Sierra Nevada Ecosystem Project
FINAL REPORT TO CONGRESS

Status of the Sierra Nevada

VOLUME II

*Assessments and Scientific Basis
for Management Options*

Wildland Resources Center Report No. 37

CENTERS FOR WATER AND WILDLAND RESOURCES
UNIVERSITY OF CALIFORNIA, DAVIS

July 1996

Sierra Nevada Ecosystem Project: Final Report to Congress
Volume II: *Assessments and Scientific Basis for Management Options*
Wildland Resources Center Report No. 37
ISBN 1-887673-01-6

Support for this research was provided by cooperative research agreement with the United States Forest Service Pacific Southwest Research Station and the University of California.

This publication is a continuation in the Wildland Resources Center Reports series. It is published and distributed by the Director's Office of the Centers for Water and Wildland Resources. The Centers sponsor projects in water and wildland resources and related research within the state of California with funds provided by various state and federal agencies and private industry. Copies of this and other reports published by the Centers may be obtained from:

Centers for Water and Wildland Resources
University of California
1323 Academic Surge
Davis, CA 95616-8750
916-752-8070

Copies of the Centers' publications may be examined at the Water Resources Center Archives at 410 O'Brien Hall, Berkeley Campus; 510-642-2666.

Please cite this volume as: *Sierra Nevada Ecosystem Project, Final Report to Congress, vol. II, Assessments and Scientific Basis for Management Options* (Davis: University of California, Centers for Water and Wildland Resources, 1996).

Photographs: Dwight M. Collins, cover, title page, sections I, II, III, V, VI; Jerry F. Franklin, section IV.

Contents

All chapters in volume II were peer reviewed.

SECTION III: BIOLOGICAL AND PHYSICAL ELEMENTS OF THE SIERRA NEVADA

SECTION IV: AGENTS OF CHANGE IN THE SIERRA NEVADA

Comparison of Alternative Late Successional Conservation Strategies
Jerry F. Franklin, David Graber,
K. Norman Johnson, Jo Ann Fites-Kaufmann,
Kurt Menning, David Parsons, John Sessions,
Thomas A. Spies, John C. Tappeiner,
Dale A. Thornburgh
(Submitted too late for publication; will be published separately later.)

Some Ecological and Economic Implications of Alternative Forest Management Policies
K. Norman Johnson, John Sessions,
Jerry F. Franklin
(Submitted too late for publication; will be published separately later.)

Preface

This volume presents the detailed scientific assessments, case studies, and background of scenarios compiled for the Sierra Nevada Ecosystem Project (SNEP). The study was conducted at the request of Congress (H.R. 5503) and was funded directly by congressional appropriation ($150,000) and by additional support ($6.5 million) from the U.S. Department of Agriculture, Forest Service. The project was managed by the University of California Centers for Water and Wildland Resources under contract with the Forest Service Pacific Southwest Research Station. A summary report of assessments, critical findings, case studies, and scenarios is published in volume I.

The chapters in volume II constitute, with few exceptions, assembly and evaluation of existing data from published and unpublished sources including expert opinion. Each chapter is authored and was prepared in response to direction from the science team and the steering committee of SNEP. The team found that much has been studied in the Sierra Nevada, although, in many areas vital to understanding the future, essential knowledge was unavailable or tests of ideas have yet to be done. Science team members were asked to draw reasonable inferences from their assessment of existing information including their own observations. They have been explicit about the bases of their knowledge and data and about where they are making assumptions or giving personal judgments.

The complete report of SNEP is contained in several volumes: Volume I offers a summary of the context for the study, the major findings from the assessments and case studies, and a presentation of alternative scenarios and their implications for the future health and sustainability of the ecosystem. Volume II contains assessments of historical, physical, biological, ecological, social, and institutional conditions in the Sierra Nevada, selected case studies, and details on the scientific bases of and methods used in scenarios. Volume III presents commissioned background reports, reports received too late for inclusion in volume II, and other supplementary materials.

In general, the study was intended to address several questions:

1. What were historic conditions, trends, and variabilities?

2. What are current conditions?

3. What are trends and risks under current policies and management?

4. What policy choices will achieve ecological sustainability consistent with social well-being?

5. What are the implications of these choices?

The first three questions were the primary focus of the assessments. The last two were the primary focus of scenarios. Some scenarios developed extensive methods and background to consider different alternatives and are presented in volume II. Several chapters are case studies where (1) existing efforts of ecosystem management were reviewed to give insight into the larger study, (2) specific locations in the Sierra were used to examine models or apply techniques of analysis, and (3) special review of the mediated settlement agreement for management of giant sequoia groves was requested by Congress.

All chapters in this volume were reviewed extensively, including anonymous peer review secured by the steering committee (see volume I). The review process involved many people who gave freely of their time and expertise and greatly improved this work. The project was conceived and executed as a scientific study by independent scientists. Thus, the reports in volume II are attributable to the authors and follow the usual standards for citation, accuracy, and statement of opinion. Throughout the study, the team fostered debate and welcomed diversity of ideas. At the end some issues remained in contention among team members and are so noted in the report. Assessment chapters, as in the journals of science, are not intended or written as consensus documents. Understanding complex ideas and recognizing areas of uncertainty come about as much by seeing different views as by studying a single, dominant perspective. But the authors have made every effort to document the basis in facts, assumptions, knowledge, and inference that they used in drawing their conclusions. Readers, by their own analyses of the information, may reach new conclusions. The team intended that the bases for its conclusions and the process of its reasoning be open and available to alternative analyses.

Authors of these chapters include the science team, special consultants, and other scientists whose expertise was solicited for assistance and coverage of various aspects of the study. Several chapters represent summaries and syntheses of larger reports or groups of reports in volume III that were commissioned for the project. In spite of the large number of reports eventually assembled, not all important components of the Sierra Nevada ecosystem could be covered by our study. The region and the scope of work is vast, and some topics must await consideration at another time.

Background data and digital databases used in geographic information systems analysis are listed in volume I, appendix 1, and are available on the Internet from the Alexandria Project at the University of California, Santa Barbara, and the California Environmental Resource Evaluation System (CERES) project of the Resources Agency of the State of California. The assistance and cooperation of the many local, state, and federal agencies, private companies, and the public in or interested in the Sierra were instrumental in conducting this study. We gratefully acknowledge them here.

DON C. ERMAN
Team Leader

SECTION I

Past Sierra Nevada Landscapes

DAVID BEESLEY
Department of History
Sierra College
Rocklin, California

Reconstructing the Landscape: An Environmental History, 1820–1960

ABSTRACT

Sierra Nevada environmental history between 1820 and 1960 can be divided into three clear phases. The first period, 1820–1900, included entrance by European-Americans, displacement of Native Americans, the discovery of gold, and the development of other resources such as timber, water, rangeland, and scenic sites. Concern for the effects of this development led to the creation of Yosemite Valley as a state park and the designation of Sequoia and Yosemite (not including the valley) as national parks. Additional Sierra Nevada forestland was included in several federal forest reserves. No effective system of regulation accompanied the creation of these areas, however. The second period of the Sierra Nevada's history, 1900–1940, saw the creation of the U.S. Forest Service and the National Park Service and the beginning of effective management of Sierra Nevada resources by these agencies. Even as federal action was being taken, private development of natural resources continued, especially logging. Automobile access stimulated recreation demand. Decisions to provide Sierra Nevada water for San Francisco, Los Angeles, and numerous hydroelectric projects affected the range as well through this period. The third period of the range's environmental history, 1940–60, was marked by accelerated use of all of its resources, especially timber and water. Improvements in automobile transportation led to increased winter uses associated with skiing. Federal and state agencies responded by trying to meet these growing demands. By 1960, many environmental groups and some elements within the federal services began to express concern over the effects of such accelerated growth. A period of environmental activism in the Sierra Nevada was about to begin.

INTRODUCTION

In December 1994 a colloquium sponsored by the Sierra Nevada Ecosystem Project (SNEP) discussed the need to describe an 1800s Sierra Nevada landscape and considered how best to summarize the region's environmental history since then. All participants agreed that no published comprehensive environmental history of the Sierra Nevada existed.

This chapter of the final report addresses one of SNEP's five fundamental questions, that is, "What were historic ecological, social and economic conditions, trends and variabilities?" (SNEP 1994).

The historians participating in the December 1994 colloquium identified three key periods in the range's history to be addressed:

1. 1820–1900: opening, unrestricted development, and early resource regulation

2. 1900–1940: establishment of agency management, continued private uses, and major environmental effects

3. 1940–60: expanding demand, agency response, and growing environmental concern

A comprehensive and detailed environmental history of the Sierra could not be expected in the short time available, nor could detailed primary source research be undertaken, because of limits set by SNEP guidelines. A more limited objective was suggested: delineate the key issues associated with ecological conditions as shaped by human actions for these three periods, and then summarize published and unpub-

Sierra Nevada Ecosystem Project: Final report to Congress, vol. II, *Assessments and scientific basis for management options.* Davis: University of California, Centers for Water and Wildland Resources, 1996.

lished secondary sources and available primary sources that bear on them. In addition, summaries of key issues and events were prepared by selected U.S. Forest Service and National Park Service personnel and other experts to supplement these sources.

Although there is no comprehensive environmental history of the Sierra Nevada, Farquhar's classic *History of the Sierra Nevada* (1965) gives a general history of the region. Many of its sections are out of date, however, and it fails to document many of the historical changes in the Sierra Nevada environment. Three environmental studies focusing on portions of the Sierra Nevada's environmental history do exist: *Challenge of the Big Trees* (Dilsaver and Tweed 1990), *Yosemite: The Embattled Wilderness* (Runte 1990), and *Tahoe: An Environmental History* (Strong 1984). Two U.S. Forest Service studies of the Sierra Nevada also exist. *The Spotted Owl: A Technical Assessment of Its Current Status* includes a historical review of Sierra Nevada forests (McKelvey and Johnston 1992), but it is limited primarily to the period from 1850 to the turn of the century. "A History of the Human Element in the Sierran Province" (Lux 1995), published as an appendix to the *Draft Environmental Impact Statement: Managing California Spotted Owl Habitat in the Sierra Nevada National Forests of California*, provides a useful extension of the time periods considered but is limited by its length as a summary to a larger report. Numerous other works, published and unpublished, deal with aspects of Sierra history. Unfortunately, no synthesis of this body of secondary sources has emerged. Many research repositories such as the Bancroft Library, the Huntington Library, the California State Library, the Yosemite Library, and the National Archives contain material relating to the range. These also have yet to be effectively synthesized.

Historians at the 1994 symposium, and others not attending, were contacted and asked to contribute. Those who responded included Richard Markley and Carmel Meisenbach, contributing information on the Tahoe National Forest; Susan Lindstrom, contributing information on the Tahoe Basin; Dana Supernowicz, contributing information on the Eldorado National Forest and other aspects of the Sierra Nevada; Pamela Conners, contributing information on the Stanislaus National Forest and Sierran water developments; Stephanie Tungate, contributing information on the Sequoia National Forest; William Tweed, contributing information on Sequoia and Kings Canyon National Parks; James Johnson, contributing information on Sierra conifer forests; William Laudenslayer, contributing information on Sierra vegetation; Linda Lux, contributing information on Sierra Nevada environmental history; Norman Wilson, contributing information on Sierran Native American patterns; Lary Dilsaver, contributing on gold rush agriculture; and Kevin McKelvey, contributing information on Sierra Nevada environmental history.

1820–1900: OPENING, UNRESTRICTED DEVELOPMENT, EARLY RESOURCE REGULATION

Opening the Sierra Nevada, 1820–55: The Impact on Natives

Prior to 1820, the Sierra Nevada was occupied by Native Americans. Anthropologists have considered the native use pattern, including the use of fire as a management tool, as resulting in a "dynamic equilibrium" (Baumhoff 1978) or even as creating a "domesticated environment" (Blackburn and Anderson 1993). Native use of the Sierra before historic contact is described in Anderson and Moratto 1996.

Between 1820 and the 1850s, native land use of the Sierra Nevada was displaced by the arrival of the Spanish and other European-Americans. Contact with the Spanish and Mexican occupants of the coastal plain and the Central Valley after 1820 changed native life patterns and the areas in the Sierra Nevada foothills occupied by natives. Spanish-Mexican pressures included conflict during exploration (Larson 1985) and raids to capture natives for mission or ranch uses. The natives responded by using armed resistance, taking livestock, and withdrawing farther into the foothills to minimize contact. Native raids in turn bred punitive actions by the Spanish and Californios, creating a cycle of disruption for the native population. All of the Sierra foothills, from the north to the south, were affected to some degree (Castillo 1978; Wilson 1995). Described by one expert as a "ripple effect," this contact would mark the beginning of the end for native dominance of the Sierra Nevada and a change to a more aggressive utilization of the range's resources (Wilson 1995).

Diseases introduced by Europeans reduced native populations. Fur trappers from the Hudson Bay Company introduced smallpox to valley natives. Contact spread the disease to Sierra Nevada foothill populations. Disease introduced to Sierra Nevada natives as a result of contact with the Spanish also had both direct and indirect effects, ranging from north to south on the western side of the range (Cook 1978; Wilson 1995; Tweed 1995).

The earliest recorded isolated crossings of the Sierra Nevada occurred in 1827 and 1833 (Smith [1827] 1977; Leonard [1833] 1978). Between 1840 and the gold rush, increasing numbers of European-Americans migrated to California. To assist this movement, federal agents such as John C. Fremont explored the northern Sierra (Goetzmann 1959). Because most of these travelers did not remain in the Sierra for any length of time, their impact on natives and the Sierra landscape was limited (Markley and Meisenbach 1995). Most of this activity was concentrated in the northern Sierra, in the modern counties of Plumas, Nevada, and El Dorado, because of their lower-elevation passes and because they were associated with more direct routes from the east. The configuration of the range, which rises in elevation as it progresses to the south, also influenced route choices. Exceptions, such as the infrequent use of Ebbetts and Sonora Passes in the central Sierra, should be noted (Stewart 1962). The hunting of wildlife, the gathering of firewood, and the grazing of travelers' animals during this crossing undoubtedly had an effect on those areas most directly contacted.

The gold rush did bring some cooperation between miners and natives (most notably in the use of natives as laborers), but contact mainly produced conflict. The most direct effect of contact during the gold rush period was upon native people living in the areas of intense mining activity or in areas adjacent to them. On the western side of the range, the contact especially affected the foothill groups, including the Konkow and Maidu in the Feather River area; the Nisenan and the Miwok in the American and Merced River drainages; and, to a lesser extent, the Yokut and Tubatulabl in the more southerly parts of the range (Castillo 1978; Larson 1985; Hurtado 1988; Markley and Meisenbach 1995; Deferrari 1995). Food resources were reduced, and game became skittish and hard to find. In one case, pressures even on such food sources as grasshoppers forced some intergroup cooperation between the Miwok and Paiutes in order to conserve a dwindling resource (Conners 1995; Deferrari 1995). Most of the impact was associated with placer mining in the northern and central foothills, although quartz developments created some negative effects as well. On the eastern side of the Sierra, the Washoe were directly affected by mining in the Comstock Lode area (present-day Virginia City, Nevada). They were forcibly excluded by loggers, commercial fishermen, and resort developers from using the Tahoe Basin, thus losing access to fish and other food resources (Lindstrom 1995). Various eastern Paiute groups were directly affected by eastern Sierra mining developments, which included the Comstock and Bodie areas, similarly losing access to subsistence resources (Malouf and Findlay 1986).

Most deaths of Sierra Nevada natives were from disease, with most mortality concentrated in the areas of most direct contact (Cook 1978). However, the effects of disease spread beyond areas of direct contact, bringing catastrophic population reductions even in the more isolated southern Sierran regions. It is estimated that nearly 100,000 Native Californians died as a result of disease (this accounted for more than 80,000 deaths, more than 80% of the total), starvation, or the effects of violence between 1848 and 1855. A great many of the dead were Sierran natives (Cook 1978; Tweed 1995; Deferrari 1995). When we consider that the number of deaths represents about one-third of the total estimated population of California natives before contact (310,000, plus or minus 10,000) (Cook 1978), it is clear that the gold rush period had catastrophic effects. Violent confrontations between natives and European-Americans were significant, producing about 12% of the native deaths of this period. While whites were killed as well, the native population bore the brunt of the casualties, at a rate of more than 50 to 1 (Cook 1978). The effects of organized and unorganized violence resulted in the displacement

of native populations from village sites and subsistence areas and a reduction in animal and plant resources. Placer mining operations reduced or eliminated fish populations (Castillo 1978; Hurtado 1988; Deferrari 1995).

The gold rush period marked the "true" opening of the Sierra Nevada to European-American occupation. A way of life marked by relatively small populations, limited technology, and seasonal limits on the use of the Sierra Nevada ecosystem was destroyed. Native fire-management practices that had shaped prehistoric forests were altered (Baumhoff 1978; Blackburn and Anderson 1993). Forest management policies developed between 1900 and 1960 specifically rejected native fire practices in their adoption of fire suppression (Show and Kotok 1923, 1924; Supernowicz 1983; Runte 1990; Tweed 1995).

Unrestricted Development, 1848–90

The discovery of gold in 1848 led to profound changes in the Sierra Nevada ecosystem. Mining promoted development in lumbering and grazing. Transportation, market hunting, tourism, and urban development followed in support of major extractive industries.

Mining

Mining history in the Sierra Nevada before 1900 can be divided into three chronological periods, recognizing that some overlap occurred. During the first of these, 1848–51, deposits of placer gold were exploited primarily by amateurs who had few skills and employed only simple technology such as pans, "rockers," and simple sluices. In the second period, between 1851 and 1859, miners exhausted most of the surface deposits and turned to the exploitation of riverbeds, veins of gold embedded in quartz, and deposits of alluvial gravel. This change required more capital, new techniques (such as the use of wing dams and ground sluicing), and larger supplies of water. In the third phase, 1860–90, Sierran mining become a capital-intensive industry employing wage-earning miners and better trained or more experienced engineers. These were employed chiefly in deep mines and gigantic hydraulic operations (Paul 1947).

Mining in the Sierra Nevada was intimately connected to the development of lumber and water resources and promoted the development of camps and towns to supply the needs of miners and loggers. Water was necessary in all gold production, and in later times it provided power for mining activities. Lumber was required to carry water in flumes, to support excavations, to provide fuel for steam engines and pumps, and to support tunnels. It was also needed for housing and business structures. Camps and towns were often consumed by fires, requiring further timber harvest. Contemporary sketches and photographs of northern and central Sierran communities show barren environments around mining settlements (Jackson 1970; Beesley 1979, 1994; Mann 1982).

Placer mining caused environmental damage. Most obvious was the mud, sand, and other debris deposited in streams.

Hillsides became pockmarked from mining operations. Channels and tunnels were cut to divert water so that streambeds could be mined. Flumes were constructed of wood to divert water from streambeds, requiring the cutting of adjoining forests. This water was used and reused farther downstream. Rivers became filled with sand. Boulders were moved out of streambeds to expose placer gold and were placed elsewhere, creating new riverine environments. Flumes leaked or collapsed, creating erosion gullies. Water storage dams burst, generating great surges of water that pushed mud, stones, and trees before the flood (Ziebarth 1984; Beesley 1994). Mercury was used to assist in the recovery of fine gold particles in placer, hydraulic, and hardrock mining during this period. Its release into stream systems stretching all the way to San Francisco Bay has to be measured in tons before 1940 (Meals 1995). The impact of this release has not been measured, and much of this metal is still in Sierran streams and soils.

The development of large-scale hydraulic mining profoundly affected the Sierra. The largest operations were located in the northern part of the range, in the Feather, American, Bear, and Yuba River drainage systems. The center of the industry was the Yuba drainage. The soil, sand, gravel, and rock displaced by hydraulic mining was immense. A report submitted by the state engineer to the California legislature in 1880 estimated that more than 680,000,000 cubic yards of debris had been washed into northern stream systems by hydraulic operations (Kelly 1959). William Brewer, a member of the state Geological Survey, made note of the immense hydraulic operations he observed in the 1860s (Brewer [1930] 1966). The impact of winter and spring floods, which carried this debris to foothill and valley communities, was disastrous. In the Sierra, hydraulic mining created areas that some contemporaries claimed could never fully recover from the damage caused (Brewer [1930] 1966; Leiberg 1902; Kelly 1959).

The direct effects of mining were not evenly distributed in the Sierra. The western foothills, ranging from Plumas County in the north to Mariposa County in the south, were most severely affected. The eastern Sierra did not produce as much gold as the western side (Clark 1980; Supernowicz 1992; Markley and Meisenbach 1995; Deferrari 1995). The Tahoe Basin was not directly affected by mining, but it was changed by the demand for lumber in the Comstock mining area (Lindstrom 1995). Mining was of very little significance in the southern third of the range, although some development occurred at Mineral King (Tweed 1995).

Logging

Logging before 1900 affected many parts of the Sierra Nevada. This industry developed primarily in support of mining activities near newly created camps and towns located on the western and eastern slopes of the northern and central parts of the range. It also provided material for the building of Central Valley towns where rail connections existed. The Sierran logging industry provided the ties, timbers, fuel, and planking necessary to build the Central Pacific and other Sierran railroads. The Central Pacific also provided timber for communities and railroads in the barren Great Basin. The Central Pacific received not only a right of way upon which to lay its track but also twenty alternate sections of land on either side. Much of the lumber to build the railroad came from these granted lands. Other private companies supplemented the railroad's cutting in nearby areas. The building of the snow sheds for the railroad near the western summit required 300 million board feet, and another 20 million board feet were required annually to keep them repaired. During the peak years of the Comstock's operation in Nevada, it consumed 70 million board feet of timbers and cordwood annually (Knowles 1942). Demand for lumber for fence posts and other uses led to the cutting of some giant sequoia near Grant Grove. At least one giant in the Kings River Basin was cut down so that it could be displayed in the East (Knowles 1942; Ayres 1958; Clar 1959; Farquhar 1965; Dilsaver and Tweed 1990).

Contemporary accounts of the Sierra timber industry written during this period reflect a dualistic perspective; most are praiseworthy, but some show concern for what are described as negative effects. On the positive side, estimates of the large volume of board feet of timber cut in Sierran forests are recorded in all contemporary accounts of the industry, reflecting pride in economic growth. Descriptions of the numerous mills are also reported in early county histories for the Sierra Nevada, for the same reason. No accurate overall total for the lumber cut for the counties included in the Sierra Nevada exists, but claims from contemporary histories range in the millions of board feet (Beesley 1984). The number of mills that operated before the turn of the century is likewise imprecise. One study written in 1924 cited 80 for the whole state in 1855 and noted that this number had increased to 320 by 1860 (Stanford 1924). Most of these mills were concentrated in the western counties of the northern and central Sierra because of the demands of gold mining (Clar 1959).

On the negative side, a California State Forestry Board report published in 1886 estimated that twenty years of cutting and fire had "consumed and destroyed" one-third of the Sierra's timber. It further estimated that if the same rate of consumption was continued, all of the range's forests would soon be cut (Clar 1959). Two turn-of-the-century U.S. Geological Survey reports detailed the impact of unregulated cutting in the northern and central Sierra, on both of its flanks. The reproduction of certain species such as sugar pine was reported to be imperiled by the wasteful high-grading practices of shake makers who took only the best parts of the large trees, leaving the rest as waste. Yellow pines were reported to have been taken in great numbers, especially in areas adjacent to mining operations, and brush and other noncommercial plant species were reported to be replacing them (Sudworth 1900; Leiberg 1902).

When confronted with conflicting reports, how can modern observers make judgments about historical logging conditions? Perhaps narrowing the view from the whole of the range to one area affected by several of the forces that dominated this period of unrestricted development can provide some perspective. The Tahoe-Truckee area was among the earliest portions of the Sierra Nevada to be contacted by European-Americans. It supported a nearby mining area and was therefore developed by loggers and railroad companies. It also was close to the route chosen for the nation's first transcontinental railroad. A study of the Tahoe-Truckee basin illustrates the intensity of logging that occurred in sections of the Sierra.

Nearly all virgin timber in the basin was cut between the 1850s and 1936, most of it between 1856 and 1880. This cutting began on the eastern side of the Sierra, near present-day Carson City, Nevada, and then moved into the Lake Tahoe Basin. From there it continued down the Truckee River corridor. Markets for this lumber included the Comstock Lode, the Central Pacific and Virginia and Truckee railroads, and cities in the Great Basin. Lumber was used to build V flumes in the steep Truckee River corridor to transport timber to Truckee and markets served by the Central Pacific (Knowles 1942). John Muir traveled up the Truckee River to Lake Tahoe in 1888 and expressed hope that eventual renewal of the forest would occur, noting that the ground was littered with "fallen burnt logs or tops of trees felled for lumber." The "best timber," he said, had been cut (Muir 1938). The forest that did return was changed, however. White fir began to dominate an area that formerly had included not only this fir but significant proportions of yellow (primarily Jeffrey) and sugar pines (Strong 1984).

Tahoe-Truckee forests at this time were also reduced by human-caused fires resulting from careless actions by logging personnel and by wasteful logging practices then common to the industry. Examples of such practices included leaving stumps as high as 3 feet because of terrain and tree girth. Felled logs were frequently cut at the point where limbs began, leaving the rest behind to serve as fuel when fires started, often damaging nearby merchantable timber. Potentially salable trees were cut to build V flumes to transport cut lumber. These V flumes consumed 135,000 board feet per mile. Steam engines called steam donkeys damaged young trees and disturbed forest soils as they dragged logs to chutes or loading pads, where they were loaded on wagons or railcars for transport to the mills. Saws at the mills generated large quantities of sawdust, which was often dumped into nearby rivers, killing fish and creating health hazards and reduced water quality for those living in cities such as Reno (Pisani 1977b). On the Truckee River, specially constructed "splash dams" were developed at one time to float logs to the mills at Truckee. The repeated scouring of the streambed that resulted when the logs and water that collected behind these dams were released altered the riparian habitat and contributed to declining fish populations (Leiberg 1902; Knowles 1942; Pisani 1977b; Beesley 1984).

This case study of one area in the Sierra Nevada contains elements that are site-specific, such as the heavy use of V flumes and the uncommon use of the Truckee River to transport logs and dispose of sawdust. It also contains elements common to other areas in the northern and central Sierra Nevada. Disturbance of soils, injury to young trees, careless fire management, wasteful cutting practices, and careless slash handling are examples that can be cited. The most significant logging areas were in the foothills of Nevada and Sierra Counties and portions of Placer County, and in association with the major mining districts stretching from Placerville to Mariposa County. Less timber harvesting occurred in the southern Sierra Nevada because of transportation difficulties associated with the terrain and because of the lack of substantial mining deposits (Supernowicz 1983; Markley and Meisenbach 1995; Lindstrom 1995; Deferrari 1995; Tweed 1995).

Grazing

Sierran meadows were heavily grazed before 1900. Most grazing involved seasonal transhumance, meaning that animals were grazed on low-elevation winter ranges and then driven to alpine areas for summer range use. This practice included cattle, sheep, horses, goats, and, in some areas, pigs. Cattle, raised for meat as well as for dairy purposes, were driven from valley or foothill areas into the Sierra for relatively lightly used summer pasture. Extreme drought in the Central Valley and in southern California in the 1860s led to increased livestock use of Sierran ranges. Meat, butter, and cheese were supplied to railroad workers, lumberjacks, miners, and town or camp dwellers in the Sierra Nevada (Edwards 1883; Sudworth 1900; Leiberg 1902; Claytor and Beesley 1979; McGlashan 1982).

The sheep industry in California developed in two distinct periods before 1900. The first, 1848–60, involved driving animals from New Mexico and southern California to mining camps and towns in the western foothills for consumption. This phase did not result in much actual grazing in the Sierra Nevada. The second phase, after 1860, depended on grazing Sierran pastures. Itinerant or "gypsy" sheep bands were driven into both sides of the Sierra Nevada from southern and central California because drought and competition for land in those areas made free range in the mountains desirable. The number of sheep that foraged on Sierran meadows before Forest Service regulation began can only be guessed at. There was no limit to the size or the number of bands that entered the Sierra before 1900, nor was there a limit on the length of time they could utilize a specific area. Undoubtedly, the number of sheep using all available meadow systems in the Sierra Nevada during this time would be in the millions (Douglass and Bilbao 1975). Some observers attribute the reduction of some native perennials and their replacement by more aggressive annual species in upper-elevation grassy hillsides and higher-elevation meadow systems to this unregulated sheep grazing (Muir 1894; Douglass and Bilbao 1975; Rowley 1985; Beesley 1985).

Sheep grazing in the Sierra Nevada before 1900 was condemned by contemporary observers. It was judged by these critics to be more destructive than cattle grazing. John Muir memorably named sheep "hoofed locusts," their being in his view more effective than fires or glaciers in destroying vegetation. Two complaints from nineteenth-century critics predominated: first, that too many animals were grazing for too long on Sierran pastures (LeConte [1875] 1930; Edwards 1883); and second, that sheepherders were starting fires to improve future range or remove barriers to sheep movement. The *First Biennial Report of the California State Board of Forestry for the Years 1885–1886*, reflecting this antisheep view, recommended that all sheep be excluded from the Sierra because of the damage they caused to soils and vegetation (Wagoner 1886; Muir 1894; Sudworth 1900; Leiberg 1902; Johnston n.d.).

Regardless of whether the contemporary observers were accurate or not in their assessment of damage caused by sheepmen, their views would shape future forest management policies. No understanding of previous native or natural fire patterns existed. In the view of contemporaries opposed to fires, the sheepmen added to naturally caused fires in a significant way. The California State Board of Forestry wanted to exclude all fires so as to improve timber production and watershed potential of Sierra Nevada forests for agricultural uses (Wagoner 1886).

Most of the Sierra Nevada was affected by grazing. Foothill, middle-elevation forests, and subalpine areas such as the upper Kern Basin were heavily impacted. Only alpine fell fields escaped impact because they had little vegetation and were in difficult terrain. Most cattle, especially those associated with dairying, were kept on lower-elevation, higher-quality, and often fenced ranges. Sheep grazed on all other rangeland (Sudworth 1900; Leiberg 1902).

Transportation

Western immigration across the Sierra Nevada into California, during its beginnings in 1840, was limited to passages across the mountain barrier. The discovery of gold in 1848 and the development of timber and rangeland resources led to a change in transportation patterns. Passes that earlier had been used for east to west movement were often abandoned for a diffuse system of routes leading into mining areas, lumber sources, or mountain pastures, as development progressed. Food and other necessities flowed from valley and foothill agricultural areas and supply towns into the Sierra. Surveys were made by private commercial interests and county governments to mark out feasible wagon routes across the mountains, and California passed legislation to create an improved wagon route through the old Carson Pass in 1858 (White 1928; Beesley 1994).

Regional differences affected transportation development. Areas with more valuable resources, such as gold or timber, were more rapidly accessed by transportation networks. The most accessible northern Sierra Nevada roads originated from valley, river, or foothill towns such as Sacramento, Yuba City, Stockton, Merced, Grass Valley, and Nevada City and extended to western Sierra Nevada mining districts and to the Comstock Lode in Virginia City, Nevada. The most used roads included the Beckwourth or Feather River route in the northern Sierra Nevada to the Comstock Lode and Virginia City; the system called Henness, which led from Marysville and Nevada City to the Comstock Lode in Nevada; and the Carson Pass route connecting the Comstock Lode to Placerville and Sacramento. Many of these were variations of earlier emigrant routes that were supplanted when the Sacramento Valley Railroad connected Sacramento to Placerville, where it joined the road that led through Carson Pass to Nevada mining areas. The Dutch Flat Toll Road from Sacramento into the northern Sierra Nevada and the Comstock later was used by the Central Pacific Railroad (Supernowicz 1983; Rice et al. 1988; Beesley 1994; Markley and Meisenbach 1995).

Economic development in the central and southern Sierra produced foothill road systems in the most accessible areas. Roads from Stockton, Modesto, Merced, and other Central Valley towns reached mining and timber areas near Jackson, San Andreas, and Sonora. Access to Yosemite was well established by the 1870s through Great Oak Flat near present-day Groveland, California. Much of the southwestern portion of the range remained isolated. Only two major roads existed, one to the Kaweah Colony timber claims and another to Mineral King. Some well-developed packing trail systems were created, however (Russell 1947; Larson 1985; Deferrari 1995; Tweed 1995).

Like road and trail development, railroad construction was confined to the more accessible areas in the northern Sierra, often along existing roads or trails. The construction of the Central Pacific, completed in 1869, used vast quantities of Sierran resources such as lumber. It also opened the way to construction of two other railroads, the Virginia and Truckee and the Nevada County Narrow Gauge, which connected mining communities to the transcontinental rails (Myrick 1962; Best 1965). Some logging operations in the Tahoe Basin used rail to transport logs (James 1915; Strong 1984). Increased access to the Sierra Nevada via the Central Pacific led to increased tourism in the Sierra, with Lake Tahoe, Weber Lake, and Independence Lake becoming tourist destinations (Beesley 1994).

Market Hunting and Fishing

A predictable food supply was needed by Sierran miners, loggers, camp residents, and city dwellers. Meat, grains, and vegetables supplied much of this need. In the earliest stages of contact, wildlife was hunted for market. Mule deer were heavily hunted for market in portions of the range (Brewer [1930] 1966). The species persisted, but their growing scarcity was noted (Deferrari 1995). Pronghorn of the eastern Sierra, bighorn sheep, and grizzly bears were also the focus of market hunting, and their numbers declined (Storer and Tevis 1978; Farquhar 1965; Beesley 1994). Market fishing, added to the effects of mining and logging, brought a reduction in fish populations, especially in the Lake Tahoe and Truckee River drainage. Commercial fishing of Lahontan trout was initiated to feed tourists and residents and later was expanded to serve consumers in San Francisco. Non-native trout species were introduced after the Lahontan variety declined in numbers (Pisani 1977b; Strong 1984).

Tourism

The Sierra Nevada provided scenery of exceptional grandeur and a variety of recreational activities. Even while rushing to get over the barrier formed by the Sierra Nevada, many immigrants noted its spectacular beauties. Some members of the Mariposa Battalion, fearfully searching for Miwok Indians, also expressed awe when they encountered Yosemite Valley. Tourism remained a secondary industry during the years of heavy resource development. But after the decline in the mining, logging, and grazing industries, recreation increased greatly in importance. Most of the focus was on the lakes of the northern Sierra Nevada, on Yosemite, and on the Big Trees in the Calaveras area. Development of the Mineral King Road opened that area to tourism in the 1880s. The Central Pacific and Southern Pacific Railroads provided access to many of these features, including the connection from the Central Valley to El Portal, near Yosemite. The publication of tourist guides to the Sierra became a regular industry by the end of the nineteenth century (Farquhar 1965; Strong 1984; Runte 1990; Beesley 1994).

Urban Development

Mining was the basis of most early European-American settlement in the Sierra. These settlements tended to be ephemeral because of the vagaries of a mining economy. Mining communities generally can be divided into short-lived camps that disappeared when the gold supply or hope of finding gold faded and more permanent towns that developed where supplies of gold persisted (Dilsaver 1985). More than three hundred mining communities existed from Plumas County in the north to Fresno County in the south during the most active period of the gold rush, from 1848 to the 1880s. Most of these camps and towns were in the northern and central portions of the western Sierra Nevada (Gudde 1975). Besides towns and cities based on gold mining, other communities, such as Truckee in the northern Sierra, developed around industries such as lumbering, railroading, or service functions for Tahoe's tourists. Lumber and tourism produced long-lived communities at Tahoe, including Glenbrook and Tahoe City. Foothill towns up and down the western side of the range supplied food and other needs for camps and towns further into the Sierra. Other towns provided government services (Hinkle and Hinkle 1949; Gudde 1975; Meschery 1978; Mann 1982; Strong 1984; Beesley 1994).

Those communities that survived the initial gold rush phase developed multiple economic activities, some of which included agricultural production. Fresh vegetables and fruit were produced in foothill communities, especially in such counties as Placer, El Dorado, Nevada, and Mariposa. In mining and timber production areas, beer and wine were produced from locally grown as well as imported ingredients (Taylor 1975; Dilsaver 1985; Beesley 1988).

The Beginning of Resource Regulation, 1864–1900

Concerns about the effects of unregulated development of the Sierra were raised by citizens, county and state officials, national politicians, and organized conservation interests before the mid-1860s. This protest led to the creation of institutionalized management of the Sierra. Five well-defined, though not coordinated, sets of actions mark the beginnings of the conservation movement in the range: (1) protection of Yosemite Valley and the Calaveras Big Trees; (2) control of hydraulic mining; (3) attempts to protect Lake Tahoe and the Truckee River; (4) attempts by the State of California to control forests; and (5) creation of national parks and forest reserves.

Protection of Yosemite Valley and the Big Trees

The discovery of the Calaveras Big Trees and Yosemite Valley quickly attracted national attention and led to actions to promote them for private gain. Prominent among these early promoters was James Hutchings, who quickly developed tourist facilities and services in Yosemite Valley (Farquhar 1965; Runte 1987). Many eastern visitors and some California citizens feared that such activities, resembling those that had commercialized Niagara Falls, could lead to similar effects in these two Sierra Nevada areas. California politicians and business interests in the Central Valley joined to demand federal protection. In 1864, Yosemite Valley and the Calaveras Big Trees area were granted to the State of California, marking the first time that federal action was taken to preserve land from development (Runte 1987).

Although California established a commission to administer Yosemite Valley and the Calaveras Big Trees, very little funding was provided. Regardless of the intent of the commission, these areas were developed. In Yosemite Valley lodging was built for tourists, commercial signs were created promoting the valley's features, trees were cut to improve views, meadows were fenced to provide pasture for livestock, orchards were planted to provide fruit for tourists, ladders were built to help people reach Vernal Falls, ferries and bridges were built to facilitate stream crossing, and Nevada Falls was altered to force more water into a central channel so that tourists would be more likely to see the waterfall in the summer (Runte 1987, 1990; Beesley 1994).

Control of Hydraulic Mining

Hydraulic mining in its most highly developed form in the late nineteenth century used the force of water collected in dams in higher Sierra Nevada elevations. It was then transported in flumes and penstocks under ever-increasing pressure to water cannons or "monitors" where it was released against gold-bearing gravel deposits. This mining technique created immense amounts of debris, which clogged stream systems and contributed to disastrous floods in the foothills and Sacramento Valley when spring flooding occurred. Irreversible damage, whole hillsides being washed away, for example, occurred in many sites. Related riparian systems were also affected by the millions of cubic yards of sand, gravel, rocks, and other debris produced by this activity. The California legislature responded to complaints from foothill and valley interests by promoting measures to control flooding. Representatives from California areas not directly affected by this problem eventually refused to appropriate state funds for flood control. Eventually farmers and cities deluged by the miners' debris turned to the courts. In 1884 an injunction against the depositing of hydraulic mining debris was granted by a federal court. In the 1890s, federal legislation responded to this injunction with legislation that allowed hydraulic mining as long as debris was contained on-site. Some small-scale operations were able to comply, and other operations continued to operate illegally for a short while. In time, however, these operations ceased, and large-scale hydraulic mining came to an end during the 1890s (Kelly 1959; Beesley 1994).

Lake Tahoe and Truckee River Protection Attempts

Concern by the public to limit resource use at Lake Tahoe (originally named Lake Bigler) and the Truckee River resulted in attempts to protect parkland around the lake and to stop the dumping of sawdust into the river drainage. In the 1860s legislation by the California legislature was considered to promote fire reduction and theft of forest resources. While most of this proposed legislation did not become law, concern was clearly demonstrated. In 1883, the California legislature created a study group, named the Lake Bigler Forestry Commission, to specifically address the problems of overcutting in the Tahoe Basin. National concern over wasteful forest practices and overcutting influenced the thought of some Californians at this time. Between 1865 and 1868 Frederick Starr's *American Forests* and George Perkins Marsh's *Man and Nature* were published. In 1873 Franklin B. Hough presented a report to the American Association for the Advancement of Sciences, which led to the creation of the American Forestry Association in 1875. The members of the Lake Bigler Forestry Commission were all familiar with the ideas generated by these books and reports (Clar 1959).

The ensuing Lake Bigler (Lake Tahoe) Forestry Commission report called for the protection of Lake Tahoe and the land around it for the use of tourists. It also called for control of the lumber operations that were rapidly harvesting the forests on its shores. A park was proposed, to be created by the transfer of state, federal, and private land to the State of California, obviously operating on the model of the earlier Yosemite Valley grant. Objections to land transfers that would bring profit to the Central Pacific Railroad prevented action to protect this area (Pisani 1977a).

The dumping of huge quantities of sawdust into the Truckee River by lumber mills was another problem noted at this time. It was the cheapest way to dispose of this bothersome by-product. But its effects, which included pollution and fish reduction, created conflict between loggers and Reno city residents, grazers, and Paiute fishermen, all of whom used the river. Finally, the California and Nevada legislatures agreed in 1889 to jointly prohibit the dumping of sawdust. Most dumping stopped. In 1894, the California Fish Commission put pressure on lumber operations persisting in dumping and secured the end of the practice (Pisani 1977b).

State Attempts at Forest Protection

Conservationists and valley irrigation interests desiring protection of watersheds combined forces to create a state Board of Forestry in 1885. Identified in 1868–69, and proposed in the Lake Bigler Forestry Commission report, this new board was intended to help manage state school lands and to promote state control over unregulated federal lands. Because most of the forestlands in California remained under private or national control, the board was capable only of studying the problem and suggesting remedies (Clar 1959). In 1886 the California Board of Forestry called for control of California forestlands by the state to reduce fire damage, prevent trespass and theft, and protect watersheds in the interest of irrigationists. Luther Wagoner, as a representative of the state Board of Forestry in 1886, called for the complete exclusion of sheep from Sierra Nevada forests for both erosion control and watershed protection reasons (Wagoner 1886; Beesley 1994).

The board published four biennial reports, collectively mentioning brush taking control of cut-over forestland, fire and erosion resulting from sheep grazing, the wasting of timber by shake makers, the composition of timber species in Sierran forests, and the need either for more effective federal management of forests or for such authority to be transferred to the State of California (Beesley 1994). Limited by funding and the fact that most California timberlands did not belong to the state, the California Board of Forestry remained weak. It was abolished following the enactment of the Federal Forest Reserve Act under President Benjamin Harrison in 1891 (Clar 1959; Beesley 1994).

The Beginning of Federal Park and Reserve Action

State measures to protect Sierran forests or monumental features such as Yosemite and the Calaveras Big Trees before 1890 dissatisfied some conservationists, who called for more protective federal action. By 1890, some forest and scenic resource issues were addressed by the creation of two national parks (Sequoia and Grant Grove, and Yosemite) and several Sierra Nevada forest reserves (Sierra, Stanislaus, and Tahoe). While federal legislation was passed, no overall policy was developed to administer these two new federal responsibilities (Runte 1987). A nongovernment group called the Sierra Club was founded at the same time to help shape policies for these areas (Jones 1965).

The park boundaries were molded through the actions of local civic interests, state politicians, railroad officials, conservationists, and national politicians who joined to protect these two specific sites (Runte 1990; Beesley 1994; Tweed 1995). They were limited in size and location by political and economic concerns. Few known commercially valuable resources were included in most Sierra Nevada park grants. Later, when mining, hydroelectric, or water resources were identified, both Yosemite and Sequoia were reduced in size or invaded. Sheep grazing in Yosemite and Sequoia was ended by the use of federal troops, who were able to exclude sheep bands. Conflict over boundaries in the Sequoia National Park led to the continued cutting of trees for several years after park designation (Runte 1987, 1990; Beesley 1994; Tweed 1995).

Forest reserves were primarily the product of national actions that came to include the Sierra Nevada (Steen 1976). Between 1893 and 1900, three forest reserves were created in the Sierra Nevada, one each in the southern, central, and northern parts of the range. The southern reserve was called Sierra, and it made up most of the current Inyo and Sequoia National Forests (Larson 1985). The drainages of the central Sierra rivers, which included the Mokelumne, Merced, Stanislaus, and Tuolumne, became the central division and were eventually included in the Stanislaus National Forest (Conners 1992). Reserved lands in the northern part of the range included elements of the Tahoe and Eldorado National Forests (Markley and Meisenbach 1995). No effective management plan or organization was proposed or developed for these reserves at this time. Trespass, unauthorized grazing, and timber theft continued. Forest reserve legislation may actually have stimulated attempts by timber speculators to file with the Government Land Office for claims before land could be placed in reserves. Railroads also relied on another federal law, called the Lieu Land Act of 1897, to trade some of their original right-of-way lands (often cut over) for forested land not in the reserves. Despite these last vestiges of uncontrolled activities, a milestone had been passed. Park and reserve lands had been withdrawn from sale. In the future, unauthorized use of forests under federal control would be reduced (Ayres 1958; Tweed 1995).

Summary

Between 1820 and 1900, the Sierra Nevada was opened to historic forces that transformed its human use patterns and changed the physical nature of the range. Native populations were decimated, and their long-established land-use practices were displaced by different technologies. European-American population densities during this development phase were greatly expanded. Large-scale extractive industries became a dominant feature of many parts of the Sierra Nevada. Mining, logging, grazing, and other activities affected many aspects of the range's ecosystem. Concern about the negative effects of the chief extractive industries and the danger to scenic resources led to the first local, state, and national actions to provide some protection. This included ending hydraulic mining and creating state and national parks and forest reserves. These actions marked a transition to a new period of Sierra Nevada environmental history.

1900–1940: ESTABLISHMENT OF AGENCY MANAGEMENT, CONTINUED PRIVATE USES, MAJOR ENVIRONMENTAL EFFECTS

The next phase of the Sierra Nevada's environmental history, taking place in the period from 1900 to 1940, featured growth of federal, state, and municipal agencies whose jurisdictions included much of the public lands and resources in the Sierra. Private ownership of Sierran resources continued to affect the most productive Sierra Nevada forestlands. Effective agency regulation in public land marked a distinct change from past national land-use practices.

Establishment of Agency Management

National Forests

The establishment of federal control over forested lands became effective with the creation of the U.S. Forest Service. Congressional action in 1905 transferred forest reserves to the Department of Agriculture under the direction of Gifford Pinchot. In 1907 these reserves were redesignated as national forests. Pinchot immediately began to professionalize the new service in line with modern forestry practices. In the Sierra Nevada, the early reserves were reorganized into eight more manageable national forests: Plumas, Tahoe, part of Toiyabe, Eldorado, Stanislaus, Sierra, part of Inyo, and Sequoia (Farquhar 1965; Steen 1976).

Most activities of Forest Service personnel before 1940 could be described as custodial. Their principal duties were establishing accurate boundaries, preventing timber theft and trespass, suppressing fires, managing special use activities such as mining and grazing, building ranger facilities, preparing and supervising timber sales, and building campgrounds (Bigelow n.d.).

Although the policy of "multiple use" of national forests was not explicitly stated until 1960, it was practiced during this early period. Under Pinchot and other Forest Service chiefs until 1945, balanced stewardship of all resources was emphasized. In practice this meant that logging would always be considered important to sound forest management where it was appropriate. Watershed protection and hydroelectric development were promoted. Grazing, mining, recreation, wildlife habitat, and hunting were seen as desirable uses, but logging was stressed as the most significant contribution to society from the national forests. Limited demand for timber on Forest Service land in the Sierra Nevada from 1907 to 1920 (the best forestland in the range in private hands was being developed heavily at this time) and reduced demand during the Great Depression, 1929 to 1939, meant that balanced use was relatively easy to maintain (Ayres 1958; Sedjo 1991; Conners 1990; Hirt 1994).

National Parks

While Yosemite Valley and the Calaveras Big Trees represented the first areas in the Sierra set aside as parks, they were under state jurisdiction. The first national parks in the Sierra Nevada (Sequoia/Grant Grove and the Yosemite high country) were established in 1890 (Runte 1987; Tweed 1995). These two parks were not integrated into any general management scheme. Protection was assigned to the U.S. Army, which used mounted patrols that were generally successful in keeping sheepherders and other trespassers at bay.

In 1905, Congress passed what is commonly called the "Right-of-Way Act," which allowed utility corridors to be created on federal lands in the West. At times portions of the designated park areas were removed because economic interests such as mining demanded them, as in 1907 when the Devil's Post Pile and Banner Mountain areas were taken away from Yosemite National Park (Russell 1947; Runte 1990). In 1913 the Minarets, a series of sharp peaks near Banner and Ritter Mountains, were also removed from Yosemite National Park for similar reasons (Albright and Cahn 1985). Hydroelectric developers secured entry into Sequoia National Park between 1905 and 1915 and cut timber, developed roads, and began construction work on a dam. The dam was not finished because of geological problems (Dilsaver and Tweed 1990).

National park policy from 1890 to 1916 was developed without any central plan to guide it. It was not clearly established just what a national park was and what could be done in one. The failure to address these two issues would lead to the flooding of Hetch-Hetchy Valley behind a dam to supply San Francisco with water. The outrage caused by this invasion of a national park was a key factor in the creation of the National Park Service in 1916 (Runte 1990; Tweed 1995).

Stephen Mather, the first head of the National Park Service, shaped national park policy in the years from 1916 to 1928. Horace Albright, a trusted associate, was chosen to replace him, thus guaranteeing a continuous policy over several decades (Albright and Cahn 1985). Some changes in park configuration in this period occurred. In 1926 Sequoia National Park was expanded to include the headwaters of the Kern River and Mount Whitney, removing them from Forest Service control. In 1940 Congress created Kings Canyon National Park, incorporating parts of the Sierra and Sequoia National Forests and all of General Grant Park into the new entity. Sequoia and Yosemite expanded tourist facilities, created roads and trails, established or expanded fire-suppression actions, and began predator control and wildlife

management (including the feeding of bears so as to reduce problems for tourists). Grazing continued in the valley until 1933, although most had been excluded by the mid-1920s. Various types of vegetation manipulation, such as mowing and some burning, were used into the 1920s in the valley (Russell 1947; Gibbens and Heady 1964; Runte 1990; Tweed 1995).

California State Actions

The history of state land management following the demise of the first Board of Forestry, as it affects the Sierra Nevada, divides into three clear periods. The first, lasting from 1890 to 1905, saw little interest by the California government in its grain, forest, or grazing lands. While private interests were concerned about watershed protection and fire dangers, no major state legislation was written. The period between 1905 and 1919 brought the establishment of a second state Board of Forestry, which tried to address the issues of fire control, reforestation of cut-over lands, and protection of state forestlands. Without much funding, the board had little impact. In 1911, under the influence of Republican Progressive reformers, a state Conservation Commission was created, largely to deal with water conservation and hydroelectric power development, reflecting the growing influence of California's urban centers. In 1927, the third period of California resource history began when Governor C. C. Young and the legislature cooperated to pass a law that created a Department of Natural Resources. Although placed under a single state agency, jurisdiction over California's forestlands was separated from that over the newly defined state parks, all of the latter of which were outside of the Sierra Nevada (Clar 1959).

Between 1922 and 1945, four major issues dominated the state's resource thinking: (1) providing sufficient funding to suppress forest fires, (2) gaining control of logged-over lands to form the basis for future state forestlands, (3) reforesting these cut-over lands, and (4) surveying and developing watersheds for irrigation and domestic water uses (Clar 1959). The water issue led the state in the 1930s to plan a major water project to utilize water from the Sacramento and American Rivers to irrigate the Central Valley. The Great Depression forced the state to abandon the idea because of funding problems, opening the way for the federal government to step in and appropriate the project (Hundley 1992).

Water Agencies

The creation of forest reserves in 1891 was based on the major premise that forests were needed for lumber, as watersheds for irrigation and domestic purposes, and for development of hydroelectricity (Conners 1992). Political pressure for water resource protection and development in the West before 1900 was based on the need of irrigation interests to preserve forests as watersheds. Unregulated logging and grazing were condemned because they threatened the forests that made possible the development of irrigated agriculture. Farming was judged more likely to produce stable societies and economies than the extractive and wasteful logging and grazing practices that existed before 1900 (Worster 1985). While small-scale irrigationists would continue to exert an influence on water policy and the protection of watersheds, the impounding of water in dams for larger-scale irrigation, hydroelectric power generation, and urban uses emerged as more important factors in molding western and Sierran water policy in the twentieth century. Water development in the Sierra Nevada from 1890 to the early 1940s would be carried out primarily by federal agencies, urban governments, public utilities such as Pacific Gas and Electric, and local irrigation agencies (Worster 1985; Frederick 1991; Hundley 1992).

The largest of all government agencies to begin development of Sierran water was the federal Bureau of Reclamation. The major actions of this bureau that affected the Sierra Nevada occurred during the Great Depression, 1929 to 1939. During that time, the federal government took over control of the state-proposed Central Valley Project (Frederick 1991). Construction of a portion of the federal project began in the 1930s, but the dams and aqueducts that constitute most of the project were not completed until the 1950s (Hundley 1992).

The quest for water by the cities of San Francisco and Los Angeles had an immediate impact on the Sierra Nevada's streams at this time. Many histories exist of these two cities' attempts to gain control of Hetch-Hetchy Valley and the Tuolumne and Owens River water (Jones 1965; Worster 1985; Reisner 1986; Hundley 1992). While water for growing populations was an important reason for seeking to use Hetch-Hetchy and the Owens River, both cities clearly wanted hydroelectric generation to be an important part of these developments (Hundley 1992). It is often stated in historical accounts that San Francisco had other options. John Muir even suggested that Lake Tahoe, its shores denuded of timber and facing degradation of its water purity anyway, be given to San Francisco (Jones 1965). Regarding this view, however, it is also possible that, given the growing population in East Bay cities, even if San Francisco had not claimed Hetch-Hetchy, other urban centers would have pressed claims for it. In the climate of Progressive politics under Presidents Roosevelt and Wilson, a conservation ethic that stressed utilization of resources in service to the public interest was likely to prevail in any struggle for power (Fox 1981; Nash 1982).

Actions taken to assure a water supply for the city of San Francisco had direct effects on the environment of the Hetch-Hetchy/Tuolumne River system, including the construction of the O'Shaughnessy Dam, the cutting of Hetch-Hetchy's forests, and the flooding of its meadows. The effects of the construction of the Los Angeles Aqueduct involved the diversion of water away from Owens Lake, converting it into a salt flat. Declining water supplies also meant that Owens Valley farmers and the remnant elements of the local Paiutes faced economic ruin as the agricultural economy died. The city of Los Angeles purchased a great deal of land in Inyo County, but the majority of the land remained as federal lands in the Inyo National Forest. Tourists from Los Angeles soon began to utilize this area as automobile travel developed (Hundley 1992).

Although large state and federal water projects had substantial environmental effects on the Sierra Nevada, the numerous medium to small-sized dams and water delivery systems built after 1900 may have had an even greater influence on the Sierra Nevada ecosystem, because of their sheer numbers. No overall study of this impact has been published. Three studies of specific portions of the range, however, serve as examples of the potential impact of these projects on its northern, central, and southern parts. The earliest significant use of water before 1900 was related to mining or agriculture. While agriculture would continue to claim Sierran water resources after 1900, generation of hydroelectric power became much more important. The Pacific Gas and Electric Company (PG&E) consolidated control over electrical generation for Placer, Nevada, and Sierra Counties between 1905 and 1913. It constructed several dams and generating facilities, the largest at Spaulding. By 1940, PG&E had become the primary producer of electricity for much of the Pacific Coast, and much of its generating capacity came from facilities within the Tahoe National Forest (Jackson et al. 1982).

In the central Sierra at various times between 1905 and 1920, several smaller companies, such as the Tuolumne Electric Company, the Main River Water Company, the Stanislaus Electric Power Company, and the Sierra and San Francisco Power Companies, constructed facilities in the Stanislaus National Forest (Conners 1992). In the southern Sierra Nevada, potential hydroelectric generation sites within Sequoia National Park and on the Kings River were identified between 1913 and 1920. Unlike the situation in the northern and central Sierra, however, in this area demands to dam or utilize streams did not produce significant effects other than exploratory construction activities (Dilsaver and Tweed 1990).

Grazing Management

Most grazing land in the Sierra Nevada came under control of the Forest Service in the period 1900–1940. A major exception was Sierra Valley, in the northern part of the range, where significant private ownership existed (Sinnott 1979). All Sierra Nevada national forests established special use permits that favored grazing by local ranchers over "gypsy" sheep bands and unauthorized cattle interests (Douglass and Bilbao 1975; Steen 1976). Local grazers in some cases resisted this new control, but many soon came to realize the advantages that came with guaranteed access by "local" interests (Bigelow n.d.; Rowley 1985). Regulation did not necessarily reduce the number of animals utilizing U.S. forests in the period after 1907. By 1917 there had been a 50% increase in the number of animals grazing in national forests, and demands during the First World War pushed usage beyond that. Better management practices and increased grazing land placed under Forest Service control have been cited as reasons for this increase (Rowley 1985). It is likely that numbers of grazing animals in the Sierra Nevada increased similarly.

Gradually during this period, cattle began to replace sheep on many Sierran ranges, resulting in more soil compaction and increased effects on vegetation in riparian zones (Lux 1995). Between 1905 and 1930 the Forest Service developed policies intended to balance grazing intensity and range conditions. These included instituting term-grazing privileges, limiting the number of animals allowed under existing climate and range conditions, and closing some areas to grazing in order to protect watersheds or to limit impacts on wild game. The Forest Service established predator control and poisonous plant reduction to serve those holding livestock permits on forest ranges (Markley and Meisenbach 1995).

In the 1930s Forest Service control of its rangeland was challenged by the creation of a rival Grazing Service in the Department of Interior following the passage of the Taylor Grazing Act of 1934. The Taylor Act did not apply to most Sierran rangeland, but it did create competition between the Interior and Agriculture departments that forced the Forest Service to modify its policies. These policy changes included offering ten-year leases, relaxing policy that previously had reduced animal usage during drought, and permitting more animals on western ranges. While these changes did not lead to grazing that was as severe in its impact on Sierra Nevada rangelands as that of the unregulated years, Sierra ranges were opened to greater usage than had occurred in the two decades between 1910 and 1930 (Rowley 1985).

Continuing Private Uses

Much of the most accessible Sierran timberland by 1900 was in private hands. Application of railroad logging techniques permitted greater amounts of lumber to be brought to market and allowed more distant areas to be logged economically. Hydraulic mining continued on a limited basis. Mining at hardrock sites continued. The 1870s mining law allowed mineral resources to be easily appropriated by private interests but did not regulate their operations adequately (Lux 1995; Markley and Meisenbach 1995).

Railroad and Other Logging in the Sierra

The use of railroads to transport lumber to distant markets began first in the northern Sierra with the construction of the Central Pacific Railroad (Edwards 1883; Knowles 1942). This railroad permitted the shipment of timber from the Sierra Nevada to Great Basin cities such as Salt Lake City. California areas along the railroad route, and national and international markets, were served before the turn of the century, when the Central and Southern Pacific railroads were integrated. Access to other markets stimulated the development of other rail logging systems to harvest Sierra Nevada timber. The railroad lumber industry of the Sierra Nevada grew most between 1890 and the 1920s, and more than eighty rail logging companies were created (Ayres 1958; Lux 1995). Logging rail systems opened formerly inaccessible privately held timberlands to intense development until the 1930s. Privately

owned rail logging systems encouraged Forest Service personnel in the Sierra to open timber sales in lands close to rail systems, thereby aiding industries and the Forest Service in reaching their separately defined goals (Conners 1990).

Railroad logging occurred in most of the Sierra, with much less taking place in the southern one-third of the range. Areas especially affected included the Tahoe-Truckee Basin, the portion of the South Yuba River drainage close to or in the Tahoe National Forest, areas located within or near the Eldorado and Stanislaus National Forests, and locations near the Sequoia National Forest. Railroad logging primarily harvested pine and redwood. Fir was generally used for fuel and pulp. Cedar was used for fuel as well. The cutting left many acres denuded. No overall figures are available, but examples can be cited to illustrate this point. Records from the Hobart Mills operation in Sierra County, only one of the three counties in the northern Sierra in which this company owned land, lists more than 105 million board feet cut between 1916 and 1919 (Knowles 1942). In the central Sierra, the West Side Lumber Company cut more than 90 million board feet in 1915 and 1916 (Conners 1990).

A report issued by S. B. Show for the Forest Service in 1926 warned that if the pine forestland in California, 80% of which was in private hands, continued to be cut at currently existing cutting rates, most of the companies involved would soon be in the "cut-over land business" (Conners 1990). Their methods of moving cut timber to the rail landings, judged to be wasteful by Forest Service standards, included flumes, steam donkeys, and chutes that caused damage or used considerable amounts of timber for construction or fuel (Brown and Elling 1981; Supernowicz 1983; Conners 1990; Markley and Meisenbach 1995). Other wasteful practices included careless slash handling and fire control practices. At times, however, cooperation between the Forest Service and private companies occurred in the area of fire control. Forest Service standards were imposed on the rail loggers when they bid for federal timber sales (Conners 1990; Markley and Meisenbach 1995).

Prices for timber from the Sierra Nevada fluctuated between 1900 and the 1930s. At times prices were low because of low demand and overproduction, and at other times they were high, especially because of wartime demand during 1914 to 1918. Rail connections reduced transportation costs and, until the late 1920s, aided rail logging operations, even though they continued to utilize older and more wasteful methods (Brown and Elling 1981; Supernowicz 1983; Lux 1995; Markley and Meisenbach 1995). During the Depression many companies operating on private land went bankrupt or adopted more aggressive cutting policies to maximize profits (Conners 1990). Economic conditions in the 1930s sometimes created advantages for the Forest Service, leading to land exchanges and extensions of cutting rights that provided short-term cash advantages to private companies and long-term gains for Sierran national forests (Conners 1990).

Sierra Nevada forests before 1940 were commonly more

open than those of today, with large, even-aged trees under which grew perennial grasses and few shrubs. The most significant effect of logging before 1940 was the removal of the largest yellow and sugar pines. Replacing these were smaller but more densely packed pines in some areas, more fir and cedar in other areas formerly dominated by pine species, and more shrubs than had existed in the earlier forests (Laudenslayer et al. 1989; Laudenslayer and Darr 1990). By 1934 more than half of the mixed conifer forestland in the north-central Sierra Nevada had been entered for harvesting, although logging was restricted primarily to ponderosa, Jeffrey, and sugar pines. Fir was less affected. The southern part of the range, where rail logging did not develop, still served primarily local markets. On Forest Service land only 7% of mixed conifer forests had regenerated second-growth stands, because of the recentness of cutting. More than 90% of the remaining unharvested conifer stands in the Sierra were located on Forest Service land (Johnston n.d.).

Mining Developments

Quartz gold mining grew in importance after 1900. Permanent communities such as Sierra City, Alleghany, Nevada City, Grass Valley, and Sonora reflected the relatively stable nature of this industry (Clark 1963, 1980; Sinnott 1976). The impact of this industry on water and other Sierran elements has not been determined. During the Second World War, most of these hardrock gold mining operations were closed so that the iron, fuel, and wood they consumed could be redirected into the war effort. Few reopened after 1945.

While gold mining declined during the period from 1900 to 1940, other types of mining developed in many areas in the Sierra. More than twenty different minerals were mined between 1900 and 1960, many having been developed before the Second World War. These minerals included copper, chromite, barite, molybdenite, and tungsten. Their development contributed to the economic viability of local communities (Jackson et al. 1982; Supernowicz 1992). Their impact on ecosystem conditions in the Sierra Nevada has not been assessed.

Other Major Environmental Effects

Agency management significantly changed Sierran use patterns after 1900, especially related to fire suppression and recreation development.

Fire Suppression

Among the Forest Service personnel in the Sierra Nevada, fire was one of the most frequently mentioned subjects: extinguishing fires, training people to fight them, establishing lookouts to spot them, establishing phone lines to report them, and requiring timber sales to limit the possibility that one would start. July and August forestry concerns were dominated by fire (Bigelow n.d.).

During the early part of the twentieth century the Forest Service identified and studied sources of fire. Fire was generally seen as a degrading force to be excluded, if possible. These attitudes were also shared by Park Service personnel. By the mid-1920s all national forests and national parks in the Sierra Nevada had fully developed policies, procedures, and organization to suppress fire in their jurisdictions; these took into consideration season, topography, and past fire histories for their special area. Regulations for timber sales required that fire control equipment be readily available, enforced brush and vegetation piling procedures, and even set limits on where loggers could smoke (Ayres 1958; Supernowicz 1983; Cermak 1988; Markley and Meisenbach 1995; Tweed 1995). All national forests in the Sierra Nevada had developed infrastructures such as lookouts and phone systems. Some observers noted that such policies did reduce fire frequency by the late 1920s, although others said that low fuel levels dating from earlier forest conditions were actually the reason for fewer fires (Ayres 1958; Cermak 1988; Lux 1995).

The decision to exclude fire from public lands came about as the result of a debate over whether to permit "light burning" or, as some called it, "Indian burning" or to use complete suppression. Studies that included field experimentation, most notably those of Forest Service personnel S. B. Show and E. I. Kotok, were established over time in several different locations, with national and state foresters coming down hard on fire-caused damage. Logging and grazing interests and even some nature writers held that light fires reduced fuel, thereby creating more open forests and lessening the danger that excessive fuels would feed destructive crown fires (Ayres 1958; Cermak 1988). Concern by the Society of American Foresters about this growing disagreement led them to offer to arbitrate the differences. A California Forestry Commission was created, with representatives from both sides of the issue appointed. Eventually, this commission supported a policy of complete suppression (Pyne 1982).

In 1923, Show and Kotok published a study that essentially settled the debate for the Forest Service. They concluded that all fires, especially repeated light fires in the same area, caused progressive damage and hence were not benign. Repeated burnings, in their view, killed young and less-fire-adapted species, creating unnatural forests that favored mature pines. While mature trees and open canopies were good for logging and grazing interests, fire discouraged effective regeneration of mixed forests. If forests were to be sources of a sustainable timber supply, fire had to be suppressed (Show and Kotok 1923). The next year they published another report that established policies to implement their conclusions (Show and Kotok 1924). In the same year, the Clarke-McNary Act was passed by Congress, and it clearly established fire exclusion as national policy. Federal money was offered to state agencies that would comply with suppression doctrine (Pyne 1982). Absolute fire suppression would form the basis of Forest Service and Park Service policy until the 1960s, when it was reconsidered (Pyne 1982; Supernowicz 1983).

In the Depression of the 1930s, declining timber sale revenues reduced fire-fighting funds. To compensate for reductions in fire-fighting personnel at this time, the Forest Service utilized the Civilian Conservation Corps (CCC), a federal employment program created for young men, to assist in fighting fires. The CCC also provided valuable help in building and improving Forest Service and national park trails and facilities (Supernowicz 1983; Markley and Meisenbach 1995; Tweed 1995).

Expansion of Recreation

After 1900, tourist revenue created by automobile access grew in economic importance in areas where roads and natural beauties existed together. Expanded demand created by recreation and tourism added another significant force to bring about further changes in environmental conditions in the Sierra Nevada (Strong 1984; Runte 1990).

Monumental Features. Before 1900 the focus of Sierran tourism was on its monumental features (Yosemite Valley State Park, the Calaveras Big Trees, Yosemite National Park, and Sequoia National Park) and its lakes (Tahoe, Independence, and Weber). Limits set by horse transportation eventually were pushed back when railroads made travel more comfortable for a growing middle class (James 1915; Hinkle and Hinkle 1949; Beesley 1979; Strong 1984; Runte 1990; Tweed 1995). Cars increased the numbers of visitors in all of these areas. Automobiles, natural beauty, and Kodak cameras acted together to stimulate increased camper and tourist use. Railroad connections to Tahoe and Yosemite rapidly declined as most tourists used the more convenient automobile. In Yosemite alone, the number of visitors using automobiles doubled every several years until the Great Depression (Schmidt 1990; Demars 1991).

Yosemite National Park, officially admitting automobiles in 1913, was affected more by automobile tourism before 1940 than Sequoia or Lake Tahoe. In 1926, a shorter all-season road was opened up on the Merced River–El Portal route, making Yosemite even more accessible. Park promotion by the service after its founding in 1916 was intense. Park Director Stephen Mather stressed that national parks were the "playgrounds of the nation." He cooperated with the National Park to Park Highway Association in its promotion of vacationing by cars. The Park Service encouraged more visitors per year and longer stays by them. In practice, this led commercial concessions to construct non-nature-oriented facilities such as bowling alleys and swimming pools. The "Fire Fall," which involved pushing burning debris over Glacier Point as an evening attraction for Yosemite Valley tourists, was reintroduced by David Curry, a park concessionaire. Thousands of car campers who visited the valley also used the Merced River for cooking and bathing. Sanitary facilities were inadequate, resulting in increased pollution. Winter use was promoted where possible. The Park Service worked closely with commercial concessionaires such as the Curry and Yosemite Park Companies. In Yosemite this meant promoting winter carni-

vals, sporting contests, snowshoeing, ice skating, and snow play. Badger Pass Ski Lodge was opened in 1935 for touring and downhill skiing (Schmidt 1990; Demars 1991).

Sequoia and General Grant National Parks had very little environmentally significant tourist impact before 1900. Appropriations by Congress allowed some low-level road and trail development in 1900, making sequoia groves in both areas more accessible. Between 1920 and 1934 the "Generals" highway was constructed, connecting Sequoia and General Grant. In the 1930s the U.S. Forest Service and the State of California cooperated to build a road from General Grant Grove to Kings Canyon, opening these spectacular areas to automobiles. By the end of the decade, Sequoia and Kings Canyon National Parks had well-developed tourist infrastructures that included roads, trails, lodges, and campgrounds (Tweed 1995).

Lake Tahoe was transformed in much the same way as Yosemite and Sequoia by automobile access and recreation. Because much of Tahoe's land was in private hands, more private recreation development occurred there than in the federal park areas. Transportation costs to Tahoe dropped considerably as adequate roads and reliable automobiles became common. Little tourism occurred in the winter, when the Tahoe Basin was isolated because of heavy snowfall (Strong 1984; Lindstrom 1995).

Highway transportation to the Tahoe Basin was built on the older network of freight and wagon roads that had developed to serve Lake Tahoe and the Comstock Lode/Virginia City, Nevada, area before 1900. Many of the engineered roads were the product of the forest highway program of the 1930s. Paved highways such as Highway 50 connected California to the lake's south shore, and state routes 89 and 28 opened the western, northern, and eastern shores as well. In 1931 Nevada relegalized gambling, and by the mid-1930s facilities on the Nevada side of the lake and in Reno began to generate more tourist and automobile travel. As a result, tourist activities changed. The railroad, lake steamer, and luxury hotel pattern of the pre-1900 days rapidly declined, with car camping, use of cabins and auto courts, cafes, and service stations replacing it. In essence, a new class of tourists came to dominate the lake (Strong 1984; Supernowicz 1983; Lindstrom 1995).

The addition of winter sports activities in conjunction with better roads and autos caused increased winter use of Lake Tahoe. In the 1920s some winter sports activities were promoted at Tahoe Tavern, including ski races, bobsled races, and sleigh rides. In 1930 a national ski jumping competition was staged. Shortly after that, a small T-bar lift was built at Spooner Summit. Large-scale winter sport development would not occur, however, until after 1945 (Strong 1984).

Other Recreation Development. Automobiles also opened areas of the Sierra that had not drawn tourist attention at earlier times. Development of county, state, and federal roads, including Highways 4, 20, 28, 40, 49, 50, 89, 108, 120, 180, and

395, opened many parts of the Sierra to Californians and other tourists. Improvement of roads within and connecting to national park areas also facilitated this process (Supernowicz 1983; Lux 1995; Lindstrom 1995; Markley and Meisenbach 1995; Tweed 1995). More camping brought new concerns to federal agencies, especially the fires caused by campers (Bigelow n.d.; Show and Kotok 1924).

Automobile access to the Sierra between 1900 and 1940 changed the range, just as other human activities had in past periods. The Tahoe, Eldorado, Stanislaus, and Sequoia National Forests offer representative examples. All saw increased use with the advent of automobile access. The forests increased the availability of improved campgrounds along major routes. Because of these campgrounds, programs to increase awareness of fire danger were instituted. Sportfishing and hunting activities were encouraged. Visitor use significantly increased in all of the national forests. During the Great Depression some consideration of camping time limits was considered because of what was deemed "squatting." In the northern Sierra Nevada, national forest land was developed for skiing and other winter sports activities, especially in the areas served by Highways 40 and 50. All of the forests encouraged the development of "summer home" sites based on terminable leases. These homesites were usually located near established highways (Markley and Meisenbach 1995), with streams or lakes acting as drawing factors (Conners 1993). Local materials were usually granted to the builders of these homes (Supernowicz 1983; Conners 1993). Homesites at Wilsonia and at Mineral King represented similar development in the southern part of the range (Dilsaver and Tweed 1990; Tweed 1995). No summary of the environmental impact of such housing exists.

Summary: Establishment of Agency Regulation

Between 1900 and 1940, the unregulated use of the Sierra Nevada came to an end. Development of the range continued, but under some form of regulation. Federal and state agencies such as the U.S. Forest Service, the National Park Service, the U.S. Reclamation Service, the California Division of Forestry, the state Natural Resources Agency, and several municipal water agencies imposed limits on the use of much of the Sierra Nevada's resources. While private land and resource development continued, notably in areas served by railroad logging operations, regulated use exerted a significant influence. Recreation emerged as a dominant force, largely because of automobiles and better roads. Water for urban, hydroelectric, and irrigation purposes was developed. The Sierra ecosystem continued to change in response to human actions, but the patterns were different from those of its opening phase.

1940–60: EXPANDING DEMAND AND AGENCY RESPONSE

The Second World War was a watershed event in California and the Sierra Nevada. After a decade of economic collapse, the 1939–45 years of war-driven economic growth put the nation's natural resources and labor power back to work. Demand for lumber alone from national forest lands rose by more than 200% compared to prewar levels, and the percentage of the nation's supply of lumber that came from Forest Service lands increased from 5% of the total to 10% (Hirt 1994). Continuing economic expansion from the end of the war to 1960 had effects on the Sierra Nevada that probably exceeded those of extraction during earlier periods (Strong 1984; Rice et al. 1988; Hundley 1992; Hirt 1994). The response of national, state, regional, and local agencies promoted growth and constitutes the third phase of this period of the range's history. Overall, there was an increase in resource demand, especially in the areas of timber and recreation. Agency actions supported attempts to offer more services and resources to meet these demands, in the belief that the Sierra Nevada was an inexhaustible resource as long as it was effectively managed. This optimistic viewpoint would be called into question in the 1960s (Hirt 1994).

Population Growth and Resource Demand, 1940–60

War-related economic expansion in the United States began in 1939. California benefited greatly from increases in defense production spending. Although California's population had grown steadily and at a faster rate than that of the nation as a whole until the 1920s, during the Depression growth declined. Following 1940, however, people relocated to the San Francisco Bay and Los Angeles areas, where war production boomed. Although some still predicted that a postwar slump would occur, the beginning of the cold war between the United States and the USSR led to continued war-related production and employment. Instead of a decline, growth in population again outpaced that of the nation as a whole (Thompson 1955; California 1970, 1979; Hart 1978; Rawls and Bean 1993).

This population growth had an immediate effect on the Sierra Nevada because of demands for timber and mineral resources. Ironically, the gold mining industry, considered by many to have been the foundation upon which the state was built, was dealt a death blow by a war-closure ruling. In order to control the use of resources such as steel and lumber, all hardrock mines were shut down. Just as federal action in the 1880s had effectively ended hydraulic mining, the mine closure order killed off most quartz mining. Although some of the largest hardrock gold mines would revive after the war, most closed by the 1950s, thus ending a significant pattern of Sierran land use. Mining of many resources, including gold, would still be a part of the range's economy and would continue to create environmental problems, but not at the levels of the past (Clark 1980; Palmer 1992).

The urbanization of California and economic growth produced increased demands for recreation facilities and use of natural places by a mobile and more affluent public. The national forests and national parks in California drew most of their visitors from urban areas within the state itself. The infrastructure that served tourism in the forests and parks did suffice in the immediate years after the war, but strains and increasing demands would exert pressure for expansion (Strong 1984; Runte 1990; Demars 1991; Tweed 1995; Lindstrom 1995).

Government agencies and private sector elements met demands for Sierra Nevada resources after the war by producing, in one word, more. Some uses, such as mining and grazing, were reduced, but other resources were exploited beyond previous levels.

U.S. Forest Service Response

In 1945, the Forest Service acknowledged that timber from the lands it managed was going to play a more important role in meeting the nation's needs than in the past. The service moved from a custodial role into a production mode (Hirt 1995). As an example, between 1902 and 1940, the total timber harvested on the Eldorado National Forest was 148.9 million board feet. From 1941 to 1945 it totaled 175.4 million board feet, reflecting wartime demand. Between 1946 and 1959, the harvest total stood at 728.9 million board feet (Supernowicz 1983), meaning that in thirteen years more than twice as much timber was harvested on the Eldorado Forest than in the preceding forty-three years.

The Forest Service was hard-pressed during World War II to meet the demand for timber while still practicing sustained-use forestry and trying to meet multiple-use ideals. After the war the service attempted to develop national regulatory standards. Private interests initially resisted, but eventually a rough sort of cooperation developed between the Forest Service and the larger private developers, who received access to increasingly valuable Forest Service sales. The Forest Service could not cut and process its own trees, but it could force private cutters to respect regulations established on public lands, because private companies needed the wood. As part of this accommodation, both public and private forestry in the postwar period moved toward "intensive timber management" practices to try to keep up with increasing public demands (Sedjo 1991; Hirt 1994).

During the 1950s, pressure from private and Congressional development interests, assisted by the Eisenhower administration, caused the Forest Service to increase its output of timber. The Republicans used the threat of reducing the amount of land administered by the service as leverage. Under Forest Director Richard McCardle, national forests allowed larger timber sales and constructed more roads. Many of the areas opened had previously been considered too remote or steep

to log. Watershed damage, erosion, and wildlife impacts resulted from this increased activity from public and private sources. Timber supplied by the national forests rose to almost one-third of the nation's supply by 1970 (White 1991). While the largest increase in logging on Forest Service lands occurred in the Pacific Northwest and portions of the Rockies, the Sierra Nevada was affected because of the market created by population growth in California. By this time demand for Sierra Nevada timber may already have exceeded timber growth (Johnston n.d.). Between 1940 and 1960, timber harvests in the state grew from 2 billion board feet to 6 billion board feet per year, with most of the rise coming after 1946 (Rice et al. 1988; Hirt 1994; Markley and Meisenbach 1995).

During the 1950s, concern about the rate at which forests were being cut and the negative effects on soil, watersheds, and wildlife resulted in resistance to the policies of the Forest Service. While the service had always stated a multiple-use approach to the lands in its control, timber had always dominated its activities. As an example of the mounting criticism, Willis Evans of the California Department of Fish and Game expressed concern to the Society of American Foresters in 1959 that the effects of high-yield production were destroying the West's forests for the sole purpose of timber production. He said that the public interest in its forests as a whole was not being served (Hirt 1994). Because of such growing public concern, Congress passed the Multiple Use–Sustained Yield Act in 1960. As a result of this policy, environmental activists began to demand changes from the Forest Service. Emphasis was placed on preservation of wilderness areas, protection of streams and watersheds, and preservation of wildlife habitat. These actions constituted the beginning of a new phase in the range's history (Strong 1988; White 1991; Hirt 1994).

Lake Tahoe and National Park Response

Lake Tahoe, Yosemite, and Sequoia/Kings Canyon, the three most important early monumental features of the Sierra, were all affected by postwar expansion. Tahoe had already been heavily logged, so its lands were not subjected to the same levels of cutting as other areas of the range following the war. Yosemite and Sequoia/Kings Canyon, long under federal resource controls, were similarly spared from such actions. They did, however, face increased recreation demands. Expansion in all tourist activities in these parks and in adjacent privately held facilities occurred following 1945 (Strong 1984; Runte 1990; Clawson and Harrington 1991; Tweed 1995).

The Lake Tahoe Area. After 1945, urbanization at Lake Tahoe occurred at a rate far exceeding that of past periods. Development would overwhelm all attempts to deal with the impact on Lake Tahoe as a whole and the adjoining areas. It would not be until the late 1950s that anyone would ask questions about problems being generated by mushrooming growth (Strong 1984).

In 1945 business and political leaders in the Tahoe area, concerned by its near desertion during the war, developed programs to draw tourists. Local booster agencies and governments staged events to draw more people. They also worked to improve transportation connections. The Placerville to Lake Tahoe Resort Owners Association exerted political pressure that led to the extension of an all-weather road to the south shore of Tahoe in 1947. With gambling in nearby Nevada, travel to the lake rose dramatically. Air travel to Tahoe between 1945 and 1949 also increased (Strong 1984; Lindstrom 1995).

Following 1955, growth accelerated along the Highway 50 corridor. Skiing resort developments such as Heavenly Valley and the opening of new gambling facilities at the lake and in Reno provided the draw. To supply services for tourists, permanent residency grew, leading to increasing pollution problems. Despite such problems, local governments promoted this transient and year-round population growth (Strong 1984). In 1960 the Olympic Winter Games were staged at Squaw Valley, dramatically increasing year-round use of the Sierra Nevada (Strong 1984). Facilitating the Olympic developments was the building of Interstate 80, which connected the area to Sacramento.

Yosemite. By 1945, park accommodations had seriously affected the Yosemite region. Wildlife had been controlled to meet visitor needs; for example, mountain lions were killed, as were aggressive bears in higher camps. Scenic and biological resources were strained, in the view of some contemporary observers, as the postwar period opened (Runte 1990).

In 1954, more than 1 million visitors came to Yosemite. By 1967 the 2 million visitor level was reached. Under pressure from concessionaires, the Park Service generally increased the number and variety of housing units, camping facilities, and different recreation activities. Some within the Park Service and the Sierra Club pushed to have all unnatural features removed from the valley floor, but to no avail (Runte 1990). One historian, speaking of the valley campgrounds in these years, described them as a rural slum (Demars 1991).

While most of the increasing use of Yosemite was concentrated in the valley, pressures also mounted to open the adjoining high country. Between 1954 and 1961, a struggle (between developers and Park Service personnel on one side and environmental interests such as the Sierra Club on the other) ensued over the construction of the Tioga Road. In the end, the road interests won, opening another area to easy tourist access (O'Neill 1984; Runte 1990; Demars 1991).

Sequoia and Kings Canyon National Parks. In comparison to the more northerly areas, Sequoia and Kings Canyon were less affected. Demand for timber had no direct effect, because logging had long been excluded. However, on adjacent land, heavy logging of all conifer species occurred, leaving the nationally protected parks as "biological islands," cut off from developed lands near them (Tweed 1995). Tourist interest led to increased use, as it did elsewhere in the range. Because the tourist and road facilities were generally overbuilt for the

needs of the 1930s, they absorbed growth in use to the 1950s. With increased use, however, a need developed during the following decade to replace and upgrade these facilities. A general National Park Service policy called "Mission 66" led to the upgrading of most facilities at Sequoia/Kings Canyon but did not increase development into new areas (Tweed 1995).

In general, park policies in the time between 1940 and 1960 continued to stress total fire suppression. Management goals stressed natural appearance and visitor safety. Transportation and urbanization remained generally at 1940 levels. Wildlife management policies were consistent with pre-1940 practices. It would be after 1960 that most of these policies would be challenged, leading to different management practices and perspectives (Tweed 1995).

Water Agency Responses

Population growth in California after the war was primarily an urban phenomenon. The Sierra Nevada supplied most of the water for the state's largest cities. San Francisco relied on Hetch-Hetchy, and Los Angeles obtained nearly 80% of its water from the Mono and Owens River drainages (Kahrl 1979). Expansion of populations in other urban areas, in the San Francisco Bay Area, in Sacramento, and in the Tahoe Basin (including Reno) would lead to further Sierran water development. In addition to being used to supply cities of California and Nevada, Sierra Nevada water was demanded by California corporate farmers for irrigation (Kahrl 1979; Strong 1984; Hundley 1992).

The Central Valley Project. In the 1930s the Bureau of Reclamation and the U.S. Army Corps of Engineers began construction of the Central Valley Project to regulate the Sacramento, American, Stanislaus, and San Joaquin Rivers and to provide water for contracted users. Most of this huge project would be completed after the war. By the 1950s, the Shasta, Keswick, Folsom, New Melones, and Friant Dams had been built, and a complex system of canals distributed more than 3 million acre-feet of water to state interests. These dams generated electricity and provided recreation for thousands of users (Hundley 1992).

The California Water Project. California lost control over development of several Sierran rivers when the federal government established the Central Valley Project. But many California agribusiness interests, angered by a federal 160-acre limit on subsidized water use, still longed for a less restrictive state water project. The recovery of the state's prosperity after the war led to political action to create a state-controlled system. The focus of water planners was on the Feather River. Flooding caused by the Feather River in the 1950s added to the justification for a dam in the eyes of many voters. Because of the high cost of the project, it was planned that not only would Central Valley farmers use the water, but much of it could be shipped south over the Tehachapis into the Los Angeles Basin. Governor Edmund Brown secured passage of legislation in 1959 to authorize funding. By 1962 Sierra Nevada water began flowing south, making it over the mountains in 1971 (Worster 1985; Hundley 1992).

Tahoe Basin Water. The need for water from the Sierra Nevada to meet urban growth demands in the Tahoe Basin and nearby Reno exceeded supply during the postwar boom. As early as the 1930s Lake Tahoe property owners and businesses at the lake were contesting with irrigators and power suppliers over Tahoe and Truckee water. The dispute was complicated because two states, several county governments on both sides of the state boundaries, and several federal agencies had conflicting jurisdictions. In 1934 a temporary solution, called the Truckee River Agreement, was cobbled together. It prohibited tunneling into Lake Tahoe or cutting its rim, as some Nevada interests had desired. Minimum and maximum lake levels were established. A reservoir was to be built by federal water agencies at Boca on the Little Truckee to store water for Nevada. After the war a second dam on Prosser Creek in the same drainage was constructed. Sierra Pacific Power and Light was allowed to build a small reservoir for power generation (Taylor 1975; Strong 1984).

Accelerated postwar growth soon made these earlier compromises unworkable. Local, county, state, and national interests began to work at cross-purposes. California and Nevada came to realize that they could not achieve any of their goals exclusively without costly court battles, something that neither really wanted. In 1955 they cooperated to create a California-Nevada Interstate Compact Commission. It took until 1963 for that group to create a report that apportioned water in the basin between the two states. It would not be until 1971 that both state legislatures ratified the agreement (Strong 1984).

The fundamental problem of growth in the basin was not addressed in any of these attempts to work toward a compromise solution. It was not just the amount of water that caused difficulty. Pollution in the Tahoe Basin, and thus eventually in the lake itself, was also an issue. Raw sewage was detected in the lake. Debris and nutrients created by runoff and development seriously affected water purity and clarity. Between 1945 and 1960, numerous studies by the city, county, and state governments were conducted. No action to address the problems would be taken until the 1970s, however (Strong 1984).

Recreation Demands

Transportation Development. The key to recreation development in the Sierra Nevada was always access. Trails, roads, rail connections, and eventually automobile connections had opened much of the Sierra to tourists by the end of the 1930s. After World War II, improved automobile transportation routes expanded tourist and recreation use. Interstate 80, in particular, changed Sierran use patterns. Instead of seasonal limits imposed by weather, the new freeway encouraged

heavier year-round use and permanent population growth. The construction of the Tioga Road increased access to Tuolumne Meadows and the Yosemite high country between 1950 and 1961. Connections from Los Angeles to ski areas such as Mammoth and June Lakes were supplied by an improved Highway 395. Two major exceptions to the response of agencies for expanded highway connections involved the projected Emerald Lake Highway at Tahoe and a trans-Sierra highway in the Banner-Ritter area. Protests from a growing number of environmentalists stopped all serious consideration of these roads (O' Neill 1984; Strong 1984; Rice et al. 1988; Runte 1990).

The Development of Skiing. The most significant expansion of recreation in the Sierra Nevada following the Second World War came from skiing. While air transport facilitated some of this expansion, notably at Tahoe-Donner, it was automobile travel that turned it from a marginal recreational use to a major industry. The earliest skiing before 1940 had been in Yosemite near Badger Pass, at the Twin Bridges area along Highway 50, or along Highway 40, especially at Cisco Grove, Norden, and Sugar Bowl. These areas would continue to develop after the war. They were joined by new areas because the Forest Service cooperated with private developers (Farquhar 1965; Fairclough 1971; Strong 1984).

Ski developments new to the Sierra or expanded after 1945 included the Mammoth and June Lakes region, Donner Summit resorts such as Sugar Bowl and Squaw Valley (stimulated by the 1960 Olympics), and Tahoe resorts such as Alpine Meadows and Heavenly Valley (Farquhar 1965; Fairclough 1971; Strong 1984).

Challenges to Growth-Oriented Policies, 1950–60

Population and economic growth after the Second World War in the nation, California, and the Sierra Nevada was supported by most of the public and all agencies of government. A nation and people scarred by a ten-year-long depression but put back to work in the war and the decade of economic growth that followed showed little concern for the effects of this growth. Access to national parks was convenient, and tourists were welcomed. Forest resources, including lumber, pasture, camping, hunting, and fishing, were all expanded to meet public demand (Strong 1984; Runte 1990; Demars 1991; Hirt 1994).

Not all who utilized the Sierra were happy with the effects of postwar growth. Groups with conflicting goals such as grazers and timber users continued to have conflicts. Conservation groups, which included the Sierra Club, the Wilderness Society, the National Wildlife Federation, and the Audubon Society, were troubled by the effects of increased postwar development. While divided on the particulars, conservation advocates were troubled by the emphasis that forest officials placed on timber production. Sierra Nevada environmental activists developed new ideas concerning wilderness and integrated wildland approaches. These were expressed most effectively by David Brower of the Sierra Club, who called for reduced demand and less impactive use on Sierra Nevada resources. Taken together, these perspectives meant a different type of environmental movement had begun (Strong 1988; Runte 1990; Hirt 1994).

Agency response to these different demands was mixed. Concern about wildlife was generally not expressed by forest and park officials. Damage to riparian areas did draw complaints from within fish and wildlife services, but their budgets did not allow them to compete strongly with silviculture. Recreation did receive more attention, but disproportionate funding for timber production continued. Some portions of national forest land, mainly "rock and ice" or higher-elevation areas, were set aside before 1940 with a primitive area designation. In the Sierra, the only example was the Desolation Valley area, which lay alongside Highway 50. No further designation of what some in the Forest Service called "wilderness reserves" occurred beyond 1940 until the passage of the Wilderness Act of 1964 (Steen 1976; O'Neill 1984; Palmer 1988; Runte 1990; Hirt 1994; Tweed 1995).

The decade of the 1950s represented a watershed in Sierra Nevada environmental history. Between 1890 and 1950, cooperation between federal agencies and environmental interests had sometimes been strained. But conservation groups compromised or made peace with these federal agencies because more access, fish and game improvements, and scenic preservation were shared values. Many conservation groups believed that not cooperating would bring more harm to the range from extractive industries such as logging and grazing. In the 1950s a new conservation movement began to address broader Sierran and national issues such as wilderness designation and wildlife protection. Their actions led to challenges to federal land-use policies, even if that meant breaking with their allies in park and forest agencies (Strong 1984; Runte 1990; Hirt 1994).

Several key issues illustrate this growing conflict. First was the Forest Service's use in the 1950s of what were defined as "salvage sales," which critics charged was a designation improperly used to justify increased timber cutting in some areas formerly designated as "primitive." Although these redesignated timber harvests were not located in the Sierra Nevada, the Sierra Club began to organize resistance to such actions because of their potential effects. Another area of concern involved the inability of federal agencies to control growth in the Tahoe Basin. Multiple jurisdictions and interests meant that growth and pollution of the lake continued without effective countermeasures. A third issue that drew widespread concern was the Tioga Road expansion within Yosemite National Park. Regardless of environmentalist complaints about the effects of the reconfiguration and the widening of this road, the National Park Service continued construction (Strong 1984; O'Neill 1984; Runte 1990; Hirt 1994).

By the late 1950s numerous conservation groups had

mounted a challenge that questioned the ability of the national forests and national parks to meet more than narrowly prescribed growth-oriented uses. The dominant agency concerns of timber production and tourist access were challenged. Wildlife policies were criticized as being oriented only toward predator control and managing animals such as deer and bears. Vegetation control was seen as focusing only on the reduction of vegetation considered harmful to grazing animals. Various political campaigns and internal protests during this period demanded that public agencies consider broader environmental health concerns (Runte 1990; Hirt 1994).

Because of this rising protest, in 1960 the Multiple Use–Sustained Yield Act was passed. For the Forest Service the law confirmed long-standing commitments dating back to the days of Gifford Pinchot to have the forests meet a broad range of uses. No interest was to have special priority (Markley and Meisenbach 1995). In reality, MUSY, as some called the act, left intensive timber harvest policies basically unchallenged. However, the law marked the opening of a new era of resistance from environmental interests. Notably absent from the groups that endorsed the law was the Sierra Club. Its rejection was based on what it described as ambiguities in the act, the prevailing timber and water priorities of the Forest Service, and an unwillingness of the service to protect and develop wilderness areas. Unquestioned cooperation between organized environmental groups such as the Sierra Club and the Forest Service had come to an end (Hirt 1994).

Summary: Agency Response to Growth, 1940–60

The unprecedented growth of the California and U.S. economies after 1940 had a tremendous impact on the Sierra. Demand for lumber, water, rangeland, and recreation access exceeded that of any other period in the range's past environmental history. All of the major resource agencies made efforts to satisfy the needs of the public. Undoubtedly, the greatest area of increased use was in timber production. Across the West, including in the forests of the Sierra, the U.S. Forest Service kept increasing production of timber. Clear-cutting became the dominant form of logging. Criticism from conservation and wildlife groups mounted. By the end of the decade a newly energized environmental movement challenged the growth-oriented policies of public agencies. In response, the agencies began to question their former policies. A new era in Sierra Nevada environmental history had begun.

REFERENCES

Albright, H. M., and R. Cahn. 1985. *The birth of the National Park Service.* Salt Lake City: Institute of the American West.

Anderson, K. 1993. *Indian fire-based management in the sequoia–mixed conifer forests of the central and southern Sierra Nevada.* Yosemite National Park: Yosemite Research Center.

Anderson, M. K., and M. J. Moratto. 1996. Native American land-use practices and ecological impacts. In *Sierra Nevada Ecosystem Project: Final report to Congress,* vol. II, chap. 9. Davis: University of California, Centers for Water and Wildland Resources.

Ayres, R. W. 1958. *History of timber management in the California national forests.* Washington, DC: U.S. Forest Service.

Baumhoff, M. A. 1978. Environmental background. In *California,* edited by R. F. Heizer, 16–24. Vol. 8 of *Handbook of North American Indians,* edited by W. C. Sturtevant. Washington, DC: Smithsonian Institution.

Beesley, D. 1979. Cornish pump. *Nevada County Historical Society Bulletin* 33 (2): 7–16.

———. 1984. Whistle punks and steam donkeys: Logging in Nevada County and the northern Sierra during the age of animal and steam power. *Nevada County Historical Society Bulletin* 38 (4): 25–30.

———. 1985. Changing land use patterns and sheep transhumance in the northeastern Sierra Nevada, 1870–1980. In *Forum for the Association of Arid Lands studies,* edited by O. Templar, 3–8. Vol. 1. Lubbock, TX: Texas Tech University, International Center for Arid and Semi-Arid Land Use Study.

———. 1988. From Chinese to Chinese American: Chinese women and families in a Sierra Nevada county. *California History* 67 (3): 168–79.

———. 1994. Opening the Sierra Nevada and the beginnings of conservation. Unpublished manuscript. Tahoe National Forest, Nevada City, California.

Best, G. M. 1965. *Nevada County Narrow Gauge.* Berkeley, CA: Howell-North Books.

Bigelow, R. N.d. History of Forest Supervisor Richard L. P. Bigelow, 1902–1936. Unpublished manuscript. Tahoe National Forest, Nevada City, California.

Blackburn, T. C., and K. Anderson. 1993. Introduction: Managing the domesticated environment. In *Before the wilderness: Environmental management by Native Californians,* compiled and edited by T. C. Blackburn and K. Anderson, 15–25. Menlo Park, CA: Ballena Press.

Brewer, W. H. [1930] 1966. *Up and down California, in 1860–1864.* Edited by F. P. Farquhar. Reprint, Berkeley and Los Angeles: University of California Press.

Brown, M. R., III, and C. M. Elling. 1981. *An historical overview of redwood logging in the Hume Ranger District, Sequoia National Forest.* Porterville, CA: U.S. Forest Service.

California, State of. 1970. *Statistical abstracts.* Sacramento: State of California.

———. 1979. *Statistical abstracts.* Sacramento: State of California.

Castillo, E. D. 1978. The impact of Euro-American exploration and settlement. In *California,* edited by R. F. Heizer, 99–127. Vol. 8 of *Handbook of North American Indians,* edited by W. C. Sturtevant. Washington, DC: Smithsonian Institution.

Cermak, R. W. 1988. Fire in the forest: Fire control in the California national forests, 1898–1955. Unpublished manuscript. Tahoe National Forest, Nevada City, California.

Clar, C. R. 1959. *California government and forestry.* Sacramento: Department of Natural Resources, Division of Forestry.

Clark, W. B. 1980. *Gold districts of California.* Bulletin 193. Sacramento: California Division of Mines and Geology.

Clawson, M., and W. Harrington. 1991. The growing role of outdoor recreation. In *America's renewable resources,* edited by K. D. Frederick and R. A. Sedjo, 249–83. Washington, DC: Resources for the Future.

Claytor, M. P., and D. Beesley. 1979. Aspen art and the sheep industry of Nevada and adjoining counties. *Nevada County Historical Society Bulletin* 33 (4): 25–30.

Conners, P. A. 1990. West Side Lumber Company contextual history. Unpublished manuscript. Stanislaus National Forest, California.

———. 1992. Influence of the Forest Service on water development patterns in the West. In *The origins of the national forests: A centennial symposium,* edited by H. K. Steen, 154–69. Durham, NC: Forest History Society.

———. 1993. Historical overview of recreational residences on the Stanislaus National Forest. Unpublished manuscript. Stanislaus National Forest, California.

———. 1995. Interpretation of Deferrari north district ecosystem analysis: First installment. Unpublished manuscript. Stanislaus National Forest, California.

Cook, S. F. 1978. Historical demography. In *California,* edited by R. F. Heizer, 91–98. Vol. 8 of *Handbook of North American Indians,* edited by W. C. Sturtevant. Washington, DC: Smithsonian Institution.

Dana, S. T., and M. Krueger. 1958. *California lands.* Washington, DC: American Forestry Association.

D'Azevedo, W. L., ed. 1986. *Great Basin.* Vol. 11 of *Handbook of North American Indians,* edited by W. C. Sturtevant. Washington, DC: Smithsonian Institution.

Deferrari, C. 1995. North district ecosystem analysis: First installment. Unpublished manuscript. Stanislaus National Forest, California.

Demars, S. E. 1991. *The tourist in Yosemite, 1855–1985.* Salt Lake City: University of Utah Press.

Dilsaver, L. 1985. After the gold rush. *Geographical Review* 75 (1): 67–88.

Dilsaver, L., and W. Tweed. 1990. *Challenge of the Big Trees.* Three Rivers, CA: Sequoia National Park History Association.

Douglass, W. A., and J. Bilbao. 1975. *Amerikanuak: Basques in the New World.* Reno: University of Nevada Press.

Edwards, W. F. 1883. *Tourists' guide and directory of the Truckee Basin.* Compiled and edited by C. D. Irons. Truckee, CA: "Republican" Job Print.

Fairclough, D. R. 1971. An administrative history of Squaw Valley, 1949–1971. Master's thesis, California State College.

Farquhar, F. P. 1965. *History of the Sierra Nevada.* Berkeley and Los Angeles: University of California Press.

Fox, S. 1981. *John Muir and his legacy: The American conservation movement.* Boston: Little, Brown.

Frederick, K. D. 1991. Water resources: Increasing demand and scarce supply. In *America's renewable resources,* edited by K. D. Frederick and R. A. Sedjo, 23–80. Washington, DC: Resources for the Future.

Gibbens, R. P., and H. F. Heady. 1964. *The influence of modern man on the vegetation of Yosemite Valley.* Berkeley: University of California, Division of Agricultural Sciences.

Goetzmann, W. H. 1959. *Army exploration of the American West, 1803–1863.* Lincoln: University of Nebraska Press.

Gudde, E. G. 1975. *California gold camps.* Edited by E. Gudde. Berkeley and Los Angeles: University of California Press.

Hart, J. D. 1978. *A companion to California.* New York: Oxford University Press.

Heizer, R. F., ed. 1978. *California.* Vol. 8 of *Handbook of North American Indians,* edited by W. C. Sturtevant. Washington, DC: Smithsonian Institution.

Hinkle, G., and B. Hinkle. 1949. *Sierra Nevada lakes*. Edited by M. Quaif. Indianapolis: Bobbs-Merrill.

Hirt, P. W. 1994. *A Conspiracy of optimism: Management of national forests since World War Two*. Lincoln: University of Nebraska Press.

Hundley, N., Jr. 1992. *The great thirst: Californians and water, 1770s–1900s*. Berkeley and Los Angeles: University of California Press.

Hurtado, A. L. 1988. *Indian survival on the California frontier*. New Haven, CT: Yale University Press.

Jackson, J. H. 1970. *Anybody's gold*. San Francisco: Chronicle Books.

Jackson, W. T., R. Herbert, and S. Wee. 1982. *History of Tahoe National Forest*. Nevada City, CA: U.S. Forest Service, Tahoe National Forest.

James, G. W. 1915. *The lake of the sky*. New York: J. F. Tapley.

Johnston, J. D. N.d. The effect of humans on the Sierra Nevada mixed conifer forests. Unpublished manuscript. Lassen National Forest, California.

Jones, H. R. 1965. *John Muir and the Sierra Club: The battle for Yosemite*. San Francisco: Sierra Club.

Kahrl, W. L., project dir. and ed. 1978. *The California water atlas*. Sacramento: State of California.

Kelly, R. L. 1959. *Gold vs. grain: The hydraulic mining controversy in California, Central Valley*. Glendale, CA: Arthur C. Clark Co.

Knowles, C. D. 1942. A history of lumbering in the Truckee Basin from 1856 to 1936. Unpublished manuscript. Forestry Library, University of California, Berkeley.

Larson, R. C. 1985. Giants of the southern Sierra: A brief history of the Sequoia National Forest. Edited by S. L. Forsberg. Unpublished manuscript. Sequoia National Forest, California.

Laudenslayer, W. F., and H. H. Darr. 1990. Historical effects of logging on the forests of the Cascade and Sierra Nevada Ranges of California. *Transactions of the Western Section of the Wildlife Society* 26:12–23.

Laudenslayer, W. F., H. H. Darr, and S. Smith. 1989. Historical effects of forest management practices on eastside pine communities in northeastern California. Paper presented at Multi-Resource Management of Ponderosa Pine Forests Conference.

LeConte, J. [1875] 1930. *A journal of ramblings through the High Sierra by the University Excursion Party*. Reprint, San Francisco: Sierra Club.

Leiberg, J. B. 1902. *Forest conditions in the northern Sierra Nevada, California*. Washington, DC: Government Printing Office.

Leonard, Z. [1833] 1978. *Adventures of a mountain man: The narratives of Zenas Leonard*. Reprint, with an introduction by M. Quaife. Lincoln: University of Nebraska Press.

Lindstrom, S. 1995. Spatial patterns of Sierra Nevada landscape change, 1820–1960—Lake Tahoe Basin. Unpublished manuscript.

Lux, L. 1995. A history of the human element in the Sierran province. Appendix G of *Draft environmental impact statement: Managing California spotted owl habitat in the Sierra Nevada national forests of California*. San Francisco: U.S. Forest Service.

Malouf, C. I., and J. Findlay. 1986. Euro-American impact. In *Great Basin*, edited by W. L. D'Azevedo, 499–516. Vol. 11 of *Handbook of North American Indians*, edited by W. C. Sturtevant. Washington, DC: Smithsonian Institution.

Mann, R. 1982. *After the gold rush*. Stanford, CA: Stanford University Press.

Markley, R., and C. Meisenbach. 1995. Historical summary: Tahoe National Forest environmental history. Unpublished manuscript. Tahoe National Forest, Nevada City, California.

McGlashan, M. N. 1982. Heritage: Early dairying. *Sierra Heritage* 2 (2): 13–17.

McKelvey, K. S., and J. D. Johnston. 1992. Historical perspective on the forests of the Sierra Nevada and the Transverse Ranges of southern California: Forest conditions at the turn of the century. In *The California spotted owl: A technical assessment of its current status*, technical coordination by J. Verner, K. S. McKelvey, B. R. Noon, R. J. Gutierrez, and T. W. Beck, 225–46. Albany, CA: U.S. Forest Service, Pacific Southwest Research Station.

Meals, H. 1995. Mercury use and gold mining: Yuba River watershed. Unpublished manuscript. Tahoe National Forest, Nevada City, California.

Meschery, J. 1978. *Truckee*. Truckee, CA: Rocking Stone Press.

Muir, J. 1894. *The mountains of California*. Garden City, NJ: Doubleday.

———. 1938. *John of the mountains: The unpublished journals of John Muir*. Edited by L. M. Wolfe. Madison: University of Wisconsin Press.

Myrick, D. F. 1962. *Railroads of Nevada and eastern California*. Vol. 1. Berkeley, CA: Howell-North.

Nash, R. 1982. *Wilderness and the American mind*. New Haven, CT: Yale University Press.

O'Neill, E. S. 1984. *Meadow in the sky: A history of Yosemite's Tuolumne Meadows region*. Groveland, CA: Albicaulis Press.

Palmer, K. 1992. Vulcan's footprint on the forest: The mining industry and California's national forests. In *Origins of the national forests*, edited by H. K. Steen, 136–53. Durham, NC: Forest History Society.

Palmer, T. 1988. *The Sierra Nevada: A mountain journey*. Washington, DC: Island Press.

Paul, R. 1947. *California gold: The beginnings of mining in the Far West*. Lincoln: University of Nebraska Press.

Pisani, D. 1977a. Lost parkland: Lumbering and park proposals in the Tahoe-Truckee basin. *Journal of Forest History* 21:4–17.

———. 1977b. The polluted Truckee: A study in interstate water quality, 1870–1934. *Nevada Historical Society Quarterly* 20 (3): 151–66.

Pyne, S. J. 1982. *Fire in America: A cultural history of wildland and rural fire*. Princeton, NJ: Princeton University Press.

Rawls, J. J., and W. Bean. 1993. *California: An interpretive history*. New York: McGraw-Hill.

Reisner, M. 1986. *Cadillac desert: The American West and its disappearing water*. New York: Viking.

Rice, R. B., W. A. Bullough, and R. J. Orsi. 1988. *The elusive Eden: A new history of California*. New York: Alfred A. Knopf.

Rowley, W. D. 1985. *U.S. Forest Service grazing and rangelands*. College Station: Texas A and M University Press.

Runte, A. 1987. *National parks: The American experience*. Lincoln: University of Nebraska Press.

———. 1990. *Yosemite: The embattled wilderness*. Lincoln: University of Nebraska Press.

Russell, C. 1947. *One hundred years in Yosemite*. Berkeley and Los Angeles: University of California Press.

Schmidt, P. J. 1990. *Back to nature: The Arcadian myth in urban America*. Baltimore, MD: Johns Hopkins University Press.

Sedjo, R. A. 1991. Forest resources: Resilient and serviceable. In *America's renewable resources*, edited by K. D. Frederick and R. A. Sedjo, 81–122. Washington, DC: Resources for the Future.

Show, S. B., and E. I. Kotok. 1923. *Forest fires in California: 1911–1920*. Washington, DC: U.S. Department of Agriculture.

———. 1924. *The role of fire in the California pine forests*. Washington, DC: Government Printing Office.

Sierra Nevada Ecosystem Project (SNEP). 1994. *SNEP update*. June.

Sinnott, J. J. 1976. *Sierra valley: Jewel of the Sierras*. Vol. 4 of *History of Sierra County*. Pioneer, CA: California Traveler.

Smith, J. S. [1827] 1977. *The southwest expedition of Jedediah S. Smith*. Reprint, edited by and with an introduction by G. R. Brooks. Lincoln: University of Nebraska Press.

Stanford, E. R. 1924. A short history of California lumbering. Master's thesis, University of California, Berkeley.

Steen, H. K. 1976. *The U.S. Forest Service: A history*. Seattle: University of Washington Press.

Stewart, G. R. 1962. *The California trail*. New York: McGraw-Hill.

Storer, T. I., and L. P. Tevis. 1978. *California grizzly*. Lincoln: University of Nebraska Press.

Strong, D. H. 1984. *Tahoe: An environmental history*. Lincoln: University of Nebraska Press.

———. 1988. *Dreamers and defenders: American conservationists*. Lincoln: University of Nebraska Press.

Sudworth, G. B. 1900. Stanislaus and Lake Tahoe Forest Reserves, California, and adjacent territory. In *Twenty-first annual report of the U.S.G.S.* Part V, *Forest reserves*. Washington, DC: Government Printing Office.

Supernowicz, D. 1983. Historical overview of the Eldorado National Forest. Master's thesis, California State University, Sacramento.

———. 1992. Draft report on mining in region five: Overview. Unpublished manuscript. Eldorado National Forest, California.

Taylor, M. 1975. Boca: A history. Unpublished manuscript. Tahoe National Forest, Nevada City, California.

Thompson, W. S. 1955. *Growth and changes in California's population*. Los Angeles: Haynes Foundation.

Tweed, W. 1995. Summary for Sequoia and Kings Canyon National Parks: S.N.E.P. Unpublished manuscript. Sequoia and Kings Canyon National Parks, California.

Wagoner, L. 1886. Report on the forests of the counties of Amador, Calaveras, Tuolumne, and Mariposa. In *First biennial report of the California State Board of Forestry for the years 1885–1886*, 39–44. Sacramento: State Office.

White, C. L. 1928. Surmounting the Sierras: The campaign for a wagon road. *Quarterly of the California Historical Society* 7 (1): 3–19.

White, R. 1991. *"It's your misfortune and none of my own": A history of the American West*. Norman: University of Oklahoma Press.

Wilson, N. L. 1995. A chronology and notes on the European discovery, exploration, and settlement in the Sierra Nevada region, 1542–1848. Unpublished manuscript.

Worster, D. 1985. *Rivers of empire: Water, aridity, and the growth of the American West*. New York: Oxford University Press.

Ziebarth, M. 1984. California's first environmental battle. *California History* 63 (4): 274–79.

SCOTT STINE
Department of Geography and
  Environmental Studies
California State University
Hayward, California

2

Climate, 1650–1850

ABSTRACT

Climate exerts a profound influence on landscape by determining the flux of both energy (solar radiation) and mass (rain, snow, and water vapor). If climate changes significantly, the landscape can be expected to respond geomorphologically, hydrologically, and biologically. These individual responses, in turn, can feed on one another, creating a cascade of landscape perturbations.

Around 1850, just as large numbers of Europeans descended on the Sierra Nevada for the first time, the region experienced a marked shift in climate, from the abnormally cool and moderately dry conditions of the previous two centuries (the "Little Ice Age"), to the relatively warm and wet conditions that have characterized the past 145 years. This climatic shift should concern land managers for two interrelated reasons: First, the landscape changes that have occurred since 1850 may not be entirely anthropogenic but rather attributable in part to the shift in climate. Second, the landscape of the immediate pre–gold rush period should not be considered an exact model for what the Sierra would be today had Europeans never colonized the region. Thus, attempts to restore "natural conditions" as part of an overall Sierra Nevada management plan should focus not on the pre-European landscape but rather on the landscape that would have evolved during the past century and a half in the absence of Europeans.

Using proxy climatic records, this chapter explores the Sierra Nevada climate of the period 1650–1850 and compares it to that of the modern (post-1850) period. The focus is on climate at the decade to century scale, rather than on individual years or meteorological events. Emphasis is placed on records from lakes, glaciers, tree lines, and tree rings that can be resolved to time scales of multiple decades or less. Other types of proxy indicators, such as pollen and pack-rat records, while indispensable for illuminating multiple-century to millennial changes in climate, are not included in this analysis.

CLIMATE GENERALIZATIONS

Climate is not a landscape component as much as a landscape determinant. It exerts an overriding influence on such landscape components as vegetation (including its type, biomass, and distribution), hydrology (including the size, distribution, fluctuations, and water quality of lakes and rivers), soils (including their thickness, stability, and nutrient capacity), and landforms (including their rates of formation and loss). It also strongly influences other landscape determinants, the most important of which may be fire (including its location, frequency, and intensity).

Climate is inherently changeable, whether by the decade, century, or millennium. It is inherently variable, with some periods characterized by frequent and/or wide departures from the average and others by infrequent and/or narrow departures. And it is inherently site-specific, differing even over small areas depending on such variables as topography, slope orientation, vegetation cover, and elevation. In a range as high, extensive in area, and complex in topography as the Sierra Nevada, the variety of local climates is too extensive to enumerate here. The following generalizations can be drawn, however:

- Temperatures generally decline with increasing elevation in the range, though drainage of cold air into valley bottoms can provide exceptions to this rule.

- Precipitation generally increases with increasing elevation in the range, though wind can keep high-elevation areas swept of snow.

- Winds generally strengthen with increasing elevation in the range.

- Snowfall composes a greater percentage of total precipitation at higher elevations in the range.

Sierra Nevada Ecosystem Project: Final report to Congress, vol. II, *Assessments and scientific basis for management options*. Davis: University of California, Centers for Water and Wildland Resources, 1996.

- At a given elevation, the western slope of the range receives greater amounts of precipitation than the eastern slope.

- At a given elevation below 3,000 m (10,000 ft), precipitation generally decreases in the southerly direction through the range.

- At a given elevation below 3,000 m (10,000 ft), temperature generally increases in the southerly direction through the range.

- At a given elevation above 1,000 m (4,000 ft), temperatures on the eastern side of the range are generally higher in the summer and lower in the winter than on the western side of the range.

CLIMATE IN THE SIERRA, 1650–1850

Climate varies not only spatially but also temporally, with some periods being relatively wet and others relatively dry, some relatively cool and others warm. Putting aside for now the year-to-year and decade-to-decade variations in climate, it is possible to characterize the period from the mid-1600s to the mid-1800s as having been, by modern standards, abnormally cool and moderately dry. This interval was preceded by several centuries of cool and wet conditions and was followed by the relatively warm and wet conditions of the past 145 years. Evidence for these generalizations comes from several sources.

Evidence from Hydrographically Closed Lakes

Under natural conditions, runoff from the eastern Sierra Nevada terminates in lakes that lack outlets. These hydrographically closed lakes (including Pyramid, Mono, and Owens) lose water only to evaporation. They thus fluctuate widely through time, rising when inflow exceeds evaporative loss and falling when the relationship is reversed. By reconstructing the past fluctuations of such lakes, it is possible to determine if a particular period was relatively wet and/or cool (i.e., "effectively wet," represented by rising lake levels and a high stand) or relatively dry and/or warm (i.e., "effectively dry," represented by declining levels and a low stand). The closed lakes of the eastern Sierra all show evidence of having been relatively high around the years 1550–1650, reflecting effectively wet conditions; of having fallen to relatively low levels between 1650 and 1850, reflecting effectively dry conditions; and of having risen to relatively high levels during "modern" (post-1850) times, reflecting a return to effectively wet conditions.

Pyramid Lake and Winnemucca Slough

Pyramid Lake, the northernmost of the hydrographically closed lakes that are fed by Sierra Nevada runoff, receives the bulk of its inflow from the Truckee River watershed. When the surface of Pyramid Lake exceeds an elevation of 1,177.4 m (3,863 ft), it overflows into Winnemucca Slough, greatly expanding the lake surface area.

Since around 1860, water has been diverted from the Truckee River for irrigation, depriving the Pyramid-Winnemucca system of a large portion of its natural inflow. As a result of these diversions, Pyramid Lake presently stands more than 20 vertical meters (65 ft) below its point of overflow. In the absence of diversions, Pyramid Lake would have spent all but a few exceptionally dry years (likely 1924–34, 1948–51, 1976–77, and 1987–92) of the modern period at the elevation of the spillway (L. Benson, telephone conversation with the author, May 1995). Winnemuca Slough would have been maintained as a lake throughout this period.

During the decades preceding 1850, prior to any significant alteration of the hydrography by humans, Pyramid Lake dropped to a level below its spillway, resulting in the desiccation of Winnemucca Slough. Hardman and Venstrom (1941) and Harding (1965) interpret Frémont's account from 1844 as indicating that in the mid-1840s the Pyramid Lake surface stood several feet below the spillway. Relict tree stumps with as many as 20 growth rings are rooted on the Truckee River delta at a yet lower elevation (about 3.35 m [11 ft] below the sill, according to Harding). For these trees to have grown, Pyramid Lake must have spent at least two decades below the elevation of 1.174 m (3,852 ft). Radiocarbon assays of these stumps (Stine, unpublished data) indicate that they date to sometime after 1750. Because the stumps are known to have existed prior to 1862 (Russell 1885b; Harding 1965), they must have become established during the late eighteenth or early nineteenth century. These lines of evidence indicate that, during the decades prior to the 1850s, Pyramid Lake stood more than 3.35 vertical meters (11 ft) below the level that it would have occupied during the past 145 years in the absence of diversions.

With Pyramid Lake more than 3.35 m below its sill, and Winnemucca Slough desiccated, the surface area of the Pyramid-Winnemucca system, and thus evaporative loss from the system, is reduced by roughly a third. The low stand that characterized Pyramid Lake prior to 1850 thus reflects a substantial diminution of effective inflow—to perhaps 66% or less of the modern natural value.

Mono Lake

Mono Lake, on the lee side of the central Sierra Nevada, receives its inflow from the Rush, Lee Vining, and Mill Creek drainages. Since 1941, much of this inflow has been diverted by the Los Angeles Department of Water and Power for domestic supply, forcing the lake to low levels. But for these diversions, the lake surface during the past century would have fluctuated within a narrow elevation interval (± 2 m [6 ft]) centered on about 1,958 m (6,423 ft) (P. Vorster, telephone conversation with the author, May 1995). This elevation is hereafter referred to as the "natural level" of the modern period.

Around 1650, Mono Lake attained a high stand at 1,967.8 m (6,456 ft), 10 m (30 ft) higher than the calculated natural level of the modern period (Stine 1987, 1990). Radiometrically dated evidence for this high stand (and thus evidence for effectively wet conditions) is seen today in the form of rooted stumps, sedimentary sequences, and geomorphic stand lines. These same lines of evidence demonstrate that between 1650 and about 1840 effectively drier conditions drove Mono Lake to 1,948.9 m (6,394 ft)—9 vertical meters (29 ft) below the calculated "natural level" of the modern period. This low level, corresponding to a surface area approximately 79% that of the modern natural value, indicates that the effective inflow to Mono Lake prior to 1850 was, on average, less than 79% of the effective inflow of the period 1937–79 (Stine 1987, 1990).

Owens Lake

Owens Lake is the natural sink for all eastern Sierra Nevada runoff south of the Mono Basin. Diversion of the Owens River (Owens Lake's main feeder stream), first by irrigation interests in the Owens Valley and subsequently by the Los Angeles Department of Water and Power, has desiccated the lake, exposing the entire playa floor. Were it not for these diversions, Owens Lake since the early 1860s would have stood within a narrow elevation interval (±1 m [3 ft]) centered on about 1,095.5 m (3,594 ft). Such a lake would cover the now-dry lake floor with up to 15.8 m (52 ft) of water (Stine 1994).

Sequences of lake-transgressive and lake-regressive deltaic sediments exposed in the walls of the stream cuts adjacent to Owens Lake provide evidence that around the half-century 1600–1650 effectively wet conditions drove the shoreline to an elevation of 1,098.8 m (3,605 ft), creating a lake with a surface area approximately 110% that of the modern natural value. Effectively dry conditions prevailed during much of the ensuing two centuries, forcing the lake to elevations below 1,088 m (3,570 ft)—more than 7 m (24 ft) lower than the natural level of the modern period. This low stand represents a surface area approximately 77% that of the natural modern value, suggesting that effective inflow to the lake was more than 23% below the natural modern value (Stine 1994; Stine, unpublished data).

In summary, the closed lakes of the eastern Sierra Nevada are consistent in being dominated by declining levels and low stands during the period from the mid-1600s to the mid-1800s. In those two centuries the lakes attained elevations lower than those of the prior century (1550–1650) and lower than the "natural levels" of the twentieth century.

Evidence from Sierra Nevada Glaciers

Following thousands of years of little or no glaciation, high-elevation cirques of the Sierra Nevada experienced ice accu-mulation for several centuries prior to 1850 (Clark and Gillespie 1995; Curry 1969). This period of minor glacier advance (typically less than 2 km), first described in the Sierra by Matthes (1939), corresponds to the "Little Ice Age"—a period of cooling over much of the globe that began in the fourteenth or fifteenth century and continued through the middle of the nineteenth century (Grove 1988).

Based on the well-documented behavior of glaciers in the European Alps, Matthes (1939), speculated that these small ice bodies of the Sierra Nevada reached their maximum extent during the period 1850–55. Maps and photos produced by Russell (1885a, 1889) show that by the early 1880s the Lyell, McClure, and Dana Glaciers had begun to retreat from their maximum positions of the Little Ice Age, with the ice front lying several hundred feet (in the case of the Lyell Glacier) up the canyons from the terminal moraines. With the exception of a few years of net positive glacier balance (Curry 1969), this shrinkage, and the loss of many small ice patches, continued into the early decades of the twentieth century. Matthes (1939, 1942a, 1942b) noted that between 1933 and 1938 the Palisades Glacier thinned by 8.2 m (27 ft); that between 1931 and 1939 the East Lyell Glacier retreated 26.5 m (87 ft); and that between 1933 and 1941 the West Lyell Glacier thinned by 3.7 m (14 ft) and the East Lyell Glacier by 6.7 to 10.4 m (22 to 34 ft). Further shrinkage of the glaciers is evident from a comparison of the U.S. Geological Survey quadrangles from the late 1940s and early 1950s with those from the 1980s.

Evidence thus indicates that the centuries prior to 1850 were abnormal in the context of Holocene climate, in that they favored ice accumulation in cirques of the high Sierra. Shortly after 1850 the glaciers began to retreat. With the exception of a few aberrant years, Sierran glaciers have experienced a net negative balance since that time.

Theoretically, this minor glaciation of the mid-sixteenth through mid-nineteenth centuries is attributable to some combination of increased precipitation (leading to greater accumulation) and decreased temperature (leading to less melting and sublimation). Since the lake level records presented earlier in this chapter are consistent in suggesting that climate was relatively dry during this period, it might be concluded, as a working hypothesis, that relatively low temperatures caused the advance of the ice. Various types of dendroclimatological evidence, presented later in this chapter, comport with this hypothesis. The dendroclimatic record, in fact, verifies that climate was both relatively cool and relatively dry during the centuries preceeding the California gold rush.

Evidence from the Hydrogen Isotope Record

According to Feng and Epstein (1994), the deuterium-to-hydrogen ratio of nonexchangeable hydrogen in the cellulose of a tree ring is systematically related to the deuterium-to-hydrogen ratio in the water used by the tree to produce that ring. At least in meteoric (i.e., atmospheric) waters, the deuterium-to-hydrogen ratio is a function mainly of air temperature at the time of condensation, with higher temperatures giving rise to greater amounts of deuterium. To the extent that the trees in question are being nurtured by meteoric waters, rather than by long-stored ground waters, the deuterium-to-hydrogen ratios in the rings can be used as a proxy for past temperatures.

Feng and Epstein applied these principles to an analysis of bristlecone pine (*Pinus longaeva*) tree rings from the White Mountains, immediately east of the Sierra Nevada. Based on that analysis, they infer that a rapid cooling started around 1600, and culminated between approximately 1750 and 1850 with the coolest temperatures of the past 8,000 years. The record indicates that, since then, temperatures have been on the rise.

Evidence from Tree Lines

The cooling of the seventeenth, eighteenth, and early nineteenth centuries inferred from the hydrogen isotope composition of bristlecone pines seems also to be reflected in the position of the upper bristlecone tree line in the White Mountains. LaMarche (1973) noted that during the past several thousand years the bristlecone tree line on both Campito and Sheep Mountains has declined from the high levels of mid-Holocene time. This recession progressed in fits and starts, with the most recent plunge, resulting in the lowest tree line of the past 7,000 and more years, occurring around 1600–1700. Bristlecone reproduction at the upper limits of this depressed tree line ceased between approximately 1700 and 1860. Since then, the pines have been reestablishing themselves at elevations nearly as high as the loftiest tree line of the Holocene epoch (LaMarche 1973, 1982).

LaMarche (1973) argued that upper tree-line elevation on Sheep Mountain is a function of temperature during the growing season, with increased temperatures leading to a higher tree line. The upper tree-line position on Campito Mountain, in contrast, reflects both temperature and precipitation, with increases in both leading to a higher tree line. Since tree lines on both mountains fell around 1600, and remained low until around 1860, LaMarche concluded that climate during this interval was both cool and dry. The warmer and wetter conditions since 1860 account for the ongoing rise in the tree lines, according to him (LaMarche 1982).

Lloyd and Graumlich (1993; Graumlich, telephone conversation with the author, June 1995) report that the foxtail pine (*Pinus balfouriana*) tree line in the southern Sierra Nevada fell between 1500 and 1600, in concert with the most recent tree-line depression in the White Mountains. Curiously, the foxtail pine tree line on Cirque Peak near the southern end of the Sierra Nevada seems to have reached low levels not around 500 years ago, as reported in the studies of LaMarche and of Lloyd and Graumlich, but around 1,400 years ago (Scuderi 1987); indeed, no further tree-line depression on Cirque Peak appears to have occurred during the Little Ice Age. The Cirque

Peak record, however, does show an upward movement of the tree line during the past century, in common with the White Mountain sites and the Sierra sites of Lloyd and Graumlich.

Evidence from Tree Rings

Graumlich's tree-ring record from subalpine conifers (*Pinus balfouriana* and *Juniperus occidentalis*) in the southern Sierra, spanning more than 1,000 years, constitutes the most recent, and arguably the most rigorous, dendroclimatic analysis from the range (Graumlich 1993). Her work permits a climatic reconstruction far more detailed temporally (to the multiple-year scale) than that derived from lakes or glaciers. It also allows the temperature factor to be isolated from the precipitation factor, an advantage that neither the lake record nor the glacial record can provide.

Employing response surfaces that relate historical summer temperature and winter precipitation to ring width, Graumlich demonstrates that

- Growing-season temperatures reached their lowest level of the past millennium around 1600 and then remained low, by modern (1928–88) standards, until around 1850.

- While the period 1713–32 was, by modern standards, characterized by relatively wet conditions, it was preceded by a century dominated by low precipitation and was followed by 130 years (particularly the intervals 1764–94 and 1806–61) of anomalous drought.

- The period 1937–86 has been the third-wettest half-century interval of the past 1,000 and more years.

Graumlich stresses that her inferred droughts and temperature variations are reflected in other tree-ring studies undertaken in and adjacent to the Sierra Nevada (Briffa et al. 1992; Michaelson et al. 1987; LaMarche 1973; Graumlich and Brubaker 1986; Fritts 1991; Hughes and Brown 1992). Such coherence indicates that her findings have applicability throughout and beyond the range, rather than just locally.

SUMMARY AND DISCUSSION

A combination of records from lakes, glaciers, tree rings, and tree lines in and adjacent to the Sierra Nevada indicates that, in general, the period from around 1650 to 1850 was characterized by anomalously cool, moderately dry conditions. After being driven to their highest levels in several millennia by effectively high Sierran runoff, Mono, Pyramid, and Owens Lakes began to fall around 1650. Physical, biotic, and historical evidence indicates that during the 1840s and 1850s these lakes stood well below their modern natural levels, leading to the inference that effective inflow was perhaps 23%–34% less than the modern value. By the early 1860s the lakes were rising toward their modern natural levels in response to increased effective runoff from the Sierra.

As the closed lakes of the eastern Sierra were falling to low levels between 1650 and 1850, glaciers were forming and advancing in the high Sierra, arguably for the first time during the Holocene (the postglacial period). (Note that the accumulation of ice in the cirques of the eastern Sierra Nevada played only an insignificant role in decreasing runoff from the Sierra during this period. The building of glaciers during the Little Ice Age is thus not, in and of itself, an explanation for the recession of the closed lakes.) Glacier increase, together with the hydrogen-isotope record from the White Mountain bristlecone pines and the tree-line depression in both the White Mountains and the Sierra, strongly suggests that the centuries prior to the California gold rush were the coolest of the past 8,000 and more years. The modern wasting of the glaciers, the upward expansion of alpine tree lines, and the hydrogen-isotope record all indicate that since 1850 the Sierra has experienced a marked warming.

Graumlich's tree-ring record from the southern Sierra provides the most detailed view of variations in the latest Holocene climate. That record confirms that the period from 1650 to 1850 was generally dry, although it points up an important exception not evident in the lake or glacial records: the interval 1713–32 was anomalously wet. Graumlich's work also provides corroboration that the period from 1650 to 1850 was, by both Holocene and modern standards, abnormally cool.

IMPLICATIONS FOR THE FIRE REGIME IN THE SIERRA NEVADA

Based on an examination of burn scars in the tree rings of giant sequoias (*Sequoiadendron giganteum*) at five groves on the west slope of the Sierra, Swetnam (1993) reconstructed a 2,000-year-long fire history. Relating fire size and fire frequency to climate series derived from tree rings of giant sequoias and bristlecone pines, he documented a close decade- to century-scale positive relationship between summer temperature and fire activity (frequency and synchrony) and a close multiple-year-scale negative relationship between precipitation and fire activity. He tentatively attributes these relationships to high-frequency (years-scale) precipitation-dependent changes in the moisture content of fuels and low-frequency (decade- to century-scale) temperature-dependent changes in fuel production.

Swetnam's record indicates that throughout the period 1650–1850 fire frequency in the groves sustained its lowest level in 900 years, a result that would be expected, given the low temperatures of the Little Ice Age.

Equally expected would be an increase in fire frequency corresponding to the increase in temperature after 1850. This modern increase in fire frequency did not occur, however, probably for three reasons: decreased fuel loads due to sheep grazing, a decrease in ignition due to the demise of Native Californian culture, and fire suppression policies. Indeed, rather than increasing, fire frequency since 1850 has decreased to its lowest value of the past 2,000 years.

These findings underline the peculiarity of the modern fire regime in the Sierra Nevada: while the disparity between fire frequency of the modern period and that of the period 1650–1850 is clearly large, the disparity between modern fire frequency and the frequency that would occur today absent European settlement is even larger.

IMPLICATIONS FOR MANAGEMENT OF THE SIERRA NEVADA

The period 1650–1850 is of great interest to Sierra Nevada land managers because it is the last interval in which Europeans exerted little if any influence on the Sierran landscape. It may be tempting to use the condition of the landscape as it existed during this two-century period as a model for what the Sierra would be today in the absence of Europeans. In a related way, it may be tempting to attribute all landscape changes that have occurred since 1850 to the agency of Europeans.

There can be no question that many of the changes that have occurred since 1850 are attributable to the activities of Europeans. But it must be borne in mind that the European incursion closely coincided with a marked shift in climate and that some of the landscape change since 1850 may be, at least in part, attributable to that climate shift. The magnitude of the shift underscores its potential importance in instigating landscape change. In temperature, the shift was from the coldest century-scale interval of the Holocene, as indicated by the tree-line and glacier records, to one of the warmest periods of the past 4,000 years, as suggested by the recent upward movement of the tree line. In moisture availability, the shift was from moderate effective drought, as evidenced by the records of tree rings and lake levels, to the relative wetness of the present century—a century that appears, from the records of lake levels, to be the fourth-wettest of the past 4,000 years (Stine 1990) and that includes the third-wettest fifty-year interval (1937–86) of the past millennium (Graumlich 1993).

Thus it seems that, even if the European incursion had never occurred, the Sierra Nevada landscape of today would differ in significant ways from that of the immediate pre–gold rush period. With this in mind, attempts to restore "natural conditions" as part of an overall management plan should focus not on the pre-European landscape but rather on the landscape that would have evolved during the past century and a half in the absence of Europeans.

REFERENCES

Briffa, K. R., P. D. Jones, and F. H. Schweingruber. 1992. Tree-ring density reconstructions of summer temperature patterns across western North America since 1600. *Journal of Climate* 5: 735–54.

Clark, D. H., and A. R. Gillespie. 1995. Timing and significance of late-glacial and Holocene glaciation in the Sierra Nevada, California. *Quaternary International* in press.

Curry, R. R. 1969. Holocene climatic and glacial history of the central Sierra Nevada, California. *Geological Society of America Special Paper* 123.

Feng, X., and S. Epstein. 1994. Climatic implications of an 8000-year hydrogen isotope time series from bristlecone pine trees. *Nature* 265: 1079–82.

Fritts, H. C. 1991. *Reconstructing large-scale climatic patterns from tree-ring data: A diagnostic analysis.* Tucson: University of Arizona Press.

Graumlich, L. J. 1993. A 1000-year record of temperature and precipitation in the Sierra Nevada. *Quaternary Research* 39: 249–55.

Graumlich, L. J., and L. B. Brubaker. 1986. Reconstruction of annual temperature (1590–1979) for Longmire, Washington, derived from tree rings. *Quaternary Research* 25: 223–34.

Grove, J. M. 1988. *The Little Ice Age.* London: Methuen.

Harding, S. T. 1965. Recent variations in the water supply of the Great Basin. Water Resources Archives Report 16. Berkeley: University of California.

Hardman, G., and C. Venstrom. 1941. A 100-year record of Truckee River runoff estimated from changes in the levels and volumes of Pyramid and Winnemucca Lakes. *Transactions of the American Geophysical Union* Part 1, 71–90.

Hughes, M. K., and P. M. Brown. 1992. Drought frequency in central California since 101 B.C. recorded in giant sequoia tree rings. *Climate Dynamics* 6: 161–67.

LaMarche, V. C. 1973. Holocene climatic variations inferred from treeline fluctuations in the White Mountains, California. *Quaternary Research* 3: 632–60.

———. 1982. Lagged response of the upper treeline ecotone to rapid changes in climate. *Seventh Biennial Conference of the American Quaternary Association, Program and Abstracts* 25.

Lloyd, A. H., and L. J. Graumlich. 1993. Abstract: Late Holocene treeline fluctuations in the southern Sierra Nevada. *Bulletin of the Ecological Society of America* 74: 334.

Matthes, F. E. 1939. Report of the Committee on Glaciers. *Transactions of the American Geophysical Union* 518–23.

———. 1942a. Glaciers. In *Hydrology,* edited by O. E. Meinzer, 149–219. New York: McGraw-Hill.

———. 1942b. Report of the Committee on Glaciers. *Transactions of the American Geophysical Union* 374–92.

Michaelson, J., L. Haston, and F. W. Davis. 1987. Four hundred years of central California precipitation variability. *Water Resources Bulletin* 23: 809–18.

Russell, I. C. 1885a. Existing glaciers of the United States. In *5th annual report for 1883–1884,* 303–55. Washington, DC: U.S. Geological Survey.

———. 1885b. *Geological history of Lake Lahontan.* Monograph 11. Washington, DC: U.S. Geological Survey.

———. 1889. Quaternary history of Mono Valley, California. In *8th annual report for 1886–1887,* 261–394. Washington, DC: U.S. Geological Survey.

Scuderi, L. A. 1987. Late-Holocene upper timberline variation in the southern Sierra Nevada. *Nature* 325: 242–44.

Stine, S. 1987. Mono Lake: The past 4000 years. Ph.D. diss., University of California, Berkeley.

———. 1990. Late Holocene fluctuations of Mono Lake, California. *Palaeogeography, Palaeoclimatology, Palaeoecology* 78: 333–81.

———. 1994. Late Holocene fluctuations of Owens Lake, Inyo County, California. Technical Report to Far Western Anthropological Research Group, Davis, California.

Swetnam, T. W. 1993. Fire history and climate change in giant sequoia groves. *Science* 262: 885–89.

WILLIAM C. KINNEY
Davis, California

3

Conditions of Rangelands before 1905

ABSTRACT

Paleoecological sources indicate that the location and extent of Sierra Nevada rangelands have varied significantly during the last 20,000 years. Modern vegetative associations are recent, with montane wet meadows appearing during the last 3,000 years. A late Pleistocene sagebrush grassland existed where montane and subalpine forests occur today. In the central Sierra a pattern of deglaciation and vegetative response was repeated at different times depending on location, from alpine grassland to a diverse mixture of conifer and shrub species in an open forest structure. The Holocene began with a decline in mesic species and increased charcoal and oak pollen, indicating a warming trend. A cooler, wetter climate followed as mesic conifers reappeared and evidence of fire decreased.

Historical accounts indicate that highly productive rangelands existed when Europeans arrived. Large ungulate populations were present, and perennial grasses dominated foothill rangelands. Numerous observers reported severe overgrazing by livestock in the late 1800s, due in part to a lack of regulation of the common rangelands. Livestock management contributed to annual grassland conversion on the west side and to juniper woodland expansion on the east side of the range.

The abundance of a diverse assemblage of large grazing mammals at the end of the Pleistocene indicates that Sierra Nevada rangelands were highly adapted to intense grazing pressure and that animal disturbance was an integral part of this highly productive system. This evidence argues for a recognition that well-managed animal disturbance is as vital as well-managed fire to ecosystem health and sustainability.

INTRODUCTION

An assessment of the health of Sierra Nevada ecosystems requires some basis from which to evaluate current conditions. That is, how similar are current ecosystems to those of the past, and to what extent do recent historical conditions represent their range of variability over longer periods of time? The purpose of this chapter is to address the second of the three basic assessment questions; that is, what were the conditions of Sierra Nevada rangeland ecosystems in the past, and how have these ecosystems varied over recent geologic and historic time intervals? Here, rangelands are defined broadly to include the grassland-, woodland-, and chaparral-dominated associations that have been used by livestock and wild ungulates within the study area in the past. A more precise definition of current Sierra Nevada rangeland ecosystems can be found in Menke et al. 1996.

This chapter reviews available historical and paleoecological information on past conditions and the natural range of variability of rangeland resources within the project study area, from the late Pleistocene to the start of public ownership at the turn of the century. The focus here is on the last 20,000 years, a period encompassing a wide range of climatic and ecological conditions, from near-maximum glacial advance, through a period of deglaciation, to more recent intervals that were evidently hotter and drier than the current climate. The first part of the chapter looks at paleoecological studies that illuminate century- and millennial-scale changes in rangeland vegetation. The second part of the chapter addresses questions concerning the conditions and use of Sierra Nevada rangelands during the initial European exploration and settlement period from 1579 to 1905, when public ownership of these lands was consolidated. The final part of

Sierra Nevada Ecosystem Project: Final report to Congress, vol. II, *Assessments and scientific basis for management options*. Davis: University of California, Centers for Water and Wildland Resources, 1996.

the chapter discusses implications of these findings for the management of Sierran rangelands. Questions relating to the use and management of these rangelands since 1905, and to their current condition and trend, are addressed in Menke et al. 1996.

Organization of the chapter is basically chronological. Following the review of paleoecological conditions is a review of historical rangeland conditions and uses. The historical review focuses primarily on foothill and montane rangelands, based on the limited coverage of early accounts. These accounts cover the period from initial European exploration and settlement to the consolidation of public ownership in 1905, with special emphasis on the period from 1860 to 1900 when rangeland resources were heavily affected by Europeans. Sources of data include accounts by explorers, trappers, settlers, naturalists, and surveyors. Although some readers may find the historical accounts "subjective," they are the best available sources of information on rangeland condition during the settlement period.

Data used by paleoecologists to reconstruct past environments of the Sierra Nevada include tree-ring chronologies; lake and meadow sediment stratigraphy and fossil content; existing and prehistoric lake levels and glacial extents; and pack-rat midden pollen and macrofossils. These proxy data, taken together, sketch an outline of Sierra Nevada rangeland ecosystems and how they have responded to climatic variability.

Unfortunately, data relating to past rangeland conditions are limited, both for the distant past and for the period of European settlement. These limitations are due in part to the lack of systematic studies of rangeland conditions during the settlement period (Burcham 1957). In addition, current paleoecological research of the Sierra Nevada suffers from limited spatial and temporal resolution. Finally, it is often difficult to interpret paleoecological data because of a diversity of ecosystem responses to changing environmental conditions. However, these sources do provide a picture of the location, general composition, and extent of broad rangeland types, and the relative productivity of past Sierra Nevada rangeland ecosystems.

In addition to assessing the health of current rangeland ecosystems and reviewing past conditions, initial rangeland research objectives included a review of evidence for a late Pleistocene megafaunal herbivory and its coevolution with rangeland vegetation; a review of adaptive livestock-management strategies that recognize these plant-herbivore relationships; evidence of successful application of these strategies by practicing land managers; and implications of this evidence for management of Sierra Nevada rangelands. After initial review of the literature in these areas, it became clear that this chapter could not adequately address all these objectives.

Considerable scientific controversy has arisen among ecologists over whether herbivory can benefit plant communities, in part because such benefits challenge established paradigms about the role of animal disturbance in wildland ecosystems.

A large body of reductionist research aimed at selected rangeland ecosystem processes has expanded our knowledge of plant-herbivore relationships but generally fails to address the full complexity of ecosystem response. In many cases, there has been a confusion over benefits to individual plants or communities versus benefits to the overall ecosystem. Key findings from these studies are presented later.

It is clear, however, that plant-herbivore interactions are very site- and time-specific and that achieving potential benefits depends on adapting rangeland management to these specific conditions. Even a complete review of this literature would not answer questions that can be addressed only through adaptive experimental designs. Given a lack of adaptive rangeland-management studies in the project area, project resources were allocated to analyzing current data on rangeland condition and trend, as addressed in Menke et al. 1996, rather than further exploring these broader questions.

LATE PLEISTOCENE PALEOECOLOGY OF THE SIERRA NEVADA

Geologists define the Quaternary period as the most recent, covering the last two million years or so of geologic time. This period is divided into the Pleistocene and the Holocene epochs, at between 12,000 and 10,000 B.P. As noted earlier, paleoecologists analyze a variety of data to infer past ecological and environmental conditions about a given site or region.

Recent investigations of paleoenvironmental conditions continue to modify the picture generated by earlier research and to call into question past theories about climate change, ecological succession and climax, and the delineation of existing plant community types. This is because it is often difficult to find modern analogs to some of the plant associations that evidently existed within the study area in the past (Davis and Moratto 1988; Grayson 1993). In addition, it appears that the Sierra Nevada has had a different climatic history than the Great Basin during the last 12,000 years (Davis and Moratto 1988).

Western-Slope Montane Vegetation

Cole (1983) analyzed one modern and six Pleistocene pack-rat middens from caves in lower Kings Canyon ranging in elevation from 920 to 1,270 meters (~3,000–4,000 ft). This area is within the present oak-chaparral woodland vegetative zone in California. The modern midden showed a high correlation with local vegetation, as determined by three 60 m (197 ft) transects within 100 m (328 ft) of the midden. The Pleistocene middens ranged in age, with five of the six falling between 20,000 and 12,500 B.P. The middens were analyzed for both pollen and plant macrofossils. Results showed that the vegetation in this zone was completely different from the present vegetation, with no evidence of the modern oak-chaparral community. Instead, the assemblages included a high abundance of *Pinus* and *Artemisia* pollens as well as macrofossils indicating a mixed coniferous forest dominated by the xerophytic conifers western juniper, ponderosa pine, sugar pine, and single-needle piñon. This mixture is similar to that found on the east side of the Sierra crest today, but 500 to 1,000 m (1,641 to 3,281 ft) higher in elevation than the Kings Canyon middens.

This assemblage seems to indicate a Pleistocene climate considerably drier than today at this site. Also present, however, are macrofossils of mesophytic taxa California nutmeg, incense cedar, and red fir, as well as a few pollen grains of giant sequoia. Although presence of these latter taxa suggests precipitation levels higher than exist today on the east side of the range, Cole (1983) felt that the presence of the xerophytic conifers indicated current east-side levels of precipitation, but with a colder climate, causing a greater proportion of moisture to fall as snow. A cooler climate would lower evapotranspiration losses and allow for more mesic microclimates in snow-filled depressions. The youngest sample (~12,500 B.P.) did not contain any fossils of the mesophytic taxa, indicating a trend to a drier climate toward the end of the Pleistocene.

Additional sources of information are pollen and plant macrofossils within and stratigraphy of montane meadow sediments. Meadow sediments have been analyzed from sediment cores (Anderson 1990; Anderson and Smith 1994; Davis and Moratto 1988; Smith and Anderson 1992) and from meadow gullies (Wood 1979).

Smith and Anderson (1992) examined a sediment core from Swamp Lake, at 1,554 m (~5,100 ft) in Yosemite National Park, that provides a record of environmental change during the last 16,000 years, based on pollen, macrofossil, and charcoal analyses. The core stratigraphy indicates Tioga-stage deglaciation between 16,000 and 13,700 B.P. During this interval, the fossil record indicates that herbs and sagebrush dominated the Swamp Lake environment, suggesting a cooler and drier climate than at present, similar to the conditions today 1,000 m (3,281 ft) higher and east of the Sierra crest.

By 12,000 B.P., pollen and macrofossil data indicate that subalpine, lower and upper montane conifers became established around the edge of Swamp Lake. These included lodgepole pine, sugar and ponderosa pine, red and white fir, mountain hemlock, incense cedar, and Sierra juniper (*Juniperus occidentalis*). Sagebrush and herbaceous pollen percentages decline throughout the interval from 13,700 to 10,400 B.P., while charcoal concentrations increase. This anomalous mix of high-elevation and lower montane conifers about 12,000 B.P. suggests a climate that was not just cooler and wetter, but fundamentally different from current conditions. Smith and Anderson (1992) offer several possible causes for this anomaly, including a greater seasonality than at present, with perhaps cooler, wetter winters and warmer summers. Alternatively, a lag in vegetation response to climatic change may have al-

lowed subalpine and upper montane species to persist in favorable microhabitats. These conditions from 12,000 to 10,000 B.P. in Yosemite are consistent with the evidence from the middens in lower Kings Canyon dated earlier than 14,000 B.P., at elevations from 920 to 1,270 m (~3,000–4,000 ft).

Davis and Moratto (1988) analyzed macrofossil and pollen samples from cores of Exchequer Meadow, at an elevation of 2,219 m (7,280 ft) in the Sierra National Forest. The base of the core was radiocarbon dated to 13,500 B.P. and was divided into three zones based on predominant vegetation in the samples. The basal *Artemisia* zone (13,500–7070 B.P.) suggests vegetation and a climate that is drier than today's on the west side of the range. It also contains giant sequoia pollen, indicating a late Pleistocene climate with temperature ranges similar to those of today.

The *Artemisia* zone was further subdivided into upper *Quercus* and lower Gramineae subzones at 10,680 B.P. The inferred vegetation in the lower subzone was an alpine grassland. High levels of spores of the dung fungus *Sporormiella* in the Gramineae subzone indicate that grazing animals were abundant during this period. These spores are present in sediments older than 11,000 B.P. in several sites in the western United States (Davis 1987) and are linked directly to extinct megafauna by their presence in fossil mammoth dung (Davis et al. 1984; Mead et al. 1986). Absence of these spores after 11,600 B.P. here may date the extinction of Rancholabrean herbivores at middle elevations in the western Sierra. Transition to the oak subzone indicates a rapid climatic warming about 11,000 B.P.

Anderson (1990) analyzed pollen and plant macrofossils from sediments in three high-elevation lakes in the Sierra, including Tioga Pass Pond at 3,018 m (9,900 ft) on the crest, Barrett Lake at 2,816 m (9,239 ft) on the east slope, and nearby Starkweather Pond at 2,438 m (8,000 ft) on the west slope, the latter two near Mammoth Lakes. His analysis indicates that deglaciation occurred later, after about 12,000 B.P. at these higher elevations, than at the middle-elevation sites cited earlier. During the remaining interval of the Pleistocene (to 10,000 B.P.), the lack of macrofossils indicates that trees were absent or poorly established at Barrett Lake and Starkweather Pond. Pollen suggests the nearby vegetation probably included pine, juniper, sagebrush, and grasses. Sediments at Tioga Pass Pond do not predate the end of the Pleistocene.

Great Basin Vegetation in the Late Pleistocene

Analysis of the paleoenvironment of the Great Basin indicates that it was considerably cooler and wetter during the late Pleistocene than at any subsequent time (Grayson 1993). The strongest indicators of this climatic condition are the levels of Great Basin lakes during this interval. Lake Lahontan at its maximum, about 14,000 B.P., covered about fourteen times more surface area than its modern remnants and extended well into California near present-day Honey Lake on the Susan River. Benson et al. (1990) show that Lake Lahontan, Lake Russell (prehistoric Mono Lake), and Lake Searles in the Owens Valley all achieved high stands between 15,000 and 13,500 B.P.

Wigand and Mehringer (1985) analyzed pollen and seeds from Hidden Cave near Carson Sink and concluded that a sagebrush steppe existed there in the late Pleistocene. Subalpine conifers occupied very low elevations at that time, with an understory of sagebrush, winterfat, and shadscale (Grayson 1993; Wigand et al. 1995).

Koehler and Anderson (1994) examined pack-rat middens 10 m (33 ft) above the highest level of Pleistocene Lake Owens, dated from about 23,000 to 14,500 B.P., to infer lakeshore vegetation during this interval. Their data suggest a woodland consisting of Utah juniper and single-needle piñon, green ephedra, wild rose, and Menodora, with Rocky Mountain juniper, which does not occur today in California, present to about 16,000 B.P. Pollen analysis indicates a xeric upland desert-scrub association nearby, with the more mesic conditions at lakeshore due to higher ground-water levels. Jennings and Elliott-Fisk (1993) analyzed middens from the White Mountains that show juniper woodlands extending 600 m (1,968 ft) lower in elevation in the late Pleistocene than at present.

Grayson (1993) argues that a great deal of evidence indicates a cool, moist climate in the late Pleistocene; such evidence includes inferred vegetation, lake levels, and human population levels and dietary patterns. As in the Sierra, plant associations with no modern analog appear in the basin during this period. Problems exist with current concepts of habitat type and condition and trend, which do not reflect recent evidence about dynamic vegetative response to disturbance and climatic change (Grayson 1993; Tausch et al. 1993). That is, the concept of "potential natural condition" of rangeland sites, resulting from successional processes in the absence of disturbance, needs review (see also Woolfenden 1996).

Late Pleistocene Herbivory

One issue in the debate about western rangeland management is the role of Pleistocene megafauna in the evolution of rangeland ecosystems. As noted earlier, initial research objectives were to review evidence for the existence of these animals in the study area and the role of large animal disturbance in rangeland ecosystems. This section reviews pertinent literature on these questions, within the limits mentioned in the Introduction.

Wagner (1989) lists twenty species of late Pleistocene grazing mammals that existed in California as late as 11,000 B.P. This list includes two or more species of horse, tapir, llama, camel, pronghorn, and bison, as well as mammoth, mastodon, shrub ox, musk ox, and the other surviving large herbivores, mule deer, elk, and bighorn sheep. Given the diversity of both herbivorous species and their likely predators, Wagner concludes that California grasslands would have sustained heavy grazing pressure, equal to or greater than what occurs in contemporary East African savannas (Wagner 1989). Edwards (1992) discusses the inferred dietary preferences of the large herbivores present in California during the late Pleistocene and their probable impact on native perennial bunchgrasses. He concludes that the native grasses would have been heavily grazed and trampled and that their ability to benefit from this kind of disturbance is still present. Edwards (1992) also compares the diversity of the California Pleistocene megafauna to that in East Africa today.

The Great Basin also had a large and diverse Pleistocene megafauna. Grayson (1993) identifies some twenty-four sites containing species of extinct Pleistocene mammals that are within or near the Sierra Nevada Ecosystem Project (SNEP) study area (see Grayson's Table 7.2). These large mammals included three species of ground sloth, giant short-faced bear, saber-toothed cat, American lion, American cheetah, several species of horse, flat-headed peccary, camel, llama, shrub ox, musk ox, mastodon, and mammoth. The diversity of large predators in this group would have caused wild herbivore behavior to differ radically from conventionally managed rangeland livestock in the western states.

Taken together, the paleontological evidence indicates that a diverse assemblage of large mammals occupied the Sierra Nevada region in the Pleistocene. As discussed earlier, the dung fungus *Sporormiella* is abundant in sediment cores of Sierra Nevada meadows before 11,000 B.P., indicating that grazing animals were abundant in the Sierra Nevada until sometime near the end of the Pleistocene (Davis and Moratto 1988). The implied diversity and abundance of this assemblage indicate that Sierra Nevada rangelands were highly adapted to grazing by large herds of wild herbivores and that this kind of severe disturbance was an integral feature of these ecosystems.

Several authors (Edwards 1992; McNaughton 1979; Savory 1988) have argued that there are in fact beneficial impacts of animal disturbance on rangeland plant communities. McNaughton (1976) showed that net primary productivity of East African rangelands was increased by the impacts of large migratory herds grazing at different times in the same areas. McNaughton (1979) later identified ten pathways through which plants benefit from grazing. These include greater photosynthetic activity from removal of senescent plant material, increased conservation of soil moisture by greater water-use efficiency, and increased nutrient recycling from dung and urine. Savory (1988) argues that many researchers ignore the role that large ungulates play in providing favorable conditions for seedling establishment through the combined effects of soil disturbance, compaction, and fertilization.

Savory (1988) developed an adaptive resource-management process that recognizes these beneficial plant-herbivore interactions. He maintains that grazing animals, when managed to mimic wild herds under predator pressure, are necessary to maintain healthy rangelands under many conditions. This model also recognizes that rangelands need to be managed as complete systems and that management must be adaptive to realize beneficial effects and avoid adverse impacts. Although many ranchers using this adaptive approach have had notable success in improving the health of their rangelands, controlled experiments on fixed rotational grazing systems have failed to duplicate these results, pointing out the difficulties in designing and conducting truly adaptive research. Similar criticisms by proponents of adaptive management have called for experimental designs that allow more flexibility and address the full complexity of ecosystem responses (Walters and Holling 1990). For additional discussions of adaptive ecosystem management, see Kusel et al. 1996.

This view of plant-herbivore relationships, which contradicts conventional wisdom about the detrimental impacts of overgrazing, has spawned a vigorous debate over the role of grazing in rangeland ecosystems (Edwards 1992; McNaughton 1979, 1983, 1986, 1993; Oesterheld and McNaughton 1991; Oesterheld et al. 1992; Savory 1988). As discussed in the Introduction, a lack of studies that faithfully duplicate the adaptive livestock-management process precluded a more thorough examination of this debate.

Nevertheless, the paleoecological record supports the view that a highly productive rangeland ecosystem existed in the project study area during the last 20,000 years and that this rangeland vegetation supported a large and diverse megafaunal assemblage before mass extinctions about 10,000 years ago. The implications of this conclusion for management of Sierra Nevada rangelands are discussed in the last section of this chapter.

CHANGING HOLOCENE PALEOENVIRONMENTS

Geologists divide the Quaternary period into Pleistocene and Holocene epochs, with the dividing date set between 12,000 and 10,000 B.P. (Grayson 1993; Wigand et al. 1995). Grayson (1993) notes that ecological changes at this time were often transitional and that the timing of these changes varies with location. It also appears that the timing inferred for these changes depends on the kind of evidence used. Changes in vegetation on the western slope of the Sierra in the late Pleistocene indicated a trend toward warmer and drier conditions than today's. The Great Basin, on the other hand, while drier than at glacial maximums, showed higher lake levels than exist today. However, the vegetative changes beginning there about 12,000 B.P. indicate a warming trend (Wigand et al. 1995).

In the Great Basin, this major change to hot and dry conditions becomes most evident by about 8000 to 7500 B.P. (Grayson 1993). In the Sierra, a shift seems to have occurred toward cooler and wetter conditions about 6500 B.P., depending on elevation and latitude, followed by another shift at about 3700 B.P., to the even cooler, wetter Neoglacial regime. A change is

also observed in the Great Basin about 4500 B.P., toward a cooler, wetter climate. This late Holocene Neoglacial interval, ending about 2000 B.P., is indicated by expansion of woodlands and a grassy sagebrush steppe in the Great Basin and by wet meadow formation in middle elevations of the Sierra (Wood 1979).

Sierra Nevada Holocene Environment

Smith and Anderson (1992) argue that an early Holocene xeric period in the Sierra began about the end of the Pleistocene, based on pollen, macrofossil, and microscopic charcoal analysis of sediments from Swamp Lake. This interval of warm, dry climate showed high oak and minimum fir pollen percentages and maximum charcoal concentrations for this site. This xeric period lasted until about 6500 B.P., when fir pollen increased and charcoal concentrations decreased, suggesting a cooler and/or wetter trend.

The zone from 10,400 to 6500 B.P. shows a steady decrease in pine pollen and a steady increase in Cupressaceae pollen (probably incense cedar). Macrofossil analysis indicates that red fir, lodgepole pine, Sierra juniper, and mountain hemlock disappeared near the beginning of the zone, while ponderosa pine, sugar pine, and incense cedar continued. At 6500 B.P., fir pollen began increasing and charcoal decreased with the cooler and wetter conditions of the mid-Holocene.

Anderson and Smith (1994) compared the sediment stratigraphies and fossil pollen samples of nine Sierra middle-elevation meadows, from 1,857 to 2,219 m (6,093–7,280 ft) elevation and spanning latitudes from Yosemite National Park to Sequoia National Forest. Their analysis suggests that the ecological, climatic, and hydrological conditions in the Sierra Nevada changed considerably during the early Holocene interval. Pollen assemblages are dominated by plants that today grow in dry microhabitats, with lesser amounts of pollen from plants that require higher soil moisture, suggesting a more open forest structure with a less effective water cycle than today's.

The sediment stratigraphies reviewed by Anderson and Smith (1994) show a basal layer of coarse sands or gravels and a transition to colluvium or finer sands generally occurring between 10,500 and 9500 B.P. These early to middle Holocene sediments consist of silty bands of varying organic content, alternating with layers of peat, coarse sand, or gravel. The authors suggest that these deposits were generated by greater erosion rates on basin slopes than exist today, perhaps due to a much sparser vegetative cover on immature soils. The stratigraphy suggests periods of alternating meadow and forest vegetation, as in Wood's analysis of Exchequer Meadow (Wood 1979).

The initiation of wet meadow development from 4500 to 3000 B.P. is indicated by a transition from predominantly colluvium to predominantly peat in the upper zone of the cores. This transition to peaty sediment represents a widespread change in hydrologic regime within the area encompassed by these meadows. Once begun, a positive feedback cycle was generated in these locations, as increasingly organic soils promoted greater water-retaining capacity, thus providing habitats for species that require very moist soil during the hot, dry summer. Wood (1979) also recognized a change in hydrologic and depositional regime, but at later dates in the meadows he examined.

Anderson and Smith (1994) also reviewed studies of pollen stratigraphy from four high-elevation and two middle-elevation lakes, noting that the same changes to wet meadow conditions occurred at higher elevations, but 1,000–1,500 years earlier. They compare these changes with other data indicating changes in climatic and ecological characteristics of the Sierra region, such as in the tree line and elevational limits of red fir and mountain hemlock.

Davis and Moratto (1988) divided the Holocene portion of Exchequer Meadow cores into a *Quercus* subzone from 10,680 to 7070 B.P., a *Pinus* zone from 7070 to 1870 B.P., and an *Abies* zone from 1870 to the present. The latter division corresponds to the point of maximum fir pollen in the meadow and may indicate meadow invasion during a brief period when wet meadows dried.

Stine (1994) analyzed drowned relic tree stumps from four sites in the Sierra Nevada that indicate two periods of extended drought between about A.D. 892 and 1350. These droughts lasted at least 220 and 140 years, respectively, and were separated by a century or so of very wet conditions.

Wood (1979) analyzed seven dissected middle-elevation meadows to investigate conditions of meadow formation in the western Sierra. He concluded that depth to water table was the sole cause for determining the presence of conifer or meadow vegetation and thus inferred a generalized sequence of meadow development. However, he discusses other studies indicating that meadow development and invasion result from a complex and dynamic interaction of fire, climatic variability, long-term climatic change, and vegetative response, and his sequence does not completely conform to the more recent evidence reviewed earlier. Wood notes that "limited chronologic control suggests that scour and fill complicate this (early and middle Holocene) unit." He also asserts that all meadows identified for his study were incised since 1900, yet only seven of his twenty-three sites had established maximum ages (Wood 1979).

Anderson's (1990) analysis of the sediments of three high-elevation lakes suggests that early Holocene, high-altitude Sierra vegetation was also structurally different from that of today, with a more open forest than currently exists at these elevations. Montane chaparral shrubs, such as mountain mahogany, manzanita, and sagebrush, dominate the pollen record. Anderson suggests that the presence of these xeric species could have resulted from poor water-retention capacity of the immature soils after deglaciation, but he felt that it was more likely the result of lower precipitation than at present.

Great Basin Holocene Environment

In contrast to the Sierra Nevada, the Great Basin in the early Holocene (11,500–7500 B.P.) was evidently cooler and wetter than it is today (Grayson 1993). This change is indicated by lake levels that were considerably higher than current levels, though lower than Pleistocene maximums. Human population in the Great Basin appears to have been higher than during subsequent intervals, and archaeological evidence indicates that food was relatively plentiful. Human sites during this interval were associated almost exclusively with these more plentiful water sources (Grayson 1993).

Wigand et al. (1995) note the existence of a grassy sagebrush steppe near Eagle Lake in the northeast Sierra Nevada about 10,200 B.P. At Connley Caves in southeast Oregon, near Paulina Marsh, twenty-two of twenty-three elk specimens were dated between 11,000 and 7200 B.P. (Grayson 1993). This evidence indicates a favorable rangeland habitat during the interval. In addition, 95% of the bird remains were deposited there during this early interval, and most of those birds were tightly associated with marshy habitats, again indicating that the marsh dried up after 7200 B.P. Other evidence indicates that humans subsequently abandoned the site during the hot, dry mid-Holocene.

At Hidden Cave near Carson Sink in western Nevada, Wigand and Mehringer (1985) have shown that at about 10,000 B.P., pine and sagebrush pollen decreased somewhat, though not to levels comparable to modern spectra until 6800 B.P. Investigations in the Mojave Desert area indicate that perennial streams with marshy edges persisted until 8,000 years ago. This regime, cooler and wetter than current conditions, then disappeared, and water tables dropped some 24 m (80 ft) (Grayson 1993).

A mid-Holocene interval that was hotter and drier than today in the Great Basin is indicated by inferred vegetation, lowered lake levels, and evidence of decreased human habitation (Grayson 1993). Mid-Holocene archaeological sites are rare, with earlier lakeshore settlements abandoned. Food sources evidently became more scarce, as the appearance of seed-grinding artifacts indicates (see also "East-Side Rangeland Conditions" in the following section). Many early Holocene water sources vanished during this interval. Inferred vegetation from pollen and pack-rat midden studies indicate a shift to more xeric plants, or changes in community elevational boundaries. Jennings and Elliott-Fisk (1993) found an upward migration in the upper boundary of piñon-juniper woodlands in the White Mountains during this time.

Stine (1990) has documented that Mono Lake has had a series of at least six significant high and intervening low stands during the late Holocene. Since about 3,500 years ago, the level of the lake has fluctuated over a range of 40 m (131 ft). He argues that these fluctuations reflect decade- to century-scale climatic change during this period and may be related to observed variations in solar activity. Grayson (1993) argues that this fine-scale change probably existed in the late Pleistocene, but that it is harder to detect in older paleoenvironmental evidence lacking the finer temporal resolution of more recent sites.

The late Holocene interval in the Great Basin showed a shift to cooler and wetter conditions about 4,500 years ago, corresponding to an expansion of human population (Grayson 1993). In the northwest part of the province, increasing grasses and sagebrush began to replace the greasewood-dominated vegetation between about 5000 and 4700 B.P. In a synthesis of studies from throughout the Great Basin, Wigand et al. (1995) identify a period of maximum areal extent of juniper-dominated woodlands during the Neoglacial (4000 to 2000 B.P.). This period was followed by a 400-year drought cycle in which desert scrub communities expanded from 1900 to 1500 B.P. Greater grass abundance in the northern Great Basin sagebrush steppe between 1500 and 1200 B.P. corresponds to dramatic increases in bison remains in the archaeological sites of this region (Wigand et al. 1995). Nowak et al. (1994) and Miller and Wigand (1994) have shown that western juniper increased in this area as effective moisture increased, and retreated during times of drought. They conclude that the more extensive expansion of juniper during the last one hundred years has been affected by both increased moisture and European settlement, especially the management of livestock.

Sierra Nevada Paleoecological Rangeland Conditions

Paleoecological sources indicate that the location and extent of Sierra Nevada rangelands have varied significantly during the last 20,000 years. Modern vegetative associations are relatively recent. At lower elevations in the southern Sierra, data from pack-rat middens indicate an anomolous mix of xerophytic and mesophytic conifers from 20,000 B.P. to at least 14,000 B.P. By 12,500 B.P. the absence of the mesophytic taxa indicates a warmer and drier interval, leaving vegetation like that at upper elevations on the east side of the Sierra today.

In the central Sierra, a pattern of deglaciation and vegetative response is repeated at different times, depending on elevation. At Swamp Lake in Yosemite National Park (~1,524 m/5,000 ft), a sagebrush grassland occurred from about 16,000 to 13,700 B.P. From 13,700 to 10,400 B.P., an anomalous mix of subalpine, lower and upper montane conifers developed there. From about 10,000 to 6500 B.P., a warm, dry interval is inferred from maximum levels of oak pollen and charcoal, disappearance of several conifer species, and an open forest structure with a shrub-sagebrush understory. At 6500 B.P., fir pollen increases and charcoal decreases, indicating a shift to cooler, wetter conditions.

In middle-elevation meadows (~1,829–2,287 m/6,000–7,500 ft) of the central Sierra, deglaciation began about 13,500 B.P., with an alpine grassland existing to about 10,400 B.P. Spores of the dung fungus *Sporormiella* in meadow sediments indicate an abundance of Pleistocene megafauna until about 11,000 B.P. A rapid warming trend is inferred at about 10,000 B.P. From

10,000 to 6500 B.P., these locations also show an open forest structure with high oak pollen and charcoal, and a shrub-sagebrush understory. The period from 7000 to 4500 B.P. shows alternating meadow and forest soil conditions, with a decrease in oak pollen and charcoal. Wet meadow development began somewhere between 4500 and 3000 B.P., depending on location. A widespread change in hydrologic regimes is indicated for this Neoglacial interval, from about 5000 to 2000 B.P.

In the Great Basin, a cool, wet late Pleistocene is indicated by maximum lake levels, a grassy sagebrush steppe in the valleys, and subalpine conifers at low elevations. Lower lake levels indicate a warming trend from 11,500 to 7500 B.P., but with conditions still cooler and wetter than they are today. Human habitation sites first appear during this interval, with population levels higher than those at subsequent periods. A hot, dry interval from 8000 to 5500 B.P. shows lowered or vanishing lakes, abandonment of human sites associated with these lakes, and a shift to desert scrub vegetation. The period from 4000 to 2000 B.P. was also cooler and wetter in the Great Basin than current conditions, with an increase in human population, while the last 2,000 years have seen a decrease in moisture and the development of modern plant associations.

HISTORICAL CONDITIONS AND IMPACTS OF EUROPEAN SETTLEMENT

This review of Sierra Nevada rangeland ecosystems during the early years of European exploration and settlement is based on the written accounts of the explorers, trappers, settlers, naturalists, and scientists who traveled the Sierra, and their descriptions of the vegetation and the grazing animals they saw there. As noted earlier, a lack of systematic surveys of rangelands during this period makes it necessary to use these sources. Most of the accounts are of western-slope foothill conditions, although John Muir, in his extensive travels and writings and with his unique powers of observation, provides some rare and invaluable descriptions of montane and Great Basin rangeland ecosystems.

Proto-historic Rangeland Conditions (1579–1850)

Conditions of Western-Slope Foothill Grasslands

McCullough (1971) cites a number of early explorers who observed the California tule elk population before Europeans settled in the state. Typically, observers recognized that the elk was not just a large deer, but they had no other animal to compare it to, so their descriptions often used this term. Bourne (1653) published an account of the landing of Sir Francis Drake on the California coast somewhere near San Francisco in the summer of 1579, based on the notes of Drake and others on the voyage. McCullough quotes their landing party:

> The inland we found to be far different from the shoare [sic], a goodly country and fruitful soil, stored with many blessings fit for the use of man: infinite was the company of a very large and fat Deer, which then we saw by thousands as we supposed in a herd.

McCullough goes on to describe the travels of Otto Von Kotzebue, the Russian sea captain who journeyed overland from San Rafael to Fort Ross in 1824. Kotzebue described the northern coast range:

> The fine, light and fertile soil we rode upon was thickly covered with rich herbage, and the luxuriant trees stood in groups as picturesque as if they had been disposed by the hand of taste. We met with numerous herds of small deer, so fearless, that they suffered us to ride fairly into the midst of them, but then indeed darted away with the swiftness of an arrow. We sometimes also, but less frequently, saw another species of stag, as large as a horse, with the branching antlers; these generally graze the hills, from whence they can see round them on all sides, and appear much more cautious than the small ones. (Thompson 1896)

McCullough (1971) estimates that the total population of tule elk in California during this period of European exploration was about 500,000 animals.

Burcham (1957) discusses the first Spanish explorers' impressions of California's range resources. He notes,

> The Spaniards entered the San Joaquin valley at least as early as 1772. . . . they noted an abundance of forage.

Jedediah Smith is believed to have been the first American explorer to cross the Sierra Nevada (Smith and Brooks 1977). Cermak and Lague (1993) describe his expedition of 1826–27. Entering the Central Valley over Cajon Pass, he skirted the western foothills of the range, crossing again from west to east in May 1827, somewhere near Ebbets Pass. Smith reported seeing oak savannas, wildflowers, and brushlands in the foothills, and a wide variety and abundance of wildlife, including elk, antelope, and wild horses.

Burcham (1957) also relates the accounts of Zenas Leonard, the cook for Joseph Walker, an early American explorer. Describing the lower reaches of the Merced River and the San Joaquin valley in 1833, Leonard observed that

> there is a level prairie on the richest soil, producing grass in abundance of the most delightful and valuable quality. (Leonard 1959)

The fur brigades undoubtedly had a significant impact on Sierra foothill ecology, through their impact on California Indian populations. The Hudson Bay Company sent expeditions to California every year from 1827 to 1843. The account of John Work, who led a Hudson Bay expedition in 1833, indicates the effect these brigades had on wildlife. His entry of February 22, referring to the Marysville Buttes area, reads:

> We have been a month here and we could not have fallen on a better place to pass a part of the dead winter season when nothing could be done in the way of trapping on account of the height of the waters. There is excellent feeding for the horses, and abundance of Animals for the people to subsist on, 395 elk, 148 deer, 17 bears, and 8 antelopes have been killed in a month which is certainly a great many more than was required. (Work and Maloney 1945)

McCullough (1971) reports the accounts of other explorers who observed the large herds of elk in the Central Valley. Wilkes (1845), reporting on federal exploration efforts of 1838–42, wrote of the American River region: "The variety of game in this country almost exceeds belief. The elk may be said to predominate." Newberry (1884), reporting on surveys to determine a route for the railroad, stated that the herds of grazing animals rivaled those of the bison of the Great Plains.

There are a number of descriptions of the Central Valley and Sierra foothills grasslands, such as Audubon's account in 1851:

> The whole country to the north and east of Stockton through to the Calaveras is most rich and splendid soil, in many places too low for farming, but the grazing was excellent, quantities of wild oats, rye grass (I think), clover and a species resembling red-top. In many places the grasses were breast high as I waded through them, but generally knee-deep. (Audubon and Audubon 1969)

From these descriptions of wild herbivore populations and their grassland habitats, it appears that the low-elevation rangelands of the project study area were very productive when the first Europeans arrived. Burcham (1957) concluded that the presettlement grasslands were composed mainly of perennial bunchgrasses, although he notes that the remains of three exotic annuals were found in the adobe bricks of the earliest missions. Burcham (1957) attributes dispersal of these species to the Spanish explorers of the 1500s.

Burcham (1957) believed that most of the conversion from perennial to exotic annual grasslands occurred in the late 1850s and early 1860s. He attributes this invasion to a variety of factors working in concert, including extended periods of drought and extremely heavy grazing pressure during this interval. However, he also notes that several early American explorers—including Leonard quoted earlier, and Audubon quoted in the previous passage—observed extensive invasion of exotic annuals in California grasslands during the middle 1800s. Thus, there was a large seed bank waiting when favorable conditions occurred.

Conditions of Montane Meadows

Descriptions of montane rangeland conditions are few, especially before 1860. Most of the accounts by early explorers seem to be focused on foothill vegetation, although it is often difficult to determine exactly where the explorers were at the time of their observations. In addition, these parties often crossed the mountains during periods of snow cover at higher elevations. It wasn't until the survey expeditions after 1860, and John Muir's accounts after 1869, that more information about rangeland conditions was provided.

Lt. John C. Fremont, U.S. Topographical Engineer, led several expeditions into California during the 1840s. Cermak and Lague (1993) discuss Fremont's crossing around Carson Pass in the winter of 1843/44. Before beginning, Fremont sent a party under Edward Kern to cross the Sierra to the south. Kern led his detachment over Walker Pass and down the South Fork of the Kern River to the location of today's Kernville in January 1844. On January 21 Kern noted,

> Among the foot-hills are beautiful groves of live and other oaks, clear from growth of underwood; the fine grass gives the country the appearance of a well-kept park.

Again, both Kern's and Fremont's crossings in January precluded any observation of rangeland ecosystems at higher elevations.

Conditions of Montane Rangelands.

John Muir arrived in California in 1868, spending his first summer in Yosemite working for a sheep rancher in 1869. In this and subsequent years he would travel extensively throughout the central Sierra Nevada, always on foot, usually without a pack animal, observing and recording in detail the geologic and ecological processes he encountered. Only a few of his observations concern rangeland vegetation, but they provide some idea of what conditions were like before the severe impacts of unregulated grazing occurred.

During his travels, Muir explored the upper areas of what is now Yosemite National Park. In 1869, he encountered a bear in a meadow near his camp above Yosemite Valley. His account also describes the meadow vegetation:

> And there stood neighbor bruin within a stone's throw, *his hips covered by tall grass and flowers*, and his front feet on the trunk of a Fir that had fallen out into the meadow, which raised his head so high that he seemed to be standing erect . . . harmonizing in bulk and color and shaggy hair with the trunks of the trees and *lush vegetation*. (Muir 1982) (emphasis added)

In September 1875, Muir undertook what was perhaps the first "ecological assessment" of the giant sequoia belt of the Sierra Nevada, traveling from Yosemite Valley south to the Kern Basin, covering more than 322 km (200 mi) in his journey (Muir 1982). He described his purpose for this reconnaissance of the giant sequoia:

> In particular, I was anxious to find out whether it had ever been more widely distributed since the glacial period; what conditions . . . were affecting it; . . . and whether, as was generally supposed, the species was near extinction.

Because this trip was planned to last for some weeks during the autumn, Muir was persuaded to take "Brownie," a small wild mule, to carry provisions and two blankets. He thus had perhaps more occasion to observe and remark upon the conditions of montane meadows he encountered along the way. Muir described the general extent of these rangeland systems, as well as their individual appearance:

> Imbedded in these majestic woods are numerous meadows, around the sides of which the Big Trees press close together in beautiful lines. (Muir 1982)

Camped along the edge of a meadow in the Kaweah Basin, Muir describes its vegetation:

> There lay the grassy, flowery lawn, three-fourths of a mile long, . . . ruffled here and there with patches of ledum and scarlet vaccinium. (Muir 1982)

Later, camping two miles downstream, he related that "Brownie had plenty of grass."

One of Muir's "secret spots" was a place then known as Shadow Lake, at about 2,225 m (7,300 ft), some 13 km (8 mi) from Yosemite Valley. He described the fall conditions around the lake during his first visit in the autumn of 1872 (Muir 1982):

> The goldenrods are in bloom; but most of the color is given by the ripe grasses, Willows, and Aspens. . . . round the shores sweeps a curving ribbon of meadow, red and brown dotted with pale yellow, shading off hear [sic] and there into hazy purple.

After years of visiting this site without seeing anyone else there, Muir relates:

> On my last visit, . . . I was startled by a human track, which I saw at once belonged to some shepherd. . . . Returning from the glaciers shortly afterward, my worst fears were realized. A trail had been made down the mountain-side from the north, and all the gardens and

meadows were destroyed by a horde of hoofed locusts, as if swept by fire. (Muir 1982)

Conditions of East-Side Rangelands. Fletcher (1987) describes the presettlement vegetation of Mono basin, based on historical accounts and ethnographic investigations. Perennial grasses were evidently abundant in the sagebrush-scrub community, especially giant wild rye (*Elymus cinereus*) and Indian ricegrass (*Oryzopsis hymenoides*). The Kuzedika Paiute, a small band centered in the Mono basin, collected seeds from these bunchgrasses and from desert needlegrass (*Stipa speciosa*) as part of their varied diet, which also included desert peach (*Prunus andersonii*), elderberry (*Sambucus mexicana*), and buffaloberry (*Shepherdia argentea*). In addition, the Kuzedika held rabbit drives every fall, setting fire to the sagebrush to flush out the animals, a practice that would have been favorable to grass growth.

John Muir also observed range conditions in the east-side type in 1869, when he encountered Kuzedika women in the Mono Basin, harvesting wild rye (probably *Elymus cinereus*) in their traditional way (Fletcher 1987):

> Four or five miles from the (Mono) lake I came to a patch of elymus, or wild rye, growing in magnificent waving clumps six or eight feet high, bearing heads six or eight inches long. The crop was ripe, and Indian women were gathering the grain in baskets by bending down large handfuls, beating out the seed, and fanning it in the wind. The grains are about five-eighths of an inch long, dark-colored and sweet. I fancy the bread made from it must be as good as wheat bread. (Muir 1911)

Vale (1975) reviewed twenty-nine journals and diaries of early travelers through the Great Basin to determine vegetative descriptions before settlement. Unfortunately, only one source indicates the type of grasses or forbs found:

> More mountainous localities . . . are covered with meadows of a tall grass resembling somewhat rye. At still more swampy points, rushes and sedge-grasses occupy the surface. Over dry, deep sandy slopes, an exceedingly nutritious grass is scattered in single bunches, bearing large sweet seeds, which are eagerly sought for by the animals and Indians. (Simpson 1876)

Unregulated Grazing of Sierra Nevada Rangelands (1850–91)

Impacts of the Livestock Industry

Burcham (1957) discusses the growth of the livestock industry in California in the 1850s in response to the tremendous surge in population after the gold rush. His data, based on U.S. Census reports, show a fivefold increase in range stock from 1850 to 1860, including about one million each of beef

cattle and sheep in 1860. He notes that these data probably underestimate true livestock populations because of census sampling biases. Wagner (1989) has adjusted the census data based on U.S. Department of Agriculture (USDA) reports, and his data indicate even larger numbers: 1,800,000 beef cattle and 1,730,000 sheep in 1860. Much of this growth was due to livestock imported from other parts of the country (Burcham 1957).

Burcham (1957) describes the slackening market demand and a series of floods and droughts in the 1860s that disrupted the California livestock industry. As the gold rush boom abated, demand for beef cattle was curtailed, just as animal numbers were peaking. As demand dropped, beef cattle numbers soared to more than 3,000,000 head in 1862 (Burcham 1957). From 1863 to 1864, severe drought devastated the state's livestock industry. As large numbers of animals died of starvation, ranchers began to drive stock into the Sierra Nevada to forage. The increase in livestock numbers, combined with periodic drought and the loss of valley rangelands to farming, contributed to the impacts of grazing on Sierra Nevada rangelands from about 1870 to 1900 (Vankat 1970). Burcham (1957) believed that the most significant damage to native California perennial grasses occurred in the 1860s, although there is some evidence that the exotic annuals had made important inroads in California even before the growth of the livestock industry.

Common Property Rangeland Resources

Beesley (1996) discusses how Sierra meadows and grasslands were apparently overused before 1900. The historical accounts quoted later seem to agree that sheep grazing, as conducted during this period, affected rangeland condition more than cattle grazing. This greater impact was mainly due to the higher numbers of sheep over a longer summer season and to the sheepherders' burning practices, which were evidently more frequent and extensive than those of the Native Californians. Beesley notes that the first report by the California State Board of Forestry included recommendations to exclude sheep grazing because of the damage it caused (Wagoner 1886).

Because there was no regulation on the number of herds or the duration of their stay in the mountains, these rangelands seem to have been severely grazed on a frequent basis (Burcham 1957). Thus, there was little or no opportunity for the vegetation to recover adequately. In addition, sheepherders set fires to remove brush and invigorate the forage, again without recognizing the recovery time needed to restore lost nutrients to the ecosystem.

The irony here is that the herd management of any individual sheepherder was evidently quite reasonable, assuming each herd would not return to any grazing site for a suitable rest period. It was the combined effect of a number of herds that created the overuse. The economic incentives of sheep and wool production were evidently so great that exploitation of the "commons" was accelerated (Burcham 1957).

Under a system of common ownership of a resource, with no regulation of individual use, the so-called "tragedy of the commons" occurs. In economic terms, the common property resource, here the public rangelands, are overused, because no individual user has any incentive to conserve or steward the resource; any reduction in his use is quickly captured by other users (Howe 1979). Sierra Nevada grazing in the late 1800s is a classic example of this type of market failure, which requires collective action to remedy.

Impacts of Sheep Grazing

John Muir was one of the most ardent critics of sheep grazing as it was practiced during this period. He wrote about

> the comprehensive destruction caused by "sheepmen." Incredible numbers of sheep are driven to the mountain pastures every summer and their course is ever marked by desolation . . . , the shrubs are stripped of leaves as if devoured by locusts, and the woods are burned. (Muir 1877)

Vankat (1970) notes that a number of travelers in the Sierra during this period agreed with Muir. The California Geological Survey, headed by Josiah D. Whitney, launched a series of scientific expeditions into the Sierra Nevada, led by William H. Brewer, Whitney's principal assistant. Clarence King first worked as a field assistant to Brewer. King's description of his second visit to the Kern Plateau in September 1873 reveals the degree of change:

> The Kern Plateau, so green and lovely on my former visit in (July of) 1864, was now a gray sea . . . no longer velveted with meadows and upland grasses. . . . shepherds . . . leaving hardly a spear of grass behind them. (King 1902)

On his ascent of Mount Whitney in 1885, Magee (1885) observed:

> Mountain meadows are abundant, but the sheep-herder and his flocks have . . . worked their ruin. . . . Each of these meadows is yearly cropped several times by various flocks of sheep, and the result is that, . . . there are now only shreds and patches. The sod and the verdure are gone—eaten and trodden out; the gravel is now in the ascendant.

William Russel Dudley, professor of botany at Stanford University, traveled throughout the Sierra in the late 1890s. In 1898 Dudley, on a trip to the Kaweah Peaks region, observed that

> the great obstacle to the explorer is not the danger from crag or chasm, but the starvation threatening his animals,

through the destruction of the fine natural meadow pasturage by sheep. (Dudley 1898)

On the east side, rangeland productivity decreased, evidently because of livestock mismanagement. Fletcher (1987) notes that geologist Israel Russell, who had visited Mono Lake in 1881, observed the effects of overgrazing there in 1887:

> There was formerly sufficient wild grass in many portions of the basin to support considerable numbers of cattle and sheep; but owing to overstocking, these natural pastures are now nearly ruined. (Russell 1889)

Early Public Ownership and Regulation (1891–1905)

Although federal ownership began with the creation of Yosemite, Kings Canyon, and Sequoia National Parks in 1890, Beesley (1996) points out that public ownership actually began in 1864, when the federal government granted Yosemite Valley and the central Sierra Big Trees to California to preserve and protect these areas. However, such efforts proved ineffective, and starting in 1881, the state began asking the federal government to protect Sierran lands from destructive wildfire and unregulated logging, mining, and grazing (Ewing et al. 1988).

In spite of federal ownership, however, including the establishment of the Sierra Forest Reserve in 1893, numerous problems of enforcement arose (Vankat 1970). Vankat (1970) notes that initial protection of the national parks was left to the U.S. Army, beginning in the summer of 1891, and was hampered by lack of adequate enforcement penalties. In addition, because the troops left in the fall, sheepherders would use the parks until driven out by winter weather. These problems were described by a number of officials and others who traveled through the Sierra during this period. In his 1893 report, the acting superintendent of the Sequoia National Park stated that the Kern River drainage was

> almost impassable to the traveler, to such an extent is every thing eaten off the face of the earth and trampled under foot by the hundreds of thousands of sheep which every year roam over that territory. (U.S. Department of Interior 1893)

In his 1894 report, the acting superintendent stated:

> For years the Kern River country has been a sheep range, and enormous numbers of sheep are driven there annually. As a consequence the country is entirely denuded of grasses and bushes and presents a barren, uninviting aspect. . . . the whole country has, from a beautiful land once covered with nice and luxuriant grass, been turned into a desert. (U.S. Department of Interior 1894)

Vankat (1970) concludes that the army had stopped summer sheep grazing in California's national parks by about 1896, a fact corroborated by Dudley (1896) in 1895, who observed:

> To pass from the trampled meadows of the reservation (Sierra Forest Reserve) to the protected meadows of the (Sequoia) National Park is a lesson in patriotism.

However, in 1898, while army troops were stationed in the Philippine Islands during the Spanish-American War, trespass and unauthorized use of national parks again increased, as noted by the acting superintendent in his 1899 report:

> It is estimated that at least 200,000 sheep roamed at will over the national preserve, a destructive fire raged in the Giant forest, and hunters frequented the parks with impunity. (U.S. Department of Interior 1899)

Army troops returned in the summer of 1899, but political pressure was also at work in trying to loosen the regulations, as John Muir complained in an 1899 letter:

> The sheep owners in particular are already giving trouble and promise more next season. I have just learned that . . . sheep invaded and desolated the reservation last summer under a concession made by Secretary Bliss. (Muir 1899)

After creation of the forest reserve system, the U.S. Geological Survey initiated systematic surveys of the new reserves (McKelvey and Johnston 1992). Although the primary purpose of these surveys was to assess timber reserves, they briefly noted range resources. Sudworth observed in his survey of 1900:

> There are practically no grasses or other herbaceous plants. The forest floor is clean. The writer can attest the inconvenience of this total lack of grass forage for in traveling over nearly 3,000,000 acres not a single day's feed for saddle and pack animals was secured on the open range. . . . Barrenness is, however, not an original sin. From a study of long-protected forest land in the same region and from the statements of old settlers, it is evident that formerly there was an abundance of perennial forage grasses throughout this territory. (Sudworth and Gannett 1900)

Many areas in the higher elevations appeared to recover rapidly after rest. Muir (1917) noted that in Yosemite "the gardens and beds of underbrush once devastated by sheep are blooming in all their glory." Dudley's observation about the contrast between the Sierra Reserve and the Sequoia National Park, quoted earlier, further supports the conclusion that many areas did recover rapidly, at least in terms of observ-

able forage. However, Burcham (1957) felt that the effects of severe grazing during this time are still evident in the plant composition of these meadows.

Historical Conditions in the Sierra Nevada Rangeland Ecosystems before 1905

Historical accounts of rangeland condition and use in the late 1800s indicate that highly productive rangeland communities existed throughout the study area when Europeans arrived. Large elk herds were present on the west side of the range. Perennial grasses were dominant in the grassland communities, although exotic annuals had begun their invasion even before the arrival of the first missions in 1769.

The first extensive use of Sierra Nevada rangelands for livestock began in the 1860s. A number of observers reported severe and repeated grazing until about 1900, due in part to a lack of regulation of the common rangelands. Livestock management, combined with extended drought, contributed to the conversion of Sierra foothills to annual grasslands and is also implicated, along with climate, in the expansion of juniper woodlands on the east side of the range.

The overall agreement of many observers about the conditions and uses of rangeland vegetation tends to corroborate their observations. Without regulated access during the late 1800s, overuse of the common rangelands of the Sierra Nevada occurred, because forage conservation by any one livestock operator would be captured by another. As a result, Sierra Nevada rangelands were overgrazed, in that forage plants did not have enough time to recover after severe, repeated grazing. As unregulated grazing was eliminated, recovery of some of the rangeland vegetation in many areas was fairly rapid, at least in terms of forage production. However, questions of long-term loss of species diversity, and other potentially adverse ecological impacts, are addressed in Menke et al. 1996.

CONCLUSIONS AND IMPLICATIONS FOR MANAGEMENT

During the last 20,000 years, the location and extent of Sierra Nevada rangeland ecosystems have changed significantly in response to major changes in climatic and environmental conditions. Modern rangeland communities became established in their current locations after the end of the Pleistocene about 10,000 years ago. In the case of many riparian meadow sites, conditions alternated between meadow and forested ecosystems during the early Holocene, with modern wet meadows occurring only within the last 2,000 to 3,000 years. During the late Pleistocene a grass-sagebrush rangeland existed where montane and subalpine forests occur today, while at lower elevations where today we see chaparral and grass-woodland communities, a mixture of xerophytic and mesophytic conifers occurred.

The sagebrush grasslands of the late Pleistocene supported a diverse ecosystem of now extinct megafauna, including a large number of herbivores and a formidable group of mammalian predators. Evidence from meadow sediments indicates that these animals were abundant until about 11,000 years ago. The abundance of this diverse assemblage indicates that these rangelands were highly productive and that animal disturbance was an integral part of ecosystem health and productivity. Like the bison on the Great Plains and the diverse herbivores of East Africa, the behavior of these animals would likely have resulted in brief episodes of severe disturbance followed by longer recovery periods, as well as sequential episodes of less severe grazing on different components of the plant community by herds of different species. This disturbance regime, quite unlike traditional livestock behavior in the United States, would have provided several crucial functions for sustaining the high productivity of that rangeland ecosystem, including the breakdown and recycling of plant materials, increasing the net productivity of rangeland vegetation, improving the rangeland water cycle, and creating conditions favorable to seed germination and seedling establishment.

It is time for resource scientists to recognize that properly managed large-animal disturbance is as natural and necessary to healthy ecosystems as properly managed wildfire. Just as fire can seriously impair ecosystem health if its timing and severity are not matched to the natural rhythms of the ecosystem, so animal impacts can also be undesirable under grazing management that does not control the timing and frequency of animal disturbance. However, when fire or animal disturbance is used adaptively, where soil conditions, plant growth rates, and community dynamics are monitored and adequate ecosystem recovery times are provided, these management tools are essential to sustaining healthy Sierra Nevada ecosystems.

What role should large herbivores play in the management of Sierra Nevada rangelands today? Is it possible to achieve this former level of productivity and diversity with existing rangeland plant communities? It's clear that animal disturbance has considerable potential to effect changes in rangeland ecosystems. Whether these changes are beneficial or undesirable depends on how animals are managed. On some sites a lack of disturbance will lead to the gradual decline of ecosystem vitality, and fire or animal disturbance is needed to restore ecosystem functioning. This restoration requires livestock-management strategies that recognize beneficial plant-herbivore relationships, monitor the impacts on vegetation closely, and adjust animal behavior to achieve desired conditions.

As stated in the Introduction, a full review of the research on plant-herbivore relationships is beyond the scope of this chapter. However, it is clear that to study these relationships

and their role in rangeland management properly, future research must be integrated with management and public involvement in an adaptive framework (see Kusel et al. 1996). To accomplish this kind of management on public lands, agencies will have to revise their regulations and guidelines, which focus too narrowly on forage utilization standards as an indicator of how to manage grazing animals. Rather, they need to focus on controlling the length of grazing and recovery periods based on site-specific monitoring of plant growth rates and on active management of animal behavior.

REFERENCES

Anderson, R. S. 1990. Holocene forest development and paleoclimates within the central Sierra Nevada, California. *Journal of Ecology* 78 (2): 470–89.

Anderson, R. S., and S. J. Smith. 1994. Paleoclimatic interpretations of meadow sediment and pollen stratigraphies from California. *Geology* 22 (8): 723–26.

Audubon, J. W., and M. R. Audubon. 1969. *Audubon's western journal, 1849–1850.* Glorieta, NM: Rio Grande Press.

Beesley, D. 1996. Reconstructing the landscape: An environmental history, 1820–1960. In *Sierra Nevada Ecosystem Project: Final report to Congress,* vol. II, chap. 1. Davis: University of California, Centers for Water and Wildland Resources.

Benson, L. V., D. R. Currey, R. I. Dorn, K. R. Lajoie, C. G. Oviatt, S. W. Robinson, G. I. Smith, and S. Stine. 1990. Chronology of expansion and contraction of 4 Great Basin lake systems during the past 35,000 years. *Palaeogeography, Palaeoclimatology, Palaeoecology* 78 (3–4): 241–86.

Burcham, L. T. 1957. *California range land: An historico-ecological study of the range resource of California.* Sacramento: California Department of Natural Resources, Division of Forestry.

Cermak, R. W., and J. H. Lague. 1993. Range of light—range of darkness: The Sierra Nevada, 1841–1905. Unpublished report. San Francisco: U.S. Forest Service, Pacific Southwest Regional Office.

Cole, K. 1983. Late Pleistocene vegetation of Kings Canyon, Sierra Nevada, California. *Quaternary Research* 19:117–29.

Davis, O. K. 1987. Spores of the dung fungus Sporormiella: Increased abundance in historic sediments and before Pleistocene megafaunal extinction. *Quaternary Research* 28:289–94.

Davis, O. K., L. Agenbroad, P. S. Martin, and J. I. Mead. 1984. The Pleistocene dung blanket of Bechan Cave, Utah. *Special Publication of the Carnegie Museum of Natural History* 8:267–82.

Davis, O. K., and M. J. Moratto. 1988. Evidence for a warm dry early Holocene in the western Sierra Nevada of California: Pollen and plant macrofossil analysis of Dinkey and Exchequer Meadows. *Madrono* 35 (2): 132–49.

Dudley, W. R. 1896. Forest reservations: With a report on the Sierra Reservation, California. *Sierra Club Bulletin* 1:254–67.

———. 1898. The Kaweah group. *Sierra Club Bulletin* 2:185–91.

Edwards, S. W. 1992. Observations on the prehistory and ecology of grazing in California. *Fremontia* 20 (1): 3–11.

Ewing, R. A., N. Tosta, R. Tuazon, L. Huntsinger, R. Marose, K. Nielson, R. Motroni, and S. Turan. 1988. *Rangelands: Growing conflict over changing uses.* Sacramento: California Department of Forestry and Fire Protection, Forest and Rangeland Resources Assessment Program.

Fletcher, T. C. 1987. *Paiute, prospector, pioneer: The Bodie–Mono Lake area in the nineteenth century.* Lee Vining, CA: Artemisia Press.

Grayson, D. 1993. *The deserts past: A natural prehistory of the Great Basin.* Washington, DC: Smithsonian Institution.

Howe, C. W. 1979. Natural resource economics. New York: John Wiley.

Jennings, S. A., and D. L. Elliott-Fisk. 1993. Packrat midden evidence of late Quaternary vegetation change in the White Mountains, California-Nevada. *Quaternary Research* 39 (2): 214–21.

King, C. 1902. *Mountaineering in the Sierra Nevada.* New York: Scribners.

Koehler, P. A., and R. S. Anderson. 1994. Full-glacial shoreline vegetation during the maximum highstand at Owens Lake, California. *Great Basin Naturalist* 54 (2): 142–49.

Kusel, J., S. C. Doak, S. Carpenter, and V. E. Sturtevant. 1996. The role of the public in adaptive ecosystem management. In *Sierra Nevada Ecosystem Project: Final report to Congress*, vol. II, chap. 20. Davis: University of California, Centers for Water and Wildland Resources.

Leonard, Z. 1959. *Adventures of Zenas Leonard, fur trader.* Norman: University of Oklahoma Press.

Magee, T. 1885. Ascent of Mt. Whitney. In *Tulare County scrapbook.* N.p.

McCullough, D. R. 1971. *The tule elk: Its history, behavior, and ecology.* University of California Publications in Zoology 88. Berkeley and Los Angeles: University of California Press.

McKelvey, K. S., and J. D. Johnston. 1992. Historical perspective on the forests of the Sierra Nevada and the Transverse Ranges of Southern California: Forest conditions at the turn of the century. In *The California spotted owl: A technical assessment of its current status*, edited by J. Verner. General Technical Report PSW-133. Berkeley, CA: U.S. Forest Service, Pacific Southwest Forest and Range Experiment Station.

McNaughton, S. J. 1976. Serengeti migratory wildebeest: Facilitation of energy flow by grazing. *Science* 191:445–67.

———. 1979. Grazing as an optimization process: Grass-ungulate relationships in the Serengeti. *American Naturalist* 113:691–703.

———. 1983. Compensatory plant growth as a response to herbivory. *Oikos* 40:329–36.

———. 1986. On plants and herbivores. *American Naturalist* 128:765–70.

———. 1993. Grasses and grazers, science and management. *Ecological Applications* 3 (1): 17–20.

Mead, J. I., L. Agenbroad, O. K. Davis, and P. S. Martin. 1986. Dung of the *Mammuthus* in the arid Southwest, North America. *Quaternary Research* 22:121–27.

Menke, J., C. Davis, and P. Beesley. 1996. Rangeland assessment. In *Sierra Nevada Ecosystem Project: Final report to Congress*, vol. III. Davis: University of California, Centers for Water and Wildland Resources.

Miller, R. F., and P. E. Wigand. 1994. Holocene changes in semiarid pinyon-juniper woodlands. *Bioscience* 44 (7): 465–74.

Muir, J. 1877. On the post-glacial history of Sequoia gigantea. *Proceedings of the American Association for the Advancement of Science* 25.

———. 1899. Letter to Theodore P. Lukens, 10 January.

———. 1911. *The mountains of California.* New York: Century Co.

———. 1917. *Our national parks.* Boston: Houghton-Mifflin.

———. 1982. *The wilderness world of John Muir.* Edited by E. W. Teale. Boston: Houghton Mifflin.

Newberry, J. S. 1884. *Notes on the geology and botany of the country bordering the Northern Pacific Railroad.* New York: New York Academy of Sciences.

Nichols, P., and N. Bourne. 1653. *Sir Francis Drake revived.* London: Nicholas Bourne.

Nowak, C. L., R. S. Nowak, R. J. Tausch, and P. E. Wigand. 1994. Tree and shrub dynamics in northwestern Great Basin woodland and shrub steppe during the late Pleistocene and Holocene. *American Journal of Botany* 81 (3): 265–77.

Oesterheld, M., and S. J. McNaughton. 1991. Effect of stress and time of recovery on the amount of compensatory growth after grazing. *Oecologia* 85:305–13.

Oesterheld, M., O. E. Sala, and S. J. McNaughton. 1992. The effect of animal husbandry on herbivore carrying capacity at a regional scale. *Nature* 355:234–36.

Russell, I. C. 1889. Quaternary history of Mono Valley, California. In *USGS Eighth Annual Report 1886–1887*, 261–394. Reprinted by Artemisia Press, Lee Vining, CA.

Savory, A. 1988. *Holistic resource management.* Washington, DC: Island Press.

Simpson, J. 1876. *Report of the explorations across the Great Basin of the territory of Utah for a direct wagon-route from Camp Floyd to Genoa, in Carson Valley, in 1859.* Washington, DC: Government Printing Office.

Smith, J. S., and G. R. Brooks. 1977. *The Southwest expedition of Jedediah S. Smith: His personal account of the journey to California, 1826–1827.* Glendale, CA: A. H. Clark.

Smith, S. J., and R. S. Anderson. 1992. Late Wisconsin paleoecologic record from Swamp Lake, Yosemite National Park, California. *Quaternary Research* 38 (1): 91–102.

Stine, S. 1990. Late Holocene fluctuations of Mono Lake, eastern California. *Palaeogeography, Palaeoclimatology, Palaeoecology* 78:333–81.

———. 1994. Extreme and persistent drought in California and Patagonia during Medieval time. *Nature* 369 (6481): 546–49.

Sudworth, G. B., and H. Gannett. 1900. Stanislaus and Lake Tahoe Forest Reserves, California, and adjacent territory. In *Twenty-first annual report of the USGS, part V, Forest reserves.* Washington, DC: Government Printing Office.

Tausch, R. J., P. E. Wigand, and J. W. Burkhardt. 1993. Viewpoint—plant community thresholds, multiple steady states, and multiple successional pathways—legacy of the Quaternary. *Journal of Range Management* 46 (5): 439–47.

Thompson, R. A. 1896. *The Russian settlement in California known as Fort Ross.* Santa Rosa, CA: Sonoma Democrat Publishing Co.

U.S. Department of Interior. 1893. *Report of the Acting Superintendent of the Sequoia and General Grant National Parks, to the Secretary of the Interior.*

———. 1894. *Report of the Acting Superintendent of the Sequoia and General Grant National Parks, to the Secretary of the Interior.*

———. 1899. *Report of the Acting Superintendent of the Sequoia and General Grant National Parks, to the Secretary of the Interior.*

Vale, T. R. 1975. Presettlement vegetation in the sagebrush-grass area of the Intermountain West. *Journal of Range Management* 28 (1): 32–36.

Vankat, J. L. 1970. Vegetation change in Sequoia National Park, California. Ph.D. diss., Department of Botany, University of California, Davis.

Wagner, F. H. 1989. Grazers, past and present. In *Grassland structure and function: California annual grassland,* edited by L. F. Huenneke and H. A. Mooney, 151–62. Tasks for Vegetation Science 20. Boston: Kluwer Academic.

Wagoner, L. 1886. Report on the forests of the counties of Amador, Calaveras, Tuolumne, and Mariposa. In *First biennial report of the California State Board of Forestry for the years 1885–1886.* Sacramento: California State Board of Forestry.

Walters, C. J., and C. S. Holling. 1990. Large-scale management experiments and learning by doing. *Ecology* 71 (6): 2060–68.

Wigand, P. E. 1987. Diamond Pond, Harney County, Oregon: Vegetation history and water table in the eastern Oregon desert. *Great Basin Naturalist* 47:427–58.

Wigand, P. E., M. L. Hemphill, S. Sharpe, and S. Patra. 1995. Great Basin semi-arid woodland and montane forest dynamics during the late Quaternary. In *Climate change in the Four Corners and adjacent regions: Implications for environmental restoration and land-use planning,* edited by W. J. Waugh. Proceedings of workshop, 12–14 September 1994. Grand Junction, CO: U.S. Department of Energy, Grand Junction Projects Office.

Wigand, P. E., and P. J. Mehringer Jr. 1985. Pollen and seed analyses. In *The archeology of Hidden Cave,* edited by D. H. Thomas. Anthropological Papers of the American Museum of Natural History 61. New York: American Museum of Natural History.

Wilkes, C. 1845. *Narrative of the United States exploring expedition during the years 1838, 1839, 1840, 1841, 1842.* Philadelphia: Lea and Blanchard.

Wood, S. H. 1979. *Holocene stratigraphy and chronology of mountain meadows, Sierra Nevada, California.* Earth Resources Monograph 4. Berkeley, CA: U.S. Forest Service, Pacific Southwest Region.

Woolfenden, W. B. 1996. Quaternary vegetation history. In *Sierra Nevada Ecosystem Project: Final report to Congress,* vol. II, chap. 4. Davis: University of California, Centers for Water and Wildland Resources.

Work, J., and A. B. Maloney. 1945. *Fur brigade to the Bonaventura: John Work's California expedition for the Hudson's Bay Company, 1832–1833.* San Francisco: California Historical Society.

WALLACE B. WOOLFENDEN
U.S. Forest Service
Inyo National Forest
Lee Vining, California

4

Quaternary Vegetation History

ABSTRACT

The geological record reveals that Sierra Nevada ecosystems are a transitory part of a dynamic and evolving landscape. The effects of climatic change and disturbance events, however, can persist in time, and many ecological processes have long time spans. Therefore, the complete understanding of the current state of an ecosystem and its potential behavior, its adaptability and resilience, may require knowing its history over many hundreds to thousands of years. The Quaternary period is a geologic name for the past 2.4 million years, during which global climate cooled and at least six successive major glacial cycles covered the Sierra Nevada with ice caps and mountain glaciers, filled lake basins in the adjacent deserts, and lowered the elevation limits of plant species. These ice ages were interspersed with shorter warm intervals when arid habitats expanded into northerly latitudes and tree lines gained elevation. Species responded individualistically to these changes, sometimes assembling into communities with no known modern analog. During the peak of the last glaciation, 18,000 to 20,000 years ago, alpine vegetation, dominated by sagebrush and herbs, grew in a dry glacial climate above 1,500 m (4,900 ft) in the Sierra Nevada where there is now montane forest. At the same time a juniper woodland occupied the lower mountain slopes and desert basins of the eastern Sierra Nevada and an expanded bristlecone and limber pine forest covered the crest of the White-Inyo Mountains. As the glaciers waned after 14,000 years ago, dense conifer forests of a species mix unlike any modern assemblage were established in wetter climatic conditions at intermediate west-slope elevations. With continual warming after around 10,000 years ago, a dry, open conifer forest with a montane shrub understory and an increased number of oaks was growing throughout the range. Greater proportions of charcoal indicate an increase in fire frequency. Higher temperatures are also documented by an upslope extension of tree lines and the expansion of xeric desert species on the east side. Effective precipitation increased after about 6,000 years ago, as indicated by an increase of subalpine conifer populations and the closure of montane forests. Subsequent to 4,000 to 3,000 years ago, the apparent downslope retreat of whitebark pine, mountain hemlock, red and white fir, and incense cedar suggests an intensification of cooler conditions and the formation of the modern Sierran forests. A brief warm-dry period between A.D. 900 and 1300 was followed by a 400-year phase with average conditions cooler and wetter than today and multiple advances of alpine glaciers. Both climatic episodes are recorded in the paleoecological record.

INTRODUCTION

Time is the issue. Why is the history of Sierra Nevada vegetation over many thousands of years important to the Sierra Nevada Ecosystem Project (SNEP)? Ecologists know that the assessment of the condition and trend of an ecosystem requires measurements on the scale at which it operates (Scholes 1990). This is as true for temporal scales as it is for spatial scales. Ecosystems are dynamic and evolving. They have a history that can be hidden, because many ecosystem processes and changes occur at timescales greater than the timescale at which they are usually observed (Magnuson 1990; Scholes 1990). Serious errors can be made, for example, in the attempt to model the behavior of a plant community if there is a time lag in the response of that community to an external disturbance and if existing community structure is a result not of present conditions but of the persistence of a past effect.

Long time intervals cannot be practically studied with biological methods of observation and experiment. Historical records of climatic variables and evidence of past conditions provided by photographs and documents are too limited. Geological records of past environments, covering time periods of centuries to millennia, are becoming increasingly avail-

Sierra Nevada Ecosystem Project: Final report to Congress, vol. II, *Assessments and scientific basis for management options*. Davis: University of California, Centers for Water and Wildland Resources, 1996.

able. Although they have lower temporal, spatial, and biotic resolution than contemporary observations, they do reveal the response of vegetation to long-term variation in climate and disturbance regimes.

The reconstruction of Sierra Nevada vegetation through such proxy records as pollen and macrofossil (visible to the naked eye) stratigraphies and plant macrofossils from pack rat middens has the potential to define natural ranges of variability and trajectories for vegetation along a broad continuum of time frequencies. Paleoecological information can also help to identify thresholds, rates of change, successional pathways, and time lags; assess community stability and resilience to disturbance; specify the effective independent and dependent variables involved in change; and provide a baseline for the assessment of the relative influence of past, current, and future human activities.

The objectives of this paper are to (1) outline the climatic chronology of the Quaternary period as a context for the vegetation history of the Sierra Nevada; (2) briefly describe the methods used in the reconstruction of past vegetation; (3) summarize findings concerning vegetation dynamics and its relation to climate; (4) review and summarize the current literature on the Quaternary vegetation history of the Sierra Nevada study area; and (5) address possible management implications of our understanding of the vegetation history of the Sierra Nevada ecosystem.

QUATERNARY CHRONOLOGY

The Quaternary vegetation history of the Sierra Nevada can be more fully understood if it is placed in the context of the global climatic chronology because climate is a major driver of vegetation dynamics. Conventional time-stratigraphic nomenclature is used in this chapter (Coleman et al. 1987). When it is convenient, chronometric ages are abbreviated *ka* and

Ma (thousand and million years, respectively, measured from the present). Thus, one thousand years ago is abbreviated 1.0 ka. Radiocarbon (^{14}C) ages are given as years before the present (*yr B.P.*), which is fixed at A.D. 1950: for example, 1200 yr B.P. Radiocarbon ages corrected for variation in the amount of ^{14}C produced in the atmosphere by calibration with the bristlecone pine tree-ring chronology is given as *cal yr B.P.* When calendar dates are precisely known from historical documents or tree-ring chronologies, they are given as B.C. or A.D.

The Quaternary period includes the Pleistocene and Holocene (or Recent) epochs. The Pleistocene was classically defined as a unique time when major glaciations occurred. This concept has somewhat lost its distinguishing characteristic with evidence of ice ages occurring throughout Earth's history since the early Precambrian over two billion years ago. Most recently, during the middle Miocene about 14 million years ago, the Antarctic ice sheet rapidly grew and mountain glaciers formed in the northern hemisphere (Flint 1971; Ghil 1991; Singh 1988). Nevertheless, the inception of the latest extensive continental glaciation is signaled by a rapid drop in levels of carbonate about 2.4 million years ago, documented in marine sediment cores from the North Atlantic Ocean (Ruddiman et al. 1990; Shackleton et al. 1984). This drop indicates the influx of ice-rafted particles from the ice sheets bordering the Atlantic that diluted the carbonates precipitating from seawater.

Chemical and microfossil analyses of marine sediments in deep ocean cores have provided continuous, high-resolution records of glacial-interglacial cycles throughout the Quaternary (Imbrie et al. 1992) (figure 4.1). Temporal variation in the distribution of foraminifera (a microscopic marine organism) assemblages reflect past temperature and salinity of ocean waters, and oxygen-isotope ratios ($\delta^{18}O$) extracted from the carbonate tests of these organisms provide estimates of temperatures and continental ice volume. The results of this research have been the resolution of what previously had been thought to be four glacial periods into at least eleven during

FIGURE 4.1

A benthic $\delta^{18}O$ record from the Atlantic Ocean for the past two million years (from Imbrie et al. 1992).

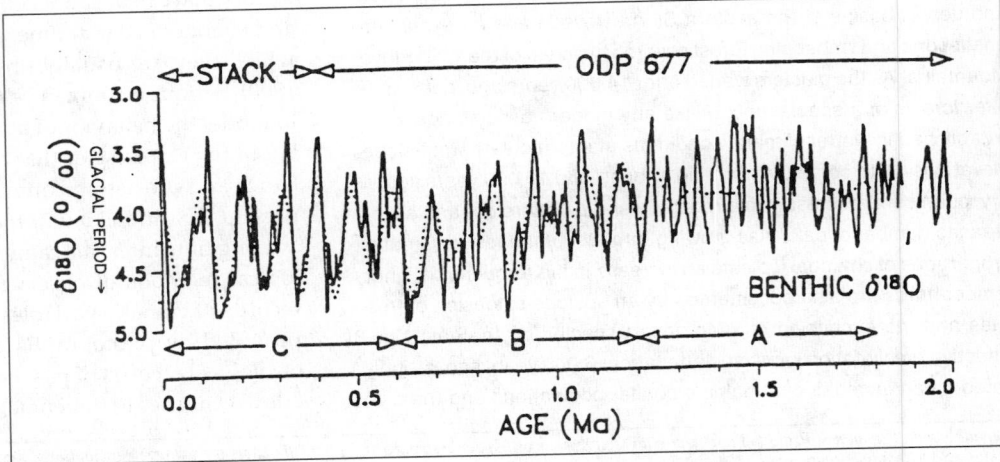

the last 900,000 years. At least six of these are recognized in the Sierra Nevada.

One of the dominant research questions in Quaternary science has been the causes of the Pleistocene ice ages. The initiation of continental glaciation has been attributed to changes in the coupled atmospheric and oceanic circulation system, which responded to such tectonic events as the closure of the isthmus of Panama and the uplift of the Sierra Nevada, Himalaya-Tibetan plateau, and other ranges (Ruddiman and Kutzbach 1989). The pattern of glacial oscillations is widely assumed to be driven by the seasonal and latitudinal distribution of incident solar radiation according to periodic variations in the earth-sun geometry (termed the Milankovitch theory, after the Yugoslav astronomer Milutin Milankovitch, who calculated and linked the orbital periodicities to climate changes [Berger 1991]). There are three major components: a 100,000-year cycle of the eccentricity (circular to elliptic) of the earth's orbit, a 41,000-year cycle of the earth's obliquity with respect to the ecliptic plane (variation in the tilt of the polar axis between 21.8° and 24.4°), and a 23,000-year precession of the equinoxes (the wobble of the earth's axis) (figure 4.2). Climatic information derived from ocean sediments, from ice cores taken from the Greenland and Antarctic ice caps, and from cores taken from terrestrial sedimentary basins has exhibited patterns that reflect these periodicities and demonstrate their influence on phenomena that are responsive to climate change. The combined effect of these geometric relationships produced the strong 100,000-year glacial-interglacial rhythm beginning about 700,000 years ago. Prior to this time the 41,000-year frequency dominated, with apparently weaker glacial expansions. Within these cycles, glacial periods averaged about 90,000 years, and warm interglacial periods averaged 10,000–20,000 years. The present Holocene climate phase, which began approximately 10,000 years ago, is the latest interglacial. Although it may be either halfway through or nearing its conclusion, the predicted global warming resulting from the injection of anthropogenic "greenhouse gases" into the atmosphere may be modifying the ancient climatic pattern.

Global ice volume grows with low summer insolation and high winter insolation and vice versa. There are also "free" oscillations generated by internal system feedback mechanisms and a large chaotic component. The presence or absence and limits of land and sea ice, sea surface temperatures, land albedo, effective soil moisture, composition of the atmosphere, and circulation of the atmosphere and oceans are other variables that can act as external forcing functions or internal variations, depending on scale. In general, external controls cause the large-scale variations over time and space, and internal controls operate on smaller scales, inducing continuous short-term variation (Bartlein 1988).

During the last deglaciation, summer solar radiation increased and winter solar radiation decreased from near present levels at 18,000 years ago to a summer maximum and winter minimum at 9,000 years ago. According to the com-

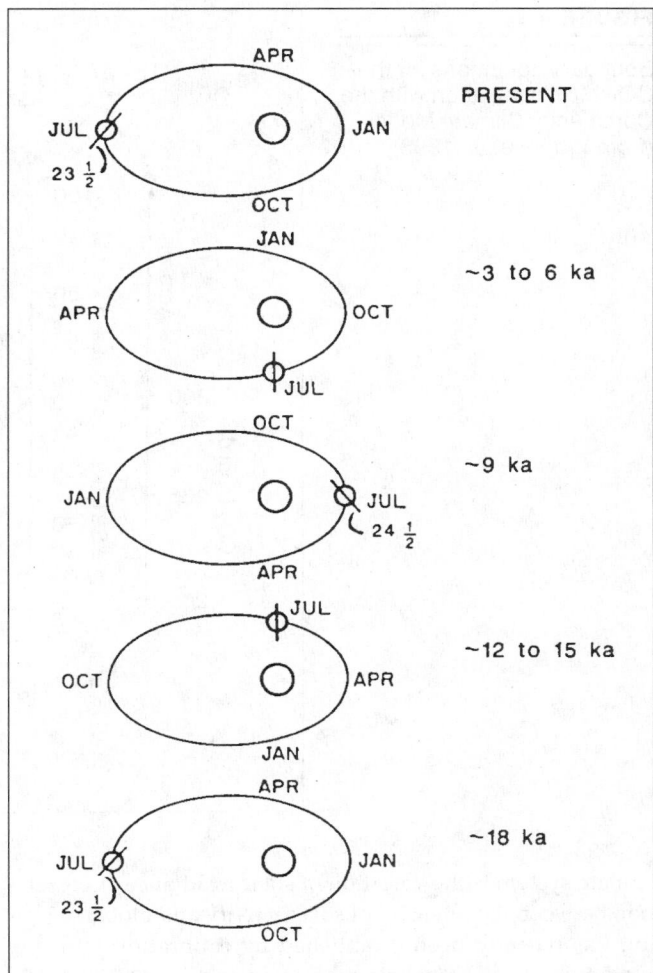

FIGURE 4.2

Earth-sun geometry at selected time intervals (from Wright et al. 1993).

puterized atmospheric general circulation models (GCMs) (COHMAP Members 1988; Kutzbach and Guetter 1986), major trends in climatic components include a decrease in land ice and an increase in sea surface temperatures, carbon dioxide, atmospheric dust, and—particularly important for vegetation change—seasonal variation in insolation (figure 4.3). Maximum summer insolation at 9,000 years ago caused the sustained above-average warm period known as the "early Holocene xerothermic" or "mid-Holocene climatic optimum," depending on the region affected. A rise in global temperature affects atmospheric circulation and, consequently, the position, orientation, and steepness of temperature gradients and precipitation patterns on a regional scale. So, while the northwestern United States became dry, the monsoons expanded and delivered increased summer rainfall to the Southwest, with resulting changes in vegetation.

At a smaller timescale another effect on the recent global

FIGURE 4.3

Boundary conditions for the
COHMAP simulation with the
Community Climate Model
(from Wright et al. 1993).

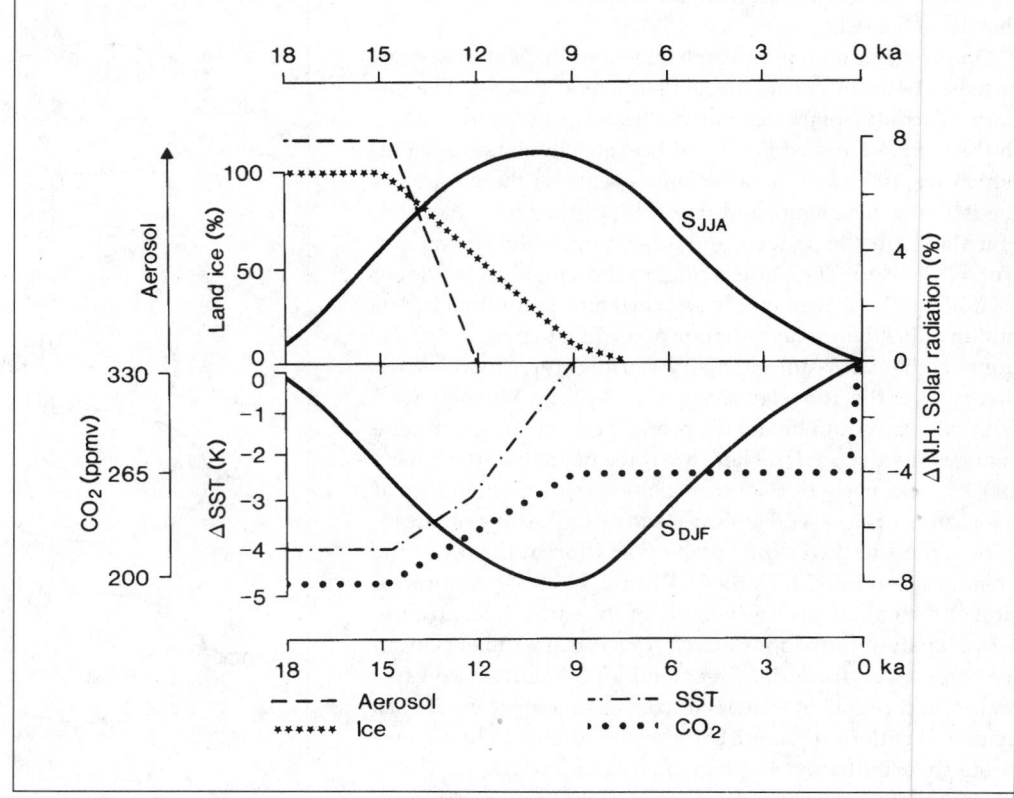

climate system is the variation of solar irradiance. A correlation between the intensity of solar activity and global warming has recently been established by calibration with the radiocarbon (^{14}C) chronology (Damon and Sonett 1991). The history of solar activity through observation of sunspot cycles since the seventeenth century and recent systematic studies of solar irradiance are linked to the radiocarbon record. The increase in solar winds as the sun becomes more active blocks the cosmic rays that produce the radioactive isotope of carbon; thus, ^{14}C production is inversely related to irradiance. Therefore, a 9,000-year radiocarbon record, taken from carbon isotopes in long tree-ring records, can be used as a proxy for solar activity and, indirectly, a proxy for global temperature variations (Davis 1994). Times of minimum solar activity (high ^{14}C production) between A.D. 1400 and A.D. 1850 correspond to the Little Ice Age. The time of maximum irradiance between A.D. 1100 and A.D. 1300 approximates the Medieval Warm Period (Hughes and Diaz 1994; Stine 1990). Patterns of climate thus vary both periodically and irregularly at all spatial-temporal scales and must be taken into account when modeling vegetation change (figure 4.4).

The surface of the earth is like a palimpsest with traces of former landscapes becoming more obscured with the passage of time. Thus, more detail is known about the most recent climatic events. We view the last glacial period (generally termed the Wisconsinan) from the vantage point of a warm interglacial peak, which we consider normal. Since the previous interglacial, which peaked about 125,000 years ago, global cooling has been accompanied by brief, warmer excursions, producing an irregular sawtooth pattern when the temperature curve is plotted. During the glacial maximum global surface temperature averaged 4°C less than that of today, ice covered one-third of the land surface of the earth, sea level was about 121 m (397 ft) below the present sea level, and atmospheric circulation patterns were modified (COHMAP Members 1988; Kutzbach and Guetter 1986). Global climate was predominantly a winter-precipitation regime. The expanded land and sea ice strengthened the latitudinal temperature gradient, displacing westerlies southward and weakening the summer monsoon. Pacific frontal storms were tracked into California farther south than today, possibly throughout much of the year due to the contraction of the northern Pacific high-pressure cell. Lower summer temperatures and greater effective moisture led to the filling of large lake basins in the Great Basin and southeast California and the altitudinal and latitudinal displacement of many plant species compared to modern distributions; elevational ranges of species were depressed more than 1,000 m in the SNEP study area. Locally in the southern Sierra Nevada, the climate was cold and dry. This greater continental climate regime may be attributed to exposure of the California continental shelf due to lowered sea levels, which captured much of the precipitation arriving onshore (Barnosky et al. 1987).

VEGETATION-CLIMATE RELATIONSHIPS

As with climate, vegetation has changed within a hierarchy of time and space. At issue here is the relationship of processes between different scales, especially which processes become dominant at higher or lower levels. Consideration of the time and space scale at which observations of vegetation change are made is important in determining how closely vegetation has tracked climate change (Cole 1985; Davis 1986; Prentice et al. 1991; Webb 1986).

On a fine scale of less than a century and within a plant community domain of about one hectare (2.47 acres) (which is also at a human scale of observation), there is an apparent stability of process, structure, and species composition and abundance classes, in spite of constant turnover through loss and recruitment. A conventional pollen record may show longer periods (100 to 1,000 years) of regional vegetation stability relative to the local dynamic equilibrium of disturbance and succession (Bradshaw 1988). Usually, however, relatively gradual as well as abrupt changes in vegetation composition are described by the higher-order timescales of regional pollen diagrams. Equilibrium and successional trend models, concepts of habitat type, and potential vegetation in ecology have been based on short-term observations. Newer, alternative concepts of thresholds and multiple, nonlinear pathways among many possible states of species composition are more compatible with the insights of vegetation dynamics gained by longer-term observations through paleoecological methods (Tausch et al. 1993).

If climate, soil genesis, and allogenic disturbances could be held stable, biotic processes alone would be sufficient to produce change, most likely a modification of species abundances. With a combination of disturbance events and climate change, a regenerating plant community that was established under a different climate may produce an altered composition and structure as it responds to new climatic conditions (Patterson and Prentice 1985, 100). An extreme example of the effect is the ghost forest of Whitewing Mountain (3,050 m [10,000 ft]) on the Inyo National Forest. Several hectares of prostrate logs litter the crest of the now bare, pumice-covered ridge, which is located southeast of the town of June Lake in the eastern escarpment of the Sierra Nevada. Below tree line, an east-side red fir–mixed conifer forest grows on the slopes of the mountain and surrounding pumice flats. According to a radiocarbon (^{14}C) date of about 900 yr B.P. from a wood sample taken from one of the logs, the forest was growing during a warm period. The source of the pumice has been identified as the nearby Glass Creek volcanic vent, which erupted about 720 yr B.P. (Miller 1984). The forest was apparently blown down by the eruption and never regenerated, perhaps because the growing season was becoming too short as the climate cooled within the following

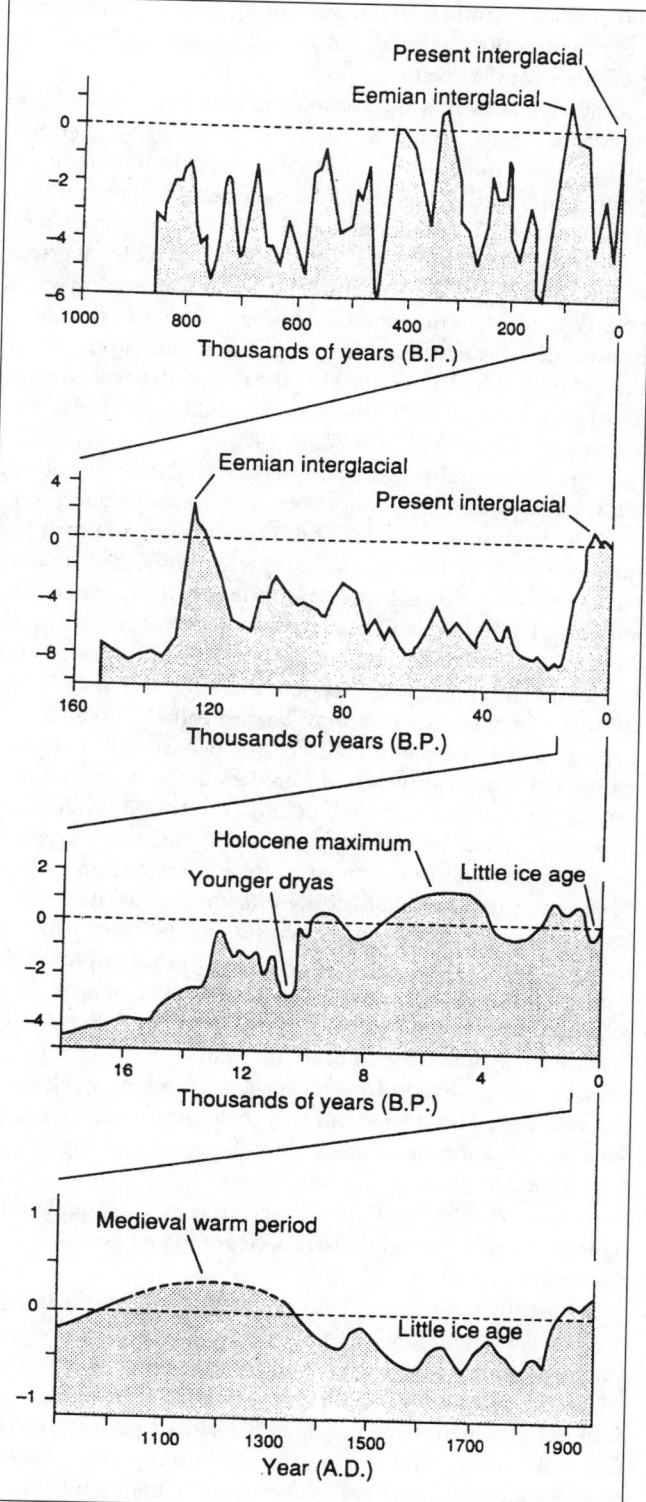

FIGURE 4.4

Global surface temperatures at four timescales (from Tausch et al. 1993; reprinted by permission of the Society for Range Management).

200 years (Graumlich 1993), preventing primary succession. At present only a few small individuals of whitebark pine are growing near the crest.

Disturbance events and variations in temperature and precipitation are by no means the only causes of vegetational change. Wide fluctuations in atmospheric carbon dioxide are being recognized as having had a major effect on ecosystems (Betancourt et al. 1993; Polley et al. 1993). For example, a recent study (Van De Water 1993) showed that a 40% increase in carbon dioxide at the last deglaciation was accompanied by a 17% decrease in stomatal density and a 14% increase in reconstructed water use efficiency in fossil limber pine. From these data it is hypothesized that the rise in atmospheric levels of carbon dioxide "influenced the ability of C3 plants to assimilate carbon" (Van De Water 1993).

Climate also alters the competitive balance of species within a stand or community because species respond individually to climate change; a shift in seasonality of precipitation, for instance, will favor some species over others, and an increase in winter drought will be differentially tolerated. The amount of suitable habitat relative to different topographic, edaphic (soil-related), and microclimatic conditions also is altered by climate change so that, even though spatial distribution of plant species may be controlled by substrate, populations expand and contract as fluctuations in climate influence their sensitivity to differences in the water-holding capacity of the soil (Graumlich and Davis 1993; Prentice et al. 1991). Global temperature variation, shifts in upper atmosphere circulation patterns, and redistribution of precipitation thus have a cascading effect down to the level of community dynamics and the physiology and life history of specific individuals, which are relatively more important at that level. But changing landscape patterns at smaller timescales reflect a transition of processual dominance that makes the effects of climatic variability unclear.

Paleoecological data are often used to estimate past climatic variables. Methods range from an ecological intuition based on a combination of intimate knowledge of the vegetation distribution in an area and precipitation and temperature data from local weather stations, lapse rates, and available autecological studies of individual species to more rigorous quantitative models.

Independent models of climate change, such as the GCMs cited earlier (Webb 1986), assist in understanding how vegetation responds to complex variation in climate. Many other types of paleoenvironmental data—the relative position and dating of glacial moraines, periglacial features, past lake levels, paleosols—contribute to a more complete representation of past climates, but they are insufficient for the identification of causes because of *convergence,* in which different processes and causes produce similar effects, and, conversely, *divergence,* in which similar processes and causes produce different effects in different regions (Schumm 1991). Synoptic to global-scale model simulations, compared with paleoenvironmental data, can help identify effective processes and causes.

PALEOECOLOGICAL METHODS

As a member of the geological sciences, paleoecology uses the consequent structures of biological and physical processes to infer past system states. The two principal methods used in the Sierra Nevada study area are pollen analysis and the analysis of plant macrofossils from sediments and pack rat middens.

Pollen Analysis

The analysis of palynomorphs (including pollen, spores, and algae colonies) can provide a continuous, high-resolution vegetation chronology. Samples of pollen and spores are preferably collected with coring apparatus from stratified wet sediments in meadows, marshes, peat bogs, ponds, and lakes. Pollen is also retrieved from soils, cave sediments, spring deposits, archaeological deposits, alluvium, glacial ice, and pack rat middens. Contemporary pollen rain is sampled by a variety of traps. A sediment core or section is subsampled at intervals in the laboratory and the pollen and spores extracted and concentrated by dissolving the organic and mineral matrix with acids. The residue is mounted on slides, and the palynomorphs are identified with a binocular microscope and tabulated. Each subsample consists of an assemblage of taxa calculated in proportional or absolute terms. The assemblages are usually plotted as a pollen diagram arranged by abundances along the horizontal axis and by depth and stratigraphic age along a vertical axis (figure 4.5). Chronological control is routinely accomplished by radiocarbon-dating organic matter or humic acids in the core. By graphically displaying variation in pollen and spore abundance through time, the pollen diagram becomes the primary instrument for interpretation of vegetation changes.

The basic assumption of pollen analysis is that pollen assemblages sampled from traps or sediments directly represent vegetation composition, with correction made for differential pollen production and dispersal among plant taxa. The pollen record of a vegetation type is dependent on a system of transportation, sedimentation, and preservation of pollen grains that is a function of mesoscale topographic winds, drainage patterns, distance to a depositional basin, basin size, and depositional environment. Areas from which pollen is derived at a collection site are conventionally divided by distance into local, extralocal, regional, and long-distance.

There are three general factors that contribute to the success of pollen analysis. (1) Distinctive morphological characteristics of shape, aperture type, and surface sculpture make pollen and spores readily identifiable to family, generic, and (rarely) species taxonomic levels. (2) The small size (10–100 microns) of pollen and spores and the enormous number produced assure that they are uniformly mixed in the atmosphere and widely distributed and therefore conducive to the appli-

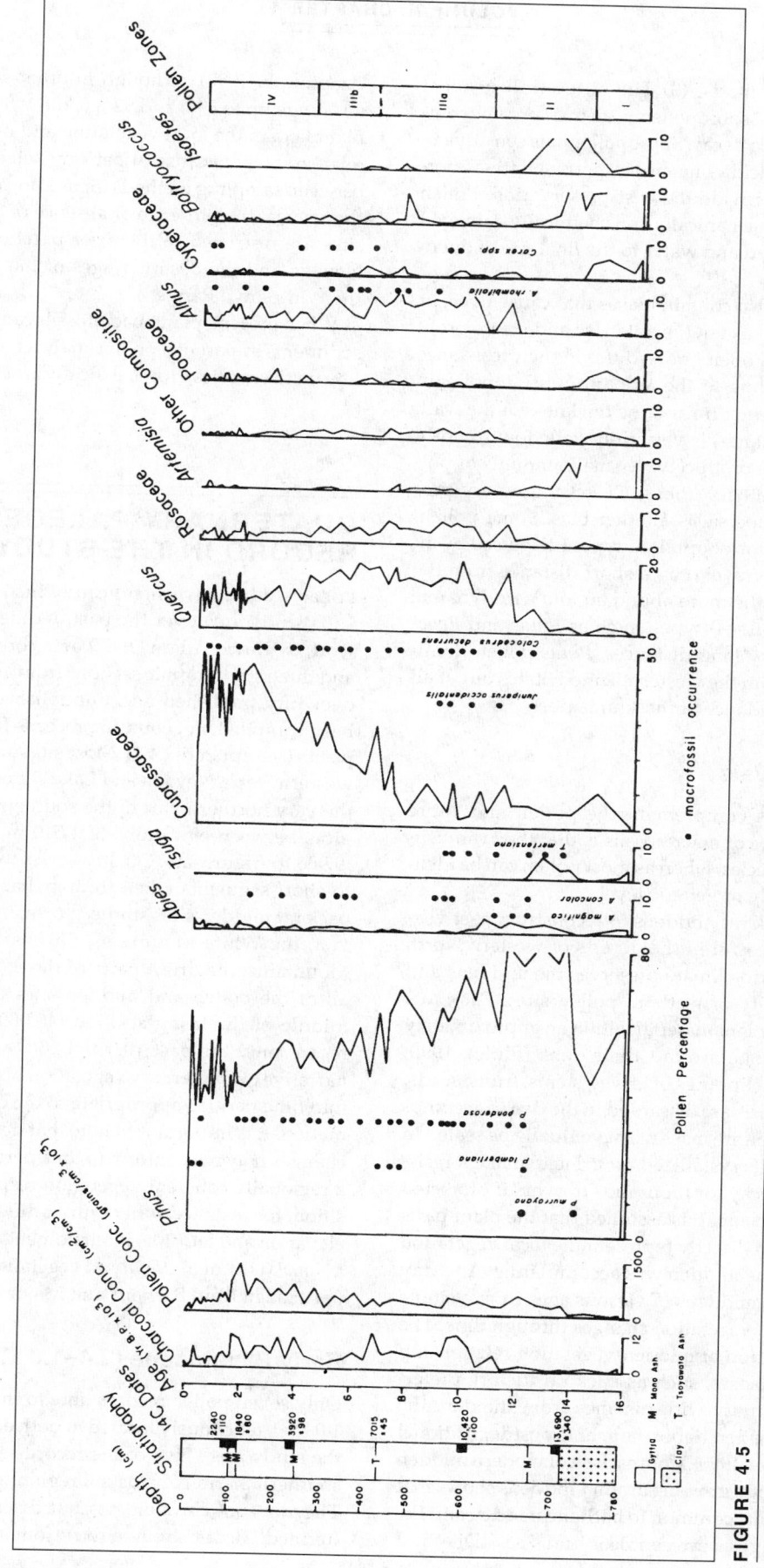

FIGURE 4.5

Summary pollen diagram with macrofossil occurrences and charcoal concentration for Swamp Lake, Yosemite National Park (from Smith and Anderson 1992).

cation of statistical methods. (3) The outer wall of most palynomorphs is made of sporopollenin, an inert organic compound highly resistant to decay. Sporopollenin is sensitive to oxidation, however, so acidic, anaerobic, permanently waterlogged environments provide the best preservation. Pollen and spores can also be mechanically and chemically damaged during transport by wind and water to the final site of deposition.

These three factors also contain biases that cause over- or underrepresentation of taxa in the database and create problems in interpretation. Pollen grains differ in their resistance to deterioration according to the amount of sporopollenin contained in the wall structure and the thickness of the wall. *Populus* pollen, for instance, is very thin walled and poorly preserved. Some plants produce very small amounts of pollen *(Linum)*, whereas others *(Pinus, Quercus)* are very prolific. Differences in dispersal is another bias. Some pollen, especially low-production zoophilous types (dispersed by insects, birds, and bats), travel only a short distance from the source plant, whereas the more abundant and aerodynamic wind-borne (anemophilous) types, such as *Pinus* and *Ephedra,* can be carried very long distances. *Pinus* pollen dominates most samples from the western United States out of all proportion to the abundance of the source trees.

Macrofossil Analysis

Preserved plant parts complement the pollen and spore records. The advantage of macrofossils is that they can usually be identified to species, whereas most pollen can be identified only to the family or generic level.

The analysis of pack rat middens to reconstruct past vegetation has been successful in the deserts of western North America, where the arid climate preserves the middens and restricts the availability of wetland pollen sites. Pack rats *(Neotoma* spp.) gather plant material within an approximately 30–50 m foraging range around their dens (Finley 1990; Vaughan 1990). These clippings of leaves, stems, fruits, seeds, flowers, and other debris are discarded in the den to accumulate as a midden. The entire mass may eventually be sealed in an indurated matrix of crystallized rat urine to make it resistant to erosion and decay for thousands of years if protected from moisture and erosion. It is assumed that the plant parts preserved in a midden directly represent the local vegetation growing at the time the midden was accumulating. An array of radiocarbon-dated middens of various ages from a single locale will then reveal vegetation changes through time. The fundamental assumption of midden-vegetation relationship is subject to several biases, such as pack rat dietary preferences, pack rat selectivity with distance from the den, the length of time for midden deposition, and postdepositional history (Spaulding et al. 1990). Comparison of modern midden contents with vegetation growing around the dens shows that the percentage of plants common to both the midden and the local vegetation ranges between 36% and 78% (Dial and

Czaplewski 1990), although in other studies similarity values range up to 89% and 92% (Cole 1983). The proportions of plant taxa in the local vegetation and in middens do not correlate reliably because of pack rat selectivity and disparities between sampling methods (Spaulding et al. 1990), but at least they reveal the presence or absence of taxa. Pack rat midden sites are restricted to the drier parts of the southern Sierra Nevada and the desert fringes of the study area, including the White-Inyo Range.

When present, plant macrofossils can also be extracted from sediments as part of a pollen analysis. They provide information on the species growing locally around the deposition site.

QUATERNARY PALEOECOLOGICAL RECORD IN THE STUDY AREA

Pollen studies have significantly increased in the Sierra Nevada study area over the past decade. Since the pioneering work by David Adam (1967) on a core from Osgood Swamp and three stratigraphic sections from Yosemite National Park, over fifty published and unpublished pollen profiles have been compiled, not counting records from archaeological deposits (see appendix 4.1). Most sites are located in the central western Sierra Nevada and Lake Tahoe Basin subregions and the very northern part of the southern Sierra Nevada subregion, between elevations of 1,510 m (4,950 ft) and 3,020 m (9,900 ft) (figure 4.6). Quality varies from deep stratigraphies to short segments taken from archaeological sites. The few pack rat midden sites studied so far are located in Kings Canyon, the White Mountains, Owens Valley, and the Scodie Mountains, the driest parts of the SNEP study area. Almost all of the pollen and midden sites together date from the middle of the last glaciation (40,000 years ago), and most record only the past 10,000–12,000 years. Accordingly, this narrative of Quaternary vegetation history will be subdivided into time periods appropriate to the detail of available information. It is important to note that descriptions of vegetation changes may not conform to time period boundaries, because a regionally coherent vegetation response to a climatic transition, for instance, will occur at different times according to elevation and latitude. Plant nomenclature follows *The Jepson Manual* (Hickman 1993), and vegetation types are, for the most part, taken from Barbour and Major (1990).

Early Pleistocene (2.4–1.0 Ma)

Only seven pollen records date to the early Pleistocene (2.4–1.0 Ma), felicitously located at both ends and in the middle of the study area. Five of the records are in the vicinity of the southern Sierra Nevada subregion (Axelrod and Ting 1962). They are not chronologies but discrete pollen assemblages (termed "floras" by Axelrod) from sedimentary units repre-

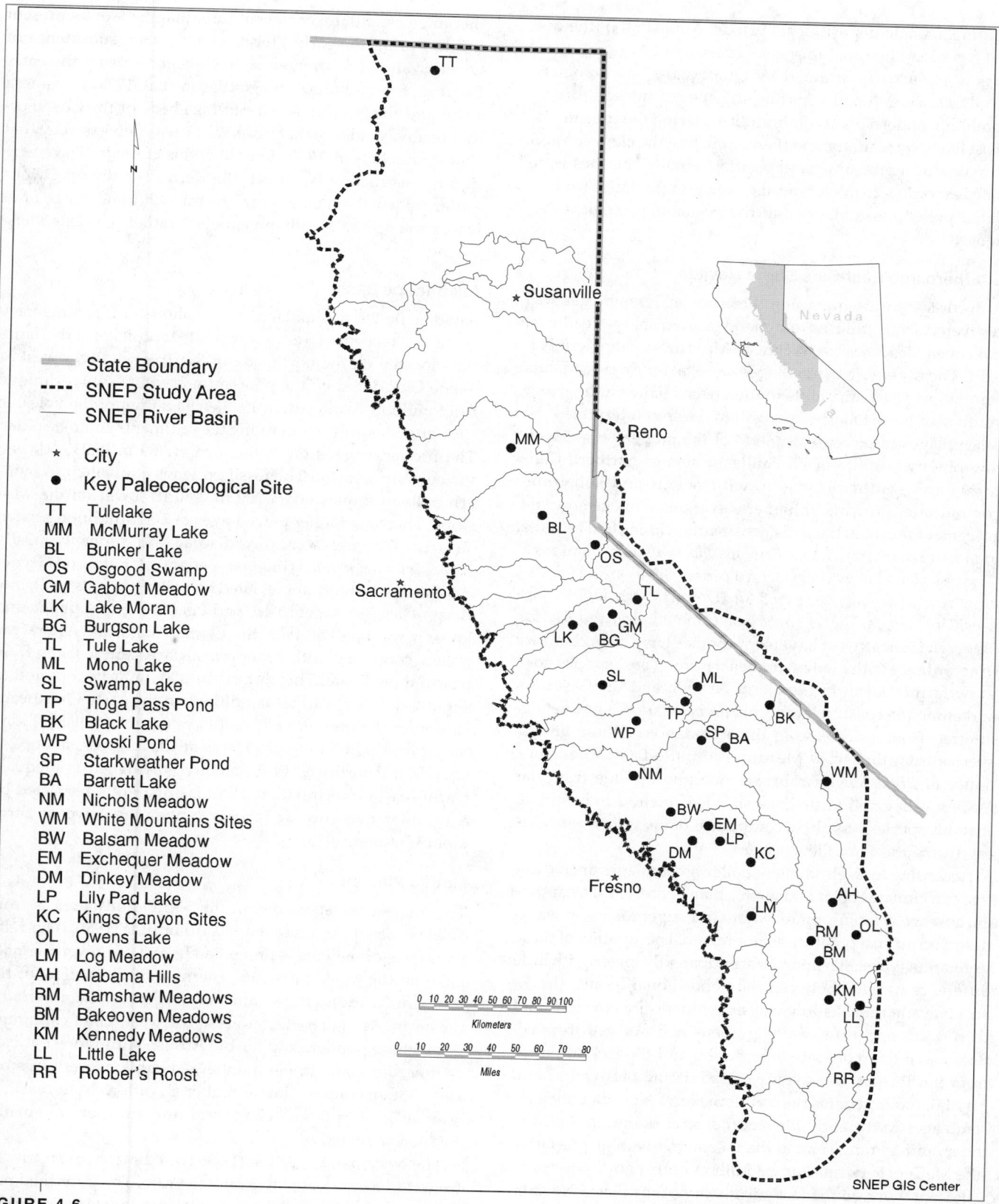

State Boundary
SNEP Study Area
SNEP River Basin

* City
● Key Paleoecological Site

TT Tulelake
MM McMurray Lake
BL Bunker Lake
OS Osgood Swamp
GM Gabbot Meadow
LK Lake Moran
BG Burgson Lake
TL Tule Lake
ML Mono Lake
SL Swamp Lake
TP Tioga Pass Pond
BK Black Lake
WP Woski Pond
SP Starkweather Pond
BA Barrett Lake
NM Nichols Meadow
WM White Mountains Sites
BW Balsam Meadow
EM Exchequer Meadow
DM Dinkey Meadow
LP Lily Pad Lake
KC Kings Canyon Sites
OL Owens Lake
LM Log Meadow
AH Alabama Hills
RM Ramshaw Meadows
BM Bakeoven Meadows
KM Kennedy Meadows
LL Little Lake
RR Robber's Roost

FIGURE 4.6

Map of key paleoecological sites in the SNEP study area.

senting a single depositional episode. Almost all of the pollen has been identified to species, a level of taxonomic resolution not usually attained by Quaternary palynologists. Analyses were done by sorting the species into groups resembling modern plant communities, placing the site into the most likely community, and then estimating the elevation and climate during the time of deposition. Finally, the sites were dated according to the composition of the taxa relative to other dated pollen assemblages and the available geological evidence.

Southern and East-Side Sierra Nevada

A north-to-south sequence of three pollen assemblages was recovered from exposed sediments at Ramshaw Meadows, Bakeoven Meadows, and Kennedy Meadows on the Kern Plateau. The other two assemblages were taken from road cuts through the Alabama Hills in the Owens Valley and farther south near Little Lake in Rose Valley. Axelrod interpreted the assemblages to be representative of the modern upper yellow pine (white fir–mixed conifer) forest of northern California and southern Oregon, with red fir and subalpine communities growing at higher elevations. This geographic placement is due to the presence of pollen identified as *Abies grandis* (grand fir), *Tsuga heterophylla* (western hemlock), *Cupressus lawsoniana* (Port Orford cedar), *Taxus brevifolia* (Pacific yew), *Pseudotsuga menziesii* (Douglas fir), *Lithocarpus densiflora* (tan oak), *Acer circinatum* (vine maple), and *Myrica hartwegii* (Sierra sweet bay) (some names have been changed in accordance with modern nomenclature). These species now grow farther north in the Sierra Nevada and the Cascades and along the coast. A latitudinal and elevational gradient is inferred from a decrease in the abundance of these humid species from Ramshaw Meadows to Little Lake. The abundance of *Pinus longaeva* (bristlecone pine) pollen from the Owens Valley and Little Lake sites led Axelrod to postulate that this species was also a component of the yellow pine forest during the early Pleistocene.

According to Axelrod, these pollen assemblages are indicative of a climate cooler and wetter than at present. The apparent absence of alpine, piñon woodland, sagebrush, and warm desert scrub taxa points to a later regional occupation of these vegetation types. Elevations were apparently lower, with little difference in relief between the depositional basins; the Sierra Nevada rain shadow was not climatically effective, and the Owens and Rose Valley grabens had not yet dropped. Because of the moderate topography and the lack of pollen from the Tertiary (a geologic period dating between 65 and 2.4 Ma), Axelrod dates the southern Sierra Nevada pollen assemblages to the early Pleistocene, after the initial uplift of the mountain range and at the inception of alpine glaciation (the McGee) between 2 and 3 million years ago.

Recent work on the tectonic history of the Sierra Nevada, however, estimates the initiation of significant uplift and westward tilting between 25 and 10 million years ago, in the late Miocene or Pliocene (Huber 1981). Uplift accelerated until

the range essentially achieved more than 80% of its present height during the early Pleistocene. Further, subsidence of Owens Valley is estimated to have begun during the early Pliocene, 5–6 million years ago (Bacon et al. 1979; Giovanetti 1979) and the 3-million-year-old lake beds of the Coso Formation give evidence that Rose Valley was depressed by that time (Duffield et al. 1980). In light of this, if Axelrod's vegetation reconstruction is correct, the Kern Plateau and Owens Valley pollen assemblages may be mid-Pliocene in age or at least greater than 3 million years old, rather than late Pleistocene.

Lake Tahoe Basin

Another pollen assemblage from a short-term sedimentary event was recovered from a clayey silt deposit beneath a latite flow located on the north side of the Truckee River west of Tahoe City, dated by the potassium-argon (K-Ar) technique to 1.9 million years (Adam 1973). The dating of the volcanic rock provides only a minimum; the sediments could be older. The present vegetation in the area (1,900 m [6,230 ft]) is a mixed conifer forest. The fossil pollen is equivalent to modern pollen samples taken 300 m (980 ft) lower on the west slope. There are higher percentages of TCT (undifferentiated Taxaceae-Cuppressaceae-Taxodiaceae taxa, probably mostly *Libocedrus decurrens* [incense cedar]), Chenopodiaceae-Amaranthus (usually abbreviated by pollen analysts to Cheno-Ams), *Artemisia* (sagebrush), and Poaceae (grass) pollen and lower percentages of *Abies* (fir), *Quercus* (oak), and Asteraceae pollen, compared with a modern surface sample taken from near the fossil site. This difference may indicate a period of slightly drier or warmer conditions in the early Pleistocene except for the presence of fossil *Picea* (spruce) pollen. Today the nearest population of *Picea* is at Clark Creek, about 240 km (150 mi) northwest of Tahoe City, where there is adequate warm-season precipitation. *Picea* is apparently restricted by a summer moisture deficiency longer than two or three months (Adam 1973).

Modoc Plateau

The longest pollen record in the SNEP study area is from Tulelake Basin (Adam et al. 1990; Adam et al. 1989). This remnant Pleistocene lake is immediately east of Lower Klamath Lake on the Modoc Plateau. The rain shadow effect of the southern Cascade Range maintains a sagebrush steppe around the basin. At higher elevations to the west, plant communities include ponderosa pine–bitterbrush shrub forest, mixed conifer forest, and upper montane–subalpine forest. In more xeric (low-moisture) sites at higher elevations to the east are manzanita-*Ceanothus* chaparral and juniper-sagebrush steppe woodlands.

The bottom of the 334 m (1,096 ft) sediment core retrieved from the basin is dated to 3 million years. The published pollen diagram presents only the most abundant taxa, including *Pinus*, *Quercus*, and TCT (probably mostly *Juniperus occidentalis* and *Calocedrus decurrens*); *Artemisia*, other

Asteraceae (composites), Cheno-*Ams*, and *Sarcobatus* (greasewood); and Poaceae and Cyperaceae (sedge and tule). Whereas these pollen types represent the surrounding terrestrial and lakeside vegetation, the presence of algae colonies and diatoms in the sediments reflects a lacustrine environment. The pollen is used by the authors primarily as a climatic record; interpretation in terms of vegetation is minimal, and the time resolution is low.

Proportional variations among these pollen taxa record changes in forest and woodland, desert, and valley bottom vegetation as they respond to climate change. Adam interprets the high percentages of *Pinus* and TCT pollen (along with temperature- and chemistry-sensitive diatom taxa) during the termination of the Pliocene as extensive coniferous forest around a warm, shallow, slowly drying lake. When the low percentages of *Artemisia* and Cheno-*Ams* pollen are also taken into account, the climate can be inferred to be more humid than it is today, although there were dry periods. At the beginning of the Pleistocene the pollen curves change from lower (smoother) to higher (spikier) frequency variation, the amplitude of TCT pollen decreases slightly, and *Artemisia* pollen becomes more abundant. An appreciable drying trend began after 1.64 million years, as indicated by a decrease in TCT pollen and coextensive increase in *Artemisia*, Asteraceae, Poaceae, and Cheno-*Ams*, along with intervals of lake desiccation. The vegetation was apparently responding to both the dryer, colder glacial climates and the increasing rain shadow as the Cascades were uplifted. The coniferous forest seems to have been opening up, and sagebrush steppe was becoming more prevalent.

Late Pleistocene (1.0 Ma–10 ka)

Modoc Plateau

The Pleistocene vegetation trends in the Tulelake Basin core continue after 1 million years with a significant decrease in TCT pollen, and *Artemisia*, other Asteraceae, Cheno-*Ams*, and Poaceae increase to the highest percentages in the core. Inverse fluctuations between pine and juniper-type pollen indicate glacial cycles imposed on the overall drying trend as the modern sagebrush steppe is established. *Quercus* pollen shows no definite trend throughout the core, but it varies in concert with TCT pollen and provides a similar signal of glacial cycles.

East-Side Sierra Nevada

The only other pollen record within the SNEP study area that spans nearly the past million years comes from Owens Lake (Litwin et al. 1995). A 323 m (1,060 ft) core recovered from the lake sediments extends back to about 800,000 years, based on the presence of the Brunhes-Matuyama magnetic-reversal boundary (780 ka) and Bishop ash (760 ka) near the bottom (Owens Lake Core Study Team 1995).

Modern vegetation surrounding Owens Lake is zoned by elevation into saltbush-greasewood scrub on the valley bottom and lake basin; creosote bush scrub, shadscale and blackbush mixed-desert scrub, and sagebrush scrub on the alluvial fans; piñon or piñon-juniper woodland above that; and upper montane, subalpine forests, and alpine communities along the crest. Riparian vegetation consists of tree and shrub forms of *Salix* spp. (willow), *Populus fremontii, P. angustifolia,* and *P. trichocarpa* (cottonwood), with *Populus tremuloides* (aspen) at higher elevations. Small stands of *Quercus kelloggii* (black oak) and *Q. chrysolepis* (canyon live oak) also grow along the drainages of the eastern escarpment.

The preliminary Owens Lake pollen diagram for the entire core has a very coarse resolution of about 5,900-year sampling intervals. General patterns of glacial cycles, however, are discernible in the inverse relation of pine and juniper pollen: as climate went into a glacial phase, *Juniperus* increased and *Pinus* decreased. Another clear periodicity is an increase in *Artemisia* that precedes *Juniperus* with the beginning of each glaciation. Long-term trends are the increase in peak abundances of juniper pollen and a slight increase in *Quercus* and *Salix*. The other taxa show no trends throughout the core.

The interval of 800,000 to about 685,000 years ago at the bottom of the core is distinguished by relatively high percentages of *Abies* and *Picea* and low percentages of *Juniperus*. *Abies* attains similar values in the upper part of the core, but *Picea* is only occasionally present. *Picea* is an anomaly. The only other place it has been identified in Sierra Nevada Quaternary sediments is in the Tahoe City assemblage (discussed earlier). It is more likely to have grown in the northern Sierra Nevada during the early Pleistocene than in the more xeric southern end of the range, even during the full glacial periods. It was not identified in the Tulelake core, although there are populations of two species (*Picea engelmanii* and *P. breweriana*) in restricted areas to the southwest and in the Siskiyou and Klamath Mountains (Griffin and Critchfield 1972). The characteristics of *Picea* pollen are distinctive enough not to be confused with *Pinus*, if preservation is good, so errors in identification are minimal. *Picea* pollen is very buoyant like *Pinus*, although nearly four times heavier and with twice as much settling velocity (Erdtman 1969), so it will not travel as far. It has been reported as a component of the modern pollen rain from Pleistocene Searles Lake, about 37 km (60 mi) southeast of the Owens Lake coring site (Leopold 1967). The nearest *Picea* populations today are in northern Arizona mountains. The pollen may have been brought into southeastern California by expanded monsoonal flow during the height of the interglacials because most of the peaks of *Picea* pollen coincide with periods of high *Pinus* and low *Juniperus*, which mark the interglacials.

Another anomalous pollen type found occasionally throughout the Owens Lake core is *Juglans* (walnut). A small stand now growing in Owens Valley may have been planted in historic times. The habitat of *Juglans hindsii*, historically present along the Sacramento River and near the delta, and *J. californica* in the southern California coastal mountains has a strong marine influence with small seasonal temperature variations, unlike the more continental regime of the eastern Sierra Nevada. It is more likely that the *Juglans* pollen of Owens Lake represents a northward expansion of *J. major* (Arizona black walnut) during the Pleistocene.

A finer-resolution (670-year intervals) pollen record from Owens Lake is being compiled for the past 150,000 years (Woolfenden 1995). This interval covers the last two major glacial cycles. As the record is interpreted so far, there was a general increase in Utah juniper (*Juniperus osteosperma*) woodland on the alluvial fans of the Sierra Nevada and Inyo Mountains and probably around the lake margins throughout the last glacial (Tioga) from about 73,000 to a maximum at about 20,000 yr B.P. *Pinus monophylla* (singleleaf piñon) and other white pine pollen have higher percentages than yellow pine pollen during this period. Most of the white pine pollen may represent the expansion of *Pinus longaeva* (bristlecone pine) and *P. flexilis* (limber pine) subalpine forests in the Inyo Mountains and the depression of the elevational range of *P. monticola* (western white pine), *P. balfouriana* (foxtail pine), *P. flexilis*, and perhaps *P. albicaulis* (whitebark pine) lower on the Sierra Nevada escarpment. Species such as *P. contorta* (lodgepole pine) and *P. jeffreyi* (Jeffrey pine) and other elements of the upper montane forest (red fir, white fir, and mountain juniper) were probably included in the forest. A component of *Pinus* pollen may have blown over from the west slopes. The increase in *P. monophylla* pollen indicates a lowering of the elevational limits of this species on the rockier portions of the bajadas as an associate of juniper, and possibly an increase in its density on the desert mountains to the south. These trends, along with the concomitant increase in *Artemisia* pollen, and higher percentages of Poaceae (signaling the expansion of sagebrush steppe) suggest a transition to a climate cooler and wetter than today's. The termination of the Tioga glaciation at about 14,000 years ago is represented by a decrease in juniper relative to pine and an increase in *Ambrosia* cf. *dumosa* (white bursage), Cheno-*Ams*, and *Cercocarpus-Purshia* (a type of Rosaceae pollen undifferentiated among *Cercocarpus ledifolius* [mountain mahogany], *Purshia tridentata* [bitterbrush], and *Coleogyne ramosissima* [blackbush]). The trend in shrub taxa indicates a spread of warm-desert species in the Owens Lake Basin. The warm and subhumid waning stages of the last interglacial, between about 110,000 to 75,000 years ago, is similarly reflected by high percentages of *Abies*—probably *Abies concolor* (white fir) and *A. magnifica* (red fir)—*Pinus*, Cheno-*Ams*, *Cercocarpus-Purshia*, and *Ambrosia* types.

Two pack rat middens collected on the northeast side of Owens Lake, 10 m (33 ft) above the highest Pleistocene lake level (1,145 m [3,757 ft]) provide evidence for the floral characteristics of lakeside vegetation between 22,900 and 14,870 yr B.P. (Koehler and Anderson 1994a). The vegetation around the site is at present a mixed shadscale-sagebrush scrub with nearby creosote bush–bursage (*Larrea tridentata–Ambrosia dumosa*) communities on the fans. The full- to late-glacial midden macrofossils are typical of a Utah juniper–singleleaf piñon woodland with an understory of xeric shrubs, cactus, and grasses such as *Ericameria cuneata* (cliff goldenbush), *Ephedra viridis* (green joint fir), *Menodora spinescens*, *Glossopetalon spinescens* (Nevada greasewood), *Opuntia basilaris* (beavertail cactus), and *Achnatherum hymenoides* (Indian ricegrass), among others. The unusual find of *Juniperus scopulorum* (Rocky Mountain juniper) in midden layers dating 20,590 and 16,010 yr B.P. is the first observation of this species west of its present westernmost location in the Charleston Mountains of southwestern Nevada. Its Pleistocene occurrence around Owens Lake may be due to high water tables or the moderating influence of the expanded lake on local climate. Pollen extracted from the midden samples reflects the macrofossil assemblage except for a high abundance of *Artemisia* pollen.

North of Owens Lake, a spectacular 31,500-year sequence of twenty pack rat middens was recovered from the Alabama Hills on three sites with southern exposures (Koehler and Anderson 1995). The macrofossils and pollen contained in the middens are congruent with the Owens Lake pollen record and further enhance the late Pleistocene vegetation history of the eastern Sierra Nevada. During the middle- to full-glacial interval from 31,450 to 20,310 yr B.P. a Utah juniper–Joshua tree (*Yucca brevifolia*) community with an understory of *Artemisia tridentata*, *Purshia tridentata*, and *Atriplex confertifolia* (shadscale) grew in the Alabama Hills. The association of *Yucca brevifolia* with *Juniperus osteosperma* has not been described and may not have a modern analog, although its association with *Juniperus californica* (California juniper) is known in the Mojave Desert (Vasek and Thorne 1990). As with *Juniperus scopulorum* on the eastern shore of Owens Lake, *Yucca brevifolia* may signal a moderate local glacial climate influenced by an expanded lake, with winter temperatures not much lower than in recent times. *Artemisia* and *Purshia tridentata*, however, imply cooler summers.

Around 19,070 yr B.P. climatically significant changes occurred with the appearance of the more xeric shrub species *Ericameria cuneata* and *Purshia mexicana* (cliffrose) and the departure of bitterbrush. Joshua trees disappeared by 17,760 yr B.P., followed by the departure of Rocky Mountain juniper and the upslope retreat of piñon from the shores of Owens Lake by 16,010 yr B.P. (Koehler and Anderson 1994a). The apparent warming and drying trend continued between 13,350 yr B.P. and 9540 yr B.P. with the arrival of blackbush, *Opuntia echinocarpa* (cholla), and *Lycium andersonii* (wolfberry) and the local departure of *Juniperus osteosperma* soon after that.

Late glacial woodland is also documented at the southern end of the Sierra Nevada (McCarten and Van Devender 1988). Three pack rat macrofossil assemblages dated at 13,800 to 12,820 yr B.P. from the 1,125 m (3,690 ft) elevation Robber's Roost site in the Scodie Mountains are dominated by *Pinus monophylla*, *Juniperus californica*, *Ceanothus greggii*, *Artemisia tridentata*, *Purshia tridentata* var. *glandulosa* (desert bitterbrush), *Quercus turbinella*, and *Yucca brevifolia*. The best analog for the assemblage is not the piñon-oak woodland 600 m (1,970 ft) just above the midden site (which is composed of *Pinus*

monophylla, *P. sabiniana*, and *Quercus chrysolepis*) but piñon-juniper woodland on the north slope of the San Bernardino Mountains. The modern vegetation aound the Robber's Roost site is a creosote bush–bursage scrub.

McCarten and Van Devender make an important point about the macrofossils at Robber's Roost that can be extended to the Alabama Hills middens. Even though these midden sites are located near the base of the eastern escarpment of the Sierra Nevada, the fossil plant assemblages do not contain Sierran montane trees that should have dispersed east into the Alabama Hills or south and east into the Scodie Mountains.

Farther north, eight pack rat middens recovered from the White Mountains and Volcanic Tablelands afforded a chronological sequence of past vegetation at six disjunct locales between 19,290 yr B.P. and 2130 yr B.P. (Jennings and Elliott-Fisk 1993). The two earliest-dated middens, from adjacent Volcanic Tablelands sites (1,341 m [4,400 ft]), span the interval from the full-glacial to the opening of the Holocene. The modern vegetation is a mixed shadscale and sagebrush scrub. Plant macrofossils of one midden show that an entirely different plant community existed 19,000 years ago. At that time Utah juniper, in association with *Purshia tridentata* var. *glandulosa* (desert bitterbrush), *Tetradymia axillaris* (cottonthorn), *T. canescens* (horsebrush), and *Ericameria cuneata*, among other species, grew around the site about 600 m (1,970 ft) below the present lower border of piñon-juniper woodland. More than 9,000 years later, as revealed by the second midden, radiocarbon-dated to 9830 yr B.P., *Juniperus osteosperma* and *Purshia tridentata* var. *glandulosa* were still in the area, but the other shrub associates were replaced by *Chamaebatiaria millifolium* (fernbush), *Prunus andersonii* (desert peach), *Ribes velutinum* (gooseberry), and *Artemisia* spp. This shift in the flora is among species now occupying shared elevation ranges on dry slopes (or, in the case of *Ribes velutinum*, moist and shaded habitats) in the White Mountains and does not necessarily signify climate change.

An 11,500-year core taken from the west side of Mono Lake contains a pollen record typical of a late-glacial Great Basin juniper-sagebrush woodland (Davis 1993). Today *Pinus monophylla* grows on the lower slopes above the lake, with a sagebrush-bitterbrush understory. There are stands of *Juniperus osteosperma* to the east on the north side of the lake. *Pinus jeffreyi* occupies the drainages. An unusual feature is the high percentage of *Sequoiadendron* pollen in the 11,500–10,000 yr B.P. interval, up to 20% near the base of the core. Studies of *Sequoiadendron* pollen dispersal at Lost Grove and Tuolumne Grove have determined that this pollen type is not widely dispersed outside the groves (Anderson 1990b). Percentages drop to about 20% at about 100 m from the edge of the groves and less than 5% within 500 m of the grove. Therefore, it is possible, although a biogeographical anomaly, that the *Sequoiadendron* pollen in Mono Lake came from nearby stands of *Sequoiadendron giganteum* (giant sequoia).

A pollen record from Tule Lake (2,080 m [6,820 ft]), at the base of the escarpment east of Sonora Pass, is undated, but time periods can be generally inferred from changes in the proportions of pollen taxa (Byrne et al. 1979). Present vegetation is sagebrush-bitterbrush scrub with open stands of *Pinus jeffreyi*, *Juniperus osteosperma*, and *Cercocarpus ledifolius* on the surrounding ridges. The lake is located behind a Tioga-age terminal moraine that gives a minimal deglaciation date for the vegetation record of about 14,000 yr B.P. The lower two-thirds of the pollen profile is dominated by *Pinus*, TCT (*Juniperus*), and *Artemisia* pollen. This probably represents a cold-dry late-Pleistocene juniper woodland with associated pine.

Just east of the Sierra Nevada crest, a high-elevation pollen record was taken from Barrett Lake (2,816 m [9,240 ft]). It is situated within the mixed mountain hemlock–red fir–lodgepole pine–limber pine forest. The pollen spectra in the 11,730–10,000 yr B.P. interval is similar to modern samples from open subalpine forest and at tree line. In addition to the lack of macrofossils in the lake sediments and very low pollen concentrations, this implies a near-treeless newly deglaciated landscape (Anderson 1987).

West-Slope Sierra Nevada

Unlike east-side Sierra Nevada and northeastern California, where the availability of deep Pleistocene lake basins and pack rat midden sites allows for extended vegetation chronologies well back into the Quaternary period, the western slope of the Sierra Nevada has produced continuous vegetation histories only since the Tioga glacial maximum (18,000–20,000 yr B.P.). (Two exceptions are discussed later.) The shorter chronologies are a consequence of the scarcity of natural lakes and wetlands at elevations below the glacial termini and conditions arid enough for the preservation of pack rat middens. Another reason may be a deficiency of fieldwork intensive enough to discover deeper sediments and middens. Before global deglaciation was initiated sometime after 17,000 years ago (Fairbanks 1989), and at least 15,000 years ago in the Sierra Nevada (Byrne et al. 1993), the 430 km (267 mi) ice cap–mountain glacier complex covered and was scouring what were to become pollen sites. The ice cap itself was limited by the balance between winter snow accumulation and summer melting. This annual snowline or equilibrium line altitude (ELA) on southern aspects ranged from about 2,400 m (8,000 ft) in the northern Sierra Nevada to about 4,000 m (13,000 ft) in the southern Sierra Nevada (Porter et al. 1983; Wahrhaftig and Birman 1965). Above an average of about 2,500 m (8,200 ft), valley glaciers fed mid-altitude ice fields created when glaciers overtopped canyon divides and coalesced. Below 1,800 m (5,900 ft) in the central Sierra Nevada, the ice fields fed valley glaciers that descended farther than 60 km (37 mi) down west-slope canyons to about 600–1,200 m (1,970–3,940 ft). On the steeper eastern escarpment, valley glaciers descended 15–27 km (9–17 mi) to elevations between 1,300 and 2,200 m (4,260–7,220 ft). It is clear that the basins from which upper-elevation pollen records were retrieved were within

reach of glaciers or that the high velocity of glacial meltwater streams prevented the accumulation of deep sediments.

The single pack rat midden record from the cismontane Sierra Nevada is a sequence of eight samples of four middens recovered from caves in lower Kings Canyon at an elevation range of 920–1,270 m (3,020–4,170 ft) and dating to a late Pleistocene time interval of greater than 40,000 yr B.P. to 12,500 yr B.P. (Cole 1983). The present vegetation type is oak–pine/chaparral woodland of Quercus chrysolepis and Pinus monophylla with an understory dominated by Cercocarpus betuloides (birch leaf mountain mahogany), Cercis occidentalis (western redbud), Umbellularia californica (California bay), Garrya flavescens (silk tassel), Ceanothus integerrimus (deer brush), Yucca whipplei (Our Lord's candle), Rhamnus crocea (spiny redberry), and Elymus trachycaulus spp. subsecundum (slender wheatgrass). The major difference between modern and midden-based Pleistocene floras is the past absence of Quercus and several xeric shrub and wheatgrass associates in middens and the presence of a very diverse mixed conifer forest during glacial times. Throughout nearly 26,000 years since 40,000 yr B.P., Kings Canyon was occupied predominantly by Juniperus occidentalis associated with Abies grandis, Calocedrus decurrans (incense cedar), Pinus lambertiana (sugar pine), P. cf. ponderosa (yellow pine), and a single occurrence (17,520 yr B.P.) of Torreya californica (California nutmeg). Juniperus macrofossils were most abundant in the sample dated to 20,000 yr B.P. Pinus lambertiana departed by 14,600 yr B.P., Abies grandis by 14,190 yr B.P., Calocedrus decurrens by 12,500 yr B.P., and Juniperus occidentalis and yellow pine after 12,500 yr B.P. Pinus monophylla was present throughout most of the interval to the present day.

Species common to Pleistocene and modern vegetation communities are Pinus monophylla, Ceanothus integerrimus, and Garrya flavescens. Sequoiadendron pollen extracted from the middens implies the presence of Sequoiadendron giganteum closer to the sites than the modern groves. Except for Abies grandis, Calocedrus decurrens, Torreya californica, and Sequoiadendron giganteum, which probably inhabited nearby shaded ravines, the composition of the Pleistocene forest is more typical of the east side than the present west-slope montane forest at higher elevations. Cole attributes this either to the marble substrate of the Kings Canyon sites, or to a cold, dry late-Pleistocene climate across the continent, or to a combination of both, with an uneven snow accumulation maintaining the mesic species (those requiring a moderate amount of moisture).

A relatively xeric late-Tioga glacial climate with a local, irregular distribution of water availability is demonstrated by cores taken from Tulare Lake in the southern San Joaquin Valley (Atwater et al. 1986; West et al. 1991). Pollen abundances show high Juniperus, Artemisia, Sarcobatus, and Sequoiadendron from about 26,000 yr B.P. to about 13,000-11,000 yr B.P. From this evidence Sequoiadendron giganteum was inferred to be more widespread and probably a component of an extended riparian woodland along the Kern River into the San Joaquin Valley. The pollen is now thought to have been transported downriver during flood events.

West-slope pollen and macrofossil stratigraphies document a consistent pattern of vegetation-climate change in the central and southern Sierra Nevada subregions, although timing and floristic details differ among sites according to latitude and elevation. The earliest of these pollen profiles, dating well into the late glacial, are from intermediate elevations, below 2,500 m (8,200 ft) in the lower and upper montane forests.

At present, Nichols Meadow provides the only full-glacial (about 18,500 yr B.P.) pollen spectra (Koehler and Anderson 1994b). The meadow is located within the Nelder Giant Sequoia Grove in the Fresno River drainage basin at an elevation of 1,510 m (4,950 ft). It is the lowest pollen site in the Sierra Nevada, near the upper margin of the ponderosa pine forest and the lower margin of the white fir–mixed conifer forest. The pollen spectra at the base of the core represent an open sagebrush-grass community or perhaps a woodland with scattered white and yellow pine and Juniperus nearby. This is indicated by very high percentages of Artemisia and Poaceae with relatively significant amounts of Pinus and Cupressaceae based on a very low pollen influx, which means the tree pollen came from a distant source or from a very few individuals in the area. Herb pollen is also common. The ratio of sagebrush to grass is equivalent to that of the late glacial sequences of the northern Rocky Mountains of Wyoming, southwestern Montana, and southeastern Idaho, and modern sagebrush steppe in the basins of northeast Wyoming (Baker 1983). The pollen spectra also resemble those of modern Sierra Nevada alpine communities east of the crest. Both reflect a cold, dry periglacial climate. The community persisted at Nichols Meadow until 12,500 yr B.P.

A subalpine sagebrush steppe–woodland, represented by high frequencies of Artemisia and (in some central Sierra Nevada sites) Juniperus and Sarcobatus pollen is recorded at other sites above 1,500 m, where the modern vegetation is montane forest.[1] This pollen zone dates from the base of the cores, between 16,000 and 11,000 yr B.P., and continues until 13,000–10,000 yr B.P., depending on elevation and latitude. It seems to have persisted into the early Holocene, until about 7000 yr B.P., at Balsam and Exchequer Meadows in the more xeric southern Sierra Nevada. Also, between 16,000 and 11,000 yr B.P., change in the hydrology is seen in a transition of the basal sediments from gravels, sands, silts, and glacial flour to organic silts, muds, and peat.

Variation in the proportions of pollen types shows that sagebrush steppe was replaced relatively rapidly by a closed conifer forest as the climate became warmer, with more effective moisture. Within about 500 years, percentages of Pinus and Abies increased as Artemisia, Poaceae, and herb pollen decreased. Pinus, especially, reached maximum values for the entire late Quaternary record.

Evidence for trees colonizing the sites is the higher percentages of arboreal pollen and the presence of macrofossils. Conifers are identified to species in the few analyses that in-

cluded macrofossils. *Abies concolor, Pinus lambertiana, Pinus ponderosa*-type, *Calocedrus decurrens,* and *Sequoiadendron* grew around Nichols Meadow. The forest at the higher Lake Moran site from 12,000 to 10,000 yr B.P. was composed of *Pinus lambertiana, P. monticola, P. contorta,* and some *Tsuga mertensiana. Pinus lambertiana* and *Abies concolor* were two species at Bunker Lake.

Similar to that of the Kings Canyon macrofossil assemblage, a very diverse forest of lower and upper montane and subalpine species was established at Swamp Lake, Yosemite National Park (YNP), after 13,700 yr B.P. and remained for about 3,300 years. Lodgepole pine, western white pine, ponderosa pine, white and red fir, incense cedar, mountain hemlock, and western juniper provided the species mix. Smith and Anderson (1992) conjecture that the upper-elevation conifers may have lagged behind in suitable microhabitats as the other trees followed the changing climate upslope and that development of a more extreme seasonality of cooler, wetter winters and warmer summers created a greater variety of habitats. The species composition at some of these paleoecological sites does not resemble any assemblage in present Sierran forest. Byrne et al. (1993) make a striking conjecture, based on pollen abundance, low frequencies of Quercus pollen, and relatively low charcoal abundance, that the late-glacial dense conifer forests had a structure similar to modern mixed conifer forests conditioned by historic fire suppression.

The steppe-forest transition took place later, after 10,000 yr B.P. at Starkweather Pond and also at Gabbott Meadow, located at 1,995 m (6,550 ft) deep into the Stanislaus River drainage basin (Mackey and Sullivan 1991). A brief maximum of *Abies* pollen percentages, accompanied by *A. concolor* and *A. grandis* macrofossils, by about 10,000 yr B.P. reveals a slight lag in arrival of these species at the site.

Meadows developed during this period. The deposition of organic-rich sediments, along with pollen of Brassicaceae, Cyperaceae *(Carex),* Asteraceae, and *Oxypolis* (its first appearance), is interpreted to signal the inception of a small meadow after 12,500 yr B.P. at Nichols Meadow (Koehler and Anderson 1994b). Similarly, after the same date, meadow conditions appear to have been established at Lake Moran until about 10,000 yr B.P. (Edlund and Byrne 1991). The evidence is a stratum of peaty sediments and a high percentage of pollen associated with meadows: Apiaceae, Liliaceae, Onagraceae, Malvaceae, and Ranunculaceae.

This cool-wet interval is marked by the high stands of pluvial lakes in southwestern deserts between 15,000 and 12,000 yr B.P. and by the maintenance of marsh and pond conditions in some Mojave Desert valleys (Benson et al. 1990; Enzel et al. 1992; Quade 1986; Wells et al. 1987). Lakes and vegetation were apparently responding to changing atmospheric circulation and water budgets over the entire region throughout deglaciation, coinciding with the northward movement of the jet stream.

Early Holocene (10–6 ka)

From 18,000 years ago average insolation over the Northern Hemisphere is computed to have increased in the summer to a July maximum of 8% higher than today's values and decreased in the winter to a January minimum of about 8% lower than today's values by 9,000 years ago (COHMAP Members 1988) (figure 4.3). The first stages of summer warming and its effect on Sierra Nevada vegetation is reflected in the pollen records discussed earlier. With an intensification of warming came a decrease in effective moisture, and the postglacial montane forests began to change rapidly. At several sites lake levels dropped, and meadows dried and were invaded by conifers (Anderson 1990a; Koehler and Anderson 1995; Wood 1975). Tree-ring dating of the remains of bristlecone pine above the present tree line on Sheep Mountain in the White Mountains has determined that the minimum altitude of the upper tree line was up to 150 m above the present tree line prior to 3,750 years ago due primarily to higher temperatures (LaMarche 1973).

West-Slope Sierra Nevada

Under a warm, dry early Holocene climate, central Sierra Nevada sites show a decrease in *Pinus* and an increase in Cupressaceae *(Juniperus* and *Calocedrus decurrens),* Artemisia, Quercus, Alnus, Pteridiam, and herb pollen. *Quercus* pollen, especially, reaches its peak frequencies during this period, exceeding modern values at several sites. *Abies* pollen also decreases after a brief, early rise in abundance. An important correlative to shifts in proportions of pollen taxa is the dramatic increase in fire frequency as indicated by high concentrations of charcoal. It is obvious that both aridity and intensification of the fire regime were opening up the denser late Pleistocene forest canopy and changing its composition. Macrofossils provide more details. Although *Pinus ponderosa* macrofossils were present at low elevations in Nichols Lake and Swamp Lake YNP sediments prior to 10,000 yr B.P., the species first appeared after this date farther upslope at Lake Moran and Bunker Lake, while *Pinus monticola, P. lambertiana,* and *P. contorta* macrofossils temporarily disappeared at Lake Moran. Mesic upper-montane and subalpine species—*P. contorta, Abies grandis, Tsuga mertensiana,* and *Juniperus*—permanently dropped out of the macrofossil records at Swamp Lake YNP.

Pollen and macrofossils from Log Meadow, in the Giant Forest of Sequoia National Park (2,948 m [6,720 ft]), records a mixed conifer forest between 10,500 and 9000 yr B.P., with an absence of *Sequoiadendron* (Anderson 1994). After 9000 yr B.P. a few individuals of *Sequoiadendron* first began to grow in proximity to the site.

High-altitude forests were already in existence after the beginning of the Holocene warm period. Macrofossils of western white pine, lodgepole pine, western juniper, and red fir appear in the sediments of Starkweather Pond at about 9300,

8500, 7300, and 6300 yr B.P., respectively. These are the associates of the modern upper montane forest in the area. An open-canopy forest structure is inferred from the abundant montane shrub pollen, such as *Cercocarpus,* Ericaceae (probably *Arctostaphylos*), *Chrysolepis,* and probably *Quercus vaccinifolia* (Anderson 1990a). The shrubs, the temporary decrease and disappearance of *Tsuga mertensiana* pollen, and the increase in charcoal frequency are all indications of early Holocene aridity.

The highest pollen site sampled so far is Tioga Pass Pond at 3,018 m (9,901 ft) (Anderson 1987, 1990a). It is situated in a subalpine forest of *Pinus contorta,* associated with a few *P. monticola* and *P. albicaulis* (whitebark pine). The pollen record begins about 9300 yr B.P. with a typical assemblage of an alpine sagebrush steppe. Trees were not growing around the pond but may have been scattered in the area.

As noted earlier, the two southern pollen sites, Exchequer and Balsam Meadows, have an early Holocene vegetation record similar to that of the high-altitude sites but resulting more from aridity than an elevational lag effect. Trees were not established around the meadows until about 7500 yr B.P. Before that date an open sagebrush steppe with local stands of *Pinus, Abies,* and *Juniperus* or *Calocedrus decurrens* apparently persisted from the late-glacial. The longer Exchequer Meadow record shows a decrease in *Artemisia* and Poaceae pollen from the higher abundances between 13,500 and 11,000 yr B.P. After 10,000 yr B.P., conditions remained fairly constant at Balsam Meadow, while a gradual trend in nearby forest development is evident at the higher-elevation Exchequer Meadow. There, *Pinus* and *Abies* pollen steadily increased throughout the interval to 7000 yr B.P. *Quercus* pollen also became more frequent in the Exchequer Meadow sediments between about 10,000 and 7000 yr B.P., and the abundance of *Sequoiadendron* pollen, centered at 10,680 yr B.P., indicates an extension of the upper-elevation range of that species.

East-Side Sierra Nevada

A lodgepole pine–limber pine forest is inferred to have been established at Barrett Lake by 10,000 yr B.P., followed by local individuals of *Tsuga mertensiana* a thousand years later. There is then no discernible trend in conifer pollen between 10,000 and 6000 yr B.P. except for a slight increase in mountain hemlock. Shrub pollen representing east-side mountain and basin vegetation also increases during this interval. Included are *Artemisia,* Cheno-*Ams,* *Cercocarpus-Purshia*-type, *Ambrosia, Chrysolepis,* and *Quercus vaccinifolia.* The trend in shrub pollen covaries with pollen and macrofossil records in Mono Lake and the Owens Valley.

The Pleistocene to Holocene transition at Tule Lake is indicated by an initial decrease in *Juniperus* followed by a rise in *Pinus* pollen at the expense of sagebrush. High fluctuating Cyperaceae (sedge) and *Typha* (cattail) percentages are also attributed to the lowering of the lake level and expansion of surrounding marshes at two intervals in the Holocene.

Pollen from Owens Lake and the Alabama Hills pack rat middens documents a continuing regional postglacial vegetation change after 10,000 yr B.P. As *Juniperus* pollen diminishes to low values, *Pinus* begins to peak with the increase of *Abies* and such xeric shrub pollen as *Cercocarpus-Purshia*-type, *Ephedra, Ambrosia,* and Cheno-*Ams.* Macrofossils date the local appearances of *Krascheninnikovia lanata* (winterfat), *Chrysothamnus teretifolius* (rubber rabbitbrush), *Grayia spinosa* (spiny hopsage), *Lycium andersonii* (wolfberry), and *Mirabilis bigelovii* (four o'clock) between 9540 yr B.P. and 8700 yr B.P. as additions to the xeric association previously established in the Alabama Hills. *Juniperus osteosperma* finally departed from the locality between 9500 and 7650 yr B.P.

An early Holocene macrofossil assemblage was taken from two pack rat middens located in Falls Canyon on the northwest slope of the White Mountains (Jennings and Elliott-Fisk 1993) at the lower range of the piñon/mountain mahogany–sagebrush *(Artemisia nova)*–desert bitterbrush woodland (1,830 m [6,004 ft]). They are radiocarbon-dated to 8790 yr B.P. and 7810 yr B.P. The middens reflect the modern community except for the lack of such xeric shrubs as *Cercocarpus ledifolius, Ephedra viridis,* and *Purshia tridentata* var. *glandulosa.* The oldest midden provides a minimum date for *Pinus monophylla* in the White Mountains; because there is no glacial-to-Holocene chronological sequence for the mountain front, it cannot be affirmed as the earliest date of occurrence. Dispersal of *P. monophylla* into the White Mountains, however, may be close to the early Holocene if the populations had retreated to the southern desert mountains as they did in the Great Basin. The 9830 yr B.P. Volcanic Tableland midden did not contain *P. monophylla,* although that area was suitable habitat during the late-glacial. The northernmost full-glacial occurrences of the species are at equivalent latitudes from 1,155 m (3,790 ft) at the western base of the southern Inyo Mountains (36°36' N) dated between 22,900 yr B.P. and 17,680 yr B.P. (Koehler and Anderson 1994a) and from a 925 m (3,035 ft) site in the Skeleton Hills in the northern Mojave Desert of southwest Nevada (36°38' N) dated to 17,900 yr B.P. (Spaulding et al. 1990). Both sites have a southeast aspect. *P. monophylla* moved north or upslope as climatic conditions became favorable, presumably when warming summers were accompanied by an expansion of monsoonal rainfall in the Southwest deserts (Spaulding and Graumlich 1986). The species arrived at the southern slopes of the Eleana Range, 50 km (30 mi) north of the Skeleton Hills and 885 m (2,900 ft) higher, by 11,700 yr B.P., so it is not unreasonable to expect it to have dispersed 106 km (66 mi) north and 675 m (2,215 ft) higher along a more continuous mountain range at a faster rate.

Farther south along the mountain front, a 5640 yr B.P. midden from near the head of Silver Canyon (3,048 m [10,000 ft]) contains *J. osteosperma* and *P. monophylla* macrofossils. This is evidence for the extension of the upper limits of these trees to elevations now occupied by bristlecone pine–limber pine woodland, presumably as a consequence of higher temperatures.

Middle to Late Holocene (6.0 ka to Present)

West-Slope Sierra Nevada

The general trends in pollen abundances at intermediate elevations over the past 6,000 years are a decrease in oak and alder from their early Holocene maxima and an increase in *Abies* and Cupressaceae *(Calocedrus decurrens)*. Charcoal concentrations also decrease. More effective moisture, from a combination of lower temperatures and higher levels of precipitation, are suggested by these data. Change in climate parallels the postulated lessening of summer insolation over the Northern Hemisphere toward present values (COHMAP Members 1988). The analysis of temperature-sensitive deuterium to hydrogen ratios (δD values) in bristlecone pine tree rings from the White Mountains has produced an 8,000-year chronology indicating optimal temperatures at 6,800 years ago followed by a continuous average cooling (Feng and Epstein 1994). There have been higher-frequency fluctuations in vegetation dynamics that the temporal resolution of the existing pollen stratigraphies either do not capture or that, when recorded, need to be further analyzed, but overall the modern vegetation communities were developed during this period.

There are some differences in details among sites. Lake Moran displays its usual lag effect, with an increase in *Quercus* pollen to a maximum frequency at about 4500 yr B.P. *Quercus* also continues to increase at Osgood Swamp until about 2800 yr B.P. High frequencies of *Alnus* pollen, along with an abundance of *Carex lenticular* and Ranunculaceae macrofossils at Nichols Meadow, are interpreted as the establishment of the modern meadow (Koehler and Anderson 1994b). At Balsam Meadow, the synchronous rise in percentages of *Abies, Calocedrus decurrens,* and *Quercus* pollen and decrease of *Pinus* pollen during the past 3,000 years are explained as a downslope shift of those species that make up the present community (Davis et al. 1985). Finally, the Log Meadow site shows an increase in *Sequoiadendron* pollen after 4500 yr B.P. from a very few to maximum abundances of the modern forest during the past several hundred years (Anderson 1994).

The high-altitude sites of Starkweather and Tioga Pass Ponds are sensitive to the response of upper montane and subalpine vegetation to Holocene climate change. They both document the variations of the upper altitudinal limits of red fir and mountain hemlock and the lower range of whitebark pine after 6000 yr B.P. At Tioga Pass Pond, *Abies* pollen increased to a more-or-less stationary level at about 4000 yr B.P., after several fluctuations during the early Holocene. Beginning with a secondary rise at 2000 yr B.P., *Abies* rapidly reached maximum values 800 years later. It just as rapidly decreased in abundance soon after, indicating an altitudinal depression. Starkweather Pond, 580 m (1,900 ft) lower, records a gentler upward trend in *Abies* to 3000 yr B.P. A macrofossil identified as *A. grandis* was recovered in sediments dated at about 2500 yr B.P. *Tsuga mertensiana* pollen became abundant just before 5000 yr B.P. as populations approached Tioga Pass Pond. The species arrived 500 years later, depositing needles in the pond

sediments, and subsequently retreated about 2500 yr B.P. At the same time, there was a steady increase in mountain hemlock pollen at Ten Lakes, located 2,743 m (9,020 ft) above the Tuolumne River in Yosemite National Park (Anderson 1987). *Pinus albicaulis* needles are relatively common in Tioga Pass Pond by 3500 yr B.P., remaining at high frequencies until 500 yr B.P.

Anderson (1990a) interprets the collective pattern from all three sites as a response to cooler conditions beginning about 3000-2500 yr B.P., depressing the upper altitudinal limits of *Tsuga mertensiana* and *Abies grandis* and the lower limits of *Pinus albicaulis*. There is widespread evidence for a cool-moist episode between 4000 and 2500 yr B.P. in the Southwest, including a high stand at Mono Lake at 3770 cal yr B.P. (Stine 1990), the lowering of the bristlecone pine upper tree line in the White Mountains between 3500 and 2500 yr B.P. (LaMarche 1973), and the existence of a shallow lake in the Silver Lake playa around 3620 yr B.P. (Enzel et al. 1992), to cite only a few examples.

Pollen and macrofossil profiles compiled from close increment sampling are important for analyzing century-scale vegetation changes. The Woski Pond diagram is one such record (Anderson 1987; Anderson and Carpenter 1991) (figure 4.7). Woski Pond is located in Slaughter House Meadow, Yosemite Valley, at 1,212 m (3,975 ft). The bottom of the core is dated to nearly 1500 yr B.P. (A.D. 1300), and samples were taken at about 50- to 150-year intervals. The lower half of the diagram represents a mixed conifer forest dominated by *Pinus ponderosa, Abies, Calocedrus decurrens,* and *Quercus* (probably *Q. kelloggii*). At about 650 yr B.P. there was an abrupt shift in pollen trends to decreasing *Pinus* and *Abies* and increasing *Quercus* and shrub pollen, primarily *Prunus-* and *Cercocarpus*-type. A very large spike of charcoal accompanying the shift indicates a fire disturbance event.

There are two lines of evidence leading to the conclusion that the disturbance was not natural but resulted from human intervention. First, around A.D. 1300 the climate began to change from a 400-year series of persistent above-average temperatures (the Medieval Warm Period) to a 500-year period with a temperature mean of 0.48°C (0.86°F) below modern levels and a precipitation mean of 6.46 cm (2.64 in) above modern averages in the southern Sierra Nevada, based on tree-ring analysis (Graumlich 1993). These conditions produced a positive glacier budget with multiple advances of alpine glaciers and a decrease in the number of fire events (Birman 1964; Burke and Birkeland 1983; Curry 1969; Gillespie 1982; Scuderi 1984, 1987; Swetnam 1993). This is the worldwide Little Ice Age, also termed the Matthes glaciation in the Sierra Nevada. The change in direction of pollen percentages in Woski Pond runs counter to what would be expected with a cooler, wetter climate phase. The initial disturbance could be explained as the cumulative effect of the previous period of droughts. The high-frequency charcoal influx between 800 and 600 yr B.P. is obvious evidence for a brief interval of intense local fires that altered the forest structure. It does not

WOSKI POND, CALIFORNIA, POLLEN PERCENT

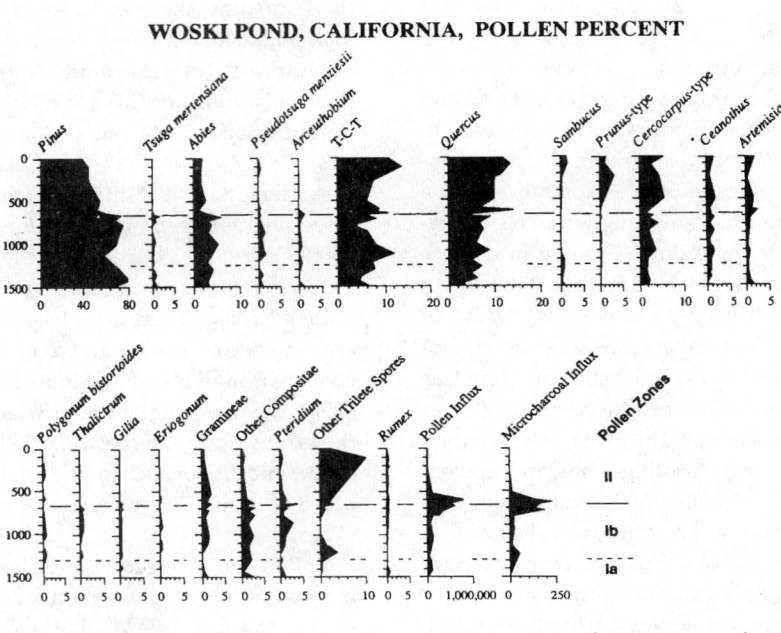

WOSKI POND, CALIFORNIA, AQUATICS

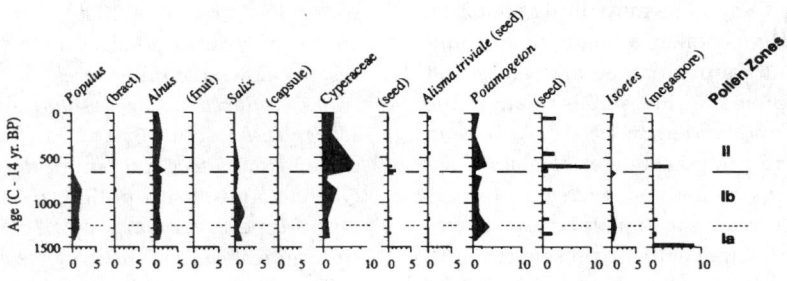

WOSKI POND, CALIFORNIA, MACROFOSSILS

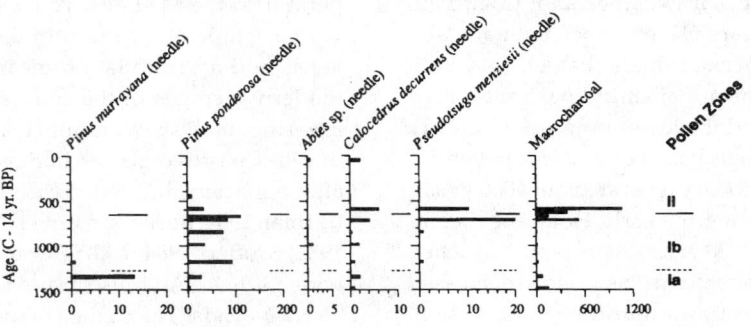

FIGURE 4.7

Pollen and macrofossil profile of Woski Pond, Yosemite National Park (from Anderson and Carpenter 1991).

explain, however, the continued decline in *Pinus* pollen and increase in *Quercus* pollen. Second, archaeological investigations document that after A.D. 1200 Yosemite Valley was permanently occupied by a large, sedentary population of Native Americans who exploited a wide variety of resources, particularly the acorn (Munday and Hull 1988). The similarity of their material culture relates this population to the historically known Central Sierra Miwok. The Miwok, like other California groups, practiced intensive vegetation management, including the use of fire (Blackburn and Anderson 1993). Correlation of the dates of human settlement, the large charcoal peak, and rapid, sustained vegetation change supports a strong case that the human inhabitants of Yosemite initially burned the forest for conversion to an open oak woodland in order to enhance the acorn crop, among other reasons. Fire may have been used periodically to maintain the vegetation type. The scope of this report, as a summary of existing studies of vegetation history, encompasses primarily the vegetation-climate relationship, but for future research it is strongly recommended that prehistoric human modification of the landscape be taken into account.

East-Side Sierra Nevada

The situation of Barrett Lake at the present lower elevational limit of *Tsuga mertensiana* and the upper limit of *Abies grandis* makes it a sensitive site for recording vegetation change. Arboreal pollen increased around the lake after 5500 yr B.P., when *A. grandis* was added to the previous lodgepole pine–limber pine–mountain hemlock forest and the *T. mertensiana* population grew. Pollen abundances peaked around 1000 yr B.P. before declining.

The remaining three middens to be considered from the southern end of the White Mountains, dating from 4510 to 2130 yr B.P., contain macrofossil assemblages that are little different from the modern flora at the sites. Similarly, only small changes are evident from the Alabama Hills midden assemblages. *Cowania mexicana* disappeared from the record, and *Echinocactus polycephalus* (cottontop cactus) was added to the flora after 2830 yr B.P.

CONCLUSIONS AND MANAGEMENT IMPLICATIONS

A great deal of information about vegetation dynamics in the Sierra Nevada region exists from paleoecological data. The information is at a fairly coarse temporal and spatial scale, however. In order to resolve the scale to a finer level, a finer network of sites in all vegetation communities and at the full range of elevations would be needed. Research is continually providing new sites: the Desert Research Institute (DRI) is analyzing cores retrieved from Eagle Lake and Pine Creek drainage in northern California; Little Valley in the Washoe

Lake area; and a series of subalpine ponds and meadows in the Desolation Wilderness, the headwaters of Lee Vining Creek, the Middle Fork of Bishop Creek, and the North Fork Kings River (Wigand 1994). A joint project between DRI and the U.S. Geological Survey is a study of macrofossils from an exposure of glacial and interglacial deposits in the Feather River canyon spanning the past 600,000 years (Wigand 1994).

The study of vegetation history is critical to management. Most important is the realization that the global climate system and climatically sensitive elements of the ecosystem, including vegetation, are in a constant state of flux. The nature and degree of climatic control of ecosystem processes, however, can be usefully conceptualized by specifying the spatial and temporal scales of interest. Change can be slow or very rapid, and the range of variability is much greater than that observed during our lifetimes or even recorded over the past century. In addition, some climatic and ecosystem changes are quasi-periodic and thus somewhat predictable; other changes are chaotic and strongly influenced by historical contingencies, frustrating any definite projections of future outcomes.

The long-term perspective has shown that plant communities we recognize today are not cohesive units but disassemble and reassemble with varying dominance patterns and composition in response to climatic variation; each species responds individualistically at its own rate, some lagging behind others, and never quite in equilibrium. Associations of species have occurred in the past for which no known modern analog can be found, implying that the full range of potential responses of taxa may not be expressed in the modern climate state. This individualistic behavior of vegetation communities (Gleason 1926) does not deny the competitive interactions between species that make any community whole "greater than the sum of its parts." It does mean that, at any one time, the landscape is a unique combination of climate, species, soils, and topography (see also Sprugel 1991).

Because the management concepts of habitat type and potential natural vegetation assume stability and a single pathway toward a specific endpoint of species composition, they need to be modified to account for uncertain changes in environmental conditions and a multiplicity of possible future states (Tausch et al. 1993). Management principles that have the achievement and maintenance of a steady-state vegetation community as an objective may be successfully applied on a small scale with concerted effort. But sustaining an entire landscape or achieving a single desired hypothetical landscape over the long term is not feasible. With insight from paleoecology, the focus of management should be "the maintenance of the dynamic capacity [of an ecosystem] to respond adaptively" (Costanza et al. 1993). This focus requires knowledge of the processes operating at present and in the past and working with those processes to allow a system enough resilience to respond to changing circumstances. The structure and function of an ecosystem may be sustained even though the species composition will vary. Close monitoring and continual adjustment in management practices is still necessary. In any case, more research is needed to acquire knowledge sufficient to understand and predict the future behavior of ecosystems and their response to climate change (Mooney 1991).

ACKNOWLEDGMENTS

This report has benefited from knowledge and ideas generously shared by David Adam, Scott Anderson, Douglas Clark, Owen Davis, Connie Millar, and Linda Reynolds and from thorough reviews by Michael Barbour, Michael Diggles, Dave Parsons, and Carl Skinner.

NOTES

1. The following sites record subalpine sagebrush steppe–woodland above 1,500 m (4,900 ft):

Swamp Lake, Yosemite National Park, 1,554 m (5,098 ft), Tuolumne River drainage, white fir–mixed conifer forest (Smith and Anderson 1992)

McMurray Lake, 1,778 m (5,832 ft), Yuba River drainage, red fir–mixed conifer forest (Byrne et al. 1993)

Swamp Lake, 1,957 m (6,420 ft), North Fork Stanislaus River drainage, white fir–mixed conifer forest (Batchelder 1980)

Burgson Lake, Stanislaus River drainage, 1,960 m (6,430 ft), red fir–mixed conifer forest (Byrne et al. 1993)

Osgood Swamp, 1,980 m (6,500 ft), Lake Tahoe Basin, mixed conifer (Adam 1967)

Lily Pad Lake, 1,980 m (6,500 ft), North Fork Kings River drainage, white fir–mixed conifer forest (Byrne et al. 1993)

Bunker Lake, 1,995 m (6,550 ft), Middle Fork American River drainage, upper margin white fir–mixed conifer forest near red fir forest (Edlund 1994)

Balsam Meadow, 2,015 m (6,611 ft), San Joaquin River drainage, upper margin white fir–mixed conifer near red fir forest (Davis et al. 1985)

Lake Moran, 2,018 m (6,621 ft), North Fork Stanislaus River drainage, red fir forest (Edlund 1991; Edlund and Byrne 1991)

Exchequer Meadow, 2,219 m (7,280 ft), Kings River drainage, Jeffrey pine forest (Davis and Moratto 1988)

Starkweather Pond, 2,438 m (8,000 ft), Middle Fork San Joaquin River drainage, red fir forest (Anderson 1987, 1990a)

REFERENCES

Adam, D. P. 1967. Late Pleistocene and recent palynology in the central Sierra Nevada, California. In *Quaternary Paleoecology,* edited by E. J. Cushing and H. E. Wright Jr. New Haven, CT: Yale University Press.

———. 1973. Early Pleistocene (?) pollen spectra near Lake Tahoe, California. *Journal of Research of the United States Geological Survey* 1:691–93.

———. N.d.a. Lake 11,100 ft. pollen analysis. Unpublished ms.

———. N.d.b. Upper Echo Lake pollen analysis. Unpublished ms.

Adam, D. P., J. P. Bradbury, H. J. Rieck, and A. M. Sarna-Wojcicki. 1990. Environmental changes in the Tule Lake Basin, Siskiyou and Modoc Counties, California, from 3 to 2 million years before present. *United States Geological Survey Bulletin* 1933:1–13.

Adam, D. P., A. M. Sarna-Wojcicki, H. J. Rieck, J. P. Bradbury, W. E. Dean, and R. M. Forester. 1989. Tulelake, California: The last 3 million years. *Palaeogeography, Palaeoclimatology, Palaeoecology* 72:89–103.

Anderson, R. S. 1987. Late-Quaternary environments of the Sierra Nevada, California. Ph.D diss., University of Arizona.

———. 1990a. Holocene forest development and paleoclimates within the central Sierra Nevada, California. *Journal of Ecology* 78:470–89.

———. 1990b. Modern pollen rain within and adjacent to two giant sequoia (*Sequoiadendron giganteum*) groves, Yosemite and Sequoia National Parks, California. *Canadian Journal of Forestry Research* 20:1289–1305.

———. 1994. *Paleohistory of a giant sequoia grove: The record from Log Meadow, Sequoia National Park.* General Technical Report PSW-151. Berkeley: U.S. Forest Service.

Anderson, R. S., and S. L. Carpenter. 1991. Vegetation change in Yosemite Valley, Yosemite National Park, California, during the protohistoric period. *Madroño* 38 (1): 1–13.

Anderson, R. S., and S. J. Smith. 1994. Paleoclimatic interpretations of meadow sediment and pollen stratigraphies from California. *Geology* 22:723–26.

Atwater, B. F., D. P. Adam, J. P. Bradbury, R. M. Forester, R. M. Mark, W. R. Lettis, G. R. Fisher, K. W. Gobalet, and S. W. Robinson. 1986. A fan dam for Tulare Lake, California, and implications for the Wisconsin glacial history of the Sierra Nevada. *Geological Society of America Bulletin* 97:97–109.

Axelrod, D. I., and W. S. Ting. 1962. Early Pleistocene floras from the Chagoopa surface, southern Sierra Nevada. *University of California Publications in Geological Sciences* 39:119–93.

Bacon, C. R., D. M. Giovanetti, W. A. Duffield, and G. B. Dalrymple. 1979. New constraints on the age of the Coso Formation, Inyo County, California. *Geological Society of America, Abstracts of Program* 11 (3): 67.

Baker, R. G. 1983. Holocene vegetation history of the western United States. In *The Holocene,* edited by H. E. Wright Jr., 109–27. Vol. 2 of *Late-Quaternary environments of the United States.* Minneapolis: University of Minnesota Press.

Barnosky, C. W., P. M. Anderson, and P .J. Bartlein. 1987. The northwestern U.S. during deglaciation: Vegetation history and paleoclimatic implications. In *North America and adjacent oceans during the last deglaciation,* edited by W. F. Ruddiman and H. E. Wright Jr., 289–321. Boulder, CO: Geological Society of America.

Bartlein, P. J. 1988. Late-Tertiary and Quaternary palaeoenvironments. In *Vegetation history,* edited by B. Huntley and T. Webb III, 113–52. Dordrecht, Netherlands: Kluwer Academic Publishers.

Batch, J. R. 1977. A post-glacial pollen record from a subalpine valley in Yosemite National Park. Master's thesis, San Francisco State University.

Batchelder, G. L. 1970. Post-glacial ecology at Black Lake, Mono County. Ph.D. diss., Arizona State University.

————. 1980. A late Wisconsinan and early Holocene lacustrine stratigraphy and pollen record from the west slope of the Sierra Nevada, California. In *Abstracts and Program, Sixth Biennial Meeting.* N.p.: American Quaternary Association.

Benson, L. V., D. R. Currey, R. I. Dorn, K. R. Lajoie, C. G. Oviatt, S. W. Robinson, G. I. Smith, and S. Stine. 1990. Chronology of expansion and contraction of four Great Basin lake systems during the past 35,000 years. *Palaeogeography, Palaeoclimatology, Palaeoecology* 78:241–86.

Berger, A. 1991. Long-term history of climate ice ages and Milankovitch periodicity. In *The sun in time,* edited by C. P. Sonett, M. S. Giampapa, and M. S. Matthews. Tucson: University of Arizona Press.

Betancourt, J. L., E. A. Pierson, A. K. Rylander, J. A. Fairchild-Parks, and J. S. Dean. 1993. Influence of history and climate on New Mexico piñon-juniper woodlands. In *Managing piñon-juniper ecosystems for sustainability and social needs,* edited by E. F. Aldon and D. W. Shaw, 42–62. General Technical Report RM-236. Fort Collins, CO: U.S. Forest Service, Rocky Mountain Forest and Range Experiment Station.

Birman, J. H. 1964. *Glacial geology across the crest of the Sierra Nevada, California.* Special Paper 75. Washington, DC: Geological Society of America.

Blackburn, T. C., and K. Anderson, eds. 1993. *Before the wilderness: Native Californians as environmental managers.* Menlo Park, CA: Ballena Press.

Bradshaw, R. H. W. 1988. Spatially-precise studies of forest dynamics. In *Vegetation history,* edited by B. Huntley and T. Webb III, 725–51. Dordrecht: Kluwer Academic Publishers.

Burke, R. M., and P. W. Birkeland. 1983. Holocene glaciation in the mountain ranges of the western United States. In *The Holocene,* edited by H. E. Wright, 3–11. Vol. 2 of *Late-Quaternary environments of the United States.* Minneapolis: University of Minnesota Press.

Byrne, R. N.d. Unpublished pollen diagram of Catfish Lake. University of California, Berkeley.

Byrne, R., D. Araki, and S. Peterson. 1979a. Report on pollen analysis of a core from Tule Lake, Mono County, California. Department of Geography, University of California, Berkeley.

————. 1979b. Pollen evidence for recent vegetation change in the Melones Dam area. Department of Geography, University of California, Berkeley.

Byrne, R., S. A. Mensing, and E. G. Edlund. 1993. *Long-term changes in the structure and extent of California oaks.* Final report to the Integrated Hardwood Range Management Program. Berkeley: University of California.

COHMAP Members. 1988. Climatic changes of the last 18,000 years: Observations and model simulations. *Science* 241:1043–52.

Cole, K. 1983. Late Pleistocene vegetation of Kings Canyon, Sierra Nevada, California. *Quaternary Research* 19:117–29.

————. 1985. Past rates of change, species richness, and a model of vegetational inertia in the Grand Canyon, Arizona. *The American Naturalist* 125:289–303.

Coleman, S. M., K. L. Pierce, and P. W. Birkeland. 1987. Suggested terminology for Quaternary dating methods. *Quaternary Research* 28:314–19.

Costanza, R., L. Wainger, C. Folke, and K. Maler. 1993. Modeling complex ecological economic systems. *BioScience* 43:545–55.

Curry, R. R. 1969. Holocene climate and glacial history of the central Sierra Nevada, California. In *United States contributions to*

Quaternary research, edited by S. A. Schuman and W. C. Bradley, 1–47. Special Paper 123. Washington, DC: Geological Society of America.

Damon, P. E., and C. P. Sonett. 1991. Solar and terrestrial components of the atmospheric ^{14}C variation spectrum. In *The sun in time,* edited by C. P. Sonett, M. S. Giampapa, and M. S. Matthews. Tucson: University of Arizona Press.

Davis, M. B. 1986. Climatic instability, time lags, and community disequilibrium. In *Community ecology,* edited by J. Diamond and T. J. Case. Cambridge, MA: Harper and Row.

Davis, O. K. 1993. Unpublished pollen diagram from Mono Lake, California. University of Arizona.

————. 1994. The correlation of summer precipitation in the southwestern U.S.A. with isotopic records of solar activity during the Medieval Warm Period. *Climatic Change* 26:271–87.

Davis, O. K., R. S. Anderson, P. L. Fall, M. K. O'Rourke, and R. S. Thompson. 1985. Palynological evidence for early Holocene aridity in the southern Sierra Nevada, California. *Quaternary Research* 24:322–32.

Davis, O. K., and M. J. Moratto. 1988. Evidence for a warm dry early Holocene in the western Sierra Nevada of California: Pollen and plant macrofossil analysis of Dinkey and Exchequer Meadows. *Madroño* 35:132–49.

Dial, K. P., and N. J. Czaplewski. 1990. Do woodrat middens accurately represent the animals' environments and diets? The Woodhouse Mesa study. In *Packrat middens: The last 40,000 years of biotic change,* edited by J. L. Betancourt, T. R. Van Devender, and P. S. Martin. Tucson: University of Arizona Press.

Dorland, D. 1980. Two post-glacial pollen records from Meyers Grade Marsh and Grass Lake, El Dorado County, California. Master's thesis, San Francisco State University.

Dorland, D., D. P. Adam, and G. L. Batchelder. 1980. Two Holocene pollen records from Meyers Grade Marsh and Grass Lake, El Dorado County, California. In *Abstracts and Program, 6th Biennial Meeting.* N.p.: American Quaternary Association.

Duffield, W. A., C. R. Bacon, and G. B. Dalrymple. 1980. Late Cenozoic volcanism, geochronology, and structure of the Coso Range, Inyo County, California. *Journal of Geophysical Research* 85:2381–404.

Edlund, E. 1994. Bunker Lake paleoecological analysis. In *Framework for archaeological research and management, national forests of the north-central Sierra Nevada,* edited by R. J. Jackson, 1–23. Placerville, CA: U.S. Forest Service.

Edlund, E. G. 1991. Reconstruction of late Quaternary vegetation and climate at Lake Moran, central Sierra Nevada, California. Master's thesis, University of California, Berkeley.

Edlund, E. G., and R. Byrne. 1991. Climate, fire, and late Quaternary vegetation change in the central Sierra Nevada. In *Fire and the environment: Ecological and cultural perspectives,* edited by S. S. Nodvin and T. A. Waldrop, 390–96. General Technical Report SE-69. Asheville, NC: U.S. Forest Service, Southeastern Forest Experiment Station.

Enzel, Y., W. J. Brown, R. Y. Anderson, L. D. McFadden, and S. G. Wells. 1992. Short-duration Holocene lakes in the Mojave River drainage basin, Southern California. *Quaternary Research* 38: 60–73.

Erdtman, G. 1969. *Handbook of palynology.* Copenhagen: Munksgaard.

Fairbanks, R. G. 1989. A 17,000-year glacio-eustatic sea level record: Influence of glacial melting rates on the Younger Dryas event and deep-ocean circulation. *Nature* 342:637–42.

Feng, X., and S. Epstein. 1994. Climatic implications of an 8000-year hydrogen isotope time series from bristlecone pine trees. *Science* 265:1079–81.

Finley, R. B., Jr. 1990. Woodrat ecology and behavior and the interpretation of paleomiddens. In *Packrat middens: The last 40,000 years of biotic change,* edited by J. L. Betancourt, T. R. Van Devender, and P. S. Martin. Tucson: University of Arizona Press.

Flint, R. F. 1971. *Glacial and Quaternary geology.* New York: John Wiley and Sons.

Ghil, M. 1991. Quaternary glaciations: Theory and observations. In *The sun in time,* edited by C. P. Sonett, M. S. Giampapa, and M. S. Matthews. Tucson: University of Arizona Press.

Gillespie, A. R. 1982. Quaternary glaciation and tectonism in the southeastern Sierra Nevada, Inyo County, California. Ph.D. diss., California Institute of Technology, Pasadena.

Giovanetti, D. M. 1979. Volcanism and sedimentation associated with the formation of southern Owens Valley, California. *Geological Society of America, Abstracts of Program.* 11 (3).

Gleason, H. A. 1926. The individualistic concept of the plant association. *Bulletin of the Torrey Botanical Club* 53:7–26.

Graumlich, L. J. 1993. A 1000-year record of temperature and precipitation in the Sierra Nevada. *Quaternary Research* 39:249–55.

Graumlich, L. J., and M. B. Davis. 1993. Holocene variation in spatial scales of vegetation pattern in the upper Great Lakes. *Ecology* 74:826–39.

Griffin, J. R., and W. B. Critchfield. 1972. *The distribution of forest trees in California.* Research Paper PSW-82. Berkeley, CA: U.S. Forest Service, Pacific Southwest Forest and Range Experiment Station.

Hickman, J. C. 1993. *The Jepson manual.* Berkeley and Los Angeles: University of California Press.

Huber, N. K. 1981. *Amount and timing of late Cenozoic uplift and tilt of the central Sierra Nevada, California—evidence from the upper San Joaquin River Basin.* Professional Paper 1197. Washington, DC: United States Geological Survey.

Hughes, M. K., and H. F. Diaz. 1994. Was there a "Medieval Warm Period," and if so, where and when? *Climatic Change* 26:109–42.

Imbrie, J., et al. 1992. On the structure and origin of major glaciation cycles, 1. Linear responses to Milankovitch forcing. *Paleoceanography* 7:701–38.

Jennings, S. A., and D. L. Elliott-Fisk. 1993. Packrat midden evidence of late Quaternary vegetation change in the White Mountains, California-Nevada. *Quaternary Research* 39:214–21.

Kilbourne, R. T. 1978. Pollen studies in the Auburn Dam area. Sacramento: California Division of Mines.

Koehler, P. A., and R. S. Anderson. 1994a. Full-glacial shoreline vegetation during the maximum high stand at Owens Lake, California. *Great Basin Naturalist* 54:142–49.

———. 1994b. The paleoecology and stratigraphy of Nichols Meadow, Sierra National Forest, California, USA. *Palaeogeography, Palaeoclimatology, Palaeoecology* 112:1–17.

———. 1995. Thirty thousand years of vegetation changes in the Alabama Hills, Owens Valley, California. *Quaternary Research* 43:238–48.

Kutzbach, J. E., and P. J. Guetter. 1986. The influence of changing orbital parameters and surface boundary conditions on climate simulations for the past 18,000 years. *Journal of the Atmospheric Sciences* 43:1726–59.

LaMarche, V. C., Jr. 1973. Holocene climatic variations inferred from treeline fluctuations in the White Mountains, California. *Quaternary Research* 3:632–60.

Leopold, E. B. 1967. Summary of palynological data from Searles Lake.

In Pleistocene geology and palynology of Searles Valley, California: Guidebook, Friends of the Pleistocene, Pacific Coast Section meeting, 23–24 September 1967, n.p.

Litwin, R. J., D. P. Adam, N. O. Frederiksen, and W. B. Woolfenden. 1995. *800,000-year pollen record from Owens Lake: Preliminary analyses.* Special Paper. Washington, DC: Geological Society of America.

Mackey, E. M., and D. G. Sullivan. 1991. Appendix E: Revised final report, results of palynological investigations at Gabbott Meadow Lake, Alpine County, California. In *Cultural resource studies, North Fork Stanislaus River Hydroelectric Development Project,* edited by A. S. Peak and N. J. Neuenschwander, 473–99. Sacramento, CA: Peak and Associates.

Magnuson, J. J. 1990. Long-term ecological research and the invisible present. *Bioscience* 40:495–501.

McCarten, N., and T. R. Van Devender. 1988. Late Wisconsin vegetation of Robber's Roost in the western Mojave Desert, California. *Madroño* 35:226–37.

Mehringer, P. J., Jr., and J. C. Sheppard. 1978. Holocene history of Little Lake, Mojave Desert, California. In *The ancient Californians, Rancholabrean hunters of the Mojave lakes country,* edited by E. L. Davis. Los Angeles: Natural History Museum of Los Angeles County.

Miller, C. D. 1984. Chronology and stratigraphy of recent eruptions at the Inyo volcanic chain. In *Holocene paleoclimatology and thphrochronology east and west of the central Sierran crest,* edited by S. Stine, S. Wood, K. Sieh, and C. D. Miller. Field trip guidebook for the Friends of the Pleistocene Pacific Cell, 12–14 October 1984, n.p.

Mooney, H. A. 1991. Biological response to climate change: An agenda for research. *Ecological Applications* 1:112–17.

Munday, W. J., and K. L. Hull. 1988. *The 1984 and 1985 Yosemite Valley archeological testing projects.* Publications in Anthropology 5. Yosemite National Park: U.S. National Park Service, Yosemite Research Center.

Owens Lake Core Study Team. 1995. An 800,000-year paleoclimatic record from Core OL-92, Owens Lake, southeast California. In *Report of 1994 Workshop on the Correlation of Marine and Terrestrial Records of Climate Changes in the Western United States,* edited by D. P. Adam, J. P. Bradbury, W. E. Dean, J. V. Gardner, and A. M. Sarna-Wojcicki, 36–55. Open-File Report 95-34. Menlo Park, CA: U.S. Geological Survey.

Patterson, W. A., III, and I. C. Prentice. 1985. Quantitative interpretation of fossil pollen spectra: Dissimilarity coefficients and the method of modern analogs. *Quaternary Research* 23:87–108.

Polley, H. W., H. B. Johnson, B. D. Marino, and H. S. Mayeux. 1993. Increase in C3 plant water-use efficiency and biomass over glacial to present CO_2 concentrations. *Nature* 361:61–64.

Porter, S. C., K. L. Pierce, and T. D. Hamilton. 1983. Late Wisconsin mountain glaciation in the western United States. In *The Late Pleistocene,* edited by H. E. Wright Jr., 71–111. Vol. 1 of *Late-Quaternary Environments of the United States.* Minneapolis: University of Minnesota Press.

Prentice, I. C., P. J. Bartlein, and T. Webb III. 1991. Vegetation and climate change in eastern North America since the last glacial maximum. *Ecology* 72:2038–56.

Quade, J. 1986. Late Quaternary environmental changes in the upper Las Vegas Valley, Nevada. *Quaternary Research* 26:340–57.

Ruddiman, W. F., and J. E. Kutzbach. 1989. Forcing of late Cenozoic Northern Hemisphere climate by plateau uplift in southern Asia and the American West. Journal of Geophysical Research 94: 18409–27.

Ruddiman, W. F., A. McIntyre, and M. Raymo. 1990. Paleoenvironmental results from North Atlantic Sites 607 and 609. *National Science Foundation, Initial Reports of the Deep Sea Drilling Project* 94:855–78.

Scholes, R. J. 1990. Change in nature and the nature of change: Interactions between terrestrial ecosystems and the atmosphere. *South African Journal of Science* 86:350–54.

Schumm, S. A. 1991. *To interpret the earth: Ten ways to be wrong.* Cambridge: Cambridge University Press.

Scuderi, L. A. 1984. A dendroclimatic and geomorphic investigation of late-Holocene glaciation, southern Sierra Nevada, California. Ph.D. diss., University of California, Los Angeles.

———. 1987. Glacier variations in the Sierra Nevada, California, as related to a 1200-year tree-ring chronology. *Quaternary Research* 27:220–31.

Sercelj, A., and D. P. Adam. 1975. A late Holocene pollen diagram from near Lake Tahoe, El Dorado County, California. *Journal of Research of the United States Geological Survey* 3:737–45.

Shackleton, N. J., et al. 1984. Oxygen isotope calibration of the onset of ice-rafting and history of glaciation in the North Atlantic region. *Nature* 307:620–23.

Singh, G. 1988. History of aridland vegetation and climate: A global perspective. *Biological Reviews* 63:159–95.

Smith, S., and R. S. Anderson. 1992. Late Wisconsin paleoecologic record from Swamp Lake, Yosemite National Park. *Quaternary Research* 38:91–102.

Spaulding, W. G., J. L. Betancourt, L. K. Croft, and K. L. Cole. 1990. Packrat middens: Their composition and methods of analysis. In *Packrat middens: The last 40,000 years of biotic change*, edited by J. L. Betancourt, T. R. Van Devender, and P. S. Martin. Tucson: University of Arizona Press.

Spaulding, W. G., and L. J. Graumlich. 1986. The last pluvial climatic episodes in the deserts of southwestern North America. *Nature* 320:441–44.

Sprugel, D. G. 1991. Disturbance, equilibrium, and environmental variability: What is "natural" vegetation in a changing environment? *Biological Conservation* 58:1–18.

Stine, S. 1990. Late Holocene fluctuations of Mono Lake, eastern California. *Palaeogeography, Palaeoclimatology, Palaeoecology* 78:333–82.

Swetnam, T. W. 1993. Fire history and climate change in giant sequoia groves. *Science* 262:813–960.

Tausch, R. J., P. E. Wigand, and J. W. Burkhardt. 1993. Viewpoint: Plant community thresholds, multiple steady states, and multiple successional pathways: Legacy of the Quaternary? *Journal of Range Management* 46:439–47.

Van De Water, P. 1993. Ecophysiological response of *Pinus flexilis* to atmospheric CO_2 enrichment during deglaciation. Master's thesis, University of Arizona.

Vasek, F. C., and R. F. Thorne. 1990. Transmontane coniferous vegetation. In *Terrestrial vegetation of California,* edited by M. G. Barbour and J. Major. Special Publication 9. Sacramento: California Native Plant Society.

Vaughan, T. A. 1990. Ecology of living packrats. In *Packrat middens: The last 40,000 years of biotic change,* edited by J. L. Betancourt, T. R. Van Devender, and P. S. Martin. Tucson: University of Arizona Press.

Wahrhaftig, C., and J. H. Birman. 1965. The Quaternary of the Pacific mountain system in California. In *The Quaternary of the United States,* edited by H. E. Wright and D. G. Frey, 299–340. Princeton, NJ: Princeton University Press.

Webb, T., III. 1986. Is vegetation in equilibrium with climate? How to interpret late-Quaternary pollen data. *Vegetatio* 67:75–91.

Wells, S. G., L. D. McFadden, and J. C. Dohrenwend. 1987. Influence of late Quaternary climatic changes on geomorphic and pedogenic processes on a desert piedmont, eastern Mojave Desert, California. *Quaternary Research* 27:130–46.

West, G. J. N.d. Ross Relles Camp pollen analysis. Unpublished ms.

West, G. J., O. K. Davis, and W. J. Wallace. 1991. Fluted points at Tulare Lake, California: Environmental background. In Contributions to Tulare Lake archaeology I, edited by W. J. Wallace and F. A. Riddell. Unpublished ms. Tulare Lake Archaeological Research Group, n.p.

Wigand, P. 1994. Letter to the author, 1 March.

Wood, S. H. 1975. *Holocene stratigraphy and chronology of mountain meadows, Sierra Nevada, California.* Earth Resources Monograph 4. San Francisco: U.S. Forest Service.

Woolfenden, W. B. 1995. Fine resolution pollen analysis of Core OL-92, Owens Lake, California. In *Report of 1994 Workshop on the Correlation of Marine and Terrestrial Records of Climate Changes in the Western United States,* edited by D. P. Adam, J. P. Bradbury, W. E. Dean, J. V. Gardner, and A. M. Sarna-Wojcicki. Open-File Report 95-34. Menlo Park, CA: U.S. Geological Survey.

_____. N.d. A late Holocene pollen record from the bog mound springs, Deep Springs Valley, Mono County, California. Unpublished research in progress.

Wright, H. E., Jr., J. E. Kutzbach, T. Webb III, W. F. Ruddiman, F. A. Street-Perrott, and P. J. Bartlein, eds. 1993. *Global climates since the last glacial maximum.* Minneapolis: University of Minnesota Press.

Zauderer, J. N. 1973. A neoglacial pollen record from Osgood Swamp, California. Master's thesis, University of Arizona.

APPENDIX 4.1

Quaternary Pollen and Macrofossil Sites in the SNEP Study Area

Site	County	Latitude	Longitude	Elevation		Reference
				m	ft	
Tulelake	Siskiyou	41°57'	121°30'	1,229	4,030	Adam et al. 1990; Adam et al. 1989
Eagle Lake	Lassen	40°35'	120°45'	1,500	4,920	Wigand 1994
McMurray Lake	Sierra	39°27'	120°39'	1,778	5,832	Byrne et al. 1993
Ross Relles Camp	Nevada	39°22'	120°57'	900	2,950	West n.d.
Tahoe City	Placer	39°10'	120°09'	1,900	6,175	Adam 1973
Bunker Lake	Placer	39°03'	120°23'	1,995	6,545	Edlund 1994
Osgood Swamp	El Dorado	38°51'	120°02'	1,980	6,500	Adam 1967; Zauderer 1973
Upper Echo Lake	El Dorado	38°50'	120°04'	2,300	7,550	Adam n.d.b
Ralston Ridge Bog	El Dorado	38°50'	120°06'	2,580	8,465	Sercelj and Adam 1975
Auburn Dam	El Dorado	38°50'	120°69'	100	330	Kilbourne 1978
Meyers Grade Marsh	El Dorado	38°50'	120°50'	2,073	6,800	Dorland 1980; Dorland et al. 1980
Grass Lake	El Dorado	38°48'	119°58'	2,347	7,700	Dorland 1980; Dorland et al. 1980
Gabbot Meadow	Tuolumne	38°25'	120°52'	1,995	6,550	Mackey and Sullivan 1991
Lake Moran	Tuolumne	38°23'	120°08'	2,018	6,621	Edlund 1991; Edlund and Byrne 1991
Swamp Lake	Tuolumne	38°22'	120°08'	1,957	6,420	Batchelder 1980
Burgson Lake	Tuolumne	38°21'	119°56'	1,960	6,430	Byrne et al. 1993
Tule Lake	Mono	38°21'	119°28'	2,079	6,820	Byrne et al. 1979
Catfish Lake	Tuolumne	38°12'	119°59'	1,850	6,070	Byrne n.d.
Mono Lake	Mono	38°00'	119°10'	1,950	6,400	Davis 1993
Swamp Lake, YNP	Tuolumne	37°57'	119°49'	1,554	5,100	Smith and Anderson 1992
Ten Lakes #3	Tuolumne	37°54'	119°32'	2,743	9,020	Anderson 1987
Tioga Pass Pond	Mono	37°54'	119°15'	3,018	9,900	Anderson 1987, 1990a
Soda Springs	Mariposa	37°53'	119°22'	2,750	9,022	Adam 1967
Polly Dome	Mariposa	37°51'	119°27'	2,650	8,700	Batch 1977
Black Lake	Mono	37°49'	118°35'	1,900	6,230	Batchelder 1970
Crane Flat	Mariposa	37°46'	119°48'	1,850	6,070	Adam 1967
Woski Pond	Mariposa	37°43'	119°37'	1,212	3,975	Anderson 1987
Hodgdon Ranch	Mariposa	37°41'	119°52'	1,400	4,590	Adam 1967
McGurk Mdw.	Mariposa	37°41'	119°38'	2,091	6,860	Anderson and Smith 1994
Starkweather Pond	Madera	37°40'	119°04'	2,438	8,000	Anderson 1987, 1990a
Barrett Lake	Mono	37°36'	119°00'	2,816	9,240	Anderson 1987, 1990a
Nichols Meadow	Madera	37°26'	119°34'	1,509	4,950	Koehler and Anderson 1994b
White Mountains	Mono	37°20' to 37°45'	118°05' to 118°25'	1,341 to 3,048	4,400 to 10,000	Jennings and Elliott-Fisk 1993
Lake 11,100 ft.	Fresno	37°12'	118°41'	3,380	11,100	Adam n.d.a
Deep Springs Valley	Mono	37°11'	118°03'	1,500	4,920	Woolfenden n.d.
Balsam Meadow	Fresno	37°09'	119°14'	2,015	6,610	Davis et al. 1985
Exchequer Meadow	Fresno	37°04'	119°06'	2,219	7,280	Davis and Moratto 1988
Dinkey Meadow	Fresno	37°00'	119°05'	1,683	5,520	Davis et al. 1985
Lily Pad Lake	Fresno	36°59'	118°59'	1,980	6,500	Byrne et al. 1993
Kings Canyon	Fresno	36°52'	119°17'	920 to 1,270	3,018 to 4,167	Cole 1983
Hightop Meadow	Fresno	36°48'	118°57'	1,908	6,260	Anderson and Smith 1994
Meadow of Honor	Fresno	36°44'	118°59'	1,857	6,090	Anderson and Smith 1994
Weston Meadow	Tulare	36°43'	118°53'	2,036	6,680	Anderson and Smith 1994
Owens Lake	Inyo	36°36'	118°05'	1,155	3,790	Koehler and Anderson 1994a
Long Meadow	Tulare	36°35'	118°44'	2,206	7,240	Anderson and Smith 1994
Circle Meadow	Tulare	36°34'	118°45'	2,085	6,840	Anderson and Smith 1994
Log Meadow	Tulare	36°33'	118°44'	2,048	6,720	Anderson 1994
Alabama Hills	Inyo	36°32' to 36°37'	118°03' to 118°07'	1,264 to 1,535	4,150 to 5,040	Koehler and Anderson 1995
Alabama Hills	Inyo	36°22'	118°07'	1,400	4,600	Axelrod and Ting 1962
Owens Lake	Inyo	36°22'	117°58'	1,100	3,600	Litwin et al. 1995
Ramshaw Meadows	Inyo	36°13'	118°13'	2,620	8,600	Axelrod and Ting 1962
Dogwood Meadow	Tulare	36°12'	118°40'	1,987	6,520	Anderson and Smith 1994
Bakeoven Meadows	Inyo	36°08'	118°10'	2,440	8,000	Axelrod and Ting 1962
Kennedy Meadows	Inyo	36°00'	118°14'	1,890	6,200	Axelrod and Ting 1962
Little Lake	Inyo	35°57'	117°54'	1,000	3,280	Mehringer and Sheppard 1978
Little Lake	Inyo	35°56'	117°53'	1,020	3,360	Axelrod and Ting 1962
Robber's Roost	Kern	35°35'	117°57'	1,215	4,000	McCarten and Van Devender 1988

CONSTANCE I. MILLAR
Institute of Forest Genetics
U.S. Forest Service
Pacific Southwest Research Station
Albany, California

5

Tertiary Vegetation History

ABSTRACT

The Tertiary period, from 2.5 to 65 million years ago, was the time of origin of the modern Sierra Nevada landscape. Climates, geology, and vegetation changed drastically in the Sierra Nevada during this time, and analyses of this period provide both context for and insight into vegetation dynamics of the current and future Sierra. During the early Tertiary, warm-humid, subtropical to tropical conditions prevailed on the low, rolling plains of the area now the Sierra Nevada. Fossil taxa with tropical adaptations and affiliations were widespread throughout the region. In the Sierra Nevada, ginkgo *(Ginkgo biloba),* avocado *(Persea),* cinnamon *(Cinnamomum),* fig *(Ficus),* and tree fern *(Zamia)* were common. At the end of the Eocene epoch, about 34 million years ago, global climates changed rapidly from warm-equable to cool-seasonal temperate conditions. In response, vegetation also shifted enormously; cool-dry-adapted conifers and hardwoods, which had been refugial during the early Tertiary in upland areas of the Great Basin–Idaho region, migrated into newly hospitable habitats of the Sierra Nevada. These floras contained many of the taxa now native to the Sierra, plus relicts from the subtropical forests of the earlier Tertiary and species adapted to temperate conditions with summer rain—mixes that seem incompatible under modern conditions. Until late in the Pliocene epoch (about 1 million years ago), adequate but diminishing rainfall distributed through the year supported many taxa now extinct in the Sierra Nevada. By the late Tertiary, in response to continued drying, winter cooling, and increasing summer drought, and to gradual uplift of the Sierra Nevada, replacement of early Tertiary floras by modern taxa and associations had occurred. With the development of a Mediterranean climate by the late Pliocene, floras of the Sierra Nevada became segregated ecologically into elevational, latitudinal, and orographic zones.

An important message for ecosystem management from a study of the Tertiary flora of the Sierra Nevada is that, although vegetation has changed drastically over 65 million years, the rate has been very slow. Human impacts in the Sierra are potentially of a similar magnitude to these evolutionary changes but can occur at rates many times faster; such changes may be more rapid than plants are likely to adapt to. Another management implication is that currently native species in the Sierra Nevada have existed in the past under drastically different climatic and environmental conditions than at present, have had very different distributions, and have occurred in mixes not seen in the recent past. Thus, assumptions about the behavior of native species in the future under unknown climates and/or novel management regimes should not be based solely on the behaviors of species in current environments. Unforeseen responses are likely, whether "positive" (population health, expansion, productivity), "negative" (population decline, extirpation), or novel. The most appropriate management action is to maintain diverse, healthy forests with conditions favoring resilience to unpredictable but changing future climates and management regimes. Plans that require landscapes to reach precise vegetation targets are likely to fail. Management programs that build flexibility, reversibility, and alternative pathways are more likely to succeed in an uncertain future.

INTRODUCTION

The Tertiary period is a slice of Earth's history, roughly defining the time between the extinction of the dinosaurs and the beginning of Northern Hemisphere continental glaciation, from 65 million years ago (denoted as 65 Ma) to about 2.5 Ma (table 5.1). This was a period of major change in global climates and of significant mountain building, overall moving from warm-mild and moist-equable regimes to seasonally dry and cool climates. The Mediterranean dry summer typical of California today was unknown until late in the Tertiary. The Tertiary was the time of initial uplift of the Sierra Nevada and volcanism in the Cascade Mountains. Accompanying these physical changes were radical transformations in the vegetation assemblages that covered the landscapes.

The human time scale for land management stretches 100–200 years into the future at its most imaginative. Why would

Sierra Nevada Ecosystem Project: Final report to Congress, vol. II, *Assessments and scientific basis for management options.* Davis: University of California, Centers for Water and Wildland Resources, 1996.

SNEP look back 65 million years? The Tertiary provides an important larger context for understanding modern landscape relationships in the Sierra Nevada. The Tertiary was the time of revolutionary development of the modern vegetation, climates, and landscape. At the onset of the Tertiary, there were humid subtropical climes in California, typical of vast periods of time prior to the Tertiary. Species such as ginkgo, avocado, figs, and palms dominated broad plains and low mountains in the area that is now the Sierra Nevada. By the mid-Tertiary, plants with affinities to modern taxa—pines, firs, oaks, and cottonwoods—appeared to be more widespread in the region of the developing Sierra Nevada. By the close of the Tertiary, most modern species and many modern vegetation assemblages were present and stratified into elevational zones. Although species and plant communities shifted in response to fluctuating conditions of the Quaternary period, which followed, these shifts were minor relative to the major evolutionary and continental-scale dynamics of the Tertiary. Thus the Tertiary sets the stage for the present.

The present flows seamlessly from the past. Knowing the origins and broad context of our flora informs our understanding and appreciation of the dynamics of current Sierra Nevada ecosystems—why species grow where they do, under what environmental and ecological conditions they have grown, what relationships have existed among plant associates, how biota respond to environmental change, and what potentials exist for rapid and dramatic natural vegetation change. Since many of our current taxa first appeared in California under different climates and evolved under very different environmental conditions, the past informs us about ecological responses that we are not able to infer from present dynamics.

OBJECTIVES

The purposes of this chapter are to:

- briefly review and assess the methods used to reconstruct the Tertiary vegetation of the Sierra Nevada

- develop a chronological overview of Tertiary geology, climate, and vegetation for the Sierra Nevada

- present floral lists and maps from published reports on Tertiary fossil deposits of the Sierra Nevada and neighboring regions and

- summarize points relevant to ecosystem management of the Sierra Nevada

The time frame for this chapter is the Tertiary period, as I define the boundaries from 65 Ma to 2.5 Ma (table 5.1), with focus on the Miocene and Pliocene epochs. Although a thor-

TABLE 5.1

Geological time chart for the Quaternary and Tertiary periods of the Cenozoic Era, showing approximate ages and durations of epochs (Odin 1982; Shackleton and Opdyke 1977; Swisher and Prothero 1990; Woodburne 1987).

Period	Epoch	Millions of Years Ago (Ma)
Quaternary	Holocene	0–0.01 (last 10,000 years)
	Pleistocene	0.01–2.5
Tertiary		
Neogene	Pliocene	2.5–7
	Miocene	7–26
Paleogene	Oligocene	26–34
	Eocene	34–54
	Paleocene	54–65

ough understanding of the biogeographic and phylogenetic origins of modern Sierra species requires studying their presence in fossil floras throughout western North America and beyond, the focus here is on what was and what was not in the Sierra Nevada during the Tertiary. Thus, the geographic focus is the greater Sierra Nevada region and parts of western and central Nevada, specifically the area defined by the fossil floras chosen for inclusion here (figure 5.1).

ASSUMPTIONS

1. This review is not intended to be exhaustive or comprehensive. Literature citations to more in-depth analyses are provided.

2. The focus is on plants and vegetation primarily, geology and climate secondarily; animals are not considered.

3. Confidence in knowledge decreases as we look further into the past; the biases of the fossil record and interpretation are discussed. Historical reconstruction is fraught with speculation.

4. Systematics and dating of the original interpretations of fossils are accepted unless subsequent revisions specific to the flora were published, or unless subsequent publications cast doubt on identifications. Other than these revisions, no modernization of nomenclature or taxonomic revision is attempted. Taxonomic revisions often lead to significant reinterpretations of biogeographic and evolutionary events. Examples of these are given to indicate the tenuousness of interpretations and the dependence on accurate taxonomy.

5. Detailed projections by original authors about paleoclimate (especially specific temperatures) and paleoaltitudes are

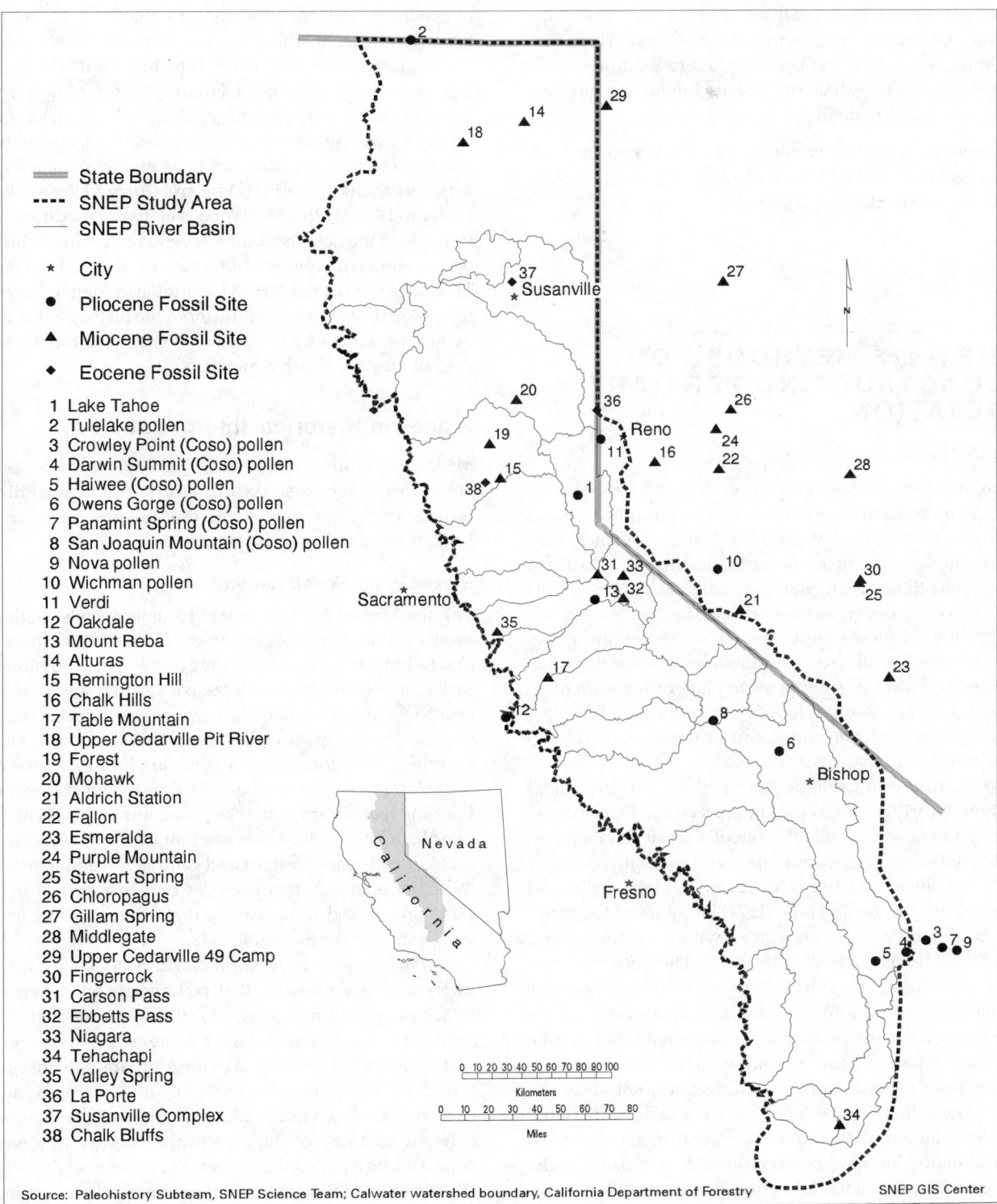

Legend

- State Boundary
- SNEP Study Area
- SNEP River Basin

- ★ City
- ● Pliocene Fossil Site
- ▲ Miocene Fossil Site
- ◆ Eocene Fossil Site

1 Lake Tahoe
2 Tulelake pollen
3 Crowley Point (Coso) pollen
4 Darwin Summit (Coso) pollen
5 Haiwee (Coso) pollen
6 Owens Gorge (Coso) pollen
7 Panamint Spring (Coso) pollen
8 San Joaquin Mountain (Coso) pollen
9 Nova pollen
10 Wichman pollen
11 Verdi
12 Oakdale
13 Mount Reba
14 Alturas
15 Remington Hill
16 Chalk Hills
17 Table Mountain
18 Upper Cedarville Pit River
19 Forest
20 Mohawk
21 Aldrich Station
22 Fallon
23 Esmeralda
24 Purple Mountain
25 Stewart Spring
26 Chloropagus
27 Gillam Spring
28 Middlegate
29 Upper Cedarville 49 Camp
30 Fingerrock
31 Carson Pass
32 Ebbetts Pass
33 Niagara
34 Tehachapi
35 Valley Spring
36 La Porte
37 Susanville Complex
38 Chalk Bluffs

Source: Paleohistory Subteam, SNEP Science Team; Calwater watershed boundary, California Department of Forestry   SNEP GIS Center

FIGURE 5.1

Distribution of Tertiary floras in the Sierra Nevada and adjacent regions of western and central Nevada. Floras are numbered in approximate order of age (from young to old). *Note:* Floras 3–10 are pollen sites, interpretations are based on a very small number of grains, and species identifications may be incorrect. These sites should be treated with question.

often not summarized here. New methods have cast doubt on the validity of some specific interpretations. Those new techniques have not yet been applied to Sierran Tertiary floras. Thus, generalizations about climate and environment are conservatively given.

6. Knowledge gained by understanding the origins of modern vegetation in the Sierra Nevada is relevant to ecosystem management of the ecoregion.

REVIEW OF METHODS FOR RECONSTRUCTING TERTIARY VEGETATION

Time Periods

The classification of time into eras, periods, and epochs (the geological time chart) is somewhat arbitrary, implying that a continuous process, time, is divisible. "Since geological time is not salami, slicing it up has no particular virtue" (Vita-Finza 1973). Despite this fundamental contradiction, historic events do tend to occur more or less periodically, lending themselves to description in pieces rather than as a continuum. Periodicities of tectonic, climatic, and biotic events are not often synchronous, however, from place to place or between plant and animal events. Boundaries of time periods are thus specific to regions, to biotas, and to causes (climate, paleomagnetic events, biotic changes).

Classification into geologic time periods was especially important in early paleontologic interpretation. Before direct dating methods were available, age of a fossil flora was assigned based on geologic stratigraphy and correlation to other local fossil floras. Radiometric dating (Dalrymple and Lanphere 1969; Steiger and Jager 1977) has relieved the temptation to date by correlation, since—within tolerances and errors—fossil floras can be directly dated. This both improves the accuracy of assigning floras to periods in the geologic time chart and relieves pressure for relying on those assignments to periods, since many floras can be discussed by direct age rather than by period. However, although radiometric methods have been available for several decades, not all of the Tertiary fossil floras in the Sierra Nevada (originally dated by stratigraphic correlation) have been confirmed radiometrically, and many that have been confirmed were dated in the early years of radiometry, when techniques were less accurate than recent methods. Quaternary sites discussed in Woolfenden 1996, by contrast, are almost all radiometrically dated. With the increased availability of accurate radiometric dating, discussion about stratigraphic definitions of boundaries, once a topic of intense debate, has subsided.

I have included radiometric dates in this report where available. I do not defend a strict view of the dates for boundaries

of epochs or eras, instead accepting that they are guidelines for orientation in the past. For convention, I adopt the Mesozoic:Cenozoic boundary at 65 Ma (Odin 1982), the beginning of the Northern Hemisphere ice ages for the Tertiary:Quaternary boundary at 2.5 Ma (Shackleton and Opdyke 1977; references in Thompson 1991), and combined North American floristic and land-mammal stages for the Tertiary epochs (table 5.1) (Odin and Curry 1985; Swisher and Prothero 1990; Wolfe 1981; Wood et al. 1941; Woodburne 1987). Workers in the field, including several reviewers of this chapter in manuscript, propose alternative dates for time periods. This underscores the fact that boundaries depend on which factors are considered significant in the history of the earth. I do not include here a review of alternative dating for the epochs of western North America.

Biases in Historical Interpretation

Misinterpretations of vegetation history occur due to inherent biases in the fossil record, errors in understanding the record, and cumulative errors due to subsequent analysis. Each of these is discussed in turn.

Biases in the Fossil Record

The single most frustrating reality about reconstructing past events is that there are gaps in the fossil record. These occur due to limited exposures in time and space of fossil-bearing rocks and sediments of successively older ages. For the Sierra Nevada, exposure of Tertiary rocks is uneven. No fossil-bearing deposits of the earliest Tertiary (Paleocene) are known; Eocene and Oligocene fossil floras are limited in extent and are present mostly in the northern Sierra; Miocene and Pliocene records are somewhat more numerous. Even for the middle to late Tertiary, however, much better representation occurs in adjacent western and central Nevada. Fortunately, these floras contain many species that later appear in the Sierra Nevada and thus provide important material from the perspective of the Sierra Nevada.

Tertiary records in western North America are primarily impression macrofossils, that is, imprints left when leaves, twigs, or fruits (macro-organs) were deposited in wetland sediments (lake bottoms, bogs, marine environments, or other wetlands). Occasionally, petrified organs and tissues are found. In these, chemical replacement of living tissues has occurred, leaving a nearly identical replica of the internal and external anatomy of the organ or tissue (usually wood or cones). From a regional perspective, macrofossil deposits bias the sampling in that they overrepresent wetland species (willows [Salix], cottonwoods [Populus], etc.) that are adjacent to depositional sites and underrepresent upland species. Sampling is assumed to be limited to plants growing about 1 km (0.62 mi) from the site of deposition (Gregory 1994). Beyond this distance, smaller leaves are preferentially preserved over large leaves, as are thick, tough leaves over fragile ones. By and large, conifer remains are readily preserved if they get into a deposit. Their usual ecological position in the uplands, however, may limit their representation in the deposit. For all these reasons, the number of specimens of a single type occurring in a fossil deposit is usually not correlated with its abundance in the environment, and many contemporaneous species may be left out of the deposit altogether. Several other biases due to preservation of individual macrofossil specimens, referred to collectively as *taphonomic bias*, distort the sampling and recovery of species from macrofossil deposits (Greenwood 1992; Spicer 1989; Wolfe and Upchurch 1986).

Macrofossils also occur in the arid parts of the Sierra Nevada region in wood rat middens (Betancourt et al. 1990). These do not date to the Tertiary and are not considered here.

The other important plant remains from the Tertiary are pollen grains and other microfossils. Wind- and waterborne pollen is preserved in wetland sediments of lakes and bogs. Pollen in these sediments is usually recovered from long cores bored through lake sediments. Sampling like this has a significant advantage over macrofossil deposits in that a continuous stratigraphic record through time may be obtained, with much better control on species mixes, stratigraphic orientation, and changes over time than lakeside macrofossil deposits can offer. Pollen samples infrequently have been taken from solid exposed sediments rather than from a core, a technique that eliminates or reduces the opportunity to analyze a continuous record.

Pollen sampling is a common method for Quaternary analysis, but only recently has it been applied to Tertiary sediments. The Tulelake core (Adam et al. 1990; Adam et al. 1989) is the only published continuous core for the Sierra Nevada region that extends into the Tertiary, although other deep cores are currently under analysis. Most notable is the Owens Lake study, which provides continuous analysis of a sediment core into the early Pleistocene (Owens Lake Core Study Team 1995).

Pollen floras suffer different kinds of systematic biases from those of macrofossil deposits. Species with abundant and wind-borne pollen grains are disproportionately represented. Of these, there is a bias related to distance, in that pollen travels different distances depending on species. For example, because of its size and shape, 95% of giant sequoia (*Sequoiadendron giganteum*) pollen falls within 500 m (1,500 ft) of a native forest source (Anderson 1990), whereas pine (*Pinus*) pollen can travel hundreds of kilometers and still be an abundant type in a pollen sample. Biases due to size of the depositional basin also occur. Pollen grains of different species degrade with time, and differential preservation is especially important in old samples, such as Tertiary pollen cores. For several of these reasons, pine, fir, and spruce may dominate the pollen record in numbers disproportionate to their representation in the original flora. Methods to calibrate these biases are routinely applied (Overpeck 1985; Prentice 1985).

Other kinds of microfossils are often identified along with pollen in Quaternary samples. These include diatoms, chrysophyte cysts and scales, radiolarians, coccoliths, ostracods, and occasionally foraminifera (in saltwater basins). Charcoal and some macrofossils (leaf tissue) may also be included in lake sediment cores. Charcoal can provide information about fire occurrence.

Biases in Reading the Record

Analysis of any fossil flora hinges critically on accurate systematic interpretation of specimens. Opportunities for misidentifying macrofossils are abundant, because of poor preservation (e.g., only part of a leaf or cone was imprinted or intact), changes in size, shape, or structure due to preservation, distant relationship to modern taxa (there is no living analog), hybridization, and natural variation in the species. Because so much interpretation depends on correct identification, old fossil floras have been reviewed and their systematics revised; these revisions have sometimes been as dramatic as assigning a specimen to a different kingdom from that in the original publication. Individual paleobotanists vary in their willingness to make identifications, with some assigning specimens confidently to species and others listing only family or genus. Methods have been developed to assess physiognomy of fossil remains independent of taxonomic identification (described below), thus circumventing the dependence on correct systematic identification for some kinds of analysis.

Microfossils also may be misidentified, but the risk is lower in part due to the lack of diagnostic characters for identifying pollen to lower taxonomic levels and the reduced temptation to try. Thus, pollen is often identified only to genus, sometimes even to a combined family level (e.g., TCT, *Taxodiaceae-Cupressaceae-Taxaceae*). The lack of species diagnostics limits the usefulness of pollen analysis in studies that require knowledge of individual species.

Some fossil floras have been independently analyzed for macrofossils and for pollen. These provide the opportunity to compare information from the two data sets and assess the relative effectiveness of one or the other method. The Chalk Bluffs fossil flora near Nevada City (figure 5.1; table 5.2; appendix 5.1, list 3), originally described by an extensive macrofossil list (MacGinitie 1941), was reevaluated for pollen taxa by Leopold (1983, 1984). This analysis revealed the biases of both approaches. Pollen did not diagnose individual species and was unable to record taxa from four families found in macrofossils, yet it added taxa from eight families not recorded in the macrofossils. The additional taxa were mostly wind-pollinated species. Despite the differences in representation of individual taxa, the vegetation and climatic interpretation of the flora was similar between the two methods, that is, that this assemblage was a rich subtropical forest in a warm, moist climate. A significant addition from the pollen was the presence of taxa from the pine family (pine [*Pinus*], fir [*Abies*], spruce [*Picea*]), with implications discussed later. Other comparisons of Tertiary pollen and macrofossil floras have yielded greater discrepancies (e.g., only 38% correlation of taxa among methods for a Washington flora [Reiswig 1983], 18% correlation for a northwestern California flora [Barnett 1983]).

TABLE 5.2

Paleogene fossil floras of the Sierra Nevada and surrounding regions, listed in approximate order of age (young to old).

Flora Name	Location	Present Elevation	Latitude and Longitude	Age	Ma	Number of Species	Reference
La Porte	La Porte, CA	1,200 m 3,900 ft	39°42' N 120°W	Early Oligocene	33[a]	41	Potbury 1935
Susanville Complex	Susanville, CA	1,500 m 4,875 ft	40°30' N 120°40' W	Middle Eocene		22[b]	Knowlton 1911; Wolfe and Hopkins 1967
Chalk Bluffs	Colfax, CA	1,000 m 3,250 ft	120°52' N 39°15' W	Early Eocene		71	MacGinitie 1941; Leopold 1983

[a]Age is radiometrically confirmed.
[b]The number of species is questionable.

Biases in Analysis

A floral list, either macrofossil or microfossil, provides the raw data for subsequent analysis. For Tertiary floras, the main analysis has been to infer paleoclimates, paleoaltitudes, and ecological relationships of vegetation assemblages. These analyses have sometimes been quite specific, attempting to define mean annual temperatures, ranges of temperatures, effective temperatures, and annual precipitation at the fossil sites, as well as elevations above sea level. There are two general approaches to environmental analysis of fossil floras: those that rely on floristic comparison (analog approach), and those that rely on morphological relationships with environment (physiognomy approach).

Floristic analysis (Axelrod 1966, 1968; Axelrod and Bailey 1969) attempts to describe Tertiary conditions (plant communities and their ecological relationships) on the basis of taxonomic composition. The approach is basically qualitative and intuitive, and it involves comparing modern species to fossil species. Modern relatives are assigned corresponding to species in the fossil deposit, and known ecological requirements of modern species are used to build a composite description of past conditions. This approach requires both high accuracy in species identification and trust in the concept of uniformitarianism—the assumption that modern species do not differ in ecological response and requirements from their fossil representatives and thus provide reliable and relatively precise indicators of paleoenvironments.

Although analogs and uniformitarian models are common in paleontology and are routinely applied in Quaternary analysis, they have been criticized for reconstruction of older environments (e.g., Bryson 1985; MacGinitie 1962; Wolfe 1971; Wolfe and Hopkins 1967). The primary bases for criticism are: (1) These models depend on accurate taxonomic identification to species level, which is doubtful for the Tertiary; (2) the potential evolutionary change within species lineages renders comparisons of ecological relationships between current and fossil groups invalid; (3) the models confuse "vegetation" with "taxon," assuming that a vegetation assemblage has a characteristic response to an environment, whereas, in fact, individual taxa respond; (4) some environmental situations have

no analog, because current climates and ecological conditions do not represent the full range of those that existed in the past; and (5) the reproducibility of results by other workers has been low.

An alternative approach based on the observation that leaf morphology varies with climate has proved useful for interpreting fossil floras that have many angiosperm taxa. This approach takes advantage of empirical relationships between leaf physiognomy and climate, notably the positive correlation of the percentage of species with entire-margined leaves in an assemblage with the mean annual temperature and equability of vegetation type (tropical rain forest, subtropical forest, deciduous oak forest, etc.). This measure has been applied to fossil assemblages since the early twentieth century (Bailey and Sinnott 1916) and has been developed more fully for Tertiary interpretations by Wolfe and his colleagues (Wolfe 1971; Wolfe and Hopkins 1967). The significance of this approach is that climatic interpretations are independent of species identifications, known empirical relationships are used, quantitative values are derived, and results are reproducible. From climate information, inferences about paleoelevations have also been made.

Univariate leaf-margin analysis is not adequate to represent fossil floras fully (Axelrod and Bailey 1969; Wolfe 1971), and other characters have been used to supplement interpretations. Recently, Wolfe (1993) developed multivariate approaches that take advantage of combined data sets to estimate temperatures and elevations of Tertiary fossil environments. These methods were applied to a critical reevaluation of the Florissant flora in Colorado, a well-known Tertiary deposit (Gregory 1994). The reevaluation indicated that climatic estimates based on multivariate approaches yield quite different values than both floristic comparisons and univariate leaf physiognomy methods (Gregory 1994).

Results at Florissant with the multivariate methods suggest that climatic interpretation of the Sierran Tertiary floras may need significant revision. For this reason, in this report I do not summarize the detailed climatic interpretations of earlier literature. Further, since the Sierra Nevada has been dominated by conifer vegetation types, which are not amenable to the leaf physiognomy methods in general, we need to take a broader view of climate implications of the deposits. The approach here is to indicate generally the ecological conditions and elevations suggested by the fossil species in the assemblages and to encourage revision of interpretations for older floras.

TERTIARY HISTORY OF THE SIERRA NEVADA

Although the Tertiary was the time of origin for modern California vegetation, this is not to say that the taxa that are distinctively Sierran originated during this period or in California (Millar and Kinloch 1991). Many lineages did undergo significant evolution in California during this period. Many taxa that are today Sierran were in other parts of North America during the early Tertiary and often occurred in assemblages with no modern analogs. Rather, the primary significances of the Tertiary were the changes in distributions of many taxa and the major environmental changes (climate and mountain building) that catalyzed these changes. The environmental changes and the resulting plant responses led to almost complete replacement of vegetation types in the Sierra Nevada and to restructuring of species mixes, geographic distribution, and elevational zones of vegetation types.

Although early work stressed the cohesiveness of vegetation assemblages (e.g., geofloras [Axelrod 1958]), biologists now almost universally accept that taxa respond individualistically to environments (Botkin 1990; Frankel and Soule 1981; Grumbine 1994; Hansen et al. 1991; Kaufman 1993). Vegetation assemblages are transient collections of taxa unified by a common environment and the intersection of biogeographic histories. This is not to say that taxa do not influence each other in space and time; ecosystem science focuses on just these interactions. At historical scales, however, migratory movements and population colonizations and extirpations are primarily related to the behavior of individual taxa.

The many plant species and complex vegetation assemblages in the Sierra Nevada are significant components of biodiversity in the region (Davis and Stoms 1996; Shevock 1996). From the historical perspective, these species and assemblages can be viewed as relicts of earlier periods and prior environmental events, with contributions from different parts of North America, exhibiting evolutionary responses to an increasing aridity and seasonality that began in the mid-Tertiary. The following sections summarize major events in the Sierra Nevada through the epochs of the Tertiary. Vegetation dynamics are inferred from Tertiary deposits in the Sierra Nevada and adjacent regions of Nevada (figure 5.1; tables 5.2–5.4). Systematic compositions of these fossil floras (using

unrevised nomenclatures) are given in appendix 5.1, lists 1–34 (in alphabetical order by flora name).

Paleocene and Eocene

During the earliest Tertiary, the region of the present Sierra Nevada was mostly low plains to low hills, dominated by old marine sediments. In the north, the region was a low plain with a river 5 km (3 mi) wide crossing near the area of Susanville today. Southward, the region consisted of low, rolling hills, with smaller rivers draining across most of the present Sierran axis (Armentrout et al. 1979). In the region of Mount Whitney, the land rose to its highest altitude, which was still quite low compared to current Sierran elevations. The western edge of the Sierra Nevada formed the Eocene Pacific Ocean coastline for all but the northern portion (figure 5.2) (Axelrod 1968; Minckley et al. 1986).

Interior to the Sierra Nevada was a large upland region that stretched throughout the northern Great Basin and intermountain areas of Idaho, western Wyoming, and western Colorado (Axelrod 1968; Ruddiman and Kutzbach 1989; Wolfe 1987). This was the only important upland region of western North America, extending to elevations over 1,225 m (4,000 ft) (Axelrod 1968) or, by Gregory's recent interpretation (1994), to 2,500 m (8,000 ft) in Colorado. This high plateau was dominated by volcanic centers and large lakes, and many of the fossil floras in the region are contained in calderas and depositional basins. This upland was unusual not only for western North America but also for temperate latitudes worldwide. As such it was an important relictual area for temperate montane taxa and a source of taxa to the Sierra Nevada in later epochs (references in Millar 1993).

Climates for the Sierra Nevada during the Paleocene and Eocene, as inferred from several sources, were different from current climates and from those in epochs before the Tertiary. Although warm-equable climates had typified the late Mesozoic (McGowran 1990; Parrish 1987; Wolfe and Upchurch 1986), temperate latitudes of the early Tertiary experienced unusually high temperatures (figure 5.3) and rainfall (references in Wolfe 1990). The trends toward increasing humidity started in the early Paleocene and continued into the Eocene, reaching maximums by the early Eocene (Savin 1977; Wolfe 1985). Major fluctuations in temperature (greater in magnitude than those of the Pleistocene) characterized the Eocene (figure 5.3), causing conditions in California to alternate between tropical and subtropical (references in Millar 1993). Truly temperate conditions (seasonally cool and dry) did not exist in California except perhaps in a few limited upland areas, and no true deserts or arid areas are known to have existed at this time in the region of the Sierra Nevada (Axelrod 1979). Except in the uplands of the northern Great Basin and Idaho, humid subtropical conditions existed in a broad zone throughout temperate latitudes in North America (Millar 1993; Wolfe 1978) throughout the early Tertiary.

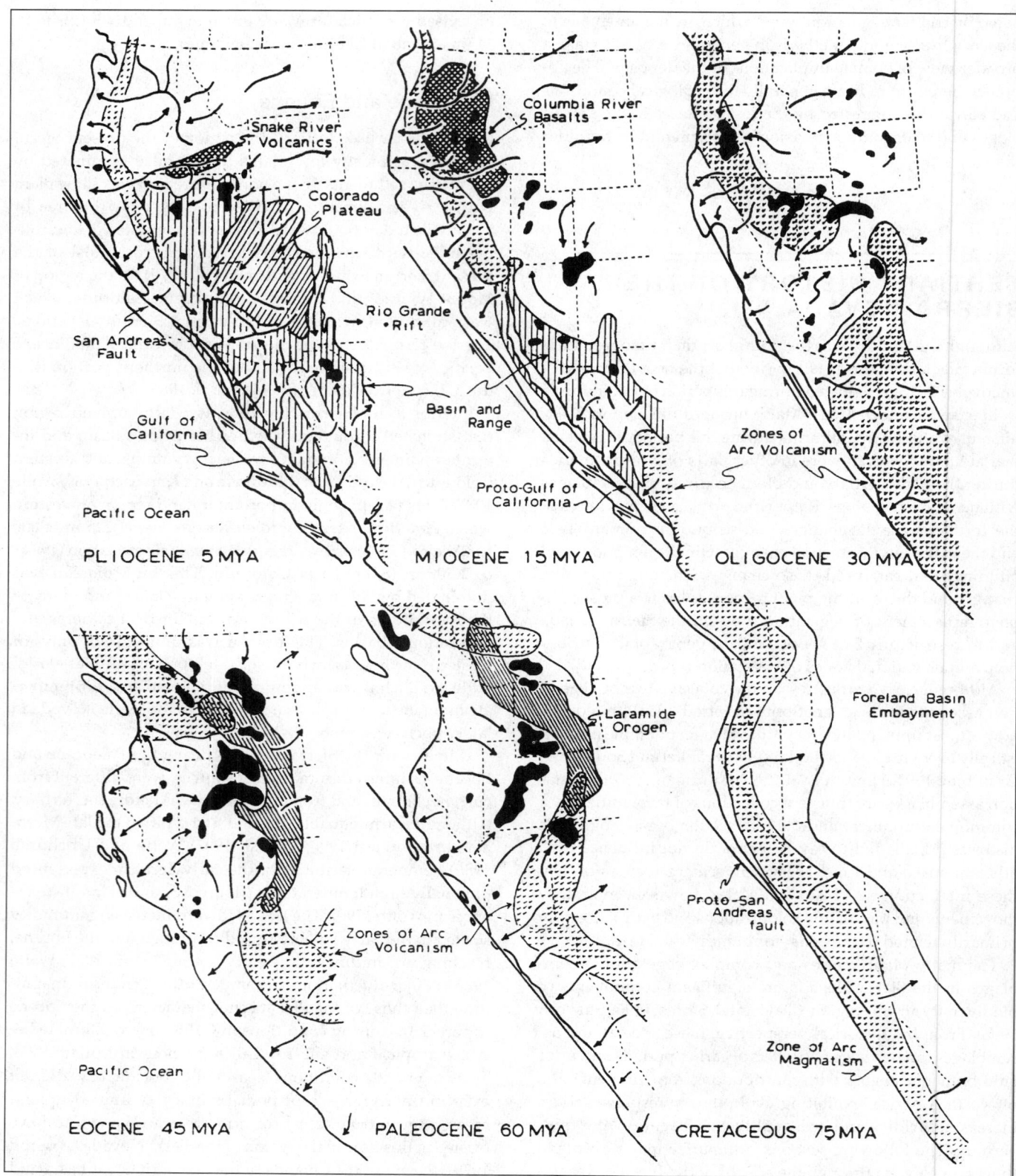

FIGURE 5.2

Tertiary coastline of western North America (from Minckley et al. 1986).

In general, the vegetation of California (like vegetation throughout most of temperate North America) reflected the tropical/subtropical climates of the early Tertiary. Starting in the earliest Paleocene, diverse tropical angiosperm species appeared with increasing representation throughout the warm-humid intervals of the Paleocene and Eocene. Subtropical assemblages are known to have expanded in temperate zones worldwide from this time, extending to 70° N latitude (Friis et al. 1987; Tiffney 1985; Wolfe 1985). Plant communities from Eocene locations were similar taxonomically and physiognomically to current rain forests of eastern Asia (e.g., Malaysia) and southern Mexico.

The Sierra Nevada has only two fossil floras from this period (table 5.2; figure 5.1). There are not many more terrestrial deposits for all of California, since most rocks of this age in California are marine sediments. These, along with the early Oligocene La Porte flora (table 5.2; appendix 5.1, list 14), occur in northern California and western central California, and they record rich, diverse, and—compared to modern California—exotic forests. Chalk Bluffs (appendix 5.1, list 3), near Colfax, California, records one of the richest Eocene floras of western North America, with seventy-one species, in families and with foliar adaptations that could only indicate humid subtropical conditions. Few taxa in the macrofossil record overlap species in the present Sierra Nevada. Rather, they

contain species of viburnum *(Viburnum)*, avocado *(Persea)*, magnolia *(Magnolia)*, fig *(Ficus)*, and many others with warm-humid affinities. Similar subtropical taxa with warm-humid physiognomic adaptations (large leathery leaves, entire margins, drip points) occur in the smaller Susanville Complex flora (appendix 5.1, list 26) and the early Oligocene La Porte flora (appendix 5.1, list 14). The only gymnosperms, represented in very low diversity in the macrofossil record, were similarly warm-humid-adapted. They included a cycad *(Zamia)*, and possibly a yew *(Taxus)* species.

The additional assessment of the Chalk Bluffs deposit for pollen added a floral component that was absent in the macrofossil record. In very small proportions, pine, spruce, and fir pollens were identified, suggesting that these species blew in from distant areas that remained habitable to conifers in the otherwise incompatible subtropical environments. Except for Chalk Bluffs, conifer taxa have not been recorded from any other Sierran sites of this age, and pines were recorded from only one other Eocene deposit in California (Axelrod and Raven 1985). Indeed, pines and possibly other conifers in general seem to have retreated to hospitable refugia during the early Tertiary. At the global level, these refugia were at very high and low latitudes and in the few upland regions that existed in temperate latitudes, such as the volcanic plateau of the Great Basin–Idaho uplift (Millar 1993). Limited

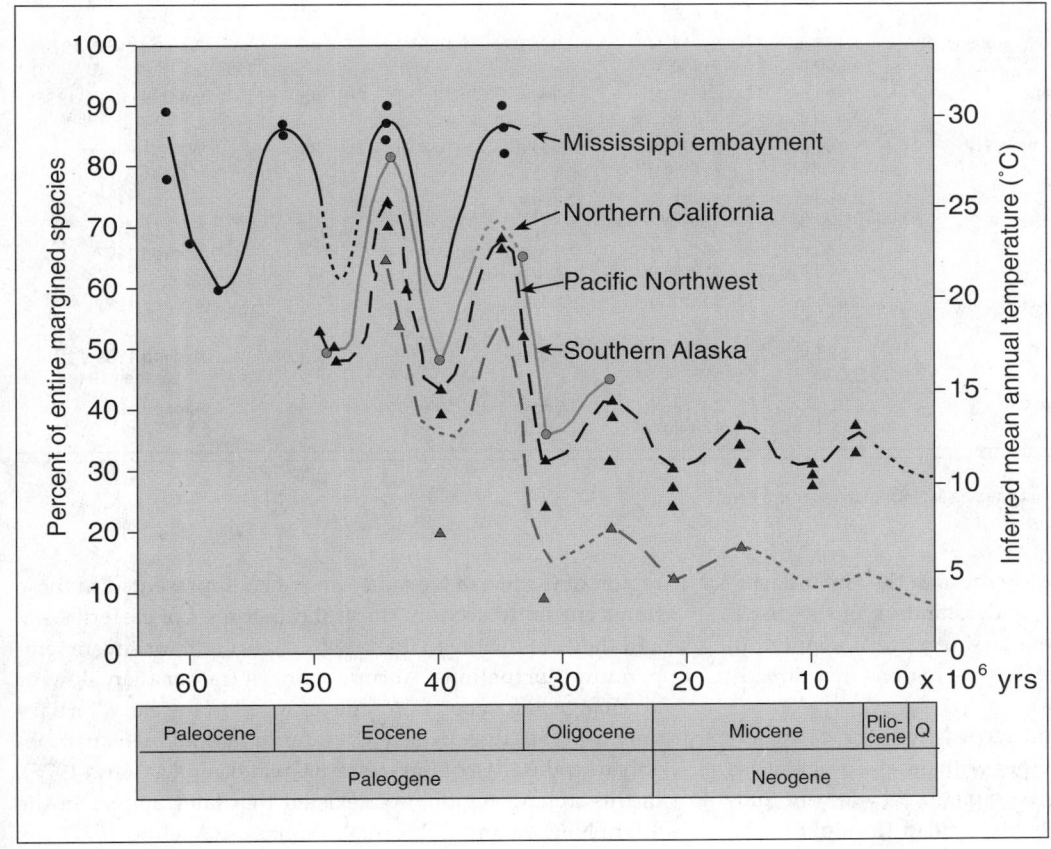

FIGURE 5.3

Tertiary paleotemperatures estimated from foliar physiognomy (from Wolfe 1978).

TABLE 5.3

Miocene fossil floras of the Sierra Nevada and surrounding regions, listed in approximate order of age (young to old).

Flora Name	Location	Present Elevation	Latitude and Longitude	Age	Ma	Number of Species	Reference
Alturas	Alturas, CA	1,350 m 4,390 ft	41°30' N 120°35' W	Late	8.1[a]	7	Axelrod 1944b
Remington Hill	Nevada City, CA	1,180 m 3,840 ft	39°17' N 120°45' W	Late	9.5	32	Condit 1944
Chalk Hills	Virginia City, NV	1,785 m 5,800 ft	39°23' N 119°33' W	Late		28	Axelrod 1962
Table Mountain	Columbia, CA	677 m 2,200 ft	38°02' N 120°23' W	Late		28	Condit 1944
Upper Cedarville Pit River	Canby, CA	1,350 m 4,390 ft	41°22' N 121°4' W	Late		21	LaMotte 1935, 1936
Forest	Sierra County, CA	1,500 m 4,875 ft	39°30' N 125°52' W	Late		9	Knowlton 1911; Chaney 1944
Mohawk	Plumas County, CA	1,500 m 4,875 ft	39°46' N 120°38' W	Middle		13	Knowlton 1911; Axelrod 1944b
Aldrich Station	Yerington, NV	?1,970 m ?6,400 ft	38°27' N 118°54' W	Middle	11[a]	35	Axelrod 1956
Fallon	Fallon, NV	1,200 m 3,900 ft	39°20' N 119°3' W	Middle	12.6	22	Axelrod 1956
Esmeralda[b]	Coaldale, NV	1,590 m 4,900 ft	38° N 117°46' W	Middle	12.7[a]		Axelrod 1940
Purple Mountain	Wadsworth, NV	1,275 m 4,140 ft	39°35' N 119°4' W	Middle	13	36	Axelrod 1976, 1995
Stewart Spring	Stewart Valley, NV	1,655 m 5,500 ft	38°36' N 117°59' W	Middle	13–14[a]	43	Wolfe 1964; Schorn 1984
Chloropagus	Hot Springs Mountains, NV	1,385 m 4,500 ft	39°42' N 118°57' W	Middle	13.9[a]	22	Axelrod 1956
Gillam Spring	Gillam Spring, NV		40°30' N 118°57' W	Middle	15	63	Axelrod and Schorn 1994
Middlegate	Clan Alpine Mountains, NV	1,415 m 4,600 ft	39°17' N 118°2' W	Middle	15.9[a]	43	Axelrod 1956, 1976, 1986
Upper Cedarville 49 Camp	49 Camp, NV	1,500 m 4,875 ft	41°36' N 119°55' W	Middle	16	43	LaMotte 1935, 1936; Chaney 1959
Fingerrock	Cedar Mountain, NV	1,754 m 5,700 ft	38°37' N 117°58' W	Middle	16.4	22	Wolfe 1964
Carson Pass[b]	Kirkwood, CA	2,550 m 8,290 ft	38°41' N 120° W	Middle			Axelrod 1977
Ebbetts Pass[b]	Markleeville, CA	2,660 m 8,645 ft	38°32' N 119°48' W	Middle			Axelrod 1977
Niagara[b]	Markleeville, CA	1,800 m 5,850 ft	38°40' N 119°48' W	Middle			Axelrod 1977
Tehachapi	Tehachapi, CA	1,538 m 5,000 ft	35°14' N 118°14' W	Middle	17.1[a]	70	Axelrod 1939; Chaney 1944
Valley Spring	Mokelumne, CA	215 m 700 ft	38°19' N 120°46' W	Middle	19.9[a]	17	Axelrod 1944b

[a]Age is radiometrically confirmed.
[b]The available literature does not give full species list information for the flora.

exposures of Eocene fossil-bearing rocks in the Great Basin–Idaho uplift area constrain our understanding of the distributions of these more temperate taxa (e.g., taxa adapted to seasonally cool and dry climates) in the regions that directly surround the Sierra Nevada.

Collectively, the limited plant record from the early Tertiary in the Sierra Nevada, together with other Eocene floras from western North America, gives a picture of warm-humid-adapted forests with high diversity, spread throughout the region of the Sierra Nevada. Many taxa represented in these floras currently occur in tropical rain forests of eastern Asia, and they are unable to tolerate frost, drought, or severe temperature fluctuations. Abrupt changes (rather than slow or no changes) in vegetation type occurred in the early Tertiary only near the edge of the major interior upland plateau, probably near the edge of the western Great Basin (Axelrod 1977). In this upland region, taxa existed that later appear in the Sierra Nevada and California mountains (Axelrod 1977).

Oligocene

The Oligocene remained a period of relative quiescence geologically for the Sierra Nevada, despite general uplift of the entire range, retreat of the coastline along the western margin of the Sierra, and local volcanic activity. Explosive volcanic activity and mountain building occurred, however, in the region of the Great Basin–Idaho uplands (McKee 1979; Stewart 1978).

The Oligocene was a time of major climatic transition in California and throughout temperate Northern Hemisphere latitudes. Climatic events that occurred during this epoch triggered the subsequent modernization of the Sierra Nevada forests. Although the dates, geographic extent, and causes remain questions of scientific interest and debate, little doubt exists that a major climatic change occurred near what is now considered the Eocene:Oligocene boundary, about 34 Ma (table 5.1) (references in McGowran 1990; Parrish 1987; Prothero 1994; Wolfe 1978). This climatic event, sometimes referred to as the *terminal Eocene event* (Wolfe 1978), was the most profound of the Tertiary. Inferred from many data sources, average temperatures in temperate latitudes at this time declined drastically (up to 13°C [23.4°F] in some areas over only one million years) (figure 5.3) and stayed low (al-

though not as low, and with minor fluctuations) throughout the rest of the Tertiary.

Accompanying this radical temperature drop was a shift to drier, more seasonal climates and to wider annual temperature ranges. The average range of temperatures is estimated to have been over 25°C (45°F), twice the present range (Wolfe 1971). Continental ice sheets occurred for the first time in the Tertiary during the Oligocene (McGowran 1990), although no glaciers are known to have formed in the Sierra Nevada until the onset of the Pleistocene.

The terminal Eocene event was marked by widespread disappearances and extirpations of the Eocene subtropical and tropical floras from middle latitudes worldwide (Wolfe 1978). Complementary expansions of cool-adapted conifers and angiosperms are recorded in many deposits of this age. These conifers may in fact have been reoccupying sites in middle latitudes that they had dominated in the late Mesozoic but had been forced out of by the pervasive warm-humid conditions of the early Tertiary (e.g., for pines, Millar 1993).

The Sierra Nevada has only one fossil deposit from the Oligocene, the La Porte flora of northern California (figure 5.1; appendix 5.1, list 14), and this is early enough to have affinities to the periods discussed earlier. The Sierra Nevada record, therefore, does not have the temporal resolution to

TABLE 5.4

Pliocene fossil floras of the Sierra Nevada and surrounding regions, listed in approximate order of age (young to old).

Flora Name	Location	Present Elevation	Latitude and Longitude	Age	Ma	Number of Species	Reference
Lake Tahoe pollen	Tahoe City, CA	1,900 m 6,175 ft	39°10' N 120°9' W	Late	1.9[a,b]		Adam 1973
Tulelake pollen	Tulelake, CA	1,240 m 4,030 ft	42° N 121°30' W	Late	3–2[a]		Adam et al. 1990; Adam et al. 1989
Crowley Point (Coso) pollen	Inyo Mountains, CA	1,180 m 3,840 ft	36°21' N 117°33' W	Late		25	Axelrod and Ting 1960
Darwin Summit (Coso) pollen	Inyo Mountains, CA	1,600 m 5,200 ft	36°17' N 117°42' W	Late		13	Axelrod and Ting 1960
Haiwee (Coso) pollen	Coso Mountains, CA	1,355 m 4,400 ft	36°14' N 117°56' W	Late	2.2[a]	37	Axelrod and Ting 1960
Owens Gorge (Coso) pollen	Bishop, CA	2,000 m 6,500 ft	37°33' N 118°37' W	Late	3.2[a]	30	Axelrod and Ting 1960
Panamint Springs (Coso) pollen	Inyo Mountains, CA	600 m 1,950 ft	36°18' N 117°26' W	Late		18	Axelrod and Ting 1960
San Joaquin Mountain (Coso) pollen	Mammoth Lakes, CA	2,954–3,052 m 9,600–9,920 ft	37°45' N 119°7' W	Late		23	Axelrod and Ting 1960
Nova pollen	Panamint Mountains, CA	800 m 2,600 ft	36°17' N 117°19' W	Late		22	Axelrod and Ting 1960
Wichman pollen	Wichman, NV	1,600 m 5,400 ft	38°42' N 119°4' W	Late		14	Axelrod and Ting 1960
Verdi	Verdi, NV	1,477 m 4,800 ft	39°31' N 119°58' W	Early	5.7[a]	19	Axelrod 1958
Oakdale	Oakdale, CA	77 m 250 ft	37°47' N 120°42' W	Early	6	16	Axelrod 1944a
Mount Reba[c]	Bear Valley, CA	2,615 m 8,600 ft	38°31' N 120°1' W	Early	7	>6	Axelrod 1976, 1977

[a]Age is radiometrically confirmed.
[b]This is a minimum age; the radiometric date is for the lava flow that overlies the pollen deposit.
[c]The available literature does not give full species list information for the flora.

trace at a fine scale the major transitional events indicated earlier. Because of this, the exact timing of climate change and vegetation response in the early mid-Tertiary cannot be determined for the Sierra Nevada.

Miocene and Pliocene

The Miocene and Pliocene epochs span the last half of the Tertiary, or about 23 million years. Fossil records are relatively abundant throughout this period, climate and mountain building are complex, and floristic relationships are significant to an understanding of modern vegetation patterns. We summarize these epochs together for this report, since no major transitions mark the Miocene:Pliocene boundary.

Uplift and mountain building of the Sierra Nevada accelerated during the Miocene, as evidenced by an increase in block faulting during this time. The range was not yet a moisture barrier to westerlies carrying ocean-laden air. The ocean that extended into the present Central Valley of California was reduced to a large inland sea bay, connected in several places to the ocean (figure 5.2). Its eastern shoreline extended along the middle third of the present Sierran axis (Axelrod 1968). By the mid-Miocene (16–18 Ma), Columbia River basalt was flowing, creating ample opportunities for fossil preservation in the still-high Great Basin–Idaho upland. Uplift of this region continued, estimated at another 920 m (3,000 ft), with higher regions in the south (Nevada) and lower in the north. Many fresh water lakes were formed during this uplift, which also proved important for fossil deposition.

Block-fault uplift of the Sierra Nevada continued through the Pliocene, with scattered volcanic activity along the eastern and, to a lesser extent, western margins. Huber (1981) estimated that, by the late Pliocene (3 Ma), the height of the central Sierra Nevada was about 2,100 m (6,825 ft). Much of the uplift of the Sierra Nevada apparently occurred after 1.9 Ma (Winograd et al. 1985), and the ranges were much less effective barriers to wet air masses from the west prior to that time. Formation of the major southeastern valleys (the Owens and Searles Valleys) may have been tied to the onset of mafic volcanism and basin-range faulting between 3 and 4 Ma (Duffield et al. 1980).

Following the major change in the climates of western North America (and elsewhere) during the Eocene-Oligocene transition, climates never again turned warm-humid and tropical in temperate western North America. There were, however, significant changes from the conditions of the Oligocene, and fluctuations in climate occurred throughout the rest of the Tertiary. From the cool-cold, strongly seasonal conditions inferred for Sierra-like regions of western North America during the Oligocene, average temperatures increased in the early Miocene, and rainfall decreased. This early Miocene warming was followed by a cooling and drying trend and increasing provincialization, which continued (with fluctuations) through the middle and late Miocene (Raven and Axelrod 1978). The late Miocene seems to have been cooler

than the middle, with evidence for dry summer climates beginning in Nevada by the late Miocene (Wolfe 1969). This is the first evidence for the Mediterranean climates and arid climates of the Sierra Nevada; prior to this, rainfall seems to have occurred year-round (Axelrod 1973, 1979; Raven and Axelrod 1978). Nevertheless, certain parts of the Sierra Nevada (e.g., near Lake Tahoe [Adam 1973]) seem to have retained more summer rain than at present until at least 1.9 Ma.

Although no major climatic, geologic, or floristic transitions mark the beginning of the Pliocene, this epoch was a period of decreased average rainfall and gradual decline in temperatures. The drying was related mostly to a change in the seasonality of precipitation, with summer drought increasing in length and severity. The full Mediterranean pattern typical of California and most of the Sierra Nevada developed only in the late Pliocene. The climate patterns may be related to increases in high-pressure areas, decreases in global temperatures, glaciation at high latitudes—all conditions during the later Pliocene that signaled the development of Pleistocene climate patterns. The middle of the Pliocene may have been the driest part of the Tertiary in California (Raven and Axelrod 1978), although some sites in the western interior of North America show evidence of greater levels of effective moisture than in modern times (Thompson 1991). Clearly there were significant fluctuations in climate during this period (e.g., Adam et al. 1990; Adam et al. 1989). By the end of the Pliocene, climates were distinctly cooler and wetter throughout the Sierra Nevada, with the estimated increase in rainfall 25–40 cm (10–15 in) above that of the mid-Pliocene (Axelrod 1977).

From the standpoint of scientific method, the Pliocene is a period that overlaps the focuses of Tertiary botany and Quaternary science, since deep pollen cores (previously restricted to the Quaternary) are increasingly penetrating Pliocene (and even Miocene) sediments. Since the methods and questions of the disciplines have differed somewhat, the Pliocene becomes rich with information from both scientific communities. When interpreting reports from the two disciplines, it is important to note that events may be described at different temporal, spatial, and conceptual scales. Quaternary scientists who investigate Tertiary phenomena focus especially on high frequency events (e.g., those with periods of 1,000 years or less) (Delcourt and Delcourt 1991, hierarchical models of time and space). Quaternary science methods are capable of looking at temporally continuous floristics and fine resolution in the temporal scale of climate; they have a strong focus on external forcing factors for detailed climate reconstruction. Quaternary scientists, however, continue to rely heavily on uniformitarian assumptions about species ecology, making the (probably valid) assumption that evolutionary change was not significant during the Holocene or even late Pleistocene. For the Pliocene, this assumption may be inappropriate, and floristic comparisons based on response-surface analyses may not be well calibrated to successively older taxa.

Analysis of continuous records of microfossils (pollen and

diatoms) in parts of California make the dating of climate events and high-frequency fluctuations more precise. An example from northeastern California, on the Modoc Plateau, is the Tulelake core, with continuous records into the late Pliocene (figure 5.1; table 5.2; appendix 5.1, list 29) (Adam et al. 1990; Adam et al. 1989). During the period from 3.0 Ma to 2.12 Ma, fluctuations in the Sierra Nevada climate are recorded from the Tulelake core. From 3.0 to 2.9 Ma, the climate was cool, whereas from 2.9 to 2.6 Ma, warmer conditions prevailed (as much as 5°C [9°F] higher than current conditions). Evidence for severe summer drought also exists. From 2.48 to 2.12 Ma, the Tulelake basin dried, and floristic compositions (e.g., sagebrush [*Artemesia*]) increased, probably reflecting colder conditions as well. A shift to cool, moist conditions around 2.0 Ma is inferred from diatoms.

Vegetation dynamics of the Miocene and Pliocene of the Sierra Nevada and adjacent regions are complex. Although this undoubtedly reflects the major environmental transitions of the time, it is probably also an artifact of the better records from younger ages. I discuss in turn the Miocene-Pliocene vegetation history under three more-or-less-chronological themes:

1. Early extinctions and vegetation replacements resulting from the Eocene:Oligocene climatic event,

2. High diversity through the middle to late Miocene correlated with summer rain, and

3. Migrations, species turnovers, increasing provincialization, zonation, and late Miocene extinctions due to decreasing summer rainfall through the Miocene-Pliocene.

Early Extinctions and Replacements

Since the Oligocene is poorly recorded in the Sierra Nevada, the response of vegetation to the major climatic transitions of the Eocene:Oligocene can be read only from the later record. Many Tertiary records from the mid-Miocene in the Sierra Nevada and western Nevada show enormous turnovers in vegetation relative to earlier (Eocene) deposits (compare table 5.2 and appendix 5.1, lists 3 and 26 to table 5.3 and appendix 5.1, lists 1, 5, 8, 9, 11, 15, 16, 22, 25, 28, 30, 32). The most dramatic changes are the loss of warm-humid-adapted angiosperms and the appearance of cool-temperate-adapted conifers and angiosperms (Axelrod 1977, 1986; Axelrod and Schorn 1994; Raven and Axelrod 1978; Schorn 1984; Wolfe 1969). Almost all of the key subtropical and tropical taxa (e.g., *Diospyros, Ficus, Engelhardtia, Magnolia, Viburnum, Cinnamomum, Persea*) that are known from Sierran and other western North American Eocene floras are missing from the fossil deposits of the mid-Tertiary. In both floristics and foliar physiognomy (percentage of entire-margined leaves), the adaptations changed from tropical to temperate (Wolfe 1969). In general, the losses of taxa from the Sierra Nevada represent major regional and even continental extirpations: many taxa

(or their nearest relatives) that occurred in California during the Eocene are found now in tropical Mexico and others only in eastern Asia (Axelrod 1977; Raven and Axelrod 1978).

Replacing these subtropical taxa, by the mid-Miocene, were abundant conifer species of the Pinaceae *(Pinus, Abies, Picea)*, Cupressaceae *(Chamaecyparis, Thuja, Juniperus)*, and Taxodiaceae *(Sequoiadendron, Sequoia)* and angiosperm species such as *Alnus, Fraxinus, Populus, Salix, Arbutus, Quercus,* and *Ceanothus* (e.g., appendix 5.1, lists 1, 5, 8, 9, 11, 15, 22). In addition to taxa currently native to the Sierra Nevada, these Miocene forests contained temperate-adapted taxa now native to other parts of North America. These included conifers and angiosperms currently native to non-Sierran provinces of California (e.g., *Chamaecyparis, Sequoia, Picea, Thuja,* coastal species of *Quercus*), as well as species that now grow in eastern North America (e.g., *Carya, Ulmus, Juglans, Liquidambar*).

Many of these temperate-adapted taxa appear to have been present during the early Tertiary on the volcanic plateau of the Great Basin–Idaho uplift, which seems to have served as a refugial island during the warm, humid phases of the early Tertiary (Axelrod 1968, 1986; Axelrod and Raven 1985; Millar 1993; Millar and Kinloch 1991). These taxa would have been closely adjacent to the Sierra Nevada, capable of migrating westward relatively rapidly into the range as Oligocene and early Miocene climates opened hospitable habitats in the Sierra Nevada (Axelrod 1977). Prior to basin and range extension, these two regions were closer together than at present (Fiero 1991).

High Diversity of Miocene Floras

One consequence of these biogeographical changes was that, although temperate taxa replaced tropical ones in the Sierra Nevada by the mid-Miocene, diversity of the new flora was high, apparently much higher even than at present in the Sierra Nevada. Not only was the total diversity of species high (many currently native species plus taxa not now in the Sierra), but also the vegetation associations were highly diverse and different from those of the present. For instance, in the Purple Mountain flora of western central Nevada (mid-Miocene) (appendix 5.1, list 22), the following species occurred together in one deposit: false cypress *(Chamaecyparis)*, red and white firs *(Abies cf. magnifica* and *A. cf. concolor)*, Santa Lucia fir *(A. cf. bracteata)*, western white pine *(Pinus cf. monticola)*, Brewer's spruce *(Picea cf. breweriana)*, giant sequoia *(Sequoiadendron cf. giganteum)*, madrone *(Arbutus)*, live oak *(Quercus cf. chrysolepis)*, cottonwood *(Populus)*, and willow *(Salix)* (Axelrod 1976). At the Upper Cedarville locality of northwestern Nevada (mid-Miocene) (appendix 5.1, list 30), false cypress, *Ginkgo*, redwood *(Sequoia cf. sempervirens)*, red fir, ponderosa pine *(Pinus cf. ponderosa)*, nutmeg *(Torreya)*, madrone, chestnut *(Castanea)*, beech *(Fagus)*, hickory *(Carya)*, ash *(Fraxinus)*, *Tilia,* and elm *(Ulmus)* grew together (LaMotte 1936). The Chalk Hills forest of western central Nevada (late Miocene) (appendix 5.1, list 4) contained false cypress, white fir, foxtail pine (*Pinus* cf. *balfouriana*), Douglas fir (*Pseudotsuga* cf. *menziesii*), giant sequoia, madrone, *Rhododendron*, hickory, and oak (*Quercus*) (Axelrod 1962).

By modern standards these and most other Sierran middle to late Miocene floras strike us as unusual in that they contain mixes of "incompatible" species. For instance, the co-occurrence of foxtail pine (subalpine), Port Orford cedar (warm-humid), giant sequoia (middle elevation, mixed conifer, fire-adapted), rhododendron (cool-mesic), hickory (continental, well-distributed rainfall), and scrub live oak (xeric) challenges our ability to imagine the Tertiary habitat. Although biases in the fossil record (such as single floras appearing contemporaneous from the stratigraphy but actually representing many years of accumulation) may skew interpretations, these diverse vegetation assemblages recur commonly enough to indicate that the associations were actually this diverse and ecologically complex by modern standards.

An explanation for the high species and association diversities is that, although climates of the Sierra Nevada and western Nevada were temperate by the Miocene, rainfall remained distributed throughout the year (Axelrod 1977; Axelrod and Schorn 1994; Wolfe 1969). A common requirement, or tolerance, of the species in these assemblages is summer rainfall. From several lines of evidence, summer rainfall appears to have persisted late into the Miocene, although the trend was toward decreasing summer rain during this period. Summer rainfall would offer permissive conditions for subtropical taxa that were remnants of the early Tertiary (e.g., *Ginkgo* at Upper Cedarville [appendix 5.1, list 30]; *Viburnum* and *Persea* at Tehachapi [appendix 5.1, list 28]; *Magnolia* at Mohawk [appendix 5.1, list 16]) as well as temperate broad-leaved species not native to California now (hickory, beech, chestnut, honey locust [*Robinia*]) and temperate conifers and angiosperms that can tolerate these conditions. Temperate climate with summer rainfall thus contributed to the high diversity of Miocene floras.

Response to Decreasing Summer Rainfall and Warmer Summers

The main catalyst for vegetation change in the late Tertiary appears to have been the trend to decreasing summer rainfall, which culminated in the well-developed Mediterranean pattern of present-day California. Lack of distribution of rainfall throughout the year, especially when associated with high temperatures during the growing season, is extremely stressful to plants. It requires specific adaptations and is apparently intolerable to many taxa. One pattern that appears from this climate trend is increasing provincialization and zonation of assemblages. Although there are few early Tertiary deposits in the Sierra Nevada, their composition parallels the many Eocene fossil floras from throughout western North America. Notable about these floras, in addition to their general subtropical or tropical adaptations, is their low degree of regional or local differentiation (Wolfe 1969). In the Sierra Nevada, little evidence of provincial development exists by the mid-Miocene. By the late Miocene and the Pliocene, floras of western North America, including those in the Sierra Nevada, had become increasingly differentiated, by region, latitude, and elevation (Wolfe 1969). Cooler winters may also have contributed to differentiation.

Although mid-Miocene floras had remained highly diverse by virtue of adequate summer rain, decreasing summer precipitation in the later Miocene and the Pliocene narrowed the adaptive zones for these species, segregating them into habitats specific to the needs of individual taxa. Thus the "incompatible mixtures" of species from the mid-Miocene started to segregate into the "compatible mixtures" of modern associations, with the result that floras became more depauperate. Compared to modern standards, the mid-Miocene floras reflect more than earlier Miocene floras the locally specific conditions (e.g., elevation, orography, local climate) currently influencing species distributions and ecosystem dynamics in the Sierra Nevada.

Another consequence of increasing summer drought was migration or shifting distributions of species. The diverse forests of the mid-Miocene in Nevada, which contained taxa that are now found in subalpine, mixed conifer, and coastal environments, appear to have contributed taxa that moved west and found suitable habitats in the Sierra Nevada by the late Miocene and the Pliocene. For example, foxtail pine, present in several early and middle Miocene floras in Nevada but absent from California in the early Tertiary, appears to have migrated westward to the Sierra Nevada and western California. It remains a relict in only two regions in California that persist in having summer rain today: the southern Sierra Nevada and the Klamath Province (Axelrod 1977). As the conifer taxa moved west and up into the Sierra Nevada, the Nevada basins and ranges lost these "montane" conifers and were colonized by piñon/juniper species, which spread rapidly across this area in the early Pliocene. Increasing aridity eventually forced even these conifers to higher elevations, and sage/bitterbrush dominated the dry basins. The effect on Sierra Nevada floras is exemplified by the Mount Reba flora of the central Sierra Nevada. Situated now at 2,625 m (8,600 ft) but probably lower in the early Pliocene, the deposit contains live oak, tan oak (*Lithocarpus* cf. *densiflora*), Douglas fir, cypress (*Cupressus*), white fir, and giant sequoia (appendix 5.1, list 17) (Axelrod 1976). By the early Pliocene, these taxa were no longer present in western Nevada, although they had been in earlier Miocene floras. Presumably many of these species were extirpated in western Nevada and had migrated westward into the Sierra, where they found more suitable habitats.

Contributing also to the modernization of the Sierra Nevada forests was gradual extinction or extirpation of many species adapted to warm winters and summer rainfall. By the late Miocene and the Pliocene, subtropical/tropical and temperate hardwoods requiring summer rain were mostly gone (e.g., Chalk Hills [appendix 5.1, list 4]; Oakdale [appendix 5.1, list 19]; Verdi [appendix 5.1, list 33]), and warm-humid conifers (Port Orford cedar, redwood, *Thuja*) and other taxa that are now coastal (Brewer's spruce [*Picea* cf. *breweriana*], Santa Lucia fir [*Abies* cf. *bracteata*]) were declining in representation. Relicts from the early Tertiary persisted longer on the west side of the Sierra Nevada than on the east side (Raven and Axelrod 1978). On the west slope, a rich oak woodland was present in the early Pliocene (e.g., Oakdale flora [appendix 5.1, list 19]) that still contained summer-rain species, such as *Celtis, Persea, Robinia, Sapindus,* and *Umbellularia,* which were gone from the eastern Sierra by that time (Axelrod 1977). These extirpations left diversity lower than in the early to middle Miocene forests and left communities both more differentiated and more adapted to summer drought. In sum, vegetation diversity was lower by the late Tertiary in part due to elimination of species now allied with taxa in eastern North America and eastern Asia and in part due to segregation into climate zones elevationally and latitudinally in the Sierra Nevada.

These patterns over the late Miocene and the Pliocene are well documented in a set of continuous sediments at Gillam Spring, northwestern Nevada (Axelrod and Schorn 1994). A rapid change in species diversity and abundance centered around 15 Ma is recorded in three stratigraphically continuous localities (appendix 5.1, list 11; figure 5.4). The compositions of the florules shift from being dominated in the oldest stratum by deciduous hardwoods allied to taxa of the eastern United States and eastern Asia to being dominated in the youngest stratum by conifers and summer-drought-adapted mountain hardwoods.

The last phase of the Tertiary was marked by climates that by some indications were warmer than at present in the Sierra Nevada and possibly not as dry. Between 2 and 3 Ma in the Tulelake samples (Adam et al. 1990), pine and *Taxodiaceae-Cupressaceae-Taxaceae* (assumed to be incense cedar or juniper) dominate, whereas the modern vegetation is sagebrush shrub. Increasing summer drought eliminated most of the sclerophyllous vegetation from western Nevada and stratified Sierran conifers both elevationally and by slope (west or east side of the crest). That the Mediterranean climate pattern had not become as intensified in the earliest Pleistocene (1.9 Ma) as at present, however, is suggested by the lingering persistence of spruce near Lake Tahoe (appendix 5.1, list 13) (Adam 1973).

The general vegetation trends described for the second half of the Tertiary can be seen in the distributions for select individual taxa (figures 5.5–5.22). Many conifers now native to the Sierra Nevada had much broader Tertiary distributions, were not stratified into the vegetation groups we now recognize, and/or were present only in western Nevada during the Tertiary. This can be seen for several species currently in the upper montane and subalpine zones (red fir [figure 5.5]; bristlecone pine [*Pinus* cf. *longaeva*] [figure 5.6]; foxtail pine

[*Pinus* cf. *balfouriana*] [figure 5.7]; western white pine [figure 5.8]; western hemlock [*Tsuga* cf. *heterophylla*] [figure 5.9]). It is also seen for conifers that currently occur in mixed conifer and primarily west-side Sierran forests (white fir [figure 5.10]; incense cedar [figure 5.11]; sugar pine [*Pinus* cf. *lambertiana*] [figure 5.12]; ponderosa [*P.* cf. *ponderosa*] and Jeffrey [*P.* cf. *jeffreyi*] pines [figures 5.13, 5.14]; Douglas fir [figure 5.15]; and giant sequoia [figure 5.16]). Many oaks similarly had broad distributions in the Sierra and western Nevada during the Miocene and Pliocene (black oak [*Quercus* cf. *kelloggii*] [figure 5.17]; white oak [*Q.* cf. *lobata*] [figure 5.18]).

Several species were broadly distributed in the middle Tertiary, eventually became extirpated from the Sierra Nevada, but now occur in coastal California habitats. These include false cypresses and Santa Lucia fir (figure 5.19), Brewer's spruce (figure 5.20), coast redwood (figure 5.21), and coast live oak (*Quercus* cf. *chrysolepis*) (figure 5.22). Blue oak (*Q. douglasii*) and scrub oak (*Q. dumosa*), both widespread now, occur in surprisingly few Tertiary deposits (figure 5.23).

The importance of taxonomic identification and revision to interpretation of biogeography and evolution can be demonstrated with two examples. Howard Schorn (letter to the author, July 1995) is revising many of the Tertiary floras from Nevada and the Sierra Nevada and provides the following information. As a generalization, the revision of taxonomy indicates no *Pseudotsuga, Larix,* or *Thuja* in the Nevada–Sierra Nevada area during the Tertiary. The many fossils that previously indicated the presence of Douglas fir (figure 5.15) are in fact now identified primarily to *Abies* and *Tsuga,* with none to Douglas fir. Thus, the biogeographic origins of Douglas fir must not parallel the history described above for many of the present-day Sierran conifers. Rather than originating on the upland plateau of Nevada, Douglas fir must have moved into the Sierra Nevada from another bioregion.

Similarly, Schorn's revisions (letter to the author, July 1995, and manuscript in preparation) indicate that, although abundant *Abies* is recorded in Tertiary deposits, no white fir (*Abies concoloroides*) is firmly identified from the Sierra Nevada or Nevada until the Pliocene about 6–8 Ma, and then only in the vicinity of Reno, Nevada. This contrasts with earlier descriptions (figure 5.10), which suggest white fir was widespread through this region during the Tertiary. Schorn identifies most of the white fir fossils as red fir (*A. magnifica*) instead. These revisions markedly change interpretations of the origins of these two taxa: white fir, like Douglas fir, apparently was not part of the upland Nevada mixed conifer forests and did not enter the Sierra Nevada from the east. Conversely, red fir was much more abundant and widespread throughout Nevada Tertiary forests than previously indicated (figure 5.5), and apparently did co-occur in the upland region as part of the diverse Miocene and Pliocene forests.

These examples underscore the importance of correct taxonomic identifications and the need for widespread revision of old fossil floras with new methods.

FIGURE 5.4

Rapid change in species diversity and abundance recorded at three stratigraphically continuous mid-Miocene sites at Gillam Spring, Nevada (from Axelrod and Schorn 1994).

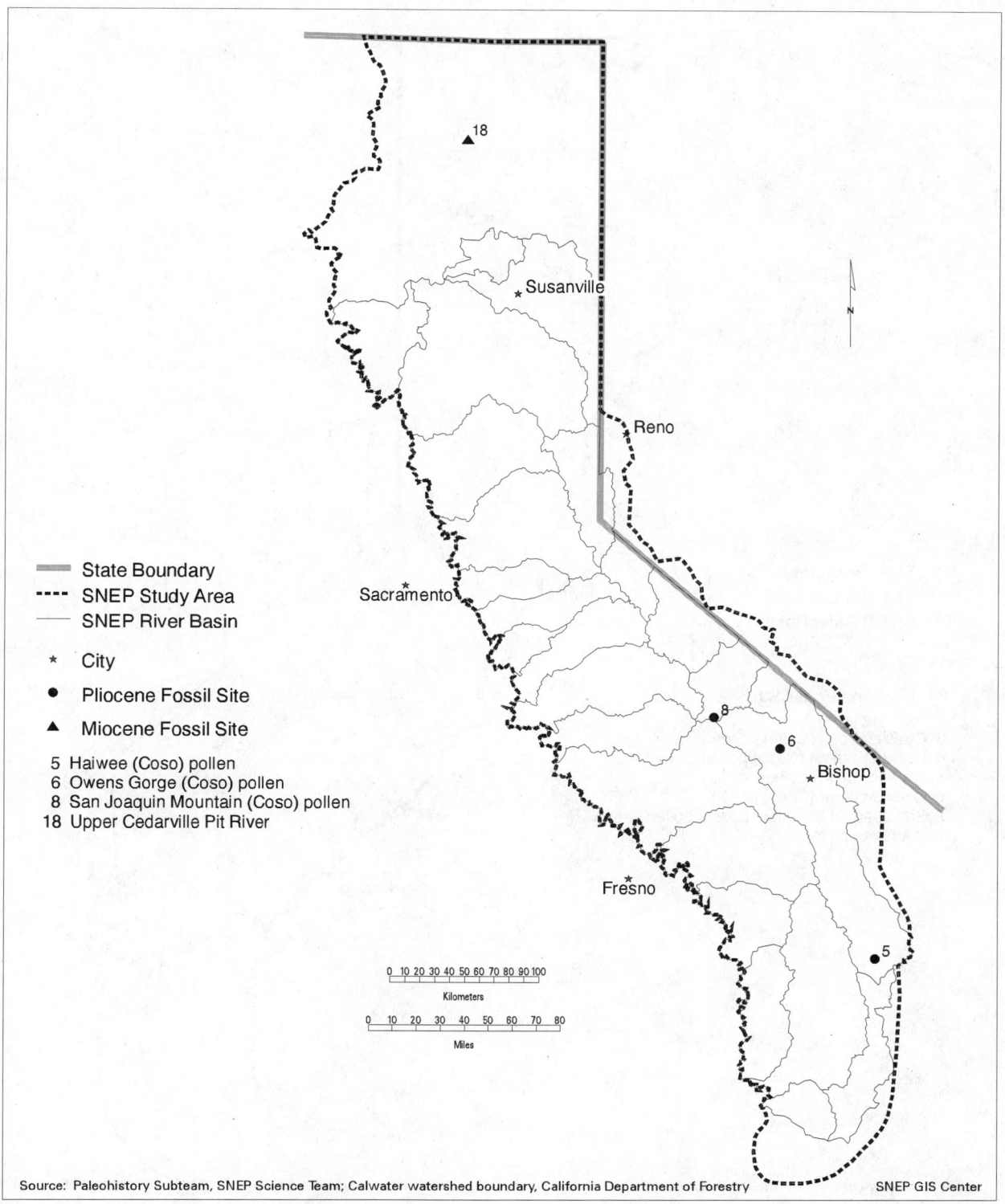

FIGURE 5.5

Distribution of red fir *(Abies* cf. *magnifica)* in the Sierra Nevada and adjacent Nevada as recorded in Tertiary fossil deposits. *Note:* Floras 5, 6, and 8 are pollen sites, interpretations are based on a very small number of grains, and species identifications may be incorrect. These sites should be treated with question. Unpublished revisions by Howard Schorn of the University of California, Berkeley, Museum of Paleontology (letter to the author, June 1995) indicate that all *Abies concolor* fossils indicated for Tertiary Nevada–Sierra Nevada are in fact red fir.

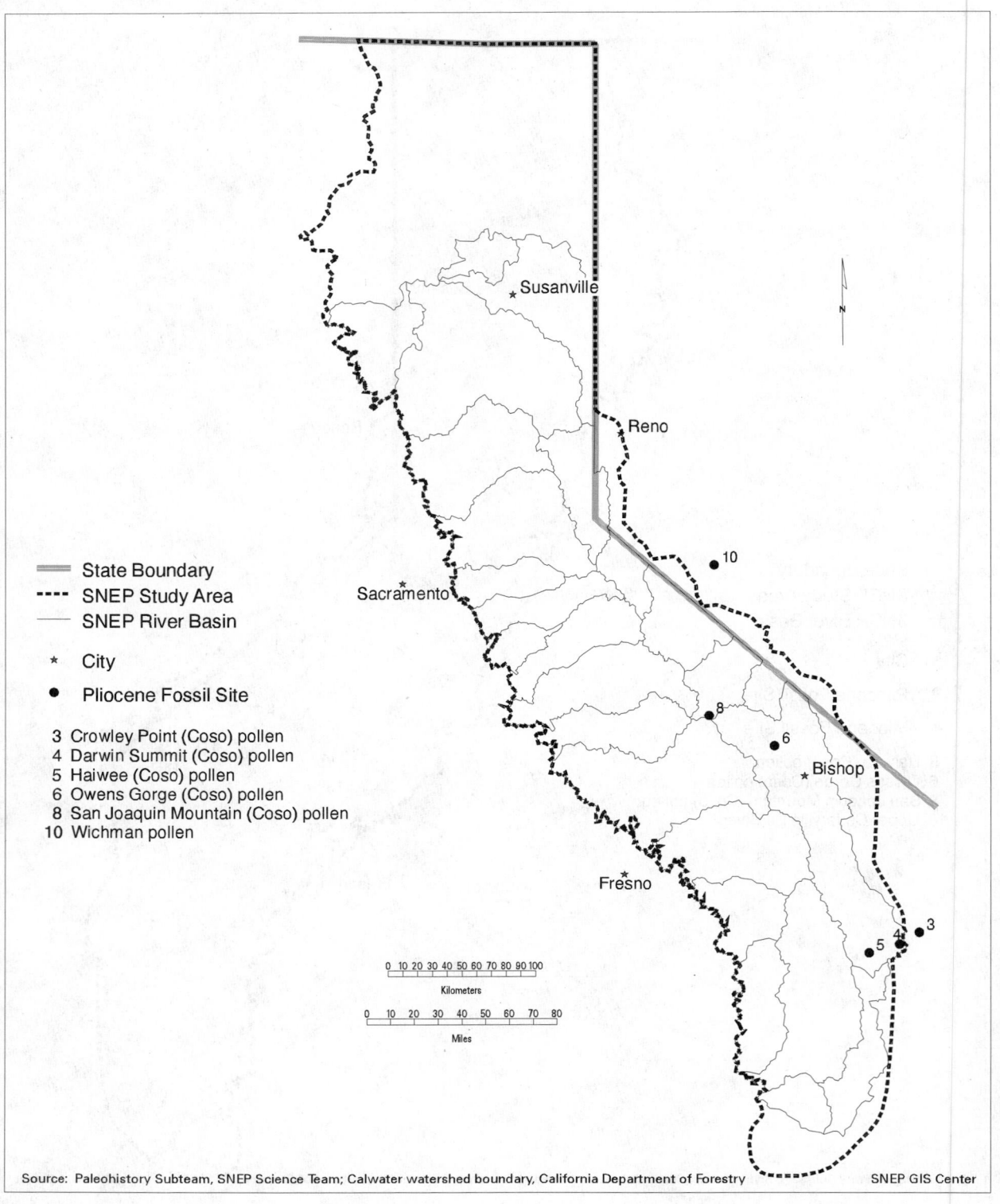

State Boundary
SNEP Study Area
SNEP River Basin

★ City

● Pliocene Fossil Site

3 Crowley Point (Coso) pollen
4 Darwin Summit (Coso) pollen
5 Haiwee (Coso) pollen
6 Owens Gorge (Coso) pollen
8 San Joaquin Mountain (Coso) pollen
10 Wichman pollen

Source: Paleohistory Subteam, SNEP Science Team; Calwater watershed boundary, California Department of Forestry   SNEP GIS Center

FIGURE 5.6

Distribution of bristlecone pine *(Pinus* cf. *longaeva)* in the Sierra Nevada and adjacent Nevada as recorded in Tertiary fossil deposits. *Note:* Floras 3–6, 8, and 10 are pollen sites, interpretations are based on a very small number of grains, and species identifications may be incorrect. These sites should be treated with question.

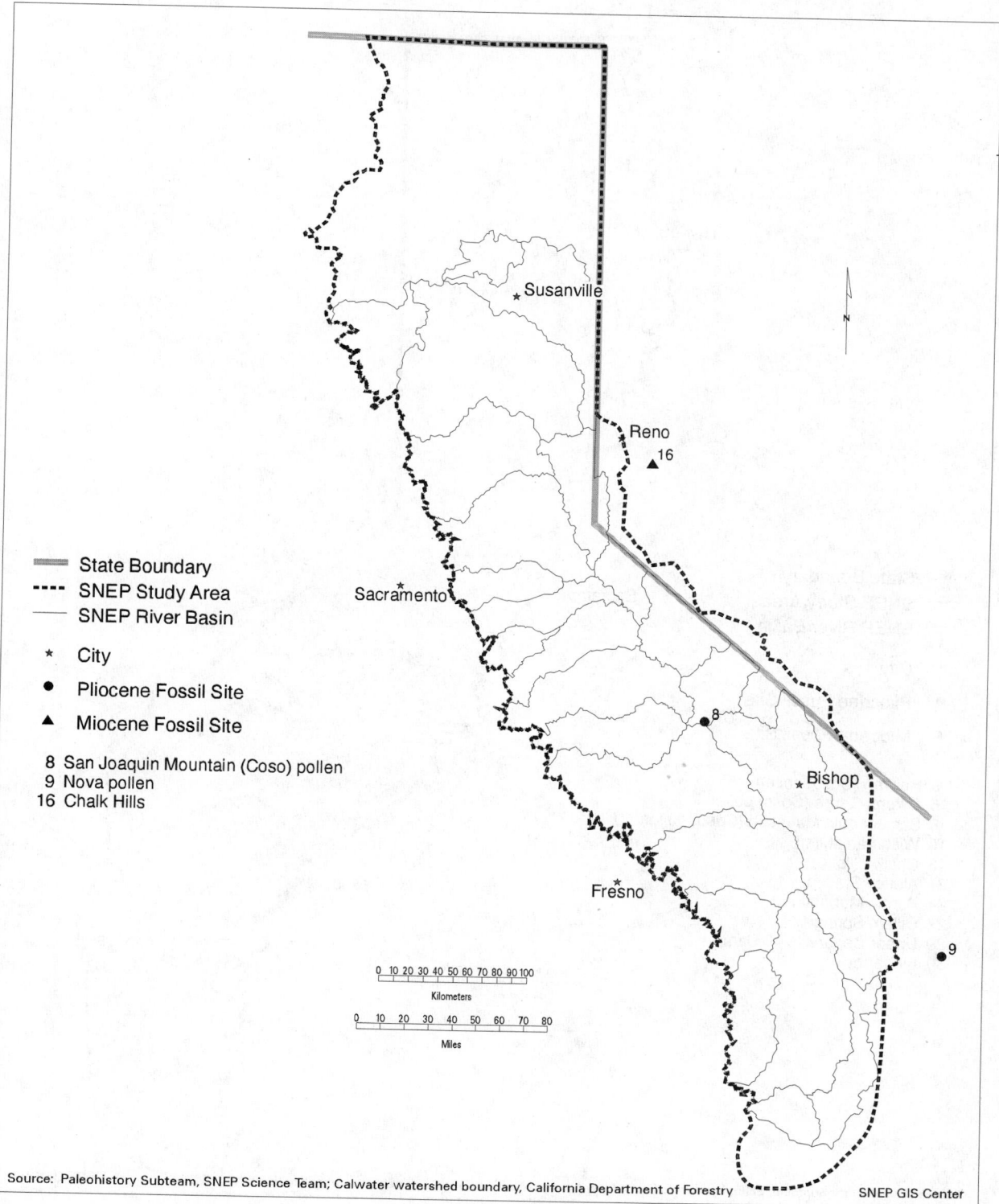

State Boundary
SNEP Study Area
SNEP River Basin

★ City

● Pliocene Fossil Site

▲ Miocene Fossil Site

8 San Joaquin Mountain (Coso) pollen
9 Nova pollen
16 Chalk Hills

Source: Paleohistory Subteam, SNEP Science Team; Calwater watershed boundary, California Department of Forestry   SNEP GIS Center

FIGURE 5.7

Distribution of foxtail pine *(Pinus* cf. *balfouriana)* in the Sierra Nevada and adjacent Nevada as recorded in Tertiary fossil deposits. *Note:* Floras 8 and 9 are pollen sites, interpretations are based on a very small number of grains, and species identifications may be incorrect. These sites should be treated with question.

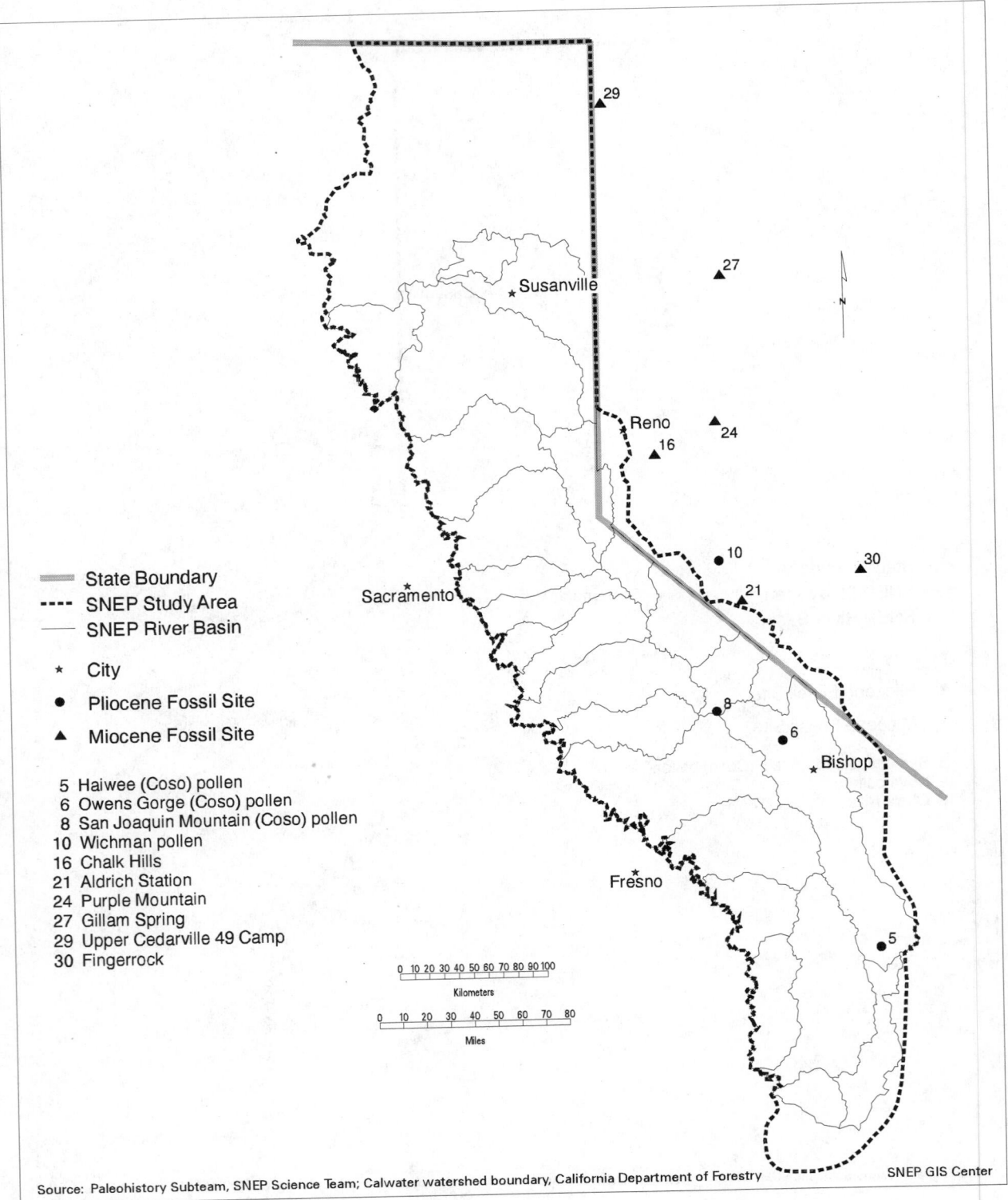

State Boundary
SNEP Study Area
SNEP River Basin

★ City
● Pliocene Fossil Site
▲ Miocene Fossil Site

5 Haiwee (Coso) pollen
6 Owens Gorge (Coso) pollen
8 San Joaquin Mountain (Coso) pollen
10 Wichman pollen
16 Chalk Hills
21 Aldrich Station
24 Purple Mountain
27 Gillam Spring
29 Upper Cedarville 49 Camp
30 Fingerrock

0 10 20 30 40 50 60 70 80 90 100
Kilometers

0 10 20 30 40 50 60 70 80
Miles

Source: Paleohistory Subteam, SNEP Science Team; Calwater watershed boundary, California Department of Forestry   SNEP GIS Center

FIGURE 5.8

Distribution of western white pine *(Pinus* cf. *monticola)* in the Sierra Nevada and adjacent Nevada as recorded in Tertiary fossil deposits. *Note:* Floras 5, 6, 8, and 10 are pollen sites, interpretations are based on a very small number of grains, and species identifications may be incorrect. These sites should be treated with question.

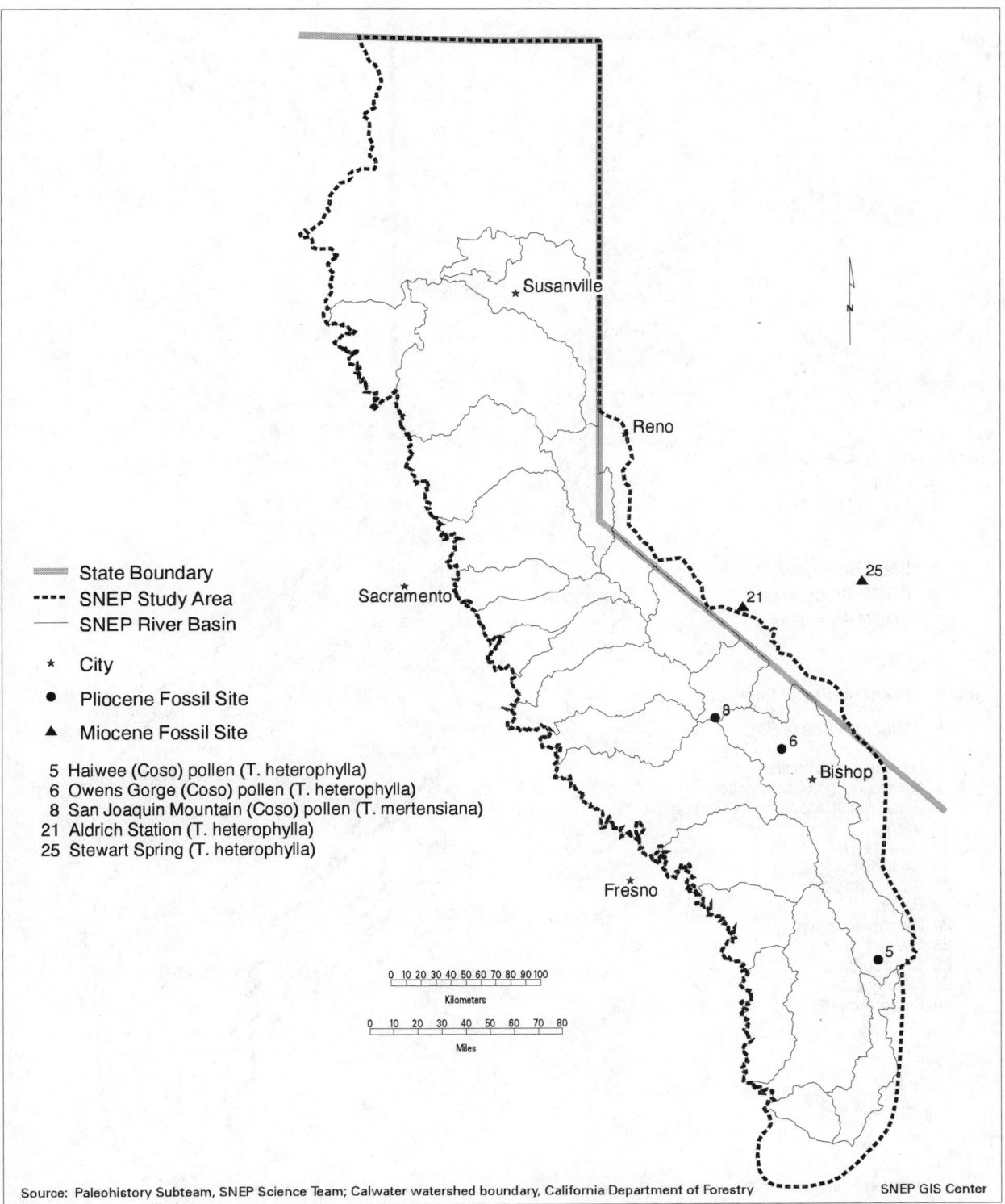

State Boundary
SNEP Study Area
SNEP River Basin

★ City
● Pliocene Fossil Site
▲ Miocene Fossil Site

5 Haiwee (Coso) pollen (T. heterophylla)
6 Owens Gorge (Coso) pollen (T. heterophylla)
8 San Joaquin Mountain (Coso) pollen (T. mertensiana)
21 Aldrich Station (T. heterophylla)
25 Stewart Spring (T. heterophylla)

Source: Paleohistory Subteam, SNEP Science Team; Calwater watershed boundary, California Department of Forestry   SNEP GIS Center

FIGURE 5.9

Distribution of hemlock *(Tsuga)* in the Sierra Nevada and adjacent Nevada as recorded in Tertiary fossil deposits. *Note:* Floras 5, 6, and 8 are pollen sites, interpretations are based on a very small number of grains, and species identifications may be incorrect. These sites should be treated with question.

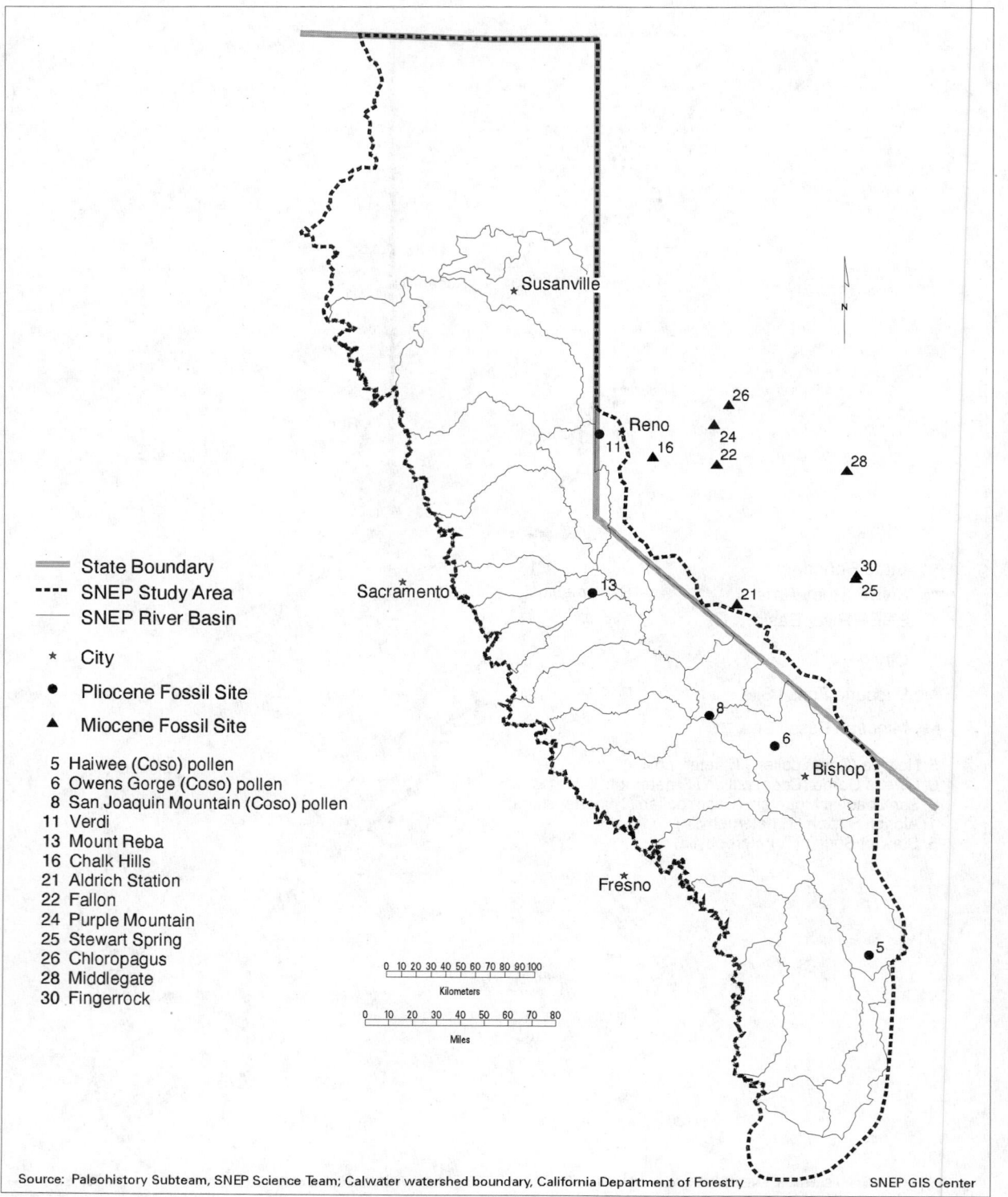

Legend:

State Boundary
SNEP Study Area
SNEP River Basin

* City
● Pliocene Fossil Site
▲ Miocene Fossil Site

5 Haiwee (Coso) pollen
6 Owens Gorge (Coso) pollen
8 San Joaquin Mountain (Coso) pollen
11 Verdi
13 Mount Reba
16 Chalk Hills
21 Aldrich Station
22 Fallon
24 Purple Mountain
25 Stewart Spring
26 Chloropagus
28 Middlegate
30 Fingerrock

0 10 20 30 40 50 60 70 80 90 100
Kilometers

0 10 20 30 40 50 60 70 80
Miles

Source: Paleohistory Subteam, SNEP Science Team; Calwater watershed boundary, California Department of Forestry   SNEP GIS Center

FIGURE 5.10

Distribution of white fir *(Abies* cf. *concolor)* in the Sierra Nevada and adjacent Nevada as recorded in Tertiary fossil deposits. *Note:* Floras 5, 6, and 8 are pollen sites, interpretations are based on a very small number of grains, and species identifications may be incorrect. These sites should be treated with question. Unpublished taxonomic revisions by Howard Schorn of the University of California, Berkeley, Museum of Paleontology (letter to the author, June 1995) indicate that there is no white fir in this bioregion except in the Reno, Nevada, area (Verdi) during the Tertiary. The fossils originally described as white fir now are primarily identified as red fir.

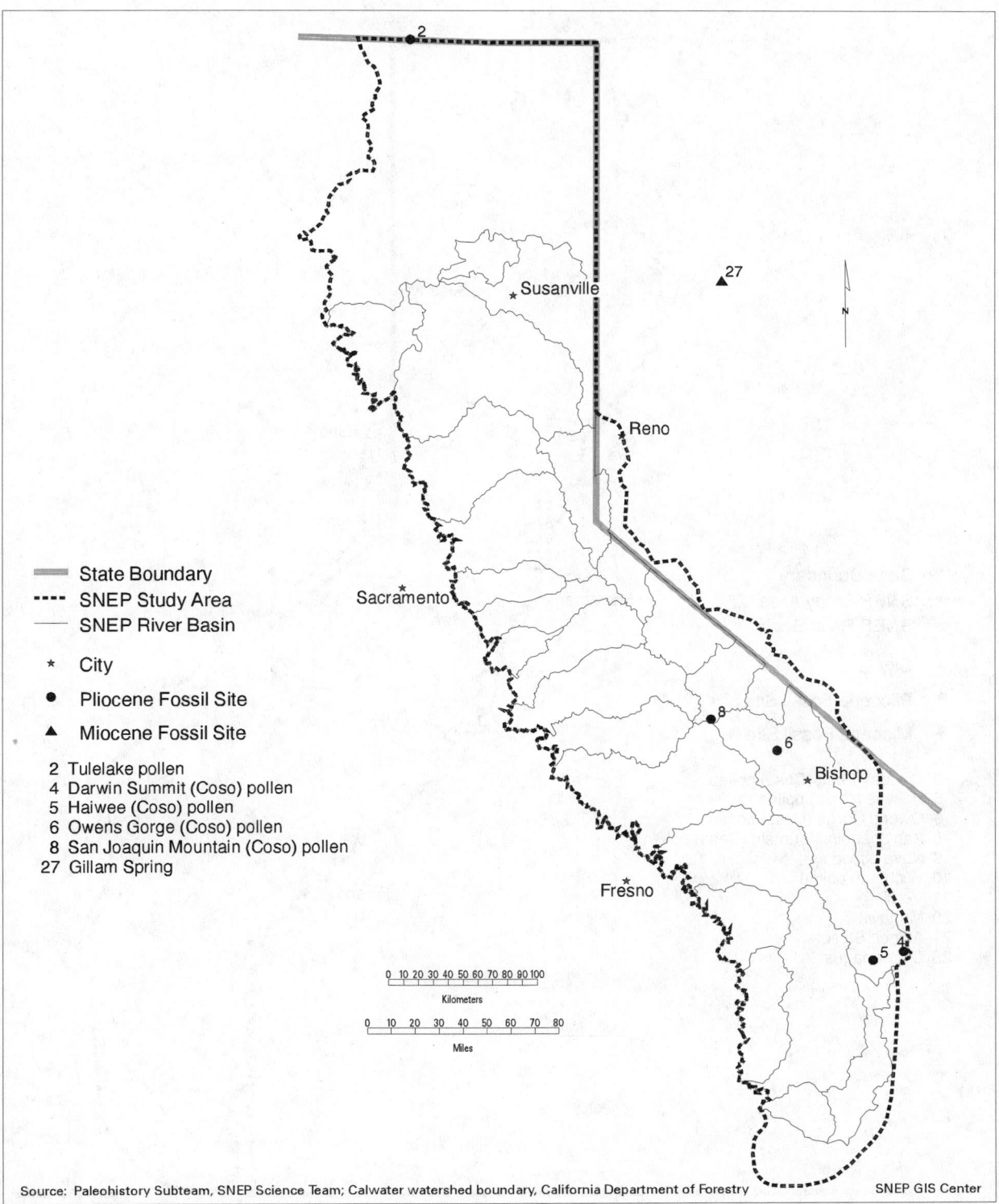

State Boundary

SNEP Study Area

SNEP River Basin

★ City

● Pliocene Fossil Site

▲ Miocene Fossil Site

2 Tulelake pollen
4 Darwin Summit (Coso) pollen
5 Haiwee (Coso) pollen
6 Owens Gorge (Coso) pollen
8 San Joaquin Mountain (Coso) pollen
27 Gillam Spring

0 10 20 30 40 50 60 70 80 90 100
Kilometers

0 10 20 30 40 50 60 70 80
Miles

Source: Paleohistory Subteam, SNEP Science Team; Calwater watershed boundary, California Department of Forestry   SNEP GIS Center

FIGURE 5.11

Distribution of incense cedar *(Calocedrus* cf. *decurrens)* in the Sierra Nevada and adjacent Nevada as recorded in Tertiary fossil deposits. *Note:* Floras 2, 4–6, and 8 are pollen sites, interpretations are based on a very small number of grains, and species identifications may be incorrect. These sites should be treated with question.

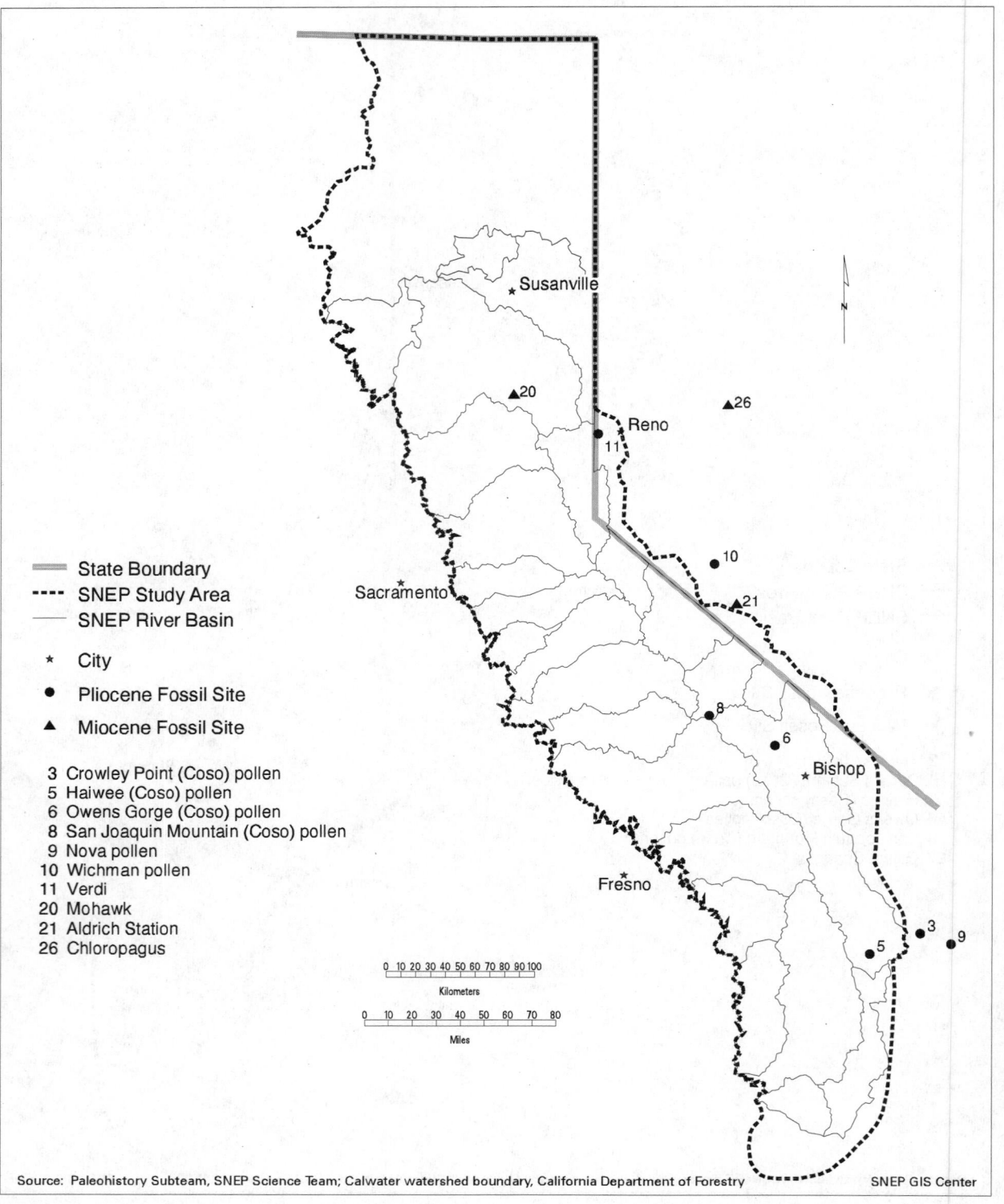

State Boundary
SNEP Study Area
SNEP River Basin

★ City
● Pliocene Fossil Site
▲ Miocene Fossil Site

3 Crowley Point (Coso) pollen
5 Haiwee (Coso) pollen
6 Owens Gorge (Coso) pollen
8 San Joaquin Mountain (Coso) pollen
9 Nova pollen
10 Wichman pollen
11 Verdi
20 Mohawk
21 Aldrich Station
26 Chloropagus

Kilometers
0 10 20 30 40 50 60 70 80 90 100

Miles
0 10 20 30 40 50 60 70 80

Source: Paleohistory Subteam, SNEP Science Team; Calwater watershed boundary, California Department of Forestry   SNEP GIS Center

FIGURE 5.12

Distribution of sugar pine *(Pinus* cf. *lambertiana)* in the Sierra Nevada and adjacent Nevada as recorded in Tertiary fossil deposits. *Note:* Floras 3, 5, 6, and 8–10 are pollen sites, interpretations are based on a very small number of grains, and species identifications may be incorrect. These sites should be treated with question.

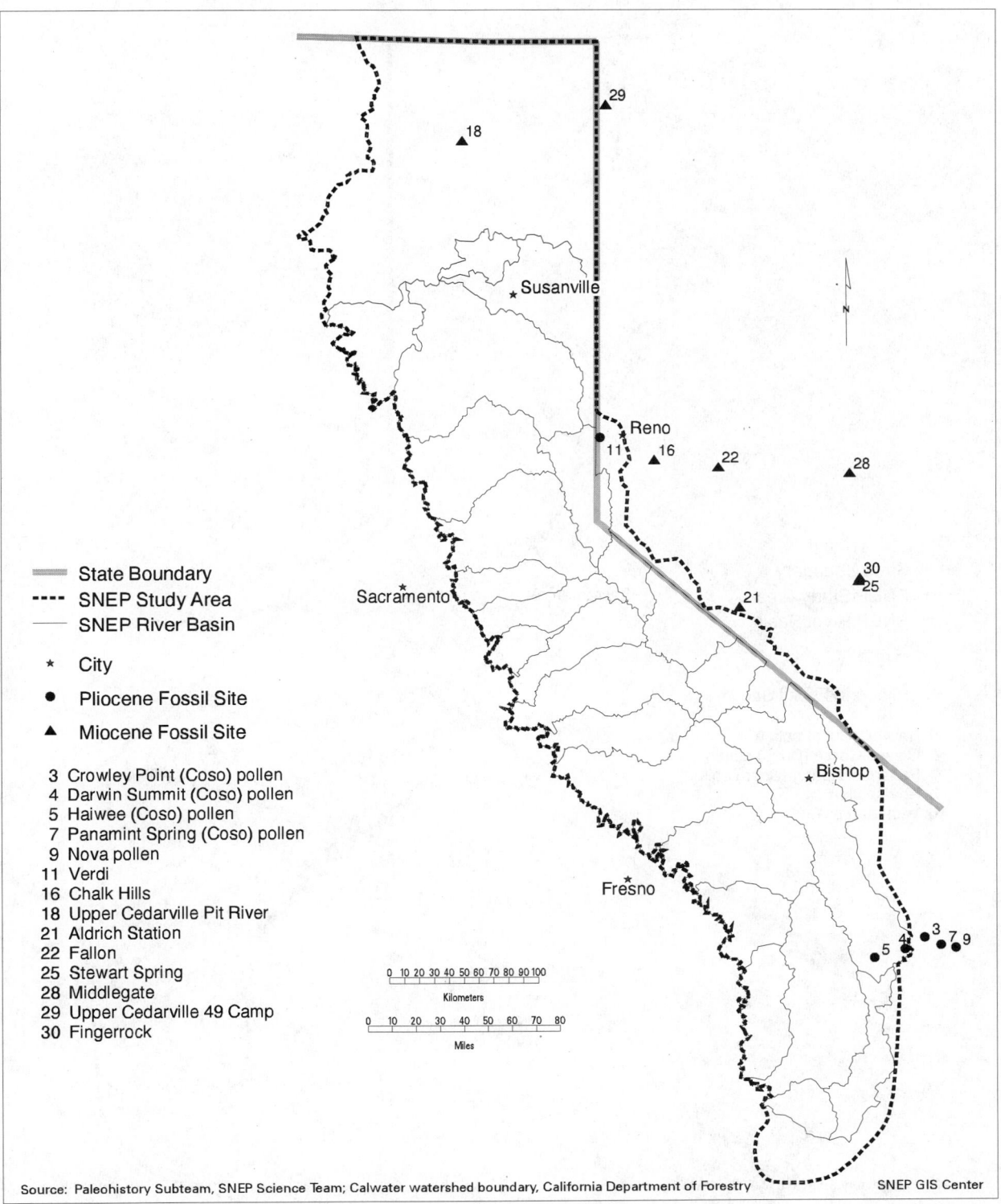

State Boundary
SNEP Study Area
SNEP River Basin

★ City
● Pliocene Fossil Site
▲ Miocene Fossil Site

3 Crowley Point (Coso) pollen
4 Darwin Summit (Coso) pollen
5 Haiwee (Coso) pollen
7 Panamint Spring (Coso) pollen
9 Nova pollen
11 Verdi
16 Chalk Hills
18 Upper Cedarville Pit River
21 Aldrich Station
22 Fallon
25 Stewart Spring
28 Middlegate
29 Upper Cedarville 49 Camp
30 Fingerrock

Source: Paleohistory Subteam, SNEP Science Team; Calwater watershed boundary, California Department of Forestry   SNEP GIS Center

FIGURE 5.13

Distribution of ponderosa pine *(Pinus* cf. *ponderosa)* in the Sierra Nevada and adjacent Nevada as recorded in Tertiary fossil deposits. *Note:* Floras 3–5, 7, and 9 are pollen sites, interpretations are based on a very small number of grains, and species identifications may be incorrect. These sites should be treated with question.

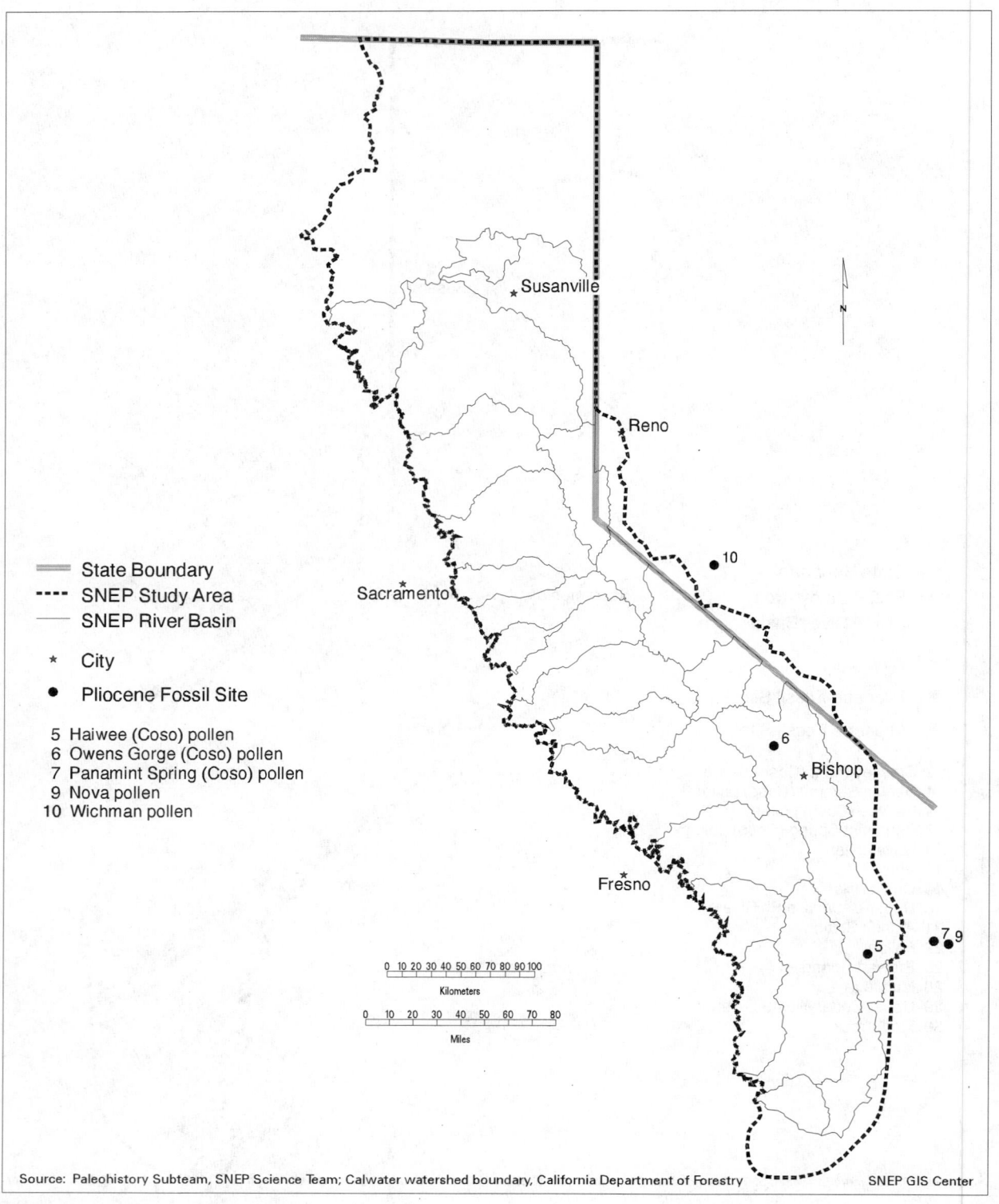

FIGURE 5.14

Distribution of Jeffrey pine *(Pinus* cf. *jeffreyi)* in the Sierra Nevada and adjacent Nevada as recorded in Tertiary fossil deposits. *Note:* Floras 5–7, 9, and 10 are pollen sites, interpretations are based on a very small number of grains, and species identifications may be incorrect. These sites should be treated with question.

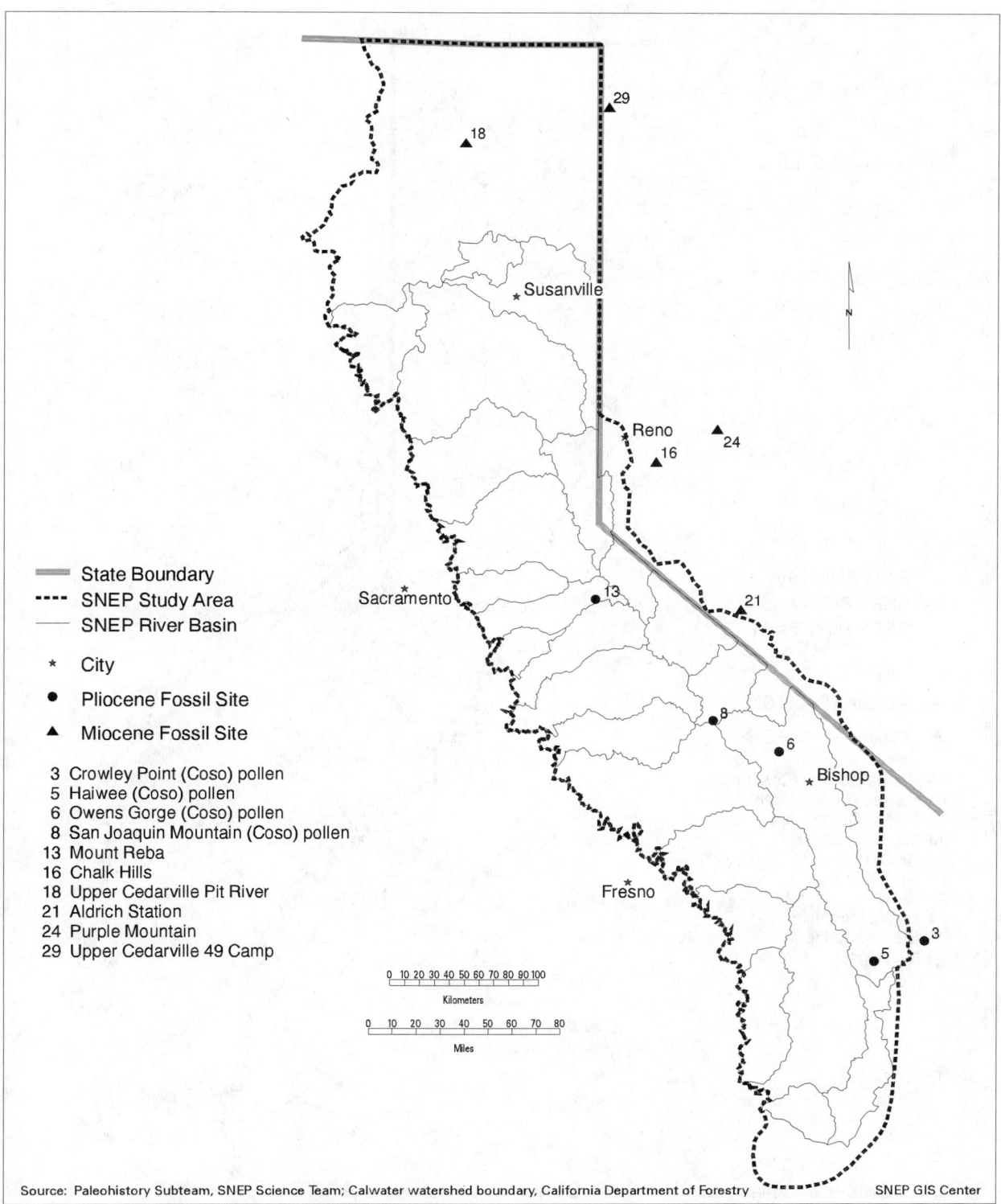

FIGURE 5.15

Distribution of Douglas fir *(Pseudotsuga cf. douglasii)* in the Sierra Nevada and adjacent Nevada as recorded in Tertiary fossil deposits. *Note:* Floras 3, 5, 6, and 8 are pollen sites, interpretations are based on a very small number of grains, and species identifications may be incorrect. These sites should be treated with question.

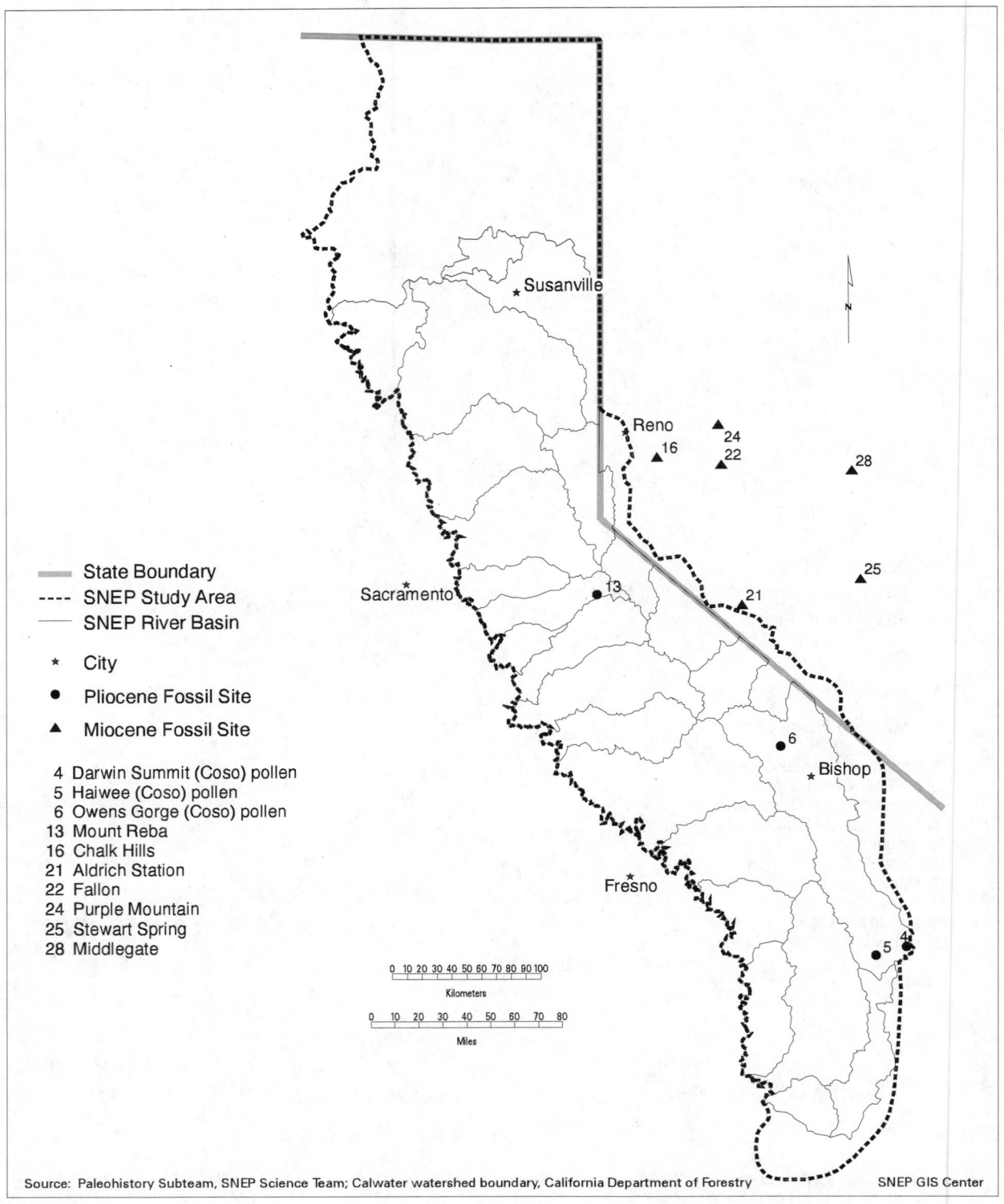

State Boundary
SNEP Study Area
SNEP River Basin

★ City
● Pliocene Fossil Site
▲ Miocene Fossil Site

4 Darwin Summit (Coso) pollen
5 Haiwee (Coso) pollen
6 Owens Gorge (Coso) pollen
13 Mount Reba
16 Chalk Hills
21 Aldrich Station
22 Fallon
24 Purple Mountain
25 Stewart Spring
28 Middlegate

0 10 20 30 40 50 60 70 80 90 100
Kilometers

0 10 20 30 40 50 60 70 80
Miles

Source: Paleohistory Subteam, SNEP Science Team; Calwater watershed boundary, California Department of Forestry   SNEP GIS Center

FIGURE 5.16

Distribution of giant sequoia *(Sequoiadendron* cf. *giganteum)* in the Sierra Nevada and adjacent Nevada as recorded in Tertiary fossil deposits. *Note:* Floras 4–6 are pollen sites, interpretations are based on a very small number of grains, and species identifications may be incorrect. These sites should be treated with question.

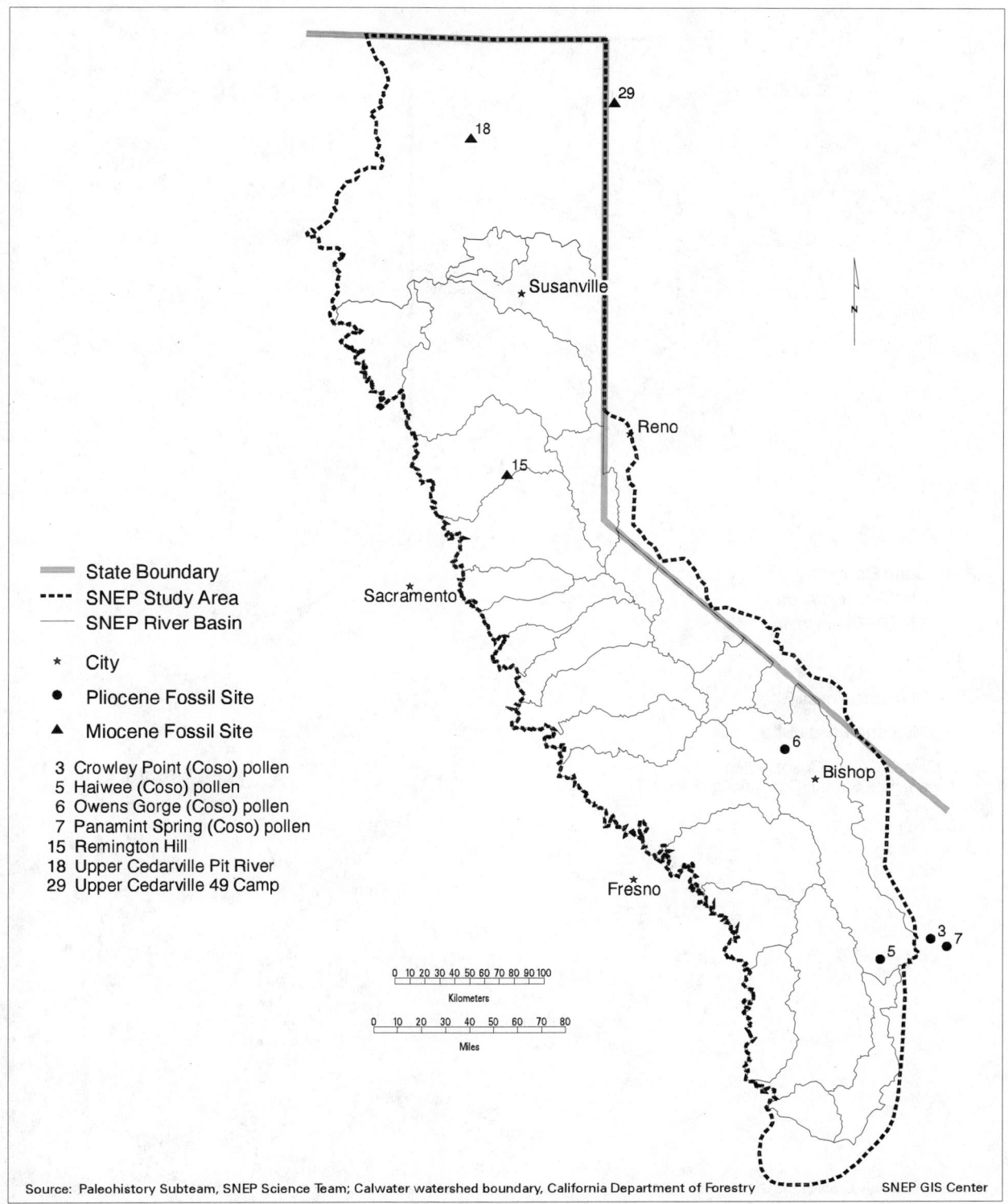

FIGURE 5.17

Distribution of black oak *(Quercus* cf. *kelloggii)* in the Sierra Nevada and adjacent Nevada as recorded in Tertiary fossil deposits. *Note:* Floras 3 and 5–7 are pollen sites, interpretations are based on a very small number of grains, and species identifications may be incorrect. These sites should be treated with question.

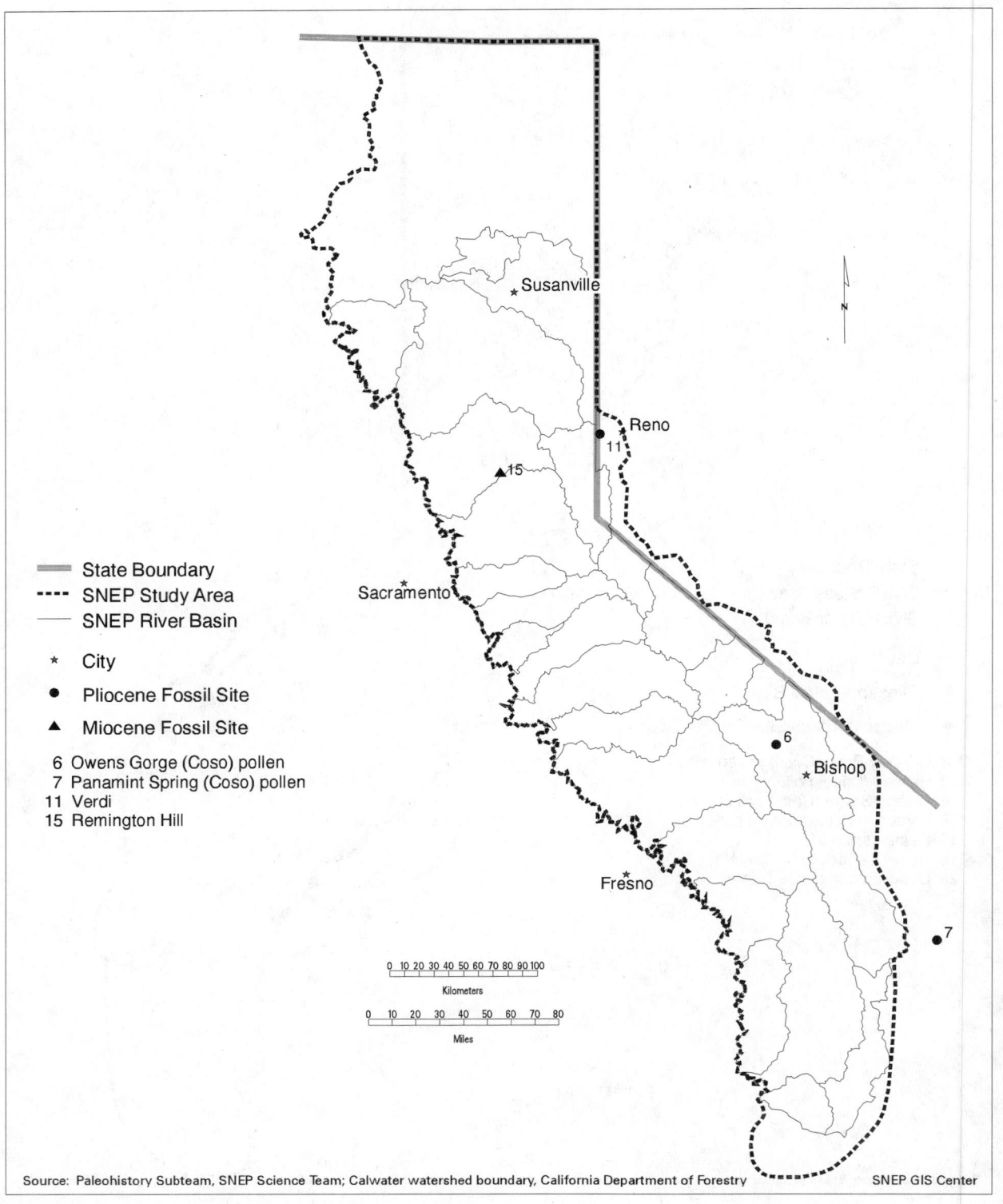

FIGURE 5.18

Distribution of white oak *(Quercus* cf. *lobata)* in the Sierra Nevada and adjacent Nevada as recorded in Tertiary fossil deposits.
Note: Floras 6 and 7 are pollen sites, interpretations are based on a very small number of grains, and species identifications may be incorrect. These sites should be treated with question. Wolfe and Tanai (1987) indicate that the Verdi site actually is an *Acer* not oak.

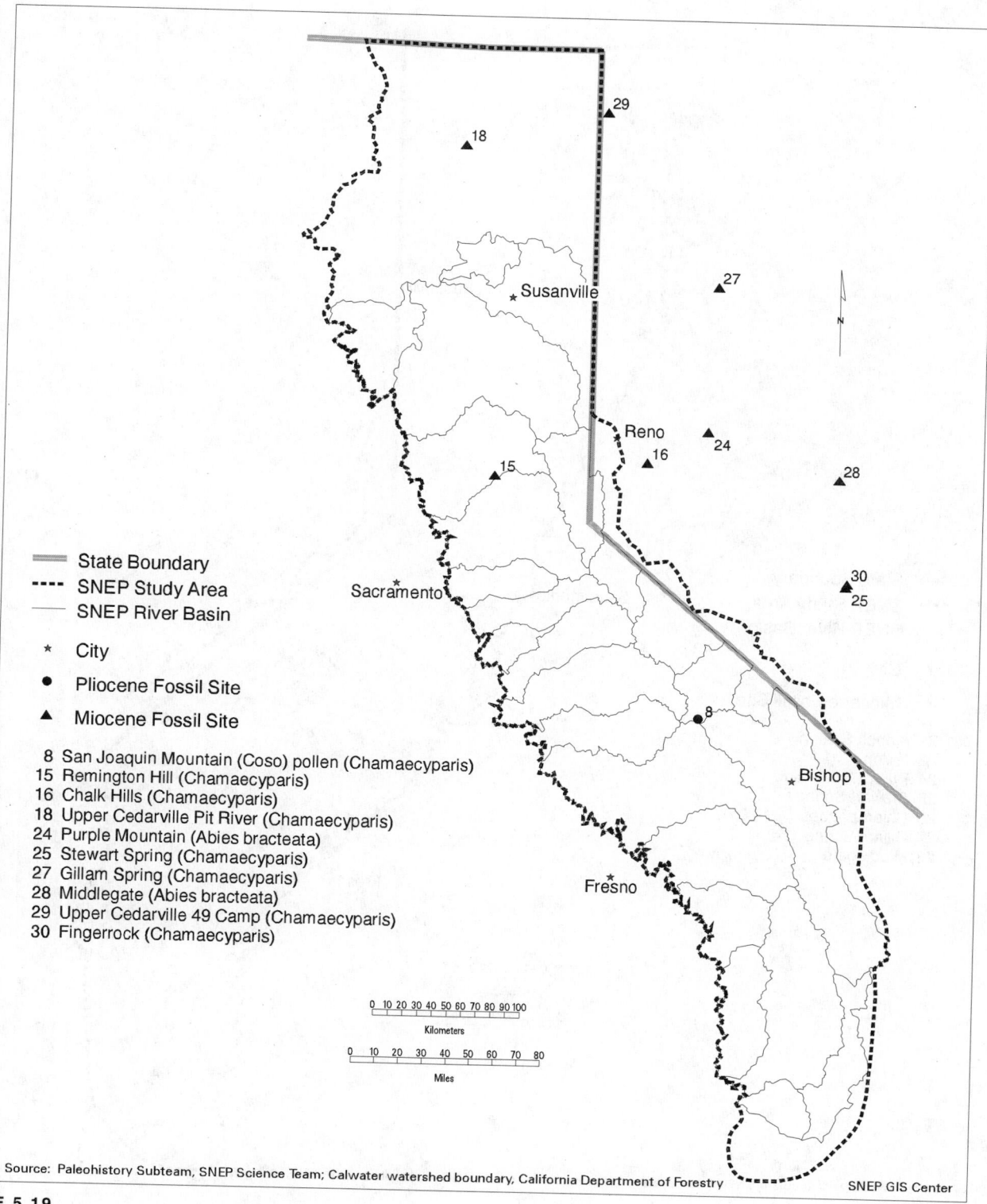

State Boundary
SNEP Study Area
SNEP River Basin

★ City
● Pliocene Fossil Site
▲ Miocene Fossil Site

8 San Joaquin Mountain (Coso) pollen (Chamaecyparis)
15 Remington Hill (Chamaecyparis)
16 Chalk Hills (Chamaecyparis)
18 Upper Cedarville Pit River (Chamaecyparis)
24 Purple Mountain (Abies bracteata)
25 Stewart Spring (Chamaecyparis)
27 Gillam Spring (Chamaecyparis)
28 Middlegate (Abies bracteata)
29 Upper Cedarville 49 Camp (Chamaecyparis)
30 Fingerrock (Chamaecyparis)

0 10 20 30 40 50 60 70 80 90 100
Kilometers

0 10 20 30 40 50 60 70 80
Miles

Source: Paleohistory Subteam, SNEP Science Team; Calwater watershed boundary, California Department of Forestry

SNEP GIS Center

FIGURE 5.19

Distribution of Santa Lucia fir *(Abies bracteata)* and false cypresses *(Chamaecyparis)* in the Sierra Nevada and adjacent Nevada as recorded in Tertiary fossil deposits. *Note:* Flora 8 is a pollen site, interpretations are based on a very small number of grains, and species identifications may be incorrect. This site should be treated with question.

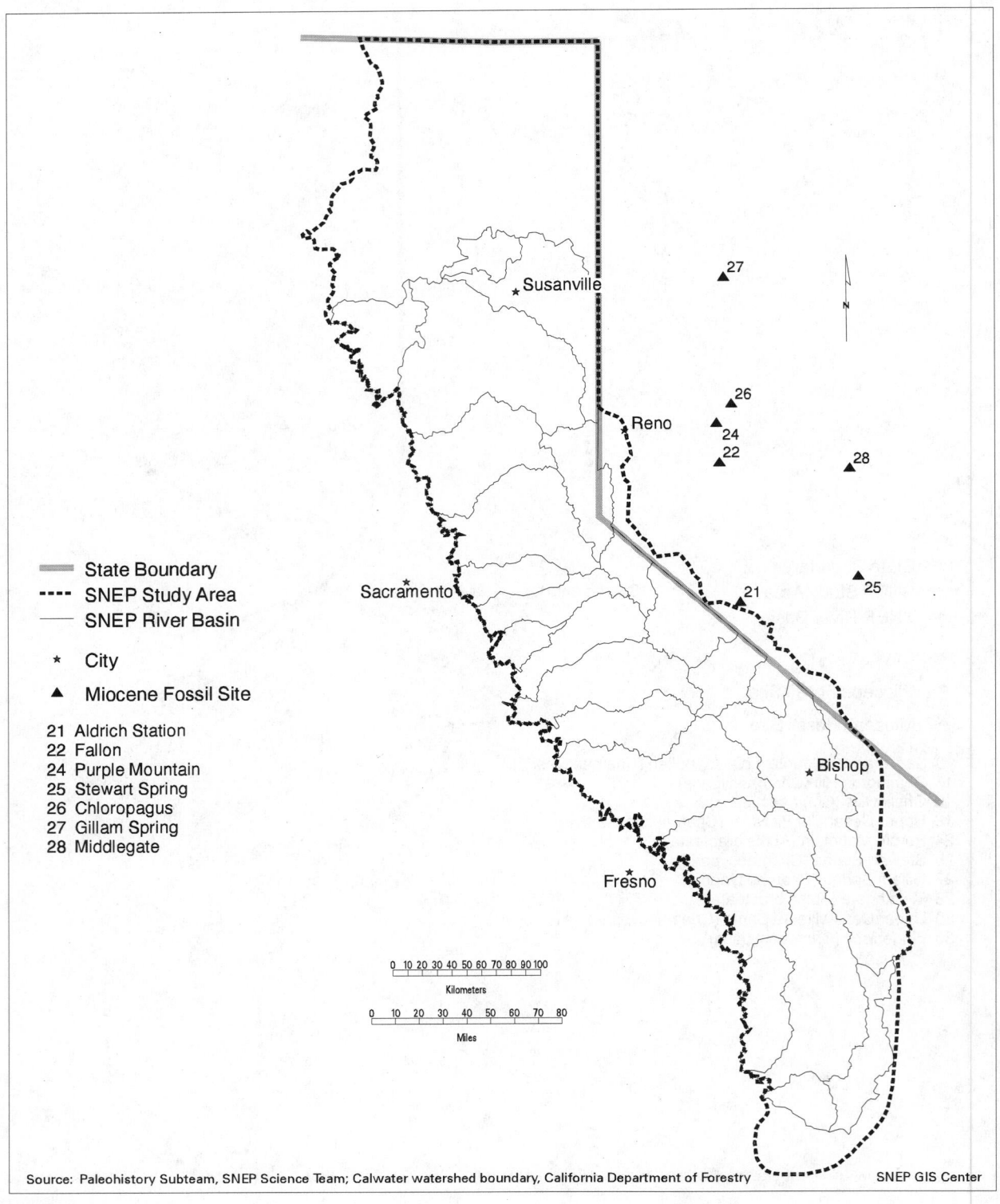

State Boundary
SNEP Study Area
SNEP River Basin

★ City
▲ Miocene Fossil Site

21 Aldrich Station
22 Fallon
24 Purple Mountain
25 Stewart Spring
26 Chloropagus
27 Gillam Spring
28 Middlegate

0 10 20 30 40 50 60 70 80 90 100
Kilometers

0 10 20 30 40 50 60 70 80
Miles

Source: Paleohistory Subteam, SNEP Science Team; Calwater watershed boundary, California Department of Forestry   SNEP GIS Center

FIGURE 5.20

Distribution of Brewer's spruce *(Picea* cf. *breweriana)* in the Sierra Nevada and adjacent Nevada as recorded in Tertiary fossil deposits.

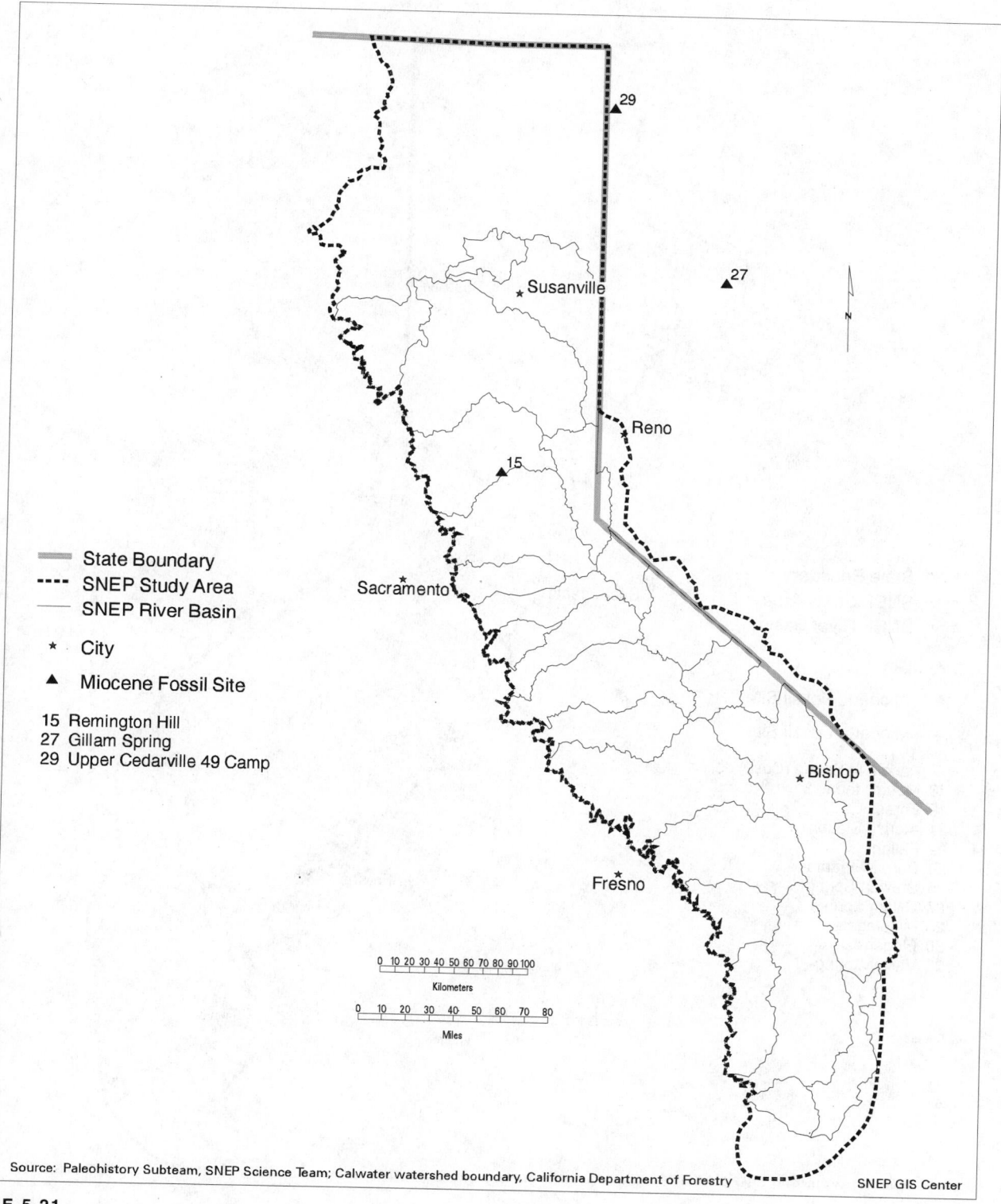

State Boundary
SNEP Study Area
SNEP River Basin
★   City
▲   Miocene Fossil Site

15 Remington Hill
27 Gillam Spring
29 Upper Cedarville 49 Camp

Susanville

Reno

Sacramento

Bishop

Fresno

0 10 20 30 40 50 60 70 80 90 100
Kilometers

0 10 20 30 40 50 60 70 80
Miles

Source: Paleohistory Subteam, SNEP Science Team; Calwater watershed boundary, California Department of Forestry

SNEP GIS Center

FIGURE 5.21

Distribution of coast redwood *(Sequoia* cf. *sempervirens)* in the Sierra Nevada and adjacent Nevada as recorded in Tertiary fossil deposits.

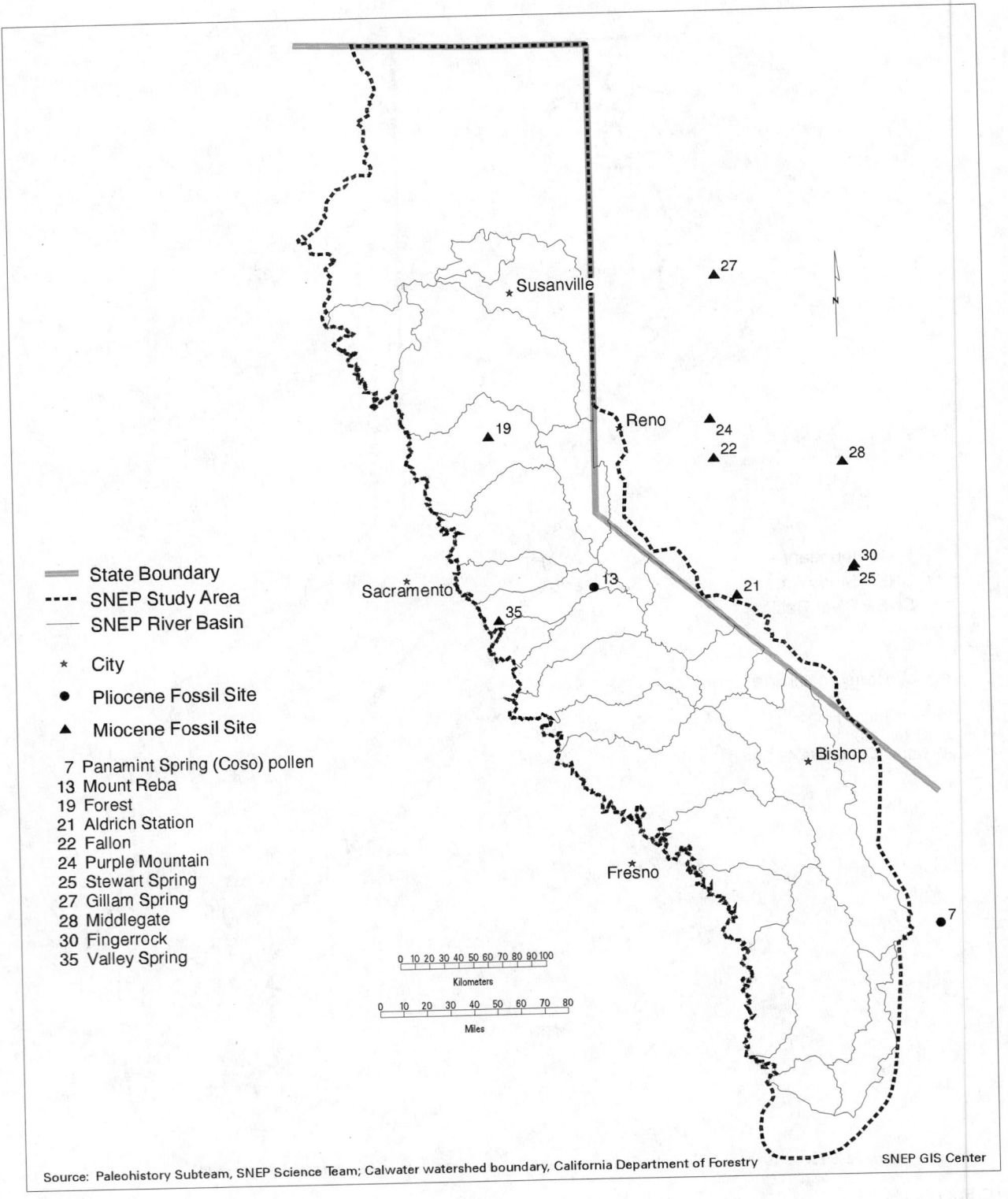

Legend:

━━━ State Boundary
----- SNEP Study Area
—— SNEP River Basin

★ City
● Pliocene Fossil Site
▲ Miocene Fossil Site

7 Panamint Spring (Coso) pollen
13 Mount Reba
19 Forest
21 Aldrich Station
22 Fallon
24 Purple Mountain
25 Stewart Spring
27 Gillam Spring
28 Middlegate
30 Fingerrock
35 Valley Spring

Source: Paleohistory Subteam, SNEP Science Team; Calwater watershed boundary, California Department of Forestry   SNEP GIS Center

FIGURE 5.22

Distribution of coast live oak *(Quercus* cf. *chrysolepis)* in the Sierra Nevada and adjacent Nevada as recorded in Tertiary fossil deposits. *Note:* Flora 7 is a pollen site, interpretations are based on a very small number of grains, and species identifications may be incorrect. This site should be treated with question.

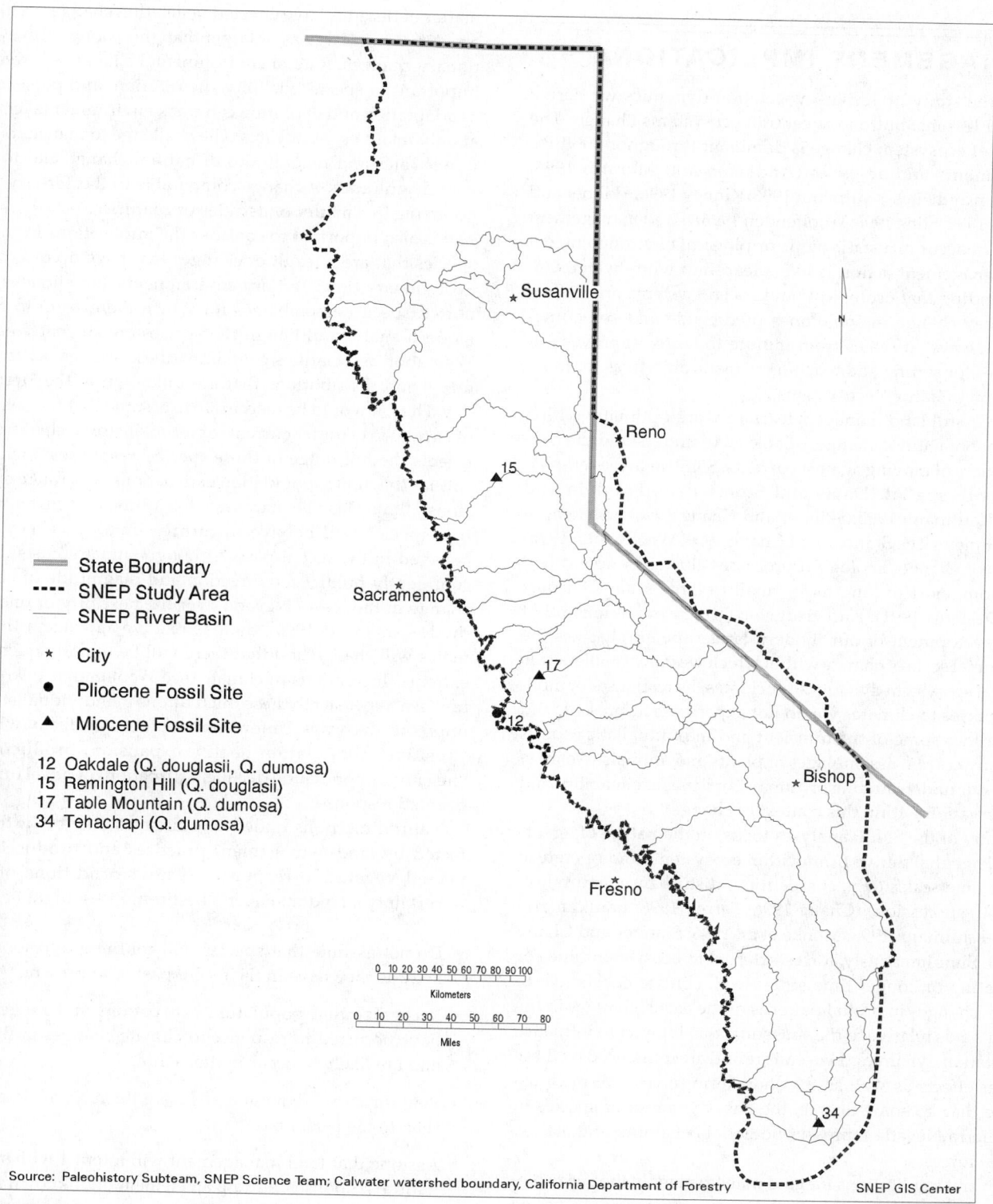

Legend:

State Boundary
SNEP Study Area
SNEP River Basin

★ City
● Pliocene Fossil Site
▲ Miocene Fossil Site

12 Oakdale (Q. douglasii, Q. dumosa)
15 Remington Hill (Q. douglasii)
17 Table Mountain (Q. dumosa)
34 Tehachapi (Q. dumosa)

Source: Paleohistory Subteam, SNEP Science Team; Calwater watershed boundary, California Department of Forestry   SNEP GIS Center

FIGURE 5.23

Distribution of blue oak *(Quercus douglasii)* and scrub oak *(Q. dumosa)* in the Sierra Nevada as recorded in Tertiary fossil deposits.

MANAGEMENT IMPLICATIONS

From the study of Tertiary vegetation dynamics we derive several lessons, but one is central: ecosystems change. The theme of ecosystem change is dominant throughout SNEP's assessments and analyses (Anderson and Moratto 1996; Franklin and Fites-Kaufmann 1996; Kinney 1996; Skinner and Chang 1996; Stine 1996; Woolfenden 1996). Land management does not occur in a static biotic or physical environment; every management action is like a leap into white water. Understanding that ecological (and social) systems are in flux, that they change in sometimes predictable and sometimes chaotic ways, at scales from minute to large, improves our chances for setting and reaching realistic objectives for management of natural ecosystems.

Successful land managers learn to work optimally within the flow of natural change, or at least to understand the consequences of moving against currents. SNEP analyses emphasize both spatial (Davis and Stoms 1996; Franklin and Fites-Kaufmann 1996; Skinner and Chang 1996) and temporal (Kinney 1996; Skinner and Chang 1996; Woolfenden 1996) scales of change. Ecological processes cluster in nested levels, from short to long term, small to large scale (Delcourt and Delcourt 1991). Knowledge of Tertiary evolution and of the development of our modern landscape emphasizes the millennial scale of change within which we view smaller-scale natural ecosystem dynamics, elucidates the responses of modern species to climates we do not experience at present, and underlines some of the transient and individualistic aspects of many current assemblages of plants and animals. We must place our understanding of human impacts, sustainability, and conservation within this context of change.

Many of the SNEP analyses focus on the nature of agents of change that derive from within ecosystems and operate at short time scales (e.g., competition, disturbance, natural selection, succession) (Chang 1996; Ferrell 1996; Franklin and Fites-Kaufmann 1996; Menke et al. 1996; Skinner and Chang 1996). Simultaneously, forces external to ecosystems and/or operating on longer time scales (e.g., climate effects like El Niño, changes in orbital patterns of the earth, changes in the earth's axis relative to the sun, sunspots) interact to influence vegetation dynamics. Taxa endure and are influenced by these larger effects as well as the short-term forces. We must assume that, to some degree, the past dynamics of species in the Sierra Nevada represent potential behaviors of these external forces.

Several implications for land management in the Sierra Nevada derive from this chapter. The most important is a scale lesson and derives from a comparison of the rate and magnitude of change that humans impose with the rate and magnitude of change on an evolutionary scale. Although 65 million years have witnessed drastic changes in Sierran ecosystems, taken at a millennial or centennial (human) scale the pace of evolutionary change is very slow. Humans, by contrast, in a matter of decades have effected vegetation changes in the Sierra Nevada as large as or larger than these long-term evolutionary changes. Rate of environmental change is extremely important to species viability, distribution, and persistence, as adaptation and evolution can track environmental change at only relatively slow rates. The challenge to managers is to mimic rates and magnitudes of natural change and to prevent magnitudes of change comparable to the Tertiary from occurring in a matter of decades or centuries.

It is also important to consider the implications for native species that are a result of changes that have occurred over evolutionary time. Tertiary environments and climates supported species assemblages for which there seem to be no modern analogs, although the species are currently native. Vegetation associations containing these species occurred in mixes and distributions that are unknown in the Sierra today. Thus, it would be inaccurate to assume that the behavior of current taxa under current (or recent historic) climates represents the full range of those species' responses (including species dynamics, population extirpations, and range expansions). It may also be inaccurate to assume that currently native species will behave in future climates as they have behaved in the recent past. Although climatologists do not confidently predict the direction and magnitude of climate change in the Sierra Nevada over the next fifty or one hundred years (Stine 1996), there is relative agreement that climates will change and that there will be more frequent and extreme fluctuations in climate (see Woolfenden 1996). Sierran taxa may react to these fluctuations in individualistic and unpredictable ways. Unforeseen responses are likely, whether "positive" (population health, expansion, productivity), "negative" (population decline, extirpation) or novel (unprecedented response).

Natural reactions under new climates may be further affected by land-management practices and produce unexpected vegetation responses. Under conditions of such uncertainty a land manager's best responses might be:

- Do not assume that species will continue to respond exactly as they have in the recent past or at present.

- Recognize that population extirpations and range shifts have occurred in response to climate changes in the past and are likely to occur in the future.

- Plan for more frequent and severe fluctuations in climate than in the recent past.

- Assume that land management will interact with species' natural responses, adding an element of unknown magnitude.

Maintaining diverse, healthy forests with conditions favoring resilience to unpredictable but changing future climates and management regimes is the most appropriate management response. Planning that depends on landscapes reaching precise vegetation targets is likely to fail. Management programs that build flexibility, reversibility, and alternative pathways are more likely to succeed in uncertain futures.

ACKNOWLEDGMENTS

I thank David Adam and Michael Diggles, both of U.S. Geological Survey, Menlo Park, CA; Owen Davis, University of Arizona, Tucson; David Parsons, U.S. Forest Service, Missoula, MT; Howard Schorn, University of California, Berkeley; and Carl Skinner, U.S Forest Service, Redding, CA, for careful, incisive, and thoughtful reviews of this chapter.

REFERENCES

Adam, D. P. 1973. Early Pleistocene pollen spectra near Lake Tahoe, California. *Journal of Research of the United States Geological Survey* 1:691–93.

Adam, D. P., J. P. Bradbury, H. J. Rieck, and A. M. Sarna-Wojcicki. 1990. Environmental changes in the Tule Lake Basin, Siskiyou and Modoc Counties, California from 3 to 2 million years before present. *United States Geological Survey Bulletin* 1933:1–13.

Adam, D. P., A. M. Sarna-Wojcicki, H. J. Rieck, J. P. Bradbury, W. E. Dean, and R. M. Forester. 1989. Tulelake, California: The last 3 million years. *Palaeogeography, Palaeoclimatology, Palaeoecology* 72:89–103.

Anderson, M. K., and M. J. Moratto. 1996. Native American land-use practices and ecological impacts. In *Sierra Nevada Ecosystem Project: Final report to Congress*, vol. II, chap. 9. Davis: University of California, Centers for Water and Wildland Resources.

Anderson, R. S. 1990. Modern pollen rain within and adjacent to two giant sequoia (*Sequoiadendron giganteum*) groves, Yosemite and Sequoia National Parks, California. *Canadian Journal of Forest Research* 20:1289–305.

Armentrout, J. M., M. R. Cole, and H. J. Terbest, eds. 1979. *Cenozoic paleogeography of the western United States*. Pacific Coast Paleogeography Symposium 3. Palo Alto, CA: Society of Economic Paleontologists and Mineralogists, Pacific Section.

Axelrod, D. I. 1939. *A Miocene flora from the western border of the Mojave Desert*. Publication 476. Washington, DC: Carnegie Institute of Washington.

———. 1940. The Pliocene Esmeralda flora of westcentral Nevada. *Washington Academy of Science Journal* 30:163–74.

———. 1944a. The Oakdale flora. In *Pliocene floras of California and Oregon*, edited by R. W. Chaney, 147–66. Publication 553. Washington, DC: Carnegie Institute of Washington.

———. 1944b. The Pliocene sequence in central California. In *Pliocene floras of California and Oregon*, edited by R. W. Chaney, 207–24. Publication 553. Washington, DC: Carnegie Institute of Washington.

———. 1956. Mio-Pliocene floras from west-central Nevada. *University of California Publications in Geological Science*, vol. 33.

———. 1958. The Pliocene Verdi flora of western Nevada. *University of California Publications in Geological Science* 34:61–160.

———. 1962. A Pliocene *Sequoiadendron* forest from western Nevada. *University of California Publications in Geological Science* 39:195–268.

———. 1966. The Eocene Copper Basin flora of northwestern Nevada. *University of California Publications in Geological Science* 59:1–124.

———. 1968. Tertiary floras and topographic history of the Snake River Basin, ID. *Geological Society of America Bulletin* 79:713–34.

———. 1973. History of the Mediterranean ecosystem in California. In *Mediterranean type ecosystems, origin and structure*, edited by F. di Castri and H. Mooney, 225–77. New York: Springer-Verlag.

———. 1976. Evolution of the Santa Lucia fir ecosystem. *Annals of the Missouri Botanical Garden* 63:24–41.

———. 1977. Outline history of California vegetation. In *Proceedings of terrestrial vegetation of California*, edited by M. Barbour and J. Major. Special Publication 9. Berkeley: California Native Plant Society.

———. 1979. Desert vegetation, its age and origin. In *Arid land plant resources*, edited by J. R. Goodin and D. K. Northington, 1–72. Lubbock, TX: International Center for Arid and Semi-Arid Land Studies.

———. 1986. Cenozoic history of some western American pines. *Annals of the Missouri Botanical Garden* 73:565–641.

———. 1995. The Miocene Purple Mountain flora for western Nevada. *University of California Publications in Geological Science* 139: 1–62.

Axelrod, D. I., and H. P. Bailey. 1969. Paleotemperature analysis of Tertiary floras. *Palaeogeography, Palaeoclimatology, Palaeoecology* 6:163–95.

Axelrod, D. I., and P. Raven. 1985. Origins of the Cordilleran flora. *Journal of Biogeography* 12:21–47.

Axelrod, D. I., and H. E. Schorn. 1994. The 15 Ma floristic crisis at Gillam Spring, Washoe County, northwestern Nevada. *PaleoBios* 16:1–10.

Axelrod, D. I., and W. S. Ting. 1960. Late Pliocene floras east of the Sierra Nevada. *University of California Publications in Geological Science* 39:1–118.

Bailey, I. W., and E. W. Sinnott. 1916. The climatic distribution of certain types of angiosperm leaves. *American Journal of Botany* 3:24–39.

Barnett, J. 1983. Palynology of the Weaverville Formation, Northwestern California. In *Proceedings of palynology of tertiary floras of western North America*, edited by L. H. Fisk, 11. San Francisco: American Association of Stratigraphic Palynologists.

Botkin, D. B. 1990. *Discordant harmonies: A new ecology for the twenty-first century*. New York: Oxford University Press.

Bryson, R. A. 1985. On climatic analogs in paleoclimatic reconstruction. *Quaternary Research* 23:275–86.

Chaney, R. W., ed. 1944. *Pliocene floras of California and Oregon*. Publication 553. Washington, DC: Carnegie Institute of Washington.

———. 1959. *Miocene floras of the Columbia Basin*. Publication 617 (part 1). Washington, DC: Carnegie Institute of Washington.

Chang, C. 1996. Ecosystem responses to fire and variations in fire regimes. In *Sierra Nevada Ecosystem Project: Final report to Congress*, vol. II, chap. 39. Davis: University of California, Centers for Water and Wildland Resources.

Condit, C. B. 1944. The Table Mountain flora. In *Pliocene floras of California and Oregon*, edited by R. W. Chaney, 57–90. Publication 553. Washington, DC: Carnegie Institute of Washington.

Dalrymple, G. B., and M. A. Lanphere. 1969. *Potassium-argon dating: Principles, techniques, and applications to geochronology*. San Francisco: W. H. Freeman Co.

Davis, F. W., and D. M. Stoms. 1996. Sierran vegetation: A gap analysis. In *Sierra Nevada Ecosystem Project: Final report to Congress,* vol. II, chap. 23. Davis: University of California, Centers for Water and Wildland Resources.

Delcourt, H. R., and P. A. Delcourt. 1991. *Quaternary ecology.* New York: Chapman and Hall.

Duffield, W. A., C. R. Bacon, and G. B. Dalrymple. 1980. Late Cenozoic volcanism, geochronology, and structure of the Coso Range, Inyo County, California. *Journal of Geophysical Research* 85:2381–404.

Ferrell, G. T. 1996. The influence of insect pests and pathogens on Sierra forests. In *Sierra Nevada Ecosystem Project: Final report to Congress,* vol. II, chap. 45. Davis: University of California, Centers for Water and Wildland Resources.

Fiero, B. 1991. *Geology of the Great Basin.* Reno: University of Nevada Press.

Frankel, O. H., and M. E. Soule. 1981. *Conservation and evolution.* Cambridge: Cambridge University Press.

Franklin, J. F., and J. A. Fites-Kaufmann. 1996. Analysis of late successional forests. In *Sierra Nevada Ecosystem Project: Final report to Congress,* vol. II, chap. 21. Davis: University of California, Centers for Water and Wildland Resources.

Friis, E. M., W. G. Chalone, and P. R. Crane, eds. 1987. *The origin of angiosperms and their biological consequences.* Cambridge: Cambridge University Press.

Greenwood, D. R. 1992. Taphonomic constraints on foliar physiognomic interpretations of late Cretaceous and Tertiary palaeoclimates. *Reviews of Palynology and Palaeobotany* 71:149–90.

Gregory, K. M. 1994. Palaeoclimate and palaeoelevation of the 35 Ma Florissant flora, Front Range, Colorado. *Palaeoclimates* 1:23–57.

Grumbine, R. E. 1994. What is ecosystem management? *Conservation Biology* 8:27–38.

Hansen, A. J., F. J. Spies, F. J. Swanson, and J. L. Ohmann. 1991. Conserving biodiversity in managed forests: Lessons from natural forests. *BioScience* 41:382–92.

Huber, N. K. 1981. Amount and timing of late Cenozoic uplift and tilt of the central Sierra Nevada, California: Evidence from the upper San Joaquin River basin. Professional Paper 1197. Washington, DC: U.S. Geological Survey.

Kaufman, W. 1993. How nature really works. *American Forests* 99:17–19, 59–61.

Kinney, W. C. 1996. Conditions of rangelands before 1905. In *Sierra Nevada Ecosystem Project: Final report to Congress,* vol. II, chap. 3. Davis: University of California, Centers for Water and Wildland Resources.

Knowlton, F. H. 1911. Flora of the auriferous gravels of California. In *The Tertiary gravels of the Sierra Nevada of California,* edited by W. Lindgren, 1–28. Menlo Park, CA: U.S. Geological Survey.

LaMotte, R. S. 1935. Studies in Tertiary paleobotany, IV: The Upper Cedarville flora of northwestern Nevada and adjacent California. Ph.D. diss., Department of Paleontology, University of California, Berkeley.

———. 1936. *The Upper Cedarville flora of northwestern Nevada and adjacent California.* Publication 455. Washington, DC: Carnegie Institute of Washington.

Leopold, E. B. 1983. Presentation at symposium, Palynology of Tertiary floras of western North America, American Association of Stratigraphic Palynologists, San Francisco.

———. 1984. Pollen identifications from the Eocene Chalk Bluffs flora, California. *Palynology* 8:8.

MacGinitie, H. D. 1941. *A middle Eocene flora from the central Sierra Nevada.* Publication 534. Washington, DC: Carnegie Institute of Washington.

———. 1962. The Kilgore flora. *University of California Publication Bulletin, Department of Geological Sciences* 35:67–158.

McGowran, B. 1990. Fifty million years ago. *American Scientist* 78: 30–39.

McKee, E. H. 1979. Ash-flow calderas: Their genetic relationship to ore deposits in Nevada. *Geological Society of America Special Paper* 180:205–11.

Menke, J., C. Davis, and P. Beasley 1996. Rangeland assessment. In *Sierra Nevada Ecosystem Project: Final report to Congress,* vol. III. Davis: University of California, Centers for Water and Wildland Resources.

Millar, C. I. 1993. Impact of the Eocene on the evolution of pines. *Annals of the Missouri Botanical Garden* 8:471–98.

Millar, C. I., and B. B. Kinloch. 1991. Taxonomy, phylogeny, and coevolution of pines and their stem rusts. *Proceedings of IUFRO rusts of pine,* edited by Y. Hiratsuka et al. Information Report NOR-X-317. Banff, Alberta: Forestry Canada.

Minckley, W. L., D. A. Hendrickson, and C. E. Bond. 1986. Geography of western North American fishes: Description and relationships to intracontinental tectonism. In *The zoogeography of North American fishes,* edited by C. H. Hocutt and E. O. Wiley, 519–613. New York: Wiley-Interscience.

Odin, G. S., ed. 1982. *Numerical dating in stratigraphy.* New York: John Wiley.

Odin, G. S., and D. Curry. 1985. The Paleogene time scale: Radiometric dating versus magnetostratigraphic approach. *Journal of the Geological Society of London* 142:1179–88.

Overpeck, J. T. 1985. Quantitative interpretation of fossil pollen spectra: Dissimilarity coefficients and the method of modern analogs. *Quaternary Research* 23:76–86.

Owens Lake Core Study Team. 1995. An 800,000-year paleoclimatic record from core OL-2, Owens Lake, southeast California. *Proceedings of report of 1994 workshop on the correlation of marine and terrestrial records of climate changes in the western United States,* edited by D. P. Adam, J. P. Bradbury, W. E. Dean, J. V. Gardner, and A. M. Sarna-Wojcicki. Open File Report 95-34. Washington, DC: U.S. Geological Survey.

Parrish, J. T. 1987. Global palaeogeography and palaeoclimate of the late Cretaceous and early Tertiary. In *The origins of angiosperms and their biological consequences,* edited by F. M. Fris, W. G. Chaloner, and P. R. Crane, 51–73. Cambridge: Cambridge University Press.

Potbury, S. S. 1935. *The La Porte flora of Plumas County, California.* Publication 465. Washington, DC: Carnegie Institute of Washington.

Prentice, I. C. 1985. Pollen representation, source area, and basin size: Toward a unified theory of pollen analysis. *Quaternary Research* 23:76–86.

Prothero, D.O. 1994. The Eocene-Oligocene transition. Vol. of *Critical moments in paleobiology and earth history,* edited by D. J. Bottjer and R. K. Bambach. New York: Columbia University Press.

Raven, P. H., and D. I. Axelrod. 1978. *Origin and relationships of the California flora.* University of California Publications in Botany 72. Berkeley and Los Angeles: University of California Press.

Reiswig, K. N. 1983. Comparisons of the megaflora and palynoflora of the Chuckanut Formation, northwest Washington. *Proceedings of palynology of Tertiary floras of western North America,* edited by L. H. Fisk, 10. San Francisco: American Association of Stratigraphic Palynologists.

Ruddiman, W. F., and J. E. Kutzbach. 1989. Forcing of the late Cenozoic Northern Hemisphere climate by plateau uplift in southern Asia and the American West. *Western Journal of Geophysical Research* 91 (D15): 18409–27.

Savin, S. 1977. The history of the earth's surface temperature during the past 100 million years. *Annual Review of Earth Planetary Science* 5:319–56.

Schorn, H. E. 1984. Palynology of the late middle Miocene sequence, Stewart Valley. *Palynology* 8:259–60.

Shackleton, N. J., and N. Opdyke. 1977. Oxygen isotope and paleomagnetic evidence for early Northern Hemisphere glaciation. *Nature* 270:216–19.

Shevock, J. R. 1996. Status of rare and endemic plants. In *Sierra Nevada Ecosystem Project: Final report to Congress*, vol. II, chap. 24. Davis: University of California, Centers for Water and Wildland Resources.

Skinner, C. N., and C. Chang. 1996. Fire regimes, past and present. In *Sierra Nevada Ecosystem Project: Final report to Congress*, vol. II, chap. 38. Davis: University of California, Centers for Water and Wildland Resources.

Spicer, R. A. 1989. The formation and interpretation of plant megafossil assemblages. In *Advances in Botanical Research*, edited by J. Callow, 96–191. New York: Academic Press.

Steiger, R. H., and E. Jager. 1977. Subcommission on geochronology: Convention on the use of decay constants in geo- and cosmochronology. *Earth and Planetary Science Letters* 36:359–62.

Stewart, J. H. 1978. *Geologic map of Nevada*. Reno: U.S. Geological Survey.

Stine, S. 1996. Climate, 1650–1850. In *Sierra Nevada Ecosystem Project: Final report to Congress*, vol. II, chap. 2. Davis: University of California, Centers for Water and Wildland Resources.

Swisher, C. C., and D. R. Prothero. 1990. Single-crystal 40Ar–39Ar dating of the Eocene-Oligocene transition in North America. *Science* 249:760–62.

Thompson, R. S. 1991. Pliocene environments and climates in the western United States. *Quaternary Science Reviews* 10:115–32.

Tiffney, B. M. 1985. Perspectives on the origin of the floristic similarity between eastern Asia and eastern North America. *Journal of the Arnold Arboretum* 66:73–94.

Vita-Finza, C. 1973. *Recent earth history*. London: Macmillan.

Winograd, I. J., B. J. Szabo, T. B. Coplen, A. C. Riggs, and P. T. Kolesar. 1985. Two-million-year record of deuterium depletion in Great Basin ground waters. *Science* 227:519–22.

Wolfe, J. A. 1964. *Miocene floras from Fingerrock Wash, southwestern Nevada*. Professional Paper 454-N. Washington, DC: U.S. Geological Survey.

———. 1969. Neogene floristic and vegetational history of the Pacific Northwest. *Madrono* 20:83–110.

———. 1971. Tertiary climatic fluctuations and methods of analysis of Tertiary floras. *Palaeogeography, Palaeoclimatology, Palaeoecology* 9:27–57.

———. 1978. A paleobotanical interpretation of Tertiary climates in the Northern Hemisphere. *American Scientist* 66:694–703.

———. 1981. A chronologic framework for Cenozoic megafossil floras of northwestern North America and its relation to marine geochronology. In *Pacific Northwest Cenozoic biostratigraphy*, edited by J. A. Armentrout, 39–46. Special Paper 184. Boulder, CO: Geological Society of America.

———. 1985. Distribution of major vegetational types during the Tertiary. *Geophysical Monographs* 32:357–75.

———. 1987. An overview of the origins of the modern vegetation and flora of the northern Rocky Mountains. *Annals of the Missouri Botanical Garden* 74:785–803.

———. 1990. Estimates of Pliocene precipitation and temperature based on multivariate analysis of leaf physiognomy. *Proceedings of Pliocene climates: Scenarios for global warming*, edited by L. B. Gosnell and R. Z. Poore, 39–42. Open File Report 90-64. Washington, DC: U.S. Geological Survey.

———. 1993. A method of obtaining climatic parameters from leaf assemblages. *United States Geological Survey Bulletin* 2041:1–71.

Wolfe, J. A., and D. M. Hopkins. 1967. Climatic changes recorded by Tertiary land floras in northwestern North America. In *Tertiary correlations and climatic changes in the Pacific*, edited by K. Hatai, 61–76. Tokyo: Eleventh Pacific Science Congress.

Wolfe, J. A., and G. R. Upchurch. 1986. Vegetation, climatic, and floral changes at the Cretaceous-Tertiary boundary. *Nature* 324:148–52.

Wood, H. E., R. W. Chaney, J. Clark, E. H. Colbert, G. L. Jepsen, J. B. Reeside, and C. Stock. 1941. Nomenclature and correlation of the North American continental Tertiary. *Geological Society of America Bulletin* 52:1–48.

Woodburne, M. O., ed. 1987. *Cenozoic mammals of North America: Geochronology and biostratigraphy*. Berkeley and Los Angeles: University of California Press.

Woolfenden, W. B. 1996. Quaternary vegetation history. In *Sierra Nevada Ecosystem Project: Final report to Congress*, vol. II, chap. 4. Davis: University of California, Centers for Water and Wildland Resources.

APPENDIX 5.1

Systematic Composition of the Tertiary Sierra Nevada Floras

The floras are listed in alphabetical order. Species names in parentheses are the closest modern affinities to fossil taxa.

LIST 1

Aldrich Station flora, western central Nevada, mid-Miocene (Axelrod 1956).

Equisetales
  Equisetaceae
    Equisetum alexanderi
Gymnosperms
  Cupressaceae
    Thuja dimorpha (plicata)
  Pinaceae
    Abies concoloroides (concolor)
    Picea magna (polita, neoveitchii)
    Picea sonomensis (breweriana)
    Pinus florissanti (ponderosa)
    Pinus wheeleri (monticola, lambertiana)
    Pseudotsuga sonomensis (menziesii)
    Tsuga sonomensis (heterophylla)
  Taxodiaceae
    Sequoiadendron chaneyi (giganteum)
Angiosperms
  Berberidaceae
    Mahonia marginata (beali)
    Mahonia reticulata (repens)
  Betulaceae
    Alnus smithiana (tenufolia)
  Caprifoliaceae
    Symphoricarpos wassukana (oreophilius)
  Celastraceae
    Pachystima nevadensis (myrsintes)
  Fagaceae
    Quercus hannibali (chrysolepis)
    Quercus simulata (myrsinaefolia)
  Hippocastanaceae
    Aesculus ashleyi (parryi)
  Leguminosae
    Amorpha oblongifolia (californica)
    Sophora spokanensis (japonica)
  Myricaceae
    Comptonia parvifolia (asplenifolia)
  Oleaceae
    Fraxinus acornia (velutina)
  Platanaceae
    Platanus paucidentata (racemosa)
  Rhamnaceae
    Rhamnus precalifornica (californica)
  Rosaceae
    Amelanchier apiculata (utahensis)
    Cercocarpus antiquus (betuloides)
  Salicaceae
    Populus alexanderi (trichocarpa)
    Populus payettensis (angustifolia)
    Populus sonorensis
    Populus subwashoensis (temula, grandidentata)
    Salix knowltonii (lemmonii)
    Salix payettensis (exigua)
  Sapotaceae
    Burmelia beaverana (lanuginosa)
  Ulmaceae
    Ulmus moorei (crassifolia)
    Zelkova nevadensis (serrata)

LIST 2

Alturas flora (macrofossil), northeastern California, late Miocene (Axelrod 1944b).

Angiosperms
  Salicaceae
    Populus alexanderi (trichocarpa)
    Populus payettensis (angustifolia)
    Populus pliotremuloides (tremuloides)
    Salix truckeana (nigra)
    Salix vanorensis (caudata)
    Salix wildcatensis (lasiolepis)
  Ulmaceae
    Ulmus moragensis (speciosa)

LIST 3

Chalk Bluffs flora (macrofossil and pollen), western central Sierra Nevada, California, early Eocene (Leopold 1983; MacGinitie 1941).

Pteridophytes
  Cyatheaceae
    Hemitelia pinnata
  Schizaeaceae
    Lygodium kaulfussi
Gymnosperms
  Cycadaceae
    Zamites californica
Angiosperms
  Aceraceae
    Acer aequidentatum
  Aesclepiadaceae
    Asclepiadites laterita
  Anacardiaceae
    Rhus mixta
  Apocynaceae
    Nerium hinoidea
    Tabernaemontana chrysophylloides

Betulaceae
Alnus peria
Burseraceae
Canarium californicum
Caprifoliaceae
Viburnum variabilis
Celastraceae
Celastrus preangulata
Cercidiphyllaceae
Cercidiphyllum elongatum
Combretaceae
Terminalia estamina
Compositae
Calycites mikanoides
Cornaceae
Cornus kelloggii
Ebenaceae
Diospyros retinervis
Euphorbiaceae
Acalypha aequalis
Mallotus riparius
Fagaceae
Castanopsis longipetiolatum
Quercus distincta
Quercus eoxalapensis
Quercus nevadensis
Hamamelidaceae
Hamamelites voyana
Liquidambar californicum
Icacinaceae
Phytocrene sordida
Juglandaceae
Carya sessilis
Engelhardtia nevadensis
Lauraceae
Cinnamomum acrodromum
Cinnamomum dilleri
Cryptocarya praesamarensis
Laurophyllum fremontensis
Laurophyllum litseaefolia
Neolitsea lata
Persea praelingue
Persea pseudo-carolinensis
Leguminosae
Dalbergia rubra
Desmodium indentum
Inga ionensis
Pongamia ovata
Strongylodon falcata
Vouapa geminifolia
Liliaceae
Smilax labidurommae
Magnoliaceae
Magnolia dayana
Meliaceae
Cedrela eolancifolia
Menispermaceae
Hyperbaena diforma
Moraceae
Artocarpus lessigiana
Ficus densifolia
Ficus goshenensis
Myrtaceae
Calyptranthes myrtifolia
Nymphaeaceae
Nelumbium lacunosum
Nyssaceae
Nyssa californica
Oleaceae
Fraxinus yubaensis
Palmae
Sabalites californicus
Platanaceae
Platanophyllum angustiloba
Platanophyllum angustiloba var. *serrata*
Platanophyllum whitneyi
Platanus appendiculata
Platanus coloradensis
Rhamnaceae
Rhamnidium chaneyi
Rhamnus calyptus
Rhamnus plenus

Rosaceae
Chrysobalanus eoicaco
Vauquelinia exigua
Sabiaceae
Meliosma truncata
Salicaceae
Salix ionensis
Sapindaceae
Cupania oregona
Thouinopsis myricaefolia
Saxifragaceae
Hydrangea californica
Simarubaceae
Ailanthus lesquereuxi
Theaceae
Gordonia egregia
Ulmaceae
Chaetoptelea pseudo-fulva
Vitaceae
Cissus pyriformus

Additions to the Chalk Bluffs flora from pollen identifications (Leopold 1983, 1984):

Gymnosperms
Pinaceae (very small fraction; inblown)
Abies
Picea
Pinus
Angiosperms
Bombacaceae
diverse genera
Juglandaceae
Juglans
Carya
Engelhardtia
Platycarya
Hamamelidaceae
Liquidambar
Icacinaceae
Phytocrene
Melicaceae
Cedrela
Platanaceae
Platanus
Schizaeaceae
Anemia
Tiliaceae
diverse genera

LIST 4

Chalk Hills flora (macrofossil), western central Nevada, late Miocene (Axelrod 1962; Howard Schorn, letter to Constance I. Millar, July 1995).

Gymnosperms
Cupressaceae
Chamaecyparis linguaefolia (*lawsoniana*)
Cupressus/Juniperus
Pinaceae
Abies concoloroides (*concolor*)
Pinus balfouroides (*balfouriana*) (per Axelrod 1986)
Pinus florissanti (*ponderosa*)
Pinus wheeleri (*monticola*)
Pseudotsuga sonomensis (*menziesii*)
Tsuga (*mertensiana*)
Taxodiaceae
Sequoiadendron chaneyi (*giganteum*)
Angiosperms
Berberidaceae
Mahonia reticulata (*nervosa, repens*)
Betulaceae
Alnus smithiana (*tenufolia*)
Betula thor (*papyrifera*)
Ericaceae
Arbutus matthesii (*menziesii*)
Rhododendron gianellana (*occidentale, albiflorum*)

Fagaceae
  Castanopsis sonomensis (chrysophylla)
  Quercus simulata (myrsinaefolia)
Juglandaceae
  Carya bendirei (ovata)
Rhamnaceae
  Ceanothus chaneyi (integerrimus)
  Ceanothus leitchii (velutinus)
  Rhamnus precalifornica (californica)
Rosaceae
  Amelanchier alvordensis (alnifolia)
  Holodiscus idahoensis (microphyllus, dumosus)
  Prunus moragensis (emarginata)
Salicaceae
  Populus eotremuloides (trichocarpa)
  Populus pliotremuloides (tremuloides)
  Populus washoensis (grandidentata, tremula var. davidiana)
  Salix knowltoni (lemmonii)
  Salix laevigatoides (laevigata)
  Salix owyheeana (hookeriana)
Saxifragaceae
  Ribes stanfordianum (sanguineum, nevadense)

LIST 5

Chloropagus flora (macrofossil), western central Nevada, mid-Miocene (Axelrod 1956).

Gymnosperms
  Cupressaceae
    Juniperus nevadensis (utahensis, californica)
    Thuja dimorpha (plicata)
  Pinaceae
    Abies concoloroides (concolor)
    Picea sonomensis (breweriana)
    Pinus wheeleri (lambertiana)
  Taxaceae
    Torreya nancyana (californica)
Angiosperms
  Berberidaceae
    Mahonia marginata (beali)
    Mahonia reticulata (repens)
  Fagaceae
    Quercus hannibali (chrysolepis)
    Quercus simulata (myrsinaefolia)
    Quercus wislizenoides (wislizenii)
  Leguminosae
    Cercis carsoniana
    Robinia californica (neo-mexicana)
  Rosaceae
    Amelanchier apiculata (utahensis)
    Cercocarpus linearifolium (ledifolius)
  Salicaceae
    Populus alexanderi (trichocarpa)
    Populus payettensis (angustifolia)
    Populus subwashoensis (tremula, grandidentata)
    Salix knowltoni (lemmonii)
    Salix payettensis (exigua)
    Salix wildcatensis (lasiolepis)
  Typhaceae
    Typha lesquereuxi (latifolia)

LIST 6

Crowley Point (Coso) flora (pollen), eastern central California, late Pliocene (Axelrod and Ting 1960). *Note:* This flora is a pollen site, interpretations are based on a very small number of grains (67), and species identifications may be incorrect. This flora should be treated with question.

Gymnosperms
  Cupressaceae
    Juniperus occidentalis
    Juniperus osteosperma

Pinaceae
    Pinus aristata or *flexilis*
    Pinus lambertiana
    Pinus cf. *monophylla*
    Pinus ponderosa
    Pseudotsuga taxifolia
Angiosperms
  Berberidaceae
    Mahonia sp.
  Caprifoliaceae
    Symphoricarpos oreophilus
  Cornaceae
    Cornus californica
    Cornus sessilus
  Fagaceae
    Quercus kelloggii
  Rhamnaceae
    Ceanothus cordulatus
    Ceanothus integerrimus
    Rhamnus crocea
  Rosaceae
    Cercocarpus betuloides
    Rosa sp.
  Salicaceae
    Populus trichocarpa
    Salix exigua
    Salix lasiandra
    Salix lasiolepis
  Saxifragaceae
    Ribes nevadense
    Ribes viscosissimum
  Sterculiaceae
    Fremontia californica
  Ulmaceae
    Ulmus alata

LIST 7

Darwin Summit (Coso) flora (pollen), eastern central California, late Pliocene (Axelrod and Ting 1960). *Note:* This flora is a pollen site, interpretations are based on a very small number of grains (17), and species identifications may be incorrect. This flora should be treated with question.

Gymnosperms
  Cupressaceae
    Calocedrus decurrens
    Juniperus osteosperma
  Pinaceae
    Pinus aristata or *flexilis*
    Pinus cf. *monophylla*
    Pinus ponderosa
  Taxodiaceae
    Sequoiadendron giganteum
Angiosperms
  Betulaceae
    Betula fontinalis
  Caprifoliaceae
    Symphoricarpos albus
  Cornaceae
    Cornus nuttallii
  Juglandaceae
    Pterocarya sp.
  Rhamnaceae
    Ceanothus cordulatus
  Salicaceae
    Salix lasiolepis
  Saxifragaceae
    Ribes roezlii

LIST 8

Fallon flora (macrofossil), western central Nevada, mid-Miocene (Axelrod 1956).

Gymnosperms
    Cupressaceae
        Juniperus nevadensis (utahensis, californica)
        Thuja dimorpha (plicata)
    Pinaceae
        Abies concoloroides (concolor)
        Picea sonomensis (breweriana)
        Pinus florissanti (ponderosa)
    Taxaceae
        Torreya nancyana (californica)
    Taxodiaceae
        Sequoiadendron chaneyi (giganteum)
Angiosperms
    Berberidaceae
        Mahonia marginata (beali)
        Mahonia reticulata (repens)
    Betulaceae
        Betula thor (papyrifera, occidentalis)
    Ericaceae
        Arbutus matthesii (menziesii)
    Fagaceae
        Quercus hannibali (chrysolepis)
        Quercus simulata (myrsinaefolia)
        Quercus wislizenoides (wislizenii)
    Leguminosae
        Sophora spokanensis (japonica)
    Oleaceae
        Fraxinus alcorni (velutina)
    Rosaceae
        Cercocarpus linearifolius (ledifolius)
    Salicaceae
        Populus eotremuloides (trichocarpa)
        Populus subwashoensis (tremula, grandidentata)
        Salix knowltoni (lemmonii)
        Salix payettensis (exigua)
    Typhaceae
        Typha lesquereuxi (latifolia)

LIST 9

Fingerrock flora (macrofossil), southwestern Nevada, mid-Miocene (Wolfe 1964).

Gymnosperms
    Cupressaceae
        Chamaecyparis nootkatensis
    Pinaceae
        Abies (concolor)
        Picea magna
        Pinus (monticola)
        Pinus (ponderosa)
    Taxodiaceae
        Glyptostrobus sp.
Angiosperms
    Aceraceae
        Acer bolanderi
        Acer (macrophyllum)
    Berberidaceae
        Mahonia reticulata (repens)
    Betulaceae
        Alnus relata
        Betula thor
    Cyperaceae
        Cyperacites sp.
    Ebenaceae
        Diospyros
    Ericaceae
        Arbutus traini (menziesii)
    Fagaceae
        Quercus (chrysolepis)
        Quercus pseudolyrata
        Quercus simulata (chrysolepis? myrsinaefolia)

    Juglandaceae
        Carya bendirei (tonkinensis)
    Platanaceae
        Platanus bendirei
    Salicaceae
        Populus lindgreni (heterophylla)
    Ulmaceae
        Ulmus newberryi
        Zelkova oregoniana

LIST 10

Forest flora (macrofossil), southern central Sierra Nevada, California, late Miocene (Chaney 1944; Knowlton 1911).

Angiosperms
    Aceraceae
        Acer arcticum
        Acer sp.
    Fagaceae
        Quercus pseudo-chrysolepis (chrysolepis)
        Quercus steenstrupiana?
        Quercus transgressus
    Liliaceae
        Smilax diforma (rotundifolia)
    Platanaceae
        Platanus paucidentata (racemosa)
    Salicaceae
        Salix hesperia (lasiandra)
    Ulmaceae
        Ulmus californica (americana)

LIST 11

Gillam Spring flora, northwestern Nevada, mid-Miocene (Axelrod and Schorn 1994). Species lists are by age of florules (upper = youngest; lower = oldest).

Upper Gillam Spring Florule
Equisetales
    Equisetaceae
        Equisetum
Gymnosperms
    Cupressaceae
        Calocedrus
        Chamaecyparis
    Pinaceae
        Abies
        Picea lahontense
        Picea magna
        Picea sonomensis
        Pinus (contorta)
        Pinus (monticola)
        Pinus (three-needle)
        Pinus tiptonia
    Taxodiaceae
        Sequoia
Angiosperms
    Aceraceae
        Acer chaneyi
        Acer scottiae
    Berberidaceae
        Mahonia reticulata
    Betulaceae
        Betula smithiana
    Ericaceae
        Arbutus idahoensis
        Vaccinium (sonomensis)
    Fagaceae
        Quercus hannibali
        Quercus simulata
    Juglandaceae
        Pterocarya
    Oleaceae
        Fraxinus

Rosaceae
    Crataegus
    Prunus chaneyi
    Robinia
Salicaceae
    Salix
    Salix boisensis
    Salix pelviga
Typhaceae
    Typha
Ulmaceae
    Ulmus
Cyperacites?
Eugenia nevadensis?

Middle Gillam Spring Florule
Gymnosperms
    Cupressaceae
        Chamaecyparis
    Pinaceae
        Picea lahontense
        Picea sonomensis
        Pinus (three-needle)
        Pinus tiptonia
Angiosperms
    Aceraceae
        Acer osmonti
        Acer schorni
    Berberidaceae
        Mahonia reticulata
    Betulaceae
        Alnus latahensis
        Betula ashley
        Betula smithiana
    Ericaceae
        Arbutus idahoensis
    Fagaceae
        Lithocarpus klamathensis
        Quercus (ovata)
        Quercus (oblonga)
    Platanaceae
        Platanus
    Rosaceae
        Cercocarpus antiquus
    Salicaceae
        Salix (two spp.)
    Typhaceae
        Typha
    Ulmaceae
        Ulmus
Pieris?

Lower Gillam Spring Florule
Gymnosperms
    Cupressaceae
        Chamaecyparis sp.
        Thuja? sp.
    Pinaceae
        Picea magna
        Pinus tiptonia
Angiosperms
    Aceraceae
        Acer (bolanderi)
        Acer chaneyi
        Acer (glabrum)
        Acer (nigra)
        Acer tyrrelli
    Berberidaceae
        Mahonia
    Betulaceae
        Betula vera
    Ericaceae
        Arbutus idahoensis
        Vaccinium
    Juglandaceae
        Fagus washoensis
        Lithocarpus
        Pterocarya
        Quercus eoprinus
        Quercus hannibali (chrysolepis)
        Quercus simulata

Oleaceae
    Fraxinus
Platanaceae
    Platanus
Rhamnaceae
    Rhamnus
Rosaceae
    Amelanchier
    Prunus cassianna
Salicaceae
    Populus eotremuloides (tremuloides)
    Salix
Saxifragaceae
    Hydrangea?
Typhaceae
    Typha
Ulmaceae
    Zelkova

LIST 12

Haiwee (Coso) flora (pollen), southeastern Sierra Nevada, California, late Pliocene (Axelrod and Ting 1960). *Note:* This flora is a pollen site, interpretations are based on a very small number of grains (123), and species identifications may be incorrect. This flora should be treated with question.

Gymnosperms
    Cupressaceae
        Calocedrus decurrens
    Pinaceae
        Abies concolor
        Abies grandis
        Abies magnifica
        Pinus aristata or *flexilis*
        Pinus jeffreyi
        Pinus lambertiana
        Pinus cf. *monophylla*
        Pinus monticola
        Pinus murrayana
        Pinus ponderosa
        Pseudotsuga taxifolia
        Tsuga heterophylla
    Taxodiaceae
        Sequoiadendron giganteum
Angiosperms
    Betulaceae
        Alnus rhombifolia
        Betula fontinalis
        Corylus californica
    Caprifoliaceae
        Sambucus glauca
    Cornaceae
        Cornus nuttallii
    Fagaceae
        Quercus kelloggii
        Quercus wislizenii
    Garryaceae
        Garrya sp.
    Juglandaceae
        Juglans nigra
    Papaveraceae
        Dendromecon rigida
    Rhamnaceae
        Ceanothus cordulatus
        Ceanothus integerrimus
    Rosaceae
        Amelanchier alnifolia
        Holodiscus dumosa
        Rubus parviflorus
    Salicaceae
        Salix exigua
        Salix lasiandra
    Saxifragaceae
        Ribes cereum
        Ribes montigenum
        Ribes nevadense
        Ribes roezlii

Sterculiaceae
    Fremontia californica
Ulmaceae
    Zelkova sp.

LIST 13

Lake Tahoe flora (pollen), central Sierra Nevada, California, late Pliocene (Adam 1973).

Gymnosperms
  Pinaceae
    Abies
    Libocedrus
    Picea
    Pinus
    Pseudotsuga
    Tsuga
Angiosperms
  Amaranthaceae
    Amaranthus?
  Betulaceae
    Alnus
    Betula
  Caryophyllaceae
  Chenopodiaceae
    Sarcobatus
  Compositae
    Artemesia
  Cornaceae
    Cornus
  Cyperaceae
  Ericaceae
  Euphorbiaceae
    Euphorbia
  Fagaceae
    Quercus
  Gramineae
  Juglandaceae
    Juglans
  Loranthaceae
    Arceuthobium
  Malvaceae
  Oleaceae
    Fraxinus
  Onagracaceae
    Oenothera?
  Polygonaceae
    Eriogonum
    Polygonum
  Ranunculaceae
  Rhamnaceae
  Rosaceae

LIST 14

La Porte flora (macrofossil), western central Sierra Nevada, California, early Oligocene (Potbury 1935).

Pteridophytes
  Polypodiaceae
    Polypodites sp.
Gymnosperms
  Cycadaceae
    Zamia mississippiensis var. *macrophylla* (spp.)
  Taxaceae
    Cephalotaxus californica (*argotaenia*)
Angiosperms
  Apocynaceae
    Tabernaemontana intermedia (*lanceolata, rupicola*)
  Aquifoliaceae
    Ilex oregona (*paraguensis*)
  Cornaceae
    Cornus kelloggii (spp.)
  Dilleniaceae
    Davilla intermedia (*aspera, rugosa*)
  Euphorbiaceae

    Acalypha serrulata (*schlechtendahliana*)
    Aleurites americana (*triloba*)
    Euphorbiophyllum multiformum (*Drypetes alba*)
    Microdesmis occidentalis (*casearifolia*)
  Fagaceae
    Quercus nevadensis (*hainanensis*)
    Quercus suborbicularia (spp.)
  Hamamelidaceae
    Liquidambar californicum (*styraciflua*)
  Lauraceae
    Cinnamomum acrodromum (*mercadoi*)
    Cinnamomum dilleri (*pedunculatum*)
    Laurophyllum intermedium (*Misanteca capitata*)
    Laurophyllum raminervum (*Ocotea* sp.)
    Ocotea eocernua (*cernua*)
    Persea praelingue (*lingue*)
    Persea pseudo-carolinensis (*podadenia*)
  Leguminosae
    Leguminosites falcatum (*Prioria copaifera*)
    Lonchocarpus coriaceus (*hondurensis*)
    Mimosites acutifolius (*Pithecolobium corymbosum*)
    Sophora repandifolia (spp.)
  Liliaceae
    Smilax goshenensis (*mexicana*)
  Menispermaceae
    Cissampelos rotundifolia (*pareira*)
    Hyperbaena diforma (*smilacina*)
  Moraceae
    Ficus goshenensis (*bonplandiana*)
  Palmae
    Sabalites rhapifolius (*Rhapis flabelliformis*)
  Rhamnaceae
    Rhamnidium chaneyi (*elaeocarpum*)
  Sabiaceae
    Meliosma goshenensis (*panamensis*)
  Sapotaceae
    Chrysophyllum conforme (*mexicanum*)
  Sterculiaceae
    Sterculia ovata (*blancoi, lanceolata*)
  Styracaceae
    Styrax curvatus (*argenteum*)
  Tiliaceae
    Columbia occidentalis (*longipetiolata*)
  Ulmaceae
    Ulmus pseudo-fulva (*mexicana*)
  Verbenaceae
    Petrea rotunda (*arborea, volubilis*)

LIST 15

Middlegate flora (macrofossil), western central Nevada, mid-Miocene (Axelrod 1956, 1976, 1986).

Equisetales
  Equisetaceae
    Equisetum alexanderi (sp.)
Gymnosperms
  Cupressaceae
    Thuja dimorpha (*plicata*)
  Pinaceae
    Abies concoloroides (*concolor*)
    Abies scherrii (*bracteata*) (Axelrod 1976)
    Picea magna (*polita, neovietchii*)
    Picea sonomensis (*breweriana*)
    Pinus (*ponderosa*) (Axelrod 1986)
  Taxodiaceae
    Sequoiadendron chaneyi (*giganteum*)
Angiosperms
  Aceraceae
    Acer alvordensis (*macrophyllum*)
    Acer arida (*grandidentatum*)
    Acer middlegatei (*saccharinum*)
    Acer minor (*negundo*)
  Anacardiaceae
    Rhus alvordensis (*glabra*)
  Berberidaceae
    Mahonia marginata (*beali*)
    Mahonia reticulata (*repens*)

Betulaceae
  Betula thor (papyrifera, occidentalis)
  Betula vera (lenta)
Ericaceae
  Arbutus prexalapensis (xalapensis)
Fagaceae
  Quercus hannibali (chrysolepis)
  Quercus simulata (myrsinaefolia)
  Quercus wislizenoides (wislizenii)
Lauraceae
  Persea coalingensis (podadenia)
Leguminosae
  Robinia californica (neo-mexicana)
Myricaceae
  Comptonia parvifolia (asplenifolia)
Nymphaeceae
  Nymphaeites nevadensis (Nymphaea spp.)
Oleaceae
  Fraxinus coulteri (oregona, americana)
  Fraxinus millsiana (anomala)
Platanaceae
  Platanus dissecta (orientalis)
  Platanus paucidentata (racemosa)
Rhamnaceae
  Ceanothus precuneatus (cuneatus)
Rosaceae
  Cercocarpus antiquus (betuloides)
  Cercocarpus holmesii (paucidentatus)
  Crataegus middlegatei (chrysophlla, erythropoda)
  Crataegus pacifica (euneata, monogyna)
  Prunus morganensis (emarginata)
Salicaceae
  Populus eotremuloides (trichocarpa)
  Populus payettensis (angustifolia)
  Populus pliotremuloides (tremuloides)
  Salix hesperia (lasiandra)
  Salix knowltoni (lemmonii)
  Salix truckeana (gooddingii)
  Salix wildcatensis (lasiolepis)
Styraceaceae
  Styrax middlegatei (californica)
Typhaceae
  Typha lesquereuxi (latifolia)

LIST 16

Mohawk flora (macrofossil), northern Sierra Nevada, California, mid-Miocene (Axelrod 1944; Knowlton 1911).

Gymnosperms
  Pinaceae
    Pinus (lambertiana)
Angiosperms
  Betulaceae
    Alnus sp.
  Ericaceae
    Arbutus matthesii (menziesii)
  Fagaceae
    Quercus payettensis
  Hamamelidaceae
    Liquidambar (styraciflua)
  Juglandaceae
    Carya egregia (ovata)
    Juglans (californica)
  Lauraceae
    Laurus californica
    Persea pseudo-carolinensis (podadenia)
  Magnoliaceae
    Magnolia dayana
  Rhamnaceae
    Berchemia multinervis (scandens)
  Salicaceae
    Populus eotremuloides (trichocarpa)
    Salix hesperia (lasiandra)

LIST 17

Mount Reba flora (macrofossil) (partial list), western central Sierra Nevada, California, early Pliocene (Axelrod 1976, 1977).

Gymnosperms
  Cupressaceae
    Cupressus (cashmeriana)
  Pinaceae
    Abies (concolor)
    Pseudotsuga (menziesii)
  Taxodiaceae
    Sequoiadendron (giganteum)
Angiosperms
  Fagaceae
    Lithocarpus (densiflora)
    Quercus (chrysolepis)

LIST 18

Nova flora (pollen), eastern central California, late Pliocene (Axelrod and Ting 1960). *Note:* This flora is a pollen site, interpretations are based on a very small number of grains (77), and species identifications may be incorrect. This flora should be treated with question.

Gymnosperms
  Cupressaceae
    Juniperus osteosperma
  Ephedraceae
    Ephedra nevadensis
  Pinaceae
    Abies sp.
    Pinus attenuata
    Pinus balfouriana
    Pinus jeffreyi
    Pinus lambertiana
    Pinus cf. *monophylla*
    Pinus ponderosa
Angiosperms
  Betulaceae
    Alnus rhombifolia
    Corylus californica
  Juglandaceae
    Juglans californica
    Juglans rupestris
    Pterocarya sp.
  Rhamnaceae
    Ceanothus cordulatus
    Rhamnus californica
  Rosaceae
    Cercocarpus betuloides
    Holodiscus sp.
  Salicaceae
    Populus angustifolia
    Salix exigua
    Salix lasiolepis
  Ulmaceae
    Ulmus alata

LIST 19

Oakdale flora (macrofossil), western central Sierra Nevada, California, early Pliocene (Axelrod 1944a).

Angiosperms
  Berberidaceae
    Mahonia marginata (fremontii)
  Ericaceae
    Arctostaphylos oakdalensis (mariposa)
  Fagaceae
    Quercus dispersa (dumosa)
    Quercus douglasoides (douglasii)
    Quercus wislizenoides (wislizenii)

Lauraceae
  Umbellularia salicifolia (californica)
Leguminosae
  Robinia californica (neo-mexicana)
Rhamnaceae
  Ceanothus precuneatus (cuneatus)
Rosaceae
  Photinia sonomensis (arbutifolia)
Salicaceae
  Populus alexanderi (trichocarpa)
  Populus paucidentata (acuminata)
  Populus pliotremuloides (tremuloides)
  Salix wildcatensis (lasiolepis)
Sapindaceae
  Sapindus oklahomensis (drummondii)
Saxifragaceae
  Ribes mehrtensis (quercetorum)
Ulmaceae
  Celtis kansana (reticulata)

LIST 20

Owens Gorge (Coso) flora (pollen), eastern central Sierra Nevada, California, late Pliocene (Axelrod and Ting 1960). *Note:* This flora is a pollen site, interpretations are based on a very small number of grains (162), and species identifications may be incorrect. This flora should be treated with question.

Gymnosperms
  Cupressaceae
    Calocedrus decurrens
  Pinaceae
    Abies concolor
    Abies magnifica
    Pinus aristata or *flexilis*
    Pinus jeffreyi
    Pinus lambertiana
    Pinus monticola
    Pinus murrayana
    Pseudotsuga taxifolia
    Tsuga heterophylla
  Taxodiaceae
    Sequoiadendron giganteum
Angiosperms
  Aceraceae
    Acer glabrum
  Betulaceae
    Corylus californica
  Caprifoliaceae
    Symphoricarpos albus
  Cornaceae
    Cornus californica
    Cornus nuttallii
  Fagaceae
    Quercus agrifolia
    Quercus breweri
    Quercus kelloggii
    Quercus lobata
    Quercus wislizenii
  Juglandaceae
    Juglans cinera
    Juglans nigra
    Pterocarya sp.
  Salicaceae
    Salix exigua
    Salix lasiandra
    Salix lasiolepis
    Salix nuttallii
  Saxifragaceae
    Ribes monteginum
  Sterculiaceae
    Fremontia californica

LIST 21

Panamint Springs (Coso) flora (pollen), eastern central Sierra Nevada, California, late Pliocene (Axelrod and Ting 1960). *Note:* This flora is a pollen site, interpretations are based on a very small number of grains (45), and species identifications may be incorrect. This flora should be treated with question.

Gymnosperms
  Pinaceae
    Pinus attenuata
    Pinus coulteri
    Pinus jeffreyi
    Pinus cf. *monophylla*
    Pinus ponderosa
Angiosperms
  Cornaceae
    Cornus nuttallii
  Fagaceae
    Castanopsis chrysolepis
    Quercus kelloggii
    Quercus lobata
  Garryaceae
    Garrya sp.
  Juglandaceae
    Juglans californica
    Juglans rupestris
  Oleaceae
    Fraxinus velutina
  Rosaceae
    Holodiscus sp.
  Saxifragaceae
    Ribes cereum
  Sterculiaceae
    Fremontia californica
  Typhaceae
    Typha latifolia
  Ulmaceae
    Ulmus alata

LIST 22

Purple Mountain flora (macrofossil), western central Nevada, mid-Miocene (Axelrod 1976, 1995).

Gymnosperms
  Cupressaceae
    Chamaecyparis sierrae (lawsoniana)
  Pinaceae
    Abies concoloroides (concolor)
    Abies klamathensis (shastensis)
    Abies scherrii (bracteata)
    Picea magna (polita)
    Picea sonomensis (breweriana)
    Pinus quinifolia (monticola)
    Pseudotsuga sonomensis (menziesii)
  Taxodiaceae
    Sequoiadendron chaneyi (giganteum)
Angiosperms
  Aceraceae
    Acer columbianum (glabrum)
    Acer middlegateii (saccharinum)
    Acer oregonianum (macrophyllum)
  Berberidaceae
    Mahonia reticulata (pinnata-insularis)
    Mahonia simplex (japonica, lomariifolia)
  Betulaceae
    Betula lacustris (papyrifera)
  Ericaceae
    Arbutus matthesii (menziesii)
  Fagaceae
    Castanopsis sonomensis (chrysophylla)
    Lithocarpus klamathensis (densiflorus)
    Quercus hannibali (chrysolepis)
  Leguminosae
    Amorpha oklahomensis (fruticosa)

Rhamnaceae
  Ceanothus leitchii (velutinus)
  Rhamnus precalifornica (californica)
Rosaceae
  Amelanchier alvordensis (alnifolia)
  Cercocarpus antiquus (betuloides)
  Cercocarpus holmesii (paucidentatus)
  Heteromeles sonomensis (arbutifolia)
  Holodiscus idahoensis (glabrescens)
  Lyonothamnus parvifolia (extinct sp.)
  Sorbus sp. *(aucuparia)*
Salicaceae
  Populus eotremuloides (trichocarpa)
  Populus payettensis (angustifolia)
  Populus pliotremuloides (tremuloides)
  Salix knowltonii (lemmonii)
  Salix sp. *(melanopsis)*
  Salix sp. *(nigra)*
  Salix wildcatensis (lasiolepis)

LIST 23

Remington Hill flora (macrofossil), western central Sierra
Nevada, California, late Miocene (Condit 1944).

Gymnosperms
  Cupressaceae
    Chamaecyparis gracilis (lawsoniana)
  Taxodiaceae
    Sequoia langsdorfii (sempervirens)
Angiosperms
  Aceraceae
    Acer negundoides (negundo)
  Berberidaceae
    Mahonia malheurensis
  Caprifoliaceae
    Viburnum platyspermum (ellipticum)
  Ericaceae
    Arbutus matthesii (menziesii)
    Arctostaphylos martzi (manzanita)
    Leucothoe sp. *(davisiae)?*
  Fagaceae
    Quercus douglasoides (douglasii)
    Quercus prelobata (lobata)
    Quercus pseudo-lyrata (kelloggii)
    Quercus remingtoni (morehus)
    Quercus simulata (myrsinaefolia)
    Quercus winstanleyi (aliena)
    Quercus wislizenoides (wislizenii)
  Hamamelidaceae
    Liquidambar pachyphyllum (styraciflua)
  Hippocastanaceae
    Aesculus preglabra (glabra)
  Juglandaceae
    Juglans pseudomorpha (nigra)
  Lauraceae
    Persea pseudo-carolinensis (borbonia)
    Umbellularia salicifolia (californica)
  Liliaceae
    Smilax diforma (rotundifolia)
  Platanaceae
    Platanus paucidentata (racemosa)
  Rhamnaceae
    Berchemia multinervis (scandens)
    Ceanothus precuneatus (cuneatus)
  Rosaceae
    Crataegus newberryi (pinnatifida)
    Prunus petrosperma (ilicifolia)
  Salicaceae
    Populus alexanderi (trichocarpa)
    Populus pliotremuloides (tremuloides)
    Populus prefremontii (fremontii)
    Salix hesperia (lasiandra)
  Sapindaceae
    Ungnadia clarki (spp.)
  Ulmaceae
    Ulmus californica (americana)
  Vitaceae
    Vitis bonseri

LIST 24

San Joaquin (Coso) flora (pollen), eastern central Sierra
Nevada, California, late Pliocene (Axelrod and Ting 1960).
Note: This flora is a pollen site, interpretations are based
on a very small number of grains (59), and species
identifications may be incorrect. This flora should be
treated with question.

Gymnosperms
  Cupressaceae
    Calocedrus decurrens
    Chamaecyparis lawsoniana
    Juniperus occidentalis
  Pinaceae
    Abies concolor
    Abies grandis
    Abies magnifica
    Pinus aristata or *flexilis*
    Pinus balfouriana
    Pinus lambertiana
    Pinus monticola
    Pinus murrayana
    Pseudotsuga taxifolia (menziesii)
    Tsuga mertensiana
  Taxaceae
    Taxus brevifolia
Angiosperms
  Aceraceae
    Acer circinatum
  Anacardiaceae
    Rhus diversifolia
  Juglandaceae
    Carya sp.
  Rhamnaceae
    Ceanothus cordulata
    Rhamnus crocea
  Salicaceae
    Salix lasiolepis
    Salix scouleriana
  Saxifragaceae
    Ribes cereum
    Ribes nevadense

LIST 25

Stewart Spring flora (macrofossil and pollen), western
central Nevada, mid-Miocene (Wolfe 1964 [macrofossil];
Schorn 1984 [pollen]).

Gymnosperms
  Cupressaceae
    Chamaecyparis (nootkatensis)
    Juniperus nevadensis (californica, utahensis)
  Pinaceae
    Abies (concolor)
    Larix (occidentalis)
    Picea (breweriana)
    Picea magna
    Pinus (edulis)
    Pinus (monticola)
    Pinus (ponderosa)
    Tsuga (heterophylla)
  Taxodiaceae
    Sequoiadendron[a]
Angiosperms
  Anacardiaceae
    Astronium mawbyi
    Rhus (integrifolia)
    Schinus savegei (gracilepis)
  Berberidaceae
    Mahonia reticulata (repens)
  Betulaceae
    Betula sp. *(lacustris)*
  Caprifoliaceae
    Sambucus sp.

Cyperaceae
    Cyperacites sp.
Eleagnaceae
    Eleagnus cedrusensis (utils)
Ericaceae
    Arbutus traini (menziesii)
    Arctostaphylos (masoni)
Fagaceae
    Quercus cedrusensis
    Quercus (chrysolepis)
Garryaceae
    Garrya axelrodi (elliptica)
Graminae
    Poacites sp.
Juglandaceae
    Carya[a]
    Juglans (major)
Oleaceae
    Fraxinus millsiana
Rhamnaceae
    Colubrina sp.
Rosaceae
    Amelanchier (cusicki)
    Cercocarpus (antiquus)
    Holodiscus fryi (dumosus)
    Lyonothamnus parvifolius
    Peraphyllum (vaccinifolium)
    Prunus sp.
    Rosa sp.
    Sorbus (acuparia)
Salicaceae
    Populus cedrusensis (brandegeei)
    Populus (tremuloides)
    Populus (trichocarpa)
    Populus washoensis (grandidentata? bonatti?)
    Salix pelviga
Sapindaceae
    Sapindus sp.
Saxifragaceae
    Philadelphus nevadensis
    Ribes webbi (cereum)
Ulmaceae
    Ulmus[a]
    Zelkova[a]

[a]Pollen taxon is additional to macrofossil list.

LIST 26

Susanville Complex flora (macrofossil), northeastern Sierra Nevada, California, mid-Eocene (Knowlton 1911; Wolfe and Hopkins 1967).

    Oreodaphne litsaeformis
Angiosperms
  Araliaceae
    Aralia lasseniana
  Fagaceae
    Quercus moorii
    Quercus olafensi
  Juglandaceae
    Juglans rugosa
  Lauraceae
    Cinnamomum scheuchzeri
    Laurus californica
    Laurus grandis
  Leguminosae
    Leguminosites sp.
  Magnoliaceae
    Magnolia hilgardiana
    Magnolia ingelfieldi

LIST 27

Table Mountain flora (macrofossil), western central Sierra Nevada, California, late Miocene (Condit 1944).

Gymnosperms
  Pinaceae
    Pinus pretuberculata (attenuata)
Angiosperms
  Aceraceae
    Acer bolanderi (acuminata)
  Anacardiaceae
    Rhus mensae (laurina)
  Aquifoliaceae
    Ilex opacoides (opaca)
  Berberidaceae
    Mahonia prelanceolata (lanceolata)
  Cornaceae
    Cornus ovalis (alternifolia)
    Nyssa elaenoides (sylvatica)
  Ericaceae
    Arbutus matthesii (menziesii)
    Rhododendron sierrae (rockii)
  Fagaceae
    Quercus bockeei (Asian type)
    Quercus convexa (engelmannii)
    Quercus dispersa (dumosa)
  Juglandaceae
    Carya typhinoides (cordiformis)
  Lauraceae
    Persea coalingensis (borbonia)
    Umbellularia salicifolia (californica)
  Leguminosae
    Cercis buchananensis (canadensis)
    Gleditsia (spp.)
    Robinia californica (neo-mexicana)
  Magnoliaceae
    Magnolia californica (grandiflora)
  Oleaceae
    Forestiera buchananensis (neo-mexicana)
  Platanaceae
    Platanus dissecta (racemosa)
  Rhamnaceae
    Berchemia multinervis (scandens)
  Rosaceae
    Cercocarpus antiquus (betuloides)
    Crataegus newberryi (pinnatifida)
  Salicaceae
    Salix californica (breweri)
  Saxifragaceae
    Philadelphus nevadensis (lewisii)
  Ulmaceae
    Celtis kansana (reticulata)
    Ulmus californica (americana)

LIST 28

Tehachapi flora (macrofossil), southern Sierra Nevada, California, mid-Miocene (Axelrod 1939; Chaney 1944).

Equisetales
  Equisetaceae
    Equisetum sp.
Filicales
  Polypodiaceae
    Pteris calabazensis
Gymnosperms
  Cupressaceae
    Cupressus mohavensis (arizonica)
  Pinaceae
    Pinus lindgreni (cembroides)
Angiosperms
  Anacardiaceae
    Rhus obovata (virens)
    Rhus preintegrifolia (integrifolia)
    Rhus sonorensis
  Berberidaceae
    Mahonia mohavensis (fremontii)

Burseraceae
    Bursera sp.
Caprifoliaceae
    Viburnum sp. *(stenocalyx)*
Ericaceae
    Arbutus mohavensis (peninsularis)
    Arbutus prexalapensis (xalapensis)
    Arctostaphylos mohavensis (sp.)
Euphorbiaceae
    Euphorbia mohavensis (hindsiana)
Fagaceae
    Quercus browni (chrysolepis)
    Quercus convexa (engelmanni)
    Quercus declinata (tomentella)
    Quercus dispersa (dumosa)
    Quercus pliopalmeri (paleri)
    Quercus turneri (arizonica)
Juncaceae
    Juncus sp.
Lauraceae
    Persea sp. *(hartwegii)*
    Umbellularia dayana (californica)
Leguminosae
    Amorpha oblongifolia (californica)
    Diphysa californica (suberosa)
    Leucanea californica (microcarpa)
    Pithecolobium miocenicum (dulce)
    Pithecolobium mohavense (mexicanum)
    Prosopis pliocenica (juliflora)
    Robinia californica (neo-mexicana)
Melicaceae
    Cedrela oregoniana (sp.)
Moraceae
    Ficus sp. *(palmeri)*
Myricaceae
    Myrica mohavensis (mexicana)
Oleaceae
    Foresteria sp. *(reticulata)*
    Fraxinus mohavensis (macropetala)
Palmaceae
    Erythea californica (brandegeei)
    Sabal miocenica (uresana)
Rhamnaceae
    Ceanothus precrassifolius (crassifolius)
    Ceanothus precuneata (cuneatus)
    Colubrina lanceolata (arborea)
    Condalia mohavensis (lycoides)
    Karwinskia californica (humboldtiana)
    Rhamnus precalifornica (californica)
Rosaceae
    Cercocarpus antiquus (betuloides)
    Cercocarpus preledifolius (ledifolius)
    Chamaebatiaria creedensis (millefolium)
    Holodiscus elliptica (discolor)
    Lyonothamnus mohavensis (floribundus)
    Photinia (Heteromeles) sp. *(arbutifolia)*
    Platanus paucidentata (racemosa)
    Prunus masoni (lyoni)
    Prunus preandersonii (andersonii)
    Prunus prefasciculata (fasciculata)
    Prunus prefremontii (fremontii)
    Rosa miocenica (mohavensis)
Salicaceae
    Populus alexanderi (trichocarpa)
    Populus lesquereuxi (angustifolia)
    Populus pliotremuloides (tremuloides)
    Populus prefremontii (fremontii)
    Populus sonorensis (monticola)
    Salix coalingensis (lasiolepis)
    Salix kernensis (bonplandiana)
Sapindaceae
    Dodonea californica (viscosa)
Sapotaceae
    Bumelia florissanti (lanuginosa)
Saxifragaceae
    Philadelphus bendirei (lewisii)
Sterculiaceae
    Fremontia trilobata (californica)
Typhaceae
    Typha lesquereuxi (sp.)
Ulmaceae
    Celtis kansana (reticulata)

LIST 29

Tulelake flora (pollen), northeastern California, late Pliocene (Adam et al. 1990; Adam et al. 1989).

Gymnosperms
    Cupressaceae
        TCT *(Taxodiaceae-Cupressaceae-Taxaceae)* (probably *Juniperus occidentalis* or *Calocedrus decurrens*)
    Pinaceae
        Pinus
Angiosperms
    Chenopodiaceae-Amaranthus
    Compositae
        Artemisia
    Cyperaceae
    Fagaceae
        Quercus
    Poaceae

LIST 30

Upper Cedarville 49 Camp flora (macrofossil), northwestern Nevada, mid-Miocene (Chaney 1959; LaMotte 1935, 1936).

Lycopods
    Lycopodiaceae
        Lycopodium prominens (obscurum)
Gymnosperms
    Cupressaceae
        Chamaecyparis gilmoreae (nootkatensis)
    Ginkgoaceae
        Ginkgo adiantoides (biloba)
    Pinaceae
        Abies laticarpus (magnifica)
        Pinus monticolensis (monticola)
        Pinus russelli (ponderosa)
        Pseudotsuga masoni (menziesii)
    Taxaceae
        Torreya bonseri (californica)
    Taxodiaceae
        Sequoia langsdorfii (sempervirens)
Angiosperms
    Aceraceae
        Acer merriami (macrophyllum)
        Acer negundoides (negundo)
        Acer osmonti (glabrum)
    Araliaceae
        Oreopanax conditi (taubertianum)
    Berberidaceae
        Odostemon simplex (nervosus)
    Betulaceae
        Alnus carpinoides (rubra)
        Ostrya oregoniana (virginiana)
    Cercidiphyllaceae
        Cercidiphyllum crenatum (japonicum)
    Ericaceae
        Arbutus matthesii (menziesii)
    Fagaceae
        Castanea lesquereuxi (pumila)
        Castanopsis chrysophylloides (chrysophylla)
        Fagus washoensis (longipetiolata)
        Quercus consimilis (myrsinaefolia)
        Quercus distincta (agrifolia)
        Quercus pseudo-lyrata (kelloggii)
    Juglandaceae
        Carya egregia (ovata)
    Lauraceae
        Umbellularia oregonensis (californica)
    Leguminosae
        Cercis spokanensis (occidentalis)
        Leguminosites vicicarpus
    Menispermaceae
        Cebatha heteromorpha (triloba)
    Oleaceae
        Fraxinus sp. *(oregona)*

Platanaceae
   Platanus dissecta (racemosa)
Rosaceae
   Crataegus newberryi (pinnatifida)
   Prunus masoni (integrifolia)
   Sorbus chaneyi (alnifolia)
Rutaceae
   Ptelea miocenica (trifoliata)
Salicaceae
   Populus eotremuloides (trichocarpa)
Sapindaceae
   Sapindus oregonianus (mukorossi)
Saxifragaceae
   Ribes sp. *(lacustre)*
Sparganiaceae
   Sparganium praesimplex (simplex)
Tiliaceae
   Tilia aspera (mandshurica)
Ulmaceae
   Ulmus speciosa (americana)

LIST 31

Upper Cedarville Pit River flora (macrofossil), northern Sierra Nevada, California, late Miocene (LaMotte 1935, 1936).

Gymnosperms
   Cupressaceae
      Chamaecyparis gilmoreae (nootkatensis)
   Ginkgoaceae
      Ginkgo adiantoides (biloba)
   Pinaceae
      Abies laticarpus (magnifica)
      Pinus russelli (ponderosa)
      Pseudotsuga masoni (menziesii)
Angiosperms
   Betulaceae
      Alnus carpinoides (rubra)
   Ericaceae
      Arbutus matthesii (menziesii)
   Fagaceae
      Castanea lesquereuxi (pumila)
      Castanopsis chrysophylloides (chrysophylla)
      Fagus washoensis (longipetiolata)
      Quercus consimilis (myrsinaefolia)
      Quercus distincta (agrifolia)
      Quercus pseudo-lyrata (kelloggii)
   Juglandaceae
      Carya egregia (ovata)
   Lauraceae
      Umbellularia oregonensis (californica)
   Oleaceae
      Fraxinus sp. *(oregona)*
   Platanaceae
      Platanus dissecta (racemosa)
   Rosaceae
      Prunus masoni (integrifolia)
   Rutaceae
      Ptelea miocenica (trifoliata)
   Sapindaceae
      Sapindus oregonianus (mukorossi)
   Tiliaceae
      Tilia aspera (mandshurica)

LIST 32

Valley Springs flora (macrofossil), western central Sierra Nevada, California, mid-Miocene (Axelrod 1944b).

Gymnosperms
   Pinaceae
      Pinus sp.
   Taxaceae
      Torreya sp.

Angiosperms
   Aceraceae
      Acer bolanderi (acuminata)
      Acer cf. *negundoides (negundo)*
   Berberidaceae
      Mahonia hollicki
   Betulaceae
      Alnus merriami
   Fagaceae
      Quercus convexa (engelmannii)
      Quercus hannibali (chrysolepis)
      Quercus simulata (myrsinaefolia)
   Lauraceae
      Persea pseudo-carolinensis (borbonia)
      Umbellularia salcifolia
   Moraceae
      Ficus microphylla
   Platanaceae
      Platanus dissecta (racemosa)
   Rhamnaceae
      Ceanothus (rigidus)
   Rosaceae
      Lyonothamnus mohavensis (floribundus)
   Salicaceae
      Salix californica (breweri)
   Sapindaceae
      Sapindus oklahomensis (drummondii)

LIST 33

Verdi flora (macrofossil), western central Nevada, early Pliocene (Axelrod 1958).

Gymnosperms
   Pinaceae
      Abies concoloroides (concolor)
      Picea (pollen, Howard Schorn personal communication cited in Adam 1973)
      Pinus florissanti (ponderosa)
      Pinus prelambertiana (lambertiana)
      Pinus pretuberculata (attenuata)
Angiosperms
   Characeae
      Chara verdiana (spp.)
   Ericaceae
      Arctostaphylos verdiana (nevadensis)
   Fagaceae
      Quercus prelobata (lobata)[a]
      Quercus renoana (engelmannii)
      Quercus wislizenoides (wislizenii)
   Grossulariaceae
      Ribes galeana (roezlii)
   Naiadaceae
      Potamogeton verdiana (spp.)
   Nymphaeceae
      Nymphaeites nevadensis
   Rhamnaceae
      Ceanothus precuneatus (cuneatus)
   Rosaceae
      Prunus moragensis (emarginata)
   Salicaceae
      Populus alexanderi (trichocarpa)
      Populus pliotremuloides (tremuloides)
      Populus subwashoensis (tremula davidiana)
      Salix boisiensis (scouleriana)
      Salix truckeana (gooddingii)

[a]Now identified as *Acer* (Wolfe and Tanai 1987).

LIST 34

Wichman flora (pollen), northern central Nevada, late Pliocene (Axelrod and Ting 1960). *Note:* This flora is a pollen site, interpretations are based on a very small number of grains (37), and species identifications may be incorrect. This flora should be treated with question.

Gymnosperms
    Pinaceae
        Pinus attenuata
        Pinus flexilis or *aristata*
        Pinus jeffreyi
        Pinus lambertiana
        Pinus monticola
Angiosperms
    Anacardiaceae
        Rhus diversilobus
    Betulaceae
        Alnus rhombifolia
    Cornaceae
        Cornus nuttallii
    Fagaceae
        Quercus breweri
    Juglandaceae
        Juglans cinerea
    Rhamnaceae
        Ceanothus cordulatus
    Rosaceae
        Cercocarpus ledifolius
        Rubus parviflorus
    Saxifragaceae
        Ribes nevadense

REFERENCES

Adam, D. P. 1973. Early Pleistocene pollen spectra near Lake Tahoe, California. *Journal of Research of the United States Geological Survey* 1:691–93.

Adam, D. P., J. P. Bradbury, H. J. Rieck, and A. M. Sarna-Wojcicki. 1990. Environmental changes in the Tule Lake Basin, Siskiyou and Modoc Counties, California from 3 to 2 million years before present. *United States Geological Survey Bulletin* 1933:1–13.

Adam, D. P., A. M. Sarna-Wojcicki, H. J. Rieck, J. P. Bradbury, W. E. Dean, and R. M. Forester. 1989. Tulelake, California: The last 3 million years. *Palaeogeography, Palaeoclimatology, Palaeoecology* 72:89–103.

Axelrod, D. I. 1939. *A Miocene flora from the western border of the Mojave Desert.* Publication 476. Washington, DC: Carnegie Institute of Washington.

———. 1944a. The Oakdale flora. In *Pliocene floras of California and Oregon,* edited by R. W. Chaney, 147–66. Publication 553. Washington, DC: Carnegie Institute of Washington.

———. 1944b. The Pliocene sequence in central California. In *Pliocene floras of California and Oregon,* edited by R. W. Chaney, 207–24. Publication 553. Washington, DC: Carnegie Institute of Washington.

———. 1956. Mio-Pliocene floras from west-central Nevada. *University of California Publications in Geological Science,* vol. 33.

———. 1958. The Pliocene Verdi flora of western Nevada. *University of California Publications in Geological Science* 34:61–160.

———. 1962. A Pliocene *Sequoiadendron* forest from western Nevada. *University of California Publications in Geological Science* 39:195–268.

———. 1976. Evolution of the Santa Lucia fir ecosystem. *Annals of the Missouri Botanical Garden* 63:24–41.

———. 1977. Outline history of California vegetation. In *Proceedings of terrestrial vegetation of California,* edited by M. Barbour and J. Major. Special Publication 9. Berkeley: California Native Plant Society.

———. 1986. Cenozoic history of some western American pines. *Annals of the Missouri Botanical Garden* 73:565–641.

———. 1995. The Miocene Purple Mountain flora of western Nevada. *University of California Publications in Geological Science* 139:1–62.

Axelrod, D. I., and H. E. Schorn. 1994. The 15 Ma floristic crisis at Gillam Spring, Washoe County, northwestern Nevada. *PaleoBios* 16:1–10.

Axelrod, D. I., and W. S. Ting. 1960. Late Pliocene floras east of the Sierra Nevada. *University of California Publications in Geological Science* 39:1–118.

Chaney, R. W., ed. 1944. *Pliocene floras of California and Oregon.* Publication 553. Washington, DC: Carnegie Institute of Washington.

———. 1959. *Miocene floras of the Columbia Basin.* Publication 617 (part 1). Washington, DC: Carnegie Institute of Washington.

Condit, C. B. 1944. The Table Mountain flora. In *Pliocene floras of California and Oregon,* edited by R. W. Chaney, 57–90. Publication 553. Washington, DC: Carnegie Institute of Washington.

Knowlton, F. H. 1911. Flora of the auriferous gravels of California. In *The Tertiary gravels of the Sierra Nevada of California,* edited by W. Lindgren, 1–28. Menlo Park, CA: U.S. Geological Survey.

LaMotte, R. S. 1935. Studies in Tertiary paleobotany, IV: The Upper Cedarville flora of northwestern Nevada and adjacent California. Ph.D. diss., Department of Paleontology, University of California, Berkeley.

———. 1936. *The Upper Cedarville flora of northwestern Nevada and adjacent California.* Publication 455. Washington, DC: Carnegie Institute of Washington.

Leopold, E. B. 1983. Presentation at symposium, Palynology of Tertiary floras of western North America, American Association of Stratigraphic Palynologists, San Francisco.

———. 1984. Pollen identifications from the Eocene Chalk Bluffs flora, California. *Palynology* 8:8.

MacGinitie, H. D. 1941. *A middle Eocene flora from the central Sierra Nevada.* Publication 534. Washington, DC: Carnegie Institute of Washington.

Potbury, S. S. 1935. *The La Porte flora of Plumas County, California.* Publication 465. Washington, DC: Carnegie Institute of Washington.

Schorn, H. E. 1984. Palynology of the late middle Miocene sequence, Stewart Valley. *Palynology* 8:259–60.

Wolfe, J. A. 1964. Miocene floras from Fingerrock Wash, southwestern Nevada. Professional Paper 454-N. Washington, DC: U.S. Geological Survey.

Wolfe, J. A., and D. M. Hopkins. 1967. Climatic changes recorded by Tertiary land floras in northwestern North America. In *Tertiary correlations and climatic changes in the Pacific,* edited by K. Hatai, 61–76. Tokyo: Eleventh Pacific Science Congress.

Wolfe, J., and T. Tanai. 1987. Systematic phylogeny and distribution of *Acer* (maples) in the Cenozoic of western North America. *Journal of the Faculty of Science, Hokkaido University* 22 (1): 1–246.

SECTION II

Human Components of the Sierra Ecosystem

PAUL F. STARRS
University of Nevada
Reno, Nevada

6

The Public as Agents of Policy

ABSTRACT

Looking at landscape alteration as the basic evidence of human activity is a far more effective technique for capturing policy change than relying on written laws or formal policy documents. Changes on the land are generally vernacular—driven from the ground up—and such changes are a sophisticated barometer of policy need and local ideology. Accepting the vernacular environment as a basic analytical unit has an added advantage; it bonds scientific investigations and expert efforts at establishing baselines to the visible and concrete expressions of human artifice in material culture. Although policy periods are often ossified into dated eras as a simple bounding device and a historical convenience, concern with the evolution of landscape—something that is much akin to "an ecosystem"—requires accepting different, especially physical and cultural, evidence. Vernacular policy is fashioned by local people to meet their everyday needs for order and resource exploitation, and to protect residents and landscape. Because policy throughout much of the Sierra is at best inchoate, the biggest contributor to policy has been official and administrative abstinence and abdication. Yet policy by inattention and inaction paradoxically is an active form of policy making, and inaction is not always picked up in a formal policy analysis or history; these are, after all, generally produced by policy analysts, economists, or historians, who are most comfortable evaluating written fact rather than changes on the land. The advantage of landscape-level studies is that they examine the substance of change, regardless of the initiator. Ultimately, there are dual authorities, producers of quite different kinds of landscape change: the influential establishment forces and the vernacular "doers." These two vessels for policy assessment, and their direct implications for the Sierra Nevada ecosystem, are examined.

INTRODUCTION

Geography is writ large in the policy past, present, and future of the Sierra Nevada ecosystem. The Sierra Nevada is seen in various lights by its diverse perceivers. Through time the Sierran realm—more than 640 kilometers (400 miles) long and up to 160 km (100 mi) wide—has been appraised as home, as impediment, as an enormous pool of natural resources awaiting exploration and exploitation, as a setting for the playing out of sundry human ambitions, as the place for parks and historic preservation, or, in the view of some, as a paragon of the pristine. There can, of course, be endless wrangling about the precise meaning of these ambitious, changeable, and contradictory visions, and a big helping of contentiousness about what exactly such cavalier classifications imply is not unexpected. Yet policy as it has evolved (and failed to develop) in the Sierra of Nevada and California owes more to geography than to an orthodox history.

Foreword—the Portents of "Policy"

Although there is room, especially in policy studies, for dispute about the neatest way to understand the Sierra Nevada—how policy-making eras and their thematic historical categories are best arranged—setting a geographical anchor is a handy way to slow down any drifting. The history of the Sierra can be treated as a phalanx of time-discrete events, but its geography is thoroughly intermixed and eventually built upon alterations that have a real physical form and a permanent and discernible effect on the land. The distinguished French author of the *Annales* school, Fernand Braudel, has put it nicely: "Landscape and panoramas are not simply realities of the present but also, in large measure, survivals from

Sierra Nevada Ecosystem Project: Final report to Congress, vol. II, *Assessments and scientific basis for management options.* Davis: University of California, Centers for Water and Wildland Resources, 1996.

the past. Long-lost horizons are redrawn and created for us through what we see; the earth is, like our own skin, fated to carry the scars of ancient wounds" (Braudel 1988, 31). The study of landscape modification is a basic stock of geography and its far younger and as yet malleable disciple discipline of environmental history. Connecting these twinned fields to the policy climate of the Sierra is hardly a stretch. But history and geography are separated by conspicuously different views of the importance and mechanics of how policy is shaped.

The founder of *Landscape* magazine, J. B. Jackson, has written about a fundamental split in American life between what he calls "establishment" and "vernacular" views, and the implications of these ways of seeing speak not only to how landscapes like those of the Sierra are regarded, but also to how they are treated, handled, and seen to have evolved. Policy is, naturally enough, the primary province of establishment figures; they make policy trying to ensure that landscapes evolve in certain ways and not in others. The wishes of elected policy makers in a democratic society are set down in debates and hearings; the results of deliberations are chronicled in judicial decisions, legislation, executive orders, and regulations.

But other important changes take place in any landscape. These changes come from the actions and inactions of less self-important and yet at least equally influential forces, representing vernacular views—the needs, desires, and deeds of the everyday residents of a landscape. But despite the distinction recognized in American law between statutory law and common practice as it can be incorporated into law, policy studies have been loath to recognize or perhaps just incapable of recognizing this difference on the ground, where policy is expressed. Vernacular or common folk "create" policy in formal terms only rarely, but they shape landscape (literally) in many and substantive ways, and do so constantly.

This contrast between vernacular and establishment forces has plenty of local and regional expression in the Sierra Nevada. Take, for example, community well-being and governance. Rural sociologists claim with glee and surprise to have discovered in the last fifteen years that western Sierra communities are clamoring for a voice, instituting efforts to create consensus and gain added power over local economies and future options. Yet it is sorely misguided to assert that these demands are anything new. As Rodman Paul and Charles Shinn have noted, the mining camps of the western slope created intricate and entirely oral and tradition-based codes that saw the communities through thick and thin during an era when "legal" and policy authority was impossibly remote in prospect or enforcement (Paul 1963; Shinn 1884).

"Miner's law," whether ruling on claim disputes, access and rights to water, or legal transgressions, was a system of justice and authority that had no formal state or federal sanction. A mining community made itself work. Not perfectly self-equilibrating, mining communities did nonetheless solve most pressing problems on their own. This home-concocted

"policy," if at times cast in a circumspect and extralegal mold, was unmistakably far removed from vigilantism: here was a direct expression of the mining (and town) community's need for order. If on its face "unofficial," the elements of informal control in mining communities were genuine, amounting to a usufructuary policy—born from use, not legal code—that governed community growth and development along the entire western Sierra front.[1] And informal rule did work, while the laws and policies of California and Nevada (to say nothing about those of local counties or the federal government) failed to engage until decades later.

Comparable unofficial forms of policy and community control have steered a multitude of activities besides mining—livestock raising, for example, had the stockgrowers association, based loosely upon the Spanish and Mexican *Mesta*; logging camps had discrete grievance boards; hunters practiced venery in the foothills when laws to control hunting, whether for sport or market sales, were at best malformed; water was administered through nonstatutory means, admittedly traditional and highly ordered, that were eventually recast into the doctrine of prior appropriation, still important in every western state. The Central Pacific Railroad conducted its sales of checkerboard land with considerable energy and spawned a swelling of forest cutting and town building that had its own private policy. And, of course, the modern-day homeowners association is nothing less than a way to control behavior that, however codified in Covenants, Conditions, and Restrictions, generally preempts law and policy—and establishes practices that are stricter (by association) than those of the surrounding society. In fact, the notion that law and policy are the best means to solve problems, especially those between neighbors, is a notably modern conceit, as Robert Ellickson has suggested with no small measure of sting (Ellickson 1986, 1991). Consistently throughout the history of the Sierra Nevada, improvisation has been the mother of policy and the father of innovation.

The contemporary plaints coming from rural, resource-dependent communities owe much of their volume and vehemence to the organizational skills of relative newcomers who do not realize that their wails of protest are old wine in new bottles (Fortmann and Starrs 1988). Asking for legal and legislative redress has been a last resort throughout much of the Sierra's Euro-American history, and not a first principle. Policy abjured by legislature or Congress, policy that on occasion went unwritten simply because there was no easy means to construct it, has usually been fashioned by local people to meet their everyday needs for order and resource exploitation and to protect residents and landscape. Without pressing into environmental determinism, it can be argued that adaptation to Sierran environments brought about a huge series of improvisatory changes in the use and practices wrought upon the Sierra and its resources.

And it is still that way. The foothills, east and west of the massif of the Sierra Nevada batholith, are changing with great speed and are growing crowded as new human populations move in (Walters 1986).[2] Now, for example, a vacuum is being filled that was first breached in the 1950s by tentative and speculative second-home developments around Lake Tahoe, in the Feather River country, and through the foothill central Sierra counties (Parsons 1972a, 1972b). These developments proliferate and pose something of a quandary, in part because there has been little discrete government effort to limit the subdivisions and second homes that John Fraser Hart has described as sitting on the "perimetropolitan fringe" (Hart 1991). Because policy on wildland fire prevention, subdivision, access, and wildlife management along the western side of the Sierra is at best inchoate, the biggest contributor to policy has actually been official and administrative abstinence and abdication. This is not necessarily bad, and it certainly has a long history. But policy by inattention and inaction paradoxically is still very much an active form of policy making, and inaction is not always picked up by a formal history. Looking at landscape alteration as the basic evidence of human activity is a more effective technique for capturing changes than relying on laws or formal policy documents. Changes on the land are generally vernacular—driven from the ground up. Accepting the vernacular environment has an added advantage; it bonds scientific investigation and attempts to establish baselines to seeing a human past that finds its concrete expression in material culture.

So, though policy eras can be ossified into bracketed dates as a historical convenience, concern with the evolution of landscape—something much akin to "an ecosystem"—is different. Landscape-level studies examine the substance of change, regardless of the initiator. Ultimately, there are two authorities, two different kinds of producers of landscape change: the influential establishment forces and the vernacular "doers."

At times the proponents of science, planning, and orderly progress can be rendered all but irrelevant. Politics, policy, and the improvisations of community and regional forces often surge ahead of science and reason, bringing change to an ecosystem before it is known or understood. That has certainly been the case in the Sierra Nevada. Vernacular forces are a basic fact of landscape change. Improvisation and imagery—how a problem or situation is perceived and how a local constituency chooses to deal with it—have had an enormous influence on the Sierra Nevada, and they still are critically important today. In fact, experts who study the material culture of everyday landscapes are increasingly comfortable with an argument that the built works of a community are frequently a remarkably effective means of understanding the ideas and ideals of the community's people; as Henry Glassie once put it, "in works are the mind." The same conclusion can be applied to the backtracking that takes place in analyzing the improvisation of vernacular policy. At a local level, policy is created when there is a significant need of either a whole population or a smaller but influential clique. Vernacular policy is still a remarkably sensitive barometer of community desire or perceived need.

John Wesley Powell ran afoul of Congress in 1878 with his notable "Report on the Lands of the Arid Region" in large measure because he insisted that surveys of arable land, water supplies, forest acreages, and ideal community sites should be completed, with appropriate land set aside into town grants, before settlers moved onto the western lands. That plan lasted a matter of months before Senator Bill Stewart of Nevada interceded to allow speculation and settlement to move forward before surveys were complete (Stegner 1953; Starrs 1988; Pisani 1992). The adventurous practice of charging ahead into a relatively unknown realm survives in American life. Changes in a landscape are made with limited regard for science, and it can take policy makers years to catch on and catch up (Pisani 1992).

The vicious infighting currently going on over the reach of the Endangered Species Act (ESA), as another case in point, is a striking example of a scientific policy in defense of biodiversity that has run up against a local desire to develop and sometimes plunder. The policy, in law and fact, is clear enough; the work-arounds on the ground, including those of the Department of the Interior, suggest that asserting something is "right" for an ecosystem, however that may be determined, generally amounts to a speed bump in the dynamic road of change.[3] Placing possibility ahead of policy and planning is in the nature of vernacular activity and, for good or bad, a deep American tradition.

Good "science" may not matter, and even "policy" can be an afterthought. The federal experience with rangeland reform and modification of the 1872 Mining Law are recent cases; California's involvement with the California Environmental Quality Act and energy policy reform is equally telling. The times when policy actually scripts an orderly landscape change are rare. Instead, landscape change is improvised, even libertarian—conducted beyond the traditional reach of establishment policy making. So studying just the stages of official policy formation turns out to be a profoundly partial choice, and one best avoided.

Policy and the Particulars of Landscape Change

The Sierra Nevada ecosystem has through time been shaped by a variety of landscape changes driven by vernacular and establishment forces. These modifications have a mappable, physical expression; most led eventually to policy changes that signal discrete historical benchmarks. But the time lag between the arrival of settlers or resource exploiters and the formulation of anything like "policy" is usually significant.

There are plenty of cases throughout the Sierra where sizable landscape changes were made (and they still can be) without any recognition at all in law that these require or can be altered by formal policy making. An epoch of sheep grazing across the Sierra; the damming of rivers for small-scale hydroelectric generation and assurance of town water supplies; and thirty years of essentially unbridled foothill subdivision are hardly trivial examples of land use and habitat modification. Policy rarely precedes land use; it is more reactive than proactive. Alas, scientific study as an influence on policy is not often farsighted either, as Donald Worster has pointed out in *Nature's Economy* (1994).

As a case in point, the long-standing influence of Native Americans was certainly a momentous fact of Sierran vegetation and fire history, and remained so until long after Ishi (Kroeber 1976). Indian management of Sierran resources was significant, and in the relatively few sites where there has been elaborate ethnographic study of foothill Indians, or in places like the Valley of the Yosemite,[4] the evidence suggests that ongoing Indian activity made for telling changes in the land.[5] But policy followed only slowly. There are good reasons for this. As Slotkin (1973, 1985) and White (1983) have suggested, to recognize the productive and landscape-changing role of Native Americans and then deal with them in law was to create a policy where there was no defense for actions that Euro-Americans were already taking. What happened instead typically involved ignoring the "problem" with a full expectation that with the passing of relatively short periods everything would be cared for (Limerick 1987).[6] The abdication of policy was a lot easier to contend with than a formal, but indefensible, policy. And so, in the case of the Sierra, it was left to settlers and disease to handle what the laws could not. Inaction is policy, as is a paralysis that prevents consensus. This is by no means unknown today—to put off making policy is itself a choice that avoids the liabilities of after-the-fact blame.

The basic task for this chapter is to look at the Sierra Nevada ecosystem from a geographer's perspective, focusing impressionistically on a single, policy-based question, "how did it get to be this way?" That is no small task, but Beesley's (1996) sizable reconstruction of the Sierra Nevada landscape precedes this effort, and he has completed some admirable spade work. Anyone who needs the dates and particulars of legislated change in the Sierra should have that account at hand. As this introduction suggests, the concerns covered here—literally changes in the land and its inhabitants—are different. The basic organization of the rest of this chapter turns on individual aspects of land and life in the Sierra, with suggestions for the major policies and facts, whether driven by establishment or vernacular forces, that have altered the Sierran landscapes.

The Sierra is divided here into component physical parts, not regionally, but topically. Because there are some resources—roads and passes, communities and property—that are social, rather than biological, they are included. Each part is taken in turn, and I comment on what I see as the most significant policy creations and omissions that have affected the Sierra's elemental resources: rangeland, wildlife, fire, minerals, forests, the hydrosphere, arable lands, movement, recreation, property (or habitat), biodiversity, and the public trust. There is room enough for other resources or issues to be introduced, but a study of the geographical aspects of policy in the Sierra needs to be more than a survey of legislative acts

and judicial opinions. It needs to include a want of policy as well, for to do nothing is, in a government sense, something very important and is itself a conscious act and a reflection of certain decisive values: in particular, it suggests satisfaction with the way things are, or a conviction that nothing should be done, or doubt that any political consensus is reachable.

An analytical frame of five windows is developed, each offering a different vista on the modes and conundrums of Sierra Nevada policy. The themes bear on almost all the resources, environments, and populations of the Sierra. *Implantation* is concerned with the nature and substance of settlements and exotic introductions in the Sierra Nevada, from Native American times until today. *Exploitation* focuses on the objective resources of the Sierra, which range from acorns to auriferous gravels, from water to wildlife and wilderness. *Tenure* examines the cadastre—how land was acquired, the patterns of ownership and use in the Sierra, and the expression in the Sierra Nevada of the familiar American obsession with land and property, including the public trust. *Intensity* takes on the relationship between core and periphery—some activities develop sizable effects in small areas; others are diffuse but enormously widespread in practice. Finally, the Sierra Nevada ecosystem assuredly does not exist in isolation, and the essential geographical relationship between the Sierra and other places is brought out in a study of *linkages*. Discussion of these concepts is incorporated into the larger text that follows, but they are as much organizational themes as analytical conveniences, which in any case reach to the heart of the resources of the Sierra Nevada ecosystem and get beyond plain science to an argument that for at least ten thousand years the Sierra has been as much and as significantly a human place as a playground for the forces of biogeography, climate, pedology, geomorphology, and hydrology to work their way.

RESOURCES

Twelve resources are examined. Many overlap, but that duplication will be glossed over. None of these surveys is intended as anything more than an illustration of the dynamic between vernacular and establishment policy forces. It is motive power and inclination that is sought here because policy is not made in straight lines but from subtly shaded or overtly self-serving zigzags of contending and opposing forces. The formal histories of the Sierra Nevada Ecosystem Project (SNEP) are the obvious places to turn for history; here only a sampling of citations for the different fields is offered. But keep in mind that each resource is posed as a talking point, an illustration of how Sierran policy is influenced. Especially important are the first clarion calls for change, because there is an almost Newtonian inclination to leave well enough alone through much of American policy history, with a few rare exceptions relating to land and agriculture, where the policy of Congress was interventionist indeed. Each locale, even within the Sierra, makes accommodations to local conditions, and the formal policy making of government came later, to broker and modify the vernacular relationship of Sierra residents to the land.

Rangeland

What is not forested or bare in the Sierra Nevada is rangeland, according to the disciplinary definitions of range managers. But even forested areas of the Sierra Nevada batholith have seen significant use by native animals and by introduced (or exotic) herbivores, primarily cattle and sheep.[7] Animals hunted by Native Californians certainly grazed, browsed, and roamed the Sierra, and were important nearly as much for ritual life as for a protein source. Domesticated animals from Spanish times on were regular visitors to the foothills, and feral horses and burros ranged in the Sierra, brought there by Spanish and, later, Mexican miners, in the early nineteenth century. Feral animals were free spirits, in some cases literal escapees from domestic use, and they formed one more layer of creatures to be exploited in California. From native wildlife to early exotic introductions, through the depredations of gold rush–era market hunting, into the present-day conflicts over water supply, the Pacific Flyway, and the always-discussed "peripheral canals," there are instructive policy lessons to be gleaned by looking at animals.

Domesticated animals in early California roamed freely and widely. Disputes over ownership were settled by "judges of the plains" who ruled with an authority granted by public stature and their informal acclaim as an elect.[8] Livestock grazed on commons and mixed freely—the concept of property was profoundly different in Hispanic Alta California than in the eastern United States, a historical incidence of a common property resource that has been insufficiently studied. In fact, almost all laws relating to resources were based upon Old World Andalusian and New World Mexican experience, with Arab North African roots, instead of English common law (except in the specific matter of water use, where appropriative doctrine ruled).[9] So the Sierra was during the Spanish era a vast potential grazing land, and it was used as such modestly, especially in the western foothill country between 300 and 1,500 m (1,000 and 5,000 ft), where there was plenty of forage and relatively easy herding. Although about 80% of *Californio* livestock grazing was centered in the Coast Ranges, the Sierra did not go untouched, especially as settlements in the southern Sacramento and northern San Joaquin valleys grew entrenched in the 1820s and 1830s.

A major trait that the Spanish-Mexican experience passed on to Californians in the Anglo period was the importance of transhumance, the seasonal movement of livestock from one foraging ground to another (Rinschede 1984, 1988). Two kinds of transhumance affected the Sierra and went unimpeded by

policy until nearly the turn of the century—and in some respects, these retain significance today.

First was the market movement of animals along the Sierra, either the east or west side, to mine sites. The east, including the Owens Valley and the precipitous eastern escarpment of the Sierra, was the main movement ground for sheep and cattle, as for the drayage animals, including horses and oxen bound for the mines of western Nevada—the superlative Comstock, Bodie, Aurora, Lida, Silver Peak, and later Goldfield, Tonopah, Rhyolite, and Bullfrog. The route available was a relatively narrow defile, because just a little to the east lay Death Valley and its known problems. These animals ate.[10] The same pattern of concentration of travel routes is true for movement (remarkably little studied) along the foothills of the Sierra, moving animals to the Mother Lode and, in later years, to the deep, hardrock mines of the western slopes. For at least forty years these animals ate and trampled, some escaped, and they proliferated; there are accounts of sizable numbers of feral animals doing well along the east and west sides of the Sierra despite the remarkably diverse corps of major predators that roamed parts of the Sierra until relatively late.

A second major effect of livestock on the Sierra came from transhumant movements. As many as six million sheep a year are documented in Inyo County tax rolls as traveling a circle route, starting near Bakersfield and moving east to the Mojave, up the east side of the Sierra, and crossing at passes—Tioga, Sonora, Ebbetts, and others. The passage of so many animals altered vegetation and established trails and use patterns that are a fact of Sierran life. None of the Sierra will ever be "pristine" again; domesticated animals and humans have been everywhere.

That is history enough; the effects, especially on the vegetation of montane meadows, were sizable. Because of the short growing season, montane environments, like deserts, show the effects of use long after the use itself is past, but change in the Sierra was nowhere near as vast as change in the Coast Ranges of California, where the vegetation mix in effect underwent nearly a perfect swap of Mediterranean grazing-adapted species for the California native plants. The upper elevations of the Sierra certainly were affected by this seasonal grazing and were geomorphologically altered, but the conversion was in no way comparable to what happened farther west. Once the major transhumant movements of livestock ceased with the creation of forest reserves in the 1890s (John Muir actually called, in a massive letter-writing campaign, for the army to be mustered to stop grazing in the Sierra), the vegetative outrage was switched from cattle and sheep to the pack animals of the Sierra Club sojourners—as many as two hundred animals, on some of the trips, resting in the same meadows that Muir had fought to see kept sheep-free.[11]

The creation of forest reserves was an important set-aside of land in some parts of the Sierra, but in others grazing continued essentially unabated into the mid-twentieth century.

The establishment of national forests proved far more important in the Sierra than the Taylor Grazing Act of 1934. In effect it was a final coffin nail, but the transhumant livestock grazers had been nearly stopped by the turn of the century. Yet grazing continued in many places. Grazing was permitted in the forest reserves, often under the watchful eyes of resident rangers, government employees who were also community members—a striking difference from post–World War II practice in the U.S. Forest Service (USFS). The checkerboard ownership pattern of the Central Pacific Railroad, which crossed the Sierra in 1868, permitted some sheep owners to continue to move their flocks across the Sierra into the 1930s and even the 1940s. That the antisheep campaign was in part racially motivated (many of the herders were Basque, Hispanic, Italian, or Irish) is hardly ever denied; John Muir was not above playing the "race card" to protect his beloved "range of light." Grazing continued in private land areas, and where grazing domesticated animals was not allowed, rangeland remained habitat.

And rangeland continued to be used in the Sierra for grazing. Its uses as wildlife habitat were self-evident; the forest reserves served rather well for those needs, although many of the areas put under USFS control were relatively high in elevation and austere and unwelcoming for up to nine months of the year. And the foothills, especially along the western slope, were and are still crucial range habitat. Private land, generally coming out of either the old checkerboard or from the cut-over and regenerating lands of private timber companies, is today grazed by ranch owners who practice a particularly precarious kind of transhumant ranching (Huntsinger 1989). Animals are wintered on grain or stubble in the Sacramento or San Joaquin valleys, with some grazing of home ranch properties, from fall through spring, and in May or June go to the mountains to work through the forests. Although these animals all wear bells, the roundup experience through the foothill forests of ponderosa and sugar pines has virtually nothing to do with *Heidi*.

Grazing of rangeland in the Sierra was most affected by the removal to preserves of vast tracts of the Sierra's land—except for the western foothills and the lower-elevation meadows of the east side, where cattle, and a few bands of sheep, were still allowed to graze. Two contemporary exceptions come to mind: Sequoia and Kings Canyon National Parks have seen a continuing imbroglio over livestock grazing into and near the park boundaries that perpetuates the long-standing debate among Park Service intelligentsia over whether United States national parks can or ought to include cultural uses. The answer for now, based upon not just the southern Sierra example but also the Great Basin National Park experience, would appear to be "no," in marked contrast to the international experience where human presence and activity are assumed. The long-term effects of this exclusion upon habitat is another matter that has already seen caustic commentary—that of Alston Chase (1986) comes to mind. Biologically, it is not possible to "shut the barn door."

Another case involves pack stock. The use of pack animals is a contentious topic that the USFS and the National Park Service are facing with distinct unease, because pack outfits are some of the oldest commercial clients in the Sierra. Government agencies work with rangeland as much by executive order and agency policy as by statute, although federal legislation has produced some crucial changes of direction—consider the National Environmental Policy Act (NEPA) and the Federal Land Management Policy Act (FLMPA), or, for that matter, the Taylor Grazing Act. It is the pleasure of the USFS, built upon the model of Bernard Frenow's German Forestry Service, to work through regulation. Increasingly, such regulation must have public input and commentary, but in the high Sierra the contending parties are relatively few. Nonetheless, the battle of titanium-frame, polarfill-bag-toting backpacker versus the mule-packed dude versus government range conservationist is an interesting test of the Forest Service's ability to cope with changing models of resource use. Otherwise, much rangeland in the Sierra is either private, and difficult to regulate, given the contemporary regulatory climate, or given over to wildlife.

There is one further observation that has significance for rangeland in the Sierra. A considerable acreage, arguably a half or more, of the western slope of the Sierra is rangeland. Especially where this is private land, it is also under tremendous pressure for parceling into ranchettes and other subdivisions. By some estimates, as much as 5% or 6% of this rangeland per decade is being turned into housing tracts, despite the remnant encouragement of proagriculture laws like the Williamson Act, which provides tax protections to landowners who keep their land in various forms of farming. Pressure is building from the 300 m (1,000 ft) contour of the Sierra up to about 1,000 m (3,000 ft) to convert land into housing for incoming residents. This pressure will not stop until, to echo Dan Luten's timeless phrase, the Sierra foothills are as repulsive as every place else in the state (Luten 1986). If not especially cheerful, that view has weight and cogency. The same thing is happening on the east side of the Sierra, from Verdi and Susanville, to Jacks Valley and Genoa, to June Lake and Olancha. Rangeland is becoming housing whenever the price is right. Establishment policy is about two decades behind reality, and unless policy changes do encourage livestock ranching, the cattle, sheep, goats, horses, and llamas will disappear, and in their wake will come rural idyllists and commuters to Fresno, Sacramento, Chico, or the capitol complex in Carson City. In more than a few places Wranglers are being replaced by Internet access for rancher wannabes; that is not innately a bad thing, but whether this change is "best for the resource" is another question. The substantive effect of this change from extractive use to neosubsistence has been little examined.

The issues relating grazing and the Sierra are gnarled and complex. Domestic animals are often nativized; so too are the exotic plants they often bring with them. Vegetation change, landscape modification at a local and rangewide scale, and hydrologic alteration of the Sierra have occurred, and occur almost anywhere animals graze in numbers. Feral animals have escaped—including domesticated cats, dogs, donkeys, and horses—and pose a sizable threat to native wildlife and to the ecosystems of the Sierra; they are treated differently from sheep, cattle, and pack stock, however. Transhumance, a historically significant activity, is still important across some acreages of the Sierra. Nonuse of public lands is a "solution" that some public-land advocates have taken to emphasizing, especially partisans of a particularly abstract formulation of pristine biodiversity. Yet battles over grazing have been a 150-year fact in the Sierra and will most likely continue with little easing. In fact, although there is some dispute about this, where livestock ranchers abandon grazing, they are likely to sell land into subdivisions, which raises a whole set of ancillary problems relating to land use and protection and to the ability of local and regional planners to cope with changes in use. Furthermore (these issues are indeed all connected), there is an ongoing stalemate between "wise-use" advocates, who suggest that public lands are to be used, and antigrazing forces, who argue that federal lands in particular are part of a national trust.

Wildlife

The initial fact about wildlife in the Sierra Nevada ecosystem is a big one, perhaps too easily ignored because it also affects wildlife across California and the United States. The first and most sizable shift that took place between the ports of southern England and Plymouth Rock in the early 1500s was a deep conviction not that religion should be free in America—it wouldn't be—but that in the colonies game and wildlife were a public property, not a private hoard.[12] Government in the United States is therefore bound to protect wildlife, which until the formation of the European Economic Community was not the case in Europe or, for that matter, almost anyplace else in the world. Determining exactly how migratory and territorial animals (and plants) are to be managed, especially where economic uses are essential for private land and frequently sought on public land, is not easy. This policy migraine has deep roots and reaches to a core of geographical and policy quandaries.

Federal and state governments are saddled with an interesting chore that has few upsides. Legally, wildlife is a statutory, government responsibility, and historically this has posed two separate bodies of problems: how should wildlife be managed, and, in particular, for whose interest are animals to be safeguarded or removed (animals and plants are rarely given standing on their own, the work of contemporary environmental ethicists notwithstanding)?[13] Second, what is to happen when the needs of private landowners, who possess about 50% of the area of California and 15% of the Sierra, are in conflict with the best management of wildlife (Ewing et al. 1988)? Opinions have changed drastically through time; Beesley (1996) has charted the eras of predator control and

animals in parks being treated like zoo residents. The fashions have shifted markedly.

Given the rather remarkable, if perhaps frail, state of the ESA, the most useful comment here probably is to note that habitat is a complex realm. Conservation biologists are still defining habitat requirements species by species and refining the nature of wildlife corridors; the basic needs of different species are less than agreed upon, and there can be little doubt that there are changes ahead, especially as the existence value of wildlife is discussed and as management priorities shift from game species to endangered species to, perhaps, all species.

Oddly enough, considering that wildlife in the Sierra preceded the arrival of Euro-Americans, many of the most important shifts in policy regarding wildlife are relatively contemporary—certainly since 1960. The influence of the Wilderness Act of 1964; NEPA of 1969, which began requiring the preparation of formal environmental impact statements; the Endangered Species Act of 1973 (and its subsequent controversies); the tenure of James Watt as Secretary of the Interior; and even the passage in 1994 of the California Desert Conservation Act are each in their own ways major federal stepping-stones toward wildlife recognition and protection. Before 1960, a generally laissez-faire attitude prevailed, and the effects of relative government indifference, or at least inattention, were certainly felt in the market hunting of the gold rush era when almost anything edible in the Sierra was hunted to provide meat for the miners. The same tendency to eat any and all meat affected livestock, of course, but though the population of domesticated animals could be, and was, replenished by the long-drive movement of animals from Mexico or the Southwest and Texas, there was no comparable restocking of wildlife. There is a notable dearth of information about the direct effects of market hunting or, for that matter, about predator control later on; the best work is in theses or is anecdotal.[14]

Wildlife policy has evolved through time from exploitation for eating to the extirpation of predators, into a contemporary era in which wildlife across the board is held to have value. Whether this "value" will extend to a reintroduction of major predators is an interesting question; the first moves of this waltz are being essayed in Montana and Idaho to less-than-happy reviews from various parties, and it may be some time before, as archenvironmentalist David Brower has suggested, grizzly bears are reintroduced to the Sierra to add a *real* element of "wilderness" to the backcountry. Because the encouragement of wildlife finds particularly vocal support in urban areas, and California and Nevada are becoming ever-more urbanized, it is true that "urban refugees" and city people are strongly partisan wildlife advocates. The existence of a constituency does not necessarily equal policy change, and the apparently intractable and extreme opposition will not break between resource users, who see wildlife and especially the ESA as an impediment, and wildlife supporters (including wildlife biologists, who tend to support the objects of their study with considerable vigor). The middle ground or compromise position is not accessible (Starrs 1994).

The linkages between Sierra wildlife populations, habitat preservation, resource use, and the desires of a relatively distant city population are strong, but the exact effects have yet to be determined. Federal and state wildlife officials face a bewildering variety of mandates that make any single policy direction difficult. This is typical of the 150-year history of official wildlife-government relations in the Sierra; the vernacular responses—hunting, trapping, habitat conversion, and use—speak more loudly.

Because habitat and resources of the Sierra often affect species at some remove, the domino effect in land use and policy is a fact: an easily understood case is the relationship between Sierra-derived water supplies and fisheries on both the east and west side—the cui-ui of Pyramid Lake, salmon and steelhead of the Sacramento–San Joaquin River systems, and the in-stream fish of the feeder streams to Mono Lake offer three notable, if notorious, cases. Vast levels of environmental engineering (like the Peripheral Canal) have been contemplated to attempt to correct a history of environmental modification like the damming of every one of the Sierra's major rivers—and the effects of California's human population rising to thirty-three million people. A similar story can be told for flyways (in both the Sierra and the once-marshy regions of the Sacramento valley), or for migratory game paths in the Sierra—animals, too, practice transhumance, and subdivisions intervene.

Because wildlife possessed in California what Native Americans did not dominate, and shared the rest, wildlife has been most influenced by human activity and change. And yet, if the policy climate for wildlife has warmed, it remains anything but friendly and tractable. The ESA attempted to put the needs of wildlife, especially endangered wildlife, first, and that act is now beleaguered.[15] When there is a concerted effort to solve problems relating to "natural" environments that have conurbations in close proximity, the results are both complex and often indifferent; Lake Tahoe's "protection," although an interesting example of intergovernmental and popular concern working in concert, is less than a complete success (Strong 1984).

Wildlife is notable in that almost all policy directions have favored one elite group or another—hunters or ranchers come to mind. Of late, a single-minded emphasis in the ESA on habitat protection is assumed, in turn, to offer a blanket protection to all local species. But that response is so out of scale that reauthorization of the act as it is now will be nearly impossible. Disproportionate policy rarely succeeds.

Fire

Whether fire is a resource or just a fact of Sierra Nevada life, it is certainly a potent motive force. Native American burning was widespread, and as active manipulators of the environment, Native Americans are receiving added recognition

(Pyne 1981, 1995). Since Indian times, policy relating to fire could hardly be less intelligent; consistently, policy on fire control has either accommodated government agencies with expansive agendas or sought to placate home owners contemplating movement into fire-threatened areas. Announcements as recently as a Saturday, 8 July 1995, *New York Times* headline to the effect that federal land managers are now considering including fire as a management tool for public-land administration come with so many caveats as to be risable. There is a certain caesura in the fire policy of the 1990s, an artful pause to allow assessment of potential costs and liabilities to catch up with the clearly stated opinions of conservation biologists and public-land managers that fire ought to be a useful tool for the sustenance and recuperation of public and private land.

The shifts in federal and state fire policy are on the record; certainly SNEP has devoted considerable effort toward gleaning a coherent fire history of the Sierra. The evolution of government fire-tolerance policies has the distinction of equaling perhaps only Bureau of Indian Affairs (BIA) changes in Native American policy for the most consecutive miscues and an equivalent contemporary state of impasse. In fact, the word "disaster" comes to mind, as several conflagrations throughout the Sierra suggest. There is no blame to be apportioned, but the evolution from let burn to moderate control to full control of any wildland fire has inevitably led to a contemporary questioning of whether zero-tolerance of fire in the Sierra has left the range ready for immolation.

The California Department of Forestry and Fire Protection (CDF) and the various fire-fighting agencies of the federal government (there are several, coordinated since the 1980s through Boise and the Interagency Fire Center) have either chosen or seen pressed upon them the role of protecting residents in the Sierra from wildland fires, on private and public land alike. Although various catastrophic California and Nevada fires have suggested that this arrangement is precarious, and the abilities of state and federal resources to protect human-inhabited wildlands is less than complete, it does bear noting that fires in Sierran settings have come nowhere near, yet, to the dollar costs and ecological damages wrought by the Oakland or Santa Monica fires of the last few years. The acreages of vegetation consumed, and the potential loss of firefighter lives, are, however, potentially huge in the Sierra, and the situation is not getting better.

Because assurances of continued fire protection exist, subdivisions and exurban movement continue to blanket the Sierra foothills—a tinderbox by nature, and especially so now that there is a sixty-year history of fire suppression. In fact, a policy of fire control has allowed growth, development, and populations to move into fire-hazard areas with some confidence that they will be protected. That can be a false faith, and the litany of CDF warnings that this particular year (fill in the blank) is potentially catastrophic because either (1) there was little precipitation so vegetation is tinder dry, or (2) it was wet and grass has grown tall, is now received as part of the yearly early summer ritual (like cleaning out the barbecue), with radio announcements and widely disseminated press releases. Because it covers the spectrum of California's Mediterranean-type climate and its fire-adapted vegetation, the warning is usually correct.

Movement of people into the Sierra foothills, often into areas of extraordinary fire hazard, is another example of vernacular activity taking place in a void left by an unwillingness, or inability, of California (and Nevada, in places like Verdi or Incline) to preclude sometimes dangerous activity. The failure to intercede is both understandable—private property rights are foremost in the United States—and in prospect deadly. There is, in essence, no one and no entity disposed to tell people that they cannot move into the Sierra foothills. Although CDF and the USFS have recently expressed reservations about claiming a fire-fighting mandate, generally in meetings rather than in explicit deliberative policy statements, they are also prisoners of an organizational style that in both cases has some paramilitary aspects—their charge is to get the job done, not to ask why (Weatherford 1981).

The current state of fire policy in the Sierra is profoundly influenced by the studies of wildfire researchers, ethnographers, cultural geographers, and resource managers; in the last three decades, knowledge has perhaps grown more in the field of fire history and ecology than in any other realm of resource management. Part of the reason for this remarkable growth is the paucity of information that existed before—the work of Harold Biswell was path-breaking and even revolutionary in its California setting (Biswell 1989).[16] And the most important point that Biswell and his students have established is simple: much of the Sierra is in natural terms a fire-evolved landscape. To expect that fire will be kept apart from so reasonable a home as the Sierra foothills is a vain hope—the big fires will come.

How policy will evolve to encompass this knowledge is an interesting question—the vernacular changes that have been made in the Sierra are dire, and almost the exact opposite of what scientific evidence suggests is desirable. With Humpty Dumpty scattered in pieces through the Sierra, it is not easy to see how accountability, self-reliance, and an awareness of fire and its hazards can be transferred from government to residents. The expectation is still that humans not only will survive, they will prevail.

Minerals

Mining in the Sierra has been a source of enormous policy change. At core, the facts of minerals and mining are geomorphology and economic geology—the study of landforms and the locations of mineral deposits. It is locating and exploiting those resources, or choosing not to, that is an ongoing problem. The innovations driven by community and miner needs in the nineteenth century have already been noted, but there are landscape changes associated with mining that deserve at least cursory mention; the substance of law is covered in

Beesley 1996. The search for gold, begun in Spanish days in the Los Angeles Basin (where gold was found), grew to obsession after 1848 and produced a pulse of migration to California that populated significant parts of the state. In fact, several of the Sierra foothill counties did not equal their 1850 and 1860 population counts until the censuses of 1980 and 1990. The ebb and flow of boom and bust is fact. And yet, the search for mineral wealth in the Sierra is a chronicle of massive and sequential changes in policy, which have reached and influenced national parks—the Mineral King Wilderness of the southern Sierra was, after all, widely held to be a potential trove of precious minerals.

Mineral exploitation and exploration in the Sierra have produced changes in community formation, property, water ownership and access, subsurface ownership rights, and the law relating to the impoundment of water, and an unceasing series of changes and innovations in the control of energy, transportation, and the technology of mining and its effects. A number of these issues are dealt with in standard histories by Kelley (1959, 1989), Paul (1963), Farquhar (1965), Rintoul (1976), Cleland (1964), and McWilliams (1949). Along with water, mining has been better treated in policy and its progressions than any other aspect of the Sierra. As a result, only a few observations really need reiteration, and these are where the connections between the commonplace and the more orthodox forces of policy making need clarification.

Mineral exploitation created numerous settlements throughout the Sierra but especially along the eastern and western margins of the range. It brought roads to the communities, including railroads in some places, and even saw water transit and riverboats reach into small Sierra tributaries. Like a shotgun blast, the gold rush in California put pellets of towns from Oroville to Yosemite and into the watershed of the Kern River. For thirty years, it was the demands of the mines that fueled the state's economy; almost everything was focused on mineral development or on meeting the secondary and tertiary needs of the merchants and mechanics, the inventors or institutional bankers who had their jobs because of mining. That is understandable enough, even though a thirty-year boom is not easily counted upon.

But the mines also produced landscape changes on an epic and problematic scale. The mine entrepôts—Sacramento, Stockton, San Francisco—were dependent upon the booms and collectively feared their slowing. The creation of financial markets, indeed, the funding of a world empire based upon investment of California mineral capital (including the financing of the Comstock Lode development in Nevada), grew from the Golden State and the Sierra. It was literally a worldwide financial market, its reach documented in journals like the *Mining and Scientific Press*, published from San Francisco (and Oakland, after 1906). These linkages to the world economy made the Sierra of more than casual worldwide importance.

The technology that developed from the Sierra's mines—placer mining on increasingly large scales, culminating in hydraulic and dredge mining and hard-rock exploration in the Southern mines—was also widely diffused and born of a boosterish climate.

Water developments associated with California mining not only shaped the water law of the West, but, linked to technology and capital, law and practice in California water were moved elsewhere. The reach of San Francisco (and later the East Bay) to the Sierra, like the stunning technological innovations of Los Angeles in the Owens Valley aqueduct, was directly tied to technical advances from the mines, including sophisticated siphon systems, turbine manufacturing, and pipeline technology. It was mineral exploitation in the Sierra that made the contemporary Bay Area and Los Angeles plausible—innovation driven from the bottom up.

And it can be said, though not without some hesitation, that the modern environmental movement owes no small debt to mining. The effluents, and in particular the sediment flows, moving downstream from the Feather, Yuba, Bear, and American River drainages as a direct result of hydraulic mining began clogging the waterways of the Sacramento valley once the river profile decreased in steepness and sediment could settle. This sediment, of course, clogged the waterways and led to the by now familiar "dike wars" of the Sacramento River system, with community after community raising its levees in hopes of excluding rising river waters that were flooding towns with increasing frequency (Kelley 1989). Towns resisted continued mine development, but their petitions were turned down repeatedly, until finally the 1884 Sawyer ruling (*North Bloomfield*) in federal court closed the door on hydraulic mining—so long as miners insisted upon dumping debris directly into streams. A comparable decision was rendered in California court in the *Gold Run* decision that same year. This one-two judicial questioning of established practice hardly signaled the cessation of hydraulic mining, but it did cap off the most extreme abuse and compelled the increasingly corporate hydraulic mining ventures to impound their waste and sediment.

This principle of accountability in mining marked a major change in western life and certainly affected the Sierra in its varied forms. No longer was environmental exploitation severed from any lasting consequences—blame and redress could be fixed. Although gold and silver were pumped into the California economy, this newly imposed responsibility was perhaps as significant in western life in general. It established a new baseline for resource development. Other changes in water law, power provision and dam ownership, the protection of urban water supplies through flumes and pipelines—even the establishment of several hundred still-extant communities throughout the Sierra—are related to mining as well. Mining was a boon to California life, if also one that literally reconfigured many of the rivers, canyons, Tertiary gravel deposits, hillsides, and not incidentally the townscapes and wallets of California.

Forests

Forests in the beneficent climate of California and Nevada still grow slowly—far more slowly, in fact, than the forests of the Southeast. The lag between forest cuts is chronologically significant. While experts praise the regrowth of the southern United States' forests, California and the Pacific Northwest are not long into their cycle of regrowth and are suffering accordingly. Policy in the California forests has been two-pronged, one prong following the course of federal control of much of the Sierra, the other tracing the changes brought by private ownership. In general, policy relating to forest use and exploitation in the Sierra is more tightly controlled than that for virtually any other activity—the apparatus of timber-harvest plans, accountability for poor harvest practices, and the requirements for replanting and, increasingly, making certain that replanted trees actually thrive are the results of a long history of forest development, even if enforcement has varied in effectiveness.

Forests were cut over hard during the nineteenth century, and regulations followed. The CDF and the USFS have split jurisdictions, CDF in charge of administering the use of private lands, including the increasingly important nonindustrial private forest lands (smaller private sales), and the USFS handling the harvest and jurisdiction over federal lands, outside of parks. In essence, much of the evolution of forest management has turned around the question of what sorts of forests are expected or planted after trees are removed—major reservations about the harvest of timber itself is a relatively recent phenomenon, although groves like the Big Trees were set aside early on for posterity (Huntsinger and McCaffrey 1995).

In the main, policy has allowed cutting of forests in the style of the day. Those styles have changed, and in the 1990s, the clear-cut, with its maximum in efficiency but low diversity in replacement species, may be permanently on the way out, in part for aesthetic reasons, but also because of increasing concerns about biodiversity and the effects of both clear-cutting and replanting of monospecific replacement trees. Other changes, including prohibition of herbicide use to knock back brush invasion and to encourage regeneration of the planted "desirable" species, may or may not last; already the herbicide rulings have been slackened.

Certainly forests are more than commercial products, and that realization creates policy implications too. Forests as a locale for wildlife and biodiversity, their importance for aesthetics and recreation, are all important alternative uses, not always in keeping with an industrial forest and policies designed to encourage maximum sustainable forest timber yield. The teeter-totter swing of federal forest harvest sales has not helped the Sierra's forests, and an ever-increasing activism of Sierra residents is placing increased emphasis on sustainable resource management instead of the more typical "cut and run" policy. But coming up with a formula for "sustainable resource management," especially given a seventy-year growing cycle and an undistinguished record of past regeneration and replanting, is easier in theory than in practice. No doubt local communities prefer steady jobs—whether the government and private forestry firms can make that happen remains to be seen.

Added to the mix should be "Hardin's Law," named after the distinguished population ecologist Garrett Hardin, which argues that no natural system can at the same time maximize both efficiency and diversity. In the generally reciprocal relationship, a trade-off always occurs; a system becomes efficient at the cost of diversity, and a stable and diverse system is not, in short-term production, particularly efficient. Many of the debates about forest health of the last thirty years have turned over this general statement, which echoes Eugene Odum's 1960s arguments about stability and diversity, arguments at the foundation of modern ecological theory that are now in some quarters being questioned by alternatives like state-transition models. Yet regardless of the scientific debate, the monocultural stand, especially as it regrows into a clear-cut site, is simply not perceived by much of the American public as a good thing. Although the forest-products industry has good reason to dispute this perception, efforts in landscape architecture theory and timber-harvest practice are strongly directed toward camouflaging, with landscape corridors and cosmetic cuts, what is pervasively seen as ugly. The effect of this perceptual resistance to what has for years been standard forest practice, like the on-again, off-again ban on the use of herbicides on federal land clear-cut sites, is cloying to forest experts, but proof that the public sometimes does care, even if "wrongly," about certain issues.

Water (the Hydrosphere)

Although Larson (1996) covers the policy implications and history of Sierra water, two essential facts deserve reiteration: California, including large parts of the Sierra, is fundamentally a semiarid realm through much of the year, and consequently the seasonality of available water is important, and, second, no state has been more highly engineered to redress this simple fact of distribution.

The water history of the Sierra has already been the subject of dozens of top-flight studies in historical geography and environmental history. Probably no aspect of life in California and Nevada has been covered as well as hydraulic history: at least thirty-six book-length studies come to mind when considering the natural and human facts of California and Nevada's waterscape. Because water is the limiting commodity in what geomorphologists refer to as a "transport-limited landscape," it is no wonder that the Sierra Nevada, on both east and west sides, is water-obsessed. Especially in matters of agriculture and urbanization, water questions are immensely complex; land and agriculture are, after all, a founding concern of the United States, dating to Jeffersonian agrarian confidence in the yeoman farmer (and distaste for urban

places). And yet, agriculture (and, increasingly, city growth) in California and Nevada is limited by water supply, so the manipulation of that water has become fundamental to the practice of agriculture. And a multiplicity of other forces contend for secure access to water.

Arable Land

Farming is important in the Sierra, and where open and appropriate land is not cultivated, it makes for splendid wildlife habitat and hay lands. But with the difficult terrain, a relative absence of open valley floors, and the very elevation of the Sierra, there is not a lot of cropland agriculture in the Sierra, certainly nothing comparable to that of South America on similar slopes. The vagaries of climate require irrigation throughout the Sierra, and in some of the areas where tree crops proliferate farmers make a steady income from orchards and nut crops.

The importance of marijuana growing in the Sierra is undetermined. The marijuana harvest is probably sizable, but no one wants to know for sure. Shifts in state and federal enforcement policy (especially relating to land seizures) have moved such cultivation off private land and into the national forests.

There is a great deal of comparable casual agriculture, in some cases by residents of long duration, in other cases by relative newcomers who savor the opportunity to farm on a five-acre parcel. Although attractive in prospect, farming small tracts is probably among the most destructive possible uses of land and a direct result of the decreasing profitability of larger-scale ranching in the foothills of the eastern and western Sierra.

Movement

Getting across the Sierra has long been a strategic necessity. Such mobility has not come cheaply. Whether the responsibility for locating routes fell to trappers, western "mountaineers," or railroad company scouts, or whether the trails followed the Native American routes across the Sierra, as had been the case in the Appalachian ranges a hundred years before, a mountain range with passes nearly 2,500 m (1.5 mi) in elevation, except at the northern and southern ends of the batholith, posed a travel hazard.

And yet, goods and people had to be moved. Initial efforts to cross the Sierra were avowedly commercial; John Sutter and other early-nineteenth-century Sacramento and San Joaquin valley landowners wanted overland visitors to whom they could sell land. They had to cross the Sierra to get to California. This movement produced industries, in the mid-nineteenth century, on both sides of the Sierra—communities on the east side for staging, to provide guides, to offer changes of stock, and to supply the travelers, and a burgeoning community on the more gently sloping west side to shepherd the arrivals to an appropriate resting place; the arrival then would have had a passing resemblance to stepping out of baggage claim in a foreign airport near the equator.

Some of the best-known trips were the unsuccessful ones; geographical information about the Sierra was not reliable, and observed data about crossing Utah and Nevada in the 1840s was still worse. Corrupt guidebooks misdirected overland travelers, some of whom perished either on the way to the Sierra, or in it. The Donner Party holds a special place in the American imagination, perhaps because it is difficult, standing near the Interstate, to imagine so much snow on the ground at Donner Lake. Yet the Donner expedition brought a dozen people to eternal repose and has fed the imagination for nearly 150 years, a traveler's (real) nightmare.

Alternative routes across the Sierra were widely sought but not often found. The best passes lay far to the north, at Beckwourth Pass, and still farther to the south, at Walker Pass. Although both routes were found relatively early, neither was especially attractive (Howard 1993; Todd 1949; McCarthy 1974; Nash 1985; Stewart 1962; Vale and Vale 1983). Routes across the central Sierra were scouted through the 1850s and 1860s, largely by groups interested in the potential of a railroad route and, of course, the speculative value of land that either came with the railroad (the "checkerboard") or land serviced by the tracks themselves. No all-weather passes were found; even the route at Donner required a vast investment in highly experimental technology, including the rotary plow, roadbeds chiseled into granite, and a vast network of snowsheds to protect the tracks from avalanche. Technological innovation was critical to transit across the Sierra. It was worth the expense; San Francisco, the Sacramento valley (and especially the booming wheat crops of Glenn and Colusa), and the San Joaquin valley beckoned.

Left behind was the entire east side of the Sierra and the pocket valleys, isolated and even today relatively sparsely settled, in the Sierra massif itself. Near Quincy, south by Sequoia at the Kaweah Colony, at John C. Frémont's Mariposa Colony, in the agricultural and irrigation colonies of the San Joaquin valley, was other potential. Once the railroad reached Quincy (late), its future along the Feather River was set; epochs of booms and busts dominated elsewhere. Genoa, on the east side of the Sierra near the foot of the Kingsbury grade to South Lake Tahoe, was the oldest community in Nevada but was deserted in the 1860s when Brigham Young grew restive, fearing an invasion from an increasingly hostile federal government looking over the Utah Territory (already dubbing itself "Deseret"). The Mormon Station at Genoa was abandoned.

The boom in trans-Sierra travel came with the Comstock exploitations. As capital shifted after 1859 from California to Nevada—sometimes literally, sometimes in the nineteenth-century equivalent of electronic banking, by telegraph—the rush was on to depart the western mines and head to the silver strike in Virginia City and, soon, elsewhere. With the railroad not yet completed, the communities between Donner Pass and Tioga Pass—at 3,003 m (9,998 ft), the highest of the main Sierra passes—grew quickly. Yet the passes had to be kept open, through sometimes difficult weather. Government monies were slow in coming to what were predominantly private ventures.

The Union Pacific–Central Pacific Railroad, finished in 1869, could run through most weather, but its access was modest, and stages and secondary lines, or livery horses, had to carry passengers once they stepped off the Zephyr in Reno, Truckee, or Rocklin. To keep the railroad open was sometimes an epic struggle and was played as such in the local newspapers. Severe storms and their effects on the trains are still recorded in photographs that dot the walls of Sierra towns. Passengers across the Sierra could be delayed for days on the east side; not bad news if they were in Reno, which was cultivating an insalubrious reputation, but decidedly inconvenient.

Railroads and transit were decisive for one industry, the mines. Railroads or the Truckee River itself transported logs and supplies from the Sierra and parts west back to the Nevada mining operations, and provided many of the luxuries that mining barons were used to, including oysters, sourdough from San Francisco, and the San Francisco daily *Call*, an early entry into that city's subpar dailies. The spillage east from California made Nevada what it was; without that rail traffic and the capital, technology, equipment, and knowledge it brought, there would have been no booms. Yet travel was arduous at best.

Roads across the Sierra were a misery. Through the 1920s, most highways were toll roads, the exception being the sometimes federally financed Lincoln Highway, ultimately completed in 1930 but in 1912 put forward as the first "complete" transcontinental road (Hokanson 1988). It was anything but. These tracks and byways have been treated by several authors—Tom Howard's (1993) dissertation is the best work to date. The routes varied in effectiveness and seasonal reliability; Sonora Pass, for example, had been identified in 1852 as a quick route, but even today its 22-degree pitch near the summit intimidates all but the boldest traveler, and it is shut with the first breath of winter. And yet, these were often the same routes that sheepherders used as a regular part of their circular routes across the Sierra; a sequential pattern of use, with passenger roads as the next alternative, was kicking in.

As the strategic importance of the West Coast grew during World War II, the pivotal role of roads was recognized. The National Defense highway system, first proposed in 1910 but not supported until the 1940s, was slowly moved over into what is now known as the Interstate highway system. Its first purpose, and raison d'être, was to move troops and matériel in the event of a West Coast front. Little wonder that the flavor of George R. Stewart's classic *Storm*, with a huge blizzard shutting down old U.S. 40 (now Interstate 80) has the crisp ring of high drama. It was vintage 1930s and '40s.

Casual travel across the Sierra was primarily by train until the 1950s. With the arrival of skiing at the Squaw Valley Winter Olympics in 1960, the Sierra became a destination resort. Even after the Winter Olympics departed, demand to see the facilities that had been spoken of so widely, and televised for the first time, built steadily. Now, the Sierra sees near-gridlock on many Friday nights and Sunday afternoons through the winter—worse than that for three-day weekends, with skiers from San Francisco to Hollister, from Mendocino to Marysville trying to get an early start and hit the slopes.

The industrial uses that once supported the railroad, and which in theory were supposed to assist the Interstate, are now regarded as bothers. Trucks jackknifing and grinding at slow speed across the battered concrete pavement are a major source of skier distaste, and with good reason. Yet the major change in transit through the Sierra is simply the replacement of private initiative with federal and state largesse—the Highway Trust Fund and state monies maintain a series of Caltrans stations along Interstate 80 that spit out a series of familiar orange trucks once the snow falls or accidents occur. Use of the other roads across the Sierra is more a matter of happenstance—Highway 50 remains open much of the year, but not always, and it is slower than its northern sibling; the roads south of Highway 50 close with the snows.

Among the great controversies every year is commerce associated with opening of the Tioga Pass road. Businesses along the east side of the Sierra are at least somewhat dependent upon traffic across the Sierra, so they eagerly await an early opening of the Tioga Pass road. Because that road goes through a national park, there is resistance to what is sometimes voiced as "accommodation" to the economic imperatives of east-side businesses, however great their hardship. And yet, each year the Tioga Pass road is plowed and the tourists flow, leaving tips and taking snapshots of scenery with them. This commerce is supported by federal monies—often by Yosemite National Park funds—but there is little choice.

Recreation

Precisely what the Sierra is supposed to be from a human perspective, what it has been through the ages, has changed greatly. The ideal of wilderness is oddly American and especially western, and this has been recognized for years. And yet this obsession with "wild" land is also often seen abroad as one of the nobler aspects of life in the United States; it draws tourists in large numbers, and there are few places on earth where so much land is given over to so little directly economically productive activity. Land and wilderness are instead said to have existence value, and for that praised. The vision of wilderness as a dark, spectral space, daunting and dismal, was by and large left to the East.[17]

And so recreation has become, especially in the post–World War II years with the development of leisure industries for the middle class, a major business for the Sierra, and that interest in re-creating (the word is significant) is the source of both profit and many policy changes for the Sierra. In essence, many of John Muir's arguments, praising preservation and existence value over economic and commercial utility, have been reversed—Gifford Pinchot, who would have won those

arguments in the 1890s, would find the situation less helpful today. No longer is utilitarianism foremost; instead, respect for the undying values of relatively raw nature is praised.

There are still exploitative uses of nature like river running, backpacking, fishing, and hunting, but they tend increasingly toward nonconsumptive use. Preservationist ideas have a certain rebound, although the number of backcountry hikers in the Sierra is down, despite a huge surge in the population of California. This, too, is part of the episodic attraction to nature; the hikers who swarmed into the Sierra in the 1970s, "loving the wilderness to death," have settled down to children, Thermarest pads, car camping, and mortgages, instead of VW microbuses, dried fruit and gorp, and cheap, cotton sleeping bags. Yet advocacy for wilderness and backcountry recreation remains high.

Among the very difficult policy topics in recreation is elitism. In majority nonwhite cities like Los Angeles and Fresno, and even in white enclaves like Chico, complaints are increasingly being tendered that government agencies like the Forest Service and the Bureau of Land Management (BLM) are not attuned to the needs or desires of minority populations. The accusation is made, whether true or not, that many of the recreational apparatchiks at a national level are, in essence, in the business of protecting a very large amount of land as the stomping ground, a kind of extended backyard, for an educated, white elite that is not interested in making that terrain available to a more diverse citizenry. Whether these agencies are elitist or not, there can be no doubt that a major policy front will involve opening up the forests, parks, and other public lands and lakes to a nonwhite population that has historically been very much in the minority of California users of such resources, largely because their needs have rarely been considered, studied, or sought out.

Parks and the uses of parks change widely through time. Galen Cranz has written about this, in particular from an urban standpoint, but her arguments are extensible to the national forests and other public lands of the Sierra Nevada ecosystems.[18] Finding what is desired, what is attractive, and what is needed will require more assiduity than has been displayed to date. The accusation of elitism is loathsome to many who grew up backpacking in the Sierra, in part because it has the ring of truth and is a trigger to volumes of white self-reproach.

The national parks are, of course, the crown jewels of the Sierra, and they have been treated in a large number of essays, enough that little reiteration is needed here. Whether the parks were chiseled from the best of the country's surface or, as Alfred Runte has argued, they were unusable lands that were placed in parks as a sop to those who wanted preservation but had no political clout to protect resource-rich land, is for the experts to debate (Orsi et al. 1993; Runte 1987). But like the urban parks, the national parks have gone through their sequence of different eras: as a lonely refuge for the very few; as a kind of botanical and scenic zoo; as a literal zoo, with regular public feedings of caged bears; as theme parks, with nightly firefalls as in Yosemite; and as safe havens for the distant-from-home, with fire and predators like mountain lions and coyotes removed. Natural features are always modified by human activity and have been in California and Nevada ever since the arrival of Native Americans. Nature is a socially constructed fact, not a biological absolute. Instead, the natural world of Yosemite, Lassen, Sequoia, Kings Canyon, Mono Lake, and other sites has been transformed into a palatable version for the visitors, many unfamiliar, now, with a "wild" nature, who are the parks' constituency. And one of the last things that anyone wanted to acknowledge through much of the history of creating American parks is that, in the wild, nature kills (Leopold 1991b).

Yet kill it does. Battles over access to parks produced safety improvements and proved the utility of parks for other purposes—like providing, in Hetch-Hetchy, a prime reservoir site for San Francisco. Roads were run in—not early, surprisingly. Yosemite passed nearly sixty years as a park without a road to the valley floor; there was no road until the Raker Act was passed in 1913, the same act that permitted San Francisco its Tuolomne water supply. Yet now there are predator-control programs, relocating "problem" bears, carefully improved trails (often trails first built by Civilian Conservation Corps or Works Progress Administration crews of the 1930s and 1940s), and even gun-toting Park Police. Parks are little different from medium-sized cities through significant parts of the year. The preservation of natural values and recreation is a matter of question, or at least, taste. Liability in the national parks is an increasing problem for the Park Service; it is difficult to overestimate the amount of trouble that city visitors can get into in a park setting.

On the other hand, there is no shortage of places where hikers and travelers can get seriously close to nature. Boating down any of the upper- and middle-range Sierra rivers, primarily on the west side, but sometimes even on the east side (the Walker, Carson, or Truckee), can be dangerous. Some choose that. Others want to drive a car through a Big Tree or pivot the satellite dish on their mobile home from the valley floor. This is the United States; forming policy to accommodate all these views is no easy matter, in particular because some of the visions (seeing pristine nature without any indication of humans and accommodating Winnebagoes, for example) are flatly contradictory and mutually exclusive.

Property (Habitat)

Land can be said to be the ultimate American sacrament, as Paul Gates taught two generations of land-tenure researchers (Gates 1960). It was the availability of land that drew settlers to North America during the seventeenth century, and it is still land and opportunities for getting land (although it is no longer "free") that draw immigrants from around the world in the late twentieth century. Acquisitiveness and opportunity for gaining land were driving forces for two hundred years of United States history, until the selling of the

public domain finally ended with FLPMA in 1976, with the recision to the BLM of land previously available to homesteading. What began in the mid-1970s was a new era, or at least a different one, in theory emphasizing government land stewardship instead of land alienation, or sale. This shift has been a boon to scientists and environmentalists, who praise the preservation of public lands (about 85% of the Sierra) as potential, if not actual, wildlife habitat and a laboratory for biological diversity. This new era is not so praised by potential resource users (who often do not own the land they exploit), or by potential settlers, who see the end of the free (or cheap) land epoch. This policy juncture is crucial, maybe the most important in the geography of the Sierra. And yet it is also a simple fact: homesteading in the Sierra is no longer possible, at least not under a government aegis, and therefore the value of land is appreciating under the exigencies of supply and demand. Land is increasingly being commodified because it is scarcer all the time—scarcer because the supply is cut off and because there are ever more people in California and Nevada who want access to the Sierra for varied purposes.

The Sierra has seen an essential contrast between different views of land and property, and understanding that land is not a fixed or universal commodity is an important facet of policy. As Joe Jorgensen has suggested, land is not the same thing to all people (Jorgensen 1984). Different visions and different attitudes toward land are shaped by cultural facts including occupation, race, ethnicity, religion, history, and language. In Jorgensen's study, the attitudes of Mormon farmers, Native Americans, ranchers, sodbusters, and environmentalists were dramatically at odds. Although his study was in microcosm, across the larger Sierra a vast panorama of views of land could be expected. These differences have produced conflict in the past and will continue to do so in the future.

Among the more intriguing developments in land policy in the Sierra is a homegrown movement in California pressing for a locally informed understanding of land-use issues. Known as the bioregional movement (or sometimes arguing for "watershed consciousness"), it presses strongly for communities to attend to local needs before becoming involved in larger or more global issues. In essence, the argument is that one's own house should be in order before the housekeeping of other groups and communities is challenged. With a number of fronts, some in the San Francisco Bay Area, some in the Sierra Nevada foothills, especially around North San Juan Ridge, this bioregional consciousness movement is anything but another "flaky" California trend; it is instead finding vast support around the United States and abroad as an example of socially and community-informed land-use planning, with extensive local involvement.[19] Yet bioregionalism has some elusive and uncertain effects upon public land and public land management, in part because the emphasis of Gary Snyder, Peter Berg, and others is on local control, not absentee ownership. The policy implications have yet to play out entirely; the movement is still building, but many groups in the Sierra are taking this bioregional ethic to heart and producing strong critiques of the federal government's distance from local issues.

The call for "community-based resource management" is loud in the environs of the Sierra Nevada and has some interesting resemblances, in fact and theory, to the wise-use movement, although with a different politic. In fact, the Sagebrush Rebellion, centered in Nevada, Utah, and other states of the intermountain West, has an expression in the Sierra as well. Although summarizing the current situation is not easy, perhaps it might best be contained within an insistence that the federal government continue to be a steward, with a higher degree of sensitivity instead of pressing agency goals that have little semblance to local needs and ideas.[20]

Existence Value (Biodiversity)

Whether humans are a part of "nature" may not necessarily sound like a policy question, but in regard to the Sierra Nevada it clearly is. Land can be preserved and conserved in many ways, and whether the human presence is to be a part of that is an important issue (Leopold 1991a). Studies of park formation and biodiversity reserves generally do not include humans as a component, an expression of a Frederick Law Olmstead tradition that still holds in the 1990s. Yet opposing views are heard, and have been in the United States since the 1920s and 1930s.[21]

Just how land is best to be preserved and sustained is always a difficult question. Among the important decisions that have to be made is whether a human presence is considered acceptable in public lands or, especially, on park lands.[22] Although the official view of such matters is clear and skeptical, the pattern in other parts of the world, where there are not such extensive tracts of "pristine" land, is much different. There, parks and government lands are often dotted with legally allowed residents; in some cases, national and regional parks actually include a variety of cultural features, sometimes entire communities, tacitly acknowledging that the human presence is "natural" there. Comparable recognition has not been achieved in the Sierra, although the "problem" of inholdings in the national forests is at times difficult and contentious and a powerful lobby supports it.

There is, furthermore, a great deal of public policy strain that develops in the legislative process between the needs and imperatives of California and the West and those of the eastern half of the country, where facts like water shortage, public lands, insecure tenure, spaciousness, a vast wildlife population, and other such fundaments of western existence are at best poorly understood. This becomes an issue, especially when views of the rights of nature and of natural features on the landscape are pressed, as in Christopher Stone's eloquent views (Stone 1987, 1974; Sax 1987). There are also some similarities between this difficult regional impasse and the public trust arguments first formed by Joseph Sax when he was at the University of Michigan—public trust arguments

are much more influential and far-reaching when they apply to half the land or more in an ecosystem.

Public Trust

The public trust argument, in general a development in law of the 1960s, holds that a number of resources amount to public goods and that it is the responsibility of government, and especially of the federal government, to preserve the quality of those resources for future generations. As Harrison C. Dunning has noted, the legal content of the public trust argument is a relative novelty in the United States, but the principles that underlie public trust date back to classic times, for it was in the public interest during Greek and Roman times to maintain navigability, preserve resource use options, and serve the public good.[23] These ideas apply especially to water, wildlife, habitat, and a variety of positive externalities such as clean air, clean surface water, unpolluted ground water, and other qualities and quantities that have a formal physical expression but that have a value not easily commodified.

Public trust doctrine explicitly puts responsibility for maintenance of several forms of environmental quality on government, and, increasingly, case law suggests that it may come to be the ultimate authority (and responsibility) of both federal and state governments to safeguard resources for posterity. The argument is still forming, however. Exactly what falls into the public trust is not always easily resolved, something that Joseph Sax recognized in 1970 (Sax 1970). But how broadly these arguments for the preservation of future options can be extended is the real policy dilemma in the Sierra. There are extremely difficult and divisive issues that tie into private property rights, and the public trust doctrine quickly runs afoul of some of these land-rights issues if habitat preservation (especially on private land) is upheld as a necessary requirement. The wise-use movement, with its own polemics, takes a Hooverian view of right and wrong ("the business of America IS business").[24] Given a 150-year history in the United States of land acts that appeared to place fee-simple ownership of 64 ha (160 acres) and up of land as a sacred right, the line between public trust preservation of wetlands, endangered species, in-stream flow, genetic resources, and other precious goods is not easy to draw. As a case in point, there is the near-total collapse of the fisheries of the Sacramento River, so reduced in vigor that any number of technical fixes, many of questionable effectiveness, keep being proposed (Black 1995).

And yet, there are arguments for the preservation of options that are impossible to gainsay. When Wallace Stegner wrote of wilderness as a part of "the geography of hope," when Starker Leopold insisted that wilderness and national parks should serve as "vignettes of primitive America," there were striking, important themes. Such extents and acreages, such reserves and refuges, may appear to residents of the developing world as an impossible luxury, but it is a luxury that many residents of the United States, at least until the

1990s, have insisted upon (Sax 1980). And they may well have a point.

CONCLUSION

Policy evolution in the Sierra can be traced through any number of different landscape elements. The course of a few are included here. The essential point is that the Sierra's landscapes have gone through a series of important evolutions driven by two forces, one essentially vernacular and improvisational, and the other perpetually reactive and "establishment" oriented. The forces of the establishment are those recognizable, even shopworn, elements that are traditional to history and policy studies but can hardly be taken as inclusive. Many of the most significant changes to the landscapes of the Sierra, maybe almost all of them, are the product of local demand and vernacular activities. Policy is after the fact—vernacular change is avowedly contemporary.

Five broad themes were introduced in this chapter, and they have framed the arguments throughout. *Implantation* has surfaced in looking at the arrival, colonization, and beginning of community entrenchment, whether in grazing, wildlife and habitat, agriculture, mining, or the use of fire as a management tool. Ideas and practices are as much subject to implantation as the scribing of town boundaries or the development of water law. *Exploitation* is concerned with the use of land, whether destructive or consumptive; in recreation or water; using the tools of fire or chainsaw or residential "ranchette" or dam. *Tenure* examines how humans have made their use and possession of the land felt, taking the land as private property, managing it as a commons of individuals or for a village or group, holding the land available to all as the seldom-seen but much discussed "open-access good." *Intensity* is perhaps the trickiest of these concepts, for the vehemence and rigor of use does vary everywhere across the face of the Sierra, and through time. Some landscapes are hard-hit by vernacular use—some have in effect been sacrificed by private or government practice, by law or by custom. A dam site would be a prime example, with the flooding of a sizable area deemed in the greater public good, with an intense use (being underwater would seem to qualify as that). Finally, *linkages* look at two different realms—the roads and power lines that bind the Sierra together or that are sundered or blocked, and therefore restrict joining. But fully as important as physical ties are the more ideological connections: economies bound together; ideas about wildlife or subdivisions; access or exclusivity that has diffused from one place to another; the expectations and aspirations that lead San Francisco Bay Area urbanites into rural retreat in the Sierra foothills, seeking some variant of "the good life." These five themes underlie everything in this chapter.

The landscapes of the Sierra show a welter of human influences. Policy has responded, sometimes appropriately, but only in the rarest of circumstances preemptively. That is an important part of the Sierra's policy story, and it is the study of geographical change, of exploitation, destructive and not, writ large.

NOTES

1. Harrison C. Dunning, of the University of California at Davis School of Law, notes that such broad statements as these require some qualification, that indeed the law is a jealous mistress and there is little that is not addressed, even by omission, in legal practice. For example, as he states, "the policy of recognizing use rights (the usufructuary policy), was in fact embedded in the establishment law as well as the vernacular practice" (H. C. Dunning, letter to the author, October 6, 1995). He is obviously correct; for government not to govern is indeed a form of government.

2. Walters's point is amply borne out by data from driver's license changes that used to be collected by the California State Department of Finance, now held by the Department of Motor Vehicles.

3. The evolution of wildlife policy is traced later in this chapter, but a cogent summary of the different American eras can be found in Dunlap 1988.

4. On the management of both Yosemite and Sequoia–Kings Canyon National Parks, there is the remarkably effective special issue of *California History*. Of special note are the essays by Dilsaver and Strong (1993) and by Runte (1993).

5. Studies of Native American roles in the Sierra are relatively numerous. Among the relatively recent works that include discussion of Native American management is Vale and Vale 1994.

6. This instructive lapse between land use and policy, especially when the group involved is a racial or ethnic minority, is discussed by Limerick (1987).

7. On the nature of environmental change, there is a great deal of discussion, much of the best from Harold Heady and James Bartolome; for an example, see Heady and colleagues 1991. For an overall picture, although dated in the light of more recent analyses, see Burcham 1970 and 1982. For details of the California social situation during Spanish-Mexican times, see Pitt 1966.

8. On the details of *Californio* ranching during the peak era, see Rojas 1979. The accuracy of the Rojas view is discussed in Haslam 1986. There are also essays by D. Hornbeck, although his voice tends to be absolutist. The writing of Jo Mora (1948) (also aptly illustrated) is a surprisingly accurate and tasteful, and therefore instructive, treatment.

9. On the overall context of the social and physical elements of land-use change after the introduction of livestock, there is much. Perhaps the best is Bishko 1981. It follows on the classic 1920 study of Klein. The great benefit of Bishko, Klein, and Butzer is that they see clearly the connections between thirteenth-century laws and practices and contemporary problems, although those are articulated with varying degrees of grace. See, as a case in point, Butzer 1992 and Jordan 1993. Commentaries on the New World roots, especially as they affect California, Nevada, and the Southwest, appear in Dusenberry 1950 and 1963.

10. See Douglass 1985. The longer, and more thorough, study is the earlier volume, Douglass and Bilbao 1975. These treatments, along

with those of Richard Harris Lane, offer a portrait of cultural assimilation and the influence of livestock grazing and the practices of transhumance on the Sierra and its ecosystems; see, for example, Lane 1985.

11. The photograph archives of the Sierra Club in the Bancroft Library show virtual equid waves, waiting to attack the Sierra front from the trailheads; photographs of meadow scenes suggest that the effect of grazing was more interspecific swap than cessation—and horses, burros, and mules are notably harder grazers, and less efficient, than sheep and cattle.

12. On wildlife and wildlife law, see Lueck 1989, Lund 1980, Smeltzer 1985, and Tober 1981. On the general status of wildlife in the United States, see Dunlap, 1988. It is matched by Mighetto 1990.

13. Stone (1974) takes one point of view, arguing for a generic equality—a view that Merchant (1992) has posited as part of a generalized ethics of nature, but which Lewis (1992) repudiates as an over-the-top form of "speciesism." Occupying an intriguing middle road is Bourjaily 1984.

14. See Stine 1980. The larger picture is prepared, in an acknowledged classic, by Dasmann (1965).

15. The ESA, its status apparently safeguarded in late June 1995 by the U.S. Supreme Court, is nonetheless the subject of much real searching into the nature of private versus public property and is an endlessly complicated issue. See Cole 1992 for a discussion that materially affects California, or Harrison 1991. For an encyclopedic treatment, see Bonnett and Zimmerman 1991.

16. The February 1994 Biswell Symposium is an apt follow-through to the ideas of Harold Biswell, nicely summarized in Biswell 1989 and Weise and Martin 1995.

17. The original treatment, novel then, was Nash 1973, followed by Nash 1989. This view found a spirited commentary from Worster (1990). The contemporary changes in wilderness theory are effectively charted in Oelschlaeger 1991. Amazingly prescient and crafted with both philosophy and wit is the early statement of Aldo Leopold, reprinted in Leopold 1991c. Aldo Leopold's views are spelled out in Flader 1974.

18. The periodization of American park making is by no means entirely ordered or rational. There is a great deal of self-aggrandized and carefully projected social engineering involved; see Cranz 1989 for a discussion.

19. An original statement of bioregional categorization and philosophy is found in Udvardy 1975. Udvardy's start has been widely discussed; see Dasmann 1976 and 1988. An eloquent discussion of bioregionalism and sense of place appears in Parsons 1985. Arguably one of the most effective workers in bioregional ideology is Gary Snyder; see Snyder 1990.

20. On the place of government tribal culture, see Weatherford 1981, which discusses Forest Service activities as a form of cult and culture that is devoted to self-preservation; it follows upon the classic study of Herbert Kaufman (1960).

21. For a reasoned argument, in historical context (the original report was prepared in the 1950s and published twelve years later), see the blue-ribbon panel discussions released by the California Public Outdoor Recreation Plan Committee (1970).

22. On the dilemmas of how culture can be incorporated into park land, see Sax 1982. Sax's skepticism is posed in more austere terms in Sax 1993.

23. See Popper and Popper 1993. The spirit of skepticism is best argued in Luten 1986. For a discussion of the problems of crowding and migration, see Starrs and Wright 1995.

24. The range of opinions about the so-called wise-use movement is blisteringly diverse. For some ideas of the scope, see Hage 1994, Helvarg 1994, Gottlieb 1989, and, finally, Echeverria and Eby 1995.

REFERENCES

Beesley, D. 1996. Reconstructing the landscape: An environmental history, 1820–1960. In *Sierra Nevada Ecosystem Project: Final report to Congress*, vol. II, chap. 1. Davis: University of California, Centers for Water and Wildland Resources.

Bishko, C. J. 1981. Sesenta años después: *La Mesta* de Julius Klein a la luz de la investigación subsiguiente. *Historia, Instituciónes, Documentos* 8:9–57.

Biswell, H. 1989. *Prescribed burning in California wildlands vegetation management.* Berkeley and Los Angeles: University of California Press.

Black, M. 1995. Tragic remedies: A century of failed fishery policy on California's Sacramento River. *Pacific Historical Review* 64 (1): 37–70.

Bonnett, M., and K. Zimmerman. 1991. Politics and preservation: The Endangered Species Act and the northern spotted owl. *Ecology Law Quarterly* 18 (1): 105–71.

Bourjaily, V. N. 1984. *The unnatural enemy.* Tucson: University of Arizona Press.

Braudel, F. 1988. Foreword. In *History and environment*, translated by S. Reynolds, vol. 1 of *The identity of France.* New York: Harper and Row.

Burcham, L. T. 1970. Ecological significance of alien plants in California grasslands. *Proceedings of the Association of American Geographers* 2:36–39.

———. 1982. *California range land: An historico-ecological study of the range resource of California.* Publication 7. Davis: University of California, Davis, Center for Archeological Research.

Butzer, K. W. 1992. The Americas before and after 1492: An introduction to current geographical research. *Annals of the Association of American Geographers* 82 (3): 345–68.

California Public Outdoor Recreation Plan Committee. 1970. *The scenic, scientific, and educational values of the natural landscape of California: Preservation of California's landscape*, edited by H. L. Mason (Chair), R. Langenheim, A. S. Leopold, M. N. Palley, C. O. Sauer, H. L. Vaughan, and F. Violich. Sacramento, CA: Department of Parks and Recreation.

Chase, A. 1986. *Playing God in Yellowstone.* Boston: Atlantic Monthly Press.

Cleland, R. G. 1964. *The cattle on a thousand hills: Southern California, 1850–1880.* San Marino, CA: Huntington Library.

Cole, C. A. 1992. Species conservation in the United States: The ultimate failure of the Endangered Species Act and other land use laws. *Boston University Law Review* 72 (2): 343–79.

Cranz, G. 1989. *The politics of park design: A history of urban parks in America.* Cambridge, MA: MIT Press.

Dasmann, R. 1965. *The destruction of California.* New York: Collier Books.

———. 1976. Future primitive: Ecosystem people versus biosphere people, and biogeographical provinces. *CoEvolution Quarterly* 11 (September): 26–37.

———. 1988. Toward a biosphere consciousness. In *The ends of the earth: Perspectives on modern environmental history*, edited by D.

Worster, 277–88. Studies in Environment and History. Cambridge and New York: Cambridge University Press.

Dilsaver, L., and D. H. Strong. 1993. Sequoia and Kings Canyon national parks: One hundred years of preservation and resource management. In *Yosemite and Sequoia: A century of California national parks,* edited by R. J. Orsi, A. Runte, and M. Smith-Baranzini, 13–32. Berkeley and Los Angeles: University of California Press.

Douglass, W. A. 1985. The Basque sheepherder. *Basque Studies Program Newsletter* 32:6–8, 12.

Douglass, W. A., and J. Bilbao. 1975. *Amerikanauk: Basques in the New World.* Reno: University of Nevada Press.

Dunlap, T. R. 1988. *Saving America's wildlife: Ecology and the American mind, 1850–1990.* Princeton, NJ: Princeton University Press.

Dusenberry, W. 1950. Constitutions of early and modern American stock grower's associations. *Southwest Historical Quarterly* 53:255–75.

———. 1963. *The Mexican Mesta: The administration of ranching in colonial Mexico.* Urbana: University of Illinois Press.

Echeverria, J., and R. B. Eby, eds. 1995. *Let the people judge: Wise use and the private property rights movement.* Washington, DC: Island Press.

Ellickson, R. 1986. Of Coase and cattle: Dispute resolution among neighbors in Shasta County [California]. *Stanford Law Review* 38:623–87.

———. 1991. *Order without law: How neighbors settle disputes.* Cambridge, MA: Harvard University Press.

Ewing, R. A., R. Tuazon, N. Tosta, L. Huntsinger, R. Marose, K. Nielson, R. Motroni, and S. Turan. 1988. *California's forests and rangelands: Growing conflict over changing uses.* Sacramento: California Department of Forestry and Fire Protection.

Farquhar, F. P. 1965. *History of the Sierra Nevada.* Berkeley and Los Angeles: University of California Press.

Flader, S. L. 1974. *Thinking like a mountain: Aldo Leopold and the evolution of an ecological attitude toward deer, wolves, and forests.* Columbia: University of Missouri Press.

Fortmann, L., and P. F. Starrs. 1988. Power plants and resource rights. In *Community and forestry: Continuities in natural resources sociology,* edited by R. G. Lee, D. R. Field, and W. R. Burch Jr., 179–94. Boulder, CO: Westview Press.

Gates, P. W. 1960. The land system of the United States in the nineteenth century. In *Proceedings of the first congress of historians from Mexico and the United States,* 222–55. México City: Editorial Cultura.

Gottlieb, A. M., ed. 1989. *The wise-use agenda: The citizen's policy guide to environmental resource issues: A task force report.* Bellevue, WA: Free Enterprise Press.

Hage, W. 1994. *Storm over rangeland: Private rights in federal lands.* 3rd ed. Bellevue, WA: Free Enterprise Press.

Harrison, G. L. 1991. The Endangered Species Act and ursine usurpations—a grizzly tale of two takings. *University of Chicago Law Review* 58 (3): 1101–24.

Hart, J. F. 1991. The perimetropolitan bow wave. *Geographical Review* 81 (1): 35–51.

Haslam, G. 1986. Arnold Rojas: Voice of the *vaqueros. Californians* 4 (5): 36–40.

Heady, H. F., J. W. Bartolome, M. D. Pitt, G. D. Savelle, and M. C. Stroud. 1991. California prairie. In *Ecosystems of the world: Natural grasslands,* edited by R. T. Coupland, vol. II, 315–337. Amsterdam: Elsevier.

Helvarg, D. 1994. *The war against the Greens: The "wise use" movement, the new right, and anti-environmental violence.* San Francisco: Sierra Club Books.

Hokanson, D. 1988. *The Lincoln Highway: Main Street across America.* Iowa City: University of Iowa Press.

Howard, T. F. 1993. Breaching the mountain barrier: The first roads over the Sierra Nevada. Ph.D. diss., Department of Geography, University of California, Berkeley.

Huntsinger, L. 1989. Grazing in California's mixed conifer forests: Studies in the central Sierra Nevada. Ph.D. diss., Department of Wildland Resource Science, University of California, Berkeley.

Huntsinger, L., and S. McCaffrey. 1995. A forest for the trees: Forest management and the Yurok environment, 1850–1994. *American Indian Culture and Research Journal* 19 (4): 155–92.

Jordan, T. G. 1993. *North American cattle ranching frontiers.* Albuquerque: University of New Mexico Press.

Jorgensen, J. G. 1984. Land is cultural, so is a commodity: The locus of differences among Indians, cowboys, sod-busters, and environmentalists. *Journal of Ethnic Studies* 12 (3): 1–21.

Kaufman, H. 1960. *The forest ranger: A study in administrative behavior.* Resources for the Future. Baltimore: Johns Hopkins University Press.

Kelley, R. L. 1959. *Gold vs. grain: The hydraulic mining controversy in California's Sacramento valley; a chapter in the decline of the concept of laissez faire.* Glendale, CA: Arthur H. Clark Company.

———. 1989. *Battling the inland sea: American political culture, public policy, and the Sacramento valley, 1850–1986.* Berkeley and Los Angeles: University of California Press.

Klein, J. 1920. *The Mesta: A study in Spanish economic history, 1273–1836.* Cambridge, MA: Harvard University Press.

Kroeber, T. 1976. *Ishi in two worlds: A biography of the last wild Indian in North America.* Berkeley and Los Angeles: University of California Press.

Lane, R. H. 1985. *Basque sheepherders of the American West: A photographic documentary.* Basque Series. Reno: University of Nevada Press.

Larson, D. J. 1996. Historical water-use priorities and public policies. In *Sierra Nevada Ecosystem Project: Final report to Congress,* vol. II, chap. 8. Davis: University of California, Centers for Water and Wildland Resources.

Leopold, A. 1991a. The conservation ethic. In *The river of the mother of God and other essays by Aldo Leopold,* edited by S. Flader and J. B. Callicott, 181–93. Madison: University of Wisconsin Press.

———. 1991b. Game and wild life conservation. In *The river of the mother of God and other essays by Aldo Leopold,* edited by S. Flader and J. B. Callicott, 164–69. Madison: University of Wisconsin Press.

———. 1991c. Wilderness as a form of land use. In *The river of the mother of God and other essays by Aldo Leopold,* edited by S. Flader and J. B. Callicott, 134–42. Madison: University of Wisconsin Press.

Lewis, M. 1992. *Green delusions: An environmentalist critique of radical environmentalism.* Durham, NC: Duke University Press.

Limerick, P. 1987. *The legacy of conquest: The unbroken past of the American West.* New York: W.W. Norton.

Lueck, D. 1989. The economic nature of wildlife law. *Journal of Legal Studies* 18 (2): 291–324.

Lund, T. A. 1980. *American wildlife law.* Berkeley and Los Angeles: University of California Press.

Luten, D. B. 1986. Empty land, full land, poor folk, rich folk. In *Progress against growth: Daniel B. Luten on the American landscape,* edited by T. Vale, 213–23. New York: Guilford Press.

McCarthy, J. 1974. The Lincoln Highway. *American Heritage* 24 (4): 32–37, 89.

McWilliams, C. 1949. *California, the great exception.* Reprint, Santa Barbara, CA: Peregrine Smith, 1979.

Merchant, C. 1992. *Radical ecology: The search for a livable world.* New York: Routledge.

Mighetto, L. 1990. *Wild animals and American environmental ethics.* Tucson: University of Arizona Press.

Mora, J. 1948. *The Californios.* New York: Scribners.

Nash, G. D. 1985. *The American West transformed: The impact of the Second World War.* Bloomington: Indiana University Press.

Nash, R. 1973. *Wilderness and the American mind.* New Haven, CT: Yale University Press.

———. 1989. *The rights of nature: A history of environmental ethics.* History of American Thought and Culture. Madison: University of Wisconsin Press.

Oelschlaeger, M. 1991. *The idea of wilderness.* New Haven, CT: Yale University Press.

Orsi, R. J., A. Runte, and M. Smith-Baranzini, eds. 1993. *Yosemite and Sequoia: A century of California national parks.* Berkeley and Los Angeles: University of California Press.

Parsons, J. J. 1972a. The California gold country. *Geographical Review* 62 (2): 269–71.

———. 1972b. Slicing up the open space: Subdivisions without homes in northern California. *Erdkunde* 26 (1): 13–17.

———. 1985. On "bioregionalism" and "watershed consciousness." *Professional Geographer* 37 (1): 1–6.

Paul, R. 1963. *Mining frontiers of the Far West, 1848–1880.* New York: Holt, Rinehart, and Winston.

Pisani, D. J. 1992. *To reclaim a divided West: Water, law, and public policy, 1848–1902.* Albuquerque: University of New Mexico Press.

Pitt, L. 1966. *The decline of the Californios: A social history of the Spanish-speaking Californians, 1846–1890.* Berkeley and Los Angeles: University of California Press.

Popper, F., and D. E. Popper. 1993. Yo, pioneers! Guess what? The western frontier is coming back strong. *Washington Post*, 5 September.

Pyne, S. J. 1981. *Fire in America: A cultural history of wildland and rural fire.* Princeton, NJ: Princeton University Press.

———. 1995. *World fire: The culture of fire on earth.* New York: Holt.

Rinschede, G. 1984. *Die Wanderviehwirtschaft im gebirgigen Western der USA und ihre Auswirkungen im Naturraum.* Eichstätter Beiträge: Schriftenreihe der Katholischen Universität Eichstätt; Abteilung Geographie, Band 10. Regensburg: Verlag Friedrich Pustet.

———. 1988. Transhumance in European and American mountains. In *Human impact on mountains*, edited by N. J. R. Allan, G. W. Knapp, and C. Stadel, 96–108. Totowa, NJ: Rowman and Allanheld.

Rintoul, W. 1976. *Spudding in: Recollections of pioneer days in the California oil fields.* San Francisco: California Historical Society.

Rojas, A. 1979. *Vaqueros and buckeroos.* Bakersfield, CA: Hall Letter Shop.

Runte, A. 1987. *National parks: The American experience.* Lincoln: University of Nebraska Press.

———. 1993. Joseph Grinnell and Yosemite: Rediscovering the legacy of a California conservationist. In *Yosemite and Sequoia: A century of California national parks*, edited by R. J. Orsi, A. Runte, and M. Smith-Baranzini, 85–96. Berkeley and Los Angeles: University of California Press.

Sax, J. L. 1970. The public trust doctrine in natural resource law: Effective judicial intervention. *Michigan Law Review* 68 (3): 471–566.

———. 1980. *Mountains without handrails: Reflections on the national parks.* Ann Arbor: University of Michigan Press.

———. 1982. In search of past harmony: In French regional parks, human inhabitants are part of the natural landscape. *Natural History* 91 (8): 42–51.

———. 1987. An advocate for nonpersons: Christopher D. Stone's *Earth and other ethics: The case for moral pluralism. Natural History* 96 (8): 60–61.

———. 1993. Nature and habitat conservation in the United States. *Ecology Law Quarterly* 20 (1): 47–56.

Shinn, C. 1884. *Mining camps: A study in American frontier government.* Reprint, Gloucester, MA: Peter Smith, 1970.

Slotkin, R. 1973. *Regeneration through violence: The mythology of the American frontier, 1600–1860.* Middletown, CT: Wesleyan University Press.

———. 1985. *The fatal environment: The myth of the frontier in the age of industrialization, 1800–1890.* Middletown, CT: Wesleyan University Press.

Smeltzer, J. F. 1985. *Wildlife law enforcement: An annotated bibliography.* Fort Collins: Colorado Division of Wildlife.

Snyder, G. 1990. The place, the region, and the commons. In *The practice of the wild*, 25–47. San Francisco: North Point Press.

Starrs, P. F. 1988. Reconsidering John Wesley Powell. *Society and Natural Resources* 1 (1): 83–85.

———. 1994. "Cattle Free By '93" and the imperatives of environmental radicalism. *Ubique: Notes from the American Geographical Society* 14 (1): 1–4.

Starrs, P. F., and J. Wright. 1995. Great Basin growth and the withering of California's Pacific idyll. *Geographical Review* 85 (4): 424–44.

Stegner, W. 1953. *Beyond the hundredth meridian: John Wesley Powell and the second opening of the West.* Reprint, Lincoln: University of Nebraska Press, 1982.

Stewart, G. R. 1962. *The California trail: An epic with many heroes.* New York: McGraw-Hill.

Stine, S. 1980. Hunting and the faunal landscape: Subsistence and commercial venery in early California. Master's thesis, Department of Geography, University of California, Berkeley.

Stone, C. D. 1974. *Should trees have standing?—Towards legal rights for natural objects.* Los Altos, CA: William Kaufmann.

———. 1987. *Earth and other ethics: The case for moral pluralism.* New York: Harper and Row.

Strong, D. H. 1984. *Tahoe: An environmental history.* Lincoln: University of Nebraska Press.

Tober, J. A. 1981. *Who owns the wildlife? The political economy of conservation in nineteenth-century America.* Contributions in Economics and Economic History, 37. Westport, CT: Greenwood Press.

Todd, C. J. 1949. The Sierra Nevada: A transportation barrier. Master's thesis, Department of Geography, University of California, Los Angeles.

Udvardy, M. D. F. 1975. *A classification of the biogeographical provinces of the world.* IUCN Occasional Paper 18. Morges, Switzerland: International Union for Conservation of Nature and Natural Resources.

Vale, T. R., and G. R. Vale. 1983. *U.S. 40 today: Thirty years of landscape change in America.* Madison: University of Wisconsin Press.

———. 1994. *Time and the Tuolomne landscape.* Salt Lake City: University of Utah Press.

Walters, D. 1986. *The new California: Facing the 21st century.* Sacramento: California Journal Press.

Weatherford, J. M. 1981. *Tribes on the hill.* New York: Rawson, Wade.

Weise, D. R., and R. E. Martin, technical coordinators. 1995. *The Biswell Symposium: Fire issues and solutions in the urban interface and wildland ecosystems.* General Technical Report PSW-GTR-158. Albany, CA: U.S. Forest Service, Pacific Southwest Research Station.

White, R. 1983. *The roots of dependency: Subsistence, environment, and social change among the Choctaws, Pawnees, and Navajos.* Lincoln: University of Nebraska Press.

Worster, D. W. 1990. The rights of nature: A history of environmental ethics, by Roderick F. Nash. *Isis* 81 (309): 798–800.

———. 1994. *Nature's economy: A history of ecological ideas.* 2nd ed. New York: Cambridge University Press.

LARRY RUTH
Department of Environmental Science,
 Policy, and Management
University of California
Berkeley, California

7

Conservation and Controversy: National Forest Management, 1960–95

ABSTRACT

In the period 1960 to 1995, policies affecting national forest management generated a variety of directions for the planning, management, conservation, and preservation of national forest lands and resources in the Sierra Nevada. The National Environmental Policy Act (NEPA) forced better disclosure of information utilized by the U.S. Forest Service, enhanced public awareness of management issues, and led to increased public involvement in agency decision making. As a result, efforts to increase timber production in Sierra Nevada national forests met with increased public scrutiny as well as political and legal opposition. The National Forest Management Act (NFMA) of 1976 mandated extensive planning to promote effective and efficient conservation of forest resources and to resolve forest management controversies. Demand for increased public timber supplies, however, conflicted with demands for increased recreation and wilderness preservation. Contestation over national forest policies did not begin with NFMA, but the broad scope of land-management planning generated remarkable public attention and controversy. Public opposition to potential impacts on wildlife habitat and other aspects of forest ecosystems and to the increased use of clear-cutting as a timber harvest method led to legal action challenging national forest plans. At present, implementation and interpretation of law and administrative policies have forced the Forest Service to revise forest plans in the Sierra to better incorporate species and habitat requirements as part of its ecosystem-management strategy.

INTRODUCTION

The origin of this research is an invitation by the Sierra Nevada Ecosystem Project (SNEP) to participate in a workshop to assist in determining the role that public policies have played in shaping the ecosystems of the Sierra Nevada. Specifically, several researchers have been asked to respond to the question "Which public policies have been most significant in shaping the ecosystems of the Sierra Nevada as they exist today?" These individuals have been asked to concentrate on the period from 1960 to the present, with the intention that a range of policies and their impacts will be analyzed. I believe that the exercise is a useful one and am delighted to contribute to this inquiry.

My effort to respond to this question will be primarily to address policies and issues associated with national forest management in the Sierra Nevada during the last thirty-five years. While I am aware that public policies have had a multitude of significant effects in many other areas of the Sierra, this chapter will not attempt to address them directly. My approach will concentrate on developing the context for policy implementation in the national forests during this period and discussing the effects and implications of these policies. As a social scientist, I will leave an authoritative determination of impacts and their significance on ecosystems to my colleagues on the SNEP Science Team. As part of the effort to organize thinking about institutional aspects of natural resource sustainability in the Sierra, I offer some thoughts on ancillary issues pertaining to policy implementation. This excursus may prove useful as the Science Team seeks to answer the ques-

Sierra Nevada Ecosystem Project: Final report to Congress, vol. II, *Assessments and scientific basis for management options*. Davis: University of California, Centers for Water and Wildland Resources, 1996.

tion posed and to consider its implications. Where feasible, I will refer to other policies that have had significant impacts on the Sierra. Generally, this will be limited to suggesting areas for further research or discussion by others. Discussion pertaining to the impact of policy in the national forests of the Sierra Nevada will focus on forces that shaped the administration of the national forests, the response to public activism surrounding national forest management, and the further implications for conservation and management of the national forests in the Sierra Nevada.

This workshop represents an initial effort to discuss with some precision the role that public policies have played in influencing the conditions of Sierra Nevada ecosystems. Ultimately, statements about policies and their effects on the ecosystem, particularly regarding the current era, must be answered by further study. This analysis should be undertaken in light of the results of the completed SNEP assessments. This point cannot be emphasized too strongly. The social scientists engaged in policy review and analysis are going to present research on various public policies and their environmental effects and impacts. This work will necessarily be based on a general knowledge of conditions at the ecosystem level, as it is presently understood. I also hasten to point out that research to date is based on an understanding of environmental conditions that is clearly imperfect and incomplete. If it were not, SNEP would not be engaged in an assessment, and we might leave to the administrative or legislative process the task of defining a policy that would be consistent with the state of scientific knowledge. After all, it was public and congressional concern about the lack of scientific knowledge, understanding, and consensus as to the ecological health of the Sierra Nevada that led to the articulation of SNEP's research mission. One may naturally anticipate that SNEP's assessments will result in better, more accessible knowledge about the ecological conditions of the Sierra. New information that may be presented in the assessments may change the key points about policy that we glean from our present discussion. Thus, it would prove of substantial value to renew the discussion of policies and their impacts in the Sierra in light of the information to be made available once the assessments are completed. Only then will it be possible to answer the question "Which public policies have been most significant in shaping the ecosystems of the Sierra Nevada as they exist today?" with any degree of accuracy or precision.

PUBLIC POLICY AND RESEARCH: CONTEXT AND IMPLICATIONS

Understanding the influence of policy is an important component of SNEP's assessment of the status of the ecosystems of the Sierra Nevada. Before discussing specific policies, I would like to express some concerns relating to the consideration of policies and their effects. I believe the limits of this kind of enterprise should be understood by the participants in this workshop. Several fundamental issues occur to me. These bear not only on the discussion about policies and their ecological effects and implications but also on the effort to discuss the broader implications of public policy in the Sierra. Ideally, an inquiry into the impact of policies will also consider the influence of social and economic dynamics on policy implementation. Describing these interactions is probably beyond the scope of this workshop. Nevertheless, full comprehension of the policy context depends on an understanding of both formal and informal social and administrative dimensions relating to policy implementation. For this reason, I particularly appreciate the inclusion of social scientists and others with practical experience in policy implementation as part of the SNEP Science Team.

A primary concern regarding the question we are asked to tackle in the policy workshop relates to the ability of individual research projects to adequately assess the role of a particular policy in shaping the ecosystem. Public policies, defined as the sum of law, regulation, administrative programs, and public projects together with their funding and implementation, affect virtually all of the land area and natural resources in the Sierra Nevada. The effect of public policies extends across both time and space, with the results of prior policies exerting an influence on the present status of resources and ecosystems. Equally, effects and implications of environmental policies may extend beyond the specific areas, issues, or programs that they were designed to affect directly. The breadth and depth of policies, and their effects on ecosystems, therefore, are likely to be substantial. As a result, the full extent to which policies have affected and influenced both the current state of the ecosystem and the present fabric of natural resource institutions will be difficult to establish. Given the task and its scope, this exercise will necessarily produce an eclectic and incomplete view of the role that policy has played in shaping ecosystems.

Another cause for concern pertains to difficulties in empirical method. The effects of public policy may appear to be the result of specific policies. Correlation of the operation of a specific policy to a particular effect may appear to be intuitively obvious, but attempting to go beyond this is a difficult task. In reality, attribution of specific causation to policies or establishing their effect with certainty is a difficult task, complicated by the operation of multiple policies and other forces that influence the same resources or ecosystems. In addition, for policy effects for which one hypothesis may be constructed, other possible explanations generally exist—some more or less likely. Specific attribution of the effects of laws, statutes, plans, programs, and projects can thus be a complicated and error-prone enterprise. Observed effects may be the indirect results of obscure policies or unintended by-products of various policy instruments. Precise attribution of effects is therefore a controversial and contentious exercise, and one that may exceed the capacities of any research design or the capability of even the most zealous researcher.

A variety of public policies that have had significant effects on the environment generally are not characterized by a specific relation to ecosystems, natural resources, or the environment. Examples of this kind pertaining to the Sierra in the last thirty-five years include the development and expansion of Interstate 80, the state highway systems, and the national forest road system. These dramatically improved access to the entire Sierra Nevada and contributed to the concentration of urban, commercial, recreational, and commodity-related development and associated environmental impacts in particular areas of the range. Individually, all of these developments have had significant environmental impacts on the ecosystems in these locations. Another example is the influence of national and state policies related to air quality and pollution control. Although evidence suggests that aerial pollutants are beginning to have profound impacts on the Sierra, the relationship between policies and these impacts is complicated (Cahill et al. 1996) and may be isolated and analyzed only with difficulty.

Additionally, attribution and discussion of public policies and their effects may be complicated by a number of factors. Several ostensibly separate policies may together contribute impacts on a resource or areas, resulting in cumulative effects that are difficult to attribute to a specific policy instrument. In other cases, funding for one of a number of interactive policies may be uneven or irregular, affecting implementation and making it difficult to draw any conclusion as to the success or failure of impact of particular public policies. Finally, in some cases the absence of policy may have implications for the state of the ecosystem that are as significant as or greater than those from policies that we can more easily define and observe. SNEP Science Team members, I believe, have already explored many individual impacts. I am hopeful that individual assessments will capture some of the effects that may be traceable to the presence or absence of certain policy phenomena.

THE NATIONAL FORESTS: POLICIES, CONTEXT, AND IMPLICATIONS

National forest management and its ecological implications in the Sierra Nevada are obvious and important sources of information regarding the impact of public policies on ecosystems in the region. Several reasons compel attention to the role of policy in the management of these lands. A large proportion of the land area of the Sierra Nevada, especially at middle to upper elevations, is national forest land. Policies and planning specifically pertaining to national forest management are the product of a number of laws and adminis-

trative policies, including the National Forest Management Act of 1976 (NFMA), *U.S. Code*, vol. 16, secs. 1600–1614 (1976); the Organic Act of 1897, *U.S. Code*, vol. 16, secs. 473–482, 551; the Multiple Use–Sustained Yield Act of 1960 (MUSY), *U.S. Code*, vol. 16, sec. 528 et seq.; the Wilderness Act of 1964; the National Environmental Policy Act of 1969 (NEPA), *U.S. Code*, vol. 42, sec. 4321 et seq.; the Endangered Species Act (ESA), *U.S. Code*, vol. 16, secs. 1531–43 (as amended in 1989); as well as other environmental laws, annual appropriations legislation, and a range of administrative policies relating to fire suppression and fuel management. These policies guide a range of activities, which necessarily are likely to have significant environmental effects. The impact of these activities is likely to be felt in the national forests, on adjacent lands, and in ecosystems beyond national forest boundaries. Another reason that national forest management policies are worth special attention pertains to the valuable information that may be obtained by reviewing the impacts and implications of policies explicitly designed to guide the conservation and management for large areas of the Sierra Nevada.

In the Sierra Nevada, national forests fulfill several important varied functions. These forests serve uses representing a wide range of natural resource–related values, including the use of forest resources to produce commodities, such as timber and forage for grazing. The national forests also contain other resources, including water, fish, wildlife, minerals, recreational opportunities, and others. Often these uses conflict or appear to conflict with one another. Natural resource management in the national forests of the Sierra Nevada has been the subject of a great deal of scrutiny and continuing controversy during the past thirty-five years. Concern and contestation regarding Forest Service resource policies did not begin with the enactment of NFMA. It is undeniable, however, that the NFMA land-management planning process, especially at certain key decision points over the last fifteen years, has generated remarkable public interest and caused considerable controversy. The reason for the intense interest in NFMA and its effects has to do with several aspects of the statutory mandate. The law's provisions were intended to reorder national forest management to develop coordinated plans for multiple use and to promote effective and efficient conservation of forest resources. The scope of the law, combined with increased demands on public timber supplies, suggested that NFMA had the potential to propose and implement management activities that would have widespread effects on management of the national forests, including those of the Sierra Nevada. Before analyzing these elements and the impact of NFMA, several earlier policies will be discussed in order to explain the culture of Forest Service administration and to provide a context for the discussion of recent forest policies and their impact on the national forests.

ADMINISTRATION OF THE NATIONAL FORESTS IN THE POSTWAR ERA

During and after World War II, the Forest Service began to focus on increased timber harvesting and other commodity considerations in its overall administration of the national forests. A housing boom had created an unprecedented demand for timber. As private timber was harvested and these supplies declined, industry pressed for expanded timber sales in the national forest to fulfill the demand (Clary 1986). The Forest Service increased timber sales, and as a result, conflicts between timber harvesting and recreation also increased, somewhat tarnishing the agency's reputation.

In the postwar era, Congress recognized that the Forest Service and other land-management agencies were being pressured to meet the needs of a diverse set of recreation users. During this period, land-management policy was still primarily guided by the Organic Act of 1897, which offered little guidance on how to reconcile administration of the national forests with changing public needs. This act stated that the forest reserves, as they were originally known, were to be managed "for the purpose of securing favorable conditions for water flows and to furnish a continuous supply of timber for the use and necessity of the citizens of the United States."

In 1958, Congress created the Outdoor Recreation Resources Review Commission (ORRRC) to review the situation and to make recommendations for meeting recreation needs in 1976 and 2000. The Forest Service supported the work of the ORRRC, because the agency had always encouraged recreational enjoyment in national forests as an adjunct to timber, range, and other uses. The ORRRC made recommendations to Congress that called for increased governmental funding for recreational development and for coordinated planning within agencies to provide better recreational opportunities.

LEGISLATIVE CHANGE

The Multiple Use–Sustained Yield Act of 1960

Throughout the postwar period, the Forest Service was confident of its ability to manage the forest for many different uses, including wilderness. The agency sought legislation that would confirm its authority to manage the expanding array of uses, enabling it to reconcile timber production and other commodity uses with public demands for more recreation opportunities and for wilderness preservation (U.S. Forest Service 1960). The Forest Service recognized that additional support for a range of other uses and activities, including the recreational goals of the ORRRC, would bring additional appropriations, allowing for development of uses that were already present in the national forests. The Sierra Club opposed

this initiative, arguing that the Forest Service commitment to timber meant that the agency would not make balanced decisions that recognized the importance of other forest resources (Dana and Fairfax 1980, 203–4). It is also likely that increased friction between the Forest Service and environmentalists over legislation proposed to designate areas of federal land, including some in the national forest, as "wilderness," contributed to this opposition. Despite the opposition, Congress enacted the Multiple Use–Sustained Yield Act in 1960. This law stated that the national forests were to be managed for "the achievement and maintenance in perpetuity of a high-level annual output or regular annual output of the various renewable resources of the national forest without impairment of the productivity of the land." The Forest Service expanded its utilization of planning in order to coordinate various forest uses, or at least to rationalize conflicting uses (Wilson 1978). A Regional Multiple Use Planning Guide was prepared for each region to guide local planning. Forest Land Use Plans were developed for each national forest to guide multiple-use integration and development (Forest Service 1973, sec. 8213). Unit Plans were then completed to tailor management specifically to the conditions of watersheds or large drainage areas ranging in size from fifty thousand acres to several hundred thousand acres. This system essentially ratified Forest Service determinations of the "greatest good for the greatest number." Planning permitted continued timber sales while also allowing the agency to claim that it had become the nation's premier provider of outdoor recreation opportunities.

The Multiple Use–Sustained Yield Act (1960, secs. 528, 529) recognized the importance of a spectrum of resource uses, including "outdoor recreation, range, timber, watershed, and wildlife and fish." The statute also recognized the value and place of wilderness in the national forests. Legislative acknowledgment of these uses permitted the Forest Service to serve the public interest by developing a variety of forest resources and activities appropriate to meet the needs of various uses and groups. Even so, the Forest Service, an organization built on compromise, began to shift its administration in response to the needs of the public. When opposition to Forest Service projects occurred in this era, it could be countered, if not diffused, by locating potentially conflicting uses in another forest area. Culhane (1981, 388–94) argued that the many different interest groups involved tended to counteract each other's power, enabling the Forest Service to pursue a middle course. Timber harvesting could take place in one area, while fishing, hiking, and other recreation uses could be located in another. Compromises allowed resource development activities to continue, with either the support or the acquiescence of interested parties and interest groups.

Forest Service policies in this period were not without critics. Congressional appropriations were primarily oriented toward timber. To some, however, Forest Service administration remained primarily attuned to the most powerful constituencies in the regions it served. Timber and other commodity interests proved to be powerful enough to compel attention. Agency policies therefore often appeared to reflect a bias toward timber production rather than attempting to serve some broader conception of the public interest (McConnell 1966) or one more closely tied to a growing constituency of national forest visitors whose interest centered on recreation. Professional foresters, who made up the bulk of agency personnel and its management, had motivations and goals different from those of the timber industry. Despite these tensions, established working relationships between forester and logger, and between the Forest Service and the timber industry, appeared to lend credence to conservationist claims that timber considerations dominated the agency's agenda. In fairness to the Forest Service, however, Congress made continuing budget appropriations in order to expand the agency's timber program. This strongly suggested that Congress believed that Forest Service timber management was consistent with the purpose of furnishing "a continuous supply of timber for the use and necessity of the citizens of the United States" and thus squarely served the public interest.

After MUSY was enacted, the Forest Service slowly began to expand its staff to bring in new kinds of professional expertise. Even so, the preponderance of foresters in the Forest Service and the agency's role in supplying timber to private industry provoked doubt about the "multiple-use" orientation of the agency. The multiple-use philosophy allowed the agency to avoid many management controversies, but as the following discussion indicates, this approach never satisfied important segments of the public.

The Wilderness Act of 1964

Advocates in support of the idea that separate areas should be set aside for wilderness preservation had always opposed the idea of multiple use as the guiding principle for all forest lands. They believed that if wilderness was accorded a status only equivalent to any other use of forest resources, wilderness would necessarily be subservient to timber and other commodity uses when the agency made determinations. Although the Forest Service had already designated many wilderness areas in the national forests,[1] many conservationists did not believe that the agency valued wilderness enough to ensure that the existing "primitive area" designations would survive the timber industry's preference for increased timber harvest levels in national forest lands. The Forest Service claimed this designation was sufficient to ensure that these lands would be managed as wilderness. Some wilderness proponents were unconvinced and sought legislation to make it impossible for the status of these lands to be altered administratively (McCloskey 1966). Some of these lands, located in national forests, national parks, and other federal lands, were to be reserved as wilderness, and the status of additional areas was to be reviewed in the following ten years.

Agency efforts to deflect attention from the wilderness issue failed to divert wilderness supporters from their goal of securing legislative protection for wilderness designations. In 1964 Congress passed the Wilderness Act, despite the opposition of the Forest Service. This statute established a National Wilderness System. Certain lands administered by the federal land-management agencies, including 2.1 million acres of land that previously had been administratively protected by the Forest Service, were designated as "wilderness" or "pristine" areas, with the status of other areas to be reviewed during the following decade. In the Sierra Nevada, a number of areas in national forests, prized for their scenic beauty and recreational value, were reserved. These areas were located mainly in the alpine and subalpine zones. The agencies retained control over wilderness areas under their administration, but the new designation limited uses on these lands. Ironically, the designation permitted no timber harvest on these lands but, subject to presidential review, continued to permit other development, including mining, grazing, and water development. Loss of the range of options that the Forest Service formerly controlled on these lands was something of a blow to agency prestige, because it implied that the agency could not be trusted to preserve this land on its own (Dana and Fairfax, 1980, 227–29).

The National Environmental Policy Act

Enactment of the National Environmental Policy Act (1969) represented a major watershed in public policy. The expression of public concern for environmental values reflected the concern of many individuals in this era. NEPA was intended to ensure that environmental factors would be considered as part of the decision-making process. A major element in the law is the requirement that an environmental impact statement (EIS) be prepared for federal actions having a "significant effect on the environment (National Environmental Policy Act 1969, sec. 102 (2)(c), *U.S. Code,* vol. 42, sec. 4332)." NEPA, however, did not require that environmentally questionable projects be abandoned, so there was no expectation that preparation of an EIS would lead to dramatic changes in Forest Service proposals or in the policy of multiple use. The EIS, however, requires that the public be provided an opportunity to comment on agency proposals. This element of the law has had a profound impact, both on the decision-making processes of all federal agencies and on the relations of these agencies with the public. Public disclosure of information also provides citizens with an opportunity to challenge these decisions in the political process and in court. Additionally, in some cases, the time required to prepare and to complete the documents required by NEPA has provided another obstacle that has deterred some project proponents. In this way, the procedural aspects of NEPA have exerted a significant influence on a wide range of Forest Service land-management activities and programs.

An early example of the procedural aspects of NEPA requirements and their far-reaching effects is amply illustrated in a celebrated controversy in the Sierra Nevada, where the proposed development of a ski resort in Mineral King (a relatively undeveloped area, then a part of the Sequoia National Forest) by Walt Disney was ultimately derailed. Although the ski resort had been approved prior to the enactment of NEPA, the Forest Service was faced with a lawsuit challenging its decision. The agency elected to prepare an EIS for the proposed ski resort. The Sierra Club sued, inter alia, to force consideration of the environmental effects on national park resources due to the expansion of the access road (*Sierra Club v Morton*, 405 US 727 [1972]). Although the lawsuit was later dropped, delays created by the lengthy administrative and legal process ultimately caused the developer to lose interest in the project (*Ecology Law Quarterly*, 1972, 1976). The demise of the proposed ski resort project meant that Mineral King would remain largely undeveloped. After several years, this area was transferred to Sequoia National Park. The ecological significance of this result was perhaps limited to the preservation of one valley and its environs. Nevertheless, it was perceived as no small victory for conservationists, who were encouraged in their struggle against what they regarded as the tendency of the Forest Service to abandon too quickly its conservation precepts as it sought some compromise in search of the public interest. In contrast, subsequent efforts by conservationists to challenge an EIS prepared for Kirkwood, another ski area in the Sierra, were unsuccessful, and the area was successfully developed.

The Forest Service, in many respects, pioneered the implementation of NEPA, but its experience has not been without some difficulties. Among other early innovations, the agency developed an environmental assessment (EA), a preliminary report used to ascertain probable environmental effects and thereby determine whether preparation of an EIS was required. Nevertheless, in the years immediately after NEPA's enactment, the Forest Service had considerable difficulty in adjusting to the law's requirements. In the first years of NEPA's existence, the early 1970s, seasoned agency managers, almost all of them foresters, generally were not fully aware of the scope and intricacies of EIS requirements, nor were they prepared to supervise a comprehensive consideration of environmental impacts (Taylor 1984, 208). Consequently, on many occasions, the agency concluded that an EA was sufficient and did not insist on preparation of an EIS for certain projects. In some of these cases, the agency's analysis established that some projects, including timber harvests (especially those where clear-cutting was employed), were unlikely to have a "significant impact" on the environment. Additionally, agency determinations of which projects required an EIS tended to exclude many projects from this requirement. Preparation of an EIS entailed an opportunity for public comment but also allowed the Forest Service opportunities to revise the project and to respond to its critics before reaching a final decision. Even when the Forest Service began its early efforts to produce environmental impact statements, several years were required to develop the skills to conduct a full analysis of environmental impacts and to produce an adequate EIS.

Reasons for the agency's inability were due in part to initial uncertainty about what preparation of an EIS entailed and also in part to lack of expertise in analyzing environmental impacts. Despite the presence of an array of forestry professionals and other interdisciplinary scientific experts in the Forest Service, the agency could not immediately deploy and utilize professionals who possessed skills appropriate for the preparation of the EIS. As a result, lawsuits successfully challenged agency decisions regarding the preparation of an EIS (*Kleppe v Sierra Club*, 427 US 390 [1976]).

After suffering losses in court, Forest Service managers recognized that the agency had to learn more about environmental impact analysis and how to prepare an EIS. It took time to recruit and to cross-train experts in disciplines not previously mastered to any significant degree by the Forest Service personnel. Resistance to formal public participation waned as the agency became more familiar with the process. As the agency developed greater interdisciplinary environmental expertise, it was gradually able to handle sensitive projects and to prepare an EIS in a professional and more defensible manner (Taylor 1984).

TIMBER SUPPLY AND THE NATIONAL FORESTS

The forest products industry was also concerned over Forest Service timber sale policies. The struggle to ensure a reliable future timber supply is essential to the stability of the industry. As the demand for timber grew in the concluding years of World War II and in the postwar years, increased cutting on private lands, particularly in the Northwest, led to shortages in the supply of mature timber on these lands. As private timber inventories were logged, some harvested lands were replanted. While these trees were growing, however, the supply of mature timber on private lands declined. In many areas, including California, a portion of the cutover lands were sold or exchanged to the Forest Service to avoid paying for replanting, fire protection, and taxes. The rotational sequence caused other operators to become dependent on the national forests for timber supplies in the middle 1970s. To accommodate this need, the forest products industry pressed for increased timber sales on the national forests (Clawson 1975; Dowdle and Hanke 1985, 85–88). Many foresters and individuals, both inside and outside of the timber industry, argued that harvest levels for the national forests have been and continue to be set substantially below what the national forests can produce on a sustained yield level (Rey n.d.; John Zivnuska, Berkeley, California, personal communication, April 4, 1986).

Industry and its supporters were frustrated by the lack of any national strategy to respond to the demand for timber. In addition to frustration over the low timber volume offered for sale, the timber industry was also somewhat concerned about the variability in sales levels from year to year. The industry was perplexed by the uncertainties of the political process that necessitates appropriations for Forest Service timber sales. The amount of timber that would be available for harvest in any one year could not be reliably predicted in advance. This was due to the time required to prepare any sizable timber sales. These efforts generally required a sustained budget that would permit an effort to continue over several years, something that was difficult to ensure in advance. Accordingly, varying levels from year to year meant that it was equally difficult for timber interests to plan capital investments to meet market demands.

Several timber supply strategies have been proposed to solve these difficulties. One idea promoted dominant uses and called for zoning public land areas that were primarily suited to a certain commodity use (such as timber, grazing, or mining). Other uses of those areas would be discouraged. This idea surfaced several times, but it never attracted the necessary support to bring it to fruition. The Public Land Law Review Commission (PLLRC), which undertook a comprehensive review of the management of public lands in 1968, recommended, among other things, that areas especially suited for timber production should be established "to manage for the dominant use (United States Public Land Law Review Commission 1970)." This was echoed in 1973 by the President's Advisory Panel on Timber and the Environment (PAPTE). The proposals were made in an earnest attempt to improve the efficiency of management of public lands. However, the emphasis of the proposals on commodities development and production was out of step with the burgeoning environmentalist sympathies that began to color public opinion at the time. As a result, no action was taken, and timber supply remained a central, if unpredictable, aspect of national forest management (Dana and Fairfax 1980, 235). Industry's needs for supplies were largely met from year to year, but its pleas for stability remained unanswered to this point. In California, in the 1970s and 1980s, the Forest Service's Region 5 harvested about 69% of the biological growth from the available timbered national forest lands. The level was higher than in other Forest Service regions. This harvesting was accomplished largely, although not entirely, through selection logging.

THE MONONGAHELA LITIGATION

A lack of willingness on the part of the Forest Service management to respond more forcefully to public concerns about clear-cutting, overcutting, and other silvicultural practices represented a significant miscalculation. In the Sierra Nevada, timber harvesting was largely accomplished by selection logging of mature trees or groups, unlike in the Pacific Northwest. Precisely because conservationists and outdoor recreation enthusiasts in other regions found themselves without recourse in the agency, they sought other means to influence agency decisions. Ultimately, a coalition of interests unhappy over plans to clear-cut an area of the Monongahela National Forest favored for hunting, fishing, and other recreational uses brought suit to enjoin further clear-cutting in the national forest. As a result of the court decision in *West Virginia Division of Izaak Walton League of America v Butz* (367 FSupp 422 [1973]), Forest Service authority to manage timber was severely impaired. This case centered around the interpretation of the Organic Act of 1897 (codified as amended at *U.S. Code*, vol. 16, secs. 473–82 and 551, in 1982). The Forest Service argued that these statutes supported its timber harvest practices. The trial court held that the Organic Act prohibited timber harvesting unless the trees were "mature" and individually "designated" and "marked" for harvest. Since the Forest Service was employing silvicultural management methods in direct contravention to this, the agency's system of timber management was effectively halted. This stunned both the Forest Service and the timber industry. It was clearly unacceptable to the timber industry, which depended on the national forests as part of their available supply. The Forest Service appealed the ruling, but the Fourth Circuit Court of Appeals upheld the District Court's opinion (*West Virginia Division of Izaak Walton League of America v Butz*, 522 F2d 945 [4th Cir 1975]). Although the effect of the ruling was confined to the Fourth Circuit, the implication was that the timber-harvesting program in the national forests, especially in its increasing reliance on clear-cutting, was in jeopardy.

Scientific and technical arguments cited by the Forest Service in support of clear-cutting as part of a properly conducted silvicultural system were of no avail in the face of public opposition and legal challenges. Finally, the Forest Service was unable to ignore or to parry the thrusts of its opponents. In the wake of the Monongahela decision, environmentalists brought similar cases in district courts in South Carolina, Texas, Tennessee, Georgia, Alaska, and Oregon (Dana and Fairfax 1980, 317). The Forest Service faced the prospect of defeat in all of these cases and, as a consequence, the deprivation of management methods on which it had come to rely. Forest Service personnel believed that these practices were essential tools for forest management and did not intend to manage the forests without them. Ultimately, congressional action was required to restore Forest Service authority to use clear-cutting to harvest stands that included immature trees (National Forest Management Act 1976).

THE FOREST AND RANGELAND RENEWABLE RESOURCES PLANNING ACT OF 1974

As the legal challenge in the Monongahela National Forest worked its way through the legal system, earlier efforts to provide for an economically stable management environment resurfaced in connection with proposals for the establishment of a strategic planning program for the nation's forest resources. This culminated in enactment of the Forest and Rangeland Renewable Resources Planning Act (RPA) in 1974 (Public Law 93-378; *Statutes at Large* 80 (1974): 476), as amended by the National Forest Management Act in 1976 (Public Law 94-588; *Statutes at Large* 90 (1976): 2949), codified at *U.S. Code,* vol. 16, secs. 1600–1614 (1982).

The RPA represents an attempt to institute a rational system of strategic planning for public and private natural resources in the United States. The statute provided a strategic framework for economic and physical planning for all forest resources and uses. These included timber, range, minerals, development, wilderness, and a host of other commodities and recreational needs. The statute directed the Forest Service to determine the aggregate national demand for all forest products. Every ten years, the agency was to inventory forest resources and public needs and to produce an "assessment" of the state of public and private forest resources in the United States.

The agency was also to develop a "program" every five years to meet those needs. Using information obtained from the national census along with other economic projections, it projects the future demand for forest resources in the United States. The program outlines the levels of commodities and other goods that can be can be supplied by the nation's forests and allocates a share of national goals to the national forest system. To meet these goals, it proposes a budget for the Forest Service for the next five years. Its recommendations are submitted to the president, who may use them as a guide for appropriation requests for the Forest Service that are submitted to Congress. The budget would reinforce the results of the program by mandating public expenditures designed to reach those goals. However, RPA was intended only to develop a strategic plan. The statute contained no explicit authority to implement the results of the RPA program, deliberately leaving this to the prerogatives of Congress and the president.

Implementation of these results remained subject to existing legislation and other political and administrative forces. Considerable agency resources were devoted to this initiative, which culminated in published reports containing RPA targets and regional disaggregation, showing how much regions such as California were expected to contribute toward achieving national goals established for various categories of forest uses and resources. During the appropriations process, congressional attention often initially focused on the RPA

documents, the assessment, and the program but was generally diverted from this focus by the politics of the budget process. As a result, the goals of the RPA were only partially realized over the years. Additionally, enactment in 1976 of the National Forest Management Act amended RPA, establishing a separate land-management planning process. This further complicated strategic planning and budgeting for the Forest Service.

As part of the review of the Forest Service land-management plans (LMPs) (discussed infra), many interested public and private parties commented on the LMPs. Remarks by the state of California and others requested that the Forest Service more strongly consider RPA goals and that the final plans for the Sierra Nevada national forests adopt as regional policy a management strategy that set timber harvest levels closer to the RPA targets. The state of California's comments on the plans for national forests in the Sierra and elsewhere in California went beyond simply discussing timber targets and sought to force the Forest Service to employ an approach that would have led to more careful consideration of a panoply of issues related to national forest planning. These comments were intended to reorient agency planning to employ more integrated views, considering the national forests and their contribution to part of a larger landscape and region. On the question of timber, for example, the state suggested that sustained yield calculations for timber should employ a regional timber inventory, using the stock of timber on both public and private land as the starting point for sustained yield calculations as opposed to that of a single national forest. State concerns also extended to a variety of noncommodity issues, seeking to draw the Forest Service more deeply into planning for watersheds and regions consisting of multiple national forests.

Sympathetic to the tenor of these comments, the Forest Service considered the requests, featuring them prominently in its response to public comment on the plans. Ultimately, however, other criteria contained in NFMA's provisions (discussed infra) and neither RPA's strategic thrust nor its targets controlled land-management planning. Without apparent irony, the final decisions of forest supervisors and other agency managers contained little more than an acknowledgment that RPA's targets called for higher harvest levels that were apparently not to be achieved under NFMA planning.

RPA's impact clearly has been decidedly less than the one intended by its sponsors. Nevertheless, it bears restating that RPA's strategic approach offered the Forest Service and others concerned with forest resources and ecosystems several valuable integrative mechanisms. The RPA related to all three Forest Service functions—the national forest system, research, and state and private forestry. While these tools were not exploited, a strategic planning approach could still prove useful in helping to realize shared goals for the conservation and management of the nation's public and private natural resources. RPA provided a broader planning authority than NFMA and is of potential relevance to emerging regional environmental planning and management initiatives. This is particularly the case in regions such as the Sierra Nevada, where significant ecological issues extend well beyond the national forests.

An RPA program designed to achieve ecological and other goals inherent in NFMA, carefully thought out and sensibly implemented, may have offered something of considerable value in the present context. This type of strategic plan might have laid the groundwork and enabled more active integration of federal and nonfederal lands in a range of cooperative ventures between different public and private landowners designed to achieve an entire spectrum of forest-related goals, not simply RPA timber targets. These ventures might include watershed and/or multiple national forest planning and regional planning. It is instructive to note that many current initiatives sponsored by the Forest Service and other groups to foster ecologically sensitive management suggest that a cooperative approach is critical, both to the success of these efforts and to the solution of a variety of national forest policy issues.

ROADLESS AREAS AND WILDERNESS, REVISITED

Under the Wilderness Act of 1964, the Forest Service was required to study certain areas to evaluate their potential for inclusion into the wilderness system or for multiple use ("primitive" areas had originally been set aside under the prior "U" regulations established by the Department of Agriculture). The agency undertook this study in 1967, projecting that the study would be completed within ten years. Of its own volition, once the initial study of these lands had been completed, the agency expanded the study to examine the larger remaining roadless areas within the national forests. The progress of this review, known as the Roadless Area Review and Evaluation (RARE), provides another example of an environmentalist challenge to agency initiatives. Conservation groups wanted the agency to pay more attention to recreation and preservation opportunities on the remaining forest land as well.

Upon the completion of the review, which indicated that approximately twelve of fifty-six million acres studied had wilderness potential, these groups remained unsatisfied. They wanted the Forest Service to increase the number of areas and the acreage recommended for wilderness. The Sierra Club immediately sued to enjoin the agency from adopting the results of the study, on the grounds that it was not accompanied by an adequate EIS (*Sierra Club v Butz*, 3 ELR 20071 [ND Cal 1972]). An out-of-court settlement restricted timber harvest in all roadless areas pending the completion of the EIS. The EIS was released in 1973, but it did not lead to legislative action. The Forest Service subsequently abandoned the first

study and embarked on a new study in 1977. This study, known as "RARE II," was released by the president in 1979. Although 65.7 million acres were recommended as potential wilderness, many areas for which the environmentalists sought protection were not included. The adequacy of the RARE II EIS was also challenged, this time by the state of California as well as the Sierra Club and other environmental groups (*California v Block*, 483 FSupp 465 [ED Cal 1980] 690 F2d 753 [9th Cir 1982]).

To end the policy stalemate, after an interagency review of RARE II, Congress took up the question of the disposition of these roadless areas. The result was a state-by-state review of the wilderness recommendations by Congress. A new series of wilderness bills proposed wilderness designations for additional acreage located in the national forests, in the national parks, and in other public land. In 1979 the first California wilderness bill was introduced by Representative Philip Burton and was subjected to five years of debate before passage in 1984. In a pattern repeated in a number of other western states, this legislation also returned other national forest lands to multiple use and reserved certain other areas for further evaluation as to their suitability as wilderness. Congress considered the Forest Service recommendations and the views of various interest groups and dealt with the decision as a political issue. Although the environmentalists compromised in Congress, accepting less acreage than they had originally sought, the bills that were passed represented further victories by the environmentalists in their struggle to force the government to permanently manage additional acreage as wilderness. This strategy worked much better for the environmentalists than did the administrative process, where they made little headway in persuading the agency to adopt their vision in either RARE or RARE II.

The timber industry, in contrast, was concerned that this initiative would result in the removal of more productive timber lands from the commercial timber base of the national forests. Industry was wary of further diminution of the timber base (Rey n.d.). The timber industry recognized that the national forests must support recreation wilderness and other nonconsumptive uses. Their position was that enough land was already preserved as wilderness, that the remaining timber should be managed as a renewable resource, and that the timberlands in question should be made available for harvest. Industry wanted to prevent more timber from being removed from the national forest's available timber base to ensure that as much timber as possible would remain available for commercial operations.

The wilderness legislation of 1984 led to additional national forest land being removed from the full spectrum of multiple uses. As before, grazing, mining (where already established), water resource development (as permitted by executive order), recreation, and other interventions such as the planting of fish were allowed to continue. In the Sierra Nevada, an additional 1.8 million acres of land in national forest were reserved by the 1984 legislation, again mainly at high elevation. Notwithstanding this victory, environmentalists remained determined to protect other roadless areas that were not reserved in this round of wilderness legislation. The new wilderness areas, along with the areas reserved by the earlier Wilderness Act of 1964, joined the national parks Yosemite, Sequoia, and Kings Canyon as the largest areas of contiguous forest land in the Sierra not subject to human intervention, save those activities associated with fire suppression.

THE NATIONAL FOREST MANAGEMENT ACT OF 1976

Natural resource management policy for the national forests of the United States has been dramatically restructured over the last twenty years. Environmentalism, public interest litigation, and internal agency and congressional initiatives together worked to force the federal government to respond to public pressure to change traditional management practices of the federal resource management agencies, including the Forest Service.

Background to New Legislation

After the Monongahela decision, Congress considered several measures to restore national forest management authority. Legislative debates reflected a continuing competition in the legislative process between two different visions of forest management. Several bills were introduced in Congress to counter the effects of the legal obstacle to the method of timber sale and harvesting that had been in use by the Forest Service. S. 2926, introduced by Senator W. Jennings Randolph of West Virginia, would have allowed timber harvesting in the national forests only with stringent prescriptions, including provisions that would have limited the size of clear-cuts to a maximum of twenty-five acres and required a 200- to 300-year "rotation," or growth period, for all trees (U.S. Forest Service 1976, 17). A competing bill, S. 3091, presented a management model more deferential to agency expertise. Sponsored by Senator Hubert Humphrey, this bill sought to restore discretionary authority to the Forest Service to employ a broad range of management practices. The bill directed the agency to develop plans that would respond to the diverse conditions encountered in each national forest and provide for management within certain limitations designed to protect the environment. This bill, far more than the others, continued reliance on agency expertise to make management decisions. After considerable debate, this version, modified by certain amendments, was adopted as the National Forest Management Act.

The National Forest Management Act called for the implementation of natural resource planning that would attempt to reconcile public demands relating to resource management

and conservation with the need for timber production and other natural resource development. Designed to resolve continuing disputes over national forest management, the new statute, together with other contemporaneous changes in the legal environment, sought to increase Forest Service responsiveness, especially to environmentalism, but also to economic efficiency criteria, through greater legalization of agency procedures. The law's emphasis on planning was intended to modify existing agency resource management policies by developing competence in a variety of scientific disciplines. Interdisciplinary analysis, once fully developed by the agency, was to provide reliable scientific information to assist in resolution of the controversies surrounding national forest management.

Under NFMA, "multiple use and sustained yield" of forest resources remained the focus of national forest management (National Forest Management Act 1976, sec. 2952, sec. 6e1, codified at U.S. Code, vol. 16, sec. 1604e1 [1982]). The premise of the planning process was that agency decisions would respond to natural conditions in the forest and to demands on the natural resources to produce fair and balanced plans. The plans were to be circulated for public comment to permit the agency to respond to public comment and to modify its decisions. Planning contemplated a range of forest management activities and land uses that was substantially the same as had existed prior to the Monongahela decision. Clear-cutting and other harvesting methods that permitted even-aged management were allowed if they could be shown to be the "optimum" silvicultural method. In this respect, the statute did not appear to represent a radical departure from prior management of the national forests. NFMA, however, did incorporate environmental protection into multiple-use planning of public natural resources.[2] Also different in NFMA were the procedures it established requiring the coordination of forest planning, environmental assessment, and public comment on management proposals prior to the initiation of management actions.

Several aspects of NFMA restructured public land management to produce more balanced plans and to reduce the likelihood of legal battles relating to management actions.[3] First, planning was undertaken pursuant to detailed statutory instructions to ensure that adequate consideration was given both to resource protection and to development. Second, building on an idea of interdisciplinary expertise already in use by the Forest Service, the statute directed the agency to develop forest plans using an "interdisciplinary team" consisting of a group of agency scientists and resource professionals with diverse scientific and professional skills. By requiring input from new kinds of "experts," NFMA intended to make certain that the sustainability of forest resources was given full consideration during agency decision making. Third, the law expanded opportunities for public involvement in the planning process, seeking to permit an unprecedented level of public participation in management decisions. These features all promoted new avenues of decision making within the agency and distinguished NFMA land-management planning from earlier Forest Service management. The impact of these provisions is worth considering because they have shaped the course of national forest management from the enactment of NFMA to the present time.

The origins of NFMA's administrative reforms have to do with the administrative culture of the Forest Service itself. The Forest Service was regarded as an example of the effort to promote expert management in administration (Clarke and McCool 1984, 41–44) and an able player in national politics (Hays 1969). Despite controversies over national forest management, the agency enjoyed an excellent reputation among politicians and social scientists as a model of effectiveness in bureaucratic management (Clarke and McCool 1984, 41–44). After World War II, many aspects of national forest management became controversial. As support for Forest Service management decisions steadily eroded, these decisions were increasingly subject to challenge. Consequently, support for management authority itself also eroded significantly. This legacy makes it particularly intriguing to study administrative change in an agency so rich in tradition and in expertise. Understanding how the reform affected the Forest Service response to NFMA's objectives also provides insights as to the difficulties in ensuring the attainment of any complex set of objectives through legislation and implementation.

NFMA accurately reflected wider political conflicts and uncertainty over goals for public land use. The statute's provisions for management reform contain less than definitive direction and emphasize planning to achieve balanced land-management plans. The implication of this arrangement is that controversies over national forest management that Congress could not resolve would remain. Land-management planning conducted pursuant to NFMA anticipated these conflicts. Planning was intended not to eliminate national forest management controversies but to provide procedures for land and resource planning that would enable conflict resolution and progress toward better management in light of conflict. Three ideas central to administrative reform are contained in NFMA's direction to the Forest Service. First, the relationship between law and administrative behavior is specified in the statutory elaboration of the planning process. Regulations further emphasized full assessment of the forests' capabilities for diverse uses and decision making consistent with that information. The law recognized that Forest Service administrators were charged with more than managing a planning process and that they were policy makers whose decisions could have a significant impact on the condition of the national forests. The statute gave the administrators general guidance in decision making but delegated to agency managers discretion to reach a decision within a range of possible outcomes that would achieve the greatest "net public benefit" (NFMA Regulations). Of course, there were other constraints on administrators. Land-management planning, like many other public programs, is conducted in a highly charged political environment. The political implications of these "administrative" decisions were closely followed by successive executive branch appointees. The Forest Service, therefore, was expected to act with both technical proficiency and sensitivity to public and political opinion.

The procedural reforms associated with planning and the NEPA process forced consideration of information that previously might easily have been undervalued or ignored. Responding to land-management planning requirements for analysis was intended to allow the agency to develop local plans in accord with the statute's substantive goals for resource development and preservation. The new procedures led to considerable changes in agency operation.

The National Forest Management Act implicitly promoted a second principle of administrative reform to ensure a stable management environment that would be responsive to changing public priorities. Even at the time of its enactment, there was considerable skepticism among scholars of public land policy concerning the power of the new law to do so. NFMA, in seeking to promote this goal, fought against an already strong tide of activism. Several concerns are worth mentioning here. First of all, establishing a comprehensive land-use planning system, as done under NFMA, which standardizes analysis and planning direction of natural resources over a very large area, represents a conceptual challenge. On top of this ambitious goal, the expectation that this system would retain flexibility sufficient to permit managers to respond to varying local needs and conditions was perhaps a forlorn hope. Early in the planning process, some doubt was expressed that any Forest Service management policy requiring assent of the public could succeed, as long as those who opposed it could find a method to block implementation of agency plans (Behan 1981, 802, 805). This statement later proved to be fairly prophetic.

Second, the National Forest Management Act explicitly recognized the continuing validity of multiple-use management. The Forest Service sought to employ this philosophy to satisfy the needs of timber and other commodity interests while also attempting to satisfy environmentalist concerns. At the same time, NFMA implicitly acknowledged that prior multiple-use management did not sufficiently accomplish this objective. The newly constructed procedures in national forest planning intended to respond to environmental constituencies without sacrificing the virtues of the established management system. To those familiar with recent public land management in the United States, who had come to view controversy as the normal condition for public land policy making, the effort to blend these conflicting aims was a formula that would achieve only added conflict and inefficient use of publicly owned natural resources (O'Toole 1988; Stroup and Baden 1983; Rosenbaum 1984). Nevertheless, the ambiguity inherent in NFMA's mission made land-management planning the subject of continuing scrutiny by public land scholars, activists, and others.

A third idea inherent in the administrative reform of the era focused on demands to increase representation and participation in government is well illustrated on NFMA's emphasis on public participation in planning. Public involvement was intended to reorient administrative decision making from a strict reliance on expert management toward decision making that resembled a political dialogue between the administrator and the public (Reich 1985; Handler 1988; Friedmann 1987). Efforts to draw the public into the planning process resulted from a tacit recognition that forest-planning decision making, although dependent on Forest Service expertise, had political implications. Out of necessity, planning required agency consideration of public opinion during all stages of the process. NFMA land-management planning employed various types of public participation so that wherever possible disagreements over administrative decisions would be settled expeditiously. Public participation allowed the public and interest groups to comment on agency proposals. The Forest Service experimented with innovative techniques, such as negotiation, that blurred distinctions between public involvement and conflict resolution in order to resolve specific policy disputes (Wondolleck 1988), drawing on collaborative approaches to settling policy questions utilized in public land management, and in other administrative and regulatory settings (Fiorino 1988; Sullivan 1984; Burton 1991).

NFMA: Impact of Land-Management Planning in the Sierra Nevada

In the middle to late 1980s, the Forest Service produced land-management plans (LMPs) calling for expanded utilization of practices, such as even-aged management, that had already generated considerable controversy. Armed with statutory language that allowed what remained to be a controversial practice, the plans clearly laid out the future of every area within the national forest. Several important elements in the plans in the Sierra illustrate some significant differences from the policies previously guiding national forest management.

Forest Service land-management planning proposed to greatly accelerate clear-cutting in many regions, including the Sierra Nevada. Overall, this reflected an apparent emphasis on enhanced productivity on national forest lands devoted to timber production. This included some related silvicultural methods, such as seed tree cutting and overstory removal. As a result of this policy, the new plans proposed significant increases in clear-cutting in the Sierra during the middle 1980s. Data from national forest timber sales reflects this increase (Verner et al. 1992, 240–41). The rationale for this increase was that many forest stands were mainly composed of mature or overmature trees, well past their peak growth period. Planning documents presented to the public suggested that as these stands were cleared and replanted, growth would increase, allowing the forest to supply more timber. Some conversion of forest types were proposed. Mixed stands, containing conifer and hardwood, were to be logged and replanted as conifer (generally pine) stands. This was apparently part of an effort to improve timber yield. Similarly, other

stands, in which white fir had increased due to the effects of earlier timber harvests and fire suppression, were to be harvested and replanted to resemble more closely the mix of species that had occurred before human intervention.

Many foresters were sympathetic to the goals of the Forest Service. Where a natural mix of species had been or was being eclipsed by the growth of white fir and the harvesting of older stands, they regarded the initiative as an effort to restore the forest landscape. This group viewed the LMPs as moving the national forests much closer to the model of a regulated forest, an ideal of scientific forestry that allows both greater productivity of forest lands and better modeling of timber growth. Environmental critics of land-management plans viewed these arguments as insufficient either to justify the increased use of clear-cutting or to increase harvest levels. Prior to the adoption of the LMPs, these practices had already been introduced in many areas in the Sierra Nevada. In the years immediately following the approval of the LMPs, utilization of these practices significantly increased.

Exceptions may be seen in the way this practice was adapted and applied to clear-cut areas around the giant sequoia trees located in Sequoia National Forest. Ironically, although justified in part by Forest Service managers as a method to leave the giant sequoias intact while promoting sequoia regeneration, these timber sales were regarded by some as proof of irresponsible stewardship in these relatively rare areas. They became one of the single most visible aspects of the changes actually implemented as a result of land-management planning. This controversy, one of apparent ecological significance, sparked appeals and lawsuits and led eventually to the mediated settlement agreement (MSA), which is being explored in greater depth elsewhere by SNEP.

Chiefly, the changes in national forest management in the Sierra had to do with increased intensity of management activities, such as clear-cutting, rather than the wholesale adoption of new forest practices. However, when the scope and intensity of these actions were laid out in the plans, it became clear that the character of many areas would change drastically under the hand of management. Predictably, some of the same individuals and organizations that the Forest Service faced in earlier struggles over clear-cutting and wilderness resurfaced to battle the Forest Service again.

OTHER NEW STATUTES AND THEIR IMPLICATIONS

During this period, several other newly enacted statutes also significantly modified the Forest Service prerogatives. Although the implications of these laws in the national forests of the Sierra Nevada will not be examined here in any depth, it is important to understand that these laws dramatically altered federal prerogatives with respect to natural resource planning and management. The Federal Land Policy and Management Act, as amended (*U.S. Code*, vol. 43, secs. 1701–84 [1976]) (FLPMA), directed the Bureau of Land Management (BLM) to undertake comprehensive land and resource planning for the public lands similar in scope to what NFMA required for the national forests. At the same time FLPMA revised and modified authorities related to the entire Forest Service lands program, altering management of rights of way, acquisition, small tracts of noncontiguous land, etc. The law also revised the administration of the minerals programs of the Forest Service and BLM.

The Federal Water Pollution Control Act (*U.S. Code*, vol. 33, sec. 1251 et seq. [Clean Water Act]) and the Clean Air Act transferred to the states the authority to regulate practices on federal lands, provided the state had obtained approval for its own program to enforce these laws. For the first time, air- and water-quality standards, as well as the authority to issue permits for a range of regulated activities, applied to federal lands, limiting federal prerogatives on public land. Forest Service discretionary authority in planning is also subject to the operation of other laws with which public forest management must also comply. The Clean Water Act requires that the activities likely to affect the quality of certain water systems must be conducted under approved procedures, or "best management practices." The Forest Service's interpretation of its responsibility in California under this law was challenged in court, resulting in the modification of certain management actions (*Northwest Cemetery Protective Association v Peterson* [764 F2d 581 (9th Cir 1985)]).

The Federal Wild and Scenic Rivers Act (*U.S. Code*, vol. 16, secs. 1271–87) is another statute with specific mandates and established standards that the Forest Service was required to take into account in the preparation of its land-management plans.

CONSERVATION OF THE CALIFORNIA SPOTTED OWL

Reshaping Resource Planning and Management in the Sierra Nevada

The present legal environment for resource management in the national forests is principally comprised of three statutes, NFMA (1976), NEPA (1969), and the Endangered Species Act (1989) (ESA). The legal requirements for protection for plant and animal species contained within these laws is significant enough to have played an important role in the evolution of federal and state resource management. More precisely, it is clear that procedural requirements for national forest planning have substantive effects. Legal protections for sensitive species contained within these laws, and their regulatory progeny, operate to modify any Forest Service land-management plan that fails to take account of the needs of the species

or to provide for its habitat requirements. Scientific analysis intended to help administrators make decisions consistent with the law may effectively dictate policy choices that have profound effects for forest management.

The California spotted owl (*Strix occidentalis occidentalis*), one of three subspecies of spotted owls, is related to the northern spotted owl (*Strix occidentalis caurina*) and the Mexican spotted owl (*Strix occidentalis lucida*) (American Ornithologists' Union 1983). The California spotted owl's range extends from the Pit River (at the northern end of the Sierra Nevada) southerly through the Sierra Nevada, along the central Coast Range south of the Golden Gate, and throughout the forested areas of southern California, including the higher montane regions. The majority of California spotted owl habitat in the Sierra Nevada is within national forests, entirely within the Pacific Southwest Region (Region 5) of the Forest Service (the range of the northern spotted owl lies mostly in the Cascade Mountain system and includes part of both the Pacific Northwest Region and the Pacific Southwest Region of the Forest Service). In the late 1980s, Forest Service management practices for both the northern and California spotted owls sought to preserve small areas of owl habitat, known as Spotted Owl Habitat Areas (SOHAs). The SOHA strategy permitted partial timber harvesting in parts of SOHAs not immediately adjacent to nest trees. Lands outside of SOHAs also were utilized for nesting, roosting, and foraging by the owls, but the SOHA policy did not affect timber harvests on the remainder of forest lands. A more complete explanation of the SOHA strategy can be found in U.S. Forest Service 1993a, III-1-2.

Research conducted on the northern spotted owl raised the possibility that the SOHA strategy did not sufficiently protect owl habitat and that the continued use of clear-cutting was detrimental to the spotted owl. As the Interagency Scientific Committee study on the northern spotted owl was nearing completion in 1989, research indicated that the existing management strategy would not sufficiently ensure the survival of the northern spotted owl and that its continued use would lead to further decline in northern spotted owl numbers (Thomas et al. 1990, 427). The Fish and Wildlife Service (FWS) listed the northern spotted owl as a "threatened" species under the ESA in June 1990. The Mexican spotted owl was also listed under the ESA (*Federal Register* 58, no. 49 [16 March 1993]: 14248–71). The research suggested that the SOHA policy and subsequent administrative actions employed to protect the habitat of the California subspecies also were inadequate and were as vulnerable to legal challenge as those employed for the conservation of the northern spotted owl.

The Forest Service had already designated the California spotted owl as a "sensitive species." This required an internal evaluation of any plans or projects to determine their effects on the spotted owl. Yet there were few demographic or ecological studies specific to the California subspecies. Accordingly, the lack of biological information made it difficult to justify any change in management guidelines or to offer guidance as to what type of habitat management should be adopted. This uncertainty caused some concern among resource managers and the public. NFMA and its regulations require that the Forest Service maintain viable populations of native and select non-native wildlife species (*Code of Federal Regulations,* vol. 36, sec. 219.19 [1988]). The Forest Service instituted a new policy, known as cumulative effects analysis (CEA), to supplement the SOHA strategy. CEA called for specific consideration as to how individual projects would affect owl habitat in relation to habitat conservation measures generally required for known or probable owl sites for pairs or resident single owls (Verner et al. 1992). Environmental groups continued to express concerns about the adequacy of the conservation measures and challenged Forest Service decisions to continue using the SOHA strategy for management of owl habitats. Eventually, environmental groups filed a number of administrative appeals challenging a number of timber sales in the region (e.g., "Appeal of the Tahoe National Forest Land Management Plan," Natural Resources Defense Council, 15 March 1991).

The decision of the Forest Service was to attempt to resolve the controversy by pursuing a remedy without waiting for the results of the administrative and legal process. This development may be regarded as the clearest signal that the land-management plans were not going to be fully implemented. The new policy precluded any large-scale use of clear-cutting or other methods intended to achieve even-aged management strategies for the Sierra. The law's mandates, especially the element of the regulations requiring the Forest Service to ensure that its plans would provide a "minimum viable population" of forest species, ultimately worked to ensure that NFMA's implementation did not lead to the wholesale changes in the forest that the Forest Service had proposed.

Developing a New Management Strategy

In June 1991, in response to growing public concern about the status of the California spotted owl, state and federal agencies convened the "California Spotted Owl Assessment and Planning Team ('Steering Committee')." The goal of this group was to assess the status of the owl and explore alternative management strategies that would conserve the subspecies and its habitat. The steering committee, co-chaired by Ron Stewart, then regional forester for the Pacific Southwest Region, and Douglas Wheeler, secretary of the Resources Agency of California, included representatives from the Resources Agency, the Forest Service, the California Department of Forestry and Fire Protection (CDF), the National Park Service (NPS), the Fish and Wildlife Service (FWS), the Bureau of Land Management (BLM), the Board of Forestry (BOF), and the Department of Fish and Game (DFG). Observers from county government, and nongovernmental observers from environmental groups, timber and forest products industries, and several other organizations were also invited to attend. Agency

representatives agreed to plan the implementation of conservation measures, especially those required if the subspecies were to be listed under the ESA. The charter for this project directed federal and state natural resource agencies to

work cooperatively . . . to assess local research, inventory and monitoring information for the . . . spotted owl [and that as] more information becomes available . . . agencies will continue to work cooperatively to incorporate other species and habitat needs into a long-term ecosystem planning strategy for the Sierra and Southern California ecosystems ("California Spotted Owl Assessment and Planning Team," ms., 14 May 1991).

The steering committee immediately created two teams, a "technical team" to provide expertise in avian biology and ecology and a "policy implementation planning (PIP) team" to provide policy and economic analysis. The project was to produce several results:

- a review by the technical team of the status of the California spotted owl, to be published as a technical report

- recommendations by the technical team for a management strategy to maintain viable populations of the owl, including an assessment of alternative measures considered

- analysis by the PIP team of socioeconomic effects resulting from the implementation of the management recommendations of the technical team, including an "evaluation of alternative institutional strategies" and regulatory applications to be considered for adoption by state and federal agencies

The technical team evaluated several alternative management strategies for the owl. The team analyzed the status of the owl and offered a set of recommendations, known as the "CASPO report," to the steering committee in May 1992. Following a period of review and comment, the report, "The California Spotted Owl: A Technical Assessment of Its Current Status," was published in late 1992 (Verner et al. 1992). A review of the impacts of this policy may be found in Ruth and Standiford (1994). The scientific analysis suggested that existing policy measures used to protect the spotted owl and its habitat were inadequate. The technical team investigated the loss of suitable habitat in the Sierra Nevada. The team noted that suitable owl habitat probably was once more extensive and concluded that habitat loss has been caused by even-aged silvicultural practices and catastrophic fire. Their research attributed further diminution in habitat to the activities of miners and sheepherders in the nineteenth century (Verner et al. 1992, 10–11, 225, 232–33, 240–41, 248–53). The team noted that in the present era current land-management policy called for increased clear-cutting and other forms of regeneration harvests. The emphasis within the plans on harvesting large-diameter trees was also viewed as detrimental by the technical team (J. K. Verner, project leader, Wildlife Monitoring and Range Research, USFS, Pacific Southwest Research Station, personal conversation with the author, July 22, 1993). These harvests removed forest structures upon which the owl was dependent. Under the LMPs for the Sierra Nevada national forests, the technical team estimated that the amount of suitable habitat would further decline at a rate of 229,000 acres per decade (Verner et al. 1992, 11 and chap. 13). The team also emphasized that suppression of fire had accelerated the accumulation of fuels and significantly increased the likelihood of fires that would destroy timber stands, including those essential to the spotted owl. The technical team concluded that these management actions had detrimental effects on spotted owl habitat and proposed an interim strategy of thinning and fuel management to begin to address these problems (Verner et al. 1992, chap. 1).

CURRENT POLICY FOR THE NATIONAL FORESTS IN THE SIERRA NEVADA

To adopt the CASPO recommendations as management policy, the agency complied with planning and public participation requirements of NFMA (1976) and NEPA that must be completed prior to making changes in management direction. Pursuant to NEPA, the Forest Service prepared the "California Spotted Owl Sierran Province Interim Guidelines Environmental Assessment" (EA) (U.S. Forest Service 1993a, III-1-2). The EA incorporated substantially all of the CASPO management recommendations into an interim management plan for the Sierra Nevada national forests. The regional forester's decision amended the regional guidelines for land management in the seven Sierra Nevada national forests. This procedure satisfied the requirements of NEPA, as these amendments were judged to be nonsignificant actions (U.S. Forest Service 1993b, DN-13-15) (Cal. Owl NOI). On January 13, 1993, the regional forester formally adopted the plan as management direction for these national forests.

This decision was not immediately accepted by the entire steering committee. A number of administrative appeals filed by timber interests, and others brought by affected counties, challenged the adequacy of the EA. In addition, members of the steering committee representing state and federal agencies alleged that the Forest Service decision to change policy abrogated the interagency agreement and departed from the exercise of shared authority they believed to be implicit in the owl assessment process. The Forest Service argued that there was no breach of this agreement, maintaining that it was clear that the long-term survival of the population of the owl could not be assured if existing policy permitting extensive clear-cutting and other forms of regeneration harvests remained in force. The Forest Service noted that the data sub-

mitted in the technical team's report left no choice: The agency was legally required to revise its management policy. The agency maintained that the steering committee's involvement in the development of policy pertaining to national forests was strictly advisory. At the same time, the agency agreed to remain part of the interagency process as it continued the preparation of the EIS.

Members of the steering committee representing the state of California found themselves in an awkward position. They fully understood and accepted the Forest Service's desire to avoid a legal challenge to its management policies. Nevertheless, state support for the California Spotted Owl Assessment had always been conditioned on the principle that in this process, the federal government had an obligation to consider the implications of conservation planning and management for national forest lands on the larger region in which they were situated. Ten years earlier, similar concerns motivated the state to comment on Forest Service plans, suggesting that the land-management plans needed to more fully consider regional effects and a more cooperative regional approach to planning. For their part, state officials had recognized the dramatic changes in theories pertaining to natural resource conservation and management. Evidence of the state's support for a more ecologically integrated, regional approach to management is reflected in the state's strong support in the drafting, adoption, and implementation of the "Memorandum of Understanding on Biological Diversity" (The Resources Agency 1992). Accordingly, state officials sought to better integrate measures to conserve habitats for multiple species while maintaining the viability of local economies into Forest Service planning and decision making. Although the state's views of how these goals should be accomplished have provoked criticism, from both environmental groups and commodity groups, the state continues to articulate these same concerns. It may be argued that the Forest Service's failure to pay heed to state concerns regarding the land-management planning process of the prior decade led to increased state support for initiatives to focus more local and regional attention on Forest Service planning. These initiatives include the California Spotted Owl Assessment, the Council on Biological Diversity, Sierra Nevada Research Planning (SNRP), and the Sierra Nevada Ecosystem Project itself.

At the present time, early in 1996, the Forest Service continues to revise the "Draft EIS: Managing California Spotted Owl Habitat in the Sierra Nevada National Forests of California" (U.S. Forest Service 1995) (DEIS). The DEIS will more fully address conservation planning for the broader suite of species living in the Sierra Nevada (Cal. Owl NOI), eventually permitting the Forest Service to revise its management policies. Currently, forest management activities in the national forests of the Sierra must continue to conform to the CASPO policies as articulated in the Cal. Owl NOI. These are subject to change upon adoption of the final EIS.

LESSONS FOR THE FUTURE

Until quite recently, environmental and natural resource–related policies operated without explicitly considering the ecosystem as a point of reference for policy formation, implementation, or evaluation. It is certainly true that some policies implemented in the Sierra Nevada by the Forest Service have had profound effects upon the landscape and, most probably, the state of Sierran ecosystems. As mentioned above, other policies—some of them largely or entirely unrelated to natural resources—have also had significant impacts on the ecosystems of the Sierra Nevada. The point of this retrospective policy analysis is not solely to establish that certain public policies positively or negatively affected ecosystems in the Sierra Nevada. Nor is it just an opportunity to call attention to those policies that the SNEP Science Team believes should be favored or avoided in the future. This retrospective, and indeed, the Sierra Nevada Ecosystem Project itself, provides a larger opportunity to review the operation of environmental and natural resource–related policies and should help to reflect on the complexity of the institutional setting surrounding the implementation of these policies. This approach will materially assist the SNEP Science Team, public officials, and the public in understanding and evaluating existing and proposed policies.

SUMMING UP NFMA'S IMPLEMENTATION AND ITS IMPACT: SUBSTANTIAL OR UNDERWHELMING?

Implementation of bold land-management initiatives proposed by the Forest Service, such as even-aged management, upon a landscape where almost any large-scale activity apparently has the potential to cause substantial environmental impacts, was unsuccessful. Examination of Forest Service policy, of the operation of NFMA, and of recent extraordinary efforts to resolve continuing controversies over natural resource management demonstrates that much of the opposition to Forest Service plans has been due to the inability of the agency to satisfy the ecological protection provisions of federal law and policy. Despite the mandate of national forest management policies to implement an ecological approach to natural resource planning, conservation, and management, the Forest Service has generally not been able to do so. Until comparatively recently, most successful agency policy initiatives tended to be responses to single-purpose functional demands, such as timber, range, or recreation. Implementation of NFMA, despite explicit requirements for the protection of ecosystem attributes and functions, has had only limited success in incorporating an ecological approach into its general

land and resource planning. Forest Service efforts to comply with these requirements continue to face political and budgetary directives that complicate resource management. Conflicting directions often appear to be supported by elements within NFMA itself.

Political and social activism related to dissatisfaction with national forest management activities remains another important force, leading to political challenges and to heightened legal scrutiny of Forest Service decisions. In many instances, activists are able to draw on elements of NFMA that tend to support a particular position. The agency, in contrast, generally must try to balance the operation of a particular provision with that of other goals within NFMA. In many instances it has done so only to find that a decision will not meet with legal or regulatory approval. Scientific and technical knowledge are essential components of decision making, but this expertise has not solved the riddle of how to meet either the demands of NFMA's several conflicting forest management goals or the conflicting demands of the public. Satisfying these objectives, especially habitat conservation and commodities production, has forced the Forest Service to promulgate policies that are effectively compromises intended to arrive at a decision that reconciles several conflicting objectives. Forest Service actions continue to face certain scrutiny, which quickly translates into opposition if the agency displays a lack of attention to legal or administrative mandates pertaining to environmental protection. The outcome—unintended by the Forest Service—is that implementation of plans and management activities poised to have a substantial impact on the ecosystem were prevented from being fully implemented.

The intense public attention focused on NFMA has been an understandable source of frustration to many Forest Service managers. Public involvement in resource planning translated into administrative and legal mechanisms to influence national forest planning, however, represents a self-correcting mechanism for policies that do not comply with the intent of the law and its attendant regulations. Imperfect and unsatisfactory as it may appear, the ongoing Forest Service's effort to conserve the California spotted owl is an example of the process at work. Simultaneously, however, experience during the recent decades with the transient nature of Forest Service policies strongly suggests that policies established to provide for management of the Sierran national forests have not yet successfully ensured long-term sustainability for the natural resources and ecosystems. Irrespective of an ongoing search for solutions, the Forest Service has been unable to implement either broad policies or land-management plans that survive longer than a few years (Yaffee 1994; Ruth 1990). No proposal for managing the national forests of the Sierra Nevada thus far has elucidated a strategy that will demonstrably satisfy the ecological, socioeconomic, legal, and political criteria by which these policies are judged. Producing land-management plans that respond to current national forest management priorities in the Sierra Nevada—providing for habitat and species conservation while promoting fuel management and commodities production—continues to be a difficult technical problem. Achieving solutions, of course, is complicated by various conflicting views among key public actors, both individuals and groups, as to what methods will best accomplish particular goals. Recognition of the difficulty in integrating ecological, technical, and social concerns into a successful management plan is part of the motivation for the Sierra Nevada Ecosystem Project. Public concerns regarding the management and conservation of national forest ecosystems will not be resolved by the SNEP Science Team or by direct application of its final report. The fruits of the project, the scientific assessment and the scenarios developed by the Science Team, however, should provide a useful step in the search for solutions.

ACKNOWLEDGMENTS

The author gratefully acknowledges the assistance of the staff of the Sierra Nevada Ecosystem Project (SNEP), members of the SNEP Science Team, and many other individuals connected with this endeavor. In particular, Professor Harrison Dunning, organizer of the SNEP Policy Retrospective, held on May 17, 1995, deserves special thanks and recognition. All those who reviewed and commented on an earlier version of this chapter provided invaluable and constructive criticism.

NOTES

1. As early as 1929, at the instigation of foresters within the agency, areas in the national forests had already been removed from harvesting and other management activities. This practice originated with the "L-20" regulation in 1929, which allowed the Forest Service to protect certain "primitive areas." This authority was expanded and more precisely defined in 1939 with the "U" regulations. Over the objections of commodity users, additional land was removed at this time. Three different types of designations were established. Regulation U-1 defined "wilderness" as unroaded, undeveloped tracts 100,000 acres or more. These areas were to be designated by the secretary of agriculture. Areas that had similar characteristics but were smaller in size could be set aside as "wild" areas by the chief forester. A third category allowed tracts of 100,000 acres or more to be designated by the chief forester as roadless areas to be managed for recreation "substantially in their natural condition." See Dana and Fairfax (1980, 157–58).

2. Senator Humphrey's original bill was amended to provide legislative assurances to conservation interests that required the Forest Service to eschew certain extractive resource policies. One provision required the use of a sustained yield forestry practice known as "nondeclining even flow." This provision mandated that timber sales from each forest were to be "equal to or less than a quantity which can be removed from such forest on a sustained yield basis: Provided, That, in order to meet overall multiple use objectives, the Secretary may establish an allowable sale quantity for any decade which departs from the projected long-term average sale quantity that would otherwise be established. . . . [S]uch planned departures must be consistent with the multiple-use management objectives of the land management plan" (sec. 11). Although this practice had already been adopted by the Forest Service in 1973 (see earlier discussion), the amendment committed the agency to plan timber harvest levels on each forest at a rate that was sustainable indefinitely.

The timber industry and many economists opposed this provision. In their view, nondeclining even flow was too restrictive because it prevented major variations in the allowable cut on a national forest that could increase economic returns while still meeting sustained timber yield goals. To accommodate this objection, the final version of the bill allowed for exceptions from the "non-declining even flow" policy in order to achieve multiple-use goals. This arrangement was emblematic of the design for national forest planning. This compromise enabled Congress to delegate discretionary authority to the Forest Service to operate within certain limits. This also allowed Congress to defer responsibility to the Forest Service for many controversial decisions regarding the determination of management priorities, land allocations, and levels of commodity development and other resource uses.

3. As a precursor to new national forest planning, NFMA contained several provisions intended to remove the threat of delays resulting from legal challenges to new planning. The statute provided that existing plans for an area would remain in force until a new land-management plan was adopted. Primarily, this meant the unit plans and timber management plans and other special use plans developed under the auspices of the Multiple Use–Sustained Yield Act of 1960. This allowed the Forest Service to continue to manage the national forests as it had before the National Forest Management Act, pending the completion of the new plans. Notwithstanding the decision in the Monongahela case, this provision tacitly permitted the use of clear-cutting on the forests, pending the release and final approval of the National Forest Management Act plans, including the period during which a new plan might be appealed. In order to remove any further doubt as to whether clear-cutting was permitted, section 11 of the statute explicitly repealed the language of the 1897 Organic Act, which had stipulated that trees could not be harvested unless the trees were "mature" and individually "designated" and "marked."

REFERENCES

American Ornithologists' Union. 1983. *Check-list of North American birds.* 6th ed. Lawrence, KS: Allen Press.

Behan, R. 1981. RPA/NFMA—time to punt. *Journal of Forestry* 79: 802–5.

Bingham, G. 1986. *Resolving environmental disputes: A decade of experience.* Washington, DC: Conservation Foundation.

Burton, L., Jr. 1991. *American Indian water rights and the limits of the law.* Lawrence: University Press of Kansas.

Cahill, T. A., J. J. Carroll, D. Campbell, and T. E. Gill. 1996. Air quality. In *Sierra Nevada Ecosystem Project: Final report to Congress,* vol. II, chap. 48. Davis: University of California, Centers for Water and Wildland Resources.

California spotted owl assessment and planning team. 1991. Manuscript, 14 May.

Clarke, J. N., and D. McCool. 1984. *Staking out the terrain*. Albany: State University of New York Press.

Clary, D. 1986. *Timber and the Forest Service*. Lawrence: University Press of Kansas.

Clawson, M. 1975. *Forests for whom and for what?* Baltimore: Johns Hopkins University Press for Resources for the Future.

Culhane, P. 1981. *Public lands politics: Interest group influence on the Forest Service and the Bureau of Land Management*. Baltimore: Johns Hopkins University Press.

Dana, S. T., and S. K. Fairfax. 1980. *Forest and range policy*. 2nd ed. New York: McGraw-Hill.

Dowdle, B., and S. Hanke. 1985. Public timber policy and the wood products industry. In *Forestlands: Public and private*, edited by R. T. Deacon and M B. Johnson, 77–102. Cambridge, MA: Ballinger.

Ecology Law Quarterly. 1972. Comment, Mineral King: A case study in Forest Service decision making. *Ecology Law Quarterly* 2:493.

———. 1976. Commentary, Mineral King goes downhill. *Ecology Law Quarterly* 5:555.

Fiorino, D. J. 1988. Regulatory negotiation as a policy process. *Public Administration Review* 48 (July/August): 764–72.

Forest and Rangeland Assessment Program (FRAP). California Department of Forestry and Fire Protection. 1988. *California's forests and rangelands: Growing conflict over growing uses*. Sacramento: California Department of Forestry.

Friedmann, J. 1987. *Planning in the public domain: From knowledge to action*. Princeton, NJ: Princeton University Press.

Handler, J. 1988. Dependent people, the state, and the modern/postmodern search for the dialogic community. *UCLA Law Review* 35:999.

Hays, S. P. 1969. *Conservation and the gospel of efficiency*. New York: Atheneum.

McCloskey, M. J. 1966. The Wilderness Act of 1964: Its background and meaning. *Oregon Law Review* 45:289.

McConnell, G. 1966. *Private power and American democracy*. New York: Knopf.

O'Toole, R. 1988. *Reforming the Forest Service*. Washington, DC: Island Press.

Reich, R. 1985. Public administration and public deliberation. *Yale Law Journal* 94:1617–40.

The Resources Agency. 1992. Memorandum of understanding, California's coordinated regional strategy to conserve biological diversity. The agreement on biological diversity. Sacramento, California.

Rey, M. N.d. (c. 1988). Albuquerque speech to Bureau of Land Management/Forest Service. Reprinted in National Forest Products Association newsletter.

Rosenbaum, K. 1984. Forest planning—bound for the courts again. *Environmental Law Reporter* 14:10195.

Ruth, L. 1990. Resolving forest planning disputes: Administrative appeals, negotiations, and continuing political conflict over national forest management. Proceedings of the 1990 annual meeting of the American Political Science Association, San Francisco, September.

Ruth, L., and R. Standiford. 1994. Conserving the California spotted owl: Impacts of interim policies and implications for the long-term. Report 33. Report of the Policy Implementation Planning Team to the Steering Committee for the California Spotted Owl Assessment. Wildland Resource Center, Division of Agriculture and Natural Resources, University of California, Davis.

Stroup, R., and J. Baden. 1983. *Natural resources: Bureaucratic myths and environmental management*. Cambridge, MA: Ballinger.

Sullivan, T. J. 1984. *Resolving development disputes through negotiation*. New York: Plenum.

Taylor, S. 1984. *Making bureaucracies think*. Stanford, CA: Stanford University Press.

Thomas, J. W., E. Forsman, J. Lint, E. Meslow, B. Noon, and J. Verner. 1990. *A conservation strategy for the northern spotted owl*. Washington, DC: U.S. Government Printing Office.

U.S. Forest Service (USFS). 1960. *Report of the chief 19*. Cited in C. F. Wilkinson and H. M. Anderson, Land and resource planning in the national forests. *Oregon Law Review* 64 (1985): 29, n. 128.

———. 1973. *Forest Service manual*. Washington, DC: USFS.

———. 1976. The National Forest Management Act of 1976. *Current information report 16*. Washington, DC: USFS.

———. 1993a. *California spotted owl Sierran province interim guidelines environmental assessment*. San Francisco: USFS, Pacific Southwest Region.

———. 1993b. *Decision notice and finding of no significant impact for California spotted owl Sierran province interim guidelines*. San Francisco: USFS, Pacific Southwest Region.

———. 1995. *Draft environmental impact statement: Managing California spotted owl habitat in the Sierra Nevada national forests of California—an ecosystem approach*. San Francisco: USFS, Pacific Southwest Region.

U.S. Public Land Law Review Commission. 1970. *One third of the nation's land*. Washington, DC: U.S. Government Printing Office.

Verner, J. K., B. Noon, K. McKelvey, R. Gutierrez, G. Gould, and T. Beck. 1992. *The California spotted owl: A technical assessment of its current status*. General Technical Report PSW-GTR-133. Albany, CA: USFS, Pacific Southwest Research Station.

Wilkinson, C. F., and H. M. Anderson. 1985. Land and resource planning in the national forests. *Oregon Law Review* 64:1.

Wilson, C. 1978. Land management planning processes of the Forest Service. *Environmental Law* 8:461.

Wondolleck, J. 1988. *Public lands conflict and resolution: Resolving forest planning disputes*. New York: Plenum.

Yaffee, S. L. 1994. *The wisdom of the spotted owl: Policy lessons for a new century*. Washington, DC: Island Press.

DAVID J. LARSON
Department of Geography and
Environmental Studies
California State University
Hayward, California

8

Historical Water-Use Priorities and Public Policies

ABSTRACT

The forces that created and maintain contemporary California's complex waterscape have exploited the Sierra Nevada for 145 years. Since the Gold Rush era, the development, manipulation, and use of its water resources has significantly modified the Sierra Nevada landscape, incalculably impacting the region's ecosystem. Focusing on selected episodes featuring the impoundment and conveyance of water and its various uses, this paper, emphasizing the historical evolution of water use priorities, seeks answers to the question: How have past public policies involving water resources—or their absence—impacted the Sierra Nevada ecosystem? Special attention is given to the scale and scope of landscape transformation in the last half of the 19th century, when technology and capital were largely unconstrained by public policies.

PROLOGUE

The constant quest for water—to use, control, and manipulate—has left deep imprints on California's history and environment. Californians have historically confronted water scarcity problems with strategies designed to augment existing supplies. Among the institutions that evolved to manage this scarce resource is a system of water rights peculiar to California (largely because it contains conflicting elements from so many traditional approaches to water rights), along with a commitment to construct large-scale storage and conveyance facilities—an attempt to physically conquer a physical problem. That problem being: plenty of water but not in the right places and the right times.

Since the Gold Rush era, the forces that created California's complex waterscape looked to exploit the resources of the

Sierra Nevada, America's longest unbroken mountain range. For 145 years, the development, manipulation, and use of its water resources has significantly modified the Sierra Nevada landscape, thereby impacting the region's ecosystem. Providing prototypes for innovative hydraulic technologies, water law, water quality, and river preservation, the Sierra is where several seminal water management issues were played out, including the first conservation versus preservation battle in United States history: John Muir's vigorous attempt to prevent a dam in the Hetch Hetchy Valley of the Tuolumne River early in this century. Ever since, major Sierran rivers and lakes have commonly known controversy. The melodramatic struggle in the 1970s to "save" the Stanislaus from the New Melones Dam, for example, was America's most publicized river conservation dispute of its time. More recently, Mono Lake, at the dry eastern base of the Sierra, has symbolized the conflicts over the allocation and use of water.

The historical evolution of water use priorities for Sierra Nevada water extends far beyond the intense battles of the past 25 years between assorted water agencies and environmental organizations. Since the 1850s, development of the Sierra Nevada's water resources has mirrored prevailing values and objectives; often specific public policies have resulted.

Focusing on selected episodes featuring the impoundment and conveyance of water and its various uses, this paper implicitly seeks answers to the question: How have past public policies involving water resources shaped the current conditions of the Sierra Nevada ecosystem? This question is not easily answered. Indeed, it may be impossible to isolate the effects of public policies on Sierran water resources, at least with any precision. However, a better understanding of several benchmark events (appendix 8.1) could possibly provide fresh insights.

While policies have unquestionably been important, especially in this century, it could be argued that the Sierra Ne-

Sierra Nevada Ecosystem Project: Final report to Congress, vol. II, *Assessments and scientific basis for management options.* Davis: University of California, Centers for Water and Wildland Resources, 1996.

vada waterscape actually experienced its heaviest impacts *before* the existence of explicit public policies. Large forces (political, social, economic) operating far from the mountains long have determined how the region's water resources would be used. And while these forces are clearly seen in a 20th century context, they were no less apparent 125 years ago, when the application of private enterprise and capital, much of it flowing from the East Coast and overseas, literally transformed the Sierra Nevada landscape. The literature on 20th century events is substantial; the cornerstone water resource issues—Hetch Hetchy, Owens Valley, the Central Valley Project and State Water Project, Mono Lake, the Stanislaus—are widely known. (They will be looked at subsequently, although not in detail). Much less known, possibly more intriguing, and therefore worthy of careful scrutiny here, is the scale and scope of landscape transformation—of waterscape impacts—in the last half of the 19th century, a time when technology and capital were largely unconstrained by public policies.

FASHIONING A HYDRAULIC LANDSCAPE

Perhaps no other area of our country of roughly equal size is so rich in the history of hydraulic engineering and technology as the northern and central Sierra Nevada. This region, bracketed by the Feather River in the north and the Merced in the south, gave rise to several hydraulic technologies that were not only the first of their kind in the West but the first of their kind anywhere in the world.

It was here that America witnessed the first large-scale development of reservoirs, ditches, and flumes for mining, irrigation, and power production. And here too was the birthplace of the Pelton Wheel (which revolutionized water power technology); the first facility in California for the generation of electric power under high heads; and the first long-distance, high voltage, power transmission line in the world.

Ditches and Flumes

Sometimes history seems unreasonably exclusive in its determination as to what shall be remembered and what largely forgotten. The extensive and elaborate water storage and transfer systems originally laid out to facilitate exploitation of California's placer gold deposits—the most widespread in North America—is an example of the latter.

More than a century ago on the summer-dry western slopes of the central Sierra Nevada more than 6,000 miles of ditches and flumes moved water from higher elevations to the foothills. Singly or in series, these man-made watercourses ran for miles along the broad east-west trending ridges, the major landform of California's Gold Country. The size of the

ditches (or canals, as they were commonly called), ranging from four feet wide and two feet deep to double those dimensions, was usually dependent on the terrain traversed. In the rugged, broken country of the higher elevations, narrow and deep ditches with steep grades were preferred: initial excavation was less costly as were repairs due to damage from snow. Larger volume ditches were more common in the lower foothills, below 2,500 feet, where the grade is gentler and snow less frequent.

By the mid-1870s, mining ditches with carrying capacities as large as 80 cubic feet per second were in operation. Where it was not possible to excavate ditches, wooden flumes were built. It was often easier (and cheaper) to cross a canyon with a flume than to follow the contour of a mountain with a ditch. For a ditch of medium capacity the cost of construction was $500 per mile; there were, however, ditches that cost upwards of $5,000 per mile. Flumes were also built on or around solid rock where the cost of blasting for a ditch would have been prohibitive. Some of these canyon-spanning flumes, sustained in the air by trestles that rose to a height of 200 feet or more, were engineering marvels.

Flumes were generally built of 1 1/2-inch plank, with a framing of four-by-four and three-by-four scantling every three feet or so. Sugar pine was the favored construction material, though flumes were built of fir and spruce as well; the forest mix closest to the flume site usually determined the wood used. To protect against strong winds, high flumes were usually anchored to trestle towers with wire or wire rope. These suspension flumes were among the most spectacular structures of Gold Rush California.

The larger Gold Country ditch and flume systems derived their water supply from dams constructed near the Sierra summit to impound water from a melting snowpack. The dams held hundreds of thousands of acre-feet of water, primarily for summer mining activity. Cement was not yet a prime construction material at the apogee of California's gold mining era (except as a component of mortar for masonry) and reinforced concrete was unknown. Water and mining company engineers thus designed dams in stone, earth, and wood, building them with crews of displaced domestic and Chinese placer miners, many of whom would later lay the Central Pacific Railroad's line over the Sierra.

The earliest water transfer systems supplied alluvial placers in the lower elevations of the Sierra foothills. Generally modest structures, they were not built to endure. Yet in each of the Gold Country counties vestiges of these waterworks can be found, and some of the ditches are still in operation after 130 years. For example, direct descendants of these ditches form the main water supply for most of present-day Tuolumne County.

The Hydraulic Mining Era

Hydraulic mining—the application of water under pressure, through a nozzle against a natural bank (as defined by State and Federal Statute)—was introduced in April of 1853. Little understood and largely ignored by historians and others, hydraulic mining, together with quartz mining, revived the declining California gold industry and set in motion the second major era of mining activity that, in terms of duration, industrial works constructed, and gold produced, dwarfed the early Gold Rush period. It was during the hydraulic mining era that water storage and transfer systems achieved a scale and scope theretofore unimagined. Water companies and mining concerns were often entwined: large ditch companies purchased claims and did their own mining, while heavily capitalized hydraulic operations eventually acquired (or built) their own reservoirs and ditch systems.

The hydraulic mining technique, a California invention and until the recent success of "heap-leaching" the greatest technological advance in the long history of alluvial placer mining, made possible economic extraction of gold from vast low-grade deposits buried deep in the earth. By the mid-1870s, giant hydraulic operations, not held accountable for reclamation, were washing gravels yielding less than five cents per cubic yard and making it pay! Without an immense water supply delivered to the mining pit by ditch and flume from sources many miles away, such results would have been impossible. When hydraulic mining was at its peak in the late 1870s *single* nozzles, up to nine inches in diameter, were discharging up to 25 million gallons of water in 24 hours! By the 1880s, in the watersheds of the Feather, Yuba, Bear, and American, where hydraulic mining achieved its greatest development, more than 150,000 acre-feet of water was stored in dammed alpine lakes and strategically located man-made reservoirs (figure 8.1).

In the 30 years hydraulic mining was actively practiced,

more than $100 million was thought to have been invested; probably $30 million was expended in the construction of ditches, flumes, and reservoirs. Hydraulic mining yielded several billion dollars in gold. But it also produced a debris flow of tidal wave proportions that, after filling mountain canyons, spilled out onto the flat Central Valley to bury thousands of acres of farmland under infertile sand and rock. This initially unforeseen consequence of hydraulic mining, after a long and bitter regional conflict, led to its eventual stoppage; Judge Lorenzo Saywer's injunction in January 1884 effectively ended large-scale hydraulic mining operations on land drained by tributaries of the Sacramento and San Joaquin rivers. Legislation which resulted from Judge Saywer's decision ("Woodruff v. North Bloomfield Gravel Mining Company") enjoined the deposit of mining debris in streams to the detriment of agricultural interests in the Sacramento Valley; hydraulic mining was practically prohibited except under the most severe restrictions, involving the permanent impounding of the debris.

It has been estimated that in 1880 alone the Sacramento and San Joaquin rivers received in excess of 46 million cubic yards of debris from hydraulic mines in the mountains. During an 18-month period in the late 1870s, the quantity of mining debris dumped into the Yuba River drainage would have entirely filled the 363-mile-long Erie Canal. Nearly a quarter of a century after Judge Sawyer's decision, the distinguished government geologist G.K. Gilbert estimated that hydraulic mining operations in the basins of the Feather, Yuba, Bear, and American, the so-called "Northern Mines," had produced nearly 1.3 billion cubic yards of debris. Total mining debris from hydraulic mining for the entire Sierra Nevada likely exceeded 1.5 billion cubic yards, or 930,000 acre-feet, almost

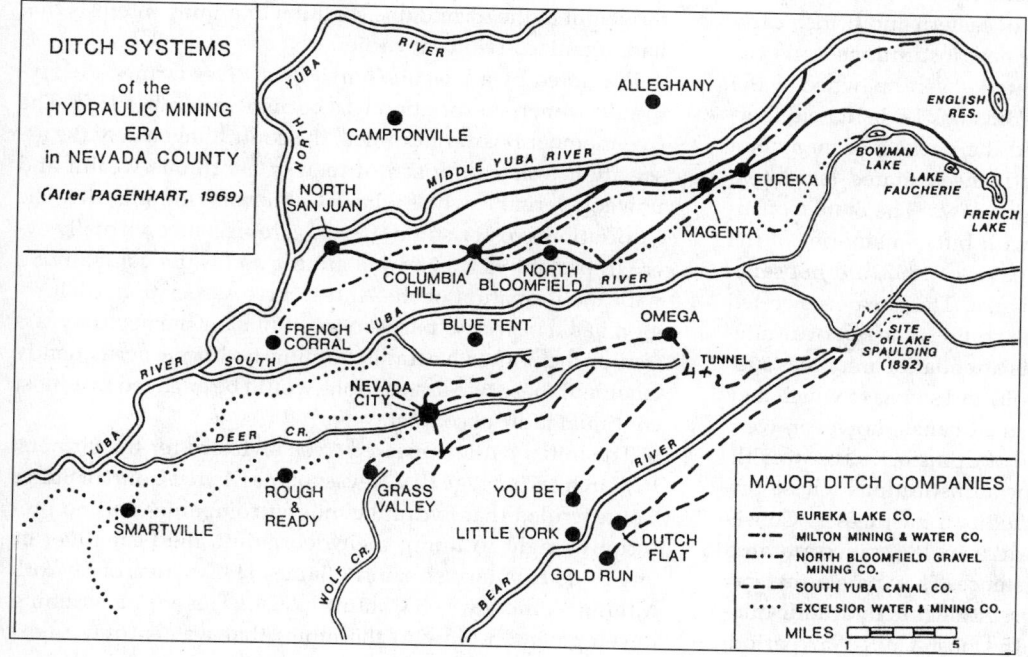

FIGURE 8.1

Ditch systems of the hydraulic mining era in Nevada County.

enough to completely fill Folsom Reservoir. Never in human history had man moved so much earth so quickly. Hydraulic mining was clearly responsible for a landscape transformation that was nothing less than geologic in nature and scope (appendix 8.2).

Why California?

Only in California, it appears, could large-scale hydraulic mining have evolved to such astonishing proportions. When gold was discovered California was a vast virgin territory without laws or precedents—or many people. The fantastically rich placer deposits happened, by chance, to be located in a region of mild climate with bountiful supplies of water and wood. The timing of the discovery of gold—roughly co-incident with major social upheavals and economic reversals abroad—coupled with its universal allure, drew to California hardy, energetic, and industrious men of all nationalities who designed and maintained laws (if not policies, per se) ideally suited to an empty land. With California courts concluding that riparian rights did not exist on federal lands where there was no private riparian claimant, a doctrine of water use by appropriation quickly evolved. River diversion, reservoir construction, and rights of way for ditches and flumes required no formal possessory title; miners had only to drive stakes along the proposed route and post notices of their intentions. A popularly elected judiciary readily adopted these mining laws and made them the fundamental laws of the State. So the physical and legal prerequisites were nicely in place.

Large-scale hydraulic mining required at least one other essential element: enormous amounts of capital. It too would be put in place. The early 1860s witnessed the sudden emergence of various financial intermediaries that, in subsequent years, would channel huge sums of eastern and British capital into the Sierra Nevada. These new institutions—and the new rules under which they operated—were spawned by the enormous wealth derived from Comstock Lode in neighboring Nevada. Silver not only lined the pockets of many San Franciscans, but, more importantly, necessitated legislative modifications in California banking laws. The State Constitution of 1849, drafted by men with bitter memories of the disastrous nationwide financial panic of 1837, did not sanction incorporated commercial banks. Their experience led them to believe that nothing good could come out of a state-chartered banking system with its attendant bank notes substituting for the "hard money" (gold and silver) to which they had grown accustomed. Commercial banks, however, were not really needed until the arrival of Comstock silver wealth, which virtually demanded financial institutions whose primary function was extending credit to businesses. Consequently, state laws were amended; in 1862, incorporated commercial banks came into existence. They paved the way for foreign as well as domestic investment in hydraulic mining, particularly after 1870. In the post–Civil War period,

America plunged into an era whose keynotes were industrialism and expansion. And hydraulic mining was one of several industrial ventures in the West that attracted the attention of American and British capitalists.

Although the first mining ditches antedated hydraulic mining, the evolution of the ditch and flume network largely reflected the changing fortunes of the hydraulic mining industry. As the hydraulic method became widespread, ditch and flume construction accelerated to meet increased water requirements. Many water companies consequently over-built their systems. Bankruptcies were common, as were consolidations. The number of ditch systems and their total mileage fell steadily throughout the 1860s. In the 1870s, however, ditch mileage stabilized, then expanded. Even more elaborate water storage facilities and distributions systems were planned for the 1880s (appendix 8.3). But then came the injunction. The story of how it came about may be of interest in the context of public policy analysis.

The Downstream Debris Dilemma: Choosing Up Sides

Devastated by repeated mining-caused flooding throughout the 1870s and disenfranchised from property and opportunity, Sacramento Valley farmers and townsfolk in 1878 launched a legal campaign that they hoped would result in nothing less than the complete abolition of hydraulic mining. Anticipating protracted litigation, their first step was to form a strong, well-funded grassroots organization: the Anti-Debris Association of the Sacramento Valley. Its mandate was to finance and prosecute lawsuits to challenge the miners' practice of filling rivers with debris. With an elected five-man board of directors, and the ability to levy assessments on members, the Anti-Debris Association was a necessary counterweight to the formidable coalition of mining interests that had organized two year earlier.

Prompted by a lawsuit from a Bear River farmer, the Hydraulic Miners Association was born in September 1876. The five-member Board of Council, the controlling arm of the association, was comprised of men of enormous wealth and power, a veritable who's who of California's mining elite. The association's roster, some 90 strong, represented virtually every important Sierra Nevada mining and water company. So with the formation of the Anti-Debris Association, each region had a powerful partisan coalition. For the next 15 years (ten of them after hydraulic mining had been permanently enjoined) these two associations would be engaged in almost continual judicial combat.

The initial battle appeared to be a victory for the farmers. In March 1879, Judge P.W. Keyser of the District Court of Sutter County ruled that hydraulic mining companies had no prescriptive right to dump their debris into the Bear River or any of its tributary streams. ("James H. Keyes v. Little York Mining Company," 53 California 724.) This was a stunning development, a blow to the mines that was entirely unexpected. The valley, jubilant, celebrated for weeks. Lawyers for the Hydraulic Minters Association quickly secured a stay of proceedings, then appealed Keyser's decision to the State Supreme Court, which issued a further stay. Angry and bewildered, farmers wondered just who had actually won the suit. Those in the bottomlands of the Bear River watched forlornly as the turbid river, displaced from its bed by an inundation of mining debris, ran riot over their fields. Then hope came in the form of crossfire from the Anti-Debris Association. In September 1879 the City of Marysville filed suite against the North Bloomfield Gravel Mining Company (and other Yuba River mining concerns) in Yuba County District Court, seeking a perpetual injunction against the company's practice of dumping debris into tributaries of the Yuba River ("The City of Marysville v. North Bloomfield Gravel Mining Co. et al.," 58 California 321).

San Francisco capitalists were outraged. The *Stock Report*, mouthpiece of San Francisco's financial community, vehemently voiced mining's position on the matter in its editorials. So too did the respected *Mining and Scientific Press*, a San Francisco weekly. Whether certain mining interests pressured the State Supreme Court is open to conjecture, but in November the district Court's injunction was overturned. The Supreme Court reasoned that it was not equitable to join all the mines of one watershed together in a single suit. The valley was stunned—while the mountains, a portrait of euphoria, rejoiced for days.

The final month of the decade wound down with the farmers deeply disappointed but not ready to give up. The abolition of hydraulic mining would simply require a broader realization that California had already passed an economic crossroads. The large loss of cropland resulting from the particularly devastating floods of 1878 brought into sharper focus the fact that agriculture, dominated by dry-farmed wheat, had become California's leading industry. Areas heavily slathered by mining debris where among the richest farmlands in the state. Thousands of acres of productive orchards and grain fields had been reduced to barren wastelands. Thousands more awaited a similar fate. History is rife with examples of short-lived societies built and based on precious metal mining; impermanence is their legacy. Was California destined to follow this path?

With the 1870s came a new vision of California's future. It lay not in gold from the Sierra Nevada, but in the splendid soils of the Central Valley. Able to produce a continuous stream of wealth, agriculture stood in sharp contrast to the hit-and-run nature of mineral extraction. And during the '70s it had replaced mining, statistically, as the leading sector of the state's economy. By late in the decade the annual value of the dry-farmed wheat crop alone had reached $40 million, more than double that of the dwindling gold output. The trend was clear and irreversible: the pivot of prosperity had shifted permanently toward the fields.

The Chess Match

As the California legislature began its 1880 session, the tension between agricultural and mining interests was almost palpable. Many lawmakers were therefore startled a few months later when the two factions jointly pushed for passage of legislation to save the Sacramento Valley. After a two-year study, the California State Engineer (an office created specifically by the debris dispute) had recommended a comprehensive flood control system consisting of brush dams in the Sierra Nevada foothills and extensive levees at key locations in the valley. The legislature subsequently passed the Drainage Act, a reclamation project to be funded by statewide taxation and the charge of one half cent on each miner's inch of water (a volumetric measure specific to Sierra Nevada mining operations) used by every hydraulic mining company. A Board of Drainage Commissioners, comprised of the governor, the state engineer, and the surveyor-general, would supervise the project. Work began almost at once as levees were thrown up along the banks of the Sacramento and dams built across the Yuba and Bear rivers. Debris-restraining dams, comprised of brush, wire, and logs, spanned canyon mouths at the edge of the Sierra foothills. Levees lined the Yuba from its junction with the Feather east to the Sierra; seven more miles of levees lined the Feather below where it is joined by the Yuba. The Bear was similarly hemmed in. Farmers and miners alike thought their debris problems were over.

The ensuing winter was wicked. Beginning in January 1881, a long parade of storms marched across northern California. By early February flooding was extensive and, according to at least one source, the most devastating ever. The integrated flood control system had failed miserably. Infuriated and frustrated, farmers and townsfolk in May revived the long-dormant Marysville v. North Bloomfield suit. Hydraulic mining, it was felt, had to be abolished. There could be no compromise solution.

In late June the Superior Court of Yuba County granted an injunction. Reluctantly, hydraulic operations in the upper Yuba basin ground to a halt. A month later, in the Superior Court of Sacramento County, the State Attorney General sought an injunction against the Gold Run Ditch and Mining Company ("The People v. The Gold Run Ditch and Mining Company," 66 California 138), whose property lay high in Placer County, on the North Fork of the American River. Two months later the California Supreme Court ruled that the Drainage Act was an unconstitutional assumption by the state of a private regional concern. All of California's citizens could not be taxed, the court reasoned, so that only the Sacramento Valley might benefit.

Throughout the winter and spring of 1882, hydraulic miners anxiously awaited the Gold Run decision. It came down in June. Judge Jackson Temple ruled that the Gold Run Company had to build restraining barriers at its mine to keep coarse debris (gravel and boulders) from entering tributaries of the American River. Once they were built, mining operations could resume.

The Anti-Debris Association, which was given new life at the time of the revived Marysville v. North Bloomfield suit, was not pleased. Judge Temple's decision, it felt, was a weak echo of the Drainage Act. Dozens of flimsy brush dams clearly were not the answer; the floods of 1881 had graphically demonstrated their limited utility. Furthermore, only one mine at a time could be enjoined in this fashion, a time-consuming and basically ineffective approach. A more sweeping injunction remained the Association's aim.

What became the decisive suit was filed in the Ninth United States Circuit Court in San Francisco in September 1882. Edwards Woodruff, a Marysville landowner, brought a suit against the North Bloomfield Gravel Mining Company and all other mines in the Yuba River watershed ("Edwards Woodruff v. the North Bloomfield Gravel Mining Company et al.," cited as 9 Sawyer 441).

Checkmate: Victory for the Valley

An air of anxiety hung over the Sierra Nevada mining communities for most of 1883. The fate of Edwards Woodruff's suit weighed heavily on the minds of many. Some miners were convinced that Judge Lorenzo Saywer, a legitimate '49er who had spent time prospecting in the Nevada City district, would come down on the side of mining interests. Others were much less confident of a favorable ruling, as anti-mining sentiments in the Sacramento Valley had never been stronger. As 1883 wore on, the farmer cause quietly gathered momentum. Some of the best hydraulic engineers left for mining ventures elsewhere around the world. Perhaps they saw the writing on the wall.

On January 6, 1884, Marysville got word that Judge Sawyer would hand down his decision the following day. Confident townsfolk prepared for a grand celebration that included an enormous bonfire. On Friday the 7th, in his San Francisco courtroom, Judge Sawyer delivered his precedent-setting perpetual injunction: The hydraulic mining companies, "their servants, agents and employees, are perpetually enjoined and restrained from discharging or dumping into the Yuba River, or any of its forks or branches...tailings, bowlders, cobble stones, gravel, sand, clay, debris or refuse matter..." When word of the decision reached the Sierra via telegraph, whole towns became immobilized with abject disbelief. The valley, though, was a scene of unabashed celebrating. The long struggle was over. The farmers had won.

Aftermath: The End of an Era

Judge Sawyer's decision dealt a death blow to the hydraulic mining industry. No longer was there any legal justification for using the rivers of the Sierra Nevada as dumping grounds for mining debris. The question was that of nuisance: damage to private property and damage to public property—in this case the navigable waterways of the state. Free and open passage on them is guaranteed by the United States Constitution. There was never any question that mining debris posed an extreme menace to navigation. Judge Saywer wrote, "So long as hydraulic mining is carried on as now pursued it will continue to be an alarming and ever-growing menace..." (9 Sawyer 441).

Hydraulic mining interests passively accepted Sawyer's judgment. No plans were made for a retrial. The largest companies were the first to concede defeat. And the mountains began to empty, not of debris but of miners themselves.

In the wake of Sawyer's decision, several other suits were filed in federal courts. Combined, they effectively shut down all remaining operations in the northern and central Sierra, except those in remote locales far from the public eye. The *Mining and Scientific Press* reported that by the end of 1886 hydraulic mining in the Sierra Nevada was virtually nonexistent. Once-giant mining companies were having their properties and apparatus auctioned off at sheriff's sales to pay back fines and court costs.

Although the hydraulic mining era appeared over, there remained a glimmer of hope. The U.S. Army Corps of Engineers, which conducted a year-long investigation of the California debris problem, recommended to Congress in 1891 that hydraulic operations be allowed to resume if adequate restraining works first were constructed. Two years later, Anthony Caminetti, a congressman from California's 2nd District, introduced legislation in the House of Representatives to create a federal agency for the purpose of regulating hydraulic mining in the Sacramento–San Joaquin drainage system. Congress subsequently passed the so-called Caminetti Act. Under supervision of the Army Corps of Engineers, permits were granted to applicants who had already built debris dams below their mine sites.

Although well intentioned, this last-gasp legislative attempt to resuscitate hydraulic mining came up short. Heavy snows in the early 1890s (the 1890 snowpack was the deepest on record until 1952) ruined many miles of flume and ditch, essential elements in any hydraulic enterprise. Financially strapped mining companies simply could not afford to rebuild the damaged water systems *and* construct restraining dams. So in spite of the Caminetti Act, there would be no hydraulic revival, no hydraulic encore. The curtain had come down for good.

Legacies: Irrigation and Power Possibilities

As early as the 1860s, irrigated agriculture was seen as the logical successor to hydraulic mining. The contemporary historian John S. Hittell predicted that many of the mining ditches and flumes would eventually be "as indispensable to the farms, orchards, and vineyards of the dry uplands as to the placer diggings." This notion was echoed years later (six months after the Sawyer decision) by a large Sacramento Valley landholder who told the *Sacramento Record Union* that "by showing that waters can be conducted anywhere, hydraulic mining has unwittingly solved a most important feature in the problem of irrigation."

Beginning in 1872, the California Legislature enacted several measures in support of local irrigation development. But these ultimately proved ineffectual in helping small farmers gain access to water resources dominated by large riparian landowners. In 1887, however, passage of the Wright Irrigation Act, authored by Senator C.C. Wright of Modesto, gave farming communities the authority to purchase, build, and operate their own irrigation systems. Irrigation districts could be created whenever a county board of supervisors approved a petition either from 50 landowners or from a majority of landowners in the area. For the first time in California history, water for irrigation was recognized as a "public use."

The complex water transfer systems abandoned by bankrupt mining companies were inherently well suited for another future use: the generation of hydroelectric power. Several years before hydraulic mining was judicially restrained, Hamilton Smith, Jr., the distinguished superintendent and chief engineer of the North Bloomfield Gravel Mining Company, foresaw hydroelectric power as an alternative use for ditch water. Yet by 1890, mining authority J.B. Hobson still could wonder "to what extent these expensive systems may yet be put in the way of furnishing water for power and irrigation cannot be very readily estimated."

The two decades prior to 1890 witnessed several important developments that, collectively, made the generation and long distance transmission of electricity possible. After discovery of the dynamo, or electric generator, in 1873, the electric motor was invented. And the generator, in conjunction with the electric arc lamp (devised by Thomas Edison in 1879) allowed a single system to light an entire city. This revolution in streetlighting was widely embraced; by the mid-1880s open arc lamps on tall wooden poles were common fixtures in America's larger cities.

While Edison and others were refining the incandescent lamp in 1880, there was an important advancement in water-power technology: the Pelton wheel, developed by Lester Pelton of Camptonville in Yuba County. Pelton's contribution consisted of placing twin buckets with split centers closely spaced around the perimeter of an impulse water wheel. This design allowed a more effective flow of water than was possible with the crude, slow, hurdy-gurdy wheel, which had single buckets around its perimeter

Pelton, a former millwright, was immediately challenged by other inventors who insisted that his idea was really theirs. But Pelton's claim was ultimately upheld, and he alone received the U.S. patent for this invention. The Pelton wheel, which saw several refinements, was the crucial first step toward making the impulse wheel an efficient prime mover. With this technological improvement, hydroelectric power generation became as economical as thermoelectric power generation. There remained a major unresolved problem,

however: electricity still could not be transmitted long distances without substantial loss.

History seems to have a way of arranging for important inventions to arrive on the scene just when they are needed most. And in the early 1880s two technological advances made electrical transmission over relatively long distances a reality: the development of transformers with the capacity to handle high voltage alternating current, and vastly improved storage batteries. The age of electric power was dawning.

Early Hydroelectric Power Generation

After the electric generator and motor had been made commercially useful, engineers looked for ways in which water rather than steam could be used to drive electric generators. And as the *Mining and Scientific Press* noted in an October 1887 editorial, the hydroelectric power potential of California was enormous and "there is no calculating the effect it will have on the industries of the State."

Two European experiments encouraged those with an eye on tapping the hydropower potential of the Sierra Nevada. In 1886 electricity was transmitted along a 2,000-volt line from a steam-driven plant in Tivoli to Rome, a distance of 17 miles. This was the first successful transmission of alternating current. Its significance to California, where transmission was a major problem, was that power finally had been carried a substantial distance. The second historic experiment took place in Germany in 1891. Electricity generated at a water-powered plant in Lauffen was sent to the International Electric Exhibition in Frankfurt, 81 miles away. Thirty thousand volts were transmitted, far more than had ever been attempted.

California's first application for water rights specifically for generating hydroelectric power was filed in 1891. Cornishman Alfonso Tregidgo, manager of a gold quartz mine in Grass Valley, was disturbed by the high cost and inefficiency of steam power for pumping and operating mining equipment. After learning of the successful application of water power and long distance transmission at a hydroelectric plant in Italy, Tregidgo decided to build a hydro plant on the South Yuba River and supply Nevada City, Grass Valley and nearby mines with power and light.

With water rights secured, the Nevada County Electric Power Company was incorporated in 1892. A powerhouse site and rights of way for transmission lines were purchased. Water diverted from the South Yuba about three miles above the powerhouse would be carried by flume to obtain a 200-foot drop to the Pelton wheels driving the electric generators. Numerous construction problems coupled with fallout from the nationwide financial depression of 1893 delayed completion of the plant, known as the Rome Power House, until February 1896, when the system was put into operation. It is of interest historically as the first plant of what came to be the Pacific Gas and Electric Company, the major utility serving most of northern and central California.

When the demand for electricity exceeded the capacity of the Rome plant, another facility, the Colgate plant, was built in 1899 on the north fork of the Yuba River. One of the most widely known plants in the history of civil engineering, the Colgate plant supplied a 60,000-volt line running out of the Sierra for 140 miles to Oakland. This was the world's first long-distance transmission line.

The unqualified success of the Nevada County hydroelectric plants spurred similar facilities throughout the Sierra Nevada. Within a few years, every major river draining the west slope of the Sierra could claim its own hydroelectric power plant. From Plumas County in the north to Fresno

County in the south, the Sierra Nevada hummed from the generation of electricity by way of falling water. Most of these plants were located such that they could tap into abandoned, but still functional, ditch systems created in the heyday of hydraulic mining, or even earlier. The Phoenix Power Plant in Tuolumne County, which took water directly from the Columbia Ditch, and the Murphy's Power Plant in Calaveras County, which used the Utica Ditch, are two examples of hydroelectric plants built in the early 20th century that were essentially dependent on ditches excavated in the early years of the Gold Rush (figure 8.2).

The initial incentive to the rapid development of hydro-

FIGURE 8.2

19th-century ditch systems in Tuolumne County.

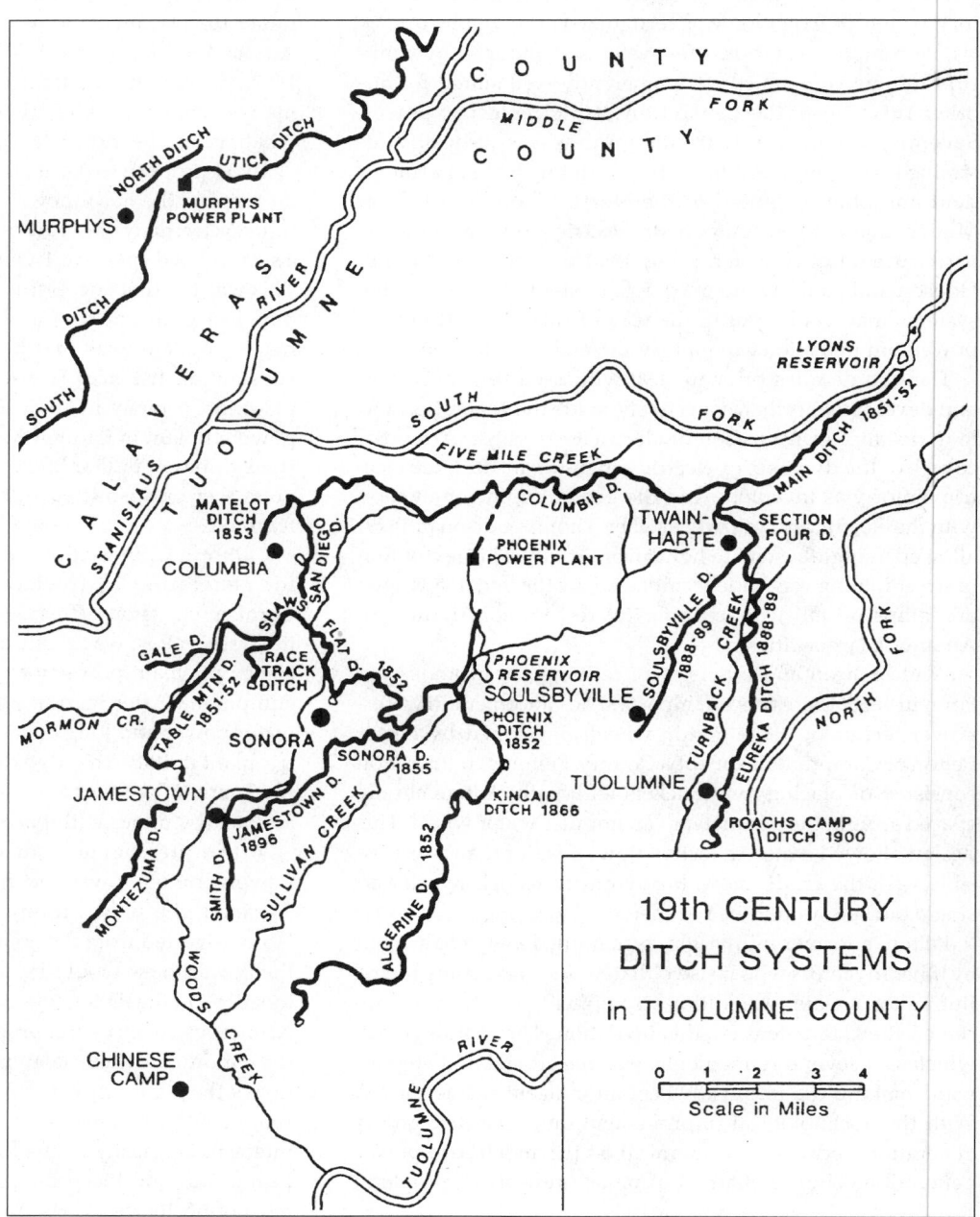

electric power in California was the high fuel costs existing around the turn of the century. Most of the electric plants in the state burned coal, which had to be imported from Australia and British Columbia. Not only was there a hefty transportation cost but an important customs duty had to be paid as well. The ability to generate electricity from falling water, clean and mechanically efficient then as now, attracted power companies, old and new, to the Sierra Nevada. The consequences for the Sierra's water resources, of course, was that an ever-increasing number of major rivers were given over to the storage of water for the purposes of hydroelectric power generation. The basins of the Yuba, Bear, American, and the eastward flowing Truckee River were especially endowed with attractive hydropower sites (figure 8.3). In most of these plants, water was used for irrigation after passing through the turbines, further underscoring the advantages of hydroelectric facilities.

Incorporated in 1905, the Pacific Gas & Electric Company was heavily capitalized by the sale of common and preferred stock. With this pool of ready capital, the utility bought most of the reservoirs, dams, ditches and flumes built by mining interests which were unable to remain commercially viable. Through its shrewd purchase of existing water storage and conveyance facilities, PG&E virtually monopolized the entire hydroelectric industry in northern California within a few years of its incorporation; however, smaller utilities such as Sierra Pacific Power Company, whose corporate roots extended all the way back to the El Dorado Canal Company, were able to maintain a niche (appendix 8.4). To this day, PG&E maintains several hundred miles of ditch and flume, whose waters turn turbines in Sierra powerhouses just as they have for nearly ninety years.

Epilogue to an Era

More than a century has passed since the roaring jets of water were turned off the high walls of the Sierra Nevada Tertiary gravels, ending an era of massive human alteration of the earth's surface. Today the cavernous hydraulic pits lie in eerie silence; their floors, studded with pines, are partially covered by rain-fed marshes. Still without a full vegetative cover, the deeply gouged walls have been further transformed by time and nature: wind and water have rounded the exposed edges, while oxidized minerals have tinted the gravels, producing multi-colored facades. The boldest landscape signature of the hydraulic mining era, these gaping red-dirt excavations endure as monuments to our unbridled assault on the earth in pursuit of precious metals.

Mining debris no longer encroaches on agricultural lands. Yet the effects of this debris and the present-day use of reservoirs, old ditch systems and other water diversions are still highly noticeable on the Sierra Nevada landscape, particularly with respect to channel morphology, channel forming processes and riparian vegetation. On steep, dry bluffs vegetation has been slow to assume control. Consequently, weath-

ering processes have wasted these barren banks, forming graded talus slopes. Upland creeks remain laden with tailings, but larger streams now run clear as glass through rocky canyons.

All but forgotten as an economic enterprise, the hydraulic mining industry in California left valuable legacies. It hastened the development of scores of engineering techniques and innovations sophisticated beyond their time and useful beyond the realm of gold mining. Important, too, was the impetus given to the young sciences of geology and hydrology. Most useful, though, were the dozens of reservoirs and thousands of miles of ditch and flume, key components of the elaborate water storage and transfer systems ready-made to help meet northern California's needs for hydroelectric power generation, irrigation, and municipal water supply.

NEW WATER-USE PRIORITIES FOR A NEW CENTURY

Private development distinguished 19th century water use in the Sierra Nevada. Beginning with the Gold Rush era and for nearly five decades, water developers supplied chiefly local needs, the mining companies and, later, irrigation districts being the principal users. In the early 20th century, however, the generation of hydroelectric power became the dominant private use of Sierra Nevada water, and electricity so generated for the first time was exported far beyond the mountains.

So too was Sierra water itself. Los Angeles and San Francisco, then California's two largest cities, effectively pioneered a water-based imperialism led by public—not private—entities with access to the public treasury. The success of these long-distance urban water grabs almost certainly opened the Sierra Nevada to exploitation by the enormous federal and state hydraulic projects of more recent decades. A better understanding of the 20th-century development of Sierra water resources requires a re-examination of the peculiar legal and legislative context out of which it grew.

Water Rights in Context

The first generation of miners were squatters on the Public Domain. Yet the federal government, preoccupied with the sectional strife that culminated in the Civil War, exerted no authority over the Sierra gold mining region until after the war. The first federal law that dealt with the disposition of mining property in California, enacted in 1866, simply recognized mining claims that the miners themselves had established.

Left to their own devices, the early miners devised a workable system of self-government to protect their property and mineral rights. Mining claims, which required "improvement"

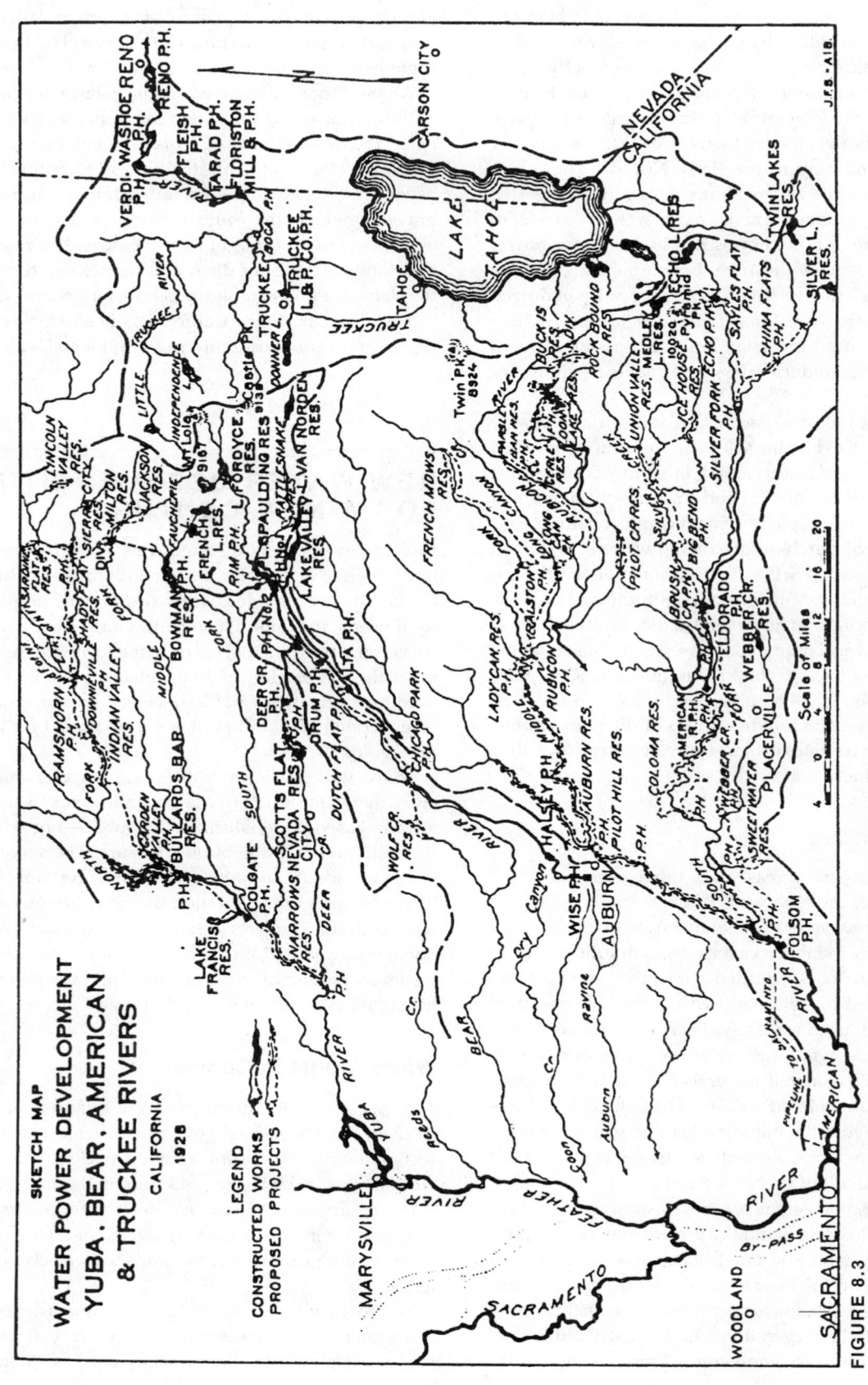

FIGURE 8.3

Water power development: Yuba, Bear, American, and Truckee Rivers (1928).

as a condition of ownership, were dispensed on a first-come-first-served basis. The system that miners used to allocate, or "appropriate," water rights closely paralleled that of land use. After the intention was recorded in county offices, water could be diverted and carried long distances away from the source, used as desired, and abandoned without concern. The only stipulation was that the water be used "diligently." Water rights were forfeited with non-use. These mining customs, particularly useful in arid lands commonly without other sources of water, form the basis of the Western doctrine of prior appropriation. Miners could not have been aware of the full significance of the legal system they were creating and the degree to which it would conflict with widely accepted practices regarding the allocation of water rights.

Circumstantial as it may have been, the miners' system of water rights allocation would become embedded in the California legal system along with the older, established concepts derived from English common law—including riparian rights, which were assigned automatically with ownership of land adjacent to a stream and were not lost through non-use. When English common law was adopted by the first California Legislature at the time of statehood (1850), the riparian system was included as part of the unexamined legal baggage even though this system of water rights had not previously been used in California. The doctrine of prior appropriation, already the quasi-legal custom throughout the Sierra Nevada, had hastened the claiming of water resources and encouraged economic development. Riparian rights, recognized by the legislature in 1872, were reaffirmed primarily in the courts, particularly the landmark 1886 California Supreme Court decision in Lux v. Haggin. The legislature's response to this decision was the Wright Act of 1887: It provided the public with power to take water and land, by act of condemnation, to create community-controlled irrigation districts.

Thus California struggled along with two contradictory systems. The riparian system was never suited to California, and the appropriative system (before 1914) did not have an orderly and effective method of administration: the rights were too easily obtained and there was no centralized system of registration. Obviously needed was legislation to set up a fair and efficient system of water rights.

In 1900, a group of progressive citizens under the name of California Water and Forest Association convinced irrigation authority Elwood Mead, then chief of the U.S. Department of Agriculture's Office of Experiment Stations, to instigate an investigation of the water rights on selected California rivers. Mead's famous report, "Irrigation Investigations in California," was published in 1901. It revealed that conditions were, if anything, worse than anyone had imagined. Virtually every stream, especially those draining the Sierra's west slope, was legally choked with oversubscribed or useless water claims.

Consequently, a bill was drawn up for California's 1903 legislative session, embodying Mead's recommendations to alleviate the obvious legal problems. The bill included a general water code to systematize water procedures. But too much opposition was encountered, and the bill failed to pass. It would take another decade and much more effort before comprehensive legislation, which placed the recording and licensing power with the state rather than counties, would be passed.

In 1911, the California Legislature established the California Conservation Commission, with George Pardee, and ex-governor philosophically opposed to the monopoly of natural resources by corporations or individuals, as chairman. The Commission's purpose: to investigate and gather data on forestry, water, and use of water, water power, electricity, mines and mining, dredging, reclamation, and irrigation; and to revise, systematize, and reform the state laws concerning those subjects. Many Commission members had been active in the California Water and Forestry Association. Thus their recommendations, contained in a report submitted in 1912, echoed those proposed nine years earlier.

As an immediate consequence, a State Water Commission was established in 1912 to administer water rights for power purposes; a comprehensive Water Commission Act covering all uses of water was enacted the following year. Opponents of the Water Commission bill successfully delayed its passage by demanding a referendum on it. With public support, the act was approved at the next general election and became effective in December 1914.

The California Water Commission Act regularized appropriative procedure. Priority was given to the earliest permittee rather than the first applicant. For the first time in the state's history, it became possible to determine just how much unappropriated water remained in California. Now long-range planning of water resources, an essential prerequisite for the new water use priorities of the new century, could be pursued.

The Water Commission Act was emblematic of a new era for California's water resources: Local control, individual ingenuity, and private enterprise was giving way to centralized control, cooperative ventures, and the use of public funds and purview. How California's two largest cities went about securing water supplies for the 20th century underscored this shift. Acting independently and exclusively in their own interests, San Francisco and Los Angeles, through their massive public water projects, initiated a process that would in time fundamentally transform the Sierra Nevada waterscape and indeed that of the entire state.

Reaching Out to the Sierra Nevada: Urban Water Grabs

As the 19th century wound down, the problem of insufficient urban water supply stood as a substantial impediment to California's 20th century prosperity. America's transition from a predominantly rural to an overwhelmingly urban society occurred in California before the country as a whole. The accelerated growth of San Francisco and Los Angeles in the last quarter of the 19th century pushed both cities up against the limits of their local water supplies. Without guarantees of additional water sources, continued prosperity would be problematic.

In 1900, San Francisco could boast of a population of 340,000, Los Angeles 100,000 and climbing rapidly; both cities had vigorous boosters, room to grow, and the transportation networks to facilitate growth; neither city, however, possessed the organizational structure required to reach far beyond city limits to tap a water source. In a departure from the national norm, California's largest cities did not control their own water franchises. Under California's riparian doctrine, water was viewed as a private resource, and private, not municipal, entities ran the water business.

Procurement and delivery of water from distant sources required substantial capital investments that private water companies were not prepared to make. Thus, municipalization of the urban water supply became the preferred solution to the "problem" created by continued urban growth: Cities were able to acquire capital through taxation and the sale of bonds. In the early 20th century, San Francisco and Los Angeles each moved to municipalize their water supplies, and each, acting independently, looked to the Sierra Nevada.

San Francisco struck first. By 1901, city leaders, moving to gain municipal control over the water franchise, commissioned an investigation of potential water projects on ten northern California rivers: the Eel, McCloud, Sacramento, Feather, Yuba, American, Consumnes, Mokelumne, Stanislaus, and Tuolumne, seven of which drain the Sierra's western slope. Reservoir sites, hydropower potential, water rights, water quality, routing considerations, and political concerns all favored a project on the Tuolumne River, but the premiere site for a reservoir in this scheme, the Hetch Hetchy Valley, lay inside Yosemite National Park.

Undeterred, San Francisco pushed for this ideal site. The city's struggle for permission from the federal government to dam Hetch Hetchy lasted a dozen years and spanned the administrations of two United states presidents and three secretaries of the Department of Interior. Advocates and adversaries were legion. Conservationists, led by Gifford Pinchot, argued that the project was an improvement upon nature: enormous water and power resources would come under municipal *public* control, and San Francisco would build roads and trails so visitors could enjoy easy access to the scenic splendors of Hetch Hetchy Valley. Preservationists, on the other hand, led by John Muir and the Sierra Club, took a jaundiced view of the dam, preferring to see it as the desecration of a natural temple. Additional attacks came from the Modesto and Turlock irrigation districts, which claimed prior rights to the Tuolumne. Despite resolute opposition, San Francisco ultimately won out.

In 1913, Congress passed the Raker Act authorizing construction of a dam across Hetch Hetchy Valley. To assure fiscal soundness while helping underwrite the project's considerable cost, the bill stipulated that a hydroelectric power system for municipal and commercial use be included. (Today, Hetch Hetchy electricity powers San Francisco's cable cars, electric trolleys and buses, and also lights and heats the airport and other municipal buildings). Confident that the dam was the best way to serve public needs without compromising the beauty of the public domain (Yosemite), President Woodrow Wilson, in December, signed the bill into law.

San Francisco's troubles were hardly over. Twenty more years of controversy, corruption, and escalating costs (to $100 million) plagued the project. O'Shaughnessy Dam was completed in 1922 but water from Hetch Hetchy Reservoir did not begin flowing into San Francisco until 1934. Hetch Hetchy quickly became a practical necessity for 20th century San Francisco: a dependable, if distant, source of water, power and revenue—a distinction it still holds. Subsequent expansion, including construction of New Don Pedro Dam in the Sierra foothills, has produced a water storage and conveyance system that delivers almost six times as much water as the original Hetch Hetchy project. Today San Francisco sells surplus water to suburban Bay Area communities and electricity to private utilities and Central Valley irrigation districts.

While San Francisco struggled for decades to secure and implement its Sierra-based municipal water system, Los Angeles, starting around the same time and eyeing a watershed adjoining the Tuolumne's, built a massive water project, a true engineering marvel, in one fifth of the time for only a quarter of the cost. In 1902, the city of Los Angeles, committed to the municipalization of the water franchise, purchased the water-distribution facilities of the private Los Angeles City Water Company following expiration of the company's lease. Spiraling population growth had convinced water department officials that local water supplies, once thought to be substantial, needed augmentation, especially if Los Angeles, as envisioned by boosters, was to become California's leading metropolis. Some 235 miles due north of Los Angeles, on the east side of the Sierra Nevada, the Owens River, fed by pristine snowmelt, seemed an ideal source for additional water.

To finance this ambitious project, whose centerpiece was a spectacular gravity aqueduct, Los Angeles floated two bonds, the first in 1905 for $1.5 million (covering the cost of surveying and land acquisitions), and the second in 1907 for $23 million (for construction). Between the two bond issues, which ran the city's legal indebtedness to the limit, Los Angeles sought and obtained from Congress in 1906 the required right-of-way for the aqueduct to pass over public domain land. Approval came with a stipulation: that no water from the project should ever be offered to private interests for resale outside the city limits. (Not coincidentally, between 1914 and 1923 Los Angeles nearly quadrupled in area as a result of annexations). Construction of the aqueduct began in 1908 and, astonishingly, was completed five years later. In November 1913, Owens River water, four times as much as the city of Los Angeles was capable of then using, arrived in the San Fernando Valley.

Like San Francisco's Hetch Hetchy, the Owens Valley aqueduct project did not escape controversy. The focus of a long running dispute involving charges of deceit and duplicity by various public officials was the fate of the Owens Valley, whose modest farming communities were totally dependent on the local water supply. As Los Angeles grew—100,000 immigrants annually in the 1920s—so too did demand for water. The city responded by acquiring additional land and water rights in the agricultural heart of the valley, thus driving out longtime farmers and ranchers, some of whom reacted by repeatedly dynamiting the aqueduct. Eventually, Los Angeles wound up purchasing virtually all of the private land in the Owens Valley, thereby becoming the largest landowner and taxpayer of Inyo County. But it was still not finished.

Los Angeles voters, perhaps fearing the repercussions of a protracted drought on an exploding population, in 1930 approved a $40 million bond issue. With funding secured, the Los Angeles Department of Water and Power extended its eastside project 105 miles further north into the Mono Basin to tap eastward-flowing Sierran streams feeding Mono Lake. Completed in 1940, the so-called Mono extension could not be operated at full capacity: Los Angeles now held rights to far more Mono Basin water than the original aqueduct could carry. A second Owens Aqueduct, roughly paralleling the original, was begun 24 years later. By 1970 it was carrying Inyo-Mono water to Los Angeles, the two aqueducts collectively supplying 80 percent of the city's annual water requirements. The Owens Valley and Mono Basin, their water supplies controlled by an absentee metropolitan landlord, had effectively become water colonies of imperial Los Angeles, just as Hetch Hetchy Valley had for San Francisco.

Similarities abound in the way San Francisco and Los Angeles, two progressive, growing cities in the early 20th century, reached out to the Sierra Nevada for water. Both cities clearly understood that "progress" meant growth, that growth was largely dependent on the availability of abundant, inexpensive water supplies. Both cities early recognized that municipal control of water (and power) would be essential for guiding future development. And both relied on the federal government for cooperation and the public treasury for funding for their water empires.

So that no other remote, sparsely populated region would suffer as the Owens Valley had, the legislature in 1931 enacted the County of Origin Statute, which authorizes counties to recapture water later needed for their development. This policy would become embedded in the 1933 legislation that gave rise to the Central Valley Project.

Impacts on the Sierra Nevada environment traceable to the two major water transfer projects are clearly visible. In the Tuolumne River watershed, they range from clear cutting and drowning Hetch Hetchy Valley behind O'Shaughnessy Dam to the constant manipulation of the Tuolumne's flows, usually for hydroelectric considerations. Changes in the Owens Valley landscape, while more subtle, are no less apparent. Diversions into the aqueduct dried up Owens Lake, caused the valley floor to drop due to subsidence, and severely lim-

ited irrigated agriculture. Moreover, land purchased by the city of Los Angeles, amounting to several hundred thousand acres, has effectively transformed this slice of the eastern Sierra into a recreational suburb of Southern California. Mammoth Mountain, for example, is the most visited ski resort in the United States. Long is the list of impacts resulting from this land use. On the other hand, Los Angeles' ownership of the Owens Valley and interest in its groundwater has likely prevented urban development that otherwise might have occurred.

Most written accounts of the Hetch Hetchy and Owens Valley water projects tend to view them a morality plays: a pristine mountain valley (within a national park, no less!) and a promising agricultural landscape are violated by the forces of greed, fraud, deceit, and duplicity perpetrated by distant avaricious metropolises flexing their epic political muscles. Obviously, there is more to it.

Conceived during the Progressive Era, these projects, promising to deliver to their cities abundant supplies of inexpensive mountain water, were veritable paragons of progressivism. Each project required and received cooperation from the federal government, whose prevailing public policy of utilitarianism—greatest good for the greatest number over the longest time—certainly validated the water transfers. The success of Hetch Hetchy and Los Angeles' aqueduct system demonstrated the enormous benefits that could be gained through public water development. And in ways that embraced philosophy as well as engineering, created a template for the colossal federal and state water delivery systems unique to California. Their impact on the Sierra Nevada spanned nearly the length of the range.

Two Great Projects for the Great Valley

During the early 20th-century agricultural transformation of the Central Valley, the waters of the Sacramento, San Joaquin and their tributaries were used without a comprehensive plan for their conservation. Valley farmers and growers, dependent on year-to-year stream flow for irrigation, lived with the specter of drought; dry years, often in succession, were facts of life in the valley. So too was destructive flooding, commonly occurring at the close of a drought cycle. To ensure against these fluctuating extremes, farmers banded together to form irrigation districts, flood control districts, reclamation districts, and other mutual aid associations. They also came to rely on the valley's enormous groundwater resource: the alluvial sands and gravels which filled the Great Valley thousands of feet in places were permeated with moisture. But decades of pumping had lowered the water table to such depths that only the most powerful pumps could tap into it.

Then in the early 1920s, the situation got worse. Severe drought significantly reduced surface flows and excessive pumping caused an alarming drop in the groundwater table. Desperate agriculturalists sought help from the state legislature, which in 1921 had begun a comprehensive study of

California's watersheds, focusing on the flood control needs and irrigation potential of the Great Valley. This investigation, which stretched out over a decade and was ultimately titled the "State Water Plan," became the basis for the Central Valley Project (CVP), a massive water system that, by transferring water from the northern Sacramento Valley south to the San Joaquin Valley, would reshape the face and future of the Great Valley—along with that of the adjacent Sierra Nevada. Authorized as a state project in 1933, the CVP, which arrived on the scene in the depths of America's worst depression, could not find financing: No market could be found for the bonds to finance construction of the dams, canals, and associated infrastructure. Called on for a bailout, the federal government officially took over the Central Valley Project in 1935. The Bureau of Reclamation was placed in charge of construction and administration of this sprawling system.

In late 1937, the Bureau broke ground on the first unit to be completed (in 1940), the Contra Costa Canal. The next year construction of Shasta Dam, keystone of the system, began. Subsequently, the Central Valley Project effectively circled the Great Valley with a necklace of dams, large and small, wedged into the canyons of the Sierra Nevada and, to a lesser extent, the Coast Ranges. The spacious reservoirs behind Folsom Dam on the American River, New Melones on the Stanislaus, and Friant on the San Joaquin were intended to capture and hold Sierra runoff during the winter and spring for agricultural use in the long, dry summer. The dams had other effects as well: flood prevention, navigation maintenance, recreation, and, of course, hydroelectric power generation. The alteration of Sierra ecosystems caused by the construction of these large dams, though difficult to quantify, was undeniably massive. (And so too has been the impacts of reservoirs, streamflow diversions and streamflow regulations on regional landscape patterns, fisheries, wildlife, riparian vegetation, groundwater supplies, channel formation and channel maintenance.)

After numerous wartime delays, in 1951, some 14 years after construction had begun, water started flowing to the San Joaquin Valley from the Sacramento drainage. But already there was discontent. As a creature of the Bureau of Reclamation, the CVP was rooted in the government regulation known as the "160-acre limitation," which specified that no single farmer may irrigate more than 160 acres with the ultra-cheap water from a federally financed reclamation project. To receive CVP water, farmers and growers, hundreds of whom owned vast acreages, had to sign contracts which included the provision that they promise to divest themselves of "surplus lands"—those in excess of 160 acres (or various legal exceptions to this figure). For many powerful agricultural interests, especially the corporation-owned ranches, the acreage limitation was an absurdity to be challenged, fought vigorously, and, ultimately, avoided. The latter option took the form of persuading the people of California to underwrite a water plan that would not only serve agribusiness in the Central Valley but would also benefit the entire state. This movement would culminate in the world's largest water transfer system: the State Water Project.

By the middle of the 20th century most of the rivers draining the long western slope of the Sierra Nevada had been plugged by dams. Huge placid reservoirs backed them up. One large river that remained wild, due primarily to its remoteness and ferocity, was the Feather, the largest tributary of the Sacramento. Subject to devastating surges, most prominently the "Christmas floods" of 1955, the Feather carried an average annual runoff of 4.5 million acre-feet. Support for a mammoth state-controlled water system, whose centerpiece would be a giant dam on the Feather (at Oroville), gained momentum, slowly, through the 1950s. The Burns-Porter Act authorizing the project cleared the state legislature in 1959. In the general election the following year, voters narrowly approved the bond measure required to finance the State Water Plan. Voting patterns were conspicuous in their regional biases: wide-spread support for the project in Southern California, while only one northern county, Butte (site of the proposed Oroville Dam), voted in favor of the $1.75 billion bond measure.

Volumes have been written on the State Water Project, one of the most scrutinized and analyzed public projects in United states history. No insights will appear here that cannot be found in the vast literature on the subject. It is worth noting, however, that this colossal water project, so grand in scale and scope, so much a product of self-serving private interests and governmental bureaucracies, is likely the last of its kind. In the past twenty years Californians in growing numbers have risen up to stifle, scuttle, or otherwise re-evaluate water projects—proposed, planned, operating—like no time in the state's history. A new era, born of fiscal austerity and widespread public support for the maintenance of a more "natural" environment, had begun. In the Sierra Nevada, this clear shift in public opinion found expression in battles to save a free-flowing river and restore a saline lake.

The Fate of the Stanislaus and Mono Lake

There is no better illustration of the fierce clash between forces representing two eras of water policy than the protracted struggle to "save" a wild and scenic stretch of the Stanislaus River, which drains the watershed immediately north of Yosemite National Park. A dam on the Stanislaus had been a foregone conclusion to most. Approved in 1944 as part of the Central Valley Project, construction of the New Melones Dam finally got underway in the early 1970s. As the fourth highest dam in the United States, it would eventually flood 26 miles of the Stanislaus, including some of the nation's heaviest traveled white water rapids. The reservoir behind New Melones would also drown archaeological and historical sites, petroglyphs and wilderness areas. Wildlife, fisheries and water quality would be adversely impacted as well.

An initiative to halt the dam, drawn up by a consortium of environmental organizations spearheaded by the specifically formed Friends of the River, appeared on the 1974 ballot as Proposition 17. Although the proposition lost, 53 percent to 47 percent, the battle had just begun. Moving to the courts, the Stanislaus case, through bureaucratic entanglements, ultimately pitted the state of California against the federal government (California v. United States). Meanwhile, construction of the dam continued throughout the decade as the controversy played out in more court action and finally in Congress. The House Interior Committee in September 1980 failed to endorse a bill that would have included the Stanislaus in the federal wild rivers system. Although the vote was close—the bill was defeated by two votes—the cause was all but lost. The extremely wet winters of 1982 and 1983, which flooded the Stanislaus Canyon behind New Melones, made any further protest moot.

On the other side of the Sierra Nevada, at the same latitude as New Melones Dam, a different kind of battle was being waged. Mono Lake, a starkly beautiful, 500,000 year-old roughly circular lake was the focus. Fed by creeks draining the Sierra's east slope, Mono Lake, with no natural outlet, had acquired a level of salinity that produced a unique ecosystem involving brine shrimp and flies and migratory waterfowl. Diversions of feeder streams by the city of Los Angeles, begun in 1941, had increased substantially after 1970 when the Mono extension of the city's Owens Aqueduct was completed. Larger diversions by Los Angeles accelerated the decline in the level of Mono Lake and increased the concentration of salts in the water, causing biologists to predict drastic consequences for the brine shrimp and flies as well as the resident visiting bird life.

Visible deterioration of the lake's ecosystem prompted the formation in 1978 of the Mono Lake Committee, which quite clearly had as its goal to "save" Mono Lake. With powerful economic and political forces lined up on the other side of the issue, the most formidable of which was the city of Los Angeles itself, environmentalists sprung an innovative offensive: In 1979, the National Audubon Society, Friends of the Earth, Mono Lake Committee, and others filed suit against the city of Los Angeles, claiming the Mono Basin diversions violated the doctrine of "public trust." Historically, this doctrine was associated with public access to navigable waters for commercial activity and for fishing. Now it was being invoked to protect an area that had documented scientific (ecological) and scenic value. The argument was that the state of California had an obligation to prevent Los Angeles from compromising the public's benefit and use of Mono Lake. The local superior court sided with Los Angeles but in 1983 the State Supreme Court, in the case of National Audubon Society v. Superior Court of Alpine County, ruled that no water can be taken from a stream, lake, or other natural source without assessing the impact on navigable waters. The public trust doctrine had been upheld.

Years of bitter engagements and legal maneuverings between the supporters of Mono Lake and the city of Los Angeles appears to have ended. In 1994, the State Water Resources Control Board voted to restore the level of Mono Lake to 6,392 feet, 18 feet above its level at the time of the ruling. Such an elevation, most studies have concluded, would preserve the ecological integrity of the lake. In a dramatic reversal of long standing policies and legal decisions favoring an ethic built around the notion that growth is good, that bigger is better, the Mono Lake case highlights the new era of limits and restrictions being imposed on a hydraulic society by a hydraulic society.

This new approach to water resources management presupposes that the status quo, especially in regard to climate, be maintained. No new large dams have been built in the Sierra Nevada in two decades and none, aside from Congressman John Doolittle's push to complete the mothballed Auburn Dam, are planned. But what happens if a prolonged cycle of drought returns? Not for six years, as in the late 1980s and early 1990s, but for 60 years, or 100? California has known them before. It will again.

SOURCE NOTES

Fashioning a Hydraulic Landscape

Ditches and Flumes

This truly astonishing aspect of Sierra Nevada history is surprisingly absent from the mainstream literature. Two unpublished works that give this subject its due are: Thomas H. Pagenhart, "Water Use in the Yuba and Bear River Basins, California," Ph.D. dissertation, 1969, University of California, and the author's own Master's thesis in geography, David J. Larson, "Ditch and Flume Systems of the Central Sierra Nevada: Evolution of a Water Transfer Network," University of California, 1982.

The Hydraulic Mining Era

Hydraulic mining as an enterprise and an agent of landscape transformation is covered in detail in the above two sources. See also the excellent monograph by Philip Ross May, *Origins of Hydraulic Mining in California* (Oakland, 1970). Another fine account of this phenomenon appears in Robert L. Kelley's *Gold vs. Grain: The Hydraulic Mining Controversy in the Sacramento Valley: A Chapter in the Decline of the Concept of Laissez-faire* (Glendale, 1959).

Two classic government reports detail the scope of damage to the landscape wrought by hydraulic mining: William H. Hall, *Report of the State Engineer to the Legislature of California*, 23 Session, 1880, Part III, The Flow of Mining Debris, (Sacramento, 1880), and Grove K. Gilbert, *Hydraulic-Mining Debris in the Sierra Nevada*, U.S.G.S. Professional Paper 105 (Washington, 1917).

Why California?

May and Larson, cited above, provide details on why California, more than any other 19th century locale, was the ideal stage on which the hydraulic mining drama could be played. Also, numerous issues of the *Mining and Scientific Press*, a San Francisco weekly newspaper, carried editorials addressing this question. Other papers consulted were the *Sacramento Union*, *Nevada Daily Transcript* (Nevada City), and the *Daily Alta California* (San Francisco).

The Downstream Debris Dilemma: Choosing Up Sides

The definitive account of this inter-regional conflict is Robert L. Kelley's *Gold vs. Grain* (1959), cited above, and his earlier "The Mining Debris Controversy in the Sacramento Valley," *Pacific Historical Review*, Vol 25 (1956), 331–346.

For an account of flooding and its effects, see: William T. Ellis, *Memories: My Seventy-Two Years in the Romantic County of Yuba, California* (Eugene, Oregon, 1939); Charles L. Brace, *The New West: or, California in 1867–1868* (New York, 1869) weighs in with a first-hand view. See also Peter J. Delay, *History of Yuba and Sutter Counties* (Los Angeles, 1924) and Charles Nordhoff, *California for Health, Pleasure and Residence* (New York, 1872). Various newspapers provided material for my description of the debris dilemma, foremost among them: *Mining and Scientific Press*, *Nevada Daily Transcript* (Nevada City), *Sutter Banner* (Yuba City), *Marysville Appeal*, and two Sacramento papers, the *Record-Union* and the *Bee*.

The full text of Judge Keyser's decision in *J.H. Keyes v. Little York Mining Company* is reprinted in the *Mining and Scientific Press*, March 22, 1879. The *Stock Report* (San Francisco) provides acid commentary after the ruling.

The rise of a wheat culture in California is explained in detail in: Horace Davis, "Wheat in California," *The Overland Monthly*, Vol. 1 (November 1868); Rodman W. Paul, "The Wheat Trade Between California and the United Kingdom." *Journal of American History*, Vol. 45 (December 1958); Lawrence J. Jelinek, *Harvest Empire, A History of California Agriculture* (San Francisco, 1979), especially Chapter 5, "Bonanza Wheat Era: 1873–1902."

The Chess Match

See William H. Hall, *Report of the State Engineer to the Legislature of California*, 23rd Session, Part III, The Flow of Mining Debris, (Sacramento, 1880).

For background on the Drainage Act, see *Statutes*, Legislature of California, 23rd Session (1880).

The floods of 1881 were termed the most devastating ever by numerous valley newspapers, specifically (and predictably) the *Marysville Appeal* and the *Bee* and *Union*.

Checkmate: Victory for the Valley

Judge Sawyer's decision is variously cited as 9 *Sawyer* 441 and *Federal Reporter*, Vol. 18, No. 14, 1884, pp. 753–813. The full text of the judgment appears in the *Mining and Scientific Press*, January 19, 1884. Other San Francisco newspapers also provided detailed coverage. The celebration in the valley is chronicled by Peter Delay, *The History of Yuba and Sutter Counties* (1924).

Aftermath: The End of an Era

For the Army Corps of Engineers Report see Thomas L. Casey, *Mining Debris, California* 51 Congress, House Exec. Doc. No. 267 (Washington, 1891). See also W.W. Harts, "The Control of Hydraulic Mining in California by the Federal Government," *Proceedings of the American Society of Civil Engineers*, Vol. 32, No. 2 (1906).

Legacies: Irrigation and Power Possibilities

J.S. Hittell's prediction appears in J. Ross Browne, *Report on the Mineral Resources of the States and Territories West of the Rocky Mountains* (Washington, D.C. 1868), p. 606; *Sacramento Record-Union*, July 19, 1884.

The early innovations in electricity generation and water-power technology are discussed in Charles M. Coleman, *P.G. and E. of California, The Centennial Story of Pacific Gas and Electric Company, 1852–1952*, (New York, 1952) and Norman Smith, "The Origins of the Water Turbine," *Scientific American*, Vol. 242 (January 1980), especially 146–147.

Early Hydroelectric Power Generation

See two obscure but illuminating works by J.W. Johnson, a professor of mechanical engineering at the University of California, Berkeley: "Engineering Highlights of the California Mining Days," *California Engineer* (May 1949) and the more detailed "Early Engineering Center in California," *California Historical Society Quarterly [California History]* Vol 29 (September 1950), 193–209. See also C.M. Coleman, *P. G. and E. of California* (1952) for the role played by companies that would eventually become a part of Pacific Gas and Electric Company.

Epilogue to an Era

Observations are based on the author's own field work in the Sierra Nevada foothills.

New Water-Use Priorities for a New Century

Water Rights in Context

For an appraisal of the relationship between early gold miners and California government see Gerald Nash, *State Government and Economic Development: A History of Administrative Policies in California, 1849–1933* (Berkeley, 1964), 38–41.

A summary history of appropriative rights is found in: Governor's Commission to Review California's Water Rights Law, *Appropriative Water Rights in California: Background and Issues* (1977), 4–5.

The hopelessly gnarled condition of water claims in turn-of-the-century California is clearly presented in Elwood Mead et al, *Irrigation Investigations in California*, Bulletin 100, U.S.

Department of Agriculture Office of Experiment Stations (Washington D.C. 1901).

Mead's recommendations resulted in Assembly Bill 735: California Legislature. Assembly, *Journal of the Assembly, 35th Session of the Legislature*, Vol II (Sacramento, 1903).

For background on the California Conservation Commission, see California Legislature, *California Statutes, 1911*, Chapter 408 (Sacramento, 1911). The recommendation of the CCC is found in: California Conservation Commission, *Report* (Sacramento, 1912).

Reaching Out to the Sierra Nevada: Urban Water Grabs

For description and analysis of the Hetch Hetchy controversy, see Warren D. Hanson, *San Francisco Water and Power: A History of the Municipal Water Department and Hetch Hetchy System* (San Francisco, 1985); Roderick Nash, *Wilderness and the American Mind*, 3rd ed. (New Haven, 1982), Chapter 10. Also consulted was Kendrick Clements, "Politics and the Park: San Francisco's Fight for Hetch Hetchy, 1908–1913," *Pacific Historical Review*, Vol. 48 (May 1979).

The preservationist versus conservationist philosophies are eloquently presented in Michael L. Smith, *Pacific Visions: California Scientists and the Environment, 1850–1915* (New Haven, 1987).

President Woodrow Wilson's views on Hetch Hetchy are found in: *Congressional Record*, 63 Cong. 2d Session (December 19, 1913), 1189.

An excellent overview of the entire Hetch Hetchy story is found in the William L. Kahrl (ed), *The California Water Atlas* (Sacramento, 1978).

The story of how the City of Los Angeles secured water rights in the Owens Valley and then engineered a transport system for that water has been the subject of many articles and several books, the latter being best represented by: William L. Kahrl, *Water and Power, The Conflict over Los Angeles' Water Supply in the Owens Valley* (Berkeley, 1982) and Abraham Hoffman, *Vision or Villainy, Origins of the Owens Valley–Los Angeles Water Controversy* (College Station, Texas, 1981). See also *The California Water Atlas* (cited above) for a fine overview of this classic case of abuses of power in the public interest. A popular history of this controversy is contained in Remi Nadeau, *The Water Seekers* (New York, 1950).

The County of Origin law (for watershed protection) is dealt with in *California Statutes*, Chapter 286 (1927). For a vivid description and analysis see Norris Hundley, Jr. *The Great Thirst, Californians and Water, 1770s–1990s* (Berkeley, 1992).

Two Projects for One (Great) Valley

An enormous volume of material has been written concerning every aspect of the Central Valley Project and the subsequent State Water Project, including numerous books devoted to either or both of these water projects. Without doubt the best source for an overview of the two projects as well as for the myriad details associated with them in Norris Hundley's *The Great Thirst, Californians and Water, 1770s–1990s*, a work

of magisterial proportion. Everything else pales by comparison. But see also *The California Water Atlas* (1978) for its superb graphics on water storage and transfer data and associated water statistics. An earlier account of water development in California, colorfully written, is Erwin Cooper, *Aqueduct Empire* (Glendale, 1968). Also worth looking at for perspective and a "big picture" approach is Marc Reisner, *Cadillac Desert, The American West and Its Disappearing Water* (New York, 1986; 1993).

The Fate of the Stanislaus and Mono Lake

For the Stanislaus River New Melones dam controversy see: Tim Palmer, *Stanislaus: The Struggle for a River* (Berkeley, 1982) and the same author's *Endangered Rivers and the Conservation Movement* (Berkeley, 1986). The definitive analysis of the legal history of this river battle is W. Turrentine Jackson and Stephen D. Mikesell, *The Stanislaus River Drainage Basin and the New Melones Dam* (Davis, June 1979). See also Samuel P. Hays, *Beauty, Health, Permanence: Environmental Politics in the United States, 1955–1985* (New York, 1987).

The pertinent governmental and legal documents are: California Water Resources Control Board, *Decision 1422* (Sacramento: April 4, 1973); *California v. United States*, 438 U.S. 645 (1978); *United States v. California Water Resources Control Board*, 694 F. 2nd 1171 (1982); California Department of Water Resources, "Management of the California State Water Project," *Bulletin 132–83* (Sacramento, November 1983) 155–156.

The Mono Lake story is told in Ron Bass, "The Troubled Waters of Mono Lake," *California Journal*, Vol. 9 (October 1979), 349–350; Daniel Chasan, "Mono Lake v. Los Angeles: Tug-of-War for Precious Water," *Smithsonian*, Vol. 11 (February 1981), 42–50; National Research Council, *Mono Lake Basin Ecosystem: Effects of a Changing Lake Level* (Washington, 1987); *The California Water Atlas* (1978); William Kahrl, *Water and Power*.

The legal scholarship on the Mono Lake case is best embodied in two works by Harrison C. Dunning: "The Significance of California's Public Trust Easement for California's Water Rights Law," *U.C. Davis Law Review*, Vol. 14 (Winter 1980), 357–398, and "A New Front in the Water Wars: Introducing the 'Public Trust' Factor," *California Journal*, Vol. 14 (May 1983); See also *National Audubon Society v. Superior Court of Alpine County*, 33 Cal. 3rd 419 (1983); denied 464 U.S. 977 (1983).

The question of whether California will experience another prolonged drought is addressed in Scott Stine, "Extreme and Persistent Drought in California and Patagonia During Mediaeval Time," *Nature* (16 June 1994).

REFERENCES

Bass, R. 1979. The troubled waters of Mono Lake. *California Journal* 9 (October): 349–50.

Brace, C. L. 1869. *The new west: Or California in 1867–1868.* New York.

Browne, J. R. 1868. *Report on the mineral resources of the states and territories west of the Rocky Mountains.* Washington, DC: Governnment Printing Office.

Casey, T. L. 1891. *Mining debris, California.* 51st Congress. House Exec. Doc. No. 267. Washington, DC.

Chasan, D. 1981. Mono Lake v. Los Angeles: Tug of war for precious water. *Smithsonian* 2 (February): 42–50.

Clements, K. 1979. Politics and the park: San Francisco's fight for Hetch Hetchy, 1908–1913. *Pacific Historical Review* 48 (2): 185–215.

Coleman, C. M. 1952. *P.G. and E. of California: The centennial story of Pacific Gas and Electric Company 1852–1952.* New York: McGraw-Hill.

Cooper, E. 1968. *Aqueduct empire.* Glendale, CA: Arthur H. Clark.

Davis, H. 1868. Wheat in California. *Overland Monthly* 1 (4): 446–49.

Delay, P. J. 1924. *History of Yuba and Sutter Counties.* Los Angeles, CA.

Dunning, H. C. 1980. The significance of California's public trust easement for California's water rights law. *UC Davis Law Review* 14 (winter): 357–98.

———. 1983. A new front in the water wars: Introducing the "public trust" factor. *California Journal* 14 (May): 189–90.

Ellis, W. T. 1939. *Memories: My seventy-two years in the romantic county of Yuba, California.* Eugene, OR.

Gilbert, G. K. 1917. *Hydraulic-mining debris in the Sierra Nevada.* USGS Professional Paper 105. Washington, DC.: U.S. Geological Survey.

Hall, W. H. 1880. *Report of the state engineer to the legislature of California.* 23rd session. Sacramento, CA.

Hanson, W. D. 1985. *San Francisco water and power: A history of the municipal water department and Hetch Hetchy system.* San Francisco: City and County of San Francisco.

Harts, W. W. 1906. The control of hydraulic mining in California by the federal government. *Proceedings of the American Society of Civil Engineers* 32 (2): 95–124.

Hays, S. P. 1987. *Beauty, health, permanence: Environmental politics in the United States 1955–1985.* New York: Cambridge University Press.

Hoffman, A. 1981. *Vision or villainy: Origins of the Owens Valley–Los Angeles water controversy.* College Station: Texas A & M Press.

Hundley, N., Jr. 1992. *The great thirst: Californians and water, 1770s–1990s.* Berkeley and Los Angeles: University of California Press.

Jackson, W. T., and S. D. Mikesell. 1979. *The Stanislaus River drainage basin and the New Melones Dam.* Davis: California Water Resources Center.

Jelinek, L. J. 1979. *Harvest empire: A history of California agriculture.* San Francisco: Boyd and Fraser Publishing Co.

Johnson, J. W. 1949. Engineering highlights of the California mining days. *California Engineer*, May, 8–10.

———. 1950. Early engineering center in California. *California Historical Society Quarterly* 29 (3): 193–209.

Kahrl, W. L., ed. 1978. *The California water atlas.* Sacramento: State of California.

———. 1982. *Water and power: The conflict over Los Angeles' water supply in the Owens Valley.* Berkeley and Los Angeles: University of California Press.

Kelley, R. L. 1956. The mining debris controversy in the Sacramento Valley. *Pacific Historical Review* 25 (3): 331–46.

———. 1959. *Gold vs. grain: The hydraulic mining controversy in the Sacramento Valley: A chapter in the decline of laissez-faire.* Glendale, CA: Arthur H. Clark.

———. 1980. *Battling the inland sea: American political culture, public policy, and the Sacramento Valley, 1850–1986.* Berkeley and Los Angeles: University of California Press.

Larson, D. J. 1982. Ditch and flume systems of the central Sierra Nevada: Evolution of a water transfer network. Master's thesis, Department of Geography, University of California, Berkeley.

May, P. R. 1970. *Origins of hydraulic mining.* Oakland, CA: Holmes Book Company.

Mead, E., et al. 1901. *Irrigation investigations in California.* Bulletin 100. Washington, DC: U.S. Department of Agriculture, Office of Experiment Stations.

Mendell, G. H. 1882. *Report upon a project to protect the navigable waters of California from the effects of hydraulic mining debris.* Washington, DC: Government Printing Office.

Nadeau, R. A. 1950. *The water seekers.* Garden City, NY: Doubleday.

Nash, G. D. 1964. *State government and economic development: A history of administrative politics in California, 1849–1933.* Berkeley and Los Angeles: University of California Press.

National Research Council. 1987. *Mono Lake Basin ecosystem: Effects of a changing lake level.* Washington, DC: National Research Council.

Nordhoff, C. 1873. *California for health, pleasure and residence.* New York: Harper and Brothers.

Pagenhart, T. H. 1969. Water use in the Yuba and Bear River Basins, California. Ph.D. diss., Department of Geography, University of California, Berkeley.

Palmer, T. 1982. *Stanislaus: The struggle for a river.* Berkeley and Los Angeles: University of California Press.

———. 1986. *Endangered rivers and the conservation movement.* Berkeley and Los Angeles: University of California Press.

Paul, R. W. 1958. The wheat trade between California and the United Kingdom. *Journal of American History* 45 (4): 391–412.

Reisner, M. 1986. *Cadillac desert: The American West and its disappearing water.* New York: Viking.

Smith, M. L. 1987. *Pacific visions: California scientists and the environment, 1850–1915.* New Haven, CT: Yale University Press.

Smith, N. 1980. The origins of the water turbine. *Scientific American* 242 (1): 138–48.

Stine, S. 1994. Extreme and persistent drought in California and Patagonia during mediaeval time. *Nature* 369 (16 June): 546–49.

Strong, D. 1984. *Tahoe: An environmental history.* Lincoln: University of Nebraska Press.

Benchmark Events Impacting the Sierra Nevada Waterscape

19th Century

1848   Discovery of gold at Sutter's Mill near South Fork of the American River.

1849   Gold Rush begins in earnest. California Constitutional Convention meets at Monterey; adopts Common Law of England with its riparian rights provisions.

1853   First documented use of hydraulic mining--the application of water under pressure, through a nozzle against a natural bank--in Nevada County.

1877   Beginning of *Lux v. Haggin* suit, a legal test of riparian versus appropriative water rights.

1878   Legislature names William Hammond Hall first "State Engineer."

1884   Sawyer Decision: Judge Lorenzo Sawyer issues perpetual injunction against hydraulic mining in the Sierra Nevada.

1886   California Supreme Court rules in favor of riparian rights in *Lux v. Haggin* decision.

1887   Legislature passes Wright Act, laying the basis for irrigation districts.

1896   Rome Power House, using water diverted from the South Yuba, is first hydroelectric plant of what came to be the Pacific Gas and Electric Company.

1897   Wright Act revised and strengthened.

1899   Colgate power plant on north fork of Yuba River supplies electricity for Oakland via the world's first long-distance transmission line.

20th Century

1912   State Water Commission established to administer water rights for power purposes.

1913   The Raker Act authorizes the damming of Hetch Hetchy Valley, allowing the City of San Francisco to invade Yosemite National Park for municipal water and power.

Owens Valley Aqueduct, carrying eastern Sierra water to southern California, is completed.

1914   California Water Commission Act regularizes appropriative procedure.

1920   Federal Power Act paves the way for a vast number of small-scale hydroelectric dams to be built in Sierra national forests.

1922   O'Shaughnessy Dam is completed in Hetch Hetchy Valley, Yosemite.

1929   EBMUD completed Mokelumne Aqueduct from the Sierra Nevada to the East Bay.

1933   Central Valley Project approved by California Legislature and, after a referendum campaign, by California voters.

1935   Bureau of Reclamation takes over construction and administration of Central Valley Project.

1941   Owens Valley Aqueduct is extended north into the Mono Basin.

1945   State Water Resources Control Board created by Legislature.

1951   Central Valley Project becomes operational as water flows through the Delta-Mendota and Friant-Kern canals.

1957   League to Save Lake Tahoe is organized.

1959   State Water Resources Development Bond Act, known as the Burns-Porter Act, passes the Legislature; ratified by voters in 1960 General Election. These actions authorize State Water Project.

1962   Commercial whitewater rafting begins on the Stanislaus River.

1964   Wilderness Act passed by Congress; includes portions of the Sierra high country.

1968   Wild and Scenic Rivers Act (Federal). The Feather River is designated a wild and scenic river.

1969   National Environmental Policy Act (NEPA). A major element of this legislation is the requirement that an Environmental Impact Statement be prepared for any federal action having a "significant effect on the environment." This deals a serious blow to the dam builders.

1970   Completion of Owens Valley Aqueduct extension into Mono Basin; The Tahoe Regional Planning Agency is organized.

1972   Wild and Scenic Rivers Act (State); includes portions of the American River along with major north coast rivers.

1973   Friends of the River is organized; a statewide initiative against New Melones Dam on the Stanislaus loses; NMD floods canyon in 1982.

1975   Construction halted on Auburn Dam on American River.

1978   North Fork of the American added to the federal wild and scenic rivers system.

1980   The lower American River is added to the federal wild and scenic rivers system.

1983   The California Supreme Court rules on the Public Trust Doctrine in favor of Mono Lake.

1984   Mono Lake Scenic Area is designated; Tuolumne River is added to federally protected rivers system.

1987   Kings, Merced, and North and South Forks of the Kern are added to the national wild and scenic rivers system.

1994   State Water Resources Control Board votes to restore the level of Mono Lake to 6,392 feet, 18 feet above its current level.

Estimates of Mining Debris Deposited, 1849–1909

	Million Cubic Yards
From Hydraulic Mining in the Basin of the:	
Upper Feather River	100
Yuba River	684
Bear River	254
American River	257
Streams tributary to lateral basins of the Sacramento River	30
Mokelumne River to Tuolumne River, inclusive	230
From Ordinary Placer Mining	60
From Quartz Mining (one-fourth in the Sacramento Basin)	50
From Drift Mining (three-fourths in the Sacramento basin)	30
Total Mining Debris:	
From Hydraulic Mining	1,555
From all Mining Tributary to the Sacramento River	1,390
From all Mining Tributary to Suisun Bay	1,665

Source: G.K. Gilbert, Hydraulic Mining Debris in the Sierra Nevada. U.S.G.S. Professional Paper 105 (Washington D.C., 1917), p. 43.

APPENDIX 8.3

Water Storage Capacities of Water and Mining Companies in the Northern Sierra Nevada

COMPANY	STORAGE CAPACITY (Millions of cubic feet)
South Yuba Canal Company	1,800
Eureka Lake & Yuba Canal Company	1,130
El Dorado Canal Company	1,070
North Bloomfield Gravel Mining Company	1,050
Milton Mining and Water Company	650
California Water Company	600
Omega and Blue Point Company	300
Spring Valley Mining Company	300
Other small reservoirs in the watersheds of the Feather, Yuba, Bear, and American Rivers	700

Source: George H. Mendell, <u>Report Upon a Project to Protect the Navigable Waters of California From the Effects of Hydraulic Mining Debris</u> (Washington, 1882), p. 13.

Sequence of Ownership: El Dorado Canal Company

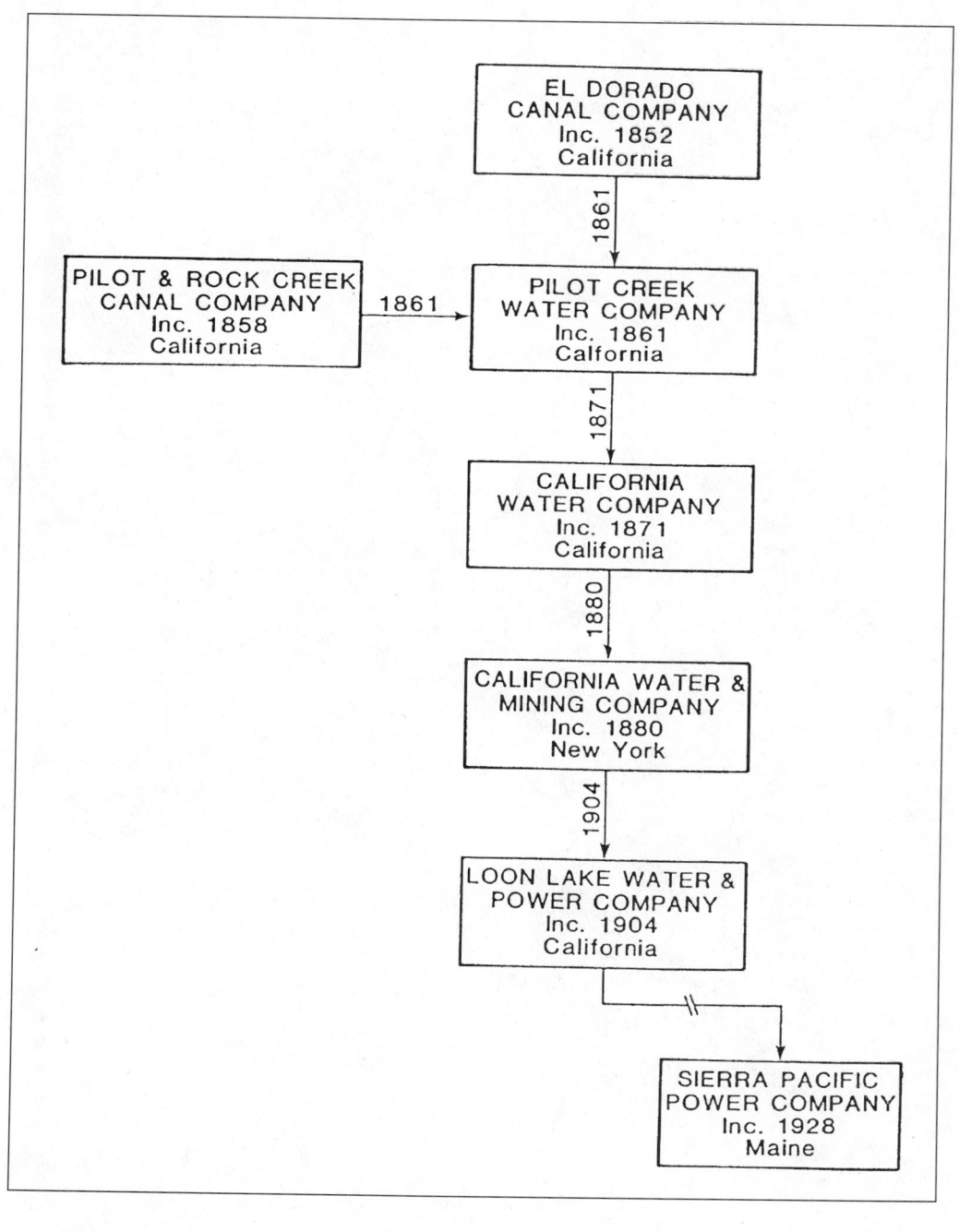

M. KAT ANDERSON
American Indian Studies Center
University of California
Los Angeles, California

MICHAEL J. MORATTO
Applied EarthWorks, Inc.
Fresno, California
and
California State University
Fresno, California

9

Native American Land-Use Practices and Ecological Impacts

ABSTRACT

During a span of 10,000 years or more, Native Americans in the Sierra Nevada were sustained by hunting and fishing, gathering, lithic quarrying, and trading. To meet their requirements for firewood, fish and game, vegetal foods, craft supplies, and building materials, the native peoples of the Sierra managed biotic resources intensively, with significant ecological and evolutionary consequences. The distribution, structure, composition, and extent of certain plant communities, for example, were shaped by burning, pruning, sowing, weeding, tilling, and selective harvesting. Numerous "protoagricultural" techniques, based upon traditional knowledge of natural processes gained over the millennia, were applied to increase the quantity and improve select qualities of focal plant species. Fire was the most important management tool, employed to clear brush, maintain grasslands and meadows, improve browse for deer, enhance production of basketry and cordage materials, modify understory species composition in forests, and reduce fuel accumulation that might otherwise sustain intense fires. Considering that the human population of the Sierra Nevada was approximately 90,000–100,000 in late prehistoric times (ca. A.D. 1300–1800), the environmental consequences of aboriginal land-use and management practices were substantial. There is currently an ecological "vacuum," or disequilibrium, in the Sierra resulting from the departure of Native American influences. The recent decline in biotic diversity, species extirpation and endangerment, human encroachment into fire-type plant communities (e.g., chaparral), and greatly increased risk of catastrophic fires are but symptoms of this disequilibrium. It is recommended, therefore, that land-managing agencies and land-use planners incorporate Native American traditional knowledge into future policies and programs for ecosystem management in the Sierra Nevada. This traditional knowledge, which permitted the adaptive success of large human populations and the maintenance of Sierran environments for more than a hundred centuries, must not be dismissed.

INTRODUCTION

Californians are faced with the growing necessity of finding ways to maintain the integrity and livability of their ecosystems (Barbour et al. 1993; Jensen et al. 1993). Increasing knowledge of the diverse ecological relationships of native peoples to their environments affords an opportunity to assess these relationships with respect to ecological principles and to assess their value for helping to solve regional and local environmental problems (Nabhan 1995). Sustainability is increasingly being defined in terms of conserving *cultural* as well as biological diversity (Manley et al. 1995). The varied past approaches of Native Americans to resource use and management in the Sierra Nevada could contribute significantly to maintaining biological and cultural diversity, and improving human livelihood (Soulé and Kohm 1989). Resource management by Native Americans in the Sierra Nevada bioregion was long term and widespread, producing ecological and evolutionary consequences in the biota (Blackburn and Anderson 1993). Therefore, many ecosystems in the Sierra are not self-maintaining islands that require only protection to remain in a "pristine" state. There is currently an

Sierra Nevada Ecosystem Project: Final report to Congress, vol. II, *Assessments and scientific basis for management options.* Davis: University of California, Centers for Water and Wildland Resources, 1996.

ecological "vacuum," or disequilibrium, in the Sierra Nevada resulting from the departure of Native Americans from managing these ecosystems.

This chapter explores why present-day management and restoration of the Sierra Nevada bioregion should be grounded in *historical* as well as ecological research. It provides an overview of the Native Americans who inhabited the Sierra Nevada during the early 1800s as well as those who live there today, their land-use activities, management practices, environmental ethics, and potential beneficial and negative ecological effects in different ecosystem types. Also, it asks a series of questions about the state of knowledge and substantiated evidence for modification of Sierran landscapes by indigenous peoples. Specific examples of production systems for three cultural use categories—basketry, foods, and cordage—are given. Ecological consequences of removing Native Americans from traditional economic and land-management roles in the Sierra Nevada are examined, and an agenda for future policy, research, and management directions as well as collaborative efforts with contemporary Native Americans is proposed.

Relevance of Native American Environmental History to Sierra Nevada Ecosystem Project Objectives

Knowledge of the history of natural systems is an essential component of scientific analysis (Crumley 1994; Smith 1994). This history influences our ability to assess the present health and condition of ecosystems in the Sierra Nevada and to predict the future (Woolfenden 1994). Plant-community organization and assemblages are expressions of species evolution and species behavior (Whittaker and Woodwell 1972), and plant adaptations are responses to past environmental conditions. Native Americans, as integral residents of the Sierra Nevada, modified environmental conditions, dispersed plant species to new areas, and created recent evolutionary modifications in the flora through human selection for particular traits. Thus, Native Americans were instigators of ecosystemic change with varying degrees of intensity during the time they inhabited the Sierra, beginning some 10,000 years ago (Elston et al. 1977; Moratto 1984; Moratto et al. 1988; Peak and Crew 1990; Rondeau 1982).

The Sierra Nevada did not fit the definition of an uninhabited, virgin wilderness at the point of Euro-American contact. Rather, it had been shaped by thousands of years of indigenous burning, pruning, sowing, selective harvesting, and tilling (Anderson and Nabhan 1991; Simms 1992). Native Americans have managed Sierran ecosystems in a nonrandom fashion, using a variety of horticultural techniques. Such management is substantiated by ethnohistoric and ethnographic records, studies of museum artifacts, paleoecological findings, fire scar studies, and ecological field studies (Anderson 1993b; Anderson and Carpenter 1991; Kilgore and Taylor 1979; Lewis 1993; Matson 1972; Roper Wickstrom 1987).

Furthermore, early humans were effective hunters, influencing the distribution, abundance, and diversity of wildlife within their tribal territories. To understand the vegetation of a particular locality or region at a specific time requires knowledge of soil, topography, climate, natural processes, *and* history of land use by Native Americans.

We are in the first stages of documenting in a detailed and intensive manner the prehistoric and historic land-management practices in the Sierra Nevada. Deliberate management of wild plant and animal resources and habitats was a major element of Native American subsistence strategies. Yet, investigation of the relationships between such land-management activities and their ecological consequences is a nascent field of study. In fact, there exists no synthesis or detailed analysis of past wildlife management by Native Americans and its potential ecological impacts in the Sierra Nevada. It is clear that Native American land-management practices had significant ecological and evolutionary consequences on the biota, but the details of these impacts will remain unknown for specific geographic regions until interdisciplinary teams conduct more comprehensive studies.

If the goal of public land-managing agencies is to preserve certain ecosystems in some semblance of their pre-contact structures and functions, then they can no longer ignore these anthropogenic effects and must investigate the possibility of simulating some of the earlier cultural practices (Anderson 1993a; Wagner et al. 1995). The most recent argument against using pre-contact vegetation as a baseline for contemporary wildland management is that it would be treating ecosystems as "living museums" rather than as dynamic systems. This argument holds that wilderness areas should be treated as places where nonhuman life and ecological processes are unimpeded (Parsons et al. 1986). Yet pre-Euro-American vegetation was far from fixed. The underlying management philosophy of Native Americans in the Sierra Nevada was to continuously introduce small disturbance regimes into various plant-community types, which created openings or clearings. These openings invited the colonization of plant species that could not grow in the surrounding dominant vegetation type. These clearings represented a series of earlier successional stages within a more homogeneous landscape. Rather than reflecting an unchanging system, these landscapes were much more dynamic under the influence of human disturbance than in their "natural" state.

The nature and intensity of human intervention varied both geographically and diachronically. For example, some areas were subject only to lightning fires; other areas experienced both lightning- and Indian-set fires; and yet other areas were shaped largely by anthropogenic forces (i.e., frequent Indian-set fires). The creation of specialized habitats intensified plant-plant, plant-animal, animal-animal, human-animal, and human-plant relationships, creating a highly interactive system that ultimately changed vegetation patterns over time. Hence, the objective is not to re-create exactly a static picture of historic landscapes, but rather to investigate and understand the native cultural *processes* that drove biological diversity and shaped various ecosystem states and to unravel the ecological principles embedded in ancient land-management systems. As Christensen (1988) has recognized, diverse disturbances play an essential role in the long-term maintenance of virtually all ecosystems.

The Study of Native American Land-Management Practices

Analysis of indigenous protoagricultural practices yields a baseline of historical ecological information about the diversity, dynamics, and functioning of plant communities in the Sierra Nevada under former disturbance regimes. It also offers other models of human cultural intervention in nature. Simulating some of these practices in long-term field experiments would elucidate the effects of aboriginal activities upon natural resources in the Sierra Nevada and disclose the extent to which ecosystem health in the areas of soil productivity, gene conservation, biodiversity, landscape patterns, nutrient cycling, and an array of ecological processes is tied to former native economic and management activities. Native Americans have influenced Sierran landscapes over many generations. Their traditional knowledge of former abundances, composition, density, and quality of plant and animal species extends to time periods long before the advent of governmental land management. Their land-use practices were successful for thousands of years in maintaining diverse and productive ecosystems. The time depth of this traditional knowledge may provide a sense of what has been lost in Sierran landscapes since aboriginal times. Contemporary native cultures still maintain some of the traditional practices, and these may serve as analogs for testing alternative wildland-management strategies, restoring endangered ecosystems and species, enhancing the productivity and biodiversity of wildlands, and maintaining culturally significant plant resources for the perpetuation of native cultural traditions (Birckhead et al. 1992; Martinez 1992). If ecologists and land managers could understand the intricacies and mechanics of how and why native people shaped ecosystems, it would enrich their inventory of management methods and enhance their ability to make informed decisions.

OVERVIEW: PRE-CONTACT NATIVE AMERICAN INTERVENTIONIST APPROACH TO NATURE

Indian Tribes of the Sierra Nevada

There were numerous, distinctive cultures in the Sierra Nevada at the time of historic contact. During the early 1800s,

this region was inhabited by approximately thirteen "tribes" (ethnic groups speaking separate languages) composed of many "tribelets" (Kroeber 1962). This variety of cultures was reflected in diverse adaptations to Sierran environments and myriad land-use and resource-management strategies. Tribes on the west side of the Sierra included the Maidu, Konkow, Nisenan, Northern Sierra Miwok, Central Sierra Miwok, Southern Sierra Miwok, Foothill Yokuts (Poso Creek, Tule-Kaweah, Kings River, and Northern Hill), Western Mono (Monache), and Tübatulabal; on the east side of the mountains were the Northern Paiute, Washoe, and Owens Valley Paiute; the Kawaiisu (and to some extent, the Washoe) held land on both sides of the range (figure 9.1).

Maidu lands included the Susan River, the Red Clover, Valley Indian, and Willow Creeks, and the upper stretches of the North Fork of the Feather River, while the Konkow occupied the watersheds of the Middle and South Forks of the Feather River and the lower stretches of the North Fork of the Feather River (Riddell 1978). The Nisenan inhabited the drainages of the Yuba, Bear, and American Rivers and the lowest reaches of the Feather River; they moved seasonally to higher elevations (Wilson and Towne 1978). The Sierra Miwok (or Me-Wuk) comprised three divisions: the Northern Sierra Miwok occupied foothills and mountains of the Mokelumne and Calaveras River drainages; the Central Sierra Miwok claimed the foothill and upland portions of the Stanislaus and Tuolumne watersheds; and the territory of the Southern Sierra Miwok embraced the upper reaches of the Merced and Chowchilla Rivers (Levy 1978). The Foothill Yokuts (or Northern Hill Yokuts) occupied the foothills from the Fresno River basin southward to the Kern River (Spier 1978a). At higher elevations were the Western Mono (Monache), with six geographic subdivisions: the Northfork Mono, Wobonuch, Entimbich, Michahay, Waksachi, and Patwisha (Spier 1978b). In the southern Sierra Nevada foothills, the Tübatulabal occupied the Kern and South Fork of the Kern River country; three Tübatulabal bands are recognized: Pahkanapïl, Palagewan, and Bankalachi (Smith 1978).

Portions of the eastern Sierra were inhabited by the Northern Paiute, Owens Valley Paiute, and Kawaiisu (figure 9.1). The Northern Paiute occupied a vast territory extending from the Sierran crest eastward to Reese River and from Mono Lake northward to the Snake River country (Fowler and Liljeblad 1986). Bordering the Northern Paiute, south of Mono Lake, the Owens Valley Paiute inhabited Owens, Round, and Long Valleys, and frequented the White and Inyo Mountains as well as the eastern slopes and crestal zone of the Sierra Nevada to obtain seasonal resources (Liljeblad and Fowler 1986). The Kawaiisu homeland was in the southeastern Sierra Nevada and adjacent portions of the Tehachapi and Piute Mountains. Settlements were focused along the Kern and South Fork of the Kern Rivers, with seasonal use of the Sierra Nevada foothills from Kelso Valley up through the Walker Pass locality (Zigmond 1981, 1986). The Washoe, linguistically unrelated to their Paiute neighbors, held the Lake Tahoe Basin, a series

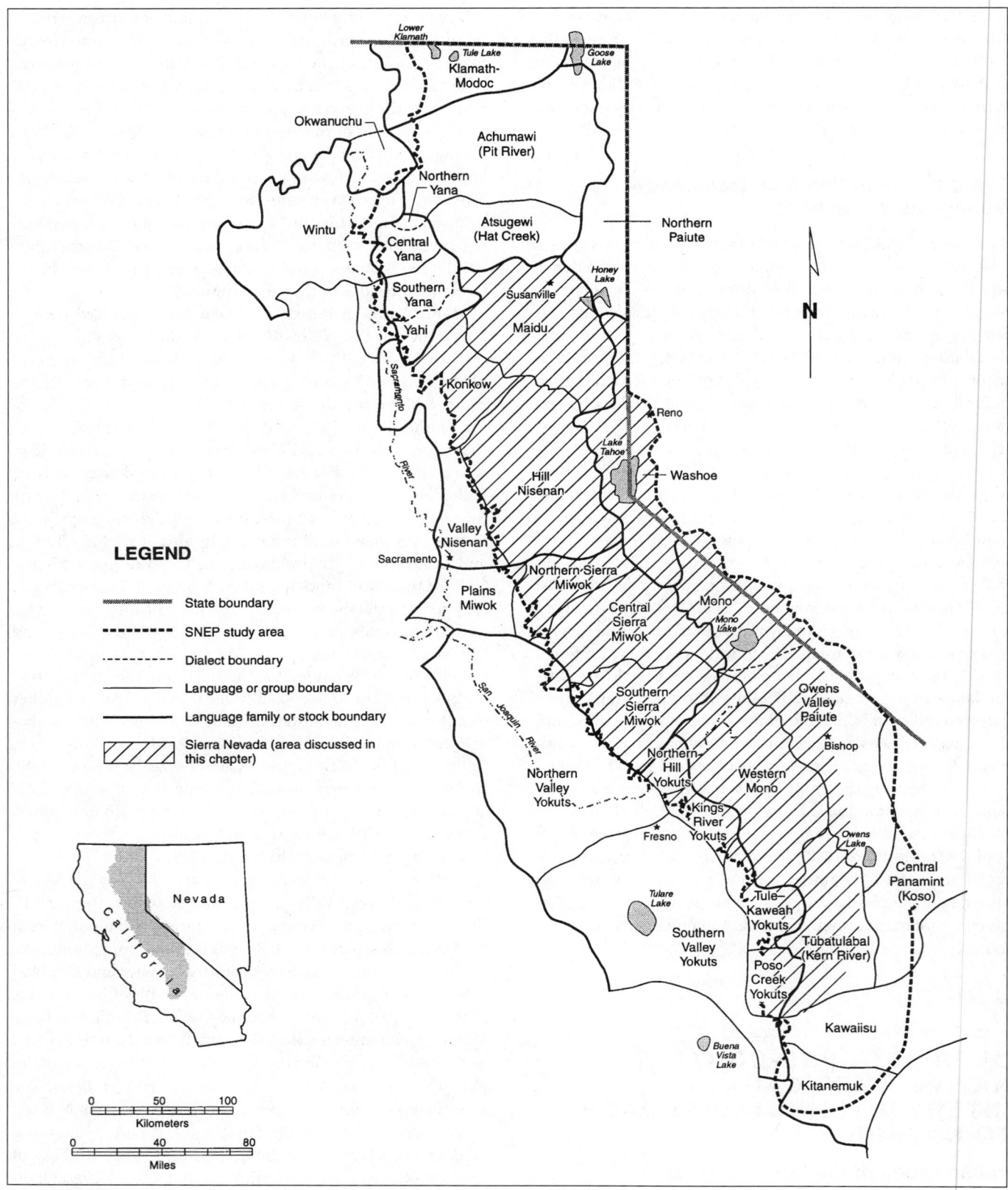

FIGURE 9.1

Tribal territories in the Sierra Nevada and adjacent regions, ca. A.D. 1800.

of montane valleys accented by Honey Lake, Washoe Lake, and Topaz Lake, and diverse biotic zones on the arid lands east of the Sierran crest, below Lake Tahoe (d'Azevedo 1986).

Native American Populations

The distribution of Native American populations in the Sierra Nevada was greatly influenced by environmental and cultural factors. On a regional scale, population densities were highest and "permanent" settlements most frequent at elevations below 1,000–1,250 m (3,300–4,100 ft); higher-altitude sites typically were occupied mostly during the warm season. Population densities tended to be substantially higher on the western side of the range than along the Great Basin rim, east of the Sierran crest. Within these broad patterns, populations were geographically diverse. In each locality such variables as terrain, biotic diversity and richness, availability of water, and access to toolstone, as well as the traditional land-use practices of the local society, affected carrying capacity of the land and thus human population levels. On a micro scale, the siting of individual camps or villages reflected such considerations as view, aspect, slope, drainage, insolation, vegetative cover, protection from wind, avoidance of cold sinks, proximity to water and economic resources, outcrops of bedrock suitable for use as mills, nearby trails, perceived flood or fire hazards, and perhaps defense, as well as the intended site function, number of residents, anticipated duration of occupation, and proximity to other settlements.

The number of residents at particular sites ranged from a few (e.g., several men in a hunting camp) to a few hundred in the larger villages. Intermediate in size were seasonal and special-purpose encampments. Late prehistoric/protohistoric Sierran peoples often were organized into "village communities," each consisting of a named, principal village under a chief or headperson and a number of smaller, tributary settlements (Kroeber 1962; Merriam 1967). The central villages of such communities were often situated near major streams in favorable settings within the lower Transition and upper Sonoran life zones.

Any estimate of aboriginal (i.e., pre-A.D. 1800) populations in the Sierra Nevada must be framed by caveats. Population levels fluctuated over time in response to paleoenvironmental changes (Moratto et al. 1978, 1988); different methods of estimation (e.g., reliance on historical accounts, ethnographic recollections, or ecologic models) yield divergent results (cf. Baumhoff 1963; Cook 1976b; Kroeber 1925; Merriam 1905); and even such "accurate" historical documents as U.S. War Department and Office of Indian Affairs records from the 1850s may not be reliable indicators of earlier population levels. As S. F. Cook (1955a, 70) has noted,

The depletion of population in the San Joaquin Valley [including the adjacent Sierra] between 1800 and 1850 was far greater than has been appreciated. . . . Warfare,

massacre, forced conversion, starvation, and exposure all took a tremendous toll of life, but the sweeping epidemics of the 1830s were even more devastating [see Cook 1955b]. Together, these forces destroyed in the aggregate fully 75 per cent of the aboriginal population.

Taking into account a wide range of information from early Spanish, Mexican, and other historical sources, Cook (1955a) estimated aboriginal populations of 7,600 for the Kaweah River drainage, 3,500 for the Merced, 9,100 for the Kings, 19,000 for the Mariposa area, Chowchilla, Fresno, and upper San Joaquin Rivers, and 4,150 for the "foothill strip," including lands of the Central and Northern Miwok. This yields a subtotal of 43,350 people in the southwestern Sierra Nevada. In the northwestern Sierra, populations of 1,050 for the Mountain Maidu and 7,400 for the combined Hill Maidu (Konkow) and Nisenan are estimated (Cook 1976b), giving a subtotal of 8,450. Adding roughly 500 for the Northern Paiute, 1,500 for the Washoe, 1,000 for the Owens Valley Paiute, and 500 for the Kawaiisu (Kroeber 1925) yields 3,500 as a subtotal for the eastern Sierra. Taken together, these estimates total 55,300.

This total, however, may be substantially lower than the actual native population of the Sierra Nevada prior to ca. 1830. Some of the estimates may fall short of the mark because of reliance on postepidemic observations. Even so, several historical accounts refer to large populations: James D. Savage, who was involved with numerous Sierran tribes before the gold rush, estimated in 1851 that 50,000–55,000 Native Americans lived in the area between the Tuolumne and the Kern Rivers; O. M. Wozencraft, a U.S. Indian commissioner, in 1852 set the native population of the area between the Yuba and the Mokelumne Rivers at 40,000, noting that old residents said the number had been twice as large in 1848; and Indian agent Adam Johnson in 1853 estimated that Sierran and Central Valley Indians totaled 80,000 (Cook 1955a). Although Cook characterized these accounts as "broad generalizations based largely upon subjective impression and applying to the years preceding 1847" (1955a, 33), they do suggest that the ethnographic population estimates are likely too low. Baumhoff's (1963) study of ecological determinants of population, showing that the "actual populations" of some groups were well below the numbers predicted on the basis of carrying capacity, also would seem to support higher estimates. The Central and Southern Sierra Miwok, for example, had predicted versus "actual" populations of 8,547 versus 4,410 and 8,503 versus 5,766, respectively (Baumhoff 1963). Moreover, the density of late prehistoric sites in many Sierran localities would suggest a level of occupational intensity greater than implied by ethnographic testimony. All things considered, 90,000–100,000 seems a reasonable approximation of the number of Native Americans living in the Sierra Nevada during the early 1800s. Ecological implications of this population level are discussed later in this chapter.

Native Americans Today

Although today their ancestral lands are occupied mostly by other peoples, each Native American polity in the Sierra Nevada has maintained a distinct ethnic identity. Their contemporary needs, goals, and worldviews have stemmed from a difficult past, including 200 years of Indian resistance to Euro-American appropriation of their land and natural resources (Cook 1976a; Heizer 1974; Hurtado 1988; Phillips 1993; Rawls 1984). Many of the native groups regulate their business and conduct financial affairs through formal tribal councils. Some tribes own and operate their own museums. Intra- and inter-tribal gatherings occur up and down the Sierra Nevada annually and are known variously as acorn festivals, Indian days, big times, and powwows. The current activities and lore of the Native Americans have emerged from a blending of ancient botanical knowledge and a sustained interest in their cultural heritage. Language is intimately tied with gathering and management knowledge. Most of the languages in the Sierra Nevada are still spoken, but some of the cultural groups have only one or two fluent speakers left (Hinton 1994). Individuals of both sexes and of all ages still gather plants. Uses of plants and animals that had been relinquished have regained importance among some families in recent years. Most of the tribes in the Sierra Nevada have an insignificant land base or none at all; consequently, they are forced to gather mostly on public lands. The loss of habitat—wetlands, overflow channels of streams, black oak–ponderosa pine mixed conifer forest, and so on—for culturally significant plants is extensive. Newly formed organizations such as the California Indian Basketweavers' Association (CIBA) have expressed concerns, on behalf of their members from different tribes, about such habitat loss and other issues facing those involved in traditional uses of the land.

Despite a turbulent history and subsequent acculturation, California Indian elders are still a substantial source of information about present and former traditional plant uses and management practices, and in some cases elders are still practicing plant management adjacent to their homes. Burning for cultural resources occurred "on the sly" on U.S. Forest Service (USFS) lands as late as the 1950s, and some traditional management is still conducted on several reservations and rancherias. Unfortunately, many plant ecologists and resource managers still distrust or discount this anecdotal information. Yet, some of the richest details of former resource-management practices have come from ethnographic interviews conducted this decade (Anderson 1993b). The accuracy of these accounts is verified through cross-referencing with testimony from other families, both within and between tribes. Oral histories are then combined with information from museum studies, ethnographic and ethnohistoric accounts, and the archaeological record to provide the most thorough reconstruction of past human activities on the land. Native American systems of knowledge about the environment have a great deal to teach resource managers. Some basket weavers have been involved in blending western and nonwestern knowledge systems in on-the-ground resource management on USFS lands (Anderson 1992; CIBA 1993; Lorri Planas, Choinumni/Mono, conversation with M. K. Anderson, 1994).

Past Economic Activities

To meet their basic needs, native peoples of the Sierra Nevada practiced such diverse subsistence activities as gathering, hunting, fishing, firewood use, and toolstone quarrying. A variety of greens, fruits, bulbs, corms, tubers, and mushrooms was gathered by each community. Staple foods included acorns from oaks (*Quercus douglasii, Q. chrysolepis, Q. kelloggii, Q. wislizenii, Q. garryana, Q. vaccinifolia, Q. berberidifolia*) and a number of small, hard seeds from native grasses and broad-leaved herbaceous plants. Additionally, native peoples procured deer, fish, small game, insects, and other animals (Barrett and Gifford 1933; Curtis 1924; DeQuille 1963; Merriam 1955; Powers 1976). This diversity of food resources was obtained by following an annual cycle of population movements that coincided with seasonal availability of specific resources; often this involved warm-season abandonment of villages in the foothills as populations dispersed to small, temporary camps at higher elevations (Barrett and Gifford 1933; Kroeber 1925). In addition to acquiring plants and animals for food, Native Americans gathered large quantities of plant material for firewood, basketry, cordage, and construction purposes. Native American relationships to the land were highly interactive. Areas were manipulated annually, biennially, triennially, or quadrennially to augment wild plant populations and create shifting mosaics of different vegetation types. Both small patches and extensive areas of vegetation were burned, and individual plants were pruned, dug, shaken, knocked, or weeded (Clara Charlie, Chukchansi-Choinimni Yokuts, conversation with M. K. Anderson, 1990; Bill Franklin, Sierra Miwok, conversation with M. K. Anderson, 1990; Grace Tex, North Fork Mono, conversation with M. K. Anderson, 1991).

Past Land-Management Practices and Ecological Consequences

Until recently, vegetation types in the Sierra Nevada were viewed as "natural," and their productivity was maintained through natural disturbance in the complete absence of human influence (Nichols 1989; Parsons et al. 1986). It is now recognized that many ecosystems in the Sierra Nevada evolved through significant human intervention (Blackburn and Anderson 1993; Lewis 1993; Wagner and Kay 1993). The ability of Native Americans to meet their economic needs was sustained not only through hunting, fishing, and gathering but also through a variety of horticultural techniques including burning, irrigating, pruning, selective harvesting, sowing, and weeding. These horticultural practices were exercised most commonly in five Sierran vegetation types: foothill woodland, chaparral, mixed conifer forest, riparian corridor, and meadow. Fire was the most important management tool, employed by the Native Americans to clear brush, maintain grasslands and meadows, improve browse for deer, enhance production of basketry and cordage materials, modify understory species composition in forests, and reduce fuel accumulation that might otherwise sustain intense fires. Ecological effects of horticultural techniques varied in time and space, depending upon the cultural objective and plant-community type. Some of the horticultural techniques used by native peoples of the Sierra Nevada and a general definition for wild-plant management are summarized as follows.

Wild-plant management is the human manipulation of native plants, plant populations, and habitats, in accordance with ecological principles and concepts, that effects a change (either beneficial or negative) in plant abundance, diversity, growth, longevity, yield, and quality to meet cultural needs (Anderson 1993a). Management techniques include:

Burning: applying fire to particular vegetation areas under specified environmental conditions and descriptors such as seasonality, fire-return interval, and dimensions to achieve select cultural purposes.

Irrigating: supplying select land areas with water by means of diversion and artificial channels.

Pruning: removing dead and living parts from native plants to enhance growth, form, or fruit and seed production.

Selective harvesting: harvesting in a discriminate, repetitive manner that leads to intended or unintended selection of traits, which in turn leads to evolutionary modifications such as enlargement of the favored plant part, reduction of the potential for reproduction by seed, or color changes in the fruit or seed.

Sowing: broadcasting seed collected from native plants onto an area, usually recently burned ground.

Tilling: removing earth in the harvest of underground perennial plant organs (e.g., roots, rhizomes, corms, bulbs), frequently followed by the subsequent dividing of these organs and leaving of individual fragments in the soil.

Transplanting: digging up a plant or a portion of a plant and moving it to another place.

Weeding: removing unwanted plant species near favored plant species.

There are three broad realms in which Native Americans acted as agents of environmental change:

1. *Dispersal agents.* Native Americans were intentional and sometimes inadvertent agents of plant dispersal that has rearranged the distribution of some species and created unusual plant distributions and polymorphisms.

2. *Agents of habitat modification.* Native Americans expanded and maintained suitable habitat in both time and space for desired species without necessarily altering character traits.

3. *Agents of genetic modification.* Native Americans modified the gene pools and genetic structures of plants through selective harvesting and transplanting. Over hundreds to thousands of years, specific genotypes of many intensively used plant species were selected by Native Americans and therefore probably still exhibit character traits that are adapted to small-scale human disturbance regimes.

The potential linkage between a culture's horticultural practices, uses of particular plant species, and selection pressures exerted on those species has not been sufficiently studied. For example, as a demonstration of combined dispersal and selection, the high variability in blue camas (Camassia quamash) in the Sierra Nevada is probably due to the trading and selective harvesting practices of different tribes (Susan D'Alcamo, conversation with M. K. Anderson, 1993). One possible study would be to compare the morphological variation in populations of a native species gathered in several different tribal territories with the differences in local harvesting and management regimes.

Past Land-Ownership Patterns

Native American societies recognized territorial boundaries and community ownership of land coupled with individual responsibility for resource conservation and use (Kroeber 1925, 1962). Increasing evidence shows that use and improvement of areas through cultivation gave a family or community exclusive use rights to that area. Thus, within each tribal territory there were numerous traditional collection sites for basketry material, acorns, clover, mushrooms, cordage fibers, and so on (Curtis 1924; Gayton 1948; Wilson 1972; Norma Turner, Mono, conversation with M. K. Anderson, 1992). Euro-American law and practice, imposed upon the Sierra Nevada during the mid-nineteenth century and subsequently, enshrined private ownership rights even at the expense of community interests. These sharp cultural differences are reflected in the land-use practices of pre- and post-contact populations.

Land-Use Ethic

Although Native American economic and management practices in the Sierra Nevada were diverse, they were nonetheless unified by a fundamental land-use ethic: to interact with nature respectfully and in ways allowing all life forms to coexist. This ethic transcended cultural and political boundaries.

It comprised spiritual, philosophical, and economic dimensions that encouraged sustained relationships between human societies and Sierran environments over spans of centuries or millennia. In Native American cosmologies, humans are viewed as part of the natural system; thus, all life forms are related to humans and must be treated with respect. Legends, ceremonies, songs, dances, and arts were and continue to be integrated parts of the spiritual systems, instructing the people in right and wrong behavior and the position and obligations of each person within the group (Swezey 1975). Land-use and land-management activities were guided by complex cultural rules, sophisticated knowledge of reproductive biology, and awareness of community ecology. Aldo Leopold's land ethic is most akin to native philosophy in that he advocated that humans should avoid both the dangers of overexploitation and the inactivity of preservation (Callicott 1990). One of the most provocative ideas found in Native American views is that human intervention in nature does not necessarily create disharmony. When Native American elders today are asked what has changed in the Sierra Nevada, they are apt to respond by saying simply, "No one is gathering or tending areas anymore." The idea that human use *ensures* an abundance and diversity of plant and animal life appears to have been an ancient one in the minds of native people, and there is very likely an ecological as well as a spiritual basis for this belief (Blackburn and Anderson 1993).

EVIDENCE FOR WIDESPREAD MODIFICATION OF SIERRAN LANDSCAPES: SPECIFIC QUESTIONS

Were the Technologies of Native Americans in the Sierra Nevada Capable of Creating Widespread Ecological Change?

Yes. Although most of the prehistoric tools (the digging stick, knocking stick, obsidian knife, seed beater, etc.) used in the Sierra Nevada appear simple and unable to affect vast areas, the fire-making kit allowed people to alter landscapes. Burning was probably the most widely employed, efficient, and significant vegetation-management tool used in the Sierra Nevada (Anderson 1994; Lewis 1993; Reynolds 1959). Knowledge and use of the slow match and torch recorded for most tribes gave native peoples the technological capability to burn either small patches or extensive tracts of vegetation in a systematic fashion. Frequent burning promoted a herbaceous understory vegetation within woodlands and coniferous forests. This continuous and sufficient fuel bed facilitated the burning of land of large areal extent. Felling trees with fire to promote type conversions was a capability of most tribes (Driver 1937; Driver and Massey 1957). Extensive trade networks in the Sierra Nevada promoted the exchange of seeds and other plant parts that could be propagated in new areas. For example, seeds were exchanged between families of Yokuts descent (Gayton 1948). In recent ethnographic interviews, Ruby Cordero (Chukchansi Yokuts) and Hector Franco (Wukchumni Yokuts) have described burning to promote seed crops. Additionally, after being burned, areas were sometimes sown with seed (Hudson 1901; Steward 1938).

Were Native American Settlement and Land-Use Patterns Repetitive and of Adequate Duration to Cause Permanent Effects on the Vegetation?

Yes. Indians have occupied the Sierra Nevada for at least 9,000–10,000 years (Moratto 1984; Moratto et al. 1988). It has been widely assumed until recently that Native Americans in the Sierra Nevada were "hunter-gatherers" who did not practice agriculture and whose environmental impacts were negligible. There is increasing archaeological, paleoecological, ethnographic, and ethnohistoric evidence, however, that human manipulations were regular, constant, and long term, causing cumulative and permanent effects in plant associations, species composition, and in the gene pools and genetic structures of species in a multitude of Sierra Nevada vegetation types (Anderson and Carpenter 1991; Blackburn and Anderson 1993; Anderson 1993a; Kilgore and Taylor 1979). This is not to say that particular land-use and resource-management activities persisted unchanged throughout the Holocene. Indeed, the archaeological record shows that population densities, land-use intensity, and specific economic practices did vary diachronically. Periods of notably intense cultural activity (e.g., ca. 7500–6000 B.C., 1000 B.C.–A.D. 700, A.D. 1300–1800) were separated by times of diminished populations and concomitantly reduced land use. During each interval, prevailing economic practices were applied over a span of centuries. The most recent period of intensive land use endured for some five hundred years before Euro-Americans entered the Sierra Nevada. This interval was long enough that Native American human activities caused substantial environmental effects.

Special areas were designated for basketry materials, bulb gathering, seed collecting, cordage-fiber harvesting, or greens picking and were shaped by continual long-term use and management (Aginsky 1943; Latta 1977; Voegelin 1938). Technologies such as basketry and cordage are extremely ancient fiber arts in North America; basketry fragments radiocarbon-dated to more than 10,000 years B.P. have been found in western North America (Adovasio 1974). These fragments demonstrate qualities that show that they were manufactured with the same techniques as those used for historic baskets. Presumably, fire was employed to stimulate the production of long shrub shoots wherever the basketry craft diffused in California. Additionally, management of gathering sites was a way of visually marking one's relationship with the area and was a signal for gaining land-use rights. Place names, ethnographic work with contemporary elders in different tribes, and the ethnohistoric and ethnographic literature all substantiate the fact that resources were gathered from long-term collection sites inherited through relatives (Gladys McKinney, Dunlap Mono, conversation with M. K. Anderson, 1992; Norma Turner, Mono, Dumna Yokuts, conversation with M. K. Anderson, 1991).

Was the Protohistoric Human Population in the Sierra of a Magnitude Sufficient to Cause Widespread Ecological Impacts?

Probably. The carrying capacity of Sierran environments for human populations varied significantly in space and time. By 1000 B.C. the west side of the Sierra Nevada was widely and intensively inhabited (Moratto 1984). The Native American population of the Sierra Nevada in A.D. 1800 was probably on the order of 90,000–100,000 (supra). There is no known archaeological evidence for larger numbers at any earlier time, although populations during the 1000 B.C. to A.D. 500 interval might have been comparable to those of late prehistoric times, ca. A.D. 1300–1800 (Moratto et al. 1988).

One measure of impact potential is population density. Kroeber (1939) calculated densities per 100 km^2 (39 mi^2) of 0–5 for the Northern Paiute; 10–25 for the Washoe, Owens Valley Paiute, and Western Mono; 25–45 for the Maidu, Konkow, and Nisenan together; 45–70 for the Sierra Miwok; and 70-plus for the Foothill Yokuts. Other estimates of population densities for Sierran tribes tend to be similar or higher (cf. Baumhoff 1963; Cook 1955a, 1976a, 1976b). Populations were not distributed evenly within any territory, but rather were concentrated near major streams in the upper Sonoran and lower Transition zones. Because of seasonal movements, a single community could affect environments in several localities at different elevations during the course of its annual cycle of dispersion and aggregation.

If one assumes a pre-contact Sierran population of about 100,000 distributed among settlements averaging, say, thirty-five residents each (five houses of seven residents each), then at any given time there would have been roughly 2,860 settlements, each of which would have required firewood, fish and game, vegetal foods, craft supplies, and construction materials for dwellings and sweat houses, ramadas, grinding booths, granaries, and, in principal villages, ceremonial lodges. Allowing for seasonal relocations and special-use camps, the number of sites occupied per year easily could have been 5,000–10,000. The magnitude of impact would have reflected not only the direct results of occupation per se (involving perhaps a few hectares per settlement) but also resource extraction, effects of predation, and intentional burning within a catchment of perhaps 5–10 km^2 (2–4 mi^2).

Not enough is known about the resource requirements and extent of land managed to meet those demands of each tribe's settlements. Quantitative models based upon detailed archaeological studies and analyses of museum specimens

gathered from fire-managed areas, as well as careful experimentation and replication, need to be developed to better understand the sustained resource needs of a typical, pre-contact Sierra Nevada community (Blackburn and Anderson 1993). One thing is clear: modern population levels and trends in the Sierra Nevada are unprecedented and already exceed those of pre-contact Native Americans by more than an order of magnitude.

Which Land-Use Activities Required the Highest Quantity of Plant Material from Managed Environments?

Basketry, cordage, firewood, and foods. These cultural use categories required gathering on a frequent, repetitive basis and demanded the collection of large amounts of plant materials from managed environments. For example, the basket-weaving industry required a large-scale effort to manage, harvest, size, cure, and weave plant materials into baskets for each village. This industry was at the very heart of Native American material culture in the Sierra Nevada. Specialized baskets were manufactured variously to serve as seed beaters, winnowing devices, burden packs, storage containers, cooking vessels for stone-boiling of mush, parching trays, bowls and cups, cradleboards, and fish traps, and for myriad other uses. Practicing the art of basketry demanded a steady, large supply of uniform plant materials for weaving. Hundreds of thousands of young shoots from different plant species were needed annually. These amounts were sufficiently large as to make opportunistic gathering (wherever one might find the right material) prohibitive. Thus, collecting basketry material was not happenstance, but was, rather, a sizable collective enterprise (table 9.1).

Great efficiency was needed to gather enough materials yearly to comply with the strict standards for the manufacture of many cultural items. Most of the basketry materials could not be used right away, but required a storage period to season them. This period varied from one to four years depending upon the plant species (Bates and Lee 1990; Margaret Mathewson, conversation with M. K. Anderson, 1992). Women had to plan ahead, gathering that year's new growth for a basket they might make two or three years later. To rely on natural fires from lightning to induce production of large numbers of desirable shoots would be risky, because lightning could strike in the wrong plant-community type, not strike in a location with suitable kinds of plant species, or hit too far away. Setting fires in the area where the plants grew was far more efficient. These facts support a burning regime that was very frequent, to keep shrubs at a young growth stage in order to obtain a continuous supply of a tremendous quantity of usable shoots for the making of many kinds of baskets.

Firewood, too, was required in large quantities. Domestic fires were used to singe game, braise meat, preheat earth ovens, heat stones for boiling acorn and other foods, raise the

TABLE 9.1

Comparison of numbers of useful shoots from unmanaged versus managed shrubs used for Western Mono basketry[a] (adapted from Anderson 1993b).

Basket Type	Plant Species Used	Shoots per Basket	Unmanaged Plants per Basket	Managed Plants per Basket
Burden	*Ceanothus cuneatus*	2	10 shrubs	1 shrub
	Rhus trilobata	1,200 (1.2 m each)	400 patches	12 patches
	Cercis occidentalis	25 (1.8 m each)	50 shrubs	1 shrub
Full-sized	*Rhus trilobata*	675	102 patches	6 patches
cradleboard	*Cercis occidentalis*	75 (1.8 m each)	150 shrubs	6 shrubs
	Ceanothus cuneatus	13	65 shrubs	1 shrub
Twined seed beater	*Ceanothus cuneatus*	2 (for rim)	10 shrubs	1 shrub
	Ceanothus cuneatus	188 (for warp and weft)	376 shrubs	15 shrubs
Seed gathering	*Ceanothus cuneatus*	2 (for rim)	10 shrubs	1 shrub
	Ceanothus cuneatus	376 (for warp and weft)	752 shrubs	31 shrubs
	Cercis occidentalis	50	100 shrubs	4 shrubs
Twined sifter	*Rhus trilobata*	1,000 (1.1 m each)	333 patches	10 patches
	Cercis occidentalis	25 (1.8 m each)	50 shrubs	2 shrubs

[a]Based on discussions with Norma Turner (Western Mono weaver). Management methods are pruning and burning.

Conversions:

Unmanaged *Cercis occidentalis*
1 1.8 m shoot/shrub
3 0.9 m shoots/shrub

Managed *Cercis occidentalis*
25 1.8 m shoots/shrub
25 0.9 m shoots/shrub

Unmanaged *Rhus trilobata*
10 short shoots/patch
3 long shoots/patch

Managed *Rhus trilobata*
100 short shoots/patch
100 long shoots/patch

Unmanaged *Ceanothus cuneatus*
1 rim shoot/5 shrubs
2 smaller shoots/shrub

Managed *Ceanothus cuneatus*
2 larger-diameter shoots/shrub (for rim)
10–15 smaller-diameter shoots/shrub (for warp and weft)

temperature in sweat houses, provide illumination after dark, and warm houses during cold weather. Fires were also used to fell and cut trees for house posts and rafters, to char post butts (as a wood preservative), and, among the Yokuts, Mono, and Paiute, to fire pottery (Gayton 1929; Liljeblad and Fowler 1986). In addition, many Sierran groups cremated their dead (Gifford 1955; Gould 1963; Kroeber 1925). Thus, firewood use was substantial. Assuming a pre-contact average of 2,860 settlements with five houses each (supra), and allowing, as a guess, 10 kg (22 lb) of daily firewood use per household, Native Americans would have burned some 143,000 kg (314,000 lb) of fuel each day. Annually, this would have amounted to 52,195 metric tons (51,165 tons avoirdupois). Further assuming that 5,000–10,000 sites were occupied each year, the average fuel consumption per settlement would have been roughly 5.8–11.5 metric tons (5.2–10.4 English tons). Some of the larger villages, with 300–500 members, might well have collected 250 metric tons or more of firewood annually. Such quantities not only would have reduced the fuels available to sustain natural fires, but also might have depleted supplies of firewood in some places sufficiently to require people to relocate.

Which Land-Use Activities Had the Greatest Impact on Sierran Plant Communities?

The single most important reason mentioned by Native American elders when asked why their ancestors burned in the Sierra Nevada was to keep the underbrush down to prevent a large, devastating fire (Clara Charlie, Chukchansis-Choinumni, conversation with M. K. Anderson, 1991; Bill Franklin, Sierra Miwok, conversation with M. K. Anderson, 1990; Ron Goode, North Fork Mono, conversation with M. K. Anderson, 1991). Tragically and ironically, many of the elders interviewed had lost their homes to fire—including precious baskets, mortars and pestles (cracked from the fire intensities), and other valuable cultural items—because of fuel accumulations on adjacent public lands due to fire-suppression policies. Accounts of past burning to keep the brush down are rich and varied:

My great aunt and mother talked about how the land was burned. If there was brush, they'd burn in the ponderosa pine and sugar pine areas. I remember there wasn't the tall brush that there is now. It's hopeless. They've let it go for so long. So when it does burn it goes and goes and kills the big trees. When they'd set the fires, it wouldn't hurt the trees.

They'd burn from the bottom of the slope. They would burn in the fall after rains. They would touch off any of the brush. It would burn some of the new needles off but it wouldn't burn way down through the duff like it does with the controlled burning today. They wouldn't burn the whole area, but anywhere it needed it. (Nellie

Williams, North Fork Mono, conversation with M. K. Anderson, 1991).

Maria Lebrado used to burn the hills on her property. The white man sure ruined this country. It's turned back to wilderness. In the old days there used to be lots more game—deer, quail, gray squirrels, rabbits. They burned to keep down the brush. The fires wouldn't get away from you. It wouldn't take all the timber like it would now. Burns were started in October, November, or December, not in January at the bottom of the slope. They burned every year. The fires wouldn't get up in the trees. There wasn't enough vegetation to get up in the trees. The plants were widely spaced. It wouldn't scorch except a few trees. They never talked about burning in the giant sequoias. They used to burn the high country in Yosemite and Crane Flat around 6,000 ft elevation. (Jim Rust, Southern Sierra Miwok, conversation with M. K. Anderson, 1989).

Clearly "burning to clear out the brush" was extremely important among native cultures. When analyzed, this purpose was perhaps the most significant reason of all for burning. Indians of the Sierra Nevada were very much aware of the double-edged sword of fire—that it could be beneficial or harmful to plant resources. For example, Native Americans actively managed vegetation patterns with fire to prevent intense fires that would promote tree scorching, which would harm valuable plant resources such as black oaks (Anderson 1993b).

A severe fire in a tribal territory would mean not only immediate loss of property, resources, and perhaps lives, but also disaster for the long-term well-being of a community. A catastrophic forest fire of the kind witnessed during the last several years, for example, could destroy hundreds or thousands of hectares of important game habitat and plant food resources. If many of the foothill pines, black oaks, sugar pines, and blue oaks were destroyed at important gathering sites, a substantial portion of the food supply would be lost. "Burning to keep the brush down" provided the environmental context within which more localized burning could then be done for specific cultural purposes. Frequent burning was the insurance policy against annihilation of important gathering and village sites.

That there were large areas of impenetrable growth in the Sierra Nevada in the middle to late 1800s is undeniable (Dudley 1896; Perlot 1985). Had the Indian burning patterns already been largely disrupted even before the arrival of the cattlemen, gold miners, and earliest settlers? Was the native population drastically reduced (because of exotic diseases), and was the brush therefore more widespread than during pre-contact times? Some scientists have argued that the population of California's Indians was not large enough and that they were not technologically capable of setting huge portions of California on fire (Burcham 1959; Clar 1959). How-

ever, if each pre-contact Indian household had burned only 10 hectares (25 acres) per year, about 143,000 ha (353,000 acres) of the Sierran landscape could have been altered annually, and many times more than this during the multiyear fire intervals. Accurate estimates of the areal extent of indigenous burning will require far more intensive studies than have yet been undertaken. Detailed studies of late prehistoric and early historic aboriginal populations in the Sierra Nevada are particularly needed.

How Were Selected Plant Species Affected by Protoagricultural Intervention?

Native Americans in the Sierra Nevada in many cases selected plant species that thrive under repeated disturbance. Cultural groups used a wide variety of plant species for many different products, but they relied heavily on a small subset of the total Sierran flora to meet their major needs. The cultural use categories that required continuous gathering of large amounts of plant parts are building construction, firewood, basketry, cordage, and foods. The understanding, exploitation, and modification of vegetative or asexual reproduction of plant species were extremely important to Indian subsistence economies. Vegetative reproductive structures have evolved with environmental disturbance in the form of flooding, fire, and small mammal and large mammal (grizzly bear, elk) activity, and human perturbations, therefore, frequently mimicked such natural disturbances. Multiplication and selection were often from clones. According to Sauer (1952), an individual plant with strong vegetative reproductive mechanisms might be divided and multiplied indefinitely. Vegetative reproduction exploited by Native Americans is of six major forms: offsets, tubers, stolons, perennial creeping root stocks, adventitious and epicormic shoots, and rhizomes. The new plant is an identical reconstitution of the parent rather than variant progeny (Sauer 1952). Native Americans gathered vegetative reproductive parts and progeny and maintained the parent plant in situ. Other gathering strategies that ensured long-term, repetitive collection in the same areas were gathering of sexual reproductive parts with maintenance of parent plant in situ and gathering of sexual reproductive parts with some seed replacement.

Indian Disturbance Regimes: Some Examples

Basketry-Production Systems. The adaptive significance of vegetative reproduction in shrubs has long been a major topic of inquiry by ecologists and evolutionary biologists (Keeley 1986; Naveh 1975; Wells 1969). Within the native flora of the Sierra Nevada are numerous species that display adventitious and epicormic sprouting capability (Kauffman and Martin 1990). All native groups in the Sierra Nevada burned and/or pruned areas in mixed conifer forests, riparian areas, oak woodlands, and chaparral to promote the growth of adventitious shoots and epicormic branches of native shrubs such as sourberry *(Rhus trilobata),* willows *(Salix* spp.), redbud *(Cer-*

cis occidentalis), and hazelnut (Corylus cornuta var. californica) (Fowler 1986; Gamble et al. 1979; McMillin 1963; Potts 1977; Clara Charlie, Chukchansi-Choinumni Yokuts, conversation with M. K. Anderson, 1990; Ruby Cordero, Chukchansi Yokuts-Sierra Miwok, conversation with M. K. Anderson, 1991; Amy Rhoan, Paiute, conversation with M. K. Anderson, 1990). All these species are believed to have displayed such growth behavior long before human management. In such cases, human management merely expanded the suitable ecological conditions that favored such growth. Most fires were set in the fall, after one or two rains, and they were set frequently (one- to several-year intervals). These fires were ignited from the bottom of the slope and were of an unknown areal extent, but probably the cumulative acreage was substantial given the density and dispersion of humans in the Sierra and the fact that large amounts of young growth were required for each village (Anderson 1993b).

Basketry was a highly developed technology in the Sierra Nevada, and the tradition is still maintained today. Historically, the use of baskets was so central to daily living that it represented 50% of the plant material culture (excluding construction materials) of the sixty or so tribes in the state (Anderson 1993a). One medium-sized cooking basket, for example, could take several thousand redbud first-year shoots to complete. The numbers of young shoots occurring "naturally" on wild shrubs are very few, justifying the need for frequent management. Native Americans set fires in ways that perpetuated native shrub species having protected, subterranean plant organs, which allowed for subsequent, in situ development. After the fires were set, hundreds of thousands of first-year shoots of various native shrubs were harvested in the following fall, winter, or early spring. Young growth was highly valued by weavers because it displayed such physiological and morphological features as anthocyanins, uniform cell density, flexibility, straightness, absence of lateral branching, and long length, which facilitated optimal construction of baskets. Additionally, young growth lacked insect or pathogen activity that would weaken basketry material (Anderson 1991).

Food-Production Systems. Leaves for greens, fruits, mushrooms, and bulbs were the edible plant parts that were managed for with fire in late summer to late fall by tribes throughout the Sierra Nevada to maintain their quality and quantity.

Fruits: Burning of chokecherries, manzanita berries, strawberries, and elderberries has been recorded among the Maidu, Foothill Yokuts, Western Mono, and Miwok tribes to increase fruit production, thin dense shrub canopies, reduce insect activity by eliminating old wood (Jewell 1971, as quoted in Roper Wickstrom 1987; Lydia Beecher, Mono, conversation with M. K. Anderson, 1991; Hector Franco, Wukchumni Yokuts, conversation with

M. K. Anderson, 1991; Avis Punkin, North Fork Mono-Miwok, conversation with M. K. Anderson, 1991).

Greens: Burning of herbage for better wild crops was recorded among the Chukchansi Yokuts; Western Mono; Southern, Central, and Northern Miwok (Aginsky 1943) to promote palatable growth, increase seed production, extend the gathering tract, and keep greens collections areas free and open. For example, clover (Trifolium spp.) patches were burned in Wukchumni Yokuts territory in October and November and in North Fork Mono territory (Rosalie Bethel, North Fork Mono, conversation with M. K. Anderson, 1991; Hector Franco, Wukchumni Yokuts, conversation with M. K. Anderson, 1992).

Mushrooms: Mushroom patches were burned by the Western Mono to improve quality and promote abundance. Species include Morchella elata, Peziza spp., Amanita spp., and Ramaria spp. (Goode 1992; Hazel Hutchins, Mono, conversation with M. K. Anderson, 1992; Nellie Williams, North Fork Mono, conversation with M.K. Anderson, 1991; Dave Bowman and Ed Bowman, Wobonuch Mono, conversation with M. K. Anderson, 1991).

Bulbs, corms, and tubers: Areas were burned by the Southern Sierra Miwok, Western Mono, and Northern Hill Yokuts to reduce competitive shrubs and grasses, recycle plant nutrients, heighten the size and quantity of underground swollen stems, and keep areas open to maintain these crops. Species included Perideridia spp., Sanicula spp., Brodiaea spp., and Allium spp. (Baxley 1865; Lydia Beecher, Mono, conversation with M. K. Anderson, 1991; Ruby Cordero, Chukchansi Yokuts-Miwok, conversation with M. K. Anderson, 1991; Ella McSwain, North Fork Mono, conversation with M. K. Anderson, 1991).

Seeds: Seed-collection sites were burned by the Western Mono, Paiute, Maidu, Nisenan, Northern Hill Yokuts, and Sierra Miwok to eliminate insects and diseases, recycle nutrients, keep open areas within forests and dry montane meadows, eliminate weed competition, augment seed production, and eliminate detritus of perennial grasses. Species included Astragalus bolanderi, Lathyrus sulphureus, Pickeringia montana, Wyethia spp., Salvia columbariae, and Calandrinia ciliata (Driver and Massey 1957; Gayton 1948; Hudson 1901; Kroeber 1925; Anonymous elder, North Fork Mono, conversation with M. K. Anderson, 1991; Hector Franco, Wukchumni Yokuts, conversation with M. K. Anderson, 1991).

Cordage-Production Systems. Cordage can be defined as "the twisting together of separate fiber strands into a single, long twined string or rope" (Mathewson 1985). Making of string or cordage is perhaps the oldest fiber art in America (Adovasio 1974). Native peoples probably brought cordage technology with them when they first entered California. The most important cordage-fiber plants used by native peoples in the Sierra Nevada were Indian hemp (*Apocynum* spp.) and milkweed (*Asclepias* spp.). These two genera contain herbaceous species with stems that are composed of excellent "bast" fibers. These fibers were collected, extracted, and manufactured into many items, including nets for fishing, deer nets, rabbit nets, netting bags, tump lines, slings, flicker feather head bands, hair nets, feather capes, feather skirts, belts, cord belts for women's aprons, and bow strings.

Herbaceous plants that contained desirable fiber were gathered primarily in the late fall or winter when the stalks had died back (Barrett and Gifford 1933). Cordage plants were periodically burned in the fall to decrease accumulated dead material, provide increased access for harvesting, allow greater sunlight to the new growth, and recycle nutrients to the soil. Plants were reputed to grow straighter and taller when burned (Peri et al. 1982; Rosalie Bethel, Mono, conversation with M. K. Anderson, 1991; Hector Franco, Wukchumni Yokuts, conversation with M. K. Anderson, 1991). Large quantities of Indian hemp and milkweed were harvested to make different cultural items (table 9.2), suggesting that the cumulative acreage burned to maintain productive collection sites was probably substantial. For example, a 12 m (40 ft) deer net made by the Sierra Miwok would require 2,134 m (7,000 ft) of string, or 35,000 plant stalks (Craig Bates, conversation with M. K. Anderson, 1992).

Native American Resource Management at Different Levels of Biological Organization

Horticultural techniques were applied at different levels of biological organization. Thus, the ecological consequences of these techniques would register at the following scales:

Organism Level. Individual plants were manipulated through spot burning and pruning to enhance production of a desired plant part. For example, single shrubs of button-

bush (*Cephalanthus occidentalis*) and willow (*Salix* sp.) were pruned for arrow-shaft material, elderberry (*Sambucus mexicana*) shrubs were coppiced for musical instruments, and brush was piled on individual shrubs of maple, redbud, and oak and set on fire to induce long shoots for basketry and looped stirring sticks (Anderson 1993b).

Population Level. Stands of bunchgrass (e.g., *Muhlenbergia rigens*) for basketry, herbaceous plants for cordage (e.g., *Apocynum cannabinum*), edible plants for greens (e.g., *Trifolium* spp.), seeds (e.g., *Madia* spp., *Wyethia* spp.), and corms, tubers, and bulbs (e.g., *Perideridia* and *Sanicula* spp.) were set afire to enhance quantity and quality, reduce plant competition, and keep surrounding vegetation from encroaching (Anderson 1993b). Populations of blue dicks (*Dichelostemma capitatum*) and yellow nut grass (*Cyperus esculentus*) were irrigated in Owens Valley by the Paiute (Lawton et al. 1976).

Plant-Community Level. Vegetation dominated by foothill woodland or coniferous forests was managed for maximum complexity of the vertical structure to encourage a variety of plant species in the understory. Disturbance, in the form of burning and digging, was frequent and of an intensity and scale to prevent monopolization of resources by one or a few species.

Landscape Level. Native Americans introduced burning to maximize plant-community diversity. Particularly important was promoting pioneer stage and fire subclimax plant communities. "Burning to keep the brush down" was a maxim adhered to by all Sierran peoples. Burning expanded special plant-community subtypes such as black oak–ponderosa pine, prolonged the life of dry meadows, and cleared out reed-choked marshlands (McCarthy 1993; Hector Franco, Wukchumni Yokuts, conversation with M. K. Anderson, 1992). Fire mosaics promoted an abundance of water in numerous springs and creeks (Duncan 1964; James Rust, Southern Si-

TABLE 9.2

Quantities of cordage material (*Apocynum* and *Asclepias* spp.) gathered for various cultural items by Native Americans of the Sierra Nevada (adapted from Lindstrom 1992; Anderson 1993a).

Tribe	Cultural Item	Use	Dimensions	Total Cordage Length	Stalks Gathered (Number)
Washoe and Northern Paiute	Gill net	Fishing	1.6 mm 2-ply 30 m x 1.4 m x 38 mm mesh (1/16" 2-ply 100' x 4.5' x 1.5" mesh)	3,665 m (12,022 ft)	60,110
Washoe and Northern Paiute	Bag net	Fishing	1.6 mm 2-ply 0.75 m x 0.75 m x 0.75 m x 25 mm mesh (1/16" 2-ply 2.5' x 2.5' x 2.5' x 1" mesh)	270 m (885 ft)	4,425
Washoe and Northern Paiute	A-frame dip/lift net	Fishing	1.6 mm 2-ply 2.1 m sq. x 1.2 m (x 4 panels) x 25 mm mesh (1/16" 2-ply 7' sq. x 4' [x4 panels] x 1" mesh)	2,405 m (7,890 ft)	39,450
Sierra Miwok	Feather cape	Ceremony	1.6 mm 2-ply 44.5 mm mesh (1/16" 2-ply 1.75" mesh)	30 m (100 ft)	500
Sierra Miwok	Deer net	Hunting	3.2 mm 2-ply 12.2 m x 1.8 m x 102 mm mesh (1/8" 2-ply 40' x 6' x 4" mesh)	2,134 m (7,000 ft)	35,000

erra Miwok, conversation with M. K. Anderson, 1989). Burning at higher elevations was for the expressed purpose of removing shrub and duff layers, promoting a more tightly assembled snowpack. This dense snowpack melted off more slowly, reducing flooding and causing ephemeral creeks and streams to run longer in the summer (Jewell 1971). Strategies for maintaining ecosystem integrity included

- hand clearing and burning detritus that might alter moisture and soil conditions—which would encourage a new array of plant species to colonize

- hand weeding and burning to maintain ecotones around special plant-community types such as meadows

- not obstructing, but rather maintaining and encouraging recurrent changes in water level and scouring along streams and marshes

Once exotic herbaceous species had begun to spread into the Sierra Nevada, they were readily incorporated into the ethnobotanies of the tribes. For example, wild mustard (*Brassica* spp.) leaves were consumed by the Maidu, Yokuts, and Tübatulabal (Duncan 1964; Gayton 1948; Latta 1977; Voegelin 1938). Fillaree (*Erodium cicutarium*) greens were eaten by the Maidu (Duncan 1964). Wild oat (*Avena fatua* and *A. barbata*) seeds were prized by the Sierra Miwok, Yokuts, and Tübatulabal (Barrett and Gifford 1933; Gayton 1948; Latta 1977). Brome (*Bromus rigidus*) seeds were added to the Miwok diet (Barrett and Gifford 1933), and *Echinochloa crusgalli* and *Polypogon* seeds were eaten by Tübatulabal (Voegelin 1938). Tribes burned areas to promote the growth and abundance of native plants for edible seeds and greens. After the introduction of exotics, burning probably continued. Because many of these exotic species thrive after periodic burning, indigenous burning perhaps contributed to expansion of the range and distribution of these aliens.

ECOLOGICAL CONSEQUENCES OF REMOVING NATIVE AMERICANS FROM TRADITIONAL ECONOMIC AND LAND-MANAGEMENT ROLES

There is a growing awareness that the decline of biodiversity in the United States may be tied directly to past fire-suppression policies of land-managing agencies (National Research Council 1992). New studies concerned with rare and endangered species in the Sierra Nevada (Boyd 1987; Verner et al. 1992) are concluding that frequent fire is necessary to the health and maintenance of habitat for certain endangered biota. Fire-suppression policies on public lands were based on a perception of fire as a destructive force without an understanding of the dynamics of fire and its ecological role; hence, those policies constituted a real threat to the very resources they were intended to protect. Fire is now a widely accepted management tool in conservation biology (National Research Council 1992). But prescribed-burning programs on public lands adjacent to urban areas are hampered by increasing fire risk, threatening human safety and valuable property. Additionally, when prescribed-burning programs are implemented, they are usually done with little or no understanding of the former role of Native Americans in setting fires and creating other kinds of human disturbances. In this light, some scientists now recognize that wildfires in the Sierra Nevada often are more severe and larger than were the wildland fires in aboriginal times and that, therefore, the wildland ecosystems are also at risk (Martin and Sapsis 1992).

Most of the plants useful to Sierran tribes are highly shade intolerant and qualify as early- to mid-successional species. That these early stages were most useful for indigenous needs has been pointed out by previous studies (Lewis 1993; Reynolds 1959). Gaps or grassy openings were created, maintained, or enlarged within diverse plant communities, resulting in many "patches" of plants in varying successional states. Human disturbance at gathering sites was a regular element of the system. For example, fire was used to maintain patches of deergrass (*Muhlenbergia rigens*) for basketry within mixed conifer forests and chaparral areas; patches of edible native grasses and forbs (*Fragaria californica, Madia* spp., *Salvia columbariae*) within oak woodlands and mixed coniferous forests; and patches of edible bulbs, corms, and tubers (*Perideridia* spp., *Sanicula* spp.) in the dry montane meadows, open understories of coniferous forests, and openings in chaparral (Anderson 1993b). The result was that plant diversity was maximized.

The heterogeneity of ecological communities was expanded through indigenous manipulations. Mixed conifer forests and oak woodlands were often managed for maximum complexity of the vertical structure to encourage a variety of plant species in the understory. Thus, woodlands and forests often exhibited widely spaced trees, giving better light interception and ultimately leading to an increase in species diversity in an area (Huston 1994). Frequent burning recycled nutrients, destroyed insects and diseases, and promoted a lush understory vegetation that provided an important food supply for Sierran tribes. A variety of understory plant species supported an abundant and diverse insect and small-mammal population, providing a valuable food source to the California spotted owl (Verner et al. 1992).

Old growth in mixed conifer forests in the Sierra Nevada featured large-diameter, healthy individuals, 12–18 m (40–60 ft) apart. The open-growth architecture made these trees more drought tolerant and disease and insect resistant than those of our overstocked forests today. Native grasses, promoted through burning, created a permeable forest soil surface that checked surface erosion. Soil fertility was enhanced by continuously decomposing feeder roots. Downed logs and snags were left intact by light surface fires and supplied nutrients, wildlife habitat, and moisture reservoirs (Martinez 1993). The tree plantations and second-growth forests in many parts of the Sierra Nevada today are structurally and biologically less diverse than natural forests under Native American burning regimes and contain impoverished faunas (cf. Mayer and Laudenslayer 1988; Verner and Boss 1980).

Ecologists hypothesize that plant communities subjected to intermediate levels of disturbance size, frequency, and intensity exhibit high levels of species diversity and high productivity (Connell 1978). The emerging subfield of "patch dynamics" in the discipline of plant ecology recognizes the key role of disturbances such as windstorms, lightning fires, lava flows, and modern human interventions in directing the successional patterns and evolution of plant populations (Mooney and Godron 1983; Pickett and White 1985). It is proposed that the Native American role in creating these "patches" in the landscape was considerable, and in the absence of native burning practices these patches are now undergoing accelerated successional changes.

Indigenous Knowledge and Rare and Endangered Plant Species

Certain plants integral to traditional cultures in the Sierra Nevada are now on rare and endangered or uncommon species lists assembled by the California Native Plant Society. These include such species as Pringle's yampah (*Perideridia pringlei*), Kaweah brodiaea (*Brodiaea insignis*), and coyote thistle (*Eryngium vaseyi*) (Zigmond 1981; Hector Franco, Wukchumni Yokuts, conversation with M. K. Anderson, 1992). The rare and endangered status of plant species is often attributed to habitat fragmentation and habitat loss due to development. Another tack worth investigating is the role that indigenous use and management played in maintaining these plant populations. In the absence of these former human disturbances, plant populations may have declined.

Cultural knowledge of native peoples may be useful in re-storing and managing other rare and endangered plants in the Sierra Nevada such as three-bracted onion (*Allium tribracteatum*) and Small's southern clarkia (*Clarkia australis*). Although we have no evidence that these species were used by Native Americans in the Sierra Nevada, we know that other species of the same genus were gathered and managed. These techniques may be transferable, across species of the same genus, and are worth investigating (table 9.3). For example, the North Fork Mono formerly burned common *Wyethia* spp. to maintain seed production (Rosalie Bethel, North Fork Mono, conversation with M. K. Anderson, 1991). Hall's wyethia (*Wyethia elata*) is uncommon, and El Dorado County mule ear (*Wyethia reticulata*) is endangered (Smith and Berg 1988). Both species occur in the Sierra Nevada in habitat types similar to those of the more common species. As fire cycles are restored to populations of these species, knowledge of Native American objectives for management of common *Wyethia* spp. and how Indians changed the frequency and intensity of fires may be integral to successful modern wildland management and restoration of these less common species.

DISCUSSION, CONCLUSIONS, AND RECOMMENDATIONS

Comparative research on how natural resources were used, maintained, and influenced by different native groups in the Sierra Nevada is useful for developing objectives and methodologies for managing, conserving, and restoring wildlands (Anderson 1993a; Gomez-Pompa and Kaus 1992). Management of nature preserves and wilderness areas will have to involve continued human intervention. Land managers need to fully understand vegetation dynamics, including the role of disturbances (Sierra Nevada Research Planning Team 1994). Native peoples have to be recognized as a contributor to the

TABLE 9.3

Possible application of Native American use and management techniques for enhancement of uncommon, rare, or endangered plant species populations.

Uncommon, Rare, or Endangered Species	Other Species in Genus Known to Be Managed	Tribe	Part Used	Use	Management Techniques
Allium tribracteatum	Common *Allium* spp. (e.g., *Allium validum*)	Western Mono	Bulb	Food	Tilling/burning
Clarkia australis	*Clarkia purpurea* ssp. *purpurea*	Central Sierra Miwok	Seed	Food	Sowing/burning
Perideridia parishii ssp. *latifolia*	*Perideridia bolanderi*; *P. gairdneri*; *P. kelloggii*; *P. parishii*	Northern Hill Yokuts; Sierra Miwok; Western Mono	Tuber	Food	Tilling/burning
Trifolium barbigerum var. *andrewsii*	Common *Trifolium* spp.	Northern Hill Yokuts	Leaf	Food	Burning
Wyethia elata and *W. reticulata*	*Wyethia helenioides*; *W. mollis*	Western Mono	Seed	Food	Burning

dynamics of ecosystem development. Similar to fires and floods, the cultivation techniques and harvesting strategies of indigenous peoples were types of disturbances that contributed to changes in structure and function of the vegetation. Understanding their past role in vegetation dynamics requires knowledge of the diversity and complexity of proto-agricultural, native land-management systems as well as the sophisticated, traditional lore upon which they are based (Soulé and Kohm 1989).

Some traditional wildland-management systems combine high species, structural, and temporal diversity, efficient nutrient cycling and energy flow, and intricate biological interactions. Such complexities have been selected over a long period in response to a wide array of cultural demands. These systems are essentially waiting to be "rediscovered" and analyzed (Anderson 1994).

Management and restoration of the Sierra Nevada for such objectives as protecting soil and water resources, maintaining wildlife habitat, and preserving biological diversity must be grounded in historical research and not rest on the illusion that the prehuman ecosystems are still intact and self-sustaining. Accurate reconstructions of interactions between native people and the natural environment in the Sierra Nevada and attempts to quantify the effects of indigenous horticultural practices on vegetation dynamics will require highly qualified, interdisciplinary teams of social, physical, and biological scientists working cooperatively with contemporary Native Americans in specific regions. Methodologies for collecting data would include oral interviews, archaeobotanical remains, and analysis of pollen, charcoal deposits, fire scar tree rings, museum artifacts, and written accounts, allowing for the independent cross-checking of conclusions (Crumley 1994). To date, these types of comprehensive studies are rare in the Sierra Nevada. One such study uses archaeological data, charcoal concentrations, and pollen cores to examine a long-term environmental change in Yosemite Valley (Anderson and Carpenter 1991). One of the biggest challenges will be to find more effective and creative ways to blend indigenous knowledge and scientific knowledge systems (DeWalt 1994). Native American experience with resource management of wildlands could be combined with theories of population biology and biogeography to develop new approaches and methods for preserving species (Primack 1993).

Future Research Priorities

To begin developing the information upon which innovative management strategies can be based, the following studies are recommended.

1. Determine whether fire and other vegetation-management methods used by Native Americans should be reintroduced. Set up a series of field experiments in the Sierra Nevada to simulate indigenous horticultural practices and harvesting strategies, and assess the interrelations and impacts of such cultural practices on individual plants, populations, communities, and ecosystem characteristics and dynamics.

2. Document knowledge systems of tribal elders. Conduct more in-depth ethnographic studies with Indian elders to ascertain details of former wildland-management practices in different plant-community types in the Sierra Nevada. Highest priority should be given to use patterns and knowledge systems that are disappearing most rapidly among the elder populations.

3. Reconstruct vegetation. Provide an accurate estimate of the understory plant species composition of late prehistoric forests in the Sierra Nevada using phytolith analysis, ethnographic interviews, early historical landscape descriptions, historical photographs, and early herbarium specimen collections.

4. Estimate indigenous populations. Develop a realistic prehistoric human-population estimate for the Sierra Nevada based upon early historical accounts, carrying-capacity estimates, archaeological site record analysis, land-use/settlement models, census data, and disease-spread models.

5. Investigate the significant prehistoric developments and their impacts. Measure and evaluate the ecological impacts of significant prehistoric developments—human entry, hunting (predation), reliance on acorns, use of bedrock mills, introduction of the bow and arrow, and exchange systems on the Sierra Nevada bioregion.

6. Calculate managed plant material quantities. Extrapolate from the numbers of adventitious shoots, flower stalks, and rhizomes needed for each basket type and herbaceous stems needed for each cordage item to the annual needs and landscape impacts for an average-sized village in three tribal territories in the Sierra Nevada.

7. Assess the importance of different plant species in historic basketry of tribes. Devise diagnostic features to accurately assess the identification of plant species used in the baskets of Sierra Nevada tribes in museum collections. Identify plant species used in different basket types. Rank the importance of plant species used in basketry by each tribe in the Sierra Nevada, and reconstruct major basketry complexes.

8. Assess the importance of different plant species in the historic diets of tribes. Reconstruct the major food complexes of different Sierran tribes through analysis of museum ethnobotanical collections, survey of existing literature, ethnographic interviews, and archaeological findings.

9. Document habitat loss of culturally significant plants. Inventory the native plant species that are useful to contemporary Native American cultures in the Sierra Nevada, and assess the habitat loss of culturally significant plants.

10. Compare life-history traits and habitat requirements of rare and endangered plant species and related common species (same genus) that were managed by Sierran tribes.

11. Investigate relationships between Native American economic practices and the faunas of the Sierra Nevada. Studies to date have emphasized Native American management of plant resources. Similar studies are needed to investigate the direct and indirect effects of traditional Native American economic practices on the nature and quantitative aspects of faunal assemblages in the Sierra Nevada.

12. Activate a regional study with an interdisciplinary team. Combine archaeological, ethnographic, paleoecological, fire-history, and museum research to yield a better understanding of the resource and management needs of prehistoric tribal villages in diverse Sierran localities. With this detailed information, it would be possible to better estimate the amount of "managed" acreage that would be needed to meet resource requirements of villages in the entire region over long periods of time.

Education, Planning, and Management Proposals

1. Establish an advisory council (or several regional councils), including Native Americans and specialists in such fields as ethnography, ethnobotany, ecology, archaeology, and ecosystems management, to assist land-managing and land-permit agencies in any general planning, zoning, and site development that could significantly affect Sierran ecosystems.

2. Recognize Indian-set fires as an integral disturbance factor in shaping Sierra Nevada ecosystems. Integrate intentional burning to simulate these former practices into overall land-use and fire-management planning.

3. Acknowledge the significance of pre-contact Native American land uses and fire-management regimes in local and regional planning and zoning, and discourage development in fire-type vegetation communities and other environmentally sensitive areas.

4. Establish traditional resource-use areas on public lands for access by Native Americans; such areas would be managed with sensitivity to traditional values.

5. Teach schoolchildren and the general public about Native American conservation ethics and traditional land-use and resource-management practices.

6. Create an ethnobiology handbook for public land managers in California that defines the field and its methodologies, major issues, research priorities, and relevance to ecosystem management and conservation biology.

7. Develop a geographic information system (GIS) database of temporally segregated archaeological site locations in the Sierra Nevada to permit modeling of past land-use patterns. Incorporate findings into modern land-use planning and zoning.

8. Systematically catalog ethnobotanical and ethnozoological information for each tribe in the Sierra Nevada into an ethnobiological database that would complement and interface with the existing Natural Diversity Data Base and SNEP's GIS.

ACKNOWLEDGMENTS

This chapter is a product of group discussions held at the SNEP workshop entitled *Reconstructing an 1800s Sierran Landscape,* in Rocklin, California, December 10–11, 1994. We gratefully acknowledge the key points made by many individuals. Special thanks are given to Michael Barbour, Richard Garcia, Ron Goode, George Gruell, Susan Lindström, C. Kristina Roper, Frederick Velasquez, and Norman Wilson. We appreciate also the efforts of Paul Rich and Lyn Meckstroth in creating the tables and map of tribal territories, and of Mary Feagins and Susan Rapp in producing several iterations of this chapter.

REFERENCES

Adovasio, J. M. 1974. Prehistoric North American basketry. *Nevada State Museum Anthropological Papers* 16:98–148.

Aginsky, B. W. 1943. Culture element distributions, XXIV: Central Sierra. *University of California Anthropological Records* 8 (4): 393–468.

Anderson, M. K. 1991. California Indian horticulture: Management and use of redbud by the Southern Sierra Miwok. *Journal of Ethnobiology* 11 (1): 145–57.

———. 1992. Restoring deer grass. *News from Native California* 6 (2): 40.

———. 1993a. The experimental approach to assessment of the potential ecological effects of horticultural practices by indigenous peoples on California wildlands. Ph.D. dissertation, Department of Environmental Science, Policy, and Management, University of California, Berkeley.

———. 1993b. Indian fire-based management in the sequoia–mixed conifer forests of the central and southern Sierra Nevada. Unpublished final report to the Yosemite Research Center, Cooperative Agreement Order Number 8027-002. Yosemite National Park: National Park Service.

———. 1994. Prehistoric anthropogenic wildland burning by hunter-gatherer societies in the temperate regions: A net source, sink, or neutral to the global carbon budget? *Chemosphere* 29 (5): 913–34.

Anderson, M. K., and G. P. Nabhan. 1991. Gardeners in Eden. *Wilderness Magazine* 55 (194): 27–30.

Anderson, R. S., and S. L. Carpenter. 1991. Vegetation change in Yosemite Valley, Yosemite National Park, California, during the protohistoric period. *Madroño* 38 (1): 1–13.

Barbour, M., B. Pavlik, F. Drysdale, and S. Lindström. 1993. *California's changing landscapes: Diversity and conservation of California vegetation.* Sacramento: California Native Plant Society.

Barrett, S. A., and E. W. Gifford. 1933. Miwok material culture. *Bulletin of the Public Museum of the City of Milwaukee* 2 (4): 119–377.

Bates, C. D., and M. J. Lee. 1990. *Tradition and innovation: A basket history of the Indians of the Yosemite–Mono Lake area.* Yosemite National Park: Yosemite Association.

Baumhoff, M. A. 1963. Ecological determinants of aboriginal California populations. *University of California Publications in American Archaeology and Ethnology* 49 (2): 155–236.

Baxley, W. H. 1865. *What I saw on the west coast of South and North America and at the Hawaiian Islands.* New York: D. Appleton and Co.

Birckhead, J., T. De Lacy, and L. Smith, eds. 1992. *Aboriginal involvement in parks and protected areas.* Canberra, Australia: Aboriginal Studies Press

Blackburn, T. C., and M. K. Anderson. 1993. Introduction: Managing the domesticated environment. In *Before the wilderness: Native Californians as environmental managers,* edited by T. C. Blackburn and M. K. Anderson, 15–25. Menlo Park, CA: Ballena Press.

Boyd, R. 1987. The effects of controlled burnings on three rare plants. In *Conservation and management of rare and endangered plants,* edited by T. S. Elias, 513–19. Sacramento: California Native Plant Society.

Burcham, L. T. 1959. Planned burning as a management practice for California wild lands. Talk given at the 59th annual meeting of the Society of American Foresters, Division of Range Management, San Francisco, 18 November.

California Indian Basketweavers Association (CIBA). 1993. Bear grass update. *CIBA Newsletter* 3:8.

Callicott, J. B. 1990. Whither conservation ethics? *Conservation Biology* 4:15–20.

Christensen, N. L. 1988. Succession and natural disturbance: Paradigms, problems, and preservation of natural ecosystems. In *Ecosystem management for parks and wilderness,* edited by J. K. Agee and D. R. Johnson, 62–86. Seattle: University of Washington Press.

Clar, R. 1959. *California government and forestry: From Spanish days until the creation of the Department of Natural Resources in 1927.* Sacramento: State of California, Division of Forestry, Department of Natural Resources.

Connell, J. H. 1978. Diversity in tropical rain forests and coral reefs. *Science* 199:1302–10.

Cook, S. F. 1955a. The aboriginal population of the San Joaquin valley, California. *University of California Anthropological Records* 16 (2): 31–80.

———. 1955b. The epidemic of 1830–1833 in California and Oregon. *University of California Publications in American Archaeology and Ethnology* 43 (3): 303–26.

———. 1976a. *The conflict between the California Indian and white civilization.* Berkeley and Los Angeles: University of California Press.

———. 1976b. *The population of the California Indians, 1769–1970.* Berkeley and Los Angeles: University of California Press.

Crumley, C. L. 1994. Historical ecology: A multidimensional ecological orientation. In *Historical ecology: Cultural knowledge and changing landscapes,* edited by C. L. Crumley, 1–16. Santa Fe, NM: School of American Research Press.

Curtis, E. S. 1924. The Maidu. In *The North American Indian,* vol. 14, edited by F. W. Hodge, 99–126. New York: Johnson Reprint Corp.

d'Azevedo, W. L. 1986. Washoe. In *Great Basin,* edited by W. L. d'Azevedo, 466–98. Vol. 11 of *The Handbook of North American Indians.* Washington, DC: Smithsonian Institution.

DeQuille, D. 1963. *Washoe rambles.* Los Angeles: Westernlore Press.

DeWalt, B. R. 1994. Using indigenous knowledge to improve agriculture and natural resource management. *Human Organization* 53 (2): 123–31.

Driver, H. E. 1937. Culture element distributions, VI: Southern Sierra Nevada. *University of California Anthropological Records* 1 (2): 53–154.

Driver, H. E., and W. C. Massey. 1957. Comparative studies of North American Indians. *Transactions of the American Philosophical Society* 47:2.

Dudley, W. R. 1896. Forest reservations: With a report on the Sierra reservation, California. *Sierra Club Bulletin* 1 (7): 254–67.

Duncan, J. W. 1964. Maidu ethnobotany. Master's thesis, Department of Anthropology, Sacramento State College (now California State University, Sacramento).

Elston, R. J., O. Davis, A. Leventhal, and C. Covington. 1977. The archaeology of the Tahoe Reach of the Truckee River. Manuscript on file, Special Collections Department, Getchell Library, University of Nevada, Reno.

Fowler, C. S. 1986. Subsistence. In *Great Basin,* edited by W. L. d'Azevedo, 64–97. Vol. 11 of *The Handbook of North American Indians.* Washington, DC: Smithsonian Institution.

Fowler, C. S., and S. Liljeblad. 1986. Northern Paiute. In *Great Basin,* edited by W. L. d'Azevedo, 435–65. Vol. 11 of *The Handbook of North American Indians.* Washington, DC: Smithsonian Institution.

Gamble, G., J. Gamble, and C. Silva. 1979. Wikchamni coiled basketry. *Journal of California and Great Basin Anthropology* 1 (2): 268–79.

Gayton, A. H. 1929. Yokuts and Western Mono pottery making. *University of California Publications in American Archaeology and Ethnology* 24 (3): 239–51.

———. 1948. Yokuts and Western Mono ethnography, I: Tulare Lake, Southern Valley, and Central Foothill Yokuts. *University of California Anthropological Records* 10 (1): 1–138.

Gifford, E. W. 1955. Central Miwok ceremonies. *University of California Anthropological Records* 14 (4): 261–318.

Gomez-Pompa, A., and A. Kaus. 1992. Taming the wilderness myth. *BioScience* 42 (4): 271–79.

Goode, R. W. 1992. Cultural traditions endangered. Report to the U.S. Forest Service, Sierra National Forest, Fresno, CA.

Gould, R. A. 1963. Aboriginal California burial and cremation practices. *University of California Archaeological Survey Reports* 60: 149–68.

Heizer, R. F., ed. 1974. *The destruction of California Indians.* Santa Barbara and Salt Lake City: Peregrine Smith.

Hinton, L. 1994. *Flutes of fire: Essays on California Indian languages.* Berkeley, CA: Heyday Books.

Hudson, J. W. 1901/n.d. Unpublished field notes on Mono/Yokuts/Yosemite/Central Miwok. Grace Hudson Museum, Ukiah, CA.

Hurtado, A. L. 1988. *Indian survival on the California frontier.* New Haven, CT, and London: Yale University Press.

Huston, M. A. 1994. *Biological diversity: The coexistence of species on changing landscapes.* Cambridge, England: Cambridge University Press.

Jensen, D. B., M. S. Torn, and J. Harte. 1993. *In our own hands: A strategy for conserving California's biological diversity.* Berkeley and Los Angeles: University of California Press.

Jewel, D. 1971. Letter to R. Riegelhuth. On file, Research Office, Sequoia and Kings Canyon National Parks, Three Rivers, CA.

Kauffman, J. B., and R. E. Martin. 1990. Sprouting shrub response to different seasons and fuel consumption levels of prescribed fire in Sierra Nevada mixed conifer ecosystems. *Forest Science* 36 (3): 748–64.

Keeley, J. E. 1986. Resilience of Mediterranean shrub communities to fires. In *Resilience in Mediterranean-type ecosystems,* edited by B. Dell, A. J. M. Hopkins, and B. B. Lamont, 95–112. Dordrecht, Netherlands: Dr. W. Junk Publishers.

Kilgore, B. M., and D. Taylor. 1979. Fire history of a sequoia–mixed conifer forest. *Ecology* 60 (1): 129–42.

Kroeber, A. L. 1925. Handbook of the Indians of California. *Bureau of American Ethnology Bulletin* 78. Washington, DC: Smithsonian Institution.

———. 1939. Cultural and natural areas of native North America. *University of California Publications in American Archaeology and Ethnology* 38:1–242.

———. 1962. The nature of land-holding groups in aboriginal California. *University of California Archaeological Survey Reports* 36:19–58.

Latta, F. F. 1977. *Handbook of the Yokuts Indians.* Santa Cruz, CA: Bear State Books.

Lawton, H. W., P. J. Wilke, M. DeDecker, and W. M. Mason. 1976. Agriculture among the Paiute of Owens Valley. *Journal of California Anthropology* 3 (1): 13–50.

Levy, R. 1978. Eastern Miwok. In *California,* edited by R. F. Heizer, 398–413. Vol. 8 of *The Handbook of North American Indians.* Washington, DC: Smithsonian Institution.

Lewis, H. T. 1993. Patterns of Indian burning in California: Ecology and ethnohistory. In *Before the wilderness: Native Californians as environmental managers,* edited by T. C. Blackburn and M. K. Anderson, 55–116. Menlo Park, CA: Ballena Press.

Liljeblad, S., and C. S. Fowler. 1986. Owens Valley Paiute. In *Great Basin,* edited by W. L. d'Azevedo, 412–66. Vol. 11 of *The Handbook of North American Indians.* Washington, DC: Smithsonian Institution.

Lindström, S. 1992. Great Basin fisherfolk: Optimal diet breadth modeling the Truckee River aboriginal subsistence fishery. Ph.D. dissertation, Department of Anthropology, University of California, Davis.

Manley, P. N., G. E. Brogan, C. Cook, M. E. Flores, D. G. Fullmer, S. Husari, T. M. Jimerson, L. M. Lux, M. E. McCain, J. A. Rose, G. Schmitt, J. C. Schuyler, and M. J. Skinner. 1995. *Sustaining ecosystems: A conceptual framework.* Berkeley, CA: U.S. Forest Service, Pacific Southwest Region and Station.

Martin, R. E., and D. B. Sapsis. 1992. Fires as agents of biodiversity—pyrodiversity promotes biodiversity. In *Proceedings of the symposium on biodiversity of northwestern California,* edited by R. R. Harris and D. C. Erman, 150–57. Berkeley: University of California, Division of Agriculture and National Resources.

Martinez, D. 1992. Native American forestry practices. In *The status and future of pesticide use in California.* Redding: California Forest Pest Council.

———. 1993. Managing a precarious balance: Wilderness versus sustainable forestry. *Winds of Change* 8 (3): 23–28.

Mathewson, M. 1985. Threads of life: Cordage and other fibers of the California tribes. Senior thesis, Department of Anthropology, University of California, Santa Cruz.

Matson, R. G. 1972. Pollen from the Spring Garden Ravine Site (4-Pla-101). In *Papers on Nisenan environment and subsistence,* edited by E. W. Ritter and P. D. Schulz, 24–27. Publication 3. Davis: University of California, Center for Archaeological Research.

Mayer, K. E., and W. F. Laudenslayer, Jr., eds. 1988. *A guide to wildlife habitats of California.* Sacramento: California Department of Forestry and Fire Protection.

McCarthy, H. 1993. Managing oaks and the acorn crop. In *Before the wilderness: Native Californians as environmental managers,* edited by T. C. Blackburn and M. K. Anderson, 213–28. Menlo Park, CA: Ballena Press.

McMillin, J. H. 1963. The aboriginal human ecology of the Mountain Meadows area in southwestern Lassen County, California. Master's thesis, Department of Anthropology, Sacramento State College (now California State University, Sacramento).

Merriam, C. H. 1905. The Indian population of California. *American Anthropologist* 7:594–606.

———. 1955. *Studies of California Indians.* Berkeley and Los Angeles: University of California Press.

———. 1967. Ethnographic notes on California Indian tribes, III: Ethnological notes on central California Indian tribes. *University of California Archaeological Survey Reports* 68 (3): 257–448.

Mooney, H. A., and Godron, M., eds. 1983. *Disturbance and ecosystems.* Berlin and New York: Springer-Verlag.

Moratto, M. J. 1984. *California archaeology.* Orlando and London: Academic Press.

Moratto, M. J., T. F. King, and W. B. Woolfenden. 1978. Archaeology and California's climate. *Journal of California Anthropology* 5 (2): 147–61.

Moratto, M. J., J. D. Tordoff, and L. H. Shoup. 1988. *Culture change in the central Sierra Nevada, 8000 B.C.–A.D. 1950.* Final report of the New Melones Archaeological Project 9. Washington, DC: U.S. National Park Service.

Nabhan, G. P. 1995. Cultural parallax in viewing North American habitats. In *Reinventing nature? Responses to postmodern deconstruction,* edited by M. E. Soulé and G. Lease, 87–102. Washington, DC: Island Press.

National Research Council. 1992. *Science and the national parks.* Washington, DC: National Academy Press.

Naveh, E. 1975. The evolutionary significance of fire in the Mediterranean region. *Vegetatio* 29:199–208.

Nichols, H. T. 1989. Managing fire in Sequoia and Kings Canyon National Parks. *Fremontia* 16 (4): 11–14.

Parsons, D. J., D. M. Graber, J. K. Agee, and J. W. van Wagtendonk. 1986. Natural fire management in national parks. *Environmental Management* 10 (1): 21–24.

Peak, A. S. 1987. *Archaeological data recovery of CA-Cal-S275, Clarks Flat, Calaveras County, California.* Cultural Resource Studies, North Fork Stanislaus Hydroelectric Development Project 3. Submitted to the Northern California Power Agency, Roseville, CA.

Peak, A. S., and H. L. Crew. 1990. *An archaeological data recovery project at CA-Cal-S-342, Clarks Flat, Calaveras County, California.* Cultural Resource Studies, North Fork Stanislaus River Hydroelectric Development Project 2. Roseville: Northern California Power Agency.

Peri, D. W., S. M. Patterson, and J. L. Goodrich. 1982. *Ethnobotanical mitigation, Warm Springs Dam—Lake Sonoma California,* edited by E. Hill and R. N. Lerner. Penngrove, CA: Elgar Hill, Environmental Analysis and Planning.

Perlot, J. N. 1985. *Gold seeker: Adventures of a Belgian Argonaut during the gold rush years,* translated by H. H. Bretnor. New Haven, CT, and London: Yale University Press.

Phillips, G. H. 1993. *Indians and intruders in Central California, 1769–1849.* Norman: University of Oklahoma Press.

Pickett, S. T. A., and P. S. White. 1985. *The ecology of natural disturbance and patch dynamics.* San Diego: Academic Press.

Potts, M. 1977. *The Northern Maidu.* Happy Camp, CA: Naturegraph.

Powers, S. 1976. *Tribes of California.* Berkeley and Los Angeles: University of California Press.

Primack, R. B. 1993. *Essentials of conservation biology.* Sunderland, MA: Sinauer Associates.

Rawls, J. J. 1984. *Indians of California: The changing image.* Norman: University of Oklahoma Press.

Reynolds, R. D. 1959. Effect of natural fires and aboriginal burning upon the forests of the central Sierra Nevada. Master's thesis, Department of Geography, University of California, Berkeley.

Riddell, F. A. 1978. Maidu and Konkow. In *California,* edited by R. F. Heizer, 370–86. Vol. 8 of *The Handbook of North American Indians.* Washington, DC: Smithsonian Institution.

Rondeau, M. 1982. The archaeology of the Truckee site, Nevada County, California. Foundation of California State University, Sacramento. Manuscript on file, North Central Information Center, California State University, Sacramento.

Roper Wickstrom, C. K. 1987. *Issues concerning Native American use of fire: A literature review.* Publications in Anthropology 6. Yosemite National Park: Yosemite Research Center.

Sauer, C. O. 1952. *Agricultural origins and dispersals.* New York: American Geographical Society.

Sierra Nevada Research Planning Team. 1994. *Critical questions for the Sierra Nevada: Recommended research priorities and administration.* Report 34. Davis: University of California, Centers for Water and Wildland Resources, Division of Agriculture and Natural Resources.

Simms, S. R. 1992. Wilderness as human landscape. In *Wilderness tapestry: An eclectic approach to preservation,* edited by S. I. Zeveloff, L. M. Vause, and W. H. McVaugh, 183–201. Reno: University of Nevada Press.

Smith, B. P. 1994. Concepts in historical ecology: The view from evolutionary theory. In *Historical ecology: Cultural knowledge and changing landscapes,* edited by C. L. Crumley, 17–41. Santa Fe, NM: School of American Research Press.

Smith, C. R. 1978. Tübatulabal. In *California,* edited by R. F. Heizer, 437–45. Vol. 8 of *The Handbook of North American Indians.* Washington, DC: Smithsonian Institution.

Smith, J. P., and K. Berg, eds. 1988. *California Native Plant Society's inventory of rare and endangered vascular plants of California.* 4th ed. Special Publication 1. Sacramento: California Native Plant Society.

Society of American Foresters. 1993. *Sustaining long-term forest health and productivity.* Task Force Report. Bethesda, MD: Society of American Foresters.

Soulé, M. E., and K. A. Kohm. 1989. *Research priorities for conservation biology.* Island Press Critical Issues Series. Washington, DC: Island Press.

Spier, R. F. G. 1978a. Foothill Yokuts. In *California,* edited by R. F. Heizer, 471–85. Vol. 8 of *The Handbook of North American Indians.* Washington, DC: Smithsonian Institution.

———. 1978b. Monache. In *California,* edited by R. F. Heizer, 426–37. Vol. 8 of *The Handbook of North American Indians.* Washington, DC: Smithsonian Institution.

Steward, J. H. 1938. *Basin-plateau aboriginal sociopolitical groups.* Bureau of American Ethnology Bulletin 120. Washington, DC: Smithsonian Institution.

Swezey, S. L. 1975. The energetics of subsistence-assurance ritual in native California. *University of California Archaeological Research Facility Contributions* 23:1–46.

Verner, J., and A. Boss, technical coordinators. 1980. *California wildlife and their habitats: Western Sierra Nevada.* Berkeley, CA: U.S. Forest Service, Pacific Southwest Forest and Range Experiment Station.

Verner, J., R. J. Gutierrez, and G. I. Gould Jr. 1992. The California spotted owl: General biology and ecological relations. In *The California spotted owl: A technical assessment of its current status,* technical coordination by J. Verner, K. S. McKelvey, B. R. Noon, R. J. Gutierrez, G. I. Gould Jr., and T. W. Beck, 55–77. U.S. Forest Service General Technical Report PSW-GTR-133. Albany, CA: Pacific Southwest Research Station.

Voegelin, E. W. 1938. Tübatulabal ethnography. *University of California Anthropological Records* 2 (1): 1–84.

Wagner, F. H., R. Foresta, R. B. Gill, D. R. McCullough, M. R. Pelton, W. F. Porter, and H. Salwasser. 1995. *Wildlife policies in the U.S. national parks.* Washington, DC: Island Press.

Wagner, F. H., and C. E. Kay. 1993. "Natural" or "healthy" ecosystems: Are U.S. national parks providing them? In *Humans as components of ecosystems,* edited by M. J. McDonnell and S. T. A. Pickett, 257–70. New York: Springer-Verlag.

Wells, P. V. 1969. The relation between mode of reproduction and extent of speciation in woody genera of the California chaparral. *Evolution* 23:264–67.

Whittaker, R. H., and G. M. Woodwell. 1972. Evolution of natural communities. In *Ecosystem structure and function,* edited by J. A. Wiens, 137–59. Corvallis: Oregon State University Press.

Wilson, N. 1972. Notes on traditional foothill Nisenan food technology. *Center for Archaeological Research Publications* 3:32–38.

Wilson, N. L., and A. H. Towne. 1978. Nisenan. In *California,* edited by R. F. Heizer, 387–97. Vol. 8 of *The Handbook of North American Indians.* Washington, DC: Smithsonian Institution.

Woolfenden, W. B. 1994. Historical ecology and the human dimension in ecosystem management. In *Draft Region 5 ecosystem management guidebook 2,* pp. I-E-4–I-E-5. Berkeley, CA: U.S. Forest Service, Pacific Southwest Region.

Zigmond, M. L. 1981. *Kawaiisu ethnobotany.* Salt Lake City: University of Utah Press.

———. 1986. Kawaiisu. In *Great Basin,* edited by W. L. d'Azevedo, 398–412. Vol. 11 of *The Handbook of North American Indians.* Washington, DC: Smithsonian Institution.

LINDA A. REYNOLDS
Inyo National Forest
Bishop, California

10

The Role of Indian Tribal Governments and Communities in Regional Land Management

ABSTRACT

Indian tribes and other Indian communities in the Sierra Nevada Ecosytem Project (SNEP) study area are its original stakeholders. Their current and future effect on land management is larger than simple demographics would suggest because federally recognized Indian tribes have a government-to-government relationship with the United States; therefore, in most matters they are not subject to state or county jurisdiction. Because this unique relationship is poorly understood by the general public, this chapter presents certain key concepts of Indian law.

Throughout the centuries of conquest and attempted assimilation, Native Californians have maintained their cultural identity and ties with the land. Today there are thirty-five recognized tribes with traditional territory in the SNEP study area, sixteen tribal communities seeking federal recognition, and two tribes seeking restoration. There are also a number of increasingly influential intertribal organizations focused around particular issues.

In the past thirty years, federal Indian policy has shifted from assimilation and termination to promotion of self-governance for tribes and, recently, toward strengthening the government-to-government relationship. Tribal efforts at economic development, reassertion of aboriginal rights, repatriation of ancestral remains and cultural items, recovery of traditional cultural lands, and protection of sacred sites have been fostered by contemporary federal legislation. Examples of involvement in land management are offered, along with a discussion of future trends.

INTRODUCTION

. . . the relationship of the Indians to the United States is marked by peculiar and cardinal distinctions which exist as nowhere else.

Chief Justice John Marshall,
Cherokee Nation v. Georgia (1831)

The Issue

The purpose of this assessment is to provide a basis for understanding the effect of Indian tribes and Indian communities on land-use management. The Indian people in the Sierra Nevada Ecosystem Project (SNEP) study area are the one social group with both ancient roots in the ecoregion and a continuing stake in its future. It goes without saying that they are culturally and historically distinct from other rural communities. Less obvious, and certainly less well understood by the general public, are the constitutional and legal distinctions that comprise the impetus for the assessment in this chapter. Federal Indian law (U.S. Code title 25) is not directly analogous to any other body of federal law. It is based on Western European international law, colonial precedents, constitutional provisions, treaties, and U.S. Supreme Court and lower court decisions (AIRI 1988; U.S. Commission on Civil Rights 1981).

In today's world the status and rights of native peoples is a global issue. Here in the United States the issue takes on its "peculiar" character through the sovereign status of tribes and the trust responsibilities of the federal government to them. It is also a local issue: in the Sierra Nevada ecoregion, where

Sierra Nevada Ecosystem Project: Final report to Congress, vol. II, *Assessments and scientific basis for management options*. Davis: University of California, Centers for Water and Wildland Resources, 1996.

over half the land base is managed by federal agencies, the unique legal status of tribal governments ensures they will be a factor in future land-management decisions.

This fact is contrary to the general public's perception. According to the congressionally chartered American Indian Policy Review, "One of the greatest obstacles faced by the Indian today in his drive for self-determination and a place in this Nation is the American public's ignorance of the historical relationship of the United States with Indian tribes and the lack of general awareness of the status of the American Indian in our society today" (U.S. Commission on Civil Rights 1981).

In addition to this pervasive lack of knowledge about the history and legal position of Native Americans in contemporary American society, there are a number of commonly held myths that often interfere in the development of working relationships between native people and other segments of society. One of the most enduring is that the Sierra Nevada was a pristine wilderness before non-Indians arrived to exploit its resources. Another is that California Indians were all "diggers" with no historical differences or cultural complexity, that they were simply small, roving bands of hunter-gatherers with, at best, a localized presence. The term *digger* itself is a demeaning racial pejorative that has been used in history books and school texts, misrepresenting Native Californian cultures and perpetuating the myth of racial inferiority. A casual conversation at any local gathering in the Sierran ecoregion will demonstrate that many of the people who were taught from these books still hold negative images of Native Californians. There are others who think that most Indians are gone from their ancestral homelands in the Sierra and have been acculturated into the "melting pot" and therefore aboriginal rights to lands and resources and ongoing traditional cultural activities need not be considered. Some people feel that Indians receive "special treatment" through minority racial status. In many Sierran communities these myths exacerbate a national problem of racial prejudice against Indian people (U.S. Commission on Civil Rights 1981).

Relevance to Other SNEP Issues

> The Indian's preservation of the land and its products for the ten thousand or more years of their undisputed occupancy was such that the white invaders wrested from them a garden, not the wilderness it salved their conscience to call it.
>
> Kroeber and Heizer 1968

Tribal governments, Indian communities, and individual Indian people must be considered separately from the general population under a suite of federal and state laws dealing with environmental analysis, religious freedom, archaeological sites, and protection of Native American human remains. This being the case, they have an effect on land management greater than simple demographics would suggest.

In California, the land base of most tribal governments is 121.5 ha (300 acres) or less, creating a reliance on federal lands for exercise of reserved rights, access to traditional resources, ceremonial use, economic development, and land acquisition that is far greater than in other states.

Another important consideration is that Indian people relate to and use the land differently than other members of society. Whether they live there or not, Indian people have a spiritual connection to their ancestral lands that derives from traditional cultural teachings about the use and management of nature. They feel that they are the original land managers and have ongoing responsibilities and rights.

The active participation of tribal governments and Indian communities and individuals in developing and implementing land-use plans is not only important, it is mandated by federal law. Tribes are not special interest groups, they are part of the family of governments in the United States. Furthermore, the well-being or lack thereof of tribal governments and Indian communities will directly affect the well-being and capacity of the larger community aggregates (Doak and Kusel 1996). As detailed in the "Future Trends" section, land acquisition/recovery plans, water rights, tribal economic development, and related issues will figure prominently in the future of the Sierra Nevada ecoregion.

Methods of Data Collection

For this chapter, government documents and published legal, anthropological, and historical sources were supplemented with input from all tribal governments and communities in the SNEP study area by means of a letter disseminated through the U.S. Forest Service's Tribal Relations Program (TRP). (The text of the letter is provided in appendix 10.1.) The letter was followed up by personal contacts, telephone calls, and presentations to tribal councils. Group presentations about SNEP and this assessment were made by the author and Connie Millar of the SNEP Science Team to twenty-four tribes and communities at a TRP workshop in Fresno and by Sonia Tamez, TRP Region 5 program manager, to the Native American Heritage Commission and to the Council on the Status of California Indians. All persons and organizations cited herein were given an opportunity to review the first draft of this chapter.

Written and verbal responses were received from members of the Kern Valley Indian Community, Chico Band of Mechoopda Indians, Big Pine Tribal Council, Pit River Tribe, Native American Heritage Commission, California Indian Basketweavers Association, Fort Independence Reservation, California Indian Legal Services, Maidu Bear Dance Committee, Tuolumne Me-wuk Tribe, Lone Pine Band of Paiute Indians, Tule River Reservation, Tyme Maidu, Commission on the Status of California Indians, Native American Heritage Preservation Council of Kern County, Holkoma Mono, Calaveras Miwok Tribal Group, and Shingle Springs Rancheria. Those individuals who were comfortable being quoted are cited in the text.

These responses clarified issues, provided examples of land uses, and identified the need for changes in terminology. In all cases where there was a preference for one term over another, for example, *unacknowledged* rather than *unrecognized*, this chapter uses the term preferred by native people. Another term that was repeatedly corrected is *traditional* insofar as respondents feel that *traditional* by itself is insufficient and needs to be accompanied by *cultural* and/or *heritage*. All respondents but one were pleased that the assessment was being written and especially that the salient legal and historical background was discussed. The idea that "people need to know these things" was expressed a number of times. Negative comments concerned timing; that is, Indian people should have been involved earlier in the SNEP study. Some respondents expressed concern that the legal and cultural separateness of individual tribes and communities would be ignored in an overview document. And one person commented that this would just be one more document that could be used against Indian people.

Information gaps exist in several areas. Demographics constitute a major gap even though Indian people and tribal membership are included in the U.S. Census: like most minority communities, they are undercounted. Another gap lies in the realm of traditional land-management practices (Anderson and Moratto 1996). This is of concern because land managers seeking a baseline "natural environment" from which to establish a desired future condition are working with the models of non-Indian researchers, who often lack knowledge of the type, extent, and duration of vegetation management practices of Native Californians. One of the most glaring gaps lies in writings about Native Californians by Native Californians. To date, most writing about American Indian culture, history, and traditions has been done by non-Indians, with resulting cultural biases and misinterpretations. This gap is closing, and we can look forward to more direct information in the future.

My knowledge of the status of tribes and future trends is largely derived from my education in anthropology and my experience as a U.S. Forest Service employee. I have worked in the Heritage Resources program since 1977—in the Sierra Nevada since 1981—and have been responsible for Native American consultation on the Inyo National Forest since 1986. In 1992, I was appointed Inyo National Forest TRP manager. Most recently, I was asked by the Timbisha Shoshone of Death Valley to coordinate a congressionally mandated interagency study of lands suitable for a reservation.

I am a non-Indian writing about sensitive topics with due respect for Native Californian peoples and cultures. I have been honored by many years of productive working relationships with Indian colleagues and friends, including consultation with native peoples from throughout the SNEP study area and beyond for this chapter. In the final analysis, however, it must be made clear that I write from my own viewpoint and do not speak for Indian people, tribes, communities, or organizations.

Organization

The next section presents key concepts of Indian law necessary for understanding the contemporary situation and future trends. The sections that follow provide an overview of the history of Native Californians and their contemporary presence in the SNEP study area and a discussion of contemporary federal policy as it relates to tribes, land use, and ecosystems management. The final section considers the relationship of tribal issues to land management, with a discussion of future trends that will affect the Sierra ecoregion.

Treatment of such a complex and little-known topic within the space allotted for this assessment necessitates that it will be general. For more information and detail, the reader is referred to the sources listed in the references, the offices of the tribal governments listed in appendix 10.2, the intertribal organizations referred to in the text, and the offices of land management agencies in the Sierra Nevada ecoregion.

KEY CONCEPTS

Allotments

Allotments are holdings of individuals or families outside a reservation; some in trust, some in fee simple. These are scattered throughout the SNEP study area. Within reservations, assignments of tribal land, as opposed to allotments, are made by the tribal government to individuals and/or families. The land so assigned remains tribal trust land.

California Indians

The term *California Indians* refers to indigenous peoples in the land now known as California. Today they continue to maintain their separate cultural identities while participating in the social and economic activities of non-Indian communities. The Indian Claims Commission defined the "Indians of California" to whom Congress has a fiduciary duty as

all Indians who were residing in the State of California on June 1, 1852, and their descendants now living, as set forth by the Act of May 18, 1928.... This identifiable group includes the descendants of members of what have sometimes been loosely described as tribes, bands, rancherias, and villages of Indians of California, and other individual Indians, who resided in California at the time of the promulgation of the Treaty of Guadalupe Hidalgo in 1848. Members of the group who were born prior to May 18, 1928, were enrolled as Indians of California by direction of the Act of Congress approved May 18, 1928.... Members of the group who were born subsequent to May 18, 1928, are to be enrolled by direction of the Act of Congress approved June 30, 1948. (Quesenberry 1993)

Federal Recognition

A tribe is federally recognized if "(1) Congress or the executive created a reservation for the group whether by treaty (1871), by statutorily expressed agreement or by executive order or other valid administrative action: and (2) the United States has some continuing political relationship with the group, such as providing services through the BIA [Bureau of Indian Affairs]" (AIRI 1988).

Indian

An Indian is a person with some amount of Indian blood who is recognized as an Indian by the person's tribe or community. . . . While membership in a federally recognized tribe is the general criteria used by the BIA for participation in most federal programs, a blood standard also is used alternatively for eligibility for some programs. In recent years Congress has not allowed the BIA to rely solely on a blood standard for federal program eligibility. (AIRI 1988)

Indian Country

The Indian Country Statute of 1948 (18 U.S. Code sec. 1151) defines Indian Country as "all land within the limits of any Indian reservation under the jurisdiction of the United States Government . . . all dependent Indian communities within the borders of the United States . . . all Indian allotments, the Indian titles to which have not been extinguished."

Indian Reservation

Federal reservations exist for several purposes. An Indian reservation is that land over which a tribe is recognized by the United States as having governmental jurisdiction (25 Code of Federal Regulations part 151). Some reservations in California have also been called rancherias or colonies.

Reserved Rights

"Tribal rights, including rights to land and to self-government, are not granted to the tribe by the United States. Rather, under the reserved rights doctrine (*United States v. Winans*, 1905), tribes retained ('reserved') such rights as part of their status as prior and continuing sovereigns" (AIRI 1988). Reserved rights are those that were not specifically extinguished by treaty or lands claim cases. In addition to land and self-government, these include hunting and fishing rights, the right to gather traditional materials, and water rights. There are many unanswered questions about reserved rights, and they are a source of conflict in many areas between states and tribes and between tribes and local communities (AIRI 1988; U.S. Commission on Civil Rights 1981).

Restoration

Tribes once federally recognized were terminated from federal recognition during the termination era. Such tribes may seek to be "restored" to their former status. One other use of the term *restoration* is as the alternative preferred by many Indian people to the term *land acquisition*. In that context herein, the phrase used is *land acquisition/restoration*.

Sovereignty

The special status of Indian tribal governments was defined by a series of United States Supreme Court decisions of the 1830s referred to as the "Marshall Trilogy" after their author, Chief Justice John Marshall. In *Cherokee Nation v. Georgia* (1831), he wrote that "it may well be doubted whether those tribes which reside within the acknowledged boundaries of the United States can, with strict accuracy, be denominated foreign nations. They may, more correctly, perhaps, be denominated domestic dependent nations. . . . Their relation to the United States resembles that of a ward to his guardian." From this decision and the two others of the trilogy—*Johnson v. M'Intosh* (1823) and *Worcester v. Georgia* (1832)—grew the concept of the sovereignty of Indian tribes and the trust responsibility of the U.S. government. (The trust relationship is discussed in a later section.)

Tribal sovereignty is the third source of sovereignty in the United States, the other two being federal and state. Indian tribes, regardless of size, are internally sovereign; external relationships with other countries are reserved to the federal government. Each tribe has a government-to-government relationship with the United States, including all agencies and bureaus, as will be discussed in the "Contemporary Federal Policy" section.

Powers that are not limited by federal law or treaty remain with tribes. These include the power to establish a form of government and to determine membership, some police powers, and the power to administer justice, to exclude persons from the reservation, to charter business organizations, and to exercise sovereign immunity. In general, state law and local law do not apply in Indian country without congressional consent. The degree to which federal statutes apply in Indian Country has been adjudicated in federal courts on a case-by-case basis.

Environmental laws, such as the National Environmental Protection Act (NEPA), apply in only some situations. These include any so-called federal action, such as the transference of other federal or private land to tribal trust, or the use of Housing and Urban Development monies for construction. Environmental laws also apply on tribal land that is held in fee simple. For example, the Big Pine Tribe of Owens Valley acquired private land from the Los Angeles Department of Water and Power. Construction of an industrial park on the land required compliance with NEPA and other federal laws because that land was owned by the tribe, not held in trust for the tribe by the federal government.

Another aspect of sovereignty was articulated at a Tribal Relations workshop in Fresno (May 2–4, 1995) by Joseph Myers, executive director of the National Indian Justice Center. In discussing the idea of sovereignty, he pointed out that it also needs to be understood in nonlegal, cultural terms, because to Indian people it also means land, inner strength, spirituality, and the wholeness of life.

Tribe

The term *tribe* has several meanings, depending upon the context. The legal meaning is discussed here, and the anthropological one is discussed in the "Native Californians" section.

> Historically, the federal government has determined that it will recognize particular groups of Indians as Indian tribes pursuant to its authority under the Indian Commerce Clause of the United States Constitution. Thus reservations variously have been set aside for ethnologically defined tribes, for bands or other subgroups of tribes, and for confederations of several tribes or bands. All are considered as tribes for legal purposes. . . . Indian groups not recognized under federal law may seek recognition through litigation, through the administrative procedures established by the BIA, or through congressional statute. (AIRI 1988)

This is codified in Code of Federal Regulations title 25, part 150, where *tribe* is defined as a tribe, band, nation, community, rancheria, colony, pueblo, or other federally recognized group of Indians. "Tribal membership requirements can be established by usage, written law, treaty, or international agreement. Today, membership typically is defined by a tribal constitution, tribal law, or a tribal roll; varying degrees of blood quantum are required by different tribes" (AIRI 1988).

Trust

As stated earlier, the trust relationship derives from the concept of tribal sovereignty. Congress has broadly construed authority over Indian tribes based on the Indian Commerce Clause of the Constitution (art. 1, sec. 8, clause 3). The executive branch has much more narrowly construed power in its relationship with Indian tribes, but, as with Congress, it has a fiduciary, that is, trustee role. Beneficiaries of the trust relationship include tribes and individual Indians.

Perhaps the most important aspect of the trust relationship is the protection of Indian landownership. The Trade and Intercourse Acts prohibited the sale of Indian land without federal consent. Indians, although not citizens at that time, held land and other property as trust beneficiaries of the United States. This arrangement, in theory at least, protected Indian landownership and allowed the federal government rather than the states to control the opening of Indian lands for non-Indian settlement. The trust relationship, therefore, enhanced federal power, but it also created federal duties relating to Indian lands and other natural resources (AIRI 1988).

The trust relationship also includes legal representation: 25 U.S. Code sec. 175 states, "In all states and territories where there are reservations or allotted Indians, the United States Attorney shall represent them in all suits at law and in equity."

Finally, we need to acknowledge that the trust relationship has been defined over time through federal court decisions, congressional actions, and executive orders:

> The trust relationship has proved to be dynamic and ongoing, evolving over time. One question that constantly arises is whether the trust relationship is permanent. Is it a perpetual relationship, or is it one that can or ought to be "terminated"? Is the purpose to protect Indian landownership and self-governing status? Or is it to give the federal government power to assimilate Indians into the larger society, to rehabilitate them as "conquered subjects" or to "civilize" them?
>
> Different eras have provided different answers to these questions. At the turn of the century the trust relationship was seen as short term and transitory. Indian land was to be protected for a brief transition period while Indians were assimilated into the "mainstream." The trust relationship was seen as the basis for congressional power to pass legislation breaking up tribal landholdings into individual allotments.
>
> More recently, the view has broadened. The trust relationship now is seen as a doctrine that helps support progressive federal legislation enacted for the benefit of Indians, such as the modern laws dealing with child welfare, Indian religion, and tribal economic development. The trust also controls contemporary interpretations of time-honored treaties and statutes. The once transitory trust relationship apparently has developed into a permanent doctrine that will serve as a benevolent influence in the future of Indian law. (Geary 1994)

NATIVE CALIFORNIANS: AN OVERVIEW

The technological and complex social organizations of California's hunter and gatherers were integrated with value systems which encouraged increased productivity and acquisition of surpluses. The abundance of plant and animal resources and the development of storage techniques and other truly skilled applications of human ingenuity allowed these people to develop beyond the normal parameters of hunting and gathering, particularly in the sociological, philosophical, and religious realms.

Bean and Lawton 1994

Scientific evidence for human presence in California extends back in time approximately 10,000–12,000 years (Chartkoff and Chartkoff 1984; Moratto 1984). Many changes in population, land-use practices, and subsistence-settlement patterns occurred during this time. The evidence that prehistoric populations in California were an interactive and effective component of their environments is substantial and is discussed by Anderson and Moratto (1996). The legacy of these millennia is found throughout the Sierra Nevada ecoregion in the form of traditional use areas, sacred places, and archaeological sites, all of which entail legal obligations that affect activities occurring on federal, state, and, in some cases, private land. Space limitations preclude a discussion of these contemporary landscape features; the focus of this chapter will be on the historical period, especially from the time the United States annexed California (1846), when the issue of tribal relations under United States law came into being.

Non-Indian Contact

The first recorded European contact with California Indians was with the voyage of Juan Rodriguez Cabrillo in 1542. Occupation followed much later, in 1769, when, in response to Russia's incursion into Alaska and British interest in the west coast of North America, Spain sent soldiers and padres north from Mexico to establish a colonial presence in California. The Spanish were followed in 1812 by the Russians, who established an outpost at Fort Ross, and the entrance—legal and otherwise—of other non-Indian peoples (Bean 1968).

At that time California was inhabited by peoples speaking many different languages and organized into myriad groupings labeled by anthropologists as tribelets, village communities, and districts (Kroeber 1925, 1962; Steward 1933). In this chapter these autonomous sociopolitical units will be referred to as tribes. Traditionally, non-Indian scholars have lumped these tribes into larger ethnolinguistic groupings that demonstrate a linguistic and cultural affiliation, not a political or corporate unity in either aboriginal or contemporary terms. (See Anderson and Moratto 1996 for a map of the territories and a listing of the ethnolinguistic groups in the Sierra Nevada circa 1800.) Concomitant with this deceptively simple political organization was a complex social web of intermarriage, trade, economic redistribution, vegetation and game management, and long-distance movement of people and resources. The Sierra Nevada itself was no obstacle to social intercourse, and people throughout the region had and still maintain connections with one another (d'Azevedo 1986; Heizer 1978a; Myers 1995).

Manifest Destiny

With the signing of the Treaty of Guadalupe Hidalgo on July 4, 1848, Mexico ceded the territory occupied by the United States during the Mexican-American War. In return, the United States guaranteed protection of the property rights and civil liberties of former Mexican citizens, including native peoples. In California this commitment was jeopardized that same year with the discovery of gold and was abrogated the following year when delegates to the California constitutional convention voted to deny citizenship to California Indians.

The difficulty of dealing with the inevitable conflicts between the gold seekers and Indian people led to protracted debate in Congress. After California was admitted to the Union on September 9, 1850, Congress authorized President Fillmore to make treaties with Native Californians; three Indian agents were named and sent to California in 1851. Between 1851 and 1852, eighteen treaties were drawn up with 138 tribes, designating land to be ceded and reservations to be established. Under urging from the California delegation, however, none of these treaties was ever ratified, and in 1852 Congress took the extraordinary measure of sealing them until 1906 (Heizer 1978b; Stewart 1978).

The Indians of California

California is thus in the unique position of being a nontreaty state; federal recognition has been gained through other actions. In the 1850s groups of Indian people were gathered onto seven former military reservations to protect them from violence by non-Indians (Heizer 1978b). Beginning in 1864, reservations were established for dispossessed Native Californians by executive order. Ultimately, 117 communities were established by the federal government on lands set aside from the public domain or purchased for the "homeless Indians of California" (Stewart 1978). The tribes who have been federally recognized by executive order have all the rights that treaty tribes have, including sovereignty and a trust relationship with the United States. Unfortunately, other tribes remain unacknowledged.

The historical circumstances in California have created a situation today in which several groups may be included in one tribe and several tribes may be located on one reservation. An individual tribe may also have several reservations intermingled with other land. In some tribes, tribal members may retain their tribal affiliation and participate in tribal affairs but reside off the reservation. Tribes and their members may also retain an interest in their aboriginal territories even if they no longer reside in the area.

Furthermore, California Indians have been treated as one group by the federal government since the Treaty of Guadalupe Hidalgo. In 1850, Congress passed the California Indians Act. "These actions taken as a whole, manifest Congress' early intent to deal with the California Indians as a single, identifiable group or community of Indians for purposes of providing federal protection and services" (Quesenberry 1993). This act was followed by a series of statutes and actions designed to provide homes, education, and other services for the "Indians of California," regardless whether the groups that benefited were federally recognized tribal governments.

In 1928, the California Indians Jurisdictional Act was passed, enabling California Indians with notice of the act to sue the federal government for the uncompensated taking of land. California Indian land claims were settled in 1950 at the 1850 price of 47 cents per acre. This rate was raised in 1968 to $1.50 per acre. Where the money was taken, land claims were extinguished; however, it is not widely recognized that the settlement was for land, not resources, so the issue of reserved rights may still be open. For those tribes and individuals who did not settle, there are still outstanding claims and questions of aboriginal rights.

General Allotment Act of 1887 (Dawes Act)

The General Allotment Act is a mighty pulverizing engine to break up the tribal mass. It acts directly upon the family and the individual.

Theodore Roosevelt (1901)

Federal policy toward Indian people in the latter half of the nineteenth century consisted of an effort to "assimilate" tribal people into Euro-American society. Children were sent away to boarding schools, traditional religion and medicine were suppressed, and through the granting of allotments, an attempt was made to break up the traditional social structure by turning native individuals into Indian versions of Jeffersonian yeoman farmers.

Some allotments from public lands had been made through individual treaties prior to the Dawes Act, but in no way did these equal its effect. Nationally, the act allowed 32 ha (80 acres) of tribal land to be allotted to individuals, 64 ha (160 acres) to a family. In California, the allotments were taken from available public lands. In central California alone, there were 1,000 public land allotments amounting to more than 2,834 ha (7,000 acres) granted between 1887 and 1984. After twenty-five years, the lands could pass from trust to fee simple, and the allottee become a citizen. Indians who became citizens under the allotment process gained the allotments in fee simple, and these could be sold to non-Indians. This practice was ended with the passage of the Indian Reorganization Act (AIRI 1988).

The Indian Reorganization Act of 1934

The Indian Reorganization Act was passed in response to the influential Merriam Report of 1928, which detailed the terrible living conditions on many reservations. One major facet was to end the parceling out of tribal lands as allotments and extend the trust period of existing allotments. Another thrust was promotion of tribal self-government. Not all tribes accepted the act, but many did and formed constitutions and corporations under its provisions (AIRI 1988).

Termination

After World War II, the federal government embarked on a policy of "mainstreaming" reservation Indians, as embodied in the Termination Act of 1953. Termination in this sense is the revocation of federal recognition. In California, this policy was implemented through the California Rancheria Act of 1958, which resulted in the termination of forty-one tribes statewide. Within ten years, 60% of the land processed for termination went to non-Indians. Another result of termination was the relocation movement, under which terminated people from all over the United States were relocated to urban areas, placing an additional burden on state services.

Several tribes sued individually for restoration of federal recognition in the 1970s, and then a class action suit, *Tille Hardwick v. United States* (1978), was filed on behalf of all terminated tribes who wished to participate. Seventeen tribes, including six in the SNEP study area, were restored to federal recognition under this lawsuit and are consequently known as *Tille Hardwick* tribes. Tribes that were dismissed from the lawsuit may still file their own suits to regain federal recognition (AIRI 1988; Slagle 1989).

Tribal Governments, Communities, and Organizations

There are thirty-five federally recognized tribal governments in or with traditional heritage lands in the SNEP study area (figure 10.1), sixteen unacknowledged tribes seeking federal recognition, and two restoration candidates. These are listed, along with addresses and other information, in appendix 10.2. There are also intertribal organizations formed around particular issues throughout the state that are gaining importance in Indian affairs in California. Some of these groups are presented in the sections that follow.

Confederation of Aboriginal Tribes of California

A group of unacknowledged central Sierra Nevada tribes, the Confederation of Aboriginal Tribes of California, was formed in 1988. Leadership was taken by the North Fork Mono, one of the unratified treaty tribes. Today, the North Fork Mono is an aboriginal nation of 600 members, with cultural and leadership systems intact and are very much involved in tribal issues. They are representative of tribes whose aboriginal rights to land and water have never been extinguished by treaty.

The confederation was established to work on federal recognition and was instrumental in bringing about the enactment of the act that established the Commission on the Status of California Indians (Goode and Franco 1995).

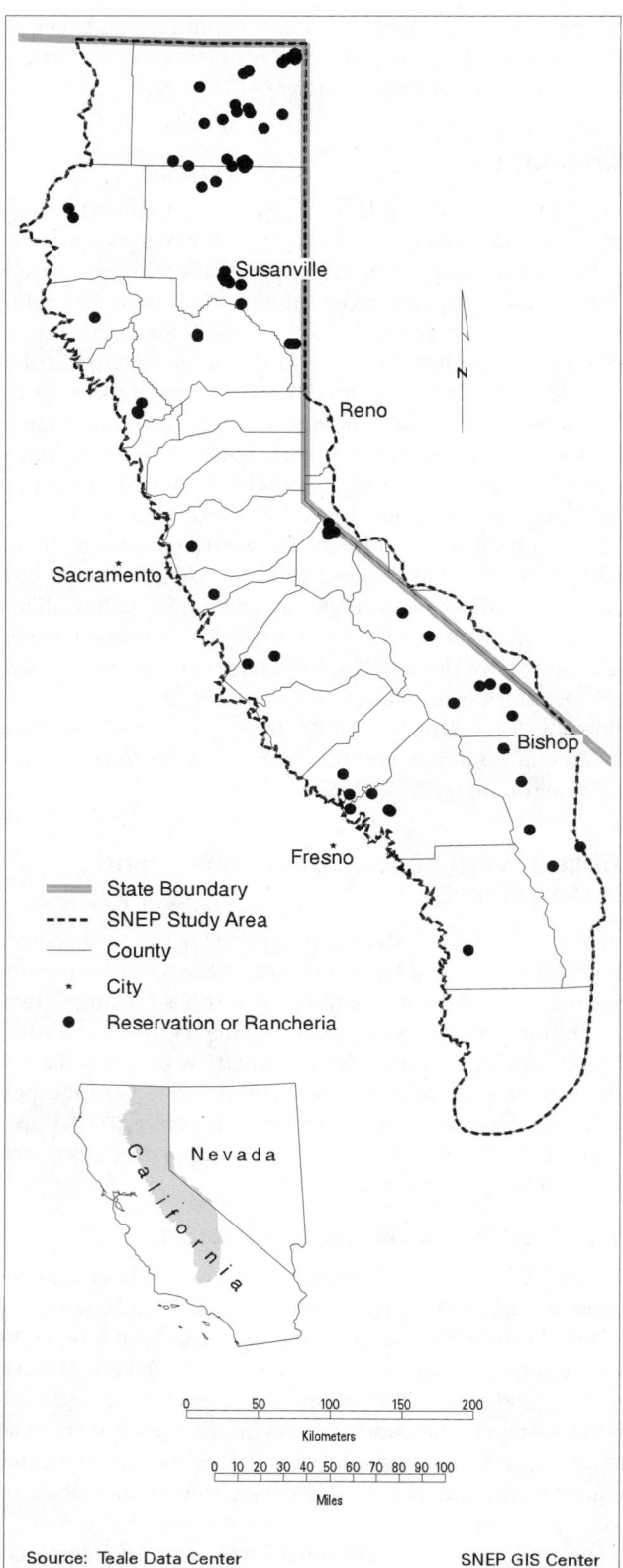

FIGURE 10.1

Reservations of groups with traditional territory in the SNEP study area.

Advisory Council on the Status of California Indians

In 1989, legislation was introduced in Congress with the purpose of clarifying the status of certain California tribes and providing recognition to others. This was passed as the Advisory Council on California Indian Policy Act of 1992. With the exception of Auburn Rancheria, specific recognition provisions in the act were deleted, and an Advisory Council of eighteen members was established. Membership is composed of representatives from sixteen tribes (acknowledged and unacknowledged), one from the BIA, and one from the Indian Health Service. They were charged with the responsibility to

1. develop a comprehensive list of California Indian tribes and a "descendancy list" for each tribe;

2. "identify the special problems confronting unacknowledged and terminated tribes";

3. "propose reasonable mechanisms to provide for the orderly and fair consideration of requests by such tribes for Federal acknowledgment";

4. assess the "social, economic, and political status of California Indians";

5. examine the effectiveness of federal policy with respect to California Indians;

6. compare services and facilities provided to California Indian tribes with those provided to Indian tribes nationwide;

7. conduct public hearings; and

8. develop recommendations.

To meet its responsibilities, the Advisory Council has held hearings throughout the state. Task forces have been established to deal with specific issues. There is a federal recognition task force, a health task force, and a legislative committee. One task force is dealing with a variety of cultural heritage issues involving sacred sites, burial grounds, archaeological sites, and ceremonial lands. The Advisory Council has prepared an executive order addressing protection of these places. In addition to land acquisition/recovery, the Advisory Council is addressing the issue of access for traditional cultural uses. Through the hearings and the work of the task forces, a tremendous amount of information has been collected. The life of the Commission on the Status of California Indians was supposed to end in October 1995, but an extension has been requested (Johnson and Manuel 1995).

The Native American Heritage Commission (NAHC)

Created by the California legislature in 1977, the Native American Heritage Commission is composed of nine members appointed by the governor, five of whom must be Native Californians. Commission meetings are held throughout the state in traditional tribal territories to make it easier for tribal members to attend.

The responsibilities of NAHC include identification of the most likely descendants when aboriginal human remains are found; liaison between Indian people and other government agencies at the federal, state, county, and city level; and maintenance of a Sacred Lands Inventory File and review of environmental documents for possible impacts to sacred lands (NAHC 1992).

California Indian Legal Services (CILS)

An offshoot of California Rural Legal Assistance, California Indian Legal Assistance formed in 1967 to deal with the special legal problems of Indian people in California. From a single office in Berkeley, it has expanded to branches in Oakland, Bishop, Escondido, and Eureka (Margolin 1993).

California Indian Basketweavers Association (CIBA)

From a gathering of northern and central California basketweavers, the California Indian Basketweavers Association has grown to a statewide organization. CIBA's purposes, as listed in its statement of purpose (CIBA 1995) are

1. to preserve, promote and perpetuate California Indian basketweaving traditions;

2. to raise awareness and provide education of Native Americans, the public, public agencies, arts, educational and environmental groups of the artistry, practices and concerns of Native American basketweavers;

3. to promote solidarity and communication between Native American basketweavers;

4. to promote and provide opportunities for Native American basketweavers to pursue the study of traditional basketry techniques and forms and to showcase their work;

5. to provide information and services to Native American basketweavers, including means of protecting their rights as artists and Native Americans;

6. to establish rapport and work with public agencies and other groups in order to provide a healthy physical, social, cultural, spiritual and economic environment for the practice of Native American basketry;

7. to increase Native American access to traditional cultural resources on public and tribal lands and traditional gathering sites, and to encourage the reintroduction of such resources and designation of gathering areas on such lands;

8. to broaden communications with other Native American traditional artists; and

9. to do all of the above in a manner which respects our Elders and Mother Earth.

Native American Heritage Preservation Council of Kern County

An example of people from different tribes working together with city and county governments to deal with local concerns is the Native American Heritage Preservation Council of Kern County, composed of representatives from the Chumash, Wachumni, Kawaisu, Paiute, Tubatulabal, Tule River Tribes, and others. It was formed in July 1991, when a development project was proposed on ancestral homelands containing burials and prehistoric archaeological sites. Since then, the council has been recognized as a credible resource by the county, by agencies, and by the private sector in dealing with Native American issues in Kern County and reviewing the adequacy of the cultural resources portions of environmental impact reports prepared under the California Environmental Quality Act (CEQA). A joint effort is currently under way between the county board of supervisors and the council to develop countywide CEQA guidelines. Other activities include burial protection, an annual gathering at California State University, Bakersfield, and other cultural activities. Currently, the council is petitioning the City of Bakersfield to name places on an eight-mile bikepath along the Kern River with native names (Gomez 1995).

California Indian Forest and Fire Management Council (CIFFMC)

The California Indian Forest and Fire Management Council is a recently constituted group comprising members from Elk Valley, Hoopa, Round Valley, Karuk, Pauman, Yurok, and Tule River Tribes. It began in 1992 as a group of concerned natural resources tribes (tribes that need to manage natural resources on their reservations). These tribes are dealing with such contemporary ecosystems management issues as prescribed burning, reintroduction of fire to the ecosystem, and cultural resources management. Their statement of purpose includes the following: "It is essential as Tribal people for us to protect and enhance the earth that we live with, especially with changes of management of natural resources compared to centuries ago" (CIFFMC 1995).

CONTEMPORARY FEDERAL POLICY

The federal agenda began to shift toward a policy supportive of tribal self-governance in the 1960s. In his 1970 message to Congress, President Nixon explicitly rejected termination policies and coined the term *self-determination*. His declaration was followed by a steady stream of legislative and executive action designed to actualize self-determination on the ground and to clarify aspects of Indian law. In addition to laws dealing specifically with Native Americans, three pieces of environmental and land-management legislation have been important in including Indian people in management activities: the National Environmental Policy Act of 1969 (NEPA), the Federal Land Policy and Management Act of 1976 (FLPMA), and the National Historic Preservation Act of 1966 (NHPA).

NEPA establishes national policy for environmental protection. Included are requirements to consider cultural values and diversity and to consult with affected tribal governments, communities, and traditional practitioners. Similar provisions are found in CEQA.

FLPMA requires coordination with Indian tribes along with other federal agencies and state and local governments in preparing and maintaining an inventory of public lands and their various resource and other values, in developing and maintaining long-range plans for the use of public lands, and in managing public lands.

NHPA requires federal agencies to take into account the effects of their activities on historic and prehistoric properties eligible for listing on the National Register of Historic Places. Where properties are eligible because of their traditional religious and/or cultural importance to Native Americans, consultation with the appropriate tribe, community, or individual is required.

The Role of Federal Agencies

On April 29, 1994, President Clinton issued an executive order to the heads of executive departments and agencies outlining the principles involved in working with federally recognized tribes as sovereign tribal governments. In it, he emphasized that each executive department and agency, "including every component bureau and office," is responsible for developing a government-to-government relationship with federally recognized tribes, for consulting with tribes on the effects of federal actions on the tribe and tribal trust resources, for removing obstacles to developing a working government-to-government relationship, and for developing methods to deal with specific tribal issues and needs. In California, the federal agencies have recently formed an Interagency Indian Policy Group with representation from the Forest Service (USFS), the Bureau of Land Management (BLM), and the National Park Service (NPS).

The United States Forest Service
(Department of Agriculture)

The *Forest Service Manual* (sec. 1563) describes the relationship between tribal governments and communities. The USFS is to

1. maintain a government-to-government relationship with federally recognized tribal governments;

2. implement programs and activities honoring Indian treaty rights and fulfill legally mandated trust responsibilities to

the extent that they are determined applicable to National Forest System lands;

3. administer programs and activities to address, respect, and be sensitive to traditional native religious beliefs and practices; and

4. provide research, transfer of technology, and technical assistance to Indian governments.

Following the president's executive order, the Inyo National Forest held a Tribal Relations workshop in October 1994 attended by representatives of ten tribes and communities from eastern California and western Nevada, USFS employees from Regions 4 and 5, NPS, BLM, California Department of Transportation (CALTRANS), the Los Angeles Department of Water and Power (DWP). In May 1995, the Sequoia, Sierra, and Stanislaus National Forests combined to hold a workshop attended by twenty-six tribes and communities, representatives of intertribal organizations, NPS, California Fish and Game, CALTRANS, and other state and local agencies.

Region 5 has a *TRP Handbook,* and many of the national forests have developed forest-specific handbooks. A working draft of the *Forest Service National Resource Book on American Indian and Alaska Native Relations* is being reviewed by agency and tribal people and will be finalized in 1996.

The Bureau of Land Management
(Department of the Interior)

The BLM is similar to the USFS in that both are multiple-use land-management agencies responsible for millions of acres of formerly tribal land and both have many opportunities to work with Indian people and tribes as partners. The BLM has established a National Native American Program Office in Santa Fe, as opposed to the USFS's more decentralized approach, in which each Region is developing policies that meet particular needs. The office in Santa Fe has recently issued national guidelines for working with tribes. Another important difference is that the BLM has identified "surplus lands" available for transfer to recognized tribes through the BIA. Transference may be accomplished by administrative action of the secretary of the interior or through legislation.

The National Park Service (Department of the Interior)

NPS has a preservation mission and has taken a correspondingly more restrictive approach toward aboriginal rights. Certain activities, like collecting traditional resources, typically are not allowed, and this often creates difficulties between the NPS and Indian people. Changes are being made in the SNEP study area, however, with draft protocols dealing with the range of government-to-government relationships by both Yosemite and Death Valley National Parks.

The Office of American Indian Trust Responsibilities under the secretary of the interior plans to issue a compilation of the Native American consultation guidelines of all the agencies in the Department of the Interior in the fall of 1995. These guidelines were drawn up in compliance with President Clinton's executive order. This, along with the USFS *National Resource Book,* should provide both agencies and the tribes with clearer and more uniform direction. The different agency approaches are as confusing and dismaying to Indian people as they are to the general public.

With the growth of tribal relations programs and consultation from all levels of government, many tribes are feeling overwhelmed by the number of requests for consultation. Many simply do not have the staff or resources to handle all of it in a timely manner, nor trained personnel to address specific issues. In this situation there is a need for personal contacts, sensitivity in communication, and technical assistance.

FUTURE TRENDS

Outstanding Aboriginal Rights, Restoration, and Federal Recognition

Tribes who were parties to the unratified treaties did not surrender any land or resources to the United States. Land claims were settled in most cases by the Land Claims Commission. However, some individuals and tribes did not accept the land settlement money, and the BIA is still holding money for those who have not extinguished their aboriginal claims. Those who did not "take the money" often express the intention of pursuing further legal action.

Though not all Indian communities want federal recognition (Velasquez 1995), eighteen unacknowledged and terminated tribes in the SNEP study area are seeking recognition or restoration. As these tribes become recognized or restored, they will be sovereign nations under federal law. As individuals and tribes continue to pursue reserved rights, local communities and state and federal land and resource management practices will be affected.

Land Acquisition/Restoration

The federal government has the authority to convey land to federally recognized tribes under different authorities: The USFS exchanges land, the BLM transfers land, and Congress may create Indian Country out of public land by legislation. A precedent for returning national park land to Indian tribes was established 1975 when the Havasupai Nation's claim to a homeland in the Grand Canyon was partially recognized. Recently, the secretary of the interior administratively transferred BLM lands adjacent to tribal governments in California, including the Utu Utu Gwaitu Tribe of Benton, the Fort Independence Reservation, the Bridgeport Indian Colony, and the XL Ranch in the SNEP study area. Lands transferred range from 16 to 101 ha (40 to 250 acres).

A precedent-setting case has arisen in Death Valley, where the Timbisha Shoshone were made homeless through the creation of a national monument in 1933. The tribe gained federal recognition in 1983, and today their temporary land base consists of 16 ha (40 acres) at Furnace Creek in Death Valley, which they hold under an expired special use permit.

The traditional homeland of the Timbisha includes the southeastern portion of the SNEP study area, on lands managed by the Inyo National Forest, the Park Service, the Ridgecrest Area Office of the BLM, and the China Lake Naval Air Weapons Station. Section 705(b) of the 1994 California Desert Protection Act mandated a study by the relevant federal agencies and the tribe to identify lands suitable for a reservation. At present, a large working group has been established along with smaller working groups focused on issues such as current land uses, legal questions, and agency mandates.

Another type of land acquisition/restoration is being sought by the Tyme Maidu, Berry Creek Rancheria, Oroville. The Tyme Maidu are a federally recognized tribe who have a small rancheria of 57 acres. The BLM had identified adjacent land to transfer to the tribe, who planned to erect a roundhouse there for traditional ceremonies. Other portions were to be used for reburial of repatriated ancestral human remains excavated during the construction of Oroville Dam. The local congressman withdrew the request for the transfer under protest from the county board of supervisors (Bjork 1995). In several areas, there is a clear conflict between tribes and counties that do not want any land to go to trust and thereby be removed from the tax base. Whether that was the case in Oroville is not known.

Self-Governance

The Indian Self-Determination and Education Assistance Act of 1975 was intended to maximize Indian people's authority to direct federal programs and services to their communities. Priorities are now set by the tribe according to its needs, laws, and individual management guidelines. In addition, authority is provided for tribal governments to acquire lands adjacent to reservations for purposes of the act.

Under current federal policy, the BIA is downsizing; in fact there are plans for the California office to be moved to Albuquerque. Through the provisions of the 1975 act, monies are being distributed to tribes rather than the BIA so that they can directly acquire necessary services. Tule River in the SNEP study area was one of the first tribes in the nation to contract programs from the government in 1980.

Water Rights

The landmark Supreme Court decision *Winters v. United States* (1908) established the Winters Doctrine, which holds that when a reservation is established under treaty, implicitly sufficient water is reserved for the tribe's present and future use. This doctrine was upheld for reservations established by executive order in *Arizona v. California* (1963). In 1952, Congress passed the McCarren Amendment, giving the states some jurisdiction to adjudicate water rights on trust lands (AIRI 1988).

An interesting situation involving five tribes exists in Owens Valley, the setting for one of the classic western water wars. In the last century, the Paiute people were displaced and their irrigation systems taken over by non-Indians. Both Indians and non-Indians were then drastically affected by claims of the City of Los Angeles to Owens River water. Indian people who had adjusted to the conquest by becoming agricultural workers were again displaced as farms were bought up to protect the watershed. A Los Angeles Department of Water and Power (DWP) study (Ford 1930) detailed the impoverished condition of the Indian people due to water diversions and lack of employment. It also pointed out the advantages to the DWP of acquiring trust lands. This report set the stage for a land exchange among the DWP, the Paiutes, and the Department of the Interior, which Congress approved in 1937. Although the legislation provided that a "fair and equal trade" be made, that is, the land exchanged should be of equal value, plus water rights, only 607 ha (1,500 acres) of DWP land was exchanged for trust land, and with no accompanying water rights.

The subsequent history is complex, with the DWP ultimately claiming that the Owens Valley tribes have no water rights. The affected tribes—Lone Pine, Fort Independence, Big Pine, Bishop, and Benton—formed the Owens Valley Indian Water Commission and engaged California Indian Legal Services to represent their interests. A federal fact-finding team composed of members from the Solicitor General's office, the BLM, and the BIA studied the case and issued a report finding that the exchange did not meet the legislative requirement for equity; therefore, the tribes have a valid claim. The secretary of the interior has requested Los Angeles to begin negotiations to settle the claim. To date, a Water Rights Negotiating Team still has to be appointed (Stidham 1995).

Another aspect of this controversy is its effect on the community as a whole. The DWP negotiated with Inyo County for years to develop a ground-water pumping agreement. When the draft environmental impact statement was completed in 1990, the tribes successfully challenged it because neither the county nor the DWP had consulted with them.

Who Owns the Past?

Two important pieces of legislation dealing with native people's rights to their cultural legacy and ancestral remains were enacted in the past twenty years: the Archaeological Resources Protection Act of 1979 (ARPA) and the Native American Graves Protection and Repatriation Act of 1990 (NAGPRA).

ARPA provides for the protection and management of archaeological resources and specifically requires notification of the affected Indian tribe if proposed archaeological investigations could result in harm to or destruction of any location considered by the tribe to have religious or cultural importance. As amended in 1992, ARPA incorporates the provisions of NAGPRA.

NAGPRA provides that federal agencies must consult with appropriate Indian tribes or individuals prior to authorizing the intentional removal of Native American human remains, funerary objects, and objects of cultural patrimony. As with other federal legislation, it applies on federal land and on any land if the project is funded with federal monies.

NAGPRA also provides for consultation over existing collections to identify and assure disposition of human remains and related cultural items in a manner consistent with the desires of lineal descendants or the appropriate tribal authorities. Since its enactment, all federal agencies and associated museums and repositories have been required to make listings of their holdings available to tribes and provide them access to view collections. Cost to both agencies and tribes has been high, and repatriation itself creates additional obligations. In the case of the Tyme Maidu mentioned earlier, additional land is needed for the tribe to inter the human remains that are being returned. In Inyo County, a state-proposed "green sticker" (off-highway vehicle) route was stopped by local opposition to the project, including by several of the Owens Valley tribes, who wish to use a portion of the area—a traditional burial ground—to perform reburial ceremonies for human remains being repatriated from the Smithsonian Institution. The Tuolumne Me-wuk Tribe has received a 1995 Economic Recovery Rural Community Assistance Program grant of $43,424 to develop a plan for an interpretive center to be used to curate repatriated human remains and associated funerary items for the five tribes of the Central Sierra Me-wuk Cultural and Historic Preservation Committee (Fuller 1995).

The protection of human remains is an important and sensitive issue for Indian people. In addition to federal legislation, California Public Resources Code, sec. 5097.99, makes it a felony to willfully, unlawfully, and knowingly obtain or possess Native American human remains taken from a Native American grave or cairn. Other state law requires immediate notification of the coroner upon discovery of human remains. Within twenty-four hours the coroner notifies the Native American Heritage Commission, which identifies the most likely descendants. The most likely descendants then make recommendations for treatment and disposition of the remains.

Tribal Courts

A lasting legacy of the termination period for California Indians is Public Law 280, enacted in 1953. PL 280 provides for partial state jurisdiction over law enforcement on some or all reservations. In some states it applies only to specific laws; in California it extends to most crimes and some civil matters. Some matters, such as zoning, taxation, and hunting and fishing rights, were specifically exempted. PL 280 does not extinguish a tribe's sovereignty or jurisdiction, and tribal courts may exist concurrently with state courts.

No monies were appropriated for PL 280, placing a financial burden on local law enforcement. At the same time, tribal expectations of effective law enforcement were not fulfilled. Neither Congress nor the BIA has really dealt with this problem, and at present the only tribal court system in the state is on the Hoopa Reservation in northern California.

There is a move toward the development of tribal circuit courts and Indian law enforcement systems in other parts of California to deal with the special cases of the Indian Child Welfare Act, voting irregularities, housing, and violations of the Indian Civil Rights Act. In Owens Valley, the Big Pine Tribe of Paiute Indians has received an Administration for Native Americans (ANA) grant from the Department of Health and Human Services to conduct a two-year demonstration project to develop a tribal court system. The system will serve tribes throughout the eastern Sierra Nevada, following a model established by the Toiyabe Indian Health Clinic, which serves tribes from Antelope Valley to Death Valley.

Environmental Justice

As roads and buildings go up, basketry material and medicine plants grow scarce, burials are disturbed and sacred places become inaccessible, if not completely destroyed. As riverbanks erode from clear-cutting, fish become scarce. Foresters use chemical herbicides to manage brush, with little thought of their effect on native food and cultural resources. Water projects flood traditional homelands, and water is diverted from rural communities to meet the needs of growing urban populations. Waste management companies, attracted by the fact that state and local regulations do no apply in Indian country, look to reservations as sites for landfills, hazardous waste sites, and recycling facilities. (Gendar 1993)

There do not appear to be any toxic waste dumps or other significant toxic problems on reservations in the SNEP study area as there are on reservations in southern California. A positive avenue for protection is available to tribes through Treatment as a State grants administered by the U.S. Geological Survey. Tribes with a land base may apply to be treated as a state for environmental purposes and may apply for monies to develop a wide range of environmental protective measures. The Utu Utu Gwaitu Tribe of Benton has achieved this status.

An important concern is protection of watersheds of all the tribes, including underground water for those reservations with no surface water. Without protection, reservation growth and development would be affected in the future.

Traditional Cultural Uses of Public Lands

Traditional tribal communities rely on the maintenance of a natural landscape and protection of key locations, plants, and animals in order to sustain their identity and exercise traditional cultural practices. They often interpret as disrespectful of nature major land alterations such as clear-cutting and road building.

Herbicide application is interpreted as harmful to the ecosystem, including the human beings who live and work on the land. As a result of consultation, some steps toward accommodating traditional cultural uses have been made in some national forests (e.g., the Eldorado, Plumas, Tahoe, and Sierra National Forests) with the establishment of herbicide-free plant collection areas. As a result of meetings by the California Indian Basketweavers Association with the chief of the USFS and the Environmental Protection Agency (EPA), the USFS and California EPA's Department of Pesticide Use are doing a joint study to develop a methodology for assessing risks to basketmakers (Greensfelder 1995).

Access to traditional heritage lands is also important for sacred and ceremonial uses. Sacred lands are designated by the Creator and cannot be "desanctified" or re-established like the churches of other religions. Often these areas are located on lands that have specific geological or other attributes that make them unique and fixed on the landscape. The American Indian Religious Freedom Act of 1978 (AIRFA) and the Religious Freedom Restoration Act of 1993 (RFRA) address this issue. AIRFA states that the policy of the United States government is to protect and preserve the rights of Native Americans to believe, express, and exercise their traditional religions, including access to religious sites. A U.S. Court of Appeals decision has determined there is a compliance element in AIRFA that requires federal consultation with Indian leaders when a proposed land use might conflict with traditional beliefs or practices so that unnecessary interference with religious practices is avoided. AIRFA was strengthened by the passage of RFRA, which restored the judicial standard that requires federal agencies to demonstrate a "compelling governmental interest" before substantially burdening a person's religious liberty. Furthermore, it must be "the least restrictive means of furthering that compelling governmental interest."

There are many sacred sites in the SNEP study area. One is Cave Rock, a prominent volcanic plug on the east shore of Lake Tahoe that can be seen as a microcosm of issues surrounding tribal rights and contemporary land-use practices. One of the first toll roads between the Nevada Comstock and the California Mother Lode wound around the outside of the rock, buttressed by rock walls constructed by Chinese workers in the 1860s. In the 1920s and again in the 1930s, tunnels were blasted through it to accommodate U.S. 50. In the 1960s a Nevada State Parks boat ramp and parking area were constructed at its base. Today, Cave Rock is a popular recreational spot offering many activities, including boating and rock climbing.

For the Washoe people of Nevada and California, Cave Rock is a focal point for spiritual power and the dwelling place of mythic beings. Indian spiritual leaders obtained their power by visiting Cave Rock, and Elders went there to pray. Despite the changes wrought by construction and recreational use, it still retains power and significance to Washoe people and continues to be treated with respect and awe. The tribe has been actively seeking assistance from managing agencies, including Nevada State Parks and the Lake Tahoe Basin Management Unit, to restore damage from rock climbers, terminate rock climbing, control boat access, and repatriate the rock (Rucks 1995).

There are also many traditional cultural gatherings and ceremonies that take place on public lands. One is the Jamani (Mountain) Maidu Wedam held early in June of each year. The Wedam is a spring ceremony popularly known as the Bear Dance. In the past, the People gathered to give thanks to Kodojapen (i.e., Earthmaker; God; Creator of Everything in This World) and to pray for protection from grizzly bears and rattlesnakes. When the private land where the Wedam had been held was sold, a proposal was made to establish a permanent campsite for the Wedam in the Eagle Lake Ranger District of Lassen National Forest. The campsite would be reserved for the Wedam fourteen days of the year, then open to the public for the rest of the time. Lassen officials reacted positively. According to a former forest supervisor, Dick Henry, "This is just one good example of the kind of partnerships that can help us realize some untapped potential in the forest. We provide the opportunity, some technical assistance, and perhaps a little funding help, and the Native American community volunteer their time and talent. They get an excellent location to gather and hold their traditional activities. The public gets a first-class campground."

Assistance in holding the Wedam has been provided by other elements of the community, including the Lassen Indian Health Center, the California Department of Fish and Game, the Lassen County Sheriff's Department, and the California Department of Forestry.

A special use permit is required for the Wedam, placing on the Bear Dance Committee the onerous burden of obtaining insurance for the event costing close to $800. They have been assisted by the Pit River Tribe in meeting this obligation, but not all groups are as fortunate. Such insurance problems are fairly common and place a burden on the free exercise of religion (Benner 1995).

Special Forest Products

Through the years there have been some conflicts between traditional and commercial uses of resources on public lands. One long-standing example involves the piñon pine nut, a traditional food staple of tribes in the eastern portion of the SNEP study area. Commercial harvesters were using machinery and methods that not only cleaned out whole groves but in many cases damaged and even killed the trees. Twenty years ago, the Inyo and Sequoia National Forests and the California BLM established a policy prohibiting commercial pine nut harvesting, while allowing collection for personal use. Increasingly, illegal harvesters have been going into the

eastern Sierra, necessitating extra efforts from law enforcement personnel during harvesting season.

Other plants, such as mushrooms, bear grass, ferns, and staghorn lichen, are being targeted by "self entrepreneurs," and the problems are exacerbated by the USFS's promotion of "special forest products," especially in areas where timber harvest has been reduced. There is a need to manage these plants for sustainability to meet all of the users' needs (Richards 1996).

Rural Economic Development

A number of avenues are open to tribes to apply for economic development grants. Two examples are presented in this section.

In 1992, the Stanislaus National Forest nominated the Tuolumne Me-wuk Tribe to participate in the Rural Development Program. The following is a summary of the grants the tribe has received in the past two years from the USFS and other sources for rural development opportunities (Montoya 1995):

1. $5,000 grant from the USFS to complete an action plan required for participation in the 1990 Farm Bill grant activities.

2. $25,000 Economic Diversification Study grant from the USFS to look in detail at economic diversification opportunities at the rancheria. The study has been completed and will be used for future grant proposals, and information will be shared with other tribes.

3. $19,500 Economic Recovery grant from the USFS to develop a business plan for a wood products manufacturing business. The business plan was completed, and future funding will be sought.

4. $20,000 grant for a native plant nursery received from the ANA. Their application was based on the action plan that was completed for the USFS Rural Development Program. The nursery opened in April 1995 and employs four tribal members. The local community is very supportive.

In the northern Sierra, a $20,000 Economic Recovery Project award through the USFS Cooperative Forestry Assistance Program has been made to develop a plan to market "A Maidu Sense of Place" in Indian Valley, one of Plumas County's most impoverished forest-dependent communities. Funds were sought by the Maidu Cultural and Development Group, a consortium of Indian Valley residents (primarily Maidu) who are interested in developing culturally based tourism and forest stewardship. Planning will focus on three areas (Ackerman 1995):

1. The Maidu Place-Name Plan, which involves youth and elders working together to add Maidu language place-names to new visitor and USFS maps of the area.

2. A Roundhouse Living Village Plan for a site on or near Highway 89 (a major tourist thoroughfare), which involves development of a cultural focus for both visitors and residents, including a replica of a traditional Maidu settlement, a craft sales area, a dance and visitors' arena, and an interpretive ethnobotany trail.

3. A Riparian and Forest Land Native American Stewardship Plan, which involves integrating Maidu resource management concepts and practices into existing USFS and community partnerships such as the Feather River Coordinating Resource Management Group.

Resource Management

The National Indian Forest Resources Management Act (NIFRMA) directs the secretary of the interior, in consultation with affected Indian tribes, to obtain an independent assessment of the status of Indian forest resources and their management. To achieve this, the secretary contracted with the Intertribal Timber Council, which in turn selected seven nationally recognized forestry experts to serve as an Indian Forest Management Assessment Team (IFMAT), which made the following findings (IFMAT 1993):

1. Tribal members emphasize different visions and goals for their forests than do BIA forestry employees.

2. Generally, a small proportion of tribal members or BIA forestry employees believe that current resource management is good or excellent.

3. The administrative relationship between the United States and each tribal government is the key factor affecting the ability of tribes to achieve their forest goals.

4. Indian forestry is seriously underfunded and understaffed compared with forestry on similar federal and private lands.

5. Managers of Indian forests are practicing more ecosystem management now than in the past.

6. The health and productivity of Indian forests are mixed and vary by forest type and geographic location.

7. Roads have contributed to a number of environmental problems.

8. Opportunities exist to substantially increase income and other benefits.

9. Forest management plans for reservation forests have the potential for meeting many tribal goals and priorities, but a narrow definition of sustained yield management, inadequate analysis in some cases, and lack of funding and personnel make attainment of goals difficult.

10. A number of issues require special planning and management, including allotments, non-Indian ownership on reservations, and off-reservations lands where tribes have treaty rights.

The Forest Division of the BIA is basically responsible for forestry operations on reservations or allotments. They must approve all Indian sales of timber. Since NIFRMA the BIA has been trying to adjust its goals to be more in synch with tribal goals. Furthermore, tribes that fall under the Self-Governance Project are now completely autonomous; they have their own budgets and make their own decisions without BIA approval (Collins 1995). In the SNEP study area, the Tule River Reservation has its own forestry program (Stewart 1996).

The Indian Mineral Development Act of 1982 gives tribal governments the authority to develop mineral resources and to enter into joint-venture agreements, operating agreements, and leases. The act conveys and extends tribal authority to regulate and cooperate with private and governmental entities in the development of tribal energy and nonenergy mineral resources.

Gaming

In a 1987 decision, the U.S. Supreme Court ruled that the tribe, not the state, has authority over gaming on reservations. The Indian Gaming Regulatory Act of 1989 created three classes of reservation gaming with varied jurisdiction:

Class 1: traditional gaming, such as hand game, which is the exclusive jurisdiction of the tribe.

Class 2: bingo and some other games, which are jointly regulated by the tribe and the federal government.

Class 3: high-stakes gambling, which requires a tribal compact with the state.

At present, there is a case before the Supreme Court on the issue of whether the state is obligated to enter into a gaming compact when requested to by a tribe.

In the SNEP study area there are several tribal governments that have applied for Class 3 compacts: Alturas Rancheria, Auburn Rancheria, Berry Creek Rancheria, Big Sandy Rancheria, Bishop Tribal Council, Chicken Ranch Rancheria, Fort Independence Reservation, Jackson Band of Mi-wuk Indians, Picayune Rancheria, Shingle Springs Band of Miwok Indians, Table Mountain Rancheria, Tule River Indian Tribe, and Utu Utu Gwaitu Tribe of Benton (Medeiros 1995).

Native American gaming is controversial, and the positive benefits to the communities where it occurs are not widely known. On December 15, 1993, the inspector general of the U.S. Department of the Interior (IGRA) issued an audit report on Indian gaming. Among other things, it found the following:

In three states, alone, benefits included dramatically increased employment levels among Indians and non-Indians; increased tax revenues, increased nongaming tourism revenues; increased housing, education, and health benefits to Indians; and reduced Government assistance to tribal and nontribal members. The IGRA . . . noted that the "striking feature" of the current debate over Indian gaming is the lack of deference to tribal views and positions. (Indian gaming news 1995)

CONCLUSION

Native Americans have a special legal relationship with the federal government unlike that of any other group of citizens. The central concepts of this relationship, tribal sovereignty and the trust responsibility, were first enunciated in Supreme Court decisions of the 1830s. Sovereignty and trust are dynamic, evolving concepts. Federal agencies are still working at the development of govenment-to-government relationships, and final guidelines have yet to be promulgated. The initial steps are being made, however, and working relationships are being forged. Because over half the land in the Sierra Nevada ecoregion is managed by federal agencies, primarily the USFS, the government-to-government relationship ensures that tribes will have a much greater role in landmanagement decisions than in the past.

The pre-contact sociopolitical organization of Native Californians was complex, with many politically autonomous groups. This fact, coupled with their treatment in the past 200 years, has produced a contemporary situation unlike that of any other group of Native Americans. In the SNEP study area alone, there are thirty-five federally recognized tribes, sixteen tribes seeking federal recognition, and two tribes seeking restoration. Not only do sheer numbers add a degree of complexity to the consultation process, but also California Indians have historically been treated as a single unit toward which Congress has fiduciary responsibilities, whether federally recognized or not.

Although the basic premise of tribal sovereignty and the trust relationship were established early in the nineteenth century, historical circumstances have prevented California Indians from effectively exercising their rights in many cases. The shift in federal policy in the past thirty years, accompanying legislation, and clear executive definition of the government-to-government relationship have all combined to empower native peoples. With new avenues of economic development such as rural economic recovery grants and gaming, tribal governments are in a position to advance their own goals and to contribute to overall rural development and well-being.

ACKNOWLEDGMENTS

I thank Jonathan Kusel, Connie Millar, and Victoria Sturtevant of the SNEP Science Team for their reviews of the initial draft and four anonymous reviewers for their review of the second draft. I additionally acknowledge Connie Millar for recognizing the need for this assessment and for her availability to tribal people and to me during the time data were compiled. Sonia Tamez of the USFS Regional Office provided an initial draft overview of tribal issues in the SNEP study area. My counterparts in the TRP—Marvin Benner, Lassen NF; Marcia Ackerman, Plumas NF; Penny Rucks, Lake Tahoe Basin Management Unit; and Jayne Montoya, Stanislaus NF—gave me some of the examples provided. Frederick Marr of the Bishop Office of California Indian Legal Services provided otherwise unavailable documents and resources. Many individuals read various outlines and drafts and made valuable comments and suggestions. I especially recognize the in-depth review and comments from Reba Fuller, NAGPRA Project Director of the Central Sierra Me-wuk Cultural and Historic Preservation Committee; Sara Greensfelder of the California Indian Basketweavers Association; Frederick Marr; Jayne Montoya; Larry Myers of the Native American Heritage Commission; and Fred Velasquez of the Calaveras Me-wok Tribe.

REFERENCES

Ackerman, M., Plumas National Forest. 1995. Conversations and e-mail with Linda Reynolds.

American Indian Resources Institute (AIRI). 1988. *Indian tribes as sovereign governments: A sourcebook on federal-tribal history, law, and policy.* Oakland, CA: AIRI Press.

Anderson, M. K., and M. J. Moratto. 1996. Native American land-use practices and ecological impacts. In *Sierra Nevada Ecosystem Project: Final report to Congress*, vol. II, chap. 9. Davis: University of California, Centers for Water and Wildland Resources.

Bean, L. J., and H. W. Lawton. 1994. Some explanations for the rise of cultural complexity in Native California with comments on proto-agriculture and agriculture. In *Before the wilderness: Environmental management by Native Californians*, edited by T. C. Blackburn and M. K. Anderson, 27–54. Menlo Park, CA: Ballena Press.

Bean, W. 1968. *California: An interpretive history.* New York: McGraw-Hill.

Benner, M., Lassen National Forest. 1995. Conversations and e-mail with Linda Reynolds.

Bjork, D., Tyme Maidu Cultural Heritage Committee. 1995. Telephone conversation with Linda Reynolds, 15 April.

California Indian Basketweavers Association (CIBA). 1995. Statement of purpose. *California Indian Basketweavers Newsletter* 10.

California Indian Forest and Fire Management Council (CIFFC). 1995. Program brochure for workshop at South Lake Tahoe, February.

Chartkoff, J. L., and K. K. Chartkoff. 1984. *The archaeology of California.* Stanford, CA: Stanford University Press.

Collins, K. 1995. Letter to Linda Reynolds.

d'Azevedo, W. L., ed. 1986. *Great Basin*. Vol. 11 of *Handbook of the North American Indians*. Washington, DC: Smithsonian Institution Press.

Doak, S. C., and J. Kusel. 1996. Well-being in forest-dependent communities, part II: A social assessment focus. In *Sierra Nevada Ecosystem Project: Final report to Congress*, vol. II, chap. 13. Davis: University of California, Centers for Water and Wildland Resources.

Ford, A. J. 1930. *Owens River Valley, California, Indian problem*. Los Angeles: Department of Water and Power.

Fuller, R., NAGPRA Project Director, Central Sierra Me-wuk Cultural and Historical Preservation Committee. 1995. Telephone conversations with Linda Reynolds, 17 and 19 July.

Geary, M. 1994. *Collaborating with first nations: Native American relationships and public land management*. Petaluma, CA: National Indian Justice Center; Washington, DC: Division of Cultural Heritage, Bureau of Land Management.

Gendar, J. 1993. Environmental issues. In *California Indian Legal Services after 25 years*. News from Native California special edition. Berkeley, CA: Heyday Books.

Gomez, R., Tubatulabal. 1995. Letter to Linda Reynolds, on file at the Inyo National Forest.

Goode, R., and L. Franco. 1995. Presentation about the Confederation of Aboriginal Tribes of California. Paper read at Tri-Forest Tribal Relations Workshop, 2–4 May, at Fresno, CA.

Heizer, R. F., ed. 1978a. *California*. Vol. 8 of *Handbook of North American Indians*. Washington, DC: Smithsonian Institution Press.

———. 1978b. Treaties. In *California*, edited by R. F. Heizer, 701–4. Vol. 8 of *Handbook of North American Indians*. Washington, DC: Smithsonian Institution Press.

Indian Forest Management Assessment Team (IFMAT). 1993. *An assessment of Indian forest management in the United States*. Portland, OR: Intertribal Timber Council.

Indian gaming news. 1995. *The Tribal Court Record* 7:26–27.

Johnson, J., and L. Manuel Jr. 1995. Presentation on the Advisory Council on the Status of California Indians. Paper read at Tri-Forest Tribal Relations Workshop, 2–4 May, at Fresno, CA.

Kroeber, A., ed. 1925. *Handbook of the Indians of California*. Bulletin 78. Washington, DC: Bureau of American Ethnology.

———. 1962. The nature of land-holding groups in aboriginal California. *University of California Archaeological Survey Reports* 36:19–58.

Kroeber, T., and R. F. Heizer. 1968. *Almost ancestors: The first Californians*. San Francisco: Sierra Club.

Margolin, M. 1993. Indian law for Indian people. In *California Indian Legal Services after 25 years*, 2–4. News from Native California special edition. Berkeley, CA: Heyday Books.

Medeiros, M., California deputy attorney general. 1995. Telephone conversation with Linda Reynolds, 28 July.

Montoya, J. 1995. Conversations and e-mail with Linda Reynolds.

Moratto, M. J. 1984. *California archaeology*. New York: Academic Press.

Myers, J. 1995. Presentation on tribal sovereignty. Paper read at Tri-Forest Tribal Relations Workshop, 2–4 May, at Fresno, CA.

Native American Heritage Commission. 1992. Memo to USFS.

Quesenberry, S. V. 1993. Letter to Francis P. McMannon. In *Federal oversight hearing on Native American legislation*. Sacramento: California Indian Legal Services.

Richards, R. T. 1996. Special forest product harvesting in the Sierra Nevada. In *Sierra Nevada Ecosystem Project: Final report to Congress*, vol. III. Davis: University of California, Centers for Water and Wildland Resources.

Rucks, P., Lake Tahoe Basin Management Unit. 1995. Conversations and e-mail with Linda Reynolds.

Slagle, A. 1989. *News from Native California*. Berkeley, CA: Heyday Books.

Steward, J. 1993. Ethnography of the Owens Valley Paiute. *University of California Publications in American Archaeology and Ethnology* 33:233–350.

Stewart, O. C. 1978. Litigation and its effects. In *California*, edited by R. F. Heizer, 705–12. Vol. 8 of *Handbook of North American Indians*. Washington, DC: Smithsonian Institution Press.

Stewart, W. C. 1996. Economic assessment of the ecosystem. In *Sierra Nevada Ecosystem Project: Final report to Congress*, vol. III. Davis: University of California, Centers for Water and Wildland Resources.

Stidham, L., California Indian Legal Services. 1995. Telephone conversation with Linda Reynolds, 30 June.

U.S. Commission on Civil Rights. 1981. *Indian tribes: A continuing quest for survival*. Washington, DC: U.S. Government Printing Office.

Velasquez, F., Calaveras Me-wuk tribal liaison with USFS. 1995. Letter to Linda Reynolds, on file at the Inyo National Forest.

Consultation Letter

United States Department of Agriculture   873 N. Main St.
Forest Service   Bishop, CA 93514
Inyo National Forest   (619) 873-2400
    TDD (619) 873-2538

Reply to: 1500

Date: 22 February 1995

To Whom It May Concern

Hello:

My name is Linda Reynolds and I am the Forest Archaeologist on the Inyo National Forest. Last year I was also made Tribal Governments Program Manager because of my years of experience working with the Paiute and Shoshone peoples whose traditional territory includes the Inyo. I am contacting you because I am coordinating the gathering of information for a report on tribal governments and communities for the Sierra Nevada Ecosystem Project (SNEP).

SNEP is an assessment of the Sierra Nevada ecoregion requested by Congress in 1992. It is not a Forest Service management evaluation or plan. In addition to a scientific evaluation of old-growth forests, watersheds and significant natural areas, it will include consideration of over-all ecological, social and economic factors. Participating scholars are from State and Federal government and academia. They will consider past conditions, the present situation, and possible future conditions based on current trends. SNEP is not a legal document like an Environmental Impact Report, nor will it recommend specific alternatives for future management. What it will do is outline for Congress a set of management options and their possible outcomes. The reports that go into compiling the reports for Congress, including this one, will be made available to the public and become part of the information base that land management agencies have access to. There is no formal post-SNEP follow-up planned. The final report will be completed in December of 1995. More detailed information on SNEP is provided in Attachment A.

Because tribal governments and communities are culturally, historically and legally distinct from other rural communities and governments a separate report will be devoted to tribal concerns. An outline of the report is provided in Attachment B. This is of course subject to modification as we get input.

The focus will be on contemporary tribal governments and communities and how they interact with other communities and environments in the Sierra Nevada ecoregion. Other reports that will deal with aspects of Native American history and concerns will include one by Kat Anderson of the University of Kansas on traditional land management practices and one by Connie Millar of the Forest Service's Pacific Southwest Research Station dealing with significant areas, including cultural sites. In addition, a section on forestry in Sequoia Kings Canyon by Bill Stewart of the University of California, Berkeley examined the forestry practices of four separate entities, including Tule River Reservation.

General community involvement workshops are being conducted in specific areas by Jonathan Kusel of the University of California, Berkeley. In February workshops were held in Bishop, Chico, and Davis. Others may be scheduled for Quincy and the central Sierra.

I have to have this report into the editorial group on the first of June. This gives us very little time to work with, but I feel the effort is justified. This assessment acknowledges the vital role of contemporary tribal governments in the Sierra Nevada ecoregion and highlights the importance of including tribal government concerns in a major assessment like this.

Because of the short timeline for completion, I am making use of the Forest Services Tribal Government Program to get the word out. I have asked colleagues in the SNEP study area to share this letter with you and invite your participation. Some of you may receive more than one notification, but I am sure we all agree it is better to have over-kill than leave any group out.

What I would like is your input on the outline, any suggestions for improvement, what topics you feel need to be discussed, and any examples of current projects. It is critical that tribal people are the ones who define their values and interests. We plan to have the Draft Final report completed by April 30, and available for review before the Final is completed and submitted to the SNEP Coordinating Committee. This summer alternative management scenarios will be developed from all the reports and you will have the opportunity to review these scenarios and their effect on tribal governments.

I am working with Kacy Collons, a graduate student at the University of California, Berkeley. Kacy will be getting a dual degree in law and urban development. She is very interested in environmental law and has worked with tribal people in the past. Kacy's contribution to the project will be a section which focuses on four individual examples. The groups chosen will not be presented as representative of all tribes and communities. They will be included as specific examples to fill in the larger picture.

Sonia Tamez, Tribal Relations Coordinator for the Forest Service Region 5, whom many of you know, is involved in the project, too. Sonia is consulting with the Native American Heritage Commission and the Committee on the Status of California Indians. Connie Millar of the Pacific Southwest Research Station, geneticist on the SNEP Science Team, and chair of the SNEP Coordinating Committee is over-seeing the project. Connie is available to answer any questions you have about SNEP and this project. Her phone number is (510) 559-6435.

My work phone is (619) 873 2423 and I can be reached evenings at (619) 387 2483. I am looking forward to hearing from you and working with you on development of this report.

Sincerely,

/s/

LINDA A. REYNOLDS
Tribal Government Program Manager
SNEP Associate

ATTACHMENT A
(Condensed from *SNEP Update* June, 1994)

The Sierra Nevada Ecosystem Project: An Introduction

The Sierra Nevada Ecosystem Project (SNEP) is an assessment of the Sierra Nevada ecoregion. Requested by Congress in 1992, this ecosystem evaluation undertakes a scientific review of late-successional forest, key watersheds, and significant natural areas on federal lands of the Sierra Nevada ecoregion. It also broadly evaluates the entire set of Sierra Nevada ecosystems, including their social, economic, and ecological components. A scientific assessment, as defined for this project, is one that is clear (e.g., uses explicit assumptions, models, criteria, and methods) and objective, carefully evaluating data quality and the validity of inferences that can be made from these data. The project will be completed in December, 1995.

Effective ecosystem management requires that elements of special concern, such as late-successional forest or intact stream systems, be examined in the context of the over-all ecosystem and its surroundings. Although it is relatively simple to manage for a single ecosystem output, such as water, single element management fails to account for other components through which the ecosystem responds. For example,

it is apparent that management for a single species may trigger changes in the ecosystem that impact other species and hence, not meet broader management goals.

With our society possessing many different values, and competition for resources in the Sierra Nevada intense, we must address all ecosystems components and processes in assessing the Sierra Nevada. For instance, we know that diversions of water for agricultural or residential use competes with preservation of organisms that require unimpeded streamflow. Forest grazing competes with use of forage resources by wildlife and may affect protection of riparian systems. The timber economy competes with organisms that depend upon the preservation of intact forests for their existence. Human settlement in the Sierra Nevada and resources that it requires compete with protection of wildlands. Development and transportation across much of California produce air pollution that stresses the forest and impedes clear, scenic vistas.

Conflict such as these are poorly resolved by single-element analysis and management, as the ecosystems that underlie these issues are far more complex and interdependent than such management allows. The more refined our understanding of ecosystem functions and interactions, and the more accurate our measures of the dynamics of internal and external ecosystem processes, the better we are able to choose ecosystem management solutions that minimize unexpected and perhaps undesirable outcomes.

Moreover, some ecosystem management scenarios present a more satisfactory combination of socially desired outcomes than do others. Desirable short-term outcomes may not be sustainable over decades or centuries. Scientists can estimate what the outcomes of a particular set of policies or actions will be, but they cannot provide a "best" choice, which is beyond the bounds of science. This decision and direction must be made by society, and the effects of such a decision monitored in the ecosystem to allow future adjustments in management practices.

The overall goal of the Sierra Nevada Ecosystem Project is to provide an accurate, multidimensional ecosystem assessment such that key structural components and functional processes can be identified and adequately described to enable the management of these systems at sustainable levels into the future. In doing this, we follow the multiple charges given to the SNEP Science Team by Congress and the SNEP Steering Committee, as well as further direction held within the correspondence between the Chief Forester and interested members of Congress. These documents direct us to address specific features of the Sierra Nevada, including "old-growth forests" and "key watersheds." Although the Science Team believes that late-successional forest and intact stream systems are indeed of appropriate concern, Congress and society will best be served by an ecosystem assessment that places these elements in context and determines the linkages among as many important components of the system as possible, so that a fuller understanding of the consequences of policy decisions may be attained.

That said, our ability to accomplish this second, far more ambitious and comprehensive goal is constrained by time, information, and the limitations of our science. We believe that of equal importance to our assessment, either of present conditions or of any predicted future state, is our development of tools, models and approaches that will be available to evaluate the potential outcomes of future management policies as they evolve over the years to come.

What SNEP is . . .

- An ecosystem management approach to assessing resource information.

- An integration of all values.

- An analysis of the landscape from three perspectives:

  - a look at past conditions and the range of variability for those conditions,

  - a look at the present situations,

  - a look at the future based on current trends of management.

- A set of management options and their possible outcomes based on a variety of goals for ecosystem sustainability.

And, what SNEP is not . . .

- Not planning or implementation of plans.

- Not a set of recommended alternatives.

- Not a foreclosure of manager's options.

- Not a definition of management objectives.

- Not a funding source for research.

- Not a delineation of political boundaries.

- Not part of a NEPA process.

The SNEP Approach

The SNEP charge as directed by Congress first asks for a scientific review of late-successional forests, key watersheds, and significant natural areas on federal lands of the Sierra Nevada ecoregion. Concurrently, the SNEP will also conduct a broad assessment of Sierra Nevada ecosystems with the goals of assessing environmental and social conditions throughout the entire Sierra Nevada and developing policy methodologies for achieving sustainable management of Sierra Nevada ecosystems. To achieve these goals, the Science Team is addressing five fundamental questions for each ecosystem issue of concern. These questions are divided into two categories:

For ecosystem assessments:

1. What are current ecological, social and economic conditions?

2. What were historic ecological, social and economic conditions, trends, and variabilities?

3. What are trends and risks under current policies and management?

For policy development and evaluation:

4. What policy choices will achieve ecological sustainability consistent with social well-being?

5. What are the implications of these choices to ecological, social and economic conditions?

The Science Team has formed workgroups and study plans to address these five questions for each of the following issues:

1. *biodiversity* (e.g., biotic community composition and structure, seral stage distribution, genetic diversity, distribution and viability of key terrestrial and aquatic plants and animals, and ecologically and culturally significant areas. Analysis of late-successional forests, key watersheds, and giant sequoia forests will receive emphasis.);

2. *natural disturbance regimes* (e.g., fire, drought, insect and pathogen effects);

3. *human disturbance regimes* (e.g., grazing, silviculture, agriculture, fire suppression, recreation);

4. *water quality and quantity* (e.g., surface waters, hydrological systems, watershed features);

5. *air quality* (e.g., natural and artificial emissions, composition, distribution and levels of airborne particulates);

6. *Sierra Nevada human communities* (e.g., rural community stability and well-being, community participation in resource policy decisions, rural economies);

7. *land development and human settlement patterns* (e.g., demographic, ecologic, economic, infrastructure and employment aspects).

These issues are of greatest public concern and of most importance to ecological sustainability, economic vitality and social health of the Sierra Nevada, as identified in proceedings and other documents from the Sierra Summit, Sierra NOW, Sierra Economic Summit, Sierra Nevada Research Planning Team, and many smaller local groups.

The team will evaluate each issue and determine historic and current environmental conditions to establish baseline conditions for future management, assessing trends and risks under current management policies, determining which policy choices will provide the highest probabilities for achieving ecosystem sustainability, and determining the implications of these choices for the economy. Recognizing that issues vary in priority regionally within the Sierra Nevada and that available data is variable in quality, we are not attempting to evaluate each issue with equal emphasis.

In addition to assessing individual elements and issues we will synthesize assessments by subregion (Modoc-Lassen; northern Sierra Nevada; central-western Sierra Nevada; central-eastern Sierra Nevada; Lake Tahoe Basin; southern Sierra Nevada), and by critical ecosystems (e.g., giant sequoia, riparian systems). Where appropriate and illustrative, we will provide case studies of assessments of environmental and social issues and of institutional approaches to ecosystem management and community participation in decision-making.

It is important to understand that each issue is not studied in isolation. Workgroups tasked to answer the five questions above are composed of an interdisciplinary cadre of Science Team members, each of whom are linked to other workgroups. Synthesis will also be achieved through simulation modeling where critical ecosystem elements and processes are quantitatively linked. Thus, results will be fully integrated for the entire system.

Development and Evaluation of Policy Choices

In developing an approach to policy analysis (questions 4 and 5), we must combine Congressional direction for the SNEP that specifically focuses on late-successional forests and key watersheds with direction we received regarding the integrated ecosystem study. To address requirements of both, we propose to:

1. *Develop and evaluate alternatives for Sierra Nevada-wide sustainable management of late-successional forests and key watersheds.* Using the extensive database and analyses produced by the initial SNEP assessment of forests and watersheds, we will develop alternative strategies for maintaining and restoring these late-successional forests and critical watersheds.

2. *Develop methodologies for subregional simulations of management alternatives that integrate diverse ecological and social inputs and outputs, and test these methodologies in representative Sierra Nevada subregions.* The subregional simulations will use inputs of many attributes (e.g., areas by different land-allocation categories, sivicultural treatments, road treatments) to provide information on forest structures and commodity outputs (e.g., commercial timber harvest volume, biomass volume, water, forage, wildlife habitat, seral stage) over time under each alternative. This information would then be used to assess risks to species and ecosystem processes, and effects on local and regional economies. At the subregional level, simulations will be implemented

in representative case studies in the western and eastern Sierra Nevada.

3. *Develop an integrated spatial model for simulating management alternatives at the watershed level, and use this in at least one representative Sierra Nevada watershed.* The objective of this analysis is to integrate and link individual models developed or used by the SNEP, such as vegetation-change models, hydrologic models, human settlement models, and forest management models, and to identify possible solutions for maintaining ecosystem sustainability across public and private ownerships. An important part of the model will be the ability to portray cumulative effects of alternatives and, in the case of alternative developments, to control cumulative effects within policy guidelines.

The SNEP will then use this model to simulate, more accurately than in the past, the implications of the many policies that prescribe spatial relationships among seral-stage patches (such as wildlife corridors), nonlinear cumulative effects (such as water quality), and simulation of non-timber activities (such as recreation, water diversions, human settlement). This model will be tested in at least a single Sierra Nevada watershed (the Cosumnes River basin), where sufficient data exist to allow such testing.

Finally, integrated ecosystem management requires that concerned publics be involved in the generation and evaluation of management strategies and alternatives. For policy analysis in the broader ecosystem study (the latter two objectives above), the Science Team will develop and evaluate methods for scientists and public representatives to cooperate in articulating issues for which alternative solutions can be considered. Public representatives will come from communities within the Sierra Nevada, and from other organizations and groups that have a vested interest in the future of the Sierra Nevada. The approach will attempt to improve upon the public participation experience of federal, state and county agencies by gathering and evaluating information from those who have intimate ecosystem knowledge, and by learning from public representatives how they perceive the benefits and costs of particular management alternatives are being distributed.

ATTACHMENT B
Draft Outline
TRIBAL GOVERNMENTS IN
THE SIERRA NEVADA REGION

I. *Introduction:* This assessment will discuss Indian tribes and communities in the SNEP study area as a distinct social group with ancient ties to the Sierra Nevada, who helped shape the historic landscape through traditional land management practices, and whose separate legal status must be taken into account in any present or future land management scenarios.
   A. Indian tribes and communities are culturally, historically and legally distinct from other rural communities.
      1. Sovereign status of recognized tribes.

2. Trust Responsibilities of the Federal Government.
3. The condition and rights of native peoples is a global issue. Indian people in the United States represent both a national and a local issue where there are tribal governments and communities, as there are throughout the Sierra Nevada. The issue is old as the United States' military annexation of California from Mexico in 1846, and it is an issue which will remain with us into the foreseeable future.

B. Relationship of Indian tribes and communities to Sierra Nevada ecosystems.
  1. Indian tribes, communities, and individuals are private land holders (reservations, rancherias, and allotments).
  2. Tribal governments are sovereign nations separate from the federal, state, and counties where they are found.
  3. Tribal governments have a government to government relationship with the federal government and federal land comprises over half of the Sierra Nevada.
  4. In some cases there are reserved rights on public lands.
  5. Indian people relate to and use the land differently than other Sierran communities.
  6. Traditional practitioners have knowledge of traditional land-use practices.
  7. Indian people have a spiritual connection to their ancestral lands and they feel they are the original land managers with on-going responsibilities and rights.

C. Public perceptions.
  1. That California was a pristine wilderness before non-indian settlement.
  2. That California Indians were all "diggers"; no understanding of differences or cultural complexity.
  3. Indians are the "invisible minority." Non-indians forest users tend to think that they are all gone.
  4. Misperception that Indians receive "special treatment" through their minority status. Lack of understanding the legal differences between Indians and non-indians.
  5. Outright prejudice in rural communities.

D. Sources of information for this assessment.
  1. Discussion of published literature.
  2. BIA
  3. Tribes and Communities. We will make use of the Forest Service's Tribal Government Program to contact groups.

II. *Native Californians*
A. Prehistory.
  1. Native California at the time of contact with European nations. There will be a brief discussion of prehistory; the focus will be on the socio-political structure at contact.
  2. Map of traditional territories.
B. History. Native Californians have been treated differently than other native peoples in the contiguous United States. A brief legal history beginning in with the military occupation of California by the US in 1846 will deal with the following topics:
  1. California is a non-treaty state.

2. The "Indians of California" are a legally distinct entity.
3. "Recognition"
  a. Non-federal "treaties"
  b. Military reservations
  c. Executive Order tribes
  d. Commission on the Status of California Indians

C. Contemporary tribes and communities within the SNEP study area. This section will also consider those groups who have traditional ties to the land.
  1. Scale of analysis: The tribe or community will be the primary level of analysis.
  2. Statistics
    a. Membership
    b. Reservation/Rancheria
    —Date Established
    —Size
    —Status (recognized, etc.)
    —Map

III. Present Conditions
A. Federal and state law as it pertains to Indian people and tribes in relation to land management.
  1. PL 280 State/law enforcement
  2. Indian Gaming Act
  3. Indian Child Welfare Act
  4. Winters Doctrine/water rights
    a. Owens Valley
  5. Etc.

B. Agencies' Tribal Relations Programs
  1. Clinton's EO
  2. Forest Service
  3. Park Service
  4. BLM

C. Traditional uses on non-tribal land
  1. Hunting/fishing/harvesting
    a. Traditional vegetation management needed to restore plants.
    b. Herbicide use; conflict between traditional uses and contemporary needs.
    c. Commercial vs. traditional use, e.g., pinyon pine nuts, mushrooms.
  2. Ceremonies, sacred sites.
  3. Access to private lands.

D. Rural development

IV. Trends
A. Land acquisition and retention rights.
B. Question of unratified treaties and retention of aboriginal rights still to be settled.
C. Increase in recognized tribes.
D. Increase in requests for land acquisition.
E. Increased interest in public lands.
F. Legislative trends.
G. Effects of environmental policy on minority communities.

APPENDIX 10.2

Indian Tribes and Communities

This appendix contains information about tribal governments and communities with traditional cultural lands in the SNEP study area. It is divided into three sections: Federally Recognized Tribes, Tribes Seeking Federal Recognition, and Restoration Candidates.

The following information is taken from Bureau of Indian Affairs *1990 Field Directory* and *1993 Tribal Directory*, supplemented with additional information learned during the course of this study.

FEDERALLY RECOGNIZED TRIBES

Alpine County

Tribe: Washoe
Tribal affiliation: Washoe
Reservation: Colonies at Woodfords, California; Dresslerville, Stewart, and Carson, Nevada
Population: 1,500 (approximate)
Land base: 33,603 ha (83,000 acres) in California and Nevada
Tribal office: 919 U.S. 395 South, Gardnerville, NV 89410

Amador County

Tribe: Buena Vista Rancheria of Me-Wuk Indians
Tribal affiliation: Me-Wuk
Reservation: Buena Vista Rancheria
Population: 1
Land base: None
Tribal office: 4650 Coalmine Road, Ione, CA 95640
Remarks: Tribe restored to federal recognition under class action suit *Tille Hardwick v. United States of America*, C-79-1910SW. Judgment filed December 22, 1983. There are no tribal trust lands; lands owned in fee status: 72 ha (67.5 acres).

Tribe: Jackson Band of Mi-wuk Indians
Tribal affiliation: Me-Wuk, Miwok, Mi-Wuk
Reservation: Jackson Rancheria
Population: Within rancheria, 20; adjacent, 8
Land base: 134 ha (330.66 acres)
Tribal office: PO Box 150, Jackson, CA 95642
Remarks: Act of March 3, 1893 (27 *Stat.* 628, c. 209) appropri-

ated $10,000 for purchase of land, etc., for the "Digger" Indians of Central California at Jackson. The rancheria was established January 7, 1895.

Butte County

Tribe: Berry Creek Rancheria of Maidu Indians
Tribal affiliation: Tyme Maidu
Reservation: Berry Creek Rancheria
Population: Within rancheria, 15; adjacent, 196
Land base: 26 ha (65 acres)
Tribal office: 1779 Mitchell Avenue, Oroville, CA 95966
Remarks: Original tract purchased March 1, 1916, by the federal government from the Central Pacific Railway Co. for the Dick Harry Band of Indians. Title to the land was vested in the United States of America with the Indians having a right only to occupancy and use of the lands, unless otherwise authorized by Congress. Approximately 13 ha (32 acres) were purchased with a HUD grant and accepted into trust pursuant to the Indian Land Consolidation Act of 1983.

Tribe: Chico Band of Mechoopda Indians
Tribal affiliation: Mechoopda
Reservation: Chico Rancheria
Population: Within rancheria, 0; adjacent, 300
Land base: 0.2 ha (0.5 acre) cemetery
Tribal office: 3006 Esplanade, Suites G and H, Chico, CA 95926
Remarks: On April 17, 1992, the status and rights of the Chico Band of Mechoopda Indians were reinstated by the federal government to the status they had before termination.

Tribe: Enterprise Rancheria of Maidu Indians
Tribal affiliation: Maidu
Reservation: Enterprise Rancheria
Population: Within the rancheria, 24; adjacent, 336
Land base: 16 ha (40 acres)
Tribal office: unknown
Remarks: One parcel sold, purchased under the Acts of 1906 and 1908. Lands purchased by authority of Act of August 1, 1914 (38 *Stat.* 58–59).

Tribe: Mooretown Rancheria of Maidu Indians
Tribal affiliation: Maidu-Concow
Reservation: Mooretown Rancheria
Population: Within the rancheria, 225; adjacent, 200

Land base: None

Tribal office: PO Box 1842, Oroville, CA 95965

Remarks: Federal recognition restored to tribe under class action suit *Tille Hardwick v. United States of America, C-79-1910SW*. Judgment filed December 22, 1983. There are no tribal trust lands; individually owned parcels remain in "fee" status.

Calaveras County

Tribe: Sheep Ranch of Me-wuk Indians

Tribal affiliation: Me-wuk, Miwok

Reservation: Sheep Ranch Rancheria

Population: Unknown

Land base: 0.37 ha (0.92 acres)

Tribal office: Unknown

Remarks: Purchase for homeless California Indians without designation of tribe on April 5, 1916.

El Dorado County

Tribe: Auburn Rancheria

Tribal affiliation: Nisenan, Southern Maidu

Reservation: Auburn Rancheria

Population: Unknown

Land base: Unknown

Tribal office: PO Box 3035, Route E, Auburn, CA 95603

Remarks: Termination: August 16, 1967; restoration 1995.

Tribe: Shingle Springs Band of Miwok Indians

Tribal affiliation: Miwok

Reservation: Shingle Springs Rancheria

Population: Within rancheria, 16; adjacent, 247

Land base: 65 ha (160 acres)

Tribal office: PO Box 1340, Shingle Springs, CA 95682

Remarks: Lands were purchased by the secretary of the interior.

Fresno County

Tribe: Big Sandy Rancheria of Mono Indians

Tribal affiliation: Western Mono

Reservation: Big Sandy Rancheria

Population: Within rancheria, 61; adjacent, 47

Land base: 48 ha (119.5 acres)

Tribal office: PO Box 337, 7302 Rancheria Lane, Auberry, CA 93602

Remarks: Pursuant to the judgment entered in *Big Sandy Band v. Watt*, the community and individually owned lands were accepted into trust.

Tribe: Cold Springs Rancheria of Mono Indians

Tribal affiliation: Mono

Reservation: Cold Springs Rancheria (Sycamore Valley)

Population: Within rancheria, 158; adjacent, 101

Land base: 63 ha (154.65 acres)

Tribal office: PO Box 209, Tollhouse, CA 93667

Remarks: Original land base established by Executive Order 2078 of November 10, 1914, which excluded lands from the Sierra National Forest, California, for the Cold Springs Band of Indians.

Tribe: Table Mountain Rancheria

Tribal affiliation: Yokut

Reservation: Table Mountain Rancheria

Population: Unknown

Land base: 25 ha (60.93 acres)

Tribal office: PO Box 177, Friant, CA 93626-0177

Remarks: The original rancheria was purchased under the authority of the act of May 18, 1916 (39 *Stat.* 123, 12), date of deed, September 27, 1916, deed in name of United States of America. The rancheria began termination under the California Rancheria Act pursuant to order in *Table Mountain v. Watt*. Lands have been restored to trust (15 ha [36.96 acres]) individually owned.

Inyo County

Tribe: Big Pine Paiute Tribe of Owens Valley

Tribal affiliation: Paiute-Shoshone

Reservation: Big Pine Reservation

Population: Within reservation, 371; adjacent, 32

Land base: 113 ha (279 acres)

Tribal office: PO Box 533, Big Pine, CA 93513

Remarks: The act of April 20, 1937, authorized the secretary of the interior to exchange Indian lands and water rights for lands owned by the City of Los Angeles and Inyo and Mono Counties (*Stat.* 50 c. 114).

Tribe: Bishop Indian Tribal Council

Tribal affiliation: Paiute-Shoshone

Reservation: Bishop Reservation

Population: Within reservation, 927; adjacent, 69

Land base: 354 ha (875 acres)

Tribal office: PO Box 548, Bishop, CA 93584

Remarks: The act of April 20, 1937, authorized the secretary of the interior to exchange Indian lands and water for lands owned by the City of Los Angeles and Inyo and Mono Counties (*Stat.* 50 c. 114).

Tribe: Fort Independence Indian Community of Paiute Indians

Tribal affiliation: Paiute

Reservation: Fort Independence Reservation

Population: Within the reservation, 83; adjacent, 40

Land base: 224 ha (552.24 acres)

Tribal office: PO Box 67, Independence, CA 93526

Remarks: Executive Order 2264 of October 28, 1915, set apart lands for this reservation. Executive Order 2375 of April 29, 1916, enlarged the reservation. Eighty-one ha (200 acres) were added in 1995 through administrative transfer of adjacent BLM lands by the secretary of the interior.

Tribe: Paiute-Shoshone Indians of the Lone Pine Community

Tribal affiliation: Paiute-Shoshone

Reservation: Lone Pine Reservation

Population: Within the reservation, 232; adjacent, 64
Land base: 96 ha (237 acres)
Tribal office: 101 S Main Street, Lone Pine, CA 93545
Remarks: The act of April 20, 1937, authorized the secretary of the interior to exchange Indian lands and water for lands owned by the City of Los Angeles and Inyo and Mono Counties (*Stat.* 50 c. 114).

Tribe: Death Valley Timbisha Shoshone Tribe
Tribal affiliation: Western Shoshone
Reservation: Timbisha Band of Shoshone Indians
Population: Within the rancheria, 55; adjacent, 145
Land base: None
Tribal office: PO Box 206, Death Valley, CA 92328
Remarks: Notice published in the *Federal Register,* November 4, 1982, acknowledged Death Valley Timbisha Western Shoshone Band as federally recognized. Notice was based on determination that the group satisfies all the criteria set forth in 25 *CFR* 83.7 (formerly 54.7). Members at present reside on a 16 ha (40 acre) site in Death Valley National Park, commonly referred to as the Indian Village. Under the California Desert Protection Act of 1995, the tribe and affected agencies have been directed to complete a study for land suitable for restoration to the tribe as a reservation.

Lassen County

Tribe: Susanville Indian Rancheria
Tribal affiliation: Paiute–Maidu–Pit River–Achomawi–Atsugewi–Washoe
Reservation: Susanville Indian Rancheria
Population: Within the rancheria, 145; adjacent, 228
Land base: 61 ha (150.53 acres)
Tribal office: PO Drawer U, Susanville, CA 96130
Remarks: Original rancheria purchased August 15, 1923, for homeless California Indians, deed in the name of United States of America. Public Law 95-459 approved October 14, 1978, provided for the United States of America to hold 49 ha (120 acres) in trust for the rancheria.

Madera County

Tribe: North Fork Rancheria of Mono Indians
Tribal affiliations: Mono
Reservation: North Fork Rancheria
Population: Within the rancheria, 75; adjacent, 205
Land base: None
Tribal office: PO Box 120, North Fork, CA 93643
Remarks: Rancheria restored to federal recognition under class action suit *Tille Hardwick v. United States of America, C-79-1910SW.* There are no tribal lands; individually owned lands restored to trust: 32 ha (80 acres).

Tribe: Picayune Rancheria of Chukchansi Indians
Tribal affiliation: Chukchansi
Reservation: Picayune Rancheria
Population: Within the rancheria, 12; adjacent, 8
Land base: 16 ha (38.76 acres)

Tribal office: PO Box 269, Coarsegold, CA 93614
Remarks: Restored to federal recognition under class action suit *Tille Hardwick v. United States of America, C-79-1910SW.* Judgment filed December 22, 1983. There are no tribal lands. One parcel, consisting of 12 ha (28.76 acres), was restored to trust for an individual.

Modoc County

Tribe: Alturas Rancheria of Pit River Indians
Tribal affiliation: Pit River–Achomawi–Atsugewi
Reservation: Alturas Indian Rancheria
Population: Within the rancheria, 8; adjacent, 0
Land base: 8 ha (20 acres)
Tribal office: PO Box 1035, Alturas, CA 96101
Remarks: Rancheria established by act of June 21, 1906, appropriating funds for purchase of lands for California Indians. Rancheria purchased by provisions of act of January 24, 1923 (43 *Stat.* L 1188); purchase date: September 8, 1924.

Tribe: Cedarville Rancheria of Northern Paiute Indians
Tribal affiliation: Northern Paiute
Reservation: Cedarville Rancheria
Population: Within the rancheria, 10; adjacent, 3
Land base: 8 ha (20 acres)
Tribal office: PO Box 126, Cedarville, CA 96104
Remarks: Rancheria established under the authority of acts of June 21, 1906, and later, appropriating funds for purchase of lands for California Indians. Purchased October 19, 1915.

Tribe: Fort Bidwell Indian Community of Paiute Indians
Tribal affiliation: Paiute
Reservation: Fort Bidwell Reservation
Population: Within the Reservation, 124; adjacent, 39
Land base: 1350 ha (3,334.97 acres)
Tribal office: PO Box 127, Fort Bidwell, CA 96112
Remarks: A joint resolution of January 30, 1879, authorized the secretary of the interior to use the abandoned Fort Bidwell Military Reserve for an Indian Training School. An act of January 27, 1913, granted land to the People's Church for a cemetery and right-of-way over the Fort Bidwell Indian School Reservation, the Indians to have right of internment therein (37 *Stat.* 652, c. 15). Executive Order 2679 of August 3, 1917, enlarged the reservation.

Tribe: Pit River Tribe of California
Tribal affiliation: Pit River
Reservation: Likely Rancheria
Population: None
Land base: 0.53 ha (1.32 acres) cemetery
Tribal office: None
Remarks: See Pit River Tribe of California, Shasta County.

Tribe: Pit River Tribe of California
Tribal affiliation: Pit River–Achomawi–Atsugewi
Reservation: Lookout Rancheria
Population: Within the rancheria, 16
Land base: 16 ha (40 acres)

Tribal office: None
Remarks: See Pit River Tribe of California, Shasta County.

Mono County

Tribe: Utu Utu Gwaitu Paiute
Tribal affiliation: Paiute
Reservation: Benton Paiute Reservation
Population: Within the reservation, 82
Land base: 166 ha (410 acres)
Tribal office: PO Box 909, Benton, CA 92512
Remarks: Executive Order July 22, 1915, recognized the tribe. One ha (2.5 acres) purchased by the tribe using HUD grant funds August 24, 1984. One hundred one hectares (250 acres) were transferred from adjacent BLM lands through administrative order of the secretary of the interior in 1995.

Tribe: Bridgeport Paiute Indian Colony
Tribal affiliation: Paiute
Reservation: Bridgeport Indian Colony
Population: Within the reservation, 53; adjacent, 26
Land base: 32 ha (80 acres)
Tribal office: PO Box 37, Bridgeport, CA 93517
Remarks: Rancheria established October 18, 1974, by Public Law 93-451. Sixteen ha of adjacent BLM land were transferred through administrative action of the secretary of the interior in 1995.

Plumas County

Tribe: Greenville Rancheria of Maidu Indians
Tribal affiliation: Maidu
Reservation: Greenville Rancheria
Population: Within the rancheria, 279; adjacent, 38
Land base: None
Tribal office: 1304 E Street, Suite 106, Redding, CA 96001
Remarks: Rancheria restored to federal recognition under class action suit *Tille Hardwick v. United States of America, C-79-1910SW*. Judgment filed December 22, 1983.

Shasta County

Tribe: Pit River Tribe of California
Tribal affiliation: Eleven Autonomous Bands—Ajumawi, Porige, Astarawi, Atsugewi, Atwamsini, Hammawi, Hewisedawi, Ilmawi, Itsatwi, Kosalektawi, Madesi
Reservation: Ajumawi-Atsugewi Nation
Land base: 3,873 ha (9,567.18 acres)
Tribal office: PO Drawer 1570, Burney, CA 96013
Remarks: The Pit River Nation comprises eleven autonomous bands. Each band head is elected by band members. The chairperson and vice-chairperson are chosen through a general election.

Tribe: Pit River Tribe of California
Tribal affiliation: Pit River–Achumawi–Atsugewi–Wintun
Reservation: Big Bend Rancheria
Population: Within the rancheria, 6
Land base: 16 ha (40 acres)

Tribal office: None
Remarks: See Pit River Tribe of California, earlier.

Tribe: Pit River Tribe of California
Tribal affiliation: Madesi Band of Pit River Indians
Reservation: Montgomery Creek Rancheria
Population: Within the rancheria, 30
Land base: 29 ha (72 acres)
Tribal office: None
Remarks: See Pit River Tribe of California, earlier.

Tribe: Redding Rancheria
Tribal affiliation: Wintun–Pit River–Yana
Reservation: Redding Rancheria
Population: Within the rancheria, 170; adjacent, 30
Land base: 13 ha (30.89 acres)
Tribal office: 2000 Rancheria Road, Redding, CA 96001-5528
Remarks: Federal recognition restored on December 15, 1985, as a result of class action suit *Tille Hardwick v. United States of America*.

Tribe: Pit River Tribe of California
Tribal affiliation: Pit River–Ajumawi–Atsugewi
Reservation: Roaring Creek Rancheria
Population: Unknown
Land Base: Unknown
Tribal Office: None
Remarks: See Pit River Tribe of California, Shasta County.

Tulare County

Tribe: Tule River Indian Tribe
Tribal affiliation: Yokut
Population: Within the reservation, 590; adjacent, 260
Land base: 22,411 ha (55,356 acres)
Tribal office: PO Box 286, Porterville, CA 93258
Remarks: An act of April 8, 1864, authorized the establishment of Indian reservations in California (13 *Stat.* 39–41 c. 48). An executive order of January 9, 1873, established this reservation and an order of October 3, 1873, canceled the order of January 9, 1873, and reestablished the reservation. (An act of May 17, 1923, changed the boundaries of the Tule River Reservation [45 *Stat.* 600–601 c. 614].)

Tuolumne County

Tribe: Tuolumne Band of Me-wuk Indians
Tribal affiliation: Me-wuk, Miwok, Yokut
Reservation: Tuolumne Rancheria
Population: Within the rancheria, 169; adjacent, 445
Land base: 144 ha (355.77 acres)
Tribal office: 19595 Miwuk Street, Tuolumne, CA 95379
Remarks: Original purchase of 177 ha (289.52 acres) on October 25, 1910, under authority of acts of June 21, 1906, and April 30, 1908. Executive Order 1517 of April 13, 1912, added 14 ha (33.58 acres), and an additional 5 ha (12.67 acres) were purchased on April 14, 1978, under authority of the act of June 18, 1934.

Tribe: Chicken Ranch Rancheria of Me-wuk Indians
Tribal affiliation: Miwok, Me-wuk
Reservation: Chicken Ranch Rancheria
Population: Within the rancheria, 3; adjacent, 3
Land base: 1 ha (2.85 acres)
Tribal office: PO Box 1699, Jamestown, CA 95327
Remarks: Tribe restored to federal recognition under class action *Tille Hardwick v. United States of America, C-79-1910SW.* Judgment filed December 22, 1983.

TRIBES SEEKING FEDERAL RECOGNITION

Tribe: American Indian Council of Mariposa County
Address: PO Box 1200, Mariposa, CA 95338

Tribe: Antelope Valley Paiute Tribe
Address: PO Box 119, Coleville, CA 96107

Tribe: Big Meadows Lodge Tribe
Address: PO Box 362, Chester, CA 96020

Tribe: Calaveras County Band of Miwok Indians
Address: Star Route 1, Bald Mountain Road, West Point, CA 95255

Tribe: Choinumni Tribe
Address: 3330 East Dakota, #113, Fresno, CA 93726

Tribe: Chukchansi Tribe
Address: PO Box 852, Oakhurst, CA 93644

Tribe: Dunlap Band of Mono Indians
Address: PO Box 126, Dunlap, CA 93621

Tribe: Ione Band of Miwok Indians
Address: Route 1, Box 191, Ione, CA 95640

Tribe: Kern Valley Indian Community
Address: PO Box 168, Kernville, CA 93238

Tribe: Maidu Nation
Address: PO Box 204, Susanville, CA 96130

Tribe: Mono Lake Indian Community
Address: PO Box 237, Lee Vining, CA 93541

Tribe: Northern Maidu Tribe
Address: 516 Grand Avenue, Susanville, CA 96130

Tribe: North Fork Band of Mono Indians
Address: PO Box 49, North Fork, CA 93643

Tribe: Plumas County Indians, Inc.
Address: PO Box 102, Taylorsville, CA 95947

Tribe: Tehatchapi Indian Tribe
Address: 219 East H Street, Tehatchapi, CA 93561

Tribe: Wukchumni Tribe
Address: 1426 W Sunny View, Visalia, CA 93291

RESTORATION CANDIDATES

Tribe: Chico Rancheria
Affiliation: Wailiki and Maidu
Address: 4237 Third Avenue, PO Box 988, Lakeport, CA 95453
Termination: June 2, 1967

Tribe: Nevada City Rancheria
Affiliation: Maidu
Address: Nevada City
Termination: September 22, 1964

TIMOTHY P. DUANE
Department of City and Regional Planning
and Department of Landscape
  Architecture
University of California
Berkeley, California

11

Human Settlement, 1850–2040

ABSTRACT

This assessment of human settlement characterizes the extent of development, its historical levels and spatial distribution, and the factors driving it; makes projections of population growth and alternative land conversion estimates for alternative human settlement patterns; assesses the likely impacts associated with development and the degree to which existing institutional mechanisms anticipate and mitigate them; and sets forth alternative growth management policies that could mitigate those impacts. The importance of human settlement as a factor in the future health and sustainability of Sierra Nevada ecosystems cannot be overstated. The human population of the Sierra Nevada is forecast to triple from 1990 to 2040, while the land area developed for human settlement could potentially quadruple if current patterns of development continue.

The population of the Sierra Nevada more than doubled from 1970 to 1990, and its current population is approximately four times the peak population during the gold rush (1849–1852). Most of the new residents have settled near the historic centers of the gold rush, but modern patterns of human settlement have resulted in much more extensive land conversion. Three out of five Sierra Nevada residents lived on less than 300 mi^2 (less than 1%) in 1990, but human settlement was spread across nearly 1,741 mi^2 at an average density of at least one housing unit per 32 acres. This constituted 5.44% of the entire Sierra Nevada, or nearly 14% of all private land (including industrial timberlands). Up to one-eighth of the entire Sierra Nevada (3,905 mi^2) may have been affected by human settlement in 1990 at an average density of at least one housing unit per 128 acres. There is no clear threshold density at which human settlement results in significant impacts on the health and sustainability of Sierra Nevada ecosystems.

The Sierra Nevada is likely to undergo significant more land conversion to accommodate continuing population growth over the next half-century. Population growth in the metropolitan centers of California is forecast to double the state's population between 1990 and 2040, leading to expansion of the emerging metropolitan centers of the Central Valley that are within commuting distance of the Sierra Nevada foothills. Metropolitan areas near the Sierra Nevada in the state of Nevada are also forecast to continue growing. This growth would create new employment opportunities on the urban edge and extend the reach of reasonable commute times into areas that have not yet faced significant settlement by commuters. The result is likely to be continuing in-migration by commuters, retirees, and former metropolitan-area residents who are seeking a rural or exurban lifestyle offering natural and social amenities. Many of these latter immigrants are likely to accept lower incomes in exchange for these amenities, but they also generally bring human and financial capital with them. They therefore have the potential to generate new employment in the Sierra Nevada.

Because these new residents are likely to have higher incomes than most existing residents, their arrival will put pressure on land and housing prices. The factors driving the exodus to exurbia over the past three decades are likely to continue, resulting in an increasingly homogeneous population of affluent, white, well-educated residents in the commuter and retiree communities near the Central Valley and the Lake Tahoe region. More isolated communities in the northern and eastern Sierra are likely to grow relatively slowly, however, with less pressure on land and housing prices. Existing patterns of human settlement are more stable in these areas, where lower land prices make significant investments in centralized infrastructure uneconomic. Large higher-density developments are likely in the Gold Country, however, where proximity to the Sacramento metropolitan area has already increased land and housing prices significantly. Nonlocal landowners have already consolidated parcels in these areas and have proposed development of several planned communities in the region. Tens of thousands of individuals and corporations own parcels in the Sierra Nevada, but relatively few landowners control most of the private land. Private industrial timber companies control the bulk of the private land in those counties where data are available.

The social, economic, and ecological ramifications of future development will depend upon specific spatial patterns of human settlement in relationship to existing communities, infrastructure services, vegetation and habitat types, and watershed boundaries. Our under-

Sierra Nevada Ecosystem Project: Final report to Congress, vol. II, *Assessments and scientific basis for management options*. Davis: University of California, Centers for Water and Wildland Resources, 1996.

standing of those relationships is still poor at this time. It is therefore impossible to characterize the specific impacts that population growth and human settlement will have in the Sierra Nevada. The range of impacts could be quite significant, however, if existing development patterns continue. Continuing the existing pattern of "sprawl" development with a high-growth scenario could result in human settlement on nearly half the private land in the Sierra Nevada (6,846 mi^2) at an average density of at least one housing unit per 32 acres. A low-growth scenario with the existing pattern of sprawl development would reduce that figure by 44%, to just 3,817 mi^2. This is still significantly greater than the 1,741 mi^2 affected by human settlement at that average housing density in 1990.

Even modified settlement patterns are forecast to result in significant land conversion from 1990 to 2040, suggesting that the scale of population growth alone could lead to significant impacts. A high-growth scenario with a more "compact" form of settlement would result in nearly a doubling of land converted to human settlement, from 1,741 mi^2 to 3,363 mi^2 at an average density of at least one housing unit per 32 acres. A low-growth scenario with a more "compact" form of settlement, on the other hand, could nearly be accommodated within the land area already converted to human settlement at an average density of at least one housing unit per 32 acres in 1990. Through infill and carefully targeted density transfers, the low population forecast for 1990–2040 would require only 1,875 mi^2 (only 8% more than in 1990). Both the scale and pattern of human settlement will therefore affect—and must therefore be considered by—local, state, and federal land and resource management agencies with responsibilities for the health and sustainability of Sierra Nevada ecosystems.

Existing institutional arrangements for land use and environmental planning in the Sierra Nevada appear inadequate for managing rapid population growth and the land conversion process associated with human settlement. Comprehensive updates of both the Nevada County and El Dorado County General Plans appear to have either significantly underestimated the likely future impacts of "buildout" or failed to mitigate significant impacts under the "overriding considerations" provision of the California Environmental Quality Act. Many of these impacts are associated with existing substandard parcels, most of which were established through subdivisions that preceded most of current state planning law.

Innovative growth management strategies to coordinate and consolidate development across these parcels may therefore be necessary if the impacts of future population growth are to be mitigated. Appropriate policies cannot be selected without a better understanding of the relationships between alternative patterns of human settlement and impacts, but creative "open space development design" through site-specific clustering could mitigate some of the likely effects. Other rural and exurban regions have adopted some of these policies, but they have not yet been embraced in the Sierra Nevada. There are a number of social, political, economic, and institutional factors that may explain why growth management has generally been ineffective in the region, but further study is necessary before specific policies are likely to be adopted. The effectiveness of those policies, in turn, will depend upon a wide range of similar factors. Some dimensions of the health and sustainability of Sierra Nevada ecosys-

tems are likely to face significant threats, however, in the absence of successful growth management. It is therefore critical that local, state, and federal land and resource management agencies assess the management implications of continuing extensive and intensive human settlement in the Sierra Nevada. This is particularly true in the western Sierra Nevada foothills, where nearly five out of every six Sierra Nevada residents lived in 1990. This fraction is expected to increase from 1990 to 2040 as regional employment centers in the Central Valley grow, increasing growth pressures in those Sierra Nevada foothill communities within commuting distance of these centers. In contrast, the more remote northern and eastern Sierra Nevada regions are forecast to have relatively slow growth.

INTRODUCTION

Human settlement in the Sierra Nevada has had and will continue to have a profound impact on Sierra Nevada ecosystems. The distribution and abundance of natural resources in the Sierra Nevada, in turn, have had an enormous effect on patterns of human settlement and the types of human activities that have taken place in the Sierra Nevada landscape. This assessment report characterizes the current pattern of human settlement, the historic pattern of human settlement from 1850 to 1990, and a range of future population projections for the Sierra Nevada and alternative scenarios of human settlement patterns. We also discuss the factors driving human settlement in the region and a range of policy alternatives to mitigate the environmental impacts of expanding human settlement.

Our analysis begins with the entire Sierra Nevada, where we describe historic population figures for the region from 1850 to 1990, which are reported for all of the California counties in the Sierra Nevada.[1] These data are not available at a subcounty level until 1970, however, so it is impossible to determine the population of the Sierra Nevada proper (as a subset of the overall population of the Sierra Nevada counties) from 1850 to 1970. More detailed data are available for selected years for some incorporated cities in the Sierra Nevada.

We then summarize subcounty population figures by county census division (CCD) from 1970 to 1990, a period in which the population of the Sierra Nevada more than doubled. That population is only about one-fourth of the population of the counties in the Sierra Nevada, however, highlighting the importance of differentiating the Sierra Nevada proper from the much larger county totals.

We follow with a discussion of the factors driving the explosion of population growth in the Sierra Nevada from 1970 to 1990. These include a wide range of factors outside the Sierra Nevada itself, linking the fate of future population growth to broader state, national, and global trends. The importance of metropolitan expansion in the Bay Area and Sacramento is highlighted. This expansion has been a key factor driving the concentration of population growth occurring in the west-central-north subregion of the "Gold Country" in Nevada, Placer, and El Dorado Counties. Highway access into and across the Sierra Nevada is also critical for higher-density development linked to metropolitan areas, along with access to water, sewers, and power.

We next present projections by the California state Department of Finance (DOF) for county-level population projections for the 1990–2040 period. We then describe a simple model for allocation of these county-level projections to the Sierra Nevada portion of each county based upon the 1970–90 share of county population growth that each Sierra Nevada CCD received. Alternative forecasting methods are discussed, and the reasonableness of the DOF forecasts is evaluated. Likely changes in the subregional distribution of the Sierra Nevada population are then described for 1990–2040. Due to data limitations, however, these changes are presented for only the large aggregate spatial units (CCDs) in the Sierra Nevada. This coarse-scale analysis is inadequate for analysis of the ecological impacts of alternative spatial patterns of human settlement. We therefore examine current and historical patterns of settlement with greater spatial resolution.

Changes in average housing densities are then reported by census block group (CBG) from 1940 to 1990 through a series of maps that graphically illustrate the expansion of human settlement throughout the Sierra Nevada over the past half-century. We then characterize the distribution of human settlement by eleven broad classes of housing density as of 1990 based upon over 50,000 census blocks, the smallest unit available for analysis. This distribution is reported here by CCD, county, and river basin.

In order to get a more detailed understanding of the processes and patterns of human settlement, we next focus on a subregional analysis of population growth and land use patterns in a five-county region that includes Amador, Calaveras, El Dorado, and Nevada Counties and portions of Placer County. This analysis focuses on the distribution of parcel sizes by frequency and area in the five-county area. More detailed analysis is then reported for Nevada and El Dorado Counties, both of which are currently updating their General Plans. Land use patterns and policies under consideration in those General Plans are evaluated in terms of social, economic, and environmental impacts as described in their associated draft environmental impact reports (DEIR's) and based upon our own independent analysis. Both Nevada and El Dorado Counties' human settlement patterns in 1990 and alternative plans for the future are then compared with prevailing patterns of land use throughout the rest of the Sierra Nevada. We also evaluate the feasibility of infrastructure investments assumed in the General Plans. The role of infrastructure is critical in determining future settlement patterns. The General Plan development process and the associated EIR analysis are then reviewed for their capacity to mitigate impacts.

The potential impacts of 1990–2040 population growth on land conversion in each of the counties and CCDs are then estimated based upon a range of alternative assumptions about future population growth and human settlement patterns. Four alternative population growth forecasts and six alternative settlement patterns are considered through four scenarios of future development based upon low- and high-population forecasts and compact versus sprawl settlement patterns. Total land area converted to human settlement under each of the resulting four scenarios is then presented for a range of threshold settlement densities. We also present the unsuccessful results of preliminary attempts to model the spatial patterns of human settlement with finer spatial resolution for the entire Sierra Nevada. Alternative modeling approaches are outlined that hold promise for future assessment. The ecological consequences of settlement are then discussed, including limitations in our present knowledge about the relationships between alternative patterns of human settlement and specific ecological consequences in the Sierra Nevada.

The assessment report concludes with a discussion of alternative policy options available to local, state, and federal land and resource management agencies to mitigate the potential impacts of conversion associated with expanding residential development. The institutional setting for adoption of those policies is then described and evaluated to determine the likelihood of alternative mitigation measures being adopted in the future. Due to significant data limitations, however, we were unable to reach firm conclusions about the efficacy of alternative policy options to mitigate the impacts of human settlement. Suggestions for further research are therefore presented to guide future assessments.

The Setting for Human Settlement in the Sierra Nevada

The Sierra Nevada core region as defined by SNEP is vast and highly heterogeneous in terms of human settlement. Some parts of it are remote and inaccessible, while others are within easy commuting distance of rapidly growing metropolitan regions. Just across the region's western boundary lies the Central Valley, where there are least six rapidly growing urban centers, each with a population greater than 100,000 in 1990. The northern and eastern boundaries of the Sierra Nevada, in contrast, are against the sparsely populated high desert of the Great Basin biogeographical province. These areas are often isolated for months every year as winter snows close the mountain passes linking these rural areas to the rest of California. There are thirty-two counties (twenty-seven in California and five in Nevada) with all or part of their territory within the SNEP study region, but only twenty-two of these counties include portions of the Sierra Nevada proper. Eighteen of these counties are in California, and four are in Nevada. Of these, only nine counties (all in California) lie entirely within the boundaries of the region (figure 11.1).[2]

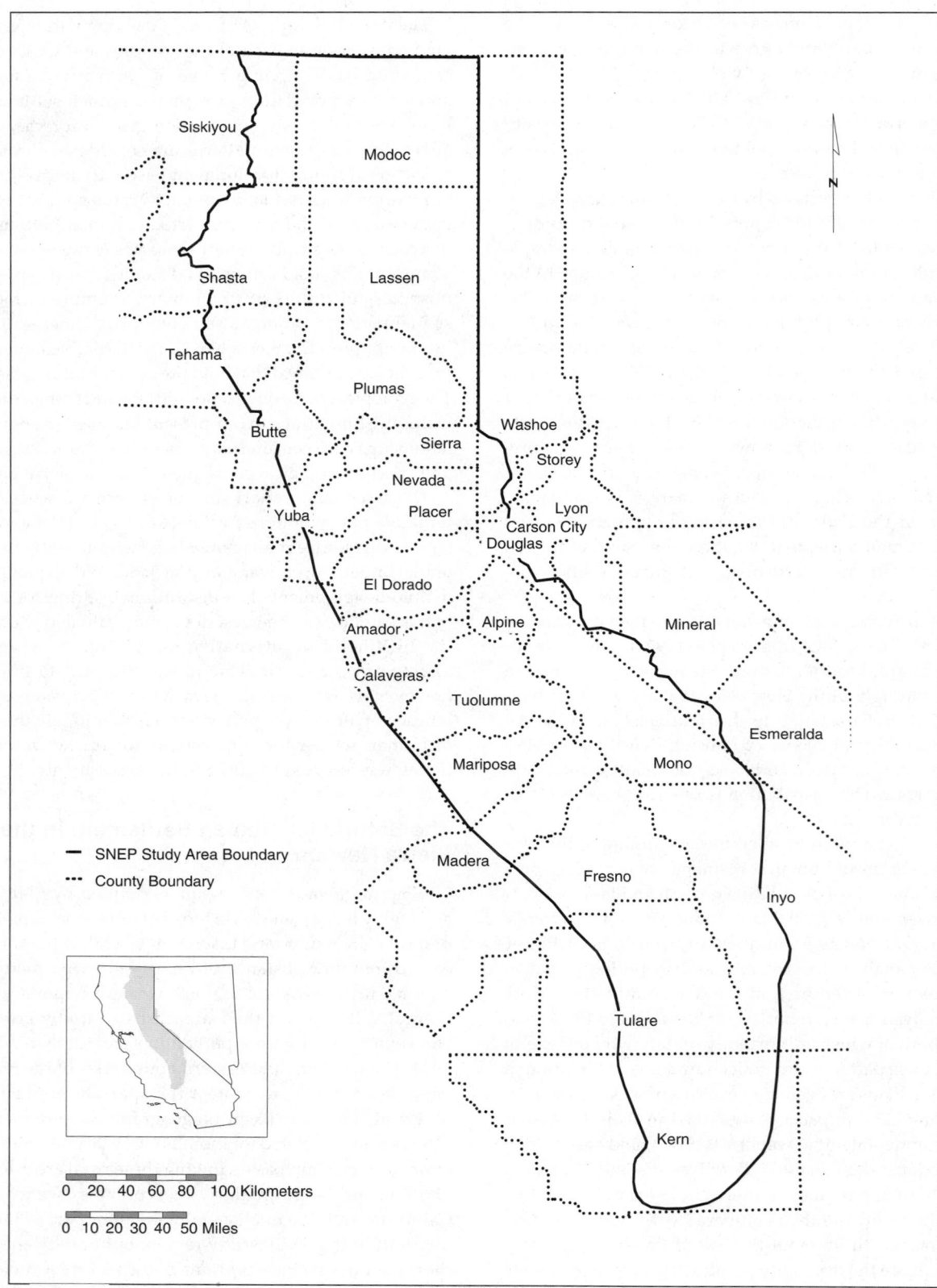

FIGURE 11.1

Counties in the Sierra Nevada region.

Beyond the Central Valley lie the major metropolitan centers of San Francisco and Los Angeles, both within 100 miles of the Sierra foothills. The only major urban centers near the eastern flank of the Sierra are Reno and Carson City, both in the vicinity of Lake Tahoe. Interstate 80 and U.S. Highway 50 connect the Reno, Carson City, and Lake Tahoe regions to the Sacramento metropolitan area and the rest of northern California. The Sierra Nevada is also crossed by state highways: from north to south, 70 (Feather River canyon), 49 (Yuba Pass), 88 (Carson Pass), 4 (Ebbetts Pass), 108 (Sonora Pass), 120 (Tioga Pass), and 178 (Walker Pass). Also providing access from the Central Valley into the Sierra Nevada are Highways 20 (Marysville to Interstate 80 via Grass Valley and Nevada City), 140 (Merced to Yosemite National Park via Mariposa), 41 (Fresno to Yosemite National Park via Oakhurst), 180 (Fresno to Sequoia and Kings Canyon National Parks), 245 and 198 (Visalia to Sequoia and Kings Canyon National Parks), and 190 (Porterville to Sequoia National Forest). Highway 49 traverses the western foothills (from Oakhurst in the south to Sierraville and Loyalton in the north), while Highway 89 cuts across Monitor Pass south of Lake Tahoe and extends north through Truckee and Quincy to Lake Almanor. U.S. Highway 395 skirts the eastern edge of the Sierra Nevada from Susanville in the north through Reno, Carson City, the Mono Basin, and the Owens Valley to southern California. Carson, Ebbetts, Sonora, Tioga and Monitor Passes are all closed seasonally in the winter, from around Thanksgiving until Memorial Day. U.S. Highway 99 connects the string of Central Valley towns west of the Sierra Nevada. Figure 11.2 shows these primary transportation corridors in and near the Sierra Nevada.

This complex pattern of road networks links the Sierra Nevada to social and economic activity throughout California and the world. It brings recreational visitors to access the wonders of the Sierra Nevada and provides for the export of the natural resources that are extracted in the range and sold as commodities in metropolitan markets. The transportation network is therefore a primary determinant of the pattern of human settlement in the Sierra Nevada. Our assessment therefore highlights the linkages between the Sierra Nevada and other parts of California.

Key Questions

This chapter attempts to answer the following key questions about the patterns of human settlement in the Sierra Nevada and the forces shaping future human settlement:

- What were the historic patterns of population growth and human settlement by county from 1850–1990?

- What were the primary factors driving exurban population growth over the past quarter-century?

- What is the likely spatial distribution of future population growth from 1990 to 2040 by county and CCD?

- What is the current spatial pattern of population distribution and housing density by density class?

- What are the relationships between development density and other 1990 Census variables?

- What are the relationships between development patterns and infrastructure access and costs?

- What are the relationships between development patterns and environmental constraints?

- What is the relationship between settlement patterns and land use designations and policies in local General Plans?

- What are the environmental impacts of land use patterns under proposed General Plans?

- What are the infrastructure needs and financing mechanisms available to support proposed General Plans?

- What is the impact of land ownership patterns on the applicability of General Plan policies to development patterns?

- What are the ecological, social, and economic impacts of population growth and alternative human settlement patterns from 1990 to 2040?

- What are the growth management policy options available for mitigating future impacts of growth or modifying its spatial pattern?

- What is the likelihood of and what are the constraints to adoption of such policy options in the current institutional setting?

- What further research is necessary to answer key questions that we have been unable to answer in this assessment of human settlement in the Sierra Nevada?

Each of these questions has implications for the degree to which human settlement and its associated activities have affected Sierra Nevada ecosystems and will continue to affect them. Each of them is also affected by the character and quality of Sierra Nevada ecosystems, which in turn affect the social and economic conditions of the human communities located in the Sierra Nevada. Answers to these questions therefore have importance for nearly every aspect of the Sierra Nevada Ecosystem Project's assessment. Human settlement per se is not necessarily of interest, but it represents a vital intermediate variable for assessment of the social, economic, and ecological state of the Sierra. This SNEP assessment focuses on the processes driving human settlement itself.

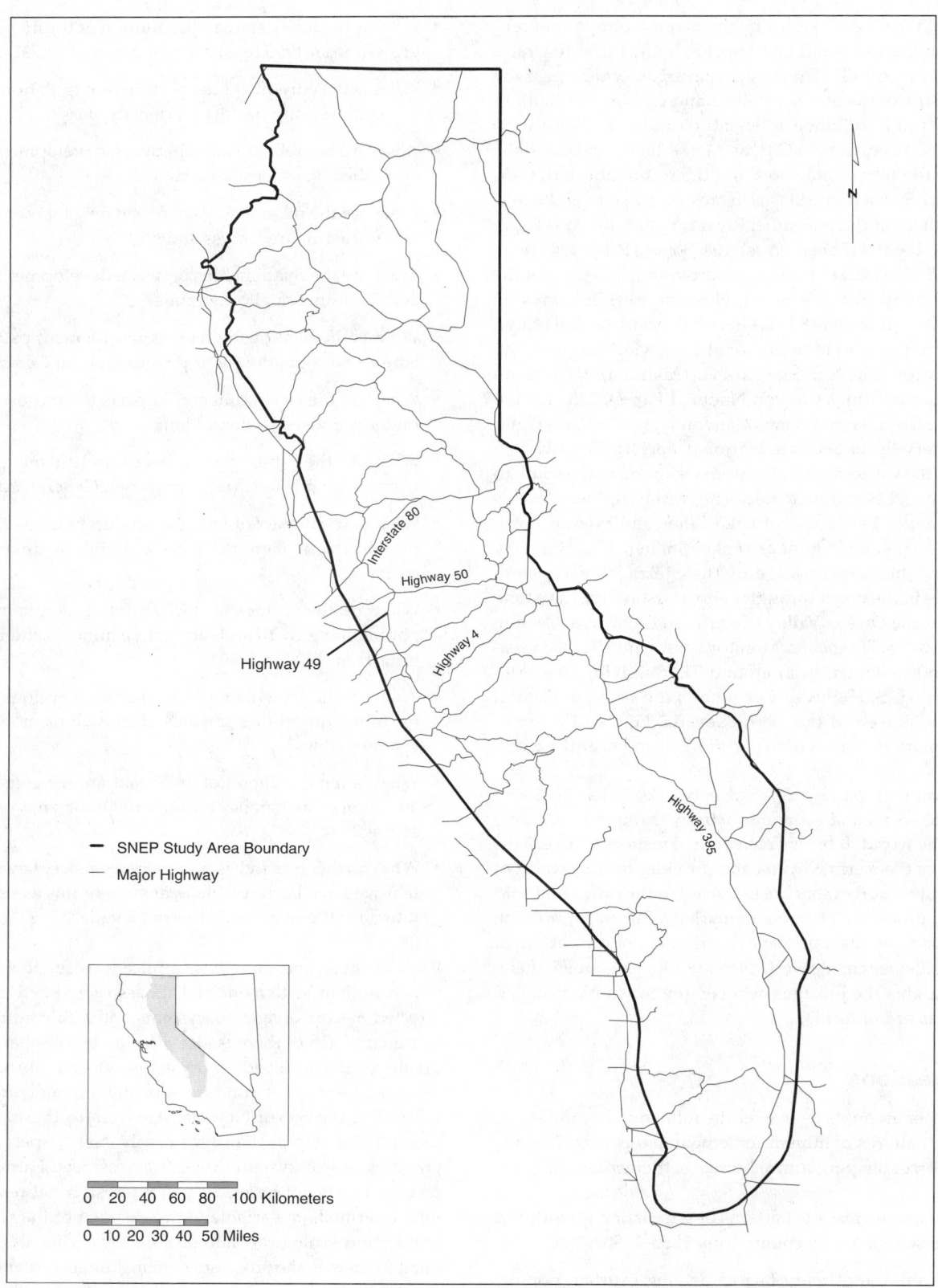

FIGURE 11.2

Major highways in the Sierra Nevada region.

BACKGROUND

Human beings have lived in and utilized the natural resources of the Sierra Nevada for millennia. The focus of this assessment, however, is on patterns of human settlement in the region since 1850. Only four ships dropped anchor in San Francisco Bay in 1848, the same year that James Marshall discovered gold at Sutter's Mill near Coloma and the South Fork of the American River. The next year brought nearly seven hundred ships through the Golden Gate. Most of their passengers unloaded in the ports of northern California and promptly set out for the gold fields of the Sierra Nevada foothills. The region has been intensely inhabited ever since, and the patterns of settlement reflect the geography of both natural and human resources. The pattern of towns, roads, waterways, and related infrastructure established by the forty-niners continues to constitute the framework within which a new wave of migration has swept over the Sierra Nevada during the past three decades. This assessment focuses on that recent migration wave and its implications for the future. The historical effects of the first century of postcontact settlement in the Sierra Nevada are described briefly in the history chapter by Beesley (1996). Our data sources provide only a broad outline of population levels by county from 1850 to 1970, so we will not attempt to delineate in fine spatial detail the historical pattern of settlement or its impacts. A more detailed analysis at the subcounty level is possible only from 1970 to 1990. We therefore focus on the factors driving recent population changes in the area. Figure 11.3 shows the eighteen California counties in the Sierra Nevada included in our analysis.

Several things stand out in the historical census record from 1850 to 1970. The first is that the overall population of the Sierra Nevada counties was relatively stable throughout the nineteenth century, although individual counties went through significant fluctuations. The population of the Sierra Nevada counties peaked in 1852 at around 150,000, which was also the peak year of gold production in California. The southernmost counties in the Sierra Nevada then began to grow rapidly throughout the twentieth century, but most of that growth took place in the Central Valley rather than the Sierra Nevada proper. Because subregional data are not available before 1970 (except for forty towns, most of which are incorporated), it is impossible to determine the precise population of the Sierra Nevada or the distribution of population within the Sierra Nevada with any accuracy from 1850 to 1970. The overall population roughly doubled between 1860 and 1960. Figure 11.4 shows the total population for all of the eighteen California counties in the Sierra Nevada from 1850 to 1990.

The population within those counties that are entirely within the Sierra Nevada grew and fell slightly as commodity prices and business cycles brought residents into and out of the range. California's population roughly doubled every two decades during this period, while it took a century for the Sierra Nevada population to double. California's population growth was primarily concentrated in the coastal regions within and near the emerging metropolitan regions of the San Francisco Bay Area, greater Los Angeles area, and San Diego County. The Central Valley towns of Sacramento, Stockton, Modesto, Merced, Fresno, Visalia, and Bakersfield also grew not far from the Sierra Nevada foothills. Reno and Carson City grew moderately in Nevada, with a drop in population following the end of the silver boom not unlike the fluctuations in the gold camps of the Sierra Nevada. Industrialization of the hardrock gold mining practices maintained population stability in the northern Sierra Nevada community of Grass Valley until the mines finally shut down in 1956. There was some increase in local gold prospecting in the foothills during the Great Depression, but most miners did not stay on. Because the census is completed only every decade, it is impossible to correlate fluctuations in population levels with annual changes in economic conditions. Figure 11.5 shows the 1850–1990 time series in more detail for Nevada County, including a breakdown for the communities of Grass Valley, Nevada City, and Truckee where available. Similar data are available for other counties and cities in the Sierra Nevada from the California Environmental Resource Evaluation System (CERES) project of the Resources Agency of the State of California (http://ceres.ca.gov/snep), and the Alexandria Project at the University of California, Santa Barbara (http://alexandria.sdc.uscb.edu/). Note that the unincorporated portion of Nevada County grew most rapidly from 1970 to 1990.

The second thing that stands out in the 1850–1970 population data is how quickly the counties of the southern Sierra Nevada grew after the turn of the century. This is in stark contrast with the other subregions of the Sierra Nevada, which did not experience rapid growth until after World War II. California became an agricultural powerhouse in the soil of these counties, which was nourished with water from the Sierra Nevada and the Sacramento River watershed in the northern Central Valley. The population growth in these counties was therefore concentrated in the San Joaquin valley rather than the Sierra Nevada proper. The data since 1900 are dominated by those southern Central Valley counties, so it is difficult to discern clear patterns for the Sierra Nevada proper during the twentieth century. The population of the Central Valley itself overwhelms the totals. Figure 11.6 shows 1850–1990 population growth by subregion.

The third feature of population patterns from 1850 to 1970 is how significant the nonwhite portion of the population was in 1850 to 1900 compared with today. In particular, the Chinese constituted a large fraction of the population of the Sierra Nevada counties from 1860 to 1900. Together with Native Americans and African-Americans ("Black" in the census), ethnic minorities accounted for over 22% of the total population of the region at their peak in 1860. This is despite the collapse of the California Indian population between the special census of 1852 and 1860 (from roughly one in eight Sierra Nevada residents to less than 4%). The collapse was most

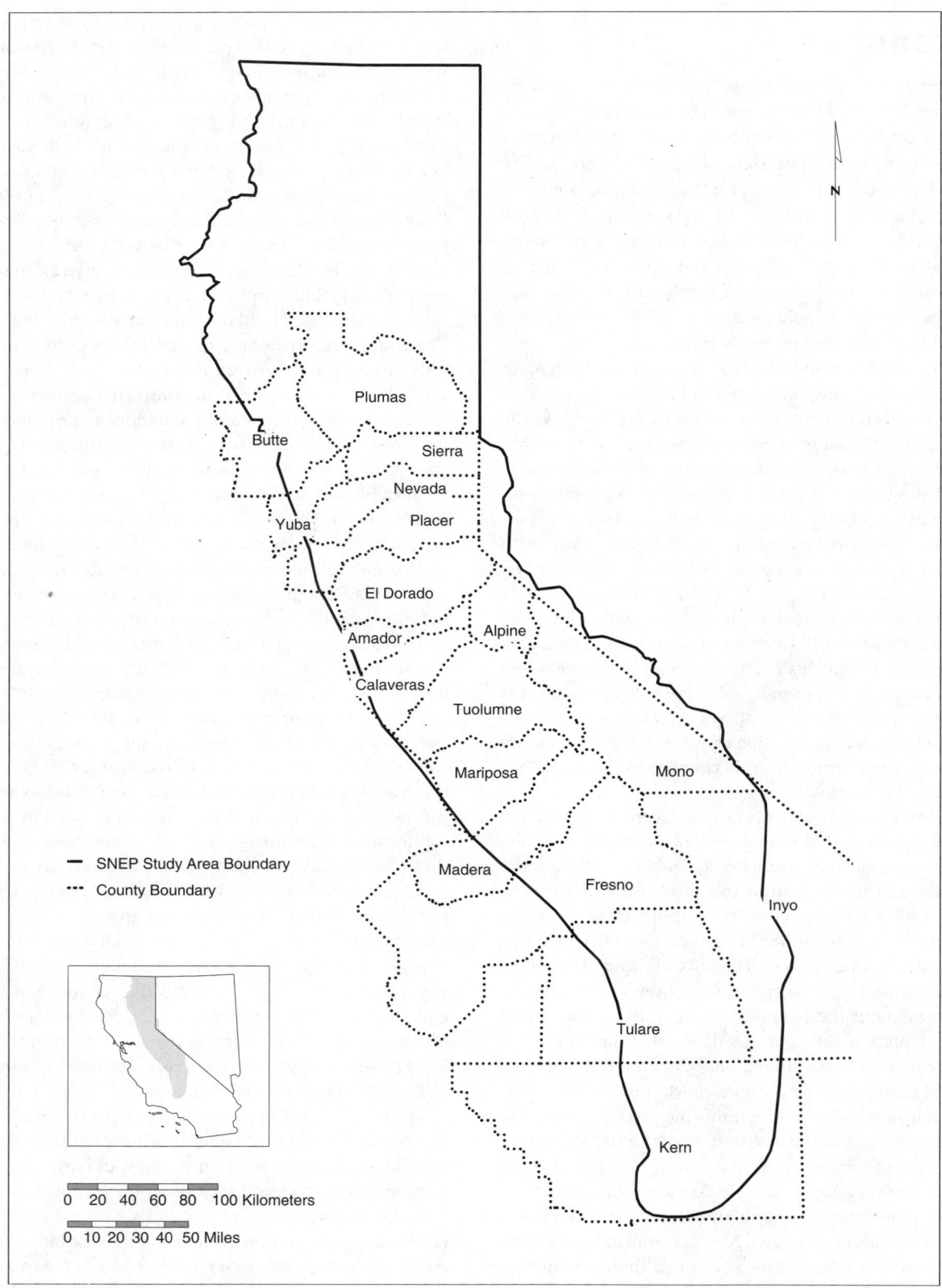

FIGURE 11.3

Counties included in historical analysis.

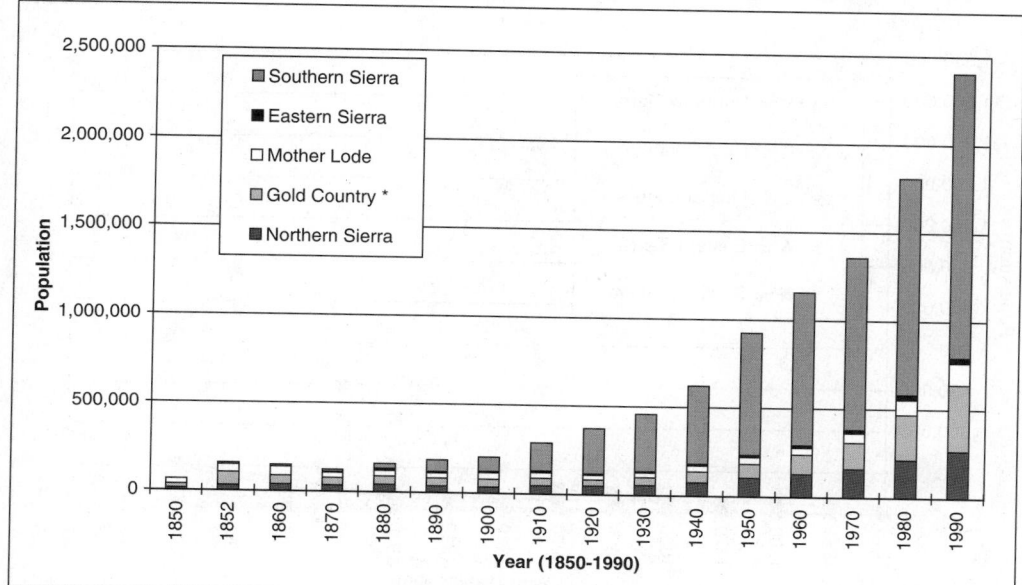

FIGURE 11.4

Population of Sierra Nevada
counties, 1850–1990.

pronounced in the Mother Lode region of the west-central Sierra Nevada, where much of the gold mining activity took place. Estimates of ethnicity are difficult during the 1850s, however, due to the high fraction of residents born in Mexican California. These residents automatically became citizens of the United States under the terms of the Treaty of Guadalupe Hidalgo. Up to one-third of the gold miners may have been foreign-born from outside pre-treaty Mexican Alta California, the new state of California, or any other territories or states of the United States of America. A large fraction of the white miners are also believed to have left the Sierra Nevada in 1859 to 1860 for the Comstock Lode of Nevada, where a silver strike presented new opportunities. This exodus may have also increased the relative share of the population by ethnic minorities in 1860. Today the population of the

Sierra Nevada is overwhelmingly white and differs significantly from the rest of California.

Chinese laborers are well known to have been a critical workforce for the transcontinental railroad (exceeding 12,000 workers at the famous "Chinese Wall" near Donner Pass)[3] and actively participated in gold mining and other activities after the initial gold rush period. They were also already present before work on the railroad began. A series of anti-Chinese activities drove many of the Chinese out of the Sierra Nevada and California around the turn of the century, however, with Nevada County's Chinese population dropping from a high of around 2,000 in 1880 to only 100 by 1910 (*Grass Valley Union* 1995a).[4] Japanese immigrants first appeared as a significant element of the population for the eighteen counties in the region during the same period. Once

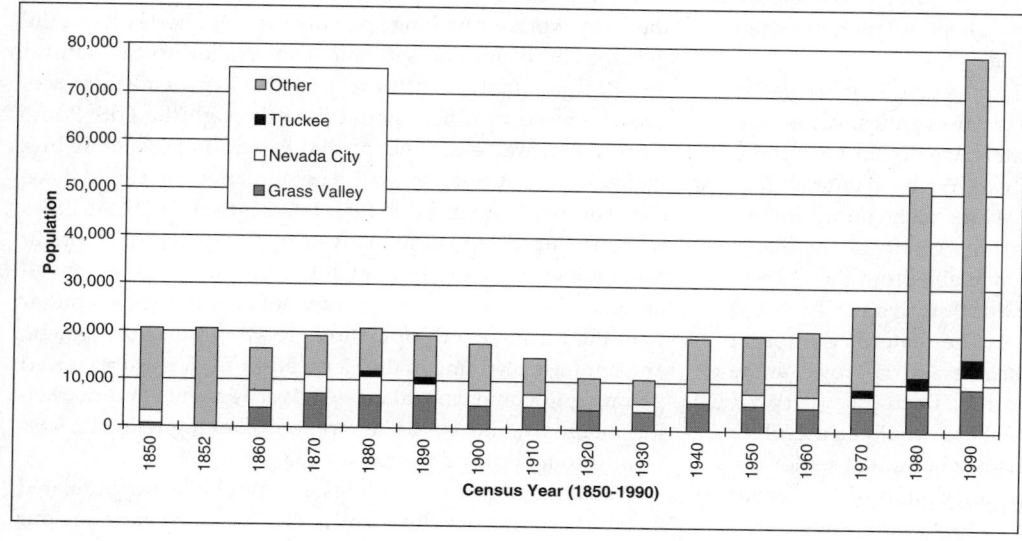

FIGURE 11.5

Population of Nevada
County, 1850–1990.

FIGURE 11.6

Population of Sierra Nevada
subregions, 1850–1990.

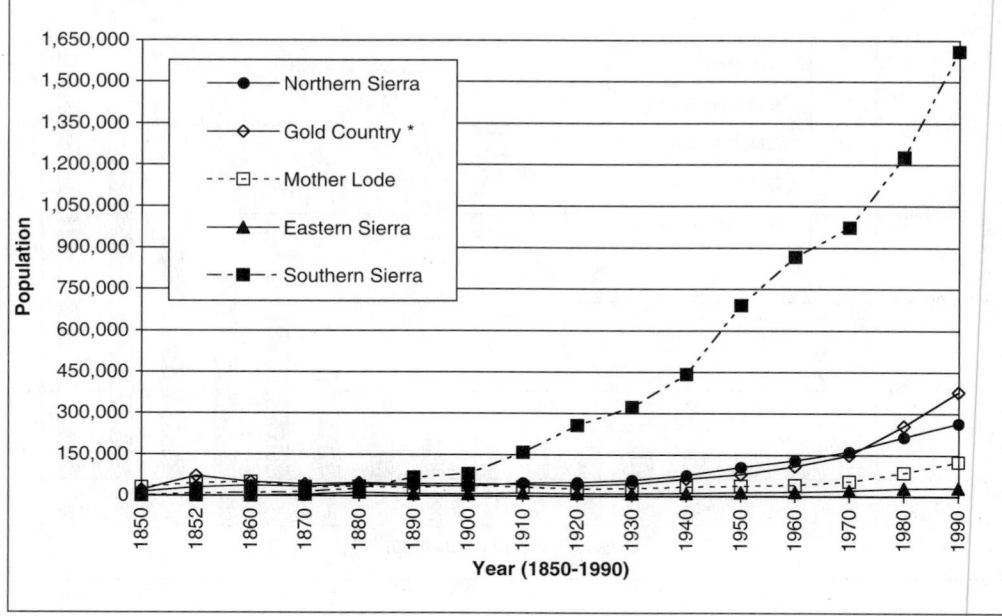

again, however, the Japanese migrants seem to have been concentrated in the southern San Joaquin valley, and it is unclear how significant they were as an element of the population in the Sierra Nevada proper. Many Japanese-Americans were also transferred to the Manzanar Relocation Camp during World War II in the Owens Valley (Inyo County), but no census was conducted between 1941 and 1945. The census record does show several thousand residents of Japanese ancestry in the Sierra Nevada throughout the twentieth century but does not capture this significant influx of forced migrants during World War II (Koda 1995). The county-level figures also show a significantly higher fraction of nonwhite residents from 1900 to 1990 than there was for the Sierra Nevada portion of the Sierra Nevada counties. In particular, the census data show a high fraction of black (African-American) residents primarily in communities in the Central Valley. Figure 11.7 shows the ethnicity of the nonwhite population recorded in the census from 1850 to 1990.

The historical records contained in the *Census of Population* are full of rich detail about individual counties, communities, and ethnic groups. Unfortunately, we do not have room to discuss those records in detail here. We have entered raw population figures from 1850 to 1990 for each county, identified community, and identified ethnic group in the Sierra Nevada into a spreadsheet that is available from the California Environmental Resource Evaluation System (CERES) project of the Resources Agency of the State of California (http://ceres.ca.gov/snep), and the Alexandria Project at the University of California, Santa Barbara (http://alexandria.sdc.ucsb.edu/). This database should be useful for more detailed queries about the history of human settlement in the Sierra Nevada. A five-page description of the census data and the history of changing county boundaries is also

included for reference. The focus of our remaining assessment of census records will be on the recent doubling of the population in the Sierra Nevada portion of the counties during the period 1970–90. We will then use that analysis as the basis for allocating county-level population forecasts to the Sierra Nevada portion of counties for the period 1990–2040.

METHODOLOGY

Our methodology for assessing patterns of human settlement in the Sierra Nevada relied upon the development of a geographic information system (GIS) on a UNIX workstation using the GIS software package Arc/Info. This GIS served as the framework for making spatially explicit queries about the distribution of human settlement in relation to the natural and human factors. Where possible, information was georeferenced to other spatial data through the GIS. Some information was either nonspatial or was not available in a digital form, however, so we did not limit our analysis to those data sources that could be integrated into the GIS. At times we relied upon statistical analysis of nonspatial data, literature review and interviews with key informants, and consultations with academic and professional colleagues familiar with the processes of population growth, human settlement, and land use planning in the Sierra Nevada. We also reviewed planning documents, real estate advertisements and marketing materials, and media reports on planning-related issues from throughout the Sierra Nevada.

We attended public meetings and public hearings related to land use and development in the Sierra Nevada for our

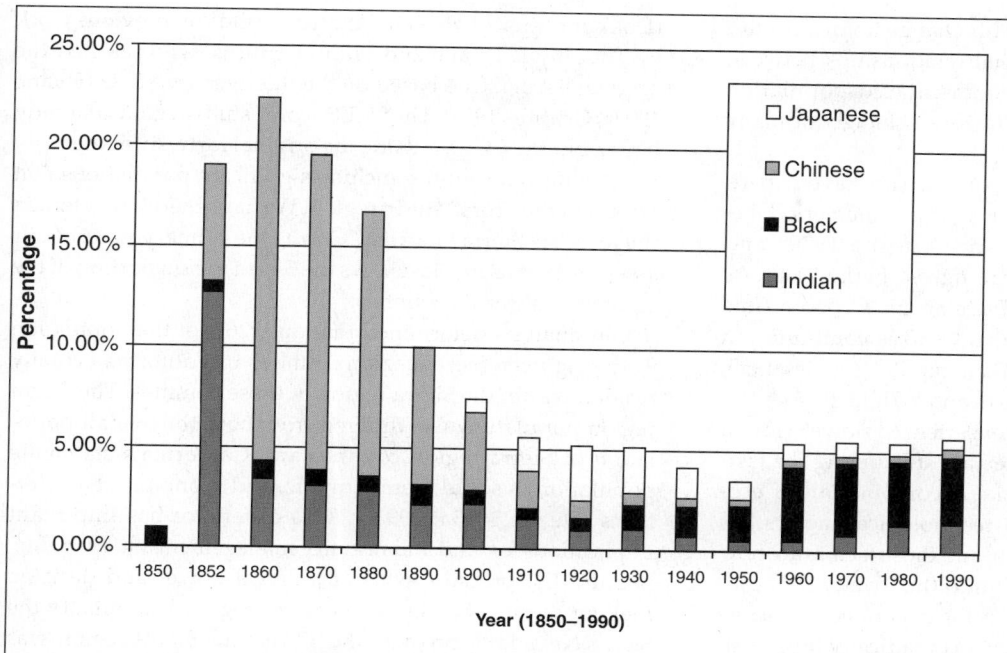

FIGURE 11.7

Minority population by ethnicity as a percentage of total Sierra Nevada population, 1850–1990.

more detailed case studies of the General Plan update processes in Nevada and El Dorado Counties. Graduate students in the Environmental Planning Studio at the University of California, Berkeley (taught by the author), studied these two General Plans in detail in January–May 1993 and January–May 1994, respectively. We relied upon specific studies completed in those classes for insights into specific aspects of the General Plans now under review. The author also worked as assistant city planner for the city of Grass Valley in 1982 and was already familiar with the 1980 Nevada County General Plan. We also reviewed General Plans for Plumas, Placer, Amador, Calaveras, Mono, and Inyo Counties and the Wildlife Habitat Management Plan for Tuolumne County, but only the General Plans for El Dorado and Nevada Counties were analyzed in any detail. These were the only two counties in the Sierra Nevada with sufficiently developed land use maps to allow GIS analysis. We also had other data available for those counties in digital form that were not generally available for the rest of the Sierra Nevada. Moreover, these two counties have experienced the greatest rates and absolute numbers of population growth in the Sierra Nevada. Assessing policies in these two counties therefore gives us some insight into how counties facing extreme growth pressures may plan for additional growth in the current institutional context. Together they have spent at least $4 million on their efforts to update their General Plans over the past five years (Rivas 1993–95; Boivin 1991–95). This presented an opportunity to build on extensive existing work rather than trying to create a database from scratch.

Specific methods are discussed in more detail in each section of this chapter.

SOURCES

Specific sources are discussed in detail in each section of the chapter.

RESULTS

The results of each of our individual analyses are described in detail in this section. The significance of the results of each analysis is then discussed in relation to the other results in the conclusion of the assessment, along with management implications.

The Second Gold Rush: 1970–90

The Sierra Nevada region grew by more than 65% in the 1970s and 39% in the 1980s, and by a total of 130% from 1970 to 1990 (an average annual rate of approximately 3.5%). This compares with overall growth of 49% for all of California and 22% for the entire United States from 1970 to 1990 (U.S. Bureau of the Census 1970, 1980, 1990). This rapid population growth boosted the population of the Sierra region from just under 273,000 in 1970 to around 618,000 in 1990.[5] More people moved into the Sierra Nevada from 1970 to 1990 than migrated into the area during the entire gold rush through the 1850s. This second gold rush resulted in a dramatic change in the social, demographic, and economic characteristics of Sierra

Nevada residents (Duane 1993a). This change in turn continues to alter the economic and social relationships between those residents and Sierra ecosystems. Rapid population growth has become the dominant factor of change for many Sierra Nevada communities.

Many rural communities in North America have experienced rapid population growth during this same period, beginning with the "rural renaissance" in the 1970s and continuing with a flood of "equity refugees" in the 1980s. As reported in a 1993 cover article in *Time* magazine, "Boom Time in the Rockies" (Bonfante 1993), this trend is continuing in the 1990s (*New York Times* 1993; Diringer 1994; Weiss 1995; *High Country News* 1993, 1994; Starrs and Wright 1994).[6] The counties of the Sierra Nevada have experienced slower growth rates in the past few years, but they are still among the fastest-growing counties in California. A combination of economic, social, demographic, and technological factors has fueled this urban-to-rural migration, and those factors are now expected to sustain the trend well into the twenty-first century. The rapid population growth being experienced in some rural areas has the potential to transform radically the physical and the social environments of those regions, including significant fragmentation of habitat and the likely loss of native biological diversity. This is certainly true in the Sierra Nevada (Duane 1993b). It is not limited to the Sierra Nevada, however, for many other nonmetropolitan communities are experiencing rapid population growth. The experience of rapid growth in the Sierra Nevada could therefore be a harbinger for the rest of the rural West.

There is no political jurisdiction with boundaries that coincide with the ecosystem or bioregional boundaries of the Sierra Nevada mountain range, but the 1991 Biodiversity Memorandum of Understanding (MOU), signed by ten state and federal land and resource management agencies, delineated rough boundaries for the Sierra Nevada bioregion that are consistent with those used by SNEP and others. Understanding the social, demographic, and economic characteristics of the Sierra Nevada population and the transformation that is occurring within the region requires a bioregional analysis of census data from 1970, 1980, and 1990. The Sierra Nevada region delineated in the MOU lies within portions of eighteen California counties and three Nevada counties, but only nine of the California counties are completely within the Sierra Nevada bioregion.[7] We therefore took the census data boundaries and included only those county census divisions (CCDs) that were largely within the Sierra Nevada bioregion, creating a composite of CCDs that was approximately coterminous with the boundaries of the Sierra Nevada bioregion and with the SNEP core area.[8] With the exception of the population within the Lake Tahoe Basin, residents of the three counties in Nevada live outside the Sierra Nevada proper. For a number of reasons discussed later, this analysis addresses only the Sierra Nevada portion of the eighteen California counties. Portions of other counties (e.g., Lassen County) were included in the social assessment work completed for SNEP

(Doak and Kusel 1996). Our analysis builds on previous work by Timothy P. Duane and Philip Griffiths, who selected the original list of CCDs based on the Biodiversity MOU (Duane 1993a; Griffiths 1993). The SNEP "core" study region is slightly larger, but the CCDs outside the original forty-six in the eighteen California counties include significant populations that are not in the "core" study region. We have therefore retained the forty-six Sierra Nevada CCDs as the primary units of our assessment. Figure 11.8 shows the Sierra region portion of the eighteen California counties.

Our analysis determined that only 26% of the population in the eighteen Sierra region counties in California actually resided within the Sierra region of those counties. The Sierra region population also differed from both the overall population of Sierra region counties and California's statewide population in social, demographic, and economic characteristics (Duane 1993a, 1993c). This difference has important implications for land and resource management and planning, because the primary locus of political power and decision making within the eighteen-county region lies outside the Sierra Nevada. Moreover, the 32,000 mi[2] Sierra region was home to only about 2% of California's population in 1990 (despite accounting for roughly 20% of the land area in the state).

Population growth in the Sierra region of the eighteen California counties in the Sierra Nevada was nearly exactly the same in absolute terms in the 1970s (175,472 people) as in the 1980s (174,101 people). In contrast, the eighteen-county region grew faster in the 1980s than the 1970s (597,935 versus 452,241). This was in the context of much greater growth in California in the 1980s than the 1970s (6,092,000 versus 3,697,000). Due to the larger base population in 1980 than 1970, however, the percentage growth rate was lower in the 1980s than in the 1970s in the Sierra region. Table 11.A1 in appendix 11.1 shows growth patterns by county and subregion of the Sierra region, which are discussed in the next section.

Population growth is not evenly distributed across individual counties, however; some areas experience more rapid growth and/or population turnover than other areas, and this has social, economic, and ecological implications for the Sierra Nevada. Figure 11.9 illustrates the pattern of county inmigration for the central Sierra Nevada region of Nevada, Placer, El Dorado, Amador, and Calaveras Counties. The western part of Placer County, including the cities of Roseville and Rocklin, is technically outside the boundaries of our study area but is included here for reference purposes. This map shows that more than 30% of the population moved to each of the respective counties between 1985 and 1990 for many of the census block groups. Note that this was not generally true around many of the established communities, such as Grass Valley and Nevada City in Nevada County, but was generally true in the unincorporated areas.

Fully 12.68% of California residents in 1990 did not live in California in 1985, but only 7.01% of Sierra region residents were from outside the state. State-level population growth is dominated by three sources: (1) natural increases; (2) foreign

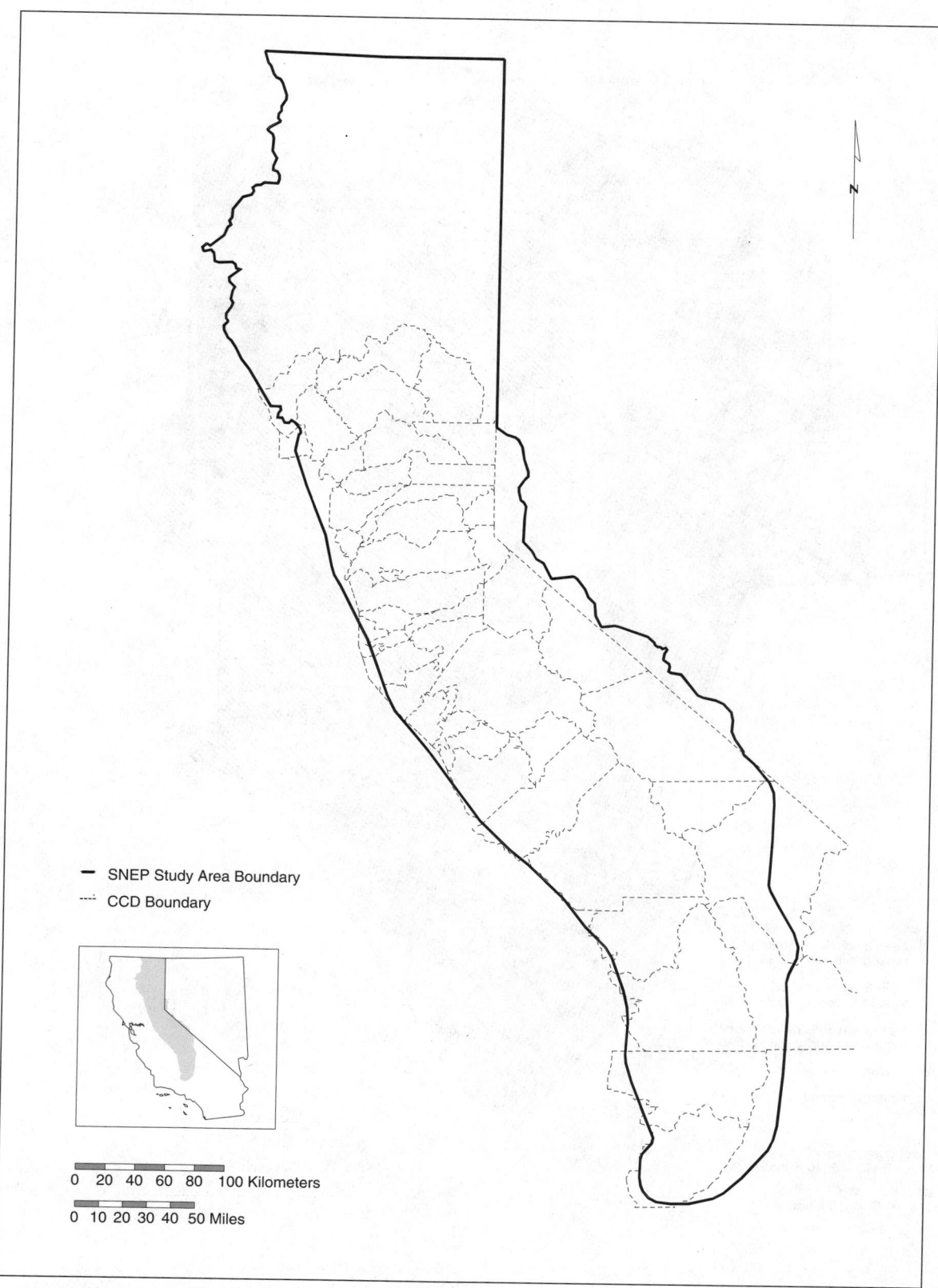

SNEP Study Area Boundary
CCD Boundary

0 20 40 60 80 100 Kilometers

0 10 20 30 40 50 Miles

FIGURE 11.8

Census civil divisions included in growth analysis.

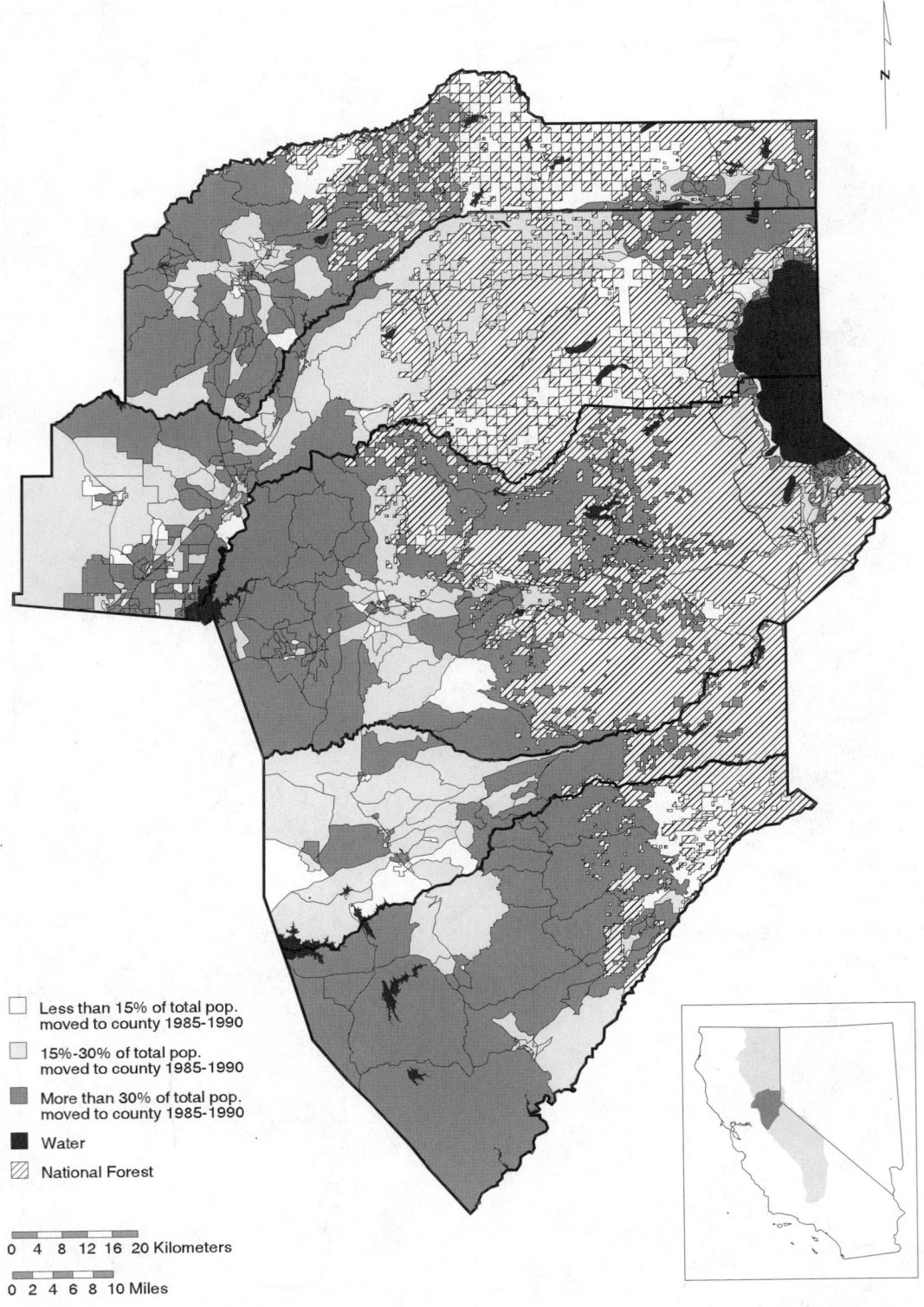

Less than 15% of total pop.
moved to county 1985-1990

15%-30% of total pop.
moved to county 1985-1990

More than 30% of total pop.
moved to county 1985-1990

Water

National Forest

0 4 8 12 16 20 Kilometers

0 2 4 6 8 10 Miles

FIGURE 11.9

County in-migration, central Sierra Nevada region (based on 1990 Census of Population Summary Tape File 3A).

immigration (both legal and illegal); and (3) domestic immigration from other states (Teitz 1990). Since the 1990 census, the state has experienced relatively high natural increases and continues to accommodate from one-fourth to one-third of the legal foreign immigration to the United States. Illegal immigrants are much more difficult to account for, but California also clearly has a disproportionate share of the nation's illegal immigration. Domestic migration has literally reversed itself: whereas the state grew by up to 453,000 people per year at the peak in 1990 through domestic migration from other states, it is believed to have had a net domestic population loss for some years from 1990 to 1993 through emigration to other states. There was a net gain of 33,000 domestic migrants in 1994, however, as lower real estate costs and a slight upturn in the California economy began to draw new immigrants. This small net gain was nevertheless overwhelmed by a net natural increase of more than 361,000 births over deaths. The age structure, birth rates, and demographic momentum of California's current population now ensures that the state will continue to grow even without significant net domestic immigration. This demographic momentum has profound implications for future population projections, which will increasingly be dominated by natural increases. Most of that natural increase is expected to occur in metropolitan areas.

Unlike the rest of California (in particular, the metropolitan areas and the Central Valley), the Sierra region experienced low natural increases, low foreign immigration, and low domestic migration from other states from 1970 to 1990. Most of the population growth in the Sierra region during this period was due to immigration from other parts of California. More than one-fourth (27%) of the Sierra region residents in 1990 lived in a different county within California in 1985. Given that the population of the entire Sierra region grew by 39% in the 1980s, we would expect that about 14% of the 1990 Sierra region population would have been nonresident in 1985 based on population growth alone (half of the 1980–90 total immigration total divided by the 1990 total). Combined with the 7% of 1990 Sierra region residents who were out of state in 1985, however, more than one-third of 1990 Sierra region residents (27% plus 7% equals 34%) were not residents of the same county just five years earlier. Some Sierra region residents may have moved across county lines and remained within the Sierra region, but these data suggest that the turnover rates among migrants are much greater than the net changes in population would suggest. A large fraction of new migrants may therefore not be staying in Sierra Nevada communities for more than five years. Fully 40% of the residents of the Tahoe Basin and Truckee areas in 1990 were not residents of the same county in 1985 (Griffiths 1993).[9]

This is partially explained by the demographic characteristics of the Sierra region population and its new migrants. The population of the Sierra Nevada in 1990 was considerably older than the population of California. The percentage of people over 55 years of age in the Sierra region (27%) was 50% greater than the percentage for the state (18%). The proportion of people 15 to 24 years of age was also lower in the Sierra region (25%) than in California as a whole (34%). The percentages of people under 15 and from 35 to 54 were similar for the Sierra region and the state. A coarse regional analysis of age-cohort changes from 1980 to 1990 suggests there is a net out-migration of young adults from 15 to 34 years of age and a net in-migration of adults 35 to 54 and over 55 years of age. Despite the common perception that it is only retirees moving into the Sierra region, therefore, the source of population growth appears to be both retirees and working-age adults. The 35 to 54 age cohort grew by 6.6% in the Sierra region from 1980 to 1990, while the over 55 age cohort remained relatively stable. This contrasts with a 3.7% increase in the 35 to 54 cohort and a 1.5% decline in the over 55 age group for the state as a whole. Despite the larger proportion of older residents, then, in-migration by additional retirees during the 1980s merely replaced those in the same cohort who had moved out of the region or died. Because the Sierra region has a disproportionately larger share of persons over 55 and a disproportionately smaller share of persons under 5 years of age, natural increase accounts for a very small fraction of annual population increases. Differences between the Sierra region and California's age structure are shown in figure 11.10 (Griffiths 1993).[10]

A more detailed cohort survival analysis of data for Nevada County shows that the working-age adults are also bringing with them young school-age children. Indeed, it appears that the arrival of kindergarten for a member of the household may be a critical factor driving migration to the Sierra region. Fewer children are projected to migrate to Nevada County by the model either under 5 years of age or between 15 and 19 than in the 5 to 9 and 10 to 14 age groups. Following graduation from high school, the young adults appear to leave the area either for school, employment, or the attractions of urban life and are not replaced by immigrants in the same age cohort. Young families in their thirties then appear to move to the area with young children who have reached school age. Similar numbers of migrants in the 30 to 34, 35 to 39, 40 to 44, 45 to 49, and 50 to 54 age cohorts are projected by the model to migrate to Nevada County. Finally, a much larger cohort of retired and semi-retired migrants over the age of 55 are projected to move into the area based upon the 1980–90 trends.[11] The projected migration patterns for Nevada County and the state as a whole are quite different, as shown by figure 11.11 and figure 11.12. The projection for California shows emigration for all age classes from 60 to 84, while immigration is strong in the 20 to 29 age class (age cohorts showing net emigration for Nevada County). In addition to migration characteristics, however, it is important to note that Nevada County's general fertility rate was only 62 per 1,000 females, compared to an average of 73 per 1,000 females for all of California. Migration is therefore a more significant factor for population growth in the Sierra region than it is for the state as a whole (Collados and Griffiths 1993).

FIGURE 11.10

Percentage of population by age, Sierra Nevada and California.

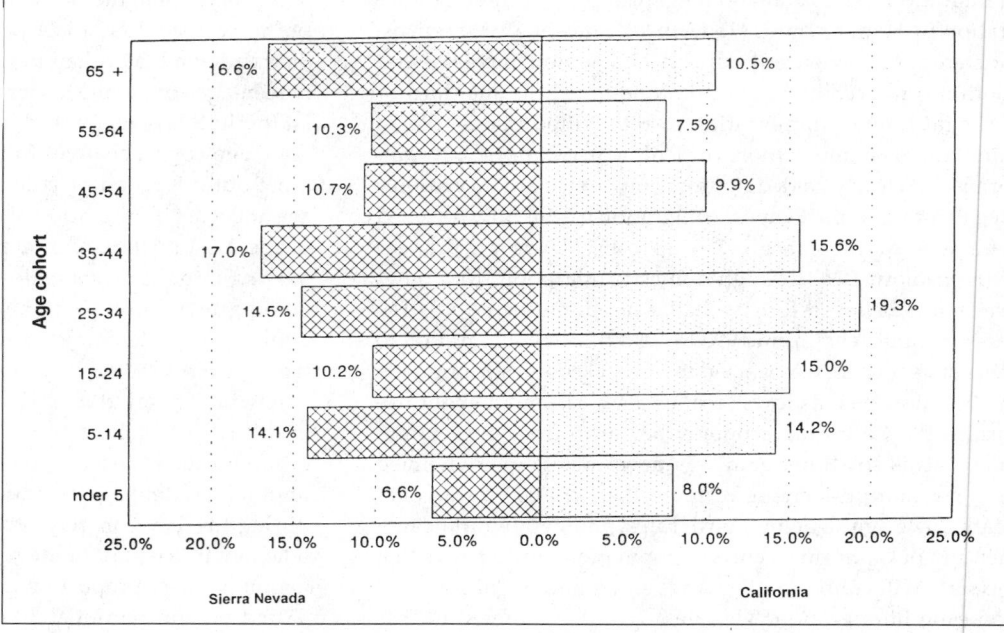

The characteristic that most distinguishes the Sierra region from the rest of California, however, is that its population is overwhelmingly white. This ethnic homogeneity of the region's population has even been cited by some as the primary reason for migration to the Sierra region (Walsh 1991). While the state of California is becoming increasingly heterogeneous in cultural and ethnic terms, approximately 92% of the Sierra region was white in 1990. The comparative figure for the state was 69%. Three of the Sierra region counties, Amador, Tuolumne, and Kern, also have state correctional facilities that account for a significant fraction of each county's population (approximately 10% of Amador).[12] The inmates at these state prisons are much more ethnically heterogeneous, so their presence tends to overstate the ethnic heterogeneity of the Sierra Nevada population.[13] Nevada County (which has no state correctional facility) was over 97% white in 1990, which made it the most ethnically homogeneous county in the entire state of California. The nonincarcerated population of the Sierra Nevada was therefore probably somewhere between 92% and 97% white in 1990.

The 1990 census is likely to have undercounted some nonwhite ethnic groups, however, and there appears to have been an increase in nonwhite residents of the Sierra region in the 1990s. The 1990 undercounting is likely to have been most significant for Hispanics or Latinos, and that is also the group that appears to have increased since 1990. South Lake Tahoe resort casinos are increasingly employing Latinos and Filipinos in low-wage kitchen and maintenance jobs instead of young, seasonal white workers. This is a phenomenon that is most evident by the predominance of Spanish behind the kitchen door or among the maids cleaning rooms on any hallway of a high-rise casino. Bilingual education has also in-

creased dramatically in South Lake Tahoe schools, while communities near both North Lake Tahoe (e.g., King's Beach) and Stateline (e.g., South Lake Tahoe) have significant pockets of poverty. An informal economy has also appeared in some areas (e.g., Truckee) where Latinos gather at a regular spot each day for day wage labor. It is unclear whether or not undocumented aliens are a significant part of this underground labor pool. Most appear to be legitimate residents, either with citizenship or a "green card" allowing work on a permanent resident visa. This is certainly true for the more formal employment sector in the tourism industry and parts of the construction industry.[14]

The dominant ethnicity of the nonwhite population of the Sierra region also varies by subregion. Portions of the Sierra

FIGURE 11.11

California projected migration by age group, 1990–2000.

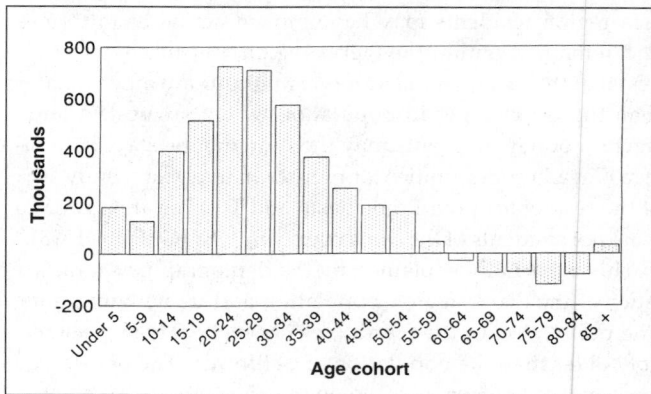

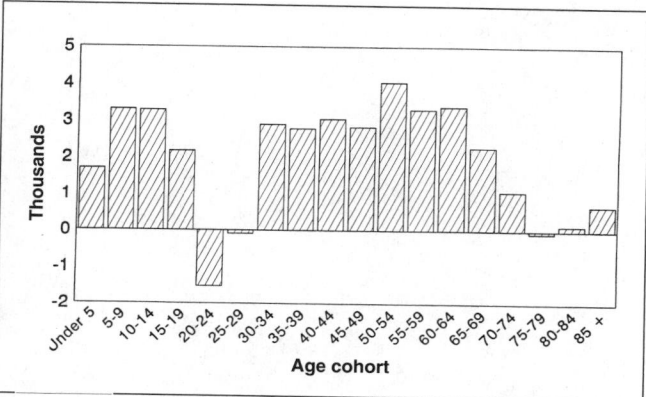

FIGURE 11.12

Nevada County projected migration by age group, 1990–2000.

Nevada have significant Native American populations, for example, as a percentage of the relatively small total subregional population. This is particularly true for the eastern Sierra subregion. The greatest ethnic heterogeneity appears in the southern Sierra subregion, with its strong ties to the agricultural communities of the southern Central Valley. There is a higher percentage of Hispanics in this portion of the Sierra region and a lower percentage of whites than in any other subregion. These differences are even more pronounced by community, as reported in Doak and Kusel (1996). Even those subregions with apparent ethnic homogeneity have communities within them that are quite different.

We use slightly different subregional boundaries and names in this chapter than those used by Doak and Kusel (1996). This difference reflects the specific emphasis of our assessment, which is human settlement and its relationship to the forces driving population growth in the Sierra Nevada. Here is a brief summary of the subregional groupings and names used by each of our respective assessments:

- *Northern Sierra* in this chapter includes Plumas, Butte, Yuba, and Sierra Counties. The social assessment chapter adds some parts of Lassen County.

- *Lake Tahoe* in this chapter includes eastern Nevada, Placer, and El Dorado Counties. The social assessment chapter uses the name *Greater Lake Tahoe Basin* and adds Alpine and parts of Washoe and Douglas Counties in Nevada.

- *Gold Country* in this chapter includes western Nevada, Placer, and El Dorado Counties. The social assessment chapter uses *west-central north* for the same area.

- *Mother Lode* in this chapter includes Amador, Calaveras, Tuolumne, and Mariposa Counties and the eastern portion of Madera County. The social assessment chapter uses *west-central south* for the same area.

- *Southern Sierra* in this chapter includes portions of Fresno, Tulare, and Kern Counties. The social assessment chapter uses *southwest* for the same area minus a small part of Tulare County.

- *Eastern Sierra* in this chapter includes Inyo, Mono, and Alpine Counties. The social assessment chapter uses *southeast* for the same area minus Alpine County and plus a small part of Tulare County.

All of the Sierra Nevada subregions used in this assessment and their relationship to one another are shown in figure 11.13. Note that the area covered in our CCD-based analysis does not coincide precisely with the SNEP core study area boundary.

Maps showing each of the subregions used in this assessment of human settlement and all of the associated CCD units appear in figures 11.14–11.19. In contrast to our CCD-based assessment of human settlement, the social assessments group developed its "community aggregations" from the 1990 "census block groups" (CBGs), which are a smaller unit of analysis than the CCD. These smaller CBG units were not delineated for the 1970 census, however, forcing us to rely upon the larger CCD units as the basis for our analysis of population growth from 1970 to 1990 below the level of the county.

These CCD units will be referred to again in our projections of 1990–2040 population and human settlement patterns in the Sierra Nevada. They are our primary units of analysis at the scale of the entire Sierra Nevada and across multiple decades. We will nevertheless translate these CCD-based estimates into more spatially explicit patterns of human settlement through analysis of the Nevada and El Dorado County General Plans.

Analysis of population growth from 1970 to 1990 shows that some subregions and some CCDs grew much faster than others. Moreover, some experienced more rapid growth in the 1970s than the 1980s and vice versa. Finally, the unincorporated areas in the Sierra Nevada accommodated the vast majority of the population growth. The dominant pattern of development was therefore beyond the service boundaries of existing water and sewer infrastructures, which are important factors influencing patterns of development. This pattern of growth also made counties (rather than incorporated cities) the dominant planning and regulatory entities with jurisdiction over land use and human settlement in the Sierra Nevada. California has a strong "home rule" tradition regarding land use, with local governments exercising planning and regulatory authority within the context of general state policies. Those state policies include specific requirements for the preparation of General Plans, consistency requirements calling for zoning to be consistent with those General Plans, and environmental review procedures under the California Environmental Quality Act (CEQA). Other state and federal regulations regarding water quality and air quality can impose constraints upon land use decisions by local governments,

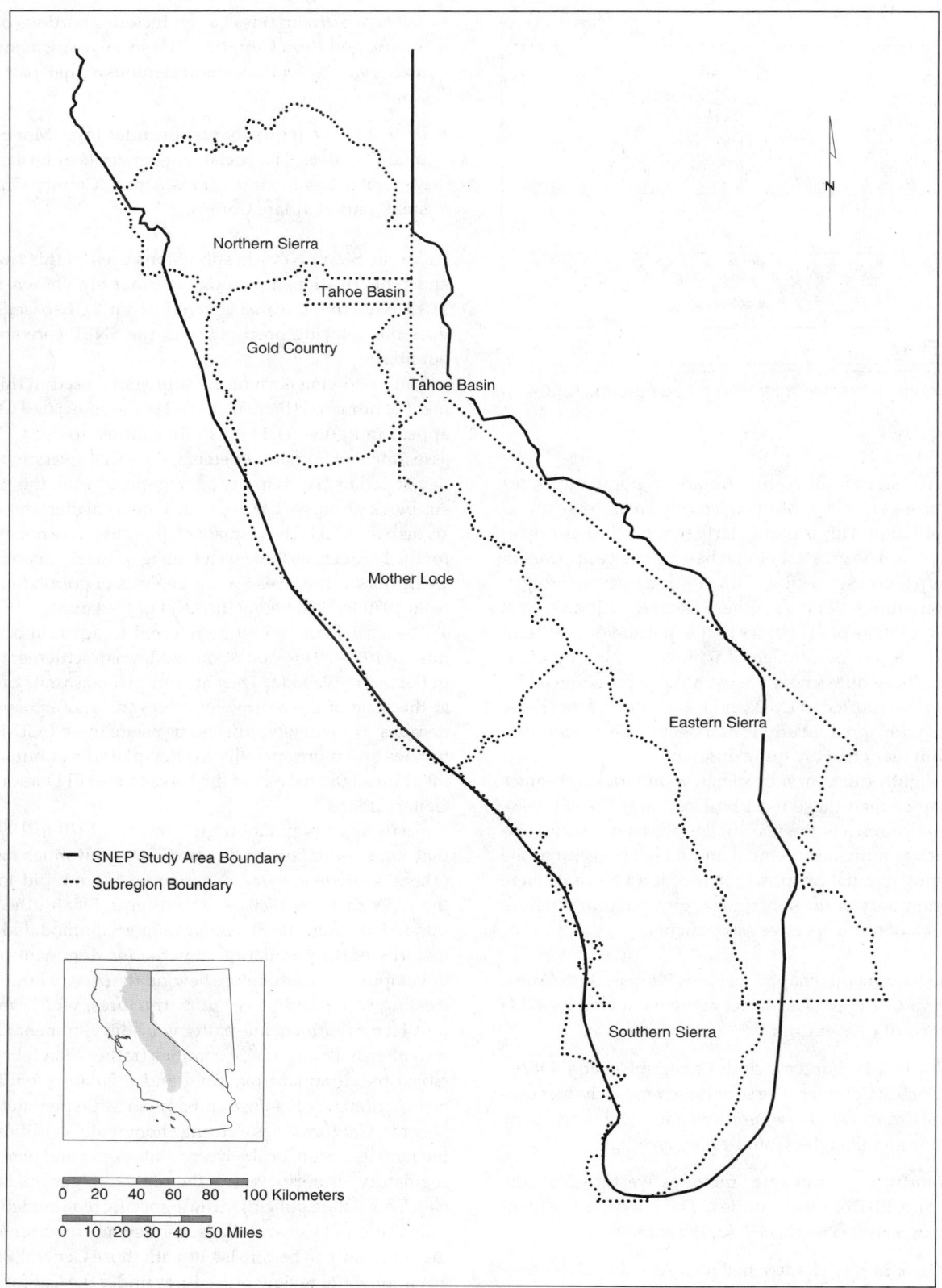

FIGURE 11.13

Subregions of the Sierra Nevada (SNEP core area).

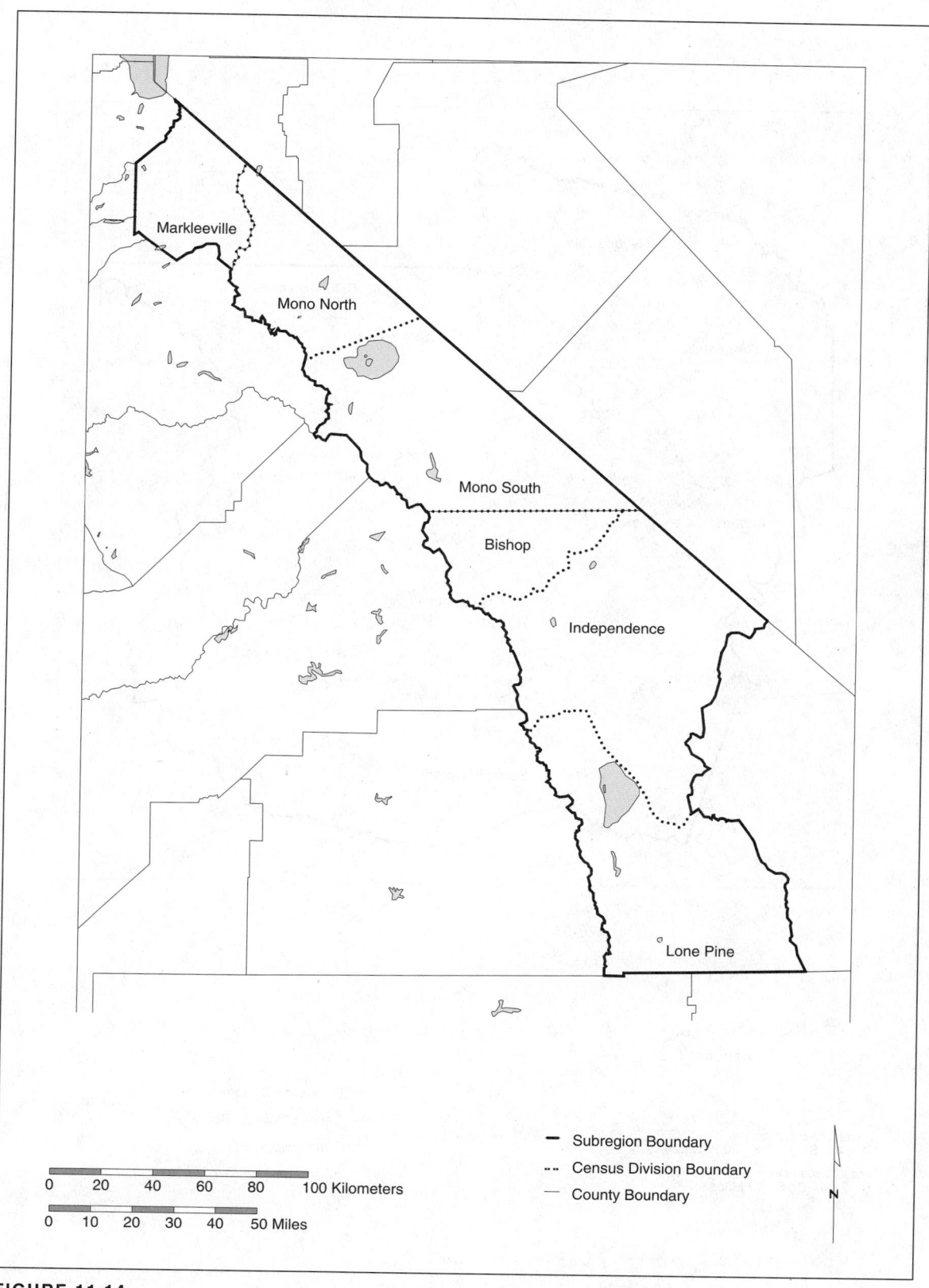

Markleeville

Mono North

Mono South

Bishop

Independence

Lone Pine

Subregion Boundary
Census Division Boundary
County Boundary

0 20 40 60 80 100 Kilometers

0 10 20 30 40 50 Miles

N

FIGURE 11.14

Eastern Sierra subregion.

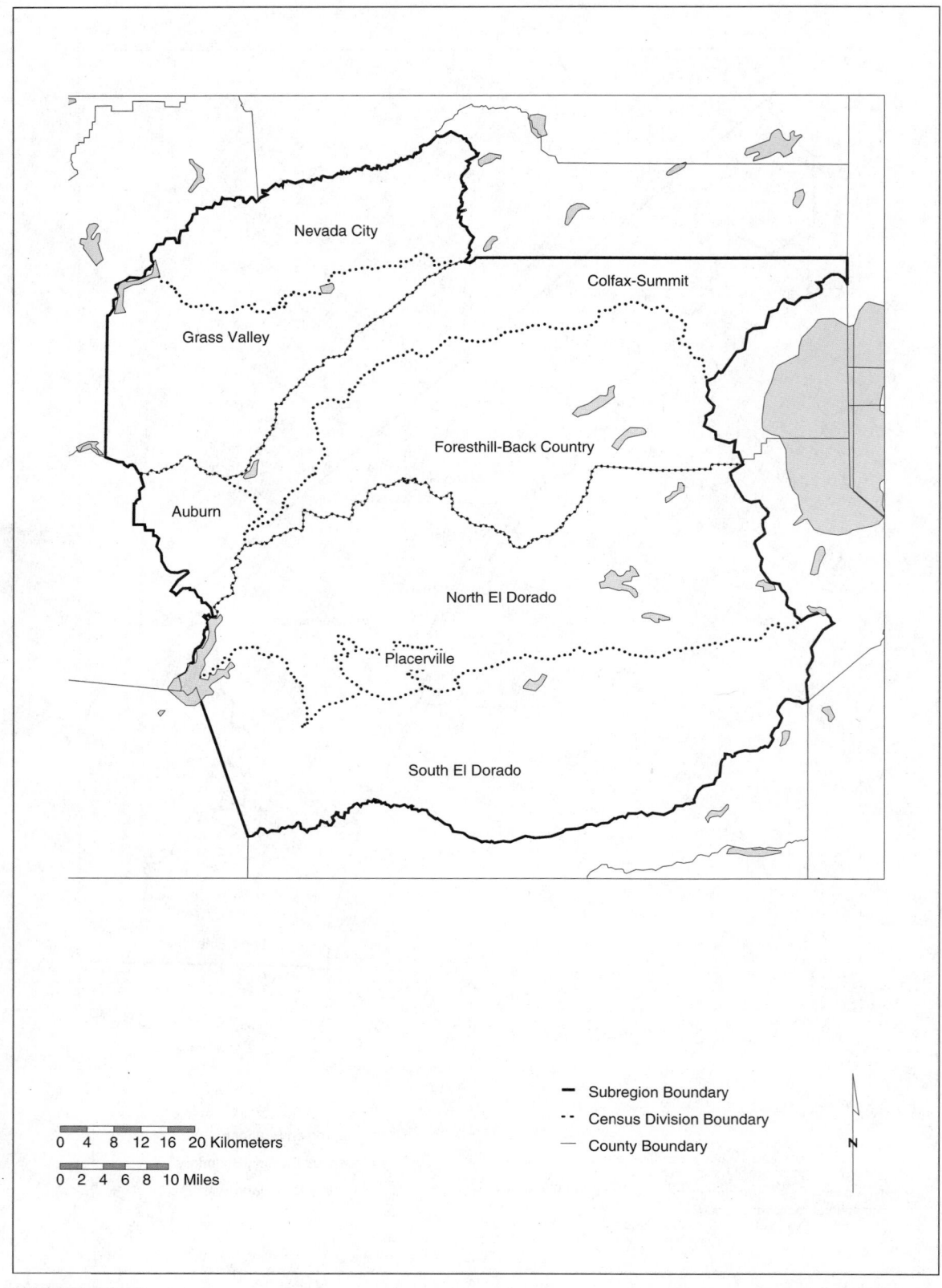

FIGURE 11.15

Gold Country subregion.

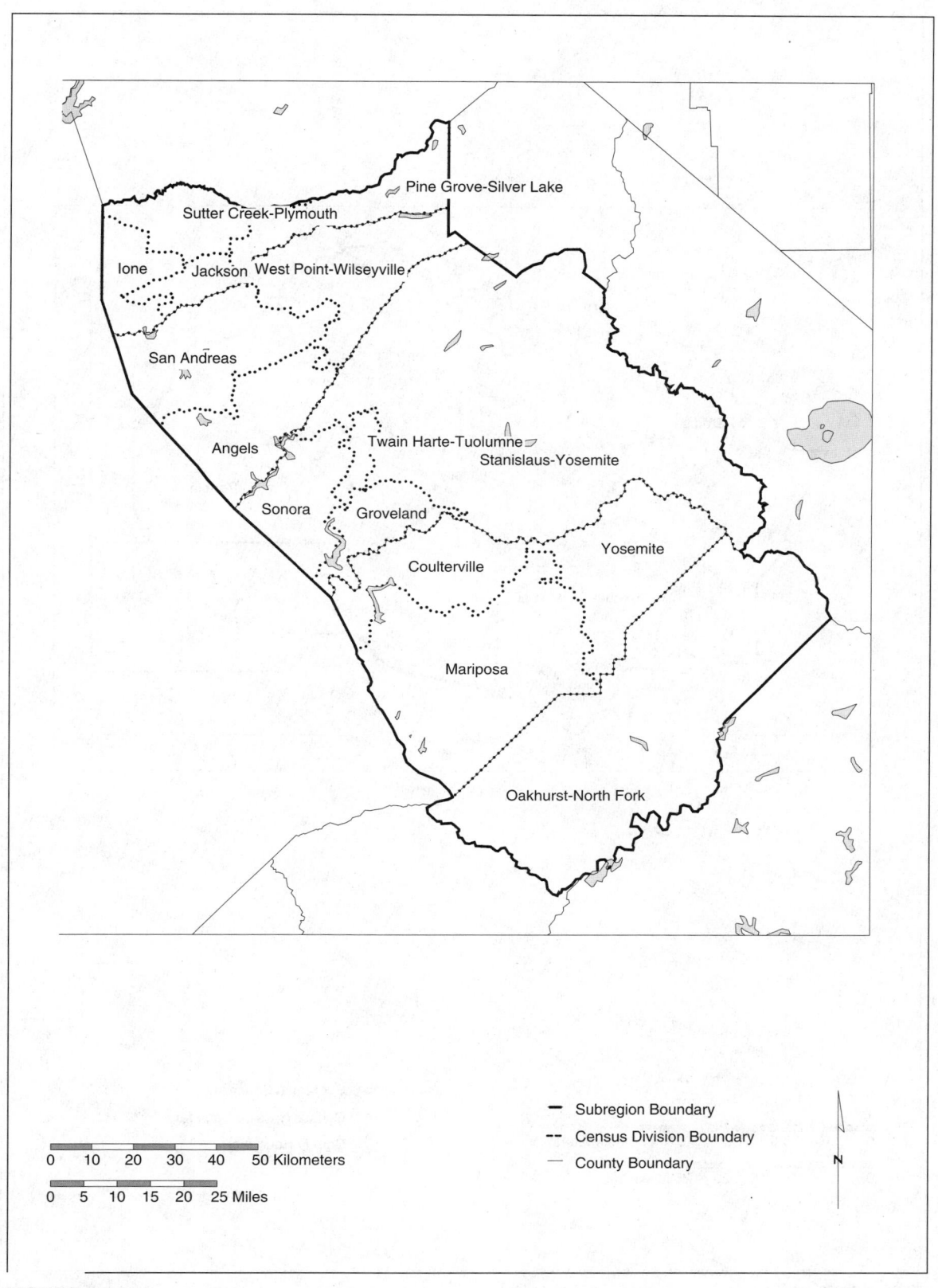

FIGURE 11.16

Mother Lode subregion.

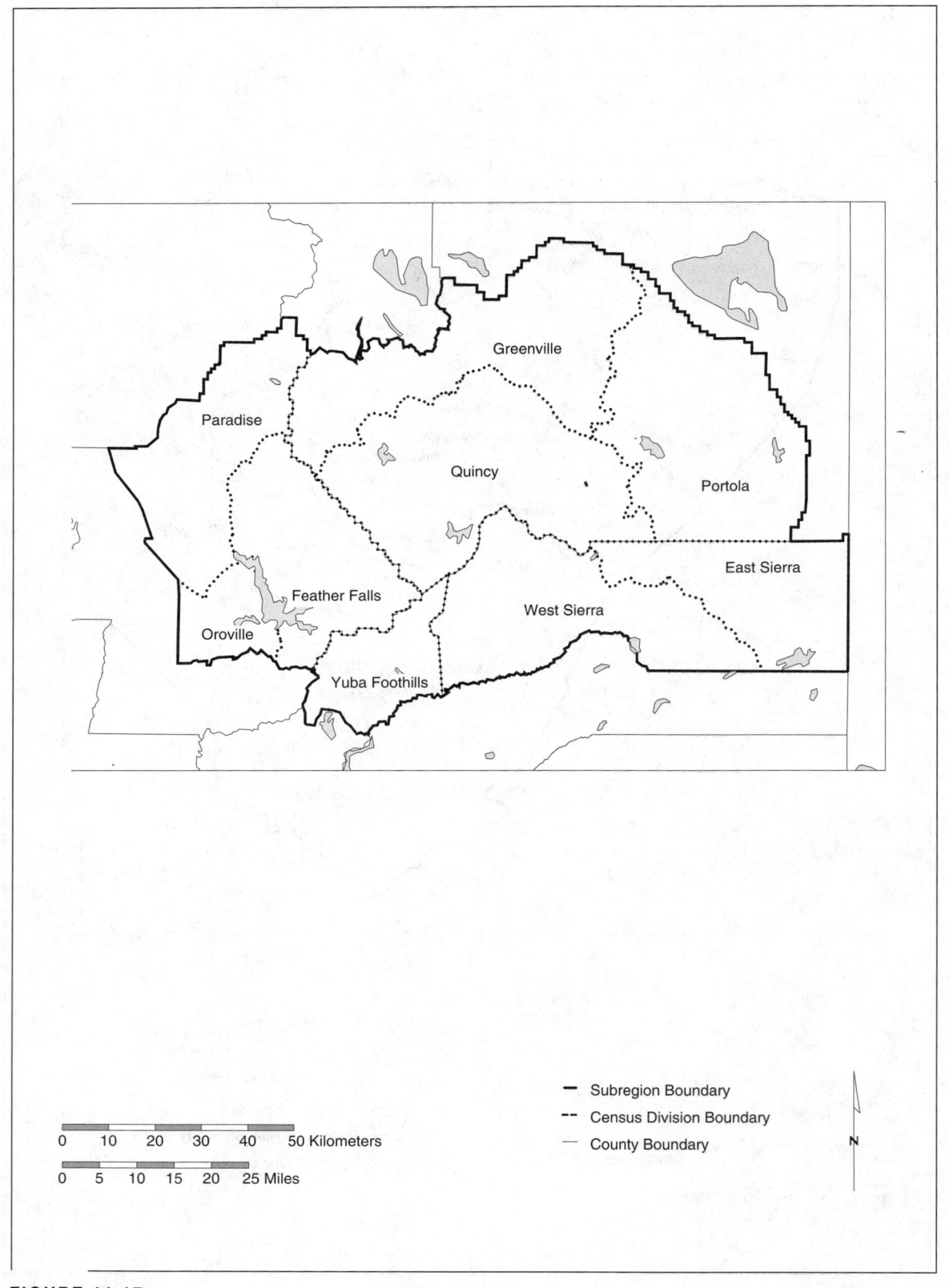

FIGURE 11.17

Northern Sierra subregion.

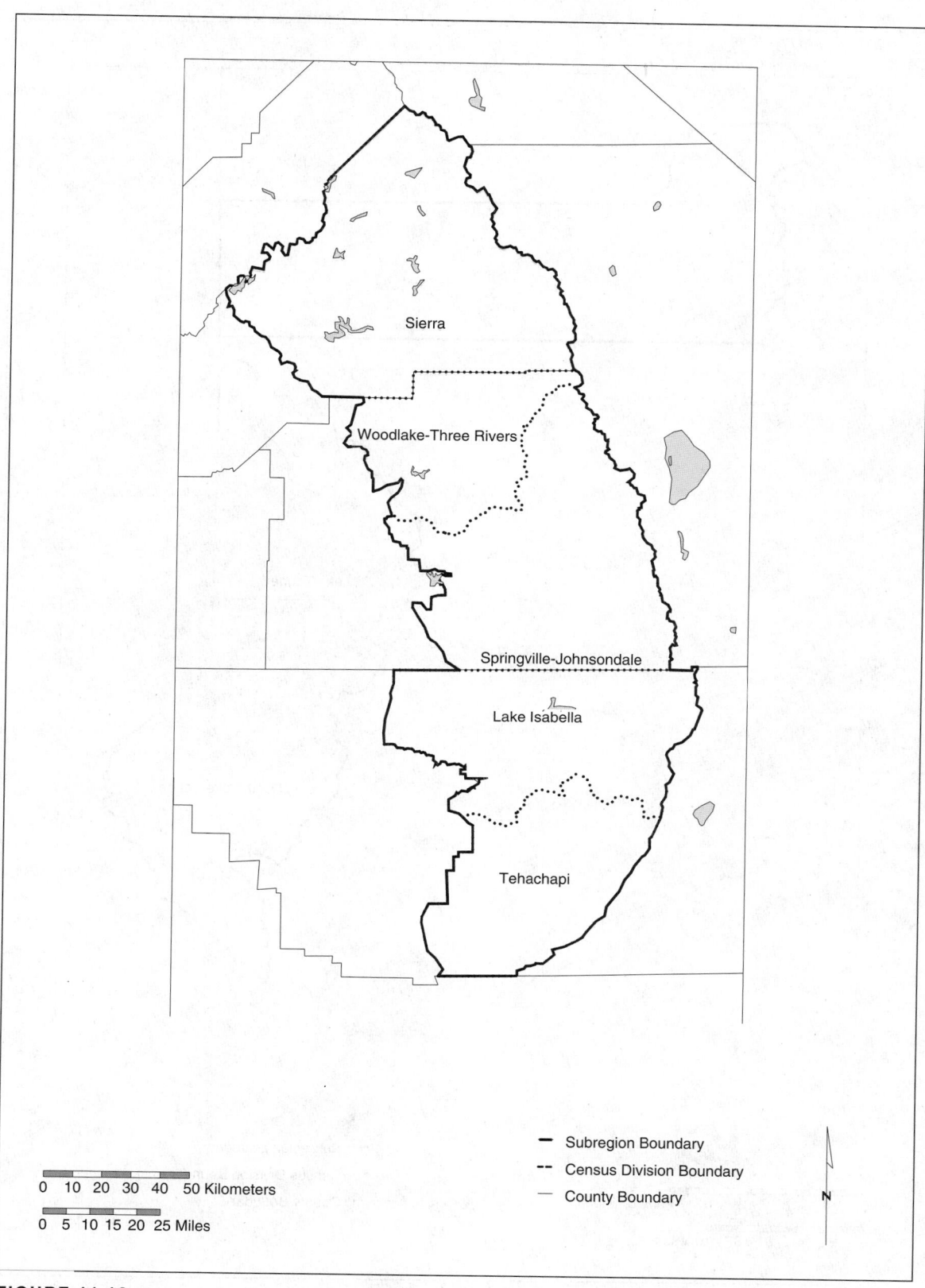

FIGURE 11.18

Southern Sierra subregion.

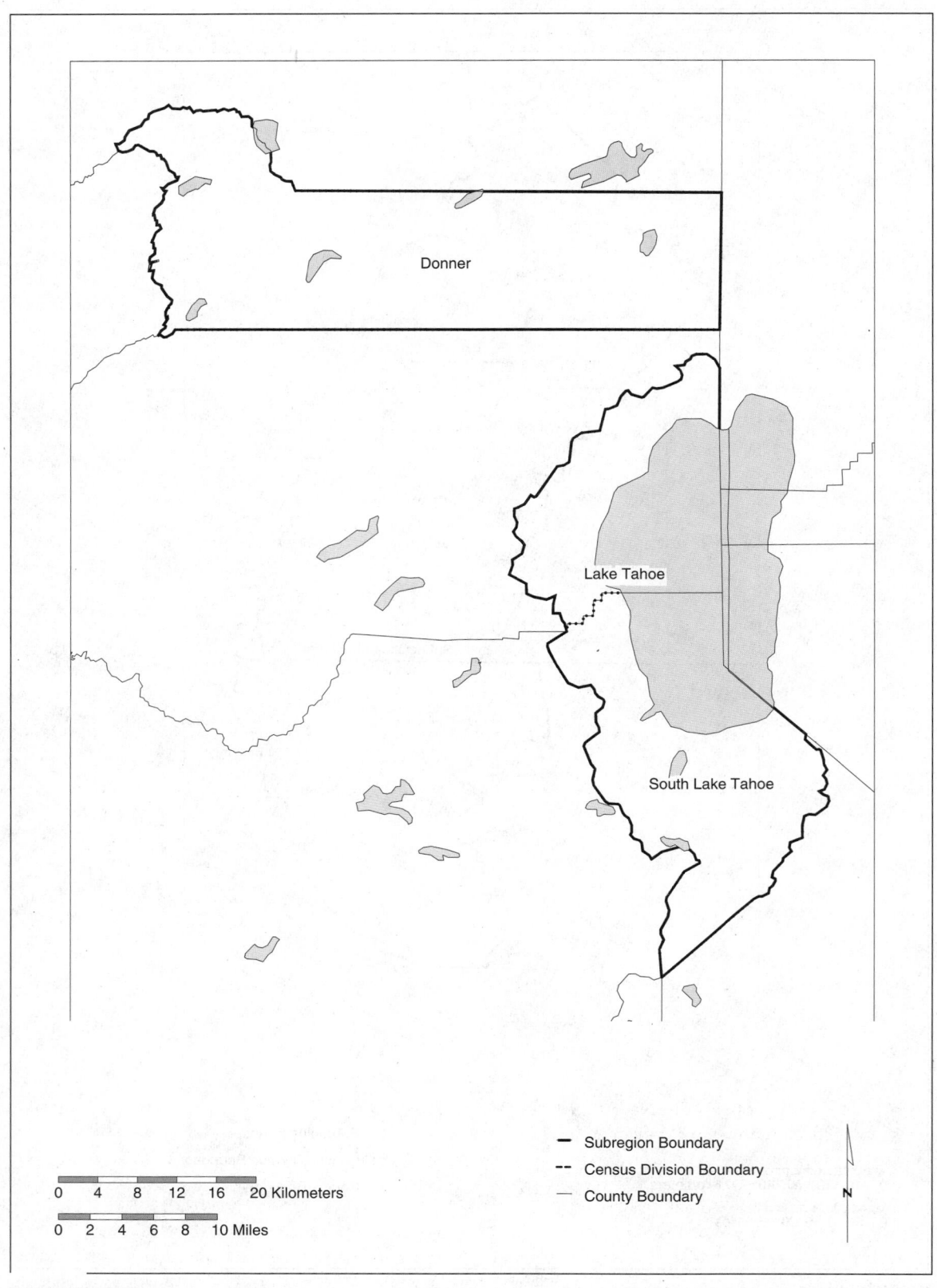

FIGURE 11.19

Tahoe Basin subregion.

and the state has recently imposed new requirements regarding fire safety near wildland areas. Primary planning authority continues to rest with local governments, however, and the county governments of the Sierra Nevada were responsible for land use where most of the development took place in the 1970s and 1980s. Incorporated cities accounted for a relatively small share of growth or land conversion associated with human settlement. As we will discuss later, however, much of the template for recent development patterns was set by policy decisions dating back to the late 1960s and early 1970s.

Nonmetropolitan Population Growth and the Exodus to Exurbia

Arthur C. Nelson offers a comprehensive discussion of exurban regions in his 1992 article "Characterizing Exurbia" (Nelson 1992). The term first appeared in a 1955 book called *The Exurbanites,* which described a group of people living farther out in the metropolitan orbit than existing rail lines (Spectorsky 1955). *Webster's New World Dictionary* offered a definition of the exurbs in 1972: "a region, generally semi-rural, beyond the suburbs of a city, inhabited by persons in the upper income group . . . commuting to the city as a business or professional person" (Nelson 1992). This is also approximately the time that rapid growth pressure began in the Sierra Nevada foothills of California. Most writers emphasize the economic dependence of exurbanites upon nearby urban areas that are accessible via a daily automobile commute (Nelson 1992; Joseph and Smit 1981). This theme is repeated in Thomas W. Sanchez and Nelson's 1994 paper "Exurban and Suburban Residents: A Departure from Traditional Location Theory?" which suggests that exurbanization may simply be an expansion and extension of suburbanization (Sanchez and Nelson 1994). Resolving this issue is of central importance to our understanding of the factors driving rapid nonmetropolitan population growth, for it influences our interpretation of the relative importance of different factors driving human settlement. Relevant data have generally been too aggregated to allow a definitive conclusion, however, because counties (especially in the West) are generally large. Many different types of communities and economic activities may therefore be dispersed throughout a county, making it a heterogenous unit. Judy Davis and her colleagues recognized this distinction in their 1994 study of the "The New 'Burbs" in the Portland, Oregon, metropolitan area, but even their analysis focused on exurban development in relationship to the metropolitan orbit (Davis et al. 1994). Their primary contribution was some hard data that distinguished between "rural," "exurban," and "small town" residents. We do not have similar data for Sierra Nevada residents, but it is clear that all three types of residents are present in the rapidly growing exurban landscape.

Due to the focus on exurban residents' need to commute to metropolitan employment on some regular basis, however,

all of these definitions of the exurban landscape imply some degree of physical proximity to metropolitan regions. This model views exurban development as an extension of the urban sphere just beyond the metropolitan periphery,[15] which reflects to some degree the metropolitan orientation of academic departments of city and regional or urban planning. This perspective is predicated in part on an industrial model of economic activity, however, that may no longer be an accurate characterization of economic relationships in a postindustrial information economy. A series of trends in the 1970s and 1980s converged to reduce the need for *physical* proximity while retaining the need for some type of *economic* integration with metropolitan areas in the 1990s. The "space of flows" (economic, social, cultural, and informational [Castells 1989]) has now become central to the cultural, and economic geography of a region: its ties to other regions (and even its residents' conception of their "sense of place") are decreasingly constrained by physical geography and their proximity to the metropolitan core. Exurban regions can therefore have economic and cultural links to metropolitan centers and the global capitalist economy even when they are well beyond commuting distance. At the same time, of course, limited access to new technologies could isolate rural areas further from the economic mainstream. The spatial structure of economic activity both within the metropolitan region and between the metropolis and the hinterlands is likely to be affected by such technological and economic changes. The forecasting challenge is to identify how those changes will translate into spatial patterns of economic activity and residential location choice. There is still considerable debate about this, of course, so it is difficult to offer definitive forecasts about exurban settlement patterns (Castells 1989).

There is nevertheless some evidence in the patterns of exurban development in the Sierra Nevada that metropolitan access is only important for a subset of individuals migrating into the region. That subset accounts for a significant fraction of overall population growth, but it is also associated with a pattern of human settlement that is higher in density and therefore results in less land conversion per housing unit. The total land area directly affected by exurban growth is therefore not dominated by commuters. The associated patterns of commuter-dependent human settlement are at present concentrated primarily along the western foothills zone in the Gold Country east of Sacramento. Other factors have dominated human settlement in other parts of the range, so we must address the full range of settlement patterns to address the impact of exurban growth on the Sierra Nevada landscape. Each of these factors is discussed in detail later.

The most important patterns of population growth in the Sierra Nevada are likely to be dominated by changing patterns of economic activity and industrial location within metropolitan areas. There is strong evidence that the boom in nonmetropolitan population growth has been most pronounced near existing metropolitan areas, for example, so we can not discount the importance of proximity to metropolitan regions as a critical driving force in the growth of exurbia (Blumenfeld 1954, 1986; Hart 1991; Nelson and Dueker 1990). The west-central Sierra Nevada counties of Nevada, Placer, and El Dorado accounted for 40% of the population growth in the Sierra region from 1970 to 1990, and they all have a significant population of commuters to the greater Sacramento metropolitan area. Expansion of that metropolitan area will tend over time to include portions of those counties, just as suburban regions in the past were once small towns or rural areas. In fact, increases in population density in these three counties is highly correlated with increases in population density in Sacramento County. Population growth in Sacramento County, in turn, is highly correlated with population growth trends in the greater San Francisco Bay Area. This reflects the phenomenal growth that has occurred in California's population in the twentieth century. The state averaged a doubling of population roughly every twenty years through 1970 and then added ten million more people from 1970 to 1990 (Teitz 1990). Those ten million people settled primarily in metropolitan areas, setting the stage for the exodus to exurbia by previous metropolitan residents.

Nelson has estimated that approximately one-fourth of all American residents lived in exurban counties in 1985, and those exurban counties accounted for nearly 30% of all population growth between 1965 and 1985. He also estimated that the land area of those counties covers nearly a third of the United States. This pattern appears to have accelerated since 1985. As Nelson notes, however, this assessment at the county level was limited by the structure of census data available at the time of his analysis. Many counties include subareas that are urban, suburban, exurban, and rural. It is therefore difficult to associate the gross land area of the county categories cited above with the net land area that may reflect an exurban pattern of settlement. This is particularly true in the western United States, where counties are generally much larger in area than those found in other parts of the country. Placer County, for example, is classified as "metropolitan" by the census because it has one city (Roseville) with a population greater than 50,000 persons. Roseville is outside of the Sierra Nevada proper, however, and it is doubtful that the presence of Roseville within the jurisdictional boundaries of Placer County makes Squaw Valley near north Lake Tahoe part of the Sacramento "metropolitan" region. Portions of nearby El Dorado County are much closer to Sacramento than Squaw Valley in terms of commuting times, yet all of El Dorado County is classified as "nonmetropolitan." Further spatial analysis of the 1990 census data is therefore necessary to determine actual patterns of density and sprawl and the relationship between growth in "urban," "rural," and "exurban" areas. Differentiation between types of exurban growth patterns is also necessary.[16] Our disaggregation of county-level data down to the CCD level has allowed us to assess the phenomenon of exurban development in the Sierra Nevada with less interference from "spillover" data from metropolitan centers within or adjacent to Sierra Nevada counties. This level

of analysis is still quite coarse, however, so it has also been necessary to analyze patterns at the block-group and block levels of the U.S. census to understand spatial patterns more accurately. It is the specific spatial pattern of human settlement on the landscape that determines the ecological consequences of development in the Sierra Nevada, so we cannot rely upon broad generalities about urban-to-rural migration patterns derived from nationwide assessments at the county level.

Based upon county-level analysis, the evidence for a broad "reverse migration" from urban to rural areas was strong throughout the United States for the 1970s. It was first identified by demographer Calvin Beale in 1975 (Beale 1975) and confirmed by the 1980 census. For the first time in American history, the 1980 census showed that nonmetropolitan areas grew faster than metropolitan areas during the previous decade.[17] This so-called rural renaissance brought great hope to residents and planners in many rural areas, which had experienced consistent decline throughout the previous century (Vining and Strauss 1977). A general sense of opportunity in rural areas came from this macro-level reading of the census data: rural areas might have more economic opportunities in the 1980s. The 1990 census showed that urban areas again grew faster than rural areas in the 1980s, however, and the serious economic difficulties of many agriculture-dependent regions highlighted how short-lived and illusory the rural renaissance had been for many areas (Barringer 1993).[18] Many planners therefore concluded that the 1970s were just an aberration. Others argued that the apparent "reverse migration" was just a statistical anomaly due to either the reclassification of counties from "rural" to "urban" between the 1970 and 1980 censuses or to "spillover" growth from metropolitan regions to adjacent nonmetropolitan counties (Nelson 1992). This interpretation argued that the historical rural-to-urban migration pattern had not been reversed but that there had simply been a shift *within* metropolitan regions to the outlying urban edge. After controlling for adjacency, however, the counterurbanization pattern was still evident for the 1970s: rural counties not adjacent to metropolitan areas also experienced net in-migration (Nelson 1992). This debate and the difficulty of differentiating "rural" from "urban" counties led to a number of recommendations for reformation of the Census Bureau's definitions of rural and urban areas to avoid the problem of "moving targets" through reclassification every ten years (Nelson 1992; Lang 1986). Nelson argued in 1992 that the Census Bureau should go even further and categorize counties as either urban, rural or "exurban" (Nelson 1992).

Whether or not the historical pattern of rural-to-urban migration in the United States has been reversed in *aggregate* for the country, however, *net* statistics for migration between "rural" and "urban" or "metropolitan" and "nonmetropolitan" regions do not reveal the uneven distribution of population growth occurring within and among rural areas. Kenneth M. Johnson made this important distinction in his 1993 article

"Demographic Change in Nonmetropolitan America, 1980–1990" (Johnson 1993). He characterizes counties as retirement, recreational, adjacent to a metropolitan area, and not adjacent to a metropolitan area. He also evaluated the importance of an urban place of at least 10,000 people within the county. As suggested by the commuter-oriented perspective on exurban growth, many of the nonmetropolitan counties showing rapid growth in the 1970-90 period were adjacent to "suburban" counties that were part of an adjacent metropolitan region. Many of the other nonmetropolitan regions that grew rapidly during this period were quite distant from metropolitan regions, but they were adjacent to large areas of contiguous public lands. These are areas generally judged to have high scenic amenities, clean air, and ready access to recreational opportunities. Many other rural areas—in particular, those whose economies continued to be exclusively dependent upon agriculture , forestry, or mineral extraction—continued to experience the historical pattern of *decline*, masking the emergence of a strong exodus from urban areas to exurban regions offering amenities. Among nonadjacent counties, those without an urban place of at least 10,000 were much more prone to decline than those with a large urban place. This factor is less influential for adjacent counties, although there was a tendency for counties with the smallest places to grow more rapidly. This latter finding suggests that other attributes, including amenity characteristics of small-town life, may be important factors driving growth.

The "exodus to exurbia" therefore appears to be associated with both a classic process of suburbanization *and* an ongoing transformation of rural economies from a commodities-oriented, natural resource–extractive industrial base to a services-oriented, amenity-driven base. Even as aggregate national statistics show a slowdown of urban-to-rural growth (from 14.1% in the 1970s to 3.7% in the 1980s [Johnson 1993]), growth continued rapidly in many desirable small towns and nonmetropolitan areas not adjacent to metropolitan areas. Moreover, as Kenneth Johnson notes, it is important to distinguish between natural increase and migration as sources of change in the total population of a region. Many agriculturally dependent regions had significantly greater gross emigration than the net emigration figures. Both gross and net migration patterns were masked by relatively high natural increases, which have historically been greater in rural than urban areas. Even in areas with net growth, however, the exodus of local residents of childbearing age (coupled with the in-migration of older retirees) meant that "natural increase" was *negative* in many areas. This means that the net increase due to migration was even greater than net population change. There also appears to be significant turnover among new residents, suggesting that the total number of immigrants is much greater than net inmigration.[19]

This is only a summary of the key literature on the processes of exurban growth. For a detailed annotated bibliography on the literature of exurban growth, nonmetropolitan employment, and rurality, see Barry and Duane (1994).

Factors Driving Population Growth in the Sierra Nevada

Most traditional approaches to economic development in rural regions would predict rapid growth in a rural region only if there is an expansion of resource extraction. This reflects a "base" view of the economy, in which exports of primary commodities are the foundation for all local economic activity. Indeed, this appears to be what drove population growth in most rural and exurban regions before the 1960s. While some economic expansion of extractive industries did occur for subperiods of the 1970s and 1980s (e.g., the western slope Rocky Mountain energy boom of the late 1970s and early 1980s; the increase in timber harvesting in the Pacific Northwest during the middle to late 1980s), these traditional rural industries generally *decreased* their employment over the 1970-90 period. Some of this decrease was associated with improved labor productivity and consolidation of operations, while some was driven by contraction of production (caused by either market forces or environmental restrictions). In either case, however, those communities that grew the fastest generally grew despite the *decline* of employment in the extractive sectors. An extraction boom was therefore *not* driving population growth. The most timber-dependent communities were the slowest-growing areas in the Sierra Nevada from 1970 to 1990, while rapidly growing areas decreased timber-sector employment. Expansion of the extractive industrial base was clearly not driving the exodus to exurbia.

If not driven by extractive industry, what was the economic foundation for this growth? What allowed people to move to these areas, and why did they choose to move there in the 1970s and 1980s? Unlike the traditional resource-extractive base of these rural areas, the base for the subtle yet profound transformation of the 1970s and 1980s has been increasing recognition of the *amenity* value of natural resources. In some situations this has made resources more valuable *in situ* than they would be if extracted and exported as commodities for sale in the urban marketplace. This new valuation reflects a broad social change in the environmental values of Americans that has simultaneously challenged traditional approaches to land and resource management over the past three decades. The new values are nevertheless not yet reflected in many of the public land and resource management policies of federal, state, or local agencies. Traditional approaches to land and resource management may therefore sometimes conflict with local social values that are newly emerging as a result of amenity-driven migration and economic diversification.

Those values are readily apparent in a detailed survey of El Dorado County residents conducted in January 1992 as part of the El Dorado County General Plan update. The survey makes it clear that exurbanites are not driven to the Sierra Nevada primarily by traditionally defined economic opportunities but instead seek a way of life (J. Moore Methods 1992). Less than one-fourth of the respondents cited "to work or to find employment" as a major reason for choosing to live in the county, while nearly half said it was to raise their family. "To get away from urban, city life" and "to live in a rural environment" were cited as major reasons by an overwhelming three-quarters of the respondents. This is the primary appeal of the Sierra Nevada. Open space, air quality, and views were cited by 72%, 65%, and 62%, respectively. At least two out of every five respondents listed recreational opportunities as a major reason, while 36% considered "affordable housing" a major reason. Just over one-fourth listed "the quality of the public schools" and fewer than one-fifth mentioned water quality as a major reason for living in the county. Nearly one-fourth of the residents specifically moved to the county to retire, and just over one-fourth mentioned the desire to be near their families as a major reason (J. Moore Methods 1992).[20]

There were no questions about specific "negatives" or "disamenities" that local residents were trying to "escape" by moving from metropolitan areas to the region. These could include several of the factors described below as important factors driving the exodus to exurbia: concern about urban crime, poor urban schools, and increasing racial heterogeneity in metropolitan areas.[21] No comprehensive survey data exist on these issues for the entire Sierra Nevada, and we were unable to undertake such a survey for this assessment. The results of the El Dorado County survey should therefore be viewed as primarily representative of Sierra Nevada residents who live in similar areas in the rapidly growing Gold Country of Nevada, Placer, and El Dorado Counties. We cannot necessarily extrapolate the results more generally to all residents of the Sierra Nevada.

Based upon our review of the literature, twenty-five years of interviews with Sierra Nevada residents, and the quantitative data in the El Dorado County General Plan survey, we believe the following factors have converged to fuel the exodus to exurbia: quality-of-life preferences, the deconcentration of metropolitan employment, information technologies, telecommunications technologies, the shift from manufacturing to services, globalization of the economy, aging of the population, equity gains of urbanites, the lower cost of living in nonmetropolitan areas, the decline of urban schools, increases in urban violence, the ethnic and racial homogeneity in nonmetropolitan areas, and recreation and tourism. This exodus has occurred within the context and against the backdrop of a transportation system that has reduced the cost and time of commuting from the Sierra Nevada foothills. In particular, the construction of Interstate 80 and improvements in U.S. 50 and U.S. 395 have made some parts of the Sierra Nevada much more accessible now to metropolitan areas.

Quality of Life Preferences

Americans have always indicated a preference for small towns and rural lifestyles in surveys, but they have generally settled in urban areas due to the greater range of economic opportunities in metropolitan regions. As Nelson notes, "The latent desire of Americans for the Jeffersonian rural life-style drives exurban development" (Nelson 1992; Carlino 1985; Elazar 1987; Wardwall 1982). Due to the factors outlined later, "the latent preference can now be expressed" (Nelson 1992; Blackwood and Carpenter 1978). It is against this prior background—of American *preferences* for rural regions—that exurbia has boomed.[22] Many Americans clearly prefer cities and metropolitan life, but they remain a minority of the population except in the case of young, single individuals. Based on survey data, most would prefer to live in a small town or rural area (Nelson 1992; Jackson 1985; Fishman 1987). Most Americans actually choose to live in metropolitan areas, however. This raises an important question about whether or not survey respondents accurately characterize their location preferences.

Deconcentration of Metropolitan Employment

The shape and extent of the American metropolis have changed dramatically ever since World War II and the initial investments in the interstate highway system in the late 1950s, and these changes have clearly affected the desirability and feasibility of exurban development. In part, the exurban growth of the 1970s and 1980s simply reflects the expansion of the American population and economy during the 1950s and 1960s, creating new opportunities to live in the "country" while working in the "city." What is more important, however, is that deconcentration of employment within metropolitan regions shifted to the periphery during the 1970s and 1980s. This put many exurban locations within commuting distance of new employment opportunities (Cervero 1986, 1993; Garreau 1991). In the Sierra Nevada, completion of Interstate 80 and expansion of U.S. 50 have had the most profound affect on expansion of the Sacramento metropolitan area and commuter relocation to Nevada, Placer, and El Dorado Counties.

Information Technologies

The microchip and the personal computer have diminished the need for traditional forms of organizational structure, eliminating the need to have a large critical mass of resources to take advantage of economies of scale in information management. What could once be done only by a large corporation with a specialized data processing department can now be accomplished by an individual with a thousand-dollar personal computer. This change has opened up the structure of business, creating new opportunities for smaller organizational structures that can respond to the need for flexible production systems. In some cases these small groups—no longer dependent upon employment in the downtown headquarters of a major corporation—have chosen to relocate based upon other criteria that reflect residential location preferences.

Telecommunications Technologies

The computer modem, facsimile machine, cellular phone, and cable television have made it possible to modify many of the historic relationships between economic activity and location. Just as the personal computer allowed individuals to analyze data without relying upon the corporate bureaucracy, these technologies allow the analysis to be completed anywhere within communications range. The establishment of overnight delivery services has made small towns in rural areas just as "close" to markets as urban areas in terms of shipping important documents and products. Overnight shipping services have also made mail-order and phone-order shopping much more convenient for exurban residents. Moreover, access to cultural material that was historically only available in urban areas—such as timely news, opera or symphony, and major-league sporting events—is now made possible with satellite-dish or cable television and same-day, West-Coast publishing of the *Wall Street Journal* and the *New York Times*. One need no longer live in a metropolitan area to gain many of its amenities and access to its wide choices and specialized consumer markets.

Shift from Manufacturing to Services

Just as globalization has reduced the relative importance of an urban location, a shift in the relative importance of services and value-added manufacturing (with relatively low material intensity, such as computer software) makes proximity to markets and/or raw materials less important than before. There are now a number of economic activities that do not have significant transportation costs associated with them. This shift is, of course, made possible partly by the other social, demographic, technological, and economic trends described here. Growth in the service sector also reflects a general trend in the structure of maturing, postindustrial economies (Powers 1996).[23] That shift increases the viability of economic activity in exurban regions.

Globalization of the Economy

The expansion of global markets and the relative decline of American dominance of the domestic market have combined to create new opportunities for business outside the United States. Because an increasing share of business is with customers who are far from domestic urban centers it is less necessary to be in those urban centers than it was when the customer base was primarily located nearby. One may need to be based in New York if all of one's customers are there, but if one's customers are spread from New York to Tokyo to London, one can just as well be in a small town in the Sierra Nevada. This is especially so for many service industries or high value-added manufacturing activities that depend on international markets. Proximity to a major airport is then an important consideration.

Aging of the Population

The aging of the U.S. population, together with the increasing wealth of urban retirees (due to both equity gains and stronger retirement savings), has created a new pool of "empty nesters" able to live wherever they want. This group then lives off the so-called mailbox economy, bringing outside income into the local economy and generating a multiplier effect as well as a demand for specialized services (e.g., health care and financial advising). This is "base" economic activity. Exurban areas experiencing rapid growth tend to have a disproportionate share of retirees, and this generally reflects in-migration rather than a natural increase for that age group (Collados and Griffiths 1993). The Sierra Nevada shows significant gains for immigrants over 55 years of age.

Equity Gains of Urbanites

Rapid population growth in urban areas during the 1970s and 1980s—particularly in California—created strong consumer demand for housing. Many of the existing homeowners were therefore able to sell their urban homes for significant capital gains based upon the difference between their investment and appreciation. These "equity refugees" were able to move from urban areas with high housing costs to rural areas with relatively low housing costs—in some cases buying new houses mortgage-free. In many cases the desire to avoid capital gains taxes compelled investment in a new home of comparable value, however, driving up the cost of housing in the rural area facing growth. The collapse of metropolitan housing markets in the early 1990s has therefore translated into a drop in demand for housing in exurbia.

Lower Cost of Living

Housing values (and the overall cost of living) are generally lower in nonmetropolitan areas than metropolitan areas. This differential therefore creates incentives to move from urban to rural regions that are consistent with strictly economic models of human behavior. Among nonmetropolitan areas, however, housing values (and therefore costs) are now generally *highest* in those amenity-oriented regions experiencing rapid growth. While they are certainly an important factor, lower housing values alone are unable to explain the migration pattern to exurban areas. Moreover, wages are also generally lower than in metropolitan areas. The specific location choices being made by exurban migrants appear to reflect other criteria. Those amenities therefore translate into significant economic value and activity.

Decline of Urban Schools

Declining quality in and public support for public schools in urban areas have led many urban families with school-age children to send their children to private schools. The increased costs of private education exacerbate the gap between the costs of housing and other services in urban areas and their costs in rural areas. Moving to rural areas is now therefore more cost-effective than it was with good urban public schools, for many families are able to get by with significantly less household income in rural areas. Private schools in urban areas can cost from $5,000 to $10,000 per student per year. Avoiding those costs alone can translate into savings for a mortgage of approximately $80,000–$170,000 for a family with two children. The mean of this range was comparable to median housing values in the Sierra Nevada in 1990.[24]

Increase in Urban Violence

Significant increases in urban violence occurred throughout the 1980s in the United States, despite more than a doubling of the incarceration rate. This violence has decreased the perception of security and well-being that many metropolitan residents still maintained throughout the 1960s and 1970s, when suburban communities were generally deemed safe from the crime of the central city. The increasing concern about safety has led some urban and suburban residents to flee to rural and exurban areas, which are generally viewed as safer. In some cases even that is not considered "safe enough," however, and many wealthy exurbanites have moved into "gated communities" that offer the perception of even greater residential safety. Many of these communities emphasize safety and security in their real estate marketing campaigns,[25] although some evidence suggests that crime rates may actually be higher within gated communities than in neighboring rural and exurban communities in the Sierra Nevada.[26]

Ethnic and Racial Homogeneity

Racism in American society may also explain some degree of the exodus to exurbia. Urban and metropolitan regions in the United States are increasingly multiethnic, multiracial, and multicultural. At the same time that white populations have reached minority status in many cities (notably in California's urban areas), the ethnic composition of the exurban areas experiencing rapid growth is overwhelmingly white. It is difficult to determine how important racism is as a determinant of migration, but the exodus to exurbia parallels the migration to suburbia of the 1950s and 1960s. Statistically valid survey data appear to be lacking on this issue (Walsh 1991). We do not believe that this factor dominates the exodus to exurbia, but we recognize it as a factor for some subset of immigrants moving to the Sierra Nevada from metropolitan areas.

Recreation and Tourism

One of the fastest-growing industries in the 1970s and 1980s was recreation and tourism, and Americans primarily travel domestically. Recreation and tourism is now a $35 billion industry in California alone, and Disney World in Florida receives more annual visitors than the nation of France. Increasing interest in outdoor recreation has in some cases focused on national parks and other public lands, but the small historic towns of the Sierra Nevada have also experienced significant increases in tourism. Increasing tourism in those areas has in turn exposed a great number of people to their associated communities, which then became the focus of residential relocation decisions when that became possible for those visitors.[27]

Many of these factors are difficult to consider explicitly in an economically based model of population migration. One quantifiable element of the Sierra region's economic ameni-

ties, however, is the availability of relatively low housing costs (Inman 1992). The weighted average median value of owner-occupied houses in the Sierra region was $128,678 in 1990, only two-thirds the median value of $195,500 for California. Despite significant growth pressures, that value increased only 80% from 1980 to 1990, while the median value for the state went up by 131%. The median value of a Sierra region owner-occupied home therefore dropped from 84% of the state median to 66% from 1980 to 1990. In addition to these relatively low housing costs, a much higher fraction of Sierra region homes are seasonal units—16.5% versus only 1.8% for the state of California as a whole. Once again, the median values of owner-occupied homes, the rate of growth in those values from 1980 to 1990, and the fraction of all housing units that are occupied only seasonally varied widely across the counties and subregions of the Sierra Nevada.

It is also important to note that housing values from the 1990 census are already outdated. The lag in housing value growth between the metropolitan areas of California and the Sierra Nevada led to a surge of price pressures and real estate speculation in the late 1980s and early 1990s. Significant increases in prices during 1990 and 1991 are not reflected in the median values reported by the (April) 1990 census data. Many urbanites "cashed out" on this significant appreciation in urban markets and built their new "equity mansions" in the Sierra Nevada. The real estate development industry also built many new homes on speculation that the price feeding frenzy in urban markets would continue to support demand for high-end custom homes in the Sierra Nevada. Figure 11.20 shows how residential construction took off during the last half of the 1980s, more than doubling in total value in Nevada County in 1985–86 alone.

This phenomenon was greatest in those areas within commuting distance of the Sacramento metropolitan area, but it also occurred in more remote areas such as the Lake Almanor

FIGURE 11.20

Nevada County residential and commercial construction, 1980–92.

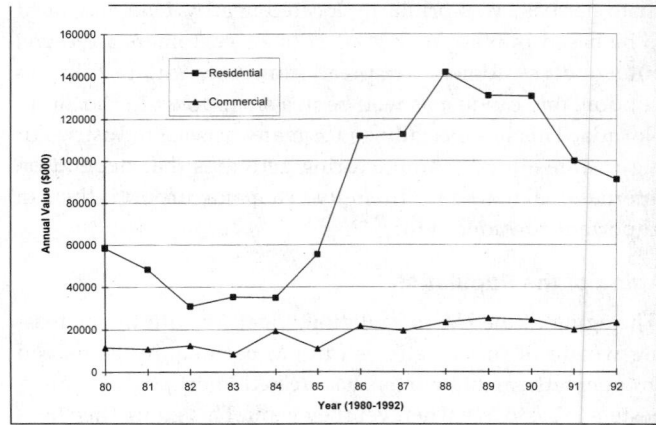

peninsula in the northern Sierra Nevada. Many of these equity refugees were not constrained by the need to maintain a job and wage or salary income. Extremely high appreciation in their urban houses gave them after-mortgage equity gains that sometimes exceeded the cost of buying a house in the lower-cost real estate market of the Sierra Nevada; without a monthly mortgage, these exurbanites could then get by with considerably less monthly income. The *wealth* of these exurban equity refugees is therefore considerably more than their income might suggest, and it has created a new crisis in affordable housing in the Sierra Nevada. There is now a "countercommute" of service workers going up Highways 50, 80, and 49 from the Sacramento metropolitan area every morning to work in the communities of the Sierra Nevada foothills. In the words of former El Dorado County Supervisor Bill Center, "The BMWs are going down the hill while the Pontiacs are going up the hill every morning," then they reverse direction and pass each other again each evening. Median housing values are now lower in the Sacramento metropolitan area than they are in the foothill communities within commuting distance (Drace 1993–95).[28] Nevada County experiences a similar two-way commute, with 30% of employed residents of the county working outside the county while 30% of the jobs in the county are held by nonresidents (Nevada County Transportation Commission 1995a; Nevada County Planning Department 1994a; Landon 1994–95). Many of these nonresident workers probably live in Placer County, however, so the pattern is not tied as directly to the Sacramento metropolitan area.

Prices dropped again slightly and average listing times on the market increased as the rest of the state dropped into recession, but they have not dropped as significantly as housing prices in the hyperinflated metropolitan markets (Marois 1995). The flood of equity refugees has subsided a bit nevertheless, demonstrating how dependent migration pressures are on the economic conditions in the metropolitan regions of California. The Rodney King riots of 1992, the Malibu fires of 1993, the Northridge earthquake of 1994, and the Los Angeles floods of 1995 all had a devastating effect upon the real estate market in southern California. What is perhaps more important, the end of the cold war and significant cutbacks in defense spending have disproportionately affected the California economy. Just as California boomed under the high level of military spending of the Reagan and Bush years, it suffered a bigger bust than most other states with the Clinton cutbacks. Part of the decrease is due to base closures (which have hit California disproportionately), but defense contractor spending cuts have been even more significant. Southern California has lost an estimated half-million jobs in the aerospace and defense sector since 1990. Many of these jobs were held by highly paid homeowners whose demand for housing helped to drive up housing prices in the metropolitan areas. That demand created the equity gains that then allowed other metropolitan residents to make their exodus to exur-

bia. The loss of those jobs has therefore helped to dry up equity gains over the past five years.

The increases in Sierra Nevada housing prices have generally held at levels comparable to 1989–90, however, and this suggests that low-cost housing may no longer be a significant draw for new migrants in the 1990s and beyond. Median home prices jumped in Nevada County from around $120,000 in 1986 to a high of nearly $200,000 in 1990, and they are now at levels (around $160,000) comparable to 1989 (Marois 1995). Moreover, significant equity gains in metropolitan California real estate markets may be dampened by the hyperinflated values that existed in the late 1980s and their subsequent collapse in the early 1990s.[29] The combination of reduced metropolitan housing costs and increased Sierra Nevada housing costs means that equity refugees have less incentive to move based simply on the economic advantages associated with housing costs. Future exurbanites may therefore be more dependent upon employment income than recent migrants in order to make the move to exurbia. This raises important issues about future commute patterns, the traffic congestion and air quality impacts of the emerging patterns of exurban development, and the economic and social mix of the emerging communities of exurbia. All of these factors will affect the future of population growth in the Sierra region, for quality of life appears to be the primary driver of population growth.[30]

Population Projections from 1990 to 2040 for the Sierra Nevada

The Demographic Research Unit of the California Department of Finance (DOF) produced county-level population projections for the period 1990–2040 in April 1993 (California Department of Finance 1993). The Center for the Continuing Study of the California Economy (CCSCE), an independent research institution in Palo Alto, California, has also produced county-level population projections for the year 2005 (Center for the Continuing Study of the California Economy 1995). The CCSCE projections are consistent with the DOF projections, but the DOF projections extend much further into the future. We therefore focus here on the DOF projections and their implications for the Sierra Nevada. The CCSCE forecast for the eighteen counties in the Sierra region for the year 2005 is 3,671,300. This figure is slightly lower than the average of the DOF forecasts for 2000 (3,421,600) and 2010 (4,356,800), which is equal to 3,889,200 (approximately 6% higher than the CCSCE forecast). The CCSCE forecast is therefore within the range of four alternative forecasts that we developed for the Sierra region based on DOF forecasts. Because the DOF forecasts are available only at the county level, we had to estimate which portion of future growth in each county would occur within the Sierra region of each county. As noted earlier, we have detailed data for population growth at the CCD level for 1970, 1980, and 1990. Those data show that the Sierra region gained approximately 175,000 each decade between 1970 and 1990. Continuing that absolute growth from 1990 to 2040 would lead to a Sierra region population of approximately 1.5 million.

We developed a simple model for allocating shares of county-level DOF population growth forecasts to each of the Sierra region CCDs based upon one of three simple factors: (1) the fraction of county-level growth in each CCD from 1970 to 1980; (2) the fraction of county-level growth in each CCD from 1980 to 1990; or (3) the fraction of county-level growth in each CCD from 1970 to 1990. Individual CCDs in the Sierra region varied, with some CCDs having a greater share of county-level growth in one decade than the other. For the entire Sierra region, however, the 1970–80 share of aggregate county-level growth for the entire eighteen-county region was around 8% greater than the 1980–90 share. The estimates we present later are based upon the highest estimate for the entire Sierra Nevada (from 1970–80 shares). These population forecasts could therefore be as much as 14% greater than the CCSCE forecasts and up to a third greater than the level that would be reached if there were continued absolute population growth of 175,000 per decade in the Sierra region from 1990 to 2040.[31] The impact on specific areas will vary widely by CCD, however, and the forecasts based upon the 1970–80 factors do not represent the highest population forecast from 1990 to 2040 for each CCD individually. The combined total population of each CCD-specific highest-growth forecast would result in a Sierra region population of approximately 2.4 million in the year 2040. This is considerably higher than the Sierra region forecasts based upon the three factors described above, which result in a total population of about 1.8 million to 2.0 million by 2040.

The DOF projections are quite daunting, for they are based on a forecast that the entire state will grow from just under 30 million persons in 1990 to nearly 49 million in 2020 and more than 63 million by 2040. Such an increase would more than double California's population in just fifty years. At least in the short term, these forecasts appear to be plausible: at the midway mark between the 1990 census and the DOF forecast for the year 2000, California's population had already exceeded 34 million by the beginning of 1995. The forecast for the year 2000 is for a statewide population of 36,444,000. Net natural increase alone accounted for approximately 361,000 people (approximately 1%) in 1994. Even without significant net domestic in-migration (which totaled only 33,000 people in 1994), continued legal international immigration and natural increases could easily exceed the DOF forecast by the year 2000. The next two most populous states, Florida and Texas, had populations of about 18 million each in 1994. California alone could therefore equal the combined total of those two states' 1990 population by the year 2000.

Although it has only one-eighth the national population, California accommodates roughly one-fourth to one-third of all legal international immigration into the United States. It

is likely to account for at least a similar amount of illegal immigration, although enforcement of Proposition 187 (passed by the voters of California in November 1994), together with stricter border patrols in recent years, could reduce the inflow of illegal immigrants. The 1990 census figures probably understate California's actual population by at least one million, however, for they significantly undercount illegal immigrants in California. The U.S. Census Bureau has also acknowledged that some ethnic groups were undercounted in 1990 within the population of legal residents. Underestimation errors are generally believed to be focused in metropolitan areas, however, so the 1990 census estimates for the Sierra region are probably not affected significantly by these errors. Other errors associated with collecting data in remote rural areas are probably more important in the Sierra region than in metropolitan areas. These include unmarked roads and many houses that are hidden from view on large parcels. The sheer inaccessibility of these residences and their low density reduces the likelihood of successful follow-up visits by census enumerators. For purposes of this analysis, however, we will assume that the 1970–90 census figures for the CCDs in the Sierra region are reliable.

Based upon those figures, the Sierra region population is forecast to more than *triple* from just over 600,000 in 1990 to nearly 2 million people (1,964,200) by the year 2040. The lower CCSCE estimation function (14% below the "high" DOF forecast) would yield a Sierra region population of 1,722,138 by the year 2040.[32] This total would be comparable to the entire San Francisco Bay Area's population in 1940 (1,734,308).[33] The overall growth rate for the Sierra Nevada from 1990 to 2040 (226%) would also be comparable to the growth rate experienced by the Bay Area from 1940 to 1990 (247%). The projections are therefore within the range of recent experience in northern California. The combined population of Sacramento and San Joaquin Counties (which include two of the primary commuter destinations for residents of the western Sierra Nevada foothills) grew by 400% from 1940 to 1990. Individual counties in the Bay Area experienced a wide range of growth rates during that same period. The densest county and employment center of the region, San Francisco, grew by only 14% (due to limited land area and existing high densities). Alameda County grew ten times as fast at 149%, Napa County grew by 289%, and Marin County's population expanded by 335%. Most of Marin County's growth occurred from 1940 to 1970, when the population jumped from 52,907 to 208,652. It grew only an additional 10% to 230,069 from 1970 to 1990. This low rate was the result of a complex set of growth management tools and a strict General Plan in the early 1970s (Teitz 1990). San Mateo County also truncated its growth after 1968, although it still grew by 481% from 1940 to 1990 (Teitz 1990). The experience of these two counties' aggressive growth management regimes from 1970 to 1990 may have relevance to other counties in the Sierra region, such as Nevada and El Dorado Counties, who are now facing rapid growth pressures and considering updates to their General Plans. This topic will be discussed in more detail in the discussion of management implications and policy options below.

With the exception of Marin County and San Mateo County, the Bay Area's suburban counties accommodated most of the growth in the region over the past fifty years and grew at rates comparable to those forecast for some Sierra region counties over the next fifty years. Napa County grew by 289%, Sonoma County by 462%, Solano County by 593%, and Contra Costa County by a remarkable 700% as the 1940 population of 100,450 mushroomed to 803,732 by 1990. Sacramento County grew by 511% during the same period, from 170,333 in 1940 to 1,041,219 by 1990. San Joaquin County, however, to the south of Sacramento County and containing the port city of Stockton on the San Joaquin River, grew at a much slower rate of 258% during this period. San Joaquin County's population of 134,207 was 79% of Sacramento County's population in 1940, but it grew to only 480,628 by 1990, only 46% of Sacramento County's population. This differential growth rate between Sacramento and San Joaquin Counties can be attributed both to the rapid growth in state government in Sacramento during this period (when California's population grew to more than 30 million), the substitution of capital and energy for labor in the agricultural sector, and the construction of Interstate 80 (I-80) through the Sacramento area. The latter effectively integrated Sacramento with the Bay Area, while Stockton and San Joaquin County remained more isolated from economic integration with the rapidly expanding Bay Area. Stockton therefore continued to function as only a regional center for the northern San Joaquin valley and the Delta region, while Sacramento emerged as not only the center of state government but also an economic extension of the Bay Area.

This latter point is critical, for it helps to explain the subregional concentration of growth within the Sierra region. Access to the Bay Area along I-80 allowed firms based in the Bay Area to locate manufacturing facilities in the greater Sacramento metropolitan area in the 1970s and 1980s, where land costs were considerably lower than in the rapidly urbanizing Bay Area. These business location decisions reflected both the economics of site development (i.e., each company's own facilities) and the economics of residential location choice (i.e., each company's employees' own residences). The former could have led to facility location decisions that shifted manufacturing activities out of the Bay Area (in particular, the Santa Clara valley for high-technology companies) to a wide range of locations with good transportation access. The latter, however, which includes both the cost of living and the amenity value of residential location, resulted in the location of manufacturing activities between Sacramento and the Sierra Nevada foothills. This location is actually *less* convenient than other relatively low-cost locations along I-80 between Sacramento and the Bay Area, because it requires additional travel time and additional risks of delays while crossing the Sacra-

mento metropolitan area in order to reach the Bay Area. It is *more* convenient, however, for employees who want to live in the Sierra Nevada foothills or at least in that part of the Sacramento metropolitan area that will provide easy recreational access to the Sierra Nevada and the American River.[34] Access to the residential amenities of the Sierra Nevada appears to have been a primary factor in the location choices of Bay Area firms relocating manufacturing facilities outside the Bay Area. The original decision to relocate those facilities, in turn, was the result of rapid growth in the Bay Area that both increased the cost of land and housing and decreased the quality of life for many employees through increased traffic delays and decreased open space. The transformation of the Bay Area landscape therefore had a direct bearing on the forces that have begun to transform the Sierra Nevada landscape over the past quarter century. The fate of the Sierra Nevada is inextricably tied to California's metropolitan centers.

Examples of these new employment centers can be found along I-80 northeast of Sacramento in Placer County and along U.S. Highway 50 on the eastern edge of Sacramento County. The former includes facilities for Hewlett-Packard and NEC, both located in Roseville and fueling nearby residential development in Rocklin and Loomis. Perhaps the most extreme example of this relocation phenomenon exists in the eastern Sacramento County town of Folsom, however, where Intel has developed a large complex of buildings that employed nearly 2,750 employees in 1994 and is home to Intel's six major product divisions as well as Intel's North and South American Sales and Marketing Operation. It is also world headquarters for the company's Information Technology organization. The corporate headquarters remains in the so-called Silicon Valley, but new technologies now allow worldwide corporate activities to be coordinated from a satellite facility located 150 miles away. That satellite facility sits on a bluff above the American River just a few miles from El Dorado County and less than a thirty-minute drive from either the gold rush town of Placerville or the state capital. The 236-acre campus had a gross payroll of about $100 million in 1994. Construction of a new 320,000-square-foot building in 1994–95 cost $52 million and will house an additional 1,750 employees. The total employment at the Intel Folsom site will then be 4,500 employees and generate a payroll of between $150 million and $200 million per year (Intel Corporation 1994). This employment base, together with the multiplier effect of the site through subcontractors and employee expenditures in the community, is likely to fuel much of the nearby Sierra region's population growth.[35] High-technology employment in the greater Sacramento area now accounts for at least 15,200 direct jobs at Packard Bell, Hewlett-Packard, Intel, NEC Corporation, and Apple Computer (*Grass Valley Union* 1995b).

Figure 11.21 shows the relationship between the Sacramento metropolitan area and these emerging employment centers along Interstate 80 and U.S. Highway 50. The Sierra Economic

FIGURE 11.21

Proximity of Gold Country communities to Sacramento and Lake Tahoe.

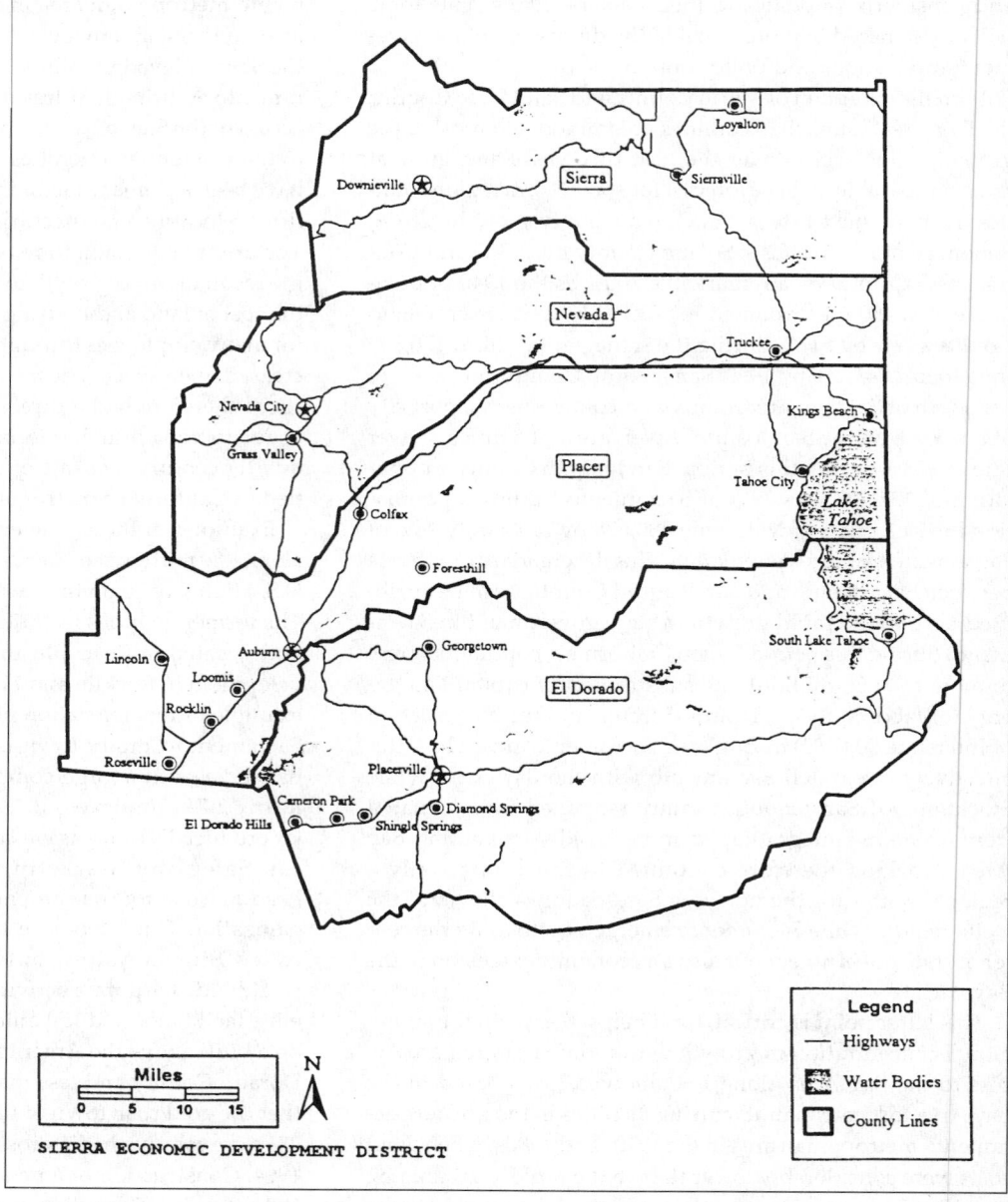

SIERRA ECONOMIC DEVELOPMENT DISTRICT

Development District (SEDD) publishes this map to show its economic links to Sacramento and its service area of Sierra, Nevada, Placer, and El Dorado Counties.

Much of the overall population growth in the Sierra region from 1970 to 1990 nevertheless appears to be "suburban" rather than "exurban" in character. This is particularly true for higher-density single-family home developments along the I-80 and U.S. 50 corridors. The literature on suburbia is therefore relevant to our understanding of both historic growth patterns and the DOF population forecasts. This literature includes a wide range of population growth models that are based on a "gravity" concept of a nested hierarchy of "urban fields." In northern California, this model indirectly links the west-central Sierra Nevada foothills to the Bay Area

through the Sacramento metropolitan area. San Francisco is the central city core of the Bay Area, with an urban field of lower-density employment centers throughout the Bay Area. Both those peripheral centers and the Sacramento metropolitan areas are effectively linked to and dependent upon the well-being of that central city core through transportation networks and economic flows. As Robert Cervero has demonstrated, however, employment on the periphery of the Bay Area metropolitan region is now comparable to that in the central cities of San Francisco, Oakland, and San Jose (Cervero 1986, 1993; Garreau 1991). This deconcentration of employment has created further opportunities for more dispersed residential locations that remain within commuting distance of Bay Area employment. A similar phenomenon in the Sacramento metropolitan area has in turn made some parts of the Sierra Foothills part of what Kenneth T. Jackson has called the "crabgrass frontier" (Jackson 1985). In that respect we have much to learn from his history of "the suburbanization of the United States." It has already had direct relevance to parts of the Sierra region over the past quarter-century.

Jackson documents how other "rural" regions on the periphery of metropolitan regions have been undergoing suburbanization since at least 1830. This process is neither a new phenomenon nor one dependent upon the automobile, although its particular form does seem to reflect the dominant transportation technologies of the time of initial development. Travel time has always been more important than travel distance or the specific means of travel in determining the maximum distance individuals are willing to commute between their residence and place of employment. The deconcentration of metropolitan employment in California, coupled with continuing population growth throughout the state, is therefore likely to continue to increase the attractiveness of the Sierra Nevada for residential location. Travel times are also reduced per unit distance traveled on less congested rural roads, so the maximum feasible commute distance is not a linear function of distance. Greater distances can be covered in the same amount of time as average settlement densities decrease and congestion delays are eliminated. Changing technologies and patterns of economic activity are, of course, highly uncertain over any fifty-year period, so we cannot assume today's structural relationships when forecasting future conditions. There is nevertheless strong historical precedent for continued settlement of the Sierra Nevada consistent with the DOF forecasts.

The rapid rates of growth forecast for the Sierra Nevada are not unusual historically and have also been experienced by other rural regions outside the San Francisco Bay Area or Sacramento. These rapid rates of growth on the suburban and exurban frontiers often accompany a slowing of growth due to density saturation in the metropolitan central city. Brooklyn, New York, was a sleepy rural village of just 7,125 people in 1820 (while nearby New York City, isolated across the East River, already had 123,706 residents). "In the next four decades, however, the town of Brooklyn was transformed. Regular steam ferry service to New York City . . . began in 1814" (Jackson 1985). Brooklyn more than doubled in population each of the next few decades, jumping to 15,384 in 1830; to 36,233 in 1840; to 96,838 in 1850; and to 266,661 in 1860. "Whether it was easy access, pleasant surroundings, cheap land, or low taxes," notes Jackson, "the suburb was growing faster than the city by 1800" (Jackson 1985). By 1890 the population of Brooklyn was 806,343, while New York City exceeded 2.5 million people. Brooklyn itself had grown from just 6% of New York City's population to 32% in seventy years. Jackson states that "one wag noted that Brooklyn 'sold nature wholesale' to real-estate developers, for sale to homeowners at retail" (Jackson 1985). This sounds quite a bit like today's Sierra Nevada real estate market. Many of the same factors that drew

people to Brooklyn in 1830 are now drawing Sacramento commuters to the west-central Sierra Nevada foothills.

These attractions have historically been accessible only to residents of a particular class, however, leading Robert Fishman to call the American suburbs "bourgeois utopias" (Fishman 1987). Here he is referring to the middle-class suburb of privilege, a residential community beyond the core of a large city. The development that he describes is more restrictive than the broader patterns and processes driving today's exurban growth, but it has many of the same roots. What is more important, it accurately describes the subset of exurban Sierra Nevada development that is most like the classic middle-class commuter suburb. Much of what we are now seeing in the Sierra Nevada is similar in intent (if not urban form) to that which first constituted a "suburb" in 1750 in London—having a house "in the country." The need for the exclusion of others from the suburb is important now, as it was two hundred years ago, in part to ensure the preservation of this idyllic setting. Establishment of successful "gated" communities with relatively high suburban densities is therefore not surprising in the context of this historic pattern of suburbanization. Fishman outlines three primary factors driving suburban development: (1) the desire for life in a picturesque space; (2) the protection of the family; and (3) the avoidance of urban problems. All three factors appear to be important considerations in the residential location decisions of recent immigrants to the Sierra Nevada and other exurban areas throughout the rural West. Historical processes of suburbanization are therefore relevant to our understanding of the processes driving exurban growth in the Sierra Nevada.

Like the "crabgrass frontier" of Brooklyn from 1820 to 1890, the west-central Sierra Nevada foothills could gain a similar share of the greater Sacramento metropolitan region's population from 1970 to 2040. And like the "bourgeois utopias" of historic London, the values of many migrants to the Sierra Nevada may reflect basic truisms about human nature and the search for the ideal as much as new technologies and lower housing costs. Suburban development certainly reflects transportation commute times, economic conditions, and land markets, but it also reflects the values, dreams, and lives of the migrants in the places they left for suburbia. Ironically, many of the exurban migrants of the past few decades have come from suburbia. This raises a fundamental question confronting planners and citizens in the Sierra Nevada today: how can we avoid a development process that will destroy the very features that make the region a desirable place to live? The historical record is not encouraging, with the recent exodus to exurbia strong evidence that the suburban ideal has not maintained itself in the face of a wide range of forces that have transformed metropolitan areas throughout the country. Meeting that challenge in the face of significant continuing population growth will not be easy in the Sierra Nevada.

Based upon the DOF forecasts, many other areas are likely to experience similar increases in commuting and

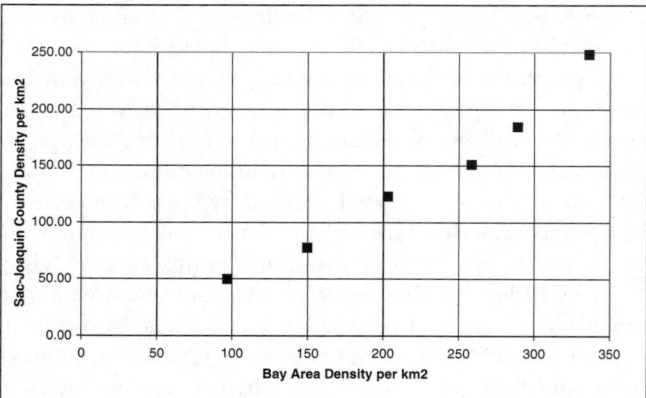

FIGURE 11.22

Density of Sacramento and San Joaquin Counties versus Bay Area, 1940–90.

suburbanization as the metropolitan centers of Stockton, Modesto, Fresno, Visalia, and Bakersfield continue to grow.[36] Fresno and Bakersfield, which are the southernmost communities in the San Joaquin valley, are emerging as significant metropolitan areas in their own right. Nearly 500,000 people lived in the Fresno CCD in 1990, and Fresno County produces more agricultural value than any other county in the United States. Bakersfield, the capital of Kern County, had a population of nearly 300,000 in its immediate CCD in 1990. Kern County produces more oil than any other county in the United States. Agriculture is also very important in Kern County, however, and oil pumps are often working away alongside farm equipment in the fields. Neither Fresno nor Bakersfield is closely tied to the Bay Area through commuting patterns. The more northern communities of Stockton and Modesto have become increasingly linked to the Bay Area economy through the development of Interstates 5, 580, and 205. These highways now link residents in Stockton, Modesto, and the

FIGURE 11.23

Gold Country versus Sacramento County density, 1940–90.

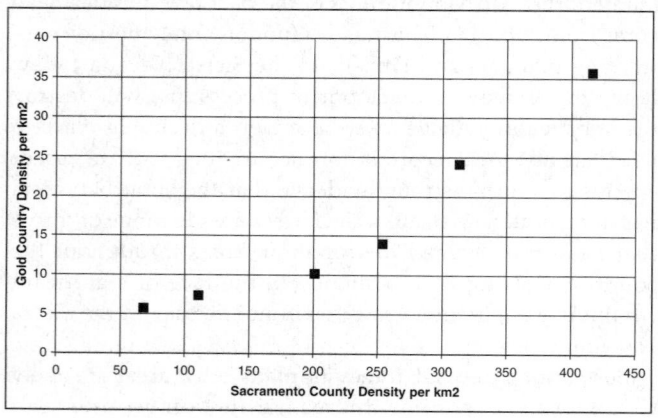

nearby towns of Tracy and Manteca to jobs across Altamont Pass in the Livermore Valley area near the intersection of Interstates 580 and 680. From there, commuters can go north to San Ramon and Walnut Creek; go south to Fremont, San Jose, and the greater Silicon Valley; or continue west to the East Bay employment centers of Oakland and Berkeley. These commutes sometimes total two hours each way, but people are willing to do them in order to have an affordable (or larger) home in a place where they feel safe. They sometimes ride in vanpools, which allows them to sleep or read each way, but they most typically ride in single-occupant automobiles that get stuck in traffic jams on Altamont Pass. The Bay Area Rapid Transit (BART) system is now constructing a feeder line out to Pleasanton and Livermore that will connect potential commuters directly to downtown San Francisco. This in turn is likely to increase commuting to bedroom communities in the Central Valley.

These bedroom communities, which were sleepy agricultural towns until recently, are now sprouting commercial centers to provide services to the commuters when they are home. Times have changed since George Lucas grew up in Modesto, which inspired his film *American Graffiti*. Freeway interchanges along Highway 99 now bustle with neon and traffic jams. Gang violence has also appeared, just as it has in Stockton, Fresno, and Bakersfield. These valley towns are now becoming suburban centers. A large mall in Modesto, growing boat sales in Stockton, and the emergence of dining establishments to feed the weary commuters have all sprung up. These in turn create new employment opportunities both for other residents in the Central Valley and for those willing to commute from the nearby Sierra Nevada foothills. There is probably relatively little commuting at this point from the foothills to these lower-paid service jobs, but the potential is there for new higher-paid employment opportunities. In this way the expansion of the Bay Area directly affects the suburbanization of the Sierra Nevada.

Figure 11.22 shows how increasing population densities in Sacramento and San Joaquin Counties have been closely tied to increasing densities in the Bay Area. A simple bivariate regression analysis of the data has an R-squared of 0.97, which means that 97% of the variation in one variable is explained by variation in the other variable. This is based on only six observations, of course, so we cannot infer much statistical significance to this finding. It nevertheless illustrates that growth in these areas appears to be closely tied.

A similar relationship appears to hold between Sacramento County and the Gold Country counties of Nevada, Placer, and El Dorado. The relationship appears weaker, with an R-squared value of only 0.92, but that appears to be due to a split in the data around 1960. This is also when Interstate 80 was completed, so we have a plausible explanatory variable that is consistent with the general theory of commuting as a primary factor in determining population growth in these counties. Figure 11.23 shows the data for 1940–90, while figure 11.24 shows the stronger relationship from 1960 to 1990.

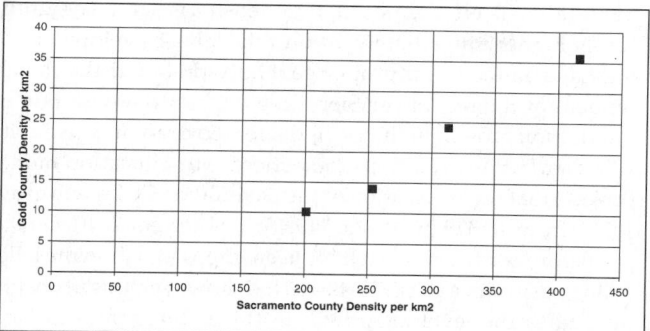

FIGURE 11.24

Gold Country versus Sacramento County density, 1960–90.

The R-squared value for the 1960–90 regression is 0.99, although there are only four data points. Again, this is simple correlation and does not necessarily indicate causation. These figures illustrate merely that the data are consistent with the urban field hypothesis of exurban growth.

Note that the average population densities in the Gold Country counties are considerably lower than the average densities in the metropolitan counties. This is not surprising, but it is exaggerated by the fact that the Gold Country counties have large areas of public land and private industrial timberland, where there is no potential for residences. Due to its relatively small land area, the average density for Sacramento County is also comparable to the average density for the Bay Area (it was lower in 1940 and 1950, almost exactly the same in 1960 and 1970, and has been higher in 1970 and 1980). Average density in San Joaquin County is midway between Sacramento and the Gold Country. Average densities in the Gold Country counties in 1990 (28 to 48 persons per square kilometer) were comparable to average densities in Marin County in 1940, Sonoma County from 1950 to 1970, Solano County until 1950, and Napa County until 1980. The growth patterns and experience in those Bay Area counties from 1940 to 1990 are therefore relevant to the future growth patterns and experience likely for the Sierra region. Even Amador and Calaveras Counties, which are within commuting distance of Sacramento and San Joaquin Counties, had average densities in 1990 (12 to 20 persons per square kilometer) that were comparable to Napa from 1940 to 1950 and Solano or Sonoma in 1940.

Under the April 1993 DOF forecast, individual counties and CCDs in the Sierra region will have widely varying rates of growth . The Sierra region of three of the most remote counties (Plumas, Sierra, and Inyo) will remain less than twice their 1990 population by the year 2040, while eight counties will experience between a doubling and a tripling of their 1990 population by 2040. Seven counties will actually see their Sierra region population more than triple. Two CCDs are forecast to increase by more than tenfold: the Yuba Foothills CCD

in Yuba County (2040 population will be 13.5 times the 1990 population) and the Lake Isabella CCD in Kern County (2040 population will be 10.4 times the 1990 population). The population for the Sierra region of Fresno County is forecast to be 9.6 times the 1990 population. Only the Yosemite CCD of Mariposa County, which lost population in 1970–80 due to the relocation of housing for National Park Service employees, is forecast to have a smaller population in 2040 than there was in 1990. Plate 11.1 compares 1990 population density to the maximum forecast population for each CCD in 2040 based upon the county-level DOF forecasts and our simple CCD population allocation methodology.

The DOF population forecasts are highly uncertain, of course, for population forecasting is a risky business. On average, the aggregate absolute population growth per decade for the Sierra Nevada under the DOF forecasts is 50% higher than the absolute population growth during either the 1970–80 or the 1980–90 period. Historic levels of population growth are therefore only about two-thirds of the DOF forecasts. The early 1990s also saw significant domestic emigration from California, taking some growth pressure off the state and the Sierra Nevada. California's net population increase in 1994 was less than half of the peak-year increase of 740,000 people in 1990 (Teitz 1990; Landis 1992). The DOF forecasts may therefore overstate growth in the Sierra region as the greater rural and exurban West absorbs an increasing fraction of the exodus from California's metropolitan areas. Reduced property values in those metropolitan areas have also reduced the opportunities for significant "cash out" by equity refugees, one of the primary factors driving the migration. This change has already reduced the rate of immigration into the Sierra Nevada in the mid-1990s. The DOF forecasts were not completed until April 1993, however, and California was already deeply mired in recession by that time. Net domestic migration is also becoming a less important determinant of overall statewide population growth as the demographic momentum of natural increase becomes more important.

Conversely, the forecasts may overstate the capacity of metropolitan areas to absorb additional growth in the state.[37] If the state-level growth forecasts are accurate (which seems to be reasonable for at least the decade of the 1990s), then this would suggest that nonmetropolitan population forecasts (such as for the Sierra region counties) are understated. Much of the Sierra region growth is also forecast to occur in those counties within commuting distance of the secondary metropolitan centers of the Central Valley. These include Sacramento, Stockton, Fresno, and Bakersfield. Those metropolitan centers may actually carry a greater fraction of their respective county's overall population growth. On the other hand, expansion of those metropolitan centers will also create new employment and service opportunities for the Sierra Nevada foothills. The bottom line is that population growth forecasts are highly uncertain over even short periods of time. Significant population growth is highly likely for most regions of the Sierra Nevada, however, even if the precise levels of such growth are difficult to predict.

Only time will tell if the April 1993 DOF forecasts will accurately predict Sierra region growth for the next fifty years. The DOF made a similarly bold fifty-year forecast in 1971 for Alpine, Amador, Butte, Calaveras, El Dorado, Nevada, Placer, Plumas, Sierra, and Yuba Counties. The 1971 forecast for those ten counties projected a doubling of the population over fifty years from 334,500 in 1970 to 465,200 in 1990 and 695,500 in 2020 (Department of Water Resources 1973). The forecast growth rate was much higher than previous rates of growth for the region, and many observers probably doubted at the time that the region would reach a population of 700,000 by the year 2020. The DOF forecast turned out to be *low,* however, for the 1990 census showed that those ten counties had already reached a population of 703,856 in just twenty years. We estimate that the Sierra region of those counties had a population of 436,426 in 1990.[38] The population of those counties by the year 2020 will undoubtedly be much higher than the 1971 DOF forecast projected. Continued growth at that absolute rate per decade for those ten counties (i.e., 184,678 persons per decade) would result in a population of 1,257,890 by the year 2020. This would be 81% greater than the original 1971 forecast population of 695,500 in the year 2020.

With a doubling of the Sierra region population from 1970 to 1990, the DOF forecast projection of a tripling of the population from 1990 to 2040 is both plausible and probable. Even if population growth in the Sierra region stayed steady at 175,000 per decade (the same absolute level as 1970–90), the total Sierra region population would increase by 140% to nearly 1.5 million people. The range of plausible population estimates for the Sierra region is therefore from just under 1.5 million people to up to 2.4 million people by the year 2040. Even the low end of the range represents a significant increase.

This growth is likely to make many areas of the Sierra Nevada look more like the Gold Country over the next fifty years. The April 1993 DOF forecast projects that the population of the Gold Country CCDs will grow 179% from 222,837 in 1990 to 621,842 by the year 2040. Due to regulatory constraints that have been imposed by the Tahoe Regional Planning Agency (TRPA) within the Lake Tahoe Basin since the 1970–80 period, however, this probably overstates the level of growth in the Lake Tahoe subregion. The balance of growth that we have forecast for the Lake Tahoe subregion would take place in the Gold Country subregion of Nevada, Placer, and El Dorado Counties. Based upon the original forecast, however, the Gold Country subregion's growth rate is relatively low compared to both the Mother Lode and southern Sierra subregions. The Gold Country subregion's share of overall Sierra region population will therefore decrease from 36% in 1990 to just 30% by the year 2040.

The average population density of more than 30 persons per km^2 in the Lake Tahoe subregion is higher than in any other subregion of the Sierra region. Because a large part of the Donner CCD in Nevada County is actually quite far from Lake Tahoe itself, the population density is actually significantly higher immediately around the lake. Development of the Lake Tahoe Basin proper generally preceded the development of the rest of the Sierra region, and the high rate of population growth in the 1970s decreased dramatically in the 1980s as TRPA regulations took effect. Our allocation model projects that the population of the Lake Tahoe CCDs will grow 147%, from 48,329 in 1990 to 119,453 by the year 2040. Due to regulatory constraints that have been imposed within the Lake Tahoe Basin since the 1970–80 period, however, this probably overstates the level of growth in this subregion (although much of that growth could occur in the spillover "bathtub ring" outside the jurisdiction of the TRPA). The balance of growth would take place in the Gold Country subregion of Nevada, Placer, and El Dorado Counties. If the Lake Tahoe subregion continues to grow according to this forecast, its share of overall Sierra region population will remain steady at 8% through the year 2040.

The Mother Lode subregion is forecast to grow 236%, from 124,795 in 1990 to 418,900 by the year 2040. Despite this rapid growth, the Mother Lode subregion's share of overall Sierra region population will increase only slightly from 20% in 1990 to 21% by the year 2040. This is due to the rapid growth forecast for the southern Sierra subregion, a remarkable 384%, from 92,366 in 1990 to 447,479 by the year 2040. This rapid growth rate means that the southern Sierra subregion's share of overall Sierra region population will increase dramatically, from 15% in 1990 to 25% by the year 2040. No other subregion increases its relative share of overall Sierra Nevada population as much. This primarily reflects the high growth forecast the DOF for the metropolitan areas of the southern San Joaquin valley. It is also slightly distorted by the growth in Kern County population due to prison construction in the Tehachapi CCD from 1970 to 1980. A similar problem exists for the Ione CCD in Amador County and the Stanislaus CCD in Sonora County, the only other CCDs in the Sierra Nevada with a state correctional facility. Both of the latter are located in the Mother Lode subregion, however, where the entire county is located within the Sierra Nevada. Only those errors associated with allocating the county-level DOF forecasts to CCD-level projections in the Tehachapi CCD in Kern County will therefore affect overall Sierra Nevada population projections.

Two of the subregions are likely to experience much slower growth than the rest of the Sierra Nevada. The eastern Sierra subregion is forecast to grow 136% under the DOF forecast, from 28,509 in 1990 to 67,418 by the year 2040. This relatively slow growth rate means that the eastern Sierra subregion's share of overall Sierra region population will decrease from 5% in 1990 to just 3% by the year 2040. The northern Sierra subregion is also forecast to grow relatively little (149%), from 102,110 in 1990 to 254,563 by the year 2040. This relatively slow growth rate means that the northern Sierra subregion's share of overall Sierra region population will decrease slightly, from 16% in 1990 to 13% by the year 2040. In both cases this reflects the relative isolation of these two subregions and their lower levels of direct economic integration with the rapidly growing metropolitan centers outside the Sierra Nevada. These subregional differences therefore highlight and reinforce the importance of metropolitan growth as a primary driver of future population growth throughout the Sierra Nevada.

The subregional summary just presented is based upon the CCD allocation factors for 1970–80, which resulted in the highest overall population forecast from 1990 to 2040 for the entire Sierra Nevada. This set of factors will not result in the highest overall population forecast for individual CCDs or subregions, however, so the spatial pattern of future growth should also be evaluated based upon the 1980–90 and 1970–90 allocation factors. Several differences between the different allocation factors for specific CCDs warrant explicit discussion here. One CCD (Yosemite, in Mariposa County) and one county (Inyo) actually lost population from 1970 to 1980. The reductions in the Yosemite CCD appear to reflect the relocation of housing for some National Park Service employees in Yosemite National Park. All of the population losses in Inyo County occurred in the Death Valley CCD, however, which is outside the Sierra Nevada region. All Sierra Nevada CCDs in Inyo County grew from 1970 to 1980. Three more CCDs (Lone Pine, in Inyo County; Greenville, in Plumas County; and Twain Harte, in Tuolumne County) lost population from 1980 to 1990. Three other CCDs (Ione, in Amador County; Tehachapi, in Kern County; and Stanislaus, in Tuolumne County) had unusually high increases in population from 1980 to 1990 due to construction or expansion of correctional facilities within the CCD. Finally, two CCDs (Yuba Foothills, in Yuba County, and Sierra, in Fresno County) had growth rates that differed by more than 10% across the two periods. Use of the allocation factors for either the low-growth or high-growth periods can therefore result in significantly different shares of overall county growth going to these CCDs during the forecast period. In both cases these CCDs account for only a small share of overall county population, so the error associated with this difference can be quite large. Estimates for all of these CCDs should therefore be viewed cautiously.

The Residential Development Process in the Sierra Nevada

The land use pattern in rural and exurban regions is mixed, and the Sierra Nevada is no exception (Nelson 1992; Yaro et al. 1988; Arendt 1994b; Davis et al. 1994). High-density clusters of structures exist in pockets at critical crossroads and in small villages and towns, but most of the landscape is uninhabited or sparsely settled. The villages and towns often have population and structure densities that are comparable to urban settings in metropolitan regions, but their *scale* (both in population and area) is significantly smaller. Rural and exurban villages and towns are typically home to 10^2 to 10^4 people, while metropolitan-area towns and cities range from 10^4 to 10^6 residents.[39] This could change if the overall level of population growth in the Sierra Nevada results in significant expansion of existing high-density urbanized areas.

Development in the Sierra Nevada is occurring primarily in the formerly rural, unincorporated areas near gold rush–era communities in the foothills zone. It is not city-centered, although the "urban" centers of the foothills often provide the essential services that the residents now demand. The result is a dominant pattern of low-density, land-intensive, large-lot exurban sprawl. More people in the Sierra Nevada live in high-density settlements than low-density settlements, but much more land area is devoted to the latter than the former. This pattern of settlement is extensive and land-intensive.

The reasons for this pattern of development are manifold. The fundamental force is the desire of new residents to live "in the country" with wooded, open spaces shielding their "homestead" from the view of neighboring homes. Depending on vegetative cover, this can be achieved at densities below approximately one unit per acre. The result is a sense of privacy and a connection with the natural world. Contact with the community comes through regular visits to the nearby town center, where daily employment and/or service needs are met. There is also limited neighborhood contact, although lower densities decrease opportunities for inadvertent interaction with neighbors. Often the center of informal social life in these areas is the post office or the grocery store. These are "sacred spaces" within the community that serve a vital social function (Hester 1990).

Some of these recent migrants are moving from metropolitan areas into "gated" communities, where reproduction of the social, physical, spatial, and market characteristics of the suburban landscapes they left may be desired. They seek the market benefits of that physical and spatial pattern without the costs of scale diseconomies associated with the larger metropolitan pattern they left. Wal-Mart is fine, because a wide variety of goods at a low price is a desirable element for them. What they want to leave behind are the traffic congestion, the crime, the graffiti, and the homelessness they faced in their daily lives in their suburban communities in California's metropolitan areas. The suburban land use pattern of sprawl is therefore only seen as a problem if and when it begins to be associated with those disamenities. Within the tighter and more homogeneous social context of exurbia, however, traffic congestion is the first disamenity they will probably experience. By then, of course, the land use pattern will be very difficult to change.

Development in the Sierra Nevada foothills has generally been through a process dominated to date by incremental construction of individual homes, unlike the "new town" subdivision process common in metropolitan real estate markets. Large parcels are often subdivided without simultaneous development of "model homes" and builder-originated construction. Instead, lots are sold to individuals without any requirement to choose a particular "model home" design or hire the subdivision developer as the home construction contractor. Many of the existing parcels in the Sierra Nevada were created through major subdivisions that were approved by local planners before the Subdivision Map Act of 1973. Many of these were intended for second homes as "recreation residences" but have subsequently been developed for year-round residences. Quarter-acre lots with on-site septic systems are common among these subdivisions from the late 1960s and early 1970s. Just three major developments from this period (Lake Wildwood, Lake of the Pines, and Tahoe Donner) accounted for two-thirds of all new home construction in Nevada County during the 1980s (Nevada County Planning Commission 1993). The land was subdivided and sold off parcel by parcel, but actual development has been incremental over nearly three decades. These rural lots are often purchased by urbanites who will hold them vacant until retirement or another personal (rather than a local real estate market) opportunity finally allows them to move to the Sierra Nevada. Each parcel is then typically developed individually in accordance with the needs of the individual lot owner. The size, style, and impact of each house on the environment therefore varies widely in exurban areas. The overall scale of the development and its eventual impact were therefore difficult to gauge during the early years of the projects. Moreover, most of these large-scale "recreational residential" developments were approved before passage of the California Environmental Quality Act (CEQA) in 1970.

Because individual landowners built their own houses (and because buildout has occurred incrementally over several decades), many of the large-scale subdivisions in the Sierra Nevada have developed without the mass-produced feeling that is more common to the large-scale suburban subdivision developments in California's metropolitan areas. (Homes built by speculators have become more common in the Gold Country since the boom of the late 1980s, although the market appears to have been overbuilt in the early 1990s. Future "spec" home activity is therefore likely to be dampened by the significant losses incurred by speculators in the early 1990s.) Most of the remaining lots have been and continue to be created through "minor" subdivisions of four or fewer parcels that are exempt from the stringent requirements of the Subdivision Map Act of 1973.[40] Concurrent subdivision and infrastructure investments (which are required under the Subdivision Map Act of 1973) have therefore been the exception rather than the rule to date in the exurban development process in the Sierra Nevada. This could change in the future, however, with several large-scale "new town" developments likely to be developed in the future. The implications of such a shift are discussed in our case studies of the General Plans for Nevada and El Dorado Counties.

Minimum lot sizes are usually set by the local government through a General Plan designation and a specific zoning ordinance.[41] The minimum for rural residential lots with on-site septic systems and on-site well water typically varies from one acre to five acres, but there is no standard policy in the Sierra Nevada. As noted earlier, many of the existing parcels were originally approved for on-site septic disposal at densities of up to four units per acre and were "grandfathered" in by the newer General Plans and zoning ordinances in the 1970s and 1980s. Two-fifths of the land in Nevada County was not yet zoned with specific density requirements as late as 1980. The General Plan updates by Nevada and El Dorado Counties now propose a minimum of 3–5 acres for on-site septic disposal systems with an on-site well water source (Nevada County Planning Department 1994a; El Dorado County Planning Department 1994). The 1980 Nevada County General Plan had a 1.5 acre minimum for the same configuration, but that has since been deemed inadequate (*Nevada County General Plan* 1980; Norman 1982; Boivin 1991–95).

Local government land use policy is usually set by a combination of five factors:

1. Existing parcelization (e.g., current land use designation)

2. Land uses on adjacent properties (e.g., typical densities)

3. Infrastructure availability (e.g., roads, water, sewers)

4. Environmental constraints (e.g., slope, soils, vegetation)

5. Philosophy, values, and ideology (e.g., the role of regulation)

Note that factors 3 and 4 (infrastructure availability and environmental constraints) could lead to land use densities that are often inconsistent with factors 1 and 2 (existing parcelization and land use on adjacent properties). Environmental constraints or a lack of infrastructure may limit the potential development density, for example, but the land may already be zoned for or adjacent to land already developed at higher densities. The final factor (philosophy, values, and ideology) seems to determine the relative weight given to the other factors and the range of alternative policies that elected and appointed officials are willing to consider (Juvinall 1995). Based upon review of the General Plan development processes in Nevada and El Dorado Counties, it appears that local officials often rely upon this existing pattern of parcelization as the primary factor in designating land uses.[42] This is the primary reason that existing General Plans and zoning designations in the Sierra Nevada are often inconsistent with the results of environmental analyses. Decisions have often been made primarily based upon adjacent land uses or existing zoning on adjacent parcels, rather than the availability of infrastructure or the environmental impacts of development. Despite CEQA, the impacts of development are therefore not fully mitigated in the county General Plan and zoning processes. "Overriding considerations" are frequently invoked under CEQA to avoid mitigation for significant effects. This is demonstrated below in our case study of the General Plan updates for Nevada and El Dorado Counties. Planning clearly takes place in a highly politicized context.

The timing, location, and degree of urbanization in metropolitan regions is often determined by major capital investments in infrastructure systems: roads, water supply, sewage collection and treatment facilities, energy supply and related systems (e.g., stormwater drainage). This policy tool—the ability of local governments to control the timing and location of investments in physical infrastructure—has significantly less influence on low-density rural and exurban land development that relies upon on-site infrastructure. It is therefore more difficult to guide development patterns in these areas, where relatively low land costs make site-specific on-site infrastructure investments economic. Indeed, most rural land development occurs without either centralized water supply or sewer systems. On-site wells and septic tanks are common. According to the 1990 census, nearly one in four of all Sierra Nevada Region housing units have private, on-site well water supplies (versus about one in twenty-five for California) and nearly three out of every five housing units have septic tanks or cesspools for waste disposal (versus less than one in ten for California as a whole).

This fact has a direct bearing on the pattern of development that occurs in exurbia. Environmental and health factors dictate that on-site well water systems and on-site septic tank systems should be separated, and therefore zoning regulations *require* low-density development patterns. This is in part due to the reliance on zoning (which is oriented toward density controls) as the primary means of regulating local land use. Local soil conditions, slopes, and the hydrologic characteristics could all be considered when determining site-specific risks and appropriate standards,[43] but comprehensive analysis of these natural factors has generally been weak in the exurban planning process. Rather than allowing development only where environmental constraints are least limiting, then, local governments have relied on large-lot zoning to increase the likelihood that there will be *some* buildable site on a given parcel. Undoubtedly, many one-acre parcels have multiple building sites and could support more than one house with an on-site septic system and on-site well water. Conversely, many one-acre parcels have poor soils, steep slopes, proximity to intermittent surface water sources, and very poor ground-water resources. Systematic analysis of environmental constraints would favor shifting development from the latter site to the former site, with less environmental impact at the same level of development. The current reliance on large-lot zoning fails to complete such analysis, however, so it promotes large-lot exurban sprawl and a landscape of fragmentation. Site-specific consideration of natural constraints tends to occur only through the building permit requirements of a percolation test (for septic systems) and minimum well water flow rates. There is rarely any site-specific evaluation of the risk of septic system failures or potential contamination of critical hydrologic resources as a result of failures.[44]

The large lot sizes dominating the prevailing pattern of exurban development are therefore a direct result of the lack of infrastructure to serve the burgeoning exurban population. This lack of infrastructure in turn is a function of both land market economics and the reliance of local land use authorities on low-density, large-lot zoning as the primary means of reducing the potential health risks associated with on-site well water and septic tank systems. These health risks in turn are a function of both on-site infrastructure technology and economics and the environmental constraints of the site. Large lots are not necessarily required to meet the market demand for homes. New residents might be just as satisfied with their "quality of life" on a half-acre lot as on a two-acre lot, for example, if the amenities they seek—privacy, clean water, wildlife habitat, and possible room for a horse—are still available to them in that alternative configuration.[45] A two-acre minimum leads to a development pattern that directly affects up to four times as much land, however, while breaking up ecosystems and habitat, through road networks, building footprints, and the influence of domestic pets, into a pattern of "islands" that are unconnected to larger habitat patches or ecological systems in the landscape. The result may be considerably more environmental damage than would be necessary if the infrastructure allowed higher densities. Higher density does not necessarily mean less environmental impact.

Patterns of Human Settlement in the Sierra Nevada

Much of the literature on rural and exurban land use has failed to distinguish between very different patterns of human settlement in the exurban landscape. Judith Davis and her colleagues made an important distinction in 1994 between "suburban," "exurban," and "small town" residents in their study of exurban counties near the Portland metropolitan area, but that study focused on social, demographic, and economic differences rather than settlement patterns per se (Davis et al. 1994). Further distinctions are necessary to understand land use and the impacts of alternative patterns of human settlement in the Sierra Nevada. Exurban development patterns generally include five distinct types of settlement:

1. Compact small towns of 10^2 to 10^4 population

2. Contiguous exurban subdivisions at suburban densities

3. Stand-alone "gated" communities at suburban densities

4. Large single-family lots with private on-site infrastructure

5. Rural agriculture, natural resource, or open space lands.

These patterns are described in the sections that follow.

Compact Small Towns of 10^2 to 10^4 Population

Communities are the core of exurban areas and the location of most commercial and service activities. In the rural and exurban West their urban form usually dates from the nineteenth century, making them "walkable" and compact in size.[46] These towns were built before the automobile had been invented and long before it had come to dominate urban form. Their architecture is usually a mixed vernacular, offering a variety of styles but relatively standard building scale of between two and four stories. Many of these towns were built around mining or other commodity extractive industries, and their architecture reflects repeated investments and an evolution from tents to shacks to wood-frame buildings to masonry brick structures.[47] Urban designer Peter Owens notes, however, that "most of these places would be illegal under current zoning codes" (Owens 1991–95). Yet they are both intuitively attractive and extremely practical forms of human settlement (Alexander et al. 1977).

Recent "neotraditional" urban designers like Andres Duany and Elizabeth Plater-Zyberk have attempted to reintroduce the spatial patterns and urban form of these traditional patterns in new developments like Seaside, Florida (Duany and Plater-Zyberk 1991). Similar proposals have been made by Peter Calthorpe for the Sierra Nevada foothills (Calthorpe 1993; Local Government Commission 1992). In theory, these neotraditional new towns promise both social and ecological benefits. In practice, the centrally planned neotraditional towns remain socially and economically segregated and lack much of the vitality of the organically developed traditional small towns (Harvey 1993). They are also limited in both population and land area, limiting their potential as a model for handling the dramatic increases in population being experienced in the exurban West. They nevertheless provide a critical social and economic function and offer important lessons for urban design that could yield significant environmental benefits.

Socially and culturally, there is daily interaction among residents in these small towns through shopping, schools, and the rural ritual of picking up mail at the post office box. Social events often revolve around participatory activities (e.g., Little League games, fund-raising pancake breakfasts) rather than professional entertainment. Volunteerism is quite common; in fact, many services that are provided by professionals in urban areas (e.g., fire fighting) are staffed primarily by volunteers in these communities. Reliance on all-volunteer fire departments is changing, however, as the Sierra Nevada grows: commuters and retirees have little interest in volunteer fire fighting, so taxes and fees must be raised to pay for more full-time professional firefighters. Population densities are "urban" within the city limits of these compact small towns—often from 2,000 to 5,000 persons per square mile (plus significant land area dedicated to commercial, industrial, and public uses). In many cases these towns are not incorporated but subject to county oversight for land use planning, regulation, and public services. Truckee was already a bustling town when the Central Pacific railroad was completed in 1869, for example, but it did not incorporate as a municipality with its own city council until 1994.[48] Until then it was "unincorporated Nevada County" and relied upon the county to provide essential public services.

Contiguous Exurban Subdivisions at Suburban Densities

Contiguous subdivisions built in the postwar period are often immediately adjacent to the pre–World War II, compact small towns. These subdivisions are usually connected to the small town's water supply and sewer system, allowing densities comparable to suburban developments in metropolitan areas (anywhere from four to eight houses per acre, or a population density of 5,000 to 10,000 per square mile of residential development after accounting for about 20% dedicated to public roads). Infrastructure access is also necessary to build much higher density multiple-family units. Infrastructure is the key element defining these developments, which have architectural features and a layout that diverge sharply from the patterns in the historic small towns. Residences are typically single-story, while they are often two levels in the historic pattern. Streets are much wider, and the houses are set back from the streets and from each other with ample yard space. The social openness of the traditional front porch has been replaced by the fenced backyard, which isolates the modern family's leisure time and diminishes opportunities for casual interaction (Jackson 1985; Fishman 1987). The garage, a small and hidden addition to the lot in the traditional small town (if it exists at all), has moved from the backyard to the front of the house. The primary means of accessing the residential space is now through the automobile. These subdivisions are designed to maximize vehicle mobility and minimize social interaction. As Michael Southworth, Peter Owens, and Eran Ben-Joseph have demonstrated, the evolution of subdivision design in America reflects a series of systematic changes by nonarchitects that date back to the 1920s and 1930s and continue to constrain urban form (Southworth and Owens 1992; Southworth and Ben-Joseph 1993; Southworth 1995).

Within the development, this settlement pattern is just like that of any other suburban subdivision; within the broader exurban context, however, it is often quite different. Its proximity to the "old town" often allows pedestrian or bicycle access to services, while its overall scale (10^1 to 10^2 acres) is usually much smaller than those developed in metropolitan regions (10^2 to 10^4 acres). This has an important social effect, for the residents retain a familiarity with their neighbors and a connection to the immediately adjacent small town that is often absent in suburbia. These higher density areas are nevertheless very different spaces from the traditional small town itself. They are often the only location in an exurban community where multiple-family housing is located. The poorest members of exurban regions tend to live either in subsidized multiple-family units or in trailers and mobile homes in the most rural (and lowest-cost) settings in the area. Gentrification of the quaint Victorian houses of the historic small towns has increased the need for this kind of housing, but state and federal funding support for affordable housing has diminished recently and is generally concentrated in declining central cities. The rapid growth of the rural and exurban West has created a new affordability crisis (Nevada County Planning

Commission 1993). These contiguous urban subdivisions are therefore becoming more important for their role as pockets of poverty in what is otherwise becoming a more affluent exurban landscape. Examples of this pattern are found in the west-central Sierra Nevada on the outskirts of the historic towns of Grass Valley, Auburn, and Placerville.

Stand-alone "Gated" Communities at Suburban Densities

The opposite condition exists in the many exurban areas that have independent "gated" communities, which are neither physically contiguous to nor socially integrated with the small towns that form the core of the exurban settlement pattern. Unlike the small towns, these "private" communities are usually homogeneous in ethnic (white), social (well-educated exurbanites), demographic (more retirees), economic (wealthy relative to the rest of the region) and political (conservative) characteristics. They are often built around significant recreational amenities (e.g., lakes and golf courses), and they generally have larger lots and more expensive homes. In some cases they have community sewer and water systems, but many older subdivisions continue to depend on private septic systems and on-site wells. This dependence on private infrastructure has not diminished densities, however, for many of these older subdivisions were approved before land use and environmental planning laws required stricter standards. Densities range from one to four houses per acre, or 1,000 to 5,000 persons per square mile. The total population of these private communities is often comparable to the compact small towns at "buildout" (5,000 to 10,000 people). They are usually unincorporated, however, and do not provide many of the service functions of the compact small towns. They are therefore "bedroom communities" that insulate themselves from the rest of the exurban region except as the rest of the region may provide necessary services (e.g., shopping and medical). Because they have assessed themselves to provide infrastructure services (e.g., road maintenance, private community sewer system), their residents often object to tax increases that will benefit the larger exurban community (e.g., for county roads or county schools).

The median assessed values of homes and median family incomes in these communities rival the highly inflated values of metropolitan California. They far exceed typical values for most of the rural and exurban West. Lake of the Pines had a median housing value of $368,500 in 1990, compared with median values of $155,685 in Nevada County, $128,678 for the Sierra region, and $195,500 for California as a whole. The median household income in the core census block group of Lake of the Pines was $55,161 in 1990, compared with medians of $32,464 for all of Nevada County, $29,595 for the Sierra region, and $35,798 for California. Lake Wildwood's median house value was $226,800, and the median household income in Lake Wildwood was $52,359 in the core census block group in 1990. The values are lower in Lake Wildwood than those in Lake of the Pines primarily because Lake of the Pines has a much higher fraction of commuters who work outside Nevada County (53%) than Lake Wildwood (23%). Lake Wildwood also has a higher fraction of retirees, with 66% of its residents at least 55 years of age ("only" 48% of the residents of Lake of the Pines are at least 55 years of age). This compares with 29% of Nevada County residents, 27% of all Sierra region residents, and only 18% of all California residents who are 55 years of age or older (U.S. Bureau of the Census 1990).[49]

These high housing values and household incomes have supported effective privatization of public services without municipal incorporation. Lake of the Pines and Lake Wildwood also rival the incorporated towns of Grass Valley, Nevada City, and Truckee as population centers in Nevada County. Unlike those three incorporated cities, however, the privatized "public" sector of the gated communities is exempt from a wide range of laws guiding public policy in California municipalities. They are exempt from open meeting laws and can structure mechanisms for controlling local land use and infrastructure decisions based upon ownership rather than equal representation. Political jurisdictions are then less relevant to infrastructure and land use decision making in privatized communities. Expansion of this pattern of "gated community" development therefore has implications for the land use planning process itself. It also has a direct bearing on the capacity to provide local infrastructure through general taxation.

Not surprisingly, many of the members of these communities see little reason to tax themselves to provide services for the rest of the county or larger community. The privatization of the public sector through the gated community structure effectively segregates the exurban landscape by class. Gated communities clearly provide a market good with a particular set of characteristics that are highly valued by many consumers in the marketplace. In that sense, they provide room for many of the equity refugees fleeing metropolitan areas for exurbia. The marketing materials for these communities emphasize personal safety and social, demographic, economic, spatial, and architectural homogeneity. Interestingly, the residents of these gated communities appear willing to accept strong restrictions on their (and their neighbor's) "private property rights" through covenants, codes, and restrictions (CC&Rs) in the title to their property. Land use regulation is often strongest in these gated communities, with the homeowners association in charge.

Large Single-Family Lots with Private On-Site Infrastructure

Most exurbanites probably live in one of the three settlement patterns just described. Most of the land area in exurban areas is probably in the fifth settlement pattern described, largely "open space." Most of the land area in exurban regions directly affected by human settlement, however, is probably a result of large single-family lots with private on-site infrastructure.[50] This is an extremely popular form of settlement, for it offers privacy as well as direct contact with the country ideal for the ex-urbanite. Ironically, it can also have significant negative impacts on the environment through habitat fragmentation and potential contamination from septic system operation. As described earlier, the large lot size is primarily a function of the public health need to separate on-site water supplies from on-site sewage disposal through septic tank and leach field systems. This requirement has resulted in minimum lots sizes of from 1 to 5 acres per dwelling unit (300 to 1,500 persons per square mile).[51] Many of the subdivisions were approved under less stringent standards that allowed development at densities up to four units per acre with on-site water and septic disposal, offering a bare minimum of adequate area for leach field drainage. Based upon experience throughout the Sierra Nevada, however, many can therefore be expected to fail under soil, slope, or hydrologic conditions that are less than optimal. In some cases this will preclude further development at the high densities allowable under current land use designations. In other cases, septic or well failures will lead to the establishment of community water supplies and/or public sewer systems, which could then lead to higher-density infill development of these substandard lots. In either case there can still be significant social, economic, and ecological impacts. Unfortunately, these impacts are not analyzed ex ante for most developments.

This pattern of development accounts for a significant fraction of the total land area developed to date in Nevada and El Dorado Counties. Parcels in the size class of 1 to 5 acres per dwelling unit accounted for 11.42% of the land area in improved parcels (6.53% of all land area) in Nevada County and 9.78% of the land area in improved parcels (45.43% of all land area) in El Dorado County in 1992. Proposed county General Plan requirements call for minimum parcel sizes in this range for on-site infrastructure, so this size class is expected to account for at least 11.87% of Nevada County's total land area and 3.03% to 3.43% of El Dorado County's total land area under "buildout" of the draft General Plan updates for each county of 1994. Parcels in the size class of 5 to 10 acres per dwelling unit, 10 to 20 acres per dwelling unit, and 20 to 40 acres per dwelling unit account for a much smaller fraction of the total parcels but a much higher fraction of total private land in 1992 in both Nevada and El Dorado Counties. Each of these size and dwelling unit density classes has different ecological impacts associated with development. Significant variation within each size class also exists due to different management practices and behavior of landowners, however, so it is difficult to generalize ecological impacts by average density or average parcel size class (Duane 1993b; Fortmann and Huntsinger 1989). McBride et al. (1996) highlight some general relationships, however, that suggest the scale of impacts. Unfortunately, we do not yet have a clear understanding of these relationships.

Due to the "grandfathered" substandard lots approved before current standards, however, any analysis based on parcel size classes alone is likely to underestimate the number and land area of parcels with on-site well water and/or septic systems. Conversely, including all smaller parcels (higher densities) is likely to overstate the dominance of this pattern, because many of the new subdivisions with treated water and/or sewage treatment facilities are being built with densities at four units per acre (quarter-acre lots). Parcel size distribution and its implications for "buildout" are discussed in more detail in the sections on the Nevada and El Dorado Counties, General Plans.

Rural Agriculture, Natural Resource, or Open Space Lands

Most residents of exurbia live in one of the four settlement patterns just described, but most of the exurban landscape is still managed primarily for agriculture or natural resources commodity extraction.[52] Its primary economic value, however, appears to be shifting in areas facing rapid population growth from a landscape of production to a landscape of visual consumption (Willis 1994; Alterman 1994). This change is consistent with the historic processes of suburbanization in metropolitan areas. Population densities on these lands are typically no more than one structure per 40–160 acres, or from 10 to 200 persons per square mile. Some exurban ranches have as few as 5 people per 10,000 acres, or less than 1 person per 3 mi[2]. Agricultural productivity is often threatened by the encroachment of exurban "ranchette" development, however, as the new exurban residents often impose new restrictions on traditional agricultural practices due to the spillover effects of those productive activities (e.g., noise, pesticides) on the consumptive enjoyment of amenities in the residential regions.[53] In response a number of rural counties have passed "right-to-farm ordinances," which limit new residents' rights to file nuisance complaints against long-standing agricultural practices.[54] The economic viability of many agricultural lands is also threatened by exurban development, however, as increasing land prices make agriculture an increasingly marginal activity when compared to the opportunity costs of subdividing and developing the land (Forero et al. 1992; Hargrave 1993). Moreover, decreasing agricultural activity in exurban regions can reduce the economies of scale in supplying the remaining farmers, increasing the cost and decreasing the availability of farm equipment and related supplies (e.g., feed).[55] A similar phenomenon can occur with natural resource management on private lands (e.g., timber), although public lands management policy is often more important to the viability of local natural resources extraction industries in the Sierra Nevada and in general throughout the West.[56] Public land managers clearly face a new and less supportive sociopolitical context for traditional commodity extraction activities as the private lands adjacent to public lands undergo rapid settlement.[57]

Both agricultural and natural resources lands function effectively as *de facto* public open space for many of the new exurban residents, offering scenic, aesthetic, recreational, and ecological benefits. The implications of human settlement on recreational opportunities for Sierra Nevada residents is discussed in Duane (1996). Recreational use of these *de facto* open spaces is probably one of the primary drivers of and values in "agricultural preservation" efforts in suburban and exurban regions. "Countryside preservation" is the real goal of many proponents of regulations maintaining large parcel sizes, while calls for "agricultural preservation" are often simply a rallying cry that invokes the self-sufficient yeoman farmer and images of Jeffersonian democracy (Kemmis 1990). The focus on agricultural production also taps into a deeply held belief in American society: that agriculture is "primary," so it should and must be protected (Powers 1996). Moreover, many environmentalists often argue that the productive soils underlying agricultural lands are a nonrenewable resource that will forever be lost if an area is "paved over" for new subdivisions.[58] This is rarely the case in the Sierra Nevada foothills, however, where the soils are not nearly as rich for agriculture as they are in the Central Valley and the Napa Valley. The rationale for agricultural preservation must therefore go beyond soils.

The increased values associated with the amenity benefits of open space lands are not easily captured by landowners, however, creating a conflict between long-term agricultural and natural resource landowners and other community members' values and interests. The *real* beneficiaries of countryside and agricultural preservation efforts are generally *not* the farmers or owners of agricultural land but the rest of the community that derives public good benefits associated with the aesthetic, recreational, and ecological goods and services provided by those private lands. Agriculturalists beyond the range of speculative development are also likely to support such efforts, for they yield marginal benefits at very low opportunity cost. Large landowners within the range of speculative development (e.g., their lands are likely to be developed within the next ten to twenty years) are likely to oppose such preservation efforts despite a long family history in agriculture and/or natural resources and a commitment to agricultural preservation. Their children often do not want to continue in this difficult line of work, and they recognize that selling their land to developers is the most effective way to transfer the value of their land to the next generation and relieve themselves of the uncertainties of agriculture. Despite their abstract support for preservation, then, their personal interest in realizing economic gains will often lead them to oppose such efforts. Social conflict is therefore likely to continue between the proponents of such efforts and the supposed beneficiaries of such efforts unless and until the true beneficiaries are willing to structure mechanisms to compensate existing landowners for reduced speculative land values. This is the fundamental challenge of growth management efforts throughout the Sierra Nevada and the rest of exurbia.

Public Land Ownership in the Sierra Nevada

Patterns of public land ownership are an important factor affecting patterns of human settlement. We used a map prepared by the Strategic Planning Program (SPP) of the California Department of Forestry and Fire Protection (CDF) to analyze patterns of land tenure. Various federal, state, and local government agencies administer more than 60% of the total land area of the Sierra Nevada region as public lands. National forests managed by the U.S. Forest Service alone account for two-thirds of publicly owned lands in the region and 40% of all lands in the region. The Bureau of Land Management manages 13% of the region, while the National Park Service administers approximately 6%. City and county governments, the state of California, the U.S. military, and other federal agencies each account for around 1%. These public lands are generally unavailable for human settlement (except for some recreational purposes and for employee housing) under current institutional arrangements. Human settlement is therefore concentrated on private land. We therefore used the land tenure overlay to reduce the error associated with using the census block group (CBG) coverage to estimate patterns of human settlement throughout the Sierra Nevada. The CBG is the next smallest unit of census data aggregation below the CCD. We intersected the two coverages to create a third coverage of "private block group" (PBG) polygons, which then allocated all population and housing units across only the private lands within the CBG. Figure 11.25 shows the distribution of public land by federal land management agency in the Sierra Nevada.

Our map shows that some counties have considerably more public land than others. Amador County is only 22.14% public land, while Inyo County is 98.34% public land. Much of the nonfederal land in Inyo County is owned by the City of Los Angeles Department of Water and Power (LADWP). Table 11.A2 in appendix 11.1 shows the area of each California county in the Sierra Nevada by tenure.

Information on population and housing unit characteristics was derived from the U.S. Census Bureau's Summary Tape File 3A, 1990 Census of Population and Housing. This publication (available in digital form) presents data from the Census Bureau's sample survey of households, including 17% of all households on average (although it may include up to 50% of all households in rural areas). Because data from STF3A are not available at the census block level (the smallest unit of analysis for census data), we used the next largest unit of aggregation, the "split block group" (SBG). The Census Bureau splits block groups that cross city and other political boundaries, providing separate data records for each block group part. This is the smallest geographic unit for which sample census data are readily available. There are 740 split block groups in the five-county central Sierra Nevada counties of Nevada, Placer, El Dorado, Amador, and Calaveras, for example, compared to approximately 17,000 census blocks and eighteen CCDs.

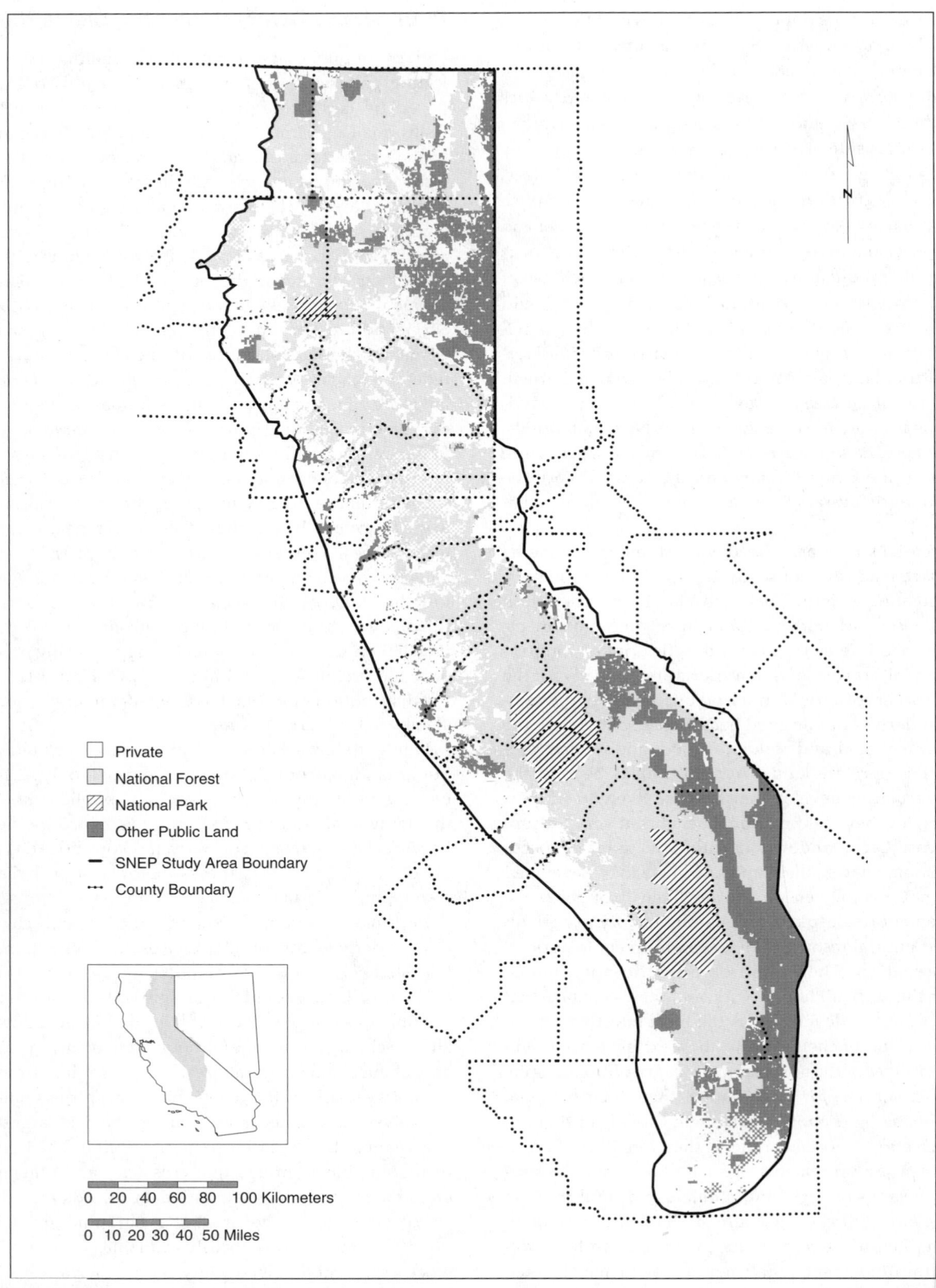

FIGURE 11.25

Public land ownership, Sierra Nevada region.

Private landownership is concentrated at lower elevations, in the foothill areas. Nevertheless, the boundary between public and private lands throughout the region is extremely complex, with numerous pockets and strips of privately owned land extending into higher elevations, particularly in the central and northern portions of the region. This pattern increases the potential for significant impacts on public lands and higher-elevation ecological conditions from development on private lands. The GAP assessment (Davis and Stoms 1996) highlights areas in the Sierra Nevada in which the public-private interface per unit of area is very high. That analysis also highlights vegetation types that are primarily represented on private lands. In particular, the following five vegetation types fall largely on private lands and are subject to settlement: nonnative grassland (88% of mapped distribution on private lands), valley oak woodland (98%), blue oak woodland (89%), interior live oak woodland (71%), and foothill pine-oak woodland (82%). In addition, there is significant human settlement on private lands in the west-side ponderosa pine forest and the lower cismontane mixed conifer–oak forest. This pattern reflects the spatial concentration of human settlement "below the green line" west of national forest boundaries up to around 1,200 m (4,000 ft) in elevation. Based on the Davis and Stoms 1996 analysis, approximately 80% of the land below 1,000 m (3,280 ft) in elevation in the northern Sierra (north of the Stanislaus River) is in private ownership.

Several higher-elevation vegetation types are also being affected by intensive recreational development activity on both public and private lands in the Lake Tahoe and eastern Sierra subregions. These include the Jeffrey pine forest, the east-side pine forest, and some subalpine meadow communities. Both ownership patterns and development patterns are more complex in the higher elevations than in the western foothills, however, so it is difficult to generalize about the relationship between human settlement and vegetation types in these areas. More detailed analysis is necessary at a vegetation-specific level in these areas.

Data limitations constrained our assessment of human settlement at the scale of the entire Sierra Nevada. We therefore focused in greater detail on the five-county central portion of the region where human settlement is already at its densest and growth pressures are high. These five counties of the central Sierra Nevada (Nevada, Placer, El Dorado, Amador, and Calaveras) are characterized by a comparatively high proportion of private land relative to their total areas, however; in all five cases, private lands cover 50% or more of the part of the county within the Sierra Nevada region. They are therefore not necessarily representative of the conditions throughout the Sierra Nevada. They nevertheless offer an interesting case study of the factors affecting human settlement.

More detailed information was available on private lands in these five counties than in any other part of the Sierra Nevada through access to a digital database developed by the Teale Data Center for CDF. The database relates county assessors' parcel records to a georeferenced coverage of the assessors' "map book pages." Each map book page contains multiple parcels, but each individual parcel's assessor's information is available in the database and can be spatially related to the map book page coverage. We assisted CDF in completion of the database through supplementary funding and a contract with Teale to incorporate ownership information into the database. This same ownership information is already publicly available from each county assessor's office, but it had not previously been related to the map book page coverage under the original contract with CDF. We therefore have no more access to parcel information than is available to the general public. Having it in digital form, however, allows us to relate ownership patterns to human settlement and to complete more detailed spatial and statistical analysis of the data than previously possible. Specific findings are discussed in more detail in sections discussing the Nevada County General Plan and the El Dorado County General Plan. All of these data are now available for public access from the California Environmental Resource Evaluation System (CERES) project of the Resources Agency of the State of California (http://ceres.ca.gov/snep), and the Alexandria Project at the University of California, Santa Barbara (http://alexandia.sdc.ucsb.edu/).

Housing Density from 1940 to 1990 in the Sierra Nevada

Social data on well-being is only available at the "census block group" (CBG) level. Census block groups are clusters of several census blocks, containing 650 people on average. We therefore had to rely on the more aggregated CBG data (685 polygons) rather than the more precise Census Blocks (over 50,000 polygons) to assess a number of factors influencing the pattern of human settlement in the Sierra Nevada. There are approximately 800 block groups in the entire SNEP study area, with 685 of those in the eighteen-county Sierra Nevada region covered by the 46 CCDs included in our assessment. Doak and Kusel (1996) used a slightly larger set of 720 census block groups that were then aggregated into the 180 "Community Aggregations" (CAs) reported in their social assessment. Once again, this reflected the different needs of our assessments. We limited our analysis of CBG data to those CBGs within the CCDs analyzed for 1970–90.

We obtained data on a variety of household characteristics from the U.S. Census Bureau's Summary Tape File 3A, 1990 Census of Population and Housing. One question in the 1990 census asked a sample of residents what year their home was built. Based upon these data, we constructed a series of coverages showing the average density of development within each private block group for each decade between 1930 and 1990. This series will tend to understate the degree of development in early years, however, for some structures built in later years may have destroyed older housing on the same site. Older houses that are unoccupied would also not be represented in the Census responses. The resulting maps nevertheless present a fascinating time-series sequence of human settlement in the Sierra Nevada. Plate 11.2 shows this pattern for each decade from 1930 to 1990.

This series of plots shows the steady expansion of human settlement throughout the Sierra Nevada during the 1930–90 period, with a rapid expansion beginning as early as 1960. Because the "private block group" (PBG) polygons are so large, however, these maps will tend to overstate the density in private industrial forest lands and understate the density in other areas within the same PBG. Human settlement was therefore actually more concentrated (and at a higher density) than that suggested by these plots. They nevertheless offer a more accurate picture than that provided by the census block group (CBG) coverage, which allocates density across both public and private land.

Note how the primary areas of increasing density are along U.S. Highway 50 and Interstate 80 between the Sacramento area and the greater Lake Tahoe Basin. It was still possible to connect large areas of low-density or unsettled land along a latitudinal gradient from north to south through the western Sierra Nevada foothills in 1930–50, but development along these highways had effectively isolated the American River drainage from the largely contiguous regions north and south of I-80 and U.S. 50 by 1980–90. Other areas of relatively high density have also appeared to fragment the landscape. The potential implications of such fragmentation are discussed in more detail later.

Access to Infrastructure Services in the Sierra Nevada

We examined a number of additional factors at a variety of spatial scales that could potentially help determine the distribution and rate of residential development in the region. These included descriptive and bivariate analyses of forty census variables for all census block groups in the SNEP study area, for the five-county study area, and for two individual counties (Nevada and El Dorado) for which we had more detailed information.

The availability of physical infrastructure is one of the factors we examined. High-density development depends on proximity to sewer, water, gas, and power lines. At lower densities, development may still be possible through use of septic systems for waste disposal and wells for water. Even low-density development and isolated rural homes almost always depend on public power, however. Unfortunately, we were not able to obtain detailed maps or data that would allow us to incorporate the location of physical infrastructure into our analysis of settlement patterns. Pacific Gas and Electric Company, the only infrastructure entity with regional responsibilities throughout the five-county area, denied our request for infrastructure network information "for competitive reasons." (Pacific Gas and Electric Company staff 1994). We received more support and cooperation from the Nevada Irrigation District (NID), the El Dorado Irrigation District

(EID), the Georgetown Divide Public Utility District (GDPUD), the El Dorado County Water Agency (EDCWA), and the El Dorado County Planning Department to convert their infrastructure data into a coverage, but their data were generally not in digital form. Only EID data were digital, and they were based on a CAD system that was not georeferenced in a system compatible with our other coverages (El Dorado Irrigation District staff 1994). NID is hoping to convert its paper-based engineering maps into a geographic information system (GIS) coverage over the next few years (Nevada Irrigation District staff 1994–95). We originally hoped to model the economic costs of infrastructure access and expansion, but we were only able to map the distribution of homes with access to some of these services through use of the private block group data in the census. This is only a proxy measure for patterns of infrastructure access and is not spatially explicit enough to allow development of an economic model of development that is directly linked to infrastructure access.

We analyzed the forty census variables for all 685 census block groups in the Sierra Nevada, but they are difficult to display in graphical form for the entire region. We therefore illustrate the spatial patterns in these variables here with a subset based upon all of the CBGs in the central Sierra Nevada counties of Nevada, Placer, El Dorado, Amador, and Calaveras. The western portion of Placer County, including the cities of Roseville and Rocklin, is outside the area of our analysis but is displayed here for reference purposes. Access to public sewage disposal varies at a county level in the Sierra Nevada from only 21% of Mariposa County residents to 83% of Mono County residents. Figure 11.26 shows the pattern of access to public sewer by CBG in the five central Sierra Nevada counties.

Our bivariate analysis of all forty census variables against one another resulted in very few strong correlations for the 685 census block groups. This result may reflect either poor associations, skewed distributions, or confounding variables not accounted for in our analysis. Population density at the CBG level is positively correlated with the distribution of access to public sewer in the central Sierra Nevada region, for example (R-squared = 0.24; t-statistic value = 14.53), but it is clear from the histogram that all CBGs above an average density of 1,000 to 1,500 persons per square kilometer have nearly 100% sewer coverage. Population density is less strongly correlated with the fraction of the CBG households with access to public water supply (R-squared = 0.157; t-statistic value = 11.09). The lower R-squared value for the public water variable reflects the fact that access to public water is more pervasive in the region. The threshold population density (POPDENS) at which a CBG had nearly complete access to public water is only 500 to 1,000 persons per square kilometer. Figure 11.27 summarizes the fit for both of these variables and shows the histogram for each variable. Note that this analysis used "No Public Sewer" (NOPSEW) and" No Public Water" (NOPWAT) as the independent variables, so there is a negative correlation with POPDENS. Figure 11.27 also shows

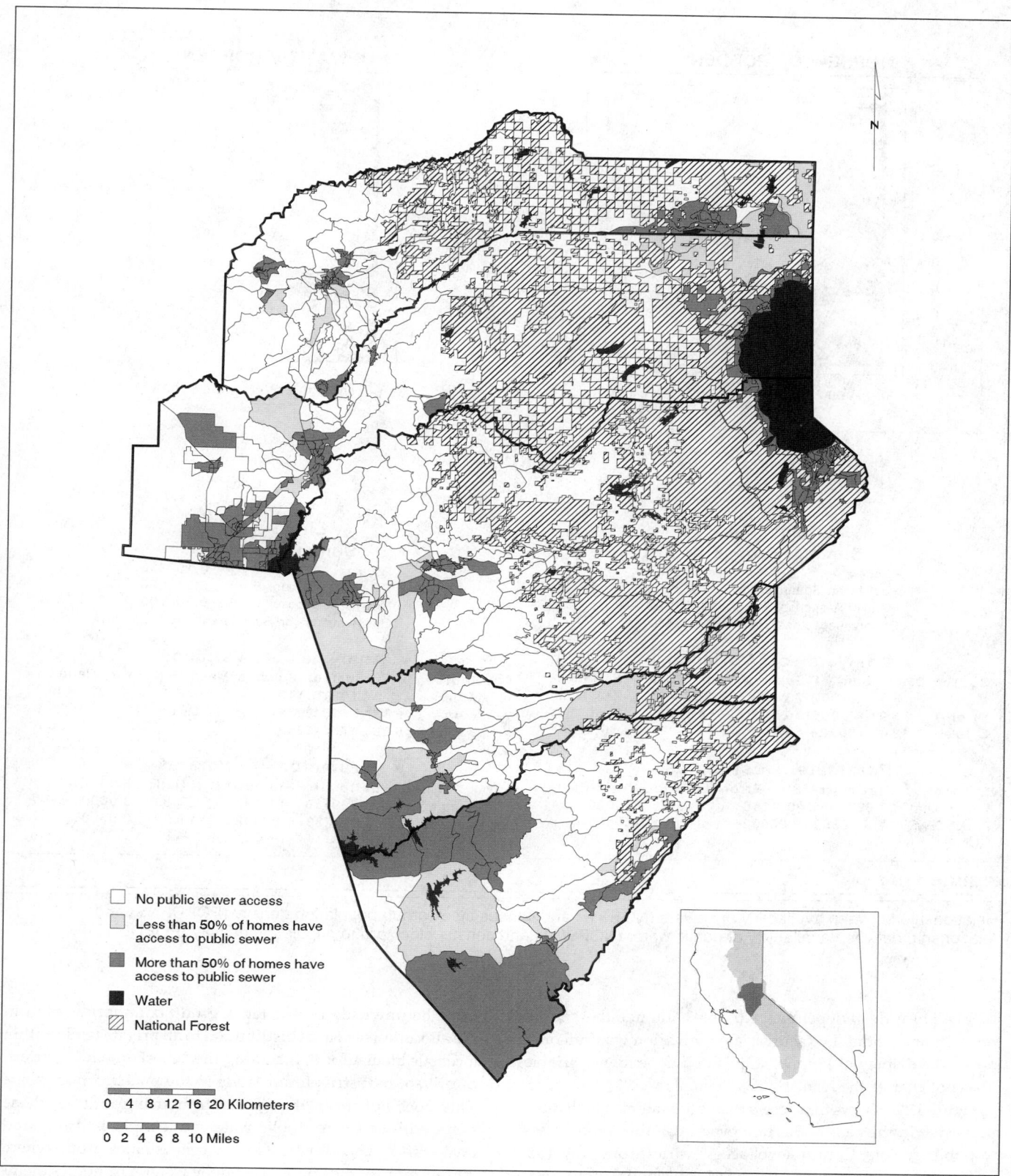

FIGURE 11.26

Access to public sewers, central Sierra Nevada region (based on 1990 Census of Population, Summary Tape File 3A).

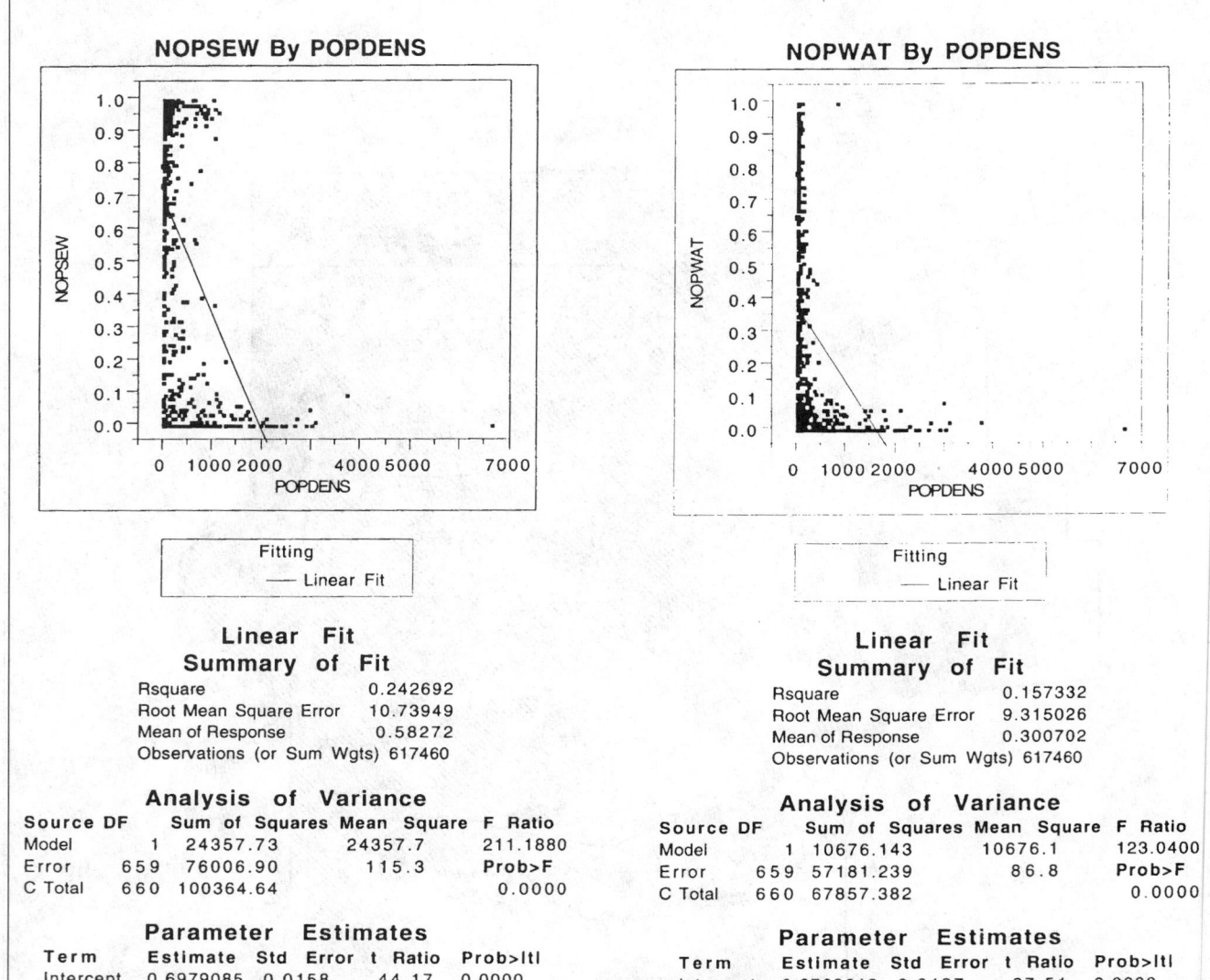

FIGURE 11.27

Relationship between availability of sewers (NOSEW) and census block group population density (POPDENS) and relationship between availability of public water (NOPWAT) and census block group population density (POPDENS).

POPDENS as the independent variable, although this simply reflects the fact that every bivariate regression was run only once. The display of POPDENS as the independent variable does not change the result.

Figure 11.28 shows that access to public water is much more pervasive in the central Sierra Nevada counties than access to public sewers. County-level access varies from only 41% of Mariposa County residents to 84% of Mono County residents. Once again, the access varies significantly within each county. Figure 11.28 highlights the spatial patterns of access in the five-county central Sierra Nevada. Note in figure 11.28

how the unwieldy census block group boundaries result in the allocation of spatial distribution to all private lands within a census block group, including the "checkerboard" pattern of private industrial forest lands in the mixed conifer zone. This does not mean that there are housing units in those areas with access to public water but simply that the land area with that shading is within a census block group where 50% or more of the homes use public water. The homes themselves are not distributed evenly throughout the census block group polygon but are concentrated within portions of each polygon. This problem led us to use the much smaller census

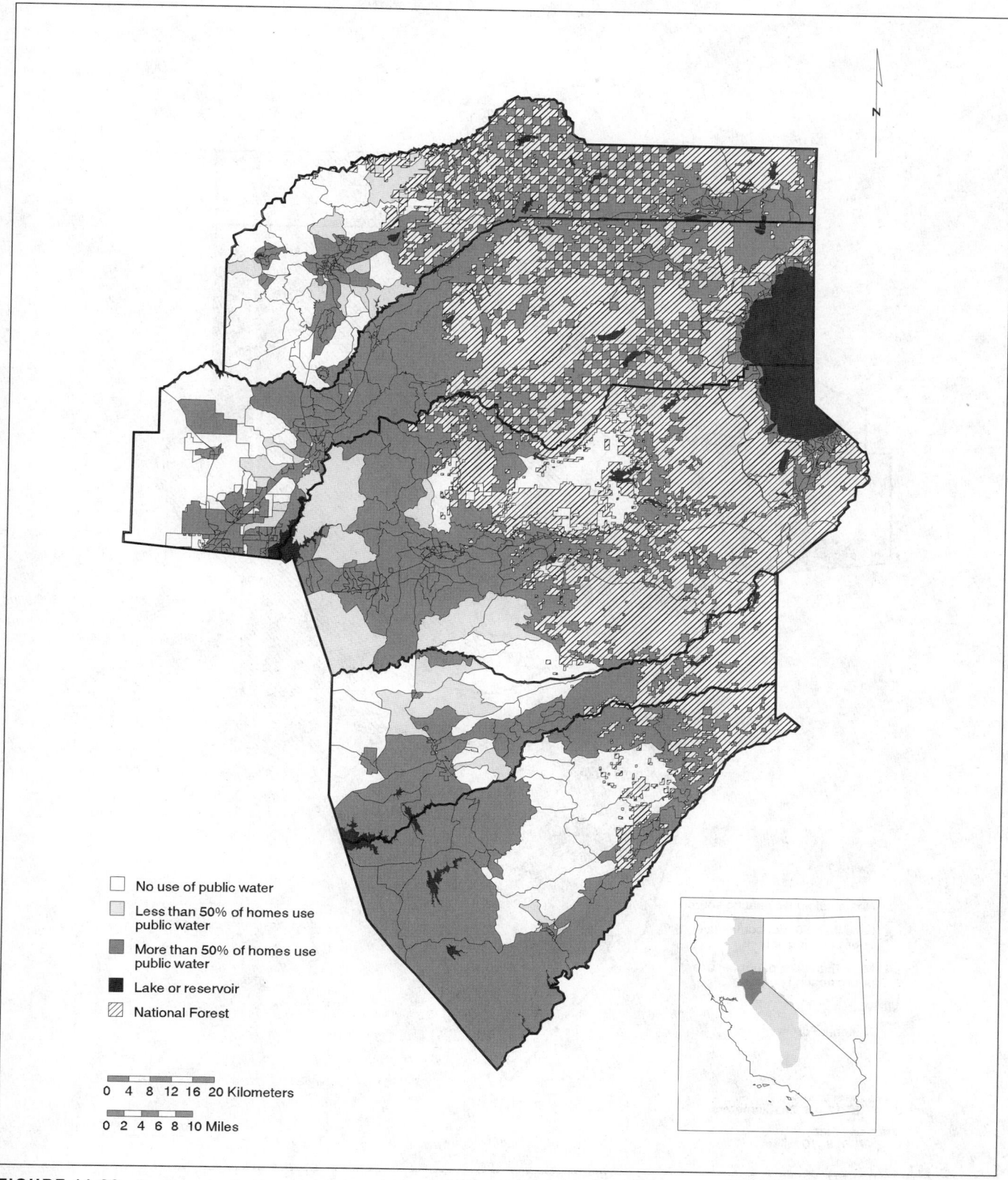

No use of public water

Less than 50% of homes use public water

More than 50% of homes use public water

Lake or reservoir

National Forest

0 4 8 12 16 20 Kilometers

0 2 4 6 8 10 Miles

FIGURE 11.28

Source of household water, central Sierra Nevada region (based on 1990 Census of Population, Summary Tape File 3A).

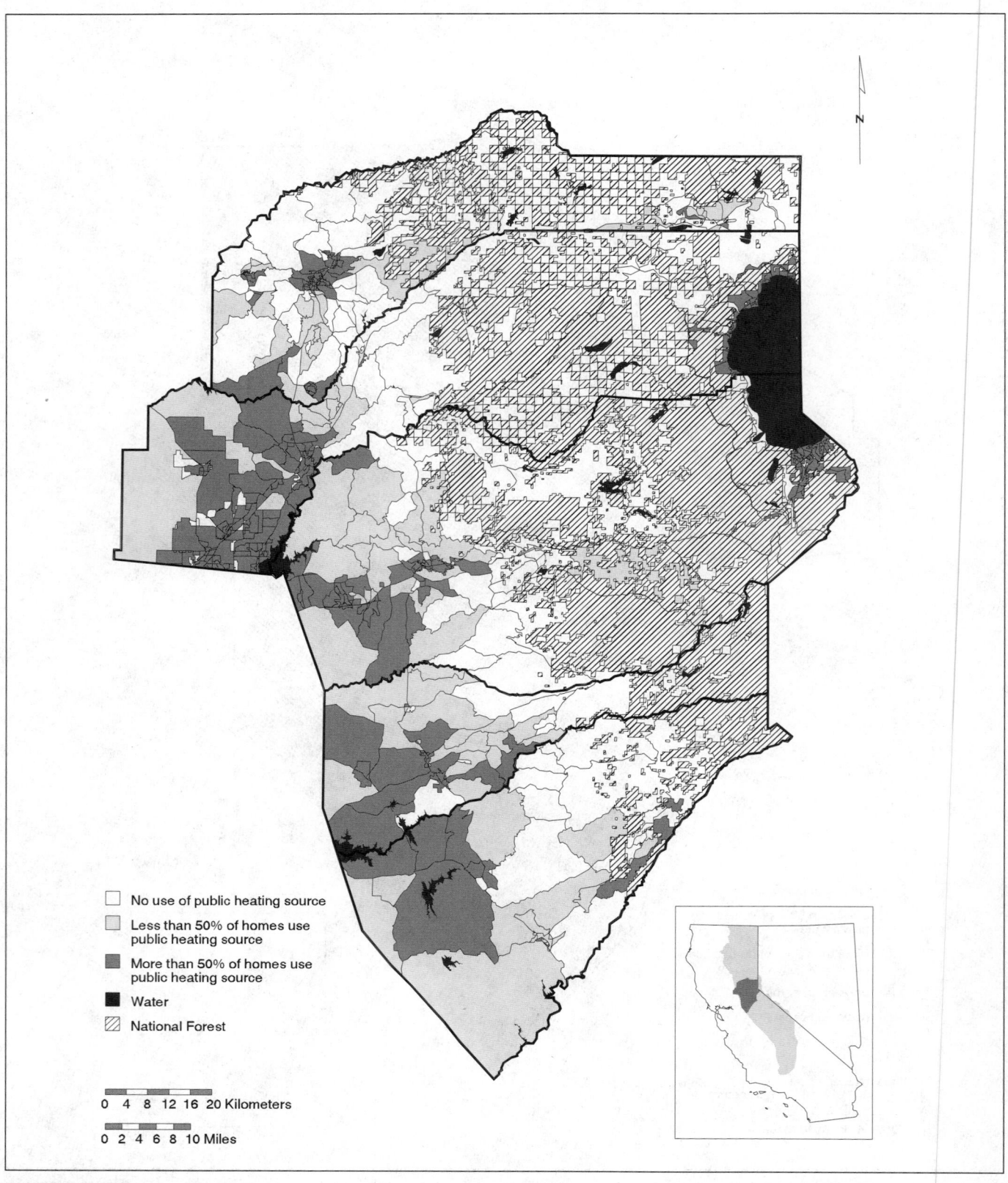

FIGURE 11.29

Source of household heat, central Sierra Nevada region (based on 1990 Census of Population, Summary Tape File 3A).

blocks when we mapped 1990 housing density rather than the less precise census block group or private block group polygons.

Access to natural gas or another public heating source is not a necessary element of development, but public utilities generally do not invest in heating infrastructure unless housing densities are high enough to allow the fixed costs to be recovered across many housing units. Figure 11.29 shows that this occurs only in a few areas: along the I-80 corridor up to Auburn in Placer County, around Grass Valley and Nevada City, along U.S. Highway 50 through El Dorado Hills, and in the area leading to Jackson and Ione in Amador County. The rest of the housing units in this area depend primarily upon a private heating source (generally wood or propane). This has important implications for air quality and market demand for fuel wood as population continues to grow throughout the Sierra Nevada. The trends in air quality discussed in Cahill et al. (1996) should be considered in the context of some reliance on wood heating at low housing densities.

Road access and proximity to major transportation corridors are additional factors influencing settlement patterns. We tested for the relationship between road density (based on the Teale coverage at 1:100,000) and census block group population density and assessor's map book page parcel density. There was no statistically significant relationship between these variables and road density, so roads could not be used as a proxy for density. Access times to employment centers in metropolitan areas outside the Sierra Nevada should be an important variable affecting development patterns, however, and this is apparent from visual inspection of the census block group data for the entire Sierra Nevada. The highest population densities are found in those CBGs proximate to four-lane highways accessing the Sacramento metropolitan area, recreational developments in the Lake Tahoe and eastern Sierra regions, and the historic nineteenth-century towns along Highway 49. We were unable to complete a more systematic analysis that modeled the relationship between commute times and settlement patterns, but we did examine the issue in more detail in our analysis of the Nevada and El Dorado County General Plans. Figure 11.30 shows the fraction of the population working outside the county in each of the census block groups in the central Sierra Nevada. Not surprisingly, these areas are concentrated within commuting distance of the Central Valley. There are also high levels of intercounty commuters at higher elevations, however, including many in eastern Nevada, Placer, and El Dorado Counties.

As the overall picture of density in the Sierra Nevada region makes clear, the spillover effect of proximity to major urban centers such as Sacramento is an important factor affecting both total levels of population and the population densities associated with patterns of human settlement. This proximity has a direct effect on land values in the western foothills of the Sierra Nevada, which in turn increases the viability of making significant investments in infrastructure (e.g., roads, sewers, water, and power) that can then allow much higher development densities. The higher densities are necessary to make such infrastructure investments economic, because they usually involve a high proportion of fixed costs. Increasing land values also make some areas that are marginal for development through on-site infrastructure (e.g., septic systems and well water) attractive for development: the relative costs of those investments declines (and the relative value of making them increases) as land values increase. The cost of drilling a 200-foot well might be a large fraction of total development costs when land is $10,000 per acre, for example, but drilling even an 800-foot well could be economical if land costs $50,000 per acre. These are the types of land value changes we have seen in Nevada, Placer and El Dorado Counties over the past decade.

As a result, many "unbuildable" lots are now being developed. Physically based models of development fail to capture this phenomenon, because the "unbuildable lot" is fundamentally an economic concept. Higher land values can therefore result in significant land development that would not otherwise occur at lower land values. The employment and income characteristics of new Sierra Nevada residents are therefore an important determinant of human settlement patterns in the region. Higher incomes associated with commuters and some retirees puts pressure on land and housing prices, which is likely to lead to development that is both more *intensive* (i.e., at higher average densities where public infrastructure is provided) and more *extensive* (i.e., across the landscape into some areas that were previously considered "unbuildable"). Higher land values are therefore unlikely to lead only to either greater density in existing areas of development or greater land area under development at existing densities. Both are likely to occur, and the two types of development have different impacts. The difficulty is predicting when and where each type of development will occur.

The economics of infrastructure in the Sierra Nevada are in part a function of federal, state, and local policies. The federal Safe Drinking Water Act imposes specific requirements for water treatment that have economic consequences for the cost of domestic water supply, for example, while Regional Water Quality Control Board requirements for sewage treatment affect the relative costs of septic versus sewer system waste disposal. Enforcement of the non-point-source (NPS) water pollution provisions of section 319 of the federal Water Quality Act of 1987 (Clean Water Act Amendments) could also lead to greater restrictions (and costs) for the use of septic systems. The California Public Utilities Commission (CPUC) also recently modified its rules and regulations for allocating the cost of power line extensions, which will increase the cost of providing power to more remote rural development sites. These new rules (effective July 1, 1995) will increase the relative attractiveness of developing those parcels that have the easiest access to existing power lines, yet could also increase pressure to develop more remote sites at higher densities (in order to allocate the fixed costs of the line extension across more housing units). Actions by a wide range

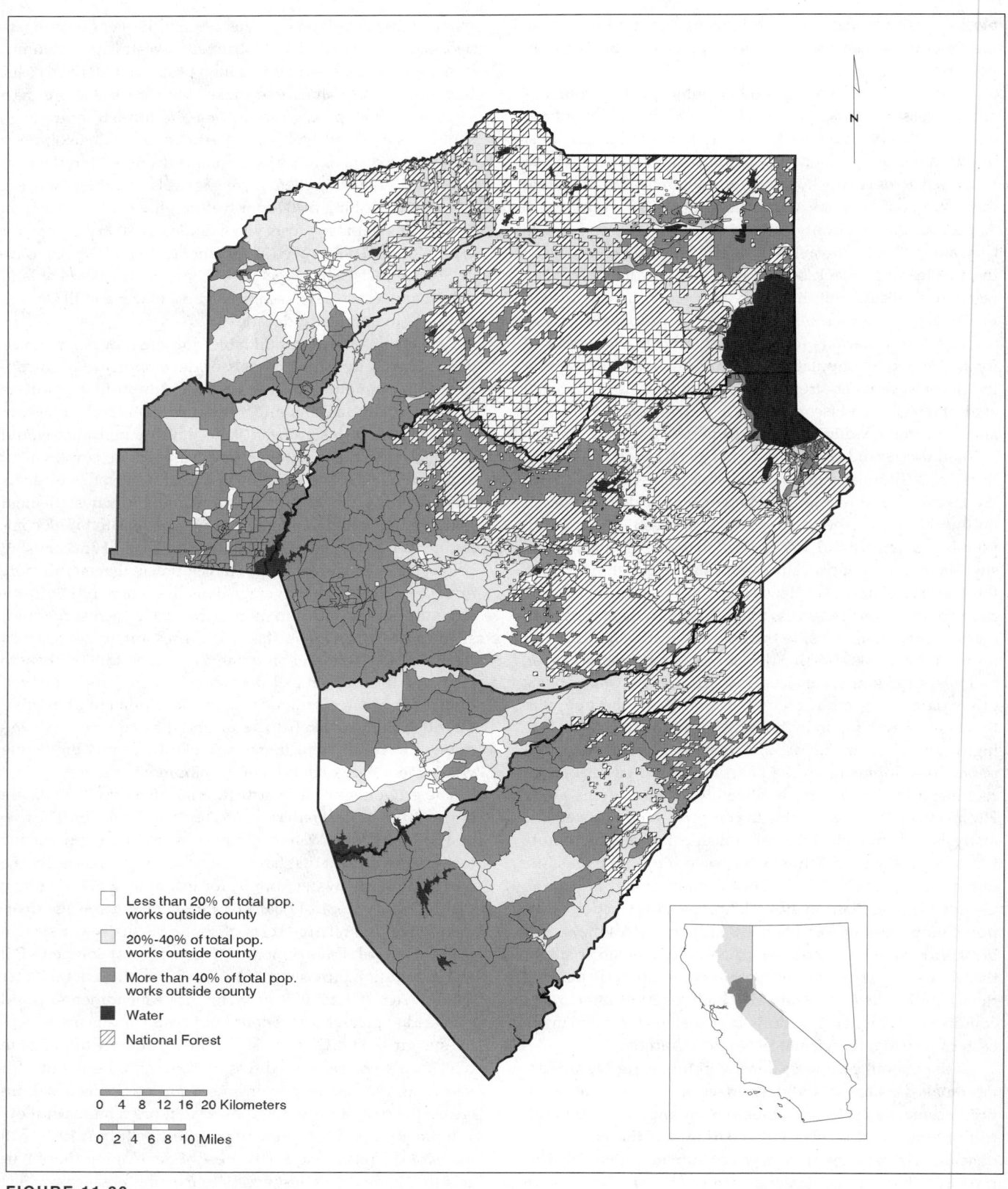

Less than 20% of total pop. works outside county

20%-40% of total pop. works outside county

More than 40% of total pop. works outside county

Water

National Forest

0 4 8 12 16 20 Kilometers

0 2 4 6 8 10 Miles

FIGURE 11.30

Place of work, central Sierra Nevada region (based on 1990 Census of Population, Summary Tape File 3A).

of public agencies therefore affect patterns of human settlement. In many cases those actions may be focused primarily on very different issues than those addressed in this assessment, but they can have profound ramifications for the Sierra Nevada.

Census Variables and Density in the Sierra Nevada

Census data were not available at the CBG level in the Sierra Nevada for 1970, but we considered development of a population forecasting model based upon data for 1980 and 1990. Our exploratory data analysis of the relationships between the forty census variables was designed to help specify such a model. Other exurban modelers have attempted similar LOGIT specifications based on a combination of census data and biophysical variables such as proximity to roads and cities (Sanchez and Nelson 1994). In particular, Ted Bradshaw and Brian Muller at the University of California, Berkeley, have developed a model of exurban growth in California's Central Valley to forecast farmland conversion for the American Farmland Trust (1995).[59] They have also developed a model of exurban development in the forested counties of California for the U.S. Forest Service Pacific Southwest Research Station at Riverside to assist modeling of fire risks (Bradshaw and Muller 1994–95).

In both cases they developed models with only limited explanatory power (Bradshaw and Muller 1994). While they did identify some statistically significant relationships, there was little theoretical basis for the relationships identified. We believe that confounding variables, together with both the inconsistent basis for establishing census block group boundaries and the limited data, make such an approach problematic at this time. We do believe this is a promising direction for future research, however, following completion of the next census in the year 2000. We will then have data for three periods (1980, 1990, and 2000) across the entire country in digital form related to a consistent geospatial reference in the census TIGER files. That data set will then allow systematic analysis of the exurban growth phenomenon throughout both the Sierra Nevada and similar areas throughout the rural western United States. Until then, however, we have chosen to rely upon the simpler modeling strategy described in the section on allocating county-level DOF population forecasts.

Despite the data limitations for forecasting, the descriptive statistical analysis and bivariate analysis were still useful. They helped to establish the distributions of particular variables across the CBG set and then helped us identify outliers (such as those associated with correctional facilities) that could otherwise confuse our analysis. They also allowed identification of potential proxies for particular settlement patterns. Finally, the bivariate analysis confirmed several of the theoretical relationships we suspected based upon the framework established earlier regarding factors driving exurban population growth in the Sierra Nevada. In particular, the bivariate analysis highlighted the relationships between income, housing values, access to water and sewer, and physical proximity to the commuting opportunities in the Central Valley (especially Sacramento). These relationships reinforce the importance of metropolitan expansion and deconcentration as critical factors driving population growth in the Sierra Nevada. They also reinforce the notion that changing land values can radically alter the housing densities that can be supported by the real estate and housing markets. This has important implications for likelihood of alternative future patterns of human settlement in the Sierra Nevada. While we have not been able to develop a more explicit economic model of infrastructure access and development densities, we know it is quite important. Future research in this area should emphasize this important relationship.

Descriptive statistics for each of the forty census variables and the bivariate analyses of relationships between each of those variables are available from the California Environmental Resource Evaluation System (CERES) project of the Resources Agency of the State of California (http://ceres.ca.gov/snep), and the Alexandria Project at the University of California, Santa Barbara (http://alexandria.sdc.ucsb.edu/). Figure 11.27 illustrates these bivariate analyses and those data.

Mapping Housing Density in the Sierra Nevada

The extent of private lands in the region establishes only the most basic template of where human settlement can expand. We decided that a more accurate representation was necessary to assess the spatial patterns of human settlement at various development densities. We therefore relied on the more detailed data from the 1990 Census of Population and Housing on the Summary Tape File 1B publication, which provides basic population and housing characteristics for individual "census blocks" (CBs). A census block is the fundamental geographic unit at which the data are originally recorded by the census following collection at the household level. In urban centers census blocks correspond with actual city blocks; in rural areas census blocks are usually delineated to correspond with logical natural or artificial boundaries, such as roads and rivers. (The relationship between the boundaries of arbitrary units such as census blocks and other geographic features becomes important when several maps are combined for analysis.) Most counties in California, even those with small populations, are divided into thousands of census blocks. They vary in size from about an acre in densely populated urban counties to hundreds of acres in more sparsely populated regions.

We calculated housing density for each census block in the Sierra Nevada by dividing the 100% housing count by the land area of the census block. The latter is reported in the STF1B file with an accuracy of one-thousandth of a square kilometer or 1,000 square meters. Conversion to units per square mile followed this formula:

$$(units/0.001 \ km^2)(1,000) \ (0.001 \ km^2/1 \ km^2)(1 \ km^2/ \\ 0.3861 \ mi^2) = (units/mi^2)$$

To create a map of census blocks in the Sierra Nevada region, we relied on the TIGER digital line files published by the U.S. Census Bureau.[60] The positional accuracy of boundary segments in these files has a maximum stated error of plus or minus approximately 51 m (167 ft); in other words, census block boundaries in our digital map should be within 51 m of their actual position. This level of accuracy is entirely adequate for a regional study such as ours. We related the polygons in our digital map to the tabular data from the STF1B publication using the census-designated labels that together uniquely identify every block in the nation: State FIPS code (2 characters), County FIPS code (3 characters), Census Tract (6 characters), Census Block (4 characters).

The resulting housing density map of the Sierra Nevada region contained over 50,000 polygons, each with a unique housing density value. For presentation purposes we aggregated individual census blocks into six broad categories based on housing density:

1. Zero housing units per square mile

2. Fewer than two units per square mile

3. Two to ten units per square mile

4. Ten to forty units per square mile

5. Forty to one hundred sixty units per square mile

6. One hundred sixty or more units per square mile

The class with the highest density therefore shows those areas where there is on average at least one housing unit for every 4 acres. In this fashion, census block clusters with relatively high densities show the actual location of communities in the region, regardless of their incorporation status. Plate 11.3 shows 1990 housing density in the Sierra Nevada based on these aggregated clusters.

Our final map of housing density in and around the Sierra Nevada strongly reflects the location of major urban centers in the Central Valley and the transportation corridors connecting the Sierra Nevada to those centers (shown in figure 11.2). Each of these centers is surrounded by areas of relatively high housing density. These areas tend to extend into and are most concentrated in the Sierra Nevada foothills in the counties of Amador, El Dorado, Calaveras, Placer, and Nevada, where the largest area of relatively high housing density in the region is found. Other areas of high-density human settlement are in high-altitude recreational centers, such as the Lake Tahoe Basin and Mammoth Lakes. This census block–based representation of human settlement in the Sierra Nevada for 1990 is more accurate spatially than either

the community aggregations used in the social assessment by Doak and Kusel (1996) or the 1930–90 coverage of housing density (based on private block groups) shown in plate 11.2. To our knowledge it is the most accurate representation ever completed for the Sierra Nevada. It is nevertheless limited by the fact that census blocks are not randomly or evenly distributed across the Sierra Nevada. Some errors are therefore likely to exist in the largest and most heterogeneous census blocks. The smallest and most homogeneous census blocks will be the most accurate.

A census block that is only 20 acres within a homogeneous subdivision developed at an average density of four units per acre will accurately show the area of the entire census block as having an average density of four units per acre. Similarly, a large census block of even 1,000 acres within an unsettled national park wilderness will accurately show the average density of zero units per acre throughout the census block. Errors are likely to occur, however, if a census block straddles the two and averages them out. A 100-acre census block that included the high-density 20-acre development and was otherwise undeveloped, for example, would assign an average density of 0.8 units per acre to the entire region. Our analysis would therefore both understate the area in high-density development and overstate the area without any development at all. We have not determined how extensive this type of error may be, but it should be low. The census block boundaries should contain relatively homogeneous units.

Based upon this analysis, we estimate the following distribution of average density of human settlement in the Sierra Nevada by land area (based on the land area within each census block and classified by average census block density). Note that these estimates are for residences only and do not include land conversion due to commercial and industrial uses, which would increase the developed area significantly.[61] These eleven density classes show a finer resolution than shown in plate 11.3, including densities up to one housing unit per acre (640 or more per mi^2). They cover all of the census blocks within the forty-six CCDs we used in the 1970–90 analysis (32,001 mi^2 or 20,481,252 acres). The total population count based on the census blocks (604,644) was slightly lower than the total for the forty-six CCDs, but we have not been able to identify the source of this error.[62] The two estimates are within 3% of each other, and the error is acceptable for an analysis of this scale. We believe the census block estimates of the spatial distribution of land area, housing units, and population density by housing unit density classes is the most accurate estimate for the region, but the CCD data are most useful for landscape-scale analysis. The errors could have come through either the census data tapes, the Arc/Info processing step, or the spreadsheet analysis we completed in Microsoft Excel. Table 11.A3 in appendix 11.1 shows the distribution of these variables by housing unit density class for the region.

Table 11.A3 shows that 1,741 mi^2 (1,114,531 acres) have an average housing density of at least one housing unit per 32 acres. This is approximately the same average housing density used by the Strategic Planning Program of the California Department of Forestry and Fire Protection (CDF) as a threshold for indicating that a wildland area has been converted to an "urban" use.[63] It is also considered a threshold below which it is difficult to practice industrial forestry. The California Department of Fish and Game considers critical deer habitat to be adversely affected when parcel sizes are 20 acres or less, while some other species may be significantly affected at higher or lower average densities. Even a threshold average density of one housing unit per 16 acres (forty housing units per square mile) affects 1,009 mi^2 (645,592 acres) of the Sierra region. A lower-density threshold of ten units per square mile (one housing unit per 64 acres) encompassed 2,632 mi^2 (1,684,189 acres) in 1990 in the Sierra Nevada. Nearly 2.5 million acres (3,905 mi^2) were affected at a threshold of five housing units per square mile (an average density of one housing unit per 128 acres). Within this settled area, nearly 300 mi^2 (190,893 acres) are settled at a density 4 acres or less per housing unit, with 89 mi^2 (56,867 acres) at a density of at least one housing unit per acre.

Actual densities can vary considerably within each census block, and the ecological effects of human settlement at higher densities can affect adjacent areas that are settled at relatively low densities. Moreover, the estimates reported here do not include any land area developed for commercial, industrial, or public uses outside the census blocks with these densities. Large commercial shopping centers and many downtown areas have relatively little housing, for example, and will typically be in the lowest-density classes. They are nevertheless the site of significant ecological impacts associated with human settlement. The total land area converted for human settlement includes nonresidential land uses. At least 1,009 mi^2 (645,592 acres) and potentially as much as 3,905 mi^2 (2.5 million acres) of the Sierra Nevada were therefore already converted to human settlement or were directly influenced by adjacent human settlement as of 1990.

It is important to note that the distribution of housing units is not a proxy for the distribution of population by housing density class in the Sierra Nevada. This is a critical factor to consider when allocating future population growth projections to housing density classes in order to estimate the total land area affected by human settlement from 1990 to 2040. Average household sizes are generally smaller in the densest class (640 or more housing units per square mile), which probably reflects the smaller household sizes typically found in multifamily housing units. This distribution could also reflect high seasonal vacancy rates in recreational residences in some of the lower-density classes, where there were vacant housing units when the census was taken in 1990. Surprisingly, some census blocks with no housing units still reported some population in 1990. These could have been temporary residents or seasonal employees. They account for only 1,630 persons, or 0.27% of the total population. Nearly two of every five Sierra Nevada residents (39.49%) lived on just 89 mi^2

(56,867 acres) in the region (0.28% of the land area) in 1990. Another fifth of the population (21.24%) lived on the 209 mi^2 (134,025 acres) settled at an average density of between 1 and 4 acres per housing unit (160–640 housing units per square mile). Three-fifths of the Sierra Nevada population (60.73%) therefore lived on less than 1% (0.93%) of its land base in 1990. This same area accounted for 64.08% of the housing units in the Sierra Nevada in 1990. Fully 80.00% of the Sierra Nevada population lives on the 1,009 mi^2 (645,592 acres) that have an average housing density of 16 acres or less per unit (forty or more housing units per square mile). Figure 11.31 illustrates the distribution of area, housing units, and population by density class.

These areas of human settlement are not distributed randomly across the Sierra Nevada landscape. Development in just two counties, Nevada and El Dorado, accounts for 30% of all the land area (791.12 mi^2 [506,317 acres]) in the Sierra Nevada that is settled at an average housing density of ten or more units per square mile (64 acres per housing unit). Those two counties account for 32% of the land area using the 32-acre-per-housing-unit threshold (559 out of 1,741 mi^2) and 35% of the land area using the 16-acre-per-housing-unit threshold (346.5 out of 1,009 mi^2). Nevada and El Dorado Counties also accounted for one-third of the population of the Sierra Nevada in 1990. This is one of the reasons our detailed case study focuses on the General Plans in these counties. Land use patterns and average densities in these more developed counties are likely to be more typical of future conditions in other parts of the Sierra Nevada as they continue to grow. The total land area affected by future population growth and human settlement is therefore likely to be less than the proportional increase expected in population. We discuss expected changes in total area for each of the density classes in our discussion of the General Plans and population forecasts for the Sierra Nevada.

We also determined the distribution of housing units by housing density class by watershed, by county, and by CCD. The twenty-four river basin boundaries in the SNEP study area do not coincide exactly with the forty-six CCDs or the counties, but we were still able to derive useful estimates of human settlement by river basin. Table 11.A4 in appendix 11.1 shows the distribution by river basin. These data, together with the distribution by county and by CCD, are available for more detailed analysis (using either dBase or Excel 5.0 for Windows) from the California Environmental Resource Evaluation System (CERES) project of the Resources Agency of the State of California (http://ceres.ca.gov/snep), and the Alexandria Project at the University of California, Santa Barbara (http://alexandria.sdc.ucsb.edu/).

Table 11.A4 shows that the American River watershed is the most populated, with 42,984 housing units and a population of 99,847 in a total watershed area of 1,887 mi^2. This is not surprising, given its location between I-80 and U.S. 50 in Placer and El Dorado Counties. The Yuba River watershed is nearly as populated, with 40,309 housing units and 90,836

FIGURE 11.31

Percentages of area, housing, and population by housing density class in the Sierra Nevada (based on 1990 census blocks in forty-six CCDs in eighteen California counties).

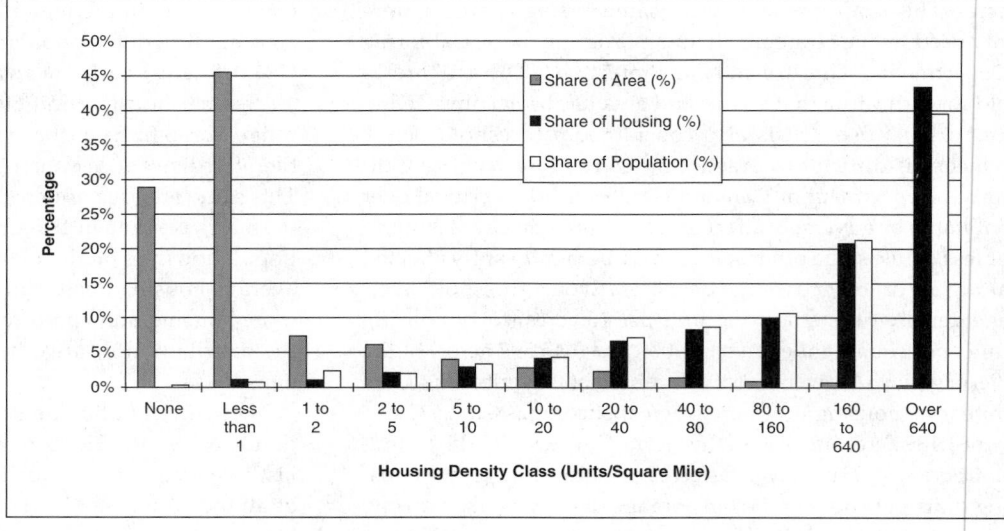

people in a total watershed area of 1,837 mi². The Yuba River watershed includes portions of Sierra, Yuba, Nevada, and Placer Counties (including the Bear River tributary to the Yuba River). The 767 mi² Truckee River watershed (including Lake Tahoe) has 42,011 housing units but only 49,767 residents. This reflects the high fraction of housing units that are seasonal. The relatively small Cosumnes River watershed (628 mi²) has 17,101 housing units and 41,700 residents, while there are 21,213 housing units and 38,681 people in the 1,710 mi² Tuolumne River watershed. The latter is largely in Yosemite National Park, however, while the former does not extend far into the higher elevations of the Sierra Nevada.

The Nevada and El Dorado County General Plans

Analysis at the level of the entire Sierra Nevada region is limited by the large size of the region, by the lack of consistent data, and by variation in local environmental conditions, demographic characteristics, and development policies and trends. For this reason, we performed two more detailed case studies at the county level that allowed us to focus on spatial patterns of development at a finer scale. These analyses focused on assessment of the General Plan update process under way for the past five years in Nevada and El Dorado Counties. Both counties adopted new General Plans in late 1995 or early 1996. Our analysis focused on the draft General Plans released in late 1994 and the draft environmental impact reports (EIRs) released in early 1995. These documents represent the most extensive and most current attempts at land use planning in the Sierra Nevada since the early 1980s. Most county General Plans in the region are now at least ten years old, and the experience of Nevada and El Dorado Counties may be useful to other local governments

as they attempt to update their General Plans in the coming decade.

We had five objectives in evaluating the General Plan update processes and land use maps in Nevada and El Dorado Counties:

1. To determine the range of spatial patterns for future human settlement and land use that represent the "official future" for Nevada and El Dorado Counties, where one-third of all Sierra Nevada residents lived in 1990.

2. To determine the range of factors considered and their relative importance in the land use planning process and the development of General Plan policies and land use maps.

3. To determine the range of environmental impacts likely to result from development under the General Plans.

4. To determine the type, timing, and costs of infrastructure investments required to achieve the objectives of the General Plans.

5. To determine the degree to which the impacts of buildout under the General Plans will be mitigated through the environmental impact review process under the California Environmental Quality Act (CEQA).

We were therefore interested in both the "product" (e.g., the General Plans, the associated land use maps and ordinances, and the mitigation measures adopted in the final EIR under CEQA) and the "process" ("planning") by which those products ("plans") were developed. Both will have a bearing on future patterns of human settlement in the Sierra Nevada.

County General Plans are important both as indicators and determinants of future development. The process that leads to the creation of a General Plan generally involves consideration of a broad range of ecological, social, and economic factors. Resulting zoning patterns therefore tend to reflect consideration of the opportunities and constraints afforded by these factors. However, a General Plan is not merely a passive document that allocates growth where it seems likely to go anyway. Zoning rules in themselves create a whole new layer of opportunities and constraints affecting an area's potential for development. General Plans are therefore a driving factor themselves.

We determined that the process currently followed by county planning agencies has several critical problems. First, the data relied upon for creating General Plan land use maps can be highly inaccurate. The most basic complication resulting from such inaccuracies is the zoning of areas at densities that have already been exceeded. Our analysis of the draft General Plans for both Nevada and El Dorado Counties indicates that tens of thousands of existing parcels are substandard under the proposed General Plans but are "grandfathered" and exempt from the new General Plan policies. The General Plans themselves therefore grossly understate the potential for new development, for they are based upon a "planimetric" analysis of future buildout under the General Plans' land use designations (rather than the underlying parcelization). The DEIRs for the General Plans therefore underestimate the scope and severity of environmental impacts associated with development under the draft General Plans.

Unfortunately, county planning departments are generally unable to complete the necessary analysis to identify the scale, severity, and spatial pattern of this problem, primarily because land use maps, zoning maps and parcel maps are typically developed on paper and lack the flexibility that would make it possible to experiment with different sets of criteria in the application of zoning. For example, a county may wish to know the effect on total housing counts of stream setbacks of various widths. The development of a series of scenarios would be an expensive, lengthy proposition if carried out exclusively through overlay of physical maps. The use of digital maps in the context of a geographic information system (GIS) allows for this type of flexibility in conducting multiple scenario analyses. Neither Nevada nor El Dorado County had this digital GIS capability during the General Plan update process. Both counties have been developing GIS capabilities, however, and the GIS was used for more limited analysis. The primary use of GIS still appears to be for plotting and presentation purposes, however, rather than geospatial analysis of human settlement and associated land use activities.

Our analysis using GIS answered some questions that were raised in the planning process in late 1994 but were not yet answered as the Nevada County Planning Commission reviewed the DEIR in early and mid-1995. These questions included the effect of existing parcelization on total buildout estimates of future population and housing units. Our results indicate that the DEIR was based upon significant underestimates of total buildout potential due to existing parcelization.

This raises questions about the reliability of the assumptions underlying the DEIR and its analysis of the environmental impacts of the draft General Plan. Not surprisingly, those impacts are at the heart of an intense local debate over both the DEIR and the General Plan itself. Consistent and reliable information about the impacts of the General Plan is generally unavailable, however, which exacerbates the conflict through disagreements about basic information. The information before the public at present appears to be erroneous.

For the purpose of this study, we analyzed the land use maps developed for one alternative in Nevada County and two alternatives in El Dorado County. We will begin our discussion of results with a comparison of the El Dorado County General Plan alternatives.

As stated in the draft text of the General Plan, complete buildout in El Dorado County, after which maximum allowable densities will have been reached throughout the county and no new development will be allowed to occur (without changes to the land use designations contained in the General Plan) will happen by the year 2040. The philosophy of the El Dorado County General Plan is also to avoid constraining the land market, however, by not limiting the total amount of land within a given land use designation only to that amount forecast to be required at buildout. The total buildout estimated based on the General Plan land use maps is therefore likely to exceed the actual buildout forecast to occur by the year 2040. Total buildout is nevertheless an accurate representation of how much development could occur without constraints under the General Plan. It is therefore the "official future" of maximum development. (As we will discuss later, however, past experience with General Plans suggests that the actual future will probably differ significantly from the "official future" of the General Plans.)

The General Plan land use maps represent two slightly different visions of how housing will be allocated spatially at buildout. One, the "Project Description," meets anticipated housing needs through a dispersed pattern of development with very limited restrictions based on infrastructure availability or environmental constraints. Plate 11.4 shows the land use map for the El Dorado County General Plan Project Description. The other option, the "General Plan Alternative," concentrates development into a more compact pattern and has greater restrictions based on infrastructure availability or environmental constraints. Plate 11.5 shows the land use map for the El Dorado County General Plan Alternative.

Both options are intended to allow roughly the same amount of total development by the year 2040, and they differ only slightly across all three of the scenarios we considered. *Significant modifications have been made to both the Nevada County General Plan and the El Dorado County General Plan since 1994, however, that are not captured in our analysis. The land use maps and associated policies relied on for our analysis were also the basis for the DEIRs released in early 1995, however, so the DEIRs are also inaccurate to the degree that the underlying assumptions in our analysis are inaccurate.*

The first step in the process was to convert the original land use maps into digital form. The county planning office originally published the maps at a scale of 1:2,000. Using a public land survey township and range section grid supplied by Teale Data Center as our base map, we digitized the zoning boundaries with an average error of 0.005% and a maximum error of 0.01%. This provided a positional accuracy comparable to the census blocks coverage also used in the analysis. Achieving a higher degree of accuracy would have been difficult for several reasons. First, township and range section coordinates may be inaccurate in rural areas, with some measurements dating back to the original land surveys of the nineteenth and early twentieth centuries. Second, some degree of warping inevitably occurs during blueprinting (we did not have access to Mylar originals). Third, and perhaps most important, the planning department itself did not intend for the maps to be used in a context requiring a high degree of accuracy and used thick lines on many of them. These thick lines reduce the accuracy of digitizing and can translate into significant boundary errors in reference to the actual location of the boundaries on the ground. Our land use maps should therefore not be used for finer-scale analysis than we have done here. In particular, they should not be used at the scale of individual parcels. Plate 11.6 shows the land use map for the Nevada County General Plan that was the basis for our analysis.

Once the digital land use maps were completed, we checked them against the originals for labeling and linework errors and assigned density ranges to each zoning area. For the purpose of our study, we used the midpoint of the stated density range for each land use classification. For example, if an area was zoned "Rural Residential" at 10–20 acres per unit, then a buildout density of 15 acres per unit was assumed. We also completed an analysis of total potential housing units at buildout using both a "low" (e.g., 20 acres per unit) and a "high" (e.g., 10 acres per unit) development density. The results of all three buildout scenarios are discussed for all of the alternative General Plans later.

The allocation of land uses by land use classification differs across the three General Plans we reviewed. Table 11.A5 in appendix 11.1 summarizes the amount of land dedicated to each land use classification under each of the land use maps. The implications of these distributions of land use classifications are discussed in more detail later.

An important point to make here is that both counties included land use designations for public lands, although neither county asserts land use jurisdiction over those lands. This stance contrasts with much more militant efforts to assert local jurisdiction in other rural counties in the West (Larson 1995).[64] El Dorado County does zone some Bureau of Land Management (BLM) lands for nonpublic purposes (including potential development) under the General Plan Alternative, however, with the apparent expectation that the BLM will release those lands from public ownership.[65] With the exception of these BLM lands, inclusion of those public lands

on the land use maps appears at this point simply to reflect prevailing land use practices under the jurisdiction of the state or federal agencies managing those lands. Some categories of land use, such as "Forest 160" or "Forest 640" in Nevada County and "Natural Resource" in El Dorado County, are dominated by federal and industrial forestlands. These lands are generally expected to continue in these resource uses under their existing ownership arrangements, although General Plan designations for private industrial timberlands can affect both market real estate values and the viability of alternative land uses. Some "checkerboard" areas near Donner Summit, for example, are at present zoned for industrial forestry and would be changed under the draft Nevada County General Plan to accommodate significant recreational, commercial, and residential development. Changing the zoning designation for those lands will change their market value, which will increase the potential cost of land exchanges or acquisition through Land and Water Conservation Act Funds.[66] Because these particular parcels include important trail access to adjacent federal lands, local land use decisions could directly impact management of federal lands.

The two alternatives under consideration by El Dorado County offer an even more dramatic example of how land use designations can affect future use of private industrial forestlands and management of adjacent public lands. The "Natural Resource" land use designation allows only one housing unit per 160 acres under the Project Description, while the same designation under the General Plan Alternative allows one housing unit per 40 acres. The latter land use map has more total area in this designation than the former, but the effective development potential under the same "Natural Resource" designation could allow up to four times as many housing units. More important than the actual density, however, is that 40-acre parcels must be accessed by a road network that would fragment existing industrial forestlands. Social constraints on harvesting could also result, along with increased restrictions on fuels management through prescribed burning. Finally, wildland fire risk could increase in these areas due to increased likelihood of ignition associated with vehicles and the presence of structures. Fire suppression in the urban-wildland intermix zone also tends to emphasize the protection of structures over natural resources. This fact could have enormous implications for fire regimes and costs of wildland resource management if it were adopted more widely in the Sierra Nevada.

This example highlights the important links between local land use planning and state or federal responsibilities for wildfire and natural resources management. Other policies within each of the General Plans have similar implications for water quality, air quality, transportation financing, and the health of local, state, and government finance. These impacts are discussed in more detail when we discuss the DEIR findings for each of the General Plans. Mitigation for these impacts is also discussed later.

Note that the total area classified in the two El Dorado

County alternatives is only 1,570.6 mi², while total area in each county is 1,791.1 mi². This reflects the fact that we did not digitize those portions of El Dorado County within the Lake Tahoe Basin, which are subject to the regulations of the Tahoe Regional Planning Agency (TRPA). Our analysis focused on development and land use on the western slope of El Dorado County. Land use within the TRPA jurisdiction must be considered in a regional context, and we did not feel that limited data for El Dorado County only would be useful. Much of this region is also within the incorporated community of South Lake Tahoe. There is a detailed discussion of the TRPA and its regulatory program (together with the associated programs of the California Tahoe Conservancy) in the SNEP case study of the Lake Tahoe subregion prepared by Elliott-Fiske et al. (1996).

We did complete a digitized coverage for all of Nevada County outside the incorporated cities of Grass Valley and Nevada City, but land use patterns in the Truckee-Donner area have been altered by the incorporation of the town of Truckee since the Draft General Plan was released in 1994 . The town of Truckee is now preparing its first General Plan, and it is expected to deviate in several respects from the Nevada County General Plan (generally resulting in lower buildout estimates) (Nevada County Planning Department 1994b). Our focus in the analysis will therefore remain on the western slope, but our county-level statistics for Nevada County do include the eastern part of the county (but do not reflect Truckee's incorporation). This focus reflects the 1994 Final Draft Nevada County General Plan.[67]

Buildout Analysis of the County General Plans

The El Dorado County General Plan land use designations include ranges of allowable density, while the Nevada County General Plan indicates fixed densities for each land use classification. El Dorado County also includes three future sites for a "Planned Community," however, at different average densities. Inclusion of specified densities for these sites allowed us to calculate the range of future housing units allowable under the General Plan maps' buildout directly from table 11.A5. These buildout estimates do not include lands within the Lake Tahoe Basin or the incorporated city of Placerville, so they understate the ultimate number of housing units at buildout in El Dorado County.

The Nevada County General Plan includes three land use classifications that are similar to the "Planned Community" designation used by the El Dorado County General Plan: "Planned Development," "Planned Residential Community," and "Special Development Area." None of these classifications has a specified allowable density of development associated with it. The "Planned Residential Community" designation applies only to four existing subdivisions, however, where we could determine existing parcelization and average buildout densities for the class. Based upon the

assessor's 1992 data, Tahoe Donner has 6,094 parcels on 3,809 acres (1.60 housing units per acre), Lake of the Pines has 2,038 parcels on 1,343 acres (1.52 housing units per acre), Lake Wildwood has 3,035 parcels on 2,189 acres (1.39 housing units per acre), and Alta Sierra has 2,855 parcels on 3,100 acres (0.92 housing units per acre). The average density for these 10,441 acres is 1.34 housing units per acre across the 14,022 parcels. We used this total of 14,022 housing units as the buildout estimate for the "Planned Residential Community" land use classification throughout all three of our scenarios.[68]

We had to develop independent estimates of future development density for lands classified as "Planned Development" (which were scattered among many parcels) or "Special Development Area" (which were focused on an area proposed for a "new town"). We analyzed all of the parcel data for any of the assessor's map book pages that intersected with lands designated as "Planned Development" to estimate average development densities. Those parcels totaled 104,802 acres, however, which is more than ten times as much land area as that designated for "Planned Development" under the General Plan. Those parcels have an average designated density of 0.53 units per acre (337 units per square mile), which is very close to the 0.62 housing units per acre (397 units per square mile) used by El Dorado County as the lowest development density for the same designation. We therefore used 0.53 units per acre for our "low" scenario of buildout of those lands designated for "Planned Development" in Nevada County.

The proposed "new town" in the land classified as a "Special Development Area" is expected to have a much higher density, however, so we estimated the future density of the SDA land use classification based on the weighted average density of all areas designated "Urban High Density" (20 units per acre on 3.33% of the land area, accounting for 23.12% of the housing units), "Urban Medium Density" (6 units per acre on 7.44% of the land area, accounting for 15.51% of the housing units), "Urban Single Family" (4 units per acre on 35.11% of the land area, accounting for 48.82% of the housing units), and "Residential" (1 unit per 1.5 acres on 54.13% of the land area, accounting for 12.55% of the housing units) under the Nevada County General Plan. The weighted average density for these areas was 1,841 housing units per square mile, or about 2.88 housing units per acre. We used this density for the "middle" scenario for the SDA designation and the "high" scenario for the "Planned Development" designation. It is within the range of allowable densities for the "Planned Community" designation in the El Dorado County General Plan and is therefore a reasonable proxy for the number of housing units that could be built under the Nevada County General Plan at buildout. We also evaluated a "low" scenario using the lowest allowable density (1.4 units per acre) and a "high" scenario using the highest allowable density (4.1 units per acre) under the "Planned Community" designation in the El Dorado County General Plan. Our "middle" scenario for the "Planned Development" lands was the average of the "low"

(0.53 units per acre) and the "high" (2.88 units per acre). "Planned Development" lands therefore had a lower average density assigned to them than the "Special Development Area" lands for each of the scenarios.

Using these assumptions, total buildout on the 10,127 acres with the "Special Development Area" designation would range from 6,278 housing units to as high as 41,519 housing units. Preliminary proposals for the Gold Country Ranch "new town" estimated only 5,249 housing units on twenty-four adjacent parcels totaling 8,232 acres (0.64 housing units per acre) (Nevada County Planning Department 1994a). Other documents from the Nevada County General Plan suggest that the "new town" site would be only 7,100 acres at an average density of 1.04 housing units per acre. This is based on 920 acres designated "Urban Single Family" (4 housing units per acre), 50 acres in "Urban Medium Density" (6 housing units per acre), 170 acres in "Urban High Density" (20 housing units per acre), and 5,960 acres without residential development. Both of these average density estimates seem very low, however, given the average density of other developments requiring centralized infrastructure. We therefore compared our estimates to densities in other "new town" projects.

The large-scale Stanford Ranch project in nearby Placer County was originally zoned for 11,000 units on 3,500 acres (3.14 units per acre) by the city of Rocklin and is now expected to be built out with around 8,000 units (2.29 units per acre). The decrease in average density is due to new wetlands restrictions and a reduced share of overall units going into multifamily units (due to overbuilding of multifamily units and low occupancy rates elsewhere in Rocklin) (Stanford Ranch Information Center 1994–95). The urban design approach employed by Peter Calthorpe for the Gold Country Ranch project generally calls for compact, "transit-oriented development" (TOD) through a "pedestrian pocket" idea (Calthorpe 1989, 1993). Because of the larger land area designated "Special Development Area,"[69] however, any "new town" projects built under the Nevada County General Plan would probably not maintain such high densities across the entire development. An average density between the "low" scenario (1.4 units per acre) and the "middle" scenario (2.88 units per acre) is therefore likely. These types of areas are very unlikely to be developed at the "high" density of 4.1 units per acre across all of the 10,127 acres in the SDA classification.

We checked these assumptions again by assigning the same average densities to the incorporated cities of Grass Valley and Nevada City, which account for another 3,424 acres that have no densities associated with them in the Nevada County General Plan. This resulted in an estimated 9,849 housing units in the two cities at an average density of 2.88 housing units per acre. This is higher than the current number, which supports a combined population of around 12,000 people. The lower density of 1.4 units per acre results in only 2,123 housing units in the incorporated cities, however, which is well below the current number of housing units. Based upon this

comparison with the existing incorporated cities in Nevada County, we believe the range of probable future buildout densities for the Nevada County General Plan is somewhere between our "low" and "middle" scenarios. We used a fixed average density of 1,000 units per square mile (1.56 units per acre) for the incorporated areas in all three scenarios, resulting in 5,350 housing units in the city limits. This allowed us to focus our analysis on the effects of different assumptions about allowable densities in the undesignated areas affected by the Nevada County General Plan. Infill could also increase average densities within the incorporated cities, of course, but that would be subject to the 1982 Grass Valley General Plan and the 1985 Nevada City General Plan. We were unable to digitize those General Plans for this analysis. The average density we assumed for the incorporated areas also reflects a high level of commercial land use within each of the incorporated cities.

Note that all of our buildout estimates are based upon "gross" acreage within a land use designation. A significant fraction of all land is likely to be dedicated to roads, however, resulting in a smaller "net" acreage available for actual development. The fraction assigned to roads and other nonbuildable uses is expected to be from 10% to 20% of all undeveloped land. General Plan policies could still allow development based upon the allowable gross acreage, however, so it is not necessarily appropriate to reduce our buildout estimates by 10–20%. The net land area allowed for development will be a function of specific General Plan policies and specific language in the zoning ordinance.

At this point we were ready to proceed with the first step in our spatial analysis, which was to simply calculate the change in housing density from current levels if complete buildout were to occur. We used the 1990 census blocks coverage as a measure of current housing density. To calculate the change in housing density from 1990 to buildout, we created a new digital map based on the intersection of the two input maps. This intersection was necessary because the boundaries of census blocks and land use designation areas do not coincide; each land use area typically consisted of several blocks, with many blocks straddling the boundary between land use areas. The output map produced by the intersection of the census blocks and land use areas contained a larger number of smaller areas, each belonging to only one census block and one land use area. Figure 11.32 illustrates this intersection for a hypothetical land use and census block.

Each General Plan reported allowable densities in terms of housing units per acre, which could be compared with the existing 1990 housing density. Because we now had a value for the current and future densities, we could calculate two different measures of change:

1. Absolute Change = (Maximum Allowable Density) – (1990 Density)

2. Relative Change = {Absolute Change} / {1990 Density} * 100

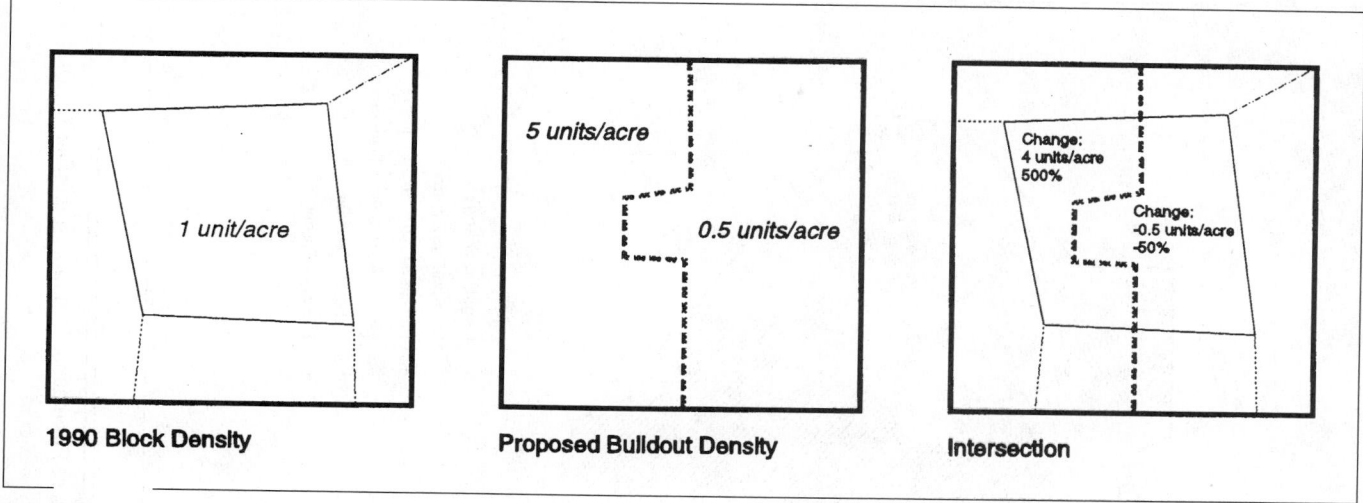

1 unit/acre

1990 Block Density

5 units/acre

0.5 units/acre

Proposed Buildout Density

Change:
4 units/acre
500%

Change:
-0.5 units/acre
-50%

Intersection

FIGURE 11.32

Illustration of method for calculating change in housing density from 1990 to General Plan buildout using census block data.

Our first analytic step was to calculate these two derived variables and add them to table 11.A5 without any further spatial analysis. The spatial patterns of absolute and relative change are also important determinants of both the ecological and economic effects of development, however. As a result, we can generate new maps illustrating the changes that occur between 1990 and buildout under the General Plan alternatives. Figure 11.33 shows the spatial pattern of the absolute changes that would occur, and figure 11.34 shows the spatial pattern of relative changes for the El Dorado County General Plan Project Description.

Figure 11.35 shows the absolute changes and figure 11.36 shows the relative changes for the El Dorado County General Plan Alternative.

As noted earlier, each of the General Plan alternatives in El Dorado County includes a range of possible densities for each land use classification. We present the "middle" scenario here (based upon the mean density allowed for each land use classification), but there is significant variation between the "low" and "high" range of possible densities. There were approximately 61,000 housing units in El Dorado County in 1990 (without Placerville or the Lake Tahoe Basin), and the "middle" scenario projects a buildout of 156,820–160,919 housing units. Based simply upon allowable density, the "low" scenario would result in only 68,065–70,574 units. These figures are clearly improbable, however, for there are more than 7,065 vacant parcels today. We therefore believe the "low" scenario is highly unlikely and should not be relied upon as the basis for evaluating the impacts of future land use. The "high" scenario would increase the total number of housing units at buildout to 243,083–253,772. This would represent more than a fourfold average increase in the total number of housing units in El Dorado County. The total number of new units above those in 1990 would increase from an average of

about 98,000 new units under the "middle" scenario to an average of around 187,000 new units under the "high" scenario. This represents a 91% increase in the absolute growth in housing units under the "high" scenario compared with that estimated under the "middle" scenario. Any estimates of future impacts associated with buildout of the El Dorado County General Plan based upon the "middle" scenario could therefore be underestimating the potential impacts by 50%. Table 11.A6 in appendix 11.1 shows these different buildout estimates for each scenario for both the Project Description and the General Plan Alternative.

We completed a similar analysis for the Nevada County General Plan, using the estimated average housing densities described for those lands designated Planned Development, Planned Residential Development, and Special Development Area. Total buildout under these assumptions results in a total of 128,265 housing units under our "middle" scenario. This compares with 37,352 housing units in 1990. The two estimated land use classifications (Planned Development and Special Development Area) accounted for 35.61% of the total, however, on only 4.17% of the total land area in the county. Our "low" scenario estimated 93,991 housing units, and our "high" scenario could result in up to 152,080 housing units. The two estimated land use designations account for 12.13% of all housing units under the "low" scenario to 45.69% of all housing units under the "high" scenario. The ultimate buildout estimates under the Nevada County General Plan are therefore highly sensitive to the allowable densities for these special land use classifications. Even without them, however, we estimate 82,588 housing units at buildout based only upon the area designated under each of the land use classifications. Table 11.A7 in appendix 11.1 shows these different buildout estimates for each of the scenarios for the Nevada County General Plan.

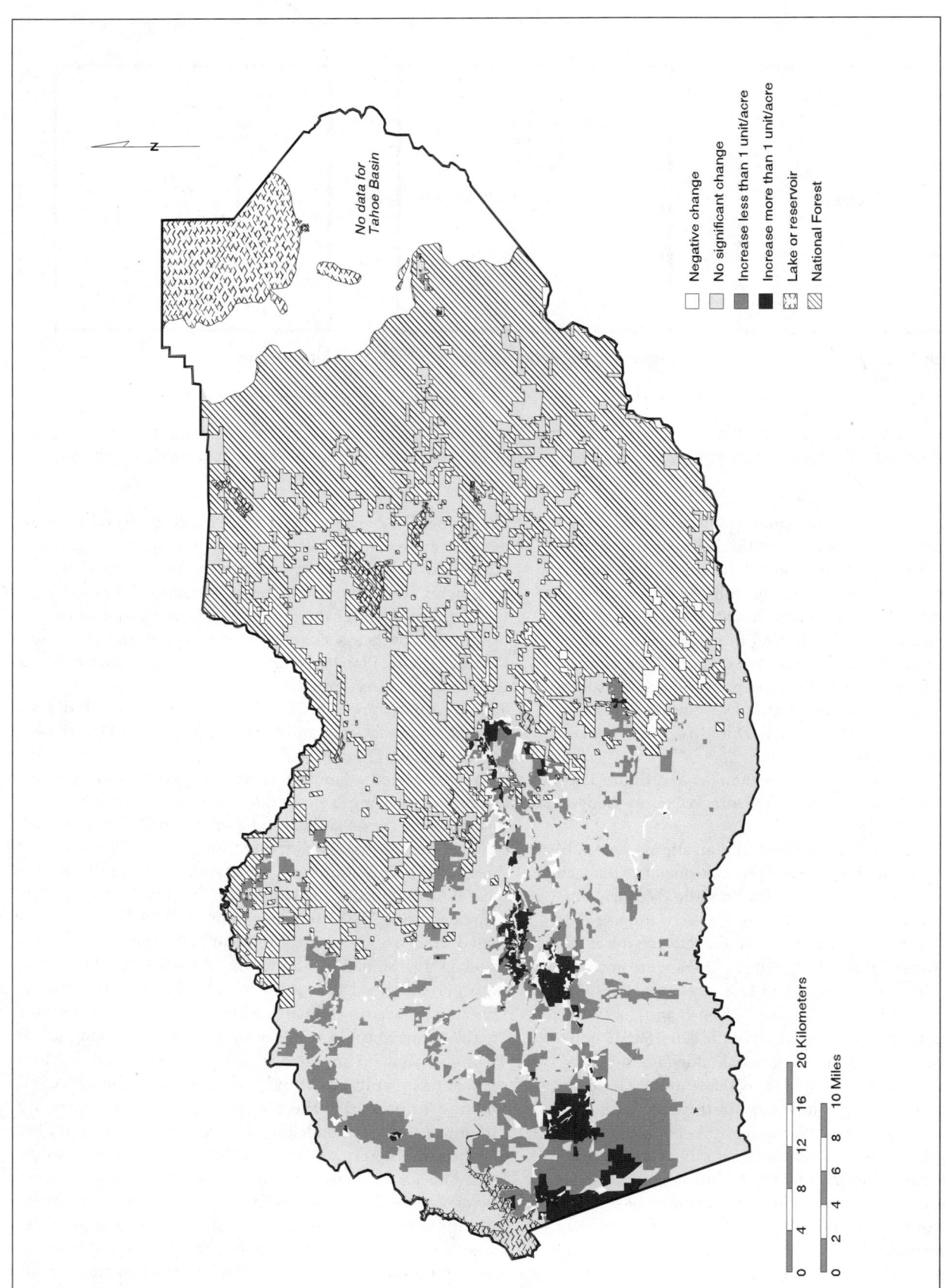

Legend:

- Negative change
- No significant change
- Increase less than 1 unit/acre
- Increase more than 1 unit/acre
- Lake or reservoir
- National Forest

No data for Tahoe Basin

0 4 8 12 16 20 Kilometers

0 2 4 6 8 10 Miles

FIGURE 11.33

Absolute change in housing density, El Dorado County, 1990 to buildout (Project Description) (from 1990 Census of Population, Summary Tape File 1B; Draft El Dorado County General Plan).

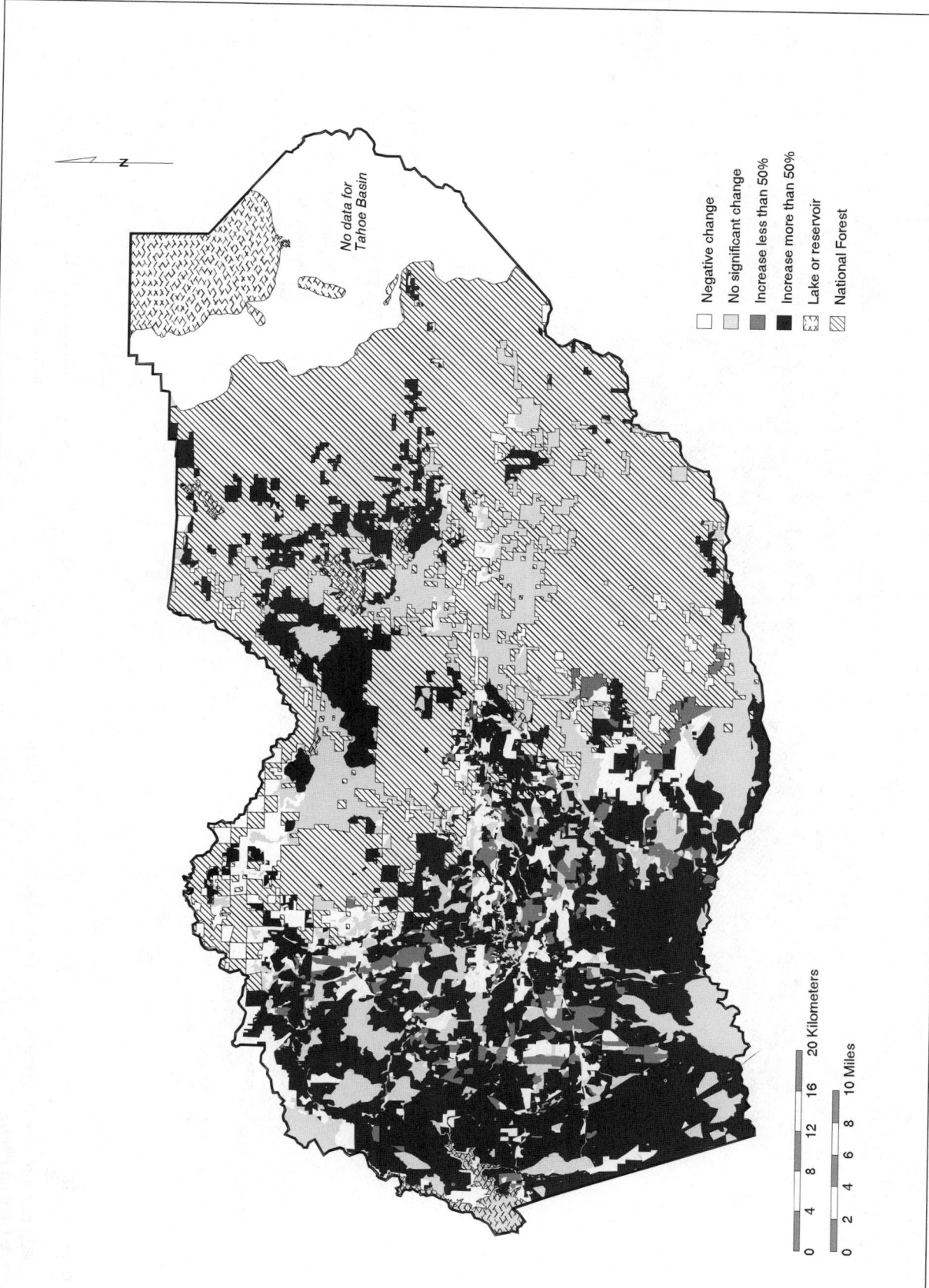

Legend:

- Negative change
- No significant change
- Increase less than 50%
- Increase more than 50%
- Lake or reservoir
- National Forest

No data for Tahoe Basin

N

0   4   8   12   16   20 Kilometers

0   2   4   6   8   10 Miles

FIGURE 11.34

Relative change in housing density, El Dorado County, 1990 to buildout (Project Description) (from 1990 Census of Population, Summary Tape File 1B; Draft El Dorado County General Plan).

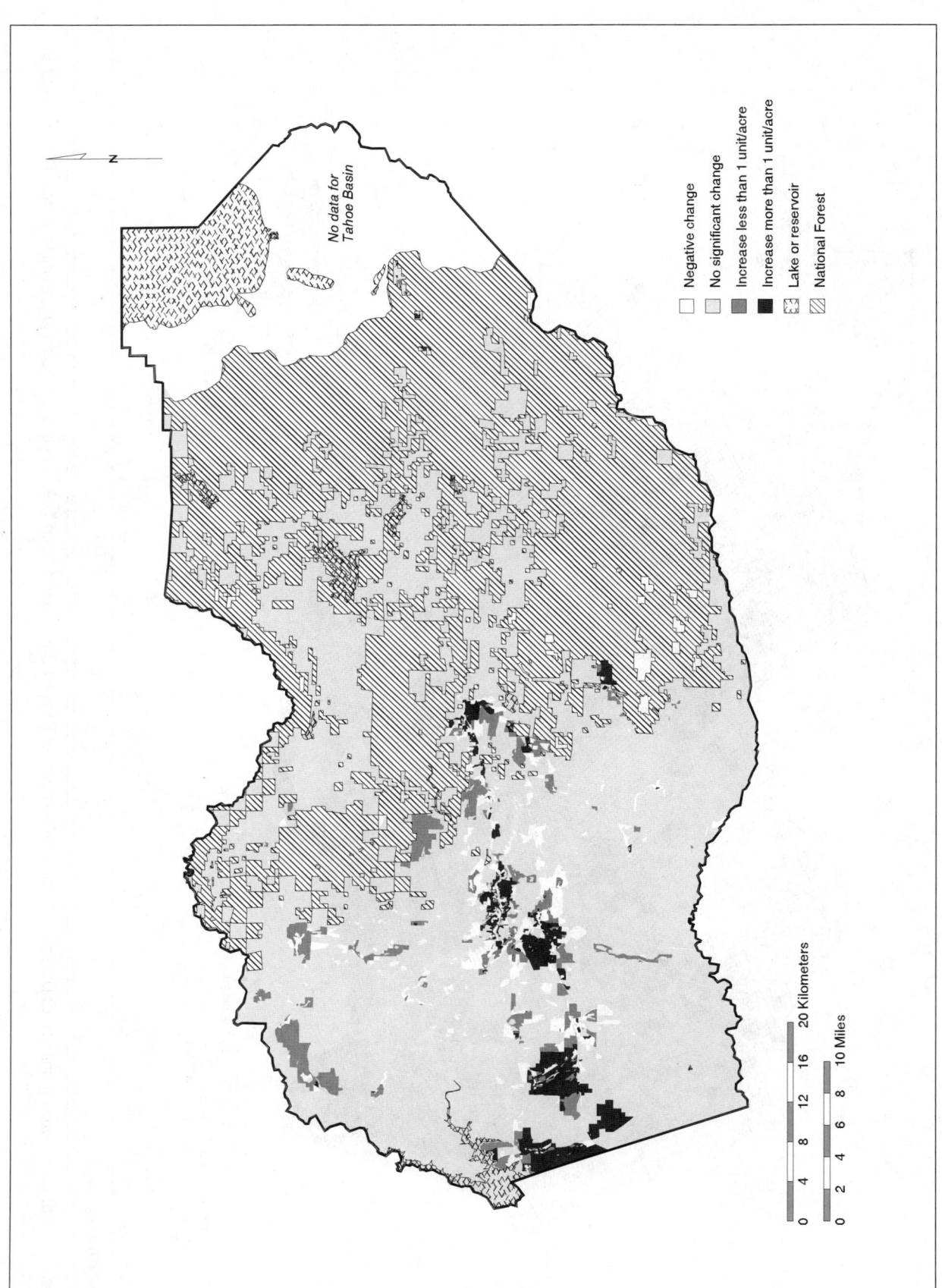

FIGURE 11.35

Absolute change in housing density, El Dorado County, 1990 to buildout (alternative) (from 1990 Census of Population, Summary Tape File 1B; Draft El Dorado County General Plan).

Legend:

- Negative change
- No significant change
- Increase less than 1 unit/acre
- Increase more than 1 unit/acre
- Lake or reservoir
- National Forest

No data for Tahoe Basin

N

0   4   8   12   16   20 Kilometers
0   2   4   6   8   10 Miles

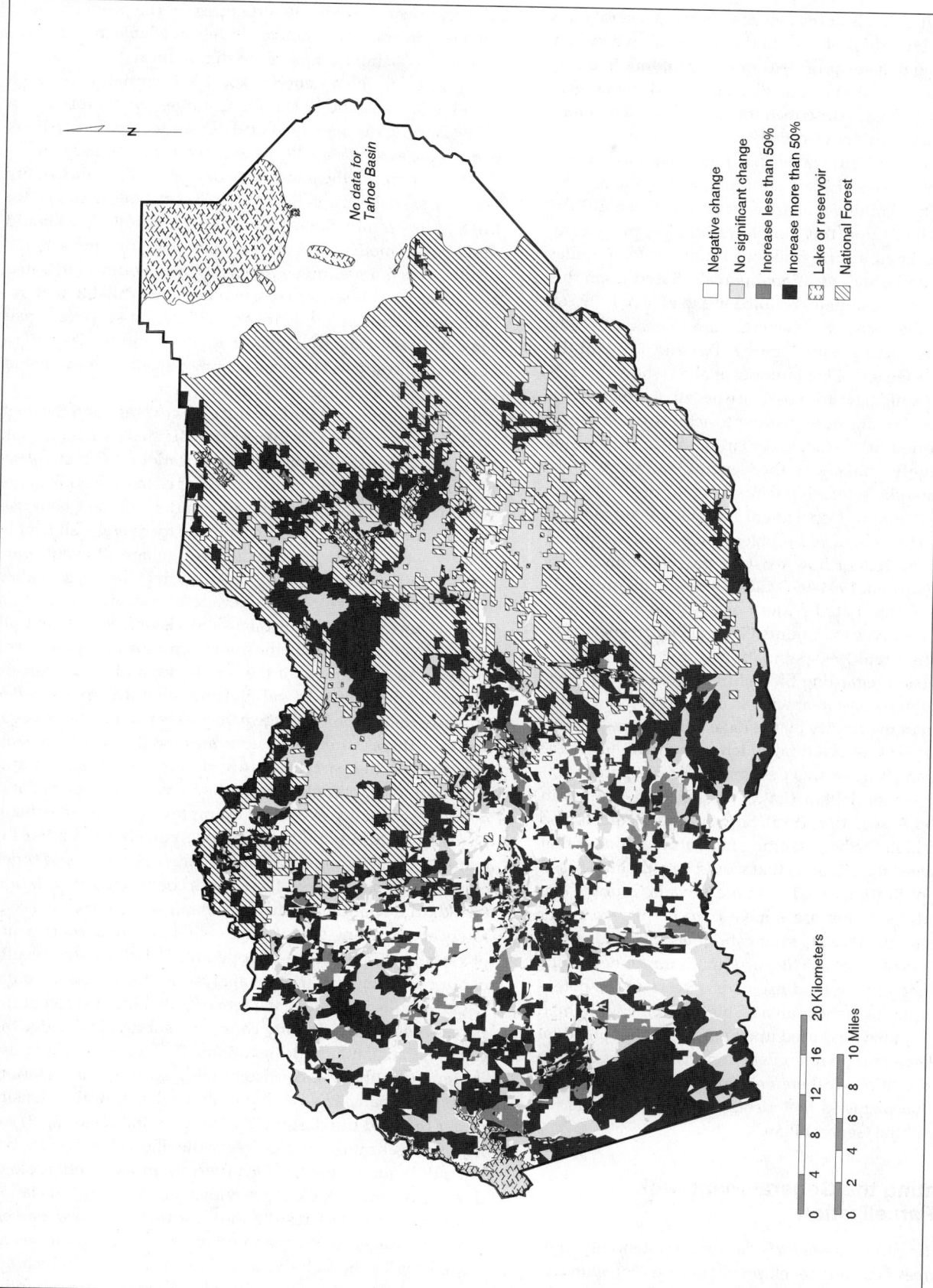

Negative change
No significant change
Increase less than 50%
Increase more than 50%
Lake or reservoir
National Forest

No data for
Tahoe Basin

N

0   4   8   12   16   20 Kilometers

0   2   4   6   8   10 Miles

FIGURE 11.36

Relative change in housing density, El Dorado County, 1990 to buildout (alternative) (from 1990 Census of Population, Summary Tape File 1B; Draft El Dorado County General Plan).

Our buildout analysis of the Nevada County General Plan identified a number of problems in the relationship between existing housing density as of 1990 and the buildout housing densities according to the General Plan land use designations. As in the "low" scenario described for the El Dorado County General Plan, any buildout estimates based only on the land use designations and allowable densities contained in the Nevada County General Plan are highly improbable. The Nevada County Planning Department and its General Plan consultants relied upon this simple estimation procedure, however, to estimate future buildout estimates for population, housing units, and additional parcels. Based upon this simplification, those estimates resulted in a total of only 29,769 additional parcels under the General Plan (Nevada County Planning Department 1994a). Figures 11.37 and 11.38 for the Nevada County General Plan buildout analysis show that significant areas would have to experience negative growth compared to actual 1990 densities in order to achieve the buildout densities reflected in the General Plan land use maps. This outcome is highly unlikely, so the General Plan maps probably understate likely future potential "build out." The Nevada County Planning Department has acknowledged that the understatement could be a problem but has not completed a systematic analysis of how existing parcelization affects "build out." (Norman 1994–95; Miller 1994–95). Figures 11.37 and 11.38 show the spatial pattern of absolute and relative changes under the Nevada County General Plan, respectively, compared with actual 1990 population.

This conclusion regarding the failure to account for existing parcelization is consistent with an analysis completed for one subregion of the county by the Lake Vera/Round Mountain Neighborhood Association, which found that there were already 476 parcels in an area designated for 232 parcels under the 1980 General Plan (Lake Vera/Round Mountain Neighborhood Association 1995). Some of those parcels also can be subdivided further, resulting in a future buildout that will vastly exceed the 232 units that would be estimated based upon the simplification relied on in the Nevada County General Plan analysis. Other areas have parcels that are larger than the minimum allowable but cannot be subdivided further (e.g., a seven-acre lot in a five-acre-minimum zone), which means the General Plan land use maps overstate the potential for development in those areas. This problem highlighted the limits to any analysis based upon land use maps and density designations that do not reflect underlying parcelization and existing densities. We therefore completed a more detailed assessment that included the effects of parcelization on the applicability of the General Plan.

Implementing the General Plans with Existing Parcelization

Parcel-specific county assessor's data are a potentially more detailed source of current development information than census data, for they include boundaries and data from the tax rolls that describe existing structures on the property. Such detailed information is not yet widely available, but the process of converting from paper to digital format has begun in many areas. In the absence of specific information for each parcel in Nevada and El Dorado Counties, we performed an analysis using the average parcel densities of each map book page in the assessor's rolls. Based on our preliminary analysis, the accuracy of the underlying data in the Nevada County assessor's database is believed to be less reliable than that for El Dorado County. Before reporting the results, we should describe our method and outline potential problems with the results. This is a preliminary analysis that should be updated as soon as parcel-based map coverages are available with reliable georeferencing.[70] Until then, it should serve as a reasonable basis for identifying the magnitude and spatial pattern of potential conflicts between existing parcelization and land use designations under the General Plans.

First we intersected the General Plan coverage and the map book page coverage. This intersection divided each map book page into one or more land use classifications. We then calculated the average allowable density for each map book page, weighted in proportion to the area under different land use designations. By inverting our estimate for average allowable density, we were able to calculate the average allowable parcel size for that map book page. Any parcel that was smaller than twice the size of the average allowable parcel was deemed unsuitable for further subdivision. We divided all parcels larger than this by the minimum allowable parcel size to arrive at an estimate of the number of additional parcels that could be created by subdivision of existing parcels. Because we relied upon average parcel sizes per map book page, however, these estimates are accurate for only those map book pages where the parcel sizes are closely distributed around the mean. A map book page with one very large parcel and many very small parcels, for example, would overestimate the capacity to subdivide the smaller parcels and underestimate the capacity to subdivide the larger parcel. These types of map book pages should have high coefficients of variation (standard deviation divided by mean), while the most accurately estimated map book pages will have small coefficients of variation. Table 11.A8 in appendix 11.1 shows the results of our coefficient of variation analysis for the assessor's data.

Our results indicate that a remarkably high fraction of the parcels within each county cannot be subdivided under the General Plan land use designations. This reflects (1) land use designations that are consistent with existing parcelization; (2) existing parcelization that is above the allowable density under the land use designation but grandfathered in; (3) existing parcelization that is below the allowable density but not subdividable further; or (4) errors in our methodology due to many map book pages with a high standard deviation in parcel sizes. The results indicate that very few parcels would be subject to the planning reviews and regulations required under the Subdivision Map Act of 1973 (which applies to all subdivisions resulting in 4 or more parcels) under

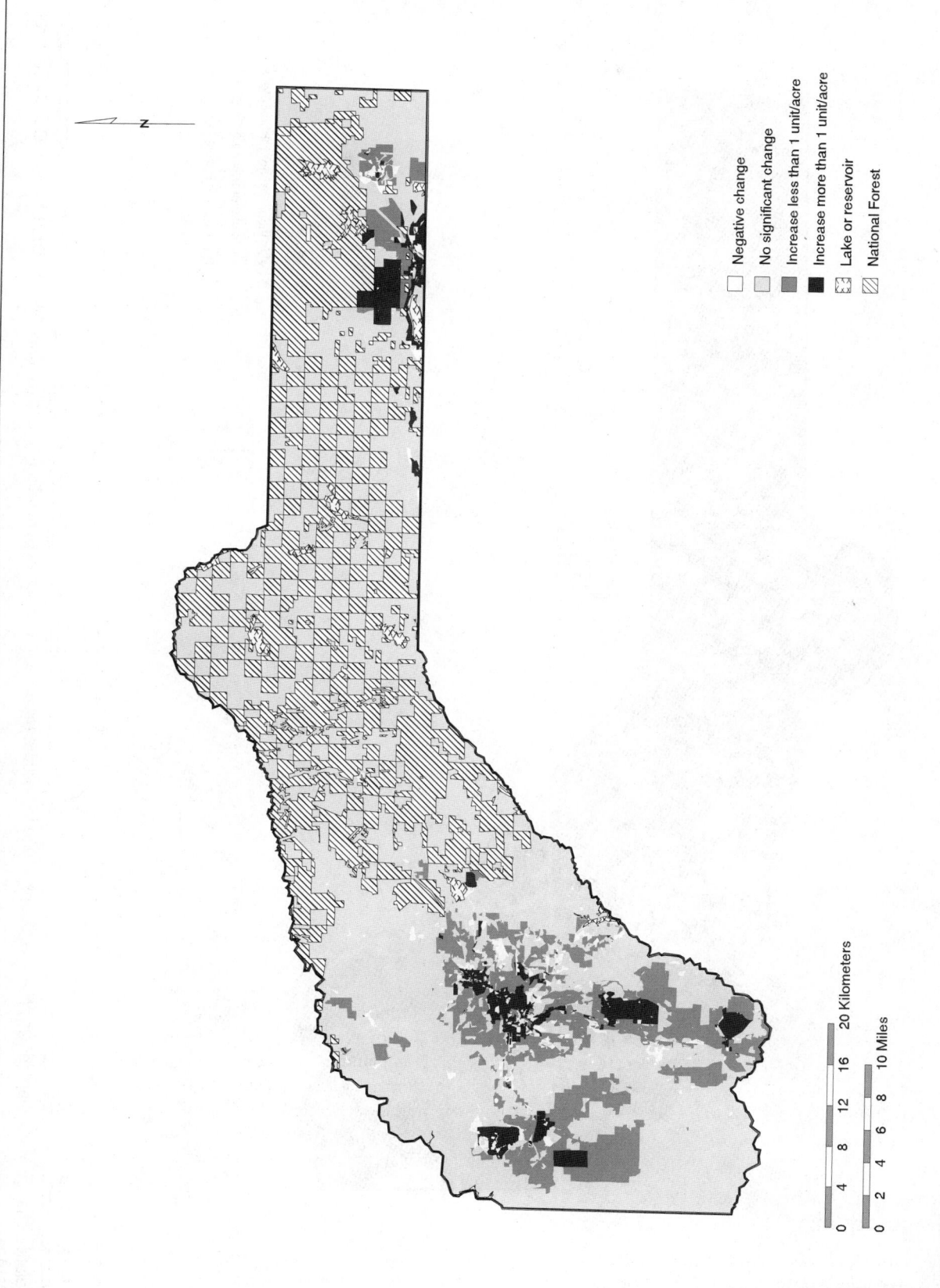

FIGURE 11.37

Absolute change in housing density, Nevada County, 1990 to buildout (from 1990 Census of Population, Summary Tape File 1B; Draft Nevada County General Plan).

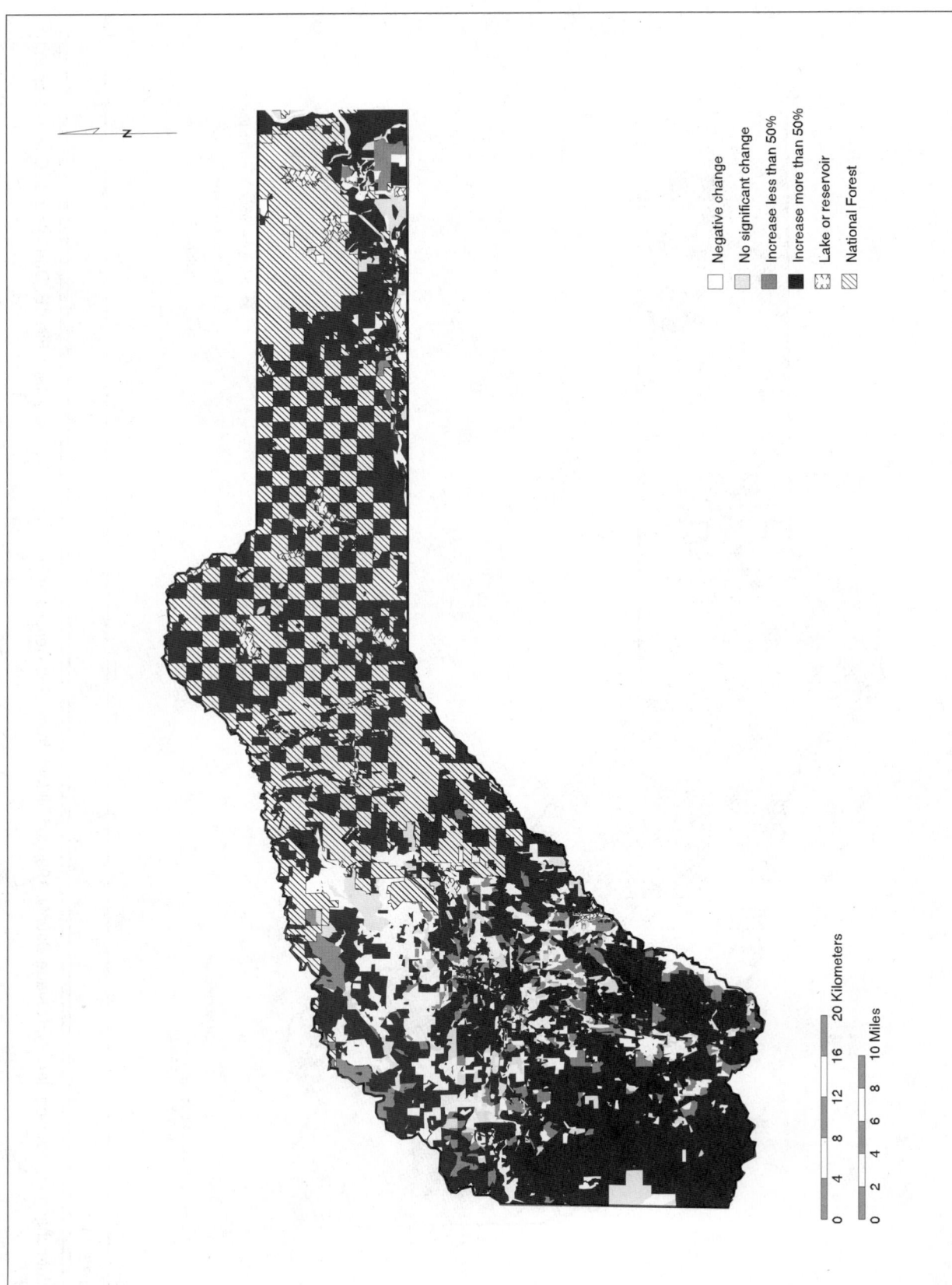

FIGURE 11.38

Relative change in housing density, Nevada County, 1990 to buildout (from 1990 Census of Population, Summary Tape File 1B; Draft Nevada County General Plan).

the new General Plans in either county: only 2,877 parcels (2.95% of the total) in El Dorado County and only 3,153 (5.44% of the total) in Nevada County. The rest of the parcels in each county could either be developed (but not subdivided further) or else be subdivided into 4 or fewer parcels through ministerial actions.[71] This is an important distinction, for ministerial actions do not generally trigger a full environmental impact report (EIR) review process under the California Environmental Quality Act (CEQA). Our estimates for El Dorado County are based upon the Project Description. Because these subdividable parcels are larger than average, however, these parcels accounted for 27.82% of the area in Nevada County that is neither state nor federal land. Fully two-thirds (67.13%) of the land area in El Dorado County that is not in state or federal ownership would be subject to the Subdivision Map Act and therefore require a specific set of requirements for development. The full distribution of parcels and their "subdividability" is summarized for both counties in table 11.A9 in appendix 11.1.

Nevada County had records for 57,963 parcels in the assessor's database in 1992 (53,314 improved and 4,649 unimproved parcels),[72] while El Dorado County had 97,681 parcels (73,780 improved and 23,910 unimproved parcels). Based simply upon existing unimproved parcels, then, there is significant potential to accommodate additional population in each county without additional subdivisions. Many of these existing parcels are within the incorporated communities of Grass Valley, Nevada City, Truckee, Placerville, or South Lake Tahoe. As noted earlier, however, many additional parcels could be created under the draft county General Plans. The primary reason is that large parcels account for a disproportionate share of total potential housing units in both of the counties. Even though only a small fraction of all existing parcels can be subdivided, those parcels tend to be large and could be subdivided into many parcels. Based upon the methodology we employed, there is a potential for the creation of another 144,470 parcels in Nevada County and 71,370 parcels in El Dorado County under the General Plans. Our methodology estimated a range of 107,796 to 181,145 potential total parcels following allowable subdivision under the Nevada County General Plan, which is much higher than the 29,769 additional parcels estimated by the Nevada County Planning Department. Based on the methodological limitations cited above, we believe the actual number lies somewhere between our respective estimates.

Our estimates are based upon the mean number of lots that could be created through subdivision for each subdivision class, with the maximum class (more than 640 parcels) assumed to equal only 640 parcels. We are fairly confident about our estimates for El Dorado County, but very cautious about the estimates for Nevada County for the methodological reasons cited. In particular, land use designations under the Nevada County General Plan are often inconsistent with underlying parcelization. Our methodology is therefore less likely to be accurate in those cases due to the high coefficient

of variation in parcel sizes for each map book page and land use designation. Due to parcelization, however, our preliminary analysis nevertheless raises serious questions about the accuracy of any projections of future parcels and housing units based only on the General Plans and their associated land use designations.

We attempted to highlight those areas with a high degree of existing parcelization by developing two related coverages for both El Dorado and Nevada Counties. The first coverage identifies the total acreage in parcels of 160 acres or more for each map book page that has any parcels of that size. All of the other map book pages have already been subdivided into smaller parcels and are represented by blank spaces in each coverage. Note that there are large contiguous areas in the western part of both counties near the historic gold rush towns of Placerville (El Dorado County) and Grass Valley and Nevada City (Nevada County) that have no large parcels. These areas then extend into areas of newer development, such El Dorado Hills and Auburn Lake Trails (El Dorado County) and Lake Wildwood, Lake of the Pines, Cascade Shores, and Chicago Park (Nevada County). There is also a large contiguous area without large parcels in the south-central part of El Dorado County, around Lake Tahoe, and near Truckee in eastern Nevada County. Figures 11.39 and 11.40 illustrate this pattern for El Dorado and Nevada Counties, respectively.

The second coverage focuses more specifically on those map book pages that have a high number of small parcels less than five acres in size. Not surprisingly, many of these areas match the map book pages without large parcels identified in the previous coverage. Figures 11.41 and 11.42 show where these small parcels are concentrated spatially.

This small parcel coverage provides a more detailed spatial representation of the pattern of parcelization, however, which allows a more accurate assessment of the potential to exceed identified buildout forecasts based only upon the General Plan land use maps. Areas with a high level of parcelization, like the Lake Vera/Round Mountain neighborhood in Nevada County, are likely to have much higher rates of both absolute and relative change in average housing densities at General Plan buildout than identified in figures 11.37 and 11.38. Other areas with lower levels of parcelization are less likely to have a problem with substandard lots that have been grandfathered in under the General Plan for higher levels of development density than would otherwise be allowed under their land use classification. Based upon our review, we believe this is a much more serious problem with the Nevada County General Plan than the El Dorado County General Plan. The El Dorado County General Plan more accurately accounts for existing parcelization in its land use designations and is therefore less likely to understate or overstate future development potential. The Nevada County General Plan often fails to recognize high levels of inconsistency between existing parcelization and land use designations for an area, so it tends to understate future development potential. There

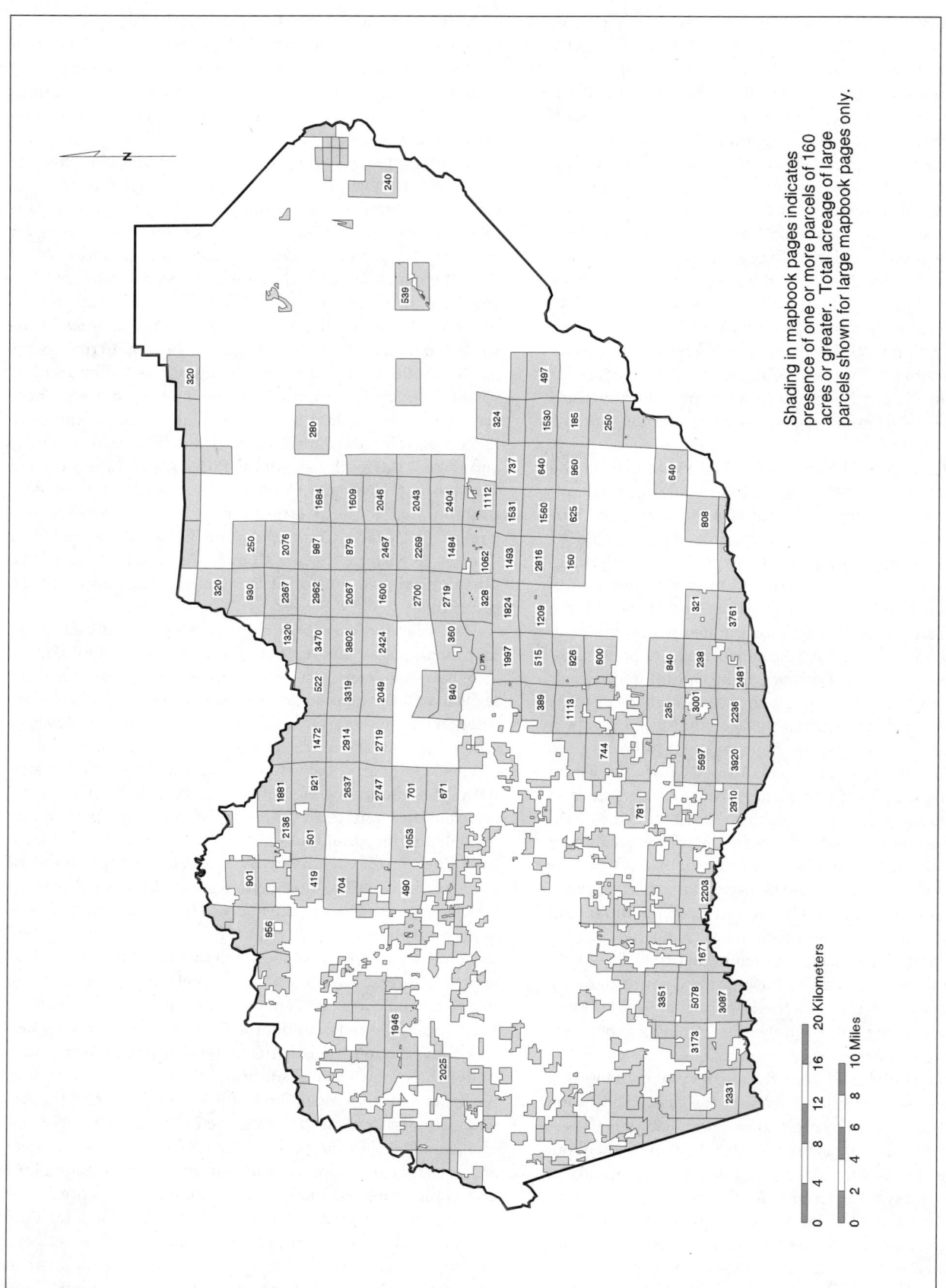

FIGURE 11.39

Acreage in large parcels, El Dorado County, 1992 (from parcel database, El Dorado County Assessor's Office).

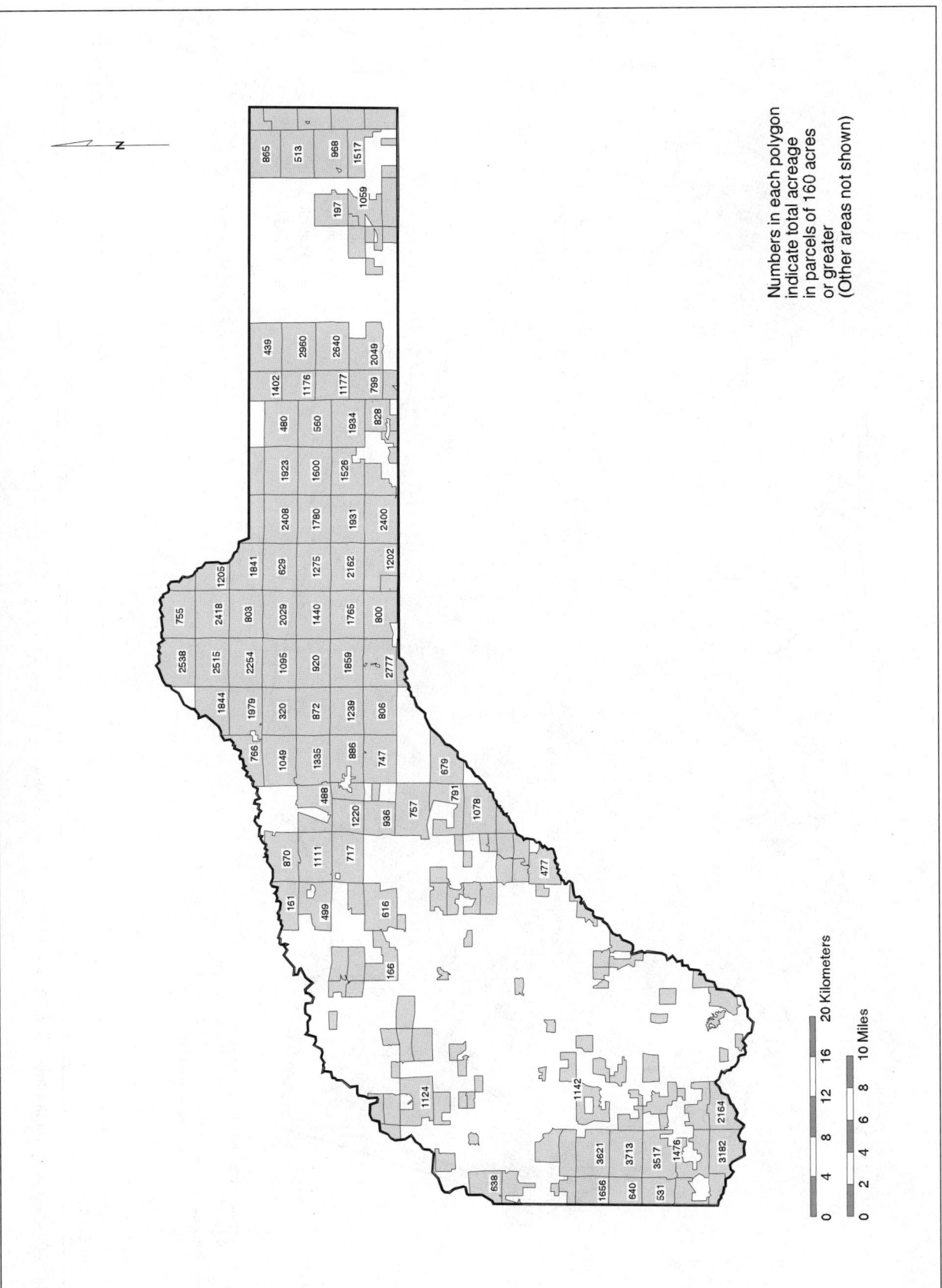

Numbers in each polygon
indicate total acreage
in parcels of 160 acres
or greater
(Other areas not shown)

FIGURE 11.40

Acreage in large parcels, Nevada County, 1992 (from parcel database, Nevada County Assessor's Office).

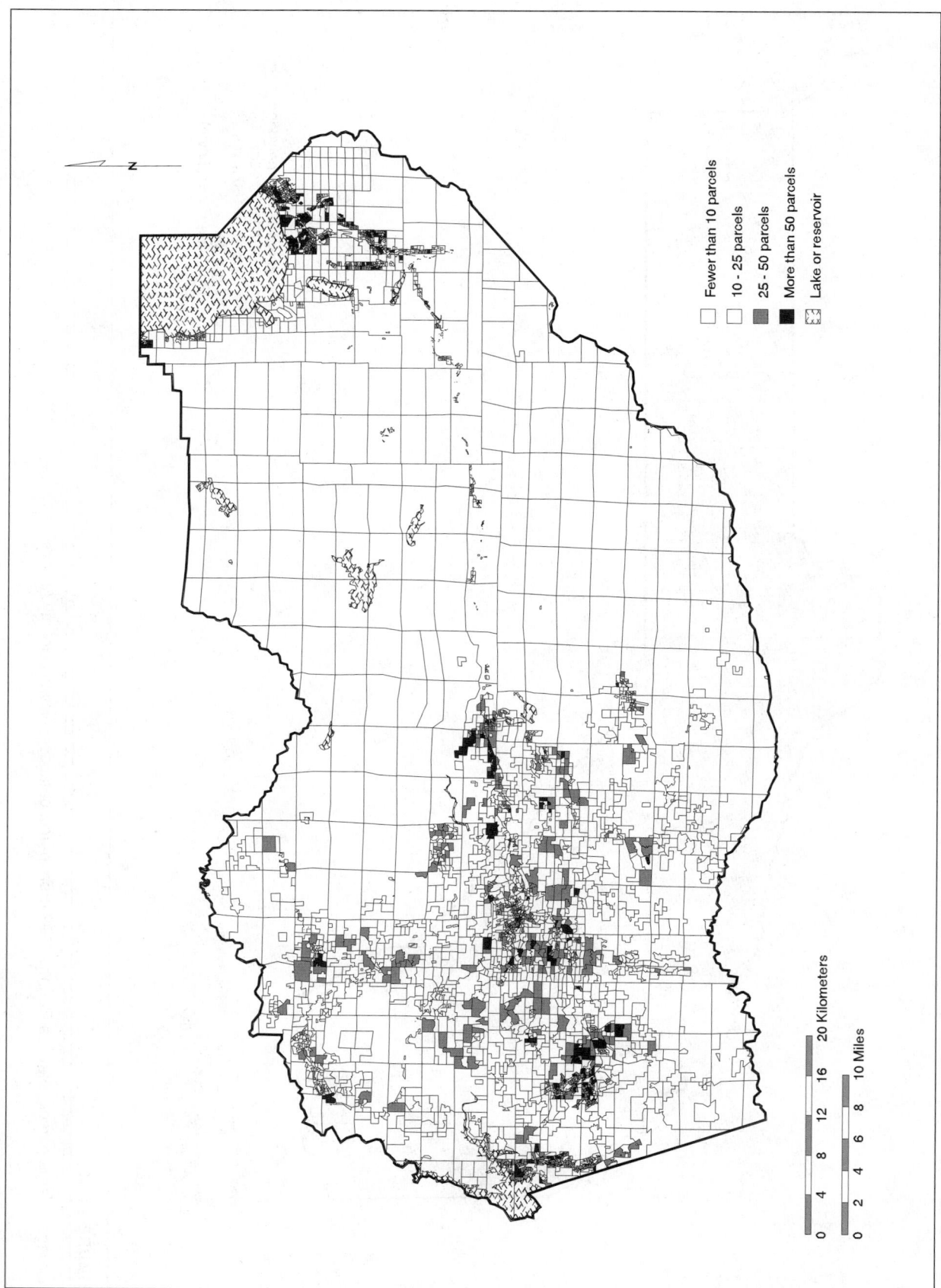

FIGURE 11.41

Concentration of small parcels, El Dorado County, 1992 (from parcel database, El Dorado County Assessor's Office).

Legend:

- Fewer than 10 parcels
- 10 - 25 parcels
- 25 - 50 parcels
- More than 50 parcels
- Lake or reservoir

0 4 8 12 16 20 Kilometers

0 2 4 6 8 10 Miles

N

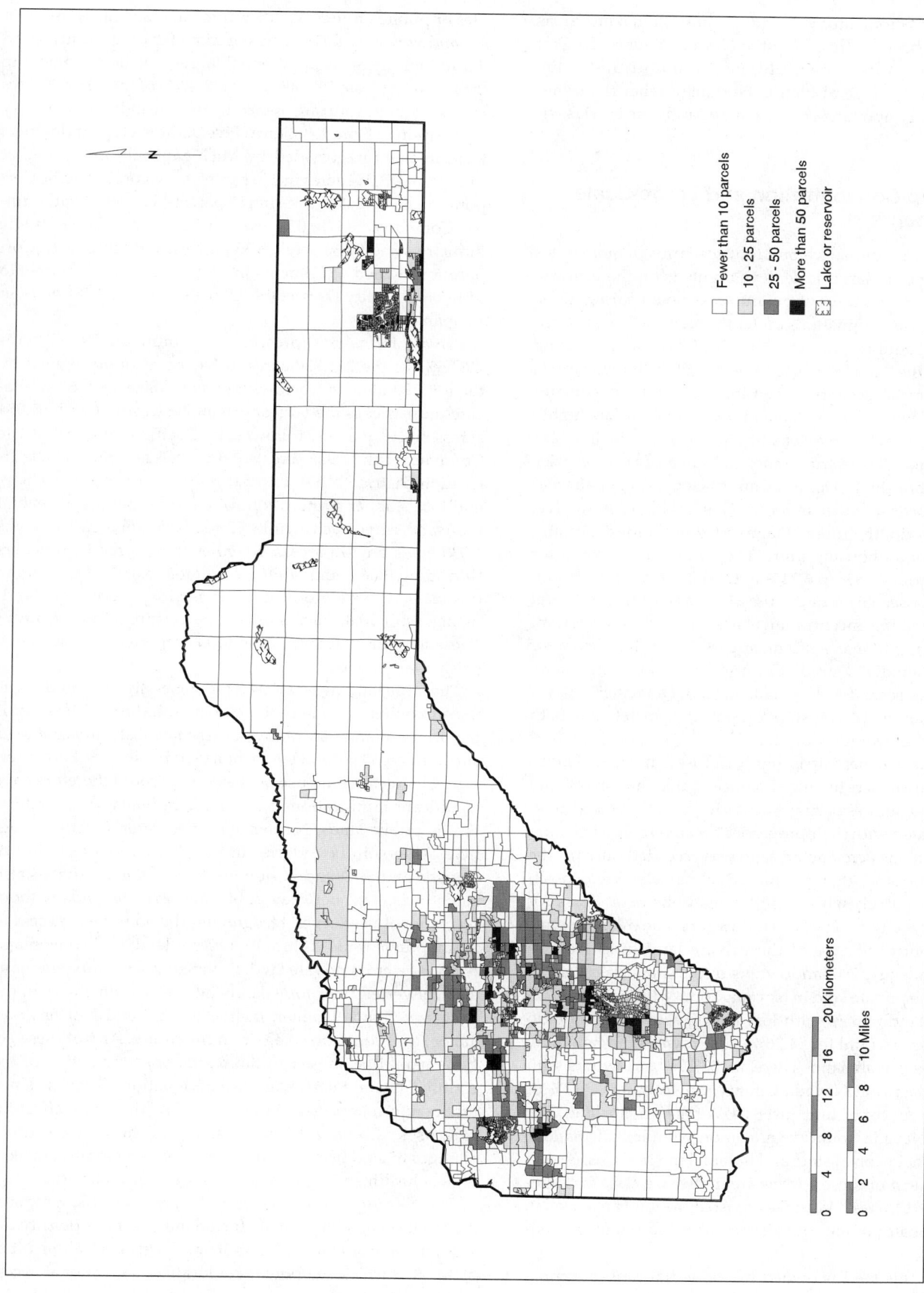

FIGURE 11.42

Concentration of small parcels, Nevada County, 1992 (from parcel database, Nevada County Assessor's Office).

is significant "momentum" for future growth in some areas that has not been addressed in the Nevada County General Plan. We therefore expect ultimate buildout under the Nevada County General Plan to be much higher than forecast based only upon the land use maps and density classifications.

Ownership Concentration and Large-Scale Developments

Our analysis of alternative buildout scenarios demonstrates the importance of large-scale developments on the ultimate number of housing units at buildout. As noted earlier, three special land use designations under the Nevada County General Plan account for anywhere from 21.33% of all housing units under the "low" scenario to 64.20% of the housing units under the "high" scenario. The ultimate buildout estimates under the Nevada County General Plan are therefore highly sensitive to the allowable densities for these special land use classifications. This is true under only one of the scenarios for buildout of the El Dorado County General Plan alternatives, the "low" scenario under the General Plan Alternative. In that scenario, the three "Planned Communities" account for 14.31% of all housing units. They otherwise account for anywhere from 4.68% to 6.71% of total housing units under the other five scenarios under the El Dorado County General Plan. Ownership concentration is nevertheless an important factor affecting human settlement patterns in the region, for a large fraction of El Dorado County's current population resides in large-scale developments (e.g., El Dorado Hills). We therefore examined ownership concentration in detail for both Nevada and El Dorado Counties.

Private land ownership is highly diffused in terms of numbers of parcel owners but highly concentrated in terms of land area in both counties. Many owners have "no reported acreage" associated with their properties because their parcels are smaller than one acre and acreage was recorded only for the subdivision (rather than the indivdual parcels). We consolidated these parcels with the "less than one acre" class for purposes of analysis. The 300 landowners who own 160 acres or more control 55.89% of the private land in El Dorado County, while just 290 landowners in the same class control 53.00% of the private land in Nevada County. In contrast, there are 66,159 landowners with less than 5 acres in El Dorado County. They account for 84.26% of the owners but hold only 6.06% of the private land. There are 35,121 landowners with less than 5 acres in Nevada County, constituting 79.13% of the owners but controlling just 8.04% of the private land. The figures for Nevada County require more complex calculations due to the inclusion of state and federal land in the assessor's database. Most of these parcels are in the size class of "owners of over 640 acres." We have adjusted the total area used in the denominator of the spreadsheet to exclude state and federal lands.

Much of this land is owned by industrial timber compa-nies or public utilities. We therefore analyzed the records for all landowners with 160 acres or more for each county in order to identify areas of potential large-scale development in the future. At least 130,500 acres (42.48%) of the 307,218 acres owned by this group are owned by private industrial forestry concerns in El Dorado County. Five of the six top landowners were in this category, led by Michigan-California Lumber Company (73,254 acres in 276 parcels).[73] Georgia-Pacific Corporation held 32,038 acres in 157 parcels, Wetsel-Oviatt Lumber Company had 10,054 acres in 82 parcels, Sierra Pacific Industries had 9,081 acres in 55 parcels, and Fibreboard Corporation owned 6,073 acres in 35 parcels. The Sacramento Municipal Utility District (SMUD) also owned 3,384 acres in 105 parcels.

Private industrial forestlands account for 59,773 acres (26.85%) of the 222,580 acres in Nevada County owned by the top 290 owners (each with at least 160 acres). Sierra Pacific Industries is the largest private landowner, with 55,054 acres on 219 parcels, followed by Pacific Gas and Electric Company with 11,926 acres on 91 parcels and the Nevada Irrigation District (NID) with 8,098 acres on 149 parcels. Several local timber owners control large properties of several thousand acres each, while Fibreboard Corporation owns 1,035 acres on 10 parcels. Together these large landowners (including PG&E and NID) account for 35.85% of the land in this class. The state of California is also a very large landowner, with 10,835 acres on 279 parcels in Nevada County. These are not included in the totals reported here for large private landowners.

The large landowners described generally own land in the mixed conifer zone and above, where industrial forestry is the primary land use. Many of these lands are adjacent to or surrounded by federal lands managed by the U.S. Forest Service. Changes in land use on these lands could therefore have significant ramifications for ecosystem management efforts on the public lands. Neither wildlife, water, nor fire recognize ownership boundaries in the checkerboard pattern of Nevada and El Dorado Counties. Recreational activities also sometimes cross between public and private lands without recreational users even recognizing the boundary. Access to many public trails is through private land, and some major trails even cross private land. Private timber companies have also converted their forestlands into recreational uses in the past (e.g., Tahoe Donner, Incline Village, and Northstar-at-Tahoe) and could do so again in the future. Any of these actions would be subject to the regulatory jurisdiction of the county General Plans. Land use designations for these lands are therefore important in terms of how they may affect future uses of adjacent public lands and how those future changes in land use may affect the social, economic, and ecological health and sustainability of the Sierra Nevada.

The Nevada County General Plan, for example, proposes designation of some private forestlands for high-density development near Donner Summit and Castle Peak along Interstate 80. Such development could affect recreational access and recreational demand if developed to the allowable density requested by the landowner in the update process. Many other forestlands could be classified to allow one housing unit on every 40 acres, which could lead to subdivision of industrial forestlands and conversion to recreational or residential uses. The presence of recreational residences on these lands could in turn affect industrial forestry operations on adjacent private and public lands. Local government action to allow increased development under the General Plan also tends to increase the market value of land, which could create a windfall for some private landowners if public acquisition of those lands is pursued later by the state or federal government (even if such efforts have been ongoing before the new land use classification). This could significantly increase the cost of land trades and land acquisition by public agencies attempting to rationalize land ownership in order to improve land and resource management in the Sierra Nevada. These efforts usually do not involve condemnation, and there are sometimes specific restrictions against the use of condemnation by public agencies (e.g., to acquire public land in a wild and scenic river corridor if more than 50% of the corridor is already owned by public agencies) (South Yuba River Citizens League 1993). The new land use designations under the General Plan would still increase the costs of land for the public agencies.

Despite these possible consequences from General Plan designations for large owners in the mixed conifer zone, however, development activity for human settlement is most likely to be concentrated "below the green line" on private lands in the west-side ponderosa pine and below in the foothills. Figure 11.43 shows a summary of the distribution of land ownership by parcel frequency and area for both counties for those areas west of the "green line" or national forest boundary in each of the counties.

Parcel size alone does not capture the potential size of future developments, however, for clustered ownerships of adjacent parcels can result in large-scale development opportunities. We therefore used the assessor's data to identify clusters of adjacent parcels with a single owner that were likely to become the site of large-scale human settlement. These sites cannot be identified based on parcel size alone, for they often involve many smaller parcels that have been aggregated to accommodate a larger development. The largest private landowner in El Dorado County that is not from the timber industry is Cook Ranch Partners of Rancho Cordova, for example, developers of the Cinnabar Ranch site south of Placerville (7,771 acres on 27 parcels). The owners of this project are primarily foreign investors who have proposed 569 housing units on 4,975 acres of the site. This represents a significant reduction in density from previous proposals, but the site design involves extensive land conversion to accommodate human settlement on a network of 5-acre parcels. There is no higher-density development on the site, but the cumulative effects of the project are still likely to be significant (Fugro-McClelland [West] 1994; McKuen 1994). Lower densities do not necessarily result in lower impacts—an important point to consider when evaluating future patterns of human settlement for the Sierra Nevada. The project averages only 46.86 housing units per square mile for the entire site and 73.20 housing units per square mile for the 4,975 acres of developed land. In contrast, the "Planned Community" designations under the two El Dorado County General Plan alternatives would have average densities of 397 to 2,624 units per square mile. They would therefore be able to accommodate the same population as the Cinnabar Ranch project on a range of 139 to 917 acres. This would take only 3–18% of the land area that will be developed by the Cinnabar design.

In Nevada County, the largest private landowner not from the timber industry is Gold Country Ranch, Inc. The land-

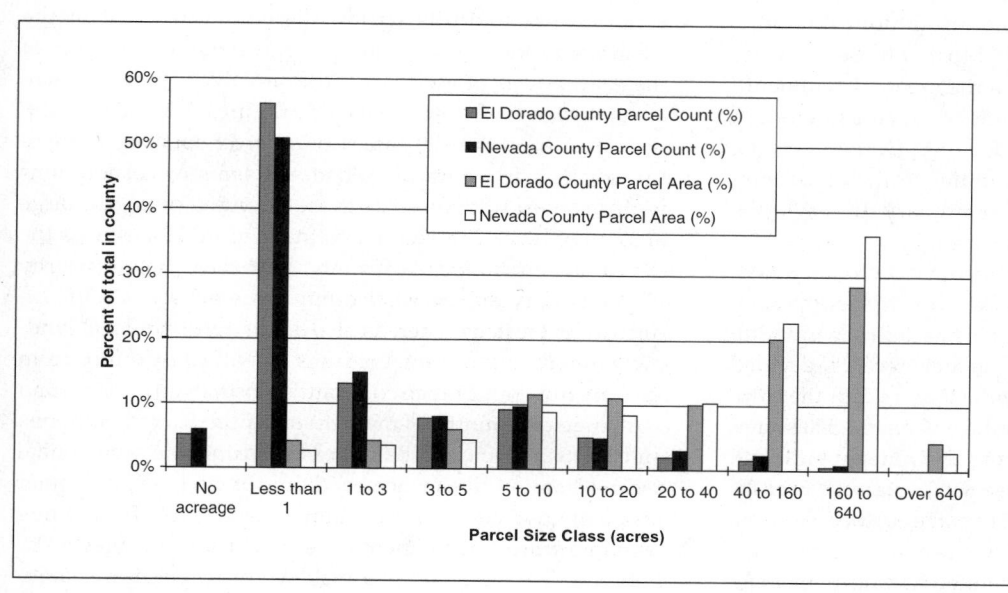

FIGURE 11.43

Number of parcels versus area by parcel size class for Nevada and El Dorado Counties "below the green line" (west of the national forests).

holdings of Gold Country Ranch cover 8,232 acres in 24 adjacent parcels. The corporation is based in Rocklin and includes several large local landowners with adjacent parcels. This project proposes a more compact development and a much higher level of total housing units and population than the Cinnabar Ranch project. Unlike the Cinnabar Ranch, however, the Gold Country Ranch project has not yet prepared an environmental impact report (EIR). It is therefore difficult to evaluate the potential impacts of the project in the absence of more specific information on the design. We therefore considered a range of future buildout scenarios for this site in our estimation of potential future housing units under the Nevada County General Plan. These buildout scenarios assume average development densities of anywhere from 397 to 2,624 units per square mile, consistent with those used for the "Planned Community" designation in the El Dorado County General Plan alternatives. Our "middle" scenario assumes an average density of 1,841 units per square mile (2.88 units per acre). These densities are considerably higher than those now expected for the Cinnabar Ranch project. Given recent changes in real estate markets, the Gold Country Ranch project proposal is likely to lie dormant for the next few years before coming forward under the new Nevada County General Plan with a more detailed site-specific proposal.[74]

This has also happened with several other "new town" proposals in nearby Yuba, Sutter, Placer, and El Dorado Counties. Only the Stanford Ranch, which broke ground in 1987, is continuing to develop at the rate and on the scale of these proposed "new town" developments. Those projects that have not yet started actual development are now delaying the significant up-front investments in infrastructure required to develop such large-scale projects. Stanford Ranch has already made those investments, so it has a strong incentive to continue the development process in order to allocate those costs to as many homeowners as possible. There are now over 2,000 homes at Stanford Ranch, with about 8,000 homes forecast to be developed on the 3,500-acre site at buildout (Stanford Ranch Information Center 1994–95). Most of those residents are expected to commute to work in the greater Sacramento metropolitan area, which now includes Roseville in western Placer County and Folsom at the edge of El Dorado County. Expansion of employment opportunities in the electronics industry and in state government are primary drivers of the growing demand for housing in this area.

Retirees are also moving to the western Sierra Nevada foothills and this portion of the greater Sacramento metropolitan area. The Del Webb corporation, which has built several "Sun City" retirement communities from scratch, recently decided to locate its first Sun City in northern California in the community of Roseville. Sun City Roseville will have 3,500 homes for an estimated 6,000 residents age 55 and older at buildout. The project was first formerly proposed in January 1993, broke ground in February 1994, and sold 629 homes in the first seven months. Unlike at Stanford Ranch, where a dozen individual developers are building homes with different price ranges and styles within the overall development, the Del Webb corporation handles every aspect of its development. Its first Sun City, near Phoenix, was started in 1960 and now has 26,000 homes. The company is now starting construction on 4 new homes every day at the Roseville site (Grass Valley Union 1995c). The residents are expected to come primarily from elsewhere in California.

Each of these developments in western Placer County represents a very different pattern of development than that which has dominated human settlement in the Sierra Nevada to date. The higher incomes of their target population represent the same sociodemographic group that has emerged as the driving force in the real estate markets of Nevada and El Dorado Counties, however, and the higher housing costs they can afford have allowed high levels of investment in centralized infrastructure. This in turn has allowed development at higher densities than those typically found in the Sierra Nevada. These developments therefore represent one important model for human settlement in the future for a significant fraction of the Sierra Nevada population. As in the private, unincorporated communities of Lake Wildwood, Lake of the Pines, and El Dorado Hills, each home owner's land use is controlled more by codes, covenants, and restrictions (CC&Rs) in the deed than by local land use regulations. Infrastructure has also been privatized, with special assessment districts that tap the development site for special property taxes that go only toward infrastructure that serves that site. The result is the effective privatization of many functions that would normally be handled by local governments. This privatization has enormous implications for the future of land use planning and infrastructure investment throughout the Sierra Nevada where these types of developments take place (Egan 1995).

The emergence of large-scale development proposals in Nevada and El Dorado Counties suggests that the land development process could be entering a new phase in the Sierra Nevada foothills within the commute orbit of the Sacramento metropolitan area (Hoge 1995). One aspect of these large-scale projects is that they involve significant ownership by landowners from outside the area. We used the ZIP code data from the 1992 assessor's parcel database to identify counties in California where residents or corporations with primary addresses in those counties owned a large amount of land within either county. Figure 11.44 shows the distribution for Nevada County, with clear ties to nearby Placer County and Shasta County (where Sierra Pacific Industries is located). There is also a pattern of nonlocal landowners with significant holdings in Nevada County from Sacramento, San Francisco, Contra Costa, Santa Clara, and Los Angeles Counties. Landowners in the Sacramento area and the San Francisco Bay Area could include recreational second-home owners who spend time in the Truckee-Donner area, but Los Angeles–area landowners are probably holding land for future development. This conclusion is suggested by both the size of the landholdings by Los Angeles–area landowners and the relative distance between Los Angeles and the land in Nevada County.

El Dorado County records show a similar pattern, although there is a higher concentration of ownership in adjacent Placer, Amador, and Sacramento Counties. There are also large land-holdings among residents and corporations based in Solano, Contra Costa, Alameda, Santa Clara, San Mateo, and Los Angeles Counties. These include both second-home owners in the Lake Tahoe region and potential land developers. Timber industry ownership shows up for Shasta (SPI), Amador (Georgia-Pacific), and Tuolumne (Fibreboard) Counties. Figure 11.45 shows the coverage for El Dorado County.

Similar analysis for the five-county central Sierra Nevada region (the only counties for which we have ownership records) show high levels of ownership by residents of the state of Nevada. Ownership by foreign interests and owners in other states was not significant. These data represent county locations for only the recorded address ZIP code in the assessor's database, however, and do not record the location of owners of partnerships or corporations with a single address. These could include significant foreign investors, such as those in the Cinnabar Ranch project. These data are therefore only a first-order measure of the spatial pattern of land-ownership in each of the counties. They do highlight large ownership in both the San Francisco Bay Area and the Los Angeles area, however, which suggests that there are significant "communities of interest" in those areas that may not be represented among the residents of each of the counties.

Incremental Development and Minor Subdivisions

The potential for large-scale developments is certainly concentrated among those landowners with large amounts of contiguous land, but the most significant effects of development under the proposed General Plans could come through the incremental development of existing parcels and minor subdivision of those existing parcels. These activities do not trigger significant environmental review under CEQA and are in most cases ministerial actions that do not require approval from the local land use authority. The existing pattern of parcelization among parcels smaller than 160 acres may therefore dominate future patterns of human settlement under the General Plans without any systematic opportunities in the future to mitigate the impacts of that development.

The existing parcels in both Nevada and El Dorado Counties are overwhelmingly small "below the green line" west of the national forest boundary. Our analysis indicates that nearly all of the parcels in the database that do not have a specific acreage associated with them are less than 1 acre and are within higher-density subdivisions. Parcels less than 1 acre in size therefore account for 61.21% of the parcels but only 4.10% of the land area in western El Dorado County below the green line. This smallest parcel class accounts for 56.75% of the parcels but only 2.59% of the land area in western Ne-

vada County outside the national forest boundary. Parcels 1–10 acres in size constitute about one-third of the total parcels in both counties (30.05% in El Dorado County and 32.51% in Nevada County), but only 21.71% of the land area in El Dorado County and 16.73% of the land area in Nevada County. These parcels are all generally on septic systems and typically get their domestic water supply from an on-site well, but many of them (especially those greater than 5 acres) could be subdivided further into several parcels that still meet the local health department's minimum lot size requirements for on-site well water with a septic system. There were 23,991 parcels in El Dorado County and 18,383 parcels in Nevada County 1–10 acres in size in 1992. In area, they accounted for 97,663 acres in El Dorado County and 74,909 acres in Nevada County.

Slightly larger parcels of 10 to 40 acres account for nearly a comparable amount of total land area (21.17% in El Dorado County and 18.77% in Nevada County) but are far fewer in number (only 6.75% of the total parcels in El Dorado County and 7.65% in Nevada County). These larger parcels could be subdivided once into two or three parcels (a "minor" subdivision), sold to several buyers while the original owner retains a home on one of the lots, then subsequently subdivided again by the purchasers of the smaller lots. In this way a 40 acre parcel may become four 10 acre parcels before each of these is ultimately split again into several parcels of 3–5 acres. What may be a single housing unit today on 40 acres could therefore easily turn into eight to thirteen housing units on septic systems and on-site well water. Multiplied across the landscape, this pattern of incremental development and minor subdivisions can increase human settlement considerably. There were 5,389 parcels in El Dorado County and 4,328 parcels in Nevada County in the 10–40 acre size class in 1992. In area, they accounted for 95,203 acres in El Dorado County and 84,018 acres in Nevada County. Combined with the totals in the 1–10 acre size class, there are therefore 192,866 acres in El Dorado County and 158,927 acres in Nevada County in parcels that are between 1 and 40 acres. All but the smallest of these parcels could potentially be subdivided through minor subdivisions and result in significant cumulative increases in housing units without the extension of sewer and/or public water infrastructure.

The final size class, which we have not yet discussed, is between 40 and 160 acres. This size class accounts for only 1,188 parcels in El Dorado County (1.49%) and 1,287 parcels in Nevada County (2.28%), but these parcels cover 91,293 acres in El Dorado County (20.30%) and 101,860 acres in Nevada County (22.75%). The mean size of these parcels was 76.8 acres in El Dorado County and 79.1 acres in Nevada County. These larger parcels could also be subdivided into four or fewer parcels without either approval from the land use authority or environmental reviews, with the same ultimate sequence of subsequent subdivisions. These larger parcels are generally more rural and remote from services, however, and they are less likely to face conversion pressures in the near term.

Specific land use designations in the General Plans could

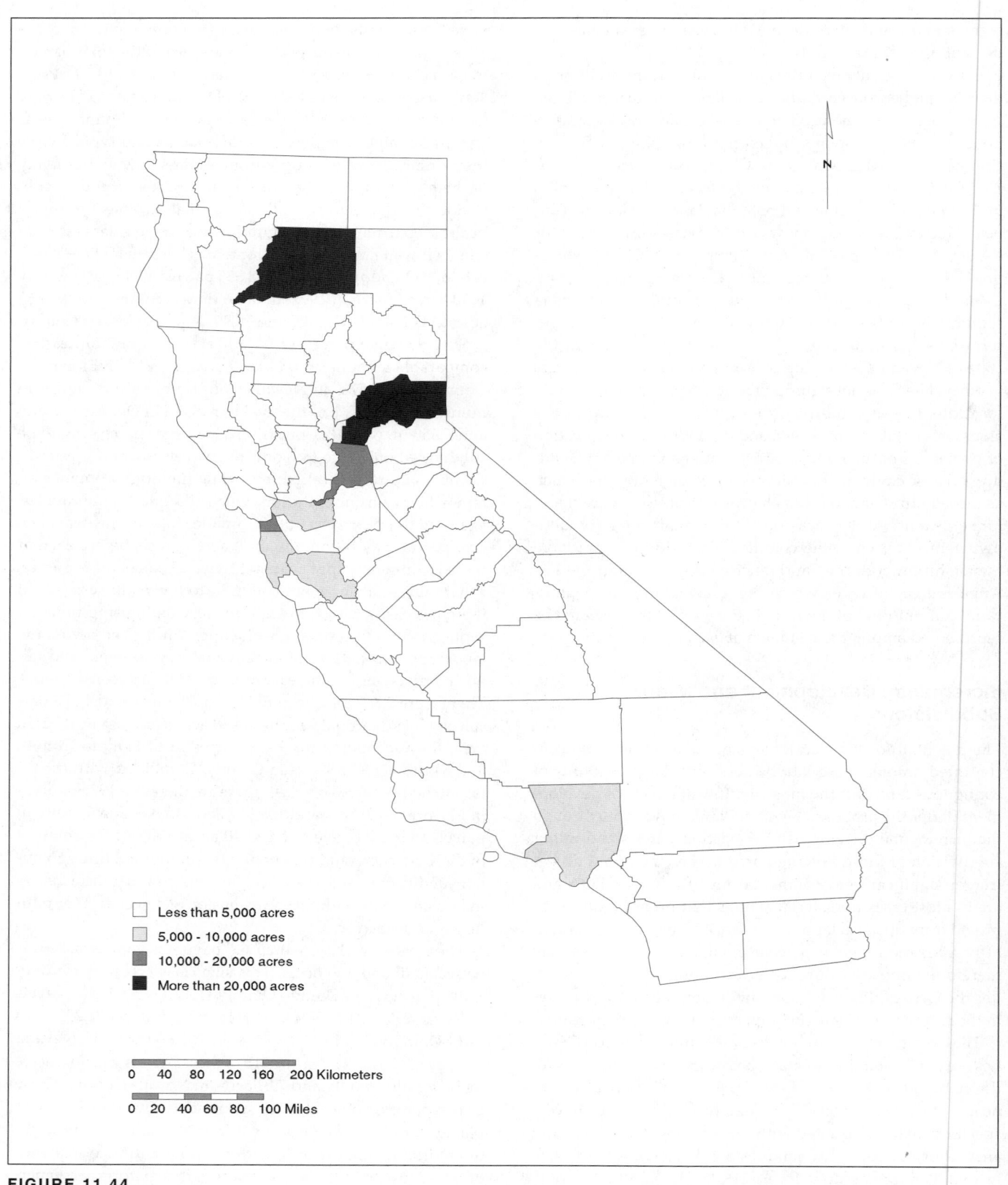

FIGURE 11.44

Distribution by county of residence for owners of land in Nevada County (from parcel database, Nevada County Assessor's Office).

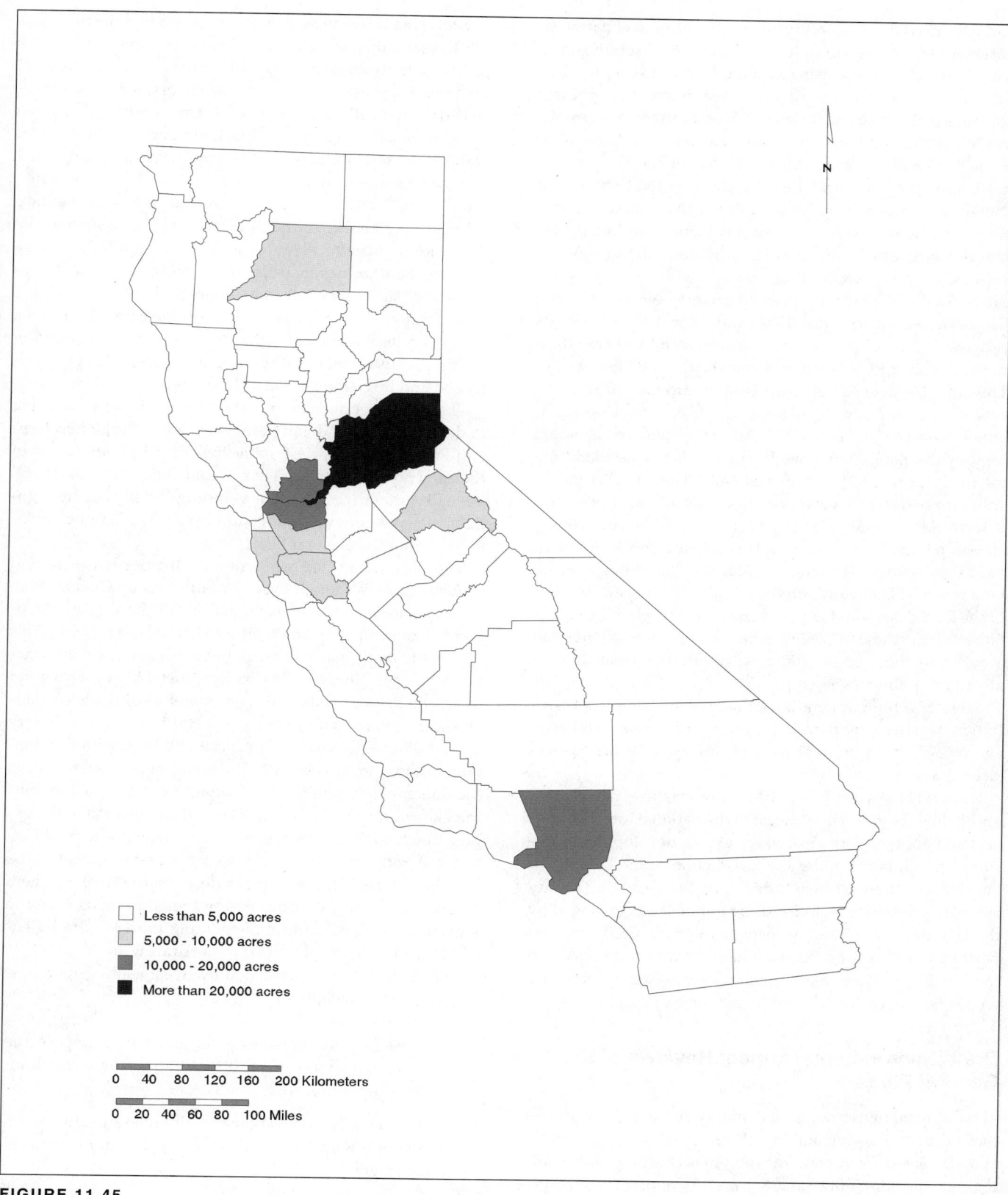

FIGURE 11.45

Distribution by county of residence for owners of land in El Dorado County (from parcel database, El Dorado County Assessor's Office).

directly affect the feasibility of significant parcelization of these larger parcels through sequential minor subdivisions. The Project Description version of the El Dorado County General Plan has a minimum parcel size requirement of 160 acres for lands in the "Natural Resource" classification, for example, while the General Plan Alternative has only a 40 acre minimum for the same classification. Subdividing a 160-acre parcel under the "Natural Resource" classification would therefore be much more difficult under the Project Description, for it would require a General Plan amendment. The landowner could easily complete a minor subdivision of a 160 acre parcel into four 40 acre parcels without needing approval under the General Plan alternative. Nevada County has designations for "Rural 40" and "Rural 160" under its proposed General Plan, with additional land use classifications of "Forest 40," "Forest 80," Forest 160," and "Forest 640." The same 160 acre parcel could be split into four 40 acre parcels without any additional approvals if it were classified as either "Rural 40" or "Forest 40," but such a subdivision would require a General Plan amendment if it were classified "Rural 160," "Forest 160," or "Forest 640." The parcel could be split into two 80 acre parcels without additional approvals if it were classified as "Forest 80." The draft Nevada County General Plan proposed classification of 36,958 acres as "Rural 40" and 34,914 acres as "Forest 40," and much of this land is probably already subdivided into 40 acre parcels. Most of the land designated in the overall "Forest" classification is either "Forest 160" (279,583 acres) or "Forest 640" (20,481 acres), but most of that land is "above the green line"— already in public ownership.

Table 11.A10 in appendix 11.1 shows the complete distribution of parcels by parcel size class (rather than total acreage owned by individual owners) for each of these parcel size classes.

Remember that table 11.A10 is only for those parcels that are "below the green line" (west of the national forest boundary) and are therefore most likely to face development pressure. Federal lands under the jurisdiction of the U.S. Forest Service are therefore excluded, but the Nevada County assessor's database includes some public lands managed by the U.S. Bureau of Land Management or the California Department of Parks and Recreation. Most of these lands are in the South Yuba River region with relatively steep slopes and would not generally face significant development pressure.

Draft Environmental Impact Reviews of the General Plans

The California Environmental Quality Act (CEQA) requires that a draft environmental impact report (DEIR) be prepared and circulated for review for the Final Draft General Plans for both Nevada County and El Dorado County. These DEIRs were circulated in early 1995, and each county held public hearings to take comment on the DEIRs, which are intended to describe the potential environmental impacts of the General Plans and to mitigate those impacts where feasible. Comments on the DEIRs were then incorporated into the final EIRs (FEIRs) by either the EIR consultants (in Nevada County's case) or county staff (in El Dorado County's case). The local planning commission and board of supervisors then evaluated the final EIRs and modified them to reflect their own independent findings regarding the impacts of the General Plan on the environment. CEQA requires that "significant" environmental effects be reduced to "less-than-significant" through mitigation measures. Significance has been established through either statute, CEQA guidelines issued by the Governor's Office of Planning and Research (OPR), or case law. Local authorities may also make findings that some impacts are "significant but unavoidable" due to "overriding considerations" that make mitigation measures infeasible. Both counties made these kinds of findings when reviewing their respective final EIRs, but those findings have been challenged and litigation has been threatened by local groups.[75]

These are large and complex documents, so we are not able to describe them in detail here. For papers on the two General Plan draft EIRs, describing the critical points found in the longer documents, see Thomas and Duane (1995a, 1995b). We will summarize the primary findings briefly here to highlight the types of environmental impacts that can be anticipated under the General Plans.

The DEIRs identified some common themes across the two General Plans, although the El Dorado County General Plan DEIR is more comprehensive and more thoroughly documented than the Nevada County General Plan DEIR. This appears to be primarily a result of both inadequate information supporting the Nevada County General Plan analysis and its greater degree of generality. It is therefore difficult to evaluate the potential environmental impacts of the Nevada County General Plan with reliability or specificity. Even with that limitation, however, it is clear that development under both of the General Plans will result in "significant and unavoidable" impacts on the environment. The critical question is the degree to which those impacts will be mitigated under CEQA. Because both counties' final EIRs were under appeal at the time of our analysis, we cannot describe the final EIR here with any certainty. Our discussion therefore focuses on the impacts associated with implementing the 1994 draft General Plans that were identified in the draft EIRs.

The draft EIRs are dominated by five general impacts associated with development:

1. Decreased wildlife habitat and recreational open space due to the conversion of land to residential, commercial, industrial, and public uses

2. Increased traffic congestion and air emissions due to development, with increased traffic delays and decreased traffic safety

3. Increased water quality problems associated with both point-source wastewater treatment facilities and non-point-source septic systems

4. Increased fire risk and safety hazards associated with settlement in the urban-wildland intermix, including emergency escape on substandard roads

5. Increased shortfalls in the capacity of local governments to provide services, including education, public safety, and parks and recreation

In addition, El Dorado County could face serious limitations in domestic water supply due to increased demand associated with development under the General Plan. Anticipation of this problem led the El Dorado County Water Agency to initiate a study of water supply and demand in 1994 that identified potential sources of future supply. This study analyzed supply and demand for all of the water suppliers in the county, which are dominated by the El Dorado Irrigation District (EID). In contrast, the Nevada Irrigation District (NID) is one of the few water districts in the state that has ample supply capacity to handle the significant growth anticipated under the Nevada County General Plan, in part due to a large fraction of agricultural users who pay relatively low prices for their water. NID therefore has the potential for significant shifts between agricultural and domestic uses. EID is more restricted in this regard, and it is negotiating to acquire water rights for the Silver Fork of the South Fork of the American River from Pacific Gas and Electric Company (PG&E) (El Dorado County Water Agency 1994; Doolittle 1994–95). Alpine and Amador Counties have opposed the water rights request, however, unless it includes guarantees to maintain lake levels at Caples and Silver Lakes. Resolution of the water supply issue is therefore dependent on resolution of that water rights controversy (Doolittle 1994–95; Brissenden 1993–1994; Center 1991–95).

Many parcels currently rely on ground water for domestic water supply, which is inconsistent and highly variable in the fractured bedrock geology of the Sierra Nevada (Swain 1994). There are very few true aquifers in Nevada and El Dorado Counties, so water supply is difficult to predict without site-specific well drilling and analysis. Limitations in groundwater availability could therefore limit future development at buildout under the General Plans. Development under the General Plans could also potentially affect existing supplies, necessitating significant expansion of the treated water supply distribution system to more remote locations currently served by wells. Due to the need to complete site-specific analysis, however, neither of the General Plans has analyzed the availability of ground water to supply the water necessary to accommodate growth. It is therefore unclear how much of the buildout population's demand could be met through on-site well water.

Land Conversion

The scale of land conversion anticipated under the General Plans is astounding. When compared with the land area in each of the density classes reported at the census block level for the 1990 census, we estimate that the Nevada County General Plan would increase the area dedicated to human settlement at an average density of one unit per acre or greater from 12 mi^2 in 1990 to 30.37 mi^2 at buildout (146%). The land area settled at an average density of 1–4 acres per unit would nearly double from 30 mi^2 in 1990 to 59 mi^2, while the 4–8 acre class would increase 76%, from 40 mi^2 to 71 mi^2. Using our most conservative estimate of 16 acres per housing unit as a threshold (forty housing units per square mile), the total land area subject to human settlement will increase from 143.43 mi^2 in 1990 to 228.37 mi^2 (a 59% increase). Another 11.0 mi^2 are dedicated to commercial, industrial, and public uses. The total land area with less than ten housing units per square mile will drop only from 662 to 528 mi^2, but more than half of that land area will move from being unsettled to being lightly settled.

El Dorado County's General Plan is similar. Once again, the density class of just two to five housing units per square mile increases the most (396%) under the General Plan Alternative. The density class of ten to twenty housing units per square mile increases the most (608%) under the Project Description. The highest-density class (one or more units per acre) roughly doubles under each alternative, with an overall increase of land settled with at least four housing units per square mile from 203.1 mi^2 in 1990 to 563.7 mi^2 (278%) under the General Plan Alternative. The area affected by at least this level of human settlement would increase to only 294.2 mi^2 (45%) under the Project Description, however, for it has greater increases in lower-density classes. The land area settled at an average density of ten to twenty housing units per square mile increases by 933% under the Project Description, from 153.4 mi^2 to 1,086.5 mi^2. The El Dorado County General Plan Alternative also has 8.6 mi^2 dedicated to commercial, industrial, and public uses, while the Project Description commits 13.4 mi^2 of land to those designations. There are also some lands with unknown densities. Tables 11.A11–11.A13 in appendix 11.1 summarize the changes in land area under each of the General Plans compared with 1990 settlement densities.

Transportation

Accommodating this level of land use change requires significant investments in new infrastructure, and the transportation sector is the one that will experience some of the greatest impacts. The El Dorado County General Plan anticipates a need for transportation improvements that would cost between $800 million and $1 billion over the twenty years (Thomas 1995, 1994). U.S. Highway 50 and local Highway E16 would need to be widened to six lanes each to handle increased traffic associated with development under the General Plan (Thomas 1994). This analysis of the El Dorado County General Plan reflects consistent and comprehensive consideration of land use changes within transportation analysis zones (TAZs) that comprise census block groups for consistent social, demographic, and economic information (Rivas 1994). Though we have not evaluated the modeling efforts used to derive the estimates, we believe they are a reasonable projection of future transportation impacts and the need for additional facilities.[76] The funding needs under the General Plan are well beyond current revenue projections, however, making many of these infrastructure investments unlikely. Further degradation of level of service (LOS) standards and significant air-quality impacts associated with a highly congested transportation system can therefore be expected in the absence of those investments.[77]

Nearby Placer County's General Plan EIR, which was released in late 1993 for a General Plan that was then adopted in August 1994, anticipated similar problems along Interstate 80. "By its own analysis," noted one newspaper article on the EIR, development under the General Plan would "induce a 20-mile-long traffic jam on Interstate 80 each weekday rush hour between Citrus Heights and Auburn" (Bowman 1994). The article also noted that "as it is now, I-80 during weekday commuting hours is generally jammed for about a two-mile stretch" (Bowman 1994). Development under the Placer County General Plan would therefore increase the highway mileage of congestion tenfold, since vehicle trips and vehicle miles traveled (VMT) were both projected to increase at a much higher rate than population. That increased congestion could in turn affect commuters from Nevada County to the greater Sacramento metropolitan area. Following the historic pattern of metropolitan deconcentration, continuing traffic congestion could then accelerate the relocation of employment opportunities to the metropolitan fringe (in order to avoid the congestion costs associated with commuting to Sacramento itself). This process of metropolitan expansion and deconcentration could then put portions of western Nevada County within commuting distance of jobs in Placer County. A similar phenomenon has already had some effect on the shift of employment to Roseville and Folsom, which are both beyond the areas of I-80 and U.S. 50 that experience daily congestion now.

Congestion within Nevada County is much more difficult to ascertain from the draft EIR. The 1994 Regional Transportation Plan (RTP) was not released until May 1995, well after the draft EIR and final EIR had been reviewed and debated by the planning commission. Transportation modeling for the Nevada County General Plan was also based upon buildout densities derived from the General Plan land use classifications, failing to account for existing parcelization and the potential for much higher buildout in some areas. Transportation modeling depends upon spatially explicit analysis of origin and destination linkages through assumed trip patterns, so it is nearly impossible to complete a reliable model of the transportation system under buildout without spatially explicit estimates of trip generation and travel patterns under buildout. Even without reliable spatially explicit data, however, the 1994 RTP anticipates significant funding short-

falls over the next twenty years. Short-term needs of $72 million, intermediate-term" needs of $84 million, and long-term needs of $54 million total $209.2 million (Nevada County Transportation Commission 1995a). As the RTP notes, however, "The regional travel demand model is not designed to analyze improvements for intersections" (Nevada County Transportation Commission 1995b). The impact of development upon intersections—which are a primary determinant of LOS and congestion—have therefore not been fully considered in the needs identification process. The $209.2 million estimate cited is therefore probably low, and only $72.5 million in likely revenues have been identified to cover those costs. The $136.7 million shortfall is nearly two-thirds the anticipated need even without additional costs for intersection improvements.

Future development under the Nevada County General Plan is therefore likely to result in significant degradation of LOS for most of the roads in the county. The 1994 RTP analyzes daily LOS standards for all of the major highways and arterials in the county both "with" and "without" identified improvements, and many of the roads have LOS ratings of F (the lowest possible) without the improvements (Nevada County Transportation Commission 1995c). Even with the improvements, however, many roads retain LOS ratings of C or D. These improvements often involve expansion of two-lane roads to four lanes, which increases road capacity and speeds. Highway 49 from Alta Sierra Drive to McKnight Way does not rise above an F rating even *with* improvements and expansion to four lanes.

Due to the uncertainty about funding for improvements, the air quality impacts associated with buildout under both the Nevada and El Dorado County General Plans is likely to be greater than that anticipated in the DEIRs. Increased congestion, particularly following "cold starts" by commuters, is likely to result in significant increases in hydrocarbons, nitrogen oxides, and carbon monoxide (especially at intersections). El Dorado County is part of the Sacramento Metropolitan Air Quality District, which is a nonattainment area and therefore subject to greater regulatory oversight. Nevada County is part of the Northern Sierra Air Quality District, however, and receives less scrutiny under both the federal Clean Air Act of 1990 and the state Clean Air Act of 1988. Modifications to air-quality regulations and/or the boundaries of the districts (especially since a significant part of Nevada County's locally generated emissions appears to be due to commuting) could result in future constraints on land use due to "indirect source" air-quality impacts.

Water Quality

Development under the General Plans must address disposal of liquid wastes through either centralized sewage treatment or on-site septic systems. As discussed earlier, the economics of infrastructure investment have led to reliance on sewage treatment only when settlement densities are high enough to allocate the high fixed costs across many users. Some of the existing sewage collection systems in Nevada and El Dorado Counties date from the nineteenth century, and most of the wastewater treatment plants (WWTPs) in the area were constructed since passage of the federal Clean Water Act in 1972. Federal and state grants financed the first round of projects throughout the 1970s, then federal and state funding sources shifted to low-interest loans rather than grants in the 1980s. Even these loan funds are now diminishing in the face of significant state and federal budgetary contractions, which could force future WWTP investments (and maintenance of depreciating existing systems) to be sustained by WWTP users. All wastewater dischargers must acquire a permit from the Regional Water Quality Control Board (RWQB) of the state of California under the National Pollution Discharge Elimination System (NPDES), which was established under the Clean Water Act in 1972 and is technically administered by the federal Environmental Protection Agency (EPA). California's Porter-Cologne Act of 1970 also establishes receiving water standards, an approach that was adopted in part by the federal Water Quality Act of 1987 (Richardson 1992–94). Section 319 of the 1987 act also establishes stricter requirements for the use of "best management practices" (BMPs) for the control of non-point-source (NPS) pollution (Thompson 1989). Possible NPS sources associated with the General Plan include erosion and sedimentation from construction activity and surface water contamination from septic systems.

El Dorado County has three major NPDES permits and four minor permits outside the Lake Tahoe Basin: Deer Creek (2.0–2.5 million gallons per day [mgd]), El Dorado Hills (1.0–1.6 mgd), and Hang Town in the city of Placerville (1.0–1.6 mgd) are major dischargers, while the Dunlap Ranch, Sierra Pacific Lumber Company, Wetzel-Oviatt Lumber Company, and the El Dorado Hills Community facilities operate under minor NPDES permits. The major permits require more frequent monitoring and are subject to somewhat greater scrutiny by the RWQCB. Nevada County has five major permits and five minor permits: Donner Summit (0.5–1.0 mgd; capacity to be upgraded to 2.0 mgd), Lake of the Pines (0.7–1.1 mgd), Lake Wildwood (1.1 mgd), Grass Valley (1.7 mgd), and Nevada City (1.3 mgd) are major dischargers, while Penn Valley (0.1 mgd), Cascade Shores (0.025 mgd), North San Juan (0.025 mgd), Mountain Lake (0.015 mgd), and Gold Creek Park (0.015 mgd) operate under minor permits. The last two cases are particularly interesting, because they service private subdivisions that are quite distant from major WWTP facilities. This highlights the potential for higher-density development in areas with limited capacity to handle septic systems even when not contiguous to existing urban areas. Higher land values will probably make this more common in the future.

The allowable flows noted for each of the NPDES permits are for "average" conditions, which are difficult to define and relatively rare in the hydrologic regime of the Sierra Nevada. In particular, many of the older systems suffer from significant "inflow and infiltration" (I & I) problems due to stormwater flows into the WWTP in the winter and spring. These flows often overflow the WWTP and result in raw sewage spills into surface waters. The California Department of Fish and Game (CDFG) is especially concerned about the impacts of WWTP operation on aquatic biota, but enforcement under the Fish and Game Code has been difficult in the absence of adequate monitoring (Lehr 1995). The RWQCB has attempted to deal with the problem through new permit requirements and selected application of "cease and desist" orders limiting additional sewer hookups (CVRWQCB 1989, 1992). In some cases, however, improvements to existing WWTPs have only occurred with state or federal financing. As noted earlier, such financing may be less likely in the future. This could become a serious problem for existing facilities as they become older and less reliable, for the fee increases necessary to renovate facilities could be extremely high. Recent funding for improvements to the Cascade Shores facility (east of Nevada City) are costing over $18,333 per parcel and over $30,000 per existing home. The state is paying $1.7 million of the costs and offering a 20-year loan of $225,000 to get the plant operating. Users will only have to repay the loan. The improvements are unlikely to have occurred without the state grant, and water-quality problems would have continued (Lauer 1995c, 1995b).

Development under the General Plans will probably be associated with some septic system failures on substandard existing parcels, which will lead to increased demands on existing WWTP capacity. These increased demands, together with discharges from new WWTPs designed and built to serve higher-density developments, could have dramatic impacts on hydrologic regimes. Dilution associated with existing natural surface-water flows could be reduced as effluent becomes a larger fraction of overall flows (EIP Associates 1995). This could result in impacts on ecological processes, recreational access, and public health in existing surface waters. Wastewater flows could significantly increase in-stream flows during drought periods of summer and autumn, while sewering existing septic systems could reduce ground-water and surface-water flows in other areas. Potential septic system failures could also result in ground-water contamination and increased demands for potable water supplies for domestic use.[78] Diversions associated with meeting that demand for domestic supply could in turn result in impacts in other watersheds.

Septic system failures have been documented in Nevada and El Dorado Counties since at least 1970 due to the poor site quality of many soils (Davis 1994; Cranmer Engineering and Halatyn 1971). Almost all classified soils in Nevada, El Dorado, and other Sierra Nevada counties have been rated with a "severe" soil limitation rating for standard conventional deep trench septic systems due to shallow depth to bedrock (less than 4 ft), steep slopes (more than 9%), slow soil permeability, rock outcroppings, and/or high shrink-swell potential (Nevada County Planning Department 1994b). The Sierra Nevada also has tremendous soil and topographic variability, however, so septic suitability is not well characterized by the large-scale (1:20,000 or 1:24,000) soil surveys prepared by the Soil Conservation Service (SCS). A more careful overlay of topography and soil types indicates that certain regions are suitable for septic tank systems, particularly if proper maintenance standards and ongoing monitoring are enforced. Higher land values also make alternative systems more feasible, making some "unbuildable lots" suitable for building with alternative systems. Standard systems cost $3,000 to $4,000, but typical installation costs range from $6,000 to $8,000 in Nevada County. Advanced sand filter systems run between $12,000 and $20,000 (Sage 1995). The failure of existing septic systems could therefore necessitate significant additional investment in on-site infrastructure if significant water quality impacts are to be avoided. Unfortunately, the potential for widespread septic system failures has not been well studied in the DEIRs for the General Plans. It is therefore difficult to estimate either the environmental or economic impact of potential septic system failures under the General Plans at buildout. The DEIR for the Nevada County General Plan called for a detailed study of this issue as a mitigation measure, but it was not used to formulate the land use designations in the General Plan itself. The background assessment work simply has not been completed.

This is only a brief overview of the ecological, technical, and economic constraints associated with water quality, water supply, and wastewater disposal issues associated with development in Nevada and El Dorado Counties. These issues are described in more detail in Megatelli and Duane (1995), which summarizes the results of our assessment of these issues.

Fire Safety

One of the most serious but least understood impacts of buildout under the General Plans is the impact on fire safety. Human settlement is associated with fire ignitions and modifies the suppression strategies for wildfire fighting in the urban-wildland intermix zone (Irwin 1987, 1989). Higher-density developments that are dominated by the built environment are less threatened by this impact, but their proximity to wildlands in an "edge" environment could still increase ignition risks (e.g., due to children playing with matches, sparks from motorcycles, etc.). Lower-density developments are both difficult to protect and difficult to evacuate. The presence of structures in the urban-wildland intermix zone alters suppression strategies and complicates sharing of fire-management responsibilities among local, state, and federal agencies. In particular, resources (e.g., firefighters, water, and equipment) are often allocated to the protection of individual structures and public safety rather than protection of wildland resources. This could result in both greater wildland resource damage and significantly greater fire-suppression costs. Finally, the presence of human settlement affects the viability of many presuppression fuel-management options. The specific patterns of human settlement that are likely to occur under the General Plans are therefore likely to have a significant impact on fire regimes in the Sierra Nevada.

The greatest risk, due to the many substandard lots and roads that have been grandfathered under the General Plans, is to public safety. Evacuation difficulties along the steep, narrow streets of Nevada and El Dorado Counties are likely to be similar to those experienced in the tragic Oakland and Berkeley hills fire of October 1991. New state standards adopted after that fire in 1992 apply only to new subdivisions, yet much of the development expected to occur under the General Plans will occur either on existing parcels or through "minor" subdivisions that are exempt from the Subdivision Map Act. The fire risk is an area that needs considerably more analysis than that in either of the DEIRs. In particular, the DEIR for the Nevada County General Plan incorrectly relies upon the General Plan land use designations to evaluate the fire risks associated with the General Plan. As noted earlier, existing parcelization makes that a dangerous assumption. El Dorado County has adopted a more stringent set of fire safety standards for new developments, but our analysis of their application in the Cinnabar Ranch project suggests that considerably more work is necessary in order to mitigate fire safety risks associated with human settlement. Alternative settlement patterns could reduce some of those risks.

Government Services

In addition to the specific funding needs identified in the DEIRs for physical infrastructure (e.g., roads, sewers, and water), development under the General Plans will affect local government revenues and local governments' capacity to provide ongoing services. These include police and fire protection, general administration, public health, planning, libraries, and the other costs of local government. The relationship between land use patterns and future revenues and costs is difficult to forecast, however, due to the instability of state and federal budget mechanisms in recent years. This has been true at least since the passage of Proposition 13 in 1978, which reduced property tax rates in California and limited the rate of increase in assessed property values. The problem has been exacerbated in the 1990s by a severe statewide recession that has resulted in greater claims by the state on local revenues. Finally, the slowdown in construction activity within the Sierra Nevada has dampened the "boost" that new construction brings to average assessed values and property taxes. General fund revenues have consequently fluctuated wildly.

Together, these conditions have made local governments increasingly reliant on growth and fees to pay for basic services. Unfortunately, growth in the cost of providing these basic services appears to have been greater than growth in local revenues. Development fees do not generally cover the full cost of providing even physical infrastructure, let alone libraries and sheriff's deputies. This situation is symbolized by the Nevada County Library, a spacious new building (built largely with state funds) that has many empty shelves. The old Nevada City library, now the county historical branch, is open only nine hours each week. Buildout under the General

Plans is not anticipated to alleviate this situation, although demand for local government services is expected to grow. Continuing degradation in the levels of service (LOS) for many of these government services is therefore a likely outcome under the General Plans.[79]

The level of these impacts has resulted in challenges to the final EIRs that were adopted by both counties in 1995. Appellants claimed that the EIRs were inadequate under CEQA due to a failure both to consider alternatives that could reduce the level of impacts and a failure to adopt specific mitigation measures that include changes to the land use maps. El Dorado County has since modified both the language and land use designations of the 1994 Final Draft General Plan and released a supplemental EIR that will include consideration of a lower-growth alternative that has less environmental impact. Most of the changes appear to increase allowable development densities and decrease requirements for comprehensive consideration of the environmental effects of development under the General Plan. In particular, the LOS standards for many of the roads in the county were reduced to a lower level. References to regional coordination for air-quality and transportation planning have also been deleted, along with many requirements for development of an integrated recreational trail system and public parks. The new board of supervisors also fired the director of the El Dorado County Planning Department following its rejection of an internally prepared "low growth alternative" to the existing General Plan. Legal counsel has been retained by the board in order to prepare for litigation on the General Plan and the final EIR, which is likely soon after the General Plan is adopted (Rivas 1995; Griffiths 1995).

The Nevada County Board of Supervisors recently rejected an appeal of its final EIR after the Nevada County Planning Commission eliminated many of the draft EIR mitigation measures in its deliberations. The board of supervisors then adopted some of those same mitigation measures as changes to the General Plan itself on October 13, 1995 (Mooers 1995). Final adoption of the El Dorado County General Plan was not expected until late 1995 or early 1996, so we have had to limit our analysis and discussion to the Final Draft General Plans released in 1994 and the DEIRs circulated for review in early 1995. Our findings would certainly be modified by the subsequent action on both of the General Plans in 1995, but we were unable to revise our analysis to incorporate those changes. In general, however, the changes made by both counties since release of the DEIRs are likely to increase the unmitigated environmental impacts of the revised General Plans. Infrastructure funding shortfalls are also likely to be exacerbated by the changes. The exception to this generalization is the decision by Nevada County supervisors to eliminate the "new town" site and to reduce average densities in some areas. They made no changes to the land use map, however, to account for existing parcelization and underestimation of future growth.

Modeling the Spatial Patterns of Future Human Settlement

Our assessment of historic population growth, projected population growth, land use planning, and the development process associated with human settlement is intended to provide the basis for estimating the ecological, social, and economic consequences of human settlement in the Sierra Nevada. The spatial pattern of future human settlement is one important determinant of these ecological, social, and economic consequences. We therefore attempted to model the spatial patterns of future human settlement for the entire Sierra Nevada through a series of relatively simple models for allocating the CCD-specific population growth forecasts for 1990–2040.

For our simplest spatial model of density-dependent population growth, we developed a series of future population counts for census blocks based upon a "contagion" model of contiguous development that was estimated as a function of two density-dependent growth factors: (1) a measure of the density of census blocks adjacent to each census block; and (2) the housing density of each census block. The first factor was calculated as a function of the area-weighted density of each adjacent census block and the fraction of the total perimeter of each census block adjacent to each adjoining census block. Larger, denser adjacent census blocks with a greater fraction of adjoining perimeter were therefore assumed to exert a greater influence on future development pressure than smaller, less dense adjacent census blocks with a smaller fraction of adjoining perimeter. The second factor, census block density-dependent growth, was then calculated based upon the positive half-period interval of a sine curve.

The general formula for a sine curve is

$$f(x) = A \sin [(2 * pi/B)(x - C)] + D$$

where

f is the height of the curve (growth rate)

A is the amplitude of the curve (maximum growth rate)

B is the period of the curve ($0.5 * B$ is the positive interval of the sine curve)

C is the horizontal shift from the origin (minimum density threshold)

D is the vertical shift from the origin (minimum growth rate)

x is the horizontal distance from the origin (density) where { x if $x <=$ threshold;

$x = \{$ threshold if $x >$ threshold and threshold $< [(0.5 * B) - C]$

In each county there were a few blocks with extremely high housing densities, usually corresponding to prisons or other institutions. To ensure that they did not skew the distribution along the interval, an additional maximum threshold was applied to the density values prior to calculation. This allowed us to use a smaller period and therefore obtain a wider spread for most of the densities.

Our use of a sine function rests on the implicit assumption that maximum growth will occur at intermediate levels of existing density. This assumption allows for noncontiguous growth to occur through "metastasis" as well as direct "contagion" through density proximity. By varying the portion of the positive half-period interval that we used, we were able to vary the density at which maximum growth would occur. We also tested alternative models that had increasing and decreasing rates of growth as a function of density. Neither linear nor nonlinear formulations of these model specifications yielded satisfactory results, however, that allowed any useful spatial differentation across the landscape for further assessment of ecological impacts. We therefore focused on the sine function model to derive results that demonstrated differential landscape changes.

To implement the sine model, we wrote a program (in the PERL computer language) in which B, C, and D were fixed for the entire region for each run and A was determined through an iterative process for each CCD such that the calculated population increment when summing across all blocks in a CCD was within 1% of the population increment determined from our analysis of the DOF forecasts. The following steps describe the process we developed for each CCD:

1. For the first iteration, an arbitrary amplitude of three times the average growth rate for the CCD was used.

2. The formula was then applied to all populated blocks in the CCD, on the empirically reasonable assumption that nonpopulated blocks are in the public domain and not likely to experience population growth.

3. A running total of the population increment of all blocks was maintained.

4. Once all blocks had been processed, the total resulting population increment was compared to the estimated population increment for the CCD from the DOF forecasts.

5. If the calculated increment was greater than the DOF estimate, the amplitude was lowered by an amount proportional to the difference between the two, or vice versa if the calculated increment was less than the DOF estimate.

6. This process was repeated until the calculated increment and estimated increment were within 1% of each other.

7. Population increments for each block were then converted to housing density increments based on the mean household size for each housing density class across the entire Sierra Nevada region.

Unfortunately, our attempts at spatial allocation of population growth based on this sine function model did not generate empirically satisfying results at the spatial scale of census block polygons. It proved impossible to generate a growth rate curve that allocated the population in a reasonable fashion such that no set of census blocks received unreasonably large shares of the growth. We were therefore unable to develop a reasonable spatial allocation of the 1990–2040 forecasts at the census block scale.

There are two characteristics of the census block polygons that help explain our difficulties. The first is the relatively nonuniform distribution of existing housing densities in the census blocks. Most clusters of housing units are contained in small, dense polgyons, with surrounding areas represented as large, sparsely settled polgyons. This situation is not apparent in our reported distribution of population by density class due to the exponential fashion in which we defined our density classes. (We suspect we would have achieved better results with an exponential scaling of housing density as the dependent variable rather than the linear scaling that we used in the sine function modeling, but we did not have adequate time to test this alternative specification.)

The degree of correlation between size and density in the census block polygons themselves represents a second complication for forecasting. Already densely settled areas are apportioned into small polygons, while less densely settled areas are apportioned into relatively large polygons. Census block polygons are therefore not randomly distributed across the landscape but are already correlated with density by size. As we added population to the polygons, the overall picture of growth would become increasingly unrealistic because additional population would appear spread throughout these polygons, while due to their large size they would register relatively minor changes in density.

These complications led us to develop an alternative conceptual model for spatial allocation of the 1990–2040 CCD growth forecasts in future modeling. Our next attempts will be based on a raster model, in which the landscape is divided into small cells of fixed size. The raster model offers several advantages over a vector model:

- Processing of raster-based models is much quicker than processing of vector-based ones, due to the simplicity of the raster data model as represented in digital form. This speed advantage will allow us to repeat a greater number of more complex permutations of our analysis scheme.

- Incorporation of natural factors such as slope, soil type, and vegetation, which tend to be highly heterogeneous over a landscape and best represented in raster form, is facilitated by the use of the raster model for the analysis itself.

- The small, fixed cell size of the raster representation eliminates the problems we experienced using the census block polygons and will allow us to allocate population and settlement growth more precisely in response to both local and adjacent factors.

The first step in implementing the raster model was to convert the vector-based census block coverage to raster form.

We chose a cell size of 30 m (98 ft) to be consistent with the elevation and slope data we will also use in the future. Figure 11.46 shows an area near Grass Valley and Nevada City in census block converted from census block polygon (vector) densities to raster-based densities using the 30 m grid size.

Part of our new allocation model stipulates that new development is dependent on existing adjacent densities (i.e., low-density areas in proximity to high-densities areas will be subjected to highest growth pressure). This assumption reflects an implicit economic model of the costs of extending infrastructure, which we will also be testing explicitly in the future. To develop a measure of adjacent density, we calculated a "focal mean" for each cell in the map. The focal mean represents an average density value for all cells within a 300 m or 10-cell radius of any particular cell. The choice of 300 m is arbitrary. We can and will also test alternative radii for derivation of focal mean density values in the future. Figure 11.47 shows the focal mean density representation (based upon the 300 m radius) for the same area near Grass Valley and Nevada City.

The simplest analysis that we can then perform is to select a threshold value for the focal mean density as defining a high degree of settlement, which yields a set of smoothly shaped areas on the map. Figure 11.48 shows the difference in density values between existing 1990 housing densities and the focal mean density value for each grid cell.

To identify preliminary areas for infill, the existing density is merely subtracted from the focal mean density. Figure 11.49 shows areas likely to experience infill using this model of proximate density as the basis for determining likely development patterns.

We were unable to apply this modeling approach to the entire Sierra Nevada, but our preliminary exploration of its specification offers promise for future application. We must therefore rely upon coarser estimates of land conversion for 1990–2040 in this assessment. Spatially explicit characterizations of future patterns of human settlement are possible at this time only for Nevada and El Dorado Counties, where General Plan land use maps have been digitized. We have therefore had to limit our estimates of land conversion for human settlement from 1990 to 2040 to the CCD-specific analysis described in the next section.

Land Conversion Estimates for Human Settlement from 1990 to 2040

The total land area converted to human settlement to accommodate 1990–2040 growth will depend upon the spatial pattern and average density of settlement, which will in turn depend upon the complex interaction of public policy, infrastructure, and land economics. Strict development controls, significant expansion of water and sewer systems, and higher land prices would likely lead to a more intensive pattern of development with less land conversion than would occur in the absence of those conditions. Continuing existing patterns

of development would consume more land than could be achieved under those conditions. The ecological implications of continuing existing patterns of development and a range of alternative potential growth management policy mechanisms for mitigating those impacts are discussed in outline later.

Without assuming any specific linkages to specific policies or market conditions, we considered six alternative distributions of future population by housing density class. These were based upon our GIS analysis of the distribution of population by housing density class under the following:

1. 1990 Sierra Nevada census blocks

2. 1990 Nevada County census blocks

3. 1990 El Dorado County census blocks

4. Nevada County General Plan

5. El Dorado County General Plan Project Description (EDCGPPD)

6. El Dorado County General Plan Alternative (EDCGPA)

The three General Plan distributions were based on the planimetric estimates of area designated for buildout at specific density classes in the General Plan land use maps but did not account for the higher levels of density that are likely due to existing parcelization. Table 11.A14 in appendix 11.1 shows the distribution of population by housing density class for each alternative.

We then considered four alternative future growth projections from 1990 to 2040 for each of the fourty-six CCDs in our analysis:

1. Based on each CCD's 1970–90 share of overall county growth (DOF7090)

2. Based on each CCD's 1970–80 share of overall county growth (DOF7080)

3. Based on each CCD's 1980–90 share of overall county growth (DOF8090)

4. A lower projection at two-thirds the DOF7090 projection, which was the approximate absolute growth rate historically 1970–90 for the entire Sierra Nevada (HISTORIC).

Combined with the six alternative population distributions by density class, these four alternative population projections for 1990–2040 result in twenty-four possible 2040 land-conversion estimates for each of the forty-six CCDs in our analysis.

The resulting 1,104 cells of land-conversion estimates are a bit overwhelming for presentation, however, and many of the population distributions by housing density class are similar to one another. We therefore simplified the set to four scenarios:

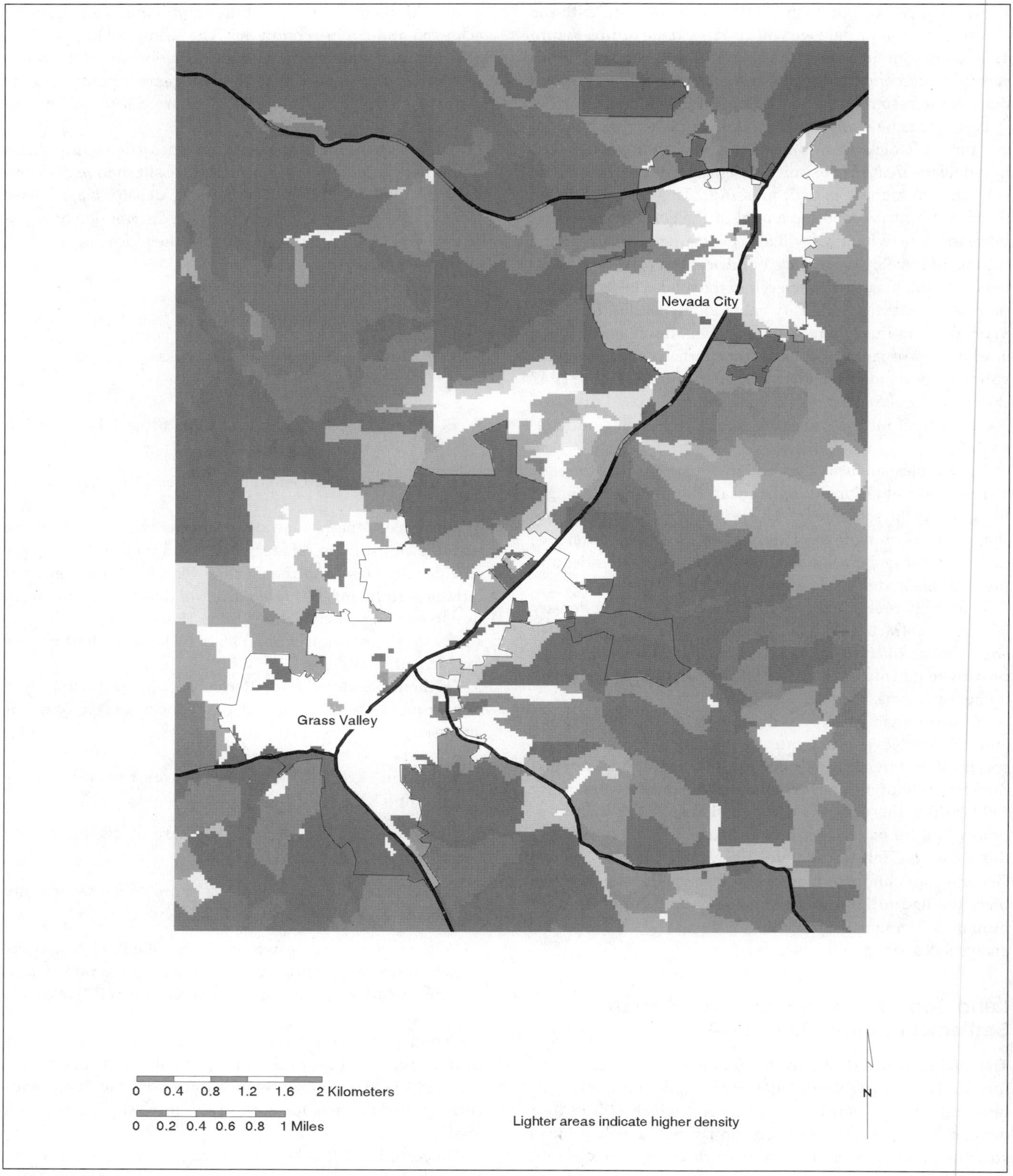

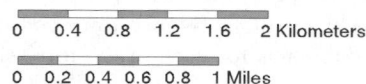

Nevada City

Grass Valley

0   0.4   0.8   1.2   1.6   2 Kilometers

0   0.2 0.4 0.6 0.8   1 Miles

Lighter areas indicate higher density

N

FIGURE 11.46

Housing density, Grass Valley–Nevada City area, 1990.

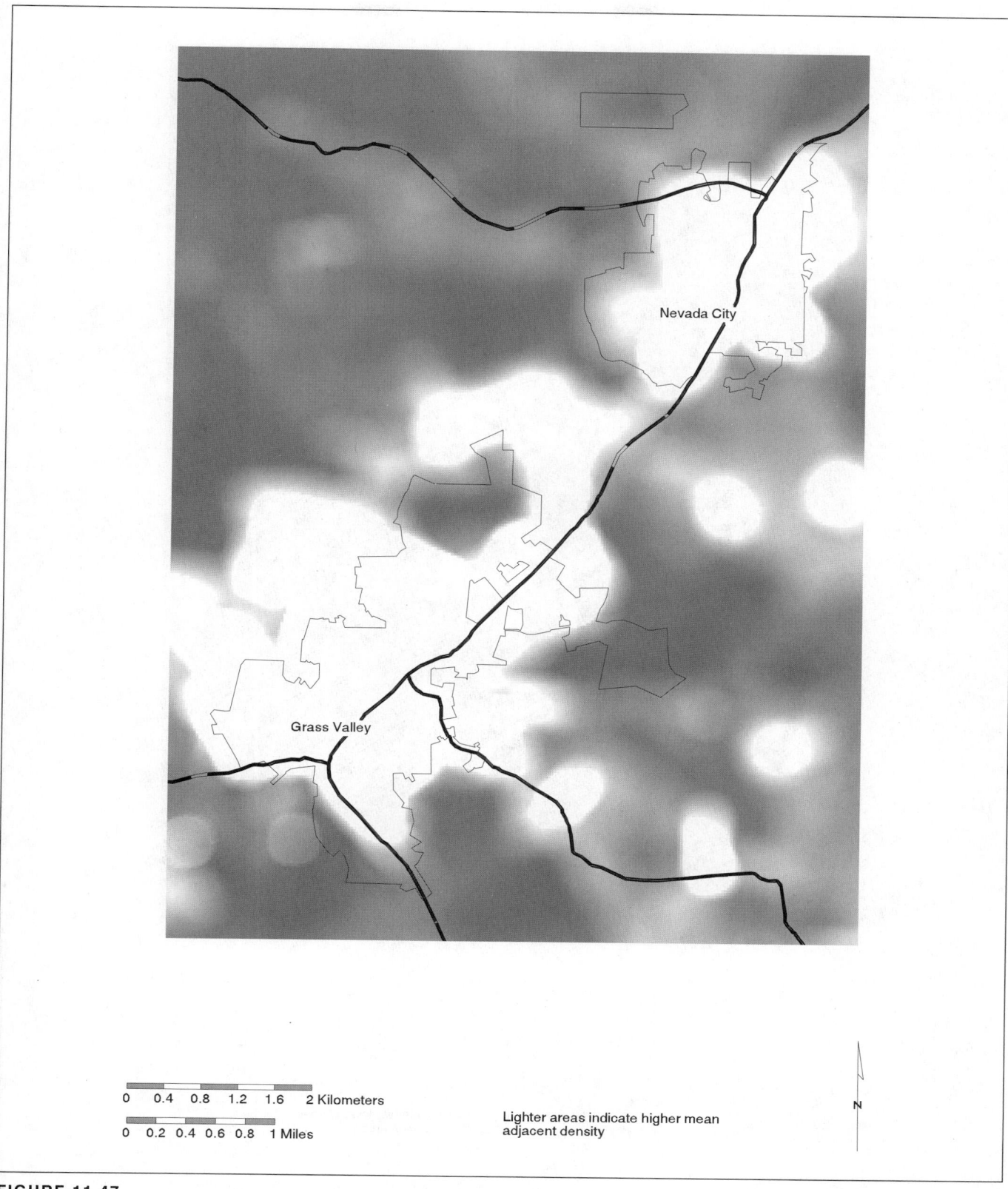

FIGURE 11.47

Focal mean housing density, Grass Valley–Nevada City area, 1990.

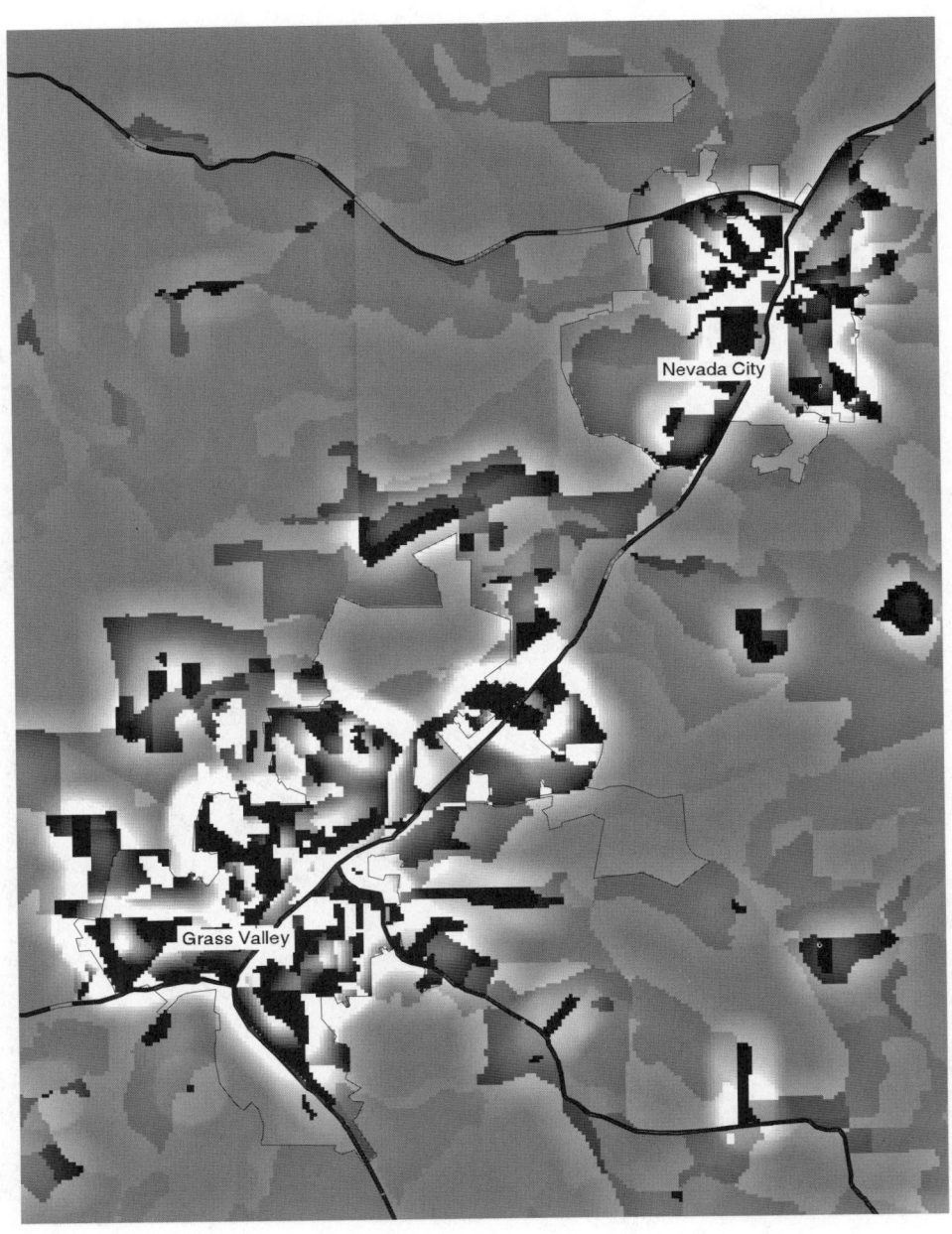

0 0.4 0.8 1.2 1.6 2 Kilometers

0 0.2 0.4 0.6 0.8 1 Miles

Lighter areas indicate lower density
relative to adjacent areas

N

FIGURE 11.48

Difference between point and adjacent housing density, Grass Valley–Nevada City area, 1990.

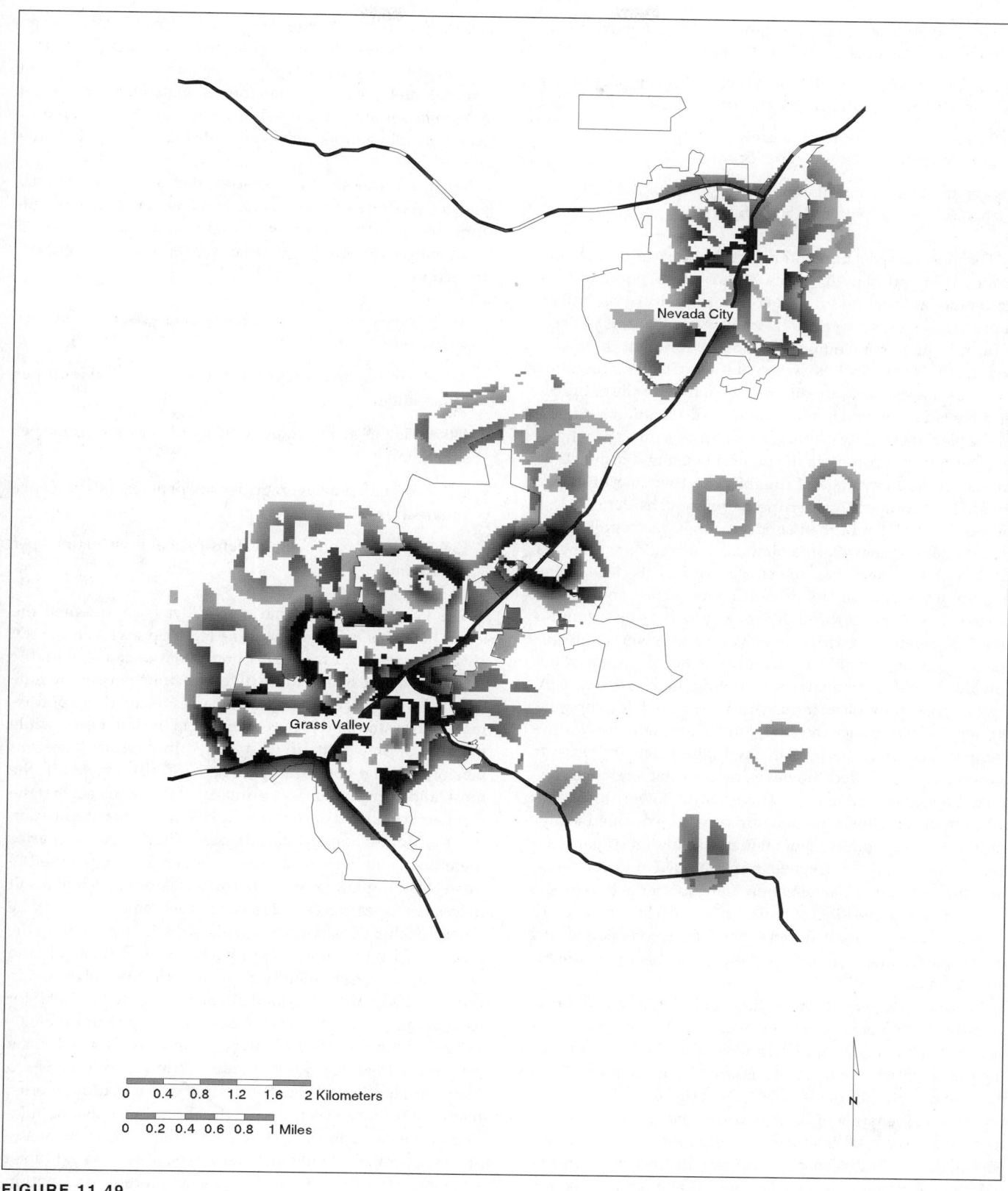

FIGURE 11.49

Opportunities for infill development, Grass Valley–Nevada City area, 1990.

Scenario *A:* low population growth with compact human settlement patterns (Low-Compact)

Scenario *B:* high population growth with compact human settlement patterns (High-Compact)

Scenario *C:* low population growth with sprawling human settlement patterns (Low-Sprawl)

Scenario *D:* high population growth with sprawling human settlement patterns (High-Sprawl)

The most compact population distribution was the Nevada County General Plan, in which 71.34% of the population is accommodated in the highest housing density class (640 or more dwelling units per square mile). Note that this is a significantly higher fraction of the population than there was living in this class in 1990, when Nevada County's distribution was not significantly different than that for the entire Sierra Nevada. The much more compact distribution assumed in the Nevada County General Plan still consumes roughly a quarter-acre per person in the highest housing density class, moreover, in an average of roughly two dwelling units per acre. This "compact" pattern is therefore considerably less dense than most suburban subdivisions in metropolitan areas. We believe this reflects a bimodal distribution within this density class, where there are clusters of parcels close to one acre in size (with on-site domestic well water and on-site wastewater disposal through septic systems) and around a quarter-acre in size (with public water and sewer). Unfortunately, we were not able to disaggregate housing density below this level for our analysis. Doubling the average density for this class (through an infrastructure-directed development strategy) could reduce the land-conversion estimates for the compact scenarios by 50% in the highest-density class. It would have little effect, however, on the total land area converted by human settlement at any of the lower thresholds for human settlement. As noted in our more detailed analysis, the Nevada County General Plan also underestimates the amount of land that is likely to be developed at lower densities due to existing parcelization. Our quarter-acre-per-person estimate for the highest housing density class is therefore a reasonable basis for estimating the land-conversion effects of compact human settlement patterns across the entire Sierra Nevada.

The most dispersed (sprawling) population distribution was the 1990 Sierra Nevada census block distribution, in which 39.49% of the population resided in the highest housing density class. We therefore assumed continuation of this existing distribution across all CCDs in the Sierra Nevada for our sprawl scenarios of human settlement. This allowed us to estimate the total land area required in each CCD to accommodate 1990–2040 population growth if existing patterns of human settlement were to continue. Land tenure relationships constrain the potential to expand the land area converted to lower housing density classes, however, so the lower housing density classes generally increase their average densities

within their density ranges rather than expand in area (e.g., land in the class of ten to twenty dwelling units per square mile might move from twelve dwelling units to eighteen dwelling units). We have therefore estimated land converted to human settlement only above the density threshold of twenty dwelling units per square mile (32 acres per dwelling unit).

Based upon these four scenarios, the range of additional land-conversion required to accommodate population growth from 1990 to 2040 (beyond the land area already converted for human settlement in 1990 that was reported earlier) is estimated to be

- 106–579 mi^2 at an average density of at least 640 units per square mile

- 299–875 mi^2 at an average density of at least 160 units per square mile

- 480–1,655 mi^2 at an average density of at least 80 units per square mile

- 477–2,957 mi^2 at an average density of at least 40 units per square mile

- 134–5,105 mi^2 at an average density of at least 20 units per square mile

The Low-Compact scenario (A) always represented the lower bound of our range and the High-Sprawl scenario (D) always represented the higher bound of our range, with the exception of the 640 or more dwelling units per square mile threshold. These two extreme scenarios resulted in approximately the same land area conversion in the latter case, while the Low-Sprawl scenario (C) resulted in the least land-conversion and the High-Compact scenario (B) resulted in the most land-conversion. This primarily reflects the fact that the compact scenarios concentrate 71.34% of the total population into the highest housing density class. The compact scenarios therefore result in more land area converted to human settlement in the highest housing density class, but they still result in less land area converted to human settlement in all of the other housing density classes. This is made clear at all of the other density thresholds. Figure 11.50 shows the total land area converted to human settlement in the Sierra Nevada in the year 2040 (including land already converted in 1990) for each of the four scenarios at each of the density thresholds.

These estimates of land conversion associated with human settlement 1990–2040 are not uniform throughout the Sierra Nevada. They reflect the distribution of population forecast by the DOF for each county and the allocation of that population by our allocation models to each of the CCDs in our analysis. In general, the land most likely to be converted to human settlement is primarily in the western foothills and within commuting distance of rapidly growing cities in the Central Valley. Some specific vegetation (Holland) types and Wildlife Habitat Relationship model (WHR) types are therefore more

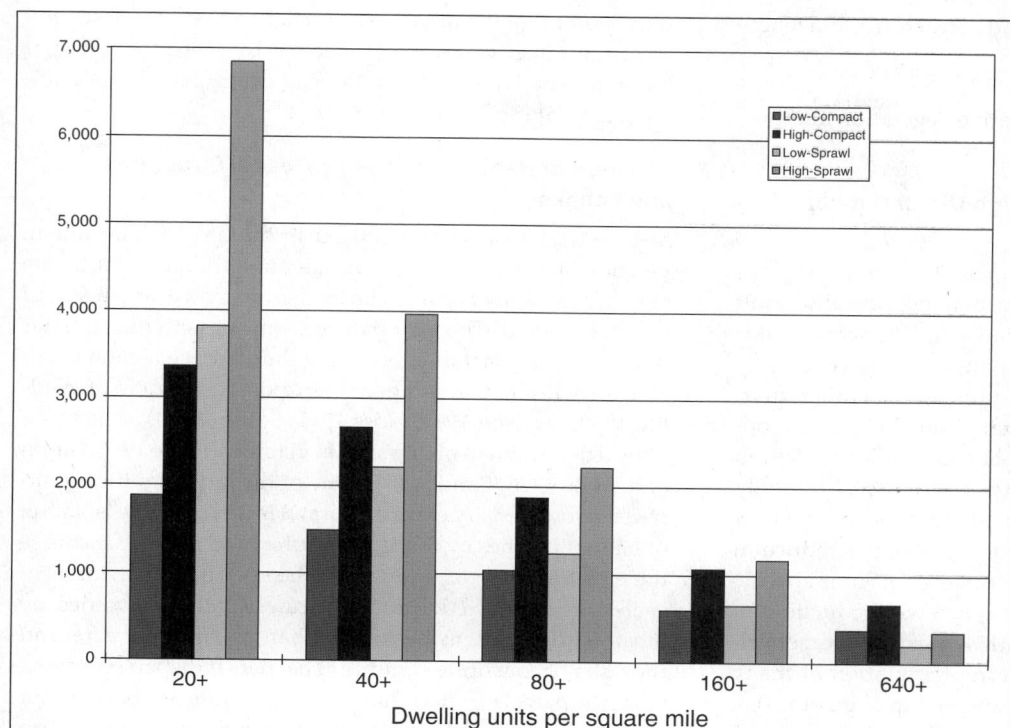

FIGURE 11.50

Land area converted to human settlement in 2040 by housing density class thresholds (square miles).

threatened by human settlement than others, reflecting the nonrandom spatial distribution of growth, private ownership, and vegetation. More spatially explicit analysis is necessary to determine the effect of specific patterns of human settlement on specific vegetation (Holland) types and WHR types. As noted earlier, we were unsuccessful in our attempt to develop a general model for spatially explicit allocation of 1990–2040 population growth.

We disaggregated vegetation (Holland) type and WHR type data by CCD in order to complete more detailed analysis of the relationship between projected land conversion and specific ecological features. Unfortunately, similar disaggregation by river basin or watershed was not possible, because the CCD boundaries often cut across watersheds. The CCDs are large analytic units, so it is impossible to infer land-conversion estimates for specific vegetation (Holland) and WHR types or watersheds without a spatially explicit model of human settlement that allocates the 1990–2040 population forecasts. We can nevertheless identify those vegetation (Holland) or WHR types that could be converted by human settlement on the private lands within each CCD with the CCD-level data. We were unable to complete such an analysis for all of the CCDs and vegetation (Holland) or WHR types for the forty-six CCDs in the Sierra Nevada, but all of our disaggregated data are available in dBase and Excel format from the California Environmental Resource Evaluation System (CERES) project of the Resources Agency of the State of California (http://ceres.ca.gov/snep), and the Alexandria Project at the University of California, Santa Barbara (http://alexandria.sdc.ucsb.edu/), for further analysis in the future.

Ecological Implications of Land Conversion for Human Settlement

The ecological implications of land conversion expected for human settlement from 1990 to 2040 in the Sierra Nevada will depend upon the spatial patterns of human settlement and the distribution of land conversion by vegetation type, wildlife habitat, watershed, slope, elevation, and a wide range of other natural factors. The specific effects of alternative patterns of human settlement are still poorly understood, although preliminary estimates of the relationship between settlement density and vegetation change have been characterized in McBride et al. (1996). Significant additional empirical work is still necessary to project the ecological impacts of future landconversion, but we can outline the range of possible impacts based upon the literature (Peck 1993). Here we offer a partial summary of the ecological implications of land conversion for human settlement.

Land conversion causes at least five direct effects on vegetation and wildlife:

1. Reduced total habitat area through direct habitat conversion

2. Reduced habitat patch size and increased habitat fragmentation

3. Isolation of habitat patches by roads, structures, and fences

4. Harassment of wildlife by domestic dogs and cats

5. Biological pollution from non-native vegetation alleles

Reduced Total Habitat Area through Direct Habitat Conversion

Reduction of habitat is the most apparent effect of development, but low-density development may not actually result in significant reduction in total habitat area. The actual building site and associated construction impacts may cover only one-fourth of an acre, for example, plus up to another quarter acre for access roads, septic system leach lines, and a domestic water well. This density level could indeed result in denaturation of up to 25%–50% of a 1 acre parcel but only 5%–10% of a 5 acre parcel. We estimate that the direct effect of low-density exurban development is probably a reduction in total habitat area by around 20% (10%–30%). Some specific habitats are disproportionately threatened with reductions in area, however, for they lie in the path of most exurban development. This fact reflects the underrepresentation of many vegetation types on land in public ownership (e.g., blue oak woodland in the Sierra Nevada foothills) and the overrepresentation of a limited number of ecosystems (e.g., "rocks and ice" of alpine wilderness preserves in the Sierra Nevada high country). The gap analysis (Scott et al. 1991) completed for SNEP (Davis and Stoms 1996) highlights specific vegetation types that are most likely to be affected by human settlement on private lands in the Sierra Nevada (most notably in the western foothills). Direct reductions in total habitat area can be significant for many rarer habitat types or those that have already suffered significant reductions in total area.

Reduced Habitat Patch Size and Increased Habitat Fragmentation

Even when reductions in total habitat area are limited, the average patch size of remaining habitat is reduced significantly with low-density exurban sprawl. Depending upon how edge effects are evaluated (e.g., distance from roads and structures), average patch size can drop from thousands of acres to less than an acre (e.g., if the entire area is fragmented into 1 acre building lots). The negative consequences of habitat fragmentation are well known theoretically and documented in a number of specific cases for both tropical and temperate regions (Harris 1984; Adams and Dove 1989; Gilpin and Soule 1986; Wilcove et al. 1986; Lovejoy et al. 1986; Soule 1991b). The distribution of patch sizes also typically shifts from a few large-sized patches to a pattern in which only a small fraction of the total number of patches for any given habitat remain large enough to support viable populations of many species. Those may be the only patches that remain "effective habitat," despite both the continued existence of many patches with similar vegetation and a relatively small

reduction in total habitat area. Total area may decrease only 20%, but effective area may decline by more than 90% at buildout. Roads are probably the single biggest source of habitat fragmentation in exurban areas.[80]

Isolation of Habitat Patches by Roads, Structures, and Fences

Neither total habitat area nor the distribution of habitat among patches of various sizes is an adequate description of the changes that may occur in the landscape matrix as a result of exurban sprawl. The way patches connect with one another is an important factor determining the effective habitat available for wildlife use and gene transport (Defenders of Wildlife 1989; Hudson 1991; Noss 1991; Soule 1991a), which is a critical determinant of population viability (Soule 1986; Gilpin and Soule 1986; Pimm 1986). One of the most significant impacts of low-density exurban sprawl is therefore the isolation of habitat patches by roads, structures, and fences. Of course the effect of each depends upon the specific life histories of each species affected. Structures can usually be avoided by most wildlife at densities of less than one unit per acre, and they do not constitute significant barriers if dispersed among adjacent parcels. Fences can serve as significant barriers for many mammals and reptiles, but they appear to constitute a relatively low barrier to the migration of birds and invertebrates or the transport of genetic material from most vegetation. Roads are probably the single most important barrier to both wildlife and genetic movement between habitat patches, just as they are the most important source of habitat fragmentation and edge effects. Unfortunately, transportation planners rarely consider the effects of transportation network design upon native biological diversity. Further research and education (of both the public and transportation planners and engineers) are necessary to develop transportation network designs that minimize these impacts.

Harassment of Wildlife by Domestic Dogs and Cats

Even if wildlife can avoid residential structures, they are often unable to avoid harassment by domestic dogs and cats. These pets extend the effective area of human settlements to a degree that development could form a significant barrier between and/or reduce the effective habitat of adjacent habitat patches. It is difficult to estimate the "dog-shed" or "cat-shed" associated with this effect, but it can be quite large. Many exurban properties have limited lawns and limited fencing, and leash laws are usually only loosely enforced. The result is that both dogs and cats are able to roam freely throughout the exurban matrix as long as they avoid conflict with humans. The range of dogs can easily be several miles in a single day, making most of the settled portion of the exurban matrix subject to the effect of dogs. Michael Soule has documented the apparent effect of cat predation on birds in the urbanizing areas of northern San Diego County (Soule 1991b), and harassment from dogs is known to affect many species that are common in areas facing rapid exurban growth.

Dogs and cats can also be a source of seed dispersal of non-native plants (discussed later) and can be a source of disease for native wildlife. Other pets or domesticated animals (e.g., cattle or sheep) can also be disease sources that can decimate native wildlife populations (e.g., ungulates). This effect could be important when seasonal migrations occur (Yuba County Community Services Department 1985; Peck 1993).

Biological Pollution from Non-native Vegetation Alleles

The risk of "biological pollution," or genetic contamination, is a concern both for non-native invasives (e.g., Scotch broom) and for nonlocal stock of species that are native to an area (e.g., Douglas fir). In the first case, the invasive species can outcompete and displace some native species, modifying the vegetative structure to a degree that affects other species and the entire landscape matrix. A sun-tolerant species may invade a recently opened forest area, for example, displacing an entire succession of species that would normally have occurred in the absence of that species. The specific species being displaced is therefore not the only one directly affected. The second instance is more subtle and much more difficult to evaluate: genetic hybridization may occur or the population with the nonlocal alleles may outcompete the local alleles. The apparent structure of the landscape matrix may not change as a result, but the genetic information contained in the resulting matrix will be different than the information in the native matrix. This change may then diminish the capacity of the entire system to respond to some significant disruption (e.g., global climate change) in the future. To the degree populations are determinants of the long-term viability of Sierra Nevada ecosystems, it may be just as important to protect against nonlocal genetic contamination as it is to minimize the risk of invasive non-natives. Considerably more research must be completed before we can confidently determine the relative importance of particular populations (Medbury 1993).

In addition to these direct effects upon vegetative composition, structure, and function (which in turn affects wildlife habitat and wildlife viability), land conversion for human settlement has several direct effects on hydrologic regimes that could be important:

6. Increased impervious surfaces and increased peak run-off

7. Increased heavy metal and oil runoff from impervious surfaces

8. Increased risk of ground-water and/or surface-water contamination through septic effluent disposal

9. Decreased ground-water flow to surface-water system due to ground-water pumping

10. Modified surface water flow due to irrigation, septic system effluent disposal, and treated wastewater discharges

Increased Impervious Surfaces and Increased Peak Runoff

Conversion of wildlands for human settlement includes the construction of roads, parking, and structures as well as soil compaction and vegetation modification. In general, these changes are likely to increase impervious surface, decrease leaf canopy and its capacity to intercept precipitation, and decrease evapotranspiration on the site. A change in the local hydrograph often results, although intervening factors may dampen the effect of these changes on sedimentation and downstream hydrological characteristics. On-site water retention timing and volume can also be affected, so it is difficult to generalize the effects of land conversion for human settlement. Changes in vegetation can also increase evapotranspiration over time as planted vegetation matures.

Increased Heavy Metal and Oil Runoff from Impervious Surfaces

Many of the impervious surfaces associated with human settlement accumulate heavy metals and oils due to the presence of transportation technologies (e.g., cars, trucks, motorcycles) and other human activities (e.g., chain saws). These substances are then likely to be removed from the site or transported from points of concentration on the site through heavy precipitation during peak runoff periods. The degree to which these materials then enter surface water systems and affect hydroecological systems depends on the characteristics of both the intervening watershed and the aquatic ecological system. It is therefore difficult to generalize these effects from human settlement, and the impact of nonresidential land uses (e.g., commercial, industrial) is likely to be greater on a per-acre basis than all but the highest density pattern of human settlement.

Increased Risk of Ground-Water and/or Surface-Water Contamination through Septic Effluent Disposal

As noted in our discussion of census data for the Sierra Nevada, the use of septic systems is significantly higher in the exurban landscape of the Sierra Nevada than it is for California as a whole. The potential risk of septic system contamination of ground-water is a function of system operation, leach field characteristics, and ground-water characteristics (Davis 1994). These are highly site-specific features in the Sierra Nevada, where both soils and ground-water characteristics are highly variable. Historic failures of septic systems have led to building restrictions, ground-water contamination, and surface-water contamination (Cranmer Engineering and Halatyn 1971; Davis 1994; Lauer 1995a, 1995c; 1995b; Lenahan 1995). All of these outcomes are possible through the failure of existing or newly developed septic systems. They may also occur even if septic systems are operating normally, however, as densities increase to the point at which soils are unable to "treat" the septic effluent to an acceptable standard (Thompson 1989; Hanson and Jacobs 1989).

Decreased Ground-Water Flow to Surface-Water System Due to Ground-Water Pumping

On-site ground-water is a primary source of domestic potable water and irrigation water for many low-density exurban households in the Sierra Nevada (Turner 1973; U.S. Bureau of the Census 1990). The ground-water system in the region is characterized by highly variable and unpredictable storage in fractured bedrock, however, rather than a clearly delineated set of ground-water aquifers (Swain 1994). This system is therefore interconnected with the surface-water system in complex and unpredictable ways. Dependence on ground-water pumping for water supply therefore has the potential to affect surface-water flows. It is unclear how significantly this may affect surface-water systems, but any effect is likely to be site-specific.

Modified Surface-Water Flow Due to Irrigation, Septic System Effluent Disposal, and Treated Wastewater Discharges

Human settlement requires access to water supplies, and the provision of water supplies usually involves either importing water through interbasin transfers or significant in-basin storage to accommodate seasonal differences between natural flow regimes and human uses (Turner 1973). Those human uses of water then result in either irrigation for outdoor uses (which can either recharge ground-water or enter the evapotranspiration cycle) or internal domestic use. Most water used internally is then discharged through either septic system disposal or sewered wastewater treatment. Septic system disposal can then affect ground-water and/or surface-water hydrology within the local watershed (Hanson and Jacobs 1989), while sewered wastewater treatment can lead to either in-basin discharges or interbasin transfer to another watershed. Wastewater can then account for a significant fraction of surface-water flow, altering both the seasonal timing and overall level of flows downstream of the point of discharge (CVRWQCB 1989, 1992).

Finally, land conversion due to human settlement can have a wide range of indirect effects on ecological structure and function. The most important of these in the Sierra Nevada is associated with impacts on the fire regime in both settled areas and adjacent wildlands. Human settlement affects the structure and level of fuel loads, the viability of presuppression fuel-management strategies, the likelihood of ignition risk, the availability of suppression resources, and the allocation of those resources through suppression efforts (Irwin 1987, 1989). Each of these will in turn affect the future risk and characteristics of fire in the Sierra Nevada. Vegetation management in the "urban forest" of areas converted to human settlement can either decrease or increase fuels in the urban-wildland intermix zone (Doyle 1995). Further research is necessary to establish empirical relationships between alternative patterns of human settlement and each of these indirect effects on the Sierra Nevada.

CONCLUSIONS

Land use in the Sierra Nevada has changed dramatically during the past fifty years (Weeks et al. 1943), beginning when California's population boomed and standards of living rose during the first two decades following World War II. The population of the Sierra Nevada has more than doubled since then, resulting in a 1990 population that is approximately four times the peak population during the gold rush. Most of the new residents have settled near the historic centers of the gold rush, but their patterns of settlement have resulted in much more extensive land conversion. Three out of five Sierra Nevada residents lived on less than 300 mi² (less than 1%) in 1990, but human settlement was spread across nearly 1,741 mi² at an average density of at least one housing unit per 32 acres to accommodate seven out of every eight Sierra Nevada residents. This constituted 5.44% of the entire Sierra Nevada, or nearly 14% of all private land (including industrial timberlands). Up to one-eighth of the entire Sierra Nevada (3,905 mi²) may have been affected by human settlement in 1990 at an average density of at least one housing unit per 128 acres.

The Sierra Nevada is likely to undergo significant land conversion through continuing population growth over the next half-century. Population growth in the metropolitan centers of California is forecast to result in a doubling of the state's population between 1990 and 2040, leading to expansion of the emerging metropolitan centers of the Central Valley that are within commuting distance of the Sierra Nevada foothills (Teitz 1990). Metropolitan areas near the Sierra Nevada are also forecast to continue growing in the state of Nevada. This growth would create new employment opportunities on the urban edge and extend the reach of reasonable commute times into areas that have not yet faced significant residential location by commuters. The result is likely to be continuing immigration by commuters, retirees, and former metropolitan-area residents who are seeking a rural or exurban lifestyle offering significant natural and social amenities. Many of these latter immigrants are likely to accept lower incomes in exchange for these amenities, but they also generally bring human and financial capital with them. They therefore have the potential to generate new employment in the Sierra Nevada.

These new residents are likely to have higher incomes than most existing residents and will put pressure on land and housing prices. The factors driving the exodus to exurbia over the past three decades are likely to continue, resulting in an increasingly homogeneous population of affluent, white, well-educated residents in the commuter and retiree communities proximate to the Central Valley and the Lake Tahoe region. More isolated communities in the northern and eastern Sierra are likely to experience relatively slow growth, however, with less pressure on land and housing prices. Existing patterns of human settlement are more stable in these areas, where lower land prices will make significant investments in centralized infrastructure uneconomic. Large higher-density developments are likely in the Gold Country, however, where proximity to the Sacramento metropolitan area has already increased land and housing prices significantly. Nonlocal landowners have already consolidated parcels in these areas and have proposed development of several planned communities in the region.

The social, economic, and ecological ramifications of future development will depend upon specific spatial patterns of human settlement in relation to existing communities, infrastructure services, vegetation and habitat types, and watershed boundaries. As discussed earlier, our understanding of those relationships is still poor at this time. It is therefore impossible for us to characterize the specific impacts that population growth and human settlement will have in the Sierra Nevada. The range of impacts could be quite significant, however, if existing development patterns continue. Continuing the existing pattern of sprawl development with a high-growth scenario could result in human settlement on nearly half the private land in the Sierra Nevada (6,846 mi^2) at an average density of at least one housing unit per 32 acres. A low-growth scenario with the existing pattern of sprawl would reduce that figure by 44%, to just 3,817 mi^2. This area is still significantly greater than the 1,741 mi^2 affected by human settlement at that average housing density in 1990.

Even modified settlement patterns are forecast to result in significant land conversion from 1990 to 2040, suggesting that the scale of population growth alone could lead to significant impacts. A high-growth scenario with a more compact form of settlement would result in nearly a doubling of land converted to human settlement, from 1,741 mi^2 to 3,363 mi^2 at an average density of at least one housing unit per 32 acres. A low-growth scenario with a more compact form of settlement, on the other hand, could nearly be accommodated within the land area already converted to human settlement at an average density of at least one housing unit per 32 acres in 1990. Through infill and carefully targeted density transfers, the low population forecast for 1990–2040 would require only 1,875 mi^2 (only 8% more than in 1990). Both the scale and pattern of human settlement will therefore affect—and must therefore be considered by—local, state, and federal land and resource management agencies with responsibilities for the health and sustainability of Sierra Nevada ecosystems.

This suggests that any factor influencing future patterns of human settlement has the potential to affect the future impacts of continuing population growth on the health and sustainability of Sierra Nevada ecosystems. One of the most important factors determining patterns of human settlement is land use policy embodied within local General Plans. These documents and associated land use maps are the legal framework within which local land use planning, infrastructure investment, and land development occur. The ecological, social, and economic effects of subsequent development under the General Plans is required to be evaluated in an environmental impact report (EIR) prepared under the California Environmental Quality Act (CEQA). Local jurisdictions are then required to mitigate environmental effects unless "overriding considerations" warrant accepting those effects. Based upon our review of the Nevada and El Dorado County General Plans and their associated EIRs, however, it appears that the current planning process fails adequately to (1) determine the scale and location of future land conversion accurately; (2) systematically determine the effects of such land conversion on a wide range of ecological, social, and economic systems; and (3) mitigate those impacts that are determined to be significant. The current General Plan and EIR process therefore appears inadequate to the task of mitigating the effects of future land conversion for human settlement in the Sierra Nevada (Johnston and Madison 1991; Bank of America et al. 1994; Governor's Interagency Council on Growth Management 1993).

MANAGEMENT IMPLICATIONS

The importance of future population growth to the future of health and sustainability of the Sierra Nevada cannot be overstated. Management implications will vary for local, state, and federal agencies, but nearly all aspects of land and resource management in the Sierra Nevada will be affected. Local agencies will be affected as specific patterns of human settlement result in specific patterns of demand for services and as that demand in turn affects the fiscal capacity of local government to provide those services. The privatization of some services through high-density, large-scale "gated" communities has very different implications than the privatization of services through low-density, large-lot exurban sprawl development relying on well water and septic systems. State agencies with responsibility for fire protection, wildlife, water quality, transportation, and air quality will also be affected directly by these different patterns of development. Federal land and resource managers are likely to be impacted by modified fire regimes, increasing social constraints on industrial timber operations, and increasing demand for local recreation and open space benefits provided by federal lands to local communities.

Alternative patterns of human settlement will affect each of these issues differently. Our evaluation of management implications must therefore address both the different patterns of human settlement that are possible and the management strategies associated with them. We will focus here on the range of growth management policies available to mitigate the impacts we have identified and the institutional setting for implementation of those policies. We are not recommending policies here, but merely outlining the potential suitability of alternative policies to mitigate the specific impacts identified in this assessment. We also believe there are significant constraints to adoption of many of those policies, however, so we discuss the ecological, economic, and social factors (including institutional factors) that influence both applicability and adoption of specific policies in the Sierra Nevada. It is clear that alternative patterns of human settlement and their implications for ecological, economic, and social systems in the Sierra Nevada are too heterogeneous to warrant a "one size fits all" policy for human settlement, so there is no "silver bullet" policy option that will mitigate the impacts of human settlement.

Growth Management Policies to Mitigate Human Settlement Impacts

There is a wide range of policies available to manage population growth and to mitigate the impacts of human settlement. The appropriateness of specific policies depends upon the impact of concern, however, as well as its specific relationship to human settlement. A particular settlement pattern might have a significant effect upon native nesting songbirds, for example, that can primarily be traced to the presence of domestic dogs and cats. Alternative settlement patterns might all have a similar impact, therefore, while alternative pet management regimes could mitigate the impact. In contrast, the effects of human settlement on hydrologic regimes may be either a linear or nonlinear function of housing density. Perhaps there is an effect that is proportional to housing density only up to (or down to) a threshold density, above (or below) which higher (or lower) density does not have an effect. The specific form of these relationships is likely to vary across impacts, so we can not make a general statement about either impacts or policies. Proper evaluation of alternative policies requires a better understanding of the relationships between alternative patterns of human settlement and a wide range of impacts.

Despite this caveat, it is still possible to hypothesize likely relationships and to evaluate the capacity of alternative policies to mitigate the likely impacts of human settlement. Growth management tools have been in use since the first case of informal urban design, when incompatible uses were separated in order to reduce the likelihood that "nuisance" uses would impact other uses (Kostof 1991). This approach has generally been formalized and institutionalized through zoning ordinances and land use planning approaches that emphasize the spatial separation of incompatible uses. Zoning has been widely used in urban areas since the landmark Supreme Court case *Euclid* in 1926,[81] but was adopted and applied to all land uses only in the 1970s and 1980s for many parts of the Sierra Nevada. Planning techniques have evolved more recently to include a complex suite of both broad and specific tools for managing growth and mitigating its impacts (Stein 1993; DeGrove 1992). We therefore have an extensive literature to draw on when discussing growth management alternatives (Innes et al. 1993; Stein 1993; DeGrove 1992). Because it is extensive, we will offer only a brief introduction here to some of the techniques that may have specific application to the rural and exurban context of the Sierra Nevada. A more systematic consideration of growth management techniques and their capacity to mitigate the effects of human settlement in the Sierra Nevada requires a better understanding of the relationship between alternative patterns of human settlement and likely impacts. Carefully targeted growth management tools can then be evaluated accordingly.

In general, growth management tools can be characterized as one of three types:

1. Spatial, in which the location of specific land uses is designated and constrained

2. Temporal, in which the timing of development is controlled by local authorities

3. Outcome, in which the activities allowed on a particular site are controlled in their timing, duration, frequency, or intensity to maintain particular outcomes or conditions

Any of these broad classes of tools might be proposed and/or adopted to address similar impacts. Conversely, different approaches are likely to be appropriate and necessary to mitigate different types of impacts. General public concern about the traffic impacts of new development, for example, could result in the following new policies:

- Limitations on new commercial development near substandard intersections

- Requirements that new commercial development can go forward only after intersections have been upgraded sufficiently to accommodate all forecast traffic flow

- Requirements that new commercial developments could be open only during certain hours in order to avoid exacerbating traffic problems at substandard intersections

These examples are simply illustrative, but they highlight how the term *growth management* can mean very different things to different people. Irving Schiffman outlines twenty-six different growth management tools in *Alternative Techniques for Managing Growth* (1989), and there are many variations on each of his themes. State-level growth management regimes have taken a variety of forms (Innes 1991), from Oregon's land use (spatial) emphasis (Knapp and Nelson 1992) to Florida's infrastructure concurrency (timing) requirements (DeGrove 1992). Local jurisdictions have also adopted a wide range of growth management approaches within California (Governor's Interagency Council on Growth Management 1993; Landis 1992; Glickfeld and Levine 1992). The effectiveness of those measures is still subject to considerable debate, for it is difficult to control study data for the specific growth management policy alone (de Neufville 1981; Landis 1988, 1992; Innes et al. 1993). Moreover, growth management policies may have the effect of increasing land prices and decreasing housing affordability (Dowall and Landis 1981; Dowall 1984, 1991). Spillover effects into adjacent jurisdictions are also difficult to capture, and many of the growth management systems have been in place only a short time or have been modified following legal challenges (Landis 1988, 1992; Glickfeld and Levine 1992). It is therefore impossible to generalize about the likely effects of growth management tools on ecological, social, or economic conditions in the Sierra Nevada.

The specific effects of human settlement in the Sierra Nevada will dictate which types of growth management tools, if any, are appropriate for impact mitigation. Urban limit lines, for example, are unlikely to make a significant contribution as long as centralized infrastructure is not a significant determinant of settlement patterns. Moreover, concerns about maintaining "rural character" and the "quality of life" in the Sierra Nevada may make a highly concentrated pattern of human settlement undesirable for many residents. Specific impacts on vegetation and wildlife, on the other hand, could call for innovative growth management approaches that have not been applied yet in other jurisdictions. These could include seasonal limitations on specific activities that could negatively affect rare and endangered or endemic native plants, for example, but would not necessarily limit the opportunities to develop adjacent areas for human settlement. The potential scope of such limitations could be identified through an overlay of local land use plans with the U.S. Geological Survey 7.5-minute quadrangle GIS database prepared by Shevock (1996). Similar analyses need to be completed in relationship to other ecosystem resources.

Other rural areas (outside the Sierra Nevada) facing rapid population growth have pursued an innovative set of policies to maintain the "rural character" and "quality of life" amenities that dominate public debate about land use planning in the Sierra Nevada. These approaches draw upon a long tradition in landscape architecture and site design that was popularized by Ian McHarg in *Design with Nature* in 1969 (McHarg 1992). A comprehensive guide to these techniques was first published regionally by the Center for Rural Massachusetts in 1988 (Yaro et al. 1988), then subsequently updated with additional examples from other locations by Randall Arendt in *Rural by Design* (1994b). Other useful guides to these techniques have been published separately both by researchers (Arendt 1994a; Pivo 1990, 1988; Wolfe 1990) and by local jurisdictions that have adopted these policies (Redman 1992; Montgomery County 1992b, 1991, 1992a; Livingston County Planning Department 1991). High-altitude mountain environments also have special design problems due to their harsh conditions (Dorward 1990), which are important considerations for many parts of the Sierra Nevada.

The basic approach to "open space development design" (Arendt 1994a) is quite simple: instead of creating a landscape of large-lot parcels, subdivision development should be clustered to protect important aesthetic, economic, and ecological resources in the community and the landscape. Protection can be achieved without restricting total development through a redistribution of the overall site development density to the most suitable locations (determined by analysis of natural factors, viewsheds, and social factors that could constitute constraints on or opportunities for development). The result is many smaller parcels for human settlement and a few larger parcels for protection of amenity values. The overall development level is not changed, however, avoiding the charge that landowners' private property rights have been violated. The pattern has simply been altered to mitigate the effects of human settlement on ecological, social, and economic concerns. (Further reductions may still be necessary in order to mitigate some impacts, however, such as transportation and air-quality impacts associated with peak levels of transportation demand. Design changes alone may therefore not be sufficient to mitigate all significant impacts.)

This approach can best be illustrated visually. Figure 11.51 shows an existing site that has a high level of potential development associated with its existing zoning designation. Figure 11.52 shows how this site would be developed under standard large-lot zoning.

This large-lot pattern has the effect of fragmenting the landscape visually, socially, ecologically, and economically. The scale, texture, and design characteristics of the landscape that brought the new residents have now been altered forever. This outcome may be acceptable, but it is not necessarily inevitable. Figure 11.53 shows how the same number of housing units can be developed through a creative open space development design that clusters the housing units at a scale appropriate to the landscape.

Most of the primary visual features of the landscape have been preserved under this design alternative. Most of the housing units also have access to more open space than they would under the conventional development pattern, and they each generally retain a sense of privacy despite being located on smaller individual parcels. Small, multiple-household septic systems have been sited under this arrangement to improve septic system operation despite the higher density of development in each of the cluster areas. The ecological and hydrological effects of this development pattern have not been analyzed systematically, but explicit design around natural features should result in lessened impacts. The scale of both the open space and clusters, however, would affect whether or not significant benefits accrue. The literature on this issue is still very weak, because most analyses of the relationships between human settlement and ecological impacts in exurban landscapes have been at a coarser scale that fails to capture subtleties of site design. This is the most significant limitation of our assessment for the Sierra Nevada: we have been unable to determine the relationships between finer-scale settlement patterns and ecological, social and economic impacts on Sierra Nevada ecosystems.

This approach was nevertheless proposed as a central mitigation strategy in the draft EIR (DEIR) for the Nevada County General Plan (Harland Bartholomew and Associates 1994). The Nevada County Planning Commission subsequently rejected a mandatory clustering requirement for all new developments, however, and the Nevada County Board of

FIGURE 11.51

Existing land use in a typical rural and exurban landscape.

Supervisors did not override that decision on appeal. The Nevada County General Plan instead relies primarily upon poorly defined "flexible" site development standards (SDSs) to mitigate some of the impacts of future development. No mechanisms were adopted to address existing parcelization, such as transfer of development rights (TDR) arrangements, and concurrent provision of infrastructure or timing limitations as a function of infrastructure capacity were also rejected. In general, the General Plan and its associated EIR were adopted despite projections of significant ecological, social, and economic effects on the grounds of "overriding consid-

erations." Those impacts are therefore still likely to occur as described earlier and in the EIR.[82]

Clustered development was also a key design feature in earlier versions of the El Dorado County General Plan, although it was not proposed as a prominent feature in either the General Plan Project Description or the General Plan Alternative released by El Dorado County in 1994. Recent changes to both the language in the General Plan and the land use designations on the General Plan land use maps will generally increase the impacts identified in the DEIR. Once again, the El Dorado County Planning Commission and El Dorado

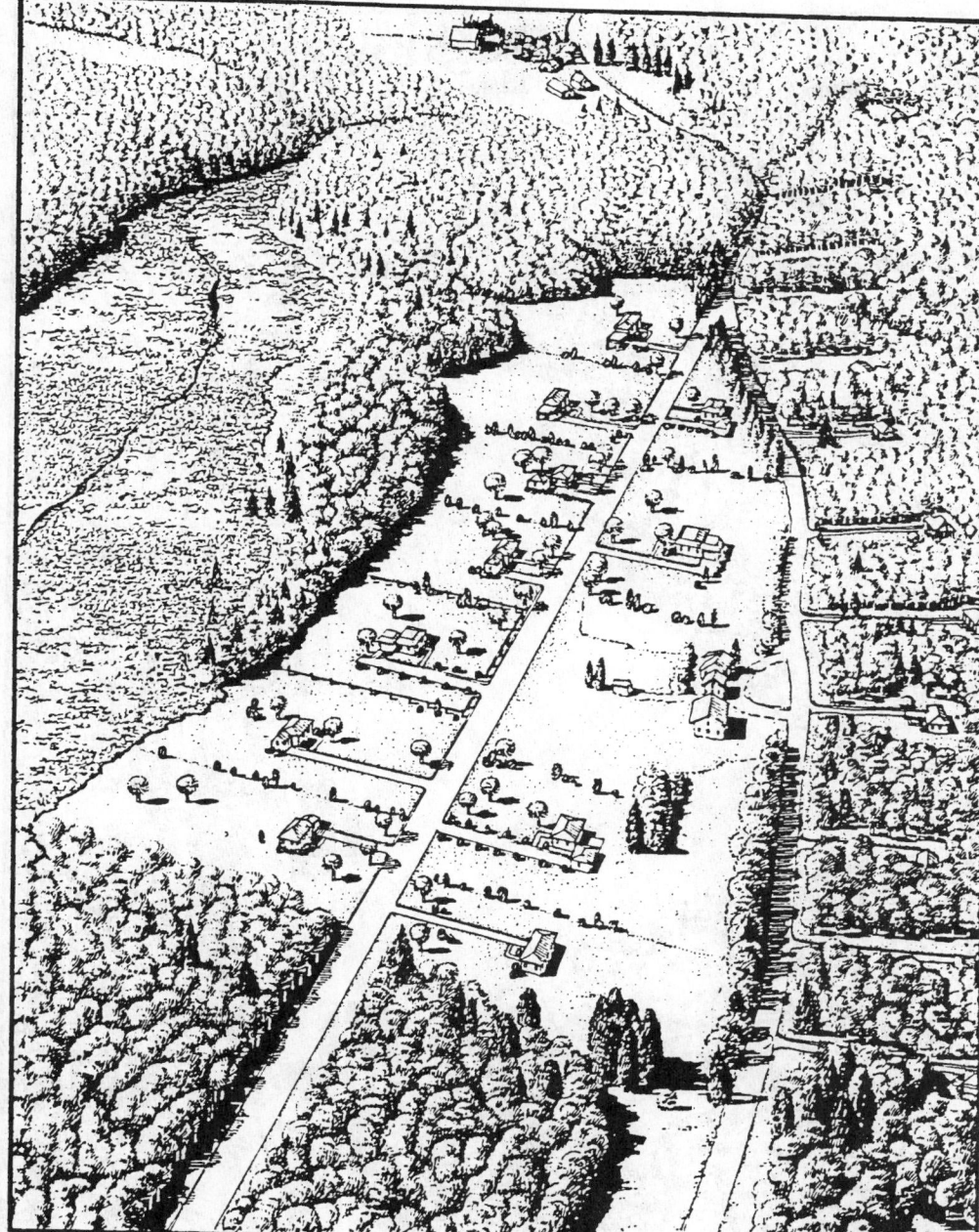

FIGURE 11.52

Development under conventional design and land use plan.

County Board of Supervisors are expected to adopt a final General Plan despite these impacts on the grounds of "overriding considerations." These include regional effects associated with air quality in the greater Sacramento metropolitan area, although the new version of the General Plan text strikes out most references to regional responsibilities.[83] Although many of those regional impacts were identified in the DEIR, the county is required only to accommodate regional housing needs under state law. The impacts of local land use policies on other resources of regional, state, or federal concern (e.g., water quality or biodiversity) must only be disclosed under CEQA. Unless other institutional mechanisms require modifications (e.g., for "listed" species under the federal Endangered Species Act of 1973 or air-quality regulations affecting transportation improvements), local agencies are not required to mitigate the impacts of their decisions.

It should be clear that the availability of appropriate growth management tools is therefore not enough to ensure their adoption and implementation. A variety of social factors have generally limited their adoption by local governments in the Sierra Nevada and other rural areas (Popper 1984). With few exceptions, those aspects of the planning process that do re-

FIGURE 11.53

Development under open
space development design
alternative.

sult in disclosure and mitiation of impacts are generally required under state law. The politics of growth management and planning are therefore critical to the success of planning and policy. It is also important to note that policy makers often lack adequate information about both the costs and the benefits of alternative patterns of human settlement and alternative policies. Establishment of basic information about these relationships may therefore be a necessary prerequisite to the adoption of more targeted policies. In some cases this information is collected through the CEQA process (Yuba County Community Services Department 1985), while some

jurisdictions have established policies as part of the general planning process (Granholm 1987). Better integration of these two processes has also been advocated widely, but it may require changes in state law (Johnston 1991; Duane 1993a). State and federal land and resource managers are generally dependent upon local planning agencies for information on future human settlement patterns, but that information is generally not well integrated with state and federal land and resource management and planning efforts (Forest and Rangeland Resources Assessment Program 1988). In response, some state agencies (e.g., the Strategic Planning Program of the California Department of Forestry and Fire Protection) have initiated efforts to improve local planning agencies' knowledge of and incorporation of regional-scale ecological concerns (Peck 1993; Giusti and Tinnin 1993).

Institutional Setting and Constraints on Effective Growth Management

Implementation of growth management policies is constrained in part by institutional considerations. One institutional issue that arises immediately for the "open space development design" approaches is who will have responsibility for management of and liability for open space. This issue has generally been dealt with through either a homeowners association or public management, although ownership in the open space lands is usually retained by the individual home owners. In either case, however, it clearly requires a greater degree of institutional coordination than that required under the large-lot conventional development model. Local residents are also often concerned that the "open space" lands could possibly be developed later, resulting in much higher levels of overall development. Complex and detailed codes, covenants, and restrictions (CC&Rs) are therefore necessary on titles and deeds to ensure maintenance of design integrity. In the case of public management, it may be necessary to establish a new entity (such as an open space district) or to empower an existing agency (such as the parks and recreation department of the local resource conservation district) with new powers. Getting the new system in place also has institutional and economic costs. Due to economies of scale, moreover, it may be more costly to accomplish this jurisdiction by jurisdiction. Many open space districts in metropolitan areas are consequently regional in nature, with responsibility for acquisition and management of lands in multiple jurisdictions (e.g., East Bay Regional Parks District, covering all of Alameda and Contra Costa Counties).

A second institutional problem emerges for application of these design concepts to areas that have already been parcelized. Coordinated planning is difficult and faces high transactions costs in these situations in the absence of other mechanisms to consolidate ownership and development rights (e.g., through TDRs). This is a common problem in the Sierra Nevada, where many parcels were created in the late 1960s and early 1970s that were later grandfathered in under stricter state land use and environmental planning requirements. Sometimes these parcels have multiple owners, and each of those owners may have purchased his or her parcel with an expectation of building a single-family home on the lot, making it extremely difficult politically to propose nondevelopment of specific lots, even if other areas might be allowed to develop at higher densities. Some mechanism must simultaneously be established to transfer development potential either from one parcel owner to another or from one parcel to another in order to rationalize land use to protect ecological, social, or economic concerns (e.g. the value of sur-

rounding properties, which could be negatively affected by the elimination of adjacent open space and views). This problem is less intractable when a single owner has retained control over multiple adjacent parcels, but that owner is still not generally under any legal obligation to modify the subdivision map and associated lot lines. In fact, the prospect of significant public opposition to a project could lead to development of a project with greater impacts simply because the existing subdivision design can avoid potential delays associated with discretionary review. This is also true for many large-acreage sites where a developer faces a choice between easy approval of a conventional development or great uncertainty associated with a more innovative open space development design. Not surprisingly, the landowner usually pursues the less risky conventional development (Arendt 1994).

Institutional considerations also affect the degree to which matters of regional, state, or federal concern are addressed in land use planning. California has a long tradition of "home rule" on matters of local land use, for example, with fierce local resistance to the imposition of state or regional controls on land use decision-making authority. Ironically, however, many of the problems associated with existing parcelization and substandard infrastructure in the Sierra Nevada reflect local land use decisions that predate stricter state requirements from the early 1970s (*Grass Valley Union* 1970a, 1970b). These state requirements include the California Environmental Quality Act (CEQA) of 1970, the Subdivision Map Act (SMA) of 1973, and General Plan consistency requirements dating from legislation in 1971. In all three cases, the state imposed new land use planning or environmental analysis requirements on the local land use planning process. Some of these requirements were substantive, such as the SMA requirement that subdivisions of more than four parcels have developed infrastructure before parcels could be sold. General Plans and zoning ordinances are also now required to be consistent under state law, and local authorities are required to have a specific set of elements in their General Plans. Many of the state-imposed requirements, however, are largely procedural. Significant impacts must be mitigated under CEQA, for example (a substantive requirement), but local authorities can avoid this substantive requirement by making findings of "overriding considerations," effectively translating the substantive requirement into a procedural requirement. The result is that many significant environmental impacts of human settlement are not mitigated.

Examples of regional or state land use control are limited and have usually involved resources of state, federal, or international significance. These have included filling restrictions in the San Francisco Bay, where the San Francisco Bay Conservation and Development Commission (BCDC) was established by state legislation in 1965; Lake Tahoe, where the bi-state Tahoe Regional Planning Agency (TRPA) was established by congressional action and the bi-state Tahoe Compact in 1970; and the California coastal zone, where the

California Coastal Commission was established through a vote of the California electorate in 1971 and subsequent legislation in 1976. Subsequent efforts to establish regional land use authorities have been rejected by the legislature in the face of strong opposition from elected local officials in the jurisdictions whose authority would be reduced. Regional institutions have been successfully established in California for the management of resources that have traditionally been managed by the state, however, such as the Bay-Delta Oversight Commission (BDOC), or where local jurisdictions have not traditionally exercised authority (e.g., ground-water management). Several regional planning efforts have also been pursued to address endangered species concerns in California. Many state and federal functions are also administered through regional offices, many of which are organized along ecological boundaries (e.g., air-quality management districts). Voluntary associations of government also act as a clearinghouse for state and federal grants administration, information generation, and coordinated planning. These "councils of government" (COGs) include the Association of Bay Area Governments (ABAG). Similar organizations in the Sierra Nevada include the Sierra Planning Organization (SPO) and the Sierra Economic Development District (SEDD). Neither plays a strong role in land use planning for the Sierra Nevada.

The California Department of Fish and Game has commented on General Plans in terms of their impacts on wildlife; the California Department of Forestry and Fire Protection has raised fire safety issues related to density, vegetation management, and access; the regional water-quality control boards have identified potential water-quality issues and the California Department of Parks and Recreation has addressed the impact of some policies on recreational activities. Despite this participation, however—which has been quite limited due to staff and budgetary limitations—local land use planners and decision makers are not required to modify General Plans to reflect the state's resource concerns. State agencies also participated in the local planning process in the late 1960s and early 1970s, sometimes raising concerns that have been realized a quarter-century later (Gerstung 1970, 1973). The linkage between state resource management concerns and local land use planning still remains weak. Local land use planning decisions continue to affect state resource management interests in a very direct way, however. State agencies have primary responsibility for wildlife, fire suppression, and water quality, for example, but the local authorities retain all control over many of the factors affecting the exercise of those state responsibilities. The state, in turn, retains some discretionary control over some resources (e.g., transportation and wastewater treatment plant funding) that directly affect local land use planning and patterns of human settlement. That discretionary control, however, has generally not been used to encourage alternative growth management policies, based on the potential impacts of settlement patterns on another state agency's area of responsibility. Coordination among state agencies on these issues appears to be relatively weak.

Finally, the linkage between the local land use planning process and federal agencies—who control roughly three-fifths of the land base in the Sierra Nevada—is also weak. Local officials have often commented on federal land management plans, but federal agencies rarely comment on or attempt to influence local land use plans. The exceptions are those cases in which local land use policies could clearly affect the capacity of the federal agencies to manage their own lands. Examples include proposed land use designations that could facilitate high-density development projects that could impede public access to recreational resources or could be critical to public use of those recreational resources (e.g., ski resorts on federal lands, where the "base" facilities are on adjacent private lands). Local land use planning and development policies can clearly have a wide range of more diffuse indirect effects, however, that could potentially constrain a wide range of management practices. Increased local settlement could increase recreational demands, alter wildlife habitat for species that travel seasonally up- and downslope, or even increase local opposition to timber harvest and fuel-management practices on federal lands.

This set of management implications is merely illustrative of the effects of human settlement on the health and sustainability of Sierra Nevada ecosystems. Considerably more research is necessary to determine the precise nature of the impacts of alternative patterns of human settlement on the system and the capacity of alternative growth management policies to mitigate those effects. Even in the absence of such definitive research, it should nevertheless be clear that human settlement in the Sierra Nevada has had and will continue to have a profound effect upon the ecological, social, and economic characteristics of the Sierra Nevada. Land and resource managers at the local, regional, state, and federal levels must now address those effects in future planning. The challenge of managing the impacts of human settlement raises a number of complex issues about the relationship between different levels of government and the relationship between private property rights and community and ecological well-being, but those conflicts are likely to be compounded in the future if institutions fail to address them proactively today (Niebanck 1984; Dubbink 1984).

ACKNOWLEDGMENTS

This research would not have been possible without the assistance of a number of graduate students, consultants, and colleagues from the University of California, Berkeley, and the Sierra Nevada Ecosystem Project. Karl Goldstein provided critical and consistent GIS analytic support throughout the project and prepared most of the maps in the report. He was clearly the most important resource I had while completing this research project. Karl worked closely with me on both the GIS analysis and the conceptual development of spatial modeling approaches described in the text. He also prepared all of the final maps.

Karl built on previous work by Phil Griffiths, who did the GIS analysis of 1970–90 population growth trends by CCD for me in 1992–93 (before SNEP). Phil was assisted during that early GIS effort by David Paradise, who volunteered his time to the research while on a brief sabbatical from Silicon Graphics, Inc. (SGI). Chris Thomas completed the analysis of the draft environmental impact reports for the Nevada and El Dorado County General Plans and the analysis of the implications of General Plan buildout for the transportation system in El Dorado County. He also prepared the initial population growth allocation spreadsheet by CCD for 1990–2040. Melody Tannam collected and input the 1850–1990 census data by county, city, and ethnicity and prepared several related charts. Dan Barry completed a literature review and annotated bibliography on the phenomena of exurban growth, demographic change, and rural communities in transition. Nabiha Megateli prepared an overview of septic systems and wastewater treatment in Nevada and El Dorado Counties. Craig Mayer completed the descriptive statistical analysis and bivariate analysis of 1990 census variables with the assistance of Karl Goldstein. Karl Goldstein, Craig Mayer, and Nabiha Megateli digitized the Nevada and El Dorado County General Plan land use maps. Miho Rahm, Jenna Lloyd, Barbara Hadenfeldt, and Carey Pelton provided administrative support through the Institute for Urban and Regional Development. Bill Mallgren assisted with the spatial analysis of the assessor's database and ownership consolidations, and Malcolm McDaniel organized the bibliography and in-text citations using EndNote. Carrie Salazar prepared final charts and graphs from my preliminary figures in Excel.

SNEP Science Team members and special consultants Greg Greenwood, Rowan Rowntree, Joe McBride, Bill Stewart, Jonathan Kusel, Dennis Machida, and Doug Leisz provided important insights on the forces driving human settlement in the Sierra Nevada. My colleagues Ted Bradshaw and Bob Twiss at the University of California, Berkeley, offered similar insights, while my colleague John Radke provided critical GIS support. Brian Muller shared his work with Ted Bradshaw on modeling exurban development in the Central Valley for the American Farmland Trust, while Bob Johnston from the University of California, Davis, shared his efforts to model transportation and land use for the Sacramento Area Council of Governments (including Placer and El Dorado Counties). Sharon Boivin and Pat Norman of the Nevada County Planning Department and Pierre Rivas of the El Dorado County Planning Department were very helpful with materials for the case studies, as was Nevada County GIS coordinator Diana Carolan. Information on water supplies and hydroelectric development issues in Nevada and El Dorado Counties was provided by Les Nicholson of the Nevada Irrigation District,

Marie Davis of the Georgetown Divide Public Utility District, Rob Alcott of the El Dorado Irrigation District, and El Dorado County water resources consultant Jim Doolittle.

Participants at our September 1994 workshop on our assessment included Betty Riley, Michael Reeves, and Lynne Purvis from the Sierra Planning Organization/Sierra Economic Development District; Bob Roan of the High Sierra Resource Conservation and Development Council; Placer County Planning Director Fred Yeager; Amador County Supervisor Stephanie D'Agostini; Amador County Planning Director Susan Grijalva; Steve Chilton and Kelly Berger from the Tahoe Regional Planning Agency; Greg Greenwood, Cathy Bleier, and Sheila Peck from the Strategic Planning Program at the California Department of Forestry and Fire Protection; and Susan Kocher of Forest Community Research (working with Jonathan Kusel on the SNEP social assessment chapters [Kusel, 1996; Doak and Kusel, 1996]). CDF ranger Jim Smith and California Department of Fish and Game employees Stafford Lehr, Daniel Hinz, and Bob Mapes also attended another workshop near Placerville in May 1995. All of their comments helped us frame the assessment and focus on the key questions.

Former El Dorado County supervisor Bill Center and former El Dorado County planning commissioner Ray Griffiths also provided insight on the development of the El Dorado County General Plan from 1991 to 1994. Dozens of participants in the Nevada County General Plan update process provided additional information on the Nevada County General Plan. These included members of the Resolution Committee (Dale Creighton, Sam Dardick, Peter Van Zant, and Laurie Oberholtzer), representatives from the Rural Quality Coalition (Peter Van Zant and Laurie Oberholtzer), and representatives from the California Association of Business, Property, and Resource Owners (Maskey Heath, Todd Juvinall, and Margaret Urke). All of these individuals and others involved in planning and land use issues in both counties provided information that assisted development of this assessment. None of them is responsible for the subsequent analysis or conclusions we have reached.

Graduate students in the Environmental Planning Studio at the University of California, Berkeley, reviewed specific aspects of the Nevada County General Plan in May 1993 (Cecilia Collados, Philip Griffiths, Scot Medbury, Andrew Partos, Leora Elazar, Steven Lewis, Perl Perlmutter, Karl Goldstein, Jeff Wutzke, Sarah Marvin, Juliet Lamont, Sheila Peck, Jennifer Knauer, Kallie Marie Kull, Trang Ngan Le, Norrie Cooper, Rebecca Coffman, Robert Faulstich, and Ming Zhao) and the El Dorado County General Plan in May 1994 (Nelia Badilla Forest, Malcolm McDaniel, Lori Tsung, Tim Hargrave, John Deck, Craig Mayer, Chris Thomas, Andrew Delaney, Rachel Arthur, Jennifer Vick, Andrea Lucas, Maria Wiseman, and Gretchen Hayes). Their individual papers and classroom discussion were invaluable, as were conversations with other graduate students and colleagues. Three anonymous peer reviewers and several members of the public also provided important comments on an earlier draft of the assessment.

Harry L. Dodson generously provided permission to use the perspective drawings in figures 11.51–11.53, which were illustrated by Kevin Wilson based on plans by Dodson Associates, Landscape Architecture and Planning, in Ashfield, Massachusetts.

My wife, Teresa McGlashan, provided patient and critical support as a "SNEP widow" during many months of late-night and weekend analysis and writing. My son, Cody Kenneth Duane-McGlashan, provided inspiration for the future of the Range of Light as the seventh generation to spend time in the Sierra Nevada. He entered the world on September 23, 1995, and he will be able to check on the validity of our analyses in the year 2040 at the early age of 45. His existence now makes our abstract, future projections of population growth and land conversion from 1990 to 2040 seem much less distant and much more concrete. Ultimately, he and his children will judge the value of the Sierra Nevada Ecosystem Project and assessments like this one.

This assessment built on work that was previously funded in part by the Committee on Research, the Beatrix Farrand Fund of the Department of Landscape Architecture, and the Townsend Humanities Center at the University of California, Berkeley. The Strategic Planning Program of the California Department of Forestry and Fire Protection also provided additional funding for the development of digital databases that were critical to the assessment. The author alone remains responsible for the contents of the assessment.

NOTES

1. We do not include some counties that border the northern boundary of the Sierra Nevada, such as Lassen.
2. Portions of other counties (e.g., Lassen in California and Washoe, Carson, and Douglas in Nevada) are in the Sierra Nevada, but the CCD boundaries for those counties extend into areas that include significant populations of non-Sierra residents. We have therefore limited our analysis to the forty-six CCDs in the eighteen California counties. The total population for the Sierra Nevada is therefore slightly larger than this.
3. Historical plaque beneath "Chinese Wall" and railroad tracks at Donner Summit and old Highway 40.
4. Based on pamphlets 25 (1) and 25 (2) of the Nevada County Historical Society, written by Patrick Tinloy.
5. Based on the Sierra Nevada portion of the population of the eighteen California counties included here.
6. An estimated 818,000 Californians moved from California to the interior West between 1985 and 1991, and many of these moved to rural or exurban regions of those states. The Sierra Nevada has also experienced continuing growth, and rural areas throughout California are growing despite net domestic emigration from California from 1990 to 1992. By comparison, only 250,000 Americans migrated westward on the Oregon Trail to California from 1843 to 1865 (others arrived by clipper ship or via other routes).

7. Plumas, Sierra, Nevada, El Dorado, Amador, Calaveras, Alpine, Mariposa, and Mono Counties.
8. Data for the state of Nevada were unavailable at the time of the analysis, so the results presented here refer only to those portions of the Sierra Nevada within California. Because the state of Nevada was not a signatory to the Biodiversity MOU, however, the focus of the policy recommendations is on California.
9. All figures cited are from our analysis of the 1990 census and cited in detail in Griffiths (1993). Note that these data are only for those persons 5 years of age and older, since younger ones were not alive in 1985!
10. This analysis is based on table 3.1 of Griffiths (1993).
11. Although the standard retirement age is 65 for many people, we have highlighted the age cohort of 55 years of age or older due both to the data on immigration and to the significant equity capital that has allowed many Sierra Nevada immigrants to "semi-retire" at an earlier age. The American Association of Retired Persons also uses 55 years of age as the basis for eligibility in the AARP, regardless of employment status.
12. The impact of these prisons on these communities is described in the social assessment by Doak and Kusel (1996). We have not calculated community-specific social and demographic characteristics.
13. This correction is most important for Amador and Tuolumne Counties' overall ethnicity estimates.
14. These characterizations of changes since 1990 are based upon personal observation and conversations with residents of the Lake Tahoe subregion (including Truckee, in Nevada County) and the eastern Sierra subregion. Employers in the tourism and construction industries commented on the increase in their utilization of a Latino workforce since 1990. This appears true for both large and small businesses. Based upon these conversations, it appears that there is very little employment of illegal aliens in the formal sector. Stricter penalties against employees and Proposition 187, together with tighter federal border controls, appear to have minimized the role of illegal aliens in the economy of the Sierra region. Bilingual education data are presented in Elliott-Fisk et al. (1996) and community poverty data are presented in Doak and Kusel (1996).
15. Nelson (1992) offers a detailed discussion of the literature on exurban development and suggests criteria for defining exurban regions within the constraints of U.S. Census Bureau data collected at the county level. His definition emphasizes the role of the central city in metropolitan regions as an employment center for exurban households. We argue that the conception of exurban development should be construed more broadly to include patterns of economic activity that are dependent upon and integrated with urban centers but not physically proximate. This means that some exurban households can be located well beyond commuting distance to cities, in areas that would otherwise be considered "rural" based on their overall appearance or their apparent physical relationship to the nearest metropolitan region.
16. Note that our definition of exurbia is broader than Nelson's (1992) and would include many areas that may be classified as "rural" by his criteria or could be classified as "metropolitan" by the census. An example of the former would be Calaveras County; an example of the latter would be Placer County.
17. The Sierra Nevada experienced its first significant population increase during this same decade.
18. Some recent data suggest that rural areas have again experienced

a small net in-migration from 1990 to 1991. Surprisingly, nonmetropolitan areas had a lower unemployment rate than metropolitan areas in fiscal year 1992 (for the first time in thirteen years, since the "rural renaissance" of the 1970s supposedly ended). The factors driving this shift are discussed in detail later.

19. The basis for this conclusion is described in our discussion of census block group data later.

20. These results are for the entire sample of 748 respondents. More detailed cross-tabs are also available by supervisorial district, age, housing status, income, school children, education, occupation, political ideology, length of residence, Sacramento commuters, area, acreage of parcel, June 1990 voter status, political party, and sex. These cross-tabs reveal some significant differences among subgroups, which highlight some of the differences in values between newcomers and oldtimers.

21. It is extremely difficult to get this information, for race is a highly charged issue in American society. Anecdotal evidence strongly supports this hypothesis, but more systematic research is needed on the issue.

22. Some writers have challenged the idyllic representation of small-town life and "the demonization of city life" (Zukin 1993) as unrealistic, and the "latent preference" cited by Nelson (1992) and Blackwood and Carpenter (1978) may indeed reflect a romantic vision of nonmetropolitan living. It is nevertheless a genuine preference for many (but certainly not all) Americans, however, and they are now able to pursue it.

23. Forthcoming in 1996. The author reviewed a final draft manuscript for Island Press in January 1995.

24. Assuming an 8%–10% interest rate and typical insurance and taxes in California, the monthly (before tax) cost of a mortgage is approximately 1% of a thirty-year, fixed-rate mortgage. Annual costs of $10,000–$20,000 for private schools equal $833–$1,667 per month. Avoiding those costs therefore frees a comparable amount for a mortgage, allowing one to acquire a mortgage of approximately $83,300–$166,700 (average = $125,000) without a reduction in net cash flow. The median owner-occupied household housing value in the Sierra Nevada was $128,678 in 1990. Lower current interest rates for mortgages translate into even greater home purchasing power for each dollar of savings from education.

25. This is a key selling feature for real estate advertisements for Lake Wildwood in Nevada County.

26. This appears to be true with Lake of the Pines in Nevada County, where local law enforcement personnel attribute a higher crime rate to high levels of "latchkey" children in two-income, commuter families. The high level of commuters in the community has reduced the daytime adult presence.

27. According to a survey conducted by the U.S. Fish and Wildlife Service, Americans spent some $14 billion on "primary nonconsumptive wildlife recreational pursuits" in 1985. The actual value of those recreational experiences was probably much higher, because there is only a limited "market" for these activities. In the jargon of economics, in other words, there was a large uncaptured consumer surplus.

28. Detailed statistics on the relative affordability of housing in Nevada County are presented in the draft 1994 Nevada County General Plan Housing Element. The comparison to costs in Sacramento is based upon materials prepared by Common Ground Communities for a Community Development Block Grant ap-

plication in 1995. The author is on the board of directors of Common Ground Communities.

29. Home buyers in the late 1980s and early 1990s purchased their homes at high prices and have seen slow or negative growth in their equity since then. This reduces the future potential for significant equity gains that could then free another wave of equity refugees to migrate to exurbia.

30. The final version of the Nevada County General Plan and the latest draft El Dorado County General Plan state that maintaining rural quality and environmental quality are essential goals for each county.

31. Assuming that the lower-bound DOF forecast is 6% greater than the CCSCE forecast and that the higher-bound DOF forecast is 8% greater than the lower-bound DOF forecast (1.08 * 1.06 =1.14).

32. The DOF estimates for 1990 differ slightly from the 1990 census figures, which are from April 1990.

33. The San Francisco Bay Area referred to here includes the nine counties, Alameda, Contra Costa, Marin, Napa, San Francisco, San Mateo, Santa Clara, Solano, and Sonoma, that are members of the Association of Bay Area Governments. Population figures are taken directly from census data summarized in published tabular form for 1940, 1950, 1960, 1970, 1980, and 1990. No digital sources were available.

34. These amenities include the Bureau of Reclamation's Folsom Lake, on the edge of the foothills.

35. Two of the three homes listed for sale in the May-June 1994 issue of the internal "Intel Folsom News" classified ads section were in El Dorado County, while the third was in nearby Placer County. We do not have any statistically valid data on the residential location of high-technology employees in Folsom.

36. The San Joaquin valley town of Merced may also have significant growth if it becomes the site for the tenth campus of the University of California. Closure of the nearby Castle Air Force Base is at present threatening to stall economic development in the area, but a new UC campus would be likely to serve as an incubator for a wide range of employment opportunities both in research and in the private sector.

37. This has been suggested as a likely problem by Ted Bradshaw at the University of California, Davis.

38. This estimate is based upon the sum of all Sierra region CCDs within the ten-county study area.

39. These values indicate the order of magnitude of the population, rather than the absolute number. Villages are usually in the 100s and up to maybe 1,000s of people, for example, while towns are in the 1,000s and possibly into the 10,000s. Rural and exurban regions usually do not have communities larger than maybe 10,000–20,000 people, but there may be some centers up to 20,000–50,000. Note that the U.S. Census would treat any county with a community larger than 50,000 in population as "metropolitan."

40. A "minor" subdivision typically involves a parcel split that results in four or fewer lots, thereby avoiding detailed planning and environmental review under the state Subdivision Map Act of 1973.

41. This has been required by California law only since the early 1970s, however, and most Sierra Nevada counties did not have zoning ordinances that were consistent with their General Plans until at least 1970.

42. This conclusion is supported by review of the 1980 Nevada County General Plan, 1982 Grass Valley General Plan, 1994–95

Nevada County General Plan update, and 1994–95 El Dorado County General Plan update. Local officials are consistently unwilling to "downzone" below existing parcelization.

43. Analysis of these environmental characteristics follows the basic framework established by Ian McHarg in his book *Design with Nature* (McHarg 1992).

44. Based upon a detailed review of General Plan documents for Nevada County and the city of Grass Valley and preliminary review of General Plan documents for Placer, El Dorado, Tuolumne, Inyo, and Mono Counties and the cities of Nevada City and Mammoth Lakes. The Lake Tahoe region is an exception due to the detailed environmental thresholds established by the Tahoe Regional Planning Agency (TRPA).

45. Existing market data on consumer preferences are limited by the lack of suitable alternatives for comparison, however, so it is difficult to ascertain how consumers would respond to the alternatives.

46. Many similar villages that dot the New England countryside date from the eighteenth century.

47. Fires often destroyed these small towns in their earliest days, with subsequent rebuilding using brick to ensure greater resistance. The current appeal of these small towns in part reflects this evolution over time.

48. There were multiple attempts to incorporate Truckee before final incorporation as a municipality.

49. All data are from the U.S. Bureau of the Census STF 3A, Population and Housing, 1990 Census. Based upon an independent analysis of Locally Adjusted Personal Income (LAPI) data for 1989, however, Stewart (1996) has determined that the 1990 census data underestimate transfer payments by a factor of 2 to 3. This error is concentrated in both the lower-income households (due to a failure to report AFDC and related social welfare payments) and the higher-income groups (due to a failure to report interest, investment, and dividend income). These estimates of median total income are therefore lower than actual median incomes.

50. The data available for this assessment did not offer a clear conclusion regarding this question.

51. Minimum lot size requirements vary widely with jurisdictions, reflecting both the high uncertainty and lack of detailed analysis of septic or well system risk associated with various soils, slopes, and other natural factors. Higher minimum parcel sizes of 10 or 20 acres per dwelling unit are often required to maintain the rural character of a place or to protect some sensitive area (e.g., steep slopes or an adjacent wetland), but those larger parcel sizes are not usually required to meet public health concerns. It is important to note that the general application of a minimum lot size requirement means that the site-specific capability to accommodate water and septic is usually not evaluated.

52. Parcel size distribution and land uses are discussed in detail later for Nevada and El Dorado Counties.

53. This has been a more significant problem to date in the Central Valley than in the Sierra Nevada.

54. Several Sierra Nevada counties have adopted these "right-to-farm" ordinances over the past decade.

55. Note that a shift in the mix of agricultural supplies may also occur rather than an overall reduction in the total value of such supplies. Many "ranchette" activities require significant supply expenditures that may actually increase the overall level of total economic activity in the agricultural supply sector.

56. Nearly half (48%) of the eleven western states and three-fifths of the Sierra Nevada is owned by the federal government, with much higher proportions of some rural counties in the Sierra Nevada.

57. For example, there were significant public protests against land trades by the Bureau of Land Management in Nevada County in 1988 due to concerns about herbicides for industrial forestry activities.

58. Materials for the People for Open Space and the Greenbelt Alliance emphasize this in the Bay Area.

59. Ted Bradshaw is now at the University of California, Davis.

60. We used the 1992 TIGER files; the 1994 files were still not available at the end of June 1995.

61. We considered use of LANDSAT Thematic Mapper satellite imagery from 1972, 1986, and 1992 for a more comprehensive analysis of land conversion, but we determined that we could not reliably differentiate changes in the TM measurements based on land conversion alone in the areas for which we had TM data. The census data is the only reliable source for human settlement for the entire Sierra Nevada region.

62. We suspect that it reflects the failure to include some blocks that are outside the SNEP core region that were otherwise included in the CCDs, because the SNEP core region does not coincide with the CCDs.

63. SPP was established following the Forest and Rangeland Resources Assessment Program (FRRAP). This threshold was used in previous studies by CDF to indicate an "urban" land conversion classification.

64. There has been a movement by some counties (notably Catron County, New Mexico, and Nye County, Nevada) to establish local land use controls over federal lands. The U.S. Justice Department filed suit in federal court in 1995 to establish exemption of federal lands from local land use regulation, and there has never been a successful case establishing local government jurisdiction over federal lands. The case was still pending as we prepared this assessment report (Larson 1995).

65. We identified this pattern through our analysis of the El Dorado County General Plan, and it was confirmed by Pierre Rivas, principal long-range planner, El Dorado County Planning Department, at our workshop with local planners from the five-county central Sierra Nevada region in September 1994.

66. Nearby private lands have been acquired through this funding source for recreational purposes.

67. The Final Draft Nevada County General Plan was adopted in March 1994 just as Truckee incorporated.

68. The Nevada County General Plan map shows only 10,141 acres in this classification (when calculated by Arc/Info), but we estimated ultimate buildout of these subdivisions at 14,022 total housing units.

69. A total of 10,127 acres are designated SDA, but Gold Country Ranch is only 8,232 acres.

70. El Dorado County had developed a parcel-based digital coverage by October 1994, but the county surveyor would not make it available for our use until the board of supervisors authorized a fee schedule for its release. Despite repeated requests, we were unable to trade our working data for use of the parcel-based coverage. We also believe there are serious errors associated with georeferencing of that coverage. Nevada County is developing a similar coverage based upon scanned images of zoning designa-

tion maps (ZDMs) and digitized centroids of each parcel, but that coverage will not allow direct calculation of area.

71. We estimated that two-thirds of the parcels in the class "Can be subdivided into 3–5 parcels" were in this class, while one-third could be subdivided into 5 parcels and were therefore subject to the Subdivision Map Act as "major" subdivisions. Note potential methodological problems cited in the text, however.

72. This statement conflicts with other reports from the Nevada County Planning Department during the General Plan process that there are over 17,000 unimproved parcels in Nevada County out of a total of only 51,000 (Boivin 1991–95).

73. These data are from 1992, before Sierra Pacific Industries acquired most of Michigan-California's assets.

74. The area designated for this "new town" on the 1994 draft Nevada County General Plan will now be designated as a development reserve zone, to be studied when a specific project is proposed.

75. The Rural Quality Coalition of Nevada County and the Federation of Neighborhood Associations have led the challenge to the Nevada County General Plan, while the El Dorado County Taxpayers for Quality Growth have led the challenge to the El Dorado County General Plan. No case has yet been filed, but it is likely that the final El Dorado County General Plan will be litigated in part on the grounds of CEQA violations.

76. Robert Johnston at the University of California, Davis, is currently modeling the impact of transportation system on land use in El Dorado and Placer Counties for the Sacramento Area Council of Governments. The results of his analysis were not available in time for inclusion or review here.

77. The most recent draft of the El Dorado County General Plan, dated August 17, 1995, establishes much lower LOS standards for many roads in comparison with the 1994 draft El Dorado County General Plan.

78. Threats to potable water supplies have been one of the primary drivers of sewering since the 1970s.

79. Both counties completed separate fiscal analyses of their respective Final Draft General Plans, but they came to very different conclusions. El Dorado County determined that development under its General Plan would not improve the fiscal condition of the county and that LOS standards would continue to decline. Nevada County determined that its General Plan would be fiscally sound for the county, but that analysis failed to consider many of the costs discussed here. It also suffered from the same inadequacy of the DEIR: reliance upon General Plan buildout estimates that fail to account for existing parcelization.

80. Their effect on the distribution of patch size (and total patch area) is a function of both their geographic distribution and the "edge effect" attributed to them from the roadway into the interior of adjoining patches. A four-lane freeway and a two-track dirt road probably have significantly different effects, and they should be considered accordingly. Much more research needs to be done on the effective edge effect of different road types and uses.

81. *Euclid v. Ambler Realty Co.*, 272 U.S. 365 (1926).

82. There are two major exceptions to this generalization in the final Nevada County General Plan adopted on October 13, 1995: (1) temporary elimination of the "new town" or special development area in western Nevada County (until further studies are completed) and (2) reductions in allowable densities in several areas that had high-density or medium-density designations. The overall buildout population estimates appear to have been re-

duced more significantly, to around 140,000 people, through other changes in assumptions about household size, net versus gross development densities, and several other more minor changes in assumptions. As noted in our analysis, however, these reductions are overwhelmed by more significant buildout underestimation errors due to failure to account for existing parcelization.

83. Compare the 1994 Draft El Dorado County General Plan to the revision released on August 17, 1995.

REFERENCES

Adams, L. W., and L. E. Dove. 1989. *Wildlife reserves and corridors in the urban environment: Guide to ecological landscape planning and resource conservation.* Columbia, MD: National Institute for Urban Wildlife.

Alexander, C., M. Silverstein, S. Ishikawa, and I. Fiksdahl-King with M. Jacobson and S. Angel. 1977. *A pattern language.* New York: Oxford University Press.

Alterman, R. 1994. *Can farmland preservation work? Lesson for the USA from a six-nation comparative perspective.* Tempe, AZ: Association of Collegiate Schools of Planning.

American Farmland Trust. 1995. *Alternatives for future urban growth in California's Central Valley: The bottom line for agriculture and taxpayers.* Davis, CA: American Farmland Trust.

Arendt, R. 1994a. *Designing open space subdivisions: A practical step-by-step approach.* Media, PA: Natural Lands Trust.

———. 1994b. *Rural by design: Maintaining small town character.* Chicago: American Planning Association.

Bank of America, Greenbelt Alliance, Resources Agency, and Alliance for Affordable Housing. 1994. *Beyond sprawl.* San Francisco: Bank of America.

Barringer, F. 1993. Population grows in rural America, studies say. *New York Times,* 25 May.

Barry, D., and T. P. Duane. 1994. Annotated bibliography of literature on exurban growth, nonmetropolitan employment, and rurality. Unpublished manuscript. Davis, CA: Sierra Nevada Ecosystem Project.

Beale, C. 1975. *The revival of population growth in nonmetropolitan America.* Washington, DC: U.S. Department of Agriculture, Economic Research Service.

Beesley, D. 1996. Reconstructing the landscape: An environmental history, 1820–1960. In *Sierra Nevada Ecosystem Project: Final report to Congress,* vol. II, chap. 1. Davis: University of California, Centers for Water and Wildland Resources.

Blackwood, L. G., and E. H. Carpenter. 1978. The importance of antiurbanism in determining residential preferences and migration patterns. *Rural Sociology* 43 (1): 31–47.

Blumenfeld, H. 1954. The tidal wave of metropolitan expansion. *Journal of the American Institute of Planners* 20 (1): 3–14.

———. 1986. Metropolis extended. *Journal of the American Planning Association* 52 (3): 346–48.

Boivin, S., Nevada County Planning Department. 1991–95. Interviews with the author.

Bonfante, J. 1993. Boom time in the Rockies: More jobs and fewer hassles have Americans heading for the hills. *Time,* 6 September.

Bowman, C. 1994. Placer plan predicts freeway tie-ups. *Lincoln News Messenger.*

Bradshaw, T., and B. Muller, University of California, Berkeley. 1994–95. Interviews with the author.

———. 1994. *Mapping the California exurbia: Patterns of non-metropolitan population growth.* Tempe, AZ: Association of Collegiate Schools of Planning.

Brissenden, J., Alpine County Board of Supervisors. 1993–94. Interviews with the author.

Cahill T. A., J. J. Carroll, D. Campbell, T. E. Gill, 1996. Air quality. In *Sierra Nevada Ecosystem Project: Final report to Congress*, vol. II, chap. 48. Davis: University of California, Centers for Water and Wildland Resources.

California Department of Finance. 1993. *Population projections by race/ethnicity for California and its counties 1990–2040.*

Calthorpe, P. 1989. *The pedestrian pocket book: A new suburban design.* New York: Princeton Architectural Press.

———. 1993. *The next American metropolis: Ecology, community, and the American dream.* New York: Princeton Architectural Press.

Carlino, G. 1985. Declining city productivity and the growth of rural regions. *Journal of Urban Economics* 18:11–27.

Castells, M. 1989. *The informational city: Information technology, economic restructuring, and the urban-regional process.* New York: B. Blackwell.

Center, Bill, El Dorado County Supervisor. 1991–95. Interviews with the author.

Center for the Continuing Study of the California Economy (CCSCE). 1995. Letter to the author.

Central Valley Regional Water Quality Control Board (CVRWQCB). 1989. *Waste discharge requirements for city of Grass Valley domestic wastewater treatment plant (Nevada County).* Order No. 89-005 (NPDES No. CA0079898).

———. 1992. *Order no. 92-112 requiring the city of Grass Valley (Nevada County) to cease and desist from discharging waste contrary to waste discharge requirements.*

Cervero, R. 1986. *Suburban gridlock.* New Brunswick, NJ: Center for Urban Policy Research.

———. 1993. Changing live-work spatial relationships: Implications for metropolitan structure and mobility. Paper presented at the Fourth International Workshop on Technological Change and Urban Form: Productive and Sustainable Cities, Berkeley, California.

Collados, C., and P. Griffiths. 1993. Implications of age composition and migration for the Nevada County General Plan. Paper prepared for T. P. Duane, University of California, Berkeley, 7–8.

Cranmer Engineering and H. Halatyn. 1971. *General water and sewer plan: Nevada County, California.* Nevada City, CA: Nevada County Planning Department.

Davis, F. W., and D. M. Stoms. 1996. Sierran vegetation: A gap analysis. In *Sierra Nevada Ecosystem Project: Final report to Congress*, vol. II, chap. 23. Davis: University of California, Centers for Water and Wildland Resources.

Davis, J. S. , A. C. Nelson, and K. J. Dueker. 1994. The new 'burbs. *Journal of the American Planning Association* 60 (1): 45–59.

Davis, M., Georgetown Divide Public Utility District. 1994. Interview with the author.

Defenders of Wildlife. 1989. *Preserving communities and corridors.* Washington, DC: Defenders of Wildlife.

DeGrove, J. M., with D. A. Miness. 1992. *The new frontier of land policy: Planning and growth management in the states.* Cambridge, MA: Lincoln Institute of Land Policy.

de Neufville, J. I., ed. 1981. *The land use policy debate in the United States.* New York: Plenum Press.

Department of Water Resources. 1973. *Sierra foothills investigation, Central District.*

Diringer, E. 1994. Crowd on the new frontier: Elbow room running out. *San Francisco Chronicle,* 15 December.

Doak, S. C., and J. Kusel. 1996. Well-being in forest-dependent communities, part II: A social assessment focus. In *Sierra Nevada Ecosystem Project: Final report to Congress*, vol. II, chap. 13. Davis: University of California, Centers for Water and Wildland Resources.

Doolittle, J., consultant to El Dorado County. 1994–95. Interviews with the author.

Dorward, S. 1990. *Design for mountain communities: A landscape and architectural guide.* New York: Van Nostrand Reinhold.

Dowall, D. E. 1984. *The suburban squeeze: Land use policies in the San Francisco Bay Area.* California Series in Urban Development. Berkeley: University of California Press.

———. 1991. Less is more: The benefits of minimal land development regulation. Working paper 531, Institute of Urban and Regional Development, University of California, Berkeley.

Dowall, D. E., and J. D. Landis. 1981. Land use controls and housing costs: An examination of San Francisco Bay Area communities. Working paper 81-24, Institute of Urban and Regional Development, University of California, Berkeley.

Doyle, J. 1995. Fear fuels Mill Valley fire plan: Vulnerable town learns from Oakland hills, Inverness. *San Francisco Chronicle,* 16 October.

Drace, R. P., President, Common Ground Communities. 1993–95. Interviews with the author.

Duane, T. P. 1993a. Managing the Sierra Nevada. *California Policy Choices* 8:169–94.

———. 1993b. *Exodus to exurbia: The threat of population growth in rural "buffer zone" regions to the conservation of biological diversity.* Tempe, AZ: Society for Conservation Biology.

———. 1993c. *Managing exurban sprawl in the Sierra Nevada foothills.* Philadelphia: Association of Collegiate Schools of Planning.

———. 1996. Recreation in the Sierra. In *Sierra Nevada Ecosystem Project: Final report to Congress*, vol. II, chap. 19. Davis: University of California, Centers for Water and Wildland Resources.

Duany, A., and E. Plater-Zyberk. 1991. *Towns and town-making principles.* Cambridge and New York: Harvard Graduate School of Design and Rizzoli.

Dubbink, D. 1984. I'll have my town medium-rural, please. *Journal of the American Planning Association* 50 (4): 406–18.

Egan, T. 1995. The serene fortress: Many seek security in private communities. *New York Times,* 3 September.

EIP Associates. 1995. *Grass Valley wastewater treatment plant expansion, draft environmental impact report.*

Elazar, D. J. 1987. *Building cities in America: Urbanization and suburbanization in a frontier society.* Lanham, MD: Hamilton Press.

El Dorado County Planning Department. 1994. *Final Draft, El Dorado County General Plan.*

El Dorado County Water Agency. 1994. *Preliminary assessment of water demand and supply balance for El Dorado County.* Placerville, CA: El Dorado County Water Agency.

El Dorado Irrigation District Staff. 1994. Interviews with the author and review of CAD-based digital files.

Elliott-Fisk, D. L., R. A. Rowntree, T. A. Cahill, C. R. Goldman, G. Gruell, R. Harris, D. Leisz, S. Lindstrom, R. Kattelmann, D. Machida, R. Lacey, P. Rucks, D. A. Sharkey, and D. S. Ziegler. 1996. Lake Tahoe case study. In *Sierra Nevada Ecosystem Project: Final report to Congress,* vol. III. Davis: University of California, Centers for Water and Wildland Resources.

Fishman, R. 1987. *Bourgeois utopias.* New York: Basic Books.

Forero, L., L. Huntsinger, and W. J. Clawson. 1992. Land use change in three San Francisco Bay Area counties: Implications for ranching at the urban fringe. *Journal of Soil and Water Conservation* 47 (6): 475–80.

Forest and Rangeland Resources Assessment Program (FRRAP). 1988. *California's forests and rangelands: Growing conflict over changing uses.* Sacramento: California Department of Forestry and Fire Protection.

Fortmann, L., and L. Huntsinger. 1989. The effects of nonmetropolitan population growth on resource management. *Society and Natural Resources* 2:9–22.

Fugro-McClelland (West), Inc. 1994. *Cinnabar planned development project (rezone 92-25), draft environmental impact report.* Placerville, CA: Planning Division, Community Development Department, El Dorado County.

Garreau, J. 1991. *Edge city: Life on the new frontier.* New York: Doubleday.

Gerstung, E. 1970. *A brief survey of the impact of subdivision activity on the fish and wildlife resources of Nevada County.* Sacramento: California Department of Fish and Game.

———. 1973. Land development and fish and wildlife protection. *Cal-Neva Wildlife,* 134–37.

Gilpin, M., and M. E. Soule. 1986. Minimum viable populations: Processes of species extinction. In *Conservation biology,* edited by M. Soule. Sutherland, MA: Sinauer Associates.

Giusti, G. A., and P. J. Tinnin, eds. 1993. *A planner's guide for oak woodlands.* Berkeley: University of California, Department of Forestry and Resource Management, Integrated Hardwood Range Management Program.

Glickfeld, M., and N. Levine. 1992. *Regional growth . . . local reaction: The enactment and effects of local growth control and management measures in California.* Cambridge, MA: Lincoln Institute of Land Policy.

Governor's Interagency Council on Growth Management (GICGM). 1993. *Strategic growth: Taking charge of the future: A blueprint for California.* Sacramento, CA: Governor's Office of Planning and Research.

Granholm, S. L. 1987. *Tuolumne County wildlife project.* Adopted by resolution 303-87. Sonora, CA: Tuolumne County Planning Department.

Grass Valley Union. 1970a. 10,000 new lots—13,000 more coming. 9 January.

———. 1970b. Planning commission rarely turns down subdivisions. 10 January.

———. 1995a. Chinatowns were immigrant havens. 15 July.

———. 1995b. Computer makers flock to California's capital. 10 October.

———. 1995c. Retirement community a success in Roseville: Sun City's new development cost-effective. 1 April .

Griffiths, P. 1993. The Sierra now: A compendium of social and economic statistics describing California's Sierra Nevada region. M.C.P. professional report, Department of City and Regional Planning, University of California, Berkeley.

Griffiths, R., Former El Dorado County planning commissioner. 1995. Interviews with the author.

Hanson, M. E., and H. M. Jacobs. 1989. Private sewage system impacts in Wisconsin: Implications for planning and policy. *Journal of the American Planning Association,* Spring, 169–90.

Hargrave, T. 1993. The impact of a federal grazing fee increase on land use in El Dorado County, California. Paper prepared for T. P. Duane, University of California, Berkeley.

Harland Bartholemew and Associates 1994. *Draft environmental impact report.* Nevada City, CA: Nevada County Planning Department.

Harris, L. D. 1984. *The fragmented forest: Island biogeography theory and the preservation of biotic diversity.* Chicago: University of Chicago Press.

Hart, J. F. 1991. The perimetropolitan bow wave. *The Geographical Review* 81 (1): 35–51.

Harvey, D. 1993. Comments at the Western Humanities Conference, Stanford, California.

Hester, R. 1990. *Community design primer.* Mendocino, CA: Ridge Times Press.

High Country News. 1993. Small towns under seige. Special issue, 5 April.

———. 1994. Grappling with growth. 5 September.

Hoge, P. 1995. Rural growth: Promise or threat? Plans causing tension in booming Sierra counties; development in El Dorado pleases some, worries others. *Sacramento Bee,* 16 October.

Hudson, W. E., ed. 1991. *Landscape linkages and biodiversity.* Washington, DC: Island Press.

Inman, B. 1992. Haven on earth. *San Francisco Focus,* April, 59–67.

Innes, J. 1991. *Group processes and the social construction of growth management : The cases of Florida, Vermont, and New Jersey.* Working paper 542, Institute of Urban and Regional Development, University of California, Berkeley.

Innes, J., J. D. Landis, and T. K. Bradshaw. 1993. *Issues in growth management: Reprints of recent growth management writings by IURD associates.* Berkeley: University of California, Institute of Urban and Regional Development.

Intel Corporation. 1994. *Intel and its Folsom, California site.*

Irwin, B. 1989. *A discussion of the county General Plan and the role of strategic fire protection planning.* Sacramento: California Department of Forestry and Fire Protection.

Irwin, R. L. 1987. Local planning considerations for the wildland-structural intermix in the year 2000. In *Symposium on Wildland Fire 2000 in South Lake Tahoe, California,* 38–46. Berkeley, CA: U.S. Forest Service, Pacific Southwest Forest and Range Experiment Station.

J. Moore Methods, Inc. 1992. *El Dorado County General Plan Survey.* Sacramento and Placerville, CA: El Dorado County.

Jackson, K. T. 1985. *Crabgrass frontier: The suburbanization of the United States.* New York: Oxford Press.

Johnson, K. M. 1993. Demographic change in nonmetropolitan America, 1980–1990. *Rural Sociology* 58 (3): 347–65.

Johnston, R. A., and M. E. Madison. 1991. *Planning for habitat protection in California: State policies and county actions to implement CEQA through improved general plans.* Contract 8CA85456. Sacramento: California Department of Forestry and Fire Protection, Forest and Rangeland Resources Assessment Program (FRRAP).

Joseph, A., and B. Smit. 1981. Implications of exurban residential development. *Canadian Journal of Regional Science* 4 (2): 207–24.

Juvinall, T. 1995. Proponents of strict zoning dishonor Constitution. *Grass Valley Union,* 15 September.

Kemmis, D. 1990. *Community and the politics of place.* Norman: University of Oklahoma Press.

Knapp, G., and A. C. Nelson. 1992. *The regulated landscape: Lessons on state land use planning from Oregon.* Cambridge, MA: Lincoln Institute of Land Policy.

Koda, C. 1995. Dancing in the borderland: Finding our common ground in North America. Unpublished manuscript.

Kostof, S. 1991. *The city shaped: Urban patterns and meanings through history.* Boston: Little, Brown.

Kusel, J. 1996. Well-being in forest-dependent communities, part I: A new approach. In *Sierra Nevada Ecosystem Project: Final report to Congress,* vol. II, chap. 12. Davis: University of California, Centers for Water and Wetland Resources.

Lake Vera/Round Mountain Neighborhood Association. 1995. Neighborhood Plan.

Landis, J. D. 1988. Estimating the housing price effects of alternative growth management strategies in the city of San Diego. Working paper 88-148, Institute of Urban and Regional Development, University of California, Berkeley.

———. 1992. *Do growth controls work? An evaluation of local growth control programs in seven California cities.* Berkeley: California Policy Seminar.

Landon, D., Executive director, Nevada County Transportation Commission. 1994–95. Interviews with the author.

Lang, M. 1986. Redefining urban and rural for the U.S. Census of Population: Assessing the need and alternative approaches. *Urban Geography* 2:118–34.

Larson, E. 1995. Unrest in the West. *Time,* 23 October 52–66.

Lauer, S. 1995a. County says it will repay state loan: Cascade Shores sewage woes could threaten properties. *Grass Valley Union,* 8 March.

———. 1995b. State wants county's financial guarantee on sewer project. *Grass Valley Union,* 6 March.

———. 1995c. Subdivision to get new sewer. *Grass Valley Union,* 22 April.

Lehr, S., Fishery biologist, California Department of Fish and Game. 1995. Interview with the author.

Lenahan, A. 1995. Cascade Shores sewer renovation under way. *Grass Valley Union,* 4 October.

Livingston County Planning Department. 1991. *PEARL: Protect Environment, Agriculture, and Rural Landscape: An open space zoning technique.* Howell, MI.

Local Government Commission. 1992. *Land use strategies for more livable places.* Sacramento, CA: Local Government Commission.

Lovejoy, T. E., R. O. Bierregaard, Jr., A. B. Rylands, J. R. Malcom, C. E. Quintela, L. H. Harper, K. S. Brown Jr., A. H. Powell, G. V. N. Powell, H. O. R. Shubart, and M. B. Hays. 1986. Edge and other effects of isolation on Amazon forest fragments. In *Conservation biology,* edited by M. Soule. Sutherland, MA: Sinauer Associates.

Marois, M. B. 1995. Home market flat: Real-estate personnel report "negative equity" in county. *Grass Valley Union,* 8 July.

McBride, J. R., W. Russell, and S. Kloss. 1996. Impact of human settlement on forest composition and structure. In *Sierra Nevada Ecosystem Project: Final report to Congress,* vol. II, chap. 46. Davis: University of California, Centers for Water and Wildland Resources.

McHarg, I. 1992. *Design with nature.* 2nd ed. New York: John Wiley.

McKuen, P. 1994. Managing partner, Cook Ranch Partners. Telephone interview with the author.

Medbury, S. 1993. Biological invasions in the cismontane Sierra Nevada: Prediction and regulation. Paper prepared for T. P. Duane, University of California, Berkeley.

Megatelli, N., and T. P. Duane. 1995. Water quality, water supply, and wastewater disposal issues in Nevada and El Dorado Counties. Unpublished manuscript. Davis, CA: Sierra Nevada Ecosystem Project.

Miller, T., Acting planning director, Nevada County Planning Department. 1994–95. Personal communications.

Montgomery County Planning Commission. 1991. *Land preservation district: Model zoning provisions.* Norristown, PA.

———. 1992a. *Land preservation district: Land development standards.* Norristown, PA.

———. 1992b. *Land preservation: Old challenge . . . new ideas.* Norristown, PA.

Mooers, J. S. 1995. At last, a General Plan. *Grass Valley Union,* 14 October.

Nelson, A. C. 1992. Characterizing exurbia. *Journal of Planning Literature* 6 (4) : 350–68.

Nelson, A. C., and K. J. Dueker. 1990. The exurbanization of America and its planning policy implications. *Journal of Planning Education and Research* 9 (2): 91–100.

Nevada County General Plan. 1980. Nevada City, CA.

Nevada County Planning Commission. 1993. *Draft Nevada County general plan. Housing element,* II-127. Nevada City, CA: Nevada County Planning Commission.

———. 1994a. Final draft general plan—additional population and parcels. Handout at October 11 Resolution Committee meeting, Nevada County Planning Department, Nevada City, CA.

———. 1994b. Resolution committee notes for October 11. Staff report. Nevada County Planning Department, Nevada City, CA.

Nevada County Planning Department. 1994a. *Final draft Nevada County general plan.* Nevada City CA.

———. 1994b. *Final draft Nevada County general plan,* vol 2, p. V-9, table 1. Nevada City, CA.

Nevada County Transportation Commission. 1995a. *Regional Transportation Plan.*

———. 1995b. *Regional Transportation Plan,* 22–26.

———. 1995c. *Regional Transportation Plan,* table A-2.

Nevada Irrigation District Staff. 1994–95. Interviews with the author.

New York Times. 1993. Eastward, ho! The great move reverses. 30 May.

Niebanck, P. L. 1984. Dilemmas in growth management. *Journal of the American Planning Association* 50 (4): 403–5.

Norman, P., Nevada County Planning Department. 1982. Interviews with the author.

———. 1994–95. Interviews with the author.

Noss, R. F. 1991. Landscape connectivity: Different functions at different scales. In *Landscape linkages and biodiversity,* edited by W. E. Hudson. Washington, DC: Island Press.

Owens, P., University of California, Berkeley. 1991–95. Interviews with the author.

Pacific Gas and Electric Company staff. 1994. Telephone conversations with the author from Auburn, Jackson, and San Francisco.

Peck, S. 1993. *Landscape conservation planning: Preserving ecosystems in open space networks.* Berkeley: University of California Cooperative Extension, Department of ESPM, Integrated Hardwood Range Management Program.

Pimm, S. L. 1986. Community structure and stability. In *Conservation biology,* edited by M. Soule. Sutherland, MA: Sinauer Associates.

Pivo, G. 1988. *Preserving ruralness through cluster housing: Problem or opportunity?* Seattle: Washington State Department of Natural Resources and Department of Urban Design and Planning, University of Washington, College of Architecture and Planning.

Pivo, G., R. Small, and C. R. Wolfe. 1990. Rural cluster zoning: Survey and guidelines. *Land Use Law,* September, 3–10.

Popper, F. J. 1984. Rural land use policies and rural poverty. *Journal of the American Planning Association* 50 (3): 326–34.

Powers, T. M. 1996. *Extraction and the environment: The economic battle to control our natural landscapes.* Washington, DC: Island Press. Forthcoming.

Redman, A D. 1992. Making rural clustering work. Easton, MD: Redman/Johnston Associates.

Richardson, D., CH2M HILL, Emeryville, CA. 1992–94. Interviews with the author.

Rivas, P., El Dorado County Planning Department. 1993–95. Interviews with the author.

Sage, L., Nevada County Environmental Health Department. 1995. Interview with the author.

Sanchez, T. W., and A. C. Nelson. 1994. *Exurban and suburban residents: A departure from traditional location theory?* Tempe, AZ: Association of Collegiate Schools of Planning. Based on a working paper prepared for the Fannie Mae Office of Housing Research.

Schiffman, I. 1989. *Alternative techniques for managing growth.* Berkeley, CA: Institute of Governmental Studies.

Scott, J. M., B. Csuti, and S. Caicco. 1991. Gap analysis: Assessing protection needs. In *Landscape linkages and biodiversity,* edited by W. E. Hudson. Washington, DC: Island Press.

Shevock, J. R. 1996. Status of rare and endemic plants. In *Sierra Nevada Ecosystem Project: Final report to Congress,* vol. II, chap. 24. Davis: University of California, Centers for Water and Wildland Resources.

Soule, M. E., ed. 1986. *Conservation biology: The science of scarcity and diversity.* Sutherland, MA: Sinauer Associates.

———. 1991a. Theory and strategy. In *Landscape linkages and biodiversity,* edited by W. E. Hudson. Washington, DC: Island Press.

———.1991b. Land use planning and wildlife maintenance: Guidelines for conserving wildlife in an urban landscape. *Journal of the American Planning* Association 57 (3): 313–23.

Southworth, M. 1995. *Walkable suburbs? An evaluation of neotraditional communities at the urban edge.* Working paper no. 639, Institute of Urban and Regional Development, University of California, Berkeley.

Southworth, M., and E. Ben-Joseph. 1993. *Regulated streets: The evolution of standards for suburban residential streets.* Working paper 593, Institute of Urban and Regional Development, University of California, Berkeley.

Southworth, M., and P. Owens. 1992. *The evolving metropolis: Studies of community, neighborhood, and street form at the urban edge.* Working paper 579, Institute of Urban and Regional Development, University of California, Berkeley.

South Yuba River Citizens League. 1993. *The South Yuba: A wild and scenic report.* Nevada City, CA: South Yuba River Citizens League.

Spectorsky, A. C. 1955. *The exurbanites.* Philadelphia: Lippincott.

Stanford Ranch Information Center. 1994–95. *Stanford Ranch: A master planned community.* Rocklin, CA: Stanford Ranch. Also interviews with representatives at the information center, November 1994 and July 1995.

Starrs, P. F., and J. B. Wright. 1994. California, out—Great Basin growth and the withering of the Pacific idyll. *Geographical Review* (October): review draft.

Stein, J. M., ed. 1993. *Growth management: The planning challenge of the 1990s.* Newbury Park, CA: Sage.

Stewart, W. 1996. Economic assessment of the ecosystem. In *Sierra Nevada Ecosystem Project: Final report to Congress,* vol. III. Davis: University of California, Centers for Water and Wildland Resources.

Swain, W., U. S. Geological Survey, Sacramento, CA. 1994. Interviews with the author.

Teitz, M. 1990. California's growth: Hard questions, few answers. *California Policy Choices* 6:35–74.

Thomas, C. 1994. Potential impacts of rural development upon habitat in El Dorado County. Paper prepared for T. P. Duane, University of California, Berkeley.

———. 1995. The once and future El Dorado County: Transportation planning beyond the edge. Paper prepared for E. Deakin, University of California, Berkeley.

Thomas, C., and T. P. Duane. 1995a. A review of the Draft Environmental Impact Report for the El Dorado County General Plan. Unpublished manuscript. Davis, CA: Sierra Nevada Ecosystem Project.

Thomas, C., and T. P. Duane. 1995b. A review of the Draft Environmental Impact Report for the Nevada County General Plan. Unpublished manuscript. Davis, CA: Sierra Nevada Ecosystem Project.

Thompson, P. 1989. *Poison runoff.* New York: Natural Resources Defense Council.

Turner, K. M. 1973. *Sierra foothills investigation, central district.* Sacramento, CA: Department of Water Resources.

U.S. Bureau of the Census. 1970, 1980, 1990. Census files compiled by the U.S. Bureau of the Census. Database on magnetic tape and CD-ROM.

———. 1990. *STF 3A, Population and Housing, 1990 Census,* compiled by the U. S. Bureau of the Census. Database on magnetic tape and CD-ROM.

Vining, D. R., and A. Strauss. 1977. A demonstration that the current deconcentration of population in the United States is a clean break with the past. *Environment and Planning A* 9:751–58.

Walsh, J. 1991. The frontiers of white flight. *San Francisco Examiner, Image Magazine,* 17 November, 36–54.

Wardwall, J. M. 1982. The reversal of nonmetropolitan migration loss. In *Rural society in the U.S.: Issues for the 1980s,* edited by D. A. Dillman and D. J. Hobbs. Boulder, CO: Westview.

Weeks, D., A. E. Wieslander, H. R. Josephson, and C. L. Hill, 1943. *Land utilization in the northern Sierra Nevada.* Special Publication of the Giannini Foundation of Agricultural Economics. Berkeley: University of California, College of Agriculture, Agricultural Experiment Station.

Weiss, P. 1995. Off the grid. *The New York Times Magazine,* 8 January.

Wilcove, D. S., C. H. McLellan, and A. P. Dobson. 1986. Habitat fragmentation in the temperate zone. In *Conservation biology,* edited by M. Soule. Sutherland, MA: Sinauer Associates.

Willis, K. G. 1994. *Preserving traditional farming practices and landscapes in the United Kingdom: An economic appraisal.* Tempe, AZ: Association of Collegiate Schools of Planning.

Wolfe, C. R. 1990. The cluster alternative: A basis for private development in the public interest. *Connecticut Planner's Journal* 3 (4): 1–2.

Yaro, R. D., R. G. Arendt, H. L. Dodson, and E. A. Brabec. 1988. *Dealing with change in the Connecticut River Valley: A design manual for conservation and development.* Boston: Massachusetts Department of Environmental Management; Amherst, MA: Center for Rural Massachusetts; Cambridge, MA: Lincoln Institute of Land Policy; The Environmental Law Foundation.

Yuba County Community Services Department. 1985. *Final environmental impact report on the cumulative impacts of rural residential development on migratory deer in Yuba County.* Marysville, CA: Yuba County Community Services Department.

Zukin, S. 1993. Comments at the Western Humanities Conference, Stanford University, Stanford, California, October.

APPENDIX 11.1

Human Settlement Data

TABLE 11.A1

Sierra Nevada population and population growth by county, 1970–90.

County	1970	1980	1990	1970–80	70–80%	1980–90	80–90%	1970–90	70–90%
Alpine	484	1,097	1,113	613	127%	16	1%	629	130%
Amador	11,821	19,314	30,039	7,493	63%	10,725	56%	18,218	154%
Butte	101,969	143,851	182,120	41,882	41%	38,269	27%	80,151	79%
Calaveras	13,585	20,710	31,998	7,125	52%	11,288	55%	18,413	136%
El Dorado	43,833	85,812	125,995	41,979	96%	40,183	47%	82,162	187%
Fresno	413,329	514,621	667,490	101,292	25%	152,869	30%	254,161	61%
Inyo	15,571	17,895	18,281	2,324	15%	386	2%	2,710	17%
Kern	330,234	403,089	543,477	72,855	22%	140,388	35%	213,243	65%
Madera	41,519	63,116	88,090	21,597	52%	24,974	40%	46,571	112%
Mariposa	6,015	11,108	14,302	5,093	85%	3,194	29%	8,287	138%
Mono	4,016	8,577	9,956	4,561	114%	1,379	16%	5,940	148%
Nevada	26,346	51,645	78,510	25,299	96%	26,865	52%	52,164	198%
Placer	77,632	117,247	172,796	39,615	51%	55,549	47%	95,164	123%
Plumas	11,707	17,340	19,739	5,633	48%	2,399	14%	8,032	69%
Sierra	2,365	3,073	3,318	708	30%	245	8%	953	40%
Tulare	188,322	245,738	311,921	57,416	30%	66,183	27%	123,599	66%
Tuolumne	22,169	33,928	48,456	11,759	53%	14,528	43%	26,287	119%
Yuba	44,736	49,733	58,228	4,997	11%	8,495	17%	13,492	30%
Total	1,355,653	1,807,894	2,405,829	452,241	33%	597,935	33%	1,050,176	77%

TABLE 11.A2

Total area and private area in Sierra Nevada by county.

County	Total Area in Sierra Nevada (square km)	Percent of Total Area	Total Private Area in Sierra Nevada (square km)	Percent of Total Private Area	Percent of County in Public Land
Alpine	1,925	1.67%	155	0.36%	91.95%
Amador	1,432	1.25%	1,074	2.51%	25.00%
Butte	2,429	2.11%	1,720	4.01%	29.17%
Calaveras	2,493	2.17%	1,941	4.53%	22.14%
El Dorado	4,639	4.04%	2,416	5.64%	47.91%
Fresno	7,135	6.21%	1,673	3.90%	76.55%
Inyo	8,677	7.55%	144	0.34%	98.34%
Kern	8,924	7.76%	5,515	12.87%	38.20%
Lassen	12,219	10.63%	4,765	11.12%	61.00%
Madera	3,439	2.99%	1,389	3.24%	59.60%
Mariposa	3,722	3.24%	1,749	4.08%	53.00%
Modoc	10,820	9.41%	3,888	9.07%	64.07%
Mono	8,074	7.02%	671	1.56%	91.69%
Nevada	2,524	2.20%	1,711	3.99%	32.20%
Placer	3,371	2.93%	1,885	4.40%	44.09%
Plumas	6,769	5.89%	2,007	4.68%	70.35%
Shasta	5,629	4.90%	3,216	7.50%	42.88%
Sierra	2,491	2.17%	743	1.73%	70.19%
Tehama	2,742	2.38%	1,738	4.06%	36.61%
Tulare	8,626	7.50%	2,285	5.33%	73.51%
Tuolumne	5,891	5.12%	1,445	3.37%	75.46%
Yuba	999	0.87%	722	1.69%	27.68%
TOTAL	114,969		42,854		62.73%

TABLE 11.A3

Area versus housing units versus population by density class for 46 CCDs based on 1990 census blocks by housing unit density class (units/sq. mi.).

Housing Density (per sq. mi.)	Area (sq. mi.)	Area (acres)	Area % (of total)	Housing Units (in class)	Housing % (of total)	Population (in class)	Population % (of total)
None	9,238	5,912,437	28.87%	0	0.00%	1,630	0.27%
Less than 1	14,547	9,309,919	45.46%	3,264	1.07%	4,298	0.71%
1 to 2	2,361	1,510,985	7.38%	3,200	1.05%	14,160	2.34%
2 to 5	1,951	1,248,638	6.10%	6,298	2.06%	11,864	1.96%
5 to 10	1,274	815,084	3.98%	8,859	2.90%	20,022	3.31%
10 to 20	890	569,659	2.78%	12,690	4.15%	25,948	4.29%
20 to 40	733	468,938	2.29%	20,317	6.64%	43,015	7.11%
40 to 80	444	284,245	1.39%	25,298	8.27%	52,369	8.66%
80 to 160	266	170,454	0.83%	29,951	9.79%	64,139	10.61%
160 to 640	209	134,025	0.65%	63,427	20.73%	128,449	21.24%
Over 640	89	56,867	0.28%	132,616	43.35%	238,750	39.49%
Grand Total	32,002	20,481,252	100.00%	305,920	100.00%	604,644	100.00%

Cumulative Totals	Area (sq. mi.)	Area (acres)	% of total	% of 80+	% of 40+	% of 20+	% of 10+
640+ (< acre/unit)	88.86	56,867	0.28%	15.74%	8.81%	5.10%	3.38%
160+ (4 acres/unit)	298.27	190,893	0.93%	52.83%	29.57%	17.13%	11.33%
80+ (8 acres/unit)	564.60	361,347	1.76%	100.00%	55.97%	32.42%	21.46%
40+ (16 acres/unit)	1,008.74	645,592	3.15%		100.00%	57.93%	38.33%
20+ (32 acres/unit)	1,741.45	1,114,531	5.44%			100.00%	66.18%
10+ (64 acres/unit)	2,631.55	1,684,189	8.22%				100.00%

TABLE 11.A4

Area by housing density class by river basin (see key to CalWaterID codes and HUD classes below).

cawatid	Data	hudclass											Grand Total
		1	2	3	4	5	6	7	8	9	10	11	
514	Sum of sq_mile	545	492	192	177	89	108	97	88	55	33	10	1,887
	Sum of hu100	0	112	212	573	642	1,587	2,881	4,810	6,307	10,356	15,504	42,984
	Sum of pop100	609	90	116	785	1,052	3,109	7,051	12,544	15,719	24,726	34,046	99,847
518	Sum of sq_mile	1,557	1,251	247	186	142	84	49	12	15	12	5	3,561
	Sum of hu100	0	466	347	587	1,043	1,163	1,278	739	1,676	3,863	6,892	18,054
	Sum of pop100	48	635	335	1,112	1,728	2,097	2,482	1,409	3,446	6,361	12,635	32,288
532.2	Sum of sq_mile	201	43	62	66	69	79	45	27	18	14	5	628
	Sum of hu100	0	18	92	201	487	1,146	1,339	1,453	1,934	4,199	6,232	17,101
	Sum of pop100	0	48	125	379	1,124	2,729	3,238	3,606	4,984	10,233	15,234	41,700
533	Sum of sq_mile	32	8	16	113	109	66	24	7	3	4	1	382
	Sum of hu100	0	3	24	327	689	925	622	402	313	980	1,020	5,305
	Sum of pop100	0	6	44	695	1,437	1,699	1,288	818	770	1,421	1,612	9,790
534	Sum of sq_mile	272	165	111	96	29	31	51	8	14	11	4	793
	Sum of hu100	0	60	193	327	207	505	1,283	469	1,561	4,001	4,368	12,974
	Sum of pop100	108	45	3,965	605	368	741	2,480	670	2,153	5,147	4,833	21,115
536	Sum of sq_mile	459	981	31	67	25	38	33	28	22	19	7	1,710
	Sum of hu100	0	161	46	238	198	542	940	1,661	2,576	5,960	8,891	21,213
	Sum of pop100	34	215	68	450	260	1,099	1,870	3,843	4,721	11,167	14,954	38,681
537	Sum of sq_mile	171	0	0	0	0	0	0	0	0	0	0	171
	Sum of hu100	0	0	0	0	0	0	0	0	0	1	0	1
	Sum of pop100	0	0	0	0	0	0	0	0	0	1	0	1
552	Sum of sq_mile	339	774	93	79	41	44	18	14	1	1	0	1,403
	Sum of hu100	0	138	125	281	255	594	501	876	69	250	307	3,396
	Sum of pop100	0	146	229	656	502	1,437	916	1,093	115	267	218	5,579
553	Sum of sq_mile	206	367	110	46	27	27	12	9	0	1	0	808
	Sum of hu100	0	140	163	154	201	379	298	487	132	292	189	2,435
	Sum of pop100	85	352	440	322	383	766	649	945	249	506	392	5,089
554	Sum of sq_mile	599	1,265	277	28	63	12	16	21	5	7	4	2,297
	Sum of hu100	0	233	349	95	404	149	418	1,253	474	1,878	4,919	10,172
	Sum of pop100	0	148	324	163	706	144	657	1,887	798	2,855	7,072	14,754
601	Sum of sq_mile	263	430	6	0	4	12	0	0	1	1	0	717
	Sum of hu100	0	163	7	1	39	185	0	21	162	165	252	995
	Sum of pop100	0	185	13	0	77	144	0	9	150	117	301	996
637.3	Sum of sq_mile	327	11	2	15	4	0	6	0	0	0	0	365
	Sum of hu100	0	3	2	39	41	2	151	15	7	288	106	654
	Sum of pop100	0	4	0	46	0	0	50	13	4	120	33	270
509521	Sum of sq_mile	520	458	30	45	25	21	18	7	6	4	4	1,139
	Sum of hu100	0	189	33	200	188	286	549	396	659	1,403	5,195	9,098
	Sum of pop100	0	235	21	466	471	637	1,210	991	1,318	3,049	10,719	19,117

cawatid	River Basin
509,521	Sacramento
514	American
516,517	Yuba
518	Feather
532.2	Cosumnes
532.4,532.6	Mokelumne
533	Calaveras
534	Stanislaus

cawatid	River Basin
536	Tuolumne
537	Merced
538-540	San Joaquin
552	Kings
553	Kaweah
554	Kern
555	Tule
556	Caliente

cawatid	River Basin
601	Mono Basin
603,552	Owens
623-625	Mojave
630,631	Walker
632,633	Carson
634-636	Truckee
637	Eagle Lake

Housing Density Classes correspond to table 11.A3 (1 = 0 units per sq. mi.).

continued

TABLE 11.A4 (continued)

cawatid	Data	1	2	3	4	5	6	7	8	9	10	11	Grand Total
							hudclass						
516517	Sum of sq_mile	579	212	166	301	158	122	128	82	47	33	10	1,837
	Sum of hu100	0	109	206	992	1,115	1,723	3,707	4,611	5,223	9,059	13,564	40,309
	Sum of pop100	157	69	165	1,668	2,472	4,306	8,753	11,064	12,736	21,227	28,219	90,836
603552.33	Sum of sq_mile	845	1,670	206	134	9	14	24	4	5	7	5	2,923
	Sum of hu100	0	327	280	372	61	160	583	223	536	2,077	11,441	16,060
	Sum of pop100	76	365	481	815	106	441	887	467	967	4,421	14,594	23,620
630631	Sum of sq_mile	172	540	112	25	24	4	7	1	1	0	0	886
	Sum of hu100	0	355	131	79	139	78	191	50	76	170	95	1,364
	Sum of pop100	72	132	309	174	385	136	261	72	126	322	165	2,154
632633	Sum of sq_mile	94	385	59	23	1	4	0	0	1	0	0	567
	Sum of hu100	0	34	101	67	4	58	10	15	77	46	80	492
	Sum of pop100	0	23	112	131	7	119	12	34	109	120	168	835
538539540	Sum of sq_mile	516	1,549	137	105	121	79	87	33	13	9	2	2,651
	Sum of hu100	0	144	183	352	832	1,193	2,340	1,809	1,516	2,424	2,208	13,001
	Sum of pop100	71	247	423	812	1,990	2,365	5,340	4,361	3,562	4,220	2,657	26,048
634635636	Sum of sq_mile	397	65	24	56	77	27	36	31	20	16	18	767
	Sum of hu100	0	10	29	193	574	320	964	1,958	2,295	5,769	29,899	42,011
	Sum of pop100	1	11	55	260	311	277	1,094	971	2,645	6,383	37,759	49,767
532.4,532.6	Sum of sq_mile	259	100	141	73	51	65	41	34	19	10	2	795
	Sum of hu100	0	17	229	247	357	902	1,164	1,986	2,140	2,639	2,678	12,359
	Sum of pop100	4	15	127	499	446	1,888	2,228	3,916	3,971	5,169	5,407	23,670
555.(1,2,4,5)	Sum of sq_mile	222	409	104	93	48	15	12	16	3	2	1	924
	Sum of hu100	0	202	160	303	378	223	373	912	332	510	770	4,163
	Sum of pop100	100	337	496	492	685	305	525	732	525	681	738	5,616
556.1,625.3	Sum of sq_mile	63	112	69	16	67	2	4	0	0	0		334
	Sum of hu100	0	40	98	49	384	34	88	25	31	1	12	762
	Sum of pop100	0	118	131	86	601	46	151	39	61	2	22	1,257
623,624(1,2), 625(1,2,4)	Sum of sq_mile	187	627	44	1	12	1	1	0	0	0	0	873
	Sum of hu100	0	227	49	49	1	21	23	2	32	14	0	417
	Sum of pop100	0	402	44	71	6	76	61	7	72	36	0	782
637.(1,2,4)	Sum of sq_mile	907	582	75	110	47	45	19	11	4	3	2	1,804
	Sum of hu100	0	131	107	424	333	619	494	621	423	889	3,669	7,710
	Sum of pop100	509	200	279	921	785	1,539	1,224	5,691	1,080	2,187	8,748	23,163
(blank)	Sum of sq_mile	11,541	14,435	1,994	2,521	1,382	994	802	416	277	237	261	34,861
	Sum of hu100	0	3,475	2,889	8,020	9,768	14,322	22,635	23,422	31,096	74,994	558,320	748,941
	Sum of pop100	9,246	7,880	14,142	21,955	33,945	41,168	63,933	67,306	89,957	207,460	1,531,369	2,088,361
Total Sum of sq_mile		21,275	26,928	4,306	4,388	2,613	1,896	1,530	852	531	426	340	65,083
Total Sum of hu100		0	6,757	6,055	14,164	18,345	27,116	42,832	48,216	59,647	132,228	676,611	1,031,971
Total Sum of pop100		11,120	11,908	22,444	33,563	49,854	67,268	106,360	122,488	150,238	318,198	1,731,895	2,625,336
SNEP Sum of sq_mile		9,734	12,493	2,312	1,867	1,230	902	728	436	253	189	78	30,222
SNEP Sum of hu100		0	3,282	3,166	6,144	8,577	12,794	20,197	24,794	28,551	57,234	118,291	283,030
SNEP Sum of pop100		1,874	4,028	8,302	11,608	15,909	26,100	42,427	55,182	60,281	110,738	200,526	536,975

cawatid	River Basin
509,521	Sacramento
514	American
516,517	Yuba
518	Feather
532.2	Cosumnes
532.4,532.6	Mokelumne
533	Calaveras
534	Stanislaus

cawatid	River Basin
536	Tuolumne
537	Merced
538-540	San Joaquin
552	Kings
553	Kaweah
554	Kern
555	Tule
556	Caliente

cawatid	River Basin
601	Mono Basin
603,552	Owens
623-625	Mojave
630,631	Walker
632,633	Carson
634-636	Truckee
637	Eagle Lake

Housing Density Classes correspond to table 11.A3 (1 = 0 units per sq. mi.).

TABLE 11.A5

Land-use designations in draft general plans (1994).

Zoning Classification	Allowable Density	Area (sq. miles)	Area (acres)	Share (%)
El Dorado County "Alternative"				
Multi-Family Residential	5-24 Units/Acre	3.1	1,972	0.20%
High Density Residential	1-7 Units/Acre	26.0	16,641	1.66%
Medium Density Residential	1 Unit/1-5 Acres	47.6	30,433	3.03%
Low Density Residential	1 Unit/5-20 Acres	211.8	135,580	13.49%
Rural Residential	1 Unit/20-40 Acres	98.3	62,900	6.26%
Natural Resource	1 Unit/40+ Acres	1010.9	646,960	64.36%
Rural Residential Low Density	1 Unit/40+ Acres	75.6	48,411	4.82%
Planned Community Three	Average 0.62 - 3.57 Units/Acre	1.1	722	0.07%
Planned Community One	Average 1.4 Units/Acre	1.6	1,028	0.10%
Planned Community Two	Average 4.1 Units/Acre	3.0	1,915	0.19%
Commercial	N.A.	6.1	3,888	0.39%
Industrial	N.A.	3.5	2,222	0.22%
Open Space	N.A.	71.9	46,045	4.58%
Public Facility	N.A.	2.5	1,577	0.16%
Research & Development	N.A.	1.4	908	0.09%
Area Plan	Unknown Density	5.5	3,545	0.35%
Unlabeled on Plan Map	Unknown Density	0.7	434	0.04%
TOTAL		1,571	1,005,183	100%
Planned Communities		6	3,665	0.36%
Total w/o PCs		1,565	1,001,518	99.64%
El Dorado County "Project"				
Multi-Family Residential	5-24 Units/Acre	3.1	2,000	0.20%
High Density Residential	1-7 Units/Acre	25.7	16,447	1.64%
Medium Density Residential	1 Unit/1-5 Acres	53.8	34,448	3.43%
Low Density Residential	1 Unit/5-10 Acres	180.3	115,390	11.48%
Rural Residential	1 Unit/10-40 Acres	295.0	188,801	18.78%
Natural Resource	1 Unit/Over 160 Acres	963.1	616,415	61.32%
Planned Community Three	Average 0.62 - 3.57 Units/Acre	3.0	1,915	0.19%
Planned Community One	Average 1.4 Units/Acre	1.6	1,028	0.10%
Planned Community Two	Average 4.1 Units/Acre	1.1	722	0.07%
Commercial	N.A.	5.2	3,345	0.33%
Industrial	N.A.	3.4	2,171	0.22%
Open Space	N.A.	31.0	19,863	1.98%
Public Facility	N.A.	2.7	1,702	0.17%
Research & Development	N.A.	1.5	932	0.09%
TOTAL		1,571	1,005,180	100%
Planned Communities		6	3,665	0.36%
Total w/o PCs		1,565	1,001,514	99.64%
Nevada County General Plan				
Urban High Density	20 Units/Acre	0.7	427	0.07%
Urban Medium Density	6 Units/Acre	1.5	955	0.15%
Urban Single Family	4 Units/Acre	7.0	4,507	0.73%
Residential	1 Unit/1.5 Acres	10.9	6,950	1.13%
Estate	1 Unit/3 Acres	32.9	21,046	3.41%
Rural 5	1 Unit/5 Acres	70.7	45,258	7.33%
Rural 10	1 Unit/10 Acres	68.3	43,703	7.08%
Rural 20	1 Unit/20 Acres	53.5	34,227	5.55%
Rural 30	1 Unit/30 Acres	25.3	16,214	2.63%
Rural 40	1 Unit/40 Acres	57.7	36,958	5.99%
Rural 160	1 Unit/160 Acres	2.8	1,786	0.29%
Forest 40	1 Unit/40 Acres	54.6	34,914	5.66%
Forest 80	1 Unit/80 Acres	1.0	665	0.11%
Forest 160	1 Unit/160 Acres	436.8	279,583	45.30%
Forest 640	1 Unit/640 Acres	32.0	20,481	3.32%
Business Park	n.a.	0.9	562	0.09%
Community Commercial	n.a.	1.2	737	0.12%
City	n.a.	3.4	2,200	0.36%
Highway Commercial	n.a.	0.2	149	0.02%
Industrial	n.a.	1.3	843	0.14%
Neighborhood Commercial	n.a.	0.4	247	0.04%
City	n.a.	1.9	1,224	0.20%
Office Professional	n.a.	0.2	138	0.02%
Open Space	n.a.	44.7	28,606	4.63%
Planned Development	n.a.	15.2	9,725	1.58%
Planned Residential Community	n.a.	15.8	10,141	1.64%
Public and Institutional	n.a.	6.6	4,252	0.69%
Rural Commercial	n.a.	0.1	60	0.01%

continued

TABLE 11.A5 (continued)

Zoning Classification	Allowable Density	Area (sq. miles)	Area (acres)	Share (%)
Recreation	n.a.	0.4	258	0.04%
Special Development Area	n.a.	15.8	10,127	1.64%
Service Commercial	n.a.	0.0	2	0.00%
Unknown	n.a.	0.3	185	0.03%
Village Business Park	n.a.	0.1	72	0.01%
Water	n.a.	10.0	6,432	(Not Included)
TOTAL		974	623,634	100%
Total in PD and SDA		31	19,852	3.22%
Total w/o PD or SDA		943	603,783	96.78%
Total in GV and NC		5	3,424	0.55%
Total w/o GV and NC		969	620,210	99.45%
Total w/o either above		938	600,358	96.23%

TABLE 11.A6

Housing units forecast under El Dorado County General Plan "buildout" scenarios.

Zoning Classification	Middle Density	Middle Housing	Middle Share (%)	Low Density	Low Housing	Low Share (%)	High Density	High Housing	High Share (%)
El Dorado County General Plan Project Description (densities are per sq. mi.)									
Multi-Family Residential	9,280	29,000	18%	3,200	10,000	11%	15,360	47,999	32%
High Density Residential	2,560	65,787	42%	640	16,447	17%	4,480	115,127	76%
Medium Density Residential	384	20,669	13%	128	6,890	7%	640	34,448	23%
Low Density Residential	96	17,309	11%	128	23,078	25%	64	11,539	8%
Rural Residential	40	11,800	8%	16	4,720	5%	64	18,880	12%
Natural Resource	4	3,853	2%	4	3,853	4%	4	3,853	3%
Planned Community Three	1,341	4,012	3%	397	1,187	1%	2,285	6,837	4%
Planned Community One	896	1,439	1%	896	1,439	2%	896	1,439	1%
Planned Community Two	2,624	2,960	2%	2,624	2,960	3%	2,624	2,960	2%
TOTAL		156,829	100%		70,574	100%		243,083	100%
Planned Communities		8,412	5%		5,587	6%		11,237	7%
Total w/o PCs		148,417	95%		64,987	94%		231,847	93%
El Dorado County General Plan Alternative (densities are per sq. mi.)									
Multi-Family Residential	9,280	28,601	18%	3,200	9,862	14%	15,360	47,340	19%
High Density Residential	2,560	66,563	41%	640	16,641	24%	4,480	116,485	46%
Medium Density Residential	384	18,260	11%	128	6,087	9%	640	30,433	12%
Low Density Residential	80	16,948	11%	32	6,779	10%	128	27,116	11%
Rural Residential	24	2,359	1%	16	1,573	2%	32	3,145	1%
Natural Resource	16	16,174	10%	16	16,174	24%	16	16,174	6%
Rural Residential Low Density	16	1,210	1%	16	1,210	2%	16	1,210	0%
Planned Community Three	1,341	1,512	1%	397	448	1%	2,285	2,577	1%
Planned Community One	896	1,439	1%	896	1,439	2%	896	1,439	1%
Planned Community Two	2,624	7,852	5%	2,624	7,852	12%	2,624	7,852	3%
TOTAL		160,919	100%		68,065	100%		253,772	100%
Planned Communities		10,804	7%		9,739	14%		11,869	5%
Total w/o PCs		150,114	93%		58,326	86%		241,903	95%

TABLE 11.A7

Housing units forecast under Nevada County General Plan "buildout" scenarios.*

Zoning Classification	Middle Density	Middle Housing	Middle Share (%)	Low Density	Low Housing	Low Share (%)	High Density	High Housing	High Share (%)
Nevada County General Plan (densities are per sq. mi.)									
Urban High Density	12,800	8,539	7%	12,800	8,539	9%	12,800	8,539	6%
Urban Medium Density	3,840	5,729	4%	3,840	5,729	6%	3,840	5,729	4%
Urban Single Family	2,560	18,030	14%	2,560	18,030	19%	2,560	18,030	12%
Residential	427	4,634	4%	427	4,634	5%	427	4,634	3%
Estate	213	7,015	5%	213	7,015	7%	213	7,015	5%
Rural 5	128	9,052	7%	128	9,052	10%	128	9,052	6%
Rural 10	64	4,370	3%	64	4,370	5%	64	4,370	3%
Rural 20	32	1,711	1%	32	1,711	2%	32	1,711	1%
Rural 30	21	540		21	540	1%	21	540	
Rural 40	16	924	1%	16	924	1%	16	924	
Rural 160	4	11		4	11		4	11	1%
Forest 40	16	873	1%	16	873	1%	16	873	1%
Forest 80	8	8		8	8		8	8	
Forest 160	4	1,747	1%	4	1,747	2%	4	1,747	1%
Forest 640	1	32		1	32		1	32	
City**	1,000	3,438	3%	1,000	3,438	4%	1,000	3,438	2%
City**	1,000	1,912	1%	1,000	1,912	2%	1,000	1,912	1%
Planned Development**	1,089	16,549	13%	337	5,125	5%	1,841	27,973	18%
Planned Residential Community**	860	14,022	11%	860	14,022	15%	860	14,022	9%
Special Development Area**	1,841	29,128	23%	397	6,278	7%	2,624	41,519	27%
TOTAL		128,265	100%		93,991	100%		152,080	100%
Total in PD and SDA		45,677	36%		11,403	12%		69,492	46%
Total w/o PD or SDA		82,588	64%		82,588	88%		82,588	54%
Total in GV and NC		5,350	4%		5,350	6%		5,350	4%
Total w/o GV and NC		122,915	96%		88,641	94%		146,730	96%
Total w/o either above		77,238	60%		77,238	82%		77,238	51%

*Note text explanation of likely errors in Nevada County General Plan forecasts due to existing parcelization at higher than densities allowable under the Plan.
**See text explanation of methods used to estimate high, middle and low average "build out" densities for these unspecified density land use classifications.

TABLE 11.A8

Coefficient of variation (C.V.) analysis of map book page data from 1992 El Dorado and Nevada County assessor's records.

C.V.	Nevada County						El Dorado County					
	Parcels	Area	Pages	%Parcels	%Area	%Pages	Parcels	Area	Pages	%Parcels	%Area	%Pages
0	959	1,654	140	2%	4%	7%	24,959	14,247	1,069	60%	41%	57%
0-0.499	21,967	27,994	1,028	51%	61%	52%	4,454	4,651	154	11%	13%	8%
0.5-0.99	11,459	12,515	517	27%	27%	26%	9,609	13,735	526	23%	40%	28%
1.0-1.99	6,705	3,022	211	16%	7%	11%	1,923	1,793	108	5%	5%	6%
2.0-2.99	918	336	29	2%	1%	1%	126	24	8			
3.0-3.99	351	96	17	1%		1%	289	262	12	1%	1%	1%
4.0-4.99	179	80	6				48	9	3			
5.0-9.99	336	65	9	1%			64	36	4			
10.0+	102	21	5				178	11	5			
Sum	42,976	45,784	1,962	100%	100%	100%	41,650	34,768	1,889	100%	100%	100%
0-0.5	22,926	29,648	1,168	53%	65%	60%	29,413	18,898	1,223	71%	54%	65%
0-0.99	34,385	42,163	1,685	80%	92%	86%	39,022	32,633	1,749	94%	94%	93%
0-1.99	41,090	45,185	1,896	96%	99%	97%	40,945	34,427	1,857	98%	99%	98%
0.5+	20,050	16,136	794	47%	35%	40%	12,237	15,870	666	29%	46%	35%
1.0+	8,591	3,621	277	20%	8%	14%	2,628	2,135	140	6%	6%	7%
2.0+	1,886	599	66	4%	1%	3%	705	342	32	2%	1%	2%
3.0+	968	263	37	2%	1%	2%	579	318	24	1%	1%	1%
4.0+	617	167	20	1%	0%	1%	290	56	12	1%	0%	1%
5.0+	438	86	14	1%	0%	1%	242	47	9	1%	0%	0%

Nevada County has 138 Map Book Pages where the C.V. is 1.0+ and the Area exceeds 10.00; El Dorado County has only 64 MBPs in this class. Nevada County has 17 Map Book Pages where the C.V. is 0.5+ and the Area exceeds 80.00; El Dorado County has 22 MBPs in this class.

TABLE 11.A9

Subdividability of existing parcels under Nevada and El Dorado County General Plans.*

Subdividability	Frequency	Area	Mean	Minimum	Maximum	Std Dev	Parcel %	Area %	Middle
El Dorado County									
At or above allowable density	78,062	0	0.00	0.00	0.00	0.00	80%		0
Lower than allowable density but cannot be subdivided further (e.g., a 7-acre parcel in a 5-acre zone)	12,462	12,462	1.00	1.00	1.00	0.00	13%	17%	0
2-3 smaller parcels possible	4,280	9,702	2.27	2.00	3.00	0.44	4%	13%	10,700
4-5 smaller parcels possible	1,079	4,776	4.43	4.00	5.00	0.49	1%	6%	4,856
6-10 smaller parcels possible	896	6,769	7.55	6.00	10.00	1.36	1%	9%	7,168
11-20 smaller parcels possible	514	7,334	14.27	11.00	20.00	2.68	1%	10%	7,967
21-40 smaller parcels possible	183	5,071	27.71	21.00	40.00	5.51		7%	5,582
40-160 smaller parcels possible	162	12,712	78.47	41.00	158.00	32.11		17%	16,200
161-640 smaller parcels possible	36	9,434	262.06	161.00	575.00	110.48		13%	14,418
640 or more smaller parcels possible	7	6,845	977.86	716.00	1,544.00	273.54		9%	4,480
TOTAL	97,681	75,105	1,375.61	963.00	2,356.00	426.61	100%	100%	71,370
Subdivision Map Act Restrictions	2,338	50,553					2%	67%	
Nevada County									
At or above allowable density	29,620	93,659	3.16	0.00	628.76	59.87	51%	16%	29,620
Lower than allowable density but cannot be subdivided further (e.g., a 7-acre parcel in a 5-acre zone)	16,934	101,723	6.01	0.09	662.76	28.14	29%	18%	16,934
2-3 smaller parcels possible	7,163	163,508	22.83	0.12	671.05	91.66	12%	29%	14,326
4-5 smaller parcels possible	1,639	80,391	49.05	0.26	760.77	142.91	3%	14%	6,556
6-10 smaller parcels possible	1,317	42,969	32.63	0.37	883.74	85.70	2%	8%	10,536
11-20 smaller parcels possible	665	32,097	48.27	0.61	638.64	111.19	1%	6%	10,308
21-40 smaller parcels possible	332	18,374	55.34	1.12	1280.00	125.65	1%	3%	10,126
40-160 smaller parcels possible	240	25,260	105.25	4.25	2081.20	200.81		4%	24,120
161-640 smaller parcels possible	50	12,113	242.27	31.00	955.00	234.13		2%	20,025
640 or more smaller parcels possible	3	1,333	444.43	219.88	560.00	194.49			1,920
TOTAL	57,963	571,429					100%	100%	144,471
Subdivision Map Act Restrictions	3,153	158,945					5%	28%	

*note text for methodological problems in analysis and likely source of errors in estimating subdividability of parcels under each of the General Plans. Higher estimate for the Nevada County General Plan than the El Dorado County General Plan reflects failure by Nevada County to address existing parcelization in land use designations.

TABLE 11.A10

Distribution of parcel sizes for Nevada and El Dorado Counties.*

Parcel Size Class	El Dorado Frequency	Nevada Frequency	El Dorado Area	Nevada Area	El Dorado Frequency %	Nevada Frequency%	El Dorado Area%	Nevada Area%
No acreage**	4,031	3,335			5%	6%		
Less than 1 acre	44,834	28,753	18,431	11,616	56%	51%	4%	3%
1 to 3 acres	10,441	8,381	18,904	15,714	13%	15%	4%	4%
3 to 5 acres	6,210	4,548	26,880	19,559	8%	8%	6%	4%
5 to 10 acres	7,340	5,454	51,880	39,636	9%	10%	12%	9%
10 to 20 acres	3,866	2,671	49,920	37,613	5%	5%	11%	8%
20 to 40 acres	1,523	1,657	45,282	46,405	2%	3%	10%	10%
40 to 160 acres	1,188	1,287	91,293	101,860	1%	2%	20%	23%
160 to 640 acres	385	451	128,046	162,647	0%	1%	28%	36%
Over 640 acres	16	14	19,116	12,597	0%	0%	4%	3%
TOTAL	79,834	56,551	449,752	447,647	100%	100%	100%	100%

*Note that these data are only for those private lands in areas within each county that are "below the green line" west of the national forest boundary.
**These parcels include parcels less than one acre in size within subdivisions for which the subdivision's area is listed in the assessor's records. They also include condominiums or time-share units, which are also generally less than one acre in size. This size class is generally < 1 acre.

TABLE 11.A11

El Dorado County General Plan Project Description at "buildout" versus 1990 census densities.

Housing Density Class	1990 Area	Project	Abs. Change	Rel. Change	Housing Units	% of Housing	Population	% of Pop
None	510.6	71.9	−438.68	−86%				
Less than 1	223.4		−223.43	−100%				
1 to 2	231.5		−231.47	−100%				
2 to 5	194.1		−194.12	−100%				
5 to 10	152.5		−152.54	−100%				
10 to 20	153.4	1,086.5	933.14	608%	17,384	11%	36,709	12%
20 to 40	122.5	98.3	−24.20	−20%	98		233	
40 to 80	84.2		−84.21	−100%				
80 to 160	62.4	211.8	149.46	240%	16,948	11%	40,833	13%
160 to 640	38.6	47.6	8.98	23%	18,260	12%	40,237	13%
Over 640	17.9	34.8	16.87	94%	105,968	67%	195,020	62%
TOTAL	1,791.1	1,551.0	−240.19	−13%	158,658	100%	313,032	100%
Commercial/Industrial/Other		13.4	11.00					

Housing Density Class is given in units per square mile and all areas are given in square miles.

TABLE 11.A12

El Dorado County General Plan Alternative at "buildout" versus 1990 census densities.

Housing Density Class	1990 Area	Alternative	Abs. Change	Rel. Change	Housing Units	% of Housing	Population	% of Pop
None	510.6	31.0	−479.6	−94%				
Less than 1	223.4		−223.4	−100%				
1 to 2	231.5		−231.5	−100%				
2 to 5	194.1	963.1	769.0	396%	3,853	2%	5,891	2%
5 to 10	152.5		−152.5	−100%				
10 to 20	153.4		−153.4	−100%				
20 to 40	122.5		−122.5	−100%				
40 to 80	84.2	295.0	210.8	250%	11,800	8%	29,293	9%
80 to 160	62.4	180.3	117.9	189%	17,309	11%	41,703	13%
160 to 640	38.6	53.8	15.3	40%	20,669	13%	45,545	15%
Over 640	17.9	34.6	16.6	93%	103,198	66%	189,923	61%
TOTAL	1,791.1	1,557.9	−233.3	−13%	156,829	100%	312,356	100%
Commercial/Industrial/Other		8.6	8.6					

Housing Density Class is given in units per square mile and all areas are given in square miles.

TABLE 11.A13

Nevada County General Plan density distribution at "buildout" versus 1990 census densities.*

Housing Density Class	1990 Area	General Plan	Abs. Change	Rel. Change	Housing Units	% of Housing	Population	% of Pop
None	365	55	−310	−85%				
Less than 1	87							
1 to 2	97	32	−65	−67%	32		28	
2 to 5	59	440	381	644%	1,759	1%	3,507	1%
5 to 10	54	1	−53	−98%	8		19	
10 to 20	79	112	34	43%	1,797	1%	4,768	2%
20 to 40	90	79	−11	−12%	2,252	2%	5,337	2%
40 to 80	61	68	7	12%	4,370	3%	10,849	4%
80 to 160	40	71	30	76%	9,052	7%	21,809	9%
160 to 640	30	59	29	97%	11,649	9%	25,669	10%
Over 640	12	9.20	−3	−25%	97,346	76%	179,153	71%
TOTAL	974	926	−48	−5%	128,265	100%	251,139	100%
Commercial/Industrial		11	11					

*Note text explanation of likely errors in Nevada County General Plan due to failure to address existing parcelization at higher than allowable densities.
Housing Density Class is given in units per square mile and all areas are given in square miles.

TABLE 11.A14

Distribution of population by housing density class.*

Housing Density Class (per sq. mi.)	1990 Census 46 Sierra CCDs	1990 Census Nevada County	1990 Census El Dorado County	Nevada County General Plan	El Dorado County General Plan Project Description	El Dorado County General Plan Alternative
None						
Less than 1	1%					
1 to 2	2%					
2 to 5	2%		1%	1%		
5 to 10	3%	1%	1%			
10 to 20	4%	4%	4%	2%	12%	
20 to 40	7%	8%	7%	2%		
40 to 80	9%	10%	9%	4%		9%
80 to 160	11%	13%	14%	9%	13%	13%
160 to 640	21%	23%	21%	10%	13%	15%
Over 640	39%	40%	44%	71%	62%	61%
TOTAL	100%	100%	100%	100%	100%	100%

*Note that General Plan estimates assume that land use designations will be accurate at "buildout" and will be unaffected by inconsistent existing parcelization.

JONATHAN KUSEL
Forest Community Research
and
Department of Environmental Science,
    Policy, and Management
University of California
Berkeley, California

12

Well-Being in Forest-Dependent Communities, Part I: A New Approach

ABSTRACT

This chapter presents a new approach to the conceptualization and assessment of well-being in forest-dependent communities. Studies of well-being in natural-resource-dependent communities (NRDCs), including agrarian communities, boomtowns (communities undergoing rapid growth), and forest-dependent communities, are examined to highlight common themes and approaches. Social indicators, which more directly address well-being, are discussed, and a five-point summary of common weaknesses is presented. The county, a commonly used unit of analysis for well-being assessment of NRDCs, is rejected in favor of a more socially relevant unit. A discussion of a new approach to well-being in forest communities begins with definitions of the terms *community* and *forest dependence;* the latter is broadened from traditional commodity-based definitions to include aesthetic and tourism-related dependence. The work of Amartya Sen, whose conceptualization of well-being focuses on the real opportunities people have and their achievements in light of their opportunities, forms the foundation of this new approach. Sen's conceptualization is further broadened by shifting analysis away from exclusive attention on the individual to include the community, which acknowledges the importance of a sense of place. Methodologically, the new approach to well-being involves collecting diverse slices of data, including secondary measures and an assessment of community capacity. Community capacity consists of three components: physical capital, human capital, and social capital. Assessment involves evaluating how community residents draw these components together to meet local needs and create opportunities. The advantage of this approach is that well-being assessment includes not only indicators suggestive of low well-being but also a measure of how communities respond and create opportunities to improve local well-being.

INTRODUCTION

Forest ecosystems in North America have recently become the focus of comprehensive and broad-scale ecosystem studies. Many of these studies have adopted an "ecosystem management" approach (see, for example, Bormann et al. 1993; Ministry of Environment 1994; and Forest Ecosystem Management Assessment Team [FEMAT] 1993, among others). Ecosystem management has been defined in diverse ways, but there is general agreement that humans and human communities are a part of ecosystems and an important area of study (Grumbine 1994; Manley et al. 1994; World Commission on Environment and Development 1987). Despite this agreement, however, no ecosystem study to date has adequately addressed the well-being of humans and human communities.

This chapter presents a new conceptual and methodological approach to assessing community well-being in communities that are dependent on natural resources, with a particular emphasis on forest-dependent communities. The focus on forest-dependent communities stems from the recent emphasis on forest ecosystem studies and from the fact that the well-being of these communities has long been narrowly discussed in the context of extractive forest management activities. Other studies involving natural-resource-dependent communities and studies using social indicators are reviewed to highlight the diversity and complexity of approaches to understanding human well-being.

This chapter is divided into two sections. The first section begins with a review of studies evaluating well-being and the lives of individuals living in natural-resource-dependent com-

Sierra Nevada Ecosystem Project: Final report to Congress, vol. II, *Assessments and scientific basis for management options.* Davis: University of California, Centers for Water and Wildland Resources, 1996.

munities. These studies narrowly define dependence in terms of commodity production, and they spend considerable energy analyzing the connection between resources and human well-being, an important though often overstated linkage. Common themes among these studies are highlighted. The first section concludes with a discussion of social indicators, which address the more basic issues of what well-being is and how it should be assessed; the limitations of social indicators; and the use of counties as the unit of analysis for understanding well-being.

The second major section of this chapter presents a new approach to the study of well-being in forest-dependent communities. Because of the confusion surrounding the terms *community* and *forest dependence*, these concepts are defined. The work of Amartya Sen, whose conceptualization of well-being focuses on the real opportunities people have and their achievements in light of their opportunities, forms the foundation of the new approach offered here. Sen's conceptualization of well-being is broadened in one important way: well-being analysis is shifted away from looking exclusively at the individual to looking at the individual and his or her community. The chapter concludes with a discussion of how these concepts and this approach can be used to develop a new methodological approach to a community well-being assessment in ecosystem management studies.

STUDIES OF WELL-BEING IN NATURAL-RESOURCE-DEPENDENT COMMUNITIES AND THE USE OF SOCIAL INDICATORS

Resource Dependency and Well-Being

The inclusion of humans and the study of human well-being in ecosystem studies is in its infancy. It is therefore useful to briefly examine empirical studies of resource dependency and human well-being in a variety of natural-resource-dependent communities (NRDCs). The objective of this section is to offer a glimpse of the diverse ways in which researchers have grappled with the linkage of resource dependency and human and community well-being. Studies of well-being in three kinds of NRDCs are reviewed: agrarian communities; boomtowns, or communities that have undergone extremely rapid growth associated with the extraction of nonrenewable resources; and forest-dependent communities.

An often implicit and underlying aspect of studies of communities that are dependent on forests and other resources is the attempt to understand the relationship between resource use (or dependence) and individual and community well-being. Yet the more basic questions of what constitutes well-being and how it might best be evaluated remain unanswered. Other research, such as the work on social indicators, has

addressed that question more directly, though still not without difficulties.

Agrarian Communities

The Jeffersonian ideal of the small, agrarian rural community forms the model against which agriculture and other resource-dependent communities are evaluated (Bealer et al. 1965; Drielsma 1984). The community in this model is stable, is small in scale, and offers the opportunity for healthy family life, independence, and entrepreneurial activity (Drielsma 1984).

The classic study of well-being in agrarian communities was conducted by Walter Goldschmidt (1947), who evaluated the structure of agriculture and its relationship to community well-being in California. The variables he examined include wages of owner-operators, industrial workers, and basic laborers; employment turnover; security in labor; social isolation of workers; labor participation in important community decisions; and the strength and diversity of community institutions and infrastructure. Goldschmidt found that an increase in the concentration of the farm sector led to a decline in rural economic and social well-being. He noted that in contrast to a community surrounded by large farms, a community surrounded by small farms had a higher percentage of self-employed and white-collar workers; a lower percentage of farm wage laborers; more business and retail trade; more schools, parks, civic and social organizations, newspapers, and churches; and a better-developed infrastructure and a more local decision-making structure. The dimensions of well-being most affected were living conditions and income.

Subsequent studies have shown that the inverse relationship between large-scale industrialized agriculture and well-being still holds true for California and nearby states in which large-scale industrial agriculture is dominant (MacCannell 1988; Swanson 1988). In raising the issue of the impact of land tenure on the well-being of agrarian communities, Goldschmidt's study raised the possibility that concentration of control of other resources, particularly in the hands of essentially absentee owners, might have similar adverse effects in other kinds of resource-dependent communities.

Boomtowns

Studies of rapid resource-related growth in small communities, commonly known as boomtown studies, generally discuss well-being in terms of population change. The focus of many of these studies is the impact of development activities. The independent variable, rapid community growth measured by population change, is associated with extractive energy projects such as oil or gas or mining development. Dependent variables have included measures of income and various aspects of employment, but the most commonly used variables by far are measures of crime (Albrecht 1982; Finsterbusch 1982; Freudenburg and Jones 1991; Gold 1982; Krannich et al. 1989; Seydlitz 1993; Wilkenson et al. 1982).

Some researchers have drawn broad conclusions suggesting that rapid development leads to the loss of integrative functions and is accompanied by a loss of local control, caused primarily by the rapid influx of outsiders overwhelming existing social services and networks (Jobes 1984a, 1984b; Gold 1985; Kennedy and Mehra 1985; Krannich et al. 1989). Gold (1985) believes disruption is caused by contrasts in lifeways and involves the replacement of close friendships and kin networks (gemeinschaft characteristics) with a less integrated social organization.

Freudenburg and Jones (1991), in an exhaustive review of boomtown studies, found that crime increased (by a factor of three, on average). This is in contrast to earlier studies, which, as Freudenburg and Jones point out, overstate the benefits of development activities. Yet, while lending support to the social disruption thesis (Finsterbusch 1982), the authors take issue with those who use grand theories and draw broad conclusions. They adopt what they term a middle-range perspective and suggest that the increase in crime associated with rapid development is due to reduced density of acquaintances.

Unlike many other boomtown researchers, Freudenburg and Jones rely on three primary data sources in their review and reanalysis of boomtown development: county-level data, survey data from communities, and case studies using crime statistics. Finally, unlike Goldschmidt's findings and the findings of researchers in forest-dependent communities, a discussion of which follows, decision making controlled by extralocal organizations was not examined or did not surface as a significant issue for researchers in these studies.

Forest-Dependent Communities

Well-being in forest-dependent communities has long been discussed in the context of community stability, a term that, for many, includes the more general notion of forest community well-being. The commonly held misconception of community stability calls for a steady flow of timber products, primarily logs, to ensure stable employment in the timber industry, which, in turn, leads to community well-being. (Community stability was once conceived in much broader terms [see Dana 1918 and Kaufman and Kaufman 1946]. Beginning in the late 1920s, however, the term became inextricably linked to timber industry employment in U.S. Forest Service discussions of sustained-yield forest management [Fortmann et al. 1989].)

One of the earliest studies of well-being in a forest community was carried out by Harold and Lois Kaufman (1946) in the Libby-Troy area of Montana. In addressing well-being, the Kaufmans used the then-popular term *stability*. But because their use of *stability* encompasses much more than employment stability, *well-being* is substituted for it in this discussion of their work.

The Kaufmans believe that creation of a prosperous economy is essential to well-being, but in addition to a concern about "what people do for a living" is a concern about "how well they live." They state, "A characteristic of the good life is that experiences in the community and of the forest are not only regarded as means but as ends in themselves—they are appreciated and enjoyed for their intrinsic worth. Also, the good life has a depth and variety of experience" (23). They point out that attainment of "the highest standard of living" can be realized only by maintaining a balance between population and natural resources. They link this concern to the limits of "timber supply, production costs and markets" (15). Like more conventional analysts, they agree that maintenance of community well-being involves the development of a stable timber industry, a diversified economy, and the practice of sustained-yield forestry. But in addition to the contribution of land use and industry to well-being, they describe five other "approaches" toward maintaining community well-being: organizing the greater community, strengthening the rural home, making religion a part of life and the church more community centered, promoting public participation in the determination of forest policy, and creating a forest-centered tradition. In these suggestions there is evidence of both the Jeffersonian tradition and a sense that the promotion of well-being involves process as well as products.

Kaufman and Kaufman question the wisdom of the Sustained-Yield Forest Management Act passed at the time of the study. They argue that it favors timber operators with large holdings, thereby concentrating economic power in the hands of a few while being "silent concerning controls that might be needed to safeguard the public interests" (71). In one of the first calls for public involvement, the Kaufmans suggest that the Forest Service involve the public in the formulation of forest policy to ensure that the concentration of economic power does not result in the abrogation of public interests and concerns. They maintain that such involvement should be "extensive" (85). The Kaufmans' study is rare in its attention to these issues.

The studies by James Fred Kelly (1974) and David Williamson (1976) demonstrate the value that loggers place on "rugged individualism" and their contempt for and resistance to the U.S. Forest Service, the agency that controls the terms of access to forest resources. Kelly's study emphasizes the importance of strong community ties and a spirit of cooperative community self-reliance for well-being, while Williamson focuses on the social organization of gyppo logging around kin networks. Carroll (1984) explores the sense of community held by loggers as an occupational group and also finds the tradition of spirited individualism firmly entrenched. (His approach is, in part, a response to a perception of the decline of community in modern society and, in part, an attempt to avoid the conundrum of locality-based definitions of community.) This individualism is empowering and plays an important role in well-being, but at the same time it binds workers to a disappearing occupation (Carroll and Lee 1990). Carroll (1984) reports that loggers and their families have powerful ties to their physical locales, although these ties do not correspond to the geographic bounds of their communities. As is the case in the earlier studies, local residents' contempt for the Forest Service is also a theme.

Marchak's (1990) study of forest-dependent towns in British Columbia emphasizes the adverse effects of uncertainty about future employment (reflected in high rates of population turnover) stemming from control of the resource base by outside firms that make decisions "without reference to the needs of workers in these communities" (99). She suggests that high turnover rates do not reflect the personal choices of workers but rather the structure of the industry. Marchak was the first researcher to note that women are particularly demoralized by the conditions in single-industry forest towns.

Kusel and Fortmann (1991) and Kusel (1991) studied forest counties and communities in California, focusing on general well-being and the capacity of forest communities to maintain and enhance local well-being. Capacity is described as "what enables communities to pull through hard times" (Kusel and Fortmann 1991, 84). Methodologically, their work comprises three separate studies: a statistical analysis of forest counties that examined indicators of well-being and explored measures of forest use, a rapid rural appraisal of seven forest communities to assess community capacity, and a long-term ethnographic study of three forest communities, examining well-being and capacity. Kusel and Fortmann also examined the relationship of ownership and control of forest resources to well-being. They found that a higher concentration of private forest landholding is associated with lower median income, and that high percentages of public timberland are associated with higher poverty rates (at the county level). They found also pockets of high poverty in low-poverty forest counties.

Kusel and Fortmann determined that communities are deeply affected by forces outside of their control, including outside employers, natural-resource decision makers, and outside money. In contrast to studies characterizing the "inevitable" culture clash between newcomers and long-standing residents (see, for example, Price and Clay 1980 and Schnaiberg 1986), Kusel and Fortmann note that recent in-migrants and women play crucial roles in mobilizing community action and increasing local capacity.

In the ethnographic study, Kusel found that extensive job loss in rural forest communities was devastating in the short and long term. Economic and social turmoil led to short-term difficulties for families and communities and to a long-term reduction in community capacity. Mill restructuring has the effect of reducing well-being through layoffs. Kusel also found that local, family-run mills contribute more to community well-being than mills owned by large, nonlocal owners.

Forest communities throughout the Pacific Northwest were included in the social assessment conducted by the Forest Ecosystem Management Assessment Team (FEMAT). (This was one of three teams created by President Clinton to "identify management alternatives that attain the greatest economic and social contribution from the forests of the region and meet all requirements of applicable laws and regulations" [FEMAT 1993, ii].) The FEMAT study is one of the first large-scale, American ecosystem studies that attempts to explicitly include and assess human communities. The objectives of the social assessment include describing the nature and distribution of social values (which were not linked to any locality), identifying the consequences of forest management alternatives for communities and individuals, and describing how alternatives affect social values and constituencies (FEMAT 1993, VII-45).

FEMAT scientists held two separate workshops with panels of community experts to assess the capacity of communities in the region. The concept of community capacity was defined in the workshops as an independent variable that in part determines community response to and the consequences of land-management alternatives. Higher-capacity communities were considered more adaptable and therefore less affected by changes in forest management.

Although a variety of secondary data were offered and used by experts, FEMAT researchers relied primarily upon experts' knowledge of communities. The concept of capacity, modified in this chapter, plays a key role in the new approach to well-being described herein and is discussed at length later.

Social Indicators and Well-Being

Two primary areas utilizing social indicators include (1) social impact assessment (SIA), which predicts and assesses the consequences of technical projects (e.g., hydroelectric projects, waste-dump siting, etc.) and specific policy actions on well-being (Interorganizational Committee 1994), and (2) broader research focused on more general well-being or life conditions (e.g., Allardt 1993; Campbell 1981). Included in this second area is an examination of the philosophical and conceptual underpinnings of well-being (see, for example, Nussbaum and Sen 1993). These two broad areas of research offer important insights to scientists studying the relationship of resource dependence to community well-being, and much of this discussion relies upon them. This section closes with a brief discussion of the county as a unit of analysis used for well-being assessment.

Terms such as *standard of living, quality of life, welfare, happiness, life satisfaction,* and others have been used in studies to characterize a good and healthy life or the critical components of one. But they may have different meanings to people and consequently have led to confusion about what well-being is and how it might best be measured. The numerous approaches to the study of well-being, such as measurement of utility, income, personal satisfaction, and happiness, to mention just a few, have yielded incommensurable and, at times, contradictory results that have only further muddied the waters of well-being assessment. Burdge (1994) states, "The field of impact assessment does not have a series of agreed upon concepts or list of variables around which to accumulate research knowledge" (3). Discussing the link between environmental planning and social assessment, he states that there is a "need to reach some tentative agreement on concepts, procedures and content."

In addition to conceptual concerns, social indicator researchers have wrestled with the problem of whether to study well-being by using subjective self-report measures or measures of external conditions (also called sociodemographic measures), considered by many to be more objective (Allardt 1993; Erikson 1993). Implicit in the debate over appropriate measures are the questions, Who should do the evaluating? and, What variables should be evaluated? Sociodemographic measures, including crime, income, employment, and poverty, are frequently the measures of choice because they are the most detailed measures available for a limited area, are easily gathered (or have already been gathered, in the case of U.S. Census Bureau data), and have more direct policy relevance for governments than other measures (De Neufville 1975). (See Burdge 1994 and Interorganizational Committee 1994 for recent discussions of categories and indices.) Yet, despite widespread use and limited researcher reflection, both sociodemographic measures and subjective self-report measures have significant limitations. Sen (1985b), in particular, and others have provided powerful critiques. A five-point summary of the limitations of social indicators is presented here, followed by a brief discussion of the problems associated with the unit of analysis used in many NRDC studies of well-being.

First, social indicators, consisting of aggregate individual data, ignore the variability of structural conditions at the level of the county or region, and of such institutional arrangements as the concentration of capital, land ownership, and power that influence well-being in a community (Kennedy and Mehra 1985; Kim 1973). Communities with greater disparities in wealth often have lower community well-being than communities with more equal distribution of wealth, even though average measures such as income may be the same. Goldschmidt's (1947) evaluation of the structure of agriculture and its relationship to community well-being is a good example of why this consideration of institutional arrangements is important.

Second, Sen (1985b) points out that sociodemographic measures of opulence, such as real income, confuse well-being with being wealthy in terms of material possessions. Measures of real income provide an indication of what an individual can buy, or his or her "commodity command," but they provide no indication of how an individual may improve his or her life with purchased commodities. Sen (1987) states that commodities provide only a means to an end and that the issue is more a "matter of the life one leads rather than of the resources and means one has to lead a life" (16). In their research, Kaufman and Kaufman (1946) expressed a similar concern with how well individuals live.

Third, and related to the previous point, is the issue of what constitutes well-being for whom. For example, sociodemographic measures of opulence do not take into account the distribution of resources within a family (Sen 1985b). For example, a male head of household may purchase luxury items for himself while other family members are inad-

equately clothed and fed. Similarly, women's concerns may differ from those of men. Nussbaum and Sen (1993) question whether the quality of female life has similar constituents to the quality of male life. Feminist research was launched out of a concern that women's perspectives and their life circumstances were not recognized. Oakley (1975) points out that women have been reduced "to a side issue from the start." The concerns of adolescents may also differ from those of adults. Freudenburg (1984) discovered that adolescents in rapidly growing communities were more likely to be dissatisfied with their locality and less satisfied with their overall quality of life than adolescents in similar towns that were not growing rapidly, whereas the same relationship did not hold for adults.

A fourth problem has to do with subjective measures and the distinction between ill-being and well-being. Subjective well-being is commonly measured with scales indicating satisfaction with the self as a person, personal freedom, personal happiness, and sense of personal control (Campbell 1981; Chamberlain 1985). Yet Headey and colleagues (1985) point out that well-being may be a different dimension than ill-being. They found that more objective measures of health and material standard of living, while contributing little to measures of well-being, significantly contributed to measures of ill-being. Bradburn and Caplovitz (1965) and Wilson (1967) found the same to be true for measures of happiness: there are positive and negative dimensions that are independent of one another. In addition to requiring measurement of positive and negative dimensions, this suggests that people may adjust their perceptions of well-being (or happiness) to the conditions they face.

Sen (1985b), studying the same issue from an economic and philosophical perspective, states that subjective measures of well-being, such as pleasure and desire fulfillment, are incomplete for two reasons: (1) they are fully based on the mental states of an individual, and (2) they lack a personal metric of value ("the mental activity of valuing one kind of life rather than another"). Sen terms these reasons "physical condition neglect" and "valuational neglect," respectively. An example illustrates the incompleteness. One who is poor, without the comfort of a home, out of work, and ill-fed but happy has obviously adjusted her expectations and taken solace in small pleasures. But fulfillment of limited desires, no matter how happy this person might be, is not suggestive of a high level of well-being. Moreover, this psychological state cannot be compared to that of another individual whose desires are greater. Sen (1984) states, "Quiet acceptance of deprivation and bad fate affects the scale of dissatisfaction generated, and the utilitarian calculus gives sanctity to that distortion. This is especially so in interpersonal comparisons" (309).

Fifth, researchers who have examined the relationship between objective and subjective measures have shown that sociodemographic indicators have little relationship to subjective measures of well-being (Barlett and Brown 1985; Campbell 1981; Gans 1962; Mastekaasa and Moum 1984;

Oppong et al. 1988; Suttles 1969). Gans (1962) reported in his study of West Enders in Boston that there existed a high satisfaction among residents of the area, yet it was declared a slum because of measures (by upper-middle-class professionals) of low physical condition and low income and was completely cleared for redevelopment. The difference between the West Enders' satisfaction with the area and the measures of the "professionals" provides a warning that not only may measures differ, but they may do so because some measures reflect the values (and power) of those who are doing the measuring more than the values of those whose well-being is being evaluated. Moum (1988) found that only 10% of the variance in quality-of-life scales is explained by sociodemographic variables.

Given gender, class, and ethnic differences and the importance of local salience, it is not surprising that numerous measures of well-being have been developed but no standard metric has emerged (Burdge 1994; Johnson 1988; Oppong et al. 1988). The arguments just expressed suggest that the "holy grail" of complete well-being assessment may indeed be unobtainable. They also demonstrate that considerable humility is necessary in any assessment and interpretation of human well-being.

The Social Unit of Analysis

In the debates over self-report measures versus sociodemographic measures and over appropriate metrics of well-being, little attention is paid to the unit of analysis or level of data aggregation used for assessment. This may lead to additional confusion about whose well-being one is discussing and the factors that influence it. Data availability (and research funding) too often determine the unit of analysis. The county has been the most common unit of analysis in studies of community stability in forest-dependent communities (Machlis and Force 1988), and its exclusive use is inadequate for several reasons (for a contrasting view, see Lobao 1990).

Perry (1986) has criticized the use of counties because they are not a unit with real social meaning. People do not generally identify with their counties, and, indeed, numerous NRDCs are alienated from their parent county. Relationships and life take place in communities, not counties.

Equally important for NRDC assessment is that only a small percentage of communities in a county may be resource dependent. County-level measures, whether they are median income, poverty, or unemployment, may have little relationship to resource activities. For example, the 1990 median income of Plumas County, a northern California county with a number of forest-dependent communities, is slightly more than $24,000. The four largest communities in the county, which are more dependent on extractive timber activities than the rest of the county, have median incomes that are well below the county median—one almost $9,000 lower and another $5,000 lower. The southern Sierra Nevada mountain communities in Tulare and Fresno Counties offer additional examples.

Most of these communities have little in common with the much larger, agriculturally dominated Central Valley communities located in the same counties. Distilling forest dependence in these communities by using county data would be difficult, if not impossible. This is not to say that the linkage between resource dependency and well-being at a county level is unimportant, but that such dependence in communities that are part of a county aggregate in which the relationship appears relatively small and insignificant will not be identified.

If one desires to understand community well-being, then, the unit of analysis must focus on and isolate community. County data alone often encompass too broad and diverse an area to be used for accurate examination of well-being in many NRDCs. Finally, a determination of the causal factors influencing community well-being more often than not requires a specificity and detail unobtainable with county-level data.

A NEW APPROACH TO FOREST COMMUNITY WELL-BEING

This section begins with a definition of *community* and a redefinition and expansion of the term *forest dependence*. The concept of community has engendered considerable debate, a debate that will not be resolved here but that nonetheless must be addressed. In sharp contrast, the concept of forest dependence has been uncritically accepted as employment generated from tree harvesting. The use of the concept here is considerably broadened from the more narrow use. A discussion of Amartya Sen's novel "capabilities and functioning" approach to well-being follows. His approach is expanded by adding an emphasis on the community, to arrive at the "capacity" approach. This section and the chapter conclude with a discussion of how the concept of capacity can be used in an assessment of community well-being.

Conceptual Clarity

The Concept of Community

Community in this paper is defined in terms of a locality-based shared identity. This definition is primarily based on Gusfield's (1975) discussion of community, which includes the intersection of two components: a relational component and a territorial component. The relational component involves "the quality or character of human relationships," which includes a sense of belonging. Selznick (1992) states that this includes shared beliefs, interests, and commitments among individuals that unite diverse groups and activities. The relational component of community is a vital part of individual well-being and is discussed further later.

Gusfield's territorial component involves what people have in common and share at their specific locale. This includes diverse institutional components: governments and law, school districts, churches, and families, among other things (Selznick 1992). Gusfield's conception of community roughly encompasses the three areas for which Hillery (1955), in a survey of the literature, found definitional agreement: social interaction, area, and common ties. (See Lee et al. 1990 for a discussion of these concepts for forest communities.) Although Gusfield does not limit his discussion to place-based communities, the focus here is primarily on geographically place-based, forest-dependent communities.

Despite this focus on locale, it is terribly important to recognize that forest communities are part of the larger society, with extensive vertical linkages, to use Warren's (1978) terminology. (Warren's [1978] observation that horizontal linkages [ties between organizations within a community] have been overwhelmed by vertical linkages [ties to organizations and institutions outside the community] is relevant here. Warren argued that the rising influence of an increasingly urban society frequently results in a decline of a community's distinctiveness, self-sufficiency, and individual interactions.) These linkages, or the lack of them, may profoundly affect a community and the opportunities it has available. A small rural community that is the home of a mill owned by a multinational corporation may have additional mill-related employment and other opportunities. This same community will also be quite sensitive to the actions of a single company (or individual in the company), which may have no local ties beyond the mill. In a somewhat similar vein, social relationships extend beyond the formal administrative and informal boundaries of a community (Selznick 1992; Strathern 1984). Individuals may hold multiple "community" identities as a result of associations at their place of work and through other organizations and institutions that are outside of their community of residence. Small NRDCs include overlapping sets of social groups, and these groups are important to local community well-being and how local communities are influenced by forest policy. The focus on place-based communities suggested here provides a clear starting point, and a critical one, for assessing well-being. Many rural NRDCs in the West, by the nature of their location and proximity to public lands, often have clear geographic boundaries.

Broadening the Concept of Forest Dependence and Recognizing the Importance of the Sense of Place

Forest-dependent communities are those immediately adjacent to forestland or those with a high economic dependence on forest-based industries, including tourism as well as timber. This broader definition is necessary to show that well-being in forest communities must focus on more than a biological resource and timber products.

First, *forest dependence* suggests that a community's primary relationship is to a biological forest, and, as the term has commonly been used, to wood products. (Machlis and Force [1988] point out that forest or timber dependency is generally determined by forest commodity production or economic measures [e.g., measures of sales by forest industries, percentage of total income from the forestry sector, and forest industry employment].) It is true that forest-dependent communities rely on the biological forest resource. However, these communities, particularly ones in which a number of residents work in the wood products industry, also depend on the economic and social structure that permits (and demands) particular uses of the forest resource. This structure mediates the terms of a community's access to the economic and social benefits of this resource. The strongest relation is to the economic and social system, not the biological one, despite its obvious importance. Thus, in a community in which many workers are employed in the wood-products industry, the ability of local residents to gain economically from the forest, as well as to create new jobs, is a function not only of the biological condition and production of the forest but also of (1) the extent to which controllers of the forest permit and promote commercial activities, (2) the extent to which those who create industry jobs make them available in or near the community, as well as the extent to which those who control wood-products jobs maintain them, and (3) the terms upon which these jobs become available. The same may be said of other forest-dependent jobs.

Second, the commodity production perspective ignores those forest-dependent communities that do not produce a single board foot of timber. Communities can be economically dependent upon the forest without any forest-commodity production whatsoever. Many communities whose raison d'être is forest tourism or retirement living are dependent on the forest, and they are increasing in number and size, particularly in the western United States.

Third, forest dependence occurs with no economic relationship to the forest resource and is based on an aesthetic, symbolic, and locality-based importance (Hester 1985; Hiss 1990; Tuan 1993; Walter 1988). The forest is a landscape and, for forest communities, part of a human sense of place. As a landscape, Relph (1976) suggests, it represents "an expression of communally held beliefs and values and of interpersonal involvements" (34). Meinig (1979) observes that "a well-cultivated sense of place is an important dimension of human well-being. Carried further, one may discover an implicit ideology that the individuality of places is a fundamental characteristic of subtle and immense importance to life on earth, that all human events *take place*, all problems are anchored in place, and ultimately can only be understood in such terms" (46).

Wendel (1987) found that a majority of the residents of a forest community in California chose the response "the trees/the forest" to a question asking what was the most important place in the community. The trees and the forest were important for many reasons: they represented a link with the residents' past tradition of logging, a connection to their present and future economic base of tourism and to aesthetic values, to mention just a few. Hester (1985) calls places that reinforce and help define the community living tradition "sacred" places. Kaufman and Kaufman (1946), using the term *stability* rather than *well-being*, state, "A meaningful tradition is always an important part of the life of a stable community. A tradition is needed . . . which magnifies the significance of the forest and portrays the relationship of forest and people" (30). Berry (1987), in a somewhat similar vein, believes community to be inseparable from its place, with community and place mutually supportive. They represent the human and natural economies, each offering the other the possibility of a lasting and livable life.

As a landscape, sacred place, or resource, the forest supports local residents and contributes to the definition they have of themselves and their understanding of who they are. The lifeways of community members and the landscape are intertwined. Thus, when discussing dependence, one must recognize that the forest provides not only the means of production, diversely defined, but sustenance to the local living tradition, economically, socially, and spiritually.

Capabilities and Functionings

Sen (1984, 1985a, 1985b, 1987, 1993) offers what he calls the capabilities and functionings approach as an alternative way of evaluating well-being. An individual's capabilities consist of the freedom one has or the opportunities from which one can choose. An individual's functionings consist of one's achievements, or what she or he "succeeds in doing with the commodities and characteristics at his or her command" (Sen 1985b, 10). Functionings vary from the more basic, which include escaping mortality and malnourishment, to the more complex, such as achieving self-respect (Sen 1993). Sen argues that these elements are part of an individual's being and must be part of a well-being assessment.

Sen's approach to well-being counters the problem of the limitation of sociodemographic measures, such as measures of opulence, by evaluating not just the goods at one's disposal or one's wealth, but *how they contribute to what a person can do.* For example, an individual who owns a bicycle, other things being equal, would be considered better off than one who does not. But if the same individual lives in a war-torn country where the roads are predominantly unridable and bicycle riders are targeted by snipers, bicycle ownership contributes little to that person's transportation functioning and may negatively affect well-being. Similarly, a job that provides an adequate income may be essential to one's (and one's family's) well-being, but if an individual cannot advance in his or her job, or if creative opportunities are desired but unavailable, diminished well-being through reduced achievement results. A job that provides adequate pay contributes to one's well-being, but the pay alone constitutes only a portion of one's achievement. These examples highlight what Sen (1984) refers to as the "capability to function" (317).

The capabilities and functionings approach addresses the subjective-indicator problem by dividing the evaluation into two parts: "(i) *specification* of the functioning achievements,

and (ii) the *valuation* of the functioning achievements" (1985b, 30). Specification requires identifying achievements for which a valuation is made. To return to the example of a poor, unemployed, ill-fed, homeless person, the specification of her functionings would clearly indicate a low level of well-being, while the personal valuation of her well-being is rendered somewhat unimportant.

What is unique about Sen's capabilities and functionings approach is that it requires an analysis of the opportunities or freedom individuals have (capabilities) *and* their achievements or successes (functionings) in light of their opportunities. For someone to have a high level of well-being, she or he not only must feel well but also must have opportunities available and be able to take advantage of them. Sen, however, restricts the analysis of well-being to the individual and avoids the sticky problem of a contextually based valuation of various capabilities and functionings, which is important for a more complete evaluation of well-being.

The Importance of Community

To allow for a complete discussion of individual opportunity, as well as to better understand the valuation of functioning achievements and well-being, requires a focus on the individual and on community. Motivations for human action spring from internalized values. Benn (1982) maintains that these flow from "traditions of behavior" that do not reflect *"individually conceived* goals, but reflect those of our culture and communities" (49–50). Selznick (1987) offers the perspective of the "implicated self," which holds that humans are dependent on others for personality development and "psychological sustenance" (447). He states, "A morality of the implicated self builds on the understanding that our deepest and most important obligations flow from identity and relatedness, rather than consent" (451). Bellah et al. (1985) and MacIntyre (1984) maintain that human identity is found in community, as a collective living tradition.

Acceptance of the perspectives of "traditions of behavior" and the "implicated self" requires a well-being assessment to examine how communities define success (or functionings), which in turn affects how individuals view success. Native American communities may define success differently than Anglo communities. Ethnically similar communities may have definitions of success that differ from one another for any number of reasons as well. Hence, beyond the most basic of functionings, proper assessment must recognize these differences. A community and its traditions must inform the evaluation of well-being. To neglect the community is to neglect context and important—indeed vital—aspects of individuals.

It is important to point out that a community of shared values does not equal a community of conformity (Lasch 1988). Lasch states that social solidarity is not "an identity of interests; it rests on public conversation. It rests on social and political arrangements that serve to encourage debate instead of foreclosing it" (178). Communal relationships, with the associated responsibilities they bring, and freedom to choose are both coveted values. Selznick (1992) points out that there must be "freedom *in* associations as well as freedom *of* association" (363). He adds that a concern for personal autonomy "assume[s] that the worth of community is measured by the contribution it makes to the flourishing of unique and responsible persons. As an attribute of selfhood and of self-affirmation, autonomy requires commitment as well as choice" (363).

The perspective of the implicated self also recognizes that taking part in the life of a community contributes to individual well-being. Humans are constituted by social relationships found in community, and there is a reciprocal and interdependent relationship between an individual and others in her or his community. Implicit in this perspective is that a collective good exists; well-being may be improved by residents working on community projects that, narrowly conceived, are of no benefit to them personally. Individual well-being is increased as a result of an increase in feelings of being a part of a community and by making the community a better place to live. This is part of the relatedness component of community discussed earlier and involves a category of individual behaviors termed *commitments*. More broadly, this behavior may be termed *civic responsiveness*. Sen would disagree with the extension of well-being analysis to include commitment behaviors. Because of the importance of his work for the approach developed here, a brief review of this disagreement and a response to it are presented in appendix 12.1. Sen (1990) nonetheless recognizes the importance of community to well-being, stating, "Some functionings are very elementary. . . . Others may be more complex but still widely valued such as achieving self-respect, or *taking part in the life of the community*" (emphasis added).

Well-Being Assessment

Adopting Amartya Sen's approach to well-being requires the assessment of individual opportunities (capacities) and achievements or successes (functionings) in light of available opportunities. Individual opportunities are shaped by conditions that individuals face personally and within the context of a community. For example, as a general rule, one who is in poverty will have fewer opportunities than one who is not. But support services and networks available for those in poverty in one community will likely lead to higher capacities compared to the capacities of those in poverty in another community without such services and networks (all other things being equal).

For large-scale ecosystem studies, it simply is not possible to evaluate opportunities and successes for each individual. Nonetheless, diverse secondary data combined with primary data about communities (including support services) can be used to develop a rudimentary understanding of conditions and opportunities. Useful secondary measures and their related functioning include but are not limited to the following: measures of poverty that indicate those who have not

secured an income adequate to escape it (escaping poverty being a very basic functioning); poverty intensity (i.e., the further one is below a poverty threshold the higher the intensity), suggestive of a lower level of functioning and greater need; and higher education levels, suggestive of higher functioning and possible opportunities. Equally important, the presence of individuals with high levels of education may lead to increased community capacity, for reasons discussed next. Other important measures that address conditions that also may address the functioning of residents include measures of crime, drug dependency, and children in families receiving public assistance. (Machlis et al. 1995 prepared a list of indices and measures for the Eastside Ecosystem Study, though no direction was provided for indicator selection or use.)

Community Capacity

The expansion of well-being analysis calls for a focus on community to assess activities (or civic involvement) that, in turn, affect opportunities for residents in a community. Capacity is more than the existence of or individual willingness to participate in voluntary organizations. It involves assessing individual commitment actions at the level of the community that, when combined with physical and human resources, determine community capacity.

Community capacity is the collective ability of residents in a community to respond (the communal response) to external and internal stresses; to create and take advantage of opportunities; and to meet the needs of residents, diversely defined. It also refers to the ability of a community to adapt to and respond to a variety of different circumstances. Community capacity depends on three broad areas: (1) *physical capital*, which includes physical elements and resources in a community (e.g., sewer systems, open space, business parks, housing stock, schools, etc.), along with financial capital; (2) *human capital*, which includes the skills, education, experiences, and general abilities of residents; and (3) *social capital*, which includes the ability and willingness of residents to work together for community goals. While physical and human capital are commonsense foundations of capacity, social capital appears to be one of the most important determinants.

Selznick (1992) discusses communities as places where people grow and flourish. He notes that a "flourishing community has high levels of participation: people are appropriately present, and expected to be present, on many different occasions and in many different roles and aspects" (364). The empirical research of Putnam (1993a) in Italy, Flora and Flora (1991) in the Midwest, and others has shown the importance of social capital and has demonstrated that it is a primary determinant of economic development and community capacity. Putnam (1993b), examining the modern-day rise of regional governments in Italy from the eleventh century, states, "The historical roots of the civic community are astonishingly deep. . . . Communities did not become civic because they were rich. The historical record strongly suggests precisely the opposite: they have become rich because they were civic. The social capital embodied in norms and networks of civic engagement seems to be a precondition for economic development, as well as for effective government." An example is offered to show the relationship between social capital and financial capital and to further explicate the role and importance of social capital. A community may have a number of residents who are quite wealthy, but if they are not involved in the community and desire little to do with it, their financial capital does nothing for the community beyond their self-interested concerns. Conversely, a community with little financial capital and high social capital may conduct numerous fund-raisers as well as reach outside the community to raise money to address local needs, thereby improving local well-being.

Measurement of community capacity can be complex. Diverse slices of data examined over time are needed for accurate assessment (see Kusel 1991, Kusel and Fortmann 1991, and Putnam 1993a). To gain a rapid understanding of community capacity of forest-dependent communities, researchers can conduct workshops with experts who are knowledgeable about diverse community issues. These experts assess the components of capacity listed previously and, more specifically, identify those most determinate of overall community capacity. Use of expert informants in workshops requires a shift in methods, a shift made considerably more difficult by the necessary addition of qualitative data collection. The selection of experts to participate in workshops is critical, as it determines the accuracy and quality of the information obtained. Experts must understand community issues, institutions, and resources and cannot be community boosters or overly partisan about issues.

The assessment of community capacity facilitates an understanding of opportunities for productive and rewarding involvement in a community and the potential for increased opportunity for individuals. Although such assessment does not allow a specification of how any single individual's well-being is affected, high community capacity itself is suggestive of higher levels of well-being for residents. High capacity suggests, too, that expansion of opportunities to meet community needs (and local well-being) is not only possible but likely. With continued shifting of responsibilities from state and federal entities to localities, and increased responsibility placed on locals for self-development and self-improvement—including those communities that have long relied on federal and state subsidies for infrastructure development and maintenance—examining the capacity of communities is an important area of well-being research.

SUMMARY: FOREST COMMUNITY WELL-BEING

In this review of the studies of well-being it should be clear that there is room for considerable improvement in the assessment of well-being in communities dependent on forests and other natural resources. Future studies of well-being cannot rely only on subjective reports of well-being, because of their incompleteness; exclusive reliance on measures of opulence such as income are equally limiting, because such measures do not address the issues of distribution of resources. Additionally, if researchers are to discuss resource dependency and well-being, they must be clear about the term *dependency:* what it means and the variety of ways in which resources can be valued. The forest as a "place" embodies a diverse array of values. If a local forest, long used as a locale for the production of wood products, is reserved exclusively for recreational use or is overcut, local well-being will decline through the diminution of socially important forest values (not to mention jobs).

Researchers also must be clear about the unit of analysis. Community well-being cannot be assessed through county-level analysis. Counties are too heterogeneous, and too often jobs associated with resources make up a small proportion of a county economy. Communities are a logical unit of study but pose methodological problems: clear identification of boundaries is often difficult, and data availability within these boundaries may be limited. Well-being analysis must often strike a balance between socially meaningful units of analysis and units for which data are available.

Communities must be thought of not only as units of analysis but also as parts of well-being assessments. Inclusion of the category of behavior termed *commitments* broadens the conception of well-being in two significant ways. First, it acknowledges that capabilities and functionings are defined *in part* by the community. A community is composed of and sustained by individuals, and individuals are shaped by their community. Viewed in this light, a community—defined here as a locality-based collection of individuals—can foster or inhibit individual thinking about capabilities and individual ability to function, and the ways in which it does so must be considered. Hence, local conditions are viewed as an influence on individual conceptualizations of well-being. A second implication of a broadened conception of well-being is that relationships within a community involve a component of responsibility to communal relationships. This involves practices of commitment that make up patterns of individual allegiance and responsibility directed toward community. These practices may profoundly affect well-being.

In broad-scale ecosystem studies, it is simply not possible to assess the resources each individual has and determine how they contribute to that individual's functioning. An assessment of community capacity allows researchers to assess in some measure the opportunities available to residents today as well as the potential for creating additional opportunities and improving well-being. A basic assumption is that the higher the capacity of a community, the more likely it is that opportunities exist or will be created to expand individual capabilities and functioning. The very act of building community capacity is not only opportunity enhancing but also leads to improved social well-being.

The emphasis on individuals, local community, and capacity does not mean that social and political arrangements beyond community boundaries should be ignored. They are an important—and in many cases a critical—component of capacity and local well-being. This is particularly true of forest communities in which local and nearby land is owned by and local jobs are controlled by outsiders. Actions originating outside of a community may contribute to or severely restrict the capabilities and functionings of local residents. For example, local capabilities may be reduced by forest management decisions that do not involve local residents and that do not take into account local needs. Good capacity assessment will identify these arrangements. Improving the ability of a community to respond to and influence decisions affecting them that are made outside community boundaries is another way of improving the well-being of community residents.

What is unique about this approach to the study of community well-being is that it involves an analysis of factors that reduce local well-being *and* an analysis of how individuals and their communities respond to these factors. Examination of capacity encompasses as well an examination of how individuals and communities create opportunities or, to use Sen's terminology, capabilities, that expand the possible functionings or achievements of community members and improve well-being. In addition to identifying general levels of individual and community well-being, one of the significant benefits of this approach is its identification of areas with low capacities and a reduced ability to self-develop and improve local well-being. It is these areas that require the most attention and will provide the most difficult challenges for ecosystem managers who have among their management goals the desire to improve the well-being of humans and human communities.

REFERENCES

Albrecht, S. L. 1982. Commentary on "Local social disruption and Western energy development: A critical review." *Pacific Sociological Review* 25 (3): 297–306.

Allardt, E. 1993. Having, loving, being: An alternative to the Swedish model of welfare research. In *The quality of life,* edited by M. C. Nussbaum and A. Sen, 88–94. Oxford: Clarendon Press.

Barlett, P. F., and P. J. Brown. 1985. Agricultural development and the quality of life: An anthropological view. *Agriculture and Human Values* 2 (2): 28–35.

Bealer, R. C., F. K. Willits, and W. P. Kuvlesky. 1965. The meaning of rurality in American society: Some implications of alternative definitions. *Rural Sociology* 30 (3): 257–66.

Bellah, R., R. Madsen, W. M. Sullivan, A. Swidler, and S. M. Tipton. 1985. *Habits of the heart.* New York: Harper and Row.

Benn, S. I. 1982. Individuality, autonomy, and community. In *Community as a social ideal,* edited by E. Kamenka, 43–62. London: Edward Arnold.

Berry, W. 1987. Does community have a value? In *Home economics.* San Francisco: North Point Press.

Bormann, B. T., M. H. Brookes, E. D. Ford, A. R. Kiester, C. D. Oliver, and J. F. Weigland. 1993. A broad, strategic framework for sustainable-ecosystem management. Prepared by the Eastside Forest Health Panel for presentation to the Chief of the U.S. Forest Service. Corvallis, OR: Forestry Sciences Lab.

Bradburn, N. M., and D. Caplovitz. 1965. *Reprints on happiness.* Chicago: Aldine.

Burdge, R. J. 1994. *A conceptual approach to social impact assessment: Collection of writings.* Middleton, WI: Social Ecology Press.

Campbell, A. 1981. *The sense of well-being in America: Recent patterns and trends.* New York: McGraw-Hill.

Carroll, M. S. 1984. Community and the northwestern logger. Ph.D. diss., University of Washington.

Carroll, M. S., and R. G. Lee. 1990. Occupational community and identity among Pacific Northwestern loggers: Implications for adapting to economic changes. In *Community and forestry: Continuities in the sociology of natural resources,* edited by R. G. Lee, D. R. Field, and W. R. Burch Jr., 107–24. Boulder, CO: Westview Press.

Chamberlain, K. 1985. Value dimensions, cultural differences, and the prediction of perceived quality of life. *Social Indicators Research* 17 (4): 345–401.

Dana, S. T. 1918. *Forestry and community development.* Bulletin 638. Washington, DC: U.S. Department of Agriculture.

De Neufville, J. I. 1975. *Social indicators and public policy.* New York: Elsevier Scientific Publishing.

Drielsma, J. H. 1984. The influence of forest-based industries on rural communities. Ph.D. diss., Yale University.

Erikson, R. 1993. Descriptions of inequality: The Swedish approach to welfare research. In *The quality of life,* edited by M. C. Nussbaum and A. Sen, 67–83. Oxford: Clarendon Press.

Etzioni, A. 1988. *The moral dimension: Toward a new economics.* New York: The Free Press.

Finsterbusch, K. 1982. Commentary on "Local social disruption and Western energy development: A critical review." *Pacific Sociological Review* 25 (3): 307–21.

Flora, J. L., and C. Flora. 1991. Local economic development projects: Key factors. In *Rural community economic development,* edited by N. Walzer, 141–56. New York: Praeger.

Forest Ecosystem Management Assessment Team (FEMAT). 1993. *Forest ecosystem management: An ecological, economic, and social assessment.* Report of the Forest Ecosystem Management Assessment Team (July). Washington, DC: U.S. Forest Service.

Fortmann, L. P., J. Kusel, and S. K. Fairfax. 1989. Community stability: The foresters' figleaf. In *Community stability in forest-based communities,* edited by D. LeMaster and J. Beuter, 44–50. Beaverton, OR: Timber Press.

Freudenburg, W. R. 1984. Boomtown's youth: The differential impacts of rapid community growth on adolescents and adults. *American Sociological Review* 49:697–705.

Freudenburg, W. R., and R. E. Jones. 1991. Criminal behavior and rapid community growth: Examining the evidence. *Rural Sociology* 56 (4): 619–45.

Gans, H. J. 1962. *The urban villagers*. New York: The Free Press of Glencoe.

Gold, R. L. 1982. Commentary on "Local social disruption and Western energy development: A critical review." *Pacific Sociological Review* 25 (3): 349–56.

———. 1985. *Ranching, mining, and human impact of natural resource development*. New Brunswick, NJ: Transaction, Inc.

Goldschmidt, W. 1947. *As you sow*. Glencoe, IL: The Free Press.

Grumbine, E. 1994. What is ecosystem management? *Conservation Biology* 8:23–38.

Gusfield, J. R. 1975. *Community*. New York: Harper and Row.

Hazard, G. C. 1988. Communitarian ethics and legal justification. *University of Colorado Law Review* 59 (4): 721–39.

Headey, B., E. Holmstrom, and A. Wearing. 1985. Models of well-being and ill-being. *Social Indicators Research* 17:211–34.

Hester, R. 1985. Subconscious places of the heart. *Places* 2 (3): 10–22.

Hillery, G. A., Jr. 1955. Definitions of community: Areas of agreement. *Rural Sociology* 20 (2): 194–204.

Hiss, T. 1990. *The experience of place*. New York: Alfred A. Knopf.

Interorganizational Committee on Guidelines and Principles. 1994. *Guidelines and principles for social impact assessment*. NOAA Technical Memorandum NMFS-F/SPO-16. Washington, DC: U.S. Department of Commerce.

Jobes, P. C. 1984a. Social structure and myth: Impacts of energy development in ranchland. Paper presented at the annual meetings of the American Sociological Association, San Antonio, Texas, August.

———. 1984b. Disintegration of a ranching community: A longitudinal case study of the impact of coal development on gemeinschaft. Unpublished manuscript. University of Michigan, Ann Arbor.

Johnson, D. J. 1988. Toward a comprehensive "quality-of-life" index. *Social Indicators Research* 20 (5): 477–96.

Kaufman, H. F., and L. C. Kaufman. 1946. *Toward the stabilization and enrichment of a forest community*. The Montana Study, University of Montana. Missoula, MT: U.S. Forest Service, Region 1.

Kelly, J. F. 1974. The Skoglund loggers. Ph.D. diss., University of California, Riverside.

Kennedy, L. W., and N. Mehra. 1985. Effects of social change on well-being. *Social Indicators Research* 17 (2): 101–13.

Kim, K. 1973. Toward a sociological theory of development: A structural perspective. *Rural Sociology* 38 (4): 462–76.

Krannich, R. S., E. H. Berry, and T. Greider. 1989. Fear of crime in rapidly changing rural communities: A longitudinal analysis. *Rural Sociology* 54 (2): 195–212.

Kusel, J. 1991. *Ethnographic analysis of three forest communities in California*. Vol. II of *Well-being in forest-dependent communities*. Sacramento: California Department of Forestry, Forest and Rangeland Assessment Program.

Kusel, J., and L. Fortmann. 1991. *Well-being in forest-dependent communities*. Vol. I. Sacramento: California Department of Forestry, Forest and Rangeland Assessment Program.

Lasch, C. 1988. The communitarian critique of liberalism. In *Community in America: The challenge of habits of the heart*, edited by C. H. Reynolds and R. V. Norman, 173–84. Berkeley and Los Angeles: University of California Press.

Lee, R. G., D. R. Field, and W. R. Burch Jr. 1990. Introduction: Forestry, community, and sociology of natural resources. In *Community and forestry: Continuities in the sociology of natural resources*, edited by R. G. Lee, D. R. Field, and W. R. Burch Jr., 3–14. Boulder, CO: Westview Press.

Lobao, L. M. 1990. *Locality and inequality: Farm and industry structure and socioeconomic conditions*. Albany: State University of New York Press.

MacCannell, D. 1988. Industrial agriculture and rural community degradation. In *Agriculture and community change in the U.S.*, edited by L. E. Swanson, 15–75. Boulder, CO: Westview Press.

Machlis, G., and J. E. Force. 1988. Community stability and timber dependent communities. *Rural Sociology* 53 (2): 220–34.

Machlis, G. E., J. E. Force, and S. E. Dalton. 1995. Monitoring social indicators for ecosystem management. Unpublished paper. Eastside Ecosystem Management Project, Walla Walla, WA.

MacIntyre, A. 1984. *After virtue: A study in moral theory*. Vienna, IN: University of Notre Dame Press.

Manley, P. N., G. E. Brogan, C. Cook, M. E. Flores, D. G. Fullmer, W. Husari, T. M. Jimerson, L. M. Lux, M. E. McCain, J. A. Rose, G. Smitte, J. C. Schuyler, and M. J. Skinner. 1995. *Sustaining ecosystems: A conceptual framework*. Version 1.0 (April). R5-Em-Tp-001. San Francisco: U.S. Forest Service Pacific Southwest Region and Station.

Marchak, P. 1990. Forest industry towns in British Columbia. In *Community and forestry: Continuities in the sociology of natural resources*, edited by R. G. Lee, D. R. Field, and W. R. Burch Jr., 95–106. Boulder, CO: Westview Press.

Mastekaasa, A., and T. Moum. 1984. The perceived quality of life in Norway: Regional variations and contextual effects. *Social Indicators Research* 14 (4): 385–419.

Meinig, D. W. 1979. Symbolic landscapes: Some idealizations of American communities (and the beholding eye: Ten versions of the same scene). In *The interpretation of ordinary landscapes*, edited by D. W. Meinig, 31–48. New York: Oxford University Press.

Ministry of Environment, Lands, and Parks (Province of British Columbia) and Environment Canada. 1994. *State of the environment report for British Columbia*. British Columbia: Minister of Environment, Lands, and Parks; Ottawa: Minister of State.

Moum, T. 1988. Yea-saying and mood-of-the-day effects in self-reported quality of life. *Social Indicators Research* 20 (2): 108–25.

Nussbaum, M. C., and A. Sen. 1993. Introduction to *The quality of life*, edited by M. C. Nussbaum and A. Sen, 1–6. Oxford: Clarendon Press.

Oakley, A. 1975. *The sociology of housework*. London: Robertson.

Oppong, J. R., R. G. Ironsides, and L. W. Kennedy. 1988. Quality of life in a centre-periphery framework. *Social Indicators Research* 20 (6): 605–20.

Perry, C. 1986. A proposal to recycle mechanical and organic solidarity community sociology. *Rural Sociology* 51 (3): 263–77.

Price, M. L., and D. C. Clay. 1980. Structural disturbances in rural communities: Some repercussions of the migration turnaround in Michigan. *Rural Sociology* 45 (4): 591–607.

Putnam, R. D. 1993a. *Making democracy work: Civic traditions in modern Italy*. Princeton, NJ: Princeton University Press.

——— 1993b. The prosperous community: Social capital and public life. *The American Prospect* 13 (spring): 35–42 .

Relph, E. 1976. *Place and placelessness*. London: Pion.

Schnaiberg, A. 1986. Reflections on resistance to rural industrialization: Newcomers' culture of environmentalism. In *Differential social impacts of rural resource development*, edited by P. D. Elkind-Savatsky, 229–58. Boulder, CO: Westview Press.

Selznick, P. 1987. The idea of a communitarian morality. *California Law Review* 75:445–63.

———. 1992. *The moral commonwealth: Social theory and the promise of community*. Berkeley and Los Angeles: University of California Press.

Sen, A. 1984. *Resources, values, and development*. Cambridge: Harvard University Press.

———. 1985a. Well-being, agency, and freedom (the Dewey lectures). *Journal of Philosophy* 82 (4): 169–221.

———. 1985b. *Commodities and capabilities*. Professor Dr. P. Hennipman Lectures in Economics, vol. 7. New York: North-Holland.

———. 1987. *The standard of living*. Edited by G. Hawthorn. Cambridge: Cambridge University Press.

———. 1990. Individual freedom as a social commitment. *New York Review of Books*, 14 June .

———. 1993. Capability and well-being. In *The quality of life*, edited by M. C. Nussbaum and A. Sen, 30–53. Oxford: Clarendon Press.

Seydlitz, R., S. Laska, D. Spain, E. W. Triche, and K. L. Bishop. 1993. Development and social problems: The impact of the offshore oil industry on suicide and homicide rates. *Rural Sociology* 58 (1): 93–110.

Society of American Foresters. 1989. *Report of the Society of American Foresters national task force on community stability*. Bethesda, MD: Society of American Foresters.

Strathern, M. 1984. The social meaning of localism. In *Locality and rurality: Economy and society in rural regions*, edited by T. Bradley and P. Lowe, 191–97. Norwich, England: Geo Books.

Suttles, G. D. 1969. *The social order of the slum*. Chicago: University of Chicago Press.

Swanson, L. 1988. *Agriculture and community change in the U.S.* Boulder, CO: Westview Press.

Tuan, Y. 1993. *Passing strange and wonderful: Aesthetics, nature, and culture*. Covelo, CA: Island Press.

Walter, E. V. 1988. *Placeways: A theory of the human environment*. Chapel Hill: University of North Carolina Press.

Warren, R. L. 1978. *The community in America*. 3rd ed. Chicago: Rand McNally.

Weeks, E. C. 1990. Mill closures in the Pacific Northwest: The consequences of economic decline in rural industrial communities. In *Community and forestry: Continuities in the sociology of natural resources*, edited by R. G. Lee, D. R. Field, and W. R. Burch Jr., 125–40. Boulder, CO: Westview Press.

Wendel, S. 1987. McCloud, California: Maintaining cherished landscapes and lifestyles. Professional report (Master of Landscape Architecture), Department of Landscape Architecture, University of California, Berkeley.

Wilkenson, K. P., J. G. Thompson, R. R. Reynolds Jr., and L. M. Ostresh. 1982. Local social disruption and western energy development: A critical review. *Pacific Sociological Review* 25 (3): 297–306.

Williamson, D. 1976. Give 'er snoose: A study of kin and work among gyppo loggers of the Pacific Northwest. Ph.D. diss., American University.

Wilson, W. 1967. Correlates of avowed happiness. *Psychological Bulletin* 67 (4): 294–306.

World Commission on Environment and Development. 1987. *Our common future*. New York: Oxford University Press.

APPENDIX 12.1

Why Civic Responsiveness (or Commitment Behaviors) Are Important: A Counterargument to Sen

Sen (1987) ignores "commitments" in the calculus of personal well-being. He does so by making a distinction between actions based on "sympathy," which are included in calculations of well-being, and actions based on "commitment," which are not. Sen includes "commitments" in a category called "agency achievement" (1987) or "agency freedom" (1985). According to Sen (1985), "agency achievement" is a more inclusive category than personal well-being, and includes "what a person is free to do and achieve in pursuit of whatever goals or values he or she regards as important." He points out (1987) that by expanding the focus of attention and including "commitments," we move from "personal well-being" to "agency achievement."

Help provided to an individual that has the effect of making the helper "feel—and indeed be—better off" is "sympathy." This increases one's personal well-being. The behavior category of "commitments," on the other hand, involves personal action (it too may be help provided to another), which Sen states, *"in the net*, [is not] beneficial to the agent himself."* Sen adds, "This would put action outside the range of promoting one's own well-being."

Sen's rejection of an action because it is "in the net" not "beneficial to the agent himself" involves an evaluation of action (and its consequences) after the fact, or a prediction of its outcomes. Sen states (1987) he is concerned with effects. He, however, does not discuss at what point this calculation should take place nor what measures should be made to determine whether a behavior is beneficial or not. Given the nature of his decision rule, Sen ignores the motivation for in-

dividual action. Both categories of behavior, "commitments" and "sympathies," may involve action for which the motivation stems from the desire to help another person. Actions to improve one's community that do not have the "effect" of contributing to one's well-being therefore are not included in Sen's well-being calculations. In this manner, Sen ignores historical, social and societal forces that not only influence action (and motivations for action) but also influence value decisions implicit in the evaluation of well-being.

Etzioni (1988) states that the category of action called "commitments" is moral behavior. This is because such action is based on intentions, not consequences or effects. Intentions may also be considered the "intrinsic character" of action, and are taken here to be the primary criteria by which to evaluate it, because the consequences of action may not be predictable. This valuational approach is central to a deontological social philosophy (Etzioni 1988; Hazard 1988).

REFERENCES

Etzioni, A. 1988. *The moral dimension: Toward a new economics*. New York: The Free Press.

Hazard, G. C. 1988. Communitarian ethics and legal justification. *University of Colorado Law Review* 59(4): 721–39

Sen, A. 1985. Well-being, agency, and freedom (the Dewey lectures). *Journal of Philosophy* 82 (4): 169–221.

———. 1987. *The standard of living*. Edited by G. Hawthorn. Cambridge: Cambridge University Press.

SAM C. DOAK
S. C. Doak & Associates
Portland, Oregon

JONATHAN KUSEL
Forest Community Research
and
Department of Environmental Science,
    Policy, and Management
University of California
Berkeley, California

13

Well-Being in Forest-Dependent Communities, Part II: A Social Assessment Focus

ABSTRACT

This chapter assesses the current state of community well-being throughout the Sierra Nevada through the analysis of a combination of socioeconomic and community capacity measures. Aggregations of census block groups were used as the primary analysis unit. One hundred and eighty "community" aggregations were identified across the study area, delineated within six regions. Information on community capacity was derived through a series of nineteen local expert workshops. A case study of community capacity was conducted in Plumas County to examine the congruence of expert capacity assessment with community self-assessments. Socioeconomic data were developed from the 1990 *Census of Population and Housing.* A socioeconomic scale was developed from a diverse set of census measures to characterize the socioeconomic status of aggregations and to highlight similarities and variation across the Sierra Nevada. Aggregations were also characterized geographically by their spatial relationships to population centers, transportation corridors, and areas dominated by public lands, and a scale of relative isolation was developed from these spatial variables.

The relationships among socioeconomic factors, community capacity, and aggregation location and proximity to other geographic features are explored. Community capacity and socioeconomic status are found to be relatively independent, suggesting that they represent different dimensions of well-being that are not strongly related to each other. They are examined together in the discussion of well-being of the 180 aggregations.

INTRODUCTION

A credible, science-based assessment of the Sierra Nevada ecosystem must include a human dimensions component that includes a focus on current socioeconomic and social dynamics that influence ecosystem use, demands, and conditions. The objectives of the Sierra Nevada Ecosystem Project (SNEP) social assessment are to assess the current state of well-being of communities throughout the Sierra through an analysis of both socioeconomic measures and community capacity. This assessment contributed to the development of SNEP policy scenarios and can prove useful in the evaluation of the consequences of policy scenarios and ecosystem management more generally.

This chapter is divided into four sections. The first section briefly describes the geographic area included in this assessment. The second section discusses the methods employed in the study and is divided into five subsections: (1) data sources and unit of analysis, (2) community socioeconomic factors, (3) community capacity, (4) spatial analysis, and (5) community self-assessment.

The third section describes and discusses the results of the assessments. Capacity assessment ratings and socioeconomic scale scores are reviewed by region, and the factors contributing to each are analyzed. The relationship of spatial variables to capacity and socioeconomic factors is discussed, as is the relationship between community self-assessment and expert capacity assessment. A summary of what is learned from

Sierra Nevada Ecosystem Project: Final report to Congress, vol. II, *Assessments and scientific basis for management options.* Davis: University of California, Centers for Water and Wildland Resources, 1996.

this integrated assessment constitutes the final section of the chapter.

The diagram in figure 13.1 demonstrates the methodological flow of the integrated SNEP social assessment and its relationship to supplementary SNEP work. The "Methods" section first describes the development of the primary data sources and the "community aggregation" analysis unit, which is based on Bureau of the Census block groups. It then discusses the methodological development of the socioeconomic scale and measurement of capacity. The socioeconomic scale is based exclusively on a diverse set of census measures. This scale includes critical components of well-being but is not in itself exhaustive of all measures of well-being. Community capacity reflects a dynamic and multidimensional component of human well-being. It was assessed through workshops held with local experts. The subsection on spatial analysis describes how "community" point data were determined and how they were used to explore the relationships among socioeconomic factors, community capacity, and aggregation location and proximity to other geographic features. The final subsection discusses the workshops devoted to a pilot community self-assessment.

STUDY LOCATION

The social assessment focuses principally on the SNEP core region, an area primarily delineated by watershed boundaries and modified by elevation and administrative boundaries. The geographic area included in the social assessment deviates slightly from the SNEP core boundary to accommodate complete Bureau of the Census block groups, units for which socioeconomic data are summarized. The social assessment study area (hereafter referred to as the "Sierra") extends from central Lassen County and Eagle Lake in the north, southward to the middle of Kern County. The eastern boundary follows the Nevada-California border to the Lake Tahoe Basin—where it includes the populated portion of Nevada in the basin—and continues south to Inyo County. The Death Valley region in Inyo County is excluded from the assessment area. The western boundary includes the foothill region of Kern, Tulare, Fresno, and Madera Counties, bisecting those counties, and follows the western boundaries of Mariposa, Tuolumne, Calaveras, Amador, and El Dorado Counties. The western boundary excludes the westernmost portion of Placer County, includes all of Nevada County, and draws in the foothill and mountain regions of Yuba and Butte Counties.

FIGURE 13.1

Social assessment methodology.

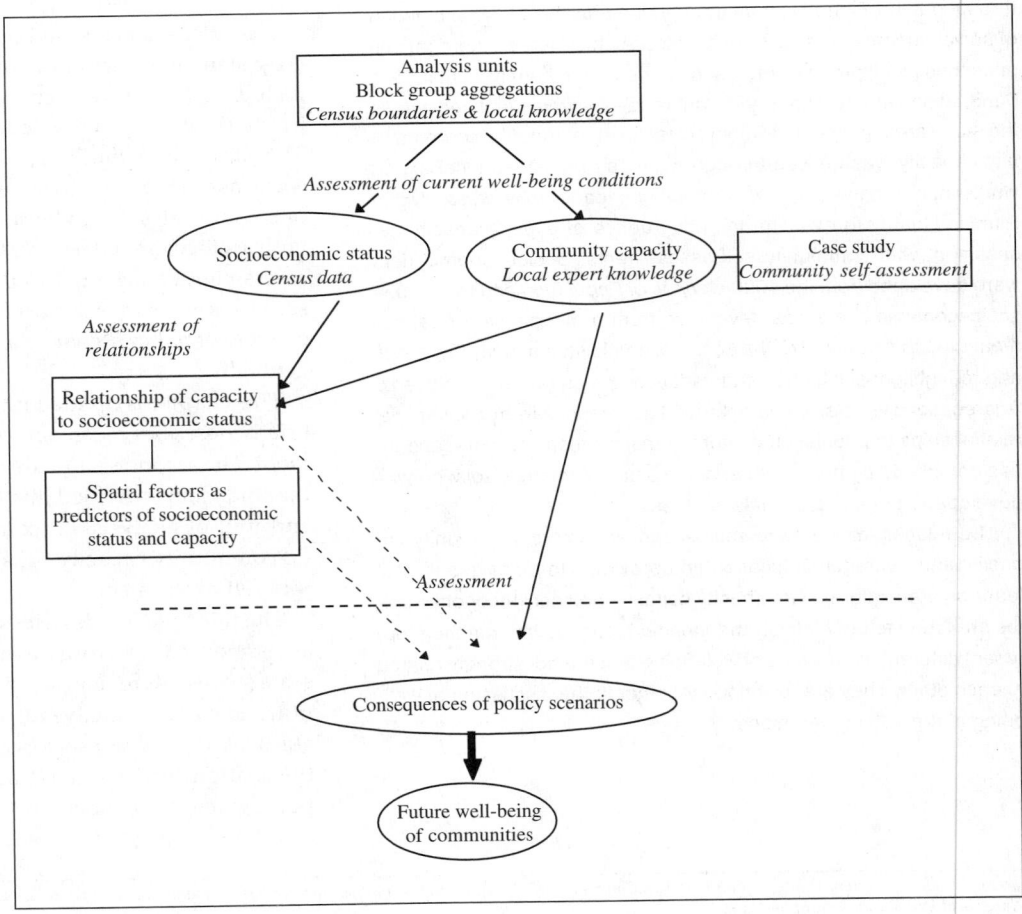

METHODS

Data Sources and Unit of Analysis

The first step in the social assessment identifies a socially relevant unit of analysis and a consistent source of data that are available for that unit and that provide information on social conditions across the Sierra. The 1990 *Census of Population and Housing* (U.S. Bureau of the Census 1990a) was selected as the principal source of secondary data. Census data include an extensive amount of socioeconomic information that is consistent throughout the Sierra, and it is accessible at several different geographic levels. Other social and economic data either are available only at a county level or are inconsistent in format. For example, statistics on crime, health, and enrollment in public support programs, measures that would be valuable to include in a well-being assessment, are available only at the level of the county or incorporated city. See Machlis et al. 1995 and Interorganizational Committee on Guidelines and Principles 1994 for listings of additional measures. Most are not included here, however, because they are not collected at a level equivalent to census block groups.

The selection of an analysis unit focuses on finding a socially recognizable unit of analysis that captures the variation in social conditions across the Sierra. The Bureau of the Census reports population and housing summary statistics in several different geographic units, including counties, tracts, places, block groups, and blocks.

The county, the most frequently used unit of analysis (see Kusel 1996), was avoided for this study because it is too gross a level and does not adequately represent social communities. County-level analyses tend to ignore within-county variation that can be observed only by focusing on smaller analysis units. Moreover, in several cases only portions of individual counties are included in the SNEP study area. It would be inappropriate, for example, to include the relatively large valley populations of Fresno and Yuba Counties in an analysis of social well-being in the Sierra Nevada. Census tracts are smaller than counties, but they are still large enough to suffer from the same deficiencies, though to a slightly lesser degree.

Places, as defined by the Bureau of the Census, include incorporated areas and census-designated places (CDPs). CDPs are the statistical counterparts of incorporated places and comprise densely settled areas that are identifiable by name but are not legally incorporated. The boundaries of both incorporated places and CDPs often omit people who may identify with the community defined by the CDP or incorporated area. Moreover, because populations in the Sierra Nevada are often dispersed, places provide only partial coverage, leaving the majority of the population in the region unaccounted for. Places were thus found to be unsuitable for the study.

Blocks and block groups are the two smallest units used by the Bureau of the Census. Blocks, the smallest unit, could not be used in this study because only limited census data are published at this level. Block groups are the smallest inclusive units for which all summary statistics are reported, including the more detailed sample data. Block group delineations, however, often do not reflect existing community lines. Blocks and block groups are delineated along major roads and other physical features and often do not coincide with communities or other meaningful social units. Many small towns and communities are split into two or more block groups.

Recognizing that many of the limitations of block groups could be overcome by combining similar adjacent areas into larger and more socially meaningful units, we selected aggregations of block groups as the primary analysis unit for the social assessment. Existing block groups were aggregated to form meaningful social units that more closely represented locally defined communities. While block groups do not cross county lines, no attempt was made to restrict aggregations to within individual counties, and in some cases aggregations do include populations in more than one county.

Block group aggregation units and names were developed through an iterative process involving extensive input from county planners and other local experts. First, discussions took place with planners in each county in the Sierra familiar with local communities and also generally familiar with census data and county demographics. These discussions led to the development of preliminary "community" block group aggregations, using the criteria just outlined. Draft county-level maps of these aggregations were then sent to each planner for review, along with a request to circulate the map to others knowledgeable about local communities. Both written and oral comments were received. Additional conversations with planners in each county clarified their suggestions regarding changes to the preliminary aggregations, and the preliminary aggregations were revised. Finally, during individual community capacity workshops, groups of local experts reviewed the revised aggregations and provided suggestions for the final aggregations.

A total of 182 aggregations were developed from 720 block groups within the core study area, although only 180 were used in the analysis. Two aggregations (Tehachapi Prison and the Correctional Center in Susanville) were excluded from all analysis since more than 99% of their populations are prisoners in correctional institutions. Two other aggregations with large prison populations, Ione and Keystone/Lake Don Pedro, were included in the analysis because the incarcerated represent a smaller percentage of the overall aggregation population. Moreover, prisoners are excluded from all but one of the socioeconomic scale measures. This is discussed further in a later section.

The following criteria were used to guide the development of aggregations.

- Aggregations are formed from one or more block groups that are spatially adjacent or linked to one another.

- Aggregations are formed from block groups in which the majority of the population is associated with a single community.

- Each aggregation should contain a minimum total population of 500. With few exceptions, all aggregations—including solitary block groups—conform to this rule. In a few areas, smaller block groups were not aggregated with others when they represented distinct communities or to preserve the heterogeneity represented by extreme social differences between adjacent areas.

- Where it is necessary to include more than one community in a single aggregation, block groups may be placed into aggregations in which populations share common service centers, common community service districts, or common school systems.

- Block groups with small, dispersed populations that conform to no single community are aggregated together when they share similar demographic characteristics, as determined by local knowledge (e.g., low-density housing, commuting patterns, or ethnicity).

- Where adjacent block group populations differ and do not fall under the previous criteria, they are kept as separate units to ensure that this diversity is captured in the analysis.

- With numerous small communities or areas with no clearly identifiable communities, aggregation determination is also based on geographic features.

Block group aggregation names are intended to be both inclusive of existing communities and descriptive of general population patterns within the aggregation. In general, aggregation names include major community names as commonly recognized by residents (e.g., Arnold/Avery/Dorrington, or Kernville/Wofford Heights). Where no definable communities exist, names are based on general geographic characteristics (e.g., Lake Oroville Area, or South County) or on relationships to nearby towns or community centers.

Socioeconomic Scale

A scale depicting variation in selected socioeconomic factors for the community aggregation units was developed from 1990 *Census of Population and Housing* data. The socioeconomic scale incorporates five primary categories: housing tenure, poverty, education, employment, and children in homes with public assistance income. These individual categories are combined into a single scale to take advantage of each individual measure while ensuring that no single one dominates. Each category is weighted equally within the scale, although the poverty category has two components. The primary assumption of the scale is that higher levels of home ownership, education, and employment indicate higher levels of

socioeconomic well-being, and higher levels of poverty and a higher percentage of children in homes receiving public assistance income indicate lower levels of socioeconomic well-being.

Components of the Socioeconomic Scale

The *housing tenure* score of the scale is the percentage of all occupied housing units that are owner occupied. Since the universe is occupied housing units, this variable includes only permanent residences that are the usual place of residence of the occupants. It excludes group quarters (e.g., military quarters, college dormitories, or prisons). The inverse of this variable is equal to the percentage of occupied housing units that are renter occupied. The housing tenure score, then, reflects the relative level of owner-occupied housing versus renter-occupied housing across the Sierra. The housing tenure component is suggestive of the relative wealth and permanence of the residents in an area and offers insight into the degree of local control of a vitally important resource.

The *poverty* score includes two equally weighted components: the percentage of all persons in poverty and a measure of poverty level and intensity. Poverty status is determined at a national level by the Bureau of the Census as a function of family income and family size. The number of persons below the poverty level is the sum of the number of persons in families with incomes below the poverty level and the number of unrelated individuals with incomes below the poverty level. Poverty status is not determined for institutionalized persons, persons in military group quarters and in college dorms, and unrelated individuals under fifteen years of age.

The first component in the poverty score, the percentage of all persons with income below the poverty level, is the ratio of persons with incomes above the poverty level to those with incomes below the poverty level. The second component of the poverty score indicates the relative intensity of poverty of those individuals with incomes below the poverty level. Three variables are combined to capture the intensity of poverty within a given area, using the following formula:

$$S = \Sigma \left[(1 * X), (3 * Y), (9 * Z) \right]$$

where: S = poverty intensity

X = percentage of persons with incomes between 75% and 99% of the poverty level

Y = percentage of persons with incomes between 50% and 74% of the poverty level

Z = percentage of persons with incomes less than 50% of the poverty level

The multiplication factors of 1, 3, and 9 are used to emphasize the intensity of poverty by placing greater weight on the highest poverty levels. These factors help to stretch out the range of numbers and create a greater distance between incomes that are just below the poverty level and those that are far below the poverty level. More linear factors of 1, 2, and 3 do not place enough emphasis on the higher levels of poverty to provide a score reflective of poverty intensity.

Education is reflected by a cumulative educational attainment score weighted toward higher levels of educational attainment. Education is assessed in the census data for all persons twenty-five years of age and older. This is the only component of the socioeconomic scale that includes the large incarcerated populations of Ione and Keystone/Lake Don Pedro. Unlike other components of the scale, the census-defined universe for educational attainment is *all* persons, including prisoners, and is reported in a way that does not allow for isolation of incarcerated populations. The education score is calculated by multiplying the percentage of persons in each of the seven census data education categories by a factor that increases by 1 at each higher level and then summing the products.

$$S = \Sigma \, [A, (B*2), (C*3), (D*4), (E*5), (F*6), (G*7)]$$

where: S = educational attainment score
A = percentage of persons with less than a ninth grade education
B = percentage of persons with a ninth to twelfth grade education, no diploma
C = percentage of persons who are high school graduates or the equivalent
D = percentage of persons with some college, no degree
E = percentage of persons with an associate degree
F = percentage of persons with a bachelor's degree
G = percentage of persons with a graduate or professional degree

The *employment* score is the percentage of the civilian labor force that is employed and is the inverse of the percentage of persons who are unemployed. All civilians sixteen years old and older are classified by the Bureau of the Census as unemployed if they (1) were neither "at work" nor "with a job but not at work" during the week of enumeration, and (2) were looking for work during the four weeks preceding enumeration, and (3) were available to accept a job. Also included as unemployed are civilians who did not work at all during the reference week and were waiting to be called back to a job from which they had been laid off. As used here, the universe for employment excludes those not "in the labor force" and those in the armed forces.

As measured by the Bureau of the Census, unemployment provides a well-defined but somewhat narrow view of the status of the labor force. Since it is limited to individuals who are actively seeking work, the measure is often inaccurate in areas of chronic unemployment where frustrated workers have dropped out of the labor force. Such workers cannot be identified using only census data. These data report employment and labor force participation for those between the ages of 16 and 64, as well as for those over 64 years of age, but the percentage of persons within these groups who are not in the labor force is not restricted to frustrated workers who have dropped out, but may include early retirees and others who are not part of the labor force by choice.

Children in households with public assistance income reflects the percentage of all children under fifteen years of age living in households that receive public assistance income. Public assistance income includes (1) supplementary security income payments by federal or state welfare agencies to low-income persons who are sixty-five years old or older, blind, or disabled; (2) aid to families with dependent children (AFDC); and (3) general assistance. It excludes payments for hospital or medical care. The percentage of children in households with public assistance income provides an indicator of families in need. Yet it is important to point out that not all families in need receive public assistance. This is true particularly in cases where cultural values limit the acceptability of public assistance.

Development of the Socioeconomic Scale

Standardized scores were calculated for each component score before they were combined into a single scale. Standardized scores, often referred to as "Z" scores, indicate the number of standard deviations above or below the mean that a particular observation falls. They are calculated by dividing the difference between a particular observation and the mean by the standard deviation. Standardization facilitates the comparison of scores from distributions. To ensure that outliers do not have undue influence on the distribution range of any score, each standardized score is then normalized to a base of 100 using two standard deviations as reference points. The individual component scores are combined into a single socioeconomic scale, which is also normalized to a base of 100.

$$X = \frac{\Sigma_S{}^{1-6} \, [((S \, / \, Z) - A) * (100/(B - A))]}{5}$$

where: X = socioeconomic scale
S = standardized scores of each of five scale components
Z = 2 if S is persons in poverty or poverty intensity; otherwise, $Z = 1$
A = −2 (two standard deviations below the mean)
B = 2 (two standard deviations above the mean)

The socioeconomic scores are reported on a seven-point categorical scale, with 1 being the lowest socioeconomic score and 7 being the highest. The ordinal scale is derived from the continuous scores, divided into categories based on the number of standard deviations from the mean of the scale. A rating of 1 is a very low socioeconomic score and includes those scores at least two or more standard deviations below the mean (i.e., standard deviation ≤−2); 2 is a low score (standard deviation > –2 and ≤ –1); 3 is a medium-low score (standard deviation > –1 and ≤ –0.5); 4 is a medium score (standard deviation > –0.5 and < 0.5); 5 is a medium-high score (standard deviation ≥ 0.5 and < 1); 6 is a high score (standard deviation ≥ 1 and < 2); and 7 is a very high score (standard deviation ≥ 2).

While income is a commonly used indicator of socioeconomic status and well-being, it is not included in the socioeconomic scale for two reasons: (1) most of the variables in the scale are closely correlated with income, and (2) income measures available from the census data are problematic. The Bureau of the Census reports income in a variety of tables and formats. Comparisons of census-reported aggregate income with other income sources indicate that census-reported income is considerably underestimated, particularly for interest and dividend and public assistance income (Stewart 1996). Adjustments can be made to compensate for these discrepancies, but they can be applied only to aggregate income, making many of the census income tables, which report income within finite categories, unusable. Due to the level of variation in interest and dividend and public assistance income among block group aggregations, these correction factors have a significant effect on the ordering of relative income among aggregations. As an additional complication, analysis of aggregate income indicates that pockets of households with extraordinarily high income throughout the Sierra can significantly distort the real distribution of income within aggregations, making *average* income measures—whether household, family, or per capita—inappropriate as relative indicators of the socioeconomic status of individuals within a particular area. Average income measures are best suited only as a means for expressing total income in an area in relative terms.

Since aggregate income, or average income derived from aggregate data, is the only measure that can be effectively adjusted for discrepancies involving interest and dividend income and public assistance income, and since average income appears to severely distort the relative ranks of aggregation by actual income, direct income measures were not included in the socioeconomic scale.

Community Capacity

Community capacity (described fully in Kusel 1996) is the collective ability of residents in a community to respond to external and internal stresses; to create and take advantage of opportunities; and to meet the needs of residents, diversely defined. It consists of three broad categories: physical capital, human capital, and social capital (see appendix 13.1).

Community capacity was assessed for the community aggregations, based on local expert knowledge. A series of local workshops was held in nineteen different locations across the Sierra. All of the workshops but one focused on aggregations falling primarily within a single county. That workshop covered aggregations in the Greater Lake Tahoe Basin, which included aggregations in six separate counties.

The number of participants in each workshop ranged from three to eighteen, depending on the area and the number of aggregations to be addressed. To ensure diverse perspectives in workshop discussions, SNEP workshop organizers selected participants from a variety of backgrounds. Included were those individuals who—by nature of their profession, local involvement, or history of residence—are knowledgeable about the physical, human, and social capital of most of the communities within each workshop's area of focus. Participants included but were not limited to planners and planning commissioners, community development professionals, current and former county supervisors, education administrators, businesspeople, health and human service providers, and long-term residents with diverse backgrounds and experiences.

To ensure consistency in the information gathered, each workshop used the following process.

1. The creation, composition, and general charge of SNEP were introduced to the group.

2. The role of the social assessment component within the SNEP process was discussed and the entire social assessment methodology reviewed, including a brief introduction to the analysis units and the concepts of well-being and community capacity.

3. The process for determining the capacity of community aggregations within the workshop's area of interest was outlined.

4. The community aggregations for the area of interest were reviewed by the group for appropriateness. In some cases, alterations were made to the aggregations.

5. The concept of community capacity was reintroduced and defined in more detail.

6. Participants were asked to indicate the various community aggregations with which they were most knowledgeable and most familiar. Based on the responses, assignments were made to individual participants to ensure that each aggregation was assessed by two different people (although limited expert knowledge and limited number of experts occasionally led to one assessment). Participants were asked to complete a separate community capacity worksheet (see appendix 13.2) for each aggregation, including a narrative assessment of capacity and a rating of capacity on a seven-point scale ranging from very low to very high. In assessing capacity, participants were asked to consider the level of physical, human, and social capital in the communities within each aggregation.

7. The individual capacity rankings for each aggregation were summarized and anonymously presented to the group for their review. During a facilitated group discus-

sion, the capacity ratings for each aggregation were discussed, there was further elaboration of issues relating to capacity for each aggregation, and a final capacity ranking was determined by the group for each aggregation.

Information gathered from each workshop includes individual community aggregation narratives and capacity rankings, final group capacity rankings, and notes from the facilitated discussion of all aggregations.

The results of each workshop were reviewed to ensure that the capacity rankings and related discussions were consistent with those of other workshops. If a group did not adequately grasp the concept of capacity or if numerical ratings generated by the experts appeared significantly different from those of other groups, a second panel was convened. Two additional panels were convened for these reasons. In these instances, the narratives of both groups were incorporated into the assessment, but the capacity ratings were selected from the group that appeared to have the best understanding of capacity and that assigned ratings consistent with those employed in other workshops.

In most workshops experts proved reluctant to apply the highest and lowest capacity ratings on the seven-point capacity scale, and very few aggregations actually received either a 1 or a 7. To ensure greater consistency in the analysis across the study, the scale was collapsed to a five-point range, with scores of 1 and 2 combined to form the lowest capacity score and scores of 6 and 7 combined to form the highest.

Spatial Analysis

A geographic point coverage was generated to represent the approximate location of the population-weighted centers of each community aggregation. The point coverage was developed to provide a population-based depiction of the aggregations and to facilitate analysis of the relationships among socioeconomic factors, community capacity, and aggregation location and proximity to other geographic features. Polygon representations of the block group aggregations do not adequately reflect the location and distribution of populations within each aggregation. (A polygon is a closed-plane figure used to represent the geographic extent of a feature on a map. Block groups are delineated by the Bureau of the Census as a series of adjacent polygons inclusive of all populated and unpopulated land and water areas in the United States.) Many aggregations include large tracts of public land and other unpopulated areas, and the physical extent of the aggregation polygons often distorts the extent of the actual populations within them.

Point representations of each aggregation were created by averaging the coordinates of the internal points of each block group within the aggregation, weighted by the population of the block group relative to the population of the entire aggregation. The internal point coordinates (latitude and longitude) of a block group, calculated by the Bureau of the Census, rep-

resent the approximate geographic center of the block group. If, due to the shape of the block group, the geographic center falls outside of the block group, the internal point is relocated within the boundaries. Likewise, if the center falls within a body of water, the internal point is relocated to a land area within the block group (U.S. Bureau of the Census 1990b). Figure 13.2 shows the point representation of the 180 block group aggregations used in the social assessment. Point locations for the aggregations could be more accurately located from block (rather than block group) center coordinates and populations; however, these data were not readily available within the time frame of the analysis.

To further characterize each aggregation by geographic location relative to infrastructure, services, public land, and other factors, we enhanced the aggregation point coverage with some basic spatial data, including

- average elevation of an area defined by a 0.5 km (0.3 mi) radius around the point
- aerial distance to the nearest federal highway or interstate
- aerial distance to the nearest state highway on major road coverage
- aerial distance to the closest major city with a population of 25,000 or greater
- aerial distance to the nearest county seat
- percentage of public land within an 8 km [5 mi] radius
- population density by aggregation area

While actual road miles or travel time may provide more explicit measures than aerial distances, time and resource limitations prohibited this type of detailed analysis. Moreover, since the aggregation center points are only representations of dispersed populations, the less precise aerial distances should suffice for this analysis. The creation of these variables permits the evaluation of spatial characterizations in developing typologies of aggregations as well as the examination of spatial relationships associated with socioeconomic factors and community capacity. These variables were selected in part due to a previous assessment of communities in the Pacific Northwest (Forest Ecosystem Management Assessment Team [FEMAT] 1993) in which a rudimentary analysis of spatial factors indicated that access to transportation corridors, density of federal land ownership, and general isolation may be related to community capacity.

The four distance measures and the percentage of public land measure were combined into a simple scale as a general proxy of isolation of each community aggregation. Standardized scores were calculated for each component of the scale and normalized to a base of two standard deviations. The normalized component variables were then combined with equal weight into a single scale. The isolation scale was created using the following formula.

FIGURE 13.2

Point locations of
aggregations.

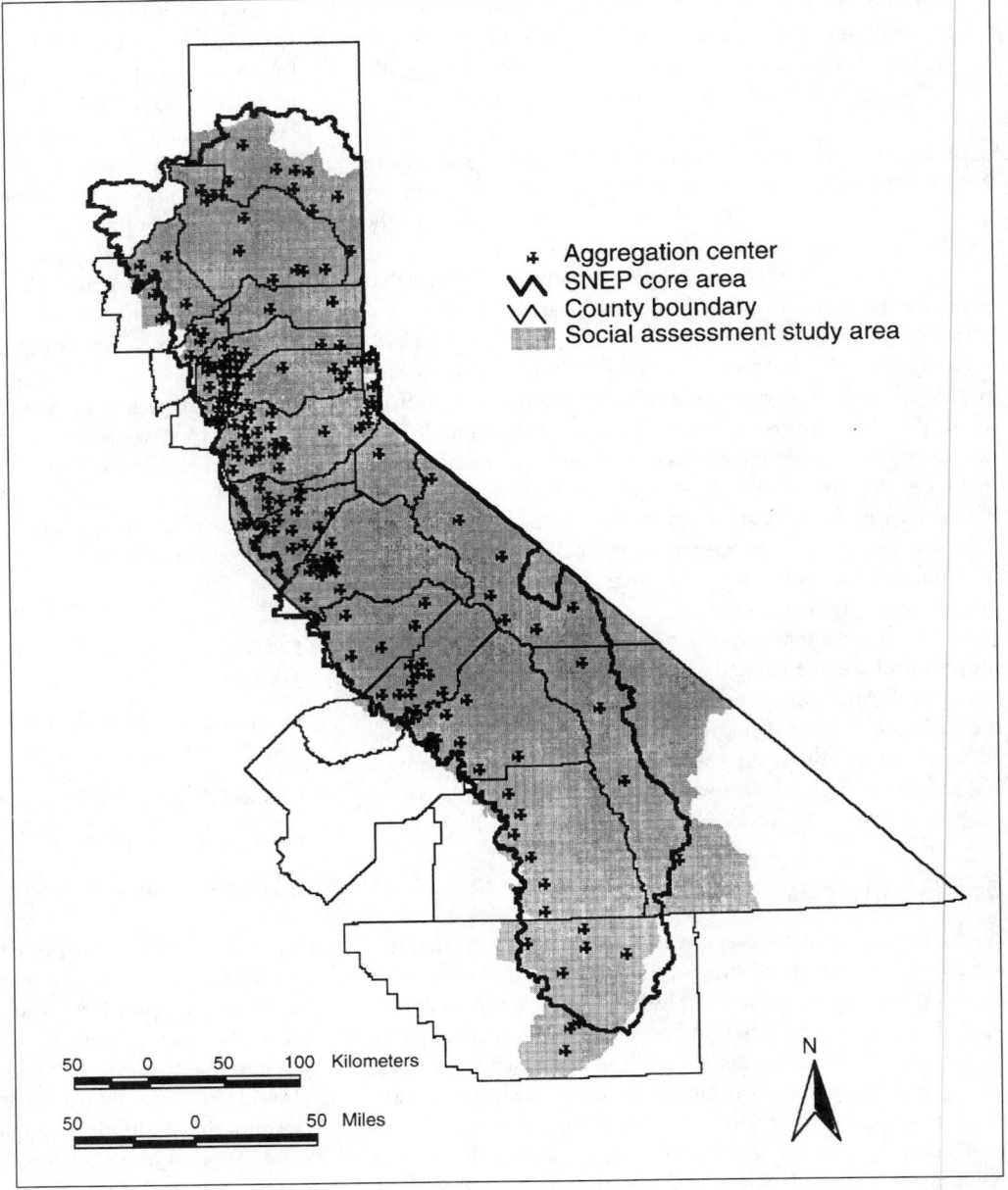

Legend:
+ Aggregation center
SNEP core area
County boundary
Social assessment study area

50   0   50   100   Kilometers

50   0   50   Miles

N

$$S = [\Sigma(A, B, C, D, E)]/5$$

where: S = isolation scale
    A = standardized distance to the nearest federal highway or interstate
    B = standardized distance to the nearest state highway
    C = standardized distance to the closest major city with a population of 25,000 or greater
    D = standardized distance to the nearest county seat
    E = standardized percentage of public land within an 8 km radius

Case Study of Capacity Assessments in Plumas County

A case study of community capacity was conducted to examine the congruence of expert capacity assessment with community self-assessments. The study of individual communities also allowed in-depth exploration of local issues. Plumas County was chosen as the case study because of the varying types of forest dependence (e.g., commodity production, a service industry associated with recreation, and a growing number of retirement communities and other communities in which the forest was important as a backdrop) and because it is the home of the Plumas Children's Network. The Plumas Children's Network, working with a grant from the Sierra

Health Foundation, was conducting community assessments to develop strategies to improve the health and well-being of children and families in Plumas County. Its staff agreed to work with the SNEP researchers because of an interest in community capacity assessment. Working with the Plumas Children's Network provided SNEP social assessment researchers access to local networks and individuals who were able to help organize community workshops and ensure higher local participation. In addition, the Plumas Children's Network was able to use SNEP research immediately for its community assessment and to help secure additional funding for a second phase of the grant. As a result, local communities benefited not only by learning about themselves but also from the advancement of the Plumas Children's Network projects.

Involving local residents in assessing community capacity required the development of a community self-assessment workshop, one quite different from the process used for the expert assessment of capacity for the aggregations. Evening workshops, which averaged two hours in length, were conducted in the towns of Chester, Graeagle, Greenville, Portola, Quincy, and Sierra Valley. The workshops followed this format:

1. Participants were introduced to researchers, to SNEP, and to the Plumas Children's Network. This was followed by a brief description of the workshop objectives and a discussion of the workshop ground rules. The workshop objectives were to identify key issues that affect local capacity and to numerically rate community capacity on a seven-point scale (1, very low; 2, low; 3, medium-low; 4, medium; 5, medium-high; 6, high; 7, very high).

2. SNEP researchers described the concept of capacity and its application to communities.

3. SNEP researchers discussed issues that define and determine community capacity.

4. Working individually, participants were asked to write on cards the most important items/issues that affect their community's capacity and to numerically rate their community's capacity.

5. Participants were individually asked to identify the three most important issues that determine capacity.

6. Working in small groups, participants shared and discussed their lists of most important items/issues with one another and determined the five or six of most importance to the small group.

7. The five or six most important items/issues from each group were posted in front of the full group.

8. Items/issues were organized into categories.

9. The large group reexamined the list, discussed it, and added any important items/issues that were missing.

10. In several workshops participants voted on the most important issues and were allotted five votes to distribute among issues they felt were most important.

11. Individually, participants rated the capacity of the community a second time.

12. A SNEP researcher and the Plumas Children's Network coordinator briefly recapped the meeting, reviewed group determinations, and thanked participants.

RESULTS AND DISCUSSION

This section is presented in four parts. The first part introduces the social assessment regions and describes the distribution of population within them. In the second part the socioeconomic status and capacity scores among the regions and individual aggregations are discussed in detail. An overview of the regions and the aggregations within them is provided, highlighting some of the variation in community capacity, socioeconomic scale scores, and other ancillary socioeconomic data. Some of the unique findings regarding individual aggregations are also discussed based on observations from the workshops and analysis of socioeconomic data. This section closes with a discussion of the concentration of populations with low socioeconomic status.

The third part focuses on relationships between diverse socioeconomic variables, including the socioeconomic status score and capacity. The discussion addresses the internal association of socioeconomic scale items; the association between capacity and socioeconomic status; the occurrence, frequency, and type of single-parent households; and the association of income to the socioeconomic scale. The relationship of capacity to socioeconomic status is described as an important determinant of overall aggregation well-being. Patterns of age distribution are described and related to socioeconomic status and capacity, along with the spatial characteristics of aggregations. The final part of this section presents the findings from the Plumas County case study.

Throughout these summaries, all discussions of socioeconomic scores and statistics related to socioeconomic variables refer to information derived from the 1990 Census of Population and Housing (U.S. Bureau of the Census 1990a). Community capacity scores discussed in the summaries reflect a five-point capacity scale—collapsed from the seven-point scale used in the workshops—where 1 indicates low and very low capacity, 2 indicates medium-low capacity, 3 indicates medium capacity, 4 indicates medium-high capacity, and 5 indicates high and very high capacity. Descriptions of individual aggregations are based largely on discussions in the local capacity workshops.

The Social Assessment Regions

Region Descriptions

Six distinct regions, somewhat different from the hydrologic and other geographic regions presented elsewhere in this volume, were identified in the social assessment. They are based on transportation corridors, commute patterns, economies, community identification, and other information collected in local workshops. These regions, while similar in some respects, are recognized as relatively distinct social and economic areas. Delineation of regions permits identification of regional patterns and trends and provides a valuable (though certainly not the only) perspective for this analysis. Figure 13.3 identifies these regions relative to the SNEP core area.

The Northern Sierra region includes the southern half of Lassen County, all of Plumas and Sierra Counties, and foothill areas on the east side of Yuba and Butte Counties. Many communities in this region are linked to the timber industry, some historically, with little or no modern-day timber-related employment, and others currently, with a significant proportion of employment in the timber industry. The area also has a growing recreation and service economy. The region is largely beyond the Sacramento and Interstate 80 commuting corridor that characterizes the area to the south.

FIGURE 13.3

Social assessment regions.

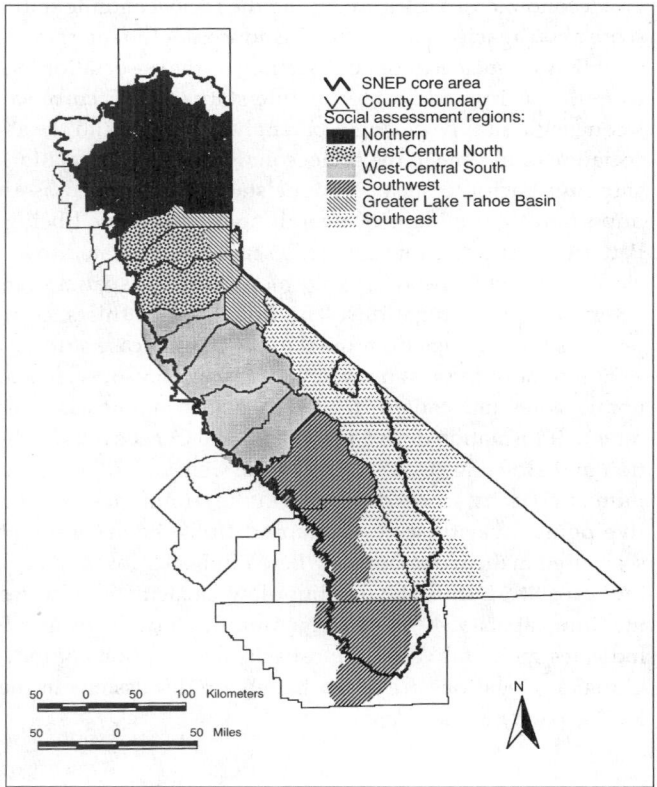

The West-Central North region includes the western portions of Nevada and El Dorado Counties and the central portion of Placer County. Each of these three counties lies along major east-west transportation routes that provide easy access to the Sacramento area and allow year-round traffic over the divide and into the Lake Tahoe Basin. This area has seen considerable growth in the last twenty years.

The West-Central South region includes Amador, Calaveras, Tuolumne, and Mariposa Counties and the eastern portion of Madera County. These five counties are linked by Highway 49, which traverses the Sierra foothills and terminates in Oakhurst in Madera County. The southern three counties are also linked by their economic reliance on Yosemite National Park.

The Southwest region includes the eastern portion of Fresno and Tulare Counties, and the north-central portion of Kern County. The Central Valley portions of these counties are outside of the SNEP core study area and are not included in this analysis. The large Central Valley cities of Fresno, Visalia, and Bakersfield, however, have a considerable and growing impact on settlement within the Southwest Sierra Nevada region.

The Greater Lake Tahoe Basin (GLTB) region consists of the easternmost portions of Nevada, Placer, and El Dorado Counties; all of Alpine County; and the southwestern portion of Washoe County and northwestern portion of Douglas County in Nevada. Alpine County and the Donner Summit and Truckee aggregations to the north are not part of the Lake Tahoe hydrologic basin, but the economies and social organization of these areas, primarily based on tourism and recreation, are similar to others in the basin and therefore are grouped for this analysis. Although a portion of Carson City extends to the shore of Lake Tahoe, this area was not included in the social assessment, because the vast majority of the population within the block group resides in the Carson Valley, a clearly separate geographic region and economy.

The Southeast region includes the east-side Sierra counties of Mono and Inyo, excluding the southeastern portion of Inyo that includes Death Valley. The small, sparsely inhabited southeastern portion of Tulare County, which drops eastward from the Sierra Divide to the Inyo County border, is included in this region. Due to the small, dispersed population across the region, some of the census block groups are extremely large and unwieldy. This led to the creation of several community aggregations encompassing small communities not closely linked to one another (e.g., Big Pine/Independence and Olancha/Cartego/Kennedy Meadows). This region is also characterized by a land ownership pattern dominated by public agencies, primarily the Los Angeles Department of Water and Power (LADWP), the U.S. Forest Service, and the U.S. Bureau of Land Management.

Population Distribution

The total population of the Sierra Nevada area considered in the social assessment is 646,769. This total excludes 5,533 persons in the Tehachapi Prison aggregation and 4,099 in the Susanville Correctional Center aggregation, and is based on 1990 data. Figure 13.4 illustrates the distribution of population in the six regions. More than one-third of the Sierra population lies in the West-Central North region, and more than one-half resides in the West-Central North and West-Central South regions combined. The two most populated aggregations in the study, however, are in the Northern region. The west-side aggregations of Oroville and Paradise/Magalia have population totals of 33,706 and 32,507, respectively, and together they make up more than half of the population in the Northern region. The Northern region has the second lowest median aggregation population of the Sierra, at 1,345.

The South Lake Tahoe aggregation, with a population of 23,319, is the third most populated in the Sierra, including more than one-third of the residents in the Greater Lake Tahoe Basin region. Another 15% of the GLTB region lies in the Truckee aggregation, with a population of 9,386, and a further 12% is in the Incline/Crystal Bay/Brockway aggregation, with 7,856. The region has a median aggregation size of 2,395.

The Auburn and Shingle Springs/Cameron Park aggregations are the fourth and fifth largest populations in the Sierra, with 23,202 and 22,270 persons, respectively. These two areas each make up approximately 10% of the total population in the West-Central North region, which has a median aggregation size of 2,888, the highest of the six regions.

The largest aggregation in the West-Central South region is Ione, with a population of 9,537. Forty-five percent of the Ione aggregation's population, however, lives in group quarters, presumably in prison. The second and third largest aggregations are Mariposa and Sonora, with populations of 8,746 and 7,418, respectively. These three largest aggregations constitute just under 18% of the region's population. The Keystone/Lake Don Pedro aggregation also has a large prison population, with 80% of the 4,812-person aggregation living in group quarters. Several aggregations have significant proportions of Native Americans; these include North Fork, O'Neals, Tuolumne, and Westpoint/Wilseyville, where Native Americans account for between 8% and 11% of the total population. The median aggregation size of the West-Central South region is 2,418.

With a population of 16,884, the Tehachapi aggregation is the largest in the Southwest region, with 28% of the population. The Lake Isabella Complex aggregation is the second largest, with a population of 8,382, and Auberry/Tollhouse/Prather/Meadow Lake/Burrough Valley, with a population of 6,940, is the third largest. Together, these three aggregations include more than 50% of the region's population. The median population size for the region's aggregations is 1,707.

The Southeast region has the smallest population of the Sierra and has the lowest median aggregation population, with 1,094. More than 70% of this region's population lies in the aggregations of Bishop (12,355), Mammoth Lakes (4,785), and Big Pine, Independence (2,531). Bishop alone contains

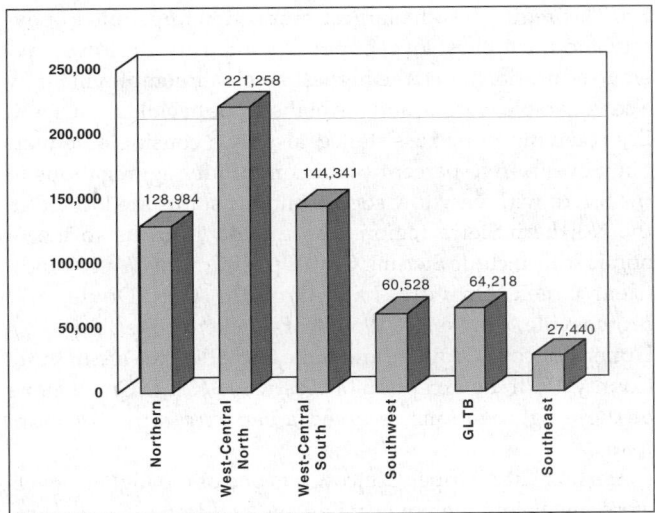

FIGURE 13.4

Population by region.

45% of the region's residents. The proportions of Native Americans and those of Hispanic origin are relatively high in the Southeast region compared to other Sierra regions. Almost 8% of the population is Native American, and 10% is of Hispanic origin.

Variation in Socioeconomic Status and Community Capacity by Region and by Aggregation

Socioeconomic conditions and community capacity vary considerably among the six social assessment regions. A summary of the regional averages of capacity scores, socioeconomic scores, and individual socioeconomic scale components is presented in table 13.1. Average regional scores are calculated from aggregation scores weighted by population.

The socioeconomic status and capacity of aggregations also vary considerably across the Sierra and within each of the regions. Table 13.2 lists each aggregation by region along with the socioeconomic score, capacity rating, and population. Low socioeconomic scores highlight a range of societal needs within aggregations. Low capacity scores indicate a reduced ability of local communities to effectively address those needs and to self-develop. The following discussion focuses primarily on individual aggregations with low socioeconomic status and low capacity by region and highlights other aggregations with unique characteristics.

Northern Region

With an average score of 2.5, the Northern Sierra region has the lowest average socioeconomic status of any region. The Northern region also has the lowest average capacity score of

2.5. This region has the largest proportion of people in poverty and the highest level of poverty intensity, the lowest average education level, the highest level of unemployment by a considerable margin, and the highest rate of children in families receiving public assistance, also by a considerable margin. Seventy-five percent of all community aggregations in the Sierra with very low socioeconomic status are located in the Northern Sierra region. Those rated 1 on the socioeconomic scale include Sterling City/Upper Concow, Westwood/Clear Creek, Oroville, Lake Oroville Area, Doyle, and Brownsville/Challenge/Woodleaf/Rackerby. Sterling City/Upper Concow, Oroville, and Lake Oroville Area are in Butte County. With the exception of Westwood/Clear Creek, none of these aggregations received a capacity rating of more than 2.

Sterling City/Upper Concow has one of the highest levels of people below the poverty line and children in families receiving public assistance of any community in the Sierra. Residents are likely to have lower incomes and to spend a greater percentage of that income on housing costs than in all other areas in the Sierra. This area has been in decline since the departure of the lumber company that built Sterling City decades ago. The threat of a very serious wildfire disaster is exacerbated by the existence of a single, limited access road.

The Oroville aggregation has many low-income residents: more than 50% earn less than 80% of the Butte County median income (a substantial portion of the county is outside the social assessment area); the community has one of the highest proportions of people in poverty and has the highest proportion of children in families receiving public assistance of any in the Sierra. The business community and town residents have recently begun to work together to improve the town.

The Lake Oroville aggregation is a large area with a dispersed and relatively poor population and no real community center. This area also includes recreation cabins and developing resorts. It has one of the highest levels of children in families receiving public assistance in the Sierra. The Brownsville/Challenge/Woodleaf/Rackerby aggregation has a high level of unemployment and a high percentage of children in families receiving public assistance. It is an extremely isolated area with few services. Doyle has one of the highest levels of children in families receiving public assistance in the Sierra and has the lowest average education level in the region. It is an insular area that recently lost its fire department and ambulance.

A large proportion of the workforce in the community of Westwood/Clear Creek is employed in the timber industry or related support services. The town has one of the lowest average educational levels and one of the highest percentages of children in families receiving public assistance. Westwood, however, has a relatively high capacity rating of 4.

Herlong/Sierra Army Depot is ranked in the middle of the regional socioeconomic scale and the capacity scale but is facing considerable uncertainty due to the proposed base closure.

Lake Almanor West is the highest rated aggregation in the region on the socioeconomic scale and one of the highest in the Sierra. It is a very small community with many high-value second homes and a well-to-do retirement community. Graeagle, with a score of 6 on the socioeconomic scale, is rated similarly to Lake Almanor West and, along with Lake Almanor Peninsula, is the second highest rated community in the region. In general, these areas have limited dependence on anything local except county road crews, emergency services, and local hospitals. Graeagle and Lake Almanor West both have capacity scores of 4, the highest level in the region.

In west-side communities in this region, such as Browns Valley and Forest Ranch/Cohasset/Butte Creek, the short commuting distance to business centers is associated with higher education and income levels. Less accessible areas, such as Sterling City/Upper Concow and Camptonville/Strawberry Valley/LaPorte, tend to have lower socioeconomic scores and lower capacity scores.

West-Central North Region

With an average socioeconomic score of 4.8, the West-Central North region has the highest overall score of the six regions in the Sierra. The average capacity score of 3.4 is second highest, next to the GLTB region. Aggregations in the West-Central North region are characterized by bedroom communities with relatively homogeneous populations of out-of-county commuters and retirees. Nonetheless, there are blue-collar and resource-extraction-dependent communities and areas in which agriculture is locally important. Despite the generally high wealth that exists in the region, there are also pockets of extreme poverty within aggregations, some of which are masked in the scale by large populations of high wealth and high education.

Aggregations with medium-low and low community capacity in this region tend to include two different types of areas: those with a correspondingly low level of socioeconomic status and those that have relatively high socioeconomic levels but that lack any community identity or structure. In the first category are the American River Canyon and Placer East, two expansive aggregations in the higher elevations of El Dorado County and Placer County, respectively. The populations in these higher-elevation areas are quite scattered and tend not to be associated with any particular communities. These areas are also characterized by generally low overall socioeconomic status and have the highest and most intense poverty rates in the region. Volcanoville/Quintette is another area with low capacity and low socioeconomic status in El Dorado County. This is a low income, rural, resource-dependent community with the highest unemployment rate and the highest incidence of children in homes receiving public assistance income in the region. In the second category are a number of aggregations, such as McCourtney/South County (Nevada), Deer Creek, Colfax/

TABLE 13.1

Regional averages of socioeconomic and capacity scores and socioeconomic scale components[a] weighted by population.[b]

Region	Total 1990 Resident Population[c]	Average Community Capacity[d]	Score on Seven-Point Socioeconomic Scale[e]	Score on Continuous Socioeconomic Scale[f]	Tenure[g]	Poverty Intensity[h]	Poverty[i]	Education[j]	Employment[k]	Families with Children Receiving Public Assistance[l]
Northern	128,984	2.5	2.5	33.87	-.21	.48	.83	-.63	-.93	1.23
West–Central North	221,258	3.4	4.8	64.75	.27	-.41	-.44	.40	.35	-.33
West–Central South	144,341	3.1	3.7	51.89	.00	-.13	-.07	-.36	.00	.23
Southwest	60,528	2.8	3.8	49.36	.02	.21	.31	-.54	-.10	.06
Greater Lake Tahoe Basin	64,218	3.9	3.5	51.52	-1.85	.07	.07	.69	.24	-.43
Southeast	27,440	3.1	3.8	46.66	-1.02	.36	.20	-.13	.10	.08
Total Sierra Nevada	646,769	3.1	3.8	52.20	-.17	-.03	.04	-.06	-.05	.15

[a]Scale components (tenure, poverty intensity, poverty, education, employment, and families with children receiving public assistance) are expressed as average standardized scores equal to the average number of deviations from the mean.

[b]Averages are weighted by the population of each aggregation relative to the total population of the region or study area.

[c]Source: U.S. Bureau of the Census 1990a. Population numbers are exclusive of 9,632 persons incarcerated in the Tehachapi Prison and Susanville Correctional Center.

[d]Average of five-point community capacity scale (1=lowest; 5=highest) determined by experts in local workshops (averages are from integer values chosen at the aggregation level).

[e]Average of seven-point socioeconomic scale developed from continuous socioeconomic scale (1=lowest; 7=highest).

[f]Average of continuous socioeconomic scale developed from 1990 census data, normalized to base of 100.

[g]Average of standardized tenure score: higher values indicate higher levels of home ownership.

[h]Average of standardized poverty intensity score: higher numbers indicate greater intensity of poverty among residents with incomes below the poverty level.

[i]Average of standardized poverty score: higher numbers indicate a higher percentage of the population with incomes below the poverty level.

[j]Average of standardized education score: higher values indicate higher overall levels of education.

[k]Average of standardized employment score: higher values indicate higher levels of employment in civilian population.

[l]Average of standardized score for families with children receiving public assistance income: higher values indicate higher percentages of families receiving public assistance income.

TABLE 13.2

Socioeconomic and capacity scores by aggregation and region.

Aggregation	Population	Socioeconomic Score (1 to 7)	Capacity Score (1 to 5)
Northern Region			
Browns Valley	1,204	4	3
Brownsville/Challenge/ Woodleaf/Rackerby	1,094	1	2
Camptonville/Strawberry Valley/La Porte	969	3	2
Central Butte	4,221	4	4
Chester	2,115	2	4
Dobbins/Challenge/ Brownsville	1,072	4	2
Downieville/North Yuba	1,289	3	2
Doyle	953	1	1
Eagle Lake	660	4	2
East Shore/Hamilton Branch	782	3	3
Forest Ranch/Cohasset/ Butte Creek Canyon	3,140	5	2
Graeagle	1,010	6	4
Greenville/Indian Valley	2,907	3	2
Herlong/Sierra Army Depot	1,534	3	3
Janesville	2,569	4	3
Lake Almanor Peninsula	997	6	4
Lake Almanor West	240	7	4
Lake Oroville Area	3,671	1	1
Milford	413	4	4
Mohawk Valley	994	2	3
Oregon House/Dobbins	1,345	4	1
Oroville	33,706	1	2
Paradise/Magalia	32,507	3	2
Portola	2,873	3	2
Quincy	6,857	3	4
Sierra Valley	879	4	4
Sierra Valley/Verdi	2,029	4	2
Standish/Litchfield	1,173	2	1
Sterling City/Upper Concow	1,547	1	1
Susanville	11,983	3	4
Westwood/Clear Creek	2,251	1	4
Total / Average	128,984	2.5	2.5
West-Central North Region			
Alta Sierra	6,389	6	3
Alta/Dutch Flat/Gold Run	1,701	5	4
American River Canyon	220	2	1
Applegate	1,497	5	3
Auburn	23,202	4	3
Auburn rural	7,001	6	3
Banner Mountain	3,744	6	4
Bowman	1,043	4	2
Camino	2,908	4	4
Cedar Grove	1,440	5	3
Cement Hill/Lake Vera	2,474	6	3
Chicago Park	3,001	5	2
Colfax	904	4	3
Colfax/Weimer	5,045	4	1
Coloma/Lotus	2,535	6	4
Cool/Pilot Hill	3,434	6	4
Deer Creek	1,395	6	2
Diamond Springs	7,179	4	3
El Dorado Hills	8,837	7	4
El Dorado/Nashville	5,273	5	2
Foresthill Divide	4,231	4	3
Garden Valley/Greenwood	878	4	5
Georgetown	2,608	4	3
Gold Hill	2,059	5	3
Grass Valley	13,573	2	4
Higgins Corner	4,699	5	4
Kelsey	1,323	3	4
Lake of the Pines	3,696	6	3

TABLE 13.2 (continued)

Aggregation	Population	Socioeconomic Score (1 to 7)	Capacity Score (1 to 5)
Lake Wildwood	5,028	6	4
Latrobe	1,323	6	4
McCourtney/South County	1,779	5	1
Meadow Vista	4,087	6	5
Mosquito	896	4	3
Nevada City	3,645	4	5
Newcastle	3,897	5	5
Newtown/Sly Park	3,721	6	3
North San Juan/French Corral/Washington	3,204	4	5
Old Auburn Road	4,503	4	2
Ophir	2,016	6	4
Penn Valley	2,208	4	4
Placer East	1,236	2	1
Placerville	14,165	4	4
Pleasant Valley (El Dorado County)	2,869	5	2
Pleasant Valley (Nevada County)	972	5	1
Pollock Pines	4,908	4	3
Rattlesnake	1,687	5	1
Red Dog/You Bet	2,666	5	3
Rescue	2,973	6	4
Rough and Ready	2,424	4	2
Scotts Flat	960	6	3
Shingle Springs/ Cameron Park	22,270	6	5
South County (El Dorado County)	3,931	4	2
Squirrel Creek	1,043	4	2
Volcanoville/Quintette	558	2	2
Total / Average	221,258	4.8	3.4
West-Central South Region			
Ahwahnee	2,921	4	4
Arnold/Avery	5,372	4	3
Bass Lake	1,393	6	4
Big Hill/Cedar Ridge	2,903	5	2
Big Oak Flat/Groveland	3,515	4	3
Camanche	847	4	1
Catheys Valley	1,472	2	4
Coarsegold	2,311	4	3
Columbia	3,403	3	2
Copperopolis/Copper Cove	2,247	4	2
Coulterville	2,250	3	2
Dorrington/Tamarack	640	6	2
East Sonora	2,479	4	4
Greater Angels Camp	2,787	3	5
Indian Lakes/ Quartz Mountain	1,424	2	3
Ione	9,537	4	4
Jackson	4,901	4	2
Jamestown Area	5,383	4	3
Jupiter	112	3	1
Keystone/Lake Don Pedro	4,812	4	2
Mariposa	8,746	3	4
Mokelumne Hill	1,349	3	4
Mono Vista/Crystal Falls	2,874	4	3
Mountain Ranch/ Sheep Ranch/Calaveritas	2,108	4	3
Murphys/Douglas Flat	3,229	4	5
North Fork	1,648	4	4
Oakhurst	4,058	4	5
O'Neals	727	4	2
Phoenix Lake	1,755	5	3
Pine Grove	3,116	4	3
Pioneer/Buckhorn	4,960	4	2
Plymouth/Fiddletown	2,868	4	3
Rail Road Flat/Glencoe	1,726	3	1
Raymond	1,499	4	3
River Pines	486	3	3
San Andreas	2,439	3	2

TABLE 13.2 (continued)

Aggregation	Population	Socioeconomic Score (1 to 7)	Capacity Score (1 to 5)
Sonora	7,418	2	5
Soulsbyville	1,382	4	2
Sutter Creek/Amador City/ Volcano	3,324	4	4
Tuolumne	3,230	2	4
Twain Harte/Strawberry	6,969	4	3
Valley Springs/ Rancho Calaveras	7,832	4	3
Wawona	302	4	4
Westpoint/Wilseyville	2,269	3	1
Yosemite Forks/ Cedar Valley	1,169	5	3
Yosemite Junction/ Wards Ferry	2,188	4	2
Yosemite Lakes	2,396	6	2
Yosemite National Park/ El Portal	1,565	4	3
Total / Average	144,341	3.7	3.1
Southwest Region			
Auberry/Tollhouse/Prather/ Meadow Lake/ Burrough Valley	6,940	4	2
Badger/other rural	2,287	2	1
Breckenridge Mountain	683	4	1
California Hot Springs	729	4	4
Cane Brake Area	1,661	2	1
Glennville	553	4	1
Hart Flat/Keene	1,904	6	2
Kernville/Wofford Heights	4,354	4	2
Lake Isabella Complex	8,382	3	3
Lemoncove	996	4	2
Lower Foothills/ Millerton Lake	1,543	6	4
Pinehurst/Miramonte/ Hume Lake	195	5	1
Shaver Lake/ Huntington Lake	855	4	3
Springville/Yokohl Valley/ Camp Nelson	2,475	4	5
Tehachapi	16,884	4	4
Tejon Ranch	1,581	5	1
Three Rivers/ National Park rural	1,752	6	5
Tule River Indian Reservation	1,812	2	1
Watts Valley Road/Foothills	739	1	4
Wonder Valley/Tivy Valley/ Squaw Valley/Dunlap	4,203	3	1
Total / Average	60,528	3.8	2.8
Greater Lake Tahoe Basin Region			
Alpine Meadows/ Ward Canyon	788	4	2
Donner Summit	733	4	3
Echo/Upper Truckee	2,425	6	3
Glenbrook	393	7	4
Incline/Crystal Bay/ Brockway	7,856	4	5
Kings Beach	2,365	1	2
Markleeville/Woodfords/ Bear Valley	1,113	2	3
Montgomery Estates/ Tahoe Paradise/Meyers	3,079	5	3
North Tahoe	2,630	5	3
South Lake Tahoe	23,319	2	4
Squaw Valley/ Olympic Valley	845	6	3
Stateline/Kingsbury	3,153	4	4
Tahoe City	2,587	4	3
Truckee	9,386	5	5

TABLE 13.2 (continued)

Aggregation	Population	Socioeconomic Score (1 to 7)	Capacity Score (1 to 5)
West Shore	1,462	4	3
Zephyr Cove/Skyland	2,084	6	2
Total / Average	64,218	3.5	3.9
Southeast Region			
Antelope Valley (Walker, Coleville, Topaz)	1,412	2	4
Big Pine, Independence	2,531	4	2
Bishop	12,355	4	4
Bridgeport/Twin Lakes/ Swauger	742	4	4
Greater Lone Pine	1,916	2	3
June Lake	607	4	2
Lee Vining/Mono Basin	415	5	4
Long Valley/ Wheeler Crest/Paradise	1,094	6	2
Mammoth Lakes	4,785	4	2
Olancha/Cartego/ Kennedy Meadows	682	2	1
Tri-Valley/Oasis	901	3	2
Total / Average	27,440	3.8	3.1

Weimer, and Newtown/Sly Park, which are characterized by somewhat dispersed populations with no real community center or identity. While community capacity is quite low, socioeconomic status ranges from medium to high.

Some of the incorporated cities of the region, including Grass Valley, Nevada City, and Placerville, have high community capacities despite relatively low socioeconomic status. Poverty levels are relatively high in these three cities. The level of home ownership in Grass Valley is the lowest in the region and in Nevada City is third lowest. The high capacity of these towns appears to be largely influenced by the strength of the business communities within them. The other cities of the region, Auburn and Colfax, have both medium capacity ratings and medium socioeconomic status, although poverty levels are also relatively high. The level of home ownership in Auburn is the second lowest in the region.

The North San Juan/French Corral/Washington aggregation is unique in the region. This high-capacity area was described in the workshops as an area "that has done more with nothing than anyone else," indicating a high level of both human and social capital. This is also an area of only moderate socioeconomic status, with high poverty levels and the third most intense poverty in the region. Workshop participants indicated, however, that the actual extent of poverty in this area may be overstated in the census data. Low reported income may be offset by unreported transfer payments as well as an active informal local economy.

West-Central South Region

The average socioeconomic score of 3.7 and the average capacity score of 3.1 for the West-Central South region closely parallel the average weighted scores for the entire Sierra (3.8

and 3.1, respectively). The aggregations in this region are diverse and range from established resource-extraction-dependent communities with long family histories to bedroom communities of commuters and retirees.

Four aggregations have a capacity of 1 in this region: Jupiter, Rail Road Flat/Glencoe, Westpoint/Wilseyville, and Camanche. All but Camanche have medium-low socioeconomic scores. Rail Road Flat/Glencoe and Westpoint/Wilseyville are both rural, isolated areas in Calaveras County. The communities in both areas appear to have a difficult time coming together to address even common issues such as fire protection. They are resource-dependent communities suffering from a lack of jobs. The Westpoint/Wilseyville aggregation used to be an economically stable area with a timber and agricultural base. Now, unemployment and poverty levels are high, and the aggregation has one of the lowest education levels in the region. Rail Road Flat/Glencoe has an even higher incidence and intensity of poverty, along with some substandard housing. This area also has some disparities in income levels, with some wealthy residents who are not assimilated into the rest of the community. The Camanche aggregation is a partially failed subdivision in Western Amador County. Residents are primarily retirees or commuters and have consistently voted down sewer and water projects that are needed to allow other lot owners to build and move into the community. The low socioeconomic status in the aggregation appears to be related to relatively high unemployment and low education levels. There is also a severe disparity in income between residents who own their own homes and those who rent.

Thirteen aggregations have a community capacity of 2. Socioeconomic scores in these areas range from medium-low to very high. Low-capacity aggregations with lower socioeconomic scores include Mariposa, Columbia, San Andreas, and Coulterville. Mariposa and Coulterville are tourism- and resource-based aggregations in Mariposa County. Coulterville is historically a mining and livestock grazing area but is focusing more on tourism with the current decline of resource-related employment. Both areas suffer the loss of their young people as they move away for work or education, and both aggregations have high proportions of retirees who demand services but in general, as was reported in workshops, contribute little to overall community capacity. The Columbia aggregation in Tuolumne County includes a community college and is an area with a mixed population and no real community center or focus. It has the highest percentage of households with children receiving public assistance income in the region. San Andreas is the county seat of Calaveras County but has little focus or sense of community. It has the fourth lowest rate of home ownership among the resident population of the region.

Aggregations with medium socioeconomic scores include Jackson (the county seat of Amador), Yosemite Junction/Wards Ferry, O'Neals, Soulsbyville, Keystone/Lake Don Pedro, and Pioneer/Buckhorn. The O'Neals aggregation has

a mixture of ranchers, commuters, and retirees and a relatively large population of Native Americans. Keystone/Lake Don Pedro is a large agricultural area with little community focus and a small retiree population. It has the lowest education level of the region, although this is quite likely due to a high prison population.

Four aggregations in the region have capacity ratings of 5. Socioeconomic scores for these areas range from medium-low to medium-high. Oakhurst and Murphys/Douglas Flat have medium socioeconomic scores. Greater Angels Camp and Sonora have socioeconomic scores of medium-low and low, respectively. The Sonora aggregation has the third lowest rate of resident home ownership in the region and the fourth highest rate of families with children receiving public assistance income.

An aggregation with one of the lowest socioeconomic scores in the region, Catheys Valley, has a community capacity rating of 4. Catheys Valley, in rural Mariposa County, was historically a mining and grazing area. The current population is largely ranchers, with some commuters and a few retirees. The residents of this aggregation were described by participants in the capacity workshop as having less wants and needs than those of other areas. While unemployment in Catheys Valley is relatively low, poverty levels are the highest in the region. The intensity of poverty here is second highest in the region.

The aggregations of Yosemite National Park/El Portal and Wawona are unique in that the ability of these communities to meet their needs is strongly influenced by the National Park Service and the single concessionaire to the park. The Yosemite aggregation has a capacity rating of 3, and Wawona has a rating of 4. These areas both have medium-low socioeconomic status. These aggregations share the lowest home ownership rates, the highest education levels (Wawona has the fourth highest education level in the Sierra), and the lowest unemployment rates in the region. The Wawona aggregation, however, has the fourth highest percentage of residents with incomes below the poverty line and the highest intensity of poverty in the region (third highest in the Sierra). At the same time, there are no families with children receiving public assistance income.

Resource-dependent communities in the region typically have relatively low socioeconomic scores, but their community capacity ratings vary greatly. Low-capacity resource-dependent communities such as Rail Road Flat/Glencoe and Westpoint/Wilseyville lack much of the history and the sense of community of higher-capacity communities such as Tuolumne and North Fork. The main sawmill in North Fork closed in 1994—and hence some of the socioeconomic indicators may be understated—but the community is working hard to develop alternative economic opportunities for its residents.

Southwest Region

The Southwest region has the second lowest average capacity score, 2.8, and a socioeconomic score of 3.8. The region has the second highest rate of poverty and the second lowest education and employment scores.

Watts Valley Road/Foothills, with a 1 on the socioeconomic scale, is the lowest scoring aggregation in the region and one of the lowest on the socioeconomic scale in the Sierra. The aggregation has one of the highest poverty levels and has the highest intensity of poverty in the Sierra. The mean education score is also low. Almost 16% of the population is of Hispanic origin and 19% is Native American. There is, however, a considerable disparity between home owners and renters in the aggregation. Census data indicate that, in general, home owners have very high incomes and renters very low incomes.

The Tule River Indian Reservation is the second lowest rated aggregation in the region on the socioeconomic scale, with low socioeconomic status. The aggregation has high poverty and high poverty intensity. It also has the third highest rate of resident home ownership in the region. Unlike Watts Valley Road/Foothills, however, it has a capacity score of 1. Management of the reservation is limited, as the County of Tulare provides law enforcement and other services, despite the tribe's sovereign nation status. Economic development activities may unite the reservation, but planning is done off the reservation. People of Hispanic origin total 9%, and 42% are Native American in the aggregation.

The Cane Brake Area, Badger/other rural, and Lake Isabella Complex aggregations all have low socioeconomic status scores. Cane Brake Area and Badger/other rural had moderately high poverty and high intensity of poverty. Cane Brake Area was identified as having a pocket of high, multigenerational poverty. Cane Brake Area and Lake Isabella Complex have very low average education levels. Both Badger/other rural and Cane Brake Area are extremely isolated, with very limited public services and a limited amount of social capital, and consequently both have low capacity. The moderately large Lake Isabella Complex aggregation received a medium capacity rating, largely because a portion of the population is responsive to community issues. This aggregation has the lowest education score in the region. High ethnic or minority group representation was found only in Badger/other rural, with 23% of the population of Hispanic origin.

The two high-capacity aggregations are Springville/Yokohl Valley/Camp Nelson and Three Rivers/National Park rural. Springville/Yokohl Valley/Camp Nelson is primarily a rural area with some ranching and a dispersed population with low income. The more "urban" center of Springville is a cohesive community. The aggregation has a moderately low socioeconomic score. Three Rivers/National Park rural has a mix of newcomers and retirees who are able to pull together despite their differences. Despite the small size of the area, there is a local newspaper and a full spectrum of community services. The aggregation has low unemployment and poverty and rates high on the socioeconomic scale.

Hart Flat/Keene and Lower Foothills/Millerton Lake are the other highly rated aggregations in the region on the socioeconomic scale. Both of these areas have very high pro-

portions of owners to renters and very low poverty. Both aggregations have growing commuter populations: Hart Flat/Keene residents commute to Bakersfield, and Lower Foothills/Millerton Lake residents commute to Fresno and Clovis. The Hart Flat portion of Hart Flat/Keene has very expensive homes and high incomes. Many of these residents are commuters. The moderately low capacity rating is primarily because the area is a bedroom community and because Keene, though small, is relatively poor. Keene is the headquarters of the United Farmworkers of America, suggesting that Hispanic capacity may be considerably different from the predominantly Anglo capacity of the aggregation as a whole. Ten percent of Hart Flat/Keene residents are of Hispanic origin. Lower Foothills/Millerton Lake has a medium-high capacity rating because residents are reportedly willing to work on community issues.

Many of the areas in the Southwest Sierra region were at one time economically dependent on the timber industry. Pinehurst/Miramonte/Hume Lake and Auberry/Tollhouse/Prather/Meadow Lake/Burrough Valley are examples, the latter of which saw the local mill close in early 1994. Economies are shifting and increasingly catering to tourism, recreation, and retirement living; an example is the Kernville/Wofford Heights area on the Kern River. A growing number of commuters to Central Valley cities are settling in Lower Foothills/Millerton Lake, Tejon Ranch, and Lemoncove, among others. These new resident commuters challenge longstanding ranching and agricultural lifestyles, though conflicts are not necessarily inevitable or intractable.

Greater Lake Tahoe Basin Region

The GLTB region has an average socioeconomic score of 3.5, the second lowest in the Sierra. The region has the highest average capacity score, 3.9. Socioeconomic status in this region, however, is actually bimodal. More than 40% of the population resides in aggregations with socioeconomic scores of 2 or less, while 47% lives in aggregations with socioeconomic scores of 5 or greater. The region has the highest level of education, the second lowest unemployment level, and the lowest proportion of children in families receiving public assistance. Yet while the GLTB region tops the Sierra in a number of scale categories, there are pockets of poverty that reflect the unequal distribution of wealth in this tourist and recreation economy. Typical of a recreation and tourist area, the GLTB has the lowest rate of home ownership among residents of any area in the Sierra by a considerable margin.

Limited affordable housing near jobs in the basin leads to increased commuter traffic and shifts the burden of housing and other service provision for workers elsewhere. For example, casinos and service employers in Incline/Crystal Bay/Brockway draw employees from Kings Beach and from Reno. Similarly, approximately one-third of South Lake Tahoe workers commute to Stateline/Kingsbury to work.

The Kings Beach community, with a score of 1, has the lowest socioeconomic scale score in the region and one of the lowest in the Sierra. With a capacity score of 2, it shares with two other aggregations the lowest capacity score in the GLTB. Reasons for low capacity in Kings Beach include little internal leadership and poor access to capital. Service workers for casinos, restaurants, and ski areas reside in Kings Beach. It has proportionally more renters than any other aggregation in the Sierra, and there is a considerable amount of substandard housing. The Kings Beach aggregation is the youngest in the Sierra, with more than two-thirds of the population under the age of thirty-five. Fifty-seven percent of the population is male, the fifth highest male population in the Sierra. Only 50% of the households in Kings Beach are family households, one of the lowest totals in the study, yet the area has the second highest percentage of families with children headed by single parents. Twenty-nine percent of the families with children are single-parent, female-headed households. Kings Beach has a very high poverty level, and poverty intensity is one of the highest in the Sierra. Unlike many other aggregations with high poverty, however, Kings Beach has relatively few children in families receiving public assistance. The aggregation also has the lowest education score for the region. A total of 37% of Kings Beach residents are of Hispanic origin, twice the next highest total for aggregations in the region, and it was reported in the capacity workshop that more than half of the elementary school population is Hispanic.

Markleeville/Woodfords/Bear Valley, which constitutes the entire populated area of Alpine County, is another aggregation in the region that ranks low on the socioeconomic scale, with a score of 2. It has a higher capacity score of 3. The Forest Service and the Bureau of Land Management control 98% of the land. This area is dependent on the ski economy for jobs and transit occupancy taxes, and on the federal forests for forest reserve funds. Most residents with full-year jobs work for the county or the schools. The area has the highest number of children in families receiving public assistance in the region. The Washoe Indian community makes up more than one-third of the aggregation population and approximately 50% of the school population.

South Lake Tahoe, with a score of 2, is another aggregation in the region with relatively low socioeconomic conditions. Like Kings Beach, this aggregation has proportionally more renters than other aggregations in the Sierra and has a relatively high percentage of residents of Hispanic origin: 17.2%. The housing stock is substandard, and many renters work in the casinos, ski areas, and related services. A large number of out-of-area residents control housing in South Lake Tahoe and do not participate in the community. Many renters also do not participate. Unlike Kings Beach, the capacity score of South Lake Tahoe is 4. South Lake Tahoe is an incorporated city and is considered to have an effective government.

Squaw Valley/Olympic Valley, with a socioeconomic score of 6 and a capacity score of 3, is an aggregation of contrasts. It has one of the highest poverty levels in the Sierra, yet there are no families with children receiving public assistance. Only

23% of the households in this aggregation are family households, by far the lowest in the Sierra. The aggregation also has one of the highest education levels in the Sierra. Nearly 60% of the population consists of young adults between the ages of eighteen and forty-four. Poverty in this area is not family poverty; it may reflect a high number of seasonal workers and may also reflect low pay. The extremely high education levels also suggest that the resident population has more choice, though local opportunities may be limited. Residents of the valley are described as very independent and extremely different from other groups in the region. The lower capacity score reflects the fact that, beyond their operations, the powerful resort companies take very limited responsibility for local issues.

Truckee and Incline/Crystal Bay/Brockway both have capacity scores of 5, the only two aggregations with this score in the basin. The Truckee aggregation includes an incorporated city with an effective government and, as a regional service center, has one of the most diversified economies. The community appears to share a common vision for the area. Incline/Crystal Bay/Brockway was rated high due to the presence of wealth, strong local representation, and a politically savvy populace.

Zephyr Cove/Skyland and Glenbrook have socioeconomic scores of 6 and 7, respectively, rating near the top of the scale for all aggregations in the Sierra. Both have very high education levels (Glenbrook has the highest in the Sierra) and no children in families receiving public assistance. The capacity of Zephyr Cove/Skyland was rated a 2; it is a bedroom community with no central core and little cohesiveness. The capacity of Glenbrook was rated a 4, primarily because of its financial and political strength.

Southeast Region

The Southeast region has an average socioeconomic scale score of 3.8, tied with the Southwest region for second highest in the Sierra. The average capacity score of 3.1 is equal to that of the Sierra as a whole. The economies of this region are based primarily on recreation and tourism, and the region has a high proportion of workers in the government and service sectors. Control of land in the Southeast region is a hotly debated issue and is frequently characterized as a double-edged sword. The Los Angeles Department of Water and Power (LADWP), U.S. Forest Service, and Bureau of Land Management control the vast majority of land. Control of the land by owners and managers who are, for the most part, outside the region leaves major landholding decisions beyond the reach of locals. Although little private land is available for local development, present management includes the retention of considerable open space and a natural landscape that is widely valued and upon which the region's tourist and recreation economy is established.

The three lowest rated aggregations on the socioeconomic scale are Antelope Valley (Walker, Coleville, Topaz), Greater Lone Pine, and Olancha/Cartego/Kennedy Meadows. These three aggregations are in the bottom 10% of the socioeconomic scale of all Sierra Nevada aggregations. Olancha/Cartego/ Kennedy Meadows is the only aggregation in the region with a community capacity rating of 1. The capacity of Greater Lone Pine is 3, and that of Antelope Valley (Walker, Coleville, Topaz) is 4. Less than half of all aggregations in the region have a capacity rating of 3 or higher, and only four received a 4. Five aggregations received a 2 for community capacity, four of which were in Mono County.

The small community aggregation of Olancha/Cartego/ Kennedy Meadows, rated low on both the socioeconomic and capacity scales, has one of the lowest education levels and highest levels of unemployment in the Sierra. A total of 14% of the population is of Hispanic origin.

Antelope Valley (Walker, Coleville, Topaz) has one of the highest percentages of people in poverty in the Sierra and a high poverty intensity score. Poverty and poverty intensity in this aggregation are the highest in the region. Almost 12% of the population is of Hispanic origin, and 10% of those over sixteen are Native American. The aggregation is one of the highest rated in capacity because, among other things, residents are quick to pool resources and pull together in times of need. In Greater Lone Pine, socioeconomic scale component scores are uniformly low. A total of 9% of the residents over sixteen are Native American, and 13% are of Hispanic origin.

The highest rated aggregation in the region on the socioeconomic score is Long Valley/Wheeler Crest/Paradise. The capacity score is 2, due primarily to a dispersed population and limited civic action.

The second highest rated aggregation in the socioeconomic scale is Lee Vining/Mono Basin. There are no children in families receiving public assistance in this aggregation. Census data indicate that both owners and renters have high incomes, with home owners being some of the wealthiest in the Sierra. Community capacity is 4, the highest capacity rating in the region. Social capital has increased as a result of a recognition of the importance of the landscape and place and consequent efforts devoted to protecting it. The area depends almost exclusively on recreation and tourism but has little control over the flow of tourists traveling over Tioga Pass and through Yosemite National Park. The National Park Service limits tourist bus volume and controls snow removal activities on the Tioga Pass road, which determine when the pass opens in the summer. The pass was described as an economic lifeline for the community.

The destination resort town of Mammoth Lakes is the center for a great many tourist-related activities associated with Mammoth Mountain. Typical of other destination resorts, Mammoth Lakes has one of the lowest proportions of home owners to renters in the entire Sierra and has one of the two highest educated populations in the Southeast region (the other is June Lake). Mammoth Lakes ranks in the middle of the socioeconomic scale for the region, rated a 4, and has a low capacity of 2. Good physical infrastructure and human capital do not offset the divisiveness between prodevelopment community members and those opposed to development. This conflict has made it difficult for people to work together. A high turnover rate due to many seasonal workers further reduces capacity.

Concentrations of Populations with Low Socioeconomic Status

While there is considerable variation in the socioeconomic status of aggregations across the Sierra, the majority of the populace at the lower end of the socioeconomic scale resides in a relatively small number of aggregations. Fifty percent of all persons in poverty in the Sierra are in 11% of the aggregations. Fifty percent of the unemployed labor force is in 12% of the aggregations, and 50% of all children in households receiving public assistance income are in only 8% of the aggregations. Included in all three of these categories are the aggregations of Oroville, Paradise/Magalia, South Lake Tahoe, Susanville, Auburn, Grass Valley, Mariposa, Placerville, Tehachapi, Bishop, Sonora, Quincy, and Lake Isabella Complex. All but three of these thirteen aggregations include incorporated cities, and all are relatively large. The Lake Isabella Complex aggregation, with 8,382 residents, and Quincy, with a population of 6,857, are the smallest; the populations of all the other aggregations exceed 13,500. These same aggregations with low socioeconomic status are also part of the nineteen aggregations that include 50% of all single-parent families within the Sierra Nevada.

Relationships

Internal Associations of Socioeconomic Scale Components

Table 13.3 shows the correlation coefficients of associations within the various scale components, Sierra-wide. With the exception of the two poverty scores, which are developed from some of the same source data, most of the components are relatively independent of one another.

Table 13.4 shows the Pearson correlation coefficients resulting from an analysis of the relationships between the socioeconomic scale and the individual components of the scale for the entire Sierra and for each of the six regions. As would be expected, since the scale is based on these components, there is a relatively strong association between each component variable and the socioeconomic scale. The poverty and poverty intensity scores have some of the highest correlation coefficients, on average. Although these two scores are closely related by nature, they each account for only one-tenth of the total scale.

Socioeconomic Status and Capacity

Socioeconomic status and capacity are both important components of well-being, but they measure different aspects of it. Correlation analysis between the two measures for the study region reveals a positive but weak relationship between socioeconomic status and community capacity (Spearman rank order coefficient 0.2371, n=180, sig=0.001). This relationship is even weaker at the regional level, except for the Northern and West-Central North regions. Social capital proved to be the most important component of capacity and is a primary reason that capacity and socioeconomic status are weakly related. A number of aggregations with medium to high socioeconomic status were rated lower in capacity because residents did not work together well. While human capital is partially reflected in the socioeconomic scale through educational attainment and income-related components, social and physical capital are not. Increasing commuter settlement appears to increase socioeconomic scale scores through higher incomes "coming to" an area, but there is no certainty that their arrival will lead to higher community capacity. New residents may add to the human capital of communities, but social capital may be negatively affected by their inability or unwillingness to contribute to community activities. Hence, not only are socioeconomic status and capacity weakly related, but it appears that the two measures assess different dimensions of well-being.

Table 13.5 shows the juxtaposition of capacity and socioeconomic status scores for the 180 aggregations. Aggregations with medium-low to very low capacity (1–2) and a very low to medium-low socioeconomic status scale score (1–3) are those considered to have the lowest level of well-being. A total of 28 aggregations, or 16% of all aggregations, fall into this group, which constitutes 18.5% of the total study population.

Aggregations with high and very high socioeconomic status (6–7) are viewed as having the highest level of well-being. Thirty-one aggregations, or 17% of all aggregations, constituting 15.5% of the study population, fall into this group. Low capacity associated with high socioeconomic status is not, in general, likely to reduce well-being as much as low capacity associated with lower levels of socioeconomic status. This is because the residents of aggregations with high socioeconomic status can and in fact do "buy" their way out of difficulties that others must work internally to overcome. For example, some of the aggregations having high socioeconomic status and a high proportion of retirees buy services such as fire protection, security, and recreation programs, whereas other communities might rely on volunteer activities, the county, or the state for provision of such services. Nonetheless, even among the aggregations with high socioeconomic status, high capacity leads to higher levels of well-being, since capacity itself is a component of well-being.

The remaining aggregations, with moderate to moderately high well-being, can be further divided into three groups. Aggregations with medium to high capacity (3–5) and very low to medium-low socioeconomic status (1–3) have a moderate level of well-being. A total of 12% of all aggregations fall into this group. Similarly, the 20% of aggregations with low to medium-low capacity (1–2) and medium to medium-high socioeconomic score (4–5) have a moderate level of well-being. While the former group has a lower socioeconomic score, the higher capacity suggests a greater ability to take

TABLE 13.3

Coefficients of correlation between components of socioeconomic scale.

	Education	Families with Children Receiving Public Assistance	Tenure	Poverty	Poverty Intensity	Employment
Education	1.0000	−.4347	.0204	−.3981	−.3547	.4316
Number of cases[a]	180	180	180	180	180	180
Significance[b]		.000	.786	.000	.000	.000
Families with Children Receiving Public Assistance	−.4347	1.0000	−.0876	.5071	.3270	−.5045
Number of cases	180	180	180	180	180	180
Significance	.000		.242	.000	.000	.000
Tenure	.0204	−.0876	1.0000	−.2709	−.3129	.0017
Number of cases	180	180	180	180	180	180
Significance	.786	.242		.000	.000	.982
Poverty	−.3981	.5071	−.2709	1.0000	.8554	−.2470
Number of cases	180	180	180	180	180	180
Significance	.000	.000	.000		.000	.001
Poverty Intensity	−.3547	.3270	−.3129	.8554	1.0000	−.1652
Number of cases	180	180	180	180	180	180
Significance	.000	.000	.000	.000		.027
Employment	.4316	−.5045	.0017	−.2470	−.1652	1.0000
Number of cases	180	180	180	180	180	180
Significance	.000	.000	.982	.001	.027	

[a]Number of cases evaluated.   [b]Level of two-tailed significance.

TABLE 13.4

Coefficients of correlation between socioeconomic scale and scale components for Sierra and by region.

	Entire Sierra	Region					
		Northern	West-Central North	West-Central South	Southwest	Greater Lake Tahoe Basin	Southeast
Socioeconomic Scale	1	1	1	1	1	1	1
Number of cases[a]	180	31	54	48	20	16	11
Significance[b]							
Education	0.6993	0.8034	0.7566	0.5141	0.6922	0.8079	0.8576
Number of cases	180	31	54	48	20	16	11
Significance	.000	.000	.000	.000	0.001	0.003	0.001
Families with Children Receiving Public Assistance	−.7552	−.8187	−.6702	−.6129	−.7888	−.7926	−.7207
Number of cases	180	31	54	48	20	16	11
Significance	.000	.000	.000	.000	.000	.000	0.005
Tenure	0.4331	0.5685	0.6192	0.3204	0.1232	0.8198	0.0557
Number of cases	180	31	54	48	20	16	11
Significance	.000	.000	.000	0.026	0.605	.000	0.871
Poverty	−.7265	−.8000	−.7293	−.6801	−.8653	−.4776	−.7491
Number of cases	180	31	54	48	20	16	11
Significance	.000	.000	.000	.000	.000	0.061	0.008
Poverty Intensity	−.6451	−.7691	−.6602	−.4273	−.8620	−.7735	−.7111
Number of cases	180	31	54	48	20	16	11
Significance	.000	.000	.000	0.002	.000	.000	0.014
Employment	0.6625	0.8141	0.5476	0.6070	0.5713	0.5179	0.6617
Number of cases	180	31	54	48	20	16	11
Significance	.000	.000	0.001	.000	0.009	0.040	0.027

[a]Number of cases evaluated.   [b]Level of two-tailed significance.

advantage of opportunities than the latter group of aggregations, which have a higher socioeconomic score. The group of aggregations with medium to high capacity (3–5) and medium to medium-high socioeconomic status (4–5) has a moderately high level of well-being. This group makes up 35% of all aggregations.

It is important to point out that the combination of a high capacity rating and a high socioeconomic status score does not mean that all residents of an aggregation enjoy a high level of well-being (though they are more likely to than if the aggregation had a low capacity and very low socioeconomic status score). Just as some families may enjoy a considerably higher level of well-being than others in the same aggregation, some groups—ethnic, occupational, or other—may collectively have considerably lower well-being. Some of these distributional effects were identified in the capacity workshops, yet some remain beyond the resolution of this analysis.

Single-Parent Households

Twenty-four percent of family households with children in the study are headed by single parents. This is low compared to a statewide figure of 35%. With the exception of Lake Almanor West, all aggregations have some single-parent households, with the percentage ranging from 9% to 45%. On the average, 69% of single-parent households are headed by a female, although this rate ranges from as low as 41% to as high as 95% of all single-parent households. The Northern region has the highest percentage of both single-parent households (28%) and female single-parent households (21%).

Thirty-nine percent of female-headed single-parent households and 21% of male-headed single-parent households in the Sierra have incomes below the poverty level. For the entire Sierra region, male-headed single-parent households are

TABLE 13.5

Number of aggregations by capacity and socioeconomic score.

Socio-economic Score	Capacity					Total	%
	1	2	3	4	5		
1	3	3	0	2	0	8	4.4
2	7	1	4	6	1	19	10.6
3	4	10	4	4	1	23	12.8
4	5	21	25	18	7	76	42.2
5	5	5	8	3	2	23	12.8
6	0	6	8	11	3	28	15.6
7	0	0	0	3	0	3	1.7
Total	24	46	49	47	14	180	
%	13.3	25.6	27.2	26.1	7.8	100	

more than three times as likely to have household incomes below the poverty level as two-parent family households. Female-headed single-parent households are more than six times as likely to have incomes below the poverty level as two-parent family households. Figure 13.5 shows average regional poverty levels of female-headed family households, male-headed family households, and family households headed by married couples.

Correlation analysis indicates an inverse relationship between socioeconomic status and single-parent households (Pearson coefficient 0.6172, n=180, p=0.000), female-headed households (Pearson coefficient 0.5931, n=178, p=0.000), and, to a lesser degree, male-headed households (Pearson coefficient 0.3636, n=180, p=0.000). This trend holds at the regional level as well. Correlations are highest in the GLTB region between socioeconomic status and single-parent households as a whole and female-headed households.

Income

The socioeconomic scale is positively correlated to proxies for median household income and median family income. Median income is reported by the Bureau of the Census at the block group level, but since the raw survey data are not available, median income cannot be calculated for the aggregation units used in the social assessment. Instead, an approximation of "average" median income was calculated from census tables reporting percentages of both families and households falling within twenty-five family and household income groups. While it is not possible to correct for documented discrepancies in income figures, as was discussed earlier, these figures do provide a general indication of the relationship between the socioeconomic scale and income.

Correlation analysis indicates that these proxies for median household and family income are closely associated with the socioeconomic scale at the Sierra level, with Pearson correlation coefficients of 0.7574 (n=180, p=0.000) and 0.7741 (n=180, p=0.000), respectively. These relatively high positive associations also persist at regional levels.

Patterns of Age Distribution

The aggregations can also be characterized by different age distribution patterns within the populations. For a number of aggregations, these patterns are associated with socioeconomic scores and capacity in revealing ways.

Cluster analysis was used as a tool to identify five types of aggregations, based on different age distribution patterns. The percentages of total population within forty-eight census-defined age group categories were used as source variables for the cluster analysis. Aggregations with populations of less than 500 were excluded from the cluster analysis, as were the Ione and Keystone/Lake Don Pedro aggregations, which have exceptionally high prison populations. Figure 13.6 shows the average age distribution of aggregations within five-year age groups for the resulting clusters (A through E).

The age patterns include three unique types that are distinguished from the general age distribution pattern in the Sierras. These can be characterized as retirement, young adult, and young family. Retirement-oriented aggregations include those with a relatively high percentage of the population in older age groups and a low percentage of young adults and youth. Young adult areas are characterized by a high percentage of the population in young adult groups, with relatively few children or older adults. Young family aggregations are those with a higher concentration of young and middle-aged adults and young children and relatively few older adults.

Retirement aggregations are typified by cluster E in figure 13.6. An average of 50% of the population of aggregations in this group is more than fifty years old. This cluster includes aggregations in four regions:

- Paradise/Magalia, Eagle Lake, Lake Almanor Peninsula, and Graeagle in the Northern region

- Lake of the Pines, Lake Wildwood, and Old Auburn Road in the West-Central North region

- Pioneer/Buckhorn, Bass Lake, and Big Oak Flat/Groveland in the West-Central South region

- Lake Isabella Complex, Cane Brake Area, and Kernville/ Wofford Heights in the Southwest region

The age distribution pattern of cluster A in figure 13.6 is indicative of young adult–dominated populations. The cluster has the youngest average population, with more than one-third of the population between the ages of fifteen and thirty-five. More than 65% of the populations in these aggregations are under forty years of age, and they have consider-

FIGURE 13.5

Regional poverty rates of family households with children, by family type.

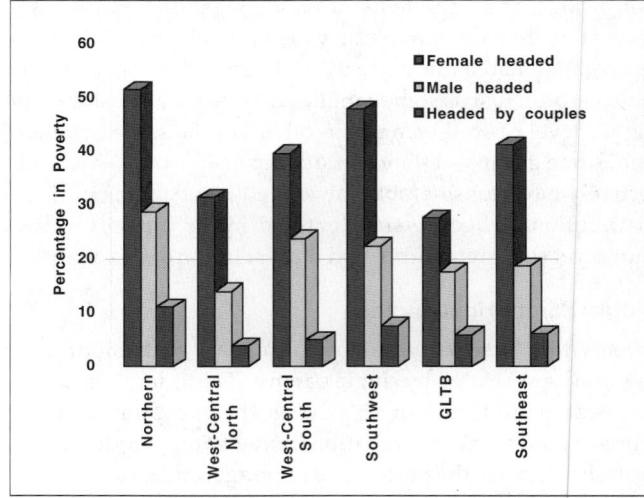

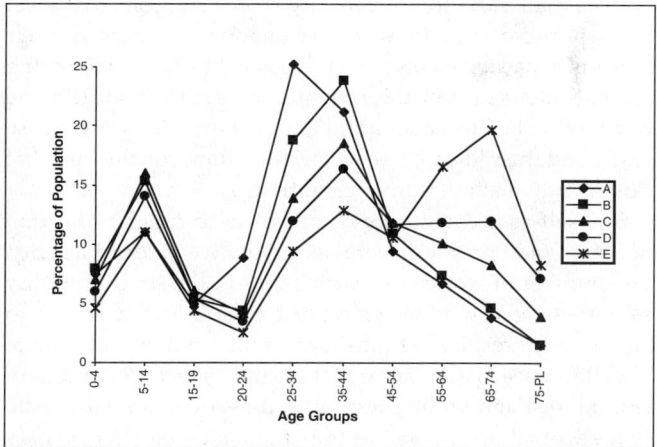

FIGURE 13.6

Age distribution by cluster.

ably fewer individuals over the age of fifty-four than the rest of the Sierra. These aggregations are typical of the young, service-oriented populations of areas with a high level of recreation use and include

- Yosemite National Park/El Portal in the West-Central South region

- South Lake Tahoe, Stateline/Kingsbury, Donner Summit, Squaw Valley/Olympic Valley, Alpine Meadows/Ward Canyon, and Kings Beach in the Greater Lake Tahoe Basin region

- Mammoth Lakes in the Southeast region

Cluster B has an age distribution pattern similar to that of cluster A with two significant differences. The aggregations in cluster B have a slightly older average adult population and considerably more children under the age of fifteen than cluster A. An average of 23% of the population is in this youth age group, higher than any other cluster or the Sierra as a whole. Like cluster A, however, this cluster has lower numbers of individuals over the age of fifty-four than the rest of the Sierra. These aggregations appear to be dominated by young families. While most are in the recreation-service-oriented Greater Lake Tahoe Basin, three aggregations are commuter communities in Western El Dorado County. The aggregations in this cluster include

- El Dorado Hills, Rescue, and Latrobe in the West-Central North region

- Echo/Upper Truckee, Montgomery Estates/Tahoe Paradise/Meyers, Truckee, Tahoe City, North Tahoe, and West Shore in the Greater Lake Tahoe Basin region

- June Lake and Long Valley/Wheeler Crest/Paradise in the Southeast region

The three age typologies just detailed are instructive in identifying unique population patterns and are suggestive of associations between these clusters and both capacity and socioeconomic status scores. The population-weighted average capacity score for the retiree cluster is 2.4, compared to 3.5 and 3.8 for the young adult and young family clusters, respectively. The weighted average socioeconomic score for the young adult group is lowest, at 2.6, reflecting the generally poorer service workers associated with these aggregations. The average socioeconomic status score of the retirement cluster is 3.7, and that of the young family cluster is 5.6. The high socioeconomic status of the young family aggregation is due to the inclusion of aggregations with highly educated, wealthy commuter populations in the West-Central North region and of some aggregations in the GLTB region having a number of professionals. Collectively, the aggregations in the three clusters represent only 23% of the total Sierra population. The majority of the aggregations fall within cluster C or cluster D, both of which have age distributions that closely mimic the average for the Sierra.

Spatial Characterization and Relationships

Based on the isolation scale, the aggregations of the West-Central North region are, on the average, the least isolated. Sixty percent of the aggregations in this region fall in the lowest (least isolated) 20% of the scale. Nearly 90% are in the lowest 40% of the scale. Only three aggregations in this region are in the highest 40% of the scale: Volcanoville/Quintette, Placer East, and American River Canyon. All three have low socioeconomic scale scores and median to very low capacity scores. The aggregations of the Southeast region are the most isolated. All but one of the eleven aggregations in this region fall within the highest 20% of the isolation scale scores. This is due in part to the high percentage of public land throughout much of this region.

The West-Central South region is the only other region with any aggregations in the lowest 20% of the isolation scale: Plymouth/Fiddletown, Ione, River Pines, and Jackson. These aggregations have medium socioeconomic scores and low to high capacity scores. More than 50% of this region lies in the highest 40% of the isolation scale. Forty-five percent of the aggregations in the Southwest region fall within the highest 20% of the isolation scale scores. Seventy-five percent are in the highest 40% of the isolation scale.

Nearly 70% of the aggregations in the Greater Lake Tahoe Basin region fall between the lowest 20% and 60% of the isolation scale. Markleeville/Woodfords/Bear Valley and Glenbrook are the only aggregations in the highest 20% of the isolation scale. They have low and very high scores on the socioeconomic scale, respectively.

The isolation scale, individual components of the scale, and other spatial variables have some associations with the socioeconomic status and capacity, at both the Sierra and regional levels. The twenty most isolated aggregations have an average socioeconomic scale score that is 0.83 standard deviations lower than the average score for the twenty least isolated. Correlation analysis indicates a weak inverse relationship between isolation and the socioeconomic scale (Pearson coefficient –0.2418, n=180, p=0.001). The direction and strength of this relationship is echoed in the relationships between the socioeconomic scale and the component parts of the isolation scale. Influenced in large part by different settlement patterns, these associations are considerably stronger in some individual regions but inverse or nearly absent in others.

The socioeconomic scale has a relatively strong inverse relationship to the isolation scale in the West-Central North region (Pearson coefficient –0.6028, n=54, p=0.000), where commuter-oriented populations predominate. Similarly, socioeconomic status is inversely related to elevation (Pearson coefficient –0.5675, n=54, p=0.000) and distance to the nearest city with a population of 25,000 (Pearson coefficient –0.4799, n=54, p=0.000) in this region. Socioeconomic status is positively, although weakly, associated with isolation, however, in regions with pockets of wealthy, isolated retirement communities such as the West-Central South region (Pearson coefficient 0.2740, n=48, p=0.059), the Northern region (Pearson coefficient 0.2257, n=31, p=0.222), and the Southeast region (Pearson coefficient 0.3250, n=11, p=0.329). Elevation is also positively related to socioeconomic status in these regions.

Findings of Plumas County Case Study

The community self-assessment case study in Plumas County provided an opportunity, albeit a somewhat limited one, to compare the results of community capacity self-assessments with expert assessments. Although it was not possible to compare numerical capacity ratings, the real value of the case study was that it offered the opportunity to identify important local issues in more detail than was possible in expert workshops. Numerical capacity ratings could not be compared, because (1) there was a small variation in community capacity scores rated by the Plumas County experts, and (2) there was very little variation in the average capacity scores for the case study communities. Expert capacity ratings for the six communities ranged from 2 (medium-low capacity) to 4 (medium-high capacity) on a five-point scale, with four of the communities rated 4 and two communities rated 2. The self-assessed communities all have mean capacity ratings that are in the middle of the capacity scale. The mean final capacity ratings in community self-assessment workshops ranged from 3.9 to 4.2 on a seven-point scale (with 1 being very low, 4 being medium, and 7 being very high capacity). The standard deviation for the means of all communities is close to 1, except for Chester, the community with the highest self-assessed score and the fewest respondents (n=11), which has a standard deviation of 1.9.

The small variation in capacity scores suggests that local workshop participants were reasonably consistent in their views of capacity and also that there may be a tendency among local residents to view their communities in the middle of the capacity scale. The small number of communities in the case study and their low degree of variation limit conclusions that can be drawn about numerical ratings.

The self-assessment workshops focused on identification of issues and items that were determinative of local capacity. The number of workshop participants in the six community self-assessment workshops ranged from a low of eleven to almost forty residents in the Greenville/Indian Valley area. The difference in number of participants reflects both the general interest and willingness of local residents to participate in the workshops as well as the organizational effectiveness of local networks of the Plumas Children's Network. In four of the six communities, priority listings of determinative issues and items of local capacity were obtained.

In Chester, participants identified the natural setting and environment, community services, and the economy as the three most important general areas determinative of local capacity. The community's beautiful natural setting and easy access to recreation resources were two subcomponents that contributed to a capacity in the natural setting and environment category. Limited county resources were identified as detracting from community services, while church groups and other local volunteer organizations and the local lumber company, which has a history of public service, led to a higher community capacity. Limited work opportunities and a weak job base detracted from local capacity in the economy category.

In Graeagle, community responsiveness and organizations, family health, and employment and economic development were the three categories identified as most important. Subcomponents of community responsiveness and organizations contributing to local capacity included a high level of community involvement and strong moral fabric, and detracting from it was the lack of a recreational center or park. A high degree of parental involvement was identified as a subcomponent of family health contributing to local capacity, and the large number of needy families was identified as detracting from it. Concerning the category of employment/economic development, limited job opportunities, particularly for teens, detracted from capacity.

In Greenville, participants identified economics, natural resources, and community as the three most important determinants of capacity. From the economics category, lack of jobs and limited economic activities were identified as detracting from overall capacity. The beauty of the valley and surrounding physical environment were identified as contributing to capacity, as were community spirit and the community in general.

At the workshop in Quincy, employment and economics, teen issues, and recreation were identified as the three most important categories. Lack of employment opportunities, with a corresponding high level of unemployment; a declining economy; declining federal, state, and county financing for jobs and services; and a school financial crisis were all identified as detracting from capacity in the first category. Teen pregnancy and drug use and a lack of prevention programs for teens and other programs for families in need detracted from capacity in the teen issues category. Subcategories of recreation included the availability of a wide spectrum of activities, which contributed to capacity, and a shortage of programs and activities, which detracted from it. A fourth category, rated just below recreation, included individual and family resources. Contributing to capacity in this category were responsiveness and willingness of community residents, recognition of the need for communication, and teamwork. Detracting from capacity was the small population base, which led to a core group being overburdened with community responsibilities.

CONCLUSIONS

The five-factor socioeconomic scale offers a useful though static perspective of socioeconomic status, while the measure of capacity provides a current and important complementary perspective to overall well-being. Low socioeconomic scores are found in areas where higher percentages of individuals and families within aggregations may lack sufficient socioeconomic resources to maintain a reasonable standard of living and, hence, experience lower well-being.

Capacity is a dynamic and multidimensional measure that provides an indication of the ability of local communities to foster an environment in which local residents can identify and address their needs and goals. Low capacity scores indicate areas that have a reduced ability to effectively address the needs of local residents and take advantage of local development opportunities that might benefit them. Low capacity, then, reflects not only lower well-being but also a reduced ability, and likelihood, of residents of aggregations to improve local well-being, including socioeconomic status.

Socioeconomic status and community capacity in the Sierra Nevada aggregations are relatively independent components of well-being, and they measure different dimensions of it. Capacity scores are positively associated with the socioeconomic scale, but this correlation is weak. The independence of these two measures appears to be due in part to the critical role of social capital, which proved to be a primary determinant of community capacity.

Capacity and socioeconomic status were combined to assess overall well-being. Aggregations with lower socioeconomic status and low capacity have the lowest level of well-being, and aggregations with high capacity and high socioeconomic status the highest. Low capacity associated with high socioeconomic status affects well-being less than low capacity associated with low socioeconomic status. This is because communities in aggregations with high socioeconomic status are considered to have fewer needs and are able to purchase or acquire services that other communities cannot afford or must work collectively to acquire. Nonetheless, aggregations with higher capacity and high socioeconomic status have higher well-being than aggregations with lower capacities and equally high socioeconomic status.

Community capacity varies widely across the Sierra Nevada. The three components of community capacity (physical, human, and social capital) sometimes appear to be in conflict with one another. That is, where human capital is perceived as being high or increasing, social capital may be low or in decline. This is particularly true where well-educated retirees or professionals fail to work cooperatively on community issues with one another or with longer-term residents. Community history is an important contributor to the human and social components of community capacity. There are a number of aggregations, particularly in the southern Sierra, in which medium-high and high capacity was linked to a long history and the continued presence of multiple old families.

Local volunteer fire departments and local schools are a common denominator in many rural communities. In areas where there is community-based support for nothing else, there is generally support for a fire department. The ability of communities to sustain such volunteer efforts is often negatively affected by increasing populations of commuters and retirees. Although residents often unify around local schools, in many aggregations school issues highlight differences in values and priorities between families and retirees or other residents without children. The needs of youth were identified as neglected in many bedroom communities where growing commuter populations lead to increasingly unsupervised youths.

Community capacity can be negatively affected by divergent values of differently aged populations. Conflicts between retirees and younger families with children were noted in a number of aggregations. Retirees often demand services but resist changes that may be necessary to provide them, and retirees are often reluctant to pay for schools and other services that seem only to benefit families with children. These clashes appear to be strongest in some of the affluent gated communities, where community capacity is negatively affected by internal strife and lack of cooperation between these two groups. In a few communities, however, the knowledge, experience, and willingness of retirees to help the community was particularly noted as a positive addition to capacity. Other volunteerism-based community services are negatively affected by populations that are aging in place, particularly in areas where youth leave communities and in bedroom communities with a large percentage of commuters.

While most of the communities in the Sierra Nevada are fairly amorphous in terms of the age distribution within the population, several communities are dominated by unique populations that can be characterized as retirement, young adult, and young family. The young adult populations are associated with service-oriented areas characterized by outdoor recreation, tourism, and gambling industries. Populations dominated by retirees are typified by high levels of natural resource amenities. Many of these areas are isolated and exclusive communities, often specifically designed to attract retirement populations. Communities dominated by young families include both bedroom commuter-oriented areas in the West-Central foothills and many of the relatively wealthy communities in the Lake Tahoe Basin.

Communities that are more isolated—in terms of distance from major cities and transportation corridors and density of nearby public land—tend to have a lower socioeconomic status, on average, than less isolated communities. This trend, however, is strongly moderated in the Sierra by certain population groups that are attracted to relatively isolated areas. Typically, residents of those isolated aggregations with higher socioeconomic status have sources of income that are more independent of location than the income sources of those with lower socioeconomic status. Several relatively isolated aggregations include affluent retirement communities whose residents are attracted in large part by the high amenity values afforded by these isolated areas.

The use of aggregations as a unit of analysis represents a significant advance in well-being assessment. Well-being is often discussed at the level of a community, but no ecosystem management study to date has actually gathered extensive and comparable community-level data for a large area. (The Columbia Basin Ecosystem Management Project, incomplete as of this writing, reported in a September 30, 1995, newsletter that it had collected economic data for numerous communities in the study area [Interior Columbia Basin 1995].) Community aggregations, based on census block groups, proved useful because in most cases they approximate meaningful social units for which comprehensive and similar data are available.

Community aggregations, however, are not without problems. In a number of instances, census block boundaries parallel main roads that are central to communities. Aggregating adjacent block groups to unify one community sometimes led to the inclusion of unrelated and unconnected communities. In some instances, single block groups cover extremely large areas that also include separate and unrelated communities. Many workshop participants who had previously worked with census data also expressed frustration with the limitations of this data. As a result, a consensus emerged from the workshops that the value of census data would be greatly improved if the Bureau of the Census would work more closely with knowledgeable local residents, such as county planners, to demarcate census geography in more consistent and socially meaningful ways. This study shows the value of the aggregate units and demonstrates how they can be identified.

Finally, the strength of this study of well-being in the Sierra Nevada region is its assessment of both socioeconomic status and community capacity for community aggregations in the Sierra. The measures of socioeconomic status and community capacity provide a comprehensive perspective of the current state of well-being of communities throughout the Sierra. The capacity workshops conducted throughout the Sierra region not only provide important information about local capacity but also offer valuable insights into the socioeconomic status of the aggregates. The measures of socioeconomic status and capacity, particularly when coupled with additional socioeconomic data pertaining to employment, can be used to evaluate the effects that various policy choices and management actions have on the residents and communities of the Sierra.

REFERENCES

Forest Ecosystem Management Assessment Team (FEMAT). 1993. *Forest ecosystem management: An ecological, economic, and social assessment.* Washington, DC: Government Printing Office.

Interior Columbia Basin Ecosystem Management Project. 1995. *Eastside Edge* (Walla Walla, WA) 2(4). September.

Interorganizational Committee on Guidelines and Principles. 1994. *Guidelines and principles for impact assessment.* NOAA Technical Memorandum NMFS-F/SPO-16. Washington, DC: U.S. Department of Commerce.

Kusel, J. 1996. Well-being in forest-dependent communities, part I: A new approach. In *Sierra Nevada Ecosystem Project: Final report to Congress,* vol. II, chap. 12. Davis: University of California, Centers for Water and Wildland Resources.

Machlis, G. E., J. E. Force, and S. E. Dalton. 1995. *Monitoring social indicators for ecosystem management.* Walla Walla, WA: Eastside Management Project.

Stewart, W. 1996. Economic assessment of the ecosystem. In *Sierra Nevada Ecosystem Project: Final report to Congress,* vol. III. Davis: University of California, Centers for Water and Wildland Resources.

U.S. Bureau of the Census. 1990a. *Census of population and housing, 1990: Summary tape file 3 (California, Nevada).* Machine-readable data files. Washington, DC: U.S. Bureau of the Census.

———. 1990. *Census of population and housing, 1990: Summary tape file 3 technical documentation.* Washington, DC: U.S. Bureau of the Census.

Definition of Capacity Presented to Local Expert Workshop Participants

COMMUNITY CAPACITY DEFINED

Community capacity in its most simple form is the ability of a community to adapt to circumstances of all sorts and to meet the needs of its residents. SNEP is interested in learning about the components of communities that affect capacity, and about the strengths and weaknesses of communities framed by the idea of capacity.

Further definition: Capacity is the ability of a community to meet local needs and expectations; to respond to internal and external stresses; and to take advantage of opportunities of all kinds. It includes the ability to adapt and to respond to changing conditions.

Community capacity can be divided into three broad areas:

Physical infrastructure includes the physical elements (e.g., sewer systems, business parks, land available for development, open space, etc.) of a community, and includes financial capital;

Human capital includes the skills, education, experiences and general abilities of residents; and

Social capital includes the ability and willingness of residents to work together for community goals (more formally defined as including networks, norms and trust that facilitate coordination and cooperation for mutual benefit).

Capacity Worksheet Used by Local Expert Workshop Participants

Sierra Nevada Ecosystem Project
Community Capacity Assessment

Worksheet

<u>Community capacity</u> is the ability of a community (or communities within a single aggregation) to adapt to circumstances of all sorts and to meet the needs of its residents. SNEP is interested in learning about the items/issues that affect community capacity, and about the strengths and weaknesses of communities in the context of capacity.

COMMUNITY AGGREGATION NAME_____

If this aggregation contains more than one distinct community, please list them.

CAPACITY NARRATIVE
Please identify the critical components of capacity (both positive and negative) for this community aggregation and describe how they are important (please refer to individual communities as appropriate)

(Please use the reverse side if you need additional space)

NUMERICAL RATING OF CAPACITY FOR THIS COMMUNITY AGGREGATE *(Please circle one number)*

1	2	3	4	5	6	7
very low	low	medium low	medium (neither low or high)	medium high	high	very high

SANDRA A. HOFFMANN
Department of Agricultural and Resource
    Economics
University of California
Berkeley, California

LOUISE FORTMANN
Department of Environmental Science,
    Policy, and Management
University of California
Berkeley, California

14

Poverty in Forested Counties: An Analysis Based on Aid to Families with Dependent Children

ABSTRACT

The Sierra Nevada Forest Counties (SNFC) have persisting problems of poverty and low incomes. Over a forty-year period the forest counties have been consistently overrepresented in the bottom third of California counties in terms of per capita income and consistently underrepresented in the top third. Only 11% of the forest counties have ever experienced what might be characterized as an economic golden age. Poverty rates in the SNFC have tended to be higher than statewide averages and, for the most part, rose between 1980 and 1990. Similarly, Aid to Families with Dependent Children (AFDC) caseloads have tended to run above statewide averages. Time-series analysis provides no evidence that the loss of timber-related employment "Granger-caused" increases in AFDC caseloads at the county level, nor that its availability would cause the decline of AFDC caseloads at the county level. Nor is there evidence to suggest that lumber and wood-products employment affects AFDC indirectly through its effects on other employment. We found that lumber and wood-products employment Granger-caused" other employment in none of the forest counties. The growth rate of lumber and wood-products employment "Granger-caused" the growth rate of other employment in only one of the forest counties. These are strong findings, particularly in light of such strongly held popular beliefs to the contrary.

INTRODUCTION

The Sierra Nevada has historically been rich—rich in timber, gold, and scenery. That richness has made some of the people who use the region's resources wealthy, some very wealthy. But other residents of the region have been poor, some very poor.

Poverty in rural areas shouldn't be news to anyone. From fictional works like *Grapes of Wrath* to the classic photography and prose of *Let Us Now Praise Famous Men* (Agee and Evans 1941) and the stark testimony in *The People Left Behind* (U.S. National Advisory Commission on Rural Poverty 1967), life in rural areas has been portrayed as short and nasty more often than pastoral and bucolic. The Sierra Nevada is no exception. Kusel and Fortmann (1991) showed that the timber counties of California, including the Sierra Nevada counties, had poverty rates that sometimes equaled or exceeded inner-city rates.

This chapter addresses the issue of poverty in the Sierra Nevada Forest Counties (SNFC). These counties include Alpine, Amador, Butte, Calaveras, El Dorado, Lassen, Madera, Mariposa, Nevada, Placer, Plumas, Sierra, Tehama, Tulare, Tuolomne, and Yuba. These are Sierra Nevada counties that in 1980 had a forest cover of more than 50% or in which 3% or more of the 1980 wages came from forest-sector industries (not including tourism) and in which timber was cut commercially. It documents the extent to which these counties, like many other areas dependent on natural resources, are

Sierra Nevada Ecosystem Project: Final report to Congress, vol. II, *Assessments and scientific basis for management options*. Davis: University of California, Centers for Water and Wildland Resources, 1996.

characterized by poverty and low incomes and explores some possible explanations.

Public Perceptions

Anecdotal explanations of economic well-being or the lack of it in the SNFC fall into four rough categories, all of which may be used by the same individuals in different contexts: The Golden Age of Timber, Environmentalism Run Amok, Corporate Greed Run Amok, and The Invasion of the Poverty Importers.

The Golden Age of Timber story suggests that when timber (and in a limited number of counties, gold) was king, towns were prosperous. The logging lifestyle was treasured:

> It's a good life to be able to work in the woods and make a good living. It really gets in your blood. It *really* does.[1]

The sequel to this story, Environmentalism Run Amok, implies that the golden age has disappeared, and its vanishing can be blamed directly on environmental regulation, which has "closed down the woods":

> And then came the spotted owl, and almost overnight the hauling jobs dried up and we had our electricity turned off and finally we received a foreclosure notice on this farm.
> —Unidentified Woman
> (California Forestry Association 1994)

> The amount of economic impact on small communities—devastating. It's going to ruin our lives to say the least. All our relatives are in the business.

> The loss is evident in the lines at the soup kitchens. And the loss is evident in the homes where unemployed workers, anxious, depressed, sunk in despair, lash out at their loved ones or find solace in alcohol or drugs. A culture, a way of life, prized and reverenced in our timber communities is dying.
> —Archbishop Thomas Murphy
> (California Forestry Association 1994)

The logical, and frequently expressed, corollary is that if increased and less-regulated timber harvesting were allowed, prosperity would once again reign in the SNFC. Interestingly, those who use these explanations are not unlikely to use a third, the Corporate Greed Run Amok story, summed up in the words of two timber fallers:

> We're just pawns in the hands of corporations—they don't care about us—you can be sure they won't lose any money. All they care about is their bonus. All they

care about is making money. (Quoted in Kusel and Fortmann 1991, 56)

> All they see is dollar signs and profits. Timber fallers are making the same money as ten years ago, but timber has gone up so the profits are going elsewhere.

The fourth anecdotal explanation for poverty in the SNFC argues that it has been imported by undesirable outsiders over time. During the 1960s and 1970s people variously stereotyped as marijuana-growing and -smoking hippies and back-to-the-landers are described as having taken up residence in the region and gone on welfare. More recently, there has been a rise in anecdotal evidence of poor urban mothers moving to rural areas for cheaper housing and greater safety as well as for all the reasons that richer in-migrants move. A subtheme is that during the more temperate summer months, urban welfare recipients "vacation" (on welfare) in the Sierra along with the more standard form of tourist.

Finally, there is anecdotal evidence about the adverse effect on the affordable housing stock of in-migration by wealthier people, which aggravates the effects of poverty.

How Does This Issue Relate to Other Sierra Nevada Issues?

Land-use choices and economic strategies are likely to affect income levels and poverty rates. These choices should be made with as clear an understanding as possible of the potential consequences for poverty and low incomes and their alleviation.

Key Questions

Our questions arise out of Kusel and Fortmann's 1991 study of well-being in forest communities, to our knowledge the only systematic statistical study of poverty in the forest counties of California. Our questions are

- What is the incidence of poverty in the SNFC?

- How persistent has it been?

- How do trends in employment and specifically in timber-industry employment affect rates of AFDC, Unemployed Parent (AFDC UP)? The lack of sufficient poverty data led us to use AFDC UP as a poverty indicator. The shortcomings of this method are discussed in detail later in this chapter.

BACKGROUND

This review of the literature begins at the national level and then addresses regional and California studies. It includes both descriptive studies of who the poor are and analysis of why poverty exists.

Rural Poverty in the United States

The most recent comparison of metropolitan and non-metropolitan poverty rates shows both declining from 1959 (when nonmetropolitan rates were just under 35%) to the early 1970s when they began a slow (albeit uneven) rise to their 1993 rates of roughly 17% for nonmetropolitan and 15% metropolitan areas. (Definitions can be found in the "Methods" section. By convention, these cumbersome terms will hereafter be shortened to *metro* and *nonmetro.)* Throughout the entire thirty-four-year period, nonmetro poverty rates have exceeded metro rates (Nord 1995). The highest nonmetro poverty rates are found in the South; the second highest, in the West (Alaska, Arizona, California, Colorado, Hawaii, Idaho, Montana, New Mexico, Nevada, Oregon, Utah, Washington, and Wyoming) (Nord 1995). In 1990, 44.4% of the U.S. nonmetro poor were in married couple families, 72.9% were white (Rural Sociological Society Task Force [hereafter RSS Task Force] 1993), and 64.7% of the families had at least one member who was formally employed (Deavers and Hoppe 1992). They were, in short, white, married, and working.

Working gets one less in nonmetro America than in metro America. McLaughlin and Perman (cited in RSS Task Force 1993) found that roughly two-thirds of the earning gap between nonmetro and metro white men is explained by the fact that education and experience result in lower incomes in nonmetro areas than in metropolitan areas. Workers in rural America are more likely to be poor than their urban counterparts with the same amount of education (Shapiro, cited in RSS Task Force 1993). The RSS Task Force on Persistent Rural Poverty (1993) concluded that "the fundamental problem resides in the low wages and inadequate employment opportunities found in rural America." We shall return to this point.

Rural Poverty in the Western Region

In 1993, among nonmetro poor households in the western region of the United States 21.8% had a full-time, full-year worker, 42.9% had part-time or part-year workers, 27.3% had no working member, and the remaining 8% had no family member of working age (that is, they were either too young or too old to work). Husband-wife families accounted for 46.8% of the nonmetro poor households, female-headed families for 32.5%, male-headed families for 1.3% and single men or women for 19.4%. Non-Hispanic whites constituted 64.9% of the nonmetro poor households in the western region. Whites made up 75% of the nonmetro poor in California. The most recent statistical data on poverty thus show that, as in the country as a whole, the nonmetro poor in the western region are likely to be white, married, and working at least part-time.

The Question of Welfare

As discussed later, we use AFDC caseloads as an indicator of poverty because of data constraints. The most exhaustive study of welfare dynamics in California (Albert 1988) does not disaggregate metro and nonmetro data. Hence, this literature offers no particular insights into rural poverty. Albert argues that expanding employment opportunities in low-skills, low-wage industries would decrease AFDC-Basic caseload. The alert reader will already have noticed that this approach is not wholly consistent with ending poverty if the RSS Task Force is correct about the causative nature of low wages.

Poverty in Natural-Resource Dependent Areas

Social scientists and economists have long since given up the search for a one-size-fits-all theory of poverty causation. We know that particular households fall in and out of poverty because of life-cycle changes such as marriage, divorce, the birth of children, the death of a breadwinner, or the onset of catastrophic illness. But we also know that systematic social and economic structures lead to prosperity for some and poverty for others. A particularly clear example is found in natural-resource dependent areas that generate substantial profits from high value products such as timber and minerals at the same time they are characterized by high rates of poverty.

The Working Group on Natural Resources of the RSS Task Force (without coming to a single, unified conclusion about causality) identified five factors affecting the creation of poverty in natural-resource dependent areas (RSS Task Force 1993):

1. rural deindustrialization (the closing of mills, employment cutbacks, the extraction of wage concessions)

2. the concentration of local political and economic power in the hands of resource-extraction firms, which may cause systematic underinvestment in human capital by restricting taxes and other measures to fund schools, and so on

3. control of state and national natural resource agencies by powerful clients (which often are large industrial concerns but, some argue, are increasingly bureaucratic national environmental interests)

4. segmented labor markets and core-periphery relations in which rural areas are the sites of low-paying, dangerous jobs while high-paying processing is located in urban areas (for a detailed discussion of this approach see Peluso et al. 1994)

5. moral exclusion from resource use through the social construction of what actually constitutes a resource (see Freudenburg and Gramling [1994] for a discussion of the moral-exclusion argument)

The group's conclusions thus simultaneously identify as poverty-generating factors both inadequate employment opportunities and low-wage employment opportunities.

Incomes and Livelihoods in California Forest Areas

Although Kusel and Fortmann's (1991) study is the only direct study of poverty in California's forest areas, the findings of studies of these areas' economies consistently suggest that the solution to poverty and low incomes is unlikely to be found in the timber industry.

Belzer and Kroll's (1986) study of the northern timber region included four forest counties (Lassen, Plumas, Sierra, and Tehama). In their argument for economic diversification in timber counties, they noted that in 1981 and 1982 timber industry employment in California was at its lowest level since the end of World War II, that California timber production had experienced an overall downward trend since 1955 despite rises in housing starts, and that from the 1950s the timber industry had tended toward concentration with smaller numbers of increasingly automated mills. They predicted permanent losses in timber employment and productive mills, with lower demands for labor.

Stewart (1993) found that significant losses of timber jobs were unrelated to changes in overall employment and that during the decline of the timber industry, per capita income in most forest counties increased because of the growing importance of public transfer payments and private capital payments in the form of interests, dividends, and rent.

Kusel and Fortmann (1991), based on a point-in-time analysis using 1980 county-level census data, found that contrary to the anecdotal "evidence" presented earlier, the greater the concentration of private timberland ownership, the lower the county median family income; the higher the percentage of public timberland, the higher the county poverty rate; and the higher the rate of in-migration between 1975 and 1980, the lower the county poverty rate.

McWilliams and Goldman (1994) tell a different story from Stewart for northern California (Butte, Del Norte, Humboldt, Lassen, Mendocino, Modoc, Nevada, Plumas, Shasta, Sierra, Siskiyou, Tehama, Trinity, and Yuba counties) where they find the forest-products industry in 1992 contributed a hefty 17.7% of the income and 22.8% of the jobs.

The Limitations of Point-in-Time Data

Piqued by the 1991 Kusel and Fortmann study, which revealed forest county poverty rates equaling or exceeding inner-city rates, we have asked a key question concerning poverty: How serious and how persistent is it in the SNFC, what causes it, and what can be done to reduce it? A preliminary attempt to update the Kusel and Fortmann study using 1990 census data suggested that the strength of the relationships found in that study had decreased considerably during the intervening decade. This change raised many questions. Did the changes reflect the declining importance of timber in the regional economy? Did in-migration act as a one-time jump start to incomes that then declined? Because income levels and poverty rates are affected by previous events, these questions cannot be answered with point-in-time data. Rather than the "snapshot" of point-in-time data, we found we needed the "movie" that time-series data can provide.

METHODS

We present two kinds of data in this study. We begin with descriptive data showing poverty-related characteristics of the SNFC. We then explore some causal relationships. Unfortunately, data on poverty rates usable at a county level are collected only once a decade, in the decennial census. The nearest surrogate for poverty rates reported monthly are AFDC caseloads. No other annual data exist. AFDC caseloads, however, are not identical to poverty rates. Not everyone who is poor receives AFDC for any number of reasons: They may not be eligible for welfare. (In 1985, California AFDC recipients who also received food stamps still fell 7% below the official poverty line [Albert 1988]. Maximum aid payments have declined 15% since 1991 [Barbara Snow, conversation with L. Fortmann, Spring 1995]). They may be eligible for welfare, but not for AFDC. They may be eligible and not know it. They may be eligible but be denied welfare nonetheless. They may be eligible but not apply because of the stigma of receiving AFDC. Thus, although AFDC is an indicator of poverty levels, it is not the same as the poverty level. Indeed, AFDC rates are likely to be below the poverty rate. Nonetheless, AFDC caseload is the best poverty indicator available across the SNFC at a frequency that was useful for time-series analysis. For this reason, we have used it.

Descriptive Statistics

Our descriptive data are taken from Nord (1995) and an analysis of the U.S. Census. The following definitions and explanations may be helpful.

Nonmetropolitan Counties

"Metropolitan statistical areas usually include an urbanized area with a population nucleus of 50,000 or more, as well as nearby communities that are economically and socially integrated with that nucleus. Nonmetropolitan counties are not linked with large cities nor with communities closely tied to large cities. This distinction is different from that between urban and rural devised by the Census Bureau (Duncan and Sweet 1992, xxvii).

Poverty Rates[2]

There are two sources of data for the 1989 poverty rates. The 1990 decennial census of population and housing is the only data source with a large enough sample to provide reliable estimates at the state and county level. The "long form" of the decennial census, filled in by about 5% of households, includes information on household composition, relationships, and income. The poverty income cutoff for each family is established based on family size and composition. The family's income for the year before the census is then compared with that poverty income cutoff level, and the people in the family are assigned the appropriate poverty status. What then shows up in the STF3C, the data file available to the public, is a total for each county of how many people had income above the poverty level, between .5 and 1.0 times poverty level, and below .5 times the poverty level. (Actually, there are a few more categories, but they are not relevant to this discussion.) These counts are also presented by race and ethnicity. The counts are population estimates based on the 5% sample. There is also a proportion of the population that is counted in the census, but for whom poverty status is not determined. The most important group is college students living in dormitories. They do not figure as either numerator or denominator in poverty-rate calculations.

The Current Population Surveys (for March 1990 and March 1994) were used to compare 1989 and 1993 data. This survey is similar to the long form of the decennial census (in terms of family and income data). It refers to income in the previous year. Like the decennial census, it calculates poverty status for each person based on the family composition and income and includes a weight variable to inflate the sample to population estimates. It is a large sample, about 55,000 households, but is not large enough to be reliable at the state level (except for states with very large populations). It is, however, done every year instead of once in ten years. It is useful, therefore, for regional estimates in the years between censuses.

Rural

"The decennial census classifies population as rural or urban ... according to the classification of the place they live. In the West, urban places include places of 2500 or more population incorporated as cities, villages, boroughs (except in Alaska), and towns, but excluding rural portions of 'extended cities.' Also included are 'Census designated places' of 2,500 or more. All other areas are classified as rural" (Nord 1995).

Timber Harvest

Total annual timber harvest data for 1949–1993 were obtained from the Strategic Planning Program, California Department of Forestry and Fire Protection.

Causal Analysis

Data Sources

Monthly data were collected on the following variables for the period 1984–1993. Databases for the primary explanatory variable of interest, monthly employment in lumber and wood-products production (Standard Industrial Classification [SIC] code 24) in California, are not consistently available for years before 1984 and therefore limit the length of the time period examined. Fortunately, from a statistical point of view, the decade from January 1984 through December 1995 saw dramatic fluctuation in the level of SIC 24 employment. This variation allows us to test the hypothesis that changes in SIC 24 employment "Granger-cause" other county employment and AFDC caseload. (The term Granger cause is defined later in this section.)

AFDC Caseload

Data on AFDC caseload are gathered by the California Department of Social Services (various years). There are two categories of AFDC cases: AFDC Unemployed Parent (AFDC UP) and AFDC Family Group (AFDC FG). AFDC UP recipients consist of two-parent households, AFDC FG of one-parent (usually the mother) households. Both programs are means tested, that is, would-be recipients must demonstrate that their income and assets fall below a certain level. Recipients can keep the first $30 they earn, plus a third of their income before aid is reduced (Snow 1995a, 1995b). We have used AFDC UP caseload in our time series analysis because it should be more sensitive to changes in timber-related employment.

County Population

Data on county population were taken from California Department of Finance reports (February 1987, July 1991, and March 1994). County population is estimated as of July 1 of each year. A monthly time series was constructed from these annual data by assuming that population changed at a uniform rate throughout the year.

County Employment

County employment data are taken from the U.S. Bureau of Labor Statistics (BLS) series on employment covered by unemployment insurance (BLS ES-202 program data, also referred to as the "Bell" series), which is collected and maintained by the California Employment Development Department. This data series was used both because very few data were missing for the period of interest and because the series is one of the few monthly employment series that is not constructed from a sample. The data series is compiled from firm-level reports, filed to comply with unemployment insurance requirements. All firms with employees covered by unemployment insurance must report the number of workers on their payroll during the pay period including the twelfth day of the month to the California Employment Development Department. "Bell" series employment data were used in this study. BLS considers the ES-202 data series to be "the most complete universe of monthly employment and quarterly wage information by industry, county, and State [available]" (U.S. Department of Labor September 1992).

BLS ES-202 categorizes employment by Standard Industrial Classification (SIC) code. In counties where confidentiality considerations do not prevent it (that is, where individual firms cannot be identified from the data), employment is reported by four-digit SIC code. SIC categories are revised periodically to reflect changes in technology and industrial structure. Pre-1988 data used in this study are classified using the 1977 edition of the SIC. Data from 1988 on were classified using the 1987 edition of the SIC (U.S. Department of Labor September 1992). Changes to the industrial categories used in this study (SIC 08 and SIC 24) were not deemed extensive enough to raise any significant issues regarding data comparability.

The other major county-level employment series available is mid-March employment reported in the U.S. Department of Commerce, Bureau of the Census, *County Business Patterns.* These data are inadequate for examining forest-related industries because forest-related employment exhibits marked seasonality, peaking in mid-summer to mid-fall.

SIC 08 (forestry employment) includes employment in "establishments primarily engaged in the operation of timber tracts, tree farms, forest nurseries, and related activities such as reforestation services." Forestry services include establishments "primarily engaged in performing, on a contract or fee basis, services related to timber production, wood technology, forestry economics and marketing, and other forestry services, not elsewhere classified, such as cruising timber, firefighting, and reforestation" (Office of Management and Budget 1987). SIC 08 employment was not used in causality tests in this study primarily because data were unavailable for most counties and most months. In addition, based on the data that are available, SIC 08 employment represents an extremely small fraction of total county employment, rarely exceeding .5% of total county employment. This level of economic activity cannot drive other activity in an economy and therefore can be safely ignored in looking for factors causing total employment or poverty.

SIC 24 (lumber and wood-products employment) includes logging, sawmills and planing mills, and production of millwork, plywood and structural members, wood containers, mobile homes, prefabricated wood buildings, and furniture and fixtures (Office of Management and Budget 1987).

This major group includes establishments engaged in cutting timber and pulpwood; merchant sawmills, lath

mills, shingle mills, cooperage stock mills, planing mills, and plywood mills and veneer mills engaged in producing lumber and wood basic materials; and establishments engaged in manufacturing finished articles made entirely or mainly of wood or related materials. Certain types of establishments producing wood products are classified elsewhere. For example, furniture and office and store fixtures are classified in Major Group 25; musical instruments, toys and playground equipment, and caskets are classified in Major Group 39. Wood working in connection with construction, in the nature of reconditioning and repair, or performed to individual order, is classified in nonmanufacturing industries. Establishments engaged in integrated operations of logging combined with sawmills, pulp mills, or other converting activity, with logging not separately reported, are classified according to the primary product shipped. . . . Independent contractors engaged in estimating or trucking timber, but who perform no cutting operations, are classified in non-manufacturing industries (Office of Management and Budget 1987).

This series is available for all study counties except Alpine for nearly all months of the study period. We use SIC 24 to represent forest-related employment because of this consistent coverage and because of its possible economic importance.

We did not include SIC 26 (paper and paper-products employment) in the study primarily because the study counties have little pulp mill activity. Furthermore, in the study counties, data on pulp and paper-mill employment, which is closely linked to timber production could not be separated from employment in the manufacture of secondary paper products like paperboard containers, coated papers, paper bags, and stationery products, which may rely on imported pulp or recycled paper rather than on California timber harvest.

"Other employment" is total county employment less employment in SIC 24. This variable was also constructed using BLS ES-202 program data.

Monthly employment data were obtained from California Employment Development Department (1994). The data were censored to protect confidentiality of county businesses.

Granger Causality Tests

Granger causality tests are widely used to investigate statistical causality over time (Cromwell 1992; Gruidle and Pluver 1991; Hoffman 1991; and Schimmelpfennig and Thirtle 1994). They have been used to investigate the impact of U.S. Forest Service policy in Oregon on forest-related employment (Burton and Berck, in press). Granger causality tests check for a very specific form of statistical causation based on two basic ideas. The first is that x can cause y only if it precedes y in time. The second is that if x does cause y, then a regression of past values of x and y on current y should predict current y significantly better than a regression of only past values of y on current y. For example, we could ask the question, Does lumber and wood-products employment "Granger-cause" AFDC caseload in a county? This is asking whether current county AFDC caseload is explained better by past values of lumber and wood-products employment and AFDC caseload in the county than by past values of AFDC caseload alone.

Granger causality is also a specific kind of causality because it is not necessarily transitive. That is, if x "Granger-causes" y, and y "Granger-causes" z, then x may or may not "Granger-cause" z. Finally, Granger causality explores causality in a purely statistical sense. By itself, it does not imply that one phenomena causes another in an economy or society. However, it does provide evidence about the plausibility of hypotheses about causation drawn from experience, observation, or theory.

More formally, y fails to Granger-cause x if for all $s > 0$ the mean squared error (MSE) of a forecast of x_{t+s} based on $(x_t, x_{t-1}, ...)$ is the same as the MSE of a forecast of x_{t+s} that uses both $(x_t, x_{t-1}, ...)$ and $(y_t, y_{t-1}, ...)$. The test is conducted by comparing two regressions: one of $(x_t, x_{t-1}, ...)$ and $(y_t, y_{t-1}, ...)$ regressed on x_{t+s}, and the other of $(x_t, x_{t-1}, ...)$ regressed on x_{t+s}. For small samples, like those used in this study, an F-statistic is used on the results of the restricted and unrestricted regressions to test the hypothesis that $(y_t, y_{t-1}, ...)$ contributes significantly to the explanation of x_{t+s} (Hamilton 1994).

Studies have found that the results of Granger causality tests can be sensitive to the number of lagged (past) values used in running the regressions and can be sensitive to the way nonstationarity (nonconstant mean or variances over time) is handled (Hamilton 1994). Said and Dickey (1984) have shown that lag lengths equal to the cube root of the number of observations used in the regressions usually provide as much information as can be obtained with greater lag lengths. The issue of whether and how to deal with nonstationarity in the underlying time series is unresolved.

In this study, transformations that increase stationarity in the observed time series materially change the interpretation of the test. Twelve-month-differencing the natural log of our observed data induces stationarity but transforms a monthly series of observations into a series made up of the annual growth rate of the variable calculated each month. This is quite different from a time series of observed past values of each variable. As a result, we have run Granger causality tests using both the raw observed time series and twelve-month differenced values of the natural log of the observed data. Granger causality tests were run using six lags of transformed series. The raw series exhibit yearly seasonal cycles. As a result, Granger causality tests on the raw observed data were run using eighteen lags (twelve months + six lags). To see how these transformations affect the interpretation of results, consider the test of whether lumber and wood-products employment Granger-causes AFDC caseload. The resulting tests on raw data can be interpreted as asking whether a combina-

tion of the past eighteen months of AFDC caseload and the past eighteen months of lumber and wood-products employment predicts current AFDC caseload significantly better than the past eighteen months of AFDC caseload alone. Tests on the transformed data ask whether the annual rate of growth in AFDC caseload in the current month is predicted significantly better by a combination of the annual growth rates of AFDC and lumber and wood-products employment for the last six months than by the annual growth rate for the past six months of AFDC caseload alone.

Granger Causality Models

The following is provided for those who desire a formal discussion of the model. Others should skip this section. Granger causality is a vector autoregressive test that defines causation so that for the time series of any two variables x and y, x fails to Granger cause y if

$$MSE\,[\hat{E}(y_t \mid y_{t-1}, y_{t-2}, ...)]$$
$$= MSE\,[\hat{E}(y_t \mid y_{t-1}, y_{t-2}, ..., x_{t-1}, x_{t-2}, ...)] \quad (1)$$

The reasoning behind this definition is that for event x to cause event y, it must precede event y. Another way to say this is that "x is exogenous to y in the time series sense" if equation (1) holds (Hamilton 1994).

We ran Granger causality tests using two types of data: raw data and twelve-month differenced natural log transformations of the raw data. With both sets of models, the number of lags was chosen because it was sufficient to induce stationarity and made sense as a representation of the information firms use for employment decisions.

We used ordinary least squares (OLS) on eighteen monthly lags of raw data to estimate

$$y_t = c + \beta_1 y_{t-1} + \beta_2\, y_{t-2} + ... + \beta_{18}\, y_{t-18} + \beta_{19} x_{t-1}$$
$$+\quad \beta_{20} x_{t-2} + ... + \beta_{36} x_{t-18} + e_t \quad (2)$$

The test for Granger causality is then simply an F-test of the null hypothesis

$$H_o: \beta_{19} = \beta_{20} = ... = \beta_{36} = 0 \quad (3)$$

For each county in the study, tests were conducted of the hypotheses that lumber and wood-products (SIC 24) employment Granger-caused non-SIC 24 employment, that SIC 24 employment Granger-caused AFDC UP caseload, and that non-SIC 24 employment Granger-caused AFDC UP caseload.

To test for Granger causality of rates of growth, we used OLS on six monthly lags of natural log transformed data to estimate

$$d(\ln(y))_t = c + \beta_1 d(\ln(y))_{t-1} + ... + \beta_6 d(\ln(y))_{t-6}$$
$$+ \beta_7 d(\ln(x))_{t-1} + ... + \beta_{12} d(\ln(x))_{t-12} + e_t.$$

where

$$d(\ln(\mathbf{y}))_t = \ln(\mathbf{y})_t - \ln(\mathbf{y})_{t-12}$$

and

$$d(\ln(\mathbf{x}))_t = \ln(\mathbf{x})_t - \ln(\mathbf{x})_{t-12} \qquad (4)$$

The annual rate of growth in a variable is the change in its natural log over a twelve month period. This is the discrete counterpart to the instantaneous rate of growth of a variable being the derivative of its natural log.

The test for Granger causality is then an F-test of the null hypothesis

$$H_o: \beta_7 = \beta_8 = \dots = \beta_{12} = 0 \qquad (5)$$

For each county in the study, tests were conducted of the hypotheses that the annual rate of growth of lumber and wood products (SIC 24) employment Granger-caused annual rate of growth of non-SIC 24 employment, that the annual rate of growth of SIC 24 employment Granger-caused the annual rate of growth of AFDC UP caseload, and that the annual rate of growth of non-SIC 24 employment Granger-caused the annual rate of growth of AFDC UP caseload.

FINDINGS

We have presented our data in two different ways. To ensure that graphs can be read and to allow geographic comparisons, graphed data are presented in small geographic clusters of counties. Tabular data are presented for the entire data set in alphabetical order.

Sierra Nevada Forest County Poverty: Descriptive Data

The Golden Age of Timber may have been a reality for some individual households. But, as can be seen in the following tables, incomes and livelihoods in the forest counties of the Sierra Nevada currently and historically compare unfavorably with the state as a whole. However, it is important to note that no forest county is included among the 24% of nonmetro counties nationally that are persistently poor, that is, counties with 20% or more of people in poverty in each of the years 1960, 1970, 1980, and 1990. Such counties are concentrated in the South, Southwest, and Alaska (Cook and Mizer 1994).

Data on the comparative rank of SNFC in terms of average per capita income are presented in table 14.1.[3] All counties in the state were ranked from one to fifty-eight. The county with the highest average per capita income is ranked number one,

and the lowest is ranked number fifty-eight. As can be seen in table 14.2, which summarizes table 14.1, the SNFC are disproportionately found among the poorer counties in California. Although they account for 28% of the counties, from 1950 to 1992 the SNFC have made up only 5% to 11% of the wealthiest third of counties, 16% to 32% of the middle third, and 45% to 60% of the poorest third. In other words, although individual households may have experienced a bonanza, since 1950, only 11% of the forest counties has experienced what might be characterized as a golden age. Furthermore, the SNFC have also been disproportionately represented among the least affluent of California counties, within the bottom third. As can be seen in table 14.1, since 1950 the SNFC have constituted between 40% and 71% of California counties with average per capita income 25% or more below the state average and (with the exception of zero in 1970) between 42% and 75% of counties averaging less than 30% of the state average.

The data presented in these tables, which show an apparent rise in the relative aggregate income levels of the SNFC while their percentage of the lowest income counties remains high, are consistent with Stewart's (1993) finding about the increasing economic importance of public transfer payments and private capital payments in the form of interest, dividends, and rent.

Data on poverty rates in SNFC presented in table 14.3 show poverty to be a persisting, indeed, increasing, problem in the forest counties. As is typical for both the United States as a whole as well as for the western region, nonmetro poverty rates exceeded metro rates. In both 1980 and 1990, half of the SNFC had poverty rates exceeding the state average. In five forest counties in 1990, nearly one in five persons fell below the poverty level. In 1990, 67% of the forest counties classified as metro had poverty rates exceeding the statewide average for metro counties, while 20% of the nonmetro counties exceeded the nonmetro average. This imbalance may, as will be seen in table 14.4, be due to the high rates of rurality in the forest metro counties, because nonmetro poverty rates are typically higher than metro rates. It is worth noting that the three foothill counties (Placer, Nevada, and El Dorado), which emerged as having relatively higher per capita income by 1992, also have some of the lowest poverty rates. (See Duane 1996 for additional insights into the foothill counties.)

Deep poverty is defined as a family income of less than 50% of the poverty level. In 1989 the average nonmetro deep poverty rate in California was 5.2% (Nord 1995). That is, 5.2% of the people in California's nonmetro areas were in deep poverty. Only two (12.5%) of nonmetro forest counties had deep poverty rates approaching or exceeding the state nonmetro average. Thus, while people in the forest counties often suffer from low incomes, most do not suffer from the deprivations of deep poverty.

The data in table 14.4 show persisting rurality in the metropolitan forest counties. As can be seen in tables 14.1 and 14.4, the 1970 census was the first to record a metro county (Placer) among the SNFC, and by 1993 six (38%) of the SNFC

TABLE 14.1

Relative rank of Sierra Nevada Forest Counties among California counties of average per capita income 1950–92 (Goldman and Hetland 1995).

County	1950	1960	1970	1980	1986	1992
Alpine	28	58[a]	52[b]	56[b]	46[b]	20
Amador	52[b]	55[a]	31	37	26	43
Butte	44	45	51	**51**[c]	**45**[b]	**44**[b]
Calaveras	57[a]	53[b]	55[b]	53	36	39
El Dorado	29	32	19	30	**17**	**18**
Lassen	31	37	48	54[b]	53[a]	53[a]
Madera	55[b]	52[b]	54[b]	27	50[a]	51[a]
Mariposa	46	24	16	52	37	41
Nevada	58[a]	56[a]	44	42	23	24
Placer	48[b]	49[b]	**29**	**16**	**14**	**12**
Plumas	22	28	23	44	34	32
Sierra	19	11	47	46	29	29
Tehama	53[b]	35	40	55	54[a]	54[a]
Tulare	51[b]	40	56[b]	45	52[a]	48[a]
Tuolumne	40	47[b]	53[b]	50	42[b]	37
Yuba	20	8	50	58[a]	56[a]	56[a]
SNFC as a percentage of counties 25% below average	64%	54%	71%	60%	47%	40%
[state total]	[11]	[13]	[7]	[5]	[17]	[15]
SNFC as a percentage of counties 30% below average	67%	75%	0%	50%	42%	45%
[state total]	[3]	[4]	[2]	[2]	[12]	[11]

[a]At least 30% below state average per capita income.
[b]At least 25% below state average per capita income.
[c]Bold indicates classification as a metro county.

were classified as metro counties. However, only one metro county, Butte, had less than 25% of its population living in rural areas. Five counties (31%) were 75%–100% rural.

Nord (1995) defines underemployment as working less than thirty-five hours a week or less than forty weeks a year. In California the average rate of underemployment among nonmetro working males in 1989 was 35.5%. As can be seen in table 14.4 in half of the nonmetro counties the underemployment rate nearly equaled or exceeded the state average.

Tables 14.1–14.4 show that the SNFC have a long history of relatively low incomes and persisting poverty. The question is why. In beginning to address this question, we have examined the relationships among AFDC Unemployed Parent caseload, lumber and wood-products employment and other employment in the SNFC.

Sierra Nevada Forest County Timber Harvest, Employment, and AFDC Caseloads: Descriptive Data

Timber Harvest

As is shown quite clearly in appendix 14.1, figure 14.A1, the timber harvest in the SNFC has been steady in comparison to the state harvest levels since the mid 1960s with a decline during the nationwide recession of the early 1980s and a recovery in the late 1980s and early 1990s. This trend contrasts with the general decline in the California timber harvest as a whole since 1955 (Belzer and Kroll 1986). Appendix 14.1, figure 14.A2 plots the SNFC timber harvest on a scale that reveals year-to-year variation more clearly. Individual county graphs are more volatile and varied (appendix 14.1, figures 14.A3–14.A6). Timber harvest is not included as a causal vari-

TABLE 14.2

Distribution of SNFC within California counties by per capita income (Goldman and Hetland 1995).

Tier	Ranking	1950	1960	1970	1980	1986	1992
1	Top 19	1 (5%)[a]	2 (11%)	2 (11%)	1 (5%)	2 (11%)	2 (11%)
2	Middle 19	5 (26%)	5 (26%)	3 (16%)	3 (16%)	6 (32%)	6 (32%)
3	Bottom 20	10 (50%)	9 (45%)	11 (55%)	12 (60%)	8 (40%)	8 (40%)

[a]Percentage rates refer to SNFC as a percentage of all California counties in the relevant tier.

TABLE 14.3

Poverty rates in Sierra Nevada Forest Counties (Bureau of the Census 1983; Nord 1995).

County	Metro/Nonmetro	Poverty Rate 1980 (State Average, 11.8%)	Poverty Rate 1990 (State Average, 12.5%; Metro Average, 12.4%; Nonmetro Average, 14.9%)	Deep Poverty 1989[a] (Nonmetro Average, 5.2%)
Alpine	Nonmetro	18.8	18.1	5–7%
Amador	Nonmetro	9.0	8.4	<5%
Butte	Metro	15.0	18.9	—
Calaveras	Nonmetro	10.1	10.1	<5%
El Dorado	Metro	8.7	7.7	—
Lassen	Nonmetro	10.3	13.3	<5%
Madera	Metro	15.7	17.5	—
Mariposa	Nonmetro	11.5	12.7	<5%
Nevada	Nonmetro	8.7	7.7	<5%
Placer	Metro	8.6	7.1	—
Plumas	Nonmetro	9.7	11.9	<5%
Sierra	Nonmetro	12.9	9.2	<5%
Tehama	Nonmetro	12.9	15.3	5–7%
Tulare	Metro	16.5	22.6	—
Tuolumne	Nonmetro	11.9	9.1	<5%
Yuba	Metro	16.1	19.5	—

[a]Nonmetro counties only. Deep poverty is defined as a family income of less than 50% of the poverty level.

able in the following statistical analysis since its effect is not direct but is mediated through employment.

Total Monthly Employment

Total monthly employment (that is, all employment, not just lumber and wood-products employment) is shown in appendix 14.2, figures 14.A7–14.A12. As in shown figure 14.A 7 (which demonstrates seasonal fluctuations), total employment in the SNFC has risen since 1984, leveling off in the 1990s well above 1984 levels. Figures 14.A8–14.A11 show employment levels in individual counties to be rising or steady.

TABLE 14.4

Rurality and underemployment of Sierra Nevada Forest Counties (Nord 1995).

County	Metro/ Nonmetro 1993	Percentage of Population Living in Rural Areas 1990[a]	Percentage of Underemployed Working Males 1989[b] (State Average, 35.5%)
Alpine	Nonmetro	75–100	<30
Amador	Nonmetro	50–75	>35
Butte	Metro	<25	—
Calaveras	Nonmetro	75–100	30–35
El Dorado	Metro	50–75	—
Lassen	Nonmetro	50–75	>35
Madera	Metro	25–50	—
Mariposa	Nonmetro	75–100	30–35
Nevada	Nonmetro	50–75	30–35
Placer	Metro	25–50	—
Plumas	Nonmetro	75–100	>35
Sierra	Nonmetro	75–100	30–35
Tehama	Nonmetro	50–75	>35
Tuolumne	Nonmetro	50–75	>35
Tulare	Metro	25–50	—
Yuba	Metro	25–50	—

[a]Overlapping categories are in the original.
[b]Nonmetro counties only.

Lumber and Wood Products Employment

SIC 24 employment includes logging, sawmills and planing mills, and production of millwork, plywood and structural members, wood containers, mobile homes, prefabricated wood buildings, wooden furniture, and fixtures. A more detailed description of this variable can be found earlier in the "Methods" section.

Absolute levels of lumber and wood-products employment are presented in appendix 14.3, figures 14.A12–14.A16. Again, seasonal fluctuations figure prominently in some of these graphs. The timber-cutting boom of the late 1980s and early 1990s also is reflected in many of these graphs. As can be seen in figure 14.A12, lumber and wood-products employment in the whole region rose during the mid to late 1980s, falling off sharply in the early 1990s. Most counties followed this general pattern, with these exceptions: Sierra was relatively stable throughout. Madera showed a much earlier drop. Plumas showed a steady decline throughout the entire period.

SIC 24 Employment as a Percentage of Total County Employment 1984–1994[4]

We have already seen that during the 1980s and early 1990s both SNFC timber harvest and SNFC lumber and wood-products employment rose and fell markedly (see appendix 14.1, figure 14.A2; and appendix 14.3, figure 14.A12). But this cycle does not appear to have had an effect on lumber and wood-products employment relative to total regional employment (see appendix 14.4, figure 14.A17). During the 1980s and early 1990s, SNFC employment in the lumber and wood products sector made up roughly 3% of total employment in the SNFC as a whole. One would not expect employment of this relative magnitude to drive a regional economy. Furthermore, while lumber and wood products employment as a percentage of SNFC total employment fell slightly relative to total SNFC employment, it fell steadily in a pattern that does not evidence the rise and fall in timber employment during the decade (see figure 14.A17). This reasonably stable regional employment picture is at variance with the dramatic stories of catastrophe with which we began this chapter. It is important to remember that stories based on real and painful individual experience are not necessarily indicative of larger trends. We shall return to this point below. It is also important to remember the point of the RSS Task Force that employment can involve low wages and poor working conditions. Our data do not address this issue.

As expected, the picture at the county level is more variable. Graphs of lumber and wood products employment as a percentage of total county employment are presented in appendix 14.4, figures 14.A17–14.A21. Again, seasonal fluctuations figure prominently in some of these graphs, as does the timber boom of the late 1980s and early 1990s. Throughout the decade, lumber and wood-products employment was consistently at or below 4% in eight counties: Butte, El Dorado, Nevada, Placer, Calaveras, Madera, Mariposa, and Tulare. These counties included all the southernmost tier of SNFC (figure 14.A21) and two of the four counties in the next tier to the north (figure 14.A20). Five of these counties (El Dorado, Nevada, Placer, Calaveras, and Mariposa) fell into the first or second tier for per capita income in 1986 and 1992. Double-digit levels of lumber and wood-products employment (roughly 10%–25%) occurred throughout most of the decade in Amador, Plumas, Sierra, and Tehama, with all but Sierra experiencing steep downward trends after 1988. Sierra experienced an upward trend, ending the decade with a slightly higher percentage of lumber and wood-products employment.

AFDC Unemployed Parent

If timber unemployment drives welfare, AFDC Unemployed Parent cases are the most likely to reflect timber employment trends. Data for AFDC UP cases per capita from 1970 to 1993 are presented in appendix 14.5, figures 14.A22–14.A26. Again, AFDC UP cases show strong seasonal fluctuations. This is consistent with Albert's (1988, 57) statewide finding that "many of the cases that open in the winter close in the spring." The SNFC are compared with California in figure 14.A22. Two features of this graph should be given particular attention. First, the per capita figures are higher for the SNFC than for the state as a whole, consistent with the income and poverty figures presented in tables 14.1–14.3. Second, at the very time that statewide AFDC caseloads were dropping and the timber industry was booming, the SNFC caseloads were rising. Why this was so will be explored in the following Granger causality analysis. Again, county trends vary. It is worth noting that four of the five counties with per capita incomes at least 30% below the state average in 1986 and 1992 (Lassen, Madera, Tulare, and Yuba) also had the highest per capita AFDC UP caseloads in 1993.[5]

Although we did not use them in the statistical analysis, data on 1970–93 AFDC Family Group caseloads are presented in appendix 14.6, figures 14.A27–14.A32. (AFDC FG households have only one parent, usually the mother, present.) In contrast to AFDC UP, AFDC FG caseloads for the SNFC briefly fell below statewide levels in the early 1980s. However, for the remainder of the period, SNFC caseloads exceeded state levels.

Tests of Causality

The remainder of this chapter explores causal relationships between variables that social science theory or popular anecdote suggest cause good or bad economic outcomes in the SNFC. We have used Granger causality tests to do this. Because of federal rules protecting confidentiality, we were not able to include data from Alpine County in these tests.

In reporting our findings we use the verb phrase "Granger-cause," a term of art in time-series analysis, because we want to be precise and clear about the limits of our findings. If we say x "Granger-causes" y, we mean that past values of x and y predict the current value of y better than the past values of y alone. Granger causality implies that the variable x_t, does not occur later in time than the variable y_t that it "Granger-causes." To repeat our caution in the Methods section, Granger causality explores causality in a purely statistical sense. By itself, it does not imply that one phenomena causes another in an economy or society. However, it does provide evidence about the plausibility of hypotheses about causation drawn from experience, observation, or theory.

Because "Granger-cause" is an awkward term, we have used "cause," within quotation marks, as a shorthand for "Granger-cause." The reader should bear in the mind the limitations on the meaning of "cause" signaled by the quotation marks.

Although we have presented diagnostic statistics for these tests, the lay reader need look only in the final column labeled Prob >F in the following tables. If there is a footnote reference beside the beside the number in that column, then, in lay terms, it is likely that x does "Granger-cause" y in that county. It is important to know that the preferred and more precise interpretation is that the question "Does x Granger-cause y" cannot be answered "no" with any statistical confidence for that county. The precision involved in the term "Granger-cause" reflects scientific method in which hypotheses can be disproved but not proved.

Tables 14.5–14.7 were calculated using eighteen months of lagged raw data. In these tables the question "Does x 'cause' y?" should be interpreted as meaning "Do the past eighteen months of x and y predict current y better than the past eighteen months of y alone?" If the answer is yes, then x "Granger-causes" y.

Table 14.5 shows that lumber and wood-products employment fails to "cause" other employment in any of the forest counties over time. More precisely, given that one can predict

TABLE 14.5

Does SIC 24 employment Granger-cause other county employment?[a]

County	Restricted R-Square	DW	F-Test Statistic	Prob >F ($F_{18, 52}$)
Amador	0.98	1.99	0.78	0.711
Butte	0.98	1.99	1.04	0.432
Calaveras	0.96	1.98	0.89	0.589
El Dorado	0.99	1.94	1.14	0.338
Lassen	0.84	2.06	1.28	0.230
Madera	0.89	1.92	1.56	0.098
Mariposa	0.93	1.91	0.55	0.924
Nevada	0.99	1.98	0.89	0.596
Placer	1.00	2.00	1.40	0.161
Plumas	0.97	1.95	0.42	0.978
Sierra	0.79	2.00	0.49	0.953
Tehama	0.97	1.98	0.62	0.869
Tulare	0.95	2.07	1.39	0.165
Tuolumne	0.96	2.01	0.93	0.552
Yuba	0.87	2.02	1.04	0.432

[a]Time series: raw data, eighteen lags.

current other employment in the county from the previous eighteen months of such employment, knowing what the lumber and wood-products employment has been during the last eighteen months will not improve ability to predict current other employment. In lay terms, employment variation in the lumber and wood-products industry over time does not cause variation in other employment. This is consistent with Stewart's (1993) finding.

Table 14.6 shows that other employment "causes" AFDC UP caseload over time in seven of the forest counties. More precisely, if one knows only what the AFDC UP caseload has been for the past eighteen months, one cannot predict the current AFDC UP caseload as well as if one also knows the past eighteen months of other employment. In lay terms,

variations in other employment causes variations in the AFDC UP caseload in seven of the fifteen counties. These findings for these seven counties are consistent with Albert's (1988) findings that employment affects aggregate levels of AFDC in California as whole.[6]

The lack of such a causal relationship in the remaining eight counties may be explained in two basic ways. First, Albert (1988) found that aggregate employment predicted AFDC case closures but not case accessions, while employment in specific industries was an accurate predictor of both. It may be that employment levels in specific industries have greater predictive power. Second, these counties may be reflecting the persisting effects of particular economic structures. Five of the counties (Amador, Calaveras, Nevada, Sierra, and Tulare) have relatively low poverty rates. It is possible that these low rates reflect poverty that persists for structural reasons, such as age distribution of the population or the wage structure of particular industries, that would not necessarily be affected by variations in employment. Similarly, since 1980 Lassen and Yuba counties have experienced average per capita incomes 30% or more below the state average. This again suggest that these low incomes persist for structural reasons and are not affected by changes in available employment. An important lesson of table 14.6 is that whatever ecological or geographical unity the Sierra Nevada may have emphatically does not translate into socioeconomic unity. These are very heterogeneous counties with differential social, political, and economic ties to state, national, and global systems.

If Albert's (1988) finding that employment levels in specific industries make a difference in predicting AFDC accession and termination, then the obvious industry to investigate in the SNFC is lumber and wood products. The data in table 14.7 show that lumber and wood-products employment fails to "cause" AFDC UP caseload over time in any of the forest

TABLE 14.6

Does other employment Granger-cause AFDC Unemployed Parent caseload?[a]

County	Restricted R-Square	DW	F-Test Statistic	Prob >F ($F_{18, 52}$)
Amador	0.86	2.05	1.08	0.389
Butte	1.00	1.97	2.05	0.019[b]
Calaveras	0.93	2.04	0.73	0.769
El Dorado	0.96	1.90	1.95	0.026[b]
Lassen	0.86	1.98	1.19	0.295
Madera	0.99	1.99	2.47	0.004[b]
Mariposa	0.94	1.94	1.82	0.041[b]
Nevada	0.96	1.99	0.99	0.480
Placer	0.99	2.08	2.08	0.017[b]
Plumas	0.91	2.04	1.63	0.078
Sierra	0.69	1.98	0.64	0.851
Tehama	0.96	1.98	2.39	0.006[b]
Tulare	1.00	1.94	2.91	0.001[b]
Tuolumne	0.94	1.99	1.51	0.115
Yuba	0.97	1.97	0.96	0.514

[a]Time series: raw data, eighteen lags.
[b]Significant at $d = .05$, that is, one cannot reject the hypothesis that other employment Granger-causes AFDC UP caseload.

TABLE 14.7

Does SIC 24 employment Granger-cause AFDC Unemployed Parent caseload?[a]

County	Restricted R-Square	DW	F-Test Statistic	Prob >F ($F_{18, 52}$)
Amador	0.86	2.03	0.91	0.566
Butte	1.00	1.93	1.62	0.081
Calaveras	0.94	2.00	1.43	0.147
El Dorado	0.96	1.99	1.89	0.033[b]
Lassen	0.87	1.94	0.14	0.170
Madera	0.99	2.04	1.06	0.411
Mariposa	0.93	1.94	1.46	0.121
Nevada	0.96	2.00	1.42	0.152
Placer	0.98	1.94	1.42	0.151
Plumas	0.93	1.96	3.21	0.000[b]
Sierra	0.75	1.94	1.58	0.101
Tehama	0.96	1.95	1.33	0.192
Tulare	1.00	1.94	0.68	0.814
Tuolumne	0.94	1.95	1.43	0.146
Yuba	0.98	2.02	1.71	0.060

[a]Time series: raw data, eighteen lags
[b]Significant at $d = .05$, that is, one cannot reject the hypothesis that SIC 24 employment Granger-causes AFDC UP caseload.

counties except El Dorado and Plumas. More precisely, if one can predict current AFDC UP caseload from the previous eighteen months of AFDC UP caseload, knowing what the lumber and wood-products employment has been during the last eighteen months will not improve ability to predict current AFDC UP per capita caseloads except in El Dorado and Plumas counties. In lay terms, employment variation in the lumber and wood-products industry over time does not cause variation in AFDC caseload except in El Dorado and Plumas counties.

Tables 14.8–14.10 address the same questions as tables 14.5–14.7, except that they use the annual difference in the natural log of raw monthly data.[7] The annual change in the natural log of a variable is the annual growth rate in the variable itself. The regression results reported in these tables use six-month lags of this transformed variable. In essence, they ask whether the annual growth rate of y from, for example, January to January is better predicted by the annual growth rate of y for the previous six months or by the annual growth rate in both x and y during the previous six months.

Table 14.8 shows that the annual growth rate of lumber and wood-products employment fails to "cause" the annual growth rate of other employment in any of the forest counties except Tulare. More precisely, if one can predict the annual growth rate of current other employment from the annual growth rate of other employment for the previous six months, then also knowing what the annual growth rate of lumber and wood products employment was during the past six months will not improve your ability to predict the current annual growth rate of other employment except in Tulare County.

In lay terms, the combination of tables 14.5 and 14.8 shows

TABLE 14.8

Does the annual growth rate of SIC 24 employment Granger-cause the annual growth rate of other county employment?[a]

County	Restricted R-Square	DW	F-Test Statistic	Prob >F $(F_{6, 89})$
Amador	0.74	2.03	1.10	0.368
Butte	0.80	1.92	1.83	0.102
Calaveras	0.60	2.00	0.60	0.733
El Dorado	0.81	1.90	0.68	0.662
Lassen	0.66	2.00	0.70	0.648
Madera	0.40	2.00	1.61	0.153
Mariposa	0.77	1.96	1.72	0.126
Nevada	0.67	1.98	0.19	0.979
Placer	0.86	1.84	1.77	0.114
Plumas	0.71	1.97	0.64	0.696
Sierra	0.58	1.95	0.27	0.948
Tehama	0.56	1.96	0.68	0.667
Tulare	0.74	2.01	2.83	0.014[b]
Tuolumne	0.72	1.97	0.64	0.699
Yuba	0.75	1.98	1.01	0.421

[a]Time series: twelve-month differences of the natural log of raw data, six lags.
[b]Significant at $d = .05$, that is, one cannot reject the hypothesis that other employment Granger-causes AFDC UP caseload.

TABLE 14.9

Does the annual growth rate of other employment Granger-cause the annual growth rate of AFDC Unemployed Parent caseload?[a]

County	Restricted R-Square	DW	F-Test Statistic	Prob >F $(F_{6, 89})$
Amador	0.79	2.01	2.02	0.071
Butte	0.71	2.01	1.69	0.134
Calaveras	0.74	2.13	1.21	0.310
El Dorado	0.85	1.88	1.94	0.084
Lassen	0.70	1.98	1.69	0.133
Madera	0.94	1.99	1.13	0.353
Mariposa	0.84	2.00	1.44	0.207
Nevada	0.90	1.99	2.00	0.074
Placer	0.92	2.03	1.05	0.397
Plumas	0.81	2.07	2.03	0.070
Sierra	0.53	2.00	1.44	0.209
Tehama	0.85	2.07	1.70	0.131
Tulare	0.97	1.91	3.93	0.002[b]
Tuolumne	0.80	2.02	3.06	0.009[b]
Yuba	0.91	1.92	1.08	0.378

[a]Time series: twelve-month differences of the natural log of raw data, six lags.
[b]Significant at $d = .05$, that is, one cannot reject the hypothesis that other employment Granger-causes AFDC UP caseload.

that lumber and wood-products employment fails to "cause" other employment in the long term (defined as eighteen months) in any of the forest counties and that annual growth in lumber and wood-products employment "causes" annual growth in other employment only in Tulare County.

In Tulare County, the ability to predict the current annual growth rate of other employment will be better if one knows the annual growth rates of both lumber and wood-products employment and other employment for the past six months. Tulare is a high poverty, persistently low-income county with the percentage of total employment accounted for by lumber and wood products employment consistently running less than 2%. The fact that results for all three Granger causality tests on annual growth rates were significant for Tulare County suggests that something distinguishes it from other forest counties.

Table 14.9 shows that the annual growth rate of other employment does not "cause" the annual growth rate of AFDC UP caseload in any of the forest counties except Tulare and Tuolumne. More precisely, if one can predict the annual growth rate of AFDC UP caseload from the annual growth rate of the AFDC UP caseload in the previous six months, then also knowing the annual rate of growth of other employment for the past six months ago does not improve ability to predict the annual growth rate in AFDC UP caseload except in Tulare and Tuolumne counties.

We have discussed Tulare County earlier. Tuolumne County has an underemployment rate of more than 35%, but a low poverty rate. Lumber and wood-products employment has accounted for roughly 2% to 6% of total employment over time.

Table 14.10 shows that the annual rate of growth of lumber

TABLE 14.10

Does the annual growth rate of SIC 24 employment Granger-cause the annual growth rate of AFDC Unemployed Parent caseload?[a]

County	Restricted R-Square	DW	F-Test Statistic	Prob >F $(F_{6, 89})$
Amador	0.78	1.99	1.39	0.226
Butte	0.69	1.98	0.71	0.645
Calaveras	0.74	2.08	1.43	0.204
El Dorado	0.86	1.90	3.87	0.002[b]
Lassen	0.70	1.99	1.62	0.152
Madera	0.94	2.05	1.75	0.120
Mariposa	0.84	1.96	0.87	0.523
Nevada	0.89	1.99	1.09	0.377
Placer	0.92	1.97	0.90	0.501
Plumas	0.81	1.99	1.86	0.096
Sierra	0.57	2.06	2.84	0.015[b]
Tehama	0.88	2.06	4.97	0.000[b]
Tulare	0.97	1.83	2.26	0.045[b]
Tuolumne	0.78	1.98	1.50	0.187
Yuba	0.90	1.93	0.67	0.673

[a]Time series: twelve-month differences of the natural log of raw data, six lags.
[b]Significant at $d = .05$, that is, one cannot reject the hypothesis that other employment Granger-causes AFDC UP caseload.

and wood-products employment "causes" the annual growth rate of AFDC UP caseload in four of the forest counties. More precisely, if one knows only the annual growth rate of the AFDC UP caseload during the last six months, one cannot predict the annual growth rate of AFDC UP caseload in the current month as well as if one also knew the annual growth rate of lumber and wood-products employment during the last six months. In lay terms, variations in the growth rate of lumber and wood-products employment in the last six months does cause variations in the growth rate of AFDC UP caseload in four of the fifteen counties.

Again, we have discussed Tulare County earlier. El Dorado County has a low poverty rate, a rising rank in average per capita income, and only 2% to 4% total employment accounted for by lumber and wood products. In Sierra County, although poverty is relatively low, lumber and wood-products employment fluctuates around 20% of total employment, which may account for the effect shown here. Tehama is a low-income county with high underemployment, high poverty, and high deep poverty. Lumber and wood-products employment has dropped from roughly 15% to 7% of total employment. In this vulnerable county, it is perhaps not surprising that short-term shocks are registered quickly.

CONCLUSIONS

The most obvious policy implication of our findings is clear-cut. Poverty and low incomes are persisting problems in the Sierra Nevada and need to be addressed. A second and equally obvious conclusion of this study is the difficulty of studying poverty.

This study had its origins in an attempt to use 1990 census data to update Kusel and Fortmann's (1991) study of poverty in California forest counties, which used 1980 census data. When the variables that predicted poverty levels in 1980 turned out to be no longer statistically significant, two inter-related explanations seemed likely. First, the descent into poverty is not necessarily instantaneous. Rather, the onset of poverty may lag behind the occurrence of a causal event, be it job loss, divorce, pay cut, or death of a spouse. Second, during the decade the structure and economic importance of the California timber industry had continued to change. Our point-in-time data could not capture the effects of these dynamic processes.

We therefore undertook a time-series analysis, which is sensitive to ongoing dynamic processes and delayed effects. We immediately encountered the data availability problems described earlier. For obvious political reasons domestic poverty is not tracked closely by the state or federal government. Lacking data on poverty suitable for time-series analysis, we used monthly AFDC UP caseloads as the closest and most accurate substitute. The limitations of this measure have been detailed earlier. Nor, although we intended to, were we able to assess the effects of employment in agriculture and tourism because reliable data suitable for county level time-series analysis were not available. We discuss alternative approaches later in the section.

Despite our inability to conduct all the analyses we had hoped to, this analysis does provide valuable policy insights. Poverty and low incomes, persisting problems in the SNFC, need to be addressed. One means of relieving poverty and increasing incomes suggested by the timber industry and supported by popular perceptions of the economy in forest counties is to increase lumber and wood-products employment in general and timber harvesting in particular.

Albert (1988) found that for California as a whole aggregate employment predicted AFDC case termination but not accessions, while employment in specific industries accurately predicted both. We therefore tested the hypothesis that lumber and wood-products employment cause AFDC UP in the SNFC either directly or indirectly by causing other employment.

Lumber and wood-products employment directly "Granger-caused" AFDC caseload in only two of fifteen forest counties. The growth rate of lumber and wood-products employment Granger-caused the growth rate of AFDC caseloads in only four of the forest counties. Although this is not as strong a finding as the absence of impact on other employment or other employment growth discussed later, it provides a marked contrast to anecdotal evidence such as the very localized stories that began this chapter. Still, these findings indicate that on a regional level a policy that attempts to increase lumber and wood-products employment or its growth rate will do little to reduce AFDC caseload or—to the extent that AFDC is a good poverty indicator—by extension, regional poverty.

Other employment appears to have more effect on AFDC caseload than does lumber and wood-products employment, whether looking at either level or growth rate. It is not possible, however, to conclude on this basis that simply increasing employment would significantly decrease AFDC caseload or poverty in the SNFC.

There is also no evidence to suggest the lumber and wood-products employment affects AFDC indirectly through its effects on other employment. We found that lumber and wood-products employment "Granger-caused" other employment in none of the forest counties. The growth rate of lumber and wood-products employment "Granger-caused" the growth rate of other employment in only one of the forest counties. These are strong findings, particularly in light of such strongly held popular beliefs to the contrary. They differ from the implications of McWilliams and Goldman's (1994) input-output analysis because input-output analysis does not account for changes in economic conditions. Input-output analysis asks what *might happen* if there were a decrease in sales by forest-related industries in the very short run and labor in all sectors were employed in fixed proportion to output. Time-series analysis reflects the adjustments that result from changes in economic conditions. That is, it asks what *actually happened in the long run.* [8]

It is clear from these findings that increasing lumber and wood-products employment is not likely to have a significant long-run impact either on other employment or on AFDC caseloads in the SNFC. That is, we have no evidence that the loss of timber-related employment "caused" increases in AFDC caseloads at the county level, nor that its availability would cause the decline of AFDC caseloads at the county level. It seems safe to conclude that policies which might increase lumber and wood-products employment in general and timber harvesting in particular would provide a crude and probably ineffective lever for addressing these issues. It must be borne in mind that this analysis has nothing to say directly about impact on household or individual income. In addition, it must be kept in mind that these are regional trends, and individual experiences in local communities may be different. Understanding what policy efforts would decrease poverty requires a broad understanding of the process that causes poverty in these counties. This is beyond the scope of the study reported in this chapter.

The analysis presented here leaves many questions unanswered. In particular, our understanding of what drives poverty in the region is not clear enough to make specific policy suggestions. In addition, many questions are beyond the scope of available data suitable for time-series analysis. This chapter does not address the dynamics of how or why people fall into poverty or the welfare system, or how they avoid doing so, or how families who lose timber-based livelihoods cope. It is important to remember that real people do lose real jobs, and to these people aggregate trends offer little consolation.

Understanding the dynamics of poverty and welfare will require systematic interviews with people in the SNFC, specifically former timber workers and former and current welfare recipients. Questions that might be asked include

- Are people staying employed by taking lower-paying jobs?

- Are people more willing to go on welfare than they used to be?

- Are people leaving the SNFC for jobs in urban areas? Who leaves? Who stays?

- Are welfare recipients moving into the SNFC? Where do they come from? Do they stay?

- Is poverty becoming "harder" as nonstandard housing becomes scarcer?

What this study does make clear is that different levels of analysis reveal different, sometimes conflicting, pictures, of poverty and economic well-being and their causes.

ACKNOWLEDGMENTS

This study has benefited from the assistance of a number of people. Professor Henry Brady, Department of Political Science, and Professor Peter Berck and Professor Irma Adelman, Department of Agricultural and Resource Economics, University of California, Berkeley, provided generous tutelage in the intricacies of time-series analysis. Professor Brady and staff at the University of California Data Archives and Technical Assistance (UC DATA) also provided poverty data and insights from their studies of poverty programs in California. Professor Keith Gilless, Department of Environmental Science, Policy, and Management; Vicky Albert, George Goldman, and Norman Hetland, Department of Agricultural and Resource Economics, University of California, Berkeley; and Russell Henly, California Department of Forestry and Fire Protection, provided insights and data. Arvis Cury and other staff of the Labor Market Information Division/Area Services Group, Employment Development Department were generous in helping us understand the limitations of California data and in providing monthly employment data. Barbara Snow, UC DATA, provided AFDC definitions. Rowan Rountree, Pacific Southwest Research Station, was helpful in many ways. Jodi Bailey helped us assemble this chapter in compliance with the guidelines of the Sierra Nevada Ecosystem Project.

NOTES

1. Quotations without attribution are from unpublished field notes. Some of the stories in Brown's (1995) sophisticated and nuanced presentation of local narratives in an Oregon timber county reveal these same themes.

2. The following discussion of the calculation of poverty rates is taken from an email communication from M. Nord, Economic Research Service, PLIB, May 24, 1995.

3. We are grateful to George Goldman and Norman Hetland, Department of Agricultural and Resource Economics, University of California, Berkeley, for making available the data on which this table is based. The actual percentages can be found in their appendix 1.

4. County SIC 24 employment does not necessarily mean that timber is being harvested in that county. A gyppo logger who lives and pays a crew from his county of residence may actually be logging elsewhere. However, milling and wood-products manufacturing would take place almost exclusively in the county.

5. Alpine County also had a high per capita caseload.

6. Albert's choice of "urban" indicators such as nonagricultural work should be borne in mind.

7. This transformation appears to increase the stationarity of each data series. There are conflicting schools of thought on whether raw or transformed data are more appropriate in tests of Granger causality (Hamilton 1994). For this reason, and because the raw and transformed data series are interpreted differently, both data types are used in this study.

8. Input-output analysis is very useful in providing a picture of the current linkages between economic sectors. However, in projecting how a change in supply or demand will affect the economy depends on a number of strong assumptions about technology and human resources. These include no substitution by firms among possible inputs, no change in relative prices, fixed proportion technologies, no labor mobility between industrial sectors, and no regional migration. In short, the input-output model does not adjust to changes in demand or supply except through unemployment and idling of production plants. As a result, input-output analysis is well known to predict higher multiplier effects than are actually experienced. Unlike input-output analysis, Granger causality makes no assumptions about production technologies or people's response to economic change. It allows the data to reveal what has occurred. Its shortcoming is that, alone, it cannot explain structurally how adjustment occurs. Its strength is that it does not base estimates of economic impact on assumptions about the structure of the economy. It measures how the economy did in fact respond.

REFERENCES

Agee, J., and W. Evans. 1941. *Let us now praise famous men.* Boston: Houghton Mifflin.

Albert, V. N. 1988. *Welfare dependence and welfare policy: A statistical study.* New York: Greenwood Press.

Belzer, D., and C. Kroll. 1986. *New jobs for the timber region: Economic diversification for northern California.* Berkeley: University of California Institute of Governmental Studies.

Brown, B. 1995. *In timber country: Working people's stories of environmental conflict and urban flight.* Philadelphia: Temple University Press.

Burton, D., and P. Berck. In press. Statistical causation: National forest policy in Oregon. *Journal of Forest Science.*

California Department of Finance. February 1987. Intercensal estimates of the population of California state and counties: 1970 to 1980. Report I70–80. Unpublished report. Sacramento, CA.

California Department of Finance. July 1991. Intercensal estimates of the population of California state and counties: 1980 to 1990. Report I80–90. Unpublished report. Sacramento, CA.

California Department of Finance. March 1994. Population estimates for California counties—July 1990–1993: Official state estimates. Report I90–93. Unpublished report. Sacramento, CA.

California Department of Social Services. (Various years). *Public welfare in California.* Sacramento, CA.

California Employment Development Department, Labor Market Information Division. 1994. *Employment and payroll data, Bell (202) data series.* Computer printout. Sacramento, CA.

California Forestry Association. 1994. *California's changing forests: A case for management.* Video. Sacramento, CA.

Cook, P. J., and K. L. Mizer. 1994. *The revised ERS county typology: An overview.* Rural Development Research Report 89. Washington, DC: U.S. Department of Agriculture, Rural Economy Division. Economic Research Service.

Cromwell, B. A. 1992. Does California drive the West? An econometric investigation of regional spillovers. *Federal Reserve Bank of San Francisco Economic Review,* no. 2:13–23.

Deavers, K., and R. Hoppe. 1992. Overview of the rural poor in the 1980s. In *Rural Poverty in America,* edited by C. M. Duncan, 3–20. New York: Auburn House.

Duane, T. P. 1996. Human settlement, 1850–2040. In *Sierra Nevada Ecosystem Project: Final report to Congress,* vol. II, chap. 11. Davis: University of California, Centers for Water and Wildland Resources.

Duncan, C. M., and S. Sweet. 1992. Introduction: Poverty in rural America. In *Rural Poverty in America,* edited by C. M. Duncan, xix–xxvii. New York: Auburn House.

Economic Development Agency of the State of California. (Various years). *California statistical abstract.* Sacramento, CA.

Freudenburg, W. R., and R. Gramling. 1994. Natural resources and rural poverty: A closer look. *Society and Natural Resources* 7 (1): 5–22.

Goldman, G., and N. Hetland. 1995. The divergence of per capita income in California metro and non-metro counties, 1950–1992. Unpublished manuscript. Cited with permission of the authors. Department of Agricultural and Resource Economics, University of California, Berkeley.

Gruidle, J. S., and G. C. Pluver. 1991. A dynamic analysis of net migration and state employment change. *Review of Regional Studies* 21 (1): 21–38.

Hamilton, J. D. 1994. *Time series analysis.* Princeton, NJ: Princeton University Press.

Hoffman, E. P. 1991. Aid to Families with Dependent Children and female poverty. *Growth and Change* 22 (2): 21–38.

Kusel, J., and L. Fortmann. 1991. *Well-being in forest-dependent communities,* vol. 1. Sacramento: California Department of Forestry and Fire Protection, Forest and Rangeland Resources Assessment Program.

McLaughlin, D., and L. Perman. 1991. Returns vs. endowments in the earnings attainment process for metropolitan and non-metropolitan men and women. *Rural Sociology* 56:339–65.

McWilliams, B., and G. Goldman. 1994. *The forest products industries in California: Their impact on the state economy.* Berkeley: University of California, Department of Agricultural and Resource Economics.

Nord, M. 1995. Data notes: Fact sheets, figures, and maps: Western poverty information. Background document prepared for the Western Rural Development Center conference Pathways Out of Poverty, Albuquerque, NM, 18–20 May.

Office of Management and Budget. 1987. *Standard industrial classification manual 1987.* Springfield, VA: NTIS.

Peluso, N., C. R. Humphrey, and L. P. Fortmann. 1994. The rock, the beach, and the tidepool: People and poverty in natural resource-dependent areas. *Society and Natural Resources* 7 (1): 23–38.

Rural Sociological Society Task Force on Persistent Rural Poverty. 1993. *Persistent poverty in rural America.* Boulder, CO: Westview Press.

Said, S. E., and D. A. Dickey. 1984. Testing for unit roots in autoregressive-moving average models of unknown order. *Biometrika* 71:599–607.

Schimmelpfennig, D., and D. Thirtle. 1994. Cointegration and causality: Exploring the relationship between agricultural R&D and productivity. *Journal of Agricultural Economics* 45 (2): 220–31.

Shapiro, I. 1989. *Laboring for less: Working but poor in rural America.* Washington, DC: Center on Budget and Policy Priorities.

Snow, B. W. 1995a. The work pays demonstration project. Unpublished manuscript.

———. 1995b. California work pays demonstration project. AFDC survey: Results from the first wave. Unpublished manuscript.

Stewart, W. C. 1993. Predicting employment impacts of changing forest management in California. Ph.D. diss., Department of Forestry and Resource Management, University of California, Berkeley.

Sullivan, B. J. 1988. Cumulative impacts of national forest timber harvests in California. Ph.D. diss., Department of Forestry and Resource Management, University of California, Berkeley.

U.S. Department of Commerce, Bureau of the Census. (Various years). *County business patterns.* Washington, DC: Government Printing Office.

U.S. Department of Labor, Bureau of Labor Statistics. 1992. *BLS handbook of methods.* Bulletin 2414. Washington, DC: Government Printing Office.

U.S. National Advisory Commission on Rural Poverty. 1967. *The people left behind.* Washington, DC: Government Printing Office.

Aggregate and County-Level Timber Harvest

FIGURE 14.A1

Timber harvest for Sierra Nevada Forest Counties and state.

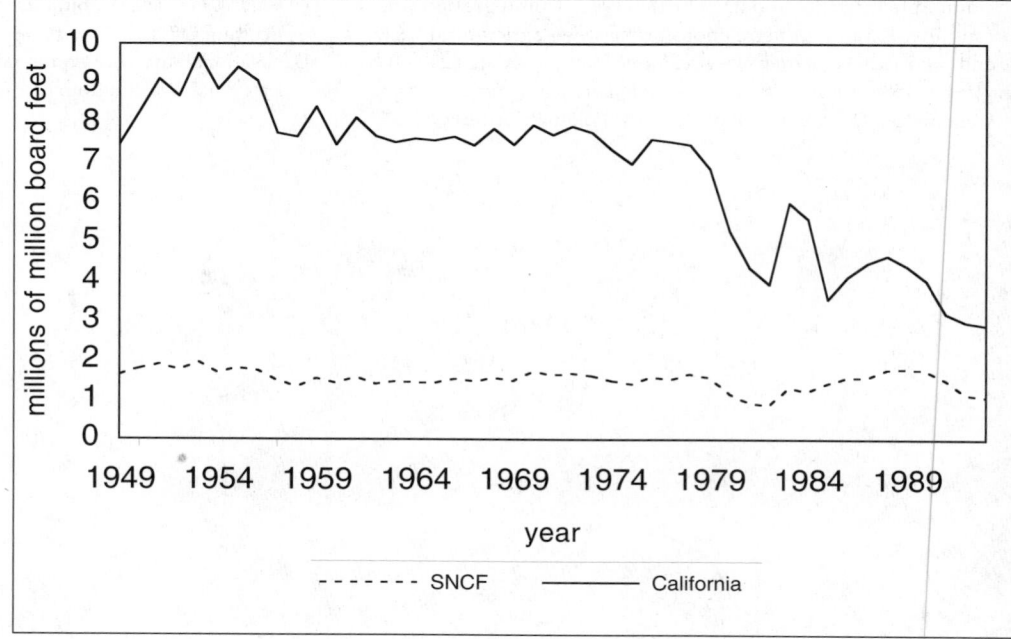

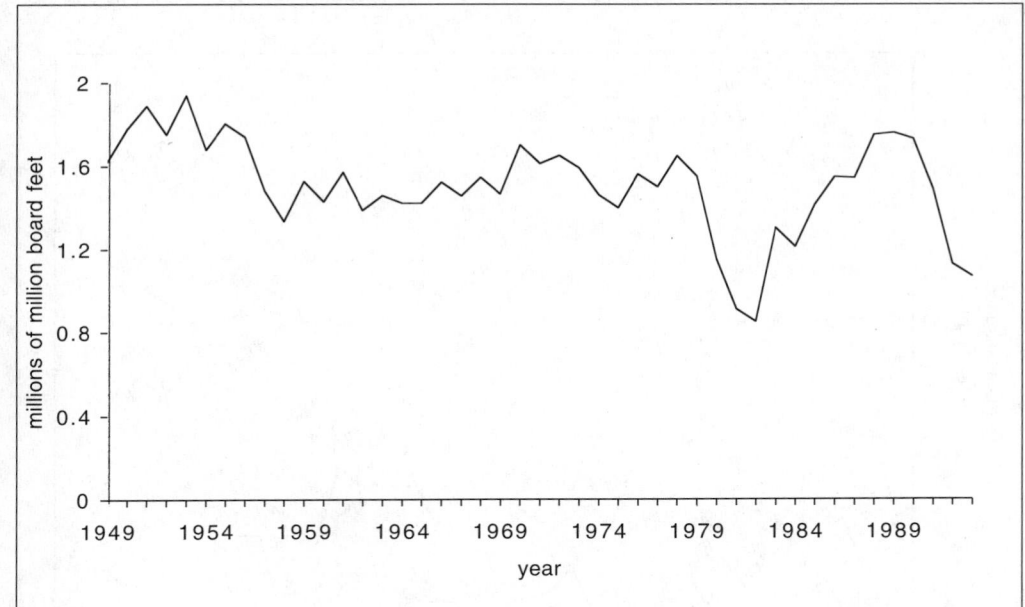

FIGURE 14.A2

Timber harvest for SNFC.

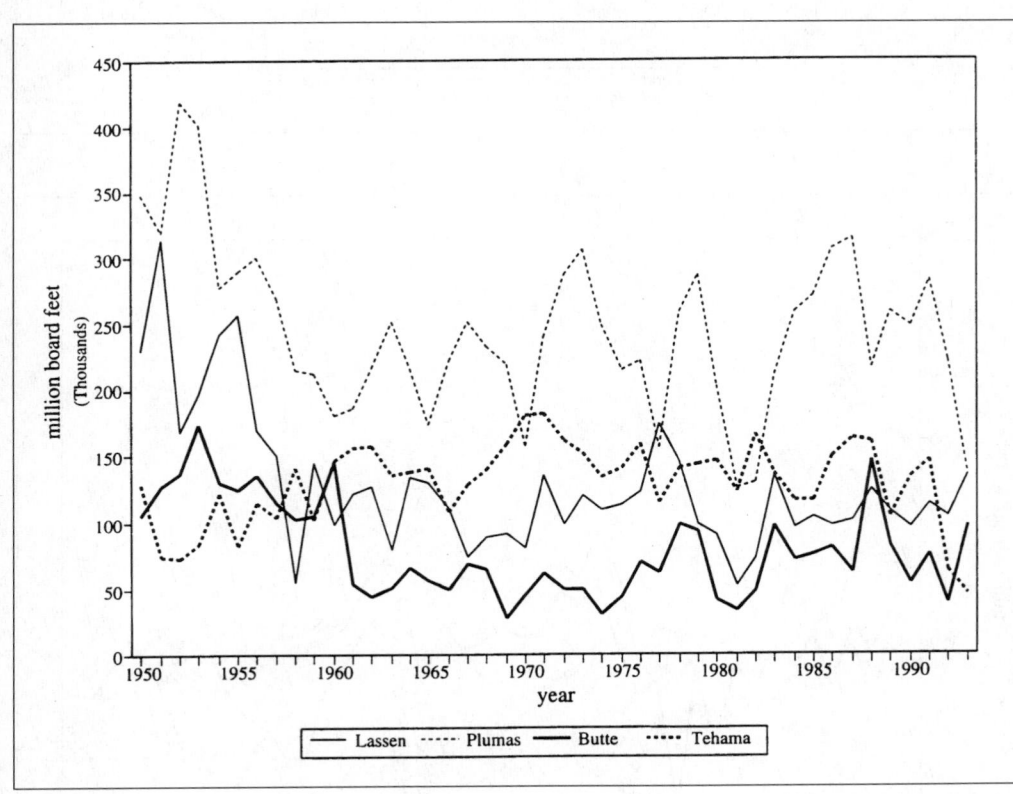

FIGURE 14.A3

Timber harvest 1949–93 for Lassen, Plumas, Butte, and Tehama counties.

FIGURE 14.A4

Timber harvest 1949–93 for
Sierra, Nevada, Yuba, and
Placer counties.

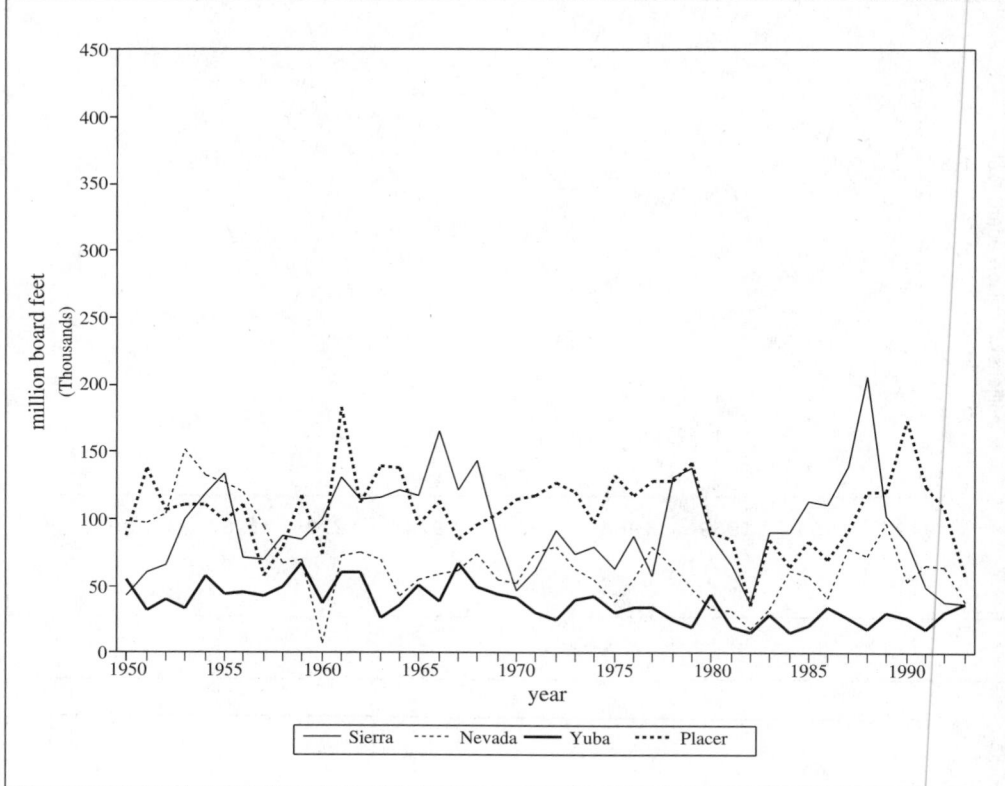

FIGURE 14.A5

Timber harvest 1949–93 for
El Dorado, Amador,
Tuolumne, and Calaveras
counties.

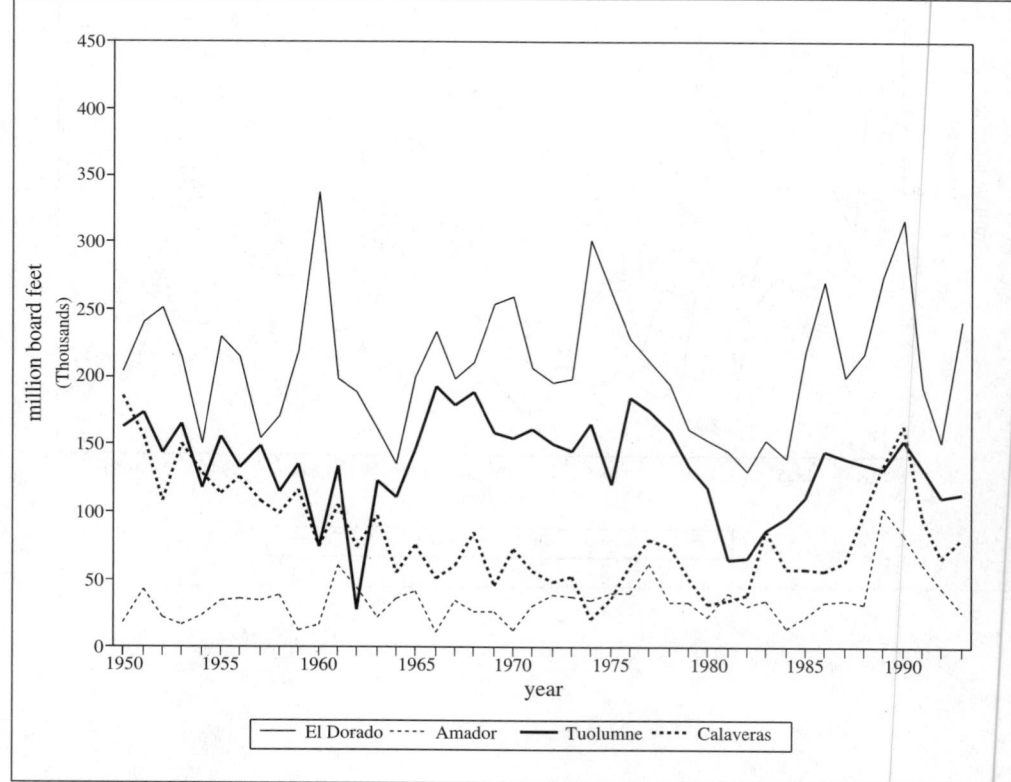

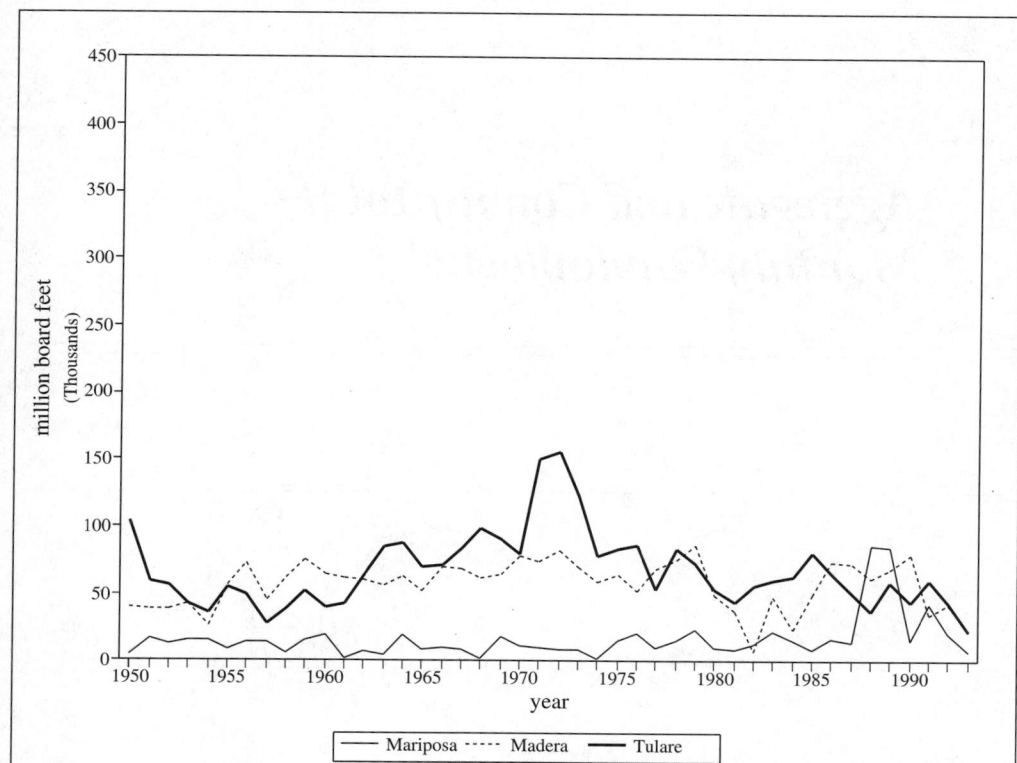

FIGURE 14.A6

Timber harvest 1949–93 for Mariposa, Madera, and Tulare counties.

APPENDIX 14.2

Aggregate and County Total Monthly Employment

FIGURE 14.A7

Total monthly employment for SNFC.

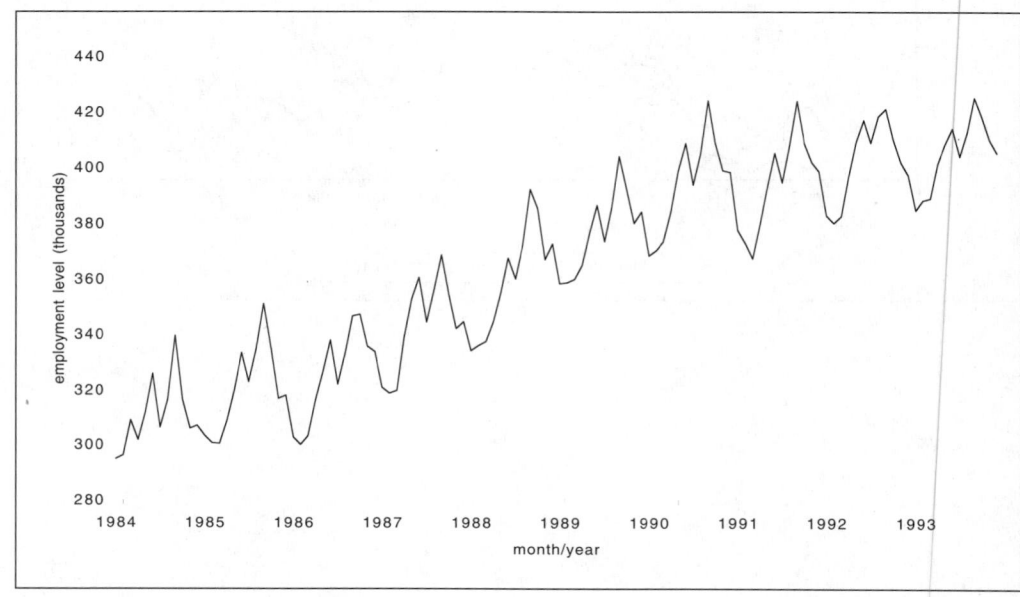

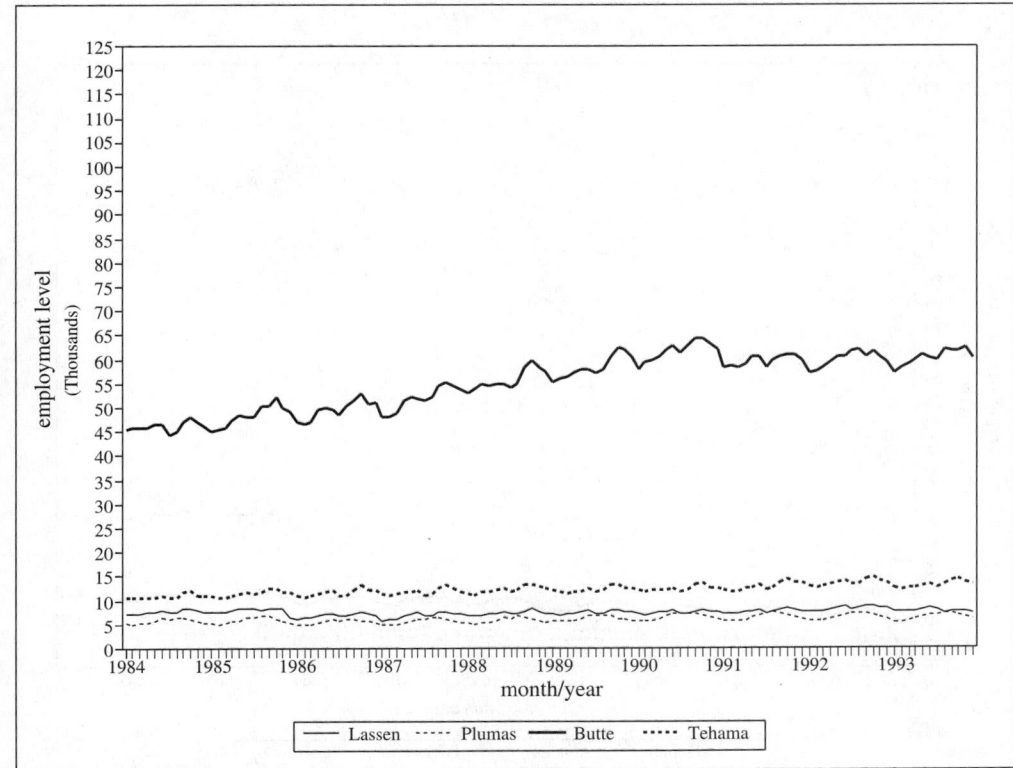

FIGURE 14.A8

Total monthly employment for Lassen, Plumas, Butte, and Tehama counties.

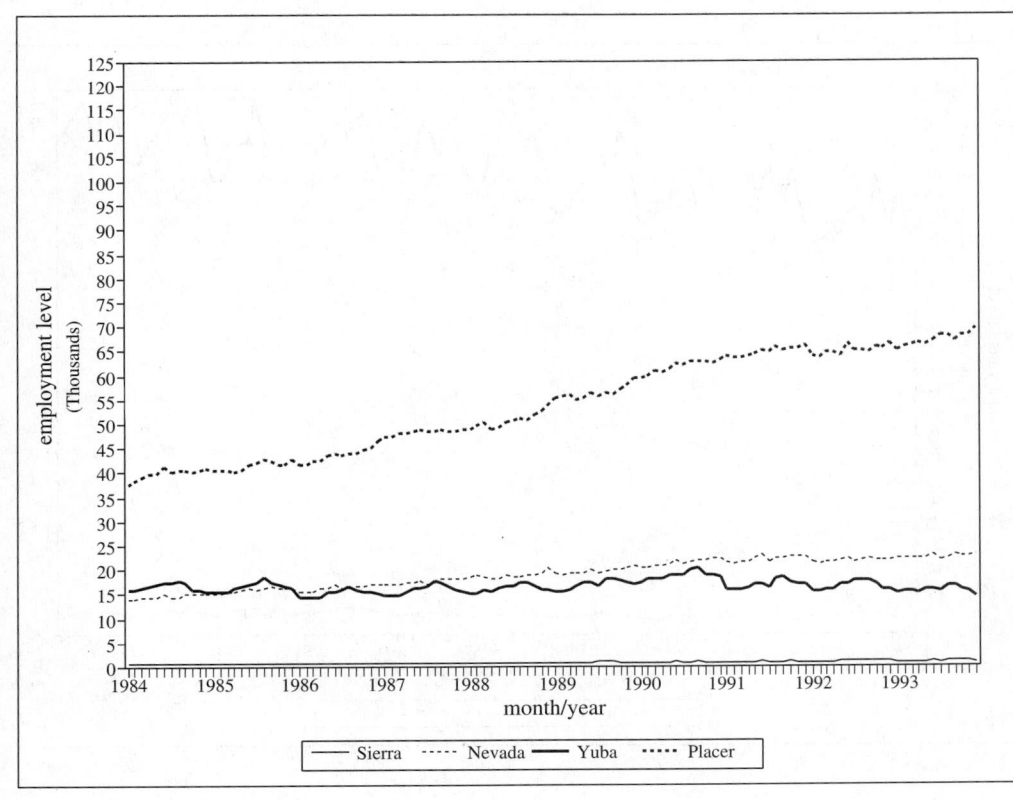

FIGURE 14.A9

Total monthly employment for Sierra, Nevada, Yuba, and Placer counties.

FIGURE 14.A10

Total monthly employment for El Dorado, Amador, Tuolumne, and Calaveras counties.

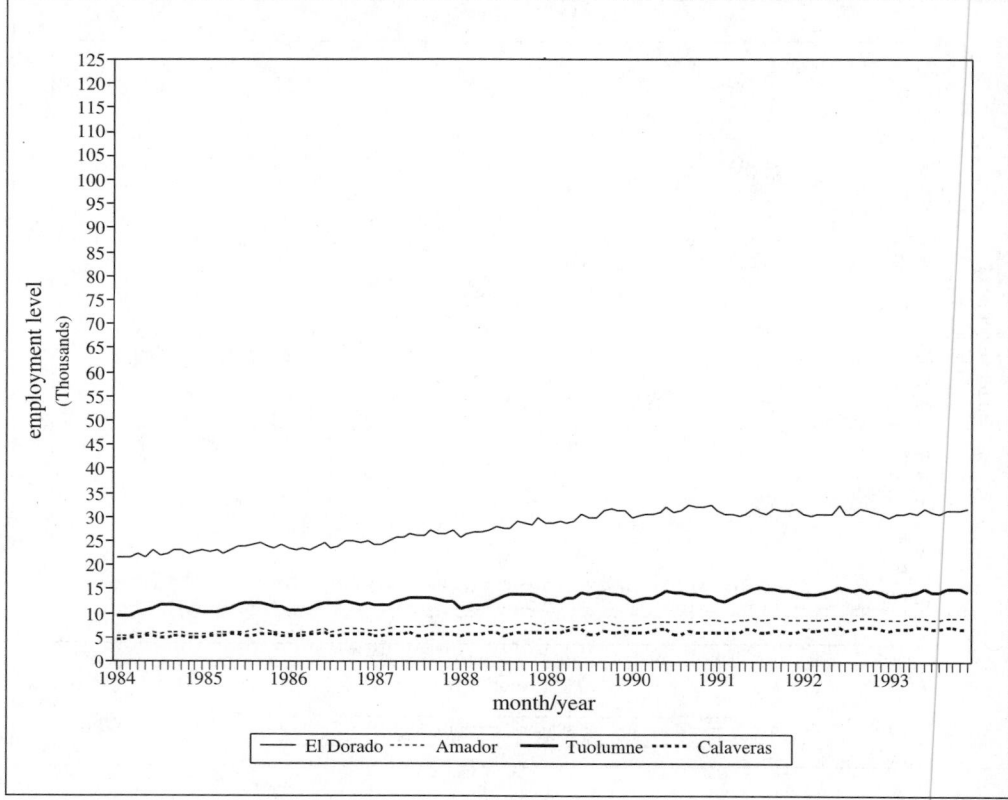

FIGURE 14.A11

Total monthly employment for Mariposa, Madera, and Tulare counties.

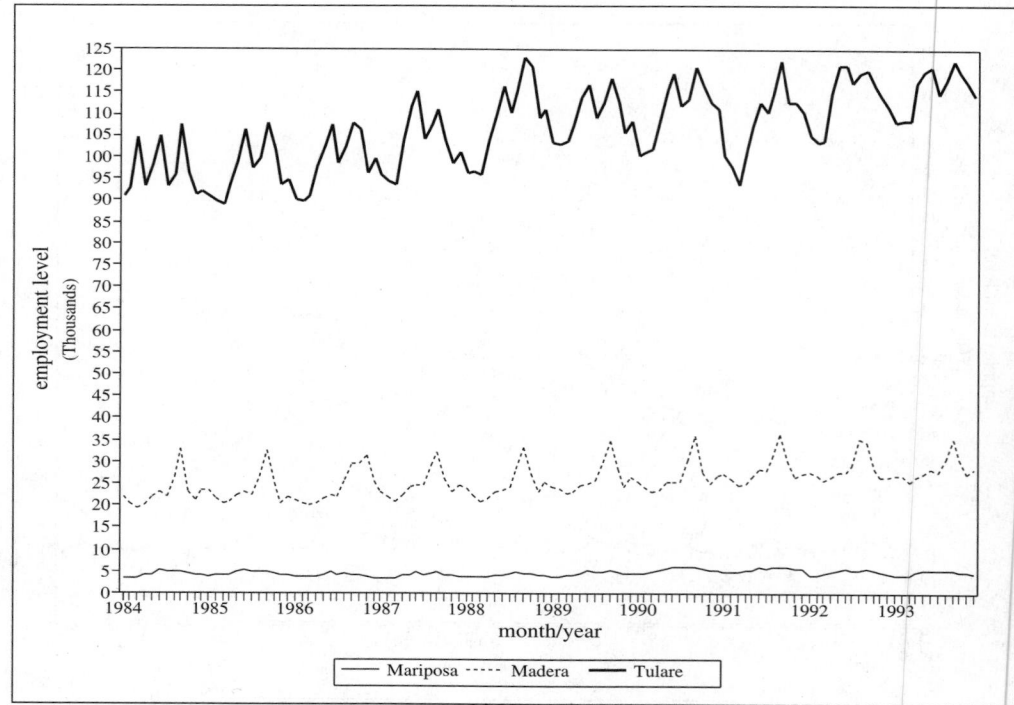

Aggregate and County Lumber and Wood-Products Employment

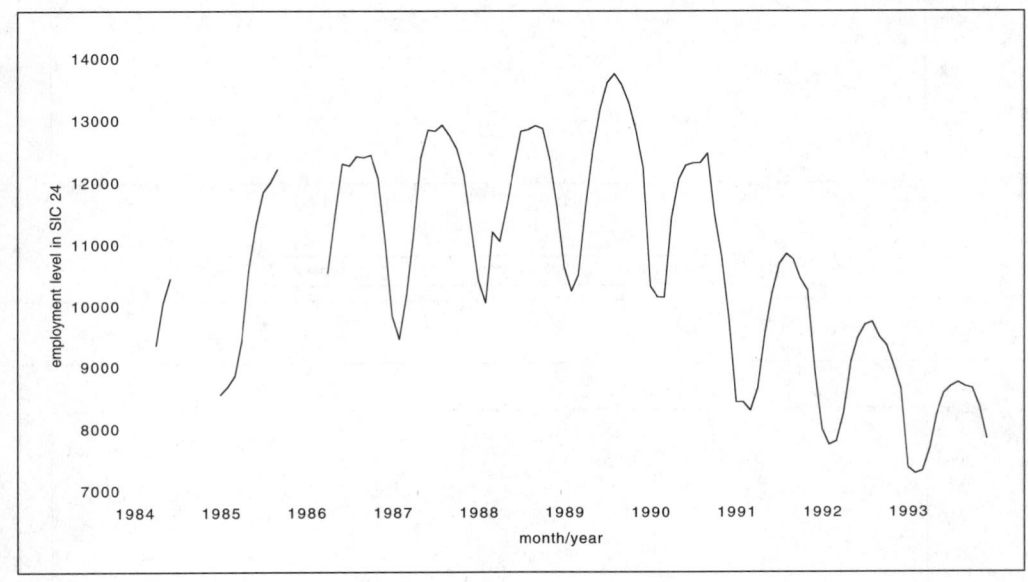

FIGURE 14.A12

Lumber and wood-products employment for SNFC.

FIGURE 14.A13

Lumber and wood-products employment for Lassen, Plumas, Butte, and Tehama counties.

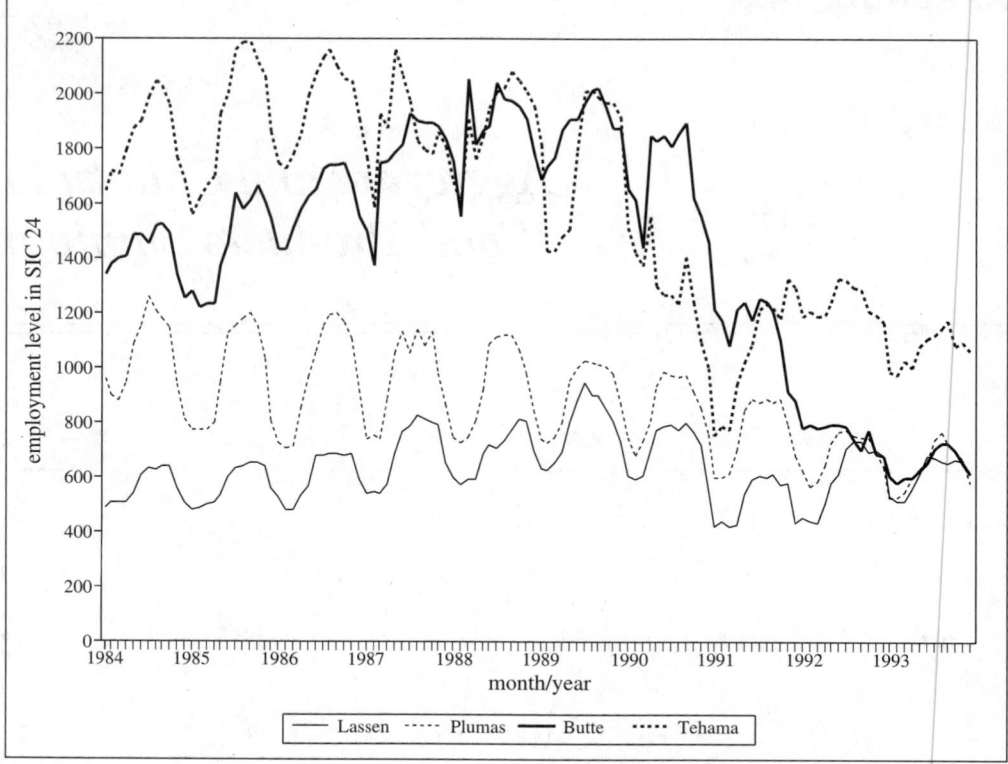

FIGURE 14.A14

Lumber and wood-products employment for Sierra, Nevada, Yuba, and Placer counties.

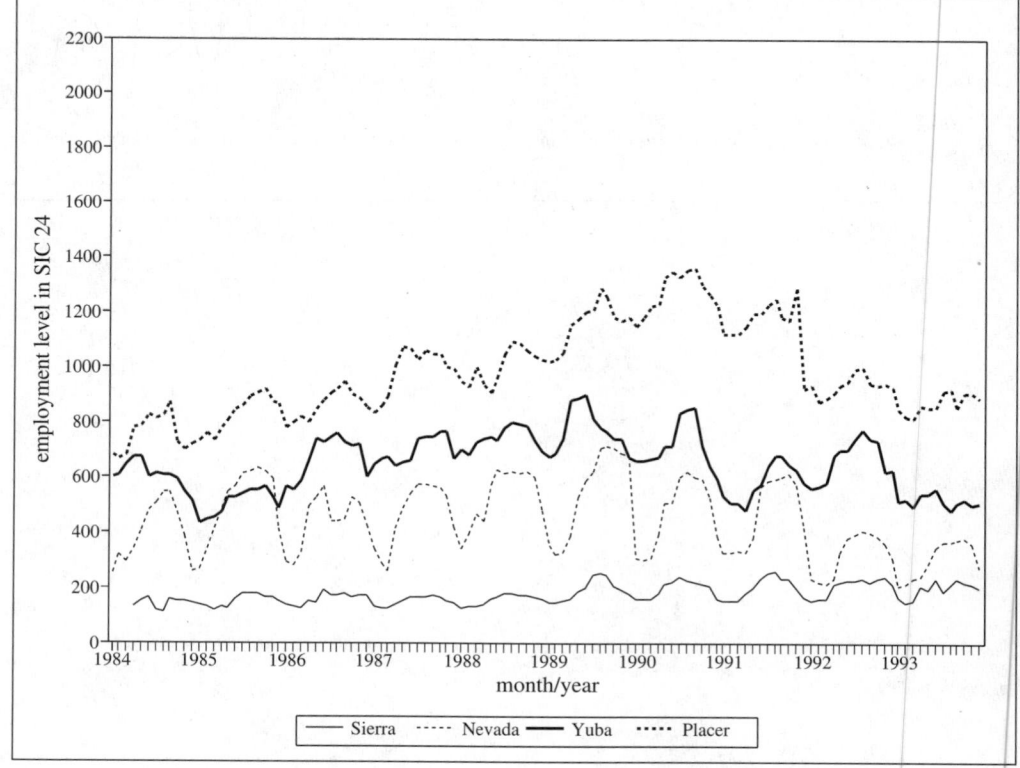

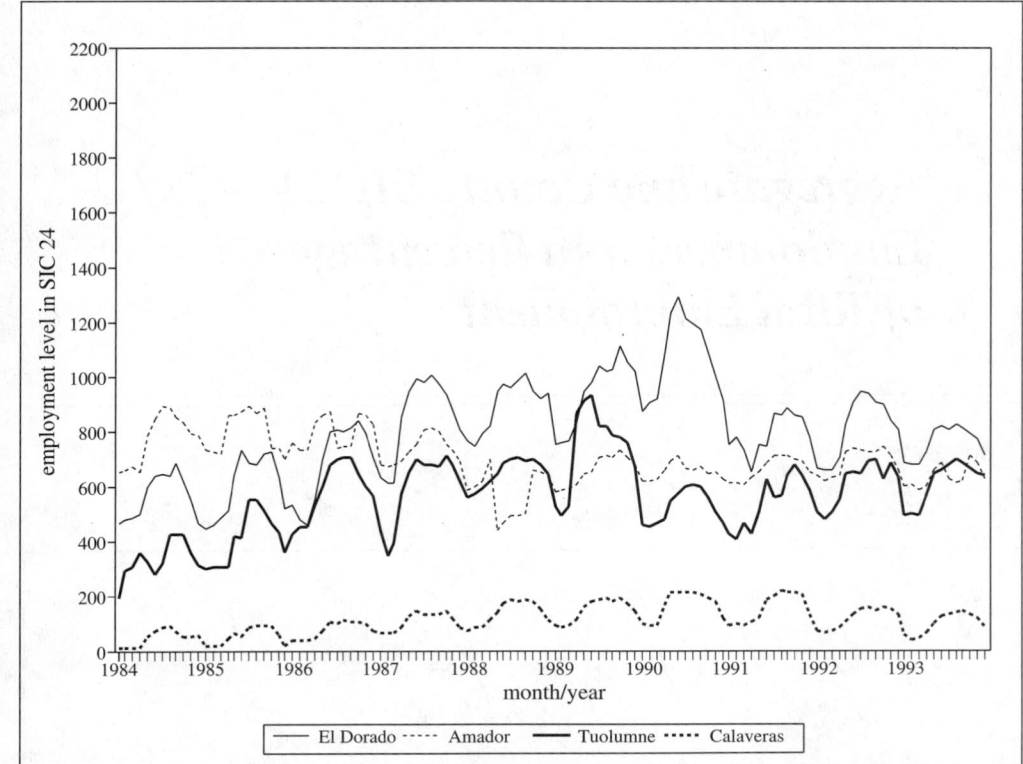

FIGURE 14.A15

Lumber and wood-products employment for El Dorado, Amador, Tuolumne, and Calaveras counties.

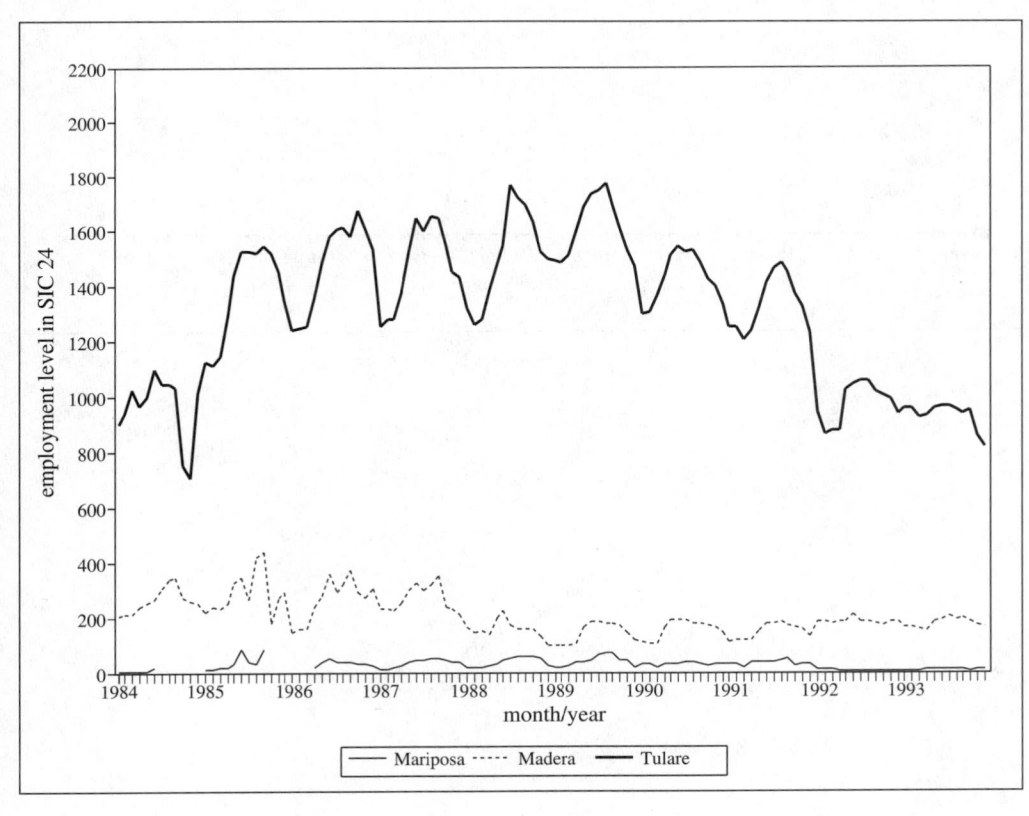

FIGURE 14.A16

Lumber and wood-products employment for Mariposa, Madera, and Tulare counties.

Aggregate and County SIC 24 Employment as a Percentage of Total Employment

FIGURE 14.A17

Total SNFC SIC 24 employment as a percentage of total SNFC employment.

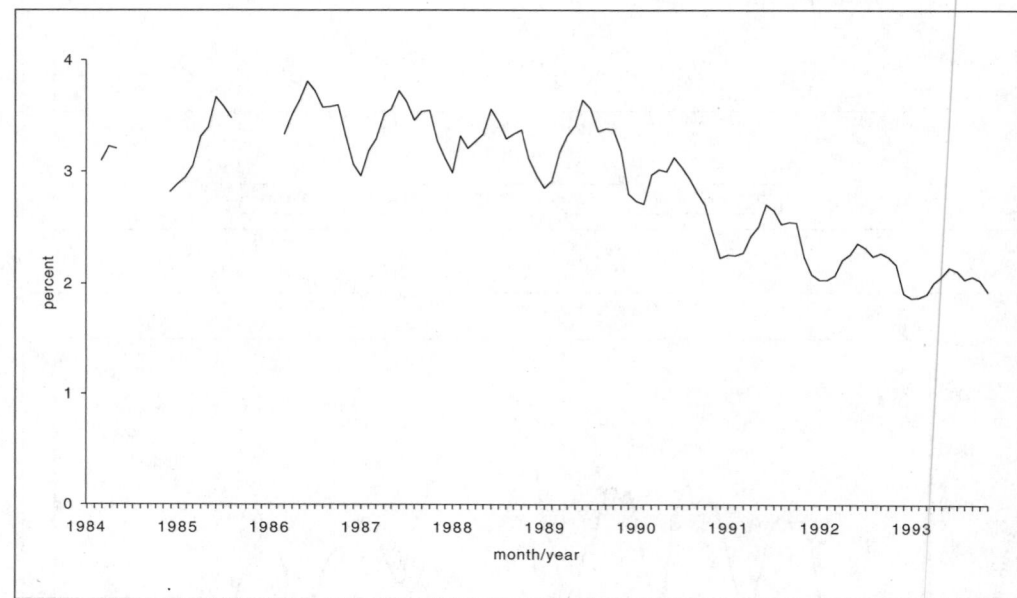

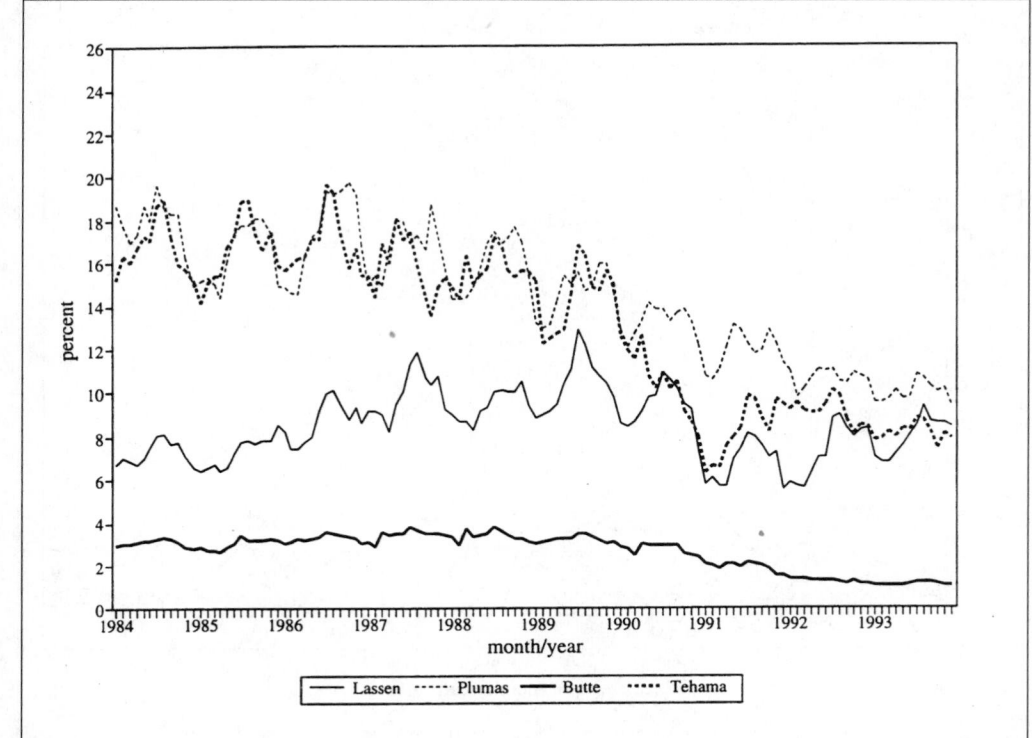

FIGURE 14.A18

SIC 24 employment as a percentage of total county employment for Lassen, Plumas, Butte, and Tehama counties.

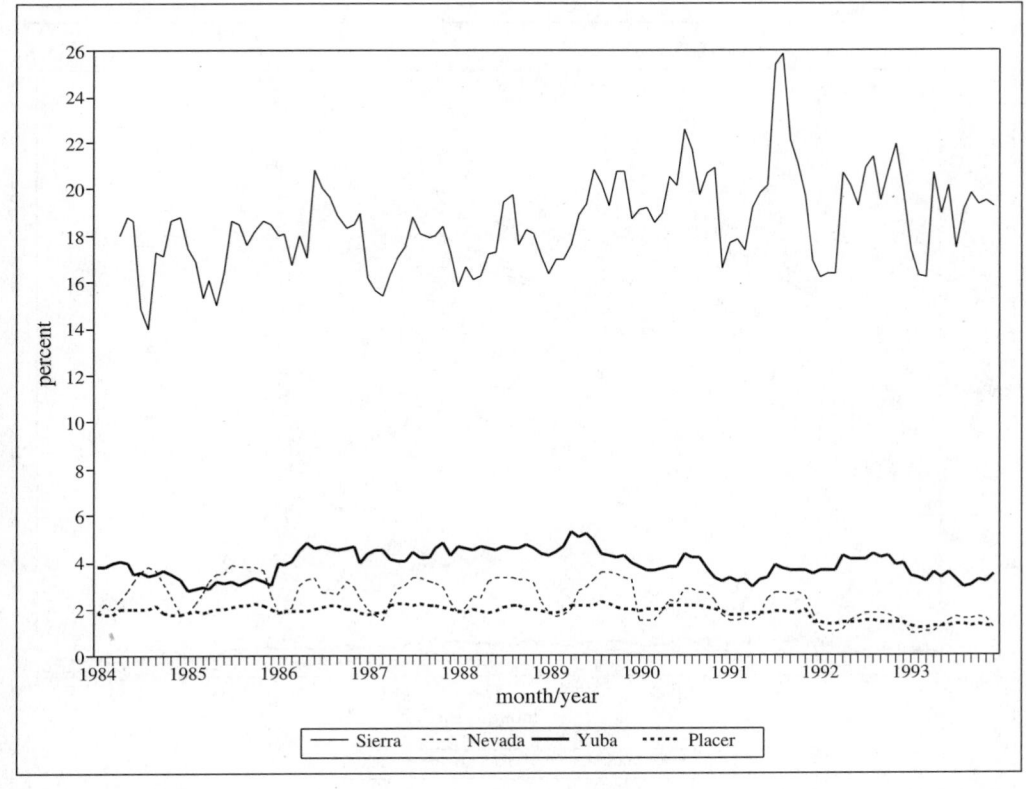

FIGURE 14.A19

SIC 24 employment as a percentage of total county employment for Sierra, Nevada, Yuba, and Placer counties.

FIGURE 14.A20

SIC 24 employment as a percentage of total county employment for El Dorado, Amador, Tuolumne, and Calaveras counties.

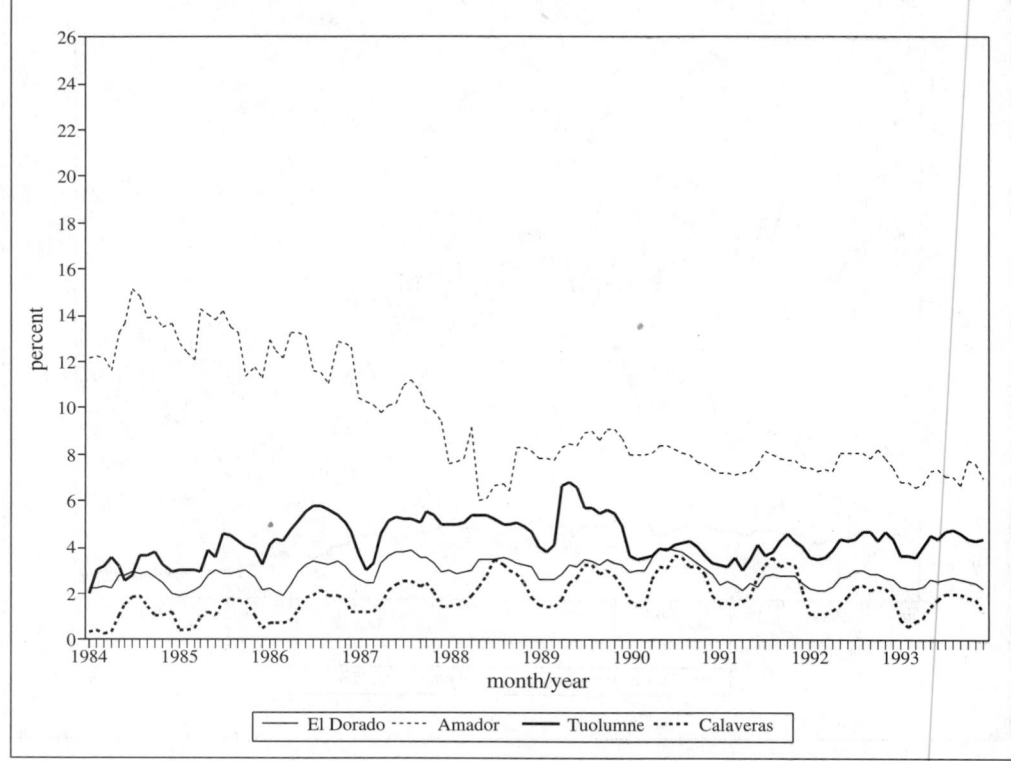

FIGURE 14.A21

SIC 24 employment as a percentage of total county employment for Mariposa, Madera, and Tulare counties.

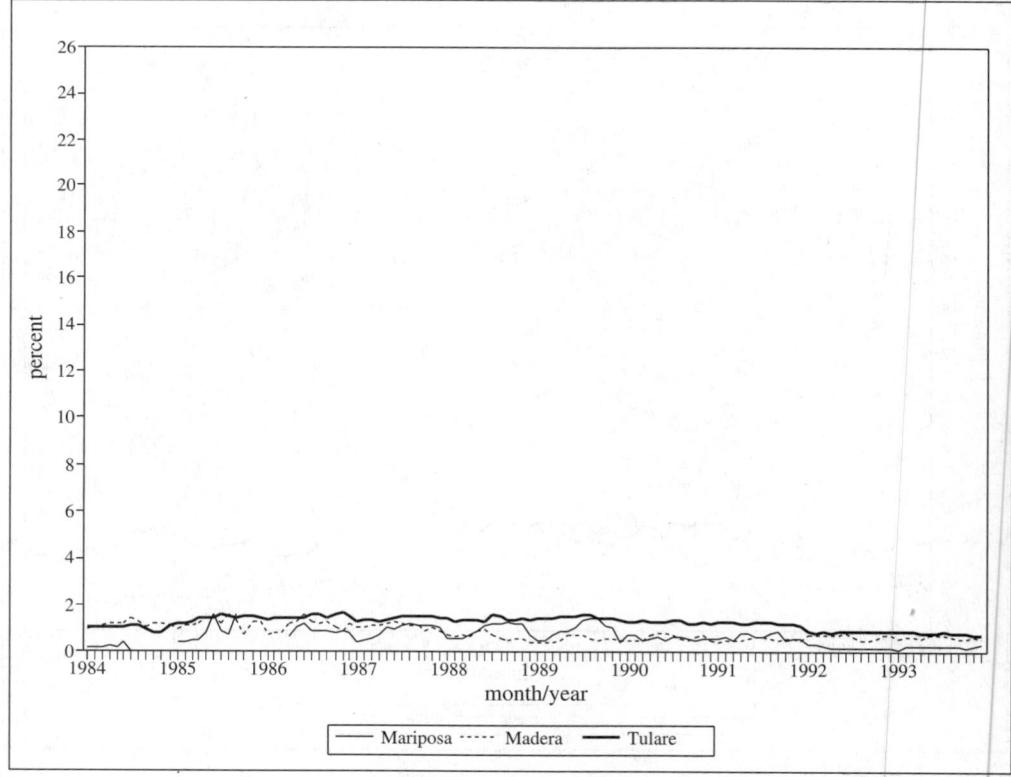

APPENDIX 14.5

Aggregate and County AFDC Unemployed Parent Caseload

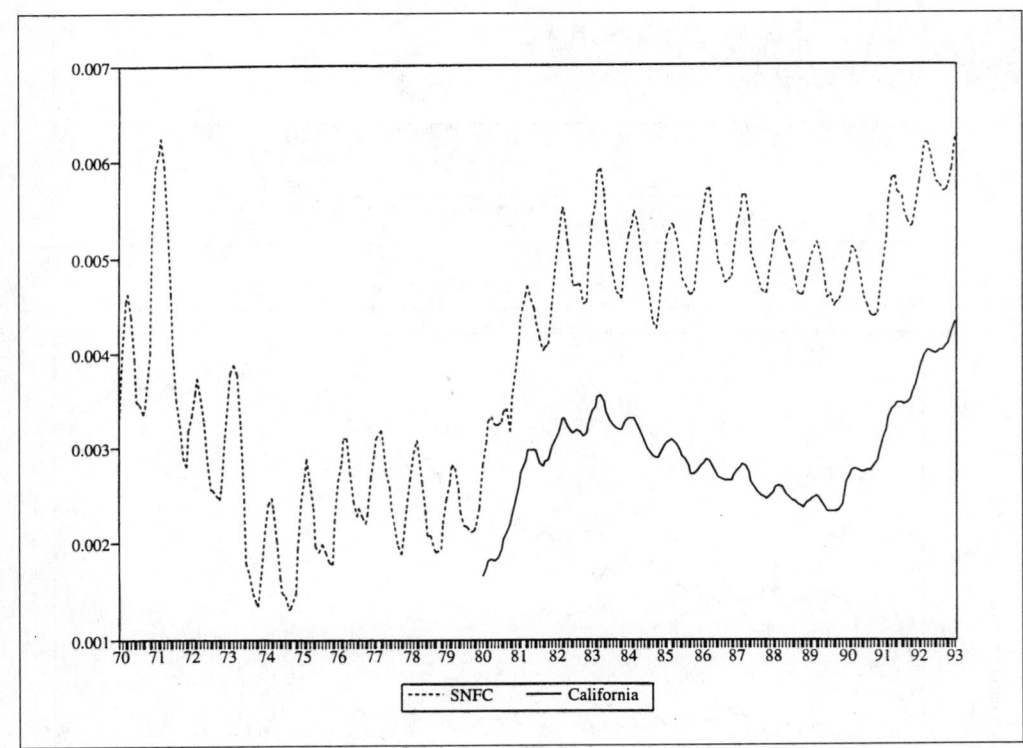

FIGURE 14.A22

AFDC Unemployed Parent program cases per capita for SNFC and state.

FIGURE 14.A23

AFDC Unemployed Parent
program cases per capita for
Lassen, Plumas, Butte, and
Tehama counties.

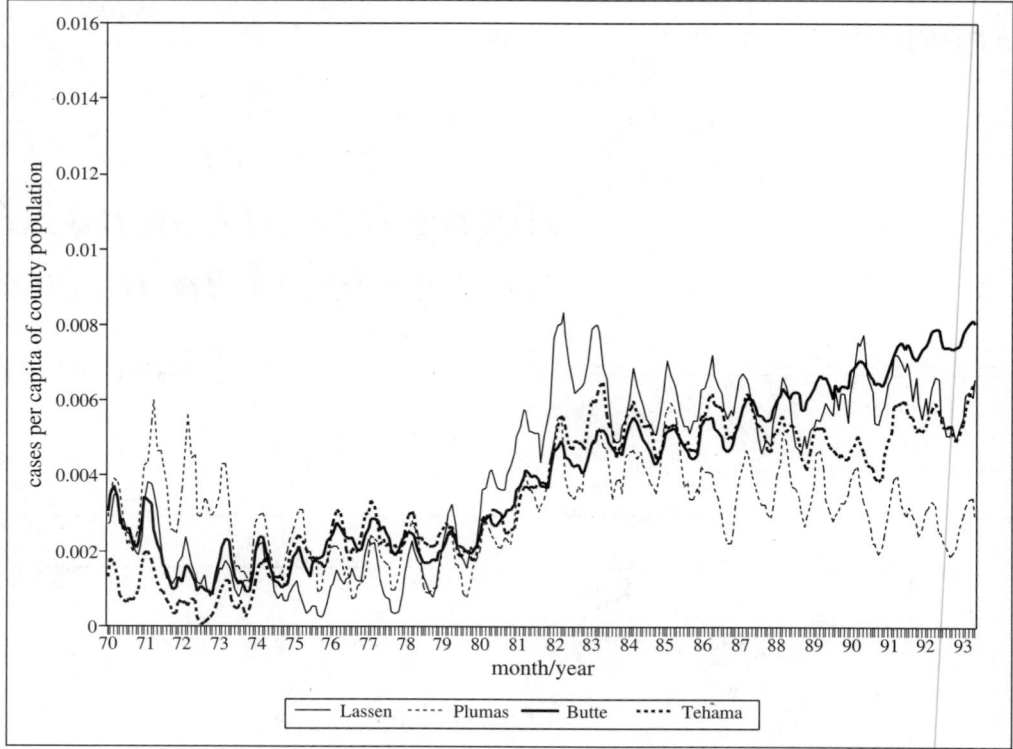

FIGURE 14.A24

AFDC Unemployed Parent
program cases per capita for
Sierra, Nevada, Yuba, and
Placer counties.

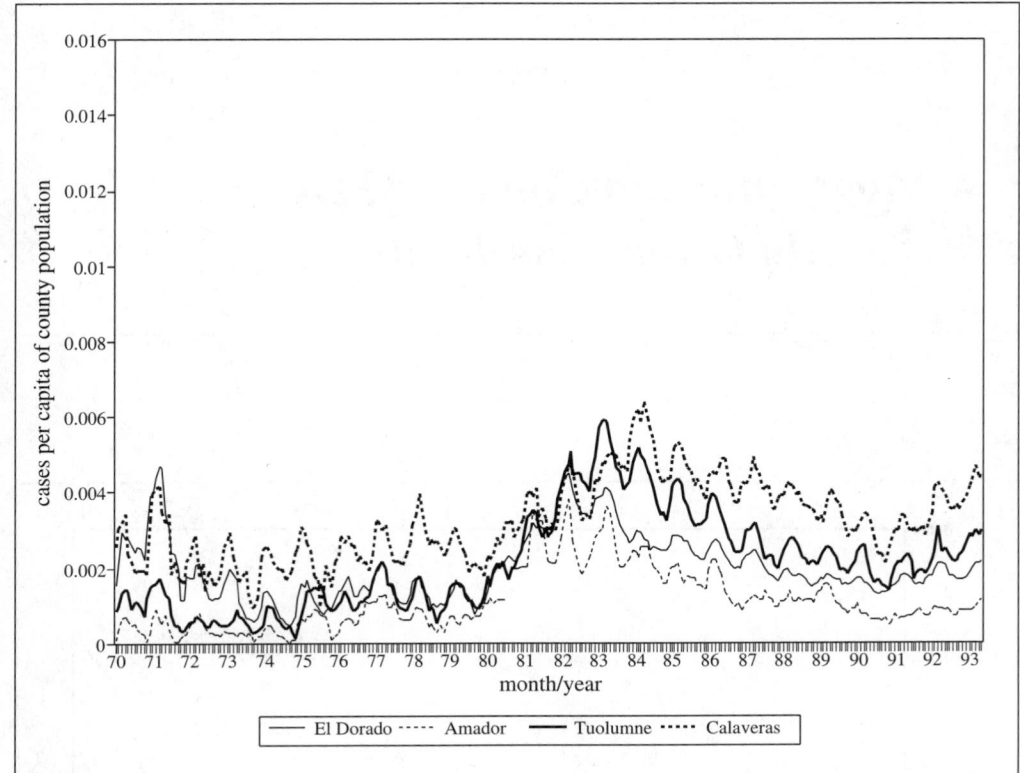

FIGURE 14.A25

AFDC Unemployed Parent program cases per capita for El Dorado, Amador, Tuolumne, and Calaveras counties.

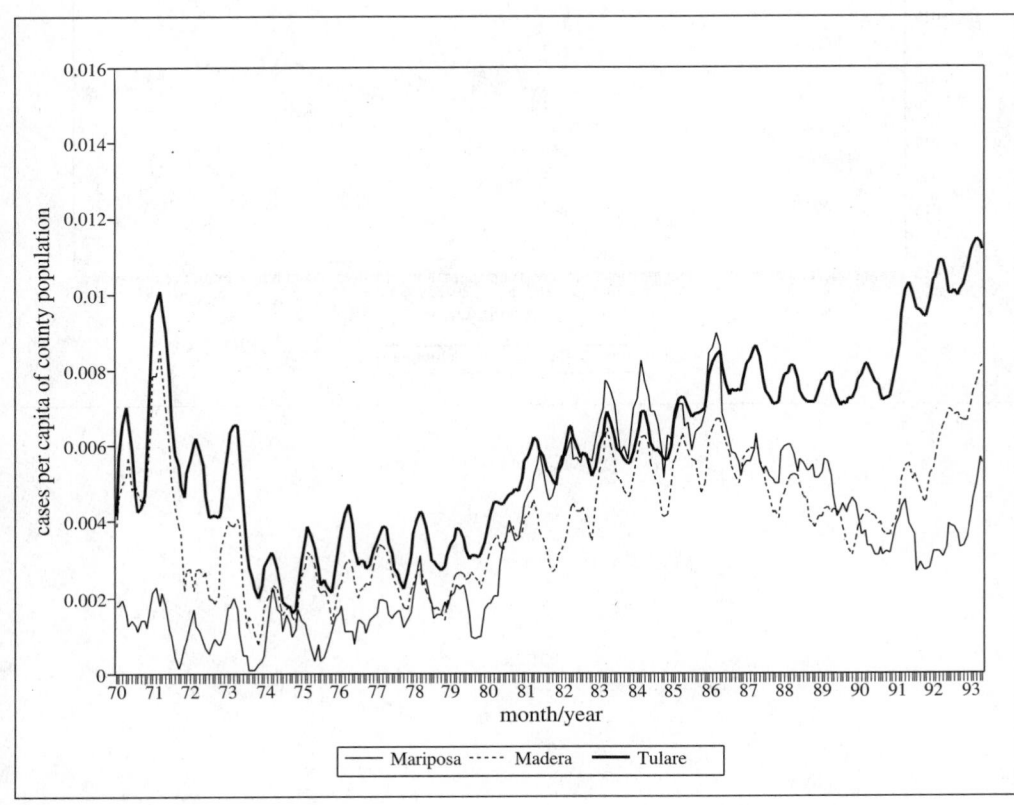

FIGURE 14.A26

AFDC Unemployed Parent program cases per capita for Mariposa, Madera, and Tulare counties.

Aggregate and County AFDC Family Group Caseload

FIGURE 14.A27

AFDC Family Group program cases per capita for SNFC and state.

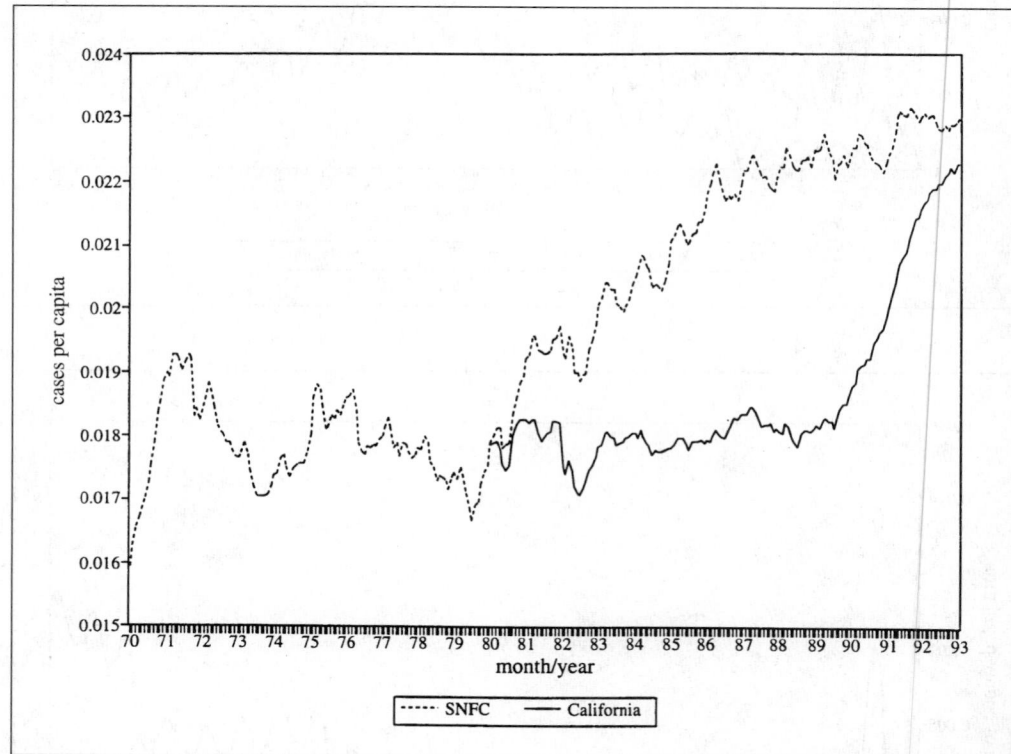

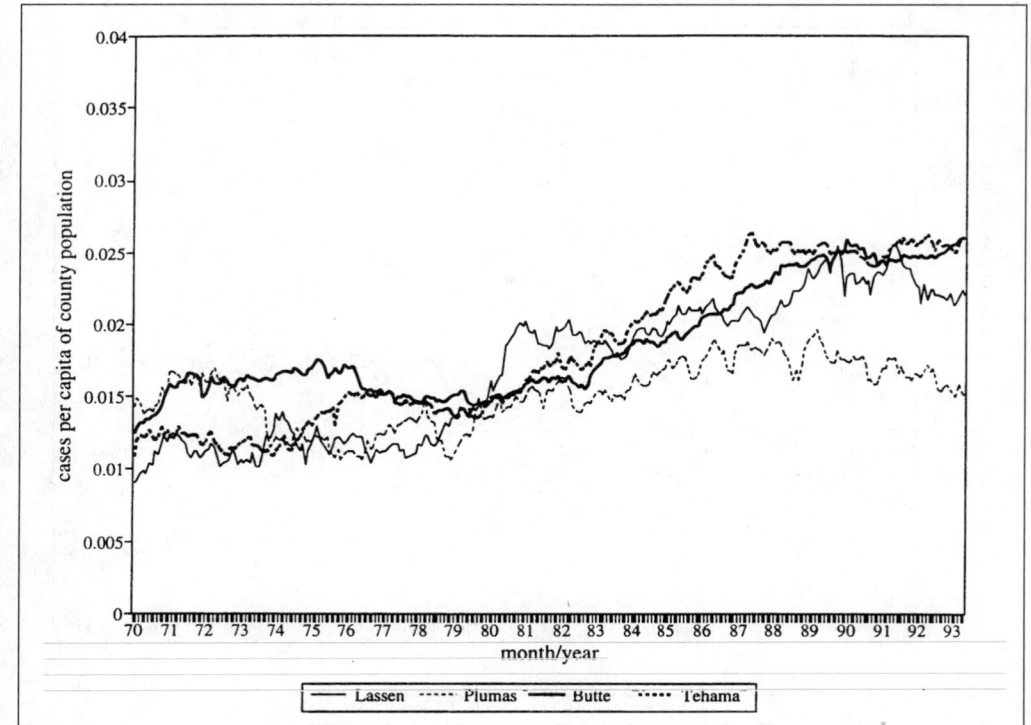

FIGURE 14.A28

AFDC Family Group program cases per capita for Lassen, Plumas, Butte, and Tehama counties.

FIGURE 14.A29

AFDC Family Group program cases per capita for Sierra, Nevada, Yuba, and Placer counties.

FIGURE 14.A30

AFDC Family Group program cases per capita for El Dorado, Amador, Tuolumne, and Calaveras counties.

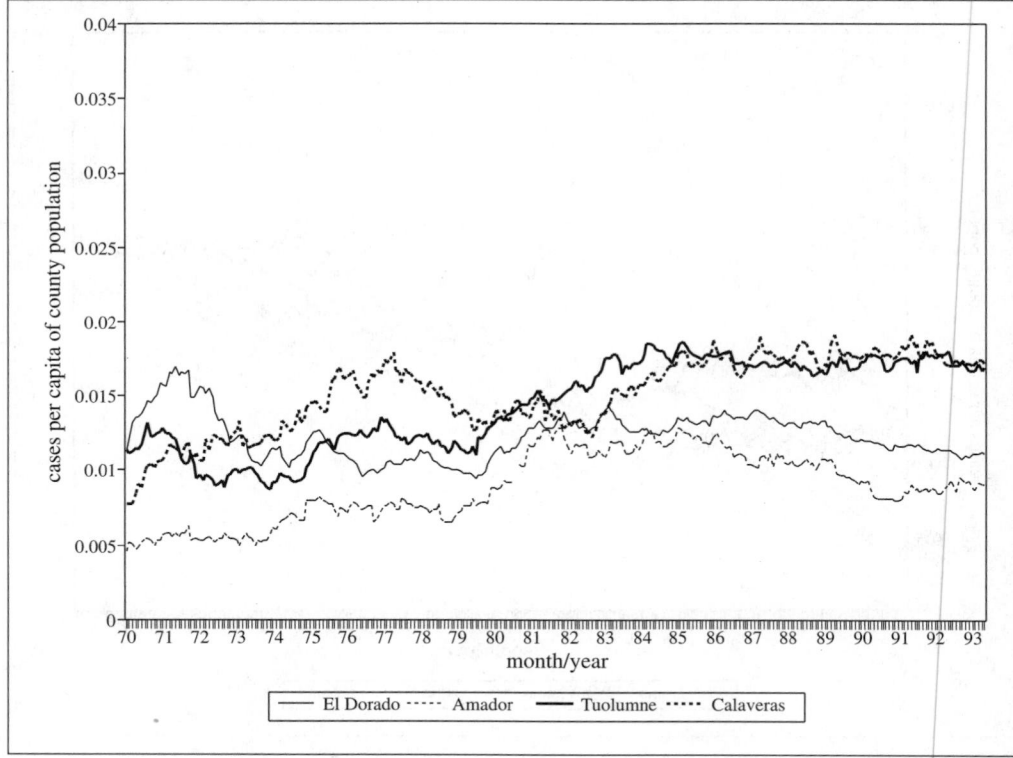

FIGURE 14.A31

AFDC Family Group program cases per capita for Mariposa, Madera, and Tulare counties.

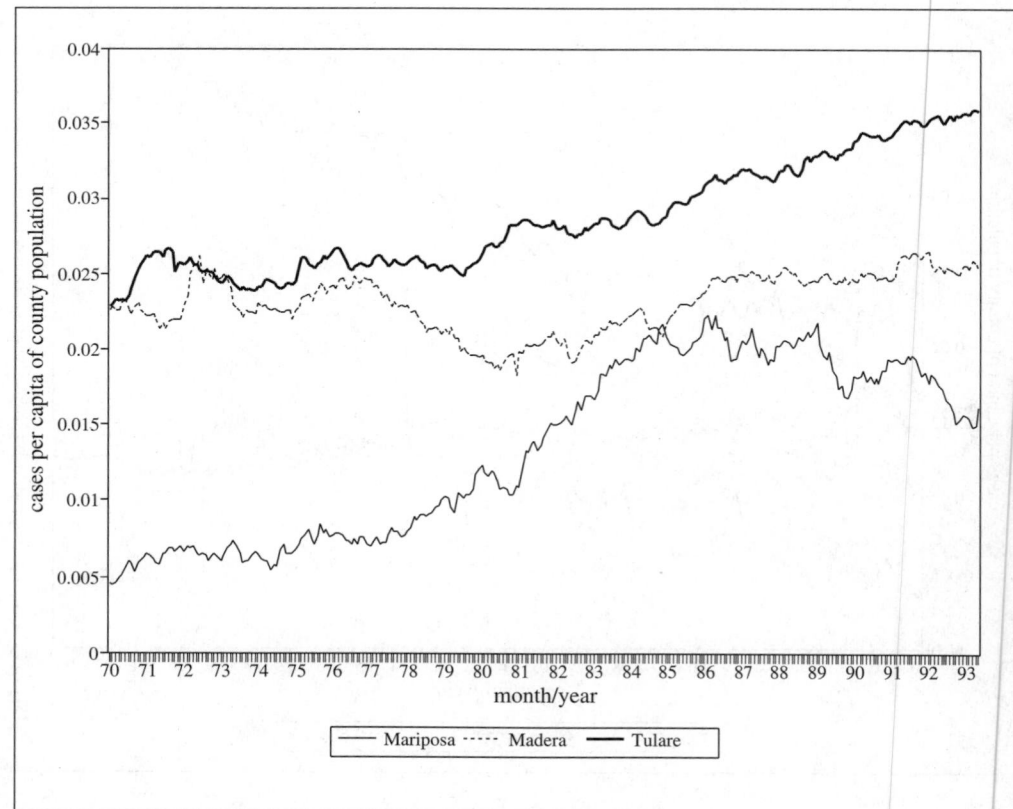

JOHN A. HELMS
University of California
Berkeley, California

JOHN C. TAPPEINER
Oregon State University
Corvallis, Oregon

15

Silviculture in the Sierra

ABSTRACT

This chapter summarizes for general readers the state of silvicultural knowledge of Sierra Nevadan forests. It has sections covering presettlement conditions, a historical overview of silvicultural and harvesting practices over the past 100 years, and discussions of silviculture and silvicultural systems. Summaries of current silvicultural practices and their effects on forest systems are based on especially commissioned reviews (attached as appendixes) by specialists in each field, including sections covering silvicultural aspects of forest soils, stand density, regeneration, vegetation management, and the current status of Forest Service plantations. A particularly important section summarizes what has been learned from three long-term studies of silvicultural treatments in the Sierra Nevada. The most significant of these is the thirty-year database from the University of California's Blodgett Forest Research Station. Additional short sections deal with silvicultural prescriptions on federal, state, and private lands. The chapter concludes with discussions of major factors and issues affecting silviculture, management, and policy analysis of Sierran forests, and identification of gaps in our current knowledge.

INTRODUCTION

This chapter summarizes the state of knowledge of Sierra Nevada silviculture for general readers. This knowledge is used by silviculturalists (1) for the science-based development of silvicultural prescriptions designed to meet a diverse array of specific management objectives within a landscape, and (2) as a basis for policy analysis and development.

Because of their mixed species composition, range of age classes, and spatial diversity, the forests of the Sierra Nevada provide broad opportunities for sustaining a wide variety of values for society and landowners. The forests of the Sierra Nevada are much more variable than the more uniform forests in the Pacific Northwest and have experienced many more harvesting and management strategies. Consequently, after nearly 150 years of use, the Sierran forests remain remarkable for their current diversity of structure and composition.

Sierran forests change in elevation from oak woodland in the foothills, to foothill pine and ponderosa pine at intermediate elevations to mixed conifer forests (consisting primarily of ponderosa pine, Douglas fir, sugar pine, California white fir, incense cedar, tan oak, and California black oak), true fir forests (red and California white fir), and lodgepole pine forests and finally to subalpine forests (whitebark, foxtail, and limber pines) at timberline (figure 15.1). The elevations at which one forest type makes a transition to another are lower in the northern Sierra than in the south.

Due to their characteristic diversity in vegetative cover, topography, and climate, Sierran forests provide a remarkable diversity of values and uses for society. They are sources of water, wildlife, timber, and aesthetic beauty, and they provide society with sites for urban communities, with jobs, and with recreation. But on a broad scale, landscape-level management and policy development are made extremely difficult by a complex mix of ownership and jurisdictional boundaries that includes federal, industrial, and small, private, nonindustrial landowners.

All forest stands are dynamic—they are continually changing in age, size, structure, and species composition due to natural and human-induced factors. Silviculture is the mixture of art and science that is concerned with the regeneration of forest stands and the development of composition and structure to provide goods, services, and values for society. Although the basic ecological requirements for the various combinations of plants, animals, insects, and diseases remain relatively stable, stands are dynamic and, as landowner and societal needs change, silvicultural practices must be tailored to meet existing stand conditions and changing management

Sierra Nevada Ecosystem Project: Final report to Congress, vol. II, *Assessments and scientific basis for management options.* Davis: University of California, Centers for Water and Wildland Resources, 1996.

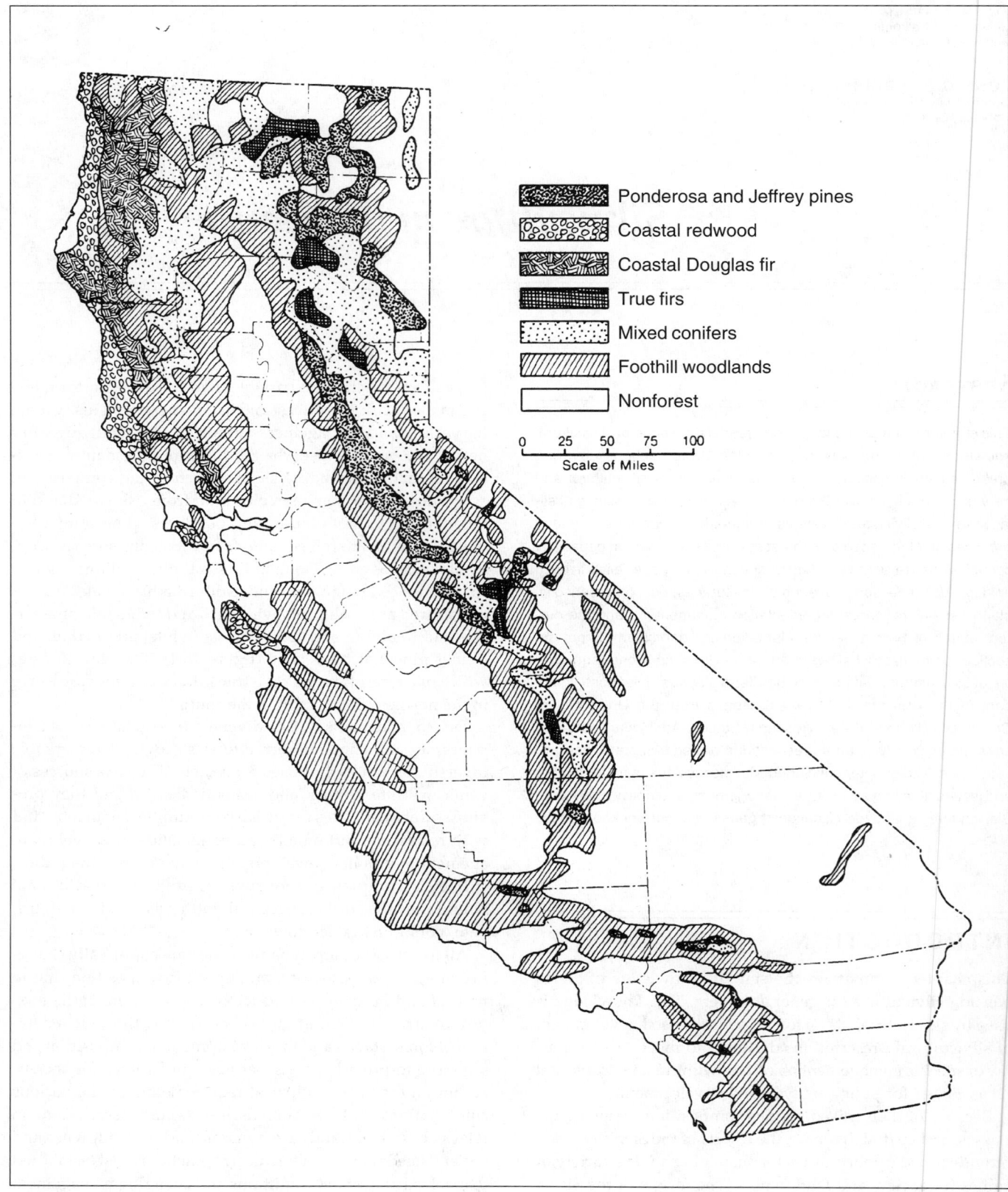

FIGURE 15.1

Distribution of forest types in California (adapted from Bolsinger 1980).

objectives. The effects of those silvicultural practices occur rapidly on sites of moderate to high productivity but take a longer time to become evident on sites of lower productivity.

There are numerous ecological problems confronting management and policy makers in the Sierra Nevada for which silviculture can offer some solutions. These include forest health; the development of a higher proportion of shade-tolerant species such as California white fir and incense cedar; the effects of wildfire suppression, including a buildup of fuels; and the sustainability of current forests and their capacity to meet the present and future needs of California and the nation.

This chapter provides historical background, presents what is known silviculturally about various forest types in the Sierra, and comments on the important current and emerging silvicultural issues. Emphasis is placed on the mixed conifer forest because of its relative importance and because of the greater abundance of knowledge regarding this forest type.

PRESETTLEMENT CONDITION OF SIERRAN FORESTS

The earliest and best-known descriptions of the mixed conifer forest were written by John Muir ([1895] 1977) who described the "inviting openness of the Sierra woods" and noted that their "park-like" condition enables one to have "little difficulty in riding on horseback through successive belts" to the peaks. The first quantitative descriptions were provided by Sudworth (1900, 1901), who measured fifty-three 0.10 ha (0.25 acre) plots in various parts of the Sierra (McKelvey and Johnston 1992). Seventeen of Sudworth's plots were located in the mixed conifer forests, and these probably provide the best description of "presettlement old-growth" forests, even though some had been subjected to partial cutting, fire, and grazing and were probably not representative of average conditions. An analysis of these plots shows that basal area (the total cross-sectional area of all trees per unit of land area) in the more southern group was considerably higher and tree size was larger (267 trees per hectare [108 trees per acre], 264 m^2/ha [1,148 ft^2/acre], quadratic mean diameter 114.3 cm [45.0 in]) than in the more northern group in the mid-Sierra (250 trees per hectare [101 trees per acre], 165 m^2/ha [719 ft^2/acre], ranging between 93 and 230 m^2/ha [406 and 1,004 ft^2/acre], and quadratic mean diameter 98.8 cm [38.9 in]). In both areas, the proportion of basal area in pines was 40% to 50%, with about 30% in either Douglas fir or California white fir and the remaining 20% in incense cedar. Additional evidence of open and parklike conditions is available in photographic records of Collins Almanor Forest taken in 1924 (Ford 1991) and from a comparison of photographs taken in the late 1800s with those taken recently (Gruell 1994). It appears from Sudworth's and Muir's descriptions and from early photo-

graphs that the original Sierran mixed conifer forests at the turn of the century, at least in some locations, were very dense and were composed of many large, old trees. These largest trees were not only pines but also Douglas fir, California white fir, and incense cedar.

The presettlement red fir forests were essentially dense monocultures of even-sized (although not necessarily even-aged) trees and were nearly devoid of understory. Because these stands were relatively inaccessible and were regarded as having lower value for wood products, they were largely ignored and unexploited. Due to their location at high elevation, the cold temperature and long periods of snow cover prevented downed woody material from decomposing as rapidly as in the lower-elevation forests. Consequently, original red fir stands were, and still are, characterized by heavy loads of dead and down fuel.

EVOLUTION OF CALIFORNIA SILVICULTURE IN THE SIERRA OVER THE PAST 150 YEARS

1850 to 1925

The first effects European settlers had on Sierran forests were associated with mining, logging, and grazing. Forests were regarded as inexhaustible, and cleared land had more value than forested land. Silviculture and forest management were irrelevant. Initially, the trees harvested were those that were most valuable, that were near rivers, or that were accessible by oxen. Later, with the advent of railroads and ground-skidding with cables, more distant stands could be harvested. Forests removed by logging or fire were replaced by either conifer regeneration or brush fields. Increasing populations of settlers greatly reduced the amount of old growth and changed the character of remaining stands through extensive "high-grade" logging (i.e., logging in which only the best trees of the most valuable species are cut), fire, and grazing.

1925 to 1979

One of the first silviculturists in California was Duncan Dunning, who was employed by the U.S. Forest Service (USFS) and had the responsibility of determining how to manage the mixed conifer forests in order to enhance their health and productivity. In 1923, Dunning reported that

the situation confronting the forester was a very difficult one. As a result of early fires, insect attacks, and grazing, the forests were usually under-stocked with a preponderance of mature and decadent timber, a deficiency of intermediate-age classes from which to select thrifty reserves, younger trees poorly distributed or stagnating in groups, and reproduction frequently absent or composed of undesirable species. The stands were often invaded by brush and "bear clover" . . . With these conditions prevailing, the forester faced the problem of improving the health of the stand, increasing the rate of growth, and securing more pine reproduction.

A major force shaping timber harvesting and regeneration on private lands in California was section 12¾ of the state constitution. This section enabled landowners who harvested at least 70% of the volume of trees on a unit of land to pay taxes only on the land, rather than on the land and timber, for forty years or until another harvest was made. This form of tax relief on private land resulted in heavy selective cutting and discouraged more modest thinnings. Section 12¾ was repealed in 1973, when the existing Forest Practice Act, which had been in place since 1943, was replaced with new legislation (described later in this chapter).

Just before World War II, technology changed with the advent of chain saws, trucks, and tractors. Changing merchantability standards increasingly permitted the removal of smaller, lower-quality trees and more species. Silviculture was centered primarily on cutting high-risk trees susceptible to insect attack or "sanitation-salvage" trees killed by frequent wildfires and outbreaks of insects and disease. Early silvicultural procedures focused on the development of risk rating systems such as those by Dunning (1928), Keen (1936), and Salman and Bongberg (1942), that identified trees for harvesting that were likely to die within a fifteen- to twenty-year period. Most harvesting in this period on both public and private lands was selective cutting in which large, old trees—particularly sugar pine, ponderosa pine, Jeffrey pine, and, to a lesser extent Douglas fir, California white fir, and red fir—were preferentially removed in order to raise stand vigor and net growth. In addition, some cutting was done to replace existing old stands with regeneration, but this activity was at a low level relative to selective cutting. It was a common practice for timber-holding companies to high-grade the pine timber from their forest lands and exchange the cut-over land with the Forest Service for trees from federal forests. Large wildfires required substantial reforestation efforts.

An early priority in forest management was to clear brush fields that had resulted from natural and human-caused forest fires and early logging, and to plant them to pine. This program required the development of a system of forest tree nurseries capable of producing seedlings that could survive the rain-free summer periods typical of the Sierra. In the 1950s, the Forest Service recognized that mixed conifer forests consisted of mosaics of age classes and structure, each requiring different silvicultural treatment. This led to a concept called unit area control, a new approach to silviculture in the Sierra that focused on prescribing treatments based on an assessment of the condition of small stands or groups rather than on individual trees.

As chair of an important committee, F. S. Baker reported to the State Board of Forestry on California's regeneration prob-

lems (Baker 1955). Faced with increasing harvesting and loss of forests due to wildfire, the committee recognized four distinct problem areas: (1) determining the most desirable regeneration density in different forest types and sites; (2) developing seedbed preparation practices that would enable conifer seedlings to survive summer drought and competition from hardwoods, grasses, and forbs; (3) combating rodent predation on conifer seed; and (4) controlling tree species composition by understanding seed producing patterns, nature of germination, and seedling establishment.

During this period, some silvicultural practices were also applied in the management of national parks. Aerial insecticides were used to control lodgepole needle miner in Tuolumne Meadows, Yosemite National Park. Extensive logging was done in Yosemite Valley to remove diseased and insect-killed high-risk trees and to restore historic vistas that were being blocked by the increased density of trees. Prescribed burning was also used extensively in giant sequoia groves in Yosemite Valley and along Highway 120 and Tioga Pass. This was done to reduce undesirable understory vegetation and to decrease fuels.

An increased desire by landowners for prompt reforestation and less reliance on the vagaries of natural regeneration led to greater use of planting, control of competing vegetation by herbicides, and control of stocking and species composition. In 1973, the old forest practice rules that had existed since 1943 were replaced by the current Forest Practice Act and its associated regulations. The old rules had no provision for enforcement of regulations, relied on natural regeneration from residual trees, and identified the seed-tree harvesting and regeneration method (table 15.1) as the preferred silvicultural practice. The new rules established minimum standards for silviculture based on methods that produce both even- and uneven-aged stands and introduced mechanisms for inspection and enforcement.

1979 to 1995

Harvesting in the late 1970s and early 1980s continued to emphasize clear-cutting on national forest lands and on some private lands. Silviculture in this period changed from relying primarily on tractors to the use of modern aerial cable logging, particularly on steeper ground. Increased harvesting in the true fir forests led to the development of a risk rating system (Ferrell 1980) that was useful not only for predicting susceptibility to insects and disease but also for characterizing trees that could provide snags or dead wood for wildlife habitat. Although reforestation often relied on natural seeding, planting was done where prompt regeneration was required. In addition, thinning the overstory to encourage the growth of young trees in the understory, known as release of advance regeneration, was used wherever possible.

Environmental concerns restricted the use of herbicides on public lands and led to hand weeding and hand cutting of

TABLE 15.1

Regeneration methods (abridged from Society of American Foresters 1995).

Even-Aged Methods
Methods used to regenerate a forest stand with a single age class.

Clear-Cutting
A method of regenerating an even-aged stand in which a new age class develops in a fully exposed microclimate after removal, in a single cutting, of all trees in the previous stand. Regeneration is from natural seeding, direct seeding, planted seedlings, and/or advance regeneration.

Seed Tree
An even-aged regeneration method in which a new age class develops from seeds that germinate in a fully exposed microenvironment after removal of all trees in the previous stand, except for a small number that are left to provide seed.

Shelterwood
A method of regenerating an even-aged stand in which a new age class develops beneath the moderated microenvironment provided by the residual trees.

Uneven-Aged Methods
Methods of regenerating a forest stand, and of maintaining an uneven-aged structure, by removing some trees in all size classes either singly, in small groups, or in strips.

Group Selection
A method of regenerating uneven-aged stands in which trees are removed, and new age classes are established, in small groups.

Single-Tree Selection
A method of creating new age classes in uneven-aged stands in which individual trees of all size classes are removed more or less uniformly throughout the stand to achieve desired stand structural characteristics.

shrubs too large to weed. In some cases, control of competing vegetation through the use of cattle was successful; sheep and goats were found to be too difficult to control to reduce vegetation successfully. Biomass harvesting to provide fuel for electric power generation permitted, for the first time, the economic thinning of dense stands of small trees. This technique the potential not only to enhance stand growth and to promote forest health but also to effectively reduce fuel loads. During this period, the effects of the wildfire suppression policy, plus preferential harvesting of pines of the preceding fifty years became evident. Sierran mixed conifer forests had changed from a mosaic with some large trees with wide spacing dominated by pines to stands with a much higher proportion of shade-tolerant California white fir and incense cedar, higher densities of young trees, and "fuel ladders" with a continuity of branches and foliage that reached from the ground to upper canopy.

In the period 1988 to 1995, two important issues have arisen. First, conservation and wildlife habitat concerns have resulted in a two-thirds reduction in the harvest of timber on public lands in California. Clear-cutting has been eliminated on public forestlands, and harvesting methods on both public and private lands have moved toward variants of single-tree and group selection. Clear-cutting on public lands has been replaced by cutting methods that include the retention of live trees. Second, public concern has led to efforts to withdraw the remaining old-growth stands from the commercial timber base. As reported in Franklin and Fites-Kaufmann (1996), 8.2% of the Sierran landscape has forests with late-seral-stage or old-growth structural characteristics. Concern about the declining amount of old-growth forest has led to interest in increasing the amount of late-seral-stage stands on federal lands.

Over the past 150 years, therefore, silviculture in the Sierra Nevada has evolved in response to changes in technology, merchantability standards, human needs for wood products, and societal concerns for the environment. This evolution is reflected in changes on both public and private forestlands from high-grading to selective cutting to regeneration cutting with an emphasis on upgrading forest conditions to the use of even- and uneven-aged silvicultural systems. Management of forests on federal lands for sustained yield of timber has been replaced by an approach focused on sustaining values, managing ecosystems, and forming strategies for protecting the spotted owl and other wildlife species. Dominating all silviculture is economics and the recognition of Sierran forest characteristics of drought, fire, insects, and disease.

WHAT IS SILVICULTURE?

Silviculture is the art and science of controlling the establishment, growth, composition, health, and quality of forests and woodlands to meet the diverse needs and values of landowners and society on a sustainable basis (Society of American Foresters 1995). A silviculturist, therefore, is an applied ecologist who has sufficient knowledge of plant growth and environmental interactions to maintain or develop forest stands or landscapes of the desired composition and structure. Stand composition is the presence and abundance of trees, shrubs, herbs, and grasses. Stand structure is the vertical and horizontal arrangement of live and dead trees, including downed woody material, and the understory species composition of shrubs, hardwoods, and grasses. Thus, silviculturists can develop forests that, like some stands in nature, are basically a single species with high density and no understory, or that consist of complex mixtures at lower density with abundant understory. To do this, silviculturists assess variables such as stand density and productivity, seed source and presence of pathogens, and environmental conditions and constraints, and prescribe various treatments, including thinning, harvesting, regeneration, weed control, prescribed burning, animal pest control, and brush-field management designed to enhance stand structure to meet society's needs.

Silviculturists use their knowledge of natural succession and stand development, as well as treatments such as planting and thinning, to achieve the desired stand conditions. For example, after a fire in a Sierran mixed conifer forest, several different stand types may result (figure 15.2) developing on

FIGURE 15.2

Four scenarios of stand development following a stand-replacement fire or other disturbance and no management or further disturbance. Species composition and structure of the future stand depend upon the species and amount of conifer seed present the first few years following the disturbance. Shrub seed (stored in the forest floor) and sprouting shrubs and hardwoods are nearly always present and will produce a dense stand after a disturbance.

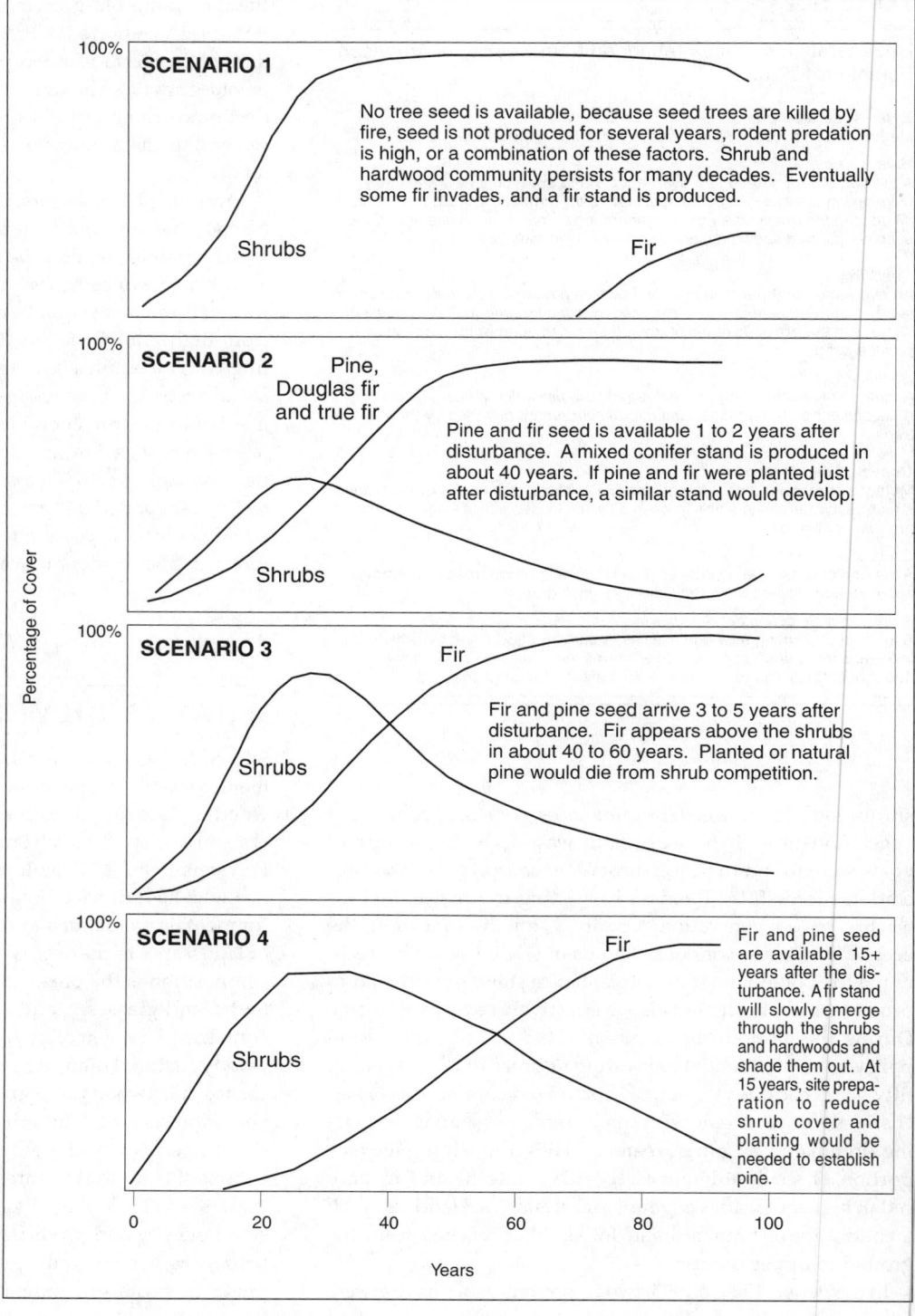

SCENARIO 1

No tree seed is available, because seed trees are killed by fire, seed is not produced for several years, rodent predation is high, or a combination of these factors. Shrub and hardwood community persists for many decades. Eventually some fir invades, and a fir stand is produced.

Shrubs

Fir

SCENARIO 2

Pine, Douglas fir and true fir

Pine and fir seed is available 1 to 2 years after disturbance. A mixed conifer stand is produced in about 40 years. If pine and fir were planted just after disturbance, a similar stand would develop.

Shrubs

SCENARIO 3

Fir

Fir and pine seed arrive 3 to 5 years after disturbance. Fir appears above the shrubs in about 40 to 60 years. Planted or natural pine would die from shrub competition.

Shrubs

SCENARIO 4

Fir

Fir and pine seed are available 15+ years after the disturbance. A fir stand will slowly emerge through the shrubs and hardwoods and shade them out. At 15 years, site preparation to reduce shrub cover and planting would be needed to establish pine.

Shrubs

Percentage of Cover

0   20   40   60   80   100

Years

different locations within the same burned-over area. If no source of conifer seed is available, the site will remain a brush field for many years, and true fir establishment may take many decades (scenario 1). However, if seed of different species of both pine and fir seed is available shortly after a disturbance, a mixed conifer forest will likely develop within about forty years (scenario 2). If seed is available after three or more years, it is likely that the next stand will slowly become stocked with fir, since pine competes poorly with established shrubs or hardwoods (scenarios 3 and 4). Assuming that no pine or fir seed is likely to be available soon after disturbance (scenarios 1, 3, and 4) and that a mixed conifer stand is the objective, a silviculturist would likely prescribe the planting of pine and fir as well as some shrub control to ensure that scenario 2 will occur. The land manager may decide that a brush field or a slowly developing fir stand would best meet management objectives, in which case no planting or shrub control would be done.

WHAT IS A SILVICULTURAL SYSTEM?

A silvicultural system is a planned sequence of operations that enables the forester to project the transition from an existing stand or landscape condition to a desired condition that conforms to a particular management objective. The process requires the articulation of management goals, the assessment of stand composition and dynamics, the capacity to modify stand treatments as conditions change, and the deliberate maintenance of options for changed objectives.

Control of Structure and Species Composition

All silvicultural treatments, in effect, influence the microenvironment, modify species composition, and thereby affect the relative capacity of grass, shrubs, and trees to grow and develop alternative stand structures. Such treatments help stands and landscapes develop structural diversity in both the horizontal and vertical dimensions, and these structures change over time as the vegetation grows and dies. Control of stand structure and species composition is effected by silvicultural prescriptions that control the interaction between species composition and the environment (temperature, water availability, light, etc.). This interaction controls the relative growth of specific forest vegetation, which, in turn, results in stands of a specific structure and composition.

Classical Systems

The traditional or classical systems of silviculture originated in Europe and were designed to upgrade stands that had been high-graded over past centuries with little regard for species mix or tree vigor. The systems result in stands having a distinctive structure representative of the range in structures in nature. The first systems developed were even-aged systems and included clear-cutting, seed tree, and shelterwood (table 15.1). Later systems developed were uneven-aged or selection systems that included single-tree and group selection. The range of techniques within both even- and uneven-aged systems facilitates the establishment and growth of both shade-intolerant and shade-tolerant tree species. The names of the systems are taken from the names of regeneration (harvest) methods. Each system represents a theoretical model of

a distinctive set of stand conditions positioned along a continuum that ranges from systems suited to the regeneration and maintenance of stands of primarily shade-intolerant species (clear-cutting, seed-tree, and group selection) to those designed primarily for the management of the most shade-tolerant species (shelterwood and single-tree selection) (Troup 1959; Daniel et al. 1979; Smith 1986; State of California 1988; USFS 1995). The choice among systems is based on their ability to create and perpetuate stands having those characteristics that best meet specific land-management and societal objectives. Overriding all technical issues in the choice and implementation of systems are the particular managerial, ecological, and societal constraints that pertain to the management area in question.

Modern Variants of Systems

In modern silviculture, management goals tend to be very complex and therefore require diverse stand structures. It is often not desirable to manage stands according to one of the five conceptually simple silvicultural systems. This is because management objectives, on both public and private lands, usually have multiple values and because the implementation of a particular system may be modified due to peculiarities of the structure and composition of the existing stand and the quality, aspect, and topography of the specific site. Diversity and complexity of stands and landscapes can best be achieved by examining the structure and composition of each stand and determining what treatments are necessary to achieve the stand structure needed to meet management objectives. It is commonly desirable to use a range of systems or approaches within one landscape or large watershed. Treatment areas may have irregular boundaries that merge into neighboring stands, and both live and dead trees might be retained in even-aged systems and group selection to provide needed structural components. In many situations, it is important that the silvicultural system chosen be aesthetically pleasing as well as ecologically sound. Consequently, in any one location, much skill is needed to develop the system used.

CURRENT SILVICULTURAL PRACTICE

Silvicultural practice must be based on a firm scientific foundation. Although much remains to be understood, substantial knowledge has been accumulated to support stand treatments designed to control species composition and stand structure. This scientific knowledge is summarized in the sections that follow.

Understanding the Forest System

Forest Soils (Based on Poff 1996)

All terrestrial forest species are dependent upon soil, which supports the health and productivity of Sierra Nevada ecosystems. Consequently, it is imperative that soil characteristics and processes and the effects of forest treatments on soils be understood.

Three important indicators of forest soil productivity are soil volume, soil porosity, and organic matter, as well as their interactions. Soil depth controls rooting depth, available water, total nutrients, and buffering capacity. Soil porosity influences rootable soil volume, water infiltration, and biological activity. Organic matter is a storehouse of plant nutrients, especially nitrogen (N), phosphorus (P), and sulfur (S), and provides a habitat for a diverse biota that carries out energy transformations and cycles nutrients. In addition, organic matter on the forest floor protects soils from erosion and retards water evaporation, and large woody material provides a water reservoir and habitat for many organisms, as well as shelter and habitat for small and medium-sized mammals (Maser and Trappe 1984; Harmon et al. 1986).

Three important processes that tend to reduce soil productivity are erosion, displacement, and mass wasting. Under a forest canopy, soil erosion is negligible (McColl and Powers 1984), but, with removal of trees, soils may erode through sheet erosion, overland flow, rill formation, gully erosion, or dry ravel. The amount of erosion depends on the extent of the disturbance, the slope, and the soil type. In the Sierra Nevada, soils developed from granitic bedrock are the most erodible, whereas soils developed from metasedimentary bedrock are the most stable.

Mass wasting results from landslides and improper road building and is more common on steep slopes (Atzet et al. 1989). Although localized mass wasting can occur in the Sierra Nevada, it is not as widespread as in northwestern California or the Pacific Northwest, because about 70% of the forested land in the Sierra Nevada is on slopes of less than 32% (Rice 1979). Soil movement might be minimized by ensuring that ground cover is maintained. Leaving large residual trees or patches of trees, as in the current practice of green tree retention, provides better protection for the soil by reducing solar insolation and slowing the rate of decomposition of forest litter. In particular, because most erosion is associated with roads, special care needs to be taken in road design, placement, construction, and maintenance.

Distribution and Cycling of Mineral Nutrients. Globally, forest soils are enormous sinks for carbon and N, with the carbon in soil equaling that of vegetation and the atmosphere combined (Powers and Van Cleve 1991, Johnson 1994). This is because as much as three-fourths of net primary production can be associated with the development of fine roots and mycorrhizae (Grier et al. 1981), especially in true fir stands (Powers and Edmonds 1992). Most nutrients are taken up by

a forest stand before crown closure, after which as much as 30% to 50% of N needs and 20% to 80% of P needs may be met by internal translocation (Powers 1979; Prescott et al. 1989; Powers and Van Cleve 1991). As a stand reaches maturity, the return of nutrients to the soil in litter fall approaches the rate of nutrient uptake (Powers 1979). Of all plant nutrients in forest ecosystems in the Sierra Nevada, N is often the most limiting (Powers and Edmonds 1991). In general, mature mixed conifer stands have about 10% of N in the forest floor, 80% in the soil, and the remaining 10% in standing biomass (Powers 1992). In true fir forests, due to colder temperatures and a slower rate of litter decomposition, the distribution of N is 40% in the forest floor, 47% in the soil, and 13% in standing biomass (Powers and Edmonds 1992).

To maintain nutrient status in forest soils, losses from leaching, biomass removal, volatilization, and soil loss must not exceed inputs from atmospheric deposition, decomposition of organic matter, and mineralization. Because the soil and forest floor on productive sites contain the vast majority of nutrients, management practices should be oriented toward maintaining their integrity and structure by minimizing compaction, erosion, and loss of soil organic matter.

Compaction. Soil compaction is a major cause of reduction in productivity. It occurs when a force is applied to a soil, particularly when it is moist (Baver 1930, Froehlich 1974, Alexander and Poff 1985), by harvesting and site preparation equipment that causes a reduction in macroporosity and an increase in density (Hatchell et al. 1970; Alexander and Poff 1985). Poor aeration, oxygen deficiency, and increased density reduce root penetration, lower water migration, and reduce mycorrhizal activity (Harmon et al. 1986).

The effects of compaction on individual tree growth are well documented (Froehlich 1978; Wert and Thomas 1981; Froehlich and McNabb 1984; Helms and Hipkin 1986a). This is because most studies have been designed to compare the growth of individual trees on compacted versus noncompacted soils on landings and skid trails. In general, the decrease in height growth of trees is directly proportional to the increase in soil density (Froehlich and McNabb 1984). Seedling height growth is commonly reduced by 20% to 40% in compacted soil (Minore et al. 1969; Murphy 1983; Froehlich and McNabb 1984). On the Foresthill divide, sixteen-year-old ponderosa pines associated with skid trails and landings having the highest amounts of compaction had a height reduction of 43% at the age of one year and 15% at fifteen years (Helms and Hipkin 1986a). In the same Foresthill study, individual tree volume growth on landings, on skid trails, and in areas adjacent to skid trails was reduced due to compaction by 22%, 29%, and 13%, respectively (Helms and Hipkin 1986b). In general, both the degree and depth of compaction are reduced by the presence of litter and various sizes of rock, and compaction increases with successive trips by machinery over a skid trail (Mace 1970; Froehlich 1978; Miles 1978; Boyer 1979; McColl and Powers 1984). Once soils are compacted, they may take up to fifty years to return to natural levels of porosity; the length of time for recovery depends on soil type, occurrence of subfreezing temperatures, and degree of compaction (Hatchell et al. 1970; Froehlich 1979; Wert and Thomas 1981; McColl and Powers 1984; Morris and Miller 1994). If compaction is deep, soils may never return to their natural state without major physical disturbance (McColl and Powers 1984).

The effects of soil compaction on the growth of an entire stand, however, are not well understood, because they are much more difficult to address. The study of the Foresthill pine plantation mentioned earlier, found that, because compaction caused substantial mortality, the reduction in stand volume per acre on landings increased from 22% due to compaction alone to a total of 69% when additional volume loss due to compaction-induced mortality, resulting in lower stocking, was considered; similarly, on skid trails, the reduction in volume increased from 29% to 55% (Helms and Hipkin 1986b). Stand growth on areas adjacent to main skid trails was reduced by 13% with no increase in mortality. Stand volume in the bulk of the stand between skid trails was not reduced. Consequently, it must be remembered that in managed stands, landings and skid trails are part of the permanent road/access system that will be used in each thinning entry and are not intended for tree growing. A reasonable approach for dealing with this issue, therefore, is to minimize that portion of the stands needed for access and harvesting, where compaction is not necessarily a potential problem, and to minimize the likelihood of causing compaction on the 90% of land devoted to tree growing.

Management strategies that can be used to limit soil compaction include using equipment that exerts a low pressure on the ground using aerial yarding systems, using winches to haul logs to the tractor on the skid trail rather than moving the tractor to each log, operating over snow or a cushion of slash or brush, avoiding operating when soils are moist, and minimizing the proportion of land area committed to skid trails.

Effects of Fire on Soils. The physical effects of fire on soils include loss of organic matter, loss of soil structure, hydrophobicity (DeBano 1979), and erosion and loss of nutrients (McNabb and Cromack 1990; Palazzi et al. 1992). The chemical effects include increase in pH, loss of the capacity to exchange cations, and loss of nutrients by volatilization, fly ash, and leaching. The biological effects include the direct mortality of soil organisms.

The effects of fire on soil productivity can be either beneficial or detrimental, depending on fire intensity, soil type, and site history (Atzet et al. 1989; Morris and Miller 1994). Adverse effects on soils increase as burn intensity increases, and the negative effects are proportional to the amount of surface duff and soil organic matter consumed (DeBano 1979; Sandberg 1980; Boyer and Dell 1980). Total N loss is almost linearly related to litter consumption and little N loss occurs

until litter consumption, exceeds 25% (Dunn and DeBano 1977; Powers 1979, Clayton and Kennedy 1985). Cations such as calcium, magnesium, sodium, and potassium released in the ash bed are susceptible to leaching, but revegetation and exchange sites in the soil usually absorb them quickly (McNabb and Cromack 1990). This generalization may not be true of coarse-textured granitic soils that are low in organic matter. Relatively cool prescribed broadcast burning that does not entirely consume the forest duff and does not totally remove effective ground cover has little effect on nutrient status except on soils of low nutrient capacity. On the other hand, the piling and burning of slash concentrates nutrients, and the high intensities can damage the soil under the piles.

Effects of Silvicultural Treatments on Soils. The effects of silviculture on soils are primarily associated with the type of treatment and the frequency and extent of disturbance. Site preparation methods used in the 1950s and 1960s, such as scalping and windrowing of topsoil aimed at reducing competition by removing sprouting species and dormant weed seeds, have now been shown to greatly reduce long-term productivity (Kittredge 1952; Morris et al. 1983; McColl and Powers 1984; Dyck and Beets 1987; Powers et al. 1988; Powers 1991; Morris and Miller 1994). Currently, site preparation treatments are designed to keep as much soil in place as possible and to retain as much cover as possible consistent with securing regeneration.

Timber harvest can potentially affect soil productivity through erosion, displacement, compaction, nutrient removal, and leaching. The percentage of bare soil following logging can range from 2% for transporting logs to the landing by helicopter to more than 75% for tractor logging (Rice 1979). In partial cutting or thinning, the disturbance is less, but the total effects depend on the proportion of biomass removed and the frequency of entry. Surface erosion from yarding is typically quite low (McColl and Powers 1984). The potential erosion from skid trails can be high but is largely eliminated by forest practice regulations that require the placement of tractor- or hand-constructed mounds of soil (water bars) across skid trails. Most erosion during timber harvest is related to roads (McColl and Powers 1984).

The amount of nutrients exported through timber harvest depends on nutrient distribution in the ecosystem, utilization standards, and rotation length. The stemwood of conifers contains about 20% of the total tree carbon (C) and about 10% of the N. Even in the most intensive harvests, less than 10% of ecosystem N would be removed (Powers and Edmonds 1992). The actual amounts of nutrients exported would be considerably less under current practices, since the amount of clear-cutting in California has been drastically reduced, and green trees, unmerchantable material, snags, and small patches of trees are typically left on the landscape. Generally, stem-only harvests of middle-aged stands have little impact on nutrient export (Powers et al. 1990), and atmospheric inputs of N, P, and probably S probably exceed harvest export

(Morris and Miller 1994). During the 1960s there was concern about nutrient loss from leaching following clear-cutting (Likens et al. 1969), but the general consensus now is that, except in very extreme cases where vegetation is absent or intentionally suppressed, nutrient losses by leaching are negligible (McColl and Powers 1984; Johnson 1994).

Stand Density (Based on Oliver et al. 1996)

Stand density is a major controller of stand growth and development. It affects the abundance and composition of ground cover, shrub development, tree growth and vigor, species composition, cover and food for wildlife, fuels and fire hazard, and the dynamics of insect and disease populations (Assmann 1970; Smith 1986; Daniel et al. 1979).

Measures of Stand Density. Stand density is measured in absolute or relative terms, depending on the purpose for which the measure is used. Absolute measures include number of trees per unit area, basal area, volume, or cover. Relative measures include comparative assessments of "normal" or "full stocking" (Dunning and Reineke 1933; Schumacher 1928; Meyer 1938). In addition, various stand density indices are used (Reineke 1933; Curtis 1970). Stand density indices have the advantage of being relatively independent of age and site quality and, as a result, are used in the development of computer simulations of stand development (Reukema and Bruce 1977; Wensel et al. 1986; Ritchie and Powers 1993). These computer models are used extensively in forest management and permit one to predict how density affects such stand characteristics as tree mortality, growth rate, crown size, and other stand and tree parameters.

Effects of Thinning on Tree and Stand Growth. As trees grow and stand density increases, trees become more crowded, and fewer resources (water, light, and nutrients) are available for maintaining individual tree growth and stand vigor. The longest time-series analysis of stand development in the Sierra Nevada (Oliver 1979; Oliver and Dolph 1992; Oliver in press) has demonstrated that, through thinning, stands can be developed that have larger, more vigorous trees with longer live crowns and that also have a vigorous understory of saplings, shrubs, and grass. The unthinned, more dense stands, in contrast, had smaller, less vigorous trees, had no understory, were a poorer habitat for large mammals, and constituted a much greater fire hazard due to the presence of dead and dying small trees with low dead branches. For example, dense forty-year-old ponderosa pine stands at Foresthill had stem diameters averaging 34.3 cm (13.5 in) and had live crown ratios of 54% of their total height. Thinned stands, on the other hand, had stem diameters averaging 53.8 cm (21.2 in) and live crown ratios of 70%, and had developed a vigorous understory (Oliver 1979; Oliver and Dolph 1992; Oliver in press).

Past studies have also quantified the extent to which thinning influences the subsequent development of stand density in Sierran stands. In ponderosa pine, the increase in stand density tends to be slower as stands increase in age and in stands on less productive sites (Oliver 1979, 1988, in press). This information is fundamentally important in determining the effects of different levels of thinning on the capacity of stands to develop the desired types of structures. For example, red and California white fir stands differ from ponderosa pine stands in that, at the age of 100 years, the fir stands reach the much higher densities of 74 to 114 m²/ha (320 to 498 ft²/acre). The effects of greater shade tolerance and vigor at these higher stocking levels are that, after thinning, trees in these old, dense, true fir stands rapidly increase their growth rate. After about ten to fifteen years, these stands grow back to their previous densities (Oliver 1988). In mixed conifer stands, the relation between density growth following thinning depends primarily on the proportion of shade-tolerant and species to shade-intolerant species in the mixture and the overall stand vigor at the time of thinning. The relationships among stand density, tree size, and stand vigor developed for the Sierra Nevada are corroborated by many other studies in temperate forests (Assmann 1970; Daniel et al. 1979).

Ponderosa pine forests maintain a relatively constant rate of biomass production over a wide range of stand densities. In mixed conifer and true fir stands, however, net productivity tends to increase as stand density increases (Daniel et al. 1979; Oliver 1988, in press). This difference in productivity/density relationships has an important bearing on prescriptions for silvicultural manipulation. In shade-intolerant pine stands, it is very important to thin to maintain stand vigor, reduce susceptibility to insect and disease attack, and reduce the potential for drought-induced mortality. In the more shade-tolerant mixed conifer and true fir stands, however, it is not as imperative to thin to maintain stand vigor, but thinning is still desirable to reduce fuel loading, increase stand value for wood products, create variability in density suitable for a variety of wildlife species, create openings for ponderosa and sugar pine seedlings as well as understory plants, and accelerate stand development toward the late seral stage.

Effect of Understory on Stand Development. In the early stages of stand development, the growth of pines on sites of low productivity (17 m [55 ft] at the age of fifty) is restricted more by shrub density than by tree density (Oliver 1984). The effect of competition is very large, as, after twenty years, trees growing without shrubs had about 40% more volume than those on plots in which shrubs were not controlled. In general, it was found that as shrub cover increased from 0% to 30%, tree growth declined rapidly. Additional shrub cover from 30% to 100% had little further effect on tree growth on these lower-quality sites (Oliver 1984). On sites of higher productivity, intertree competition may be more important than tree-shrub competition.

These findings are important not only for timber management but also, for example, in deciding the tradeoff between providing dense layers of shrubs for wildlife cover versus the potential loss in wood production. Competing brush can substantially lengthen the time to first commercial thinning, regardless of tree spacing or site productivity, especially on lower-quality sites (Oliver 1979, 1984; Fiske 1982).

An additional issue concerns the contribution of shrubs to maintaining nutrient stability within forest systems. In the Sierra, *Ceanothus* species on poorer sites are commonly believed to contribute as much as 75 kg of nitrogen to the site per hectare per year (67 lbs nitrogen/acre/year). The proportion of this amount that is available to associated trees is questionable. However, the presence of nodulating shrubs is undoubtedly of value in adding nitrogen to the system and in cycling nutrients.

Shrubs affect the nutrient cycle in forest stands by increasing the concentrations and amount of nutrients, particularly cations, in the annual litter fall and by increasing nutrient turnover rates (Tappeiner and Alm 1975; Fried et al. 1989). In these studies, the increased annual input to the forest floor did not result in additional nutrients in the soil organic layers or the upper layer of soil. Compared to our knowledge of conifers, our understanding of the effects of shrubs or hardwoods on soil properties and long-term site productivity are not well understood. We know of no studies on the effects of understory shrubs on nutrient cycling and soil properties in Sierra Nevada forests.

Stand Density, Insects, and Disease. Considerable evidence exists that the susceptibility of stands to attack by a variety of insects is related to the decline in stand vigor with increasing density (Ferrell 1974; Ferrell and Smith 1976; Berryman and Ferrell 1988; Waring and Pitman 1985). Also, as stands become more dense they become more susceptible to root diseases, storm damage, and drought (Oliver 1985; Powers and Oliver 1970). A prime example occurred during the major drought in the late 1980s when Ferrell and colleagues (1994) showed that 98% of the variation in mortality in true fir could be explained by stand density and that individual tree characteristics were unimportant. This close relationship between density and insect attack has been observed in the Sierra, where ponderosa and Jeffrey pine stands that had been thinned had very low levels of insect attack, whereas about half of the trees in the unthinned stands were killed by bark beetles. On the other hand, there are some examples of apparently vigorous trees being killed by bark beetles. Insect populations can readily become epidemic because, once insects commence breeding, they are likely to spread to adjacent stands, even though these stands may be more vigorous.

The principal insect pests in Sierran forests are the tussock moth, western spruce budworm, and bark beetles which attack Douglas fir, and the fir engraver beetle and roundheaded fir borer which attack true firs. The insects most damaging to pines are the western pine beetle, the mountain pine beetle, the red turpentine beetle, and the California five-spined ips, which breeds in downed logs and slash. Diseases of particular concern include dwarf mistletoe (which is very prevalent but species-specific, so that the mistletoe associated with one tree species cannot infect another) and several root diseases, especially the annosus root disease, black stain root disease, and shoestring fungus (Burns and Honkala 1990; Scharpf 1993).

Of particular importance is a recognition of the interaction among bark beetles, fungi, and host trees that results in the death of pine trees. Wood 1993 provides three examples resulting from long-term collaborative research among forest entomologists and pathologists in the Sierra:

- As the symptoms of needle injury due to oxidant air pollution (ozone) increase, the incidence of infestation of ponderosa pine by the western pine beetle and mountain pine beetle increases.

- Attacks of black stain root fungus and the annosus root rot predispose ponderosa pine to infestation by pine beetles and white fir to infestation by the fir engraver beetle, particularly at low beetle populations:

- Tunneling by bark beetles, especially the red turpentine beetle, into ponderosa pine has been shown to introduce pathogenic fungi that cause the death of pines by interrupting water conduction in the xylem.

Sugar pine is a particularly noteworthy tree species of the Sierra Nevada. It is well known for its beauty, large-diameter stems, spreading crowns, large cones, and wood quality. It also has a remarkable capacity to continue rapid growth well into maturity (150 years or more). Unfortunately, it has been severely affected by white pine blister rust, which was introduced into North America from Asia at the turn of the century and extended into California in about 1930. This disease, which depends on an alternate host (*Ribes* spp.) for development, severely damages and kills smaller sugar pines in the cooler, moister northern Sierra. Although larger trees are less affected by the disease, the future of sugar pine as a component of northern Sierran forests is uncertain. Forest Service geneticists and pathologists at the Pacific Southwest Station are working on a program to produce rust-resistant stock for planting. Three different types of resistance to white pine blister rust have been identified, all of which are strongly inherited. One form of resistance appears to be controlled by a dominant gene that occurs with low frequency, another is a race of trees having slower rates of infection, and a third is a type of age-dependent resistance (Kinloch and Scheuner 1990).

Currently, some landowners in the Sierra are collecting cones from particular sugar pine trees and, by exposing seedlings to the rust, determining whether or not the parent trees are potentially resistant to the rust. Seed from sugar pine trees identified as being potentially rust resistant are collected and used for reforestation. In addition, normal silvicultural practices might also help retain sugar pine in Sierran stands. When groups of seed-bearing sugar pine trees with no (or slight) infection are found, thinning to favor the best sugar pines would reduce their risk of being destroyed by fire and possibly increase seed production. Soil scarification during thinning would prepare a seedbed and encourage natural regeneration. Thus, cohorts of new seedlings with potential rust resistance might be established throughout the Sierra, augmenting the rust resistance being developed by research.

In 1995, issues related to mortality caused by drought, insects, and disease reached major biological and political proportions due to the extremely high rate of mortality stimulated by a sustained drought in the Sierra since 1987. Most mortality occurred on relatively dry sites in stands having relatively high stocking levels, and consisted of the more shade-tolerant California white fir species. It appears that the trees initially lost vigor due to the drought and were finally killed by the fir engraver beetle. In portions of the Tahoe Basin, for example, 50% of the trees within certain stands died, raising complex issues regarding the extent to which these dead and dying trees should be salvaged for wood products or left to provide wildlife habitat and to be replaced by natural succession. The issue is made more complex by considerations of the extent to which dead trees contribute to fire hazard.

Management of Stand Density. In order to maintain diverse uses and values from Sierran forests it is generally necessary to provide a mosaic of stands of varying age and density throughout the landscape. In addition, mixed conifer stands have diverse species composition, whereas stands of lodgepole pine, true fir, and east-side pine are more pure. As individual stands develop within a landscape, the location of stands having particular characteristics changes over time. From past studies we know that, in general, stands with relatively high density have

Higher levels of:
Biomass production
Tree mortality
Slowed tree growth, with smaller branches
Susceptibility to insect attack
Storm damage
Insectivorous birds
Fuel loading and ladder fuels
Susceptibility to drought
Shade-tolerant species

Lower levels of:
Tree and stand vigor
Tree size
Rapid tree growth and larger branches
Understory cover and regeneration
Wildlife dependent on open space and
large trees
Shade-intolerant species
Stand diversity

Consequently, to attain diverse land-management objectives, particularly on public lands, a mosaic of stands of varying

density should be maintained, particularly if they are associated with a continuum of conditions from early to late seral successional stages.

Stand Growth Models. Several computer growth simulators are being used to project the growth of conifer forest vegetation types in the Sierra Nevada:

• CACTOS (University of California, Berkeley, Agricultural Experiment Station), the California Conifer Timber Output Simulator, is an interactive, FORTRAN 77–based computer program designed to simulate the growth and partial harvest of conifer forest stands in northern California. It is a distance-independent, individual-tree model. It predicts uninterrupted growth or the effects of various silvicultural prescriptions. Of particular value is the "free" harvest subroutine, which gives the user considerable flexibility in harvesting by diameter classes of varying width or species groups. The heaviness of cutting can be set by basal area limits or by a fraction of the trees present. Cutting limits can be set to prevent overharvesting.

• SYSTUM-1 (U.S. Forest Service, Pacific Southwest Research Station, Redding, California) is a computer program designed to simulate the growth of very young conifer stands in inland northern California and southern Oregon from three years after planting until the trees reach a diameter compatible with entry into CACTOS, ORGANON (from Oregon), and other models. The simulator is a distance-independent, individual-tree model. Trees are grown on a plot-by-plot basis to allow for varying densities. The growth reduction effects of different levels of up to six species of competing vegetation can be simulated. Precommercial thinning can be introduced.

• PROGNOSIS (U.S. Forest Service, Intermountain Research Station, Missoula, Montana) is a generic computer program that has two variants, SORNEC and WESSIN (U.S. Forest Service, WO-TM Service Center, Fort Collins, Colorado), that are used in northeastern California and in the western Sierra Nevada. These variants are designed to simulate the development of forest stands and silvicultural treatments that include stocking control, regeneration methods, site preparation, and thinning. The simulators can be linked to other models that predict the effects of pest outbreaks on forest stands and the production of other forest on. The combined outputs provide a basis for multiresource planning.

Each of these simulators has been used extensively in the Sierra to predict the likely outcome of alternative silvicultural prescriptions and management strategies. They do not predict the establishment of understory or the effects of fire, but they do model the growth of established stands.

Management goals for Sierran forests are commonly complex and diverse. Of greatest importance is the maintenance of forest health and the reduction of potentially destructive stand-replacement wildfires. Also, depending on specific goals, there is a need to regulate species composition, maintain tree growth and vigor, enhance wood production, and maintain wildlife habitat (Assmann 1970; Smith 1986; Daniel et al. 1979). Diverse stand conditions associated with different management goals can be obtained by controlling stand density and species composition through thinning combined with underburning. The absence of these two silvicultural treatments is evident today in Sierran forests, which are generally overly dense and unhealthy. Controlling density is also important to provide wildlife habitat. In general, populations of California spotted owl are associated with a mixture of denser, older stands for nesting and roosting and more open, younger stands for hunting prey. In terms of wood production, even though biomass production may be similar for stands growing at a wide range of densities, merchantable yields of wood are much higher in thinned stands composed of fewer, larger trees than in unthinned stands composed of many smaller trees. Further, greater financial efficiencies, from the standpoint of wood production, are generally obtained in stands maintained at relatively lower stocking levels.

It should be apparent that managing stand density by thinning requires considerable technical knowledge of the effects of thinning on growth and yield and on diverse wildlife species, as well as knowledge of fire behavior in stands of varying density. Knowledge is also needed regarding how to treat slash to avoid unacceptable fire hazard and insect buildup and regarding the potential for soil compaction and nutrient depletion. Trade-offs need to be made between the frequency and heaviness of thinning as well as among timing, tree size, choice of equipment, markets for products, and cost-effectiveness. Managing stand density is therefore a complex issue requiring the combined talents of silviculturists and specialists in other natural resource disciplines.

Regeneration (Based on McDonald and Tappeiner 1996)

The ability to regenerate stands after disturbances such as fire and timber harvest, as well as within the urban/wildland interface is an important part of forest management in the Sierra Nevada. Methods of regeneration include natural seedling establishment, planting, release of advance regeneration, and coppicing of hardwoods. Aerial seeding is not now being used in the Sierra, because (1) the hot, dry summers preclude the spring application of stratified seed; (2) fall seeding is not successful due to excessive losses of seed over the winter; (3) distribution of untreated seed results in most seed being consumed by rodents and birds, and coating seeds with repellents to avoid predation is no longer environmentally acceptable; (4) seed is now very expensive, so aerial seeding is not economically feasible. All regeneration methods (except aerial seeding) are appropriate in specific circumstances, and their use is determined by an evaluation of management objectives and such variables as availability of a seed source, microclimate, soil characteristics, potential competition from shrubs and grass, vigor and distribution of advance growth, and the desired species composition and structure of the future stand. In practice, combinations of methods are commonly used.

On many sites, federal and state regulations set standards for regeneration. Generally, they require that sites be stocked with 250 to 741 seedlings/ha (100 to 300/acre) within three to five years after any harvest that removes most of the trees, depending on tree species and site quality.

The need to develop reliable methods to regenerate Sierran forests was recognized well over half a century ago (Dunning 1923). Large fires from lightning, mining, logging, and railroads had resulted in hundreds of thousands of acres of brush fields. Because it would commonly take at least sixty years before these brush fields returned to forests—and these would consist of shade-tolerant fir and cedar— foresters embarked on a large program of brush-field clearing and planting of ponderosa pine, which was a dominant species of the presettlement forest. In 1955, Baker began a concerted effort to develop regeneration practices based on an evaluation of the interaction among the physiological condition of seedlings, genetic considerations of seed source, and the microclimate of the site to be regenerated (Baker 1955). This work resulted, in 1961, in improved seed zone maps in which the State is divided into eighty-three seed zones (Buck et al. 1970), supplanting the older maps developed by Fowells in 1946 that had only twelve zones. These new seed zones provide a mechanism by which seedlings can be planted in the same zones from which the seed was collected. The work also resulted in guidelines for seed collection and handling, nursery stock production, site preparation, planting techniques, and vegetation control (Schubert and Adams 1971; Jenkinson 1980).

Natural Regeneration. Sierran forests are, of course, generally capable of being established naturally from seed. In some stands, such as those consisting of east-side ponderosa pine, this might not be the case, as reported by Show (1926):

> In the virgin forest, particularly in pure yellow pine, when a good crop of seed is produced (and this only occurs at intervals of from five to eight years), the chances are slight that an equally good crop of seedlings will result. The long dry season and the severe frosts typical of the region will destroy the vast majority of the young seedlings during the first years of their lives. Only once in every 10 to 25 years, on the average, does a satisfactory stand of seedlings become established.

In nature, the timing and success of regeneration depends on the coincidence of favorable factors that include the availability of seed, presence of a suitable seedbed, limited predation (insects, rodents, birds, and disease), favorable microclimate (limited frost, moderate temperature, and low evaporative stress), adequate precipitation, and limited competing vegetation (Baker 1952; Tevis 1953; Schubert 1956; Beetham 1963; Stark 1963,; Tappeiner and Helms 1971; Ustin et al. 1984; Laacke and Tomascheski 1986). The success of regeneration under specific conditions is also influenced by the relative shade tolerance of the species in question. Another factor adding to the uncertainty of natural regeneration is that conifers typically do not produce a good crop of seeds every year. The interval between good seed crops varies throughout the Sierra but in true firs is generally about one to four years, whereas the interval in ponderosa pines and Douglas fir is about two to eight years (Fowells and Schubert 1956; Tappeiner 1966; Gordon 1986; Burns and Honkala 1990; McDonald 1992). California black oak has a periodicity of heavy acorn crops every eight years (McDonald 1992). Recognizing all these uncertainties, the silviculturist attempts to create microclimates suitable for natural regeneration by choosing the most appropriate harvesting and site preparation methods (Gordon 1970, 1979; Laacke and Tomascheski 1986; McDonald 1976, 1983; Dunlap and Helms 1983).

Release of Advance Regeneration. In the understory of both natural and managed stands, there is commonly advance regeneration—particularly of California white fir, red fir, and incense cedar, as well as California black oak and tan oak—which develops in openings in the canopy that are caused by thinning or by natural events. This advance growth grows slowly and may persist for decades in the understory. When the canopy is opened up through treatment or natural gaps, this advance growth is often released and becomes an important component of the stand structure (Dunning 1923; Von Althen 1959; Gordon 1973; Oliver 1985; Tesch and Korpella 1993). The rate of growth of advance regeneration after release can be predicted from measurement of the prerelease live crown ratio and annual height growth (Helms and Standiford 1983). Under partial harvesting, the capacity of understory seedlings to respond to increasing levels of light is in accordance with generally accepted shade-tolerance rankings (Minore 1979), in which California white fir is the most tolerant and shows the most rapid growth of all species with increasing light; Douglas fir, incense cedar, and sugar pine are similar and somewhat less tolerant than California white fir; and ponderosa pine is the least tolerant and exhibits the least growth—half that of other species—under all partial light conditions (Oliver and Dolph 1992).

A survey of literature and personal experience with natural regeneration indicates that

- Shade-tolerant species such as California white fir, incense cedar, and red fir reproduce well in microsites ranging from exposed to shaded. The fir species will seed into shrub communities and eventually overtop them and produce pure stands of conifer.

- Ponderosa pine, Douglas fir, and sugar pine will not reproduce reliably in dense stands nor in brush fields. Well-

timed disturbance to the forest floor, an open environment, and seed availability are necessary for their establishment. In addition, competition from shrubs and grasses must be low for at least the first year and often for several years.

- Light shade aids the survival and early establishment of all species. Too much shade retards growth. Once seedlings have become established, all species grow best in full sunlight, regardless of tolerance.

- Natural regeneration of Sierra Nevada conifers can be used to regenerate sites after logging, providing there is sufficient seed source and a favorable microenvironment. Foresters must be able to use appropriate site preparation treatments such as well-timed scarification of the forest floor or prescribed burning that coincides with a seed crop and controls grass and shrub competition.

- Advance regeneration is an important component of regeneration. Careful logging is needed, and it may be desirable to interplant with shade-intolerant species to ensure well-stocked mixed conifer stands.

Planting. In today's forests, where prompt regeneration is a prime concern, the likelihood that natural regeneration will be a failure in any one year is such that forest managers often prefer to supplement natural regeneration with planting. Planting ensures that the desired mix and density of species can be established before competing weeds dominate the site. This is particularly important when it is necessary to regenerate a site after a disturbance or when it is desired to keep shade-intolerant species such as ponderosa pine, sugar pine, and giant sequoia in mixed conifer forests.

In the early years of planting conifers in California, a high mortality rate was common, primarily because of the characteristic summer drought in the Sierra Nevada. It was realized that seedlings had to be physiologically capable of becoming established rapidly during the short period in the spring when the soil is fully charged with water. The prime need was to produce seedlings capable of producing new roots immediately after being planted.

Research on seedling physiology, nurseries, and planting procedures has provided a scientific basis for successfully establishing Sierra Nevada conifers. The several state, federal, and private nurseries in California that produce conifer planting stock have utilized this knowledge and now produce seedlings with a high capacity for survival. The original research on the physiology of pine seedlings showed that nursery practices, especially seed source, time of lifting, and method of seedling storage, greatly influence seedling capacity to produce new roots (Stone and Schubert 1959; Stone et al. 1963; Stone and Jenkinson 1970, 1971). Measures of root growth capacity as an index of seedling vigor were shown to be directly related to the survival of field-planted seedlings (Jenkinson 1976, 1980). Other factors affecting seedling vigor are sowing schedules in the nursery and infection with beneficial mycorrhizal-forming fungi (Jenkinson and McCain 1993; Jenkinson et al. 1993). However, even seedlings in the best physiological condition will have lower survival rates if they are planted when soils are less than approximately 10°C for pines and less than approximately 3° to 4°C for red and California white fir (J. L. Jenkinson, U.S. Forest Service, Pacific Southwest Research Station, conversation with the author, 1995) and if they are planted improperly by inexperienced or poorly supervised crews.

Genetic Considerations in Reforestation. The growth and development of trees depends on the interaction between their genetic makeup and environment for growth. In any reforestation program it is important to use seeds from trees having desirable phenotypic characteristics. In addition, genetic makeup varies from tree to tree, from stand to stand, and at larger geographic or provenance levels. Consequently, to ensure adaptation to the site, as well as high survival and productivity, it is safest to use seeds or seedlings from a local seed source. As was mentioned earlier, to assist land managers in obtaining seed or seedlings from known sources, California is divided into approximately eighty seed zones (Buck et al. 1970; Arvola 1978). All seed or seedlings that are used for reforestation purposes are identified as to the zone from which they originated.

Because of genetic diversity within tree populations, there is an opportunity to develop seed sources or planting material that will ensure that reforested areas have broad genetic variation that will buffer the stands from changes in the environment. Opportunities for tree improvement are available at many levels of management, including by leaving better phenotypes as seed trees and shelterwood for natural regeneration, and by upgrading stands through selective cutting. In addition, genetic improvements can be made in the manner in which seed is collected or through the development of seed orchards. The first formal tree improvement plan in California was developed by the USFS in 1963. Which it established goals for hybridization of pines, seed production areas, sugar pine resistance to blister rust, and superior tree selection and seed orchards (Fowler 1963). This program was broadened into a tree improvement cooperative in 1971 involving the California Division of Forestry, several major forest industries, and the University of California, Berkeley (Kitzmiller 1976). This cooperative developed six seed orchards and a number of progeny testing sites in the northern Sierra and in other forest regions. The tree improvement plan had two areas of focus: (1) a base-level program aimed at ensuring the maintenance of at least the status quo with respect to native populations on all forestland while striving for gains in volume growth of up to 10%, and (2) a high-level program aimed at achieving sustained high genetic gains in adaptability and volume growth in major species through repeated cycles of selection and breeding. The second, longer-term program involved delineating breeding zones for each of the four major species, selecting 200 superior trees from each zone,

testing their progeny, and establishing seed orchards of parents propagated by seed or by grafting (Kitzmiller 1976). The long-term aim of this cooperative program is to produce genetically improved trees for regeneration of both federal and private forestland in California that will ensure adaptability, be less susceptible to insects and disease, and have higher productivity.

Vegetation Management

Throughout the Sierra Nevada, almost all disturbed sites rapidly become invaded by shrubs, herbs, grasses, or sprouting hardwoods. Characteristically, soil surface layers contain millions of seeds per acre of ceanothus, manzanita, and other shrubs and forbs, some of which can remain dormant but viable in the soil for a hundred years or more (Quick 1956). Dormancy is broken by fire or by abrasion from machines, and seed rapidly germinates. Other particularly aggressive plants are bear clover and bracken fern, which sprout dense networks of underground rhizomes that produce dense foliar cover.

Shrubs are an important component of forest stands because they provide habitat for various wildlife species, help cycle soil nutrients and prevent erosion and because some fix nitrogen. However, depending on density, they can be severe competitors that prevent the establishment of conifers, particularly shade-intolerant pines (Harrington et al. 1991; McDonald and Fiddler 1989). The effect of competition on reducing conifer growth is well understood (Bolsinger 1980; Radosevich et al. 1976; Conard and Radosevich 1982b; Lanini and Radosevich 1986; Walstad and Kuch 1987; Tappeiner et al. 1992; McDonald and Oliver 1984; Oliver 1984; Conard and Sparks 1993). In some cases, survival of pine seedlings has been reduced by 80% as shrub biomass increased to about 7,000 kg/ha (6,245 lb/acre) (Hughes et al. 1987; McDonald and Radosevich 1992). The effect of shrub removal can be long lasting. Conard and Sparks (1993) showed that California white fir saplings maintained a continued response eight years after treatment. In general, research has supported the common observation that shrub canopy in the Sierra must be reduced to below approximately 30% before detectable increases in growth in conifer saplings will occur (Conard and Radosevich 1982a; Conard and Sparks 1993; Oliver 1984).

Options for controlling competing vegetation in the Sierra can include mulching, manual methods (cutting or grubbing), grazing, and herbicides (McDonald and Fiddler 1993). A review of forty studies has produced the conclusion that all of these methods will release conifer seedlings; however, herbicides continue to be the most effective and least costly method of control. Use of this method, however, is often constrained by societal concerns. Cattle browsing is very effective in some areas (Allen 1987; Huntsinger 1988). Mulching is often expensive, and hand grubbing is effective only when applied to herbaceous plants and small shrubs that have germinated from seed. Several cuttings may be needed to control vigorous sprouters such as tan oak, manzanita, and *Prunus* spp.

The need to release conifers from weed competition must be considered in all regeneration methods. This is true for both even-aged and uneven-aged methods because in both, bare mineral soil must be provided to permit conifer seedlings or germinants to become established, and this also favors the rapid growth of shrubs and herbs. All release treatments should be prescribed at the minimum level needed to provide desired plants with a temporary advantage relative to competing plants. Once the desired plants are established, the desired proportion of ground cover vegetation can be maintained.

In summary, forest managers have several methods for regenerating Sierran stands that are based on research and have been successfully applied. Experienced silviculturists, with firsthand knowledge of their sites and the information outlined here, are a key part of any successful regeneration project. Application of this information varies throughout the Sierra and must be adapted differently to east-side and west-side forests, depending on specific stand conditions. The potential natural regeneration of both trees and shrubs, methods for shrub control if needed, and the operational aspects of tree planting and evaluating seedling survival and growth vary from site to site.

Productivity of Sierran Forests

Except for stands on serpentine soils or on east-side volcanics, Sierran forests are productive, with growth rates varying within each forest type. Very productive mixed conifer sites can produce 17.5 m^3/ha/year (250 ft^3/acre/year) (Oliver in press), with trees that are 42.6 m tall trees (140 ft) in 50 years (Biging and Wensel 1985). Red fir forests at high elevation are often pure, sometimes mixed with California white fir, and can be very dense and highly productive. They are capable of producing 15.0 to 16.4 m^3/ha/yr (214 to 235 ft^3/acre/year), with trees that are 36.6 m (120 ft) tall in fifty years (Schumacher 1928). They also can attain high basal area densities of 114 m^2/ha (498 ft^2/acre) (Oliver 1988). In the northeastern California and on the east side of the Sierra crest, site quality is much lower due to low precipitation and shallow soils. East-side ponderosa pine forests, for example, reach densities of 32 m^2/ha (140 ft^2/acre) but have a productive capacity of only 5.6 m^3/ha/year (80 ft^3/acre/year) (Oliver 1972).

Knowledge Gained from Research Forests

The University of California Blodgett Forest Research Station: Mixed Conifer (Based on Olson and Helms 1996)

Blodgett Forest Research Station is a 1,214 ha (3,000 acre) property in the mixed conifer forest type at an elevation of between 1,188 and 1,463 m (3,900 and 4,800 ft) near Georgetown, California. It was given as a gift to the University of California, Berkeley, in 1933 after it had been logged of most high-grade commercial timber. Over time, new conifers became established and young-growth stands and residual large trees grew in size until, in 1958, the university was able to begin

annual timber sales. Blodgett Forest is dedicated to research and the demonstration of forestry. In order to provide diverse stand structures, a management plan was initiated that divided the 109 compartments among even-aged management, uneven-aged management, and unmanipulated reserves. Regeneration methods used in Blodgett Forest include even-aged methods (clear-cutting and shelterwood) and uneven-aged methods (single-tree selection and group selection). In addition a treatment called overstory removal was used; this involves removing large trees to release existing conifer regeneration. All forest operations are done within the context of a Timber Harvest Plan approved by the State Department of Forestry and Fire Protection and meet the requirements of the California Forest Practice Act. Regeneration is accomplished with a mixture of natural seeding and planting. Because of sequential forest inventories, the records at Blodgett Forest permit an analysis of sixty years of forest growth and thirty-six years of comparative silviculture. These records are the longest time-series data set in the mixed conifer forest and provide the most extensive information on the long-term outcome of alternative silvicultural practices in the Sierra Nevada. Details of inventories, growth and yield, and outcomes from the use of alternative silvicultural treatments are provided in Olson and Helms 1996.

Growth in the Forest from 1899 to 1994. As was mentioned in an earlier section, Sudworth (1900, 1901) measured one of his fifty-three 0.10 ha (0.25 acre) plots in the Sierra Nevada in what is now Blodgett Forest. This single plot had, in 1899, 108 trees per acre, the average height of which was 45.7 m (150 ft), with clear stems 9.1 to 10.6 m (30 to 35 ft). The average area was 214 m^2/ha (930 ft^2/acre), which is twice that shown as maximum in Dunning and Reineke's 1933 tables for young-growth mixed conifer stands at age 150 years. Of the total basal area, 49% was in California white fir, 19% was in ponderosa pine, 13% was in incense cedar, 10% was in sugar pine, and 9% was in Douglas fir. The largest eight trees per acre, those between 121.9 and 137.2 cm (48 to 54 in) diameter at breast height (dbh), were all California white fir and ponderosa pine, but all five species were represented in diameters greater than 111.7 cm (44 in). No trees reported were less than 40.6 cm (16 in) dbh, and all had fire scars, with the most recent having occurred fifteen years previously. There was abundant regeneration of all species, aged from one to twelve years old. With a historic ground fire cycle of between seven and twenty years, most regeneration would presumably be repeatedly destroyed, giving rise to the commonly described open, parklike stand conditions of the presettlement period. Over the first decade of the 1900s the forest was extensively logged by railroad. The first inventory of the forest was made in 1934, and subsequent forestwide inventories were made in 1955, 1973, and 1994 (table 15.2).

These data show that growth in Blodgett Forest from 1934 to 1994 was 265,700 m^3 (56.3 million board ft). From 1955 to 1994, total harvest in the forest was 318,100 m^3 (67.4 million

TABLE 15.2

Growth over the period 1934 to 1994.

Year	Forest Stocking (Million Board Ft)[a]	Stocking (Board Ft/Acre)[b]	Growth (Board Ft/Acre/Year)
1934	26.18	12,700	
1955	37.82	18,350	257
1973	63.94	22,680	918
1994	82.47	29,270	959

[a]1 board ft = 0.02832/6 m^3.
[b]1 board ft/acre = 0.06997/6 m^3/ha.

board ft). Total growth was therefore 583,400 m^3 (123.7 million board ft), and the accumulated harvest through 1994 (adjusted for changes in forest area over time) represents 68% of total growth. As Blodgett Forest became more fully stocked, growth acre/year increased from 3.0 m^3/ha/year (257 board ft/acre/year) in 1934 to 11.2 m^3/ha/year (959 board ft/acre/year) in 1994.

Finding 1: Mixed conifer forests on high-quality sites are capable of having an increased stocking and growth rate while at the same time sustaining a harvest equivalent to 68% of growth. Concurrently, the diameters of trees can increase by as much as 76.2 to 101.6 cm (30 to 40 in), and overall forest structure can become more diverse.

Effects of Silvicultural Treatments on Growth. Productivity in terms of growth per acre per year depends both on silvicultural treatment and on the unit of measure (table 15.3). Because of the absence of harvesting, growth in compartments left as reserves was among the highest of all treatments. The highest-volume growth (including harvest volume) occurred in group selection areas, but this was due to relatively high harvest levels in areas with that treatment. The highest growth in basal area occurred in the clear-cut and group selection areas, reflecting the high growth rates of pine regeneration, the fact that all trees in a young planted stand grow rapidly, and the fact that areas with other treatments contain suppressed and intermediate trees that grow relatively slowly. Growth in single-tree selection areas was very similar to that in areas managed under even-aged methods with thinning. Stands that were clear-cut and replanted had lower average growth than continually stocked uneven-aged stands, due to the period during which clear-cut stands are being regenerated.

Finding 2: The board foot productivity of areas with group selection, single-tree selection, and even-aged treatment was similar. In time, it is expected that clear-cut areas will also produce about 1,000 board ft/acre/year. Differences in productivity and individual tree growth are due to the residual stocking levels after harvest (or the heaviness of the cutting).

Effects of Silvicultural Treatments on Regeneration. Regeneration surveys done in all treatment areas in 1994 show that the total number and distribution by species of seedlings 0 to 1.2 m (0 to 4 ft) tall and saplings 0 to 10.1 cm (0 to 4 in) dbh is not markedly different among silvicultural methods of regeneration (tables 15.4 and 15.5). The data show that the most abundant species of conifer regeneration on all treatment areas are California white fir and incense cedar. Least abundant are the shade-intolerant pines (ponderosa pine and sugar pine, particularly in areas where cutting has been aimed at thinning as opposed to establishing regeneration. Douglas fir is intermediate in abundance. California black oak and other hardwoods are abundant in all areas. Other softwoods (Pacific yew and giant sequoia) occur in small numbers. Regeneration in all treatment areas (including those using clear-cutting and group selection) is adequate to maintain representative species composition and control of quality by precommercial thinning. Additionally, observations show that seedlings planted in areas on which the shelterwood, group selection, and clear-cutting treatments were used grow much more rapidly than natural regeneration in these treatments.

Finding 3: Natural regeneration of all species is adequate to provide well-stocked stands in all silvicultural methods, providing that bare mineral soil is available as a seedbed. Growth of regeneration is enhanced with exposed microsites, lower stocking levels, and control of competing vegetation. Planting ensures prompt conifer regeneration that can more readily compete with grass and shrubs.

Effects of Silvicultural Treatments on Understory Vegetation. Over the past thirty years of management of Blodgett Forest, the understory vegetation in many compartments has been manipulated to provide a temporary advantage to regenerating conifers. Treatments have included underburning, herbicide applications, grazing, and mixtures of treatments. For the past twenty years, records have been kept on the

TABLE 15.3

Growth by silvicultural treatment, 1980 to 1994.

Treatments and Reserves	Growth + Harvest (All Species)		Basal Area
	Volume		
	(Ft^3/ Acre/Year)[b]	(Board Ft/ Acre/Year)	(Ft^2/ Acre/Year)
Reserves[a]	194	1,510	3.9
Group selection	216	1,090	8.4
Single-tree selection	123	710	4.5
Even-aged thinning	136	890	3.8
Overstory removal	70	400	3.1
Clear-cutting (0 to 14 years)	77	190	9.6

[a]No harvesting occurred in reserves.
[b]1 ft^3/acre/year = 0.06997 m^3/ha/year.

TABLE 15.4

Number of seedlings per acre[a] by species and treatment in young-growth and old-growth reserve areas.

Treatment	Ponderosa Pine	Sugar Pine	Douglas Fir	White Fir	Incense Cedar	California Black Oak	Other Hardwoods	Other Softwoods	Total
Group selection	140	170	98	398	423	1,525	121	0	2,875
Single-tree selection	355	90	138	379	445	1,271	140	0	2,818
Overstory removal	139	43	166	334	546	1,456	123	0	2,807
Even-aged thinning	68	45	189	482	450	1,637	447	3	3,321
Young-growth reserve	6	36	53	369	142	272	478	25	1,381
Old-growth reserve	26	50	200	647	1,185	479	612	82	3,281

[a]1 seedling/acre = 2.471 seedlings/ha.

amount of ground cover vegetation present in the forest. Table 15.6 shows the average amount of ground cover, by silvicultural treatment and method of control of ground vegetation when remeasurements were taken in 1993-94. Table 15.6 shows that without any treatment, the amount of understory vegetation in managed compartments is approximately 16% to 30%. This amount of cover can be compared with the level of 30% that has often been cited (Oliver 1984 and others) as the threshold level above which growth of conifer saplings is markedly reduced. The relatively low level of 7.5% cover in untreated reserve compartments is due to their having high stocking levels of 67 to 83 m²/ha (290 to 360 ft²/acre). The effectiveness of herbicides is shown by the fact that compartments with this treatment have understory vegetation cover of between 5% and 10%. As might be expected, cattle grazing results in the most variable amount of control, with cover ranging from 4% to 52%. The use of a combination of grazing plus spot herbicide application kept understory cover to 7% to 9%. Underburning, with an average ground cover of 14%, was not as effective as other methods of control (except grazing) but was still satisfactory. It should be noted that, in all methods of shrub and herb control, substantial amounts of understory vegetation have been retained. The management goal in vegetation control is not to eliminate all ground cover but to reduce it below the 20% to 30% level.

Effects of Silvicultural Treatments on Fuels. Surveys show that the amount of fuels in any given treatment area depends primarily on the regeneration method (table 15.7). Clear-cutting, site preparation, and planting resulted in one-third less fuel than other treatments. The least fuels were removed in the single-tree selection method.

Finding 4: Clear-cutting and planting removed the most fuels, and single-tree selection retained the most fuels. There was little difference among group selection, overstory removal, and thinned areas. Overstory removal and clear-cutting resulted in less accumulation of material greater than three inches in diameter. All treatments retained similar amounts of the nutrient-rich duff layer that is also critical to protect the soil from erosion.

Summary of Major Findings at Blodgett Forest. Records of 60 years of forest development have shown that mixed conifer forests respond well to diverse treatments:

- Various kinds of stand structures can be sustained without affecting stand growth.

- The amount of regeneration amount and the species composition are relatively insensitive to the cutting method used.

TABLE 15.5

Number of saplings per acre[a] by species and treatment in young-growth and old-growth reserve areas.

Treatment	Ponderosa Pine	Sugar Pine	Douglas Fir	White Fir	California Incense Cedar	Black Oak	Other Hardwoods	Softwoods	Total
Group selection	8	2	19	83	63	2	64	0	241
Single-tree selection	22	2	26	112	100	5	10	0	277
Overstory removal	7	10	10	49	44	10	23	2	155
Even-aged thinning	0	3	50	58	55	37	26	0	229
Young-growth reserve	8	3	19	108	53	3	47	6	247
Old-growth reserve	3	9	6	85	79	0	26	9	217

[a]1 sapling/acre = 2.471 saplings/ha.

TABLE 15.6

Control of understory vegetation under different silvicultural regimes (data from forty-four compartments).

	Understory Cover (%)		
	Shrubs	Herbs	Total
No Shrub Treatment			
Group selection	21.7	1.6	23.3
Single-tree selection	11.4	4.4	15.8
Thinning	23.3	6.3	29.6
Reserve	6.0	1.5	7.5
Herbicide			
Group selection	6.2	1.8	7.9
Single-tree selection	0.8	4.5	5.3
Overstory removal	4.9	5.4	10.3
Grazing			
Group selection	18.5	8.0	26.5
Single-tree selection	2.9	0.7	3.6
Overstory removal	33.0	18.5	51.5
Shelterwood	2.2	3.7	5.9
Grazing + Herbicide			
Group selection	7.7	1.9	9.2
Single-tree selection	4.9	1.7	6.6
Overstory removal	7.6	3.0	7.2
Underburning			
Overstory removal	11.8	2.0	13.8

- Growth and yield depend primarily on the number of trees per hectare.

- Clear-cutting and overstory removal had the least amount of fuels remaining after treatment; individual tree selection and reserves had the most.

Experience at Blodgett Forest suggests that in environments without stand-replacing fires, the mixed conifer forests of the Sierra are amenable to a wide variety of silvicultural treatments without a loss of productivity. All silvicultural treatments are applicable, ranging from those using even-aged through uneven-aged structures and those aimed at retaining few to many species. Stands and landscapes can be managed to provide a wide array of forest values and uses. This experience forms the basis of figures 15.3 and 15.4, which show the range of stand structures that can be attained from mixed conifer stands by silvicultural treatment. Both figures show stands that differ in the age and condition of existing trees, and the manner in which stand structure can change over time through the use of different even- and uneven-aged silvicultural systems. In both cases, it is important to recognize that a wide variety of stand conditions can be maintained and that, if desired, a similar long-term stand condition can be arrived at even though stands were harvested or regenerated using different treatments.

It must be recognized that Blodgett Forest is situated on a relatively level and very productive site. The management opportunities and rates of response to treatments are, therefore, wider and more rapid than those of forests on sites of lower productivity. In general, however, sites of all levels of productivity have a similar range of opportunities for stand development, the difference being that the rates of stand development are slower as site quality decreases. In addition, as site quality decreases, the silviculturist becomes increasingly concerned about the effect of treatments on the maintenance of soil nutrient status.

U.S. Forest Service Blacks Mountain Experimental Forest—East-Side Ponderosa Pine

The east-side pine type of northeastern, central, and southeastern California developed in the presence of relatively frequent fires that burned with variable intensities (Martin and Dell 1978; Weatherspoon 1983). These fires, coupled with heavy grazing by sheep, maintained surface fuels at low levels and kept understories relatively free from tree regeneration. Old-growth trees were two hundred to three hundred years old and approximately 76 cm (30 in) in diameter, and stands varied in density from being overstocked and having a high level of mortality due to beetle infestation to being more open, averaging about 12 trees/ha (5 trees/acre) (Dolph et al. 1995). With the cessation of extensive grazing and the

TABLE 15.7

Amount of fuels (tons per acre)[a] remaining after application of treatments.

Treatment	Duff[b]	Size Class[c]			Total
		0–1 in	1–3 in	>3 in	
Group selection	17.5	2.1	3.4	12.5	35.4
Single-tree selection	25.5	2.1	3.4	15.0	46.0
Even-aged overstory removal	15.0	3.9	3.4	7.9	30.2
Even-aged clear-cutting/regeneration	14.9	1.0	1.5	4.8	22.2
Even-aged—thinning from below	18.1	2.0	2.3	15.4	37.8
Old-growth reserve	26.9	3.1	2.4	14.6	47.0
Young-growth reserve	19.7	1.8	2.0	11.6	35.2

[a]1 ton/acre = 2.511 tonnes/ha.
[b]Decomposed leaves and branches.
[c]Undecomposed branches and stems.

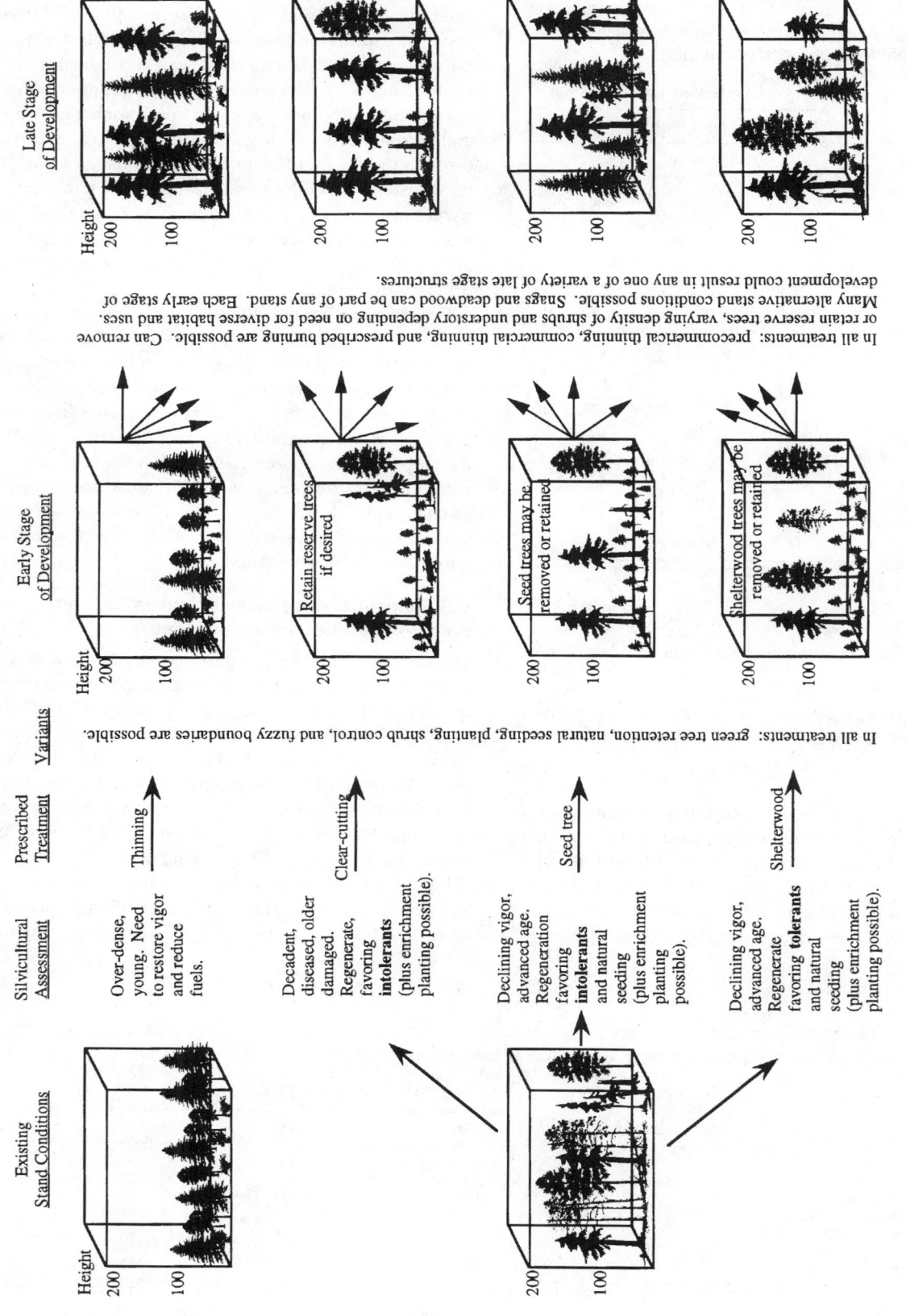

FIGURE 15.3

Hypothetical development of even-aged stand structures following differing silvicultural treatments in mixed conifer stands of differing initial condition.

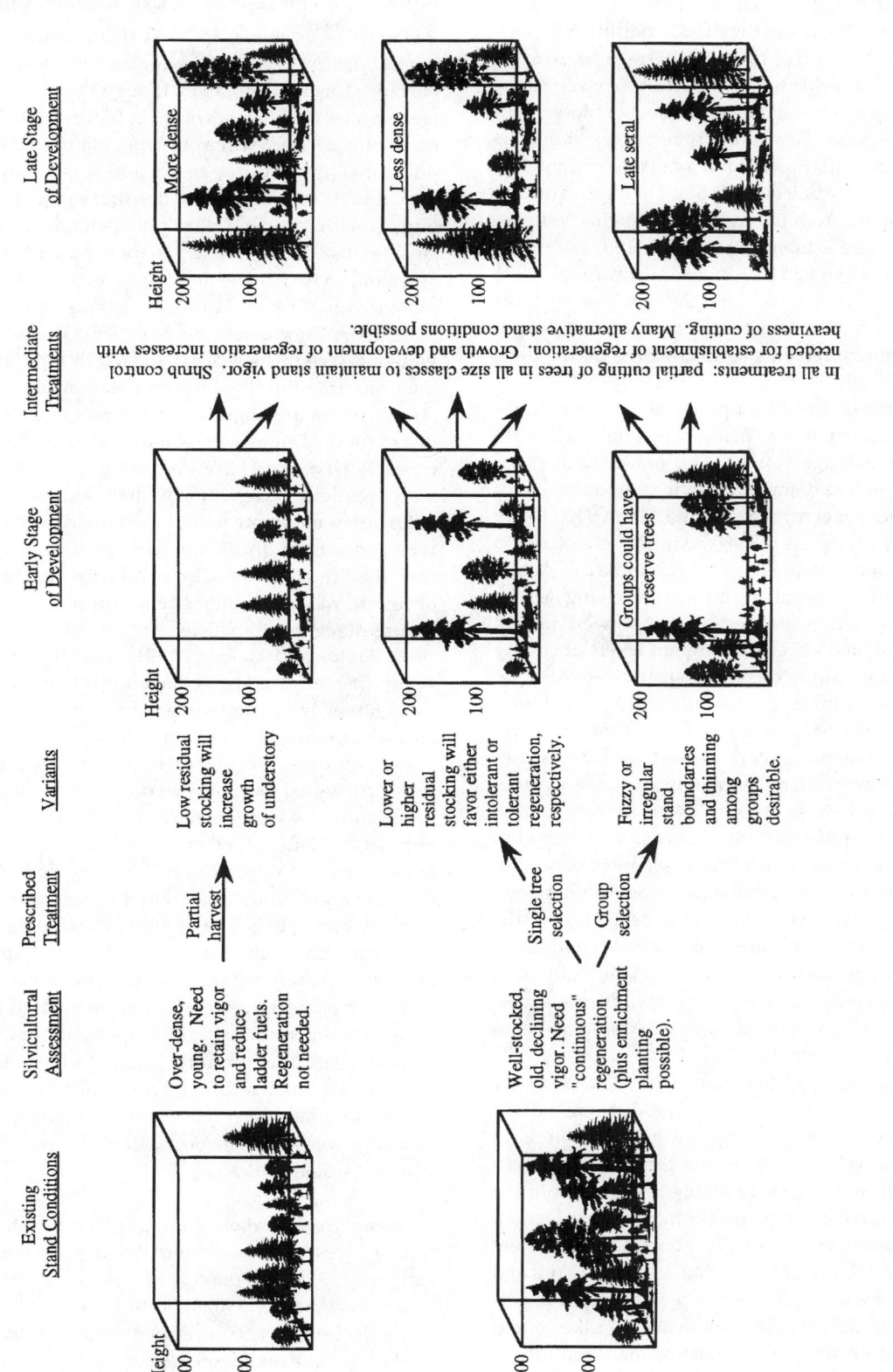

FIGURE 15.4

Hypothetical development of uneven-aged stand structures following differing silvicultural treatments in mixed conifer stands of differing initial conditions.

introduction of a fire-suppression policy, much of this forest type has developed abundant conifer saplings and poles.

In 1938 a study was begun on the Blacks Mountain Experimental Forest at an elevation of 1,706 to 2,103 m (5,600 to 6,900 ft) in northeastern California to test the effects of six levels of harvest on the growth and stand development of old-growth interior ponderosa pine. This forest type covers about 2.3 million acres, nearly 14% of the total available commercial forest area in California (Bolsinger 1983). The six treatments consisted of removing from 0% to 95% of existing volume. Forty-seven plots were established that covered, with their buffer strips, about 485.6 ha (1,200 acres). Because the plots were installed over a period of years, the database contains records ranging from forty-four to fifty-two years in length and consists of remeasurements of 13,274 trees on the plots (Dolph et al. 1995).

Diameter growth was lowest for pine and highest for California white fir, with actual rates being higher on stands with lowest residual stocking. Many large, old-growth pines showed zero or negative diameter growth due to the loss of bark. Initial volumes per acre ranged from 172.8 m^3/ha (16,010 board ft/acre) to 205.0 m^3/ha (18,480 board ft/acre). Growth ranged between 1.5 to 2.2 m^3/ha/year (22 and 32 ft^3/acre/year) with higher growth occurring on areas receiving heavier partial harvests. Growth responses in all plots were in proportion to the combined effects of different levels of cutting and competition from numerous small-sized trees. Also, ingrowth volume (the volume of trees that, since the previous inventory, have grown past the lower limit of measured diameter) increased, and mortality decreased with intensity of cutting. The loss of larger-diameter trees in all plots was probably due to increased stress brought about by competition from the large number of small trees. This stress resulted in reduced vigor and eventual mortality caused by bark beetles.

An evaluation of trends in plant succession found no evident relationships between changes in the percentage of the pine and California white fir portion of the pole component and intensity of cutting. Because the stands on the study plots developed under a Forest Service policy of wildfire suppression, the number of stems in the smaller size classes have increased dramatically since 1930.

The study results showed that if, as was the case in the 1930s, the objective is to maximize timber production, the best course is to convert slow-growing old-growth stands to young, faster-growing stands. Some old-growth characteristics have been lost in all plots, probably resulting from the wildfire-suppression policy. Perhaps the major lesson learned from this long-term study was a verification of the statement made by DeBell and Franklin (1987) that "characteristics and functions of old-growth stands cannot be guaranteed in perpetuity by simply preserving old-growth tracts." Like young-growth stands, old-growth stands must be managed for the desired attributes (Dolph et al. 1995).

U.S. Forest Service Swain Mountain Experimental Forest—Red Fir (Based on Laacke and Tappeiner 1996)

Red fir (*Abies magnifica*) and the varietal Shasta (*Abies magnifica* var. *shastensis*) occur in the higher elevations of the Sierra Nevada (Laacke 1990). Even though there are several associated species (California white fir, Jeffrey pine, incense cedar, sugar pine, western white pine, mountain hemlock, and lodgepole pine), much of the red fir forest is a natural monoculture. Red fir forests have an unusual capacity for sustaining high growth rates and developing stands having high density. Because of this, red fir forests commonly lack an understory and contain fewer and less diverse flora and fauna than more-diverse forests at lower elevations (Barbour and Woodward 1985; Gordon and Bowen 1978). The total numbers of vertebrate wildlife species in the red fir type, as indicated by the California Wildlife Habitat Relationships model (WHR) and database are slightly more than one-half of those predicted for the Douglas fir, ponderosa pine, and mixed conifer types (R. J. Laacke, U.S. Forest Service, Pacific Southwest Research Station, conversation with the author, 1995). Although wildfires of moderate to high severity do occur in the red fir type and can produce large patches of even-aged regeneration, the fire potential is low, and fires are characteristically of low to moderate intensity and result in small scattered groups of regeneration (Kilgore 1971, 1973; Agee 1990; Taylor 1993; Taylor and Halpern 1991). Red fir forests consist of stands of even-sized trees that appear to be of even age even though they were commonly regenerated over a long period and are thus actually of diverse ages.

With the exception of local impacts such as the narrow corridors harvested for fuelwood during the building of the railroads and the concentrations of livestock around new population centers (Leiberg 1902), incursion into the red fir forests was late in coming. Logging had begun in 1943 (Oosting and Billings 1943), but it wasn't until the mid 1950s that harvesting began to be significant. As recently as 1962, little was known about the silvical characteristics of red fir, and the type was regarded as too small, variable, and secondary in value to warrant management. Red fir forests continue to be, however, critically important from the standpoint of watersheds that produce much of California's water. Because of its inaccessibility, red fir has not been logged as heavily as other forest types and, after subalpine and lodgepole pine forests, is one of the least altered of the Sierra Nevada ecosystems.

Natural Regeneration. Natural regeneration is common in red fir forests (Barbour and Woodward 1985; Taylor and Halpern 1991), with good seed years occurring every one to four years (Laacke 1990). In a study of different regeneration methods, Gordon (1979) verified the importance of abundant seed and shade for adequate natural regeneration. Seedlings are most abundant in the most narrow clear-cuts, in small group selection cuttings, and under shelterwood stands having the largest number of trees in the overstory. The growth of seedlings is inversely related to the density of overtory (Laacke and Tomascheski 1986). Existing advance regeneration releases well following removal of the overstory (Gordon 1973), with the rate of release depending on the physiological condition of the regeneration (Gordon 1978; Oliver 1985). Initially, making large openings in stands resulted in little competition from grass and sedge. However, recent observations indicate that in later, adjacent openings the density of this vegetation increased due, probably, to an increased capacity for development of propagules. There is evidence that the increased cover of grasses, sedges, and herbaceous plants is providing opportunities for increased populations of gophers, which cause severe mortality to red fir regeneration. Although resistance to gopher damage increases with tree size, even large saplings can be killed when gopher populations are high (Gross and Laacke 1984).

Planting. Early reforestation efforts focused on planting cleared areas with Jeffrey pine, due to a high mortality rate among planted red fir. In many cases, natural red fir seedlings became established under the planted pines and became dominant before the pines reached maturity. However, on some sites, Jeffrey pine is now the dominant species. In the last two decades, studies of seed production (Gordon 1978); collection (Oliver 1974); and nursery production, storage, and handling (Jenkinson 1980) have provided the knowledge needed to raise vigorous seedlings with a high potential for survival after planting.

Silvicultural System. Past studies on Swain Mountain in the northern Sierra have demonstrated that red fir can be managed using either even- or uneven-aged systems. Harvested stands can be regenerated through natural seeding, providing the harvested areas do not exceed 61m (200 ft) in diameter. If the stands surrounding an area of this size are at least 45.7 m (150 ft) in height, they will offer sufficient shade and seed supply to provide regeneration. Larger blocks with a width of 106 m (350 ft) have about a ten-year delay in becoming fully stocked by natural seeding but, in time, develop into vigorous stands. With the current availability of high-quality seedling stock, it is expected that harvested blocks on favorable sites can readily be regenerated by planting. On more difficult sites, shelterwoods would enhance the survival of either natural or planted stock. A concern in even-aged management and group selection is the tendency for cut blocks to develop grasses and sedges that compete with seedlings and provide habitat for gophers. It has been demonstrated that red fir stands can readily be managed through the use of the shelterwood system. In stands of trees averaging 76 to 91 cm (30 to 36 in) in diameter, shelterwood cuttings left a range of cover represented by basal areas ranging from 12 to 48 m^2/ ha (50 to 210 ft^2/acre). Most regeneration was obtained in the densest stands, but the highest growth rate was obtained in the least dense stands. Maximum growth of red fir, once established, occurred in fully exposed sites. Consequently, once

seedlings have become established, the best growth is obtained by prompt removal of the overstory.

Because of the tolerant nature of the species and the response in its growth rate to partial cutting, there is considerable opportunity for the use of single-tree and group selection systems. This is indicated by abundant regeneration in natural gaps. The major constraint in uneven-aged approaches is the prevalence of dwarf mistletoe, which, where it occurs, makes it risky to regenerate stands using single-tree selection. The exception is on the most productive sites, where the growth rate of young trees will outpace the effects of the mistletoe. An additional difficulty with single-tree and group selection, particularly on steep ground, which requires the use of cable systems, is the potential for butt rot caused by logging damage. A risk rating system has been developed (Ferrell 1980) that provides quantitative guidelines based on crown characteristics for assessing the vigor of fir trees and their potential for mortality. These guidelines can be used to predict which trees are likely to die or produce snags and cavities and which have the potential to grow rapidly into larger size classes.

In all silvicultural treatments in true fir, it is necessary to avoid damage to trees, which often leads to infection by diseases (Aho et al. 1983, 1989). To ensure prompt natural regeneration, adequate soil disturbance produced by logging is needed to expose sufficient bare mineral soil as a suitable seedbed. Designated skid trails should be used to restrict soil compaction to permanent landings and access systems not used for growing trees. Red fir forests are noted for erodible soils and a high proportion of forest nutrient capital in the litter and down woody material on the forest floor (Powers and Edmonds 1992). Consequently, the use of practices such as broadcast burning of slash require more care in red fir forests than in other parts of the Sierra.

Even though red fir forests are dominated by one species, there is considerable silvicultural flexibility in their treatment, with many options available (Gordon 1970). Site-specific prescriptions will normally vary from the northern to the southern Sierra. For example, in the southern Sierra, lodgepole pine may regenerate with red fir. Providing the ground is not too steep, both even- and uneven-aged systems would work well, and variants of these methods that incorporate irregular-sized openings, thinnings, and the retention of groups of large trees, down logs, and snags would provide any structure needed for wildlife habitat, wood production, and watershed protection. Steep ground that requires the use of cable logging would limit treatment options.

U.S. Forest Service Challenge Experimental Forest—Mixed Conifer and Hardwood

The Forest Service Experimental Forest at Challenge, California, is the site of a long-term analysis of growth of natural regeneration under five different cutting methods (McDonald 1976). The site is highly productive, and the mixed conifer stands averaged 42.6 m (140 ft) tall. The study was installed

TABLE 15.8

Seedling density in study area, by cutting method.

Cutting Method	Seedling Density at Nine Years (per Acre)[a]						
	Ponderosa Pine	Sugar Pine	Douglas Fir	White Fir	Incense Cedar	Hardwoods	Shrubs
Single-tree selection	860	111	308	400	44	1,330	-
Group selection	1,500	185	134	565	16	807	-
Shelterwood	3,620	240	80	192	470	2,225	-
Seed tree	2,100	75	174	66	67	2,937	-
Clear-cutting	1,115	51	157	166	-	746	6,523

[a]1 seedling/acre = 2.471 seedlings/ha.

in the period 1960 to 1963, and either natural or artificial seeding was provided one to two years after cutting. Silvicultural methods tested were single-tree selection (20% of volume removed); group selection with 9.1, 18.2, and 27.4 m (30, 60, and 90 ft) diameter openings; shelterwood with 30 trees/ha (12 trees/acre), seed tree with 10 to 20 trees/ha (4 to 8 trees/acre), and clear-cutting. Table 15.8 shows the seedling density for each treatment at the end of the ninth growing season. These data show that pine was the most abundant conifer in all cutting methods, due to the predominance of pine seed sources in the overstory. In all cutting methods, the seedlings of all species were more than sufficiently abundant to develop into a mixed conifer forest, providing competing overstory was removed to allow seedling growth. Table 15.9 shows height growth for the site.

All species grew substantially faster in the clear-cut area; however the tallest species in all cutting methods were hardwoods and shrubs. Without exception, the height of every species increased as the availability of light and water increased, with growth being the least in single-tree selection.

Currently, approximately thirty years later (P.M. McDonald U.S. Forest Service, Pacific Southwest Station, conversation with the authors, 1995), reproduction in the group selection openings is about 4.6 m (15 ft) tall and constitutes releasable regeneration. The best growth is on trees in the center of the largest openings. Abundant conifers and hardwoods, especially those tolerant to shade, are present in all openings. All of the original species are present, but the growth of shrubs is limited.

Knowledge gained from the Challenge Experimental Forest includes the following:

- Each cutting method creates a specific microenvironment in terms of bare mineral soil, light, moisture, and soil surface temperature.

- The growth and development of species in these various microenvironments is directly related to their shade tolerance, with pine growing fastest and California white fir growing slowest as the amount of tree density and shade decreases.

- Hardwoods compete strongly with conifers and need to be controlled, initially, if conifer regeneration is to develop adequately.

- Tree growth is greatest in the centers of the largest openings.

- Densities of conifer seedlings and saplings densities need to be reduced if conifer regeneration is to develop adequately.

Silvicultural Prescriptions

Federal Lands

Silvicultural approaches in Sierran forests are based on policy, ownership, and existing stand conditions. Until the 1980s, decisions as to silvicultural treatments (such as harvesting, regeneration, and stand treatments) on federal lands were

TABLE 15.9

Seedling height in study area, by cutting method.

Cutting Method	Seedling Density at Nine Years (ft)[a]						
	Ponderosa Pine	Sugar Pine	Douglas Fir	White Fir	Incense Cedar	Hardwoods	Shrubs
Single-tree selection	0.5	1.0	0.7	1.4	0.9	2.4	2.1
Group selection	1.0	1.5	1.5	1.5	0.9	2.4	2.9
Shelterwood	2.7	2.6	2.1	2.9	1.9	2.9	3.1
Seed tree	3.7	3.4	3.1	3.7	2.7	4.3	3.7
Clear-cutting	6.2	5.5	4.2	4.0	-	12.6	7.8

[a]1 ft = 0.305 m.

based primarily on a goal of multiple use with an emphasis on timber management to support land-use plan goals. Assessments were made of stocking, health, and growth rate relative to equivalent well-stocked and healthy stands. Stands were then ranked in priority of need for treatment. Stands that were well-stocked, healthy, and growing well were bypassed, other stands were thinned, and the highest priority for treatment were those understocked, unhealthy, or slow-growing stands that could be made more productive by harvesting and regenerating. Culmination of mean annual volume increment (which indicates the maximum rate of biological productivity and commonly occurs in mixed conifer stands at the age of about 120 years when using merchantable measures of volume) was commonly used as an indicator of need to harvest and regenerate. In general, however, it was common for decisions regarding any one stand to be made incrementally, without a consideration of their implications —for example, a decision on regeneration might not be deliberately related to later needs for weed control or precommercial thinning.

Recognizing the need for better linkage between sequential decisions, the U.S. Forest Service in the mid-1970s instituted an advanced training program designed to teach silviculturists to develop an analytical and defensible approach to defining stand treatments. This approach centered on the development of a silvicultural prescription, which is a technical document supporting a recommendation on how to maintain or change existing stand conditions to meet desired conditions consistent with management objectives. It generally consists of (1) a description of stand location; (2) an assessment of current stand conditions (soil characteristics, site productivity, topography, climate, species, age, stocking, cover, structure, health, growth, understory composition, and cover); (3) a statement of management objectives; (4) a statement of ecological, managerial, and social constraints; (5) a list of proposed treatments; and (6) a description of the desired stand structure. The main objective of prescription writing is to formalize a procedure for assessing the ecological, managerial, and social characteristics of a particular situation, and to use this procedure to replace the more qualitative and subjective approaches used previously. Prescriptions vary in length and complexity, depending on the ecological and social sensitivity of the stand in question.

Stands that were diseased, poorly stocked, or slow growing due to past fires, harvesting, or insect and disease damage, were commonly clear-cut and planted with pines; the more shade-tolerant species native to the site were expected to seed in after the pines were established to re-form the mixed species composition. During the past decade, planting of mixtures of conifer species has become common. The sizes of clearcut patches were small, averaging 2.0 to 8.1 ha (5 to 20 acres). If adequate advance regeneration was present and could be released, the treatment was called overstory removal. Other stands might be thinned (i.e., harvested selectively) to maintain health and to accelerate growth. This approach led to the

maintenance of a mosaic of stand conditions on the landscape. Uneven-aged methods (single-tree selection and group selection) were seldom used in stands that are part of the "timber base" on federal lands.

Private Lands

On industrial private lands, with a goal of sustained timber production and consideration of other diverse land values, approaches similar to those used by the Forest Service were used to identify which stands were growing acceptably and which had a higher priority for harvesting or regeneration. Because existing stands had previously been salvage-cut or selectively cut, the diversity of size classes present provided a basis for more careful selective cutting (or thinning) designed to upgrade stand quality. Clear-cutting and planting was done in those stands that had a relatively uniform structure and where most trees were of economic size. In more structurally diverse stands, clear-cutting was not used, because of the need to dispose of the sometimes substantial amount of small-diameter, unmerchantable material and the probable need to plant. Approaches to regeneration varied considerably, depending on ownership and harvesting methods used. Natural regeneration was commonly relied on and, because of shaded conditions, the species that became established were primarily California white fir and incense cedar. This led to a change in species composition in the Sierra from the previously dominating pines to a preponderance of California white fir, incense cedar, and, in substantial areas, the dominance of released hardwoods. The past use of individual tree cutting on some private lands has led to an easy transition to adoption of the selection system.

Fire and Forest Management

There has been considerable experience with prescribed burning in the Sierra Nevada on park service and national forest lands. However, no comprehensive synthesis or summary of this experience has been undertaken to gain an understanding of the effects of the programmatic use of prescribed fires to limit wildfires, reduce fuels, or control the understory of established stands. Two important issues require research and application:

1. How to develop and maintain stand structures, densities and species composition that can be burned periodically to maintain fuelbreaks and forest health.

2. How can fire be introduced into stands that have not been underburned naturally or had prescribed fire for many years.

With regard to the first issue, there are currently no standard prescriptions. Perhaps a suitable goal would be to create wide spacing such that the crowns do not touch (about 40% crown closure). Maintaining this density in rapidly growing young stands (about thirty years of age) may require thinning every ten to fifteen years. Maintaining low levels of surface fuels (fallen trees, branches, shrubs, etc.) may require repeated prescribed burning at similar intervals. Pruning might be beneficial in these stands because it would limit the likelihood that wildfire and underburning will carry fire into the large crowns that develop at wide spacing and would also improve the quality of wood removed during thinning. If wood quality is not an issue, satisfactory reduction in crown length can be achieved through careful prescribed burning. Alternatively, maintaining a closed canopy might be beneficial on some sites, since this would encourage self-pruning, maintain a relatively cooler microclimate in the understory, and minimize the development of a shrub/hardwood/conifer understory. However, as trees age it may be necessary to encourage some conifer and hardwood regeneration to offset normal mortality. California black oak, tan oak, and Pacific madrone may be good fuel-break species, since they will sprout if their tops are damaged by prescribed fire.

The second issue involves several factors. First, on many sites where fire has not occurred for decades—especially productive sites—abundant fuel has accumulated from dying trees or limbs, growth of understory, and logging slash. To prevent controlled burns from doing unacceptable damage, these fuels should probably be reduced by either a combination of thinning from below and chipping followed by prescribed fire or if these treatments are not appropriate, a very careful initial burn followed by one or more reburns to consume the new fuels created by the initial burn. Second, fire commonly stimulates germination or sprouting of shrubs and hardwoods from seed banks and from buds on root crowns and rhizomes. Thus, controlled burning may produce more fuel than before the treatment. Reburning, herbicide applications, or mechanical treatments may therefore be needed. Burning in relatively dense stands can reduce sprouting vigor and kill most shrub seedlings. Third, areas that have not been burned for long periods commonly have deep accumulations of organic debris at the base of trees. Near the soil surface, the decomposing organic matter is generally rich in nutrients, holds considerable water, and therefore has abundant fine roots. Fire at the base of these trees tends to become very hot and, as well as killing or damaging the stem cambium, tends to completely remove this organic matter and associated fine roots. This is especially true for older ponderosa pine and sugar pine trees, since complete consumption of litter at the base of trees is likely, regardless of the season of burning. For California white fir, damage to the bases of the stems can be fairly easily controlled by burning in spring, when the lower duff is wet.

The important objective in fuels and fire management is, therefore, to design prescriptions to make the transition from the existing condition of wildfires that often burn all organic matter down to mineral soil to fires that are more frequent, cooler, and controlled. Considerable experimentation and research will be necessary before effective burning prescriptions can be developed that take into account the interrelationships

among fire intensity, soil and nutrient characteristics, soil water availability, insect and disease dynamics, and forest growth. Important social issues are those associated with the risk of fires escaping, smoke management, and cost. The liability associated with escaped fires is a major issue limiting the use of prescribed fire. As a result the costs of burning associated the liability resulting from an escaped fire are prohibitive. Current regulations and air-quality standards have limited the number of days in which prescribed burning can be done to such an extent that prescribed burning probably cannot accomplish the required fuels reduction in all the areas.

Logging Systems

Future management of Sierra forests will require innovative logging systems because of the more complex stand structures needed to meet diverse management objectives. Use of cable systems and low-pressure ground systems (used in other western forest types) should become more common. Efficient, low-impact systems that remove small material can play an important part in reducing fuels in the understory of older stands, in thinning young stands to reduce fuel loading, and in lowering the potential for future insect outbreaks. New yarding systems can be quite efficient if planning is done prior to their use. In addition, they require less road building and cause less soil disturbance than many ground-based systems.

Role of Planted Stands

Planted stands have always been a part of Sierran silviculture. Initially, this activity focused on the clearing and planting of brush fields that had increased in extent due to the combined effects of lightning-caused wildfire and fires associated with mining, logging, and grazing. Initially, planting was restricted to pines because of the historic value and preference for this species and also because the exposed microclimatic conditions were suited to pine establishment. Currently, wildfires still consume thousands of acres in the Sierra Nevada annually. On private lands, which are mostly at lower elevations, these burns are commonly replanted to ponderosa pine. On federal lands, which are more commonly at higher elevations, mixed conifer species are commonly planted. Because wildfires will continue into the future, planted forests will continue to have an important role in the Sierra. The important question to address, then, is, what is the appropriate seed source and species composition, structure, and size and location of these planted forests on the landscape? Forest sites in the Sierra can grow productive stands in a wide variety of forms. Depending on land-management goals and ownership, planted stands can vary from dense, productive monocultures with high timber yields to mixed, uneven-aged, more open stands that could be virtually indistinguishable from natural stands. Similarly, planted stands can vary in age before harvest from short-rotation, biomass stands grown for pulpwood production to stands that can take 100 years or more to reach the stage at which they contain uneven-sized trees, contain dead and dying trees for habitat, and have many of the characteristics of late-seral-stage stands.

Performance of Young USFS Plantations (Based on Landram 1996)

The Forest Service has recently completed an analysis of plantation information to determine the programmatic success of artificial reforestation. New analytical techniques using the most current information were employed.

The USFS reports that forestland on national forests in the Sierra Nevada covers approximately 3.1 million ha (7.6 million acres). Four percent of this land, 123,000 ha (300,000 acres), consisted of plantations in 1991. Another 2% consisted on nonstocked areas not yet planted or scheduled for replanting, most of which resulted from the large fires on the Plumas and Stanislaus National Forests that occurred in 1987. There are about 9,400 plantations in the SNEP study area. About 500 of these were established after fire and the remainder after timber harvest. However, because of the large size of the burns, about half of the plantation area resulted from burns and half from timber harvest. The average size of plantations following timber harvest is 6 ha (15 acres), and the average size following a fire is 138 ha (340 acres).

Using site-specific automated stand records, the Forest Service has estimated that about 90% of the area where planting was done between 1988 and 1992 (inclusive) was stocked at more than 371 trees/ha (more than 150 trees/acre) at the time of the last survey. About 10% of the area had not been successfully regenerated, and work is continuing on these areas.

Recent inventories of older plantations (average age twenty-three years) to evaluate growth and species composition have shown that

- 65% of trees are either ponderosa or Jeffrey pine.

- 2% of trees are sugar pine.

- California white fir and incense cedar are becoming established underneath the planted pines in most forest types.

- Shrub cover averages about 50% and tree cover about 25%.

- Average stocking in 1991 was 680 trees/ha (277 trees/acre) and 10.8 m^2 basal area/ha (47 ft^2/acre). Thinning to reduce intertree competition and to maintain tree vigor will probably be desirable in about ten to fifteen years in many plantations.

- The volume at age fifty-five years (the probable time of first commercial thinning) was forecast to be 252 m^3/ha (3,600 ft^3/acre), which would meet timber management objectives in the current national forest plans.

In developing this assessment of plantation status, it is recognized that additional inventories of older mapped plantations will provide better estimates of species composition,

stocking, and growth rates. Random spot checks of actual plantations have been done to verify the general characterizations. The new development of GIS-based maps for each national forest, showing the location of plantations and other forestlands provides a sound basis for monitoring the results of silviculture projects both in plantations and in other vegetation types.

SILVICULTURE OF OAK WOODLANDS

As described in McDonald 1995, California's indigenous hardwood resources can be divided into groups: foothill woodlands at lower elevations and forest-zone hardwoods at higher elevations.

The hardwood rangeland in the foothills throughout California covers about 4 million ha (10 million acres), 18% of which is publicly owned. Four percent is in reserves (Bolsinger 1988; Greenwood et al. 1993). Historically, these lands have been regarded as of limited value. Currently, however, they are recognized as being important for wildlife habitat, as a watershed, for rangeland, for conserving biological diversity, and for retaining open space (Standiford and Tinnin 1992; Helms 1994). A survey of county residents showed that the most important management issues are retaining oaks for shade, aesthetics, wildlife habitat, and fuelwood; improving oak regeneration; and offsetting the effects of land development (LeBlanc et al. 1989). It has been estimated that, from 1945 to 1985, oak woodlands were reduced by approximately 485,600 ha (1.2 million acres) (Bolsinger 1988). Earlier in this period, losses of oaks were due primarily to clearing for rangelands. In later decades, the primary causes of cutting were residential and commercial development and road and freeway construction. Since 1985, however, oaks have been valued much more than previously for wildlife habitat, soil protection, and enhanced property values (Standiford and Tinnin 1992). Consequently, there is less tendency for oaks to be cut and an increased interest in management and conservation (Standiford and Tinnin 1992).

Perhaps the most important silvicultural issue in the management of oak woodlands is lack of regeneration. In recent years, statewide studies have been conducted through the University of California's Integrated Hardwood Range Management Program (Standiford and Tinnin 1992) that have evaluated how various environmental and management factors affect oak regeneration. Studies have been made of site preparation, the use of augers to break up compacted soil, fertilizing to ensure adequate nutrient availability, top pruning to develop a more satisfactory shoot-to-root ratio to offset loss of water to evapotranspiration, weed control, methods of protecting seedlings from animal damage, and the role of introduced grasses and fire. These studies have shown that establishing oak seedlings is not as difficult as was once thought. Naturally regenerated oak seedlings tend to grow slowly, however, planted oaks will grow rapidly (0.6 m/year [2 ft/year]) if adequate soil moisture is available, especially if competing grasses are removed by clearing areas 1.2 to 1.8 m (4 to 6 ft) in diameter for each oak seedling (McCreary 1991; Standiford and Tinnin 1992). Where livestock browsing is a problem, tree shelters have been shown to be particularly effective—although somewhat expensive ($5 per shelter)—both in preventing browsing and in enhancing the microclimate for rapid growth (Standiford and Tinnin 1992).

Stocking control and thinning have received little attention in oak woodlands. A valuable contribution in this area is a report on the results of a five-year thinning study in ten stands of coast live oak (Pillsbury and Joseph 1991). The stands were sixty to eighty years old and averaged 25 to 27 cm (10 to 11 in) in diameter, 741 to 865 trees/ha (300 to 350 trees/acre), and 34 to 37 m²/ha (150 to 160 ft²/acre), and prior to thinning averaged about 4.9 m³/ha/year (70 ft³/acre/year). Stands were thinned to 23 and 12 m²/ha (100 and 50 ft²/acre). Five years after thinning, growth in the basal area of trees in the 23 and 12 m²/ha (100 and 50 ft²/acre) plots exceeded that of trees in control plots by ratios of 9:1 and 11:1, respectively. In terms of volume growth, trees in the 23 and 12 m²/ha plots grew 74% and 128% more, respectively, than those in the controls. The authors suggest that landowners can conduct an economically viable thinning operation and enhance stand vigor while promoting sound land stewardship.

Hardwood types in the forest zone occur at elevations from about 180 m (600 ft) in the north to about 2,400 m (7,800 ft) in the southern Sierra (McDonald and Huber 1995). Almost all the hardwoods are of sprout origin following fire or cutting. They include tan oak, California black oak, giant chinquapin, Pacific madrone, canyon live oak, and red alder. They are commonly found as single trees or clumps and in association with Sierran conifers and shrubs such as manzanita, deer brush, coffeeberry, *Prunus* spp., chamise, and forbs and grasses. The important uses of this zone are for wildlife, aesthetics, wood products, and water. The role of silviculture in this zone is to maintain, create, and sustain the species composition, density, and structure of vegetation desired by society and landowners to meet their diverse needs (McDonald 1992). A particular issue in this zone is increasing urbanization, with its associated problems of fragmentation, increased wood use, and greater fire hazard.

CALIFORNIA STATE FOREST PRACTICE REGULATIONS

Silviculture on private forestland in the Sierra Nevada is largely constrained by policies and regulations contained in the state Forest Practice Act. California has had a Forest Prac-

tice Act since 1943. The original Act, which was in effect until 1973, identified the seed tree system as the basic harvesting method. It encouraged private landowners to use "due diligence" in avoiding damage and protecting other resource values, but it had no budgetary provisions to enable administration or inspection. This act was declared unconstitutional in 1973 on the basis that it was administered by a Board of Forestry that was dominated by the timber industry.

In 1973, the existing Z'berg-Nejedly Forest Practice Act was approved, which changed the membership of the Board of Forestry such that a majority now consists of members of the public and environmental groups. Requirements for submission of a timber harvest plan and minimum stocking standards are defined in the act, and detailed constraints on land management are defined in accompanying regulations. The stated intent of the act is to ensure that

> The goal of maximum sustained production of high-quality timber products is achieved while giving consideration to values relating to recreation, watershed, wildlife, range and forage, fisheries, regional economic vitality, employment, and aesthetic enjoyment.

The minimum stocking standards are defined in the act as follows:

> Within five years after completion of timber operations, the harvested area must be acceptably stocked by:
> a) having a point count of 741/ha (300/ac) where i) each countable tree that is not more than four inches in diameter at breast height counts as one; ii) each countable tree over 10.1 cm (4 in) and not more than 30.5 cm (12 in) in diameter at breast height counts as three; and iii) each countable tree over 30.5 cm (12 in) in diameter at breast height counts as six.
> b) The average residual basal area, measured in stems one inch or larger in diameter, is at least 20 m²/ha (85 ft²/ac) on Site I, or 12 m²/ha (50 ft²/ac) on Sites II or lower.

The rules pertaining to silviculture in the Sierra Nevada are defined in two districts, the Northern Forest District and the Southern Forest District. A registered professional forester is required to select silvicultural methods or alternatives that achieve maximum sustained production of high-quality timber products. The harvest plan must designate one or a combination of regeneration methods, prescriptions, or intermediate treatments defined in the rules, or a defensible alternate method. Detailed specifications are provided for all even- and uneven-aged regeneration methods. If clear-cutting is used, block size is limited to 8.1 ha (20 acres). Provisions are made for wildlife protection practices that cover nest sites for many species of birds and habitats for animals. Protection is provided for riparian areas and buffer strips bordering streams. Timber harvest plans are reviewed by the

Department of Forestry and Fire Protection (CDF), the Department of Fish and Game, and other agencies before to CDF grants approval to harvest the designated trees.

SILVICULTURE AND FOREST POLICY ANALYSIS: PRESCRIPTIVE VERSUS GOAL-ORIENTED POLICIES

Recently, several major forest assessments and analyses have provided bases for major shifts in forest management. These include FEMAT (Forest Ecosystem Management Assessment Team) in the Pacific Northwest, the CASPO (California Spotted Owl) report and the U.S. Forest Service DEIS (draft environmental impact statement) for implementing CASPO and SNEP (Sierra Nevada Ecosystem Project).

In evaluating any policy report, it is important to determine whether the intent is actually to regulate what can be done on the ground or simply to provide general guidance to achieve the overall forest structure that will provide for multiple values and uses. Both FEMAT and the CASPO report contain quite prescriptive language that specifies silvicultural systems or practices and provide limited flexibility to respond to varying ecological conditions within and among stands. The DEIS is less prescriptive and provides limited ranges of opening sizes and canopy cover. The difficulty with this approach is that the more prescriptive the silvicultural policy the greater the difficulty in making it meet multiple resource objectives, because of the wide variability in existing stand conditions. Implementing overly prescriptive policy leads to a reduction in landscape diversity. For example, specifying a diameter limit on cutting, the amount of basal area to be retained, continuous cover, and the number of snags over broad ranges of stand types could benefit certain wildlife species but would drive stand composition to shade-tolerant species and would also likely lead to stand-replacement fires, loss in stand vigor, and increased susceptibility to insects. On the other hand, general, goal-oriented policy statements that specify the need for reduced fire potential and certain types of habitat on a certain percentage of the forest and that allow forest managers flexibility in choosing the methods to implement the policy are much more likely to attain the desired objectives. For example, the "green tree retention" (GTR) policy in the DEIS would provide for better habitat, stand productivity, and visual quality by suggesting a broad range of trees to be left. Similarly, the group selection method should encourage a variable opening size that enables the silviculturist to obtain regeneration of varying levels of shade tolerance, develop "fuzzy" or more natural, irregular boundaries, address fire and fuels issues, and use safe logging practices. There is no single, universal silvicultural method or prescription that will result in sound management practices in all situations. Given the wide variability in existing stands in Sierran forests, flexibility is needed in prescribing treatments that best meet management and policy objectives.

In the development and analysis of forest management policy, evaluations are made of the long-term implications of various management alternatives. This means that the effects of alternative treatments on variables such as forest structure, wildlife habitat, timber and water yields, and fire potential must be projected over time. These projections commonly assume that certain "standard" silvicultural prescriptions will be implemented throughout the forest and that the vegetation will respond predictably. The standardized prescriptions used for these projections may be reasonable for policy analysis but should be used only as general guidelines for policy implementation. Sierran forests are so diverse in species mix, site productivity, density, and vigor that it is impossible to attain desired goals using a prescriptive approach. Also, it must be recognized that over the last two decades, some goals have undergone radical change. The alternative is to define goals and intent and to charge professional silviculturists with the responsibility and accountability of meeting these goals using appropriate, site-specific treatments.

The use of a prescriptive or regulatory approach is probably motivated by lack of trust. Resource specialists, special interest groups, or policy makers may not trust an agency's ability or an industry's willingness to implement a policy in the manner intended. However, rather than include prescriptive language in policy statements, it is far better to state desired outcomes and mechanisms (including monitoring procedures) for ensuring that the desired end results are achieved. In that way, the agencies can best use their professional expertise to devise workable solutions to complex forest management and silvicultural issues, and policy makers can ensure that their intent is carried out.

MAJOR ISSUES AFFECTING SIERRAN SILVICULTURE: IMPLEMENTATION OF NEW POLICIES

The major issues affecting the silviculture of mixed conifer forests are as follows:

The enormous buildup of small-diameter understory trees and fuels. One hundred years of selective forest harvesting in the Sierra Nevada have, in contrast to the Pacific Northwest where clear-cutting was the predominant practice, retained a heterogeneous forest of mixed species that still includes substantial amounts of large-sized trees, particularly on federal lands. The major change in forest structure and composition is the growth of an enormous quantity of small-sized, shade-tolerant trees in the understory, which has resulted from a combination of sixty years of wildfire suppression, logging that stimulated new growth of trees and shrubs, and the lack of markets for small-sized material. Ten years ago, this problem began to be addressed by a rising market for biomass thinnings that were burned to generate electricity. This thinning operation has covered about 40,469 ha (100,000 acres) per year for the past ten years. However, the subsidy for fuel chips has been removed, and use of woody biomass for power generation is no longer economical. Accompanying the rise in log stumpage prices, several sawmills have been built in California that are designed specifically to harvest small-sized trees for the production of 5 by 10 cm (2 by 4 in) studs, lumber, and fencing material. These operations only begin to address the enormous supply of small-sized trees that have accumulated in the Sierra. More plants of this kind will be needed if the problem is to be adequately addressed and potential fuels removed in a significant way. Prices for pulp chips fluctuate markedly, but even when high, additional markets for small material are needed in order to make possible sustained thinnings of small-sized understory trees, particularly of shade-tolerant species.

The high probability of catastrophic fire in the Sierra Nevada makes the leaving of unmanaged reserves for parks or wildlife habitat silviculturally untenable. Experience in Yellowstone and elsewhere has demonstrated the critical need to manage fuels on a landscape basis. Fuels management has a large silvicultural component, including controlling stand densities where appropriate.

The fact that California is currently importing about 65% of the wood fiber it needs. Sierran forests are generally very productive. If policies are developed that aim at increasing the proportion of wood grown for industrial consumption, silvicultural knowledge is already available that could double the current average production of 5.2 m^3/ha/year (75 ft^3/acre/year). Some stands could be designated primarily for sustained production of wood. The structure and species composition of these could be quite simple and of relatively short rotation or quite complex, uneven-aged, long in rotation, and "natural-looking." Silvicultural knowledge is already available to develop and manage both types of planted forests.

The need for long-term sustainability of forests. Silvicultural knowledge is available, and is currently being applied, that ensures long-term sustainability of forest stands from the standpoints of soil stability and nutrient demands. What needs to be developed is a new approach to the application of silvicultural systems that preserves the integrity of forests at the landscape or watershed level and results in the desired spatial distribution of diverse stand structures. A critical issue is the buildup of stand densities, with the associated increase in risk from fire and insects. In addition, better collaboration is needed

among landowners, logging engineers, and silviculturists to ensure that operations are conducted in an ecologically sensitive manner.

Forest health. Forest health is a complex subject that involves the interaction of many factors, including stand density, vigor, and susceptibility to forest pests. What makes the concept difficult is that forest health can be interpreted at various spatial scales that include individual trees, stands, watersheds, and landscapes. In addition, forest health has a temporal characteristic that necessarily changes as one views it in terms of a year, a decade, or a millennium—that is, on human or ecosystem time scales. It must be recognized that healthy forests must periodically experience and be able to withstand drought, fires, and epidemic levels of insects and disease. Thus, a healthy forest is characterized by having long-term resilience to disturbance. In this context it is reasonable to include within "disturbance" the effects of timber harvesting, including limited clear-cutting, where these disturbances are used within the goals of sustaining long-term forest productivity.

The need for wildlife habitat for threatened and endangered species, in the form of old-growth, late-seral-stage forests. Increased awareness of the need to address forest fragmentation and to enhance the proportion of late-seral-stage forests does not require new silvicultural knowledge but the application of existing knowledge in a new way. Experience gained by both publicly and privately employed silviculturists has demonstrated a capacity to control stand structure and composition. It is a small extrapolation of this knowledge to anticipate that silviculturists can manipulate stands to provide many of the characteristics of old-growth forests even though they may be of relatively young age. Stands having large, well-spaced trees with suitable understory and downed woody material and with snags and dead trees in the canopy layer can probably be produced on lands of high site quality within 120 years, starting with bare ground. Existing seventy-year-old stands could be transformed into ones having many of the late-seral-type characteristics in fifty years. These manipulations would depend on markets for small material and opportunities for operations such as prescribed burning and the use of herbicides. Because silvicultural prescriptions are increasingly needed to meet complex wildlife habitat requirements, a particular need is to link silvicultural expertise in manipulating stand composition and structure to wildlife habitat as defined by the Wildlife Habitat Relations (WHR) models, which are used to predict the species of wildlife that are likely to be present in a stand of given characteristics. The approach used in these models of defining stands in terms of classes of canopy closure and diameter is increasingly being used for land classification and management planning.

In addressing wildlife habitat needs in the Sierra Nevada, it is probably inappropriate to design silvicultural treatments across landscapes and watersheds aimed at solving the problems of an individual target species. In general, the best way to ensure desirable habitat for diverse species of wildlife is by sustaining a diversity of stand structures and habitat. This can be done only by retaining the capacity within watersheds to use a diversity of silvicultural treatments and regeneration methods; the feasibility of doing this has been demonstrated at Blodgett Forest. Past history in California has shown the detrimental effects of the overuse of clear-cutting that was intended to overcome problems created by the previous overuse of high-grade harvesting. Similarly, rather than focusing on solving wildlife problems by a pattern of unmanaged reserves interconnected with corridors, it would seem preferable to emphasize management of lands to provide diverse habitats and values.

The need to protect of riparian areas and improve stream habitats. The most sensitive areas in mixed conifer forests are the areas adjacent to streams and wet meadows. It is essential that streams and aquatic ecosystems be maintained or restored to provide high-quality habitat. Existing forest practice rules provide stringent restrictions on management practices within streamside buffer zones and riparian areas. These rules must continue to be used on private lands and at least equivalent practices must be used on public lands. However, to maintain and enhance these areas, it is necessary to permit appropriate manipulation designed to meet these objectives. Precluding all management may lead, as has been shown in Yellowstone National Park and other ecosystems, to an unstable system, due largely to the elimination of natural disturbances such as fire.

The need to ensure sustained production of high-quality water. Considerable work was done in earlier decades aimed at evaluating the effects of different harvesting methods on water yields in the Sierra (Kittredge 1953; Anderson and Gleason 1959; West 1962; Turner 1985, 1986). This area of research must receive continued attention, since water is the most valuable product from Sierran watersheds. Forests play an important role in watersheds in protecting the soil and in ensuring that water is incorporated into the soil and released slowly into streams. Evapotranspiration and interception of water by forests is largely influenced by leaf area and canopy architecture—both of which are readily manipulated by silvicultural treatment.

The increasing pressure on urban/wildland interface forests. The growing population in California, with its associated increase in societal demands and expectations for goods and services, is placing increasing strain on Sierran forests. In particular, the rapid increase in urban de-

velopment in the mixed conifer forests is creating a new zone of land use in the Sierra (USFS 1995). Silvicultural treatments are needed in these areas to ensure the health of trees that are in close proximity to structures and paving. Regeneration is needed to provide replacements as trees mature and die. And special treatments are needed to reduce fire hazards and risks and to retain forest health.

Fragmented forests resulting from checkerboard patterns of forest ownership and differing land ownership objectives. Ecosystem, watershed, and regional approaches to land management are requiring a new type of silviculture that transcends traditional stand-level and independent ownership approaches. This can already be seen in USFS approaches to the management of land by watershed or other landscape-level analysis as opposed to the traditional approach in which a given watershed might have several independent plans based on separate federal ownerships. The situation is made much more complex where watersheds contain private landholdings, which introduces major issues of private property rights. Silviculture could be the common language in these complex situations where collaborative land-management planning is becoming increasingly necessary.

In addressing these issues, silviculturists are redefining their field in the context of new approaches to the management of public lands:

Ecosystem management. Ecosystem management is also a complex subject. As envisioned by the Chief of the Forest Service at its inception, it is "an ecosystem approach" to management rather than the management of ecosystems. Although defined in various ways, the U.S. Forest Service defines it as "the skillful, integrated use of ecological knowledge at various scales to produce desired resource values, products, services and conditions in ways that also sustain the diversity and productivity of ecosystems" (Hazelhurst et al. 1995). This approach adds an expanded dimension to standard silvicultural prescriptions that traditionally have been used at the stand level. Ecosystem management requires a broader responsibility from the silviculturist and requires landscape-level planning.

Watershed analysis. Federal agencies have adopted a new approach to land management and interagency cooperation in the Pacific Northwest. Because agencies such as the Forest Service and the Bureau of Land Management often share responsibilities in the same watershed, a new concept of watershed analysis has developed as an outcome of FEMAT (Reid et al. 1994). Although currently being tested in north coast California forests, this concept may have relevance to Sierran forests. Watersheds have been adopted as the geographic unit for evaluating habitat needs and physical and socioeconomic conditions. The objective of watershed analysis is to summarize existing information, to identify information gaps, and to describe large-scale and interdisciplinary relationships. The analysis is commonly expected to be completed within a two-month period. Following this analysis will come detailed site inventory, analyses, and project planning. Watershed analysis is expected to aid in the management of riparian areas and to be a means of understanding ecosystems and cumulative effects (Grant 1994; Reid et al. 1994). It may also result in reducing the frequency of silvicultural entry within a watershed.

Adaptive management. An adaptive management strategy strengthens and formalizes a concept used by the Forest Service that encourages feedback and adjustment in forest management. It consists of a series of steps structured to promote rapid learning and to modify management in responsive to changing societal objectives and evolving knowledge of ecosystems (USFS 1994). It is a strategy designed to increase the probability of attaining desired outcomes. It links actions through time and is seen as an important component of ecosystem management. The concept of adaptive management has traditionally been used by silviculturists, particularly those who have had the benefit of long-term association with a given land base. Adaptive management formalizes this approach which will be a positive influence in the development of sound silvicultural prescriptions on public lands.

CONCLUSIONS

A substantial amount of silvical and silvicultural knowledge is available to guide the management of Sierra Nevadan forests. Silviculturists have the knowledge and capacity to maintain or modify forest composition and structure to meet the diverse needs of society. All coniferous forest types in the Sierra—mixed conifer, true fir, and ponderosa pine—respond well to treatment and can be manipulated to provide both even- and uneven-aged structures having diverse mixtures of early, middle, and late seral stages. Both even- and uneven-aged assemblages of forest vegetation can be developed relatively quickly that have many of the characteristics of old-growth stands. Stands and watersheds can be managed for a variety of species mixes and vertical and horizontal structures. Thus, managers have considerable flexibility in meeting diverse management goals. Results from Blodgett Forest Research Station and elsewhere suggest that, given adequate time, the differences in productivity between even- and uneven-aged systems are likely to be small. Whether this is true

across sites of varying quality remains to be determined. Differences may occur in stands on steep ground and on sites of lower quality, where there will be practical limitations on the choice of logging and silvicultural options. Also, experience in uneven-aged silviculture is very limited, and there might well be programmatic problems, such as economics, practicality of record keeping, contracting for work, and workforce organization and supervision, applying the system to large land areas. However, in the absence of particular biological, managerial, or societal constraints, all systems are potentially applicable and, over a landscape, can have the same range of density, size classes, frequency of entry, road density, and species mixes. Sites of higher quality will respond more rapidly and will be more amenable to a wider range of treatments than will sites of lower quality. Past history in California and elsewhere has shown that it is important to ensure that no one or two harvesting methods dominate the landscape. Silviculturists must learn to merge systems and take advantage of existing heterogeneity through the use of "fuzzy" variants that leave green and dead trees, develop understories, have vertical and horizontal structural diversity, and use irregular boundaries.

The critical silvicultural challenges are to keep forests healthy, to reduce the potential for catastrophic crown fires, and to keep soil in place on hillsides. The basic knowledge to do this is available; all that is required are the will and finances to do it. Two major deterrents are regulations, such as air-quality standards that limit use of prescribed burning, and policies and regulations that limit silvicultural choice of harvesting and regeneration methods. In a broader context, regulatory problems that affect silviculture and increase costs are more complex because they involve nonsilvicultural issues such as planning ordinances affecting residence construction in forested areas and road access in forest zones. Because of the diversity of conditions in the Sierra, it is preferable to define the desired outcome and require professional silviculturists to prescribe and use the most effective methods to achieve the desired forest composition and structure at the landscape and watershed levels. To encourage the long-term use of sound silvicultural practices, markets are needed for small-sized trees, and stability is needed in the policy and regulatory environments.

Research Needs

More knowledge is needed to support the development of new silviculture aimed at addressing current and emerging issues. The following areas are particularly important. Current research is addressing many of these, but sustained funding and greater effort are needed to ensure adequate knowledge covering the broad range of forest types and situations in the Sierra.

- Fire-risk rating systems need to be developed for stands and fuel profiles over landscapes.

- We need to know how to establish and maintain effective fuel breaks.

- We need to find ways in which to encourage the widespread use of low-impact logging systems to selectively remove fuels and small trees from the understory of older stands and to thin young stands.

- We need to know the relative regeneration and growth rates of conifers, hardwoods, shrubs, and grasses in the understory under varying levels of overstory density, as well as their relative capacities for release in uneven-aged systems and their relation to wildlife habitat.

- We need to understand the effects of fire and biomass removal on soils and nutrient status on sites of differing levels of productivity.

- Better information is needed on the interrelationships among site quality, species composition, stand structure, weather (climate), wildlife, and insects and pathogens.

- Improved site-specific, GPS-based, multiresource inventory information is needed.

- Models are needed that project the likely development of a variety of forest structures at the landscape level under alternative management scenarios.

- We need to know how to culture hardwood sprouts.

- An understanding is needed of the applicability of silviculture to enhance the development of late-seral-stage stands.

- An understanding is needed of the applicability of uneven-aged systems at large-scale levels, on sites of diverse productivity, to meet diverse societal needs.

- We need to know the likely dynamics of forest composition and structure in the context of possible climate change and increasing air pollution.

REFERENCES

Agee, J. K. 1990. The historical role of fire in Pacific Northwestern forests. In *Natural and prescribed fire in the Pacific Northwest*, edited by J. D. Walstad, S. R. Radosevich, and D. V. Sandberg, 25–38. Corvallis: Oregon State University Press.

Aho, P. E., G. Fiddler, and G. M. Filip. 1989. *Decay losses associated with wounds in commercially thinned true fir stands in northern California*. Research Paper PNW-RP-403. Portland, OR: U.S. Forest Service, Pacific Northwest Experiment Station.

Aho, P. E., G. Fiddler, and M. Srago. 1983. *Logging damage in thinned, young-growth true fir stands in California and recommendations for prevention*. Research paper PNW-304. Portland, OR: U.S. Forest Service, Pacific Northwest Experiment Station.

Alexander, E. B., and R. J. Poff. 1985. *Soil disturbance and compaction in wildland management*. Earth Research Monograph 8. San Francisco: U.S. Forest Service, Pacific Southwest Region.

Allen, B. H. 1987. Livestock grazing as a vegetation management tool. In *Proceedings, eighth annual forest vegetation management conference*, 43–55. Redding, CA: Forest Vegetation Management Conference.

Anderson, H. W., and C. H. Gleason. 1959. Logging effects on snow, soil moisture, and water losses. In *Proceedings, Western Snow Conference*, 57–63. Fort Collins: Colorado State University.

Arvola, T. F. 1978. *California forestry handbook*. Sacramento: Department of Forestry, Resources Agency.

Assmann, E. 1970. *The principles of forest yield study*. New York: Pergamon Press.

Atzet, T., R. F. Powers, D. H. McNabb, M. P. Amarantus, and E. R. Gross. 1989. Maintaining long-term forest productivity in southwest Oregon and northern California. Chap..10 in *Maintaining the long-term productivity of Pacific Northwest forest ecosystems*, edited by D. A. Perry, R. Meurisse, B. Thomas, R. Miller, J. Boyle, J. Means, S. R. Perry, and R. F. Powers, 185–201. Portland, OR: Timber Press.

Baker, F. S. 1952. Reproduction of ponderosa pine at low elevations in the Sierra Nevada. *Journal of Forestry* 110:401–4.

———. 1955. *California's forest regeneration problems*. Sacramento: Department of Natural Resources, Division of Forestry, State Board of Forestry, Regeneration Committee.

Barbour, M. G., and R. A. Woodward. 1985. The Shasta red fir forest of California. *Canadian Journal of Forest Research* 15:570–76.

Barclays. 1995. *Barclays official California code of regulations*. Title 14, Natural Resources, Division 1.5, Department of Forestry. South San Francisco: Barclays Law Publishers.

Baver, L. D. 1930. The Atterberg consistency constants: Factors affecting their values and a new concept of their significance. *Agronomy Journal* 22:935–48.

Beetham, N. M. 1961. The ecological tolerance range of the seedling stage of *Sequoia gigantea*. Ph.D. diss., Duke University. Abstract in *Dissertation Abstracts* 24:479–80.

Berryman, A. A., and G. T. Ferrell. 1988. The fir engraver beetle in western United States. In *Dynamics of forest insect populations: patterns, causes, implications*, edited by A. A. Berryman, 555–77. New York: Plenum Press.

Biging, G. S., and L. C. Wensel. 1985. *Site index equations for young-growth mixed conifers of northern California*. Research Note 8. Berkeley: University of California, Northern California Forest Yield Cooperative.

Bolsinger, C. L. 1980. *California forests: Trends, problems, and opportunities*. Research Bulletin PNW-89. Portland, OR: U.S. Forest Service, Pacific Northwest Research Station.

———. 1983. An overview of current forest resources and trends in the eastside pine type of California. In *Proceedings, symposium on the management of the eastside pine type in northeastern California*, edited by T. F. Robson and R. B. Standiford, 29–38. SAF 83-06. Arcata, CA: Northern California Society of American Foresters.

———. 1988. *The hardwoods of California's timberlands, woodlands, and savannas*. Research Bulletin PNW-RB-148. Portland, OR: U.S. Forest Service, Pacific Northwest Research Station.

Boyer, D. 1979. *Guidelines for soil protection and restoration for timber harvest and post-harvest activities*. Portland, OR: U.S. Forest Service, Pacific Northwest Region, Division of Watershed Management.

Boyer, D. E., and J. D. Dell. 1980. *Fire effects on Pacific Northwest forest soils*. WM-040. Portland, OR: U.S. Forest Service, Pacific Northwest Region.

Buck, J. M., R. S. Adams, J. Cone, M. T. Conkle, W. J. Libby, C. J.

Eden, and M. J. Knight. 1970. *California tree seed zones.* San Francisco: U.S. Forest Service, California Region; Sacramento: California Division of Forestry.

Burns, R. M., and B. H. Honkala, eds. 1990. *Confires.* vol. 1 of *Silvics of North America.* Agricultural Handbook 654. Washington, DC: U.S. Forest Service.

Clayton, J. L., and D. A. Kennedy. 1985. Nutrient losses from timber harvest in the Idaho batholith. *Soil Science Society of America Journal* 49:1041–49.

Conard, S. G., and S. R. Radosevich. 1982a. Growth responses of white fir to decreased shading and root competition by montane chaparral shrubs. *Forest Science* 28:309–20.

———. 1982b. Postfire succession in white fir *(Abies concolor)* vegetation of the northern Sierra Nevada. *Madroño* 29 (1): 42–56.

Conard, S. G., and S. R. Sparks. 1993. *Abies concolor growth responses to vegetation changes following shrub removal, northern Sierra Nevada California.* Research Paper. PSW-RP-218 Riverside, CA: Pacific Southwest Research Station.

Curtis, R. O. 1970. Stand density measures: An interpretation. *Forest Science* 16:402–14.

Daniel, T. W., J. A. Helms, and F. S. Baker. 1979. *Principles of silviculture.* New York: McGraw-Hill.

DeBano, L. F. 1979. Effects of fire on soil properties. In *California forest soils: A guide for professional foresters and resource managers and planners,* edited by R.J. Laacke, 109–18. Agricultural Science Publication. Berkeley: University of California.

DeBell, D. S., and J. F. Franklin. 1987. Old-growth Douglas-fir and western hemlock: A thirty-six year record of growth and mortality. *Western Journal of Applied Forestry* 2 (4): 111–14.

Dixon, G. E. 1992. *The South Central Oregon/Northeastern California Vegetation Simulator (SORNEC).* Fort Collins, CO: U.S. Forest Service, WO-TM Service Center.

———. 1994. Western Sierra Nevada prognosis geographic variant of the forest vegetation simulator. Fort Collins, CO: U.S. Forest Service, WO-TM Service Center.

Dolph, K. L., S. R. Mori, and W. W. Oliver. 1995. Long-term response of old-growth stands to varying levels of partial cutting in the eastside pine type. *Western Journal of Applied Forestry* 10 (3): 101–8.

Dunlap, J. M., and J. A. Helms. 1983. First-year growth of planted Douglas-fir and white fir seedlings under different shelterwood regimes in California. *Forest Ecology and Management* 5:255–68.

Dunn, P. H., and L. F. DeBano. 1977. Fire's effect on the biological properties of chaparral soils. Paper read at Symposium on Environmental Consequences of Fire and Fuel Management in Mediterranean Ecosystems, Palo Alto, CA.

Dunning, D. 1923. *Some results of cutting in the Sierra forests of California.* Bulletin 1176. Washington, DC: U.S. Department of Agriculture.

———. 1928. A tree classification system for the selection forests of the Sierra Nevada. *Journal of Agricultural Research* 36:755–71.

Dunning, D., and L. H. Reineke. 1933. *Preliminary yield tables for second-growth stands in the California pine regions.* Technical Bulletin 354. Washington, DC: U.S. Department of Agriculture.

Dyck, W. J., and P. N. Beets. 1987. Managing for long-term site productivity. *New Zealand Forestry* 32 (3): 23–26.

Ferrell, G. T. 1974. Moisture stress and fir engraver attack in white fir infected by tree mistletoe. *Canadian Entomology* 106:315–18.

———. 1980. *Risk-rating systems for mature red fir and white fir in northern California.* General Technical Report PSW-39. Berkeley, CA: U.S. Forest Service, Pacific Southwest Research Station.

Ferrell, G. T., W. J. Otrosing, and J. C. J. Demars. 1994. Predicting susceptibility of white fir during a drought associated outbreak of the fir engraver in California. *Canadian Journal of Forest Research* 24:302–5.

Ferrell, G. T., and R. Smith. 1976. Indicators of *Fomes annosus* and bark beetle susceptibility in sapling white fir. *Forest Science* 22: 365–69.

Fiske, J. N. 1982. Evaluating the need for release from competition from woody plants to improve conifer growth rates. In Proceedings, third annual forest vegetation management conference, 25–44. Redding, CA: Forest Vegetation Management Conference.

Ford, B. 1991. *Collins Almanor Forest: fifty-one years of forest management.* Chester, CA: Collins Pine Co.

Fowells, H. A., and G. H. Schubert. 1956. *Seed crops of forest trees in the pine region of California.* Technical Bulletin 1150. Washington, DC: U.S. Forest Service.

Fowler, C. 1963. *Tree improvement program for U.S. Forest Service California region.* San Francisco: U.S. Forest Service, R-5.

Franklin, J. F., and J. A. Fites-Kaufmann. 1996. Analysis of late successional forests. In *Sierra Nevada Ecosystem Project: Final report to Congress,* vol. II, chap. 21. Davis: University of California, Centers for Water and Wildland Resources.

Fried, J. S., J. R. Boyle, J. C. Tappeiner, and J. K. Cromack. 1989. Effects of bigleaf maple on soils in Douglas-fir forests. *Canadian Journal of Forest Research* 20:259–66.

Froehlich, H. A. 1974. Soil compaction: implication for young-growth management. In Managing young forests in the Douglas-fir region, edited by A. B. Berg, 47–62. Corvallis: Oregon State University, School of Forestry.

Froehlich, H. A. 1978. *Soil compaction from low ground-pressure, torsion-suspension logging vehicles on three forest soils.* Research Paper 36. Corvallis: Oregon State University Forest Research Lab.

Froehlich, H. A., and D. W. McNabb. 1984. Minimizing soil compaction in Pacific Northwest forests. Paper read at Forest Soils and Treatment Impacts. Sixth North American Forest Soils Conference, Knoxville, TN.

Gordon, D. T. 1970. *Natural regeneration of white and red fir: Influence of several factors.* Research Paper PSW-58. Berkeley, CA: U.S. Forest Service, Pacific Southwest Research Station.

———. 1973. *Released advanced reproduction of white and red fir... growth, damage, and mortality.* Research Paper PSW-95. Berkeley, CA: U.S. Forest Service, Pacific Southwest Research Station.

———. 1979. *Successful natural regeneration cuttings in California true firs.* Research Paper PSW-140. Berkely, CA: U.S Forest Service, Pacific Sorthwest Research Station.

———. 1978. *White and red fir cone production in northeastern California: Report of a sixteen-year study.* Research Note PSW-330. Berkeley, CA: U.S. Forest Service, Pacific Southwest Research Station.

Gordon, D. T., and E. E. Bowen. 1978. *Herbs and brush on California red fir regeneration sites: A species and frequency sampling.* Research Note PSW-329. Berkeley, CA: U.S. Forest Service, Pacific Southwest Research Station.

Grant, G. 1994. Introduction to watershed analysis: A retrospective. Watershed Management Council Newsletter 6 (2): 16–17.

Greenwood, G. B., R. K. Marose, and J. M. Stenback. 1993. *Extent and ownership of California's hardwood rangelands.* Sacramento: California Department of Forestry and Fire Protection, Strategic and Resources Planning Program.

Grier, C. C., K. A. Vogt, M. R. Keyes, and R. L. Edmonds. 1981. Biomass distribution and above-and below-ground production in young and mature *Abies amabilis* zone ecosystems of the Washington Cascades. Canadian Journal of Forest Research 11:155–67.

Gross, R., and R. J. Laacke. 1984. Pocket gophers girdle large true firs in northeastern California. Tree Planters Notes 35 (2): 28–30.

Gruell, G. E. 1994. *Understanding Sierra Nevada forests: Historical overview of Sierra Nevada forests provides insights to future.* Sacramento: California Forest Products Commission.

Harmon, M. E., J. F. Franklin, P. Sollins, S. V. Gregory, J. D. Lattin, N. H. Anderson, N. G. Aumen, J. R. Sedell, G. W. Lienkaemper Jr., K. Cromack, and K. W. Cummins. 1986. Ecology of coarse woody debris in temperate ecosystems. *Advances in Ecological Research* 15:133–302.

Harrington, T. B., J. C. Tappeiner, and T. F. Hughes. 1991. Predicting average growth and size distributions of Douglas-fir saplings with sprouts of tanoak or Pacific madrone. *New Forests* 5:109–30.

Hatchell, C. E., C. W. Ralston, and R. R. Fiol. 1970. Soil disturbance in logging. Journal of Forestry 68:772–75.

Hazelhurst, S., F. Magary, and K. S. Hawke eds. 1995. *Sustaining ecosystems: A conceptual framework.* Version 1.0. R5-EM-TP-001. San Francisco: U.S. Forest Service, Pacific Southwest Region and Station.

Helms. J. A. 1994. The California region. In *Regional silviculture of the United States.* 3rd ed. Edited by J. W. Barrett, 441–97. New York: John Wiley.

Helms, J. A., and C. Hipkin. 1986a. Effects of soil compaction on height growth of a California ponderosa pine plantation. *Western Journal of Applied Forestry* 1 (4): 104–108.

———. 1986b. Effects of soil compaction on tree volume in a California ponderosa pine plantation. *Western Journal of Applied Forestry* 1(4):121–24.

Helms, J. A., and R. B. Standiford. 1983. Predicting release of advance reproduction of mixed conifer species in California following overstory removal. *Forest Science* 31 (1): 3–15.

Hughes, T. F., C. R. Latt, J. C. Tappeiner, and M. Newton. 1987. Biomass and leaf area estimates for varnish-leaf ceanothus, deerbrush, and white leaf manzanita. *Western Journal of Applied Forestry* 2:124–28.

Huntsinger, L. 1988. Grazing in California's mixed conifer forests: Studies in the central Sierra Nevada. Ph.D. diss., University of California, Berkeley.

Jenkinson, J. L. 1976. Effects of nursery conditioning and cold storage on root growth capacity and survival of 2-0 Douglas-fir and noble-red fir planting stock. Unpublished Humboldt Nursery Administrative Study. U.S. Forest Service, Pacific Southwest Research Station, Albany, CA.

———. 1980. *Improving plantation establishment by optimizing growth capacity and planting times of western yellow pine.* Research Paper PSW-154. Berkeley, CA: Pacific Southwest Research Station.

Jenkinson, J. L., and A. H. McCain. 1993. *Winter sowings produce 1-0 sugar pine planting stock in the Sierra Nevada.* Research Paper PSW-RP-219. Berkeley, CA: U.S. Forest Service, Pacific Southwest Research Station.

Jenkinson, J. L., J. A. Nelson, and M. E. Huddleston. 1993. *Improving planting stock quality—The Humboldt experience.* General Technical Report PSW-GTR-143. Albany, CA: U.S. Forest Service, Pacific Southwest Research Station.

Johnson, D. W. 1994. Reasons for concern over impacts of harvesting. In *Impacts of forest harvesting on long-term site productivity,* edited by W. J. Dyck, D. W. Cole, and N. B. Comerford, 1–12. London: Chapman and Hall.

Keen, F. P. 1936. Relative susceptibility of ponderosa pine to bark-beetle attack. *Journal of Forestry* 34:919–27.

Kilgore, B. M. 1971. The role of fire in managing red fir forests. *Transactions, North American Wildlife* and *Natural Rresources Conference* 36:405–16.

———. 1973. The ecological role of fire in Sierra conifer forests: Its application to national park management. *Quaternary Research* 3:496–513.

Kinloch, B. B., and W. H. Scheuner. 1990. *Pinus lambertiana,* Doug., Sugar pine. In *Conifers,* edited by R. M. Burns and B. H. Honkala, 370–79. vol. 1 of *Silvics of North America.* Agricultural Handbook. Washington, DC: U.S. Forest Service.

Kittredge, J. 1952. Deterioration of site quality by erosion. *Journal of Forestry,* 50:554–56.

———. 1953. Influences of forests on snow in the ponderosa–sugar pine–fir zone of the central Sierra. *Hilgardia* 22 (1): 1–96.

Kitzmiller, J. H. 1976. *Tree improvement master plan for the California region.* San Francisco: U.S. Forest Service.

Laacke, R. J. 1990. *Abies magnifica* A. Murr. California red fir. In *Conifers,* edited by R. M. Burns and B. H. Honkala, 71–79. Vol. 1 of *Silvics of North America.* Agricultural Handbook 654. Washington, DC: U.S. Forest Service.

Laacke, R. J., and J. C. Tappeiner. 1996. Red fir ecology and management. In *Sierra Nevada Ecosystem Project: Final report to Congress,* vol. III. Davis: University of California, Centers for Water and Wildland Resources.

Laacke, R. J., and J. H. Tomascheski. 1986. *Shelterwood regeneration of true fir: Conclusions after eight years.* Research Paper PSW-184. Berkeley, CA: U.S. Forest Service, Pacific Southwest Research Station.

Landram, M. 1996. Status of reforestation on national forest lands within the Sierra Nevada Ecosystem Project study area. In *Sierra Nevada Ecosystem Project: Final report to Congress,* vol. III. Davis: University of California, Centers for Water and Wildland Resources.

Lanini, W. T., and S. R. Radosevich. 1986. Response of three conifer species to site preparation and shrub control. *Forest Science* 32 (1): 61–77.

LeBlanc, J. W., K. Churches, R. Standiford, R. Logan, and D. Irving. 1989. Attitudes abour oaks in Calaveras County. California Agriculture 43 (14): 11–13.

Leiberg, J. B. 1902. *Forest conditions in the northern Sierra Nevada, California.* Professional Paper 8, Series H, Forestry 5. Washington, DC: U.S. Geological Survey.

Likens, G. E., F. H. Borman, and N. M. Johnson. 1969. Nitrification: Importance to nutrient losses from a cut-over forest ecosystem. *Science* 163:1205–6.

Lilieholm, R. J., L. S. Davis, R. C. Heald, and S. P. Holmen. 1990. Effects of single tree selection harvests on stand structure, species composition, and understory tree growth in a Sierra mixed conifer forest. *Western Journal of Applied Forestry* 5 (2): 43–46.

Mace, A. C. 1970. *Soil compaction due to tree length and full tree skidding with rubber-tired skidders.* Minnesosota Forest Research Note 214.

Martin, R. E., and J. D. Dell. 1978. *Planning for prescribed burning in the inland Northwest.* General Technical Report PNW-76. Portland, OR: U.S. Forest Service, Pacific Northwest Research Station.

Maser, C., and J. M. Trappe. 1984. *The seen and unseen world of the fallen tree.* Portland, OR: U.S. Forest Service and U.S. Bureau of Land Management.

McColl, J. G., and R. F. Powers. 1984. Consequences of forest management on soil-tree relationships. Chap. 14 In *Nutrition of plantation forests,* edited by G. D. Bowen and E. K. S. Nambia. New York: Academic Press.

McCreary, D. D. 1991. Seasonal growth patterns of blue and valley oaks established on foothill rangelands. In *Proceedings, Symposium, oak woodlands and hardwood rangeland management,* technical coordination by R. B. Standiford, 36–40. General Technical Report PSW-126. Berkeley, CA: U.S. Forest Service, Pacific Southwest Research Station.

McDonald, P. M. 1976. *Forest regeneration and seedling growth from five major cutting methods in north-central California.* Research Paper PSW-115. Berkeley, CA: U.S. Forst Service, Pacific Southwest Research Station.

———. 1983. *Clearcutting and natural regeneration . . . Management implications for the northern Sierra Nevada.* General Technical Report PSW-70. Berkeley, CA: U.S. Forest Service. 70.

———. 1992. Estimating seed crops of conifer and hardwood species. *Canadian Journal of Forest Research* 22:832–38.

McDonald, P. M., and G. O. Fiddler. 1989. *Competing vegetation in ponderosa pine plantations: Ecology and control.* General Technical Report PSW-113. Berkeley, CA: U.S. Forest Service, Pacific Southwest Research Station.

———. 1993. Feasibility of alternatives to herbicides in young conifer plantations in California. *Canadian Journal of Forest Research* 23:2015–22.

McDonald, P. M., and D. W. Huber. 1995. *California's hardwood resource: Managing for wildlife, water, pleasing scenery, and wood products.* General Technical Report PSW-GTR-154. Berkeley, CA: U.S. Forest Service, Pacific Southwest Research Station.

McDonald, P. M., and W. W. Oliver. 1984. Woody shrubs retard growth of ponderosa pine seedlings and saplings. In Proceedings, fifth annual Forest Vegetation Management Conference, 65–89. Redding, CA: Forest Vegetation Management Conference.

McDonald, P. M., and S. R. Radosevich. 1992. General principles of forest vegetation management. In *Silvicultural approaches to animal damage in Pacific Northwest forests,* edited by H. C. Black 67–91. General Technical Report PNW-GTR-287. Portland, OR: U.S. Forest Service, Pacific Northwest Research Station.

McDonald, P. M., and J. C. Tappeiner. 1996. Regeneration of Sierra Nevada forests. In *Sierra Nevada Ecosystem Project: Final report to Congress,* vol. III. Davis: University of California, Centers for Water and Wildland Resources.

McKelvey, K. S., and J. D. Johnston. 1992. Historical perspectives on forests of the Sierra Nevada and the Transverse Ranges of southern California: Forest conditions at the turn of the century. Chap. 13 in *The California spotted owl: A technical assessment of its current status,* technical coordination by J. Verner, K. S. McKelvey, B. R. Noon, R. J. Gutiérrez, G. I. Gould Jr., and T. W. Beck, General Technical Report PSW-GTR-133. 225–46. Albany, CA: U.S. Forest Service, Pacific Southwest Research Station.

McNabb, D. H., and J. K. Cromack. 1990. Effects of prescribed fire on nutrients and soil productivity. In *Natural and prescribed fire in Pacific Northwest forests,* edited by J. D. Walstad, S. R. Radosevich, and D. V. Sandberg, Corvallis: Oregon State University Press.

Meyer, W. H. 1938. *Yield of even-aged stands of ponderosa pine.* Technical Bulletin 630. Washington, DC: U.S. Department of Agriculture.

Miles, J. A. 1978. Soil compaction produced by logging and residue treatment. *American Society of Agricultural Engineers Transactions* 21:60–62.

Minore, D. 1979. *Comparative autecological characteristics of northwestern tree species . . . A literature review.* General Technical Report PNW-87. Portland, OR: U.S. Forest Service.

Minore, D., C. Smith, and R. Woollard. 1969. *Effects of high soil density on seedling root growth of seven Northwestern tree species.* Research Note PNW-112. Portland, OR: U.S. Forest Service, Pacific Northwest Research Station.

Morris, L. A., and R. E. Miller. 1994. Evidence for long-term productivity change as provided by field trials. In *Impacts of forest harvesting on long-term site productivity,* edited by W. J. Dyck, D. W. Cole, and N. B. Comerford, 41–80. London: Chapman and Hall.

Morris, L. A., W. L. Pritchett, and B. F. Swindel. 1983. Displacement of nutrients into windrows during site preparation of a pine flatwoods forest. Soil Science Society of America Journal 47: 591–94.

Muir, J. [1895] 1977. *The Mountains of California.* Reprint, Berkeley, CA: Ten Speed Press.

Murphy, B. 1983. Pinus radiata survival, growth, and form four years after planting off and on skidtrails. New Zealand Journal of Forestry 28 (2): 184–93.

Oliver, W. W. 1972. *Growth after thinning ponderosa and Jeffrey pine pole stands in northeastern California.* Research Paper PSW-85. Berkeley, CA: U.S. Forest Service, Pacific Southwest Research Station.

———. 1974. *Seed maturity in white fir and red fir.* Research Paper PSW-99. Berkeley, CA: U.S. Forest Service Pacific Southwest Research Station.

———. 1979. *Fifteen-year growth patterns after thinning a ponderosa–Jeffrey pine plantation in northeastern California.* Research Paper PSW-141. Berkeley, CA: U.S. Forest Service, Pacific Southwest Research Station.

———. 1984. *Brush reduces growth of thinned ponderosa pine in northern California.* Research Paper PSW-172. Berkeley, CA: U.S. Forest Service, Pacific Southwest Research Station.

———. 1985. *Growth of California red fir advance reproduction after overstory removal and thinning.* Research Paper PSW-180. Berkeley, CA: U.S. Forest Service, Pacific Southwest Research Station.

———. 1988. Ten-year growth response of a California red and white fir saw timber stand to several thinning intensities. *Western Journal of Applied Forestry* 3:41–43.

———. In press. Growth and yield of planted ponderosa pine repeatedly thinned to different stand densities. *Western Journal of Applied Forestry.*

Oliver, W. W., and K. L. Dolph. 1992. Mixed-conifer seedling growth varies in response to overstory release. *Forest Ecology and Management* 48:179–83.

Oliver, W. W., G. T. Ferrell, and J. C. Tappeiner. 1996. Density management of Sierra Nevada forests. In *Sierra Nevada Ecosystem Project: Final report to Congress,* vol. III. Davis: University of California, Centers for Water and Wildland Resources.

Olson, C. M., and J. A. Helms. 1996. Forest growth and stand structure at Blodgett Forest Research Station, 1933–1995. In *Sierra Nevada Ecosystem Project: Final report to Congress,* vol. III. Davis: University of California, Centers for Water and Wildland Resources.

Oostings, H. J., and W. D. Billings. 1943. The red fir forest of the Sierra Nevada: *Abietum magnificae. Ecological Monographs* 13 (3): 260–74.

Palazzi, L. M., R. F. Powers, and D. H. McNabb. 1992. Geology and soils. Chap 3 in *Reforestation practices in southwestern Oregon and northern California, edited by* S. D. Hobbs, S. D. Tesch, P. W. Owston, R. E. Stewart, J. C. Tappeiner, and G. W. Wells, 49–72. Corvallis: Oregon State University, Forest Research Lab.

Pillsbury, N. H., and J. P. Joseph. 1991. Coast live oak thinning study in the central coast of California: Fifth-year results. In Proceedings, symposium, oak woodlands and rangeland management technical coordination by R. B. Standiford, 320–32. General Technical Report PSW-126. Berkeley, CA: U.S. Forest Service, Pacific Southwest Research Station.

Poff, R. J. 1996. Effects of silvicultural practices and wildfire on productivity of forest soils. In *Sierra Nevada Ecosystem Project: Final report to Congress*, vol. II, chap. 16. Davis: University of California, Centers for Water and Wildland Resources.

Potter, D. 1994. *Guide to the forested communities of the upper montane in the central and southern Sierra Nevada.* San Francisco: U.S. Forest Service, Pacific Southwest Region.

Powers, R. F. 1979. Mineral cycling in temperate forest ecosystems. In *California forest soils: A guide for professional foresters and resource managers and planners, edited by* R. J. Laacke, 89–108. Agricultural Science Publication. Berkeley: University of California.

———. 1990. Do timber management practices degrade long-term site productivity? What do we know and what do we need to know? In *Proceedings, eleventh annual forest vegetation management conference*, 87–106. Redding, CA: Forest Vegetation Management Conference.

———. 1991. Are we maintaining the productivity of forest lands? Establishing guidelines through a network of long-term studies. Paper presented at workshop, Management and Productivity of Western-Montane Forest Soils, Boise, ID.

Powers, R. F., and R. L. Edmonds. 1992. Nutrient management of subalpine *Abies* forests. Chap. 4 in *Forest fertilization: Sustaining and improving nutrition and growth of western forests, edited by* H. N. Chappell, G. F. Weetman, and R. E. Miller. Contribution 73. Seattle: University of Washington, Institute of Forest Research.

Powers, R. F., and W. W. Oliver. 1970. *Pole breakage in a pole size ponderosa pine plantation: . . . More damage at high densities.* Research Note PSW-218. Berkeley, CA: U.S. Forest Service, Pacific Southwest Research Station.

Powers, R. F., and K. Van Cleve. 1991. Long-term ecological research in temperate and boreal forest ecosystems. *Agronomy Journal* 83: 11–24.

Powers, R. F., S. R. Webster, and P. H. Cochran. 1988. Estimating the response of ponderosa pine forests to fertilization. Paper read at Future Forest of the Mountain West, a stand culture symposium, Missoula, MT.

Prescott, C. E., J. P. Corbin, and D. Parkinson. 1989. Biomass, productivity, and nutrient-use efficiency of aboveground vegetation in four Rocky Mountain coniferous forests. *Canadian Journal of Forest Research* 19:309–17.

Quick, C. R. 1956. Viable seeds from the duff and soil of sugar pine forest. *Forest Science* 2 (1): 36–42.

Radosevich, S. R., P. C. Passof, and O. A. Leonard. 1976. Douglas-fir release from tanoak and Pacific madrone competition. *Weed Science* 24:144–45.

Reid, L. M., R. R. Ziemer, and M. J. Furniss. 1994. *Watershed analysis in the federal arena.* Unpublished Manuscript, Interagency workshop, Humboldt Interagency Watershed Analysis Center, 20 April.

Reineke, L. H. 1933. Perfecting a stand-density index for even-aged forests. *Journal of Agricultural Research* 46:627–38.

Reukema, D. L., and D. Bruce. 1977. *Effects of thinning on yield of Douglas-fir: Concepts and some estimates obtained by simulation.* General Technical Report Portland, OR: U.S. Forest Service, Pacific Northwest Research Station.

Rice, R. M. 1979. Sources of erosion during timber harvest. In *California forest soils: A guide for professional foresters and resource managers and planners, edited by* R. J. Laacke 59–68. Agricultural Science Publication. Berkeley: University of California.

Ritchie, M. W., and R. F. Powers. 1993. *A user's guide for SYSTUM-1 (Version 2.0): A simulator for young stand trends under management in California and Oregon.* General Technical Report PSW-GTR-147. Albany, CA: U.S. Forest Service.

Salman, K. A., and J. W. Bongberg. 1942. Logging high-risk trees to control insects in the pine stands of northeastern California. *Journal of Forestry* 40:533–39.

Sandberg, D. V. 1980. *Douglas-fir duff reduction.* Research Paper PNW-272. Portland, OR: U.S. Forest Service, Pacific Northwest Research Station.

Scharpf, R. F. 1993. *Diseases of Pacific Coast conifers.* Agricultural Handbook 521. Albany, CA: U.S. Forest Service, Pacific Southwest Research Station.

Schubert, G. H. 1956. *Early survival and growth of sugar pine and white fir in clear cut openings.* Research Note 117. Berkeley, CA: U.S. Forest Service, California Forest and Range Experiment Station.

Schubert, G. H., and R. S. Adams. 1971. *Reforestation practices for conifers in California.* Sacramento: State of California, Division of Forestry.

Schumacher, F. X. 1928. *Yield, stand, and volume tables for red fir in California.* Bulletin 456. Berkeley: University of California, Agricultural Experiment Station.

Show, S. P. 1926. *Timber growing and logging practice in California pine region.* Bulletin 1402. Washington, DC: U.S. Department of Agriculture.

Smith, D. M. 1986. *The practice of silviculture.* New York: John Wiley.

Society of American Foresters. 1995. *Silviculture terminology.* Bethesda, MD: Society of American Foresters.

Standiford, R. B., and P. Tinnin, 1992. *Integrated hardwood range management program. Fifth progress report, July 1991 to December 1992.* Berkeley: University of California, College of Natural Resources, Department of Forestry and Resource Management.

Stark, N. 1963. Natural regeneration of Sierra Nevada mixed conifers after logging. *Journal of Forestry* 63:456–60.

State of California. 1988. *California's forests and rangelands: Growing conflict over changing Uses.* Sacramento: California Department of Forestry and Fire Protection.

Stone, E. C., R. W. Benselor, F. J. Baron, and S. L. Krugman. 1963. Variation in the root regeneration potential of ponderosa pine from four California nurseries. *Forest Science* 9 (2): 217–25.

Stone, E. C., and J. L. Jenkinson. 1970. Influence of soil water on root growth capacity of ponderosa pine transplants. *Forest Science* 16:288–97.

———. 1971. Physiological grades for ponderosa pine nursery stock based on predicted root growth capacity. *Journal of Forestry* 69 (1): 31–33.

Stone, E. C., and G. Schubert. 1959. Root regeneration by ponderosa pine nursery stock based on predicted root growth capacity. *Journal of Forestry* 5:322–32.

Sudworth, G. B. 1900. Stanislaus and Lake Tahoe forest reserves, California, and adjacent territory. In *Annual Report of U.S. Geological Survey*, part 5, 505–61. Washington, DC: Government Printing Office.

———. 1901. *Notes on regions in the Sierra Forest Preserve: 1898–1900.* Washington, DC: U.S. Bureau of Forestry.

Tappeiner, J. C. 1966. Natural regeneration of Douglas-fir (*Pseudotsuga menziesii* (Mirb.) Franco) on Blodgett Forest in the mixed conifer type in the Sierra Nevada of California. Ph.D. diss., Department of Forestry, University of California, Berkeley.

Tappeiner, J. C., and A. A. Alm. 1975. Undergrowth vegetation effects on the nutrient content of litterfall and soils in red pine and birch stands in northern Minnesota. *Ecology* 56:1193–1200.

Tappeiner, J. C., and J. A. Helms. 1971. Natural regeneration of Douglas-fir and white fir on exposed sites in the Sierra Nevada of California. *American Midland Naturalist* 86 (2): 358–70.

Tappeiner, J. C., M. Newton, P. M. McDonald, and T. B. Harrington. 1992. Ecology of hardwoods, shrubs, and herbaceous vegetation: Effects on conifer regeneration. In *Reforestation practices in southwestern Oregon and northern California,* S. D. Hobbs, S. D. Tesch, P. W. Owston, R. E. Stewart, J. C. Tappeiner, and G. E. Wells, 137–64. Corvallis: Oregon State University, Forest Research Lab.

Taylor, A. 1993. Fire history and structure of red fir (*Abies magnifica*) forests, Swain Mountain Experimental Forest, Cascade Range, northeastern California. *Canadian Journal of Forest Research* 33: 1672–78.

Taylor, A., and C. B. Halpern. 1991. The structure and dynamics of *Abies magnifica* forests in the southern Cascade Range, USA. *Journal of Vegetation Science* 2:180–200.

Tesch, S. D., and E. J. Korpella. 1993. Douglas-fir and white fir advanced regeneration for renewal of mixed conifer forests. *Canadian Journal of Forest Research* 23:1427–37.

Tevis, L. P. 1953. Effect of vertebrate animals on the seed crop of sugar pine. *Journal of Wildlife Management* 17:128–31.

Troup, R. S. 1959. *Silvicultural systems.* Oxford: Clarendon Press.

Turner, P. E. 1985. Water salvage through vegetation management in California. Paper presented at Chaparral Ecosystems Research: Meeting and field conference, University California, Santa Barbara, 16–17 May.

———. 1986. Estimates of annual streamflow from precipitation and vegetation cover data. Paper presented at California Watershed Management Conference, West Sacramento, CA, 18–20 November.

U.S. Forest Service (USFS). 1994. Adaptive management process report. Unpublished manuscript. U.S. Forest Service Federal Working Group, Washington, Oregon, California.

———. 1995. *Draft environmental impact statement: Managing California spotted owl habitat in the Sierra Nevada National Forests of California — An ecosystem approach.* San Francisco: U.S. Forest Service, Pacific Southwest Region.

Ustin, S. L., R. A. Woodward, M. G. Barbour, and J. L. Hatfield. 1984. Relationships between sunfleck dynamics and red fir seedling distribution. *Ecology* 65:1420–28.

Von Althen, F. W. 1959. *A contribution to the study of edge effects on the regeneration of small forest openings in the Sierra Nevada.* MF Professional Paper. Berkeley: University of California.

Walstad, J. D., and P. J. Kuch, eds. 1987. *Forest vegetation management for conifer production.* New York: John Wiley.

Waring, R. H., and G. B. Pitman. 1985. Modifying lodgepole pine stands to change susceptibility to mountain pine beetle attack. *Ecology* 66:889–97.

Weatherspoon, C. P. 1983. Residue management in the eastside pine type. In *Proceedings, symposium on management of eastside pine type in northeastern California,* edited by T. F. Robson and R. B. Standiford, 114–21. SAF 83-06. Arcata, CA: Northern California Society of American Foresters.

Wensel, L. C., P. J. Dougherty, and W. J. Meerschaert. 1986. *CACTOS user's guide: The California Conifer Timber Output Simulator.* Bulletin 1920. Berkeley: University of California, Agricultural Experiment Station.

Wert, S., and B. R. Thomas. 1981. Effects of skid roads on diameter height, and volume growth in Douglas-fir. *Soil Science Society of America Journal* 45:629–23.

West, A. J. 1962. Interrelationships between production and water supply. In *Research and land management in the Upper Sierra: Conference on interrelated problems of natural resources conservation,* edited by J. P. Gilligan 38–48. Berkeley: University of California, Agricultural Experiment Station, Wildland Research Center.

Wood, D. L. 1993. Bark beetles, fungus, and host interactions involved in the death of pines in California. Founder's Award lecture. In Proceedings, forty-forth meeting, Western Forest Insect Work Conference, 20–26.

Wycoff, W. R., N. L. Crookston, and A. R. Stage. 1982. *User's guide to the stand prognosis model.* General Technical Report INT-133. Ogden, UT: U.S. Forest Service, Intermountain Experiment Station.

ROGER J. POFF
R. J. Poff and Associates
Nevada City, California

16

Effects of Silvicultural Practices and Wildfire on Productivity of Forest Soils

ABSTRACT

This chapter reviews the research literature on the effects of timber harvest, site preparation, and cultural treatments on Sierra Nevada forest soils. It is not an assessment of Sierra Nevada soils.

Silvicultural activities alter soil productivity as they alter soil volume, soil porosity, and soil organic matter. Because most of the nutrient capital in Sierra Nevada forest ecosystems is contained in the upper soil layers and in the forest floor, and not in the standing biomass, activities that displace or compact the surface soil have the greatest potential to alter site productivity. The potential for altering the surface soil is greatest during site preparation, and somewhat lower during timber harvest. Intermediate silvicultural treatments generally have low impacts on soil productivity. Wildfire has a much greater impact on soil productivity than prescribed fire, especially during postwildfire salvage and recovery operations. Although there is anecdotal evidence of locally severe losses in soil productivity, the extent of degraded soils in the Sierra Nevada is unknown.

INTRODUCTION

Soil—along with air, water, and sunlight—is a basic building block of ecosystems. Like air, water, and sunlight, soil is so common that it is nearly invisible. Unlike air, water, and sunlight, however, soil is a nonrenewable resource because it accrues so slowly. Soil provides vegetative growth and clean water and buffers the effects of major disturbances. Not including microorganisms, more than 75% of the species in forest ecosystems reside in the soil. Directly or indirectly, all terrestrial forest species, and many aquatic species as well,

are dependent on the soil. The health and productivity of Sierra Nevada ecosystems are strongly affected by the potential and condition of their forest soils.

Forest soil processes are commonly misunderstood, even by practicing resource professionals. Many people think that extraction of valuable resources such as timber carries the cost of some soil degradation, without fully understanding the long-term consequences of this degradation or its real costs. Others have the misconception that removal of any biomass from the forest will decrease productivity.

Silvicultural activities and wildfire have the potential to alter significantly the long-term productivity of forest soils. This chapter addresses the effects of timber harvest, silvicultural treatments, and wildfire on forest soils in the Sierra Nevada.

INDICATORS OF SOIL PRODUCTIVITY

Soil Productivity

The productivity of forest soils is difficult to assess directly. Commercial wood growth, net primary productivity (NPP), and soil properties are all indicators of soil productivity. Commercial wood growth is affected by stocking, stand age, genetics, weed competition, and plant pests, and the rate of wood growth changes over time (Powers 1991). Also, the full potential of a site is rarely reached, and yield tables tend to underestimate potential site quality (Powers 1991). NPP, or biomass production, may be a better reflection of inherent productivity, but it is not easily estimated. NPP is usually re-

Sierra Nevada Ecosystem Project: Final report to Congress, vol. II, *Assessments and scientific basis for management options*. Davis: University of California, Centers for Water and Wildland Resources, 1996.

ported as above-ground biomass. However, bolewood ranks relatively low in allocation of fixed carbon compared to leaves, roots, and reproductive parts, and as much as 75% of NPP may be allocated below ground (Powers 1990b). Also, NPP can change as stands age. Using soil properties to assess site productivity is not without its problems, either. Soil variables are surrogates for productivity, but they have not been fully calibrated against stand productive potential, much less against NPP (Powers et al. 1990a). Yet changes in soil properties and measurements of tree growth are the only tools currently available, although more rigorous efforts are under way (Powers et al. 1990a). Although imperfect, commercial wood growth remains the best-documented index of productivity and is a useful index of ecosystem health (Powers et al. 1990b). The studies cited in this chapter express changes in soil productivity in terms of tree growth.

The productivity of forest soils is a function of both soil potential and soil condition. Soil potential is defined by physical, chemical, and biological properties such as depth, amount of rock, organic matter content, texture, porosity, clay mineralogy, and temperature and moisture regime. Properties such as texture, clay mineralogy, and temperature and moisture regime are not readily altered. Soil condition is defined by readily altered surface properties such as thickness of surface soils (soil volume), porosity, and soil organic matter. These three properties are also integrators of many soil processes. In this chapter, the effects of silvicultural activities and wildfire on soil productivity will be examined by evaluating how they affect these three indicators of soil condition.

Soil Volume

Soil volume, or soil depth, controls rootable volume, plant-available water, and total nutrient storage, as well as hydrologic function and buffering capacity. A loss of soil depth is nonrenewable because soils form so slowly. Soil volume can be reduced by surface erosion, displacement, and mass wasting.

Soil Porosity

Forest productivity is highly correlated with soil porosity. Rootable soil volume, water infiltration and retention, gas exchange, and biological activity all depend on soil porosity. Forest species are dependent on aerobic mycorrhizal associations. The very high porosity—especially macroporosity—typical of healthy forest soils is the result of biological activity. Coarse, relatively indigestible needles and twigs deposited on the forest floor depend on a succession of macroinvertebrates to break them down to sizes microorganisms can decompose. The large pores created in this process are fragile and readily compressed by heavy equipment.

Organic Matter

The organic matter of forest soils can be grouped into three types: (1) soil organic matter, (2) the forest floor, sometimes referred to as duff and litter, and (3) large woody material, or decaying logs.

Soil Organic Matter

Although it makes up only 5% to 10% of the soil volume, soil organic matter profoundly affects soil properties. Soil organic matter is a storehouse of plant nutrients and is the primary source of plant-available nitrogen, phosphorus, and sulfur. Soil organic matter provides habitat for the diverse soil biota that carries out energy transformations and cycles nutrients and that is responsible for the strong granular structure and high porosity of healthy forest soils. Soil organic matter increases water-holding capacity and infiltration and, by promoting soil structure, protects the soil from erosion. Soil organic matter is composed of two fractions: (1) an active, rapidly recycled labile fraction consisting of plant roots, soil organisms and their feces, and recently dead plant and animal materials; and (2) a recalcitrant humus fraction consisting of the end products of decay, humic acids. Recalcitrant organic matter, the dominant form, may take hundreds to thousands of years to recycle.

The Forest Floor

Litter fall from conifers is highly resistant to decay by soil microbes and thus accumulates under closed forest canopies. Accumulated duff and litter modulate extremes of temperature and moisture, providing an environment favorable for the macroinvertebrates that recycle litter fall. The forest floor is the source of soluble organic ligands that chelate, dissolve, and leach aluminum and iron from mineral soil, thus providing a buffer against metal toxicity, particularly in sites subject to acid deposition (Powers et al. 1990a). The forest floor protects soils from erosion and enhances infiltration and hydrologic function.

Large Woody Material

Large woody material decomposes slowly (Harmon et al. 1986). As it decays, such material provides structural habitat for organisms that fix N nonsymbiotically and acts as refugia for many organisms, particularly the mycorrhizal fungi so critical to the health of forest species. Large woody material provides habitat for small mammals that inoculate openings with the spores of hypogeous fungi (Maser and Trappe 1984).

Much of the research on large woody material and soil wood has been conducted in the Pacific Northwest and Intermountain regions. In the Pacific Northwest decaying logs can persist for two hundred years or more (Sollins et al. 1987). But Harmon and colleagues (1987) found it took only sixty years for large logs to decay in the southern Sierra Nevada. The rate of decay also varies by species. Because sapwood decays rapidly and heartwood decays more slowly, trees with a high heartwood to sapwood ratio persist the longest. These results point out the hazards of extrapolating research information from one ecosystem type to another.

Interactions

Soil porosity, organic matter, and surface soil volume are highly interdependent. For example, soil organic matter and the forest floor foster the biological activity that produces soil structure, increasing porosity. Porosity enhances gas exchange and creates an aerobic environment favorable to soil organisms. Strong structural aggregates increase resistance to erosion and loss of soil volume. Loss of porosity reduces infiltration, increasing erosion and loss of volume. Loss of organic matter reduces soil structure, increasing soil erosion. These effects may be synergistic, not merely additive, and have yet to be quantified.

SOIL PROCESSES

Soil Erosion, Displacement, and Mass Wasting

Soil productivity is reduced by soil loss from erosion, displacement, and mass wasting. Water quality can also be affected by soil erosion and sedimentation from silvicultural activities. It is important not to confuse soil productivity and water quality. For example, displacing topsoil can severely reduce productivity, but if the displaced soil does not reach a watercourse, it has little effect on water quality. In contrast, erosion on roads and skid roads can deliver sediment directly to streams, but this erosion has a minor effect on overall site productivity. The following discussion is on the effects of erosion, displacement, and mass wasting on long-term productivity, not on the impacts of soil erosion on water quality.

Surface Erosion

The effects of erosion on site productivity are difficult to assess. The magnitudes of inputs, outputs, and components are not generally known, and rates of soil formation are not measured directly but are estimated by differences in mass-balance equations (McColl and Powers 1984). It is difficult to measure precisely amounts of soil loss significant to long-term productivity. This is partly because soil loss may not be reflected in sediment measured at the watershed mouth due to considerable on-slope and in-channel storage (Clayton and Kennedy 1985; McColl and Powers 1984), and partly because most erosion occurs during large, episodic events. Where a forest floor has developed under a forest canopy, erosion rates are near zero (McColl and Powers 1984), but rates can increase to ten or more times soil formation rates for short periods following disturbance (Clayton and Kennedy 1985).

Soil productivity will decline if soil is removed at a rate faster than it is replenished, even though reductions may not be measurable over short periods (Alexander 1988). This loss can result from surface erosion, soil displacement, or mass wasting.

Using an elemental balance equation and data from eighteen watersheds with noncarbonate lithology, Alexander (1988) estimated that soils form at a rate of about 0.02 to 1.9 t/ha per year. He suggested loss tolerance limits should be lower than 2.24 t/ha per year (1 ton/acre per year) for shallow and moderately deep soils on plutonic rocks. One ton per acre is equivalent to the thickness of two sheets of paper.

Surface erosion can be caused by overland flow, or it can occur as sheet erosion, rill and gully erosion, or dry ravel. Sheet erosion is the nearly imperceptible loss of soil through the action of falling raindrops. Sheet erosion is greatest on steep slopes because the splash of each raindrop has a greater probability of moving downslope. Overland flow rarely occurs on undisturbed forest soils with surface litter because infiltration is high. Rill and gully erosion results when runoff is concentrated by an impervious surface such as a road, skid trail, landing, or area of rock outcrop or shallow soils. Dry ravel is downslope movement by gravity alone. Dry ravel occurs on very steep slopes, particularly on sandy soils or after intense wildfire. Rill and gully erosion is the most common type of surface erosion on forest soils (Rice 1979).

In the Sierra Nevada, soils developed from granitic bedrock are the most susceptible to rill and gully erosion and to dry ravel; soils developed from metasedimentary bedrock are the most stable. Mature soils with high contents of iron oxides appear to be more resistant to erosion, perhaps because they tend to form stable soil aggregates.

The most effective cover type is a forest canopy with a well-developed forest floor. This cover type not only reduces sheet erosion, but also improves hydrologic function by improving lateral infiltration and movement of water in near-surface soil layers.

The amount of surface erosion approaches zero in undisturbed forests. McColl and Powers (1984) report losses of nitrogen and calcium by erosion in undisturbed forests to be about 100 g/ha per year (.09 lb/acre per year), and about 25 g/ha per year (.02 lb/acre per year) for magnesium and potassium. These losses by surface erosion are significantly less than losses by deep leaching.

Even with extreme disturbance, the loss from surface erosion does not appear large compared to that from mass wasting. In a study of 80% gradient slopes bare of vegetation following slash burning in western Oregon and Washington, Fredriksen and colleagues (1975) report an annual surface erosion rate of 3.6 m³/ha (1.9 yd³/acre), and a rate of 21 m³/ha (11.1 yd³/acre) from mass wasting. On a helicopter-logged and broadcast-burned clear-cut in the Idaho Batholith, Clayton and Kennedy (1985) reported erosion rates of 1.8, 13, 4, and 4 t/ha per year (0.8, 5.8, 1.8, and 1.8 tons/acre per year) for the first winter, first summer, second winter, and second summer after treatment, respectively. Even at their maximum, these rates are only six times the maximum soil loss tolerance limit suggested by Alexander (1988), and in the second year they were only two times the tolerance limit and decreasing rapidly.

Frequency of disturbance is a major factor in determining how surface erosion affects forest soil productivity. After a major disturbance, erosion rates exceed the rate of soil formation for only a few years, until a new forest floor accumulates and provides effective cover, assuming that reforestation is swift and successful. If the disturbance is not unusually severe, disturbance intervals of eighty to one hundred years—a normal rotation—should not lead to a decline in productivity. In the final analysis, there are many unknowns, and surface erosion effects on long-term productivity can only be inferred (Powers 1991). However, because the loss of surface soil is irreversible, even very small losses should not be taken lightly. Over many rotations even small losses could result in a significant decline in productivity.

Soil Displacement

The biggest threat to soil productivity is direct mechanical displacement of the surface soil. Practices that manipulate the top layer of the soil, and particularly those that remove it, degrade productivity by any standard (Atzet et al. 1989) and inevitably lead to a decline in site quality (Powers and Edmonds 1992).

In the 1950s and 1960s, topsoil was intentionally stripped during site preparation and pushed into windrows to reduce competition by removing sprouting species and dormant weed seeds in the surface soil and litter (McColl and Powers 1984). This is no longer done. The justification for this practice was the overwhelming research results that show early growth and survival of trees is greatest where mineral soils are most disturbed and the most biomass is removed (Morris and Miller 1994). This short-term growth response is due to increased N availability and reduced competition.

Soil may also be displaced unintentionally when slash is machine-piled for burning. More soil is displaced when machines grub out shrubs at the same time. Extensive and relatively "clean" machine piling tends to concentrate nutrients (McNabb and Cromack 1990) and may disrupt the natural decay cycles of large, rotting logs.

Another form of displacement is dusting. When very dry, fine-textured soils—especially soils high in volcanic ash—are subjected to heavy traffic, airborne soil drifts short distances, which can lead to entrenchment of skid roads.

Mechanized removal of slash into piles or windrows has a very high potential to reduce productivity because it can displace large quantities of organic matter, soil, and associated nutrients on much of the site (Morris and Miller 1994; Powers 1991). Productivity is reduced because soil organic matter and readily available nutrients are usually concentrated near the soil surface and decline rapidly with depth (Powers 1990a). In a windrowed slash pine site Morris and colleagues (1983) reported the P, K, Ca, and Mg in windrows represented displacements of between 15% to 40% of the total organic plus soil-extractable nutrient reserves of the ecosystem.

Growth reductions associated with nutrient loss by displacement do not become apparent until after crown closure (Morris and Miller 1994). But when they do finally appear, the growth reductions can be enormous. After only one treatment, the following losses have been reported: a 20% decrease in site index with 25% removal of surface soil (Kittredge 1952); a 30% loss in volume with displacement of about 2 cm (0.8 in) of topsoil (Dyck and Beets 1987); trees within 3 m (10 ft) of windrows produced two times the volume of those farther from windrows (Atzet et al. 1989); and scalped plantations produced three times as much volume after N fertilization as unscalped ones (Powers et al. 1988).

Compared to nutrient export from timber harvest, the N displaced in windrows can be six times that removed by harvest (Morris et al. 1983), and N and P losses can be two to three times those of whole-tree harvesting (Powers et al. 1990b). Nutrients besides N are also affected. Morris and colleagues (1983) report that windrow displacements of P, K, Ca, and Mg can represent displacements of 15%–40% of the total organic plus soil-extractable reserves of the ecosystem.

Forest soils in fir ecosystems are disproportionately vulnerable to productivity loss by displacement, because a much higher proportion of nutrients is in the forest floor (Powers and Edmonds 1992), and nutrients in the underlying mineral soil are typically concentrated in the upper 5–10 cm (2–4 in).

Although productivity losses by displacement can be very high, over many rotations this soil loss may be less serious than the smaller amounts of soil lost by erosion, because displaced soil, if it has not left the site, can be respread. However, respreading soil displaced by dusting may not be an option because the displaced soil is not concentrated.

Mass Wasting

Mass wasting can result from roads, increased pore water pressure, and loss of root strength by decay. Because landslides expose less-fertile subsoils, productivity can be reduced. Miles and colleagues (1984) found that Douglas fir regenerating on landslides in western Oregon averaged 25% less stocking and 62% less height growth than on nearby clearcuts.

Landslides are much more common on steep slopes. Atzet and colleagues (1989) report that landslides on the Siskiyou National Forest are twenty times more likely on slopes steeper than 70% than on slopes with 50% to 70% gradients, and two hundred times more likely than on slopes less than 50% in gradient.

Although mass wasting associated with silvicultural activities occurs locally in the Sierra Nevada, it is not as widespread as in northwestern California or the Pacific Northwest. Where it does occur, mass wasting is usually associated with roads or with geologic contact zones such as the base of the Mehrten formation. Evidence of past shallow debris flows is also common on the steep slopes of canyon inner gorges. The relative lack of widespread mass wasting in the Sierra Nevada may be because the steeper slopes have not been heavily impacted by road building and timber-harvest activities. Rice (1979) reports that about 70% of the forested land in the Sierra Nevada is on slopes with gradients lower than 32%.

Mass wasting has not had a major impact on soil productivity in the Sierra Nevada, but shallow debris flows and other forms of mass wasting could become more common if activity on steeper slopes increases.

Management Strategies to Minimize Soil Loss

Sheet erosion can be effectively mitigated by maintaining effective ground cover. Soil displacement and loss of ground cover can be eliminated by selectively manipulating slash and fuels with special equipment such as small excavators. The careful use of dozers with brush rakes is also effective. Displacement of topsoil into windrows is no longer considered necessary to reforest burns or clear-cut areas. Leaving residual trees or patches of trees can reduce insolation and slow the rate of decomposition of forest litter to protect the site until a needle cast creates a new forest floor. Cool prescribed burns that do not entirely consume the forest duff can reduce fuels and prepare sites for planting without totally removing effective ground cover. Carefully placing and designing roads and maintaining some live trees on unstable sites can reduce the risk of mass wasting. Silvicultural systems that cause minimal disturbance to the forest floor and do not entirely remove the forest canopy will cause little loss of soil to erosion, displacement, or mass wasting.

Soil Compaction

Compaction, Porosity, Density, and Strength

When a force or load is applied, soil will compact until it has enough strength to bear that load or force. As soil compacts, porosity decreases and density increases. Therefore, compaction can be thought of as a decrease in porosity with an increase in density and strength, the result of reduced pore space as air is expelled (Alexander and Poff 1985). Porosity is expressed as a percentage of volume; density as weight per unit volume; and strength as resistance to deformation, usually in kPa of pressure (Alexander and Poff 1985). Soil strength is highly dependent on moisture content.

The natural variability in density of forest soils is quite high (Alexander and Poff 1985). Such variability is not surprising, considering the effects large trees have on the soil in redistributing organic matter and nutrients (Zinke 1962) and the soil mixing that results from windthrows.

Effects on Plant Growth

The penetrating abilities of roots are reduced by poor aeration; oxygen deficiency and excess carbon dioxide both reduce root penetration (Alexander and Poff 1985). Roots do not enter rigid pores smaller than their diameters but, because of their axial pressure, can create their own pores in friable surface soils with low density. Roots are dependent upon extension in dense subsoils. When soil porosity is lost because of

compaction, less soil volume is available for roots to occupy, and plant nutrients are relatively immobile. Under such conditions, even water cannot migrate through the soil rapidly enough to supply plant transpiration needs when plants are under extreme moisture stress. Less rootable volume thus equates to less plant growth. This problem is further compounded in forest ecosystems where conifer growth is dependent on mycorrhizal fungi, which are highly aerobic (Harmon et al. 1986). The effects of compaction on tree growth are well documented. Wert and Thomas (1981) found that trees in skid roads produced 74% less bolewood than trees in an adjacent undisturbed area. Because skid roads occupied only part of the area, stand growth was reduced by only 11.8%. On the Foresthill divide, Helms and Hipkin (1986) reported a volume reduction of 59% on soils with the highest amounts of compaction; the volume of an average tree was 21% less on the most compacted soils compared to the least compacted. The decrease in height growth of trees is nearly a linear function of the increase in soil density (Froehlich and McNabb 1984).

Soils are most compactible when moist but not saturated (Baver 1930). Water reduces frictional forces between particles, decreasing the resistance of soil to deformation (Alexander and Poff 1985). Compaction causes a greater reduction in macropores in moist soils than in drier soils (Hatchell et al. 1970). However, when soils are saturated and pores filled with water, which is relatively incompressible, a process called puddling occurs. Puddling destroys soil structure and reduces macroporosity (Alexander and Poff 1985) and can cause a greater loss of porosity and infiltration than compaction, without an increase in density (Hatchell et al. 1970). Soil compaction penetrates deeper under wet conditions than under dry conditions (Froehlich 1974). Because forest vegetation transpires moisture, thereby drying the soil, timing of vegetation removal can be used to manage soil compaction (McNabb 1981).

Soil organic matter has a strong effect on compaction. Free and colleagues (1947) found that soils with the most organic matter were compacted less by a given compactive effort at a given soil moisture content than were soils with the least organic matter. In their study of California forest soils, Howard and colleagues (1981) found a high content of organic matter reduced the effects of soil compaction. Organic matter increases resistance to compaction partly because it increases soil structure (Boyer 1979). Organic matter may also increase resiliency or rebound after cessation of stresses, even though organic matter per se is not considered to be an elastic material (Stone and Larson 1980).

When bare soil is exposed in logging, some soil disturbance and some compaction occurs, but organic litter cushions these effects (Alexander and Poff 1985). Froehlich (1978) found that both the degree and the depth of compaction were reduced by the presence of a litter layer; densities increased with successive trips and as the litter was removed. At the Blodgett Experimental Forest, Miles (1978) attributed the relatively small amounts of compaction on minor skid roads to the presence of 6 to 8 cm (2.4 to 3 in) of organic litter cover, which was no longer present after several trips on primary skid trails. Mace (1970) found that the amount of slash was important in reducing compaction, and Boyer (1979) suggests a surface layer 5 cm (2 in) or greater in thickness will provide protection from compaction at moisture contents approaching field capacity and may support up to two trips with equipment before compaction occurs. Avoiding or minimizing disturbance of the litter layer and surface soil high in organic matter is key to solving soil-compaction problems (McColl and Powers 1984).

Once soil is compacted, it takes decades for porosity to return to natural levels. The length of time to recover from compaction varies with soil type and the degree of compaction (McColl and Powers 1984). Effects of compaction have been reported to persist unchanged for sixteen, eighteen, thirty-two, and fifty years (Froehlich 1979; Hatchell et al. 1970; McColl and Powers 1984; Wert and Thomas 1981). It may take forty to fifty years or more for some soils to recover from compaction (Hatchell et al. 1970; Morris and Miller 1994). If compaction is deep, soils may never return to their original state without major physical disturbance (McColl and Powers 1984). Because the origin of porosity is essentially biological, it follows that recovery from compaction is also biological. Thus, both the depth of the compaction and the amount of organic matter in the compacted soil affect how long it takes to recover. Where compaction occurs in horizons high in organic matter, recovery will be more rapid than that in subsoils lower in organic matter.

Management Strategies to Minimize Compaction

Although the processes associated with soil compaction are complex, management strategies can be developed by thinking of compaction as the result of two opposing forces: a compacting force (or load) applied to the soil and the resistance of the soil (strength) to deformation by that force (Alexander and Poff 1985). Management strategies can then be expressed in terms of manipulating these opposing forces:

- Avoid compactive forces—for example, use aerial yarding systems such as helicopters, balloons, or cables, or yard material by end-lining.

- Reduce compactive forces—for example, use low ground-pressure equipment.

- Absorb compactive forces—for example, operate on a cushion of slash with cut-to-length forwarding equipment, or operate over snow.

- Operate when soil strength is high—for example, when soil moisture is low or when the soil is frozen.

- Confine compactive forces—for example, limit the area compacted by designating skid roads, and either restore porosity by tillage or accept compaction in the skid roads as a cost of resource extraction.

Distribution and Cycling of Mineral Nutrients

Forests are sinks for carbon and nitrogen, and vast amounts are stored above and below ground, especially in the soil. Although varying by biome, amounts of organic carbon in the forest floor plus soil exceed that of the standing forest (Powers and Van Cleve 1991). Globally, soil carbon equals that of vegetation and the atmosphere combined (Johnson 1994).

Tree Biomass Accumulation

A major portion of site productivity is directed below ground, and as much as 75% of net primary productivity can be below ground as fine roots and mycorrhizae (Grier et al. 1981). The proportion of primary productivity directed below ground is much higher in true fir than in mixed conifer forests (Powers and Edmonds 1992).

The rate of nutrient accumulation by forests also changes over time. Early in the life of a stand, crown and bole weights accumulate at similar rates, but after crown closure, crown weight remains constant while bole weight continues to accumulate. Therefore, because most nutrients are in foliage, early in the life of a stand crowns contain the most nutrients, but after crown closure, boles accumulate an increasing proportion of nutrients (Powers 1979). Most nutrients are taken up before crown closure. After crown closure trees internally translocate phloem-mobile nutrients from senescing parts to actively growing sites (Powers and Van Cleve 1991). As much as 30% to 50% of N and 20% to 80% of P may be translocated internally before leaf senescence (Prescott et al. 1989). As a stand reaches maturity, return of nutrients to the soil in litter fall approaches the rate of nutrient uptake (Powers 1979).

Litter Accumulation

In a young stand, the rate of litter accumulation on the forest floor is initially low because open crowns have light litter fall and allow high surface soil temperatures that encourage decomposition. With crown closure, the rate of litter fall increases, decomposition slows, and litter accumulates more rapidly. It may take one hundred years or more to reach an equilibrium between litter accumulation and litter decay (Powers 1979).

Nitrogen

Of all plant nutrients in forest ecosystems, N is often the most limiting—particularly in western forests (Powers and Edmonds 1992). Nitrogen is added to the ecosystem in rainfall, by symbiotic and nonsymbiotic N fixation, and in negligible amounts by rock weathering (Powers 1979). Most N accumulates in soil organic matter and in the forest floor in forms unavailable to plants. In their classic study of a thirty-six-year-old Douglas fir stand, Cole and colleagues (1968) found that 84.8% of total ecosystem N was in the soil, 5.3% in the forest floor, 0.2% in the understory, and 9.7% in the trees. In general, mature mixed conifer stands have about 10% of total ecosystem N in standing biomass, 10% in the forest floor, and 80% in the soil (Powers 1991). Of the standing biomass, at least half of the N is in foliage and branches. In true fir forests, a higher proportion of N is in the forest floor: standing biomass, 13%; forest floor, 40%; soil, 47% (Powers and Edmonds 1992). Although there are significant differences, the distribution of other nutrients generally follows a similar pattern.

Most soil N is unavailable to plants and must be mineralized by bacteria to ammonia or nitrate for plant uptake. Mineralization is regulated by moisture and temperature, with the highest rates occurring at middle elevations in the mixed conifer zone. Cold temperatures limit rates of mineralization at higher elevations; lack of summer moisture limits rates at lower elevations (Powers 1990b). In true fir forests, N is mineralized under cold, moist conditions by cold-loving microbes, although rates are very low, suggesting that increases in mineralization following timber harvest will be less in true fir forests than at lower elevations (Powers and Edmonds 1992).

Management Strategies to Minimize Nutrient Export

Nutrients are lost from forested sites by leaching, biomass removal, volatilization, and soil loss. Because the soil and forest floor hold the vast majority of nutrients on a forest site, they buffer the impacts of ecosystem disturbances such as fire, insects and disease, storm damage, and timber harvest (McColl and Powers 1984). Management practices that maintain the integrity and structure of surface organic matter will have the least impact. Useful strategies to minimize nutrient losses include harvesting boles only, using specialized equipment to selectively manipulate slash without disturbing the forest floor, and using cool prescribed burns that do not consume the lower half of the duff layer.

EFFECTS OF FIRE ON SOILS

Fire can have physical, chemical, or biological effects on soils. Physical effects include loss of soil organic matter, loss of soil structure, hydrophobicity, erosion, and, in extreme cases, destruction of soil clay minerals. Chemical effects include an increase in pH, a loss of cation exchange capacity, and the loss of nutrients by volatilization, in fly ash, or by leaching. Biological effects include direct mortality of soil organisms and loss of their habitat. Fire effects on soil productivity can be either beneficial or devastating, depending on fire intensity, soil type, and site history.

Adverse fire effects on soils increase as burn intensity increases, and the effects are proportional to the amount of surface duff and soil organic matter consumed (DeBano 1979). High amounts of moisture in the soil, particularly in the lower half of the duff layer, reduce organic matter consumption (Sandberg 1980). Maximum temperatures reached, even if only in pulses of short duration, govern the magnitude of effects (DeBano 1979).

Soil temperatures above 50°C are lethal for fungi, and nitrifying bacteria are killed at 100°C (Boyer and Dell 1980). Destructive distillation of organic compounds begins at about 200°C, and organic matter ignites at 260 to 425°C (DeBano 1979). Below 200°C organic matter is not destroyed, but it can be distilled and moved within the soil and affect wettability. Above 200°C, N and S are oxidized rapidly, and at 500°C most N has been volatilized (DeBano 1979). Temperatures of 760°C have been measured at the soil surface during fires in chaparral (DeBano 1979).

Fire may temporarily sterilize soils. Hot burns on moist soils may increase the mortality of soil organisms by driving steam into the soil. Changes in populations of soil organisms usually last only a year or two but vary with fire intensity. After a fire, invertebrates decline, fungi decrease, and bacteria increase. Where fires create very large openings, the loss of host plants for mycorrhizal fungi can lengthen the time it takes to reinoculate the site (Borchers and Perry 1990).

Soil organic matter has a high cation exchange capacity. When organic matter is burned, a flush of cations such as Ca, Mg, Na, and K is released and made more readily available to plants (DeBano 1979). Hotter burns may produce bicarbonate anions, further mobilizing cations in the soil solution (McColl and Powers 1984). Cations released in the ash bed are potentially susceptible to leaching, but revegetation and exchange sites in the soil usually absorb cations quickly, preventing this type of nutrient loss (McNabb and Cromack 1990). Leaching loss could be significant under very intense burns on coarse-textured soils low in organic matter.

Under intense burns, all surface litter may be removed, making soils highly susceptible to erosion. Debris movement and dry ravel may also increase when small organic-debris dams are burned out (DeBano 1979). The formation of hydrophobic layers may accelerate soil erosion. Temperature gradients near the soil surface can be very steep; for example, 760°C at the surface, but 200°C at 5 cm (2 in). Hydrophobic layers form when organic compounds volatilized in the surface litter are driven into the soil and condense on the underlying, cooler soil particles (DeBano 1979). Generally, hydrophobic layers occur deeper, and are more water repellent, in sandy soils because these soils have high macroporosity and low surface area. Strongly hydrophobic layers create an effectively very shallow soil, making the wettable surface soil very vulnerable to erosion. Hydrophobic layers may also form at the surface if soils are moist or clayey, or where fire intensity is low. Surface hydrophobicity protects the soil from erosion but can greatly increase channel scour by causing rapidly accelerated runoff. The formation of hydrophobic layers in forest soils of the Sierra Nevada is quite variable.

As burn intensity increases, increasing amounts of N, and, to a lesser extent, P and S, are volatilized and lost to the atmosphere. In large fires of high intensity, other nutrients may be lost in the smoke plume as convective fly ash (Clayton and Kennedy 1985).

The plant nutrient most affected by fire is nitrogen. Nitrogen loss is almost linearly related to litter consumption, and little N is lost until more than 25% of the litter has been consumed (Dunn and DeBano 1977). McColl and Powers (1984) summarized N losses for different burn intensities. Under severe burns, N losses ranged from 72% to 99%, but under moderate-intensity burns, losses ranged from 11% to 38%. Fire can also increase soil nitrogen. Heating and combustion increase ammonia (Dunn and DeBano 1977), making it readily available for plant uptake. Nitrification is also stimulated by reduction of repressive tannins and by increases in ammonium (Powers 1979).

Losses of sulfur and phosphorus are proportional to nitrogen losses, though smaller—only about 5% to 9% of nitrogen loss. Sulfur is important in decomposition of organic matter and in nitrogen metabolism. Sulfur is of concern because it is not fixed, but is added to the ecosystem abiotically through precipitation and mineral weathering. The origin of atmospheric S includes fossil-fuel consumption, acid deposition, and volcanic eruptions (McNabb and Cromack 1990). Sulfur losses have been detected indirectly in the Pacific Northwest, and as much as 50% of total S may be oxidized at 800°C (Boyer and Dell 1980). Sulfur deficiencies are readily overcome by small amounts of S in fertilizer.

Fire frequency, in the context of a site's natural fire regime, has a major impact on soil productivity (McNabb and Cromack 1990). Frequent, low-intensity fires, on sites where the vegetation has adapted to them, will increase soil productivity over the long term (Klemmendson et al. 1962). On the other hand, frequent high-intensity fires, except on sites adapted to such a fire regime, are likely to reduce nutrient reserves and to initiate long-term productivity decline. Intense fires that consume all the forest floor are particularly damaging to fir forests, where a high proportion of nutrients is contained in the forest floor (Powers and Edmonds 1992).

It is important to distinguish between prescribed fire and wildfire. Wildfires have a far greater potential to affect long-term soil productivity than does prescribed fire (McNabb and Cromack 1990). In contrast to prescribed fires, wildfires are more intense, consume more organic matter, burn longer, occur when soils are drier, and have higher levels of volatilization and convective losses. An intense wildfire may volatilize the equivalent of two hundred years of N input from precipitation (Powers 1979). Soils that are subjected to intense wildfire more frequently than every one hundred years may experience productivity decline (McNabb and Cromack 1990).

An indirect effect of wildfire is the sequence of activities associated with fire suppression, timber salvage, and reforestation that follows major wildfire. The effects of these activities are discussed later in this chapter.

Management strategies to reduce the negative impacts of prescribed fire on soils involve reducing fire intensity (DeBano 1979). They include burning under high humidity, low temperatures, and low wind speeds, and burning smaller areas. Reducing fuel loading before burning, burning when the soil and duff are moist, and burning downslope with a less intense backing fire can also reduce fire intensity.

EFFECTS OF SILVICULTURAL TREATMENTS

In the following discussion, the reader should keep two things in mind. First, silvicultural practices have changed dramatically during the past two or three decades. Site preparation methods of the 1950s, 1960s, and 1970s were especially damaging to soils. Many of the clear-cut and broadcast-burn practices of the 1970s and 1980s were also harsh. From the mid-1980s to the present silvicultural practices have shifted away from large clear-cuts and bare-ground site preparation, to smaller openings and more residual trees. Logging equipment available and wood products considered merchantable have also changed.

Second, because forests grow slowly, most of the citations in the following sections are for retrospective studies on treatments made from the 1950s to 1970s. In spite of these limitations, this research expanded our knowledge of forest soil processes and certainly provides us a historical perspective on the impacts of past treatments. Some effort is needed, however, to interpret these research findings for the issues facing the Sierra Nevada today.

Timber Harvest

Timber harvest can affect soil productivity through erosion, displacement, compaction, biomass export, and leaching. The effects vary with the type of harvest—for example, clear-cutting versus partial removal—and with the degree of disturbance. This section examines only the removal of timber. Site preparation is covered in the following section.

Erosion

The amount of soil erosion caused by timber harvest is directly related to the degree of soil disturbance, which in turn is related to logging method. Percentage of bare soil following logging can range from less than 2% for helicopter yarding to more than 75% for tractor logging (Rice 1979). In clear-cutting, about 6% to 19% of bare soil is exposed using aerial yarding systems, and 15% to 30% or more with ground-based systems. In uneven-aged systems disturbance is less at each entry, but frequency of entry may be higher than in even-aged systems. Although considerable erosion can occur on the skid roads of ground-based systems, surface erosion from just the yarding is typically quite low (McColl and Powers 1984), due in part to surface roughness and to the slash left on the site, which tends to trap sediment and prevent its movement off-site. Most of the erosion during timber harvest operations is related to roads (McColl and Powers 1984).

Compaction

Porosity may be reduced when timber is harvested with ground-based logging systems. The impact on productivity is directly related to the area in skid roads. Uncontrolled skidding in clear-cuts typically results in 20% to 40% of the area in skid trails. Skid-road area may be only 8% to 10% in selection-cutting, but this is per entry. With repeated entries, skid roads under selection-cutting can occupy more than 64% of the harvested area if skid-road locations are not controlled (Dyrness 1965).

Compaction in uneven-aged systems can be difficult to mitigate with tillage. Although specifically designed tillage implements such as the forest cultivator can be used, tillage may damage the roots of residual trees, increasing their susceptibility to disease.

These impacts on soil productivity are not a necessary cost of timber harvest. Modern harvesting equipment, such as cut-to-length processors and forwarders, does not compact soil, even when operating on moist soils. Compaction can also be reduced to acceptable levels using conventional ground-based equipment if designated skid trails and end-lining are used (Froehlich et al. 1981).

Forest Floor

Disturbance of duff and litter during timber harvest may slightly increase the rate of decomposition, but the changes in temperature and moisture resulting from increased insolation and lack of litter fall have the greatest effect on decomposition. In clear-cuts the forest floor disappears in less than a decade; significant losses can also occur under partial cutting (McColl and Powers 1984).

Biomass Export

The amount of nutrients exported through timber harvest depends on nutrient distribution in the ecosystem and utilization standards. About 20% of carbon and 10% of nitrogen are in bolewood in young, mature forests. Even in the most intensive harvests, less than 10% of ecosystem N would be removed (Powers and Edmonds 1992). Actual amounts exported would be considerably less under current practices, even for clear-cutting, because unmerchantable material, snags, and small patches of green trees are typically left. The general consensus is that stem-only harvests of mid-age stands have little impact on nutrient export. Atmospheric inputs of N, P, and probably S exceed harvest export, and soil reserves of K, Ca, and Mg are high, even without weathering inputs (Morris and Miller 1994). However, whole-tree harvesting, where slash and unmerchantable boles are also exported, could be of concern on less productive soils if rotation length is short (Johnson 1983; Zinke et al. 1982).

Nitrogen

Timber harvest can increase ammonification and nitrification by raising summer temperature, increasing moisture, and by adding labile organic matter to the soil (Frazer et al. 1990). On the Challenge Experimental Forest, N mineralization remained elevated for seventeen years after clear-cutting, but the additional N was incorporated into rapidly growing vegetation. Such increases should be considerably less under partial cutting, because forest litter has a strong repressive effect on nitrification (Frazer et al. 1990).

Leaching

During the 1960s there was concern about nutrient loss from leaching following clear-cutting, in part triggered by misinterpretation of the classic Hubbard Brook study (Likens et al. 1969). The current consensus is that, except in extreme cases where vegetation is absent or intentionally suppressed, nutrient losses by leaching are negligible (Johnson 1994; McColl and Powers 1984). In their study of nutrient leaching on a high-porosity, low cation exchange capacity soil, Cole and Gessel (1965) found that nearly all elements released from the forest floor were retained within the rooting zone or taken up by vegetation. Under uneven-aged management, the effects of residual vegetation could be expected to eliminate leaching losses entirely.

Rotation Length in Even-Aged Management

The effects of timber harvest on soil productivity are exaggerated by short rotations, or as frequency of disturbance is increased (Johnson and Todd 1987; Morris and Miller 1994; Powers 1991; Powers et al. 1990b; Switzer et al. 1981). Rate of nutrient uptake is greatest at about the point of crown closure, so short rotations place a greater drain on nutrients than do long rotations (Powers 1990a). Short rotations also forgo the nutrient accretion that occurs in mature stands (Sollins et al. 1980), because of more frequent periods with less crown protection. In general, with normal harvests, rotations greater than sixty to eighty years should not export nutrients faster than they accrue (Powers et al. 1990a). Longer rotations or lighter harvests may be necessary on low-quality sites to avoid productivity decline.

Site Preparation

The potential for impacts on long-term soil productivity is greatest during site preparation. At that time the forest floor and surface soil are most subject to manipulation and most vulnerable to damage (McColl and Powers 1984). The amount of soil and forest floor manipulated varies with type of harvest, clear-felling being the most severe. Partial cutting under uneven-aged systems generates less slash and requires less manipulation of the forest floor and topsoil.

Displacement

The effects of soil displacement on soil productivity are great and well documented. Most research on soil displacement has been conducted on clear-cuts and plantations created from the 1950s to 1970s. Although harsh site-preparation treatments, such as scalping and windrowing, are no longer done, the research on them provides valuable insights on soil processes and the importance of the forest floor and surface soil in soil productivity. The effects of soil displacement, discussed earlier in this chapter, will not be repeated here.

Compaction

Most of the research on soil compaction has been conducted on skid roads. However, the general principles governing soil compaction discussed earlier in this chapter apply to site preparation as well. Conditions during mechanical site preparation make soils highly vulnerable to soil compaction. Typically, soils are moist, bare soil is exposed, and multiple equipment passes are made. Also, a far greater proportion of the treated area can be affected by site preparation. Timing makes a big difference. For example, when mechanical site preparation follows winter logging, subsoils stay moist well into the summer because transpiring trees have not "pumped" moisture out. The compaction that occurs under these conditions is insidious because it goes unnoticed and because it does not readily recover without tillage.

The use of modern equipment drastically reduces, or avoids entirely, soil displacement and compaction during mechanical site preparation. Small excavators equipped with grapple heads, for example, are used to selectively pile logging slash without disturbing the forest floor, without compaction, and even without disturbing decaying logs. The resulting piles contain no soil and few nutrients, and burn clean. Even conventional equipment, used prudently and under the right conditions, can be used to pile slash with minimal soil impacts.

Prescribed Burning

Two general types of prescribed burning are used in site preparation: broadcast burning, and piling and burning. Piling and burning slash concentrates nutrients, and the high temperatures reached under burned piles damage soils. Piling and burning may also cause displacement and compaction as discussed earlier in this chapter.

Broadcast burning has many of the impacts described for fire (see "Effects of Fire on Soils," earlier in this chapter), but the effects are usually less extreme because ignition can be limited to periods when soil moisture is high enough to prevent complete consumption of the duff and litter. Heavy fuels are often removed before burning to reduce the intensity of broadcast burns. Common techniques are yarding unmerchantable material to landings, harvesting material as chips for "hogfuel," or various forms of partial "whole-tree" harvesting, such as yarding and chipping some crowns.

Given these complexities, not to mention weather conditions at the time of burning, it is not surprising there is little published literature on the effects of broadcast burning. Studies that have been done are of complex situations and are often confounded by other factors (Palazzi et al. 1992).

In general, the effects of broadcast burning are related to the condition of duff and litter prior to burning and to what remains afterward. Consumption of the forest floor is a function of its moisture content at the time of burning (Sandberg 1980). Forest floors less than 2 cm (0.8 in) thick generally do not hold enough moisture to withstand a broadcast burn (Boyer and Dell 1980).

The amount of N lost is proportional to duff consumption. Surface erosion is also related to duff consumption, and erosion of the ash bed and surface soil after broadcast burns may be the primary mechanism of nutrient loss (McNabb and Cromack 1990). Accelerated surface erosion is commonly observed after broadcast burns, but surface erosion is difficult to measure, and real data are rare. In the Idaho Batholith, rates of 1.8 to 13 t/ha (0.8 to 5.8 tons/acre) per year were measured the first two years after fire on a broadcast-burned clear-cut (Clayton and Kennedy 1985), about six times the estimated soil formation rate. On many national forests in California, broadcast burning is not allowed, or is severely limited, on soils derived from granitic bedrock because experience has shown that erosion rates are consistently high. The formation of hydrophobic layers under prescribed burning has not been reported and is probably rare.

The effects on soil productivity that have been reported are variable. One severely burned clear-cut had one-third less mineralizable soil N than adjacent unburned areas; but paired burned and unburned units on the Six Rivers National Forest showed no differences in most soil properties (Atzet et al. 1989). Most studies have not found consistent differences in growth between burned and unburned areas, but this result may be confounded because burning can reduce total nutrients while increasing nutrient availability (Morris and Miller 1994).

In summary, the effects of broadcast burning on soil productivity range from minimal, or even beneficial, to extremely severe depending on site conditions. As with timber harvest, the frequency and intensity of biomass removal are probably major factors in determining nutrient loss.

Intermediate Cultural Treatments

After site preparation and planting, a number of treatments can be applied to maintain stocking and growth, and to protect the stand from fire. These treatments have variable effects on soil volume, porosity, and organic matter and nutrient cycling.

Clipping and Hand Grubbing

Weed and brush control is important to stand survival. Treatments are usually applied during the first few years after planting. Clipping, sometimes combined with herbicide applied to sprout stumps, is beneficial to the soil because it increases effective ground cover, protecting the soil from erosion. Grubbing, essentially hoeing brush and weeds around tree seedlings, has variable impacts. The area grubbed is bare and susceptible to erosion. The actual area grubbed depends on stocking level and grubbing radius. For example, grubbing to a 0.8 m (4 ft) radius where stocking is 3 m by 3 m (10 ft by 10 ft), can result in 50% or more of the site in bare soil. On steep slopes, especially on a site that was burned hot, grubbing and the foot traffic associated with it can cause dry ravel and expose the soil to severe erosion, at least for a season or two. In general, treatments that leave the majority of slash in place have little effect on soil productivity (Morris and Miller 1994).

Herbicides

The types and amounts of herbicides normally used in forestry have negligible impacts on long-term soil productivity. Herbicides may be used to control weeds early in the life of a stand. Herbicides may alter biological populations in soil, but very little foliar-applied herbicide reaches the forest floor, and herbicide levels in soils seldom exceed toxic levels for long periods (McColl and Powers 1984). The impacts vary by type of herbicide. Ammonium sulphamate was found harmful to collembola, isopods, and millipedes, affecting the breakdown of litter; asulam reduced nitrate production, reducing N leaching (Norris 1983). Generally, the rate of degradation and mobility in the soil determine the relative hazard of herbicides in the environment (Norris 1983). Because most herbicides are strongly sorbed onto organic particles in the surface soil, the greatest risk of herbicide movement is by erosion. The overall health and condition of the soil—porosity, organic matter, surface duff and litter—control how well the soil will buffer the effects of herbicides.

Grazing

Grazing by cattle or sheep can be used to control brush and weeds early in the life of a stand. The effects are variable, depending on specific site conditions and on how the stock are herded and managed. On gentle slopes, impacts can be negligible. When forced onto steep slopes, grazing cattle can accelerate dry ravel and erosion.

Thinning

Generally the effects of thinning on the soil are minor. Soil compaction in thinning operations is generally insignificant but depends on the type of equipment used. Slash and the forest floor cushion the impact of ground-based equipment, and actively growing trees transpire moisture, creating periods when the soil is dry. Smaller equipment, or equipment with low ground pressures, is often used. Mastication of brush to release young stands, however, has a greater potential for compaction because less duff and litter have accumulated. Equipment used in mastication is highly variable.

Opening the forest canopy can raise soil temperatures, increase biological activity, and accelerate decomposition of the forest floor. Mobile nutrients including N and K contained in foliage and bark can be concentrated in through fall and stem flow, as compared to precipitation in the open, and opening a

stand can double the rate of mobile elements leached from the forest floor (McColl and Powers 1984). These nutrients, however, are rapidly absorbed by vegetation or the soil. The foliage added to the forest floor is richer in nutrients than normal needle cast, increasing the substrate needed for ammonification, which may lead to temporary nitrifier activity (McColl and Powers 1984).

If biomass is removed in thinning, the effect will depend on what is removed. If 5% of ecosystem N is bolewood and half the trees are thinned and the boles exported, a maximum of 2.5% of the N would be removed. Because the thinned trees are generally smaller in diameter, the actual removal would be somewhat less. If boles and crowns are removed, however, the impact could be somewhat greater because nutrients are concentrated in actively growing crowns. This loss of nutrients could be of concern on heavily impacted sites, for example, intense wildfires followed by heavy site preparation, such as scalping.

Fertilization

Fertilization with N can restore productivity. In general, soils most responsive to fertilization have more than 10 cm (4 in) of available water-holding capacity and a site index of less than 30 m (95 ft) in fifty years (Miles and Powers 1988). Resources other than N limit growth on sites with lower available water capacity. Nitrogen is not limiting on the more productive sites. The effect of fertilization in increasing stand growth lasts about a decade. Nitrogen fertilization is most effective when combined with other silvicultural treatments such as thinning, and is most effective on stands near crown closure (Powers et al. 1988). Once trees have reached crown closure, N is recycled internally in the crowns and no further nitrogen is necessary. No operational fertilization is being done in the Sierra Nevada.

Fire Protection

Underburning

Prescribed burns are carried out under defined conditions, with high soil and duff moisture. They are much less intense than wildfires, and their effects are quite different. Underburns are also typically less intense than broadcast burns and tend to be more patchy. Because fire intensity is low, nutrient losses to the atmosphere through fly ash are negligible. The pruning and scorching of lower crowns add a needle cast to the forest floor, compensating for the lost duff and providing protection from erosion.

Underburns essentially oxidize the forest floor more rapidly than biological processes, removing organic matter and releasing the more rapidly recycled nutrients (McColl and Powers 1984). Plant growth is stimulated by the nutrients released into the ash layer in forms readily available to plants. Small but measurable gains in soil N occur after light underburns (Klemmendson et al. 1962). Nitrification is stimulated by underburning, possibly by elevated levels of ammonium and the reduced amount of repressive tannins in the duff and litter (Powers 1979).

Manipulation of Fuels

Fuels may be manipulated mechanically to reduce the risk of fire. The effects on soil productivity will depend on site history, the type of equipment used, and whether material is left on-site or exported.

Loss of porosity by soil compaction depends on the type of equipment used, the amount of area affected, and the thickness of slash and litter. Equipment operations could break up decaying logs, interrupting the decay process and forgoing the benefits from it.

If fuels are manipulated and left on-site, as chips for example, the result could be decreased soil temperatures, suppressed nitrification, and increased soil moisture. Bolewood contains phenols that can suppress the activity of microorganisms. Chipped material that is returned to the forest floor is unlikely to alter the C:N ratio of soils unless it is finely divided and well mixed into the soil (McColl and Powers 1984).

If fuels are removed from the site, the result could be increased soil temperatures and, with disturbance, a more rapid oxidation of the forest floor. Export of biomass and nutrients will depend on what material and how much of it is removed: if primarily poles and saplings, the amount will be low, because smaller-diameter materials contain less biomass and nutrients (Zinke et al. 1982).

Fuel Breaks

Shaded fuel breaks can be maintained by underburning or by cultivation. Those maintained by underburning have the same effects as described earlier. Fuel breaks maintained by cultivation have a higher risk of erosion, although generally only from summer precipitation. Cultivation increases oxidation of organic matter, and the benefits of a forest floor are forgone. Compaction may occur, although it is mitigated by cultivation. Considering the small land area involved and the benefits in preventing or controlling a major fire, this dedicated land use benefits soil productivity.

Soil Restoration

Tillage

Implements specially designed for forest soils can be used to recover porosity lost to compaction. Where used correctly and under the right conditions, tillage of compacted forest soils can be quite effective. Tillage breaks compacted soils into smaller aggregates, increasing porosity and surface area, allowing water to penetrate and biological activity to resume, and renewing the natural biological processes that are the source of forest soil porosity.

Tillage must be done with care in residual stands. Where root pathogens are present, damage to the roots of trees in stands can lead to root diseases such as black stain or annosus

(Kliejunas 1995). Mechanisms of infection differ, so the type of tillage implement used is important.

Respreading Topsoil

Where surface soils have been scalped and piled into windrows, practices common in the 1950s and 1960s but no longer done, lost productivity can be recaptured by respreading the topsoil. Five-year productivity gains of 37% have been reported from respreading topsoil (R. F. Powers, U.S. Forest Service, letter to the author, June 16, 1995).

Forest Roads

Although forest roads are essential for forest management, they also have both direct and indirect effects on soil productivity. The direct effect is removal of land area from the growing base. Indirect effects include landslides, gullies, and side-cast material. Roads can also disrupt the subsurface flow of water, drying out sites downslope or ponding water upslope, thus changing soil moisture regime and productivity. Roads can be restored only with difficulty and at great expense, but restoration efforts have been successful in Redwood National Park (Steensen and Spreiter 1992).

POSTWILDFIRE SALVAGE AND REFORESTATION

Loss of protective ground cover and deterioration in soil structure following wildfire increase the risk of soil loss through erosion. The use of prescribed fire to reduce fuel loading after salvage logging operations carries a high risk of increased erosion through deterioration of soil structure and further reductions in cover.

Periods following wildfire are especially critical for soil compaction. Intense wildfires can remove all surface duff and litter and may even consume surface-soil organic matter. Because trees that would normally transpire moisture are dead or removed, soil moisture levels remain critically high for several seasons following the fire. Soil moisture often is high in the subsoil, creating a situation ideal for compaction during salvage logging and subsequent site preparation activities.

Where biomass in the forest floor, crowns, and fine fuels has already been consumed by wildfire, the amount of biomass and nutrients removed in postfire salvage operations can be relatively low. The value of leaving large amounts of severely charred large woody material is questionable. It may provide some wildlife habitat, but such material adds little to soil productivity. Charring disrupts the normal decay processes of large woody material, which adds nitrogen by nonsymbiotic fixation. Burned logs potentially can trap sediment, but unless they have good soil contact and are aligned on the contour, they may actually accelerate gully erosion.

Salvage logging can generate slash, adding ground cover to reduce erosion. In some cases, salvage operations can be used to break up hydrophobic soil layers near the surface, further reducing erosion (Poff 1988). However, this benefit may be offset by other soil disturbance associated with salvage logging. Depending on the site history, soil disturbance during salvage logging may stimulate brush species by bringing viable seeds to the surface, which can have either desirable or undesirable consequences for soil productivity.

Large openings created by wildfires may create opportunities for soil restoration. Topsoil piled into windrows can be respread, and tillage can be used to restore porosity.

There is a common misperception that allowing natural succession to reforest areas following major wildfire builds soil. This idea assumes that the intense wildfires the Sierra Nevada has experienced recently are natural phenomena, and not the result of fuel buildup as a consequence of fire protection and fire exclusion. It also assumes that the soils on these burns contain sproutable roots or a seed bank of desirable species that will revegetate the site.

Ceanothus species and western mountain mahogany have root nodules with nitrogen-fixing capability (Biswell 1974). However, not all shrubs fix nitrogen, nor do they provide the same degree of soil protection. For example, Zinke (1969) has shown that soil nitrogen decreases over time under chamise. Biswell (1974) reports that chamise has extremely poor soil-protecting qualities. Research from southern California chaparral ecosystems is not transferable to forest ecosystems of the Sierra Nevada, but it does show that the ecology of shrub communities is complex and that generalizations are risky.

Erosion rates are substantially higher under shrub vegetation than under forest cover, due in part to the lack of stable soil aggregates (Perry et al. 1987) under most shrub species, and in part to the lack of effective soil cover. Runoff is more rapid under shrubs, and sheet flow is more common than where a forest floor is present. Shrub species differ in their ability to protect soil from accelerated erosion. Manzanita, for example, does not form a surface mulch that protects soils from erosion.

Nutrient retention by herbaceous communities is low in early successional stages, because little biomass is accumulated (Johnson and Swank 1973). In contrast, young vigorous forests accumulate biomass and immobilize large quantities of nutrients. Rapid reinvasion by shrubs and herbs after a fire may be important in preventing leaching of nutrients released into the ash bed. However, its importance will depend on how well buffered the soil is. For example, deep, clayey soils, high in organic matter, will allow less nutrient leaching than will shallow, coarse-textured soils, low in organic matter.

Another argument for natural succession is that reforestation with one species, commonly pine, will lead to a monoculture with low diversity. This situation rarely occurs, even when attempted in plantations, because it is difficult to exclude invading shrubs and shade-tolerant species. In a study of California plantations McDonald and Fiddler (1993) found considerable diversity, particularly in shrub species. Natural succession after fire may lead to a thick fir stand, a cover type even less resistant to fire.

The more quickly a site reaches crown closure and the more quickly a forest floor develops, the sooner soil productivity will be stabilized. However, that does not justify the severe activities used in the past such as windrowing, scalping, and intensive grubbing on steep slopes. These kinds of activities are likely to cause more degradation of soil productivity than allowing an extended period of shrub cover.

The multiple successional pathways following a wildfire will vary with ecological type, site history, burn intensity, soil type, and soil condition. Which pathway to follow will depend on resource objectives, but maintaining long-term soil productivity must be an objective common to all choices. In terms of lost site productivity, the true cost of allowing a previously forested site to remain in brush for decades could be unacceptably high.

NEEDS FOR RESEARCH AND INVENTORY

Although a great deal is known about the effects of silvicultural activities on forest soils in the Sierra Nevada, much remains to be done. Some of the published research on basic processes was done in other regions and must be extrapolated; older studies were often done on practices no longer used, and existing information is not organized into forms readily accessible to land managers.

Research Needs

Basic Productivity

There is need for basic research on how changes in soil porosity and soil organic matter affect long-term soil productivity, and on how they interact. The U.S. Forest Service studies on long-term soil productivity (Powers et al. 1990a) are noteworthy in pursuing this goal. Eight long-term soil productivity (LTSP) installations now are operating in the Sierra Nevada mixed conifer forest and are part of the world's leading research effort on the subject (R. F. Powers, U.S. Forest Service, letter to the author, June 16, 1995).

Large Woody Material

Much emphasis is being placed on preserving large woody material in Sierra Nevada forests. Although decaying logs do provide wildlife habitat, little is known of their significance to long-term soil productivity in Sierra Nevada ecosystems. Most research on large woody material has been done in the Pacific Northwest and Intermountain regions, which have ecosystems quite different from those of the Sierra Nevada. However, research has been under way since 1993 at Blacks Mountain Experimental Forest (R. F. Powers, U.S. Forest Service, letter to the author, June 16, 1995).

Postwildfire Plant Succession

There is a common misperception that after a major wildfire the best treatment is to allow sites to reforest by natural plant succession. The true costs in terms of soil productivity gains or losses under such a strategy are not known. Much work has been done on natural succession in chaparral in southern California, but knowledge from these ecosystems may not apply to forested ecosystems of the Sierra Nevada. Similarly, little is known about the use of native versus non-native plants to control erosion in postfire emergency watershed treatments.

Soil Biology

Very little is known about the soil macrofaunal populations of the Sierra Nevada and their role in soil processes. Most of the research on soil biology has been done in the Pacific Northwest and is not readily extrapolated to the Sierra Nevada. Research specific to the Sierra Nevada has just begun (Moldenke 1992).

Soil Erosion and Rates of Soil Formation

Soil erosion and formation rates are not well documented for the Sierra Nevada. Although rates of erosion are typically low after timber harvest, they could be potentially serious after intense wildfire or severe site preparation. More knowledge about soil erosion and formation rates would assist in determining appropriate postfire strategies for reforestation and for emergency treatments for burned areas. Limited work is under way at the LTSP installations.

Alternative Fuel Treatments

The effects of the mechanical treatment of fuels on forest soil processes are not well understood. Fine surface organic matter and large woody material both have structural functions that affect soil biology beyond their nutrient content. Limited work on the effects of chipping has been done on the Foresthill Divide (Lanini and Radosevich 1986), and research on chipping, fungal inoculation, and N fixation has begun.

Riparian-Terrestrial Ecosystem Linkages

With the current focus on protection of riparian ecosystems, a better understanding of the linkages between terrestrial and aquatic ecosystems is needed. Geomorphologic relationships suggest that the linkages vary considerably from site to site.

Forest Soils Extension

There is a wealth of information on Sierra Nevada forest soils in research, inventory, and practical experience that is not being fully used in planning, modeling, or designing and implementating projects. Better technology transfer is needed between researchers and practicing land managers in the field.

Researchers have many insights into forest soil processes that should be shared with field resource specialists and resource managers. Although the experiences of resource managers and field specialists are often anecdotal and unverified, these resource specialists have years of field observations and experience that are of much value. If the results of research are to be implemented, there must be a stronger link from researcher, to field specialist, to interdisciplinary team member, to decision maker and implementer.

Inventory Needs

Soil Survey

Except for a few isolated foothill areas, soil inventories have been completed for all of the Sierra Nevada at Order 3 or Order 2 levels (Order 4 in wilderness areas). Soils have been classified using *Keys to Soil Taxonomy* (Soil Survey Staff 1992). This information is adequate for small watershed and regional planning but should be verified in the field for project-level work. Only portions of this soil information are in a geographic information system (GIS) database that can be readily accessed and utilized.

Soil Analysis

Comprehensive laboratory analyses have been conducted on Sierra Nevada forest soils (Zinke et al. 1982). However, the soils analyzed have not been correlated and classified using *Keys to Soil Taxonomy*, making extrapolation of the results difficult. These laboratory data should also be correlated with newly developed ecological plant associations.

Inventory of Soil Condition

A comprehensive inventory of the condition of Sierra Nevada forest soils is needed. Such an inventory is essential to assess watershed condition, to identify areas needing restoration, and to identify areas at risk. This inventory could be carried out using information on the history of land treatments that is already available. History of past land disturbances could help identify areas most likely to have lost surface soil by erosion or displacement, and where soil compaction is most likely. Soil condition has been altered most severely on old burns of the 1950s that were subjected to timber salvage and severe site preparation, usually scalping to remove the topsoil and its content of weed seeds.

Soils most at risk would be soils with initially low resiliency that had been subjected to the most disturbance. Sites with more robust soils that had received modest levels of disturbance would be less at risk; sites with high-potential soils with high levels of disturbance would be candidates for restoration because there is more opportunity for recovery.

A model predicting soil condition could be quickly developed and field-tested with random sampling. After initial field-testing, predictions could be made and checked in the field to evaluate the accuracy of the model.

Stand-Record Card System

The U.S. Forest Service has a system of stand-record cards that contains nearly fifty years of detailed historical information on timber stands and their treatments. This information should be captured in an electronic database to prevent its being lost and to facilitate its use. It could be invaluable in assessing the condition of Sierra Nevada forest soils, discussed earlier in this chapter.

Soil Interpretations

Interpretations of how Sierra Nevada forest soils respond to management treatments are inconsistent. Except for the Soil Erosion Hazard Rating system—an interdisciplinary effort sponsored by several state and federal agencies and universities—there is no unified system for interpreting the response of Sierra Nevada soils to use and management. Many good soil interpretations are available, but they are scattered and occur in many forms.

Soil Erosion in the Mediterranean Region

The destruction of forests and extensive erosion in the Mediterranean region is frequently cited as an example of what not to allow in the Sierra Nevada. The lessons of the Mediterranean should not be taken lightly. It took several thousand years and hundreds of harvests to reach the level of destruction in the Mediterranean (Thirgood 1981), whereas there are already areas of serious soil erosion in the Sierra Nevada after barely 150 years of activity. We have developed machinery capable of major soil impacts much more rapidly than we have acquired knowledge of what these impacts mean (Powers et al. 1990b).

The Sierra Nevada and the Mediterranean have similarities in climate, soils, and ecosystems. Yet, there are also important differences. The destruction of the Mediterranean forest has been well documented by Thirgood (1981). Although there were large wildfires and periods of heavy harvest for ship-building, much of the forest destruction and soil erosion occurred incrementally as a result of overgrazing, especially by nomadic herds of goats. Destruction was accelerated during periods of political instability. Intensive agriculture was also practiced. Although heavy grazing has occurred in the Sierra Nevada, the area is not subject to nomadic grazing, agricultural impacts have been relatively minor, and much of the land base is in highly regulated public ownership. Private forest lands are managed under some of the most restrictive forest practice rules in the world, although it could be argued that soil-management issues have not received enough emphasis under these rules. Most aspects of forest management in the Sierra Nevada receive a high level of public scrutiny.

Although the lessons of the Mediterranean are sobering, it is unlikely that these sequences of soil degradation will be repeated in the Sierra Nevada. Perhaps the most important lesson from the Mediterranean experience is that soil losses too small to observe or measure can, if allowed to continue for a long time, result in a severe decline in forest soil productivity.

REFERENCES

Alexander, E. B. 1988. Rates of soil formation: Implications for soil-loss tolerance. *Soil Science* 145:37–45.

Alexander, E. B., and R. J. Poff. 1985. Soil disturbance and compaction in wildland management: *Earth Resources Monograph* 8. San Francisco: U.S. Forest Service, Region 5.

Atzet, T., R. F. Powers, D. H. McNabb, M. P. Amaranthus, and E. R. Gross. 1989. Maintaining long-term forest productivity in southwest Oregon and northern California. In *Maintaining the long-term productivity of Pacific Northwest forest ecosystems*, edited by D. A. Perry, R. Meurisse, B. Thomas, R. Miller, J. Boyle, J. Means, C. R. Perry, and R. F. Powers, 185–201. Portland, OR: Timberland Press.

Baver, L. D. 1930. The Atterberg consistency constants: Factors affecting their values and a new concept of their significance. *Agronomy Journal* 22:935–48.

Biswell, H. H. 1974. Effects of fire on chaparral. In *Fire and ecosystems*, edited by T. T. Kozlowski and C. E. Ahlgren, 321–65. New York: Academic Press.

Borchers, J. G., and D. A. Perry. 1990. Effects of prescribed fire on soil organisms. In *Natural and prescribed fire in Pacific Northwest forests*, edited by J. D. Walstad, S. R. Radosevich and D. V. Sandberg. Corvallis: Oregon State University Press.

Boyer, D. 1979. Guidelines for soil protection and restoration for timber harvest and post-harvest activities. Portland, OR: U.S. Forest Service, Pacific Northwest Region.

Boyer, D. E., and J. D. Dell. 1980. Fire effects on Pacific Northwest forest soils. Portland, OR: U.S. Forest Service, Pacific Northwest Region.

Clayton, J. L., and D. A. Kennedy. 1985. Nutrient losses from timber harvest in the Idaho Batholith. *Soil Science of America Journal* 49:1041–49.

Cole, D. W., and S. P. Gessel. 1965. Movement of elements through forest soil as influenced by tree removal and fertilizer additions. In *Forest soil relationships in North America*, edited by C. T. Youngberg, 95–104. Corvallis: Oregon State University Press.

Cole, D. W., S. P. Gessel, and S. F. Dice. 1968. Distribution and cycling of nitrogen, phosphorus, potassium, and calcium in a second-growth Douglas-fir ecosystem. In *Primary productivity and mineral cycling in natural ecosystems*, edited by H. E. Young, 197–232. Orono: University of Maine Press.

DeBano, L. F. 1979. Effects of fire on soil properties. In *California forest soils: A guide for professional foresters and resource managers and planners*, edited by R. J. Laacke, 109–18. Berkeley: University of California, Agricultural Sciences Publications.

Dunn, P. H., and L. F. DeBano. 1977. Fire's effect on the biological properties of chaparral soils. General Technical Report WO–3. Washington, DC: U.S. Forest Service.

Dyck, W. J., and P. N. Beets. 1987. Managing for long-term site productivity. *New Zealand Forestry* November:23–26.

Dyrness, C. T. 1965. Soil surface conditions following tractor and high-lead logging in the Oregon Cascades. *Journal of Forestry* 63:272–75.

Frazer, D. W., J. G. McColl, and R. F. Powers. 1990. Soil nitrogen mineralization in a clear-cutting chronosequence in a Northern California conifer forest. *Soil Science Society of America Journal* 54 (4): 1145–52.

Fredriksen, R. L., D. G. Moore, and L. A. Norris. 1975. The impact of timber harvest, fertilization, and herbicide treatment on streamwater quality in western Oregon and Washington. Paper presented at the Fourth North American Forest Soils Conference, Montreal, Quebec, August 1973.

Free, G. R., J. Lamb, and E. A. Carleton. 1947. Compatibility of certain soils as related to organic matter and erosion. *Journal American Society of Agronomy* 39:1068–76.

Froehlich, H. A. 1974. Soil compaction: Implications for young-growth management. In *Managing young forests in the Douglas-fir region*, edited by A. B. Berg, 49–62. Corvallis: Oregon State University Press.

———. 1978. Soil compaction from low ground-pressure, torsion-suspension logging vehicles on three forest soils. Oregon State University Forest Research Laboratory Research Paper 36, Corvallis, OR.

———. 1979. Soil compaction from logging equipment: Effects on growth of young ponderosa pine. *Journal of Soil and Water Conservation* 34:276–78.

Froehlich, H. A., D. E. Aulerich, and R. Curtis. 1981. Designing skid trail systems to reduce soil impacts from tractive logging machines. Oregon State University Forest Research Laboratory Research Paper 44, Corvallis, OR.

Froehlich, H. A., and D. W. McNabb. 1984. Minimizing soil compaction in Pacific Northwest forests. Paper presented at Sixth North American Forest Soils Conference, Knoxville, TN, June 1983.

Grier, C. C., K. A. Vogt, M. R. Keyes, and R. L. Edmonds. 1981. Biomass distribution and above- and below-ground production in young and mature *Abies amabilis* zone ecosystems of the Washington Cascades. *Canadian Journal of Forest Research* 11:155–67.

Harmon, M. E., K. Cromack Jr., and B. G. Smith. 1987. Coarse woody debris in mixed-conifer forests, Sequoia National Park, California. *Canadian Journal of Forest Research* 17:1265–72.

Harmon, M. E., J. F. Franklin, F. J. Swanson, P. Sollins, S. V. Gregory, J. D. Lattin, N. H. Anderson, S. P. Cline, N. G. Aumen, J. R. Sedell, G. W. Lienkaemper, K. Cromack Jr., and K. W. Cummins. 1986. Ecology of coarse woody debris in temperate ecosystems. *Advances in Ecological Research* 15:133–302.

Hatchell, G. E., C. W. Ralston, and R. R. Foil. 1970. Soil disturbance in logging. *Journal of Forestry* 68:772–75.

Helms, J. A., and C. Hipkin. 1986. Effects of soil compaction on tree volume in a California ponderosa pine plantation. *Western Journal of Applied Forestry* 1:121–24.

Howard, R. F., M. J. Singer, and G. A. Frantz. 1981. Effects of soil properties, water content, and compactive effort on the compaction of selected California forest and range soils. *Soil Science Society of America Journal* 45:231–36.

Johnson, D. W. 1983. The effects of harvesting intensity on nutrient depletion in forests. In *IUFRO symposium on forest site and continuous productivity, Seattle, WA, 22–28 August 1983*, edited by R. Ballard and S. P. Gessel, 157–66. General Technical Report PNW-163. Portland, OR: U.S. Forest Service, Pacific Northwest Forest and Range Experiment Station.

———. 1994. Reasons for concern over impacts of harvesting. In *Impacts of forest harvesting on long-term site productivity*, edited by W. J. Dyck, D. W. Cole and N. B. Comerford, 1–12. London: Chapman and Hall.

Johnson, D. W., and D. E. Todd. 1987. Nutrient export by leaching and whole-tree harvesting in a loblolly pine and mixed oak forest. *Plant and Soil* 102:99–109.

Johnson, P. L., and W. T. Swank. 1973. Studies of cation budgets in the southern Appalachians on four experimental watersheds in contrasting vegetation. *Ecology* 54:70–80.

Kittredge, J. 1952. Deterioration of site quality by erosion. *Journal of Forestry* 50:554–56.

Klemmendson, J. O., A. M. Schultz, H. Jenny, and H. H. Biswell. 1962. Effect of prescribed burning of forest litter on total soil nitrogen. *Soil Science Society of America Proceedings* 26:200–202.

Kliejunas, J. 1995. *Pathogens as disturbance agents in California forest ecosystems*. San Francisco: U.S. Forest Service, Pacific Southwest Region.

Lanini, W. T., and S. R. Radosevich. 1986. Response of three conifer species to site preparation and shrub control. *Forest Science* 32:61–77.

Likens, G. E., F. H. Borman, and N. M. Johnson. 1969. Nitrification: Importance to nutrient losses from a cut-over forest ecosystem. *Science* 163:1205–6.

Mace, A. C. 1970. Soil compaction due to tree length and full tree skidding with rubber-tired skidders. Minnesota Forest Research Notes No. 214. St. Paul, MN: University of Minnesota.

Maser, C., and J. M. Trappe. 1984. The seen and unseen world of the fallen tree. Portland, OR: U.S. Forest Service and Bureau of Land Management.

McColl, J. G., and R. F. Powers. 1984. Consequences of forest management on soil-tree relationships. In *Nutrition of plantation forests*, edited by G. D. Bowen and E. K. S. Nambiar, 379–412. New York: Academic Press.

McDonald, P. M., and G. O. Fiddler. 1993. Feasibility of alternatives to herbicides in young conifer plantations in California. *Canadian Journal of Forest Research* 23:2015–22.

McNabb, D. H. 1981. Managing soil moisture to control soil compaction. Workshop notes, Managing Forest Stands to Minimize Soil Compaction, March 25–26. FIR Program, Medford, OR.

McNabb, D. H., and K. Cromack Jr. 1990. Effects of prescribed fire on nutrients and soil productivity. In *Natural and prescribed fire in Pacific Northwest forests*, edited by J. D. Walstad, S. R. Radosevich, and D. V. Sandberg, 125–42. Corvallis: Oregon State University Press.

Miles, D. W. R., F. J. Swanson, and C. T. Youngberg. 1984. Effects of landslide erosion on subsequent Douglas-fir growth and stocking levels in the western Cascades. *Soil Science Society of America Journal* 48:667–71.

Miles, J. A. 1978. Soil compaction produced by logging and residue treatment. *American Society of Agricultural Engineers Transactions* 21:60–62.

Miles, S. R., and R. F. Powers. 1988. Ten-year results of forest fertilization in California. *Earth Resources Monograph* 15. San Francisco: U.S. Forest Service, Region 5.

Moldenke, A. R. 1992. Non-target impacts of management practices on the soil arthropod community of ponderosa pine plantations. Paper presented at the Thirteenth Annual Forest Vegetation Management Conference, Eureka, CA, January 14–16.

Morris, L. A., and R. E. Miller. 1994. Evidence for long-term productivity change as provided by field trials. In *Impacts of forest harvesting on long-term site productivity*, edited by W. J. Dyck, D. W. Cole, and N. B. Comerford, 41–80. London: Chapman and Hall.

Morris, L. A., W. L. Pritchett, and B. F. Swindel. 1983. Displacement of nutrients into windrows during site preparation of a pine flatwoods forest. *Soil Science Society of America Journal* 47:591–94.

Norris, L. A. 1983. Behavior of chemicals in the forest environment. Paper presented at Chemistry, Biochemistry, and Toxicology of Pesticides, a shortcourse, Pendleton, OR.

Palazzi, L. M., R. F. Powers, and D. H. McNabb. 1992. Geology and soils. In *Reforestation practices in southwestern Oregon and northern California,* edited by S. D. Hobbs, S. D. Tesch, P. W. Owston, R. E. Stewart, J. C. Tappeiner II, and G. E. Wells, 48–72. Corvallis: Oregon State University, Forest Research Laboratory.

Perry, D. A., R. Molina, and M. P. Amaranthus. 1987. Mycorrhizae, mycorrhizospheres, and reforestation: Current knowledge and research needs. *Canadian Journal of Forest Research* 17:929–40.

Poff, R. 1988. Resource recovery: Compatibility of timber salvage operations with watershed values. Paper presented at symposium on Fire and Watershed Management, Watershed Management Council, Sacramento, CA, October.

Powers, R. F. 1979. Mineral cycling in temperate forest ecosystems. In *California forest soils: A guide for professional foresters and resource managers and planners,* edited by R. J. Laacke, 89–108. Berkeley: University of California, Agricultural Sciences Publications.

———. 1990a. Do timber management practices degrade long-term site productivity? What we know and what we need to know. In *Proceedings, 11th annual forest vegetation management conference, November 7–9, Sacramento, CA,* 87–106. Forest Vegetation Management Conference, Redding, CA..

———. 1990b. Nitrogen mineralization along an altitudinal gradient: Interactions of soil temperature, moisture, and substrate quality. *Forest Ecology and Management* 30:19–29.

———. 1991. Are we maintaining the productivity of forest lands? Establishing guidelines through a network of long-term studies. In *Management and productivity of western-montane forest soils, April 10–12, 1990. Boise, ID,* 70–81. General Technical Report INT-280. Ogden, UT: U.S. Forest Service, Intermountain Research Station.

Powers, R. F., D. H. Alban, R. E. Miller, A. E. Tiarks, C. G. Wells, P. E. Avers, R. G. Cline, R. O. Fitzgerald, and N. S. Loftus Jr. 1990a. Sustaining site productivity in North American forests: Problems and prospects. Paper presented at the Seventh North American Forest Soils Conference, Vancouver, B.C., July 1988.

Powers, R. F., D. H. Alban, G. A. Ruark, and A. E. Tiarks. 1990b. A soils research approach to evaluating management impacts on long-term soil productivity. Paper presented at Impact of Intensive Harvesting on Forest Site Productivity, IEA/BE A3 workshop, South Island, New Zealand.

Powers, R. F., and R. L. Edmonds. 1992. Nutrient management of subalpine *Abies* forests. In *Forest fertilization: Sustaining and improving nutrition and growth of western forests,* edited by H. N. Chappell, G. F. Weetman, and R. E. Miller, 28–42. Seattle: University of Washington Institute of Forest Resources.

Powers, R. F., and K. Van Cleve. 1991. Long-term ecological research in temperate and boreal forest ecosystems. *Agronomy Journal* 83:11–24.

Powers, R. F., S. R. Webster, and P. H. Cochran. 1988. Estimating the response of ponderosa pine forests to fertilization. In *Proceedings of the future forests of the mountain west: A stand culture symposium,* edited by W. C. Schmidt, 219–25. General Technical Report INT-243. Ogden, UT: U.S. Forest Service.

Prescott, C. E., J. P. Corbin, and D. Parkinson. 1989. Biomass, productivity, and nutrient-use efficiency of aboveground vegetation in four Rocky Mountain coniferous forests. *Canadian Journal of Forest Research* 19:309–17.

Rice, R. M. 1979. Sources of erosion during timber harvest. In *California forest soils: A guide for professional foresters and resource managers and planners,* edited by R. J. Laacke, 59–68. Berkeley: University of California, Agricultural Sciences Publications.

Sandberg, D. V. 1980. Douglas-fir duff reduction. Portland, OR: U.S. Forest Service, Pacific Northwest Forest Range Experiment Station.

Soil Survey Staff. 1992. *Keys to soil taxonomy.* 6th ed. Washington, DC: U.S. Department of Agriculture, Natural Resources Conservation Service.

Sollins, P., S. P. Cline, T. Verhoeven, D. Sachs, and G. Spycher. 1987. Patterns of log decay in old-growth Douglas-fir forests. *Canadian Journal of Forest Research* 17:1585–95.

Sollins, P., C. C. Grier, F. M. McCorison, K. Cromack Jr., and F. Fogel. 1980. The internal element cycles of an old-growth Douglas-fir ecosystem in western Oregon. *Ecological Monographs* 50:261–85.

Steensen, D. L., and T. A. Spreiter. 1992. Watershed rehabilitation in Redwood National Park. Paper presented at the national meeting of the American Society for Surface Mining and Reclamation, Duluth, MN, June 1992.

Stone, J. A., and W. E. Larson. 1980. Rebound of five one-dimensionally compressed unsaturated granular soils. *Soil Science Society of America Journal* 44:819–22.

Switzer, G. L., L. E. Nelson, and L. E. Hinesley. 1981. Effects of utilization on nutrient regimes and site productivity. Paper presented at Forest Fertilization Conference, Institute of Forest Resources, Contribution 40, Union, WA, September 1979.

Thirgood, J. V. 1981. *Man and the Mediterranean forest: A history of resource depletion.* New York: Academic Press.

Wert, S., and B. R. Thomas. 1981. Effects of skid roads on diameter, height, and volume growth in Douglas-fir. *Soil Science Society of America Journal* 45:629–32.

Zinke, P. J. 1962. The pattern of individual forest trees on soil properties. *Ecology* 43:130–33.

———. 1969. *Biology and ecology of nitrogen.* Washington, DC: National Academy of Science.

Zinke, P. J., A. G. Stangenberger, M. J. Fox, B. Parker, and R. Stone. 1982. Elemental drain of fertility from a Sierra mixed-conifer forest site due to intensive harvest of fuels. California Forest Note 82. Sacramento: California State Department of Forestry.

Forest Soils of the Sierra Nevada

Figure 16.A1 is a generalized soil map showing the distribution of Sierra Nevada forest soils. The map was made using a November 1, 1994, 1:650,000 map of Cal-Veg types, and a 1:650,000 map of soil associations from the STATSGO soils data base for California as reference. A transparency of the STATSGO soil associations was placed over the Cal-Veg map, and soil areas were hand-drawn for small scale reproduction. A map of California soil temperature regimes and personal knowledge of the Sierra Nevada provided additional guidance in making the soil groupings.

The map is intended to show the distribution of forest soils in the Sierra Nevada. Forest soils occur primarily in map areas 1–8 and 12–13. No attempt was made to differentiate soils in areas 9–11 and 14. These latter areas are either outside the Sierra Nevada proper or contain relatively few forest soils. Dominant soil types, Cal-Veg types, parent material, and soil temperature regime are listed for each soil area in table 16.A1. Soils are classified according to soil taxonomy (Soil Survey Staff 1992).

REFERENCE

Soil Survey Staff. 1992. *Keys to soil taxonomy*. 6th ed. Washington, DC: U.S. Department of Agriculture, Natural Resources Conservation Service.

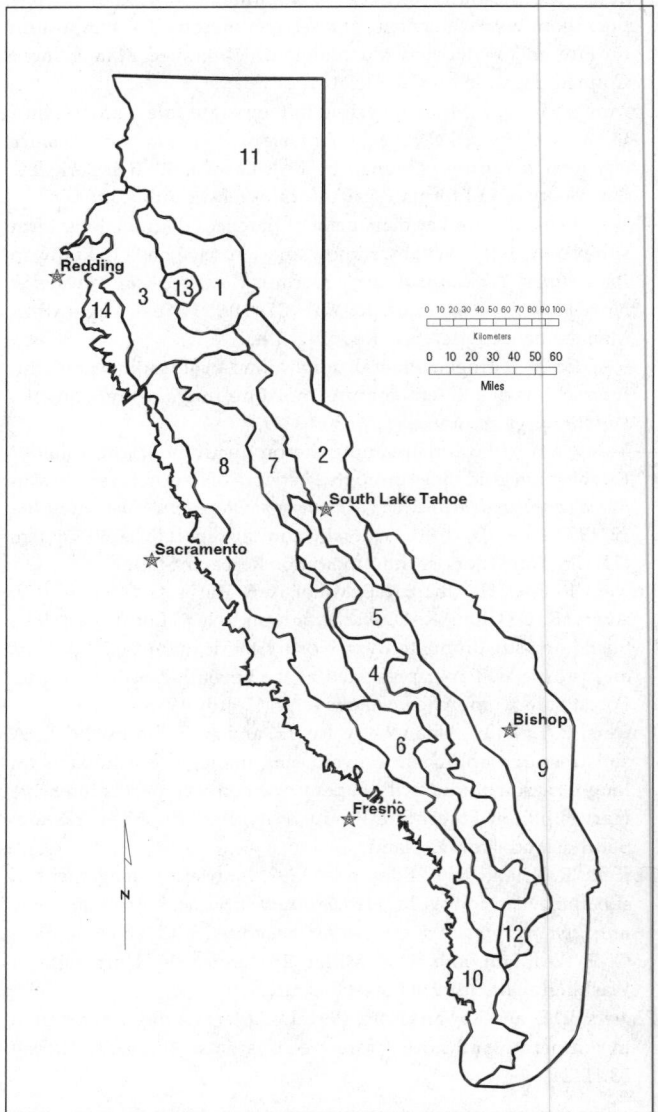

FIGURE 16.A1

Soils areas within the SNEP study area.

TABLE 16.A1

Descriptions of forest soil areas on map.

Map Area	Parent Material	Cal-Veg Type	Dominant Soils	Soil Temperature	Comments
1	Volcanic Jeffrey Pine Basin Sagebrush	Mixed Conifer–Fir Ultic Argixerolls Andic Haploxeralfs Andic Xerochrepts	Ultic Haploxeralfs	Frigid	
2	Volcanic Granitic Basin Sagebrush	Mixed Conifer–Fir Jeffrey Pine Typic Xeropsamments	Ultic Argixerolls Ultic Haploxeralfs	Frigid	
3	Volcanic Mixed Conifer–Pine Manzanita	Ultic Haploxeralfs Andic Xerumbrepts Xeric Haplohumults	Mixed Conifer–Fir	Mesic and foothills	Eastside mountains
4	Granitic	Red Fir Dystric Xeropsamments Andic Haplumbrepts Andic and Lithic Cryumbrepts	Typic Xerumbrepts	Frigid	
5	Granitic Glacical drift Mountain Hemlock Barren/Cushion Plant	Red Fir Lodgepole Pine Dystric Xeropsamments Typic Xerumbrepts	Andic Cryumbrepts Andic Haplumbrepts	Frigid Cryic	Crest of Sierra Nevada
6	Granitic Mixed Conifer–Fir Chamise	Ponderosa Pine Dystric Xerochrepts Dystric Xeropsamments	Ultic Haploxeralfs	Mesic	
7	Volcanic Glacial drift Metasediments	Red Fir Mixed Conifer–Fir	Andic Xerumbrepts Typic Xerumbrepts Ultic Haploxeralfs	Frigid	
8	Medisediments Volcanic Granitic	Mixed Conifer–Pine	Typic Haploxerults Ultic Haploxeralfs Xeric Haplohumults Andic Xerumbrepts	Mesic	Mountains of western slope of Sierra Nevada
9	(not mapped)	Basin Sagebrush Piñon–Juniper Creosote	(not mapped)	Hyperhermic to Cryic	Eastern Sierra Nevada foothills and basin and range
10	(not mapped)	Blue Oak Annual Grassland Interior Live Oak Chamise	(not mapped)	Thermic mesic	Western foothills and edge of Great Central Valley
11	(not mapped)	Basin Sagebrush Ponderosa Pine Western Juniper	(not mapped)	Mesic frigid	Modoc Plateau
12	Granitic	Jeffrey Pine Chamise	Dystric Xeropsamments Typic Xerothents Entic Haploxerolls	Mesic frigid	
13	Granitic	Red Fir Ceanothus	Andic Xerumbrepts Ultic Haploxeralfs	Frigid	
14	Mixed	Blue Oak Manzanita Mixed Conifer–Pine Typic Rhodoxeralfs	Lithic-Ruptic-Xerorthentic Xerochrepts Pachic Argixerolls	Thermic mesic	Foothills and valleys

JANET HENSHALL MOMSEN
Department of Human and Community
   Development
University of California
Davis, California

Agriculture in the Sierra

ABSTRACT

In the five intercensal years 1987 to 1992, the Sierra Nevada's proportion of California's farms increased marginally, from 6.99% to 7.08%, while the region's share of the state's farmland decreased markedly, from 16.1% to 11.2%. The income from these farms was relatively low, with Sierra Nevada farms contributing just over 2% of the state's gross farm income in 1994. The region is characterized by two major agro-ecosystems: a foothill pattern of irrigated specialized crop and animal production and a system of extensive stock grazing in the drier high-altitude rangelands with some cultivation in better-watered areas. Both these agro-ecosystems have come under pressure as the rapidly growing valley urban population looks increasingly to the Sierra Nevada for rural residential, recreational, and hobby farming opportunities. This study looks at the historical background to these agro-ecosystems and the contemporary agricultural land use and land management of the region.

INTRODUCTION

Commercial agriculture in the Sierra Nevada originated and changed in response to the demands of other resource-utilizing activities in the region: first, gold mining, followed by timber extraction and, more recently, recreation. The pattern of exploitation of these natural resources of minerals, vegetation, and landscape was largely determined by accessibility. Altitude, as it affects length of growing season, and the availability of irrigation water have been primary influences in shaping contemporary agricultural production patterns.

Two major agro-ecosystems have developed in the Sierra Nevada: a foothill pattern of irrigated specialized crop and animal production and a system of extensive stock grazing in the drier high-altitude rangelands with some cultivation in better-watered areas. Both these agro-ecosystems have come

under pressure as the rapidly growing valley urban population looks increasingly to the Sierra Nevada for rural residential, recreational, and hobby farming opportunities. This influence has been concentrated along major access routes such as Interstate 80 and is spreading north and south from these infiltration points. In order to fully represent these agro-ecosystems, this chapter includes the full SNEP study area, extending northward to the Oregon border beyond the SNEP ecoregion boundary.

Mountainous areas are generally marginal for agricultural production. Under the Least Favoured Areas (LFAs) directive that the European Community adopted in 1975 as its first common instrument of regional agricultural structural policy, mountain areas are identified as the main type of LFA. The European Community defined LFAs as areas where agriculture is hampered by permanent natural handicaps (Bertrand and Hulot 1990). Recent research in European LFAs suggests that in the course of economic growth, natural conditions become increasingly important in determining the level of agricultural income because, with an improvement in regional economy, the gap in agricultural income between LFAs and normal areas widens (Terluin et al. 1995). California may well be one of the best non-European examples of these findings. As California's agriculture has become more and more efficient and productive, the farms of the Sierra Nevada have become relatively more marginal. Yet at times in the past, agriculture in the Sierra Nevada had statewide importance, and it has been a major employer in the study area since the 1860s.

In 1994 the cash farm income in Sierra Nevada counties was among the lowest in the state. Of the thirty-seven non–Sierra Nevada counties of California, only eight had gross farm incomes from crops and livestock below $100 million in 1994 (California Farmer 1995). In the Sierra Nevada, all counties had 1994 gross farm incomes of less than $100 million, even with income from sales of timber included with that from crops and livestock. The region's counties recording the larg-

Sierra Nevada Ecosystem Project: Final report to Congress, vol. II, *Assessments and scientific basis for management options*. Davis: University of California, Centers for Water and Wildland Resources, 1996.

est gross farm incomes in 1994 were Lassen ($94,900,950), Modoc ($94,788,800), and El Dorado ($82,264,000) while Inyo, Plumas, and Sierra Counties all had farm incomes below $10 million (California Farmer 1995). Overall, thirteen Sierra Nevada counties (figures for Alpine and Placer were not available) produced a mere 2.24% of the state's gross farm income in 1994 (California Farmer 1995). Even among those farms with farm sales of over $10,000 in 1992, only Shasta County had an average farm net cash return above the state mean and all but Mono had average net cash farm returns of less than half that for California, according to the 1992 Census of Agriculture (Bureau of the Census 1994).

Only a small part of the total area of the Sierra Nevada is in farms, with but two foothill counties, Amador and Calaveras, with 62.6% and 37.7% respectively of their area in farmland, having more than the county mean (28.9%) for the whole state in 1992 (Bureau of the Census 1994). Furthermore, Amador has gone against the state trend and increased its proportion of farmland between 1987 and 1992 by almost 5%. Alpine, El Dorado, Inyo, Mono, Plumas, Sierra, and Tuolumne Counties all had less than 10% of their total area in farmland, although all but El Dorado and Alpine showed an increase between 1987 and 1992 (Bureau of the Census 1994). Overall, in the five intercensal years 1987 to 1992 the Sierra Nevada's proportion of California's farms increased marginally, from 6.99% to 7.08%, while the region's share of the state's farmland decreased markedly, from 16.1% to 11.2% (Bureau of the Census 1994).

Between 1987 and 1992, farm size declined in the foothill counties (Placer, El Dorado, Calaveras, Mariposa, Sierra, and Tuolumne), while it increased in the northern counties of Lassen, Modoc, Plumas, and Shasta and in the eastern counties of Mono and Inyo, reflecting the opposing pressures of suburbanization and economies of scale. At the same time, the proportion of nonresident farmers fell in the foothills (Amador, Calaveras, El Dorado, Nevada, and Placer Counties) and increased in the more isolated counties of Inyo, Modoc, Plumas, and Sierra. In California as a whole, two-thirds of farm operators have farming as their principal occupation, but in the Sierra Nevada, nine counties have less than half their farmers in this category, and only Calaveras and Modoc have more of such farmers than the state average. Clearly the nature of farming and the direction of change is not consistent across the whole Sierra Nevada.

This chapter looks at the historical development of agriculture in the region and at patterns of agrodiversity and land use. The counties of Modoc and El Dorado are considered in more detail as the most valuable agricultural producers in the Sierra Nevada and as examples of different agro-ecosystems. The changing nature of part-time farming is examined. Agro-ecosystems are defined and described, and patterns of crop-livestock mix are outlined. Emphasis is placed on spatial variations within the region.

METHODOLOGY

The focus is on agro-ecosystems seen as a subset of the general ecosystems of the region. The identification of agro-ecosystems and the allocation of counties to individual agro-ecosystems is based on principal components analysis (PCA). The dynamics of key structural and functional features of agro-ecosystems are defined using time series analysis. Considered within agro-ecosystems is agrodiversity, by which is meant the many ways farmers exploit the natural diversity of the bio-geosphere. More specifically, agrodiversity includes the maintenance of both biotic and management diversity within agro-ecosystems and responses to natural ecosystem diversity and dynamics (Brookfield and Padoch 1994). Agrodiversity of farming practices ensures that a range of "ecotypes" exist in close proximity and often succeed one another through time. Commercialization and monoculture lead to reduced agrodiversity, but the small, semisubsistence farms of the Sierra Nevada have been instrumental in maintaining agrodiversity in the region. The practices of these small and part-time farms in relation to the dynamics of agrodiversity, as developed by Zarin (1995), are examined. We also consider intensification and innovation in relation to land-management practices, the sensitivity and resilience of an ecosystem and the role of the creation of landesque capital, and population and production pressure on land management.

DATA SOURCES

We have not generated any new information through primary research in this assessment. All of the information on which this report is based is publicly available but has not been accessible in an integrated form. The main sources of written data are the agricultural censuses for California from 1860 to 1992 and the annual reports of the county agricultural commissioners. These data sources are supplemented by other publications from various county authorities and by historical studies of local areas. The mapped data comes from land-use data supplied by the United States Geological Survey (USGS) for the year 1970 and from data for 1988 and 1992 supplied by the Farmland Mapping and Monitoring Program (FMMP) of the Office of Land Conservation of the California Department of Conservation. The USGS 1970 data was obtained from the Geography Department at the University of California, Santa Barbara. 1992 land-use data downloaded in ARC-export files from the Environmental Protection Agency (EPA) Web site on the Internet. We had some mapping problems with this 1992 data and eventually discovered that it is based on 1976 USGS mapping corrected by the EPA using a statistical model to predict 1992 land use. Thus, it is not a very solid basis for measuring actual change in land-use categories over time and so we abandoned this data set.

The statistical data is published for whole counties, whose boundaries do not coincide with those of the SNEP study area. Some counties, such as Kern, Fresno, Tulare, Madera, Merced, Stanislaus, Yuba, Butte, Tehama, and Siskiyou, have only a small, predominantly nonagricultural part of their area lying within the Sierra Nevada, so it was decided to omit these counties from the analysis. Another problem with agricultural census data is that when the number of farms in a category is so few that individual farms could be identified, the information is left out because of the need to maintain confidentiality. This was a particular problem in small counties such as Alpine and Sierra. The census allows the development of time series analysis, but the length of time between censuses varies. The first United States Census of Agriculture was taken in 1840, but the first available for California was in 1860. The next census we have for California was taken in 1880 and after that in 1910 and 1920. Censuses were then taken every five years until 1950, then next in 1954 and every five years until 1974, then at four-year intervals until 1982, returning to five-year intervals in 1987 and 1992. The content of the censuses is also not directly comparable over time.

Questions asked in the census, definitions, and county boundaries have changed. In 1860 the number of farms in each county was not recorded, and Alpine, Inyo, Lassen, Modoc, and Mono Counties did not exist. The number of acres in improved and unimproved farmland was given but in 1880 only improved land acres were recorded (U.S. Census Office 1864, 1883). The 1910, 1920, and 1930 agricultural censuses were very limited in scope (Bureau of the Census 1913, 1922, 1932) and the 1930 and 1935 censuses had a more restricted definition of cattle than in other censuses (Bureau of the Census 1936). County boundaries also changed: part of El Dorado County was annexed to Placer and part of Placer County annexed to El Dorado in 1913. More recently, in the 1974 census there was a major change in the definition of a farm, so earlier censuses are not strictly comparable with those from 1974 onward. Since 1850, when minimum criteria defining a farm for census purposes were first established, the definition of a farm has been changed nine times. In 1959 a farm was defined on the basis of the number of acres in the place, that is, the land on which agricultural operations were conducted. Farms with less than ten acres were counted if the estimated value of sales of agricultural products for the year amounted to at least $250. Farms of more than ten acres had to have an estimated minimum annual value of production of $50 (Bureau of the Census 1961, xiv–xv). In 1976 the definition was changed: the number of acres criterion was abolished and the minimum value of sales criterion raised to $1,000 per year (Bureau of the Census 1977, ix). This change had its greatest effect on small part-time farms, such as many of those in the Sierra Nevada (Bureau of the Census 1977, B1). From 1969 the census has been based on mailed questionnaires; previously it had been carried out by direct enumeration in the field. There have also been several changes in the time of year at which the census was taken. In order to minimize the impact of these internal census changes, most of the data used here are presented as percentages or related to other data from the same census.

The annual reports of the county commissioners are even more varied than the censuses. The series of reports starts in different years for the various counties, and some counties, such as Inyo and Mono, are combined. The variables for which information is presented vary from county to county and from year to year, influenced by the specific changes in each county, the availability of data, and the interests of the individual county commissioners. This variety gives a freshness and vitality that adds color and explanation to the information gleaned from the census. It also allows an appreciation of short-term changes within the intercensal period. However, these reports do not start until after the Second World War, and the publication years for each county differ.

Because data in the census and county commissioner's reports are aggregated for the whole county, it is impossible to know where within the county specific crops are grown. However, the data from the United States Geological Survey (USGS) does provide this information in great detail, giving us a snapshot of the land use of the Sierra Nevada for the year 1970. This has been supplemented by land-use data provided by the Farmland Mapping and Monitoring Program (FMMP) of the California Office of Land Conservation. The FMMP compiles two kinds of farmland maps: Important Farmland maps for those areas that have modern soil surveys and Interim Farmland maps for those areas lacking modern soil survey information and for which there is expressed local concern on the status of farmlands. Consequently, much of the agriculturally marginal land of the Sierra Nevada is not mapped by the FMMP. Only forty counties of California are mapped, excluding Alpine, Calaveras, and Tuolumne in the central Sierra; Lassen in the north; and Mono and Inyo on the east. Sierra Valley is the only part of Plumas and Sierra Counties that is mapped. These maps show land capability based mainly on the physical and chemical qualities of the soils, plus growing season and moisture availability, but they also broadly reflect current land use.

We found major problems when trying to integrate the mapped data and the county-level statistical data. Despite the considerable amount of land-use mapping undertaken by various agencies, inconsistencies in the coverage and changes in classifications make it impossible to draw meaningful conclusions about changes in land use over time for the whole SNEP region. Interrelationships between land use, production and input levels, economic returns, and farm structure and population can be obtained only from the census data. In addition, the categories used for land use differ from one source to another and sometimes vary from year to year and from county to county as the Farmland Mapping and Monitoring Program (FMMP) responds to changes in definitions made by other agencies, such as the Soil Survey (California

Department of Conservation 1992). Consequently, the mapped information has been used to illustrate and supplement the census data but, only to a limited extent, to inventory the conversion of agricultural land. The census data have been utilized for time series analysis and for the identification of agricultural ecosystems using principal components analysis (PCA).

HISTORICAL BACKGROUND TO AGRO-ECOSYSTEMS

The history of agriculture in the Sierra Nevada can be seen as consisting of four main periods, beginning nearly a century and a half ago with the gold rush. Prior to the discovery of gold in El Dorado County in 1848, agriculture and ranching had been confined to the more accessible and fertile parts of the state, and there had been very little European settlement of the Sierra Nevada. Native American occupation of the region was based on hunting and gathering, with long-term settlement occurring wherever local food resources were plentiful, as with the seeds of a water lily, *Nuphar advena*, at Tulelake (Pease 1965, 44). In these areas of denser settlement, conflict between American Indian and European settlers over land took place. The early boom period of the gold rush was followed by one of adjustment to loss of local markets caused by declining mining activity and technological change. In the third period large-scale lumbering, power industries, and specialized agriculture developed. The fourth period is distinguished by rural residential expansion, agricultural pluriactivity, and the suburbanization of agriculture, especially in the foothills. These stages occurred first in the central foothills and somewhat later in the higher and more isolated areas of the Sierra Nevada.

The Boom Period of the Gold Rush, 1848–60

By the end of 1848, an estimated 10,000 to 12,000 men from California, Oregon, Central and South America, and the Pacific Islands had arrived in the foothills. Within five years some one-third of a million persons had migrated to the gold camps and the boom towns of the Sierra Nevada from all over the world. El Dorado County, where Marshall's eventful discovery of gold was made, rapidly became the most populous county in the state. By 1852 it had a diverse population of 40,000, while Nevada County contained approximately 20,000 people. In addition to the mining camps, settlements such as Placerville, Gold Run, and Nevada City sprang up throughout the gold-producing regions of the foothills (Weeks et al. 1943).

This growing population generated a demand for various support activities, such as lumbering, hauling of supplies, and food production. Farming developed in the foothills during the 1850s to meet the needs of the mining camps. Many disillusioned miners moved on to new discoveries elsewhere in the western United States and Canada, but some settled as farmers in the Sierra Nevada. They cleared extensive areas of timber and brushland for the production of barley, wheat, oats, and hay to meet the heavy demands of the horse teams that transported food, lumber, mining equipment, and passengers to the gold mining areas. Peach and apple orchards were established and vegetables and potatoes grown on lands irrigated with water from ditches built by mining companies. By 1860 the value of orchard produce from El Dorado County was the highest in the state (U.S. Census Office 1864), and the Sierra Nevada counties were producing about one-third of the state's orchard fruit. The three foothill counties of Mariposa, El Dorado, and Tuolumne produced 11.7% of the state's wine, and the wine output of Mariposa County alone was greater than that of Napa County. Some 35% of the state's market-garden (i.e., truck-farm) crops by value were produced in the Sierra Nevada by 1860.

The livestock industry also expanded into both foothill and mountain regions. Dairying grew rapidly, with milk, butter, and cheese finding local markets in the foothill towns and mining camps. All the Sierra Nevada counties produced butter contributing 14% of the state's total, but only seven of the region's counties made cheese (U.S. Census Office 1864). Sierra farms had less than 3% of the state's dairy cows, concentrated in the northern part of the region and in El Dorado County, and transportation difficulties clearly encouraged a concentration on butter and some cheese rather than on fresh milk. Meat production was very important: El Dorado County had by far the highest value of animals slaughtered of any county in the state and, when combined with the figures for Amador, Sierra, and Tuolumne Counties, accounted for almost one-quarter of the total for California. Thus at this period the farms of the Sierra Nevada were among California's major producers of food for local consumption using relatively intensive methods.

The effect of summer drought on pastures at low elevations soon led to the practice of driving dairy and beef cattle from foothill ranges to meadows in the high mountains. The sight of large flocks of sheep moving between winter ranges in the Sacramento Valley and summer grazing lands in the mountains also became commonplace following the introduction of stock from eastern states. As a contemporary account noted, "Here is a succession of grassy meadows—one called the Big Meadows is several miles in extent—and some men have cut a trail in and have driven up a few hundred cattle that were starving in the plains" (Brewer quoted in Burcham 1957, 153). Plumas County was the major producer of hay in 1860, with the Sierra as a whole growing 18% of the total for the state (U.S. Census Office 1864). In 1860 the Sierra Nevada contained 14% of the state's livestock by value, with Siskiyou (then including present-day Modoc County) as the leading county in the region.

Adjustment to the Decline in Gold Mining, 1860–1910

The dramatic boom period of gold mining was relatively brief and followed by a period of bust. Exhaustion of the more accessible surface placers was rapid, and California's gold production declined sharply from a peak in 1852. The attraction of new mining discoveries in Nevada, Idaho, British Columbia, and Alberta (Momsen 1990) from the 1860s onward initiated a long decline in foothill population. However, considerable local agriculture was maintained in the foothill region to supply the remaining local markets and the booming mining operations at Virginia City and other towns in Nevada. Population movement to the state as a whole continued, and many new immigrants, finding the fertile lands of the valleys in Spanish grants or other large holdings, turned to the foothill region, where they acquired land for farms by patent or homesteading (Weeks et al. 1943). The acreage of improved farmland in the foothills increased steadily, reaching a peak about 1880, with the greatest expansion occurring away from the early gold mining areas. Sierra Valley was settled by Swiss and Italians, who produced food for the silver miners of western Nevada. From 1860 to 1880, according to the agricultural censuses, acreage of improved land in Sierra County increased by 579% and in Placer County by 414%, but in El Dorado County it declined by 20%.

At higher altitudes, settlement and the related development of agriculture came later than in the foothills. American Indians were a strong deterrent to the settlement of the northeastern uplands in the 1850s and 1860s, but many of the displaced miners and ranchers remembered the fertile meadows they had seen there along the wagon trail on their journey to California (Pease 1965). In the 1850s mining in Shasta Valley had attracted enough people to provide mutual protection. Agricultural settlement on the tableland east of the Cascade-Sierra volcanic ridge was a different matter, however, because individual ranches were far apart and therefore vulnerable to attack. Military posts were set up to protect the settlers, but they also protected the American Indian peoples from roving vigilante groups of whites. The Modoc Indians were removed northward onto a reservation in Oregon in 1863, and by 1867 farm settlements had been established in the Honey Lake, Fall River, and Shasta Valleys. In 1864 the first settlement in what is now Modoc County was established, and by 1865 there were 300 residents in Surprise Valley (Pease 1965, 75). The route to Idaho ran through the valley, and farming developed to supply the wagon traffic.

Modoc County, formerly the eastern part of Siskiyou County, was established in 1874 in the northeastern corner of the state. By 1880 the initial phase of settlement of the region had been completed and the contemporary pattern of population distribution established. Two factors made the spread of farms and ranches possible during this period: subjugation of the Native American population and the availability of free or very low cost land (Pease 1965, 79).

The last hostilities in the region, the Modoc War of 1872–73, did not deter settlement on the Lost River meadows, although several ranchers were killed (Pease 1965, 79). The Treaty of Round Valley in 1868 opened the way for white settlement of the Big, South Fork, Warm Springs, and Goose Lake Valleys (Pease 1965, 79). Although this treaty only assured good conduct and did not extinguish Indian title, lands adjacent to the Pit River were immediately assumed to be public domain and so open to European settlement. The Native Americans of Round Valley had been granted a reservation, but this did not protect their land and many died of starvation and disease. Not until 1959 were courts to decide that this land had been taken from the Native American peoples illegally (Pease 1965, 80; see Reynolds 1996).

The Homestead Act of 1862 was the most common method of land acquisition. Where land had not yet been surveyed, settlers could protect themselves under the Preemption Act of 1841, which allowed the settlement of unsurveyed land with preference for eventual purchase at $1.25 an acre. Surveys took place in time for initial settlement of the northeast to be made under the Homestead Act. This act limited the amount of land that could be acquired by free patent to 160 acres, which was not enough to support a family in the higher, more remote areas of the Sierra, but the Desert Land Act of 1877 allowed up to 640 acres of land to be patented if part was irrigated within three years. The proportion to be irrigated was ambiguous in 1880, although it was later fixed at one-eighth, so much of the land acquired under this act was never irrigated (Pease 1964, 80–81). The Swamp Act of 1850 also made possible acquisition of land at low cost. Land covered by this act had to be swampy or liable to seasonal flooding, which in much of the Sierra included the highly desirable meadowland. The land was made available at $1.00 per acre, of which 80 cents could be on credit. All that was required for land to be designated swamp was for a local official to swear that the land in question was subject to flooding. This situation led the state surveyor in 1870 to complain that many were trying to "seek shelter under State laws and gain land from the Swamp Land Act" and that many who desired large holdings hoped to see their land classified as swamp (U.S. General Land Office Report, 1 August, 1870, page 461, quoted in Pease 1965, 81). Mountain meadow wetland that could be classified as swampland was especially valuable in the late 1860s, when hay commanded a high price in the mines of western Nevada. After 1873, when the Spanish doctrine of appropriation of riparian rights was legalized under the state Civil Code, water rights became an important factor in land acquisition. By 1880, 141,000 acres, 22% of total farmland in 1992, were in private ownership in Modoc County. Only 1.1% of this land was in harvested cropland. Two-thirds of the cropland was meadow on which hay was grown, and the rest of the land was in dry-farmed wheat and barley (Bureau of the Census 1883).

The numbers of dairy and beef cattle grew considerably between 1860 and 1880, in the higher parts of the Sierra Nevada, and the number of sheep almost doubled, indicating a move into livestock farming (figures 17.1 and 17.2). Movement of cattle from the valley was encouraged by the heavy rains of 1861/62 which led to the death of many cattle, and by drought in 1863 and 1864 during which many herds died of hunger and lack of water (Burcham 1957, 152). This trend was reinforced by the passing of the "no fence laws" by the state legislature in 1866, which made cattlemen liable for damage done to unfenced crops by their animals.

The demand for hay and feed grains declined after 1869, when the overland railroad was completed, and by 1880 a branch line of the Central Pacific Railroad had reached as far north as Redding. Local markets for food gradually fell as the mines in Nevada were depleted, and a further reduction occurred in the 1880s when hydraulic mining ceased as a result of the 1884 decision prohibiting the uncontrolled washing of debris into the rivers. Dry farming in the foothills became still more unprofitable following the replacement of teams by trucks and tractors during the early decades of the twentieth century. Competition from the more fertile valley farms became more intense as transportation costs were reduced by the construction of highways from the Sacramento Valley into the foothills. Except in areas where irrigation water was available, many farms were abandoned to brush and second-growth timber or utilized for extensive livestock production. By the close of the nineteenth century, much of the mountain and foothill rangeland was severely overgrazed in a struggle for forage among the numerous cattle and sheep outfits (Weeks et al. 1943).

With the inclusion of large parts of the higher parts of the Sierra Nevada in national forest reserves after 1891 and the establishment of the United States Forest Service in 1905, un-

desirable seasonal use of ranges was gradually brought under control. The Modoc Forest was established in 1903 to protect the livelihood of local ranchers (Menke et al. 1996). Grazing preference for national forest ranges was granted to small, owner-operated ranches located on foothill land adjoining the national forests in an effort to foster community prosperity and stability. Many of these small ranches, however, became uneconomic and were abandoned, and most of the private rangeland was consolidated into larger units. In 1880 the 13,000 acres of South Fork meadowland in Modoc County were controlled by only a few families, and in 1886 two of them joined to create the Modoc Land and Livestock Company, with an area of more than 11,000 acres (Duke 1939).

There was also an increase in the area of irrigated pastures used for beef and dairy cattle and for spring lambs. At higher altitudes, livestock farming was based on the use of summer range. The number of cattle in Modoc County increased from 16,000 in 1880 to 44,000 in 1909, while the number of sheep grew from 23,000 to 76,500 over the same period. The existence of large flocks of sheep in the county was an outgrowth of negotiations between local ranchers and transient flockmasters. Basque shepherds began to enter the county from 1880 as they moved their flocks from the Sacramento Valley to the mountains for summer pasture, and at about the same time Irish flockmasters moved south from Oregon into Modoc County. By the turn of the century these sheep threatened to destroy the rangeland for cattle. After 1905 permits had to be obtained for the use of rangeland and "the number of transient sheep bands was significantly reduced throughout the Sierra" (Menke et al. 1996). Shepherds could obtain these permits by settling in the area, so a number of sheep farms were established. These sheep, known as "resi-

FIGURE 17.1

Number of cattle in Amador, Calaveras, and Sierra Counties, 1860–1992.

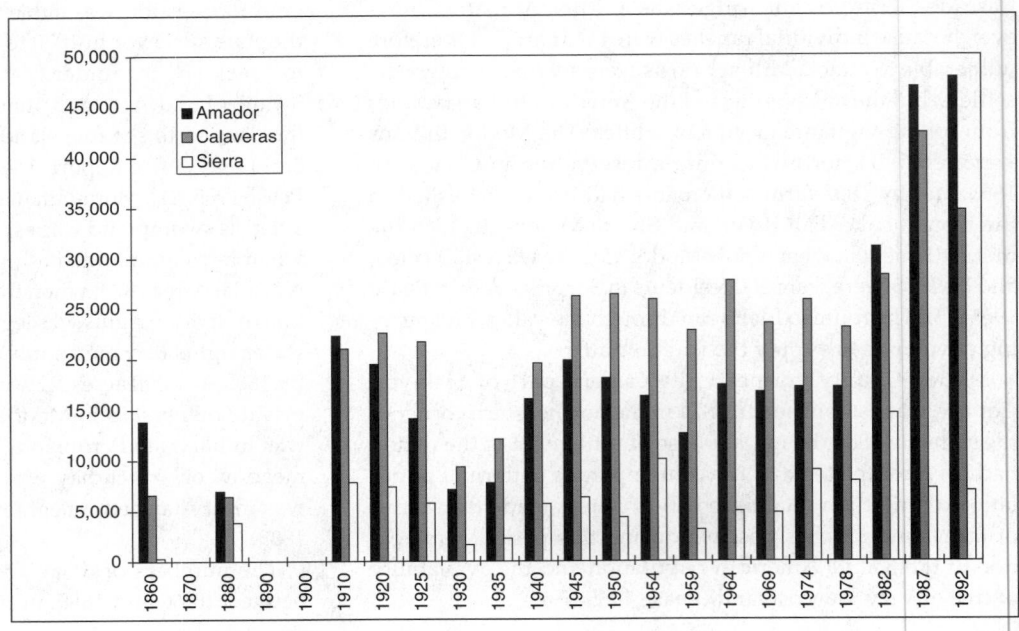

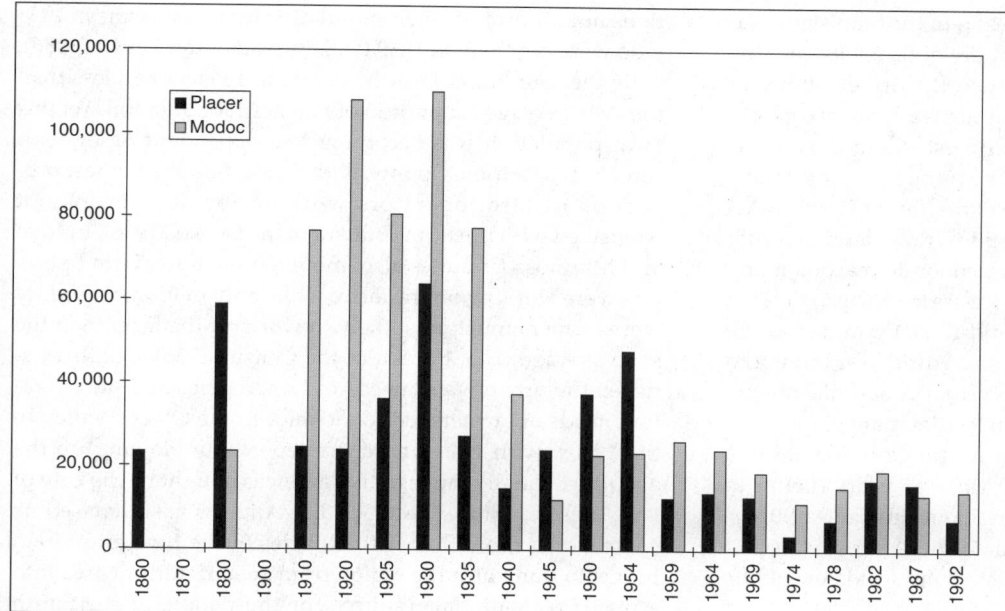

FIGURE 17.2

Number of sheep in Placer and Modoc Counties, 1860–1992.

dent sheep," were allotted rangeland too dry for cattle, so beef production was protected. In order to grow enough hay to feed the increased numbers of horses and cattle, swampy meadows were drained, and over 80,000 acres of other land in Modoc County was irrigated by 1912 (Pease 1965, 105).

New Activities, 1910–50: Lumbering, Hydroelectric Power, and Specialized Agriculture

After the turn of the century, large-scale logging became the dominant economic activity in the region (Weeks et al. 1943). The network of flumes and ditches built by the earlier hydraulic miners was gradually taken over and adapted for power, irrigation, and domestic uses. The first application for water rights for generating hydroelectric power was filed in 1891 by the Cornish manager of a gold mine in Nevada County (Larson 1996). An expansion of crop agriculture during and following the agricultural boom of the First World War brought renewed prosperity to the region. In the foothills the area in orchard crops expanded rapidly in response to the organization of irrigation districts and the rehabilitation or new construction of irrigation facilities. Pears and other fruit trees were planted on a number of ridge areas where fertile soils and water were available, as in the regions adjoining Placerville, Auburn, Grass Valley, Oroville, and Paradise. By 1924 Placer and El Dorado Counties had 15% of the state's pear trees (U.S. Bureau of the Census 1927).

The average size of farms increased from 299 acres in 1880 to 755 acres in 1925 in Modoc County, although both the number of farms and the area of farmland peaked in 1920 (U.S. Census Office 1883; U.S. Bureau of the Census 1922, 1927). In

Alpine County both the number of farms and the amount of farmland fell between 1880 and 1925, as it did in Sierra and Nevada Counties, although Shasta, Mono, Inyo, El Dorado, and Tuolomne Counties increased their number and acreage of farms (U.S. Census Office 1883; U.S. Bureau of the Census 1927). In Placer County the number of farms increased rapidly, from 514 in 1880 to 1,448 in 1925, although the amount of total farmland declined: average farm size fell from 267 acres in 1880 to 233 acres in 1910 and to 157 acres in 1925. Woodland was still being cleared for agriculture, and in the counties of Placer, El Dorado, and Shasta nearly 10,000 acres (18% of the state total) on almost 900 farms (24% of the state total) was brought into agricultural production. By 1925 Modoc County was second only to San Joaquin for acres of hay grown (U.S. Bureau of the Census 1927). Clearly, several processes were going on, with a retreat from marginal land in many areas following the end of the First World War accompanied by an expansion into new areas as accessibility improved.

Increased demand during the world wars and widespread poverty during the 1930s also affected use of public grazing lands. During these periods of national crises, there was increased livestock use of national forests and other public lands throughout the West, and often inappropriate stocking levels were disregarded. During the First World War demand for wool and mutton was high and so sheep grazing increased, while during World War II cattle usage rose. The foot-and-mouth disease epidemic of 1924–25 permanently reduced grazing in the Stanislaus National Forest, where all livestock for that season were slaughtered (Menke et al. 1996). Sonora Pass was closed to transient sheep to limit the spread of the disease, and so grazing in the eastern Sierra was also affected

(Menke et al. 1996). After 1925 stocking in the Stanislaus National Forest was reduced to 66% of previous levels, and the closure of Sonora Pass to sheep ended the driveway use of the forest (Menke et al. 1996). In many areas a series of drought years between 1919 and 1935 and overstocking during the First World War led to depletion of public grazing lands. However, it was only after passage of the Taylor Grazing Act in 1934 that much attention was paid to rangeland carrying capacities (Menke et al. 1996). For economic reasons many grazing allotments changed livestock class from sheep to cattle in the interwar years (figures 17.1 and 17.2). Permitted usage of public lands rose during the Second World War but not to the pre-1920 levels. In some areas actual usage did not increase as cattlemen concentrated on feedlot management because of the shortage of manpower for range riding and the high cost of transportation during the war years (Letter to Tuolumne County Supervisor from Stanislaus Forest Supervisor, 1965, quoted in Menke et al. 1996). After this period there was a permanent decline in stocking levels on public grazing lands (Menke et al. 1996).

Tenant farming became less popular throughout California between 1910 and 1925, but in all the Sierra Nevada counties by 1925 it was below the state average of 14.7% and Mono County had the lowest proportion of tenant farmers in California. Farm values also fell. In 1880 no county in the Sierra Nevada had farms with an average value less than twice that of California farms as a whole (U.S. Census Office 1883). By 1925 the agricultural census recorded only Alpine, Lassen, Mono, and Sierra County farms as having values above the state average. However, rankings among the region's counties had changed little. Farms in Mono County were still the most valuable, with an average value exceeded by only five other counties in California, while the farms in Tuolumne County had become the poorest in the state.

The agricultural boom also resulted in a revival of population growth in the foothill region. The population of El Dorado County, which had fallen steadily from its peak of 40,000 persons in 1852 to only 6,400 in 1920, began to rise again. However, the more isolated rural areas continued to lose people while population became concentrated in the towns, suburban areas, and fruit-producing districts of the foothills. In Modoc County "farmers and ranchers immediately adjacent to the towns frequently chose town residences" (Pease 1965, 97). Small concentrations of population developed in scattered mining districts as renewed gold mining during the depression years of the 1930s once more attracted people to the foothills. Two gold-mining districts were active in Modoc by 1912, and at the peak of the boom seventy mines employing several hundred men were in operation. This new gold mining activity was short lived but stimulated production so successfully that in 1939 the output of gold from California exceeded that of any year since 1862.

For many people the Sierra Nevada became "a last refuge from unemployment" (Weeks et al. 1943, 8). Nevada County, with 38.5%, and Alpine, with 35.4%, had the highest propor-

tions in California of their population living on farms in 1935 who were not there in 1930 (U.S. Bureau of the Census 1936). Only the counties of Lassen, Sierra, and Placer had less than the state average proportion of new farm population. Yet this farm population was becoming less dependent on agriculture. In 1935 Alpine County, with 58.8%, had the highest proportion of farm operators working for pay at jobs not connected with the farm (Bureau of the Census 1936). In Inyo and Mariposa Counties also, more than half the farm operators were working off the farm, while only in Plumas County were farmers much less likely to work off the farm than the state average (U.S. Bureau of the Census 1936). Conflict between the urban water needs of Los Angeles and the irrigation needs of farmers led to violence in the Owens Valley in the 1920s, with valley ranchers repeatedly blowing up the aquaduct. Finally much of the land was bought by the City of Los Angeles. "Since the 1930s, Los Angeles has exercised its control over more than 300,000 acres of the Inyo and Mono basins to transform the region from an agricultural area into a major recreational resource for the people of the South Coast" (Kahrl et al. 1978, 33). Only in 1995 were rural interests able to force the City of Los Angeles to reduce the amount of water it took from Mono Lake in order to preserve this lake's unique features.

Highway construction encouraged the use of the Sierra Nevada for recreation by the urban population of the state. Summer homes were built around Lake Tahoe and along streams, hunting of game became popular, and interest in historical sites grew (Weeks et al. 1943). These summer visitors stayed for relatively long periods, since the journey from the Bay Area to Lake Tahoe usually took two days, and so provided a new market for local farm produce (Trussel 1989).

In the Sierra Nevada during the first half of the twentieth century, more people were dependent on agriculture for their livelihood than on any other single economic activity. Much of this agriculture, however, involved part-time "subsistence" farming, with farmers producing some livestock and crops mainly for home consumption, while deriving supplementary income from lumbering, mining, road maintenance, activities related to recreation, or work with water or power companies. At the same time, these nonagricultural occupations did create a local market for agricultural produce.

The Suburbanization of Agriculture, 1950 to the Present

Sierra Nevada agriculture in the second half of the twentieth century is characterized by increased specialization of production, greater diversity of products, increased use of chemical inputs and integrated pest management (IPM) from the 1970s, and the development of organic farming. The role of the state in the restructuring of agriculture through subsidies for marketing and production or nonproduction of commodities and new trade, credit, and migration policies have brought many changes to the Sierra Nevada. Rural residential development and hobby farming, often involving "equity refugees" (Starrs 1996) and more women farm operators have encouraged diversity of management strategies and changed Sierran transhumance patterns and agricultural activities. There has been much discussion of the socioeconomic impact on the Sierra Nevada of the proliferation of "ranchettes," that is, holdings of less than ten acres, but in 1992 only Placer County had a higher proportion of such farms than the state average (Bureau of the Census 1994). Indeed, the number of ranchettes declined for the SNEP area as a whole between 1987 and 1992 (U.S. Bureau of the Census 1994) although the counties of Calaveras, El Dorado, Mariposa, Mono, Nevada, and Sierra recorded small increases.

Improved communications and new counterurbanization flows (Champion 1989) have reduced the differences between rural and urban communities, and the static concept of the rural-urban continuum has become an inadequate analytical framework for the study of rural communities (Smith 1991). The influx of former urbanites, most of whom are better educated and richer than the traditional rural populace, has introduced new social divisions into many rural communities. Hobby farmers maintain the land in agriculture but have different interests from traditional farmers. Younger "in-comers" to the foothills often commute to work during the week, increasing traffic congestion and air pollution on rural roads, and have little time for community activities. Retirees moving onto small holdings may contribute by volunteering for community services but may be resented because of the new ideas and attitudes they bring. If we define nonmetropolitan counties as those not linked with large cities or with communities tied to large cities (Hoffmann and Fortmann 1996) then all the Sierra Nevada counties were nonmetropolitan until 1970, when Placer County became metropolitan. Placer was joined in this category by El Dorado and Shasta Counties in 1980. Many commuters from Sacramento and even the Bay Area have moved into El Dorado, Placer, and Calaveras Counties, while Sierra Valley has attracted Reno commuters.

Although average household incomes remain generally low in the Sierra Nevada, as in most rural areas, some counties have shown remarkable variation over time. In 1950 Mono County had the third highest average income in California and was the only county in the study area to have an average family income above the state norm. By 1992 it ranked twenty-third, with an average household income 12% below the mean for the state (Hoffmann and Fortmann 1996). On the other hand, Placer County has changed its rank from forty-seventh in 1950, when its average family income was 25% below the state norm, to twelfth in 1992, when it became the first Sierra Nevada county for more than three decades to have an average household income above the mean for California (Hoffmann and Fortmann 1996), emphasizing the increasing suburbanization of this county.

There is a worldwide trend toward an increase in farm size in order to take advantage of economies of scale as levels of mechanization and commercialization increase, and Califor-

nia has been a leader in the United States. However, this trend is not so clear in the Sierra Nevada. Between 1974 and 1982 farm size increased, and the number of farms declined throughout the study area, as was expected (U.S. Bureau of the Census 1977, 1984). But during the 1980s the direction of change became more confused: average farm size continued to increase in the higher-elevation counties of Shasta, Modoc, Lassen, Plumas, and Sierra and the eastern counties of Inyo and Mono, but it fell in the foothill counties of Placer, Nevada, Mariposa, and El Dorado (figure 17.3). In Amador and Calaveras, both farm numbers and farm size increased, reflecting an expansion of farmland acreage, while in Alpine County mean farm size was almost halved because farmland fell from 7,352 acres in 1982 to 4,768 acres in 1992 (U.S. Bureau of the Census 1984, 1994). The increase in the number of farms in the foothills during the 1980s indicates the widespread impact of the growth of rural residences and hobby farms in this part of the study area, while those counties farther from major urban centers were less affected by counterurbanization trends.

Taking the counties of Modoc and El Dorado as examples of these two recent trends in Sierra Nevada agriculture, it is possible to examine the differences in greater detail. In Modoc County the number of farms increased from 1974 to 1982 and then began to decline, as did the number of farm workers, tractors, and farms in individual ownership, while the proportion of full-time farmers increased from 44% in 1982 to 48% in 1992. Concentration on livestock production grew, with the proportion of livestock farms increasing from 53% in 1982 to 64% in 1992. This change suggests the substitution of family labor for hired labor and the growth of corporate extensive farming. In El Dorado the number of farms grew very rapidly during the 1980s as rural residential lots spread, and the average farm size was only one-tenth that of Modoc farms. Yet the proportion and number of commercial farms (farms with sales of over $10,000 per year) increased in El Dorado between 1987 and 1992, suggesting intensification of production. Crop production became the dominant activity on 43% of farms in 1992, as compared with only 30% in 1982. The number of tractors and hired workers peaked in 1982, but as in Modoc, the number of full-time farmers increased between 1982 and 1992. The more rapid turnover of farmers in El Dorado than in Modoc can be seen in the average years on farm statistics for 1992: 13.9 for El Dorado versus 17.4 for Modoc County (U.S. Bureau of the Census 1984, 1994)

Clearly, change has accelerated since 1980. Intensification, commercialization, and specialization have become widespread. Urbanization of foothill counties has occurred. This trend was most marked in Placer County between 1988 and 1992, when 15% of farmland was developed (California Department of Conservation 1992).

FIGURE 17.3

Average farm size in 1992 by county.

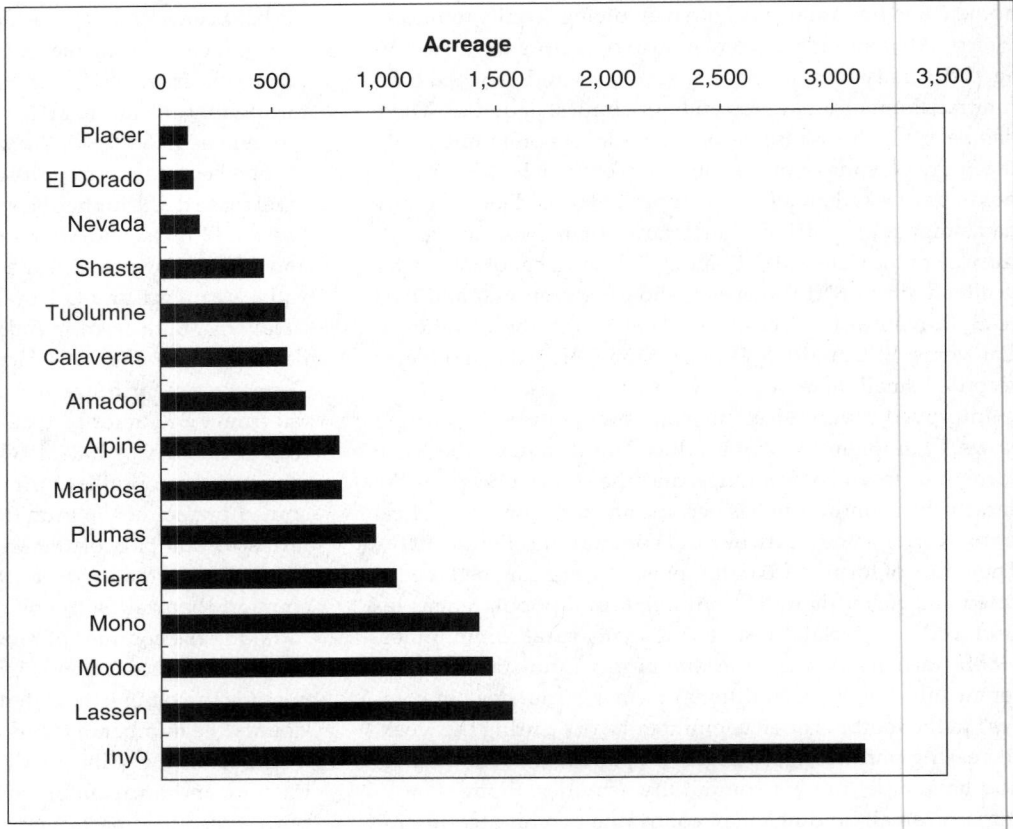

LAND MANAGEMENT

The purpose of land resource management is to ensure immediate and future production, not environmental conservation. Future production is valued over a time span that varies greatly according to the circumstances of the farmer. Conservation, in these circumstances, will arise only where future livelihood is threatened by perceived degradation or where the values of both the community and the farmer include the preservation of natural landscape and biota and a rejection of cultivation methods seen as damaging to the environment. The way individual farmers manage their land is influenced by personal perceptions that are a product of the stage in the life cycle of the farmer and the time spent on the farm, in terms both of years of experience and of labor time available daily. These perceptions will in turn influence the adoption of innovations by farmers. All farmers are faced by uncertain weather, diseases of plants and animals, and unpredictable market conditions. The strategies of risk-aversion include mixed farming, holding land across a range of resource types, and taking out insurance. Innovators will tend to be the most financially secure farmers, who are prepared to take risks in the hope of future gain.

In 1982 California farms had one hired worker per forty acres of farmland; this proportion rose to one per forty-three acres in 1992 (U.S. Bureau of the Census 1984, 1994). On Sierra Nevada farms, the ratio of workers to acres in 1982 varied from 1:86 in Placer County to 1:2131 in Inyo County. Over the next decade only five counties (Alpine, Amador, El Dorado, Nevada, and Lassen) went against the state trend with a reduction in the number of acres per worker. In El Dorado the number of acres per hired worker fell from 110 in 1982 to 63 in 1992, indicating increasing intensity of production. In both 1982 and 1992 the SNEP counties employed only 1.8% of all the hired farm workers in California (U.S. Bureau of the Census 1984, 1994).

Another way to measure intensification of production by increased inputs is to measure use of agricultural chemicals. Herbicide application was chosen as an indicator because herbicides can be used on both cropland and pastureland and because their use measures a certain level of sophistication and might be expected to decline as interest in organic production grows. In the state as a whole, herbicides were used on 15.2% of farmland in 1982, rising to 22.3% in 1992. In the SNEP study area, farms in only three counties (Mono, Modoc, and Placer) used herbicides on more than 2% of their farmland in 1982. During the 1980s herbicide use increased in all counties except Calveras, Inyo, Mariposa, and Mono so that by 1992 Placer County, with 11.2% of farmland utilizing herbicide, Modoc (4.5%), and Shasta (2.1%) were the leading counties. This ranking reflects the position of Placer and

Modoc as the counties with the highest proportions in the study area of cropland harvested in 1992.

Gender and Farm Management

There is empirical evidence for the United States that a high proportion of organic farmers are women; thus, the gender of the farm operator becomes an important element in farm management strategies.

The Sierra Nevada has long had a high male sex ratio, with men outnumbering women 12:1 in the early mining days. By 1925 the farm population in the study area was still more male than that of the state as a whole, varying from 45 women per 100 men in Alpine to 74 per 100 in Calaveras. (Inyo, where the sexes were numerically almost balanced, was the exception). Even today women rarely constitute more than 10% of farm operators in the industrialized world and even less in extensive ranching areas. Data on the number of women farmers in California have been published only since 1978 in the agricultural census, but the recorded increase in the proportion of women has been steady since then. In California as a whole, women farm operators made up only 7.9% of total farmers in 1978 but had increased to 12.4% by 1992. However, among the Sierra Nevada counties, only Modoc and Mono were below the state norm in 1992. Even more amazing is that in El Dorado, Nevada, Placer, Plumas, Shasta, and Tuolumne Counties women made up almost one-fifth of farm operators in 1992. Unfortunately, this high rate is probably more a reflection of the marginality of agriculture in the study area than of the skills of the local women, although affirmative action policies, especially in relation to farm credit, may have been a factor.

As is commonly found throughout the world, farms operated by women were much smaller than those operated by men across the region. In Mariposa, Modoc, and Plumas, the reverse was true in 1992 but appeared to be due to the effect of small numbers of large farms changing hands, possibly because of the death of a husband, as it was not true for earlier years.

Part-Time Farming

Another characteristic of farming in less-favored areas is the importance of part-time farming, which has been widespread in the study area since gold rush days. Farming was often seen as a stopgap activity until something more profitable turned up or as a source of subsistence to supplement low incomes from other activities. Today it may also be seen as a hobby for professionals who can work from their rural homes, for people taking early retirement, or for those choosing to commute from the countryside to urban employment. Part-time farming affects farm management strategies in various ways. Shortage of time may lead to the substitution of equipment and chemicals for labor to an extent that normally would not be economic. On the other hand, income from another job may support high levels of capital investment and provide the financial security that allows for innovation and risk taking. This nonfarm income may also reduce the pressures for high productivity and maximizing income; part-time farmers may be farming for pleasure rather than livelihood. Such perceptions and management strategies lie behind many of the specialized animal holdings, such as those for Arabian horses, llamas, and ostriches, found in El Dorado County and elsewhere in the foothills.

In 1982 a majority of farmers in the SNEP study area were not in full-time farming. The highest proportions of full-time farmers were found in those areas farthest from urban settlement, where non-farm jobs were not easily available as in Mono, Modoc, and Alpine Counties. The trend toward part-time farming is found in most countries, especially in environmentally marginal areas, yet unexpectedly, in the SNEP region by 1992 the proportion of farmers with no other occupation had increased in all counties except Alpine and had reached 58% in Mono County. Farmers who worked off the farm more than 200 days per year were most prevalent in those counties associated with smaller farms but with good road accessibility to urban employment opportunities. The highest proportions of such farmers (over 36% in 1992) were in the central foothill counties of Amador, Calaveras, El Dorado, and Placer (U.S. Bureau of the Census 1994). Many of these farmers could be classified as hobby farmers.

Landesque Capital and Equipment

Landesque capital is defined as physical works, created for the purpose of improving or sustaining production, that have a useful life well beyond that of a single season, crop, or crop cycle. It includes irrigation, drainage, and water-control works and tree crops such as orchards or Christmas trees. The creation of landesque capital is an important element in the management of land, biota, and water. In the foothill counties the expansion of irrigation works has been going on for the last half-century, although the number of acres under irrigation declined temporarily in the late 1950s (figure 17.4). Orchard crops have been grown on foothill farms since the gold rush days and continue to be important. In Sierra and Plumas Counties in 1964, land improvement in the form of leveling of land took place (Plumas-Sierra Counties Agricultural Commissioner 1965). At higher altitudes wetlands have been drained, as have lakes in Modoc County. The Tulelake basin is a high montane valley 4,200 feet above sea level extending from Modoc County into Oregon. Starting in the early 1900s the flat valley bottomland covered by shallow Tulelake and the surrounding marshlands were drained for irrigated agriculture and between 1917 and 1948 were opened for homesteading. Concerns for conservation of wetland habitat for wildlife led to the passage of the Kuchel Act in 1964, which enforced the coexistence of waterfowl management and agriculture in the area. Both these activities are now facing serious problems, stimulating the development of new man-

FIGURE 17.4

Irrigated acres for Placer and Mono Counties, 1944–92.

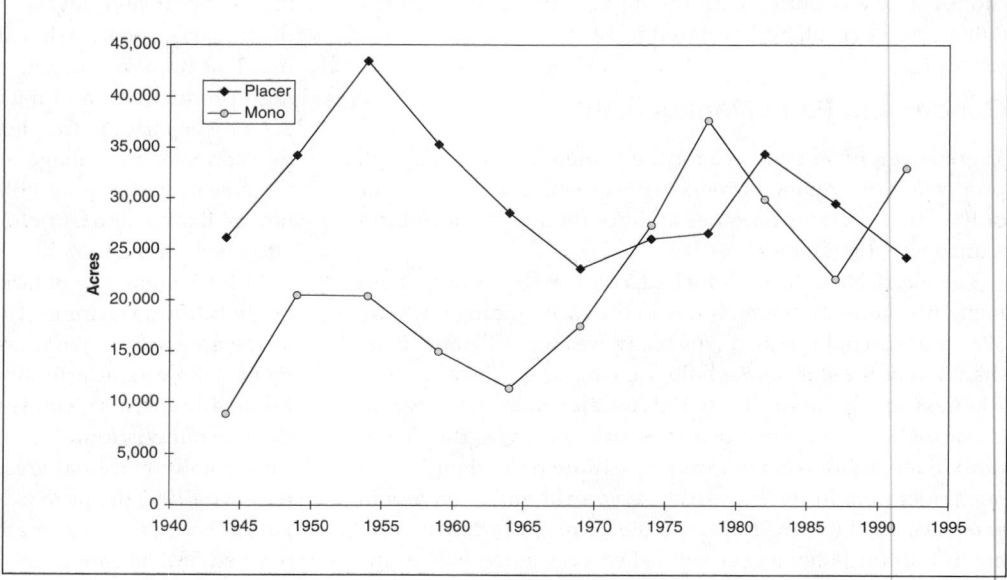

agement plans (Modoc County Agricultural Commission 1942–). Agricultural productivity has been declining because of soil infestation with nematodes and fungal pathogens. Agricultural chemicals have led to eutrophication and threaten wildlife. Researchers are now seeking nonchemical control methods for soil-borne pests and experimenting with seasonal flooding. "Sump rotation," that is, rotating areas of existing wetland into drained cropland in conjunction with flooding areas of existing cropland to create new areas of wetland, is being tried in pilot projects (Shannon 1995), an example both of the unexpected problems associated with landesque capital and of innovative management strategies.

Investment in machinery and equipment also influences management strategies. For much equipment there is a minimum size of holding below which a unit of equipment is uneconomic. Thus small farms tend to have less equipment than large farms, although part-time farms may be relatively overequipped. A farmer's investment in expensive, specialized equipment often tends to reduce flexibility in crop or livestock selection. Many of California's farms have high levels of capital investment in the most modern equipment, which is used to replace the scarce and expensive production factor of labor. Sierra Nevada farms differ from the rest of the state in this management strategy.

LAND USE

Any discussion of land use is limited by the lack of data on land use for the region as a whole. Only for 1970 is wide coverage available, so analysis of change in land use is impossible. In 1970 the Sierra Nevada presented a land-use pattern largely determined by elevation and moisture availability (figure 17.5). The SNEP western boundary delimited the upper reach of orchards and vineyards except for a few small outliers in El Dorado, Placer, Amador, and Tulare Counties (figure 17.6). Between the rich agricultural lands of the Central Valley and the forests of the mountains lay a band of herbaceous rangeland (figure 17.5). On the eastern slopes of the Sierra Nevada the forests graded into shrub rangeland, which is the dominant land use of Mono and Inyo Counties. Within the mountains and on the volcanic plateau to the north, better-watered fertile basins formed islands of cultivation and pastureland (figure 17.7).

In 1970 the USGS mapping revealed that only in Kern, Yuba, Lassen, and Inyo Counties was less than half the land in forest (table 17.1), while in Butte, Nevada, Plumas, and Shasta Counties more than four-fifths of the land within the SNEP boundaries was forested. Rangeland was most widespread in Inyo County and least in Shasta, Fresno, and Placer Counties. Only in the foothill counties of Placer, Yuba, and Fresno SNEP areas did agriculture occupy more than one-fifth of the land (figure 17.6), while in the central Sierra Nevada counties of Alpine, Inyo, Madera, Mariposa, and Tuolumne less than 1% of the land was in agriculture.

Land Capability

According to the *Atlas of California* (Donley et al. 1979), land in capability Classes I and II, defined as good cultivable land (USDA 1950), in the SNEP study area is confined to "a few places east of the mountains" (Donley et al. 1979, 73), which according to the map are in Modoc, Lassen, Alpine, and Mono Counties only. In 1974 virtually all this land was irrigated (Donley et al. 1979, 66). The California Office of Land Conservation in its Farmland Mapping and Monitoring Program

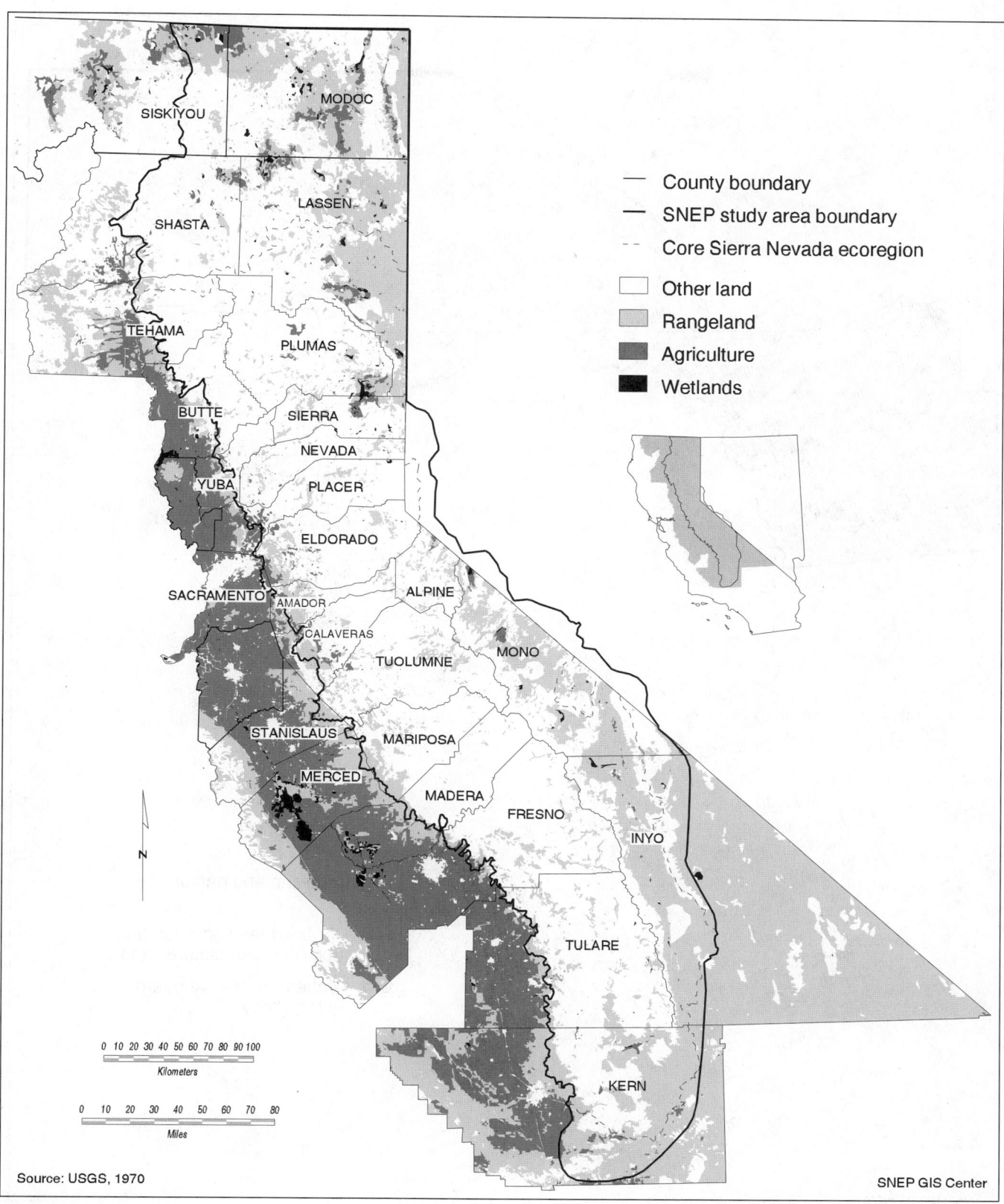

County boundary

SNEP study area boundary

Core Sierra Nevada ecoregion

Other land

Rangeland

Agriculture

Wetlands

0 10 20 30 40 50 60 70 80 90 100
Kilometers

0 10 20 30 40 50 60 70 80
Miles

Source: USGS, 1970

SNEP GIS Center

FIGURE 17.5

Land use of the Sierra Nevada, 1970.

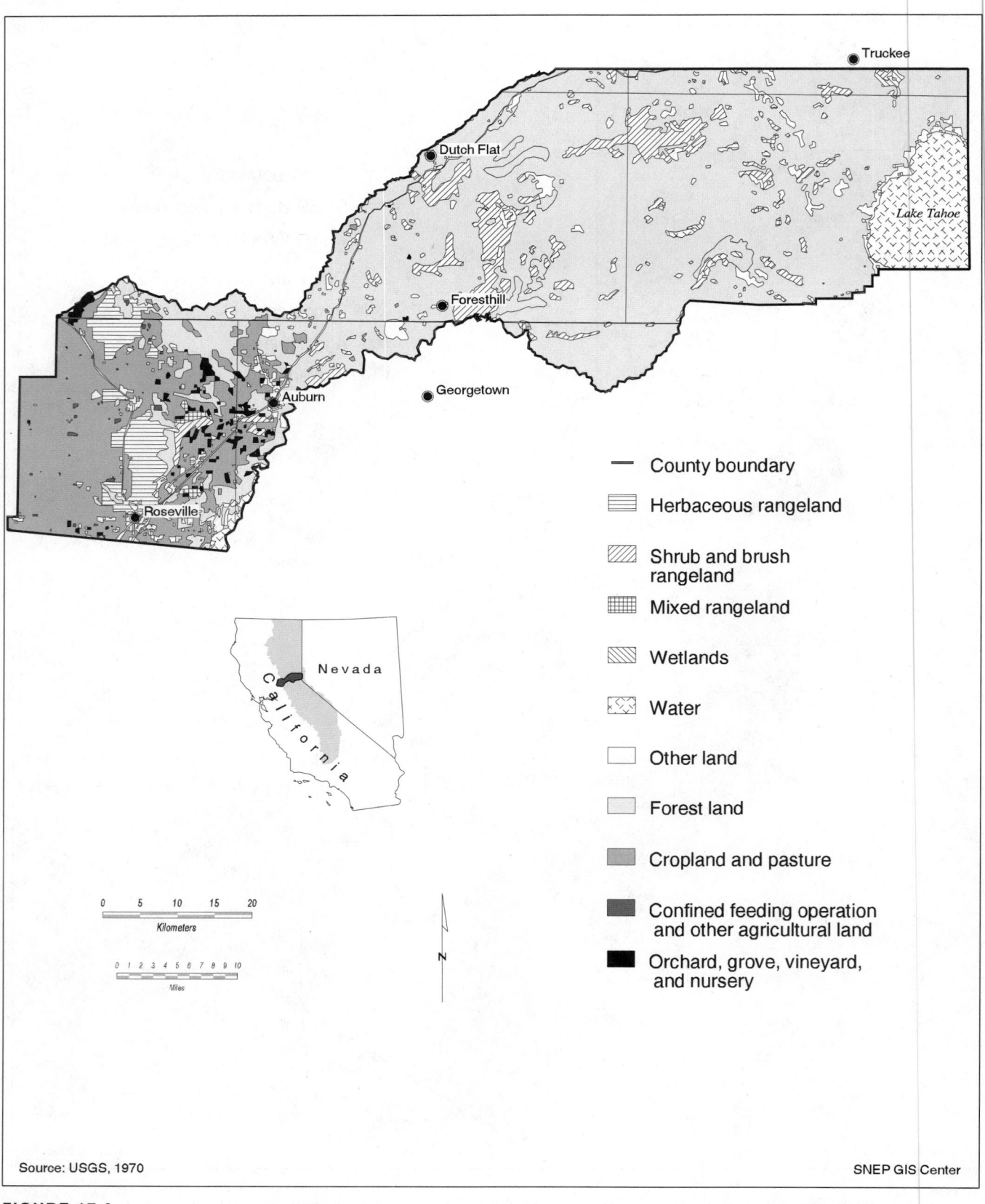

FIGURE 17.6

Placer County land use, 1970.

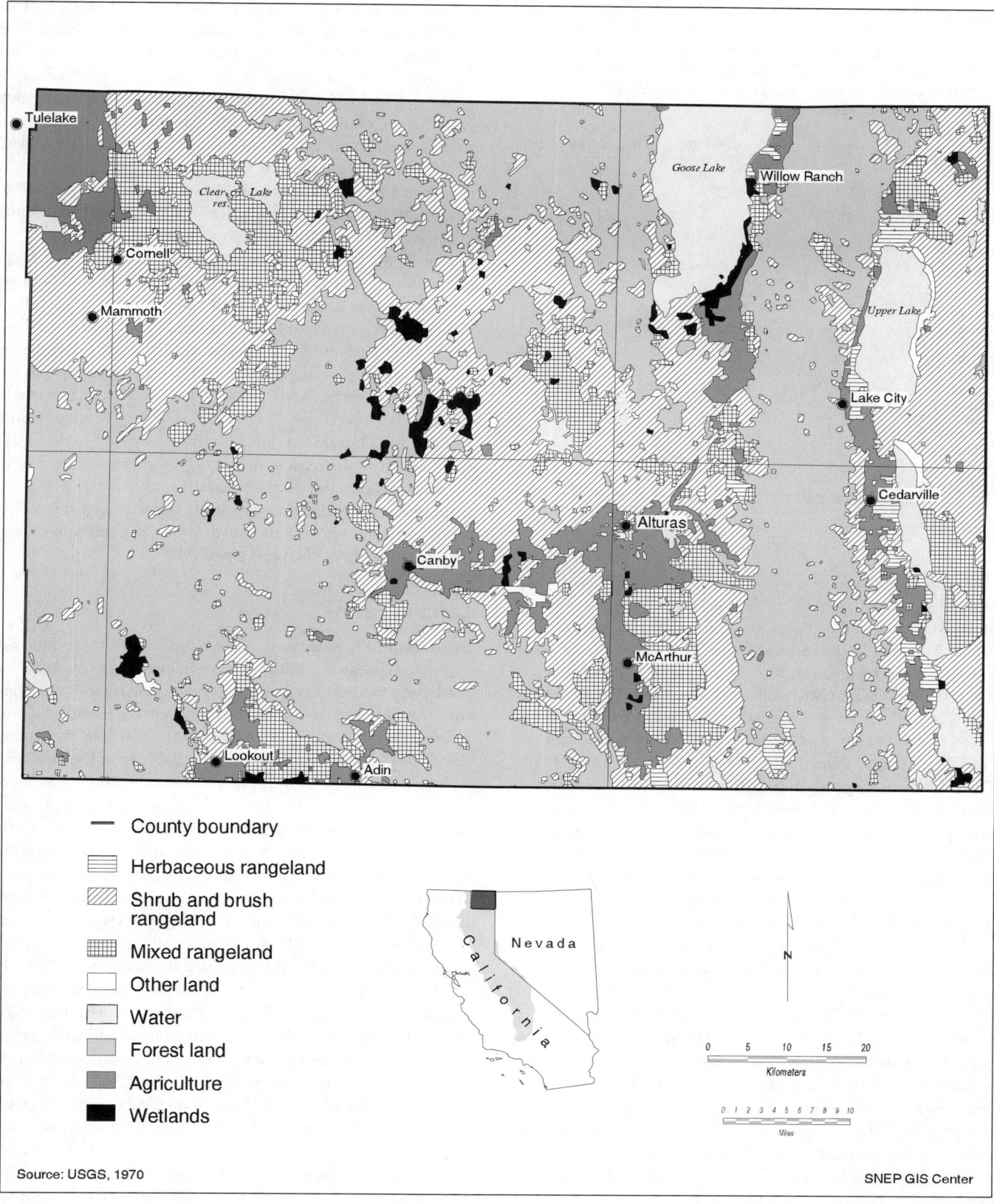

Legend:

— County boundary

▦ Herbaceous rangeland

▨ Shrub and brush rangeland

▦ Mixed rangeland

☐ Other land

Water

Forest land

Agriculture

Wetlands

Source: USGS, 1970

SNEP GIS Center

FIGURE 17.7

Modoc County land use, 1970.

TABLE 17.1

Percentage of agricultural, forested, and range land in the SNEP area by county, 1970 (USGS land-use mapping). Only those parts of counties lying within the SNEP boundary are included.

County	Forest	Range	Agriculture
Alpine	71.8	20.0	0.9
Amador	55.3	36.3	3.5
Butte	83.5	10.0	1.8
Calaveras	63.9	17.0	12.4
El Dorado	73.7	13.0	4.1
Fresno	61.4	7.2	24.7
Inyo	8.2	89.2	0.2
Kern	21.9	64.8	11.9
Lassen	47.7	43.6	3.8
Madera	73.7	23.2	0.2
Mariposa	79.1	19.7	0.1
Modoc	50.4	41.3	7.7
Nevada	82.3	9.6	1.1
Placer	62.0	7.5	20.9
Plumas	83.5	10.4	2.2
Shasta	92.8	4.1	1.3
Sierra	76.6	16.7	4.3
Siskiyou	75.4	12.8	5.5
Tulare	54.3	24.8	16.2
Tuolumne	79.7	11.8	0.4
Yuba	47.9	15.9	29.6

(FMMP) recognized two main areas classified as Prime Farmland in the SNEP area. The largest zone of such land is in Modoc County in the six basins of lava-derived alluvium of which Surprise Valley, Goose Lake Valley, and South Fork Valley are the most important (figure 17.8). Sierra Valley, a glacial lake bed, also has some Prime Farmland in the south in Sierra County. There are small, scattered patches of Prime Farmland identified and mapped by FMMP in Amador, El Dorado, Nevada, Placer, and Shasta Counties (figure 17.9). The minimum mapping unit is ten acres. Prime Farmland is land with the best combination of physical and chemical features able to sustain long-term production of agricultural crops. To be included on the map by the FMMP, this land must have been used for production of irrigated crops at some time during the two update cycles prior to the mapping date (California Department of Conservation 1992).

Farmland of Statewide Importance is similar to Prime Farmland but with minor shortcomings such as steeper slopes or soil with a lower capacity for moisture storage. There is some of this in Surprise Valley (figure 17.8) and in the northern part of Sierra Valley. The third category is Unique Farmland, which is poorer than the previous two categories and not always irrigated. There are examples of this category in Sierra Valley and western Placer County (figure 17.9). The fourth cropland category is Farmland of Local Importance, and it is determined by each county's board of supervisors and a local advisory committee (figures 17.8, 17.9). In general it includes land that is capable of agricultural production but often does not have irrigation water. It is more extensively distributed than the other cropland types. However, this land can be re-

classified from year to year, which makes identifying true land-use change difficult: For example, between 1988 and 1992 a large area of Other Land in Shasta County was reclassified as Farmland of Local Importance. The last agricultural category is Grazing Land, which is defined as "land on which existing vegetation, whether grown naturally or through management, is suitable for grazing or browsing of livestock" (California Department of Conservation 1992, 14). This is mapped in eastern Modoc County and along the western edge of the SNEP study area in a buffer zone between forestlands and the cultivated farmlands (figures 17.8, 17.9).

Not all areas are yet included in this mapping system, so analysis of changes between 1988 and 1992 is limited to those areas that were mapped in both years. The SNEP counties with at least partial coverage are Amador, El Dorado, Mariposa, Modoc, Nevada, Placer, and Shasta. Most counties showed an increase in urban land, although Modoc and Mariposa actually recorded decreases. There were small increases in the acreage of Prime Farmland except in Modoc (figure 17.8) and El Dorado Counties. Both Placer and Shasta Counties had large decreases in grazing land acreage, much of which seemed to have been reclassified as Farmland of Local Importance. Overall it does not appear from this evidence that there has been any serious loss of prime farmland to urbanization in the Sierra Nevada since 1988.

Land in Farms

Settlement in frontier areas always involves a period of trial and error. The first settlers did not in every case identify immediately the best agricultural lands in the region, nor were they familiar with the vagaries of the climate. Some land was cleared that eventually proved uneconomic and was abandoned, while other land became profitable with the availability of irrigation water. In El Dorado County there was more land in farms in 1860 than in 1992 (figure 17.10) (U.S. Bureau of the Census 1864, 1994). In 1860 only 28% of California farmland was improved, but in El Dorado and Plumas Counties, and in Siskiyou (which at that time included Modoc County), about four-fifths of the land in farms was improved. This comparison reminds us of the relative importance of agriculture in the Sierra Nevada at this early period.

In the state as a whole, farmland declined from 28.9% of total land in 1900 to 27.6% in 1925, except for a brief expansion during the First World War (Bureau of the Census 1913, 1927). However, in the Sierra Nevada some counties displayed quite different patterns: the counties of El Dorado and Shasta saw steady growth in farm acreage from 1900 to 1925, the eastern counties of Inyo and Tuolumne began to increase their acreage after 1910, but Mono and Plumas experienced a decline in acreage during the war years (Bureau of the Census 1913, 1922, 1927). In Alpine County farm acres almost doubled between 1900 and 1910 and thereafter declined (U.S. Bureau of the Census 1913). Although the population of California increased by almost two-thirds between 1900 and 1910, the

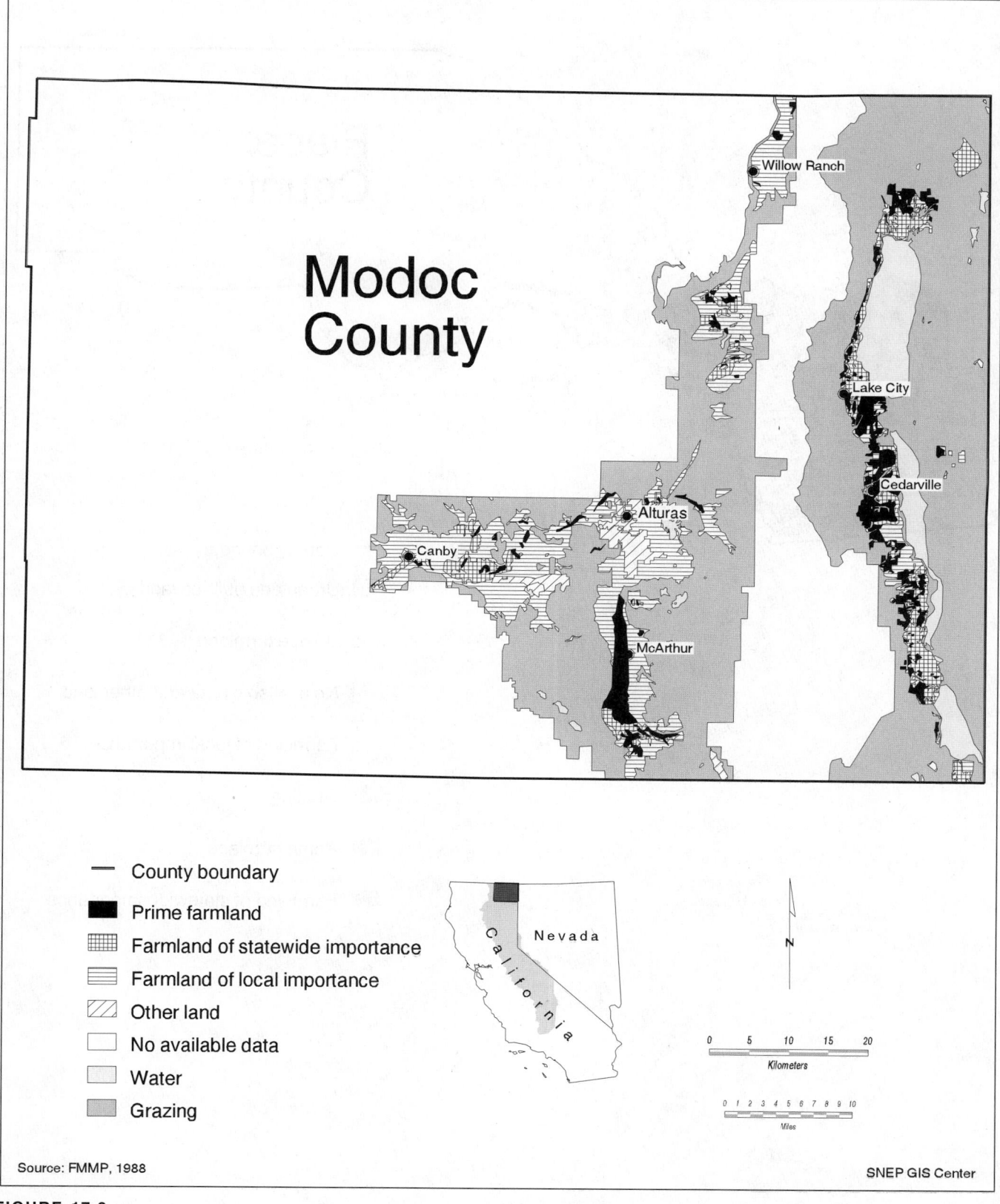

Modoc
County

Willow Ranch

Lake City

Cedarville

Alturas

Canby

McArthur

— County boundary

Prime farmland

Farmland of statewide importance

Farmland of local importance

Other land

No available data

Water

Grazing

California

Nevada

N

0   5   10   15   20
Kilometers

0 1 2 3 4 5 6 7 8 9 10
Miles

Source: FMMP, 1988

SNEP GIS Center

FIGURE 17.8

Modoc County land use, 1988.

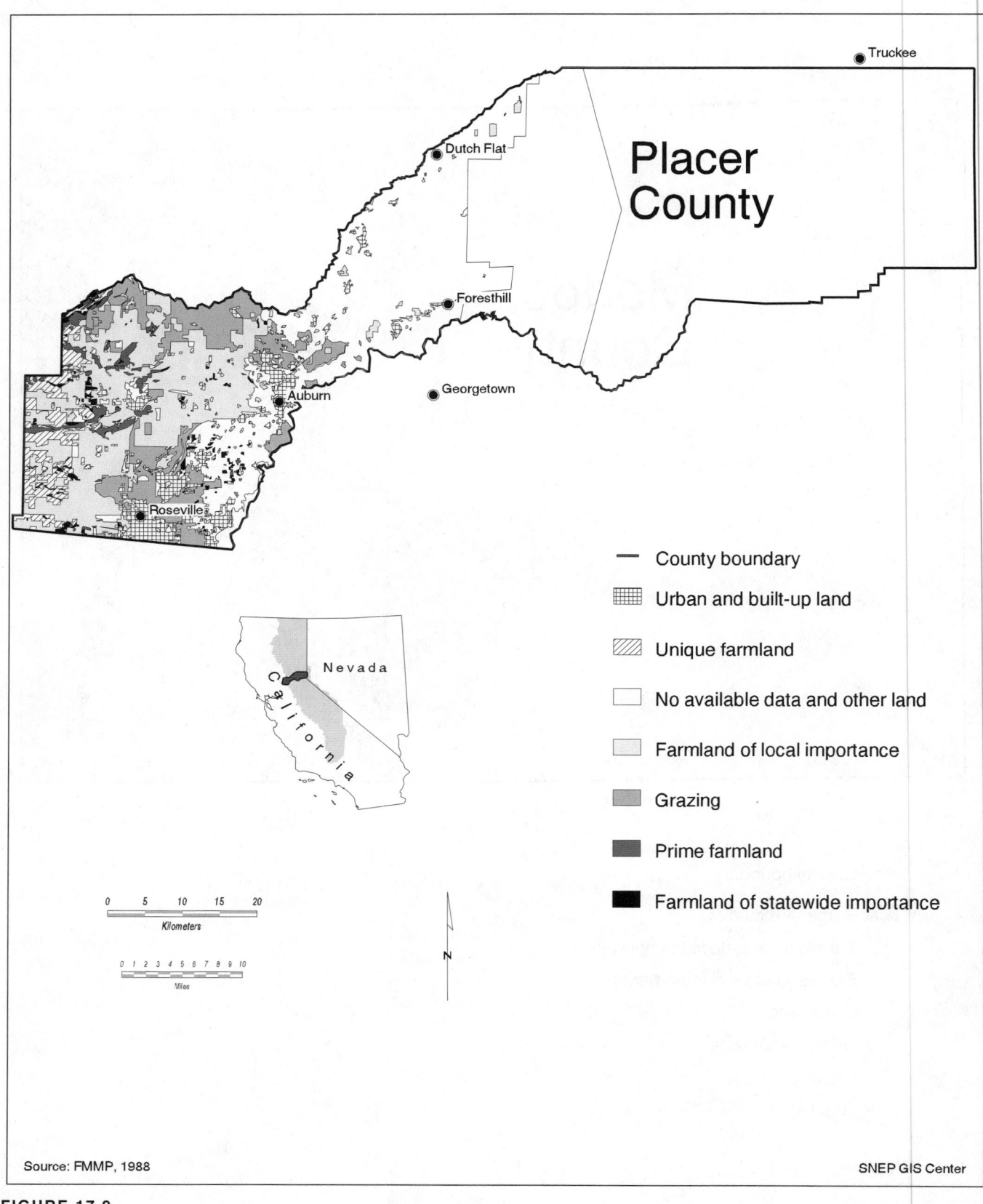

Truckee

Dutch Flat

Placer County

Foresthill

Georgetown

Auburn

Roseville

Nevada

California

— County boundary

▦ Urban and built-up land

▨ Unique farmland

☐ No available data and other land

▢ Farmland of local importance

▨ Grazing

▩ Prime farmland

■ Farmland of statewide importance

0 5 10 15 20
Kilometers

0 1 2 3 4 5 6 7 8 9 10
Miles

N

Source: FMMP, 1988

SNEP GIS Center

FIGURE 17.9

Placer County land use, 1988.

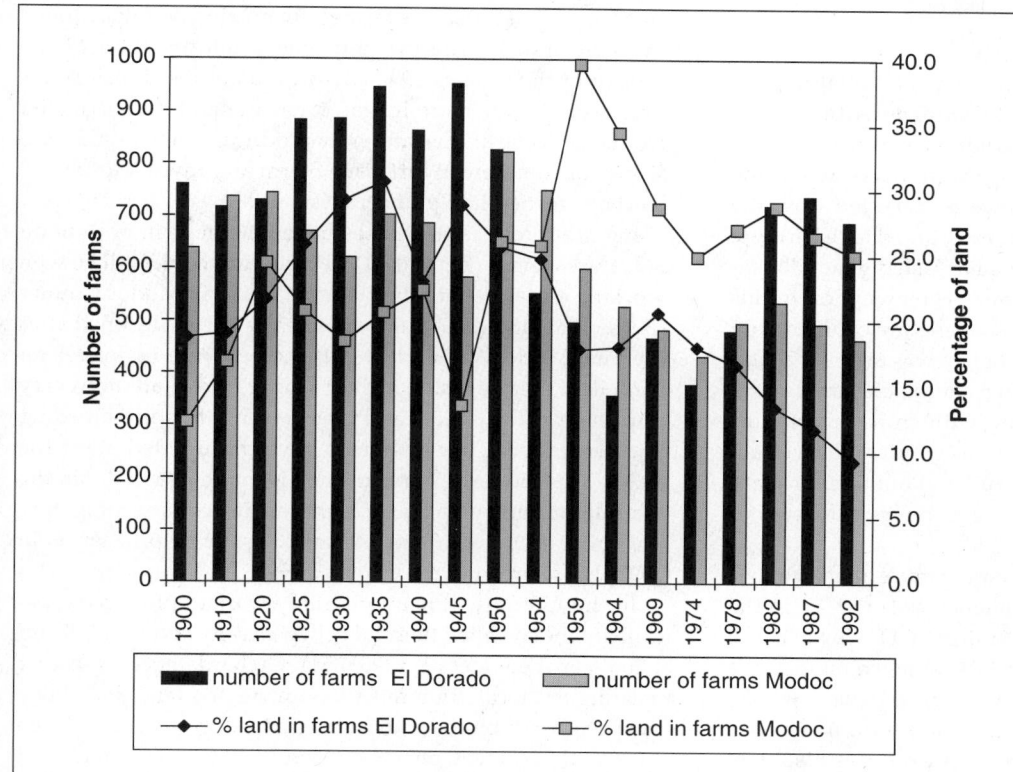

FIGURE 17.10

Number of farms and proportion of land in farms, 1900–92.

counties of Alpine, Amador, Calaveras, El Dorado, Mariposa, Mono, Nevada, and Tuolumne lost population. Only the northern counties of Modoc, Plumas, Siskiyou, Shasta, Lassen, Sierra, and Placer, plus Inyo, had a population increase during this period. Land values more than doubled for California farmland between 1900 and 1910 but declined for the Sierra Nevada as a whole, reflecting the growth of new economic opportunities outside the region (U.S. Bureau of the Census 1913), although in four counties land values increased enormously during this decade: in Modoc and Inyo land values more than quadrupled, and in Lassen and Placer they almost trebled. Only Amador, Calaveras, and Nevada Counties had a higher proportion of their land in farms than the state as a whole, and only Lassen was above the norm in proportion of improved farmland. These variations within the Sierra Nevada at the beginning of the twentieth century suggest an increasing concentration on agriculture as the dominant economic activity in many foothill counties and an expansion of rural settlement in the more isolated northern and eastern counties (figure 17.10) (U.S. Bureau of the Census 1913).

The amount of farmland in the state stabilized during the 1930s, expanded again during the Second World War, reaching a peak of 37.7% of total land in 1954, and declined in the face of competition from urban uses to 29% in 1992 (U.S. Bureau of the Census 1927, 1952, 1994). If we look at these trends at a county scale, however, a considerable amount of variation is noticeable. In the northeast of the Sierra Nevada region, in the counties of Modoc, Lassen, Plumas, Sierra, and Mono, farmland decreased during the Second World War, probably reflecting the lack of accessibility and the high transportation costs of this region. Farm acreage peaked in 1920 in Lassen County, in 1925 in Mono and Tuolumne Counties, and in 1935 in Amador, Calaveras, and El Dorado Counties. For the other Sierra Nevada counties, the greatest expansion of farmland occurred after the Second World War: in 1945 for Mariposa County; in 1954 for Placer, Plumas, Shasta, and Sierra Counties; in 1959 for Nevada County; in 1964 for Modoc; and as recently as 1969 for Inyo County (figure 17.10) (U.S. Bureau of the Census 1952, 1957, 1961, 1967, 1972).

Sierra Nevada farmland constituted 8.6% of California's total in 1860 (U.S. Bureau of the Census 1864), rose to 12.4% in 1959 (U.S. Bureau of the Census 1961), and then gradually declined to 11.1% in 1992 (U.S. Bureau of the Census 1994). Within the SNEP study area, farmland has long constituted a higher proportion of the total land area than in the state as a whole despite its relatively low productivity. In 1945 the percentage of farmland in the SNEP area was 18.2% (U.S. Bureau of the Census 1947), rising to 20.1% in 1950 (U.S. Bureau of the Census 1952) and falling from this peak to 14.3% in 1992 (U.S. Bureau of the Census 1994). By 1992 only the central foothill counties of Amador and Calaveras had a higher proportion of their land in farms than the state average, indicating an expansion of agricultural activities as urbanization pressures pushed farmers out of the Central Valley into the foothills.

Harvested Cropland

Much of Sierra Nevada farmland, that is, the total land in farm holdings, is not cropped, and even land with crops is not always harvested. The census definition of harvested cropland includes land from which crops were harvested in the census year. If two or more crops were harvested from the same land during the year, the acres are counted for each crop; therefore, the total acres of all crops harvested generally exceeds the acres of cropland harvested. The exception to this procedure is that for hay crops, whose acres are counted only once even if more than one cutting of hay was taken. If a crop was planted but not harvested, then these acres are not reported as harvested cropland. Land with crops grown purposefully for grazing is also reported as cropland harvested. Acres with bearing or nonbearing fruit and nut trees or vines are counted as harvested whether the crop was harvested or failed.

Agricultural production was becoming more intensive, and between 1924 and 1944 there was an increase of 31.7% in the amount of cropland harvested in California. However, in the mountains only Lassen County, at 61%, showed an increase in harvested acres greater than that of the state as a whole, and many of the foothill farms harvested crops from a smaller portion of their land. Mono and Inyo Counties also saw marked declines in cropland harvested because of the special situation of competition for water resources with Los Angeles. In the immediate postwar five years, as wartime demand for food disappeared, harvested cropland declined throughout the Sierra Nevada, except in Modoc, where the draining of Tulelake opened up fertile new land for homesteading. The greatest declines occurred in Shasta and Inyo Counties. At the same time, California as a whole increased its harvested cropland by 5.6%. The end of the Second World War marked the point at which the Sierra Nevada became most clearly marginalized in terms of agricultural development compared with the rest of the state. By 1992, in California 27% of farmland was in harvested cropland. The proportion of harvested cropland in the SNEP study area remained fairly constant between 1949 and 1992 at just over 7% of farmland. Placer county had 17% of its farmland in harvested cropland and Modoc 15% but all other counties harvested crops from less than 10% of their farmland.

Crop and Livestock Specialist Areas

Beef production has been important since the gold rush days, with Modoc producing the most in 1992. Calaveras and Amador Counties were the leading cattle counties in 1992 in the SNEP ecoregion, and the number of cattle increased in the 1980s (figure 17.1). In the foothills, overgrazing and repeated burnings have reduced the value of some pastures, and chaparral has engulfed many abandoned farms. Several smaller foothill properties combine feeding and grazing, producing dry-farmed grains and hay. The Sierra Nevada study area has 37% of the state's farms with grazing permits. Modoc has more of such farms than any other California county (U.S. Bureau of the Census 1992). Two-thirds of the fifteen counties studied have more irrigated pastureland than irrigated cropland. Some of this land now produces grass more intensively for commercial turf, as in Sierra and Inyo Counties, or has been turned into golf courses.

In addition to cattle, horses have long been important on SNEP area farms (figure 17.11). Percherons could still be seen working on farms into the 1960s, as noted in Modoc County (Pease 1965, 154). The number of heavy horses declined after the First World War, and by 1964 horses were no longer recorded in the census. However, riding horses are now very important in the region, and ranches for both horse breeding and recreational use of horses have proliferated since the 1970s. This modern "horsiculture" is found especially in the foothill counties, with Placer as the leading county. Altogether the SNEP study area had 14% of the state's horse farms in 1992 (U.S. Bureau of the Census 1994).

In the foothills, climate permits a variety of crops, especially in the so-called thermal belt, generally between 200 and 1,200 feet (Peters et al. 1995, 351). Orchards needing cooler weather are located on mostly nonirrigated land just above this level, with plums and cherries generally grown between 1,000 and 2,500 feet, pears between 1,500 and 3,500 feet, and apples between 2,000 and 4,000 feet (El Dorado County Agricultural Commissioner 1968). Pears were hit by disease in 1960, with Placer County losing most of its trees and El Dorado almost half (El Dorado County Agricultural Commissioner 1961, Placer County Agricultural Commissioner 1961). El Dorado County has been an important orchard area since the gold rush days, but Calaveras saw a large increase in fruit-tree planting in 1956 (Calaveras Agricultural Commissioner 1957). Harvest seasons are extended by planting many varieties of these fruit trees (El Dorado County Chamber of Commerce 1994). In 1992 the SNEP study area had 16% of California's pear farms, 15% of the plum farms, and 15% of the apple-producing farms, mainly in El Dorado County.

Proximity to Sacramento and the Bay Area and good road access mean that farmers in El Dorado County can focus on direct farmgate and U-pick sales and on on-farm value added to the product through bottling of wine, drying of fruit, and making of pies, jams, and preserves. Cut flowers, plants, and Christmas trees are also produced for direct sale. This on-farm marketing has led to the development of a combination of agriculture and recreation on many foothill farms. This concept was first developed in the Apple Hill area in 1966 and extended into the Somerset and Georgetown areas in 1983 and is now widespread (El Dorado County Agricultural Commissioner 1963). In 1992 14% of Sierra Nevada farms were involved in direct sales, more than twice the state average (U.S. Bureau of the Census 1994).

Grape growing began in El Dorado County in the early 1970s and was focused on the tourist market from the beginning. In 1981 five wineries were recorded, and by 1987 the

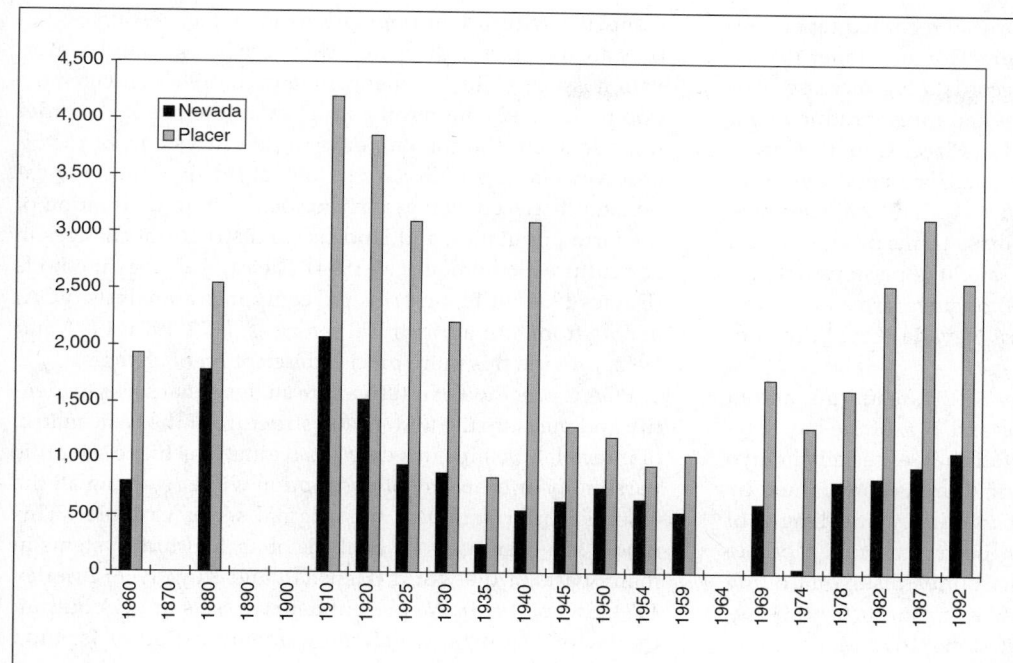

FIGURE 17.11

Number of horses in Nevada and Placer Counties, 1860–92.

county had fourteen wineries, mainly near Placerville and Somerset in the southern part of the county (El Dorado County Agricultural Commissioner 1963–).

Christmas trees are grown in the northern part of El Dorado County, with thirty-eight specialist farms listed by the Chamber of Commerce in 1994 (El Dorado County Agricultural Commissioner 1963–). Christmas tree production started in

1967 in El Dorado, followed by Sierra, Plumas, Calaveras, and Nevada Counties in 1969 (Calaveras County Agricultural Commissioner 1950–; Nevada County Agricultural Commissioner 1962–; Plumas-Sierra County Agricultural Commissioner 1960–). El Dorado and Plumas are the leading counties, with El Dorado increasingly concentrating on "choose and cut" customers (figure 17.12). Christmas trees were grown as

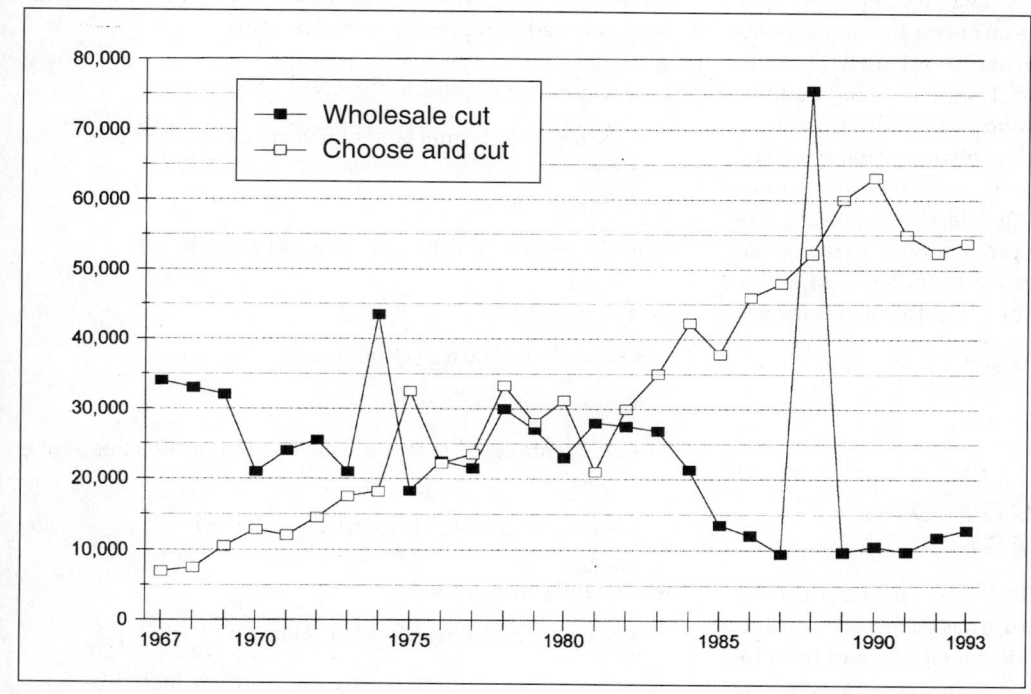

FIGURE 17.12

Number of Christmas trees grown in El Dorado County, 1967–93.

early as 1953 in Placer County but are not recorded separately after 1971 (Placer County Agricultural Commissioner 1946–). The 1992 Census of Agriculture records a big increase over 1987 in the value of Christmas trees and forest products sold and the number of farms involved in Placer County. Christmas trees were grown in Mono County for a short period in the mid-1980s (Inyo-Mono Counties Agricultural Commissioner 1974). Very few (1.3%) California farms produce forest products and Christmas trees for sale. It is perhaps not surprising that almost one-fifth (19%) of such farms are found on the forested slopes of the Sierra Nevada (U.S. Bureau of the Census 1994).

Hay is a major product on many SNEP area farms, and in 1992 the area had 10% of the farms and 11% of the hay acreage in California. Much of this is alfalfa hay, especially in Inyo and Modoc. At this elevation, agriculture is constrained by climatic limits, particularly frost frequency and length of growing season. Modoc has long been a specialist potato and onion producer, with Siskiyou County (Siskiyou County 1961–); potatoes are also grown in Shasta (Shasta County Agricultural Commissioner 1949) and Mono (Inyo-Mono Counties Agricultural Commissioner 1974–) and were introduced into Inyo County in 1979 (Inyo-Mono Counties Agricultural Commissioner 1980). Today Inyo and Mono also grow onions and garlic (Inyo-Mono Counties Agricultural Commissioner 1974–), while Modoc grows horseradish and sugar beet in the Tulelake basin (Shannon 1995).

Agrodiversity may be measured by both crop-livestock diversity and diversity of management strategies. On both measures Sierra Nevada agriculture is becoming more diverse. As California agriculture has grown more specialized, the Sierra Nevada has remained an area of semisubsistence and part-time farms producing a great range of crops and livestock, including many exotics. Off-farm employment has enabled many marginal small farms to survive. Farmers practicing organic agriculture have been increasing in number over the last decade. Isolation and inaccessibility are now being used to advantage by producers of marijuana, and some people think this illegal crop now may well be one of the most valuable in the region. However, the Sierra Nevada is probably less important for marijuana production than the northern coastal area of the state because of its less hospitable climate. Recently the production of hemp for fiber has been legally permitted in the Sierra Nevada.

FACTORIAL ECOLOGY OF AGRO-ECOSYSTEMS

The discussion so far has described how the structure and land use of Sierra Nevada agriculture has changed over time. Agriculture is marginal in much of the region, and farm income is lower than in the rest of the state. Farmers have long competed with urban populations for water resources, but they are now facing pressures from both housing and conservation issues. Counterurbanization in the 1980s changed the population and community structure in the region. In order to understand the contemporary agro-ecosystems of the Sierra Nevada, it is necessary to look at the quantitative relationship between changes in the socioeconomic situation of the farm population and land-use and structural changes in agriculture (Schulman et al. 1994). Factorial ecology methods (Davies 1984) utilizing principal components analysis (PCA) of data from four agricultural censuses, 1959, 1974, 1982, and 1992, provide this more precise description of change.

PCA is a method of multivariate analysis that seeks to identify and measure the underlying structure of the basic matrix of interrelationships in a data set. It aims to achieve scientific parsimony or economy of description while retaining all the essential information of the original set of variables. This model has been used for analysis of agricultural systems in many parts of the world (Henshall and King 1966; Brierley 1974; Swope 1995). We present here two sets of PCA output: the factor loadings, which allow identification of farming types based on the associations between the variables and the factors, and the factor scores, which measure how each county is related to the factors. In this way we are able to identify both farming types and regional farming systems.

Because this research was limited to preexisting information, the analysis is based on census data at the county level. We have included the fifteen counties that lie wholly or mostly within the SNEP study area. Because the model requires fewer variables than observations, we were able to input a maximum of fourteen variables. A further restriction was the need to choose variables that were available for all fifteen counties for all four study years. Within these parameters, variables were selected to represent six major aspects:

1. Farmer characteristics

   - Number of farmers over 65 years old (variable 12)

2. Labor input

   - Percentage of full-time farmers (variable 13)

3. Farm structure

   - Number of farms (variable 1)

   - Average size (variable 3)

   - Percentage of farms of ten to forty-nine acres (variable 8)

   - Percentage of land tenanted (variable 11)

4. Intensity of production

   - Farms using irrigation (variable 7)

   - Percentage of farms using fertilizer (variable 14)

5. Livestock

- Number of cattle (variable 9)

- Number of farms with sheep (variable 10)

6. Land use

- Acres of farmland (variable 2)

- Acres of woodland on farms (variable 6)

- Number of farms with cropland and pastureland (variable 5)

- Number of farms with harvested cropland (variable 4)

These data are shown in tables 17.2, 17.3, 17.4, and 17.5.

The study years were chosen to highlight key stages in the recent development of Sierra agriculture. In 1959 there was a change in the definition of a farm, and Sierra agriculture had not yet been influenced by the major road building that opened up the Sierra Nevada. In 1974 the definition of a farm was changed once again and data from this year is directly comparable with that for later years; this year also illustrates the early period of diversification, modernization and pluriactivity. The decade spanned by the censuses of 1982 and 1992 was one of rapid rural residential development.

Farm Systems

In each of the four years considered, three factors were extracted that, taken together, explained 82.3% of the total variance between counties in 1959, 87.7% in 1974, 84.25 in 1982, but only 77.4% in 1992, indicating growing complexity in the system since 1974. In all four years the first factor, which explained more than half the total variation in every year except 1992, was associated with intensive crop and livestock production on smaller farms, while the second factor was identified with large holdings raising cattle. The third factor varied from year to year but was always linked to an aspect of farm structure. The identification of these factors is based on the presence of high negative or positive "loadings" for the variables on each factor, as shown in figures 17.13–17.16. The analysis of these factor loadings allows two underlying dominant elements in the structure of Sierra Nevada farming to be recognized. These elements may be considered basic agro-ecosystems. One is associated with relatively intensive crop and livestock production and one with extensive cattle ranching.

Although there is general stability over time of the major agro-ecosystems, the weight of variables associated with these factors varies from year to year. In all years the first factor, which is associated with the dominant type of farming in the SNEP region, is identified by high loadings for the total number of farms, the number of farms of ten to forty-nine acres, the number of farms with harvested acreage and irrigated land, and the percentage of farmland fertilized and the number of sheep farms. The negative loading for average farm size declines over time, indicating a weaker relationship between small farms and intensive agriculture in 1992 than in 1959. The woodland acres variable is linked strongly and positively to Factor 1 in 1959 and 1982 and to Factor 2 in 1992 but quite weakly to all three factors in 1974, suggesting that there is no stable relationship between woodlots and other types of agricultural enterprise on farms in the SNEP area.

The number of sheep farms variable loads on both Factors 1 and 2 in 1959 but in the following three study years it is very strongly linked to Factor 1. The distribution of hogs and pigs was found, in a separate analysis, to follow a similar pattern. This association indicates an increasing division between cattle ranching and mixed farming with small stock and horses from 1974 onward.

Factor 2, which measures the second most important farming type, has high loadings for the number of acres in farmland in a county and the number of cattle in all four analyses. In the 1959, 1974, and 1982 studies, the percentage of acres fertilized is also positively linked to this factor, but in 1992 the link is negative. The addition of a high loading for woodlots and for older farmers to this factor in 1992 suggests declining intensity of production on cattle ranches.

Factor 3, associated with a minor farming type, in 1959 has high positive loadings for the proportion of farmland tenanted in the county and the proportion of farmers over 65 years of age but negative loadings for full-time farmers. Thus it seems to identify counties with older, part-time tenant farmers. In 1974, 1982, and 1992, there is a negative relationship between tenancy and older farmers, suggesting that younger farmers have been taking up tenant farms in recent years. In 1974, 1982, and 1992, Factor 3 is associated with larger farms. In 1992 this factor is identified with large full-time farms with a secondary link with cattle, and the earlier association with tenancy and older farmers has disappeared. The overall proportion of elderly farmers and of tenant farms has increased in the Sierra Nevada between 1959 and 1992 so that by 1992 it is no longer seen as being distinctive to any particular type of farming.

Full-time farming is strongly negatively linked to elderly and tenant farmers in 1959. In the 1974 study, full-time farming is closely and positively associated with small-farm, mixed-farming areas. By 1982 this former link has become negative, and full-time farming is associated with cattle ranching in counties with a higher proportion of farmland. In the 1992 study, full-time farming is seen as being associated only with large farms, suggesting the increasing importance of economies of scale for profitable farming.

Farming Areas

The factor scores for all four study years show the changing importance of types of farming in different parts of the SNEP study area (tables 17.2–17.5). Thus factor scores provide a spatial grouping of types of farming based on the analysis of

FIGURE 17.13

Rotated factor loadings, 1959 analysis.

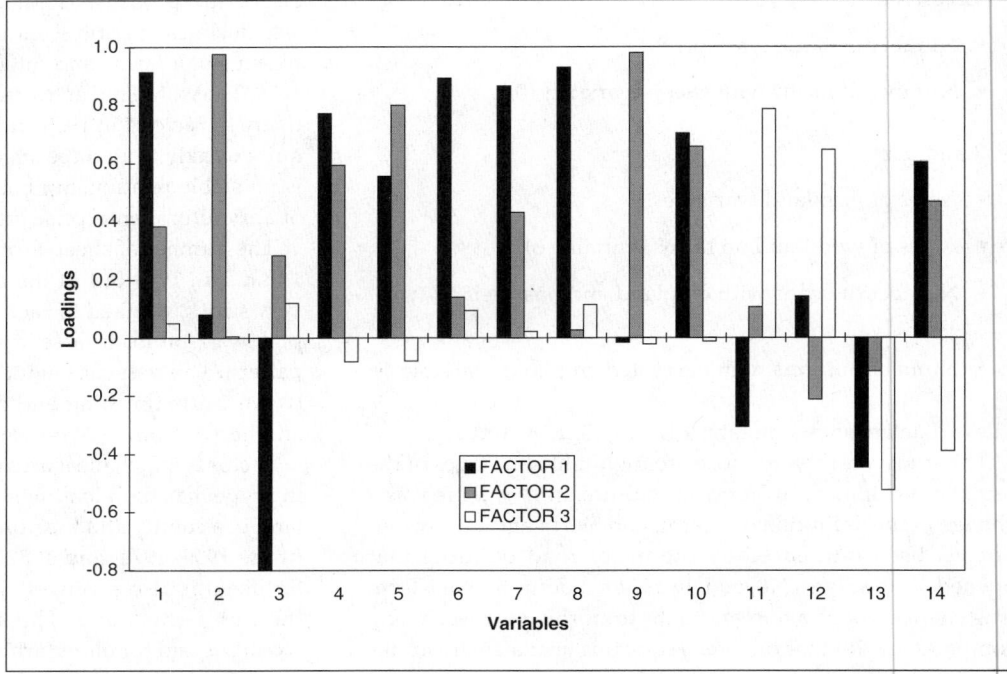

the original fourteen variables for the fifteen counties. In 1959 scores on Factor 1 suggest that the major spatial division in the region's farming is between central and northern foothill counties and eastern counties, with scores on Factor 2 recognizing a secondary division between the counties of Lassen, Modoc, and Shasta in the north and the rest of the study area. This identification of a separate farming area in the north appears for each year studied but is less distinctive in 1992, when the northern counties are linked on the third factor with Inyo and Mono for the first time, perhaps indicating the grow-

FIGURE 17.14

Rotated factor loadings, 1974 analysis.

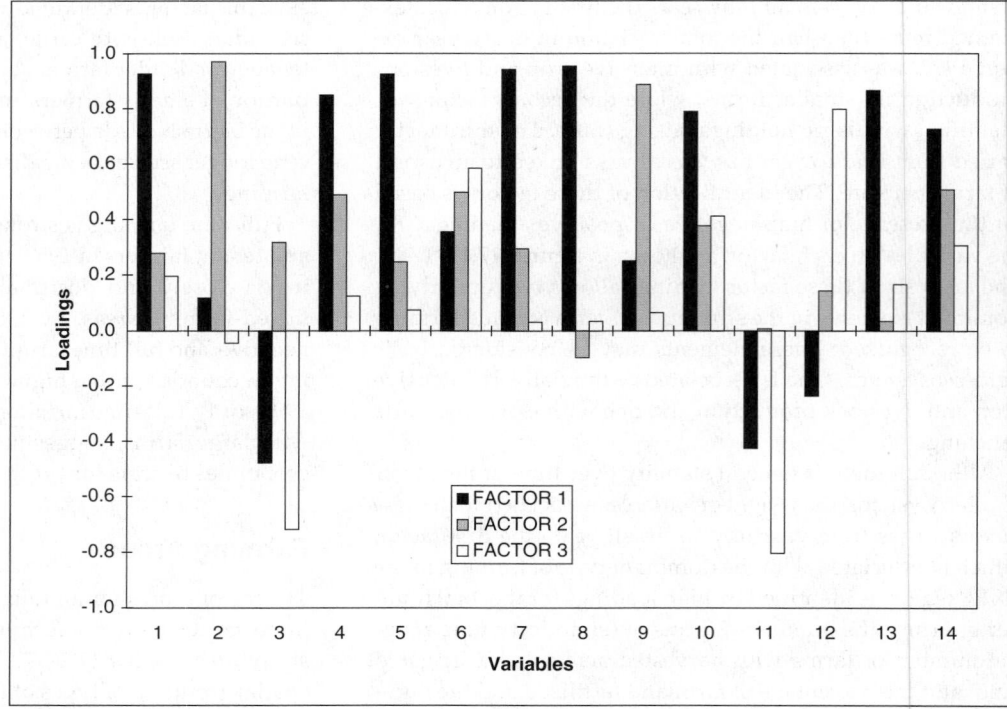

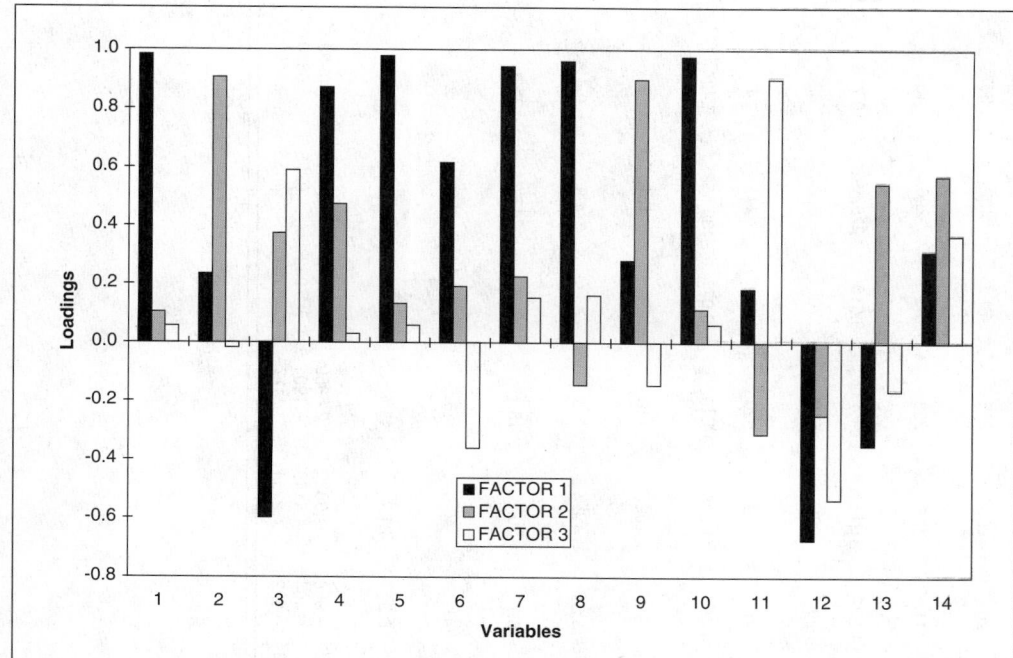

FIGURE 17.15

Rotated factor loadings, 1982 analysis.

ing importance of relative accessibility for agriculture. The third factor in 1959 identifies a minor division between the high-altitude north and the south.

By 1992 the major division is between the counties with many irrigated farms, that is, the northern county of Shasta and the central foothill counties of El Dorado and Placer, and the more rugged southern counties of Mariposa and Tuolumne and the high-elevation counties. The secondary division in 1992 is between the counties with large farms (Modoc, Lassen, Mono, and Inyo) and those with many small

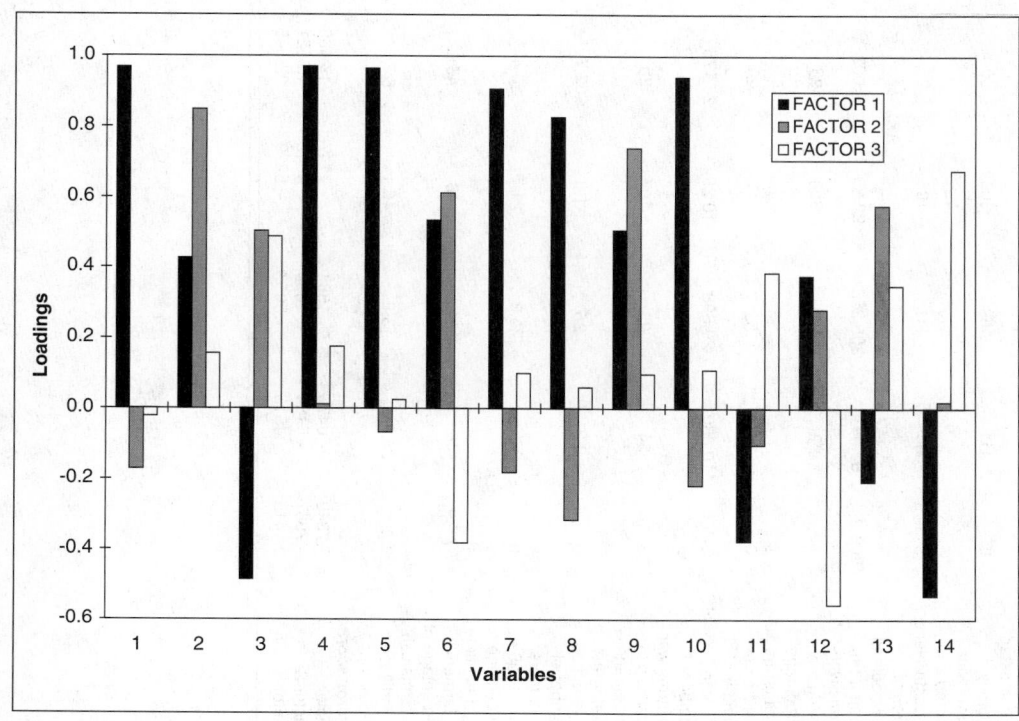

FIGURE 17.16

Rotated factor loadings, 1992 analysis.

TABLE 17.2

Analysis of farming areas, 1959.

Variables	County														
	Alpine	Amador	Calaveras	El Dorado	Inyo	Lassen	Mariposa	Modoc	Mono	Nevada	Placer	Plumas	Shasta	Sierra	Tuolumne
Number of farms	7	263	356	497	104	323	276	600	32	448	1213	103	880	49	308
Farmland acres	12,050	219,968	342,867	194,770	329,782	692,053	204,200	1,035,911	89,938	227,152	258,825	95,465	603,166	92,189	194,233
Average size of farm	1,721	836	963	392	3,171	2,413	1,030	1,727	2,811	507	213	927	686	1,881	631
Farms with harvested crops	5	111	159	306	60	266	59	545	20	257	685	71	402	36	98
Farms with crop and pastureland	5	65	111	105	21	199	47	295	11	115	208	38	296	18	54
Acres of woodland on farms	860	63,491	205,252	98,857	9,002	102,839	87,085	86,140	1,527	81,157	644,432	9,518	275,562	10,718	84,882
Farms with irrigation	7	82	163	309	93	198	55	485	29	336	972	70	688	32	107
Percentage of farms 10–49 acres	0	34	52	117	20	28	56	12	0	139	523	12	241	4	62
Number of cattle	1,315	10,880	20,598	10,394	26,176	57,775	15,991	80,467	4,456	11,875	16,490	13,077	38,552	5,494	10,533
Number of farms with sheep	3	64	93	107	23	98	49	161	9	102	140	16	161	9	52
Percentage of farms tenanted	0	6.2	2.7	1.6	32.7	6.7	3.6	1.5	3.2	1.8	7.3	7.1	4.9	9.4	3.5
Number of farmers over 65	0	26.2	25.3	21.7	20.4	15.3	19.3	11.0	6.7	20.1	15.1	23.5	17.3	22.9	23.8
Percentage of full-time farmers	57	46	44	48	43	53	49	52	47	40	46	50	39	66	49
Percentage of farms using fertilizer	28.6	24.7	13.8	45.1	8.7	24.8	10.5	47.3	6.3	21.7	45.4	21.4	30.3	4.1	5.8
Factor scores															
Factor 1	−.7349	.0475	.1584	.7147	−1.2563	−.6736	−.2420	−.1283	−.9582	.5929	2.6859	−.3689	1.1816	−.8932	−.0305
Factor 2	−.7967	−.5184	−.1869	−.4045	.5821	1.5476	−.5700	2.6460	−.4646	−.4149	−.1901	−.7092	.9275	−.6680	−.7599
Factor 3	−2.1415	.4515	.5657	−.5229	2.5505	−.2483	.0072	−1.1587	−.5225	.2254	.1117	.1361	.4685	−.2160	.2934

TABLE 17.3

Analysis of farming areas, 1974.

									County							
Variables	Alpine	Amador	Calaveras	El Dorado	Inyo	Lassen	Mariposa	Modoc	Mono	Nevada	Placer	Plumas	Shasta	Sierra	Tuolumne
Number of farms	4	194	275	379	79	296	161	433	43	169	813	89	679	28	172
Farmland acres	6,525	197,910	254,626	197,619	455,078	631,268	231,755	653,185	74,419	91,483	167,705	133,033	386,479	77,511	122,532
Average size of farm	1,631	1,020	926	521	5,760	2,133	1,439	1,509	1,731	541	206	1,495	569	2,768	712
Farms with harvested crops	3	81	102	196	39	233	23	370	26	69	435	58	379	20	26
Farms with crop and pastureland	3	85	96	155	25	177	46	156	16	67	383	54	349	14	74
Acres of woodland on farms	100	53,748	18,715	52,857	145	54,164	9,270	37,528	1,113	27,642	18,325	5,891	63,313	3,360	11,383
Farms with irrigation	4	44	88	199	68	185	37	335	36	108	608	45	450	22	62
Percentage of farms 10–49 acres	1	25	41	129	18	41	15	21	6	60	399	12	218	0	30
Number of cattle	1,990	17,190	25,769	12,732	23,355	62,399	27,144	122,715	6,405	5,189	30,819	13,781	46,385	8,982	16,050
Number of farms with sheep	1	35	46	51	6	41	27	65	7	34	80	9	64	2	32
Percentage of farms tenanted	25.00	11.30	10.20	7.40	34.20	6.40	12.40	9.90	11.60	8.90	5.80	7.90	10.20	21.40	8.10
Number of farmers over 65	1.00	32.47	23.64	21.90	10.13	22.64	24.84	17.78	9.30	25.44	15.50	19.10	15.17	28.57	19.19
Percentage of full-time farmers	2.00	95.00	119.00	139.00	21.00	127.00	71.00	127.00	227.00	74.00	297.00	48.00	237.00	15.00	65.00
Percentage of farms using fertilizer	1.00	26.29	18.18	41.16	17.72	29.59	20.63	52.19	23.26	27.81	46.86	17.05	43.22	7.14	17.44
Factor scores															
Factor 1	-.4279	-.6561	-.1990	.4513	-.7111	-.2259	-.6450	.2382	-.0414	-.2881	2.7888	-.5588	1.6551	-1.0419	-.3882
Factor 2	-1.1282	.0851	-.1434	-.1703	.9585	1.5841	-.1021	2.4500	-1.0105	-.6383	-.7453	-.5800	.5305	-.4167	-.6737
Factor 3	-1.5541	1.3949	.5394	.8826	-2.5466	.4793	.3848	.0270	-.6030	.9374	-.4196	.1808	-.0039	-.1719	.4550

TABLE 17.4

Analysis of farming areas, 1982.

							County								
Variables	Alpine	Amador	Calaveras	El Dorado	Inyo	Lassen	Mariposa	Modoc	Mono	Nevada	Placer	Plumas	Shasta	Sierra	Tuolumne
Number of farms	5	359	405	721	97	418	250	536	70	363	1,335	112	990	61	254
Farmland acres	7,352	198,135	214,881	146,644	300,594	555,958	231,183	747,787	77,731	79,402	182,792	103,289	405,180	53,373	110,680
Average size of farm	1,470	552	531	203	3,099	1,330	925	1.395	1,110	219	137	922	409	875	436
Farms with harvested crops	1	147	121	309	39	306	53	449	42	140	514	70	529	38	51
Farms with crop and pastureland	2	126	147	229	25	189	62	206	23	141	550	54	411	27	90
Acres of woodland on farms	0	13,988	31,335	13,439	1	22,051	19,461	12,360	1	14,538	13,977	5,958	87,105	3,886	13,639
Farms with irrigation	4	99	121	371	64	292	53	430	54	234	936	66	707	32	97
Percentage of farms 10–49 acres	0	96	128	309	21	108	45	62	21	158	581	22	327	12	67
Number of cattle	1,108	31,016	28,232	12,707	19,489	64,629	29,906	98,802	8,093	8,717	29,767	15,139	47,564	14,577	14,033
Number of farms with sheep	0	41	70	112	10	75	29	80	9	60	192	22	131	8	32
Percentage of farms tenanted	20.00	5.04	10.84	20.62	40.88	5.98	6.08	4.46	11.32	21.53	32.95	4.48	11.51	15.19	7.59
Number of farmers over 65	20.00	23.96	22.47	17.06	18.56	16.51	24.00	18.66	27.14	20.11	14.46	23.21	16.16	19.37	23.23
Percentage of full-time farmers	40.00	31.48	29.38	27.18	25.77	36.60	26.40	43.84	48.57	31.40	27.12	28.57	30.40	31.15	31.10
Percentage of farms using fertilizer	40.00	16.16	15.06	30.37	19.59	21.77	25.20	45.71	22.86	22.87	33.56	16.07	34.34	22.95	3.15
Factor scores															
Factor 1	-1.0419	-.1944	.0912	.9134	-1.0231	.0563	-.4471	-.0132	-1.0722	.1573	2.3653	-.5843	1.7696	-.6690	-.3079
Factor 2	-.0424	-.2544	-.4330	-.6907	.0156	1.3901	-.1345	2.8981	.0123	-.8386	-.6214	-.5535	.6624	-.4923	-.9176
Factor 3	1.0253	-1.0287	-.8368	.4514	2.5767	-.1804	-.6382	-.0313	-.5199	.0769	1.2452	-.6434	-.6596	.2448	-1.0820

TABLE 17.5

Analysis of farming areas, 1992.

Variables	Alpine	Amador	Calaveras	El Dorado	Inyo	Lassen	Mariposa	Modoc	Mono	Nevada	Placer	Plumas	Shasta	Sierra	Tuolumne
								County							
Number of farms	6	367	438	690	79	312	256	466	73	415	1,125	125	844	53	249
Farmland acres	4,768	236,222	246,077	102,028	247,550	487,499	206,138	686,876	103,294	72,471	137,723	119,514	388,084	55,446	137,530
Average size of farm	795	644	562	148	3,134	1,562	805	1,474	1,415	175	122	956	460	1,046	552
Farms with harvested crops	2	157	118	348	28	176	40	320	44	161	423	53	396	19	49
Farms with crop and pastureland	4	122	127	157	22	134	56	201	21	148	444	72	363	26	74
Acres of woodland on farms	0	24,482	14,124	0	423	23,188	17,197	27,795	1	9,698	10,157	7,888	36,289	6,650	20,176
Farms with irrigation	4	125	106	364	56	6	37	338	58	275	783	65	594	29	94
Percentage of farms 10–49 acres	1	106	121	308	14	356	63	46	14	182	469	22	272	3	52
Number of cattle	1,213	47,812	34,658	11,355	17,837	10,381	26,410	92,986	10,402	9,630	27,990	16,627	45,050	6,909	13,685
Number of farms with sheep	0	34	45	82	9	46	21	57	6	48	147	15	74	3	11
Percentage of farms tenanted	.00	6.50	12.30	12.50	30.80	11.80	7.50	3.50	46.40	15.70	16.20	6.80	12.60	47.80	9.70
Number of farmers over 65	.00	33.50	26.90	27.40	22.80	24.70	31.30	27.30	23.30	29.20	28.00	30.40	29.10	24.50	31.70
Percentage of full-time farmers	33.30	35.40	35.80	34.50	39.20	41.40	42.60	48.10	57.50	39.50	33.10	34.40	38.20	35.80	36.90
Percentage of farms using fertilizer	33.30	26.40	12.60	40.40	21.50	39.70	3.90	36.90	20.50	21.00	30.80	18.40	35.30	17.00	5.20
Factor scores															
Factor 1	-.3655	-.3202	-.3536	1.2620	-.5930	.4624	-1.0847	.4739	-.5697	.0925	2.3504	-.8077	1.3310	-.8061	-1.0716
Factor 2	-1.7712	1.1195	.6162	-.8748	-1.0498	-.2076	1.0030	1.4146	-1.2601	-.0845	-.2572	.2269	1.0103	-.8294	.9440
Factor 3	-.2565	-.2144	-.5386	-.6965	1.2380	1.1247	-.4801	2.2946	1.1447	-.8807	-1.0191	-.6712	.2807	-.2835	-1.0419

farms, such as Placer County. Thus in 1992 there is some indication that variation in farm structure is a more important basis for the identification of farming areas than environmental differentiation.

The PCA study provides a quantitive measure of the changes in farming types and areas in the Sierra Nevada between 1959 and 1992. There is some indication that over time, accessibility and farm structure have become more important than altitude in differentiating within the region. The basic division between extensive ranching and intensive crop agriculture can be identified in all four years studied, but small stock are increasingly associated with mixed farming, while the link between small farms and intensive agriculture has weakened over time. Farmer characteristics such as age and part-time farming have become less important as differentiating features among the region's counties.

CONCLUSION

The restructuring of agriculture came later to the Sierra Nevada than to many parts of California but is now well advanced. Most areas show increased intensity of inputs, diversity of products, and farm operator pluriactivity. Farming in the Sierra Nevada is often integrated with other economic activities, such as the timber industry and tourism, but farm incomes are generally lower than elsewhere in the state.

The Sierra Nevada is environmentally marginal for agriculture. It has become economically more marginal to the state but socially more integrated. A century ago it was a substantial contributor to the state's agricultural output, but today production from the region is of minor importance. There are still many small farms in the region, and the increasing specialization of agriculture is maintaining agrodiversity. Regional differences in the structure of agriculture, such as the prevalence of part-time farming and elderly and tenant farmers, have become less marked.

Several changes have occurred in the use of land, and it appears that these changes may be accelerating. Rapid urbanization elsewhere in the state is pushing farmers into the Sierra Nevada, where they must concentrate on intensive production of high-value items in order to have a viable farm. The value of farmland and farm buildings in the foothills increased more rapidly than in most parts of California between 1987 and 1992, especially in Calaveras and Amador Counties. Such high valuations reflect amenity values rather than a value based on returns from agriculture and are indicative of the demand for rural residential sites in these areas. A smaller proportion of farmland was lost to urbanization between 1988 and 1992 in the foothills of the Sierra Nevada than in the Central Valley and the Los Angeles conurbation (California Department of Conservation 1992). Unfortunately, it is impossible to document the rate or even the direction of change over a longer period or a wider area because of the incompatibility of the existing land-use surveys. A comparison of figures 17.6 and 17.9 for Placer County shows the major urbanization that occurred around Roseville and Auburn between 1970 and 1988, much of it at the expense of cropland. The agricultural census provides the only long-term source of comparable land-use data, but it is based on an aggregation of individual farm data, some of which is omitted because of the need for confidentiality, and only deals with land use on farms. A new land-use survey of the whole area is vital for comparison with the 1970 situation. Such changes as urbanization and deforestation may be more or less extensive than popularly believed, but if environmental damage is to be minimized it is necessary to identify the location and the rate of change.

REFERENCES

Amador County Agricultural Commissioner. 1948, 1950–. *Amador County report of agriculture.* Jackson, CA: Office of Agricultural Commissioner, Director of Weights and Measures.

Bertrand, J.-M., and Hulot, J-F. 1990. *Farms in mountain and less-favoured areas of the community.* Luxembourg: Commission of the European Communities.

Brierley, J. S. 1974. *Small farming in Grenada, West Indies.* Manitoba Geographical Studies 4. Winnipeg: University of Manitoba.

Brookfield, H. C., and Padoch, C. 1994. Appreciating agrodiversity: A look at the dynamism and diversity of indigenous farming practices. *Environment* 36 (5): 6–11, 37–45.

Burcham, L. T. 1957. *California range land: an historical-ecological study of the range resource of California.* Sacramento: California Department of Natural Resources, Division of Forestry.

Butte County Agricultural Commissioner. 1946–. *Butte County agricultural crop report.* Oroville, CA: Office of Agricultural Commissioner, Weights and Measures.

Calaveras County Agricultural Commissioner. 1950–. *Calaveras County report of agriculture.* San Andreas, CA: Department of Agriculture and Weights and Measures.

California Department of Conservation. Office of Land Conservation. 1992. *A guide to the Farmland Mapping and Monitoring Program.* Sacramento.

California Farmer. 1995. *California at a glance. Crop year: 1994.* 2300 Clayton Road, Suite 1360, Concord, CA 94520.

Champion, A. G., ed. 1989. *Counterurbanization: The changing pace and nature of population deconcentration.* London: Edward Arnold.

Coy, O. C. 1923. *The genesis of California counties.* Sacramento, CA: State Printing Office.

Davies, W. K. D. 1984. *Factorial ecology.* Aldershot, Hants: Gower.

Donley, M. W., S. Allan, P. Caro, and C. C. Patton. 1979. *Atlas of California.* Portland, OR: Academic Book Center.

Duke, V. 1939 *The South Fork Valley and other stories.* Monterey, CA: W. T. Lee (printed 1946).

El Dorado County Agricultural Commissioner. 1946–61, 1963–. *El Dorado County agricultural crop report.* Placerville, CA: El Dorado County Department of Agriculture.

El Dorado County Chamber of Commerce. 1994. *Ranch marketing.*

Agriculture and Tourism Councils of the El Dorado County Chamber of Commerce, Placerville, California.

Farquhar, F. P., ed. 1949. *Up and down California in 1860–1864: The Journal of William H. Brewer.* Berkeley and Los Angeles: University of California Press.

Henshall, J. D., and L .J. King. 1966. Some structural characteristics of peasant agriculture in Barbados. *Economic Geography* 38: 183–95.

Hoffmann, S. A., and L. Fortmann. 1996. Poverty in forested counties: An analysis based on Aid to Families with Dependent Children. In *Sierra Nevada Ecosystem Project: Final report to Congress,* vol. II, chap. 14. Davis: University of California, Centers for Water and Wildland Resources.

Inyo-Mono Counties Agricultural Commissioner. 1974–. *Inyo-Mono Counties annual crop and livestock report.* Bishop, CA: Inyo-Mono Department of Agriculture.

Kahrl, W. L., W. A. Bowen, S. Brand, D. L. Fuller, D. A. Ryan, and M. L. Shelton. 1978. *The California water atlas.* Sacramento: Governor's Office of Planning and Research and the California Department of Water Resources.

Larson, D. J. 1996. Historical water-use priorities and public policies. In *Sierra Nevada Ecosystem Project: Final report to Congress,* vol. II, chap. 8. Davis: University of California, Centers for Water and Wildland Resources.

Lassen County Agricultural Commissioner. 1946–. *Lassen County crop report.* Susanville, CA: Lassen County Department of Agriculture.

Madera County Agricultural Commissioner. 1941–. *Agricultural crop report.* Madera, CA: Department of Agriculture.

Mariposa County Agricultural Commissioner. 1978–. *Agricultural crop report: Mariposa County.* Mariposa, CA: Mariposa County Department of Agriculture and Weights and Measures.

Menke, J., C. Davis, and P. Beesley. 1996. Rangeland assessment. In *Sierra Nevada Ecosystem Project: Final report to Congress,* vol. III. Davis: University of California, Centers for Water and Wildland Resources.

Modoc County Agricultural Commissioner. 1942–. *Modoc County annual crop report.* Alturas, CA: Modoc County Department of Agriculture.

Momsen, J. D. 1990. Gender, class, and ethnicity in western Canadian mining towns. *Journal of Women and Gender Studies* 1: 119–34.

Nevada County Agricultural Commissioner. 1962–. *Nevada County annual crop and livestock report.* Grass Valley, CA: Nevada County Department of Agriculture.

Pease, R. W. 1965. *Modoc County: A geographic time continuum on the California volcanic tableland.* Berkeley and Los Angeles: University of California Press.

Peters, G. L., D. W. Lantis, R. Steiner, and A. E. Karinen. 1995. *California.* Dubuque, IA: Kendall/Hunt Publishing.

Placer County Agricultural Commissioner. 1946–. *Placer County agricultural crop report.* Auburn CA: Department of Agriculture.

Plumas-Sierra Counties Agricultural Commissioner. 1960–. *Plumas-Sierra Counties annual crop and livestock report.* Quincy, CA: Plumas-Sierra Counties Department of Agriculture.

Reynolds, L. A. 1996. The role of Indian tribal governments and communities in regional land management. In *Sierra Nevada Ecosystem Project: Final report to Congress,* vol. II, chap. 10. Davis: University of California, Centers for Water and Wildland Resources.

Schulman, M. D., C. Zimmer, and W. F. Danaher. 1994. Survival in agriculture: Linking macro- and micro-level analyses. *Sociologia Ruralis* 34 (2–3): 229–51.

Shasta County Agricultural Commissioner. 1947, 1949–. *Shasta County crop report.* Redding, CA: Shasta County Department of Agriculture.

Shannon, C. 1995. Rotational management of wetlands and cropland in the Tulelake Basin. Unpublished project proposal.

Siskiyou County. 1961–. *Siskiyou County annual crop and livestock report.* Yreka, CA: Siskiyou County Department of Agriculture.

Smith, M. P. 1991. The new geography of land and the end of rural society. In *Frontière et frontières dans le monde anglophone,* edited by J. R. Rougé, 88–104. Paris: Presses de l'Université de Paris—Sorbonne.

Starrs, P. F. 1996. The public as agents of policy. In *Sierra Nevada Ecosystem Project: Final report to Congress,* vol. II, chap. 6. Davis: University of California, Centers for Water and Wildland Resources.

Swope, L. H.. 1995. *Factors influencing rates of deforestation in Lijiang County, Yunnan Province, China: A village study.* Master's thesis. Department of Geography, University of California, Davis.

Terluin, I. J., F. E. Godeschalk, H. von Meyer, J. H. Post, and D. Strijker. 1995. Agricultural income in less favoured areas of the EC: A regional approach. *Journal of Rural Studies* 11 (2): 217–28.

Trussel, M. E. 1989. *Sierra summers: Fireside tales to share with young and old.* Bodega Bay, CA: Talking Mountain Publishing.

Tulare County Agricultural Commissioner. 1941–44, 1946–. *Tulare County agriculture crop and livestock report.* Visalia, CA: Agricultural Commissioner/Sealer.

Tuolumne County Agricultural Commissioner. 1941, 1946–. *Tuolumne County crop and livestock report.* Sonora, CA: Tuolumne County Department of Agriculture.

U.S. Bureau of the Census. 1913. *Thirteenth census of the United States taken in California.* Washington DC: Government Printing Office.

———. 1922. Agriculture reports for states with statistics for counties. In *Fourteenth census of the United States taken in the year 1920,* vol. 6, pt. 3. Washington DC: Government Printing Office.

———. 1927. The western states. In *Reports for states with statistics for counties,* pt. 3. Vol. 19 of *United States census of agriculture: 1925.* Washington DC: Government Printing Office.

———. 1932. The western states. In *Agriculture,* pt. 3. Vol 2 of *Fifteenth census of the United States: 1930.* Washington DC: Government Printing Office.

———. 1939. California: Statistics by counties. In *Census of agriculture: 1935.* Washington DC: Government Printing Office.

———. 1942. Statistics for counties. In *State reports,* pt. 6. Vol. 1 of *Sixteenth census of the United States: 1940. Agriculture.* Washington DC: Government Printing Office.

———. 1952. California. In *Counties and state economic areas,* pt. 33. Vol. 1 of *United States census of agriculture: 1950,* prepared by the Agriculture Division. Washington, DC: Government Printing Office.

———. 1956. California. In *Counties and state economic areas,* pt. 33. Vol. 1 of *United States census of agriculture: 1954,* prepared by the Agriculture Division. Washington, DC: Government Printing Office.

———. 1961. California. In *Counties,* pt. 48. Vol. 1 of *United States census of agriculture: 1959,* prepared by the Agriculture Division. Washington, DC: Government Printing Office.

———. 1967. California. In *State and county statistics,* pt. 48. Vol. 1 of

United States census of agriculture: 1964, prepared by the Agriculture Division. Washington, DC: Government Printing Office.

———. 1972. California. In *Area reports*, pt. 48. Vol. 1 of *1969 census of agriculture*, prepared by the Agriculture Division. Washington, DC: Government Printing Office.

———. 1977. California. In *State and county data*, pt. 5. Vol. 1 of *1974 census of agriculture*, prepared by the Agriculture Division. Washington, DC: Government Printing Office.

———. 1981. California. In *State and county data*, pt. 5. Vol. 1 of *1978 census of agriculture*, prepared by the Agriculture Division. Washington, DC: Government Printing Office.

———. 1984. California state and county data. In *Geographic area series*, pt. 5. Vol. 1 of *1982 census of agriculture*, prepared in the Agriculture Division. Washington, DC: Government Printing Office.

———. 1994. California state and county data. In *Geographic area series*, pt. 5. Vol. 1 of *1992 census of agriculture*, prepared by the Agriculture Division. Washington, DC: Government Printing Office.

U.S. Census Office. 1864. *Agriculture of the United States in 1860*. Washington DC: Government Printing Office.

———. 1883. *Report on the productions of agriculture as returned at the tenth census (June 1, 1880); embracing general statistics and monographs*. Washington DC: Government Printing Office.

U.S. Department of Agriculture (USDA). Soil Conservation Service. 1950. *Generalized classification of land according to its capability for use*. Washington, DC.

U.S. General Land Office. 1870. *Report of the commissioner of the General Land Office, August 1st*. Washington DC.

Weeks, D., A. E. Wieslander, H. R. Josephson, and C. L. Hill. 1943. *Land utilization in the northern Sierra Nevada*. Berkeley: University of California, College of Agriculture, Agricultural Experiment Station.

Yuba County Agricultural Commissioner. 1941, 1946–. *Yuba County agricultural crop report*. Marysville, CA: Yuba County Department of Agriculture.

Zarin, D. J. 1995. Diversity measurement methods for the PLEC clusters. *PLEC News and Views* 4 (March):11–21.

MICHAEL F. DIGGLES AND
NINE OTHER AUTHORS
James R. Rytuba
Barry C. Moring
Chester T. Wrucke
Dennis P. Cox
Steve Ludington
Roger P. Ashley
William J. Pickthorn
C. Thomas Hillman
Robert J. Miller

18

Geology and Minerals Issues

ABSTRACT

This chapter consists of maps and text describing possible undiscovered and known ore deposits, mineral-development issues, public health and safety issues such as asbestos, mercury, and open-pit mines, the role of calcium, as well as the hazards posed by volcanoes and earthquakes in the Sierra Nevada ecosystem. Maps of likely undiscovered and known deposits of gold, copper, sulfide, and others show the present distribution of such ores and are useful in metal-supply issues, water quality studies, and planning for avoiding environmental hazards. Information on exploration shows potential for development and is important to rural communities, the general economy, employment, and mineral supply. Mercury has important effects on plant communities and water systems in the Sierra Nevada ecosystem, with possible consequences for humans. Asbestos occurs naturally in the study area, though its health risk for humans is controversial. Calcium from carbonate rocks is essential for metabolic processes of aquatic life and in the buffering system of waters in the area. Volcanic activity and earthquakes have played a major role in shaping the environment of the study area.

EARTH SCIENCES OVERVIEW

Geology

As part of the U.S. Geological Survey (USGS) National Mineral Resource Assessment Program, an overview of California minerals was produced by Dellinger (1989), excerpts of which are reproduced here. The Sierra Nevada extends 644 km (400 mi) from the Modoc Plateau in the north to the Mojave Desert in the south; it varies in width from 64 to 161 km (40 to 100 mi). The range is highest and most rugged along much of its east side, and overall elevation gradually decreases to the west. The range can be divided into western and eastern belts that have different geological characteristics. The western belt,

along the west edge of the north half of the range, consists of Paleozoic to Mesozoic oceanic, island-arc, and composite terranes that are intruded by Mesozoic quartz dioritic to granodioritic plutons. These three terranes form narrow bands that parallel the trend of the range; generally the oceanic terrane band is closest to the core of the range, and the island-arc terrane band lies along the west edge. The eastern belt of the Sierra Nevada is batholithic terrane, composed almost entirely of Mesozoic granodioritic to granitic plutons that enclose remnants of older Mesozoic, deep-marine volcanic and clastic rocks in its western part and Paleozoic, shallow-marine quartzose and carbonate rocks in the eastern part. The different geologic characteristics of the eastern and western belts are reflected in the different mineral deposit types found in these two regions.

The western belt is historically important as the center of the California gold rush in the middle nineteenth century. The Mother Lode is in the western belt, and about a third of California's total gold production was taken from gold-quartz veins in this region; the belt contains nearly all the large lode-gold deposits in California (Albers, 1981). Virtually the entire western belt is considered geologically permissive for mineral deposits containing gold, chromium, nickel, copper, zinc, manganese, or mercury, and about a third of the belt has numerous known deposits of one or more of these metals. The large disseminated-gold deposit at Jamestown, southwest of Sonora, lies within the western belt. Gold-quartz vein deposits are present in all three terranes of this belt, but the existence of other mineral deposit types is more restricted. Podiform chromite and laterite nickel deposits are associated with ultramafic rocks that occur typically within or at the margins of oceanic and composite terranes. Chert-associated manganese deposits occur mostly in oceanic and composite terranes, but they also occur in a small area of island-arc terrane that lies northeast of the oceanic terrane at the north end of the range. Pyrite-rich massive sulfide deposits are restricted to the island-arc band.

Sierra Nevada Ecosystem Project: Final report to Congress, vol. II, *Assessments and scientific basis for management options*. Davis: University of California, Centers for Water and Wildland Resources, 1996.

Internet Access to SNEP Earth Sciences Data

The files produced by the USGS and the U.S. Bureau of Mines for the Sierra Nevada Ecosystem Project (SNEP) are on the Internet and available via anonymous file transfer protocol (FTP) from mojave.wr.usgs.gov/pub/mdiggles/snep. This area also contains a text file, *contents.txt*, that describes the files in the directory. The maps were produced by Barry C. Moring and Robert J. Miller of the USGS and Charles Bishop of the U.S. Bureau of Mines. Users may address questions to Michael F. Diggles at USGS via e-mail at mdiggles@mojave.wr.usgs or phone (415) 329-5404. A summary of U.S. Geological Survey and U.S. Bureau of Mines work for SNEP is available via the Internet's World Wide Web (WWW) in a Home Page whose Universal Resource Locator (URL) ("address") is http://caldera.wr.usgs.gov/mdiggles/SNEP-USGS.html.

MINERAL DEPOSITS IN THE SIERRA NEVADA

The major minerals issues within the SNEP study area concern metallic deposits of gold, silver, copper, lead, and zinc. Tungsten and industrial minerals such as sand and gravel are also potential issues. The assessments of undiscovered resources that are part of the USGS Mineral Resource Surveys program provide useful tools for addressing minerals issues. Some areas that are permissive for undiscovered mineral resources are areas where, were a deposit discovered and developed, an environmental impact could take place. Knowledge of where such deposits are likely to be prepares land-use managers to help minimize impacts.

For mineralized rock in a deposit to be considered "ore," it must be possible to mine it at a profit. The quantitative assessment presents numerical estimates of amounts of copper, zinc, lead, silver, and gold in undiscovered mineral deposits. It does not include additions to resources that can be made by extending known deposits. That an undiscovered deposit might exist does not mean it will be discovered or, if discovered, if it will be mined.

Quantitative Resource Assessment Technique

This resource assessment is part of a pilot project for an ongoing assessment of all mineral resources in the United States. The purpose of the assessment is to maintain a consistent, minimum level of current mineral-resource information so that such information can be considered in planning for the optimum use of public lands and for obtaining secure, long-term mineral supplies from domestic and international sources. Further details are given in McCammon and Briskey 1992. An assessment is an estimation or evaluation in this instance of undiscovered resources of base and precious metals within specific volumes of rock. This assessment is quantified in that the result is expressed in numbers. Because of the uncertainty inherent in such an assessment, the obtained results are presented probabilistically.

Methods used in this study are based on mineral deposit models. Mineral deposit models are collections of data in a convenient form that describe a group of deposits that have similar characteristics and origins. They are based on worldwide literature and observation. They contain information on the common geologic attributes of the deposits and the environments in which they are found. Grade and tonnage models consist of information on the grade and size of the individual deposits, which serve as examples for that deposit type. To begin an assessment, we review the geology of the area and select appropriate deposit models. We then delineate permissive tracts for each type of deposit. The permissive tract is defined by the environments of formation described in the deposit model. Geologic maps and maps showing location and type of mineral deposits and occurrences, if any exist, are used in outlining these permissive tracts. Geophysical and geochemical maps as well as satellite images, are also useful, and the exploration history may be important. Estimates of undiscovered resources are made to a depth of 1 km (0.6 mi) beneath the surface of the permissive tract. If an area of permissive rock is covered by more than 1 km (0.6 mi) of rock or sediment, it is excluded from the tract. Then we review the worldwide data on grade and tonnage for each model. Reasoning by analogy, the undiscovered deposits that are estimated in the area should be similar in grade and tonnage to known examples. For many deposit types, these data are available in the form of grade and tonnage models in USGS Bulletins 1693 (Cox and Singer, 1986) and 2004 (Bliss, 1992).

We review the grade and tonnage data for known deposits, if there are any, in the tract, decide whether the worldwide models are appropriate for the tract, and modify them if necessary. In well-exposed areas the largest deposits are often discovered first, and we consider how this might affect the expected tonnage of undiscovered deposits.

The Mineral Resource Data System (MRDS) of the USGS is a computerized database containing nearly 110,000 records of mines, prospects, and occurrences throughout the world. Records can contain detailed information on the name(s), location, commodities, exploration and development, deposit description, geology, references, production, and reserves and resources. Most of the data have been collected from published literature and unpublished files maintained by USGS geologists. Additionally, some data have come from the field observations of USGS geologists and, to a lesser degree, contributions from other federal agencies (such as the U.S. Bureau of Mines, U.S. Forest Service, U.S. Bureau of Land Management [BLM], U.S. Department of State), state geological surveys, mining companies, and private consultants. Plans are under way to release MRDS and its updates on CD-ROM in the near future. For access to MRDS in the meantime, contact Nancy Milton, Acting Eastern Regional Geologist, or Joseph S. Duval, Eastern Minerals Chief Scientist, at the USGS, 12201 Sunrise Valley Drive, Reston, Virginia 22092. For small requests, contact the senior author, Michael F. Diggles.

Finally, we estimate the number of undiscovered deposits of each type in the permissive areas. The estimates are subjective and are expressed in terms of percentage chance of X or more deposits. Estimates are made by teams of geoscientists who know about the deposit type or about the area, preferably both. The result of the estimation process is a probability distribution of numbers of undiscovered deposits. Details regarding the deposit estimation procedure are given in Root et al. 1992, which include a an explanation of the Mark 3 simulation program.

The deposits estimated should be consistent with the grade and tonnage model. That is, if ten deposits are estimated, five of them are visualized to be larger than the median tonnage and five of them (not necessarily the same five) are visualized to have a higher grade than the median grade. If the grade and tonnage model is based on district data, then the number of undiscovered districts is estimated.

There are many geologic, geochemical, and geophysical characteristics useful in estimating undiscovered mineral deposits. Estimates can be guided by counting mineral occurrences, geochemical anomalies, or exploration "plays" and assigning to each a probability of its being a member of the grade and tonnage distributions. Estimates can also be guided by analogy with well-explored areas that contain known numbers of deposits and that are geologically similar to the study area. Areas with geologic or morphologic structures commonly associated with mineralization are likely permissive.

The probability distribution of numbers of undiscovered deposits is then combined with probability distributions for tonnage and grades. A Monte Carlo simulation technique is used to select randomly, from each distribution, a number, a tonnage, and a grade. This procedure is done iteratively by computer many thousands of times, and a new probability distribution is generated that shows the distribution of contained metal.

Qualitative Descriptions

Quantitative assessments were made for lead, copper, zinc, gold, and silver. In addition to these assessments, treated in separate sections later, qualitative descriptions and discussion of issues for some other deposit types are included here.

Placer Gold and Recreational Prospecting

Information on placer gold and recreational prospecting is useful to rural communities and also to studies of spawning gravels in riparian habitat and in water-clarity work. Anecdotal information from the Gold Prospectors Association of America suggests that about 50,000 to 100,000 Californians a year engage in recreational prospecting in the Sierra Nevada.

An additional 10,000 to 20,000 nonresidents are thought to participate as well. The average recreational prospector spends from three to ten days a year in this activity and spends perhaps $50 a day in the local areas. Under the Mining Law of 1872, placer deposits can be claimed, and the claim holder acquires the mineral rights. In 1994, to increase public access to placer deposits for recreational prospecting, the BLM has removed from Mining Law coverage the 11 km (7 mi) stretch of the Mokolumne River from Highway 49 to Electra and now issues two-week permits for prospecting in placer deposits there. The U.S. Bureau of Mines has summarized placer locations on maps elsewhere in this chapter that show areas with the greatest number of occurrences.

Tungsten Mineralization

Tungsten, with associated gold, silver, copper, and molybdenum, is found in narrow bands of highly altered rock ("tactite" or "skarn") at contacts between granitic intrusions and metamorphosed calcareous sedimentary rocks (Einaudi and Burt 1982). Most of these deposits occur at the margins of metamorphic rock pendants generally located in the eastern part of the central Sierra Nevada.

In 1916, deposits of scheelite, a tungsten-bearing mineral, were discovered in the Pine Creek area (Bateman 1965). The Pine Creek mine began operation in 1918; it has been among the free world's largest tungsten producer. Scheelite has also been discovered and mined in the nearby Mount Morrison area and in numerous small bodies of calcareous metamorphic rocks elsewhere in the east-central Sierra Nevada. The Pine Creek and Mount Morrison pendants contain most of the known tungsten resources in the Sierra Nevada (Newberry 1982). The Pine Creek mine and associated deposits, which lie just outside the eastern boundary of the John Muir Wilderness in the Pine Creek pendant, form the largest, most productive tungsten reserves in the United States (du Bray et al. 1982), with proven reserves of 1.5 million tons of mineralized rock as of 1995. The Pine Creek mine shut down when President Carter, in reaction to the former Soviet Union's invasion of Afghanistan, put an embargo on such high-tech exports to the former USSR as deep-drilling technology for Siberia. As a result, Hughes Tool Company required less tungsten for the production of drill bits. By the time the embargo was lifted, United States trade relations with China—the country with the largest tungsten deposit in the world—had been normalized. Availability of low-cost Chinese tungsten and the lower demand for bomb casings since the end of the Vietnam War have kept the world price at a level too low for the Pine Creek mine to be reopened. A joint venture of Strategic Minerals Corporation of Connecticut and Avocet Ventures, Inc., of London hopes to mine Pine Creek reserves in the near future (R. Kattelmann, letter to M. Diggles, August 1995). In the long run, the world's known tungsten reserves can do nothing but decrease. For mineralized rock to be considered "ore," it must be possible to mine it at a profit, something that has not been done at Pine Creek recently. Because the Pine Creek mine is located on the up-thrown side of the Sierra Nevada front fault, water drains from the mine without the use of pumps. The tendency of the mine to drain naturally and the fact that the ore here falls easily into the crusher enhance the potential that operators will reopen the mine sometime in the future.

North of Lake Tahoe, the Sierra Nevada batholith contains a few roof pendants that have deposits of, and are geologically permissive for, tungsten or gold deposits. Much of the area between Lake Tahoe and Bridgeport is permissive for gold, tungsten, mercury, manganese, or uranium in skarn deposits in roof pendants and in hot-spring, vein, and other deposits in Cenozoic volcanic rocks, but known deposits are sparse (Dellinger, 1989). South of Bridgeport, the batholithic terrane contains numerous roof pendants; collectively, these bodies have numerous deposits of, and are considered to be geologically permissive for, tungsten and molybdenum in skarn deposits, gold in quartz veins, iron in epigenetic magnetite deposits, or chromium in podiform chromite deposits that occur in roof pendants of oceanic affinity along the west edge of the batholith. Batholithic rocks along the central east edge of the Sierra Nevada, from Mono Lake to Independence, contain numerous roof pendants and are considered to be geologically permissive for deposits of tungsten and gold. Deposits of tungsten, gold, or molybdenum occur locally, but known deposits are sparse or absent in most of this area. Most of the Sierra Nevada south of the thirty-sixth parallel is considered to be geologically permissive for gold, tungsten, lead, zinc, antimony, or mercury in roof-pendant skarn deposits or in gold-quartz and other vein deposits. Known deposits of gold, tungsten, antimony, or mercury occur in parts of the area, but most of the area lacks known deposits.

Sand and Gravel

Sand and gravel production in California had a total value 1.2 times as much as gold in 1994. Sand and gravel deposits are abundant in alluvium in the lower parts of drainage basins. Development of these deposits is limited by the high cost of transporting them to markets, but as sources closer to the markets are depleted and Sierra Nevada population centers grow, these resources will become more important. As the highway systems widen, expand, and need repair or replacement, the need for sand and gravel deposits will increase. A consideration in developing these deposits is the potential for a significant impact on riparian habitat and siltation of streams needed for clean water supplies. A source of sand and gravel available in some parts of southern California is closed military bases, particularly low-lying alluvial areas occupied by airfields. These areas, however, sometimes contain dump sites for fuels and solvents that have permeated the gravels (L. Darlene Batatian, conversation with Michael F. Diggles, December 1991). With these issues in mind, such deposits may be available to fill the needs in parts of the Sierra Nevada. Land-use planners need to be able to adjust from heavy-industry sand and gravel production during the life of

the deposit to some other land use after the deposit is depleted (Kockelman 1990). The sites need to be reclaimed for other uses, particularly for urban buildout that is moving toward these deposit sites.

Environmental problems in the sand and gravel industry faces include damage to riparian habitat and fisheries, water quality, air emissions of small particulate matter, diesel-engine emissions, and blasting activities. Commercial and housing developments are being built ever closer to these operations, and as potential issuers of complaints against ground vibrations, their needs must be met (Clark 1992). Dust can be controlled by watering down the operation or other more sophisticated methods (Scherer, 1992). Solutions are apt to increase water-supply needs and introduce runoff considerations. Congress has been concerned with visibility around national parks and may enact new visibility regulations that could affect the industrial-mineral industry. Mining operations may be in conflict with wetlands programs in which a no-net-loss policy is common. There may be cases where industry has the opportunity to create wetlands.

LOW-SULFIDE GOLD-QUARTZ VEIN DEPOSITS

The Mother Lode held the world record for gold deposits for many decades. It still produces some gold today and has its own health and safety issues.

Descriptive Model

Low-sulfide gold-quarts vein deposits contain gold in massive persistent quartz veins mainly in shear zones in regionally metamorphosed volcanic rocks and volcanic sedimentary rocks (Model 36A, Berger 1986).

Rationale for Model Choice

Gold-bearing mesothermal quartz veins are localized along major deep-seated, through-going structural features in low- to moderate-grade marine metasedimentary and metavolcanic rocks. Many of the type examples used by Berger (1986) in the low-sulfide gold-quartz descriptive model (Model 36A) are in the Pacific Coast region.

Rationale for Tract Delineation

The permissive tract was defined principally by the location of low- to moderate-grade regionally metamorphosed marine sedimentary and volcanic rocks of Jurassic and older age, based on the state geologic maps for California and Oregon (Jennings 1977; Walker and MacLeod 1991) and the personal knowledge of the assessors (figure 18.1). Geophysical evidence

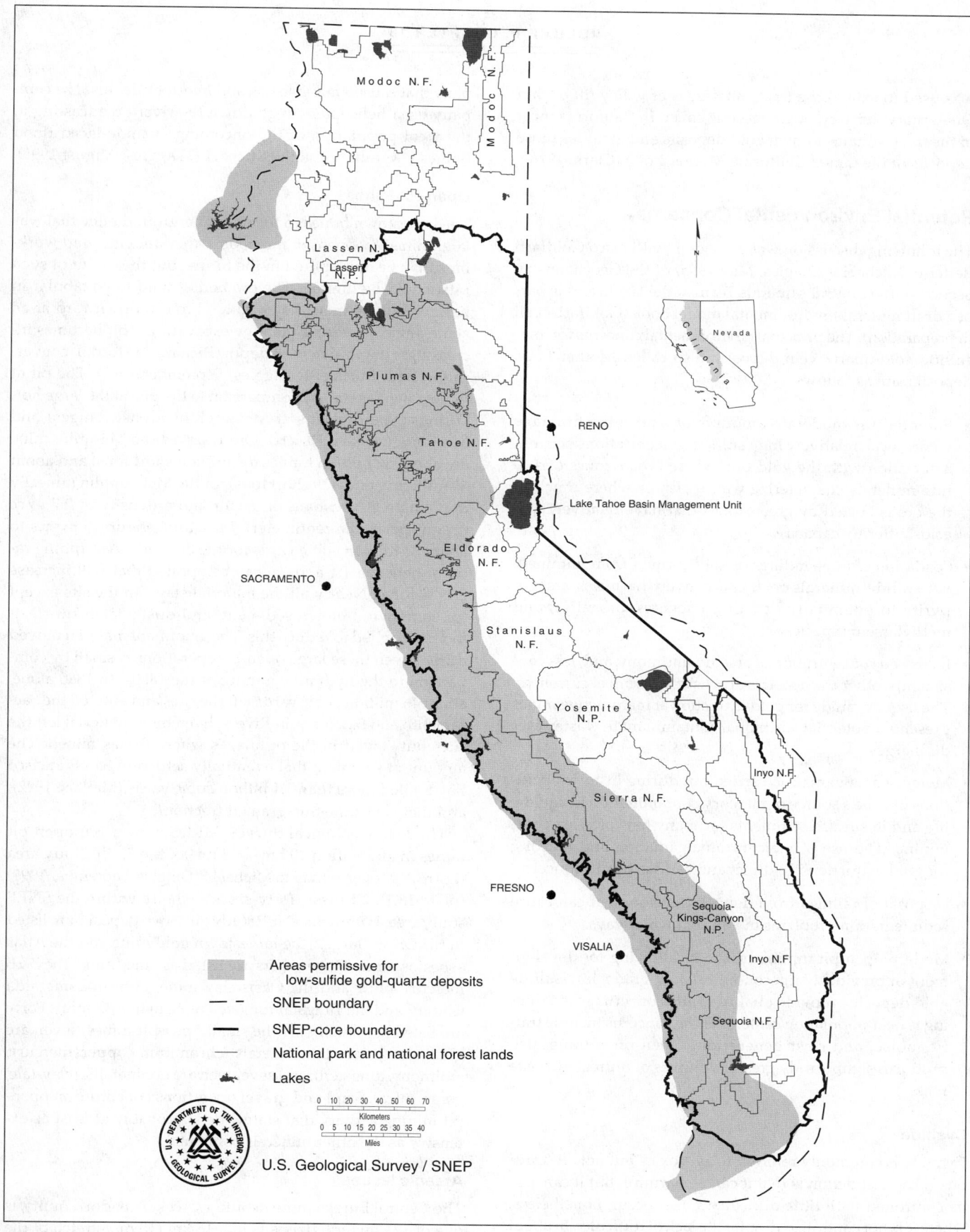

FIGURE 18.1

Tracts permissive for low-sulfide gold-quartz deposits.

was used to extend the tracts into areas of valley fill or thin Quaternary, Tertiary, or Cretaceous cover. In California most of the tract contains known gold deposits and includes those deposits in the famed California Mother Lode (Clark, 1976).

Potential Environmental Concerns

The following discussion was provided by Richard Goldfarb (letter to Michael F. Diggles, May 1995) of the Geochemistry Section of the Central Minerals Team at the USGS and is part of a draft mineral-environmental model book (Goldfarb et al. in preparation). The principal environmental concerns for low-sulfide gold-quartz vein deposits, also called Mother Lode deposits, are as follows:

- Potential for moderate amounts of acid mine drainage where local, relatively high sulfide concentrations occur in association with the gold ore, where broad zones of sulfide minerals characterize wall rocks, or where much of the ore is hosted by greenstones that have relatively low acid-buffering capacity.

- Oxidation of mine tailings or soils formed from unmined, yet sulfide minerals rock can convert harmless arseno-pyrite to potentially harmful arsenates, arsenites, and methylarsenic species.

- Increased concentration of arsenic, antimony, and, less consistently, other trace metals downstream from occurrences. The use of cyanide for gold extraction at many active mines presents a potential additional contaminant in wastewater discharges.

- Mercury amalgamation carried out during historic operations may be a source of mercury contamination in aquatic life and in surface sediments for many tens of years post-mining. The use of mercury amalgamation and roasting for gold extraction is a direct and serious health concern.

- Disposal of tailings from deposit development can cause sedimentation problems in adjacent waterways.

- Modern open-pit mining methods, allowing for development of previously uneconomical, low-grade, low-sulfide gold deposits, provide quality-of-life concerns when mining is near population centers. These concerns include traffic, noise, and dust generation. Open-pit mining also produces significantly greater volumes of untreated waste rock.

Cyanide

Cyanide is commonly referred to as "toxic" but not "hazardous." The distinction is that it can be harmful, but it can also be controlled with little difficulty. Cases occur, nonetheless, where it is not controlled despite the lack of difficulty. In about August 1995, there was a major breach of a cyanide-laced tailings pond in South America by a United States gold company that caused massive fish and livestock deaths. The company used helicopters to go down the river broadcasting to the local population of the oncoming cyanide-laced flood (James J. Rytuba, e-mail to Michael F. Diggles, August 1995).

Open-Pit Mines

The Jamestown mine is a former underground mine that was later mined by open-pit methods. Other underground workings may be open-pitted in the future, but the extent of such pits would be limited. The ore bodies tend to be tabular in nature, and, because they also tend to dip downward at an acute angle, accessing them by excavating a pit becomes increasingly expensive with depth (Richard M. Tosdal, conversation with Michael F. Diggles, September 1995). The pit at Jamestown is small by comparison to the pits at the large hot-springs-type deposits at McLaughlin (Coast Ranges) and Mesquite (Mojave Desert). Ore reserves the Mesquite mine are about 56 million tonnes (62 million short tons) and about 35 million tonnes (39 short tons) for the McLaughlin mine. By applying a 20% expansion factor and a density of 2.7 g/cc (grams per cubic centimeter), Mesquite's figure converts to about 33 million cubic yards for the ore body. A stripping ratio of between 1:1 and 1:6 can be applied that will increase this volume. Nearly all the material stays on the site except for some wind-blown sediment, and dust-abatement treatment is applied to reduce this. The amount of material moved during even these large open-pit operations is small by comparison to the hydraulic mining of the 1880s. In 1880 alone, about 46 million cubic yards left the sites and entered the Sacramento and San Joaquin Rivers. Even more material left the sites but stayed in the drainages where it was mined. The amount of sediment that eventually settled in San Francisco Bay totaled more than 1.1 billion cubic yards (McPhee 1992), and that is just the fine-grained fraction.

The U.S. Geological Survey MRDS shows 106 open-pit mines in and within 10 km^2 (6.2 mi^2)of the SNEP study area (Lorre A. Moyer, e-mail to Michael F. Diggles, September 1995) (plate 18.1). Of these, forty-six are wholly within the SNEP study area. Within the SNEP study area, two deposits are listed in MRDS as "large," the Jamestown gold mine and the Atlas asbestos mine. Three others are listed as "medium," the B&B mercury mine and the Oasis clay mine, both in Esmeralda County, and the Tungstar tungsten mine in Inyo County. Forty are listed as "small." Of the small open-pit mines, seven are gold, sixteen are other metals (chromium, copper, mercury, antimony, tungsten), and seventeen are nonmetallic (clay, talc, gems, mica). Sand and gravel operations constitute an open-pit mining activity that statewide is probably at least as extensive as the other minerals combined.

Arsenic Issues

The Central Eureka mine in Sutter Creek, Amador County, is one of 123 mineral deposits in a 16 km (10 mi) stretch of the Mother Lode belt that are in the USGS MRDS database. The Mesa de Oro housing project is situated on the mine lands and is the topic of an extensive U.S. Environmental Protection Agency (EPA) project addressing arsenic content. Other such sites may be of interest for buildout as the population of the Sierra Nevada increases and a workable, effective management strategy for dealing with arsenic in the Mother Lode becomes established.

Until arsenic-rich soils are covered, it is difficult to sell property because potential buyers cannot get mortgages. The community consists of one-third finished houses, one-third deserted houses, and one-third vacant lots. According to interviews with home owners, (M. F. Diggles, unpublished data), when two feet of topsoil are in place to cover the arsenic-rich surface that is there now, an attachment to each home owner's deed will state that the work has been done, so that banks will lend on the properties again. Work is also being done to stabilize the sides of the "mesa" because when it gets wet, the finely ground material runs off. Popular opinion is mixed, but generally it is felt that the EPA thresholds are fair. Measurements are 1,000 to 1,200 parts per million (ppm) arsenic compared to about a tenth that in Jackson. Many homes are built around Mother Lode tailings heaps in the foothills, but Mesa de Oro is the only housing project built on one.

Important Examples of Deposit Type

The low-sulfide gold-quartz grade and tonnage model is based on deposits containing 100 metric tons or more of gold (Bliss 1986). More than 50% of the deposits in the worldwide model (Bliss and Jones 1988) are located within this tract and include those of the California Mother Lode and the Grass Valley district in the Sierra Nevada foothills, as well as deposits in the Klamath, Siskiyou, and Trinity Mountains in northern California and southern Oregon. Several major mines in the tract are currently in operation or have been active recently. Most notable are the Harvard mine near Jamestown and the Royal Mountain King mine near Copperopolis, both in the California Mother Lode. Reserves at the Harvard mine are estimated to be in excess of 100 metric tons of gold (Bliss and Jones 1988).

Rationale for Numerical Estimate

The team concluded that nearly all deposits that are exposed at the surface have been discovered. The part of the tract most favorable for undiscovered deposits is that part that is covered by less than 1 km (0.6 mi) of Cretaceous sedimentary and Tertiary volcanic rocks between the Klamath Mountains and the Sierra Nevada. Estimates for this area were made using deposit density data. The shallow-covered region extending southeast from the Klamath Mountains has an area of 490 km^2 (191 mi^2). The shallow-covered region extending northwest from the Sierra is 1,700 km^2 (664 mi^2) in area. The density of distribution of low-sulfide gold-quartz vein deposits in the Klamath Mountains is four deposits per 1,000 km^2

(390 mi^2); for the Sierra Nevada, the density is 4.6 according to Bliss et al. (1987). Inasmuch as the concentration of deposits tends to drop off with distance from the zones of greatest mineralization, we expect that the density of deposits in the areas beyond the known deposits is half that given by Bliss. These densities (2 and 2.3 per 1,000 km^2 [390 mi^2]) multiplied by the areas of covered tract results in an expected value of about one deposit for the Klamath Mountains and about four for the Sierra Nevada.

The tract considered includes not only the Sierra Nevada but the Klamath Mountains, areas near Monterey, and areas in southern California . Of the 113,300 km^2 (43,745 mi^2) considered, the 51,800 km^2 (20,000 mi^2) in the SNEP study area are thought to contain nearly all the deposits. The estimates, therefore, should be treated as minimums but only slightly low. In the exposed part of the tract, the team estimated that only about one undiscovered deposit could exist giving a total expected value of about six undiscovered deposits. For the ninetieth, fiftieth, tenth, fifth, and first percentiles, the team estimated to, six, nine, twelve, and fifteen or more deposits consistent with the grade and tonnage model of Bliss (1986). The "expected value" for this set of estimates is 5.8 deposits. Deposits of this type worldwide have a mean tonnage of 0.03 million tonnes (0.033 million short tons) and a mean grade of 1.6 grams per tonne (ppm) gold and 2.5 g/t (ppm) silver. The mean amount of contained metal is 9.54 tonnes (10.54 short tons) gold and 1.22 tonnes (1.35 short tons) silver.

MASSIVE SULFIDE DEPOSITS, SIERRAN KUROKO TYPE

Base and precious metals produced from kuroko types of deposits are important to the nation's metal supply. Acid drainage is a major environmental consideration. These deposits contain copper and zinc in massive sulfide deposits in intermediate to felsic marine volcanic rocks. Triassic and Jurassic deposits have significantly lower tonnages than worldwide kuroko-type deposits (Model 28A.1, USGS Bull. 2004).

Rationale for Model Choice

Volcanogenic massive sulfide deposits are located in and near the foothills of the Sierra Nevada, in belts of intermediate to felsic marine volcanic rock. Volcanogenic massive sulfide deposits have been a historically important source of copper, zinc, silver, and gold in this region. Relatively high grades of polymetallic ores, simple metallurgy, and potential for large deposits make these deposits attractive exploration targets; exploration for, and development of, these deposits continues. Although both kuroko (Singer 1986) and Cyprus types of massive sulfide deposits are possible in this tract, the team assessed only the kuroko type because it is the only type with significant known deposits. The Sierran kuroko model, which is defined to be restricted to deposits of Triassic and Jurassic age (Model 28A.1) (Singer, 1992), was selected because the known deposits, many of which are included in the Sierran kuroko model, are in Jurassic rocks.

Potential Environmental Concerns

These types of deposits, with their high sulfide content, have been areas with acid mine-drainage problems. The Penn mine superfund cleanup site is from one such deposit. Acid-buffering maps can be made for mine-drainage and water-clarity work once a digital geologic map of the Sierra Nevada is released. A draft had been completed by July 1995. As Mount (1995) pointed out, base metals become more toxic to aquatic life in water that has a higher temperature, contains reduced, dissolved oxygen, and is acidic. Where high-carbonate-content rocks crop out downslope or downstream from kuroko-type deposits, sulfuric acidformed from the weathering of pyrite could be buffered. Conversely, where such buffering is not available, managers and mineral producers can be made aware of potential acid mine-drainage issues and take mitigating steps. A treatment of mine-drainage issues and mineral-environmental models is given by Plumlee et al. 1994.

Durkin (in press) describes techniques of mitigation in an acid mine drainage area. His paper is also available at: http://www.info-mine.com/technomine/enviromine/ case_hist/ch1.html on the World Wide Web. Action includes removing reactive rock from the dump and leach pads, backfilling, grading slopes, capping with a multimedia cover, and revegetation. The cover consists of "6 inches of onsite crushed limestone, 18 inches of compacted low-permeability manufactured soil, 4.5 feet of nonreactive crushed waste material for thermal/frost/root protection of the manufactured soil layer, and 4 to 6 inches of topsoil." The revegetation is done with "a mixture of aggressive grass species to limit the establishment of deeply rooting woody species and trees that could damage the integrity of the soil liner." The cap included a riprap-lined channel to manage runoff and control erosion. The reactive ore left on the leach pad was treated by mixing limestone with it. Postclosure monitoring is included in the plan.

Rationale for Tract Delineation

All map units in the foothills of the Sierra Nevada that contain sequences of submarine volcanic rocks have been included and define the permissive tract for volcanogenic massive sulfide deposits (figure 18.2). The tract extends westward under the Great Valley where the depth to Jurassic basement is no more than 1 km (.62 mi), based on drillhole data (Wentworth et al., in press).

Important Examples of Deposit Type

The Penn mine is one of the larger examples of volcanogenic massive sulfide deposits from the Sierra foothills metavolcanic terranes. It produced 38,000 metric tons of copper, 10,000 metric tons of zinc, 66 metric tons of silver, and 2 metric tons of gold. The Blue Moon and Western World deposits have been actively explored in recent years.

Rationale for Numerical Estimate

The tract considered includes not only the SNEP study area, but areas nearby. Of the 13,500 km^2 (5,300 mi^2) considered, the 9,240 km^2 (3,600 mi^2) in the SNEP study area are thought to contain nearly all the deposits. The estimates, therefore, should be treated as minimums but only slightly low. This region has been explored extensively in the past, focusing on easily observed surface gossans. There are seven known deposits and nineteen smaller occurrences in the area, and we judged that about a quarter of those occurrences could be deposits falling within the grade and tonnage models with further exploration and development. This observation, coupled with consideration of substantial amounts of concealed potential host rocks, guided our estimate for the fiftieth percentile. The substantial concealed area also guided our estimate for the tenth percentile. For the ninetieth, fiftieth, and tenth percentiles, the team estimated two, thirteen, and twenty-five or more deposits consistent with the Sierran kuroko grade and tonnage model of Singer 1992. The "expected value" for this set of estimates is 13.2 deposits. Deposits of this type worldwide have a mean tonnage of 0.31 million tonnes (0.34 million short tons and a mean grade of 0.37% copper, 2.9% zinc, and 2.4% lead. The mean amount of contained metal is 1.51 tonnes (1.66 short tons) gold, 81.59 tonnes (89.95 short tons) silver, 25,987 tonnes (28,651 short tons) zinc, and 11,733 tonnes (12,936 short tons) lead.

PORPHYRY COPPER DEPOSITS

The generalized porphyry copper deposit model includes various subtypes, all of which contain chalcopyrite in stockwork veinlets in hydrothermally altered porphyry and adjacent country rock (Model 17, Cox 1986a).

Rationale for Model Choice

The Lights Creek porphyry copper deposit (Storey 1978) is located in the northernmost part of the Sierra Nevada, in the Plumas County copper belt of Knopf (1935). Lights Creek consists of two mineralized zones 3 km (1.9 mi) apart. It differs from typical porphyry copper systems in that magnetite is more abundant than pyrite and chlorite is the main alter-

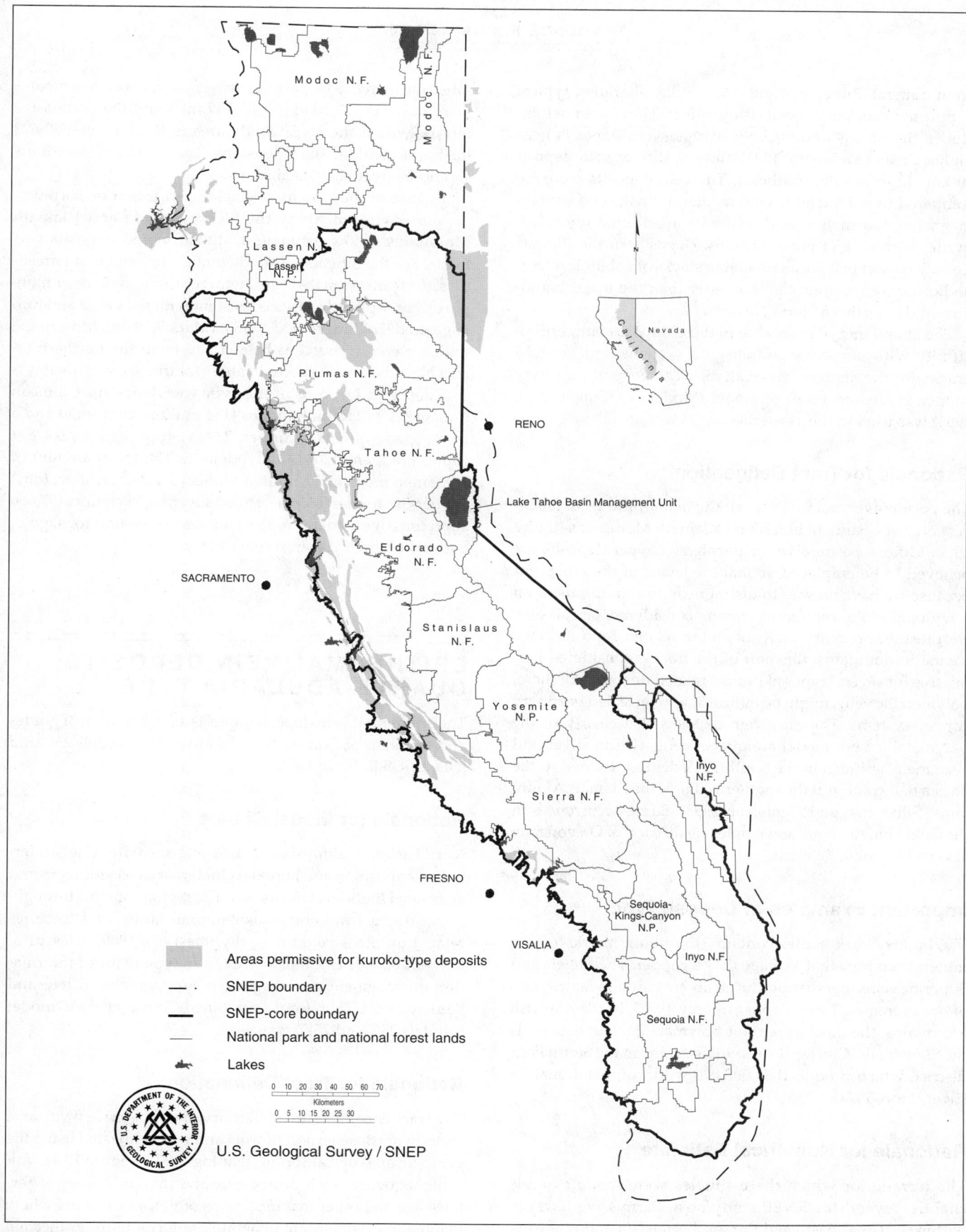

FIGURE 18.2

Areas permissive for kuroko-type deposits.

ation mineral. Zoned potassic and phyllic alteration typical of porphyry copper deposits is indistinct. The copper belt includes the Engels and Superior copper vein deposits near Lights Creek (Anderson, 1931) and the Walker vein deposit 20 km (12 mi) to the southeast. The vein deposits are large compared to polymetallic vein median tonnage and contain magnetite, tourmaline, and actinolite in addition to chalcopyrite, bornite, and other sulfides. Deposits in the Plumas County copper belt are all related to stocks of gabbroic to granodioritic composition that are older than the major batholiths in the northern Sierra Nevada.

The abundance of magnetite in these deposits suggests an affinity with porphyry copper-gold systems, but no gold grades for the deposits are available. The general porphyry copper grade and tonnage model (Model 17) (Singer et al. 1986) was used in the assessment.

Rationale for Tract Delineation

The permissive tract includes all the major plutons of the Sierra Nevada, Salinian block, and Klamath Mountains (figure 18.3). Plutons permissive for porphyry copper deposits are believed to be emplaced at shallow levels in the crust, but because we have no way to distinguish this environment on a regional scale, the entire tract is considered permissive. Despite the apparent scarcity of pluton-related deposits in the Klamath Mountains, this part of the tract is considered permissive for several types of pluton-related deposits. Of these, polymetallic veins might be indicators of concealed porphyry copper systems. The only vein deposits that clearly fit the polymetallic vein model are quartz veins rich in silver and base metal sulfides in the South Fork district, located in the Shasta Bally pluton at the southern edge of the Klamath Mountains (Silberman and Danielson 1993). Some occurrences in the Gold Hill, Ashland, and Applegate districts of Oregon may also be polymetallic veins.

Important Examples of Deposit Type

The Lights Creek bodies contain 315 million metric tons of mineralized rock that average 0.34% copper. The Engels and Superior veins together produced about 4 million metric tons of ore averaging 1.79% copper (Storey 1978). In the Klamath Mountains, the most important polymetallic vein deposit is the Silver Falls–Chicago Consolidated mine in the South Fork district, which produced $1,000,000 worth of metal, mainly silver (Hotz 1971).

Rationale for Numerical Estimate

The terrane for which the estimates were made extends slightly beyond the SNEP study area. Permissive terranes within the Great Basin and the southern basin and range extend slightly into the SNEP study area. Those extensions are about as unfavorable for these types of deposits as the terrane considered. The permissive terrane for which estimates were made is 143,100 km^2 (42,749 mi^2) and the permissive terrane within the SNEP study area is 109,438 km^2 (32,693 mi^2) and includes the extensions. The estimates, therefore, should be treated as maximums.

Because of the scarcity and restricted extent of porphyry copper environments in the Sierra Nevada and Klamath Mountains, a low estimate of undiscovered deposits was made. For the ninetieth, fiftieth, tenth, fifth, and first percentiles, the team estimated zero, zero, zero, one, and one or more porphyry copper deposits consistent with the grade and tonnage model of Singer et al. (1986) (Mark 3, no. 4). Most of the undiscovered resource is believed to be in the northern Sierra Nevada. The "expected value" for this set of estimates is 0.08 deposits. Deposits of this type worldwide have a mean tonnage of 140 million tonnes (154 million short tons) and a mean grade of 0.54% copper, 2.6 g/t (ppm) silver, 0.4 g/t (ppm) gold, and 0.03% molybdenum. The mean amount of contained metal is 3.1 million tonnes (3.4 million short tons) copper, 592 tonnes (653 short tons) silver, 88,586 tonnes (97,666 short tons) molybdenum, 28.83 tonnes (64.86 short tons) gold, and 593 tonnes (654 short tons) silver.

EPITHERMAL VEIN DEPOSITS, QUARTZ-ADULARIA TYPE

The epithermal vein deposit model is a combination of grades and tonnages of Comstock and Sado types (Models 25C and 25D, USGS Bulletin 1693).

Rationale for Model Choice

Northeastern California contains intermediate to felsic Tertiary composition and bimodal Quaternary volcanic rocks and associated high-level intrusions. The region contains through-going fracture systems, major normal faults, and fractures related to intrusive doming (Rytuba 1988; 1989). Classification of deposits as Comstock or Sado type requires information on basement geology that is not available (Klein and Bankey 1992). Therefore, a combined Comstock-Sado model (Models 25C and 25D) was used.

Rationale for Tract Delineation

The tract encompasses all Tertiary and Quaternary volcanic rocks in northeastern California and was delineated using the geologic map of California (Jennings 1977) (figure 18.4). Volcanic sequences include andesite and rhyolite domes of Tertiary age and other manifestations of volcanic centers where magmatic events might generate ore-forming hydrothermal systems.

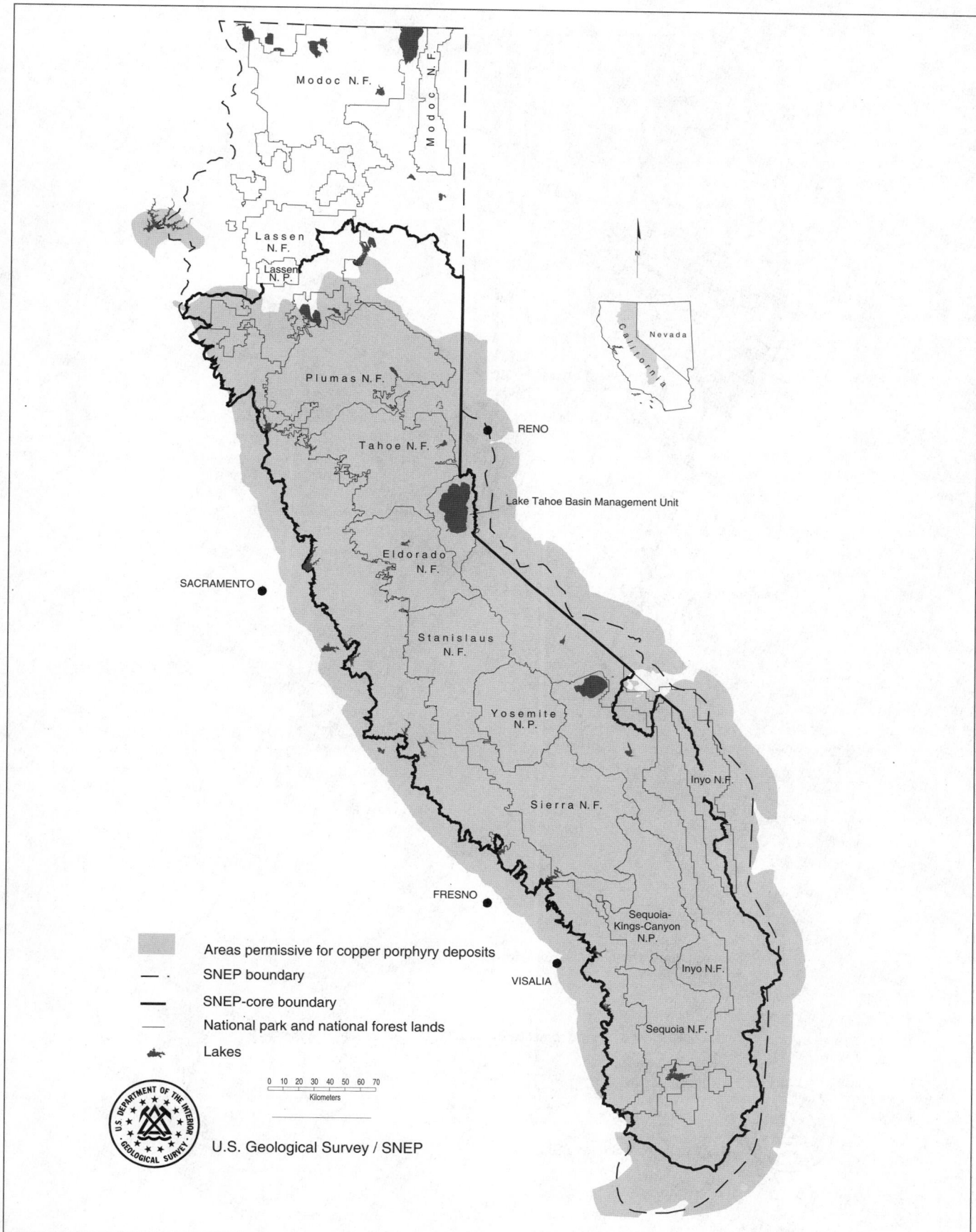

FIGURE 18.3

Areas permissive for copper porphyry deposits.

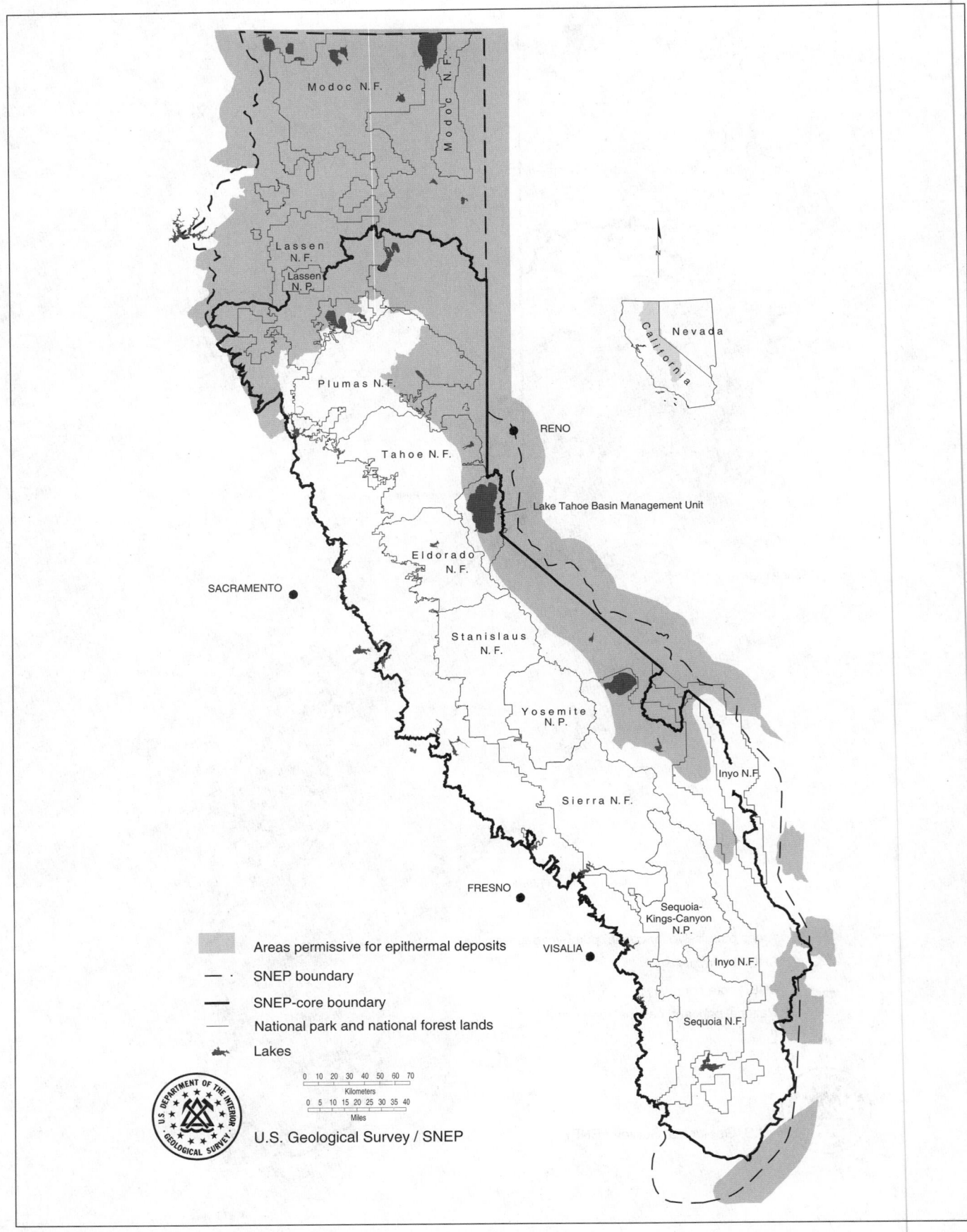

FIGURE 18.4

Areas permissive for epithermal deposits.

Important Examples of Deposit Type

There are no known quartz-adularia deposits in the area. The Skedaddle Mountain Wilderness Study Area on BLM land east of Susanville had more than 280 lode claims for epithermal deposits (Diggles et al. 1988; Munts and Peters 1987).

Rationale for Numerical Estimate

The terrane for which the estimates were made does not include the entire permissive tract within the SNEP study area, but rather a large subset of it. The permissive terrane within the SNEP study area is 68,495 km^2 (26,756 mi^2) and the terrane within it for which the estimates were made is 47,760 km^2 (18,656 mi^2). The remainder is in five subsets of larger terranes that include much of the Great Basin and Oregon. The estimates, therefore, should be treated as minimums but only slightly low. The estimators thought that there was a 50% chance of one or more undiscovered districts. For the ninetieth, fiftieth, and tenth percentiles, the team estimated zero, one, and two or more quartz-adularia epithermal-vein districts. The "expected value" for this set of estimates is one deposit. Because the size of any undiscovered deposit will depend on its deposit type, more information about the target area is needed to list an expected mean tonnage and grade; Comstock-type deposits worldwide have a mean contained amount of silver of 1.6 million tonnes (1.8 million short tons) while Sado-type deposits have just over 10% as much.

COPPER SKARN

Copper-skarn deposits contain chalcopyrite in calc-silicate metasomatic rocks near contacts with weakly mineralized igneous intrusive rocks (Model 18B, Cox and Theodore 1986).

Rationale for Model Choice

Although copper skarn deposits (Model 18B) are much less common in this tract than lead-zinc dominated systems, widespread igneous intrusions into carbonate rocks represent permissive conditions for copper skarn formation. Copper skarn mineral assemblages were observed in ores from the Gold Bottom–Copper Queen mine (Dennis Cox, unpublished data).

Rationale for Tract Delineation

The permissive tract is a nearly continuous area marked by small plutons that intrude Precambrian metamorphic and Precambrian and Paleozoic sedimentary rocks in a highly faulted part of the Mesozoic continental margin (figure 18.5). There are few areas that truly lack carbonate or other reactive rocks, so the entire sedimentary section is included. Small areas of basin fill more than 1 km (0.6 mi) in depth are excluded.

Rationale for Numerical Estimate

The terrane for which the estimates were made does not include the entire permissive tract within the SNEP study area, but rather a subset of it. The permissive terrane within the SNEP study area is 63,147 km^2 (24,667 mi^2) and the terrane within it for which the estimates were made is 17,700 km^2 (6,900 mi^2). The remainder is in four subsets of larger nearby terranes. The additional terrane is thought to be mostly much less favorable for deposits of this type. The estimates, therefore, should be treated as minimums but only slightly low. For the ninetieth, fiftieth, tenth, and fifth percentiles, the team estimated zero, one, one, and two or more copper skarn deposits consistent with the grade and tonnage model of Jones and Menzie (1986) (Mark 3, no. 8). The "expected value" for this set of estimates is 0.7 deposits. Deposits of this type worldwide have a mean tonnage of 0.56 million tonnes (0.62 million short tons) and a mean grade of 1.7% copper, 2.8 g/t (ppm) gold, and 36 g/t (ppm) silver. The mean amount of contained metal is 55,496 tonnes (61,184 short tons) copper, 0.91 tonne (1.00 short ton) gold, and 8.20 tonnes (9.0 short tons) silver.

POLYMETALLIC SKARN AND REPLACEMENT DEPOSITS

Polymetallic skarn and replacement deposits are hydrothermal, epigenetic deposits that contain silver, lead, zinc, and copper minerals in massive lenses in limestone, dolomite, or other reactive rocks, with or without calc-silicate minerals, near igneous intrusive contacts (Models 19A and 18C, USGS Bulletin 1693).

Rationale for Model Choice

Mesozoic plutons on the southeast flank of the Sierra Nevada batholith show a consistent association with numerous zinc-lead skarn and polymetallic replacement districts where they intrude Paleozoic carbonate rocks (figure 18.6). These districts include Darwin, Cerro Gordo, Modoc, Santa Rosa (MacKevett 1953), and Ubehebe (McAllister 1955). Darwin, the largest, produced more than a million metric tons of ore containing about 6% lead, 6% zinc, 0.2% copper, 200 g/t (ppm) silver and recoverable gold (Hall and MacKevett 1962; Newberry et al. 1991).

The Shoshone (Tecopa) district (Carlisle et al. 1954) produced about 600,000 metric tons of lead-zinc-silver ore and is the largest of a group of similar districts that includes Queen of Sheba, Honolulu, Ashford, Paddy's Pride, and Blackwater.

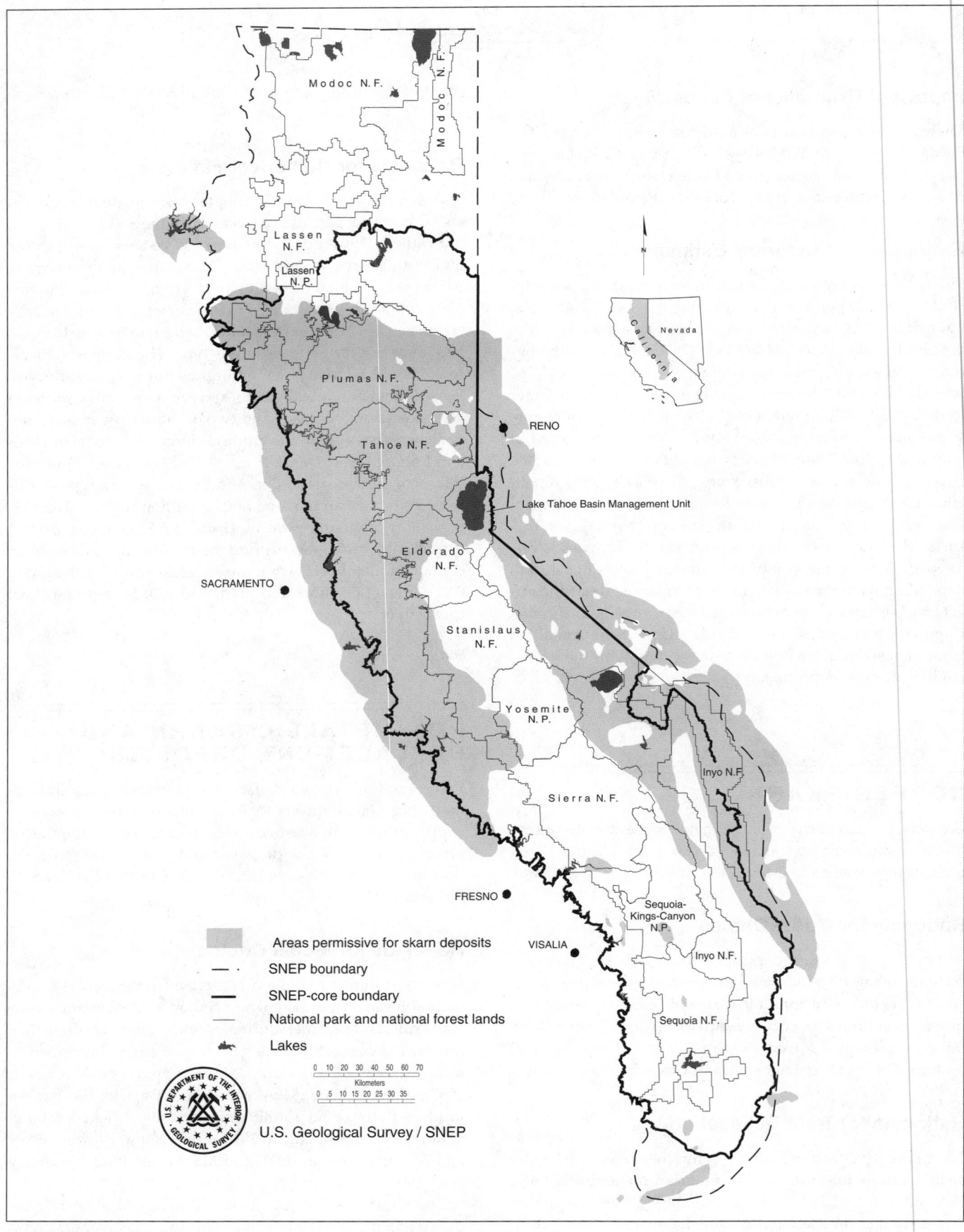

FIGURE 18.5

Areas permissive for skarn deposits.

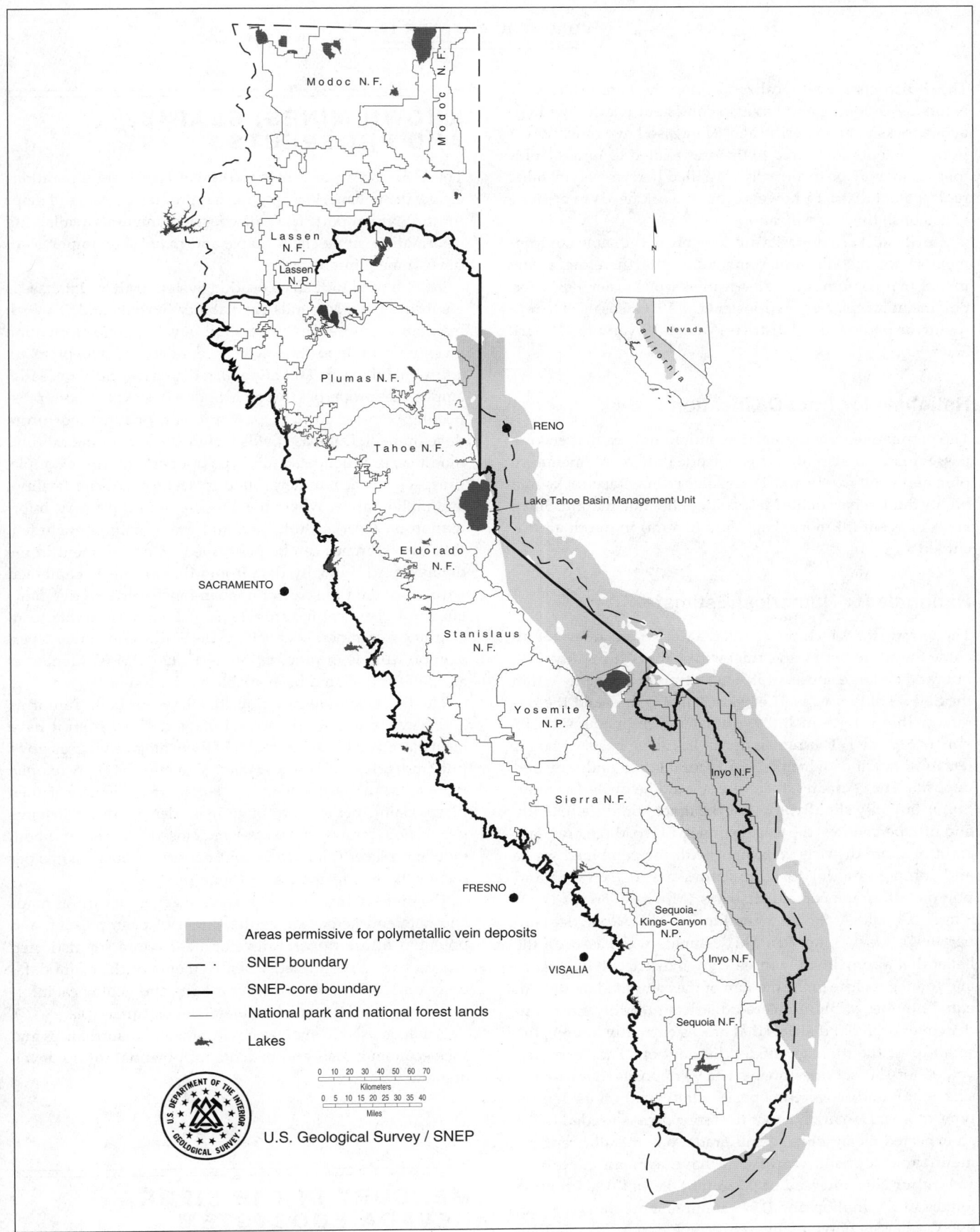

FIGURE 18.6

Areas permissive for polymetallic skarn and replacement deposits.

These districts are all localized within the Late Proterozoic Noonday Dolomite and have no consistent relation with igneous rocks of any specific age. They also have much lower pyrite contents compared to the ores related to Jurassic plutons. They may be incorrectly classified here as polymetallic replacement districts; however, more work needs to be done to establish their classification.

Zinc-lead skarn deposits are possible in the same environment as polymetallic replacement deposits; therefore, a new model that combines the zinc-lead skarn (Mosier 1986) and polymetallic replacement (Mosier et al. 1986) was used to represent the undiscovered districts (Models 19A and 18C) (Mark 3, no. 92).

Rationale for Tract Delineation

The permissive tract is a nearly continuous area that encompasses many small plutons that intrude Proterozoic metamorphic and Proterozoic and Paleozoic sedimentary rocks in a highly faulted part of the Mesozoic continental margin. Small areas of basin fill more than 1 km (0.6 mi) in depth are excluded.

Rationale for Numerical Estimate

The terrane for which the estimates were made does not include the entire permissive tract within the SNEP study area, but rather a large subset of it. The permissive terrane within the SNEP study area is 31,680 km^2 (12,375 mi^2) and the terrane within it for which the estimates were made is 19,505 km^2 (7,619 mi^2). The remainder is in two subsets of larger terranes, much of which is less favorable for undiscovered deposits. The estimates, therefore, should be treated as minimums but only slightly low. For the ninetieth, fiftieth, tenth, and fifth percentiles, the team estimated zero, three, five, and eight or more districts consistent with the combined grade and tonnage model for zinc-lead skarn (Cox 1986) and polymetallic replacement districts (Morris 1986) of D. A. Singer (Donald A. Singer, e-mail to David H. Root, September 1994) (Mark 3, no. 92). This estimate was based on the belief that approximately three known districts have grades and tonnages close to the median of the combined model. An equal number of undiscovered districts probably exists under cover or in old districts that are, as currently known, too small to fit the tonnage model distribution. The "expected value" for this set of estimates is 2.93 deposits. Because the size of any undiscovered deposit will depend on its deposit type, more information about the target area is needed to list an expected mean tonnage and grade; polymetallic replacement-type deposits worldwide have a mean contained amount of zinc and lead of 0.336 million and 0.344 million tonnes (0.370 million and 0.379 million short tons), respectively, whereas polymetallic skarn-type deposits have 0.113 million and 0.256 million tonnes (0.125 million and 0.282 million short tons) respectively.

KNOWN MINES, CLAIMS, AND PROSPECTS

The U.S. Bureau of Mines at the Western Field Operations Center in Spokane, Washington, has produced a series of maps for the SNEP report. The data consist of minerals availability information, mine-claim maps, and maps showing areas of development interest.

The Mineral Industry Location System (MILS) database is a subset of the Minerals Availability System (MAS) which maintains comprehensive minerals development information on significant deposits as well as less-detailed data on many additional deposits and mineralized areas worldwide. MILS emphasizes properties that have undergone exploration/prospecting or some form of past, present, or intended future development. Database fields include location, mineral commodities, production status, type of operation (for example, prospect, surface, underground, placer, processing facility), and bibliography. Where the data are not proprietary, information on reserves, tonnage, and grade may be included. Database searches can be performed on any element in the database. Additionally, data from MILS are easily combined with other data to screen on abandoned and inactive mine lands for chemical hazards. Plate 18.2 shows the total concentration of mineral deposits in the SNEP study area. A version of MILS is in preparation as a CD-ROM (Oddenino et al., 1995) and should be available by the end of 1995.

The BLM claim recordations database for both California and Nevada was used. Several maps derived from it have been placed on the anonymous FTP site mojave.wr.usgs.gov/pub/mdiggles/snep. These maps show the diversity of open lode claims per section (640 acres) (plate 18.3), density of open placer claims per section (plate 18.4), density of total claims per section, both open (active) and closed (former) and both lode and placer (plate 18.5), and density of open claims per section, both lode and placer (plate 18.6).

The mine-claim database is used to generate various maps and lists and help analyze the extent of past, present, and potential future mineral development. More sophisticated analyses were performed when elements of the claim database were combined with other mining and geological information to delineate known mineral deposit areas (plate 18.7). This map is used to analyze environmental disturbances and socioeconomic costs and benefits for potential future development.

MERCURY IN THE SIERRA NEVADA ECOSYSTEM

Both anthropogenic and natural sources of mercury affect the Sierra Nevada ecosystem. Natural sources of mercury include mineral deposits containing anomalous concentrations of mercury, undeveloped hot springs and thermal gas vents, recently active faults, passively degassing volcanoes, and mercury derived from the natural global atmospheric mercury cycle and added to the ecosystem through wet and dry deposition (plate 18.8). Anthropogenic sources of mercury include mercury introduced during gold mining and processing (Lindberg et al. 1979), developed geothermal areas, and mercury, derived from the atmosphere, primarily from coal combustion (Mason et al., 1994), and added to the ecosystem by precipitation. Plant communities in the ecosystem are important in concentrating and redistributing mercury (Lindberg et al. 1995).

Mercury from the Global Atmospheric Mercury Cycle

The primary source of mercury to the Sierra Nevada ecosystem is from the global atmospheric mercury cycle. Both anthropogenic and natural sources of mercury are of equal importance to the global mercury cycle, adding about 6,000 to 7,500 metric tons of mercury to the atmosphere per year (Nriagu and Pacyna 1988). The most important natural source is from evasion of elemental mercury vapor, Hg^o, from the world's oceans; the major anthropogenic source comes from coal combustion (Fitzgerald 1989; Rasmussen 1994). Soil gas emission is the third most important source of Hg^o to the atmosphere. The atmospheric mercury flux to the terrestrial environment has increased substantially since the beginning of the industrial period (about 1850) because of anthropogenic release of mercury into the atmosphere.

In California, sources of anthropogenic mercury emission to the atmosphere include waste incineration, electric utility power plants (including coal, oil, and geothermal powered plants), coal and petroleum combustion and uses, industrial-commercial sources, paint emissions, mine roasting and smelters, mobile sources (vehicles, tractors, locomotives), fugitive emissions from mine tailings and waste, landfills, and, in the past, mercury catalysts for caustic soda (now nearly all replaced). Natural sources of mercury to the atmosphere in decreasing importance include oceans, erupting volcanoes, soil vapor flux (Lindberg et al. 1995), geothermal systems and hot springs (Janik et al. 1994; Rytuba and Miller 1994), passively degassing volcanoes (Varekamp and Buseck 1986), fumaroles, soil vapor from mineralized areas, and active faults.

Wet Deposition of Mercury

Pacific weather fronts that move through the Sierra Nevada ecosystem are the primary source of wet deposition of mercury from the atmosphere in either rain or snow. The mercury species in precipitation is primarily in the ionic, oxidized form, Hg^{++}, which is derived from photooxidation of elemental mercury vapor, Hg^o, and Hg^o aerosols, primarily by ozone (Gill and Bruland 1990). The flux of wet deposition of mercury to the Sierra Nevada ecosystem is about 13 mg/m^2/yr

and may vary by several micrograms near regional point sources that vent mercury to the atmosphere.

Dry Deposition of Mercury

Dry deposition of mercury is important in the terrestrial part of the atmospheric mercury cycle. Elemental mercury vapor, Hg^o, and Hg^o aerosols from the atmosphere are taken up by plant communities in the Sierra Nevada ecosystem. Dry deposition accounts for about 6 mg/m^2/yr of mercury added to the ecosystem.

Effect on Plant Communities of Elevated Mercury Content

Mercury in Ambient Air

Typical mercury concentrations in ambient atmosphere in the Sierra Nevada ecosystem range from 2 to 3 mg/m^3. Ambient concentration of mercury in the atmosphere is important in determining whether plant communities in the Sierra Nevada ecosystem take up mercury into their leaves. Mercury uptake through the plant's root system is low relative to uptake through foliage. At low ambient mercury concentrations, plants give off mercury to the atmosphere through their foliage, and the foliage does not increase in mercury concentration during its growth cycle. At high ambient concentrations of mercury, plants take up mercury through their leaves directly from the air. Thus, for example, in a contaminated mine site where ambient air concentration of mercury is high, the mercury content of leaves increases throughout the growth cycle. Litter fall from plants and wash off from leaves are major components of mercury added to soils, creeks, and lakes.

Elemental mercury vapor, Hg^o, is the dominant mercury species in the atmosphere at normal ambient concentrations (Seigneur et al. 1994) as well as in geologic environments where elevated concentrations of mercury are present in air. Hg^{++} is present at low levels in the atmosphere. Particulate Hg may be a significant component in emanations from volcanoes such as Lassen Peak. In volcanic and geothermal environments and contaminated mine sites, the ambient air concentration can increase to several orders of magnitude above normal.

Natural geologic sources of mercury having the greatest potential regional impact on ambient atmospheric concentration include, in decreasing importance, active volcanoes, passively degassing volcanoes, geothermal areas, mineralized areas, and active faults. Permissive tract maps for epithermal deposits (figure 18.4), low-sulfide gold quartz (Mother Lode gold) (figure 18.1), and kuroko (figure 18.2) (listed in decreasing order of importance) delineate those areas with the potential for elevated ambient mercury concentrations in air where these deposits are exposed at the surface or covered by residual soils and alluvium. Deeply buried, these deposits will not contribute mercury to the ambient atmosphere. Elevated fluxes of soil gas emission from these areas, combined with the atmospheric mercury flux, may cause above-normal concentrations of mercury in plant foliage. Plant material falling to the forest floor leads to litter-fall deposition flux of mercury exceeding the normal level of 20 mg/m²/yr. Plants in these tracts are an important source of mercury contamination to creeks and lakes.

Mercury Concentrations at Geothermal Sites

Natural point sources of mercury that persist over geologically long periods may have a regional impact on the ambient atmospheric concentration of mercury in the atmosphere. More importantly, these sources will affect the uptake of mercury by plant communities adjacent to these areas. Geothermal areas in the Sierra Nevada ecosystem such as the Lassen, Long Valley, and Coso volcanic centers are point sources for mercury emission into the atmosphere. Ambient mercury concentrations in air are elevated in these areas, and adjacent plant communities will take up mercury through their leaves and concentrate mercury in foliage. Mercury concentrations in wash off and litter fall from these plant communities will have elevated and will exceed the normal flux of mercury from litter fall, about 20 mg/m²/yr, and wash off, about 6 mg/m²/yr, to the soil and into drainage basins and lakes.

Mercury Concentrations at Mine Sites and Ore Deposits

In areas with kuroko, low-sulfide gold-quartz (Mother Lode gold), and epithermal ore deposits (see figures 18.1, 18.2, and 18.4), elevated mercury levels can be expected in soils above and adjacent to the deposit. In mine sites where mercury was used in the gold amalgamation process, elevated mercury levels are present in the soils and tailings. In these areas, native mercury was also released into nearby creeks.

Gas emission from soils above and adjacent to ore deposits and from contaminated soils in mine sites can be expected to exceed the natural soil gas flux of mercury of about 7 mg/m²/yr. Elemental mercury vapor flux from contaminated soils, soils from mineralized areas, and soils in areas with naturally elevated background mercury is controlled primarily by volatilization of Hg^o and secondarily by biotic and abiotic reduction of Hg^{++} in pore water to Hg^o. The Hg^o flux from soil increases exponentially with increases in soil temperature. Thus, during the summer, soil gas emission will be greatest, and uptake of mercury into plant leaves will be highest.

Typical Hg^o flux from contaminated soils in mine sites and measured above contaminated soils range from 10,000–22,000 mg/m²/h, compared to background values of 1–10 mg/m²/h. Mercury concentration in air over contaminated soils ranges from 3–4 mg/m³, compared to background concentrations of 1–2 mg/m³. The yearly soil emission flux of mercury from a contaminated soil or mine site ranges from 10–100 kg of Hg^o/km².

Mercury in Water and Sediment of Seepage Lakes

In seepage lakes within the Sierra Nevada ecosystem, the global increase in the mercury flux is recorded in lake sediments (Hurley et al. 1994). Before about 1850, baseline mercury concentrations in lake sediments are relatively constant and range from 10–50 mg/m²/yr. From about 1850 until about 1960–70, anthropogenic mercury associated with industrialization added to the atmospheric mercury cycle and increased mercury content in lake sediments. Maximum recorded mercury concentrations range from 55–2,000 mg/g. Because of restriction of mercury release to the atmosphere since the 1960–70 period, sediments deposited in lakes since that time show decreased mercury concentration. However, this mercury content is still above initial baseline concentration.

The ratio of the maximum mercury content in the sedimentary record at about 1960–70 to the baseline concentration before evidence of anthropogenic input is termed the *enrichment factor*. Enrichment factors for seepage lakes are a function of the lake's location and range from two to about six. Little is known about the exact magnitude of enrichment factors for lakes in the Sierra Nevada ecosystem or about the effect of regional point source mercury sites on these lakes.

Methylmercury concentration in lake sediments shows a similar pattern of enrichment and then decline in last two decades, but the concentrations of methylmercury are considerably lower. The enrichment factors are much higher however, ranging from twelve to fifteen. In preindustrial (pre-1850) sediments, methylmercury typically constitutes 0.15%–0.3% of the total mercury present, and, in the sediments reflecting high anthropogenic mercury input, methylmercury constitutes from 0.4%–1.4% of the total mercury present. Methylmercury concentration in sediment corese before the industrial period is about 0.2 mg/g and in modern-day sediments ranges up to 1.5 mg/g.

In the lakes, biologically mediated reduction of Hg^{++} in large part controls the Hg^o concentration and evasion rate of Hg^o from the lake to the atmosphere. Removal of mercury through sedimentation is relatively low and about equivalent to the fluvial flux of mercury into lakes. Evasion is the primary factor in removal of mercury from seepage lakes in the Sierra Nevada ecosystem. The total mercury flux added to lakes is about 50 mg/m²/yr.

Methylmercury in Lakes

Formation of methylmercury in lakes is an important environmental factor. Methylmercury (CH_3Hg^+) concentration increases in organisms that are higher in the food chain resulting in significant biomagnification of methylmercury in species such as fish. This high concentration results principally because methylmercury has a long biological half-life in fish. Low levels of methylmercury in lake water may result in very high levels in fish and affect the wildlife that consume fish, as well as the quality of fisheries for human consumption. This problem is the primary environmental concern with respect to methylmercury.

Formation of methylmercury in lakes and wetlands is favored by low pH of the water and high levels of dissolved organic carbon. Mercury methylation is a co-metabolic reaction, and biotic methylation accounts for nearly all the methylmercury in most lakes. Sulfate-reducing bacteria are the most important mediators in the biotic methylation process, a process inhibited by the absence of sulfate. Methylmercury from precipitation accounts for less than 1% of total mercury in a lake system.

Wetlands play a major role in forming methylmercury because of their low pH and high levels of dissolved organic carbon. In the wetland environment, abiotic methylation of Hg^{++} most likely occurs in the presence of humic acids and metal catalysts.

Demethylation is primarily mediated by enzymes in single cells in a variety of organisms within lakes. Detrital particles such as clays and organic matter absorb Hg^{++} from water, and sedimentation of these particles reduces the methylation process. Methylmercury is also partly removed from the water during sedimentation and sequestered in sediments. The sediment trap flux is typically an important factor in the methylmercury cycle.

Nearly all methylmercury is formed in the lake ecosystem. For many lakes the primary concentration of methylmercury resides in the fish population. Methylmercury in the water column is relatively low and only somewhat higher than that present in zooplankton.

SERPENTINE AND LIMESTONE AS HOSTS FOR RARE PLANTS

A preliminary digital version of the geologic map of California (Jennings 1977) is currently being revised by the USGS and the California Division of Mines and Geology. We have used this draft digital file to produce a map showing areas underlain by ultramafic rocks, such as serpentine (unit um) (Jennings 1977) (plate 18.9). This map is useful in assessing the extent of areas where rare plants are likely to be present and in assessing the completeness of the rare-plant lists, because the plants that have adapted to serpentine soils are uncommon. Plants that grow in such soils include, but are not limited to, Lewisia, pitcher plant, Indian paintbrush, Mariposa lily, and lady's slipper (Kruckeberg 1984). Soils developed on carbonate rocks also host rare plants. Therefore, we plotted those areas underlain by carbonate rocks, such as limestone and for other map units that contain carbonate rocks. The units that contain carbonate rocks are further divided into those dominated by carbonate rocks and those dominated by clastic rocks. On Jennings's (1977) map, the units are ls (limestone), Pm (Permian rocks that may contain limestone and dolomite), Tr (Triassic units that likewise may include carbonate rocks), and D (Devonian units that likewise may include carbonate rocks).

SERPENTINE-HOSTED ASBESTOS

Chrysotile (white asbestos) is found in veins in serpentinized ultramafic rocks. Protoliths include ophiolite and stratiform complexes. The veins commonly originate from fractures developed in response to stresses in shear zones, notably near changes in rock competency, as near margins of serpentinite bodies and contact aureoles of later intrusive rocks, but some deposits may result from more uniformly distributed stresses. Minor asbestos veins can be found in unaltered ultramafic rocks adjacent to serpentinite. General references on the geology of these deposits are Virta 1989; Ross 1981; Shride 1969, 1973; Chidester et al. 1978; and Anhaeusser 1986. References on the environmental geology include Coleman in press and Derkies 1985.

The following discussion is part of a draft mineral-environmental model book that is in preparation by the Minerals Teams at the USGS (Wrucke in preparation).

Economic Geology

Serpentine-hosted asbestos deposits do not show significant correlation between tonnage and grade (Orris 1992). The mean tonnage is 17 tonnes (19 short tons) with the range at one standard deviation of between 2 and 149 tonnes (2.2 and 164 short tons). The mean grade asbestos is 4.6% with the range at one standard deviation of between 1.9% and 11%. These figures are reported with no distinction between types of asbestos (cross and slip fibers).

Potential Environmental Impact

Reserves of chrysotile fiber at the asbestos deposit near Copperopolis, Calaveras County, are reported to be about 1.2 million tonnes (1.3 million short tons) (Rice 1966). The deposit, now closed, is used as an asbestos waste dump. The impact of deposits of this type consist of the following

1. Natural exposures of asbestos-bearing rock, particularly serpentinite derived from ultramafic rocks, are readily eroded by natural agents and human activities because most serpentinite is composed of incompetent, highly fractured rock. A few serpentinite bodies are highly resistant to erosion.

2. Asbestos from sedimentary deposits and debris slides from asbestos-bearing rocks is redistributed by water and wind.

3. Vehicles driven across serpentinite and mine waste can dislodge asbestos, adding it to dust or runoff. Roads also produce channels that aid runoff. The surface area of roads in the southern half of the chrysotile-bearing New Idria serpentinite in San Benito County exceeds the area disturbed by the three largest asbestos mines (Woodward-Clyde Associates 1989).

4. Waste generated from asbestos mining and milling operations exposes asbestos to erosion by natural agents. The EPA considers mine waste containing more than 1% asbestos to be hazardous (Derkies 1985). The California Air Resources Board considers asbestos contents of mine waste greater than 5% as a potential toxic hazard (Resolution 91-27, April 1990).

5. Asbestos fibers can be picked up by surface drainage in areas of asbestos-bearing rocks and mines. In central California, water in the California aqueduct system contains asbestos (Kanarek et al. 1980; Coleman 1995). However, the EPA has concluded that there is no significant risk of cancer from the ingestion of asbestos fibers (EPA 1991).

6. Chrysotile deposits may contain small amounts of fibrous tremolite, which the EPA classifies as asbestos and a risk to human health (OSHA 1975).

7. Health risks to humans from exposure to small quantities of chrysotile asbestos in the environment are controversial. The controversy results from the EPA assumption that any amount of asbestos is potentially hazardous.

Environmental Considerations

Host Rocks and Surrounding Geologic Terrane

Serpentine-hosted deposits are present in massive serpentinite, commonly highly sheared and widely exposed, that has largely replaced the host protolith. Associated ultramafic rocks locally host asbestos veins. Most serpentine-hosted deposits have developed in ophiolite complexes, which are composed of oceanic crustal fragments consisting of a basal peridotite (that becomes serpentinized) overlain in sequence by cumulate gabbro, sheeted dikes, and pillow basalt, often capped by deep-oceanic pelagic sedimentary deposits. Accreted ophiolite often is dismembered into structurally complicated relationships.

Mineral Characteristics

Chrysotile is one of six mineral species called asbestos because of their fibrous habit (Skinner et al. 1988). Of these, chrysotile is the only fibrous serpentine mineral. The five other asbestos minerals belong to the amphibole group; these are grunerite asbestos (commonly referred to as amosite), riebeckite asbestos (commonly referred to as crocidolite), anthophyllite asbestos, tremolite asbestos, and actinolite asbestos.

Hydrology

Debris slides along the flanks of serpentinite bodies and in drainage channels can contain asbestos-bearing material available for removal and dispersion by streams (Cowan 1979). During flood stage, streams flowing into the San Joaquin Valley from the New Idria mass have introduced sediments into the California aqueduct (Coleman 1995). Asbestos fibers have been found in the water supply for San Francisco (Kanarek et al. 1980).

Asbestos Mobility from Mining Operations

Mine dumps and mill tailings at chrysotile deposits are sources of asbestos in runoff water and are more easily eroded than outcrops. The principal concerns associated with mineral processing include dust and tailings from asbestos milling operations. Asbestos fibers in tailings are available for airborne and fluvial transport.

Controversy Regarding Health Risks to Humans from Chrysotile Asbestos

The risk of asbestos to human health has been known since at least 1906 when deaths of workers in an asbestos weaving mill in France were noted (D'Agostino and Wilson 1993). In the 1920s, a death from asbestosis (fibrosis of the lung) was reported, and the disease was named (Sawyer 1987). However, not until the 1960s were the biologic effects of asbestos fibers documented in great detail and a relationship clearly established between exposure to asbestos and lung disease, including cancer. As a result of increasing concern about the health hazards of asbestos minerals, the U.S. Occupational Safety and Health Administration (OSHA) in 1971 issued regulations restricting airborne asbestos in the workplace, and in 1987 OSHA established the current standard of 0.2 fibers per cubic centimeter of air in the workplace. However, the regulations do not discriminate between fibers of the different asbestos minerals. Since the establishment of these regulations, numerous studies have supported the conclusion that significant differences exist in the health effects of the various asbestos minerals. For example, the few cases of mesotheliomas (cancer of the pleura or peritoneum) in Canadian chrysotile miners and mill workers appear to be not from chrysotile but perhaps from small amounts of tremolite asbestos, and a study of British workers manufacturing friction materials using only chrysotile showed no excess of deaths from lung diseases (Mossman et al. 1990). Chrysotile has been found to have the least health effects of any asbestos mineral in occupational exposure and to produce no excess lung cancer in people exposed to chrysotile alone in amounts more than ten times higher than recommended by the EPA (Coleman in press).

Risk Assessment

Hazards resulting from the inhalation of asbestos fibers have been documented by the EPA and have been the topic of considerable scientific inquiry (McDonald and McDonald 1995; D'Agostino and Wilson 1993; Mossman et al. 1990; Skinner et al. 1988; Ross 1981, 1987; Ross and Skinner 1994). Although asbestos as a cause of lung disease is well documented, debate continues regarding the risk from low levels of exposure to asbestos fibers. In a report prepared for the California Environmental Protection Agency, risk from asbestos was not ranked because data on low-level exposure were considered inadequate (California Comparative Risk Project 1994). Studies show that important factors to be considered in evaluating risk from the inhalation of asbestos are the type of asbestos mineral, length and diameter of the asbestos fibrils, amount of asbestos inhaled, and the duration of the exposure. Yet, despite conclusions that risks from chrysotile asbestos are almost certainly lower than for other asbestos minerals, uncertainty in the degree of risk from exposure to chrysotile remains and results in part from disagreement on the threshold below which which inhalation can be considered safe or if such a threshold level even exists (D'Agostino and Wilson 1993).

The EPA has concluded that inhalation of any amount of asbestos is potentially hazardous, that a single asbestos fiber can be lethal (Abelson 1990). This conclusion results from belief that the relationship between asbestos dose and health risk is linear, such that there is some risk even at very low levels of exposure (D'Agostino and Wilson 1993). According to this theory, there is no safe level of asbestos exposure. In a nonlinear relationsh, risk from exposure decreases rapidly at low levels, and a threshold value can be reached below which the risk is zero. Recent studies suggest that low-level exposure to chrysotile asbestos in the environment has generated unwarranted concern (D'Agostino and Wilson 1993) and that the single-fiber view is unproved (Abelson 1990). Other studies suggest that there is a threshold value below which exposure to chrysotile asbestos causes no measurable health effects (Ross 1987).

Estimates of risk to human health from numerous activities have been quantified, including everyday risks, and a few attempts have been made to quantify risk of exposure to chrysotile asbestos under different environmental conditions (D'Agostino and Wilson 1993; Coleman 1995). For example, data show that risks from inhaling asbestos during recreational activities at the chrysotile-bearing New Idria serpentinite or from exposure to asbestos in schools are low. Coleman (in press) concluded that "the apparent risk in making one trip by automobile to New Idria is 300 times greater than inhaling [chrysotile] fibers during a lifetime of recreation in this area." Risks from occupying schools containing chrysotile fibers is even lower and has been categorized as harmlessly small (Abelson 1990).

CALCIUM CONCENTRATIONS

James Shevock, a botanist with the U.S. Forest Service in San Francisco, suggested the usefulness of a calcium-concentration map for SNEP. Such a map is also being produced by Tom Frost, Gary Raines, and Lynn Decker for the Interior Columbia Basin Ecosystem Management Project. The following text is excerpted and adapted from Frost, Raines, and Decker (Thomas P. Frost, letter to Michael F. Diggles, February 1995).

Rock Types and Calcium Content

A bedrock calcium-concentration map (see plate 18.9 for simplified version) might serve as a tool for both addressing stream productivity and providing a measure of acid mine drainage and acid rain buffering capacity. We produced such a map by grouping rock types into calcium content classes. The base map from which the classes were defined is the digital version of the geologic map of California at 1:750,000 scale (Jennings 1977). Calcium carbonate, or calcite, is the dominant carbonate phase or mineral.

Rocks have different calcium contents and the easiest way to classify elemental abundances is by the weight percentage of the equivalent oxide, in this case CaO. Calcium rarely occurs as a mineral CaO, but it occurs complexed in distinct minerals. For example, the mineral calcite ($CaCO_3$) is the main component of limestone and calcic marble or calcite. Calcite is very soluble in water relative to other common minerals and is typically the dominant mineral dissolved to form hard water. Other calcium-rich minerals are less soluble under normal surficial conditions and thus may not be as readily available to the aquatic habitat under otherwise similar conditions. Plagioclase feldspar is one of the most common minerals in the earth's crust, but there are variations in the calcium and sodium contents of plagioclases found in different rock types. Calcium in basalts is localized in calcic plagioclase, clinopyroxene, and glass, which are relatively unstable under surficial conditions and readily alter to clays, calcite, and other soft minerals. The calcium, once soluble, is available for biotic uptake.

Rocks such as granites have low calcium content. In areas such as that underlain by the Sierra Nevada batholith, relatively little Ca is available. Some of the folded, faulted, and locally metamorphosed sedimentary rocks of the Sierra Nevada roof pendants are high in Ca and are included in plate 18.9 with other carbonate-rich rocks.

Carbonate rocks are soluble, but where climate is arid, as on the east side of the Sierra Nevada, carbonate rocks tend to be the rock types most resistant to erosion. In arid areas, abundant calcium may not be available to the aquatic environment because there is no mechanism to dissolve it.

Role of Calcium in Aquatic Systems

The calcium content of waters varies from region to region, reflecting both local geography and climate. Calcium is essential for metabolic processes in all living organisms. It plays an important role in its effect on pH and is essential in the main buffering system of natural waters.

Virtually all vertebrates, mollusks, and certain other invertebrates require large quantities of calcium (in the form of $CaCO_3$) as a major skeletal-strengthening material. Because animals need large amounts of calcium, their growth and, ultimately, size may be limited by the lack of it. Where calcium is plentiful, for example in what are commonly referred to as "chalk streams," one will potentially find large macroinvertebrates and large fish.

Most lakes have a pH of 6 to 9. When the pH of a lake falls below 4 or 5, the species diversity is likely to be severely restricted. Most fish species can live in waters with pH ranging from 5 to 9. In general, the higher the pH, the greater the productivity of the waters and the faster the growth rates of the fish and macroinvertebrates. The most productive waters (in general) have a pH of about 8. However, only a few fish are adapted for living in water of pH 8.5 or greater. One example of a fish adapted for life in alkaline waters is the Eagle Lake trout in Lassen County, where pH levels may reach 9.6. In general, water of pH greater than 10 or less than 4 will be fatal to fish.

There are at least two obvious difficulties with attempting to work out ratios of various minerals in solution in a drainage basin. One is the differential solubilities of minerals and the other is that the aquatic biota exert selective effects on dissolved substances.

VOLCANIC HAZARDS

The principal volcanic-hazard area in the SNEP study area is in the vicinity of Mammoth Lakes in the Long Valley caldera. Outside the SNEP core study area and within the greater SNEP study area, volcanic centers of the Cascade Range are important as well. These include Lassen Peak, Mount Shasta, and Medicine Lake Volcano.

Long Valley Caldera

Long Valley and Mono Craters have been the site of volcanic activity for millions of years. Bailey (1989) presents a geologic map of the region that shows extent, ages, and descriptions of volcanic rocks in the area. He includes a comprehensive treatment of the formation of the caldera, which was the resulted from massive eruptions more than 700,000 years ago.

Earthquake activity in the Long Valley region began to in-

crease in 1978 and peaked in 1980. This activity was interpreted to be the result of magma movement beneath Long Valley caldera. Miller et al. (1982) show potential hazard zones in the region, taking into account common wind directions and topographic barriers.

A comprehensive response plan for volcanic hazards in the Long Valley caldera and Mono Craters area was prepared by Hill et al. (1991) of the USGS in cooperation with the California Division of Mines and Geology. In their report, Hill et al. (1991) state that recurring earthquake swarms in Long Valley caldera through the 1980s and associated inflation of the resurgent dome in the caldera emphasize that this geologically youthful volcanic system is capable of further volcanic activity. Specific response actions under their plan are keyed to a five-level status ranking of activity level. The activity levels are eruption likely within hours to days, and intense strong, moderate, or weak unrest. The USGS continuously monitors volcanic activity in Long Valley caldera and vicinity by means of a seismic network and deformation monitoring networks (dilatometers [strainmeters], tiltmeters, and magnetometers).

If activity levels indicate that an eruption is likely, the response plan states that an eruption will most likely produce small to moderate volumes of silicic lava similar to the eruptions that occurred 650 years ago at the north end of Mono Craters and 550 years ago at the Inyo Domes. In this case, we may expect to see

- phreatic eruptions as the magma interacts with the shallow ground water producing steam blasts that can throw large rocks several hundred meters from the vent (the "eruption" could stop at this point as it did with the phreatic blasts that formed the Inyo Craters)

- an explosive magmatic phase during which hot pumice and ash would be ejected thousands of feet into the air producing thick pumice accumulations near the vent, extensive deposits of fine ash hundreds of kilometers downwind, and destructive pyroclastic flows that may reach distances as great as 5 to 10 km (3–6 mi) from the vent

- a final phase that involves the slow extrusion of lava to form steep-sided flows and domes.

Like the eruptions 550 and 650 years ago, eruptions may occur from several separate vents in succession with the vents spaced over a distance of 5 to 10 km (3 to 6 mi). Individual eruptions may be separated in time by days or perhaps weeks. Larger, more destructive eruptions following the same basic pattern are possible, but less likely. Also possible, but less likely, is a small to moderate eruption of basaltic lava similar to the eruptions that produced the Red Cones several thousand years ago. This lava could travel at speeds ranging from a few meters per hour to several kilometers per hour. The resulting lava flows may extend 10 km (6 mi) or more from the vents depending on the vigor and duration of the eruption.

Miller et al. (1982) include a hazard zone for the unlikely event of an eruption as large as that which took place 700,000 years ago. Devastation within 120 km (75 mi) would be severe to total. Pyroclastic flows would move at speeds of several hundreds of kilometers per hour. Deposits of ash 15 cm (6 in) thick would fall as far away as 500 km (300 mi) with appreciable thickness deposited all across North America. Such an event has not taken place anywhere on the earth in historic times.

Cascade Volcanoes

The three volcanic centers of the Cascade Range within the SNEP study area are Medicine Lake Volcano, Lassen Peak, and Mount Shasta (USGS 1994). Information about Cascade volcanoes is available from the Cascade Volcano Observatory of the USGS on the WWW Home Page whose URL is http://vulcan.wr.usgs.gov/home.html. Lassen Peak erupted early this century, and Mount Shasta's last eruption (Miller 1980) was about two hundred years ago. Medicine Lake Volcano has not erupted in nearly a thousand years.

The following text is published on the WWW at http://vulcan.wr.usgs.gov.

Lassen Peak

This text is excerpted from Hoblitt et al. 1987. The Lassen volcanic center consists of a chain of vents aligned roughly north-south that extends about 8 km (5 mi) north from Lassen Peak. Although volcanism began between about 600,000 and 350,000 years ago, events of the last 35,000 years are the most thoroughly studied and form the basis for assessing hazards from future eruptions in the region. The stratigraphic record of late Pleistocene and Holocene eruptions in this region contains evidence of many eruptions during the last 35,000 years

35,000 years ago: Eruptions produced two pyroclastic flows from a vent east of Sunflower Flat near the north end of the chain. These eruptions were followed by extrusion of one or more domes at vents in the same area.

25,000–35,000 years ago: Eruptions at Hat Mountain produced andesitic lava flows that reached up to 6 km (4 mi) from their vents. About the same time, eruptions at a vent now buried by the Lassen Peak dome produced at least four pyroclastic flows and several short rhyolite lava flows.

20,000 years ago: Eruptions formed an ancestral dome, now buried by the Lassen Peak dome, which is thought to have erupted shortly before 11,000 years ago. During late Wisconsin deglaciation, lahars formed on the slopes of Lassen Peak and flowed at least several kilometers, primarily to the northeast.

1,000–1,200 years ago: The Chaos Crags eruptive episode began with eruption of a pumiceous tephra. At least two pyroclastic flows traveled west down Manzanita Creek about 4 km (2.5 mi) and a similar distance north down Lost Creek. Explosive activity generated pyroclastic flows that extended down Manzanita, Lost, and Hat Creeks. Shortly thereafter, extrusion of five dacite domes formed the Chaos Crags.

300 years ago: Three or more rockfalls from the Chaos Crags generated high-velocity avalanches of rock debris that traveled as far as 4.3 km (2.7 mi) westward from the Chaos Crags. The falls may have resulted from earthquakes, steam explosions, or intrusion of a dome into the central part of the Chaos Crags.

A.D. 1914–1917: The most recent eruption at Lassen Peak took place early in this century, when a small phreatic eruption occurred on May 30, 1914, at a new vent near the summit of the peak. More than 150 explosions of various sizes occurred during the following year. A vertical eruption column resulting from the pyroclastic eruption rose to an altitude of more than 9 km (5.6 mi) above the vent and deposited a lobe of pumiceous tephra that can be traced as far as 30 km (19 mi) to the east-northeast. The fall of fine ash was reported as far away as Elko, Nevada, more than 500 km (300 mi) east of Lassen Peak. Intermittent eruptions of variable intensity continued until about the middle of 1917.

The record of late Pleistocene and Holocene eruptive activity at the Lassen volcanic center suggests that the most likely hazardous future events include pyroclastic eruptions that produce pyroclastic flows and tephra. The Lassen volcanic center is one of the principal candidates in the Cascade Range for future silicic, probably explosive, eruptions. Based on its history, pyroclastic flows could endanger areas within several tens of kilometers of an active vent. Lahars and floods could affect low-lying areas even farther from the vent, particularly if eruptions occur during periods of thick snow cover. Eruptions that produce lava flows are generally less dangerous, although both lava flows and domes can become unstable and produce pyroclastic flows and rockfall avalanches that could affect areas as far as several kilometers away. Mixing of hot debris with snow can generate lahars that could inundate valley bottoms for tens of kilometers as in 1915.

Mount Shasta

Future eruptions like those of the last ten thousand years will probably produce deposits of lithic ash, lava flows, domes, and pyroclastic flows and could endanger works of humans that lie within several tens of kilometers of the volcano. Lava flows and pyroclastic flows may affect low areas within about 15–20 km (9–13 mi) of the summit of Mount Shasta or any satellite vent that might become active. Lahars could affect valley floors and other low areas as far as several tens of kilometers from Mount Shasta. Owing to great relief and steep slopes, a part of the volcano could also fail catastrophically and generate a very large debris avalanche and lahar. Such events could affect any sector around the volcano and could reach more than 50 km (32 mi) from the summit. Explosive lateral blasts could also occur as a result of renewed eruptive activity, or they could be associated with a large debris avalanche; such events could affect broad sectors to a distance of more than 30 km (19 mi) from the volcano. On the basis of Holocene behavior, the probability is low that Mount Shasta will erupt large volumes of pumiceous ash in the future. The distribution of Holocene tephra and the prevailing wind directions suggest that areas most likely to be affected by tephra are mainly east and within about 50 km (32 mi) of the volcano's summit. However, the andesitic and dacitic composition of its products suggests that Mount Shasta could erupt considerably larger volumes of tephra in the future. Moreover, Christiansen (1982) has suggested that because it is a long-lived volcanic center and has erupted only relatively small volumes of magma for several thousand years, Mount Shasta is the Cascade Range volcano most likely to produce a very large explosive eruption (10^1–10^2 km^3 [24.7–24.9 mi^3]). Such an event could produce tephra deposits as extensive and as thick as the Mazama ash deposits and pyroclastic flows that could reach more than 50 km (32 mi) from the vent. The annual probability for such a large event may be no greater than 10^{-5}, but it is finite.

Medicine Lake Volcano

Eruptions occurring during the past ten thousand years form a reasonable basis for assessing hazards from future eruptions. Similar eruptions of silicic magma are likely from vents within and just outside the summit caldera, which is thought to be underlain by silicic magma (Heiken 1978; Eichelberger 1981), part of which could still be molten. These eruptions probably will produce tephras that could fall as much as several hundred kilometers downwind and mostly east of the volcano (Christiansen 1982; Miller, in press). Such eruptions could also produce pyroclastic flows that could endanger areas within about 10 km (6 mi) of the active vent, although such phenomena are not known to have occurred during Holocene time. Silicic eruptions are likely to culminate with eruption of dacite to rhyolite lava flows or domes that could reach as far as several kilometers from their vents. Eruptions of basalt and basaltic andesite lava may also occur from vents on the flanks of the Medicine Lake volcano (Christiansen 1982). Such eruptions may begin by forming cinder cones and dispersing mafic tephra as far as 20 km (13 mi) from the active vent and culminate with the production of lava flows that may extend for tens of kilometers downslope from their vents. Eruptions of both mafic and silicic magma may be fed by dikes. As a consequence, eruptions of basalt and rhyolite may occur simultaneously, or nearly so, from multiple, probably aligned vents. Eruptions of volumes larger than those of Holocene time are possible, including a caldera-forming eruption (Christiansen, 1982), because of the inferred existence of a large body of silicic magma beneath the Medicine Lake vol-

cano (Heiken 1978; Christiansen 1982). Future eruptions of this type could deposit thick accumulations of tephra over wide regions and produce pyroclastic flows that could affect areas more than 50 km (32 mi) from the vent. Debris avalanches and laterally directed blasts are not known to have occurred in this region in the past. Owing to the limited relief of the Medicine Lake volcano, debris avalanches are not considered likely in the future. Because of the absence of permanent snow and ice, future eruptions are not likely to generate large-volume lahars and floods, although lahars and floods of moderate volumes are possible if eruptions occur when snow covers the ground.

EARTHQUAKE HAZARDS

The alignment of epicenter clusters along the east side of the Sierra Nevada branches northward from the south end of the San Andreas fault system in the Salton Trough and bends back toward the north terminus of the San Andreas fault system at the Mendocino triple junction in northern California (Hill et al. 1990). A dense cluster of epicenters in the Mammoth Lakes area represents an episode of intense earthquake activity in Long Valley caldera that began in 1978 (Hill et al. 1985). To make a map showing twentieth-century seismicity in the Sierra Nevada (plate 18.10), we downloaded the latest data (as of January 16, 1996) from the Northern California Earthquake Data Center at the University of California, Berkeley, using their WWW Home Page whose URL is http://quake. geo.berkeley.edu/ and imported the data into Arc/Info software. A total of 3,321 earthquakes with magnitudes of at least 3.0 that occurred since 1910 are plotted on the map. From magnitude 3.0 to 3.9, there are 2,614 points; from magnitude 4.0 to 4.9, there are 616 points; from magnitude 5.0 to 5.9 there are 79 points; and from magnitude 6.0 to 6.3 there are 12 points. No earthquakes greater than magnitude 6.3 have occurred in the SNEP area during this century.

The largest historic earthquakes in the region were in Owens Valley in 1872. Between twenty-three and twenty-nine people died in the magnitude 7.6 earthquake in Lone Pine on May 26 (Sharp 1972), and the town was virtually leveled when the entire 100–110 km (63–69 mi) length of the Owens Valley fault ruptured (Ellsworth 1990). A magnitude 6.75 aftershock occurred later that same day, and another magnitude 6.75 aftershock occurred north of Bishop several days later (Goter et al. 1994). Adobe and brick buildings in Owens Valley sustained most of the damage. John Muir experienced the shaking from Yosemite Valley where he witnessed a rockfall triggered by the event. Ellsworth (1990) points out that the first long-term seismic forecast was made by G. K. Gilbert in 1883 when he noted that rebuilding Independence with wood was an extravagance because "the spot which is the focus of an earthquake . . . is thereby exempted [unlikely to have an earthquake] for a long time" (Gilbert 1884). It has now been a long time, and attention to seismic building standards is wise and encouraged.

A more general map showing all of California and Nevada was produced by Goter et al. (1994). It covers all events up to its publication date, including those with magnitudes below 1.0 and those that occurred in the last century. In that report, the authors point out that many faults capable of producing large earthquakes are quiescent for long periods between such events. Therefore, the faults indicated by seismicity on maps do not represent all faults in the region that have seismic-hazard potential.

ACKNOWLEDGMENTS

The authors of the different parts of this chapter are Michael F. Diggles (U.S. Geological Survey, Menlo Park, California), "Earth Sciences Overview" and "Mineral Deposits in the Sierra Nevada"; William J. Pickthorn (Isochem Co., Palo Alto, California) and Michael F. Diggles, "Low-Sulfide Gold-Quartz Vein Deposits"; Dennis P. Cox (U.S. Geological Survey, Menlo Park, California), Steve Ludington (U.S. Geological Survey, Menlo Park, California), and Michael F. Diggles, "Massive Sulfide Deposits, Sierran Kuroko Type"; Dennis P. Cox and Roger P. Ashley (U.S. Geological Survey, Menlo Park, California), "Porphyry Copper Deposits"; Michael F. Diggles, "Epithermal Vein Deposits, Quartz-Adularia Type"; Dennis P. Cox, "Copper Skarn" and "Polymetallic Skarn and Replacement Deposits"; C. Thomas Hillman (U.S. Bureau of Mines, Spokane, Washington) and Michael F. Diggles, "Known Mines, Claims, and Prospects"; James R. Rytuba (U.S. Geological Survey, Menlo Park, California), "Mercury in the Sierra Nevada Ecosystem"; Michael F. Diggles and Barry C. Moring (U.S. Geological Survey, Menlo Park, California), "Serpentine and Limestone as Hosts for Rare Plants"; Chester T. Wrucke (U.S. Geological Survey, Menlo Park, California) and Michael F. Diggles, "Serpentine-Hosted Asbestos"; Michael F. Diggles, Barry C. Moring, and Robert J. Miller (U.S. Geological Survey, Menlo Park, California), "Calcium Concentrations"; Michael F. Diggles, "Volcanic Hazards"; and Michael F. Diggles, "Earthquake Hazards."

Information for the discussion of undiscovered precious and base metal deposits (gold, silver, copper, lead, and zinc) is from work in review on the resource assessment of the United States. Dennis P. Cox and Steve Ludington were the national team leaders for this work. Barry C. Moring, Dan L. Mosier, and Paul C. Schruben were the software engineers. Michael F. Diggles was the regional team leader; David H. Root and William Scott were the statisticians and programmers. Present and former USGS staff who helped make estimates or otherwise contributed to the process are as follows: George V. Albino, Roger P. Ashley, Byron R. Berger, Richard J. Blakely, Joseph A. Briskey Jr., William F. Cannon, Stanley E. Church, Dennis P. Cox, Michael F. Diggles, Lawrence J. Drew,

Susan H. Garcia, Donald F. Huber, Robert C. Jachens, M. Dean Kleinkopf, Steve Ludington, W. David Menzie, J. Thomas Nash, Steven G. Peters, Jocelyn A. Peterson, Willaim J. Pickthorn, James J. Rytuba, Michael G. Sawlan, Donald A. Singer, Gregory T. Spanski, Richard M. Tosdal, and Robert A. Zierenberg. Compilations and other work on data for known deposits were done by Russell C. Evarts, Steve Ludington, Dan L. Mosier, Lorre A. Moyer, and Miles L. Silberman of the USGS and Charles Bishop, J. Douglas Causey, Thomas Gunther, C. Thomas Hillman, and Paul C. Hyndman of the U.S. Bureau of Mines.

David Oppenheimer at USGS suggested the earthquake data-retrieval and display strategy. Douglas Neuhauser at Northern California Earthquake Data Center at University of California, Berkeley, set up Diggles's account on their machine and helped perform the data retrieval. Barry C. Moring at USGS combined the data with standard SNEP layers in Arc/Info software and produced the plot.

REFERENCES

Abelson, P. H. 1990. The asbestos removal fiasco.: *Science* 247:1017.

Albers, J. P., 1981, A lithologic-tectonic framework for the metallogenic provinces of California. *Economic Geology* 76 (4):765–90.

Anderson, C. A. 1931. Geology of the Engels and Superior Mines, Plumas County, California (with a note on the ore deposits of the Superior mine): *California University Geological Sciences Bulletin* 20 (8): 293–330.

Anhaeusser, C. R. 1986. The geological setting of chrysotile asbestos occurrences in southern Africa. In *Mineral deposits of southern Africa,* edited by C.R. Anhaeusser and S. Mask, 359–75. Johannesburg:, Geological Society of South Africa.

Bailey, R. A. 1989. Geologic map of the Long Valley caldera, Mono-Inyo Craters volcanic chain, and vicinity, eastern California. U.S. Geological Survey Miscellaneous Geologic Investigations Map I-1933, scale, 1:62,500.

Bateman, P. C. 1965. *Geology and tungsten mineralization of the Bishop district, California.* U.S. Geological Survey Professional Paper 470.

Berger, B. R. 1986. Descriptive model of low-sulfide Au-quartz veins. In *Mineral deposit models,* edited by D. P. Cox and D. A. Singer, 239. U.S. Geological Survey Bulletin 1693.

Bliss, J. D. 1986. Grade and tonnage model of low-sulfide Au-quartz veins. In *Mineral deposit models,* edited by D. P. Cox and D. A. Singer, 239–43. U.S. Geological Survey Bulletin 1693.

Bliss, J. D., ed. 1992. *Developments in mineral deposit modeling.* U.S. Geological Survey Bulletin 2004.

Bliss, J. D., and G. M. Jones. 1988. *Mineralogic and grade-tonnage information on low-sulfide Au-quartz veins.* U.S. Geological Survey Open-File Report 88-0229.

Bliss, J. D., W. D. Menzie, G. J. Orris, and N. J Page. 1987. Mineral deposit density: A useful tool for mineral-resource assessment. *U.S. Geological Survey research on mineral resources—1987 program and abstracts,* edited by J. S. Sachs, 6. U.S. Geological Survey Circular 995.

California Comparative Risk Project. 1994. *Toward the 21st century: Planning for protection of California's environment.* Sacramento: California Environmental Protection Agency.

Carlisle, D., D. L. Davis, M. B. Kildale, and R. M. Stewart. 1954. *Base metal and iron deposits of southern California.* California Division of Mines and Geology Bulletin 170:41–49.

Chidester, A. H., A. L. Albee, and W. M. Cady. 1978. *Petrology, structure, and genesis of the asbestos-bearing ultramafic rocks of the Belvidere Mountain area in Vermont.* U.S. Geological Survey Professional Paper 1016.

Christiansen, R. L. 1982. Volcanic hazard potential in the California Cascades. *California Division of Mines and Geology, Special Publication* 63:41–59.

Clark, D. 1992. Future actions to meet industrial-mineral needs. In *Industrial minerals in the Basin and Range region—workshop proceedings,* edited by E. W. Tooker, 96–98 U.S. Geological Survey Bulletin 2013.

Clark, W. B. 1976. *Gold districts of California.* California Division of Mines and Geology Bulletin 193.

Cowan, D. S. 1979. Serpentinite flows on Joaquin Ridge, southern Coast Ranges, California. *Geological Society of America Bulletin* 81:2615–28.

Cox, D. P. 1986a. Descriptive model of porphyry Cu deposits. In *Mineral deposit models,* edited by D. P. Cox and D. A. Singer, 76. U.S. Geological Survey Bulletin 1693.

———. 1986b. Descriptive model of Zn-Pb skarn deposits. In *Mineral deposit models,* edited by D. P. Cox and D. A. Singer, 90. U.S. Geological Survey Bulletin 1693.

Cox, D. P., and D. A. Singer, eds. 1986. *Mineral deposit models.* U.S. Geological Survey Bulletin 1693.

Cox, D. P., and T. G. Theodore. 1986. Descriptive model of Cu skarn deposits. In *Mineral deposit models,* edited by D. P. Cox and D. A. Singer, 86. U.S. Geological Survey Bulletin 1693.

D'Agostino, J. R., and R. Wilson. 1993. Asbestos: The hazard, the risk, and public policy. In *Phantom risk,* edited by K. R. Foster, D. E. Bernstein, and P. W. Huber, 183–210. Cambridge: Massachusetts Institute of Technology.

Dellinger, D. A., 1989. *California's unique geologic history and its role in mineral formation, with emphasis on the mineral resources of the California Desert region.* U.S. Geological Survey Circular 1024.

Derkies, D. 1985. *Wastes from the extraction and beneficiation of metallic ores, phosphate rock, asbestos, overburden from uranium mining, and oil shale,* prepared for Environmental Protection Agency. National Technical Information Service Report PB 88-162631.

Diggles, M. F., J. G. Frisken, D. Plouff, S. R. Munts, and T. J. Peters. 1988. *Mineral resources of the Skedaddle Mountain Wilderness Study Area, Lassen County, California, and Washoe County, Nevada.* U.S. Geological Survey Bulletin 1706-C.

du Bray, E. A., D. A. Dellinger, H. W. Oliver, M. F. Diggles, M. F. Johnson, F. L. Thurber, H. K. Morris, T. J. Peters, and D. S. Lindey. 1982. Mineral resource potential map of the John Muir Wilderness, Fresno, Inyo, Madera, and Mono counties, California. U.S. Geological Survey Miscellaneous Field Studies Map MF-1185-C, scale 1:125,000, 2 sheets.

Durkin, T. V. In press. Acid mine drainage: Reclamation at the Richmond Hill and Gilt Edge mines, South Dakota. *Proceedings EPA seminar series on managing environmental problems at inactive and abandoned metals mine sites, Anaconda, MT, Denver, CO, Sacramento, CA, 1994.*

Eichelberger, J. C. 1981. Mechanism of magma mixing at Glass Mountain, Medicine Lake Highland volcano, California. In *Guides to some volcanic terranes in Washington, Idaho, Oregon, and northern California,* edited by D. A. Johnston and J. M. Donnelly-Nolan, 183–89. U.S. Geological Survey Circular 838.

Einaudi, M. T., and D. M. Burt. 1982. Introduction—terminology, classification, and composition of skarn deposits: *Economic Geology* 77:745–54.

Ellsworth, W. L. 1990. Earthquake history, 1769 - 1898. In *The San Andreas fault system, California,* edited by R. E. Wallace, 153–88. U.S. Geological Survey Professional Paper 1515.

Fitzgerald, W. F. 1989. Atmospheric and oceanic cycling of mercury. *Chemical Oceanography* 10:151–86.

Gilbert, G. K. 1884. A theory of the earthquakes of the Great Basin, with a practical application. *American Journal of Science* ser. 3, 27 (157): 49–53.

Gill, G. A., and K. W. Bruland. 1990. Mercury speciation in surface freshwater systems in California and other areas: *Environmental Science and Technology* 24 (13): 1392–1400.

Goldfarb, R. J., B. R. Berger, T. L. Klein, W. J. Pickthorn, M. L. Silberman. In preparation. Low sulfide Au quartz veins (Model 36a, Cox And Singer, 1986). *Mineral-environmental models,* edited by E. A. Du Bray. U.S. Geological Survey Bulletin.

Goter, S. K., D. H. Oppenheimer, J. J. Mori, M. K. Savage, and R. P. Massé. 1994. Earthquakes in California and Nevada. U.S. Geological Survey Open-File Report 94-647, scale 1:1,000,000.

Hall, W. E., and E. M. MacKevett Jr. 1962. *Geology and ore deposits of the Darwin quadrangle, Inyo County California.* U.S. Geological Survey Professional Paper 368.

Heiken, G. 1978. Plinian-type eruptions in the Medicine Lake Highland, California, and the nature of the underlying magma. *Journal of Volcanology and Geothermal Research* 4:375–402.

Hill, D. P., J. P. Eaton, and L. M. Jones. 1990. Seismicity, 1980–86. In *The San Andreas fault system, California,* edited by R. E. Wallace, 115–52. U.S. Geological Survey Professional Paper 1515.

Hill, D. P., M. J. Johnston, J. O. Langbein, S. R. McNutt, C. D. Miller, C. E. Mortensen, A. M. Pitt, and S. A. Rojstaczer. 1991. *Response plans for volcanic hazards in the Long Valley Caldera and Mono Craters area, California.* U.S. Geological Survey Open-File Report 91-270.

Hill, D. P., R. E. Wallace, and R. S. Cockerham. 1985. Review of evidence on the potential for major earthquakes and volcanism in the Long Valley-Mono Craters-White Mountains regions of eastern California. *Earthquake Prediction Research* 3 (3-4): 571–93.

Hoblitt, R. P. C. D. Miller, and W. E. Scott. 1987. *Volcanic hazards with regard to siting nuclear-power plants in the Pacific Northwest.* U.S. Geological Survey Open-File Report 87-0297.

Hotz, P. E. 1971. *Geology of lode gold districts in the Klamath Mountains, California and Oregon.* U.S. Geological Survey Bulletin 1290.

Hurley, J. P., D. P. Krabbenhoft, C. L. Babiarz, and A. W. Andren. 1994. Cycling of mercury across the sediment-water interface in seepage lakes. In *Environmental chemistry of lakes and reservoirs,* edited by A. Baker, 425–49. American Chemical Society Series No. 237.

Janik, C. J., F. Goff, and J. J. Rytuba. 1994. Mercury in waters and sediments of the Wilbur Hot Springs area, Sulphur Creek Mining District, California: *EOS Transactions American Geophysical Union* 75 (44): 243.

Jennings, C. W. 1977. Geologic map of California: California Division of Mines and Geology Geologic Map of California, scale 1:750,000.

Jones, G. W. and W. D. Menzie. 1986. Grade-tonnage model of Cu skarns. In *Mineral deposit models,* edited by D. P. Cox and D. A. Singer, 86–89. U.S. Geological Survey Bulletin 1693.

Kanarek, M. S., P. M. Conforti, L. A. Jackson, R. C. Cooper, and J. C. Murcho. 1980. Asbestos in drinking water and cancer incidence in the San Francisco Bay area: *American Journal of Epidemiology* 112: 54-72.

Klein, D., and V. Banky. 1992. Geophysical model of Creede, Comstock, Sado, Goldfield and related epithermal precious metal deposits, Cox and Singer models 25b Creede, 25c Comstock, 25d Sado, and 25e quartz-alunite Au. In *The geophysical expression of selected mineral deposit models,* edited by D. B. Hoover, W. D. Heran, and P. L. Hill, 98–106. U.S. Geological Survey Open-File Report 92-557.

Knopf, A. 1935. The Plumas County copper belt, California. In *Copper Resources of the World* 1:241–45. Sixteenth International Geologic Congress, Washington DC.

Kockelman, W. J. 1990. Land-use planing and reclamation. In *Industrial minerals in California: Economic importance, present availability, and future development,* edited and compiled by E. W. Tooker and D. J. Beebe, 43–44. U.S. Geological Survey Bulletin 1958.

Kruckeberg, A. R. 1984. *California serpentines, flora, vegetation, geology, and soils.* Berkeley and Los Angeles: University of California Press.

Lindberg, S. E., D. R. Jackson, J. W. Huckabee, S. A. Janzen, M. J. Levin, and J. R. Lund. 1979. Atmospheric emission and plant uptake of mercury from agricultural soils near the Almaden mercury mine: *Journal of Environmental Quality* 8:572–78.

Lindberg, S. E., K. Kim, T. P. Meyers, and J. G. Owens. 1995. Micrometeorological gradient approach for quantifying air/surface exchange of mercury vapor: tests over contaminated soils: *Environmental Science and Technology* 29 (1): 126–35.

Lindberg, S.E., T. P. Meyers, G. E. Taylor, R. R. Turner, and W. H. Schroeder. 1992. Atmosphere-surface exchange of mercury in a forest: Results of modeling and gradient approaches: *Journal of Geophysical Research* 97 (D2): 2519–28.

MacKevett, E. M., Jr. 1953. *Geology of the Santa Rosa lead mine, Inyo County California.* California Division of Mines and Geology Special Report 34.

Mason, R. P., W. F. Fitzgerald, and F. M. M. Morel. 1994. The biogeochemical cycling of elemental mercury: Anthropogenic influences: *Geochimica et Cosmochimica Acta* 58 (15): 3191–98.

McAllister, J. F. 1955. *Geology and mineral deposits of the Ubehebe Peak quadrangle.* California Division of Mines and Geology Special Report 42.

McCammon, R. B., and J. A. Briskey, Jr. 1992. A proposed national mineral-resource assessment: *Nonrenewable Resources* 1:259–66.

McDonald, J. C., and A. D. McDonald. 1995. Chrysotile, tremolite, and mesothelioma. *Science* 267:778–79.

McPhee, J. 1992. Annals of the former world: Assembling California— I. *The New Yorker,* 7 September, 36–68.

Miller, C. D. 1980. *Potential hazards from future eruptions in the vicinity of Mount Shasta volcano, northern California.* U.S. Geological Survey Bulletin 1503.

———. In press. *Potential volcanic hazards from future volcanic eruptions in California.* U.S. Geological Survey Bulletin.

Miller, C. D., D. R. Mullineaux, D. R. Cradell, and R. A. Bailey. 1982. *Potential hazards from future volcanic eruptions in the Long Valley-Mono Lake area, east-central California and southwest Nevada—a preliminary assessment.* U.S. Geological Survey Circular 877.

Morris, H. T. 1986. Descriptive model of polymetallic replacement deposits. In *Mineral deposit models,* edited by D. P. Cox and D. A. Singer, 99–100. U.S. Geological Survey Bulletin 1693.

Mosier, D. L. 1986. Grade-tonnage model of Zn-Pb skarns. In *Mineral deposit models,* edited by D. P. Cox and D. A. Singer, 90–93. U.S. Geological Survey Bulletin 1693.

Mosier, D. L., H. T. Morris, and D .A. Singer. 1986. Grade-tonnage model of polymetallic replacement deposits. In *Mineral deposit models,* edited by D. P. Cox and D. A. Singer, 101–4. U.S. Geological Survey Bulletin 1693.

Mossman, B. T., J. Bignon, M. Corn, A. Seaton, and J .B. L. Gee. 1990. Asbestos: Scientific developments and implications for public policy. *Science* 247:294–301.

Mount, J. F. 1995. *California rivers and streams: The conflict between fluvial process and land use.* Berkeley and Los Angeles: University of California Press.

Munts, S. R., and T. J. Peters. 1987. *Mineral resources of the Skedaddle study area, Lassen County, California and Washoe County, Nevada.* U.S. Bureau of Mines Open-File Report MLA 22-87.

Newberry, R. T. 1982. Tungsten-bearing skarns of the Sierra Nevada. I. The Pine Creek mine. *Economic Geology* 77:823–44.

Newberry, R. J., M. T. Einaudi, and H. S. Eastman. 1991. Zoning and genesis of the Darwin Pb-Zn-Ag skarn deposit, California: A reinterpretation based on new data. *Economic Geology.* 86(5): 960–82.

Nriagu, J. O., and J. M. Pacyna. 1988. Quantitative assessment of worldwide contamination of air, water and soils by trace metals: *Nature* 333:134–39.

Oddenino, C. L., L. V. Coppa, and J. Dillon. 1995. *Spatial data extracted from the Mineral Availability System/Mineral Industry Location System (MAS/MILS).* U.S. Bureau of Mines Special Publication 12-95, CD-ROM. Available from the U.S. Government Printing Office, Superintendent of Documents, Mail Stop SSOP, Washington, DC, 20402–9328, Telephone order: (202) 783-3238.

Orris, G. J. 1992. Grade and tonnage models of serpentine-hosted asbestos, model 8d. In *Industrial minerals deposit models: Grade and tonnage models,* edited by G. J. Orris and J. D. Bliss, 2–4. U.S. Geological Survey Open-File Report 92-437.

Plumlee, G. S., K. S. Smith, and W. H. Ficklin. 1994. *Geoenvironmental models of mineral deposits and geology-based mineral-environmental assessments of Public lands.* U.S. Geological Survey Open-File Report 94-203.

Rasmussen, P. E. 1994. Current methods of estimating atmospheric mercury fluxes in remote areas: *Environmental Science and Technology* 28 (13): 2233–41.

Rice, S. J. 1966. Asbestos. In *Mineral resources of California,* 86–92. California Division of Mines and Geology Bulletin 191.

Root, D. H., W. D. Menzie, and W. A. Scott. 1992. Computer Monte Carlo simulation in quantitative resource assessment. *Nonrenewable Resources* 1:125–38.

Ross, M. 1981. The geologic occurrences and health hazards of amphibole and serpentine asbestos. In *Amphiboles and other hydrous pyriboles—mineralogy,* edited by D. R. Veblen, 279–323. Mineralogical Society of America Reviews in Mineralogy 9A.

———. 1987. Minerals and health: the asbestos problem. In *Aggregates to zeolites (AZ) in Arizona and the Southwest: Proceedings of the 21st forum on the geology of industrial minerals, April 9-12, 1985,* edited by W. H. Peirce, 101–15. Tucson: Arizona Bureau of Geology and Mineral Technology.

Ross, M., and C. W. Skinner. 1994. Minerals and cancer. *Geotimes* 39: 13–15.

Rytuba, J. J. 1988. Volcanism, extensional tectonics, and epithermal systems in the northern Basin and Range, CA, NV, OR, and ID [abs.]. *Geological Society of Nevada Newsletter* May, 1–2.

———. 1989. Volcanism, extensional tectonics, and epithermal mineralization in the northern basin and range province, California, Nevada, Oregon, and Idaho. *U.S. Geological Survey research on mineral resources—1989 program and abstracts, fifth annual V. E. McKelvey forum on mineral and energy resources* [abs.], edited by K. S. Schindler, 59–61. U.S. Geological Survey Circular 1035.

Rytuba, J. J., and W. R. Miller. 1994. Environmental geochemistry of active and extinct hot spring mercury deposits in the California Coast Ranges. In *USGS research on mineral resources—1994 part A—program and abstracts*, edited by L. M. H. Carter, M. I. Toth, and W. C. Day, 90–91. U.S. Geological Survey Circular 1103-A.

Sawyer, R. N. 1987. Asbestos exposure: health effects update. *National Asbestos Council Journal* 5 (2): 25–29.

Scherer, J. 1992. Environmental concerns for land, air, and water. In *Industrial minerals in the Basin and Range region—workshop proceedings*, edited by E. W. Tooker, 94–96. U.S. Geological Survey Bulletin 2013.

Seigneur, C., J. Wrobel, and E. Constantinou. 1994. A chemical kinetic mechanism for atmospheric mercury: *Environmental Science and Technology* 28:2433–40.

Sharp, R. P., 1972, *Geology field guide to southern California*. Dubuque, IA: Kendall/Hunt.

Shride, A. F. 1969. Asbestos. In *Mineral and water resources of Arizona*, 63–73. Arizona Bureau of Mines Bulletin 180.

———. 1973. Asbestos. In *United States mineral resources*, edited by D. A. Brobst and W. P. Pratt, 63–73. U.S. Geological Survey Professional Paper 820.

Silberman, M. L., and J. Danielson. 1993. Gold-bearing quartz veins in the Klamath Mountains in the Redding 1° x 2° quadrangle, northern California. *California Geology* 46:35–44.

Singer, D. A. 1986. Descriptive model of kuroko massive sulfide. In *Mineral deposit models*, edited by D. P. Cox and D. A. Singer, 189. U.S. Geological Survey Bulletin 1693.

———, 1992. Grade and tonnage model of Sierran kuroko deposits. *Developments in mineral deposit modeling*, edited by J. D. Bliss, 29–32. U.S. Geological Survey Bulletin 2004.

Singer, D. A., D. L. Mosier, and D. P. Cox. 1986. Grade-tonnage model of porphyry copper. In *Mineral deposit models*, edited by D. P. Cox and D. A. Singer, 77–81. U.S. Geological Survey Bulletin 1693.

Skinner, H. C. W., M. Ross, and C. Frondel. 1988. *Asbestos and other fibrous materials, mineralogy, crystal chemistry, and health effects.* New York: Oxford.

Storey, L. O. 1978. Geology and mineralization of the Lights Creek stock, Plumas County, California. *Proceedings of the porphyry copper symposium: Arizona Geological Society Digest* 11:49–58.

U.S. Environmental Protection Agency (EPA). 1991. Final national primary drinking water rules: 56 Federal Register 3578 (30 January, 1991).

U.S. Geological Survey (USGS). 1994. *Preparing for the next eruption in the Cascades.* U.S. Geological Survey Open-File Report 94-585. Available from http://vulcan.wr.usgs.gov/Vhp/OFR_nexteruption.html.

U.S. Occupational Safety and Health Administration (OSHA). 1975. Occupational exposure to asbestos. Federal Register 47652, 47660 (9 October).

Varekamp, J. C., and P. R. Buseck. 1986. Global mercury flux from volcanic and geothermal sources: *Applied Geochemistry* 1:65–73.

Virta, R. L., 1989, Asbestos. *Metals and minerals.* U.S. Bureau of Mines Minerals Yearbook 2:127–34.

Walker, G. W., and N. S. MacLeod. 1991. Geologic map of Oregon. U.S. Geological Survey, 2 sheets, scale 1:500,000.

Wentworth, C. M., G. R. Fisher, P. Levine, and R.C. Jachens. 1995. *The surface of crystalline basement, Great Valley and Sierra Nevada, California: A digital map database.* U.S. Geological Survey Open-File Report 95-0096.

Woodward-Clyde Associates. 1989. Draft regional study of mining disturbances and exploration-related disturbances of asbestos-bearing material in the New Idria–Coalinga–Table Mountain study region. Report prepared for U.S. Environmental Protection Agency, EPA contract 68-01-6939.

Wrucke, C. T. In preparation. Serpentine-hosted asbestos (Model 8e; Cox and Singer 1986) and carbonate-hosted asbestos (Model 18d; Cox and Singer 1986), geoenvironmental model. 1995. In *Mineral-environmental models*, edited by E. A. Du Bray. U.S. Geological Survey Bulletin.

TIMOTHY P. DUANE
Department of City and Regional Planning
   and Department of Landscape
   Architecture
University of California
Berkeley, California

19

Recreation in the Sierra

ABSTRACT

Recreation is a significant activity in the Sierra Nevada, which serves as a center for a wide range of recreational activities. The Sierra contains some of the world's outstanding natural features, and they attract visitors from throughout the country and the world. Lake Tahoe, Yosemite Valley, Mono Lake, and the Sequoia Big Trees attract millions of visitors each year. Recreational activities on public lands alone account for between 50 and 60 million recreational visitor days (RVDs) per year, with nearly three-fifths to two-thirds of those RVDs occurring on lands administered by the U.S. Forest Service. The California Department of Parks and Recreation has the second greatest number of RVDs, followed by the U.S. Bureau of Reclamation, the National Park Service, and the U.S. Bureau of Land Management. Additional recreational activities on private lands account for millions more RVDs that are currently not accounted for by any agency in a consistent or reliable format that would allow direct comparisons with public land recreational use data. Inconsistency in the data classification and collection methodologies of the various public agencies also limits the usefulness of the recreational activity data that are available. This report brings the available data together into a common digital format and makes it available for analysis. The role of state and federal agencies in providing recreational opportunities in the Sierra Nevada is summarized, and more specific data provided about the types of recreational activities pursued under each agency's jurisdiction. There is significant variation by subregion and recreational activity class, moreover, which makes some agencies more important than others for specific types of recreation in specific areas. These differences by subregion and recreational activity class must be accounted for in any assessment of policy scenarios for the Sierra Nevada that might affect the availability of future opportunities for recreation. A more detailed assessment of recreational activities in the eastern Sierra subregion is also described to illustrate how subregional assessments can provide critical information on user characteristics and activities at a finer level of disaggregation.

INTRODUCTION

The Sierra Nevada region is a popular destination for recreationists. Year-round local residents and California residents and nonresidents pursue a wide variety of recreational activities. These pursuits occur throughout the entire region, from the bottom of steep river canyons to the top of the highest mountain peaks. The mountain range is the natural infrastructure that supports wilderness backpackers, skiers, fishing enthusiasts, off-road vehicle users, naturalists, and many others. All individuals who pursue outdoor activities within the Sierra Nevada rely upon the natural world for an enjoyable experience. The ecological conditions of the Sierra Nevada are therefore important factors influencing patterns of recreational activity. The frequency, duration, timing, and spatial pattern of recreational activities will in turn affect those ecological conditions.

Ecological, social, and economic conditions for many Sierra Nevada communities and residents are closely intertwined in the recreation sector. Tourism activity in the region, of which recreation constitutes a significant part, is also dependent in part upon the condition of Sierra Nevada ecosystems. The assessment in this chapter focuses exclusively on recreational activities on the public lands and public waters in the Sierra Nevada. This recreational activity may be either local in origin or involve tourism, which is in turn a subset of all activity related to the travel industry. Tourism that does not involve recreational activities utilizing the natural resources of the Sierra Nevada are not addressed in this report. Tourism throughout the Sierra Nevada is nevertheless conducted against the backdrop of the Sierra Nevada's recreational opportunities, so the two are closely intertwined and include most of the economic activity described by Stewart (1996) in the tourism and developed recreation sectors.

Sierra Nevada Ecosystem Project: Final report to Congress, vol. II, *Assessments and scientific basis for management options.* Davis: University of California, Centers for Water and Wildland Resources, 1996.

Unfortunately, there is very limited quantitative information linking specific levels and types of recreational activity to specific levels of tourism for the Sierra Nevada as a whole. The potential impact of changing ecological conditions on recreational activity and specific levels of tourism is therefore poorly understood. This is true despite several decades of work on the topic (Knight and Gutzwiller 1995), including several studies that were specifically focused on conditions in the Sierra Nevada (Parmeter 1976; Foin 1977). We therefore do not attempt to infer specific responses by the recreation sector to alternative management actions or changes in ecological conditions. Recreational activity is a function of many factors, and for most types of recreation ecological conditions are not necessarily the dominant factor. The availability of developed facilities and a wide range of behavioral considerations, including cultural factors, are probably equally important. The institutional arrangements for the provision of recreational opportunities (e.g., whether they are public or private and whether or not there is a fee for the activity) also influence recreational activity. Finally, aesthetic considerations are important for many types of outdoor recreation in the Sierra Nevada. Aesthetic appeal is not necessarily consistent with ecological well-being, however, so ecological well-being is not necessary to support many types of recreational activities that are dominated by aesthetic considerations.

This report has several significant deficiencies: (1) it has no reliable information about recreational activities on private lands; (2) it does not address the qualitative dimensions of different types of recreational experiences or activities; (3) it does not address the impact of recreation on ecological conditions; (4) it does not address the impact of public land policies and alternative institutional arrangements upon future recreational opportunities; (5) it does not discuss the vast literature on recreation or recreation's important historical role in determining the institutional arrangements for land and resource management in the Sierra Nevada. Limited resources required us to focus on available data in a common format, which touches on only one dimension of recreation in the Sierra Nevada. Each of these other dimensions is also important, but we were unable to address them within the framework of this assessment.

Significant recreational resources in the Sierra Nevada are located on private lands, however, and a significant level of recreational activity occurs on those lands. Much of the shoreline of the lakes and rivers of the Sierra Nevada is on private land, and recreational activities in the region are often focused around water resources. The high Sierra resort communities of Truckee and Donner Lake, much of Lake Tahoe, Lee Vining, and the town of Mammoth Lakes are all situated on private land. Access to the Sierra Nevada's spectacular national parks is primarily through the western gateway communities of Groveland, El Portal, Oakhurst, and Three Rivers. Other communities, such as Lone Pine and Bishop in the Owens Valley, are important centers for recreation on nearby public lands.

There is also a significant level of locally based recreational activity occurring within the Sierra Nevada. Finally, indoor recreation by both Sierra Nevada residents and visitors occurs largely on private lands. Our estimates of recreational activity in the Sierra Nevada are therefore conservative: overall social, economic, and ecological importance of recreation and tourism is much greater than indicated by the activity figures reported here. The impact of recreation on social and ecological well-being remains largely unexplored in the Sierra Nevada. These linkages are now being explored by local groups and recreation providers in recreation-dependent communities, however, such as the Tahoe Coalition of Recreation Providers (TCORP) and the Coalition for Unified Recreation in the Eastern Sierra (CURES).

KEY QUESTIONS

This assessment has attempted to answer five basic questions about recreation in the Sierra Nevada. The answers may help policymakers, citizens, agencies, and others to evaluate how various future policy alternatives (or other trends) may affect recreational activities in the region:

1. What are the current levels and types of recreational activities in the Sierra Nevada?

2. What is the spatial and temporal distribution of those recreational activities?

3. Who participates in these recreational activities (e.g., age, gender, residence)?

4. Who provides the opportunities for these recreational activities (e.g., agency)?

5. What changes are likely to occur in the future in recreational activities and users?

We are also interested in how the answers to these questions have changed over time, although we have not attempted to complete a comprehensive historical analysis of recreation in the Sierra Nevada; our focus has been on the recent past and its implications for the future. David Beesley (1996) discusses some of the more important historical events regarding recreation in the Sierra Nevada, including the designation of several national parks. These events continue to affect land and resource management in the region, so it is clear that recreation has played an important role in determining the present social, economic, ecological, and institutional context for management of the Sierra Nevada. This role is widely recognized by both the public and government officials at the federal, state, and local levels. It is most apparent when considering differences between the various federal land and resource management agencies. Nearly every agency is involved with or affected by recreation within its jurisdiction, so this assessment has relevance throughout the Sierra Nevada. Nearly every human community within the Sierra Nevada is also affected by recreational activity in some way. This assessment's primary challenge is therefore to clarify the current state of our knowledge regarding recreational activities across multiple jurisdictions. As noted, however, this report addresses only a few of the dimensions of that knowledge and is therefore not a comprehensive treatment of the subject.

BACKGROUND

We have not generated any new information through primary research. All of the information that we present here has already been publicly available, but it has generally been inaccessible to anyone interested in an overview of recreation in the Sierra Nevada. Before this research, it was well known that recreation was significant throughout the range. It was also well known that recreational activities within the Sierra Nevada included a wide range of users, jurisdictions, and activities. Although local residents participate in those recreational activities, it was also clear that most recreational activity in the Sierra Nevada has been by nonresidents. Those nonresidents were known to be primarily Californians but included both other Americans and foreign visitors. The non-Californians were believed to be drawn primarily to the "world-class" recreational resources of the Sierra Nevada, such as Yosemite Valley. It was unknown to what degree other parts of the Sierra Nevada were visited by non-Californians.

What we didn't know was how to answer the five questions we raised. Estimates of different types of recreational activity varied by agency and interest group, with no common basis for discussion of the relative importance of and conflicts surrounding different types of recreational activities. Due to limitations in available information, some of that remains highly uncertain. There is very little accounting consistency between recreational providers, and not all public land management agencies keep records. Private landholders also have very few incentives to maintain records on recreational use of their lands, and the records they do keep are generally unavailable. We have nevertheless helped to close some of the gaps and to identify where the remaining gaps may be. This assessment should therefore much improve our understanding of those aspects of recreation in the Sierra Nevada for which we have detailed records that are generally comparable. Considerably more research is necessary to develop a comprehensive understanding of recreational activity in the region.

METHODOLOGY

Recreation providers within the SNEP study area maintain use records of variable quality, consistency, and reliability. For this assessment, existing recreational use information was collected from all agencies that would provide it and manually entered into spreadsheet files. These agencies were contacted from June 1994 to December 1994 for information about current and historical recreation activity. Data were then compiled and analyzed for each recreation provider within the study area. Interagency data analysis was generally limited due to inconsistent data collection methodologies, varying units of measurement, and data gaps. Whenever possible, however, recreational use information was converted to recreational visitor days (RVD), a measurement unit employed by the U.S. Forest Service (USFS). One RVD equals one twelve-hour visit to a site or twelve hours of an activity. Four hours of an activity would equal, therefore, one-third of an RVD; participating in an activity for two hours per week for six weeks would equal one RVD. We have attempted to use RVDs as a standard measure here in order to allow comparisons across multiple jurisdictions and data sources. It is difficult to convert USFS estimates from RVDs into comparable units for comparison with other agency measures, but we have been able to convert most other agency measures into RVDs.

We did not have RVD data from the California Department of Parks and Recreation (CDPR), which reported numbers of visitors rather than the length of time those visitors spent on specific recreational activities. Visitation data from Yosemite National Park for 1981–91 shows that visitation is strongly correlated with RVDs (R = 0.998). Figure 19.1 shows this relationship. The average visitor generated 2.24 RVDs, ranging from a low of 2.20 to a high of 2.29. We have therefore converted visitor figures to RVD figures using a ratio of 2.24 wherever·necessary. Yosemite National Park is clearly not a "typical" recreational destination in the Sierra Nevada, however, for it receives a higher level of day visitors than most areas. We nevertheless believe it is an appropriate proxy for visitation at other sites where developed recreation is the dominant use (as a fraction of overall visitation). These areas also have a higher level of day visitors than many USFS areas. Use of the Yosemite ratio will tend to result in conservative estimates of overall recreational activity at most other sites. Both Sequoia and Kings Canyon National Parks had higher RVDs per visitor than Yosemite, while Lassen Volcanic National Park (LVNP) had a much lower RVD per visitor ratio (1.03). Figure 19.2 shows annual RVD per visitor ratios for all four national parks from 1981 to 1993. Because the LVNP ratio is so much lower than the Yosemite ratio, we estimated RVDs for the CDPR and USBOR using both conversion factors.

The RVD accounting methodology itself has several significant weaknesses, however, which include (1) variable and inconsistent accounting practices between administrative

FIGURE 19.1

Recreational visitor days (RVDs) versus visitors for Yosemite National Park, 1981–93.

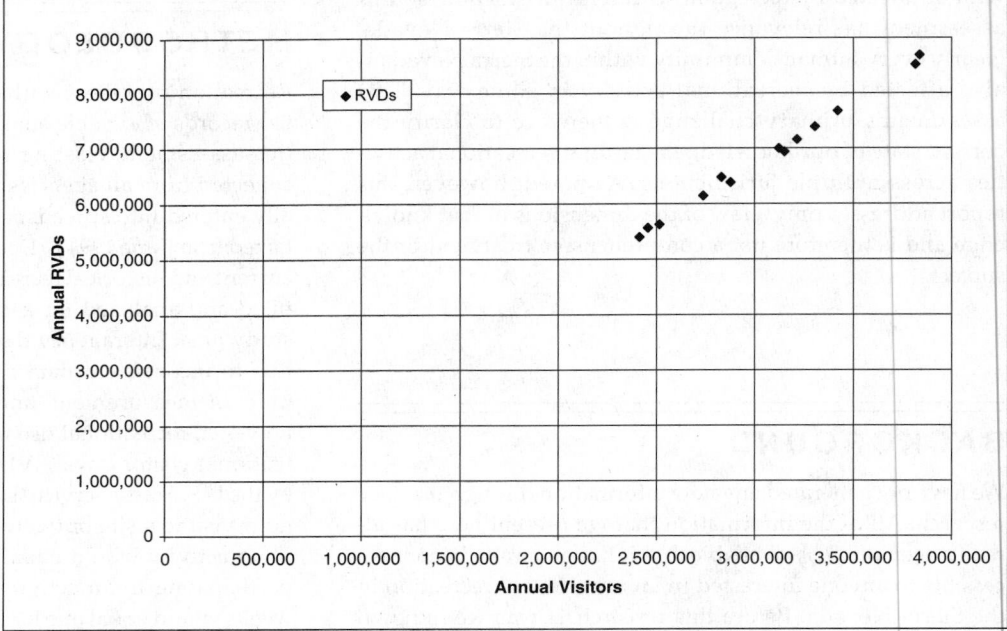

units (e.g., the National Park Service and the U.S. Forest Service; different national forests; different ranger districts within a single national forest) and over time due to changes in personnel and/or methods; (2) poorly defined RVD accounting classifications, resulting in inconsistent classification of some activities (especially new recreational activities as they first emerge); and (3) highly subjective accounting procedures that exacerbate problems of both classification and accounting. Systematic sampling procedures are generally poorly defined and rarely applied consistently enough to generate a statistically reliable basis for analysis. Together these flaws result in inconsistent data both within individual agency units (e.g., a

FIGURE 19.2

RVDS per visitor for all national parks, 1981–93.

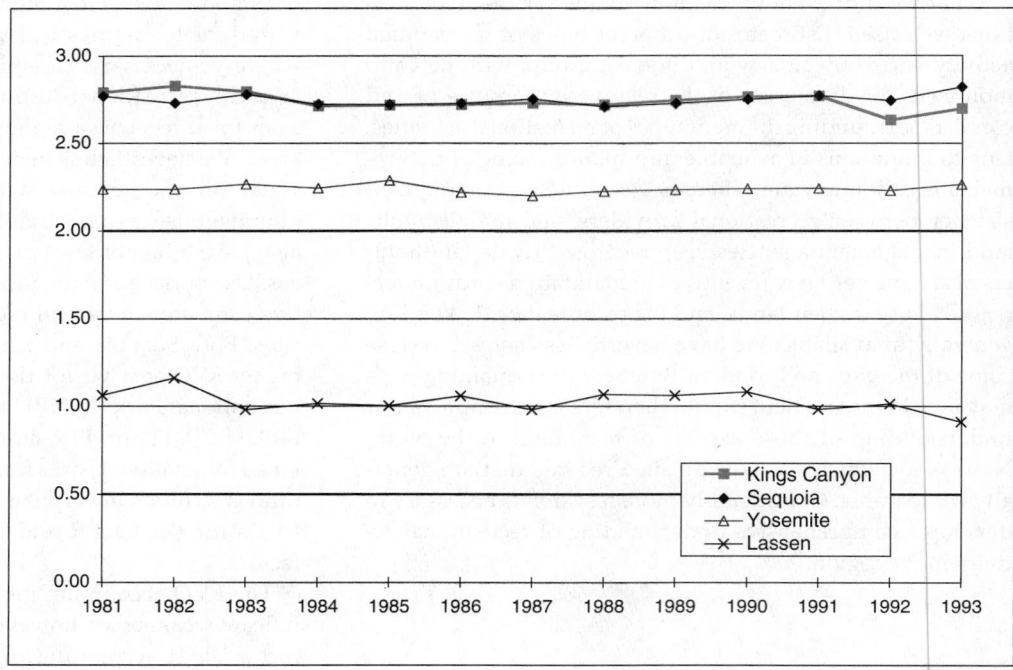

single national forest) and across agency units and agencies. Even direct quantitative comparisons are therefore uncertain.

Perhaps the most significant limitation of RVD accounting practices, however, is that the information generated is not detailed enough to guide recreational site planning and resource management activities within various subregions of a given agency or administrative unit. It is therefore often necessary to collect additional site-specific information about recreational activities and visitor preferences in order to evaluate land and resource management alternatives (Duane and Knauer 1996). This additional information can then complement the existing RVD information to assist land and resource managers and the public as they consider management alternatives. In particular, this additional information can address the quality of the visitors' recreational experiences rather than merely the number of hours spent on an activity. Likely visitor responses to management alternatives can then be assessed through a variety of methods, including focus group sessions, key informant interviews, and field surveys of recreational users. An example of this type of more detailed analysis is included for the Inyo National Forest at the conclusion of the subregional study of the eastern Sierra later in the chapter.

RECREATION ON PRIVATE LANDS

The availability of accessible recreational opportunities on both public and private lands appears to be an important factor for many residents of the Sierra Nevada in choosing where to live. The direct and indirect social and economic effects of locally based recreation can therefore be significant. Community capacity (Kusel 1996) may be both enhanced by and reflected in community recreational activities, for example, while residential location decisions based upon access to recreational opportunities may bring both construction employment and income from retirees or commuters into the local economy. Further work is needed to assess the relative importance of this access, but a survey of El Dorado County residents (J. Moore Methods 1992) found that 41% listed the recreational opportunities as a major reason they've chosen to live in the county. Another 23% listed it as a moderate reason, while 34% listed it as a minor reason (2% expressed no opinion). It can probably be assumed that those listing it as a major or moderate reason for living in the county participate in recreational activities at least fairly regularly.

If we assume that a similar proportion of Sierra Nevada residents participate in recreational activities throughout the Sierra Nevada, we can derive an approximate estimate of the number of RVDs associated with locally based recreation that is not likely to be recorded in our records here. If 41% of the region's 700,000 residents[1] recreate locally for an average of three hours per week during the year and 23% recreate lo-

cally for an average of one hour per week, this activity alone accounts for 4.4 million RVDs. This figure is likely to be conservative, yet it is nevertheless greater than the combined RVDs for Yosemite, Sequoia, and Kings Canyon National Parks in 1993 (which had a total of 3,352,667 RVDs). Assuming just one hour of recreational activity per week for all Sierra Nevada residents yields more than 3 million RVDs per year. Based on informal review of local recreation plans (for seven counties in the Sierra Nevada: Nevada, Placer, El Dorado, Amador, Calaveras, Mono, and Inyo), this is probably a conservative assumption. Total RVDs for locally based recreation, although it is widely dispersed, is therefore probably significant in the Sierra Nevada. Some of that activity is likely to be accounted for in other agencies' recreational use data (e.g., hiking or fishing on USFS land), but any of it that occurs on private land is not recorded in the results that we report.

It is also important to note that the relatively low density of human settlement in the Sierra Nevada is accompanied by large areas of open space that are privately owned. Much of this land is fenced and posted against trespass, but other land remains generally accessible for informal public recreational activities of a dispersed, low-intensity nature. These activities include running, walking, mountain biking, cross-country skiing, snowmobiling, and nature study. Similar activities occur on large private land holdings at higher elevations, especially those that are interspersed with public lands. Recreational users often cross between public and private lands on a single trail, for example, without even knowing whether they are on federal, state, local, or private land at a given time. Recreational use estimates for the public agencies described in this chapter record only those activities that occur on those lands or resources within the management jurisdiction of those public agencies. Additional recreational activities occur on private lands, and the potential for conflicts over trespass are highest at the public-private land interface. Moreover, reductions in informal public access to privately owned open space are also likely as human settlement increases parcelization and population density on large blocks of private land. The implications for trends in human settlement and public lands management for recreation are discussed in more detail later.

SOURCES

We contacted the following sources and evaluated recreational activity and visitor information (when available). In many cases, these organizations either had no data or their data duplicated other data provided by public land and resource management agencies. Detailed data sets and records of our data collection are available from the California Environmental Resource Evaluation System (CERES) project of the Resources Agency of the State of California (http://ceres.ca.gov/snep), and the Alexandria Project at the University of California, Santa Barbara (http://alexandria.sdc.ucsb.edu/).

- Federal agencies: Forest Service, National Park Service, Bureau of Reclamation, Bureau of Land Management, Army Corps of Engineers

- State agencies: Department of Parks and Recreation, Department of Fish and Game, Department of Water Resources, State Lands Commission

- Public utilities: East Bay Municipal Utility District, Hetch Hetchy Water and Power, Los Angeles Water and Power, Sacramento Municipal Utility District, Placer County Water Agency, El Dorado Irrigation District, Nevada Irrigation District

- Utility companies: Pacific Gas and Electric Company, Southern California Edison Company, Sierra Pacific Power Company

- Local government: twenty-one county parks and recreation departments, several special/community service districts

- Nongovernmental organizations: Ducks Unlimited, Friends of the River, Sierra Club, Nature Conservancy, Trust for Public Land

- Private camps: American Campground Association, Christian camps and camping centers, twenty-one county health departments

- Miscellaneous: two wildland skill schools, Recreational Equipment, Inc., other recreation researchers

Our reported results for public agencies are biased toward those agencies that have kept reliable records and reported them to us. These include the Forest Service, the National Park Service, the California Department of Parks and Recreation, the Bureau of Reclamation, the California Department of Fish and Game, Pacific Gas and Electric Company, the East Bay Municipal Utility District, and El Dorado County. It is therefore not a random sample of recreational activities in the Sierra Nevada. This list covers all of the major land and resource management agencies in the region, however, so it should be an accurate approximation of the degree and types of recreational activities on public lands and waters in the Sierra Nevada. Recreational activities are most underrepresented in our data and analysis for foothill-area water sports and local parks.

Development of a common framework for sampling, recording, reporting, and analyzing recreational activity information for public agencies would assist future efforts at analysis. The State of California's Outdoor Recreation Plan (SCORP) is the only effort currently directed toward systematic evaluation of recreational activities for the entire Sierra Nevada region.

RESULTS

Results will be described here for each of the individual data sources and agency providers of recreational opportunities. Potential problems with data and preliminary interpretations of the data are described here, although the primary product of this assessment is the integrated provision of the data itself in digital form. Considerably more analysis of individual data sets will offer additional insights into specific policy questions. All of the data sources are available in Excel 5.0 for Windows spreadsheets from the California Environmental Resource Evaluation System (CERES) project of the Resources Agency of the State of California (http://ceres.ca.gov/snep), and the Alexandria Project at the University of California, Santa Barbara (http://alexandria.sdc.ucsb.edu/).

Recreation is a significant activity in the Sierra Nevada, which serves as a center for a wide range of recreational activities. The Sierra contains some of the world's outstanding natural features, and they attract visitors from throughout the country and the world. Lake Tahoe, Yosemite Valley, Mono Lake, and the Sequoia Big Trees attract millions of visitors each year. Recreational activities on public lands alone account for between 50 and 60 million recreational visitor days (RVDs) per year, with nearly three-fifths to two-thirds of those RVDs occurring on lands administered by the U.S. Forest Service (USFS). The California Department of Parks and Recreation (CDPR) has the second greatest number of RVDs, followed by the U.S. Bureau of Reclamation, the National Park Service, and the U.S. Bureau of Land Management. Our range of estimates for total RVDs is a function of the RVD per visitor ratio we assumed for the California Department of Parks and Recreation, for which only visitor data are available. Figure 19.3

FIGURE 19.3

Agency shares of RVDs (assuming CDPR = 2.24 RVD ratio).

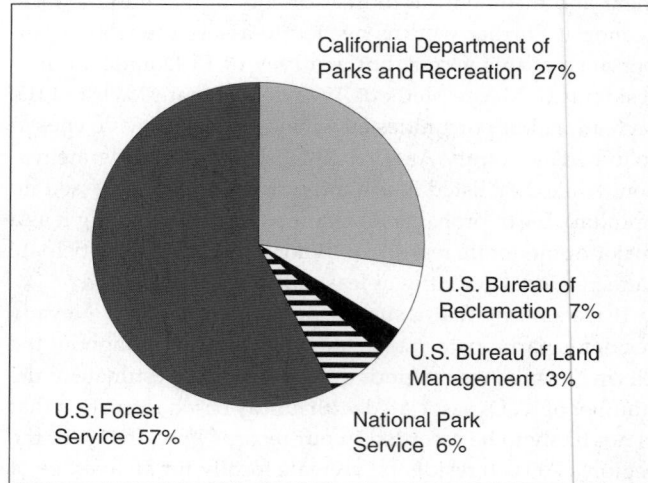

California Department of Parks and Recreation 27%

U.S. Bureau of Reclamation 7%

U.S. Bureau of Land Management 3%

National Park Service 6%

U.S. Forest Service 57%

and table 19.1 show that the USFS contributes 57% and the CDPR contributes 27% of all public RVDs in the Sierra Nevada if the Yosemite National Park ratio of 2.24 RVDs per visitor is assumed.

Figure 19.4 and table 19.2 show that the CDPR contribution drops considerably, to less than 15%, if the Lassen Volcanic National Park ratio of 1.03 RVDs per visitor is assumed. This increases the USFS share to nearly 67% of all public RVDs in the Sierra Nevada. It also decreases the total number of RVDs from nearly 59 million to about 50 million per year.

These alternative sets of assumptions do not alter either the rank order or magnitude of RVDs for other public recreation providers in our database. They do affect their relative shares of total RVDs, however, with each of the other agencies holding a higher share of the smaller total under the Lassen RVD assumption of 1.03 RVDs per visitor for CDPR visitors. The U.S. Bureau of Reclamation has about 3.9 million RVDs per year, the National Park Service has about 3.4 million RVDs per year, and the Bureau of Land Management has about 1.7 million RVDs per year. Recreational activity on reservoirs and lands of the East Bay Municipal Utility District, and Pacific Gas and Electric Company and commercial rafting through private land on the South Fork of the American River totals about 0.5 million RVDs per year. We will now describe the results for each of these agencies in greater detail.

FEDERAL AGENCIES

United States Forest Service

The USFS is the largest land manager in the Sierra Nevada and accounts for the majority of total RVDs on public lands in the region. Nine national forests or USFS administrative units are located within the SNEP study area (figure 19.5): Eldorado, Inyo, Plumas, Sequoia, Sierra, Stanislaus, Tahoe,

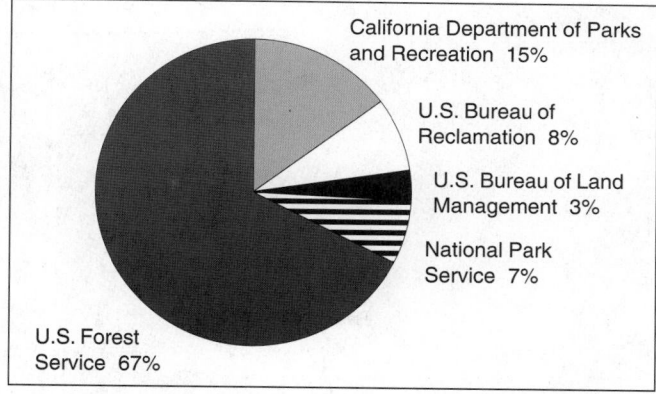

FIGURE 19.4

Agency shares of RVDs (assuming CDPR = 1.03 RVD ratio).

Toiyabe, and the Lake Tahoe Basin Management Unit (LTBMU). All of these except the Toiyabe National Forest are part of Region 5 of the USFS (headquartered in San Francisco) and have common procedures and accounting systems for collecting recreational use data. We acquired data for each of these forests from the regional office in San Francisco. Toiyabe National Forest is headquartered in Sparks, Nevada, and is part of Region 4 (headquartered in Ogden, Utah). Only the Carson and Bridgeport Ranger Districts of the Toiyabe National Forest are within the Sierra Nevada, and data for the Toiyabe were reported in a different format and for a different period than those for Region 5. We have therefore estimated mean annual RVDs for the USFS units in Region 5 based on 1987–93 data and mean annual RVDs for the Sierra Nevada portion of the Toiyabe National Forest based on 1987–91 data. The data are otherwise aggregated into the same recreational activity classes.

We have aggregated historical data from 1966 to 1993 for all national forest units in Region 5, but these are not disag-

TABLE 19.1

Agency shares of Sierra Nevada RVDs if CDPR = Yosemite (2.24 RVDs per visitor).

Agency	Annual RVDs	Percentage of Total
Pacific Gas and Electric Company	97,292	0.17
South Fork of the American River	118,000	0.20
East Bay Municipal Utilities District	306,106	0.52
California Department of Parks and Recreation	15,868,723	26.99
U.S. Bureau of Reclamation	3,881,000	6.60
U.S. Bureau of Land Management	1,660,033	2.82
U.S. National Park Service	3,352,607	5.70
U.S. Forest Service	33,500,739	56.99
Total	58,784,500	100.00
Public	58,569,208	99.63

TABLE 19.2

Agency shares of Sierra Nevada RVDs if CDPR = Lassen (1.03 RVDs per visitor).

Agency	Annual RVDs	Percentage of Total
Pacific Gas and Electric Company	97,292	0.19
South Fork of the American River	118,000	0.24
East Bay Municipal Utilities District	306,106	0.61
California Department of Parks and Recreation	7,296,779	14.53
U.S. Bureau of Reclamation	3,881,000	7.73
U.S. Bureau of Land Management	1,660,033	3.31
U.S. National Park Service	3,352,607	6.68
U.S. Forest Service	33,500,739	66.72
Total	50,212,566	100.00
Public	49,997,264	99.57

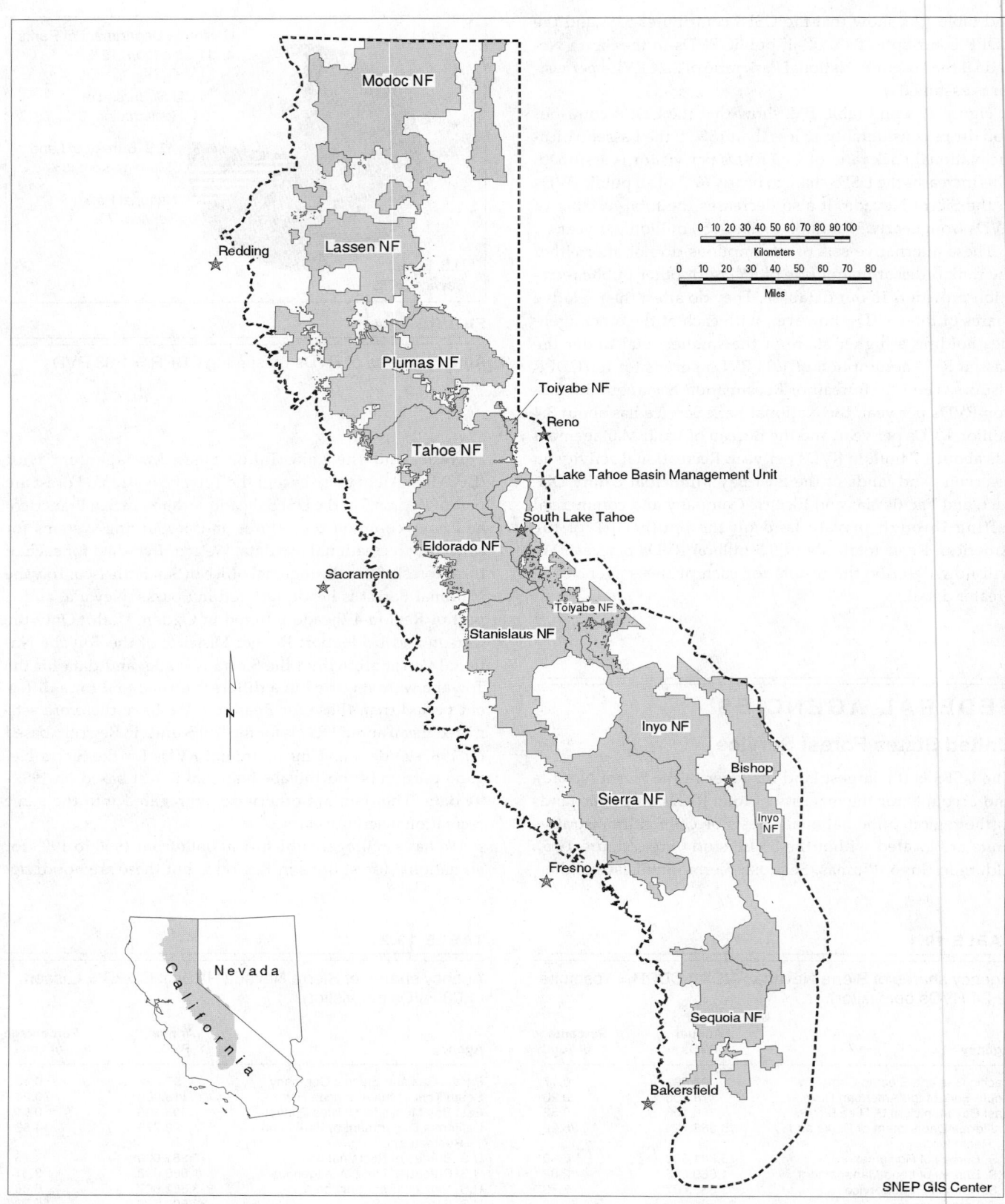

FIGURE 19.5

National Forests within study area.

gregated by individual USFS unit. Time-series analysis of that data shows that most activities were basically flat during that period, however, with the exception of three major categories: auto travel, sightseeing, and miscellaneous. All three of these categories began to grow rapidly in the mid-1980s, around the same time that the national forests in the Sierra Nevada completed their land and resource management plans under the National Forest Management Act of 1976. It is unclear what could have accounted for this increase. Because of this shift, our USFS data from 1987–93 is higher than the historical data from 1966–86. Most of that increase is in these unspecified and difficult-to-count categories, however, so the estimates may be inflated for the 1987–93 mean RVDs.

Recreational activity on the national forests in Region 5 totalled 31.9 million RVDs in 1993. The Toiyabe National Forest averaged over 1.6 million RVDs per year in the Sierra Nevada from 1987 to 1991. Total USFS RVDs are therefore more than two-and-one-half times the total RVDs for California state parks within the Sierra Nevada, ten times the total RVDs for the national parks within the Sierra Nevada, a dozen times the total RVDs for the Bureau of Reclamation, and two-dozen times the total RVDs for the BLM. Overall, the 33.6 million RVDs on USFS lands accounted for 57% of the 58.6 million RVDs reported here by public agencies for the 1987–93 period. These totals for the Sierra Nevada do not include RVDs for the Lake Tahoe State Park in Nevada, parks and/or reservoirs operated by local and regional agencies, or recreational activities on private lands in the region. They nevertheless illustrate the importance of the USFS as a provider of recreational opportunities. Many of the RVDs that occur on private lands in the Sierra Nevada are also associated with activities on the public lands, however, when recreationists spend the night on private lands but recreate during the day on public lands. The RVDs on public lands are therefore likely to be tied to total recreation-related activities and expenditures in the region.

The two largest national forests are the Inyo and the Sierra (figure 19.6). The four national forests with the highest proportion of their land base designated as wilderness are the Sierra, the Inyo, the Sequoia, and the Lake Tahoe Basin Management Unit (figure 19.7). This distribution of designated wilderness has important implications for the spatial distribution of specific recreational activities across the USFS land base in the Sierra Nevada. The southern Sierra Nevada is the only place in the contiguous forty-eight states where one can draw a straight line on a map for more than 150 miles and not cross a road, which occurs near the John Muir Trail between Tuolumne Meadows in Yosemite National Park and Monache Meadows just south of the Golden Trout Wilderness. The area between these two points includes portions of Yosemite National Park, the Ansel Adams Wilderness, the John Muir Wilderness, Kings Canyon National Park, Sequoia National Park, and the Golden Trout Wilderness. A comparison of total acreage to wilderness acreage indicates that the Sierra, Inyo, and Sequoia National Forests (all located in the southern Sierra Nevada and adjacent to national parks) proportionally contain the most wilderness within their boundaries (table 19.3).

Time-series RVD data were available for the national forests from 1966 to 1993, but uncertainty about changes in RVD accounting practices led us to focus on more recent data for consistency. We used the 1987–93 period to derive mean RVD figures for each of the USFS units on Region 5. This also allowed comparisons with other agency data, which were generally available in detail only for the more recent period. As

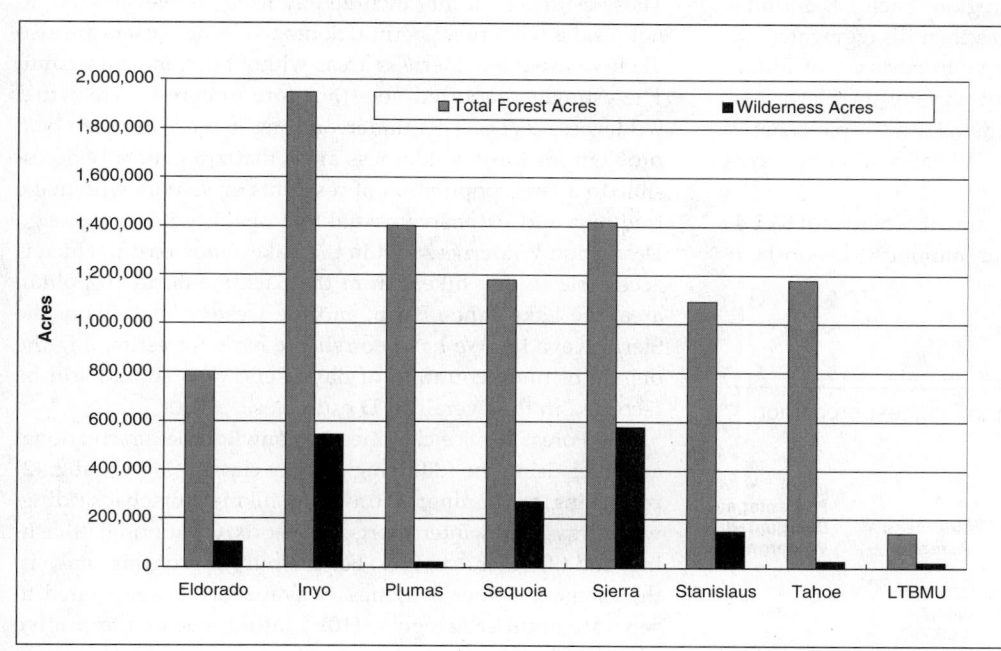

FIGURE 19.6

Size attributes of individual national forests.

FIGURE 19.7

Wilderness share of national forests.

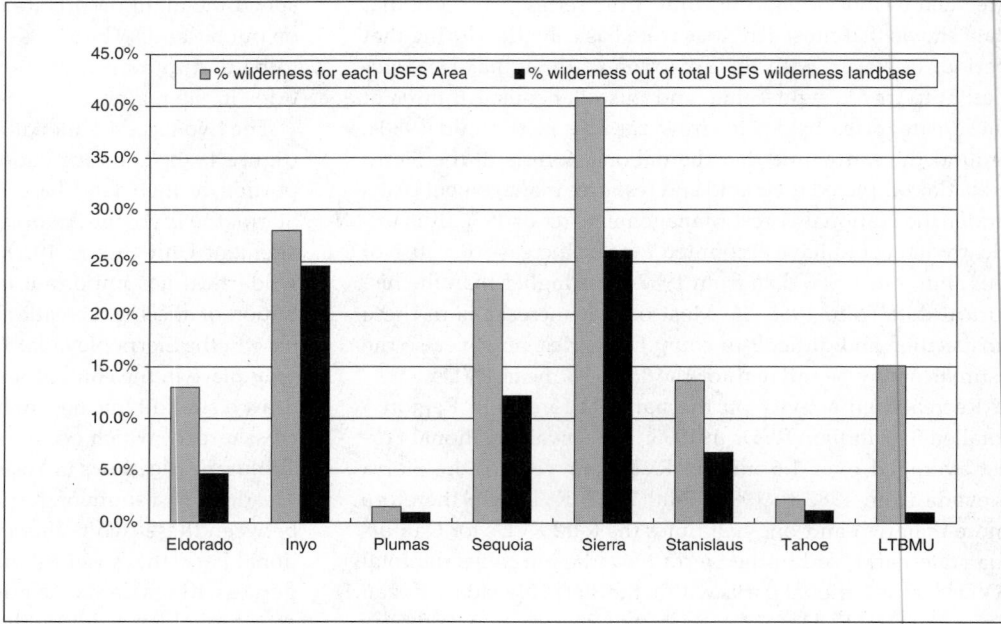

noted earlier, we used 1987–91 data for the Toiyabe National Forest. The more detailed breakdown of RVDs for the entire 1966–93 period is available from the California Environmental Resource Evaluation System (CERES) project of the Resources Agency of the State of California (http://ceres.ca.gov/snep), and the Alexandria Project at the University of California, Santa Barbara (http://alexandria.sdc.ucsb.edu/). Mean annual RVD data were therefore compiled for each national forest in the Sierra Nevada and then aggregated across the region to determine the relative contribution of each USFS unit to overall USFS RVDs in the region. Each USFS unit's total mean annual RVD estimate was then disaggregated by activity class to determine the relative importance of different recreational activities in different parts of the Sierra Nevada. Each of these is reported in detail, and a pie chart is included showing each USFS unit's RVD allocation by recreational activity class.

The Inyo National Forest is the dominant provider of RVDs, accounting for 23% of overall mean annual RVDs on land managed by the USFS in the Sierra Nevada. The Tahoe National Forest provided 16%, followed by the Sierra National Forest (13%). All of the other national forests each accounted for less than 10% of the total USFS RVDs in the Sierra Nevada (figure 19.8).

Each USFS unit estimates its total number of wilderness RVDs based upon the number of wilderness permits issued. The Inyo National Forest accounted for 36% of the total wilderness RVDs among Sierra Nevada national forests, followed by the Sequoia (17%) and the LTBMU (12%) (figure 19.9). These estimates do not include day users, however, who do not need a wilderness permit. Some wilderness users are also likely to use the wilderness areas without obtaining a permit. The estimates reported here therefore underestimate actual wilderness RVDs. This undercounting is most likely to be a problem for those wilderness areas that are generally accessible to a large population of residents or visitors who make frequent and extensive casual use of the wilderness (e.g., Desolation Wilderness within the Lake Tahoe Basin, which is accessible to day hikers from the Sacramento metropolitan area, the Lake Tahoe Basin, and the western foothills of the Sierra Nevada). We have no reliable basis for estimating the degree of undercounting of day users, who should still be recorded in the overall RVD estimates.

The Forest Service classifies its nonwilderness recreational activities using the following activity classes (1) camping; (2) picnicking, swimming; (3) travel; (4) hiking, horseback riding, water travel; (5) winter sports; (6) resorts; (7) hunting; (8) fishing; and (9) other activities. For the purposes of this analysis, the category "other activities" was further disaggregated to separate another category, (10) "nature study/interpretive activities." (This particular subcategory might be important

TABLE 19.3

Sierra Nevada national forests with the highest proportion of wilderness.

National Forest Service Area	Total Acreage	Wilderness Acreage	Percentage Designated Wilderness
Sierra National Forest	1,417,355	577,654	41
Inyo National Forest	1,944,710	544,667	28
Sequoia National Forest	1,179,193	269,790	23

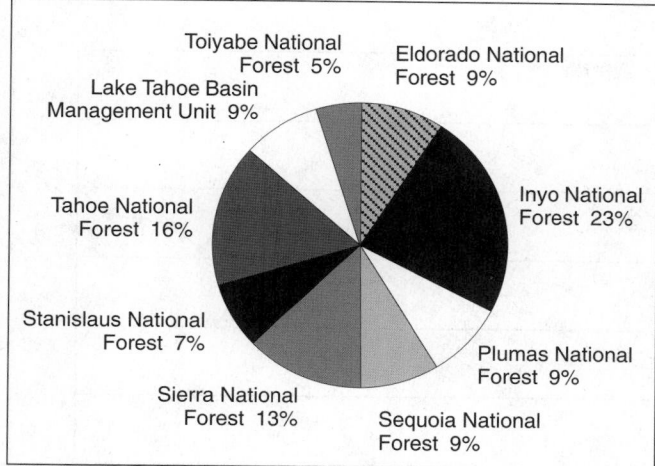

FIGURE 19.8

Distribution of mean total annual RVDs for USFS units, 1987–93.

for more detailed assessment of the policy scenarios.) Using variable methodologies, Forest Service personnel annually estimate the number of recreational visitor days (RVDs) for each activity class by ranger district. More detailed analysis of data at the ranger district level was possible only for the Inyo National Forest and is reported in a section on the eastern Sierra Nevada later in the chapter. Note that some of the activity classes include seemingly unrelated activities, while some related activities (e.g., swimming, water travel, and fishing) are recorded across multiple activity classes. This makes it difficult to disaggregate recreational activities on USFS lands based upon physical characteristics (e.g., access to water). This in turn limits our capacity to analyze the impact of various policy scenarios on recreation. Figure 19.10 shows historical patterns in wilderness and primitive area use under USFS jurisdiction from 1964 to 1993 as reported by the administering USFS units.

Mean annual RVDs were calculated for each activity class by individual Forest Service unit and across all Sierra Nevada national forests. The most popular recreational activities, as measured by RVDs, within each of the nine Forest Service areas were the two activity classes of "automobile travel" (32%) and "camping, picnicking, and swimming" (29%) (figure 19.11). Together with resorts (11%), these three general classes of recreational activity accounted for nearly three-quarters (72%) of all RVDs on USFS units in the Sierra Nevada.

Approximately 18% of the total number of RVDs for the activity class "camping, picnicking, and swimming" were attributed to the Inyo National Forest, followed by the Sierra (15%), Sequoia (13%), and Tahoe (12%) National Forests (figure 19.12). The Inyo National Forest also accounted for 28% of the Sierra Nevada RVDs in USFS units in activities related to the "travel" activity class, while the Tahoe National Forest provided 21% (figure 19.13). The Inyo and Sierra National

Forests each accounted for one-fourth of the "hiking, horseback riding, and water travel" activity class on USFS lands (figure 19.14).

Tahoe National Forest received the most RVDs (36%) for the category "winter sports," with the Inyo National Forest comprising 26% of the total for that activity class. The LTBMU accounted for another 16% and the Eldorado National Forest provided 11% of the "winter sports" RVDs (figure 19.15). The figures for the greater Lake Tahoe area (totaling 63% of all winter sports RVDs in the Sierra Nevada) reflected activity at a number of major ski resorts, while the Inyo National Forest RVDs were almost exclusively due to the presence of a single large ski area, Mammoth Mountain and nearby June Mountain.

One-fifth (19%) each of the total number of RVDs in activities in the "resorts" class occurred on the Inyo National Forest and the Sierra National Forest. Another 16% of "resort" activity occurred on the Stanislaus National Forest and 14% on the Eldorado National Forest (figure 19.16). Hunting was also most popular on the Eldorado National Forest, which accounted for 25% of total hunting activity on USFS lands in the Sierra Nevada. Other USFS units with more than a 10% share of total hunting RVDs were the Inyo National Forest (16%), the Sierra National Forest (15%), the Plumas National Forest (14%) and the Tahoe National Forest (12%). The LTBMU reported no RVDs for this class (figure 19.17). The Inyo National Forest received the most fishing RVDs (20% of the total for the Sierra Nevada on USFS lands), while the Plumas National Forest had the second highest (15%). Sierra National Forest provided 14% of all fishing RVDs and the Eldorado and Sequoia National Forests provided 10% and 11%, respectively (figure 19.18). Nearly two-thirds (65%) of the total num-

FIGURE 19.9

Distribution of mean total annual wilderness RVDs for USFS units, 1987–93.

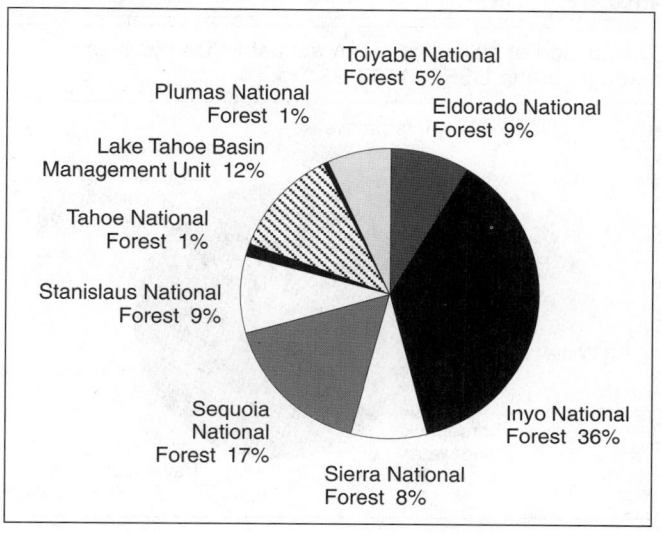

FIGURE 19.10

USFS wilderness and primitive area use by RVD, 1965–91.

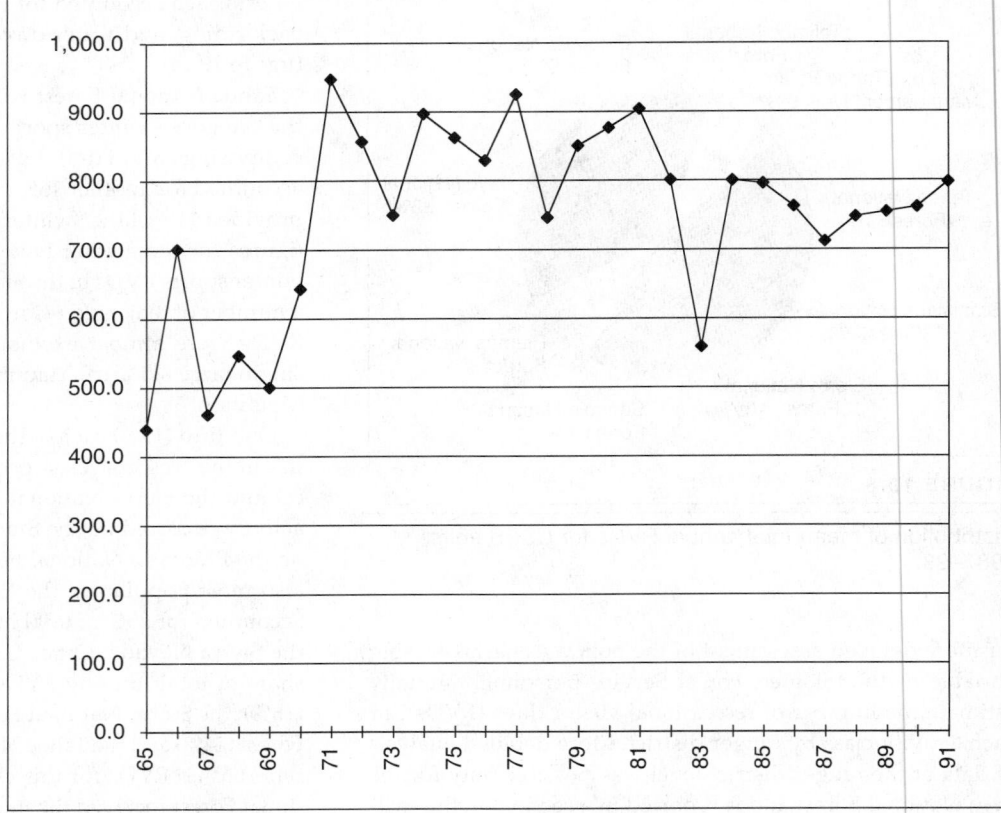

ber of hours spent on nature study or interpretive activities occurred on the Inyo National Forest, which stands out in comparison to the other USFS units in the Sierra Nevada within that activity class (figure 19.19). The Toiyabe National Forest reported no RVDs for this activity class.

Time-series trends from 1966 to 1993 for several aggregated recreational activity classes are shown in figure 19.20 for all

FIGURE 19.11

Distribution of activity in mean annual RVDs by class throughout the USFS, 1987–93.

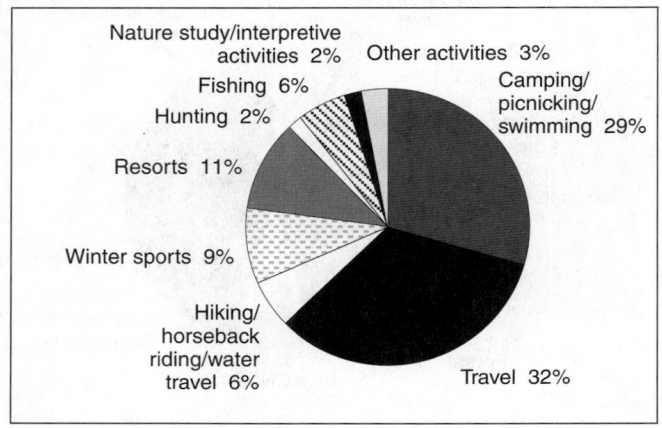

USFS units in the Sierra Nevada (except Toiyabe National Forest). Note that two major classes, "travel/sightseeing" and "miscellaneous," account for most of the growth in USFS RVD estimates. Winter sports RVDs increased significantly in the late 1970s but have been relatively flat since the early 1980s. Camping RVDs have also been flat since the late 1970s, when they dropped after climbing quickly from 1968 to 1974. Hunting and fishing RVDs have also been relatively flat since the early 1980s after a significant decline. RVDs for the "hotels, resorts, cabins and camps" class have also remained flat after a decline from 1966 to 1972. Shares of total USFS RVDs in the Sierra Nevada are shown for each of these classes in figure 19.21. Note that "travel/sightseeing" now accounts for as many RVDs as camping, which is a significant change over the time series. "Miscellaneous" now exceeds all other activity classes.

Trends for the components of the "miscellaneous" class are broken down in figure 19.22. The class as a whole has been climbing steadily since 1968, but there has been a significant jump in the "miscellaneous" class within our "miscellaneous" aggregation since 1986. Figure 19.23 shows that this jump has resulted in declining or steady shares of this aggregated class for other activities. These include "hiking, biking, and horses," "other water sports," "off-highway vehicles," and "picnicking." The "miscellaneous" component of the "miscellaneous" RVD class is not defined with any specificity by the USFS.

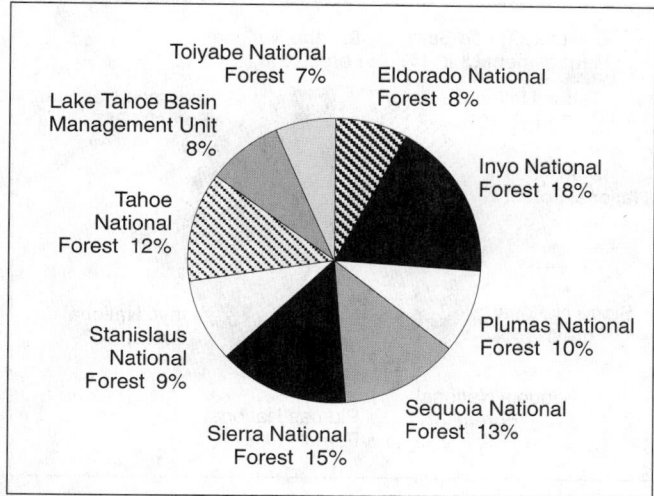

FIGURE 19.12

Distribution of the "camping, picnicking, and swimming" class in mean annual RVDs throughout the USFS, 1987–93.

The increase in RVDs for the "travel/sightseeing" and "miscellaneous" classes since 1986 has dramatically increased the total RVD estimates for USFS lands in the Sierra Nevada. The *increase* alone in "miscellaneous" RVDs since 1986 totals nearly as many RVDs (3.1 million) as occur annually in the *combined* areas of Yosemite, Sequoia, and Kings Canyon National Parks (3.4 million RVDs per year). Total "travel/sightseeing" RVDs on USFS lands are estimated to be around 8 million RVDs per year, which approaches the combined total of all park service, BLM, and Bureau of Reclamation RVDs (8.9 million RVDs per year). "Travel" now accounts for 23%

of all USFS RVDs. This figure seems to be exceptionally high, and there is no clear reason for USFS "travel/sightseeing" RVDs to have increased so dramatically from 1986 to 1993. The "miscellaneous" category can be explained in part by the recent popularity of new forms of recreation, such as mountain biking and snowboarding. We nevertheless suspect that the USFS data probably overstates the total RVDs due to unsubstantiated "travel/sightseeing" RVDs.

The time-series data for 1966–93 show that recreational activity varies substantially from year to year, so we have averaged seven years of data (1987–93) to reduce the likelihood of significant errors due to selection of an unusual year for analysis. These averages then serve as the basis for our estimates of total USFS RVDs and the share of those totals attributable to specific recreational activity classes or administrative units. Figures 19.24–19.32 summarize the proportion of RVDs by activity class for each Forest Service administrative unit. These pie charts show the fraction of total RVDs within each unit, whereas the figures reported earlier compare total RVDs by activity class across units.

Downhill ski area RVD information is accounted for within the Eldorado, Inyo, Sequoia, Sierra, Stanislaus, and Tahoe National Forests and the LTBMU. Between 1967 and 1991, total mean annual ski area RVDs increased by 79%, with an increase of more than 200% between 1967 and 1986 and a 46% decrease between 1986 and 1991 during the drought (figure 19.33). Figures 19.34–19.40 detail the total ski area RVDs for each Forest Service area. Between 1967 and 1991, only two areas showed a consistent increase in annual ski area RVDs, the Tahoe National Forest and the LTBMU. Ski resorts within these two USFS units, together with ski areas within the Inyo National Forest, have made significant investments in both snow-making equipment and new high-speed, detachable quad chairlifts during the past decade. Much of the increase

FIGURE 19.13

Distribution of the "traveling" class in mean annual RVDs throughout the USFS, 1987–93.

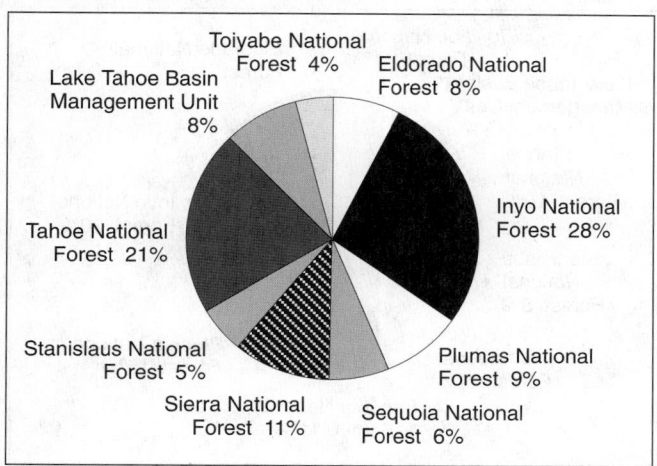

FIGURE 19.14

Distribution of the "hiking, horseback riding, and water travel" class in mean annual RVDs throughout the USFS, 1987–93.

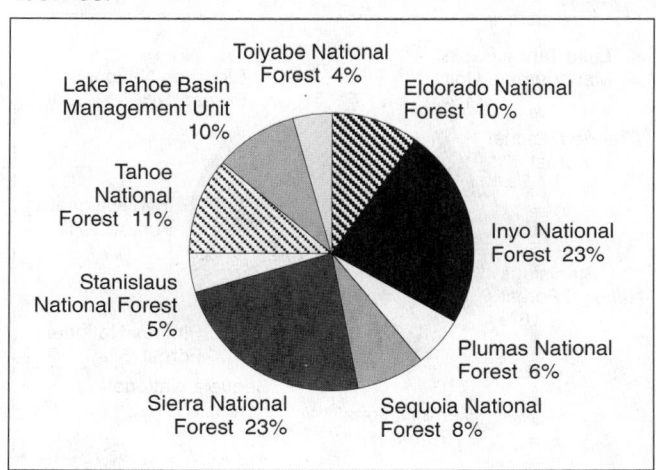

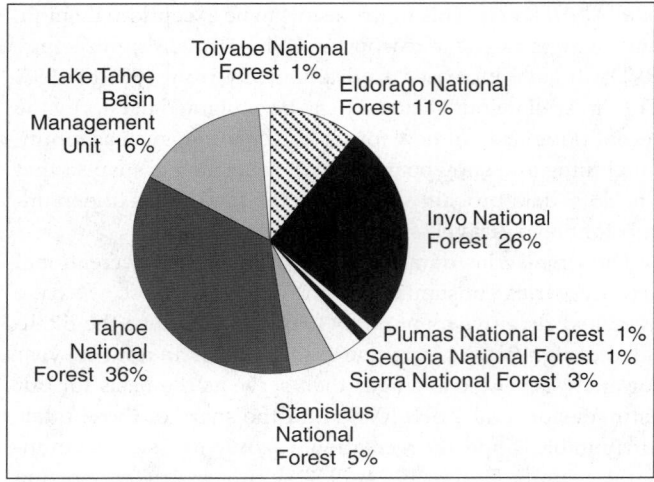

FIGURE 19.15

Distribution of the "winter sports" class in mean annual RVDs throughout the USFS, 1987–93.

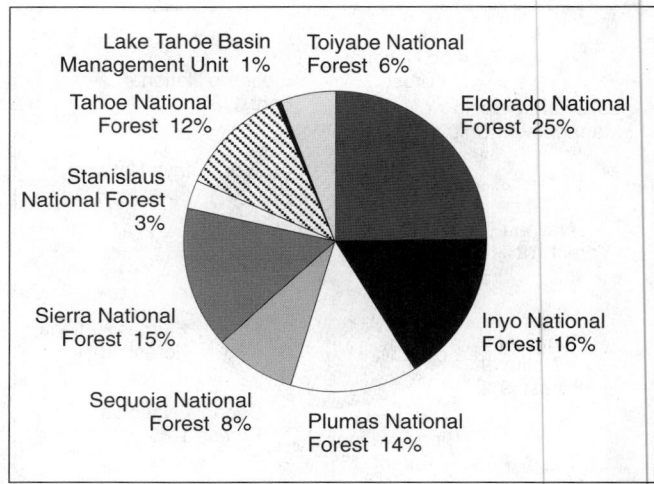

FIGURE 19.17

Distribution of the "hunting" class in mean annual RVDs throughout the USFS, 1987–93.

is therefore attributable to more efficient management of skiers at these ski areas. There has also been some limited expansion of ski runs (e.g., at Northstar, Sugar Bowl, and Diamond Peak). Ski areas in the western Sierra Nevada south of the greater Lake Tahoe area have not expanded their capacity or visitation levels. Continued population growth in the southern San Joaquin valley could create growing demand for additional skiing opportunities in this region in the future. There was no information available to indicate where skiers from that area now ski in the Sierra Nevada. Ski resorts in the Lake Tahoe area currently dominate the market

for skiing among residents of the San Francisco Bay Area and the Sacramento metropolitan area, while Mammoth Mountain dominates the market for southern California skiers. Competition from out-of-state ski areas has also been intense recently, with low-cost air fares and inexpensive package trips to resorts in Utah, Colorado, Idaho, and British Columbia. The ski resorts in the Sierra Nevada appear to rely primarily on California skiers and are not "destination" resorts relying primarily on out-of-state skiers.

The USFS data offer a clear picture of the spatial distribution of RVDs by activity class at the coarse scale of the na-

FIGURE 19.16

Distribution of the "resorts" class in mean annual RVDs throughout the USFS, 1987–93.

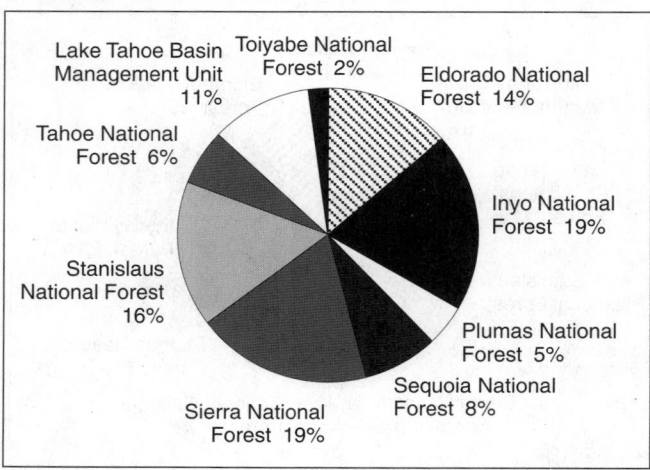

FIGURE 19.18

Distribution of the "fishing" class in mean annual RVDs throughout the USFS, 1987–93.

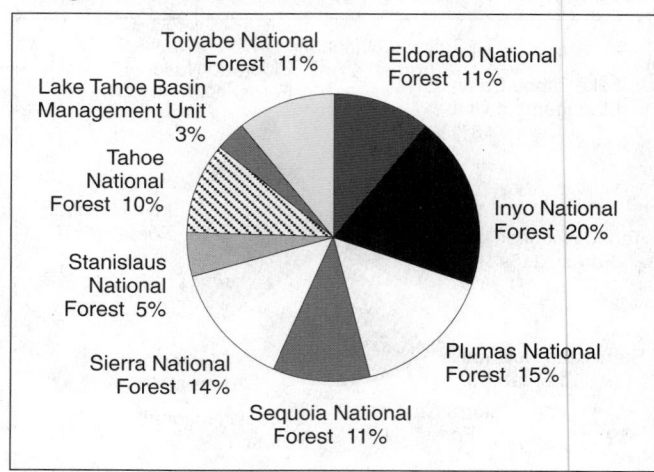

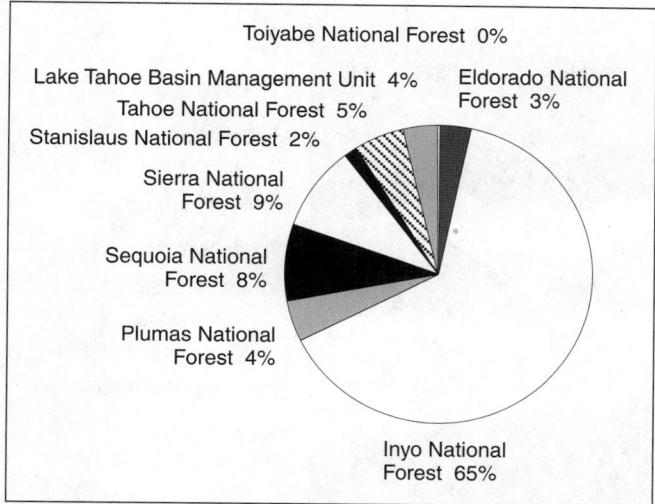

FIGURE 19.19

Distribution of the "nature study and interpretive activities" class in mean annual RVDs throughout the USFS, 1987–93.

tional forest, but these data are limited by the high level of aggregation both spatially and by activity class. It is difficult to evaluate RVDs for individual watersheds, for example, or by county. This limits our ability to relate these coarse-scale data to ecological, social, and economic conditions. It also limits our ability to evaluate policy scenarios in terms of their impacts on RVDs. There are further limitations in these data

due to the ambiguity (and consequent uncertainty) associated with some specific activity classes. Approximately 11.2 million of the 33.5 million mean annual overall RVDs (33%) for the USFS in the Region 5 national forests of the Sierra Nevada, for example, are in the "travel" class. Without these "travel" RVDs, the USFS would account for only 22.3 million of the 38.8–47.4 million RVDs (57% and 47%, respectively) reported in our data here for the entire Sierra Nevada. Other agencies (e.g., BLM) also report "travel" RVDs, but the USFS "travel" class seems to be disproportionately high. It is unclear whether these figures may include some visitors who are simply traveling through the national forests on their way to other destinations (e.g., on Interstate 80, U.S. 50, and U.S. 395). Recreation planners with each USFS unit estimate these RVDs from a variety of sources, but there is little empirical support for allocating a specified fraction of overall travel through USFS lands to the "travel" RVD class. Though this travel certainly does constitute an important activity on national forest lands (with a variety of impacts), it is arguable whether it should all be counted as "recreational" activity and included in the RVD estimates. We were unable to determine with any consistency how the "travel" class is counted by the USFS, so it is unclear at this time whether there are problems with the "travel" RVD estimates.

National Park Service

There are three national parks within the SNEP core area: Sequoia, Kings Canyon, and Yosemite (figure 19.41). The National Park Service (NPS) also operates the Devils Postpile

FIGURE 19.20

RVD trends for all USFS units (except Toiyabe), 1966–93.

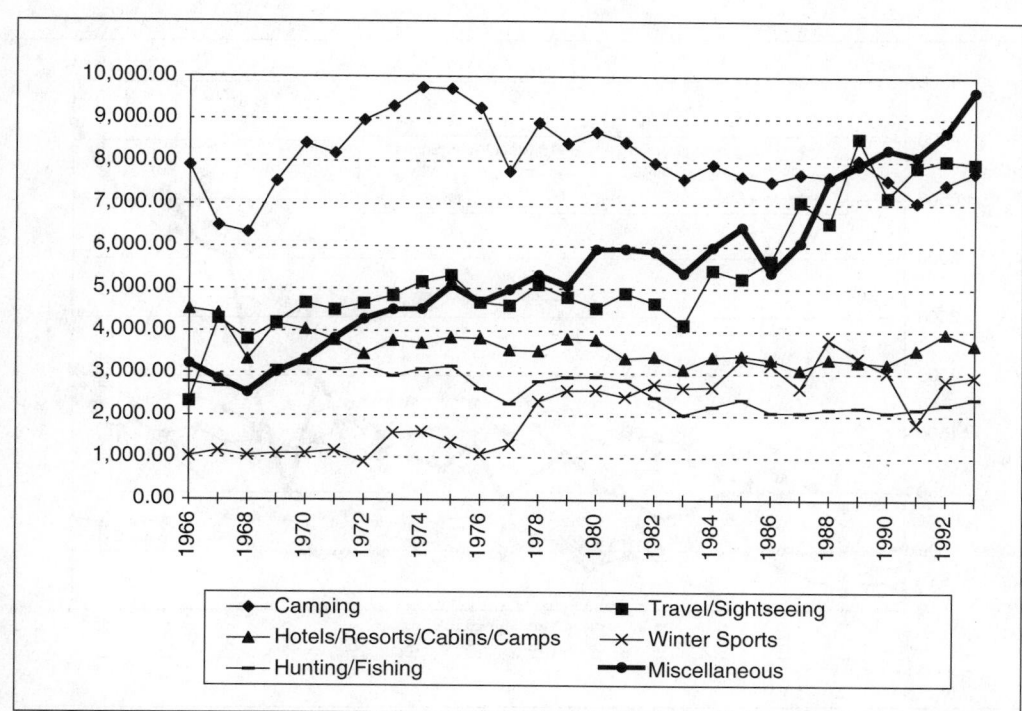

FIGURE 19.21

RVD activity class shares of all USFS RVDs (except Toiyabe National Forest), 1966–93.

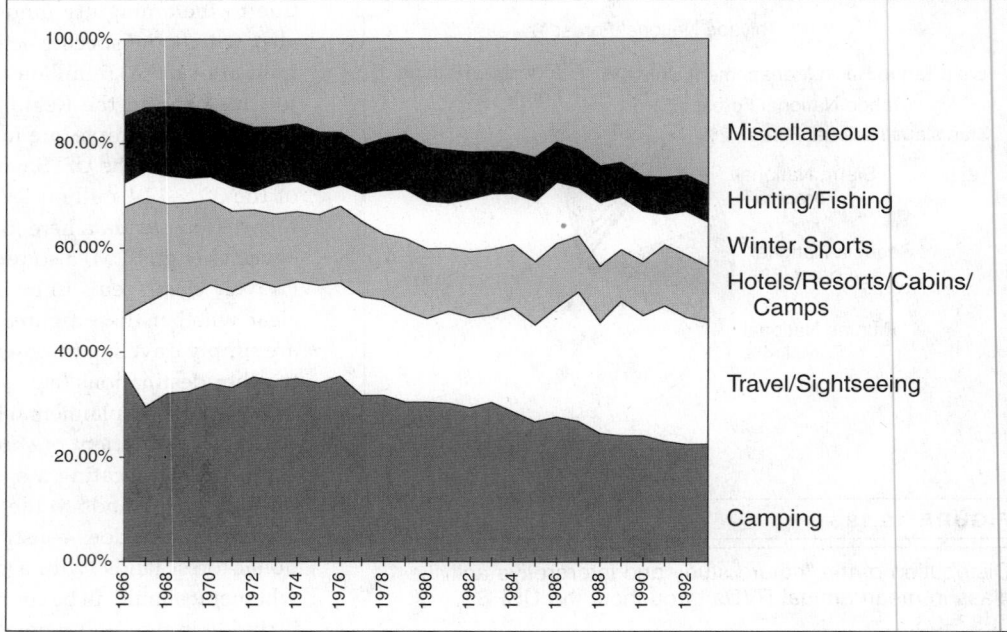

National Monument, whose visitation figures are reported by the administration of Sequoia–Kings Canyon National Parks. Lassen Volcanic National Park is located just north of the Sierra Nevada in the Cascade Range, but figures for Lassen were not included in our analysis.[2] The NPS has visitation records for 1971–93 (figure 19.42) but RVD estimates only from 1981 to 1993. Between 1981 and 1993, the total annual RVDs increased by 24% at NPS units for the region, from 2,697,634 to 3,352,607 (figure 19.43). Sequoia and Kings Canyon National Parks maintained very similar mean annual RVD rates during the twelve-year period, despite a significant decrease in RVDs for Kings Canyon between 1991 and 1993 (figures 19.44 and 19.45). During the twelve-year period, Yosemite National Park maintained the highest rate of visitation of the three national parks, averaging more than double the mean annual RVDs for either Sequoia or Kings Canyon alone. The com-

FIGURE 19.22

USFS RVDs trends within the "miscellaneous" RVD activity class, 1966–93 (Toiyabe National Forest not included).

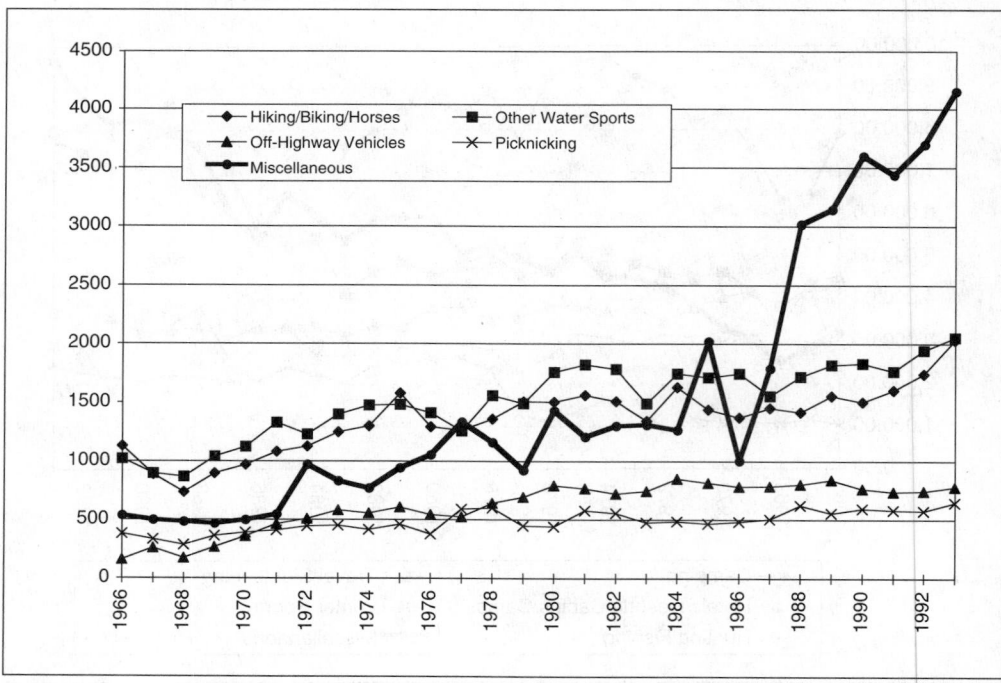

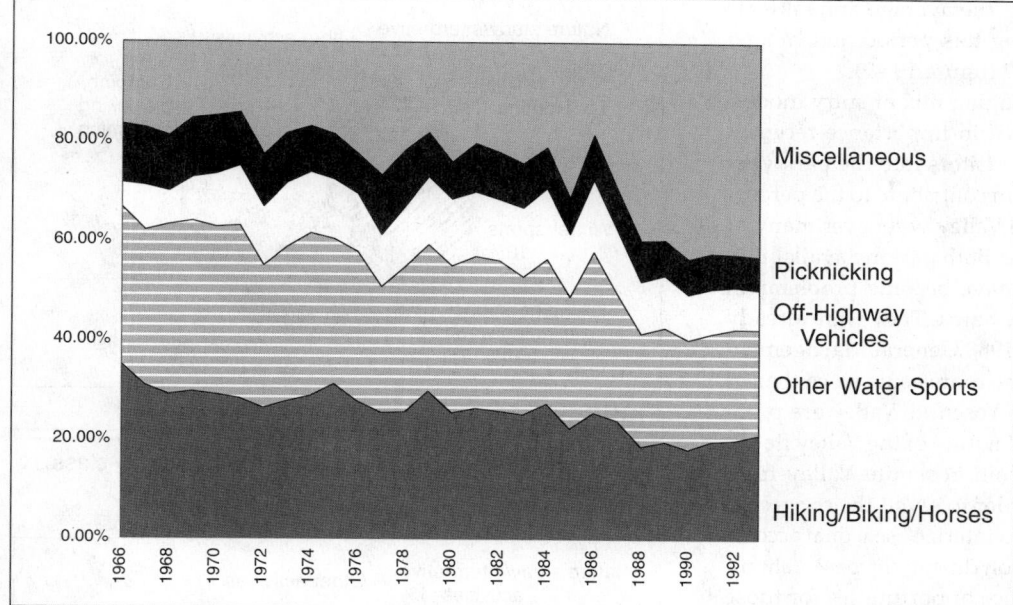

FIGURE 19.23

Shares of "miscellaneous" class of USFS RVDs, 1966–93 (Toiyabe National Forest not included).

bined total for Sequoia and Kings Canyon National Parks averaged 1,435,153 annual RVDs, however, approximately 77% of Yosemite National Park's average annual RVD figure of 1,853,237. Yosemite National Park's annual RVDs increased by 54% between 1981 and 1993 (figure 19.46), however, and recently, high levels of congestion in Yosemite Valley have required temporary closure of park entrances on weekends. Table 19.4 summarizes the RVD rates for the three individual parks during the twelve-year period 1981–93.

The change in the number of visitors to the park (who may visit for either more or less than one RVD) has been even more dramatic from 1971 to 1993. This time-series highlights Yosemite's continued growth in popularity while Sequoia and Kings Canyon remained relatively stable. Yosemite received only 2.3 million visitors in 1971, but that number had grown by 64% in 1993 to 3.8 million visitors. In contrast, Sequoia and Kings Canyon each had just under 900,000 visitors in 1971, and each had climbed to an annual visitation rate of around 1.1 million by 1991 (growing by 28% and 21%, respectively, while Yosemite grew by 46% during the same period). Because of a precipitous drop in Kings Canyon visitation in 1991–93, 1993 visitation levels were only 72% of 1971 levels, while Sequoia achieved an overall increase of 21% from 1971 to 1993. Yosemite National Park increased its share of total NPS RVDs in the Sierra Nevada during that time to nearly half (figure 19.47). The 1991–93 drop in Kings Canyon visits appears to be explained primarily by significant declines in tent camping and recreational vehicle RVDs from 1991 to 1993. It is unclear whether this is a result of changes in park management policies regarding camping, but it is somewhat surprising, given the other trends for greater demand for "front country" activities. Disaggregated by activity, RVD trends

show a slight decline in backpacking (with significant variation year-to-year) and a slight increase in concessionaire accommodations (with only slight variations year-to-year). Figure 19.48 shows the pattern by recreational activity class for Kings Canyon.

Visitation is highly correlated with RVD values for all three of the national parks during the 1981–93 period, suggesting that the historical 1971–93 visitation data are a good proxy for RVDs from 1971 to 1981. Average visitation appears to be fairly steady, with no clear trend in the RVD versus visitation

FIGURE 19.24

Eldorado National Forest mean annual RVDs by activity class, 1987–93.

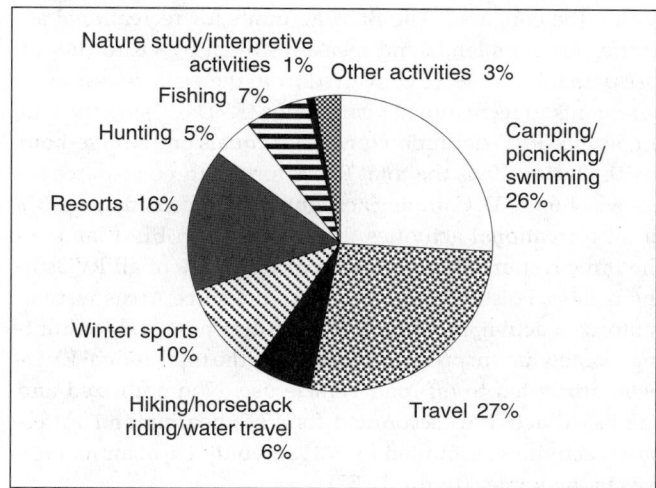

data for 1981–91. As noted earlier, the average annual RVD ratio was 2.24 for Yosemite during this period and ranged from a low of 2.20 to a high of 2.29 (figure 19.49).

Yosemite has been facing a changing mix of entry modes, however, as bus tours have grown in importance recently. Buses transport large numbers of visitors into the park very efficiently, but those same buses can contribute to the perception of crowding within Yosemite Valley whenever many of them arrive at about the same time. Both parking availability and congestion at popular sites have become problems in management of the Yosemite Valley area. They were already identified as problems when the 1980 General Management Plan was adopted but have become more acute since then.

Many of the current visitors to Yosemite Valley are probably comfortable with the "urban" nature of the Valley floor,[3] many other potential visitors avoid Yosemite Valley from Memorial Day to Labor Day in order to avoid the congested conditions.[4] Recent proposals to "winterize" seasonal accommodations and to increase visitation during off-peak "shoulder" seasons could therefore reduce opportunities for those potential visitors to experience Yosemite Valley under the conditions they prefer. This could reduce opportunities for visitors to experience Yosemite Valley under the conditions they prefer even as total visitation and RVDs increased. This scenario raises concerns about conflicts between similar uses and the impact of congestion on the quality of recreational experiences, which is discussed in more detail later. It is a problem that could apply generally to the Sierra Nevada under future conditions.

Bureau of Land Management

The Bureau of Land Management (BLM) manages land scattered throughout the Sierra Nevada that is located primarily along the periphery of the national forests (figure 19.50). Three BLM resource areas are located within the Sierra Nevada core area: Bishop, Folsom, and Eagle Lake (figure 19.51). Portions of other BLM resource areas, such as Caliente and Redding, intersect the Sierra Nevada region but are not predominantly within the core area. The BLM accounts for recreational activities upon its lands and measures them by visitor hours. Use data for 1992 were converted from the visitor-hour measurements to recreational visitor days (RVDs), using the Forest Service RVD definition (one RVD equals one twelve-hour visit). During 1992, the total RVDs for the three resource areas was 1,660,033. Camping accounted for approximately 43% of all recreational activities that occurred on BLM lands in the three resource areas. Approximately 19% of all RVDs in the Bishop, Folsom, and Eagle Lake Resource Areas were in motorized activities, while 15% were in fishing and/or hunting. Somewhat surprisingly, only 4% of the total annual RVDs were attributed to off-road vehicle use. Nonmotorized and site-based activities accounted for 13%, boating and water-based activities accounted for 6%, and only 1% of annual use was in the winter (figure 19.52).

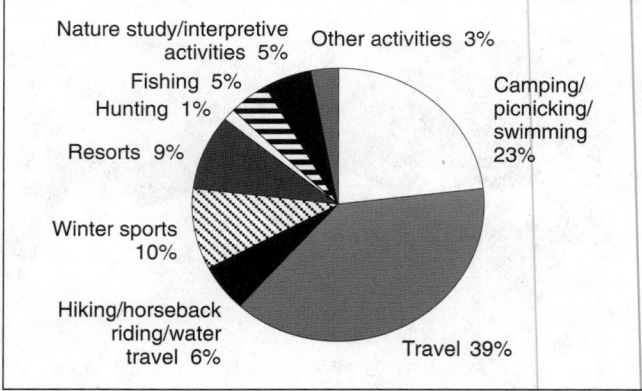

FIGURE 19.25

Inyo National Forest mean annual RVDs by activity class, 1987–93.

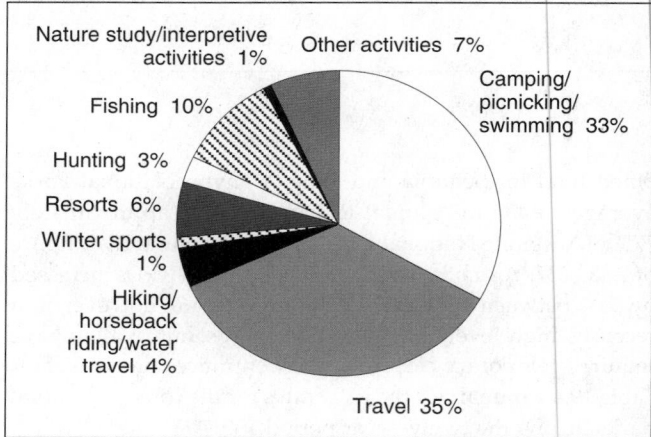

FIGURE 19.26

Plumas National Forest mean annual RVDs by activity class, 1987–93.

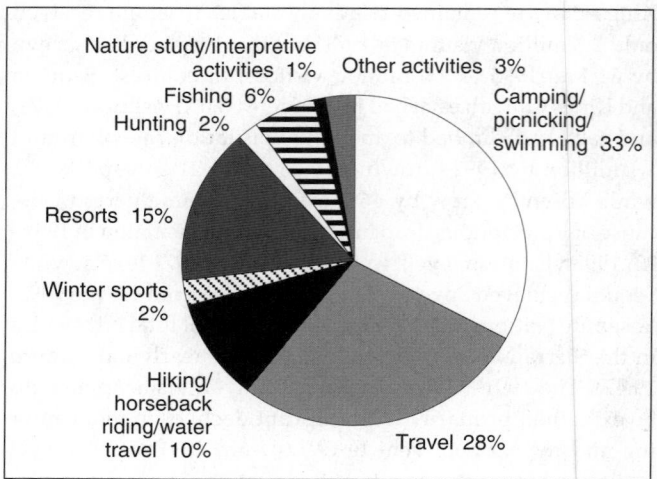

FIGURE 19.27

Sierra National Forest mean annual RVDs by activity class, 1987–93.

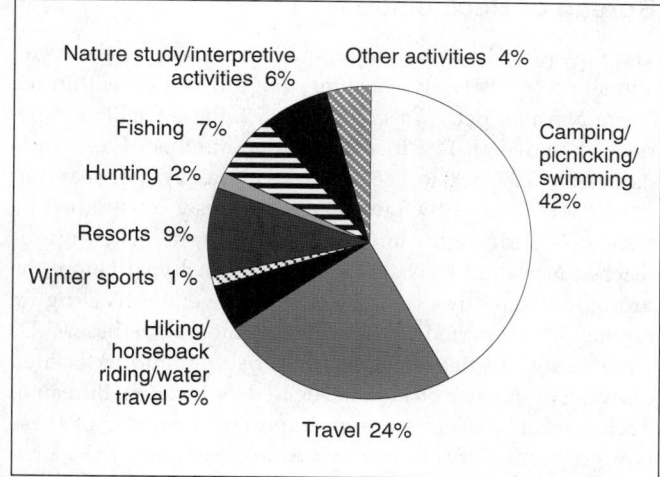

FIGURE 19.28

Sequoia National Forest mean annual RVDs by activity class, 1987–93.

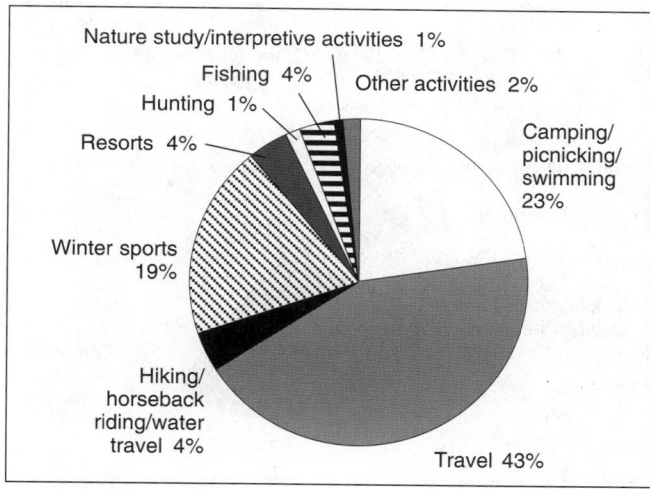

FIGURE 19.30

Tahoe National Forest mean annual RVDs by activity class, 1987–93.

Folsom Resource Area had approximately three times more total RVDs than either Bishop or Eagle Lake Resource areas (figure 19.53). This higher level of use probably reflects the proximity of the BLM lands in the Folsom Resource Area to the urban population of the Sacramento metropolitan area and the rapidly growing western Sierra Nevada foothills. Population is much more sparse in both the Bishop and Eagle Lake Resource Areas. Table 19.5 summarizes fishing and hunting RVD information for each resource area. More people fished at the Bishop Resource Area than Folsom or Eagle Lake; it had approximately 80% of the total fishing RVDs. This probably reflects the high-quality fly-fishing resource of the eastern Sierra Nevada in Mono and Inyo Counties. Hunting RVDs

were highest at the Eagle Lake Resource Area, which drew nearly 90% of the total hunting RVDs for the three areas. The Eagle Lake data includes many lands outside the core area, however, so much of this activity may have occurred on the Modoc Plateau.

This breakdown by BLM resource area for the hunting and fishing classes highlights that each resource area (and the subareas within that administrative unit) has very different RVD profiles by activity class. Geographically specific assessments must be made to evaluate the implications of the policy scenarios for recreational activities on BLM lands in the Sierra Nevada. A more detailed breakdown by geographic area and activity class is available from the California Environ-

FIGURE 19.29

Stanislaus National Forest mean annual RVDs by activity class, 1987–93.

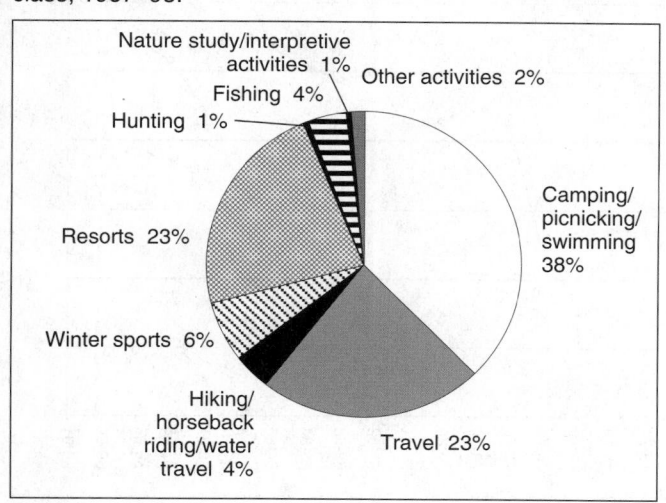

FIGURE 19.31

Lake Tahoe Basin Management Unit mean annual RVDs by activity class, 1987–93.

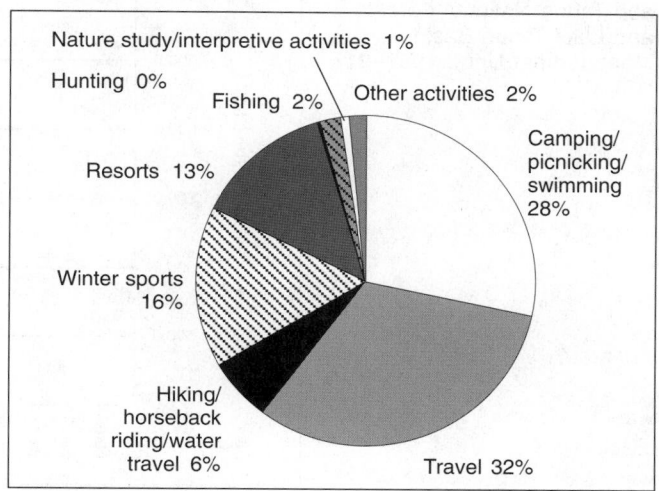

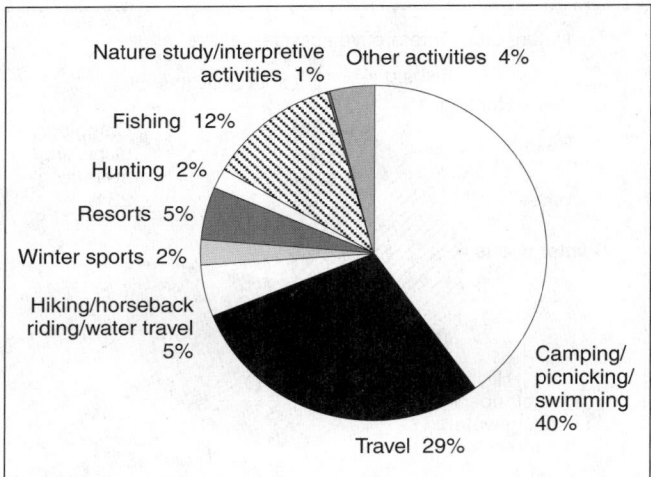

FIGURE 19.32

Toiyabe National Forest mean annual RVDs by activity class, 1987–92.

mental Resource Evaluation System (CERES) project of the Resources Agency of the State of California (http://ceres.ca.gov/snep), and the Alexandria Project at the University of California, Santa Barbara (http://alexandria.sdc.ucsb.edu/). The RVDs by activity class in the Bishop Resource Area are described in more detail later.

Bureau of Reclamation

The Bureau of Reclamation annually accounts for the recreational use of its twelve facilities that are located within the Sierra Nevada. Between 1970 and 1992, these facilities experienced an overall 14% increase in total number of recreation days, from 3,392,000 to 3,881,000 recreation days per year (figure 19.54). These RVD figures were seriously constrained by the 1987–94 drought, however, for the total RVD figures reached a peak of 6,566,000 in 1987 (the last year before the drought affected reservoir levels). The average RVD figure for the 1970–92 period was 3,917,000. Almost all of these RVDs were related to flat-water boating, fishing, and associated camping or day use on shore. Due to their location, Bureau of Reclamation reservoirs are an important provider of these types of recreational opportunities for residents of the Central Valley and the Reno metropolitan area and for visitors to the Lake Tahoe–Truckee area. Folsom Reservoir alone accounted for 42% of the average 1970–92 total Bureau of Reclamation RVDs. It serves as an important recreational resource for both the Sacramento metropolitan region and the rapidly growing Sierra Nevada foothill regions of Placer and El Dorado Counties (figure 19.55). In addition to the drought, however, increased concerns about flooding in the Sacramento area following the February 1986 floods resulted in modified reservoir operation. Further modifications may result following the 1995 floods or due to significant development of the floodplain north of Sacramento. The future capacity of Folsom Reservoir to provide recreational opportunities at pre-drought levels is therefore in question. Future RVD activity is therefore likely to approximate the average 1970–92 levels rather than return to the unusually high RVD levels of 1987. Devel-

FIGURE 19.33

Annual ski area RVDs on USFS land, (Eldorado, Inyo, Sequoia, Sierra, Stanislaus, and Tahoe National Forests, and Lake Tahoe Basin Management Unit), 1967–91.

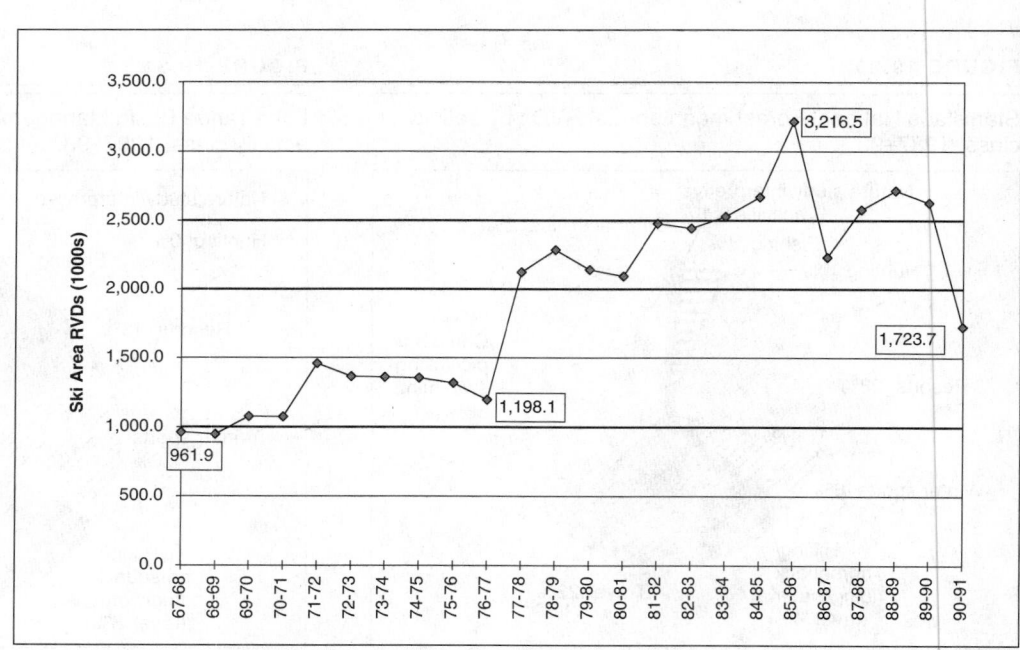

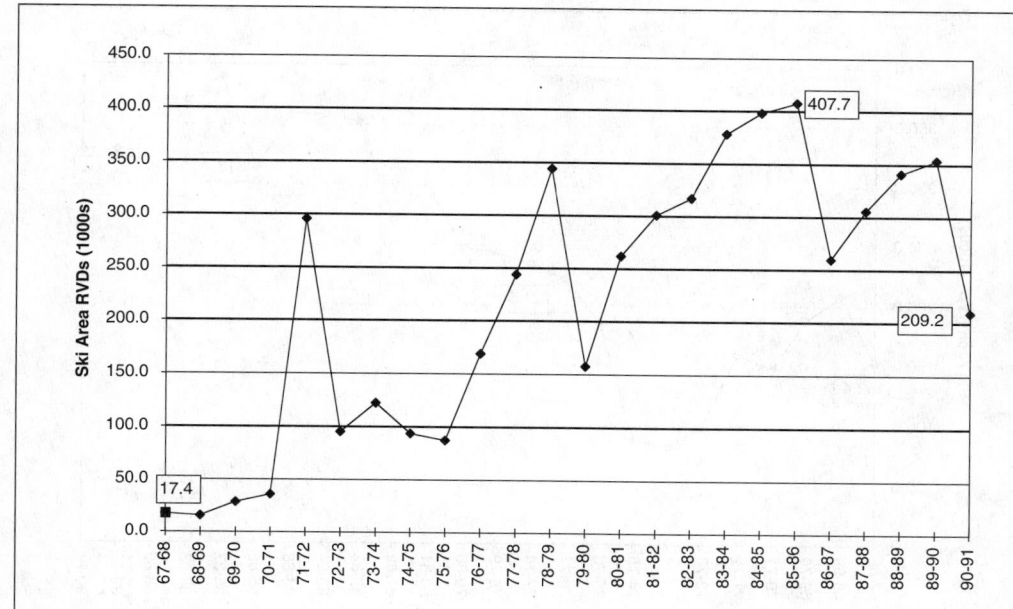

FIGURE 19.34

Annual ski area RVDs on the
Eldorado National Forest.

opment of an upstream Auburn Dam for flood control would reduce restrictions on operation, however, while potentially increasing competitive opportunities for flat-water recreation.

Army Corps of Engineers

The Army Corps of Engineers operates several reservoirs within the Sierra Nevada, but no data were available from the California offices of the Corps regarding recreational use levels or trends at those facilities. Relative to recreational use on other public lands within the Sierra Nevada, the Corps of Engineers facilities are believed to be little visited. Our RVD records nevertheless underestimate RVDs for those activities occurring at Corps facilities. An example of a Corps facility not reflected in our data is Englebright Reservoir, located on the Yuba River along the boundary of Yuba and Nevada Coun-

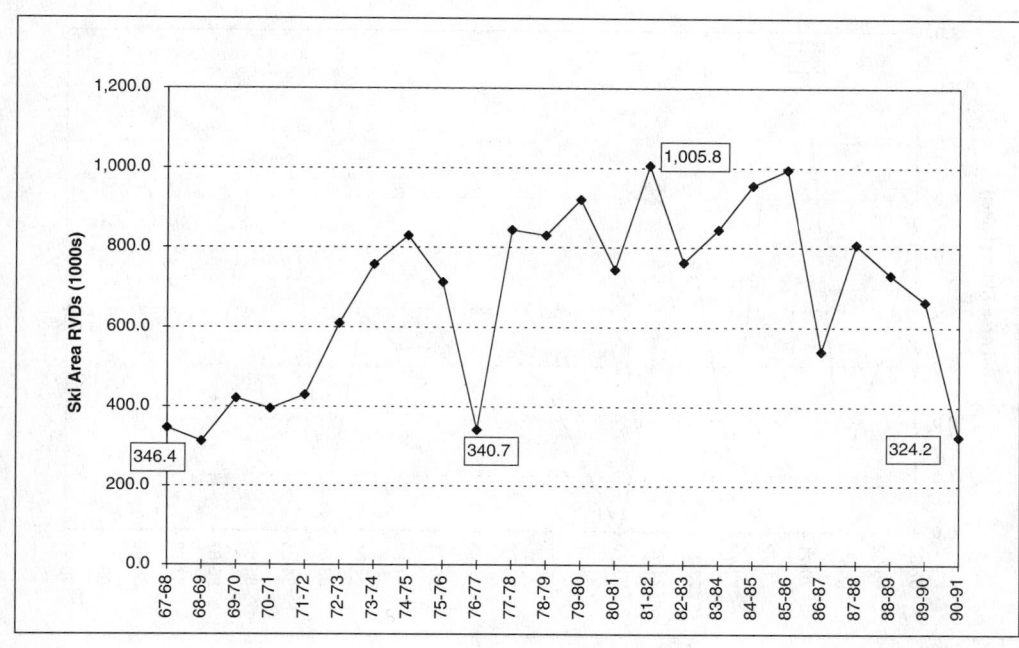

FIGURE 19.35

Annual ski area RVDs on the
Inyo National Forest.

FIGURE 19.36

Annual ski area RVDs on the Lake Tahoe Basin Management Unit.

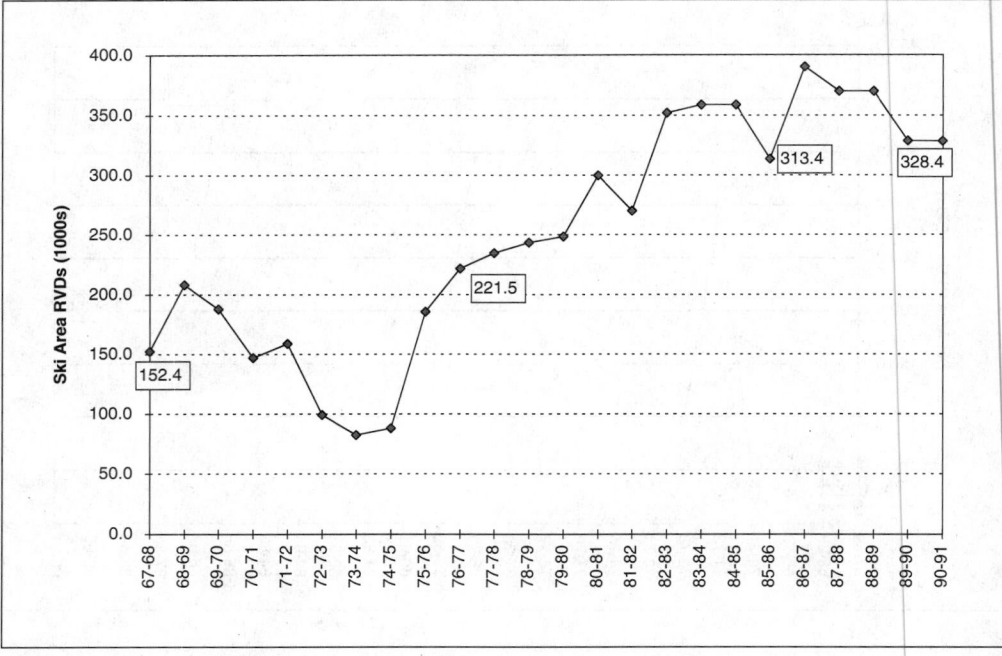

ties. Almost all of these recreation days would be related to flat-water boating, fishing, and associated camping or day use on shore.

STATE AGENCIES

California Department of Parks and Recreation

The State of California administers parks throughout the Sierra Nevada (figure 19.56), and maintains very reliable an-

FIGURE 19.37

Annual ski area RVDs on the Sequoia National Forest.

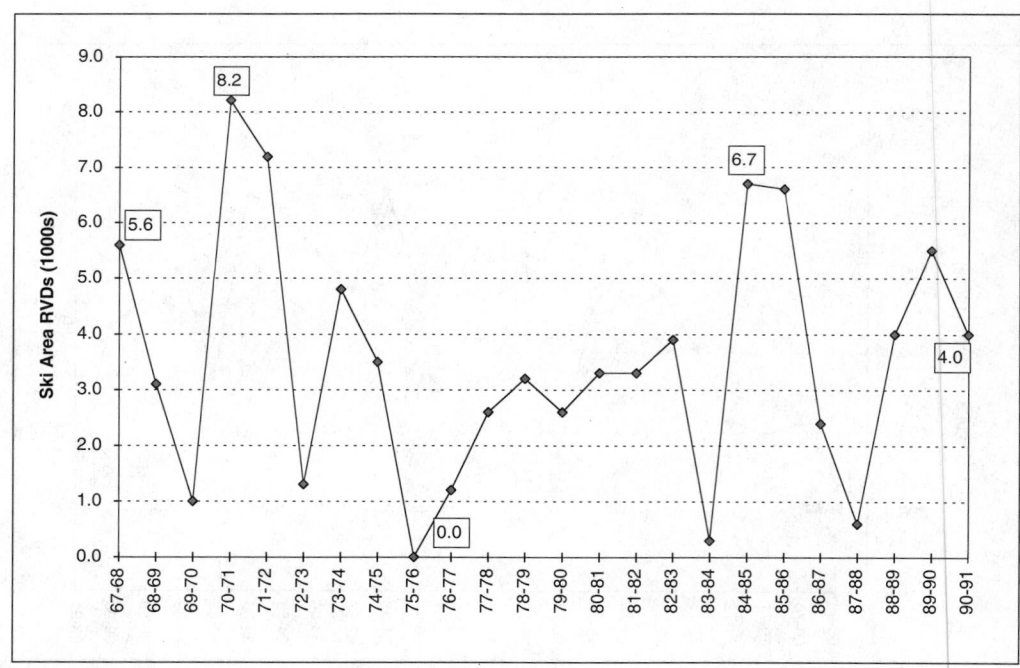

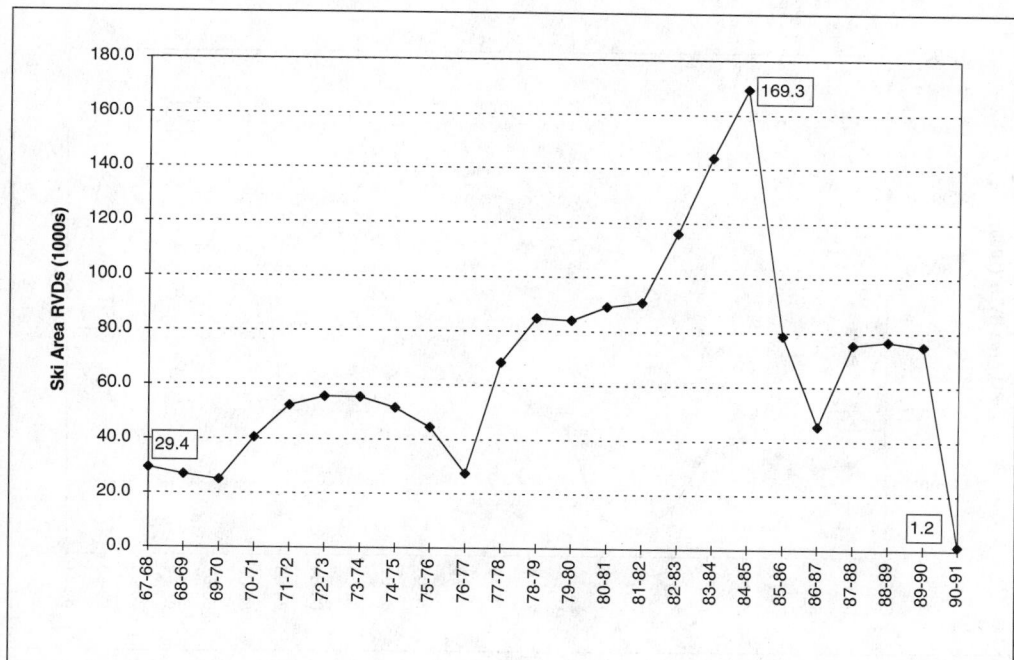

FIGURE 19.38

Annual ski area RVDs on the
Sierra National Forest.

nual use figures based upon entrance fees collected. These parks are part of one of the best state park systems in the United States (Ostertag 1995). We analyzed twenty-nine years of state park use data for Alpine, Butte, Calaveras, El Dorado, Fresno, Madera, Mono, Nevada, Placer, and Tuolumne Counties. Between 1963 and 1992, the total number of visitors per year decreased by 9%, from 7,984,899 to 7,241,246 individu-

als per year (figure 19.57). This occurred despite an increase in the state's population from 10 million in 1960 to nearly 31 million in 1990. Visitation may be either constrained by available capacity (e.g., campground reservations are usually required throughout the summer, and requests typically exceed spaces on weekends) or negatively affected by the relative cost of admission to the state units (generally higher than

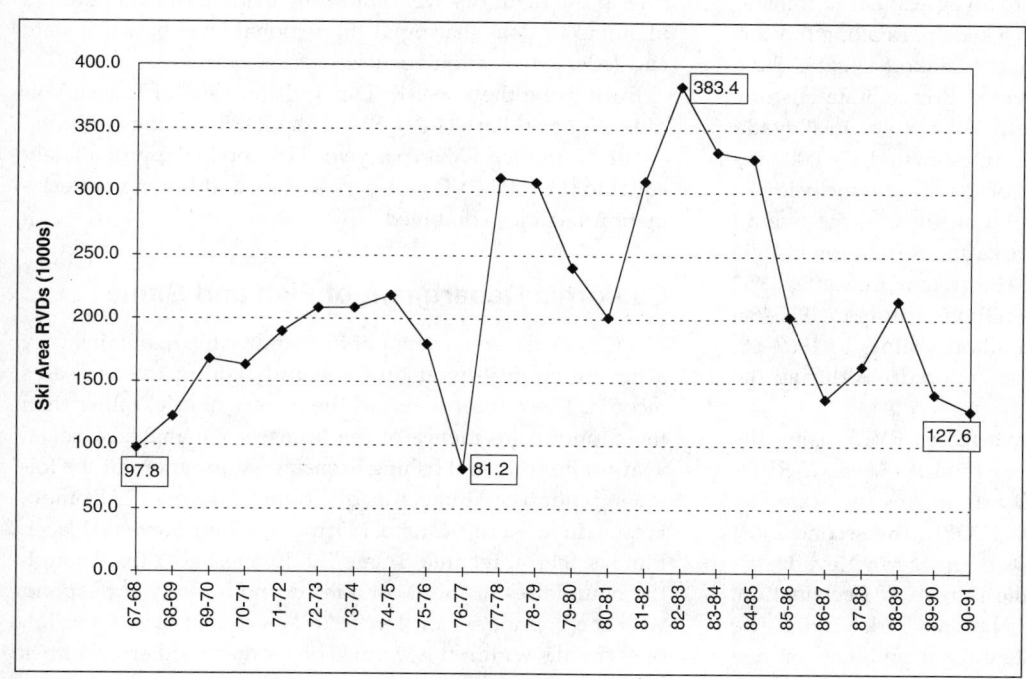

FIGURE 19.39

Annual ski area RVDs on the
Stanislaus National Forest.

FIGURE 19.40

Annual ski area RVDs on the Tahoe National Forest.

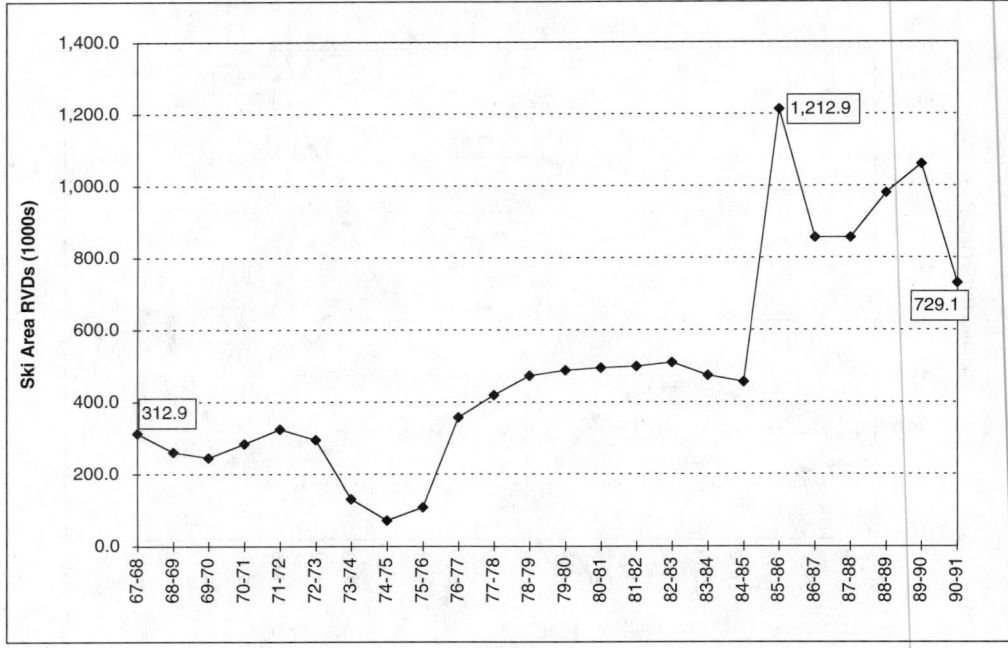

nearby facilities on federal lands). The mean annual total number of visitors for that twenty-nine-year period was 6,474,592 individuals per year and 6,413,253 for the more recent 1987–93 period.

These figures do not include the South Yuba River Project in Nevada County, which had an estimated 671,000 visitors in 1991–92 (South Yuba River Citizens League 1993) on lands either owned by the state (but not yet formally designated as a state park) or managed by the federal Bureau of Land Management. The BLM has entered into an agreement to transfer those lands to the state to develop a state park along the corridor that would eventually connect Malakoff Diggins State Historic Park and Bridgeport Covered Bridge State Historic Park downstream along the South Yuba River in Nevada County. These user figures include an estimated 170,000 visitors at the Highway 49 crossing alone. With the inclusion of the South Yuba River Project visitation figures, the overall visitation for the Department of Parks and Recreation in 1992 were comparable to those in 1963. The system grew after 1963 and reached a peak of roughly 9 million visitors in 1965–66, however, dropping to around 6 million visitors in 1967–68. The historic low of 4.9 million visitors occurred during the 1976–77 drought.

These visit estimates were converted to RVDs using the Yosemite RVD ratio of 2.24 to derive a total of 15,868,723 RVDs in the system in 1993. This RVD rate makes the state Department of Parks and Recreation (CDPR) the second most important public provider of RVDs in the Sierra Nevada, exceeding the combined totals of the Bureau of Reclamation, Bureau of Land Management, and National Park Service. The potential impact of California's state fiscal problems on rec-

reational activity in the Sierra Nevada can therefore be significant if it results in further strains on the state parks in the region. Evaluation of policy scenarios designed to manage recreational activities on the public lands in the Sierra Nevada must also clearly include careful consideration of state recreational policies. The relationship between state and federal recreation policy has generally been weak, with most of the state's recreation planning capacity eliminated through recent state budget cuts. Site-specific planning may be occurring in the field, but we found little evidence of cooperative planning or data sharing at the regional level between state and federal recreation agencies.

Even using the lower RVD per visitor ratio of Lassen Volcanic National Park (1.03), RVDs for CDPR facilities still total about 7.3 million RVDs per year. This total is approximately equal to the total RVDs for park service and Bureau of Reclamation facilities combined.

California Department of Fish and Game

The California Department of Fish and Game maintains very accurate county-level hunting and fishing license sales records. These records record the county of sale, rather than the county of residence of the licensee. We analyzed seven years of hunting and fishing licensing information for the following counties: Alpine, Amador, Butte, Calaveras, El Dorado, Fresno, Inyo, Kern, Madera, Mariposa, Mono, Nevada, Placer, Plumas, Sierra, Tehama, Tulare, Tuolomne, and Yuba. Though the boundaries of these counties do not exactly correspond with the boundaries of the Sierra Nevada, at least a portion of each falls within the Sierra. Three other northern counties

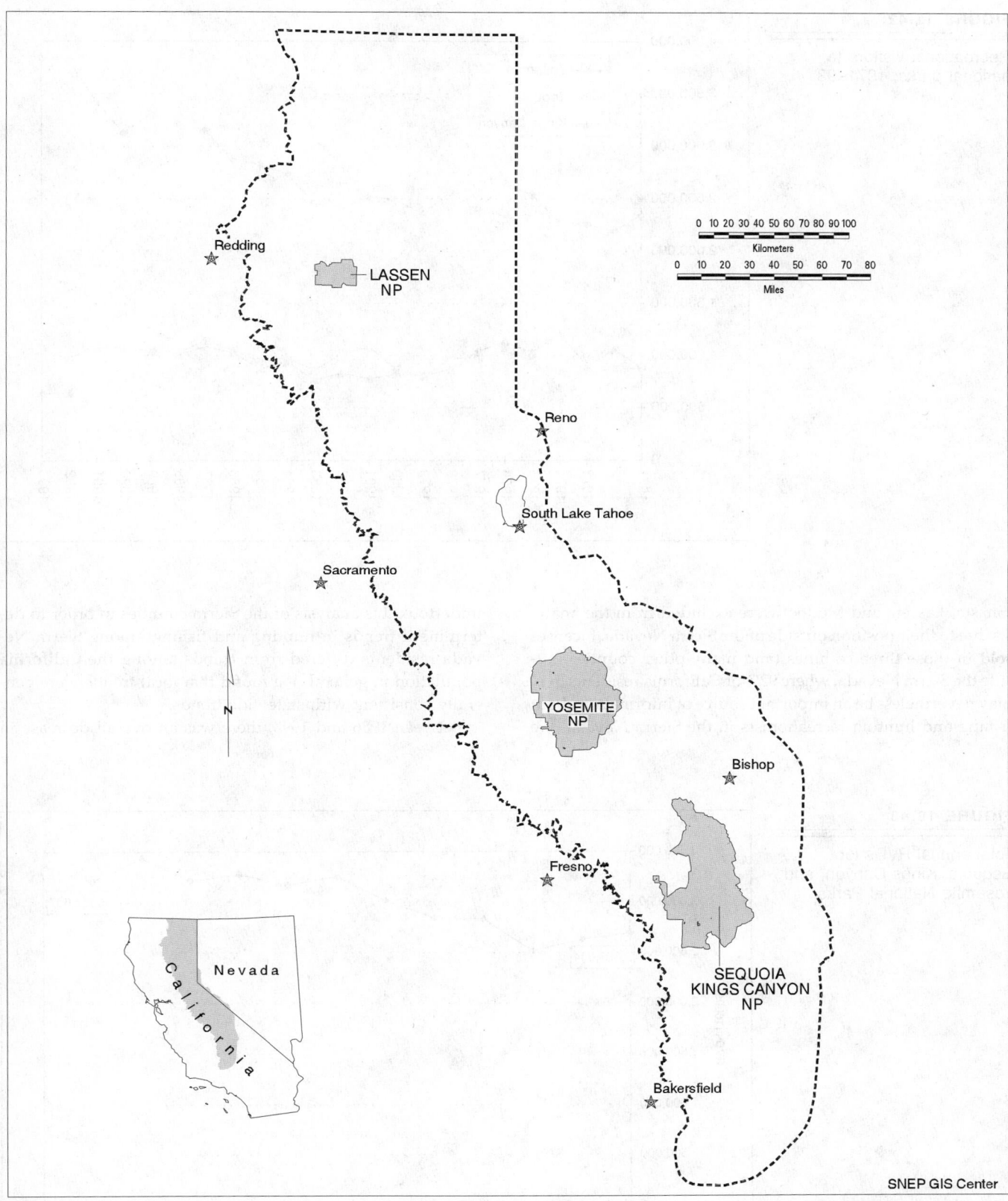

FIGURE 19.41

National parks within study area.

FIGURE 19.42

Recreational visitors to
national parks, 1971–93.

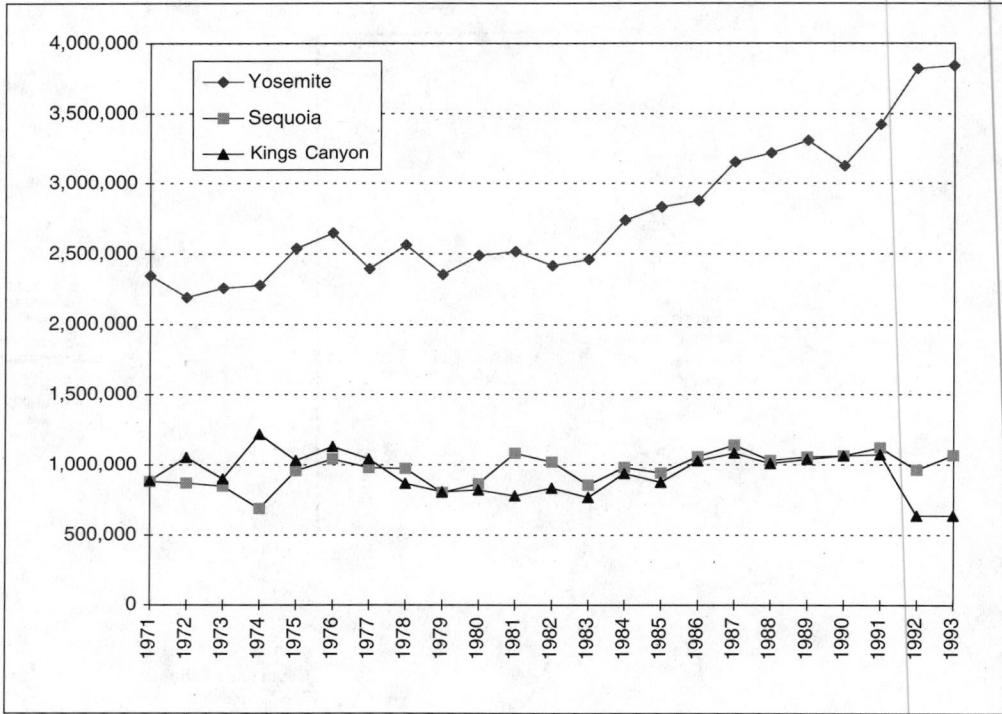

(Shasta, Lassen, and Modoc) were excluded from the analysis due to their position outside of the Sierra Nevada. Licenses sold in those three counties (and many other counties outside the Sierra Nevada, where 92% of California residents live) may nevertheless be an important source of information about fishing and hunting recreationists in the Sierra Nevada. We undertook this analysis of the Sierra counties in order to determine if trends in hunting and fishing among Sierra Nevada residents differed from trends among the California population in general. We found that local trends were generally consistent with statewide trends.

Between 1986 and 1993, there was an overall decrease in

FIGURE 19.43

Total annual RVDs for
Sequoia, Kings Canyon, and
Yosemite National Parks.

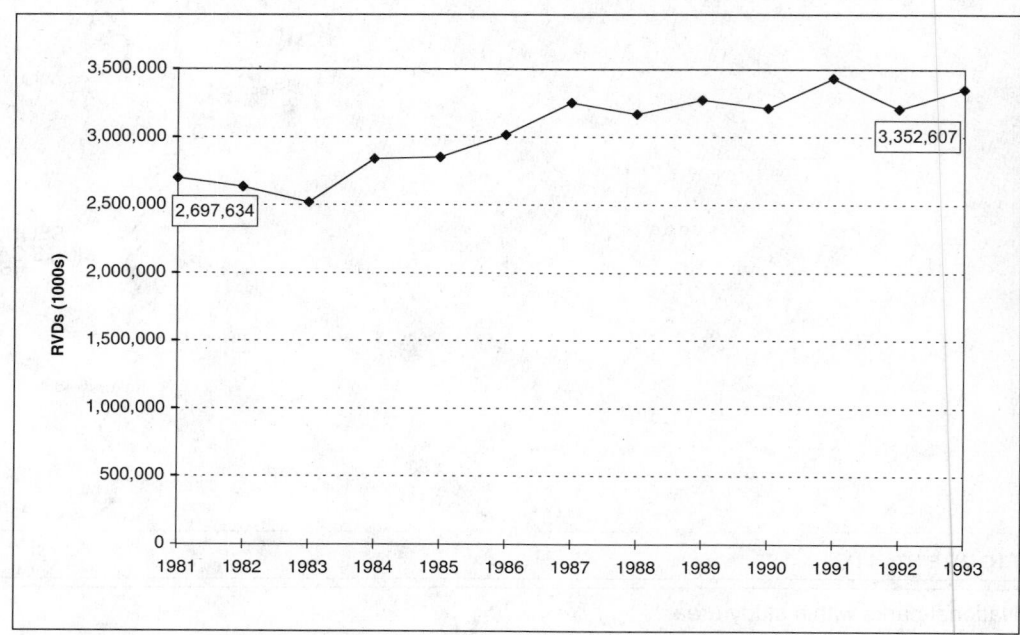

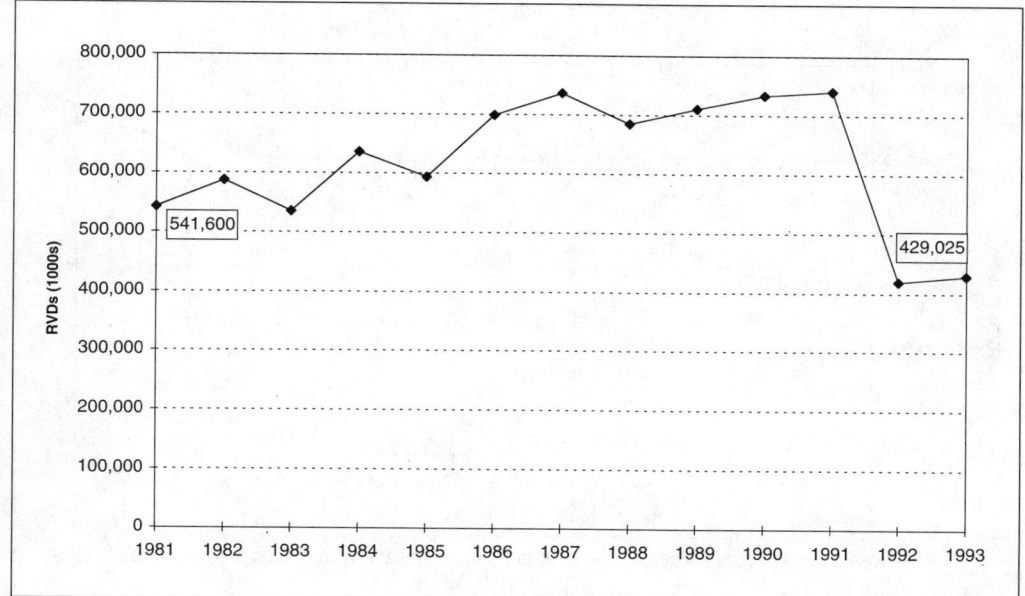

FIGURE 19.44

Total annual RVDS for Kings
Canyon National Park.

the total number of fishing and hunting licenses issued. During the 1986–87 fishing season, 298,939 fishing licenses were issued, dropping by 4% to 282, 341 in 1992–93 (figure 19.58). The total number of issued hunting licenses decreased by 15% between the 1986–87 and 1992–93 seasons, from 73,712 to 62,955 (figure 19.59). The total mean annual number of licenses issued during the seven-season period is summarized in table 19.6, showing that fishing is considerably more popular than hunting in the counties in the Sierra Nevada (by more than four-to-one).

The five counties within the Sierra Nevada region that issued the most fishing licenses were Fresno, Kern, Butte, Mono, and Inyo Counties. An analysis of USFS recreational visitor day (RVD) data during the years 1987–93 supported the county-level analysis of fishing licenses: Inyo National Forest contained the highest proportion of fishing RVDs for Forest Service areas within the Sierra Nevada (figure 19.18). These RVDs on USFS lands were probably dominated by nonresident recreationists, however, including residents in the Central Valley portions of Fresno, Kern, and Butte Counties, which

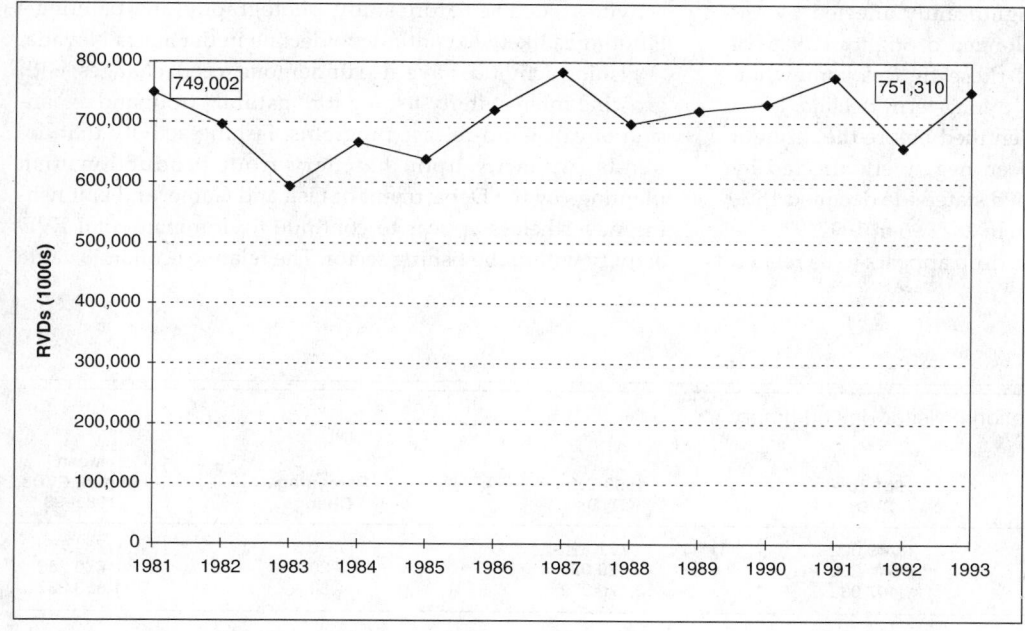

FIGURE 19.45

Total annual RVDs for
Sequoia National Park.

FIGURE 19.46

Total annual RVDs for
Yosemite National Park.

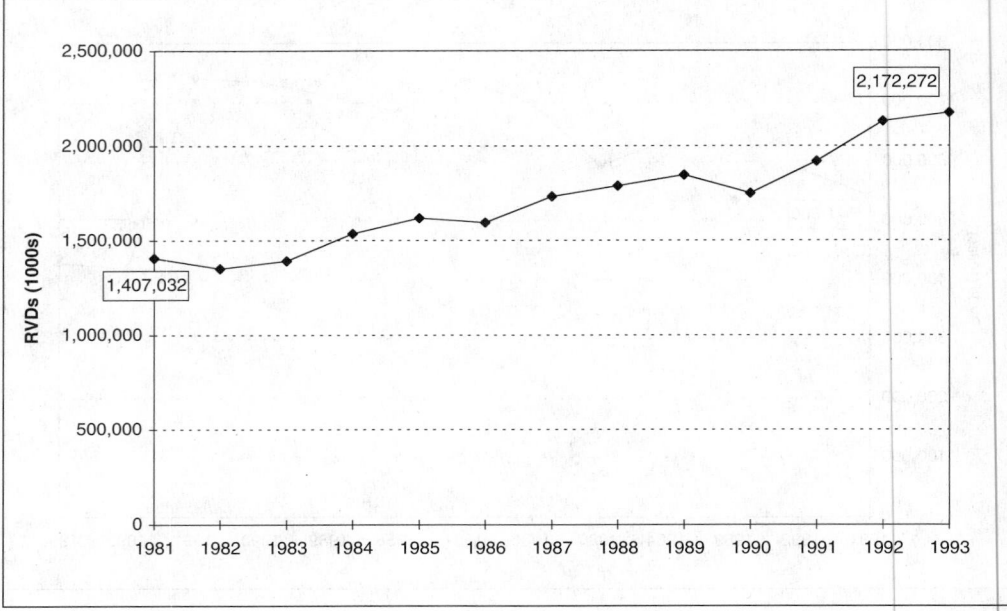

issued the largest number of hunting licenses (figure 19.60). These three counties extend into the Central Valley and have much higher populations than most of the other counties in the Sierra Nevada region.

Hunting and fishing licenses issued in other counties in California are often used in the Sierra Nevada. Those recreational activities that occur on public lands would generally be captured in agency-specific recreational use data, but hunting and fishing on private lands is generally unaccounted for in our data. The total number of fishing licenses issued by the state dropped 16%, from 1,708,900 in 1986 to 1,430,646 in 1993. The seven-year mean was 1,532,787 fishing licenses. Fishing, like whitewater rafting, is also significantly affected by annual weather variations. The prolonged drought of 1986–94 could therefore have dampened these figures somewhat. Hunting licenses appear to be on a long-term decline, however, which is a trend that was identified before the drought and generally would not have been negatively affected by the drought. Hunting licenses issued statewide declined 18%, from 351,389 in the 1986–87 season to 287,096 in 1992–93. The seven-year mean was 319,198. This drop appears to be related

to the continuing urbanization of California's population and changing social values regarding hunting. Fewer than 1% of Californians now hunt.

Restrictions on some types of hunting (e.g., mountain lions) may also have reduced the number of hunting licenses, but hunting licenses are predominantly issued for deer or waterfowl. Deer hunting is also on the decline. Land-use changes and the impact of habitat alteration on the probability of a successful hunt may also be reducing the relative attractiveness of deer hunting in the Sierra Nevada. This decline in hunting activity has also been accompanied by an apparent increase in nonconsumptive wildlife-related recreational activities, such as nature study, photography, and painting.[5] Hunting is likely to continue to decline in the Sierra Nevada.

Fishing activities have also undergone recent changes, with growing interest in fly-fishing for "natural" trout and expansion of catch-and-release programs. Fishing activity that depends primarily upon hatchery trout production, fish plantings by the Department of Fish and Game, and bait fishing nevertheless appear to continue to dominate total RVD activity within the fishing sector. The relative economic value

TABLE 19.4

National Park Service area recreational visitor day summary.

National Park Service Area	1981 Total RVDs	1993 Total RVDs	Percentage Change	Mean Annual RVDs, 1981–93
Sequoia National Park	749,002	751,310	+ 0.3	764,770
Kings Canyon National Park	541,600	429,025	−20	670,383
Yosemite National Park	1,407,032	2,172,272	+ 54	1,853,237

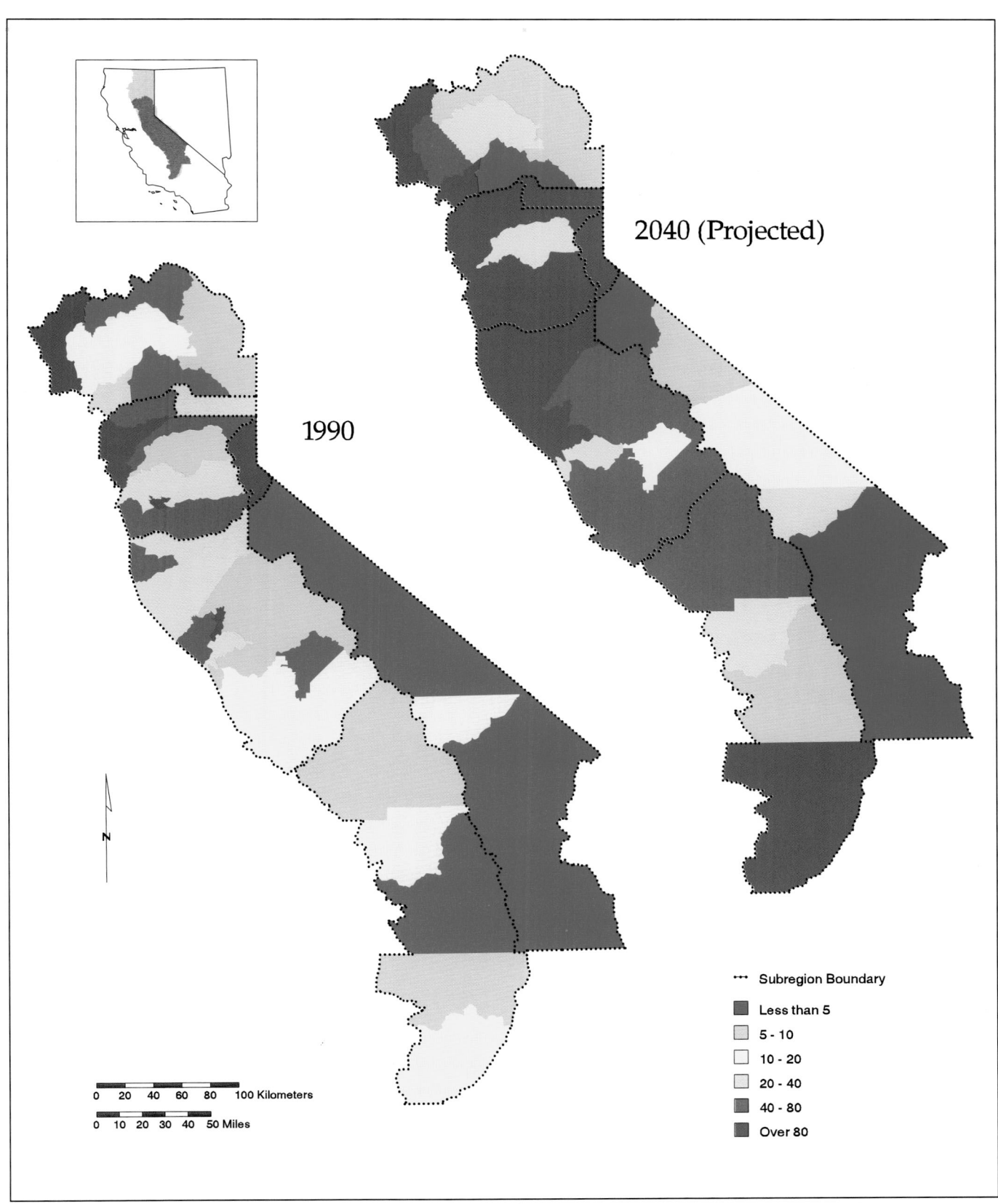

2040 (Projected)

1990

Subregion Boundary

Less than 5

5 - 10

10 - 20

20 - 40

40 - 80

Over 80

0   20   40   60   80   100 Kilometers

0   10   20   30   40   50 Miles

PLATE 11.1

Current and projected population density (persons per square mile) in the Sierra Nevada.

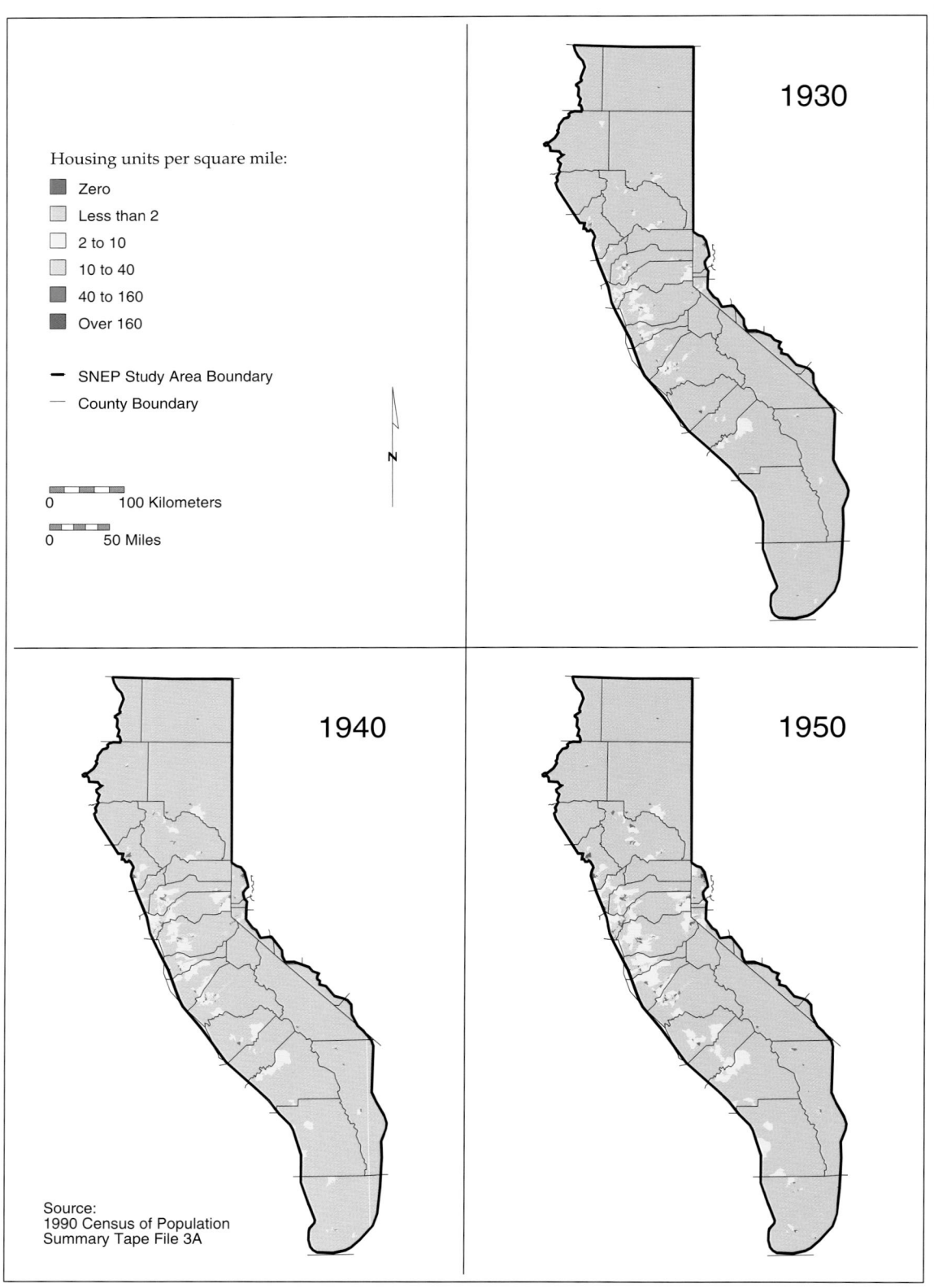

Housing units per square mile:

- Zero
- Less than 2
- 2 to 10
- 10 to 40
- 40 to 160
- Over 160

— SNEP Study Area Boundary
— County Boundary

0 ——— 100 Kilometers

0 ——— 50 Miles

N

1930

1940

1950

Source:
1990 Census of Population
Summary Tape File 3A

PLATE 11.2

Change in housing density in the Sierra Nevada region, 1940–90 (based on 1990 Census of Population, Summary Tape File 3A).

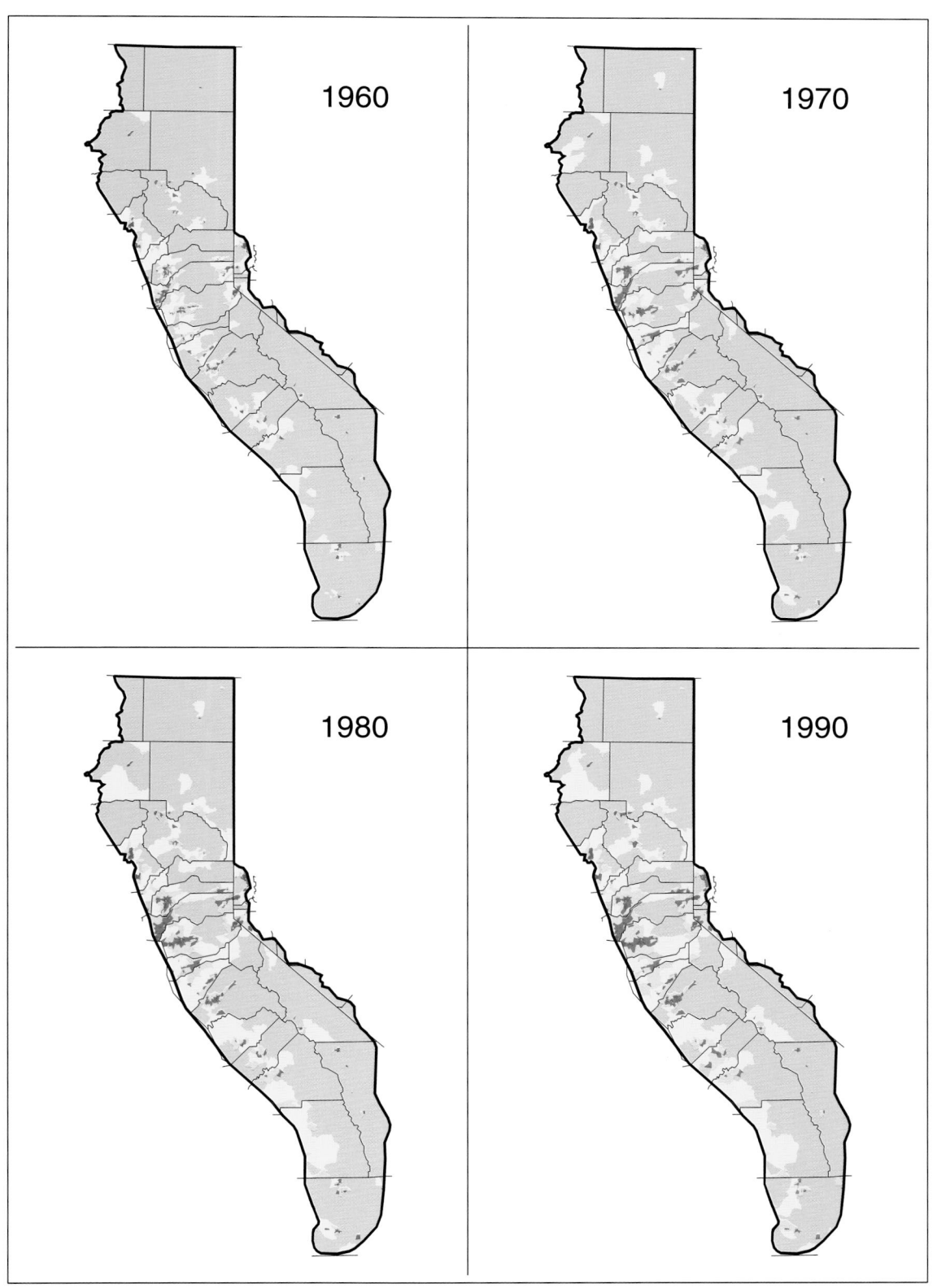

PLATE 11.2 (continued)

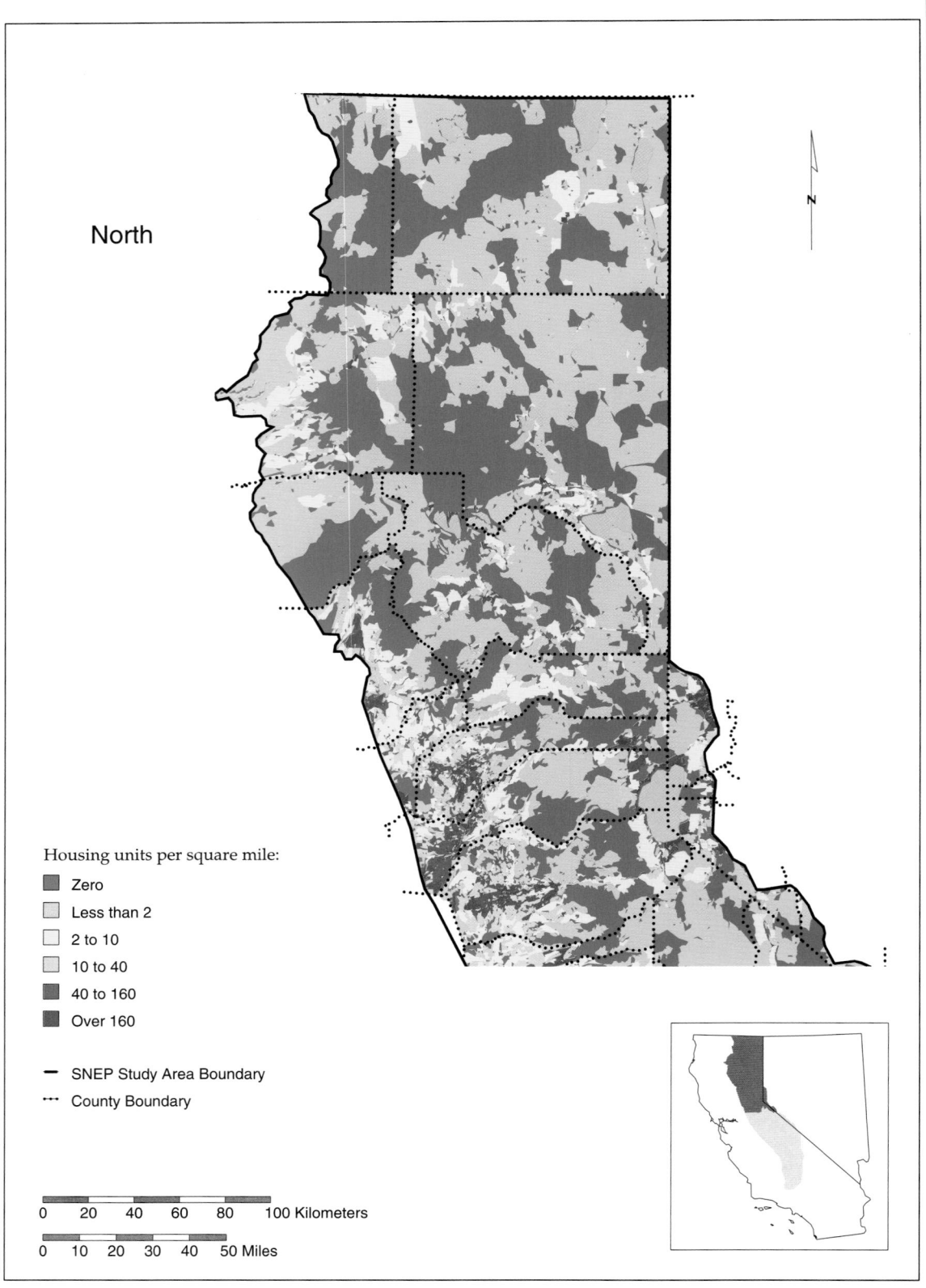

North

Housing units per square mile:
- Zero
- Less than 2
- 2 to 10
- 10 to 40
- 40 to 160
- Over 160

— SNEP Study Area Boundary
··· County Boundary

0   20   40   60   80   100 Kilometers

0   10   20   30   40   50 Miles

PLATE 11.3

Housing density in the Sierra Nevada region (from 1990 Census of Population, Summary Tape File 1B).

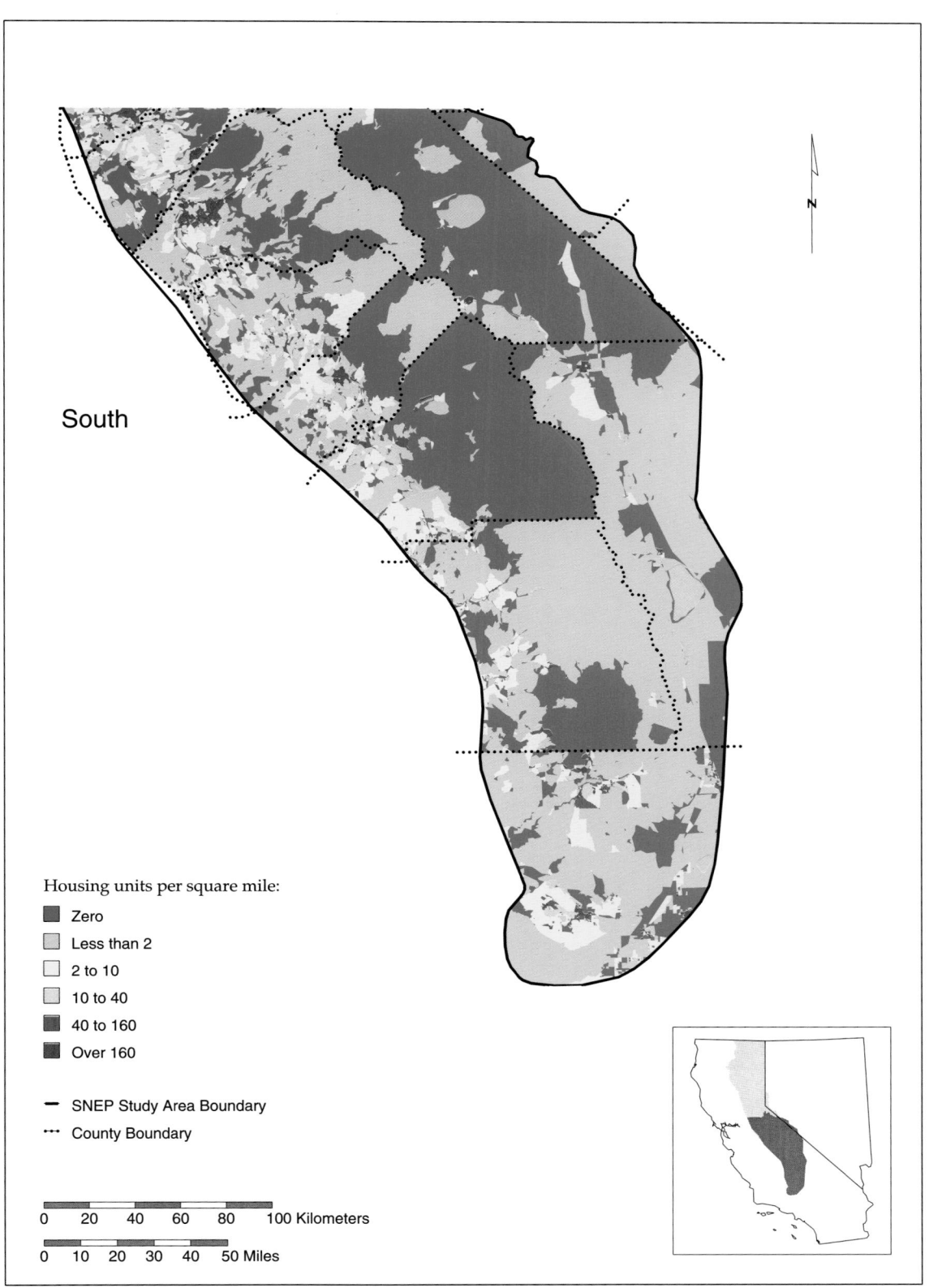

South

Housing units per square mile:

■ Zero
■ Less than 2
□ 2 to 10
□ 10 to 40
■ 40 to 160
■ Over 160

— SNEP Study Area Boundary
⋯ County Boundary

0 20 40 60 80 100 Kilometers

0 10 20 30 40 50 Miles

PLATE 11.3 (continued)

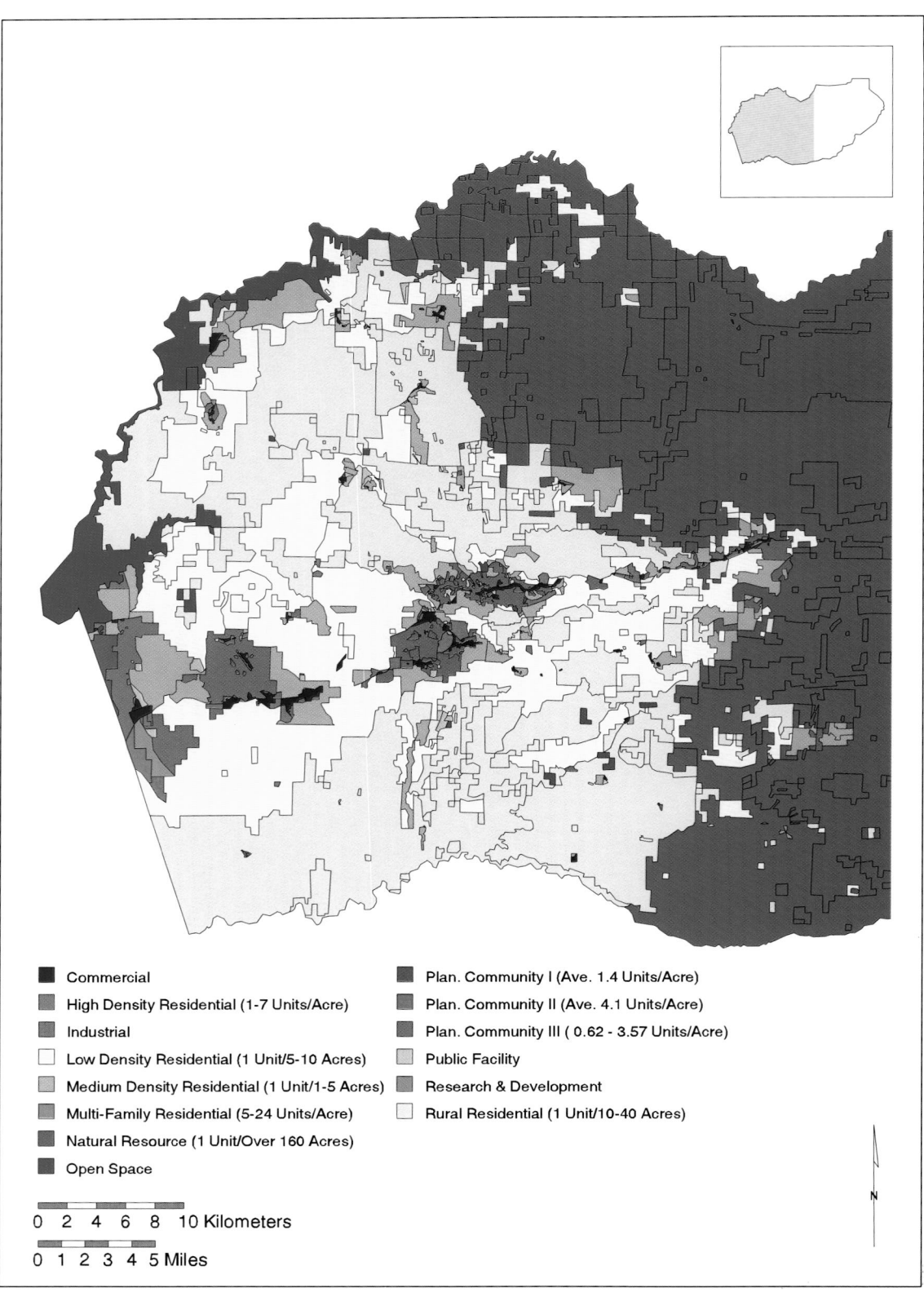

Commercial

High Density Residential (1-7 Units/Acre)

Industrial

Low Density Residential (1 Unit/5-10 Acres)

Medium Density Residential (1 Unit/1-5 Acres)

Multi-Family Residential (5-24 Units/Acre)

Natural Resource (1 Unit/Over 160 Acres)

Open Space

Plan. Community I (Ave. 1.4 Units/Acre)

Plan. Community II (Ave. 4.1 Units/Acre)

Plan. Community III (0.62 - 3.57 Units/Acre)

Public Facility

Research & Development

Rural Residential (1 Unit/10-40 Acres)

0 2 4 6 8 10 Kilometers

0 1 2 3 4 5 Miles

PLATE 11.4

Draft General Plan, Project Description, El Dorado County (western portion) (El Dorado County Planning Department).

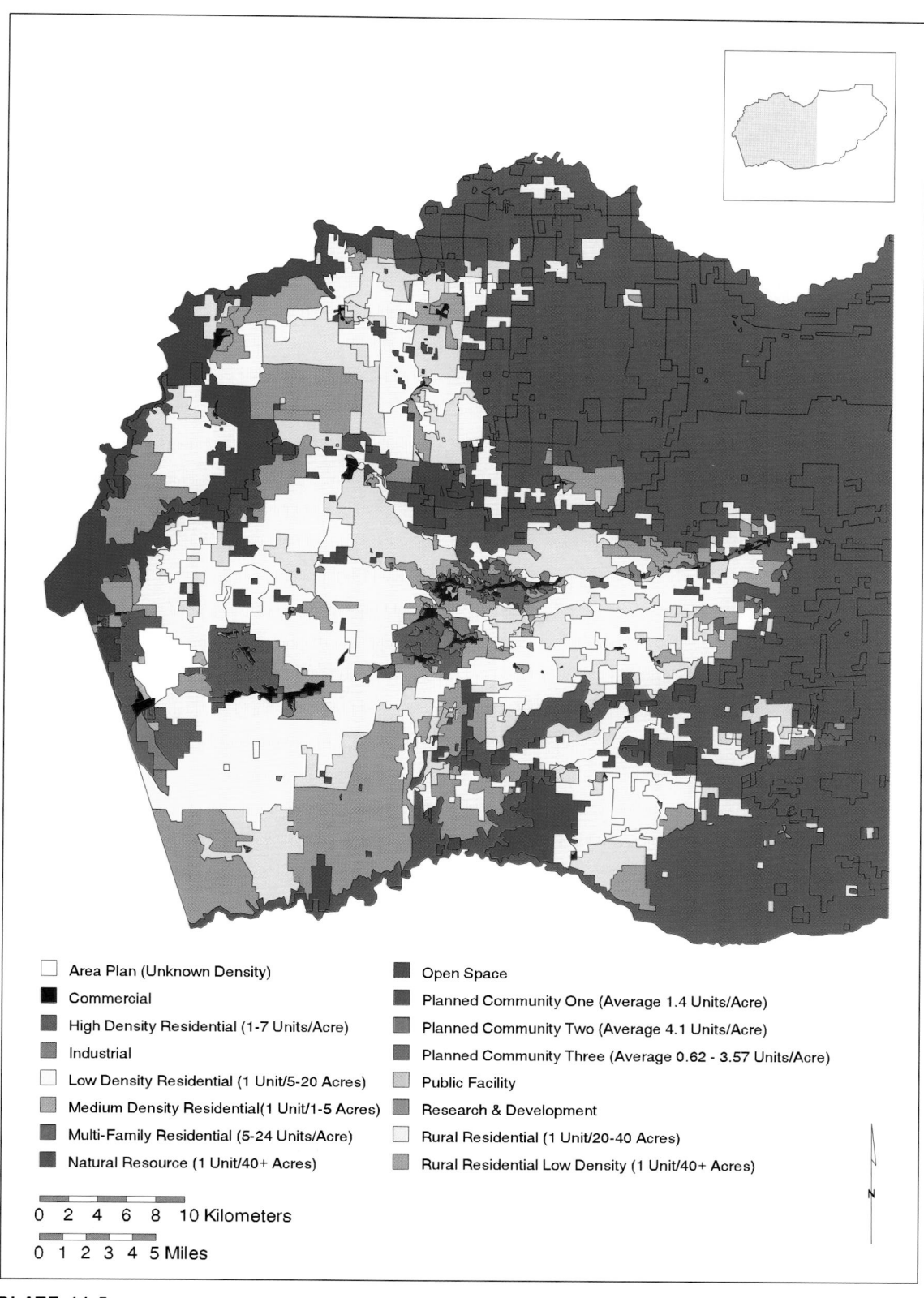

Area Plan (Unknown Density)

Commercial

High Density Residential (1-7 Units/Acre)

Industrial

Low Density Residential (1 Unit/5-20 Acres)

Medium Density Residential(1 Unit/1-5 Acres)

Multi-Family Residential (5-24 Units/Acre)

Natural Resource (1 Unit/40+ Acres)

Open Space

Planned Community One (Average 1.4 Units/Acre)

Planned Community Two (Average 4.1 Units/Acre)

Planned Community Three (Average 0.62 - 3.57 Units/Acre)

Public Facility

Research & Development

Rural Residential (1 Unit/20-40 Acres)

Rural Residential Low Density (1 Unit/40+ Acres)

0 2 4 6 8   10 Kilometers

0 1 2 3 4 5 Miles

N

PLATE 11.5

Draft General Plan, Alternative, El Dorado County (western portion) (El Dorado County Planning Department).

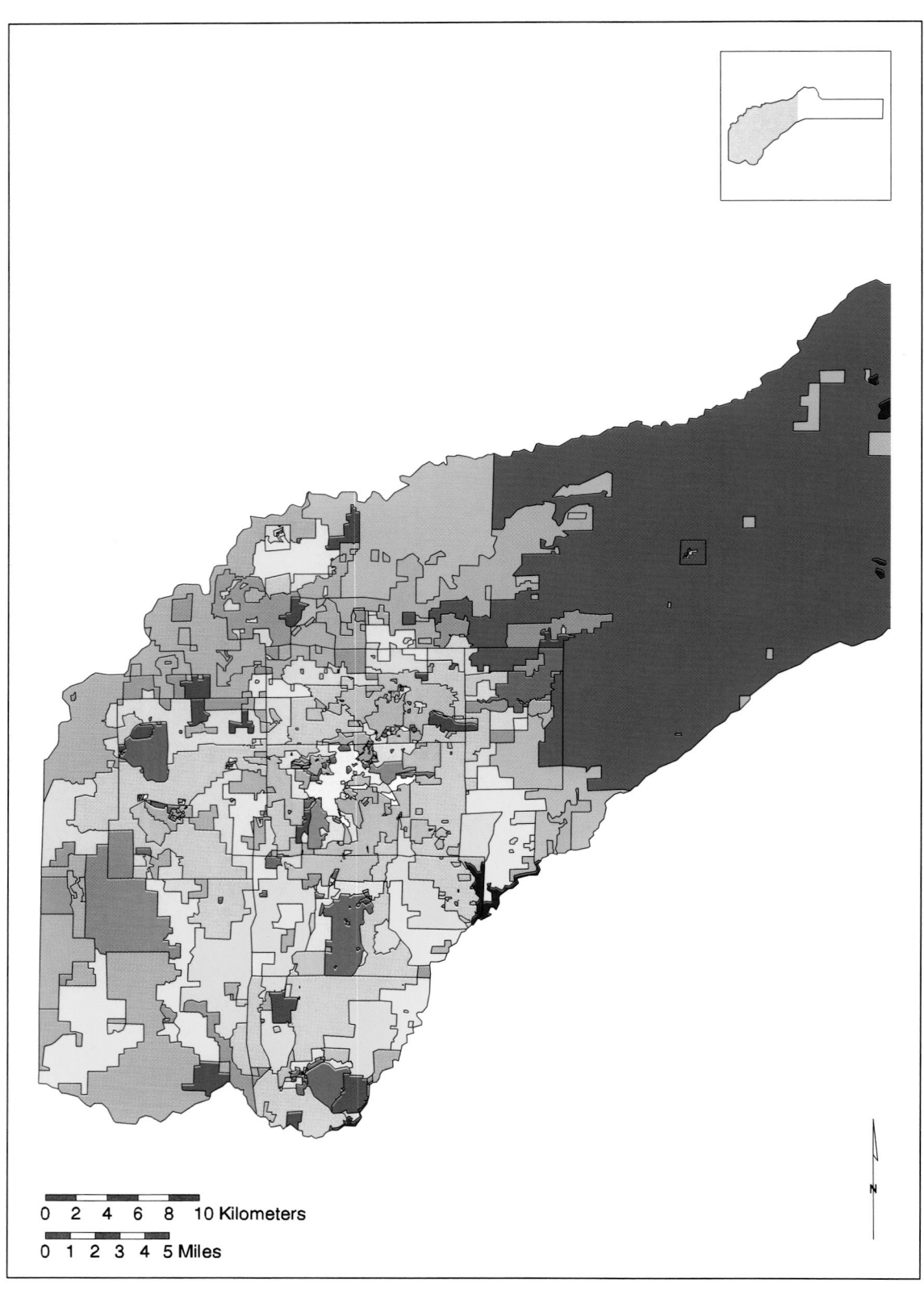

PLATE 11.6

Draft General Plan, Nevada County (western portion) (Nevada County Planning Department).

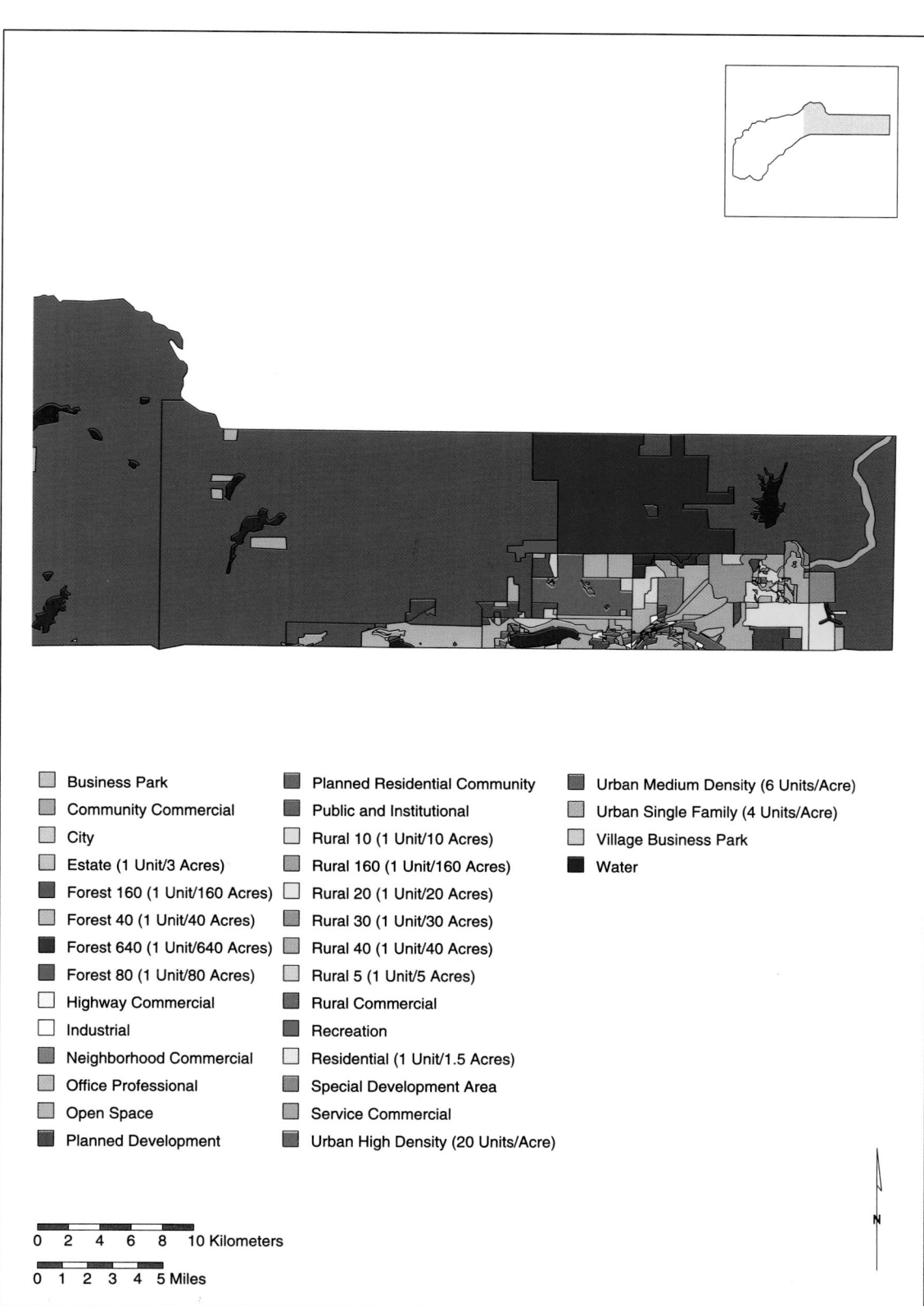

Business Park

Community Commercial

City

Estate (1 Unit/3 Acres)

Forest 160 (1 Unit/160 Acres)

Forest 40 (1 Unit/40 Acres)

Forest 640 (1 Unit/640 Acres)

Forest 80 (1 Unit/80 Acres)

Highway Commercial

Industrial

Neighborhood Commercial

Office Professional

Open Space

Planned Development

Planned Residential Community

Public and Institutional

Rural 10 (1 Unit/10 Acres)

Rural 160 (1 Unit/160 Acres)

Rural 20 (1 Unit/20 Acres)

Rural 30 (1 Unit/30 Acres)

Rural 40 (1 Unit/40 Acres)

Rural 5 (1 Unit/5 Acres)

Rural Commercial

Recreation

Residential (1 Unit/1.5 Acres)

Special Development Area

Service Commercial

Urban High Density (20 Units/Acre)

Urban Medium Density (6 Units/Acre)

Urban Single Family (4 Units/Acre)

Village Business Park

Water

0 2 4 6 8 10 Kilometers

0 1 2 3 4 5 Miles

N

PLATE 11.6 (continued)

Draft General Plan (eastern portion).

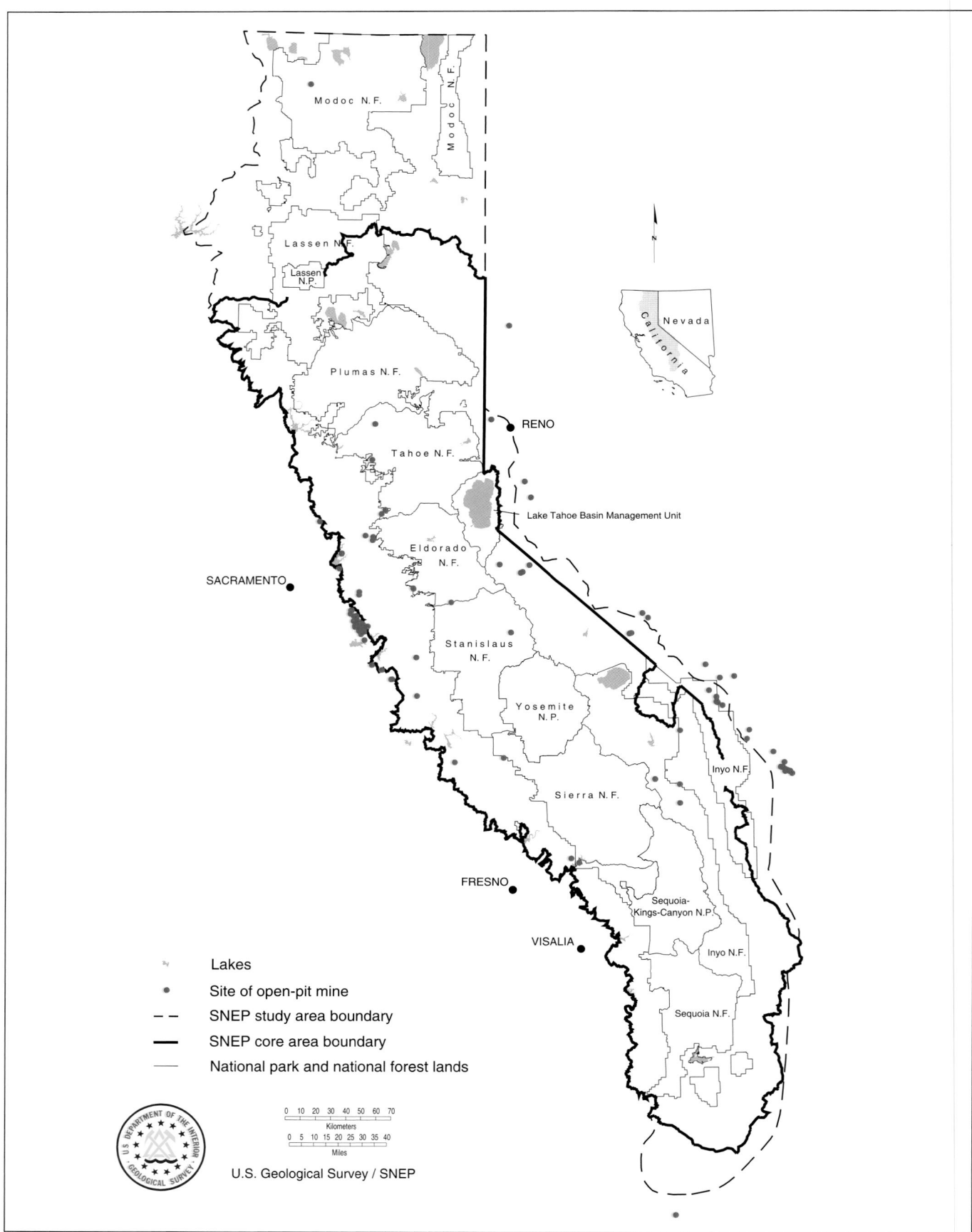

Modoc N.F.

Modoc N. F.

Lassen N.F.

Lassen
N.P.

Plumas N.F.

Nevada

California

RENO

Tahoe N.F.

Lake Tahoe Basin Management Unit

Eldorado
N. F.

SACRAMENTO

Stanislaus
N. F.

Yosemite
N. P.

Inyo N.F.

Sierra N.F.

FRESNO

Sequoia-
Kings-Canyon N.P.

VISALIA

Inyo N.F.

Sequoia N.F.

Lakes

Site of open-pit mine

SNEP study area boundary

SNEP core area boundary

National park and national forest lands

| 0 | 10 | 20 | 30 | 40 | 50 | 60 | 70 |
Kilometers
| 0 | 5 | 10 | 15 | 20 | 25 | 30 | 35 | 40 |
Miles

U.S. Geological Survey / SNEP

U.S. DEPARTMENT OF THE INTERIOR
GEOLOGICAL SURVEY

PLATE 18.1

Locations of open-pit mines within and near the SNEP study area.

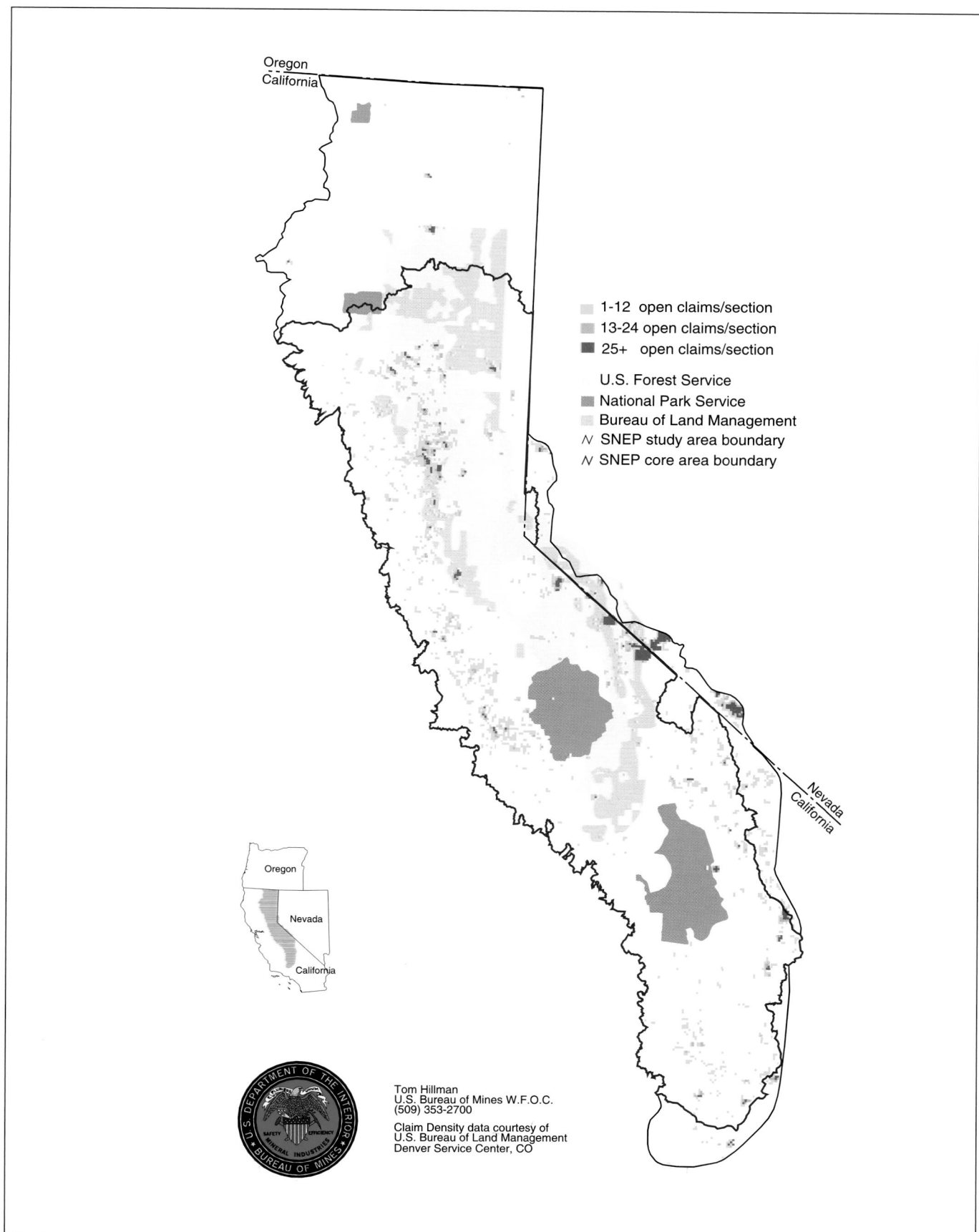

1-12 open claims/section
13-24 open claims/section
25+ open claims/section

U.S. Forest Service
National Park Service
Bureau of Land Management
N SNEP study area boundary
N SNEP core area boundary

Oregon
California

Oregon

Nevada

California

Nevada
California

Tom Hillman
U.S. Bureau of Mines W.F.O.C.
(509) 353-2700

Claim Density data courtesy of
U.S. Bureau of Land Management
Denver Service Center, CO

PLATE 18.3

Density of open lode claims per section in the SNEP study area.

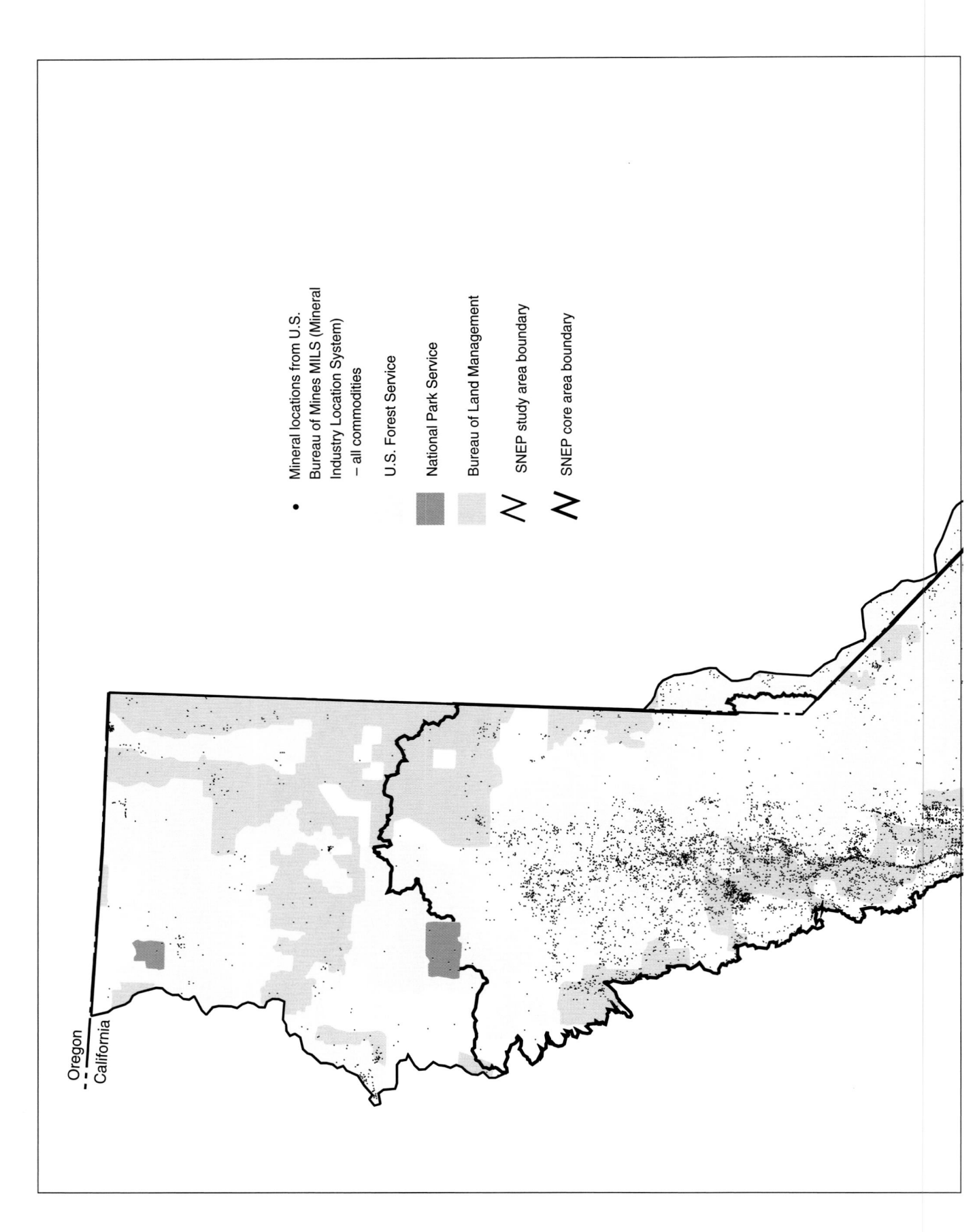

Mineral locations from U.S.
Bureau of Mines MILS (Mineral
Industry Location System)
– all commodities

U.S. Forest Service

National Park Service

Bureau of Land Management

SNEP study area boundary

SNEP core area boundary

Oregon
California

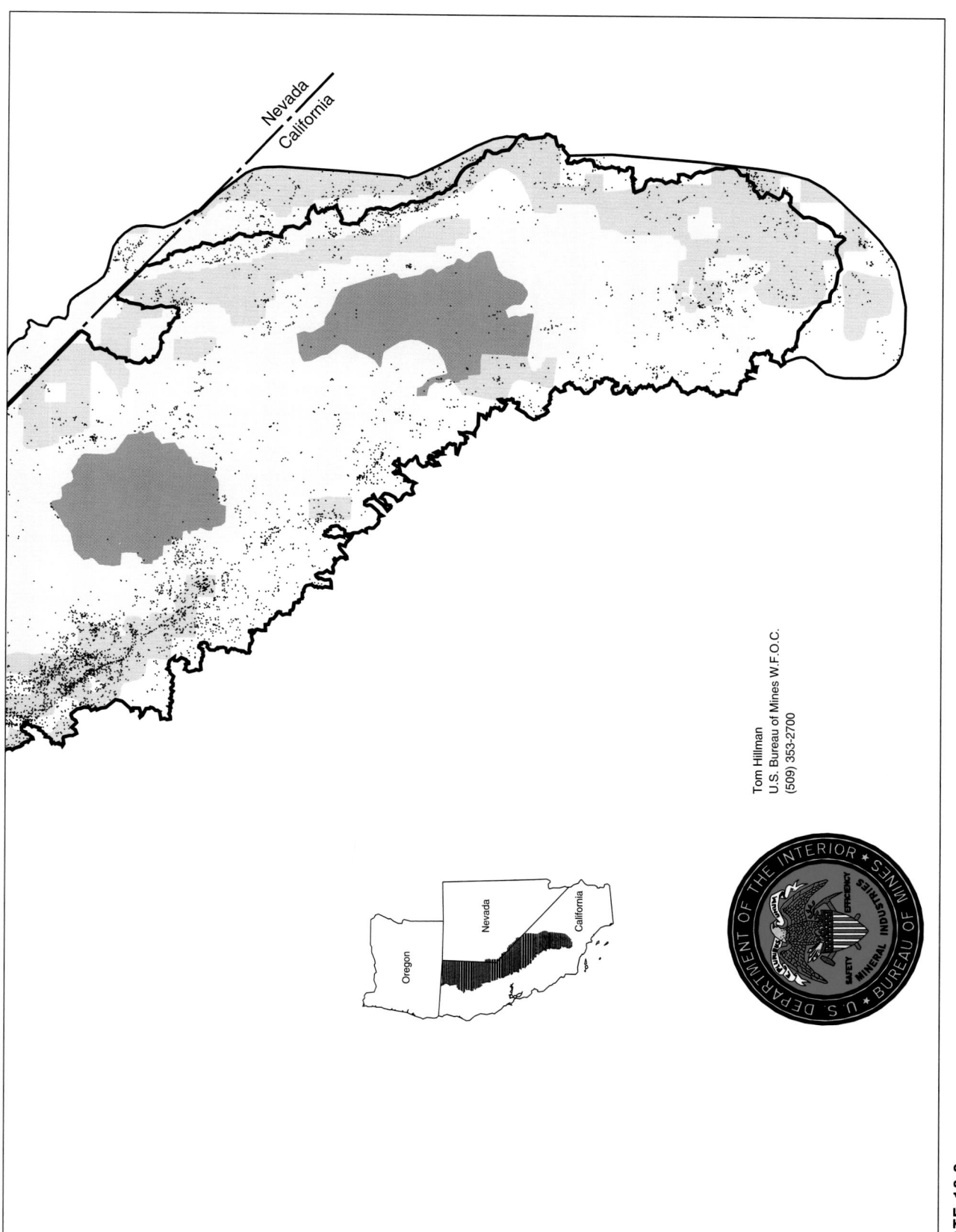

PLATE 18.2

Location of mineral deposits within the SNEP study area.

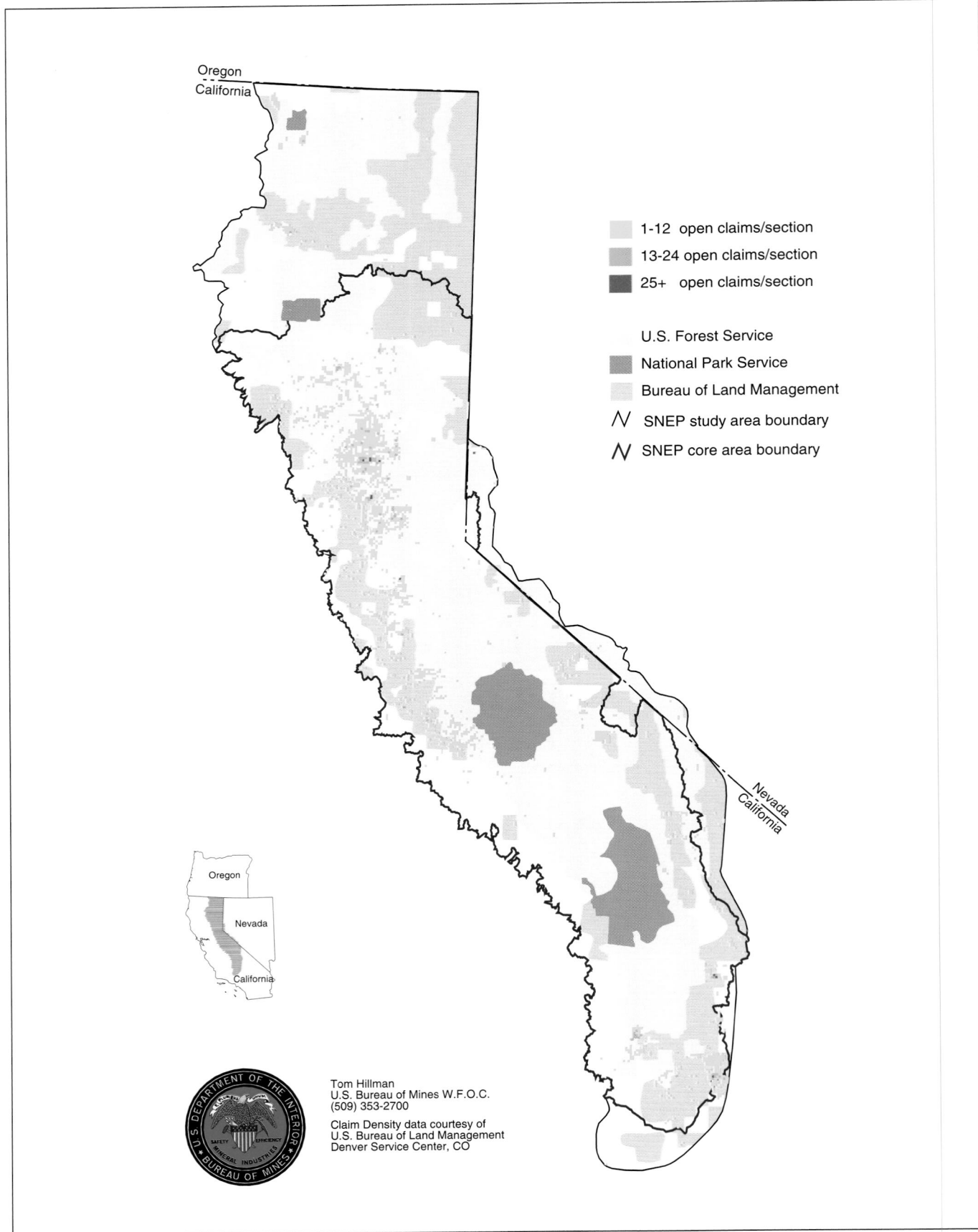

Oregon
California

1-12 open claims/section

13-24 open claims/section

25+ open claims/section

U.S. Forest Service

National Park Service

Bureau of Land Management

N∕ SNEP study area boundary

N∕ SNEP core area boundary

Oregon

Nevada

California

Nevada
California

Tom Hillman
U.S. Bureau of Mines W.F.O.C.
(509) 353-2700

Claim Density data courtesy of
U.S. Bureau of Land Management
Denver Service Center, CO

PLATE 18.4

Density of open placer claims per section in the SNEP study area.

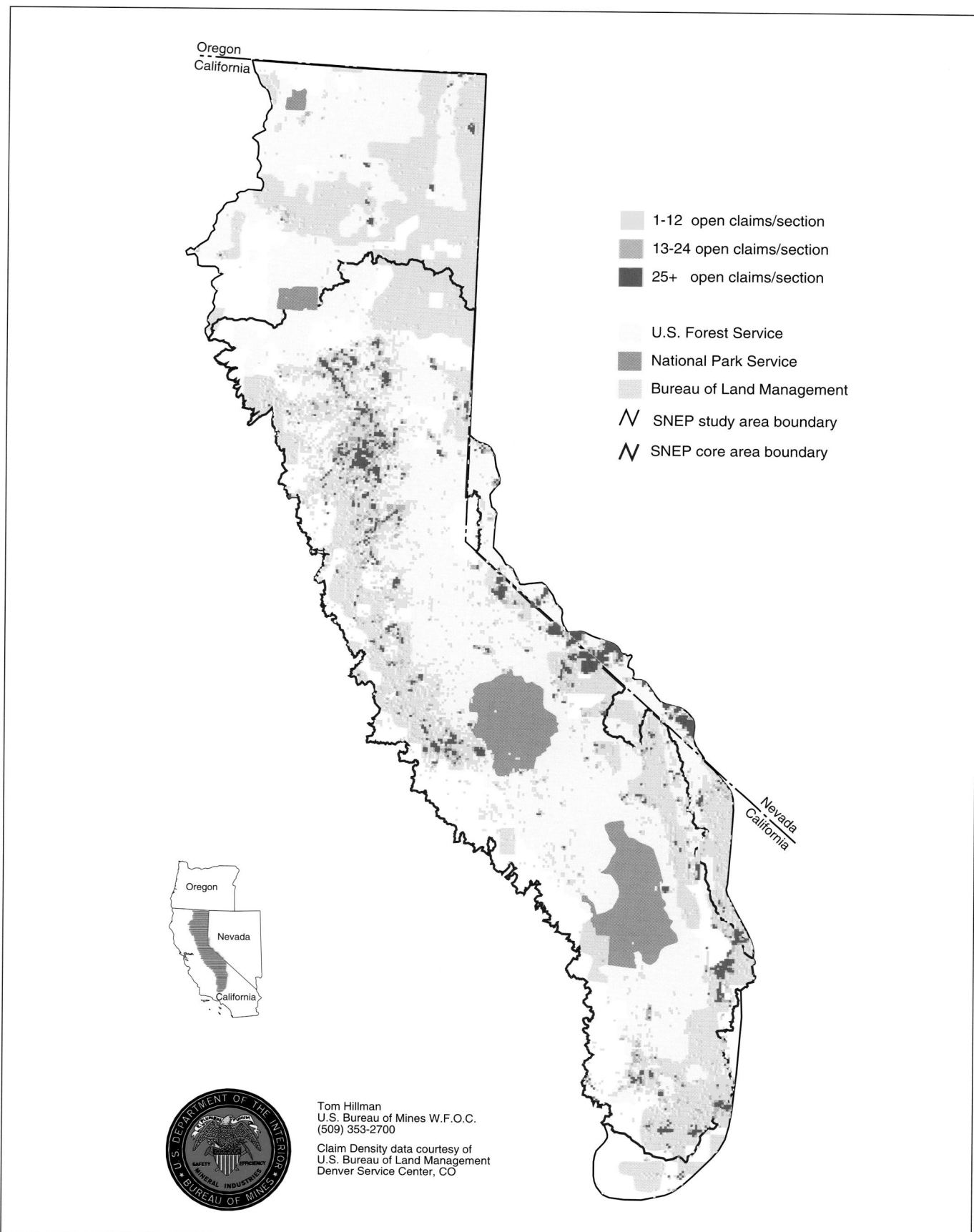

Legend:

1-12 open claims/section

13-24 open claims/section

25+ open claims/section

U.S. Forest Service

National Park Service

Bureau of Land Management

N SNEP study area boundary

N SNEP core area boundary

Oregon
California

Nevada
California

Oregon

Nevada

California

Tom Hillman
U.S. Bureau of Mines W.F.O.C.
(509) 353-2700

Claim Density data courtesy of
U.S. Bureau of Land Management
Denver Service Center, CO

PLATE 18.5

Total claim density per section in the SNEP study area.

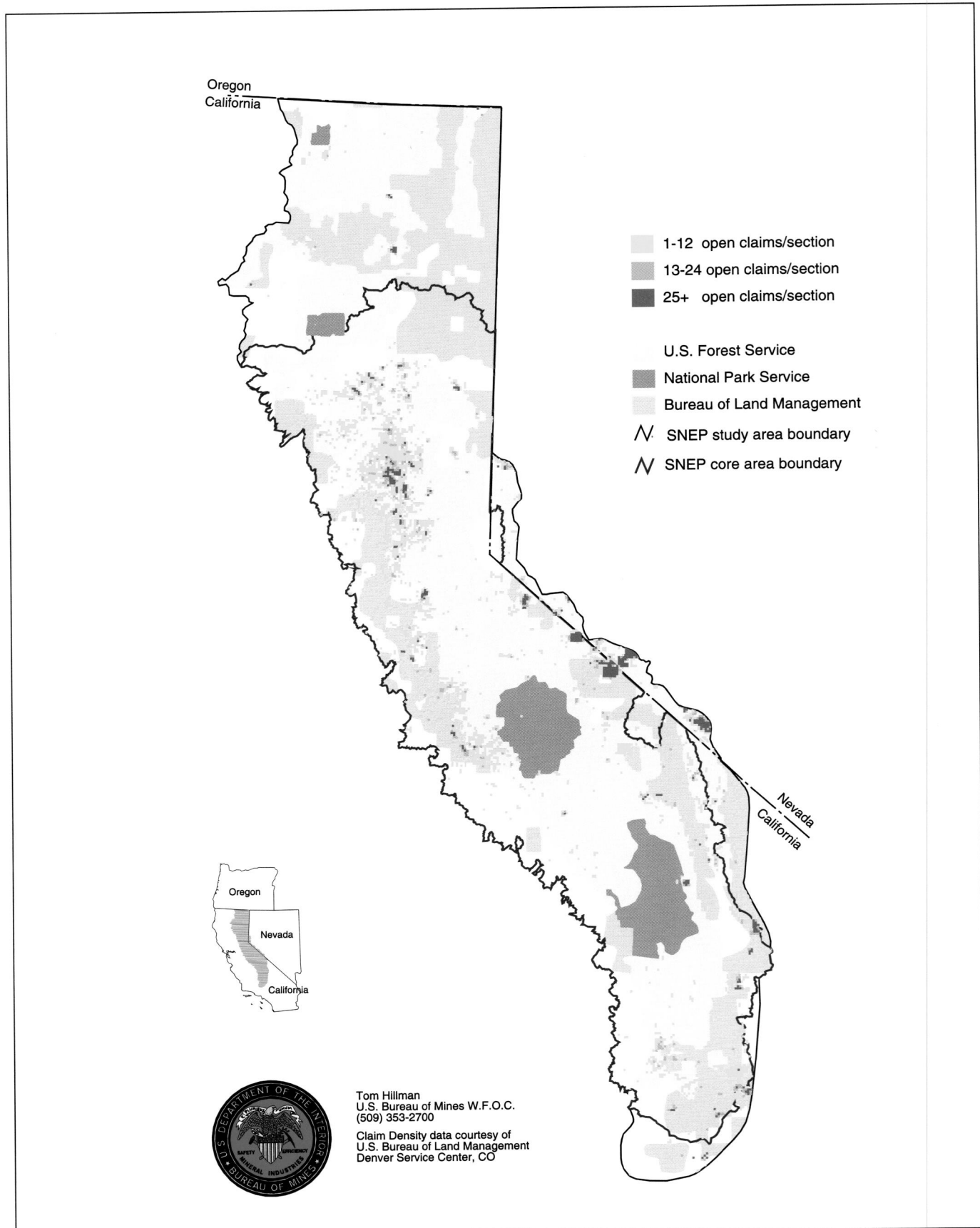

Legend:

1-12 open claims/section

13-24 open claims/section

25+ open claims/section

U.S. Forest Service

National Park Service

Bureau of Land Management

N. SNEP study area boundary

N SNEP core area boundary

Oregon
California

Nevada
California

Oregon

Nevada

California

Tom Hillman
U.S. Bureau of Mines W.F.O.C.
(509) 353-2700

Claim Density data courtesy of
U.S. Bureau of Land Management
Denver Service Center, CO

PLATE 18.6

Open claim density per section in the SNEP study area.

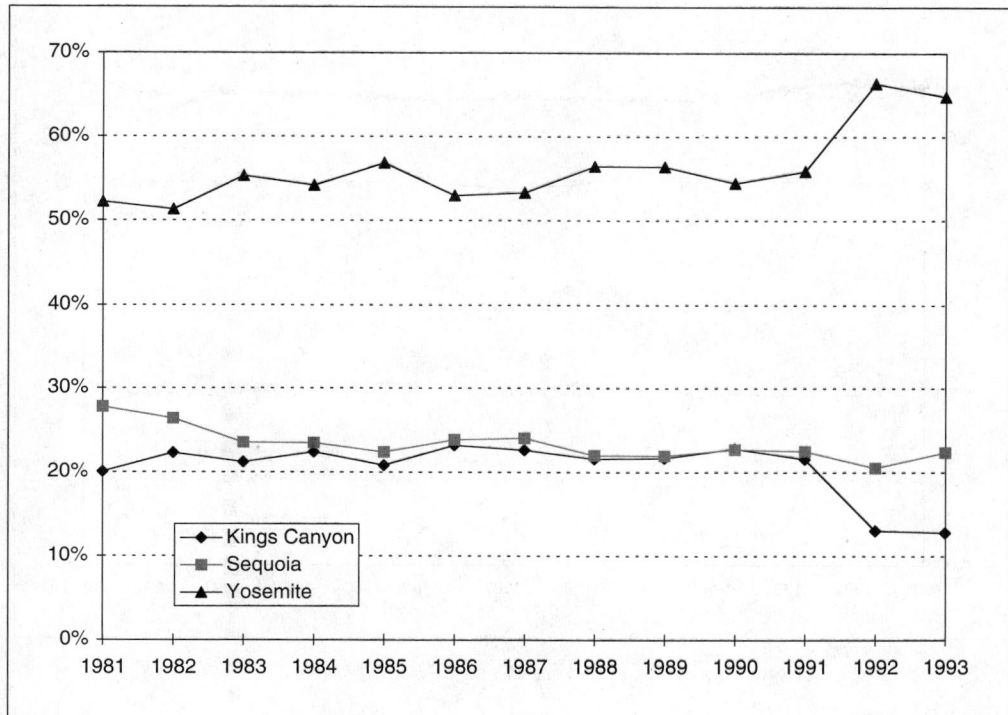

FIGURE 19.47

Park shares of National Park
Service RVDs, 1981–93.

of these different types of fishing appears to be significantly different, however, as are the implications of trends within the fishing activity class for land and resource management in the Sierra Nevada. Unfortunately, there are no good quantitative data available for an accurate estimate of specific activities within the fishing activity class for the entire Sierra

Nevada. More detailed results are presented in Knauer and Duane (1994) for the eastern Sierra Nevada.[6] This data weakness seriously limits SNEP's ability to analyze the policy implications of various land and resource management scenarios. We project fishing demand to remain relatively stable, as growth in California's population overcomes any declines

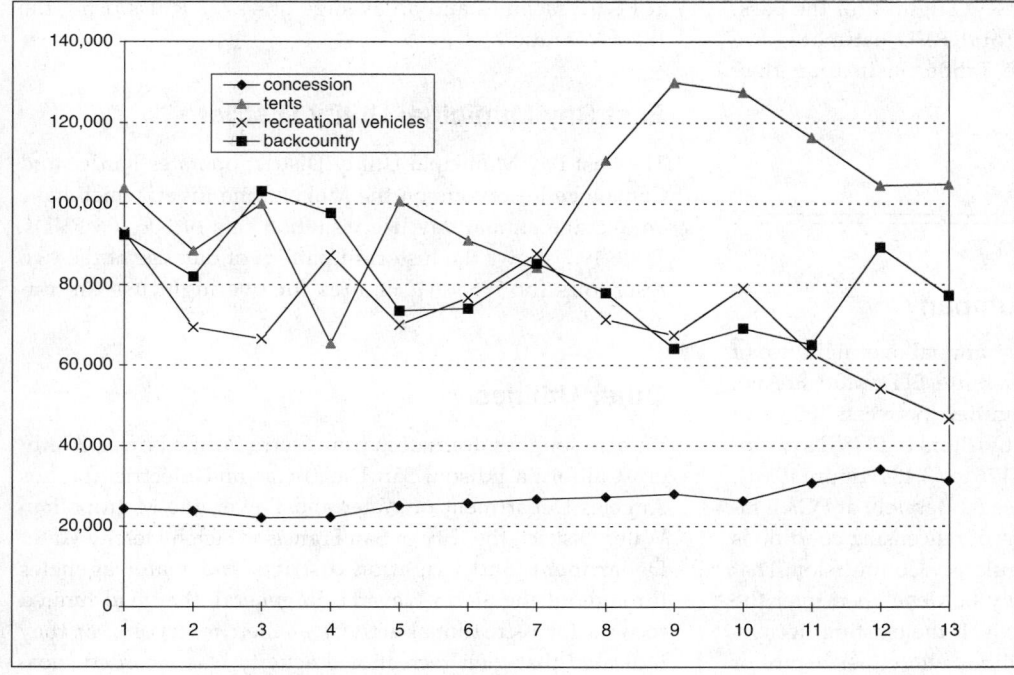

FIGURE 19.48

Kings Canyon RVDs by
activity, 1981–93.

FIGURE 19.49

RVDs per visitor in national parks, 1981–93.

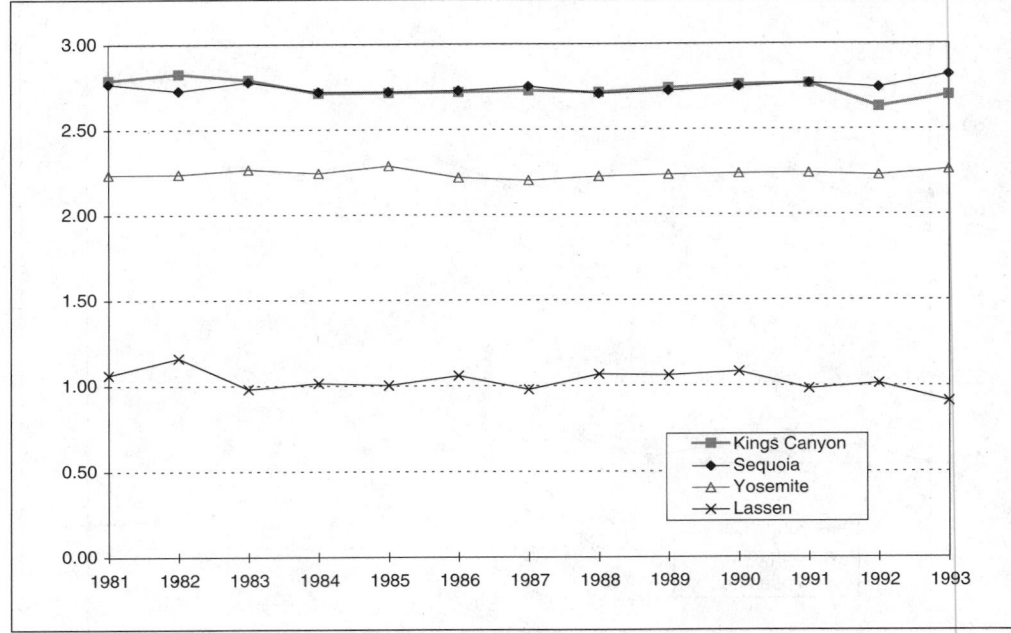

in per-capita fishing rates. We also believe that drought is primarily responsible for the 1986–93 decline, although other social, demographic, economic, and resource availability factors are also important.

Nevada Department of Parks

The state of Nevada Department of Parks operates the Lake Tahoe State Park on the east shore of Lake Tahoe, which is the most popular park in the Nevada system. Unfortunately, we were unable to obtain visitation or RVD figures for the park. This omission means that our total RVD estimates are considerably lower for the Lake Tahoe Basin than they should be.

PUBLIC UTILITIES

Pacific Gas and Electric Company

Pacific Gas and Electric accounts for annual overnight use of its campgrounds in Alpine, Amador, Butte, El Dorado, Fresno, Madera, Nevada, and Plumas Counties. Between 1985 and 1993, the total number of overnight visitors to PG&E's campgrounds increased by 60%, from 27,176 to 43,434 (figure 19.61). Camping capacity may also increase moderately at PG&E facilities in the future under the terms of relicensing conditions required by the Federal Energy Regulatory Commission. That additional capacity would probably be developed over the next five to fifteen years, during which the existing licenses for many PG&E hydroelectric facilities will either expire or

be renewed. There is also a possibility that some PG&E facilities will be acquired by other parties, however, with unknown consequences for the future operation of PG&E campgrounds. PG&E also operates a small number of facilities for the use of its employees, but visitation figures for these were unavailable.

The visitation figures for PG&E indicate numbers of overnight visitors, who can be assumed to have participated in more than one RVD for each of their visits. Using the 2.24 RVD ratio described earlier, we estimate 97,292 RVDs in 1993 at PG&E facilities and an average of 44,737 RVDs using the 1.03 RVD ratio.

East Bay Municipal Utility District

The East Bay Municipal Utility District operates Pardee and Camanche Reservoirs on the Mokelumne River, which have an average annual day use visitation rate of 306,106 RVDs. Table 19.7 shows the historical pattern of day use at the two reservoirs for 1988–94. Figures for overnight use are unavailable.

Other Utilities

We also sought information from Sierra Pacific Power, Southern California Edison, San Diego Gas and Electric, the Los Angeles Department of Water and Power, the Metropolitan Water District, the City of San Francisco Hetch Hetchy Water Department, and irrigation districts and water agencies throughout the Sierra Nevada. In general, they had limited records for recreational activity at their reservoirs, or they indicated that their recreational activity was recorded sepa-

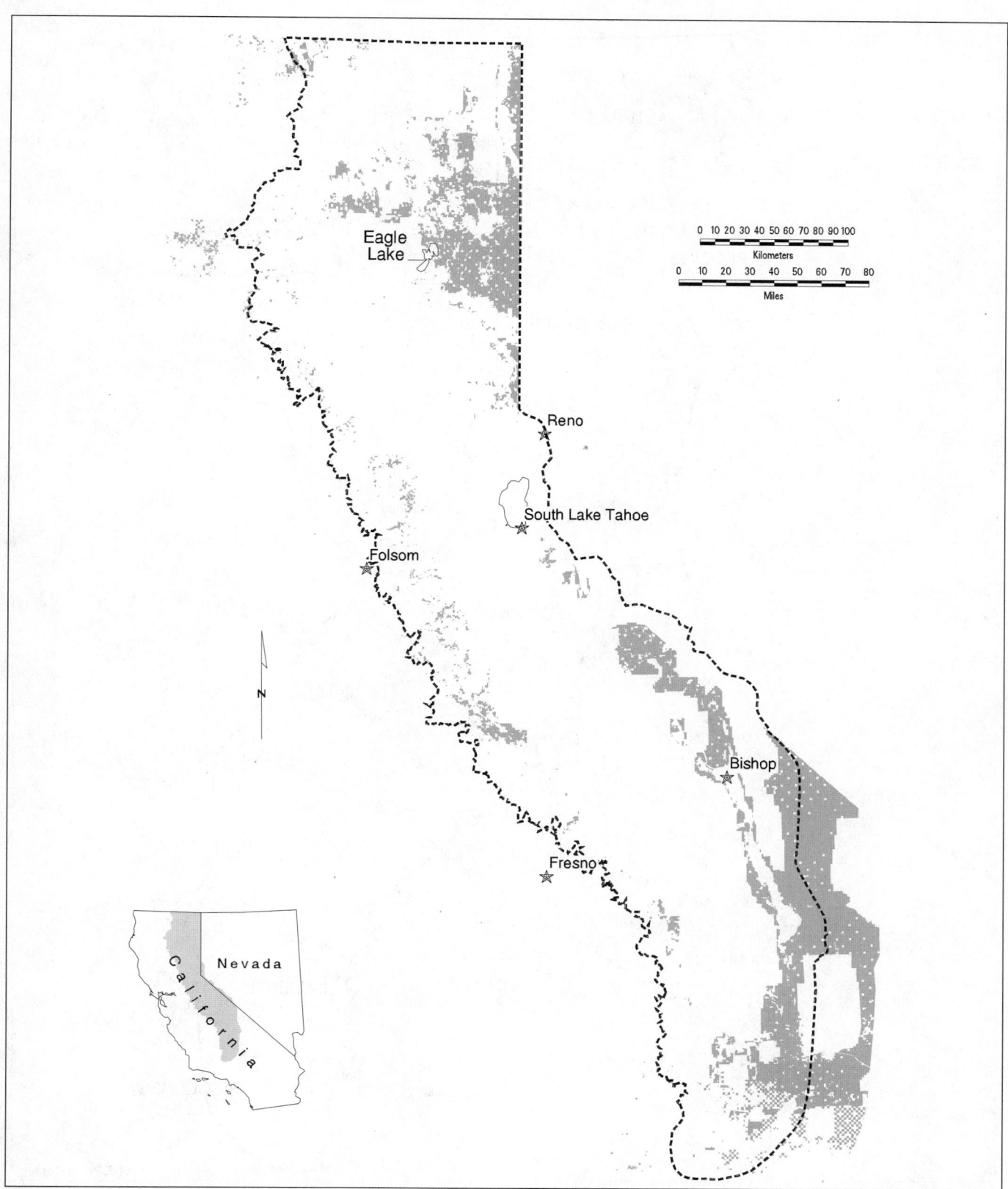

FIGURE 19.50

Bureau of Land Management lands within study area.

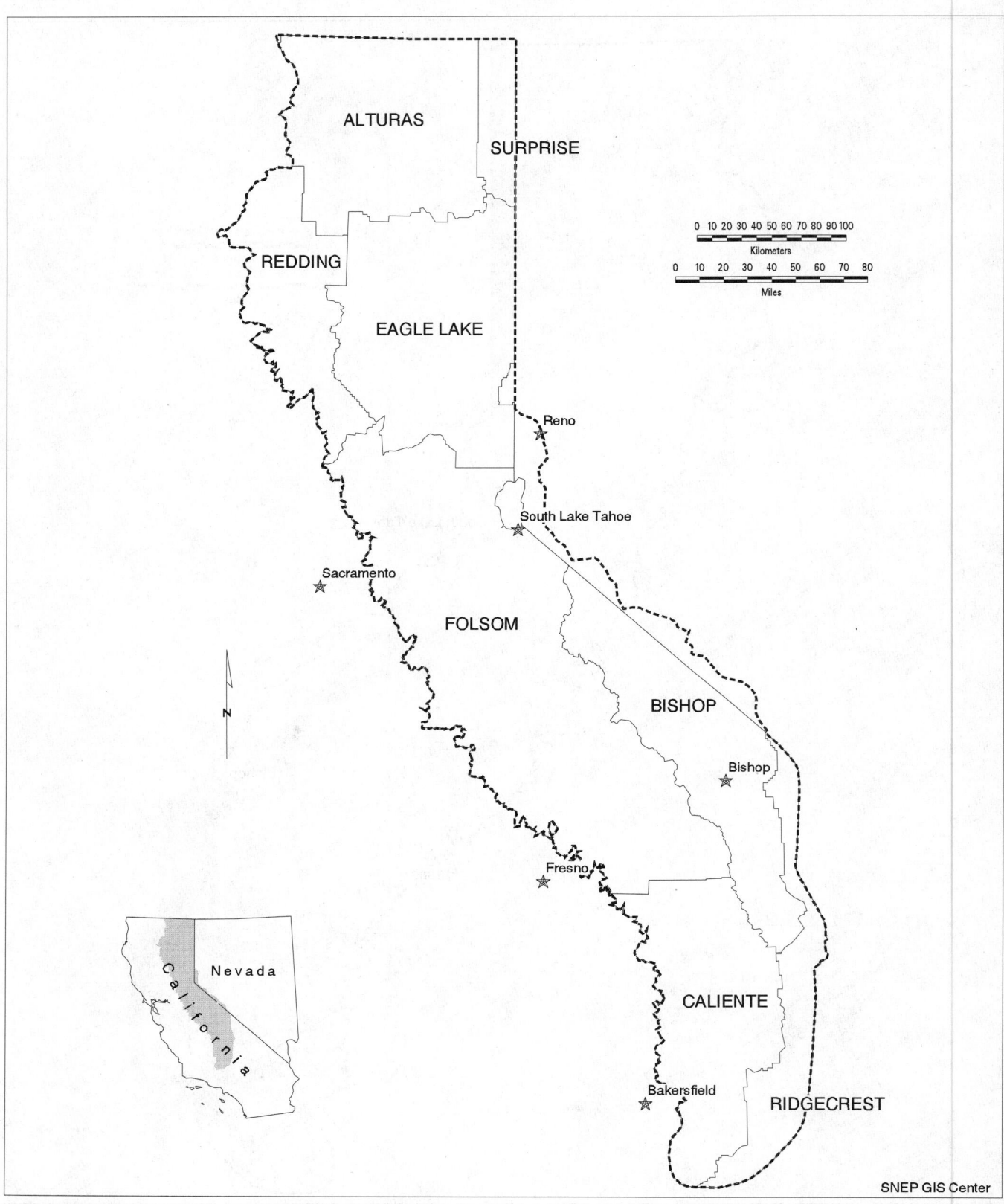

FIGURE 19.51

Bureau of Land Management resource areas.

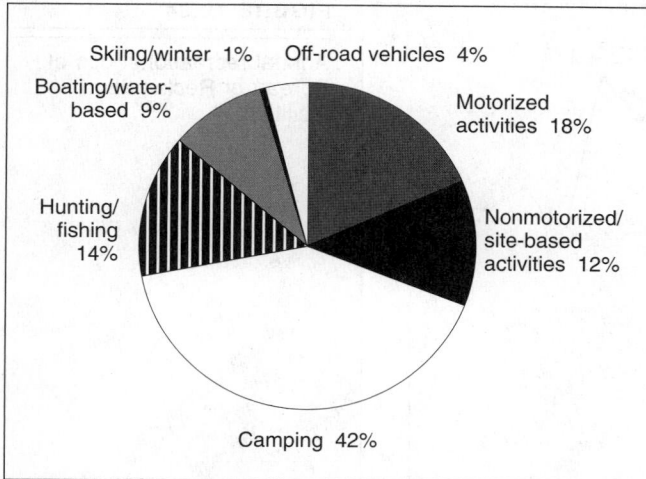

FIGURE 19.52

1992 RVDs by activity type for BLM lands.

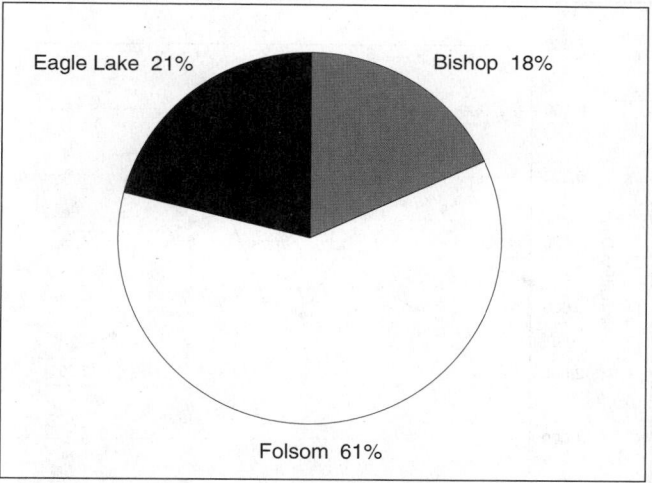

FIGURE 19.53

Mean percentage of total BLM RVDs for Bishop, Folsom, and Eagle Lake BLM Resource Areas.

rately by another entity (e.g., the U.S. Forest Service). In many cases this reflected the fact that their facilities were licensed by the Federal Energy Regulatory Commission and built on federal lands; PG&E was exceptional in its private ownership of most of its reservoir sites and recreational facilities. We therefore have no RVD estimates for other utilities. The lower-elevation reservoirs are generally not on federal land, however, and activities at those sites are not adequately reflected in our overall RVD estimates for the Sierra Nevada.

Privately Operated Camps

Many cities, counties, and nonprofit associations (e.g., the Boy Scouts, Girl Scouts, 4-H, and Campfire Girls) operate resorts and/or camping facilities in the Sierra Nevada. We were unable to get reliable or consistent data for these activities, however, across the Sierra Nevada. Individual visitation figures for a limited number of organizations are summarized in the California Environmental Resource Evaluation System (CERES) project of the Resources Agency of the State of California (http://ceres.ca.gov/snep), and the Alexandria Project at the University of California, Santa Barbara (http://alexandria.sdc.ucsb.edu/). Many of these facilities also operate under special use permits from the Forest Service, and their figures are included in USFS estimates of RVDs.

LOCAL AGENCIES
El Dorado County

River-based recreational activities, such as kayaking, rafting, and canoeing, have increased in popularity. Though there are

significant levels of recreational use on many of the major rivers located within the study area, these activity types are often difficult to quantify either seasonally or annually. El Dorado County administers river-based recreation for the South Fork of the American River (SOFAR), one of the most popular destinations for river enthusiasts in the United States (Wilderness Conservancy 1989). The county has an eight-year record of annual user days for private and commercial use (figure 19.62). Commercial, or professionally guided, river trips require permits and have decreased by 10% between 1987 and 1994. With the exception of 1993, every year during this period was a drought year in California. Privately led river trips do not require permits, thus the county's annual user-day record is less reliable for these than for commercial trips. El Dorado County accounts for privately led river trips during the summer season, from May to September 1, but its sampling procedure is not defined. These river-based recreational activities without permits have increased by 150% during the eight-year period. Following a lawsuit in 1994, El Dorado County is now planning to assess all types of recreational use on the river and revise its permit system (with an accompanying environmental impact report) in the next three years. Changes to the permit system could lead to requirements that private trips also get permits, which could in turn be either

TABLE 19.5

Hunting and fishing RVDs for Bishop, Folsom, and Eagle Lake Resource Areas during 1992.

	Bishop Resource Area	Folsom Resource Area	Eagle Lake Resource Area
Hunting RVDs	17,042	1,408	139,267
Fishing RVDs	72,242	6,567	11,683

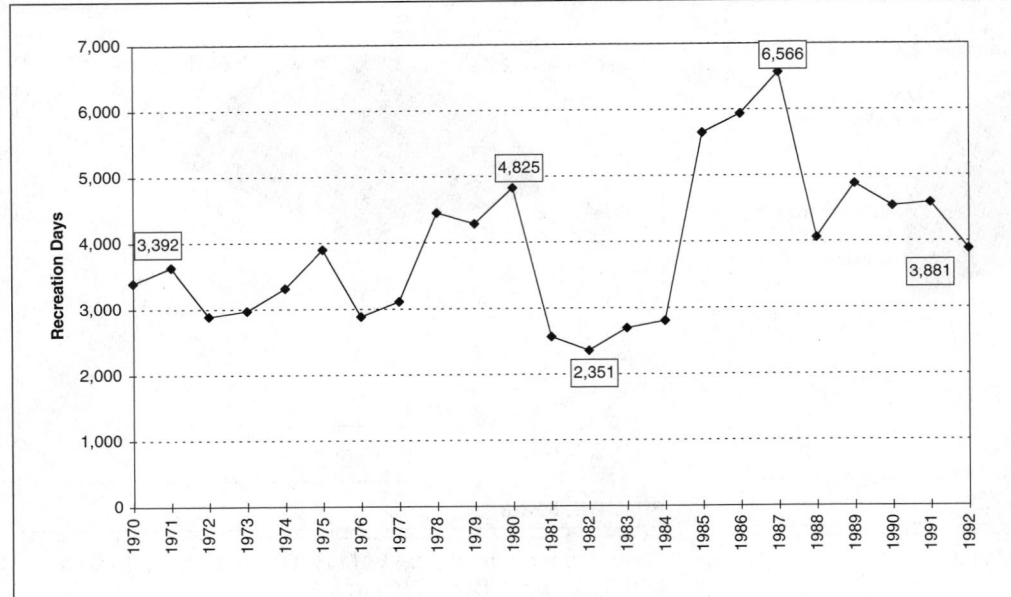

FIGURE 19.54

Annual recreational use of Bureau of Reclamation facilities.

limited or expanded under future county policies. Table 19.8 summarizes use on the South Fork of the American River between 1987 and 1994.

Note that all of this visitation data follows the construction of the New Melones Dam in the early 1980s, which eliminated the second most popular whitewater run in the United States on the Stanislaus River (Palmer 1982). Because there are no data for the South Fork American River before this period, it is difficult to estimate how much of the current recreational activity there formerly took place on the Stanislaus. There is also considerable whitewater recreation on the Middle Fork and North Fork American River, but there is no permit system in place to ensure reliable data collection for those rivers. Proposals by the U.S. Army Corps of Engineers to build the Auburn Dam could affect this activity on the Middle Fork and North Fork of the American River if the Auburn Dam floods those whitewater runs.

Other popular whitewater rivers in the Sierra Nevada include the North Yuba River, the Tuolumne River, the Merced River, and the Kern River. The Stanislaus River also saw considerable whitewater use again during the drought, when New Melones Reservoir was low enough to expose the free-flowing whitewater run temporarily. No estimates are available for whitewater recreation on these rivers outside of federal or state lands cited earlier.

THE EASTERN SIERRA

The data summarized above were generally supplied by public land and resource management agencies and are therefore generally available in a consistent format for the entire Sierra Nevada. There is a significant "gray literature" of unpublished reports and studies of recreational activity in the Sierra Nevada, however, that is not available in a consistent format. This literature includes unpublished reports, theses, dissertations, and surveys administered by agencies, academics, or local organizations with an interest in recreation and tour-

FIGURE 19.55

Folsom Lake and surrounding area.

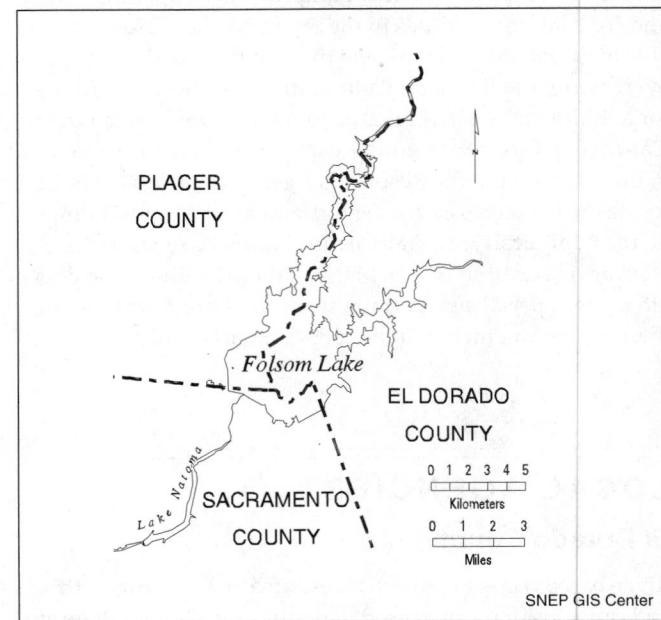

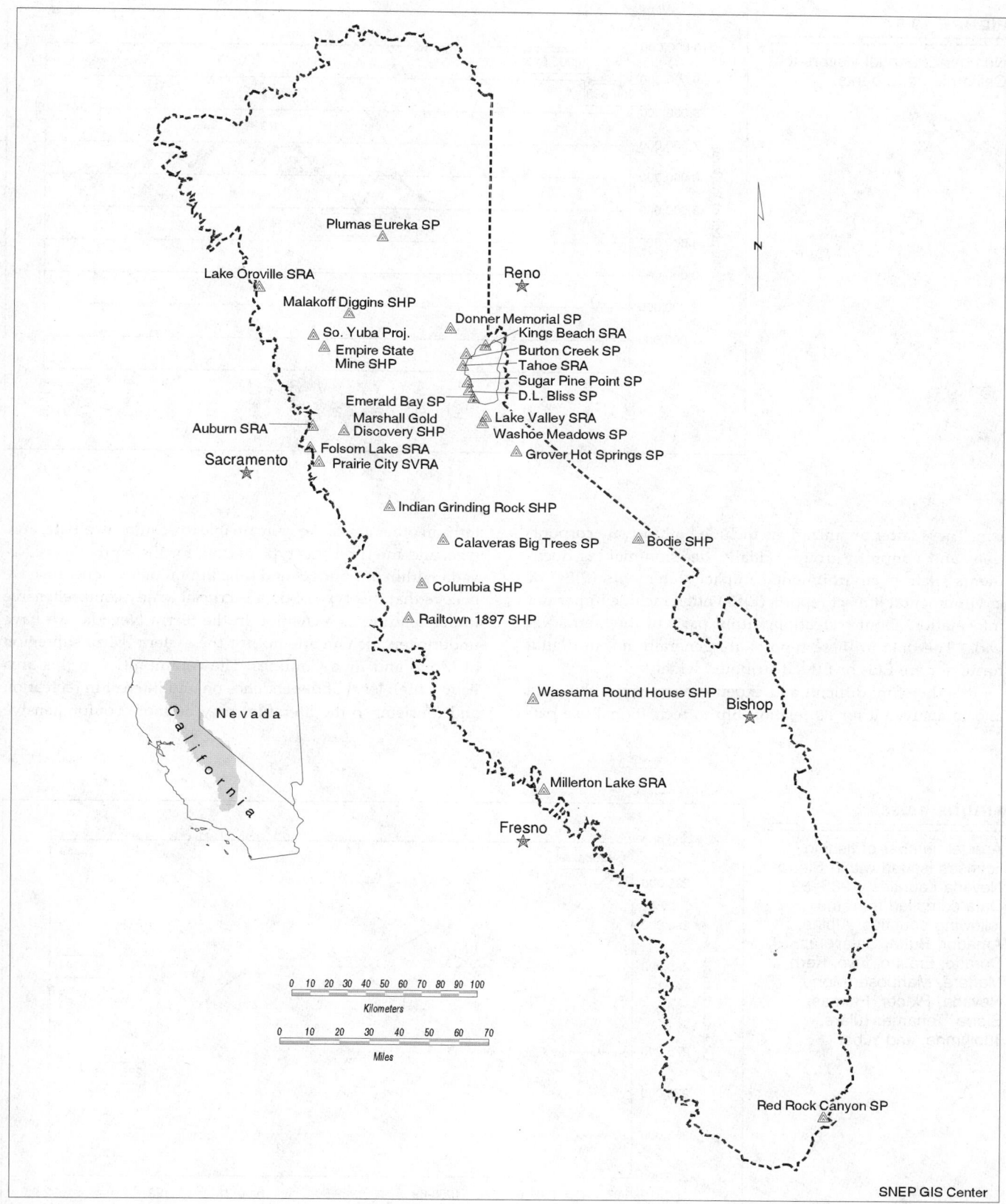

Plumas Eureka SP

Reno

Lake Oroville SRA

Malakoff Diggins SHP

So. Yuba Proj.
Empire State
Mine SHP

Donner Memorial SP
Kings Beach SRA
Burton Creek SP
Tahoe SRA
Sugar Pine Point SP
D.L. Bliss SP

Emerald Bay SP
Marshall Gold
Discovery SHP

Lake Valley SRA
Washoe Meadows SP

Auburn SRA

Sacramento

Folsom Lake SRA
Prairie City SVRA

Grover Hot Springs SP

Indian Grinding Rock SHP

Calaveras Big Trees SP

Bodie SHP

Columbia SHP

Railtown 1897 SHP

Nevada

California

Wassama Round House SHP

Bishop

Millerton Lake SRA

Fresno

Red Rock Canyon SP

0 10 20 30 40 50 60 70 80 90 100
Kilometers

0 10 20 30 40 50 60 70
Miles

SNEP GIS Center

FIGURE 19.56

California state park units.

FIGURE 19.57

Number of annual visitors to California state parks.

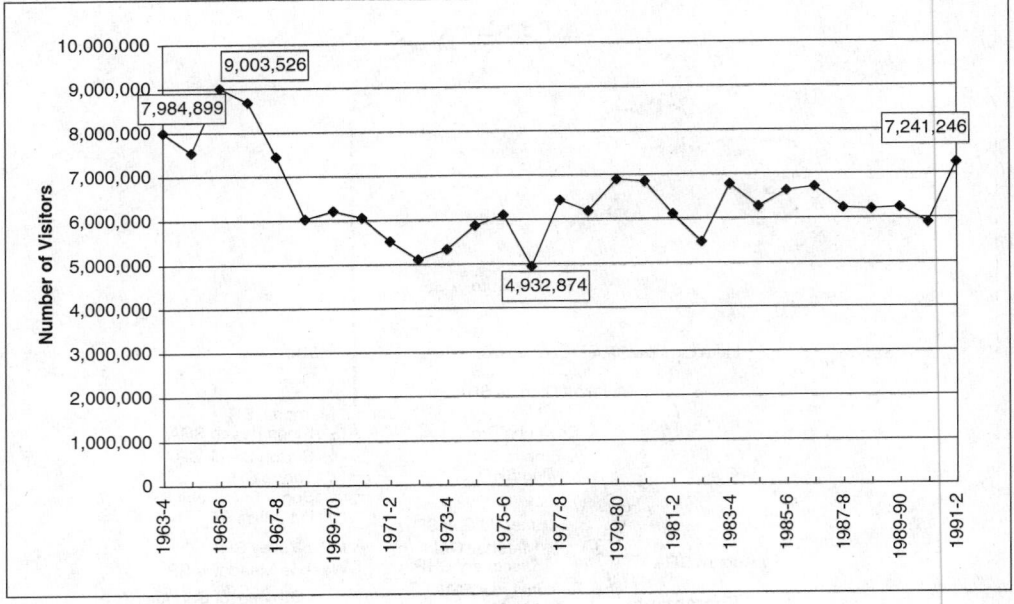

ism. These latter organizations include both private corporations and nonprofit groups. Finally, decision-making documents such as environmental impact statements (EISs) or environmental impact reports (EIRs) often include important information about recreation within parts of the Sierra Nevada. The data in these reports are generally not in digital form, nor are EISs or EIRs distributed widely.

It is therefore difficult and expensive to acquire this data and to analyze it for its relationship to recreational use patterns in other areas. We were unable to acquire, evaluate, analyze, and interpret this type of data for the entire Sierra Nevada within our budget and time limitations. But because we believe that this type of data is crucial to any comprehensive understanding of recreation in the Sierra Nevada, we have undertaken such an attempt for the eastern Sierra subregion of Mono and Inyo Counties. The communities in this area have a high level of dependence on and interest in recreation and tourism, so the literature may be more comprehensive

FIGURE 19.58

Annual number of fishing licenses issued within Sierra Nevada counties, 1986–93. Data compiled from the following counties: Alpine, Amador, Butte, Calaveras, El Dorado, Fresno, Inyo, Kern, Madera, Mariposa, Mono, Nevada, Placer, Plumas, Sierra, Tehama, Tulare, Tuolumne, and Yuba.

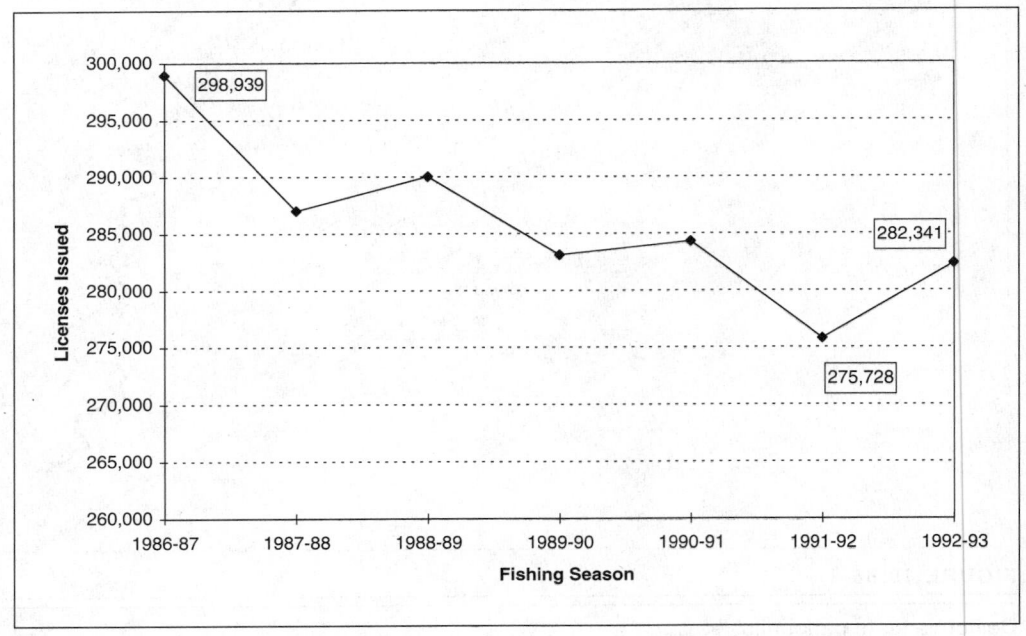

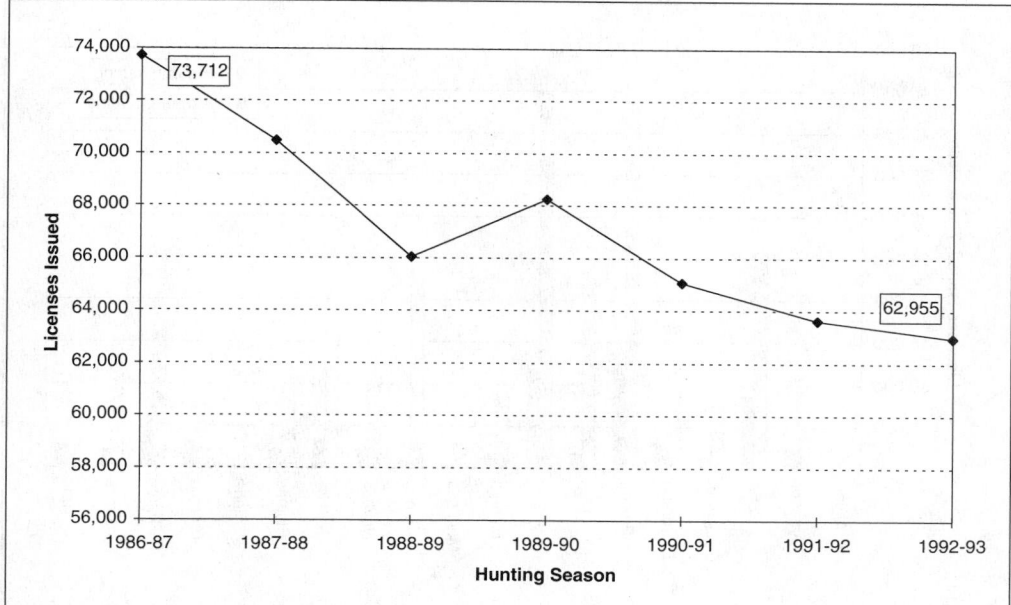

FIGURE 19.59

Annual number of hunting licenses issued within Sierra Nevada counties, 1986–93. Data compiled from the following counties: Alpine, Amador, Butte, Calaveras, El Dorado, Fresno, Inyo, Kern, Madera, Mariposa, Mono, Nevada, Placer, Plumas, Sierra, Tehama, Tulare, Tuolumne, and Yuba.

for this area than any other subregion in the Sierra Nevada outside the greater Lake Tahoe area. We nevertheless believe that this more detailed case study is illustrative of the types of data that are available and the types of analyses that can be completed at a subregional level.

The eastern Sierra is dramatic desert and mountain landscape where the Mojave Desert, the Sierra Nevada, and the Great Basin meet. This landform juncture is characterized by extraordinary topographical features, a rich diversity of natural communities, and sparse human settlement. The subregion's striking beauty is in its rugged extremes: arid desert valleys (e.g., the Owens Valley) are flanked by two of the highest mountain ranges in the continental United States. The eastern Sierra is a descriptive term that refers to the region along the eastern escarpment of the Sierra Nevada, bounded roughly by Mount Whitney to the south and Yosemite National Park and the Bodie State Historic Park to the north. This 125-mile stretch of the Sierra Nevada bounds the region to the west, with the White-Inyo Range forming the eastern boundary. From south to north, the intervening valleys comprise four distinct basins: the Owens Valley, Long Valley, Mono Basin, and Bridgeport Valley. The distance from the top of Mount Whitney and a dozen other peaks over 14,000 feet to the floor of the adjacent valleys is nearly two vertical miles. This area is therefore one of the most important and active mountain-climbing regions in the world (Porcella and Burns 1991). Mount Whitney is also less than one hundred miles from Badwater Point in Death Valley National Park, the lowest point in the contiguous forty-eight states at 282 feet below sea level. The second-largest roadless area in the contiguous forty-eight states is also in the eastern Sierra, which is the only place in the country outside Alaska where one can draw a line on a map for 150 miles and not cross a road (Foreman and Wolke 1992). The John Muir Trail and the Pacific Crest Trail draw backpackers, hikers, runners, and equestrians from around the world to the high-country wilderness of the eastern Sierra (Winnett 1978; Schaffer et al. 1989).

Most of the population in the eastern Sierra lives within the basins in the towns of Lone Pine, Independence, Big Pine, Bishop, Mammoth Lakes, June Lake, Lee Vining, and Bridgeport. The population of the region swells when visitors enter the area on winter weekends for skiing and all summer long for outdoor recreation. The local economy, in turn, is heavily dependent upon this influx of recreational visitors. With the exception of water resources, public land and resource management policy in the region emphasizes recreational activities and associated values.

Visitors to the eastern Sierra are treated to vast expanses of open, undeveloped space. During the last century, human settlement patterns have been constrained by the limited amount of private land available for development (Kahrl 1982; Walton 1992). Water for local development has also been limited by control of water rights by the Los Angeles Depart-

TABLE 19.6

Total mean annual fishing and hunting licenses issued, 1986–93.

License Type	Mean Annual Licenses Issued (1986-93)
Fishing	285,921
Hunting	67,157

FIGURE 19.60

Mean number of fishing and hunting licenses issued within Sierra Nevada counties, 1986–93.

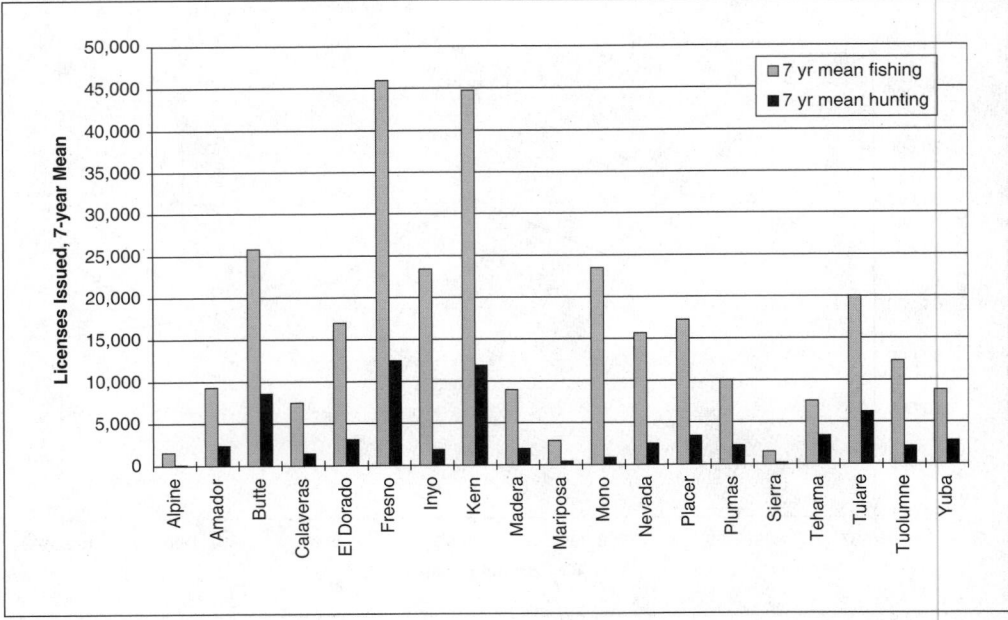

ment of Water and Power (LADWP). Most of the region's land is either publicly held or owned by the LADWP. Recreation and tourism is a mainstay of the local economies. Visitors travel from within the state of California, from out-of-state, and from other countries to participate in a wide variety of active and passive recreational activities. Many of those activities occur on public lands in the region. The dominant land manager and recreation provider in the area is the Inyo National Forest (figure 19.63). The California Department of Fish and Game plays an important role in local recreation through its management of fish and wildlife that serve as a critical draw to the ribbons of water that thread from the high escarpment of the Sierra Nevada crest down to the high desert on the valley floors.

FIGURE 19.61

Annual overnight use of Pacific Gas and Electric Company campgrounds.

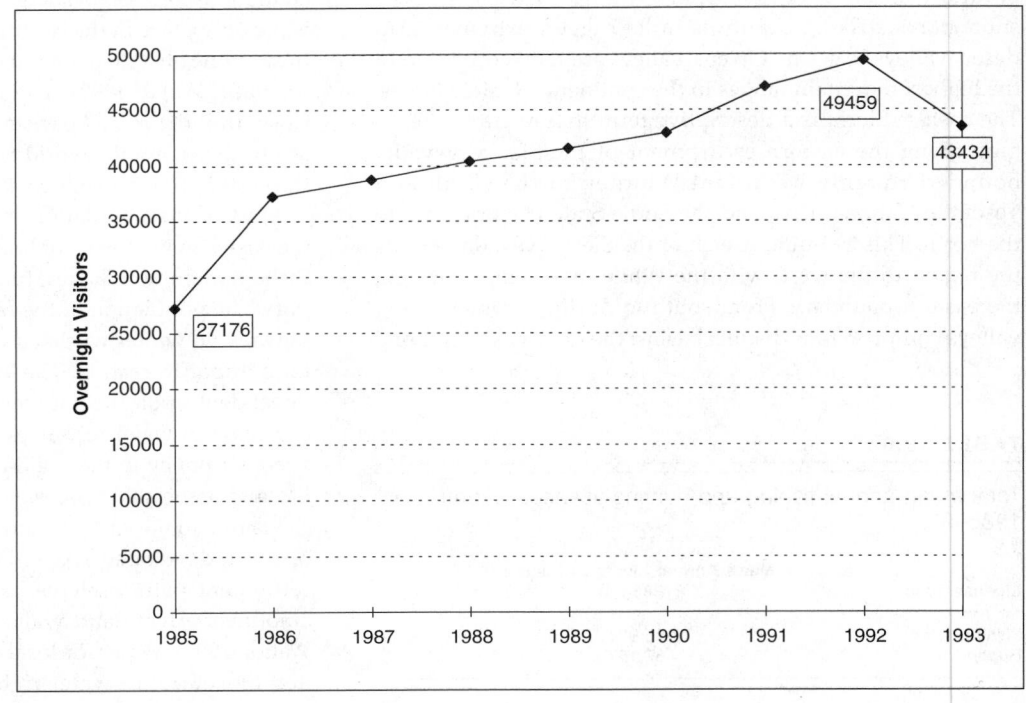

TABLE 19.7

EBMUD recreational day use.

Year	Camanche	Pardee	Total
1988	107,157	209,886	317,043
1989	258,190	83,149	341,339
1990	338,401	139,965	478,366
1991	318,456	103,909	422,365
1992	387,001	95,240	482,241
1993	388,090	99,247	487,337
1994	345,447	100,795	446,242
Sum	2,142,742	832,191	2,974,933
Mean	306,106	118,884	424,990

It is primarily the magnificent natural landscape that draws visitors to the eastern Sierra. Local economic well-being is therefore directly related to the condition of that landscape. The Coalition for Unified Recreation in the Eastern Sierra (CURES), a local group in Inyo and Mono Counties, is now pursuing a marketing theme for the area that calls for potential visitors to "Experience the Wild Side of California." CURES and the Inyo National Forest provided unpublished materials to SNEP in 1994 to allow us to examine recreational activity and user data at a less aggregated level than that which we reported for the entire Sierra Nevada. These more detailed data (summarized in Knauer and Duane 1994) allow us to gain some insight into the complexity of the coarse-grain information we reported on recreational activities by agency or subregion. Although many individual studies have either empirically or descriptively documented the social and demographic characteristics of visitors to the region, none were

multijurisdictional in approach. The objective of our study was to synthesize the results of existing studies and create a database that would be useful to policy makers, private citizens, special interest groups, and regional groups such as CURES. It can also serve as a model for subregional investigations in other parts of the Sierra Nevada.

Methodology

We summarized, analyzed, and synthesized over thirty existing secondary resources in the eastern Sierra. Resource types included formal empirical studies, land management agency use statistics, informal visitor surveys, and qualitative research conducted by agencies and academics. We also observed and monitored several ongoing recreation planning and management processes and interviewed key informants involved in recreational activities in the eastern Sierra.

The integrity of information was not uniform between secondary resources, however, due to differing research methods, varying temporal and spatial extents of study areas, and inconsistent research documentation. Secondary resources were first summarized to identify research methods, data integrity, and study results. Relevant socioeconomic data from each study were entered into Excel 4.0 for Macintosh spreadsheets. Data were then thematically pooled across studies into three information classes, and all values were normalized to account for rounding within and between studies. The three main categories of recreation and tourism information were (1) visitor information, (2) trip information, and (3) visitation. Visitor information refers to social demographic characteristics of individual visitors, such as age. The trip information class contains information relevant to the visitor's trip to the

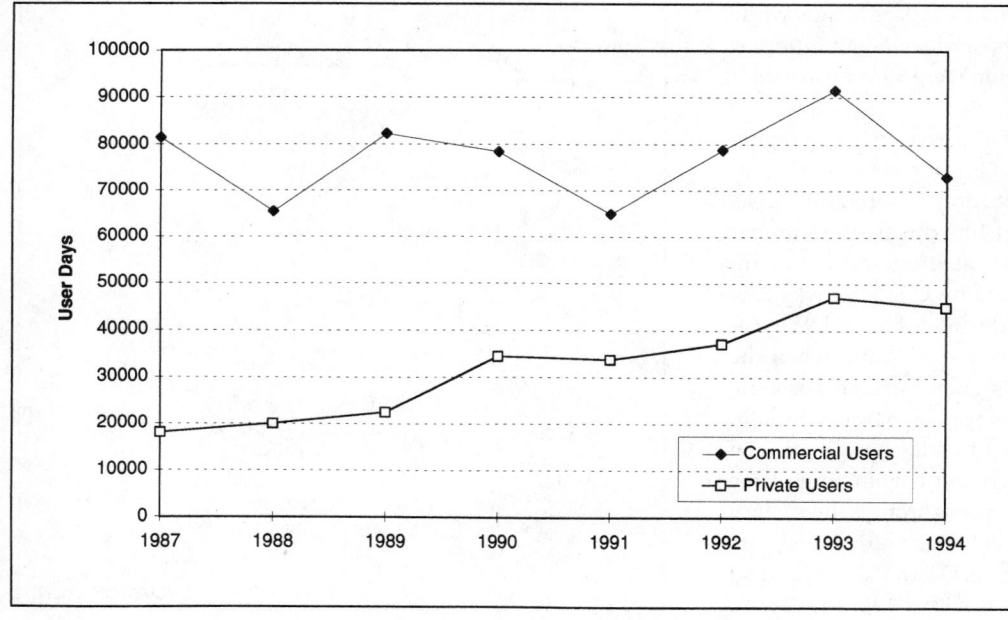

FIGURE 19.62

Annual commercial and private users on the South Fork of the American River.

TABLE 19.8

User-day summary for the South Fork of the American River.

User Type	1987 User Days	1994 User Days	Percentage Change
Commercial	81,466	73,021	- 10
Private	18,000	45,000	+ 150
Total	99,500	118,000	+ 19

eastern Sierra, such as lodging type. The visitation class comprises visitor records from specific institutions, such as the Forest Service or museums. Three data matrixes were created to summarize the specific types of information that were contained within each secondary resource. Information gaps in the recreation and tourism database were then identified, and key informant interviews were conducted to gather additional information to fill in some of the gaps.

Sources

All of the reports that we reviewed are summarized in Knauer and Duane (1994), an unpublished report that is available from the California Environmental Resource Evaluation System (CERES) project of the Resources Agency of the State of California (http://ceres.ca.gov/snep), and the Alexandria Project at the University of California, Santa Barbara (http://alexandria.sdc.ucsb.edu/).

Visitor Information

Visitors to the eastern Sierra during the summer season are generally younger than visitors in the winter. Sixty-nine percent of summer visitors were less than 45 years of age, while 57% of winter visitors were 45 years or older. Forty-four percent of all winter visitors to the region were 55 years or older (figure 19.64).

Approximately 80% of visitors to the eastern Sierra region reside within the state of California. Out-of-state visitors comprised about 15%, while the proportion of foreign tourists averaged 2% during the winter and 10% during the summer season. Most of the California visitors are from southern California, reflecting the access provided by U.S. Highway 395. Access to northern and central California is cut off by the Sierra Nevada except in the summer and autumn, when the Tioga Pass road (state route 120) is open through Yosemite National Park. Many foreign visitors appear to travel through the region as part of more extended holidays originating in either Los Angeles or San Francisco and terminating at the other. These "open jaw" trips often pass through the eastern Sierra as part of a larger trip that will include either Yosemite National Park or the Lake Tahoe region, Death Valley National Park, and Las Vegas, Nevada (often with a flight to Grand

Canyon, Arizona, from Las Vegas). Many domestic out-of-state travelers visit the area as part of a longer trip to the Sierra Nevada, California, and/or the national parks in the western United States.

The gender of summer visitors to the eastern Sierra was 35% female and 65% male. A narrower and more specific study of mountain bicyclists on the Inyo National Forest found that only 25% of this user group was female and 75% male. Winter visitors appear to be more balanced by gender.

Existing studies have not assessed the racial composition of visitors throughout the entire eastern Sierra region. Two comprehensive empirical studies of visitors to the Inyo and Toiyabe National Forests indicated that few minority tourists travel to the eastern Sierra region. Close to 90% of visitors to the eastern Sierra are Caucasians; approximately 5% are Hispanics; and slightly more than 3% are Asians. The remaining 2% of visitors are of Native American or African American descent. Discussions with Forest Service officials suggest that visitors of non-Caucasion racial backgrounds comprise a greater proportion of overall visitation than was documented within the two surveys. Officials also indicate that visitors' trip preferences and trends often differ according to race and ethnicity. Despite a lack of empirical studies that document

FIGURE 19.63

Inyo National Forest and surrounding area.

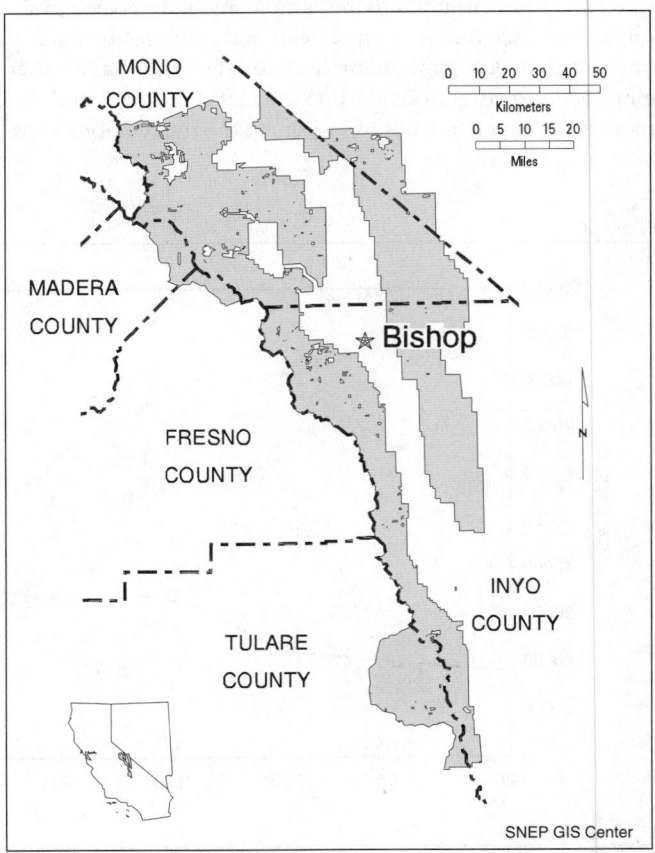

SNEP GIS Center

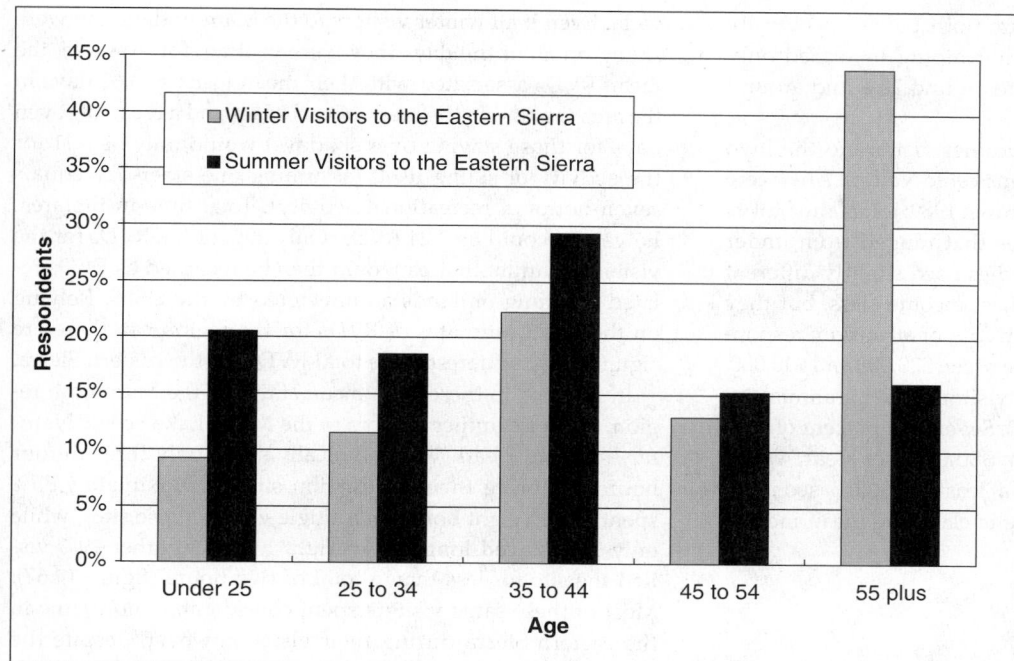

FIGURE 19.64

Age class of visitors to the eastern Sierra.

these trends in the Inyo National Forest specifically, USFS officials believe that many minority visitors prefer to recreate in and visit more developed, accessible sites, as opposed to less developed, remote sites. This theory is consistent with research findings by Deborah Chavez and others in southern California forests near the Los Angeles area (Laidlaw 1992; Chavez 1992, 1993a, 1993b; Chavez et al. 1993a, Ewert et al. 1993) , where the majority of visitors to the eastern Sierra reside.

A large proportion of visitors to the eastern Sierra have at-

tended college, with approximately 25% holding a bachelor's degree and a little more than 15% possessing a graduate degree. Summer visitors to the Mammoth Lakes area tended to have less formal education than winter visitors (figure 19.65) (Sports Research 1989, 1990). This difference may in part be explained by the higher incomes of winter skiers. Access to camping facilities during the summer also increases the feasibility of travel to the area for travelers with lower incomes. Even in the winter, moreover, about twice the fraction of visitors to the eastern Sierra subregion in general (20%) camped

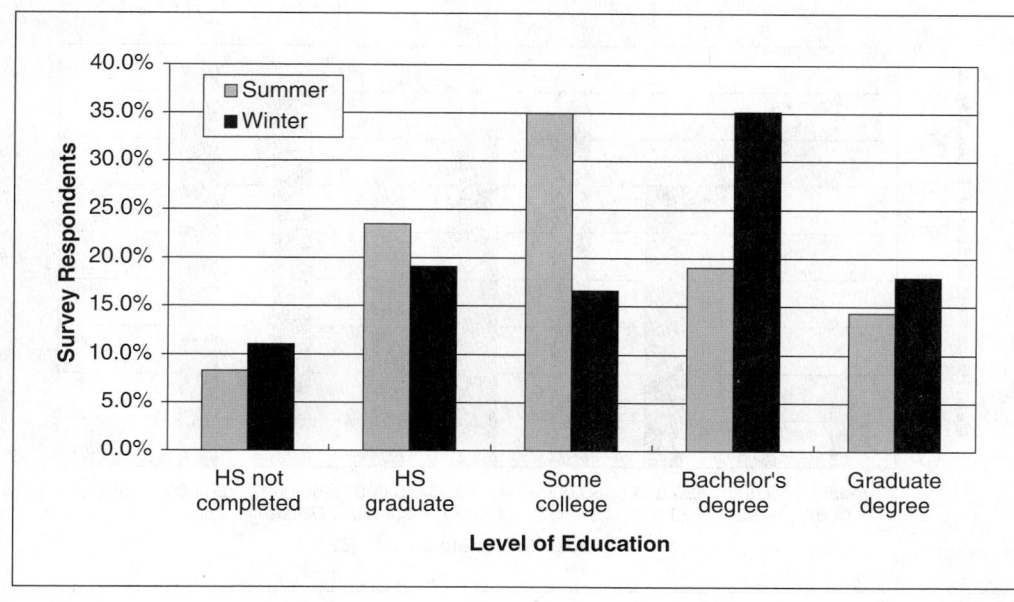

FIGURE 19.65

Mean level of education of visitors to the eastern Sierra.

as compared with visitors to Mammoth Lakes specifically (10%). The shares of total visitors camping increased only slightly in the summer, however, to around 25% and around 12% of visitors, respectively.[7]

The mean annual income of summer visitors to the Inyo National Forest and the Mono Basin Scenic Natural Area were reported in two studies (Lee and Brown 1989; Jones and Stokes Associates 1993) in income classes that ranged from under $10,000 to over $80,000. These studies used slightly different income classifications for the highest income class, but they are comparable. Slightly more than 25% of all survey respondents reported an annual income between $20,000 and $40,000. Approximately 40% of surveyed visitors had an annual income between $40,000 and $80,000. Seventeen percent of visitors to these areas earned over $80,000 per year, while one-third had an annual income of at least $60,000. Based upon the mean values within each income class, the mean income was over $55,000 (figure 19.66).

Trip Information

There is no existing study that comprehensively examines the trip characteristics of visitors to the eastern Sierra region. Existing studies were not designed to assess the travel patterns, spending patterns, or activity patterns of visitors who travel throughout the region. Several surveys collected such trip-related information within specific subregions of the eastern Sierra, however. Typically, 60% of winter visitors to the Mammoth region stayed two to three days, 25% stayed four to five days, and 12% stayed over six days. These data highlight the important relationship between public land RVD estimates and recreational activities on private land that have social, economic, and ecological importance to the Sierra Ne-

vada. Even if all winter visitors to the Mammoth region were skiing on all of the days they were visiting, for example, the mean RVDs associated with their mean visits of 3.62 days in the area (assuming the mean of each class and a mean of seven days for those staying over six days) would only be 2.11 for the activity of skiing itself (assuming one skier-day equals seven hours of recreational activity). Total time in the area, however, would be 7.24 RVDs. Only 29% of the RVDs for the visits to Mammoth area would then be recorded as RVDs related to skiing on lands administered by the USFS. Relying on the USFS estimates of RVDs for the region can therefore significantly underestimate total RVDs for the eastern Sierra.

In contrast to these high mean stays for the Mammoth region, 25% of summer visitors to the Mono Lake Scenic Natural Area and Death Valley typically spent only three to four hours exploring their destination site. Approximately 35% spent five to eight hours on a single visit to those sites, while only 21% stayed longer than eight hours. Another 19% visited the site for less than a total of two hours (figure 19.67). Most of these same visitors spent considerably more time in the eastern Sierra during their visits, however, despite the relatively brief stops at these two highly scenic attractions. Note that visitors' stays are also probably considerably longer in the spring and autumn, when the lower temperatures at these two desert sites encourage longer stays.

Visitors to the eastern Sierra stay at hotels, condominiums, campgrounds, private residences, and other types of accommodations (figure 19.68) (Mammoth Mountain Ski Area 1984, 1989; Sports Research 1989, 1990, Pisarowicz 1991; Littlejohn 1991; Klages and Associates 1992). Approximately 40% of summer visitors choose to stay at hotels, with camping being the second most favored accommodation choice during the summer (24% across multiple studies). A 1989 summer sur-

FIGURE 19.66

Annual household income of summer visitors to the Inyo National Forest and Mono Basin Scenic National Recreation Area.

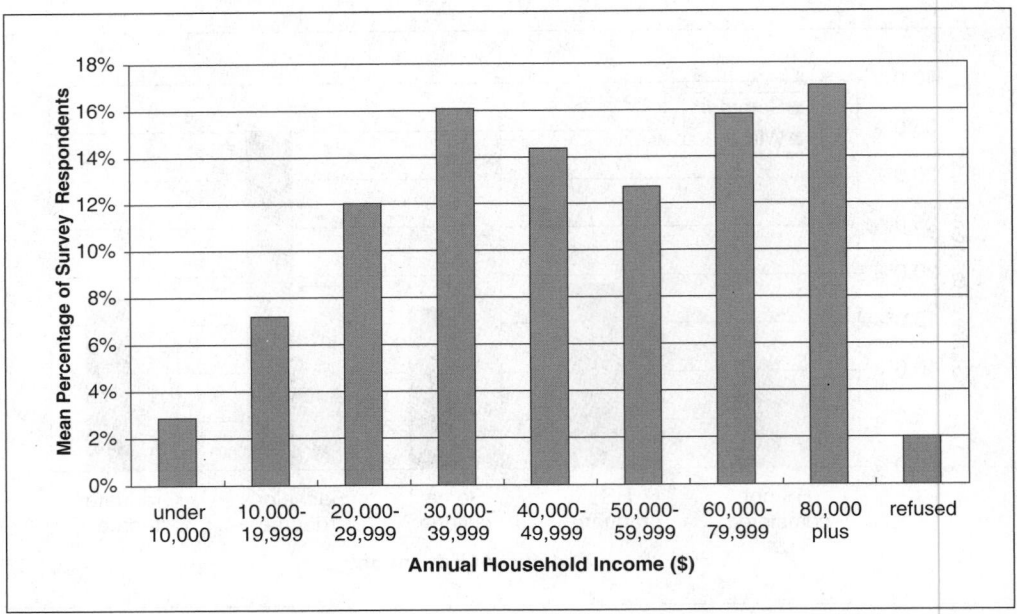

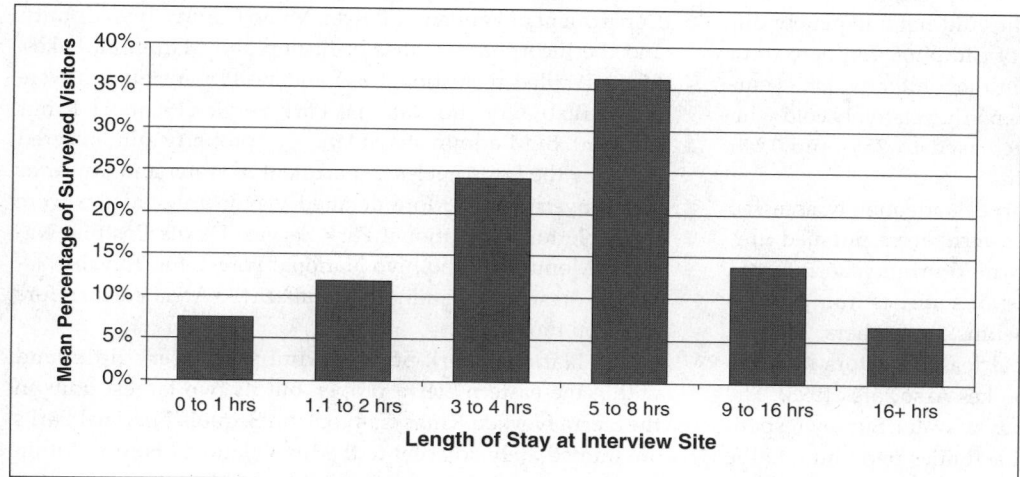

FIGURE 19.67

Single-visit length of stay for summer visitors to Mono Lake and Death Valley.

vey of visitors in the Inyo National Forest found considerably higher rates of camping, however, with 68% of respondents camping (Lee and Brown 1989). Both the timing and location of visitor surveys influence the results, so it is important to avoid coarse generalizations from these limited studies. Summaries and analysis of the individual studies should be consulted in Knauer and Duane (1994) to determine which studies are most applicable in specific circumstances and local conditions.

Surveys of winter visitors in the eastern Sierra are dominated by the detailed market research conducted by the Mammoth Mountain Ski Area. These surveys show that slightly less than 40% of winter visitors rent condominiums during their stay in the eastern Sierra, and about 30% choose to stay in a hotel. The preference for condos most likely reflects seasonal trends in the Mammoth subregion, since many skiers opt to rent a condominium rather than stay in a hotel room. As noted earlier, 20% of all winter visitors to the eastern Si-

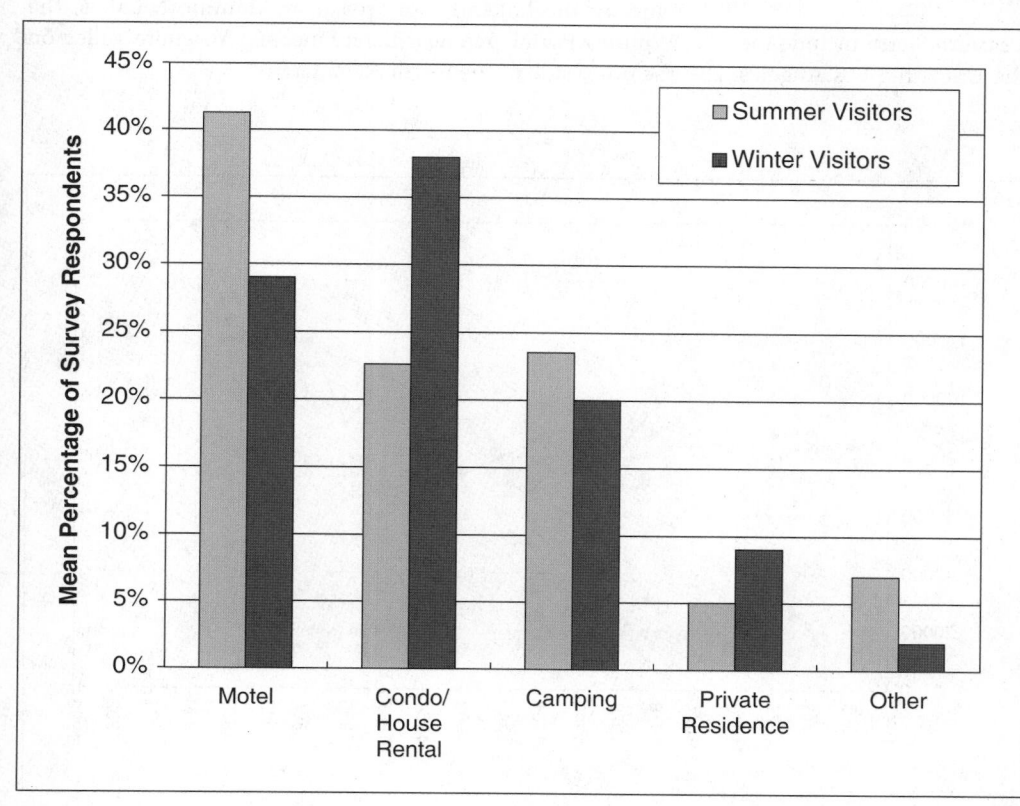

FIGURE 19.68

Type of accomodations.

erra camp, although existing studies did not adequately differentiate between various types of camping. Ten percent of Mammoth visitors even camped in the winter, which is surprisingly high given the elevation and the relatively cold winter conditions. These figures increased to 25% and 12%, respectively, in the summer months.

None of the existing visitor surveys adequately assessed the activities that visitors to the eastern Sierra pursued during their stay. This type of information may be inferred through an examination of the visitor statistics from various land management agencies, museums, and others. Two recent surveys assessed the principal reason visitors travel to the Mono Basin area (Jones and Stokes Associates 1993). The most favored activity varied by body of water. Survey respondents at Mono Lake reported that activities pertaining to the site's unique natural history were most favored, such as sightseeing (59%), viewing the tufa towers (12%), photography (10%), and bird-watching (9%). The majority of visitors surveyed at Grant Lake and Crowley Lake, on the other hand, responded that the principal reason for their visit pertained to fishing. This reflects the unusual features of Mono Lake and its distinctiveness as a natural feature. Both the Grant Lake and Crowley Lake visitor responses are probably typical for other artificial reservoirs in the region, where fishing is often the primary activity of visitors. Many associated activities, such as camping, may also be dependent upon these fishing opportunities.

Visitation and Use Information

Land management agencies in the eastern Sierra include the BLM, the National Park Service, the USFS, the Los Angeles Department of Water and Power, Mono County, Inyo County, and the incorporated cities of Bishop and Mammoth Lakes. More detailed recreational use and visitation statistics were not available for the National Park Service (Sequoia–Kings Canyon), BLM-administered lands, or property administered either by the Los Angeles Department of Water and Power or local governments. More detailed visitor-use statistics were available for the National Park Service Devils Postpile National Monument, the Inyo National Forest, the Toiyabe National Forest, and Mammoth Mountain Ski Area. We therefore focus on those here.

The National Park Service administers very little land within the eastern Sierra proper, but its two largest units in the Sierra Nevada Kings Canyon and Sequoia National Parks are immediately adjacent to the Inyo National Forest. Within the Inyo National Forest in the Mammoth subregion is the Devils Postpile National Monument. In 1979, the park service limited automobile traffic into Devils Postpile and initiated a bus service to alleviate traffic problems along the narrow road leading down into the canyon of the San Joaquin River. The road is generally open for private vehicles early each morning and late each evening, and parties with campground or resort reservations can enter the area during the day. Following the introduction of the bus service, there has been an 18% increase in the overall number of visitors and a 22% decrease in the mean number of cars traveling into Devils Postpile (figure 19.69). The results of this sixteen-year experiment could have management implications for other high-demand areas experiencing transportation problems, including the Lakes Basin area near Mammoth Lakes, the Whitney Portal area near Lone Pine, and Yosemite Valley on the other side of the Sierra Nevada.

FIGURE 19.69

Historical visitor use of Devils Postpile National Monument.

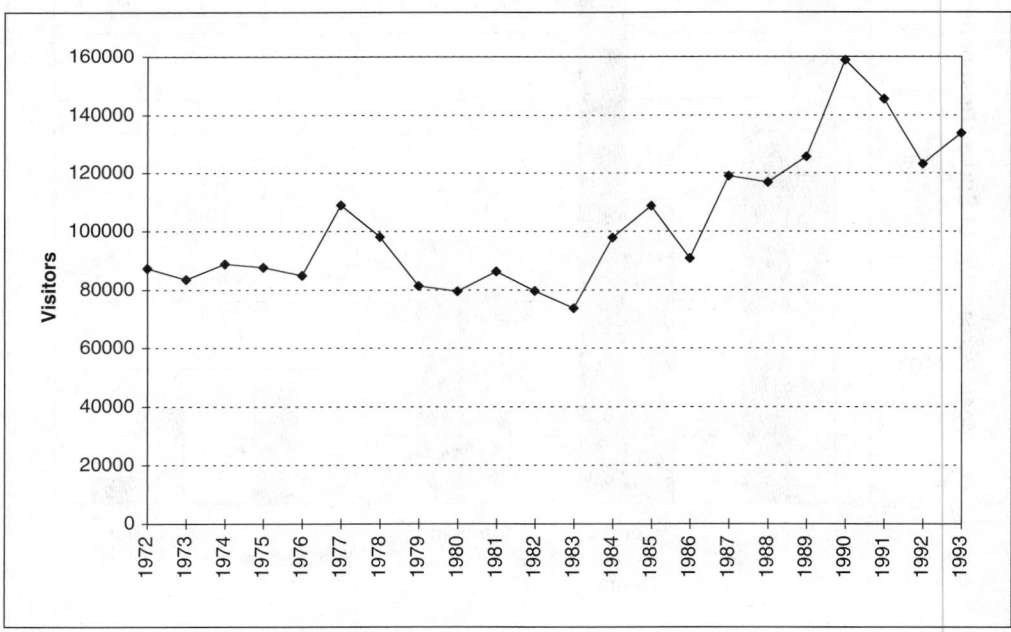

Inyo National Forest

The Inyo National Forest is the primary provider of recreational opportunities within the eastern Sierra and maintains annual records of visitor use of its land at the ranger district level. These records are then aggregated into the forestwide records that are reported to the regional office in San Francisco, which are the data we reported on earlier for the entire Sierra Nevada. There are four ranger districts in the Inyo National Forest: the Mount Whitney, White Mountain, Mammoth Lakes, and Mono Lake Ranger Districts.[8] Recreation and tourist activities on the Inyo National Forest are diverse, ranging from technical climbing of Mount Whitney (the tallest mountain in the contiguous United States) to sightseeing from automobiles. Some areas, such as the Mammoth Lakes region, have intensive developed "front country" uses that sometimes conflict. Other areas, such as the John Muir Trail through the Ansel Adams Wilderness and the John Muir Wilderness, are among the most popular dispersed backcountry recreational sites in the country. These competing demands for different types of recreational experiences (and the high demand for recreation in the area) sometimes lead to conflict, both between prospective users competing for wilderness permits (e.g., backpackers camp out all night to get the first-come, first-served permits issued each morning) and between different types of backcountry users (e.g., between large, commercial pack trips and smaller, private backpacking trips). More detailed analysis of the recreational activities on the Inyo National Forest therefore offers a useful window into the problems of recreational use management in the Sierra Nevada.

Similar issues confront land and resource managers throughout the range.

Detailed RVD statistics from 1991, 1992, and 1993 were assessed forestwide (figure 19.70). As noted earlier, however, our detailed analysis of both the data and data collection practices in the field raised general questions about the integrity of USFS RVD statistics for three reasons: (1) the accounting practices for RVDs vary among ranger districts and over time due to changes in personnel and/or methods; (2) RVD classes are ambiguous and not clearly defined; and (3) it is nearly impossible to sample and therefore account for all visitor activities throughout such a vast, dispersed geographic area. Some ranger districts are conservative in their visitor counts, reporting only those RVDs for which an actual sample was taken. Other ranger districts seem quite liberal in their estimates of some activities, perhaps in recognition of the importance of recreation to the local economy and internal Forest Service management incentives. Despite the possible data incongruities, several interesting trends were apparent within the Inyo National Forest when data were disaggregated to the ranger district level.

As noted earlier, the "travel" RVD class accounted for the highest number of RVDs on the Inyo National Forest. Due to the remote location of the Inyo National Forest relative to population centers, most visitors tour the region by automobile. Other, less popular modes of transportation used to access the Inyo National Forest include motorcycles, buses, and bicycles. Visitors prefer tent camping to other types of camping (e.g., trailer camping). Hiking, walking, and horseback riding are the most common ways of exploring the interior of

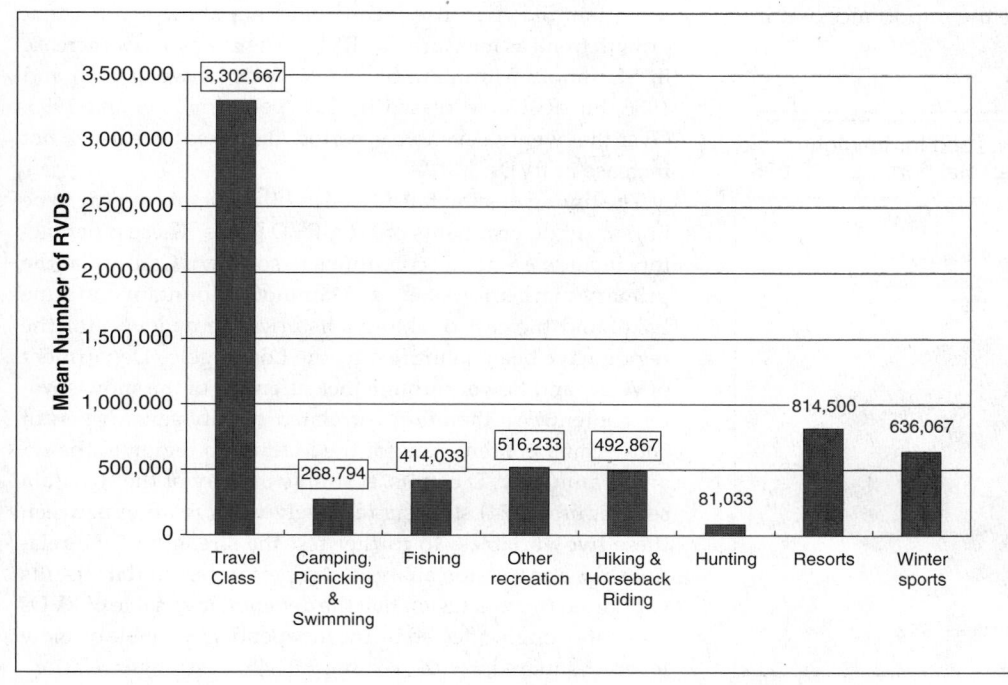

FIGURE 19.70

Mean number of RVDs per activity class for the Inyo National Forest, 1991–93.

the Inyo National Forest. Big-game hunting and cold-water fishing occur more than small-game hunting, bird hunting, or ice fishing. Skiing, both cross-country and downhill, are the most popular winter sports on the Inyo National Forest.

The Inyo National Forest includes portions of the John Muir Wilderness and the Ansel Adams Wilderness, and many day-hikers, backpackers, rock climbers, mountain climbers, and stock (e.g., horses, mules) users travel to these wilderness areas using trailheads that are located on the Inyo National Forest. Overnight visitors to these wilderness areas are required to fill out a backcountry permit, thus allowing the Inyo National Forest to maintain records regarding party size and length of stay. Most backcountry travelers obtain permits for the Ansel Adams Wilderness through either the Mono Ranger District or the Mammoth Ranger District. Backcountry permits for the John Muir Wilderness are most often acquired from the Mammoth Ranger District, Mono Lake Ranger District, Whitney Ranger District, and White Mountain Ranger District. Over 12,000 wilderness permits were issued during 1993, 7% to stock users and 93% to individuals traveling on foot (figure 19.71). Only fourteen percent of the 847 stock permits were noncommercial; while 86% of those using stock entered the wilderness with a commercial guide. Commercial permittees "write their own permits," however, so there is no independent confirmation of the usage figures reported. The backpacking permits, by contrast, are issued directly by USFS personnel. Some prospective backpackers who are unable to get permits enter on commercial stock permits and then continue their trips backpacking. The steep eastern escarpment of the Sierra Nevada has also created a good business in carrying backpacks up to the high country on the backs of mules, so some trips are "assisted" by stock.

These raw estimates of permits issued for stock and foot access to the wilderness understate the importance and im-

pact of stock access. In 1993, there were approximately 89% more stock users on a given permit than backpackers in both the Ansel Adams and the John Muir Wildernesses. The mean number of backpackers per wilderness permit was 3.19, while stock users averaged 6.02 individuals per wilderness permit (figure 19.72). Visitors with noncommercial foot-access permits spent an average of four days in either the Ansel Adams or John Muir Wilderness. Stock users typically spent about 36% more time in the John Muir Wilderness than backpackers. Backpacker trips in the Ansel Adams Wilderness during 1993 were, on average, about 11% shorter than those taken by stock users (figure 19.73). Taken together, the effects of both larger group size and longer trip length for stock users resulted in stock users' accounting for 13% of wilderness permit RVDs even though they were issued only 7% of the wilderness permits. Approximately 80% of wilderness-permit RVDs on the Inyo National Forest were for the John Muir Wilderness, with the remaining 20% for the Ansel Adams Wilderness. There were a total of 39,870 visitors and 371,122 RVDs in wilderness use in the Inyo National Forest in 1993. Note that the average RVD ratio of 9.31 for these visitors is more than four times the average RVD ratio for Yosemite National Park visitors.

Downhill skiing is an activity requiring a permit on the Inyo National Forest and occurs primarily in the Mammoth and June Lakes subregion. There is a fairly reliable RVD record for the Mammoth Mountain ski area, because the concessionaire submits annual ski ticket sales records to the Forest Service. Forest Service officials subsequently convert the ticket sales records into RVD units. Unfortunately, RVD counts for the Mammoth Mountain Ski Area reflect all four seasons; disaggregated data was not available to assess the ratio of winter RVDs to annual totals. The twenty-seven-year Mammoth Mountain Ski Area use record does not show a consistent growth trend as measured by RVDs. There was a 64% increase in Mammoth Mountain Ski Area RVDs between 1966 and 1986, but RVDs decreased by 33% between 1986 and 1993. Over the twenty-seven-year period, there has been a 46% net increase in RVDs.

The timing of snowfall as well as other factors appear to be important determinants of skier RVD levels. These other factors include economic conditions in southern California (the primary market for skiers at Mammoth Mountain and June Lake) and the cost of skiing. Historical snow levels for the region have been quantified by the Los Angeles Department of Water and Power through measurement of the snow's water content. We therefore correlated twenty-seven years of snow water content data for the Mammoth region to the ski area's annual RVD counts. A simple overlay of the two data sets (figure 19.74) shows a relatively weak relation between these two variables. To further test the strength of the relationship, a regression analysis was performed, and its results supported the conclusion that the dependent variable of RVDs was not strongly affected by the independent variable of snow levels (r-squared = 0.027; t-value = 0.845). A stronger relation-

FIGURE 19.71

Proportion of 1993 permits by type issued for the John Muir and Ansel Adams Wilderness Areas. Total permits = 12,095.

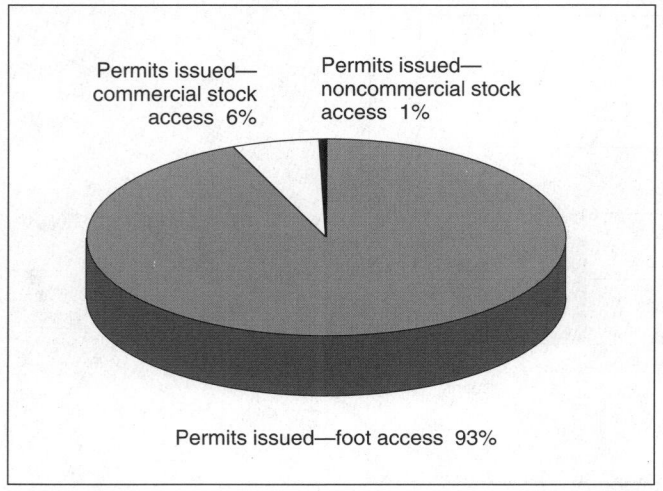

Permits issued—commercial stock access 6%

Permits issued—noncommercial stock access 1%

Permits issued—foot access 93%

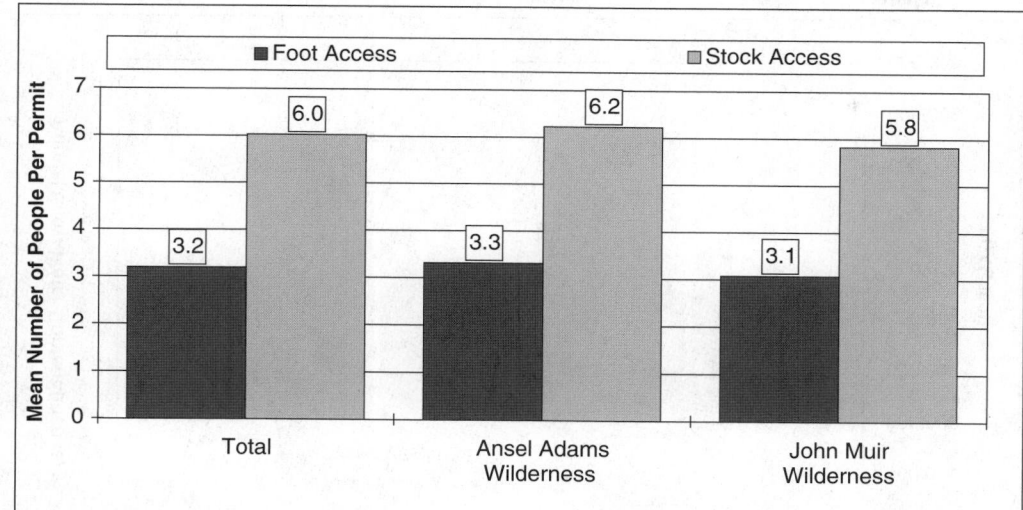

FIGURE 19.72

Mean number of people per wilderness permit issued by the Inyo National Forest, 1993.

ship between the amount of snow and RVDs may have emerged if the Mammoth Mountain Ski Area RVD data was disaggregated by season, but overall snow levels are still unlikely to be significant. Other factors appear to be more important.

The drought of 1987–94, together with changes in the Southern California economy, appears to have put a long-term damper on growth in skier RVDs at Mammoth Mountain. Even the good snow year of 1992–93 did not restore RVDs to their pre-drought level, and young snowboarders are now estimated to account for 25% of current "skier" RVDs. Broader demographic changes and economic changes in the southern California area make it unlikely that the Inyo National Forest skier RVDs will continue to grow as fast as they did before 1987. This has important implications for future land and resource management in the region, for at least two new ski developments have been proposed for the Inyo National Forest. Increased competition from destination resorts in Utah and Colorado, together with local accessibility problems within the town of Mammoth Lakes, appear likely to continue to be as important as federal land management policy to the health of the local ski industry.

Mammoth Mountain Ski Area also offers multiseasonal, nonskiing recreational activities through its "Adventure Connection." Currently, the Adventure Connection coordinates a wide variety of recreational opportunities, including a mountain-bike park, the largest organized mountain-bike event in the country, an artificial rock-climbing wall, a ropes course, orienteering courses, guided fishing, guided hiking, guided mountain biking, dog sledding, and snowmobiling. There were no statistics available as to how many recreationists, overall, have taken part in the activities coordinated by the

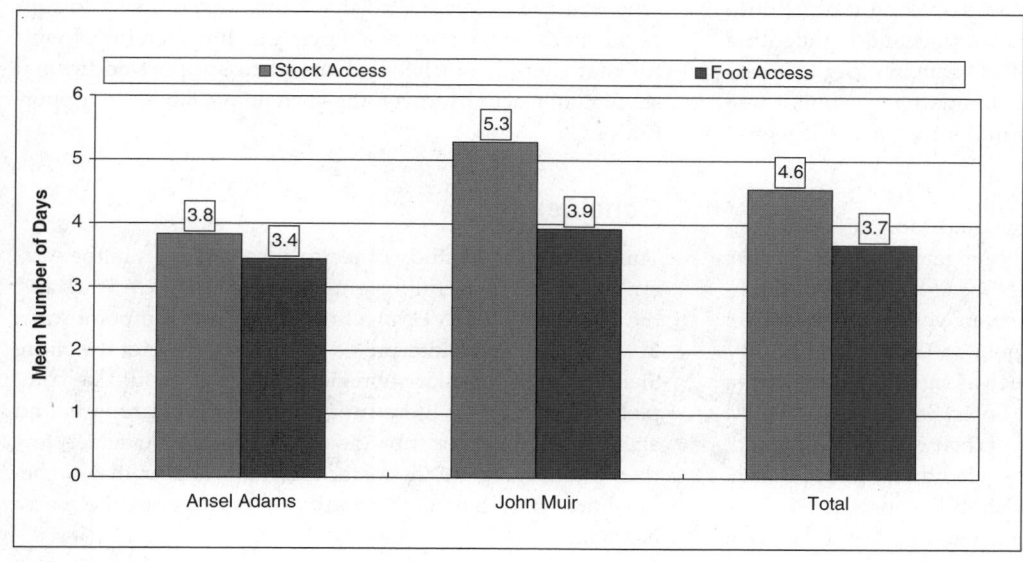

FIGURE 19.73

Mean number of days per wilderness permit issued by the Inyo National Forest, 1993.

FIGURE 19.74

Comparison of Mammoth Mountain Ski Area RVDs and annual water content of snow for the Mammoth region.

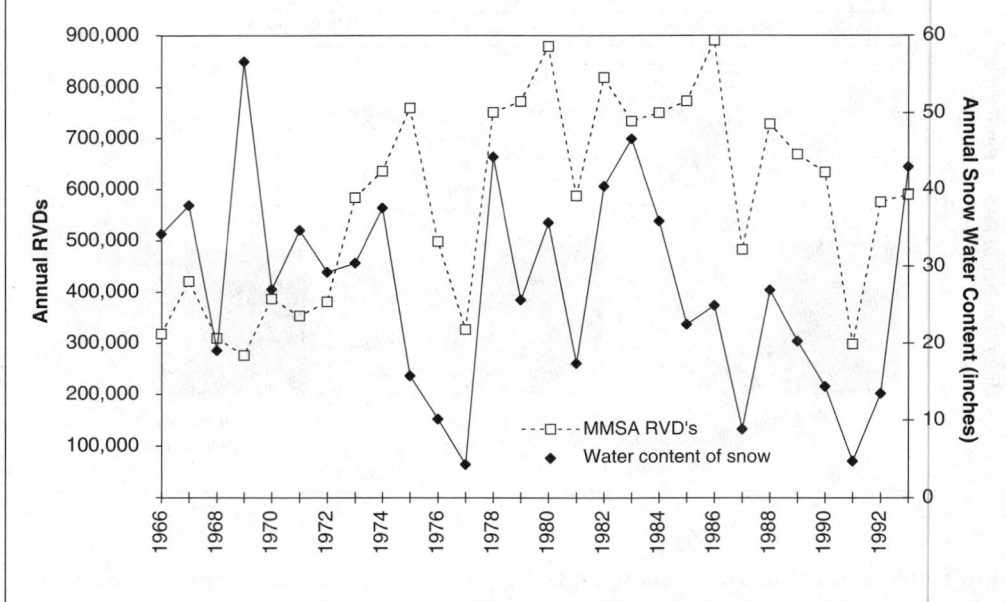

Adventure Connection. Between 1990 and 1993, however, there was an overall 160% mean increase in ticket sales for the mountain-bike park. Over two thousand riders participate in the annual mountain-bike races, and between 1993 and 1994 there was a 14% increase in the number of race participants. The ropes course, outdoor artificial wall, and orienteering courses are administered through a subcontract with Adventure Associates, a firm based in Berkeley, California. During the 1994 summer and fall season, over six thousand individuals used the artificial climbing wall and ropes/challenge course. These numbers appear relatively small compared with skier RVDs, but they represent significant new areas of growth in recreational activity on the Inyo National Forest. They are also activities that do not fit neatly into the traditional recreational activity classes, making it difficult to track them across jurisdictions or over time and highlighting the need for more disaggregated RVD data in order to assess the relative importance of emerging trends in recreation. These activities would generally appear under the "miscellaneous" class in USFS data.

Off-highway vehicle (OHV) or off-road vehicle (ORV) use on public lands occurs in both Inyo and Mono Counties, although minimal data was available regarding levels of use, user profiles, or seasonality of use. Specific types of off-highway vehicles used include all-terrain vehicles, four-wheel drives, dune buggies, and snowmobiles. Certain OHV/ORV activities are seasonal in nature, such as snowmobiling within the Mammoth subregion. Both the Forest Service and the BLM allow OHV/ORV use on portions of their eastern Sierra land, but the BLM data is more disaggregated than the USFS data. Surprisingly, BLM data for the Bishop Resource Area show only 42,450 RVDs for off-road vehicle use (5% of the total of

860,875 RVDs) and 6,175 RVDs for snowmobiling (1% of the total RVDs). Local OHV/ORV advocates have argued through CURES that OHV/ORV use is quite significant to the local economy, but it appears to be a relatively small activity. Camping accounted for 60% of the Bishop Resource Area RVDs, fishing accounted for another 9%, and nonmotorized recreation accounted for 6% of the total RVDs. Some of these other activities may have been conducted in conjunction with OHV/ORV activity, but OHV/ORV activity itself appears to be a relatively small part of overall recreational activity even on those lands that appear most suitable for such activities. Detailed information about snowmobiling activity on the Inyo National Forest was unavailable, but snowmobile permittees appear to be having a difficult time financially. Forest Service officials have also indicated that demand is unlikely to support expansion of concession operators. It is therefore doubtful that there is sufficient demand to support additional snowmobile activity, given the current availability of opportunities.[9]

Conclusions

This more detailed study of recreational activities in the eastern Sierra offers useful insights into user characteristics and the distribution of RVDs at a finer spatial and temporal scale, but it should not be interpreted as representative of the entire Sierra Nevada. Similar subregional analysis should be completed of the "gray literature" of unpublished reports and agency data for other subregions of the Sierra Nevada. A few themes do emerge in the eastern Sierra, however, that we believe are consistent with conditions throughout the Sierra Nevada:

- Recreational users are primarily Californians who live outside the Sierra Nevada.

- Recreational users are primarily Caucasians and do not represent the ethnic diversity of the rest of the state of California, where most of the recreationists live.

- Recreational users are primarily traveling to the area via private automobile.

- Recreational users are primarily male (especially in the summer).

- Recreational users in the winter are more affluent and well educated than users in the summer.

- Each wilderness permit issued for a stock user results in nearly twice as many RVDs as each wilderness permit issued for backpackers, with the additional impact of stock use on those permits and the associated impacts on riparian zones and subalpine meadows.

- Recreational activity in isolated portions of the Sierra Nevada is highly dependent upon access to urban centers (e.g., the Tioga Pass road closes for part of each year).

- Recreational activities in particular areas are often linked to recreational activities in other areas in the Sierra Nevada, California, and the West for out-of-state visitors.

- The relative importance of skiing in traditional ski resort communities is declining as spring, summer, and autumn activities continue to grow in importance and skiing stays flat.

- Levels of recreational activity vary widely on a seasonal and annual basis in response to many factors outside either the Sierra Nevada or resource management policy.

Even this limited set of conclusions suggests some important issues for consideration throughout the Sierra Nevada. Combined with the coarser-scale data we have for the entire Sierra Nevada, our more detailed assessment of recreation in the eastern Sierra Nevada raises a number of issues. It is clear that additional information must be considered in order to evaluate those issues, however, for the data that we worked with in this assessment is inadequate for policy formulation.

CONCLUSIONS AND MANAGEMENT IMPLICATIONS

Recent management experience in Yosemite Valley (where congestion led to closure of several entrances to Yosemite National Park on a series of summer weekends in 1995), together with other high-intensity-use areas such as Mammoth Lakes and Mount Whitney, raises the important point that there are often conflicts over desired conditions for recreational opportunities even within a given activity class, institution and/or management unit.

These types of conflicts may become more important for many recreation providers in the Sierra Nevada if California's population continues to grow, as forecast, to 63 million people by the year 2040 (California Department of Finance 1993). If the population of the state of Nevada is 2–3 million, roughly 65–66 million people will be within a one-day drive of the Sierra Nevada in 2040. Our assessment suggests that current use on public lands in the Sierra Nevada for the agencies reported here is slightly less than two RVDs per year for each resident of these two states. Recreational activity has been relatively steady, however, despite a doubling of the population in both the Sierra Nevada itself and California and Nevada during the 1966–93 period. A doubling of the population of California and Nevada between 1990 and 2040 will therefore not necessarily double total demand for recreational activity in the region and increase conflicts between different types of recreational activities. Growth in demand for recreational opportunities exceeded population growth as American incomes grew rapidly and the "baby boomers" were born and raised during the two to three decades following World War II, but demand has been stagnant since then. This shift coincided with stagnating personal incomes per capita and smaller families following the 1946–64 "baby boom."

The growing population of California also has quite different social, demographic, economic, and ethnic characteristics than the dominant recreational users in the Sierra Nevada today. The state's emerging population is therefore likely to have different needs and demands for recreational opportunities in the Sierra Nevada in the future. Anticipating the character of those needs and demands is a challenge. In general, the current recreational activities of this emerging population appear to be directed more toward "developed" and "front-country" activities than many of the traditional wilderness-type uses that have been so important in the Sierra Nevada throughout the past three decades. We should not project that recreational demand profile into the future without caution, however, for recreational activities are influenced by many social forces. Increased affluence, together with decreased access to other open space, could change those patterns within a single generation. It is impossible to say how the groups that are minorities in California in 1995 will value the wilderness landscape when they constitute a majority of the population fifty years from now. What is clear is that they will be among the recreational users of the Sierra Nevada then, and potential differences in their use patterns will therefore be relevant.

Even without a proportionate doubling of demand, however, conflicts are likely to increase between recreational activities and other uses of public lands and resources. Significant population growth in California would diminish access to and availability of open space and other recreational opportunities on private lands throughout the state, increasing the importance of the public lands in less populated regions such as the Sierra Nevada. The result would probably be increasing pressure to manage the remaining public lands in the Sierra Nevada to provide recreational opportunities for non-local Californians. The social, demographic, and economic characteristics of those non-local Californians will therefore be a critical determinant of what types of recreational activities will be demanded. The population of the Sierra Nevada is overwhelmingly Caucasian and projected to remain that way (Duane 1996). Potential conflicts could therefore emerge between the local population and the recreational needs of the larger urban populations to the degree that any of the future residents of the Sierra Nevada moved to the region to escape California's growing ethnic and cultural diversity. Public agencies providing recreational opportunities in the Sierra Nevada need to address this potential conflict proactively in planning today. Good research is being conducted on these issues, but we still have very little information about how different populations in California view recreational opportunities in the Sierra Nevada specifically or the role of public and private lands in providing it. Primary research is necessary to identify potential trends, evaluate their management implications, and formulate strategies to meet future recreation demand. Those strategies should explicitly account for the role of private recreation in the Sierra Nevada, which we were not able to address systematically in this chapter.

Analysis of policy scenarios must also consider the aesthetic impacts of various land and resource management activities on recreational activities. These aesthetic impacts include visual quality, noise levels, and general perceptions of human disturbance. Each of these elements affects the experience of recreationists—even if the source of the impact is outside the jurisdiction of the administrative unit on which the recreational activity occurs. Wilderness areas are affected by adjacent uses on other public lands, for example, and recreation on public lands is affected by development and use patterns on private lands. Similarly, competing recreational activities within a given area will affect the quality of the recreational experience for other recreationists in the area. User perceptions must therefore be integrated into recreation planning in a systematic way that addresses changes in the quality of the experience. The Recreational Opportunity Spectrum (ROS) system of the USFS attempts to establish a common framework for characterizing different types of recreational experiences, but it has not been integrated systematically across all recreation providers in the Sierra Nevada with empirical field research to determine how management actions affect visitors' perceptions of the recreational experience. This is especially true when considering the impact of nonrecreation activities on recreational experiences.

The potential for conflict between competing uses is most apparent for the USFS and the BLM, which both face a multiple-use mandate for management. Agencies such as the park service and California's Department of Parks and Recreation already manage primarily for recreational activities and the preservation of natural resources that have unique natural qualities. Together they have responsibility for management of some of the most spectacular recreational resources in the Sierra Nevada. The Forest Service remains the most important provider of recreational opportunities in the Sierra Nevada, however, with 57–67% of all RVDs on public lands in the Sierra Nevada taking place on USFS lands.

Expansion of recreational activities on USFS and BLM lands could constrain other management activities, however, including some types of commodity production. But as demonstrated in the SNEP economic analysis by Stewart (1996), recreation is already a more important economic activity in the Sierra Nevada than commodity production. The economic value of recreation and tourism is likely to increase significantly as the population with easy access to the mountain range continues to grow. The recreational resources of the Sierra Nevada are limited and not infinitely substitutable. Land and resource management agencies will therefore need to consider how their management actions today will affect the recreational opportunities of tomorrow for a rapidly growing population.

Our analysis of USFS data shows that the spatial distribution of RVDs is not random. The Inyo, Sequoia, and Sierra National Forests—each of which is adjacent to at least one of the national parks in the southern and central Sierra Nevada—account for 45% of all RVDs on the USFS lands in the Sierra Nevada. Together with the national parks, this portion of the Sierra Nevada probably represents one of the highest levels of recreational activity in the entire world. Over 18.5 million recreational visitor days occur in the national parks and national forests of the southern Sierra Nevada. This is also the region of the Sierra Nevada forecast to experience the greatest population growth in nearby urban centers (especially Fresno and Bakersfield) in the next few decades. As noted in the SNEP air quality assessment by Cahill et al. (1996), the area is therefore threatened by degradation of air quality that could diminish vistas and heighten ecological stress, which could in turn diminish the quality of recreational experiences in the area. The wilderness areas of this region constitute the second-largest roadless area in the contiguous United States, with some of the most spectacular scenery in the world. The Lake Tahoe Basin represents a similar focal point for recreation in the Sierra Nevada, with much of the recreational activity on the Tahoe National Forest, the Eldorado National Forest, and the Toiyabe National Forest occurring in association with activities in the Lake Tahoe Basin Management Unit. The Lake Tahoe Basin is also threatened by diminishing air quality near urbanizing areas such as Truckee and South Lake Tahoe.

Recreational activity is the engine that drives the social, economic, and ecological conditions and management policies in the region. The long-term viability of recreation and tourism in the region may nevertheless be negatively affected by reduced visibility and scenic value just as it has historically been threatened by reduced water clarity. The landscape of the Sierra Nevada is the primary economic asset underlying recreational activity in the Sierra Nevada. Without that asset, many of the recreation-dependent communities of the Sierra Nevada face social and economic difficulties due to their isolation. Marketing brochures for many of the resorts in the Sierra Nevada emphasize the natural resources of the area, including resorts that themselves offer highly urbanized and developed recreational experiences.[10]

Much of the Sierra Nevada functions as California's outdoor playground. As noted in the SNEP economic assessment by Stewart (1996), however, very little of the economic value associated with recreational activity in the Sierra Nevada now goes back into resource management in the region. Local communities also capture very little of the economic value of that activity under existing institutional arrangements. Both the human communities and the land and resource managers of the Sierra Nevada at present make important decisions about future management of the Sierra Nevada without adequate information about the social and economic importance of recreational activity on the public lands in the region. This chapter is a small step toward accounting for the level of and importance of recreation in the Sierra Nevada, but there are still significant gaps in both our knowledge and the ways in which we use that knowledge.

Fully recognizing the potential value of recreational activities in the management of both public and private lands in the region could require significant institutional innovation. There is strong evidence that the existing institutional structure does not adequately reflect the value and significance of recreation in either long-range planning processes or on-the-ground land and resource management actions. Recognizing the level and types of recreational activities in the region is a first step toward such institutional innovation, but it is only a first step. This report only begins to account for the recreational activities in the Sierra Nevada in a consistent way. Institutional arrangements in the Sierra Nevada are generally still a long way from recognizing the value of those activities in management decisions.

Perhaps the biggest gap lies in the relationship between recreation on public lands and related activities on private lands. The estimates of RVDs presented in this report are almost exclusively limited to activities that take place on the public lands. Those activities are the driving force behind considerable related activity on private lands, however, that are not accounted for in our assessment. Many of those public RVDs, in turn, depend upon the provision of services on private lands to support public land recreation. Efforts such as CURES in the eastern Sierra and the Tahoe Coalition of Outdoor Recreation Providers (TCORP) constitute important efforts to improve cooperation and integration across the public and private sectors in the communities that most depend upon recreation for their economic lifeblood. A similar effort could be useful throughout the Sierra Nevada to promote improved data collection, analysis, and consistent policy direction by both public land and resource management agencies and the private sector dependent upon public-sector recreation. This effort could be coordinated by the state Resources Agency as part of the State of California Outdoor Recreation Plan (SCORP), which is updated every five years or so.

Such efforts must be broadly inclusive, however, to ensure that all of the values of all who are interested in the resources of the Sierra Nevada are incorporated into policy decisions that affect recreation in the Sierra Nevada. These values include social, economic, and ecological concerns that are both long-term and short-term, both local and nonlocal. As the custodians for all Californians and all Americans, state and federal agencies must also ensure that the broadest public interest is served. This is a challenge in a rapidly changing world. The tension between local economic concerns (which often call for expanding recreational activity in the short term) and other social values (which often call for limiting recreational activity over the long term to protect ecological or aesthetic values) is not a new one. It was at the heart of many policy decisions over the past century-and-a-half in the Sierra Nevada that still define the parameters for today's policy debates.

The land and resource management institutions of the twenty-first century will continue to face conflicts over these issues as long as noncommodity uses of the public lands are not valued explicitly. Alternative institutional arrangements may therefore be necessary to create incentives for both public agencies and private landowners to manage in ways that are consistent with the full range of public values. These alternatives may involve anything from incremental steps (such as coordinated data collection among public agencies) to much more comprehensive innovations, such as recreation fees or permit systems for the public lands. We make no specific recommendations here, but we urge careful consideration of a wide range of alternatives.

ACKNOWLEDGMENTS

This report was prepared with the extensive assistance in 1994 of Darla Guenzler, who collected most of the information and entered it into the Excel 5.0 spreadsheets, and in 1995 of Jennifer Knauer, who completed data entry and prepared the tables and charts presented in the report. Jennifer Knauer also assisted in final report preparation and completed the more detailed study of recreational activity and visitor characteristics in the eastern Sierra Nevada in 1994, which was supported in part by the Inyo National Forest, the Coalition for Unified Recreation in the Eastern Sierra Nevada (CURES), and the White Mountain Research Station of the University of California. Bill Bramlette, deputy supervisor and former recreation officer of the Inyo National Forest, was especially helpful with the more detailed study of recreational activities in the eastern Sierra Nevada. Other members of CURES and the Inyo National Forest also helped with this effort. Bob Hawkins deserves special appreciation for catching a significant error in the July 1995 draft report. Three anonymous peer reviewers also offered useful criticism of that draft report. Carrie Salazar prepared the final figures from draft figures prepared by the author and Jennifer Knauer. The author alone is responsible for the contents of this report and any errors.

NOTES

1. This is an approximate average of estimates used by Bill Stewart (Duane 1996) and Doak and Kusel (1996); it is used here only to approximate RVDs for local residents.
2. Recreational activities at Lassen Volcanic National Park clearly have an impact on several communities in the northern Sierra Nevada, but the park itself is located outside the Sierra Nevada proper and the SNEP core area.
3. Comments made by the Yosemite National Park superintendent at the Association for Environmental Professionals conference, Yosemite National Park, March 1993.
4. Comments made by representatives from the Yosemite Committee of the Sierra Club at the Sierra Now conference, Sacramento, July 1992.
5. This conclusion is based upon a review of marketing materials for workshops and guided activities in the Sierra Nevada for these activities. We found no systematic data for quantitative estimation of these activities or trends.
6. This qualitative characterization of trends within the fishing activity class is based upon our more detailed assessment of data for the eastern Sierra Nevada, which is where most of the fly-fishing in the Sierra Nevada occurs.
7. As explained in Knauer and Duane 1994, this comparison is based upon pooled studies for summer and winter.
8. Reorganization of the Inyo National Forest in 1995 instituted new landscape zones and eliminated ranger districts.
9. This discussion of OHV/ORV activities is based upon analysis of the data and key informant interviews in the area.
10. See marketing materials for the conference titled Competition and Change: Creating and Economic Vision for Lake Tahoe, Stateline, Nevada, October 1992, and comments by the author at that conference.

REFERENCES

Baker, M., and W. Stewart. 1996. Ecosystems under four different public institutions: A comparative analysis: In *Sierra Nevada Ecosystem Project: Final report to Congress*, vol. II, chap. 51. Davis: University of California, Centers for Water and Wildland Resources.

Beesley, D. 1996. Reconstructing the landscape: An environmental history, 1820–1960. In *Sierra Nevada Ecosystem Project: Final report to Congress*, vol. II, chap. 1. Davis: University of California, Centers for Water and Wildland Resources.

Cahill T. A., J. J. Carroll, D. Campbell, and T. E. Gill. 1996. Air quality. In *Sierra Nevada Ecosystem Project: Final report to Congress*, vol. II, chap. 48. Davis: University of California, Centers for Water and Wildland Resources.

California Department of Finance. 1993. *Population projections for California state and counties 1990–2040 with age/sex and race/ethnicity*. Report 93 P-3. Sacramento, California Department of Finance, Demographic Research Unit.

Chavez, D. J. 1992. Telephone conversation with the author. June.

———. 1993a. Hispanic recreationists in the urban-wildland interface. Unpublished report. U.S. Forest Service, Pacific Southwest Research Station, Riverside, California.

———. 1993b. *Visitor perceptions of crowding and discrimination at two national forests in Southern California*. PSW-216. Riverside, CA: U.S. Forest Service, Pacific Southwest Research Station.

Chavez, D. J., J. M. Bass, and P. L. Winter. 1993a. *Mecca Hills: Visitor research case study*. U.S. Bureau of Land Management, Sacramento, in cooperation with U.S. Forest Service, Pacific Southwest Research Station, Riverside, California.

Chavez, D. J., P. L. Winter, and J. M. Bass. 1993b. Recreational mountain biking: A management perspective. *Journal of Park and Recreation Administration* 11 (3). Fall.

Doak, S. C., and J. Kusel. 1996. Well-being in forest-dependent communities, part II: A social assessment focus. In *Sierra Nevada Ecosystem Project: Final report to Congress*, vol. II, chap. 13. Davis: University of California, Centers for Water and Wildland Resources.

Duane, T. P. 1996. Human settlement, 1850–2040. In *Sierra Nevada Ecosystem Project: Final report to Congress*, vol. II, chap. 11. Davis: University of California, Centers for Water and Wildland Resources.

Duane, T. P., and J. L. Knauer. 1996. Recreational activities and visitor characteristics in the Mount Whitney complex and the Mammoth Lakes basin of the Inyo National Forest. Unpublished report. Inyo National Forest, Bishop, California.

Ewert, A. W., D. J. Chavez, and A. W. Magill, eds. 1993. *Culture, conflict and communication in the wildland-urban interface*. Boulder, CO: Westview Press.

Foin, T. C., Jr., ed. 1977. *Visitor impacts on national parks: The Yosemite ecological impact study*. Publication 10. Davis, CA: University of California, Institute of Ecology.

Foreman, D., and H. Wolke. 1992. *The big outside: A descriptive inventory of the big wilderness areas of the United States*. New York: Harmony Books.

J. Moore Methods, Inc. 1992. *El Dorado County General Plan survey*. Placerville, CA: El Dorado County Planning Department, and Sacramento, CA: J. Moore Methods, Inc.

Jones and Stokes Associates. 1993. Mono Basin draft environmental impact report. Sacramento.

Kahrl, W. 1982. *Water and power: The conflict over Los Angeles' water supply in the Owens Valley*. Berkeley and Los Angeles: University of California Press.

Klages and Associates. 1992. Mammoth Lake visitors study. Unpublished report. Mammoth Lakes Visitors Bureau, Mammoth Lakes, California.

Knauer, J. L., and T. P. Duane. 1994. *Recreation and tourism in the eastern Sierra Nevada: Visitors and activities data*. Unpublished draft Report. Sierra Nevada Ecosystem Project, Davis, California.

Knight, R. L., and K. J. Gutzwiller, eds. 1995. *Wildlife and recreationists: Coexistence through management and research*. Covelo, CA: Island Press.

Kusel, J. 1996. Well-being in forest-dependent communities, part I: A new approach. In *Sierra Nevada Ecosystem Project: Final report to Congress*, vol. II, chap. 12. Davis: University of California, Centers for Water and Wildland Resources.

Laidlaw, R. M. 1992. Interview with the author. July.

Lee, M. E., and P. J. Brown. 1989. An analysis, interpretation, and report of recreational user data collected on the Inyo National Forest. Unpublished report. Inyo National Forest, Bishop, California.

Littlejohn, M. 1991. *Visitors services project—Death Valley National Monument.* Moscow: University of Idaho, Cooperative Park Studies Unit.

Mammoth Mountain Ski Area. 1984. Mammoth Mountain Ski Area summer survey. Unpublished report. Mammoth Mountain Ski Area, Mammoth Lakes, California.

———. 1989. Analysis of 1989 World Mountain Bike Races: Mammoth Mountain. Unpublished report. Mammoth Mountain Ski Area, Mammoth Lakes, California.

Ostertag, R., and G. Ostertag. 1995. *California state parks: A complete recreation guide.* Seattle: The Mountaineers.

Palmer, T. 1982. *Stanislaus: The struggle for a river.* Berkeley and Los Angeles: University of California Press.

Parmeter, J. R. 1976. *Ecological carrying capacity research, Yosemite National Park.* Final report, vols. 1–4. University of California, College of Natural Resources, Berkeley: Department of Plant Pathology.

Pisarowicz, J. 1991. *Death Valley National Monument visitor services survey.* Death Valley Natural History Association (prepared by the National Park Service).

Porcella, S. F., and C. M. Burns. 1991. *California's fourteeners: A hiking and climbing guide.* Modesto, CA: Palisades Press.

Schaffer, J. P., B. Schifrin, T. Winnett, and R. Jenkins. 1989. *The Pacific Crest Trail: Volume 1: California.* Berkeley, CA: Wilderness Press.

Secor, R. J. 1992. The *high Sierra: Peaks, passes, and trails.* Seattle: The Mountaineers.

South Yuba River Citizens League. 1993. *The South Yuba: A wild and scenic river report.* Nevada City, CA: South Yuba River Citizens League.

Sports Research, Inc. 1989. Unpublished report. National ski opinion survey: Mammoth Mountain. Mammoth Mountain Ski Area, Mammoth Lakes, California.

———. 1990. Unpublished report. National ski opinion survey: Mammoth Mountain. Mammoth Mountain Ski Area, Mammoth Lakes, California.

Stewart, W. C. 1996. Economic assessment of the ecosystem. In *Sierra Nevada Ecosystem Project: Final report to Congress,* vol. III. Davis: University of California, Centers for Water and Wildland Resources.

Walton, J. 1992. *Western times and water wars: State, culture, and rebellion in California.* Berkeley and Los Angeles: University of California Press.

Wilderness Conservancy. 1989. *The American River: North, Middle and South Forks.* Auburn, CA: Protect American River Canyons.

Winnett, T. 1978. *Guide to the John Muir Trail.* Berkeley, CA: Wilderness Press.

JONATHAN KUSEL
Forest Community Research
and
Department of Environmental Science,
    Policy, and Management
University of California
Berkeley, California

SAM C. DOAK
S. C. Doak and Associates
Portland, Oregon

SUSAN CARPENTER
Susan Carpenter and Associates
Riverside, California

VICTORIA E. STURTEVANT
Department of Sociology
Southern Oregon State College
Ashland, Oregon

20

The Role of the Public in Adaptive Ecosystem Management

ABSTRACT

The role of the public in adaptive approaches to natural resource management is reviewed and discussed. Two approaches to adaptive management are observed: participation-limited, where the public is generally excluded from active involvement; and integrated adaptive management, where the public plays an active role along with managers and scientists. Integrated adaptive management is a process where the public works iteratively and continuously with managers and scientists, and public input is genuinely integrated into the process and evaluated on a par with other information. Implementation of integrated adaptive management is explored with a focus on identifying subgroups of the public and describing appropriate methods for providing active roles for the public. The public participation process used in Sierra Nevada Ecosystem Project is described to illustrate elements of a successful integrated adaptive process. Guidelines for integrated adaptive management emphasizing active roles for the public are presented and discussed. Integrated adaptive management offers opportunities to effectively involve stakeholders in the development of reasoned solutions to resource management problems.

INTRODUCTION

Adaptive ecosystem management is increasingly discussed by scientists and managers as a new approach to resource management. Those who use this approach acknowledge uncertainty in ecosystem management, embrace an experimental perspective to learning, and understand that it requires new roles and relationships to be developed for scientists and managers. Adaptive ecosystem management is important because it addresses incomplete knowledge and uncertainty associated with ecosystem processes and has the potential to address the political and social components of management. Compared to traditional approaches to resource management, adaptive management is more responsive to changing conditions of and demands on ecosystems. Relationships and responsibilities of the public in adaptive ecosystem management, however, have only recently received attention, and the importance of the public's role has yet to be fully realized.

We begin by describing adaptive management and argue that successful adaptive ecosystem management must include a new and active role for the public that approaches that of scientists and managers, within legal and practicable limits. Moreover, public input must be considered on a par with other information and genuinely integrated into the adaptive process. In our discussion of adaptive management we describe the roles and relationships of scientists, managers and the public and how they differ in traditional—primarily federal—land management, and in two general models of adaptive processes. To further illustrate integration of the public into an adaptive process, we describe public participation in the Sierra Nevada Ecosystem Project (SNEP) which was based on adaptive principles and resulted in a new relationship between the public and project scientists.

Based both on the SNEP experience and research in public involvement in natural resource management, we conclude with guidelines for integrated adaptive processes. Much of the research and experiences in public involvement cited in this paper are based in federal land management because of mandated public involvement requirements. Adaptive pro-

Sierra Nevada Ecosystem Project: Final report to Congress, vol. II, *Assessments and scientific basis for management options.* Davis: University of California, Centers for Water and Wildland Resources, 1996.

cesses, however, can and have been successfully applied in private as well as public land management.

WHAT IS ADAPTIVE MANAGEMENT?

Adaptive management is a term widely used and one with at least several forms and interpretations. Everett and colleagues (1993) present it as an essential component of "ecosystem management." Much of the recent discussion of adaptive processes in natural resources has evolved from the work of C. S. Holling and others in their description of an adaptive framework for environmental assessment and management. Holling (1978) characterizes the adaptive approach as "an interactive process using techniques that not only reduce uncertainty but benefit from it."

An underlying premise of adaptive management is that knowledge of the system managed is not only incomplete but elusive (Walters and Holling 1990). This is particularly relevant in management of ecosystems such as those in the Sierra Nevada where complexity is both highly dynamic and scientifically daunting. Within this uncertain environment, management actions must be designed not only to meet specific objectives but to also yield knowledge and address social goals (Walters and Holling 1990; Lee 1993). By focusing on the refinement of knowledge through management, learning is achieved through the experience of management itself rather than solely through basic research or theory development (Walters 1986; Lee 1993). Bormann and colleagues (1994) describe this as "learning to manage by managing to learn." Thus, at its core, an adaptive process both focuses and accelerates learning to create more effective management.

Critics may argue that adaptive management is a new name for traditional approaches to resource management. Certainly, natural resource management and policy have always been revised based on past successes and failures. Yet changes are frequently crisis motivated and reactive in nature. Bormann and colleagues (1994) assert that learning associated with reactive change is too slow to deal with rapidly changing issues. Consequently, a shift to an adaptive approach in resource management suggests a fundamental change in how learning takes place and how the "system" under management is approached. Walters and Holling (1990) identify three approaches to adaptations in resource management: incremental, passive, and active adaptive management. Incremental management approaches evolve from a reduced set of previously tried techniques, or trial and error. Passive adaptations are based on historical information that form a single best approach along a linear path assumed to be correct. Both of these approaches are typical of traditional management in their linear approach to change and failure to plan to learn.

As described by Walters (1986) and Walters and Holling (1990), "active" adaptive management differs from traditional management approaches through the purposeful integration of experimentation into policy and management design and implementation. Policies and management activities designed in an adaptive framework are specifically treated as experiments and opportunities for learning (Lee 1993). Establishment of procedures for evaluating a range of management actions is a critical component of adaptive management and must be part of the original design, not simply an afterthought (Holling 1978). Consequently, a shift to an adaptive approach in resource management suggests a fundamental change in how learning takes place and how the "system" under management is approached. In summary, adaptive management is defined here as a process for acting deliberately under uncertainty by increasing opportunities to develop new information and redirecting management actions in a timely manner. Management actions are designed not only to meet specific objectives but also as learning experiences that focus on the constant re-evaluation of goals, objectives and perceptions of processes as new information is developed.

PRIMARY GROUPS, ROLES, AND RELATIONSHIPS IN ADAPTIVE PROCESSES

The purposeful integration of experimentation into policy and management design and the creation of sustained learning processes through adaptive management require, among other things, a redefinition of the relationships among three primary groups of participants: scientists, managers, and the public. We begin this section by describing these three groups. We then explain how the public was addressed within three subgroups in the Sierra Nevada Ecosystem Project to offer one example of how the public might be more effectively integrated into an adaptive process.

The relationships between scientists, resource managers and the public highlight fundamental distinctions between traditional natural resource management and adaptive processes, and, additionally, are used to further identify two types of adaptive management: participation-limited and integrated adaptive management. *Traditional natural resource management* processes, even those with public involvement components, typically do not foster working relationships between the public and managers and make no attempt to encourage relationships between the public and scientists (figure 20.1). Adaptive processes, on the other hand, are fundamentally about changing the relationships between these three groups. *Participation-limited adaptive management* focuses principally on the relationship of scientists and managers (figure 20.2), while *integrated-adaptive management*, as we define it here, requires a new role for the public that includes the establish-

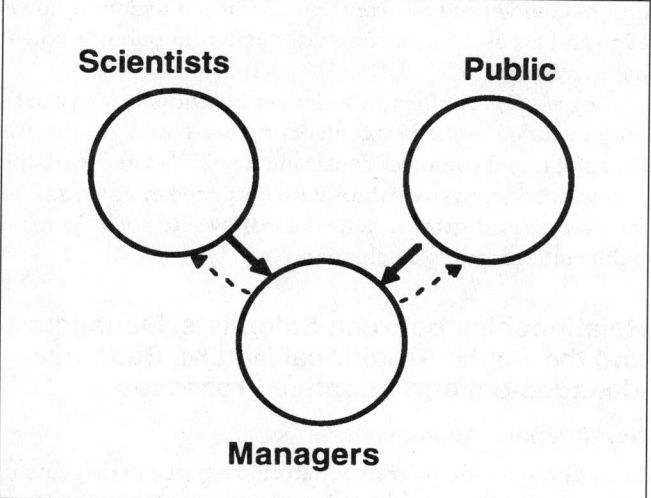

FIGURE 20.1

Traditional management.

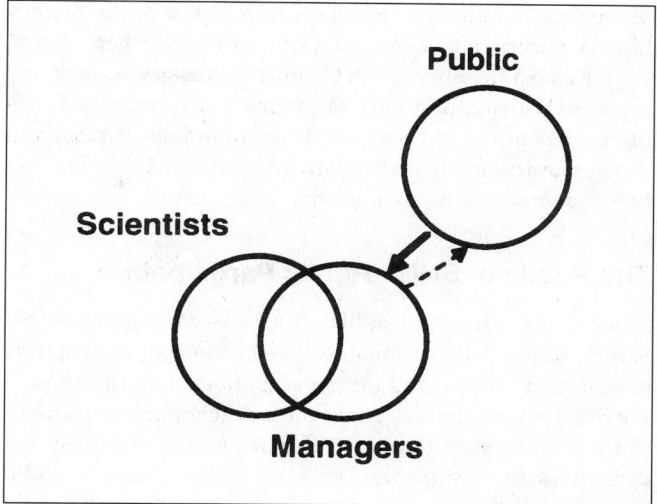

FIGURE 20.2

Participation-limited adaptive management.

ment of active working relationships between managers, scientists and the public (figure 20.3). We conclude this section with a discussion of SNEP public participation activities that helped build a new collaborative relationship between scientists and the public.

The Primary Groups in Adaptive Processes

Scientists, managers and the public make up the three primary groups in this discussion of adaptive management. *Scientists* include individuals and organizations that typically engage in scientific research. The scientific community includes individuals with training and expertise in experimental design and methods testing. Scientists may be found within management organizations as well as within separate institutions of learning, research and development. *Managers* include individuals and organizations endowed with the responsibility and authority to manage or regulate the land and resources that are under consideration for management action. This group includes higher level policy makers as well as field managers and related staff of federal, state and private organizations with resource management responsibilities. *The public* includes other individuals and organizations that are not included in the groups of managers or scientists. The public, as a group, includes a diverse array of interests.

It is useful to further identify the primary subgroups within the public in order to address better their needs and interests as well as the various skills and knowledge they might bring to a particular process. One useful dichotomy distinguishes two principal communities within the public: communities-of-place and communities-of-interest. Communities-of-place include members of the public who may be affected by or interested in management decisions and actions by nature of

their residency within or near management activities. For example, residents of the community of Lee Vining in the Eastern Sierra, organized to affect management of Mono Lake, represent a place-based group. Communities-of-interest include groups with a focused interest in (often accompanied by organized efforts to influence) management of resources unrelated to their member residence. Such groups are "communities" through this shared interest. Groups in this category include regional and national-level organizations with broad constituencies, many of whom often reside in urban areas. Examples of groups with interests in the Sierra Nevada include the Sierra Club, Women in Timber, California Forestry Association, and Audubon Society.

It is possible for an individual to be a member of both a community-of-interest and a community-of-place, an example being a resident of Lee Vining who is also a member of the

FIGURE 20.3

Integrated adaptive management.

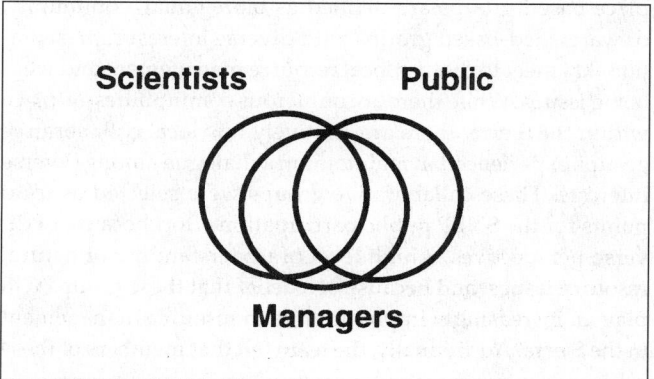

Mono Lake Committee. Similarly, a member of the Quincy Library Group, a place-based group in the Northern Sierra, who is also a member of the California Forestry Association is part of both a community-of-interest and a community-of-place. Depending on how broadly communities-of-place and -interest are defined, this public dichotomy—while quite useful—may not be entirely inclusive.

The Public in SNEP Public Participation

Three distinct types of public groups were targeted in the SNEP public participation effort: key contacts; collaborative place-based groups; and the general public. Identification of the first two groups was based on the dichotomy of communities-of-interest and communities-of-place, respectively. Because these two groups were more narrowly defined by SNEP and due to the broad nature of the SNEP study, a third more general category for the public was also identified. Distinct activities were developed for each of these groups.

The charter for the Sierra Nevada Ecosystem Project recommended that the team rely on a group of key contacts to help accomplish project objectives. The initial key contact group consisted of individuals who participated in previous planning and evaluations of the Sierra Nevada (e.g., the Sierra Summit Steering Committee and Sierra Nevada Research Planning Team). Additional individuals were added to the group as the team identified regions or areas of interests that were not represented. Communities-of-interest participated in SNEP as members of the key contact group and independently in team meetings with the general public. The key contact group totaled approximately 100 individuals representing various interest groups, scientific or other perspectives within the communities-of-interest of the Sierra Nevada. A sub-group of about a dozen key contacts assisted the SNEP public participation team in planning meetings with the general public and key contacts during the final year of the project. A diversity of interests were also represented in this sub-group, including recreation groups, public agencies, timber industry, and the environmental community.

Place-based collaborative groups, which focus efforts in communities "placed" in the Sierra, were selected as focal points for SNEP's local public participation activities. Collaborative place-based groups are defined as bioregional, community, or watershed-based groups with diverse interest representation that meet to discuss local resource management and well-being issues. While there are numerous communities-of-place within the Sierra, there are relatively few local collaborative groups experienced in maintaining a dialogue among diverse interests. These collaborative groups were selected as focal points for the SNEP public participation effort because of diverse perspectives, a high level of understanding of natural resource issues, and because of a belief that these groups will play an increasingly important role in resource management in the Sierra. Additionally, the team felt that members of these groups could effectively contribute local and regional knowledge and act as catalysts for additional local public involvement in SNEP.

The general public includes all other individuals not specifically included in the key contact or place-based groups. Although limited resources constrained SNEP's work with the public, working relationships with both the key contacts and the place-based groups helped to draw and focus general public interest and participation.

Relationships between Scientists, Managers and the Public: Traditional Natural Resource Management and Adaptive Processes

Non-Adaptive Approaches

In traditional (non-adaptive) natural resource management, an approach common with federal land management and particularly national forest planning in the 1970s and 1980s, managers operate in relative isolation and bear the responsibilities of planning, decision-making, and implementation within a closed system. Plans and decisions are based on traditional practices and available scientific knowledge from the research community. The scientific community is generally left largely on its own to design and engage in research that may or may not contribute to future management and decision-making. Though some research may be specifically linked to land management activities, scientists and managers generally do not directly interact.

Although many traditional management approaches include "public involvement" programs, the rigidity and formality of these programs typically preclude active involvement of the public in resource planning, management or decision-making. Over the last three decades, public involvement in federal land management has been generally treated as both a mandated activity and a necessary evil rather than as part of an active process integral to learning and successful management. While a major thrust of opening federal and state environmental decision-making to the public was to increase government accountability and responsiveness to citizens, there is considerable evidence that public input too often was not taken seriously or was integrated inadequately into decisions (Blahna and Yonts-Shepard 1989; Rosenbaum 1976; Wondolleck 1988). Agency responses to mandated requirements were often formal and characterized by rigid compliance with procedures, since satisfying legal requirements frequently held higher priority than meeting the spirit of the law. Staff hired to coordinate public involvement were separated from resource specialists and managers, and in the interests of "fairness" and "objectivity," public comments were analyzed, synthesized, coded, counted, and removed from their vital contexts. Agencies often took the role of neutral arbiter in processes that, by design, promoted adversarial relations among various interests (Wondolleck 1988). Moreover, the only opportunity for public comment was with draft decisions, too late to permit real integration of public input—which otherwise should have taken place during formulation of project objectives, approach and design—as these decisions were "draft" in name only (Krannich et al. 1994; Wondolleck 1988).

Adaptive Approaches

Inherent in adaptive processes is a general recognition of the need to redefine traditional roles and relationships of the primary players in natural resource management. Redefinition is unclear, however, because the public's role in adaptive management and the relationship of the public to scientists and managers has not been well defined. Two general models of adaptive processes for ecosystem management can be distinguished based on who is actively involved: a more commonly recognized participation-limited approach focusing principally on the activities and interactions of scientists and managers; and an integrated adaptive management approach that creates active roles for the public and the research community and requires new and active relationships between managers, scientists and the public.

Participation-Limited Adaptive Management. Participation-limited adaptive management focuses on the collaborative relationship between managers and scientists in the design, implementation and monitoring of adaptive policies. Describing the use of workshops in an adaptive assessment process, for example, Holling (1978) states that it is critical to have all "prime actors" present, yet he includes only scientists, managers, and "policy people" in his discussion. Lee (1993) describes adaptive management as a process where managers work closely with scientists to collect and analyze information and improve understanding. Bormann and colleagues (1994) refer to a need to "blur" the institutional boundaries between research and management. To design management plans as experiments, Holling (1978) calls for a "more elaborate and productive interplay" between scientists and decision-makers. For Holling, scientists provide a certain level of rigor and understanding of fundamental processes, and managers provide a balance to "scientist's penchant for exquisite detail and excessive resolution."

Participation-limited adaptive management includes adaptive processes and demands a new relationship between scientists and managers, but, similar to traditional resource management, it does not actively include the public in the process. The merging of science and management is a productive step, but by limiting active participation to scientists and managers, it becomes a technocratic approach, that alone is inadequate. Walters (1986) points out that traditional scientific programs tend to focus on previously validated tools and methods, leaving many research paths untouched. In a case study analysis of three adaptive management processes

in the Northwest, McLain and Lee (1994) indicate that the effectiveness of these adaptive approaches is hampered by a failure to address the role of diverse stakeholders, and, it is important to add, diverse values. They point out that little attention has been given to the types of institutional procedures that can facilitate the incorporation of social values and processes in adaptive management.

Integrated Adaptive Management. In integrated adaptive management the public is included in the adaptive process and provided an active role that approaches that of managers and scientists. Bormann and colleagues (1994) suggest that successful adaptive management of ecosystems will require collaboration, coordination and information sharing among all interested parties, including the public. They point out that adjacent landowners and various social communities will be more accepting of management and will be more willing to take responsibility if they are given a greater role in shaping experimental management decisions (Bormann et al. 1994). Everett and colleagues (1993) propose that adaptive ecosystem management strategies should be developed through the collaborations of all relevant social communities, managers and the research community. Despite interest expressed in the literature in elevating the role of the public in adaptive processes, this role has been neither adequately elaborated or widely implemented.

In an integrated adaptive process, the public, managers and scientists iteratively work together to design, implement, monitor, evaluate and jointly assess management options in an adaptive process. The degree to which the public participates in each stage of the adaptive process will vary depending on the interests, skills and knowledge that the public brings to any one situation, as well as any legal and practicable constraints to their participation. More significantly, however, public input must be genuinely integrated into the adaptive process and assessed on a par with other information. Successful application of an integrated adaptive approach thus not only permits public access and actively encourages it at each stage, but also promotes earnest exchange of information among scientists, managers, and the public and the sincere integration of diverse stakeholder knowledge and social values into the process.

The advancement of the public's role in integrative adaptive management has the advantage over participation-limited approaches of recognizing the importance of social values in goal setting and the role and importance of an active public for learning. The emphasis of a participation-limited approach is often on resolving tensions between scientists and managers. Integrated adaptive management, on the other hand, recognizes the political nature of natural resource management and the necessity of involving the public in an active process of contributing and learning.

Why Integrated Adaptive Management?

We have characterized adaptive management as a process embodying deliberate management actions designed to increase opportunities for learning in order to redirect management with improved information. The process of adaptive management has also been described as a deliberate cycle of planning, acting, monitoring, evaluating, and adjustment (Forest Ecosystem Management Assessment Team 1993), and similarly, as a continuous system involving adjustment, linked actions, feedback, and information synthesis (Bormann et al. 1994). However the various steps in adaptive management might be described, the entire process moves forward through the advancement of ecological *and* social learning. As a mechanism for learning and acting, it is the creation, modification, interpretation and flow of information that connects each step with the next and provides a critical basis for communication between the various participant groups. Information and knowledge are both the foundation and a product of adaptive management. Bormann and colleagues (1994) recognize the unique information needs of adaptive management systems by elevating information to a status equal to or greater than traditional resources.

Integrated adaptive management emphasizes active public participation in this evolution and flow of information. Integrated adaptive management requires a shift away from public participation as a discrete activity for involving the public to a process-oriented approach that calls for maximizing opportunities for the public to gather, modify, synthesize, evaluate, design, learn from and avail themselves of information related to the ecosystem processes and management decisions at hand.

Transcending both traditional management and participation-limited adaptive management, integrated adaptive management offers the public both a significant role and a permanent one. The public is not brought in occasionally simply to bless or critique a plan, rather it is invited to help design, monitor and evaluate in an iterative process. Given clear roles in each phase, the public becomes a partner with managers and researchers in a continuous process of experimentation, learning and management. There are three fundamental reasons why adaptive ecosystem management must include active roles for the public.

1. *As sources of information and knowledge.* The collection and interpretation of information is fundamental to the adaptive management process (Lee 1993). The research community is the primary source for scientific information and interpretation, and managers offer experiential knowledge. The public, however, also has a critical role as a source of information. First, landowners, local residents and communities-of-place offer a wealth of knowledge regarding local attributes and ecological and social processes affecting an ecosystem. Second, the public is a source of information on social values and can also act as filters for interpreting information on social values that are collected at a broader scale.

2. *To build trust and broaden support.* The dynamic nature of adaptive management necessitates that the public be allowed to actively participate in the process rather than simply be informed of it. Some may argue that adaptive management is too complicated for the public to understand. If this is the case, however, excluding them from the process will only result in greater confusion and distrust. "Adaptive management must take into account the broad array of public communities, and it must be interactive with the scientific discovery process rather than react to it" (Everett et al. 1993). Managing in an adaptive framework will not further social learning if the public is not part of the process. Additionally, excluding the public from adaptive ecosystem management means ignoring the political nature of land management, thereby increasing the likelihood that the process itself will not be supported.

3. *To generate ideas and question paradigms.* A primary theme of adaptive management is to "reject recipes and rituals in favor of a search for better processes to promote imagination and learning" (Walters 1986). Both scientists and managers tend to favor well-trodden paths. The public is more likely to question these favored approaches and can act as a catalyst for devising new ones. Greater societal participation can ultimately lead to a wider array of learning opportunities in natural resource management.

SNEP Public Participation as an Adaptive Process

Public participation in SNEP provides an example of how the public can be integrated in an adaptive process. The public was informed about the study and involved in SNEP through an adaptive exchange as the team iteratively sought and responded to public input. As an adaptive ecosystem management process, however, SNEP was unusual because there was no explicit role for managers, quite the opposite of what would be the case if it was a land-based adaptive process. Numerous managers did, however, participate in the SNEP adaptive process, but they did so by attending general public meetings or workshops.

Public participation in SNEP did not begin as part of an integrated adaptive process. Initially, SNEP public involvement bore closer resemblance to a narrow public participation process in traditional resource management. The first public meeting provides a striking example of this. At this meeting, following the release of the interim SNEP Progress Report, the public participated as listeners and were allowed to submit only written questions. The entire science team was seated facing the SNEP Steering Committee seated in the front of the room, with team members' backs to the public. The only interactions in the meeting were between members of the science team and members of the steering committee, except for the written questions. Maintaining a distance between the science team and the public was supported by traditional views of many team scientists. They felt that public involvement would be messy and, worse, would compromise the independent scientific study.

Through a fairly dramatic evolution, over a period of approximately six months, public involvement in SNEP grew into an adaptive process. During this period, there were numerous conversations among team members in which the potential benefits as well as the risks of an adaptive process were discussed, and an outline of a public involvement plan was developed. One of the explicit goals of SNEP public participation, as it was redefined, was to promote mutual learning and a feedback process between scientists and the public.

The first SNEP team meeting with the key contacts was held after the first public meeting as a new public involvement strategy was being developed. At this meeting the team learned the importance of public access to scientists. Key contacts were able to question and engage scientists in discussions following individual presentations. Meeting breaks in many instances proved more valuable than the presentations themselves, as participants took these opportunities to discuss issues with scientists directly. During this meeting several key contacts volunteered to assist the SNEP public participation team in planning public meetings during the final year of the project. The key contact work group, as it came to be called, was instrumental in providing ideas on how to ensure productive interactions between SNEP scientists and the public in these additional meetings.

As redefined, the SNEP public participation strategy employed a diverse array of methods to provide meaningful roles and maintain continuous involvement with the public. The core of the public involvement strategy consisted of a multi-stage approach involving newsletters providing general information on project activities and preliminary findings, an open letter to the public requesting information and calling for public contributions to scenario development, meetings and workshops, and focused public reviews of draft assessments. The open letter resulted in forty-one public submissions, all of which were reviewed by the science team.

Meetings and Workshops

A series of public meetings and workshops were conducted with collaborative groups (and communities-of-place), key contacts (largely communities-of-interest) and the general public.

Two public meetings were held in succession within the geographical areas of each of the two collaborative groups. These meetings were co-hosted by the collaborative groups which made arrangements and ensured that the broader public was invited. The local meetings were attended by a subgroup of the science team representing a diverse range of disciplines that were of particular interest to the collaborative groups. Each meeting had a different complement of scientists. Brief presentations were made by members of the science team on approaches and progress within their individual assessments. The last half of these two- to three-hour meetings was dedicated to informal questions and answers and to open discussion between scientists and attendees. The second meetings with the collaborative groups also included interactive demonstrations of some of SNEP's computer-based geographic information system (GIS) data. Notes were taken at all meetings to ensure that questions, concerns and suggestions from the public were captured, and these notes were later shared with the full team.

Two large public meetings were centrally held along with a separate workshop with the key contacts. Considerable time was allocated for interaction between scientists and the public at each of these meetings. The first public meeting under the integrated public involvement strategy focused on introducing the approaches used by scientists in the assessment and on discussions of preliminary findings. The full-day meeting began with formal presentations by some of the scientists, including questions from the public. Nearly three hours of the meeting were dedicated to an open workshop format where attendees were able to engage in discussions with scientists in small groups at tables organized by resources and disciplines. Included in this arrangement was an area dedicated to interactive demonstrations of some of SNEP's GIS data. Note-takers were again stationed at each table to capture the questions and suggestions offered by the public.

A special workshop was held with the key contacts to specifically solicit ideas regarding the development of policy scenarios. During this workshop, participants were briefed on a list of possible scenarios based on ideas drawn from responses to the open letter, previous public meetings, scientific models, and the team's resource assessments. Attendees were then divided into small groups composed of both scientists and key contacts. Led by SNEP facilitators, the groups discussed concerns and offered suggestions regarding scenario development. Notes were taken on poster sheets. Representatives from each group summarized their discussions to the full group. The dialogue captured in this workshop was used by the science team to both expand and refine the development of a suite of scenarios.

The final public meeting was scheduled to provide sufficient time to incorporate public comment gained during the meeting into the development of scenarios. This meeting offered an opportunity for the public to understand and evaluate the range of scenarios developed up to that point, and for the SNEP scientists to listen to the public's concerns, insights and suggestions. Scheduling additional time to incorporate public comment allowed the SNEP team greater opportunity to fashion scenarios which incorporated local expertise and better reflected public concerns. The round-robin type of interaction in which the public conversed with scientists face-to-face was repeated in this meeting. Following a few formal presentations by SNEP scientists, participants were given the opportunity to discuss scenarios and findings directly with scientists at tables organized by scenario focus and general resource area. As with previous meetings, notes were recorded at each table to capture the questions and suggestions offered by the public, and circulated to the team shortly thereafter.

Reviews

Key contacts and place-based groups also participated in *a focused review of SNEP assessment reports*. Draft assessments were subjected to a blind peer review process and were simultaneously sent to key contacts and place-based groups, on request, for their review and comment. Key contacts and place-based groups coordinated public review of these drafts, taking responsibility for summarizing responses and returning them to the team within the same time period allotted peer reviewers. Similar to a formal peer review process, team scientists used comments received from the key contacts and place-based groups to inform subsequent revision of their assessments.

Other Interactions

The formal public involvement strategy detailed above was supplemented by a variety of other interactions between the individual scientists on the SNEP team and the public. Interactions were often intended to either inform the public of SNEP or gather specialized knowledge, but often accomplished both. These included meetings with agencies, private industry, county supervisors, and interest groups, a series of workshops with local experts to assess community capacity and well-being, and other workshops to identify and map late successional forest types. Scientists were encouraged throughout the process to meet with individuals and groups who had information and ideas that would assist project assessments.

The Success of SNEP's Integrated Adaptive Management Process

SNEP public participation succeeded as an adaptive process largely due to the development of an active role for the public and the advancement of an interactive relationship between the public and scientists. The scientists as well as the public benefited. Most SNEP scientists, including those who were initially skeptical of interaction with the public, found the public involvement process both instructive and valuable. Many scientists were influenced in a variety of ways by public interaction, and near project end virtually all scientists were positive about exchanges with the public. It is important to reiterate, however, that the SNEP process represents a limited example of adaptive ecosystem management because the study did not expressly include managers or address specific land management projects.

The incorporation of public ideas into the science team's work represented another success. Though it is impossible to pinpoint specific "public" ideas that influenced the scientists, it is clear that public involvement did influence the work of the team. Numerous times in SNEP team meetings a scientist would reference a public comment to reinforce a point or make clear that the issue under discussion must be addressed to respond to public concern. Interaction with the public often influenced how data were presented and conclusions drawn, as well as influencing the development of scenarios.

Finally, the adaptive process itself had significant effects on the public as a group. Through their involvement, the public's perceptions of the science project itself changed. Individuals who initially felt the project was a waste of time later expressed a genuine concern that the best possible science be used to address the complex social and resource issues in the Sierra Nevada. Perhaps most importantly, people who had long been sitting on opposite sides of issues agreed that resolution of complex resource management issues would only be achieved with them working together and not against one another. There appeared to be broad agreement among these participants to continue the dialogue begun in this adaptive process after SNEP.

INTEGRATED ADAPTIVE MANAGEMENT: LESSONS FROM THE PAST, GUIDELINES FOR THE FUTURE

The complexity of resource issues, institutional environments, and diversity of stakeholders' values and knowledge, among other things, make clear that no single formula is sufficient to involve the public in integrated adaptive management. Guidelines for effective public involvement in an integrated adaptive management process, however, can be identified. The following guidelines are drawn in part from lessons learned from the successes and failures of public involvement in traditional federal land management activities and the fields of conflict resolution, conflict mediation and collaborative problem solving. Experiences in these fields reflect a trend in public participation towards adaptive processes and are discussed in more detail in appendix 20.1. It is, in fact, many of these lessons that informed the approaches and activities used to involve the public in SNEP.

Involve the Public Early

Public participation in natural resource management has proven to be most effective if it involves the affected public early in the process (Blahna and Yonts-Shepard 1989; McMullin and Nielsen 1991; Gericke and Sullivan 1994). Early involvement ensures that the knowledge, concerns, and values of the public are incorporated in the design stage, rather than driving reaction to later decisions or activities. Similar to SNEP public involvement, in an adaptive framework early involvement logically includes seeking the public's input in designing the involvement process itself.

Maintain Continuous Involvement

The public should be kept continuously informed and involved throughout the process. Adaptive processes are by definition continuous and regularly produce information that must be absorbed, evaluated and integrated into decisions. Daneke and colleagues (1983) and Howell and colleagues (1987) note the importance of continued public involvement, especially during periods when key decisions are made. Strategies for continuous involvement include providing opportunities for members of the public to become partners in the process by assuming specific roles, such as those associated with monitoring activities, that facilitate sustained participation.

Use Diverse Involvement Methods

The use of diverse involvement methods enhances the inclusiveness of public participation and provides opportunities to learn which methods are most effective. The effectiveness of different techniques for involving the public varies with the population of interest, the stage of the process, and the issues and activities at hand. Experiences in public involvement in natural resource planning and management indicate that various segments of the public respond differently to different formats. Cortner and Shannon (1993), for example, report that local wood-products workers favor informal settings and oral communication, rather than formal hearings and written comments, and Syme and Nancarrow (1992) observed that the more highly educated sectors of the public tend to participate in surveys, formal hearings and workshops. In the SNEP study, different activities were specifically structured for communities-of-place, communities-of-interest and the general public due to the unique knowledge and skills each group was able to bring to the process.

Emphasize Small Group Activities

Small group activities are the most effective public participation technique in natural resource management (Gericke et al. 1992; Blahna and Yonts-Shepard 1989). Most citizens prefer dialogue in small groups and other methods that involve two-way communication and shared decision-making (Cortner and Shannon 1993). Gericke and Sullivan (1994) note that meetings where individuals are seated as equals around a table in an informal setting have different results from those where uniformed agency personnel stand at the front of the room.

Be Inclusive

To be effective and democratic, public involvement must be inclusive and representative of all stakeholder groups, including communities-of-interest and communities-of-place. Inclusive public involvement requires that specific involvement techniques should be targeted to specific populations and activities. Moreover, the success of individual techniques should be evaluated based on their effectiveness in reaching target groups rather than in mere numbers of participants. In gathering local knowledge, for example, workshops may attract a large number of participants, but the few individuals with the greatest knowledge to offer—usually long-term local residents—are often the least likely to respond to open houses or workshops. Inviting people personally and providing forums with which they are comfortable reaches beyond those most motivated to participate. Greater societal participation in the design of adaptive management projects will lead to a wider potential array of treatments for scientific analysis as the diverse views and experiences offered by the public challenge both scientific paradigms and management dogma.

Daneke and colleagues (1983) found that public involvement practitioners were likely to be more concerned with organized interest groups and political power than inclusiveness, favoring public meetings that were often influenced by organized groups. Cortner and Shannon (1993) note that groups that feel alienated, due in part to perceived threats to their resource-based lifestyle, are likely to extend their disaffection towards planning processes. Yet, as Priscoli (1983) points out, it is these groups that may be most directly affected that should be actively encouraged to participate.

Recognize and Incorporate Local Knowledge

The active development and incorporation of local, or indigenous, information can reveal critical knowledge about resources, patterns and processes, and even management actions that are not part of the common scientific information base. People who live and work within or near ecosystems know much about them, particularly those who have done so for several decades or have multi-generational ties to certain areas. Local experience with resource management on private lands may provide valuable insight and historical information. Sturtevant and Lange (1995) provide an example of agency foresters expanding their understanding of forest stand dynamics and thinning regimes upon examining privately owned forests and discussing management techniques with private landowners. Local involvement "facilitates learning from local knowledge and reflects local concerns" (Slocombe 1993).

The incorporation of local knowledge into the adaptive management process may challenge the notions of researchers and managers of scientifically "valid" data. Local knowledge, however, should augment, not supplant scientific information. Instituting a formal process for documentation provides an opportunity to identify the strengths and limitations of this information as well as identifying inconsistencies and competing understandings.

Rely on the Public to Define Social Values

Sustainability in ecosystem management is defined by the interaction of social values and ecological conditions. Managing successfully for sustainability depends upon public awareness and the adaptation of behavior to knowledge about an ecosystem. Management units include people, their social and economic activities and their shared and individual values (Slocombe 1993). Information development, therefore, must include knowledge of social well-being within ecosystems as well as social values affecting management. Social values may change more rapidly than ecological conditions, but it is important to recognize that they may do so in response to changes in the ecosystem, particularly when restoration is needed and/or the supply of goods and services are disrupted.

The public offers critical input in identifying expected outcomes and indicating evaluation procedures necessary to maintain public trust. In recommending a proactive public involvement strategy for the then intact AT&T Corporation, Toffler (1985) noted that timely information about changes in social structure and values can only come from the public, "whose members are, in fact, involved in these changes." Assessments of social values often rely on statistical surveys and aseptic analyses of secondary data designed and conducted by researchers removed from subject populations. While these are useful information gathering mechanisms, their veracity and acceptability are greatly increased by providing opportunities for the affected public to assist in both the design and the interpretation of results.

Make Information Accessible

Participatory, collaborative processes in adaptive management can only be achieved if all information necessary for effective decision-making is equally accessible to all participants, including the public. Knowledge and information are central to the process of adaptive management. Greater information access increases opportunities for public participation and trust, and can lead to further exchange and improvement of information as it is shared, evaluated and compared to local observations. Processes for sharing existing information, however, are often limited. Local communities and interest groups generally do not have access to the range of data and information that is available to management agencies and research organizations, thereby constraining dialogue in the adaptive process. Improvements in information accessibility must also address the capacity of the participating public to fully use and understand data and other information. Communities, then, must also have access to appropriate information tools.

Critical information technologies such as geographic information systems (GIS) must be readily shared in order to make information truly accessible. Moreover, the presentation and format of available information must be appropriate to the needs and capabilities of all potential users, not only those with access to and understanding of the latest technology.

Foster Positive Working Relationships

Integrated adaptive management requires developing and maintaining healthy relationships among the primary players in the processes. Hostile relationships are lethal to adaptive processes; people distrust each other, are unable to communicate productively, and ultimately are incapable of reaching decisions and implementing them. Healthy relationships develop when participants are open and clear with each other about actions or steps which are taken, differences are acknowledged, and all parties work together to find solutions (Carpenter and Kennedy 1988).

Clearly Define Roles, Responsibilities, and Realistic Expectations

Participants must be clear about roles and responsibilities for effective participation (Cortner 1995). The role of the public cannot be limited to expressing preferences and values, nor should individuals enter the process with greater aspirations for affecting the final outcome than is politically, scientifically or legally possible. Nothing dissolves public support and trust more than having the process of collaboration build up expectations and then hit barriers which prohibit implementation. Administrative and scientific opportunities and limitations must be made clear at the beginning and throughout the process. Decision-makers, in particular, must be clearly identified. As a continuous cycle, adaptive management provides opportunities for participants to assume different roles and responsibilities at different stages. Clear communication of the extent of these roles is essential.

Promote Facilitative Leadership

Multifaceted leadership must be recognized and encouraged in order to stimulate an atmosphere of creativity and inclusiveness. Sirmon (1993) describes the ideal agency leader as a facilitator and guide who is more than a conveyer of the community of interests, but also an effective intervener and one who actively participates in dialogue. A leader, Sirmon adds, should also be an educator, a provider of data, a developer of viable alternatives, an interpreter of law and regulation, and a representative of those not able to participate. These traits are important to adaptive management on federal land where public agencies dominate the landscape. In general, however, the leadership necessary to sustain adaptive management processes may not always come from government agencies, even when agencies are instigators of the process. Integrated adaptive management benefits from leaders who promote the process itself rather than any individual project or goal. Leaders may emerge as those participants in adaptive management—whether managers, researchers or members of the public—who act as catalysts to the process, encouraging participation and innovation and fostering a climate of trust. It is these individual leaders who should be encouraged, to the extent possible, to continue their facilitative efforts in order to foster a vital and sustainable integrated adaptive process.

Be Flexible

Adaptive processes must be flexible ones, particularly during the critical creative phase of designing management experiments. A conscious effort must be made to develop a process where rules, regulations, dogma and political agenda are not allowed to impede or distort the process of social learning. Flexibility has proved to be an important element in the success of public involvement and conflict resolution activities. Plans developed with an understanding that changes might occur in the time frame, scope of issues, type of activities and number and type of participants as the project unfolds are more likely to succeed in the long run. Rigidly adhered to processes are an additional source of conflict for the manager (Carpenter and Kennedy 1988). Walters (1986, viii) exhorts, "An essential feature of dealing adaptively with uncertainty is to reject recipes and rituals, in favor of a search for better processes to promote imagination and learning." As with any activity in a adaptive framework, public participation approaches should be evaluated and adjusted.

Provide Open Dialogue for Information Synthesis and Evaluation

At the critical stage of synthesis and evaluation of information, where science, experience and social values mix, open dialogue is essential. As information is generated through adaptive management, the public collaborates with managers and scientists in synthesizing new information, evaluating how well management activities meet their objectives, and generating alternative hypotheses (the next experiment) and new management goals. If a set of measures can be agreed to, participants in the evaluation process can determine the overlap between what is biologically possible and what is socially desirable.

Ravetz (1986) notes that as society's knowledge increases, our relevant ignorance increases even more rapidly. He states, "Coping with ignorance in the formation of policy for science, technology, and environment is an art which we have barely begun to recognize, let alone master." Whatever knowledge may be gathered through adaptive management, then, will never be enough to make "pure" scientific decisions. Evaluation of information derived from adaptive management involves considerable interpretation of data couched in a framework of experience, scientific knowledge and social

values. Inclusion of the public in this process broadens the scope of the evaluation dialogue, and is critical to the social learning process. Noting that the principal of self-discovery is an important element of adaptive policy design, Walters (1986) suggests that "people only change their basic attitudes when they devise the arguments to do so for themselves."

CONCLUSION

In this paper we offer an integrated adaptive management approach because the role of the public in adaptive management has not yet been clearly articulated. Also, based on lessons from traditional resource management and public involvement over the past three decades, adaptive processes that do not provide an active role for the public will likely fail. Integrated adaptive management explicitly recognizes the role of the public alongside that of managers and scientists, and the value of information that is generated and considered through the interaction of these principal groups.

Integrated adaptive management challenges current institutional arrangements and requires support from community stakeholders and broader communities-of-interest, as well as the scientific community and managers. It requires opening up decision processes and, within legal and administrative constraints, sharing responsibility among managers, scientists and the public for decisions and for the development of information upon which they are based. This inclusive approach advances the social learning objectives of adaptive management by providing a forum to generate ideas and improve knowledge and understanding, and offers opportunities to build trust and broaden support for natural resource management activities. Allowing people to own science by making it accessible through adaptive management permits stakeholders to question assumptions, re-interpret findings, seek applications and develop better process and solutions.

The integrated adaptive public participation process used in SNEP proved time consuming, but by actively involving the public in an adaptive process, fears about the study were reduced and better integration of public knowledge and values into the study was achieved. Perhaps most importantly, diverse groups now view the SNEP study as a valuable source of information and ideas rather than an inflexible plan. Hence the study is just one part of a long-term adaptive process—one in which the public is now actively and cooperatively engaged—which will lead to improved management of the Sierra Nevada.

The emerging paradigm of adaptive management parallels the evolution of the public's role in natural resource management. While neither adaptive management nor increased active public participation are panaceas, the union of the two into an integrated adaptive management process offers opportunities to craft reasoned solutions to resource management problems while advancing society's knowledge and understanding of ecological processes. Integrated adaptive management presents a real opportunity to strike a balance between scientific management of natural resources and democratic involvement in decision-making.

REFERENCES

Blahna, D. J., and S. Yonts-Shepard. 1989. Public involvement in resource planning: Toward bridging the gap between policy and implementation. *Society and Natural Resources* 2 (3): 209–27.

Bormann, B. T., P. G. Cunningham, M. H. Brookes, V. W. Manning, and M. W. Collopy. 1994. *Adaptive ecosystem management in the Pacific Northwest.* General Technical Report PNW-GTR-341. Portland, OR: U.S. Forest Service, Pacific Northwest Research Station.

Carpenter, S., and W. J. Kennedy. 1988. *Managing public disputes: A practical guide to handling conflict and reaching agreements.* San Francisco: Jossey-Bass.

Cortner, H. J. 1995. Legal and institutional considerations in public participation in the United States. Paper presented at the International Symposium on Public Participation and Environmental Conservation, Tokyo.

Cortner, H. J., and M. A. Shannon. 1993. Embedding public participation in its political context. *Journal of Forestry* 91 (7): 14–16.

Creighton, J., J. D. Priscoli, and M. Dunning, eds. 1983. *Public involvement techniques: A reader of ten years experience at the institute for water resources.* Fort Belvoir, VA: U.S. Army Corps of Engineers.

Daneke, G. A. 1983. Introduction. In *Public involvement and social impact assessment,* edited by G. A. Daneke, M. Garcia, and J. Priscoli, 11–33. Boulder, CO: Westview Press.

Daneke, G. A., M. Garcia, and J. Priscoli, eds. 1983. *Public involvement and social impact assessment.* Boulder, CO: Westview Press.

Everett, R., C. Oliver, J. Saveland, P. Hessburg, N. Diaz, and L. Irwin. 1993. *Adaptive ecosystem management,* vol. II, *Ecosystem management: Principles and applications,* 340–53. General Technical Report PNW-GTR-318. Portland, OR: U.S. Forest Service, Pacific Northwest Research Station.

Forest Ecosystem Management Assessment Team (FEMAT). 1993. Implementation and adaptive management. In *Forest ecosystem management: An ecological, economic, and social assessment,* chap. 8. Washington, DC: Government Printing Office.

Gericke, K. L., and J. Sullivan. 1994. Public participation and appeals of Forest Service plans—an empirical examination. *Society and Natural Resources* 7:125–35.

Gericke, K. L., J. Sullivan, and J. D. Wellman. 1992. Public participation in national forest planning: Procedures, perspectives, and costs. *Journal of Forestry* 90 (2): 35–38.

Gray, B. 1989. *Collaborating: Finding common ground for multiparty problems.* San Francisco: Jossey-Bass.

Gusman, S. 1981. Policy dialogue. *Environmental Comment,* November, 14–16.

Holling, C. S. 1978. *Adaptive environmental assessment and management.* New York: John Wiley.

Howell, R. E., M. E. Olsen, and D. Olsen. 1987. *Designing a citizen involvement program: A guidebook for involving citizens in the resolution of environmental issues.* Corvallis: Oregon State University, Western Rural Development Center.

Krannich, R. S., M. S. Carroll, S. E. Daniels, and G. B. Walker. 1994. Incorporating social assessment and public involvement processes into ecosystem-based resource management: Applications to the east side ecosystem management project. Unpublished report prepared for the U.S. Department of Agriculture, Eastside Ecosystem Management Project, Walla Walla, WA. October.

Lee, K. 1993. *Compass and gyroscope: Integrating science and politics for the environment.* Washington, DC: Island Press.

McLain, R. J., and R. G. Lee. 1994. Adaptive management: Promises and pitfalls. Paper presented at the Rural Sociological Society annual meeting, Portland, OR.

McMullin, S. L., and L. A. Nielsen. 1991. Resolution of natural resource allocation conflicts through effective public involvement. *Policy Studies Journal* 19:553–59.

Priscoli, J. 1983. The citizen advisory group as an integrative tool in regional water resources planning. In *Public involvement and social impact assessment,* edited by G. Daneke, M. Garcia, and J. Priscoli. Boulder, CO: Westview Press.

Ravetz, J. R. 1986. Usable knowledge, usable ignorance: Incomplete science with policy implications. In *Sustainable development of the biosphere,* edited by W. C. Clark and R. E. Munn. Cambridge: Cambridge University Press.

Rosenbaum, N. M. 1976. *Citizen involvement in land use governance: Issues and methods.* Washington, DC: Urban Institute.

Sirmon, J. M. 1993. National leadership. In *Environmental leadership: Developing effective skills and styles,* edited by J. C. Gordon and J. K. Berry, 165–84. Washington, DC: Island Press.

Slocombe, D. S. 1993. Implementing ecosystem-based management. *BioScience* 43 (9): 612–22.

Sturtevant, V. E., and J. I. Lange. 1995. Applegate partnership case study: Group dynamics and community context. Report submitted to the U.S. Forest Service, Pacific Northwest Research Station. Southern Oregon State College, Ashland, Oregon.

Syme, G. J., and B. E. Nancarrow. 1992. Predicting public involvement in urban water management and planning. *Environment and Behavior* 24 (6): 738–58.

Toffler, A. 1985. *The adaptive corporation.* New York: McGraw-Hill.

Walters, C. 1986. *Adaptive management of renewable resources.* New York: Macmillan.

Walters, C. J., and Holling, C. S. 1990. Large-scale management experiments and learning by doing. *Ecology* 71:2060–68.

Wondolleck, J. 1988. *Public lands conflict and resolution: Managing national forest disputes.* New York: Plenum Press.

Process-Oriented Public Involvement Techniques

Public involvement techniques such as consultations, conflict resolution and conflict management, and collaborative problem solving are important tools for engaging the public in integrated adaptive management of public land. These process-oriented techniques have been increasingly and successfully used in public participation activities associated with land management activities over the last two decades. Agencies and other management organizations have been increasingly turning to consultative activities and process-oriented techniques drawn from the rapidly evolving fields of conflict resolution, conflict management and collaborative problem solving. They supplement standard newsletters, questionnaires, interviews, surveys, polls, and public meetings traditionally used to gather and disseminate information. Recognizing the need for more active and inclusive forms of public involvement, and in response to a failure of past public involvement activities, agencies and management organizations have turned to these techniques.

CONSULTATIONS

One way that agencies and other organizations widen communication is through the use of advisory groups, workshops, focus groups, dialogue groups and open houses. Face-to-face dialogues and discussions, termed consultations, are used to identify issues, explore options and, to a lessor degree, develop recommendations. In many cases the exchange of information, clarification of issues, and discussion of options through these various consultative efforts can reduce public concerns sufficiently to permit projects or policies to move forward (Creighton et al. 1983).

CONFLICT RESOLUTION AND CONFLICT MANAGEMENT

Conflict resolution and conflict management activities have been used to expedite the resolution of issues by drawing together representatives of conflicting parties and other affected stakeholder groups and identifying solutions that all parties can support. The goal of conflict resolution is to reach agreements on conflicting issues among affected parties while the goal of conflict management is to handle conflict productively. In either case, managing a conflict outside the legal arena permits the parties, and subsequently the public, to examine a broader range of issues.

Conflict resolution in natural resource management initially evolved from two separate backgrounds. Mediators with experience in labor-management negotiations worked with conflicting parties to structure a process to facilitate understanding of the range of issues and interests, develop acceptable options, and reach agreements. Following the labor management model, many mediators intervened only after negotiating parties reached deadlocks characterized by highly polarized conflict. Mediators with backgrounds in peacemaking and organizational development convened disputing parties earlier in the process to improve communication, exchange critical information, or clarify their particular issues and concerns. While some interventions result in joint options or recommendations, others focus on conflict management rather than agreement, and are known as policy dialogues (Gusman 1981), workshops, and information sharing sessions. In each of these efforts, mediators help conflicting parties work more effectively with each other.

Negotiation is the principal tool of conflict resolution, and these efforts are often called roundtables, mediations or negotiations. Negotiations typically rely on face-to-face exchanges between affected parties, but when direct meetings are psychologically or logistically difficult mediators may use shuttle diplomacy or single-text negotiations. In shuttle diplomacy mediators move back and forth between parties in conflict to achieve agreement. In single-text negotiation a mediator discusses the case and possible solutions with parties, creates a draft of an agreement and then circulates the draft among parties asking for ways to improve it. The draft continues to be circulated and revised until all parties are satisfied with the document.

In contrast to more general public involvement activities, conflict resolution activities in natural resource management have involved the public as decision-makers, or as integral players in a decision-making process, along with representatives from a responsible management organization and other affected parties. Conflict resolution shifts the dynamics from adversarial behavior, where the winner-takes-all, to joint problem solving where parties work together to produce a mutually acceptable solution. Significantly, much of the responsibility for making decisions shifts to the primary stakeholders.

COLLABORATIVE PROBLEM SOLVING

"Consensus building" initiatives or "collaborative problem solving" programs are aimed at reaching agreements earlier in the planning cycle before parties become deeply entrenched in their positions. Similarly, "consensus decision-making" engages the public in discussions about development of policy and plans and their implementation, and seeks the consensus of stakeholders in reaching acceptable solutions.

The underlying premise of collaborative problem solving assumes that if the right people are brought together in a process that encourages learning and joint exploration of solutions, they will identify good solutions and produce results. Going beyond the search for compromise that characterizes conflict resolution, collaborative processes promote mutual education among all stakeholders and encourage the development of new options. As Gray (1989) notes, collaboration is a process by which "parties who see different aspects of a problem can constructively explore their differences and search for solutions that go beyond their own limited vision of what is possible." She adds that collaboration creates "a richer, more comprehensive appreciation of the problem among the stakeholders than any one of them could construct alone."

Most collaborative processes are initiated in response to particular problems or issues. Managers or facilitators establish agendas, suggest and enforce ground rules, offer process suggestions, and in general, manage group discussion. The collaborative process ends once a plan or proposal has been approved by participating groups.

As with conflict resolution, collaborative problem solving allows the public to become a decision-maker, sharing power with managers and other traditional decision-makers. Management organizations consequently become a player, one voice among many stakeholders. As the World Bank (1995) noted in a recent study on participation agencies must "work with representatives of key interests to identify issues, jointly generate options, and seek solutions." Moreover, because stakeholders collaborate to forge options and decisions in collaborative processes there generally is little opposition to implementation. Chrislip and Larson (1994) note that collaboration is a mutually beneficial relationship between parties "who work toward common goals by sharing responsibility, authority and accountability for achieving results."

Collaborative processes are limited, however, by their typically ad hoc nature. The process ends once an issue has been addressed, and the mutual learning and working relationships that were advanced during the collaboration cease on a formal basis.

REFERENCES

Chrislip, D., and C. Larson. 1994. *Collaborative leadership.* San Francisco: Jossey-Bass.

Creighton, J., J. D. Priscoli, and M. Dunning, eds. 1983. *Public involvement techniques: A reader of ten years experience at the institute for water resources.* Fort Belvoir, VA: U.S. Army Corps of Engineers.

Gray, B. 1989. *Collaborating: Finding common ground for multiparty problems.* San Francisco: Jossey-Bass.

Gusman, S. 1981. Policy dialogue. *Environmental Comment,* November, 14–16.

World Bank, Environment Department. 1995. *World Bank participation sourcebook.* Washington, DC: World Bank.

SECTION III

Biological and Physical Elements of the Sierra Nevada

JERRY F. FRANKLIN
College of Forest Resources
University of Washington
Seattle, Washington

JO ANN FITES-KAUFMANN
U.S. Forest Service
Plumas National Forest
Quincy, California

21

Assessment of Late-Successional Forests of the Sierra Nevada

ABSTRACT

Late-successional, including old-growth, LS/OG, forest conditions were assessed for the Sierra Nevada using stand structural criteria as measures of the level of LS/OG forest function, such as in providing habitat for LS/OG-related species. Larger landscape units (polygons) which were relatively uniform in type and distribution of vegetation patches were mapped using available imagery, maps, ground-based information and the expert interpretations of resource specialists. Characteristics of the major patch types in each polygon were identified and tabulated and a composite late-successional structural ranking was calculated for each polygon on a scale that extended from 0 (no contribution to LS/OG forest function) to 5 (very high level of contribution to LS/OG forest function). Maps and databases were used in assessing current LS/OG forest conditions in the Sierra Nevada and, in other SNEP exercises, constructing and evaluating alternative management scenarios. Forests with high LS/OG structural rankings are currently uncommon in the Sierra Nevada; only 8.2% of the mapped polygons had structural rankings of 4 or 5. Commercially important forest types—such as the mixed conifer and east-side pine forests—are particularly deficient relative to their potential as a result of past timber harvesting. Key structural features of LS/OG forests—such as large-diameter trees, snags, and logs—are generally at low levels. On the positive side, the forest cover in most areas is not highly fragmented by clear-cutting and stands have sufficient structural complexity to provide for at least low levels of LS/OG forest function. National parks provide the major concentrations of high-ranked LS/OG forest with about twice as many polygons in moderate to very high rankings as adjacent National Forest lands. Furthermore, much of the remaining highly-ranked LS/OG forest on national forests is unreserved and potentially available for harvest. Forest health is generally good in the Sierra Nevada; areas of epidemic mortality are localized in subregions such as the eastern face of

the Sierra Nevada. The current extent of high-quality LS/OG forest is believed to be far below levels that existed prior to western settlement; based upon several lines of evidence, the majority of commercial forestlands were probably occupied by such forests at that time. If maintenance of high-quality LS/OG forest ecosystems is adopted as public policy, a program needs to be initiated that will 1) maintain existing high-quality LS/OG forests; 2) restore such conditions where existing LS/OG forests are insufficient to achieve objectives; 3) restore fire as an important process and to reduce risks of catastrophic loss; and 4) restore structural complexity in the matrix. Elimination of timber harvest within existing high-quality LS/OG forests for at least an interim period and restoration of low- to moderate intensity fire to existing and prospective LS/OG forest ecosystems are probably the most important immediate actions. Larger management units, called Areas of Late-successional Emphasis, are proposed as an approach which incorporates both reserves and areas managed intensively to reduce the potential for catastrophic fire. Restoration of LS/OG structures and functions in the matrix is also very important and can be achieved by developing and applying silvicultural prescriptions which restore and maintain key LS/OG structures, such as large-diameter trees and the snags and logs derived from them.

INTRODUCTION

Late-successional, including old-growth, forests are an important resource on federal lands in the Sierra Nevada. Late-successional forests are typically forests that have developed over one to many centuries without a major disturbance—i.e., a disturbance which destroyed much or all of the stand. Interest in late-successional forests reflects the fact that they

Sierra Nevada Ecosystem Project: Final report to Congress, vol. II, *Assessments and scientific basis for management options.* Davis: University of California, Centers for Water and Wildland Resources, 1996.

fulfill many important functions for human society such as by 1) providing critical habitat for many wildlife species as well as other elements of biological diversity; 2) performing important ecological functions as part of the carbon (or energy), nutrient, hydrologic and other material cycles in the Sierra Nevada, North America, and the globe; and 3) providing important inspirational, recreational, and cultural resources. The broader category of late-successional forests, incorporating both old-growth and mature forests, is used here since mature forests (stands with tree dominants 100 to 200 years old) often provide some habitats and services comparable to truly old forests; the acronym LS/OG (late-successional, including old growth) is used hereafter to refer to this array of forest ecosystems.

The significance of LS/OG forests is emphasized by the specific charge to Sierra Nevada Ecosystem Project (SNEP) to provide an assessment of the condition and distribution of these forests. The United State Congress provided this direction in language that was a part of two bills in the House of Representatives in 1992 (see appendix A in Sierra Nevada Ecosystem Project 1994): HR 5503 (passed) called for a "scientific review of the remaining late-successional forest in the National Forests of the Sierra Nevada" including production of "maps identifying the old-growth forest ecosystems"; HR 6013 (proposed) identified six tasks including an inventory of "watersheds and late-successional forests" and recommendation of alternative management strategies [for] watersheds and late-successional forests". The congressional direction was incorporated in the charges provided by the SNEP Steering Committee.

An assessment of the distribution and condition of forests contributing to late-successional forest function on federal lands in the Sierra Nevada is the subject of this paper. We begin by briefly reviewing the diversity of forest types and conditions and the methodology adopted; the mapping and characterization exercise made extensive use of resource specialists familiar with on-the-ground conditions. Subsequently, we report our findings regarding quality and distribution of forests contributing to late-successional forest function including differences among major forest types and lands administered by different agencies. Our purpose is to provide both an information base on late-successional forests and some interpretation of the findings for interested individuals, including resource managers and decision makers.

FOREST ECOSYSTEMS OF THE SIERRA NEVADA

Primer on Forest Ecosystems

An introduction to ecosystem, disturbance, and succession concepts is useful in understanding both the late-successional assessment and its implications so we begin with a primer on aspects of forest ecosystems.

The ecosystem is a holistic concept which incorporates both organic and physical components—biotic (e.g., organisms) and abiotic (e.g., climatic conditions)—and their relationships. There are numerous perspectives on how ecosystems are structured and how they work (see, e.g., Likens 1992); one useful approach is to recognize that ecosystems have three primary attributes—composition, function, and structure (Franklin in press). It is important to recognize that ecosystems are dynamic rather than static and that much of this dynamism involves responses to disturbances, both natural and human. The ecosystem concept is also very flexible as to spatial dimensions; the scale of an ecosystem is defined by the particular application—i.e., the functions that are of interest.

Composition, Function, and Structure of Ecosystems

Composition refers to the organisms which are present in an ecosystem and their relative proportions. The complement of species that are present is one common measure of biodiversity but there are others, such as measures of equitability among species and of functional diversity. The bulk of the species are, of course, small, inconspicuous organisms, such as insects, fungi, and bacteria, but size is not a measure of importance since many of these species are critical elements in maintaining the productivity of the ecosystem such as by decomposing organic material and releasing nutrients or, in the case of mycorrhizal-forming fungi, assisting vascular plants in acquisition of moisture and nutrients.

Function refers to the work carried out by ecosystems; it is important to understand that all forests are working ecosystems providing a variety of goods and services to human kind not just forests managed for timber production. Productivity, through the capture of the sun's energy by photosynthesis and its conversion to various organic materials, is an important example. Primary production by green plants is, of course, the energetic basis or basic source of "food" for most life forms as well as providing the marvelous structures that we harvest for wood. Other examples of important ecosystem functions are conservation of nutrients and soil, regulation of the hydrologic cycle, and provision of habitat for wildlife and other organisms.

Structure refers to the numbers, sizes, and kinds of "pieces" of the ecosystem and their spatial arrangement. Forest ecosystems may have a wide variety of organic structures such as live trees of various species, sizes, and conditions as well as snags (standing dead trees) and logs which are, of course, derived from live trees. "Logs" are defined here as tree boles or stems or pieces of such boles present on the forest floor primarily through natural processes and not pieces of felled trees created in logging activities. Structurally diverse ecosystems have a wide variety of life forms and structures.

Spatial patterns are as important as the diversity of individual structures in describing and understanding stand structure. What is the spatial arrangement of the individual structures? Are they uniformly distributed throughout the ecosystem or do they have an irregular or clustered distribution?

Examples of important stand-level structural patterns in forests are openings (gaps) in overstory canopies and development of multiple or vertically continuous canopy layers.

Structure is commonly emphasized in ecological and forestry analyses because 1) structure can function as a surrogate or indicator for species or processes (functions) that are difficult to measure directly and 2) structure is what we commonly manipulate through management. Many organisms are difficult to observe directly as are many processes, such as productivity. Structural measurements are, therefore, often used as surrogates or substitute measures of such organisms or processes, since they are relatively easy to make. For example, structures required as habitat for specific wildlife species (such as large-diameter, hard snags for pileated woodpeckers) are often used as indicators of suitable habitat for dependent organisms. The Wildlife Habitat (WHR) guidelines utilize such structural indicators. Similarly, productivity is often measured by observing changes (increases or decreases) in structures (e.g., tree dimensions) in stands rather than through laborious measurements of photosynthesis and other processes.

Structure is the attribute of ecosystems that humans generally manipulate either directly, as in logging, or indirectly, as through fire control or prescribed burning programs. Silviculture is based primarily on structural manipulation—although humans may also directly manipulate composition through removal or addition of specific species.

Succession and Disturbances

Ecosystems are constantly undergoing changes in composition, structure and function as a result of interactions among the organisms and changes in abiotic conditions. Some of these involve long-term directional (as opposed to cyclical) changes which are typically referred to as succession. Succession is most commonly thought of in terms of species changes, such as from an early colonizing or pioneer species to late arriving species; while common such compositional changes are not universal. However, succession in forests always involves structural and functional changes. Ecosystems that have reached a point where changes in composition and structure are very slow or imperceptible are often referred to as late-successional. Late-successional ecosystems may still be very dynamic, such as in turnover of individual trees, but with very little net change in conditions in the stand. In the Sierra Nevada many late-successional forests typically incorporated periodic, light to moderate intensity wildfires which helped maintain high levels of structural diversity within small areas.

Severe disturbances, such as wildfire, windstorm, insect outbreaks, or timber harvest, can disrupt the gradual, internally-driven successional changes in ecosystems; the level of disruption depends upon the type, intensity, and frequency of the disturbance. Most disturbances leave behind large numbers of surviving organisms and organic materials, which are sometimes referred to as biological legacies; most disturbances do not kill all organisms present, let alone sterilize the site. Such legacies can be very important because it means that much of the recovery will be based on organisms and materials already in place rather than requiring recolonization of the site from outside. Since most forest disturbances kill trees but consume little of the wood, legacies of particular significance to forest ecosystems are dead trees in the form of snags and down logs. The numerous legacies, including living trees as well as snags and logs, are a primary reason why clear-cutting is not like most natural disturbances.

Ecosystems at the Landscape Level

Much of ecology and resource management, including ecosystem science and forestry, has focused at the scale of the single stand or patch. For example, activities have often been planned and conducted at the level of a forest stand or a stream or river reach without regard for conditions in the surrounding area.

Problems invariably arise when activities or interpretations lack a larger spatial or landscape context. For example, unacceptable cumulative effects on water quality can result when a management activity, such as road building, is not considered in relation to other management activities that have been carried out or are planned within a watershed. Extensive dispersed patch clear-cutting can result in a landscape condition known as forest fragmentation in which large blocks of continuous forest are broken into small patches with significant changes in their ecological properties (Franklin and Formann 1987).

Boundaries between ecosystems, or edges, are a very important consideration in landscape ecology. Adjacent patches have significant reciprocal influences on each other which are sometimes known as edge effects. Extent of edge effects varies with the parameter of interest; for example, whether the issue of interest is tree mortality, predation on songbird nests, or air temperature (Chen et al. 1992, 1993). The greater the contrast in structural conditions between the two patches, the more intense the interaction and depth of edge effects. Maximum interactions or edge effects occur where conditions are extremely contrasting, such as along an edge between a recent clear-cut and an old-growth forest; for example, on a hot, dry summer afternoon the clear-cut may affect relative humidity and wind speed for 400 m or more into an old-growth stand (Chen et al. 1993).

Landscape-level perspectives are critical in understanding and managing ecosystems to achieve desired objectives. Recognizing the scale and pattern of patches of different ecosystem types and conditions and its relationship to the spatial patterns in the intensity of disturbances can be very important. Landscape-level perspectives also help in recognizing the linkages between terrestrial and aquatic ecosystems.

Major Forest Types and Their Characteristics

The forests of the Sierra Nevada are very complex in composition, structure, and function. This complexity reflects 1) the wide variations in environmental conditions on both a local and a regional scale, 2) a rich flora, and 3) a highly varied history of natural and human disturbances. The importance of environmental diversity in creating a complex template is easily understood: forests occupy a large elevational and latitudinal range and a wide range of geological substrates, landforms, and soil types. Hence, moisture, temperature, and nutrient regimes are extremely varied and often contrast over very short distances, such as on adjacent opposing aspects. Disturbances then interact with the flora on this template to produce truly complex mosaics of forest and other plant communities. Plant community classifications developed by agency and academic ecologists (e.g., Barbour 1988; Sawyer and Keeler-Wolf 1995) quantify much of this richness.

In this report we aggregate forests into a relatively few major forest type groupings in order to simplify the analysis while still recognizing some of the important variability. These groups differ in a variety of important factors including species composition, function (including productivity), structure, environment, and disturbance patterns. The forest type groupings are

- Foothills pine and oak

- Westside mixed conifer

- White fir

- Red fir

- Jeffrey pine

- Subalpine

- Eastside pine

- Eastside mixed conifer and white fir

- Piñon and juniper

- Riparian hardwood

Aspects of the distribution, composition, and disturbance regimes of these type groups are provided in table 21.1. Forest type groups are illustrated in figures 21.1 through 21.23. Westside mixed conifer forests (figures 21.2–21.8) are found at middle elevations on the western slopes of the Sierra Nevada. The subalpine forest type (figures 21.14–21.19) includes a diversity of forest conditions found at high elevations throughout the Sierra Nevada and the White Mountains. These forests vary widely in density and structural complexity but include large areas of low tree density. Lodgepole pine–dominated forests are a major component of the subalpine forest type group and display a broad array of structural conditions; figures 21.16–21.19 are all from the same polygon in Yosemite

National Park. Pure or mixed stands of piñon pine and juniper are characteristic of the eastern margins of the Sierra Nevada, White Mountains, and Modoc Plateau (figures 21.21–21.23). The general distribution of the type groups is illustrated in plate 21.1.

METHODS USED IN ASSESSMENT

Major Issues and Assumptions

The Sierra Nevada provides a very challenging region for assessing late-successional, including old-growth, LS/OG forest conditions. Most fundamental is the necessity to define a set of criteria by which you can recognize late-successional forests. Such assessments, while still difficult, are relatively straightforward in regions like the Pacific Northwest where infrequent disturbances typically produced large patches with dense overstory canopies dominated by cohorts of relatively shade-intolerant trees. Dominant age classes—and, hence, old-growth—is relatively easily identified by both age and structural analyses (e.g., Johnson et al. 1991; Franklin and Spies 1991a, 1991b). Furthermore, timber harvest has been almost exclusively by clear-cutting. Hence, in the Coastal Ranges and the Cascade Range, the high level of structural contrast makes distinctions between natural and managed stands particularly easy.

The forests of the Sierra Nevada are quite varied in their intrinsic structure and in the structural and compositional characteristics induced by natural, as well as human, disturbances. Although many old, uncut forest stands in the Sierra Nevada are dense, closed-canopy old-growth forests like those found in the Pacific Northwest many LS/OG forests are not of this type. Many high-quality LS/OG forests in the Sierra Nevada have low to moderate overstory tree densities, moderate canopy cover, and gaps of sufficient size for successful reproduction of the relatively shade-intolerant pioneers, such as pines and a variety of brush species (figures 21.24–21.25). Wildfires of light to moderate intensity and moderate to high frequency have been important in creating and maintaining this structure. Periodic, localized extensive mortality from bark beetles has also been important.

Selective timber harvest—the dominant approach in the Sierra Nevada—has helped maintain much of this structural complexity. Late in the nineteenth and early in the twentieth century, timber harvest approximating clear-cutting did occur on private lands in the Sierra Nevada (some of which were later incorporated into the national forests) although partial cutting was probably the more common practice. On the majority of federal timberlands clear-cutting has occurred only during the last several decades, however. Even-aged management, including shelterwood, seed tree, and clear-cutting, was initially utilized on stands with moderate and low stocking and did not preclude continued selective harvest in sig-

TABLE 21.1

Characteristics of the major forest type groups of the Sierra Nevada.

Forest Type	Dominant Trees		Landscape Patterns	Primary Disturbances	Presettlement Fire Regime	
	Northern Sierra	Southern Sierra			Northern Sierra	Southern Sierra
Foothill Pine & Oak	Foothill pine, ponderosa pine, blue oak, live oak, Douglas fir	Foothill pine, ponderosa pine, blue oak, live oak	mostly open structure, limited patches of dense forest, frequent natural openings (chaparral & outcrops)	fire, insects, pathogens, drought	low severity regime: frequent, low intensity fires	same
Westside Mixed Conifer	Douglas fir, ponderosa pine, sugar pine, white fir, incense cedar, black oak, tanoak	ponderosa pine, sugar pine, incense cedar, black oak, giant sequoia, Jeffrey pine	primarily continous forest with few extensive natural openings (eg outcrops)	fire, insects, pathogens, drought	low to moderate severity regimes: areas >50" annual ppt likely mixture of low and moderate intensity fires in complex mosaic with sufficient variability in interval to perpetuate Douglas fir; areas <50" annual ppt likely more dominantly low intensity fires. Infrequent large-scale high severity fires.	low severity regime: dominantly low intensity fires
White Fir	white fir	same	same as westside mixed conifer	insects, pathogens, fire, drought	moderate severity regime: frequent but variable extent or frequency, variable intensity with small patches of moderate to high intensity	same?
Red Fir	red fir, lodgepole pine, western white pine	same	fine to moderate scale high patch diversity of natural openings (meadows, outcrops) and open or closed forest; large extensive patches limited	insects, pathogens, fire, drought, wind, avalanche	moderate severity regime same as white fir	moderate severity regime same as white fir
Jeffrey Pine (upper montane)	Jeffrey pine	same	generally extensive uniform patches of very open forest or woodland interspersed with small pockets of denser forest	insects, pathogens, fire, drought	low severity regime: low intensity and/or small extent of fires due to discontinuous fuels	low severity regime
Subalpine	lodgepole pine, mountain hemlock, western white pine, whitebark pine	lodgepole pine, mountain hemlock, western white pine, whitebark pine, limber pine, foxtail pine, western juniper	highly variable patterns but generally diverse patch mosaic with large meadows, small patches of dense forest embedded in a large matrix of open forest or scattered trees and rock outcrops	avalanches, wind	low severity regime: low intensity and/or small extent of fires due to discontinuous fuels and infrequent ignitions (due to precipitation) associated with lightning	

continued

TABLE 21.1 (continued)

Forest Type	Dominant Trees		Landscape Patterns	Primary Disturbances	Presettlement Fire Regime	
	Northern Sierra	Southern Sierra			Northern Sierra	Southern Sierra
Eastside Mixed Conifer & White Fir	white fir, ponderosa pine, Jeffrey pine (some Douglas-fir, sugar pine & incense cedar)	white fir, Jeffrey pine	variable patterns, most often occur in a coarse-scale mosaic with eastside pine related to aspect	fire, insects, pathogens, drought	low to moderate severity regime: frequent low intensity fires, but with variable intervals enabling recruitment of Douglas fir and white fir to large sizes; greater proportion of moderate intensity fires than eastside pine due to greater productivity and fuel accumulations from variable intervals	same
Eastside Pine	ponderosa pine, Jeffrey pine, lodgepole pine	Jeffrey pine, lodgepole pine	large, continous patches of open forest that are often interspersed with large meadows, grasslands/shrublands	fire, insects, pathogens, drought	low severity regime: dominantly frequent, low severity fires	same
Pinyon & Juniper	western juniper	Utah and western juniper, pinyon pine	large continuous savannas and woodlands	fire, grazing, woodcutting	low severity regime: frequent low intensity fires	same
Riparian Hardwood	black cottonwood, aspen	water birch, black cottonwood, aspen	streamside strips	flood, debris flow	low severity regime: infrequent fire	same

FIGURE 21.1

Representative stand of foothill pine and oak forest, found at lower elevations on the western slopes of the Sierra Nevada, near Johnsville, California, on Sequoia National Forest; late-successional structural ranking of 1 (rangewide standard).

FIGURE 21.2

Young plantation of ponderosa pine on the Sierra National Forest; late-successional structural ranking of 0 (rangewide standard).

nificant parts of the forest. Earlier partial cutting retained at least modest levels of forest cover and many structures (e.g., trees, snags, and logs) from previous stands. Forest conditions are, of course, highly modified over much of the mid- and low-elevation range, compared with pre-Euro-American arrival (McKelvey and Johnston 1992). However, the extreme contrasts that exist between managed and natural forests in the Pacific Northwest are currently not the norm in the Sierra Nevada, despite centuries of human activity.

Hence, it was necessary in this assessment to develop a process which recognized and accounted for the role of a much broader range of forest stands and conditions than in previous assessments of LS/OG forests. We began by considering potential bases for assessing late-successional quality. For several reasons, including the uneven-aged nature of many stands, tree ages were rejected as a primary criterion. After extensive discussion with other scientists, both inside and outside SNEP, our decision was to utilize structural features as the basis for

FIGURE 21.3

Young mixed conifer stand on Plumas National Forest; late-successional ranking of 1 (rangewide standard).

assessing the level of late-successional attributes in a forest (LS/OG quality) or, more accurately, the contribution of a stand or landscape segment to late-successional forest function in the Sierra Nevada.

Several specific challenges were present once structural features were adopted as the primary criteria:

- Identifying the appropriate structural features to use as criteria in the analysis;

- Developing a gradient or continuous scale (based on structural features) for rating the contribution of forests to late-successional forest function in the Sierra Nevada; and

- Dealing with the complex, fine scale-mosaic of stands characteristic of many Sierra Nevada landscapes.

High levels of spatial heterogeneity of structure over a broad range of scales is common throughout much of the Sierra Ne-

vada. Structural complexity is typical of many of the forest stands on both the vertical and horizontal dimensions (figures 21.24–21.25). For example, late-successional, mixed conifer stands commonly incorporate a full range of tree species, sizes and conditions, including a component of large-diameter trees—producing the vertical complexity. These are not uniformly distributed throughout the stand, however, so small openings or semi-openings exist where tree reproduction is successfully established. Indeed, many high-quality late-successional mixed conifer and pine forests in the Sierra incorporate all tree stages within a single stand, from seedlings and saplings to large, decadent trees. Hence, it is important to recognize that structural complexity in both the vertical and horizontal dimensions is characteristic of many Sierran LS/OG stands.

FIGURE 21.4

Moderate density old-growth stand dominated by sugar and Jeffrey pine and incense cedar with understory of bear clover. Open stands of this type have high fire resilience and respond well to prescribed burning. Located in Yosemite National Park; late-successional structural ranking of 4 (rangewide standard).

FIGURE 21.5

Dense old-growth mixed conifer stand with high-canopy coverage and abundant large-diameter trees on Eldorado National Forest; late-successional structural ranking of 5 (rangewide standard).

FIGURE 21.6

External view of mixed conifer stand dominated by sugar pine and white fir in Yosemite National Park; late-successional structural ranking of 5 (rangewide standard).

FIGURE 21.7

Dense pine reproduction in small opening created by prescribed burning. Same stand as in figure 21.4.

Many forests with a varied history of management practices and disturbances have highly simplified structures but still make at least some contribution to late-successional forest function and many could potentially, under appropriate management, make additional contributions over time. Highly simplified structural conditions are typical of stands developed following clear-cutting which dramatically limit their current contribution to late-successional forest function in the Sierra Nevada.

A gradient of forest structural conditions was developed and utilized in this assessment to incorporate the variable contribution made by forest stands to late-successional forest functions in the Sierra Nevada. This gradient is partitioned into six classes recognizing a range from ecosystems which have no structural attributes characteristic of LS/OG forests and make no contribution to LS/OG function in the Sierra Nevada

FIGURE 21.8

Stands with giant sequoia have been incorporated into the westside mixed conifer type group in this analysis. Such stands typically have very high structural complexity and late-successional rankings of 4 and 5 (this stand) unless previously logged; Log Creek drainage, Sequoia-Kings Canyon National Park.

FIGURE 21.9

Representative stand of the white fir type group, which is typically found on cooler and moister habitats than the westside mixed conifer type; unknown location, structural ranking of 4 (rangewide standard).

There is a great deal of variability in the levels of structural complexity of Sierran forests, however, due to both natural and human disturbances. Absolute distinctions between "old-growth and non-old-growth", "late-successional and early successional", and managed and natural stands are not possible. There are stands which make extraordinarily high contributions to late-successional forest functions and are clearly the best remaining examples of old-growth forests (e.g., large old trees with medium to high decadence and large snags and logs); these we refer to as "high-quality" LS/OG forests. Some old and undisturbed stands, such as pine forests associated with a bear clover understory, have low tree densities and open canopies; these provide different, but important, habitat conditions than LS/OG forests which have high densities, especially of shade-tolerant species such as white fir.

FIGURE 21.10

Reproduction stand of California red fir developed following natural decline of old-growth stand; Tahoe National Forest, late-successional structural ranking of 1 (rangewide standard).

FIGURE 21.11

Old-growth California red fir stand; Yosemite National Park, late-successional structural ranking of 4 (rangewide standard).

FIGURE 21.12

Low-density stand dominated by Jeffrey pine typical of those associated with granite outcrops and shallow soils; Sequoia–Kings Canyon National Park, successional structural ranking of 2 (rangewide standard).

(structural ranking of 0) to ecosystems which incorporate high levels of structural features characteristic of LS/OG forests and make high contributions to LS/OG function in the Sierra Nevada (structural ranking of 5). This gradient makes it possible to recognize the measurable contributions made by managed as well as recently disturbed forests.

Many Sierran landscapes are intricate mosaics of different forest and nonforest conditions (figures 21.26–21.28). One common pattern in forested landscapes can be described as a "fine-scale, low-contrast mosaic." *Fine-scale* refers to the size of patch (an area distinguishable from adjacent areas by composition and structure) which is often very small (e.g., from a fraction of a hectare to several hectares). *Low contrast* refers to the small structural differences between adjacent forest patches. Coarser-textured landscapes dominated by relatively few large patches of continuous closed canopy old-growth forest do exist in the Sierra Nevada, but are not as common. There are also many

plify the variability which might occur in a hypothetical polygon, a polygon might contain patches of: 1) Extensive, selectively cut, mixed conifer stands on gentle topography with remnant large trees and logs; 2) unlogged riparian forests with high levels of late-successional structures; and 3) interspersed open rock outcrops.

Mapping and Polygon Characterization

Successfully meeting the challenges outlined in the preceding sections, including delineation and characterization of landscape units, required full utilization of all available information. Resource specialists were the most critical, single source of information; these specialists provided the extensive, personal, on-the-ground knowledge of forest conditions. Approximately 100 resource specialists participated, gathering in Sacramento, California in March 1994 to develop the initial

FIGURE 21.13

Mixed stand of Jeffrey and lodgepole pines and juniper on moderately deep soils; Yosemite National Park, late-successional structural rating of 4 (rangewide standard).

FIGURE 21.14

Open subalpine forest of California red fir and western white pine; Stanislaus National Forest, late-successional rating of 2 (rangewide standard).

landscapes which are fine-scale, high-contrast mosaics but these typically represent situations where forest stands form a patchwork with rock outcrops or other nonforested conditions. These mosaics are the result of both highly localized variations in environment (e.g., soil depth and microclimate) and a complex disturbance history.

We did not feel that it would be possible to deal effectively with tens of thousands of small patches when assessing conditions and developing management scenarios over the entire Sierra Nevada. To perform an effective analysis we decided to map and analyze larger land areas (e.g., 500 to 5000 acres) which were uniform with regards to landscape pattern, i.e., type, proportion and spatial distribution of different vegetational patches. These landscape units are hereafter referred to as "polygons"; they incorporated many, often hundreds, of individual vegetational patches. The polygons represent areas which are specifically judged to be uniform in the amount, type, distribution and functional level of LS/OG forests. To exem-

maps and data base; these specialists were subsequently involved in field review and revision of the late-successional maps and database developed by SNEP. Collectively these individuals represented two thousand years of professional, on-the-ground experience in such diverse areas as ecology, wildlife, silviculture, fire management and timber management. They included employees of the USDA Forest Service, USDI National Park Service, USDI Bureau of Land Management, and California Department of Parks and Recreation. To help insure that the focus was on resource conditions and not influenced by current project plans (such as for timber sales), the personnel involved were staff specialists and not line managers.

The resource specialists were provided with all available information about forest conditions in the Sierra Nevada: aerial and satellite photographs at a variety of scales; orthophotos of quadrangles; maps of forest and wildlife habitat conditions; geologic and topographic maps; inventory data; and maps and information provided by stakeholder groups. We believed that by combining these data with the collective knowledge and wisdom of the field experts, we would achieve the most comprehensive and accurate mapping of late-successional forests possible without initiating new, extensive, and expensive field data collection efforts. Since inventory data on LS/OG structural features (such as large-diameter snags) are very limited for Sierra Nevada forests, a complex synthesis of existing data with knowledge of on-the-ground conditions was essential.

Initial Mapping and Characterization

Major steps in the initial exercise consisted of 1) identifying polygons which were internally consistent in ecological fea-tures, including LS/OG function; 2) delineating the polygons on orthophotos or maps; 3) characterizing the ecological conditions of each of the homogenous patch types within each polygon; and 4) determining an overall ranking for each polygon based on its level of late-successional forest attributes.

All steps, criteria, and procedures were pilot tested on the Eldorado National Forest, and suggestions from participating forest staff were incorporated.

1. Polygons which were logical landscape units or groups of patches from the standpoint of function and characteristics were identified and mapped (figure 21.29). The objective was to map polygons that were relatively uniform throughout in terms of the major landscape elements (patch types) and their spatial relationships and which contrasted in one or more mapping criterion from adjacent areas. The size range suggested to the mappers for the polygons was 500 to 2,500 ha (roughly 1250 to 6625 acres). However, smaller polygons were allowed to distinguish unusual and important forest conditions, and much larger polygons were allowed where forest conditions were uniform over very large areas.

2. Conditions within each polygon were characterized utilizing a standard document form (see appendix F in Sierra Nevada Ecosystem Project 1994). Critical aspects of the characterization were identification of the major patch types found within a polygon, their relative importance, and a quantitative characterization of late-successional attributes (such as density and size of large diameter trees and snags) within the forested patch types in each polygon. Hence, the major patch types or ecosystems within each polygon

FIGURE 21.15

Forest landscape representative of many subalpine forest areas in the Sierra Nevada; Yosemite National Park, late-successional structural rating of 2 (rangewide standard).

FIGURE 21.16

Recently burned lodgepole pine stand; late-successional structural rating of 1 (rangewide standard).

FIGURE 21.17

Low-density stand of lodgepole pine associated with granite domes; late-successional structural rating of 1 (rangewide standard).

FIGURE 21.18

Dense, old-growth lodgepole pine stand; late-successional structural rating of 3 (rangewide standard).

FIGURE 21.19

Riparian lodgepole pine and willow community; late-successional structural rating of 2 (including adjacent dense forest, rangewide standard).

were identified and characterized even though their spatial locations within the polygon were not mapped. Also, information was initially developed (and subsequently expanded) on disturbances within the polygon, including disturbance by logging, mining, grazing, recreation, and wildfire. Thus, polygons could be classified based on different ecological or social goals. Mapping was done on Mylar laid over either standard base maps (1/2"=1 mile) or orthophoto quadrangles.

3. Patch types and polygons were ranked with regards to their contribution to late-successional forest function according to a six-point scale:

   0 No contribution

   1 Very low contribution

   2 Low contribution

3 Moderate contribution

4 High contribution

5 Very high contribution level

Quantitative standards to guide resource specialists in their ranking of patch types and polygons were provided (see appendix 21.1). Polygon rankings were ultimately based upon area-weighted averages of the ratings for the patch types within the polygon; hence, although only six classes are used in this presentation, finer scale distinctions are possible. Structural conditions found in the most productive forest types of the Sierra Nevada—Westside Mixed Conifer, White Fir, Red Fir, and Eastside Pine—provided the standards for the six-point scheme for structural complexity. Photographic examples of patch types representative of the six levels of contribution are provided in figures 21.1 to 21.23.

FIGURE 21.21

Young Jeffrey pine stand developed following complete harvest of the overstory trees; Inyo National Forest, late-successional structural rating of 1 (rangewide standard).

FIGURE 21.20

Representative stand of eastside mixed conifer and white fir type group; Plumas National Forest, structural rating of 3 (rangewide standard).

FIGURE 21.22

Unlogged old-growth Jeffrey pine stand; Indiana Summit Research Natural Area, Inyo National Forest, late-successional structural rating of 4 (rangewide standard).

4. Standardization among management areas (national forests, national parks, BLM lands, and state parks) was accomplished using a variety of mechanisms including frequent plenary discussions of issues and approaches among the resource specialists and the SNEP team leaders. Most important were the directions and standards provided to the resource specialists, and a continuing review of the mapping and characterization activity by SNEP mapping team leaders. A final review of each management unit was conducted by at least two SNEP team members to assure comparability in mapping, characterization, and ratings; the last part of this review was a check-out procedure for each resource specialist team performed by SNEP staff members to insure that maps and data sheets were complete and ready for digitizing.

FIGURE 23.23

Dense piñon pine stand on the Inyo National Forest that has a late-successional structural rating of 2 (rangewide standard).

5. Mapping delineations and polygon characterizations were entered into computerized data bases by the SNEP Geographic Information Systems (GIS) laboratory and used to produce maps and overlays of the late-successional polygons at the scale of 1/2"=1 mile as well as Sierra-wide maps at the scale of 1:633,000.

Initial polygon delineations and characterizations were completed for forests on federal and state park lands in the project study area, including the Sequoia, Sierra, Stanislaus, Eldorado, Tahoe, Plumas, Modoc, Lassen, Inyo, and Toiyabe National Forests, Lake Tahoe Basin Management Unit, and Lassen Volcanic, Yosemite, and Sequoia-Kings Canyon National Parks.

Subsequent Steps in Mapping and Characterization

An extensive series of steps to review and revise the Mark I maps and characterizations was undertaken following completion of the initial exercise in March 1994. These included: 1)

FIGURE 21.24

Vertical heterogeneity: westside mixed conifer forest transect based on a stand near Aspen Valley, Yosemite National Park (drawn by Robert VanPelt).

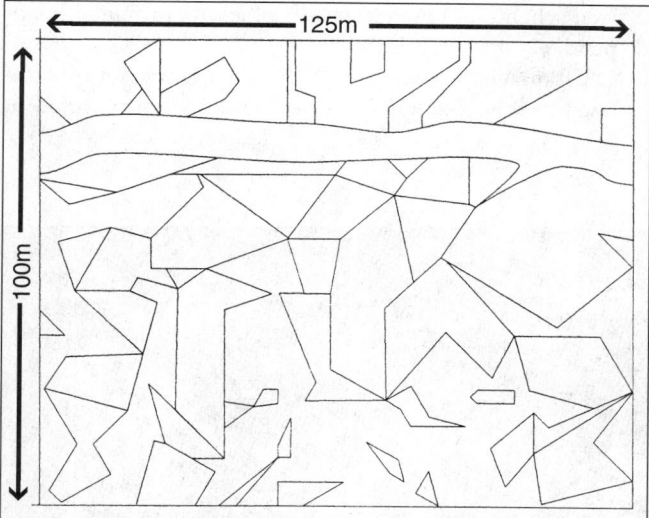

FIGURE 21.25

Horizontal heterogeneity: spatial pattern of reproduction, mature, and old tree groups, Suwanee Creek mixed conifer reference stand, Sequoia–Kings Canyon National Park; note fine-scale mosaic (based on maps from Riegel et al. 1988).

extensive field reviews of the Mark I maps by mappers and SNEP team members and revision of these maps based upon additional data; 2) independent quantitative sampling of patches and polygons to assess the quality of the mappers' characterizations; 3) development of separate structural rating scales for each major forest type group ("series-normalized" scale) to supplement the single ("Sierra-wide") scale utilized in the initial mapping effort and application of this scheme to patches and polygons; 4) external review of the map and database products by individuals knowledgeable regarding forest conditions in portions of the Sierra Nevada; and 5) final revision of the LS/OG maps and polygon characterizations by the authors.

1. Mappers and SNEP team members devoted thousands of hours to field review of the Mark I maps during the 1994 and 1995 field seasons. Mappers concentrated their field efforts in 1994 and concentrated on polygons for which they had lacked adequate information about on-the-ground conditions or about which they had questions of interpretation during the mapping exercise. During 1994 and 1995 SNEP team members conducted field checks on conditions and rankings of polygons with the objectives of visiting all management units (parks, forests, and other lands) that were part of the assessment; field examinations included the relevant mappers as well as other resource personnel for the management unit whenever possible. Priorities for the SNEP review teams were to examine:

   • any polygons about which mappers had questions (which were less than 5% of all polygons)

   • polygons representing a complete range of ratings (from 0 to 5)

   • polygons with a rank of 4 or 5.

   Field examinations by the SNEP team members ultimately included all federal land management units as well as two state parks and covered approximately 20 percent of all of the polygons and 80 percent of all polygons given a structural rank of 4 or 5. Extensive low-elevation flights were also carried out over the entire Sierra Nevada to collect additional information and photographs of conditions in the polygons not readily obtained by ground visits.

   The field review process did result in minor revision of the Mark I product: boundaries or ratings were revised on less than 5% of the polygons. SNEP team reviews judged that mapper polygon ratings followed guidelines and standards in the vast majority of cases. Less than 1% of the polygons were judged to deviate by more than 1 structural class.

2. An independent validation exercise was conducted by Dr. Philip G. Langley under contract from SNEP to quantitatively assess the quality of the late-successional patch ratings (Langley 1996). Unfortunately it was limited in size and geographic scope involving only 400 plots in mixed conifer forest. Following random selection of polygons for sampling, the relevant mappers were asked to map the distribution of patches within these polygons. Plots were than located within patch types to determine whether mappers' delineations showed distinct differences in structural criteria. Results of this validation exercise are provided by Langley (1996) and provide one measure of the quality of portions of the Mark I mapping and characterization effort; the classification accuracy ranged between 44 and 78 percent at the patch level but 82% of the patches were identified within +/- one ranking unit.

3. Late-successional structural-rating schemes—named the "series-normalized" ratings—were developed for each major forest type group during the winter of 1994-95 to supplement the single, rangewide standard used in the initial mapping effort. The original scale was based upon the conditions associated with productive westside mixed conifer, yellow pine, and red fir forests. However, there are clearly major difference in the levels of structural complexity that can be achieved among the different forest types. Subalpine lodgepole pine forests, for example, can not achieve the same level of structural richness as mixed conifer forests because of tree species and site limitations; hence, the most structurally complex lodgepole forest might only be rated as a 3 in a structural scheme based upon mixed-conifer forest.

   Consequently, it was decided to also develop structural rating scales for each forest type group in which the highest level of structural complexity that could be developed would constitute a series-normalized rating of 5. A tabular "crosswalk" was developed which allowed cross comparisons and conversions between the series-normalized and range wide structural rating schemes and all polygons were supplemented with ratings based upon the series-normal-

ized schemes. These crosswalk tables are provided in appendix 21.1.

Ultimately, the series-normalized rankings were not utilized in this LS/OG assessment and are not reported in

FIGURE 21.26

Fine-scale, low-contrast forest mosaics are typical of much of the Westside Mixed Conifer Zone; Plumas National Forest near Quincy, California.

FIGURE 21.27

Highly fragmented landscapes with fine- to medium-scale, high-contrast patch mosaics are common at middle to high elevations where granite domes and outcrops are common; Tahoe National Forest.

FIGURE 21.28

Fragmentation of forest landscapes by clear-cutting, as illustrated here, is not nearly as common as on federal lands in Oregon and Washington; Sierra National Forest.

detail here because of their potential to mislead audiences as to the structural conditions which currently exist in the Sierra Nevada. For example, much of the subalpine forest area in the Sierra Nevada is close to their potential with regards to LS/OG condition and structural complexity and would, therefore, be ranked as 4 or 5 in a series-normalized standard; such forests by the standards of productive forest sites, such as in the mixed-conifer zone, would only have structural rankings of 2 or 3. Furthermore, such forests cannot provide the structures or structurally complex stands necessary to provide for many LS/OG-related species, such as the California spotted owl. Hence, reporting LS/OG conditions based upon a series-normalized standard could cause individuals and institutions using the assessment to grossly overestimate the amount of structurally complex LS/OG forest actually present in the Sierra Nevada.

4. External review of the maps and database products was conducted during the summer of 1995 to provide some independent assessment of the accuracy and overall quality of the products; reviewers were provided access to the maps for the individual national forests and parks (1/2" = 1 mile) as well as to the Sierra-wide maps and complete polygon database with characterizations of the patch conditions. Reviews were particularly sought from individuals that were believed to have detailed knowledge regarding on-the-ground conditions. Included were individuals associated with universities, nongovernmental organizations and forestry groups, and community organizations.

This review process provided additional detailed information regarding late-successional forest conditions. One common concern of external reviewers was the occurrence of small patches of high-quality late-successional forest (4s

FIGURE 21.29

Division of landscape into polygons that vary in patch pattern and contribution to late successional forest function.

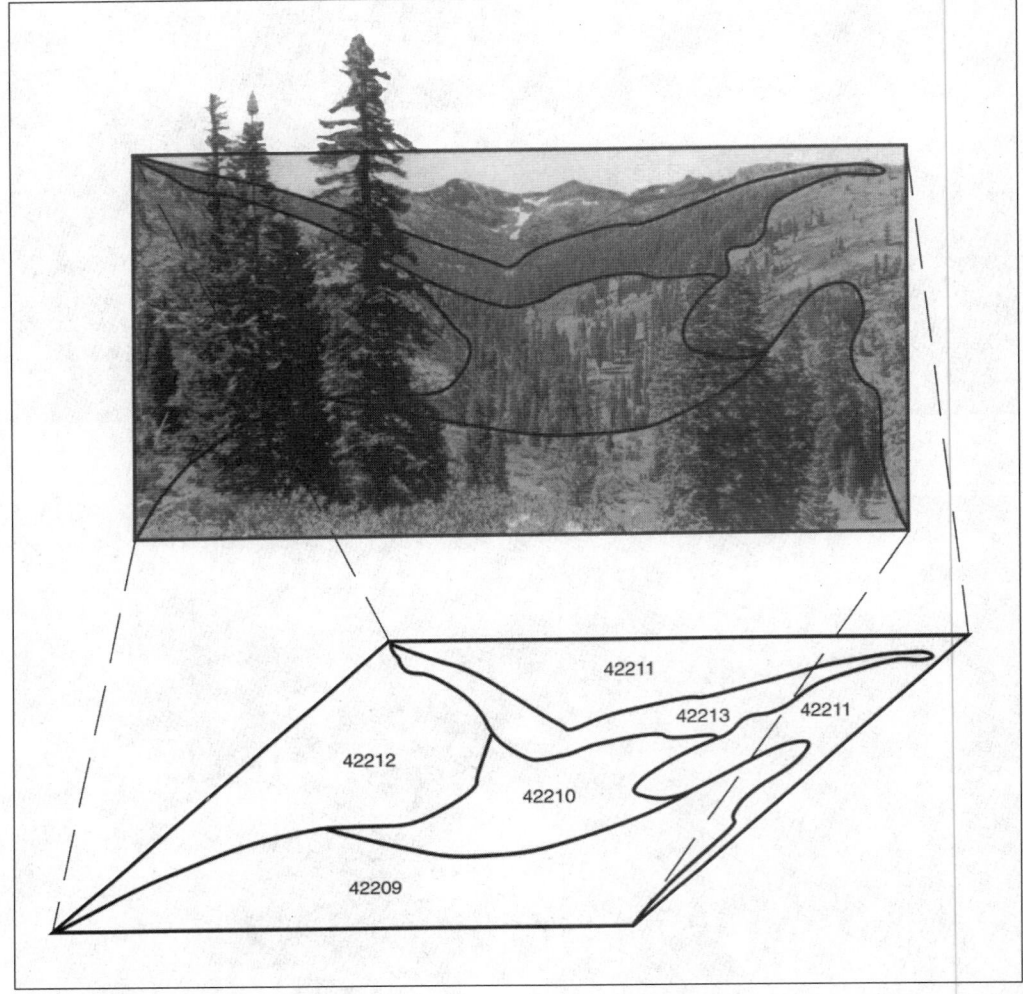

or 5s) within large polygons that had low to medium ratings (1s to 3s) in terms of their contribution to late-successional forest conditions. Such occurrences provided the possibility that important but small areas of high-quality late-successional forest could be "lost" from view. As a result of these comments, additional analyses were conducted to determine how many 3-ranked polygons that had significant patches (e.g., greater than 10 or 25% by area) of higher-rated forest conditions and where they were located. This information already existed in the database and was ultimately utilized in development of alternative management scenarios (Franklin et al. 1996).

5. Final revision of the maps was conducted by SNEP team members following these review processes. Individuals with primary responsibility for this activity were Franklin and Fites-Kaufmann.

PRODUCTS OF THE ASSESSMENT

Primary products of the assessment of late-successional and old-growth (LS/OG) forest conditions in the Sierra Nevada are maps showing the boundaries and ratings for the landscape units (LS/OG) polygons and a database providing information on the patch types and their characteristics for each polygon. Both of these products are available in digitized form.

LS/OG maps have been produced for both individual management units and for the entire Sierra Nevada. Maps for the major public land units (national forests and parks) have been produced at the scale of 1/2"=1 mile and provide more detailed information on polygon boundaries. This information has also been aggregated to produce generalized maps at the scale of the entire Sierra Nevada. Two versions of these Sierra-wide maps are reproduced as plates 21.2–21.6. Both versions are based upon the Sierra-wide structural standard; plate 21.2 provides an overview of conditions throughout the Sierra Nevada while plates 21.3–21.6 provide a more detailed, enlarged

TABLE 21.2

Number and average size of polygons by administrative unit.

Administrative Unit	Acres Mapped	Number of Polygons	Acres per Polygon
Eldorado NF	774,929	179	4,329
Inyo NF	2,108,445	302	6,982
Lassen NF	1,454,149	219	6,640
Lake Tahoe Basin MU	340,306	26	13,089
Modoc NF	1,602,733	140	11,448
Plumas NF	1,474,914	214	6,892
Sequoia NF	1,185,689	209	5,673
Sierra NF	1,409,418	148	9,523
Stanislaus NF	957,588	281	3,408
Tahoe NF	1,127,435	152	7,417
Toiyabe NF	1,041,264	38	27,402
Lassen Volcanic NP	105,400	33	3,194
Sequoia-Kings Canyon NP	863,025	179	4,821
Yosemite NP	751,592	280	2,684
Bureau of Land Management	133,907	412	325
California State Parks	23,857	49	487
All Units	15,354,652	2861	5,367

view. A Sierra-wide map illustrating distribution of polygons utilizing series-normalized ratings is provided in plate 21.7.

Databases include information for each polygon on the percentage of each major patch type and on tree composition, structural conditions, disturbance history, level of fragmentation due to natural (e.g., rock outcrops) and human (e.g., clearcutting) causes, and other attributes recorded by the mappers.

The polygon mapping and characterization exercise resulted in identification of 2,861 polygons with an average size of 5,367 acres over the entire study area. Results appear to be reasonably comparable across the management units in terms of the size of polygon (table 21.2) with several caveats. Polygons are generally smaller on national and state park lands than on national forest lands probably due to more detailed knowledge of vegetative conditions. In contrast, the small size

of Bureau of Land Management polygons reflects the highly fragmented nature of public domain lands. Polygons are generally much larger on the eastern than on the western side of the Sierra Nevada; this was expected since forest stand conditions tend to be much more uniform over large areas east of the Sierra Nevada due to both topographic conditions and management history; good examples are the Modoc National Forest and northeastern halves of the Lassen and Plumas National Forests.

LATE-SUCCESSIONAL, INCLUDING OLD-GROWTH, CONDITIONS

This analysis of LS/OG conditions utilizes the database created in the LS/OG mapping and characterization exercise. Most of the analysis is at the level of the polygons and only federal lands and state parks are considered. The reader is reminded that the scale of structural classes runs from 0 (no contribution to late-successional forest function) to 5 (high level of forest structural complexity and late-successional forest function). The following analysis utilizes only the range-wide structural ranking scheme in order to minimize the potential for misunderstandings with regards to the current extent of structurally complex forest stands in the Sierra Nevada; concerns with regards to the series-normalized structural scale were noted in the earlier section on mapping and polygon characterization. Basic data for the LS/OG assessment are tabulated by structural class and forest type group in table 21.3.

Structural ratings provide the basis for judgments regarding the ability of a polygon to provide for types and levels of late-successional forest functions. One example of such a forest function would be a polygon's ability to provide habitat for species which prefer or require LS/OG forest conditions,

TABLE 21.3

Percentage of polygons by major forest type group and late-successional forest ranking (rangewide structural standard) for all national forest, national park, state park, and Bureau of Land Management lands.

Forest Type Group	Total Acres Ranked	Percent by Rank					
		0	1	2	3	4	5
Foothills Pine & Oak	238,720	14	24	54	8	0	0
Westside Mixed Conifer	3,344,960	4	12	33	31	15	5
White Fir	217,583	3	16	34	33	7	7
Red Fir	1,476,390	0	9	28	34	17	13
Jeffrey Pine	339,759	1	7	28	55	9	0
Subalpine	2,025,003	5	27	32	32	4	<0.5
Eastside Pine	2,776,024	9	24	45	14	5	2
Eastside Mixed Conifer & White Fir	711,982	4	22	39	26	9	0
Pinyon & Juniper	1,461,157	19	75	5	1	0	0
Riparian Hardwood	314,197	7	47	33	7	6	0
All Forest Types	12,905,775	5	20	32	28	10	4

such as the California spotted owl, pine marten, or fisher. Regulation of hydrogic regimes is another ecosystem function which varies with both forest structure and composition. For example, stands with high LS/OG structural ratings are likely to contribute less runoff during rain-on-snow flood events than forests on comparable sites but with lower structural ratings, resulting in lower peak flood flows. This is because of several processes, including patterns of snow interception and melt and protection of snowpack within forest stands in old-growth forests (Harr 1986, Harr, Coffin, and Cundy 1989).

The important findings with regards to LS/OG forest conditions in the Sierra Nevada are as follows:

There is relatively little high-quality late-successional forest remaining in the Sierra Nevada, particularly in the commercial forest zones. Forests which have high levels of the structural complexity expected of fully functional LS/OG forests are not common. Only 14% of the polygons have a structural ranking of 4 or 5 in the current Sierran landscape (table 21.3).

Many more polygons—28% of the total—have a structural ranking of 3. The large percentage of polygons with this ranking reflects several factors. In the case of the major commercial types, polygons rated as structural class 3 represent several different circumstances: 1) forests that have been selectively logged at light to moderate levels; 2) productive forest sites on which structurally complex, mature forests have regrown following earlier logging; or 3) naturally fragmented landscapes in which structurally complex stands are interspersed with nonforested areas. About half of the 3-ranked polygons have a significant percentage of their acreage (>25%) in patches with a structural rank of 4 or 5 (table 21.4); these are mostly polygons in productive Westside Mixed Conifer, White Fir, or Red Fir forests.

Several of the major forest type groups do not have the potential to produce stands structurally comparable to those found on the most productive forest sites and therefore make little contribution to highest ranked polygons although they do contribute to the 3-ranked polygons (table 21.3). These include the Foothills Pine & Oak, Subalpine, and Piñon & Juniper type groups. Stands belonging to these forest types may contain trees which are large or old or both and both the stands and trees are important natural features of the Sierra Nevada

even if they do not display the structural complexity characteristic of high-quality late-successional forest stands on more productive sites. The low ratings for the Riparian Hardwood type group in table 21.3 are not representative of structural conditions found in this type; most riparian stands were actually included within polygons assigned to other types.

Commercially important forest types—such as the westside mixed-conifer and eastside pine forests—are most deficient in high-quality late-successional forest relative to their potential (table 21.3) and to presettlement conditions. Among the commercial forest types structurally complex forest stands are rarest in eastside ponderosa and Jeffrey pine forests; 78% of the Eastside Pine polygons have ratings of 0, 1, or 2 and less than 7% are rated at structural class 4 or 5. About 14% of Westside Mixed Conifer polygons are ranked as 4 or 5. Both type groups have been foci for commercial forest harvest activities but differ in that the Westside Mixed Conifer is well represented in national parks while Eastside Pine is not; the significance of this is discussed below.

There are several important factors contributing to the current condition of the Eastside Pine forests. Many of these stands were logged very heavily between 1860 and 1900 to support mining and railroad activities on the eastern slope of the Sierra Nevada—Comstock, Silver Mountain City, Bode, etc. Since commercial forests were much less extensive than on the western slopes of the Sierra Nevada, these activities essentially stripped many areas, such as the Lake Tahoe basin. Furthermore, most of the Eastside Pine stands are on gentle topography that is readily accessible to logging. Consequently, repeated selective harvest has been widespread throughout eastside pine forests for much of the last century.

The Westside Mixed Conifer type has also been subjected to a long history of timber harvest utilizing both selection and clearcut methods. Although 14% of the polygons assigned to this type had structural rankings of 4 or 5, two-thirds of these are found in national parks.

Among the forest types traditionally subject to significant timber harvest, polygons assigned to the Red Fir type had the highest proportions of structurally complex forest. Thirty percent of the Red Fir polygons had a structural ranking of 4 or 5 with another 34% in 3-ranked polygons (table 21.3). This probably reflects the less intensive history of logging in Red Fir than in lower-elevation forest types.

Key structural features of late-successional forests—such as large diameter trees, snags, and logs—are generally at low levels in the forests of the Sierra Nevada. The low structural ratings for many polygons reflects the widespread absence of key structural features of LS/OG forests, many of which are critical in providing for late-successional functions such as provision of habitat for many elements of biological diversity (Graber 1996). A logical inference from both the rankings and the tabulated characterizations of the patches developed in the mapping exercise is that large-diameter decadent trees and their derivatives—large snags and logs—are generally absent or at greatly reduced levels in accessible, unreserved forest areas throughout the Si-

TABLE 21.4

Proportion of 3-ranked polygons with varying percentages of included 4- and 5-ranked patches for several forest type groups (rangewide structural standard).

Forest Type Group	Percentage of polygon in 4- or 5-ranked patches			
	<5	5–25	25–50	>50
Westside Mixed-Conifer	59	9	18	14
Red Fir	48	20	13	19
Eastside Pine	60	23	12	5

erra Nevada. This reflects the selective removal of the large trees in past timber harvest programs as well as the removal of snags and logs to reduce forest fuels due to wildfire concerns. Snag removal programs have been underway on both public and private lands for over 60 years and log reduction programs have been underway for about half that period.

The inferences drawn from the SNEP LS/OG database are consistent with those of McKelvey and Johnston (1992) who compared current stand structure data from Westside Mixed Conifer forests with data collected from sample plots on the forest reserves at the beginning of the 20th century.

On the positive side, *the forest cover of the Sierra Nevada is relatively intact and most forest stands have sufficient structural complexity to provide for at least low levels of late-successional forest function.* Despite nearly 150 years of significant activity by western man, there is still a high level of continuity in the forest landscapes (figures 21.26–21.28). Of course, there is substantial natural fragmentation of forests at higher elevations (figure 21.13). However, high-contrast human fragmentation due to settlement clearing, mining, logging, and other activities is relatively low, particularly in comparison with the highly fragmented conditions on federal forestlands in Washington, Oregon, and northwestern California. Differences between the Sierran and Cascadian regions are clearly due to relatively recent introduction of modern clearcutting techniques on public lands in the Sierra Nevada. Partial cutting has been the most common harvest system on forestlands in the Sierra Nevada up until the 1970s and some ranger districts initiated extensive clearcutting operations as recently as the early 1980s. In contrast, dispersed patch clearcutting has been used almost exclusively on federal timberlands in the Pacific Northwest since the 1950s.

Other important factors contributing to the continuity of forest cover in the Sierra Nevada are the significant recuperative ability of forests, especially on the productive mixed-conifer sites, and active and successful reforestation programs which converted many nonstocked and understocked lands to fully stocked forest stands.

While forest continuity is high in the Sierra Nevada, as noted above, the forest structure has been greatly simplified relative to presettlement conditions so that these forests do not provide the same level of wildlife habitat and other ecological functions characteristic of high-quality LS/OG forests. Nevertheless, the majority of forests do have sufficient structural complexity to provide for at least low to moderate levels of LS/OG function. For example, about 75% of the forested polygons have a structural ranking of at least 2 (table 21.3). This rating indicates stand structural complexity comparable to 1) a maturing forest stand in the Westside Mixed Conifer zone which developed following clearcutting or 2) a stand which has been subjected to intensive selective harvest of larger-diameter trees. Forests of these types—with trees of at least moderate diameter and moderate to high canopy coverage or forests with scattered large-diameter trees—are widspread in the Sierra Nevada and provide at least some habi-

TABLE 21.5

Differences between adjacent national forests and national parks in proportions of polygons with high rankings for late-successional forest function (rangewide structural standard).

| | Percentage of Polygons | |
Administrative Unit	3+4+5	4+5
Lassen National Forest	31	5
Lassen Volcanic National Park	64	35
Stanislaus and Sierra National Forests	33	9
Yosemite National Park	54	22
Sequoia National Forest	24	9
Sequoia–Kings Canyon National Park	38	18
All national forests	32	9
All national parks	50	21
All federal lands	34	11

tat and other ecosystem functions characteristic of late-successional forests.

High-quality late-successional forest areas (structural classes 4 and 5) do exist throughout the federal landbase. However, there are significant differences in the amounts of such forest among the federal management units (table 21.5).

National parks provide the major concentrations of high-quality late-successional forests, especially at the landscape level, and, on a percentage basis, have about twice as much highly-rated forest as adjacent national forests (table 21.5). The major concentrations of high-quality LS/OG forests associated with Yosemite, Sequoia-Kings Canyon, and Lassen Volcanic National Parks is apparent in plates 21.2, 21.3, and 21.4. When these properties are compared with adjacent national forest lands the percentage of structural class 4 and 5 is over twice as great on the national park lands (table 21.5); similarly, the percentages of polygons ranked as 3 or better is nearly twice as high on national park lands. Timber harvest activities on the national forest lands are almost certainly a major factor in these differences in structural complexity between adjacent national forests and parks.

The national park forests provide an important reference point for presettlement levels of high-quality late-successional forest in the Sierra Nevada. The percentage of polygons ranked as structural classes 3, 4, and 5 found in national parks probably represents conditions that were general throughout the Sierra Nevada in comparable forest types prior to initiation of timber harvest and other modern human activities. There are some factors that confound such an interpretation. For example, it can be argued that fire control programs on national parks have moved forests toward a more structurally complex state (therefore reflecting an "unnatural" condition); i.e., an assertion that wildfire control has inflated the percentage of structurally complex forests. There are counter arguments, however. For example, fire control programs have had as much effect on national forest lands. More important is the fact that Sierran stands subject to frequent to moderate fire regimes typically display high levels of structural complexity. Hence, unless there was substantially more high-intensity stand-replacement fire in the presettlement landscape then is currently believed, fire control should not have significantly altered the collective percentage of polygons ranked as 3s, 4s, and 5s. Fire control could have altered stand densities and shifted the average rating toward a higher value, however.

The distribution of national forest polygons with high-quality late-successional forest is not uniform; many high-ranked areas in the northern and central Sierra Nevada are associated with major river canyons. Many of the remaining high-quality LS/OG areas on the Stanislaus, Eldorado, Tahoe, Plumas and Lassen National Forests occur within the canyons of major river drainages along the western edges of the national forests. Most remaining high-quality LS/OG forest was expected to occur in more remote locations within the center of the range. However, many steep and relatively inaccessible canyons and canyon walls have escaped significant logging and contain good to excellent examples of structurally complex forest habitats. Polygons of this type were mapped in the American, Feather, Yuba, Cosumnes, Rubicon, Mokelumne, and Stanislaus River and the Mill and Deer Creek drainages. Since such areas are often at the interface with rural and urban environments, wildfire is a major concern if the high-quality LS/OG forest condition is to be maintained.

Much of the highly-ranked late-successional forest on national forest lands is unreserved and potentially available for harvest. About half of the remaining structurally-complex forest on national forest lands is unreserved, i.e., within the landbase potentially available for harvest. For example, 46% of the Westside Mixed Conifer polygons ranked as structural classes 4 and 5 are in the "suitable" land class under current national forest plans and therefore available for harvest. The comparable figure for polygons assigned to the Red Fir type is 30% of the 4- and 5-ranked polygons in the suitable landbase. Conversely, there is very little high-quality Westside Mixed Conifer forest found within congressionally reserved areas, such as Wilderness, except for the national parks. The percentage of high-quality LS/OG forests available for timber harvest would be less under the preferred alternative in the California spotted owl EIS (USDA Forest Service Pacific Southwest Region 1995).

We conclude this section on results with some general observations on forest health in the Sierra Nevada. These are based upon several sources of information including the LS/OG database, current research by the second author, and observations and photographs made during evaluations of the LS/OG maps, including extensive low-elevation flights over the Sierra Nevada.

Forest health in the Sierra Nevada Range is generally good; problem areas are localized. Most forest stands in the Sierra Nevada appear to be healthy; i.e., levels of mortality due to insects and disease are at levels that are normal or near normal for natural stands. Catastrophic mortality of trees in forest stands is found in particular localities many of which are close to the margins of the forested zones, i.e., near the lower elevation transitions between forests and savannas or nonforested vegetation. This

is predictable since greater physiological stresses occur at such locations during periods of drought such as was recently experienced. The ecotonal areas are also the sites where some of the greatest shifts in stand density and composition have occurred as a result of fire suppression.

Forests along the eastern face and forest margins of the Sierra Nevada are the most common locale of stands which are undergoing (or have already undergone) catastrophic mortality. Examples of such stands are found on portions of the Inyo, Toiyabe, Plumas, and Lassen National Forests and in the Lake Tahoe Basin Management Unit. In many of these stands the bulk of the mortality has been in the white fir component (figure 21.30) although other species have also undergone significant mortality.

Many forests and woodlands along the western boundary of the southern Sierra Nevada have also undergone high levels of mortality, particularly of pine trees. Air pollution is an important factor in this part of the Sierra Nevada (Cahill et al. 1996) in addition to stresses associated with drought cycles. Conditions of stand collapse are not widespread at this time, however.

DISCUSSION OF FINDINGS AND IMPLICATIONS FOR MANAGEMENT

Limitations in Local Application of Database

There are limitations in how the information from this assessment can be applied. Users need to recognize that the objective of the assessment is to provide information for use in development and evaluation of policy scenarios at the scale of entire Sierra Nevada, not as a basis for site specific projects. Consequently, the databases need to be utilized with caution in interpreting localized conditions. *Databases and maps should not be utilized for local management purposes without additional ground-based measurements.* Detailed on-the-ground examinations are important in assessing the appropriateness of polygon boundaries, patch characterizations, and overall rating of forest structure and function from the standpoint of late-successional species and processes. The validation exercise (Langley 1996) confirms the appropriateness of the assessment at the larger scale and identifies problems that may be encountered in trying to apply the assessment within local areas without further checking.

Individuals using the LS/OG database should note that the analysis actually utilized a continuous scale of structural complexity. This perspective is sometimes lost since polygons were assigned to one of six discreet grades (0 to 6). In fact, polygon ratings were based on weighted averages which provided gradations (such as rankings of 3.1, 4.2, etc.) but the rankings were rounded to the nearest whole number on the maps and in most analyses.

Reviewers identified several problems with the continuous structural scale and the use of a weighted average to calculate overall ratings for large polygons. One of the greatest concerns was the potential for small patches of structurally complex, high-ranked LS/OG forest to get "lost" in polygons dominated by forests with low structural rankings. Polygons with overall late-successional rankings of 3 were particular problems because these polygons were numerous and many contained patches of forest with structural ranks of 4 or 5. A subsequent analysis of the 3-ranked polygons confirmed that this is a significant issue (table 21.4). For example, about 1/3 of the 3-ranked Mixed Conifer polygons had more than 25% of their area occupied by patches ranked 4 or 5. The LS/OG database can, of course, be queried to identify polygons which contain higher-ranked forest patches should subsequent users wish to do so.

Presettlement Extent of Late-Successional Forest Ecosytems

The original extent of high-quality LS/OG forests in the Sierra Nevada and its relation to current forest conditions is an issue of interest. It is our conclusion that the current extent of high-quality LS/OG forest ecosystems in the Sierra Nevada is far below levels that existed prior to western settlement. This comment is intended simply to put the current situation in a historical context, not to propose that these levels should be recreated or are necessary to maintenance of late-successional forest function in the Sierra Nevada.

Several lines-of-evidence support the conclusion that LS/OG forests were once much more extensive. Descriptions of forests in early surveys of forest reserves, such as those by Leiberg (1902), Sudworth (1900), Fitch (1900a, 1900b), and Marshall (1900), indicate that structurally-complex forests dominated by large-diameter trees were very widespread except where stands had been affected by logging or catastrophic fire. McKelvey and Johnston (1992) provide an excellent review of this information, including an evaluation of human

FIGURE 21.30

Major forest health problems are currently located along the margins of the forested zones, particularly at the eastern ecotone with the sagebrush and grasslands. Illustrated here is extensive mortality of white fir in a dense stand on the east slope of the Sierra Nevada, Toiyabe National Forest.

impacts between approximately the mid-19th century and the present.

The widespread condition of structurally-complex LS/OG forest ecosystems can also be inferred from the fire regime currently believed to have been characteristic of the presettlement Sierra Nevada landscape. A regimen of frequent, light- to moderate-intensity fires would result in the dominance of structurally and compositionally heterogenous forests incorporating the major structural features characteristic of high-quality LS/OG forests: large-diameter pine trees, snags, and logs and areas with low overstory density (gaps) dominated by tree reproduction and shrub communities (figures 21.24, 21.25). Structural simplification would generally occur only following more extensive, high-intensity fires, a circumstance currently believed to have been uncommon.

Current conditions in the national parks as identified in this assessment provide a third basis for drawing inferences about presettlement conditions (table 21.5). The estimate that 50% of the national park landscape is in moderate- to high-quality (structural ranks 3 through 5) LS/OG forest includes all polygons and not just those within the commercial forest types. For polygons within the national parks identified only with the five mid-elevation forest types, the percentages of various structural ranks are:

	Rank	
	3 + 4 + 5	4 + 5
All national forests	42	13
All national parks	82	55

Hence, current forests on productive sites in the national parks are overwhelmingly dominated by structurally complex conditions. Even assuming that densities and compositions have increased in these forests as a result of fire control programs, it is still reasonable to infer that most of these forest types were in stands of moderate to high structural complexity in presettlement times based upon the presumed fire regimes.

The collective inference from all lines of evidence is that stands with moderate to high levels of LS/OG-related structural complexity occupied the majority of the commercial forestlands in the Sierra Nevada in presettlement times.

Maintaining High-Quality Late-Successional Forest Ecosystems

The discussion in this section assumes that the maintenance of structurally-complex, LS/OG forest ecosystems is an objective of public policy in the Sierra Nevada and, further, that the intention is to maintain sufficient amounts of and linkages between LS/OG forests so as to provide a high probability of the long-term persistence of viable LS/OG ecosystems and associated organisms. Such a policy has *not* been adopted but an analysis of issues related to implementation of such a policy was a part of the SNEP assignment. We further assume that any LS/OG strategy will be integrated with other objectives

including maintenance of riparian and aquatic ecosystems and activities to reduce risks of catastrophic events to acceptable levels. The discussion is focused upon the major commercial forest types of the Sierra Nevada (Mixed Conifer, Eastside Pine, White Fir, and Red Fir); most Subalpine forests are already reserved and the Foothill Pine & Oak forests of the western slopes and Piñon-Juniper woodlands of the eastside generally do not provide structurally complex forests of the type found in the densely forested zones.

The Working Group on Late-successional Conservation Strategies (Franklin et al. 1996) has identified and discussed issues associated with the development and evaluation of conservation strategies for late-successional forests in the Sierra Nevada. We rely heavily upon their conclusions as a basis for this discussion and refer the reader to their paper for more complete information. Some of the key elements of an LS/OG conservation strategy which they identify include: 1) retaining existing high-quality LS/OG forests; 2) providing for large, contiguous blocks of LS/OG forests; 3) spatially explicit planning; 4) designating reserves where maintenance of high-quality LS/OG forests is the primary objective; 5) restoring fire as an important component of management; and 6) restoring conditions in the matrix. Available information on LS/OG forest ecosystems, processes, and organisms is an important limitation in devising conservation strategies resulting in more conservative approaches than might be necessary with a larger information base.

If maintenance of high-quality LS/OG forest ecosystems is adopted as a policy objective, the goals of that program need to be defined and management programs initiated which will: 1) maintain existing high-quality LS/OG forests; 2) restore such conditions where the existing LS/OG forests are insufficient to achieve objectives; 3) restore fire as an important process in maintaining and protecting LS/OG forest ecosystems; and 4) restore structural complexity in the matrix.

If maintenance of high-quality LS/OG forests is adopted as policy on federal forestlands in the Sierra Nevada further timber harvest within existing high-quality LS/OG forest areas should be halted for at least an interim period of planning and assessment. The desirability of maintaining existing high-quality LS/OG forests in the Sierra Nevada is based upon their limited extent and a high level of uncertainty regarding our ability to fully recreate comparable stands silviculturally (Franklin et al. 1996).

The appropriate areal extent of high-quality LS/OG forests needed to achieve specific purposes is not clear from existing information. However, the current level of high-quality LS/OG forests is far below levels that existed in the presettlement landscape and as well as the natural range-of-variability. Hence, restoration of LS/OG conditions in structurally simplified stands is likely to be an important part of achieving desired amounts of LS/OG forests in some localities, particularly where levels are currently very low, such as in much of the Eastside Pine type.

Regardless of the acreage objective and of the management strategy ultimately adopted for LS/OG forest ecosystems, the interim reservation of existing high-quality LS/OG forests from further timber harvest would maintain the largest set of options out of a relatively small existing set.

Active management to restore low- to moderate-intensity fire to existing and prospective LS/OG forest ecosystems is the most important management action needed to restore more natural conditions and reduce risks of loss to catastrophic disturbances, i.e., intense stand-replacement wildfires. Such programs are an essential element in a reserve-based conservation strategy and must be carried out at sufficient scale and frequency to be effective. Passive or lassize faire approaches to management may result in unacceptable losses of such forests. Current prescribed and managed fire programs in the national parks provide a model for active management of LS/OG forest ecosystems although the scale of the national park programs may not be adequate to achieve objectives. It is probably not possible (or, perhaps, desirable) to completely eliminate the potential for stand replacement fire; rather the overall goal should presumably be to reduce the probability of such fires to levels that would allow some desired level of high-quality LS/OG forest to be maintained in the Sierra Nevada over long time periods.

Active management to reduce risks of catastrophic fire are particularly critical at the interfaces between LS/OG forests and suburban, rural, and recreational developments. The LS/OG mapping identified a number of polygons which are outstanding examples of high-quality LS/OG forests at interfaces with urban developments along the western boundaries of the national forests. Some eastside forests, such as those in the Lake Tahoe Basin, also exhibit this juxtaposition of forest and human development.

Planning for maintenance of LS/OG forest ecosystems should be at larger spatial scales—i.e., scales of hundreds to thousands of acres. One reason is to make fire management programs practical. Activities such as the development of fuel breaks cannot be designed and implemented at the level of individual small patches. Planning at larger spatial scales is also necessary to insure availability of the large contiguous blocks of high-quality LS/OG forests which may be important to some LS/OG organisms and processes (Franklin et al. 1996).

Active management programs for maintenance of high-quality LS/OG forests need to recognize the near-natural processes, structures, and populations which are a primary value of such forests. Hence, treatment of identified high-quality LS/OG areas should emphasize prescribed fire and minimize mechanical disturbances. Intensive management activities, such as creation of shaded fuel breaks, removal of small- to moderate-size trees, and other fuel reduction activities should generally be located in areas adjacent to the high-quality late-successional forests rather than within them.

Larger management units, known as Areas of Late-Successional Emphasis (ALSEs), are proposed as one zoned,

landscape-level approach to maintaining concentrations of high-quality late-successional forest function. Using the ALSE approach, landscape-level (multi-polygon) areas have been identified for the western slopes of the Sierra Nevada using existing high-ranked polygons (4s and 5s) as cores (figure 21.31). Management plans for the ALSEs recognize two primary zones: 1) LS/OG reserves covering 60 to 80% of the ALSE within which prescribed burning and other less intrusive management practices are utilized and 2) intensively managed areas where activities such as shaded fuel breaks and "biomassing" can be carried out. Objectives in the intensively managed areas would be to: a) reduce the potential for catastrophic fire within the core LS/OG stands, b) facilitate movement of organisms between the core stands, and, c) produce forest products consistent with the first two provisions. Completely eliminating fire from the ALSEs is not a management objective but reducing the potential for intense, stand-replacement wildfires is a management objective.

ALSE strategies are discussed further in chapters on the SNEP policy analysis. A representative, well-distributed system of such areas for the western Sierra Nevada is illustrated in plate 21.5. Except for the Lake Tahoe Basin and Sequoia National Forest) the eastern slopes of the Sierra Nevada are not included because existing areas of high-quality LS/OG forests are insufficient to provide the core for a system of ALSEs.

Restoration of LS/OG Conditions in the Matrix

Late-successional management strategies for the Sierra Nevada must also address restoration of structural complexity in the managed forests or matrix. Forests on both sides of the Sierra Nevada have undergone significant structural simplification as a result of timber harvest. This is particularly notable in dramatically reduced numbers (or complete absence) of large-diameter trees and their derivatives (large snags and logs). High levels of structural complexity are needed in the matrix to provide for more of the functions of natural forests as outlined by Franklin et al. 1996.

The importance of matrix-based strategies for conservation of biological diversity are receiving increasing attention because of their importance in sustaining diversity, including species and processes essential to the long-term productivity of the matrix forest itself, and in improving overall landscape connectivity for organisms (Franklin 1993, 1996, Franklin et al. in press). Structural diversity within the matrix can provide refugia which will sustain species immediately following harvest and allow displaced species to repopulate or inoculate the area following stand recovery. Some of the processes and species—such as the array of fungi which can form mycorrhizae with trees and other plans—are of significant direct importance in maintaining the long-term productivity of the site.

Silvicultural harvest systems which provide for retention and long-term maintenance of structures from the existing stand—including large-diameter trees and their derivatives—would produce a struc-

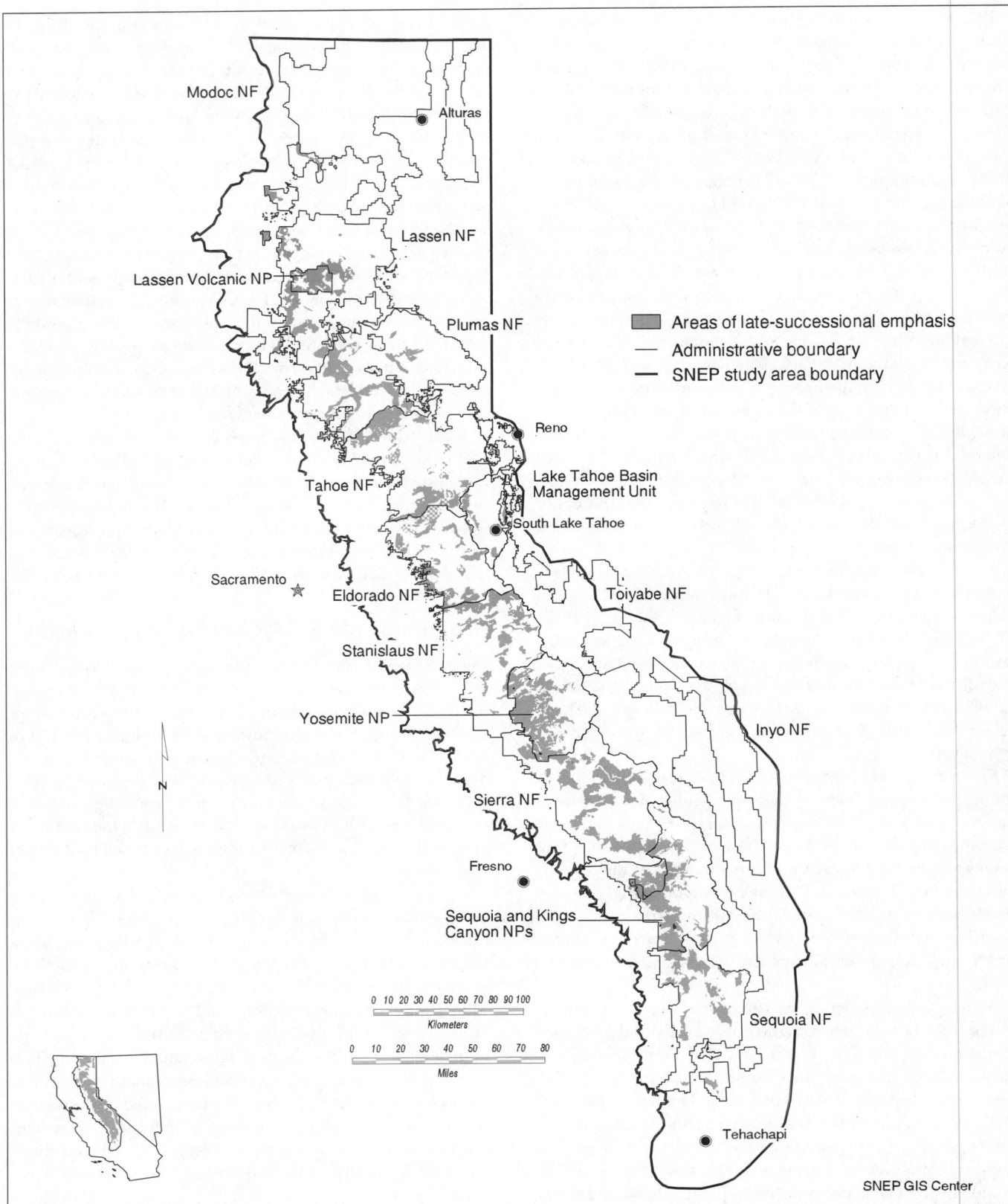

FIGURE 21.31

Distribution of proposed Areas of Late-Successional Forest Emphasis (ALSEs) in the Sierra Nevada.

turally complex managed forest matrix. Various forms of partial cutting can be designed to achieve the objective of maintaining structures at the time of harvest (Franklin et al. in press). Structural goals, such as the numbers, minimum size, and distribution of large-diameter trees, should vary according to management objectives for the stand. The importance of large-diameter snags and logs, as well as large-diameter trees, must be recognized as a part of the silvicultural design. Large-diameter trees and their derivatives fulfill many important ecosystem functions in addition to provision of wildlife habitat. For example, large trees are most likely to survive major fires to provide a legacy of live green trees in the postburn stand; i.e., the large trees substantially improve fire resiliency in the forest stands of which they are a part.

Silvicultural prescriptions for Sierra Nevada forests should also incorporate tree species as well as structural objectives. Where appropriate to site conditions, maintenance of a pine component is an important ecological objective as they provide distinctive tree, snag, and log structures. Maintaining and restoring sugar pine to mixed-conifer stands is of particular concern since this species once dominated the mixed-conifer forests in many areas (see, e.g., Sudworth 1900) and provides a unique structural resource. Sugar pine has been preferentially harvested for nearly 150 years and is currently subject to mortality from the introduced disease, white pine blister rust. Fortunately there is increasing evidence that pines can successfully reproduce under conditions of partial shade (e.g., Oliver and Dolph 1992).

Two silvicultural prescriptions have been proposed for the Sierra Nevada which will maintain or restore higher levels of late-successional forest structures. Group selection is one of these approaches. The scale of selected group that is often proposed—1 to 2 acres—is larger than the scale of mosaic of structural patches found in many natural mixed-conifer and yellow pine stands, however. Moreover, some structural retention within the groups selected for harvest may be desirable to maintain certain features (such as very large decadent trees and snags, for example) which could not be created in adequate numbers within the selected rotation period. Another approach would be to permanently reserve some groups or a portion of the matrix from harvest in order to maintain those structural features (Franklin et al. in press).

Silvicultural prescriptions which maintain or restore specific levels of structures—such as large diameter trees, snags, and logs—have not yet been extensively developed and applied. The interim CASPO guidelines (Verner et al. 1992) are a significant step toward demonstrating the practicality of prescriptions which maintain a high level of late-successional forest function while providing for significant timber harvest. Simple diameter-limit guidelines are not adequate to achieve long-term objectives, however; goals identifying the desired density, size, species composition, and distribution of large trees are needed along with multiple-entry prescriptions which systematically provide for replacements and insure that the large snags and logs derived from these trees are retained on site.

ACKNOWLEDGMENTS

Key individuals in completion of this assessment were Douglas R. Liesz, who participated in all phases of the activity and the personnel of the SNEP Geographic Information Systems Laboratory, especially John and Karen Gabriel. Several members of the SNEP team were significant participants in the mapping exercise and subsequent field evaluations and analyses: David Graber, David Parsons, Don Erman, Erin Fleming, Kay Gibbs, Connie Millar, Chris Riper, K. Norman Johnson, John Sessions, Michael Barbour, Phil Witherspoon, and Rusty Kaufmann. The authors wish to acknowledge the mappers who provided the major source of expertise in mapping and characterizing the late-successional forest conditions: Lisa Acree, Andy Aldrich, Lori Allessio, Steve Anderson, Ed Armenta, Jim Barry, Elizabeth Bergstrom, John Bradford, Bart Bloom, Frank Brassell, Chuck Bredesen, Amedee Brickey, Mike Brown, Ann Carlson, Bob Carroll, Bob Cary, Dick Castaldini, Dominic Cesmat, Dennis Clemens, Chris Click, Rich Coakley, Artie Colson, Beth Corbin, Peggy Cranston, Teri Drivas, Dan Duriscoe, Don Errington, Dale Evans, Pat Farrell, Eric Fischer, Liz Fisher, Mary Flores, Larry Ford, Charis Genter, Gerald Grevstad, Dennis Haas, Wayne Harrison, Sylvia Haultain, Karen Hayden, Terry Hicks, Tom Higley, Andy Hosford, Mike Jablonski, Bob Jennings, Dale Johnson, Lew Jump, Marie Kennedy, Dan Leedy, Jerry Kent, Ron Keil, Chuck Loffland, Tina Mark, Dan Marlatt, Mike Martini, Mike Mateyka, Brian Miller, Paul Miller, Peggy Moore, Mike Newman, Kathy Noland, Ginelle O'Connor, Ron O'Hanlon, Joe Oden, Bea Olson, Erik Ostly, Richard Perloff, Don Potter, Julie Rechtin, Ramiro Rojas, Gary Rotta, Stephanie Sager, Jim Schmidt, Roland Shaw, Joe Sherlock, Dave Sinclear, Dave Smith, Sydney Smith, Nate Stephenson, Slim Stout, George Stundinski, Pat Stygar, Neil Sugihara, John Sweetman, Linda Tatum, Boyd Turner, Steve Underwood, Jan van Wagtendonk, Kathy Van Zuuk, Al Vazquez, Rod Vineyard, Russ Volkle, Tom Warner, Steve Weaver, Judy Welles, Jerry Westfall, and Genny Wilson.

REFERENCES

Barbour, M. G. 1988. Californian upland forests and woodlands. In *North American Terrestrial Vegetation,* edited by M. G. Barbour and W. D. Billings, 131-64. Cambridge University Press: New York.

Cahill, T. A., J. J. Carroll, D. Campbell, and T. E. Gill. 1996. Air quality. In *Sierra Nevada Ecosystem Project: Final report to Congress,* vol. II, chap. 48. Davis: University of California, Centers for Water and Wildland Resources.

Chen, J., J. F. Franklin, and T. A. Spies. 1992. Vegetation responses to edge environments in old-growth Douglas-fir forests. *Ecological Applications* 2:387-96.

———. 1993. Contrasting microclimates among clearcut, edge, and interior of old-growth Douglas-fir forest. *Agricultural and Forest Meteorology* 63:219-37.

Fitch, C. H. 1900a. Sonora quadrangle, California. In *Annual reports of the Department of Interior, 21st annual report of the U.S. Geological Survey,* part 5, 569-71.

———. 1900b. Yosemite quadrangle, California. In *Annual reports of the Department of Interior, 21st annual report of the U.S. Geological Survey,* part 5, 571-74.

Fites-Kaufmann, J. 1996. Landscape patterns of pre-fire suppression forest structure: Integration of fire regime and environmental influences in westside mixed-conifer forests of the northern Sierra Nevada. Ph.D. thesis, University of Washington.

Franklin, J. F. 1993. Preserving biodiversity: Species, ecosystems, or landscapes? *Ecological Applications* 3:202-5.

———. In press. Ecosystem management: An overview. In *Ecosystem management: Applications for sustainable forest and wildlife resources.* Yale University Press.

Franklin, J. F., D. R. Berg, D. A. Thornburgh, and J. C. Tappeiner. In press. Alternative silvicultural approaches to timber harvest: Variable retention harvest system. In *Creating a forestry for the 21st century,* edited by K. Kohm and J. F. Franklin. Island Press.

Franklin, J. F., and R. T. T. Formann. 1987. Creating landscape patterns by forest cutting: Ecological consequences and principles. *Landscape Ecology* 1:5-18.

Franklin, J. F., D. M. Graber, K. N. Johnson, J. A. Fites-Kaufmann, K. Menning, D. Parsons, J. Sessions, T. A. Spies, J. C. Tappeiner, and D. A. Thornburgh. 1996. Comparison of alternative late-successional conservation strategies. In *Sierra Nevada Ecosystem Project: Final report to Congress.* Davis: University of California, Centers for Water and Wildland Resources.

Franklin, J. F., and T. A. Spies. 1991a. Composition, function, and structure of old-growth Douglas-fir forests. In *Wildlife and vegetation of unmanaged Douglas-fir forests,* edited by L. F. Ruggiero et al., technical coordinators. General Technical Report PNW-GTR-285. Portland, OR: U.S. Forest Service Pacific Northwest Region.

———. 1991b. Ecological definitions of old-growth Douglas-fir forests. In *Wildlife and vegetation of unmanaged Douglas-fir forests,* edited by L. F. Ruggiero, et al., technical coordinators. General Technical Report PNW-GTR-285. Portland, OR: U.S. Forest Service Pacific Northwest Region.

Graber, D. M. 1996. Status of terrestrial vertebrates. In *Sierra Nevada Ecosystem Project: Final report to Congress,* vol. II, chap. 25. Davis: University of California, Centers for Water and Wildland Resources.

Harr, R. D. 1986. Effects of clearcutting on rain-on-snow runoff in western Oregon. *Water Resources Bulletin* 18:785-89.

Harr, R. D., B. A. Coffin, and T. W. Cundy. 1989. Effects of timber harvest on rain-on-snow runoff in the transient snow zone of the Washington Cascades. Interim final report submitted to Timber, Fish, and Wildlife (TFW) Sediment, Hydrology, and Mass Wasting Steering Committee for Project 18 (rain-on-snow). U.S. Forest Service Pacific Northwest Forest and Range Experiment Station, Portland, Oregon.

Johnson, N., J. Franklin, J. Gordon, and J. W. Thomas. 1991. Alternatives for management of late-successional forests of the Pacific Northwest.

A report to the Agriculture Committee and the Merchant Marine Committee of the U.S. House of Representatives. College of Forest Resources, Corvallis, Oregon.

Langley, P. G. 1996. Quality assessment of late seral old-growth forest mapping. In *Sierra Nevada Ecosystem Project: Final report to Congress,* vol II, chap. 22. Davis: University of California, Centers for Water and Wildland Resources.

Leiberg, J. B. 1902. Forest conditions in the northern Sierra Nevada, California. U.S. Geological Survey Professional paper 8, series H, forestry 5.

Likens, G. E. 1992. *The ecosystem approach: Its use and abuse.* Oldendorf/Luhe, Germany: Ecology Institute.

Marshall, C. H. 1900. Mount Lyell quadrangle, California. In *Annual Reports of the Department of the Interior, 21st annual report of the U.S. Geological Survey,* part 5, 574-76. Washington, DC: Government Printing Office.

McKelvey, K. S., and J. D. Johnston. 1992. Historical perspectives on forests of the Sierra Nevada and Transverse Ranges of southern California: Forest conditions at the turn of the century. In *The California spotted owl: A technical assessment of its current status,* edited by Jared Verner et al., 225-46. General Technical Report PSW-GTR-133. San Francisco: U.S. Forest Service Pacific Southwest Region.

Minnich, R. A., M. G. Barbour, J. H. Burk, and R. F. Fernau. 1995. Sixty years of changes in Californian conifer forests of the San Bernardino Mountains. *Conservation Biology* 9:902-14.

Oliver, W. W., and K. L. Dolph. 1992. Mixed-conifer seedling growth varies in response to overstory release. *Forest Ecology and Management* 48:179-83.

Riegel, G. M., S. E. Greene, M. E. Harmon, and J. F. Franklin. 1988. Characteristics of mixed conifer forest reference stands at Sequoia National Park, California. University of California Cooperative National Park Resources Studies Unit Technical Report 32. University of California Davis.

Sawyer, J. O., and T. Keeler-Wolf. 1995. *A manual of California vegetation.* Sacramento, CA: California Native Plant Society.

Sierra Nevada Ecosystem Project. 1994. Unpublished progress report. University of California, Davis, Centers for Water and Wildland Resources.

Sudworth, G. B. 1900. Stanislaus and Lake Tahoe Forest Reserves, California, and adjacent territory. In *Annual reports of the Department of Interior, 21st annual report of the U.S. Geological Survey,* part 5, 505-61.

U.S. Forest Service. 1995. Draft environmental impact statement. Managing California spotted owl habitat in the Sierra Nevada national forests of California (an ecosystem approach). 2 vols., various pagination. U.S. Forest Service Pacific Southwest Region, San Francisco.

Verner, J., K. S. McKelvey, B. R. Noon, R. J. Gutierrez, G. I. Gould, Jr., and T. W. Beck. 1992. General Technical Report PSW-GTR-133. San Francisco: U.S. Forest Service Pacific Southwest Region.

Guides to Structural Analysis and Rating of Late-Successional Forests

Examples for Westside Mixed Conifer, Red Fir, White Fir, and Subalpine forest groups of the tables used for guides in ranking late-successional structural complexity for forest patches and for crosswalking between the Sierra-wide and series-normalized structural standards. Patch codes are those used by the U. S. Forest Service Pacific Southwest Region in timber inventories and are provided only as a cross-reference to that system. Major structural criteria utilized were: size and number of large-diameter trees; coverage of overstory (OS) and intermediate (Int.) canopy levels; significant decadence in large live trees (yes or no); levels of coarse woody debris; and disturbance history of the patch. Ranking columns refer to ranking of 1) current conditions by rangewide (column A) and series-normalized (column C) standards and 2) maximum potential ranking (based on site productivity) by rangewide (column B) and series-normalized (column D) standards.

Forest Grouping: Mixed Conifer/Westside (WMC)

Patch Code	Large Trees dbh, in.	Large Trees Trees/ac	Canopy OS	Canopy Int.	Decadence of Live Large Trees	Coarse Woody Debris Snags	Coarse Woody Debris Logs	Patch History Grazing	Patch History Harvest	Other	Ranking A	Ranking B	Ranking C	Ranking D
WMC5a	>40"	>10	>60%	Y	Y	C	C	little or none	none		5	5		
WMC4a	>40"	6-10	40-60%	na	Y	C	C	little or none	little or none		4	4		
WMC4b	>40"	2-10	>60%	Y	Y	C	C	little or none	little or none		4	4		
WMC4c	>40"	>10	>60%	na	Y	F/O	F/O	little or none	little or none		4			
WMC4d	same as 5a but							+/-	1-10%		4	5		
WMC3a	>40"	>6	20-40%	na	Y	F/O	F/O	little or none	little or none		3	3		
WMC3b	>40"	2-6	40-60%	na	Y	F/O	F/O	little or none	little or none		3	3		
WMC3c	>30"	>6	>40%	na	Y	F/O	F/O	little or none	little or none		3	3		
WMC3d	same as 4a-c but							+/-	2-30%	harv. areas w/ 0 or no LS	3	4		
WMC3e	same as 5a but							+/-	1-03-%	"	3	5		
WMC2a	>40"	2-6	20-40%	na	Y	na/N	na/N	little or none	little or none		2	2		
WMC2b	>30"	>2	>20%	na	Y	na/N	na/N	little or none	little or none		2	2		
WMC2c	same as 4a-c but								30-60%	"	2	4		
WMC2d	same as 5a but								30-60%	"	2	5		
WMC2e	same as 3a-c but								2-30%	"	2	3		
WMC2f	>24"	>20	>60%	na	na	na	na	little or none	little or none	no signif. LS	2	4-5		
WMC1a	>30"	0.5-2	>10%	na	Y	F/O	F/O	little or none			1	1		
WMC1b	same as 4a-c, 5a but								>60%	harv. areas w/ scattered LS	1	3-5		
WMC1c	same as 3a-c but								>30%	harv. areas w/ little or no LS	1			
WMC1d		0	0	na	na	A	C+	+/-	little or mod	major burned area	1	3-5		
WMC1e	>24"	>20	>40%	na	na	na	na	little or none		no signif. LS structure	1	3-4		

A Range-wide Standard - Current
B Range-wide Standard - Potential
C Group Standard - Current
D Group Standard - Potential

M = many, >4/ac
C = common, 2-4/ac
F = few, 1/2 - 2/ac
O = none, 0/ac

LS= late-successional forest structure (e.g. large live trees)

Forest Grouping: Red Fir (RF)

Patch Code	Large Trees dbh, in.	Large Trees Trees/ac	Canopy OS	Canopy Int.	Decadence of Live Large Trees	Coarse Woody Debris Snags	Coarse Woody Debris Logs	Patch History Grazing	Patch History Harvest	Other	Ranking A	B	C	D
RF5a	>40"	>10	>60%	Y	Y	A	A	little or none		superlative; usu w/ mixed conifer	5	5	5	5
RF5b	>40"	>10	>60%	N	Y	A	A	little or none	none			5	5	5
RF4a	>40"	>6	>40%	na	Y	C	C	little or none	none		4	4	4	4
RF4b	>30"	>6	>40%	na	Y	C	C	none	none		3	4	4	4
RF4c	like RF5a, 5b, but								10%		3		4	5
RF3a	>30"	2-6	20-40%		Y	F/O	F/O	little or none			2	3	3	
RF3b	like RF5a, 5b, 41, 4b, but					F/O	F/O		10-30%		3	3	3	4
RF2a	like RF5a, 5b, 4a, 4b								30-50%		1		2	
RF2b	like RF3a								10-30%		1		2	
RF2c	>24"	>20	>40%	na	na	na	na		none	no signif. LS characteristics	1		2	
RF1a	like RF5a, 5b, 41, 4b but								>50%		1		1	
RF1b	like RF3a								>30%		1		1	
RF1c	>16"	>20	>40%	na	na	na	na		none	no signif. LS characteristics	1		1	
RF1d	0		na	na	na	A	C+	little or mod		major burn area	1		1	

A Range-wide Standard - Current
B Range-wide Standard - Potential
C Group Standard - Current
D Group Standard - Potential

M = many, > 4/ac
C = common, 2-4/ac
F = few, 1/2 - 2/ac
O = none, 0/ac

LS = late-successional forest structure (eg, large live trees)

Forest Grouping: White Fir/Eastside (WF)

Patch Code	Large Trees dbh, in.	Large Trees Trees/ac	Canopy OS	Canopy Int.	Decadence of Live Large Trees	Coarse Woody Debris Snags	Coarse Woody Debris Logs	Patch History Grazing	Patch History Harvest	Other	Rank A	Rank B	Rank C	Rank D
WF5a	>40"	>10	>60%	Y	Y	C	C	little or	none		5	5	5	5
WF5b	>40"	>10	>60%	N	Y	C	C	"	"		4	4	5	5
WF4a	>40"	>6	>40%	na	Y	C		little or none	none		4	4	4	4
WF4b	in 5a or 5b but							little or none	1-10%		4	5	4	5
WF4c	>30"	>10	>40%	na	Y	C	C	little or none	none		3	3	4	4
WF3a	>30"	6-10	>40%	na	Y	C	C	little or	none		3	3	3	3
WF3b	>40"	2-6	>40%	na	Y	F	F	"	"		3	3-4	3	3-4
WF3c	same as 5a, 5b but								10-30%		3	5	3	5
WF3d	same as 4a, 4c but								1-10%		3	4	3	4
WF2a	>30"	2-6	>20%	na	Y	F/O	F/O	little or none			2	2	2	2
WF2b	same as 5, 4a, 4c, but								30-60%		2	2	2	2
WF2c	>24"	>20	>40%	na	na	na	na		l or none	no sig. LS	2		2	
WF2d	same as 3a but								10-30%		2		2	
WF1a	0	0	0	na	na	A	C+	+/-	little/mod major burn		1		1	
WF1b	same as 3a but								>30%		1		1	
WF1c	same as 5, 4a, 4c but								>60%		1		1	

A Range-wide Standard - Current
B Range-wide Standard - Potential
C Group Standard - Current
D Group Standard - Potential

M = many, >4/ac
C = common, 2 - 4/ac
F = few, 1/2 - 2/ac
O = none, 0/ac

LS = late-successional
forest structure
(e.g. large live trees)

Forest Grouping: Subalpine (Includes High Elevation Lodgepole Pine) (SA)

Patch Code	Large Trees dbh, in.	Large Trees Trees/ac	Canopy OS	Canopy Int.	Decadence of Live Large Trees	Coarse Woody Debris Snags	Coarse Woody Debris Logs	Patch History Grazing	Patch History Harvest	Other	Ranking A	Ranking B	Ranking C	Ranking D
SA5a	>30"	>10	>40%	na	Y	F	F	little or none	none	eg. mtn hemlock	3		5	5
SA4a	>30"	6-10	>20%	na	Y	F	F/O	little or none	none		2		4	4
SA4b	>24"	>10	>40%	Y	Y	C	C			eg, moist lodgepole pine	2		4	4
SA3a	>30"	2-6	>10%	na	Y	F/O	F/O	little or none	none		2		3	3
SA3b	>24"	2-10	?20%	na	Y	F	F/O	little or none	none		1		3	3
SA3c	same as 5a, 4a, 4b but							heavy &/or 1-30%			2		3	###
SA2a	>30"	0.5-2	>2%	na	Y	F/O	F/O	little or none	none		1		2	2
SA2b	>24"	0.5-2	10-20%	na	Y	F/O	F/O	little or none	none		1		2	2
SA2c	same as 5a, 4a, 4b but							+/-	30-60%		1		2	4-5
SA2d	same as 3a, 3b but								1-30%		1		2	3
SA1a	>24"	0.5-2	>2%	na	Y	F/O	F/O	little or none	none		0		1	1
SA1b	>30"	scattered trees, >0.5	na	na	Y			little or none	none		0		1	1
SA1c	same as 5a, 4a, 4b but								>60%		0		1	4-5
SA1d	same as 3a, 3b but								>30%		0		1	3

M = many, >4/ac
C = common, 2-4/ac
F = few, 1/2 - 2/ac
O = none, 0/ac

A Range-wide Standard - Current
B Range-wide Standard - Potential
C Group Standard - Current
D Group Standard - Potential

PHILIP G. LANGLEY
Forest Data Corporation
Walnut Creek, California

22

Quality Assessment of Late Seral Old-Growth Forest Mapping

ABSTRACT

A program was undertaken to assess the consistency of the late seral old-growth (LSOG) classification maps developed for the Sierra Nevada Ecosystem Project. The effort included sampling to gather data appropriate for assessing the accuracy of LSOG ratings assigned to specific mapped areas and for making such assessments. The assessment focuses on the correlation observed between patch ratings assigned by the LSOG mapping team and structural characteristics of the forest as observed on the ground. In the mixed conifer forest type, the classification accuracies ranged between 44% in the higher LSOG rating classes and 78% in the lower classes at the patch level. Consequently, the LSOG maps prepared by the SNEP team can serve as a basis for stratifying the Sierra Nevada into broad groups of late successional forest structural patterns. However, a high level of variation in structural components should be expected for any given LSOG class when making statements concerning the structural composition of patch ratings specific to given mapped polygons.

INTRODUCTION

This report describes the results of a quality assessment of late seral old-growth (LSOG) patch ratings assigned by the Sierra Nevada Ecosystem Project (SNEP) mapping team early in 1994. The assessment focuses on (1) the structural characteristics of biological material (e.g., live trees, snags, and down matter) characteristic of each assigned patch rating as observed on the ground and (2) the occurrence of human entry or management activity at some time in the past in sample patches.

To construct the maps, a team was assembled in one place for several days. Working in groups, they constructed LSOG maps by drawing on group members' special knowledge of each area, previously constructed resource maps, aerial photographs and other data. Large polygons, up to several thousand acres in size, were constructed encompassing land areas of apparently similar forest characteristics. Each polygon was more finely characterized as consisting of one to five patch types, depending on the heterogeneity within-polygons. In addition, each polygon was stratified into a major forest type. Revisions were made to patch proportions later, with different people on some national forests.

The attributes used by the mapping team for identifying patches having various levels of late successional characteristics include such items as forest type, number of large trees, number of snags, dominant species, and canopy closure. After the patch types were identified, an LSOG rating was assigned to each patch type within a polygon. The sum of the products of the proportion of patch area times the patch rating (estimated on a scale of 1 to 5) yielded the LSOG rating for a whole polygon.

The quality assessment described in this chapter was undertaken to learn about the reliability of the assigned polygon ratings as they relate to late seral forest structures observed on the ground. Because the polygon ratings depend on the proportion and rating of patch types within polygons, the problem can be addressed at the patch level to bring the scale of the assessment task within reasonable bounds.

Knowledge about the reliability of assigned patch ratings, and thus polygons, is important to assure the proper charac-

Sierra Nevada Ecosystem Project: Final report to Congress, vol. II, *Assessments and scientific basis for management options.* Davis: University of California, Centers for Water and Wildland Resources, 1996.

terization of the LSOG rating levels and spatial distribution of late seral forests in the Sierra Nevada. It is also important for any large-scale stand projections that may be attempted in the future. The key questions this quality assessment attempts to answer, therefore, are

- How consistent are the patch ratings assigned by the mapping team based on structural characteristics of the forest as observed on the ground?

- What are the structural characteristics of each assigned LSOG rating when measured on a scale of 1 to 5?

BACKGROUND

While preparing the LSOG maps, the mapping team had access to previously prepared orthophotos (e.g., rectified, scaled, and mosaicked aerial photographs), data files, and each mapper's unique individual experience in different parts of the study area. With such a diversity of input, it is natural to expect that the LSOG maps in different regions of the Sierra Nevada would exhibit a high degree of variability when compared with ground conditions. Furthermore, the LSOG maps are based on site-specific, current predictions of forest parameters that can be reliably assessed only on the ground and not from aerial photographs, previously prepared maps, or even human memory.

To objectively address the two questions posed, it would seem prudent to employ sample data that have been collected during the normal course of other ongoing forest surveys, such as the U.S. Forest Service's Forest Inventory and Analysis (FIA) program or their management-plan mapping and inventory programs. However, to assess the quality of any particular map set, it is necessary to have data that are specific to a known geographical area. Furthermore, the accompanying test data should be consistent with the categorical partitioning of the maps being tested.

None of the maps made during the course of other surveys, nor their accompanying field data, conform to the scale, range, or structure of the SNEP maps. Therefore, to evaluate the current LSOG maps, it became necessary to obtain at least some new data describing the forest structure and late seral stage at specific sample sites within the areas covered by these maps. Problems arose, however, as we endeavored to devise a sampling plan to gather data for the quality assessment.

First, in order to assess quality at the patch level efficiently, it is necessary to target the patches directly for possible inclusion in a sample. Unfortunately, though the mapping team estimated the proportion of each polygon occupied by each patch type, they made no attempt to delineate patch boundaries or otherwise locate patches within polygons. Therefore, sampling patches directly on a global basis, would have required an expensive data organizational and field sampling

procedure. The most practical alternative we saw was to draw a random sample of polygons, with probability proportional to their area, and then have the mappers delineate the patch boundaries within the selected polygons. Obviously, this raised the possibility that bias might enter the process, because sample patches would be spatially clustered within sample polygons and not drawn completely at random from the whole population of polygons at a known relative frequency. This was a risk we nevertheless had to accept.

Second, selecting polygons according to a basic randomization scheme provided no control over the number of patches in each type or rating that would appear in the sample. The result is that, although we would have a variable probability random sample of polygons, we would have a cluster sample of patch types within polygons thus compromising, to an unknown degree, the validity of any statistical tests that assume a complete randomization of observations over the entire project area. Given the geographic scope of the LSOG program, the short time span available for completion, and the exploratory nature of the results expected, we elected to proceed with the sampling plan.

METHODS

Sampling Plan

A plan was devised for obtaining the sample data necessary to assess the quality of the LSOG maps in conformance with the two questions we posed. The strategy we used is called stratified two-stage sampling with variable probabilities of selection in the first stage. In this plan, a stratum contains all the mapped LSOG polygons in a major forest type as determined by the mapping team. However, because of time and cost constraints, only the mixed conifer forest type was sampled sufficiently for evaluation purposes. Even in the mixed conifer type, we were unable to obtain a representative distribution of data throughout the entire SNEP area. To help remedy this deficiency, plot data from the U.S. Forest Service Forest Inventory and Analysis (FIA) program were used to expand the geographical coverage over more of the mixed conifer forest type.

Data from the FIA field plots were reformatted so that the same structural components could be extracted from them as were obtained from the SNEP plots. Then, if feasible, we planned to combine the two data sets to obtain quality assessments that could be extrapolated to more forest types in the SNEP area. It turned out to be not feasible, however, to combine the two data sets because of significant differences between several pairs of common structural variables. Therefore, separate analyses were done for each set.

For SNEP sampling, the first-stage sample units consist of the mapped polygons within a stratum (major forest type). Polygon selection was random, with probability proportional

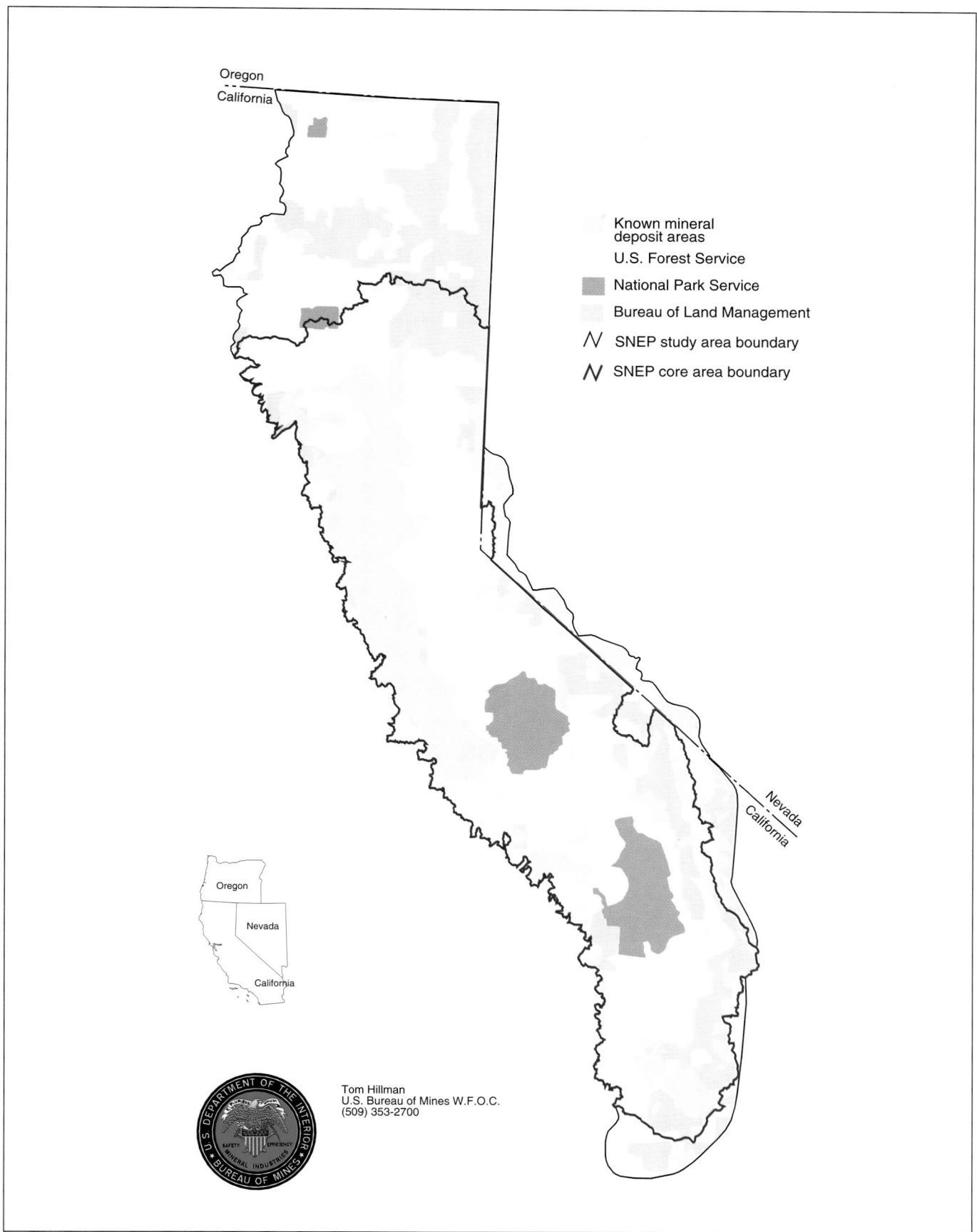

Oregon
California

Known mineral
deposit areas
U.S. Forest Service
National Park Service
Bureau of Land Management
SNEP study area boundary
SNEP core area boundary

Nevada
California

Oregon

Nevada

California

Tom Hillman
U.S. Bureau of Mines W.F.O.C.
(509) 353-2700

PLATE 18.7

Known mineral deposit areas in the SNEP study area.

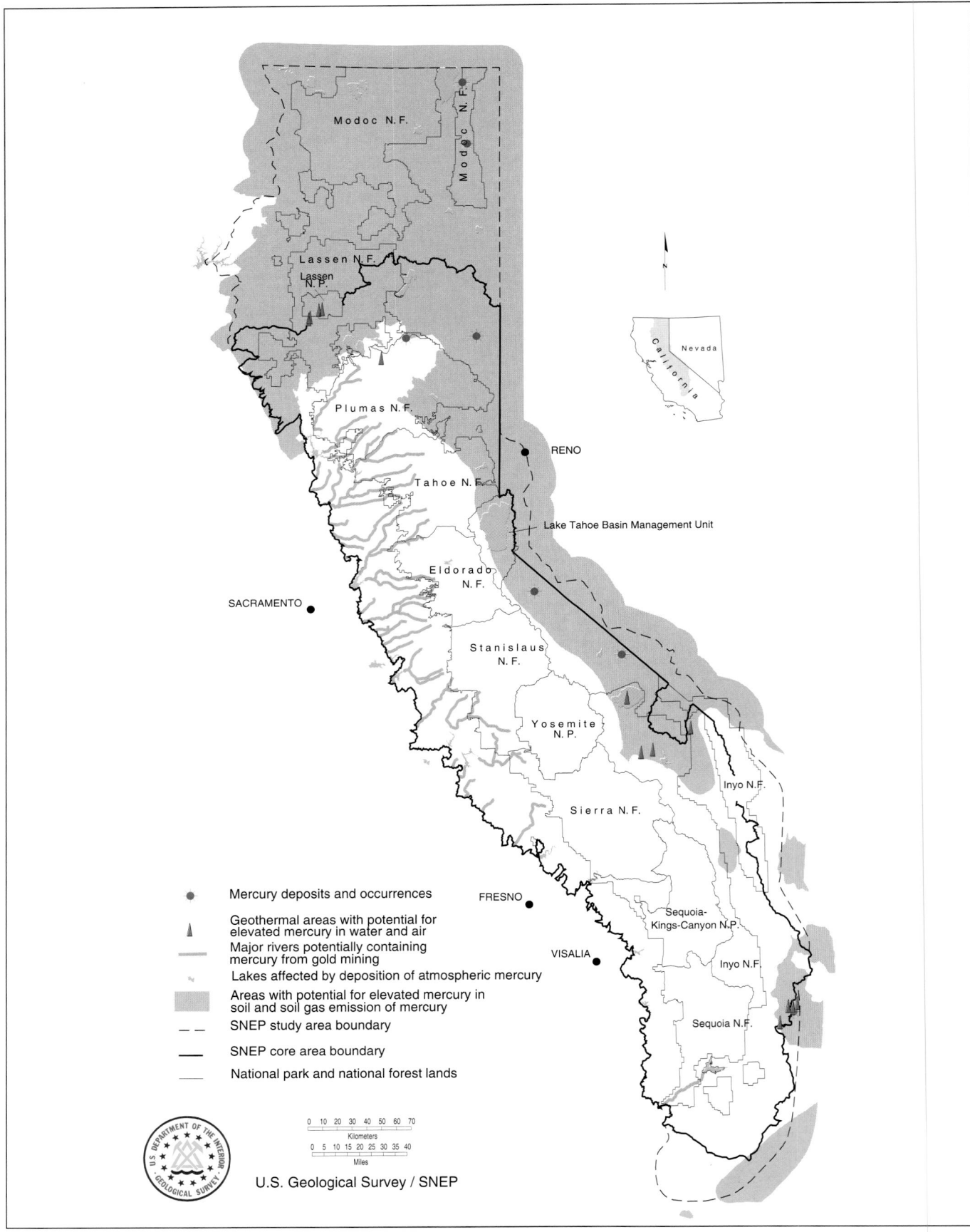

Mercury deposits and occurrences

Geothermal areas with potential for
elevated mercury in water and air

Major rivers potentially containing
mercury from gold mining

Lakes affected by deposition of atmospheric mercury

Areas with potential for elevated mercury in
soil and soil gas emission of mercury

SNEP study area boundary

SNEP core area boundary

National park and national forest lands

U.S. Geological Survey / SNEP

PLATE 18.8

Areas of possibly elevated mercury concentration.

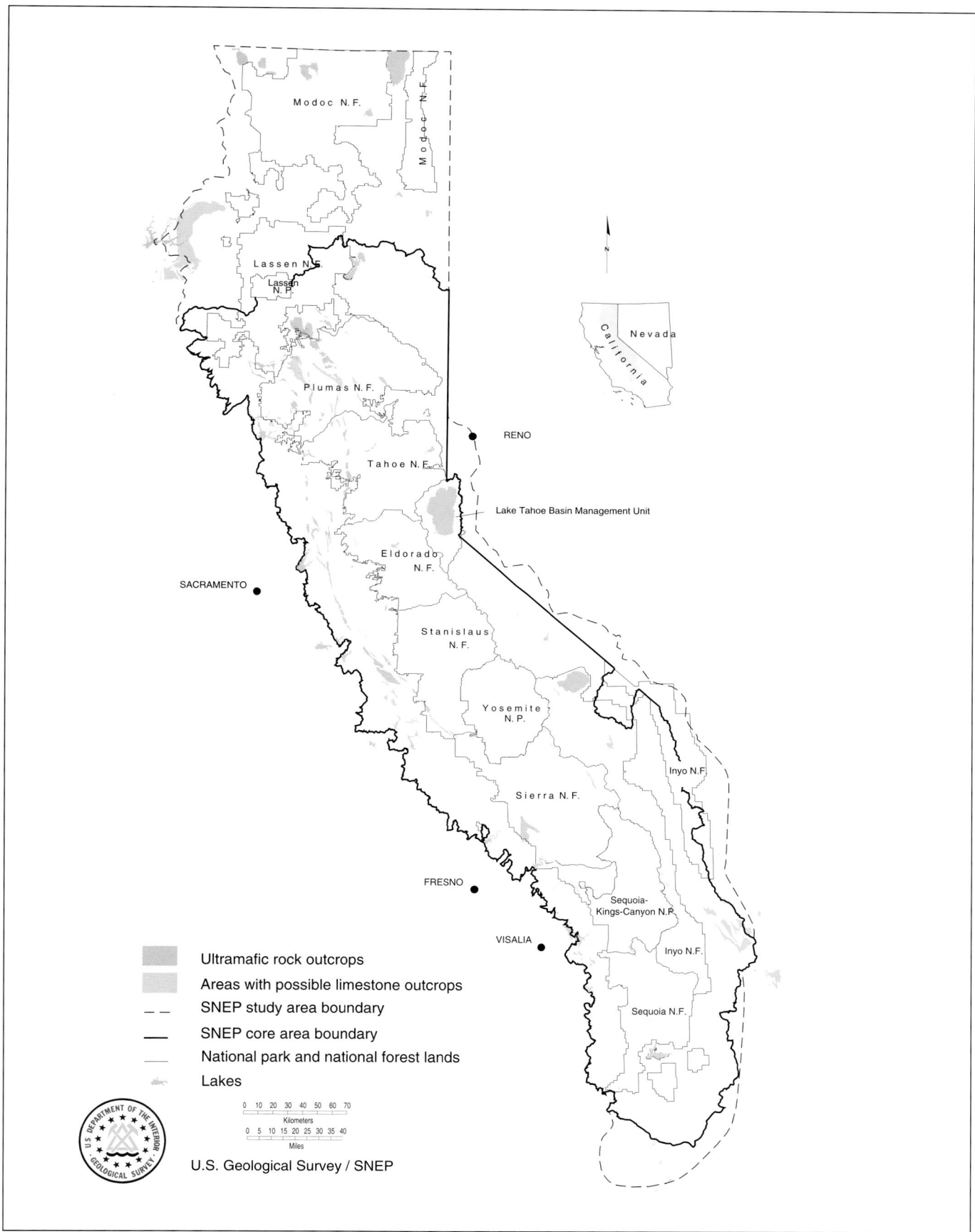

Ultramafic rock outcrops

Areas with possible limestone outcrops

--- SNEP study area boundary

— SNEP core area boundary

National park and national forest lands

Lakes

0 10 20 30 40 50 60 70
Kilometers
0 5 10 15 20 25 30 35 40
Miles

U.S. Geological Survey / SNEP

PLATE 18.9

Serpentine and carbonate-rich rocks.

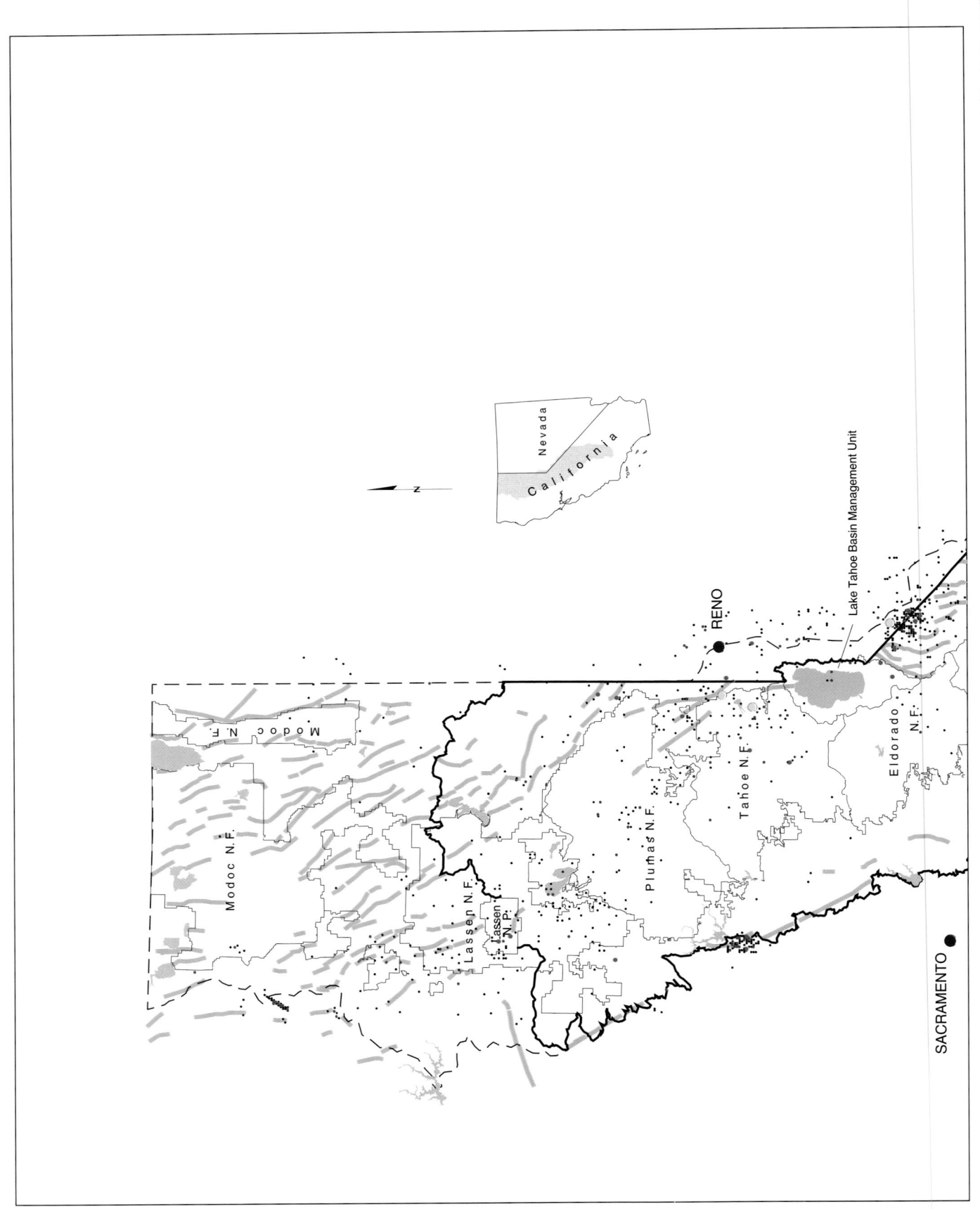

Nevada

California

RENO

SACRAMENTO

Lake Tahoe Basin Management Unit

Modoc N.F.

Modoc N.F.

Lassen N.F.

Lassen N.P.

Plumas N.F.

Tahoe N.F.

Eldorado N.F.

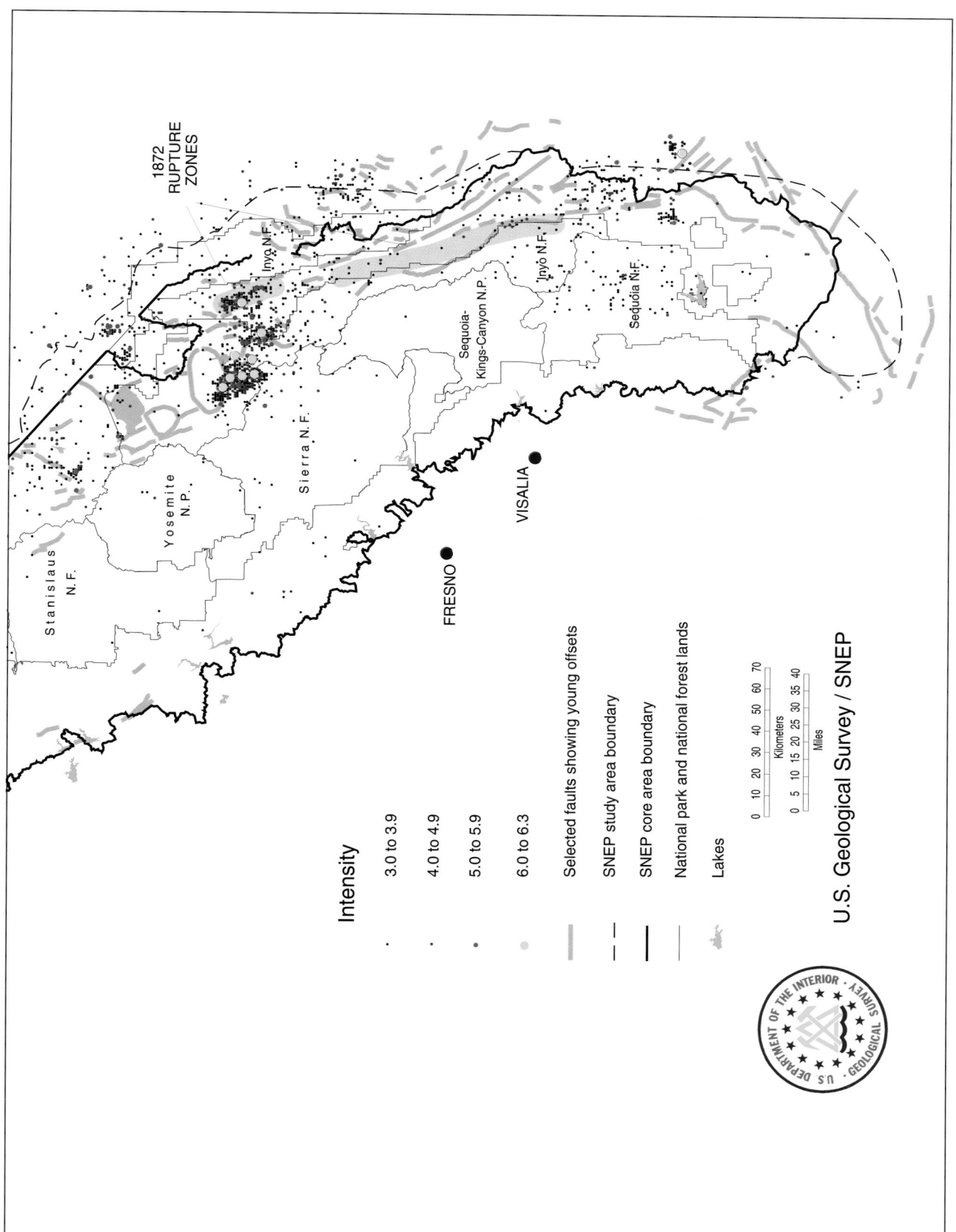

PLATE 18.10

Earthquake epicenters from 1910 to the present.

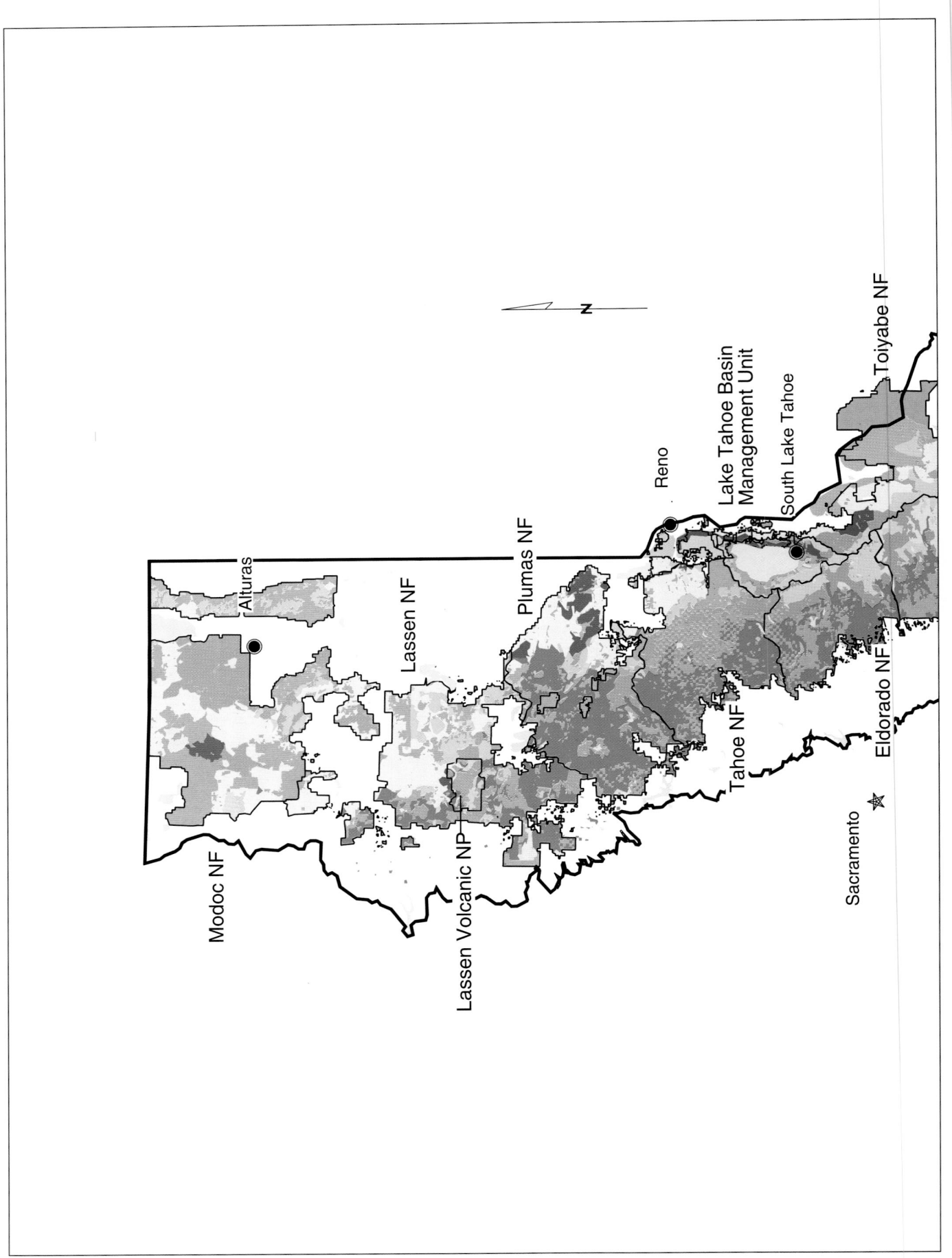

Modoc NF

Alturas

Lassen NF

Lassen Volcanic NP

Plumas NF

Reno

Lake Tahoe Basin
Management Unit

South Lake Tahoe

Tahoe NF

Toiyabe NF

Eldorado NF

Sacramento

N

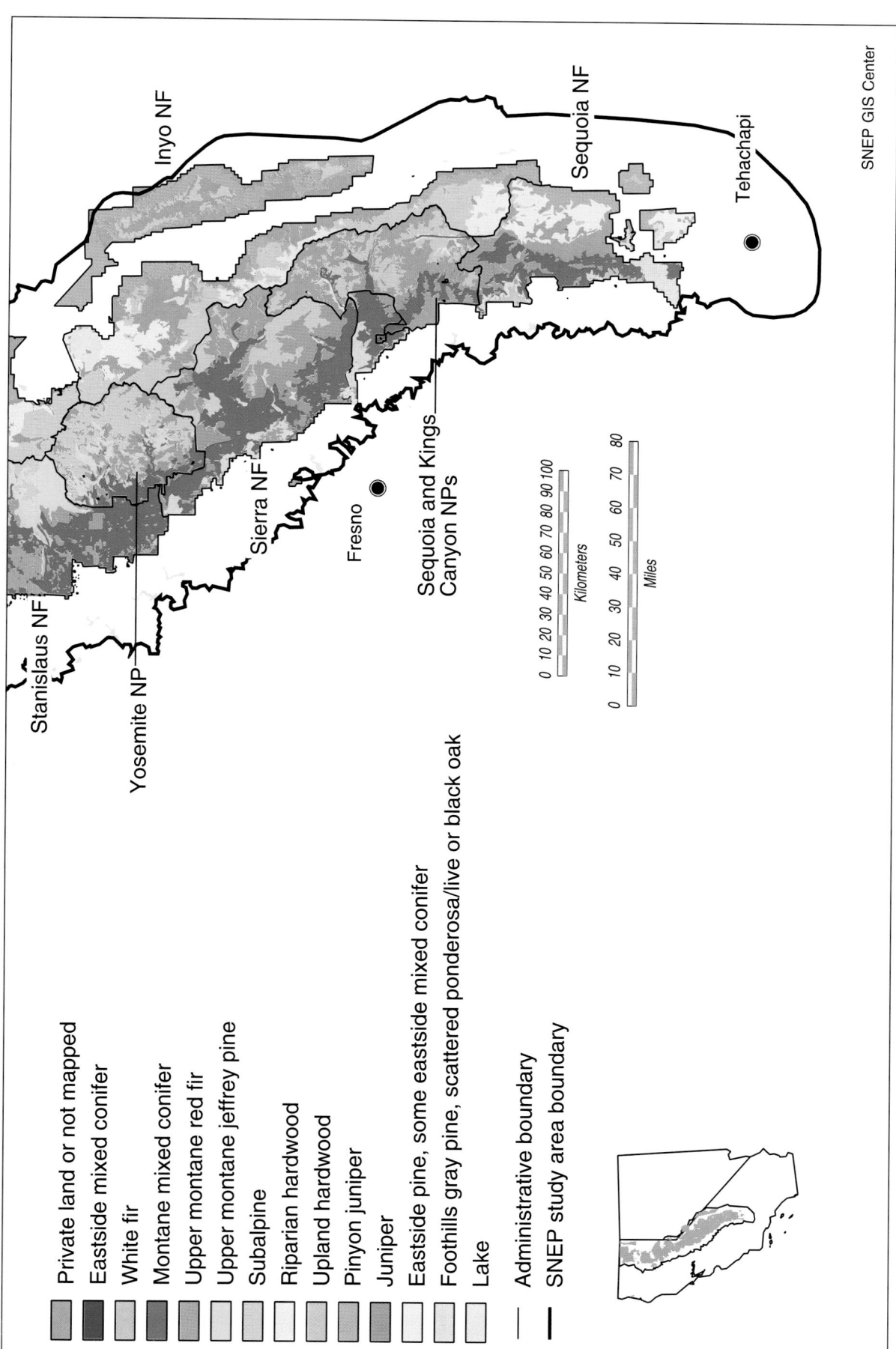

Legend:
- Private land or not mapped
- Eastside mixed conifer
- White fir
- Montane mixed conifer
- Upper montane red fir
- Upper montane jeffrey pine
- Subalpine
- Riparian hardwood
- Upland hardwood
- Pinyon juniper
- Juniper
- Eastside pine, some eastside mixed conifer
- Foothills gray pine, scattered ponderosa/live or black oak
- Lake
- —— Administrative boundary
- ▬▬ SNEP study area boundary

Stanislaus NF
Yosemite NP
Sierra NF
Fresno
Sequoia and Kings Canyon NPs
Inyo NF
Sequoia NF
Tehachapi

Kilometers
0 10 20 30 40 50 60 70 80 90 100

Miles
0 10 20 30 40 50 60 70 80

SNEP GIS Center

PLATE 21.1

Distribution of the major forest type groups in the Sierra Nevada.

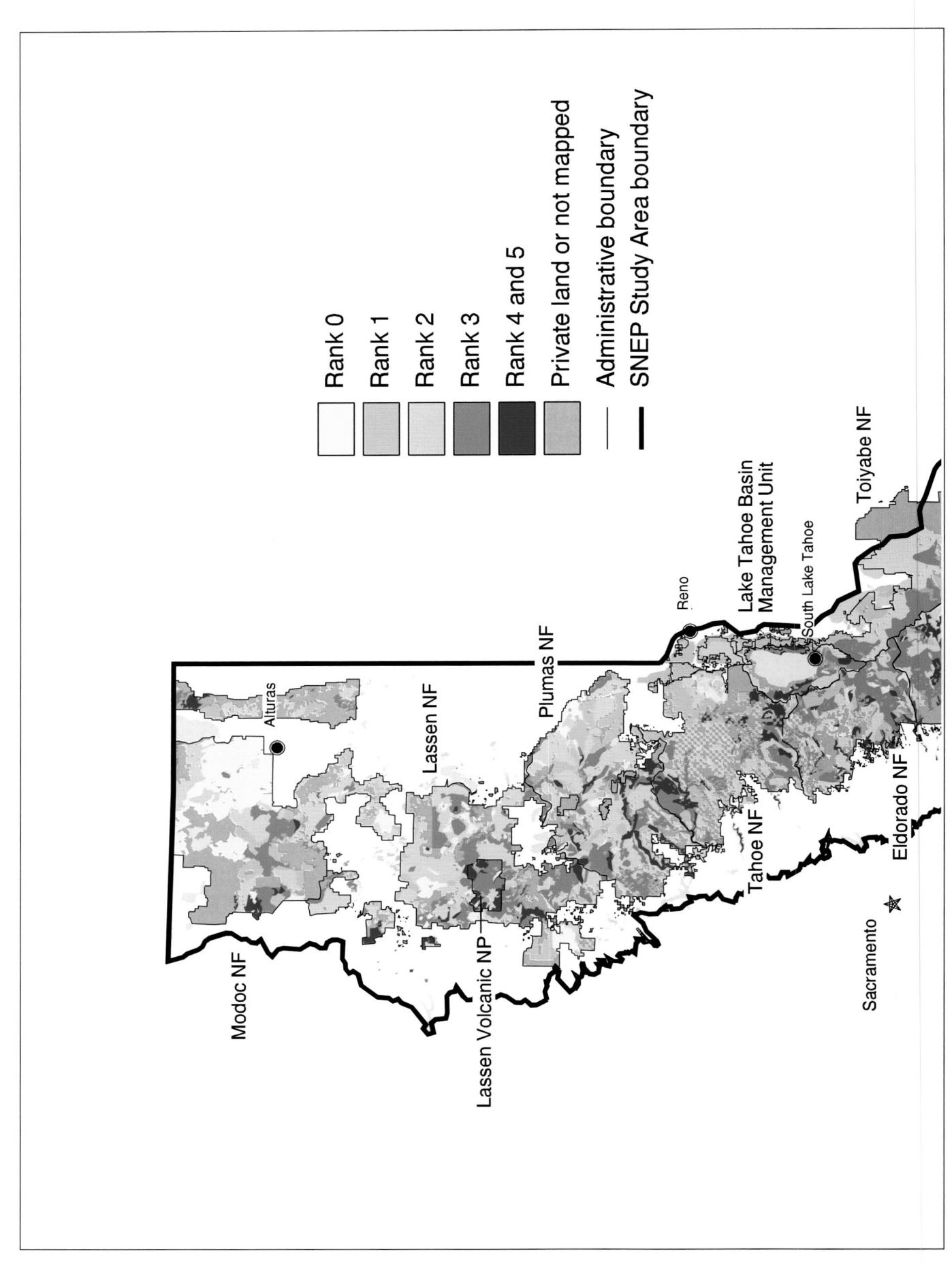

Rank 0
Rank 1
Rank 2
Rank 3
Rank 4 and 5
Private land or not mapped
Administrative boundary
SNEP Study Area boundary

Modoc NF

Alturas

Lassen NF

Lassen Volcanic NP

Plumas NF

Reno

Lake Tahoe Basin
Management Unit

South Lake Tahoe

Toiyabe NF

Tahoe NF

Sacramento

Eldorado NF

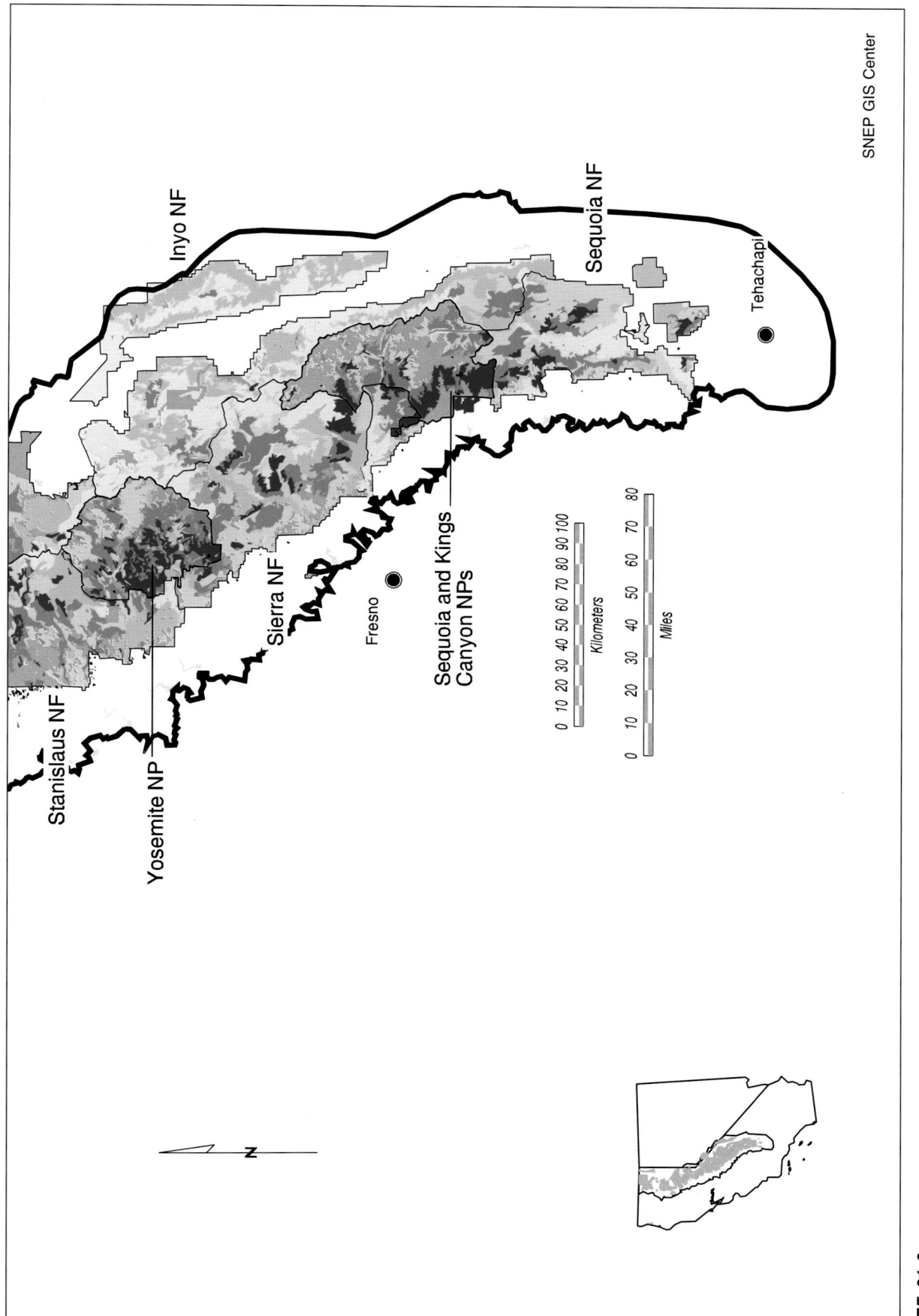

Stanislaus NF

Yosemite NP

Sierra NF

Inyo NF

Sequoia NF

Fresno

Sequoia and Kings
Canyon NPs

Tehachapi

0 10 20 30 40 50 60 70 80 90 100
Kilometers

0 10 20 30 40 50 60 70 80
Miles

N

PLATE 21.2

Distribution of polygons on federal lands throughout the Sierra Nevada color-coded as to their degree of LS/OG structural complexity and contribution to late successional forest function (rangewide structural standard).

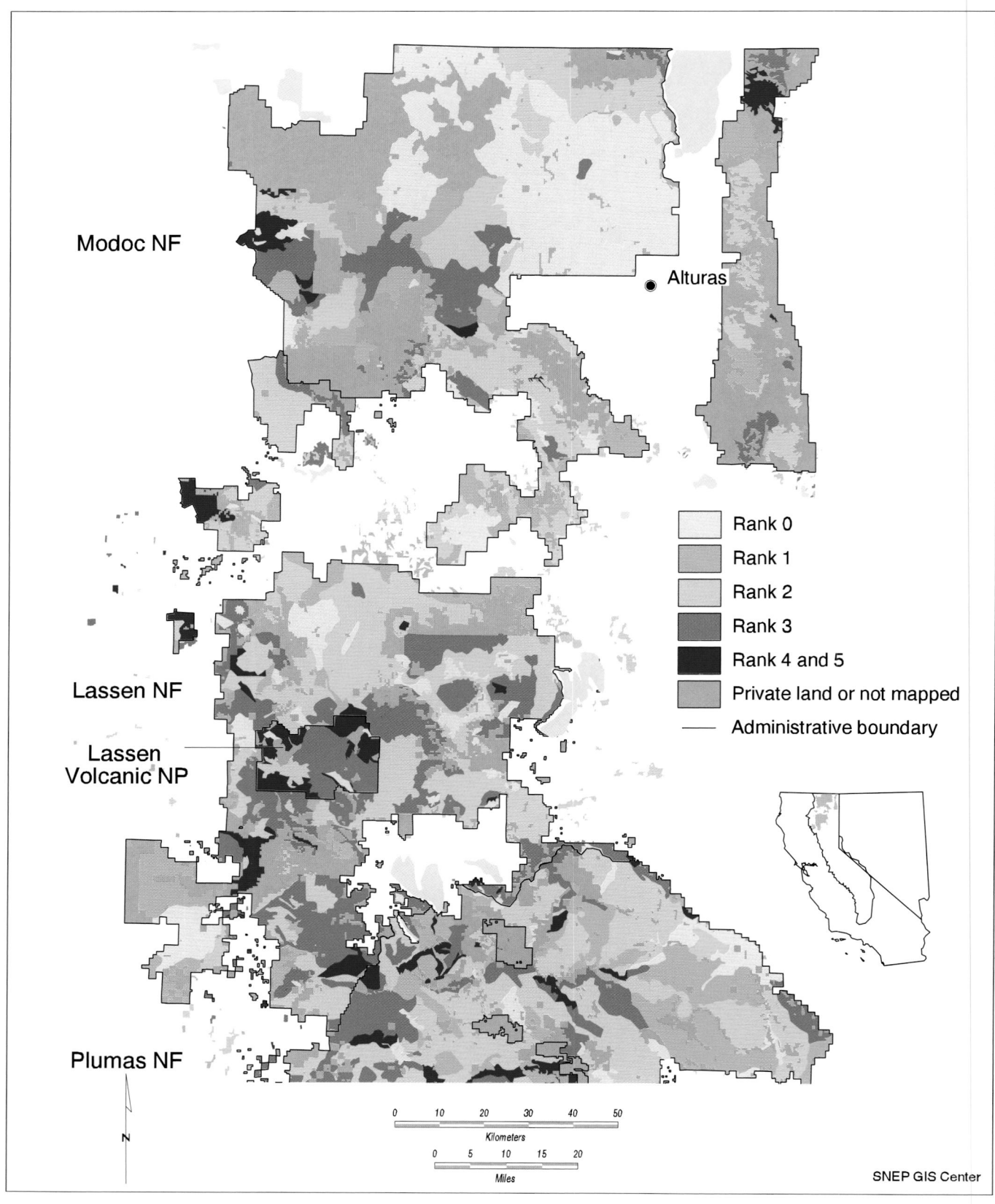

Rank 0
Rank 1
Rank 2
Rank 3
Rank 4 and 5
Private land or not mapped
Administrative boundary

Modoc NF

Alturas

Lassen NF

Lassen
Volcanic NP

Plumas NF

0   10   20   30   40   50
Kilometers

0   5   10   15   20
Miles

SNEP GIS Center

PLATE 21.3

Map of the northern section of the Sierra Nevada showing the distribution of polygons on federal lands color-coded as to their degree of LS/OG structural complexity and late successional forest function (rangewide structural standard).

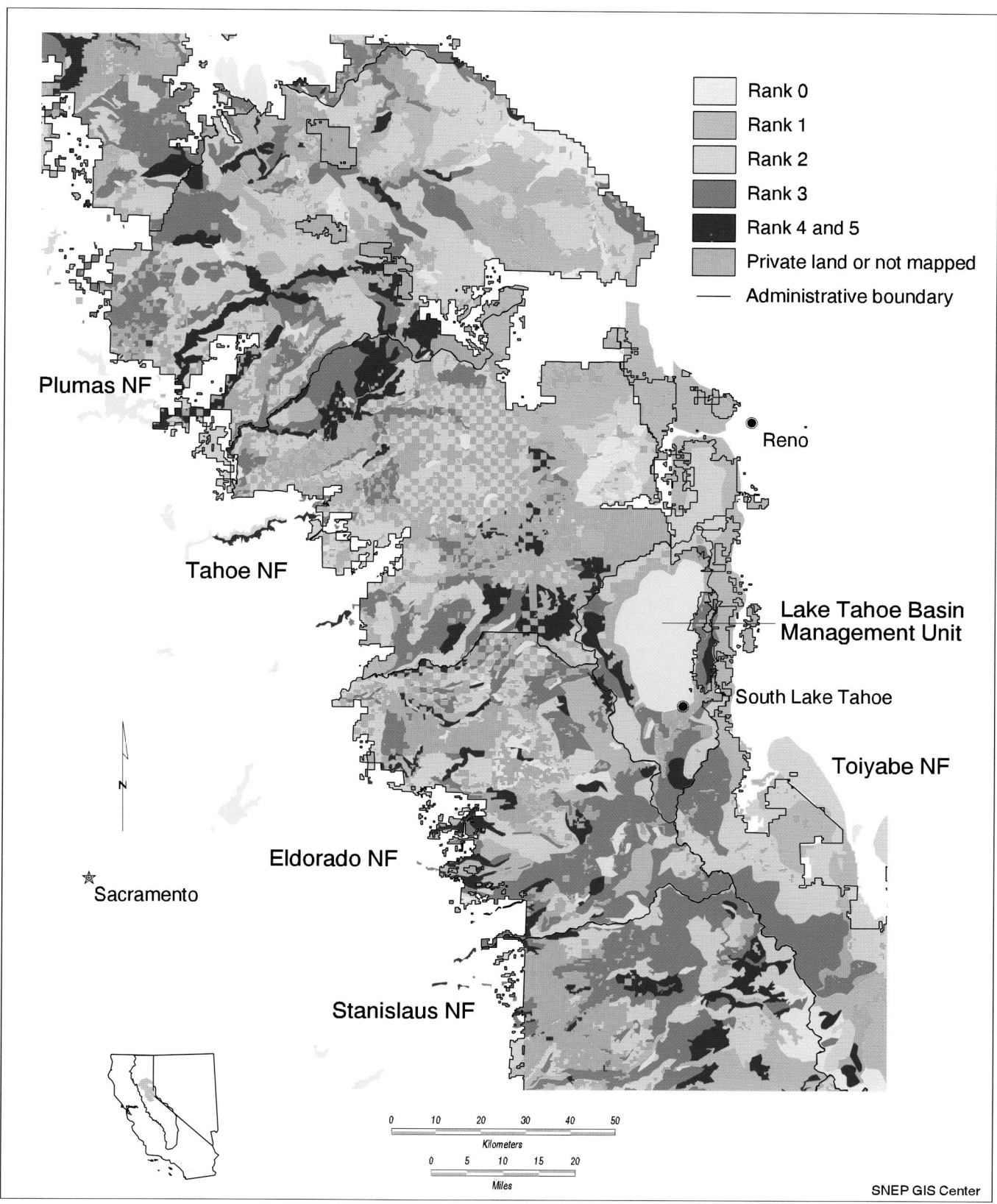

Rank 0
Rank 1
Rank 2
Rank 3
Rank 4 and 5
Private land or not mapped
Administrative boundary

Plumas NF

Reno

Tahoe NF

Lake Tahoe Basin
Management Unit

South Lake Tahoe

Toiyabe NF

Eldorado NF

Sacramento

Stanislaus NF

0   10   20   30   40   50
Kilometers
0   5   10   15   20
Miles

SNEP GIS Center

PLATE 21.4

Map of the north-central section of the Sierra Nevada showing the distribution of polygons on federal lands color-coded as to their degree of LS/OG structural complexity and late successional forest function (rangewide structural standard).

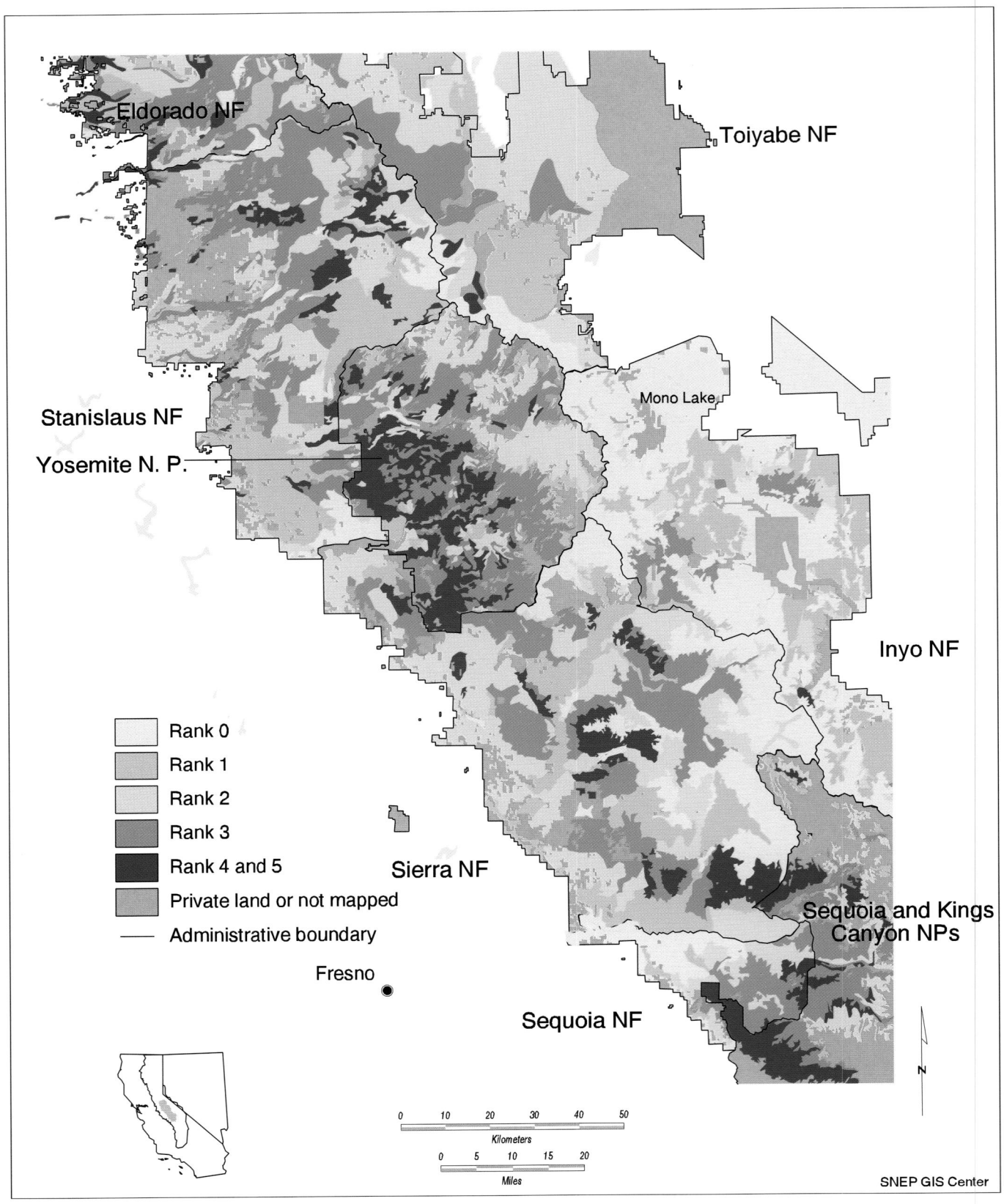

Rank 0
Rank 1
Rank 2
Rank 3
Rank 4 and 5
Private land or not mapped
Administrative boundary

Eldorado NF

Toiyabe NF

Stanislaus NF

Yosemite N. P.

Mono Lake

Inyo NF

Sierra NF

Sequoia and Kings Canyon NPs

Fresno

Sequoia NF

0   10   20   30   40   50
Kilometers

0   5   10   15   20
Miles

N

SNEP GIS Center

PLATE 21.5

Map of the south-central section of the Sierra Nevada showing the distribution of polygons on federal lands color-coded as to their degree of LS/OG structural complexity and late successional forest function (rangewide structural standard).

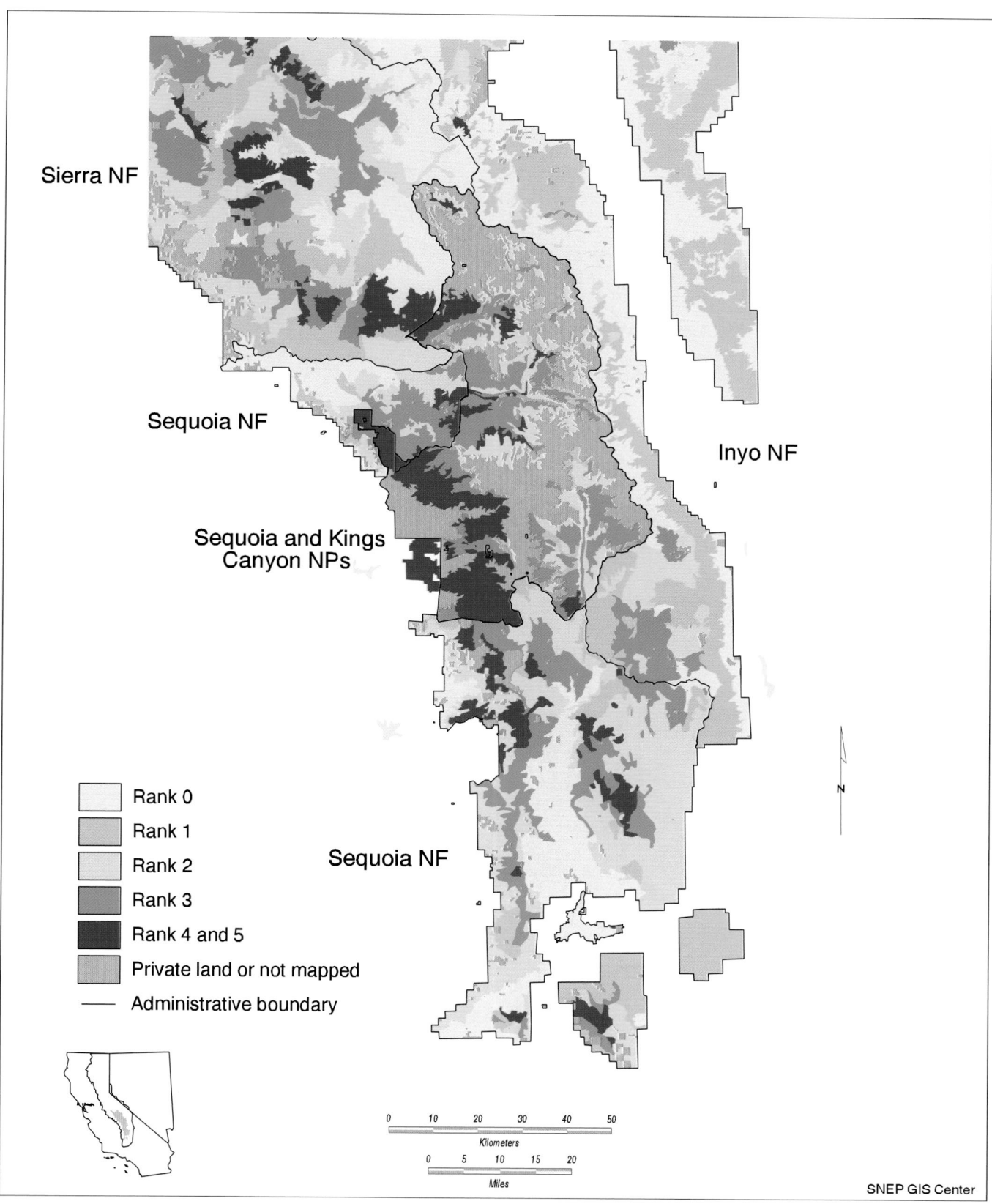

Sierra NF

Sequoia NF

Inyo NF

Sequoia and Kings
Canyon NPs

Sequoia NF

Rank 0
Rank 1
Rank 2
Rank 3
Rank 4 and 5
Private land or not mapped
— Administrative boundary

0   10   20   30   40   50
Kilometers

0   5   10   15   20
Miles

SNEP GIS Center

PLATE 21.6

Map of the southern section of the Sierra Nevada showing the distribution of polygons on federal lands color-coded as to their degree of LS/OG structural complexity and late successional forest function (rangewide structural standard).

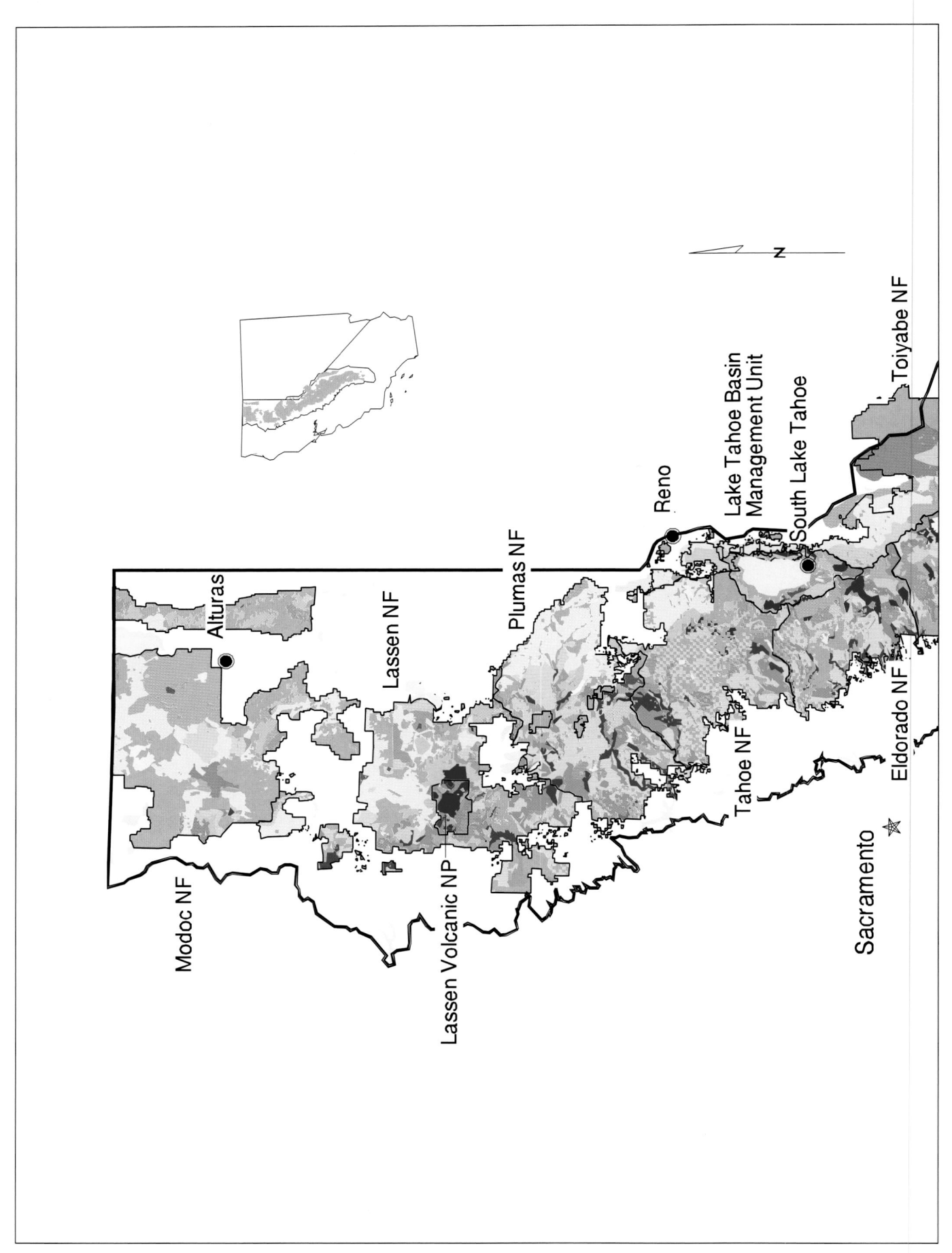

Modoc NF

Alturas

Lassen NF

Plumas NF

Reno

Lassen Volcanic NP

Lake Tahoe Basin
Management Unit

South Lake Tahoe

Tahoe NF

Toiyabe NF

Sacramento

Eldorado NF

N

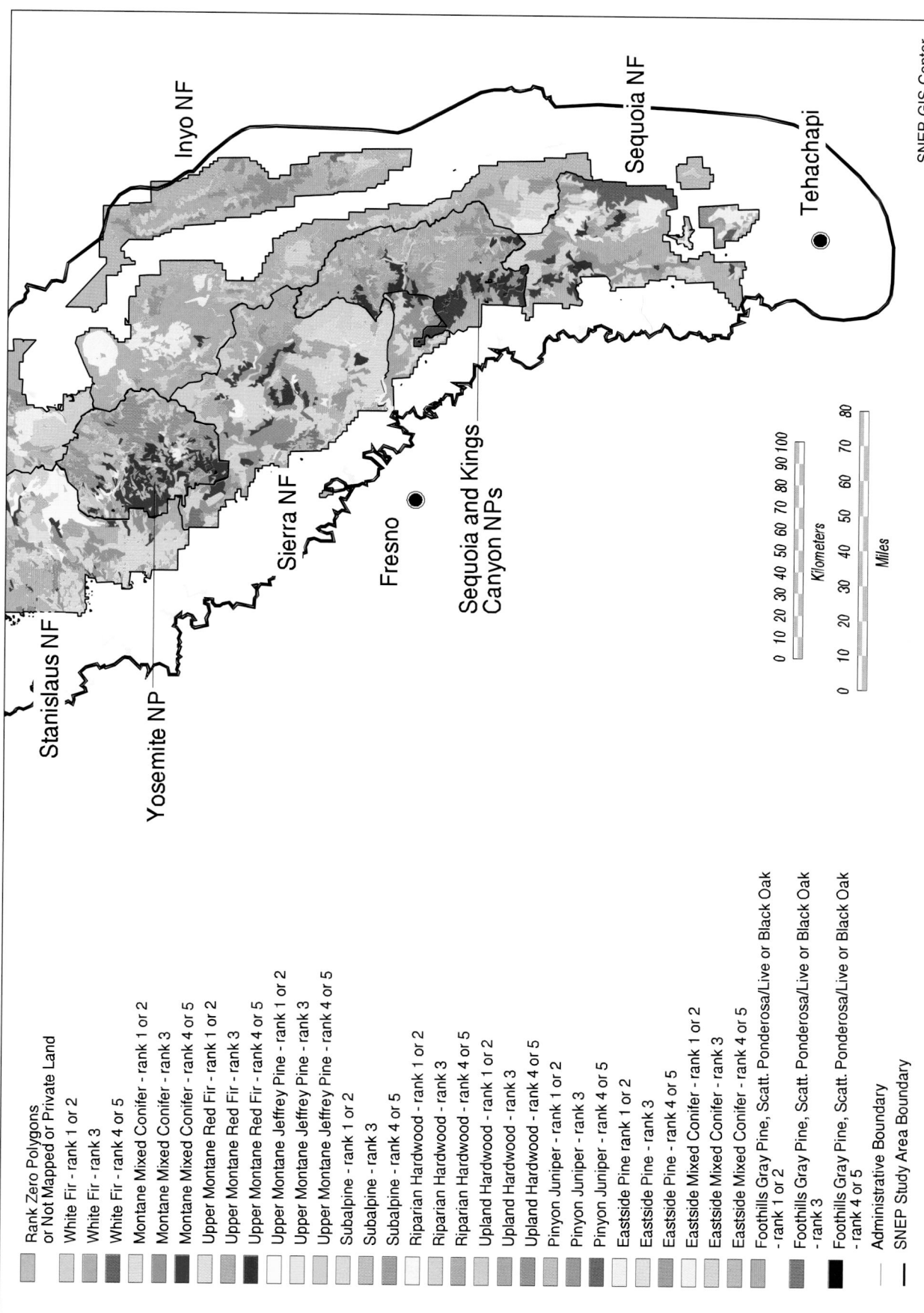

Legend

- Rank Zero Polygons or Not Mapped or Private Land
- White Fir - rank 1 or 2
- White Fir - rank 3
- White Fir - rank 4 or 5
- Montane Mixed Conifer - rank 1 or 2
- Montane Mixed Conifer - rank 3
- Montane Mixed Conifer - rank 4 or 5
- Upper Montane Red Fir - rank 1 or 2
- Upper Montane Red Fir - rank 3
- Upper Montane Red Fir - rank 4 or 5
- Upper Montane Jeffrey Pine - rank 1 or 2
- Upper Montane Jeffrey Pine - rank 3
- Upper Montane Jeffrey Pine - rank 4 or 5
- Subalpine - rank 1 or 2
- Subalpine - rank 3
- Subalpine - rank 4 or 5
- Riparian Hardwood - rank 1 or 2
- Riparian Hardwood - rank 3
- Riparian Hardwood - rank 4 or 5
- Upland Hardwood - rank 1 or 2
- Upland Hardwood - rank 3
- Upland Hardwood - rank 4 or 5
- Pinyon Juniper - rank 1 or 2
- Pinyon Juniper - rank 3
- Pinyon Juniper - rank 4 or 5
- Eastside Pine rank 1 or 2
- Eastside Pine - rank 3
- Eastside Pine - rank 4 or 5
- Eastside Mixed Conifer - rank 1 or 2
- Eastside Mixed Conifer - rank 3
- Eastside Mixed Conifer - rank 4 or 5
- Foothills Gray Pine, Scatt. Ponderosa/Live or Black Oak - rank 1 or 2
- Foothills Gray Pine, Scatt. Ponderosa/Live or Black Oak - rank 3
- Foothills Gray Pine, Scatt. Ponderosa/Live or Black Oak - rank 4 or 5
- — Administrative Boundary
- — SNEP Study Area Boundary

Map labels: Stanislaus NF, Inyo NF, Yosemite NP, Sierra NF, Fresno, Sequoia and Kings Canyon NPs, Sequoia NF, Tehachapi, SNEP GIS Center

Scale: Kilometers 0 10 20 30 40 50 60 70 80 90 100; Miles 0 10 20 30 40 50 60 70 80

PLATE 21.7

Distribution of polygons on federal lands throughout the Sierra Nevada color-coded as to major forest type group and degree of LS/OG structural complexity (series-normalized structural standard).

to polygon area. The sample in the second-stage consists of ground plots deployed within each sample first-stage unit (polygon). Before deploying field plots, however, each sample polygon was exhaustively partitioned into patches. This partitioning was performed by members of the original mapping team and was necessary to help ensure that all designated patch types within primary sample polygons would be sampled in the field. Also, it was specified that a minimum of two field plots be measured in each patch type occurring in a sample polygon so that within-patch averages could be estimated. The field plots were rectangular in shape and oriented with the long side parallel to the slope. A 2-chain-by-4-chain plot, 0.8 acres in size, was used to sample snags and down material. A 0.4-acre subplot was used to measure live trees greater than 24 in dbh (diameter breast height).

The FIA field plots were deployed before this study was contemplated and, therefore, independently of the SNEP polygon structure. Hence the data extracted from them were used solely in the context of single plot locations.

Data Items

During the field phase of the SNEP data collection effort, plots were deployed in the mixed conifer types on the Eldorado, Lassen, Sequoia, and Stanislaus national forests. The variables recorded on the SNEP field plots total 30 data items: 24 of these are structural components and 6 pertain to past occurrences of human intrusions and site quality.

The 6 specific items for intrusions and site quality are presence of an intermediate canopy, site class on a scale of 1 to 5, salvage harvest, selection harvest, tree thinning, and other human intrusion. All but site class are binary values.

The structural variables comprise number of live trees, number of snags, and down material. The 9 variables for number of live trees by 2 in classes are hardwoods 24–28 in dbh, 30–38 in, and greater than or equal to 40 in; true firs 24–28 in, 30–38 in, and greater than or equal to 40 in. The variables for number of snags by 2 in classes are hardwoods, with or without bark, 24–28 in dbh, 30–38 in, and greater than or equal to 40 in; true firs 24–28 in, 30–38 in, and greater than or equal to 40 in; and other conifers 24–28 in, 30–38 in, and greater than or equal to 40 in. The variables for down material of irregular length are true firs 20–28 in, measured at the large end, 30–38 in, and greater than or equal to 40 in and other conifers 20–28 in, 30–38 in, and greater than or equal to 40 in.

The data items extracted from the FIA database include the same structural variables as the SNEP plots. However, the human intrusion and site data comprise three instead of six variables:

1. presence of an intermediate canopy

2. site class on a scale of 1 to 5

3. a history code that is not compatible with the SNEP items concerning human intrusions

Analyses of SNEP Data

As the plot data came in, they were processed through a specially prepared program that screened for omissions, obvious mistakes, and internal consistency. Then, each variable was expanded to a per-acre basis and written to a new data file in a format compatible with our statistical analysis programs.

Discriminant Analysis

Discriminant analysis (DA) is a useful tool for obtaining a better understanding of the relationships among a set of independent variables and the population groups to which they belong. It is also used to classify individual entities, such as field plots, into unique groups based on those variables.

The specific problem that we address is how well it is possible to cluster the structural variables found on each field plot into discrete LSOG ratings as specified by the mapping team. We used discriminant analysis in three situations: (1) at the plot-level using SNEP data, (2) at the patch level using SNEP data, and (3) at the plot-level using FIA data.

Table 22.1 shows how the discriminant analysis distributed the SNEP plots for each of the mappers' assigned ratings into new ratings based on structural characteristics, including the intrusion and site variables. For example, in our set of 400 sample plots, the LSOG mappers assigned 58 plots to LSOG rating 1. Based on the structural characteristics measured in the field, however, the DA assigned 69.0% of those 58 plots to rating 1, 17.2% to rating 2, 10.3% to rating 3, and 1.7% each to ratings 4 and 5. The percentage of plots in each rating class for which the mappers' ratings correlate consistently with structural characteristics are shown in the diagonal elements of the classification matrix. These are the boldface values in table 22.1.

There are 10 variables that appear to be significant to the classification process at a probability level less than 0.10 (e.g., 90% level of confidence). These are:

1. intermediate canopy; probability of F .0003

2. site class; probability of F .0380

3. select harvest; probability of F .0002

4. other intrusions; probability of F .0677

5. live hardwood, 24 in–28 in; probability of F .0006

6. live hardwood, greater than or equal to 40 in; probability of F .0792

7. other conifer, 30 in–38 in; probability of F .0647

8. other conifer, greater than or equal to 40 in; probability of F .0000

9. hardwood snags, 30 in–38 in; probability of F .0000

10. down conifer, 30 in–38 in; probability of F .0101

Discriminant analysis is subject to the assumptions of normality in the independent (structural) variables, class by class, although the requirements for DA are less stringent than those for other statistical procedures. The data used in our analyses adhere to the normality assumptions in various degrees, that is, some variables appear to be close to normal (based on observing histograms), and others are definitely non-normal. The canopy, site class, and intrusion responses are especially susceptible to the non-normality condition when the sample size is small because, in reality, they are discrete binomial or multinomial variables rather than continuous variables. Therefore, while the probability levels reported for the significance tests may indicate the relative importance of specific variables, they are not absolute and must be viewed with some skepticism.

Table 22.1 shows the results of a discriminant analysis wherein the intrusion and site variables are included with the structural variables. Table 22.2 shows the results of a similar analysis that includes the structural variables only. A comparison of these two tables suggests that human intrusion factors strongly influence the LSOG rankings. Thus, the rankings should not be taken to reflect primarily seral stages, because they combine existing structural characteristics with those reflecting human influences.

The smaller set of variables used to obtain table 22.2 is roughly 5.5% less effective in reducing the classification error than the full set used for table 22.1. The largest difference between tables 22.1 and 22.2 shows up in rating class 3, where only 34% of the plots correspond to the structural characteristics observed on the ground compared with 46% as shown in table 22.1. The comparison indicates also that the canopy, human intrusion, and site variables, taken together, account for about 11% of the total classification accuracy in table 22.1.

When incorporating structural variables only in the DA, There are eight variables that appear to be significant to the classification process at a probability level less than 0.10:

1. live hardwood, 24 in–28 in; probability of F .0002

2. live hardwood, 30 in–38 in; probability of F .0880

3. live true firs, 24 in–28 in; probability of F .0451

4. other conifers, 30 in–38 in; probability of F .0171

5. other conifers, greater than or equal to 40 in; probability of F .0000

6. hardwood snags, 30 in–38 in; probability of F .0000

7. down conifers, 20 in–28 in; probability of F .0430

8. down conifers, 30 in–38 in; probability of F .0083

It is important to note that the values shown in tables 22.1 and 22.2 reflect the structural characteristics of the forest at the plot-level, not at the patch level, thus encompassing both within- and between-patch variability. Table 22.3 shows the results of a discriminant analysis at the patch level using the same plots as those used in tables 22.1 and 22.2 but averaged at the patch level. Only structural components are used for generating table 22.3, however, because the human intrusion variables, being discrete, cannot be averaged among plots within a patch. Hence table 22.3 should be compared with table 22.2.

According to the classification statistics computed by the discriminant analysis, there is a 49.5% reduction in classification error due to the structural components when averaged at the patch level. In the plot-level analysis of table 22.2, there is only a 30.3% reduction, indicating that 19.2% of the reduction in classification error is due to the within-patch averaging of structural components. The main differences among the LSOG rating assignments can be seen by comparing the two classification tables, especially the diagonal elements.

When incorporating structural variables at the patch level, we obtain five variables that appear to be significant to the classification process at a probability level less than 0.10:

1. live other conifers, 30 in–38 in; probability of F .0093

2. live other conifers, greater than or equal to 40 in; probalbility of F .0048

3. hardwood snags, 30 in–38 in; probability of F .0000

4. down other conifers, 20 in–28 in; probability of F .0537

5. down other conifers, 30 in–38 in; probability of F .0036

TABLE 22.1

SNEP plot-level data, mixed conifer forest type, classification matrix for all variables. Total number of correct classifications = 199 (49.8%).

| Mappers' Rating | Number of Plots | Rating Assignments from DA% | | | | | |
		1	2	3	4	5	Total
1	58	**69.0**	17.2	10.3	1.7	1.7	100
2	114	13.2	**63.2**	18.4	5.3	0.0	100
3	94	9.6	36.2	**45.7**	7.4	1.1	100
4	128	7.8	26.6	29.7	**32.8**	3.1	100
5	6	16.7	33.3	16.7	0.0	**33.3**	100
Total	400						

TABLE 22.2

SNEP plot-level data, mixed conifer forest type, classification matrix for all variables. Total number of correct classifications = 177 (44.3%).

Mappers' Rating	Number of Plots	Rating Assignments from DA%					
		1	2	3	4	5	Total
1	58	**63.8**	24.1	8.6	1.7	1.7	100
2	114	15.8	**62.3**	16.7	5.3	0.0	100
3	94	11.7	44.7	**34.0**	8.5	1.1	100
4	128	8.6	35.2	25.8	**27.3**	3.1	100
5	6	0.0	33.3	33.3	0.0	**33.3**	100
Total	400						

These results seem to indicate that the LSOG classifications assigned by the mapping team are more consistent at the patch level than the plot-level, because there is obviously a substantial within-patch variability in structural components that is absorbed by averaging plots within patches. On the other hand, the overall differences among LSOG ratings are less significant at the patch level than at the plot-level.

Analyses of FIA Data

In terms of structural components and intrusion variables, the FIA data set yielded classification results similar to the SNEP data when subjected to discriminant analyses. However, when the SNEP and FIA data sets were made factors in a two-way multivariate analysis of variance (MANOVA), the two sets tested as being significantly different at the .05 probability level. Because of this, it was inappropriate to combine the two sets of data directly in one large discriminant analysis. Instead, the results from the two data sets were reported separately.

When employing the FIA data set, we again concentrated on the mixed conifer forest type with data obtained from the Lassen, Plumas, Sequoia, and Stanislaus national forests. Only plot-level data were used, of course, because the FIA plots were not deployed in concert with SNEP polygons or patches. Both structural components and intrusion factors were used

in analyzing the FIA data. However, the intrusion variables were different here than in the SNEP data, as noted earlier.

In table 22.4, it appears that the rating assignments made by the DA are clustered somewhat more tightly around the diagonal elements than for any of the SNEP analyses. Also, the results seem to be somewhat more accurate, because correct classifications were obtained 53.0% of the time compared with the 49.8% correct classification rate shown in table 22.1, a modest increase. No explanation is available for these differences.

On the other hand, the averages of many of the structural components are significantly different when comparing the SNEP and FIA data sets. This indicates either that they represent different geographical areas or that the measurement standards and/or definitions for structural components mean different things to different people. There is credence to the notion that the FIA plots represent different geographical areas than the SNEP plots. We obtained no FIA plots from the Eldorado National Forest, where many SNEP Plots are located, and conversely, there are no SNEP plots in the mixed conifer type in the Plumas, Sierra, or Tahoe National Forests, where FIA plots are located. If, indeed, there are significant differences among the structural variables between geographical subareas, then each major forest type should be further stratified into smaller spatial units such as national forests, counties, or natural watersheds.

TABLE 22.3

SNEP patch-level data, mixed conifer forest type, classification matrix for all variables. Total number of correct classifications = 87 (59.6%).

Mappers' Rating	Number of Plots	Rating Assignments from DA%					
		1	2	3	4	5	Total
1	14	**64.3**	28.6	7.1	0.0	0.0	100
2	46	6.5	**78.3**	15.2	0.0	0.0	100
3	40	2.5	40.0	**52.5**	5.0	0.0	100
4	43	2.3	23.3	30.2	**44.2**	0.0	100
5	3	0.0	33.3	0.0	0.0	**66.7**	100
Total	146						

Only three variables in the FIA data set are sufficiently significant to seriously affect the results if removed from the analysis:

1. Intermediate canopy; probability of F .0794

2. Live hardwoods, greater than or equal to 40 in; probability of F .0655

3. True fir snags, 30 in–38 in; probability of F .0620

Structural Characteristics

From the analyses of both the SNEP and FIA data, we obtained table 22.5, showing the average value (e.g., number of pieces per acre) of each structural component by LSOG rating class. This table may be useful because it shows the average structure of the forest in each LSOG rating class as these ratings are currently thought of by the mapping team. By working with this table, it should be possible to construct definitions for the expected structural composition of LSOG rating classes and, perhaps, correlate the structural composition of patches to the variables used by the mappers to assign ratings in the first place.

CONCLUSIONS

Because of limitations in the geographical scope of the data we were able to obtain, this discussion is limited to portions of the mixed conifer type as defined by the LSOG mapping team.

The LSOG maps prepared by the SNEP team can serve as a tool for stratifying the Sierra Nevada into broad groups of late successional forest structural patterns. Our analyses indicate, however, that there is a high level of structural diversity for any given assigned patch rating.

The amount of diversity in stand structures within patches of different LSOG ratings is exemplified in tables 22.2 and 22.3. In table 22.2, "correct" assignments of patch rating at the plot-level ranged from 27% to 64% with an average classification accuracy of only 44.3%. For the patch level averages shown in table 22.3, "correct" assignments of LSOG ratings were made between 44% and 78% of the time, with an average accuracy of 59.6%; this would indicate that considerable smoothing takes place when averaging at the patch level.

When accuracies such as these are combined at the polygon level, substantial variations are likely to occur, particularly because different combinations of patch ratings occur in different polygons having similar overall LSOG ratings. Within polygons, the classification accuracies vary considerably among rating classes at the patch level. When assessing polygons, therefore, it may be useful to note the patch ratings within the polygons being evaluated to obtain an indication of reliability at the polygon level.

The magnitude of the classification errors shown in this report indicate that it would be dangerous to attempt detailed site-specific predictions of forest structure at the plot, patch, or even polygon levels directly from the LSOG maps. On the other hand, it may be feasible to use the LSOG ratings with structural values to simulate average stand development over larger land areas, such as national forests, counties, or large watersheds. The average values of structural components, such as those shown in table 22.5, illustrate the kind of data that might be used for this purpose. There is a problem, however, in that the values in table 22.5 were derived from plot-level data. To be utilized properly, the SNEP LSOG maps require data that are averaged at the patch level. Unfortunately, no large quantity of data exists to satisfy this condition over the range of the Sierra Nevada Ecosystem Project. Also, it would be difficult to obtain such data because, except for a few sample polygons, no patch boundaries within polygons are defined on the maps.

Returning to the primary questions posed in the introduction, how consistent are the patch ratings assigned by the mapping team based on structural characteristics of the forest? At the patch level, the mappers were consistent in making rating assignments about 60% of the time overall, at least

TABLE 22.4

FIA plot data, mixed conifer forest type, classification matrix for all variables. Total number of correct classifications = 98 (53.0%).

| Mappers' Rating | Number of Plots | Rating Assignments from DA% | | | | | |
		1	2	3	4	5	Total
1	10	**50.0**	50.0	0.0	0.0	0.0	100
2	77	14.3	**63.6**	19.5	1.3	1.3	100
3	49	12.2	26.5	**53.1**	6.1	2.0	100
4	43	4.7	27.9	27.9	**34.9**	4.7	100
5	6	0.0	16.7	33.3	0.0	**50.0**	100
Total	185						

TABLE 22.5

SNEP and FIA plot-level data, average values of structural variables pieces per acre.

| | LSOG Rating Class | | | | | | | | | |
| | 1 | | 2 | | 3 | | 4 | | 5 | |
Variable	SNEP	FIA	SNEP	FIA	SNEP	FIA	SNEP	FIA	SNEP	FIA
Live trees										
Hardwood, 24"–28"	.34	.00	.59	.32	.37	.67	1.00	.51	.83	.39
Hardwood, 30"–38"	.23	.00	.24	.14	.13	.21	.23	.15	.83	.00
Hardwood, >=40"	.00	.00	.13	.03	.08	.01	.01	.06	.00	.00
True firs, 24"–28"	.09	2.21	2.39	2.73	3.03	3.77	3.34	3.79	6.25	4.26
True firs, 30"–38"	.17	.74	1.45	1.31	2.42	2.09	3.16	2.26	3.33	2.69
True firs, >=40"	.26	.32	.46	.49	1.14	.94	1.21	.69	2.08	.89
Other conifers, 24"–28"	2.16	3.15	3.36	3.64	3.59	4.65	3.98	5.88	2.50	4.80
Other conifers, 30"–38"	1.16	1.77	2.48	2.33	2.13	3.86	3.85	4.06	.42	3.68
Other conifers, >=40"	.26	1.12	1.23	.90	1.36	1.73	3.22	1.81	3.33	2.67
Snags										
Hardwood, 24"–28"	.11	.00	.03	.02	.05	.04	.09	.01	.42	.00
Hardwood, 30"–38"	.02	.00	.00	.01	.01	.03	.04	.04	.83	.00
Hardwood, >=40"	.02	.00	.00	.01	.00	.01	.00	.01	.00	.00
True firs, 24"–28"	.02	.24	.36	.25	.51	.33	.62	.62	.21	.60
True firs, 30"–38"	.11	.24	.19	.13	.29	.22	.56	.43	.62	.87
True firs, >=40"	.00	.00	.08	.10	.25	.19	.25	.23	.21	.47
Other conifers, 24"–28"	.17	.20	.31	.22	.48	.20	.28	.20	.21	.40
Other conifers, 30"–38"	.17	.28	.36	.31	.37	.30	.40	.30	.00	.20
Other conifers, >=40"	.11	.12	.23	.05	.33	.17	.47	.15	.00	.33
Down Material										
True firs, 20"–28"	.19	.40	1.41	.85	1.64	.93	1.84	1.12	2.08	2.80
True firs, 30"–38"	.04	.24	.55	.50	.73	.33	.85	.45	1.46	1.73
True firs, >=40"	.04	.00	.43	.11	.65	.20	.62	.02	1.46	.53
Other conifers, 20"–28"	1.53	2.08	1.96	2.32	2.47	2.14	2.54	1.60	2.71	1.60
Other conifers, 30"–38"	.62	1.20	1.46	.77	1.17	.57	1.31	.47	3.12	.80
Other conifers, >=40"	.28	.08	.90	.27	.84	.16	1.06	.15	1.67	.13

in the mixed conifer forest type. The reliability of these assignments is higher in the lower rating classes (about 65% for classes 1 and 2) and lowest in the higher ratings (about 44% for class 4).

What are the structural characteristics of each assigned LSOG rating when measured on a scale of 1 to 5? Table 22.5 summarizes these results to the extent they are known at present.

REFERENCES

Hair, J. F. 1992. *Multivariate data analysis with readings.* New York: Macmillan.

James, M. 1985. *Classification algorithms.* New York: John Wiley.

Manly, B. F. J. 1994. *Multivariate statistical methods: A primer.* New York: Chapman and Hall.

Number Cruncher Statistical System. Product 5.3, Advanced Statistics. Jerry L. Hintze, Kaysville, Utah. This is the statistical software package used for this study.

FRANK W. DAVIS
Institute for Computational Earth System
  Science
University of California
Santa Barbara

DAVID M. STOMS
Institute for Computational Earth System
  Science
University of California
Santa Barbara

23

Sierran Vegetation: A Gap Analysis

ABSTRACT

Gap analysis assesses the distribution of plant community types among land classes defined by ownership and levels of protection of biodiversity. Gap analysis helps to identify which plant communities and species might be especially vulnerable to different human activities that can lead to habitat conversion or degradation.

This chapter presents a gap analysis of plant community types for the Sierra Nevada region, an area of 63,111 km^2 (24,367 mi^2). Ownership of the region is 37% private, 47% national forests, 10% national parks, 5% Bureau of Land Management, and less than 2% other public lands. Land ownership and land management patterns contrast sharply between the northern Sierra Nevada and the central and southern subregions. Parks and reserve lands constitute less than 2% of the northern region versus 27% of the central/southern.

We mapped eighty-eight natural plant community types within the region. Sixty-seven types were mapped over areas greater than 25 km^2 (9.65 mi^2). The ownership profiles of Sierran plant communities systematically reflect the concentration of private lands at lower elevations and of national parks in the central and southern portion of the range. Less than 1% of the foothill woodland zone of the Sierra Nevada is in designated reserves or other areas managed primarily for native biodiversity, and over 95% of the distribution of most foothill community types is available for grazing. Low- to middle-elevation Sierran forests are not well represented in designated reserves, especially in the northern Sierra Nevada. However, large areas of most of these forest types on U.S. Forest Service lands have been administratively withdrawn from intensive timber management based on current forest plans. Many high-elevation forest and shrubland community types are well represented in parks and ungrazed wilderness areas. Our analysis identifies thirty-two widespread community types whose conservation status warrants concern and twelve types that appear well protected based on their present distributions.

INTRODUCTION

Because land ownership and administrative designation establish the kinds of human activities that can occur in an area, they are usually strongly related to biodiversity status and trends. A map showing how native species and communities are distributed with respect to categories of ownership and conservation management helps to identify which elements of biodiversity might be especially vulnerable to habitat conversion or degradation. Gap analysis makes such an assessment by overlaying maps of land ownership and management onto maps of the distributions of plant community types (Scott et al. 1993). Community types and species whose distributions fall largely outside the areas whose primary management objective is to conserve native biodiversity are identified as "gaps" in biodiversity conservation.

The gap analysis of the Sierra Nevada described in this chapter represents a collaboration between the Sierra Nevada Ecosystem Project (SNEP) and the National Biological Service Gap Analysis Program (GAP). The goals of GAP are (1) to identify vegetation types and vertebrate species that are underrepresented in areas managed primarily for native biodiversity, and (2) to locate sites for new management areas where additional conservation measures could efficiently reduce the vulnerability of native biodiversity (Scott et al. 1993). This chapter focuses on the first goal and is confined to a gap analysis of vegetation types.

By quantifying broad patterns of land ownership/management in relation to vegetation, the gap analysis of the Sierra Nevada contributes one piece to SNEP's overall assessment of the region's biodiversity. It is not our objective in this chapter to provide a detailed description of Sierran vegetation, to analyze its past or current ecological condition, or to address

Sierra Nevada Ecosystem Project: Final report to Congress, vol. II, *Assessments and scientific basis for management options*. Davis: University of California, Centers for Water and Wildland Resources, 1996.

specific alternatives pertaining to vegetation management and conservation. These questions are addressed by other chapters. We describe and apply a model for siting new management areas based on the results of gap analysis in volume 1 of this report.

ASSESSMENT AREA AND QUESTIONS

The gap analysis of California is being conducted on a regional basis (Davis et al. 1995) using the ten major physical regions of California as defined in *The Jepson Manual of Higher Plants of California* (Hickman 1993). The Sierra Nevada Region encompasses 63,111 km² (24,367 mi²) extending from Tejon Pass at the southern end to the North Fork of the Feather River at the north. That region overlaps 73% of the SNEP core area. The remainder of the SNEP core area falls within other Jepson regions: Mojave Desert (2.7%), Great Basin East of Sierra Nevada (11.3%), Modoc Plateau (4.2%), and Cascades Region (8.3%) (figure 23.1). The gap analysis reported here pertains only to the Jepson Sierra Nevada Region. The remaining areas will be treated in subsequent regional analyses as part of the statewide gap analysis.

Because of the size and biological heterogeneity of the Sierra Nevada, we also conducted gap analyses for a northern versus a central/southern subregion divided at the Stanislaus River. In *The Jepson Manual* the Stanislaus River divides the northern from the central and southern Sierra Nevada.

The following digital geospatial data were compiled for this analysis:

- topography (100 m [328 ft] grid)

- vegetation (classified to Holland types using a 100 ha [247 acre] minimum mapping unit [mmu]. The mmu is the nominal extent of the smallest mapped feature.)

- dominant plant species (100 ha [247 acre] mmu)

- land ownership and administrative designation in terms of conservation (200 ha [494 acre] mmu)

- U.S. Forest Service (USFS) grazing allotment boundaries (1 ha [2.47 acre] grid)

- USFS land suitability classes (1 ha [2.47 acre] grid)

These data were analyzed to address four specific questions:

1. How do land ownership and land management vary among elevation zones?

2. What are the sizes and locations of existing parks, wilderness areas, and reserves?

3. How is each terrestrial plant community type distributed with respect to land ownership and conservation management?

4. Which major terrestrial plant community types may be vulnerable to degradation of habitat and which types appear to be relatively well protected based on their current management profile?

METHODS

Detailed descriptions of the gap analysis approach and methods can be found in Scott et al. 1993, Beardsley and Stoms 1993, and Davis et al. 1995.

Land Ownership and Land Management

GAP classifies land ownership and management into four categories intended to capture the degree to which the land is managed to maintain biodiversity (Scott et al. 1993). We depart slightly from the GAP categories by distinguishing lands based on permitted use. We assume that the most pervasive land uses affecting the status and trends of terrestrial biodiversity in the Sierra Nevada are grazing, fire suppression, timber harvest, and urban, residential, and agricultural development. Other activities, such as recreation, trapping, and mining, are certainly important but more localized and/or less readily mapped. Thus we have distinguished five ownership/management classes based on fire policy and on potential for development, timber harvest, or grazing.

Class 1: public or private land formally designated for conservation of native biodiversity and within which economic activities such as development, grazing, and timber harvest are precluded. Natural disturbance events are generally allowed to proceed without interference or are mimicked through management. The areas may be used for primitive recreational activities. Examples include national parks, national monuments, ungrazed lands within USFS wilderness areas, USFS research natural areas, USFS wild and scenic rivers, Blue Ridge National Wildlife Refuge, The Nature Conservancy preserves, and state parks and ecological reserves. (See appendix 23.1 for a listing of Class 1 areas.)

Class 2: national forest land that is generally managed for its natural values but is not formally designated for conservation of native biodiversity. Development and grazing are excluded, and timber harvest is generally excluded because it conflicts with other multiple-use objectives. Wildfires are generally suppressed. The distribution of recreational activities on Class 2 lands is

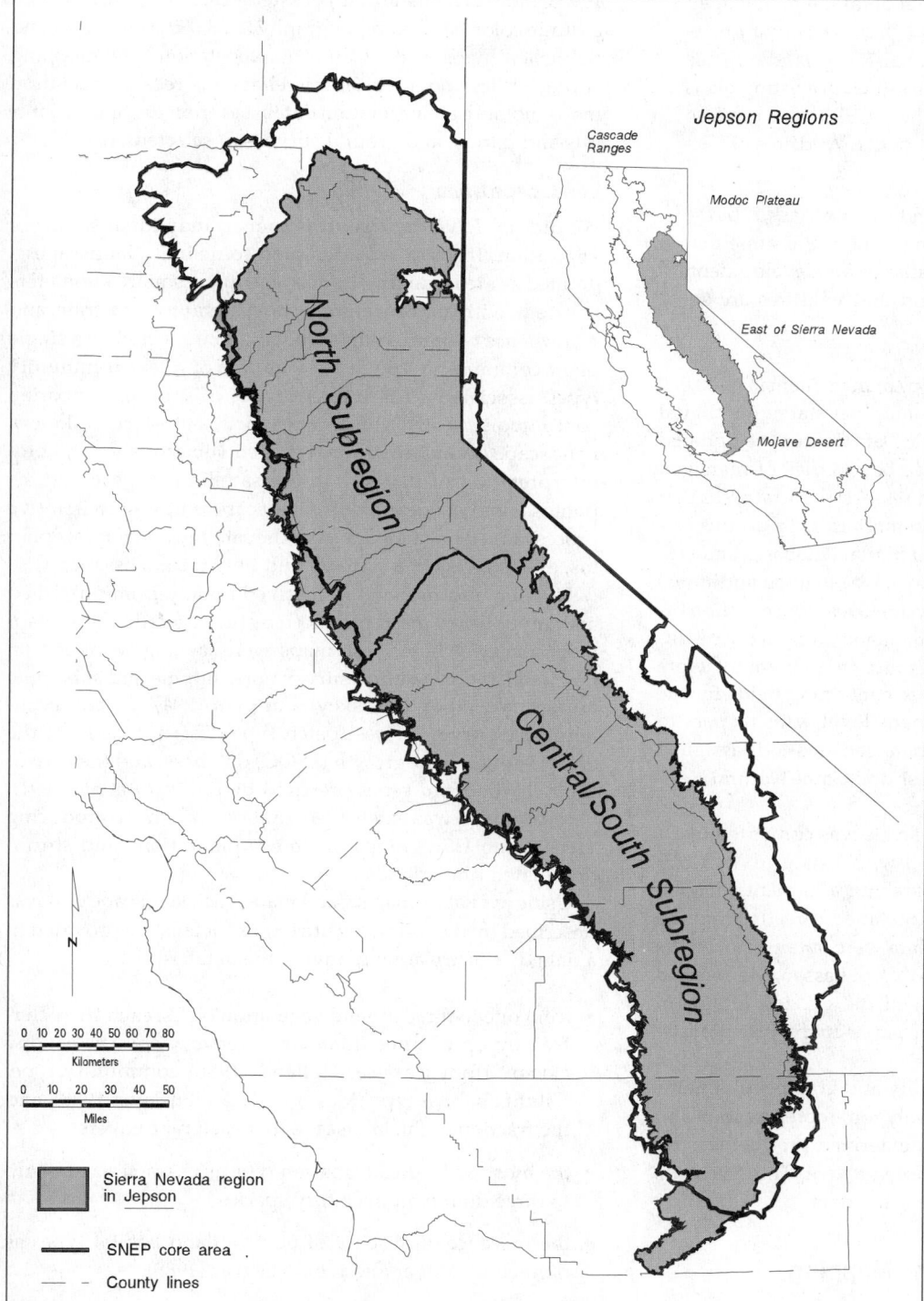

FIGURE 23.1

Regional location map of
Jepson Sierra Nevada
Region in relation to the
SNEP core area.

unknown, but a small fraction of the land is developed for recreational facilities.

Class 3: public land that is generally managed for its natural values, is treated in existing management plans as unsuitable for timber harvest, and may be grazed. Wildfires may be actively suppressed. Examples include grazing allotments within USFS wilderness areas, grazing allotments on national forest lands classified as unsuitable for timber harvest, the San Joaquin Experimental Range, Bureau of Land Management (BLM) areas of critical environmental concern, and BLM wilderness areas.

Class 4: other public lands not included in Classes 1

through 3, mainly multiple-use federal lands managed by the Bureau of Indian Affairs (BIA), Bureau of Reclamation, BLM, and USFS. National forest lands in this category include areas that are classified in existing plans as suitable for timber harvest. These USFS areas can also be within existing grazing allotments. Wildfires are actively suppressed.

Class 5: private lands other than those in Class 1. In the absence of more detailed zoning data, we assume that these lands are potentially available for development, timber harvest, and grazing and that wildfires are actively suppressed.

The base map for land ownership/management is 1:100,000 BLM surface management status maps. A statewide digital coverage was provided by the Teale Data Center. We updated and enhanced this map to include boundaries of managed areas such as wilderness areas and research natural areas that do not coincide with ownership boundaries. To do this, we consulted national forest maps and digital databases and U.S. Geological Survey topographic maps. We obtained additional maps and information from many agencies, conservation organizations, and land trusts. All managed areas in the resulting regional map of land ownership/management were described in an associated database containing fields for the managing agency, the management level with respect to biodiversity conservation, and a managed area code assigned by the California Department of Fish and Game Natural Heritage Division.

The map of land management levels was converted to a 1 ha (2.47 acre) grid and intersected with 1 ha grids of USFS land suitability class maps and grazing allotments. Digital land suitability class maps were obtained directly from the USFS. Digital grazing allotment data were obtained from the USFS for all of the national forests except Lassen, Modoc, and the Lake Tahoe Basin. We digitized the grazing allotment boundaries on these forests from paper maps provided by USFS range conservation staff.

Maps of timber harvest suitability and grazing allotments were converted back to a vector (polygon) representation and overlaid with land ownership. The derived product was reclassified into the five classes defined above. This five-class map was then overlaid with vegetation data.

Vegetation Classification and Mapping

Vegetation types were classified based on overstory structure, cover, and dominant species composition. The overstory is described by one to three species, each contributing more than 20% of the relative canopy cover. These species assemblages (Davis et al. 1995) were subsequently reclassified into natural plant community types used by the California Department of Fish and Game Natural Heritage Division (Holland 1986).

Maps of actual vegetation were produced using summer 1990 Landsat Thematic Mapper satellite imagery, 1985–90 high altitude color infrared photography (1:58,000 scale), draft and published maps of the California vegetation type mapping survey (Wieslander 1946), miscellaneous recent vegetation maps (notably the vegetation databases from the national forests and parks), and ground surveys of selected areas.

Landscape Units

We did not have the resources to map individual stands of vegetation. Instead, we attempted to delimit "landscapes," defined as areas ranging from one to many square kilometers in extent, with uniform climate, physiography, substrate, and disturbance regime. A landscape could be covered by a single plant community type or by a mosaic of a few community types associated with different types of sites (e.g., riparian zones, moist north-facing slopes, dry south-facing slopes). Landscape boundaries were mapped subjectively by photo-interpretation of patterns in the satellite imagery and air photos. Final delineation of a landscape unit was an iterative process based on evidence from the satellite imagery, air photos, existing vegetation maps, and field reconnaissance.

Floristic information was derived mainly from published and unpublished maps produced by the vegetation type mapping survey. Where these maps were lacking we relied on USFS soil and vegetation survey notes (alpine and subalpine areas surveyed by R. Taskey), our own 1994/95 field reconnaissance surveys, forest patch type descriptions from the SNEP late seral old-growth (LSOG) database, and the map of foothill woodland types prepared by Pillsbury et al. (1991). Our draft map was extensively updated in timber-producing areas using USFS maps of timber plantations and shrub-dominated timberlands.

Using available imagery and maps, each landscape unit was described by the following attributes (details are provided in a data dictionary accompanying the database):

- from one to three upland vegetation types, each characterized by up to three dominant overstory species, canopy closure (four classes), Holland (1986) community type, wildlife habitat type (Mayer and Laudenslayer 1988), and the fraction of the landscape that each type covers

- the most widespread riparian type as characterized by up to three dominant overstory species

- the presence or absence of nine wetland habitat types as defined by Mayer and Laudenslayer (1988)

- miscellaneous data, including evidence of disturbance in the landscape, occurrence of species of special interest, air photo identification number, information sources, University of California, Santa Barbara (UCSB) analyst, and comments

The draft database for the Jepson Sierra Nevada Region consists of 6,724 landscape units providing distributional information on 189 dominant species, 88 plant community types, and 35 wildlife habitat types. Analysts can query the database to retrieve distribution data on individual species, unique combinations of species, or vegetation types defined by physiognomy and/or composition.

Vegetation Map Accuracy

Because source information ranged widely in date and reliability, the current database is uneven in both level of detail and accuracy. We did not have the resources to assess the statistical accuracy of the vegetation map and associated database. However, we have appraised the product using less formal methods that have guided our use of the product. Based on UCSB field surveys in 1994 and 1995 and on comparisons with independent sources of vegetation data, the vegetation map probably overestimates the extent of conifer forest types and underestimates the extent of shrubland and middle-elevation hardwood forest types. Floristic information is more reliable in the northern and central subregions than in the southern subregion, which was only partially covered by the mapping survey of vegetation types. Floristic information is also more reliable on public lands than on private lands and better for the national parks than for the national forests. The data on upland community types and wildlife habitat types are more reliable than information on individual species or on wetland or meadow habitats. We will continue to revise the vegetation data based on review and testing by interested parties.

ASSUMPTIONS AND LIMITATIONS OF GAP ANALYSIS

Gap analysis provides a regional overview of the distribution and ownership profile of major terrestrial plant communities and vertebrate species habitats. It is not a substitute for a detailed biological inventory. Our assessment focuses on floristically defined plant community types and does not account for variations in stand age or physical stature within a type. For example, we do not distinguish late seral old-growth forest from younger forest of the same general community type.

The extent and spatial scale of the input maps of vegetation, wildlife habitat, and land management make a formal, statistical analysis of map accuracy impractical for both financial and logistical reasons. As a result, we cannot with confidence place error terms on our estimates of area or management status of plant communities.

The method that we used to map vegetation is not suited to the analysis of most wetland types or other communities that are restricted to very local environments. The mapping method is well suited to analysis of shrubs and trees, but it provides little or no information on the distribution of herba-

ceous species. Our analyses assume that the vegetation types attributed to a map unit (polygon) are dispersed uniformly throughout the unit.

Estimates of area made from maps are very sensitive to map scale and mapping methods. For example, vegetation types that typically occur in small patches may be overlooked or their extent underestimated using a vegetation map with relatively coarse spatial resolution. Our vegetation map is less sensitive to spatial resolution than traditional paper maps, because we maintain database records of secondary and tertiary vegetation types that are too fine to map using a 100 ha (247 acre) mmu. The point to remember is that our estimates of the acreage and distributions of species and types may differ considerably from areal estimates and from distributions of the same types derived from maps prepared at a finer or coarser resolution.

Land ownership/management profiles provide a crude measure of risk of development or resource overexploitation. We assume that native species are at risk in areas that have no legal or legislative mandate to protect and maintain self-sustaining natural ecosystems. Species and communities can also be at risk due to climatic change, introduced competitors and pathogens, and many other ecological factors. Furthermore, there is wide variation in land management practices within each of our five ownership/management classes. Some private lands are well managed for the maintenance of plant diversity, and some reserves are managed in a way that threatens some native species. Private land management also depends heavily on zoning status. Data on county zoning are needed for a fuller analysis of present and future management of private lands.

The static nature of the gap analysis data also limits their utility in assessing conservation risks. Our database provides a snapshot of a region in which land cover and land ownership are both very dynamic.

MAJOR FINDINGS

Results for the Jepson Sierra Nevada Region as a whole are presented first, followed by analyses of northern versus central/southern subregions.

Sierra Nevada Region as a Whole

We mapped the Jepson Sierra Nevada Region over an area of 63,111 km^2 (24,367 mi^2). We classified 56,587 km^2 (21,848 mi^2) (89.7%) of this area as vegetated (table 23.1). Non-vegetated areas included urban areas, lakes, reservoirs, rock outcrops, and alpine areas with little or no vascular plant cover.

Thirty-seven percent of the region is privately owned. The remainder, in public lands, is largely national forests (47%) and national parks (10%). The Bureau of Land Management

TABLE 23.1

Ownership and area of plant community types of the Sierra Nevada.

Type of Plant Community (Holland 1986)	Holland (1986) Code	Percentage of Mapped Distribution by Ownership									Total Mapped Distribution Area (km²)
		Private	Nongovernmental Organization(s)	County and Regional	State	Department of Defense	Other U.S. Department of the Interior[a]	National Park Service	Bureau of Land Management	U.S. Forest Service	
Scrub											
Mojave creosote bush scrub	34100	50							39	11	7
Mojave mixed scrub and steppe	34200	28							71	2	261
Mojave mixed woody scrub	34210	71							19	9	8
Blackbush scrub	34300	35							61	4	164
Great Basin mixed scrub	35100	17			4	< 1	< 1		17	61	303
Big sagebrush scrub	35210	22				1		1	16	59	183
Low sagebrush scrub[b]	35211	10			2				8	79	156
Silver sagebrush scrub[b]	35212	10							1	89	16
Subalpine sagebrush scrub	35220	36			5		1		7	51	25
Sagebrush steppe	35300	23			1	< 1	< 1		31	45	822
Rabbitbrush scrub	35400	7								93	46
Cercocarpus ledifolius woodland[b]	35500	4			1			1	1	94	252
Wyethia mollis[b]	35600	27							< 1	73	30
Chaparral											
Upper Sonoran mixed chaparral	37100	39								61	6
Northern mixed chaparral	37110	29			< 1	< 1		4	15	51	176
Chamise chaparral	37200	52			< 1	2	< 1	3	24	18	820
Semidesert chaparral	37400	39						8	26	27	109
Mixed montane chaparral	37510	21			1		< 1	3	1	73	1,388
Montane manzanita chaparral	37520	44			< 1	< 1	1	< 1	5	49	457
Montane ceanothus chaparral	37530	26			1			< 1	1	72	195
Deer brush chaparral	37531	75								25	2
Shin oak brush	37541	60			< 1			22	3	15	42
Huckleberry oak chaparral	37542	22			< 1			12		66	179
Bush chinquapin chaparral	37550	11			1			2	1	85	77
Buck brush chaparral	37810	22			1	< 1		1	10	65	155
Scrub oak chaparral	37900	71			4		3		15	8	48
Interior live oak chaparral	37A00	71	< 1		< 1		1	1	8	19	203
Upper Sonoran manzanita chaparral	37B00	18						2	33	47	163
Ione chaparral	37D00	96							4		1
Mesic north-slope chaparral	37E00	16					4	3	11	65	132
Upper Sonoran subshrub scrub	39000	50				1			22	26	42
Herbaceous											
Valley needlegrass grassland	42110	80						3		17	31
Non-native grassland	42200	88	< 1		2	1	< 1	< 1	2	7	1,923
Montane meadow	45100	14						38	< 1	48	127
Wet subalpine or alpine meadow	45210	20			2	< 1		13	3	61	201
Dry subalpine or alpine meadow	45220	10								90	4
Great Basin montane meadow[b]	45230	100								< 1	1
Alkali meadow	45310	82	5			7			6		2
Transmontane alkali marsh	52320	92	1			5			2		17
Riparian Woodland											
Great Valley cottonwood riparian forest	61410	73	8			5			2	13	20
Great Valley mixed riparian forest	61420	63			2	4			< 1	31	12
Great Valley valley oak riparian forest	61430	97			1	2			< 1		16
White alder riparian forest	61510	34								66	5
Aspen riparian forest	61520									100	0.1
Montane black cottonwood riparian forest	61530	40						56	2	1	6
Montane riparian scrub	63500	38			< 1			16	3	43	119

[a]Includes the Bureau of Reclamation, the Bureau of Indian Affairs, and the U.S. Fish and Wildlife Service.
[b]Addition to the standard Holland classification.

TABLE 23.1 (continued)

Type of Plant Community (Holland 1986)	Holland (1986) Code	Percentage of Mapped Distribution by Ownership									Total Mapped Distribution Area (km²)
		Private	Nongovernmental Organization(s)	County and Regional	State	Department of Defense	Other U.S. Department of the Interior[a]	National Park Service	Bureau of Land Management	U.S. Forest Service	
Broad-Leaved Woodland											
Oregon oak woodland	71110	43							2	56	21
Black oak woodland	71120	55			< 1	< 1	3	1	8	32	460
Valley oak woodland	71130	98			2	< 1			1	< 1	340
Blue oak woodland	71140	89	< 1		1	1	2	1	3	3	5,430
Interior live oak woodland	71150	71			1	< 1	1	1	4	22	1,299
Conifer Woodland											
Open foothill pine woodland	71310	58			1	1	< 1		21	19	441
Nonserpentine foothill pine chaparral	71322	43			1	1	5	7	22	21	249
Foothill pine–oak woodland	71410	82	< 1		1	1	1		4	10	2,976
Cismontane juniper woodland[b]	71500	< 1			< 1			14		86	155
Oak-piñon woodland[b]	71600	8						4	62	27	117
Northern juniper woodland	72110	3			2			2	7	85	182
Great Basin piñon-juniper woodland	72121	11							42	47	404
Great Basin piñon woodland	72122	21			< 1		< 1	1	25	53	863
Great Basin juniper woodland and scrub	72123	4			15				9	72	9
Mojavean juniper woodland and scrub	72220	67				< 1			27	5	63
Joshua tree woodland	73000	9							91		73
Broad-Leaved Forest											
Canyon live oak forest	81320	24			< 1		3	13	5	55	916
Interior live oak chaparral	81330	76	< 1		2	< 1	2	1	3	17	1,545
Black oak forest	81340	28			1	< 1	2	3	5	60	1,087
Tan oak forest	81400	77								23	24
Aspen forest	81B00	6			1			2	3	89	99
Conifer Forest											
Knobcone pine forest	83210	50			1		< 1		23	26	17
Southern interior cypress forest	83330	23							24	53	3
West-side ponderosa pine forest	84210	35			1	< 1	< 1	8	3	53	4,402
East-side ponderosa pine forest	84220	18			1				4	76	1,614
Sierran mixed conifer forest	84230	32			< 1		< 1	5	1	62	5,935
Sierran white fir forest	84240	23			1			6	< 1	70	540
Big tree forest	84250	5			11		1	52	1	31	71
Jeffrey pine forest	85100	9			1		< 1	13	2	75	1,961
Red fir–western white pine forest[b]	85120	7			< 1			18	< 1	75	1,594
Jeffrey pine–fir forest	85210	11			< 1			9	< 1	80	2,956
Red fir forest	85310	9			< 1			30	< 1	61	3,395
Lodgepole pine forest	86100	3			< 1			42	< 1	55	2,156
Whitebark pine–mountain hemlock forest	86210	1						37		62	378
Whitebark pine–lodgepole pine forest	86220	1		< 1	< 1			12		86	373
Foxtail pine forest	86300	1						77		21	238
Whitebark pine forest	86600	1			< 1			31	< 1	68	219
Limber pine forest	86700	< 1						3		97	21
Lower cismontane mixed conifer–oak forest[b]	87100	45			1	< 1	1	4	4	46	4,234
Upper cismontane mixed conifer–oak forest[b]	87200	20	< 1		< 1	3		14	14	48	261
Alpine Habitats											
Sierra Nevada fell field	91120	1						27		72	122
Alpine dwarf scrub	94000	< 1						1		99	394
Total Area											
Vegetated lands											56,587
Vegetated and unvegetated lands		37	< 1	< 1	1	< 1	1	10	5	47	63,111

[a]Includes the Bureau of Reclamation, the Bureau of Indian Affairs, and the U.S. Fish and Wildlife Service.
[b]Addition to the standard Holland classification.

FIGURE 23.2

Frequency of Class 1 areas by size class (bars) and cumulative area (curve) in the Sierra Nevada.

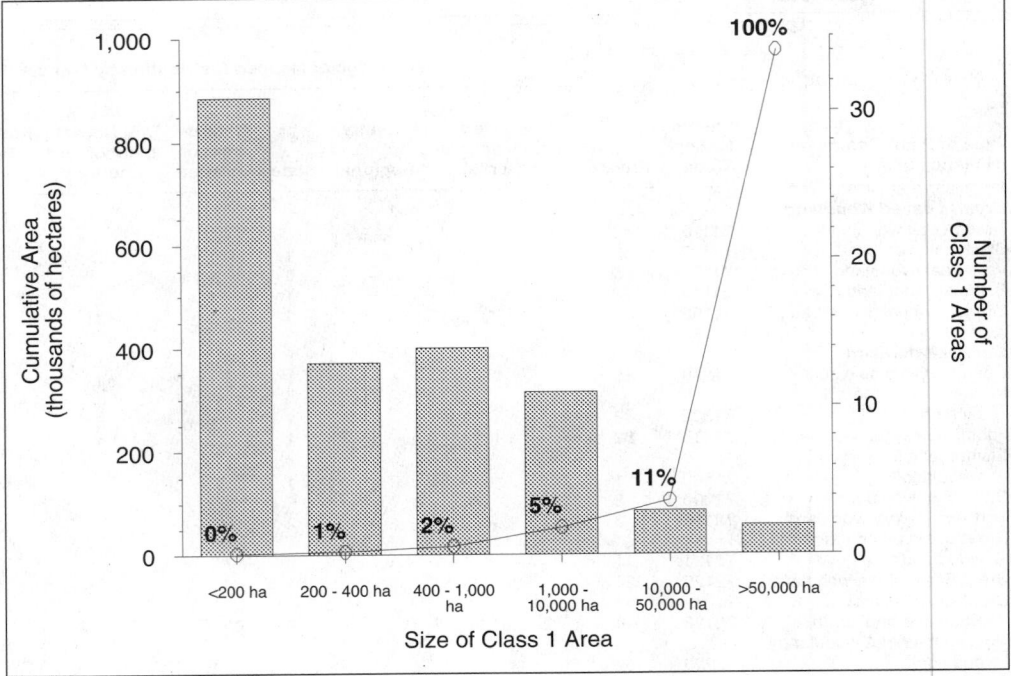

administers 5% of the region. The Bureau of Indian Affairs, other Department of Interior agencies, and the state oversee the remaining 2% of the region's land base.

We found that 15% of the region is in Class 1 management status. Yosemite and Sequoia–King's Canyon National Parks account for 89% of the Class 1 area. The size distribution of Class 1 areas is strongly skewed toward parcels of less than 200 ha (494 acres) (figure 23.2). These account for nearly half of the Class 1 parcels but contribute less than 1% of the total Class 1 area.

An additional 7% of the Sierra Nevada region is in Class 2 lands in national forests. By summing Classes 3, 4, and 5, we estimate that roughly 80% of the region is available for graz-ing (89% of vegetated lands). Summing Classes 4 and 5, we estimate that 56.5% of the land area (63.3% of vegetated lands) is available for timber harvest, although not all of this land is actually timberland.

Based on our system for converting dominant species com-binations to natural community types, we mapped eighty-eight natural plant community types within the region. Sixty-seven types were mapped over an area greater than 25 km^2 (9.65 mi^2). Sierran mixed conifer forest and blue oak woodland are the most extensive types, covering 5,933 km^2 (2,290 mi^2) and 5,426 km^2 (2,094 mi^2), respectively. Eleven community types collectively contribute 65% of the region's total vegetated acreage (table 23.2).

The ownership profiles of Sierran plant communities sys-tematically reflect the concentration of private lands at lower elevations and of national parks in the central and southern portion of the range. Many of the foothill community types fall largely on private lands, notably non-native grassland (88% of mapped distribution on private lands), valley oak woodland (98%), blue oak woodland (89%), interior live oak woodland (71%), and foothill pine–oak woodland (82%). These percentages differ somewhat from the statewide esti-mates of private ownership provided by Bolsinger (1988). His estimates are lower for valley oak woodland (86% private ownership) and blue oak woodland (75%) and higher for in-terior live oak woodland (82%). Our estimates of private own-ership and conservation of blue oak and blue oak–foothill pine community types are comparable to those of Greenwood et al. (1993).

A number of relatively widespread community types fall disproportionately on national forest lands, notably low sage-

TABLE 23.2

Eleven widespread vegetation types that collectively cover 65% of the vegetated portion of the Jepson Sierra Nevada Region.

Plant Community Type (Holland 1986)			
Name	Code	Area (km^2)	Percentage of Total Area
Mixed conifer forest	84230	5,933	10.5
Blue oak woodland	71120	5,426	9.6
West-side ponderosa pine forest	84210	4,406	7.8
Lower cismontane mixed conifer–oak forest	87100	4,231	7.5
Red fir forest	85310	3,395	6.0
Foothill pine–oak woodland	71410	2,975	5.3
Jeffrey pine–fir forest	85210	2,956	5.2
Lodgepole pine forest	86100	2,156	3.8
Jeffrey pine forest	85100	1,961	3.5
East-side ponderosa pine forest	84220	1,614	2.9
Non-native grassland	42200	1,922	2.8

TABLE 23.3

Upland rangeland plant community types in areas that can be grazed. These are types with areas greater than 25 km^2 (9.65 mi^2) with more than 90% of their mapped distribution potentially grazed.

Plant Community Type	Holland (1986) Code	Percentage of Mapped Distribution by Land Management Class			Total Mapped Distribution Area (km^2)
		Class 1 (Protected)	Classes 1–2 (Ungrazed)	Classes 3–5 (Potentially Grazed)	
Shrubland Types					
Mojave mixed scrub and steppe	34200	<0.1	<0.1	100.0	261
Blackbush scrub	34300	<0.1	<0.1	100.0	164
Chamise chaparral	37200	5.8	7.7	92.3	820
Scrub oak chaparral	37900	0.0	6.4	93.6	48
Upper Sonoran subshrub scrub	39000	3.9	6.6	93.4	42
Woodland Types					
Black oak woodland	71120	1.5	9.0	91.0	460
Valley oak woodland	71130	0.0	0.1	99.9	340
Blue oak woodland	71140	1.2	1.8	98.2	5,426
Interior live oak woodland	71150	1.0	6.2	93.8	1,299
Open foothill pine woodland	71310	1.4	6.3	93.7	441
Foothill pine–oak woodland	71410	0.4	5.6	98.4	2,975
Oak–piñon woodland[a]	71600	3.6	7.7	92.3	117
Northern juniper woodland	72110	5.8	10.7	89.3	182
Great Basin piñon woodland	72122	2.3	3.0	97.0	863
Mojavean juniper woodland and scrub	72220	1.3	1.8	98.2	63
Joshua tree woodland	73000	0.0	0.0	100.0	73
Forest Types					
Interior live oak forest	81330	1.8	4.2	95.8	1,545
East-side ponderosa pine forest	84220	0.9	8.5	91.5	1,614

[a]Addition to the standard Holland classification.

brush scrub (79%), rabbitbrush scrub (93%), *Cercocarpus ledifolius* woodland (94%), mixed montane chaparral (73%), montane ceanothus chaparral (72%), bush chinquapin chaparral (85%), cismontane juniper woodland (86%), northern juniper woodland (85%), aspen forest (89%), east-side ponderosa pine forest (76%), Jeffrey pine forest (75%), Jeffrey pine–fir forest (80%), red fir–western white pine forest (75%), whitebark pine–lodgepole pine forest (86%), and alpine dwarf scrub (99%).

Foxtail pine forest is the only type whose distribution falls mainly inside the national parks (77%). The BLM controls the largest portion of the distribution for a few community types that are marginal to the Jepson Sierra Nevada Region, notably Mojave mixed scrub and steppe (71%), blackbush scrub (61%), oak–piñon woodland (62%), and Joshua tree woodland (91%).

The mapped community types display a wide range of land management profiles. We would call special attention to four distribution types:

1. Upland rangeland plant community types mainly in areas that can be grazed. Table 23.3 lists 18 out of 67 types with areas greater than 25 km^2 (9.65 mi^2) and with more than 90% of their distribution in Classes 3–5 and therefore potentially grazed. These types merit special attention for grazing management and conservation. The main distribution for several of the types lies outside of the Jepson Sierra Nevada Region (e.g., Mojave mixed scrub and steppe, Joshua tree woodland, blackbush scrub, and the sagebrush types). While we have less confidence in our mapping of riparian and wetland types, we should note that all riparian types and most wetland habitats were also mapped with more than 90% of their distribution in Classes 3–5.

2. Forest plant communities mainly located in unprotected areas. Table 23.4 lists six types with areas greater than 25 km^2 and with less than 10% of their distribution in Class 1 land, which is designated for conservation of native biodiversity. These types are of special management concern related to timber harvest and/or fire suppression. However, except for interior live oak forest, these types are widely distributed on national forest lands that are classified in current forest plans as unsuitable for timber harvest (Class 2).

3. Chaparral community types mainly located in unprotected areas. Table 23.5 lists eight types with areas greater than 25 km^2 and with less than 10% of their distribution on Class 1 land. The policy of suppressing wildfire on Class 2–5 public and private lands and the widespread conversion of chaparral to grasslands on private ranchlands raise concern for the long-term sustainability of these fire-adapted plant communities. A similar concern arises for knobcone pine forest, a fire-dependent community that is also very poorly represented in Class 1 areas.

TABLE 23.4

Forest plant community types mainly located in unprotected areas. These are types with areas greater than 25 km^2 (9.65 mi^2) with less than 10% of their mapped distribution in areas formally designated for conservation (Class 1 land).

Plant Community Type	Holland (1986) Code	Class 1 (Protected)	Classes 1–3 (Not Available for Timber Harvesting)	Classes 4–5 (Available for Timber Harvesting)	Total Mapped Distribution Area (km^2)
Interior live oak forest	81330	1.8	18.6	81.3	1,545
Black oak forest	81340	6.5	44.4	55.6	1,087
East-side ponderosa pine forest	84220	0.9	27.6	72.4	1,614
Sierran mixed conifer forest	84230	8.1	32.9	67.1	5,933
Sierran white fir forest	84240	7.7	38.1	61.9	540
Lower cismontane mixed conifer–oak forest[a]	87100	4.9	29.9	70.1	4,231

[a]Addition to the standard Holland classification.

4. Plant community types that are well protected. Table 23.6 lists twelve types with areas greater than 25 km^2 and more than 25% of their distribution in Class 1 areas. These types are of relatively low priority for additional land acquisition or redesignation to reserve status.

Northern Sierra Subregion

The northern subregion totals 27,483 km^2 (10,611 mi^2) in area and is largely national forest or private land. Only 2.1% of the land in this subregion is in Class 1 areas (appendix 23.2). An additional 10.1% is Class 2. Potentially grazed lands (Classes 3–5) account for 87.8% of the area, while 71% is eligible for intensive timber harvesting (Classes 4–5). Private lands constitute 45.3% of the total area.

Ownership and management vary systematically by elevation zone. More than 80% of the land below 1,000 m (3,280 ft) is unreserved private land (Class 5), while less than 0.1% is in Class 1 (figure 23.3). In contrast, Class 5 constitutes less than 10% of areas above 2,000 m (6,560 ft).

Vegetation was mapped into 3,869 polygons with a median polygon size of 371 ha (916 acres). Of the sixty-eight community types mapped, forty-six had mapped distributions greater than 25 km^2 (9.65 mi^2) in extent. Sierran mixed conifer was mapped over 4,523 km^2 (1,746 mi^2) or 17.5% of vegetated lands. Other widespread types include west-side ponderosa pine forest (9% of vegetated lands), lower cismontane mixed conifer–oak forest (9%), east-side ponderosa pine forest (6%), foothill pine–oak woodland (5%), red fir forest (5%), Jeffrey pine–fir forest (4%), and Jeffrey pine forest (4%). These eight community types make up roughly 60% of the total vegetation. Only eight of the forty-six types with areas greater than 25 km^2 have more than 5% of mapped distribution in Class 1 land.

Many of the rangeland types are largely on land available for grazing, notably big sagebrush scrub (93% of distribution), rabbitbrush scrub (96%), chamise chaparral (99%), nonnative grassland (98%), black oak woodland (93%), valley oak woodland (99%), blue oak woodland (99%), interior live oak woodland (99%), open foothill pine woodland (99%), foothill

TABLE 23.5

Chaparral plant community types mainly located in unprotected areas. These are types with areas greater than 25 km^2 (9.65 mi^2) with less than 10% of their mapped distribution in areas formally designated for conservation (Class 1 land).

Plant Community Type	Holland (1986) Code	Class 1 (Protected)	Class 5 (Private—Available for Timber Harvesting, Grazing, or Urban Development)	Total Mapped Distribution Area (km^2)
Chamise chaparral	37200	5.8	51.6	820
Montane manzanita chaparral	37520	4.9	44.3	457
Montane ceanothus chaparral	37530	1.5	25.8	195
Bush chinquapin chaparral	37550	6.0	11.0	77
Buck brush chaparral	37810	1.1	22.1	155
Scrub oak chaparral	37900	0.0	70.6	48
Interior live oak chaparral	37A00	4.3	70.6	203
Upper Sonoran manzanita chaparral	37B00	4.9	18.2	163

TABLE 23.6

Well-protected plant community types. These are types with areas greater than 25 km² (9.65 mi²) with more than 25% of their mapped distribution in areas formally designated for conservation (Class 1 land).

Plant Community Type	Holland (1986) Code	Percentage of Mapped Distribution in Class 1	Total Mapped Distribution Area (km²)
Montane meadow	45100	54.0	127
Cismontane juniper woodland[a]	71500	31.4	155
Big tree forest	84250	51.6	71
Red fir–western white pine forest[a]	85120	28.8	1,594
Red fir forest	85310	33.2	3,395
Lodgepole pine forest	86100	53.5	2,156
Whitebark pine–mountain hemlock forest	86210	61.7	378
Whitebark pine–lodgepole pine forest	86220	56.1	372
Foxtail pine forest	86300	92.6	238
Whitebark pine forest	86600	58.0	219
Sierra Nevada fell field	91120	27.5	122
Alpine dwarf scrub	94000	89.5	394

[a]Addition to the standard Holland classification.

pine–oak woodland (99%), northern juniper woodland (93%), and Great Basin piñon woodland (99%).

Of the major forest types, interior live oak forest is distinctly concentrated on private lands (90%). Over half of the area in west-side ponderosa pine forest is privately held. Ponderosa pine may have previously dominated much of what we classified as lower cismontane mixed conifer–oak forest, a low-elevation type that is also predominantly on private land (63%). The middle-elevation forest types are more concentrated in the national forests (60% to 90% on public lands).

Treating the five major low- to middle-elevation conifer timber types (west-side ponderosa pine, east-side ponderosa pine, Sierran mixed conifer, Sierran white fir, and lower cismontane mixed conifer–oak forests) collectively, we estimate that 22.5% of lower montane timberlands are in reserve status or are on national forest land classified as unsuitable for intensive timber harvest.

The five high-elevation conifer types that may be used for timber production include red fir–western white pine, red fir, Jeffrey pine, Jeffrey pine–fir, and upper cismontane mixed conifer–oak forests. Currently 50% of the total area in these types is reserved or withdrawn from intensive timber harvesting.

Central and Southern Sierra Subregion

We mapped a total of 35,620 km² (13,753 mi²) as the Jepson central and southern Sierra Nevada subregion. Because both Yosemite and Sequoia–Kings Canyon National Parks fall within this area, its land management profile is strikingly different from that of the northern subregion. Class 1 areas and private lands are roughly equal in extent, respectively 25.7% and 29.8% of the area. Like those in the northern subregion, Class 1 lands are concentrated at higher elevations (figure 23.4).

Approximately 12% of the region was classified as nonvegetated (mainly land at high elevation with little or no ground cover). Vegetation was mapped into 3,143 polygons with a median size of around 500 ha (1,235 acres). The central/southern polygons are larger than their northern counterparts mainly because much of the region was not mapped by vegetation type mapping crews, and thus we relied more heavily on USFS timber type maps and on our own field vis-

FIGURE 23.3

Proportion of land in each management class by elevation zone in the northern Sierra Nevada subregion.

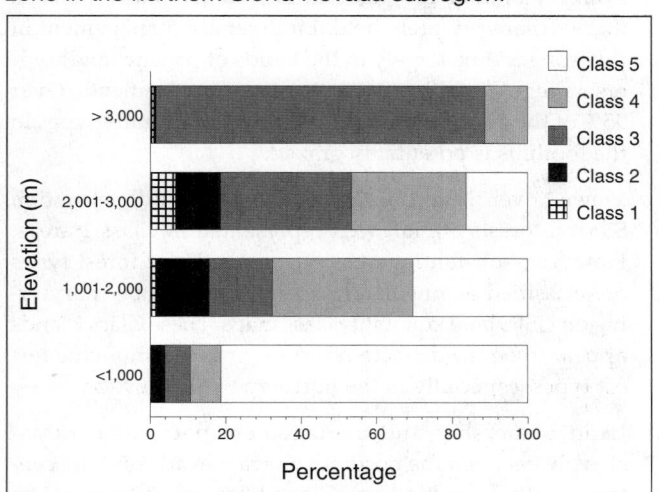

FIGURE 23.4

Proportion of land in each management class by elevation zone in the central/southern Sierra Nevada subregion.

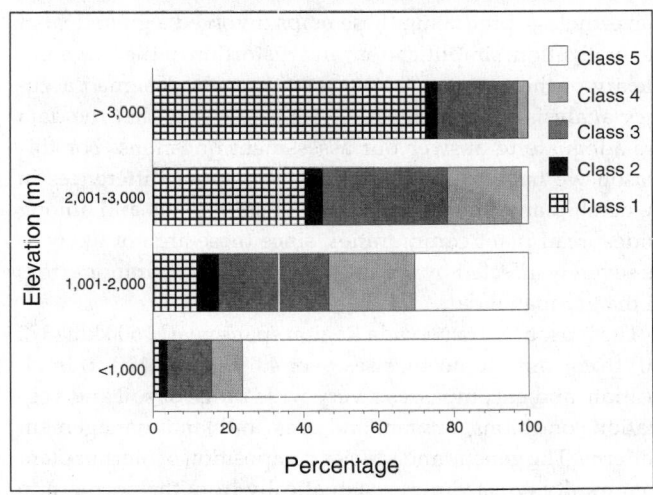

its to about 700 polygons to define polygon boundaries and composition.

Of the seventy-nine mapped community types, fifty-nine are greater than 25 km² (9.65 mi²) in extent. Taken together, blue oak woodland, foothill pine–oak woodland, and non-native grassland occupy 6,974 km² (2,692 mi²), or 22.3% of the vegetated portion of the subregion. The other extensive community types include red fir forest (7% of vegetated area), west-side ponderosa pine forest (7%), lower cismontane mixed conifer–oak forest (7%), lodgepole pine forest (6%), Jeffrey pine–fir forest (6%), and Sierran mixed conifer forest (4.5%).

Private lands and public grazing allotments cover roughly three-fourths of the vegetated area. Thus, practically the entire distribution of many plant community types is potentially grazed here, as it is in the northern subregion. Especially noteworthy are the foothill woodland and grassland types (more than 97% of mapped area available for grazing), Mojavean scrub and woodland types (98%), blackbush scrub (99%), and Great Basin piñon woodland (96%).

The largest difference between the northern and the central/southern subregions lies in the management profiles of the major forest types. With the exception of the lower cismontane mixed conifer–oak forest and the black oak forest, virtually all of the timber-producing community types have at least 20% of their distribution on Class 1 land.

A number of community types are very well represented in Class 1 areas. Twenty-three of fifty-nine extensive communities show at least 25% of their mapped distribution on Class 1 land, notably montane chaparral types, mixed conifer forest types, and subalpine woodland types.

SUMMARY AND DISCUSSION

The databases used in this gap analysis comprise the most spatially and taxonomically detailed land management and vegetation maps ever assembled for the region as a whole. Nevertheless, producing these maps involved a great deal of generalization, simplification, and distortion of the true complexity of the region. Without a statistically designed accuracy analysis we cannot state with confidence that the data are adequate to answer our assessment questions. For this reason we have tried to focus on very gross differences in ownership and management among subregions and among widespread plant communities, since these are not likely to be severely affected by the mapping scale or by minor errors in the geospatial data.

The Jepson Sierra Nevada Region spans nearly 600 km (372 mi) from south to north, rises over 4,000 m (13,120 ft) in elevation, and encompasses a very wide range of soil and vegetation conditions, human land uses, and land management patterns. The genetic and species composition of Sierran plant community types varies systematically from the northern to the southern end of the range (e.g., Taylor 1977; Walker 1992). For example, Walker (1992) estimated the average plant species turnover in Sierran mixed conifer forest to be one species per kilometer along the long axis of the range. The mixed conifer flora of the far northern Sierra Nevada shares only half of its plant species with its southern counterpart. Many plant taxa are endemic to one subregion. For this reason, the status of plant community types of the Sierra Nevada is best viewed on a subregional basis. Similarly, strategies for maintaining native Sierran biodiversity must account for the systematic and often profound differences, both administrative and biological, between the northern and the central/southern subregions, as well as between the foothill zone and higher elevations, between lower- and middle-elevation mixed hardwood-conifer and conifer community types, and between community types with predominantly west-side versus east-side distributions.

Our general conclusions are:

1. Fifteen percent of the Sierra Nevada is in designated conservation lands. An additional 7% is in national forest lands that are not grazed and/or are deemed unsuitable for timber production.

2. More than 80% of designated Class 1 areas are less than 200 ha (494 acres) in size. These small parcels collectively contribute less than 1% of total Class 1 area. Yosemite and Sequoia–Kings Canyon National Parks contribute 89% of Class 1 lands. Most remaining Class 1 areas are high-elevation, ungrazed parcels within wilderness areas in the national forests.

3. Eighty-nine percent of the vegetated area of the Sierra Nevada is privately held or is public land where grazing is legally permitted.

4. Less than 1% of the foothill zone of the Sierra Nevada is in designated reserves or other areas managed primarily for native biodiversity.

5. Roughly 80% of the lands at elevations below 1,000 m (3,280 ft) are privately held. Biodiversity management in this zone is thus largely in the hands of private landholders as regulated by state and county governments. Over 95% of the distribution of most plant community types in the foothills is potentially grazed.

6. Viewed over the entire range, low- and middle-elevation Sierran forests are not well represented in Class 1 areas. However, substantial areas of most of these forest types are classified as unsuitable for intensive timber harvesting on USFS land suitability class maps. These Class 2 lands appear to be the de facto reserves for lower montane forest types, especially in the northern Sierra Nevada.

7. Land ownership and management patterns contrast sharply between the northern Sierra Nevada and the central/southern subregion. Class 1 lands contribute less

than 2% of the northern region versus 27% of the central/southern.

8. Based on our land management classification, biodiversity of the lower montane forests of the northern Sierra Nevada is considerably more vulnerable than forest biodiversity elsewhere in the range.

9. Many high-elevation forest and shrubland types are well represented in parks and ungrazed wilderness areas. In the central/southern subregion, twenty-three of fifty-nine widespread community types are especially well protected, with over one-quarter of their distribution on Class 1 lands.

ACKNOWLEDGMENTS

Financial support for this research was provided by the USFS Sierra Nevada Ecosystem Project, the National Biological Service Gap Analysis Program, and the California Department of Fish and Game. Computing support was provided by a grant from the IBM Corporation Environmental Research Program.

Mike Bueno provided technical support and administered the computing system used in the analysis. The following University of California, Santa Barbara, staff and student research assistants worked long and hard to prepare the vegetation and land ownership maps and databases: David Court, Josh Graae, Violet Gray, Nicole Griffin, Allan Hollander, Curtice Jacoby, Paul Mills, Dennis Odion, Daniel Sarr, Laurie Schwalm, Yvonne Thompson, Jim Thorne, Rich Walker, Eric Waller, Joe Walsh, Katherine Warner, and Dan Wolnick.

We gratefully acknowledge accounting and administrative support from the staff of the Institute of Computational Earth System Science and the Center for Wildlands and Water Resources, University of California, Davis.

Geographical Information System data and support were provided by the staff of the SNEP GIS lab. Special thanks to Karen Gabriel, John Gabriel, and Russ Jones for prompt handling of our requests for data.

USFS personnel provided field data, advice, and support to our field crews. We would especially like to thank Ralph Warbington, JoAnn Fites, Jim Shevock, Connie Millar, Lenea Hansen, Beth Corbin, Stacey Scott, Terry Hicks, Bob Rogers, Lou Jump, Joanna Clines, Ron Taskey, and Neil Sugihara.

The draft manuscript benefited from the careful and constructive reviews of Zipporah Collins, Michael Barbour, Laurel Ames, William Stewart, and an anonymous reviewer.

REFERENCES

Beardsley, K., and D. M. Stoms. 1993. Compiling a digital map of areas managed for biodiversity in California. *Natural Areas Journal* 13:177–90.

Bolsinger, C. L. 1988. *The hardwoods of California's timberlands, woodlands, and savannas*. Portland, OR: U.S. Forest Service, Pacific Northwest Research Station.

Davis, F. W., P. A. Stine, D. M. Stoms, M. I. Borchert, and A. D. Hollander. 1995. Gap analysis of the actual vegetation of California: 1. The southwestern region. *Madroño* 42:40–78.

Greenwood, G. B., R. K. Marose, and J. M. Stenback. 1993. *Extent and ownership of California's hardwood rangelands*. Technical Report. Sacramento: California Department of Forestry and Fire Protection.

Hickman, J. C. (ed.). 1993. *The Jepson manual: Higher plants of California*. Berkeley and Los Angeles: University of California Press.

Holland, R. F. 1986. *Preliminary descriptions of the terrestrial natural communities of California*. Sacramento: California Department of Fish and Game.

Mayer, K. E., and W. F. Laudenslayer Jr. 1988. *A guide to wildlife habitats of California*. Sacramento: California Department of Forestry and Fire Protection.

Pillsbury, N. H., M. J. de Lasaux, R. D. Pryor, and W. D. Bremer. 1991. *Mapping and GIS database development for California's hardwood resources*. Technical Report. Sacramento: California Department of Forestry and Fire Protection, Forest and Rangeland Resources Assessment Program.

Scott, J. M., F. Davis, B. Csuti, R. Noss, B. Butterfield, C. Groves, H. Anderson, S. Caicco, F. D'Erchia, T. C. Edwards Jr., J. Ulliman, and R. G. Wright. 1993. Gap analysis: A geographic approach to protection of biological diversity. *Wildlife Monographs* 123:1–41.

Taylor, D. U. 1977. Floristic relationships along the Cascade-Sierran axis. *American Midland Naturalist* 97:333–49.

Walker, R. E. 1992. Community models of species richness: Regional variation of plant community species composition on the west slope of the Sierra Nevada, California. Master's thesis, Department of Geography, University of California, Santa Barbara.

Wieslander, A. E. 1946. *Forest areas, timber volumes, and vegetation types in California*. Forest Survey Release 4. Berkeley: California Forest and Range Experiment Station.

List of Designated Biological Reserves in the SNEP Core Region

Agency/Organization, Area Name	Area (ha)	Subtotal (ha)
Private		**12,683**
The Nature Conservancy Preserves		12,683
Dye Creek	10,312	
Kern River	337	
Mary Elizabeth Miller	152	
Table Mountain	1,882	
State		**58,549**
Department of Parks and Recreation		
State Parks and Reserves		11,670
Burton Creek	789	
Calaveras Big Trees	2,450	
Donner Memorial	155	
Emerald Bay/D. L. Bliss	646	
Grover Hot Springs	182	
Plumas-Eureka	1,885	
Red Rock Canyon	4,833	
Sugar Pine Point	730	
Department of Fish and Game		
Ecological Reserves		846
Blue Ridge Condor	623	
Fish Slough	74	
Limestone Salamander	48	
Pine Hill	101	
Wildlife Areas		46,033
Antelope Valley	1,781	
Bass Hill	1,312	
Biscar	226	
Coon Hollow	212	
Crocker Meadows	730	
Daugherty Hill	967	
Doyle	5,734	
Fay Canyon	159	
Hallelujah Junction	2,630	
Heenan Lake	524	
Honey Lake	2,963	
Hope Valley	1,199	
Red Lake	315	
Slinkard/Little Antelope	4,706	
Smithneck Creek	607	
South Fork (Corps of Engineers)	535	
Spenceville	3,463	
Tehama	16,618	
Warner Valley	277	
Willow Creek	1,075	
Federal		**1,732,984**
Bureau of Land Management		
Areas of Critical Environmental		
Concern and Wild and Scenic Rivers		84,399
Blue Ridge Condor	1,299	
Bodie Bowl	2,427	
Conway Summit	724	
Crater Mountain	2,325	
El Dorado Manzanita	42	
Fish Slough	13,986	
Fossil Falls	664	

Agency/Organization, Area Name	Area (ha)	Subtotal (ha)
Jawbone Butterbredt	52,827	
Last Chance Canyon	774	
Limestone Salamander	641	
Merced River	106	
Red Hills	2,917	
Sand Canyon	1,269	
Slinkard Valley	4,259	
Tuolumne River (Wild and Scenic River)	139	
Wilderness Areas		124,053
Bright Star	3,244	
Chimney Peak	5,081	
Coso Range	19,744	
Domeland	951	
Golden Valley	3,917	
Inyo Mountains	17,287	
Ishi	77	
Kiavah	15,843	
Malpais Mesa	8,294	
Owens Peak	29,530	
Piper Mountain	7	
Sacatar Trail	20,078	
U.S. Fish and Wildlife Service		
National Wildlife Refuge		457
Blue Ridge	457	
National Park Service		
National Monuments and Parks		666,120
Devils Postpile	326	
Lassen Volcanic	15,370	
Sequoia and Kings Canyon	348,473	
Yosemite	301,951	
U.S. Forest Service		
Research Natural Areas		18,461
Babbitt Peak	541	
Backbone Creek	164	
Bell Meadow	273	
Big Grizzly Mountain	310	
Bishop Creek	660	
Bourland Meadow	210	
Church Dome	592	
Clark Fork	946	
Cub Creek	1,545	
Graham Pinery	351	
Grass Lake	130	
Green Island Lake	445	
Harvey Monroe Hall	1,579	
Indiana Summit	422	
Indian Creek	1,481	
Jawbone Ridge	316	
Last Chance Meadow	249	
Long Canyon	954	
Lyon Peak/Needle Lake	306	
McAffee Meadow	1,408	
Moses Mountain	383	
Mount Pleasant	581	
Mountaineer Creek	678	

Agency/Organization, Area Name	Area (ha)	Subtotal (ha)	Agency/Organization, Area Name	Area (ha)	Subtotal (ha)
Mud Lake	183		Kaiser	8,977	
Peavine Point	453		Kiavah	17,709	
Secate Ridge	1,689		Mokelumne	40,843	
Sentinel Meadow	277		Monarch	17,862	
Snow Canyon	327		South Sierra	24,530	
Soda Ridge	467				
Station Creek	287		*Special Interest Areas*		22,224
Sugar Pine Point	254		Ancient Bristlecone Pine Forest	2,097	
			Bodfish Piute Cypress	237	
Wilderness Areas[a]		803,488	Butterfly Valley	192	
Ansel Adams	94,973		California Bighorn Sheep	14,377	
Bucks Lake	8,737		Carpenteria	180	
Caribou	7,593		Feather Falls	3,643	
Carson-Iceberg	63,980		Kings River	178	
Desolation	25,504		Little Last Chance Canyon	546	
Dinkey Lakes	12,280		McKinley Grove	183	
Dome Land	35,197		Neider Grove	591	
Emigrant	45,587				
Golden Trout	121,416		*Wild and Scenic Rivers*		13,782
Granite Chief	9,857		Feather River	7,447	
Hoover	19,484		Kern River	2,025	
Inyo Mountains	7,966		Merced River	1,656	
Ishi	15,903		Tuolumne River	2,654	
Jennie Lakes	4,257				
John Muir	220,833		***Total Area***		**1,804,216**

[a]Wilderness areas include grazing allotments.

Management Status (Classes 1–5) by Subregion for Plant Communities of the Jepson Sierra Nevada Region

Percentage of Mapped Distribution by Land Management Class

Type of Plant Community (Holland 1986)	Holland (1986) Code	Class 1			Class 2			Class 3			Class 4			Class 5			Total Mapped Distribution (km²)		
		North	South	Total	North	South	Total	North	South	Total	North	South	Total	North	South	Total	North	South	Total
Scrub																			
Mojave creosote bush scrub	34100		0.1	0.1					50.0	50.0					49.9	49.9		7	7
Mojave mixed scrub and steppe	34200								69.5	69.5		2.9	2.9		27.6	27.6		261	261
Mojave mixed woody scrub	34210		<0.1	<0.1		0.1	0.1		9.3	9.3		19.2	19.2		71.4	71.4		8	8
Blackbush scrub	34300		<0.1	<0.1					63.2	63.2		1.9	1.9		34.9	34.9		164	164
Great Basin mixed scrub	35100	0.1	13.3	2.7	18.8	22.0	19.5	20.8	14.0	19.5	39.1	48.6	41.0	21.1	2.2	17.4	243	60	303
Big sagebrush scrub	35210	2.2	14.1	11.2	5.0	1.5	2.3	59.8	43.3	47.2	29.3	13.0	16.9	3.8	28.1	22.4	43	140	183
Low sagebrush scrub[a]	35211	0.7	0.2	0.7	15.0	39.5	15.5	44.9		43.9	29.0	59.6	29.6	10.4	0.8	10.2	153	3	156
Silver sagebrush scrub[a]	35212				48.1		32.2	30.4	19.4	26.8	7.2	78.3	30.6	14.3	2.3	10.3	11	5	16
Subalpine sagebrush scrub	35220	10.3		10.3	11.0		11.0	21.5		21.5	20.8		20.8	36.4		36.4	25		25
Sagebrush steppe	35300	1.0	1.9	1.6	26.3	2.5	12.1	29.8	47.4	40.3	30.6	18.1	23.2	12.4	30.0	22.9	333	489	822
Rabbitbrush scrub	35400	0.9		0.9	3.6		3.6	23.6		23.6	65.0		65.0	7.0		7.0	46		46
Cercocarpus ledifolius woodland[a]	35500	0.5	42.5	18.6	4.0	40.8	19.9	35.1	7.9	23.4	54.8	7.7	34.5	5.6	1.1	3.7	143	109	252
Wyethia mollis[a]	35600	6.8		6.8	13.1		13.1	30.2		30.2	23.2		23.2	26.7		26.7	30		30
Chaparral																			
Upper Sonoran mixed chaparral	37100				44.1		44.1				16.6		16.6	39.3		39.3	6		6
Northern mixed chaparral	37110	1.7	15.1	13.8		11.6	10.8	0.9	23.3	21.4	15.4	25.9	25.0	82.0	24.0	28.9	15	161	176
Chamise chaparral	37200	0.8	10.4	5.8		2.7	1.9	3.4	24.1	14.9	14.4	34.9	25.8	81.3	27.9	51.6	364	456	820
Semidesert chaparral	37400		15.2	15.2		6.1	6.1		11.7	11.7		28.4	28.4		38.7	38.7		109	109
Mixed montane chaparral	37510	6.2	30.5	12.1	16.8	13.8	16.1	21.7	34.4	24.8	29.8	13.6	25.9	25.5	7.8	21.1	1,048	340	1,387
Montane manzanita chaparral	37520	0.1	9.0	4.9	9.2	7.6	8.4	8.5	20.4	14.9	16.7	37.3	27.6	65.5	25.6	44.3	214	243	457
Montane ceanothus chaparral	37530	1.2	8.2	1.5	6.0	14.8	6.4	20.6	13.3	20.3	46.0	48.7	46.1	26.2	15.0	25.8	187	8	195
Deer brush chaparral	37531							25.3		25.3	74.7		74.7				2		2
Shin oak brush	37541		23.4	23.4		4.9	4.9		7.4	7.4		4.0	4.0		60.3	60.3		42	42
Huckleberry oak chaparral	37542	<0.1	42.8	23.4	26.6	0.7	12.5	17.8	45.0	32.7	6.7	11.4	9.3	48.8	0.1	22.2	81	98	179
Bush chinquapin chaparral	37550	4.6	48.9	6.0	8.1		7.9	26.2	48.5	26.9	49.7	2.6	48.2	11.4		11.0	75	3	77
Buck brush chaparral	37810		1.2	1.1	0.9	17.7	15.3	15.9	50.4	45.6	14.9	16.0	15.9	68.3	14.7	22.1	21	133	155
Scrub oak chaparral	37900			0.0	39.8	3.9	6.4	36.6	4.0	6.2	18.8	16.7	16.8	4.9	75.4	70.6	3	45	48
Interior live oak chaparral	37A00		4.4	4.3		7.0	6.8	18.4	8.1	8.4	14.4	9.8	9.9	67.2	70.7	70.6	5	199	203
Upper Sonoran manzanita chaparral	37B00		5.2	4.9	1.5	4.9	4.7		25.6	24.3	27.3	49.0	47.8	71.2	15.2	18.2	9	154	163
Ione chaparral	37D00										3.8		3.8	96.2		96.2	1		1
Mesic north-slope chaparral	37E00		11.5	10.2	8.2	9.9	9.7	6.8	49.3	44.6	8.9	20.5	19.2	76.1	8.8	16.3	15	118	132
Upper Sonoran subshrub scrub	39000		3.9	3.9		2.7	2.7		28.0	28.0		15.4	15.4		50.0	50.0		42	42
Herbaceous																			
Valley needlegrass grassland	42110		64.7	12.7	<0.1	6.0	1.2	0.1	29.3	5.9	0.2		0.1	99.7		80.1	25	6	31
Non-native grassland	42200	<0.1	0.8	0.4	2.4	1.2	1.8	6.2	8.4	7.2	2.0	2.7	2.3	89.5	86.9	88.3	1,026	897	1,922
Montane meadow	45100		73.0	54.0	8.4	4.3	5.4	28.5	17.8	20.6	19.6	1.5	6.2	43.5	3.5	13.9	33	94	127
Wet subalpine or alpine meadow	45210	1.9	33.1	14.8	10.8	1.8	7.1	35.9	52.9	42.9	20.2	7.4	14.9	31.2	4.8	20.3	118	83	201
Dry subalpine or alpine meadow	45220		0.7	0.7		14.8	14.8		69.3	69.3		5.3	5.3		9.9	9.9		4	4
Great Basin montane meadow[a]	45230							0.5		0.5				99.5		99.5	1		1
Alkali meadow	45310		4.9	4.9					8.8	8.8		3.9	3.9		82.4	82.4		2	2
Transmontane alkali marsh	52320		3.9	1.4					17.3	6.3	0.6	0.4	0.5	99.4	78.4	91.7	11	6	17

[a]Addition to the standard Holland classification.

Percentage of Mapped Distribution by Land Management Class

Type of Plant Community (Holland 1986)	Holland (1986) Code	Class 1			Class 2			Class 3			Class 4			Class 5			Total Mapped Distribution (km²)		
		North	South	Total	North	South	Total	North	South	Total	North	South	Total	North	South	Total	North	South	Total
Riparian Woodland																			
Great Valley cottonwood riparian forest	61410		19.8	9.6	1.2	0.8	1.0	0.3	31.9	15.6	1.1	0.5	0.8	97.4	46.9	73.0	10	10	20
Great Valley mixed riparian forest	61420								37.0	37.0		0.5	0.5		62.5	62.5		12	12
Great Valley valley oak riparian forest	61430								2.6	2.6		0.4	0.4		97.0	97.0		16	16
White alder riparian forest	61510		65.3	65.3					0.3	0.3		<0.1	<0.1		34.4	34.4		5	5
Aspen riparian forest	61520							85.0		85.0	15.0		15.0				0.1		0.1
Montane black cottonwood riparian forest	61530	0.7	100.0	56.7	0.6		0.3	1.4		0.6	5.9		2.6	91.3		39.9	2	3	6
Montane riparian scrub	63500	3.2	37.7	23.5	3.9	1.8	2.6	22.2	14.0	17.3	17.8	19.8	19.0	52.9	26.8	37.5	49	70	119
Broad-Leaved Woodland																			
Oregon oak woodland	71110	<0.1	5.4	1.5		<0.1	<0.1	40.9	70.3	53.6	4.3	3.2	3.8	54.8	26.5	42.6	12	9	21
Black oak woodland	71120				6.9	9.2	7.5	9.5	21.0	12.7	21.8	28.0	23.5	61.8	36.5	54.7	332	128	460
Valley oak woodland	71130					0.1	0.1	6.5	0.1	1.8	0.9	0.4	0.5	92.6	99.5	97.6	92	248	340
Blue oak woodland	71140		1.4	1.2		0.7	0.6	5.3	6.8	6.5	2.6	3.3	3.2	92.2	87.7	88.6	1,031	4,395	5,426
Interior live oak woodland	71150		1.4	1.0	0.1	7.1	5.1	6.4	17.8	14.6	4.4	9.8	8.3	89.1	63.9	71.0	364	935	1,299
Conifer Woodland																			
Open foothill pine woodland	71310	0.7	1.9	1.4	0.7	6.4	4.9	2.6	21.2	16.4	5.1	24.3	19.3	91.7	46.3	58.1	114	327	441
Nonserpentine foothill pine chaparral	71322	1.8	12.9	8.4	1.8	4.5	3.5	19.0	31.1	26.9	9.4	23.4	18.5	69.8	28.2	42.7	87	162	249
Foothill pine–oak woodland	71410	0.2	0.7	0.4	0.2	2.0	1.2	5.4	15.6	11.2	3.3	6.8	5.2	91.2	74.9	82.0	1,293	1,682	2,975
Cismontane juniper woodland[a]	71500	3.4	34.9	31.4	3.4	0.6	1.0	75.4	62.3	63.9	14.5	2.1	3.6	0.1	0.2	0.2	20	135	155
Oak–piñon woodland[a]	71600		3.6	3.6		4.1	4.1		82.1	82.1		2.6	2.6		7.6	7.6		117	117
Northern juniper woodland	72110	2.4	17.6	5.8	5.0	4.6	4.9	46.1	12.6	38.7	42.7	64.7	47.5	3.8	0.6	3.1	142	40	182
Great Basin piñon–juniper woodland	72121	0.4	12.3	12.0		4.8	4.7	14.7	45.6	44.8	81.8	26.1	27.5	3.6	11.4	11.2	10	394	404
Great Basin piñon woodland	72122		2.6	2.3		0.9	0.8	2.1	50.6	43.8	82.6	24.5	32.7	14.9	21.4	20.5	122	741	863
Great Basin juniper woodland and scrub	72123				27.3		27.3	48.6		48.6	20.4		20.4	3.8		3.8	9		9
Mojavean juniper woodland and scrub	72220		1.3	1.3		0.5	0.5		4.4	4.4		26.8	26.8		66.9	66.9		63	63
Joshua tree woodland	73000								85.8	85.8		4.9	4.9		9.3	9.3		73	73
Broad-Leaved Forest																			
Canyon live oak forest	81320	5.4	22.8	17.9	22.4	9.6	13.2	23.5	33.0	30.3	15.1	14.1	14.4	33.7	20.6	24.3	258	658	916
Interior live oak forest	81330		3.2	1.8	0.9	3.7	2.4	4.3	22.3	14.4	5.4	6.0	5.7	89.5	64.9	75.7	676	870	1,545
Black oak forest	81340	2.3	12.2	6.5	17.8	9.8	14.4	22.2	25.2	23.5	23.9	32.3	27.4	33.9	20.5	28.2	624	463	1,087
Tan oak forest	81400				9.5		9.5	3.5		3.5	9.8		9.8	77.3		77.3	24		24
Aspen forest	81B00	1.9	27.4	18.0	13.3	23.3	19.6	30.6	25.3	27.3	44.1	20.8	29.4	10.2	3.3	5.8	37	63	99
Conifer Forest																			
Knobcone pine forest	83210		0.8	0.6	40.3	2.1	12.6		5.2	5.9		34.1	30.9		57.8	50.0	5	12	17
Southern interior cypress forest	83330		3.5	3.5		1.1	1.1		39.1	39.1		33.4	33.4		22.8	22.8		3	3
West-side ponderosa pine forest	84210	1.4	19.9	10.3	6.6	4.4	5.5	7.9	26.8	17.7	22.4	34.7	31.8	29.5	14.2	34.7	2,286	2,120	4,406
East-side ponderosa pine forest	84220	0.9		0.9	7.4	42.8	7.6	19.1	26.8	19.2	54.0	29.8	54.0	18.6	0.6	18.5	1,605	9	1,614
Sierran mixed conifer forest	84230	1.3	29.8	8.1	10.8	4.3	9.2	12.9	24.4	15.6	37.4	29.5	35.5	37.6	12.1	31.6	4,523	1,411	5,933
Sierran white fir forest	84240	0.5	28.9	7.7	16.0	5.1	13.2	17.8	15.1	17.1	38.5	40.5	39.0	27.2	10.3	22.9	403	138	540
Big tree forest	84250		51.6	51.6		2.6	2.6		24.4	24.4		16.7	16.7		4.7	4.7		71	71
Jeffrey pine forest	85100	2.7	34.0	16.9	10.5	5.6	8.3	33.8	27.0	30.7	40.6	28.1	34.9	12.4	5.3	9.2	1,073	888	1,961

[a] Addition to the standard Holland classification.

Percentage of Mapped Distribution by Land Management Class

Type of Plant Community (Holland 1986)	Holland (1986) Code	Class 1			Class 2			Class 3			Class 4			Class 5			Total Mapped Distribution (km²)		
		North	South	Total	North	South	Total	North	South	Total	North	South	Total	North	South	Total	North	South	Total
Red fir–western white pine forest[a]	85120	12.4	52.4	28.8	10.7	3.2	7.6	48.1	33.7	42.2	17.9	10.4	14.8	10.9	0.3	6.5	942	653	1,594
Jeffrey pine–fir forest	85210	3.1	22.9	15.6	13.2	5.0	8.0	28.1	36.3	33.2	30.1	33.2	32.1	25.5	2.6	11.1	1,095	1,861	2,956
Red fir forest	85310	2.4	49.1	33.2	14.0	2.0	6.1	24.0	32.7	29.7	35.9	15.2	22.2	23.7	1.0	8.7	1,153	2,241	3,395
Lodgepole pine forest	86100	4.1	60.0	53.5	11.6	3.1	4.1	39.8	30.8	31.9	23.2	5.6	7.7	21.4	0.5	2.9	252	1,904	2,156
Whitebark pine–mountain hemlock forest	86210	20.8	72.0	61.7	5.7	1.9	2.6	61.5	19.8	28.1	8.6	6.4	6.8	3.5	0.1	0.7	76	303	378
Whitebark pine–lodgepole pine forest	86220	3.1	65.3	56.1	3.8	20.4	18.0	68.6	6.9	16.1	16.6	7.0	8.4	7.8	0.4	1.5	55	317	372
Foxtail pine forest	86300		92.6	92.6		1.0	1.0		5.2	5.2		0.1	0.1		1.1	1.1		238	238
Whitebark pine forest	86600	9.5	67.7	58.0	7.2	7.3	7.3	35.2	2.9	8.3	43.4	22.1	25.6	4.8	<0.1	0.8	37	182	219
Limber pine forest	86700		5.4	5.4		16.9	16.9		77.6	77.6		0.1	0.1		<0.1	<0.1		21	21
Lower cismontane mixed conifer–oak forest[a]	87100	0.6	9.7	4.9	8.6	8.6	8.6	8.9	24.8	16.4	19.2	32.4	25.4	62.8	24.6	44.7	2,229	2,002	4,231
Upper cismontane mixed conifer–oak forest[a]	87200	18.0	20.6	20.1	8.6	6.0	6.6	16.0	33.4	29.7	50.1	16.5	23.6	7.4	23.4	20.0	55	205	261
Alpine Habitats																			
Sierra Nevada fell field	91120		27.7	27.5	0.7		<0.1	16.5	54.1	53.9	82.8	17.7	18.2		0.5	0.5	1	121	122
Alpine dwarf scrub	94000		89.5	89.5		7.1	7.1		3.1	3.1		0.2	0.2		0.2	0.2		394	394
Total Area																			
Vegetated lands																	25,381	31,198	56,580
Vegetated and unvegetated lands		2.1	25.7	15.4	10.1	4.1	6.7	16.8	24.9	21.4	25.7	15.6	20.0	45.3	29.8	36.5	27,483	35,619	63,102

[a]Addition to the standard Holland classification.

JAMES R. SHEVOCK
U.S. Forest Service
Pacific Southwest Region
San Francisco, California

24

Status of Rare and Endemic Plants

ABSTRACT

The Sierra Nevada represents nearly 20% of the California land base yet contains over 50% of the state's flora. Approximately 405 vascular plant taxa are endemic to the Sierra Nevada. Of this total, 218 taxa are considered rare by conservation organizations and/or state and federal agencies. In addition, 168 other rare taxa have at least one occurrence in the Sierra Nevada. Five monotypic genera are endemic to the Sierra Nevada (*Bolandra, Carpenteria, Orochaenactis, Phalacoseris,* and *Sequoiadendron*). Information on rarity and endemism for lichens and bryophytes for the Sierra Nevada is very speculative and fragmentary due to limited fieldwork and the small number of available collections. Two mosses are endemic to the Sierra Nevada. Parameters obtained for each rare and/or endemic taxon include habitat type and distributions by county, river basin, and topographic quadrangle. Distribution information for many taxa remains incomplete based on limited field studies and vouchered specimens, especially in the more unroaded and rugged areas of the Sierra Nevada. Rare and endemic species are not evenly distributed throughout the Sierra Nevada. The Kern, Kings, Merced, San Joaquin, Tuolumne, and Feather River Basins contain the largest concentrations of rare and endemic taxa in the Sierra Nevada. For the eastern slope of the Sierra Nevada, the Owens River Basin is rich in species composition as well as rare and endemic taxa. Of the three geographical subunits, the northern, central, and southern Sierra, the southern Sierra is extremely rich in endemics, rare species, and total floristic composition. Adverse impacts to some Sierran rare plants are occurring along the western fringe of the range adjacent to the Central Valley, where conversion of lands to agriculture and urbanization may greatly restrict or alter essential habitat for some Sierran endemics and/or rare species.

INTRODUCTION

For more than 100 years, the flora of the Sierra Nevada has fascinated botanists even beyond the borders of the United States. Visions of Yosemite, giant sequoias, and extensive mixed conifer forests have added to an awareness of this magnificent mountain range. The Sierra Nevada, part of the California Floristic Province, is characterized by high rates of plant endemism (Stebbins and Major 1965; Raven and Axelrod 1978; Messick 1995). For most of this century, plant collecting and floristic research remained the pursuits of professional botanists with ties to major scientific and educational centers (Shevock and Taylor 1987). Floristic studies have as one of their primary goals documentation of all the taxa (species, subspecies, varieties) for a particular geographic region and determination of their distribution and abundance within that study area (Palmer et al. 1995). Rare, endemic, and disjunct taxa have a special place in such studies because they contribute to the diversity and uniqueness of a flora.

Remarkably, the Sierra Nevada lacks a comprehensive floristic treatment. Portions of the range are covered by a great variety of floristic studies, ranging from detailed floras to florulas and checklists. Floristic studies generally fall into four categories: county floras (Clifton 1994; Oswald 1994; True 1973; Twisselmann 1967), floristic studies by watershed (Henry 1994; Lavin 1983; Palmer et al. 1983; Savage 1973; Shevock 1978; Smith 1973, 1983; Taylor 1981), studies based on park or preserve boundaries (Gillett et al. 1961; Knight et al. 1970; Potter 1983; Pusateri 1963; Rice 1969; Showers 1982), and studies by specific topographical features and habitats (Forbes et al. 1988; Howell 1951; Hunter and Johnson 1983; Sharsmith 1940; Smiley 1921; Tatum 1979; Williams et al. 1992).

Sierra Nevada Ecosystem Project: Final report to Congress, vol. II, *Assessments and scientific basis for management options.* Davis: University of California, Centers for Water and Wildland Resources, 1996.

Much acreage remains in the Sierra Nevada that is not botanically surveyed or systematically vouchered, especially in unroaded or relatively rugged areas.

With the passage of the federal Endangered Species Act of 1973 (ESA, as amended) came a distinct shift in plant collecting and subsequent conservation efforts toward a focus on those taxa believed to be candidates for threatened or endangered status. These distribution data were increasingly obtained by plant enthusiasts, botanical consultants, and various state and federal agency botanists rather than traditional academic botanists with ties to major educational institutions (Ertter 1995; Shevock and Taylor 1987). Initially, the information available to determine which taxa were in fact rare and/or endemic was fragmentary, with most information restricted to a handful of herbarium specimens (Powell 1974). With efforts directed at rediscovery of old herbarium records, along with systematic and focused fieldwork to document new occurrences, understanding of the distribution of rare and endemic species has greatly improved (Smith et al. 1980; Smith and York 1984; Smith and Berg 1988; Skinner and Pavlik 1994).

Floristic inventories are becoming ever more important as a method of documenting the plant diversity of a specific land base. However, many of the currently available floras and checklists lack citations of representative vouchered specimens to validate each of their entries. Without references to vouchered plant material deposited in major herbaria, these floras and checklists have reduced value because material on which the catalogue of names is based is not available for future study and taxonomic review (Palmer et al. 1995; Ferren et al. 1995). Of course, many floras are based on a review of herbarium records, but again, representative specimens are rarely cited in the publication of floristic studies. Wilken (1995) provides a convincing case for continued floristic studies in California that emphasize comparative analyses based on biogeographical patterns of diversity at both regional and local levels.

It may come as a surprise to many not familiar with the California flora that vascular plants are still being discovered and described as new to science for the Golden State. The majority of these newly published species are both endemic and rare. The period 1968–86 yielded over 220 newly described vascular plant taxa for California; sixty-five of these occur in the Sierra Nevada (Shevock and Taylor 1987). With publication of *The Jepson Manual* (Hickman 1993), ongoing floristic analysis by Shevock and Taylor (in preparation) will document that since 1986 this trend of discovery and publication of new vascular plant taxa continues. The southern Sierra Nevada in particular, along with other areas of carbonate and serpentinite geology, remains an area of the state worthy of continued floristic study and research (Norris 1987; Shevock 1988). During the past few years several new species have been discovered in the Sierra Nevada. Because many of these new taxa are rare and/or endemic to a single river basin, they are incorporated in this assessment with the specific epithet

"sp. nov.," "ssp. nov.," or "var. nov." until the names have been effectively published according to the International Code of Botanical Nomenclature.

This assessment was developed to determine the distribution of both endemic and rare plant taxa in the Sierra Nevada, primarily at the river basin level. For the core study area, the Sierra Nevada was divided into twenty-four river basins (figure 24.1) ranging in size from the Feather at 971,611 ha (2,399,878 acres) to the Calaveras at 94,018 ha (231,285 acres). River basin boundaries are useful because they are easy to determine both in the field and on maps, in contrast to political boundaries such as counties, forests, and parks, which have the potential to change through time. Furthermore, river basins provide a biogeographical context in which to evaluate floristic components such as rare and/or endemic species. Size, elevation range, geology and soils, vegetation types, and geographical location of each river basin are factors used to speculate why river basins vary widely in total number of taxa, including rare and endemic species. As a general overview, the northern Sierra is predominantly volcanic in origin, and the central and southern Sierra are both mainly granitic, with several areas of metamorphic and metasedimentary parent materials.

Lum (1975) may have been the first to address broad patterns of vascular plant species diversity based on *A California Flora* and *Supplement* (Munz and Keck 1959; Munz 1968). Lum also evaluated the distribution of all taxa displayed in this floristic treatment at the county level and further divided the counties to address physiographic provinces. This approach subdivided California from fifty-eight counties into ninety-four geographical units for her diversity analysis of the flora. Lum evaluated each entry in the flora (5,902 taxa) in a database with several parameters obtained per taxon. Although this approach provided many insights into the distribution and diversity within the flora, it appears that Lum made several "taxonomic decisions" by aggregating varietal and subspecific taxa to the species level, which reduced the number of highly localized endemics that could be analyzed. Nonetheless, Lum's contribution toward an understanding of vascular plant diversity in California is noteworthy. Walker (1992) added to the study of plant diversity in the Sierra Nevada by analyzing species richness and variation by plant community. Both of these studies were used to evaluate distribution patterns for rare and/or endemic vascular plants within the study area.

Nonvascular plants for this study include both lichens and bryophytes. Lichens are actually not plants but photosynthetic associations consisting of dense populations of green algae or cyanobacteria within the fungal tissue (Ahmadjian 1995). The process of this unusual association forms a "plant body" technically called a thallus, which has little resemblance to either an alga or a fungus (Hale and Cole 1988). Botanists and ecologists have historically treated lichens as a group of nonvascular plants primarily because lichens can colonize

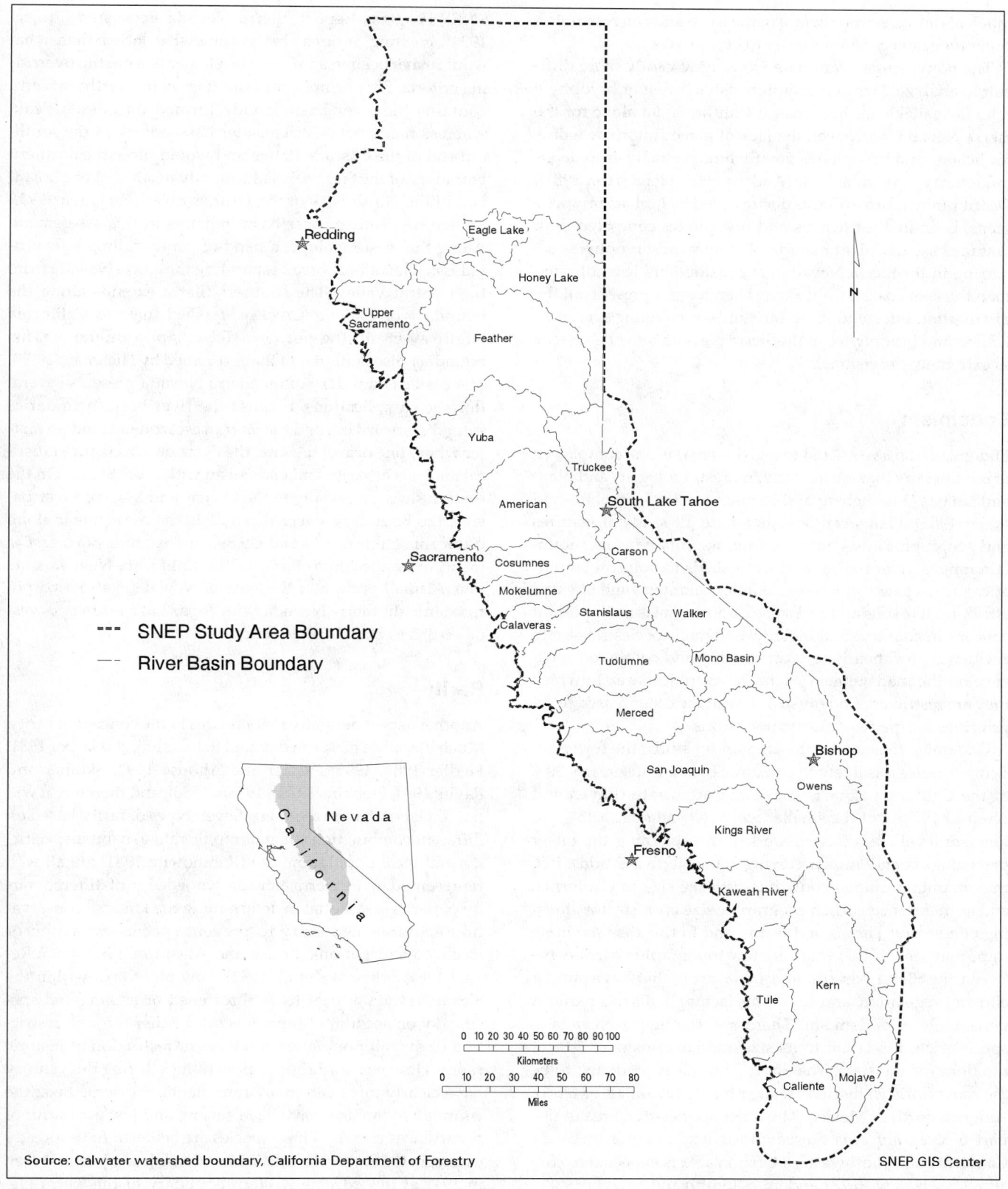

FIGURE 24.1

SNEP study area with river basin boundaries.

much of the terrestrial surface of the earth and can resemble a cover type of vegetation similar to bryophytes.

The nonvascular plant data were significantly more difficult to analyze. First, no comprehensive lichen or bryophyte flora is available at this time for California, let alone for the Sierra Nevada. Moreover, the lack of general floristic works for lichens and bryophytes greatly hinders analysis to determine rarity and endemism (Andrus 1990; Henderson 1981). Distribution information, as documented by herbarium specimens, is limited for lichens and bryophytes compared with that for vascular plant material. Lichens and bryophytes occurring in the Sierra Nevada are considerably less collected than those of coastal California. Therefore, at present, all the distribution information on rare and/or endemic species of lichens and bryophytes in the Sierra Nevada must be viewed as extremely provisional.

Endemism

Endemism (taxa restricted to a given area) is one component of biodiversity that particularly interests biologists and plant enthusiasts (Kruckeberg and Rabinowitz 1985; Stebbins and Major 1965). Plants can be endemic to all kinds of features and geographic areas, ranging from mountain peaks, mountain ranges, river basins, and watersheds to political boundaries such as parks, preserves, counties, and states and physical attributes like soils and rock types. Endemism is an important concept in conservation biology. Endemic species must rely exclusively for their long-term viability and continued existence on the management of the geographical area to which they are restricted. Endemism is one of the criteria used to set priorities for species conservation efforts.

Generally, the smaller the area under study, the fewer endemic species it is likely to contain. For example, nearly 30% of the California flora is endemic to the state (Raven and Axelrod 1978), but the smaller Sierra Nevada has a 15% endemism level. Two factors are key in explaining the lower percentage of endemics occurring in the Sierra Nevada. First, this mountain range is only one-fifth the size of California, and as mentioned earlier, geographic size is one factor affecting endemism. The second factor, and in this case the more important one, is that there are few topographic barriers between the Sierra Nevada and other areas that have similar climate, vegetation, and soils. This factor results in a reduced percentage of endemism. Therefore, the Sierra Nevada is species-rich in relation to its size and contains over 50% of the flora for the state. The Sierra Nevada is predicted to be the most floristically diverse area for its size in all of North America north of Mexico. The other species-rich area is also part of the California Floristic Province. The northwestern California and southwestern Oregon area is expected to contain 3,500 taxa with 281 endemics (Smith and Sawyer 1988).

The boundary selected for this study affects which species are viewed as "endemic" to the Sierra Nevada. The boundary used follows the primary river basins as identified in the SNEP Progress Report (Sierra Nevada Ecosystem Project 1994), creating an area that is somewhat larger than what would have been used if floristic elements were the overriding criteria. For example, the boundary in the northern Sierra contains the river basin divide formed on Lassen Peak, whereas many botanists consider Lassen Peak as the southern end of the Cascade Range and would place the northern boundary of the Sierra Nevada slightly northwest of the canyon of the North Fork of the Feather River (Hickman 1993). The entire Feather River Basin remains in this assessment. Along the western slope, a band of gently rolling hills provides an elevational break separating the Sierra Nevada from the Central Valley. The southern Sierra extends along the boundary of Caliente Creek watershed (next to California Highway 58) in the northern Tehachapi Mountains. This boundary also is similar to that expressed by Hickman (1993). The eastern boundary of the Sierra Nevada presents several floristic complications, because the river basin boundaries extend eastward beyond the Sierran escarpment, and no matter where one draws the line, the decision affects the statistical analysis of rarity and endemism within the Sierra. On the eastern slope, especially in the Owens and Mojave River Basins, the boundary selected parallels the escarpment along California Highway 14 and US 395 and extends northeast to incorporate the Mono Basin to the California-Nevada state line. A small portion of the state of Nevada is also incorporated into the Sierra Nevada from Topaz Lake north and west of US 395 to Hallelujah Junction.

Rarity

Another aspect of biodiversity relates to the concept of rarity. Much literature has been devoted to this subject (Harper 1981; Fiedler 1986, 1995; Fiedler and Ahouse 1992; Skinner and Pavlik 1994; Stebbins 1978a, 1978b, 1980), and therefore it will not be elaborated further here. Seven types of rarity, based on different combinations of geographic range, habitat specificity, and local population size (Rabinowitz 1981) are all well represented in the Sierra Nevada. Knowledge of different rarity patterns is essential in determining the kinds of conservation activities necessary to prevent species extinction or localized extirpations (Lesica and Allendorf 1992, 1995; Reveal 1981; Schemske et al. 1994). Few plant taxa within the Sierra Nevada appear to be threatened or endangered specifically on account of human actions either through restriction of overall population numbers or restriction of historic range. However, anthropogenic activities during this century have clearly impacted many rare plants. For some taxa, the cumulative impacts have been severe, and just a handful of occurrences remain. Three species are believed to be already extinct in the Sierra Nevada. *Monardella leucocephala*, last seen in 1941 at the extreme western boundary of the Sierra Nevada in Merced County, is presumed to be extinct. This extinction is attributed to human activities this century that changed valley grassland habitat to agricultural land (Skinner and Pavlik 1994). Two other species, also in the central Sierra Nevada, are viewed as possibly extinct: *Erigeron mariposanus*, last seen in 1900, and *Mimulus whipplei*, last seen in 1854. *Erigeron mariposanus* was collected several times between 1892 and 1900. It is suspected that this species was restricted to a specialized habitat that may have been altered this century. *Mimulus whipplei* was described from a single herbarium collection. Botanists are uncertain whether it merely represents an aberrant collection of a more widespread species or actually warrants taxonomic recognition. Focused surveys and field studies may yet rediscover these presumed extinct taxa (Skinner et al. 1995).

Another problematic species is *Sedum pinetorum*. This species appears to be a good taxon. In 1925, Jepson thought it was so different from *Sedum* (in having solitary flowers and tuberous roots) that he published the monotypic genus *Congdonia* to accommodate it. This taxonomic circumscription was carried forward by Munz and Keck (1959). *Congdonia*, however, proved to be an invalid name due to prior use, and this taxon returned to the genus *Sedum*. The concern, therefore, lies not with its taxonomic distinctiveness but rather its presumed location. Collected in 1913 by Katherine Brandegee and described in 1916 by T. S. Brandegee, *Sedum pinetorum* is believed to have been collected along the eastern slope of the Sierra Nevada in the vicinity of Mammoth, Mono County. However, the type population has never been relocated, nor has the species been collected since.

Based on examination of seeds within the fragment packet with the type specimen (at the University of California, Berkeley collection), there is speculation that the specimen probably came not from the Sierra Nevada but more likely from Mexico (Moran 1950). *Sedum pinetorum* was subsequently deleted from the third edition of the California Native Plant Society *Inventory* (Smith and Berg 1988), the California Natural Diversity Database, and most recently *The Jepson Manual* (Hickman 1993), even though no one has found *Sedum pinetorum* in Mexico either. I included the species in this assessment until another collection of *Sedum pinetorum* becomes available to resolve this mystery. Again, focused surveys may yet rediscover this inconspicuous plant of the eastern Sierra Nevada in the Mammoth area.

Gaps and Caveats

Clearly there are gaps in the known distribution information of many Sierran rare and endemic plants. One factor may be that some river basins have been explored more systematically than others, primarily as a function of access by roads. This assessment acknowledges that the data sets are far from complete. It is hoped this assessment will encourage general botanical collecting (especially at the river basin level) with the objective of filling in the gaps in distribution data with vouchered material and perhaps discovering plants that have heretofore remained undescribed.

The perceived rarity or endemic status for some rare and/or endemic Sierran taxa is expected to change as systematic fieldwork continues. Also, further taxonomic studies (mainly detailed monographic works) should clarify some of the taxonomic uncertainties that currently exist (Skinner et al. 1995). For rare taxa, this study follows the taxonomic circumscriptions used by Skinner and Pavlik 1994 and California Department of Fish and Game 1995 and does not dismiss difficult taxa or those submerged in more recent floristic treatments (Hickman 1993; Skinner and Ertter 1993). This approach and rationale is based in part on my own field experience, in which ecological differences and growth forms may not readily be observed solely from a review and evaluation of an herbarium specimen.

METHODS

For the analyses reported here, a database was developed to query various Sierra Nevada rare and endemic plant records. The data fields are divided into four broad categories: taxonomic, rarity, endemism, and distribution information. Taxonomic fields include plant code, vascular or nonvascular, family, genus, and specific and infraspecific epithets. Rarity fields include federal listing, state listing, California Natural Diversity Database list, California Native Plant Society list, and U.S. Fish and Wildlife Service list. Endemism fields include Sierra Nevada rare, Sierra Nevada endemic, and plants rare but located beyond the Sierra Nevada. Distribution fields include counties, river basins, habitat types, and topographic quadrangles.

For California rare plant species, the California Native Plant Society (CNPS) publication *Inventory of Rare and Endangered Vascular Plants of California* (Skinner and Pavlik 1994) was the primary reference consulted, together with data maintained by the rare plant component of the California Natural Diversity Database (CNDDB), California Department of Fish and Game (1995). Rare plants for the state of Nevada were obtained from the Nevada Heritage Program (Morefield and Knight 1991; Morefield 1994). For each rare and/or endemic taxon, the range of distribution by primary river basin within the Sierra Nevada was plotted by analyzing these data, including the analysis conducted by Lum (1975). For plants that are endemic to the Sierra Nevada but not rare or endangered, these data were laboriously gathered by species distribution information in *A California Flora* (Munz and Keck 1959) and *Supplement* (Munz 1968) plus taxa newly described since 1968 (Shevock and Taylor 1987). *The Jepson Manual* (Hickman 1993) was also analyzed, by Dean Taylor, for additional rare and/or endemic taxa occurring in the Sierra Nevada. I obtained additional distribution information for nonrare Sierran vascular plant endemics by reviewing numerous floras, florulas, and checklists (both published and unpublished) within the

Sierra Nevada. An extensive herbarium review to obtain supplemental distribution information for vascular plants was not conducted as part of this study.

For this study, lichen and bryophyte data sets were provided by leading professional authorities in these two fields of taxonomic study. Bruce Ryan at Arizona State University, Tempe, offered field knowledge and expertise on Sierran lichen distributions, and Daniel Norris at the University of California, Berkeley, provided the data sets for Sierran rare and endemic bryophytes. Selected California herbaria with important lichen and bryophyte collections (California Academy of Sciences [CAS], San Francisco State University [SFSU], and University of California, Berkeley [UC]) were visited to obtain distribution information necessary to plot occurrences within Sierran river basins.

RESULTS AND DISCUSSION

Nonvascular Plant Taxa

Of the nearly 3,330 species of lichens known to occur in the continental United States and Canada (Egan 1987), over 1,000 species have been documented to occur in California (Hale and Cole 1988; Tucker and Jordan 1978). Knowledge of lichen distributions is rather fragmentary compared with that of vascular plants. There are relatively few lichen floristic studies within the Sierra Nevada. In these available floras and checklists, 207 lichens have been documented for a portion of Tulare County (Smith 1980; Wetmore 1985, 1986; Sequoia National Park 1995), and 85 lichen taxa were reported from a mixed conifer forest at Calaveras Big Trees State Park (Pinelli and Jordan 1978). Herbert and Meyer (1984) documented 76 lichen species within a small area dominated by a blue oak woodland at the U.S. Department of Agriculture San Joaquin Experimental Range in Madera County. For the east slope of the Sierra Nevada, Herre (1911) reported 59 lichen species in the vicinity of Reno, of which he described two species as new. Ryan and Nash (1991) collected over 100 species in the Eastern Brook Lakes watershed in Inyo County, of which 30 species were new lichen records for California.

From the data sets that are at present available, it appears that no lichens currently can be considered endemic to the Sierra Nevada. However, endemism cannot be determined with any reliability for lichens in the Sierra Nevada primarily because so few lichenologists are available to conduct taxonomic work or study them systematically. For example, there are several crustose lichens, such as *Lecidea truckeei*, that may actually be rare, but this group is taxonomically very difficult to identify. Many crustose lichens may have much wider distributions than is currently understood, while others, perceived to be more widespread, could indeed be rare or endemic. At this time, however, it is not possible to determine accurate distributions, because the majority of specimens in most lichen herbaria have not yet been properly identified or are labeled only to the genus level. This situation is even more acute for the crustose group of lichens (Ryan 1995a).

Lichens display different rarity and distribution patterns than those commonly observed in vascular plants. Many lichens have wide-ranging distributions in North America but within their geographic range display a pattern of very localized occurrences restricted to specific habitats and/or narrow microenvironments. Rarity in lichens is therefore characterized by small, disjunct populations. The species richness of the lichen flora is greater in the maritime and coastal-fog-influenced areas of California (Hale and Cole 1988). The drier interior of the state, a Mediterranean climate with continental influences, has a different assemblage of lichens, with more crustose lichen taxa and fewer foliose and fruticose lichen taxa.

For this assessment, eight lichen species are considered rare in the Sierra, and two of these represent the only known occurrences in California: *Dermatocarpon moulinsii*, *Dimelaena oreina* (atypical forms in the Sierra), *Hydrothyria venosa*, *Hypogymnia metaphysodes*, *Rhizoplaca glaucophana*, *Rhizoplaca marginalis*, *Umbilicaria torrefacta*, and *Waynea stoechadiana* (Ryan 1995). *Hydrothyria venosa* is very unusual as it is the only aquatic foliose lichen species. It is restricted to rocks in clear, unpolluted streams. Within California, this species is currently known from only a few streams, ranging from Calaveras Big Trees State Park, Calaveras County, south to a tributary of the Kern River Basin in Sequoia National Forest, Tulare County.

Lichens as a taxonomic group are often used in air-quality monitoring studies because many species are sensitive to air pollution (Ryan 1990b). Air-quality degradation in the Sierra Nevada has adverse effects on some lichens. A few species may become extirpated in portions of their range within the Sierra Nevada if air quality continues to deteriorate. Extirpations of lichens in other mountains of California have already occurred. In their study of the San Gabriel and San Bernardino Mountains, Sigal and Nash (1983) were not able to relocate eight species of lichens that were collected there in conifer forests by Hasse in his *Lichen Flora of Southern California* published in 1913. Three Sierran baseline monitoring studies of lichens and air quality have been conducted. Ryan (1990a, 1990b) identified over 90 lichen species for both the Desolation Wilderness in the Lake Tahoe area and the Emigrant Wilderness, Stanislaus National Forest; the third study, by Wetmore (1985, 1986), obtained 207 lichens for Sequoia National Park and the Grant Grove section of Kings Canyon National Park.

The bryophytes (mosses, liverworts, and hornworts) are a diverse group of nonvascular plants (Crosby 1980) with nearly 23,000 described species worldwide. In North America, more than 1,220 species have been documented (Schofield 1980). Endemism in bryophytes approaches 23% for North America. For the west coast of North America, Lawton (1971) identified nearly 600 mosses in her treatment of the Pacific Northwest, the largest moss flora of a geographic region within

North America. Bryophyte taxa generally have much wider distribution ranges than vascular plants. Endemism at smaller geographic scales such as the state of California drops markedly (Koch 1950a, 1954).

Though no modern-era (post-1950s) bryophyte flora for California currently exists, such a flora is in preparation by Dan Norris and Brett Mishler at the University of California, Berkeley. Surprisingly, California and the adjacent southwestern United States have the least studied bryophyte floras in North America. At this time, Lawton 1971 and Flowers 1973 are the two floras available to provide identification of Sierran mosses. Based on herbarium records and fieldwork to date, the California bryophyte flora contains 508 species of mosses, 116 species of liverworts, and 6 species of hornworts (Mishler 1995). This record shows a significant increase in the number of taxa in California over earlier checklists of 317 mosses by Koch (1950) and 86 liverworts by Howe (1899). Of the thirty-five thallose liverworts recorded for California by Whittemore 1982, twenty-seven occur in the Sierra Nevada.

Bryophytes are well developed in the Pacific Northwest, especially in the temperate rain forests of Sitka spruce and western hemlock from southern Alaska and British Columbia to the coast redwood zone of northwest California (Lawton 1971). The bryophyte flora for the Sierra Nevada is predicted to be less diverse because this area is significantly more xeric than the Pacific Northwest. Showers (1982) documented 149 moss taxa for Lassen Volcanic National Park. Koch (1958) listed 72 mosses for the Harvey Monroe Hall Research Natural Area and vicinity toward Lee Vining along the eastern escarpment of the Sierra Nevada in Mono County. However, too few studies have been conducted to allow comparisons between the Sierra Nevada and the overall California bryophyte flora. In fact, Thiers and Emory (1992) listed the sixteen master's theses on California bryophytes for 1969–90, and only one floristic study was conducted within the SNEP boundary, the master's thesis work by Showers completed in 1978 and subsequently published in Showers 1982. There appear to be no other bryophyte floras for any smaller geographic areas within the Sierra Nevada.

Bryophytes share many of the distribution patterns observed in lichens. They tend to be highly localized to specific microenvironments defined by factors related to water availability, temperature, light, and substrate chemistry (Mishler 1995). For this assessment, seventeen mosses are considered rare in the Sierra Nevada (Norris 1995a, 1995b; Showers 1995). *Grimmia hamulosa* and *Orthotrichum spjutii* are endemic to the Sierra Nevada. *Mielichhoferia tehamensis* is endemic to Lassen Volcanic National Park. The remaining fourteen mosses, which are distributed beyond California, are *Andreaea nivalis, Bruchia bolanderi, Campylium stellatum, Distichium inclinatum, Grimmia moxleyi, Hydrogrimmia mollis, Lescuraea pallida, Mnium arizonicum, Myurella julacea, Orthotrichum euryphyllum, Polytrichum sexangulare, Racomitrium hispanicum, Tayloria serrata,* and *Tortula californica.* Several of these mosses are Holarctic (northern hemisphere) in distribution or have

widely scattered occurrences in North America. No liverworts or hornworts are considered rare or endemic to the Sierra Nevada.

Based on the distribution ranges of all Sierran plant endemics, clearly *Orthotrichum spjutii* is the rarest of all. This moss is known from a single rock face in the spray of a waterfall on the eastern slope of the central Sierra within the Walker River Basin. *Mielichhoferia tehamensis* occurs in the northernmost portion of the Sierra Nevada, with most occurrences within the headwaters of the Feather River Basin. This rare and endemic moss is restricted to deeply shaded, north-facing, steep canyon walls where winter snows remain well into July (Showers 1982).

One of the principal findings of this assessment is that there is a great need for systematic collecting and taxonomic study of lichens and bryophytes to aid in understanding species endemism, rarity, and distribution. Besides the small number of specimens available for study, there are also few botanists trained to study lichens and bryophytes. The importance of lichens and bryophytes to ecosystem function has been largely overlooked by many conservation biologists (Ahmadjian 1995; U.S. Forest Service 1995).

Vascular Plant Taxa

The Sierra Nevada plant database developed for this assessment contains 572 vascular plant entries, of which 383 taxa are being tracked by the CNDDB (California Department of Fish and Game 1995) and by the CNPS (Skinner and Pavlik 1994). Rare plants comprise 386 taxa, of which 168 extend beyond the Sierra Nevada. There are 405 vascular plant taxa endemic to the Sierra Nevada. Of this total, 223 are considered rare, and the remaining 182 plant taxa appear to be distributed across the landscape in such numbers and occurrences that their likelihood of their becoming rare, threatened, or endangered is low at this time. The distributions of these taxa by river basins provides a basis for analysis and interpretation of the flora, species richness, and diversity, with a focus on both rarity and endemism. Forty Sierran endemics are widespread, occurring in all three geographical subdivisions (northern, central, and southern). Another 133 endemic taxa occur in two subdivisions. Within these subdivisions, 55 taxa are endemic to the northern Sierra, 53 are endemic to the central Sierra, and 124 are endemic to the southern Sierra.

California has twenty-six endemic genera of vascular plants (Howell 1957; Raven and Axelrod 1978). The Sierra Nevada contains five endemic genera that are also monotypic. These were all described as "new to science" as a result of early botanical explorations in the Sierra Nevada during the middle to late 1800s. *Sequoiadendron* and *Carpenteria* were described in 1853, *Bolandra* and *Phalacoseris* in 1868, and *Orochenactis* in 1883. There have been no new vascular plant genera recognized in California since *Dedeckera* was discovered in the adjacent White Mountains in 1976. The probability that a new vascular plant genus will be discovered in the Sierra Nevada seems unlikely based on the level of fieldwork conducted in this mountain range during the past hundred years.

Of the seven river basins within the boundary of the northern Sierra Nevada, those of the Feather and American Rivers have the greatest number of taxa, including endemic and rare taxa (table 24.1). Of the 140 study taxa located within the Feather River Basin, 79 are endemic to the Sierra Nevada, and 95 are rare. The large number of rare taxa is primarily a result of the location of this river basin adjacent to the Cascade Range; several rare plants from the Klamath Mountains and Cascade Range reach their southernmost distribution limits within the Feather River Basin. Based on this study, eleven taxa are endemic to this river basin. The American River Basin shares a similar pattern except that it has a higher proportion of Sierran endemics, but only seven taxa are endemic to this river basin. The Truckee River Basin has the second highest number of endemics, with ten for the northern Sierra, of which two are endemic to the portion of the range in Nevada.

The central Sierra, which contains ten river basins, displays a completely different pattern from the one observed in the northern Sierran river basins (table 24.1). Three adjacent river basins (Tuolumne, Merced, and San Joaquin) are nearly identical in total numbers of taxa, Sierran endemics, and rare elements. They also share high numbers of endemic and rare

taxa as compared with the other river basins within the central Sierra Nevada. For taxa that are endemic to a single river basin, the San Joaquin leads, with six taxa, and the Merced and Tuolumne have four each.

The southern Sierra, with seven river basins, has a greater number of taxa, Sierran endemics, and rare elements than the northern or central Sierra (figure 24.1). The southern Sierra contains the highest elevation within the Sierra Nevada (the Mount Whitney area), and at the same time the range is narrower in width and thus steeper than the central and northern Sierra. This portion of the Sierra Nevada contains extensive alpine and subalpine areas that provide habitat for many Sierra endemic and rare taxa (Stebbins 1982). The greater aridity of the southern Sierra Nevada is also part of the reason for the high number of endemics compared with the northern Sierra. The Kern River Basin has 200 study taxa, of which 167 are endemic to the Sierra Nevada and 91 are rare. This single river basin has twenty-two plants endemic within its boundary; the Owens River Basin has eight endemics, and the Kings River Basin has six. The Owens River Basin, which contains the steep and rugged east face of the Sierra Nevada, dominates transmontane river basins in total taxa, Sierran endemics, and rare elements. Table 24.2 provides the catalogue of taxa endemic to individual river basins in the Sierra.

The geographic unit traditionally used to evaluate plant distributions in California is the county (Munz and Keck 1959; Skinner and Pavlik 1994). Table 24.3 displays rare and endemic plant information by county in a format suitable for comparison with river basins (table 24.1). Though the boundaries of the northern, central, and southern Sierra Nevada are altered to follow county lines (figure 24.2), the boundary differences do not in general alter the overall trends of the data presented. In the northern Sierra Nevada, El Dorado and Plumas Counties contain the greatest number of taxa, Sierran endemics, and rare taxa, and these two counties correspond closely to the boundaries of the American and Feather River Basins. Within the central Sierra Nevada, Mariposa County dominates all categories, followed closely by Madera County. The four counties represented in the southern Sierra Nevada all contain high concentrations of endemic and rare taxa. Tulare and Fresno Counties lead in all categories and are the counties with the largest number of endemics and rare species for the entire Sierra Nevada. Table 24.4 provides the catalogue of taxa endemic to the Sierra Nevada by counties.

The use of topographic quadrangles is another geographic approach for addressing the distribution of species. In this assessment, only rare taxa were recorded to the topographic quadrangle level, based on data obtained from Skinner and Pavlik (1994). Data sets for CNPS list 4 taxa (plants of limited distribution) lack quadrangle information. For these taxa, I attempted to review all CNPS and CNDDB manual files to obtain these data sets where available. However, many distribution gaps remain at the quadrangle level for CNPS list 4 plants.

TABLE 24.1

Distribution of rare and endemic plants by Sierran river basins.

River Basin[a]	Number of Taxa from Database	Sierran Endemics	Rare Taxa	Endemic to River Basin
Northern Sierra				
American	104	85	46	7
Eagle Lake	5	1	5	0
Feather	140	79	95	11
Honey Lake	20	6	20	1
Truckee	49	38	29	10
Upper Sacramento	46	12	42	2
Yuba	91	69	45	2
Central Sierra				
Calaveras	63	61	16	1
Carson	22	15	12	0
Consumnes	72	66	21	0
Merced	153	140	53	4
Mokelumne	90	80	31	2
Mono Lake	65	45	32	1
San Joaquin	149	135	57	6
Stanislaus	100	93	31	0
Tuolumne	152	133	59	4
Walker	33	18	23	4
Southern Sierra				
Caliente	29	22	18	1
Kaweah	112	104	37	3
Kern	200	167	91	22
Kings	160	150	56	6
Mojave	28	19	21	4
Owens	104	71	59	8
Tule	87	83	31	2

[a]Several river basins extend beyond the study area boundary. Rare and endemic species are recorded only for those taxa within the study area.

TABLE 24.2

Sierra Nevada endemics at the river basin level.

Northern Sierra Nevada	Central Sierra Nevada	Southern Sierra Nevada
American River Basin Ceanothus roderickii Draba asterophora var. macrocarpa Galium californicum ssp. sierrae Lewisia serrata Navarretia prolifera ssp. lutea Phacelia stebbinsii Wyethia reticulata	**Calaveras River Basin** Mimulus whipplei	**Caliente Creek Basin** Clarkia tembloriensis ssp. calientensis
Feather River Basin Astragalus lentiformis Astragalus webberi Calamagrostis sp. nov. Ceanothus sp. nov. Clarkia mosquinii ssp. mosquinii Clarkia mosquinii ssp. xerophylla Erigeron lassenianus var. deficiens Monardella stebbinsii Penstemon personatus Sedum albomarginatum Senecio eurycephalus var. lewisrosei	**Merced River Basin** Clarkia lingulata Eriophyllum congdonii Plagiobothrys torreyi var. torreyi Viola adunca var. kirkii	**Kaweah River Basin** Eriogonum nudum var. murinum Mimulus norrisii Ribes tularense
	Mokelumne River Basin Eriogonum apricum var. apricum Eriogonum apricum var. prostratum	**Kern River Basin** Abronia alpina Astragalus ertterae Astragalus shevockii Camissonia integrifolia Castilleja praeterita Ceanothus pinetorum Clarkia xantiana ssp. parviflora Cordylanthus rigidus ssp. brevibracteatus Crythantha incana Delphinium purpusii Erigeron multiceps Eriogonum breedlovei var. breedlovei Eschscholzia procera Galium angustifolium ssp. onycense Githopsis tenella Heterotheca shevockii Horkelia tularensis Mimulus microphyllus Mimulus shevockii Nemacladus twisselmannii Streptanthus cordatus var. piutensis Swertia tubulosa
Honey Lake River Basin Scutellaria holmgreniorum	**Mono Lake Basin** Arabis tiehmii	
Truckee River Basin Arabis rigidissima var. demota Arabis rigidissima var. simulans Astragalus austinae Astragalus whitneyi var. lenophyllus Berberis sonnei Elatine gracilis Eriogonum robustum Ivesia aperta var. canina Rorippa subumbellata Tonestus eximus	**San Joaquin River Basin** Calyptridium puchellum Collomia rawsoniana Erigeron mariposanus Erythronium pluriflorum Lupinus citrinus var. deflexus Phacelia ciliata var. opaca	
Upper Sacramento River Basin Calystegia atriplicifolia ssp. buttensis Rupertia hallii	**Tuolumne River Basin** Allium tuolumnense Brodiaea pallida Senecio clevelandii var. heterophyllus Verbena californica	**Kings River Basin** Arabis sp. nov. Cordylanthus tenuis var. barbatus Eriogonum nudum var. regirivum Gilia australis ssp. nov. Heterotheca sp. nov. Streptanthus fenestratus
Yuba River Basin Plagiobothrys glyptocarpus var. modestus Sidalcea stipularis	**Walker River Basin** Draba incrassata Orthotrichum spjutii Plagiobothrys glomeratus Senecio pattersonensis	**Mojave River Basin** Chamaesyce vallis-mortae Eriogonum kennedyi var. pinicola Hemizonia arida Lomatium shevockii
		Owens River Basin Astragalus sepultipes Galium hypotrichium ssp. inyoense Lomatium rigidum Lupinus pratensis var. eriostachys Penstemon papillatus Phacelia inyoensis Sedum pinetorum Sidalcea covillei
		Tule River Basin Clarkia springvillensis Dudleya cymosa ssp. costifolia

Rare plants were documented on 623 of the 7.5-minute topographic quadrangles covering the Sierra Nevada. Figure 24.3 provides a visual representation of the concentration and distribution of rare taxa ranging from one to fifteen rare plants per quadrangle. There are several areas throughout the Sierra Nevada where concentrations or ensembles of rare and endemic species are located. Quads with five or more rare plants generally represent the presence of an ensemble area.

Many of these ensembles are located on unusual substrates or soils, occur in areas with high plant species diversity, or occur in uncommon habitats or vegetation types. Examples of ensemble areas include the Ione Formation in Amador County, the Red Hills in Tuolumne County, the serpentinites of the Feather River Canyon in Butte and Plumas Counties, and the sandy granitic meadow borders of the Kern Plateau in Tulare County.

TABLE 24.3

Distribution of Sierran rare and endemic plants by county.

County[a]	Number of Taxa from Database	Sierran Endemics	Rare Taxa	Endemic to County
Northern Sierra				
Butte	89	58	52	6
El Dorado	103	89	45	5
Lassen	34	6	34	0
Nevada	75	59	55	4
Placer	86	71	35	0
Plumas	104	61	66	5
Shasta	29	3	27	0
Sierra	53	36	34	1
Tehama	31	3	30	0
Yuba	34	29	11	0
Central Sierra				
Alpine	36	24	19	2
Amador	77	72	23	2
Calaveras	83	80	23	1
Madera	131	123	37	1
Mariposa	154	143	55	6
Merced	8	5	6	0
Mono	103	60	65	6
Tuolumne	154	135	61	7
Southern Sierra				
Fresno	176	160	70	7
Inyo	113	81	56	6
Kern	111	84	71	17
Tulare	215	190	98	21

[a]Several counties extend beyond the study area boundary. Rare and endemic species are recorded only for those taxa within the study area.

Table 24.5 provides another perspective for assessment of speciation and distribution patterns. The genera with the largest number of rare and/or endemic taxa in the Sierra Nevada are *Eriogonum* (27), *Astragalus* (22), and *Mimulus* (19). When compared with the largest genera within the California flora (Smith and Noldeke 1960; Noldeke and Howell 1960), *Eriogonum* ranks fifth, *Astragalus* second, and *Mimulus* sixth. *Astragalus*, with over 2,000 species, is well known for its worldwide level of speciation, being among the largest genera of flowering plants. *Eriogonum*, with approximately 250 species, most of which are in the western United States, is also known for its high number of rare and endemic taxa. *Mimulus*, with nearly 150 species, has its center of distribution in North America. The Sierra Nevada provides a diversity of habitats in which endemic annual *Mimulus* species have evolved. The fourth-ranked genus with rare and endemic taxa in the Sierra Nevada is *Clarkia* (18). With forty-one species, of which thirty-nine occur in California, this genus appears to be speciating into ecological niches, with thirteen taxa being both rare and endemic to the Sierra Nevada. Thirty percent of all taxa that are either rare and/or endemic for the Sierra Nevada are distributed within the eleven genera displayed in table 24.5.

There are few endemic tree and shrub species in the Sierra Nevada as compared with the Coast Ranges of California. The Sierra Nevada has not been an important center for speciation of woody taxa as compared with other parts of California. Three endemic tree species and twenty-two endemic

shrub species occur in the Sierra Nevada. The three endemic trees (Sierra foxtail pine, Piute cypress, and giant sequoia) are found within the southern Sierra, with only giant sequoia occurring in all three geographical subunits of the Sierra Nevada. The twenty-two endemic shrubs are distributed among the following genera: *Arctostaphylos* (5), *Ceanothus* (5), *Chrysothamnus* (2), *Ribes* (2), *Tenestus* (2), *Berberis* (1), *Carpenteria* (1), *Fremontodendron* (1), *Myrica* (1), *Pyrrocoma* (1), and *Salix* (1). Several of these shrubs have narrow distribution ranges. *Arctostaphylos* and *Ceanothus*, two of the largest genera of shrubs in California, have few endemic representatives in the Sierra Nevada. Both genera are common components of various types of chaparral vegetation, and in the Sierra Nevada they also occur in montane environments as either a seral stage of a coniferous forest series or as a component of a montane chaparral series.

The diversity of habitat types that occur in the Sierra Nevada also explains the great richness of endemic and rare species within this mountain range. The broad plant communities used by Munz and Keck 1959, along with those used in Skinner and Pavlik 1994, provide the basis for recording habitat preferences for rare and/or endemic taxa in this study. Several additional habitat types not based on vegetation were recorded, because many rare taxa seem to be more dependent on them than on the surrounding vegetation type. For example, seventy taxa, or nearly 12% of rare and/or endemic taxa, can be found on rock outcrops. Many endemics and/or rare taxa are located exclusively on a particular rock type, such as carbonate, serpentinite, basalt, or granite. Other taxa have distributions that correspond more closely with elevation zones or that span several habitat types.

Distribution patterns for rare and/or endemic species differ considerably from river basin to river basin, and the distribution of these elements between habitat types is also varied. There are five dominant habitat types: Jeffrey and ponderosa pine forest types contain 211 taxa, the largest concentration of rare and endemic elements in the Sierra; the second largest, the foothill woodland, contains 139 taxa; subalpine forests contain 124 taxa; meadows have 116 taxa; chaparral, 90 taxa.

Of the five habitat types that contain the most rare and endemic taxa, the foothill woodland and chaparral are receiving the greatest increase in impacts and/or fragmentation by urbanization along the western slope of the Sierra Nevada. In chaparral vegetation types, the frequency of fire has been altered to protect other resource values, such as timber and homes. An example of this change in the Sierra Nevada is occurring at the residential development areas of Cameron Park, Pine Hill, and Salmon Falls in El Dorado County. Several rare taxa are restricted or locally endemic to gabbro soils in this area and are impacted by a direct loss of habitat by development. Those taxa that are dependent on fire as part of their life history and ecology may be negatively impacted by long-term changes in the management of chaparral vegetation. The changes may include a shift from fall to spring burning, mechanical treatments, or alteration of the fire frequency

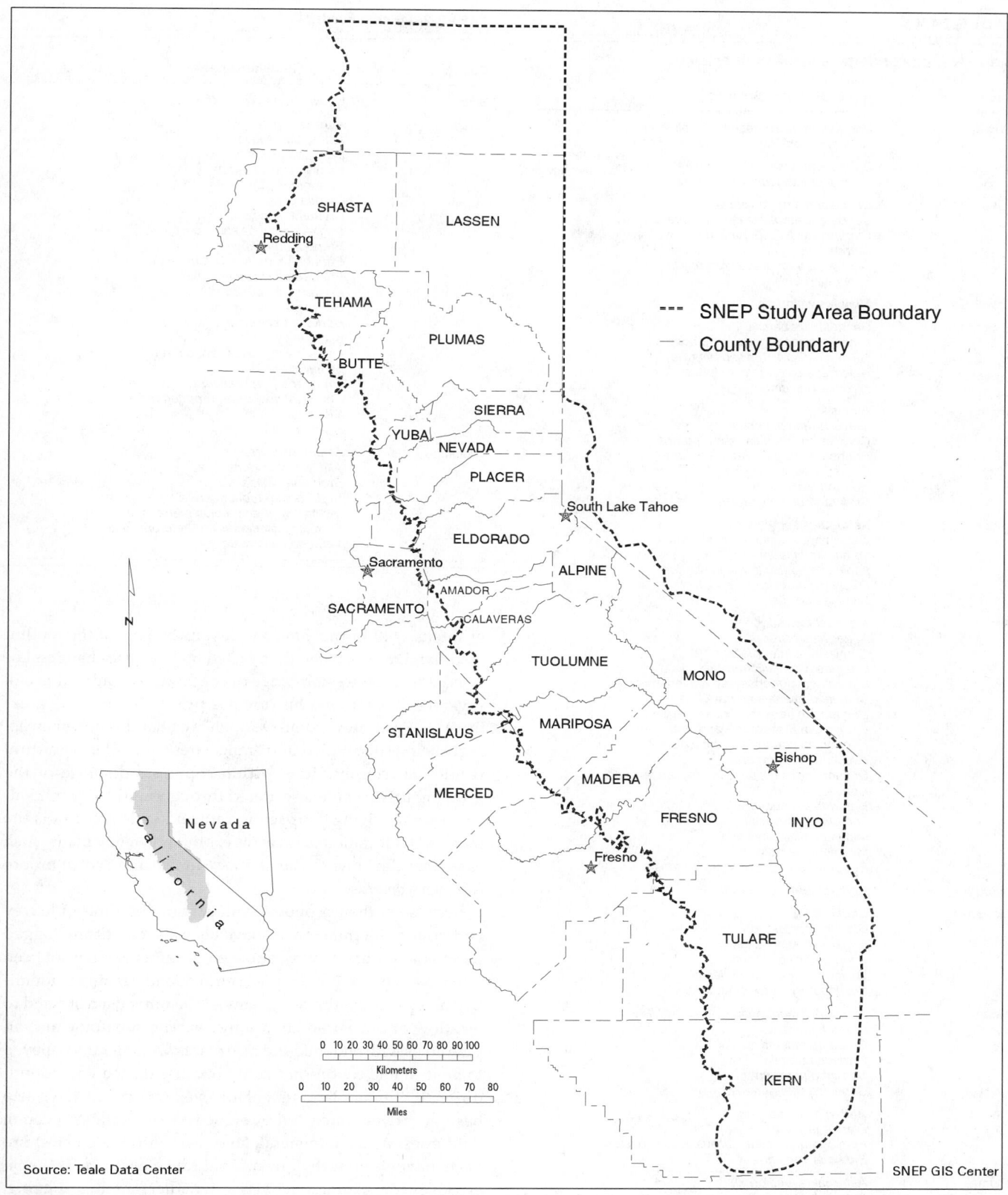

Nevada

California

SNEP Study Area Boundary
County Boundary

SHASTA
Redding
LASSEN
TEHAMA
PLUMAS
BUTTE
SIERRA
YUBA
NEVADA
PLACER
South Lake Tahoe
ELDORADO
Sacramento
AMADOR
ALPINE
SACRAMENTO
CALAVERAS
TUOLUMNE
MONO
STANISLAUS
MARIPOSA
MADERA
Bishop
MERCED
FRESNO
INYO
Fresno
TULARE
KERN

0 10 20 30 40 50 60 70 80 90 100
Kilometers

0 10 20 30 40 50 60 70 80
Miles

Source: Teale Data Center

SNEP GIS Center

FIGURE 24.2

SNEP study area with county boundaries.

TABLE 24.4

Sierra Nevada endemics at the county level.

County	Endemic Species
Alpine	*Eriogonum microthecum* var. *alpinum* *Galium hypotrichium* ssp. *ebbettsense*
Amador	*Eriogonum apricum* var. *apricum* *Eriogonum apricum* var. *prostratum*
Butte	*Calycadenia oppositifolia* *Calystegia atriplicifolia* ssp. *buttensis* *Clarkia gracilis* ssp. *albicaulis* *Clarkia mosquinii* ssp. *mosquinii* *Clarkia mosquinii* ssp. *xerophylla* *Sidalcea robusta*
Calaveras	*Mimulus whipplei*
El Dorado	*Ceanothus roderickii* *Draba asterophora* var. *macrocarpa* *Galium californicum* ssp. *sierrae* *Navarretia prolifera* ssp. *lutea* *Wyethia reticulata*
Fresno	*Arabis* sp. nov. *Carpenteria californica* *Cordylanthus tenuis* ssp. *barbatus* *Eriogonum nudum* var. *regirivum* *Gilia* sp. nov. *Heterotheca* sp. nov. *Streptanthus fenestratus*
Inyo	*Astragalus sepultipes* *Galium hypotrichium* ssp. *inyoense* *Lomatium rigidum* *Lupinus magnificus* var. *hesperius* *Lupinus pratensis* var. *eriostachys* *Sidalcea covillei*
Kern	*Allium shevockii* *Astragalus ertterae* *Camissonia integrifolia* *Chamaesyce vallis-mortae* *Clarkia tembloriensis* ssp. *calientensis* *Clarkia xantiana* ssp. *parviflora* *Delphinium hanseni* ssp. *kernense* *Eriogonum breedlovei* var. *breedlovei* *Eriogonum kenedyi* var. *pinicola* *Eschscholzia procera* *Galium angustifolium* ssp. *onycense* *Hemizonia arida* *Heterotheca shevockii* *Lomatium shevockii* *Mimulus microphyllus* *Mimulus shevockii* *Streptanthus cordatus* var. *piutensis*
Madera	*Erythronium pluriflorum*
Mariposa	*Clarkia biloba* ssp. *australis* *Clarkia lingulata* *Erigeron mariposanus* *Eriophyllum congdonii* *Lupinus citrinus* var. *deflexus* *Plagiobothrys torreyi* var. *torreyi*
Mono	*Astragalus monoensis* *Carex tiogana* *Draba incrassata* *Lupinus duranii* *Sedum pinetorum* *Senecio pattersonensis*
Nevada	*Berberis sonnei* *Elatine gracilis* *Plagiobothrys glyptocarpus* var. *modestus* *Sidalcea stipularis*
Plumas	*Astragalus lentiformis* *Astragalus webberi* *Calamagrostis* sp. nov. *Ceanothus* sp. nov. *Erigeron lassenianus* var. *deficiens* *Monardella stebbinsii*

TABLE 24.4 (continued)

County	Endemic Species
Sierra	*Ivesia aperta* var. *canina*
Tulare	*Abronia alpina* *Astragalus shevockii* *Brodiaea insignis* *Castillega praeterita* *Ceanothus pinetorum* *Clarkia springvillensis* *Crythantha incana* *Dudleya cymosa* ssp. *costifolia* *Erigeron multiceps* *Eriogonum nudum* var. *murinum* *Eriogonum twisselmannii* *Erythronium pusaterii* *Geranium coccinnum* *Horkelia tularensis* *Iris munzii* *Lotus oblongifolius* ssp. *cupreus* *Mimulus norrisii* *Oreonana purpurascens* *Phacelia eisenii* var. *brandegana* *Ribes tularense* *Silene aperta*
Tuolumne	*Allium tribracteatum* *Allium tuolumnense* *Brodiaea pallida* *Erythronium tuolumnense* *Iris hartwegii* ssp. *columbiana* *Senecio clevelandii* var. *heterophyllus* *Verbena californica*

or intensity of burns. Blue oak savannas (part of the foothill woodland) are also being impacted by land-use changes. Located along the western edge of the Sierra Nevada, blue oak savannas have a long historic use primarily for cattle grazing. What was once extensive open rangeland is now increasingly being subdivided into "ranchettes" and other semirural residential communities within commute distance of the growing urban centers scattered throughout the Central Valley. Another change impacting rare and endemic taxa is the increased infestation of invasive exotic and weedy plants such as yellow star-thistle (*Centaurea solstitalis*) and Scotch broom (*Cytisus scoparius*).

Because of their economic value, many hectares of Jeffrey and ponderosa pine forests have been systematically logged for over a century. Few meadow environments have not been intensively grazed in the past century, and grazing is identified as a potential threat to many of the rare taxa restricted to meadow and riparian environments. The subalpine and alpine areas with endemic and rare plants are generally believed to be in stable condition mainly because difficult access traditionally limited land uses. However, rare plants have also been negatively impacted by some recreational uses even in wilderness areas. In general, land uses with the greatest impacts have been at the low and middle elevations of the Sierra Nevada. Human activities, whether grazing, logging, mining, or recreation, by and of themselves may not threaten species. Rather, it is the interactions between the timing, intensity, frequency, and distribution of these various activities

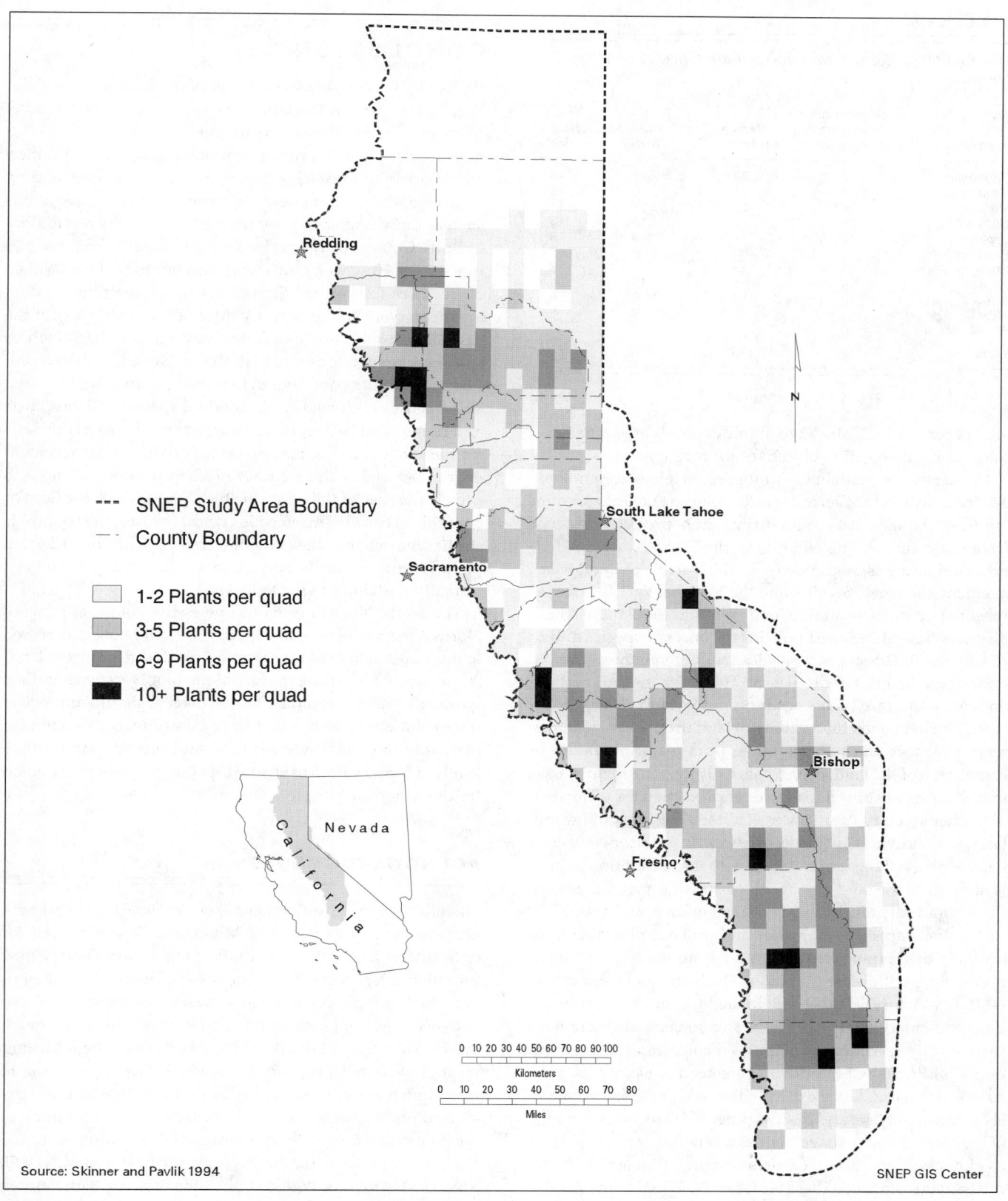

FIGURE 24.3

Distribution of rare plants by topographic quadrangle (Skinner and Pavlik 1994).

TABLE 24.5

Distribution of Sierran rare and endemic plants by largest genera.

Genera	Number of Taxa from Database	Sierran Endemics	Rare Taxa	Rare and Endemic
Eriogonum	27	23	18	14
Astragalus	22	14	16	8
Mimulus	19	16	15	11
Clarkia	18	18	13	13
Lupinus	17	17	9	9
Phacelia	15	13	9	7
Erigeron	14	13	9	8
Carex	14	5	13	4
Ivesia	12	10	8	6
Allium	11	9	8	5
Streptanthus	10	7	7	4
Totals	179	145	125	89

and practices that can have significant adverse effects on long-term conservation objectives for rare species.

Conservation guidelines, strategies, or plans for endemic and rare taxa in the Sierra Nevada need to take into account land-use changes that are occurring or are projected to occur in the near future. The majority of the Sierra Nevada is federal land administered chiefly by the Bureau of Land Management, the Forest Service, and the National Park Service. To meet the intent of several federal laws and regulations, these agencies have developed policies to conserve species and to reduce the likelihood that species will become threatened or endangered under federal law. However, not all Sierran endemics and rare plants occur on public land. Many are located on land zoned for a variety of land uses, where management practices and policies range from major alteration or conversion of the landscape by agriculture or residential uses to utilization of natural resources to protection for watershed and other amenity values (Messick 1995). Besides zoning and land uses, individual plant occurrences are at considerable risk if the ecology of the species along with its distribution pattern is not part of the land management decision-making process (Lesica and Allendorf 1992, 1995; Schemske et al. 1994).

Adverse impacts to many rare and endemic plants could be lessened by an improved analysis within the required environmental public review statutes. The National Environmental Policy Act (NEPA) for federal actions and the California Environmental Quality Act (CEQA) for nonfederal actions have to date addressed rare plant issues inconsistently. As a minimum, NEPA and CEQA documents need to clearly state the analysis used to assess the distribution and population dynamics of rare plant taxa and discuss potential threats, the ecology of the species, and management recommendations considered as either conservation measures or mitigation actions to reduce adverse impacts. These NEPA and CEQA documents also need to be viewed in a larger context to evaluate the cumulative impacts as well as monitor the effectiveness of conservation and species protection actions.

CONCLUSION

We can only estimate that over half of the California vascular flora occurs within the boundary of the Sierra Nevada because no detailed floristic treatment exists. The situation for nonvascular plants is even more problematic. Several smaller-scale floristic studies have contributed to our understanding of gross distribution patterns in recent years: localized floras such as those undertaken for master's theses, the recent flora of Butte County, or the revised flora of Lassen Volcanic National Park. However, much work remains to be done (Wilken 1995). Even for nonrare Sierran endemics, distribution data are incomplete, and this study does not resolve many of the distribution questions. Our collective understanding of which species are endemic or rare to the Sierra Nevada will be modified as fieldwork continues in more remote areas and focused surveys are conducted for individual species. All checklists and floras need to document voucher specimens as the basis for the catalogue of names presented so that future researchers can determine the accuracy of identification and need to provide a sense of the depth of study upon which the flora or checklist is based. This need is critical because monographic studies may change the taxonomic circumscription of certain plant groups or clarify species that traditionally have been difficult to identify in the field.

The Sierra Nevada remains one of the botanical gems of North America. New plant species are still being discovered in this range, and land managers across this magnificent landscape need to be aware of the unique biodiversity contained within the Sierra Nevada. Land managers should appreciate the evolutionary forces that have contributed to such a remarkable rare and endemic flora and provide appropriate levels of conservation to ensure that this resource is sustained for the American people.

ACKNOWLEDGMENTS

I thank SNEP Science Team members Michael Barbour, Frank Davis, David Graber, Connie Millar, and David Parsons for critically reviewing the first draft of this paper. Their review provided a key focus for subsequent editions. Dean Taylor and Mark Skinner provided expert technical review that also enhanced the final form of this paper. Their efforts are much appreciated. Special thanks to Dan Norris and Brent Mishler (bryophytes) and Bruce Ryan (lichens). They helped me to learn much about these two ecologically important and neglected plant groups. I also acknowledge the contribution of the many field botanists who provided distribution information on rare plants in the Sierra Nevada to CNPS and CNDDB. Thanks also to John Willoughby, Anne Bradley, Beth Painter, and Emily Roberson for helpful comments and suggestions.

REFERENCES

Ahmadjian, V. 1995. Lichens are more important than you think. *BioScience* 45(3):124.

Andrus, R. E. 1990. Why rare and endangered bryophytes? In *Ecosystem management: Rare species and significant habitats: Proceedings of the fifteenth annual Natural Areas Conference,* edited by R. Mitchell, C. Sheviak, and D. Leopold, 199–201. Bulletin 471. Albany, NY: New York State Museum.

California Department of Fish and Game (CDFG). 1995. Special plants list. In *Natural diversity data base.* Sacramento: CDFG.

Clifton, G. 1994. Plumas County and Plumas National Forest flora, 1994 draft. Unpublished file report.

Crosby, M. R. 1980. The diversity and relationship of mosses. In *The mosses of North America,* edited by R. J. Taylor and A. E. Leviton, 115–29. Lawrence, KS: Allen Press.

Egan, R. S. 1987. A fifth checklist of the lichen-forming, lichenicolous, and allied fungi of the continental United States and Canada. *The Bryologist* 90(2):77–173.

Ertter, B. 1995. The changing face of California botany. *Madroño* 42(2):114–22.

Ferren, W. R., Jr., D. L. Magney, and T. A. Sholars. 1995. The future of California floristics and systematics: Collecting guidelines and documentation techniques. *Madroño* 42(2):197–210.

Fiedler, P. L. 1986. Concepts of rarity in vascular plant species with special reference to the genus *Calochortus* Pursh (Liliaceae). *Taxon* 35:502–18.

———. 1995. Rarity in the California flora: New thoughts on old ideas. *Madroño* 42(2):127–41.

Fiedler, P. L., and J. J. Ahouse. 1992. Hierarchies of cause: Toward an understanding of rarity in vascular plant species. In *Conservation biology: The theory and practice of nature conservation, preservation, and management,* edited by P. L. Fiedler and S. K. Jain, 23–47. New York: Chapman and Hall.

Flowers, S. 1973. *Mosses: Utah and the West.* Provo, UT: Brigham Young University Press.

Forbes, H. C., W. R. Ferren, and J. R. Haller. 1988. The vegetation and flora of Fish Slough and vicinity, Inyo and Mono Counties, California. In *Plant biology of eastern California,* edited by C. Hall and V. Doyle-Jones, 99–138. Los Angeles: University of California, White Mountain Research Station.

Gillett, G. W., J. T. Howell, and H. Leschke. 1961. A flora of Lassen Volcanic National Park. *Wasmann Journal of Biology* 19(1):1–185. Revised and expanded edition by V. Oswald, D. Showers, and M. Showers, 1995. Sacramento: California Native Plant Society.

Hale, M. E., Jr., and M. Cole. 1988. *Lichens of California.* Berkeley and Los Angeles: University of California Press.

Harper, J. L. 1981. The meanings of rarity. In *The biological aspects of rare plant conservation,* edited by H. Synge, 189–303. New York: John Wiley and Sons.

Henderson, D. M. 1981. The conservation of lower plants: Report from a panel discussion. In *The biological aspects of rare plant conservation,* edited by H. Synge, 125–37. New York: John Wiley and Sons.

Henry, M. A. 1994. A checklist of plants from Short Canyon. Unpublished file report based on fieldwork 1970–94 along the eastern slope of the Sierra Nevada, Kern County.

Herbert, J. R., and R. W. Meyer. 1984. Lichens of the San Joaquin Experimental Range, California. *The Bryologist* 87(3):251–54.

Herre, A. W. C. T. 1911. The desert lichens of Reno, Nevada. *Botanical Gazette* 51:286–97.

Hickman, J. C., ed. 1993. *The Jepson manual: Higher plants of California.* Berkeley and Los Angeles: University of California Press.

Howe, M. A. 1899. The hepaticae and anthocerotes of California. *Memoirs of the Torrey Botanical Club* 7:1–208.

Howell, J. T. 1951. The arctic-alpine flora of three peaks in the Sierra Nevada. *Leaflets of Western Botany* 6(7):141–56.

———. 1957. The California floral province and its endemic genera. *Leaflets of Western Botany* 8(5):138–41.

Hunter, K. B., and R. E. Johnson. 1983. Alpine flora of the Sweetwater Mountains, Mono County, California. *Madroño* 30(4): supp. 89–105.

Knight, W., I. Knight, and J. T. Howell. 1970. A vegetation survey of the Butterfly Valley Botanical Area. *Wasmann Journal of Biology* 28(1):1–46.

Koch, L. F. 1950a. Mosses of California: An annotated list of species. *Leaflets of Western Botany* 6(1):1–40.

———. 1950b. The distribution of Californian mosses. Ph.D. diss., University of Michigan, Ann Arbor.

———. 1954. Distribution of California mosses. *American Midland Naturalist* 51(2):515–38.

———. 1958. Mosses of California, VI. Hall Natural Area and Mono County. *Madroño* 14(6):206–11.

Kruckeberg, A. R., and D. Rabinowitz 1985. Biological aspects of endemism in higher plants. *Annual Reviews of Ecological Systematics* 16:447–79.

Lavin, M. 1983. Floristics of the headwaters of the Walker River, California and Nevada. *Great Basin Naturalist* 43:93–130.

Lawton, E. 1971. *Moss flora of the Pacific Northwest.* Nichinan, Japan: Hattori Botanical Laboratory.

Lesica, P., and F. W. Allendorf. 1992. Are small populations of plants worth preserving? *Conservation Biology* 6(1):135–39.

———. 1995. When are peripheral populations valuable for conservation? *Conservation Biology* 9(4):753–60.

Lum, K.-L. 1975. Gross patterns of vascular plant species diversity in California. Master's thesis, Department of Ecology, University of California, Davis.

Messick, T. 1995. California floristic province U.S.A. and Mexico. In *The Americas,* edited by S. D. Davis, V. H. Heywood, O. H. MacBryde, and A. C. Hamilton. Vol. 3 of *Centers of plant diversity: A guide and strategy for their conservation.* Covelo, CA: Island Press.

Mishler, B. D. 1995. Bryophytes. Unpublished file report on weekend workshop, 18–19 March, University of California, Berkeley.

Moran, R. 1950. Whence *Sedum pinetorum* Brandegee? *Leaflets of Western Botany* 6(3):62–63.

Morefield, J. D. 1994. *Updated supplement to endangered, threatened, and sensitive vascular plants of Nevada.* Carson City: Nevada Natural Heritage Program.

Morefield, J. D., and T. A. Knight. 1991. *Endangered, threatened, and sensitive vascular plants of Nevada.* Carson City: Nevada Natural Heritage Program.

Munz, P. A. 1968. *Supplement to "A California flora."* Berkeley and Los Angeles: University of California Press.

Munz, P. A., and D. Keck. 1959. *A California flora.* Berkeley and Los Angeles: University of California Press.

Noldeke, A. M., and J. T. Howell. 1960. Endemism and a California flora. *Leaflets of Western Botany* 9(8):124–27.

Norris, D. H. 1995a. Letter to author identifying rare mosses, 13 July.

———. 1995b. Endemism and rarity in the California bryoflora. Unpublished file report. 15 August.

Norris, L. L. 1987. Status of five rare plant species in Sequoia and Kings Canyon National Parks. In *Conservation and management of rare and endangered plants*, edited by T. Elias, 279–82. Sacramento: California Native Plant Society.

Oswald, V. H., in collaboration with L. Ahart. 1994. *Manual of the vascular plants of Butte County, California*. Sacramento: California Native Plant Society.

Palmer, M. W., G. L. Wade, and P. Neal. 1995. Standards for the writing of floras. *BioScience* 45(5):339–45.

Palmer, R., B. L. Corbin, R. Woodward, and M. Barbour. 1983. Floristic checklist for the Headwaters Basin area of the North Fork of the American River, Placer County, California. *Madroño* 30(4):52–66.

Pinelli, J. J., and W. P. Jordan. 1978. Lichens of Calaveras Big Trees State Park, California. *The Bryologist* 81(3):432–35.

Potter, B. R. 1983. A flora of the Desolation Wilderness, El Dorado County, California. Master's thesis, Department of Biology, Humboldt State University, Arcata, CA.

Powell, W. R., ed. 1974. *Inventory of rare and endangered vascular plants of California*. Special Publication 1. Berkeley: California Native Plant Society.

Pusateri, S. J. 1963. *Flora of our Sierran national parks: Yosemite, Sequoia, and Kings Canyon*. Tulare, CA: Carl and Irving Printers.

Rabinowitz, D. 1981. Seven forms of rarity. In *The biological aspects of rare plant conservation*, edited by H. Synge, 205–217. New York: John Wiley and Sons.

Raven, P. H., and D. I. Axelrod. 1978. Origin and relationships of the California flora. *University of California Publications in Botany* (Berkeley) 72:1–134.

Reveal, J. L. 1981. The concepts of rarity and population threats in plant communities. In *Rare plant conservation: Geographical data organization*, edited by L. Morse and M. S. Henifin, 41–47. Bronx: New York Botanical Garden.

Rice, B. 1969. Plant checklist for Mineral King, California. Unpublished file report.

Ryan, B. D. 1990a. Lichens and air quality in the Desolation Wilderness, California: A baseline study. Unpublished file report, U.S. Forest Service, San Francisco.

———. 1990b. Lichens and air quality in the Emigrant Wilderness, California: A baseline study. Unpublished file report, U.S. Forest Service, San Francisco.

———. 1990c. Lichens and air quality in the San Gabriel Wilderness, California: A baseline study. Unpublished file report, U.S. Forest Service, San Francisco.

———. 1995a. Letter to author identifying rare lichens, 2 March.

———. 1995b. Letter to author updating Sierran lichens, 28 August.

Ryan, B. D., and T. H. Nash III. 1991. Lichen flora of the Eastern Brook Lakes Watershed, Sierra Nevada Mountains, California. *The Bryologist* 94(2):181–95.

Savage, W. 1973. Annotated checklist of vascular plants of Sagehen Creek Drainage Basin, Nevada County, California. *Madroño* 22(3):115–39.

Schemske, D. W., B. C. Husband, M. H. Ruckelshaus, C. Goodwillie, I. M. Parker, and J. G. Bishop. 1994. Evaluating approaches to the conservation of rare and endangered plants. *Ecology* 75(3):584–606.

Schofield, W. B. 1980. Phytogeography of the mosses of North America (north of Mexico). In *The mosses of North America*, edited by R. J. Taylor and A. E. Leviton, 131–70. Lawrence, KS: Allen Press.

Sequoia National Park. 1995. *Electronic database for lichen collections of Blakeman and Wetmore 1984–85 within Sequoia National Park, Tulare County, California*. Three Rivers, CA: Sequoia National Park.

Sharsmith, C. W. 1940. A contribution to the history of the alpine flora of the Sierra Nevada. Ph.D. diss., Department of Botany, University of California, Berkeley.

Shevock, J. R. 1978. Vascular flora of the Lloyd Meadows Basin, Sequoia National Forest, Tulare County, California. Master's thesis, Department of Biology, California State University, Long Beach.

———. 1988. New rare and geographically interesting plants along the crest of the southern Sierra Nevada. In *Plant biology of eastern California*, edited by C. Hall and V. Doyle-Jones, 161–66. Los Angeles: University of California, White Mountain Research Station.

Shevock, J. R., and W. D. Taylor. 1987. Plant explorations in California: The frontier is still here. In *Conservation and management of rare and endangered plants*, edited by T. Elias, 91–98. Sacramento: California Native Plant Society.

Showers, D. W. 1982. The mosses of Lassen Volcanic National Park, California. *The Bryologist* 85(3):324–28.

———. 1995. Conversation with author identifying rare mosses, 28 July.

Sierra Nevada Ecosystem Project. 1994. *Progress Report to Congress*. Davis: University of California, SNEP.

Sigal, L. L., and T. H. Nash III. 1983. Lichen communities on conifers in southern California mountains: An ecological survey relative to oxidant air pollution. *Ecology* 64(6):1343–54.

Skinner, M. W., and B. Ertter. 1993. Whither rare plants in *The Jepson Manual? Fremontia* 22(3):23–27.

Skinner, M. W., and B. M. Pavlik, eds. 1994. *Inventory of rare and endangered vascular plants of California*. 5th ed. Special Publication 1. Sacramento: California Native Plant Society.

Skinner, M. W., D. Tibor, R. Bittman, B. Ertter, S. Boyd, T. Ross, A. Sanders, J. Shevock, and D. Taylor. 1995. Research needs for conserving California's rare plants. *Madroño* 42(2):211–41.

Smiley, F. J. 1921. A report on the boreal flora of the Sierra Nevada. *University of California Publications in Botany* (Berkeley) 9:1–423.

Smith, D. W. 1980. A taxonomic survey of the macrolichens of Sequoia and Kings Canyon National Parks. Master's thesis, Department of Biology, San Francisco State University.

Smith, G. L. 1973. A flora of Lake Tahoe Basin and neighboring areas. *Wasmann Journal of Biology* 31(1):1–231.

———. 1983. Supplement to "A flora of Lake Tahoe Basin and neighboring areas." *Wasmann Journal of Biology* 41(1–2):1–46.

Smith, G. L., and A. M. Noldeke. 1960. A statistical report on a California flora. *Leaflets of Western Botany* 9(8):117–23.

Smith, J. P., and K. Berg, eds. 1988. *Inventory of rare and endangered vascular plants of California*. 4th ed. Special Publication 1. Sacramento: California Native Plant Society.

Smith, J. P., Jr., R. C. Cole, and J. O. Sawyer, eds. 1980. *Inventory of rare and endangered vascular plants of California*. 2nd ed. Special Publication 1. Berkeley: California Native Plant Society.

Smith, J. P., and J. O. Sawyer Jr. 1988. Endemic vascular plants of northwestern California and southwestern Oregon. *Madroño* 35(1):54–69.

Smith, J. P., and R. York, eds. 1984. *Inventory of rare and endangered vascular plants of California*. 3rd ed. Special Publication 1. Berkeley: California Native Plant Society.

Stebbins, G. L. 1978a. Why are there so many rare plants in California? I. Environmental factors. *Fremontia* 5(4):6–10.

———. 1978b. Why are there so many rare plants in California? II. Youth and age of species. *Fremontia* 6(1):17–20.

———. 1980. Rarity of plant species: A synthetic viewpoint. *Rhodora* 82:77–86.

———. 1982. Floristic affinities of the high Sierra Nevada. *Madroño* 29(3):189–99.

Stebbins, G. L., and J. Major. 1965. Endemism and speciation in the California flora. *Ecological Monographs* 35:1–35.

Tatum, J. W. 1979. The vegetation and flora of Olancha Peak, southern Sierra Nevada, California. Master's thesis, University of California, Santa Barbara.

Taylor, D. W. 1981. *Plant checklist for the Mono Basin, California.* Mono Basin Research Group Contribution 3. Lee Vining, CA: Mono Lake Committee.

Thiers, B. M., and K. S. G. Emory. 1992. Bryology in California. *The Bryologist* 95(1):68–78.

True, G. H. 1973. The ferns and seed plants of Nevada County, California. Unpublished file report.

Tucker, S. C., and W. P. Jordan. 1978. A catalog of California lichens. *Wasmann Journal of Biology* 36(1–2):1–105.

Twisselmann, E. C. 1967. A flora of Kern County, California. *Wasmann Journal of Biology* 25(1–2):1–395. Reprinted with a key to vascular plant species of Kern County by M. Moe, 1995. Sacramento: California Native Plant Society.

———. 1971. A preliminary checklist of the flowering plants of the Kern Plateau of the Sierra Nevada of California. Unpublished file report.

U.S. Forest Service. 1995. Lichens and bryophytes. In *Draft environmental impact statement: Managing California spotted owl habitat in the Sierra Nevada national forests of California (an ecological approach)*, edited by J. Gauthier, vol. 3, pp. 213–23. Sacramento, CA: USFS.

Walker, R. E. 1992. Community models of species richness: Regional variation of plant community species composition on the west slope of the Sierra Nevada, California. Master's thesis, Department of Geography, University of California, Santa Barbara.

Wetmore, C. M. 1985. Lichens and air quality in Sequoia National Park: Final report. National Park Service, Air Quality Division, Denver.

———. 1986. Lichens and air quality in Sequoia National Park and Kings Canyon National Park: Supplementary report. National Park Service, Air Quality Division, Denver.

Whittemore, A. 1982. The thallose liverworts of California. Master's thesis, Department of Biology, Humboldt State University, Arcata, California.

Wilken, D. H. 1995. Flowers in the garden: What next for California floristics? *Madroño* 42(2):142–53.

Williams, M., J. T. Howell, G. H. True Jr., and A. Tiehm. 1992. A catalogue of vascular plants on Peavine Mountain, in Peavine Mountain, Nevada, edited by J. L. Nachlinger, F. F. Peterson, and M. Williams. *Mentzelia* 6(2):3–84.

DAVID M. GRABER
National Biological Service
Sequoia and Kings Canyon Field Station
Three Rivers, California

25

Status of Terrestrial Vertebrates

ABSTRACT

The terrestrial vertebrate wildlife of the Sierra Nevada is represented by about 401 regularly occurring species, including three local extirpations in the 20th century. The mountain range includes about two-thirds of the bird and mammal species and about half the reptiles and amphibians in the State of California. This is principally because of its great extent, and because its foothill woodlands and chaparral, mid-elevation forests, and alpine vegetation reflect, in structure and function if not species, habitats found elsewhere in the State. About 17% of the Sierran vertebrate species are considered at risk by state or federal agencies; this figure is only slightly more than half the species at risk for the state as a whole. This relative security is a function of the smaller proportion of Sierran habitats that have been extensively modified. However, foothill species and those associated with riparian habitats have been substantially reduced. Continuing appropriation of native foothill communities, damage to riparian systems, and compromise of remaining late-successional forests appear to pose the greatest potential risks to Sierran wildlife. The California Wildlife Habitat Relationships System will become an increasingly critical tool in wildlife habitat management and policy decisions because it is an expert system that offers the potential for predicting the outcome on wildlife of proposed land-use changes. However, poor information on the past and present distribution, abundance, population trends, and micro-habitat requirements of most vertebrate species, and consequently the models derived from these data, presently weakens conservation efforts because agencies are likely unable to detect many real problems while overstating or seeking the wrong solutions to others.

INTRODUCTION

There are approximately 401 species of terrestrial vertebrates that use the Sierra Nevada now or in recent times according to the California Wildlife Habitat Relationships System (CWHR) (California Department of Fish and Game 1994) (appendix 25.1). Of these, thirteen are essentially restricted to the Sierra in California (one of these is an alien; i.e. not native to the Sierra Nevada); 278 (eight aliens) include the Sierra in their principal range; and another 110 (six aliens) use the Sierra as a minor portion of their range. Included in the 401 are 232 species of birds; 112 species of mammals; thirty-two species of reptiles; and twenty-five species of amphibians (appendix 25.1). By comparison, CWHR lists 643 vertebrates as regularly occurring in the State. The Sierra Nevada includes range for 68% of the birds, 62% of the mammals, 43% of the reptiles, and 54% of the amphibians in the State. There is proportionately less mesic amphibian habitat and warm-xeric reptile habitat in the Sierra Nevada than for the State as a whole (Mayer and Laudenslayer 1988). The distributions of species in the Sierra are, for the majority of non-avian species, based upon point samples taken over many years and for the most part constitute scientific best guesses. Among the species listed are those whose principal ranges are Great Basin, Central Valley, or Mojave Desert, but which appear to lap into the foothills of the Sierra Nevada on the east or west sides. Both the paucity of sample data and lack of a precise natural and generally agreed-upon boundary for the mountain range means that the figure of 401 species contains an uncertainty of about 10%.

Sierra Nevada Ecosystem Project: Final report to Congress, vol. II, *Assessments and scientific basis for management options*. Davis: University of California, Centers for Water and Wildland Resources, 1996.

KEY QUESTIONS

This report addresses the following questions:

- What species of vertebrates amphibians, birds, mammals, and reptiles presently occur in the Sierra Nevada?

- What is the present status of Sierran vertebrates, and how has that status changed in recent times?

- What are the factors that influence the status and trend of terrestrial vertebrates in the Sierra Nevada?

- What is the present state of scientific knowledge of the status, distribution, trend, and habitat requirements of Sierran terrestrial vertebrates?

- What are the factors relevant to effective conservation and management of Sierran vertebrates in the future?

METHODS

This assessment is a summary synthesis of terrestrial vertebrate status and trends derived from scientific literature, agency reports, and several publicly-accessible databases, including the California Wildlife Habitat Relationships System (CWHR) and the California Natural Diversity Database. CWHR and its associated publications was used as the starting point to develop a list of Sierra Nevada vertebrate species, their status, and their habitat dependencies (appendix 25.1). Because amphibians are treated in detail in other Sierra Nevada Ecosystem Project assessments, they are included in this assessment only schematically, and to the extent that they provide insight into the overall condition of vertebrates in the Sierra Nevada.

For purposes of this report, the study area is the California portion of the Sierra Nevada *sensu strictu*, approximately bounded by the Central Valley to the west, Owens Valley and the Great Basin to the east, the Cascade Mountains to the north, and Lake Isabella and the Kern River to the south. Because sources of distribution information vary somewhat in their terminology and geographic precision, there is some imprecision in boundaries among different groups of animals.

SPECIES AT RISK

Species considered at risk in the Sierra, through listing as endangered or threatened by State or federal government, special concern by California, or sensitive by federal land

managers (but not those locally at risk only elsewhere) include thirty-three birds, nineteen mammals, four reptiles, and thirteen amphibians: 17% of the Sierran terrestrial fauna. These figures are based upon listing either of the species as a whole, or listing of a Sierran subspecies. For the State as a whole, ~30% are so listed in the CWHR database (California Department of Fish and Game 1994), although this number is fluid. Thus based on this *administrative* criterion alone, Sierran terrestrial vertebrates proportionately are nearly twice as secure under present conditions as the full State fauna.

Three species well-distributed in the range at the time of European settlement are now extirpated from the Sierra Nevada entirely: Bell's vireo *(Vireo bellii)*, California condor *(Gymnogyps californianus)*, and grizzly bear *(Ursus arctos)*, 0.7% of the Sierran vertebrate fauna. Evidence that the gray wolf *(Canis lupus)* regularly occupied the Sierra in recent centuries is unpersuasive (Schmidt 1991). For California as a whole, although no terrestrial species is extinct, seven (~1.4%) are extirpated from California (Steinhart 1990). Except for the North American Breeding Bird Survey (BBS), there has been no systematic and widespread monitoring to measure declines in population density or contraction of range. The California Department of Fish and Game has monitored some game species, particularly mule deer *(Odocoileus hemionus)* herds in the Sierra, but variability in methodologies has made inferences from these efforts uncertain at best. Thus the information leading to listing a species or population to be of special concern, or sensitive as the U.S. Forest Service and the Bureau of Land Management use these terms, usually means there are local indications of problems: Either population numbers appear to be low, or habitat believed to be required by the species in question is declining over a significant portion of the species range. While these are reasonable and pragmatic measures of risk, the lack of broad geographic data over a significant period of time for most species means that the term as used here assuredly misses some (perhaps many) species and probably incorrectly targets others.

There is not a close match between species officially judged at risk and those for which direct population data indicate serious and progressive declines as opposed to simple rarity. Great gray owl *(Strix nebulosa)* and limestone salamander *(Hydromantes brunus)*, for example, are quite rare and local, but there is no compelling evidence of population declines. Contemporary loss of a significant portion of habitat is the most frequent cause for initially assigning a species to a risk category. However, once listed, studies to generate demographic data are often funded (California Department of Fish and Game 1991; Thelander 1994). In the case of breeding landbirds for which there are Breeding Bird Survey (BBS) data (Davidson and Manley 1993), DeSante (1995) identifies six species clearly and significantly declining: band-tailed pigeon *(Columba fasciata)*, red-breasted sapsucker *(Sphyrapicus ruber)*, American robin *(Turdus migratorius)*, chipping sparrow *(Spizella passerina)*, white-crowned sparrow *(Zonotrichia* leucophrys), and olive-sided flycatcher (*Contopus borealis*) Of these six, only the last is a listed species, while the other five are notably widespread and common. There is evidence from BBS for the probable contemporary decline of another twelve species (DeSante 1995). Based on historical descriptions of their abundance, harlequin duck (*Histrionicus histrionicus*) and yellow-breasted chat (*Icteria virens*) are greatly reduced today, while Barrow's goldeneye (*Bucephala islandica*) no longer breeds in the Sierra Nevada, and willow flycatcher (*Empidonax traillii*) has nearly vanished there.

The California red-legged frog (*Rana aurora draytonii*) appears to have reached the point of virtual extinction in the Sierran western foothills on the margins of its range (Jennings and Hayes 1994). The Yosemite toad (*Bufo canorus*), foothill yellow-legged frog (*Rana boylii*), and mountain yellow-legged frog (*Rana muscosa*) all appear to be declining perilously in recent decades (Jennings and Hayes 1994; Fellers 1995) based on recent field re-examinations of historic museum collection sites. The California horned lizard (*Phrynosoma coronatum frontale*) has disappeared from most of its limited western foothill historic sites, while the western pond turtle (*Clemmys marmorata*), also a Sierran foothill fringe species, is still present at most sites but appears to be suffering perilous population declines because of poor survival of young (Jennings and Hayes 1994). In very recent years, the recovery of mountain sheep (*Ovis canadensis*) in the Sierra Nevada through reintroduction has suffered a severe reversal, and that species is in dangerous decline (Wehausen 1995).

PREHISTORIC AND RECENT FAUNA OF THE SIERRA NEVADA

During the Pleistocene, California's fauna included camels, horses, giant ground-sloths, mammoths, bison, and saber-toothed cats, all of which became extinct by the early Holocene, about 10,000 years ago. This megafauna largely occupied the valleys and coastal plains, but undoubtedly lapped into the foothills of the Sierra Nevada on both sides, although only a very few remains have been found there. Because the vegetation of the range, and the extent of its glaciation, varied considerably on a millennial scale (Anderson 1990), both now-extinct and presently extant vertebrates, particularly the large herbivores, may have occupied Sierran ranges in the past that are now unsuitable (Wagner 1989; Grayson 1993). Although the extinction of megafauna throughout North America is associated with the change in climate at the end of the Pleistocene, this is only several thousand years after the time when people crossed into North America from Asia. Human predation may well have played a role in this transformation of the faunal landscape. At the time of European settlement, large herds of tule elk (*Cervus*

elaphus) and pronghorn (*Antilocapra americana*) were still present, especially in the interior valleys, while mule deer (*Odocoileus hemionus*) dominated the foothills and mountain sheep (*Ovis canadensis*) occupied the crest and eastern slopes. All four of these ungulates were hunted heavily by Spanish and Anglo settlers for their own needs and for city markets. This greatly reduced populations, while prime habitats were converted to use by domestic livestock. During the 19th and early 20th centuries, fur trapping for beaver (*Castor canadensis*), mink (*Mustela vison*), otter (*Lutra canadensis*), red fox (*Vulpes vulpes*), marten (*Martes americana*), and fisher (*Martes pennanti*), and trapping and shooting wolverines (*Gulo gulo*) as vermin, greatly reduced all of these species in the Sierra Nevada.

MODERN EXTIRPATIONS IN THE SIERRA NEVADA

Only three vertebrates are known to have been lost from the Sierran fauna in historic times.

Grizzly Extinction

The last California grizzly bear (*Ursus arctos*) identified with reasonable certainty was killed by cattleman Jesse B. Agnew near Horse Corral Meadow, Sequoia National Forest, in August, 1922; identification by lower canine tooth was made by C. Hart Merriam (Storer and Tevis 1955). A large bear that may well have been a grizzly was spotted by road crews in Sequoia National Park several times in the spring of 1924; in October of that year a Three Rivers cattleman named Alfred Hengst observed a bear in Cliff Creek (Sequoia N. P.) that "was the biggest bear I've ever seen, bigger than any cow, and looked as though sprinkled with snow. I had a close view of the beast which was undoubtedly a grizzly" (Fry 1924). That is the last likely sighting in California. Grizzly bears were well-distributed in California at the time of Spanish settlement, recorded everywhere but for the Great Basin, deserts, and eastern Modoc Plateau; they were concentrated in the open country of the valleys and coastal plains, especially in the riparian zones. In the Sierra they were reported most frequently in the foothill savannahs, woodlands and chaparral, but they appear to have been distributed throughout the range, selecting open country including montane meadows and the alpine zone during the snow-free months. Although largely herbivorous, grizzlies preyed upon cattle and other stock; Spanish and later Anglo settlers set out systematically to exterminate them, using large-bore rifles and steel-jawed traps as large as 5 feet in width. The closest surviving grizzly populations are in northeastern Washington and in the northern Rocky Mountains.

Least Bell's Vireo Extirpation

The least Bell's vireo (*Vireo bellii pusillus*) was historically distributed widely in riparian habitat of the San Joaquin Valley, southern Coast Range, and southwestern California, as well as the lower foothills of the Sierra Nevada. This bird still persists in small numbers in a few locations in southern California and the central coast, where it is listed as endangered by both state and federal governments. The decline of Bell's vireo parallels the spread of brown-headed cowbirds (*Molothrus ater*) in California, and in fact local control programs of this brood parasite have significantly increased nesting success (Goldwasser et al. 1980; Small 1994; S. A. Laymon personal communication). However, the destruction of willow-dominated riparian habitat has played a substantial role in the vireo's decline and has isolated remaining populations in small habitat islands.

California Condor Extinction in the Wild

The last wild California condor (*Gymnogyps californianus*) was captured in Kern County in 1987, one of 27 birds removed to captivity in the 1980s in an effort to save the species from extinction through captive breeding. The condor is a forager of open plains and savannahs, where it once apparently utilized the carcasses of Pleistocene megafauna, the surviving ungulates of the Holocene, and finally the cattle and sheep that replaced them. In the 20th century it ranged over the southern San Joaquin Valley, southern and central Coast Range, and as far south as the Transverse Range of Ventura and Los Angeles counties. However, condors selected cavities in cliffs, and even giant sequoias (*Sequoiadendron giganteum*), as nest sites, which brought them well into the western slope of the Sierra, as far north as Tuolumne County in recent times. In the 19th century, condors ranged from Canada to Baja California (Koford 1953). It is most likely that the decline of the vast herds of Pleistocene grazing animals upon which condors fed had made it a rare bird by the time of European exploration. In recent years the final decline of condors appears to have been accelerated by ingestion of lead shotgun pellets, collisions with power lines, eggshell thinning from DDT, and other largely-anthropogenic factors (Wilbur 1978). Experimental reintroductions from captive-bred zoo populations are now beginning, but it is uncertain whether the Sierra foothills and adjacent valley provide sufficient habitat quantity and quality, and whether known hazards can be mitigated sufficiently to reestablish a viable population of California condors.

SPECIES RESTRICTED TO THE SIERRA NEVADA IN CALIFORNIA

Thirteen vertebrates are essentially restricted to the Sierra Nevada in California. Six of these are amphibians, including the Yosemite toad (*Bufo canorus*), three species of salamanders in the genus *Hydromantes,* and two in the genus *Batrachoseps.* All of these are fully confined to locales in the Sierra Nevada. Montane and Great Basin endemism in amphibians is likely related to population isolation and subsequent speciation that took place during the great Holocene climatic changes. The Yosemite toad is closely related to the widely-distributed western toad (*Bufo boreas*), while the Kern Canyon slender salamander (*Batrachoseps simatus*) and the relictual slender salamander (*B. relictus*) are among a generous handful of extremely localized slender salamander species occupying moist micro-sites within generally xeric habitats in California. Of the three web-toad salamanders of the genus *Hydromantes* that occur in California, all of them restricted to the Sierra Nevada, only the Mount Lyell salamander (*Hydromantes platycephalus*), found in the alpine and subalpine zones of the Yosemite to Kings Canyon Sierra, is relatively widely distributed. The limestone salamander of the Merced Canyon (*H. brunus*), and the as-yet unnamed Owens Valley web-toed salamander (*Hydromantes* sp.), found in some riparian areas on the lower eastern slope, are quite restricted (Jennings and Hayes 1994). Both slender and web-toed salamanders belong to the family Plethodontidae, which are lungless salamanders that do not require free water for reproduction.

There are four mammals restricted to the Sierra Nevada (Zeiner et al. 1990b), and a fifth, the heather vole (*Phenacomys intermedius*), which also has a very localized population on Mount Shasta, and is well-distributed in the Pacific Northwest (Ingles 1965). Two are chipmunks (*Tamias*), another highly speciose genus in the west. Sierran endemics include the alpine chipmunk (*Tamias alpinus*) and the long-eared chipmunk (*T. quadrimaculatus*). The Mount Lyell shrew (*Sorex lyelli*) has been found only a few times, in riparian areas near Mount Lyell in the Yosemite Region. The yellow-eared pocket mouse, (*Perognathus xanthonotus*) has been recorded only in the vicinity of Walker Pass, Kern County, at the junction of the Sierra Nevada and the Tehachapi ranges. It is closely-related to the more widely distributed great basin pocket mouse (*P. parvus*) and likely has similar chaparral and desert scrub habitat preferences (Zeiner at al. 1990b).

In California, pine grosbeaks (*Pinicola enucleator*) reside only in the upper montane and subalpine forests of the Sierra Nevada, where they are restricted to wet meadows and other riparian habitat. Beyond California, pine grosbeaks range widely through the moist forests of the Rocky Mountains, the Northwest, and the Northeast. White-tailed ptarmigan (*Lagopus leucurus*) have been introduced to the Sierra Nevada from the Rocky Mountains.

ALIEN SPECIES AND THEIR EFFECTS

Of the fifteen terrestrial vertebrate species now established in the Sierra but not native to the region, seven are birds, seven are mammals, and one is an amphibian. Several of these were intentionally introduced into the Sierra Nevada by the California Department of Fish and Game as game species. These include wild pig (*Sus scrofa*), chukar (*Alectoris chukar*), white-tailed ptarmigan (*Lagopus leucurus*), and wild turkey (*Meleagris gallopavo*). White-tailed ptarmigan is native to the Rocky Mountains and the Pacific Northwest, where it uses open, alpine habitats dominated by willow. It was introduced to the Mono Pass region of the Sierra in 1971–72 by the California Department of Fish and Game as a prospective game species, and this alien has since expanded its range from Sonora Pass in the north to northeastern Kings Canyon National Park (Small 1994; National Park Service files). Muskrat (*Ondatra zibethicus*) was introduced for commercial purposes, as was bullfrog (*Rana catesbiana*). Virginia opossum (*Didelphis virginiana*) spread into the Sierra from an introduction in San Jose, and possibly elsewhere in California, early in the century, while brown-headed cowbird (*Molothrus ater*) was first recorded breeding in California in 1870 and spread progressively throughout much of the State; it is described here as alien, although likely self-introduced, because its establishment and spread in California is closely connected to anthropogenic habitat disturbance. House sparrow (*Passer domesticus*), European starling (*Sturna vulgaris*), and rock dove (*Columba livia*) spread into California from intentional introductions in the eastern United States; in the Sierra Nevada they remain close to areas of human settlement and agriculture. House mouse (*Mus musculus*), brown rat (*Rattus norvegicus*), and black rat (*Rattus rattus*) are notorious pests in urban and some suburban areas including parts of the Sierra Nevada; they have been inadvertently introduced from Eurasia many times. Feral cats (*Felis domesticus*) prey on small vertebrates and compete with small native carnivores adjacent to settlements. Although truly feral dogs (*Canis domesticus*) are unusual, roaming packs of pet dogs have impacts on wildlife, especially ungulates, and domestic stock as well. Lastly, the domestic goat (*Capra hircus*) has escaped and established feral populations in a few locations in the central Sierran foothills, as have horses (*Equus caballus*) and cows (*Bos taurus*), the latter two usually only on a local and temporary basis in the Sierra, but nonetheless occasionally damaging wetlands and riparian habitats in particular.

Several of these species have had a significant impact on the ecology of the Sierra Nevada and its native species. The most serious effects have been produced by the brown-headed cowbird. The spread of this brood parasite in the Sierra Nevada (and the West in general) has mirrored farming, livestock grazing, clear-cut logging, and suburban development. (Gaines 1977; Rothstein et al. 1980; Verner and Ritter 1983;

Airola 1986; Coker and Capen 1995). Preferred foraging habitats in the Sierra include heavily grazed meadows, recent clear-cuts, especially those that are grazed, open forest with short grass understory, pack stations and stables, picnic areas and campgrounds, lawns and golf courses, and residential areas with bird feeders. Closed-canopy and multi-layered forests, forests with shrub understory, tall-grass meadows, and clear-cuts after shrubs and trees are established do not provide cowbird foraging habitat (Laymon 1995). Brown-headed cowbirds were first reported in the Sierra foothills by Grinnell and Storer (1924) west of Yosemite in Snelling in 1915, and at Mono Lake in 1916. The species is now widespread throughout the lower and middle elevations. Cowbirds travel as far as 7 km from feeding areas to host nests (Rothstein et al. 1984; Airola 1986). The greater the area of disturbed landscape within 7 km, the greater the likelihood that a nest will be parasitized (Coker and Capen 1995). Cowbirds are implicated in or directly charged with the decline of a variety of songbirds in the Sierra Nevada, especially willow flycatcher, Bell's vireo, yellow warbler, chipping and song sparrow (DeSante, 1995).Most passerine birds are susceptible, but parasitism and its effects can be highly local (Laymon 1987). Parasitism rates in excess of 10% are cause for concern, and those in excess of 30% are a serious problem (Laymon 1995).

European starlings and house sparrows are largely restricted in the Sierra to the foothills in or adjacent to urban or agricultural lands. They compete aggressively for nest sites with a number of native birds, and starlings in particular may have a significant impact on the nesting success of cavity nesters: western bluebird, ash-throated flycatcher, woodpeckers, and swallows especially purple martin (Small 1994). Thus some settlement patterns lead to reductions or local disappearances of some native species, less through the loss of habitat than the introduction of alien competitors.

Bullfrogs, native to the eastern United States, are now widely distributed in ponds and slow-moving streams in California, including the foothills of the Sierra Nevada; they have been recorded at elevations as high as 2,500 m in Sequoia National Forest. Bullfrogs have almost completely replaced red-legged frogs and foothill yellow-legged frogs in many locations, and are undoubtedly a factor in the precipitous declines of the native Ranid frog species (Moyle 1973; Hayes and Jennings 1986). Bullfrogs also prey on young western pond turtles, where they may be a significant factor in the decline of this species, as well as ducklings and other aquatic and riparian vertebrates.

Wild pigs compete with mule deer, black bears, band-tailed pigeons, squirrels, and many other native species for mast, mushrooms, and other food items. They destroy herbaceous vegetation and root extensively, making them pests in agricultural as well as park lands. Pigs are increasing in numbers and range in California, including the Sierra Nevada foothills. They are the second most hunted big game species in the State (Barrett 1977; Wood and Barrett 1978).

Chukar, white-tailed ptarmigan, and turkey are all local and uncommon residents of the Sierra, where they appear to be providing hunting opportunities with little obvious ecological impact on native species.

SIERRAN MAMMALS

The 112 species of mammals that regularly use the Sierra Nevada, 62% of the State's mammals, are dominated in species richness by the smallest of them: shrews (7), bats (17), rabbits (7), and rodents (56) (appendix 25.1). Among the rich assemblage of rodents are seventeen squirrels and chipmunks as well as a variety of pocket gophers, pocket mice, kangaroo rats, white-footed mice, and voles, as well as larger rodents including mountain beaver, yellow-bellied marmot, [true] beaver, and porcupine. Most of these are nocturnal and seldom-observed except for the squirrels. Distribution records depend largely on museum specimens collected from a limited number of locales. Most mammalian data sets emphasize species of economic importance as game, pelts, or pests, and a small number of charismatic species that attract public attention, such as cougar, coyote, and mountain sheep.

Bats

Seventeen species of bats are believed to use the Sierra Nevada. Of these, seven have been nominated for listing under the Endangered Species Act. Three of those and one additional species have been listed as sensitive or special concern. Concerns began to be raised about many bat species when numbers using known historic roosts were noticeably smaller or had disappeared entirely. One obvious potential culprit in these declines has been pesticides, since bats are insectivorous and like birds have very high metabolisms. But habitat requirements of most bat species have been based on a very small number of sites. Recent work by Pierson (1995) and others in California suggests that the large, old trees and snags associated with late-successional forests may be quite important to long-eared myotis (*Myotis evotis*), long-legged myotis (*M. volans*), and fringed myotis (*M. thysanodes*) as healthy populations have been found only in late-successional forests. The large trees and snags of conifers possess cavities and crevices that provide thermal protection for these bats. The presence of spotted bat (*Euderma maculatum*), Brazilian free-tailed bat (*Tadarida brasiliensis*), and western mastif bat (*Eumops perotis*) is correlated with meadows, while many if not most Sierran bats forage over water, especially riparian corridors. As bats use lower elevations for part of the year, loss of high-quality riparian habitat there may be factor in the apparent decline of so many species. Relatively high densities of spotted bats and western mastif bats have been found only in the vicinity of the substantial cliffs afforded by large river drainages such as the Kings, Kaweah, Merced, and Tuolumne rivers (Pierson 1995)

Forest Carnivores

This group of species, typically referring to red fox (*Vulpes vulpes*), fisher (*Martes pennanti*), marten (*Martes americana*), and wolverine (*Gulo gulo*) has been the subject of considerable attention for the past several decades, particularly after publication of *Status of Six Furbearer Populations in the Mountains of Northern California* by Schempf and White (1977). Its title reflects a preoccupation with the former economic importance of these species, their present apparently-reduced numbers, and factors affecting recovery. In recent years, substantial efforts have been made to assess the status of fishers and martens in the Sierra Nevada as well as elsewhere in California (Zielinski et al. 1996a, 1996b) using systematic grids of baited track and camera stations. Although unable to assess trends from only a few years of data, the authors found martens to occupy much of their historic range in the Sierra Nevada. However, while they found significant fisher populations in the southern Sierra Nevada west-side mixed conifer zone, they were unable to detect fishers north of Yosemite National Park, despite reports of their presence there by Grinnell et al. (1937), and scattered reports from the 1960s collected by Schempf and White (1977). Because the northern Sierra Nevada habitat of fisher (i.e., late-successional forest) has been extensively modified by timber harvest and other resource-extractive activities, and heavy fisher trapping also took place there, more than one factor may be involved. The red fox subspecies (*V. v. necator*), found principally in the northern Sierra Nevada in California (as well as Cascades), has been seldom detected and almost unstudied. Much the same can be said of wolverine, although it is regularly but infrequently reported from one location or another throughout the Sierra Nevada from montane forests into the alpine zone. Schempf and White (1977) reported an increase in sightings in the 1970s in the southern Sierra.

Mountain Sheep

As in other places in the west, mountain (bighorn) sheep (*Ovis canadensis*) populations in the Sierra Nevada were decimated following the arrival of Europeans in the mid-19th century (Buechner 1960). Sheep populations in the Sierra were originally scattered along the crest and east slope from Sonora Pass south, and along the Great Western Divide of what is now Sequoia National Park; there was also a population in the Truckee River drainage (Jones 1950; Wehausen 1988). Likely causes for the precipitous population decline include market hunting, severe overgrazing by domestic stock, and probably most importantly the transmission of respiratory bacteria from domestic sheep to bighorn that were fatal to the latter (Wehausen 1980).

Bighorn sheep were gone from the Yosemite region before the turn of the century (Grinnell and Storer 1924). By the 1970s, only two populations remained in the Sierra Nevada: in the vicinity of Mount Baxter (ca 220 individuals) and Mount Williamson (ca 30 individuals), west of Independence. The Mount Baxter herd was increasing during the 1970s (Wehausen 1980). From 1979 until 1988, the Mount Baxter population was used by the California Department of Fish and Game, in cooperation with the U.S. Forest Service and the National Park Service, to successfully reestablish herds near Wheeler Ridge, Mount Langley, and Lee Vining Canyon. Some cougars were removed from the Lee Vining Canyon area to reduce significant losses while that herd was getting established. By 1990, the three introduced herds were all increasing, and the overall Sierra bighorn population was at least 300 (Bleich et al. 1990).

Between 1977 and 1987, cougar (*Felis concolor*) depredation reports in Inyo and Mono counties, as well as for California as a whole, increased dramatically (Foley et al. 1995). During that period, fifty predation losses to the Mount Baxter herd were discovered on its escarpment-base winter range. Losses by cougar predation were detected in the other herds as well. During the extended drought of the late 1980s and early 1990s, the herds gradually abandoned their low elevation winter ranges for much higher elevation sites that, while inferior from the standpoint of forage and protection from cold, were relatively snow-free during the drought and afforded protection from predation. This profound behavior change is attributed by Wehausen (1995) to heavy cougar predation pressure on the traditional low-elevation ranges. Concurrent with this change in behavior has been a steady decline in population. The Mount Baxter population had 108 ewes in 1978; no more than twenty were counted in 1995. Twelve sheep died in a single avalanche on Wheeler Ridge in 1995; only ten ewes remain as its reproductive base. The Lee Vining Canyon population declined from approximately thirty-six ewes in 1993 to fourteen in 1995. Whether from accidents or an inferior energetic balance, the new situation is distinctly pessimistic, with the Sierra Nevada population probably well below the 250 recorded when reintroduction began in 1979.

There is no reason to assume cougar populations were smaller than at present prior to settlement, although they may well have fluctuated significantly over time. But whereas sheep were widespread in the Sierra at settlement, presently they only persist in scattered small pockets of high elevation habitat where snow depths are tolerable and cougars absent. One possible explanation is that in the past, sheep herds were sufficiently well-distributed and large that herds in decline on account of heavy predation or weather were supplemented by colonists from other thriving herds, thus providing a regional buffer for local perturbations as well as maintaining genetic diversity. The small and isolated populations now present can no longer provide either function.

Management of the Sierran bighorn is facilitated by the Sierra Interagency Bighorn Sheep Advisory Group, which in-cludes technical representatives from participating agencies. This group is now considering a recommendation that a captive breeding program be established as insurance against complete collapse of the Sierran populations, and as a source for future reintroduction. However, domestic sheep and cattle allotments on the public lands of the eastern slope and Sierra crest, with their well known potential for disease introduction into bighorn, greatly restrict the number of potential sites available for reintroduction. So long as populations are relatively small and disconnected, some controls on predation, especially through cougar removals, may also be necessary.

SIERRAN BIRDS

The avifauna of the Sierra Nevada is still reasonably intact. Only Barrow's goldeneye (as a breeding species), Bell's vireo, and California condor have been wholly extirpated from the Sierra, but several species, including harlequin duck, great gray owl, and willow flycatcher, appear to be at great risk in the Sierra. The latter two are California endangered species, the only two species of Sierran landbirds (as defined here) that are currently officially listed, although bald eagle and peregrine falcon, which also breed in the Sierra, are on both state and federal threatened or endangered lists. Several additional Sierran landbirds are federal candidates for listing or California Species of Special Concern: Federal candidates include California spotted owl, olive-sided flycatcher, Bell's sage sparrow, as well as non-landbirds: harlequin duck, northern goshawk, and western sage grouse. Seven Sierran landbirds have been included on the California State Department of Fish and Game's list of "Species of Special Concern": Long-eared owl, black swift, Vaux's swift, purple martin, loggerhead shrike, yellow warbler, and yellow-breasted chat. Other Sierran "Species of Special Concern" include: osprey, sharp-shinned hawk, cooper's hawk, golden eagle, and prairie falcon. A new list of California Species of Special Concern is currently in draft stage. Thirteen additional Sierran landbirds are being considered for this new list.

Breeding Landbirds

(This section is extensively adapted from DeSante 1995.)

With the exception of game species, only breeding landbirds have been monitored both systematically and over the length of the range for multiple decades: The North American Breeding Bird Survey (BBS) routes began in 1966. However, only seventeen routes have been established in the entire Sierra Nevada physiographic region. As at least fourteen routes are necessary to establish trends with certainty by providing sufficient sample size, the majority of Sierra breeding landbirds are assigned to insufficient sample size; trend may be stable or unknown (Davidson and Manley 1993). In particular, routes in the lower west-slope foothills have been inadequate to establish trends with confidence for the bulk of species breeding there.

Nonetheless, because of these systematic data, breeding landbirds provide the most useful group of vertebrates to examine as an indicator of Sierran vertebrate status and trend. DeSante (1995) has analyzed the literature of birds breeding in the Sierra Nevada, as well as BBS data for the period 1966–91, and the Monitoring Avian Productivity and Survivorship (MAPS) data for twelve stations operated in the Sierra during 1990–94. DeSante's report to the Sierra Nevada Ecosystem Project is summarized here.

Twenty-six years (1966–91) of BBS data indicate that only six breeding landbird species are definitely decreasing in the Sierra Nevada physiographic region (according to the classification system described above): band-tailed pigeon –5.5% per year, red-breasted sapsucker –7.5%, olive-sided flycatcher –3.2%, American robin –2.7%, chipping sparrow –5.0%, and white-crowned sparrow –9.7%. More species, however, likely would be found to be decreasing were it not for the paucity of BBS routes in the Sierra. In fact, twelve other species appear to be decreasing by amounts ranging from 1.2% to 8.5% per year: Mourning dove, belted kingfisher, western wood-pewee, Steller's jay, mountain chickadee, golden-crowned kinglet, Swainson's thrush, black-headed grosbeak, dark-eyed junco, brown-headed cowbird, house finch, and lesser goldfinch. These eighteen decreasing species have little apparent in common except that many of them are among the commonest, most widely distributed, and most characteristic landbird species in the Sierra. It's important to note that change detection in either direction is most likely to occur for species that occur on many transects.

Only four of these eighteen definite or likely-decreasing species are true neotropical migrants: olive-sided flycatcher, western wood-pewee, Swainson's thrush, and black-headed grosbeak. Marshall (1988) previously documented the disappearance of olive-sided flycatcher and Swainson's thrush from an area of the southern Sierra and suggested that it was caused by tropical deforestation on the species wintering grounds. Except for nighthawks and various swallows, olive-sided flycatcher, western wood-pewee, and Swainson's thrush are the three longest distance migrants among the Sierra's neotropical migrant landbirds, and relatively few or no individuals of these species winter in western Mexico where the majority of the Sierra's neotropical migrants are assumed to winter. An additional eleven species may be decreasing in the Sierra: flammulated owl, white-throated swift, northern rough-winged swallow, scrub jay, American crow, chestnut-backed chickadee, white-breasted nuthatch, blue-gray gnatcatcher, Townsend's solitaire, pine grosbeak, and evening grosbeak. Only six of twenty-nine definitely, likely, or possibly decreasing species are neotropical migrants. In contrast, fourteen of these twenty-nine decreasing species are short-distance migrants or short-distance/neotropical migrants, and nine are resident or resident/short-distance migrants. This suggests

that local influences may be having a more significant negative effect than tropical deforestation on landbird populations.

Marshall (1988) also documented the disappearance of mountain quail, flammulated owl, northern pygmy-owl, spotted owl, and hairy woodpecker from his study area in the southern Sierra. The BBS data shows flammulated owl as possibly decreasing and northern pygmy-owl with a decreasing tendency, but shows hairy woodpecker as probably relatively stable (–0.8% per year). BBS data also show mountain quail as likely relatively stable (–0.6% per year). BBS data are insufficient to provide reliable trend information for spotted owl. Overall, the disappearances recorded by Marshall in the southern Sierra seem to be reflected in other parts of the Sierra as well. However, intensive work on the spotted owl, in the region (Sequoia and Kings Canyon National Parks) that abutted Marshall's (1988) study area, did detect the species there but was unable after four years of censuses to determine if the population was declining (Verner et al. 1992). Interestingly, calling for spotted owls by investigators also generated more locations for flammulated owl in the two national parks than had been recorded in all previous years (Sequoia and Kings Canyon National Parks files), suggesting that detection of this and species with similar habits may be quite poor.

On the other side of the ledger, four species were found to be definitely increasing in recent decades: White-headed woodpecker +3.4% per year, cliff swallow +26.3% per year, common raven +9.1%, and fox sparrow +3.2%. DeSante (1995) suspects that all of these increases result directly from human activities and adaptive responses on the part of the birds to these activities: Cliff swallow from increased nesting locations afforded by bridges and buildings, common ravens from increased human traffic on roads and a resulting increase in road kills that ravens have learned to utilize, fox sparrow from increased amounts of upland brushy habitat resulting from logging operations, and white-headed woodpecker from selective harvest practices (thinning) which white-heads favor. Eight other species are likely increasing by amounts ranging from +1.7% to +5.5% per year: Hammond's flycatcher, black phoebe, house wren, solitary and warbling vireos, and yellow, yellow-rumped, and MacGillivray's warblers. Six of these twelve definite or likely increasing species are true neotropical migrants: Hammond's flycatcher, cliff swallow, solitary and warbling vireos, and yellow and MacGillivray's warblers. Except for cliff swallow, which winters in South America, Sierran populations of these other five neotropical migrants probably winter primarily in western Mexico. An additional seven species are possibly increasing in the Sierra: tree swallow, hermit thrush, black-throated gray warbler, hermit warbler, western tanager, rufous-sided towhee, and Brewer's sparrow. Five of these (all but the towhee and sparrow) are true neotropical migrants that winter primarily in western Mexico. Thus, eleven of nineteen definitely, likely, or possibly increasing species are true neotropical migrants. These data, taken together with data on decreasing species presented above, provide no indication that neotropical migrants as a group are decreasing in the Sierra at any greater rate than other species. More species seem to be decreasing in the Sierra (29) than increasing there (19), although the difference is doubtfully significant (DeSante 1995).

DeSante (1995) was able to identify four species as having relatively stable population trends in the Sierra: Northern flicker, pileated woodpecker, Bewick's wren, and Cassin's finch. Two of these (pileated woodpecker and Bewick's wren) are resident or resident/short-distance migrants. Ten other species were identified as having probably relatively stable population trends in the Sierra: Hairy woodpecker, red-breasted nuthatch, brown creeper, orange-crowned, Nashville, and Wilson's warblers, green-tailed towhee, Brewer's blackbird, purple finch, and pine siskin. Three of these (the woodpecker, nuthatch, and creeper) are also resident or resident/short-distance migrant species. Finally, eight additional species were identified as having possible relatively stable population trends in the Sierra: Common nighthawk, Anna's hummingbird, downy woodpecker, barn swallow, bushtit, wrentit, song sparrow, and northern oriole. Three of these are also resident or resident/short-distance migrant species. Thus, eight of the relatively stable species seem to be resident or resident/short-distance migrant species.

DeSante tested patterns regarding the number of species showing decreasing, increasing, or relatively stable population trends among the various migratory groups by means of a contingency table and Chi-square tests. The mean population trends for species with various migration strategies provide a further indication that neotropical migrants are not declining in the Sierra more than residents or short-distance migrants. The mean population trends for the twenty-one species of residents and resident/short-distance migrants having definite, likely, or possible population trends was −1.2% per year; for the twenty-six species of short-distance migrants and short-distance/neotropical migrants having definite, likely, or possible population trends it was −1.7% per year; and for the twenty-three species of neotropical migrants having definite, likely, or possible population trends, it was +1.9% per year (when cliff swallow with a +26.3% population trend was eliminated from the neotropical migrants, the mean population trend for the remaining twenty-two species was +0.8% per year). Limiting this analysis only to species showing definite or likely population trends, the results were: residents and resident/short-distance migrants +0.7%; short-distance migrants and short-distance/neotropical migrants −2.4%; and neotropical migrants +2.6% (when cliff swallow was eliminated this value became +0.6%).

Short-distance migrants, as a group, may be faring the worst among landbirds in the Sierra. DeSante's (1995) findings agree with Hutto (1988), who questioned the decline of neotropical migrants wintering in western Mexico, and with DeSante and George (1994) who found that neotropical migrants generally showed fewer and smaller decreasing trends than short-distance migrants over western United States as a whole. This should not be interpreted as indicating that problems do not exist among neotropical migrants nor that tropical deforestation is not a problem for Sierran landbirds, but merely that gross generalizations regarding massive declines in neotropical migratory landbirds in western North America in general, and the Sierra in particular, may be unfounded based upon available data.

Moreover, a few misclassifications of migratory behavior or population trend in the contingency table could alter the results so that they were not significant. Indeed, the data as presented have rather poor statistical power or robustness. There are other cautions: Much of this analysis is based on species that were recorded on less than fourteen routes in the Sierra. Such data are generally considered inadequate for detecting reliable regional trends. Thus the results presented here must be viewed as suggestive rather than conclusive. Second, the analysis presented here utilized twenty-six years of BBS data from 1966–91. Because trends for the more recent thirteen years were not separated from trends in the early thirteen years, the situation in the Sierra could have begun to deteriorate in recent years. However, DeSante and George (1994) found the reverse to be true. Populations of both short-distance and neotropical migrants tended to fare better during the more recent thirteen years than during the earlier thirteen years. Third, BBS results are based on roadside surveys and may not be valid for areas away from roads. A variety of habitat conditions may exist adjacent to the road that are unrepresentative of the area as a whole. And, because the locations of the transects were chosen in part for accessibility, landscape development or other modifications there may likewise be unrepresentative of the region.

Potential Risks Faced by Sierran Landbirds

Grazing

Grazing of Sierran habitats, particularly montane meadow and montane riparian habitats, may constitute a significant threat to Sierran landbirds. Grazing of montane meadows has been implicated as a major cause of the drastic decline of willow flycatchers in the Sierra; Gaines (1988) claims that willow flycatchers do not nest in willows whose lowermost foliage has been denuded by livestock. Grazing has also been implicated in the decline of great gray owls outside of Yosemite National Park; great gray owls do not forage in grazed meadows, perhaps because grazed meadows are attractive to great horned owls which exclude them (Gaines 1988), or because of changes in prey populations.

The major deleterious effects of grazing on montane meadows are decrease in the density and height of herbaceous growth in the meadow. Many of the landbird species utilizing these meadows depend upon insects that either live on the herbaceous growth or depend upon the primary productivity of the herbaceous growth for sustenance. (The dense concentrations of aphids on lupines and corn lilies in these meadows is one example.) A decrease in the quantity of this herbaceous growth will result in a decrease in the food resources of landbirds that use the meadow. A decrease in the quantity of herbaceous vegetation may also lead to a concomitant proportional increase in the amount of shrubby woody vegetation. However, the increase in shrubby vegetation does not always translate into an increase in the quality of the willows that are usually present in montane meadows, as livestock often extensively browse and effectively defoliate the lowermost foliage of willows, thereby greatly reducing the usefulness of this resource to landbirds. Grazing also tends to destroy the banks of the streams flowing through the meadow which both widens and deepens the stream channels and thus increases the rate of channelization and lowers the water table (Ohmart 1994). All of these effects tend to cause a drying out of the meadow and to hasten its demise. Grazed riparian habitats, even without associated meadows, seem to be affected in a similar manner (Ohmart 1994). And finally, the grazing of montane meadows promotes contact between cowbirds (which are attracted to the grazing livestock) and a high density of nearby nests of many host species, including both those that nest in the meadow itself and those that nest, often in higher than average numbers, in the adjacent forest.

Montane meadows and montane riparian habitats are extremely important for Sierran birds. Not only is there a substantial subset of species that are dependent upon these habitats, the population densities of many forest-inhabiting species are often highest on the edges of montane meadows (DeSante 1995). Moreover, they are often used as important supplemental habitat for a variety of species, including the rapidly-declining red-breasted sapsucker, which depends upon willows in montane meadows for a steady supply of sap during the breeding season, and a number of finch species which require a daily water supply. Finally, montane meadows serve as a critical molting and pre-migratory staging area for the young and, to a lesser extent, the adults of many Sierran landbirds. Montane meadows in mid-summer may be the single most critical Sierran habitat requirement for many species that do not even utilize this habitat during the actual breeding season (DeSante 1995). Species such as orange-crowned and Nashville warblers fall into this category. The effects of grazing on other Sierra habitats are also likely deleterious to landbirds, but probably to a lesser extent than grazing in montane meadows. In all cases grazing tends to decrease the amount of herbaceous plant growth present in forest, woodland, and brushland habitats, thereby negatively affecting the food resources of many granivorous and some insectivorous species, and tends to increase the contact between cowbirds and their host species.

While the extent of cowbird parasitism in the Sierra may be increased by grazing, grazing itself may not be the basic cause for the increase in cowbirds in the Sierra. The fundamental cause for the increase in cowbirds in the Sierra may be related to agricultural practices and feedlots in the major valleys both east and west of the Sierra. The large populations of cowbirds that inhabit these valleys may serve as source populations for cowbirds that parasitize landbirds in the Sierra. Widespread, comprehensive cowbird control programs in the Sierra may be ineffective for reducing the overall problem; however, local cowbird control programs at certain critical meadows and riparian habitats may be necessary for protecting remnant populations of certain very rare species, such as willow flycatchers. The amount of grazing in the Sierra, at least at mid- and higher elevations, has been decreasing in recent years (Menke et al. 1996). Perhaps related to this, BBS indicates cowbird populations seem to be decreasing as well (DeSante 1995). However, at the present time grazing and its secondary effects may well be the single most significant negative factor in the maintenance of native Sierran landbird populations.

Logging

Forestry management practices, particularly logging and fire suppression, can have a profound effect on landbird populations in the Sierra and elsewhere (Hejl 1994). Extensive clearcutting is obviously detrimental to most forest-inhabiting species because it removes large areas of forest habitat. The even-aged forests that tend to result from planting after clearcuts often lack the tree species diversity and, apparently more importantly, structural diversity that seems to permit large and diverse bird populations to persist. Selective logging that preserves multi-aged stands and the structural diversity of the forest, may offer a better forest management prescription from an avifaunal standpoint than even-age forestry practices. Selective cutting, however, can also be detrimental if it removes or modifies important components or characteristics of the forest that are critical for certain species such as large snags and logs. Considerations of forest fragmentation are also important with regard to the management of Sierran forests. Fragmentation increases the ratio of forest edge to forest interior and has been implicated in the loss of bird species diversity in eastern forests, apparently primarily through increased rates of cowbird parasitism and nest predation (e.g., Coker and Capen 1995). It is possible that similar effects could be occurring in the Sierra, although perhaps to a lesser degree since Sierran forests naturally feature fine-scale fragmentation mosaics (Franklin and Fites-Kaufmann 1996).

A sufficient amount and distribution of old-growth and mature forests can serve as locations for source populations for species dependent upon such habitats. It also includes a sufficient quantity of the snags, logs, and other dead wood that are required by both primary and secondary cavity nesters, and used by such species as great gray owls (Hayward and Verner 1994) and spotted owls (Verner et al. 1992). These two critical aspects of forest management, providing a sufficient amount and distribution of late-successional forests, and providing a sufficient quantity and distribution of snags and other dead wood in forests of all ages with all degrees of canopy cover and tree densities, appear to be crucial for the continued existence of an intact and healthy Sierran forest avifauna. There are, of course, species including some possibly declining that prefer open stands or forest openings as would have occurred in many places under aboriginal fire regimes; these conditions could be simulated with appropriate forestry practices.

Fire Suppression

Fire suppression in the Sierra Nevada has led to forest and chaparral stand conditions inimical to many Sierra landbirds because of loss of micro-habitat elements. These include dense ingrowth of shade-tolerant tree species in place of forest openings containing herbs and shrubs, and decadent stands of chaparral with low productivity instead of mosaics of various seral conditions. And, of course, the high fuels associated with suppression can lead to large, stand-destroying fires that eliminate large, old trees, snags, and logs.

Development of the Sierra and the Loss of Breeding Habitats

Development pressures throughout the Sierra, but especially in the foothills and lower elevations of the west slope, are becoming an increasingly important threat to the viability of Sierran landbird fauna, and to the ecological integrity of the Sierra as a whole. Two habitat types stand out as most endangered by this development, the arborescent riparian habitat along the west slope's rivers and streams, and oak woodland and forest. Chaparral, however, is also threatened by this development. The risks that these habitats face from development come from a number of sources. Dam building, water diversions, and agriculture have had massive negative effects on the riparian habitats and other wetlands of the west slope, especially in the lower foothills (Kattelmann 1996; Moyle and Randall 1996). Not only have forests of typical riparian species, such as willows, cottonwoods, and sycamores, been reduced to remnants, riparian valley oak communities have disappeared from all but a handful of locales. As most of the original riparian forest habitat in the Central Valley is gone, the remaining riparian habitat in the lower foothills becomes essential to a number of species with limited habitat and critically low population levels in the Sierra, such as black-chinned hummingbird, common yellowthroat, yellow-breasted chat, and blue grosbeak.

Low-density foothill and mid-elevation developments ("ranchettes"), can produce subtle but significant problems. Grazed paddocks and large expanses of mown grass provide centers for cowbird parasitism problems. Agricultural, residential, and commercial development of the Sierran foothills increases the number of starlings inhabiting those areas which negatively affects cavity nesters by usurpation of their nest holes (Small 1994). Pets, especially house cats, prey on many bird species, while they reduce the numbers of reptiles and small mammals that serve as prey for many birds. On the other hand, ponds, orchards, and some ornamental plantings may actually increase local native diversity, depending on size, management practices, and surrounding habitat (Mayer and Laudenslayer 1988).

Little information exists regarding the population trends of the landbirds of the Sierran oak woodland (interior live, blue, canyon, and black) and chaparral habitats, despite the fact that these habitats represent areas of high vertebrate species diversity, including landbirds, for California and the Sierra (Barrett 1980; Block and Morrison 1990; Garrison 1996). Many, perhaps most, Sierran species that specialize or reach high densities in oak woodland habitats seem to be decreasing in the Sierra (band-tailed pigeon, Lewis' and acorn woodpeckers, scrub jay, plain titmouse, blue-gray gnatcatcher, western bluebird, lesser and Lawrence's goldfinches). DeSante (1995) suspects that a number of rare or uncommon chaparral-inhabiting species (such as greater roadrunner and rufous-crowned, black-chinned, and sage sparrows) are likewise decreasing, although BBS data are too sparse for most of these species to provide population trends. Grassland species of the lower foothills (such as western kingbird, horned lark, and lark and grasshopper sparrows) might also be declining as a result of increased development of the foothills (DeSante 1995), but adequate transect data are lacking. It would appear that the foothill areas and lower west slopes of all the Sierra are the areas that are now in critical need of avian research and monitoring efforts.

Increased Recreational Use of the Sierra

Increased recreational use of the Sierra and the increased vehicular traffic associated with it may present a serious threat to certain species that specialize in, or are limited to, areas of high recreational use. Montane meadows and montane riparian areas, including those at high altitudes, stand out as being most vulnerable because of their great popularity with campers, hikers, and equestrians. On the other hand, national forest and national park rules governing these areas have become much more restrictive than in former years. Increased accessibility in the Sierra will bring more humans into contact with wildlife, especially in relatively remote areas, but should have little effect on most landbirds (except possibly game birds including band-tailed pigeons, mourning doves, quail, and grouse). Annual revisions of hunting regulations and bag limits to reflect trend data and new knowledge of species biology can significantly ameliorate any effects of taking.

Pesticide Use

Pesticide use could be having serious deleterious effects on Sierran bird populations, and may provide an explanation for otherwise unexplained declines where habitats appear to be intact, but there is little direct evidence of such effects. Pesticides can potentially affect landbird populations in two ways: (1) by directly reducing the prey base available to the birds; and (2) by chemical contamination of the birds via pesticide accumulation up the food chain. Recent work suggests that exposure by the developing zygote to even extremely dilute concentrations of some common pesticides the so-called estrogen mimics may ultimately reduce fertility. Two situations in which the direct depletion of the prey base could occur are: (1) heavy pesticide use on forest insect outbreaks such as those of bark beetles at mid- and higher elevations in the Sierra; and (2) heavy pesticide use in the Central Valley that could negatively affect those flying insects that are wind-drifted to higher elevations in the Sierra and that may provide a major food source for swifts, nighthawks, olive-sided flycatchers, and even, perhaps, gray-crowned rosy finches that feed extensively on wind-drifted insects precipitated on snow banks. Pesticide contamination of birds via accumulation up the food chain is likely to be most important for diurnal raptors and owls (and also, perhaps, kingfishers and other waterbirds) but could also possibly affect most insectivorous species to some degree. Heavy pesticide use on the tropical wintering grounds of Sierran species could also be exerting a negative effect through either or both of the above-mentioned mechanisms. Considerably more research on all aspects of pesticide accumulation and its effects are needed before this potential risk to Sierran birds can be dismissed.

Habitat Destruction and Degradation of Wintering Grounds

Habitat loss on wintering grounds has been implicated as an important factor causing decreases of a number of forest-inhabiting neotropical landbirds of eastern North America (Robbins et al. 1989; Terborgh 1989), but the extent to which it is the major factor is unknown. Marshall (1988) suggested that tropical deforestation was also the major factor involved in the declines of olive-sided flycatchers and Swainson's thrushes in the southern Sierra. While this may be a correct assessment for these two rapidly disappearing species, it is doubtful that habitat loss and degradation on tropical wintering grounds can be implicated as the general overriding cause of population declines in Sierran landbirds since the Sierran BBS data do not indicate that neotropical migrants in the Sierra are faring worse that resident or short-distance migrants in the Sierra. Similarly, MAPS data from the Sierra provide no indication that Sierran neotropical migrants as a class have lower annual adult survival rates than Sierran resident or short-distance migrant species. Individual species, however, such as olive-sided flycatcher and Swainson's thrush and, perhaps western wood-pewee and black-headed grosbeak, may be adversely affected by this problem. Clearly, additional data on the relative productivity and survivorship of Sierran landbirds is needed. On the other hand, habitat loss or degradation of the temperate (southern U.S. or northern Mexico) wintering grounds of a number of relatively short-distance migrants may be a more serious problem. BBS data suggest that relatively more species of Sierran short-distance migrants may be declining than either resident or neotropical migrants. Moreover, a number of the declining short-distance migrants seem to be species that winter in grassland, brushland, or riparian habitat in the Southwest. Degradation or loss of these habitats caused by adverse agricultural and grazing practices, residential and commercial development, and pes-

ticide use may be having a strong negative affect on the landbird avifauna of the Sierra. Obviously, more work is needed in this regard.

Large-Scale Climate Change

Landbird productivity data from constant-effort mist netting in a California coastal scrub habitat in a Mediterranean climate suggests that productivity is at a maximum under relatively average weather conditions and that productivity decreases both when weather conditions are drier or wetter than average (DeSante and Geupel 1987). If such a relationship exists in the montane environment of the Sierra, then the pattern of extreme weather conditions that has characterized Sierran weather during the past two decades may have depressed the productivity of landbirds. The mechanism for this effect could be concomitant changes in primary productivity in general, or changes in the production of critical food resources for the birds, including acorn, berry, and insect production. This could, perhaps, be the overriding reason why more species of Sierran landbirds seem to show population decreases than population increases. Consequently, a future period of more extreme weather than generally characterized the Sierra during the years from 1900–1980, when we as a society developed our notions of normal Sierran climate, may result in long-term lower productivity during a time when other risk factors are also increasing.

SIERRAN REPTILES

There are 32 species of reptiles occurring in the Sierra Nevada; all are native there. Four of these are considered presently at risk: western pond turtle (*Clemmys marmorata*), blunt-nosed leopard lizard (*Gambelia silus*), California horned lizard (*Phryonosoma coronatum frontale*), and California legless lizard (*Anniella pulchra*). All of these are species that are largely found elsewhere and only minimally lap into the western Sierran foothills, although western pond turtles and California legless lizards occasionally range above 1850 m in appropriate habitat. Jennings and Hayes (1994) found that western pond turtles continue to be extant in all but a few southernmost Sierran historical sites, although they have been eliminated from many southern San Joaquin Valley and south coastal sites. Population structure of this long-lived species indicates recruitment failure in many locales, likely stemming from some combination of aquatic nesting habitat damage and predation by alien bullfrogs and bass (*Micropterus* sp.). California legless lizards, which live underground in loose soil in mostly open country, may well have been inadvertently introduced into some of the higher Sierra Nevada sites through nursery operations, transported in the roots of shrubs and trees (Jennings and Hayes 1994).

Blunt-nosed leopard lizards once ranged fairly widely in

the southern San Joaquin Valley and adjacent low foothills, but have lost most habitat to urbanization and agriculture (California Department of Fish and Game 1991). The California horned lizard (*P. c. frontale* subspecies of the coast horned lizard) is largely a coastal and valley creature of central California which ranges into the western Sierran foothills in appropriate habitat. Although widely distributed, this horned lizard has disappeared from more than a third of its range, almost certainly as a result of habitat alteration by agriculture and development. Jennings and Hayes (1994) believe that reductions may be more severe than they appear because, like the western pond turtle, this species is long-lived and may persist for years when recruitment is no longer occurring. In the Sierra, principal threats appear to be urbanization (including domestic cats) with concomitant modification of the exposed substrate and open habitat preferred by horned lizards. Pesticides, especially those that mimic estrogen, may also be a factor for those species in or adjacent to croplands.

Of the remaining Sierran reptiles (appendix 25.1), most are valley and foothill animals that range into the warm, xeric portions of the western or eastern Sierra foothills, some of them only marginally. On the other hand, several are truly montane animals in whole or in part, regularly occurring at elevations above 2,000 m or more. These include the western rattlesnake *(Crotalus viridis)*, rubber boa *(Charina bottae)*, California mountain kingsnake *(Lampropeltis zonata)*, western terrestrial garter snake *(Thamnophis elegans)*, western fence lizard *(Sceloporus occidentalis)*, sagebrush lizard *(S. graciosus)*, and northern alligator lizard *(Gerrhonotus coeruleus)*.

As with amphibians and the smaller mammals, status and trend have been crudely estimated by revisiting the sites where museum specimens often are very old and sometimes of uncertain provenance. This procedure, the only one available without more extensive contemporary recording of distributions, poses the risks both of failing to detect losses in locations where no collecting has been done, and of mistaking disappearance in a few revisited sites as representing widespread decline.

SIERRA NEVADA WILDLIFE HABITATS

The principal predictor of the presence of a particular vertebrate is appropriate habitat. Appropriate habitat for a wildlife species may vary by season or even activity. Wildlife habitats are largely equivalent to vegetation types or biological communities, but may also require the presence of abiotic elements such as cliffs, caves, lakes and streams, or sandy soils, and of biotic structural elements such as shrubs or trees at a particular seral stage, size, or density (e.g., large, decadent trees), snags, sufficient canopy cover, logs, litter, and duff important for some aspect of a vertebrate species life cycle. A

system for classifying wildlife habitats in California, including the Sierra Nevada, has been developed by Mayer and Laudenslayer (1988), and others. Habitats in the Sierra Nevada can be thought of as features that, at their grossest level, run parallel to the axis of the Sierra at different elevations, either on the west slope or the east. Examples are blue oak woodland, mixed chaparral, mixed-conifer forest, alpine dwarf shrub, or piñon-juniper woodland. Within these macrohabitats, however, a particular species may be confined to specific locales, including meso- or micro-habitats such as riparian corridors, canyon cliffs, or wet meadows adjacent to late-successional forest. Many of these finer-scale habitat features tend to run perpendicular to the axis of the range, along river drainages.

Changes in Sierran Habitats

In general, most habitat types that occur in the Sierra are generously distributed there, and most of these types are also reasonably well approximated by similar types in the White Mountains, Coast Range, Cascades, or Klamath Mountains. This is the likely explanation for the relatively low level of vertebrate endemism in the Sierra, given its large land area.

On the other hand, habitat elements associated with river and stream systems are far scarcer, and these have suffered proportionately greater reduction through human modification or appropriation. These modifications include water diversions, drowning of bottom lands by reservoirs, long-term grazing in the riparian zones, timber harvest, and human settlement.

The factors that make riparian habitats key to so many Sierran species include not only the availability of water itself in a region with six-eight months of drought, but lower temperatures during summer, shade, higher productivity of riparian plants for food, hiding cover, increased availability of insect prey, special plant structures (e.g., willow thickets). East and west trending riparian corridors provide food and protection for animals that move locally or seasonally migrate to different elevations.

Similarly, oak savannahs and woodlands, and foothill chaparral on the western slopes have been extensively modified. The native herbaceous understory in these communities was virtually replaced by introduced Eurasian grasses and dicots in the mid-nineteenth century. Most of these areas with an extensive herb understory have been grazed heavily for many years, leading to progressive loss of shrub cover, or converted to agriculture; some former chaparral has been converted to grazing land and much of the remainder has become decadent or even succeeded to conifer forest owing to fire suppression (Cheatham and Haller 1975). On the other hand, local burning and firewood collection have reduced the availability of large, old trees, snags, and fallen logs in some woodlands. The foothill communities, especially along those streams where bank slopes are gentle, have also been extensively settled. Foothill savannah, woodland, chaparral, and riparian habitats on the west slope of the Sierra offer mild winter conditions and comparatively higher productivity when the remainder of the range is cold and under snow, attracting migratory birds and wintering mammals that spend their summers at higher elevations. These habitats support species requiring open grassland, or grassland with scattered trees for nesting, perching, and feeding.

In the conifer forests, habitat has been less extensively and severely modified. Timber harvest combined with fire suppression, especially in west-side pine, east-side pine, Sierran mixed conifer, and red fir forests has modified the distribution of tree size and density, as well as large logs and snags. Late-successional conifer forests are important to species requiring moderated climates produced by the high, relatively closed canopy. Furthermore, multiple tree layers, large snags or logs provide sites for cover, nesting, feeding, and roosting. Clear-cut or burned areas produce montane chaparral or early-successional hardwood and conifer habitats that is converted quickly or quite slowly to conifer stands. Biological communities and structural elements that were present in aboriginal times have persisted, although some floristic components, size and spatial distribution of each habitat component may be different to varying degrees. (e.g., Minnich et al. 1995) Summer grazing in montane and subalpine meadows and grassy patches within forests appropriates highly-quality forage from wildlife to domestic stock, but the qualitative and quantitative effects on biodiversity of this nutrient removal with associated trampling in these locales are poorly known.

Habitat Dependency

In the Sierra, eighty-three terrestrial vertebrate species are considered dependent upon riparian (including wet meadow or lakeshore) habitat to sustain viable Sierran populations; 24% of these are at risk. Seventeen species are similarly dependent upon late-successional forests; 24% of these are at risk. There are eighty-six species that require west-slope foothill savannah, woodland, chaparral, or riparian habitats (some double-counted with riparian above) for Sierran population viability; 16% of these are listed as at risk (appendix 25.1) (California Department of Fish and Game 1994). This latter number is misleadingly low because many of these species are more widely distributed elsewhere, such as the Coast Range.

WILDLIFE MANAGEMENT

State of Knowledge of Sierran Wildlife

For only a few groups of vertebrates are there both longitudinal (over time) and geographic (over space) knowledge of status and trends in the Sierra Nevada: These include breeding land birds—there are seventeen long-term transects of the North American Breeding Bird Survey in the Sierra—and the

most popular game species, such as mule deer and waterfowl. Birds, in general, are relatively easy to observe and popular with amateurs who keep records, such as the annual Christmas Bird Counts, so a fair amount of unsystematic data exist. For all other species, range maps have been developed by summing together dozens to hundreds of museum specimens for which good location data are presumed, and extrapolating on a map. These few vouchers have been collected over a span of as much as a century, and many or most locations are represented by specimens collected in the early or middle part of this century, the golden age of collecting for California.

The file drawers of national forest, national park, and Department of Fish and Game biologists, some county agencies, as well as many private land managers and landowners, often contain records of observation of rare or unusual wildlife, often with behavioral or habitat use attached. While of far less value than systematic scientific surveys, longitudinal studies, or investigations of species-habitat relations, they can be invaluable at improving the resolution of distribution information, and for making correlative inferences about habitat preferences and other ecological attributes of the species. At present there is no efficient way to locate these data. The California Natural Diversity Database, managed by the California Department of Fish and Game, keeps site records of agency-listed plant and animal species in the State. It has the potential to serve as a clearinghouse and manager of data on all species throughout California. However, to be effective at this function, a budget many times its present one would be required. This would be an invaluable service to land managers, landowners, and government agencies throughout California.

Trend, which is the change in abundance or distribution over time, is typically based on a handful of sites for which a handful of sampling time-points exist. From these it is a dubious business to infer status and trend of a species over its range in the Sierra Nevada. For the Breeding Bird Survey of seventeen routes in the entire Sierra Nevada, there are twenty-five years of data, although not all routes for all years. Similar data exist for waterfowl on refuges, and mule deer herds in some locations. A promising synthesis of the Breeding Bird Survey and the California Wildlife Habitat Relationships System (CWHR) has been produced by U.S. Forest Service, Pacific Southwest Region. Entitled *Avesbase*, (Davidson and Manley 1993), this computer database and analytical engine combines information about population trend and habitat distribution for neotropical migrant birds as an aid in assessing their risk in California.

California Wildlife Habitat Relationships System

The California Wildlife Habitat Relationships System (CWHR) was initiated in the early 1980s to provide a formalized and generally agreed-upon compendium of knowledge about the distribution and habitat preferences of all the amphibians, reptiles, birds, and mammals in California as a means toward both improved wildlife management and improved land management practices. The immediate inspiration for CWHR came from the work of J. W. Thomas and his associates in a pioneering effort for the Blue Mountains of Oregon and Washington (Thomas 1979), but reflected an emerging national interest in such information (Nelson and Salwasser 1982). Although CWHR was intended by its developers to be a tool for professional wildlife biologists as the starting point in assessing the wildlife response to land management practices, in recent years the database has been used as an expert system both to calculate present wildlife species presence on land units, and response to habitat modifications.

From the start, CWHR was a cooperative effort that included biologists from government agencies, universities, and private industry. Principal cooperators were the U.S, Forest Service, California Department of Fish and Game, University of California, U.S. Bureau of Land Management, Southern California Edison, and the California Department of Forestry and Fire Protection. While the California Department of Fish and Game is the manager of the CWHR system including database, program, manuals, and user support, other agencies provided publication outlets for support documents, such as *California Wildlife and Their Habitats: Western Sierra Nevada* (Verner and Boss 1980). An informal technical support group, The California Interagency Wildlife Task Group, acts as steward of CWHR by maintaining quality control, assuring its continued development, and encouraging its use in resource management. Initially, existing publications and species experts were used to synthesize distribution and habitat preferences for each separate species, a significant step beyond Thomas (1979), which used wildlife guilds as units of analysis. Unifying the many existing California vegetation classification systems required a distinct, parallel effort to develop a wildlife habitat classification system (Mayer and Laudenslayer 1988) which brought wildlife and vegetation experts together to examine biological communities.

The objectives of the CWHR program are (Airola 1988):

- develop a system that can predict the potential of habitat to support wildlife species and to predict the effects of habitat changes on them

- provide easy access to a vast array of wildlife and habitat information through preparation of published volumes and a computerized database system

- encourage an ecosystem orientation that considers all wildlife species that may occur in an area

- foster consistency of analysis, so that impacts of different projects may be more readily compared, and understood by decision-makers and the public

The information in CWHR was ported to a computer database within a supporting menu-driven program (Timossi, Sweet, et al. 1994); while it was intended as an expert system, codifying the findings and opinions of species and habitat experts, it was designed for use by natural resource professionals who had been provided training in its use. The CWHR computer program and support documents (Mayer and Laudenslayer 1988; Timossi, Sweet, et al. 1994; Zeiner et al. 1988, 1990a, 1990b) are now widely in use in California.

In brief, the CWHR database has four levels of habitat suitability by a vertebrate species to each habitat/seral stage: optimum, suitable, marginal, or unsuitable. These levels are in turn differentiated into breeding, feeding, and resting values. Where special habitat elements are required by a species (e.g., large snag, mud flat, etc.), these are specified. The database provides species distribution information (contemporary and recent range) in a variety of ways, including biological province, county, national forest or BLM district, Fish and Game region, hydrologic region, and latitude/longitude.

In the most typical use of CWHR one develops a list of vertebrate species potentially present on a site, based upon additional information about the site such as location and habitat elements present (or absent), which further restricts the output list of candidate species present. CWHR was designed to minimize errors of omission—species present on a site but not listed by the program. Thus it tends to produce errors of commission—species listed but not present on a site. If one wishes to assess the change in wildlife species composition or habitat values produced by a land management action (e.g., timber harvest), field investigation is necessary to trim the initial list (Garrison 1994).

The CWHR is widely acknowledged by its developers, users, and critics to be highly imperfect. Present and potential distribution, the nature and distribution of habitats, and suitability of particular habitats to most terrestrial vertebrate species are poorly known. However, from the start CWHR was designed to evolve in response to new scientific knowledge and feedback from its users. Thus CWHR Version 5 is a substantial improvement over earlier versions.

Presently, developers are working to produce regional CWHRs for more precision, and an effort is underway to build geographic information system (GIS)-based habitat suitability models that account not only for the presence of appropriate habitat, but its extent and spatial arrangement as well (Timossi, Woodard, et al. 1994a, 1994b), and ultimately minimum viable population requirements. This has promise to provide more realistic population viability predictions, but is profoundly constrained at present both by lack of mechanistic habitat models for most vertebrates, and lack of detailed habitat information for all but a few locales.

Because CWHR uses relatively gross-scale habitat components (e.g., dominant vegetation and seral stage), it is least successful for small vertebrates, including many amphibians and reptiles, that key in to much finer-scale habitat requirements within broad types, such as particular kinds of prey, cover types, or aquatic conditions. Efforts are now in their initial stages at the U.S. Forest Service, Pacific Southwest Forest Research Station at Arcata, to develop pattern recognition models for some of these species. Again, this admirable effort faces the dual difficulty of limited information on habitat requirements for many species, and even more limited availability of the mapped distribution of required habitat elements once determined.

Gap Analysis Program

A promising new scientific strategy for habitat conservation is the Gap Analysis Program (GAP) managed by the U.S. Department of the Interior, National Biological Service (Scott et al. 1993). Gap analysis uses geographic information systems (GIS) to overlay map of plant communities with those of land ownership and land use. This system facilitates the identification of biological communities, and thus the vertebrates that depend upon those habitats, that are vulnerable to conversion and degradation. The Sierra Nevada Ecosystem Project collaborated with GAP to complete a model for the Sierra Nevada (Davis and Stoms 1996). The power of GAP promises to be greatly enhanced by new models designed to use GAP and other GIS-based data as a starting point in identifying the most efficient sites for conservation strategies. One such strategy of note is the Biodiversity Management Area (BMA) strategy developed by Davis et al. (1996)

CONCLUSION

Compared to the more intensively-developed regions of California, the terrestrial vertebrate fauna of the Sierra Nevada is relatively intact. There have been few extinctions and most species appear to retain an approximation of their aboriginal geographic extent. The most important factor in population viability for nearly all species has been and continues to be habitat quantity and quality. Habitats that have suffered the greatest reductions in extent and integrity, and therefore the greatest losses of vertebrate biodiversity, appear to be the western-slope foothills, riparian habitats, and late-successional forests. The greatest threat to the preservation of viable populations of native wildlife in the Sierra may well be the poor quality of information about status, distribution, trend, and species biology especially species-habitat relationships. This uncertainty, unless corrected, will continue to lead to inefficient conservation strategies and unpredictable outcomes from land use changes, as well as public dissatisfaction with conservation policies and their outcomes. Once the quality of data is improved, however, models such as CWHR and GAP can be effectively applied to sound ecosystem management practices.

REFERENCES

Airola, D. A. 1986. Brown-headed cowbird parasitism and habitat disturbance in the Sierra Nevada. *Journal of Wildlife Management* 50:571–75.

———. 1988. *Guide to the California wildlife habitat relationships system.* Sacramento: California Resources Agency, Department of Fish and Game.

Anderson, R. S. 1990. Holocene forest development and paleoclimates within the Sierra Nevada, California. *Journal of Ecology* 78 :470–89.

Barrett, R. H. 1977. Wild pigs in California. In *Research and management of wild hog populations*, edited by G. W. Wood, 111–13. Georgetown, SC: Clemson University, Baruch Forest Science Institute.

———. 1980. Mammals of California oak habitats management implications. In *Proceedings of the symposium on the ecology, management, and utilization of California oaks*, technical coordination by T. R. Plumb. General Technical Report PSW-44. Berkeley, CA: U.S. Forest Service.

Bleich, V. C., J. D. Wehausen, K. R. Jones, and R. A. Weaver. 1990. Status of bighorn sheep in California, 1989, and translocations from 1971 through 1989. *Desert Bighorn Council Transactions* 34:24–26.

Block, W. M., and M. Morrison. 1990. Wildlife diversity of the central Sierra foothills. *California Agriculture*, March-April, 19–22.

Buechner, H. K. 1960. The bighorn sheep in the United States, its past, present, and future. *Wildlife Monographs*, no. 4:1–74.

California Department of Fish and Game. 1991. *1990 annual report on the status of California's state listed threatened and endangered plants and animals.* Sacramento: California Department of Fish and Game.

———. 1994. *California Wildlife Habitat Relationships Database system*, version 5.0. Computer database and program. Sacramento: California Department of Fish and Game.

Cheatham, N. H., and J. R. Haller. 1975. An annotated list of California habitat types. Unpublished manuscript. University of California Natural Land and Water Reserve System.

Coker, D. R., and D. E. Capen. 1995. Landscape-level habitat use by brown-headed cowbirds in Vermont. *Journal of Wildlife Management* 59:631–37.

Davidson, C., and P. Manley. 1993. *Avesbase: A conservation database for California birds*, version 1.0, San Francisco: U.S. Forest Service, Pacific Southwest Region.

Davis, F. W., and D. M. Stoms. 1996. Sierran vegetation: A gap analysis. In *Sierra Nevada Ecosystem Project: Final report to Congress*, vol. II, chap. 23. Davis: University of California, Centers for Water and Wildland Resources.

Davis, F. W., D. M. Stoms, R. L. Church, W. J. Okin, and N. L. Johnson. 1996. Selecting biodiversity management areas. In *Sierra Nevada Ecosystem Project: Final report to Congress*, vol. II, chap. 58. Davis: University of California, Centers for Water and Wildland Resources.

DeSante, D. F. 1987. Landbird productivity in central coastal California: The relationship to annual rainfall, and a reproductive failure in 1986. *Condor* 89:636–53.

———. 1995. The status, distribution, abundance, population trends, demographics, and risks of the landbird avifauna of the Sierra Nevada mountains. Unpublished file report to Sierra Nevada Ecosystem Project, Davis, CA.

DeSante, D. F., and T. L. George. 1994. Population trends in the landbirds of western North America. In *A century of avifaunal change in western North America*, edited by J. R. Jehl Jr. and N. K. Johnson, 173–90. Studies in Avian Biology 15. Lawrence, KS: Allen Press.

DeSante, D. F., and G. R. Geupel. 1987. Landbird productivity in central coastal California: The relationship to annual rainfall, and a reproductive failure in 1986. *Condor* 89:636–53.

Fellers, G. 1995. Conversation with D. M. Graber, National Biological Service, Point Reyes National Seashore, California.

Foley, J. E., P. Foley, and S. Torres. 1995. *Mountain lion depredation in California, 1972–1992.* Unpublished file report, California Department of Fish and Game, Sacramento.

Franklin, J. F., and J. A. Fites-Kaufmann. 1996. Analysis of late successional forests. In *Sierra Nevada Ecosystem Project: Final report to Congress*, vol. II, chap. 21. Davis: University of California, Centers for Water and Wildland Resources.

Fry, W. 1924. The California grizzly. *Sequoia National Park, Historic Series, Nature Guide Service.* Bulletin 2, 4 December.

Gaines, D. 1977. *Birds of the Yosemite Sierra: A distributional survey.* Oakland: California Syllabus.

———. 1988. *Birds of Yosemite and the east slope.* Lee Vining, CA: Artemisia Press.

Garrison, B. A. 1994. Determining the biological significance of changes in predicted habitat values from the California Wildlife Habitat Relationships system. *California Fish and Game* 80:150–60.

———. 1996. Vertebrate wildlife species and habitat associations. In *Guidelines for managing California's hardwood rangelands*, edited by R. B. Standiford and P. Tinnin. Berkeley: University of California, Division of Agriculture and Natural Resources . In press.

Goldwasser, W., D. Gaines, and S. R. Wilbur. 1980. The least Bell's vireo in California: A de facto endangered race. *American Birds* 34:742–45.

Grayson, D. 1993. *The deserts past: A natural prehistory of the Great Basin.* Washington, DC: Smithsonian Institution Press.

Grinnell, J., J. S. Dixon, and J. M. Linsdale. 1937. *Fur-bearing mammals of California.* 2 vols. Berkeley and Los Angeles: University of California Press.

Grinnell, J., and T. I. Storer. 1924. *Animal life in the Yosemite.* Berkeley and Los Angeles: University of California Press.

Hayes, M. P., and M. R. Jennings. 1986. Decline of Ranid frog species in western North America: Are bullfrogs (*Rana catesbiana*) responsible? *Journal of Herpetology* 20:490–509.

Hayward, G. D., and J. Verner, eds. 1994. *Flammulated, arboreal, and great gray owls in the United States: A technical conservation assessment.* General Technical Report RM-253. Fort Collins, CO: U.S. Forest Service.

Hejl, S. J. 1994. Human-induced changes in bird populations in coniferous forests in western North America during the past 100 years. In *A century of avifaunal change in western North America*, edited by J. R. Jehl Jr. and N. K. Johnson, 232–46. Studies in Avian Biology 15. Lawrence, KS: Allen Press.

Hutto, R. L. 1988. Is tropical deforestation responsible for the reported decline in neotropical migrant populations? *American Birds* 42: 375–79.

Ingles, L. G. 1965. *Mammals of the Pacific states.* Stanford, CA: Stanford University Press.

Jameson, E. W., Jr., and H. J. Peeters. 1988. *California mammals.* Berkeley and Los Angeles: University of California Press.

Jennings, M. R., and M. P. Hayes. 1994. *Amphibian and reptile species of special concern in California.* Sacramento: California Department of Fish and Game.

Jones, F. L. 1950. A survey of the Sierra Nevada bighorn. *Sierra Club Bulletin* 35:29–76.

Kattelmann, R. 1996. Hydrology and water resources. In *Sierra Nevada Ecosystem Project: Final report to Congress,* vol. II, chap. 30. Davis: University of California, Centers for Water and Wildland Resources.

Koford, C. B. 1953. *The California condor.* Research Report 4. New York: National Audubon Society.

Laymon, S. A. 1987. Brown-headed cowbirds in California: Historical perspectives and management opportunities in riparian habitats. *Western Birds* 18:63–70.

———. 1995. Brown-headed cowbirds in the Sierra Nevada. Unpublished file report, Sierra Nevada Ecosystem Project, Davis, CA.

Marshall, J. T. 1988. Birds lost from a giant sequoia forest during fifty years. *Condor* 90:359–72.

Mayer, K. E., and W. F. Laudenslayer Jr., eds. 1988. *A guide to wildlife habitats in California.* Sacramento: California Department of Forestry and Fire Protection.

Menke, J., C. Davis, and P. Beesley. 1996. Rangeland assessment. In *Sierra Nevada Ecosystem Project: Final report to Congress,* vol. III. Davis: University of California, Centers for Water and Wildland Resources.

Minnich, R. A., M. G. Barbour, J. H. Burk, and R. F. Fernau. 1995. Sixty years of change in Californian conifer forests of the San Bernardino Mountains. *Conservation Biology* 9:902–14.

Moyle, P. B. 1973. Effects of introduced bullfrogs, *Rana catesbiana*, on the native frogs of the San Joaquin valley, California. *Copeia* 1973(1):18–22.

Moyle, P. B, and P. J. Randall. 1996. Biotic integrity of watersheds. In *Sierra Nevada Ecosystem Project: Final report to Congress,* vol. II, chap. 34. Davis: University of California, Centers for Water and Wildland Resources.

Nelson, R. D., and H. Salwasser. 1982. The Forest Service wildlife and fish habitat relationships program. *Transactions of the Northern American Wildlife and Natural Resources Conference* 47:174–83.

Ohmart, R. D. 1994. The effects of human-induced changes on the avifauna of western riparian habitats. In *A century of avifaunal change in western North America*, edited by J. R. Jehl Jr. and N. K. Johnson, 273–85. Studies in Avian Biology 15. Lawrence, KS: Allen Press.

Pierson, E. D. 1995. Letter to D. M. Graber, National Biological Service, Sequoia and Kings Canyon Field Station, Three Rivers, California, 16 October.

Robbins, C. S., J. R. Sauer, R. S. Greenberg, and S. Droege. 1989. Population declines in North American birds that migrate to the neotropics. *Proceedings of the National Academy of Sciences (USA)* 86:7658–62.

Rothstein, S. I., J. Verner, and E. Stevens. 1980. Range expansion and diurnal changes in dispersion of the brown-headed cowbird in the Sierra Nevada. *Auk* 97:253–67.

———. 1984. Radio-tracking confirms a unique diurnal pattern of spatial occurrence in the parasitic brown-headed cowbird. *Ecology* 65:77–88.

Schempf, P. F., and M. White. 1977. *Status of six furbearer populations in the mountains of northern California.* San Francisco: U.S. Forest Service.

Schmidt, R. H. 1991. Gray wolves in California: Their presence and absence. *California Fish and Game* 77(2):79–85.

Scott, J. M., F. Davis, B. Csuti, R. Noss, B. Butterfield, C. Groves, H. Anderson, S. Caicco, F. E'Erchia, T. C. Edwards Jr., J. Ulliman, and

R. G. Wright. 1993. Gap analysis: A geographic approach to protection of biological diversity. *Wildlife Monographs* 123:1–41.

Small, A. 1994. *California birds: Their status and distribution.* Vista, CA: Ibis Publishing.

Steinhart, P. 1990. California's wild heritage: *Threatened and endangered animals in the golden state.* San Francisco: Sierra Club Books.

Storer, T. I., and L. P. Tevis. 1955. California grizzly. Berkeley and Los Angeles: University of California Press.

Terborgh. J. 1989. *Where have all the birds gone? Essays on the biology and conservation of birds that migrate to the American tropics.* Princeton, NJ: Princeton University Press.

Thelander, C. G., ed. 1994. *Life on the edge.* Santa Cruz, CA: Biosystems Books.

Thomas, J. W., technical ed. 1979. *Wildlife habitats in managed forests: The Blue Mountains of Oregon and Washington.* Agricultural Handbook 553. Washington, DC: U.S. Forest Service.

Timossi, I., A. Sweet, M. Dedon, and R. H. Barrett. 1994. *User's manual for the California Wildlife Habitat Relationship microcomputer database,* version 5.0. Sacramento: California Department of Fish and Game.

Timossi, I. C., E. L. Woodard, and R. Barrett. 1994a. Habitat suitability models for use with Arc/Info: Blue grouse. Contribution to the Wildlife Habitat Relationships system. Unpublished file report, California Department of Fish and Game, Sacramento.

———. 1994b. Habitat suitability models for use with Arc/Info: Pileated woodpecker. Contribution to the Wildlife Habitat Relationships system. Unpublished file report, California Department of Fish and Game, Sacramento.

Verner, J., and A. S. Boss. 1980. California wildlife and their habitats: Western Sierra Nevada. General Technical Report PSW-37. Berkeley, CA: U.S. Forest Service, Pacific Southwest Forest and Range Experiment Station.

Verner, J., K. S. McKelvey, B. R. Noon, R. J. Gutierrez, and G. I. Gould, technical coordinators. 1992. *The California spotted owl: A technical assessment of its current status.* General Technical Report PSW-GTR-133. Albany, CA: U.S. Forest Service.

Verner, J., and L. V. Ritter. 1983. Current status of the brown-headed cowbird in the Sierra National Forest. *Auk* 100:355–68.

Wagner, F. H. 1989. Grazers, past and present. In *Grassland structure and function: California annual grassland,* edited by L. F. Huenneke and H. A. Mooney. Boston: Kluwer Academic.

Wehausen, J. D. 1980. *Sierra Nevada bighorn sheep: History and population ecology.* Ph.D. diss., University of Michigan, Ann Arbor.

———. 1988. The historical distribution of mountain sheep in the Owens Valley region. In *Mountains to desert: Selected Inyo readings,* 97–105. Independence, CA: Friends of the Eastern California Museum.

———. 1995. Letter to David Graber, National Biological Service, Sequoia and Kings Canyon National Parks, 6 October.

Wilbur, S. R. 1978. *The California condor, 1966–1976: A look at its past and future.* North American Fauna 72. Washington, DC: U.S. Department of the Interior, Fish and Wildlife Service.

Wood, G. W., and R. H. Barrett. 1978. Status of wild pigs in the United States. *Wildlife Society Bulletin* 7:237–46

Zeiner, D. C., W. F. Laudenslayer Jr., and K. E. Mayer, compiling eds. 1988. *California's wildlife,* vol. I, *Amphibians and reptiles.* Sacramento: California Department of Fish and Game.

———. 1990a. *California's wildlife,* vol. II, *Birds.* Sacramento: California Department of Fish and Game.

———. 1990b. *California's wildlife,* vol. III, *Mammals.* Sacramento: California Department of Fish and Game.

Zielinski, W. J., T. E. Kucera, and R. H. Barrett. 1996a. The current distribution of fishers (*Martes pennanti*) in California. *California Fish and Game,* in press.

———. 1996b. The current distribution of martens (*Martes americana*) in California. *California Fish and Game,* in press.

APPENDIX 25.1

Sierra Nevada Vertebrate Species

WHR Code[a]	Common Name[b]	Scientific Name[c]	Sierra Nevada Use[d]	LSOG Habitat[e]	Western Foothills Habitat[f]	Riparian Habitat[g]	Risk[h]	Native[i]
A001	California tiger salamander	Ambystoma californiense	3	3	1	1	2	T
A003	Long toed salamander	Ambystoma macrodactylum	2	3	3	1	2	T
A006	Rough-skinned newt	Taricha granulosa	3	2	2	1	3	T
A007	California newt	Taricha torosa (sierrae)	2	2	1	1	3	T
A012	Ensatina	Ensatina eschscholtzi	2	2	3	3	3	T
A014	California slender salamander	Batrachoseps attenuatus	2	2	2	3	3	T
A015	Black-bellied slender salamander	Batrachoseps nigriventris	2	2	2	3	3	T
A016	Pacific slender salamander	Batrachoseps pacificus	2	3	2	3	3	T
A017	Kern canyon slender salamander	Batrachoseps simatus	1	3	1	3	2	T
A017	Relictual slender salamander	Batrachoseps relictus	1	3	2	2	2	T
A020	Black salamander	Aneides flavipunctatus	3	2	1	2	3	T
A022	Arboreal salamander	Aneides lugubris	2	2	2	2	3	T
A023	Mount lyell salamander	Hydromantes platycephalus	1	3	3	3	2	T
A023	Owens valley web-toed salamander	Hydromantes sp.	1	3	3	1	2	T
A025	Limestone salamander	Hydromantes brunus	1	3	3	3	2	T
A028	Western spadefoot	Scaphiopus hammondi	3	3	1	1	2	T
A029	Great basin spadefoot	Scaphiopus intermontanus	3	3	3	1	3	T
A032	Western toad	Bufo boreas	2	3	2	2	3	T
A033	Yosemite toad	Bufo canorus	1	3	3	1	2	T
A039	Pacific treefrog	Pseudacris regilla	2	3	2	3	3	T
A040	Red-legged frog	Rana aurora draytonii	2	3	1	1	2	T
A043	Foothill yellow-legged frog	Rana boylei	2	3	1	1	2	T
A044	Mountain yellow-legged frog	Rana muscosa	2	3	3	1	2	T
A045	Northern leopard frog	Rana pipiens	3	3	3	1	2	T
A046	Bullfrog	Rana catesbeiana	3	3	2	1	3	F
B003	Common loon	Gavia immer	3	3	2	3	2	T
B006	Pied-billed grebe	Podilymbus podiceps	3	3	2	2	3	T
B007	Horned grebe	Podiceps auritus	3	3	3	3	3	T
B009	Eared grebe	Podiceps nigricollis	2	3	2	3	3	T
B010	Western grebe	Aechmophorus occidentalis	3	3	2	3	3	T
B010	Clark's grebe	Aechmophorus clarkii	3	3	2	2	3	T
B042	American white pelican	Pelecanus erythrorhynchos	3	3	3	3	2	T

[a]CWHR code is the California Wildlife Habitat Relationships System code.
[b]Common name is the CWHR appellation.
[c]Scientific name is the CWHR appelation
[d]Sierra Nevada use:
    1 indicates that all of species range in California is in Sierra Nevada.
    2 indicates that principal range includes Sierra Nevada.
    3 indicates that Sierra Nevada is peripheral range only.
[e]LSOG (late succesional and old-growth) habitat:
    1 indicates that population viability in Sierra requires LSOG habitats.
    2 indicates that species uses LSOG but is not dependent upon it.
    3 indicates that species does not use LSOG habitat significantly.
[f]Western foothills habitat:
    1 indicates that population viability (in Sierra) requires western foothills habitat.
    2 indicates that species uses western foothills habitat but is not dependent upon it.

    3 indicates that species does not use western foothills habitat significantly.
[g]Riparian habitat (including lakeshores and wet meadows):
    1 indicates that population viability requires riparian habitat.
    2 indicates that species uses riparian habitat but is not dependent upon it.
    3 indicates that species does not use riparian habitat significantly.
[h]Risk:
    1 indicates extirpated in Sierra Nevada.
    2 indicates on state or federal list as endangered, threatened, or special concern either for species as a whole, or Sierra portion.
    3 indicates not known to be at risk.
[i]Native:
    T indicates native to Sierra Nevada.
    F indicates non-native.

WHR Code[a]	Common Name[b]	Scientific Name[c]	Sierra Nevada Use[d]	LSOG Habitat[e]	Western Foothills Habitat[f]	Riparian Habitat[g]	Risk[h]	Native[i]
B049	American bittern	Botaurus lentiginosus	3	3	2	1	3	T
B051	Great blue heron	Ardea herodias	2	3	2	1	3	T
B053	Snowy egret	Egretta thula	3	3	2	1	3	T
B058	Green-backed heron	Butorides striatus	3	3	1	1	3	T
B059	Black-crowned night heron	Nycticorax nycticorax	3	3	1	1	3	T
B062	White-faced ibis	Plegadis chihi	3	3	3	1	2	T
B067	Tundra swan	Cygnus columbianus	3	3	2	1	3	T
B075	Canada goose	Branta canadensis	3	3	2	1	3	T
B076	Wood duck	Aix sponsa	3	3	2	1	3	T
B077	Green-winged teal	Anas crecca	3	3	1	1	3	T
B079	Mallard	Anas platyrhynchos	2	3	2	1	3	T
B083	Cinnamon teal	Anas cyanoptera	3	3	1	2	3	T
B084	Northern shoveler	Anas clypeata	3	3	1	1	3	T
B085	Gadwall	Anas strepera	3	3	3	2	3	T
B087	American wigeon	Anas americana	3	3	1	1	3	T
B089	Canvasback	Aythya valisineria	3	3	1	3	3	T
B090	Redhead	Aythya americana	3	3	3	3	3	T
B091	Ring-necked duck	Aythya collaris	2	3	2	2	3	T
B094	Lesser scaup	Aythya affinis	2	3	1	3	3	T
B096	Harlequin duck	Histrionicus histrionicus	3	3	3	1	2	T
B101	Common goldeneye	Bucephala clangula	2	3	3	2	3	T
B102	Barrow's goldeneye	Bucephala islandica	3	3	3	1	2	T
B103	Bufflehead	Bucephala albeola	3	3	2	2	3	T
B105	Common merganser	Mergus merganser	2	3	2	1	3	T
B107	Ruddy duck	Oxyura jamaicensis	3	3	2	1	3	T
B108	Turkey vulture	Cathartes aura	2	3	1	3	3	T
B109	California condor	Gymnogyps californianus	2	3	2	3	1	T
B110	Osprey	Pandion haliaetus	2	3	2	1	2	T
B111	White-tailed kite	Elanus caeruleus	3	3	1	3	3	T
B113	Bald eagle	Haliaeetus leucocephalus	3	3	2	1	2	T
B114	Northern harrier	Circus cyaneus	3	3	2	2	2	T
B115	Sharp-shinned hawk	Accipiter striatus	2	2	2	2	2	T
B116	Cooper's hawk	Accipiter cooperii	2	3	2	2	2	T
B117	Northern goshawk	Accipiter gentilis	2	1	3	2	2	T
B119	Red-shouldered hawk	Buteo lineatus	2	3	1	1	3	T
B123	Red-tailed hawk	Buteo jamaicensis	2	3	2	3	3	T
B124	Ferruginous hawk	Buteo regalis	3	3	3	3	2	T
B125	Rough-legged hawk	Buteo lagopus	3	3	3	3	3	T
B126	Golden eagle	Aquila chrysaetos	2	3	2	3	2	T
B127	American kestrel	Falco sparverius	2	3	1	3	3	T
B128	Merlin	Falco columbarius	3	3	2	2	2	T
B129	Peregrine falcon	Falco peregrinus	2	3	2	3	2	T
B131	Prairie falcon	Falco mexicanus	2	3	2	3	2	T
B132	Chukar	Alectoris chukar	3	3		3	3	F
B134	Blue grouse	Dendragapus obscurus	2	2	3	3	3	T
B135	White-tailed ptarmigan	Lagopus leucurus	1	3	3	3	3	F
B137	Sage grouse	Centrocercus urophasianus	3	3	3	3	2	T
B138	Turkey	Meleagris gallopavo	2	3	2	2	3	F
B140	California quail	Callipepla californica	2	3	2	3	3	T
B141	Mountain quail	Oreortyx pictus	2	3	2	3	3	T
B145	Virginia rail	Rallus limicola	2	3	2	1	3	T
B146	Sora	Porzana carolina	3	3	2	1	3	T
B148	Common moorhen	Gallinula chloropus	3	3	1	1	3	T
B149	American coot	Fulica americana	3	3	2	1	3	T
B158	Killdeer	Charadrius vociferus	3	3	2	2	3	T
B164	American avocet	Recurvirostra americana	3	3	3	1	3	T
B168	Willet	Catoptrophorus semi-palmatus	3	3	3	1	3	T
B170	Spotted sandpiper	Actitis macularia	2	3	2	1	3	T

[a]CWHR code is the California Wildlife Habitat Relationships System code.
[b]Common name is the CWHR appellation.
[c]Scientific name is the CWHR appelation
[d]Sierra Nevada use:
    1 indicates that all of species range in California is in Sierra Nevada.
    2 indicates that principal range includes Sierra Nevada.
    3 indicates that Sierra Nevada is peripheral range only.
[e]LSOG (late succesional and old-growth) habitat:
    1 indicates that population viability in Sierra requires LSOG habitats.
    2 indicates that species uses LSOG but is not dependent upon it.
    3 indicates that species does not use LSOG habitat significantly.
[f]Western foothills habitat:
    1 indicates that population viability (in Sierra) requires western foothills habitat.
    2 indicates that species uses western foothills habitat but is not dependent upon it.

    3 indicates that species does not use western foothills habitat significantly.
[g]Riparian habitat (including lakeshores and wet meadows):
    1 indicates that population viability requires riparian habitat.
    2 indicates that species uses riparian habitat but is not dependent upon it.
    3 indicates that species does not use riparian habitat significantly.
[h]Risk:
    1 indicates extirpated in Sierra Nevada.
    2 indicates on state or federal list as endangered, threatened, or special concern either for species as a whole, or Sierra portion.
    3 indicates not known to be at risk.
[i]Native:
    T indicates native to Sierra Nevada.
    F indicates non-native.

WHR Code[a]	Common Name[b]	Scientific Name[c]	Sierra Nevada Use[d]	LSOG Habitat[e]	Western Foothills Habitat[f]	Riparian Habitat[g]	Risk[h]	Native[i]
B185	Least sandpiper	Calidris minutilla	2	3	2	1	3	T
B199	Common snipe	Gallinago gallinago	2	3	2	1	3	T
B200	Wilson's phalarope	Phalaropus tricolor	3	3	2	3	3	T
B214	Ring-billed gull	Larus delawarensis	3	3	2	3	3	T
B215	California gull	Larus californicus	3	3	2	3	2	T
B216	Herring gull	Larus argentatus	3	3		2	3	T
B227	Caspian tern	Sterna caspia	3	3	3	2	3	T
B233	Forster's tern	Sterna forsteri	3	3	3	1	3	T
B235	Black tern	Chlidonias niger	3	3	3	1	3	T
B250	Rock dove	Columba livia	2	3	2	3	3	F
B251	Band-tailed pigeon	Columba fasciata	2	2	2	3	3	T
B255	Mourning dove	Zenaida macroura	2	3	1	3	3	T
B260	Greater roadrunner	Geococcyx californianus	3	3	1	3	3	T
B262	Common barn owl	Tyto alba	2	3	1	3	3	T
B263	Flammulated owl	Otus flammeolus	2	2	3	3	3	T
B264	Western screech owl	Otus kennicottii	2	3	2	2	3	T
B265	Great horned owl	Bubo virginianus	2	3	2	3	3	T
B267	Northern pygmy owl	Glaucidium gnoma	2	2	2	2	3	T
B269	Burrowing owl	Athene cunicularia	3	3	1	3	2	T
B270	Spotted owl	Strix occidentalis	2	1	2	2	2	T
B271	Great gray owl	Strix nebulosa	2	1	3	2	2	T
B272	Long-eared owl	Asio otus	2	2	1	2	2	T
B273	Short-eared owl	Asio flammeus	2	3	1	3	2	T
B274	Northern saw-whet owl	Aegolius acadicus	2	2	2	3	3	T
B276	Common nighthawk	Chordeiles minor	2	3	2	2	3	T
B277	Common poorwill	Phalaenoptilus nuttallii	2	3	2	3	3	T
B279	Black swift	Cypseloides niger	2	3	2	3	2	T
B281	Vaux's swift	Chaetura vauxi	2	1	2	3	3	T
B282	White-throated swift	Aeronautes saxatalis	2	3	2	3	3	T
B286	Black-chinned hummingbird	Archilochus alexandri	2	3	1	3	3	T
B287	Anna's hummingbird	Calypte anna	2	3	2	3	3	T
B289	Calliope hummingbird	Stellula calliope	2	3	2	2	3	T
B290	Broad-tailed hummingbird	Selasphorus platycercus	3	3	3	2	3	T
B291	Rufous hummingbird	Selasphorus rufus	3	3	2	2	3	T
B292	Allen's hummingbird	Selasphorus sasin	3	3	2	3	3	T
B293	Belted kingfisher	Ceryle alcyon	2	3	2	1	3	T
B294	Lewis' woodpecker	Melanerpes lewis	2	3	2	3	3	T
B296	Acorn woodpecker	Melanerpes formicivorus	2	3	1	3	3	T
B298	Red-naped sapsucker	Sphyrapicus nuchalis	3	2	3	1	3	T
B299	Red-breasted sapsucker	Sphyrapicus ruber	2	2	2	2	3	T
B300	Williamson's sapsucker	Sphyrapicus thyroideus	2	2	3	3	3	T
B302	Nuttall's woodpecker	Picoides nuttallii	2	3	1	2	3	T
B303	Downy woodpecker	Picoides pubescens	2	3	2	1	3	T
B304	Hairy woodpecker	Picoides villosus	2	2	2	2	3	T
B305	White-headed woodpecker	Picoides albolarvatus	2	1	3	3	3	T
B306	Black-backed woodpecker	Picoides arcticus	2	2	3	3	3	T
B307	Northern flicker	Colaptes auratus	2	3	2	3	3	T
B308	Pileated woodpecker	Dryocopus pileatus	2	1	3	3	3	T
B309	Olive-sided flycatcher	Contopus borealis	2	2	2	3	2	T
B311	Western wood-pewee	Contopus sordioulus	2	3	2	2	3	T
B315	Willow flycatcher	Empidonax traillii	2	2	2	1	2	T
B317	Hammonds' flycatcher	Empidonax hammondii	2	2	2	3	3	T
B318	Dusky flycatcher	Empidonax oberholseri	2	3	2	3	3	T
B319	Gray flycatcher	Empidonax wrightii	3	3	3	3	3	T
B320	Cordilleran flycatcher	Empidonax difficilis	3	3	2	3	3	T
B320	Pacific-slope flycatcher	Empidonax occidentalis	2	3	2	2	3	T
B321	Black phoebe	Sayornis nigricans	2	3	2	1	3	T
B323	Say's phoebe	Sayornis saya	3	3	2	3	3	T

continued

aCWHR code is the California Wildlife Habitat Relationships System code.
bCommon name is the CWHR appellation.
cScientific name is the CWHR appelation
dSierra Nevada use:
  1 indicates that all of species range in California is in Sierra Nevada.
  2 indicates that principal range includes Sierra Nevada.
  3 indicates that Sierra Nevada is peripheral range only.
eLSOG (late succesional and old-growth) habitat:
  1 indicates that population viability in Sierra requires LSOG habitats.
  2 indicates that species uses LSOG but is not dependent upon it.
  3 indicates that species does not use LSOG habitat significantly.
fWestern foothills habitat:
  1 indicates that population viability (in Sierra) requires western foothills habitat.
  2 indicates that species uses western foothills habitat but is not dependent upon it.

  3 indicates that species does not use western foothills habitat significantly.
gRiparian habitat (including lakeshores and wet meadows):
  1 indicates that population viability requires riparian habitat.
  2 indicates that species uses riparian habitat but is not dependent upon it.
  3 indicates that species does not use riparian habitat significantly.
hRisk:
  1 indicates extirpated in Sierra Nevada.
  2 indicates on state or federal list as endangered, threatened, or special concern either for species as a whole, or Sierra portion.
  3 indicates not known to be at risk.
iNative:
  T indicates native to Sierra Nevada.
  F indicates non-native.

WHR Code[a]	Common Name[b]	Scientific Name[c]	Sierra Nevada Use[d]	LSOG Habitat[e]	Western Foothills Habitat[f]	Riparian Habitat[g]	Risk[h]	Native[i]
B326	Ash-throated flycatcher	Myiarchus cinerascens	2	3	2	3	3	T
B333	Western kingbird	Tyrannus verticalis	2	3	2	3	3	T
B337	Horned lark	Eremophila alpestris	2	3	2	3	3	T
B338	Purple martin	Progne subis	2	3	2	3	2	T
B339	Tree swallow	Tachycineta bicolor	2	2	2	1	3	T
B340	Violet-green swallow	Tachycineta thalassina	2	3	2	3	3	T
B341	Northern rough-winged swallow	Stelgidopteryx serripennis	2	3	2	2	3	T
B343	Cliff swallow	Hirundo pyrrhonota	2	3	2	2	3	T
B344	Barn swallow	Hirundo rustica	2	3	2	2	3	T
B346	Steller's jay	Cyanocitta stelleri	2	2	3	3	3	T
B348	Scrub jay	Aphelocoma coerulescens	2	3	1	3	3	T
B349	Pinyon jay	Gymnorhinus cyanocephalus	3	3	2	3	3	T
B350	Clark's nutcracker	Nucifraga columbiana	2	3	3	3	3	T
B351	Black-billed magpie	Pica pica	3	3	3	1	3	T
B352	Yellow-billed magpie	Pica nuttalli	3	3	1	2	3	T
B353	American crow	Corvus brachyrhvnchos	3	3	2	3	3	T
B354	Common raven	Corvus corax	2	3	2	3	3	T
B356	Mountain chickadee	Parus gambeli	2	3	3	3	3	T
B357	Chestnut-backed chickadee	Parus rufescens	2	2	3	3	3	T
B358	Plain titmouse	Parus inornatus	2	3	1	3	3	T
B360	Bushtit	Psaltriparus minimus	2	3	1	3	3	T
B361	Red-breasted nuthatch	Sitta canadensis	2	1	3	3	3	T
B362	White-breasted nuthatch	Sitta carolinensis	2	2	2	3	3	T
B363	Pygmy nuthatch	Sitta pygmaea	2	1	3	3	3	T
B364	Brown creeper	Certhia americana	2	1	3	2	3	T
B366	Rock wren	Salpinctes obsoletus	2	3	1	3	3	T
B367	Canyon wren	Catherpes mexicanus	2	3	2	2	3	T
B368	Bewick's wren	Thryomanes bewickii	2	3	1	3	3	T
B369	House wren	Troglodytes aedon	2	3	2	2	3	T
B370	Winter wren	Troglodytes troglodytes	2	1	3	1	3	T
B372	Marsh wren	Cistothorus palustris	2	3	1	1	3	T
B373	American dipper	Cinclus mexicanus	2	3	2	1	3	T
B375	Golden-crowned kinglet	Regulus satrapa	2	2	3	2	3	T
B376	Ruby-crowned kinglet	Regulus calendula	2	3	2	3	3	T
B377	Blue-gray gnatcatcher	Polioptila caerulea	2	3	1	3	3	T
B380	Western bluebird	Sialia mexicana	2	3	2	3	3	T
B381	Mountain bluebird	Sialia currucoides	2	3	3	3	3	T
B382	Townsend's solitaire	Myadestes townsendi	2	3	2	3	3	T
B385	Swainson's thrush	Catharus ustulatus	2	3	2	2	3	T
B386	Hermit thrush	Catharus guttatus	2	2	2	2	3	T
B389	American robin	Turdus migratorius	2	3	2	2	3	T
B390	Varied thrush	Ixoreus naevius	2	2	2	3	3	T
B391	Wrentit	Chamaea fasciata	2	3	1	3	3	T
B393	Northern mockingbird	Mimus polyglottos	3	3	1	2	3	T
B394	Sage thrasher	Oreoscoptes montanus	3	3	3	3	3	T
B398	California thrasher	Toxostoma redivivum	2	3	1	2	2	T
B404	American pipit	Anthus rubescens	2	3	2	3	3	T
B407	Cedar waxwing	Bombycilla cedrorum	2	3	2	3	3	T
B408	Phainopepla	Phainopepla nitens	2	3	1	3	3	T
B409	Northern shrike	Lanius excubitor	3	3	3	3	3	T
B410	Loggerhead shrike	Lanius ludovicianus	2	3	1	3	3	T
B411	European starling	Sturnus vulgaris	2	3	1	3	3	F
B413	Bell's vireo	Vireo bellii	2	3	1	1	1	T
B415	Solitary vireo	Vireo solitarius	2	3	2	3	3	T
B417	Hutton's vireo	Vireo huttoni	2	3	1	3	3	T
B418	Warbling vireo	Vireo gilvus	2	3	2	2	3	T
B425	Orange-crowned warbler	Vermivora celata	2	3	2	2	3	T

[a]CWHR code is the California Wildlife Habitat Relationships System code.
[b]Common name is the CWHR appellation.
[c]Scientific name is the CWHR appelation
[d]Sierra Nevada use:
 1 indicates that all of species range in California is in Sierra Nevada.
 2 indicates that principal range includes Sierra Nevada.
 3 indicates that Sierra Nevada is peripheral range only.
[e]LSOG (late succesional and old-growth) habitat:
 1 indicates that population viability in Sierra requires LSOG habitats.
 2 indicates that species uses LSOG but is not dependent upon it.
 3 indicates that species does not use LSOG habitat significantly.
[f]Western foothills habitat:
 1 indicates that population viability (in Sierra) requires western foothills habitat.
 2 indicates that species uses western foothills habitat but is not dependent upon it.

 3 indicates that species does not use western foothills habitat significantly.
[g]Riparian habitat (including lakeshores and wet meadows):
 1 indicates that population viability requires riparian habitat.
 2 indicates that species uses riparian habitat but is not dependent upon it.
 3 indicates that species does not use riparian habitat significantly.
[h]Risk:
 1 indicates extirpated in Sierra Nevada.
 2 indicates on state or federal list as endangered, threatened, or special concern either for species as a whole, or Sierra portion.
 3 indicates not known to be at risk.
[i]Native:
 T indicates native to Sierra Nevada.
 F indicates non-native.

WHR Code[a]	Common Name[b]	Scientific Name[c]	Sierra Nevada Use[d]	LSOG Habitat[e]	Western Foothills Habitat[f]	Riparian Habitat[g]	Risk[h]	Native[i]
B426	Nashville warbler	Vermivora ruficapilla	2	3	2	3	3	T
B427	Virginia's warbler	Vermivora virginiae	2	3	3	3	3	T
B430	Yellow warbler	Dendroica petechia	2	2	2	2	2	T
B435	Yellow-rumped warbler	Dendroica coronata	2	3	2	3	3	T
B436	Black-throated gray warbler	Dendroica nigrescens	2	3	2	3	3	T
B438	Hermit warbler	Dendroica occidentalis	2	1	2	3	3	T
B460	Macgillivray's warbler	Oporornis tolmiei	2	3	2	1	3	T
B461	Common yellowthroat	Geothlypis trichas	2	3	2	1	2	T
B463	Wilson's warbler	Wilsonia pusilla	2	3	2	1	3	T
B467	Yellow-breasted chat	Icteria virens	2	3	1	1	2	T
B471	Western tanager	Piranga ludoviciana	2	3	2	3	3	T
B475	Black-headed grosbeak	Pheucticus melanocephalus	2	3	2	3	3	T
B476	Blue grosbeak	Guiraca caerulea	3	3	1	1	3	T
B477	Lazuli bunting	Passerina amoena	2	3	2	2	3	T
B482	Green-tailed towhee	Pipilo chlorurus	2	3	3	3	3	T
B483	Rufous-sided towhee	Pipilo erythrophthalmus	2	3	2	3	3	T
B484	California towhee	Pipilo crissalis	2	3	1	3	3	T
B487	Rufous-crowned sparrow	Aimophila ruficeps	2	3	1	3	3	T
B489	Chipping sparrow	Spizella passerina	2	3	2	3	3	T
B491	Brewer's sparrow	Spizella breweri	2	3	3	3	3	T
B493	Black-chinned sparrow	Spizella atrogularis	2	3	3	3	3	T
B494	Vesper sparrow	Pooecetes gramineus	3	3	1	3	3	T
B495	Lark sparrow	Chondestes grammacus	2	3	1	3	3	T
B496	Black-throated sparrow	Amphispiza bilineata	3	3	1	3	3	T
B497	Sage sparrow	Amphispiza belli	2	3	1	3	3	T
B499	Savannah sparrow	Passerculus sandwichensis	3	3	1	2	3	T
B501	Grasshopper sparrow	Ammooramus savannarum	2	3	1	3	3	T
B504	Fox sparrow	Passerella iliaca	2	3	2	3	3	T
B505	Song sparrow	Melospiza melodia	2	3	2	1	3	T
B506	Lincoln's sparrow	Melospiza lincolnii	2	3	2	1	3	T
B509	Golden-crowned sparrow	Zonotrichia atricapilla	2	3	2	2	3	T
B510	White-crowned sparrow	Zonotrichia leucophrys	2	3	2	1	3	T
B512	Dark-eyed junco	Junco hyemalis	2	3	2	3	3	T
B519	Red-winged blackbird	Agelaius phoeniceus	2	3	2	1	3	T
B520	Tricolored blackbird	Agelaius tricolor	3	3	1	1	2	T
B521	Western meadowlark	Sturnella neglecta	2	3	1	3	3	T
B522	Yellow-headed blackbird	Xanthocephalus xanthocephalus	3	3	1	1	3	T
B524	Brewer's blackbird	Euphagus cyanocephalus	2	3	2	3	3	T
B528	Brown-headed cowbird	Molothrus ater	2	3	2	2	3	T
B532	Northern oriole	Icterus galbula	2	3	1	2	3	T
B534	Rosy finch	Leucosticte arctoa	2	3	3	3	3	T
B535	Pine grosbeak	Pinicola enucleator	1	3	3	1	3	T
B536	Purple finch	Carpodacus purpureus	2	1	2	2	3	T
B537	Cassin's finch	Carpodacus cassinii	2	1	3	2	3	T
B538	House finch	Carpodacus mexicanus	2	3	1	3	3	T
B539	Red crossbill	Loxia curvirostra	2	3	3	3	3	T
B542	Pine siskin	Carduelis pinus	2	3	2	3	3	T
B543	Lesser goldfinch	Carduelis psaltria	2	3	2	1	3	T
B544	Lawrence's goldfinch	Carduelis lawrencei	3	3	2	2	3	T
B545	American goldfinch	Carduelis tristis	2	3	1	2	3	T
B546	Evening grosbeak	Coccothraustes vespertinus	2	1	2	3	3	T
B547	House sparrow	Passer domesticus	2	3	2	3	3	F
M001	Virginia opossum	Didelphis virginiana	2	3	1	2	3	F
M002	Mt. Lyell shrew	Sorex lyelli	1	3	3	1	2	T
M003	Vagrant shrew	Sorex vagrans	2	2	3	1	3	T
M004	Dusky shrew	Sorex monticolus	2	2	3	1	3	T
M006	Ornate shrew	Sorex ornatus	2	2	1	1	3	T
M008	Inyo shrew	Sorex tenellus	2	3	3	2	3	T

continued

[a]CWHR code is the California Wildlife Habitat Relationships System code.
[b]Common name is the CWHR appellation.
[c]Scientific name is the CWHR appelation
[d]Sierra Nevada use:
   1 indicates that all of species range in California is in Sierra Nevada.
   2 indicates that principal range includes Sierra Nevada.
   3 indicates that Sierra Nevada is peripheral range only.
[e]LSOG (late succesional and old-growth) habitat:
   1 indicates that population viability in Sierra requires LSOG habitats.
   2 indicates that species uses LSOG but is not dependent upon it.
   3 indicates that species does not use LSOG habitat significantly.
[f]Western foothills habitat:
   1 indicates that population viability (in Sierra) requires western foothills habitat.
   2 indicates that species uses western foothills habitat but is not dependent upon it.

   3 indicates that species does not use western foothills habitat significantly.
[g]Riparian habitat (including lakeshores and wet meadows):
   1 indicates that population viability requires riparian habitat.
   2 indicates that species uses riparian habitat but is not dependent upon it.
   3 indicates that species does not use riparian habitat significantly.
[h]Risk:
   1 indicates extirpated in Sierra Nevada.
   2 indicates on state or federal list as endangered, threatened, or special concern either for species as a whole, or Sierra portion.
   3 indicates not known to be at risk.
[i]Native:
   T indicates native to Sierra Nevada.
   F indicates non-native.

WHR Code[a]	Common Name[b]	Scientific Name[c]	Sierra Nevada Use[d]	LSOG Habitat[e]	Western Foothills Habitat[f]	Riparian Habitat[g]	Risk[h]	Native[i]
M010	Water shrew	Sorex palustris	2	2	3	1	3	T
M012	Trowbridge's shrew	Sorex trowbridgii	2	2	3	3	3	T
M018	Broad-footed mole	Scapanus latimanus	2	3	2	2	3	T
M021	Little brown myotis	Myotis lucifugus	2	2	3	2	3	T
M023	Yuma myotis	Myotis yumanensis	2	2	2	2	2	T
M025	Long-eared myotis	Myotis evotis	2	3	3	2	2	T
M026	Fringed myotis	Myotis thysanodes	2	3	3	2	2	T
M027	Long-legged myotis	Myotis volans	2	2	2	3	2	T
M028	California myotis	Myotis californicus	2	3	2	3	3	T
M029	Western small-footed myotis	Myotis ciliolabrum	2	3	2	2	2	T
M030	Silver-haired bat	Lasionycteris noctivagans	2	2	2	2	3	T
M031	Western pipistrelle	Pipistrellus hesperus	2	3	1	2	3	T
M032	Big brown bat	Eptesicus fuscus	2	3	2	2	3	T
M033	Western red bat	Lasiurus blossevillii	2	3	1	2	3	T
M034	Hoary bat	Lasiurus cinereus	2	2	2	3	3	T
M036	Spotted bat	Euderma maculatum	2	3	2	2	2	T
M037	Townsend's big-eared bat	Plecotus townsendii	2	3	2	2	2	T
M038	Pallid bat	Antrozous pallidus	2	3	1	3	2	T
M039	Brazilian free-tailed bat	Tadarida brasiliensis	2	3	1	3	3	T
M042	Western mastiff bat	Eumops perotis	2	3	2	3	2	T
M043	Pika	Ochotona princeps	2	3	3	3	3	T
M045	Brush rabbit	Sylvilagus bachmani	2	3	1	3	3	T
M046	Nuttall's cottontail	Sylvilagus nuttallii	2	3	3	3	3	T
M047	Desert cottontail	Sylvilagus audubonii	3	3	1	3	3	T
M049	Snowshoe hare	Lepus americanus	2	3	3	3	2	T
M050	White-tailed hare	Lepus townsendii	2	3	3	3	2	T
M051	Black-tailed hare	Lepus californicus	2	3	2	3	3	T
M052	Mountain beaver	Aplodontia rufa	2	2	3	1	3	T
M053	Alpine chipmunk	Tamias alpinus	1	3	3	3	3	T
M054	Least chipmunk	Tamias minimus	2	3	3	3	3	T
M055	Yellow-pine chipmunk	Tamias amoenus	2	3	3	3	3	T
M057	Allen's chipmunk	Tamias senex	2	3	3	3	3	T
M060	Merriam's chipmunk	Tamias merriami	2	3	2	3	3	T
M062	Long-eared chipmunk	Tamias quadrimaculatus	1	3	3	3	3	T
M063	Lodgepole chipmunk	Tamias speciosus	2	3	3	3	3	T
M064	Panamint chipmunk	Tamias panamintinus	2	3	3	3	2	T
M065	Uinta chipmunk	Tamias umbrinus	2	3	3	3	3	T
M066	Yellow-bellied marmot	Marmota flaviventris	2	3	3	2	3	T
M067	White-tailed antelope squirrel	Ammospermophilus leucurus	3	3	3	3	3	T
M070	Belding's ground squirrel	Spermophilus beldingi	2	3	3	2	3	T
M072	California ground squirrel	Spermophilus beecheyi	2	3	2	3	3	T
M075	Golden-mantled ground squirrel	Spermophilus lateralis	2	3	3	3	3	T
M077	Western gray squirrel	Sciurus griseus	2	2	2	3	3	T
M079	Douglas' squirrel	Tamiasciurus douglasii	2	2	3	3	3	T
M080	Northern flying squirrel	Glaucomys sabrinus	2	1	3	3	3	T
M081	Botta's pocket gopher	Thomomys bottae	2	3	2	3	3	T
M083	Northern pocket gopher	Thomomys talpoides	2	3	3	2	3	T
M085	Mountain pocket gopher	Thomomys monticola	2	3	3	2	3	T
M086	Little pocket mouse	Perognathus longimembris	3	3	1	3	3	T
M088	Great basin pocket mouse	Perognathus parvus	3	3	3	3	3	T
M090	Yellow-eared pocket mouse	Perognathus xanthotus	1	3	2	3	3	T
M091	Long-tailed pocket mouse	Chaetodipus formosus	3	3	3	3	3	T
M095	California pocket mouse	Chaetodipus californicus	2	3	1	3	3	T
M097	Dark kangaroo mouse	Microdipodops megacephalus	3	3	3	3	3	T
M104	Heermann's kangaroo rat	Dipodomys heermanni	2	3	1	3	3	T

aCWHR code is the California Wildlife Habitat Relationships System code.
bCommon name is the CWHR appellation.
cScientific name is the CWHR appelation
dSierra Nevada use:
   1 indicates that all of species range in California is in Sierra Nevada.
   2 indicates that principal range includes Sierra Nevada.
   3 indicates that Sierra Nevada is peripheral range only.
eLSOG (late succesional and old-growth) habitat:
   1 indicates that population viability in Sierra requires LSOG habitats.
   2 indicates that species uses LSOG but is not dependent upon it.
   3 indicates that species does not use LSOG habitat significantly.
fWestern foothills habitat:
   1 indicates that population viability (in Sierra) requires western foothills habitat.
   2 indicates that species uses western foothills habitat but is not dependent upon it.

   3 indicates that species does not use western foothills habitat significantly.
gRiparian habitat (including lakeshores and wet meadows):
   1 indicates that population viability requires riparian habitat.
   2 indicates that species uses riparian habitat but is not dependent upon it.
   3 indicates that species does not use riparian habitat significantly.
hRisk:
   1 indicates extirpated in Sierra Nevada.
   2 indicates on state or federal list as endangered, threatened, or special concern either for species as a whole, or Sierra portion.
   3 indicates not known to be at risk.
iNative:
   T indicates native to Sierra Nevada.
   F indicates non-native.

WHR Code[a]	Common Name[b]	Scientific Name[c]	Sierra Nevada Use[d]	LSOG Habitat[e]	Western Foothills Habitat[f]	Riparian Habitat[g]	Risk[h]	Native[i]
M105	California kangaroo rat	Dipodomys californicus	2	3	1	3	3	T
M107	Panamint kangaroo rat	Dipodomys panamintinus	3	3	3	3	3	T
M109	Desert kangaroo rat	Dipodomys deserti	3	3	3	3	3	T
M110	Merriam's kangaroo rat	Dipodomys merriami	3	3	3	3	3	T
M112	Beaver	Castor canadensis	2	3	2	1	3	T
M113	Western harvest mouse	Reithrodontomys megalotis	2	3	2	2	3	T
M116	California mouse	Peromyscus californicus	2	3	2	3	3	T
M117	Deer mouse	Peromyscus maniculatus	2	3	2	2	3	T
M118	Canyon mouse	Peromyscus crinitus	3	3	3	3	3	T
M119	Brush mouse	Peromyscus boylii	2	3	2	3	3	T
M120	Pinyon mouse	Peromyscus truei	2	3	1	3	3	T
M121	Northern grasshopper mouse	Onychomys leucogaster	3	3	3	2	3	T
M122	Southern grasshopper mouse	Onychomys torridus	3	3	2	3	2	T
M126	Desert woodrat	Neotoma lepida	3	3	2	3	3	T
M127	Dusky-footed woodrat	Neotoma fuscipes	2	2	2	2	3	T
M128	Bushy-tailed woodrat	Neotoma cinerea	2	3	3	2	3	T
M129	Western red-backed vole	Clethrionomys californicus	3	1	3	2	3	T
M130	Heather vole	Phenacomys intermedius	1	3	3	2	3	T
M133	Montane vole	Microtus montanus	2	3	3	1	3	T
M134	California vole	Microtus californicus	2	3	2	2	3	T
M136	Long-tailed vole	Microtus longicaudus	2	3	3	2	3	T
M138	Sagebrush vole	Lemmiscus curtatus	3	3	3	3	3	T
M139	Muskrat	Ondatra zibethicus	3	3	1	1	3	F
M140	Black rat	Rattus rattus	2	3	2	3	3	F
M141	Norway rat	Rattus norvegicus	3	3	2	3	3	F
M142	House mouse	Mus musculus	2	3	2	3	3	F
M143	Western jumping mouse	Zapus princeps	2	2	3	2	3	T
M145	Porcupine	Erethizon dorsatum	2	2	2	2	3	T
M146	Coyote	Canis latrans	2	3	2	3	3	T
M147	Sierra nevada red fox	Vulpes vulpes	2	2	3	3	2	T
M149	Gray fox	Urocyon cinereoargenteus	2	3	2	2	3	T
M151	Black bear	Ursus americanus	2	2	2	2	3	T
M151	Grizzly bear	Ursus arctos	2	3	2	3	1	T
M152	Ringtail	Bassariscus astutus	2	3	2	2	3	T
M153	Raccoon	Procyon lotor	2	3	2	1	3	T
M154	Marten	Martes americana	2	2	3	3	2	T
M155	Fisher	Martes pennanti	2	1	3	3	2	T
M156	Ermine	Mustela erminea	2	2	3	3	3	T
M157	Long-tailed weasel	Mustela frenata	2	3	2	3	3	T
M158	Mink	Mustela vison	2	3	2	1	3	T
M159	Wolverine	Gulo gulo	2	2	3	3	2	T
M160	Badger	Taxidea taxus	2	3	2	3	2	T
M161	Western spotted skunk	Spilogale gracilis	2	3	2	3	3	T
M162	Striped skunk	Mephitis mephitis	2	3	2	3	3	T
M163	River otter	Lutra canadensis	2	3	2	1	3	T
M165	Mountain lion	Felis concolor	2	3	2	3	3	T
M166	Bobcat	Felis rufus	2	3	2	3	3	T
M176	Wild pig	Sus scrofa	3	3	2	2	3	F
M181	Mule deer	Odocoileus hemionus	2	3	2	2	3	T
M183	Mountain sheep	Ovis canadensis	2	3	3	3	2	T
M186	Feral goat	Capra hircus	3	3	2	3	3	F
R004	Western pond turtle	Clemmys marmorata	3	3	1	1	2	T
R019	Blunt-nosed leopard lizard	Gambelia silus	3	3	2	3	2	T
R022	Western fence lizard	Sceloporus occidentalis	2	3	2	3	3	T
R023	Sagebrush lizard	Sceloporus graciosus	2	3	3	3	3	T
R024	Side-blotched lizard	Uta stansburiana	3	3	2	3	3	T
R029	California horned lizard	Phrynosoma coronatum frontale	3	3	2	3	2	T

continued

[a]CWHR code is the California Wildlife Habitat Relationships System code.
[b]Common name is the CWHR appellation.
[c]Scientific name is the CWHR appelation
[d]Sierra Nevada use:
  1 indicates that all of species range in California is in Sierra Nevada.
  2 indicates that principal range includes Sierra Nevada.
  3 indicates that Sierra Nevada is peripheral range only.
[e]LSOG (late succesional and old-growth) habitat:
  1 indicates that population viability in Sierra requires LSOG habitats.
  2 indicates that species uses LSOG but is not dependent upon it.
  3 indicates that species does not use LSOG habitat significantly.
[f]Western foothills habitat:
  1 indicates that population viability (in Sierra) requires western foothills habitat.
  2 indicates that species uses western foothills habitat but is not dependent upon it.

3 indicates that species does not use western foothills habitat significantly.
[g]Riparian habitat (including lakeshores and wet meadows):
  1 indicates that population viability requires riparian habitat.
  2 indicates that species uses riparian habitat but is not dependent upon it.
  3 indicates that species does not use riparian habitat significantly.
[h]Risk:
  1 indicates extirpated in Sierra Nevada.
  2 indicates on state or federal list as endangered, threatened, or special concern either for species as a whole, or Sierra portion.
  3 indicates not known to be at risk.
[i]Native:
  T indicates native to Sierra Nevada.
  F indicates non-native.

WHR Code[a]	Common Name[b]	Scientific Name[c]	Sierra Nevada Use[d]	LSOG Habitat[e]	Western Foothills Habitat[f]	Riparian Habitat[g]	Risk[h]	Native[i]
R030	Desert horned lizard	Phrynosoma platyrhinos	3	3	3	3	3	T
R036	Western skink	Eumeces skiltonianus	2	3	2	3	3	T
R037	Gilbert's skink	Eumeces gilberti	2	3	1	3	3	T
R039	Western whiptail	Cnemidophorus tigris	2	3	2	3	3	T
R040	Southern alligator lizard	Gerrhonotus multicarinatus	2	3	2	3	3	T
R042	Northern alligator lizard	Gerrhonotus coeruleus	2	3	3	3	3	T
R043	California legless lizard	Anniella pulchra	3	3	1	3	2	T
R046	Rubber boa	Charina bottae	3	2	3	2	3	T
R048	Ringneck snake	Diadophis punctatus	3	3	1	3	3	T
R049	Sharp-tailed snake	Contia tenuis	2	3	2	2	3	T
R051	Racer	Coluber constrictor	2	3	2	3	3	T
R052	Coachwhip	Masticophis flagellum	3	3	1	3	3	T
R053	California whipsnake	Masticophis lateralis	3	3	1	3	3	T
R054	Striped whipsnake	Masticophis taeniatus	3	3	3	3	3	T
R057	Gopher snake	Pituophis melanoleucus	2	3	1	3	3	T
R058	Common kingsnake	Lampropeltis getulus	2	3	1	2	3	T
R059	California mountain kingsnake	Lampropeltis zonata	2	2	3	2	3	T
R060	Long-nosed snake	Rhinocheilus lecontei	3	3	1	3	3	T
R061	Common garter snake	Thamnophis sirtalis	2	3	1	1	3	T
R062	Western terrestrial garter snake	Thamnophis elegans	2	3	3	1	3	T
R063	Western aquatic garter snake	Thamnophis couchi	2	2	2	1	3	T
R069	Southwestern black-headed snake	Tantilla hobartsmithi	3	3	1	3	3	T
R071	Night snake	Hypsiglena torquata	2	3	1	3	3	T
R074	Speckled rattlesnake	Crotalus mitchelli	3	3	3	3	3	T
R075	Sidewinder	Crotalus cerastes	3	3	3	3	3	T
R076	Western rattlesnake	Crotalus viridis	2	2	2	2	3	T

[a]CWHR code is the California Wildlife Habitat Relationships System code.
[b]Common name is the CWHR appellation.
[c]Scientific name is the CWHR appelation
[d]Sierra Nevada use:
    1 indicates that all of species range in California is in Sierra Nevada.
    2 indicates that principal range includes Sierra Nevada.
    3 indicates that Sierra Nevada is peripheral range only.
[e]LSOG (late succesional and old-growth) habitat:
    1 indicates that population viability in Sierra requires LSOG habitats.
    2 indicates that species uses LSOG but is not dependent upon it.
    3 indicates that species does not use LSOG habitat significantly.
[f]Western foothills habitat:
    1 indicates that population viability (in Sierra) requires western foothills habitat.
    2 indicates that species uses western foothills habitat but is not dependent upon it.

    3 indicates that species does not use western foothills habitat significantly.
[g]Riparian habitat (including lakeshores and wet meadows):
    1 indicates that population viability requires riparian habitat.
    2 indicates that species uses riparian habitat but is not dependent upon it.
    3 indicates that species does not use riparian habitat significantly.
[h]Risk:
    1 indicates extirpated in Sierra Nevada.
    2 indicates on state or federal list as endangered, threatened, or special concern either for species as a whole, or Sierra portion.
    3 indicates not known to be at risk.
[i]Native:
    T indicates native to Sierra Nevada.
    F indicates non-native.

LYNN S. KIMSEY
Department of Entomology
University of California
Davis, California

26

Status of Terrestrial Insects

ABSTRACT

Insects are potentially powerful indicators of habitat richness, pollution, and environmental perturbations. The most informative groups for these measures are insect herbivores and detritivores (debris feeders). Predators and parasites or parasitoides (insects whose larvae develop on and eventually kill a single host) are less informative because they are far less influenced by environmental conditions than by prey or host availability. Unfortunately, the insect species of California, including those of the Sierra Nevada ecoregion, are not well known. There are two basic reasons for this: (1) collections of California insects are incomplete or lacking for many regions, and (2) we lack the systematic entomology expertise to identify the California insect fauna. However, based on the limited information available, there are estimated to be about 100,000 insect species in California. Of these, 10% may be new to science, 12% are endemic to California, and only 0.9% are endemic to the Sierra Nevada. Most Sierran terrestrial endemic insect species belong to the families of wasps and bees, and of grasshoppers. Examining thirty-five families in six orders of terrestrial insects reveals two major areas of endemicity: the Owens Valley and high-altitude areas above 2,000 m (6,000 ft). Changes to these environments can make them unsuitable for the endemic species.

INTRODUCTION

California has a great variety of ecosystems and environmental conditions. Accordingly, California also has a rich and highly diverse insect fauna. In fact it may have the largest percentage of endemic insect species in America north of Mexico. There are thirty-two orders of insects; thirty-one of these occur in California (the exception is the order Zoraptera), and within these orders there are roughly five hundred families of insects in California.

One of the more recognizable geographic features in California is the Sierra Nevada Mountains. These mountains cover roughly one-sixth of the state and consist of a number of ecological zones, ranging from high-altitude tundra to low-altitude riparian sites, black oak forest to Great Basin sagebrush desert. Large numbers of insect species occupy all these zones, although some ecological zones have far more insect diversity than others. The majority of California insects are widespread and have distributions that fit five major patterns: Pacific Northwest, southern California coastal, Central Valley, Great Basin, and Mojave/Sonoran desert.

Because of the enormous number of insect species and the many different biological adaptations, insects should prove sensitive indicators of habitat quality and richness. Unfortunately, the incomplete state of our knowledge about insects makes it difficult to use them as environmental indicators or to make many generalizations about the insect fauna of California. Even though many entomologists have been working in California during the past hundred years, we still simply do not know, with any accuracy or completeness, what species occur in the state, or where they occur. For the majority of insect species it is impossible to determine distributions without having specimens collected from each locality in hand. Few species, if any, can be identified in the field.

Museum collections are inadequate in their representation of the California insect fauna. Most collections have been put together for teaching programs, specific research projects, or pest survey programs, or because of proximity to tourist sites. Each of these sources of collection materials influences known distributions of insects and results in some interesting information gaps. These gaps are caused by a number of phenomena. One is the *teaching effect,* one of the best examples of which occurs in the vicinity of the Sagehen Creek Field Station, run by the University of California. Entomology classes have collected at this site for decades, and diversity probably reflects the amount of collecting done there rather than the comparative richness of the fauna, because data for other sites in the state are inadequate. Based on collection records, this alpine valley has the highest recorded family diversity in Califor-

Sierra Nevada Ecosystem Project: Final report to Congress, vol. II, *Assessments and scientific basis for management options*. Davis: University of California, Centers for Water and Wildland Resources, 1996.

nia. A total of 396 families of insects has been recorded at this site, as opposed to substantially fewer numbers observed elsewhere (personal observation). In fact, most of the state has never been examined. Published distribution maps of insect species tend to show some amusing patterns. One of the most noticeable is perhaps best termed *casino distributions,* where the largest number of insects was collected along and between Interstate Highways 80 and 50 through the Sierra to Reno. Then there are *landmark distributions.* These occur in various national parks, such as Yosemite, Death Valley, and Lassen. Fourth is a *seasonal component.* Most sites are visited during only a small part of the year, usually in the summer or spring months. Clearly, not all insects are active at these times. Finally, there is *collector artifact.* Every collector "sees" different insect species. Typically a researcher can take a group of students to the same site at the same time to collect insects and have each come back with different species. In addition, it is physically impossible for one person to collect everything at one time. Different guilds of insects require different, sometimes incompatible, collecting techniques.

Next we come to the third part of the equation—species identification. Despite rumor to the contrary, it is impossible to identify 99% of insects to the species level in the field. Most insects are tiny, less than 5 mm (0.25 in) in length. Even professional systematists are probably unable to identify more than about 2,000 species in their groups of study. Assuming there are 100,000 insect species in California, at least fifty specialists would be required to identify everything just based on numbers, particularly because many features that distinguish insect species are subtle and can be seen only under the microscope. It is safe to say that there are not fifty taxonomic specialists currently studying species that occur in California. In 1985 there were probably thirty professional insect systematists in the state. By 1995 that number had dropped to twenty-two. We are losing taxonomic expertise at an alarming rate. Thus our ability to pinpoint areas of endemicity is waning as well. It is also important to realize that many species in California are still undescribed. Particularly in certain groups such as moths, wasps, beetles, and flies, at least 10% of the species may be new to science.

However, despite these data gaps some general patterns based on published monographs and museum collections appear. Overall, few terrestrial insect species are endemic to Sierra Nevada ecosystems, and of these most are found at high altitudes or in the Owens Valley, for reasons discussed later in this chapter.

METHODS

Distributions of insect species were taken from studies published in the *Bulletin of the California Insect Survey* (Bohart and Grissell 1975; Bohart and Horning 1971; Bohart and Schlinger 1957; Bright and Stark 1973; Camras and Hurd 1957; Foote and Blanc 1963; Grigarick and Stange 1968; Hardy 1961; Huckett 1971; Hurd 1951, 1955; Hurd and Linsley 1951; Hurd and Michener 1955; Linsley and MacSwain 1951; Menke 1979; Merritt and James 1973; Middlekauff 1960, 1969; Strohecker et al. 1968; and Wasbauer and Kimsey 1985), from other monographs (Bohart and Kimsey 1982; Otte 1981, 1984), and from specimens in the Bohart Museum of Entomology, University of California, Davis. Characteristics of species treated in these studies are outlined in tables 26.1 and 26.2.

These studies cover only a small percentage of the California insect fauna, but the group of families treated represents a broad enough range of terrestrial habits (Borror et al. 1989; Richards and Davies 1977) to make some generalizations. However, it must be emphasized that species that appear to be endemic to California, and, more specifically, to the Sierra Nevada region, may not actually be so. Collections of material from within California and from surrounding states are inadequate to make this determination. Another problem inherent in relying on older taxonomic studies is that the species names used may no longer be valid. In Strohecker et al. (1968) three of the sixteen species of grasshoppers listed as endemic to California are actually California populations of much more widespread species (Otte 1981, 1984). Thus some so-called California endemics may actually be much more widespread than indicated in the literature. Conversely, some California populations may represent unrecognized endemic species. It is also true, particularly in California, that old collection localities may have changed drastically over time, becoming, for example, shopping districts, ski slopes, or parking lots. To further confuse the picture, there are also new species in the state that remain to be described. I estimate that roughly 2% (or 2,000) of the insect species are undescribed and new to science. The point in discussing these problems is that we lack sufficient information to evaluate the California insect fauna in any detail. Underlying these problems is a severe shortage of taxonomic expertise.

FINDINGS

Insects as Indicators of Habitat Richness

It would be natural to assume that the large number of insect species would provide a very fine-grained picture of habitat diversity. This assumption is undoubtedly valid if insect groups are carefully chosen, based on their biological attributes. Most insect species (between 60% and 80%) are parasites or parasitoids of plants, animals, or other insects. These groups tend to be poor indicators because they are restricted to particular habitats only indirectly or not at all. Additionally, some species are host specific, and these are typically found in a subset of the host's distribution. Their absolute distribution within the host's range is more likely to be dic-

TABLE 26.1

Characteristics of the insect orders, families, genera, and species considered in this chapter, with a rough estimate of the number of world species in each family.

Insect Order	Family	Number of California Species	Number of California Endemic Species	Number of Sierran Endemic Species	Total Number of Genera Treated	Percentage of Total Species Endemic to California	Percentage of Total Species Endemic to Sierra	Percentage of Total Endemic California Species That Are Sierran	Estimated Number of Total World Species
Orthoptera[a]	Tetrigidae[c]	7	1	1	2	14.3	14.3	100.0	500
	Acrididae	143	49	13	40	34.3	9.1	26.5	2,000
	Eumastacidae	6	2	1	2	33.0	16.7	50.0	100
	Tanaoceridae	4	2	0	2	50.0	0	0	500
Hemiptera[b]	Hebridae	6	0	0	2	0	0	0	100
	Mesoveliidae	2	0	0	1	0	0	0	100
	Hydrometridae	1	0	0	1	0	0	0	50
	Macroveliidae	3	1	0	2	33.0	0	0	100
	Veliidae	10	0	0	2	0	0	0	200
	Gerridae	7	0	0	3	0	0	0	200
	Nepidae	3	1	0	1	33.0	0	0	200
	Belostomatidae	8	2	0	3	25.0	0	0	500
	Corixidae	30	4	1	7	13.3	3.3	25.0	1,000
	Ochteridae	1	0	0	1	0	0	0	100
	Gelastocoridae	4	0	0	2	0	0	0	100
	Naucoridae	10	3	0	2	30.0	0	0	500
	Notonectidae	11	0	0	2	0	0	0	1,000
Diptera	Micropezidae[c]	7	1	0	2	14.3	0	0	500
	Anthomyidae[c]	204	25	2	18	12.3	1.0	8.0	10,000
	Conopidae[d]	48	4	0	7	8.3	0	0	1,000
	Stratiomyidae[c]	61	3	1	12	4.9	1.6	33.3	2,000
	Tephritidae[a]	123	19	0	32	15.4	0	0	5,000
	Bibionidae[c]	23	6	3	3	26.1	13.0	50.0	1,000
Coleoptera	Platypodidae[a]	1	0	0	1	0	0	0	500
	Scolytidae[a]	185	16	0	43	8.6	0	0	1,000
	Rhipiphoridae[d]	22	7	0	2	31.8	0	0	200
Hymenoptera	Megachilidae[e]	200	53	5	12	26.5	2.5	9.4	5,000
	Melectidae[e]	15	1	0	3	6.7	0	0	100
	Mutillidae[d]	28	3	0	4	10.7	0	0	2,000
	Siricidae[a]	14	1	0	1	7.1	0	0	100
	Anthophoridae[e]	9	1	0	1	11.1	0	0	5,000
	Cephidae[a]	5	0	0	3	0	0	0	100
	Sphecidae[b]	112	13	1	9	11.6	0.9	7.7	30,000
	Pompilidae[b]	75	3	0	15	4.0	0	0	1,000
	Scoliidae[d]	8	0	0	3	0	0	0	500
	Chrysididae[d]	196	18	2	17	9.2	1.0	11.1	2,000
TOTALS (36 families):		1,592	239	30	263	15.0	1.9	12.6	73,750

[a]Herbivores. [b]Predators. [c]Scavengers/detritivores. [d]Parasitoids. [e]Pollinators.

TABLE 26.2

Biological characteristics of some insect orders found in California, with percentages for each order found in the Sierra Nevada ecoregion and the number of endemic species in that group found in California in general.

Insect Order	California Endemic Species					Immatures in Different Habitats Than Adults	Total Number of Species Considered
	Herbivores	Predators	Parasitoids	Scavengers	Pollinators		
Orthoptera		0	0	0	0	no	160
Percentage	100						
Number of Species	54						
Hemiptera	0		0	0	0	no	96
Percentage		100					
Number of Species		11					
Diptera		0			0	yes	466
Percentage	14		14	72			
Number of Species	19		4	43			
Coleoptera		0		0	0	yes	208
Percentage	67		33				
Number of Species	16		7				
Hymenoptera				0		yes	662
Percentage	20	20	30		30		
Number of Species	1	16	5		224		

tated by historical accident and dispersal capabilities than by absolute environmental factors. Other species may attack a number of host species, and this is also not usually environmentally informative. Then there is the question of dispersal. Small-bodied parasitoids, such as aphids or mymarid wasps, may be poor fliers but are readily dispersed as aerial plankton by air currents. Predators are also poor indicators because they are generally effective dispersers, and their distributions are more often determined by the availability of appropriately sized hosts than by environment. Primary consumers, such as herbivores and detritivores, are very likely the most valuable environmental indicators because they are directly linked to plant communities and are, therefore, directly susceptible to environmental conditions.

Insects often have complicated life cycles, which allow them to finely subdivide the environment and avoid competing with their own offspring for resources. In the majority of species, adults occupy one habitat, larvae another, as discussed later. An extreme example of this habitat partitioning can be seen in some wasps, such as the tiphiids, where even males and females occupy different habitats, except during mating. In certain groups, therefore, occupying different niches during different life stages also makes them sensitive environmental indicators.

Orthoptera (Grasshoppers)

All grasshoppers are plant feeders. Those in immature stages are simply smaller, flightless versions of adults and occupy the same habitats. Some adults are also flightless. There are apparently thirteen species endemic to the Sierra Nevada, found primarily at high altitudes on, for example, granite slopes, or in the Owens Valley (Otte 1981, 1984; Strohecker et al. 1968).

Hemiptera (True Bugs)

As with the Orthoptera, immature true bugs are smaller, flightless versions of the adults. All species considered here are predators of other insects in aquatic habitats. Less than 10% of these species are endemic to California, and apparently only one, *Caenocorixa kuiterti* (Corixidae), is endemic to the Sierra Nevada (Menke 1979).

Diptera (Flies)

Flies have complex life cycles. Adults are winged, and many are powerful fliers. Larvae are soft-bodied and limbless, occupying cryptic, sheltered habitats. Larvae of Micropezidae, Anthomyidae, Stratiomyidae, and Bibionidae are generally scavengers or detritivores (debris feeders) in terrestrial habitats. Most California endemic species are in these families, particularly in the Anthomyidae, although few are endemic to the Sierra Nevada. None of the phytophagous Tephritidae (fruit flies) or Micropezidae are endemic to the Sierra, nor are the conopids, which are parasitoids on wasps, bees, and grasshoppers (Camras and Hurd 1957; Foote and Blanc 1963; Hardy 1961; Huckett 1971; Merritt and James 1973).

Coleoptera (Beetles)

Only three groups of beetles have been monographed for California, which is a poor representation for this huge order of insects. The Platypodidae and Scolytidae are bark boring in trees, and the parasitoid family Rhipiphoridae attacks primarily larval wasps and bees in their nests. None of these has species endemic to the Sierra Nevada, even though at least the Scolytidae are primary pests of pine and fir trees of the Sierra (Bright and Stark 1973; Linsley and MacSwain 1951).

Hymenoptera (Wasps, Bees, and Ants)

Hymenoptera have biologies very different from the other orders considered in this study. The majority of sawflies have free-living herbivorous larvae and winged, nectar-, pollen-, or insect-feeding adults. Bees are the most important animal pollinators of flowering plants. Wasps are either parasitoids or predators of other insects. Both bees and wasps have helpless larvae that are placed either in or on a host or in a nest by the adult female. Of these groups the largest number of endemic species, to both California in general and the Sierra Nevada specifically, are some of the bees (Megachilidae) and predatory sphecids and their parasites (Chrysididae) (Bohart and Grissell 1975; Bohart and Horning 1971; Bohart and Kimsey 1982; Bohart and Schlinger 1957; Grigarick and Stange 1968; Hurd 1951, 1955; Hurd and Linsley 1951; Hurd and Michener 1955; Middlekauff 1960, 1969; Wasbauer and Kimsey 1985).

Areas of Endemicity

Few insect species are endemic to the Sierra Nevada. However, of the groups considered in table 26.1 certain generalizations can be made (figure 26.1). Most endemic species are found in the Owens Valley in Inyo County or are high montane, occurring at elevations above 2,000 m (6,000 ft), particularly above the tree line. Most of the endemic Orthoptera (grasshoppers) occur at high elevations, generally above the tree line in the Sierra. In the Hymenoptera (wasps, bees, and ants) and nonaquatic Diptera (flies) the majority of endemic species occur in the Owens Valley. Finally, a smaller number of species in these orders and the others detailed earlier occur at middle elevations in the western-slope of the Sierra.

CONCLUSIONS

Reasons for Low Sierran Endemicity

Rough estimates made by the author suggest that 100,000 insect species exist in California, about 12% of which are endemic to the state. However, only 0.9% of the terrestrial species are endemic to the Sierra Nevada. Why there are so few Sierran endemic species of insects relative to California as a whole is a complex story that can be broken down into two components: ecological and historical. Ecologically, at least from the perspective of the majority of insect families, the Sierra Nevada offers few unique habitats. Most are shared with other western states and Mexico. Alpine and western-slope Sierra Nevadan habitats between 700 and 1,000 m (2,000 and 5,000 ft) form a nearly continuous corridor with the northern California Coastal, Siskiyou, and Cascade Mountains. The eastern slope of the Sierra Nevada shares the majority of species with the Great Basin habitats of Nevada and Utah. The only parts of these mountains that function as habitat islands are the high-altitude, boreal sites and perhaps the far southern end of the mountain range.

From a historical perspective there are other factors as well. The geological structure of California that we see today is relatively recent. The uplift of the Sierra Nevada and Coastal Mountains began less than ten million years ago. The entire western coastal region of North America was uplifted, forming a continuous corridor with the Cascades and Rocky Mountains. As a result new species have had relatively little time to develop in the Sierra Nevada, particularly because most of the Sierran habitats are not physically isolated from other regions. Insects do not speciate rapidly. Personal observations of insects found in Ice Age (Pleistocene) deposits, such as cave deposits and tar pits, indicate that these were essentially modern species. These remains, as well as insects fossilized in amber, copal, mudstone, and other materials, suggest that speciation generally takes many tens or even hundreds of thousands of years. Many species are found in both the Coastal Mountains of California and the Sierra. However, the farther south in the Sierra, the more isolated populations have become, and, as expected, the higher the number of endemic species. Based on current information, the Owens Valley in Inyo County is the site of highest endemicity of terrestrial insects in the Sierra Nevada region. Insect species there appear to have been isolated for a considerable period from both the rest of the Sierran species to the west, by the precipitous eastern slope, and the Great Basin species to the east, by the White Mountains.

Management Implications

The implications of land use in these areas of concentrated endemism must be considered. High-montane species often occur in sites ideal for ski slopes. Removing vegetation from these sites damages the site for species such as flightless grasshoppers. Clearly, water is an issue of importance on the east side of the Sierra Nevada and particularly for the Owens Valley. Drastically changing drainage and flow patterns in this habitat clearly changes the suitability of the area for many of these endemic species.

ACKNOWLEDGMENTS

This study was supported by the Sierra Nevada Ecosystem Project as authorized by Congress (HR 5503) through a cost-reimbursable agreement No. PSW-93-001-CRA between U.S. Forest Service, Pacific Southwest Research Station, and the Regents of the University of California, Wildland Resources Center.

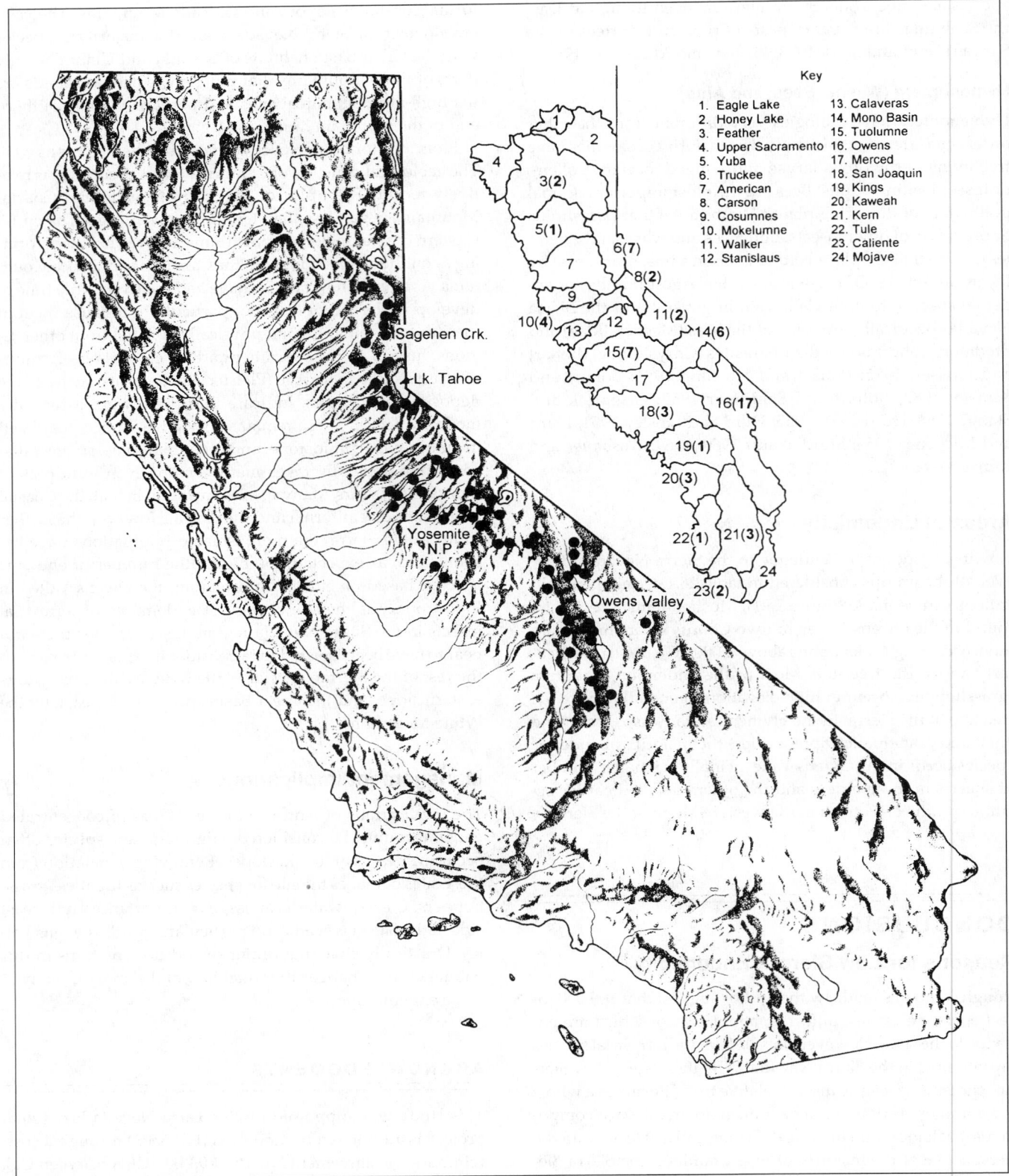

Key

1. Eagle Lake	13. Calaveras
2. Honey Lake	14. Mono Basin
3. Feather	15. Tuolumne
4. Upper Sacramento	16. Owens
5. Yuba	17. Merced
6. Truckee	18. San Joaquin
7. American	19. Kings
8. Carson	20. Kaweah
9. Cosumnes	21. Kern
10. Mokelumne	22. Tule
11. Walker	23. Caliente
12. Stanislaus	24. Mojave

FIGURE 26.1

Distribution of endemic species in the Sierra Nevada ecoregion. Numbers of endemic species, in parentheses, found in each of the major river basins in the study area, numbered to the left of the parentheses.

REFERENCES

Bohart, R. M., and E. E. Grissell. 1975. California wasps of the subfamily Philanthinae. *Bulletin of the California Insect Survey* 19: 1–92.

Bohart, R. M., and D. S. Horning Jr. 1971. California bembicine sand wasps. *Bulletin of the California Insect Survey* 13:1–49.

Bohart, R. M., and L. S. Kimsey. 1982. A synopsis of the Chrysididae in America north of Mexico. *Memoirs of the American Entomological Institute* no. 33:1–266.

Bohart, R. M., and E. I. Schlinger. 1957. California wasps of the genus *Oxybelus. Bulletin of the California Insect Survey* 4 (4): 103–34.

Borror, D. J., C. A. Triplehorn, and N. F. Johnson. 1989. *An introduction to the study of insects.* Philadelphia: Saunders College.

Bright, D. E., Jr., and R. W. Stark. 1973. The bark and ambrosia beetles of California: Scolytidae and Platypodidae. *Bulletin of the California Insect Survey* 16:1–169.

Camras, S., and P. D. Hurd Jr. 1957. The conopid flies of California. *Bulletin of the California Insect Survey* 6 (2): 19–40.

Foote, R. H., and F. L. Blanc. 1963. The fruit flies or Tephritidae of California. *Bulletin of the California Insect Survey* 7:1–117.

Grigarick, A. A., and L. A. Stange. 1968. The pollen-collecting bees of the Anthidiini. *Bulletin of the California Insect Survey* 9:1–113.

Hardy, E. 1961. The Bibionidae of California. *Bulletin of the California Insect Survey* 6 (7): 179–95.

Huckett, H. C. 1971. The Anthomyiidae of California exclusive of the subfamily Scatophaginae. *Bulletin of the California Insect Survey* 12:1–121.

Hurd, P. D., Jr. 1951. The California velvet ants of the genus *Dasymutilla* Ashmead. *Bulletin of the California Insect Survey* 1 (4): 89–114.

———. 1955. The carpenter bees of California. *Bulletin of the California Insect Survey* 4 (2): 35–72.

Hurd, P. D., Jr., and E. G. Linsley. 1951. The melectine bees of California. *Bulletin of the California Insect Survey* 1 (5): 119–35.

Hurd, P. D., Jr., and C. D. Michener. 1955. The megachiline bees of California. *Bulletin of the California Insect Survey* 3:1–247.

Linsley, E. G., and J. W. MacSwain. 1951. The Rhipiphoridae of California. *Bulletin of the California Insect Survey* 1 (3): 79–87.

Menke, A. S. 1979. The semiaquatic and aquatic Hemiptera of California. *Bulletin of the California Insect Survey* 21:1–166.

Merritt, R. W., and M. T. James. 1973. The Micropezidae of California. *Bulletin of the California Insect Survey* 14:1–27.

Middlekauff, W. W. 1960. The siricid wood wasps of California. *Bulletin of the California Insect Survey* 6 (4): 59–77.

———. 1969. The cephid stem borers of California. *Bulletin of the California Insect Survey* 11:1–19.

Otte, D. 1981. *The North American grasshoppers. I. Acrididae: Gomphocerinae and Acridinae.* Cambridge, MA: Harvard University Press.

———. 1984. *The North American grasshoppers. II. Acrididae: Oedipodinae.* Cambridge, MA: Harvard University Press.

Strohecker, H. F., W. W. Middlekauff, and D. C. Rentz. 1968. The grasshoppers of California. *Bulletin of the California Insect Survey* 10:1–177.

Wasbauer, M. S., and L. S. Kimsey. 1985. California spider wasps of the subfamily Pompilinae. *Bulletin of the California Insect Survey* 16:1–130.

ARTHUR M. SHAPIRO
Center for Population Biology
University of California
Davis, California

27

Status of Butterflies

ABSTRACT

The Sierra Nevada has an unusually rich butterfly fauna that, however, is distinguished by little endemism at either species or subspecies levels. This may change soon, as more taxonomic subspecies are named. The fauna is structured altitudinally, latitudinally, and between east and west slopes. Maximum species richness occurs at middle elevations on the west slope and around lower passes. Endemism and relictualism are concentrated at high elevation (subalpine and alpine) and on unusual soils at lower elevations. Some patterns of endemism and relictualism suggest a very dynamic biogeography in the Quaternary period, further supported by phylogeographic (genetic) studies. The historic butterfly record is so poor that the effects of land use and management on the fauna can only be guessed at. Taxa of special concern are mostly relicts, especially on ultramafic soils; one is found in marshes and wet meadows on the east slope (Speyeria nokomis apacheana).

INTRODUCTION

The middle-elevation Sierra Nevada has one of the richest butterfly faunas in temperate North America; its only close competitor is the Colorado Front Range (Scott and Epstein 1987). On June 8, 1992, I observed sixty-two species on the wing along four miles of Old Highway 40 (Donner Pass Road). To put this in perspective, only about sixty-five species have ever been recorded in the British Isles, and only fifty-eight occur there today (Dennis 1992). There are few sites outside the tropics where one could see sixty-two butterfly species in one day.

This is especially striking when one considers that butterflies are uncommon in old-growth forests. The Sierran fauna is overwhelmingly adapted to successional and edaphic, nonforest habitats (meadows, barrens, riparian corridors, and alpine fell fields). Most of the Sierra is forested, yet most of its butterfly diversity is not found in the forest—a fact first noted by Emmel and Emmel (1963b).

Butterflies are important for biodiversity and conservation biology because they are diverse enough that patterns in distribution and diversity are demonstrable; their taxonomy is in relatively good shape, at least compared with that of most other invertebrates; they include both ecological generalists and specialists, with some of these specialists tied to unusual and/or endangered habitats; they often have close and potentially coevolved relationships with larval host plants and sometimes with adult nectar sources; they are relatively easy to study and are large enough to be marked individually (and are identifiable as individuals without recapture); and perhaps most importantly they are *pretty,* often charismatically so. Their appeal thus extends beyond professionals to a larger number of dedicated amateurs and to the public at large (Pollard and Yates 1993).

Not only are people interested in butterflies in an abstract sense, but they also like butterflies and want them as part of their environment, even though butterflies have no perceived economic importance. In the Sierra Nevada, butterflies are often a prominent part of the landscape. I know of no formal studies of public opinion about butterflies, but twenty-four years' experience as a Sierran field naturalist has taught me that people are interested in butterflies and like to talk about them. The most common remark I hear—with or without foundation—is that numbers of butterflies have declined "since [my interlocutor] was a kid."

The existence of so many butterfly enthusiasts has both good and bad aspects. It generates pro- and anticollecting tensions, support and hostility for endangered species regulation and enforcement, reliable and spurious identifications and distribution records, good and bad taxonomy. All of these impinge on both perception and management of the Sierran butterfly fauna.

Given how popular butterflies are, it is remarkable that the

Sierra Nevada Ecosystem Project: Final report to Congress, vol. II, *Assessments and scientific basis for management options.* Davis: University of California, Centers for Water and Wildland Resources, 1996.

Sierran fauna is so poorly documented. Butterfly studies in California date to 1849, when a Frenchman, Pierre Joseph Michel Lorquin, began collecting butterflies for the distinguished Parisian entomologist J. B. A. Boisduval (Emmel and Emmel 1973). Lorquin's notes are lost, but Boisduval records that Lorquin collected in the Sierra between 1850 and 1856 and ascended at least the Sacramento and Feather Rivers. Boisduval published the first faunistic treatment of California butterflies in 1868, with occasional references to the Sierra. In his introductory remarks he praises Lorquin for "braving the tooth of the bear and the fangs of the rattlesnake." Boisduval described many common and a few scarce California butterflies, none of them endemic to the Sierra. The absence of high-altitude taxa (species) indicates that Lorquin never reached the alpine zone.

Although California butterflies continued to be described—eventually by American authors—the next faunistic paper on Sierran butterflies (really the first, because Boisduval attempted to cover the entire state) took twenty-two years to appear (Behr 1890). It was a report of twenty butterfly species from Yosemite, followed closely by two more Yosemite lists (Dyar 1892; Van Dyke 1892). Van Dyke enumerated thirty-eight identified species, with accurate ecological notes. Newcomer (1910) produced an excellent paper, containing a credible sketch of the vegetation, climate, and topography of the Tahoe Basin as well as records of seventy-five taxa. Unsurprisingly, the species-level taxonomy has not fared well in eighty-five years. Nonetheless, the portrait of the Tahoe fauna is recognizable; the biological information (such as host plants) is quite accurate.

Unfortunately, the same cannot be said for the next paper (McGlashan 1914), which was privately published and, perhaps fortunately, has rarely been referenced since. It is misleading and highly inaccurate but of interest as an example of "local color."

It was ostensibly written by Ximena McGlashan, the young daughter of Truckee's most prominent citizen, Charles Fayette McGlashan. Actually the elder McGlashan, an indefatigable promoter, probably wrote it himself—just as he did Ximena's subscription magazine, *The Butterfly Farmer*. The paper purports to be a list of Truckee butterflies. It also served as a sales list (all the species were said to be available by mail for 5 cents each). The paper included ninety-one species, a reasonable number for Truckee, but explicitly disclaimed completeness and does not define "Truckee" at all. The list contains species of the western foothills, the high desert, and the alpine zone as well as species that might reasonably be expected at Truckee. Clearly the McGlashans had a very broad concept of Truckee boundaries, and the list is useless for any study of faunal turnover—illustrating the pitfalls of taking historical documents at ecological face value (S. Smallwood and A. M. Shapiro in preparation).

It took sixteen years for the next faunistic paper on Sierran butterflies to appear (Martin and Ingham 1930), listing eighty-three taxa for the Huntington Lake area, Fresno County. This paper suffers from some archaic taxonomy and is tainted by appalling "game hog" collecting data, but it is nonetheless useful. It was quickly followed by two papers published by John S. Garth (1935a, 1935b) on Yosemite butterflies, also in the *Bulletin of the Southern California Academy of Sciences*. In 1959 James W. Tilden published another Yosemite paper, "The Butterfly Associations of Tioga Pass," in the *Wasmann Journal of Biology*. These three papers set many precedents, which came to fruition in 1963 when the authors combined to publish a ninety-six-page monograph, *Yosemite Butterflies*, that in turn set the standard for such studies (Garth and Tilden 1963).

Garth and Tilden use the Merriam life-zone concept, which had great currency then—it was also used by Storer and Usinger in *Sierra Nevada Natural History* in the same year. Using thirty years of records, Garth and Tilden cross-characterize each species' distribution in terms of Merriam's altitudinal zones and Dice's (1943) biotic provinces, which reflect the role of the Sierra as a climatic divide. They also list each species' usual plant community associations.

One year earlier, in 1962, two young brothers, John and Thomas Emmel, published a preliminary fauna for Donner Pass. Though they missed the autumn fauna by leaving too early, they did a sophisticated study of butterfly activity as a function of weather and climate (Emmel and Emmel 1963a) and a study of biodiversity as a function of community and landscape that was decades ahead of its time (Emmel and Emmel 1963b). Also published in 1962 was a minor commentary on the high-altitude fauna (Eriksen 1962). Shields (1966) published a list of seventy-four species from a middle-elevation, west-slope locality in Tuolumne County, embracing diverse habitat types. The next Sierran faunistic paper was my own on Castle Peak, Nevada County, one of the last alpine areas in the northern Sierra (Shapiro 1978). This is, to my knowledge, the last faunistic paper on Sierran butterflies, though my group has published on various components of the northern Californian montane faunas. We have twenty-four years of Sierran faunistic and phenological data, as yet unpublished but constituting one of the largest butterfly data sets of its sort in existence.

These data, however, have nearly all emanated from the Interstate 80/Highway 20 corridor, where I have maintained a permanent transect. There is a much more extensive data set not only for the Sierra but for all of California. The Emmel brothers and Sterling O. Mattoon have been working on a definitive butterfly fauna of California, which when published will be the most complete and detailed such work ever done in the United States (and perhaps the world). They have assembled most of the data from private and institutional collections, filling gaps by targeted collecting of their own. They have also reared and photographed the early stages of virtually the entire fauna, most of which was previously unreared.

Had their book appeared before the Sierra Nevada Ecosystem Project (SNEP), this chapter would be very different. Because their data set is not yet available for analysis, this chapter is based on a much smaller data set, along with frequent conversations with the Emmels and Mattoon. A by-product of their project is a taxonomic work (Emmel 1995) that will have major repercussions for our perception of the Sierran fauna. I will refer repeatedly to this potential problem/opportunity.

THE SIERRAN FAUNA

Butterfly taxonomy has been singularly unstable recently, which makes it difficult to quantify the Sierran butterfly fauna. Philosophers and biologists may argue over the nature and reality of species, but to discuss faunistics one must have a taxonomic touchstone. Using the 1981 *Catalogue/Checklist* of the North American butterflies by Miller and Brown (because it is the most copiously documented and widely circulated of the possible lists), I estimate that there are 155 species of butterflies (excluding rare casuals) in the Sierra Nevada. I define the Sierra as extending from Buck's Lake, Plumas County, to Tehachapi Valley, Kern County. Buck's Lake is the approximate northern limit of a number of characteristic Sierran butterflies, for example, *Polites sabuleti tecumseh, Phyciodes campestris montana*, and *Anthocharis stella*. To the north are increasing numbers of Cascadian and Klamath taxa, absent from the Sierra proper. The estimate of 155 is conservative. Garth and Tilden (1963) recorded 134 species in Yosemite; I have counted 115 at Donner Pass in twenty-four years. Both counts include strays. I am not appending a taxonomic list to this chapter because imminent nomenclatorial changes are sure to render any such effort obsolete within a year. Table 27.1 gives the distribution among families.

Intuitively, this seems a rich fauna, but proving it is not easy. Faunal richness is related not only to area but also to topographic, climatic, and vegetational diversity—which are not simple functions of area. It is also related to history. It is not self-evident how to make appropriate comparisons, and there are few reliable species lists for well-defined montane

TABLE 27.1

Taxonomic composition of the Sierra Nevada and North Coast Range butterfly faunas (approximated from Miller and Brown 1981).

Family	Sierra Nevada		North Coast Range	
	Species	Subspecies	Species	Subspecies
Hesperiidae	30	31	30	31
Pieridae	20	22	14	16
Papilionidae	7	8	6	6
Lycaenidae	52	57	38	47
Nymphalidae	36	44	31	35
Satyridae	8	9	4	4
Danaidae	2	2	1	1
Totals	155	173	124	140

TABLE 27.2

California counties with more than 135 butterfly species recorded. The two smallest counties are entirely Sierran and are the top-ranking counties in terms of species per area.

County	Land Area (mi²)	Number of Species[a]	Species per mi²	Rank in Species per mi²
San Bernardino	20,119	153	0.0076	8
Inyo[b]	10,130	150	0.0148	7
Tulare[c]	4,844	150	0.0310	4
San Diego	4,262	146	0.0343	3
Riverside	7,176	143	0.0200	5
El Dorado[c]	1,726	140	0.0811	2
Kern[b]	8,152	139	0.0171	6
Nevada[c]	975	138	0.1415	1

[a]According to Ray E. Stanford, phone call, June 1995. Includes strays and casual records.
[b]Counties with significant Sierran area.
[c]Mainly Sierran counties.

areas in western North America. Stanford and Opler (1993) collate distribution data for the western butterflies by county. Interpretation of data arranged in this way is difficult, because counties seldom correspond to physiographic or vegetational units. Arizona has 142.5 species per county versus 103.1 for California; but Arizona has fourteen counties with an average area of 13,065 km² (8,120 mi²), while California has fifty-eight counties with an average area of 4,344 km² (2,700 mi²), plus a much greater variance in county area. On a prorata basis California has the richer fauna, but statewide Arizona has more species per mi². Table 27.2 shows the most species-rich California counties. Although the top three are all non-Sierran, the Sierra counties far surpass these southern desert counties in richness. A multivariate analysis of butterfly diversity patterns in California akin to the plant study by Richerson and Lum (1980) is in progress in my lab. It is already evident that butterflies and plants do not respond identically to environmental factors; the best diversity predictors are likely to be different, though topography is important to both.

Using Miller and Brown (1981) and treating every species as represented by at least one subspecies, the Sierran fauna contains 173 subspecies; Garth and Tilden (1963) recorded 151 subspecies in Yosemite.

Distribution Patterns within the Sierran Fauna

Some butterfly species extend completely across the Sierra, transcending Merriam life zones and Dice biotic provinces. These species usually are seasonal or altitudinal migrants (*Colias eurytheme, Vanessa* spp., *Plebeius acmon*) (Shapiro 1980), or they have distinctive ecotypes in different climates (*Papilio zelicaon*) (Shapiro 1995). Many species, however, are confined to either the west (Californian biotic province) or east (Artemisian) slope. The west-slope species include mixed mesic (characterized by moderate moisture) forest endemics like *Pieris napi* and *Amblyscirtes vialis;* the east-slope ones are high plains–high desert–steppe species such as *Colias alexandra, Satyrium behrii,* and *Pontia beckerii*. In other cases different subspecies of the same complex replace each other on the two slopes (*Anthocharis s. sara* and *s. thoosa, Euchloe h. hyantis* and *h. lotta, Coenonympha tullia california* and *t. ampelos*), or the situation is taxonomically ambiguous (species or subspecies?—The Cupressaceae-feeding *Mitoura*). As with the vegetation, elements of the high-desert fauna penetrate the alpine (*Pontia occidentalis, P. sisymbrii, Cercyonis oetus, Lycaena heteronea,* etc.). Altitudinal stratification of the fauna overall is much more pronounced on the west than on the east slope.

Many west-slope species have well-defined altitudinal ranges or are confined to particular vegetation belts (or Merriam zones). Not all of these extend the entire length of the range, but those extending over more than a few counties usually show a pronounced north–south altitudinal gradient. Some species listed by Garth and Tilden (1963) as "Hudsonian" or "Arctic-Alpine" at Yosemite dip down to middle elevation in the northern Sierra, often in very un-Hudsonian plant communities. Thus *Lycaena cupreus* and *L. editha,* as well as *Polites sabuleti tecumseh,* occur on mesic meadows at 1,500 m (4,950 ft) north of Interstate 80. Garth and Tilden's ecological typology thus does not hold for the entire range.

It did, however, characterize the Merriam life-zone spread for every species at Yosemite, permitting a comparison of faunal diversity on a zonal transect across the region. (I treat their "unrestricted" species as occurring in all zones, though this is usually not true.) The distribution (table 27.3) shows a species maximum in the Transition Zone (1,200–2,100 m [3,950–6,900 ft], according to Garth and Tilden), corresponding to mixed mesic forest, which also has the greatest precipitation and the greatest floristic diversity (Barbour and Major 1977). The cumulative faunas for the Sierran stations on my Interstate 80 transect (table 27.4) reach their maximum at Donner Pass (2,100 m [6,900 ft]). The vegetation here is in Merriam's Canadian Zone, with some Hudsonian elements. The difference is probably an artifact of the topography. Donner Pass is low enough for many east- and west-slope species to pass into the other's territory in at least some years, partially dou-

TABLE 27.3

Distribution of butterfly species by Merriam life zones at Yosemite National Park according to Garth and Tilden 1963. "Unrestricted" species are included in the totals for all zones.

Merriam Zone	Number of Species
Lower Sonoran	19
Upper Sonoran	81
Transition	86
Canadian	72
Hudsonian	53
Arctic-Alpine	35

TABLE 27.4

Distribution of butterfly species on a transect parallel to Interstate 80 across the north-central Sierra, based on 1972–94 data by A. M. Shapiro, unpublished.

Station	Elevation	Number of Species (Breeding Residents)
Washington, Nevada County	803 m	73(55)
Lang Crossing, Nevada County	1,500 m	99(80)
Donner Pass, Nevada-Placer Counties	2,100 m	115(85)
Castle Peak, Nevada County	2,730 m	58(45)
Sierra Valley, Sierra County	1,500 m	72(62)

bling the fauna. The passes at Yosemite are higher (Tioga Pass, 3,010 m [9,930 ft]) and more difficult to cross. The middle-elevation maximum is consistent with various other insect studies, though the causes of the pattern remain controversial (McCoy 1990).

The most famous of the "unrestricteds" are the mass migrants, the California Tortoiseshell (*Nymphalis californica*) and the Painted Lady (*Vanessa cardui*). Both migrate by the millions in favorable years; the Tortoiseshell has often tied up July traffic over the summits. Its larvae cause spotty mass defoliation of foothill *Ceanothus* in spring and of *C. velutinus* (tobaccobrush) in the high country in summer.

Anthocharis lanceolata is predominantly a west-slope species (also in the Coast Range and northwestern California) that, however, also occurs locally on the east slope, for example, in the Carson Range, in canyons east of Monitor Pass, and in southern Inyo County, and thence (as subspecies *australis*) into the desert ranges of southern California. It is associated with rocky canyon walls, feeding on various *Arabis,* but on the west slope is commonly found in mesic forest openings on the rather weedy *Arabis glabra*. This may be a non-native plant in the Sierra, and the presence of *A. lanceolata* in mesic forest may be a recent phenomenon. It is a mobile species whose disjunctions do not necessarily imply relictualism, as witnessed by its dramatic movement upslope in the Donner Pass area during the late 1980s–early 1990s under drought conditions.

In several cases species pairs occur in which one is a weedy ecological generalist and the other a narrow specialist; the generalist is highly dispersive and regularly transgresses zonal and community boundaries, while the specialist is philopatric ("stay-at-home"). Examples are *Plebeius acmon* and *P. lupini,* and *Lycaena helloides* and *L. nivalis*. In one case three specialists are stratified altitudinally, from west (foothill) to east (subalpine, alpine): *Thorybes pylades, diversus,* and *mexicana. T. diversus,* a rare and poorly known species, seems more common in the Coast Range and Trinities.

The most unusual pattern in the Sierran fauna (Shapiro 1992a) has two species or subspecies altitudinally stratified with a "no-man's land" between them, occupied permanently by neither. In the pairs *Anthocharis sara sara* and *A. stella,* and

Phyciodes c. campestris and *P. c. montana*, the low-elevation entity is widespread beyond the Sierra, but the high-altitude one is endemic. In the *Pontia protodice* and *P. occidentalis* pair both species have immense ranges, but neither is a permanent resident at middle elevation. Some other species (*Euchloe ausonides, E. hyantis, Everes amyntula*) have unnamed ecological races or ecotypes differing dramatically in altitudinal and ecological distribution and host plants. *Papilio zelicaon* has already been mentioned.

North of Mount Lola the crest dips below the subalpine zone, and the high-country biota disappears. Of the Sierran, truly alpine butterflies, only *Callophrys lemberti* jumps to Mount Lassen. Species diversity is lower in the north than in the central and southern Sierra, presumably reflecting the lessened topographic diversity. In the far north the subspecies of *Coenonympha tullia*, isolated by the crest elsewhere, intergrade (merge into one another) near Portola and again in the Pit River drainage (Porter and Geiger 1988).

Endemism in the Sierran Fauna

How much of the Sierran fauna is endemic, and at what taxonomic levels? The taxonomic level of endemism is significant in that, in general, higher-level endemism indicates greater antiquity. A Sierran endemic is here defined as any taxon found only in the Sierra, as previously defined. Again, the analysis is tied to Miller and Brown (1981). Raw endemism figures are meaningful only in some comparative context; how else can we say if endemism is "high" or "low"? Good data are available for Ball Mountain (Siskiyou County), the Trinity Alps, the Eddies and the Trinity Divide, and the North Coast Ranges. All but the last have strong Cascadian affinities that reduce the usefulness of the comparison unless the Cascades are included. I will restrict my comparisons here to the North Coast Ranges, defined as extending from Marin County to Highway 299 from Eureka to Weaverville, to avoid this problem.

As noted earlier, the Sierran fauna is estimated at 155 species and 173 subspecies, or 1.12 subspecies per species. This figure suggests little differentiation *within* the Sierra, hence little endemism on a fine scale. At the species level there are only three endemics, and two of them are problematic; to list one, I had to deviate from Miller and Brown (1981) and adhere to more recent biological information. They list *Anthocharis stella* as a subspecies of *A. sara*. In 1986 Geiger and I convincingly demonstrated that *A. stella* was specifically distinct, a conclusion now supported by several localities where the two are sympatric (occur together) with no evidence of hybridization (unpublished findings by Shapiro, T. C. Emmel, J. F. Emmel, S. O. Mattoon, and G. Austin). *Oeneis ivallda* is treated as a full species both by Miller and Brown and by Stanford and Opler (1993), but its relationship to what has been called *Oe. chryxus stanislaus* has been in doubt since the latter was discovered. Porter and Shapiro (1991) found no genetic evidence for speciation, and subsequent unpub-

lished work by C. C. Nice and Shapiro supports their conspecificity (condition of belonging to the same biological species). Thus *ivallda* may have to be demoted to a subspecies-level endemic. Only *Colias behrii* is transparently both a species and endemic.

This seems like very low endemism. There are many endemics in the Rockies, but it is unclear how that huge montane area should be subdivided for purposes of comparison; the Sierra, we must recall, is one continuous range. Austin and Murphy (1987) conclude that there are *no* species-level endemics in the entire Great Basin.

The North Coast Ranges, including the high country of the Yolla Bollys, cover roughly 35%–40% the area of the Sierra Nevada. They have no alpine and only tiny amounts of subalpine habitat, but their topography is often very rugged. They have 124 species and 140 subspecies, for a ratio of 1.13 subspecies per species—indistinguishable statistically from the Sierra.

Because the area involved is so much smaller, however, an equal ratio of subspecies per species implies that subspecific differentiation has occurred on a finer geographic scale in the Coast Ranges than in the Sierra (the average ranges of subspecies are smaller). There are two broad sets of explanations for this: either the habitat mosaic is finer in the Coast Ranges (as perceived by butterflies), or the fauna is older. Both may be true. The percentage of endemism at the subspecies level is identical for the two ranges (20%; table 27.5). Again, this equivalence really translates into more endemism in the Coast Ranges relative to area, some of which is surely due to the distinctive climates of the immediate coastal fog belt.

Behr (1890) commented on the apparent lack of Sierran midaltitude endemism, going so far as to declare that "there is much more affinity between [the California North Coast] and the Sierra Nevada, up to 4000 or 5000 feet, than there is for instance between the insect fauna of the Andalusian Coast and that of the Sierra Nevada of Granada [or] between Marseilles

TABLE 27.5

Taxonomic composition of the endemic Sierra Nevada and North Coast Range butterfly fauna (approximated from Miller and Brown 1981), and frequency of endemics at the species and subspecies levels in both faunas.

Family	Sierra Nevada		North Coast Range	
	Species	Subspecies	Species	Subspecies
Hesperiidae	0	3	0	3
Pieridae	2	0	0	2
Papilionidae	0	3	0	1
Lycaenidae	0	7	0	13
Nymphalidae	0	19	0	9
Satyridae	1	3	0	0
Totals	3	35	0	28

Sierra Nevada: Species 3/155 = 0.019
Subspecies 35/173 = 0.202
North Coast Range: Species 0/124 = 0
Subspecies 28/140 = 0.200

and the upper valley of the Rhône." He also noted the floristic affinities of the North Coast and the Sierra Nevada. (Ironically, a recent paper underscores the lack of endemism in the Sierra de Javalambre, central Spain [Sánchez-Rodríguez and Baz 1995].)

Subjectivity and the Recognition of Endemism

Interpretation of endemism at the subspecies level is complicated by the lack of a "biological" concept of the subspecies; subspecies are inherently subjective. There is neither a phenotypic nor a genomic criterion available to objectify them. There is no particular reason to suspect that taxonomists have been more reluctant to name subspecies in Sierran versus Coast Range butterflies. However, I suspect a statistical investigation would reveal that the California montane butterfly fauna is less "split" at the subspecies level than other western North American montane faunas. Does the relatively low endemism reflected in table 27.5 indicate a genuine lack of differentiation in the fauna, or taxonomic conservatism?

This point is important, because the faunistic work of Emmel, Emmel, and Mattoon has generated a long list of new subspecific taxa to be published in Emmel (1995). With this book, the appearance of a poorly subspeciated California fauna will vanish. About 150 new subspecies are being named statewide (nearly one for every two species). Of these, twenty to twenty-five will be Sierran (J. F. Emmel, personal communication, August 1995). Once the new names are validated we will be able to assess whether the Sierra is still relatively low in endemism, in comparison with the North Coast Ranges, for example. Of course, whatever the answer, nothing will have changed about the biology—only our perception of it. With that change may come both collecting pressure on the newly recognized subspecies and moves to protect some of them under federal or state legislation. Perceptions of endemism are important not only for understanding how faunas evolve but also for their potential economic and political consequences.

Large-Scale Biogeographic Affinities of the Fauna

Although butterfly biogeography was addressed by European and American workers in the nineteenth century, the first major analysis of paleoprocesses on butterfly distributions was done by Kostrowicki (1969) for the Palearctic region. Kurentsov (1986) analyzed the role of Beringia in Northern Hemisphere insect biogeography. A formal biogeographic analysis akin to Kostrowicki's has yet to appear in North America, but some patterns are already evident.

Shapiro, Palm, and Wcislo (1981) discuss the derivation of the Trinity-Eddy faunas, placing them in a phytogeographic context. Because butterfly fossils are very rare (and none is from California), butterfly paleogeography must be inferred

from other types of evidence, mainly paleobotanical. We must assume that host relationships have been stable during this time frame and that plant associations, even on a very coarse scale, are useful as butterfly indicators. These are risky assumptions (Dennis 1977, 1992). Ecotypes often transgress perceived ecological associations, threatening our ability to extrapolate butterfly ranges based on vegetation; and some paleocommunities existed in climates without modern analogs, leaving us clueless as to potential butterfly faunas.

The oldest butterfly fossils are mid-Tertiary; some belong to extant genera. The antiquity of the major families is unknown, but clearly butterflies were widely distributed and fairly diverse by the mid-Oligocene. The western montane fauna can be interpreted as derived from Madro-Tertiary and Arcto-Tertiary sources (Raven and Axelrod 1978). Among Arcto-Tertiary elements the most extraordinary is the Golden Oak Hairstreak, *Habrodais grunus,* which with its recently discovered sister species *H. poodyi,* restricted to Baja California, clearly represents a relict (persistent remnant) of a stock otherwise confined to the Old World and best developed in East Asia.

The Sierran component of the Gray-Veined White (*Pieris napi*) complex is indistinguishable from inner Coast Range populations. This is a group with strong indications of multiple invasions across Beringia (and perhaps also across the North Atlantic). The Californian populations may ultimately be found most closely related to warm-temperate East Asian ones, representing one of the older episodes of dispersal. A preliminary outline of the biogeography of this group appears in Geiger and Shapiro (1992).

The rest of the western cordillera has high-altitude taxa with arctic or subarctic affinities. Chabot and Billings (1972) noted that the circumpolar-boreal relict element in the Sierran alpine flora was unusually poor. The same is true in the butterflies. *Lycaena phlaeas* occurs in Yosemite and in the White Mountains, and then skips to the northern Rockies and northeastern Oregon. *Colias behrii* is a very localized subalpine endemic in the central Sierra that is obviously derived from either the circumpolar *C. palaeno* or (less likely) the Nearctic boreal *C. pelidne.* It is thus presumably of Quaternary origin. A strikingly convergent endemic, probably of similar age, exists in the Andes between Santiago de Chile and Mendoza, Argentina *(C. mendozina)* (Shapiro 1991).

One of the strangest relictual patterns, currently being studied genetically in our lab, concerns the Greenish Blue, *Plebeius saepiolus.* Virtually all Californian populations of this cordilleran-boreal species are unique in having only a brown morph in the female. Most Rocky Mountain and boreal populations have only blue females. Fixed blue-female populations occur along the far north coast of California and northward, and in the subalpine and alpine zones of the White Mountains. The Sierran populations across from the Whites have brown females only. This situation hints broadly of a double invasion.

The characteristic arctic and alpine Satyrid genus *Erebia* does not occur in California at all. Its associate *Oeneis* is represented by *Oe. c. stanislaus* and *Oe. ivallda* in the Sierra, which are discussed later but do not seem to be strandings of tundra species, and by *Oe. nevadensis,* which barely reaches the Sierra and represents a low-elevation species group found in cool, mesic forest around the Northern Hemisphere. *Neominois ridingsii* is a subalpine-steppe species in the Sierra, Warners, and Whites, disjunct from the Rockies and Great Basin ranges. *Neominois* appears to be the sister-genus of the large Central Asian *Karanasa* and appears to be derivative from Pleistocene steppe-tundra, as may also be *Pontia occidentalis, P. beckerii, Euchloe ausonides, Lycaena cupreus,* and perhaps others. These distributions are related to that of the Crucifer *Stroganowia* (Rollins 1982). Rigorous phylogeographic studies (Avise 1994) may strengthen these scenarios.

Relations of the Sierra Fauna to Northern and Northwestern California Faunas

Shapiro (1992b) provided outlines of a biogeographic scenario relating the various montane butterfly faunas in California. It was based on a nested pattern of distributions, interpreted in terms of Holocene climatic change. The broad outlines of this process are now emerging in the light of recent progress in paleoclimatic reconstruction, and the forthcoming Emmel, Emmel, and Mattoon book may provide distributional data that are detailed enough to allow the scenario to be fine-tuned for at least some of the most interesting butterflies. Prospects seem particularly good for the North Coast Range disjunctions and serpentine relics discussed later.

I identified four components in the nested pattern. Group I species are found in northwestern California and northward but are unknown in the Sierra Nevada (e.g., *Colias occidentalis,* a species frequently misrecorded from the Sierra—as in Garth and Tilden 1963—based on confusion with female *Zerene eurydice*). Group II is similar but has at least one known population in the northern Sierra (*Oeneis nevadensis, Carterocephalus palaemon*). Group III includes many montane species widely distributed in both northwestern California and the Sierra Nevada. Some of these, such as the red fir forest specialist *Chlosyne hoffmanni,* are subspecifically differentiated between the two ranges. The two subspecies of *Parnassius phoebus* (*sternitzkyi* in the northwest, *behrii* in the Sierra) are phenotypically very different and may not be closely related. Group IV consists of the strict Sierran endemics.

Shapiro, Palm, and Wcislo (1981) attempted to use Trinity Alps butterfly distributions to test competing scenarios for colonization of the high Sierra from the north (Cascades) versus the east (Rockies, across the Great Basin [Major and Bamberg 1963, 1967]). None of the Rocky Mountain–related high Sierran taxa were found in the Trinities, but because the Trinity climate was more severe than the Sierran in the Pleistocene, the result was not definitive. Since then, the newly discovered (as yet undescribed) species of *Agriades* (Lycaenidae) has been found in both the high Sierra and northwest California, the only such distribution seen to date (J.

Emmel and S. O. Mattoon, personal communication). The presence of relict populations of *Oeneis nevadensis* and *Carterocephalus palaemon* in the northern Sierra supports the idea that the cool, moist-adapted fauna retreated and mostly disappeared from the low northern Sierra in the Hypsithermal/Xerothermic. The presence of relicts such as *Parnassius clodius* (in the Yolla Bollys, formerly in Santa Cruz County) and the unnamed Cascade-Trinity subspecies of *Polites sabuleti* (in Colusa and northern Lake Counties) shows a similar northward retreat in the North Coast Ranges. The Mormon Fritillary, *Speyeria mormonia,* has relict populations in the Eddies, on Ball Mountain north of Mount Shasta, and in the Warners. These populations seem to suggest its route between the Cascades and the Sierra Nevada, but as a meadow species it might well have been able to cross a cool, pluvial (characterized by abundant rain) Great Basin. Phylogeography offers a concrete hope of resolving its regional history.

Phyciodes orseis is one of the rarest butterflies in northern California. It has a Group III distribution, with different subspecies (*P. orseis orseis* in northwest California, formerly south to Marin County [although there is some ambiguity about the authenticity of Marin County due to the age of the records]; *P. o. herlani* in a small area centered on the Lake Tahoe basin). However, both "subspecies" may actually be stabilized hybrid swarms between *P. mylitta* and the corresponding geographic subspecies of *P. campestris* (*P. c. campestris* in the northwest, *P. c. montana* in the Sierra). If so, they arose independently as epiphenomena of the distribution of the parental species. *P. mylitta* and *P. campestris* are widely sympatric at low elevations today, with no apparent hybridization. Scott (1994) ignores the hybrid hypothesis.

The lack of relicts of northern affinity in bogs in the Sierra Nevada is striking, because they are relatively common in the higher North Coast Ranges. At least one North Coast Range butterfly, *Lycaena xanthoides,* gives hints of ancient hybridization with its close relative *L. editha,* now confined to the Sierra and south Cascades in California. A probable stabilized hybrid population between these two exists in far northern California, from Dunsmuir to near Yreka, and there is a smaller apparent hybrid zone in canyons in the White Mountains (Shapiro and Geiger in preparation).

At the southeast end of the Sierra a fairly small faunistic element derived from the desert enters on the east side. One of the most spectacular Sierran butterflies, the Nokomis Fritillary (*Speyeria nokomis apacheana*), belongs to this element; it is confined to wet, east-slope meadows. The species as a whole is in decline, and its presence in the Sierra at all reflects wetter times in the Pleistocene. For its genetics see Britten et al. 1994.

Another pattern of disjunction that we are studying phylogeographically is associated with serpentine soils in the North Coast Range and Sierra. Once again the butterfly ranges are nested, but the pattern is asymmetrical. *Mitoura muiri* and *Hesperia columbia* are found mostly on serpentine. *M. muiri* has never been recorded from the Sierra, while *H. columbia* was reported from two sites (Mariposa and Kern Counties) by Shields (1978). *H. columbia* is a bunchgrass feeder with undetermined preferences; *M. muiri* feeds on Sargent and MacNab cypresses; both thus could occur in the Sierra. *Hesperia lindseyi*, another bunchgrass feeder, and *Erynnis brizo lacustra*, which in northern California feeds only on *Quercus durata*, were formerly thought to be absent from the Sierra. They are now known to occur on a number of ultramafic barrens in Nevada, Placer, and El Dorado Counties (both) and Mariposa County (*brizo*). Both extend through the Central Coast Ranges and the Transverse Ranges, so that there is a potential dispersal route into the Sierra from the south—perhaps followed by extinction south of El Dorado County as the climate became hotter and drier along the east flanks of the San Joaquin Valley. The strangest component of this pattern is an unnamed subspecies of the *Hesperia comma* complex. The normal west-slope Sierran entity is *H. c. yosemite*, which occurs at middle elevations with a single brood in June and July. A phenotypically different entity, but apparently also a member of the *comma* complex, is known from a serpentine barren in Nevada County, where it flies in the third week of September and into early October. Normal *yosemite* occurs on nonserpentine soils both above and below this site. Apparently the same entity is widespread in the south Yolla Bollys on nonserpentine soils; it also flies in late September and October. Additional Sierran localities are reported from El Dorado and Mariposa Counties. The biogeography of the *comma* complex in California is extremely difficult, and many "blend zone" populations are known that mix characteristics of named subspecies (MacNeill 1964). Only phylogeography is likely to clarify this confusion.

Genetics and the Subspecies Problem

The "subspecies problem" has bedeviled lepidopterists for decades. Porter and Geiger (1988) focus on the problem in their revision of the *Coenonympha tullia* complex, based on electrophoretic genetics (a technique used to make concealed genetic variability in populations visible in the lab). We have traditionally named subspecies based on color and pattern, but molecular-phylogeographic techniques, including both enzyme electrophoresis and various DNA-based methods (Avise 1994), now give us new access to the genetic architecture of populations and species complexes. Baughman et al. (1990) attempted to work out the history of the *Euphydryas editha* complex in western North America. This species is highly colonial and breaks down into a number of ecotypes associated with particular host plants on a geographic basis, but its genomic architecture was remarkably uninformative, with most populations very similar and a few anomalously, and idiosyncratically, distinct. The result of Baughman et al. is similar to that of Tong and Shapiro (1989) on the physiologically very distinct, but electrophoretically nearly identical, ecotypes of Californian *Papilio zelicaon*. Porter and Geiger (1988) and Porter and Shapiro (1991) found that the Satyrids *Coenonympha tullia ampelos/california* and *Oeneis ivallda/chryxus*

stanislaus were less differentiated electrophoretically than phenotypically. Recent unpublished work by C. C. Nice and Shapiro on various Lycaenids (*Lycaeides idas* and *melissa* complexes, *Mitoura nelsoni/muiri/siva*) points in the same direction. If parallelism is common in butterfly ecotypes, discordance with phylogeography is to be expected.

Recently, Ball and Avise (1992) reviewed phylogeographic versus phenotypic differentiation in avian subspecies. Predictably, they found that some taxonomic subspecies corresponded to well-defined genetic entities, while others did not. Their conclusions bear on butterflies perhaps even more than on birds:

> Recognition of deep historical separations may not be the only rationale for subspecies descriptions . . . . Any mutations serving as genetic markers of breeding populations (including those underlying particular morphological or behavioral traits) can be of great utility . . . even if the mutations are of recent origin and do not reflect long-term population separations or genome-wide patterns of differentiation . . . . We have argued that short-term population separation should not be sufficient to justify formal taxonomic recognition of subspecies (in part, because sensitive and refined genetic assays will likely reveal significant structure even at deme and family levels in most species) . . . . Subspecies names should be reserved for the major subdivisions of gene-pool diversity within species . . . concordant subdivisions at multiple independent loci . . . therefore, some other means of cataloging geographic distributions of individual markers should be implemented. Overall, an enlightened perspective on intraspecific differentiation would recognize the great variety of evolutionary breadths and patterns likely to be represented among populations, and the various taxonomic and population applications to which these levels of genetic separation might be applied.

Overall, genetic studies of Sierran butterflies—both published and in progress—point to a predominance of stasis at the genomic level. Yet, in many species, ecotypic differentiation (genetically based ecological races) is obvious; it involves phenology, diapause (developmental arrest; a time of seasonal dormancy), and host-plant use, but seldom color and pattern (which seem to vary in other, nonconcordant ways). These attributes are under genetic control, and to the extent they have been studied in our fauna and others, that control tends to be simple and Mendelian. Presumably we are seeing strong selection on life-history traits, superimposing the resultant variation on an otherwise nearly invariant background. We simply do not see the level of genetic differentiation in butterflies that we would expect in so large and ecologically complex an area as the Sierra, were the fauna old. (The single deviant case, *Speyeria nokomis apacheana*, involves extremely small effective population size and is consistent with drift

[Britten et al. 1994]). Genetics thus dovetails with geography in suggesting that the existing fauna only quite recently attained its current distribution.

CONSERVATION AND MANAGEMENT ISSUES

Some Basic Natural History

Butterflies are holometabolous (having complete metamorphosis) insects. This fact has tremendous implications for butterfly ecology and management. To succeed, a butterfly population must have access to appropriate resources in all life stages. Butterflies are diverse enough in their life histories that it is difficult to generalize about them. This section is largely abstracted from Dennis 1992 and Scott 1986 and adapted to a Sierran faunal context. Rather than burden this chapter with dozens of references to basic (non-Sierran) butterfly biology, I refer the reader to the bibliographies of these works. In addition, I will not attempt to cite the hundreds of publications touching on the natural history of species found in the Sierra. Emmel, Emmel, and Mattoon will inventory this literature in their forthcoming book. Meanwhile, the Dennis book in particular is extremely useful.

As far as is known, the larvae of all Sierran butterflies are phytophagous (feeding on plants). Host-plant adaptations may be very strict (some *Euphilotes* ecotypes on single species or even races of *Eriogonum*) or extremely broad (*Vanessa cardui* on many unrelated plant families). Most species lie somewhere in between, feeding on a few plants typically sharing their defensive chemistry: *Pieris*, *Euchloe*, and their close relatives eat only plants containing glucosinolates (Cruciferae and Capparidaceae), but not those having an additional line of defense, such as *Thlaspi* and *Erysimum*; *Junonia coenia* eats Scrophulariaceae, Plantaginaceae, and the Verbenaceous genus *Lippia*, which share iridoid glycosides. The more specialized the relationship, in general, the more vulnerable to disruption. Multiple-brooded species may have different host plants in successive generations; *Nymphalis californica* tracks young foliage of *Ceanothus* upslope as the season advances. Some species feed on particular parts of the plant or distinct seasonal stages (phenophases): *Euchloe* and *Anthocharis* eat only buds, flowers, and fruit of Cruciferae; many blues, such as *Everes amyntula*, are seed feeders. The plants used as larval hosts by butterflies are a small and very nonrandom selection from the total flora. Certain genera and families are particularly important; these include willows *(Salix)*, oaks *(Quercus)*, wild buckwheats *(Eriogonum)*, Malvaceae, and Papilionaceous legumes *(Vicia, Lathyrus, Astragalus, Lupinus, Trifolium)*. All the Californian Satyrids and Hesperiine skippers feed on grasses or sedges, but their preferences in the wild are largely unknown, largely because few people can identify graminiforms well. These butterflies rarely feed on naturalized annual species. Annuals in general are infrequently used as butterfly hosts. Most associations are with herbaceous perennials and woody plants, but because of the clustered taxonomic preferences noted earlier and because of climatic correlations, species diversity of these plant groups is not a very useful predictor of butterfly diversity. Many Lycaenid and Riodinid larvae have mutualistic relationships with ants, here as elsewhere. These relationships may be obligate or facultative.

Most butterflies are heliotherms (they depend on incoming solar radiation to heat their bodies to temperatures sufficient for flight), and thus are animals of sunny climates. Coastal fog-belt climates have high floristic diversity but few butterflies. Cloudiness is seldom if ever limiting to Sierran butterflies during the flight season, except perhaps in the alpine zone (but compare Emmel and Emmel 1963a).

The number of broods per year may be fixed or variable. Univoltine (single-brooded) species tend to be thus throughout their ranges, while multivoltine species rarely persist where they can produce only one brood per year. There is a predictable seasonal succession of adult butterflies in a given locality. The flight periods of different species tend to have a constant seasonal relationship even though the actual flight dates may vary greatly with the weather. At higher elevations the timing of snowmelt is critical to initiating the flight season. *Philotes sonorensis*, typically the earliest-flying butterfly in the Sierra that does not hibernate as an adult, has emerged at the same 1,500 m (4,950 ft) site as early as late February (in drought years) and as late as May to mid June (in years of late snowmelt).

All breeding residents in the alpine zone are univoltine, and some, such as *Oeneis ivallda*, *Oe. chryxus stanislaus*, and *Neominois ridingsii*, require two years to complete a generation (diapausing twice as larvae). This phenomenon also occurs in alpine butterflies in Eurasia and in some arctic species. Both alpine species and those of stressful foothill habitats, especially serpentine, often show multiple-year diapause as larvae or pupae. This pattern appears to be a hedge against short-term catastrophe, analogous to a soil seed bank in plants.

Most foothill butterflies are either spring-univoltine or spring-bivoltine, a cycle that allows them to exploit the combination of lush vegetation and sunny days that occurs only then. At middle elevation about half of the fauna is univoltine, increasing to 80% at 2,100 m (6,900 ft). Univoltines may fly at any season, but very few (*Apodemia mormo, Neophasia menapia, Ochlodes sylvanoides*) fly in the last third of the season. Like crops, butterflies are highly vulnerable to density-independent catastrophes, including unusual or severe weather and fire. Management practices with the potential to produce catastrophic mortality (logging, pesticide use, prescribed burning, grazing) need to figure butterfly life cycles into their scheduling if sensitive species are at risk.

The peak of butterfly adult diversity occurs in the lower foothills in May, moving gradually upslope with the advancing season. There may be striking differences in butterfly seasonality with slope and exposure, especially at high elevations. At Carson Pass, for example, the dry, south-facing slopes of Red Lake Peak and Little Round Top may be three weeks advanced relative to the north-facing slopes of Round Top at the same elevation, just across Highway 88.

Adult butterflies feed on nectar. Some species also or primarily visit sap fluxes or rotting fruit. Although most flower-visiting butterflies have preferences (which typically reflect the geometry of the flower relative to the tongue and leg lengths of the animals), there are apparently no tightly co-evolved pollinator-butterfly systems in the Sierra. Flowers particularly important to butterflies in general include *Chrysothamnus, Aster* and *Solidago* (Compositae), *Eriogonum* (Polygonaceae), *Rhamnus* (Rhamnaceae), *Aesculus* (Hippocastanaceae), *Agastache* and *Monardella* (Labiatae), *Spraguea* (Portulacaceae), *Apocynum* (Apocynaceae), and *Asclepias* (Asclepiadaceae). The only butterfly that visits turpentine weed and vinegar weed (*Trichostema*, Labiatae) is the skipper *Ochlodes sylvanoides;* it is unclear if it is an effective pollinator.

In addition to the vicinity of nectar sources, large butterfly aggregations are observed on mud puddles and on rocky summits. The puddling groups consist almost entirely of young males, with each species tending to cluster separately. Among frequent participants are various Blues (*Lycaeides, Celastrina, Everes, Euphilotes, Plebeius*), *Pieris napi, Papilio zelicaon, indra,* and *eurymedon,* and *rutulus, Adelpha bredowii, Chlosyne* and *Euphydryas* spp., *Erynnis* spp., *Thorybes* spp., and *Hesperia nevada.* These animals seem to be collecting minerals that may be physiologically necessary for reproductive activity, but the phenomenon is still poorly understood.

Hilltop aggregations are interpreted as an epigamic (mate-locating) strategy for low-density populations spread over difficult terrain. On the summits males may either perch or patrol a territory, and females are only transient visitors, coming to mate and then departing. Among hilltopping species commonly observed in the Sierra are *Pontia occidentalis, P. sisymbrii, Euchloe hyantis, Papilio zelicaon, indra,* and *eurymedon, Parnassius phoebus behrii, Oeneis* spp., *Vanessa* and *Nymphalis* spp., *Speyeria egleis, Thorybes mexicana nevada,* and *Hesperia nevada.*

It is important to remember that hilltop aggregations may draw on a large area and cannot be interpreted as representative of overall population density. Hilltopping species are highly vagile (free-moving) and may cover several kilometers a day in routine upslope and downslope flights between breeding and mating sites. On the other hand, some species (such as many *Euphilotes*) are intensely philopatric ("stay-at-home") and may spend their entire lives within a few meters of the plant they fed on as larvae. Hilltopping promotes gene flow and may prevent population differentiation. From a genetic standpoint, philopatric butterflies are likely to show much more population substructuring than others; most of the likely candidates are Blues (Lycaenidae).

The adverse season is typically spent in diapause. Each species or lineage has a characteristic phase of the life cycle capable of diapause, and there is usually little or no flexibility in this regard. In subalpine and alpine climates unseasonable snow may make it difficult to complete development in one season. In most species natural selection has made the cycle more conservative than it need be in most years, as "insurance" against unusual weather events. In a few species, such as *Papilio zelicaon,* normally univoltine populations may produce a partial second brood in unusually favorable years.

Historical Changes in the Sierran Fauna

Because there are no pre–gold rush records and very few precise records before 1930 anywhere in the Sierra, it is almost impossible to say anything about historic change unless we extrapolate from vegetation or community-level change to the probable butterfly fauna. This practice is risky.

Fire suppression in the Sierra Nevada has undoubtedly changed the environment for butterflies. Because butterflies are rare in continuous closed-canopy forest, the more open landscapes maintained by fire may have been more conducive to butterflies than are current ones, but this depends in part on the distribution of understory resources, including larval host plants and adult nectar sources. The middle-elevation forests of the inner North Coast Range may approximate historic Sierran conditions better than Sierran forests do today. In these forests we may travel long distances and see few if any butterflies, only to encounter tremendous concentrations along streams or anywhere such butterfly flowers as *Apocynum* or *Agastache* are abundant. Fritillaries (*Speyeria* spp.), which are strong fliers, tend to be abundant in these forests. The openness of the forest floor also favors their host plants, *Viola* spp.

Insofar as it opens up the forest, selective logging may mimic the beneficial effects of natural fire for butterfly breeding. In much of the Sierra, bull thistle (*Cirsium vulgare*) is virtually the signature of logging disturbance, and it is a valuable nectar source in habitats where native flowers are uncommon in late summer.

Because few butterflies occur in mature forest but many species are associated with successional vegetation, both clear-cutting and fire, even in its current form, are likely to enhance the richness of butterfly species in the short to medium term. The butterfly fauna of montane chaparral (*Ceanothus velutinus, Arctostaphylos* spp., *Prunus emarginata, Quercus vacciniifolia,* etc.) is small but consistent, including *Nymphalis californica, Incisalia iroides,* and *Celastrina argiolus echo* and usually *Callophrys (dumetorum* or *lemberti*) and *Hesperia (comma* complex and *juba*). The richest butterfly fauna in the Sierra are found in vegetationally diverse successional habitats with many perennial herbs and in canyon bottoms with rock faces in close proximity to riparian vegetation (especially on west- and southwest-facing slopes).

Experience suggests that butterfly species richness may peak in the early stages of tree establishment, when the community is becoming multilayered but is still strongly insolated.

Fire suppression leads to enhanced shade, and a few mesic-adapted species such as *Pieris napi, Parnassius clodius,* and *Amblyscirtes vialis* may have expanded their ranges beyond their normal riparian corridors at middle elevation under these conditions.

There is no reason to think that either logging or fire has added to or subtracted from the Sierran butterfly fauna. Undoubtedly there have been local changes in distribution and abundance, but no direct information is available.

Nor is direct information available on grazing effects on Sierran meadow and riparian butterflies. Intuitively, we would expect selective grazing to reduce plant diversity and thus injure butterflies through depletion of host plants and/or nectar resources. We might also expect injury from trampling and the disturbance of wet, peaty soils. In fact, there is no evidence known to me that this has occurred in the Sierra, and many meadow butterflies use disturbed mud and animal tracks as puddling sites, enhancing their apparent abundance.

Climatic Instability in Recent Decades

The climatic instability in the past twenty-five years has been correlated with both conspicuous and subtle changes in Sierran butterfly distribution and abundance as tracked by my Interstate 80 transect. In no case is there solid proof of causal mechanisms, but plausible mechanisms exist in many cases.

The transect study, initiated in 1972, embraces the 1975–77 drought, the 1982/83 "year of the big snow," the December 1990 cold wave, and the 1994/95 snowy winter, as well as less dramatic but in some cases even more significant climatic perturbations in other years. The short but intense 1975–77 drought had few effects, but the less intense but more prolonged drought of the 1980s through early 1990s coincided with many changes in butterfly distribution and abundance. *Polites sabuleti tecumseh* disappeared from the lower part of its range at 1,450–1,500 m (4,785–4,950 ft), at the same time becoming more abundant at 2,100–2,400 m (6,900–7,920 ft). Its disappearance from Bear Valley (Nevada County) coincided with the removal of grazing, however, and this may also have been a factor. *Euchloe ausonides* and *Thorybes pylades,* both resident at my 800 m (2,625 ft) site, colonized and bred repeatedly at 1,500 m (but apparently disappeared over the winter of 1994/95). *Anthocharis lanceolata,* common at and below 1,500 m, colonized and bred at 2,100 m. Several species declined precipitously at 2,100 m, including *Lycaena arota,* which had been abundant in the 1970s and early 1980s, and *Plebeius shasta,* which was widespread at Donner Pass when the Emmels worked there thirty years ago but may now be extinct there below 2,300 m (7,590 ft). The data are not entirely unambiguous, but there are hints that the Castle Peak (2,700 m [8,900 ft]) fauna was systematically enriched from below during the drought years, with previously incidental species beginning to breed at tree line. This coincided with a decline in the resident fauna nearby at 2,100 m.

Although species numbers have fluctuated strikingly at Donner Pass since 1971, the most dramatic and causally explicit changes occurred in 1992. The 1991/92 snowpack was unusually light and melted very early. May 1992 was warmer than a normal June; at Donner (as elsewhere in the Sierra) both species numbers and individual abundances hit record highs, and there was an influx of subtropical strays from the desert (Shapiro 1993). A sudden snowstorm June 10–11 and accompanying cold wave dropped the number of species flying from sixty-two on June 8 to twenty-nine a week later. The fauna has not yet recovered from this event and what followed. Much of the reproduction achieved before June 10 was undoubtedly lost. Stragglers of many species continued to emerge after the cold wave, but in very low numbers. Immatures that survived the storm were then subjected to very severe drought conditions in late summer, with the vegetation senescing (drying up) 4–7 weeks early. Almost the entire fauna at Donner is univoltine, and any given species typically can diapause only in a particular life-history stage. Diapause is usually initiated in summer and continues until late winter. The 1992 diapausers were subjected to about a month more of hot, dry weather than normal, which must have imposed a tremendous physiological burden on them. There may also have been significant losses when hosts senesced before the larvae were done feeding. The snowy winter of 1992/93 seems to have abetted over-winter survival, but in summer 1994 the vegetation again dried early enough to cause larval mortality.

The alpine zone at nearby Castle Peak was not sufficiently advanced at the time of the June 1992 storm to be seriously affected, and diversity there was not harmed. Drought conditions in summer 1994 were very severe above tree line at Kit Carson Pass, but 1995 flights were mostly good after a very snowy winter.

As dramatic as these events have been, they are only moderately unusual on the scale of historic weather records and have been greatly surpassed in both intensity and duration over the Holocene as documented by palynological, dendrochronological, and other proxy climatic records. I return to climatic lability and butterfly faunistics in the conclusions to this chapter.

Weedy Ecotypes and Anthropogenic Range Extensions

Much of the low-elevation California butterfly fauna now eats exotic weeds (Shapiro 1984); one species, *Pieris rapae,* is itself an animal "weed" introduced from Europe in the nineteenth century. Relatively few montane or alpine butterflies appear to breed on introduced plants at this time, but some of those that do are widespread and common and may owe some of their success to the use of weedy hosts. Table 27.6 lists some examples known to me. Of these, only *Lycaena xanthoides* currently seems to be expanding its range in the Sierra, and it suffered at least temporary reversals in 1995.

TABLE 27.6

Some native Sierran butterflies now using weedy host plants. (Several other Sierran butterflies use weeds commonly in the Central Valley but not in the mountains.) Species with distinct weed-adapted ecotypes are not included.

Species	Native Hosts	Weedy Hosts
Pontia protodice, P. occidentalis	Cruciferae	Cruciferae: Cardaria, Lepidium, Descurainia, Sisymbrium, etc.
Colias eurytheme	Astragalus?	Legumes: Alfalfa (Medicago), Sweet Clover (Melilotus), etc.
Plebeius saepiolus	Native Trifolium	Naturalized Trifolium
Lycaena cupreus	Oxyria digyna, Native Rumex?	Rumex acetosella
Lycaena editha	Polygonum phytolaccoides, Native Rumex	Weedy Rumex, incl. R. acetosella
Lycaena xanthoides	Native Rumex	Weedy Rumex, incl. R. acetosella
Phyciodes mylitta	Native thistles	Cirsium vulgare, other weedy spp.

A special case is the formation of disturbance-associated ecotypes feeding on weeds. The species in question had not been associated with disturbed or weedy environments in the past. There have been several spectacular instances of this phenomenon in the eastern United States, including the shift of the skipper *Poanes viator* from aquatic grasses and sedges to common reed (*Phragmites communis*) in the Philadelphia–New York corridor and of *Erynnis baptisiae* from its native, scarce, and local host *Baptisia* to the introduced vetch *Coronilla varia*, planted for erosion control on highway embankments in Pennsylvania (Shapiro and Shapiro 1973; Shapiro 1979); in both cases an obscure species became very abundant. (See also Thomas et al. 1987.)

In California the Anise Swallowtail, *Papilio zelicaon*, has evolved a multitude of host-specialist ecotypes with appropriate phenology and diapause, from sea level to tree line (Shapiro 1995). The multivoltine, weedy ecotype that feeds on sweet fennel (*Foeniculum vulgare*), which was already widespread near sea level, has followed its host along freeway embankments into much of the Gold Country, up to at least 1,300 m (4,250 ft) (1995), and is still expanding. It now ranges above univoltine and partially bivoltine ecotypes on native hosts on serpentine soils and in rocky canyons. We are rearing increasing numbers of wild nondiapausers from populations where we formerly rarely encountered any, suggesting gene exchange and the dilution of the strong diapause strategy.

We do not know when the Silvery Blue (*Glaucopsyche lygdamus*) discovered weedy annual vetches along freeway embankments. Its native hosts are perennial vetches and lupines. Populations using introduced *Vicia* (*villosa, benghalensis*) may have been present on the floor of the Sacramento valley

twenty-five years ago (Shapiro 1974), but they, and the vetches, only recently appeared in the Sierran foothills; *Vicia*-feeding *lygdamus* first appeared at Colfax, Placer County, in 1991 and by 1994 had moved down to a disturbed site in the American River canyon, where it now flies within 1 km (0.62 mi) of nonweedy, native populations. Curiously, the same phenomenon is occurring simultaneously in the same species in the northeastern United States (Dirig and Cryan 1991). These ecotypes reinforce the apparent climatic trend by carrying the low-elevation fauna higher into the foothills.

Is the Fauna in Danger?

There are no federal or California threatened or endangered butterfly taxa in the Sierran fauna—at least not yet. Endemic taxa, especially those with very small ranges or those with very narrow ecological specializations, are at the highest risk. The low level of observed endemism would suggest little ground for concern, but once many new subspecies are named, perceptions of threats and pressures for protection are likely to follow.

Endemic Sierran taxa are concentrated in the subalpine and alpine zones. High-altitude taxa are most at risk from climatic change, though no data suggest that any are actually in peril. Anecdotally, we know that the numbers of alpine butterflies have fluctuated wildly with the climatic instability of the past two decades. We do not know if that is a problem; nor do we know the capacity of these insects for multiple-year diapause. Dennis (1992) addresses some of these issues (see also Botkin et al. 1991).

Another group of species of special concern is the relicts restricted (as far as we know) to a few edaphic (soil-determined) barrens—*Erynnis brizo lacustra* and *Hesperia lindseyi*, in particular, and the unnamed North Coast Range *Hesperia comma* entity flying in autumn amidst early-summer Sierran races. Some of the best-known edaphic barrens in the Sierra (e.g., the Ione clays) have no special butterflies. A systematic survey of Sierran serpentine and similar sites needs to be done to document the full extent of the distribution of the relict skippers and to assess the need for protection.

There have been recent changes in the butterfly fauna of the Sierra Valley, north of Truckee, that may have been climate mediated. These include the apparent loss of the once-abundant *Phyciodes c. campestris–c. montana* intergrade (phenotypically intermediate) populations and of *Colias philodice* and a drastic decline of *Cercyonis pegala boopis*. There is no basis for assessing whether such phenomena are purely local or reflect regional processes. The Sierra Valley is an area of special interest because of the intimate interdigitation of the Sierran and Great Basin biota. The most dramatic change observed there has been the replacement of *Lycaena arota arota* by *L. a. virginiensis* in 1994/95, apparently representing a climate-driven extinction and colonization at the subspecies level. Populations of the *Eriogonum*-specialist genus *Euphilotes* in the nearby hills adjoining Dog Valley and Sardine Valley appear to be evolutionarily active. Some of these may have been affected by the August 1994 fires in the area.

Like many other organisms, butterflies are vulnerable to habitat fragmentation, which prevents reinforcement or recolonization of local populations, inhibits gene flow, and decreases species' ability to rebound from climatic or other natural disasters and to track geographically any directional shifts in climate. Although no specific threats can be adduced, it should be self-evident that rapid development of the western foothills may put at least some butterflies at risk in the region.

One specific threat that can be dealt with by prudent planning is the use of microbial insecticides for the suppression of forest-defoliator outbreaks. *Bacillus thuringiensis* (BT, Dipel®) is a nonselective lepidopteran larvicide. To date, the U.S. Forest Service has been sensitive to this threat and has solicited input from lepidopterists on potential consequences for nontarget species. It is important that this attitude be maintained and that adequate lists of lepidopterists exist.

Recreational and scientific collecting, done in moderation, has not been identified as a potential problem for any Sierran butterfly except perhaps *Speyeria nokomis apacheana*. Because of their low reproductive capacity, Parnassians could be vulnerable here, as in Europe, where nearly all populations are now legally protected. However, Californian populations have not been finely subdivided taxonomically as have those in Europe, and there seems to be very little pressure on them. Highly specialized Lycaenid populations (Arnold 1983), having low vagility and very exacting environmental requirements, will always be the butterflies most vulnerable to overcollecting or habitat alteration. Few Sierran Lycaenids have been recognized subspecifically to date, however, and none is known to have been lost.

CONCLUSIONS

1. There is no historic record of Sierran butterflies before 1849. Hardly any serious ecological or faunistic work was done before the 1930s. Faunal change can thus be inferred only by the use of risky assumptions or, more reliably, from relatively short-term and local data sets.

2. Changing concepts of subspecific butterfly taxonomy are likely to change our perceptions of the Sierran fauna in the near future.

3. For a charismatic, popular group, the butterflies are remarkably underdocumented. This will change when the Emmel, Emmel, and Mattoon book on California butterflies appears.

4. The Sierran fauna is rich in species. Some of the richest butterfly faunas in temperate North America occur there,

and the richest California counties (corrected for area) are Sierran.

5. Although it is difficult to demonstrate rigorously, the Sierran butterfly fauna as currently understood taxonomically has a low degree of endemism. There are only three endemic species and surprisingly few subspecies, given the area of the range. The North Coast Ranges have higher subspecies-level endemism, corrected for area.

6. Butterflies as a group are not adapted to old-growth, closed-canopy forest and are thus irrelevant to conservation decisions about such habitats. Most butterflies in the Sierra occur in successional, or climatically or edaphically treeless, environments, steppe or savanna, or in riparian corridors.

7. Fire suppression has probably altered butterfly ranges and abundance, but there are no hard data. Neither fire nor logging is necessarily inimical to butterflies, and both may even be beneficial. The impact of grazing in the Sierra is not understood.

8. With the possible exception of *Speyeria nokomis apacheana*, no Sierran butterflies are at serious regional or global risk. Continuing climatic instability or systematic climatic change can be expected to cause (possibly major) faunal changes. Habitat fragmentation and destruction, especially loss of edaphic barrens with disjunct relicts, could become a problem, especially in the western foothills.

9. Both biogeographic and genetic studies suggest that the existing Sierran butterfly fauna is young. Both the geography of the fauna and existing ecotypes probably date only from the Holocene. Although there are few relict butterfly populations in the Sierra, statewide and regional patterns of relictualism point to northward regression of the fauna in the Hypsithermal, consistent with current understanding of the paleovegetational sequences. There is no basis to consider either historic or current butterfly faunas to be in "equilibrium."

ACKNOWLEDGMENTS

This report could not have been prepared without information supplied informally by John F. Emmel, Sterling O. Mattoon, Greg Kareofelas, Carol Whitham, Christopher C. Nice, Adam H. Porter, Ray E. Stanford, and George Austin. Given its preliminary nature, errors and omissions are to be expected, and for these I take full responsibility. Field work since 1972 has been supported in part by several California Agricultural Experiment Station projects and currently by National Science Foundation grant DEB93-06721.

REFERENCES

Arnold, R. A. 1983. Ecological studies of six endangered butterflies: Island biogeography, patch dynamics, and design of habitat preserves. *University of California Publications in Entomology* 99:1–161.

Austin, G. T., and D. D. Murphy. 1987. Zoogeography of Great Basin butterflies: Pattern of distribution and differentiation. *Great Basin Naturalist* 47:186–201.

Avise, J. C. 1994. *Molecular markers, natural history, and evolution.* New York: Chapman and Hall.

Ball, R. M., Jr., and J. C. Avise. 1992. Mitochondrial DNA phylogeographic differentiation among avian populations and the evolutionary significance of subspecies. *The Auk* 109:626–36.

Barbour, M. G., and J. Major. 1977. *Terrestrial vegetation of California.* New York: John Wiley.

Baughman, J. F., P. F. Brussard, P. R. Ehrlich, and D. D. Murphy. 1990. History, selection, drift, and gene flow: Complex differentiation in checkerspot butterflies. *Canadian Journal of Zoology* 68:1967–75.

Behr, H. H. 1890. Yosemite Lepidoptera. *Zoe* 1:177–79.

Boisduval, J. B. A. de. 1868. Lépidoptères de la Californie. *Annales de la Société Entomologique de Belgique* 12:5–95.

Botkin, D. B., R. A. Nisbet, S. Bicknell, C. Woodhouse, B. Bentley, and W. Ferren. 1991. Global climate change and California's ecosystems. In *Global climate change and California: Potential impacts and responses*, edited by J. B. Knox and A. F. Scheuring, 123–49. Berkeley and Los Angeles: University of California Press.

Britten, H. B., P. F. Brussard, D. D. Murphy, and G. T. Austin. 1994. Colony isolation and isozyme variability of the Western Seep Fritillary, *Speyeria nokomis apacheana* (Nymphalidae) in the western Great Basin. *Great Basin Naturalist* 54:97–105.

Chabot, B. F., and W. D. Billings. 1972. Origins and ecology of the Sierra alpine flora and vegetation. *Ecological Monographs* 42:163–99.

Dennis, R. L. H. 1977. *The British butterflies: Their origin and establishment.* Faringdon, U.K.: E. W. Classey.

———. 1992. *Butterflies and climate change.* Manchester, U.K.: Manchester University Press.

Dice, L. R. 1943. *The biotic provinces of North America.* Ann Arbor: University of Michigan Press.

Dirig, R., and J. F. Cryan. 1991. The status of silvery blue subspecies (*Glaucopsyche lygdamus lygdamus* and *G. l. couperi*: Lycaenidae) in New York. *Journal of the Lepidopterists' Society* 45:272–90.

Dyar, H. G. 1892. Collecting butterflies in the Yosemite Valley. *Entomological News* 3:30–33.

Emmel, T. C., ed. 1995. *Systematics of western North American butterflies.* Gainesville, FL: Mariposa Press. In press.

Emmel, T. C., and J. F. Emmel. 1962. Ecological studies of Rhopalocera in a high Sierran community—Donner Pass, California. I. Butterfly associations and distributional factors. *Journal of the Lepidopterists' Society* 16:23–44.

———. 1963a. Ecological studies of Rhopalocera in a high Sierran community—Donner Pass, California. II. Meteorological influences on flight activity. *Journal of the Lepidopterists' Society* 17:7–20.

———. 1963b. Composition and relative abundance in a temperate zone butterfly fauna. *Journal of Research on the Lepidoptera* 1:97–108.

———. 1973. *The butterflies of southern California.* Science Series 26. Los Angeles: Natural History Museum of Los Angeles County.

Eriksen, C. 1962. Further evidence of the distribution of some boreal Lepidoptera in the Sierra Nevada. *Journal of Research on the Lepidoptera* 1:89–93.

Garth, J. S. 1935a. Butterflies of the Boundary Hill Research Reserve, Yosemite National Park, California. *Bulletin of the Southern California Academy of Sciences* 33:131–35.

———. 1935b. Butterflies of Yosemite National Park. *Bulletin of the Southern California Academy of Sciences* 34:37–75.

Garth, J. S., and J. W. Tilden. 1963. Yosemite butterflies. *Journal of Research on the Lepidoptera* 2:1–96.

Geiger, H. J., and A. M. Shapiro. 1986. Electrophoretic evidence for speciation within the nominal species *Anthocharis sara* Lucas (Pieridae). *Journal of Research on the Lepidoptera* 25:15–24.

———. 1992. Genetics, systematics, and evolution of Holarctic *Pieris napi* species-group populations (Lepidoptera: Pieridae). *Zeitschrift für zoologische Systematik und Evolutionsforschung* 30:100–122.

Kostrowicki, A. S. 1969. *Geography of the Palearctic Papilionoidea (Lepidoptera).* Krakow: Panstwowe Wydawnictwo Naukowe.

Kurentsov, A. I. 1986. Significance of Beringian links in Holarctic insect zoogeography. In *Beringia in the Cenozoic Era*, edited by V. L. Kontrimavichus, 529–54. Rotterdam: A. A. Balkema.

MacNeill, C. D. 1964. The skippers of the genus *Hesperia* in western North America with special reference to California. *University of California Publications in Entomology* 35:1–230.

Major, J., and S. A. Bamberg. 1963. Some cordilleran plant species new for the Sierra Nevada of California. *Madroño* 17:93–109.

———. 1967. Some cordilleran plants disjunct in the Sierra Nevada of California and their bearing on Pleistocene ecological conditions. In *Arctic and alpine environments*, edited by H. E. Wright and W. H. Osborn, 171–88. Bloomington: Indiana University Press.

Martin, L. M., and C. H. Ingham. 1930. An annotated list of the diurnal Lepidoptera of Huntington Lake region, Fresno County, California. *Bulletin of the Southern California Academy of Sciences* 29:115–34.

McCoy, E. D. 1990. The distribution of insects along elevational gradients. *Oikos* 58:313–22.

McGlashan, X. 1914. List of Truckee butterflies. *The Butterfly Farmer* 1:125–27.

Miller, L. D., and F. M. Brown. 1981. *A catalogue/checklist of the butterflies of North America north of Mexico.* Memoir 1. Sarasota, FL: Lepidopterists' Society.

Newcomer, E. J. 1910. The butterflies of the Lake Tahoe region. *Entomological News* 21:274–317.

Pollard, E., and T. J. Yates. 1993. *Monitoring butterflies for ecology and conservation.* London: Chapman and Hall.

Porter, A. H., and H. J. Geiger. 1988. Genetic and phenotypic population structure of the *Coenonympha tullia* complex (Lepidoptera: Nymphalidae: Satyrinae) in California: No evidence for species boundaries. *Canadian Journal of Zoology* 66:2751–65.

Porter, A. H., and A. M. Shapiro. 1991. Genetics and biogeography of the *Oeneis chryxus* complex (Satyrinae) in California. *Journal of Research on the Lepidoptera* 28:263–76.

Raven, P. H., and D. I. Axelrod. 1978. Origin and relationships of the California flora. *University of California Publications in Botany* 72:1–134.

Richerson, P. J., and K.-L. Lum. 1980. Patterns of plant species diversity in California: Relation to weather and topography. *American Naturalist* 116:504–36.

Rollins, R. C. 1982. A new species of the Asiatic genus *Stronganowia* (Cruciferae) from North America and its biogeographic implications. *Systematic Botany* 7:212–20.

Sánchez-Rodríguez, J. F., and A. Baz. 1995. The effects of elevation on the butterfly communities of a Mediterranean mountain, Sierra de Javalambre, central Spain. *Journal of the Lepidopterists' Society* 49:192–207.

Scott, J. A. 1986. *The butterflies of North America*. Stanford, CA: Stanford University Press.

———. 1994. Biology and systematics of *Phyciodes*. *Papilio* 7:1–120.

Scott, J. A., and M. E. Epstein. 1987. Factors affecting phenology in a temperate insect community. *American Midland Naturalist* 117:103–18.

Shapiro, A. M. 1974. The butterfly fauna of the Sacramento valley, California. *Journal of Research on the Lepidoptera* 13:78–82, 115–22, 137–48.

———. 1978. The alpine butterflies of Castle Peak, Nevada Co., California. *Great Basin Naturalist* 37:443–52.

———. 1979. *Erynnis baptisiae* (Hesperiidae) on crown vetch (Leguminosae). *Journal of the Lepidopterists' Society* 33:258.

———. 1980. Mediterranean climate and butterfly migration: An overview of the California fauna. *Atalanta* 11:181–88.

———. 1984. Geographical ecology of the Sacramento valley riparian butterfly fauna. In *California riparian systems: Ecology, conservation, and productive management*, edited by R. E. Warner and K. M. Hendricx, 934–41. Berkeley and Los Angeles: University of California Press.

———. 1991. The zoogeography and systematics of the Argentine Andean and Patagonian Pierid fauna. *Journal of Research on the Lepidoptera* 28:137–238.

———. 1992a. Twenty years of fluctuating parapatry and the question of competitive exclusion in the butterflies *Pontia occidentalis* and *P. protodice* (Lepidoptera: Pieridae). *Journal of the New York Entomological Society* 100:311–19.

———. 1992b. Genetics and the evolution and biogeography of some Klamath butterflies in a regional context. In *Proceedings of symposium on biodiversity of northwestern California, Oct. 28–30, 1991*, edited by H. M. Kerner, 237–48. Berkeley: Wildland Resources Center, University of California.

———. 1993. Long-range dispersal and faunal responsiveness to climatic change: A note on the importance of extralimital records. *Journal of the Lepidopterists' Society* 47:242–44.

———. 1995. From the mountains to the prairies to the oceans white with foam: *Papilio zelicaon* makes itself at home. In *Genecology and biogeographic races*, edited by A. R. Kruckeberg, R. B. Walker, and A. E. Leviton, 67–99. San Francisco: Pacific Division of the American Association for the Advancement of Science.

Shapiro, A. M., C. A. Palm, and K. L. Wcislo. 1981. The ecology and biogeography of the butterflies of the Trinity Alps and Mount Eddy, northern California. *Journal of Research on the Lepidoptera* 18:69–152.

Shapiro, A. M., and A. R. Shapiro. 1973. The ecological associations of the butterflies of Staten Island (Richmond County, New York). *Journal of Research on the Lepidoptera* 12:65–128.

Shields, O. 1966. The butterfly fauna of a yellow pine forest community in the Sierra Nevada, California. *Journal of Research on the Lepidoptera* 5:127–28.

———. 1978. *Erynnis brizo lacustra* and *Hesperia columbia* in the Sierra Nevada. *Journal of the Lepidopterists' Society* 32:61–62.

Stanford, R. E., and P. A. Opler. 1993. *Atlas of western United States butterflies*. Denver and Fort Collins, CO: Authors.

Storer, T. I., and R. L. Usinger. 1963. *Sierra Nevada natural history*. Berkeley and Los Angeles: University of California Press.

Thomas, C. D., D. Ng, M. C. Singer, J. L. Mallet, C. Parmesan, and H. L. Billington. 1987. Incorporation of a European weed into the diet of a North American herbivore. *Evolution* 41:892–901.

Tilden, J. W. 1959. The butterfly associations of Tioga Pass. *Wasmann Journal of Biology* 17:249–71.

Tong, M. L., and A. M. Shapiro. 1989. Genetic differentiation among California populations of the Anise Swallowtail, *Papilio zelicaon* Lucas (Papilionidae). *Journal of the Lepidopterists' Society* 43:217–28.

Van Dyke, E. C. 1892. Notes on some of the butterflies of the Yosemite Valley and adjacent region. *Zoe* 3:237–41.

DEBORAH L. ROGERS
Institute of Forest Genetics
Pacific Southwest Research Station
U.S. Forest Service
Berkeley, California

CONSTANCE I. MILLAR
Institute of Forest Genetics
Pacific Southwest Research Station
U.S. Forest Service
Berkeley, California

ROBERT D. WESTFALL
Institute of Forest Genetics
Pacific Southwest Research Station
U.S. Forest Service
Berkeley, California

28

Genetic Diversity within Species

ABSTRACT

Based on our review of literature and survey of geneticists working on California taxa, we find genetic information lacking for most species in the Sierra Nevada. This situation is likely to remain in the future, with specific groups of taxa or occasional rare or high-interest species receiving specific study. Where we do have empirical information, we find few generalities emerging, except occasionally within closely related or ecologically similar taxa. Despite these difficulties in assessing genetic diversity, we direct attention to situations estimated to be most deserving of attention from a genetic standpoint.

Severe wildfire: With the significantly increased risk of severe fires currently facing the Sierra Nevada, large, stand-replacing fires present significant risks to gene pools of most middle- and low-elevation Sierran forests, with direct and indirect consequences to the genetic diversity of plants and animals that live in them.

Habitat alteration: For most taxonomic groups evaluated in the Sierra Nevada, the major threat to genetic diversity is habitat destruction, degradation, or fragmentation. Estimated effects involve not only direct losses of population-level genetic structural diversity but also changes in genetic processes (gene flow, selection), effective population sizes, and genetically based fitness traits. High-priority areas would be the foothill zone on the west slope, several of the trans-Sierran corridors (especially in the central Sierra Nevada), and scattered locations of concentrated development elsewhere.

Silviculture: Management actions that are extensive across the landscape yet intensive in manipulating individuals and populations have the greatest theoretical potential (but limited if no empirical evidence) for direct and significant genetic effects. As such, silvicultural activities, including tree improvement programs, operational forest regeneration (artificial and natural), and timber harvest, potentially affect gene pools of target spe-cies. Fortunately, tree improvement programs in the Sierra Nevada (both public and private cooperatives) have long used sophisticated and ecologically appropriate genetic diversity and genetic conservation guidelines. Similarly, in operational forest regeneration, federal, state, and local regulations regarding genetic diversity in planting mostly have high standards and are backed by a fair amount of research. Seed banks exist for public and private reforestation that maintain high standards of seed origin and genetic diversity, although exigencies presented by potentially large, severe wildfire may not be adequately met. The focus for seed banking is the commercial conifers, and only slowly has seed banking emphasized other species with storable seeds. These programs, which have histories of several decades in the Sierra Nevada, serve as models for other taxa where similar activities occur (e.g., fish stocking).

Research is inconclusive about the long-term genetic consequences of timber harvest on commercial tree species. Nevertheless, traditional silvicultural practices, which were designed primarily to maximize growth of the target species, tended to result in spatial patterns of harvest and live-tree retention that acted in concert with genetic conservation guidelines. By contrast, some new forestry practices, which combine fiber production with ecological stewardship for wildlife and nontimber species, may have potential for minor dysgenic effects on native timber species. For instance, leaving clumps of trees, especially suppressed individuals (e.g., for wildlife protection) may promote inbreeding or lowered fitness if the members of the clumps are related, as they appear to be.

Ecological restoration: Practitioners of ecological restoration have only recently become aware of genetic concerns in planting. Although many programs focus on restoring correct native species, an understanding of the appropriate genetic material within species, its origin, diversity, and collection, remains missing or rudimentary in many programs. Thus, genetic contamination problems may be more severe than if exotic species

Sierra Nevada Ecosystem Project: Final report to Congress, vol. II, *Assessments and scientific basis for management options.* Davis: University of California, Centers for Water and Wildland Resources, 1996.

had been planted. The significance of this genetic threat in the Sierra Nevada is lowest in projects of ecological community restoration and highest in postfire erosion control projects. Frequently these involve grass species and occasionally forb mixes. Although exotic grasses (especially rye grass) previously were used routinely, native grasses are increasingly becoming favored. There is often little understanding of the potential genetic consequences of planting seeds of native species but unknown (often commercial nursery) origins.

Fish management: Management of fish species and genetic diversity within species in the Sierra Nevada is done in a way that potentially disrupts many native gene pools. The introduction of hatchery, nonlocal, and genetically altered genetic stocks of native fish species has had the direct effect of creating conditions for intraspecific hybridization, gene contamination, and gene pool degradation. Indirectly, the introduction of exotic fishes has large effects on biodiversity through displacement of native fish species and impacts on aquatic invertebrates and amphibia, which affects gene pools through loss of populations.

Range improvement: Similar to fish management, although lesser in effect in the Sierra Nevada, is the direction and intent of range improvement projects. In past decades, range shrubs, particularly bitterbrush, were widely planted in Great Basin areas (on the border of the Sierra Nevada) to improve rangelands for cattle. Very little of the shrub germ plasm planted in the past derived from local seed zones or followed genetic diversity guidelines that maintain native genetic structure. More recently, shrubs have been planted for wildlife habitat enhancement. These are increasingly falling under seed transfer and genetic diversity guidelines, with the result that native local seeds are being collected and planted.

Exotic pathogens: Exotic pathogens create direct and indirect genetic threats in the Sierra Nevada. For example, white pine blister rust is fatal to sugar pines that carry the susceptible gene. The resistant gene exists naturally in very low frequencies in sugar pine. Although a well-funded and genetically sophisticated program exists for developing and outplanting sugar pine that is resistant to white pine blister rust, there has been limited recognition of the genetic consequences of the current federal harvest practices for the species. At present, known resistant old-growth sugar pines are not cut, but susceptible trees may be harvested, and in areas where resistance is unknown, harvest proceeds without genetic testing. The potential loss of genetic diversity, through harvest, of traits other than the resistance loci is significant. Indirect genetic effects occur when populations are so devastated as to drastically decline in size or become extirpated. An example is the exotic pathogen that moves from domestic to native bighorn sheep This pathogen causes a disease that is extremely serious and usually fatal to bighorn sheep, exterminating populations.

Taxon-specific issues: Many activities theoretically have significant genetic effects on specific taxa in the Sierra Nevada. Examples of these include the sport collecting of butterflies, the harvesting of special forest products (especially mushrooms and other fungi, ladybird beetles, lichens, etc.), the use of biocides with wide action against native insects, and forest-health practices whose goals are to reduce or eliminate populations of native insects and pathogens.

Land management: Most human-mediated (as well as natural) activities have some genetic consequences. The question is not whether we create genetic change, but which effects are significant enough to warrant altering our behavior. In general, there has been a pervasive lack of awareness of the potential genetic consequences of land management, from local practices to regional landscape plans. Genetic awareness, evaluation, prescription, mitigation, monitoring, and restoration have generally been very low in public and private management and have been concentrated in a few land-use programs (e.g., tree regeneration). Although it is broadly recognized that most management actions have effects on wildlife, there are few instances where environmental analyses—for instance in National Environmental Policy Act (NEPA) contexts—have considered genetic effects. Land-management agencies do not place geneticists broadly throughout the Sierra Nevada, and genetic knowledge usually resides centralized (e.g., with tree improvement headquarters) or within silviculture staffs, where it is focused mostly on the already established genetic management programs of commercial timber species.

What is needed is a general awareness that genetic consequences must be considered and evaluated for land-management activities in general, and a framework and strategy for doing so. It is not enough to lump these concerns under general biodiversity evaluation, since this often takes into account only immediate effects on the population or species viability of a few indicator species. This chapter proposes some management guidelines and standards for preserving and enhancing genetic diversity in the Sierra Nevada.

INTRODUCTION

Genetic diversity is not a front-page, public issue. Whereas species extinctions, loss of old-growth forests, and degradation of air and water quality are readily grasped and easily comprehended, to many people gene pool integrity remains arcane, invisible, and dismissable as academic. Yet genes are the fundamental unit of biodiversity, the raw material for evolution, and the source of the enormous variety of plants, animals, communities, and ecosystems that we seek to protect and use. Genetic variation shapes and defines individuals, populations, subspecies, species, and ultimately the kingdoms of life on earth. The gene pool of widespread species is spread throughout many populations; for a rare species it may consist of a single population. From one species to the next, the composition and structure of individual gene pools vary. Each has a unique relationship to the viability and long-term survival of the population and species.

Human actions on the landscape almost always have some genetic effect. While many changes in genetic diversity occur naturally (genetic change is the basis of evolution), human activities in the Sierra Nevada, as elsewhere, may accelerate or change the direction of evolution in undesired ways. Genetic erosion, genetic engineering, genetic contamination, and extinctions of populations and species are potential effects or sources of genetic change mediated by humans. What are the responsibilities of SNEP and of decision makers in the Sierra Nevada for addressing genetic concerns in policy development and land management? As is the case with other biodiversity issues, the main questions regarding genetic diversity are

- What important compositional, structural, and functional genetic diversity exists in Sierra Nevada taxa?

- How much, what kind, and what distribution of genetic diversity is desired or enough?

- How do human activities affect, both directly and indirectly, genetic diversity detrimentally, and what actions can be taken to prevent or mitigate undesired consequences?

Although these questions are reasonable theoretically, our ability to answer them is extremely limited by lack of information. If we consider that the genes of all organisms from all species (known and unknown) of the Sierra Nevada collectively make up the gene pool of the range, we begin to see why even a basic inventory of genetic diversity is impossible to obtain practically. Genetic diversity is difficult to measure; cannot be observed, counted, or monitored directly in the field; and requires the use of either elaborate laboratory methods or long-term field trials for detection. Genetic interpretation depends on information from proxies and markers that don't necessarily reflect traits of interest to managers. Ultimately, it is unknowable today what genes will be important as raw material for the evolution of adaptations to meet unknown environmental challenges of the future.

With one significant exception, genetic conservation concerns in land management have for the most part been lumped into the category of biodiversity management and not directly tackled in regional land-management policy or practice. Forest genetic programs have long made use of sophisticated genetic conservation and management policies and practices, both in operational forest regeneration and in tree improvement programs. Beyond the scope of commercial forest trees, however, the ecological consequences of genetic changes brought about by land management have only begun to be addressed programmatically. The U.S. Forest Service (USFS), for example, has expanded its forest genetics programs to provide guidance to all taxa (Hessel 1992). In 1992, a scientific roundtable convened in Wisconsin to develop regional management recommendations for ecosystem management of the Chequamegon and Nicolet National Forests. This is one of the few bioregional efforts where genetic diversity concerns pertaining to many aspects of land management were addressed (Crow et al. 1994).

Objectives

The inherent nature of genetic diversity and its recalcitrance to measurement and interpretation make the task of assessing genetic diversity in the Sierra Nevada quite different from assessments of other biodiversity attributes. Notwithstanding practical barriers, genetic theory is very well developed and has been tested and confirmed in extremely successful genetic manipulations in medicine, agriculture, and animal husbandry. This theory, along with the direct genetic studies that have been done for some taxa in the Sierra Nevada, provide the basis for both our genetic assessments and our suggestions for genetic management. Since geneticists tend to focus on specific taxonomic groups rather than working across taxa, there has been little sense of how much information is actually available in total, what the genetic patterns are for various taxa (whether the patterns are concordant or conflicting), or what the implications of this information for management might be. Information on genetic diversity of the Sierra Nevada is scattered in the literature and has not previously been compiled under a common theme. The objectives for this chapter, therefore, also differ somewhat from other SNEP assessments:

- Inform the public and land managers about pertinent questions and priorities regarding genetic diversity and its role in ecosystem health and sustainability; bring them to a broader awareness and understanding of the concerns and opportunities of genetic diversity.

- Compile information collectively about genetic diversity for major taxonomic groups in the Sierra Nevada, summarizing patterns of within- and among-population genetic composition and structure relevant to the long-term health, sustainability, and management of populations.

- Assess genetic diversity in the few cases where information is available, recognizing that general trends cannot be developed from these specific cases.

- Assess genetic diversity indirectly, using inferential tools as available. In many cases, the best that can be done is to develop conceptual frameworks to guide future individual, local, and case-specific assessments.

- Suggest approaches for integrating genetic diversity concerns and opportunities into land-management planning and practice.

This report documents efforts to address the SNEP assessment and management questions as they pertain to genetic variation within species of the Sierra Nevada:

- What are the current conditions? We develop here a summary overview of what is known about the gene pools of major taxonomic groups within the Sierra Nevada—the amount and pattern of genetic variation, which species are best genetically studied, and which are least well understood. We further attempt to identify, at a broad level, genetic significance in terms of rich, rare, or representative portions of the gene pools and any evidence of the factors underlying the genetic patterns observed.

- What were historical conditions, trends, and variabilities? Very little historical information exists on genetic variation, and even less exists that is specific to the Sierra Nevada. Many of the tools currently in use for measuring and monitoring genetic variation are relatively recent (e.g., allozymes and DNA techniques). The few species that have been the subject of temporal genetic studies (e.g., a few insect and fish species) have brief life cycles, and the studies investigated less than a decade in the lifetime of the species. The extinction of species and the expansion and contraction of their ranges is frequently a subject for study through the pollen record (e.g., Anderson 1990), and the impact on levels of genetic variation is inferred (e.g., Critchfield 1986). However, the changes in genetic variation over the lifetime of extant species are rarely assessed directly.

    Researchers frequently analyze historical relationships among taxonomic groups by studying current levels of genetic variation and inferring the time since divergence of these species based on the amount of genetic dissimilarity or distance. However, assumptions, rather than direct evidence, form the basis for this type of study, and these assumptions are built on tenuous theoretical or empirical foundations. Further, they are more often directed at relationships among species rather than relationships among subspecies or populations within a species. This type of study has not been included in this chapter.

    Thus, the genetic answer to the question regarding historical conditions relies mainly on theoretical, rather than empirical, evidence. Any evidence of historical trends, including any apparent relationships with climatic or geographic factors, has been reported in the section "Inferences of Genetic Significance."

- What are the trends and risks under current policies and management? Threats to the genetic integrity of Sierra Nevada species can be either direct (e.g., genetic contamination of native gene pools of fish by hybridization with introduced exotics or non-native populations) or indirect (e.g., increased inbreeding leading to inbreeding depression of certain species due to fragmentation of their habitats via land-conversion practices). We address these threats with empirical evidence or specific examples where available, and with implications of theoretical consequences in the absence of such data. Particularly vulnerable areas or species are identified. We identify specific policies and practices that historically, currently, or potentially affect genetic composition and/or structure, as well as the nature of the effects on gene pools (e.g., increases or decreases in genetic diversity). The difference between a positive outcome and a negative one is one of context: the species targeted and the specific quality of the populations affected, the temporal and spatial context, and the manner and scale in which the policy or practice is applied dramatically affect whether an action is a genetic threat or not.

- What are the genetic management options for the future? We summarize some specific ongoing programs in the Sierra Nevada, suggest generic guidelines that could be more broadly applied in land management and policy situations, and offer general strategies for integrating genetic diversity considerations into land management.

Assumptions

We made the following assumptions in the preparation of this chapter:

- We assume that the goal of land management is to maintain and promote ecosystem health and sustainability. This also becomes the goal of genetic conservation, as explicitly assumed in this chapter, and the standard by which we evaluate the status and trends of genetic diversity in the Sierra Nevada.

- Genetic diversity is fundamental to, and thus critically important for, the short- and long-term viability of Sierra Nevada taxa and to the integrity of the ecosystems they compose. Most traits of interest in managing taxa, populations, and ecosystems have genetic bases, although environmental variation plays an important role in determining phenotypic plasticity and response.

- Changes in gene pools occur naturally and continuously, in response to natural selection and stochastic effects (e.g., gene flow, mutation, genetic drift).

- Human actions affect genetic diversity. Some kinds of genetic change mimic natural change or are negligible, acceptable, or desirable; others are undesired and warrant preventative actions or mitigation.

- Direct genetic data are extremely limited, and interpretations regarding the ecological and evolutionary significance of genetic changes are limited.

- In the absence of direct data, genetic and genecological theory is strong enough to support cautious inferences regarding assessments of genetic diversity, to evaluate management effects, and to suggest practical management and monitoring guidelines.

- Case-by-case assessments, evaluations, and management prescriptions are essential and are not developed here other than to provide examples. Few generalizations are robust across taxa and situations. When offered, they are tentative.

BACKGROUND AND METHODS

Genetic variation is not readily measured in organisms from native habitats because of the confounding effect of genetic and environmental influences on phenotypic variation. To measure genetic variation nearly always requires removing individuals from their native habitats and either growing them in experimentally controlled environments or using laboratory analyses to assess traits whose expression is not greatly influenced by the environment. Because these analyses are neither field based nor particularly intuitive, we give background information on these methods here to aid the reader in understanding and interpreting their results.

Genetic Hierarchies

Because genes are basic building blocks of biological organisms (e.g., populations and species) and biological assemblages (e.g., communities and ecosystems), their diversity is expressed at hierarchical levels. Different processes are more important at the various levels, and thus assessments and management considerations must take these scalar issues into account, remaining cognizant of the relevant context. Genetic diversity is manifested as differences within and among gametes or embryos (haploid/diploid), individuals within populations, interbreeding populations (also called demes), ecotypes, local races or strains, geographic races and subspecies, and species.

Factors that influence and determine the structure of genetic diversity play different roles at the various levels (table 28.1). Mutation occurs within individuals and is expressed among individuals within populations; gene flow (geographic migration and exchange of genes) occurs among individuals within populations, among populations, and occasionally among ecotypes or races or at higher levels (e.g., interspecific hybridization). Natural selection exists on all levels but gains in ecological and evolutionary significance at the level of individuals within populations and among populations. Genetic drift (stochastic changes in genetic diversity due to sampling phenomena of small population size) occurs within and among populations. Inbreeding is another genetic process that is significant within small populations of some species.

The genetic effects of natural and human actions also vary in significance at the different levels. At the level of individual organisms, change in phenotype (including death) is the most obvious effect of mutation. At the population level, effects are observed as changes in allele, genotype, and phenotype frequencies and, correspondingly, in population expansions

TABLE 28.1

Structural, compositional, and functional levels of genetic diversity (only functions that are most significant for the level are given).

Structure	Composition	Function
Gametes, embryos	Sperm, pollen, spores, seeds	Mutation, fertilization
Individuals	Individual plants, animals, etc.	Mutation, mate selection
Populations, demes	Interbreeding individuals	Gene flow, natural selection, drift, inbreeding
Ecotypes, local races	Genetically distinct demes	Natural selection, drift, gene flow
Geographic races and subspecies	Genetically distinct groups of ecotypes and local races	Natural selection, drift
Species	Genetically distinct groups not regularly interbreeding	Natural selection, drift

and population extinctions. At the species level, genetic effects result in the creation of new species (speciation) or extinction. Although genetic diversity is relevant from the molecular to the ecosystem level, the focus of this chapter is on aspects of genetic biodiversity within species.

Although genetic variation can be considered at various spatial scales, it is difficult to standardize across species, since the relevant domain varies according to the size, mobility, and genetic structure of the species. For example, more-local scales might be appropriate for discussing genetic variation in small, relatively immobile (e.g., nonmigratory, having a restricted gene flow), and/or locally adapted species, while broader scales would be more appropriate for larger, highly vagile (e.g., migratory, having pollen flow or seed dispersal across long distances), and/or broadly adapted species. The scale at which genetic information is available for any given species is also a function of the sampling design of available studies. Thus, genetic variation is discussed at the level most appropriate and for which there is information available across species groups.

To summarize, we recognize the following aspects to genetic diversity in regard to ecosystem health and sustainability:

- Genetic diversity, both structure and process, is both input to ecological systems (i.e., it influences and determines fitness, viability, and evolution) and output (i.e., ecological and environmental effects determine genetic structure and process) at many ecosystem levels (individual to population to species to community).

- Genetic diversity is but one factor that contributes to the status and health of ecological systems. The significance of genetic diversity relative to other factors (e.g., demography, reproduction, and stochastic events such as disturbance) varies by taxon, location, season, and so on.

- Because each species has unique life histories and unique ecological relationships, the resulting genetic architectures and the importance of genes to viability and sustainability are unique.

- Human actions in ecosystems are analogous in their potential impacts to natural forces and are potent in their ability to alter genetic structures and processes, with diverse effects on viability and sustainability. Human actions can take place in concert with natural processes or can run counter to them.

The standards against which we implicitly measure status and trends of genetic diversity in the Sierra Nevada in this chapter are those that exist in those native Sierran ecosystems that have been minimally disturbed by human activities relative to their historical condition. For example, many wilderness and noncommercial forestlands in public ownership would be regarded as reference conditions, despite the effects of fire suppression, grazing, air pollution, and so on. From a genetic standpoint, even many manipulated lands (e.g., those used for timber) may not be far from an original condition, depending on the silvicultural treatments used. We recognize that these conditions are not "natural" or "pristine" in being uninfluenced by humans, either prehistorically or historically, directly or indirectly. We assume that effects on genetic diversity of prehistoric human use and of many recent human activities, however, are relatively minor at the broad scale in the Sierra Nevada, although local effects may have been intense.

Accepting such conditions as a reference standard does not imply that they are optimally adapted, inherently ideal, or naturally in balance. We accept this reference for evaluating genetic diversity for several reasons:

- Near-natural systems are the closest analogs to sustainable systems that we can describe—the dynamism and change in such systems are of a level that we accept in management.

- In many cases we cannot actually measure either the quantitative status of genetic diversity or absolute values of the contribution of genetic diversity to individual, population, and species viability. We can, however, make an assessment of whether particular actions have caused or might cause deviation from the present state, we can predict what the genetic consequences might be, and we can project what the impacts of such actions on health and sustainability might be. Thus, because our ability to evaluate impacts is limited to relative change, we take the present, minimally disturbed state as the standard for comparison.

- By accepting minimally disturbed conditions as reference, we do not risk the arrogance of assuming that humans can predict, understand, or create optimal conditions better than natural structures and processes can; instead we simply compare results between the reference and the manipulated lands.

Where systems are highly altered and the current conditions obviously do not serve as adequate reference standards, standards must derive either from adjacent or analogous minimally disturbed locations or from an analysis of historic ranges of variability.

Measurement of Genetic Variation

Several kinds of data are used to measure levels and patterns of genetic variation. The longest-standing method is to describe readily observable attributes of individuals, such as various metric traits, color, time required to reach certain developmental stages, and so on. (The collective attributes of an individual, the result of both genetic and environmental influences, is a phenotype.) This is typically called morphological data, or morphometric if quantitative. However, to detect heritable variation in these traits, the environment in which the individuals are raised or grown must be uniform and subject to experimental controls (i.e., replication and randomization); otherwise, observed differences could simply be environmental effects rather than genetic differences. For obvious reasons, plants are more amenable to this type of genetic assessment. This type of data has historically been used for genetic studies and does not necessarily require any sophisticated equipment. Whereas common-garden studies identify genetic differences within the same generation, the complementary type of analysis more common for animals—pedigree analysis—requires several generations.

A second type of commonly collected genetic data is biochemical traits, including enzymes, terpenes, flavonoids, blood groups, and other physiological markers. The most commonly studied of these traits are differences in isoenzymes—called allozymes. Such enzymes can readily be extracted from tissue samples, such as leaves or blood. In general, these enzymes have been shown to be under strict genetic control, and differences in the forms of each enzyme can be interpreted as genetic differences between the individuals sampled. This technique has been in common use for a broad range of organisms since the 1970s. Because the enzymes are typically not modified by the environment of the organism, samples can be taken from individuals in diverse geographic areas and reliably given a genetic interpretation.

The third type of genetic data, and the most recent to become available, is derived from a collection of techniques that assess molecular traits, such as DNA (from the nucleus, mitochondria, and/or chloroplast) and RNA. For example, DNA can be extracted from tissue samples from individuals, and variation between individuals can then be assessed either by comparing random segments of the DNA or by sequencing the DNA directly. Like the allozyme data, most DNA data are free from environmental influences and can be directly interpreted as genetic differences. This type of data has been gaining in popularity since the late 1980s.

The relationships among these kinds of genetic data (morphometric, allozyme, and DNA) are not well understood: sometimes they present concordant patterns, sometimes not. The genetic basis and freedom from environmental influence of biochemical and molecular traits makes them attractive for genetic studies. However, the adaptive and ecological importance of the genetic variation they reflect is uncertain. To some extent, observed differences among the results from the three data types may reflect study parameters and assumptions. For example, morphometric data for animal taxa are often confounded by environmental influences. Plant species are more amenable to being grown in common gardens; the morphometric traits measured in such situations therefore reflect genetic differences, the environmental component having been controlled. When morphometric data are obtained in common-garden studies in native habitats, or when correlations are found with environments, they are often considered to reflect some aspect of adaptively significant genetic variation.

Allozyme data, often considered to reflect neutral genetic variation (i.e., variation that is not under selective pressure and therefore represents time-dependent divergence based on reduction or lack of gene flow rather than adaptively significant genetic variation), sometimes show levels and patterns of genetic variation that are concordant with the morphometric data, especially when they are analyzed by multivariate methods. Often, however, they overemphasize certain patterns (e.g., within-population versus among-population genetic variation) relative to morphometric data. In some instances, certain genetic markers have been shown to be under the influence of strong selection, challenging the assumption that they are basically neutral. The significance of allozyme data thus seems to depend on a variety of factors, including the number and type of allozymes studied, the taxon under consideration, and the type of data treatment.

DNA studies sometimes agree with the results from one or both of the other two data types and sometimes present a different perspective. Current and common understandings of DNA studies suggest that, since plastid (chloroplast or mitochondrial) DNA evolves more quickly than the nuclear DNA (which is the basis of morphometric and allozyme data, as well as some molecular—e.g., PCR—data), studies of plastid DNA reflect more recent influences on genetic variation. Under this model, for example, a recently colonized species might show little genetic variation based on allozyme data but more variation in plastid DNA within and/or among populations, suggestive of recent and local processes, including adaptation.

Clearly, we are still learning about the relationships among the questions we are asking, the types of genetic data we collect, and the adaptive, evolutionary, management, and conservation implications of these data. In general, we recognize the effects of study design (including amount of data, scale at which data were collected, etc.) in determining this relationship. Life history characteristics are also likely to figure greatly into this relationship. There may also be some general trends according to taxonomic group, again largely a function of life history characteristics. Thus, our presentation of genetic information includes reference to the type of data and how the results agree or disagree among data types. The genetic significance is then discussed relative to the type of data, type of studies, level of representation of the available studies, and taxonomic group. If there are generalizations that can be made within a taxonomic group, these are also presented.

Conventional measures of genetic variation often refer to the hierarchical organization described earlier. Within-population genetic variation, when measured with allozymes, is often expressed as *heterozygosity,* an estimate of the percentage of individuals in a population who have two alternate forms of a gene, averaged over all the genes considered in the study. A statistic represented as F_{ST} reflects the amount of genetic differentiation within a species or the relative amount of variation among populations. The various forms of each (studied) gene present, and their frequencies, are often compared among populations, subspecies, or even closely related species, using indices that reflect the amount of genetic similarity or, conversely, genetic difference. The latter is more commonly known as genetic distance. Standards for measuring genetic diversity and population differences have changed over the twenty years that allozyme data have been collected, from similarity measures to the various genetic distances (Nei's and Rogers' are most commonly cited) to measures of population differentiation—F_{ST} (Wright 1978).

Information Acquisition

The information contained in this chapter was assembled in a three-stage process. First, the SNEP Genetics Workshop was convened in Placerville, California, on September 22 and 23, 1994. On that occasion, the attending individuals offered their research findings, general understandings, and educated opinions regarding genetic variation in various taxa within the Sierra Nevada. Approximately twenty-five scientists participated (appendix 28.2). The information gained from this workshop was captured in a preliminary report. Second, literature searches were conducted to fill in information gaps from the preliminary report and to include available references and context for the information offered during the workshop. Finally, the expanded report was circulated to all workshop participants, and also to approximately fifteen others, for review and comment. This chapter incorporates all review comments.

CURRENT CONDITIONS

Genetic Variation of Major Taxonomic Groups in the Sierra Nevada

Our general objective is to summarize the available information on genetic variation in species of the Sierra Nevada ecosystems. This is not an exhaustive inventory but rather a summary of the types of information available, highlighting the taxa that are best and least studied in each major taxonomic grouping (table 28.2). Life forms have been included and categorized according to the availability of data. Thus, we present information organized within seven taxonomic divisions: plants, mammals, birds, amphibians and reptiles, fish, insects, and fungi.

Information is included from many available sources, not only the published literature, and may refer to data from morphological, allozyme, and DNA studies. In general, the summary is limited to intraspecific genetic variation rather than the taxonomic or systematic studies at the species level, although this is noted when only the latter types of information are available. Patterns in genetic architecture, when known, are noted. Associations between geographic and genetic patterns are described, either as generalizations or by specific examples if data do not support generalizations. Further, any relationships between life history characteristics and patterns of genetic variation for that taxonomic group are presented. Finally, major research needs or concerns, as expressed by workshop participants, are noted.

Plants

There are approximately 3,500 species of vascular plants in the Sierra Nevada, representing approximately 50% of the taxa in California (Shevock 1996). Of those, approximately 45 are

tree species (Griffin and Critchfield 1972). Among Sierran plant species, most genetic studies have been on tree species, and most of these have focused on the widespread and commercially important conifer species (table 28.3). Best studied are several pine species, most notably ponderosa pine (*Pinus ponderosa*) (e.g., Conkle and Critchfield 1988) and sugar pine (*P. lambertiana*) (e.g., Harry et al. 1983), followed closely by Douglas fir (*Pseudotsuga menziesii*) and white fir (*Abies concolor*) (e.g., Hamrick 1976). For these species, there are comprehensive accounts of genetic variation among and within populations in the Sierra, based on both morphological and allozyme data. Most other pine species of the Sierra, as well as other conifers such as giant sequoia (*Sequoiadendron giganteum*) (Fins and Libby 1982) and incense cedar (*Calocedrus decurrens*) (Harry 1984) have been studied at the morphological and/or allozyme levels. Some genetic information is also available for Jeffrey pine (*Pinus jeffreyi*), Washoe pine (*P. washoensis*), whitebark pine (*P. albicaulis*), foxtail pine (*P. balfouriana*), grey or foothill pine (*P. sabiniana*), knobcone pine (*P. attenuata*), lodgepole pine (*P. contorta*), red fir (*Abies magnifica*), and cypress species (*Cupressus* spp.) (see appendix 28.1). Recently, interest stemming from conservation concerns has prompted genetic studies of several oak (*Quercus*) species, resulting in data on allozyme variation within and among populations (Millar et al. 1990a). There is some population-level genetic information for a few other angiosperm species, but much of it is inferred from studies outside of the Sierra Nevada (as is the case with *Populus trichocarpa*) (Dunlap et al. 1994).

The substantial amount of information on the genetic architecture of many Sierran tree species is based largely on morphological and/or allozyme data. Because common-garden studies were the early standard for genetic studies of forest tree species (begun over fifty years ago in the Sierra Nevada), the "morphological" variation referred to here is

TABLE 28.2

Characterization of information on genetic variation for the biota of the Sierra Nevada.

Taxon Division	Research Emphasis	Least Studied
Plants	Gymnosperms: commercially significant tree species, pines in particular	Most angiosperm species, especially geographically restricted and/or endemic species
	Angiosperms: Exotic annuals, low-elevation species, herbaceous dicots, oaks, woody shrubs, widespread species	
Mammals	Pre-1940s distributions of species and subspecies	Current distributions of species and subspecies
	Population studies for a few species of ground squirrels, gophers, rats and mice	Populations of most species Bats
Birds	Population studies for approximately 15 species in Sierra Nevada	Raptors
Reptiles and amphibians	In general, commercial species and rare or endemic species Turtles (one species); some information on lizards Salamanders; some toads and frogs	Widespread, common species All snake species
Fish	Salmonids, commercial species	Non-game-fish species Salmonids in central/western Sierra Nevada
Insects	Human-interest species, including insect pests to crops and trees, disease-vector species, and butterfly species (aesthetics)	Most species, including such major taxonomic groups as aquatic species and (specialist) species with rare host plants
Fungi	Fungal pathogens of commercially significant plants (e.g., *Heterobasidium, Verticicladiella, Peridermium* spp.)	Most fungal species, especially those in the Zygomycota and Chytridiomycota

TABLE 28.3

Genetic variation (heterozygosity) within populations of Sierra Nevada plants.

Species	Number of Loci	Heterozygosity[a]	F_{ST}	Sampling Range	Reference
Gymnosperms					
Abies concolor		0.24		Central Sierra Nevada	Conkle 1992
Calocedrus decurrens	20	0.21	0.042	California	Harry 1984
Pinus attenuata		0.14	0.12	California	Millar et al. 1988
Pinus jeffreyi	20	0.185	0.068	Central Sierra Nevada	Furnier and Adams 1986
	20	0.255	0.092	Southern Sierra Nevada	Furnier and Adams 1986
	25	0.137	0.004	Mono County	Millar et al. 1993
Pinus lambertiana		0.25		Oregon and California	Conkle 1992
Pinus ponderosa		0.21		Sierra Nevada	Conkle 1992
Pinus sabiniana		0.14		California	Conkle 1992
Pinus washoensis	26	0.15		Oregon and California	Niebling and Conkle (1990)
Pseudotsuga menziesii		0.28		?	Conkle 1992
Sequoiadendron giganteum	8	0.143		Sierra Nevada	Fins and Libby 1982
Taxus brevifolia	11	0.170	0.107	Continental U.S. and Alaska	Doede et al. in press
Angiosperms					
Antennaria corymbosa	19	0.078	0.173	Rocky Mountains	Bayer 1988
Antennaria media	19	0.058	0.07	Sierra Nevada	Bayer 1989b
Antennaria rosea	19	0.114	0.378	Western U.S.	Bayer 1989a
Avena spp.					Allard et al. 1968
Bromus tectorum	25	0.012	0.478	U.S.	Novak el al. 1991
Calochortis minimus	16	0.168		Sierra Nevada	Ness et al. 1990
Calochortis nudus	16	0.095		Sierra Nevada	Ness et al. 1990
Clarkia speciosa	17	0.183	0.043	Kern County	Soltis and Bloom 1986
Elymus glaucus	20	0.086	0.549	Washington, Oregon, California	Knapp and Rice in press
Ipomopsis aggregata	23	0.099		Sierra Nevada	Wolf et al. 1991
Lewisia spp.	22		0.541	Sierra Nevada	Carroll et al. n.d.
Lewisia cantelovii	22	0.208		Yuba and Sacramento Rivers	Carroll et al. n.d.
Lewisia congdonii	22	0.234		Yosemite and Kings Canyon National Parks	Carroll et al. n.d.
Lewisia serrata	22	0.148		American River	Carroll et al. n.d.
Salix exigua	15	0.122	0.258	Southwestern U.S.	Brunsfeld et al. 1991
Salix melanopsis	15	0.147	0.180	Northwestern U.S.	Brunsfeld et al. 1991
Scutellaria bolanderi	18	0.023	0.720	Sierra Nevada	Olmstead 1990
Scutellaria californica	18	0.129	0.288	California	Olmstead 1990
Scutellaria nana	18	0.117	0.327	Northern California and Nevada	Olmstead 1990
Scutellaria siphocampyloides	18	0.042	0.628	California	Olmstead 1990
Wyethia					Ayers (SNEP Genetics Workshop)
Ferns					
Cheilanthes gracillima			0.286	Eldorado County	Soltis et al. 1989

[a]Heterozygosity is the proportion of heterozygous genotypes per locus per individual.

genetically based, without the problems of plasticity that typify studies where field observations of morphology are made without controlled tests. Some common-garden studies, taking advantage of vegetative propagation techniques, have examined genetic architecture from the population to the clonal level of variation (e.g., Fins and Libby 1982). The general trend seen in most genetic architecture studies is one of substantial genetic variation residing within populations, although the amount of such variation ranges widely among traits and can change with age (Namkoong and Conkle 1976).

Genetic diversity within populations (as measured by allozymes) tends to be relatively high in Sierran conifer populations (table 28.3) in comparison with that found in gymnosperm species in general (e.g., mean heterozygosity is 0.151 [Hamrick et al. 1992]). Differences among populations tend to be fairly low ($F_{ST} < 0.10$) (table 28.3), consistent with gym-

nosperms in general and with woody plant species that are wind dispersed and wind pollinated (Hamrick et al. 1992). As is the case with metric traits, differentiation can be greater along elevational gradients in the Sierra than along latitudinal ones. For example, the F_{ST} among four elevational transects in sugar pine, which ranged from the Eldorado National Forest in the north to the Sequoia National Forest in the south, was 0.015, whereas differences from low to high elevation along each transect averaged 0.038 (Westfall 1995). Among congeneric species, the distribution of genetic variation fluctuates. Among the pines, there tend to be large genetic differences among species, as is the case among the closed-cone pines (Millar et al. 1988). In the cypresses, the differences are much more modest (Millar and Delany n.d.). This pattern is not uniform among all genetic markers. For example, differences among the closed-cone pine species are much lower in mitochondrial DNA (Strauss et al. 1993) than in allozymes.

As these patterns are based largely on studies of the widespread conifer tree species—outcrossing, wind-pollinated gymnosperms—it is not known whether they will hold true for species with different life forms, specific habitats (e.g., riparian), or different mating systems, dispersal systems, or modes of reproduction (e.g., clonal species). Also, some replicated common-garden experiments point to the existence of genotype x environment interactions in some species. However, these data are neither abundant nor consistent enough to permit generalizations about the significance of these interactions.

The taxa best represented in the genetic knowledge of nontree plants in the Sierra Nevada are characterized as exotic annuals, especially low-elevation species. For example, Novak and colleagues (1991) found substantial rangewide differentiation in *Bromus tectorum*. Beyond the exotics, native taxa have been studied in a nonordered fashion, with scattered representation from both herbaceous dicots and shrubs (table 28.3), and much of this work has focused on taxonomic issues rather than intraspecific genetic structure. Comprehensive information on the genetic architecture of species is scarce. The studies of widespread species have focused on among-population variation, while the little information available for more restricted species is usually reflective of within-population variation.

Levels of genetic diversity within populations of many of the Sierran angiosperms are nearly as high as those of the gymnosperms (table 28.3). In some species, especially those limited in range or habitat, such as Bolanders skullcap (*Scutellaria bolanderi*), diversity is low (Olmstead 1990). Genetic differentiation among populations of many of the angiosperm species studied is quite high compared to that of gymnosperms, suggesting genetic isolation among populations. For example, differences among Sierran willow (genus *Salix*) populations isolated by river drainages tend to be high (Brunsfeld et al. 1991). In recent studies of the native grasses *Nassella* and *Danthonia*, Knapp (1994) and Knapp and Rice (1994b), respectively, have noted that allozyme variation is correlated with geographic distance. However, this is not the case for blue wild rye (*Elymus glaucus*) (Knapp and Rice 1995). Genetic drift (in small populations) and interruption of gene flow have been suggested as the genetic processes responsible for this lack of relationship (Slatkin 1993). Few direct studies of gene flow between species have been conducted. Nason and colleagues (1992) found evidence for first-generation gene flow between two Sierran species of manzanita (genus *Arctostaphylos*), one occupying dry sites and the other moist ones.

Few studies are available that have investigated clonal diversity in asexually reproducing plants in the Sierra Nevada. In a study of *Antennaria rosea* over much of its range in western North America, it was observed that populations tend to be composed of one or a few rather localized clones (Bayer

1990). Samples from the Sierra Nevada had some of the highest numbers of clones per population (i.e., eleven clones). The proportion of polyclonal populations detected (73%) is similar to the average reported for a wide range of clonal plant species (77%) (Ellstrand and Roose 1987).

Differentiation in other biochemical traits has also been studied in Sierran species, both gymnosperms and angiosperms. Desrochers and Bohm (1993) found greater complexity in flavonoid compound (responsible, in part, for color) profiles in southern *Lasthenia californica* populations than in northern ones. In the monoterpene profiles (monoterpenes being components of pine pitch), Zavarin and colleagues (1993) found differentiation between the Sierran and the Great Basin and Rocky Mountain populations.

A major constraint in assessing the genetic variation in plants of the Sierra Nevada other than trees is the lack of basic biological information, such as fine-scaled species-distribution maps and approximate population sizes. As it is not feasible to conduct common-garden studies for all plant species, plasticity poses a problem for genetic studies based on variation in morphological characters observed in the field. One approach to widespread genetic assessments is to attempt to link certain life history characteristics, such as mating system, with patterns of genetic variation (e.g., Hamrick et al. 1992). However, there might be many exceptions to such generalizations, and even basic life-history characteristics (e.g., mating system) are not known for many species and cannot necessarily be inferred from morphology. Allozyme studies, as we mentioned earlier, are convenient, but the relationship between this level of genetic variation and adaptive or evolutionarily significant variation is unclear. The best hope for gaining comprehensive information on genetic variation in many species lies in defining some morphological markers under known genetic control. Patterns in such simply inherited characters have been similar to those in allozymes. For example, in stem-surface phenotypes in Sierran pondersosa pine, one of which confers resistance to the gouty pitch midge (*Cecidomyia piniinopsis*), Ferrell et al. (1989) show a complex geographic pattern in the Sierra Nevada very similar to that in allozymes (Westfall and Conkle 1992; Westfall et al. n.d.). In trees, a suite of independently varying characters, such as growth, bud break, and cold hardiness, has been used in forming seed-transfer guidelines (Campbell 1986; Rehfeldt 1990). It is possible that such approaches will be necessary in other plant groups.

Mammals

About 110 species of mammals are found in the Sierra Nevada (Zeiner et al. 1990b). Most of the available information consists of distributions of species and subspecies. However, much of this was collected before 1940 and hence may not reflect the current situation. Modern studies cover a small and scattered sample (about 25% to 30%) of all taxa, and most have focused on non-Sierran and arid-land populations.

The genetic information available on Sierran mammals is

mostly based on allozyme data (i.e., genetically based variants of certain enzymes) and nuclear and mitochondrial DNA (mtDNA); there is some information on chromosomal data and a small amount on morphological characters that distinguish species and subspecies. Early studies focused on taxonomic issues; more recently they have tackled population structures.

Few genetic studies have been conducted on populations within the Sierra Nevada. Exceptions include the well-studied pocket gopher *(Thomomys bottae)*, some kangaroo rats (genus *Dipodomys*), some mouse species (genus *Peromyscus)*, and the kit fox *(Vulpes macrotis)*. Within-population genetic diversities in allozymes (i.e., observed heterozygosity) for the few Sierran mammals surveyed tend to be low, generally less than 0.05 (table 28.4). Heterozygosities in allozymes for some of the mouse *(Peromyscus)* species, at 0.07–0.09, are among the highest for mammals (Avise et al. 1979). In contrast, allozyme variation within species of the kangaroo rat *(Dipodomys* spp.) is very low and approaches zero for one population of Panamint kangaroo rat *(D. panamintinus)* sampled within the Sierra Nevada (Johnson and Selander 1971).

In addition, differentiation among populations of many Sierran species is also low (table 28.5), suggesting high

amounts of gene flow or recent history (Slatkin 1993). Genetic similarities among populations within species of kangaroo rats (genus *Dipodomys*) are very high, even in the wide-ranging species (Johnson and Selander 1971). Although no data exist for Sierran populations, low levels of differentiation are observed in such wide-ranging species as the mule deer *(Odocoileus hemionus)* among populations ranging across the western United States ($F_{ST} = 0.048$) (Cronin 1991). Differences were very low ($F_{ST} = 0.004$) between deer populations that were adjacent but separated by the Continental Divide (Cronin et al. 1991b). In the Brazilian free-tailed bat *(Tadarida brasiliensis)*, southwestern populations that include those occupying distinct migrational groups show low F_{ST} values (0.05), even though band and recapture data suggest low exchange among migratory groups (Svobda et al. 1985). Intercolony differences in the bat species are even lower ($F_{ST} = 0.008$) (McCracken et al. 1994).

Patterns in some rodent species that occupy the Sierra contrast greatly with those just described. Although rangewide population differences are relatively low in the deer mouse, *P. maniculatus* ($F_{ST} = 0.16$), this value is relatively high in comparison to that for the coyote ($F_{ST} = 0.09$), which is equally wide-ranging. Californian populations of this deer mouse

TABLE 28.4

Genetic variation (heterozygosity) within populations of Sierra Nevada mammal species.

Taxonomic Name	Common Name	Number of Loci	Heterozygosity[a]	Sampling Range	Reference
Thomomys bottae	Pocket gopher	23	0.093	Southwestern U.S., including Sierra Nevada	Patton and Yang 1977
Dipodomys ordii	Ord's kangaroo rat	17	0.008	Western U.S.	Johnson and Selander 1971
Dipodomys microps	Chisel-toothed kangaroo rat	17	0.007	Western U.S.	Johnson and Selander 1971
Dipodomys agilis	Pacific kangaroo rat	17	0.040	Western U.S.	Johnson and Selander 1971
Dipodomys heermanni	Hermann's kangaroo rat	17	0.042	Western U.S.	Johnson and Selander 1971
		22	0.00	Butte County	Patton et al. 1976
Dipodomys panamintinus	Panamint kangaroo rat	17	0.000	Western U.S., including Sierra Nevada	Johnson and Selander 1971
		22	0.05	Kern County	Patton et al. 1976
Dipodomys deserti	Desert kangaroo rat	17	0.010	Western U.S.	Johnson and Selander 1971
Dipodomys merriami	Merriam's kangaroo rat	17	0.051	Western U.S.	Johnson and Selander 1971
Dipodomys nitratoides	San Joaquin kangaroo rat	17	0.040	Western U.S.	Johnson and Selander 1971
Peromyscus maniculatus	Deer mouse	22	0.091	U.S. and Canada, including Sierra Nevada	Avise et al. 1979
Peromyscus californicus	California mouse	31	0.027–0.124	Coastal California, Baja California and foothills of Sierra Nevada	Smith 1979
Microtus californicus	California vole	4[b]	0.24	California Coast Range	Bowen 1982
Canus latrans	Coyote	53	0.050	Zoo	Fisher et al. 1976
Canus latrans	Coyote	10[c]	0.50	Southern California	Roy et al. 1994
Vulpes macrotis	Kit fox	24	0.025–0.111	Western U.S., including Sierra Nevada	Dragoo et al. 1990
Ursus americana	Black bear	33	0.015	Yosemite National Park	Manlove et al. 1980
Martes americana	Marten	24	0.170	Wyoming	Mitton and Raphael 1990

[a] Heterozygosity is the proportion of heterozygous genotypes per locus per individual.
[b] Variable loci only.
[c] Microsatellite (repeated) DNA.

TABLE 28.5

Population differentiation (F_{ST}) among populations of Sierra Nevada mammal species.

Taxonomic Name	Common Name	Number of Loci	F_{ST}[a]	Sampling Range	Reference
Tadarida brasiliensis	Brazilian free-tailed bat	38	0.05	Southwestern U.S.	Svoboda et al. 1985
Marmota flaviventris	Yellow-bellied marmot	8[b]	0.07	East River Valley, Colorado	Schwartz and Armitage 1980
Peromyscus maniculatus	Deer mouse	22	0.16	U.S. and Canada, including Sierra Nevada	Avise et al. 1979
Microtus californicus	California vole	4[b]	0.04	California Coast Range	Bowen 1982
Odocoileus hemionus hemionus	Rocky Mountain mule deer	9[b]	0.048	Western U.S.	Cronin 1991
Odocoileus hemionus hemionus; Odocoileus hemionus columbianus	Mule and black-tailed deer	9[b]	0.38	Western Canada	Cronin 1991
Canus latrans	Coyote	10[c]	0.09	Southern California	Roy et al. 1994

[a] F_{ST} is the amount of genetic differentiation among populations of a species.
[b] Variable loci only.
[c] Microsatellite (repeated) DNA.

diverged significantly from the rest of the western populations (based on variation in mitochondrial DNA) and also showed substantial genetic diversity within the populations. Differentiation among populations was greater for the pinyon mouse (*P. truei*), which occupies more restricted habitats (Avise et al. 1979).

One of the most intensively studied species, the pocket gopher (*Thomomys bottae*), also has relatively high genetic diversities within populations (heterozygosity of 0.09) and geographic structuring that follows chromosomal patterns (Patton and Yang 1977; Smith and Patton 1980). This species shows more genetic variation among than within populations. In areas where populations are small and widely spaced, genetic drift has resulted in considerable genetic homogeneity within populations. In the Sierra Nevada, Patton and Smith (1990) list major geographic subdivisions in the pocket gopher as northern and southern Sierra foothills, the Yosemite Valley, the Kern River Plateau, the Inyo-White Mountains, and the Mount Whitney complex. Dispersal is limited and effective population sizes low, which results in substantial population structuring, but dispersal among more distant populations is sufficient to maintain genetic diversity (Daly and Patton 1990; Patton and Feder 1981). Similar dispersal patterns and geographic structures are suggested in the marmot, *Marmota flaviventris* (Schwartz and Armitage 1980), and in some non-Californian species of ground squirrel, *Spermophilus* (e.g., van Staaden et al. 1994).

Genetic differences among species are also low, even among morphologically distinct and geographically separated taxa. In an allozyme and morphological study of the kit fox (*Vulpes macrotis*, sampled in the northeastern Sierra) and the swift fox (*V. velox*), Dragoo et al. (1990) found little differentiation between the two species (Nei's distances ranged 0.000 to 0.013) and concluded that the two should be reclassified as subspecies. Johnson and Selander (1971) found similarly low levels of differentiation among species of the kangaroo rat (including some Sierran and near-Sierran populations), another wide-ranging arid-land taxonomic group in the West.

Variation in mitochondrial DNA (mtDNA) has also been studied in taxonomic relationships in Sierran and proximal-Sierran species. Based on this type of data, Cronin (1992) found little genetic variation in elk (*Cervus elaphus*), both within and between the two Californian subspecies. The same study found substantial genetic variation in the mule deer subspecies of *O. hemionus* (populations east of the Sierra Nevada), but none in the black-tailed deer (populations west of the Sierran and Cascade crests). Cronin and colleagues (1991a) also found substantial genetic diversity in mitochondrial DNA clones in black bear populations of the Pacific Northwest, suggesting maternally based structuring. Although Cronin (1991b) claimed to find gene flow between mule deer and black-tailed deer in a contact zone between the subspecies (in British Columbia), mtDNA and allozyme F_{ST} values (0.56 and 0.378) suggest very little gene flow (equivalent to 0.2 to 0.41 individuals per generation). In a study of interspecific and intraspecific mitochondrial DNA variation in grasshopper mice (genus *Onychomys*), Riddle and Honeycutt (1990) found regional subdivisions that conform to existing taxonomic subdivisions, with relatively little differentiation within regions (based on single-individual samples). The northern grasshopper mouse (*O. leucogaster*) was clearly differentiated from the southern (*O. torridus*); the data suggested that the Great Basin populations of *O. leucogaster* had become isolated from the populations to the east during the Pleistocene.

In summary, many of the Sierran mammalian species sampled show fairly low levels of within-population genetic variation as well as high levels of gene flow among populations. However, there are notable exceptions, such as the pocket gopher, with its well-differentiated groups in the Sierra Nevada, and the California vole, with its relatively high level of heterozygosity. Most of the mammalian species native to the Sierra Nevada have not been genetically studied, and fewer still have been studied in Sierra Nevadan populations. Thus, it is unknown how well these trends represent the complete group of Sierran populations.

A widely communicated need at the SNEP Genetics Workshop is for population-level studies on most mammalian species. Also expressed was the need to study the temporal context for normal ranges of variation in populations, as demographic and genetic structures may vary over time (as in Patton and Feder 1981).

Birds

There are approximately two hundred bird species with winter and/or summer ranges in the Sierra Nevada bioregion (Zeiner et al. 1990a). Little genetic information is available for most species at the among-population or within-population level. Approximately fifteen species of birds in the Sierra Nevada have been the subject of population genetics studies (mostly allozyme-level studies, some of DNA data). Subspecies designations, based on morphometric characteristics, within bird species are common. Sierra Nevada examples of subspecies structuring include the fox sparrow (*Passerella iliaca*), northern flicker (*Colaptes auratus*), Hutton's vireo (*Vireo huttoni*), and spotted owl (*Strix occidentalis*). Recent mitochondrial DNA studies have even suggested species designations for some subspecies (e.g., the fox sparrow and Hutton's vireo). Thus, some subspecies or even populations within the Sierra Nevada may actually be distinct species rather than intraspecific genetic variants.

Bird species have several life history characteristics that are likely to be relevant to the amount and structure of genetic variation they exhibit. First is the mating system: most, but not all, bird species are monogamous (Lack 1968), allowing for more genetic variation than if the offspring were largely the result of a few males who mated with many females. Second is their pattern of distribution. Two distribution models are common: colonial and continuously dispersed. Colonial species tend to have greater population differentiation than do evenly distributed species (Barrowclough 1980). Third is the tendency to be either migratory or sedentary. One would expect sedentary populations to be less panmictic (i.e., less likely to exhibit random mating and hence less likely to have gene pools that are thoroughly mixed) than migratory populations, since there is less opportunity for mating among populations (e.g., Johnson and Marten 1992).

Another factor that has been demonstrated to influence genetic structure in avian species is directional bias in gene flow (i.e., immigration and hybridization between two populations may occur more frequently in one direction than in the other). Peterson (1991) studied gene flow between two groups of scrub jays (*Aphelocoma coerulescens*) that are strongly differentiated morphologically and are physically separated by geographic barriers, chiefly deserts. Using morphological criteria, he estimated that gene flow east to west across the Mojave Desert, from the *woodhouseii* populations to the *californica* populations, was two to seven times stronger than

west-to-east movement. The two forms approach one another closely (within about 20 km [12 mi]) in the Owens Valley. Here, 16% of the individuals studied showed eastern influence and 4% were apparently first-generation immigrants. Peterson hypothesized that the bias in the direction of gene flow in this case was due to habitat differences in the two subspecies, and that a stronger psychological barrier to entering desert habitats exists for the *californica* jays. Their normal habitat is oak woodlands, which are structurally distinct from the desert, whereas the *woodhouseii* subspecies occupies a more diverse pinyon-juniper-woodland habitat that seems to grade directly into desert habitats.

Morphological differentiation in birds, the usual basis of subspecies designations, usually reflects both environmental and genetic influences, captive-rearing studies in birds (i.e., those where birds are raised in common environments so that genetic effects can be distinguished from environmental effects) being difficult to administer and rarely occurring in the literature. One exception is an egg transplant study in the red-winged blackbird (*Agelaius phoeniceus*), a widespread species that also inhabits the Sierra Nevada. Northern to southern and reciprocal transplants of eggs revealed that a large component of morphometric variation was indeed environmental (James 1983). This result may help explain why morphometric studies in bird species may show different patterns of variation than those revealed by allozyme or DNA studies.

Population differentiation (F_{ST}) in birds of the Sierra Nevada, mainly based on estimates from allozyme studies, is generally low (table 28.6). This is consistent with previous findings that North American avian species generally consist of populations of moderate to large effective size with moderate to high levels of gene flow (i.e., successful mating) among them (Barrowclough 1980; Barrowclough and Johnson 1988). Only four of the F_{ST} values for the Sierra Nevadan bird species listed are over 0.10 (this small value suggests little population differentiation, due to migration and mating among populations), and the larger values are for species with two or more subspecies in the study sample. For example, the California subspecies of Hutton's vireo (*Vireo huttoni huttoni*) and the interior subspecies in Arizona (*V. h. stephensi*) have a mean F_{ST} of 0.614, a value more characteristic of interspecific than intraspecific differentiation. This species is highly sedentary, and the two subspecies exist in different habitats. Indeed, these two taxa were probably isolated even prior to the Wisconsin glacial maximum (approximately 18,000 years BP) and are definitely approaching, or have already reached, species level (Cicero and Johnson 1992). The few studies showing F_{ST} values greater than 0.10 are also often based on very few polymorphic loci.

This pattern of little geographic structuring and moderate levels of gene flow among populations is reflected in Hammond's flycatcher (*Empidonax hammondii*), a species that nests in boreal forests and woodlands of western North America. Samples from breeding localities at the extremes of

TABLE 28.6

Genetic population structure (F_{ST}) for bird species of the Sierra Nevada.[a]

Taxonomic Name	Common Name	Number of Loci[b]	F_{ST}[c]	Sampling Range	Reference
Riparia riparia	Bank swallow	—	0.051	North America	Barrowclough 1980
Cistothorus palustris	Marsh wren	—	0.061	North America	Barrowclough 1980
Callipepla californica	California quail	37(16)	0.032	California and Baja California	Zink et al. 1987
Larus californicus	California gull	35(8)	0.004	California and Utah	Zink and Winkler 1983
Sphyrapicus ruber	Red-breasted sapsucker	39(7)	0.019	California and Oregon	Johnson and Zink 1983
Pipilo erythrophthalmus complex	Rufous-sided towhee	16(1)	0.229	Maine to California	Sibley and Corbin 1970
Passerella iliaca	Fox sparrow	38(14)	0.014	California, Oregon, Nevada	Zink 1986
Passerella iliaca	Fox sparrow	13(13)	0.013	California and Nevada	Burns and Zink 1990
Zonotrichia leucophrys	White-crowned sparrow	19(3)	0.047	California and Colorado	Baker 1975
Branta canadensis	Canada goose	35(24)	0.065	North America	Van Wagner and Baker 1986
Colaptes auratus	Northern flicker	3(3)	0.098	U.S.	Fletcher and Moore 1992
Amphispiza belli	Sage sparrow	41(17)	0.112	California and Nevada	Johnson and Marten 1992
Icterus galbula	Northern oriole	2(2)	0.027	North America	Corbin et al. 1979
Vireo huttoni	Hutton's vireo	33(6)	0.614	Arizona and California	Cicero and Johnson 1992
Strix occidentalis	Spotted owl	23(1)	0.55	Oregon, New Mexico, California	Barrowclough and Gutiérrez 1990
Empidonax hammondii	Hammond's flycatcher	36(16)	0.043	Western North America, including Sierra Nevada	Johnson and Marten 1991

[a] All species listed are native to (i.e., have summer and/or winter ranges in) the Sierra Nevada.
[b] The first two entries are based on dispersal data; all others are based on allozyme data. The number of loci assayed is followed by the number of polymorphic loci in the sample, in parentheses.
[c] F_{ST} is the amount of genetic differentiation among populations of a species.

its nesting distribution, including the southern Sierra Nevada, showed that only 4.3% of the genetic variation present (based on allozyme data) was distributed among populations (Johnson and Marten 1991). This indicates moderate to high genetic heterogeneity among populations, a pattern reflected in the bird's high degree of morphological homogeneity over its entire nesting distribution.

In general, bird species of the Sierra Nevada and elsewhere have low levels of individual genetic variation when sampled at a specific set of genes. For example, it is possible for a bird to have two forms of each gene; the number of times this occurs, averaged over all the genes sampled and all the individuals sampled in a population, leads to an estimate of the genetic variation in the average individual, a value referred to as observed heterozygosity. Observed heterozygosity values are available for some of the bird species (and from the same studies) listed in table 28.6: they are all low. Average observed heterozygosity for the Canada goose (*Branta canadensis*) is 0.051 (range: 0.031–0.083) (Van Wagner and Baker 1986); for the sage sparrow (*Amphispiza belli*) is 0.042 (range: 0.03–0.55) (Johnson and Marten 1992); for Hutton's vireo (*Vireo huttoni*) is 0.014 (Cicero and Johnson 1992); and for Hammond's flycatcher (*Empidonax hammondii*) is 0.026 (range: 0.012–0.039) (Johnson and Marten 1991).

Amounts and patterns of geographic variation in birds differ among studies using allozyme, mitochondrial DNA, and morphometric evidence. Usually, allozyme evidence is more conservative, perhaps because it is less influenced by the environment, showing less population structure than is suggested by the other two types of data. For example, the Canada goose (*Branta canadensis*) shows spectacular amounts of mor-

phometric differentiation across its range, yet a rangewide isozyme study (including California samples) found very little population differentiation, as suggested by the low F_{ST} value of 0.065 (Van Wagner and Baker 1986). Similarly, in the northern flicker (*Colaptes auratus*), three subspecies are recognized in this widespread bird species, yet very little population differentiation is evident from an allozyme study ($F_{ST} = 0.098$) (Fletcher and Moore 1992). These allozyme values indicate much less population differentiation than that indicated by morphometrics.

To further understand these surprising differences, a third source of genetic material can be studied. It has been suggested (e.g., Zink and Dittman 1991) that, since mitochondrial DNA evolves more quickly than nuclear DNA, the former might be more likely to reveal geographic patterns of variation than the latter. Morphometric and allozyme data largely reflect products from nuclear DNA. However, among studies of birds of the Sierra Nevada, there appear to be more exceptions than trends in the relationships among patterns from mtDNA, morphometric, and allozyme studies. Two examples were presented previously of morphometric patterns that were not supported by allozyme variation. However, in a study of scrub jays (*Aphelocoma coerulescens*), Peterson (1991) found that the allozyme data agreed with the morphometric designation of five subspecies (and much gene flow among populations within subspecies). In a study of the brown towhee complex (*Pipilo* spp.), a group of four currently recognized species mainly inhabiting the southwestern United States, Zink and Dittman (1991) found that mtDNA and allozyme data revealed similar evolutionary and geographic patterns. Similarly, allozyme, morphometric, and mtDNA evidence showed strong and similar differentiation between two groups in samples of the sage sparrow complex (*Amphispiza belli*) taken from California and Nevada (Johnson and Marten 1992).

Mitochondrial DNA data do not necessarily reflect geographic patterns. For example, in a study of the chipping sparrow (*Spizella passerina*), a widespread migratory North American passerine species that also inhabits the Sierra Nevada, no geographic differentiation was observed in mitochondrial DNA at all, in spite of the fact that a large part of the range was sampled (Zink and Dittman 1993b). Thus, the three named subspecies for this sparrow have no support from mtDNA data. In fact, the lack of mtDNA geographic structure over relatively large distances is typical of several passerine bird species that inhabit areas that were recently glaciated (Zink 1994). In spite of considerable geographic variation in size and plumage color in the song sparrow (*Melospiza melodia*) across its continental U.S. range (including six California samples), mtDNA did not reveal any geographic structure (Zink and Dittman 1993a). Similarly, mitochondrial data from thirty-nine locales (including several in the Sierra Nevada) of the fox sparrow (*Passerella iliaca*) complex failed to show any geographic variation within four major taxonomic groupings, despite marked morphometric clines (i.e., gradual changes in species characteristics that parallel some geographic or environmental trend) within these groups (Zink 1994). In these cases, it has been postulated that isolation by distance has not been very important in shaping population genetic structure as measured by mitochondrial DNA. Rather, historical isolating events may be more important (Zink 1994).

In summary, most of the bird species in the Sierra Nevada for which there are genetic data show weak population differentiation based on allozyme data. Subspecies may be differentiated at the allozyme level if they have been isolated for a long time. Morphometric variation, even at the subspecies or rangewide level, is often not accompanied by allozyme or mitochondrial DNA variation. Together with information from limited captive-rearing studies, this may indicate that much morphometric variation does not have a genetic basis. Although the life history traits of mating system, pattern of dispersal, and migratory tendencies seem logically related to the amount and pattern of genetic information, there are too few empirical studies to verify these theories.

Reptiles and Amphibians

Of the approximately forty-six species of reptiles and thirty species of amphibians listed as occurring in the Sierra Nevada (Zeiner et al. 1988; Mark Jennings, e-mail communication with the authors, 1994), approximately half have genetic information available. Most of this information is on amphibians, particularly salamanders and frogs. Among reptiles, lizard species have been best studied genetically. Snakes appear to be the least well-studied group: except for some morphological and behavioral studies, there is virtually no genetic information for the twenty-seven snake species in the Sierran

and proximal-Sierran regions (Zeiner et al. 1988). For amphibians and reptiles, it is interesting to note that, in general, the most common, widespread, and abundant species have been least studied, and the endemics and rare species have been studied most. This is postulated to be a consequence of both the current public focus (and associated funding opportunities) on rare or endangered species and the early distributional and morphological studies of the more common species, leaving only the less rewarding (genetic) increments of information to be gained.

Within the amphibian and reptile populations studied, genetic variability in allozymes is low to moderate (table 28.7). For example, recent studies (Wake and Yanev 1986; Wake et al. 1989) report that the invading population *Ensatina eschscholtzii xanthoptica* in the Sierra Nevada has the lowest heterozygosity of any population in the genus, in line with theoretical expectations. Heterozygosity values are generally less than 0.100. In addition, genetic differences among populations within some species of salamanders are often extremely high, sometimes approaching or exceeding genetic differences among related species (Hedgecock and Ayala 1974; Wake and Yanev 1986). Even in a species that occupies a limited area, such as the Inyo Mountains salamander (*Batrachoseps campi*), which occurs only in specific areas along a 32 km length of the Inyo Mountains, heterozygosity can vary somewhat from one population to another (its heterozygosity range is 0.04–0.08) (Yanev and Wake 1981). The limitation of these groups to moist, sometimes riparian habitats (Zeiner et al. 1988) restricts their dispersal and, consequently, gene flow between populations. This, along with extinction and recolonization with climatic fluctuations can result in random loss of genetic variability. Heterozygosities ranged from 0.01 to 0.09 among seventeen populations of the Pacific tree frog (*Hyla regilla*) (Case et al. 1975) and from 0.02 to 0.25 among sixteen populations of the lungless salamander (*Ensatina eschscholtzii*) (Wake and Yanev 1986). Differences in heterozygosity from one population to another are present, though to a lesser extent, in western frog species (Case 1978a, 1978b; Case et al. 1975).

Few studies have addressed genetic differentiation among Sierran populations of reptile or amphibian species. The available data, which are almost exclusively for amphibians, show a pattern of strong genetic differentiation among populations (table 28.8). In a study of nineteen populations of the lungless salamander (*Ensatina eschscholtzii*), including populations in the Sierra Nevada, Wake and Yanev (1986) noted "profound" allozymic differentiation among populations ($F_{ST} = 0.705$) (see also Jackman and Wake 1994; Wake et al. 1994). Similarly, a high level of genetic differentiation was demonstrated among populations of the Inyo Mountains salamander (*Batrachoseps campi*) (Yanev and Wake 1981). A high level of genetic differentiation exists among California populations of the black salamander (*Aneides flavipunctatus*), suggesting that there has been little gene flow among populations (since the Pliocene or late Pleistocene epochs) (Larson 1980). Another example

TABLE 28.7

Mean heterozygosity values for reptile and amphibian species of the Sierra Nevada.

Taxonomic Name	Common Name	Number of Loci	Heterozygosity[a]	Sampling Range	Reference
Amphibians					
Ambystoma macrodactylum	Long-toed salamander	21	0.066	Oregon and Idaho	Howard and Wallace 1981
Ambystoma tigrinum	Tiger salamander	32	0.055	California[b]	Shaffer 1984[c]
Aneides flavipunctatus	Black salamander	21	0.103	California	Larson 1980
Batrachoseps campi	Inyo Mountains salamander	33	0.060	Inyo Mountains, California	Yanev and Wake 1981
Ensatina eschscholtzii	Ensatina	26	0.112	California	Wake and Yanev 1986
Hydromantes brunus	Limestone salamander	18	0.180	Mariposa County, California[b]	Wake et al. 1978
Hydromantes platycephalus	Mount Lyell salamander	18	0.080	Tuolumne County, California[b]	Wake et al. 1978
Hydromantes shastae	Shasta salamander	18	0.080	Shasta Lake, California	Wake et al. 1978
Hyla regilla	Pacific tree frog	14	0.007–0.093	Oregon and California	Case et al. 1975
Rana aurora	Red-legged frog	15	0.039	California Coast Range	Case 1978a
Rana boylei	Foothill yellow-legged frog	15	0.038	California	Case 1978a
Rana boylei	Foothill yellow-legged frog	15	0.045	California	Case 1978b
Rana cascadae	Cascade frog	15	0.037	Lassen County, California	Case 1978a
Rana catesbeinana[d]	Bullfrog	15	0.000	California[b]	Case 1978a
Rana muscosa	Mountain yellow-legged frog	15	0.060	Sierra Nevada	Case 1978a
Rana muscosa	Mountain yellow-legged frog	15	0.070	Sierra Nevada	Case 1978b
Taricha torosa	California newt	18	0.094	Sierra Nevada	Hedgecock and Ayala 1974
Reptiles					
Anniella pulchra	California legless lizard	27	0.022	Coastal California	Bezy et al. 1977
Elgaria coerulea	Southern alligator lizard	34	0.063		Good 1988
Elgaria multicarinata	Northern alligator lizard	34	0.013		Good 1988
Elgaria panamintina	Panamint alligator lizard	34	0.015		Good 1988
Sceloporus graciosus	Sagebrush lizard	19	0.030	Five western states	Thompson and Sites 1986
Uta stansburiana	Side-blotched lizard	18	0.053	California	McKinney et al. 1972

[a]Heterozygosity is the proportion of heterozygous genotypes per locus per individual.
[b]One population.
[c]Recalculated in Shaffer and Breden 1989.
[d]Introduced from eastern United States.

of this trend is found in a widely occurring reptile species sampled outside of the Sierra Nevada. Thompson and Sites (1986) measured an average F_{ST} of 0.231 among western steppe populations of the sagebrush lizard (*Sceloporus graciosus*).

Little work has been done on genetic variation in morphological characteristics in Sierran populations of reptiles; most recent studies have been done on non-Sierran populations of the Pacific Coast. Seeliger (1945) surveyed morphological variation in the western pond turtle (*Clemmys marmorata*) and found that individuals in Sierra Nevadan populations had morphological characteristics that were intermediate to those in north coastal and south coastal areas of California. Although the genetic basis for these differences has not been established, such polymorphisms are the basis for further study. Bechtel and Whitecar (1983), in controlled breeding of Californian samples, have established inheritance of color patterns in the gopher snake (*Pituophis melanoleucus*). However, no studies have been done on the population structure or the ecological context of this variation in color pattern. In a study of geographic variation in feeding preferences in Cali-

fornian populations of the terrestrial garter snake (*Thamnophis elegans*), Arnold (1980a, 1980b) established that differences between coastal and interior populations are inherited and affect feeding (or avoidance) responses to slugs, amphibians, and the toxic newts (*Taricha* spp.). Although Sierran populations were not included in the study, they do indicate adaptation to available food and suggest habitat-based population differences in Sierran garter snakes (as in Jennings et al. 1992). Sinervo and colleagues have done extensive work on physiological and biophysical genetics in lizards of the genera *Uta* and *Sceloporus*, comparing populations from contrasting environments in the western United States. They have found differences among species and among and within Oregon and southern California populations in thermal physiology and growth (Sinervo 1990; Sinervo and Adolph 1994); differences among and within Coast Range populations in the trade-offs between reproductive numbers, egg size, and survival (Sinervo et al. 1992, 1991); and heritability of running performance and leg and tail size (Tsuji et al. 1989).

Although population structure and gene flow are critical to genetic assessments of these species, it is not certain how the well-studied species groups may serve as models for the less-studied majority. There is little information on how life history traits might relate to fitness-relevant genetic variation, but this is considered to be a worthy topic for further research. Recent losses in amphibian populations require that Case's early work on frogs be reassessed and expanded. There is also very little information on local adaptation. Other high-priority research needs are comparisons of mitochondrial and nuclear DNA patterns to assess variation in sex-biased gene flow (as in Wade et al. 1994).

Fish

In the Sierra Nevada bioregion, there are approximately forty fish species native to the inland lakes and rivers, and another thirty or thirty-one species have been introduced (Moyle 1976). The overwhelming majority of genetic information (morphological, allozyme, and mtDNA) is for the commercially significant salmonid (family Salmonidae) species. Although only a few salmonid species are native to the Sierra Nevada (e.g., whitefish [*Prosopium williamsoni*], cutthroat trout [*Oncorhynchus clarki*], golden trout [*O. mykiss whitei* and *O. mykiss aguabonita*], and rainbow trout [*O. mykiss*]), several others are anadromous visitors (e.g., chinook salmon [*Oncorhynchus tshawytscha*], steelhead salmon [*O. mykiss irideus*], and Pacific lamprey [*Lampetra tridentata*]), and others have been introduced (e.g., brook trout [*Salvelinus fontinalis*], lake trout [*Salvelinus namaycush*], brown trout [*Salmo trutta*], and kokanee [*Oncorhynchus nerka*]). California, and to a lesser extent the Sierra Nevada, contains the southernmost populations of most of the anadromous fish of the Pacific coast of North America (Moyle 1994) and thus have the potential for harboring rare and/or genetically significant populations (Nielsen et al. 1994).

Of the life-history traits that most affect the amount and pattern of genetic variation in fish species, perhaps the most significant is the spawning migration typical of many of the species discussed here. Returning to their natal streams to reproduce would tend to restrict gene flow among the residents of various streams and thereby impose a certain geographic structure on genetic variation (e.g., Bartley et al. 1992).

Although species and subspecies designations in fish, like those in other taxa, have often relied heavily on morphometric characteristics, it has been suggested that the environment may play a larger role in determining phenotype in fish than in many other animal species (e.g., Allendorf et al. 1987). For example, certain physiological traits of fish (including indeterminate growth capacity, greater sensitivity to variation in temperature than other homoisothermic [cold-blooded] vertebrates, and greater flexibility in traits associated with reproductive success) may allow great phenotypic plasticity, thereby widening the potential gap between genetic and phenotypic variation.

In general, fish species of the Sierra Nevada for which there is some genetic information show rather low to moderate levels of within-population genetic variation and low levels of population differentiation (tables 28.9 and 28.10). In a review of eight salmonid taxa, Allendorf and Leary (1988) reported mean heterozygosity values of 0.013 to 0.095. The importance of within-population genetic variation is emphasized in studies comparing hatchery populations (which may have undergone a severe bottleneck—that is, may have been generated from a fairly small sample of the original population) with wild populations. In such cases (e.g., Bartley and Gall 1990), there is little evidence that hatchery populations have reduced amounts of genetic variation compared with natural populations. This suggests that if the species has low levels of genetic (or, at least, allozyme) variation in natural populations, this variation is not reduced significantly by taking a sample and using this as breeding stock.

However, the comparison of hatchery and wild trout and salmon populations in California shows different results when different genetic markers are used. With mitochondrial DNA (maternally inherited), there is an increase in diversity in hatchery stocks due to the introduction of geographically di-

TABLE 28.8

Population differentiation values (F_{ST}) for reptile and amphibian species of the Sierra Nevada.

Taxonomic Name	Common Name	Number of Loci	F_{ST}[a]	Sampling Range	Reference
Amphibians					
Ambystoma macrodactylum	Long-toed salamander	21	0.350	Oregon and Idaho	Howard and Wallace 1981
Aneides flavipunctatus	Black salamander	21	0.470	California	Larson 1980
Batrachoseps campi	Inyo Mountains salamander	33	0.470	Inyo Mountains, California	Yanev and Wake 1981
Ensatina eschscholtzii	Ensatina	26	0.705	California	Wake and Yanev 1986
Reptiles					
Sceloporus graciosus	Sagebrush lizard	19	0.231	Five western states	Thompson and Sites 1986

[a] F_{ST} is the amount of genetic differentiation among populations of a species.

TABLE 28.9

Mean (observed) heterozygosity values for fish species of the Sierra Nevada.

Taxonomic Name	Common Name	Number of Loci[a]	Heterozygosity[b]	Sampling Range	Reference
Oncorhynchus kisutch	Coho salmon	45(23)	0.027	Northern and central California	Bartley et al. 1992
Oncorhynchus tshawytscha	Chinook salmon	53(21)	0.038	Coastal and inland northern California	Bartley and Gall 1990
O. tshawytscha	Chinook salmon	78(47)	0.053	Northern California and southern Oregon	Gall et al. 1992
Gasterosteus aculeatus	Three-spined stickleback	18(7)	0.100	Northern California	Haglund et al. 1992
Catostomus tahoensis	Tahoe sucker	~60(12)	0.023	Pyramid Lake, Nevada	Buth et al. 1992
O. mykiss whitei	Golden trout	12(6)	0.134[c]	Southern Sierra Nevada	Gall et al. 1976
Oncorhynchus mykiss	Rainbow trout	32(24)	0.092	Northern and coastal California	Berg and Gall 1988

[a] Number of loci assayed is followed, in parentheses, by the number of loci polymorphic and scorable.
[b] Heterozygosity is the proportion of heterozygous genotypes per locus per individual.
[c] Excludes the heterozygosity estimate for the rainbow trout sample in the study.

vergent maternal lineages in the process of egg and fry transfers (Nielsen et al. 1994). With microsatellite DNA (paternally inherited), the opposite trend is noted, with hatchery rainbow trout in southern California showing greatly reduced levels of genetic diversity when compared with wild populations (J. L. Nielsen, Pacific Southwest Research Station, U.S. Forest Service, e-mail communication with the authors, 1995).

Probably the best studied of the western salmonids is the rainbow trout (Oncorhynchus mykiss). Although native to a relatively small range on the Pacific coast of Canada and the United States, including the Sierra Nevada, it has been distributed around the world, and its high commercial value has prompted genetic investigation, particularly on heritability of morphometric characteristics (e.g., Elvingson and Johansson 1993; Gjedrem 1992; Gjerde and Schaeffer 1989). Even within its native range, it displays a wide range of phenotypes, including being freshwater and anadromous, inhabiting great ranges in water temperature and flow rate, and even displaying some variation in chromosome number (Hershberger 1992). However, the extent to which the phenotypic variation is due to genetic variation is still under investigation.

Since we have only limited information on Sierra populations, we can gain insight by examining studies of the same species in nearby regions. In a study of thirty-two rainbow trout populations in Idaho, Oregon, and Washington, allozyme analysis revealed a high degree of polymorphism. The mean heterozygosity value of 0.059 is high for a salmonid (Allendorf 1975). A more recent study of coastal California rainbow trout populations (including several in the Sierra Nevada—for example, the Middle Fork of the Feather River) also revealed high levels of heterozygosity (Berg and Gall 1988). The mean heterozygosity value for these populations (0.092, table 28.9) is even higher than the estimate for the more northern populations. Analysis showed little evidence of geographic structuring in these populations and suggested moderate to high levels of gene flow (table 28.10). Although a few populations were distinguished by the presence of a few uncommon alleles, Berg and Gall (1988) suggested that this could be due to temporal fluctuations in allele frequencies rather than stable geographic structure.

Recent studies based on mitochondrial DNA and microsatellite alleles of the southern steelhead, the anadromous form of rainbow trout, show significant differences in genetic frequency among three biogeographic zones in California: northern, from Humboldt Bay to Gualala Point; cen-

TABLE 28.10

Genetic population structure (F_{ST}) for fish species of the Sierra Nevada.

Taxonomic Name	Common Name	Number of Loci[a]	F_{ST}	Sampling Range	Reference
Oncorhynchus kisutch	Coho salmon	45(23)	0.158[b]	Northern and central California	Bartley et al. 1992
Oncorhynchus tshawytscha	Chinook salmon	53(21)	0.177[b]	Coastal and inland northern California	Bartley and Gall 1990
O. tshawytscha	Chinook salmon	78(47)	0.106[b]	Northern California and southern Oregon	Gall et al. 1992
Gasterosteus aculeatus	Three-spined stickleback	18(7)	0.163	Western North America	Haglund et al. 1992
Oncorhynchus mykiss	Rainbow trout	32(24)	0.127	Northern and coastal California	Berg and Gall 1988

[a] Number of loci assayed is followed, in parentheses, by the number of loci scorable and polymorphic.
[b] Actually calculated as G_{ST}.

tral, from the Russian River to Point Sur; and southern, from San Simeon Point to Santa Monica Bay (Nielsen et al. 1994).

An allozyme study of thirty-five populations of chinook salmon (*Oncorhynchus tshawytscha*) from inland and coastal waters of California (including samples from the Yuba River, Bear Creek, Merced River, Stanislaus River, and Tuolumne River in the Sierra Nevada) revealed a mean heterozygosity value of 0.038, typical for salmonid species (table 28.9). The lowest heterozygosity values were found in the Klamath–Trinity River drainage (0.008–0.022), the authors speculating that this may be due to effects from relatively recent volcanic activity (Bartley and Gall 1990). Ash from volcanoes kills fish through suffocation and mechanical abrasion, thereby reducing population sizes and removing some of the naturally occurring genetic variation within populations. Somewhat higher population differentiation exists among these California populations ($F_{ST} = 0.177$) than has been reported in Alaska populations, the authors suggesting that this resulted from the longer time since glaciation in California. Further, although there is some evidence of coastal-inland genetic structuring in certain other salmonids, such as cutthroat and rainbow trout, there is no allozyme evidence for such a distinction in chinook salmon in California (Bartley and Gall 1990).

More population differentiation is observed in chinook salmon when mitochondrial DNA, rather than isozymes, is studied. Four groups in the Sacramento–San Joaquin Basin, which have historically been recognized as discrete units based on the seasonal distribution of their peak spawning times, were also recognized as genetically divergent based on mitochondrial data (Nielsen et al. 1994).

Two inland subspecies of the cutthroat trout are recognized as native to the Sierra Nevada, the Lahontan cutthroat (*Oncorhynchus mykiss henshawi*) and the Paiute cutthroat (*O. m. seleneris*). An allozyme study of populations within the range of the Lahontan subspecies in northeastern California and northern Nevada showed much population differentiation, with 45% of the observed allozyme variation accounted for as genic diversity among populations (Loudenslager and Gall 1980). This is unusual among Sierra Nevada fishes and not the standard pattern for cutthroat; other cutthroat subspecies show less population differentiation (Loudenslager and Gall 1980). The authors suggest that the population differentiation observed may be due to the fact that this subspecies inhabits the large lakes and headwater tributaries of the Humboldt, Truckee, Carson, and Walker Rivers, drainages that are presently isolated from one another. Further, since the final desiccation of glacial Lake Lahontan occurred 8000 years BP, the populations may have been isolated for a long time. More recent (unpublished) molecular studies of the Humboldt and Lahontan cutthroat trout populations show significant loss of genetic diversity in fragmented populations, regardless of geological time of isolation (J. L. Nielsen, e-mail communication with the authors, 1995).

Golden trout, endemic to the southern Sierra Nevada, has been the subject of recent genetic studies. Concern has focused on this species due to its extremely narrow range, human disturbance of its fragile habitat, and the high probability of hybridization between endemic goldens and rainbow trout, which have been introduced into significant portions of the golden's range, diluting and contaminating natural populations (Gall et al. 1976). Also, golden trout has been widely planted outside its native range. The estimate for mean heterozygosity for the two subspecies (*S. a. aguabonita* and *S. a. whitei*) is 0.134, indicating considerably more within-population allozyme diversity than most other studied fish species of the Sierra Nevada. Indeed, even the Cottonwood Creek population of *S. a. aguabonita*, which was a planted population started with only twelve or thirteen trout, has a high heterozygosity estimate (0.126) (Gall et al. 1976). The estimate for the two populations of *S. a. whitei*, while somewhat lower than the other populations (0.088), is still reflective of considerable genetic variation and leads the authors to suggest that the so-called "threatened" Little Kern golden trout "did not appear to be in immediate danger of extinction through lack of adaptive capability" (Gall et al. 1976).

Some genetic information is also available for the threespine stickleback (*Gasterosteus aculeatus*), a species with a widespread, circumboreal distribution that includes coastal California. It was introduced to some inland California waters, including the Mono Basin (Moyle 1976). A recent allozyme study of widespread populations in this species estimated the heterozygosity of the northern California sample to be 0.100, a value higher than that of most salmonid species and comparable to values for other teleosts (Haglund et al. 1992).

In summary, fish species of the Sierra Nevada show very modest levels of genetic variation within their populations. Although there is much apparent phenotypic variation within most species, a great deal of this may be due to environmental influences rather than genetic differences. In species that are anadromous or that occupy large river drainages, there is surprisingly little population differentiation. Striking counterexamples of much population differentiation occur in a few species that have been isolated in lakes or in nonconnected rivers for a long time. Most of the genetic studies on species of the Sierra Nevada are based on allozyme data, thereby limiting the opportunity to compare genetic patterns based on various different types of data. Fish introductions, transplants, and the release of hatchery-raised fish complicate the assessment of "natural" levels and patterns in genetic variation of fish species, perhaps more than in other taxonomic groups.

Insects

In the state of California, there are approximately 28,000 species of insects (Powell and Hogue 1979), many of them represented in the Sierra Nevada bioregion. In spite of their vast species representation, genetic information for this taxonomic group is both limited and sporadic (table 28.11). Most intraspe-

TABLE 28.11

Genetic population structure (F_{ST}) for insect species of the Sierra Nevada.

Taxonomic Name	Common Name	Number of Loci	F_{ST}	Sampling Range	Reference
Pieris rapae[a]	Cabbage butterfly	4	0.014	United States	Pashley et al. 1985
Cydia pomonella	Codling moth	4	0.066	Africa, Europe, U.S., Australia	Pashley 1980
Euphydryas chalcedona	Chalcedona checkerspot	8	0.090	Central California	McKechnie et al. 1975
Euphydryas editha	Checkerspot	8	0.118	Central California	McKechnie et al. 1975
Chrysomela aeneicollis	Montane leaf beetle	5	0.135	Sierra Nevada	Rank 1992
Coenonympha tullia	[Satyrine butterflies]	21	0.051	Northern California, southwestern Oregon, northern Nevada	Porter and Geiger 1988
Oeneis chryxus	[Satyrine butterflies]	16	0.081	Sierra Nevada	Porter and Shapiro 1989
Speyeria nokomis apacheana	Western seep fritillary	25	0.022	Sierra Nevada	Britten et al. 1994b

[a] Introduced species.

cific studies have been driven by human-interest factors such as health issues (e.g., mosquitoes), agricultural or forest crop concerns (e.g., budworms, bark beetles, and grasshoppers), or aesthetic interests (butterflies). For example, some studies have investigated genetic variation in response to pesticides in defoliating insect species (e.g., *Choristoneura* spp.) (Stock and Robertson 1980). A few species of particular phylogenetic interest or scientific value (e.g., fruit flies) have received considerable attention. Aquatic insects, including stoneflies, mayflies, and water striders, have received relatively little attention (White 1988).

Beyond the insects affecting health concerns and agricultural and timber production, and species of special interest, most insects have not been genetically studied, even at the species level. Thus, there are major gaps in the knowledge of intraspecific genetic variation. In particular, there is little genetic information for species with rare host plants. As there is much variation in breeding systems and other life history traits among insect species, and because the available genetic information is so sporadic, generalizations across these taxa are difficult.

Of those few insects of the Sierra Nevada that have been studied, the most common trend is for them to have genetic architectures displaying little population differentiation and high levels of gene flow among populations (table 28.11). In a study of twenty-one populations of a satyrine butterfly complex (*Coenonympha tullia*) in the northern Sierra Nevada, southwestern Oregon, and northern Nevada, Porter and Geiger (1988) found a high degree of polymorphism within populations but little interpopulation differentiation (mean F_{ST} = 0.051; 35%–59% polymorphic loci; expected heterozygosity = 13%–20%).

In an allozyme study of forty-one populations of the checkerspot butterfly (*Euphydryas editha*), little genetic variation was found among populations despite great geographic distances and ecological differences within the range of the

species in the western United States (Baughman et al. 1990). An analysis of nineteen isozyme loci showed six major groupings, three of which have representative populations in the Sierra Nevada.

Considerable genetic research has been conducted on the bark beetles (family Scolytidae), largely owing to their destructive effects on pine forests in North America. More than 170 species of bark beetles occur in California (Powell and Hogue 1979), many of them in the Sierra Nevada. Until recently, most genetic research on bark beetles had been directed toward understanding the evolution of the various species (Hayes and Robertson 1992). Allozyme studies of the mountain pine beetle (*Dendroctonus ponderosae*) have revealed fairly high levels of heterozygosity (0.17 in a California population) and moderate differentiation among geographically separated populations (Stock et al. 1992). For both this species and the pine engraver (*Ips pini*), the observation has been that heterozygosity levels tend to be higher in populations that inhabit severe environmental conditions, a situation opposite to that found in coniferous trees.

Morphological, allozyme, and DNA data have recently been compared in one termite genus that is restricted to western North America, the dampwood termites (*Zootermopsis* spp.). Two of the three currently recognized extant species of this genus, *Z. nevadensis* and *Z. angusticollis*, are distributed sympatrically along the Pacific Coast from British Columbia to Baja California, Mexico, including the Sierra Nevada. Analysis of cuticular hydrocarbon (i.e., a phenotypic characteristic of unknown genetic basis) had been shown to distinctly identify all three species as well as to suggest two subspecies within *Z. nevadensis* (Korman et al. 1991). However, the two putative subspecies were not confirmed by allozyme differences. Allozyme variation within *Z. nevadensis* (expected heterozygosity 0.080), with samples included from the Sierra Nevada, was somewhat lower than that found in *Z. angusticollis* (expected heterozygosity 0.199) and close to an average reported for insects. More recent genetic studies, based on mitochondrial DNA, also showed slightly higher genetic variation within *Z. angusticollis* than within *Z. nevadensis*. Sierra Nevada samples for the latter species were included in this study. The several putative subspecies within *Z. nevadensis*, proposed on the basis of morphological data, were neither confirmed nor disputed by the mitochondrial data.

Of those species having been studied intraspecifically, genetic information has mainly been based on phenotypic or allozyme data, and the study objectives have most commonly been gene flow. However, studies of phenotypic variation often have not been genetically controlled and are usually confounded by environmental plasticity. Thus, it is uncertain how the allozyme data relate to genetic variation of adaptive significance. For the previously mentioned reasons, available gene-flow information is difficult to extrapolate to less-studied species. One possible generalization among insects is based on species mobility. Insects, like birds, can be considered to consist of two subgroups: one is colonial, highly specialized in its needs, and sedentary; the other is highly vagile and more generalized in resource utilization. In the former group, one would expect to find higher rates of inbreeding and endemism; in the latter, less. Few data exist, however, on insects in the Sierra Nevada to substantiate this.

In a direct examination of the relationship between vagility and population structure, Zera (1981) examined two species of water striders that differ in degree of winglessness. One species, *Gerris remigis,* has a widely distributed range, including the Sierra Nevada, and is nearly wingless. The other species, *Limnoporus canaliculatus,* occurs in the eastern United States and is wing-polymorphic. *G. remigis,* the wingless species, exhibited strong population structuring, fixation of alleles within populations, and relatively low heterozygosity within populations. The more vagile *L. canaliculatus,* in contrast, showed little population structuring and four times as much heterozygosity as *G. remigis.*

In summary, much of the genetic research on insects of the Sierra Nevada has been concerned with evolutionary history and species relationships. The relatively small number of species that have been genetically studied at the intraspecific level tend to show fairly high levels of genetic variation (e.g., high heterozygosity). The degree of population differentiation may be related to the migratory behavior and level of specialization of the species: species that are highly vagile and broadly associated with hosts might show less population differentiation than the converse.

Fungi

The number of fungal species in the Sierra Nevada cannot defensibly be estimated. This is a very large collection of species including the Basidomycota, Ascomycota, Zygomycota, and Chytridiomycota. Only a small fraction of the species in these groups have been studied genetically anywhere. Those that have been studied often have some economic importance

(e.g., *Armillaria* [Basidomycota]) or have value as an experimental organism (e.g., *Neurospora* [Ascomycota]). Heritability estimates have been provided for some characteristics of an ectomycorrhizal fungus *(Pisolithus tinctorius)* that is beneficial to a commercially important tree species in the southeastern United States, slash pine *(Pinus elliottii)* (Rosado et al. 1994). In general, species in the Zygomycota and Chytridiomycota have received very little attention (Bruns et al. 1991), and taxa in general from the Sierra Nevada are scarcely even described, let alone studied, for intraspecific variation.

There may be more levels of intraspecific variation in fungal species than in other taxa. One reason relates to the biological species concept. This concept pertains to the definition of species based on reproductive barriers: individuals that can interbreed belong to the same species; those that can't belong to different species. This concept of a species doesn't fit fungal behavior well. There are often intersterility groups, or ISGs (groups in which the individuals from one group cannot reproduce sexually with individuals from another group), within fungal species that by definition could be classified as separate species. Other factors leading to potentially exceptional levels of within-species diversity relate to morphological or life history characteristics. For example, in endomycorrhizal fungi (Zygomycota), the mycelium can be partitioned morphologically and metabolically, the nuclei migrate, and germ tubes start new individuals that must adapt to host-soil conditions at the same or different locations. Somatic mutations have major consequences, since all cells of a fungal organism are totipotent (that is, each cell maintains the potential to develop into a complete organism). Accumulation of deleterious effects of these mutations can be averted if those nuclei are partitioned in nongerminating structures (Morton and Bentivenga 1994). There are potentially high rates of mutation in fungal species. For example, one race of wheat rust *(Puccinia graminis)* gave rise to more than fifty mutational variants during the 1960s (Burdon and Roelfs 1985b).

For these and other reasons, morphometric characters are seldom considered representative of genetic variation in fungal species. Often there may be very little variation in morphology within a species (Gardes et al. 1991; Otrosina et al. 1992). Alternatively, life cycle morphs complicate the study of intraspecific genetic variation by the examination of morphometric traits. For example, rust fungi—one of the largest groups of obligately parasitic plant pathogens—frequently produce five morphologically distinct spore stages and alternate between two unrelated plant hosts, making morphological identification difficult (Gardes and Bruns 1993).

Intraspecific genetic variation in fungal species is typically measured by DNA analysis, allozyme analysis, or virulence studies. (For a virulence study, fungal spores are collected and inoculated with the appropriate host plant. After remaining in conditions suitable for germination for a week or two, evidence of infection can be noted.) In some cases, the genetic patterns revealed by different methods are not consistent. For example, in a study of wheat rust *(P. graminis)* in Australia, a virulence analysis detected sixteen different races, while an isozyme survey detected no intraspecific variation (Burdon and Roelfs 1985b). In contrast, a study of eastern United States wheat rust that used these same two methods showed complementary patterns. The authors suggest that the difference may be due to life history characteristics. In the latter study, the populations had undergone sexual reproduction until the 1920s, when the alternate host, the barberry, was eradicated. The former (Australian) study was based on populations that had never possessed a functional sexual cycle (Burdon and Roelfs 1985b). A study using both isozyme and RFLP (i.e., fragments of DNA that indicate genetic differences based on how well they match a control library of DNA fragments) analysis of the fungal pathogen *Rhizoctonia solani* showed complementary genetic patterns, revealing five genetically distinct intraspecific groups (Liu and Sinclair 1992).

Genetic variability in plant pathogens has often been studied on a large geographic scale but seldom at the level of individual populations (McDonald and Martinez 1990). Since many fungal species occur worldwide, the scale of sampling can have significant consequences for the pattern of genetic variation observed. For example, in an isozyme study of wheat stem rust *(Puccinia graminis)* in thirteen countries, most of the alleles were widespread, yet the author observed that considerable variation could occur at more local levels (Burdon 1986). Indeed, in a hierarchical study (locations in field, stems within locations, lesions within a leaf, etc.) of a haploid fungus *(Septoria tritici)* sampled in one field near Davis, California, the authors found considerable genetic variation. The pattern observed was of a population highly subdivided into a mosaic of independent clones without significant migration between different locations in the field. The authors hypothesized that genetically diverse founding populations had provided the initial inoculum, and reproduction via asexual spores had resulted in localized clusters of clones. They concluded that most genetic variation in this fungal species may be distributed on a local, rather than broad, geographic scale (McDonald and Martinez 1990).

There are only a few examples of genetic studies based on samples of fungal species within the Sierra Nevada. An allozyme study has been conducted on the pathogen *Heterobasidion annosum (= Fomus annosus)*, which causes root rot of coniferous tree species in temperate forests worldwide, including the Sierra Nevada. Two ISGs within this species have been described in North America. In the western United States the two groups occur sympatrically (i.e., in the same or overlapping areas) on a variety of host species. Samples from Oregon and California (including Yosemite National Park, Sequoia National Park, Plumas National Forest, etc.) showed isozyme patterns that largely concurred with the intersterility groups (Otrosina et al. 1992). Very few alleles were shared by the two groups, and the study results suggested that little or no gene flow occurs in nature between the two ISGs. Within each ISG there was a high degree of allele fixation. Given that

there are differences in host preferences by the two ISGs, paleoecological factors that influence host species distributions may be major forces driving the genetic differentiation in this species.

An allozyme study of the western gall rust fungus *(Pteridermium harknessii)* in the western United States (including samples from the Sierra Nevada) showed considerable genetic variation (six of fifteen isozyme loci were polymorphic) (Vogler et al. 1991). Further, the isozyme profiles separated the samples into two distinct groups ("zymodemes"), each of which had a characteristic electrophoretic profile. Populations from both zymodemes were found in the Sierra Nevada, whereas only one of the zymodemes was found in the populations sampled in southern Oregon and coastal California. In forests in the Sierra Nevada where both zymodemes occurred, there was no evidence of gene flow between the two. Despite the great range in pine host species and the geographic area covered, most of the genetic variation in the fungus was between the zymodemes, with persistent heterozygosity in one group. Thus, this fungus presents a very different pattern of genetic variation than that of its pine hosts.

Another example of slow rates of gene flow is found in a study of the fungus that causes white pine blister rust *(Cronartium ribicola)*, which infects North American white pines, including sugar pine *(Pinus lambertiana)*. It is an introduced pathogen, probably arriving in western North America in about 1910. There is a major gene for resistance to the fungus in sugar pine (Kinloch 1992). However, a virulent race of the blister rust fungus has been discovered that can completely overcome the resistance normally conferred by the gene. This virulent race was first discovered in 1978 near Happy Camp in the Siskiyou Mountains of northern California. A virulence study by Kinloch and Dupper (1987) of samples of rust from Washington to California provided no evidence of the virulent race except in the vicinity of the initial discovery site. Thus, the authors concluded that if the gene is moving it is migrating slowly, and that possibly the fungus is largely inbreeding, thereby slowing the rate of gene flow.

In conclusion, based on the genetic studies reviewed, it seems that a primary factor in the amount and pattern of genetic variation in fungal species is the mode of reproduction. The presence or absence of intersterility groups, asexual or sexual reproduction, and time since cessation of sexual reproduction in currently asexual species or populations have all been found to have profound impacts on genetic variation.

Conclusions

The overview of genetic information for species of the Sierra Nevada highlights the fact that species have been studied sporadically, leaving large information gaps and making generalizations difficult. Certain groups have been well studied at the population level, often due to human-interest factors. Examples include salmonid fishes, plethodontid salamanders, butterflies, and commercially important forest tree species.

For many species, even basic taxonomic information or recent species-distribution maps are not available. Although some generalizations can be offered for the well-studied species concerning the relationship between genetic variation and geographic patterns or life history characteristics (e.g., low F_{ST} values for most but not all organisms studied at the population level), the sporadic information base makes it difficult to extrapolate these patterns to the unknown majority. One often-noted geographic trend at the SNEP Genetics Workshop was a stronger genetic relationship with elevation (i.e., an east-west trend) than with latitude. In terms of research needs, a common refrain was the need for more studies on gene flow and genetic structuring at the population level, prerequisites for understanding genetic processes and enabling better extrapolation and prediction. A common concern was the relative lack of information on adaptive variation. Because molecular markers are often employed in genetic studies, an important concern is the extent to which these markers reflect genetic variation of adaptive significance. There was much discussion about the conditions under which these data would reflect fitness-related genetic variation of particular interest to land management.

Patterns of Genetic Significance in the Sierra Nevada

In view of the need to set priorities for natural resource management, it is important not only to discuss the nature of genetic variation but also to attempt to define what is genetically significant. Such a definition addresses three issues: (1) What are attributes of significance? (2) What levels in the genetic continuum are significant? and (3) What is the physical scale or standard for defining genetic significance?

One set of attributes for significance includes genetic variation that is rare, rich, or representative (see Millar et al. 1996). Rarity implies those genetic entities that are unusual in some respect, often, but not always, in an ecological context. Examples include disjunct populations in an otherwise contiguous species, marginal populations, and organisms displaying unusual adaptations, such as those occupying unusual soil types or elevational extremes. Rarity can also be used to describe those entities that, while not necessarily rare in an ecological sense, are evolutionarily significant or phylogenetically rare. Examples could include monotypic species or those with unusual phylogenetic histories. Another type of rarity involves rare alleles or rare genotypes. In these cases, there may be a geographic or adaptive significance that is not associated with an obvious environmental gradient. Rarity, as it is used here, does not refer to any legal or policy-oriented interpretations.

Richness implies high levels of genetic diversity. Richness criteria can be applied at any level of genetic variation, from the levels at which it originates (DNA base pairs to individual genes to chromosomes to genotypes) to the levels at which it is manifested or structured (populations to subspecies to races). Richness is often associated with three arbitrary levels of organization: within populations, among populations, and among species. Although within the context of genetic significance it is desirable to understand richness at the population level or within populations, information is often not available to allow more than a definition of species richness. Hybrid zones are another example that frequently show a high degree of genetic richness.

Representativeness is perhaps the most significant type of genetic variation. Representative individuals or populations typify the genetic composition and structure (allelic and genotypic frequencies) of a reference group. Such a group most likely is the locally adapted ecotype but could be a race or subspecies. This concept is often discussed nongenetically at the plant community or vegetation association level, where one might refer to an area, for example, that is representative of the mixed conifer forest. Representative genetic variation reflects species structure, geographic subdivisions, and large-scale adaptiveness within species.

A second issue when considering genetic significance is the biological level under consideration. Genetic variation is a continuum, spanning the levels at which it originates (diversity among alleles, chromosomes, or individual genotypes) through those levels at which it can be manifested (i.e., within individuals, among individuals in a population, among individuals among various populations, among individuals in geographic or ecological regions, etc.). The value of one level of genetic variation to another—for example, the value of individual diversity to ecosystem health—is assumed but not well understood. The level at which an attribute is described more often reflects the level at which there is some current information rather than the level that is perhaps most responsible for the attribute. As such, attributes will most often be described as pertaining to the species complex, species, or population. It is also understood that genetic significance pertains to levels that are hierarchically arranged and natural in origin. Thus, human-vectored gene introductions, exotic species, and "contrived" ecosystems (e.g., Monterey pine "ecosystems" in New Zealand) all represent levels of genetic variation that lie outside the scope of this chapter.

The geographic domain or context within which to assess genetic significance must also be addressed. For the purposes of this chapter, genetic significance (rarity, richness, or representativeness) is assessed relative to the Sierra Nevada bioregion and, where possible, relative to watershed domains.

The sections that follow address genetic significance for each taxonomic group in response to two questions:

1. What are the key factors in this taxonomic group that contribute to geographic patterns of genetic variation?

2. What are some locations of genetic significance—by attribute and level?

TABLE 28.12

Key factors underlying geographic patterns of intraspecific genetic variation for conifers of the Sierra Nevada.

Priority[a]	Factor
1	East-west subdivision: this reflects climate, elevation, biogeographic history, soil moisture availability, and biotic interactions such as competition. (This is also a key factor for species-level diversity.)
2	Elevation (primary) and aspect (minor): from the north down to the Sierra National Forest on the east side and on the west side, respectively. Glacial history: from the Sierra National Forest south (Note: This distinction of key factors is due to a correlation between latitude and elevation.)
3	Glacial/tectonic history. (This is also a key factor for species-level diversity.)
4	Species-specific gene flow factors: barriers to or corridors for gene flow, e.g., riparian areas, mountains, canyons.
5	Unusual or modifying factors, e.g., rare substrate, aspect.

[a]1 is the most influential factor, etc.

Plants

A relatively large database for commercial conifer tree species and some others supports generalizations regarding key factors that contribute to geographic patterns of genetic variation (table 28.12). The available data indicate that the most notable trend in intraspecific genetic variation in coniferous

FIGURE 28.1

Influence of seed source elevation on two-year-old seedling height of sugar pine (Harry et al. 1983), ponderosa pine (Mirov et al. 1952), and white fir (Hamrick 1976) from an elevational transect in the Sierra Nevada. The common-garden study elevation is approximately 800 m. (From Harry et al. 1983.)

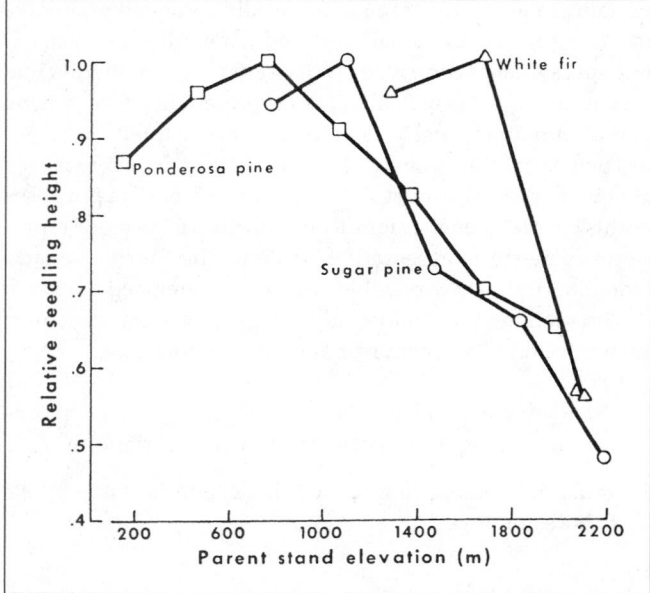

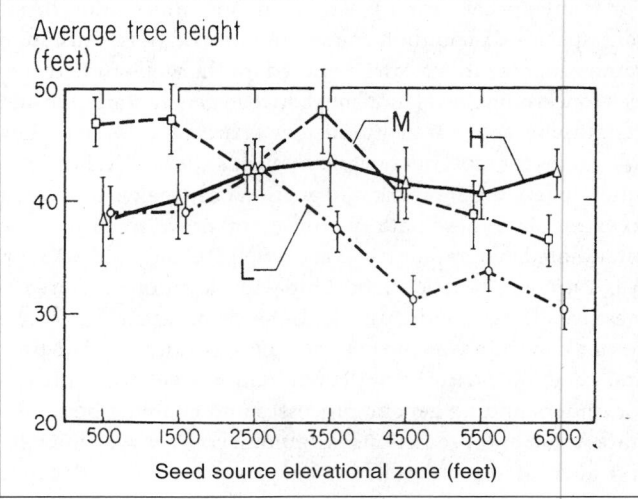

FIGURE 28.2

Height of ponderosa pine from a transect study, after twenty-nine years, at three common-garden sites differing in elevation: L = low elevation, 960 ft (about 300 m); M = middle elevation, 2,730 ft (about 800 m); and H = high elevation, 5,650 ft (about 1,700 m). (From Conkle 1973.)

trees of the Sierra Nevada is an east-to-west transition. Rather than being related to one specific factor, this longitudinal pattern is related to several integrated factors, most notably elevation, and reflects changes in such ecosystem elements as climate (especially temperature and precipitation), soil moisture availability, biotic interactions, and biogeographic history. This east-west gradient of genetic variation has been revealed in several transect studies of such species as white fir (*Abies concolor*) (e.g., Hamrick 1976), sugar pine (*Pinus lambertiana*) (e.g., Harry et al. 1983), and ponderosa pine (*P. ponderosa*) (e.g., Conkle 1973). Seedlings from each of the species were collected along a transect along the western slope of the Sierra Nevada, encompassing an elevational range of approximately 2,100 m (7,000 ft). Seedlings were grown in a common-garden environment at a site near Placerville (about 800 m [2,640 ft]) and measured periodically. Height growth of seedlings after two years shows similar trends among the three species in the influence of source elevation (figure 28.1).

This genetic relationship between tree height and source elevation is complex. The common-garden study just described was replicated with ponderosa pine at three sites of varying elevation. Conkle (1973) analyzed the results after twenty-nine years of growth and found evidence of genotype × environment interactions (figure 28.2). While height differences are relatively minor among seed sources at the high-elevation site, they vary dramatically among seed sources at the low-elevation site. Thus, the greatest risk in seed transfer is from high-elevation sites to low-elevation ones.

Multilocus analysis of allozyme data has confirmed these east-west trends for several species, including ponderosa pine,

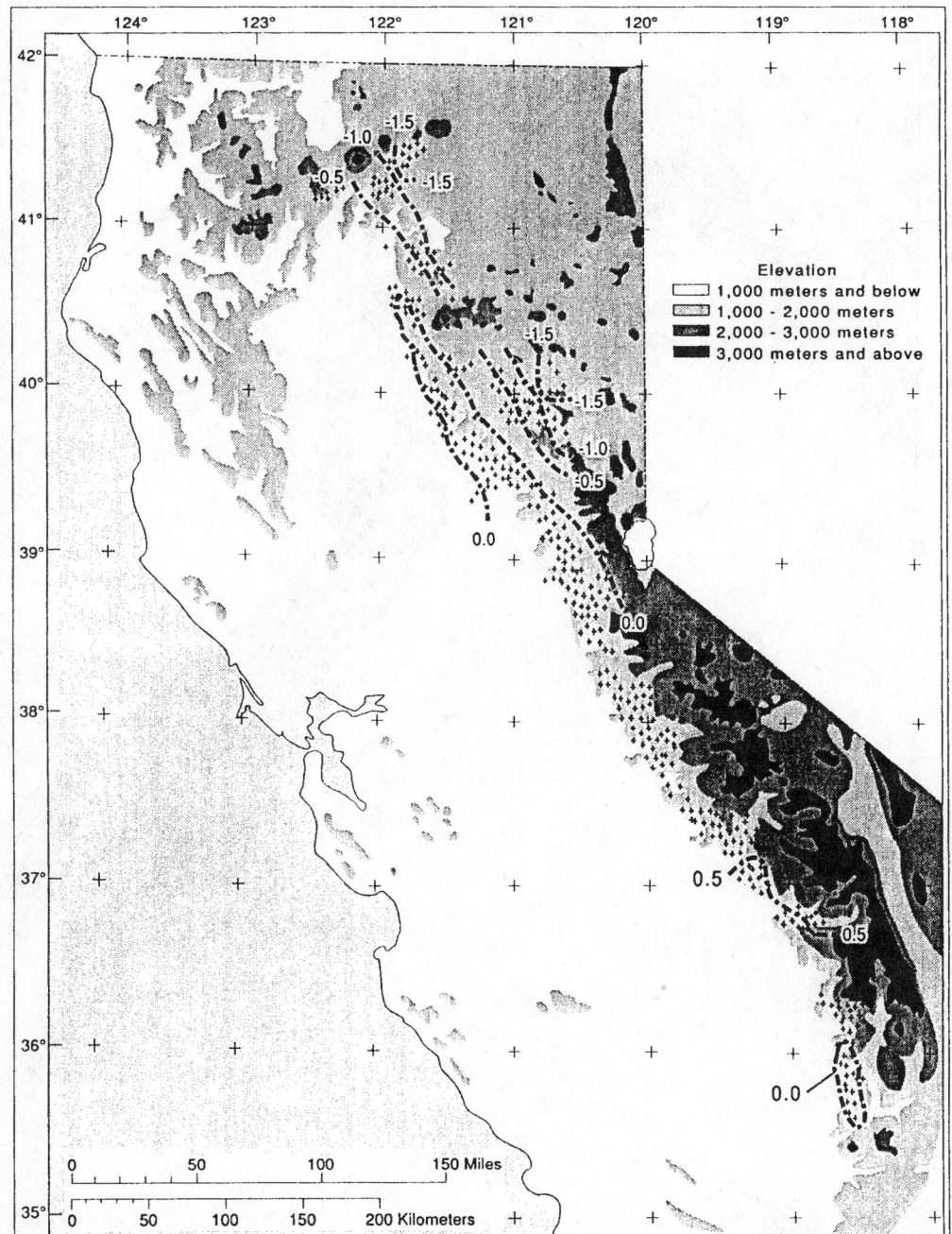

FIGURE 28.3

Multilocus contour plot of the first vector from a canonical trend-surface model for ponderosa pine. (From Westfall and Conkle 1992.)

in the Sierra Nevada (Westfall and Conkle 1992). Contour plots (based on the first canonical vector of a geographical trend-surface equation, $R^2 = 0.25$) for ponderosa pine show regions of similarity that are largely differentiated in an east-to-west direction (figure 28.3). When the next two vectors are added ($R^2 = 0.40$), elevation becomes more influential in determining the areas of genetic similarity (figure 28.4).

Geographic subdivisions recognizing these trends were proposed twenty-five years ago (Buck et al. 1970) and applied to seed transfer in commercially important conifers (figure 28.5) (Kitzmiller 1976). Subsequent common-garden and allozyme studies have modified the intitial geographic patterns and associated guidelines in minor ways. For example, latitudinal zones are now considered to be larger than those recommended originally, although the 152 m (500 ft) elevational zonation is confirmed by genetic studies (Kitzmiller 1990; J. H. Kitzmiller, Regional Office, U.S. Forest Service, conversation with R. D. Westfall, 1994).

A recent synthesis of allozyme and morphological (common-garden, early-expressed or juvenile traits) data for five

FIGURE 28.4

Multilocus contour class intervals of the first three vectors from a canonical trend-surface model for ponderosa pine. (From Westfall and Conkle 1992.)

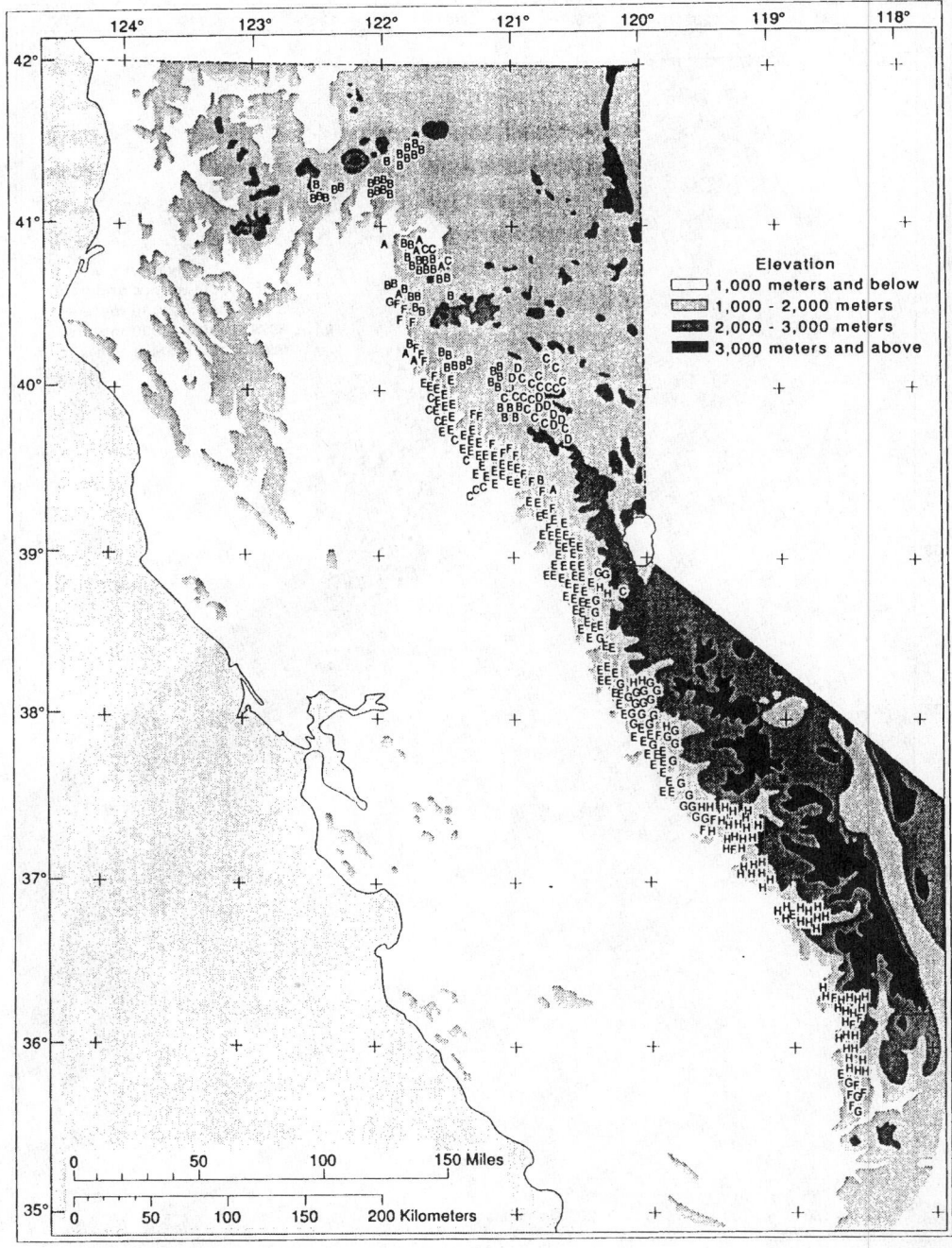

commercial conifers in the Sierra Nevada (ponderosa pine, sugar pine, white fir, Douglas fir, and incense cedar) shows much agreement with the original patterns of geographic variation (Millar et al. 1991). Both types of data were evaluated according to the percentage change in genotypes along an elevational transect. This calculation provided an assessment of the risk in transferring genotypes in reforestation processes; expressed as the percentage change in genotypes per 1,000 ft of elevational change, this is an index of transfer risk (table 28.13). This analysis provided three conclusions.

First, it confirmed that genetic variation in the Sierra Nevada for these species changes much more rapidly with elevation than with latitude (data for latitude are not presented here). Second, the northern low-elevation populations within species were genotypically similar to southern populations at higher elevations. Third, for this kind of risk analysis, allozyme and morphological data showed similar trends.

From the northern Sierra Nevada south to the Sierra National Forest, elevation appears to be the predominant factor affecting patterns in genetic variation; south of there, the patterns seem more reflective of glacial, topographic, and tectonic history than of elevation. These variables partition species differentiation more than known intraspecific variability. Factors reflecting corridors for or barriers to gene flow, such as riparian zones or mountain ranges, may be significant influences in the Sierra for speciation. Finally, some particular conditions, such as unusual soil types, may provide another level of genetic partitioning for some species (although there is little empirical data). Associations with serpentine soils have perhaps been best studied (e.g., *P. sabiniana* [Griffin 1965] and *P. ponderosa* [Jenkinson 1977]).

An extensive series of allozyme and common-garden studies for several commercial species has provided some insight regarding genetically (intraspecifically) rich areas for trees of the Sierra Nevada. First, examples of genetically rich areas are known from both spatially heterogeneous and homogeneous areas, indicating either that selection is not always the major force in shaping the genetic composition or that selec-

tive agents are not obvious. Second, for species with wide ranges, such as ponderosa pine, the highest levels of intraspecific genetic variation are found in the midsections of the major vegetation zones that they cross. Third, some species of this group exist at the boundaries of the major vegetation zones they traverse (e.g., seemingly localized populations of ponderosa pine exist above and below the mixed conifer forest type).

Table 28.14 presents some examples of species- and population-level genetic richness and rarity. These may occur at the margins of species ranges, as in the case of the unusual and northernmost population of giant sequoia (*Sequoiadendron giganteum*) in Placer County, California, which is distinct and low in genetic diversity relative to other giant sequoia populations (Fins and Libby 1982). Genetic variation within populations sometimes shows an increasing trend toward the south of a species' range. (This is discussed further in the section "Inferences of Genetic Significance.") For species whose

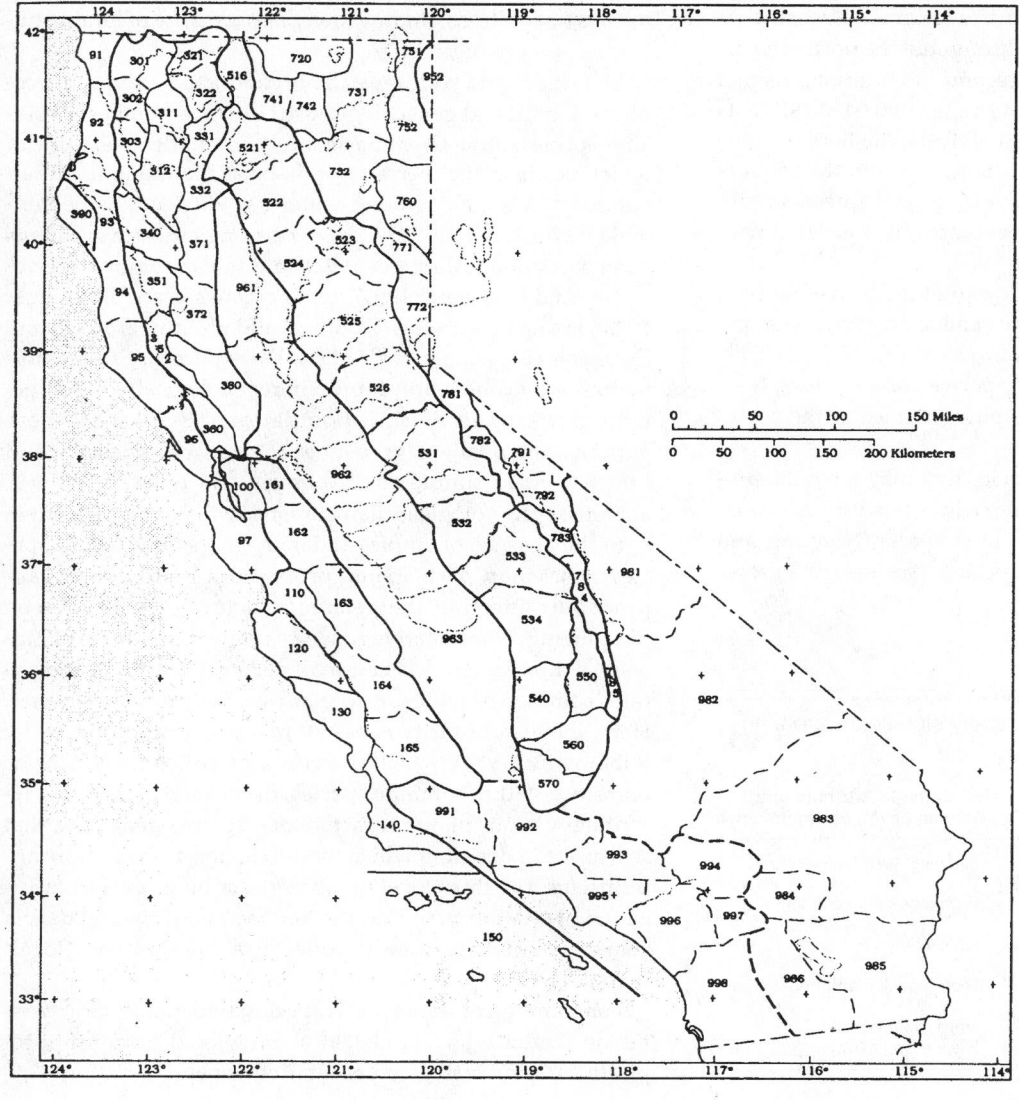

FIGURE 28.5

California tree seed zone map. (From Kitzmiller 1976.)

TABLE 28.13

Transfer risk index for elevation transfer of five central Sierran conifers.[a]

Species	Type of Data	Elevational Transfer Risk Index[b]
White fir	Allozyme	0.18
	Nursery	0.19
Ponderosa pine	Allozyme	0.25
	Nursery	0.20
Incense cedar	Allozyme	0.30
	Nursery	0.30
Sugar pine	Allozyme	0.35
	Nursery	0.27
Douglas fir	Allozyme	0.52

[a] From Millar et al. 1991.
[b] Transfer risk index is the proportion of genotypes in one population that do not match those present in another location. Risk is expressed here as proportion mismatched per 1,000 feet of elevational change.

ranges have their southern limits in the Sierra Nevada, these southernmost populations may be genetically rich or rare. One example of this pattern is found in western white pine (*P. monticola*) (Steinhoff et al. 1983). Throughout the northern part of its distribution, western white pine populations have a mean heterozygosity of 0.13 (with a range of 0.04–0.19). In its southern populations in the Sierra Nevada, the heterozygosity values range from 0.26 to 0.32. Sugar pine in the Sequoia National Forest exhibits high levels of allozyme diversity among populations; these high levels are not associated with elevation.

Genetic richness or rarity may occur at the convergence of major biophysical regions (e.g., the genetically rich, at the species and population levels, area of the Sierra Nevada–Cascade transition) and as unusual species compositions (e.g., the rich and rare Washoe–Jeffrey pine complex in the northeastern Sierra).

One example of a rare tree species that may be in the process of being assimilated into a more widespread species (ponderosa pine) is Washoe pine (*P. washoensis*) (Niebling and Conkle 1990). A high-elevation species, Washoe pine grows

TABLE 28.14

Some specific examples of genetically significant areas in the Sierra Nevada for trees.

Geographic Area or Subdivision	Genetic Attribute (and Genetic Level) of Significance
Northeastern Sierra: Washoe–Jeffrey pine complex	Rich and rare (species)
Plateau at Sierra-Modoc-Cascade convergence: ecotypes of many species, including red fir, white fir, ponderosa pine, and western juniper	Rich (species, ecotype)
Southern Sierra/Kern Plateau: rich in species of five-needled pines	Rich (species and population)
Southern Sierra: foxtail pine	Rare (population)
Placer County: giant sequoia	Rare (population)

primarily in stands on the eastern edge of the Sierra Nevada and the western edge of the Great Basin. Only three populations are well documented, all of which are in the greater Sierra Nevada region—the Warner Mountains, Babbitt Peak, and Mount Rose.

Factors underlying genetic variation within other (i.e., nonconifer) plant species in the Sierra Nevada are more conjectural. Beyond the early reciprocal transplant/elevational transect studies of Clausen, Keck, and Hiesey (e.g., Clausen et al. 1948), little information exists on adaptive variation in native species. In a morphologically based common-garden study in *Nassella*, E. E. Knapp (1994) suggests that genetic differentiation is much more pronounced on an east-west gradient, from coastal to interior populations, than from north to south. This has been interpreted as a response to climatic patterns. However, this geographic pattern is based on analysis of widespread species; it is not known to what extent this pattern might hold true for more narrowly defined species or those specific to the Sierra Nevada. Rice and Mack (1991) have conducted a reciprocal transplant study of intermountain *Bromus* populations, finding lower performance of local populations at more mesic sites.

Ehleringer and colleagues have conducted a series of ecophysiological and genecological studies on a number of arid-land species of the Great Basin and western deserts, some of which occur in the Sierra Nevada or at the montane-desert boundary. They have found water-use efficiency to be heritable (Schuster et al. 1992), to vary along climatic gradients (Comstock and Ehleringer 1992), and to vary among species (Evans and Ehleringer 1994), gender (Dawson and Ehleringer 1993), life history classes (Donovan and Ehleringer 1992), and life forms (Ehleringer et al. 1991). We expect that such information would be helpful in defining ecologically based genetic patterns in Sierran populations. Even with these patterns, the relationship between water-use efficiency and fitness is not a simple one (Donovan and Ehleringer 1994), and other physiological characteristics will have to be added.

In the absence of empirical data for nonconifer plant species, workshop participants proposed a more general approach to inferring the genetic structure of, and thereby recognizing representative genetic variation in, nontree plants, using a three-tiered decision-making key. The first level separates common or widespread plant species from rare species. Here, the idea of rarity is based on a structure proposed by Rabinowitz (1981), which uses the concept of "seven forms of rarity." Within common species, the next level involves describing various plant characteristics, such as gene flow, that will assist in deciding which model (in tier three) is more appropriate. The third level involves describing representative genetic variation based on the key factors of two models: a coarse-grained (regional) model or a fine-grained (local) model.

For common or widespread species, two main characteristics determine which model, local or regional, is more appropriate for structuring representative genetic variation. One of the most important characteristics is gene flow. There is much variation among widespread plant species in factors affecting gene flow, including breeding system, mode of reproduction (asexual, sexual, or mixed), and pollen and seed-dispersal systems. Gene flow is strongly related to genetic architecture and thus to the choice of a regional or local model for describing representative genetic variation. The second characteristic germane to the choice of model is the spatial pattern of population distribution. For example, continuous distributions might be most appropriately described with a regional model, while "patchy" or disjunct distributions might indicate a local model, even though the species range is broad.

For common species, different key factors (table 28.15) underly representative genetic variation in the local and regional models, a reflection of their relative scales. The regional model is mainly structured by climatic variables, both those correlated with east-west distance (e.g., temperature regime) and those correlated with north-south distance (e.g., day length). The local model is more finely structured by both physical (e.g., edaphic) and biotic (e.g., local patterns of competition) factors. In summary, then, two models or sets of factors have been hypothesized as descriptors for patterns of representative genetic variation in plants. The choice of which model is more appropriate will depend mainly on gene flow and the distribution characteristics of the plants. In some cases, such as a widespread species with much local differentiation, a mixed or hierarchical use of the models would be indicated.

In structuring representative genetic variation for rare or restricted plants, a Rabinowitz (1981) model for identifying rarity might be useful. For example, plants that are highly restricted in their distribution would have their resident site as the basis of representative genetic variation. For plants that are few in number but widely distributed (e.g., "sparse"), the factors underlying patterns in widespread or common species might be an appropriate way to structure their variation. For plants with intermediate numbers or distribution ranges, structuring by "specialized communities" might adequately capture their representative genetic variation.

Beyond the structuring of representative genetic variation,

TABLE 28.15

Key factors underlying the structure of representative genetic variation for common or widespread (nonconifer) plant species in the Sierra Nevada, according to regional and local models.

Regional Model	Local Model
Elevational factors (primarily east-west gradient)	Physical factors
Temperature regime, growing season, etc.	Edaphic factors
Moisture, rain shadow effect, etc.	Aspect
	Slope
Latitudinal factors (primarily north-south gradient)	Biotic factors
Day length	Local herbivores
Annual precipitation	and pathogens
	Competitors

another issue is the occurrence and pattern of rare genetic variation. Although rarity in plants is difficult to illustrate geographically, there appear to be several correlates or indicators of rarity. First, rare species or populations often occur as edaphic endemics—for example, on gabbro and serpentine soils. More than a dozen serpentine endemics are known to exist in the Sierra Nevada (Kruckeberg 1987). Second, biogeographical history may point to rare forms—for example, nonglaciated refugia. Third, hybrid zones, at the species or population levels, may be areas of evolutionary significance. Finally, peripheral populations may have rare characteristics relative to more centrally located populations.

In conclusion, it is difficult to provide more detail or make specific recommendations with respect to describing or inferring significant genetic variation in nonconifer plants of the Sierra Nevada, because of the lack of information. Even detailed species-distribution maps with rough estimates of population sizes are unavailable. Hence, the workshop participants developed a more generalized approach to structuring variation within this taxonomic group, and we are not able to identify specific geographic areas of genetic significance other than for a few well-studied species.

Mammals

Areas or instances of rich or rare genetic variation are generally unknown for most mammalian species, due to the few genetic studies of Sierran populations. However, the well-studied pocket gopher (*Thomomys bottae*) provides some examples. In a study comparing cranial features and allozyme variation of samples from thirty-one geographic locations in eastern California, Smith and Patton (1988) report two areas of genetic significance in the Sierra Nevada. The first is on the northern and eastern shores of Owens Lake. Here, the gophers have allozyme similarity to other pocket gophers in the general area, yet have distinct cranial features. For all other gophers sampled in this study, allozyme and cranial data showed concordant patterns. The second unusual area is in the vicinity of Lone Pine. Here, evidence was noted of apparent intergradation (i.e., interbreeding with the result of hybrid progeny) between two of the three putative subspecies, *T. perpes* and *T. melanotis*.

Of special genetic interest may be subspecies that are peripheral isolates of more widespread species. Such populations may be centers for evolutionary change (W. Z. Lidicker, University of California, Berkeley, note to D. Rogers, September 1994). Examples of locations where such peripheral isolates of several species occur are the Kern River Plateau, Sierra Valley, and Mono Basin. Also, for at least one widespread mammal (*Phenacomys intermedius*, the heather vole), the main Sierran cordillera contains such an isolate.

Areas in which congeneric species or conspecific subspecies meet may represent another kind of genetic richness or uniqueness. For example, the ranges of the Columbian black-tailed deer (*Odocoileus hemionus columbianus*) and the Rocky Mountain mule deer (*O. h. hemionus*) meet in a narrow contact zone in the Sierra Nevada and Cascade Range (Cronin 1991).

Overall, little genetic information is available that identifies factors underlying mammalian genetic patterns in the Sierra Nevada. With the exception of the pocket gopher, with its known population differentiation into six regions of the Sierra Nevada, most known geographic patterns reflect subspecies, rather than population, associations. At this level, most mammalian data support the major biogeographic subdivisions in the Sierra of east side and west side and north and south, with second-order structuring according to major vegetation type. Particularly strong and recurring patterns of subspecies divisions occur between east and west, both along mountain crests and in the foothills. Certain north-south divisions (e.g., south of Yosemite and south of Tahoe) are also described on the basis of phenotypic variation.

Birds

There is little evidence in the available literature of rare or endemic bird species in the Sierra Nevada. However, there are several examples of rich or unusual geographic areas with respect to avian genetics—areas that are implicated as hybrid zones among subspecies, as harboring multiple and noninterbreeding subspecies, or as being habitat for unusual populations or subspecies.

FIGURE 28.6

Probable areas of intersubspecific gene flow (arrows) in cowbirds *(Molothrus ater)* in the Sierra Nevada (shaded area). Number and letter designations refer to sampling sites. (From Fleischer et al. 1991.)

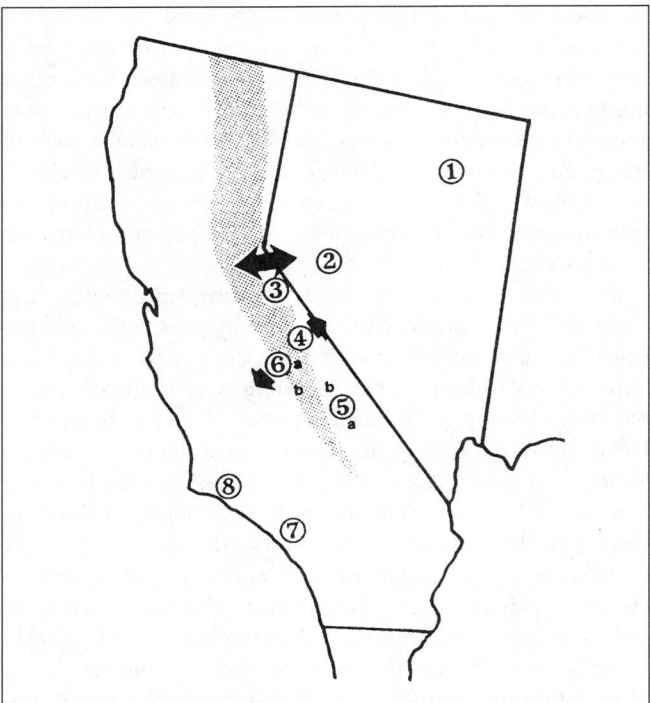

Two areas in the Sierra Nevada are probable corridors for gene flow between two subspecies of the brown-headed cowbird *(Molothrus ater obscurus* and *M. a. artemisiae).* This species is a relatively recent inhabitant of the Sierra Nevada, and contact between the two subspecies in the Sierra Nevada is even more recent, probably during the last twenty to fifty years. The subspecies *M. a. artemisiae* has historically inhabited the Great Basin, strictly east of the Sierra Nevada. The subspecies *M. a. obscurus* historically inhabited southern Arizona to Texas but has expanded its range gradually northward, initially invading southern California from the lower Colorado River in about 1900. Samples from ten sites in California and Nevada provided mitochondrial DNA evidence that gene flow is occurring between the two subspecies along the Sierran crest at two points: the Mammoth Lakes area and Lake Tahoe (figure 28.6) (Fleischer et al. 1991).

Two well-differentiated putative species of fox sparrow, *Passerella megarhyncha* and *P. schistacea,* appear to have a narrow zone of contact and hybridization along the interface of the Great Basin and the Sierra Nevada. Samples taken from the White Mountains, Warner Mountains, and Mono Lake show individuals from both groups as well as mitochondrial DNA evidence of hybridization. In this contact zone, and in the White Mountains in particular, there are a high number of unusual (mtDNA) haplotypes found nowhere else in the group of species collectively referred to as fox sparrows and apparently due to the genetic consequences of hybridization (Zink 1994).

In spite of the low level of population differentiation, on average, within sage sparrows *(Amphispiza belli),* there is a strongly differentiated population of sage sparrows *(A. b. nevadensis)* near Chalfant Valley. This population was discovered to have an unusually high frequency of a single unique allele in an allozyme study of twenty-two populations of the three subspecies in this complex in California and Nevada (Johnson and Marten 1992). The subspecies *A. b. nevadensis* is highly migratory. This general geographic area is also of interest as it is the only area where the ranges of two subspecies, *A. b. nevadensis* and *A. b. canescens,* meet (figure 28.7). The two subspecies are strongly differentiated on both morphological and isozyme data. No genetic evidence of hybridization has been found.

Although there is no direct evidence for this based on Sierra Nevada species, there is reason to believe that some elevation-related adaptations may be present in some of the species or subspecies that inhabit mountainous regions. Among finches, there is evidence of interspecific variation in blood properties related to elevation of native habitat. Rosy finches *(Leucosticte arctoa),* native to altitudes above 3,500 m (11,550 ft) have been shown to have much higher blood O_2 affinity than house finches *(Carpodacus mexicanus),* native to low altitudes, when both are measured at similar elevations (Clemens 1990).

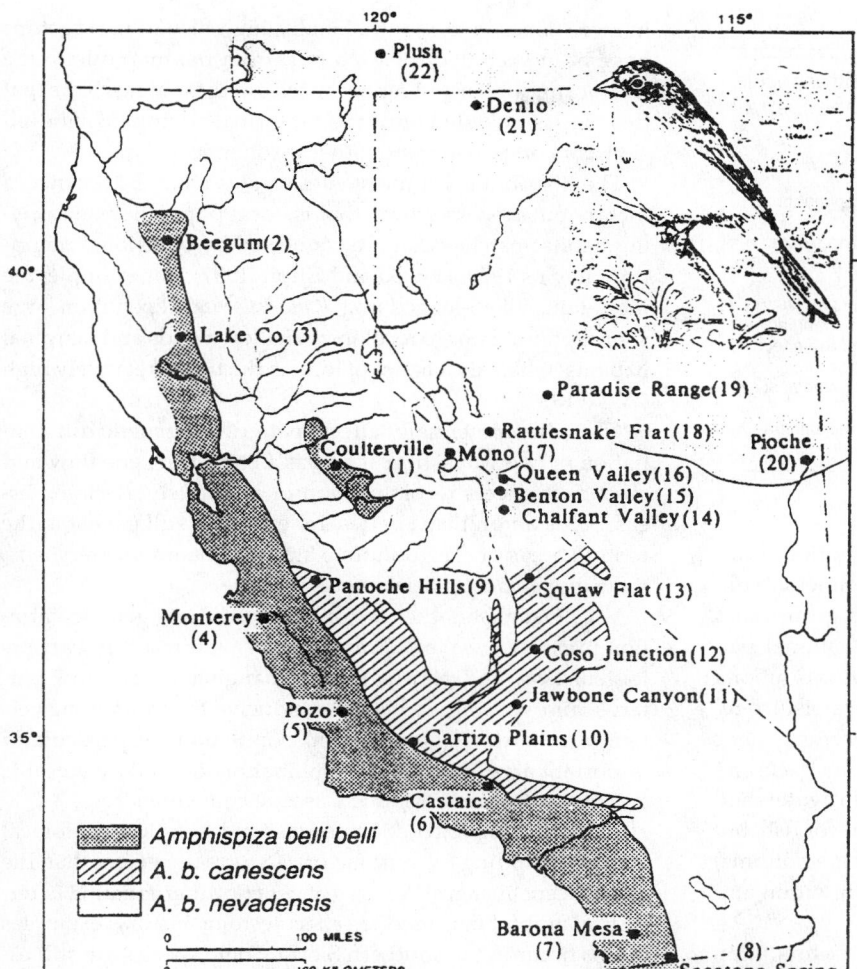

FIGURE 28.7

The known nesting distribution of the sage sparrow *(Amphispiza belli)*, showing the areas of contact between two subspecies *(A. b. canescens* and *A. b. nevadensis)* in the Sierra Nevada. (From Johnson and Marten 1992.)

Reptiles and Amphibians

Several life history characteristics pertain to the subject of genetic significance in reptiles and amphibians. First, amphibian species tend to hybridize (interspecifically) more than reptiles (e.g., Wake et al. 1978). As such, genetic richness in amphibians may be represented by hybrid zones more often than in reptiles. Second, due to the extreme phylogenetic ages of some reptile and amphibian lineages (i.e., relative to mammals), another attribute of genetic significance for this group is phylogenetic significance or age of lineage. For example, one species might be assigned more significance than another because it has no phylogenetic relatives or has fewer than the comparison species. Third, species that metamorphose may be more genetically variable, in general, than those that don't (Shaffer and Breden 1989).

For reptiles, the areas of highest species richness occur where the warm-adapted species from the south converge with the cool-adapted species from the north. Generally, they meet in areas in Kern and Tulare Counties, with the more southern species occupying the foothills and the more northern species the higher-elevation areas.

For amphibians, the notions of genetic significance are based largely on extensive salamander data. As such, there appears to be more species richness in the south for amphibians, with some notable ancient phylogenetic relicts in the Inyo Mountains and Kern Plateau. Indeed, the highest levels of species richness for salamanders in the state of California are in Kern County. (This is somewhat counterintuitive due to the moisture requirements of this taxonomic group.) Frogs are one exception to the pattern of southern species richness, with somewhat higher levels of species richness in the northwestern than the southwestern Sierra. Rare or highly restricted species occur not infrequently in the Sierra Nevada. Examples include the black toad *(Bufo exsul)*, occurring only in Deep Springs Valley between the White and Inyo Mountains in Inyo County, and the Yosemite toad *(Bufo canorus)*, restricted to the central high Sierra from El Dorado County south to near Kaiser Pass, Fresno County (Zeiner et al. 1988).

Table 28.16 lists key factors affecting the spatial structure of significant genetic variation for these three taxonomic groups in the Sierra Nevada. Because the most significant factors vary somewhat among, and even within, the three groups,

TABLE 28.16

Key factors underlying geographic patterns of intraspecific genetic variation for reptiles and amphibians in the Sierra Nevada (no priority inherent in presentation).

Elevation and its associated temperature regimes.
Rainfall, including seasonal and longer-term patterns of precipitation.
Snowpack, which reflects longer-term moisture availability.
Watershed boundaries—populations exist on cool, north-facing slopes, not on dry, south-facing slopes; thus, ridgetops demark areas of distinction.
Metapopulation structure as a result of habitat patchiness and life history.
Glaciation, which both forms a barrier and affects areas of recolonization.
Volcanism—important due to its effect as a barrier, its role in changing stream-flow patterns, its areas of multiple boundaries, and the effects (mainly inhospitable) of volcanic soils.
Tectonic effects—older phylogenetic lineages tend to reflect geological events to a greater extent than more recent species.
Edaphic/geological factors—e.g., rare populations of terrestrial salamanders in the Inyo Mountains occur in limestone creeks.

the order in which the factors are listed does not reflect priority. Elevation and its associated temperature regimes are understandably meaningful to this ectothermic group, and intraspecific genetic variation often shows elevational patterns. Moisture is another critical factor in patterns among and within species. Three major factors reflect moisture regime, and each may be correlated with intraspecific genetic variation. The first is precipitation; the second, snowpack; and the third, watershed boundaries. Not only do watershed boundaries have obvious gene-flow implications for fish, but ridgetops are often important (more so than stream bottoms) in influencing patterns of genetic differentiation within amphibian species (e.g., the Great Western Divide).

Elevation and moisture may be underlying factors in an apparent geographic trend of genetic variation in the Sierra Nevada. In some reptiles and amphibians, including plethodontid (lungless) salamanders, ranid (true) frogs, and iguanid lizards, greater population subdivision (among and within species) has been observed in the southern Sierra than in the northern Sierra. This trend appears to be independent of life history characteristics. From their extensive experience with *Ensatina* and other salamanders (e.g., Jackman and Wake 1994) and other amphibians, Wake and colleagues in the SNEP Genetics Workshop have proposed a preliminary and informal genetic zone map, whereby the subdivisions are at ridgetops between watersheds (figure 28.8). One difference between this map and the seed zone map that attempts to recognize genetic patterns in commercial tree species (figure 28.5) (Kitzmiller 1976) is the border definition. For trees, borders may often coincide with river boundaries, whereas for amphibians and reptiles, the ridgetops become more appropriate dividing lines. Part of the reason for this is that terrestrial amphibians are favored in the closed-canopy forests of north-facing slopes but are often excluded by the more open forests and chaparral on south-facing slopes. The ridgetops per se are thus not barriers, but they effectively mark barrier areas. Also, the map portrays the higher levels of interspecific and intraspecific genetic richness in the south, resulting

in more zones in the south. Finally, high-elevation ridgetops have been excluded (blacked out) from the map due to the lack of amphibian and reptile species in these areas. Note that this map is intended only as an informal attempt at subdivision, for comparison with other taxonomic groups.

The distribution of moist sites creates what is known as a metapopulation structure, that is, local populations occupying habitat patches that are connected by occasional migration (Levins 1970; Hanski and Gilpin 1991). For example, the mountain yellow-legged frog, *Rana muscosa*, depends on large source populations to recolonize shallow ponds and marginal habitats, where the chance of local extinction is relatively high (Wake 1994).

Glaciation potentially affects patterns of genetic differentiation in two ways: first, it acts as a barrier to gene flow and second, it affects recolonization of previously glaciated areas. Some amphibian species, for example, still persist in the foothill areas of the Tuolumne and San Joaquin watersheds, reflecting ancient glacial effects.

Volcanism is another factor that influences genetic variation in various ways. Volcanic activity can result in barriers to gene flow and plays a role in changing stream-flow patterns, and volcanic soil is not conducive to maintaining certain (e.g., terrestrial amphibian) populations. A particularly important area of genetic differentiation defined by volcanic activity is the Sierra-Modoc-Cascade convergence.

The older phylogenetic lineages tend to reflect geological history, including tectonic factors, to a greater extent than the more recent lineages. Patterns of genetic differentiation in terrestrial amphibians tend to reflect tectonic history; examples of this occur in the southern Sierra. Some species are still established along major fault lines.

Finally, edaphic or geological factors such as the occurrence of limestone have important implications for certain species, especially those in the genus *Hydromantes* (e.g., *H. brunus*, a highly restricted species occurring along the Merced River in Mariposa County). Another example is the recently discovered populations of a (rare) terrestrial salamander (*Batrachoseps campi*) in the limestone creeks of the Inyo Mountains (Zeiner et al. 1988).

Fish

Many species of fish are endemic to the Sierra Nevada (see Moyle et al. 1996). The widespread species (especially those within drainages that reach the coast), however, may show little population differentiation, although they vary from moderate to high levels of gene flow. An isolated drainage system can completely arrest gene flow and thus contribute to population differentiation. The length of time of isolation of the Lahontan drainage system (on the California side) is perhaps responsible for its rich array of distinctive taxa, including the endemic Tahoe sucker (*Catostomus tahoensis*) and distinctive Lahontan forms of speckled dace (*Rhinichthys osculus*), tui chub (*Gila bicolor*), mountain sucker (*Catostomus platyrhynchus*), and cutthroat trout (*Salmo clarki*). Another well-known

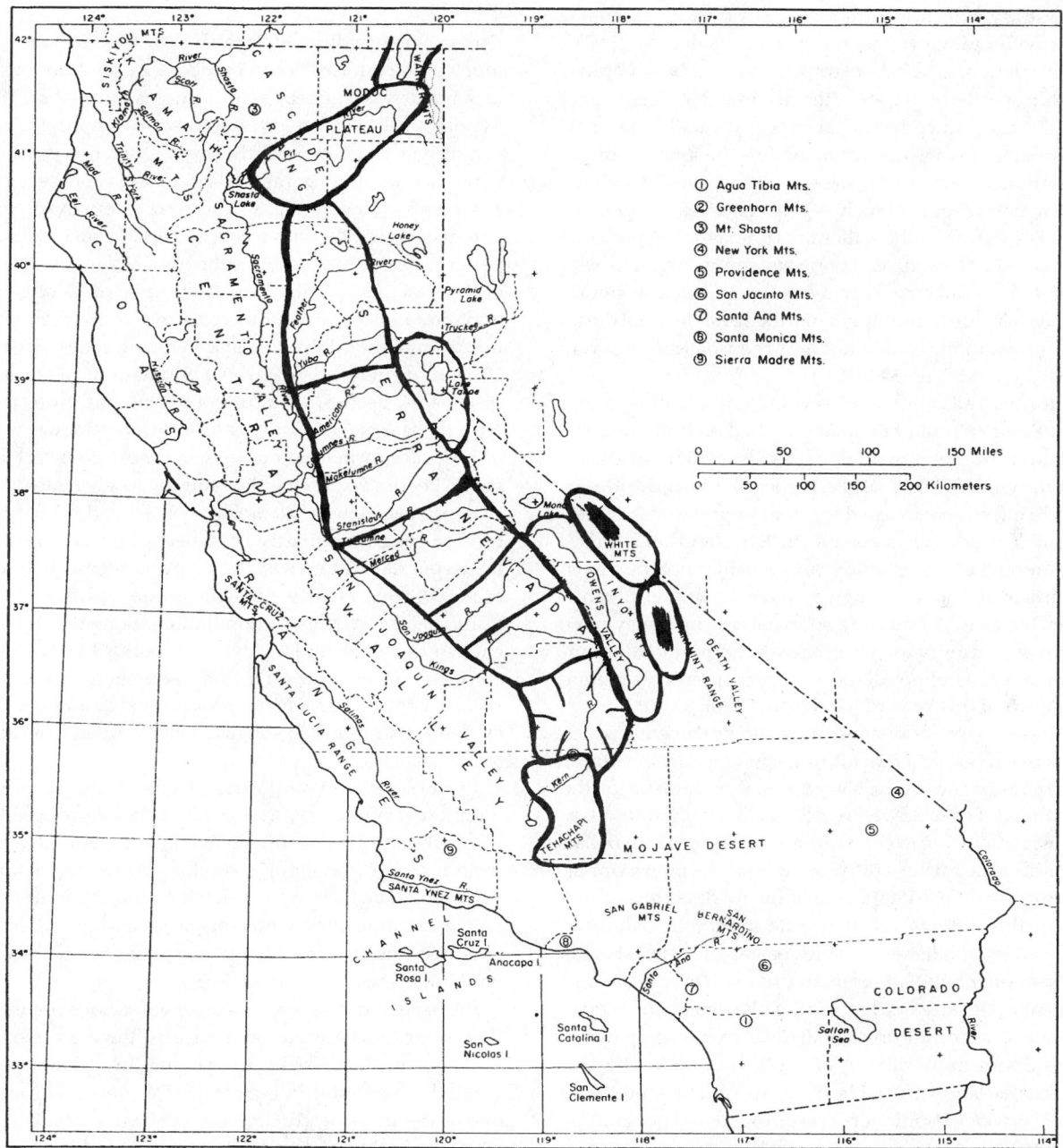

FIGURE 28.8

Patterns of geographic subdivision of significant genetic variation in reptiles and amphibians of the Sierra Nevada. Blackened areas on some ridgetops indicate areas with little or no occurrence of reptile or amphibian species. (From Wake 1994.)

example of a Sierran endemic is the golden trout, *Oncorhynchus mykiss whitei*, with its two known subspecies in the Kern River and Little Kern River drainages.

Levels of gene flow among migratory fish species are often high, and geographic patterns are weak. In a hierarchical study that considered the effects of regions (i.e., coastal versus inland), river drainages, rivers within drainages, and samples within rivers, Bartley and Gall (1990) found only

weak geographic patterns in allozyme variation among thirty-five samples of chinook salmon (*Oncorhynchus tshawytscha*) from northern California, including the Sierra Nevada ($F_{ST} = 0.177$). Most of the among-population differentiation was due to river differences (within drainages), with drainages having the second greatest degree of differentiation. Very little was due to coastal versus inland location. This pattern was reinforced by the results of a more recent and extensive allozyme study, which included samples from Oregon and assayed a wider range of enzyme systems (Gall et al. 1992).

Genetic relationships between species and among populations within species of fish are often discussed in a geological context, the time since (physical or, occasionally, thermal) separation often being directly related to the amount of genetic variation. In most cases, the separation appears to have been due to natural events such as glaciers retreating or lakes receding. For example, the Lahontan drainage subspecies of the cutthroat trout (Salmo clarki henshawi) is considered to have differentiated not only from other subspecies but also among its populations, due to the length of time since the final desiccation of pluvial Lake Lahontan and the isolation of rivers (Loudenslager and Gall 1980).

A Californian, although non-Sierran, example illustrates a similar but more recent phenomenon based on human-mediated isolation. The construction of the Chabot Dam in Contra Costa County in 1875 and of the Upper San Leandro Reservoir in 1926 effectively isolated resident populations of steelhead trout (Oncorhynchus mykiss) in Redwood and Kaiser Creeks. A recent allozyme study suggests that not only have they been isolated long enough to have become differentiated from the coastal founding source, but, since they have not had an opportunity to hybridize with domestic trout (unlike the extant coastal populations), they represent a unique "pure" source of this species (Gall et al. 1990).

In addition to river drainages and geological events, which are interrelated, one other factor underlying genetic patterns in fish species of the Sierra Nevada may be the size of the river inhabited. For example, Bartley et al. (1992) found less geographic structuring in coho salmon in California than had been found in an earlier study of chinook salmon populations (Bartley and Gall 1990). One of the reasons the authors suggest for this difference is that coho salmon in California are restricted to smaller, less stable coastal streams, whereas chinook salmon inhabit larger inland rivers. The smaller and more unstable the stream, the greater the chance of "straying"—fish not returning to their natal streams to spawn—which results in more mixing of the gene pools. Thus, in general, smaller and less stable rivers would promote more gene flow between populations, resulting in less geographic structuring.

One final and specific factor that has been linked with geographic patterns of genetic variation in fish is selection for certain allele frequencies. Certain transferrin (TFN) genotypes (i.e., individuals with certain variations of the enzyme transferrin) have been shown to have increased resistance to bacterial kidney disease in specific stocks of coho salmon. A north-south cline in the frequency of the TFN-(103) allele was found to exist between samples of coho salmon from California and Oregon. The authors contend that "the fact that this cline exists, in spite of the homogenizing effects of stock transfers, may indicate a selective advantage for certain transferrin genotypes in California" (Bartley et al. 1992).

Insects

Due to the very limited database for this taxonomic group, much of the information provided regarding underlying factors and areas of genetic significance is inferential. The factor hypothetically expected to be the most important is the coevolutionary history with the host. (Often this is a host plant; in the case of insect parasitoids the host is another insect species.) Among insect species there are many examples of apparent coevolution with a host species. In such cases, genetic patterns of the insect species theoretically might reflect those of the host. Empirically, however, many cases of geographic variation in host affiliation are merely consequences of local host availability (Futuyma and Peterson 1985). For example, in an allozyme study of twelve populations of seven species of Ips bark beetles, including a population from the Sierra Nevada near Nevada City, California, no evidence was found of host race formation on the seven host pine species in the study. Beetles occupying the same or closely related pine species, or species with oleoresin similarity, did not show any greater genetic similarity than beetles on widely divergent host species (Cane et al. 1990). A similar conclusion was reached for a study of a montane willow leaf beetle (Chrysomela aeneicollis) in populations along three river drainages in the eastern Sierra Nevada (Rank 1992). The genetic composition of the beetles was homogeneous across both willow species hosts (Salix orestera and S. boothi). Thus, no evidence was found to suggest genetic divergence according to host species.

Examples of coevolution may be more common among the highly specialized, relatively sedentary subgroups of insects, which the species just mentioned do not exemplify. Examples include alpine grasshoppers (belonging to several families of Arthoptera), which are restricted to mountaintops, and the Euphilotes butterflies, with subspecies and population differentiation closely related to the phenology of their host plants, wild buckwheat (Eriogonum spp.).

Regardless of the paucity of genetic evidence for coevolution or host race formation in insects, the converse—genetic selection in the host by insect pressure—has been documented. The western pine beetle (Dendroctonus ponderosa) is one of the most destructive insect species attacking ponderosa pine in the western United States. Studies of ponderosa pine populations in northern California (including a Plumas County population) and southern Oregon showed monoterpene profiles (i.e., frequency distributions of the various monoterpenes present) that suggest a coevolutionary relationship between the tree and insect species. For example, the ponderosa pines in the Plumas County population and in other northern California populations, which have a continuous history of western pine beetle predation, are characterized by high concentrations of limonene relative to adjacent populations. Limonene is toxic to the western pine beetle (Sturgeon 1979). Populations without a history of predation have lower levels of limonene, suggesting that the beetle may have exerted selection pressure on its host species.

High-elevation ridges have been shown to be factors affecting gene flow in some insects of the Sierra Nevada. For example, allozyme analysis of populations of a montane willow leaf beetle *(Chrysomela aeneicollis)* in the eastern Sierra Nevada showed genetic subdivision among the three river drainages in which it was sampled (Big Pine Creek, Bishop Creek, and Rock Creek). The drainages are separated by high-elevation ridges. Although the F_{ST} among river drainages was only 0.135, this is higher than the F_{ST} values across broad geographic scales for many flying insects, including bark beetles *(Drosophila* spp.) and several lepidopterans (Rank 1992). Interestingly, although high-elevation ridges apparently are barriers to gene flow among populations, gene flow also does not occur through low-elevation connections. Although the drainages are connected by nearly continuous stands of willow at lower elevations, this is apparently unsuitable habitat for the beetle, and any gene flow that occurs among the populations occurs over the ridges instead of along the streams (Rank 1992). The low-elevation connections, however, occur outside of the Sierra, in very different habitat conditions.

Substrate or soil parent material has sometimes been suggested as a factor underlying genetic differentiation in insect species, but this is not upheld in genetic tests. Two alpine butterfly "species" endemic to the Sierra Nevada *(Oeneis ivallda* and *O. chryxus stanislaus)* occur on different substrates. In the southern Sierra, the lighter *ivallda* type occurs mainly on granitic substrates and the darker *stanislaus* type on andesite. However, in the northern Sierra, this relationship is less clear, with an increase in frequency of *ivallda* types but on a mainly andesite substrate. Porter and Shapiro (1989) found minor allozyme differentiation between these "species" (F_{ST} = 0.081), which are mainly characterized by wing color. They recommend classification of these two color types as a single species, given the lack of evidence for interruption of gene flow between them (figure 28.9).

One seldom-considered factor that is correlated with patterns of genetic variation in one insect species is time of day when populations are active, which is presumably related to ambient temperature. In some species, nested within spatial levels of variation is a genetically based temporal array of genotypes, as demonstrated in *Colias* butterflies (Watt et al. 1983). Samples from Tracy, California, and Gunnison, Colorado, showed correlations between flight patterns (e.g., time of day of flight initiation) and distinct allozyme patterns. It is not clear that similar phenomena occur in other systems (i.e., in other animals or in other loci not related to the flight muscle metabolism).

Two other geographic subdivisions or associations relate to representative genetic variation in Sierra Nevada insects. The first is the natural plant community, or biome. For example, three biotypes of *Apodemia mormo* occur in three distinct plant communities in the southeastern Sierra Nevada and the western Mojave Desert. They differ in the larval host plants to which each is adapted and may even deserve spe-

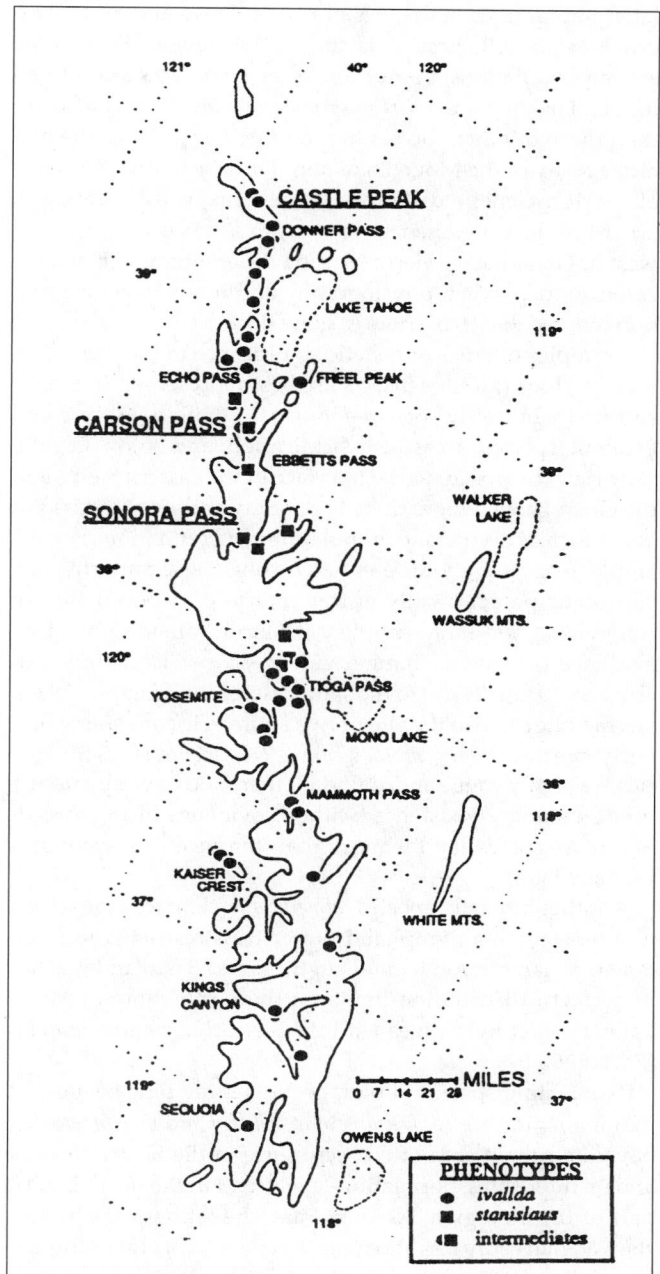

FIGURE 28.9

Distribution of alpine habitats in the Sierra Nevada with known localities of *Oeneis ivallda* and *O. chryxus stanislaus* (satyrine butterflies). Sampled populations are in the larger-sized font. (From Porter and Shapiro 1989.)

cies status (Pratt and Ballmer 1991). The association between the insects and the plant communities may be related to natural selection. The second factor is the physiographic subdivisions within biomes (for example, a Sierra Nevada/Great Basin subdivision, or watershed subdivisions). The western seep fritillary *(Speyeria nokomis apacheana)* exists in small, isolated populations in the western Great Basin and eastern Sierra Nevada. Allozyme data suggest low levels of gene flow among populations, unique alleles in some (e.g., Round Valley), and mean population heterozygosity levels that are lower than those of other species in the same family (e.g., the heterozygosity of the Mono Lake population is 0.016) (Britten et al. 1994b). Genetic distances suggest major differentiation according to watershed areas and, to a lesser extent, east/west or Great Basin/Sierra Nevada differentiation. Here, the association between physiography and insect genetic variation may be due to barriers to gene flow.

Examples of areas of genetic significance in the Sierra Nevada (rich or rare) for insects are best represented in the literature by butterfly species—unusual populations, rare and endemic species, areas of hybridization, and so on. Genetically rich areas occur at the interface of the eastern Sierra and the Great Basin, where there is not only species richness but also much ecotypic and population differentiation. For example, two strongly differentiated congeneric butterfly species occur parapatrically at this interface. A population of *Anthocharis sara sara*, sampled at Sierra Valley, showed no evidence of gene exchange with *A. sara stella* sampled at Truckee, 40 km (24 mi) to the south (Geiger and Shapiro 1986). Sierra Valley is also the site of two sympatric congeneric butterfly species (*Pontia protodice* and *P. occidentalis*). Both species are highly vagile and abundant there and have apparently been in stable coexistence, without evidence of interbreeding, in Sierra Valley for more than ten years (Shapiro and Geiger 1986).

Another butterfly species, *Limenitis lorquini weidemeyerii*, a middle-elevation nymphalid butterfly, is restricted to montane riparian canyon habitats in the Great Basin and reaches its western distribution limits on the north shore of Mono Lake. Here, it hybridizes with the Sierran *L. lorquini lorquini* (Porter 1989).

Two sibling species of bark beetle, Jeffrey pine beetle and mountain pine beetle (*Dendroctonus jeffreyi* and *D. ponderosae*), have their main area of co-occurrence in the Sierra Nevada and an interesting population-level trend in the northeastern part of the bioregion. At Yuba Pass, the Jeffrey pine beetles showed markedly less allozyme diversity than other sampled populations of that species, and at nearby Sattley, the mountain pine beetles showed considerably more allozyme diversity than other sampled populations of that species (Higby and Stock 1982).

A hybrid zone within the giant silk moths (*Saturniidae*) has been studied at Monitor Pass in Alpine County (Collins 1984).

A lone population of the checkerspot butterfly (*Euphydryas editha*) near Big Meadow (Tulare County, California) was found to be genetically distinct from forty other sampled populations within the species range across the western United States (Baughman et al. 1990).

Numerous examples exist of butterfly species, semispecies or subspecies, endemic to the Sierra Nevada, including *Phyciodes montana* and *Anthocharis stella* (Shapiro 1992), and

Oeneis ivallda and *O. chryxus stanislaus* (Porter and Shapiro 1989). For the latter two species, there is a genetically diverse area near Tioga Pass where an abrupt transition zone occurs between the two color types that distinguish the species. This zone has individuals spanning the full range in wing coloration among both species (Porter and Shapiro 1989) (figure 28.9). A specimen resembling *Xyleborus californicus* was found near Georgetown, California, in the late 1980s. This species may be either a recent introduction from South America or Southeast Asia or an extremely rare endemic species (Hobson and Bright 1994).

The gene-flow corridors provided by the transmontane rivers, such as the Pit and North Feather Rivers, are another example of species-rich areas. A complex cline involving three subspecies of the *Coenonympha tullia* group of satyrine butterflies (*C. california, C. eryngii,* and *C. ampelos*) occurs in the Pit River drainage (Porter and Geiger 1988). Rare species are often found on unusual soil types (e.g., serpentines) and on wetlands and bogs of Pleistocene origin.

Although introduced species may often have low genetic variation due to the bottleneck experienced during their introduction (e.g., *Holocnemus pluchei*) (Porter and Jakob 1990), the Sierra Nevada may hold unusual populations of these often widespread insect species. For example, the introduced European cabbage butterfly (*Pieris rapae*) was sampled over much of its current distribution in the United States. Although the species has existed for a longer time in the eastern United States (since about the 1860s), the eastern populations show little allozyme differentiation, while the population sample from Reno, Nevada, is distinctive (Vawter and Brussard 1984). Indeed, the western populations appear to have diverged not only from the eastern populations but also from one another. This genetic differentiation in the West is attributed to fragmentation of suitable habitat. The "suitable habitat" found in the West consists of agricultural or urban areas interspersed with inhospitable desert or montane natural areas.

In summary, the key factors underlying genetic differentiation due to natural selection in insects of the Sierra Nevada are, theoretically, coevolution with host plant species, climate, elevation, and substrate. However, there are few studies of such relationships and even fewer that confirm a genetic basis for the morphological differences observed. Differentiation due apparently to restriction of gene flow is more readily apparent, as in the case of populations differentiated due to ridgetops, geographic distance, or sedentary habit.

Fungi

The most important factor underlying patterns of genetic variation in fungal species, beyond life history characteristics, is theoretically the relationship with the plant host (M. Garbelotto, University of California, Berkeley, conversation with D. Rogers, September 1994). Implications based on this assumption have been made—for example, that forest pathogens are potentially more genetically diverse than fungi of domesticated crops and may have complex population structures that reflect the heterogeneity of their hosts and environments (Vogler et al. 1991). Studies documenting this pattern, however, are scarce.

Although there is no evidence from Sierra Nevada fungi for this relationship, an elegant study by Burdon and Roelfs (1985a) demonstrates a specific kind of host-pathogen relationship, namely, the genetic consequences of eradication of an alternate plant host. Wheat stem rust *(Puccinia graminis)* populations in the eastern United States have been asexually reproducing since the late 1930s, coincident with the eradication of the alternate plant host for this species, the common barberry. Populations from this area were compared with sexually reproducing populations (due to the presence of the barberry) from the Pacific Northwest. The two groups showed striking genetic differences, the sexual populations being more genetically diverse in all variables measured. The structure of genetic diversity between the two groups also differed. The sexual populations portrayed a pattern consistent with random mating, and isozyme alleles and virulence genes were unrelated. In contrast, the asexual populations were strongly subdivided along clonal lines, and there was close agreement between isozyme and virulence structure (Burdon and Roelfs 1985a).

Elevation and climate, paleohistory and recent history (mainly anthropogenic disturbances) are potentially important to fungal patterns (M. Garbelotto, conversation with D. Rogers, September 1994), although specific examples are rare. Substrate or parent material is likely to be important, especially for mycorrhizal fungi.

There is an interesting example of a genetic pattern in the western gall rust *(Pteridermium harknessii)* in the Sierra Nevada, for which there is no definitive underlying causation. In an area just east of Lake Tahoe, an allozyme analysis revealed two strongly differentiated forms of the fungus (called "zymodemes") that coexist without apparently interbreeding (figure 28.10). South of this area, populations of zymodeme II were found almost exclusively, while north and west of this area, zymodeme I was almost exclusively present. The authors interpreted this pattern, together with the lack of recombinant genotypes, to be indicative of asexual reproduction. However, the reason for the change from one zymodeme to the other in the Lake Tahoe area remains elusive (Vogler et al. 1991).

Conclusions

Genetic significance is described by attributes of rarity, richness, and representative genetic variation, as well as other attributes. For example, some species may be highly valued due to their phylogenetic significance. Significant genetic variation is also described as that portion of the total variation that preserves evolutionary potential, which is theoretically defensible but nearly impossible to recognize. Furthermore, genetically significant units vary from the subpopulation to family level. Thus, both the attributes of ge-

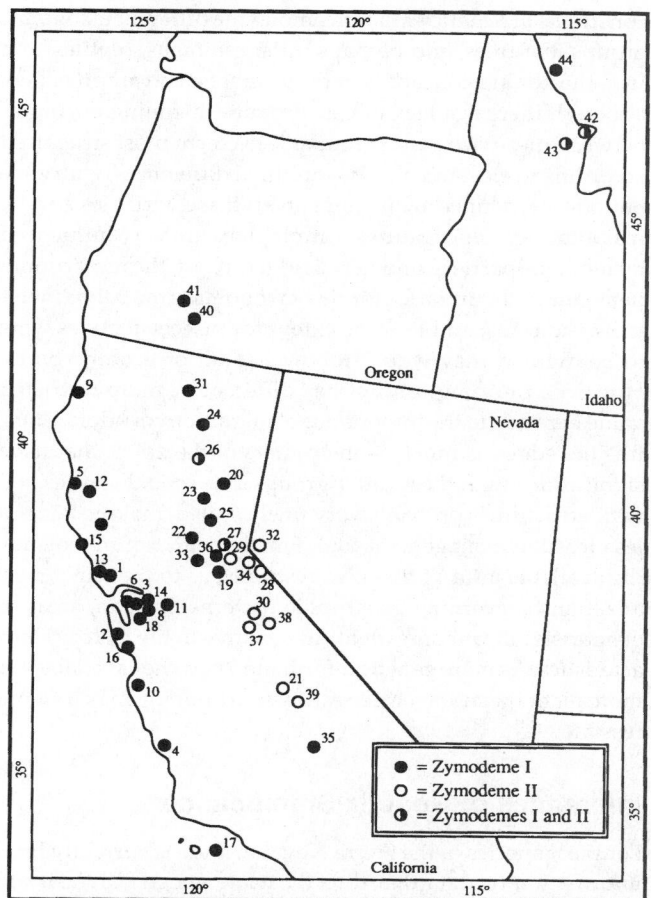

FIGURE 28.10

Geographic sources of *Pteridermium harknessii* isolates collected from the western United States. Numbers refer to collection sites. (From Vogler et al. 1991.)

netic significance and the level at which they are described are closely interconnected and dependent on the overall (management or research) objective.

The ability to define areas of genetic significance in the Sierra Nevada is greatly hampered by the lack of information, even of range distributions, for many species and by the corresponding overrepresentation of a few species in the body of current knowledge. Key factors underlying genetic subdivisions vary among taxonomic groups, with perhaps the only generalization being that east-west gradients, with their associated temperature and elevational factors, are more determinant of genetic variation, in general, than the north-south gradients. However, even this generalization has caveats. For example, although there is an east-west dichotomy in species richness for amphibians, within the species-rich west side the structure of genetic variation more accurately reflects north-south factors.

The idea of constructing zones of representative genetic variation across the Sierra Nevada is challenged by differing constructs of genetic variation among the different taxonomic groups. For trees, and perhaps for amphibians, reptiles, and fish, the zoning concept is perhaps more easily approached, although there is a lack of concordance in zoning attributes between these two major groups. Trees seem most structured according to elevational and latitudinal differences, with zone boundaries defined by increments in these variables and by major topographic features. Amphibians are very influenced by montane barriers, and hence ridgetops are the most prominent genetic boundaries for this taxonomic group. For mammals and birds, and to some extent for insects, if zones were to be defined they would follow habitat or ecotone types. However, the zone concept has different or more restricted value here due to the importance of migration corridors. Thus, it is the edges as much as the centers of the zones that have significance with these latter groups. For fungal species, genetic structure is probably very finely scaled, making zones a less feasible management tool. Finally, with nontree plants, the available information is so restricted as to prevent generalizations concerning geographic patterns. Instead, areas of genetic significance are identified relative to life history characteristics. As more genetic information becomes available for more plant species, such geographic partitioning may be more feasible.

Inferences of Genetic Significance

For most species in the Sierra Nevada, there is currently little information or data that directly relate genetic diversity to adaptive significance. Genetic significance may be particularly germane to conservation and management as a means of establishing priorities. In the absence of such information, some idea of the extent and nature of genetic significance may be available from concepts based on correlations or associations that have generalized genetic variation within a certain context. These concepts are neither mutually exclusive nor necessarily independent of the need for genetic assessments. Rather, they are additional perspectives from which genetic structure, and thus significance, may be inferred. These concepts are, with a few exceptions, preliminary suggestions developed by workshop participants for the purposes of the current discussion.

Life Form and Life History Associations

Sufficient genetic information is available from taxa beyond the Sierra Nevada within certain taxonomic divisions (e.g., plants) for researchers to have realized correlations between the level and pattern of genetic variation and certain life form and life history characteristics (Hamrick and Godt 1990; Hamrick et al. 1992, 1979). This correlation has been investigated statistically for plants (Hamrick and Godt 1990). In a review of more than four hundred plant species, correlations were noted between amount and structure of genetic variation (as measured by allozymes) and such features as the spe-

cies' geographic range, longevity, seed-dispersal mechanisms, and breeding system.

The relationship between allozyme variation and plant characteristics was investigated at three levels—genetic variation within the species as a whole, genetic variation among populations, and genetic variation within populations (tables 28.17–28.19). For example, widespread plant species tend to have greater amounts of allozyme variation than do narrowly distributed species (table 28.17). Long-lived perennials generally have higher levels of genetic variation than short-lived perennials. Breeding system is highly associated with gene flow: self-pollinating, or selfing, species tend to have relatively high amounts of variation among populations, while outcrossing species have relatively little. Population differentiation shows very different correlations (table 28.18). Selfing species have more population differentiation (i.e., more genetic variation among populations), in general, than do outcrossing, wind-pollinated species. Finally, the amount of genetic variation within plant populations shows a set of correlated traits that is similar, but not identical, to those correlated with species-level genetic variation (table 28.19). Widespread species again have higher levels of genetic variation within populations than do endemics; however, breeding system characteristics are even more highly correlated with within-population levels of variation than is geographic range. Mixed-mating, wind-pollinated species have considerably higher levels of genetic variation within populations than do selfing species.

For certain California tree species, a generalization has been noted concerning latitude and genetic variation within populations (Ledig 1987). Genetic variation within populations in the south of a species' range tends to be the highest, decreasing toward the north. A striking example of this trend occurs in a non-Sierran tree species, coulter pine (*Pinus coulteri*). In this species, heterozygosity increases from 0.11 in the northernmost populations near Mount Diablo, California, to 0.19

TABLE 28.17

Correlates with genetic variation within plant species.[a]

Trait[b]	Highest Level	Lowest Level
Geographic range	Widespread	Endemic
Life form	Long-lived, woody perennials	Short-lived perennials
Breeding system	Mixed mating, wind pollinated	Mixed mating, animal pollinated
Seed-dispersal mechanism	Attached	Explosive
Taxonomic status	Gymnosperms	Dicots
Regional distribution	Boreal-temperate	Tropical or temperate
Mode of reproduction	Sexual	Sexual and asexual
Successional status	Late successional	Mid successional

[a] Derived from Hamrick and Godt 1990.
[b] Traits are arranged in approximate order of their strength of correlation with genetic variation. Thus, geographic range is very strongly correlated with genetic variation within a species, but successional status is almost insignificant.

TABLE 28.18

Correlates with genetic variation among populations within plant species.[a]

Trait[b]	Highest Level	Lowest Level
Breeding system	Self-pollinated	Outcrossing, wind pollinated
Life form	Annuals	Long-lived, woody perennials
Seed-dispersal mechanism	Gravity dispersed	Gravity attached
Successional status	Early successional	Late successional
Taxonomic status	Dicots	Gymnosperms
Regional distribution	Temperate	Boreal-temperate

[a] Derived from Hamrick and Godt 1990.
[b] Traits are arranged in approximate order of their strength of correlation with genetic variation. Thus, breeding systems are very strongly correlated with genetic variation among populations, but regional distribution is not well associated.

in the southernmost populations in Baja California (Ledig 1987). This pattern is also seen in such Sierran species as giant sequoia (*Sequoiadendron giganteum*), Jeffrey pine (*P. jeffreyi*), western white pine (*P. monticola*), sugar pine (*P. lambertiana*), and Douglas fir (*Pseudotsuga menziesii*). One explanation offered for this apparent trend is glacial history: as species migrated northward following glacial retreat, the new and more northerly populations might have arisen from only a small sample of the original species—those that dispersed and successfully colonized northward. This smaller sample would have contained only a fraction of the species' original gene pool. If the process was an iterative one, the populations migrating northward would have originated from a smaller and smaller gene pool sample, manifested today as lower levels of within-population genetic variation. Conversely, the southern populations would have been refugial, presumably harboring genetic diversity. Yet another possibility is selection rather than historic condition—that is, that higher tempera-

TABLE 28.19

Correlates with genetic variation within populations of plant species.[a]

Trait[b]	Highest Level	Lowest Level
Breeding system	Mixed mating, wind pollinated	Self-pollinated
Geographic range	Widespread	Endemic
Life form	Long-lived, woody perennials	All others
Taxonomic status	Gymnosperms	Dicots
Seed-dispersal mechanisms	Attached	Explosive
Regional distribution	Boreal-temperate	Temperate or tropical
Successional status	Late successional	Early successional

[a] Derived from Hamrick and Godt 1990.
[b] Traits are arranged in approximate order of their strength of correlation with genetic variation. Thus, breeding systems and geographic range are very strongly correlated with genetic variation within populations, but successional status is only mildly related.

tures and summer rainfall favor higher allelic diversity levels than in northern latitudes.

Similar correlations between life history characteristics and allozyme variation were noted (at the SNEP Genetics Workshop) for insects, on a more informal basis. For example, sedentary insects with strong host-plant relationships tend, or are expected, to show stronger differentiation among populations than highly vagile insects with less specialized trophic relationships (Shapiro 1994).

This type of generalization is useful in that life history characteristics such as breeding system and geographic range are known for some species and can often be inferred for others by field observations. However, there are some constraints and caveats in the application of this approach. Correlations with genetic variation are weak at best, may change across broad taxonomic groups, and, with the exception of plants, have either little direct evidence or have been entirely inferred on theoretical grounds for other groups. Further, the correlations have so far been demonstrated only for allozyme genetic variation and not other types. Thus, assessments are often inferred, and many exceptions to generalizations occur.

Evolutionarily Significant Units

Evolutionarily significant units (ESUs) were suggested not so much as a way of replacing genetic information with a proxy, but rather as a way of placing the emphasis on a different genetic measurement—that of genetic distance between populations or groups of populations, such as golden trout. One definition of an ESU is a historically isolated set of populations; the genetic criteria for recognizing an ESU are currently under discussion (e.g., Moritz 1994). The rationale is that the more historically isolated the groups, the more likely they are to have distinct genetic attributes and different evolutionary potential. Within this framework, an ESU could be a population, a group of populations, a species, or a grouping of species. For example, weakly differentiated species might be grouped together as an ESU; in contrast, strongly differentiated populations of one species might each be an ESU. This concept suggests a way to guide decisions about biodiversity protection. One does a phylogenetic analysis of genetic (e.g., allozyme or DNA sequences) and morphological data and than recognizes clusters of close relatives and progressively more distantly related forms (this has been called phylogenetic ranking in Moritz 1994). This helps inform decisions about conservation priorities; that is, when one must make choices, it may be more important to maintain major lineages, not necessarily all the minor phylogenetic branches.

This means of assessing (and valuing) genetic variation is attractive in that it recognizes the dynamic nature of gene pools; we want to conserve not only short-term adaptations but also longer-term evolutionary potentials. However, this assessment still requires species-specific genetic data and still relies on the same kinds (allozyme or DNA) of genetic measurements.

Environmental Correlates

As was discussed in previous sections, often an environmental variable (or variables) can be identified that correlates with patterns in genetic variation. Even in the absence of genetic data to confirm these patterns, there are some features, particularly geographic features that would tend to limit gene flow, that might at least be reasonably good indicators of population differentiation. For example, ridgetops have previously been discussed as presenting gene-flow barriers for some amphibian species, and thus genetic differentiation according to this geological feature could be expected. Similarly, fish populations in nonconnected river drainages might be predicted to be more strongly differentiated than those in continuous drainages. Much of this is inferential and has not been tested or confirmed, and many generalizations will occur.

Species Richness and Distribution Patterns

In lieu of species-specific genetic information, species richness provides some measure of biodiversity. Species diversity is not the primary focus of this chapter; however, genetic significance might be inferred to some extent not only by species presence but also by distribution patterns. For example, if the species tends to be distributed as disjunct populations, or if the area under consideration has marginal populations (i.e., populations that are near the edge or limits of the species' natural range), the populations might be more genetically distinct (as compared with midrange populations of a widespread, contiguously distributed species). For example, the dusky shrew (*Sorex monticolus*) is a widespread species, yet has two areas with isolated populations—one in the Sierra Nevada and one in the San Gabriel Mountains of southern California (Zeiner et al. 1990b). The fact that these populations are isolated might be used to infer genetic significance.

This approach to assessing genetic variation has the advantage of requiring only census data, rather than sampling and genetic analysis. The number of species expected within a given geographic area can usually be obtained from species range maps, which are readily available for most mammals, birds, fish, and trees, although to a lesser extent for nonwoody plants, insects, and fungi.

ASSESSMENT OF CONDITIONS AND TRENDS

Sources of Genetic Threats and Consequences

An essential step in linking the conservation of significant genetic diversity with management practices is the identification of threats to diversity. In this context, a threat could be defined as anything that potentially or actually reduces or changes genetic diversity by a significant amount relative to the standard in situ at any of the levels at which diversity is recognized, from genes through ecosystems. However, as there is a continuum from *action* to *threat* to *consequence*, it is often difficult to identify, with objectivity, the threat. For example, the following hypothetical sequence of events could occur: urban development, leading to population fragmentation, leading to an increase in inbreeding, leading to inbreeding depression, leading to a decrease in fitness, leading to population extinction. Urban development is an action, and population extinction is a potential genetic consequence, but a threat could be defined as any event from urban development through decrease in fitness, whereas a consequence could be any event from population fragmentation through population extinction.

The definition of genetic threat is further complicated by its interactive nature with the biological attributes involved. For example, if the taxon is already depauperate in genetic variation as a result of its evolutionary (e.g., *Pinus resinosa*, red pine) or recent (e.g., North American gypsy moth populations) history, then certain actions may not pose the same genetic threat as they would with a different, genetically diverse species. In other words, species context is very important in predicting whether imposed changes in genetic diversity are significant or not and whether they are potentially detrimental or not. Differences in genetic architecture among species help geneticists to determine which actions might be threats. For example, the importance of interpopulation gene flow would influence the likelihood of population fragmentation posing a threat. Further, the cumulative nature of threats means that some actions become threats only if they occur in concert with or compound other potential threats.

An obvious factor influencing the identification of threats is the amount of information available. Specific information on the amount and structure of genetic variation for a taxon, and the interrelationships between it and ecological processes, is almost never available. Even when genetic information is available, and a genetic consequence identified, the long- and short-term *adaptive* consequences are rarely clear. SNEP Genetics Workshop participants felt that an appropriate if conservative approach in the face of minimal empirical knowledge yet a strong theoretical foundation was to assume that detrimental effects on gene pools may occur when large changes in gene diversity are predicted.

Given the variability and subjectivity of defining threats and consequences, it is desirable to address the question, What constitutes an action that may significantly alter genetic diversity? by taxonomic group. This is done in the sections that follow. In addition, those taxa and geographic areas most implicated in the Sierra Nevada are identified to the extent they are known.

Management opportunities to mitigate threats and consequences become possible with the development of standards. Standards reflect the idea of a threshold: When does an activity become a problem? When is it time to take action? Like the concept of threats, standards are fraught with the problems of inadequate information, interactions and complexities in systems, arbitrariness, and subjectivity. Time lags between cause and effect, or action and threat, or threat and consequence, are a major issue. Genetic patterns, in particular, reflect conditions in the recent or distant past, further driving a wedge between present management options and future desirable conditions. Further, standards must somehow embrace the dynamic nature of populations and their genetic attributes (the natural ranges of genetic variation). The sections that follow address the issue of standards for each taxonomic group.

Commercial Tree Species

Spectrum of Actions Likely to Cause Major Changes in Native Gene Pools. While, as is the case with other taxonomic groups, the major threats to tree species in the Sierra Nevada are habitat loss and fragmentation, forest-management practices involving these species, particularly commercial conifers, have potential genetic effects across much of their native ranges in the Sierra Nevada. These potential management- and development-related actions and their consequences are presented in more detail in table 28.20. Mitigating actions that have been taken in forest-management programs are discussed in a later section. Potential consequences may be the result of cumulative effects of multiple concurrent or sequential actions. For example, depending on their implementation, commercial regeneration practices for pines and Douglas fir could have genetic consequences of inbreeding depression, outbreeding depression, reduction in genetic diversity, and/or alteration of genetic architecture, although these effects are undocumented.

Taxonomic Groups and Geographic Areas Involved. By virtue of their being harvested, bred, and planted, the commercial conifers that are the subject of intensive timber management are species that deservedly receive the most attention. Other conifers that are planted and harvested in operational forestry practices are also subject to activities that could potentially alter the gene pool. These include the mixed conifer species, yellow pines, and true firs in the Sierra Nevada (table 28.21). On U.S. Forest Service lands, all of these species are subject to rigorous genetic diversity standards developed to maintain broad adaptability and local adaptations. Some concerns remain that, although standards are in place, they might not be maintained adequately, due to local negligence, urgency, or practical operational realities. Since most of the genetic information available pertains to the commercial tree species, it is possible to be more specific about the nature of the activities used on these species and their consequences, as well as the specific populations and geo-

TABLE 28.20

Types of potential threats and their possible genetic consequences to trees in the Sierra Nevada.

Threat	Possible Genetic Consequences
Artificial-selection pressures	Conditions imposed on tree seedlings for regeneration and restoration efforts at the nursery may not prepare them for planting. Location (climate, soils) and management (soil moisture and fertility, freedom from competition, etc.) regimes at the nursery may not mimic natural selection pressures, leaving seedlings maladapted to the planting site.
Genetic bottlenecks	In out-planting programs, the amount of genetic variation in planted seedlings may be decreased relative to that expected in natural regeneration, due to the initial sampling procedures and subsequent selection in seed orchards and nurseries.
Inbreeding depression	Many tree species have outcrossing mating systems and are susceptible to inbreeding depression. Inbreeding may be a problem in seed collections from wild stands or from seed orchards, especially if not monitored or mitigated. The effects may not manifest themselves immediately in terms of mortality of inbreds, but lowered viability or fitness may occur. Thus, individuals with low fitness may be included in out-planting programs.
Introduction of maladapted genes	Primary introduction of maladapted genes may occur as the result of introducing inbred or nonadapted genotypes to a site. Secondary introduction may occur when the introduced trees reach reproductive maturity and begin combining with the local gene pool. "Outbreeding depression" in the hybrid generation may occur if the introduced genotypes were not adapted to the site or if the introgressed progeny have lower fitness due to hybridization of dissimilar genomes.
High-grading	Selection and removal of certain phenotypes or ecotypes, often the most fit or vigorously growing, known as high-grading, occurred historically on private and public lands but may now be largely controlled by forest practice standards. However, this may continue to be a consideration in some areas, especially where such standards are not in place or enforced. For example, in some "commercial" clear-cuts, trees of commercial size and quality are removed, leaving a small number of various defective and suppressed trees for stocking, seed production, and visual acceptance. Although these cuts may have higher visual acceptance, they likely are very dysgenic.
Attack by exotic insects and pathogens	If resistance to the insect or pathogen is found at very low frequencies in the host populations, as it is in the case of resistance in white pines to the white pine blister rust, then genetic diversity within the host population may be severely lowered, resembling a bottleneck process. Out-planting of nursery-grown stock may also contribute to this situation if the stock has been infected with nursery-found pathogens and serves as a vector to wild populations.
Introgression	In some cases, such as Washoe pine, naturally occurring but sparse or rare species may be in danger of being swamped by the more widespread and co-occurring species (in this case, Jeffrey pine) if the two hybridize. In the case of Washoe pine, some populations potentially at risk in this regard might be those that were apparently heavily logged in the early settlement period, such that population size (already small) shrank drastically in a short time.
Ecological displacement of native populations by introduced species	The introduced species, such as *Eucalyptus* spp., may cause displacement both by being a better short-term competitor for resources and by modifying the site so that it becomes less amenable to the original native species in the longer term.

graphic areas that may be most important to monitor. With less-studied tree species that are nonetheless targets of some kind of manipulative management (e.g., certain oak species), one of the main concerns, based on experience from commercial conifers, is the possibility of inappropriate management. For sparsely distributed species, for which we have little or no genetic information, a major concern is lack of genetic awareness in management or inappropriate genetic management that has relied on inferences from other, genetically dissimilar species.

Standards or Thresholds for Evaluating Effects. If quantitative standards were to be developed, they would have to apply to the genetic consequence (e.g., the allowable change in rate of inbreeding) rather than the threat (e.g., fragmentation), as the latter may be cumulative, qualitative, or not obvious, or may not be the source of the problem. However, due to the lag time between activity and genetic consequence, which is especially pronounced in long-lived species, basing standards on current genetic parameters may mean that action comes too late to mitigate some situations. Thus, it seems that a quantitative approach to standards is ill suited to this taxonomic group. Instead, qualitative statements about levels of tolerance within different species to various kinds of activities must be developed for individual cases. Tolerance could be further related to specific attributes or values such as within-population diversity or genetic architecture. Qualitative standards could then be developed based on the resilience that the value or attribute of interest has to the threat. Also, current and potential management activities could be described in terms of their likelihood of enhancing or lowering tolerance.

As an example, consider within-population genetic diversity for the commercial conifers. Potential threats (i.e., events that could lead to undesired genetic consequences) include severe wildfires, extended climatic extremes, loss of seed-dispersal agents, and excessive or inappropriate seed transfers. Depending on the number of threats at any time, a threat

might be considered acceptable or unacceptable. Also, as more information becomes available, the relative nature of the threats (or specific levels of tolerance to each) may play a role in the decision-making process. Further, natural and management-provided processes that increase resilience can be identified and balanced with the perceived threat. In this example, resilience to loss of within-population diversity is provided by the relatively high frequency of reproductive effort in these species and by the relatively high levels of genetic information and its incorporation into management activities.

Other Plants

Spectrum of Actions Likely to Cause Major Changes in Native Gene Pools. Three main classes of events potentially pose serious threats to plant populations or species (table 28.22). The first is habitat destruction or degradation. Only events that are large or genetically significant relative to natural disturbances are considered here. These include alterations in natural cycles, including fire, hydrological, and mineral or nutrient cycles; land development and its associated effects; and human-initiated biotic disturbances such as exotic weed invasions and pesticide-driven loss of pollinators. The second inferred threat is genetic contamination, both within species as a result of inappropriate human movement of genetic material and between species due to hybridization with human-introduced exotics. The third threat is posed by activities that seriously fragment population structure, which can reduce effective population size and reduce gene flow. In the short term, such fragmentation may result in an increase in inbreeding depression. In the long term, genetic diversity in the population may be lost due to genetic drift (random loss of genetic diversity from a population due to sampling effects of small population size) and loss of exchange of genetic diversity. Potential threats become more serious when they either are widespread and cumulative or affect small populations of rare or sparse species. However, any consequences vary as a function of genetic architecture and breeding system.

TABLE 28.21

Examples of trees and tree habitats potentially at risk in the Sierra Nevada.

Taxa	Area and Threat
Sugar pine	Threatened throughout its range, particularly in areas where it is affected by both harvesting and white pine blister rust.
Whitebark pine, limber pine, other white pines	Possibly threatened in high-elevation areas by white pine blister rust. Resistance-oriented planting programs are less likely to reach these less-accessible sites. Also at risk in southern Sierra where (pine) species are common.
Giant sequoia	Historic harvest threatened some populations. Possible genetic consequences in areas where regeneration practices have involved transfer of seedlings among groves (populations). Potential inbreeding within isolated groves.
Ponderosa pine	Harvesting, water diversion, and regeneration practices threaten this species, particularly near urban areas, in low-elevation sites, and on private lands.
Oak species	Oak regeneration problems may result in changed genetic diversity.
Sargant cypress	At risk in west-central Sierra Nevada foothills, where urban encroachment has eliminated several isolated, disjunct populations.
Washoe pine	Especially at risk on Babbit Peak and Mount Rose, due to loss of habitat from wildfire and genetic swamping from ponderosa pine.
Riparian species	At the species level, willow and poplars throughout the Sierra Nevada are affected due to water diversion practices, loss of habitat, and lack of specific information on their patterns of genetic variation to guide restoration practices.

TABLE 28.22

Categorization of threats to plants in the Sierra Nevada.

Threat	Importance to Common Species	Importance to Rare Species
Habitat Destruction or Degradation Altered natural cycles (fire regimes, hydrological cycles, etc.) Human development • Erosion caused by road building or clear-cutting • Air pollution • Effects due to livestock presence Biotic disturbance (usually human caused) • Invasion by and competition from weeds • Presence of feral animals • Loss of pollinators • Introduction of new pests and pathogens	More important for genetically subdivided species (e.g., rare alleles may be lost from peripheral populations)	*Always Important!* (May be somewhat less important for "sparse" widespread species) Species extinction is more likely
Genetic Contamination Native species: both intraspecific and interspecific effects Exotics: interspecific hybridization (especially with exotic congeners)	Human-caused contamination more likely; usually results in loss of genetic architecture rather than extinction	Species extinction is more likely
Population Fragmentation Short term: increased inbreeding depression due to reduced effective population size Long term: loss of genetic diversity due to genetic drift Increased probability of local population extinction via demographic factors	Can be important, depending on breeding system and genetic architecture	Usually more important than in common species; impact depends on breeding system Species extinction is more likely

Taxonomic Groups Involved, Geographic Areas Affected, and Standards or Thresholds for Evaluating Effects. Due to the lack of genetic studies for many of the plant species in the Sierra Nevada, it is difficult to identify specific taxonomic groups or geographic areas that might be threatened. Also, because plants are a diverse taxonomic group in terms of distribution patterns and life history traits, species-specific interactions between these features and the threats will determine the consequences. Similarly, appropriate standards are hard to generalize, given the complexity of these interactions. Instead, a general approach to defining groups and areas at risk has been taken (table 28.23). Two factors, level of gene flow and spatial distribution, have been selected as being among the most important indicators of genetic consequence, given certain threats. Thus, according to the attributes of these two factors, the likely severity of genetic consequences is assessed relative to the threats listed in table 28.22. Those plant species that are most highly threatened, then, could be identified or conjectured according to their generalized patterns of gene flow and spatial distribution. Similarly, geographic areas that are most threatened could be projected by their high concordance of plants with certain spatial distributions. Standards are best approached as qualitative assessments of the likely severity of a consequence, given the species' features and type of threat.

Gene flow has been broadly classified according to only two levels: high and low. The level used here incorporates both intrapopulation and interpopulation gene flow. Spatial

distribution of plants has been subdivided into four categories: common or widespread species and three types of rarity (e.g., few plants on few sites). The rarity categories have been based roughly on part of the classification system proposed by Rabinowitz (1981), with one important difference: adaptive processes or degree of habitat specificity have not been addressed here. They are another important layer of consideration affecting genetic consequences, but their inclusion was considered too complicated to accommodate here.

Several examples from table 28.23 illustrate these concepts. In general, habitat destruction has more severe consequences for species or populations with low levels of gene flow than for those with high levels. This assumption is based on the generalization that high levels of gene flow lead to more mixing of the species' gene pool, and loss of one population would probably not mean the loss of many unique alleles or a large portion of the total genetic diversity of the species. Conversely, low levels of gene flow are often associated with substantial local differentiation, and the loss of even one population might represent a loss of genetic diversity not found elsewhere in the species. Rare species, regardless of the type of rarity, are more likely to be affected by destruction of their habitat than are more common species, as any loss of habitat will represent a larger proportion of the total genetic diversity of a rare species.

The consequences of human-caused genetic contamination are more complicated. Here, the relationship between gene flow and consequence depends on levels of both intrapopu-

TABLE 28.23

Relationships between threats and consequences to plants in the Sierra Nevada.

Genetic Threat	Gene Flow[b]	Severity of Consequences[a] Spatial Distribution of Plants			
		Common	Few Plants/ Few Sites	Many Plants/ Few Sites	Few Plants/ Many Sites
Habitat destruction or degradation (rapid removal of genotypes)	Low	**	*****	*****	***
	High	*	****	***	**
Human-caused genetic contamination[c]	Low	**	****	***	***
	High	*	****	***	***
Fragmentation—inbreeding depression	Low	*	****	***	**
	High	**	****	****	***
Fragmentation—genetic drift	Low	**	****	***	***
	High	*	****	**	**

[a] Rating system is based on degree of severity of consequence, from least (*) to most (*****) severe.
[b] Gene flow refers to both spatial and temporal gene flow and incorporates gene flow both within and among populations.
[c] For common plants, the consequences of genetic contamination are not so much a function of the biological situation as of the frequency of the threat. These plants are often the target of restoration or revegetation projects, so the threats are more common here, even if individually the consequences may not be severe. For rare species, the distinction between species with high and low gene flow is complicated here. The consequences will vary depending on the rate of interpopulation and intrapopulation gene flow, as well as other factors.

lation and interpopulation gene flow and other situation-specific factors (such as differences in gene frequencies between introduced and native groups, differences in number of breeding individuals, differences in adaptive capacities, etc.). For example, if a plant species possessed high levels of within-population gene flow, the introduced ("contaminating") genes might quickly circulate and swamp the natural diversity of a population; however, if the interpopulation levels of gene flow were low, the contamination would remain somewhat localized, mitigating the genetic consequence. Another aspect of genetic contamination is that common species are generally quite resilient to this type of threat. However, they are portrayed here as being subject to somewhat severe consequences, due to the frequency of occurrence of this type of threat. Common or widespread species are often used in restoration or revegetation efforts. Thus, while each individual event in itself may not have large consequences, the probabilities of ultimate genetic consequences may be additive.

The consequences of fragmentation should be considered in both short-term (inbreeding depression) and long-term (genetic drift) contexts. In the short term, species with low levels of gene flow might suffer fewer consequences than those with high levels. Low levels of gene flow are often associated with a largely inbreeding mating system, and thus there would be less likelihood of inbreeding depression. Conversely, outbreeding species would be more susceptible to inbreeding depression. If the species or population survived the short-term genetic consequences, it might be challenged by genetic drift. Here, the consequences would be felt more strongly in the species with inherently low levels of gene flow, which would remove or minimize any opportunities for bolstering levels of genetic diversity by incorporating pollen or seed from other populations.

One generalization that can be drawn from table 28.23 is that for habitat fragmentation, the main distinction in severity of consequences is between those species with few individuals and those with many. For habitat destruction, the major difference in consequences lies between species that are broadly distributed and those that are narrowly distributed (i.e., a function of spatial distribution).

Mammals

Spectrum of Actions Likely to Cause Major Changes in Native Gene Pools.
Threats to mammalian species in the Sierra Nevada fall into five categories, four of which are anthropogenic. One general category includes management that generally fails to support wildlife habitat. This results from both misinformation (inappropriate use or interpretation of information, poor or inadequate studies, etc.) and lack of information (impediments to research, lack of specific information on genetic architecture of species, etc.). Often missing is the context for interpretation provided by long-term studies and those that help determine the relationship between common measures of genetic variation (e.g., heterozygosity and allozymes in general) and population fitness.

A second threat is change in natural metapopulation structure (i.e., the structure of the group of populations that interbreed, if only occasionally). Two aspects are important: fragmentation events that decrease gene flow among populations or that subdivide previously contiguous populations, and activities that connect previously disjunct populations, thereby increasing gene flow above normal levels. Severe fragmentation or habitat loss may prevent the occasional gene transfers that are critical to adaptations and long-term resilience.

A third type of threat is loss of migratory routes, including winter and summer habitats and elevational corridors. This is a critical concern for birds and mammals.

Competition and predation from exotics and gene swamping from nonlocal transplants are another major threat. Domestic grazing species, in particular, are major competitors with natural populations.

A fifth type of threat is disease. Although disease is a natural component of the ecosystem, populations that are already stressed may succumb to diseases that would not otherwise be significant. For example, the mountain sheep *(Ovis canadensis)*, native to the southern Sierra Nevada and recently reintroduced into Inyo County and into the South Warner Wilderness of Modoc County, is extremely sensitive to disease. Diseases, particularly those transmitted from livestock, could likely be a major factor in the decline and loss of mountain sheep populations (Zeiner et al. 1990b). Loss of populations potentially leads to loss of ecotypic genetic variation in the species.

Taxonomic Groups Involved. The species most at risk are those with migratory patterns or highly specialized niches (riparian species, localized endemics, etc.) (table 28.24).

Geographic Areas Affected. Areas with easy access, lands where species and habitats have little protection, and areas of species richness are all identified as high-risk areas (table 28.24). Examples are the Sierra Valley and the Kern River Plateau. Also threatened are areas of significance to migratory species or aquatic-dependent species.

Standards or Thresholds for Evaluating Effects. In the absence of empirical data for defining standards to evaluate genetic threats in mammalian species of the Sierra Nevada,

TABLE 28.24

Taxonomic groups and geographic areas potentially most threatened in genetic diversity among vertebrate species of the Sierra Nevada.

Taxonomic Group	Threatened Taxa	Threatened Geographic Area
Amphibians	Terrestrial plethodontid salamanders (TPS), most native frogs and toads.	All areas where TPSs are found. The Kern River Plateau is *very* sensitive. All high-elevation areas of the Sierra Nevada, e.g., above 6,000 ft, and the southwestern Sierra Nevada at all elevations.
Birds	Neotropical migrants, aquatic-dependent species.	Migration stopover areas and staging areas.
Mammals	Bats, top carnivores, localized endemics, aquatic-dependent species.	Kern River Plateau, foothill areas, Sierra Valley, riparian areas.

theoretical guidelines should be considered. If genetic parameters or proxies are to be used effectively to evaluate threats or as indicators of resilience or health, there must be a context within which to interpret them. There must be a temporal context, that is, knowledge of the normal or perhaps cyclical range of variation and how the parameters relate to demographic trends. There must also be a spatial context—a means of interpreting at the management unit level how local parameters relate to regional or larger-scale patterns and trends. Further, choices must be made as to which taxa can serve as representative units, as it is unrealistic to develop standards for every species. The information needed to establish such a context is not currently available. As such, it is perhaps most appropriate to discuss the approach to developing standards. At the community level, two aspects are important, monitoring and ecosystem indicator species.

An important initial step in monitoring the health of biotic communities is to inventory the community or communities within the management unit. Although the initial inventory is key, it is important to inventory regularly. After the broadscale inventory, detailed monitoring should be provided for all indicator species (criteria for choosing these species are listed later). Minimum information collected for these species would include densities of local populations, numbers of local populations, and degree of genetic differentiation among populations, these three parameters providing some measure of metapopulation functioning. Other essential data include the appearance of new alleles and changes in the distribution of private alleles (i.e., alleles unique to a population). One or both of these could be a warning sign of disturbance. The point here is that changes in any of these features could indicate reason for concern and would eventually provide the connection, currently lacking, among threat, genetic consequence, and demographic consequence that would allow development of standards.

For choosing indicator species, an appropriate scale for reference might once again be a biotic community level or management unit. The number of species chosen will depend upon the complexity of the management unit. Choices should be locally appropriate.

SNEP Genetics Workshop participants suggested that representative or indicator species for a management unit include the following:

- Species with varying life history characteristics—short and long individual life spans, wide and narrow distributions, and so on

- Species representing various trophic levels

- Species that are strongly associated with the management unit (e.g., endemics)

- Keystone species

- Representative specialist and generalist species

- Species that are net transporters of nutrients and/or energy in or out of system (e.g., bats, certain fish)

- Species that opportunistically use the management unit and thus may help to monitor the health of the system

- Recent and ancient phylogenetic taxa

- Recent and historical residents

- Species that represent soil microfauna and microflora

Birds

Spectrum of Actions Likely to Cause Major Changes in Native Gene Pools. Given the high degree of gene flow observed in many bird species of the Sierra Nevada, an obvious genetic threat would be any action that resulted in the loss of natural metapopulation structure. Threats would be fragmentation events that decreased gene flow among populations or subdivided previously contiguous populations Severe fragmentation or habitat loss may prevent the occasional but critical gene transfers between metapopulations.

As most of the species in this bioregion are migratory, loss of migratory routes, including winter and summer habitats and elevational corridors, is a critical concern.

Activities that change the quality of habitat will have direct consequences for the resident bird species, although it is unclear whether these are detrimental, neutral, or beneficial. For example, chestnut-backed chickadees *(Parus rufescens)*, prior to 1940, inhabited mainly coastal areas of northern and north-central California. Shortly after that time, they started to move inland, and today their distribution includes both their former range and much of the Sierra Nevada (Brennan and Morrison 1991). One explanation presented for this range expansion is that successional patterns following widespread logging in the Sierras caused an increase in the proportion of Douglas fir in the mixed conifer forest, which subsequently provided habitat favorable to chestnut-backed chickadees. Although in this case the range expansion of one bird species was apparently not accompanied by a range decrease in another, this is potentially a risk with such (management) activities.

Taxonomic Groups Involved. Taxa with very specific and narrow habitat requirements might be most threatened. Specific examples were not known by workshop participants (table 28.24).

Geographic Areas Affected. No specific examples are available.

Standards or Thresholds for Evaluating Effects. Little information is directly available from which to develop standards for evaluating genetic threats to avian species. A useful avian example exists, however, of a more general issue regarding the use of genetic information for evaluating threats. In 1990 an allozyme study was reported of seven populations of the spotted owl *(Strix occidentalis)*, covering the three currently recognized subspecies from Oregon, California, and New Mexico (Barrowclough and Guttiérrez 1990). Twenty-three allozyme loci were scored; all were monomorphic for the two subspecies in Oregon and California, *S. o. caurina* and *S. o. occidentalis*. This implies both zero heterozygosity (for those loci) and no population differentiation between those subspecies based on these data. The New Mexico subspecies *(S. o. lucida)* was differentiated from the other two at only one polymorphic locus. However, in spite of the current conservation concerns regarding this species, the low heterozygosity values were not interpreted as evidence of a genetic risk or as evidence of higher than expected levels of inbreeding. Rather, it was suggested that the low values are the result of a historical bottleneck, low effective population sizes (small populations generally tend to have low levels of heterozygosity), and/or the inherently low levels of variation in many of the genes sampled in the study (across all species). The certain consequence of low heterozygosity values is that monitoring the species for genetically significant changes will be more difficult. Generally, this example illustrates that genetic information, like all other data, must be carefully interpreted within the proper biological context.

Reptiles and Amphibians

Spectrum of Actions Likely to Cause Major Changes in Native Gene Pools. The most significant threats to reptiles and amphibians in the Sierra Nevada are all related to direct (management) or indirect (e.g., urbanization and forest management) human effects. Population fragmentation and habitat loss not only result in direct removal of individuals or populations but also can have the secondary effect of increasing inbreeding depression. For terrestrial species, management activities are a threat to suitable habitat. Some specific examples of this type of threat are known for salamanders. Clear-cutting or the removal of significant and contiguous portions of the overstory on the western slopes in the Sierra Nevada is a serious threat to salamanders in areas where it dries the understory and substrate below adequate moisture levels. Similarly, controlled burns that are too intense, or severe wildfires, dry out even moisture-laden logs that otherwise could serve as temporary refugia for these species. Habitat improvement for other species, such as mammals, can pose a serious threat to amphibious species. For example, improving access for large-game mammals to a spring-fed pond in Sequoia National Forest destroyed the only known population of an undescribed species of *Batrachoseps*, a lungless salamander (Wake 1994).

Significant to amphibians are activities that have disrupted metapopulation phenomena. For example, *Rana muscosa* was once widespread in the Sierra Nevada, but its range has significantly contracted. Frogs have a natural metapopulation structure and depend on large source populations for recolonization of shallow ponds and marginal habitats, where local natural extinction rates are high. With the introduction of non-native trout into most large ponds and lakes of the high Sierra, the source populations have been devastated. Thus, local extinction continues to occur in the peripheral sites, but there is no recolonization, and the once-temporary absences become permanent extinctions. The situation may be even more extreme with *Rana cascadae* in the Lassen area (D. B. Wake, University of California, Berkeley, e-mail to the authors, 1995).

Introduced exotics are a threat to amphibians, primarily as predators. Non-native fishes have been stocked in most of the lakes in the Sierra Nevada, nearly all of which were previously fishless (Christenson 1977; R. A. Knapp 1994, 1996). Subsequently, fish dispersed throughout the streams that interconnect these lakes. Such introduced fishes appear to have nearly eliminated the mountain yellow-legged frog *(Rana muscosa)* where they co-occur (Bradford et al. 1993), and the same probably happened to *R. boylei, Hyla regilla* (Pacific tree frog), and *Ambystoma macrodactylum.* These eliminations are believed to have isolated many of the remaining populations of highly aquatic species such as *R. muscosa* (Bradford et al. 1993) and thus have been significant factors in causing habitat fragmentation and disruption of metapopulation structure, resulting in persistent local extinctions (D. B. Wake, e-mail to the authors, 1995).

In summary, urbanization activities such as land conversion, expansion of housing, development of recreation areas, road building, and dam construction are historical and continuing threats to reptiles and amphibians in the Sierra Nevada, causing habitat loss, disruption of metapopulation structure, and population fragmentation. More is known about the specific threats and consequences for amphibians than for reptiles, due to the void of genetic information for the latter group. Many of the most serious, current, and specific threats to amphibians (at least) can be categorized as management activities targeting nonamphibians without regard for amphibian habitat needs. Currently, the most profound human impacts on aquatic communities in the high Sierra appear to be related to historical and ongoing stocking of exotic fish species in high Sierra waters (Bradford et al. 1994).

Taxonomic Groups and Geographic Areas Involved. The lack of genetic information for many species makes it difficult to target specific taxonomic groups or geographic areas. However, it is known that some populations of plethodontid salamander species are threatened (table 28.24). The Kern River Plateau, because of its species richness, is an especially sensitive area; logging and other management or development activities threaten many species here.

Standards or Thresholds for Evaluating Effects. Again, due to the lack of information for many species and the complexity of genetic-ecosystem relationships, it is not currently possible, and perhaps not appropriate, to define quantitative standards. Natural patterns of genetic structure may provide

some guidelines regarding the relative resilience of different taxonomic groups to various threats. For example, species with high levels of among-population diversity (e.g., high F_{ST} values) might be more resilient to the effects of fragmentation and inbreeding depression. The most appropriate approach to conserving genetic diversity in these taxa is to focus on protection of their habitat and the health of their biotic communities.

Fish

Spectrum of Actions Likely to Cause Major Changes in Native Gene Pools. Actions that potentially alter natural genetic levels and patterns of genetic variation in Sierra Nevada fish populations include natural phenomena, human activities affecting water flow and quality, introduction of non-native fish species, direct manipulation of native gene pools, and genetic technologies.

Natural if rare phenomena such as volcanic activity can indirectly reduce fish populations via suffocation and gill abrasion and perhaps have historically removed some fish species from waters in the Mono Lake basin (Moyle 1976). Volcanic activity can alter geographic structuring of genetic variation by increasing the incidence of straying (i.e., the change in water quality confuses fish and/or dissuades them from returning to their natal waters to spawn). For example, the high ash content of rivers near Mount Saint Helens increased straying in chinook salmon, as the fish tended to avoid ash-laden water (Bartley and Gall 1990).

Human activities that may affect water quality for fish and thus potentially affect genetic structure include hydraulic mining, water diversion projects, hydroelectric projects, roadbuilding, wildfire, and logging. Habitat degradation associated with these activities has been linked to the decline of populations of chinook salmon (Bartley and Gall 1990). These natural phenomena and human activities that degrade fish habitat have two main genetic impacts. First, fish population sizes decline rapidly, pushing the remaining populations through a genetic bottleneck with possible loss of genetic variation. Second, the decrease in water quality (e.g., from turbulence, mud slides, and volcanic ash) can alter natural migration patterns, thereby affecting geographic structure in genetic variation. Further, alteration of watercourses can bring previously isolated species or populations into contact with one another, causing artificial mixing of gene pools. For example, alteration of traditional salmon spawning routes after construction of Lewiston Dam on the Trinity River may have led to natural hybridization between chinook and coho salmon in Deadwood Creek, California (Bartley et al. 1990).

Introduced species potentially have many deleterious effects on genetic structure through their impacts on existing fish populations and habitats, including potentially displacing native species (e.g., Ferguson 1990). For example, introduced brook trout *(Salvelinus fontinalis)* and brown trout *(Salmo trutta)* are thought to have displaced many populations of bull trout *(Salvelinus confluentus),* a species that historically existed in the Upper Sacramento River drainage and is now thought to be extinct in California (Leary et al. 1993). Non-natives may also hybridize with native species, diluting native gene pools and perhaps reducing fitness due to the creation of sterile or less fit interspecific hybrids. Evidence from allozyme studies suggests that several golden trout (*Oncorhynchus mykiss whitei*) populations from the Little Kern River basin have hybridized with introduced rainbow trout (*Oncorhynchus mykiss*) (Gall et al. 1976).

Manipulation of native gene pools occurs through hatchery practices that do not recognize the importance of maintaining large founder population sizes, of keeping fish stock separate by location of origin, or of returning local populations to their origins. The use of nonlocal stock in hatcheries and the practice of transferring stock between populations potentially can reduce local levels of adaptation by swamping and hybridizing with native gene pools (called outbreeding depression, which is not empirically documented). Use of only a few founder fish in hatcheries acts as a genetic bottleneck, potentially reducing the amount of genetic variation in the subsequent populations. Inbreeding depression may become a threat when levels of genetic variability become reduced below naturally occurring levels. Evidence of inbreeding depression has been described for the well-studied rainbow trout, including increased mortality of eggs, alevins, and fry; decreased growth rate of fingerlings; and decreased body weight for adult rainbow trout (Gjedrem 1992).

Hatchery fish may also affect geographic structure in genetic variation due to increased straying. It is thought that hatchery-reared fish, in some cases, may be less imprinted on their natal river or stream than wild fish due to particular management practices in their hatchery environment. For example, the increased level of gene flow observed among chinook salmon populations in the San Joaquin–Sacramento River system has been interpreted as being at least partly due to the hatchery practices in this drainage. Hatchery-released fish may not have an opportunity to imprint properly, due to limited hatchery residence time and/or water differences between the hatchery and local areas. This may lead to increased straying, increased gene flow among populations, and consequent changes in genetic structure (Bartley and Gall 1990).

Genetic technologies present both an opportunity and a potential threat to natural patterns of genetic variation. Rainbow trout has had many extreme technologies successfully applied, such as chromosome manipulation (e.g., production of triploid fish), induction of androgenesis and gynogenesis (all-paternal and all-maternal inheritance, respectively), and gene transfers (Thorgaard 1992). This facility of fish in general, and the rapid increase in knowledge of the rainbow trout genome in particular, are potentially valuable for both research and commercial fisheries interests. However, uninformed or accidental release of manipulated stock could pose a new spectrum of genetic threats to native populations.

Taxonomic Groups and Geographic Areas Involved. The aforementioned threats to genetic variation in fish species are pervasive over the Sierra Nevada, although genetic effects have rarely been directly measured. The Little Kern River golden trout is one specific example. Evidence exists for hybridization between one subspecies, *Oncorhynchus mykiss whitei*, and rainbow trout in the Little Kern River (Gall et al. 1976). However, there are also apparently pure populations remaining in upper Soda Springs Creek and Deadman Creek, both apparently physically isolated from the introgressed populations and possessing high levels of within-population genetic variation (heterozygosity).

Standards or Thresholds for Evaluating Effects. Heterozygosity has been used as an indicator of adaptive potential. For example, in the previous description of the viability of golden trout populations, the resident levels of heterozygosity in the pure populations were interpreted (by the study authors) as indicative of adaptive capability (Gall et al. 1976). However, these allozyme data must be interpreted within their spatial and temporal context. There is some evidence that sampling methods in genetic studies may bias results due to naturally existing temporal variation in allele frequencies. Between the two sample periods of 1984–86 and 1987–88, Gall and colleagues (1992) found, for twelve allozyme loci, significant differences in allele frequencies among eighteen populations of chinook salmon. This suggests that fish populations have a genetic structure that may also be related to the season or year.

In general, the task of accumulating the desirable genetic baseline data for fish is perhaps more complicated than for many other taxa, due to the high degree of manipulation of natural populations prior to genetic sampling. This is well expressed in Bartley et al. 1992: "The excessive and often undocumented transplants of coho salmon throughout the Pacific Northwest may obscure natural patterns of genetic variability and make geographical identification of stock difficult."

Insects

Spectrum of Actions Likely to Cause Major Changes in Native Gene Pools. The greatest potential threat to insects is the loss of habitat, leading to fragmentation of populations, loss of corridors for gene flow, and ultimately the alteration and loss of genetic variability. Although no direct studies on the effects of grazing have been done in the Sierra Nevada, habitat loss from trampling by grazing domestic sheep was implicated in the initial decline in populations of the Uncompahgre fritillary butterfly (*Boloria acrocnema*) in the Rocky Mountains (Britten et al. 1994a). This species, now limited to one population, is close to extinction. However, its intolerance of Holocene climates may have caused its demise anyway—it was already restricted to extremely cool, moist alpine slopes (A. M. Shapiro, University of California, Davis, e-mail to the authors, 1995).

A second, related threat is land-management activities that affect and alter forest successional stages. Many insect species are dependent on a particular seral stage. Management that affects the seral development or that arrests succession at a subsequent or earlier stage may lead to habitat loss or fragmentation. An example close to the Sierra Nevada is the apparent demise of the last population of an endangered butterfly species restricted to a Pacific Gas and Electric power line corridor in Mendocino County, California. Because the butterfly was federally listed (as threatened or endangered), the company was not allowed to cut vegetation in the corridor, and thus succession proceeded. The species was not well adapted to later successional stages, and, due to extreme fragmentation in its habitat, was not able to "escape" to a more favorable, earlier successional area. Thus, the population, and presumably the species, was driven to extinction by normal successional processes, in combination with fragmentation (Shapiro 1994). This relationship with a successional stage occurs in many butterfly species; in general, they are not adapted to old-growth habitats. Thus, successional processes might lead to genetic depletion and population extinction if seral diversity is not maintained at appropriate landscape scales and mixes.

A third potential threat may be the swamping of natural populations by human-associated new ecotypes of native (insect) species. Increasing in occurrence are ecotypes of native insect species that have adapted to feed on introduced (exotic) weeds. These are particularly prevalent in disturbed situations such as railway embankments and roadsides. As these ecotypes spread and contact the local native populations, they threaten to swamp them genetically, resulting in a loss in genetic variability that is relevant to the natural (herbaceous) host species. One example in northern California is the silvery-blue butterfly (*Glaucopsyche lygdamus*). It has a recently evolved, genetically distinct ecotype that feeds on introduced annual vetches (Shapiro 1995). This new ecotype has now spread to the point of contact with native populations that feed on native and local legumes. As such, the native populations may be threatened (Shapiro 1994). The same phenomenon is occurring independently in the northeastern United States in the same species (Dirig and Cryan 1991).

A fourth potential threat is overcollection by hobbyists. Distinctive phenotypes, especially in butterfly species, are sought and removed by sport and commercial collectors. The rarer the phenotype, the more avidly it is sought, and the larger the proportionate reduction in its species when it is removed. Once rare species or populations have been taxonomically recognized, they may be eligible for listing under state or federal legislation as threatened or endangered. However, this recognition may also increase their vulnerability to collectors. This phenomenon is associated with other charismatic species as well, including damselflies, tiger beetles, and longhorn beetles. Ladybugs are collected in great numbers in some forests by Laotians and other groups.

A fifth category of threat is posed by overuse or misuse of wide-spectrum biocides. For example, *Bacillus thuringiensis* (Bt) is a broad-spectrum biocide that is often employed against spruce budworm. However, it is an effective agent against all members of the Lepidoptera, and thus may kill all caterpillars in the application or drift area. Thus, its indiscriminate use could lead to losses of populations or races of nontarget species. While Bt provides a particularly striking example, this is a typical description of many biocides.

A last example of a type of threat that is particularly relevant to insects is the potential effects of homogenization of host species due to management activities. Standardizing, or reducing the amount of genetic variability within, host plant species (e.g., the use of standardized genotypes or clones in a forest regeneration effort) may be followed by depletion of genetic variability within the dependent insect species, although this is undocumented in the Sierra Nevada among native taxa.

Taxonomic Groups Involved. While data are not available to provide a comprehensive list of threatened species or populations for the Sierra Nevada, the Lycaenid butterflies possess the classic characteristics of a vulnerable taxon. These characteristics include having well-defined and local populations; in this case, ecotypes are differentiated according to host plants (A. M. Shapiro, e-mail to the authors, 1995). The life history characteristics of the ecotypes appear to coevolve with those of the host. The species are highly localized, specialized, and fragmented, and usually disperse only locally. They are thus vulnerable and cannot "escape" if threatened. Most of the endangered butterflies listed at either the state or federal level are members of the Lycaenidae (Arnold 1983a).

Other groups of insects may also be genetically threatened or vulnerable. These include taxa isolated on mountaintops, such as alpine grasshoppers and beetles. Parasitoids of narrow, specialist organisms are possibly threatened. Specialized roaches, such as wood roaches, might be vulnerable. Cave crickets and damselflies might also be included in this group of threatened species.

Geographic Areas Affected. For insects, three types of environments have high concentrations of potentially vulnerable organisms, from a genetic perspective. One is alpine environments, for example, the White Mountains. Insects in these environments tend to be restricted and unique, and small changes in climate or habitat threaten these organisms. Another is edaphic islands. The biota of islands of serpentine, gabbro, or Ione clay, and of sand dune areas of the Sierra Nevada, is potentially endangered. A third environment is riparian zones. Insect taxa may depend on these habitats as corridors for gene flow. Increasing fragmentation and isolation make their resident biota more vulnerable.

Another type of area that may be vulnerable for insects is described as a suture zone. This is an area consisting of multiple, overlapping hybrid or integration zones. Historically, they are often the result of refugial species meeting after expanding following major climatic shifts. These zones are often rich in unique genetic diversity due to these features of secondary contact and hybridization. Examples of this suture zone phenomenon in the Sierra Nevada include the Sierra Valley, Warner Mountains, riparian blend zones such as the Upper Feather and Pit River drainages, and parts of the Plumas National Forest (Collins 1984; Porter 1989; Porter and Geiger 1988; Shapiro and Nice n.d.).

There are several other examples of specific areas that, while not fitting the aforementioned generalizations, are host to apparently vulnerable organisms. The first is a low-elevation area of Oregon oak–juniper communities on Ball Mountain in Siskiyou County, California. This area supports a population of a butterfly that normally occupies only wet or even inundated meadows at moderate to high elevations (A. M. Shapiro, e-mail to the authors, 1994). Its existence on Ball Mountain is tenuous due to the dry summer condition of the area, and any additional stress might overwhelm the population's resilience. A second example of a threatened area is a specialized community on Goat Mountain in Colusa County, California. Here, a local population of the Mormon metalmark (*Apodemia mormo*) butterfly feeds on a local ecotype of wild buckwheat, *Eriogonum wrightii*. Similar situations may exist in the Sierra Nevada.

Standards or Thresholds for Evaluating Effects. Although specific and quantitative standards are beyond the scope of currently available data, several issues are germane to insects. First, it is unlikely that direct genetic data will suffice for developing insect standards. Because most insect sampling for genetic purposes is (currently) necessarily destructive, such studies for many species will not be undertaken, due to the rarity of many target taxa, legal protection of species, and the morality of such studies as electrophoresis when the interpretation and biological meaningfulness of the data are uncertain. At present, not enough is known about the correlation of insect population health with habitat attributes to suggest ecological or community-level proxies as a standard.

A second issue is historical sequence. Standards must be chosen so as to take into account the taxon's history. For example, low intraspecific variability would not necessarily be indicative of viability problems if that species had largely been inbreeding or homozygous for a long period of time. Fitness and viability problems are more likely to occur when there is a rapid depletion of genetic diversity, as occurs in an anthropogenic bottleneck process.

The challenge, then, is to develop standards based on trends and to have the means to examine trends in genetic attributes. One recently initiated approach with insects involves taking small samples from museum specimens, assaying them with PCR-based techniques (PCR, or polymerase chain reaction, is a means of detecting genetic variation with high sensitivity even in very small samples of DNA), and comparing the results with extant samples. One such project is currently underway in regions adjacent to the Sierra Nevada with an endangered species of butterfly, the Oregon silverspot (R. VanBuskirk, communication with the authors, 1995). This approach is not ideal, due to the limitations of museum specimens (i.e., they are few in number, are usually derived from a restricted geographic area, have the potential for being mislabeled, etc.) and the statistical challenges of comparing the historical and extant samples. However, at least qualitative results should be possible.

Plasticity is also problematic in developing standards. The role of environment in modulating phenotype may preclude the development of reliable morphological standards. For insects, the concern is the recognition of eco-phenotypes that may not have a genetic basis for differentiation. For example, silvering on the underside of hind wings in the fritillary butterflies is used for taxonomic classification. However, silvering is a highly plastic characteristic that may be greatly affected by humidity and have low heritability (Arnold 1983b, 1985; Hammond 1986).

Fungi

Relatively little information is available concerning specific effects to fungal species. This is partly a function of the general lack of (population-level) genetic information and partly due to the classification of some fungal species as pathogens. Their harmful nature to commercially significant plants has usually led to a desire to lower their populations, not conserve them. The so-called beneficial fungal species (e.g., mycorrhizae) are sometimes cultured domestically and cultivated as clones or races, making genetic variation in natural populations less of a concern. Nevertheless, it is possible that breeding commercial tree species with genes for resistance to fungal pathogens and incorporating these trees in large numbers in forests may have a negative impact on genetic variation in the target fungal species (e.g., the planting of sugar pine trees with a major gene for resistance to exotic white pine blister rust).

One fungal species that may suffer genetic consequences as a result of human activity is the edible and commercially valuable North American matsutake or tan oak mushroom (*Tricholoma magnivelare*). Although the possible genetic impacts are unknown at present, and ecological studies are only now underway, harvesting of this species in areas such as the Klamath National Forest has rapidly expanded since 1990 (Richards 1994). In recent years, the limited and traditional gathering of the mushrooms by local Native Americans and hobbyists has been outscaled by commercial harvesting (Richards 1996).

Recent and increasing attention has focused on the development of techniques to distinguish not only among fungal species but also among strains within species. These tools include random amplified polymorphic DNA (e.g., Garbelotto et al. 1993) and PCR in combination with RFLP and/or sequencing techniques (e.g., Gardes et al. 1991).

Assessment Conclusions

Because of the nature of genetic variation, its measure, and interpretation, it is extremely difficult to arrive at firm synoptic conclusions about threats to genetic diversity in the Sierra Nevada. Nevertheless, several specific issues can be singled out as being of high priority, and several others are general categories of concern.

Severe Wildfire

SNEP assessments clearly indicate the changed nature of fire regimes over the last century in the Sierra Nevada (see McKelvey et al. 1996). The risk of severe fires is higher than during any other period that has been evaluated in the Holocene. Large, stand-replacing fires such as are likely now present significant risks to gene pools of forest trees and plant communities, with direct and indirect consequences to other plants and animals that live in them.

Habitat Alteration

For most taxonomic groups, the major known threats to genetic diversity are habitat destruction, degradation, and fragmentation. These not only result in direct losses of genetic structural diversity at the population level, but also change genetic processes (gene flow, selection), affect effective population sizes, and contribute to changes in genetically based fitness. Habitat alteration and loss have both trickle-down and trickle-up effects, in that lower-level genetic diversity is affected (within and among individuals and within and among populations) and there are potential effects to species viability. Although we have not emphasized in this chapter the assessment of which geographic and taxonomic locations are most affected by habitat alteration in the Sierra Nevada, information from other chapters confirms that high-priority areas would be the foothill zone on the west slope, several of the trans-Sierran corridors (especially in the central Sierra Nevada), and scattered locations of concentrated development elsewhere.

Silviculture

Management actions that are extensive across the landscape yet intensive in manipulating individuals and populations have the greatest potential for direct and significant genetic effects. As such, silvicultural activities, including tree improvement programs, operational forest regeneration (artificial and natural), and timber harvest, potentially affect the gene pools of target species. Fortunately tree improvement programs in the Sierra Nevada (both public and private cooperatives) have long used sophisticated and ecologically appropriate genetic diversity and genetic conservation guidelines. Similarly, in operational forest regeneration, most federal, state, and local regulations regarding genetic diversity in planting have high standards and are backed by significant amounts of research. Seed banks for public and private reforestation exist that maintain high standards for seed origin and genetic diversity. These programs, which have histories dating back several decades in the Sierra Nevada, serve as models for other taxa where similar activities occur (e.g., fish stocking).

Although these programs and guidelines are genetically sophisticated and widely practiced throughout the Sierran forests, there is room for implementation error. For instance, about half of the trees planted by the U.S. Forest Service in the Sierra Nevada are in unanticipated plantations in areas burned by forest fire (Landram 1996). The seed banking program of the U.S. Forest Service bases the quantity of seeds it procures and stores primarily on a determination of planting needs. National forests are required to maintain a ten-year supply of seed for the relevant seed zones (the actual supply quantity varies from 5 to 12 years, depending on species and zone). This supply is based on estimates of planting needs determined from harvest plans and an estimation of the amount of seed needed for replanting following wildfires. Since national forests must pay for cold storage space and periodic seed testing, there is no incentive to maintain extra quantities to handle exigencies of severe and large wildfire. After large wildfires occur, there is often pressure internally in the agencies, as well as from the public, to reforest rapidly. If local and appropriate seed is not in the seed bank, the pressure to use seeds from nonlocal seed zones, low-diversity seed lots, or old seed collections may be high, despite the awareness of seed-transfer and genetic diversity guidelines. In practice, when local seed is not available, seed from adjacent zones is sought (as directed by policy), and consultations with geneticists occur when transfers are necessary.

Also important is the fact that seed is collected on national forests primarily from the timber forest types, and primarily from the commercial tree species, although this has been changing in recent years. Ability to reforest high-elevation, high-stress sites or noncommercial species within timber zones would be hampered by inadequate seed supplies.

Research studies are inconclusive about the long-term genetic consequences to commercial tree species of timber harvest, as well as about the ecological and evolutionary significance of those consequences. Nevertheless, traditional silvicultural practices, which were designed primarily to maximize growth of the target species, tended to result in spatial patterns of harvest and live-tree retention that acted in concert with genetic conservation guidelines. By contrast, some new forestry practices, which combine fiber production with ecological stewardship for wildlife and nontimber species, may have potential for dysgenic genetic effects on the native timber species. For instance, leaving clumps of trees, especially suppressed individuals (as, for example, for wildlife protection), may promote inbreeding or lowered fitness if the members of the clumps are related, as they appear to be. Similarly, leaving large, isolated live individuals as snag recruits or perch trees may lead to inbred seed if these act as seed sources. On a case-by-case basis, these effects are probably minor relative to background natural genetic diversity, although cumulative effects should be considered.

A specific taxon of concern in this regard is sugar pine. Although a well-funded and genetically sophisticated program exists for developing and out-planting sugar pine that is resistant to white pine blister rust, there has been limited recognition of the genetic consequences of the current federal harvest practices for the species. At present, known resistant old-growth sugar pines are not cut, but susceptible trees may be harvested, and in areas where resistance is unknown, harvest proceeds without genetic testing. Although leaving resistant trees and harvesting susceptible ones may seem genetically appropriate, it causes a significant loss of genetic diversity in traits other than the resistance loci and may seriously impede sugar pine's ability to pass through the pending blister rust bottleneck (Millar et al. in press). In the case of sugar pine, all mature trees should be left unharvested, especially in areas where the rust is not presently a major problem, unless the reasons for harvest are carefully evaluated and justified.

Ecological Restoration

Although tree improvement and regeneration programs have followed genetic diversity guidelines for decades, practitioners of ecological restoration have only recently become aware of genetic concerns in planting (Millar and Libby 1989). Although many programs focus on restoring correct native species, an understanding of the appropriate genetic material within species, its origin, diversity, and collection, is missing from many programs. Thus, genetic contamination problems may be more severe than if exotic species had been planted. The significance of this genetic threat in the Sierra Nevada is lowest in projects of ecological community restoration (primarily because in the Sierra Nevada such projects are highly limited in number and extent and are conducted by knowledgeable users) and highest in postfire erosion control projects. These frequently involve grass species and occasionally forb mixes. Although exotic grasses (especially ryegrass) were previously used routinely, native grasses are increasingly becoming favored. There is often little understanding of the potential genetic consequences of planting seeds of native species but unknown (often commercial nursery) origin. Further, even where there is awareness, the lower cost of commercial seeds (of unknown origin and diversity) compared to that of local, custom-picked seeds and the pressure to plant rapidly following a fire encourage the use of inappropriate genetic stock. Similar situations may arise in watershed restoration projects, where the genetic implications of activities are often not considered or evaluated.

Fish Management

Management of fish species and genetic diversity within species in the Sierra Nevada is done in a way that potentially disrupts many native gene pools. Fish raised in hatcheries and introduced into native Sierran waters are not managed to maintain or promote natural genetic architecture. Selection is for endurance and resilience to both hatchery conditions and a wide range of natural conditions, regardless of native genetic architecture. The introduction of hatchery, nonlocal, and genetically altered genetic stocks of native fish species has had the direct effect of creating conditions—and of continuing to create the potential—for intraspecific hybridization, gene contamination, and gene pool degradation. Indirectly, the introduction of exotic fishes has enormous effects on biodiversity through the displacement of native fish species as well as through impacts on aquatic invertebrates and amphibia, which affect gene pools through loss of populations (i.e., the introduction of exotic fishes is another example of habitat alteration).

Range Improvement

Similar to fish management, although of lesser effect in the Sierra Nevada, is the direction and intent of range improvement projects. In past decades, range shrubs, particularly bitterbrush, were widely planted in Great Basin areas (on the Sierra Nevada border) to improve rangelands for cattle. Germ plasm of these shrubs was almost invariably nonlocal, often from distant states. Some stock was derived from shrub improvement programs, which genetically bred stock for tolerance to wide conditions and resistance to stress and disease, but which did not promote maintenance of native genetic architecture. Thus, very little shrub germ plasm planted in the past derives from local seed zones or follows genetic diversity guidelines that maintain native genetic structure. More recently, shrubs have been planted to enhance wildlife habitat. These projects are increasingly falling under seed-transfer and genetic diversity guidelines similar to those of tree regeneration programs, with the result that native local seeds are now being collected and planted in many instances.

Exotic Pathogens

Exotic pathogens create direct and indirect genetic threats in the Sierra Nevada. White pine blister rust is fatal to sugar pines that carry the susceptible gene. The resistant gene exists in very low frequencies naturally in sugar pine; thus, the pending bottleneck from the disease epidemic will have significant and pervasive genetic effects throughout sugar pine's range in the Sierra Nevada. In other taxa and disease situations, resistance, if it exists at all, is often not simply inherited but is a combination of genetic and environmental effects. Indirect genetic effects occur when populations are so devastated as to drastically decline in size or become extirpated. An example is the exotic pathogen that moves from domestic to native bighorn sheep (which are being reintroduced into the Sierra Nevada) (see Kinney 1996). This pathogen causes a disease that is extremely serious and usually fatal to bighorn sheep, exterminating entire populations, with consequent genetic impacts.

Taxon-Specific Issues

In addition to the high-priority issues just described, there are many activities that have serious effects on specific taxa in the Sierra Nevada. Examples of these include the sport collecting of butterflies, the harvesting of special forest products (especially mushrooms and other fungi, ladybugs, lichens, etc.) (see Richards 1996), the use of biocides with wide action against native insects, and forest-health practices whose goals are to reduce or eliminate populations of native insects and pathogens. Beyond these, indirect impacts on the gene pools of specific taxa are numerous and are categorized with those that alter habitats of specific taxa. Examples include the effects of fire suppression on plant species whose seeds require fire for germination, the decline of amphibians due to the stocking of exotic fish, the displacement of native grasses by exotic perennials, the decline of taxa that depend on old-growth habitat, and so on.

Land Management

The specific activities listed in the previous sections soon grade into the comprehensive set of human activities that have some effect on gene pools. As we have noted before, most human-mediated (as well as natural) activities have some genetic consequences. The question is not whether we create genetic change but which effects are significant enough to be worthy of altering our behavior. In general, there has been a pervasive lack of awareness of the potential (theoretically inferred) genetic consequences of land management, from local practices to regional landscape plans. Levels of genetic awareness, evaluation, prescription, mitigation, monitoring, and restoration have generally been very low in public and private management, and they have been concentrated in a few land-use programs (e.g., tree regeneration). Although it is broadly recognized that most management actions have effects on wildlife, there are few instances where environmental analyses—for instance in National Environmental Policy Act (NEPA) contexts—have considered genetic effects. Land-management agencies do not place geneticists broadly throughout the Sierra Nevada, and genetic knowledge is usually centralized (e.g., with tree improvement headquarters) or resides within silvicultural staffs, where it is focused mostly on the already established genetic management programs of commercial timber species.

What is needed is a general awareness that genetic consequences must be considered and evaluated for land-management activities in general, and a framework and strategy for doing so. It is not enough to lump these concerns under general biodiversity evaluation, since this often takes into account only immediate effects on the population or species viability of a few indicator species.

The identification of taxa most at risk in the Sierra Nevada is difficult, due to the lack of specific genetic information for most species. Because the threatened taxa are widely distributed, there are few trends that point to specific geographic areas that are most threatened. For certain taxa, for which there is considerable genetic information, it may be possible to define genetic standards, such as levels of inbreeding. However, as the interpretation of genetic measures is very much affected by species' characteristics (mating system, genetic architecture, etc.), standards would be difficult to generalize. At a minimum, they would need to be structured according to basic life history characteristics. In general, taxa and their resident levels and patterns of genetic diversity and evolutionary potential are best protected by standards aimed at the biotic community level. For many species, standards based on genetic parameters (e.g., levels of heterozygosity) may be ineffective and misleading due to the time lag between threat and the genetic consequence. Addressing the following two basic needs would assist in the development of either standards or alternative approaches to risk assessments:

1. There is a need for research into the relationships between genetic parameters and the fitness of a species, and for inspecting such relationships for patterns related to life history characteristics.

2. There is a need to establish long-term monitoring programs that are systematically organized. Information on normal ranges of variation (both spatial and temporal) is essential to the development of biotic standards.

MANAGEMENT OPTIONS FOR GENETIC CONSERVATION

Although genetic effects due to land use, land management, and other human-mediated actions (e.g., air pollution) occur pervasively in plant, animal, and fungal populations of the Sierra Nevada, we cannot hope to, nor is there reason to, directly manage the entire Sierra Nevada gene pool. One responsibility of genetic conservation policy is to reduce the scope to one that is manageable. Management actions most likely to have significant genetic consequences can be prioritized, allowing management attention to be effectively focused. Certain taxa, actions, and situations are more likely to result in undesired genetic consequences than others. Thus, we recognize that (1) time and money are not available—nor is it practical—to gather genetic information that would allow all management decisions to be made wisely or defensibly; (2) in some taxa and conditions, a little genetic information incorrectly interpreted is actually misleading; ecological commonsense, knowledge of life histories and past land use, and application of sound genetic reasoning are best; and (3) some actions are more significant than others in their genetic consequences and ecological impacts; conservation efforts should be tailored to focus on high priorities but to be aware of detrimental genetic consequences both averted and caused. Baseline standards that broadly maintain the health of diverse taxa and promote the maintenance of the ecological process will provide a safety net for maintaining genetic diversity.

The following sections briefly summarize specific ongoing programs in the Sierra Nevada that address genetic diversity management and give general guidelines for genetic conservation as well as strategic approaches for integrating genetic diversity perspectives into regional planning, landscape analysis, and project implementation.

Existing Genetic Management Programs and Guidelines for Genetic Conservation

Outside of research, the longest ongoing operational program in the Sierra Nevada with direct genetic resource management and conservation objectives is the Tree Improvement and Regeneration Program of the U.S. Forest Service (Kitzmiller 1976, 1990) and cooperators in the state (the California Department of Forestry and Fire Protection) and the timber industry. The focus of the genetic conservation aspects of this program traditionally was on a small number of commercial forest tree species, at two levels: intensive tree breeding (the high-level program) and operational forest regeneration following timber harvest (the base-level program). The geographic scope is federal lands for the Forest Service program and state and private lands on which timber harvest has occurred for the other cooperators.

Within the high-level program, genetic conservation efforts traditionally were directed at maintaining broad genetic adaptedness and natural levels of genetic diversity within the families being bred for increased fiber production and other desired traits for the wood industry. This approach was counter to the prevailing agricultural and animal husbandry models, in which pedigrees were iteratively bred for reduced genetic diversity, favoring the desired traits in homogeneous lines (monocultures). It was recognized early in forest genetics that populations that lack diversity would not be stable in long-lived species and under the uncontrollable and highly variable environments and climates of natural forestlands. Diversity is maintained in the improved lineages in several ways while they are being bred for desired traits: breeding zones are used to develop locally adapted improved strains (Kitzmiller 1976), whereby parents are chosen from within certain areas of the Sierra, breeding is among parents only from within a zone, and improved progeny are out-planted within the same zone as their parents. Selection in the intensive tree improvement program is for a mix of traits, focusing on general adaptability of trees and retention of diversity. Improved stock developed in this way takes many years to become available and, with one notable exception, has not yet contributed to production of seed for regeneration in forestlands in the Sierra Nevada.

The exception in the level of production is the Blister Rust Resistance Program for sugar pine undertaken by the U.S. Forest Service and cooperators. Research on genetic resistance to white pine blister rust caused by *Cronartium ribicola* (Kinloch

1992) has transferred in the last decade to intensive operational resistance breeding. Although the target for breeding sugar pines is more focused on genetic resistance than in the general tree improvement program, the philosophy of maintaining and selecting general adaptedness and of maintaining adherence to local breeding zones remains. This program is highly productive and produces a large annual volume of resistant (having major gene resistance) sugar pines. Other landowners (industrial forest owners and state forestry) cooperate in similar resistance breeding of sugar pine for their land.

Much more extensive in its effects on lands and forests in the Sierra Nevada is the base-level program of the Forest Service tree improvement program (and its analogs in the state program) (Kitzmiller 1976, 1990). The base-level program is basically a genetic conservation approach integrated into operational forest regeneration and plantation management activities. The key elements are a focus on maintaining adaptedness, local genetic variation, and high genetic diversity while applying mild selection for desirable tree and stand traits (Kitzmiller 1990). This is accomplished through use of the seed zone map discussed previously (Buck et al. 1970; Kitzmiller 1976, 1990), which defines zones of genetic structure (presumably adaptedness) throughout California. Seed transfer, collection, and stock management guidelines maintain local and broad genetic diversity throughout the reforestation program, from selecting trees for seed collection through nursery operations and plantation management to out-planting.

This program, and similar ones throughout the timber industry and agencies, traditionally focused only on commercial tree species and had little influence on other aspects of land management or land use that might have genetic consequences. More recently, the tree improvement programs throughout the Forest Service have been broadened in scope to include all wildland taxa and any situations involving genetic management (Hessel 1992). The strategic plan focuses on developing policies for genetic adaptability, guidelines for genetic reserves, genetic policies for rare and endangered taxa, and management strategies for maintaining natural genetic diversity. In California, the former tree improvement program just described has thus expanded to become the Genetic Resource Management Program for the Forest Service Pacific Southwest region (Kitzmiller 1993), serving all aspects of genetic concerns in ecosystem management for Forest Service programs of the Sierra Nevada.

The Pacific Southwest (PSW) region of the Forest Service, at the recommendation of the PSW Genetic Resource Management Program, developed and issued a directive, *Use of Native Vegetative Material on National Forests, and Genetic Guidelines for Native Plant Collections* (Stewart 1993), that extends the genetic approach in the area of tree improvement and regeneration to all activities that deploy seeds or nursery stock into wildland situations.

Another component of the PSW Genetic Resource Management Program is the National Forest Genetic Electrophoresis Laboratory (Kitzmiller 1990; USFS 1994). Established in 1988, it was created to generate genetic information rapidly to support timber management programs, using biochemical and molecular genetic markers. In recent years, this laboratory has also expanded to focus on the assessment of genetic variation in all aspects of ecosystem management, including genetic analyses pertinent to the management of rare and endangered plants, ecological restoration, and postfire reclamation.

Given the relatively long history of genetic conservation in tree improvement and timber programs, genetic management issues have been slow to be incorporated into other programs and activities where genetic manipulation is explicit or implicit. The community concerned with ecological restoration has been the most active in considering genetic guidelines. In the last several years, the California Native Grass Association has developed guidelines similar to those described for trees (California Native Grass Association 1993), emphasizing use of native species and local germ plasm for all grass planting. Although conceptually these guidelines were developed from genetic theory and experience in forest genetics, increasing study of grass genetics is allowing the refinement specific to native grass taxa (Knapp and Rice 1994a). Similarly, as genetic studies expand for other Sierra Nevada taxa that may be threatened genetically, specific guidelines are being developed, as in the case of oaks (Millar et al. 1990a; Millar and Guinon 1990).

Several programs specifically focused on genetic aspects of Sierra Nevada wildland taxa are excluded from the scope of this chapter, as they address genetic manipulation and breeding with goals that do not include maintenance of native genetic diversity. These programs occur, for instance, in range shrub and browse species improvement programs and in sport fishery programs. Genetic manipulation focuses on developing strains that are genetically resistant to disease or to specific environmental challenges (e.g., to mine spoils or hatchery environments), but maintenance of local, native diversity is not a prerequisite goal.

At a broader level, several general policies apply to Sierra Nevada activities and impose forest and wildland management guidelines that implicitly include genetic conservation measures. The National Forest Management Act of 1976 contains specific language to maintain native diversity on national forests of the United States. This has been interpreted primarily at the species and community level (e.g., ensuring that reforestation restores the same mix of species as was in the forest prior to harvest), although it would be a natural extension to direct this language to genetic diversity within species. The California Forest Practice Act, which includes strong regulatory action for reforestation, has a genetic policy that applies to "commercial species from a local seed source or a seed source which the registered professional forester determines will produce trees physiologically suited for the area involved." Several California regional agency regulations

advocate the use of native species and local germ plasm in reintroduction or restoration projects (e.g., the National Park Service and the Soil Conservation Service), but little detail is given and the guidelines are very general.

Broader still are guidelines on how to incorporate genetic considerations into ecological restoration and reintroduction in general. These do not focus specifically on Sierra Nevada situations or taxa but are applicable to these specific conservation situations. Examples include Falk and Holsinger 1991, Falk et al. 1995, and Millar and Libby 1989. Similarly, both general approaches (Cheatham et al. 1977) and specific approaches (Millar et al. 1993, 1991; Wilson 1990) have been developed that have been applied to Sierra Nevada situations (Millar et al. 1996). Many papers provide examples of specific programs and general guidelines on genetic conservation, which would apply to the diverse situations in the Sierra Nevada (Falk and Holsinger 1991; Falk et al. 1995; Millar 1993; Millar et al. 1990b; Millar and Westfall 1992; Schonewald-Cox et al. 1983).

Genetics in Policy Criteria, Standards, and Monitoring

Only a few examples exist of instances in which genetic policy or guidelines have been developed systematically for land management and land use across a set of specific ecosystems (e.g., Crow et al. 1994, for Chequamegon and Nicolet National Forests, Wisconsin). In the Pacific Northwest, as a project of the president's forest plan, a model framework for genetic conservation planning is being developed to guide genetic management of forest resources. This project is just beginning and will focus primarily on forest trees. The PSW region of the U.S. Forest Service recently developed a conceptual framework for California national forests that provides an analysis process for implementing ecosystem management (Manley et al. 1995). As part of the analytical process, genetic diversity is considered a key ecosystem element with specific environmental indicators to be addressed at hierarchical domains of landscape scale. This process is intended to guide ecosystem management throughout the national forests of the Sierra Nevada (e.g., see Millar 1996) and provides a valuable analytical approach for incorporating genetic considerations into land-management planning and project implementation.

Based on our review of literature and survey of geneticists working on California taxa, we find genetic information lacking for most species in the Sierra Nevada. This situation is likely to remain in the future, with specific groups of taxa or occasional rare or high-interest species receiving specific study. Where we do have empirical information, we find few generalities emerging, except occasionally within closely related or ecologically similar taxa. As an attempt to provide guidance on how to manage genetic diversity in the face of diverse situations and genetic architectures and of limited empirical knowledge about most genetic architectures but strong theoretical foundations, we offer the following approaches for incorporating genetic concerns into land use and management.

Theoretical Standards for Genetic Management

The following standard and corollary for genetic management derive from population-genetic theory, based on a goal of maintaining locally adapted genetic diversity, short-term population viability, and long-term species sustainability (adaptability and resilience). In most cases, it is explicit that we do not have direct information on these variables, are unable to provide numeric baseline values for these standards, and must respond via proxies, preventative actions, inferences, and so on (see "Best Management Practices" later in this chapter). We propose the following single primary standard, along with the corollary standards listed after it.

Maintain natural levels of genetic diversity and genetic process at local to regional scales (individual to subspecies to species diversity). Natural levels are defined in an appropriate historical context, with the understanding that population and species extinction and creation, as well as other abrupt and gradual changes in gene pools, occur as an integral part of evolution. Some change is expected and appropriate; other levels and types of change are undesired.

- Avoid significant losses in genetic diversity. At the local level, avoid reductions in population sizes that would result in inbreeding depression or declines in viability that would lead to increased probabilities of population extinction.

  - Avoid losses of genes known or suspected to confer resistance to insects or pathogens, especially to exotic pests.

  - At the landscape and regional level, avoid losses in ecotypic, racial, or subspecies diversity.

- Avoid incorporating nonlocal genes into natural populations or disrupting genotypic combinations adapted to local environments. Through either direct effects on viability or indirect effects on coadapted gene complexes, genes that have not evolved locally or gene combinations novel to a population may lead to declines in individual, and thus population, viability. Avoid unnatural intraspecific or interspecific hybridizations.

- Promote natural levels and patterns of genetic process, including gene flow, natural selection, and drift (e.g., stochastic effects).

- Promote natural spatial patterns of genetic architecture, from local to regional levels.

Management Activities of Highest Concern

Generally, activities of greatest concern are those that significantly add individuals to or remove individuals from natural populations or areas adjacent to natural populations; activities that translocate individuals among locations; and activities that directly and significantly affect sex ratios, number of breeding individuals, fecundity, population establishment, viability of individuals, or mortality of different age classes in natural populations.

Specifically, the following management activities or situations involve manipulations that potentially have significant genetic consequences:

Timber harvest

Tree, shrub, or grass breeding

Wildlife habitat improvement

Land settlement

Fire suppression

Ecological restoration

Fish and other wildlife reintroductions

Forest tree planting

Air and water pollution

Recreation projects

Livestock (including wild horse) grazing

Range improvement

Habitat conversion

Prescribed and "unnatural" wildfire

Reclamation

Fish and other wildlife stocking control

Forest-health control

Biological control

Watershed restoration

Road and dam construction

Best Management Practices (BMPs)

Despite defensible theoretical standards (given earlier), it is usually impossible in practice to determine how much genetic diversity is enough or when changes in diversity are significant. Even monitoring and adaptive management approaches are difficult, due to the difficulties of successfully partitioning genetic effects from other ecological factors. Thus, except in obvious cases, it is rarely practical to determine whether standards are being met or violated. The conservative approach is to rely on preventative policy through BMPs.

Two approaches are suggested—coarse and fine filter. Coarse-filter approaches apply when no major changes in the standards are anticipated and/or none of the management activities listed in the previous section are implicated. Coarse-filter approaches focus on maintaining species, habitat, and community integrity, such as plant and animal species composition, vegetation structure and fragmentation, and disturbance regimes. The assumption is that the maintenance of these functioning systems maintains the genetic standard.

When specific management activities occur that may cause significant changes in the theoretical standards, fine-filter (intensive or case-specific) approaches are needed. Because management actions vary so much, and because of the many different effects on genetic diversity, these cases should be handled individually, with specific guidelines determined by a genetics specialist.

For fine-filter situations, some general guidelines or criteria pertain, with many exceptions:

1. When introducing individuals into natural habitats (through tree planting; plant, fish, or animal reintroductions; wildlife habitat or range improvement; reclamation; biological control; etc.):

   • Maintain local native germ plasm. Use germ plasm from donor populations that are geographically close and ecologically similar to those of the introduction site. The meaning of *local* depends on the species and context, but is related to the size of genetic neighborhoods, selection gradients, and historic events. Specific detailed guidelines ("transfer rules") have been developed for some taxa.

   • Collect donor germ plasm from local populations that are also relatively large, viable, uncontaminated (by nonlocal genotypes of the same species or interspecific hybridization), and healthy.

   • Do not use germ plasm of uncertified origin. (This guideline is exceptionally defensible.)

   • Maintain natural sex ratios and demographically appropriate age-class structures.

   • Maintain high effective population sizes through the germ plasm collection-to-introduction phases. Within the guidelines listed here, maximize the number and diversity of distinct founding genotypes, and maintain equal contributions from each donor individual through to the out-planting or introduction phase. No general rules exist (although detailed guidelines for specific taxa are available) except that larger is safer.

   • Introduce healthy founders; avoid, when possible, introducing disease with founders.

   • Favor rapid early population growth.

   • Choose introduction sites that match the habitat requirements of the species (both physical and ecological, e.g., metapopulation structure).

   • Avoid sites surrounded by or adjacent to (i.e., within significant gene-flow distance of) populations of nonlocal genotypes or races of the same species capable of contaminating the introduced populations. If necessary to accomplish this, flag the area for special concern.

   • Choose sites that are geographically large enough to accommodate large effective population sizes, unless metapopulation structure suggests otherwise.

   • Minimize inbreeding (in species that naturally outbreed) by maintaining large population sizes, minimizing relatedness in founders (avoid using clones), equalizing sex ratios, maintaining age-class stocking, and maximizing diversity (within above standards).

   • Promote reproduction and dispersal through the maintenance of ecological functioning; that is, favor natural pollinators, seed dispersers, corridors, disturbance regimes, and habitat availability (safe sites) for sexually reproducing species.

   • For asexually reproducing species, maintain high numbers of clones, as they will determine the amount and distribution of resident genetic diversity.

2. When removing individuals from natural populations (through timber harvest, fishing [native species], livestock grazing, stocking control, prescribed fire, etc.):

   • Avoid significant reductions in effective population size (i.e., reductions that bring the population size below naturally expected N_e's) (e.g., significant reductions in census number of individuals, unequal sex ratios, unequal contributions from parents, unequal numbers of offspring, drastically fluctuating population sizes).

   • Avoid actions that may lead to unnatural changes in mating systems (e.g., isolated seed trees or clumps of leave trees [trees left standing following harvest] may promote inbreeding; none of the new silvicultural or prescribed fire practices have been evaluated in this regard).

   • Avoid actions that may lead to increases in undesired intraspecific or interspecific hybridization (e.g., changes in gene-flow corridors, fragmentation).

   • Mimic natural structural patterns (spatial distributions, age-class distributions, etc.) and processes, especially disturbance regimes.

   • Mimic natural patterns and intervals of mortality.

   • Mimic inferred natural selection regimes.

3. When monitoring: Because of the nature of measuring genetic diversity and the difficulty of interpreting its significance, genetic monitoring has not been—and is unlikely to become—a routine activity in Sierra Nevada ecosystems. Rather, genetic monitoring remains somewhat of a research, or at least highly specialized, activity. Incorrect or misleading interpretations based on genetic analysis of,

for example, monitoring of marker genes could lead to imprudent management or policy.

In some cases, however, where a genetic researcher or specialist can be involved (e.g., from the National Forest Genetic Electrophoresis Laboratory), it will be useful to monitor specific aspects of genetic diversity and genetic attributes directly and plan management accordingly.

Because of these limitations, SNEP Genetics Workshop participants and geneticists generally are advocating more reliance on management decisions inferred from sound genetic theory rather than relying on the monitoring of direct genetic trends. Genetic data (i.e., direct monitoring) provide supplemental tools to inform monitoring.

A more generalized monitoring approach relying on combined empirical and theoretical insights could be as follows: Monitor levels and trends of overall genetic diversity in the population of concern, together with other proxy data, to interpret genetic status. Proxies may be ecological traits that would otherwise be part of habitat monitoring, such as demographic attributes and life history parameters of population growth and viability. If results from the monitoring of these traits indicate a population decline or significant drop in viability, and allelic or genotypic diversity similarly has dropped significantly, then it is possible that genetic factors have contributed to the decline and should be addressed. Conversely, if overall levels of genetic diversity are maintained or increase, and the population is viable and healthy, it can conservatively be assumed that genetic diversity is adequate. Abrupt changes in allele frequencies (i.e., the appearance of unique alleles) may indicate gene contamination or interspecific hybridization and should be followed by careful inspection of neighboring populations.

ACKNOWLEDGMENTS

We thank the SNEP Genetics Workshop participants for their enduring commitment to contribute to this report, from comments on structuring the workshop to contributions at the workshop to extensive and intensive reviews of the draft manuscript. We also thank the additional reviewers who contributed valuable comments on the manuscript. The names of workshop participants and additional reviewers are listed in appendix 28.2. We thank Jackie Diedrich for her help in facilitating the workshop, Chris Nelson and Erin Fleming for material support during the workshop, and the Institute of Forest Genetics, PSW Research Station, for providing the venue for the workshops.

REFERENCES

Allard, R. W., Jain, S. K., and Workman, P. L. 1968. The genetics of inbreeding populations. *Advances in Genetics* 14: 55–131.

Allendorf, F. W. 1975. Genetic variability in a species possessing extensive gene duplication: Genetic interpretation of duplicate loci and examination of genetic variation in populations of rainbow trout. Ph.D. diss., University of Washington, Seattle.

Allendorf, F. W., and R. F. Leary. 1988. Conservation and distribution of genetic variation in a polytypic species, the cutthroat trout. *Conservation Biology* 2:170–184.

Allendorf, F. W., W. Rymar, and F. M. Utter. 1987. Genetics and fishery management. In *Population genetics and fishery management*, edited by N. Rymar and F. Utter. Seattle: University of Washington Press.

Anderson, R. S. 1990. Holocene forest development and paleoclimates within the central Sierra Nevada, California. *Journal of Ecology* 78:470–89.

Arnold, R. A. 1983a. *Ecological studies of six endangered butterflies: Island biogeography, patch dynamics, and design of habitat preserves.* University of California Publications in Entomology, no. 99.

———. 1983b. *Speyeria callippe* (Lepidoptera: Nymphalidae): Application of information-theoretical and graph-clustering techniques to analyses of geographic variation and evaluation of classifications. *Annals of the Entomological Society of America* 76: 929–41.

———. 1985. Geographic variation in natural populations of *Speyeria callippe* Boisduval (Lepidoptera: Nymphalidae). *Pan-Pacific Entomology* 61:1–23.

Arnold, S. J. 1980a. Behavioral variation in natural populations. 1. Phenotypic, genetic, and environmental correlations between chemoreactive responses to prey in the garter snake, *Thamnophis elegans. Evolution* 35 (3): 489–509.

———. 1980b. Behavioral variation in natural populations. 2. The inheritance of a feeding response in crosses between geographic races of the garter snake, *Thamnophis elegans. Evolution* 35 (3): 510–15.

Avise, J. C., M. H. Smith, and R. K. Selander. 1979. Biochemical polymorphism and systematics in the genus *Peromyscus.* 7. Geographic differentiation in members of the *truei* and *maniculatus* species groups. *Journal of Mammalogy* 60 (1): 177–92.

Baker, M. C. 1975. Song dialects and genetic differences in white-crowned sparrows (*Zonotrichia leucophrys*). *Evolution* 29:226–41.

Baldwin, B. G. 1993. Molecular phylogenetics of *Calycadenia* (Compositae) based on its sequences of nuclear ribosomal DNA: Chromosomal and morphological evolution reexamined. *American Journal of Botany* 80 (2): 222–38.

Ball, C. T., J. Keeley, H. Mooney, J. Seeman, and W. Winner. 1983. Relationship between form, function, and distribution of two *Arctostaphylos* species (Ericaceae) and their putative hybrids. *Acta Oecologica* 4 (2): 153–64.

Barrowclough, G. F. 1980. Gene flow, effective population sizes, and genetic variance components in birds. *Evolution* 34 (4): 789–798.

Barrowclough, G. F., and R. J. Guttiérrez. 1990. Genetic variation and differentiation in the spotted owl (*Strix occidentalis*). *Auk* 107: 737–44.

Barrowclough, G. F., and N. K. Johnson. 1988. Genetic structure of North American birds. In *Acta nineteenth congress of international ornithology,* edited by H. Ouellet, 1630–38. Vol. II. Ottawa, Quebec: University of Ottawa Press.

Bartley, D. M., et al. 1992. Population genetic structure of coho salmon (*Oncorhynchus kisutch*) in California. *California Fish and Game* 78 (3): 88–104.

Bartley, D. M., and G. A. E. Gall. 1990. Genetic structure and gene flow in chinook salmon populations of California. *Transactions of the American Fisheries Society* 119:55–71.

Bartley, D. M., G. A. E. Gall, and B. Bentley. 1990. Biochemical genetic detection of natural and artificial hybridization of chinook and coho salmon in northern California. *Transactions of the American Fisheries Society* 119:431–37.

Baughman, J. F., P. Brussard, P. R. Ehrlich, and D. D. Murphy. 1990. History, selection, drift, and gene flow: Complex differentiation in checkerspot butterflies. *Canadian Journal of Zoology* 68:1967–75.

Bayer, R. J. 1988. Patterns of isozyme variation in western North American *Antennaria* (Asteraceae: Inuleae). 1. Sexual species of sect. *Dioicae*. *Systematic Botany* 13 (4): 525–37.

———. 1989a. Patterns of isozyme variation in the *Antennaria rosea* (Asteraceae, Inuleae) polyploid agamic complex. *Systematic Botany* 14 (3): 389–97.

———. 1989b. Patterns of isozyme variation in western North American *Antennaria* (Asteraceae: Inuleae). 2. Diploid and polyploid species of section *Alpinae*. *American Journal of Botany* 76 (5): 679–91.

———. 1990. Patterns of clonal diversity in the *Antennaria rosea* (Asteraceae) polyploid agamic complex. *American Journal of Botany* 77 (10): 1313–19.

Bechtel, E. R., and T. Whitecar. 1983. Genetics of striping in the gopher snake, *Pituophis melanoleucus*. *Journal of Herpetology* 17 (4): 362–70.

Berg, W. J., and G. A. E. Gall. 1988. Gene flow and genetic differentiation among California coastal rainbow trout populations. *Canadian Journal of Fisheries and Aquatic Science* 45 (1): 122–31.

Bezy, R. L., G. L. Gorman, Y. J. Kim, and J. W. Wright. 1977. Chromosomal and genetic divergence in the fossorial lizards of the family Anniellidae. *Systematic Zoology* 26:57–71.

Bohm, B. A., and L. D. Gottlieb. 1989. Flavonoids of the annual *Stephanomeria* (Asteraceae). *Biochemical Systematics and Ecology* 17 (6): 451–53.

Bowen, B. S. 1982. Temporal dynamics of microgeographic structure of genetic variation in *Microtus californicus*. *Journal of Mammalogy* 63 (4): 625–38.

Bradford, D. F., D. M. Graber, and F. Tabatabai. 1994. Population declines of the native frog, *Rana muscosa*, in Sequoia and Kings Canyon National Parks, California. *Southwestern Naturalist* 39 (4): 323–27.

Bradford, D. F., F. Tabatabai, and D. M. Graber. 1993. Isolation of remaining populations of the native frog, *Rana muscosa*, by introduced fishes in Sequoia and Kings Canyon National Parks, California. *Conservation Biology* 7 (4): 882–88.

Brennan, L. A., and M. L. Morrison. 1991. Long-term trends of chickadee populations in western North America. *Condor* 93: 130–37.

Britten, H. B., P. F. Brussard, and D. D. Murphy. 1994a. The pending extinction of the Uncompahgre fritillary butterfly. *Conservation Biology* 8 (1): 86–94.

Britten, H. B., M. B. Brussard, D. D. Murphy, and G. T. Austin. 1994b. Colony isolation and isozyme variability of the western seep fritillary, *Speyeria nokomis apacheana* (Nymphalidae), in the western Great Basin. *Great Basin Naturalist* 54 (2): 97–105.

Bruns, T. D., T. J. White, and J. W. Taylor. 1991. Fungal molecular systematics. *Annual Review of Ecology and Systematics* 22:525–64.

Brunsfeld, S. J., D. E. Soltis, and P. S. Soltis. 1991. Patterns of genetic variation in *Salix* section *longifoliae* (Salicaceae). *American Journal of Botany* 78 (6): 855–69.

———. 1992. Evolutionary patterns and processes in *Salix* sect. *longifoliae*: Evidence from chloroplast DNA. *Systematic Botany* 17 (2): 239–56.

Buck, J. M., R. S. Adams, J. Cane, M. T. Conkle, W. J. Libby, C. J. Eden, and M. J. Knight. 1970. *California tree seed zones*. Miscellaneous Publication. San Francisco: U.S. Forest Service, Regional Office.

Burdon, J. J. 1986. Isozymic variation in *Puccinia graminis* f.sp. *tritici* detected by starch-gel electrophoresis. *Plant Disease* 70:1139–41.

Burdon, J. J., and A. P. Roelfs. 1985a. The effect of sexual and asexual reproduction on the isozyme structure of populations of *Puccinia graminis*. *Phytopathology* 75 (9): 1068–73.

———. 1985b. Isozyme and virulence variation in asexually reproducing populations of *Puccinia graminis* and *P. recondita* on wheat. *Phytopathology* 75 (8): 907–13.

Burns, K. J., and R. M. Zink. 1990. Temporal and geographic homogeneity of gene frequencies in the fox sparrow (*Passerella iliaca*). *Auk* 107:421–24.

Buth, D. G., T. R. Haglund, and W. L. Minckley. 1992. Duplicate gene expression and allozyme divergence diagnostic for *Catostomus tahoensis* and the endangered *Chasmistes cujus* in Pyramid Lake, Nevada. *Copeia* 1992 (4): 935–41.

California Native Grass Association. 1993. Genetic guidelines for native grasses. *Newsletter of the California Native Grass Association*, pp. 1–3.

Campbell, R. K. 1986. Mapped genetic variation of Douglas-fir to guide seed transfer in southwest Oregon. *Silvae Genetica* 35 (2–3): 85–96.

Cane, J. H., M. W. Stock, D. L. Wood, and S. J. Gast. 1990. Phylogenetic relationships of *Ips* bark beetles (Coleoptera: Scolytidae): Electrophoretic and morphometric analyses of the *grandicollis* group. *Journal of Systematics and Ecology* 18 (5): 359–68.

Carroll, E., M. Foster, and V. Hipkins. N.d. Unpublished data on *Lewisia* spp.

Case, S. M. 1978a. Biochemical systematics of members of the genus *Rana* native to western North America. *Systematic Zoology* 27: 299–311.

———. 1978b. Electophoretic variation in two species of ranid frogs, *Rana boylei* and *R. muscosa*. *Copeia* 1978:311–20.

Case, S. M., P. G. Haneline, and M. F. Smith. 1975. Protein variation in several species of *Hyla*. *Systematic Zoology* 24:281–95.

Cheatham, N. H., W. J. Barry, and L. Hood. 1977. Research natural areas and related programs in California. In *Terrestrial vegetation of California*, edited by M. G. Barbour and J. Major, 75–109. Special Publication 9. Davis: California Native Plant Society.

Chong, D. K. X., R. C. Yang, and F. C. Yeh. 1994. Nucleotide divergence between populations of trembling aspen (*Populus tremuloides*) estimated with RAPDs. *Current Genetics* 26 (4): 374–76.

Christenson, D. P. 1977. History of trout introductions in California high mountain lakes. In *Proceedings of symposium on the management of high mountain lakes in California's national parks*, edited by A. Hall and R. May, 9–16. California Trout, Inc., and American Fisheries Society.

Cicero, C., and N. K. Johnson. 1992. Genetic differentiation between populations of Hutton's vireo (Aves: Vireonidae) in disjunct allopatry. *Southwestern Naturalist* 37 (4): 344–48.

Clausen, J., D. D. Keck, and W. M. Heisey. 1948. *Experimental studies on the nature of species.* 3. *Environmental responses of climatic races of Achillea.* No. 581. Carnegie Institute of Washington.

Clemens, D. T. 1990. Interspecific variation and effects of altitude on blood properties of rosy finches (*Leucosticte arctoa*) and house finches (*Carpodacus mexicanus*). *Physiological Zoology* 63 (2): 288–307.

Collins, M. M. 1984. Genetics and ecology of a hybrid zone in *Hyalophora* (Lepidoptera: Saturniidae). *University of California Publications in Entomology* 104:1–93.

Comstock, J. P., and J. R. Ehleringer. 1992. Correlating genetic variation in carbon isotopic composition with complex climatic gradients. *Proceedings of the National Academy of Sciences USA* 89 (16): 7747–51.

Conkle, M. T. 1973. Growth data for twenty-nine years from the California elevational transect study of ponderosa pine. *Forest Science* 19:31–39.

———. 1992. Genetic diversity—seeing the forest through the trees. *New Forest* 6:5–22.

———. N.d. Unpublished allozyme data on sugar pine.

Conkle, M. T., and W. B. Critchfield. 1988. Genetic variation and hybridization of ponderosa pine. In *Ponderosa pine: The species and its management,* 27–43. Pullman: Washington State University Cooperative Extension.

Corbin, K. W., C. G. Sibley, and A. Ferguson. 1979. Genic charges associated with the establishment of sympatry in orioles of genus *Icterus. Evolution* 33:624–33.

Critchfield, W. B. 1956. Morphological and physiological variation in *Pinus contorta* Dougl. Ph.D. diss., University of California, Berkeley.

———. 1986. Impact of the Pleistocene on the genetic structure of North American conifers. In *Proceedings of eighth North American Forest Biology Workshop,* edited by R. M. Lanner, 70–118. Logan, UT.

Cronin, M. A. 1991. Mitochondrial and nuclear genetic relationships of deer (*Odocoileus* spp.) in western North America. *Canadian Journal of Zoology* 69 (5): 1270–79.

———. 1992. Intraspecific variation in mitochondrial DNA of North American cervids. *Journal of Mammalogy* 73 (1): 70–82.

Cronin, M. A., S. C. Amstrup, G. W. Garner, and E. R. Vyse. 1991a. Interspecific and intraspecific mitochondrial DNA variation in North American bears (*Ursus*). *Canadian Journal of Zoology* 69 (12): 2985–92.

Cronin, M. A., M. E. Nelson, and D. F. Pac. 1991b. Spatial heterogeneity of mitochondrial DNA and allozymes among populations of white-tailed deer and mule deer. *Journal of Heredity* 82 (2): 118–27.

Crow, T. R., A. W. Haney, and D. M. Waller. 1994. *Report on the scientific roundtable on biological diversity convened by the Chequamegon and Nicolet National Forests.* General Technical Report NC 166. Rhinelander, WI: U.S. Forest Service, North Central Forest Experiment Station.

Daly, J. C., and J. L. Patton. 1990. Dispersal, gene flow, and allelic diversity between local populations of *Thomomys bottae* pocket gophers in the coastal ranges of California. *Evolution* 44 (5): 1283–94.

Dawson, T. E., and J. R. Ehleringer. 1993. Gender-specific physiology, carbon isotope discrimination, and habitat distribution in boxelder, *Acer negundo. Ecology* 74 (3): 798–815.

Delany, D. L. N.d. Unpublished allozyme data on *Pinus monophylla* and *P. edulis.*

Desrochers, A. M., and B. A. Bohm. 1993. Flavonoid variation in *Lasthenia californica* (Asteraceae). *Biochemical Systematics and Ecology* 21 (4): 449–53.

Dirig, R., and J. F. Cryan. 1991. The status of silvery blue subspecies (*Glaucopsyche lygdamus* and *G. l. couperi:* Lycaenidae) in New York. *Journal of the Lepidopteran Society* 45 (4): 272–80.

Dodd, R. S., Z. A. Rafii, and E. Zavarin. 1993. Chemosystematic variation in acorn fatty acids of Californian live oaks (*Quercus agrifolia* and *Q. wislizenii*). *Biochemical Systematics and Ecology* 21 (2): 279–85.

Doede, D. L., E. Carroll, R. Westfall, R. Miller, H. J. Switzer, and R. M. Snader. In press. Variation in allozymes, taxol, and propagation by rooted cuttings in Pacific yew. In *Proceedings of international yew resources conference.* Berkeley, CA: Native Yew Society.

Donovan, L. A., and J. R. Ehleringer. 1992. Contrasting water-use patterns among size and life-history classes of a semi-arid shrub. *Functional Ecology* 6 (4): 482–88.

———. 1994. Carbon isotope discrimination, water–use efficiency, growth, and mortality in a natural shrub population. *Oecologia* 100 (3): 347–54.

Dragoo, J. W., J. R. Choate, T. L. Yates, and T. P. O'Farrell. 1990. Evolutionary and taxonomic relationships among North American arid-land foxes. *Journal of Mammalogy* 71 (3): 318–32.

Dunlap, J. M., et al. 1993. Intraspecific variation in photosynthetic traits of *Populus trichocarpa. Canadian Journal of Botany* 71 (10): 1304–11.

Dunlap, J. M., P. E. Heilman, and R. F. Stettler. 1994. Genetic variation and productivity of *Populus trichocarpa* and its hybrids. 7. Two-year survival and growth of native black cottonwood clones from four river valleys in Washington. *Canadian Journal of Forest Research* 24 (8): 1539–49.

Ehleringer, J. R., S. L. Phillips, W. S. F. Schuster, and D. R. Sandquist. 1991. Differential utilization of summer rains by desert plants. *Oecologia* 88 (3): 430–34.

Ellstrand, E., J. M. Lee, J. E. Keeley, and S. C. Keeley. 1987. Ecological isolation and introgression in an *Arctostaphylos* (Ericaceae) population. *Acta Oecologica* 8 (4): 299–308.

Ellstrand, N. C., and M. Roose. 1987. Patterns of genotypic diversity in clonal plant species. *American Journal of Botany* 74 (1): 123–31.

Elvingson, P., and K. Johansson. 1993. Genetic and environmental components of variation in body traits of rainbow trout (*Oncorhynchus mykiss*) in relation to age. *Aquaculture,* no. 118: 191–204.

Evans, R. D., and J. R. Ehleringer. 1994. Water and nitrogen dynamics in an arid woodland. *Oecologia* 99 (3–4): 233–42.

Falk, D., and K. Holsinger, eds. 1991. *Genetics and conservation of rare plants.* Cary, NC: Oxford University Press.

Falk, D. A., C. I. Millar, and P. Olwell, eds. 1995. *Restoring diversity: Reintroduction of rare and endangered plants.* Washington, DC: Island Press.

Ferguson, M. M. 1990. The genetic impact of introduced fishes on native species. *Canadian Journal of Zoology* 68: 1053–57.

Ferrell, G. T., W. D. Bedard, and R. D. Westfall. 1989. Geographic variation in *Pinus ponderosa* susceptibility to the gouty pitch midge, *Cecidomyia piniinopsis,* in the Sierra Nevada and southern Cascade Mountains of California. In *Insects affecting reforestation: Biology and damage,* edited by R. I. Alfano and S. G. Glover, 205–12. Victoria, BC: Forestry Canada, Pacific Forestry Centre.

Fiedler, P. L. 1985. Heavy metal accumulation and the nature of edaphic endemism in the genus *Calochortus* (Liliaceae). *American Journal of Botany* 72 (11): 1712–18.

Fins, L., and W. J. Libby. 1982. Population variation in *Sequoiadendron:* Seed and seedling studies, vegetative propagation, and isozyme variation. *Silvae Genetica* 31 (4): 102–9.

Fisher, R. A., W. Putt, and E. Hackel. 1976. An investigation of the products of fifty-three gene loci in three species of wild Canidae: *Canus lupus, Canus latrans,* and *Canus familiaris. Biochemical Genetics* 14: 963–74.

Fleischer, R. C., S. I. Rothstein, and L. S. Miller. 1991. Mitochondrial DNA variation indicates gene flow across a zone of known secondary contact between two subspecies of the brown-headed cowbird. *Condor* 93:185–89.

Fletcher, S. D., and W. S. Moore. 1992. Further analysis of allozyme variation in the northern flicker, in comparison with mitochondrial DNA variation. *Condor* 94:988–91.

Flowers, L., and K. J. Rice. 1994. Ecotypic variation in California brome (*Bromus carinatus*). Paper and abstract presented at the fourth annual meeting of the California Native Grass Association, Sacramento, 4 November.

Freeman, D. C., W. A. Turner, E. D. McArthur, and J. H. Graham. 1991. Characterization of a narrow hybrid zone between two subspecies of big sagebrush (*Artemisia tridentata*, Asteraceae). *American Journal of Botany* 78 (6): 805–15.

Furnier, G. R., and W. T. Adams. 1986. Geographic patterns of allozyme variation in Jeffrey pine. *American Journal of Botany* 73 (7): 1009–15.

Furnier, G. R., P. Knowles, M. A. Clyde, and B. P. Danik. 1987. Effects of avian seed dispersal on the genetic structure of whitebark pine populations. *Evolution* 41 (3): 607–12.

Futuyma, D. J., and S. C. Peterson. 1985. Genetic variation in the use of resources by insects. *American Review of Entomology* 30: 217–38.

Gall, G. A. E., D. Bartley, B. Bentley, R. Gomulkiewicz, and M. Mangel. 1992. Geographic variation in population genetic structure of chinook salmon from California and Oregon. *U.S. Fishery Bulletin* 90:77–100.

Gall, G. A. E., B. Bentley, and R. C. Nuzum. 1990. Genetic isolation of steelhead rainbow trout in Kaiser and Redwood Creeks, California. *California Fish and Game* 76 (4): 216–23.

Gall, G. A. E., C. A. Busack, R. C. Smith, J. R. Gold, and B. J. Kornblatt. 1976. Biochemical genetic variation in populations of golden trout, *Salmo aguabonita:* Evidence of the threatened Little Kern River golden trout, *S. a. whitei. Journal of Heredity* 67: 330–35.

Garbelotto, M., T. D. Bruns, F. W. Cobb, and W. J. Otrasina. 1993. Differentiation of intersterility groups and geographic provenances among isolates of *Heterobasidion annosum* detected by random amplified polymorphic DNA assays. *Canadian Journal of Botany* 71:565–69.

Gardes, M., and T. D. Bruns. 1993. ITS primers with enhanced specificity for basidiomycetes: Application to the identification of mycorrhizae and rusts. *Molecular Ecology* 1993 (2): 113–18.

Gardes, M., T. J. White, J. A. Fortin, T. D. Bruns, and J. W. Taylor. 1991. Identification of indigenous and introduced symbiotic fungi in ectomycorrhizae by amplification of nuclear and mitochondrial ribosomal DNA. *Canadian Journal of Botany* 69:180–90.

Geiger, H., and A. M. Shapiro. 1986. Electrophoretic evidence for speciation within the nominal species *Anthocharis sara* Lucas (Pieridae). *Journal of Research in Lepidopterans* 25 (1): 15–24.

Gjedrem, T. 1992. Breeding plans for rainbow trout. *Aquaculture,* no. 100:73–83.

Gjerde, B., and L. R. Schaeffer. 1989. Body traits in rainbow trout. 2. Estimates of heritabilities and of phenotypic and genetic correlations. *Aquaculture,* no. 80:25–44.

Good, D. A. 1988. Allozyme variation and phylogenetic relationships among the species of *Elgaria* (Squamata: Anguidae). *Herpetologica* 44 (2):154–62.

Griffin, J. R. 1965. Digger pine seedling response to serpentinite and non-serpentinite soil. *Ecology* 46: 801–7.

Griffin, J. R., and W. B. Critchfield. 1972. *The distribution of forest trees in California.* Research Paper PSW-82. Berkeley, CA: U.S. Forest Service, Pacific Southwest Research Station.

Haglund, T. R., D. G. Buth, and R. Lawson. 1992. Allozyme variation and phylogenetic relationships of Asian, North American, and European populations of the threespine stickleback, *Gasterosteus aculeatus. Copeia* 1992 (2): 432–43.

Hammond, P. C. 1986. A rebuttal to the Arnold classification of *Speyeria callippe* (Nymphalidae) and defence of the subspecies concept. *Journal of Research in Lepidopterans* 24:197–208.

Hamrick, J. L. 1976. Variation and selection in western montane species. 2. Variation within and between populations of white fir on an elevational transect. *Theoretical and Applied Genetics* 47 (1): 27–34.

Hamrick, J. L., and M. J. W. Godt. 1990. Allozyme diversity in plant species. *Plant population genetics: Breeding and genetic resources,* edited by H. D. Brown, M. T. Clegg, A. L. Kohler, and B. S. Wier, 43–63. Boston: Sinnauer Associates.

Hamrick, J. L., M. J. W. Godt, and S. L. Sherman-Broyles. 1992. Factors influencing levels of genetic diversity in woody plant species. *New Forestry* 6:95–124.

Hamrick, J. L., and W. J. Libby. 1972. Variation and selection in western U.S. montane species. Part 1: White fir. *Silvae Genetica* 21 (1–2): 29–35.

Hamrick, J. L., Y. B. Linhart, and J. B. Mitton. 1979. Relationships between life history characteristics and electrophoretically detectable genetic variation in plants. *Annual Review of Ecology and Systematics* 10:173–200.

Hanski, I., and M. Gilpin. 1991. Metapopulation dynamics—brief history and conceptual domain. *Biological Journal of the Linnean Society* 142 (1–2): 3–16.

Harry, D. E. 1984. Genetic structure of incense-cedar (*Calocedrus decurrens*). Ph.D. diss., University of California, Berkeley.

Harry, D. E., J. L. Jenkinson, and B. B. Kinloch. 1983. Early growth of sugar pine from an elevational transect. *Forest Science* 29 (3): 660–69.

Hayes, J. L., and J. L. Robertson. 1992. An (ecologically biased) view of the current status of bark beetle genetics and future research needs. In *Proceedings of workshop on bark beetle genetics: Current status of research,* edited by J. L. Hayes and J. L. Robertson, 1–2. Berkeley, CA: U.S. Forest Service.

Hedgecock, D., and F. J. Ayala. 1974. Evolutionary divergence in the genus *Taricha* (Salamandridae). *Copeia* 1974 (3): 738–47.

Hershberger, W. K. 1992. Genetic variability in rainbow trout populations. *Aquaculture,* no. 100:51–71.

Hessel, D. L. 1992. *Genetic resource program strategic plan.* Washington, DC: U.S. Forest Service, National Office, Timber Management.

Higby, P. K., and M. W. Stock. 1982. Genetic relationships between two sibling species of bark beetle (Coleptera: Scolytidae), Jeffrey pine beetle and mountain pine beetle, in northern California. *Annals of the Entomological Society of America* 75:668–74.

Hobson, K. R., and D. E. Bright. 1994. A key to the *Xyleborus* of California, with faunal comments (Coleoptera: Scolytidae). *Pan-Pacific Entomology* 70 (4): 267–68.

Hodges, S. A., and M. L. Arnold. 1994. Floral and ecological isolation between *Aquilegia formosa* and *Aquilegia pubescens. Proceedings of the National Academy of Sciences USA* 91 (7): 2493–96.

Holsinger, K. E., and L. D. Gottlieb. 1988. Isozyme variability in the tetraploid *Clarkia gracilis* (Onagraceae) and its diploid relatives. *Systematic Botany* 13 (1): 1–6.

Hong, Y. P., V. D. Hipkins, and S. H. Strauss. 1993. Chloroplast DNA diversity among trees, populations, and species in the California closed-cone pines *(Pinus radiata, Pinus muricata,* and *Pinus attenuata). Genetics* 135 (4): 1187–96.

Howard, J. H., and R. L. Wallace. 1981. Microgeographic variation of electrophoretic loci in populations of *Ambystoma macrodactylum columbianum* (Caudata: Ambystomatidae). *Copeia* 1981 (2): 466–71.

Jackman, T. R., and D. B. Wake. 1994. Evolutionary and historical analysis of protein variation in the blotched forms of salamanders of the *Ensatina* complex (Amphibia, Plethodontidae). *Evolution* 48 (3): 876–97.

James, F. C. 1983. Environmental component of morphological differentiation in birds. *Science* 221:184–87.

Jelinski, D. E., and W. M. Cheliak. 1992. Genetic diversity and spatial subdivision of *Populus tremuloides* (Salicaceae) in a heterogeneous landscape. *American Journal of Botany* 79 (7): 728–36.

Jenkinson, J. L. 1977. *Edaphic interactions in first–year growth of California ponderosa pine.* Research Paper PSW-127. Berkeley, CA: U.S. Forest Services, Pacific Southwest Research Station.

Jennings, W. B., D. F. Bradford, and D. F. Johnson. 1992. Dependence of the garter snake *Thamnophis elegans* on amphibians in the Sierra Nevada of California. *Journal of Herpetology* 26 (4): 503–5.

Johnson, N. K., and J. A. Marten. 1991. Evolutionary genetics of flycatchers. 3. Variation in *Empidonax hammondii* (Aves: Tyrannidae). *Canadian Journal of Zoology* 69: 232–38.

———. 1992. Macrogeographic patterns of morphometric and genetic variation in the sage sparrow complex. *Condor* 94:1–19.

Johnson, N. K., and R. M. Zink. 1983. Speciation in sapsuckers *(Sphyrapicus).* 1. Genetic differentiation. *Auk* 100:871–84.

Johnson, W. E., and R. K. Selander. 1971. Protein variation and systematics in kangaroo rats (genus *Dipodomys). Systematic Zoology* 20 (4): 372–405.

Kinloch, B. B. 1992. Distribution and frequency of a gene for resistance to white pine blister rust in natural populations of sugar pine. *Canadian Journal of Botany* 70 (7): 1319–23.

Kinloch, B. B., Jr., and G. E. Dupper. 1987. Restricted distribution of a virulent race of the white pine blister rust pathogen in the western United States. *Canadian Journal of Forest Research* 17: 448–51.

Kinney, W. C. 1996. Conditions of rangelands before 1905. In *Sierra Nevada Ecosystem Project: Final report to Congress,* vol. II, chap. 3. Davis: University of California, Centers for Water and Wildland Resources.

Kitzmiller, J. H. 1976. *Tree improvement master plan for the California region.* San Francisco: U.S. Forest Service.

———. 1990. Managing genetic diversity in a tree improvement program. *Forest Ecology Management* 35 (1–2): 131–49.

———. 1993. *Genetic resource management in the USFS PSW region.* San Francisco: U.S. Forest Service, Regional Office.

Knapp, E. E. 1994. Genetic architecture of purple needlegrass: Implications for restoration. Paper and abstract presented at the fourth annual meeting of the California Native Grass Association, Sacramento, 4 November.

Knapp, E., and K. Rice. 1994a. Starting from seed: Genetic issues in using native grasses for restoration. *Restoration and Management Notes* 12 (1): 40–45.

———. 1994b. Isozyme variation within and among populations of *Danthonia californica.* Unpublished report to U.S. Forest Service. University of California, Davis.

———. 1994c. Morphological and allozyme variation within and among populations of *Nassella pulchra.* Unpublished report to the Nature Conservancy. University of California, Davis.

———. In press. Genetic architecture and gene flow in *Elymus glaucus* (blue wildrye): Implications for native grassland restoration. *Restoration Ecology.*

Knapp, R. A. 1994. The high cost of high Sierra trout. *Wilderness Record* 19:1–3.

———. 1996. Non-native trout in natural lakes of the Sierra Nevada: An analysis of their distribution and impacts on native aquatic biota. In *Sierra Nevada Ecosystem Project: Final report to Congress,* vol. III. Davis: University of California, Centers for Water and Wildland Resources.

Korman, A. M., D. P. Pashley, M. I. Haverty, and J. P. LaFage. 1991. Allozymic relationships among cuticular hydrocarbon phenotypes of *Zootermopsis* species (Isoptera: Termopsidae). *Annals of the Entomological Society of America* 84 (1): 1–9.

Kruckeberg, A. K. 1987. Serpentine endemism and rarity. In *Conservation and management of rare plants: Proceedings of a California conference on the conservation and management of rare and endangered plants,* edited by T. S. Elias, 121–28. Sacramento: California Native Plant Society.

Lack, D. 1968. *Ecological adaptations for breeding in birds.* London: Methuen.

Landram, M. 1996. Status of reforestation on national forest lands within the Sierra Nevada Ecosystem Project study area. In *Sierra Nevada Ecosystem Project: Final report to Congress,* vol. III. Davis: University of California, Centers for Water and Wildland Resources.

Larson, A. 1980. Paedomorphosis in relation to rates of morphological and molecular evolution in the salamander *Aneides flavipunctatus* (Amphibia, Plethodontidae). *Evolution* 34 (1): 1–17.

Leary, R. F., F. W. Allendorf, and S. H. Forbes. 1993. Conservation genetics of bull trout in the Columbia and Klamath River drainages. *Conservation Biology* 7 (4): 856–65.

Ledig, F. T. 1987. Genetic structure and the conservation of California's endemic and near-endemic conifers. In *Conservation and management of rare and endangered plants: Proceedings of a California conference on the conservation and management of rare and endangered plants,* edited by T. S. Elias, 587–94. Sacramento: California Native Plant Society.

Levins, R. 1970. Extinction. In *Some mathematical problems in biology,* edited by M. Gesternhaber, 77–107. Providence, RI: American Mathematical Society.

Libby, W. J., K. Isik, and J. P. King. 1980. Variation in flushing time among white fir population samples. *Annales Forestales* 8 (6): 123–38.

Liu, Z. L., and J. B. Sinclair. 1992. Genetic diversity of *Rhizoctonia solani* anastomosis group 2. *Phytopathology* 82: 778–87.

Loudenslager, E. J., and G. A. E. Gall. 1980. Geographic patterns of protein variation and subspeciation in cutthroat trout, *Salmo clarki*. *Systematic Zoology* 29: 27–42.

MacDonald, S. E., and V. J. Lieffers. 1991. Population variation, outcrossing, and colonization of disturbed areas by *Calamagrostis canadensis*: Evidence from allozyme analysis. *American Journal of Botany* 78 (8): 1123–29.

Mahalovich, M. F. 1985. A genetic architecture study of giant sequoia: Early growth characteristics. Master's thesis, University of California, Berkeley.

Manley, P. N., G. E. Brogan, C. Cook, M. E. Flores, D. G. Fullmer, S. Husar, T. M. Jimerson, L. M. Lux, M. E. McCain, J. A. Rose, G. Schmitt, J. C. Schuyler, and M. J. Skinner. 1995. *Sustaining ecosystems: A conceptual framework.* San Francisco: U.S. Forest Service, Pacific Southwest Region and Station.

Manlove, M. N., R. Baccus, M. R. Smith, and D. Graber. 1980. Biochemical variation in the black bear. In *Proceedings of bears— Their biology and management: Third Bear Biology Association conference,* edited by C. J. Martinka and K. L. McArthur, 37–41.

Mastrogiuseppe, R. J. 1972. Geographic variation in foxtail pine, *Pinus balfouriana* Grev. & Balf. Master's thesis, Humboldt State University, Arcata, CA.

Mastrogiuseppe, R. J., and J. D. Mastrogiuseppe. 1980. A study of *Pinus balfouriana* Grev. & Balf. (Pinaceae). *Systematic Botany* 5: 86–104.

McArthur, E. D., C. L. Pope, and D. C. Freeman. 1981. Chromosomal studies of subgenus Tridentatae of *Artemisia*: Evidence for autopolyploidy. *American Journal of Botany* 68 (5): 589–605.

McCracken, G. F., M. K. McCracken, and A. T. Vawter. 1994. Genetic structure in migratory populations of the bat *Tadarida brasiliensis mexicana*. *Journal of Mammalogy* 75 (2): 500–14.

McDonald, B. A., and J. P. Martinez. 1990. DNA restriction fragment length polymorphisms among *Mycosphaerella graminicola* (Anamorph *Septoria tritici*) isolates collected from a single wheat field. *Phytopathology* 80: 1368–73.

McKechnie, S. W., P. P. Erlich, and C. L. Hogue. 1975. Population genetics of *Euphydryas* butterflies. Part 1: Genetic variation and the neutral hypothesis. *Genetics* 81:571–94.

McKelvey, K. S., C. N. Skinner, C. Chang, D. C. Erman, S. J. Husari, D. Parsons, J. W. van Wagtendonk, and C. P. Weatherspoon. 1996. An overview of fire in the Sierra Nevada. In *Sierra Nevada Ecosystem Project: Final report to Congress,* vol. II, chap. 37. Davis: University of California, Centers for Water and Wildland Resources.

McKinney, C. O., R. K. Selander, W. R. Johnson, and S. Y. Yang. 1972. Genetic variation in the side-blotched lizard (*Uta stansburiana*). In *Studies in genetics,* vol. 7, 307–18. University of Texas Publication 7213.

Meyer, S. E., and S. B. Monsen. 1991. Habitat-correlated variation in mountain big sagebrush (*Artemisia tridentata* ssp. *vaseyana*) seed germination patterns. *Ecology* 72 (2): 739–42.

Meyer, S. E., S. B. Monsen, and E. D. McArthur. 1990. Germination response of *Artemisia tridentata* (Asteraceae) to light and chill: Patterns of between-population variation. *Botanical Gazette* 151 (2): 176–83.

Millar, C. I. 1993. Conservation of germplasm in forest trees. In *Clonal forestry II: Conservation and application,* edited by W. J. Libby and R. Ahuja, 42–65. Berlin: Springer-Verlag.

———. 1996. The Mammoth-June Ecosystem Management Project, Inyo National Forest. In *Sierra Nevada Ecosystem Project: Final report to Congress,* vol. II, chap. 50. Davis: University of California, Centers for Water and Wildland Resources.

Millar, C. I., M. Barbour, D. L. Elliott-Fisk, J. R. Shevock, and W. B. Woolfenden. 1996. Significant natural areas. In *Sierra Nevada Ecosystem Project: Final report to Congress,* vol. II, chap. 29. Davis: University of California, Centers for Water and Wildland Resources.

Millar, C. I., and D. L. Delany. N.d. Unpublished allozyme data on the California cypresses. Berkeley, CA: U.S. Forest Service, Pacific Southwest Research Station.

Millar, C. I., D. L. Delany, and L. A. Riggs. 1990a. Genetic variation in California oaks. *Fremontia* 18 (3): 20–21.

Millar, C. I., D. L. Delany, and R. D. Westfall. 1993. Effects of silvicultural treatments on genetic diversity in Jeffrey pine (*Pinus jeffreyii*). Unpublished manuscript. U.S. Forest Service, Berkeley, CA.

Millar, C. I., and M. Guinon. 1990. Planting oaks: Don't forget the genes. *Newsletter of the California Oak Foundation* 2 (2): 1, 8–9.

Millar, C. I., B. K. Kinloch, and R. D. Westfall. In press. Conservation of genetic biodiversity in sugar pine. In *Proceedings of sugar pine: Its biology, management, and conservation,* edited by B. K. Kinloch. Davis, CA: U.S. Forest Service, Pacific Southwest Research Station.

Millar, C. I., F. T. Ledig, and L. A. Riggs. 1990b. Conservation of diversity in forest ecosystems: Introduction. *Forest Ecology and Management* 35 (1–2): 1–4.

Millar, C. I., and W. J. Libby. 1989. Restoration: Disneyland or native ecosystem? *Fremontia* 17 (2): 3–10.

Millar, C. I., S. M. Strauss, M. T. Conkle, and R. D. Westfall. 1988. Allozyme differentiation and biosystematics in the California closed-cone pines (subsection *Oocarpae* Little & Critchfield, genus *Pinus*). *Systematic Botany* 13 (3): 351–70.

Millar, C. I., and R. D. Westfall. 1992. Allozyme markers in forest genetic conservation. In *Population genetics of forest trees,* edited by R. Adams, S. M. Strauss, D. L. Copes, and A. R. Griffin, 347–72. New York: Kluwer Academic Publishers.

Millar, C. I., R. D. Westfall, and D. Nelson. 1991. Genetic conservation areas on the Placerville Ranger District, Eldorado National Forest. Unpublished report. U.S. Forest Service, Pacific Southwest Research Station, Berkeley, CA.

Mirov, N., J. W. Duffield, and A. R. Liddicoet. 1952. Altitudinal races of *Pinus ponderosa*: A twelve-year progress report. *Journal of Forestry* 50: 825–31.

Mitton, J. B., and M. G. Raphael. 1990. Genetic variation in the marten, *Martes americana*. *Journal of Mammalogy* 71 (2): 195–97.

Mooring, J. S. 1994. A cytogenetic study of *Eriophyllum confertiflorum* (Compositae, Helenieae). *American Journal of Botany* 81 (7): 919–26.

Moritz, C. 1994. Defining "evolutionarily significant units" for conservation. *Trends in Ecology and Evolution* 9 (10): 373–75.

Morton, J. B., and S. P. Bentivenga. 1994. Levels of diversity in endomycorrhizal fungi (Glomales, Zygomycetes) and their role in defining taxonomic and non-taxonomic groups. *Plant and Soil* 159:47–59.

Moyle, P. B. 1976. *Inland fishes of California.* Berkeley and Los Angeles: University of California Press.

———. 1994. The decline of anadromous fishes in California. *Conservation Biology* 8 (3): 869–70.

Moyle, P. B., R. M. Yoshiyama, and R. A. Knapp. 1996. Status of fish and fisheries. In *Sierra Nevada Ecosystem Project: Final report to Congress,* vol. II, chap. 33. Davis: University of California, Centers for Water and Wildland Resources.

Namkoong, G., and M. T. Conkle. 1976. Time trends in genetic control of height growth in ponderosa pine. *Forest Science* 22 (1): 2–12.

Nason, J. D., N. C. Ellstrand, and M. L. Arnold. 1992. Patterns of hybridization and introgression in populations of oaks, manzanitas, and irises. *American Journal of Botany* 79 (1): 101–11.

Ness, B. D. 1989. Seed morphology and taxonomic relationships in *Calochortus* (Liliaceae). *Systematic Botany* 14 (4): 495–505.

Ness, B. D., D. E. Soltis, and P. S. Soltis. 1990. An examination of polyploidy and putative introgression in *Calochortus* subsection *nudi* (Liliaceae). *American Journal of Botany* 77 (12): 1519–31.

Niebling, C. R., and M. T. Conkle. 1990. Diversity of Washoe pine and comparisons with allozymes of ponderosa pine races. *Canadian Journal of Forest Research* 20:298–308.

Nielsen, J. L., C. Gan, and W. K. Thomas. 1994. Differences in genetic diversity for mitochondrial DNA between hatchery and wild populations of *Oncorhynchus*. *Canadian Journal of Fisheries and Aquatic Science* 51 (S1): 290–97.

Novak, S. J., R. N. Mack, and D. E. Soltis. 1991. Genetic variation in *Bromus tectorum* (Poaceae): Population differentiation in its North American range. *American Journal of Botany* 78 (8): 1150–61.

Olmstead, R. G. 1990. Biological and historical factors influencing genetic diversity in the *Scutellaria angustifolia* complex (Labiatae). *Evolution* 44 (1): 54–70.

Otrosina, W. J., T. E. Chase, and F. W. Cobb Jr. 1992. Allozyme differentiation of intersterility groups of *Heterobasidion annosum* isolated from conifers in the western United States. *Phytopathology* 8:540–45.

Pashley, D. P. 1980. Genetic comparisons between native and introduced populations of the codling moth, *Laspeyresia pomenella* (Tortricidae) and among other tortricid species. Ph.D. diss., University of Texas, Austin.

Pashley, D. P., S. J. Johnson, and A. N. Sparks. 1985. Genetic population structure of migratory moths: The fall armyworm (Lepidoptera: Noctuidae). *Annals of the Entomological Society of America* 78: 756–762.

Patterson, R., and B. D. Tanowitz. 1989. Evolutionary and geographic trends in adaptive wood anatomy in *Eriastrum densifolium* (Polemoniaceae). *American Journal of Botany* 76 (5): 706–13.

Patton, J. L., and J. H. Feder. 1981. Microspatial genetic heterogeneity in pocket gophers: Non-random breeding and drift. *Evolution* 35 (5): 912–20.

Patton, J. L., H. MacArthur, and S. Y. Yang. 1976. Systematic relationships of the four-toed populations of *Dipodomy heermanni*. *Journal of Mammalogy* 57 (1): 159–62.

Patton, J. L., and M. F. Smith. 1990. *The evolutionary dynamics of the pocket gopher Thomomys bottae, with emphasis on California populations*. University of California Publications in Zoology, vol. 123. Berkeley and Los Angeles: University of California Press.

Patton, J. L., and S. Y. Yang. 1977. Genetic variation in *Thomomys bottae* pocket gophers: Macrogeographic patterns. *Evolution* 31 (4): 697–720.

Peterson, A. T. 1991. Gene flow in scrub jays: Frequency and direction of movement. *Condor* 91:926–34.

Porter, A. A., and E. M. Jakob. 1990. Allozyme variation in the introduced spider *Holocnemus pluchei* (Araneae, Pholcidae) in California. *Journal of Arachnology* 18:313–19.

Porter, A. H. 1989. Gene flow statistics and species level systematics of butterflies. Ph.D. diss., University of California, Davis.

Porter, A. H., and H. Geiger. 1988. Genetic and phenotypic population structure of the *Coenonympha tullia* complex (Lepidoptera: Nymphalidae: Satyrinae) in California: No evidence for species boundaries. *Canadian Journal of Zoology* 66:2651–65.

Porter, A. H., and A. M. Shapiro. 1989. Genetics and biogeography of the *Oeneis chryxus* complex (Satyrinae) in California. *Journal of Research in Lepidopterans* 28 (4): 264–76.

Powell, J. A., and C. L. Hogue. 1979. *California insects*. Berkeley and Los Angeles: University of California Press.

Pratt, G. F., and G. R. Ballmer. 1991. Three biotypes of *Apodemia mormo* (Riodinidae) in the Mojave Desert. *Journal of the Lepidopteran Society* 45 (1): 46–67.

Rabinowitz, D. 1981. Seven forms of rarity. In *The biological aspects of rare plant conservation*, edited by H. Synge, 205–17. New York: John Wiley.

Rank, N. E. 1992. A hierarchical analysis of genetic differentiation in a montane leaf beetle *Chrysomela aeneicollis* (Coleptera: Chrysomelidae). *Evolution* 46 (4): 1097–111.

Rehfeldt, G. E. 1990. Genetic differentiation among populations of *Pinus ponderosa* from the upper Colorado River Basin. *Botanical Gazette* 151 (1): 125–37.

Rice, K. J., and R. N. Mack. 1991. Ecological genetics of *Bromus tectorum*. 3: The demography of reciprocally sown populations. *Oecologia* 99:91–101.

Richards, R. T. 1994. Wild mushroom harvesting in the Klamath bioregion: A socioeconomic study. Unpublished report. University of California, Davis.

———. 1996. Special forest product harvesting in the Sierra Nevada. In *Sierra Nevada Ecosystem Project: Final report to Congress*, vol. III. Davis: University of California, Centers for Water and Wildland Resources.

Riddle, B. R., and R. L. Honeycutt. 1990. Historical biogeography in North American arid regions: An approach using mitochondrial DNA phylogeny in grasshopper mice (genus *Onychomys*). *Evolution* 44 (1): 1–15.

Rogers, D. L., D. E. Harry, and W. J. Libby. 1994. Genetic variation in incense-cedar *(Calocedrus decurrens)*. 1. Provenance differences in a twelve-year-old common-garden study. *Western Journal of Applied Forestry* 9 (4): 113–16.

Rogstad, S. H., H. Nybom, and B. A. Schaal. 1991. The tetrapod DNA fingerprinting M-13 repeat probe reveals genetic diversity and clonal growth in quaking aspen (*Populus tremuloides*, Salicaceae). *Plant Systematics and Evolution* 175 (3–4): 115–23.

Rosado, S. C. S., B. R. Kropp, and Y. Piche. 1994. Genetics of ectomycorrhizal symbiosis. 2. Fungal variability and heritability of ectomycorrhizal traits. *New Phytologist* 126:111–17.

Roy, B. A. 1993. Patterns of rust infection as a function of host genetic diversity and host density in natural populations of the apomictic crucifer, *Arabis holboellii*. *Evolution* 47 (1): 111–24.

Roy, M. S., E. Geffen, D. Smith, E. Ostrander, and R. Wayne. 1994. Patterns of differentiation and hybridization in North American wolflike canids, revealed by analysis of microsatellite loci. *Molecular Biology and Evolution* 11 (4): 553–70.

Schierenbeck, K. A., G. L. Stebbins, and R. W. Patterson. 1992. Morphological and cytological evidence for polyphyletic allopolyploidy in *Arctostaphylos mewukka* (Ericaceae). *Plant Systematics and Evolution* 179 (3–4): 187–205.

Schonewald-Cox, C. M., S. M. Chambers, B. MacBryde, and L. Thomas, eds. 1983. *Genetics and conservation: A reference for managing wild animal and plant populations*. Menlo Park, CA: Benjamin Cummings.

Schuster, W. S., D. L. Alles, and J. B. Mitton. 1989. Gene flow in limber pine: Evidence from pollination phenology and genetic differentiation along an elevational transect. *American Journal of Botany* 76 (9): 1395–403.

Schuster, W. S. F., S. L. Phillips, D. R. Sandquist, and J. R. Ehleringer. 1992. Heritability of carbon isotope discrimination in *Gutierrezia microcephala* (Asteraceae). *American Journal of Botany* 79 (2): 216–21.

Schwartz, O. A., and K. B. Armitage. 1980. Genetic variation in social mammals: The marmot model. *Science* 207:665–67.

Seeliger, L. M. 1945. Variation in the pacific mud turtle. *Copeia* 1945 (3): 150–56.

Shaffer, H. B. 1984. Evolution in a paedomorphic lineage. 1. An electrophoretic analysis of the Mexican ambystomic salamanders. *Evolution* 38 (6): 1194–206.

Shaffer, H. B., and F. Breden. 1989. The relationship between allozyme variation and life history: Non-transforming salamanders are less variable. *Copeia* 1989 (4): 1016–23.

Shapiro, A. M. 1992. Twenty years of fluctuating parapatry and the question of competitive exclusion in the butterflies *Pontia occidentalis* and *P. protodice* (Lepidoptera: Pieridae). *Journal of the New York Entomological Society* 100 (2): 3111–319.

———. 1994. Observations presented at the SNEP Genetics Workshop, Placerville, CA, 22–23 September.

———. 1995. Unpublished data. University of California, Davis.

Shapiro, A. M., and H. Geiger. 1986. Electrophoretic confirmation of the species status of *Pontia protodice* and *P. occidentalis* (Pieridae). *Journal of Research in Lepidopterans* 25 (1): 39–47.

Shapiro, A. M., and C. Nice. N.d. Unpublished data. University of California, Davis.

Shevock, J. R. 1996. Status of rare and endemic plants. In *Sierra Nevada Ecosystem Project: Final report to Congress*, vol. II, chap. 24. Davis: University of California, Centers for Water and Wildland Resources.

Sibley, C. G., and K. W. Corbin. 1970. Ornithological field studies in the Great Plains and Nova Scotia. *Discovery* 6:3–6.

Sinervo, B. 1990. Evolution of thermal physiology and growth rate between populations of the western fence lizard *(Sceloporus occidentalis)*. *Oecologia* 83 (2): 228–37.

Sinervo, B., and S. C. Adolph. 1994. Growth plasticity and thermal opportunity in sceloporus lizards. *Ecology* 75 (3): 776–90.

Sinervo, B., P. Doughty, R. B. Mury, and K. Zamudis. 1992. Allometric engineering: A causal analysis of natural selection on offspring size. *Science* 258 (5090): 1927–30.

Sinervo, B., R. Hedges, and S. C. Adolph. 1991. Decreased sprint speed as a cost of reproduction in the lizard *Sceloporus occidentalis:* Variation among populations. *Journal of Experimental Biology* 155 (January): 323–36.

Slatkin, M. 1993. Isolation by distance in equilibrium and non-equilibrium populations. *Evolution* 47 (1): 264–79.

Smith, M. F. 1979. Geographic variation in genic and morphological characters in *Peromyscus californicus*. *Journal of Mammalogy* 60 (4): 705–22.

Smith, M. F., and J. L. Patton. 1980. Relationships of the pocket gopher *(Thomomys bottae)* populations in the lower Colorado River. *Journal of Mammalogy* 61:681–96.

———. 1988. Subspecies of pocket gophers: Causal bases for geographic differentiation in *Thomomys bottae*. *Systematic Zoology* 37 (2): 163–78.

Smith, R. H., and H. K. Preisler. 1988. Xylem monterpenes of *Pinus monophylla* in California and Nevada. *Southwestern Naturalist* 33 (2): 205–14.

Smith-Huerta, N. L. 1986. Isozymic diversity in three allotetraploid *Clarkia* species and their putative progenitors. *Journal of Heredity* 77:349–54.

Snajberk, K., E. Zavarin, and D. Baily. 1979. Systematic studies of *Pinus balfouriana* based on volatile terpenoids from wood and needles and on seed morphology. *Biochemical Systematics and Ecology* 7:269–79.

Soltis, D. E., M. S. Mayer, P. S. Soltis, and M. Edgerton. 1991. Chloroplast DNA variation within and among genera of the *Heuchera* group (Saxifragaceae): Evidence for chloroplast transfer and paraphyly. *American Journal of Botany* 78 (8): 1091–112.

Soltis, D. E., P. S. Soltis, and J. N. Thompson. 1992. Chloroplast DNA variation in *Lithophragma* (Saxifragaceae). *Systematic Botany* 17 (4): 607–19.

Soltis, P. S., and W. L. Bloom. 1986. Genetic variation and estimates of gene flow in *Clarkia speciosa* subsp. *polyantha* (Onagraceae). *American Journal of Botany* 73 (12): 1677–82.

Soltis, P. S., D. E. Soltis, and B. D. Ness. 1989. Population genetic structure in *Cheilanthes gracillima*. *American Journal of Botany* 76 (8): 1114–18.

Soltis, P. S., D. E. Soltis, and P. G. Wolf. 1990. Allozymic divergence in North American *Polystichum* (Dryopteridaceae). *Systematic Botany* 15 (2): 205–15.

Steinhoff, R. J., D. G. Joyce, and L. Fins. 1983. Isozyme variation in *Pinus monticola*. *Canadian Journal of Forest Research* 13:1122–32.

Stewart, R. E. 1993. *Use of native vegetative materials on national forests, and genetic guidelines for native plant collections*. San Francisco: U.S. Forest Service, Pacific Southwest Region.

Stock, M. W., G. D. Ammar, and B. J. Bentz. 1992. Isozyme studies of bark beetle population genetics and systematics. In *Proceedings of workshop on bark beetle genetics: Current status of research*, edited by J. L. Hayes and J. L. Robertson, 7–9. Berkeley, CA: U.S. Forest Service.

Stock, M. W., and J. L. Robertson. 1980. Inter- and intraspecific variation in selected *Choristoneura* species (Lepidoptera: Tortricidae): A toxicological and genetic survey. *Canadian Entomology* 112:1019–27.

Strauss, S. H. 1987. Heterozygosity and developmental stability under inbreeding and crossbreeding in *Pinus attenuata*. *Evolution* 41 (2): 331–39.

Strauss, S. H., and M. T. Conkle. 1986. Segregation, linkage, and diversity of allozymes in knobcone pine. *Theoretical and Applied Genetics* 72:483–93.

Strauss, S. H., Y. P. Hong, and V. D. Hipkins. 1993. High levels of population differentiation for mitochondrial DNA haplotypes in *Pinus radiata, muricata,* and *attenuata*. *Theoretical and Applied Genetics* 86 (5): 605–11.

Sturgeon, K. B. 1979. Monoterpene variation in ponderosa pine xylem resin related to western pine beetle predation. *Evolution* 33 (3): 803–14.

Svobda, P. L., J. R. Choate, and R. K. Chesser. 1985. Genetic relationships among southwestern populations of the Brazilian free-tailed bat. *Journal of Mammalogy* 66 (4): 444–50.

Thompson, P., and J. W. Sites Jr. 1986. Comparison of population structure in chromosomally polytypic and montypic species of *Sceloporus* (Sauria: Iguanidae) in relation to chromosomally mediated speciation. *Evolution* 40 (2): 303–14.

Thorgaard, G. H. 1992. Application of genetic technologies to rainbow trout. *Aquaculture,* no. 100:85–97.

Tsuji, J. S., R. B. Huey, F. M. Vanberkum, T. Garland, and R. G. Shaw. 1989. Locomotor performance of hatchling fence lizards *(Sceloporus occidentalis):* Quantitative genetics and morphometric correlates. *Evolutionary Ecology* 3 (3): 240–52.

U.S. Forest Service (USFS). 1994. *National Forest Genetic Electrophoresis Laboratory: A lab for monitoring genetic diversity.* San Francisco: U.S. Forest Service, Pacific Southwest Region.

van Staaden, M. J., R. K. Chesser, and G. R. Michner. 1994. Genetic correlations and matrilineal structure in a population of *Spermophilus richardsonii. Journal of Mammalogy* 75 (3): 573–82.

Van Wagner, C. E., and A. J. Baker. 1986. Genetic differentiation in populations of Canada geese *(Branta canadensis). Canadian Journal of Zoology* 64:940–47.

Vasek, F. C. 1977. Phenotypic variation and adaptation in *Clarkia* section *Phaeostoma. Systematic Botany* 2:251–79.

Vawter, A. T., and P. F. Brussard. 1984. Allozyme variation in a colonizing species: The cabbage butterfly *Pieris rapae* (Peiridae). *Journal of Research in Lepidopterans* 22 (3): 204–16.

Vogelmann, J. E., and G. J. Gastony. 1987. Electrophoretic enzyme analysis of North American and eastern Asian populations of *Agastache* sect. *Agastache* (Labiatae). *American Journal of Botany* 74 (3): 385–93.

Vogler, D. R., B. B. Kinloch, F. W. Cobb, and T. L. Popenuck. 1991. Isozyme structure of *Pteridermium harknessii* in the western United States. *Canadian Journal of Botany* 69:2434–41.

Wade, M. J., M. L. McKnight, and H. B. Shaffer. 1994. The effects of kin–structured colonization on nuclear and cytoplasmic genetic diversity. *Evolution* 48 (4): 1114–20.

Wake, D. B. 1994. Observations presented at the SNEP Genetics Workshop, Placerville, CA, 22–23 September.

Wake, D. B., L. R. Maxson, and G. Z. Wurst. 1978. Genetic differentiation, albumin evolution, and their biogeographic implications in plethodontid salamanders of California and southern Europe. *Evolution* 32 (3): 529–39.

Wake, D. B., and K. P. Yanev. 1986. Geographic variation in allozymes in a "ring species," the plethodontid salamander *Ensatina escholtzzii* of western North America. *Evolution* 40 (4): 702–15.

Watt, W. B., R. C. Cassin, and M. S. Swan. 1983. Adaptation at specific loci. 3. Field behavior and survivorship differences among *Colias* PGI genotypes are predictable from *in vitro* biochemistry. *Genetics* 103:725–39.

Westfall, R. D. 1995. Unpublished allozyme data on sugar pine. U.S. Forest Service, Pacific Southwest Research Station, Berkeley, CA.

Westfall, R. D., and M. T. Conkle. 1992. Allozyme markers in breeding zone designation. *New Forestry* 6:279–309.

Westfall, R. D., M. T. Conkle, and F. T. Ledig. N.d. Unpublished nursery data on ponderosa pine.

White, E. E. 1990. Chloroplast DNA in *Pinus monticola.* 2: Survey of within–species variability and detection of heteroplasmic individuals. *Theoretical and Applied Genetics* 79 (2): 251–55.

White, M. M. 1988. Age class and population genic differentiation in *Pteronarcys proteus* (Plecoptera: Pteronarcyidae. *American Midland Naturalist* 122:242–48.

Wilson, B. C. 1990. Gene pool reserves of Douglas–fir. *Forest Ecology Management* 35 (1–2): 121–30.

Wolf, P. G., and P. S. Soltis. 1992. Estimates of gene flow among populations, geographic races, and species in the *Ipomopsis aggregata* complex. *Genetics* 130 (3): 639–47.

Wolf, P. G., P. S. Soltis, and D. E. Soltis. 1991. Genetic relationships and patterns of allozymic divergence in the *Ipomopsis aggregata* complex and related species (Polemoniaceae). *American Journal of Botany* 78 (4): 515–26.

Wright, S. 1978. *Evolution and the genetics of populations.* Chicago: University of Chicago Press.

Yanev, K. P., and D. B. Wake. 1981. Genetic differentiation in a relict desert salamander, *Batrachoseps campi. Herpetologica* 37 (1): 16–28.

Zavarin, E., L. G. Cool, and K. Snajberk. 1993. Geographic variability of *Pinus flexilis* xylem monoterpenes. *Biochemical Systematics and Ecology* 21 (3): 381–87.

Zavarin, E., W. B. Critchfield, and K. Snajberk. 1978. Geographic differentiation of monoterpenes from *Abies procera* and *Abies manifica. Biochemical Systematics and Ecology* 6:267–78.

Zavarin, E., Z. Rafii, L. G. Cool, and K. Snajberk. 1991. Geographic monoterpene variability of *Pinus albicaulis. Biochemical Systematics and Ecology* 19 (2): 147–56.

Zavarin, E., et al. 1982. Variability in essential oils and needle resin canals of *Pinus longaeva* from eastern California and western Nevada in relation to other members of subsection *Balfourianae. Biochemical Systematics and Ecology* 10 (1): 11–20.

Zavarin, E., K. Snajberk, and L. Cool. 1990a. Chemical differentiation in relation to the morphology of the single–needle pinyons. *Biochemical Systematics and Ecology* 18 (2–3): 125–37.

———. 1990b. Monoterpene variability of *Pinus monticola* wood. *Biochemical Systematics and Ecology* 18 (2–3): 117–24.

Zeiner, D. C., W. F. Laudenslayer Jr., and K. E. Mayer, eds. 1988. *Amphibians and reptiles.* Vol. 1 of *California's wildlife.* Sacramento: California Department of Fish and Game.

Zeiner, D. C., W. F. Laudenslayer Jr., K. E. Mayer, and M. White, eds. 1990a. *Birds.* Vol. 2 of *California's wildlife.* Sacramento: California Department of Fish and Game.

———. 1990b. *Mammals.* Vol. 3 of *California's wildlife.* Sacramento: California Department of Fish and Game.

Zera, A. J. 1981. Genetic structure of two species of waterstriders (Gerridae: Hemipetera) with differing degrees of winglessness. *Evolution* 33 (3): 218–25.

Zink, R. M. 1986. *Patterns and evolutionary significance of geographic variation in the Chistaceae group of the fox sparrow* (Passerella iliaca). Ornithological Monographs 40. Washington, DC: American Ornithologists' Union.

———. 1994. The geography of mitochondrial DNA variation, population structure, hybridization, and species limits in the fox sparrow *(Passerella iliaca). Evolution* 48 (1): 98–111.

Zink, R. M., and D. L. Dittman. 1991. Evolution of brown towhees: Mitochondrial DNA evidence. *Condor* 93:98–105.

———. 1993a. Gene flow, refugia, and evolution of geographic variation in the song sparrow *(Melospiza melodia). Evolution* 47 (3): 717–29.

———. 1993b. Population structure and gene flow in the chipping sparrow and a hypothesis for evolution in the genus *Spizella. Wilson Bulletin* 105 (3): 399–413.

Zink, R. M., D. F. Lott, and D. W. Anderson. 1987. Genetic variation, population structure, and evolution of California quail. *Condor* 89:395–405.

Zink, R. M., and D. W. Winkler. 1983. Genetic and morphological similarity of two California gull populations with different life history traits. *Biochemical Systematics and Ecology* 11:397–403.

APPENDIX 28.1

Population Genetic Studies for Plant Species Native to the Sierra Nevada

Taxon	Type of study[1]	Sampling range	Source/author(s)
PLANTS			
Trees			
Gymnosperms			
Abies concolor	M[a]	Elev. transect, Eldorado Co.	Hamrick (1976)
	M[a]	Range-wide, including SN	Hamrick & Libby (1972);
			Libby et al (1980)
			PSW Region Genetic Regeneration Program (n.d.)
	M[a]	Northern & central SN	Jenkinson (n.d.-b)
	A[a]	Northern & central SN	Westfall and Conkle (1992)
A. magnifica	B	Sierra Nevada	Zavarin et al. (1978)
	M[a]	Northern & central SN	Sorensen et al. (1990)
			Jenkinson (n.d.-b)
Calocedrus decurrens	A & M[a]	Includes Sierran populations	Harry (1984)
	M[a]	Includes Sierran populations	Rogers et al. (1994)
Cupressus spp.		Includes Sierran species	Millar and Delany (n.d.)
Juniperus osteosperma	B*		Adams (1994)
Pinus albicaulis	A*		Furnier et al. (1987)
	B	Rangewide, including SN	Zavarin et al. (1991)
P. attenuata	A	Includes Sierran populations	Strauss and Conkle (1986)
	A & M[a]	Includes Sierran populations	Strauss (1987)
	A	Rangewide, including SN	Millar et al. (1988)
	D	Includes Sierran populations	Hong et al. (1993);
			Strauss et al. (1993)
P. balfouriana	B & M	Rangewide, including SN	Snajberk et al. (1979)
	M	Rangewide, including SN	Mastrogiuseppe (1972);
			Mastrogiuseppe and Mastrogiuseppe (1980)
P. contorta	M & P	Rangewide, including SN	Critchfield (1956)
P. flexilis	A*		Schuster et al. (1989)
	B	Rangewide, including SN	Zavarin (1993)
P. jeffreyi	A	CA, including SN	Furnier and Adams (1986)
	A	Mono Co.	Millar et al. (1993)
P. lambertiana	A	Rangewide, including SN	Conkle (n.d.)
	A[a]	Sierra Nevada	Westfall and Conkle (1992)
	M[a]	Elevational transect	Harry et al. (1983)
	M[a]	Rangewide, including SN	Jenkinson (n.d.-a)
	M[a]	Northern SN	Kitzmiller and Stover (in press)
P. longaeva	B & M	Includes Californian populations	Zavarin et al. (1982)
P. monophylla	A	Includes Californian populations	Delany (n.d.)
	B	California	Smith and Preisler (1988)
	B	Includes Californian populations	Zavarin et al. (1990a)
P. monticola	A	Rangewide, including SN	Steinhoff et al. (1983)
	B	Rangewide, including SN	Zavarin et al. (1990b)
	D*		White (1990)
P. ponderosa	A & M	Includes Sierran population	Linhart et al. (1989)
	M	Includes Sierran population	Grant et al. (1989)
	P	Includes Sierran population	Monson and Grant (1989)
	A[a]	Sierra Nevada	Westfall and Conkle (1992)
			PSW Region Genetic Resources Program n.d.
	M[a]	Elevational transect	Conkle (1973)
	M[a]	Sierra Nevada	Westfall et al. (n.d.)
P. sabiniana	M		Griffin (1965)
P. washoensis	A	California and Oregon	Niebling and Conkle (1990)

continued

Taxon	Type of study[1]	Sampling range	Source/author(s)
Pseudotsuga menziesii	A[a] M[a]*	Sierra Nevada	Westfall and Conkle (1992) Campbell (1986)
	M[a]	Northern SN	N. Sierra Tree Improvement Assoc. (n.d)
Sequoiadendron gigantea	A & M[a]	Sierra Nevada	Fins and Libby (1982)
	M	Sierra Nevada	Mahalovich (1985)
Taxus brevifolia	A & B	Rangewide (in USFS locations), includes SN	Doede et al. (1995)
Angiosperms			
Acer negundo	P*		Dawson and Ehleringer (1993)
Salix spp.	A	Includes Sierran spp.	Brunsfeld et al. (1991);
	D		Brunsfeld et al. (1992)
Populus tremuloides	A*		Jelinski and Cheliak (1992)
	D*		Rogstad et al. (1991);
			Chong et al. (1994)
P. trichocarpa	M[a]*		Rogers et al. (1989); Dunlap et al. (1994)
	P*		Dunlap et al. (1993)
Quercus chrysolepsis			Riggs?
Q. douglasii	A	Rangewide, including SN	Millar et al. (1990)
Q. kelloggii			Riggs?
Q. wislizenii	A*		Nason et al. (1992)
	B	Includes SN	Dodd et al. (1993)
Non-Tree Angiosperms			
Achillea	M[a]	Sierran populations	Clausen et al. (1948)
Agastache spp.	A	Includes Sierran species	Vogelmann and Gastony (1987)
Antennaria corymbosa	A	Sierran populations	Bayer (1988)
A. media	A		Bayer (1989b)
A. rosa	A		Bayer (1989a; 1990);
Arctostaphylos spp.	M	Yosemite	Ball et al. (1983)
	A	Yosemite	Ellstrand et al. (1987)
A. mewukka	M & M[b]	Sierra Nevada	Schierenbeck et al. (1992)
Arabis holbolellii	A & M[2]*		Roy (1993)
Artemisia spp.	M[b]	Includes Sierran species	McArthur et al. (1981)
A. tridenatat	M/P*		Meyer et al (1990); Meyer and Monson (1991)
	B & M*		Freeman et al. (1991)
Aquilegia spp.	D	Sierra Nevada	Hodges and Arnold (1994)
Bromus carinatus	M		Luedke (as cited at SNEP wksp)
	M[a]	Includes Sierran populations	Flowers and Rice (1994)
B. tectorum	A	Includes Sierran populations	Novak et al. (1991)
	M[a]*		Rice and Mack (1991a; 1991b; 1991c)
Calamagrostis canadensis	A*		MacDonald and Lieffers (1991)
Calchortus spp.	P	Includes Sierran spp.	Fiedler (1985)
	M	Includes Sierran spp.	Ness (1989)
	A	Includes Sierran spp.	Ness et al. (1990)
Calycadenia	D & M/M[b]	Includes Sierran spp.	Baldwin (1993)
Carpenteria californica	A & M		Clines (1994)
Ceanothus			
Clarkia spp.	M	Includes Sierran spp.	Vasek (1977)
	M		Baldwin (as cited at SNEP wksp)
	A	Includes Sierran spp.	Smith-Huerta (1986); Holsinger and Gottlieb (1988)
C. speciosa	A	Kern Co.	Soltis and Bloom (1986)
Danthonia californica	A	Includes Sierran populations	Knapp and Rice (1994a)
Elymus glaucus	A	Includes Sierran populations	Knapp and Rice (1995)
Eriastrum densifolium	M	Includes Sierran populations	Patterson and Tanowitz (1989)
Eriophyllum confertiflorum	M[b]	Includes Sierran populations	Mooring (1994)
Hymenoclea salsola	P*		Comstock and Ehleringer (1992)
Ipomopsis aggregata	A	Includes Sierran populations	Wolf et al. (1991); Wolf and Soltis (1992)
Lewisia spp.	A	Sierra Nevada	Carroll et al. (n.d.)
Lithophragma spp.	D	Includes Sierran spp.	Soltis et al. (1992)
Lupinus	M		Harding (as cited at SNEP wksp)
Plantago	M		Stebbins (as cited at SNEP wksp)
Nassella pulchra	M & A	Includes Sierran populations	Knapp and Rice (1994b)
Polemonium	M		Pritchet (as cited at SNEP wksp)
Potentilla	M		Knapp, Rice (as cited, SNEP wksp)
Stephanomeria spp.	B	Includes Sierran spp.	Bohm and Gottlieb (1989)
Scutellaria bolanderi	A	Includes Sierran populations	Olmstead (1990)
S. californica	A	Includes Sierran populations	Olmstead (1990)
S. nana	A	Includes Sierran populations	Olmstead (1990)
S. siphocampyloides	A	Includes Sierran populations	Olmstead (1990)
Tellima grandiflora	D	Includes Sierran population	Soltis et al. (1991)
Vulpia microstachys	M	Includes Sierran populations	Kannenberg and Allard (1967) Allard and Kannenberg (1968)
Wyethia	A, M		Ayers (as cited at SNEP wksp)

Taxon	Type of study[1]	Sampling range	Source/author(s)
Ferns			
Cheilanthes gracillima	A	Includes Sierran populations	Soltis et al. (1989)
Polystichum spp.	A*		Soltis et al. (1990)
ANIMALS			
Mammals			
Tadarida brasilliensis	A	SW US (not including SN)	Svobda et al. (1985)
	A	SW US (not including SN)	McCracken et al. (1994)
Marmota flaviventris	A	East R.Valley, CO	Schwartz and Armitage (1980)
Thomomys bottae	A	SW U.S.A., including SN	Patton and Yang (1977)
	A	Lower Colorado River	Smith and Patton (1980)
	A & M	California, including SN	Patton and Smith (1990)
	D[3]	Not given (single sample)	Hafner et al. (1994)
Dipodomys agilis	A	Western U.S.A.	Johnson and Selander (1971)
D. deserti	A	Western U.S.A.	Johnson and Selander (1971)
D. heermanni	A	Western U.S.A.	Johnson and Selander (1971)
D. merriami	A	Western U.S.A.	Johnson and Selander (1971)
D. microps	A	Western U.S.A.	Johnson and Selander (1971)
	A	Butte Co.	Patton et al. (1976)
D. nitratoides	A	Western U.S.A.,	Johnson and Selander (1971)
D. ordii	A	Western U.S.A.	Johnson and Selander (1971)
D. panamintinus	A	Western U.S.A., including SN	Johnson and Selander (1971)
	A	Kern Co.	Patton et al. (1976)
Peromyscus maniculatus	A	U.S.A. & Canada, including SN	Avise et al. (1979)
	D	U.S.A. & Canada, including SN	Lansman et al. (1983)
	D	U.S.A. & Canada, including SN	Neigel and Avise (1993)
	M	Arizona & Nevada	Thompson (1990)
P. californicus	A	Coastal CA, Northern Baja CA and foothills of SN	Smith (1979)
Onychomys spp.	D	Western, including CA populations	Riddle et al. (1990)
Microtus californicus	A	Calif. coast range	Bowen (1982)
	M		Lidicker and Ostfeld (1991)
Canus latrans	A	Zoo	Fisher et al. (1976)
	A[4]	Not known	Wayne and O'Brien (1987)
C. latrans	D	Southern Calif.	Roy et al. (1994)
Vulpes macrotis	A	Western U.S.A., including SN	Dragoo et al. (1990)
Ursus americana	A	Yosemite NP	Manlove et al. (1980)
	D	E & NW US	Cronin et al. (1991a)
Martes americana	A	Wyoming	Mitton and Raphael (1990)
Odocileus hemionus			
hemionus	A*	Colorado	Scribner et al. (1991)
	A & D	Western US	Cronin (1991)
		Montana	Cronin et al. (1991b)
	A	Western US (including SN)	Derr (1991)
	D	Western US	Cronin (1992)
O. h. columbianus	A & D	AK & OR, US; BC, Canada	Cronin (1991)
	A	California (one population)	Derr (1991)
	D	Pacific Coast	Cronin (1992)
Amphibians			
Ambystoma macrodactylum	A*	Oregon and Idaho	Howard and Wallace (1981)
A. tigrinum	A[4]	California	Shaffer (1984)
Aneides flavipunctatus	A & M	California	Larson (1980)
A. lugubris		Sierra Nevada	Jackman (n.d.)
Batrachoseps campi	A	Inyo Mountains, CA	Yanev and Wake (1981)
Other *Batrachosep* spp		S. Sierra Nevada	Yanev 1978
Elgaria		Sierra Nevada	Good (1988)
Ensatina eschsholtzii	A	California, including SN	Wake and Yanev (1986)
			Jackman and Wake (1994)
			Wake et al. (1989)
Hydromantes brunus	A	Mariposa Co., California	Wake et al. (1978)
H. platycephalus	A	Toulumne Co., California	Wake et al. (1978)
H. shastae	A	Shasta Lake, California	Wake et al. (1978)
Hyla regilla	A	Oregon and CA	Case et al. (1975)
Rana aurora	A	Californian coastal ranges	Case (1978a)
R. boylei	A	California	Case (1978a)
R. boylei	A	California	Case (1978b)
R. cascadae	A	Lassen County, California	Case (1978a)
R. catesbeinana	A	California	Case (1978a)
R. muscosa	A	Sierra Nevada	Case (1978a)
R. muscosa	A	Sierra Nevada	Case (1978b)
Taricha torosa	A	Sierra Nevada	Hedgecock and Ayala (1974)
			Tan (1995)

continued

Taxon	Type of study[1]	Sampling range	Source/author(s)
Reptiles			
Anniella pulchra	A	Coastal CA	Bezy et al. (1977)
Elgaria coerulea	A		Good (1988)
E. multicarinata	A		Good (1988)
E. panamintina	A		Good (1988)
Sceloporus graciosus	A*	Five western states	Thompson and Sites (1986)
Suaromalus obesus	D*	Southwestern deserts	Lamb et al. (1992)
Uta stansburiana	A	California	McKinney et al. (1972)
Xerobates agassizi	D	Southwestern deserts	Lamb et al. (1989)

[1]Type of study:
M = morphological (morphological and/or phenological characteristics) data; M[a] = based on common-garden studies (and therefore NOT including plasticity); M[b] = based on cytological data.
A = allozyme data, single-locus data analysis; A[a] = allozyme data analyzed as multi-locus phenotypes.
B = biochemical data.
D = DNA data (RFLP, RAPD or PCR-based).
P = physiological studies.
[2]Infections by pathogens.
[3]Host-parasite systematics.
[4]Systematic study
Data include samples taken from the Sierra Nevada, unless otherwise indicated.An asterisk (*) indicates that samples were mostly or entirely outside of the Sierra Nevada portion of the species' range.

REFERENCES

Adams, R. P. 1994. Geographic variation in the volatile terpenoids of *Juniperus monosperma* and *J. osteosperma*. *Biochem. Syst. Ecol.* 22(1):65-71.

Allard, R. W., and L. W. Kannenberg. 1968. Population studies in predominantly self pollinated species.XI.Genetic divergence among the members of the *Festuca microstachys* complex. *Evolution* 22:517–528.

Allendorf, F. W. 1975. *Genetic variability in a species possessing extensive gene duplication; genetic interpretation of duplicate loci and examination of genetic variation in populations of rainbow trout.* Ph.D. dissertation, University of Washington, Seattle.

Allendorf, F. W., and R. F. Leary. 1988. Conservation and distribution of genetic variation in a polytypic species, the cutthroat trout. *Conservation Biol.* 2:170–184.

Allendorf, F. W., W. Rymar, and F. M. Utter. 1987. Genetics and Fishery management. In: N. Rymar and F. Utter (eds.). *Population Genetics and Fishery Management.* University of Washington Press, Seattle.

Anderson, R. S. 1990. Holocene forest development and paleoclimates within the central Sierra Nevada, California. *Journal of Ecology* 78:470-489.

Anderson, W. T. 1987. *To Govern Evolution: Further Adventures of the Political Animal.* Harcourt Brace Jovanovich, Orlando, Florida.

Arnold, R. A. 1983a. Ecological studies of six endangered butterflies: Island biogeography, patch dynamics, and design of habitat preserves. *Univ. Calif. Pub. Entomol.* 99:1-161.

Arnold, R. A. 1983b. *Speyeria callippe* (Lepidoptera: Nymphalidae): Application of information-theoretical and graph-clustering techniques to analyses of geographic variation and evaluation of classifications. *Ann. Entom. Soc. Am.*76:929-941.

Arnold, R. A. 1985. Geographic variation in natural populations of *Speyeria callippe* Boisduval (Lepidoptera: Nymphalidae). *Pan-Pac. Ent.* 61:1-23.

Arnold, S. J. 1980a. Behavioral variation in natural populations. I. Phenotypic, genetic and environmental correlations between chemoreactive responses to prey in the garter snake, *Thamnophis elegans. Evolution* 35(3):489-509.

Arnold, S. J. 1980b. Behavioral variation in natural populations. II. The inheritance of a feeding response in crosses between geographic races of the garter snake, *Thamnophis elegans. Evolution* 35(3):510-515.

Avise, J. C., M. H. Smith, and R. K. Selander. 1979. Biochemical polymorphism and systematics in the genus *Peromyscus*. VII. Geographic differentiation in members of the *truei* and *maniculatus* species groups. *J. Mammal.* 60(1):177-192.

Baker, M. C. 1975. Song dialects and genetic differences in white-crowned sparrows (*Zonotrichia leucophrys*). *Evolution* 29:226-241.

Baldwin, B. G. 1993. Molecular phylogenetics of *Calycadenia* (Compositae) based on its sequences of nuclear ribosomal DNA: Chromosomal and morphological evolution reexamined. *Am. J. Botany* 80(2):222-238.

Ball, C. T., et al. 1983. Relationship between form, function, and distribution of two *Arctostaphylos* species (Ericaceae) and their putative hybrids. *Acta Oecologica* 4(2):153-164.

Barrowclough, G. F. 1980. Gene flow, effective population sizes, and genetic variance components in birds. *Evolution* 34(4):789-798.

Barrowclough, G. F., and R. J. Guttiérrez. 1990. Genetic variation and differentiation in the spotted owl (*Strix occidentalis*). *Auk* 107:737-744.

Barrowclough, G. F., and N. K. Johnson. 1988. Genetic Structure of North American Birds. pp. 1630-1638. In: H. Ouellet (ed.). *Acta XIX Congr. Int. Ornithol.* II. Univ. Ottawa Press.

Bartley, D. M., et al. 1992. Population genetic structure of coho salmon (*Oncorhynchus kisutch*) in California. *California Fish and Game* 78(3):88-104.

Bartley, D. M., and G. A. E. Gall. 1990. Genetic structure and gene flow in Chinook salmon populations of California. *Trans. Amer. Fish. Soc.* 119:55-71.

Bartley, D. M., G. A. E. Gall, and B. Bentley. 1990. Biochemical genetic detection of natural and artificial hybridization of Chinook and coho salmon in northern California. *Trans. Amer. Fish. Soc.* 119: 431-437.

Baughman, J. F., et al. 1990. History, selection, drift, and gene flow: complex differentiation in checkerspot butterflies. *Can. J. Zool.* 68:1967-1975.

Bayer, R. J. 1988. Patterns of isozyme variation in western North American *Antennaria* (Asteraceae: Inuleae). I. Sexual species of sect. *Dioicae. Syst. Bot.* 13(4):525-537.

Bayer, R. J. 1989a. Patterns of isozyme variation in the *Antennaria rosea* (Asteraceae, Inuleae) polyploid agamic complex. *Syst. Bot.* 14(3):389-397.

Bayer, R. J. 1989b. Patterns of isozyme variation in western North American *Antennaria* (Asteraceae: Inuleae). II. Diploid and polyploid species of section *Alpinae. Am. J. Botany* 76(5):679-691.

Bayer, R. J. 1990. Patterns of clonal diversity in the *Antennaria rosea* (Asteraceae) polyploid agamic complex. *Am. J. Botany* 77(10):1313-1319.

Bechtel, E. R., and T. Whitecar. 1983. Genetics of striping in the gopher snake, *Pituophis melanoleucus. J. Herpetol.* 17(4):362-370.

Berg, W. J., and G. A. E. Gall. 1988. Gene flow and genetic differentiation among California coastal rainbow trout populations. *Can. J. Fish. Aquat. Sci.* 45(1):122-131.

Bezy, R. L., et al. 1977. Chromosomal and genetic divergence in the fossorial lizards of the family Anniellidae. *Syst. Zool.* 26:57-71.

Bohm, B. A., and L. D. Gottlieb. 1989. Flavonoids of the annual *Stephanomeria* (Asteraceae). *Biochem. Syst. Ecol.* 17(6):451-453.

Bowen, B. S. 1982. Temporal dynamics of microgeographic structure of genetic variation in *Microtus californicus. J. Mammal.* 63(4):625-638.

Brennan, L. A., and M. L. Morrison. 1991. Long-term trends of chickadee populations in western North America. *Condor* 93:130-137.

Britten, H. B., P. F. Brussard, and D. D. Murphy. 1994a. The pending extinction of the Uncompahgre fritillary butterfly. *Conservation Biol.* 8(1):86-94.

Britten, H. B., et al. 1994b. Colony isolation and isozyme variability of the western seep fritillary, *Speyeria nokomis apacheana* (Nymphalidae), in the western Great Basin. *Gr. Basin Nat.* 54(2):97-105.

Bruns, T. D., T. J. White, and J. W. Taylor. 1991. Fungal molecular systematics. *Ann. Rev. Ecol. Syst.* 22:525-564.

Brunsfeld, S. J., D. E. Soltis, and P. S. Soltis. 1991. Patterns of genetic variation in *Salix* section *longfoliae* (Salicaceae). *Am. J. Botany* 78(6):855-869.

Brunsfeld, S. J., D. E. Soltis, and P. S. Soltis. 1992. Evolutionary patterns and processes in *Salix* sect *longifoliae*—evidence from chloroplast DNA. *Syst. Bot.* 17(2):239-256.

Buck, J. M., et al. 1970. California tree seed zones. Misc. Pub.USDA Forest Service.

Burdon, J. J. 1986. Isozymic variation in *Puccinia graminis* f.sp. *tritici* detected by starch-gel electrophoresis. *Plant Disease* 70:1139-1141.

Burdon, J. J., and A. P. Roelfs. 1985a. The effect of sexual and asexual reproduction on the isozyme structure of populations of *Puccinia graminis. Phytopathology* 75(9):1068-1073.

Burdon, J. J., and A. P. Roelfs. 1985b. Isozyme and virulence variation in asexually reproducing populations of *Puccinia graminis* and *P. reconditaon* wheat. *Phytopathology* 75(8):907-913.

Burns, K. J., and R. M. Zink. 1990. Temporal and geographic homogeneity of gene frequencies in the fox sparrow (*Passerella iliaca*). *Auk* 107:421-424.

Buth, D. G., T. R. Haglund, and W. L. Minckley. 1992. Duplicate gene expression and allozyme divergence diagnostic for *Catostomus tahoensis* and the endangered *Chasmistes cujus* in Pyramid Lake, Nevada. *Copeia* 1992(4):935-941.

Campbell, R. K. 1986. Mapped genetic variation of Douglas-fir to guide seed transfer in southwest Oregon. *Silvae Genet.* 35(2–3):85-96.

Cane, J. H., et al. 1990. Phylogenetic relationships of *Ips* bark beetles (Coleoptera: Scolytidae): Electrophoretic and morphometric analyses of the *grandicollis* group. *Journal of Systematics and Ecology* 18(5):359-368.

Carroll, E., M. Foster, and V. Hipkins. n.d. Unpublished data on *Lewisia* spp.

Case, S. M. 1978a. Biochemical systematics of members of the genus *Rana* native to western North America. *Syst. Zool.* 27:299-311.

Case, S. M. 1978b. Electophoretic variation in two species of ranid frogs, *Rana boylei* and *R. muscosa. Copeia* 1978:311-320.

Case, S. M., P. G. Haneline, and M. F. Smith. 1975. Protein variation in several species of *Hyla. Syst. Zool.* 24:281-295.

Chong, D. K. X., R. C. Yang, and F. C. Yeh. 1994. Nucleotide divergence between populations of trembling aspen (*Populus tremuloides*) estimated with RAPDs. *Current Genetics* 26(4):374-376.

Cicero, C., and N. K. Johnson. 1992. Genetic differentiation between populations of Hutton's vireo (Aves: Vireonidae) in disjunct allopatry. *SW Nat.* 37(4):344-348.

Clausen, J., D. D. Keck, and W. M. Heisey. 1948. Experimental studies on the nature of species. III. Environmental responses of climatic races of *Achillea.*No. 581. Carnegie Institute of Washington.

Clemens, D. T. 1990. Interspecific variation and effects of altitude on blood properties of rosy finches (*Leucosticte arctoa*) and house finches (*Carpodacus mexicanus*). *Physiological Zoology* 63(2):288-307.

Clines, J. M. 1994. *Reproductive ecology of Carpenteria californica.* M.A. thesis, California State Univ., Fresno.

Clines, J. M. 1995. Data mailed to R. Westfall.

Collins, M. M. 1984. Genetics and ecology of a hybrid zone in *Hyalophora* (Lepidoptera: Saturniidae). *Univ. Calif. Pub. Entomol.* 104:1-93.

Comstock, J. P., and J. R. Ehleringer. 1992. Correlating genetic variation in carbon isotopic composition with complex climatic gradients. *Proc. Nat. Acad. Sci. USA* 89(16):7747-7751.

Conkle, M. T. 1973. Growth data for 29 years from the California elevational transect study of ponderosa pine. *For. Sci.* 19:31-39.

Conkle, M. T. 1992. Genetic diversity—seeing the forest through the trees. *New For.* 6:5-22.

Conkle, M. T. n.d. Unpublished allozyme data on sugar pine.

Conkle, M. T., and W. B. Critchfield. 1988. Genetic variation and hybridization of ponderosa pine. pp. 27-43. *Ponderosa Pine: The Species and its Management.* Washington State Univ. Coop. Ext., Pullman.

Corbin, K. W., C. G. Sibley, and A. Ferguson. 1979. Genic charges associated with the establishment of sympatry in orioles of genus *Icterus. Evolution* 33:624-633.

Critchfield, W. B. 1956. *Morphological and physiological variation in Pinus contorta Dougl.* Ph.D. dissertation, University of California, Berkeley.

Critchfield, W. B. 1986. Impact of the Pleistocene on the genetic structure of North American conifers. pp. 70-118. In: R. M. Lanner (ed.) *Proceedings of 8th North Amer. Forest Biol. Workshop.* Logan, UT.

Cronin, M. A. 1991. Mitochondrial and nuclear genetic relationships of deer (*Odocoileus* spp) in western North America. *Can. J. Zool.* 69(5):1270-1279.

Cronin, M. A. 1992. Intraspecific variation in mitochondrial DNA of North American cervids. *J. Mammal.* 73(1):70-82.

Cronin, M. A., et al. 1991a. Interspecific and intraspecific mitochondrial DNA variation in North American bears (Ursus). Can. J. Zool. 69(12):2985-2992.

Cronin, M. A., M. E. Nelson, and D. F. Pac. 1991b. Spatial heterogeneity of mitochondrial DNA and allozymes among populations of white-tailed deer and mule deer. J. Heredity 82(2):118-127.

Crow, T. R., A. W. Haney, and D. M. Waller. 1994. Report on the scientific roundtable on biological diversity convened by the Chequamegon and Nicolet National Forests. General technical report.NC 166. U.S. Dept. of Agriculture, Forest Service, North Central Forest Experiment Station.

Daly, J. C., and J. L. Patton. 1990. Dispersal, gene flow, and allelic diversity between local populations of Thomomys bottae pocket gophers in the coastal ranges of California. Evolution 44(5):1283-1294.

Dawson, T. E., and J. R. Ehleringer. 1993. Gender-specific physiology, carbon isotope discrimination, and habitat distribution in boxelder, Acer negundo. Ecology 74(3):798-815.

Delany, D. L. n.d. Unpublished allozyme data on Pinus monophylla and P. edulis.

Derr, J. N. 1991. Genetic interactions between white-tailed and mule deer in the southwestern United States. J. Wildlife Manag. 55(2):228-237.

Desrochers, A. M., and B. A. Bohm. 1993. Flavonoid variation in Lasthenia californica (Asteraceae). Biochem. Syst. Ecol. 21(4):449-453.

Dirig, R., and J. F. Cryan. 1991. The status of silvery blue subspecies (Glaucopsyche lygdamus and G.L. couperi: Lycaenidae) in New York. J. Lepid. Soc. 45(4):272-280.

Dodd, R. S., Z. A. Rafii, and E. Zavarin. 1993. Chemosystematic variation in acorn fatty acids of Californian live oaks (Quercus agrifolia and Q. wislizenii). Biochem. Syst. Ecol. 21(2):279-285.

Doede, D. L., et al. 1995. Variation in allozymes, taxol, and propagation by rooted cuttings in Pacific yew. In: Proceedings of Int. Yew Resources Conf. Berkeley, CA. In press.

Donovan, L. A., and J. R. Ehleringer. 1992. Contrasting water-use patterns among size and life-history classes of a semi-arid shrub. Funct. Ecol. 6(4):482-488.

Donovan, L. A., and J. R. Ehleringer. 1994. Carbon isotope discrimination, water-use efficiency, growth, and mortality in a natural shrub population. Oecologia 100(3):347-354.

Dragoo, J. W., et al. 1990. Evolutionary and taxonomic relationships among North American arid-land foxes. J. Mammal. 71(3):318-332.

Dunlap, J. M., et al. 1993. Intraspecific variation in photosynthetic traits of Populus trichocarpa. Can. J. Bot. 71(10):1304-1311.

Dunlap, J. M., P. E. Heilman, and R. F. Stettler. 1994. Genetic variation and productivity of Populus trichocarpa and its hybrids. 7. Two-year survival and growth of native black cottonwood clones from four river valleys in Washington. Can. J. For. Res. 24(8):1539-1549.

Ehleringer, J. R., et al. 1991. Differential utilization of summer rains by desert plants. Oecologia 88(3):430-434.

Ellstrand, E., et al. 1987. Ecological isolation and introgression in an Arctostaphylos (Ericaceae) population. Acta Oecologica 8(4):299-308.

Ellstrand, N. C., and M. Roose. 1987. Patterns of genotypic diversity in clonal plant species. Am. J. Botany 74(1):123-131.

Elvingson, P., and K. Johansson. 1993. Genetic and environmental components of variation in body traits of rainbow trout (Oncorhynchus mykiss) in relation to age. Aquaculture 118(1993):191-204.

Evans, R. D., and J. R. Ehleringer. 1994. Water and nitrogen dynamics in an arid woodland. Oecologia 99(3-4):233-242.

Falk, D., and K. Holsinger (eds.). 1991. Genetics and Conservation of Rare Plants. Oxford University Press, Cary, NC.

Ferguson, M. M. 1990. The genetic impact of introduced fishes on native species. Can. J. Zool. 68:1053-1057.

Ferrell, G. T., W. D. Bedard, and R. D. Westfall. 1989. Geographic variation in Pinus ponderosa susceptibility to the gouty pitch midge, Cecidomyia piniinopsis, in the Sierra Nevada and southern Cascade Mountains of California. pp. 205-212. In: R. I. Alfano and S. G. Glover (eds.). Insects Affecting Reforestation: Biology and Damage Forestry Canada, Pacific Forestry Centre, Victoria, BC.

Fiedler, P. L. 1985. Heavy metal accumulation and the nature of edaphic endemism in the genus Calochortus (Liliaceae). Am. J. Botany 72(11):1712-1718.

Fins, L., and W. J. Libby. 1982. Population variation in Sequoiadendron: Seed and seedling studies, vegetative propagation, and isozyme variation. Silvae Genetica 31(4):102-109.

Fisher, R. A., W. Putt, and E. Hackel. 1976. An investigation of the products of 53 gene loci three species wild Canidae: Canus lupus, Canus latrans, and Canus familiaris. Biochemical Genetics 14:963-974.

Fleischer, R. C., S. I. Rothstein, and L. S. Miller. 1991. Mitochondrial DNA variation indicatesgene flow across a zone of known secondary contact between two subspecies of the brown-headed cowbird. Condor 93:185-189.

Fletcher, S. D., and W. S. Moore. 1992. Further analysis of allozyme variation in the northern flicker, in comparison with mitochondrial DNA variation. Condor 94:988-991.

Flowers, L., and K. J. Rice. 1994. Ecotypic variation in California brome (Bromus carinatus). Paper and abstract presented at the IV Annual Meeting, California Native Grass Assoc., Nov 4, 1994, Sacramento.

Freeman, D. C., et al. 1991. Characterization of a narrow hybrid zone between 2 subspecies of big sagebrush (Artemisia tridentata, Asteraceae). Am. J. Botany 78(6):805-815.

Furnier, G. R., and W. T. Adams. 1986. Geographic patterns of allozyme variation in Jeffrey pine. Am. J. Botany 73(7):1009-1015.

Furnier, G. R., et al. 1987. Effects of avian seed dispersal on the genetic structure of whitebark pine populations. Evolution 41(3):607-612.

Futuyma, D. J., and S. C. Peterson. 1985. Genetic variation in the use of resources by insects. Ann. Rev. Entomol. 30:217-238.

Gall, G. A. E., et al. 1992. Geographic variation in population genetic structure of Chinook salmon from California and Oregon. Fishery Bulletin, U.S. 90:77-100.

Gall, G. A. E., B. Bentley, and R. C. Nuzum. 1990. Genetic isolation of steelhead rainbow trout in Kaiser and Redwood creeks, California. California Fish and Game 76(4):216-223.

Gall, G. A. E., et al. 1976. Biochemical genetic variation in populations of golden trout, Salmo aguabonita. Evidence of the threatened Little Kern River golden trout, S. a. whitei. J. Heredity 67:330-335.

Garbelotto, M. 1994. Conversation with D. Rogers, SNEP Genetic Workshop, Sept. 1994.

Garbelotto, M., et al. 1993. Differentiation of intersterility groups and geographic provenances amongisolates of Heterobasidion annosum detected by random amplified polymorhic DNA assays. Can. J. Bot. 71:565-569.

Gardes, M., and T. D. Bruns. 1993. ITS primers with enhanced specificity for basidiomycetes - application to the identification of mycorrhizae and rusts. Molec. Ecol. 1993(2):113-118.

Gardes, M., et al. 1991. Identification of indigenous and introduced symbiotic fungi in ectomycorrhizae by amplification of nuclear and mitochondrial ribosomal DNA. Can. J. Bot. 69:180-190.

Geiger, H., and A. M. Shapiro. 1986. Electrophoretic evidence for speciation within the nominal species *Anthocharis sara* Lucas (Pieridae). *J. Res. Lepid.* 25(1):15-24.

Gjedrem, T. 1992. Breeding plans for rainbow trout. *Aquaculture* 100(1992):73-83.

Gjerde, B., and L. R. Schaeffer. 1989. Body Traits in rainbow trout II. Estimates of heritabilities and of phenotypic and genetic correlations. *Aquaculture* 80(1989):25-44.

Good, D. A. 1988. Allozyme variation and phylogenetic relationships among the species of *Elgaria* (Squamata: Anguidae). *Herpetologica* 44(2):154-162.

Grant, M. C., Y. B. Linhart, and R. K. Monson. 1989. Experimental studies of ponderosa pine. II. Quantitative genetics of morphological traits. *Amer. J. Bot.* 76(6):1033-1040.

Griffin, J. R. 1965. Digger pine seedling response to serpentinite and non-serpentinite soil. *Ecology* 46:801-807.

Griffin, J. R., and W. B. Critchfield. 1972. The distribution of forest trees in California. Research Pap. PSW-82. US Forest Service.

Hafner, M. S., et al. 1994. Disparate rates of molecular evolution in cospeciating hosts and parasites. *Science* 265(5175):1087-1090.

Haglund, T. R., D. G. Buth, and R. Lawson. 1992. Allozyme variation and phylogenetic relationships of Asian, North American, and European populations of the threespine stickleback, *Gasterosteus aculeatus*. *Copeia* 1992(2):432-443.

Hammond, P. C. 1986. A rebuttal to the Arnold classification of *Speyeria callippe* (Nymphalidae) and defense of the subspecies concept. *J. Res. Lepid.* 24:197-208.

Hamrick, J. L. 1976. Variation and selection in western montane species. II. Variation within and between populations of white fir on an elevational transect. *Theor. Appl. Genet.* 47(1):27-34.

Hamrick, J. L., and M. J. W. Godt. 1990. Allozyme diversity in plant species. pp. 43-63. In: H. D. Brown, et al. (eds.). *Plant Population Genetics, Breeding and Genetic Resources* Sinnauer Associates, Inc., Mass. U.S.A.

Hamrick, J. L., M. J. W. Godt, and S. L. Sherman-Broyles. 1992. Factors influencing levels of genetic diversity in woody plant species. *New For.* 6:95-124.

Hamrick, J. L., and W. J. Libby. 1972. Variation and selection in western U.S. montane species. I. White fir. *Silvae Genetica* 21(1-2):29-35.

Hamrick, J. L., Y. B. Linhart, and J. B. Mitton. 1979. Relationships between life history characteristics and electrophoretically-detectable genetic variation in plants. *Ann. Rev. Ecol. Syst.* 10:173-200.

Harry, D. E. 1984. *Genetic structure of incense-cedar (Calocedrus decurrens)*. Ph.D. dissertation, Univ. California, Berkeley.

Harry, D. E., J. L. Jenkinson, and B. B. Kinloch. 1983. Early growth of sugar pine from an elevational transect. *For. Sci.* 29(3):660-669.

Hayes, J. L., and J. L. Robertson. 1992. An (ecologically biased) view of the current status of bark beetle genetics and future research needs. pp. 1-2. In: J. L. Hayes and J. L. Robertson (eds.). *Proceedings of Workshop on Bark Beetle Genetics: Current Status of Research.* Berkeley, CA. US Forest Service.

Hedgecock, D., and F. J. Ayala. 1974. Evolutionary divergence in the genus *Taricha* (Salamandridae). *Copeia* 1974(3):738-747.

Hershberger, W. K. 1992. Genetic variability in rainbow trout populations. *Aquaculture* 100(1992):51-71.

Higby, P. K., and M. W. Stock. 1982. Genetic relationships between two sibling species of bark beetle (Coleptera: Scolytidae), Jeffrey pine beetle and mountain pine beetle, in northern California. *Ann. Entomol. Soc. Am.* 75:668-674.

Hobson, K. R., and D. E. Bright. 1994. A key to the *Xyleborus* of California, with faunal comments (Coleoptera: Scolytidae). *Pan-Pac. Ent.* 70(4):267-268.

Hodges, S. A., and M. L. Arnold. 1994. Floral and ecological isolation between *Aquilegia formosa* and *Aquilegia pubescens*. *Proc. Nat. Acad. Sci. USA* 91(7):2493-2496.

Holsinger, K. E., and L. D. Gottlieb. 1988. Isozyme variability in the tetraploid *Clarkia gracilis* (Onagraceae) and its diploid relatives. *Syst. Bot.* 13(1):1-6.

Hong, Y. P., V. D. Hipkins, and S. H. Strauss. 1993. Chloroplast DNA diversity among trees, populations and species in the California closed-cone pines (*Pinus radiata, Pinus muricata* and *Pinus attenuata*). *Genetics* 135(4):1187-1196.

Howard, J. H., and R. L. Wallace. 1981. Microgeographic variation of electrophoretic loci in populations of *Ambystoma macrodactylum columbianum* (Caudata: Ambystomatidae). *Copeia* 1981(2):466-471.

Jackman, T.R. n.d. Thesis in Museum of Vertebrate Zoology, University of California, Berekley.

Jackman, T. R., and D. B. Wake. 1994. Evolutionary and historical analysis of protein variation in the blotched forms of salamanders of the *Ensatina* complex (Amphibia, Plethodontidae). *Evolution* 48(3):876-897.

James, F. C. 1983. Environmental component of morphological differentiation in birds. *Science* 221:184-187.

Jelinski, D. E., and W. M. Cheliak. 1992. Genetic diversity and spatial subdivision of *Populus tremuloides* (Salicaceae) in a heterogeneous landscape. *Am. J. Botany* 79(7):728-736.

Jenkinson, J. L. 1977. Edaphic interactions in first-year growth of California ponderosa pine. Research Paper. PSW-127. US Forest Service.

Jenkinson, J. L. n.d.-a. Unpublished common garden data on sugar pine.

Jenkinson, J. L. n.d.-b. Unpublished seed source data on red and white fir.

Jennings, W. B., D. F. Bradford, and D. F. Johnson. 1992. Dependence of the garter snake *Thamnophis elegans* on amphibians in the Sierra Nevada of California. *J. Herpetol.* 26(4):503-505.

Johnson, N. K., and J. A. Marten. 1991. Evolutionary genetics of flycatchers.III. Variation in *Empidonax hammondii* (Aves: Tyrannidae). *Can. J. Zool.* 69:232-238.

Johnson, N. K., and J. A. Marten. 1992. Macrogeographic patterns of morphometric and genetic variation in the sage sparrow complex. *Condor* 94:1-19.

Johnson, N. K., and R. M. Zink. 1983. Speciation in sapsuckers (*Sphyrapicus*): I. Genetic differentiation. *Auk* 100:871-884.

Johnson, W. E., and R. K. Selander. 1971. Protein variation and systematics in kangaroo rats (*Genus Dipodomys*). *Syst. Zool.* 20(4):372-405.

Kannenberg, L. W., and R. W. Allard. 1967. Population studies in predominantly self-pollinating species. VIII. Genetic variability in the *Festuca microstachys* complex. *Evolution* 21(227-240).

Kinloch, B. B. 1992. Distribution and frequency of a gene for resistance to white pine blister rust in natural populations of sugar pine. *Can. J. Bot.* 70(7):1319-1323.

Kinloch, B. B., Jr., and G. E. Dupper. 1987. Restricted distribution of a virulent race of the white pine blister rust pathogen in the western United States. *Can. J. For. Res.* 17:448-451.

Kitzmiller, J. H. 1976. Tree Improvement Master Plan for the California Region. US Forest Service, San Francisco.

Kitzmiller, J. H. 1990. Managing genetic diversity in a tree improvement program. *For. Ecol. Manage.* 35(1-2):131-149.

Kitzmiller, J. H. 1994. Conversation with R. D. Westfall.

Knapp, E. E. 1994. Genetic architecture of purple needlegrass: Implications for restoration. Paper and abstract presented at the IV Annual Meeting, California Native Grass Assoc., Nov 4, 1994, Sacramento.

Knapp, E., and K. Rice. 1994a. Isozyme variation within and among populations of *Danthonia californica*. Unpublished Report to US Forest Service. Univ. California, Davis.

Knapp, E. E., and K. J. Rice. 1994b. Morphological and allozyme variation within and among populations of *Nassella pulchra*. Report to The Nature Conservancy. University of California, Davis.

Knapp, E. E., and K. J. Rice. 1995. Genetic architecture and gene flow in *Elymus glaucus* (blue wildrye): Implications for native grassland restoration. *Restoration Ecology*. Submitted.

Korman, A. M., et al. 1991. Allozymic relationships among cuticular hydrocarbon phenotypes of *Zootermopsis* species (Isoptera: Termopsidae). *Ann. Entomol. Soc. Am.* 84(1):1-9.

Kruckeberg, A. K. 1987. Serpentine endemism and rarity. pp. 121-128. In: T. S. Elias (ed.). *Conservation and Management of Rare Plants.Proceedings of a California conference on the conservation and management of rare and endangered plants, Nov 5-8, 1986*. California Native Plant Society, Sacramento.

Lack, D. 1968. *Ecological Adaptations for Breeding in Birds* Methuen, London.

Lamb, T., J. C. Avise, and J. W. Gibbons. 1989. Phylogeographic patterns in mitochondrial DNA of the desert tortoise *(Xeribates agassuzi)*, and evolutionary relationships among the North American gopher tortoises. *Evolution* 43(1):76-87.

Lamb, T., T. R. Jones, and J. C. Avise. 1992. Phylogeographic histories of representative herpetofauna of the southwestern United States—mitochondrial DNA variation in the desert iguana *(Dipsosaurus dorsalis)* and the chuckwalla *(Sauromalus obesus)*. *Journal of Evolutionary Biology* 5(3):465-480.

Lansman, R. A., et al. 1983. Extensive genetic variation in mitochondrial DNA's among geographic populations of the deer mouse, *Peromyscus maniculatus*. *Evolution* 37(1):1-16.

Larson, A. 1980. Paedomorphosis in relation to rates of morphological and molecular evolution in the salamander *Aneides flavipunctatus* (Amphibia, Plethodontidae). *Evolution* 34(1):1-17.

Leary, R. F., F. W. Allendorf, and S. H. Forbes. 1993. Conservation genetics of bull trout in the Columbia and Klamath river drainages. *Conservation Biol.* 7(4):856-865.

Ledig, F. T. 1987. Genetic structure and the conservation of California's endemic and near-endemic conifers. pp. 587-594. In: T. S. Elias (ed.). *Conservation and Management of Rare and Endangered Plants: Proceedings of a California Conference on the Conservation and Management of Rare and Endangered Plants*. California Native Plant Society, Sacramento, CA.

Libby, W. J., K. Isik, and J. P. King. 1980. Variation in flushing time among white fir population samples. *Annales Forestales* 8(6):123-138.

Lidicker, W. Z. 1994. Written notes to D. Rogers at the SNEP Genetics Workshop, Sept. 1994.

Lidicker, W. Z., and R. S. Ostfeld. 1991. Extra-large body size in California voles—causes and fitness consequences. *Oikos* 61(1):108-121.

Linhart, Y. B., M. C. Grant, and P. Montazer. 1989. Experimental studies in ponderosa pine.I.Relationship between variation in proteins and morphology. *Amer. J. Bot.* 76(7):1024-1032.

Liu, Z. L., and J. B. Sinclair. 1992. Genetic diversity of *Rhizoctonia solani* anastomosis group 2. *Phytopathology* 82:778-787.

Loudenslager, E. J., and G. A. E. Gall. 1980. Geographic patterns of protein variation and subspeciation in cutthroat trout, *Salmo clarki*. *Syst. Zool.* 29:27-42.

MacDonald, S. E., and V. J. Lieffers. 1991. Population variation, outcrossing, and colonization of disturbed areas by *Calamagrostis canadensis*—evidence from allozyme analysis. *Am. J. Botany* 78(8):1123-1129.

Mahalovich, M. F. 1985. *A genetic architecture study of giant sequoia: Early growth characteristics*. M.S. thesis, University of California, Berkeley.

Manlove, M. N., et al. 1980. Biochemical variation in the black bear. pp. 37-41. In: C. J. Martinka and K. L. McArthur (eds.). *Proceedings of bears—their biology and management*. Third Bear Biology Association Conf.

Mastrogiuseppe, R. J. 1972. *Geographic variation in foxtail pine, Pinus balfouriana* Grev. & Balf. M.S. thesis, Humboldt State Univ., Arcata, CA.

Mastrogiuseppe, R. J., and J. D. Mastrogiuseppe. 1980. A study of *Pinus balfouriana* Grev. & Balf. (Pinaceae). *Syst. Bot.* 5:86-104.

McArthur, E. D., C. L. Pope, and D. C. Freeman. 1981. Chromosomal studies of subgenus Tridentatae of *Artemisia*: Evidence for autopolyploidy. *Am. J. Botany* 68(5):589-605.

McCracken, G. F., M. K. McCracken, and A. T. Vawter. 1994. Genetic structure in migratory populations of the bat *Tadarida brasiliensis mexicana*. *J. Mammal.* 75(2):500-514.

McDonald, B. A., and J. P. Martinez. 1990. DNA restriction fragment length polymorphisms among *Mycosphaerella graminicola* (Anamorph *Septoria tritici*) isolates collected from a single wheat field. *Phytopathology* 80:1368-1373.

McKechnie, S. W., P. P. Erlich, and C. L. Hogue. 1975. Population genetics of *Euphydryas* butterflies: I. Genetic variation and the neutral hypothesis. *Genetics* 81:571-594.

McKinney, C. O., et al. 1972. Genetic variation in the side-blotched lizard *(Uta stansburiana)*. pp. 307-318. In: *Studies in Genetics VII*, Univ. Texas Publ. 7213.

Meyer, S. E., and S. B. Monsen. 1991. Habitat-correlated variation in mountain big sagebrush *(Artemisia tridentata* ssp *vaseyana)* seed germination patterns. *Ecology* 72(2):739-742.

Meyer, S. E., S. B. Monsen, and E. D. McArthur. 1990. Germination response of *Artemisia tridentata* (Asteraceae) to light and chill—patterns of between-population variation. *Botanical Gazette* 151(2):176-183.

Millar, C. I., and D. L. Delany. n.d. Unpublished allozyme data on the California cypresses. US Forest Service.

Millar, C. I., D. L. Delany, and L. A. Riggs. 1990. Genetic variation in California oaks. *Fremontia* 18(3):20-21.

Millar, C. I., D. L. Delany, and R. D. Westfall. 1993. Effects of silvicultural treatments on genetic diversity in Jeffrey pine *(Pinus jeffreyii)*. Unpublished manuscript. US Forest Service.

Millar, C. I., et al. 1988. Allozyme differentiation and biosystematics in the California closed-cone pines (Subsection *Oocarpae* Little & Critchfield, Genus *Pinus*). *Syst. Bot.* 13(3):351-370.

Millar, C. I., R. D. Westfall, and D. Nelson. 1991. Genetic conservation areas on the Placerville Ranger District, Eldorado National Forest. Unpublished report.USDA Forest Service.

Mirov, N., J. W. Duffield, and A. R. Liddicoet. 1952. Altitudinal races of *Pinus ponderosa*—a 12-year progress report. *Journal of Forestry* 50:825-831.

Mitton, J. B., and M. G. Raphael. 1990. Genetic variation in the marten, *Martes americana*. *J. Mammal.* 71(2):195-197.

Monson, R. K., and M. C. Grant. 1989. Experimental studies of ponderosa pine. III. Differences in photosynthesis, stomatal conductance, and water-use efficiency between two genetic lines. *Amer. J. Bot.* 76(7):1041-1047.

Mooring, J. S. 1994. A cytogenetic study of *Eriophyllum confertiflorum* (Compositae, Helenieae). *Am. J. Botany* 81(7):919-926.

Moritz, C. 1994. Defining 'evolutionarily significant units'for conservation. *Tr. Ecol. Evol.* 9(10):373-375.

Morton, J. B., and S. P. Bentivenga. 1994. Levels of diversity in endomycorrhizal fungi (Glomales, Zygomycetes) and their role in defining taxonomic and non-taxonomic groups. *Plant and Soil* 159:47-59.

Moyle, P. B. 1976. *Inland Fishes of California* University of California Press, Berkeley.

Moyle, P. B. 1994. The decline of anadromous fishes in California. *Conservation Biol.* 8(3):869-870.

Namkoong, G., and M. T. Conkle. 1976. Time trends in genetic control of height growth in ponderosa pine. *For. Sci.* 22(1):2-12.

Nason, J. D., N. C. Ellstrand, and M. L. Arnold. 1992. Patterns of hybridization and introgression in populations of oaks, manzanitas, and irises. *Am. J. Botany* 79(1):101-111.

Neigel, J. E., and J. C. Avise. 1993. Application of a random walk model to geographic distributions of animal mitochondrial DNA variation. *Genetics* 135:1209-1220.

Ness, B. D. 1989. Seed morphology and taxonomic relationships in *Calochortus* (Liliaceae). *Syst. Bot.* 14(4):495-505.

Ness, B. D., D. E. Soltis, and P. S. Soltis. 1990. An examination of polyploidy and putative introgression in *Calochortus* subsection *nudi* (Liliaceae). *Am. J. Botany* 77(12):1519-1531.

Niebling, C. R., and M. T. Conkle. 1990. Diversity of Washoe pine and comparisons with allozymes of ponderosa pine races. *Can. J. For. Res.* 20:298-308.

Novak, S. J., R. N. Mack, and D. E. Soltis. 1991. Genetic variation in *Bromus tectorum* (Poaceae)—population differentiation in its North American range. *Am. J. Botany* 78(8):1150-1161.

Olmstead, R. G. 1990. Biological and historical factors influencing genetic diversity in the *Scutellaria angustifolia* complex (Labiatae). *Evolution* 44(1):54-70.

Otrosina, W. J., T. E. Chase, and F. W. Cobb, Jr. 1992. Allozyme differentiation of intersterility groups of *Heterobasidion annosum* isolated from conifers in the western United States. *Phytopathology* 8:540-545.

Pashley, D. P. 1980. *Genetic comparisons between native and introduced populations of the codling moth, Laspeyresia pomenella (Tortricidae) and among other tortricid species*. Ph.D. dissertation, University of Texas, Austin.

Pashley, D. P., S. J. Johnson, and A. N. Sparks. 1985. Genetic population structure of migratory moths: The fall armyworm (Lepidoptera: Noctuidae). *Ann. Entomol. Soc. Am.* 78:756-762.

Patterson, R., and B. D. Tanowitz. 1989. Evolutionary and geographic trends in adaptive wood anatomy in *Eriastrum densifolium* (Polemoniaceae). *Am. J. Botany* 76(5):706-713.

Patton, J. L., and J. H. Feder. 1981. Microspatial genetic heterogeneity in pocket gophers: Non-random breeding and drift. *Evolution* 35(5):912-920.

Patton, J. L., H. MacArthur, and S. Y. Yang. 1976. Systematic relationships of the four-toed populations of *Dipodomys heermanni*. *J. Mammal.* 57(1):159-162.

Patton, J. L., and M. F. Smith. 1990. *The evolutionary dynamics of the pocket gopher Thomomys bottae, with emphasis on California populations*. University of California Publications in Zoology; V. 123. University of California Press, Berkeley.

Patton, J. L., and S. Y. Yang. 1977. Genetic variation in *Thomomys bottae* pocket gophers: Macrogeographic patterns. *Evolution* 31(4):697-720.

Peterson, A. T. 1991. Gene flow in scrub jays: Frequency and direction of movement. *Condor* 91:926-934.

Porter, A. A., and E. M. Jakob. 1990. Allozyme variation in the introduced spider *Holocnemus pluchei*(Araneae, Pholcidae) in California. *J. Arachnol.* 18:313-319.

Porter, A. H. 1989. *Gene flow statistics and species level systematics of butterflies*. Ph.D. dissertation, University of California, Davis.

Porter, A. H., and H. Geiger. 1988. Genetic and phenotypic population structure of the *Coenonympha tullia* complex (Lepidoptera: Nymphalidae: Satyrinae) in California: No evidence for species boundaries. *Can. J. Zool.* 66:2651-2765.

Porter, A. H., and A. M. Shapiro. 1989. Genetics and biogeography of the *Oeneis chryxus* complex (Satyrinae) in California. *J. Res. Lepid.* 28(4):264-276.

Powell, J. A., and C. L. Hogue. 1979. *California Insects* University of California Press, Berkeley, CA.

Pratt, G. F., and G. R. Ballmer. 1991. Three biotypes of *Apodemia mormo* (Riodinidae) in the Mojave Desert. *J. Lepid. Soc.* 45(1):46-67.

Rabinowitz, D. 1981. Seven forms of rarity. pp. 205-217. In: H. Synge (ed.). *The Biological Aspects of Rare Plant Conservation* John Wiley & Sons, New York.

Rank, N. E. 1992. A hierarchical analysis of genetic differentiation in a montane leaf beetle *Chrysomela aeneicollis* (Coleptera: Chrysomelidae). *Evolution* 46(4):1097-1111.

Rehfeldt, G. E. 1990. Genetic differentiation among populations of *Pinus ponderosa* from the upper Colorado River Basin. *Bot. Gaz.* 151(1):125-137.

Rice, K. J., and R. N. Mack. 1991a. Ecological genetics of *Bromus tectorum*.I. A hierarchical analysis of phenotypic variation. *Oecologia* 99(1):77-83.

Rice, K. J., and R. N. Mack. 1991b. Ecological genetics of *Bromus tectorum*.II. Intraspecific variation in phenotypic plasticity. *Oecologia* 99(1):84-90.

Rice, K. J., and R. N. Mack. 1991c. Ecological genetics of *Bromus tectorum*.III. The demography of reciprocally sown populations. *Oecologia* 99(1):91-101.

Richards, R. T. 1994. Wild mushroom harvesting in the Klamath Bioregion: A socioeconomic study. Draft Report.University of California at Davis.

Riddle, B. R., and R. L. Honeycutt. 1990. Historical biogeography in North American arid regions - an approach using mitochondrial DNA phylogeny in grasshopper mice (Genus *Onychomys*). *Evolution* 44(1):1-15.

Rogers, D. L., D. E. Harry, and W. J. Libby. 1994. Genetic variation in incense-cedar *(Calocedrus decurrens)*: I. Provenance differences in a twelve-year-old common-garden study. *Western Journal of Applied Forestry* 9(4):113-116.

Rogers, D. L., R. F. Stettler, and P. E. Heilman. 1989. Genetic variation and productivity of *Populus trichocarpa* and its hybrids. III. Structure and pattern of variation in a 3-year field test. *Can. J. For. Res.* 19:372-377.

Rogstad, S. H., H. Nybom, and B. A. Schaal. 1991. The tetrapod DNA fingerprinting M-13 repeat probe reveals genetic diversity and clonal growth in quaking aspen (*Populus tremuloides*, Salicaceae). *Plant Systematics and Evolution* 175(3-4):115-123.

Rosado, S. C. S., B. R. Kropp, and Y. Piche. 1994. Genetics of ectomycorrhizal symbiosis. II. Fungal variability and heritability of ectomycorrhizal traits. *New Phytol.* 126:111-117.

Roy, B. A. 1993. Patterns of rust infection as a function of host genetic diversity and host density in natural populations of the apomictic crucifer, *Arabis holboellii. Evolution* 47(1):111-124.

Roy, M. S., et al. 1994. Patterns of differentiation and hybridization in North American wolflike canids, revealed by analysis of microsatellite loci. *Mol. Biol. Evol.* 11(4):553-570.

Schierenbeck, K. A., G. L. Stebbins, and R. W. Patterson. 1992. Morphological and cytological evidence for polyphyletic allopolyploidy in *Arctostaphylos mewukka* (Ericaceae). *Plant Systematics and Evolution* 179(3-4):187-205.

Schuster, W. S., D. L. Alles, and J. B. Mitton. 1989. Gene flow in limber pine: Evidence from pollination phenology and genetic differentiation along an elevational transect. *Amer. J. Bot.* 76(9):1395-1403.

Schuster, W. S. F., et al. 1992. Heritability of carbon isotope discrimination in *Gutierrezia microcephala* (Asteraceae). *Am. J. Botany* 79(2):216-221.

Schwartz, O. A., and K. B. Armitage. 1980. Genetic variation in social mammals: The marmot model. *Science* 207:665-667.

Scribner, K. T., et al. 1991. Temporal, spatial, and age-specific changes in genotypic composition of mule deer. *J. Mamm.* 72(1):126-137.

Seeliger, L. M. 1945. Variation in the Pacific mud turtle. *Copeia* 1945(3):150-.

Shaffer, H. B. 1984. Evolution in a paedomorphic lineage: I. An electrophoretic analysis of the Mexican ambystomic salamanders. *Evolution* 38(6):1194-1206.

Shaffer, H. B., and F. Breden. 1989. The relationship between allozyme variation and life history: Non-transforming salamanders are less variable. *Copeia* (4):1016-1023.

Shapiro, A. M. 1992. Twenty years of fluctuating parapatry and the question of competitive exclusion in the butterflies *Pontia occidentalis* and *P. protodice* (Lepidoptera: Pieridae). *J. New York Emtomol. Soc.* 100(2):3111-319.

Shapiro, A. M. 1994. Observations presented at the SNEP Genetics Workshop, Sept. 1994.

Shapiro, A. M. 1995. Communication to D. Rogers.

Shapiro, A. M. n.d. Unpublished data. University of California, Davis.

Shapiro, A. M., and H. Geiger. *1986*. Electrophoretic confirmation of the species status of *Pontia protodice* and *P. occidentalis* (Pieridae). *J. Res. Lepid.* 25(1):39-47.

Shapiro, A. M., and C. Nice. n.d. Unpublished data. University of California, Davis.

Shevock, J. 1995. Message to D. Rogers.

Sibley, C. G., and K. W. Corbin. 1970. Ornithological field studies in the Great Plains and Nova Scotia. *Discovery* 6:3-6.

Sinervo, B. 1990. Evolution of thermal physiology and growth rate between populations of the western fence lizard (*Sceloporus occidentalis*). *Oecologia* 83(2):228-237.

Sinervo, B., and S. C. Adolph. 1994. Growth plasticity and thermal opportunity in *Sceloporus* lizards. *Ecology* 75(3):776-790.

Sinervo, B., et al. 1992. Allometric engineering—a causal analysis of natural selection on offspring size. *Science* 258(5090):1927-1930.

Sinervo, B., R. Hedges, and S. C. Adolph. 1991. Decreased sprint speed as a cost of reproduction in the lizard *Sceloporus occidentalis*— variation among populations. J. Exp. Biol. 155(Jan):323-336.

Slatkin, M. 1993. Isolation by distance in equilibrium and non-equilibrium populations. *Evolution* 47(1):264-279.

Smith, M. F. 1979. Geographic variation in genic and morphological characters in *Peromyscus californicus. J. Mammal.* 60(4):705-722.

Smith, M. F., and J. L. Patton. 1980. Relationships of the pocket gopher (*Thomomys bottae*) populations in the lower Colorado River. *J. Mammal.* 61:681-696.

Smith, M. F., and J. L. Patton. 1988. Subspecies of pocket gophers: Causal bases for geographic differentiation in *Thomonys bottae. Syst. Zool.* 37(2):163-178.

Smith, R. H., and H. K. Preisler. 1988. Xylem monterpenes of *Pinus monophylla* in California and Nevada. *SW Nat.* 33(2):205-214.

Smith-Huerta, N. L. 1986. Isozymic diversity in three allotetraploid *Clarkia* species and their putative progenitors. *J. Heredity* 77:349-354.

Snajberk, K., E. Zavarin, and D. Baily. 1979. Systematic studies of *Pinus balfouriana* based on volatile terpenoids from wood and needles and on seed morphology. *Biochem. Syst. Ecol.* 7:269-279.

Soltis, D. E., et al. 1991. Chloroplast DNA variation within and among genera of the *Heuchera* group (Saxifragaceae) - evidence for chloroplast transfer and paraphyly. *Am. J. Botany* 78(8):1091-1112.

Soltis, D. E., P. S. Soltis, and J. N. Thompson. 1992. Chloroplast DNA variation in *Lithophragma* (Saxifragaceae). *Syst. Bot.* 17(4):607-619.

Soltis, P. S., and W. L. Bloom. 1986. Genetic variation and estimates of gene flow in *Clarkia speciosa* subsp. *polyantha* (Onagraceae). *Am. J. Botany* 73(12):1677-1682.

Soltis, P. S., D. E. Soltis, and B. D. Ness. 1989. Population genetic structure in *Cheilanthes gracillima. Am. J. Botany* 76(8):1114-1118.

Soltis, P. S., D. E. Soltis, and P. G. Wolf. 1990. Allozymic divergence in North American *Polystichum* (Dryopteridaceae). *Syst. Bot.* 15(2):205-215.

Sorensen, F. C., R. K. Campbell, and J. F. Franklin. 1990. Geographic variation in growth and phenology of seedlings of the *Abies procera/ A. magnifica* complex. *For. Ecol. Manage.* 36:205-232.

Steinhoff, R. J., D. G. Joyce, and L. Fins. 1983. Isozyme variation in *Pinus monticola. Can. J. For. Res.* 13:1122-1132.

Stock, M. W., G. D. Ammar, and B. J. Bentz. 1992. Isozyme studies of bark beetle population genetics and systematics. pp. 7-9. In: J. L. Hayes and J. L. Robertson (eds.). *Proceedings of Workshop on Bark Beetle Genetics: Current Status of Research*. Berkeley, CA. US Forest Service.

Stock, M. W., and J. L. Robertson. 1980. Inter- and intraspecific variation in selected *Choristoneura* species (Lepidoptera: Tortricidae): A toxicological and genetic survey. *Can. Entomol.* 112:1019-1027.

Strauss, S. H. 1987. Heterozygosity and developmental stability under inbreeding and crossbreeding in *Pinus attenuata. Evolution* 41(2):331-339.

Strauss, S. H., and M. T. Conkle. 1986. Segregation, linkage, and diversity of allozymes in knobcone pine. *Theor. Appl. Genet.* 72:483-493.

Strauss, S. H., Y. P. Hong, and V. D. Hipkins. 1993. High levels of population differentiation for mitochondrial DNA haplotypes in *Pinus radiata, muricata, and attenuata. Theor. Appl. Genet.* 86(5):605-611.

Sturgeon, K. B. 1979. Monoterpene variation in ponderosa pine xylem resin related to western pine beetle predation. *Evolution* 33(3):803-814.

Svobda, P. L., J. R. Choate, and R. K. Chesser. 1985. Genetic relationships among southwestern populations of the Brazilian free-tailed bat. *J. Mammal.* 66(4):444-450.

Tan, An-ming. n.d. Thesis in Museum of Vertebrate Zoology, University of California Berkeley.

Thompson, D. B. 1990. Different spatial scales of adaptation in the climbing behavior of *Peromyscus maniculatus*: Geographic variation, natural selection, and gene flow. *Evolution* 44(4):952-965.

Thompson, P., and J. W. Sites, Jr. 1986. Comparison of population structure in chromosomally polytypic and montypic species of *Sceloporus* (Sauria: Iguanidae) in relation to chromosomally-mediated speciation. *Evolution* 40(2):303-314.

Thorgaard, G. H. 1992. Application of genetic technologies to rainbow trout. *Aquaculture* 100(1992):85-97.

Tsuji, J. S., et al. 1989. Locomotor performance of hatchling fence lizards *(Sceloporus occidentalis)*—quantitative genetics and morphometric correlates. *Evolutionary Ecology* 3(3):240-252.

Van Buskirk, R. 1995. Personal communication to D. Rogers.

van Staaden, M. J., R. K. Chesser, and G. R. Michner. 1994. Genetic correlations and matrilineal structure in a population of *Spermophilus richardsonii*. *J. Mammal.* 75(3):573-582.

Van Wagner, C. E., and A. J. Baker. 1986. Genetic differentiation in populations of Canada geese *(Branta canadensis)*. *Can. J. Zool.* 64:940-947.

Vasek, F. C. 1977. Phenotypic variation and adaptation in *Clarkia* Section *Phaeostoma*. *Syst. Bot.* 2:251-279.

Vawter, A. T., and P. F. Brussard. 1984. Allozyme variation in a colonizing species: The cabbage butterfly *Pieris rapae* (Peiridae). *J. Res. Lepid.* 22(3):204-216.

Vogelmann, J. E., and G. J. Gastony. 1987. Electrophoretic enzyme analysis of North American and eastern Asian populations of *Agastache* sect. Agastache (Labiatae). *Am. J. Botany* 74(3):385-393.

Vogler, D. R., et al. 1991. Isozyme structure of *Pteridermium harknessiiin* the western United States. *Can. J. Bot.* 69:2434-2441.

Wade, M. J., M. L. McKnight, and H. B. Shaffer. 1994. The effects of kin-structured colonization on nuclear and cytoplasmic genetic diversity. *Evolution* 48(4):1114-1120.

Wake, D. B. 1994. Observations made at the SNEP Genetics Workshop, Sept. 1994.

Wake, D. B., L. R. Maxson, and G. Z. Wurst. 1978. Genetic differentiation, albumin evolution, and their biogeographic implications in plethodontid salamanders of California and southern Europe. *Evolution* 32(3):529-539.

Wake, D.G., Yanev, K., and Frelow, -.1989. (Complete citation not available.) In Otte and Endler (eds.) Speciation and its Consequences.Sinauer Associates.

Wake, D. B., and K. P. Yanev. 1986. Geographic variation in allozymes in a "ring species," the plethodontid salamander *Ensantia eschscholtzii* of western North America. *Evolution* 40(4):702-715.

Watt, W. B., R. C. Cassin, and M. S. Swan. 1983. Adaptation at specific loci. III. Field behavior and survivorship differences among *Colias* PGI genotypes are predictable from *in vitro* biochemistry. *Genetics* 103:725-739.

Wayne, R. K., and S. J. O'Brien. 1987. Allozyme divergence within the Canidae. *Syst. Zool.* 36(4):339-355.

Westfall, R. D. 1995. Unpublished allozyme data on sugar pine. US Forest Service.

Westfall, R. D., and M. T. Conkle. 1992. Allozyme markers in breeding zone designation. *New For.* 6:279-309.

Westfall, R. D., M. T. Conkle, and F. T. Ledig. n.d. Unpublished nursery data on ponderosa pine.

White, E. E. 1990. Chloroplast DNA in *Pinus monticola*. 2. Survey of within-species variability and detection of heteroplasmic individuals. *Theor. Appl. Genet.* 79(2):251-255.

White, M. M. 1988. Age class and population genic differentiation in *Pteronarcys proteus* (Plecoptera: Pteronarcyidae). *Am. Mid. Nat.* 122:242-248.

Wolf, P. G., and P. S. Soltis. 1992. Estimates of gene flow among populations, geographic races, and species in the *Ipomopsis aggregata* complex. *Genetics* 130(3):639-647.

Wolf, P. G., P. S. Soltis, and D. E. Soltis. 1991. Genetic relationships and patterns of allozymic divergence in the *Ipomopsis aggregata* complex and related species (Polemoniaceae). *Am. J. Botany* 78(4):515-526.

Wright, S. 1978. *Evolution and the Genetics of Populations* Univ. Chicago Press, Chicago.

Yanev, K.1978. Ph.D. Dissertation.University of California, Berkeley.

Yanev, K. P., and D. B. Wake. 1981. Genetic differentiation in a relict desert salamander, *Batrachoseps campi*. *Herpetologica* 37(1):16-28.

Zavarin, E., L. G. Cool, and K. Snajberk. 1993. Geographic variability of *Pinus flexilis* xylem monoterpenes. *Biochem. Syst. Ecol.* 21(3):381-387.

Zavarin, E., W. B. Critchfield, and K. Snajberk. 1978. Geographic differentiation of monoterpenes from *Abies procera and Abies magnifica*. *Biochem. Syst. Ecol.* 6:267-278.

Zavarin, E., et al. 1991. Geographic monoterpene variability of *Pinus albicaulis*. *Biochem. Syst. Ecol.* 19(2):147-156.

Zavarin, E., et al. 1982. Variability in essential oils and needle resin canals of *Pinus longaeva* from eastern California and western Nevada in relation to other members of subsection *Balfourianae*. *Biochem. Syst. Ecol.* 10(1):11-20.

Zavarin, E., K. Snajberk, and L. Cool. 1990a. Chemical differentiation in relation to the morphology of the single-needle pinyons. *Biochem. Syst. Ecol.* 18(2-3):125-137.

Zavarin, E., K. Snajberk, and L. Cool. 1990b. Monoterpene variability of *Pinus monticola* wood. *Biochem. Syst. Ecol.* 18(2-3):117-124.

Zeiner, D. C., W. F. Laudenslayer, Jr., and K. E. Mayer (eds.). 1988. *California's Wildlife. I. Amphibians and Reptiles*. California Dept. Fish and Game, Sacramento.

Zeiner, D. C., et al. (eds.). 1990a. *California's Wildlife. III. Mammals*. California Dept. Fish and Game, Sacramento.

Zeiner, D. C., et al. (eds.). 1990b. *California's Wildlife. II. Birds*. California Department of Fish and Game, Sacramento.

Zera, A. J. 1981. Genetic structure of two species of waterstriders (Gerridae: Hemiptera) with differing degrees of winglessness. *Evolution* 33(3):218-225.

Zink, R. M. 1986. *Patterns and evolutionary significance of geographic variation in the chistaceae group of the fox sparrow (Passerella iliaca)*. Ornithological Monographs. American Ornithologists' Union, Washington D.C.

Zink, R. M. 1994. The geography of mitochondrial DNA variation, population structure, hybridization, and species limits in the fox sparrow *(Passerella iliaca)*. *Evolution* 48(1):98-111.

Zink, R. M., and D. L. Dittman. 1991. Evolution of brown towhees: mitochondrial DNA evidence. *Condor* 93:98-105.

Zink, R. M., and D. L. Dittman. 1993a. Gene flow, refugia, and evolution of geographic variation in the song sparrow *(Melospiza melodia)*. *Evolution* 47(3):717-729.

Zink, R. M., and D. L. Dittman. 1993b. Population structure and gene flow in the chipping sparrow and a hypothesis for evolution in the genus *Spizella*. *Wilson Bull.* 105(3):399-413.

Zink, R. M., D. F. Lott, and D. W. Anderson. 1987. Genetic variation, population structure and evolution of California quail. *Condor* 89:395-405.

Zink, R. M., and D. W. Winkler. 1983. Genetic and morphological similarity of two California gull populations with different life history traits. *Biochem. Syst. Ecol.* 11:397-403.

Genetics Workshop Participants and Report Reviewers

WORKSHOP PARTICIPANTS

Diane Elam
Natural Heritage Division
1220 S Street
Sacramento, CA 95814

Deborah L. Elliott-Fisk
Natural Reserve System
University of California
Office of the President
300 Lakeside Dr., 6th floor,
Oakland, CA 94612
Current address:
Department of Wildlife, Fish, and Conservation Biology
University of California
Davis, CA 95616

Gary M. Fellers
National BiologiCA l Service
Point Reyes National Seashore
Point Reyes, CA 94956

George Ferrell
USDA Forest Service
Silviculture Laboratory
2400 Washington Ave.,
Redding, CA 96601

Matteo Garbelotto
Department of Environmental Science,
Policy and Management
108 Hilgard Hall
University of California,
Berkeley, CA 94720

Graham A.E. Gall
Department of Animal Science
2237 Meyer Hall
University of California
Davis, CA 95616

Jay H. Kitzmiller
USDA Forest Service
Chico Genetic Resource Center
2741 Cramer Lane
Chico, CA 95928

Eric E. Knapp
Department of Agronomy and
Range Science
161 Hunt Hall
University of California
Davis, CA 95616

Bohun B. Kinloch
Institute of Forest Genetics
USDA Forest Service
Pacific Southwest Research Station
P.O. Box 245
Berkeley, CA 94701

F. Thomas Ledig
Institute of Forest Genetics
USDA Forest Service
Pacific Southwest Research Station
P.O. Box 245
Berkeley, CA 94701

William Z. Lidicker
Museum of Vertebrate Zoology
3101 Valley Life Sciences Building
University of California
Berkeley, CA 94720

Marjorie Matocq
Department of Biology
University of California
Los Angeles, CA 90024

Constance I. Millar
Institute of Forest Genetics
USDA Forest Service
Pacific Southwest Research Station
P.O. Box 245
Berkeley, CA 94701

Chris Nice
Section of Evolution and Ecology
University of California
Davis, CA 95616

James L. Patton
Museum of Vertebrate Zoology
3101 Valley Life Sciences Building
University of California
Berkeley, CA 94720

CA lvin O. Qualset
Genetic Resources Conservation Program
133 University Extension Building
University of California
Davis, CA 95616

Kevin Rice
Department of Agronomy and Range Science
University of California
Davis, CA 95616

Larry Riggs
Biosphere Genetics Inc.
P.P. Box 9528
Berkeley, CA 94709

Deborah L. Rogers
Institute of Forest Genetics
USDA Forest Service
Pacific Southwest Research Station
P.O. Box 245
Berkeley, CA 94701

H. Bradley Shaffer
Section of Evolution and Ecology
3208 Storer Hall
University of California
Davis, CA 95616

Arthur M. Shapiro
Section of Evolution and Ecology
6347 Storer Hall
University of California
Davis, CA 95616

Thomas B. Smith
Department of Biology
San Francisco State University
1600 Holloway Ave.,
San Francisco, CA 94132

David B. Wake
Museum of Vertebrate Zoology
3101 Valley Life Sciences Building
University of California
Berkeley, CA 94720

Robert D. Westfall
Institute of Forest Genetics
USDA Forest Service
Pacific Southwest Research Station
P.O. Box 245
Berkeley, CA 94701

Randy Zebell
Department of Biology
San Francisco State University
1600 Holloway Ave.,
San Francisco, CA 94132

REPORT REVIEWERS

Don Buth
Department of Biology
University of California
Los Angeles, CA 90095

David Bradford
U.S. Environmental Protection Agency
Las Vegas, NV 89193

Diane Elam
Natural Heritage Division
1220 S Street
Sacramento, CA 95814

Norm Ellstrand
Department of Botany and Plant Sciences
University of California
Riverside, CA 92521

Gary Fellers
National BiologiCA l Service
U.S.D.I., Pt. Reyes National Seashore
Point Reyes, CA 94956

John Helms
Department of ESPM
University of California
Berkeley, CA 94720

Ned Johnson
Museum of Vertebrate Zoology
University of California
Berkeley, CA 94720

Eric Knapp
Department of Agronomy and Range
University of California
Davis, CA 95616

Bohun Kinloch
Institute of Forest Genetics
USDA Forest Service, PSW Research Station
Berkeley, CA 94701

Bill Lasley
Institute of Toxicology and Environmental Health
University of California
Davis, CA 95616

Bill Libby
28 Valencia
Orinda, CA 94563

Jennifer Neilsen
Hopkins Marine Station
Stanford University
Pacific Grove, CA 93950

James Patton
Museum of Vertebrate Zoology
University of California
Berkeley, CA 94720

Kevin Rice
Department of Agronomy and Range Science
University of California
Davis, CA 95616

Art Shapiro
Section of Evolution and Ecology
University of California
Davis, CA 95616

David Wake
Museum of Vertebrate Zoology
University of California
Berkeley, CA 94720

Philip Ward
Entomology Department
University of California
Davis, CA 95616

Under SNEP Central review process:
John Hopkins
RANGE WATCH

CONSTANCE I. MILLAR
Institute of Forest Genetics
U.S. Forest Service
Pacific Southwest Research Station
Berkeley, California

MICHAEL BARBOUR
Department of Environmental Horticulture
University of California
Davis, California

DEBORAH L. ELLIOTT-FISK
Natural Reserve System
University of California
Oakland, California

JAMES R. SHEVOCK
Fish, Wildlife, and Botany
U.S. Forest Service
Pacific Southwest Region
San Francisco, California

WALLACE B. WOOLFENDEN
Historical Ecology
U.S. Forest Service
Inyo National Forest
Lee Vining, California

29

Significant Natural Areas

ABSTRACT

The Sierra Nevada Ecosystem Project mapped 945 areas in the Sierra Nevada of ecological, cultural, and geological significance. These areas contain outstanding features of unusual rarity, diversity, and representativeness on national forest and national park lands. More than 70% of the areas were newly recognized during the SNEP project. Local agency specialists familiar with local areas mapped 553 ecological areas (average size 1,359 ha [3,349 acres]), 198 cultural areas (average size 2,371 ha [5,804 acres]), and 194 geological areas (average size 3,822 ha [9,443 acres]) during workshops held throughout the Sierra Nevada. Ecological and cultural areas are concentrated primarily in the southern Sierra, especially in the national parks, and secondarily in the northern and eastern Sierra. Geological areas concentrate somewhat at high elevations and along river corridors. Although more than a third of these areas are in "protected" designations (wilderness, natural reserves, parks, etc.), more than half were recorded as having had past impacts to biodiversity values from recreation and other intensive human uses. Forty percent have had impacts from grazing. The areas with these impacts are scattered through the Sierra Nevada. Both of these activities are permitted in many significant areas within "protected" designations, which suggests that land designation per se may not adequately maintain the biodiversity values for which these areas are recognized. Timber harvest and associated impacts were noted on about a quarter of the areas, concentrated in a few primarily west-side forests. Mining and pollution were minor and local impacts to a small per-

centage of sites. Collectively these areas represent a network of sites identified for superlative values across the Sierra Nevada. Site-specific evaluation and coordinated management with adjacent and matrix lands at the landscape level would most likely promote the greatest maintenance of biodiversity values over the range.

INTRODUCTION

This chapter reports on areas of natural diversity in the Sierra Nevada, which SNEP refers to as significant areas. SNEP defines *significant areas* as "lands in the Sierra Nevada that contain special features of ecological, cultural, or geological diversity; a feature is special if it is unusually rare, diverse, or representative of natural diversity." SNEP distinguishes significant areas from natural areas primarily on the basis of management implications. *Natural areas* are "lands that may contain special features but, more importantly, are managed to maintain or restore a state of naturalness or wildness" (Bonnicksen 1988; UNESCO 1974, 1984; World Resources Institute 1991). Some level of human use has occurred on most of these lands, and in SNEP's context, naturalness implies less the absence of humans than the dominance of nonhuman ecological processes and structures (Diedrich et al. 1994; Hoerr 1993). Management of natural areas, as old-growth areas, criti-

Sierra Nevada Ecosystem Project: Final report to Congress, vol. II, *Assessments and scientific basis for management options.* Davis: University of California, Centers for Water and Wildland Resources, 1996.

cal watersheds, and wildlife habitat, often centers on the concept of reserve management. A *reserve management strategy* "promotes protection of natural habitats by restricting human use and access." Many categories of natural areas and reserve management exist (e.g., World Resources Institute 1991). In the context of SNEP, significant area, as a category of land, does not a priori imply a certain type of management. As discussed later, the special features within SNEP significant areas are heterogeneous, not only in their identifying attributes, but in the ways they are assessed and in their management needs.

OBJECTIVES

The SNEP significant areas project was primarily an inventory effort to map and compile information about features in the Sierra Nevada that have special ecological, geological, and cultural significance. The project did not attempt to be exhaustive but rather to contribute to the list of areas already known for the Sierra Nevada. By inventorying these areas and their special attributes, SNEP highlighted their existence, general condition, and potential management needs.

Specific objectives in some cases overlap, and are accomplished by, other SNEP efforts. Collectively, the work in SNEP to identify habitats and areas of high ecological value for late successional forests, for watersheds, for endemic plants and animals, for genetic diversity, and for significant areas have a common goal of inventorying biodiversity in the Sierra Nevada. Collective objectives of SNEP projects that involve in situ biodiversity areas are to

- Compile, in GIS format, map and attribute information about previously designated natural areas, for example, wilderness, national parks, and research natural areas (Davis et al. 1996).

- Standardize approaches to selection, size, and coverage of significant areas in the Sierra Nevada; expand criteria beyond rare elements; include ecological, geological, and cultural features. Map new areas on the national forests and national parks of the Sierra Nevada, and enter them into the SNEP GIS, achieving a broad coverage of landscapes. Collect standard attribute data (this chapter).

- Inventory, map, and assess aquatic significant natural areas (Moyle 1996).

- Assess areas of concentration and management of significant plant communities and botanical resources in the Sierra Nevada (Davis et al. 1996; Shevock 1996).

- Evaluate conditions of resources broadly within the SNEP significant areas, recognizing relationships with past management, trends for the future, and management options.

These objectives derive from the five SNEP assessment and policy questions.

The following assumptions underlie our analysis of natural areas:

- Significant areas make up a heterogeneous class. Definitions of significance are arbitrary and relative to geographic scale, to biodiversity values, and to human values.

- Previous and ongoing efforts exist in the Sierra Nevada to identify natural and significant areas. SNEP's work adds to, and does not replace, these efforts.

- Because of the nature of significant areas, the SNEP significant areas mapping effort does not try to be exhaustive. Many more areas exist in the Sierra Nevada that fit the criteria and were not inventoried, either because they are unknown or because SNEP did not reach an expert who knew about them. SNEP's goal was to add to existing inventories in a systematic way.

- The significant areas inventory was based on expert-opinion knowledge; on-the-ground evaluation of significant areas was not undertaken.

- Many categories of mappers could have been used (academic, public, agency). Each has its own type of knowledge of the landscape and biases about what significant areas are. SNEP used local agency specialists, who have intimate, broad knowledge of Sierra Nevada places and who have not been systematically queried in past natural area inventories.

- SNEP focused more on locating special *features* of the Sierra Nevada and less on mapping *areas* that contain them. Thus, boundaries are rough, indicating general locations of features, and are not intended to be formal management boundaries. Site-specific management (not within SNEP's scope) would address appropriate boundaries.

- Uniform management of significant areas in the Sierra Nevada is not implied by SNEP's recognition of an area as special. SNEP assumes that the diverse features mapped in significant areas have varying management needs and priorities for protection, and that appropriate management would not automatically lead to set-aside areas of exclusive or restrictive use.

BACKGROUND

The concepts of specialness and diversity are inherent to SNEP's criteria for choosing significant areas. A brief background is developed here to explain the logic that underlies SNEP's significant areas effort.

Significant areas attempt to inventory certain types of

biodiversity in situ. Biodiversity can be considered along three "dimensions": biological organization, space, and time.

Biological Organization

Biotic diversity is hierarchic and scalar. Increasingly complex levels of biological organization are recognizable along a continuum from molecules to biomes. *Genes* are fundamental units of biodiversity and are packaged within *individuals;* interbreeding individuals compose *populations;* populations of potentially interbreeding individuals define *species;* species interact within *communities* and *ecosystems;* and related communities and ecosystems evolve in response to regional environments and climatic regimes as *biomes* (Frankel and Soule 1981; Keystone Center 1991; Salwasser 1991; U.S. Congress 1987; Wilson 1988).

Because it is more practical often to inventory or measure composition and structure (e.g., numbers and types of species, stand structures, landscape patterns), we tend to bias our thinking toward this aspect of biodiversity. In fact, process (e.g., fire, nutrient cycling, reproductive functions) may be the most important focus for sustainable land management. Because composition, structure, and function are related, one may act as a proxy or indicator for another, allowing us to infer from the more practical aspects some of the more hidden or complex aspects of the system. Process, for instance, may most easily be interpreted by analyzing changes in state over time or space.

Space

Diversity at any level of biological organization is played out in space. On the geographic scale, biological diversity is recognized relative to microsites, watersheds, landscapes, regions, or continents (Crow 1991; Diaz and Apostol 1992; Forman and Godron 1986; Interagency Team 1994). Although these levels are arbitrarily defined, they reflect a real hierarchic or nested order. Often, different processes occur and patterns emerge at different geographic scales (Crow 1991; Harris 1984).

Time

Although an intangible dimension of consequence only, time has a practical significance in that we observe different compositions, structures, and processes occurring as a function of years, decades, centuries, or millennia (Delcourt and Delcourt 1991). Further, from a biodiversity perspective, relative time is important: the past is meaningful to the present and to the future, because biotic systems evolve cumulatively through time (Woolfenden 1996; Millar 1996b). Traditionally, land managers and policy makers have short time horizons for planning and have not looked far to the past for information or considered that futures they manage may be different from the present. Recognizing that biodiversity acts on long

scales as well has opened the door for managers to view natural systems in their evolutionary context as dynamic and individualistic (Stine 1996; Kinney 1996; Woolfenden 1996; Millar 1996b; Botkin 1990; Delcourt and Delcourt 1991; Kaufman 1993).

Cultural Diversity

SNEP adopts the broadest view of biodiversity to include humans. Like any other species, humans have levels of biological organization and possess habitat attributes in space and time. Humans have lived in the Sierra Nevada for nearly 10,000 years (Bettinger 1991; Blackburn and Anderson 1993; Anderson and Moratto 1996) in compositions (ethnicities, demographics), structures (settlement groups and economic classes), and processes (trade, diet, hostilities, land use and conversion) that have changed dramatically over past millennia and will certainly change over the next decades. Distinctions may be made between ancient and modern cultures, between cultures that practice traditional, extensive land use and husbandry and those that introduce intensive, industrial technologies, and between cultures of native and introduced ancestry. These elements are as much part of the SNEP charge to inventory and assess as are the nonhuman components.

Physical Diversity

Geological, hydrological, lithological, soil, and climatic factors define ecosystems and govern the expression of biodiversity within them. The Sierra Nevada's more than several hundred million year history of uplift, erosion, volcanism, and glaciation has produced a broad suite of rock types, including many kinds of igneous, sedimentary, and metamorphic rocks, with a wide range of ages from Cambrian to Quaternary (Huber 1981; McPhee 1993; Norris and Webb 1990). Soils that have weathered from these rocks range from shallow, residual soils developed over bedrock at high altitudes to deep, depositional soils in valley floors developed over river alluvium. With varying parent materials, land stabilities, and climates, the soils of the Sierra Nevada are even more diverse than their geologic substrates.

METHODS

Criteria for Significance and Guidelines for Selection of SNEP Significant Areas

SNEP's significant area project defined significance as extending to the broad range of biological, cultural, and physical diversity. Many institutional programs for natural or significant areas in the Sierra Nevada have focused on specific aspects of biodiversity, such as the species or vegetation communities level (e.g., old-growth forests, giant sequoia groves, rare species), while other levels are ignored, such as genetic and biome levels, large spatial scales, and ecosystem processes such as nutrient and water cycles.

A first general criterion for significance is that a feature or element is *rare, rich, or representative* (Bonnicksen 1988; Hoshovsky 1994; Wilson 1988). Rarity, the quality of being uncommon or unusual, is widely discussed in ecology and conservation biology literature for its significance (Bonnicksen 1988; Frankel and Soule 1981; Hoshovsky 1994; Wilson 1988). Rarity is classified and recognized according to evolutionary origin, ecological condition, and geographic position (Fiedler and Ahouse 1992; Rabinowitz 1981; Schoener 1987). For mapping purposes, rarity was standardized to mean features that exemplify significant rare genetic, species, community, ecosystem, cultural, or geological elements. *Rare* means fewer than about five occurrences on a national forest or national park, or that the national forest or park was the only place where an element occurred in the Sierra Nevada, even if more than five occurrences existed on a national forest or park (i.e., local endemic). Distinctly unusual features were sought for the significant areas inventory. This meant, for instance, that the emphasis was on distinct or unusual phylogenetic elements (e.g., monotypic species), unusual disjunctions (e.g., disjunct population far from main ranges), extreme assemblages (unusual mix of species), unexpected landforms, and so on. Thus, in this category, the inventory sought primarily examples of rare and unusual phenomena and, secondarily, rare examples of common phenomena.

Richness, or diversity, is widely classified and debated for its meaning in ecological and evolutionary contexts (May 1973; Pimm 1986; Turelli 1978). Richness implies a larger than expected number of parts, structures, or processes occurring within an area. For SNEP's mapping, richness was standardized to mean features on national forest or park lands that best exemplify high or unusual genetic, species, community, ecosystem, cultural, or geological diversity. Candidates were considered if there were fewer than about five occurrences of equal diversity on the national forest or park.

The attribute *representative* is often not as widely acknowledged as being special as rarity or diversity. From the standpoint of ecological role, conservation importance, and human utility, however, the common situations—widespread species (e.g., ponderosa pine, Douglas fir), common vegetation types (e.g., mixed conifer), routine functions (e.g., water, nutrient cycling, fire), "central" ecological niches (optimum habitats)—make up the essence of ecosystems, ecosystem services, and natural resources. Further, in temperate latitudes, including the Sierra Nevada, these common, widespread elements often receive high human impact (Beesley 1996; Duane 1996; Franklin and Fites-Kaufmann 1996; Davis and Stoms 1996; Moyle and Randall 1996). Representatives of common types often contain much of the diversity of rare or rich situations, although possibly in lower frequencies.

Representativeness was standardized as national forest or park lands that best represent common genetic, community,

TABLE 29.1

Example from Tahoe National Forest of vegetation types developed ad hoc for mapping representative significant areas on the national forest. These types were listed by the local mappers to reflect common conditions on the forest. At least one significant area per national forest or national park was chosen to represent each type.

Red fir	Aspen–alder–cotton willow riparian
Mountain hemlock	Aspen (slope)
Mixed conifer	Canyon live oak
East-side pine	Montane chaparral
Big sagebrush–mountain mahogany	Madrone–tan oak
Western juniper	Foothill pine
Knobcone pine	Serpentine
Black oak	Giant sequoia
Subalpine shrub	Lodgepole pine
Montane meadow	Blue oak–white oak
Bog	Western white pine
Fen	White fir

ecosystem, geological, or cultural diversity. Common diversity was interpreted as meaning the best representative elements of a standard classification system (e.g., vegetation series, geological classification, cultural phase). Classifications for representative category were ecological, cultural, and geological.

Ecological

Although initially SNEP planned to use a standard vegetation type classification to select representative features, it became clear from pilot mapping sessions that this inappropriately limited choices. The current or pending classifications (e.g., USFS 1992; Allen 1987; Cheatham and Haller 1975; Parker and Matyas 1979; Holland 1986; or Sawyer and Keeler-Wolf 1996) were either too general or too specific for the scale of a national forest or national park and did not represent the mix of types that was specific to each forest or park. Instead, we chose to develop *ad hoc* vegetation lists that reflected the conditions on each forest or park. Representative areas were then chosen to exemplify each type (e.g., table 29.1).

Cultural

For purposes of choosing representative significant cultural areas, cultural diversity was classified into four categories, based primarily on time:

1. Historic (last 200 years) Indian

2. Historic (last 200 years) non-Indian

3. Archaic (200–6,000 years) Indian

4. Paleo-Indian (more than 6,000 years)

Geological

Representative significant geological areas were classified by age, landform, and rock type as

- Geological age

  - Quaternary: Holocene or Pleistocene

  - Tertiary

  - Mesozoic

  - Paleozoic

- Landform (list not specified; left open)

- Rock type

  - Sedimentary

  - Metamorphic

  - Igneous: Volcanic or Plutonic

A second criterion for significance is type of diversity. SNEP's significant areas project considered biological (ecological), cultural, and physical aspects of diversity in the Sierra Nevada (table 29.2). In the following discussions, if not specified, biological or ecological diversity or significance includes cultural aspects. Together, the two levels of criteria provide a matrix that guided SNEP's selection of special features and areas. Cells within the matrix became targets for finding significant areas in the different parts of the Sierra Nevada (figure 29.1).

In each case, "best exemplify" or "best represent" refers to that element (population, species, plant community, cultural or geological site) among the pool of qualifying areas that is (1) most stable from an ecological, cultural, or geological context (unless the element is obviously a dynamic one) and (2) most viable (population, species, vegetation assemblage, restoration status), largest, most diverse, and has contained within the area the most environmental variability. Size ranges for choosing sites were given as

- less than about 400 ha (1,000 acres) for genetic, species, or cultural features

- less than about 4,000 ha (10,000 acres) for plant communities

- less than about 20,000 ha (50,000 acres) for geological features

These size ranges were merely guidelines and were meant to standardize the relative types of significant areas that SNEP inventoried and to aid mappers in selecting appropriate categories of candidates.

Past or current administrative status was not considered a primary factor in selecting SNEP significant areas. Areas currently designated for reserve status (e.g., research natural areas, special interest areas) were mapped and inventoried by the SNEP Gap Analysis Program (GAP) (Davis and Stoms 1996) and kept in separate GIS layers. Past and current management status affected candidacy when past actions had sig-

FIGURE 29.1

Matrix of criteria that defined targets for SNEP significant area project. Each cell was considered a potential target for selecting significant areas within a specific national forest or national park. For the representative criterion, classifications were developed for each type of diversity.

nificantly and detrimentally impacted the special feature for which an area might have been chosen. If the feature persisted despite inappropriate management, and also met other criteria, it could have been included. Selection was intended to be relatively blind to management and administrative status, unless the feature was so impacted by these aspects that it did not function in its natural condition. Although these aspects of management were not considered essential in selection, they were noted in the attribute database and became part of the assessment.

Geographic scale in general was an important aspect defining SNEP's choice of significant areas. Whereas the Sierra Nevada as a whole is considered a significant feature on the continental and global scales, and Yosemite and Lassen National Parks are considered significant features at the Sierra Nevada and national scales, the SNEP significant areas project tried to standardize areas by limiting their sizes, as described. At the sizes suggested, the candidate areas were chosen if they were significant relative to an individual national forest or national park. We did not seek exhaustive lists, for example, of every archaeological site or every rare plant population on a national forest or national park. We attempted to stress elements that were "most special" along the guidelines and at the scales described.

Methods for Selecting and Mapping SNEP Significant Areas

The SNEP significant areas project used an expert opinion and target elements approach. The matrix of criteria (figure 29.1) created the basis for target cells, and the geographic focus was primarily the Sierran national forests and national parks. Mapping sessions were held at central offices on each national forest or park, and an interdisciplinary group of lo-

TABLE 29.2

Examples of criteria for significance used in the SNEP significant areas project at four levels of biological organization, the physical environment, and cultural diversity. Descriptive criteria area considered relative to spatial and temporal scales.

1. Genetic
High diversity
Unique diversity
Rare or threatened genetic
    types
Important hybrid diversity
Relictual
Refugial
Ecotones/clines/ecotypes
Chromosomal races

2. Species
Rare, threatened, endangered
    species
Marginal or unusual
    distributions
Keystone, critical, indicator
    species
Representative populations or
    species

3. Community
High diversity of species
Marginal location for type
Important disturbance regimes
Relictual, refugial
Contains rare, endemic species
Unique edaphic situation
Critical role in ecoregion
Pristine, undisturbed
Sharp or unusual ecotones

4. Ecosystem
High diversity
Endangered
High endemism
Critical role in bioregion

5. Physical
Relictual, ancient
Fossil-bearing or otherwise
    significant paleoecologically
Rare or distinctive
Exemplary of landforms, geologic
    eras, rock types

6. Cultural
Unique or representative
    archaeological elements
Highly valued socially
Traditional use
Historic value
Scenic

cal agency staff was convened for each session. The group of mappers was chosen for (1) individual knowledge of the local area, (2) diversity of disciplinary knowledge, and (3) geographic coverage. Staff areas included geology, hydrology, soils, lands and resources, landscape ecology, fire, archaeology, ecology, wildlife biology, botany, recreation, range, and land-management planning. Most sessions had at least fifteen mappers present for a national forest or park.

Specialists mapped on planimetric national forest maps (0.5 in:1 mi or 1:125,000), which SNEP had prepared with registered Mylar overlays. Separate maps were provided for mapping geological, cultural, and ecological areas. Mappers first selected areas of rarity and richness for each category, then developed a forest-appropriate list of vegetation types to serve as targets for the representative categories (e.g., table 29.1); the standard classifications described earlier were used to select representative sites of cultural and geological significance. Some cultural sites were considered too sensitive to release location data. For security, these were (1) mapped very generally with boundaries that would not be detailed enough to locate the specific cultural site (which was often very small), (2) described but not mapped, or (3) excluded entirely.

Mappers were instructed to locate polygons by drawing general boundaries on Mylar overlays. Boundaries were not intended to reflect suggested management units but rather to signify geographic locations of the special feature on the landscape.

For each area mapped, attribute data were collected about area name and location, reason for selection, significant attributes, short- and long-term management (current and future), and past impacts to the resources for which the area was identified (figure 29.2). If an area was already in a protection category such as wilderness, research natural area, or botanical area, this was noted.

Only minimal mapping was done on nonfederal lands. Federal mappers did identify areas of nonfederal public lands (e.g., state parks) if they felt them to be important relative to other federal lands in their region. Agency mappers were instructed to map special features on public lands. Boundaries drawn to identify special features on public lands may have shown pieces of private lands. This was especially the case where private lands are intermixed (e.g., Tahoe National Forest). In these cases, the actual boundaries indicated were done purposely at a coarse level to show the general region of the special features.

A few large industrial landowners in the western Sierra Nevada were contacted directly by SNEP for information on many aspects of SNEP inventories. Mapped information on significant areas with attribute data was contributed by Fiberboard Corporation.

Mapped areas were entered in the SNEP GIS, with attribute data attached in a database.

Due to the exigencies of schedules, the Plumas National Forest could not be mapped. The ecological map for the Eldorado National Forest and the cultural map for the Stanislaus National Forest were not completed in time to be included.

RESULTS

Current Conditions

Designated Reserve Areas

Many areas in the Sierra Nevada have been formally designated by public agencies, academic institutions, and nongovernmental organizations for natural significance. These areas are characterized by histories of low human disturbance and by primary management objectives of resource preservation and reserve management strategy. Many programs for the preservation of natural diversity exist in California (Cochrane 1986; Davis and Stoms 1996; Davis et al. 1996) These types of areas include (administrative authority and number of areas designated in the Sierra Nevada are in parentheses):

- Areas of critical environmental concern (BLM: 11)

- Biosphere reserves (UNESCO: 2)

Polygon Attribute Form
Tahoe NF Significant Areas
Sierra Nevada Ecosystem Project
3/95

Date: _MARCH 6, 1995_

Mapper(s) name: _____

Polygon code: _14_

Area name (e.g. Devil's Post Pile): _Duncan_

Watershed: _____

Reason for selection as a significant area
Ranking is based on four categories: genetic/species significance, ecosystem/community significance, physical/geologic, and cultural/scenic importance and 3 significant criteria rare, rich and representative. If area is selected for representative criteria also indicate what type it represents (e.g. vegetation series or geologic era). Rank the polygon based on these four characteristics, you may use numbers 1-4, with 1 being the most important and 4 the least.

Category	Rare	Rich	Representative	Type
genetic/species				
ecosystem/community	1	2	3	
physical/geologic				see code form
cultural/scenic				see code form

List of attributes that make the area significant
(e.g. deepest lake in North America, petroglyphs, etc.) Do not exceed five responses and 30 characters per response.

1. _Large block of undisturbed mixed conifer old growth_
2. _high density spotted owls / goshawks_
3. _Rich botanical_
4. _Rich ecological_

Management

Primary Emphasis (See code sheet)
76

Secondary Emphasis (See code sheet)
51, 52

Long Range Management Goal (see code sheet and cite planning documents)
34, 73

Activities within past 25 years and current impacts to significant value of area
check all that apply

Activities	Impact occurred	Category
		Logging/timber harvest
✓		Grazing
		Road building
		Mining
		Pollution
✓		Recreation (describe, e.g. mountain biking)
		other 1 (describe)
		other 2 (describe)
		other 3 (describe)

FIGURE 29.2

Example of attribute form used to describe information about new significant areas mapped through SNEP.

- Special interest areas (USFS: botanical [23], cultural [8], geological [19], zoological [1], and scenic [5])

- Ecological reserves (California Department of Fish and Game: 7)

- National parks and monuments (NPS: 4)

- Natural preserves (California Department of Parks and Recreation: 6)

- Cultural preserves (California Department of Parks and Recreation: 4)

- The Nature Conservancy preserves (5)

- National scenic areas (USFS: 1)

- Research natural areas (USFS: 21 established, 22 candidate; NPS: 8)

- State historic parks (California Department of Parks and Recreation: 8)

- State parks and reserves (California Department of Parks and Recreation: 10)

- State recreation areas (California Department of Parks and Recreation: 8)

- University of California Natural Reserve System (UC: 5)

- World heritage sites (UNESCO: 1)

- Wilderness areas (USFS: 20, NPS: 4)

- Wild trout waters (California Department of Fish and Game: 15)

- Wild and scenic rivers (state and federal: 9)

Areas within most of these categories are located and identified nonsystematically, with designation *posthoc* after informal recognition of the area's significance. A few categories have an a priori target system to identify specific elements of significance and systematically search for areas to fit these targets. Special interest areas, for example, are designated to protect significant botanical, cultural, geological, paleontological, scenic, and zoological resources (Cochrane 1986). The Research Natural Areas Program of the U.S. Forest Service systematically surveys areas that represent ecological types administered by the Forest Service throughout the country (Cheatham et al. 1977; Federal Committee on Ecological Reserves 1977). Target matrices based on plant community types (tree, shrub, and understory) have been developed for broad regions of the Sierra Nevada, and exemplary areas are sought to fill the cells (Keeler-Wolf 1985).

The SNEP Gap Analysis Program has digitized maps for designated natural areas in the Sierra Nevada into a GIS, and has completed a database with administrative information on these areas (see Davis and Stoms 1996).

SNEP Significant Areas Inventory

In all, 945 natural areas were mapped and attributed in the Sierra Nevada by SNEP, including 553 ecological areas, 198 cultural areas, and 194 geological areas (table 29.3). These include areas mapped on the Eldorado (ecological areas map not completed), Inyo, Lassen, Lake Tahoe Basin (cultural areas map not completed), Tahoe, Sequoia, Sierra, Stanislaus (cultural areas map being redrawn), and Toiyabe National Forests (Bridgeport District); the Sequoia–Kings Canyon and Yosemite National Parks; and one BLM resource area (figures 29.3, 29.4, and 29.5).

Ecological areas are distributed across the Sierra Nevada (figure 29.3), with concentrations of larger areas in Sequoia–Kings Canyon and Yosemite National Parks and in the southern Sierra and with smaller areas widely distributed in the eastern Sierra. Some regional differences may be due to the interpretations of mappers in different sessions. Cultural areas (figure 29.4) also tend to be concentrated in and around the national parks, in the southern Sierra Nevada, and in the far northern Sierra Nevada. Geological areas (figure 29.5) are more clustered around the high peaks and river corridors.

Mappers often designated sites for several "primary significance" aspects, although we requested only one. Thus, many of the categories overlap in one area, and attributes indicated are not mutually exclusive (table 29.3). Ecological areas were smaller on average than cultural areas, which were smaller than geological areas. The size ranges originally given to the mappers as guides only vaguely matched what they felt to be the actual landscape areas. Geological areas were smaller than we had suggested, and cultural areas were much larger.

Current management of significant areas mapped by SNEP is given in table 29.4. About a third of the areas are currently in some form of land designation intended to protect the biodiversity values, most in significant areas mapped in the national parks, with the remainder in wilderness and a very small proportion in other designations (e.g., research natural areas, special interest areas, wild and scenic rivers, state parks). Among the national forests, the Bridgeport District of the Toiyabe was unusual in mapping many areas in wilderness. About a third of the areas receive some form of intensive human activity (including utility corridors, multiple resource areas, recreation, transportation, administrative sites, experimental forests). About equal proportions of the areas have grazing (14%) and timber (12%) activities present. The east-side forests (Inyo and Toiyabe) had more areas in grazing designations than elsewhere and, with the exception of the national parks, fewer areas where timber activities occur. The Tahoe National Forest had the largest number of significant areas (40%) in timber zones.

Assessment and Trends in Protection of Special Features

Past impacts to significant areas are summarized in several categories (table 29.5). Sierra-wide, over half the areas were determined to have impacts from recreation, and 40% had impacts from grazing. Recreation impacts were distributed in significant areas broadly over the Sierra Nevada. Less than a third of the areas were noted as having past impacts from logging or road construction. Mining and pollution affected only a small proportion of areas (less than 10%).

Considering past management impacts and current management together suggests in general that areas traditionally considered protected for biodiversity values in fact may not

TABLE 29.3

Sierra-wide summary statistics of geographic attributes for significant natural areas mapped by SNEP.

Statistic	Ecologically Significant Areas	Culturally Significant Areas	Geologically Significant Areas
Number chosen primarily for ecological/cultural/ geological significance	553	198	194
Average size	1,359 ha (3,349 acres)	2,349 ha (5,804 acres)	3,922 ha (9,443 acres)
Number of sites containing signficant richness	127	207	68
Average size of sites chosen for richness	1,648 ha (4,060 acres)	2,371 ha (5,840 acres)	2,229 ha (5,491 acres)
Number of sites containing significant rarity	68	108	110
Average size of sites chosen for rarity	1,480 ha (3,647 acres)	2,249 ha (5,540 acres)	4,454 ha (10,971 acres)
Number of sites containing significant representativeness	253	144	168
Average size of sites chosen for representativeness	995 ha (2,453 acres)	1,936 ha (4,770 acres)	2,673 ha (6,584 acres)

Total number of areas mapped in the Sierra Nevada: 945
Total average size of areas mapped in the Sierra Nevada: 1,355 ha (3,348 acres)

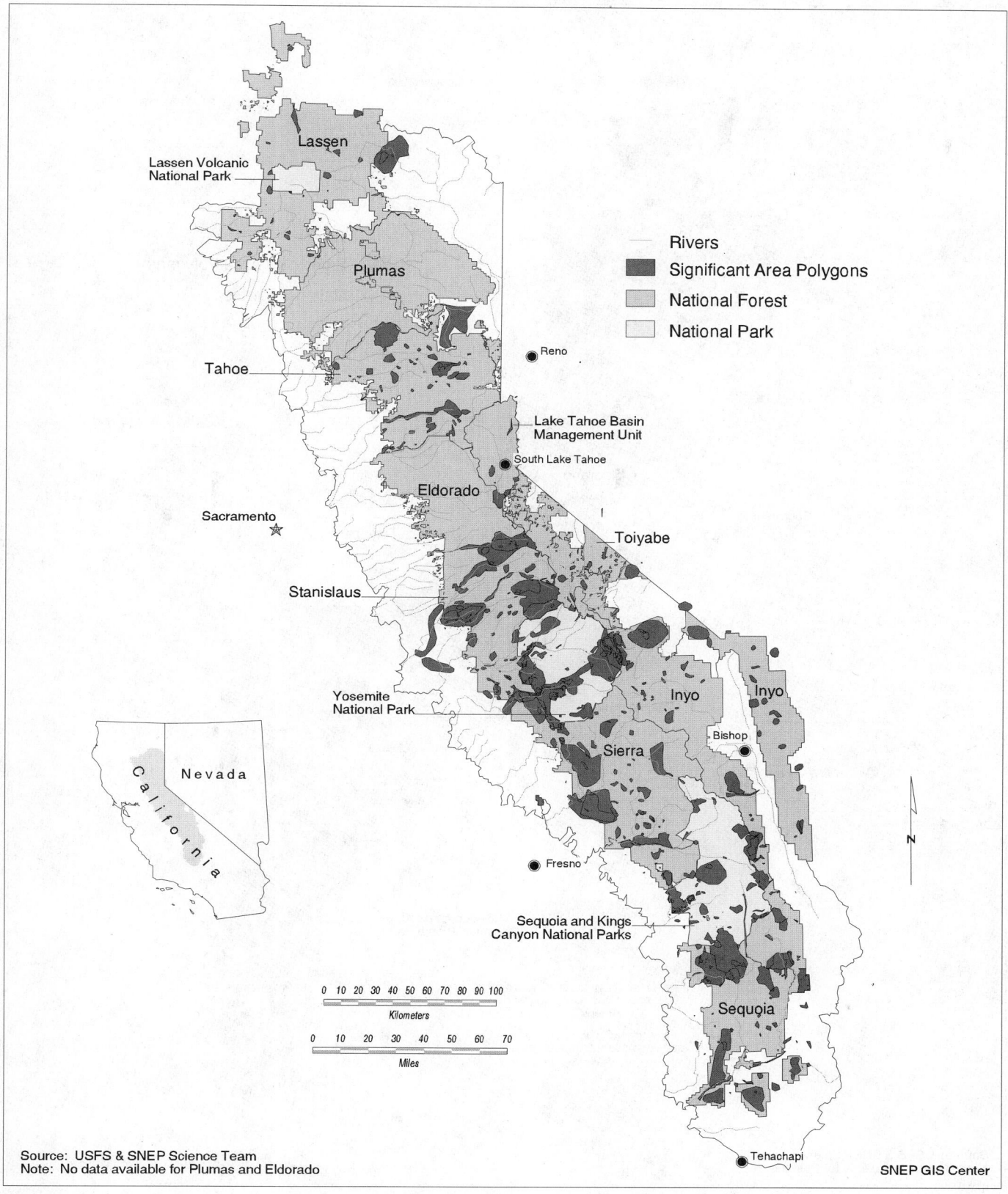

Rivers
Significant Area Polygons
National Forest
National Park

Lassen

Lassen Volcanic
National Park

Plumas

Reno

Tahoe

Lake Tahoe Basin
Management Unit

South Lake Tahoe

Eldorado

Sacramento

Toiyabe

Stanislaus

Yosemite
National Park

Inyo   Inyo

Bishop

Sierra

Nevada

California

Fresno

Sequoia and Kings
Canyon National Parks

0 10 20 30 40 50 60 70 80 90 100

Kilometers

0   10   20   30   40   50   60   70

Miles

Sequoia

Tehachapi

Source: USFS & SNEP Science Team
Note: No data available for Plumas and Eldorado

SNEP GIS Center

FIGURE 29.3

Significant ecological areas mapped by SNEP (Plumas and El Dorado not included).

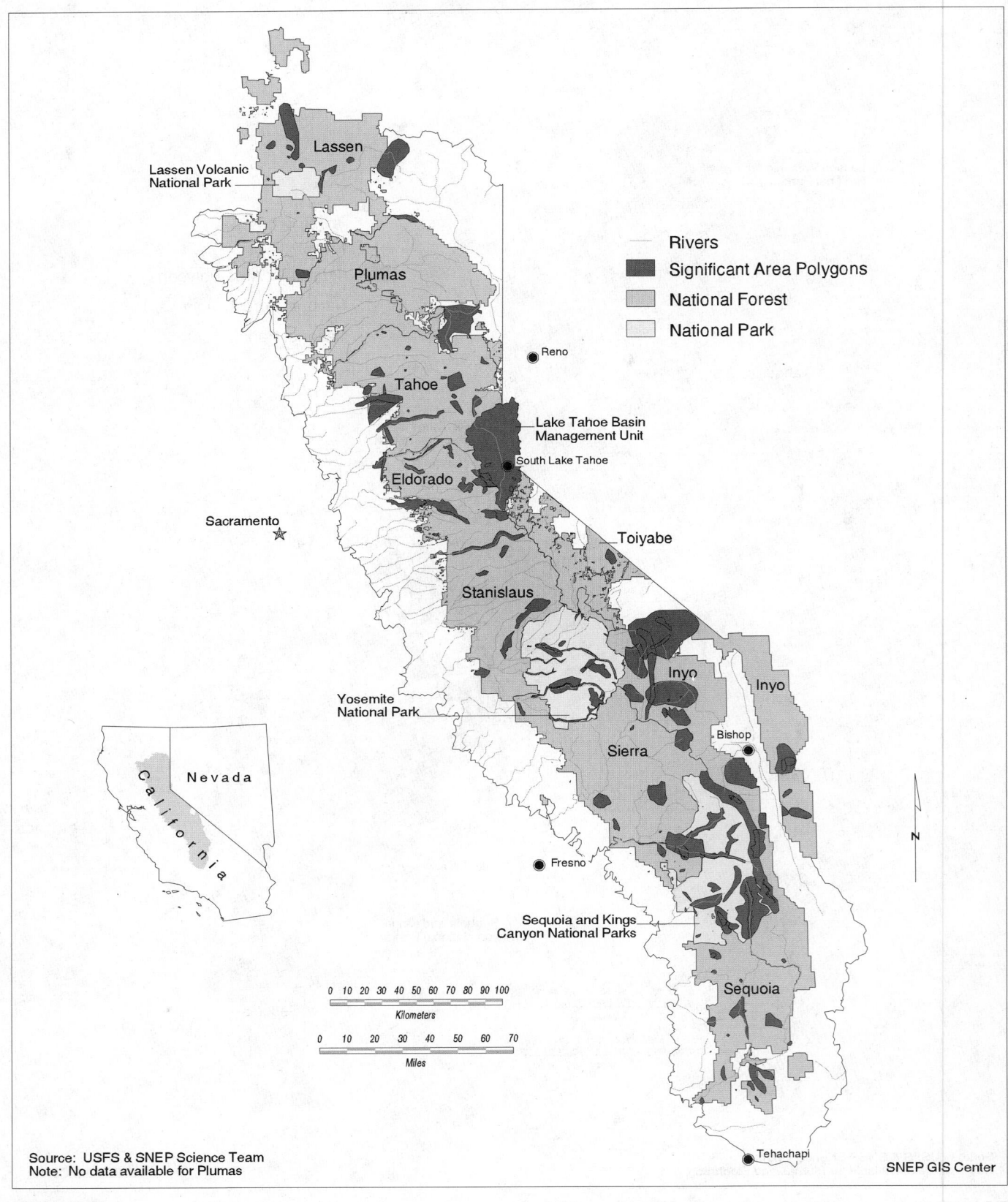

FIGURE 29.4

Significant geological areas mapped by SNEP (Plumas not included).

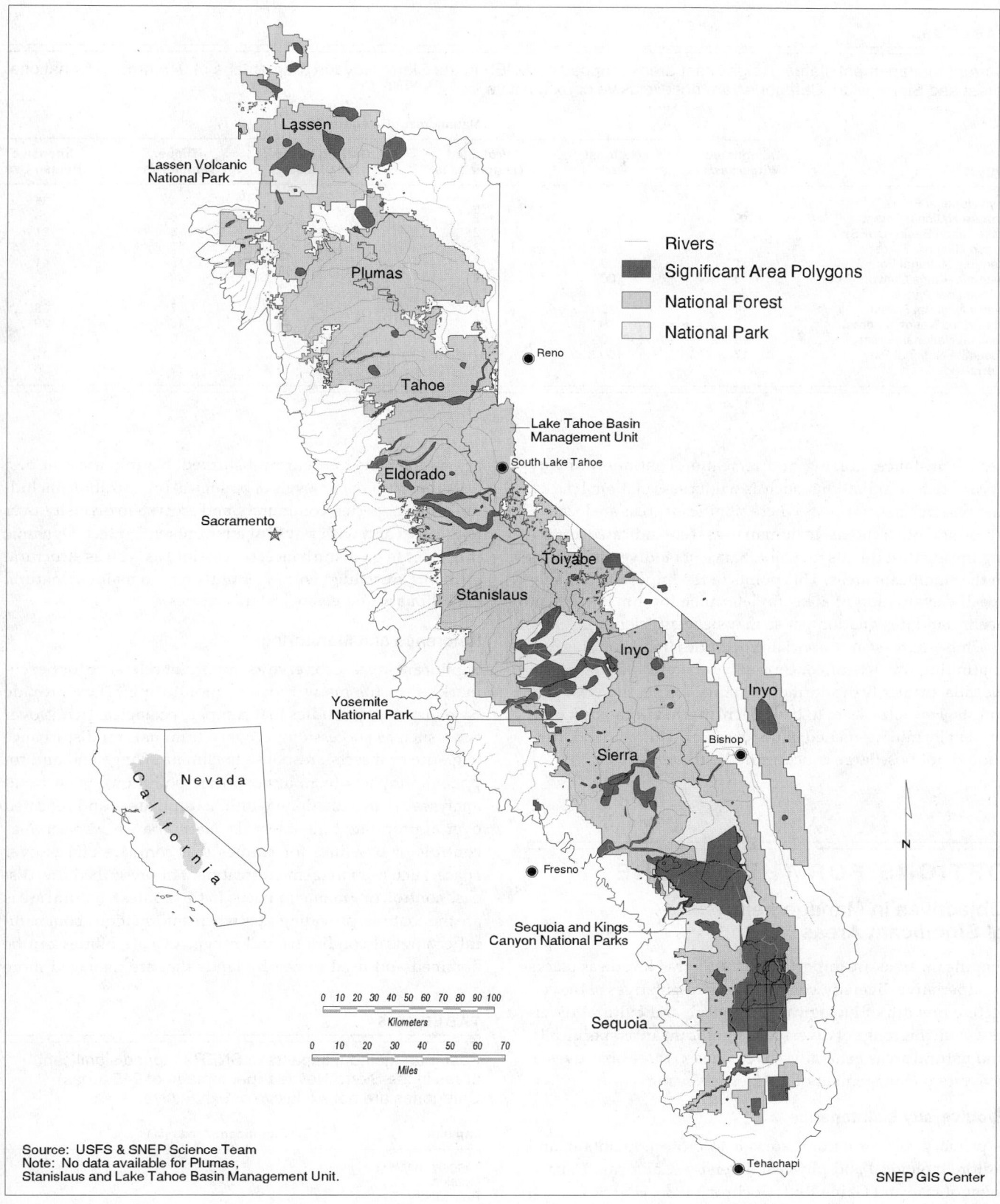

Rivers
Significant Area Polygons
National Forest
National Park

Lassen

Lassen Volcanic
National Park

Plumas

Reno

Tahoe

Lake Tahoe Basin
Management Unit

South Lake Tahoe

Eldorado

Sacramento

Toiyabe

Stanislaus

Inyo

Inyo

Yosemite
National Park

Bishop

Nevada

California

Sierra

Fresno

Sequoia and Kings
Canyon National Parks

0 10 20 30 40 50 60 70 80 90 100
Kilometers

0 10 20 30 40 50 60 70
Miles

Sequoia

Source: USFS & SNEP Science Team
Note: No data available for Plumas,
Stanislaus and Lake Tahoe Basin Management Unit.

Tehachapi

SNEP GIS Center

FIGURE 29.5

Significant cultural areas mapped by SNEP (Plumas and Stanislaus Counties and Lake Tahoe Basin Management Unit not included).

TABLE 29.4

Current management status of significant areas mapped by SNEP in the Sierra Nevada (percentage of 945 areas), by national forest and Sierra-wide. Categories are not exclusive or exhaustive.

| | Management Category (%) | | | | | |
Forest	Designated Wilderness	National Park	Protected Designations	Grazing Range Emphasis	Timber Emphasis	Intensive Human Use
Inyo National Forest	4	3	42	28	4	35
Lassen National Forest	6	1	42	18	6	25
Lake Tahoe Basin (USFS)	0	0	12	12	8	21
Tahoe National Forest	0	0	2	19	40	51
Sequoia National Forest	1	0	17	13	16	41
Sequoia–Kings Canyon National Park	9	100	48	2	0	8
Sierra National Forest	6	0	44	17	7	38
Stanislaus National Forest	1	0	23	16	16	29
Toiyabe National Forest	19	0	32	26	0	39
Yosemite National Park	17	100	31	8	0	17
Sierra-wide	4	26	32	14	12	36

be. For instance, grazing and recreation continue in many "protected" designations, such as wilderness, national parks, and research natural areas where significant areas are located. These and other intensive human uses were indicated as posing the greatest threats to maintenance of biodiversity values in the significant areas. This points to the importance of site-specific evaluation of areas to determine local management needs; land designation per se may be insufficient.

Timber harvest and associated activities (road building) are continuing impacts of concern in certain areas of the Sierra Nevada, primarily the northern national forests and west-side forests. Few sites seem to have been affected in the past or at present by mining-related issues or toxic pollutants, and these should not pose large concerns in the future.

OPTIONS FOR THE FUTURE

Objectives in Management of Significant Areas

Significant areas are important to the Sierra Nevada as places of superlative diversity, containing representatives of the collective breadth of biological, geological, and cultural diversity. With this range of diversities, significant areas specifically, and natural areas generally, have (at least) three objectives in ecosystem management of the Sierra Nevada.

Biodiversity Maintenance

A primary role for natural areas is to protect, maintain, and restore biological and physical diversity (California Department of Fish and Game 1991; Cochrane 1986; UNESCO 1974, 1984; World Resources Institute 1991). As examples of in situ conservation, these areas ideally would be sites of sufficient size and condition to enable natural ecological and evolution-

ary interactions to occur undisturbed. Natural areas at best would protect many levels of biological organization, including genetic, species, community, and ecosystem diversity, over long times and with physical environments intact. Dynamic processes (e.g., disturbance, succession) as well as structural elements are valued within natural areas to maintain natural evolutionary and ecological trajectories.

Reference and Monitoring

Significant areas also serve as important reference (or benchmark) sites for many types of monitoring. They provide baseline sites for studies that compare ecological trends over time, such as succession, recovery from natural disturbance (e.g., fire or insects), response to climate change and anthropogenic impacts (e.g., atmospheric pollution). Such trend analyses are usually done within natural areas and repeated over appropriate time intervals. Natural areas also serve as controls or baselines for studies that compare effects over space, such as management treatment of prescribed fire, disease control, or grazing impacts. In this context, natural areas are the controls, providing sites where information about natural ecological conditions and ranges of variabilities can be obtained and used in nearby lands that are managed more

TABLE 29.5

Past management impacts to SNEP-mapped significant areas in the Sierra Nevada (percentage of 945 areas). Categories are not exclusive or exhaustive.

Impacts	Significant Areas (%)
Logging, harvest	28
Grazing	40
Recreation	51
Roaded areas	27
Mining	10
Pollution	4

intensively. This latter role is increasingly important to ecosystem management, where objectives for landscapes often include reintroduction of natural structure and process. Natural areas provide places to observe and compare natural structure and process.

Research and Education

In addition to studies that provide information directly relevant to ecosystem management, significant areas can be important natural laboratories for research in general. Depending on management, such aspects as minimally disturbed conditions, longtime security, concentration of research studies, and protection of research equipment and experimental plots from vandalism are desirable to research. Similarly, natural areas are important outdoor classrooms for firsthand observation and study.

Integrated Management

SNEP's main assumption about significant area management is that special features for which the areas were selected are worthy of maintaining. Thus, a primary management objective for SNEP significant areas would be to maintain and safeguard the special features (elements) for which each area was recognized, including both short- and long-term needs for viability within the ecological and environmental contexts.

The goals of biodiversity protection, reference, monitoring, and research are best achieved with integrated management. Management of natural areas per se is critical, but equally important is coordinating with other areas managed for similar objectives, with management of adjacent matrix lands, and ultimately within watershed and landscape management (Salwasser 1991).

Natural Area Networks

It would be a major advance in achieving the goals of biodiversity protection, monitoring, and research if natural areas in the Sierra Nevada were better coordinated and managed jointly as part of a bio-geodiversity network (Diedrich et al. 1994; Noss 1983; Noss and Harris 1986; USFS 1992). The large and heterogeneous collection of lands managed for maintenance of special features and natural systems could represent in its cumulative nature a core web over the Sierra Nevada. At present this level of collective network is not achieved in the Sierra Nevada, although some individual programs are administered as integrated networks with regional-to local-level planning, targets, and goals (e.g., USFS Research Natural Areas Program). There is little integrated planning among programs within agencies, however (e.g., USFS programs on wilderness, research natural areas, and special interest areas are not coordinated), and only some interagency collaboration at any level. An interagency natural areas coordinating committee, which functioned to provide communication among agencies on natural areas efforts, has been superseded by the California Executive Biodiversity Council, which does not maintain the natural areas communication function that the original committee attempted.

Communication and functional coordination both among programs within agencies and among agencies necessitates analysis at the ecoprovince level, irrespective of administrative ownership. This would include evaluation of elements (species to ecosystems, cultural to geological) present across the landscape, representation by natural areas of different category and administration, and gaps in representation, and development of a coordinated planning and management strategy among the programs and agencies. The SNEP GAP data (Davis and Stoms 1996) primarily, with the SNEP significant areas inventory and other SNEP assessments, provides inventory and analysis of the first points. It remains an option for agencies and landowners in the Sierra Nevada to coordinate lands and programs into networks so as to achieve higher levels of integration and improved efficiency in conservation functions (Pressey et al. 1993).

Bioregional Integration

In addition to integrating natural areas into regional networks, integrating natural areas with management of matrix and adjacent lands both locally and regionally adds efficiency to achieving conservation goals (Dyer and Holland 1991; Noss 1983). An immediate opportunity is to analyze needs and uses of all lands within a local watershed or landscape (e.g., landscape or watershed analysis [Interagency Team 1994; Manley et al. 1995]), and then manage accordingly (e.g., Mammoth-June case study [Millar 1996a]). Bioregional analyses consider "two-way management," that is, reciprocal needs of adjacent units. For instance, opportunities to provide functionally large habitat areas for organisms that use a natural area may be promoted in multiple-use lands adjacent to natural areas. Depending on the organisms, relatively intensive uses might be applied on lands adjacent to significant areas, as long as specific needs for organisms and processes are provided. Conversely, information obtained within natural areas may serve to inform managers about best practices on adjacent multiple-use lands. For instance, natural areas can provide information that is useful when developing silvicultural prescriptions on adjacent lands, such as number of snags per acre, or size and age-class distribution of dominant trees, density and use of nesting trees by birds, or historic fire intervals (Millar 1996a). High-quality natural areas, where natural processes predominate, can be important places to show managers and public about conditions that exist in minimally disturbed states and thus avert conflict over what "might have been" under no-disturbance management.

SNEP has developed a set of tools for developing scenarios of regional integration for biodiversity protection and restoration (Davis et al. 1996). This model uses preexisting reserves, recognized natural areas, and areas of known value, such as significant areas, as a starting point to build a network. Significant areas as part of this scheme are considered in that report.

As discussed earlier (and in Davis et al. 1996; Davis and Stoms 1996), significant areas specifically and natural areas generally vary in their status regarding biodiversity protection. For objectives of reference/monitoring and research, however, most natural areas in the Sierra Nevada, both those designated and those just recognized as containing special features, are vastly underused. Some areas have become well known for research, based on attributes of high or exemplary diversity, ecological, cultural, or geological integrity, research protection, on-site research facilities, cumulative knowledge gain, and publication familiarity. Areas that currently attract and receive research attention are not necessarily those located in common or widespread ecosystems, nor are they in community types of high interest to managers. Thus, many existing sites could benefit by concerted programs that focus on research in basic biological or physical mechanisms. The University of California Natural Reserve System is exemplary. Monitoring for baseline ecological trends or management treatment comparison is greatly underutilized on natural areas in the Sierra Nevada. Natural scientists and managers alike would benefit by programmatic approaches to monitoring that take advantage of the benchmark conditions offered by natural areas in the Sierra Nevada.

ACKNOWLEDGMENTS

We acknowledge with appreciation all the individuals who contributed to identifying and mapping significant areas throughout the Sierra Nevada. Without their extensive, intensive knowledge and willing contributions, the inventory would never have been compiled. Kay Gibbs, John Gabriel, Karen Gabriel, and Russ Jones in the SNEP GIS center developed the database and map layers for this project and conducted analyses. We thank them all for their hard work. Of these, we especially thank Kay, whose dedicated efforts and care to detail ensured high-quality maps and information.

REFERENCES

Allen, B. H. 1987. *Ecological type classification for California: The Forest Service approach*. Berkeley, CA: U.S. Forest Service, Pacific Southwest Research Station.

Anderson, M. K., and M. J. Moratto. 1996. Native American land-use practices and ecological impacts. In *Sierra Nevada Ecosystem Project: Final report to Congress*, vol. II, chap. 9. Davis: University of California, Centers for Water and Wildland Resources.

Beesley, D. 1996. Reconstructing the landscape: An environmental history, 1820-1960. In *Sierra Nevada Ecosystem Project: Final report to Congress*, vol. II, chap. 1. Davis: University of California, Centers for Water and Wildland Resources.

Bettinger, R. L. 1991. Native land use: Archaeology and anthropology. In *Natural history of the White-Inyo Range, eastern California*, edited by C. A. J. Hall. Berkeley and Los Angeles: University of California Press.

Blackburn, T. C., and K. Anderson, eds. 1993. *Before the wilderness: Environmental management by Native Californians*. Menlo Park, CA: Ballena Press.

Bonnicksen, T. M. 1988. Standards of naturalness: The national parks management challenge. *Landscape Architecture* 78 (22): 120–34.

Botkin, D. B. 1990. *Discordant harmonies: A new ecology for the twenty-first century*. New York: Oxford University Press.

California Department of Fish and Game. 1991. Fact sheets for the Sierra region study area maps. Sacramento: Resources Agency.

Cheatham, N. H., W. J. Barry, and L. Hood. 1977. Research natural areas and related programs in California. In *Terrestrial vegetation of California*, edited by M. G. Barbour and J. Major. Davis: California Native Plant Society.

Cheatham, N. H., and J. R. Haller. 1975. An annotated list of California habitat types. Berkeley: University of California.

Cochrane, S. 1986. Programs for the preservation of natural diversity in California. Sacramento: California Department of Fish and Game, Nongame Heritage Program.

Crow, T. R. 1991. Landscape ecology: The big picture approach to resource management. In *Challenges in the conservation of biological resources: A practitioner's guide*, edited by D. J. Decker, M. E. Karsny, G. R. Geoff, C. R. Smith, and D. W. Gross. Boulder, CO: Westview Press.

Davis, F. W., and D. M. Stoms. 1996. Sierran vegetation: A gap analysis. In *Sierra Nevada Ecosystem Project: Final report to Congress*, vol. II, chap. 23. Davis: University of California, Centers for Water and Wildland Resources.

Davis, F. W., D. M. Stoms, R. L. Church, W. J. Okin, and N. L. Johnson. 1996. Selecting biodiversity management areas. In *Sierra Nevada Ecosystem Project: Final report to Congress*, vol. II, chap. 58. Davis: University of California, Centers for Water and Wildland Resources.

Delcourt, H. R., and P. A. Delcourt. 1991. *Quaternary ecology*. New York: Chapman and Hall.

Diaz, N., and D. Apostol. 1992. Forest landscape analysis and design: A process for developing and implementing land management objectives for landscape patterns. Portland, OR: U.S. Forest Service, Pacific Northwest Region.

Diedrich, J., A. Evenden, S. Greene, D. Harmon, M. Peterson, and S. Sater. 1994. The role of natural areas in the Columbia River Basin assessment and planning. U.S. Forest Service Internal Report.

Duane, T. P. 1996. Human settlement, 1850–2040. In *Sierra Nevada Ecosystem Project: Final report to Congress*, vol. II, chap. 11. Davis: University of California, Centers for Water and Wildland Resources.

Dyer, M. I., and M. M. Holland. 1991. The biosphere-reserve concept: Needs for a network design. *BioScience* 41:319–25.

Federal Committee on Ecological Reserves. 1977. *A directory of research natural areas on federal lands of the United States of America*. Washington DC: U.S. Forest Service.

Fiedler, P. L., and J. J. Ahouse. 1992. Hierarchies of cause: Toward an understanding of rarity in vascular plant species. In *Conservation biology: The theory and practice of nature conservation and management*, edited by P. L. Fiedler and S. K. Jain. New York: Chapman and Hall.

Forman, R. T., and M. Godron. 1986. *Landscape ecology*. New York: John Wiley.

Frankel, O. H., and M. E. Soule. 1981. *Conservation and evolution*. Cambridge: Cambridge University Press.

Franklin, J. F., and J. A. Fites-Kaufmann. 1996. Analysis of late successional forests. In *Sierra Nevada Ecosystem Project: Final report to Congress,* vol. II, chap. 21. Davis: University of California, Centers for Water and Wildland Resources.

Harris, L. 1984. *The fragmented forest: Island biogeography theory and the preservation of biotic diversity.* Chicago: University of Chicago Press.

Hoerr, W. 1993. The concept of naturalness in environmental discourse. *Natural Areas Journal* 13:29–32.

Holland, R. F. 1986. Preliminary descriptions of the terrestrial natural communities of California. Sacramento: California Department of Fish and Game, Nongame Heritage Program.

Hoshovsky, M. 1994. Biodiversity considerations for natural areas. Sacramento: California Department of Fish and Game.

Huber, N. K. 1981. Amount and timing of late Cenozoic uplift and tilt of the central Sierra Nevada, California: Evidence from the upper San Joaquin River Basin. *U.S. Geological Survey Professional Papers* 1197:1–28.

Interagency Team. 1994. A federal agency guide for pilot watershed analysis. Portland, OR: Federal Interagency Report.

Kaufman, W. 1993. How nature really works. *American Forests* 99 (2, 3): 17–19, 59–61.

Keeler-Wolf, T. 1985. *Inventory of research natural areas of California.* Vol. GTR-125. Berkeley, CA: U.S. Forest Service, Pacific Southwest Research Station.

Keystone Center. 1990. Biological diversity on federal lands. Report of a Keystone policy dialogue. The Keystone Center, Keystone, Colorado.

Kinney, W. C. 1996. Conditions of rangelands before 1905. In *Sierra Nevada Ecosystem Project: Final report to Congress,* vol. II, chap. 3. Davis: University of California, Centers for Water and Wildland Resources.

Manley, P. N., G. E. Brogan, C. Cook, M. E. Flores, D. G Fullmer, S. Husari, T. M. Jimerson, L. M. Lux, M. E. McCain, J. A. Rose, G. Schmitt, J. C. Schuyler, and M. J. Skinner. 1995. *Sustaining ecosystems: A conceptual framework.* San Francisco: U.S. Forest Service, Pacific Southwest Region and Station.

May, R. M. 1973. *Complexity and stability.* Princeton, NJ: Princeton University Press.

McPhee, J. 1993. *Assembling California.* New York: Farrar, Straus, and Giroux.

Millar, C. I. 1996a. The Mammoth-June Ecosystem Management Project, Inyo National Forest. In *Sierra Nevada Ecosystem Project: Final report to Congress,* vol. II, chap. 50. Davis: University of California, Centers for Water and Wildland Resources.

Millar, C. I. 1996b. Tertiary vegetation history. In *Sierra Nevada Ecosystem Project: Final report to Congress,* vol. II, chap. 5. Davis: University of California, Centers for Water and Wildland Resources.

Moyle, P. B. 1996. Potential aquatic diversity management areas. In *Sierra Nevada Ecosystem Project: Final report to Congress,* vol. II, chap. 57. Davis: University of California, Centers for Water and Wildland Resources.

Moyle, P. B., and P. J. Randall. 1996. Biotic integrity of watersheds. In *Sierra Nevada Ecosystem Project: Final report to Congress,* vol. II, chap. 34. Davis: University of California, Centers for Water and Wildland Resources.

Norris, R. M., and R. W. Webb. 1990. *Geology of California.* 2nd ed. New York: John Wiley.

Noss, R. 1983. A regional landscape approach to maintaining diversity. *BioScience* 33:700–706.

Noss, R., and L. Harris. 1986. Nodes, networks, and MUMs: Preserving diversity at all scales. *Environmental Management* 10:299–309.

Parker, I., and W. J. Matyas. 1979. Vegetation mapping and classification in California. San Francisco: U.S. Forest Service.

Pimm, S.L. 1986. Community stability and structure. In *Conservation biology: The science of scarcity and diversity,* edited by M. E. Soule. Sunderland, MA: Sinauer Associates.

Pressey, R. L., C. J. Humphries, C. R. Margules, R. I. Vane-Wright, and P. H. Williams. 1993. Beyond opportunism: Key principles for systematic reserve selection. *Trends in Ecology and Evolution* 8: 124–28.

Rabinowitz, D. 1981. Seven forms of rarity. In *The biological aspects of rare plant conservation,* edited by H. Synge. New York: John Wiley.

Salwasser, H. 1991. Roles for land and resource managers in conserving biological diversity. In *Challenges in the conservation of biological resources: A practitioner's guide,* edited by D. J. Decker, M. E. Krasny, G. R. Goff, C. R. Smith, and D. W. Gross. Boulder, CO: Westview Press.

Sawyer, J., and T. Keeler-Wolf. 1996. *A manual of California vegetation: Series-level descriptions.* Sacramento: California Native Plant Society.

Schoener, T. W. 1987. The geographic distribution of rarity. *Oecologia* 74:161–73.

Shevock, J. R. 1996. Status of rare and endemic plants. In *Sierra Nevada Ecosystem Project: Final report to Congress,* vol. II, chap. 24. Davis: University of California, Centers for Water and Wildland Resources.

Stine, S. 1996. Climate, 1650-1850. In *Sierra Nevada Ecosystem Project: Final report to Congress,* vol. II, chap. 2. Davis: University of California, Centers for Water and Wildlsnd Resources.

Turelli, M. 1978. A reexamination of stability in random versus deterministic environments with comments on the stochastic theory of limiting similarity. *Theoretical Population Biology* 13: 244–67.

U.S. Congress. 1987. *Technologies to maintain biological diversity.* Washington, DC: U.S. Government Printing Office.

UNESCO. 1974. Task force on criteria and guidelines for the choice and establishment of biosphere reserves. *Man and the Biosphere.* Report 22, Paris: UNESCO.

———. 1984. Action plan for biosphere reserves. *Nature and Resources* 20 (4):11–22.

U.S. Forest Service (USFS). 1992. National strategy for U.S. Forest Service research natural areas. Washington, DC: U.S. Forest Service.

Wilson, E. O. 1988. *Biodiversity.* Washington DC: National Academy Press.

Woolfenden, W. B. 1996. Quaternary vegetation history. In *Sierra Nevada Ecosystem Project: Final report to Congress,* vol. II, chap. 4. Davis: University of California, Centers for Water and Wildland Resources.

World Resources Institute. 1991. Biodiversity conservation strategy for North America The Keystone Center, Keystone, CO.

RICHARD KATTELMANN
University of California Sierra Nevada
 Aquatic Research Laboratory
Mammoth Lakes, California

30

Hydrology and
Water Resources

ABSTRACT

Water is a critical component of the resource issues and conflicts of the Sierra Nevada. Almost every environmental dispute in the range involves water as principal or secondary concern. Most human activities have some potential to influence the quantity, distribution, or quality of water.

Rivers of the Sierra Nevada appear to have shown remarkable resiliency in recovering from the gold mining era; however, so few channels were left untouched by historic disturbances that reference streams in a completely natural state may not exist for comparison. Water management structures developed concurrently with hydraulic mining and have since come to dominate the flows of water from the Sierra Nevada. Few river systems in the range have natural flow regimes over much of their length. In most river basins, this active management of the water itself affects the annual water balance, temporal distribution, flood hydrology, minimum flows, and water quality much more than any human disturbance of the landscape. Ironically, the primary benefits to society of water from the Sierra Nevada cause the primary impacts. By trying to serve the so-called highest beneficial uses, domestic water supply and production of food and power, we have caused the greatest impacts.

Watershed disturbance in the form of mining, road building, logging, grazing, fire, residential development, and other uses has altered vegetation and soil properties in particular areas. Where these disturbances have altered a large fraction of a watershed, including areas near stream channels, flows of water and sediment may be changed significantly. Nevertheless, major changes in hydrologic processes resulting from watershed disturbance have been noticed in only a few streams. More extensive changes are suspected, but they have not been detected because of the minimal monitoring network that is in place. Proposed programs for reducing the amounts of fuels in forests have potential for significant aquatic impacts; however, catastrophic wildfire carries far greater risks of grave damage to aquatic systems.

INTRODUCTION

> *Water, in all its forms, is indeed the crowning glory of the Sierra. Whether in motion or at rest, the waters of the Sierra are a constant joy to the beholder. Above all, they are the Sierra's greatest contribution to human welfare.*
>
> Farquhar 1965, 1

Water is central to the resource issues and conflicts of the Sierra Nevada. Changes in water availability, stream-flow quantity and timing, flooding, quality of surface and ground water, aquatic and riparian habitat, soil erosion, and sedimentation have occurred throughout the range as results of land disturbance and resource management (Kattelmann and Dozier 1991). However, the magnitude of such changes, their relative importance, and the ability of natural and human communities to adapt to or recover from alterations in hydrologic processes in the Sierra Nevada are largely unknown. Concern about degradation of water quality is widespread in public reaction to past and proposed resource management activities. Californians need to know whether their primary water source, the Sierra Nevada, is functioning well in general and what problems need attention.

The Sierra Nevada generates about 25 km^3 (20 million acre-feet [AF]) of runoff each year out of a total for California of about 88 km^3 (71 million AF) or about 28% (Kahrl 1978; California Department of Water Resources 1994). This runoff accounts for an even larger proportion of the developed water resources and is critical to the state's economy. The rivers of the Sierra Nevada supply most water used by California's cities, agriculture, industry, and hydroelectric facilities. The storage and conveyance systems developed to utilize the water resources of the Sierra Nevada are perhaps the most

Sierra Nevada Ecosystem Project: Final report to Congress, vol. II, *Assessments and scientific basis for management options*. Davis: University of California, Centers for Water and Wildland Resources, 1996.

extensive hydrotechnical network in the world. Major water supply systems have tapped the Tuolumne River for San Francisco, the Mokelumne River for Alameda and Contra Costa Counties, eastern Sierra streams for Los Angeles, and the Feather River for the San Joaquin valley and other parts of southern California. Irrigated agriculture throughout California consumes more than the annual runoff of the Sierra Nevada and accounts for more than 90% of consumptive use in the state (U.S. Geological Survey 1984; California Department of Water Resources 1994). More than 150 powerhouses on Sierra Nevada rivers produce about 24 million megawatt-hours of electricity per year (see Stewart 1996). Operations of most of the water projects are quite sensitive to fluctuations in climate over periods of a few years. Sierra Nevada rivers support extensive aquatic and riparian communities and maintain the Sacramento–San Joaquin Delta estuary ecosystems (see Jennings 1996; Moyle 1996; Moyle and Randall 1966; Moyle et al. 1996; Erman 1996).

Perhaps the most common perception of water from the Sierra Nevada is no perception at all, merely benign ignorance. For many, water is something that appears at the kitchen faucet, showerhead, garden hose, or is a choice among bottled beverages. Water rarely makes the general news except in times of serious shortage or excess. Agricultural and urban communities of the Central Valley that are dependent on water from Sierra Nevada rivers probably have the greatest direct interest in water issues, but they are chiefly concerned about the amount delivered and how fisheries policies might affect those deliveries. Most residents of the Sierra Nevada are probably knowledgeable and concerned about local water supplies and ground water but are not known to harbor any common misperceptions about the local resource, just a shared hope that there always will be enough water available. People in cities benefiting from water supplies exported from the Sierra Nevada are concerned about quantity and quality of water at the tap, but many are unsure about the source of their water. Visitors to the Sierra Nevada are usually concerned about the aesthetic qualities of water that they see. Environmentally conscious segments of the public may believe the water resources of the Sierra Nevada are substantially degraded. Serious water problems in parts of the Sierra Nevada and throughout the country may be extrapolated and perceived as occurring throughout the Sierra Nevada. For example, if poor logging practices in the Pacific Northwest are initiating landslides and ruining fish habitat, then some people may assume the same things are happening within the Sierra Nevada. Water issues highlighted in popular books (e.g., Reisner 1986; Postel 1992; Doppelt et al. 1993; Palmer 1994) are often assumed to apply to the Sierra Nevada but may not be of similar severity.

Water flowing from the Sierra Nevada has far-reaching effects. On the western slope, runoff naturally flowed through the Central Valley of California and San Francisco Bay to the Pacific Ocean or, in the south, contributed to Tulare and Buena Vista Lakes. On the eastern slope, streams flowed toward the terminal lakes of the western Great Basin. In all cases, the waters of the Sierra Nevada enriched the lands through which they flowed. In the past century, the fluid wealth of the mountains has been extended well beyond natural hydrographic boundaries through engineering projects to distant agricultural and urban areas. Electricity generated from falling water in the Sierra Nevada and distributed through the western power grid affects distant communities. Crops grown with and containing water precipitated over the Sierra Nevada are sold around the world. The recreational and aesthetic qualities of Sierran rivers and lakes attract visitors from throughout the United States and the world. Artwork portraying water in the Sierra Nevada is found around the globe; for example, a watercolor mural in traditional Chinese style of waterfalls in Yosemite Valley hangs in the Taipei airport as an example of Chinese scenery.

Water has played a critical role in Euro-American affairs in the Sierra Nevada since the discovery of gold in a channel leading to a water-powered sawmill in 1848. Water was essential to large-scale gold mining and processing. Water development for mining led to one of the nation's earliest major decisions in environmental law (that halted hydraulic mining) and to our intricate network of hydrotechnical structures that transfer water from the Sierra Nevada to farms, cities, and powerhouses. Conflicts over water from the Sierra Nevada are likely to be a continuing part of the California scene. Water is simply too valuable to society and all forms of life to be anything but a high priority for resource policies and management. Water eventually emerges in almost all environmental disputes, even when the debate starts on some other distinct issue. All parties to the dispute can usually agree that water is an influence on or is influenced by the original issue. Water is tied to all other issues considered by SNEP, with some links more obvious than others, but it is literally an integral component of the ecosystem approach.

GENERAL STATE OF KNOWLEDGE

Despite the importance of water to California, there have been remarkably few integrative studies of water resources in the state or the Sierra Nevada. State agencies have issued reports about statewide water matters for more than a century (e.g., Hall 1881; Conservation Commission 1913; California Department of Public Works 1923). The first *California Water Plan* was released by the Department of Water Resources in 1957. Originally a description of proposed water projects, updates to the *California Water Plan* have evolved into a more thorough evaluation of water supply, demand, and management (e.g., California Department of Water Resources 1994). Comprehensive descriptions of water in the state appear in books by Harding (1960), Seckler (1971), and the Governor's Office of Planning and Research (Kahrl 1978). The history of water development in California is treated by Hundley (1992). The condition of California's rivers is assessed by the California State Lands Commission (1993). Possible scenarios of the future of water resources in California have been developed by the California Department of Water Resources (1994) and the Pacific Institute (Gleick et al. 1995). Although all these books deal with the Sierra Nevada as a critical part of the California waterscape, and books devoted to the Sierra Nevada (e.g., Peattie 1947; Lee 1962; Johnston 1970; Webster 1972; Bowen 1972; Palmer 1988) at least mention water resources, a thorough treatment of water in the Sierra Nevada has yet to be written. Thousands of articles, chapters, and reports address the various aspects of hydrology and water resources in the Sierra Nevada, but there has been little synthesis of this vast work. The isolated, topical work provides a wealth of information about specific details but does not inform society about the context of that work at the scale of the mountain range or even of a river basin. In addition, there are serious gaps in the collection of information about Sierra Nevada waters. Although knowledge is far from complete for most aspects of the water-resource situation, the most troubling gap is the virtual absence of experimental research on hydrologic impacts of land management activities. Because of this near lack of local research, we usually had to infer the likely consequences of disturbance from studies done outside the Sierra Nevada. In addition, the state does not have a thorough description of each river basin that would be adequate for environmental assessments. Comprehensive lists of environmental problems in each river basin do not exist. There is no consistent method for characterizing watersheds. Absence of consistent criteria for evaluating ecological conditions along streams or in watersheds inhibits assessment of management consequences or need for restoration (California State Lands Commission 1993).

OVERALL APPROACH AND SOURCES

This assessment was primarily a literature review augmented with the author's experiences throughout the Sierra Nevada over the past two decades and a few weeks of specific field checking during the SNEP period. The libraries of the University of California and the Water Resources Center Archives at Berkeley, in particular, were critical to the effort. Offices of the national forests in the Sierra Nevada also provided a wealth of documents. Other materials were provided by dozens of agencies and individuals. Newspapers were essential sources of current information. Interviews with agency personnel and private parties augmented the written word. The primary challenges were to compile and synthesize the diversity of material. The quantity and quality of information

gathered varied widely between river basins throughout the range. Important sources were undoubtedly overlooked because of ignorance of their existence and inability to actually locate all known sources. One of the critical assumptions of this assessment was that the reported material was indeed reliable. Multiple sources of information that were consistent provided greater confidence in most material. Information was organized by resource, by impact, and geographically by river basin. Use of natural hydrologic areas was a central tenet of this effort. Consideration of nested catchments from headwaters to large river basins provides a logical hierarchy that makes physical and ecological sense. Watersheds are becoming a more common unit of analysis and planning. The California Resources Agency is organizing many of its programs on a watershed basis and has adopted a watershed delineation scheme called Calwater. This system was used in this study and by other parts of SNEP. River basins and major streams of the study area are identified in figures 30.1–30.3.

Attributes of Water

There are several attributes of water and streams that are impacted by management activities. The physical attributes are briefly described in the following paragraphs (see Moyle 1996; Moyle and Randall 1996; Erman 1996; Moyle et al. 1996; Jennings 1996 for biological impacts). The present study did not perform any systematic analyses of these attributes. Such analyses (within the constraints of readily available data) would not provide clear indications of the health of the hydrologic system of the Sierra Nevada. Instead, synthesis of existing analyses originally performed for various other purposes provided the basis of this assessment.

Stream flow (or stream discharge) is the most fundamental aspect of watershed hydrology. Stream flow will usually be addressed in this assessment just as a concept: volume of water passing by a point on a stream over some period of time. Fortunately, this concept is also measured at hundreds of sites within the SNEP study area. However, the number of sites with data useful to this study is much more limited, numbering in the dozens. Most stream-flow measuring stations are located in association with some water management project rather than for scientific study. Therefore, most information is available on highly regulated streams that suggest little about hydrologic response to changes in the landscape other than the direct manipulation of water in the channel. Gauges on unregulated (often called unimpaired) streams often have short or incomplete records or are sited in locations inappropriate for any particular after-the-fact study. The number of such gauges in the Sierra Nevada has decreased with time as costs have risen. Stream gauging stations are operated by the U.S. Geological Survey, utilities, irrigation districts, and a few other public agencies. Many of the records are published as daily values in annual volumes by the U.S. Geological Survey (USGS). Most of these records are now available on CD-

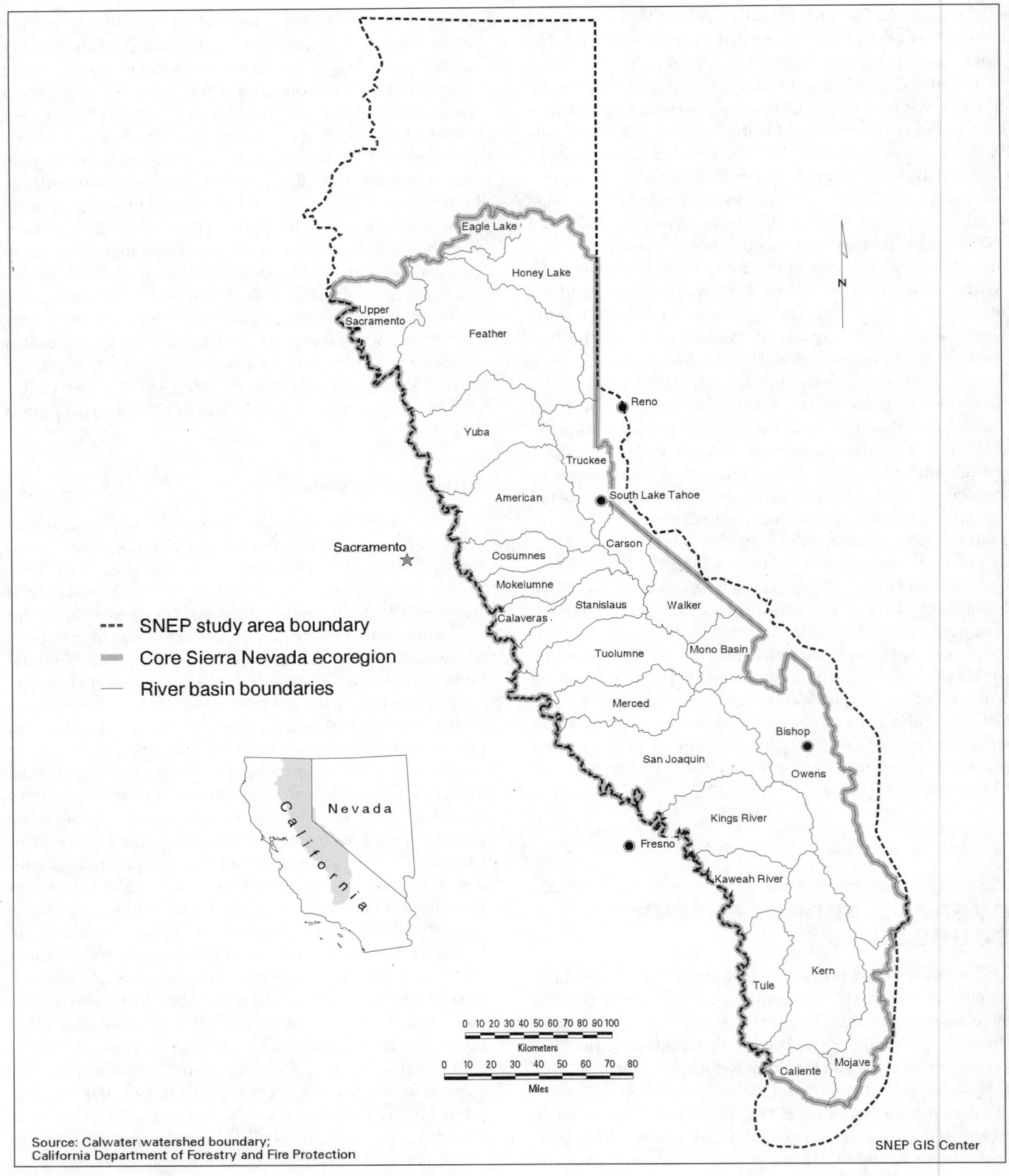

FIGURE 30.1

Major river basins of the SNEP core study area.

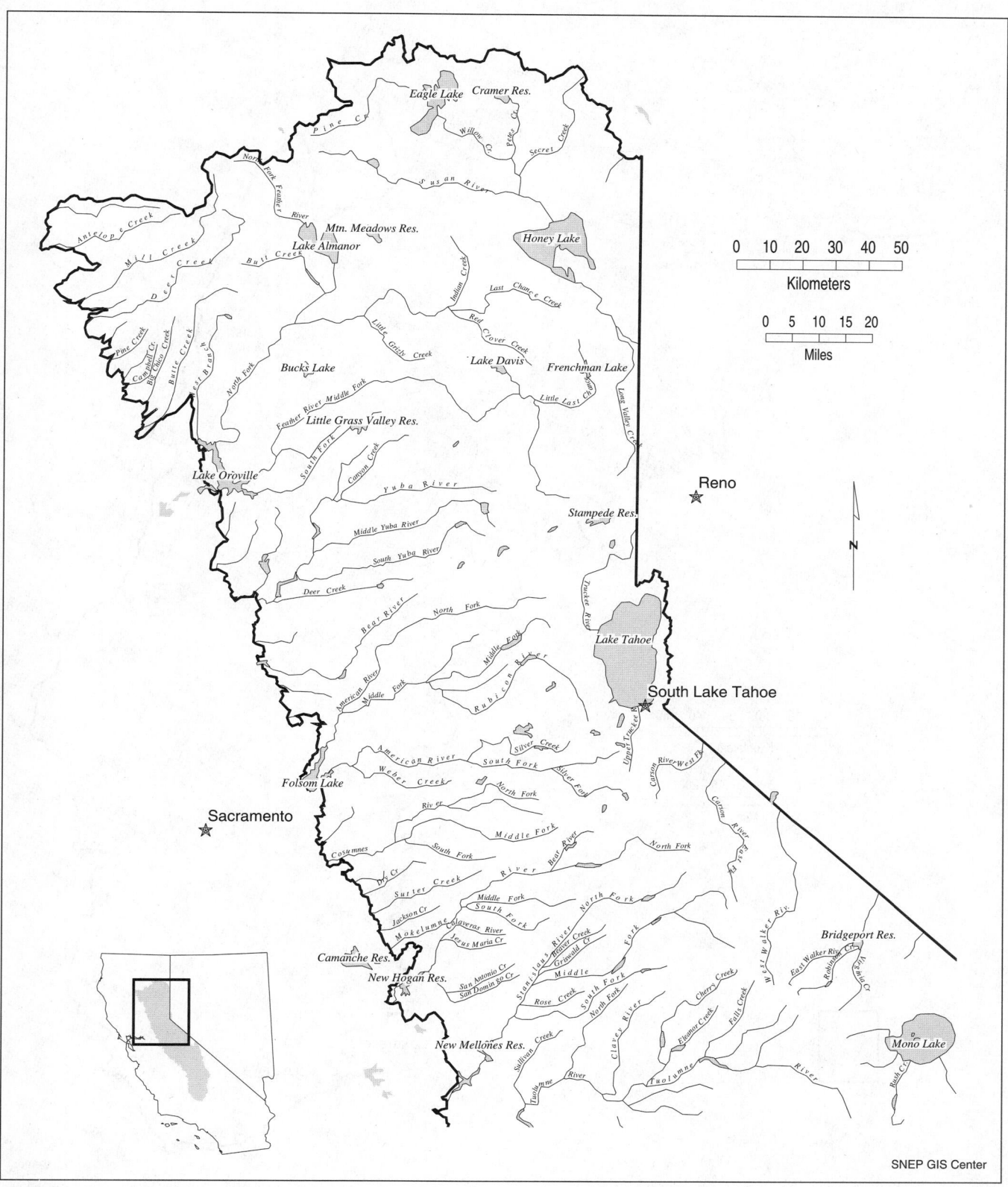

FIGURE 30.2

Principal rivers and streams of the northern half of the SNEP core study area.

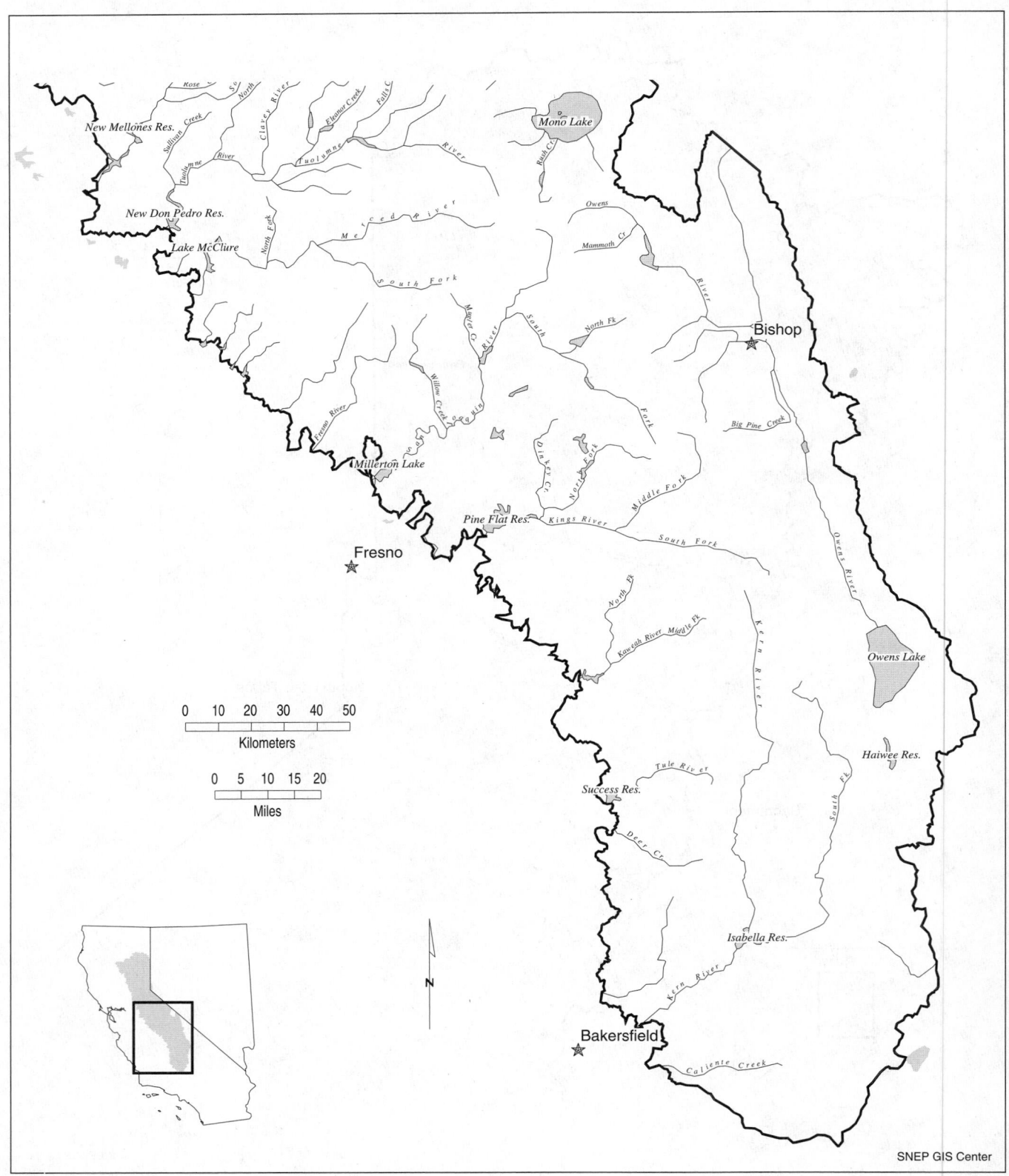

FIGURE 30.3

Principal rivers and streams of the southern half of the SNEP core study area.

ROM from private firms and the USGS. The usual characteristics of stream flow that are studied are the annual volume, distribution over time (i.e., annual hydrograph), maximum flows, and minimum flows. Measurement of flows in natural channels is not a trivial exercise, and errors can exceed 10% to 20% of the measured value (Herschy 1985). Natural variability in all aspects of stream flow can be quite high.

Sediment is the other main constituent of the fluid flowing in streams that we casually call water. Changes in sediment removal, transport, and deposition affect the general nature of stream channels and riparian areas and their biota, as well as affecting human uses of water. Mineral particles eroded from the land surface and transported mostly by flowing water gradually (or sometimes suddenly) move downslope from the mountains. These particles of various sizes are either suspended in the water or bounce along the channel as bedload. Both types of sediment move episodically as the capacity of streams to transport sediment varies with flow velocity. Sediment can be stored in a channel for years (or even centuries) before conditions are right to dislodge and transport it. Individual particles have a discontinuous journey downstream, with intermittent advances of varying lengths interrupted by temporary storage of varying duration. Suspended sediment is sampled at few stations throughout the Sierra Nevada and is a marginal measure of total sediment load. Bedload moving past a point is not routinely measured anywhere in the Sierra Nevada, but it has been measured in special studies (e.g., Andrews and Erman 1986). Repeated surveys of the bottom topography of natural and artificial lakes and calculation of the change in volume over the time interval is the best means of estimating total sediment transport (Dunne and Leopold 1978). However, this technique integrates sediment production from a large area and duration and, therefore, is difficult to associate with particular land-use activities. It is also very expensive.

Most other materials that are found in flowing water constitute the dissolved load of streams. A variety of ions occur naturally in streams, although the waters of the Sierra Nevada tend to have relatively low amounts of dissolved constituents compared with other rivers of the world (e.g., California State Water Resources Control Board 1992a). The chemical quality of streams is routinely measured at only a few gauging stations in the Sierra Nevada. There is also a biotic component of streams, ranging from viruses and bacteria to invertebrates and fish (see Moyle 1996; Moyle et al. 1996; Erman 1996).

Water temperature is another important attribute of streams, particularly with respect to suitable conditions for aquatic life. Some creatures can tolerate only relatively narrow ranges in temperature at different stages in their life cycles. The amount of dissolved oxygen also varies with temperature, decreasing as temperature increases. As with other water quality parameters, temperature is measured at only a few river gauging stations. Stream temperature varies primarily with stream discharge, original temperature of water in-

puts, exposure to sunlight, geothermal conditions, and temperature of reservoir releases.

Attributes of ground water that are of primary interest and are subject to change are the amount and quality of water in storage. Ground-water conditions in the Sierra Nevada are not routinely monitored in the manner of stream flow. Public utilities that pump ground water for water supply monitor their own wells but do not systematically report their results. Most of the publicly available information about ground water in the Sierra Nevada is from a handful of special studies.

Scale

Consideration of the scale of hydrologic impacts is crucial to understanding how water resources are affected by disturbance. A point of reference is necessary. Usually, some particular point along a stream where measurements are made of flow and/or quality parameters provides the geographic context. Impacts upstream of that point may have some measurable effect at the reference site. Activities in the river channel itself are likely to produce the greatest noticeable impacts in the channel at the point of reference. Activities near the channel in areas with occasional hydraulic connection to the channel will also have direct impacts (i.e., change in water or sediment yield) in the channel under consideration. Such areas may be surface-runoff contributing zones (sometimes called variable source areas) that yield water as sheet flow or near-surface pipe-flow in response to rainfall. At greater distances from main channels or ephemeral tributaries, water resulting from rainfall or snowmelt moves slowly downslope through soil or subsoil. Alteration of hillslope properties at locations distant from the stream channel simply has less opportunity to make a difference at the downslope and downstream point of reference. To restate the typical effect of geographic location on hydrologic impacts, a given disturbance matters less on a ridgetop than adjacent to a channel.

Cumulative Effects

One must also consider the combined or cumulative effects of activities on attributes of water at a point of reference. Altering the local water balance of a small fraction of a watershed or even adding a small quantity of pollutant to a stream usually will not result in any detectable change at some distant downstream point of reference. However, altering many small fractions of the watershed or adding the small quantity of pollutant at many places along the stream will cause a detectable change downstream. Even though each individual impact is insignificant with respect to the whole watershed, their cumulative effects may be dire. An instructive example of cumulative watershed effects occurs in the Lake Tahoe Basin, an easily visualized hydrologic unit in which the lake is the item of reference. Construction of roads, houses, casinos, parking lots, ski runs, septic systems, and so on initially affected only the immediate area of the particular development.

However, at some time, perhaps in the 1960s, there were so many individual disturbances that the nutrient balance of the lake was profoundly changed and algal production increased, with a consequent decrease in lake clarity (e.g., Goldman 1974 and 1990). Similarly, while a small diversion from a stream for irrigation might not be detectable at a downstream point, hundreds of small diversions can totally dry up a stream. Ground-water overdraft is typically the cumulative result of hundreds of small extractions. Therefore, when considering the potential impact of some activity on water resources, one must examine the intensity of the impact, how extensive it is (what fraction of the watershed is affected), the proximity of the activity to a stream channel, what other impacts in the watershed it is adding to, and the degree of recovery from past impacts. These questions of scale are implicitly addressed throughout this chapter.

HISTORY OF IMPACTS ON WATER RESOURCES

Examination of past impacts to streams and rivers helps us understand their current condition. Impacts of Native Americans on the hydrologic system appear to have been minor, largely because of the comparatively small population in the mountains and limited technology. Their deliberate use of fire as a vegetation-management tool would have been the primary agent in altering local hydrology. To the extent that intentional fires removed vegetation, evapotranspiration was reduced, water yields were increased, and surface erosion was increased. The geographical extent, intensity, and frequency of such fires cannot be quantified. Therefore, about all we can say concerning the hydrologic consequences of this activity is that there were some. Areas near to population centers were probably impacted to a greater degree than remote areas. Little is known about water development by Native Americans. Perhaps the best documented projects occurred on Bishop and Big Pine Creeks. Starting perhaps 1,000 years ago, the Paiute built dams and large irrigation canals to irrigate areas exceeding 5 km^2 (2 mi^2) in the bottomlands of the Owens Valley to enhance the growth of native vegetation (Steward 1934; Lawton et al. 1976). More modest water impoundments and diversions were built in the Tahoe basin by the Washoe (Lindstrom 1994).

The discovery of gold in 1848 had swift and dramatic consequences for streams and rivers of the Sierra Nevada. Streams were dammed, diverted, dewatered, excavated, polluted, and filled with debris from enormous hydraulic mines. Removal of trees over large areas for flumes, mine timbers, buildings, and fuel resulted in soil loss, augmentation of downstream sedimentation, and major changes in vegetative cover. Gold mining also led to many innovations in water institutions and engineering. Miners established the principle of priority in determining water rights just as in mining claims. The resultant doctrine of prior appropriation has far-reaching effects in the allocation of water resources throughout the western United States. Acquisition of water for hydraulic mines developed engineering technology and physical works that have had lasting impacts on California's water distribution system. Generation of power for mines and mills led to one of the world's most extensive hydroelectric networks.

Initially, miners worked as individuals on small claims with simple implements. Shallow gravels were excavated and washed with water in pans, rockers, long toms, and other crude devices (Silva 1986). Virtually all streams on the central western slope of the Sierra Nevada were prospected (Averill 1946; Clark 1970). Although the depth of disturbance was limited, these excavations destabilized channel beds and banks and devastated riparian vegetation over a vast area. As the surface gravels were exhausted, more intensive methods required cooperation and consolidation of the miners. Flumes were constructed to carry the summer flows of streams so that beds could be blasted and excavated. Small dams were built so that several hours of discharge could be stored and released suddenly to disaggregate the gravels hydraulically and carry away lower-density sediments in a practice known as booming or gouging. Diversions and flumes were also built to supply water to off-channel claims for separating gold and for ground sluicing where diverted water was used to erode ancient stream deposits (Averill 1946). Natural channels were often totally dewatered to supply maximum flow in an artificial waterway (Pagenhart 1969).

The erosive power of water was marshaled to great effectiveness by containing water within pipes and hoses under high pressure and then directing it at hillslopes composed of gold-bearing gravels (Stanley 1965; May 1970). As an example of the power and water use of hydraulic techniques, flumes and pipes with 120 m (400 ft) of head could deliver about 3.8 million liters (1 million gal) of water per hour through a 25 cm (10 in) nozzle at a speed of about 200 kph (120 mph) (Logan 1948). Sediment-laden runoff from the eroded hillslopes was directed into long sluice boxes, often in tunnels, to extract the gold and then discharged into the nearest creek.

At the peak of hydraulic mining, there were more than four hundred hydraulic mines in operation (Wagner 1970). Hydraulic mining was most prevalent from the Feather River to the North Fork American River (Gilbert 1917; Averill 1946). The largest quantities of material were found in the South Yuba, lower Yuba, Bear River, and North Fork American River (Gilbert 1917; James 1994). Collapse of the English Dam on the Middle Fork of the Yuba (Ellis 1939; McPhee 1993) in June 1883 released almost 18 million m^3 (15,000 AF) of water suddenly and cleaned out much of the stored mining debris in that channel (James 1994). Several of the individual pits excavated more than 75 million m^3 (60,000 AF) of material and flushed it downstream (Gilbert 1917; Senter 1987; McPhee 1993). Channels immediately downstream of the hydraulic pits were usually overwhelmed by the enormous sediment loads and stored the sediments until high-flow events flushed some of the material downstream. Surveys of the 1870s showed accumulations of 30 to 60 m (100 to 200 ft) depth in tributaries to the Bear River (Pettee in Whitney 1880, cited by James 1994). Debris was redeposited throughout the channels, but often formed tailings dams at confluences where channel gradients lessened (James 1994). Temporary reservoirs formed behind these debris accumulations, which occasionally failed catastrophically, releasing large volumes of sediment, perhaps as hyperconcentrated flows. In the early years of hydraulic mining, the upper gravels of the Tertiary river channels were attacked first. After 1870, the lower gravels, which were more strongly cemented than those above, were mined by more powerful methods that moved even more of the landscape (Lindgren 1911). This second phase of hydraulic mining produced coarser sediments and greater quantities of debris than the first period (James 1988).

As sediments moved downstream, valley rivers aggraded dramatically, and coarse sediments were deposited on farms and fields. Thousands of acres of farmland became inoperable under annual deposits of unnaturally coarse sediments (Hundley 1992). As the farmers of the Central Valley gained economic and political power, they were able to successfully challenge the mining interests (Kelley 1959). In 1884, after eighteen months of deliberation in the case of *Woodruff v. North Bloomfield,* Judge Lorenzo Sawyer of the Ninth U.S. Circuit Court in San Francisco issued an injunction against further discharge of mining debris. This decision held that release of mining waste inevitably damaged the property of others and destroyed the navigability of the Sacramento and Feather Rivers, violating both common and statutory law and interfering with commerce (Hundley 1992).

After hydraulic mining was halted, some of the debris created earlier continued to move through the rivers, largely in pulses during peak flows. Debris that was not entrained during the phase of active stream incision continues to erode into channels and perpetuates the enhanced sediment delivery of the affected streams (James 1988). Many of the small debris dams intended to stabilize mining sediment failed and released the stored material. Large competent dams have effectively stopped transport of upstream sediment to the lower reaches of the main rivers. Even after a century, exposed surfaces in the pits continue to erode through mass failures, gullying, rainsplash, and rill erosion and produce substantially elevated sediment concentrations downstream of the old mine sites (i.e., Senter 1987).

The total volume of mining debris delivered to the Central Valley has been estimated at about 1.1 billion m^3 (900,000 AF) from five rivers, with the Yuba contributing about 40% of that quantity (Gilbert 1917; Mount 1995). Gilbert (1917) also estimated that mining sediment was produced at rates about ten times greater than natural sediment yield from the Sierra Nevada, although these estimates of background rates were highly uncertain.

Mercury used in ore processing is another legacy of the mining era remaining in stream channels. The amount of mercury used in gold extraction in the Sierra Nevada and largely lost to soils and streams has been estimated at 3.4 million kg (7.6 million lb) (Central Valley Regional Water Quality Control Board 1987). Much of this mercury has moved downstream, and some of it may have contaminated mudflats of San Francisco Bay. Large amounts of mercury are still found in stream sediments throughout the Gold Country and are also trapped in reservoir sediments (Slotten et al. 1995). The cyanide process for extracting gold from powdered rock was introduced about 1896 (Clark 1970; Shoup 1988). The degree of water pollution resulting from its use and the earlier chlorination process is unknown.

Underground mining, also called hard-rock, quartz, or lode mining, began shortly after the discovery of gold in streambeds, with the Argonaut mine near Jackson opening in 1850. The Sixteen to One mine in the Yuba River Basin persisted as the main gold mine in California until 1965 and was reopened a few years ago. Hundreds of quartz mines were operated throughout the Mother Lode of the western slope (Jenkins 1948; Clark 1970). The main mining districts of the eastern Sierra Nevada were at West Walker River, Bodie, Green Creek, Virginia Creek, Lundy Canyon, Tioga Pass, Mammoth Creek, Pine Creek, Bishop Creek, and Independence Creek (De Decker 1966; Clark 1970). Both lode and placer deposits were mined in the Kern River drainage beginning in 1851 (Troxel and Morton 1962). Disposal of tailings, mine water, and ore-processing effluent were the main impacts of the underground mines on streams. Although perhaps significant locally, these impacts were minor compared with those of the surface operations.

Dredging was an important source of gold and a major impact on the lower reaches of the main rivers where the Sierra Nevada meets the Central Valley. Large-scale river dredging began in 1897 along the Yuba near Marysville (Logan 1948) and lasted until 1967 (Clark 1970). The largest dredging operations were at Hammonton on the lower Yuba and near Folsom on the lower American. Dredging was also practiced on Butte Creek, Honcut Creek, the lower Feather, the Bear near Lincoln, the Cosumnes at Michigan Bar, the Calaveras at Jenny Lind, the Mokelumne at Camanche, the Tuolumne at La Grange, and the Merced at Snelling (Aubury 1910; Clark 1970). Between 1900 and 1910, dredge capacity increased from about 20,000 m^3 (25,000 yd^3) to 200,000 m^3 (250,000 yd^3) per month (Aubury 1910). Reclamation and revegetation of dredge spoils were concerns as early as 1910 (Aubury 1910).

The development of mining towns put great pressure on local resources, which probably had consequent impact on local streams. Towns sprang up quickly when new strikes were rumored and were successively rebuilt after surprisingly frequent fires. Some towns like Elizabethville in Martis Valley and Summit City near Cisco grew to several thousand people before suddenly collapsing. Development of trails, roads, railroads, and agriculture to support the towns converted forests to bare and compacted soil, which was susceptible to erosion. Overgrazing for food production further altered plant cover and degraded riparian zones. Harvesting of fish and mammals for food and loss of habitat decimated wildlife populations and altered ecological processes (Hinkle and Hinkle 1949; Strong 1984). Demand for wood for shelter and fuel quickly depleted the forests closest to the new towns and then progressively expanded the circle of destruction. Lumber was also needed for underground mine supports, railroad ties, and flumes. In a few cases such as Bodie, lumber and fuel wood were imported from considerable distances. Forests of the Tahoe basin were cut extensively to supply wood for the Comstock silver mines near Virginia City. The Bonanza mines alone consumed 28,000 m³ (12 million board feet) of lumber and 145,000 m³ (40,000 cords) of fuel wood per year. An extensive network of skid trails, haul roads, railroads, tug boats, and flumes efficiently removed the forests of much of the Tahoe basin. An estimated 600 million board feet of lumber were buried in the Comstock mines (Hinkle and Hinkle 1949; Strong 1984). More than twenty sawmills in the Middle and South Forks of the American River produced lumber for buildings and replacement flumes for those destroyed by annual floods (Lardner and Brock 1924, cited by James 1994).

The first known miner's ditch was a V-shaped flume about 3 km (2 mi) long built at Coyote Hill near Nevada City in March 1850 (Pagenhart 1969). Later that year, a 14 km (9 mi) long ditch was built by the Rock Creek Water Company, which recovered its investment in just six weeks from the sale of water (Wagner 1970). Natural lakes in the upper Yuba basin were augmented and regulated with crude dams as early as 1850 (Pagenhart 1969). Acquisition and delivery of water to mines became a huge industry that was probably more profitable than mining. If ditches were important to the mining of surficial placer gold, they became critical to the hydraulic mining industry. Large companies built vast networks of reservoirs and waterways acquired through purchase, filing on abandoned claims, court challenges to water rights, and real and implied violence (Hundley 1992).

The levels of investment, labor, and engineering skill devoted to the miners' ditches were impressive. The main supply ditches were 2.4–4.6 m (8–15 ft) wide at the top, 1.2–1.8 m (4–6 ft) wide at the bottom and at least 1 m (3 ft) deep (Wagner 1970). Water was conveyed across valleys and rock outcrops in wooden flumes or iron pipes mounted on trestles (Logan 1948). By 1857, $13.5 million had been invested in mine water systems, with $3 million of that total in Calaveras and Tuolumne Counties (Langley 1862, cited by Shoup 1988). In the 1860s, more than 8,500 km (5,300 mi) of main canals and about 1,280 km (800 mi) of branch ditches had been constructed (Browne 1868; Logan 1948; McPhee 1993). By 1884, the total length of ditches, flumes, and pipelines built for mining purposes reached 12,800 km (8,000 mi) (Wagner 1970). (This figure was probably for all of California.) The South Yuba Canal Company maintained 720 km (450 mi) of waterways at its peak, and the Auburn and Bear River Canal operation included 460 km (290 mi) of ditches. The dam at Meadow Lake

constructed by the South Yuba Canal Company in 1858 was 12 m (42 ft) tall and 350 m (1,150 ft) long (Hinkle and Hinkle 1949). By 1880, the California Water Company had twenty-one reservoirs and 400 km (250 mi) of flumes and ditches between the Middle Fork and the South Fork of the American River. At the peak of hydraulic mining activity, there were more than 1,600 km (1,000 mi) of ditches in Nevada County (Kahrl 1978). By the 1870s, artificial reservoirs in the Sierra Nevada stored more than 185 million m³ (150,000 AF) of water (Pisani 1984). The Eureka Lake and Yuba Canal Company operated four high-elevation reservoirs to supply water to mines near North San Juan, 100 km (65 mi) away (Wagner 1970). In the same region, the North Bloomfield Gravel Company used 55 million m³ (45,000 AF) of water annually at up to 110,000 liters (30,000 gal) per minute (McPhee 1993) or 227,000 m³ (184 AF) per day and had reservoir storage capacity of 28 million m³ (23,000 AF) (Pisani 1984). The company's Bowman Dam was 22 m (72 ft) tall in 1876 and was raised another 7 m (23 ft) to increase storage as the mine's water demand increased. The Cherokee mine in Butte County used up to 150,000 m³ (123 AF) of water per day (Hundley 1992). Abandoned ditches have become naturally revegetated but can still affect runoff processes today (Pagenhart 1969). Occasional failure of both maintained and abandoned ditches can cause local debris flows and gully erosion.

Water Development

Water was also sold for domestic use and for water power for lumber and stamp mills, air compressors, and Pelton-wheel electric generators after 1890. Increasing scarcity of wood for fuel led to the use of high-pressure water for mechanical power. By the mid-1880s, most of the large hard-rock mines were using water power instead of steam power. The first known use of electrical generation for operation of mining and milling equipment in California occurred in El Dorado County in February 1890 (Logan 1948). After hydraulic mining was halted in 1884, many of the canals were acquired by irrigation districts and later by power companies. The Nevada Irrigation District still relies on reservoirs and canals built for mines in Nevada County. The Pacific Gas and Electric Company eventually took over 520 separate ditch enterprises and their water rights and facilities. By the 1890s, the log-and-brush and earth-filled dams of the miners were replaced by more substantial concrete structures (Pisani 1984). Irrigated agriculture in the foothills occupied about 36 km² (14 mi²) around Auburn and Placerville in 1880 (Pisani 1984) and grew substantially in the following decades (see Momsen 1996).

The vast network of artificial channels built for mining allowed the hydroelectric industry to take off as soon as water-powered generating technology became available. A dam on the American River at Folsom begun in 1866 that was originally intended for hydromechanical power later provided water for the first transmission of hydroelectricity out of the

Sierra Nevada. This project at Folsom began supplying power for an electric railroad in Sacramento in 1895 (Fowler 1923). After its first dam failed in 1892, a hydroelectric power plant on the South Yuba was completed in 1896 and supplied electricity for the Grass Valley and Nevada City area (Pacific Gas and Electric Company 1911). In the next two decades, dozens of hydroelectric facilities were completed throughout the Sierra Nevada: Knight's Ferry-Stanislaus, 1895; Electra-Mokelumne, 1897; Kern 1897; Newcastle-Bear, 1898; Colgate-Yuba, 1899; Farad-Truckee, 1899; Phoenix-Stanislaus, 1901; American, 1903; De Sabla-Butte 1903; Bishop Creek No. 4, 1905; San Joaquin No. 3, 1906; Kittredge-Merced, 1906; La Grange-Tuolumne, 1907; Big Creek No. 1, 1913; Kaweah No. 3, 1913 (Pacific Gas and Electric Company 1911; Fowler 1923; Coleman 1952). Independent companies were quickly merged and integrated, and multiunit projects were developed by the two companies that emerged from the consolidation battles, Pacific Gas and Electric and Southern California Edison. The Crane Valley project involving Bass Lake and Willow Creek was developed between 1900 and 1920. On the North Fork of the Feather, Pacific Gas and Electric was filling Lake Almanor and Bucks Lake by 1928 (Coleman 1952). The Big Creek project, started in 1911 by the Pacific Power and Light Corporation, was completed by Southern California Edison in 1929 and included three large reservoirs, eight tunnels, and five powerhouses (Redinger 1949).

In addition to the dozens of hydroelectric projects taking advantage of the mining waterways, three immense municipal-supply projects began as mining faded out. A scheme to develop Lake Tahoe as a water supply for San Francisco was proposed even earlier, in 1866, but failed to find support (Strong 1984). The city of San Francisco itself began prospecting for water in the Sierra Nevada as early as 1886 (Kahrl 1978). The city remained focused on the Tuolumne River with a dam at Hetch Hetchy Valley despite other feasible alternatives (Freeman 1912; Jones 1965). Largely because the project was in a national park, the proposal generated enormous controversy; however, the city prevailed with congressional approval of the Raker Act in 1913. Hydroelectric generation on a subsidiary portion of the project began in 1918, but water deliveries to San Francisco did not begin until 1934. The Owens Valley project of the city of Los Angeles was constructed more rapidly. After the general concept arose in the 1890s, construction bonds were approved in 1907 and work began in 1908. The project was operational in 1913, when Owens Valley water reached the San Fernando Valley. An extension into the Mono basin was built between 1934 and 1940. A second aqueduct was completed in 1970, enabling greater export of surface water and pumped ground water. The controversies created by the Owens Valley diversions have been described by dozens of authors (i.e., Chalfant 1922; Nadeau 1950; Kahrl 1978; Hoffman 1981; Kahrl 1982; Reisner 1986; Walton 1992; Davis 1993; Sauder 1994). By comparison, political conflict was almost absent in the Mokelumne project of the East Bay Municipal Utility District. Work began in 1923,

and Pardee Reservoir began filling in 1929, with water deliveries to Alameda and Contra Costa Counties that same year (Harding 1960). These systems deliver large volumes of water to distant communities with a large net production of electricity. Hydroelectric power production has been a key source of revenue in the financing of water projects in the Sierra Nevada.

The federal government's involvement with water in the Sierra Nevada began with the Newlands Reclamation Act of 1902, which authorized the Truckee-Carson Project. Preexisting dams that raised the level of Lake Tahoe were reconstructed to provide 1.8 m (6 ft) of controllable storage. The newly created Bureau of Reclamation assumed operation of the Tahoe dam in 1913 for irrigation of lands near Fallon, Nevada. The interstate and tribal conflicts created by this project have maintained a steady stream of litigation for eight decades (Jackson and Pisani 1973; Jones 1991; Chisholm 1994). Early in the twentieth century, the state government began considering large-scale water development (Kahrl 1993). A report by the Conservation Commission (1913) devoted half of its 500 pages to water resources. The first comprehensive plan for water development in California was prepared by R. B. Marshall in 1919. A few years later, the California Department of Public Works (1923) released the first statewide hydrographic survey, which examined 1,270 potential reservoir sites and recommended dams at 260 of them. That report led to another comprehensive development plan (Bailey 1927). After California voters approved the concept in 1933 as a state project, California was unable to sell the bonds required for financing. The U.S. Congress stepped in, federalized the proposal, and authorized the Bureau of Reclamation to begin construction of the Central Valley Project in 1935 (Harding 1960; Kelley 1989). Most of the project's water originates outside the Sierra Nevada in the upper Sacramento and Trinity Rivers. The main pieces of the Central Valley Project in the Sierra Nevada, the Friant, Folsom, and New Melones Dams, took decades to complete.

After World War II, other big water projects got under way in the Sierra Nevada, with major dams constructed on the San Joaquin, Kern, Kings, and American before 1960. The big-dam era continued at full speed through the sixties, with projects completed on the San Joaquin, Kaweah, Bear, Mokelumne, Calaveras, American, Merced, Tuolumne, and Yuba Rivers (Kahrl 1978; California Department of Water Resources 1994). The Feather River Project (later named the State Water Project) was approved by the California legislature in 1959 and by the voters in 1960. The centerpiece of the project, Oroville Dam, was completed in 1967.

Although mining in stream channels and water development have been the overwhelming impacts on hydrologic processes in the Sierra Nevada, other human activities in the past 150 years have also altered the hydrology and streams of the range. Unfortunately, there is relatively little information about the extent of these various impacts. We are left, therefore, to a few broad inferences and generalizations.

Grazing

Grazing was perhaps the most ubiquitous impact, as cattle and sheep were driven virtually everywhere in the Sierra Nevada that forage was available (see Menke et al. 1996; Kinney 1996). Anecdotal accounts describe vast herds and severe overgrazing (Sudworth 1900; Leiberg 1902). Overgrazing has been blamed for accelerated erosion beginning in the late 1800s and massive gullying of meadows in the decades that followed (Wagoner 1886; Hughes 1934). Widespread deterioration of meadows led to efforts by the U.S. Forest Service to reduce the degradation (Kraebel and Pillsbury 1934). However, continuing presence of large herds did not allow riparian vegetation to recover enough to reduce erosion of stream banks.

Timber Harvesting

Timber harvesting in the nineteenth century certainly impacted local streams but perhaps mainly because of its typical location: near streams. We can assume that riparian and near-channel forests were targeted during the mining era because they grew on gold-bearing stream deposits and wood was needed where most of the activity was: along streams. Rivers were also used for log transport. As early loggers got farther away from streams, their impacts presumably diminished. Because transportation of the logs was difficult, large amounts of slash were apparently left in the woods. Such material could reduce erosion. In addition, loggers of the 1800s simply lacked the heavy equipment that can grossly disturb hillsides. The advent of railroads had two major impacts on Sierra Nevada forests. Railroad construction consumed vast quantities of lumber for ties, trestles, and snowsheds, and the steam engines burned wood. Railroad logging caused a change in harvesting practices: economics favored removal of almost all trees near the tracks instead of taking individual trees selected for wood quality and relative ease of transportation. Where railway networks allowed large fractions of a watershed to be harvested, local yields of water and sediment could be expected to have increased. Because the degree of ground disturbance from these early logging operations is unknown, their hydrologic effects are difficult to infer. However, because early harvests did not involve road construction and persistent ground skidding to centralized landings, they may be assumed to have had lower impacts than those following World War II.

Wildfire

Fire suppression policies that began early in this century may have caused extensive and persistent changes in the water balance of the forest zone. If forest density significantly increased beyond that generally maintained under a pre-1850 fire regime, then we may assume that evapotranspiration has been maximized in the absence of harvesting. Therefore, the presumably denser forests resulting from fire suppression may have reduced water yields in many basins of the western slope. Quantifying such a reduction is not possible without knowing something about the water relations of forests before the gold rush. Regionally, changes probably do not amount to more than a few centimeters (inches) of areal water depth at most. However, the local effects of denser stands in some instances could be sufficient to reduce the flow of springs and headwater creeks. The thick ground cover resulting from the lack of fires has probably decreased surface erosion as well.

Roads

Following the Second World War, timber production increased markedly, as did construction of forest roads necessary to serve emerging techniques of log removal. The road-building boom of the 1950s through the 1970s was the greatest disturbance of the Sierra Nevada landscape since the gold rush. Initially, forest roads were just built, rather than properly engineered to minimize the risk of mass failure and surface erosion. Stream crossings were particular problems when fords or cull-logs covered with dirt were the preferred means of crossing water. Inadequate road drainage and undersized culverts were common causes of road failure and sediment production. With time, road engineering improved, but total mileage increased as well. At the extreme, up to a tenth of the land area of some catchments became road surface, with a large number of stream crossings.

Point-Source Water Pollution

The first known water pollution by industry other than mining in the Sierra Nevada involved the sawmills near Truckee. Mill waste was disposed of in the nearest de facto sewer, the Truckee River. The large loads of sawdust filled pools in the river, clogged the gravels, and probably removed oxygen from the water, killing fish in the river. Acts of both the California and Nevada legislatures in 1890 and continued enforcement by the California Fish Commission were required to halt the pollution (Pisani 1977). Construction of a pulp and paper mill at Floriston in 1899 added chemical pollutants to the Truckee. This pollution continued until the mill closed for economic reasons in the 1930s (Pisani 1977). Growth of communities in the Sierra Nevada led to water quality problems relating to solid waste and sewage disposal. All known problems were local and relatively minor. Technology for centralized sewage treatment has been both improved and widely deployed throughout the range. Bacteriological water quality around the mining camps may have been poor, as inferred by common intestinal ailments of Euro-American miners that spared the boiled-tea-drinking Chinese laborers (Johnson 1971).

SURFACE WATER QUANTITY

The Sierra Nevada annually yields a large but variable amount of water. Continuous stream-flow records began to be maintained in the mountains less than one hundred years ago and are of short duration with respect to longer-term natural variability. Based on this recent historical record, the Sierra Nevada generates about 25 km^3 (20 million AF) of runoff each year, on average, out of a total for California of about 88 km^3 (71 million AF). Stream flow in the Sierra Nevada is generated by seasonal rainfall and snowmelt. About half of average annual precipitation occurs during winter, about a third in autumn, about 15% in spring, and generally less than 2% in summer (Smith 1982). About 50% of annual precipitation falls as snow at 1,700 m (5,600 ft) at a latitude of 39° N (Kahrl 1978). Stream flow generated below 1,500 m (4,900 ft) is usually directly associated with storms, while stream flow above 2,500 m (8,200 ft) is primarily a product of spring snowmelt. Between these approximate bounds, stream flow is generated both by warmer storms and by melt of snow cover in spring. Of course, the major rivers collect inputs throughout their elevation range with a mix of events. Cayan and Riddle (1993) calculated the seasonal distribution of runoff of six Sierra Nevada rivers (table 30.1), which illustrates that snowmelt runoff becomes more important and midwinter rainfall runoff becomes less important with increasing elevation. In the American River Basin, less than half of annual runoff occurs from April through July in the lower two-thirds of the basin. In small catchments of the American adjoining the Sierra Nevada crest, more than two-thirds of annual runoff occurs during this period (Elliott et al. 1978).

Disposition of Precipitation

Overall, about half the precipitation in the major river basins of the west slope of the Sierra Nevada becomes stream flow (table 30.2) (Kattelmann et al. 1983). Stream flow, both in absolute magnitude and as a proportion of precipitation, increases with elevation. In the American River Basin, stream-flow data from twenty-five subbasins (Armstrong and Stidd

TABLE 30.1

Seasonal distribution of stream flow in selected rivers (from Cayan and Riddle 1993).

River Basin	Mean Elevation (m)	Percentage of Mean Annual Stream Flow			
		Aug–Oct	Nov–Jan	Feb–Apr	May–Jul
Cosumnes	1,120	1	21	59	18
American	1,430	2	19	46	33
Stanislaus	1,770	3	13	38	46
San Joaquin	2,290	6	9	29	56
East Carson	2,490	7	11	24	58
Merced	2,740	4	5	21	70

TABLE 30.2

Approximate disposition of precipitation in major rivers (from Kattelmann et al. 1983).

River (Gauging Station)	Precipitation (cm)	Stream Flow (cm)	Losses (cm)
Feather (Lake Oroville)	120	60	60
Yuba (Smartville)	160	100	60
American (Folsom Lake)	135	65	70
Cosumnes (Michigan Bar)	105	35	70
Mokelumne (Pardee Reservoir)	120	65	55
Stanislaus (Melones Reservoir)	115	75	40
Tuolumne (Lake Don Pedro)	110	55	55
Merced (Exchequer Reservoir)	115	45	70
San Joaquin (Millerton Lake)	110	50	60
Kings (Pine Flat Reservoir)	95	50	45
Area weighted average	120	60	60

1967) indicate an increase in stream flow of about 3 cm per 100 m (3.6 in per 1,000 ft) gain in elevation. Also, in the American River Basin, runoff efficiency increases from about 30% in the foothills to more than 80% near the crest (Elliott et al. 1978). In four small catchments in the Kings River Basin at 1,900 to 2,500 m (6,300 to 8,100 ft), about half the precipitation became stream flow on average. However, there was considerable variation among the nine years of record, depending on total precipitation. Runoff efficiency in the four years with more than 120 cm (47 in) of stream flow ranged from 63% to 75%, while in the five years when stream flow was less than 30 cm (12 in), runoff efficiencies ranged from 21% to 33% (Kattelmann 1989a). A stream gauge on the North Fork of the Kings River at 2,480 m (8,130 ft) is the highest long-term station on the western slope of the Sierra Nevada. This basin of 100 km^2 (39 mi^2) extends above 3,700 m (12,100 ft). The average annual stream flow of 74 cm (29 in) is 70% to 80% of the estimated annual precipitation. About 85% of the annual flow in this basin occurs from April to July (Kattelmann and Berg 1987). In a 1 km^2 (250 acre) research basin in Sequoia National Park at 2,800–3,400 m (9,200–11,100 ft), 75% to 90% of the annual precipitation became stream flow (Kattelmann and Elder 1991). The high-elevation portion of the Sierra Nevada, which covers approximately 3% of California and produces an average of 90 cm (35 in) of annual runoff, contributes about 13% of the state's annual stream flow (Colman 1955). This contribution amounts to an even higher proportion of the state's developed water supply because of its persistence into summer.

Snow

Snow plays a dominant role in the overall hydrology of the Sierra Nevada. Storage of frozen precipitation in winter as snow cover and its subsequent release during the spring snowmelt period controls the seasonal distribution of flow in most major rivers. Snow cover is measured at about 400 index locations (300 manually measured snow courses and 100 telemetered snow sensors) in the Sierra Nevada that are used for river forecasting. Basinwide means of April 1 water equivalence for snow courses above 2,500 m (8,200 ft) suggest that peak snowpack water equivalence for the high Sierra Nevada averages 75 to 85 cm (30 to 33 in), decreases from north to south, and is lower on the east side of the crest than on the west side (Kattelmann and Berg 1987). Snow courses between 1,800 m and 2,500 m (5,900 and 8,200 ft) have an average peak water equivalence of about 60 cm (24 in).

Flow Variability

Flow in Sierra Nevada rivers is highly variable in time, both within and between years. Peak flows can be up to five orders of magnitude greater than minimum flows. Annual volumes can be twenty times greater in very wet years than in very dry years. Some smaller streams cease flowing during prolonged dry periods.

Floods

High water levels are an integral feature of Sierra Nevada rivers and have a variety of effects on aquatic biota as well as channel morphology (Erman et al. 1988). Peak flows in the Sierra Nevada result from snowmelt, warm winter storms, summer and early-autumn convective storms, and outbursts from storage (Kattelmann 1990). In rivers with headwaters in the snowpack zone, snowmelt floods occur each spring as periods of sustained high flow, long duration, and large volume. However, they rarely produce the highest instantaneous peaks. The magnitude of a snowmelt flood depends on the spatial distribution of both snow and energy input to the snowpack. The largest volumes occur when all or almost all the basin is contributing high rates of snowmelt runoff. In basins spanning hundreds of meters of elevation with varied aspects, such situations are rare. Snow usually disappears from south-facing slopes and low elevations long before melt rates peak on north aspects and high elevations. Large snowmelt floods occurred in many river basins of the Sierra Nevada during 1906, 1938, 1952, 1969, and 1983. In all cases, snow deposition was more than twice average amounts and persisted into April and May even at low elevations. In basins of less than 100 km^2 (39 mi^2) within the snow zone, maximum specific discharges during snowmelt have ranged from 0.2 to 0.8 m^3 per second per km^2 (18 to 73 ft^3/s/mi^2) on the western slope and 0.1 to 0.2 m^3/s/km^2 (9 to 18 ft^3/s/mi^2) on the eastern slope.

Midwinter rainfall on snow cover has produced all the highest flows in major Sierra Nevada rivers during this century (Kattelmann et al. 1991). The most important factor in rain-on-snow floods is probably their large contributing area. During these warm storms, most of a basin receives rain instead of snow, generating short-term runoff from a much larger proportion of the basin than during cold storms. However, even during the warmest storms, snowpacks above 2,500

m (8,200 ft) rarely melt much because temperatures are close to 0°C (32°F). If snow cover extends to low elevations prior to a warm storm, there can be a substantial snowmelt contribution from those areas. In basins that are largely above 2,000 m (6,600 ft), the highest peaks also tend to be caused by rain-on-snow events. For example, in the Merced River in Yosemite National Park, the four highest floods were caused by rain on snow and were 1.5 to 1.8 times greater than the maximum snowmelt peak of record in 1983. In the past sixty years, six large-magnitude floods (peak flows greater than twice the mean annual flood) have occurred in almost all rivers draining the snow zone: December 1937, November 1950, December 1955, February 1963, December 1964, and February 1986. Specific discharges of these largest floods ranged from 0.2 to 4 m^3/s/km^2 (18 to 360 ft^3/s/mi^2). The largest flood in California history occurred in January 1862. Following hundreds of millimeters (tens of inches) of antecedent rainfall and snowfall down to the floor of the Central Valley, 250 to 400 mm (10 to 16 in) of rain fell in Sacramento (and undoubtedly higher amounts in the Sierra Nevada) between January 9 and 12. High-water marks on the American River near Folsom were 3.5 m (11 ft) above those observed in 1907, the third highest flood measured on the American. In the eastern Sierra Nevada, Owens Lake rose 3 to 4 m (10 to 13 ft) during that winter.

When subtropical air masses move into the Sierra Nevada in summer and early autumn, sufficient moisture is available to generate extreme rainfall. Intense convective storms occurring over a period of three or four days can generate local flooding. These convective storms can generate the greatest floods in some alpine basins that are high enough to avoid midwinter rain-on-snow events. For example, the four highest floods of Bear Creek (gauged near Lake Thomas A. Edison) were generated by summer rainfall. The peak discharge was more than twice that of the largest snowmelt flood in this basin of 136 km^2 (52 mi^2) with a mean elevation of about 2,850 m (9,300 ft). The greatest recorded floods in several east-side streams occurred in late September 1982 when 150 to 200 mm (6 to 8 in) of rain fell in two days.

In limited areas, the greatest floods occur during a sudden outburst from storage because of avalanche-induced displacement of lake water or failure of a natural or man-made dam or aqueduct. Peak flows generated by such mechanisms can be several times greater than those produced by meteorological events.

Droughts

At the other extreme, stream flow in Sierra Nevada rivers can become quite low during intense and/or extended droughts. For example, during 1977 when average snow water equivalence in early April was only 25% of the long-term mean, stream flow as a proportion of average annual flow ranged from 0.08 to 0.26. Basins with most of their area at low elevations generally had the lowest proportions of average volumes. Dry periods may last for several years. From 1928 through 1937, runoff was below average in each year. The past two decades have included record droughts for one year (1977), two years (1976–77), three years (1990–92), and six years (1987–92). The recent six-year drought was similar to the 1929–34 dry period. Total stream flow averaged across many rivers was about half of average in each case (California Department of Water Resources 1994). Other indications of past climate suggest that severe droughts in the Sierra have persisted for periods from decades to more than two centuries (Graumlich 1993; Stine 1994, 1996; Millar 1996; Woolfenden 1996). The presence of tree stumps well below modern lake levels in Lake Tahoe and Lake Tenaya and elsewhere provides strong evidence for very arid conditions in the past (Stine 1994). The period 1937 through 1986 was an anomalously wet period in a 1,000-year-long reconstruction of precipitation from dendrochronological evidence (Graumlich 1993). However, our water resources infrastructure and institutions were largely developed during this period. Inferences about the climate of the past 100,000 years (e.g., Broecker 1995) suggest that great variability in temperature has been common and the temperature of the last 10,000 years was anomalously stable. Any resumption of such a variable climate would be challenging to California's water resource system and society in general. Dramatic shifts in climate could alter the distribution of vegetation over decades to centuries and could interact with a changed precipitation regime to alter runoff generation (Beniston 1994; Melack et al. in press).

Trends

In both extremes of wet and dry conditions, there do not appear to be any strong trends in water becoming more or less available in the recent past. Concern was raised a few years ago that the proportion of annual runoff occurring in the months of April through July in the Sacramento, Feather, Yuba, and American Rivers had declined since about 1910 (Roos 1987). However, this trend appears to be a result of increased runoff for the remainder of the year and no change in absolute amounts during spring (Wahl 1991; Aguado et al. 1992). On the western slope of the Sierra Nevada, there have been no obvious trends in flood magnitude or frequency over the historical period. In rivers of the eastern slope, clusters of events at both extremes have been evident in recent years (Kattelmann 1992). Five of the largest eight to eleven snowmelt floods (in terms of volume) since the 1920s occurred from 1978 to 1986. Five of the smallest thirteen or fourteen snowmelt floods since the 1920s occurred from 1987 to 1991. Instantaneous peak flows have a similar distribution. For example, in Rock Creek, four of the ten largest annual floods and three of the six smallest annual floods occurred during the 1980s. These events support theories of some climatologists that extreme events are becoming more common in the western United States (Granger 1979; Michaelson et al. 1987).

Variability in flow remains a defining characteristic of Sierra Nevada rivers.

Even the limited variability in precipitation and runoff that occurred in this century caused water managers to attempt to augment supplies through deliberate weather modification. Soon after the theoretical basis for cloud seeding to increase precipitation was established, the world's first operational program began in the eastern Sierra Nevada in 1948. Within the next few years, cloud seeding programs were started in the San Joaquin, Kings, Mokelumne, and Feather Rivers (Henderson 1995). A dozen programs were active in the Sierra Nevada in 1994 and 1995. Despite dozens of studies, the effectiveness of cloud seeding remains uncertain. Conventional wisdom suggests that a well-designed cloud seeding program may yield up to 6% additional stream flow (American Meteorological Society 1992). Hundreds of papers have been written on environmental effects of cloud seeding (e.g., Berg and Smith 1980; Parsons Engineering Science 1995), but major impacts have not been found, perhaps because of the uncertainty in the amount of precipitation augmentation. The amounts of the primary seeding agent, silver iodide, released in a typical year (7–18 kg [15–40 lb]) over a large river basin are several orders of magnitude less than quantities naturally present in soil.

SURFACE WATER QUALITY

The Sierra Nevada is generally regarded as producing surface water of excellent quality, meaning the water is suitable for almost any use and contains lower amounts of contaminants than specified in state and federal standards. Most of the runoff would be suitable for human consumption except for the risk of pathogens. Very little of the water of the Sierra Nevada can be considered highly polluted (i.e., contaminated with materials having potential adverse effects at concentrations above natural background). Areas of lower water quality correspond to those areas with greater human activities and access. Headwater streams are particularly sensitive to pollution because of low flow conditions and nutrient limitations. The relatively few point sources of pollution throughout the range are mostly associated with inactive mines, dumps, and towns. Many contaminants that enter Sierra Nevada streams can be considered non-point-source pollutants because they are generated over large areas. Livestock waste is an example of non-point-source pollution. Sediment is the most pervasive pollutant because its production may be increased above natural background levels by almost any human activity that disturbs the soil or reduces vegetation cover. Sediment augmented above natural levels usually impairs some beneficial uses of streams. Erosion and sediment are discussed separately in another section of this chapter. Ground-water quality is discussed in the section about ground water. Water temperature is treated in Kondolf et al. 1996.

Human activities in the watershed have the potential to alter nutrient cycling. A classic study in New England provided some of the first measurements of changes in nutrient budgets as a result of complete killing (but not removal) of trees in a small catchment (Likens et al. 1970). This study at Hubbard Brook found that loss of nitrates in stream flow increased by forty times in the first year following devegetation, and export of other nutrients increased several times. Studies in Oregon (Fredriksen 1971; Brown et al. 1973) suggested that typical harvesting procedures that impact less than half of a watershed with deep soils will not significantly contaminate small streams or risk serious declines in soil productivity (Brown 1980). However, frequent harvesting of large portions of catchments with shallow soils and low cation exchange capacity can result in substantial nutrient losses from soils to streams. Elevated concentrations of nitrates and phosphates may be expected in catchments with agriculture, fish farms, and residences. Most of the work on nutrient cycling in the Sierra Nevada has been done in the Lake Tahoe area (e.g., Coats et al. 1976; Coats and Goldman 1993). In one catchment in the Tahoe basin, biological processes effectively prevented release of nitrogen in nitrate form in surface water or ground water (Brown et al. 1990). These authors cautioned that creation of impervious surfaces allows nitrates to bypass potential sinks. Human activities that decrease residence time of water in soils have potential to increase nitrate export. Nitrate concentrations sampled in seventy-seven streams of the eastern Sierra Nevada were less than 1 mg/l in all cases and usually less than 0.1 mg/l, demonstrating that there is usually little export of nitrates in streams (Skau and Brown 1990).

Point-Source Pollutants

There are very few known localized sources of water pollution in the classic outfall-into-the-stream sense in the Sierra Nevada because of the virtual absence of industries that process chemicals and continuing abatement of the few existing sources. Point-source pollution has also been reduced very effectively under the Clean Water Act of 1972 and subsequent amendments. Municipal and industrial discharges are controlled through National Pollutant Discharge Elimination System permits. Most pollution of that general nature is associated with active and abandoned mines and is discussed in the section on mining. Industrial-type pollutants may also be found in the vicinity of many cities and towns and abandoned lumber mills. However, serious problems of this nature are not known to exist (Central Valley Regional Water Pollution Control Board 1957; Central Valley Regional Water Quality Control Board 1991; Lahontan Regional Water Quality Control Board 1993). Over the entire western slope, there are only ten "municipal and industrial discharger groups": Chester, Quincy, Paradise, Portola, Nevada City, Auburn, Placerville, Jackson, Sonora, and Bass Lake (Central Valley

Regional Water Quality Control Board 1991). Water quality was considered impaired in streams receiving wastewater from Nevada City, Grass Valley, Placerville, Jackson, and the Columbia-Sonora area (Central Valley Regional Water Quality Control Board 1991).

Sewage

Most communities with a centralized population in the Sierra Nevada have common sewage collection and treatment systems. Discharges from treatment facilities are regulated by the regional water quality control board; however, short-term failures are a persistent difficulty. Disposal of treated wastewaters on land instead of directly into streams is encouraged where practicable (Central Valley Regional Water Quality Control Board 1991). An experiment in Tuolumne County demonstrated several problems with spraying treated effluent on hillsides: the soil became overloaded with nutrients, salts, and water, and algal growth effectively sealed the soil surface, minimizing infiltration (California Division of Forestry 1972). Effluent from a sewage treatment plant in the Lake Tahoe Basin was sprayed over a 40 ha (100 acre) area from 1960 to 1965. Even five years after application ceased, substantial amounts of nitrates were entering a creek downgradient from the site. A stand of Jeffrey pine at the site was also killed by the persistent high level of soil moisture (Perkins et al. 1975).

A significant fraction of the residences in the Sierra Nevada are too dispersed to allow connection to community sewage facilities and rely on individual septic systems (Duane 1996a). Septic systems in Nevada County have led to significant bacteriological contamination in streams below unsewered subdivisions (California Department of Water Resources 1974). Septic tank and leach field systems on individual lots provide a good example of cumulative watershed effects. The soils of a particular catchment have sufficient capacity to treat a particular quantity of sewage under a particular set of conditions. When the soil system is overloaded, some fraction of the waste or its derivatives is discharged to streams. Each residential septic system contributes only a small fraction of the total, but the community as a whole has polluted the catchment. Recreational developments such as ski areas and campgrounds also generate significant quantities of sewage and may have their own treatment facilities if geographically isolated. In the 1950s, Yosemite Valley was the most significant wastewater source in the upper-elevation parts of the San Joaquin River Basin (Central Valley Regional Water Pollution Control Board 1957).

Urban storm water runoff can add a variety of contaminants directly to streams. Pet waste can be a significant source of fecal coliform bacteria in some areas. Street runoff in the Lake Tahoe Basin is beginning to be routed into publicly owned lots to allow for some pollutant removal.

Even in the backcountry, inadequate disposal of human waste from dispersed recreationists has contaminated enough of the streams in remote areas of the Sierra Nevada to make consumption of any untreated water somewhat risky. Although the level of risk is unknown, pathogens including coliform bacteria, campylobacter, and *Giardia* have been found in many areas throughout the range (Hermann and McGregor 1973; Suk et al. 1986). In a survey of seventy-eight backcountry locations with varying levels of recreational use, *Giardia* cysts were found in 44% of water samples collected downstream of heavily used areas and 17% of samples from areas of relatively low use (Suk et al. 1987). *Giardia* cysts have also been detected in fecal matter of cattle grazing in backcountry areas (Suk et al. 1985). Recreational pack stock contribute to nutrient and bacterial pollution. Heavily used trails (e.g., Mt. Whitney) have had sufficient problems with human waste to warrant the installation of backcountry toilets. Low-level release of nutrients from wilderness campers have stimulated increased plant growth on lake bottoms (Taylor and Erman 1979).

Non-Point-Source Pollution

When non-point-source pollution gained widespread recognition as a critical water quality problem in the 1970s, administrative and regulatory approaches were lacking. Eventually, Congress (in the Clean Water Act of 1977 and Water Quality Act of 1987) and the Environmental Protection Agency adopted the concept of best management practices (BMP). This general concept can be stated as doing the best one can to minimize water pollution and meet water quality standards while still conducting the intended activities. Different approaches to developing and applying BMPs have been tried in different states. Ideally, BMPs should reflect the most cost-effective approach to minimizing water pollution in a specific area using practical technology (Dissmeyer 1993; Brown and Binkley 1994). Determining what is most effective and efficient in a particular region should be an iterative process of applying a practice, monitoring its effectiveness, evaluating the cost and impact, modifying the practice in its next application, and so on. Unfortunately, monitoring has been limited, so there is often little basis for improving techniques. However, the learning and refinement process has led to continual improvements in BMPs on national forests in California and on all lands in the Lake Tahoe Basin (U.S. Forest Service 1992; Tahoe Regional Planning Agency 1988). A recent review of forest management impacts on water quality concluded that the use of BMPs in forest operations was generally effective in avoiding significant water quality problems (Brown and Binkley 1994) . However, this report cautioned that proper implementation of BMPs was essential to minimizing non-point-source pollution and that ephemeral channels were often overlooked in the application of BMPs. Additionally, further development work is necessary for BMPs with respect to grazing, maintenance of slope stability, and avoiding losses of nitrates from soils (Brown and Binkley 1994). Much can be done to protect water quality simply by avoiding activities in sensitive areas, such as riparian zones, areas susceptible to mass movement, and areas where soils may become saturated and produce overland flow (Megahan and King 1985). The Tahoe Keys development in a former marsh on the upper Truckee River is an outstanding example of a major failure to respect such areas.

Forest Chemicals

Following the example of agriculture, forest management incorporated the use of fertilizers, pesticides, and herbicides in its operations during the 1960s and 1970s. As concerns about the environmental hazards of such chemicals have grown, their use appears to have decreased (Norris et al. 1991). Even at its peak, the use of silvicultural chemicals was tiny compared with that of agricultural chemicals. On the average, less than 1% of commercial forest land in the United States received any chemical treatment in a year (Newton and Norgren 1977). By contrast, most agricultural land receives multiple treatments every year.

Chemicals have been used in forest management for a variety of purposes (see Helms and Tappeiner 1996). Herbicides limit competition from other species so as to enhance opportunities for conifer regeneration and growth. Herbicide use has declined markedly since the early 1980s, when legal decisions in the Pacific Northwest limited their use and Region 5 of the U.S. Forest Service halted aerial applications of herbicides. However, chemical use now seems to be increasing again under new regulations. The use of insecticides has varied widely between years, depending on insect outbreaks (Norris et al. 1991). Fungicides and soil fumigants can control certain diseases and have been used mostly in tree nurseries. Rodenticides limit damage from gophers and other rodents, and animal repellents have been used to reduce damage to trees from porcupines and rodents. Fertilizers are used to enhance productivity by selectively compensating for nutrient deficiencies (Allen 1987). Fire retardants are the only class of forest chemicals that do not have a parallel in agriculture. They are used at margins of wildfires to slow the rate of fire spread.

Because pesticides, by definition, are toxic to some organisms, they pose hazards to some components of ecosystems. They have long been regarded as a particular threat to water quality and aquatic life (Brown 1980). Their use assumes that managers have decided that the pest that is the object of control efforts really should be eliminated or reduced in number. Therefore, the ecological risk associated with pesticides involves the consequences of that decision and the impacts on nontarget species. In general, the hazard to nontarget organisms depends on the exposure to significant doses and the toxicity of the chemical (Brown 1980). However, some groups of organisms, such as butterflies, are at risk from exposure to certain chemicals (see Shapiro 1996). Toxicological studies of forest chemicals in common use are reviewed by Norris et al. (1991).

Pesticides have the greatest potential to contaminate streams by direct (presumably unintentional) application and wind-borne drift into water courses. Toxicants used in fisheries management are applied intentionally to streams but may have a variety of unintended consequences (see Erman 1996). Spraying by ground crews is much more effective at placing all the pesticide where desired. The greatest potential for pesticides to appear in runoff exists when substantial precipitation occurs soon after the pesticide is applied. Opportunities for a chemical to reach a stream via overland flow depend on the distance from the stream to the closest point of chemical application, infiltration properties of soil and litter, the rate of flow toward the stream, and adsorptive characteristics of soil and organic matter (Brown 1980). Chemicals that reach streams may be removed through volatilization, adsorption on sediments, adsorption by aquatic biota, degradation by chemical, photochemical, or biological processes, and simple dilution with downstream movement (Norris et al. 1991).

Current practice generally limits insecticide and fungicide use to well-defined problems over relatively limited areas, such as insect-outbreak zones and nurseries. By contrast, herbicides can have rather broad application in forestry, and there is public concern about the potential for indiscriminate use. The Record of Decision on the California Region Final Environmental Impact Statement for Vegetation Management and Reforestation (U.S. Forest Service 1988) contains language prohibiting the use of hexazinone and similar herbicides "when they are expected to enter ground water or surface water, such as when soils are very sandy or have low clay or organic matter contents." A letter of October 30, 1990, to forest supervisors from the regional forester suggested that a margin of safety be established so that expected dose levels should be 100 times less than the dose level for which no adverse effects have been detected by laboratory studies. The standard that the Central Valley Regional Water Quality Control Board has established follows EPA practice as 200 parts per billion (ppb) for hexazinone (Stanislaus National Forest 1993). Monitoring for hexazinone in streams has been conducted on the Eldorado National Forest and Sierra National Forest after fall applications between 1991 and 1993. On the Eldorado, fifteen samples out of ninety contained hexazinone ranging from 1 to 19 ppb. No hexazinone has been detected and reported yet on the Sierra National Forest (Stanislaus National Forest 1993). However, a news media account suggested that hexazinone had killed riparian vegetation downstream of an application area on the Sierra National Forest in 1993.

Glyphosphate and triclopyr are two other herbicides that are being used more widely in the Sierra Nevada. Herbicide monitoring programs (Frazier and Carlson 1991) on three national forests in the Sierra Nevada in 1992 and 1993 found trace amounts of the two chemicals in only 3 of more than 120 samples, and those samples testing positive were suspected of being contaminated (Stanislaus National Forest 1993). In studies throughout the United States, chronic entry of herbicides into streams has not been observed (Norris et al. 1991). Artificial alteration of vegetation composition and cover has some potential for alteration of nutrient cycling. We are not aware of research concerning this issue at an operational scale. Pesticides are also widely used in residential areas in the Sierra Nevada and could cause localized contamination.

Fire retardants are applied during crisis situations without the opportunity for careful planning or management. Therefore, their impacts must be considered well before the time they are actually deployed. When aerial application of fire retardants was first used, the main active ingredient was sodium-calcium borate. After a few years, this material was noticed to have a tendency to sterilize the soil and restrict growth of new vegetation following the fire. In recent years, ammonium phosphate and ammonium sulfate have become the primary retardants in active use. Nitrogen in several forms is released as a breakdown product of these chemicals. Non-ionized ammonia (NH_3) is the only reaction product that is highly toxic to fish. A series of experiments relating to environmental impacts of ammonium fire retardants found that the compounds had little adverse effect on soil fertility, contributed a short-duration pulse of ammonia to streams, and moderately elevated levels of nitrates in receiving waters (Norris et al. 1978). The quantity of nutrients released by burning is likely to overwhelm any signal of those resulting from retardant application.

Forest chemicals may have a variety of unintended indirect effects on ecosystems by performing more or less as intended but in the wrong places. Insecticides may kill aquatic insects and reduce food supplies for fish. Herbicides can kill aquatic plants and disrupt the food chain at higher levels. Herbicides can also kill riparian vegetation, thereby reducing cover and shade benefits for fish and possibly increasing sediment yields. Death of riparian vegetation can add much organic debris to streams over a relatively short time and possibly deplete dissolved oxygen as it decomposes and also reduce the longer-term supply of organic matter until vegetation is reestablished on the banks. Fertilizers can contribute to eutrophication if the receiving waters are nutrient limited. To restate the obvious, minimizing the adverse impacts of forest chemicals on aquatic ecosystems requires that the chemicals be kept away from the streams and riparian zones.

Atmospheric Deposition

During the 1980s, concerns about the potential effects of atmospherically derived pollutants on aquatic ecosystems (Roth et al. 1985; Schindler 1988) focused attention on high-elevation lakes of the Sierra Nevada (Tonnessen 1984; Melack et al. 1985). The California Air Resources Board initiated a comprehensive study of the sensitivity of a small alpine lake basin in Sequoia National Park as part of a statewide acid-deposition program (Tonnessen 1991). This study explored the hydrochemical processes and biotic responses of this high-elevation system to possible shifts in precipitation chemistry (e.g., Williams and Melack 1991; Kratz et al. 1994). Hydrology and water chemistry of six other high-elevation lakes have been monitored over the past few years (Melack et al. 1993), and deposition has been monitored at several sites (Melack et al. 1995). These studies indicate that the loading rates of hydrogen, sulfate, nitrate, and ammonia are relatively low in the Sierra Nevada compared with rates in other parts of the country. However, snowpack processes can produce a distinct ionic pulse in the early part of the snowmelt season that temporarily lowers the pH of streams and lakes in high-elevation catchments with little buffering capacity (e.g., Williams and Melack 1991). Such surface waters may be at risk of acidification if air pollution and acidic deposition increase (see Cahill et al. 1996). A comprehensive state-of-knowledge review of aquatic impacts of acidic deposition by the University of California at Santa Barbara and the California Air Resources Board should be completed in 1996.

Monitoring

Obtaining adequate knowledge of water quality conditions throughout the Sierra Nevada on a continual basis is challenging at best. Frequent and long-term sampling from dozens to hundreds of sites is necessary to respond to sudden events, detect long-term trends, enforce regulations on discharges, improve the effectiveness of best management practices, and assess overall status. Sampling methodologies and analytical techniques are now fairly well developed (Stednick 1991; MacDonald et al. 1991). Bioassessment techniques using aquatic invertebrates as an integrative index or screening tool of water quality conditions is gaining widespread acceptance (U.S. Environmental Protection Agency 1989). However, broad strategies and philosophies for deciding what parameters to measure in what locations for what purpose have yet to be refined. Interpretation of water quality data to provide a sound basis for management or regulatory actions remains problematic (Ward et al. 1986). Most agencies and individuals concerned with water issues probably find the scarcity of monitoring data frustrating and inadequate to meet their needs. Additions to the present monitoring network will require implementation of creative mechanisms to provide substantial funding. No single agency can accomplish all the necessary monitoring independently. Interagency coordination is needed to maximize efficiency from available funds.

Evaluations of Water Quality in Streams

Assessments of water quality are made by the Department of Water Resources, the Central Valley and Lahontan Regional Water Quality Control Boards, the Environmental Protection Agency, the U.S. Geological Survey, the U.S. Forest Service, reservoir operators and proponents, and various other agencies. Every other year, the State Water Resources Control Board compiles water quality data from the regional water quality control boards and presents its findings to the Environmental Protection Agency under section 305(b) of the federal Clean Water Act. The 1992 Water Quality Assessment listed twenty-one streams draining the west slope of the Sierra Nevada as having serious quality problems. The principal problems in more than half these cases were degradation of fisheries habitat and inadequate flow. Mine drainage was noted in four cases, and sedimentation was recognized as a problem in tributaries of the Feather River and Little Butte Creek. Recreational impacts were mentioned as an additive problem in some cases (California State Water Resources Control Board 1992a). More rigorous criteria were used on the eastern slope, where almost all streams had some impairment of water quality, usually from water diversion or overgrazing. A subset of those streams (Blackwood Creek, Bryant Creek, Carson River, Heavenly Valley Creek, Monitor Creek, and Ward Creek) had more serious problems where violations of water quality objectives had occurred either from sedimentation or mine drainage. A list of thirty streams throughout the Sierra Nevada with various kinds of toxic contamination appeared in a companion report (California State Water Resources Control Board 1992b). Unfortunately, this listing does not rank the problems in terms of severity, and some problems on the list are known to be much more significant than others. What is worse, there is no information available for the majority of streams in the Sierra Nevada.

The Central Valley Basin plan summarizes water quality in Sierra Nevada streams above 300 m (1,000 ft) as "excellent" in terms of mineral content (Central Valley Regional Water Quality Control Board 1991). In general, concentrations increased from east to west (downslope and downstream). The Chowchilla and Fresno Rivers had the highest levels of total dissolved solids among western-slope rivers, but those amounts were still much lower than for streams in the Central Valley. A major assessment of water quality in the Sacramento River Basin was started by the U.S. Geological Survey in 1994 and will continue through 1998.

An evaluation of water quality in ten rivers in the central Sierra Nevada was carried out from 1975 to 1987 (California Department of Water Resources 1989). Nine of the rivers had very low levels of total dissolved solids (less than 150 mg/l—adequate for most industrial applications and well below a state criteria for drinking water of 500 mg/l). The tenth river in the survey, the East Walker, occasionally had high levels of total dissolved solids (up to 800 mg/l). High-elevation lakes in the Sierra Nevada as a group had the lowest ionic concentrations of any region sampled in the United States (Landers et al. 1987).

Several studies have focused on the Truckee River. Because of the high public value of the clarity of Lake Tahoe, water quality in the Lake Tahoe Basin is more thoroughly monitored than that in any other river basin in the Sierra Nevada. Water quality in most of the tributaries to the lake would be considered fine if not for the high sensitivity of the lake to nutrient additions. Downstream of Lake Tahoe, the Truckee River has largely recovered from the intense insults to water quality of the 1870s to 1930s (log transportation on artificial floods, sawdust dumping from lumber mills, and chemical waste from a pulp and paper mill) (Pisani 1977). Today, the principal problem in the Truckee River above Reno is elevated temperature resulting from water storage in Martis Creek, Prosser, Boca, and Stampede Reservoirs (Bender 1994). Total dissolved solids have been in the 6 to 210 mg/l range. Naturally occurring uranium is found in Sagehen Creek, and iron is high in a few places within the Truckee River system (Bender 1994). Water quality problems have been identified on Leviathan/Bryant Creeks (bacteria, nutrients), Little Truckee (nutrients), and Trout Creek (total dissolved solids, suspended sediments) (California State Water Resources Control Board 1984).

Although the surface waters of the Sierra Nevada are no longer pristine in terms of quality or other attributes, most streams could rank as excellent or outstanding compared with conventional standards or water elsewhere in the state, nation, or world. However, water quality in the Sierra Nevada, as elsewhere, is intimately connected to water quantity. Reduction in natural flows because of diversions is perhaps the most widespread water quality problem. Water remaining in the stream must support the same habitat needs and dilute whatever material and heat loads that arrive downstream of the points of diversion. For these reasons, what is usually considered a quantity problem is also a problem of quality. Additionally, there are persistent problems in different river basins. In the Lake Tahoe Basin, nutrient loads that would be considered small anywhere else are accelerating eutrophication of the lake. Within many parts of the Feather River Basin, unstable stream banks resulting from long-term overgrazing and roads are producing sediment yields at the basin scale that are up to four times greater than natural yields. Throughout much of the Sierra Nevada, a few problem mines continue to leach heavy metals into streams, and mercury remains in the beds of many streams from a century ago. Isolated problems such as poorly designed and located septic systems and roads impact local portions of streams and should be correctable by moderate investments for improved water quality.

EROSION AND SEDIMENTATION

Soil erosion, mass wasting, channel erosion, and sedimentation are natural processes that alter the landscape and streams. They are important disturbance mechanisms in terrestrial and aquatic ecosystems. These geomorphic processes are critical in nutrient cycling, transport of organic matter, and creation of fresh surfaces for colonization (Naiman et al. 1992). The rates at which they occur are highly variable across the landscape and over time. These processes operate most intensely

in association with major rainstorms and so can be considered episodic in nature. Nevertheless, streams tend to adjust their form to accommodate the long-term sediment supply. Processes that detach and transport particles of soil and rock downslope and downstream can be lumped together as erosion. Sedimentation occurs when these particles come to rest in transitory or long-term storage.

Aquatic Effects

Alteration of stream sediments can seriously impact populations of fish and other aquatic organisms. Aquatic ecosystems have developed in response to a particular regime of water and sediment flows and channel conditions. When conditions change, such as when annual floods cease because of a dam or the proportion of silt-size sediments increases because of a road built next to the stream, some organisms will benefit and some will suffer. Trout and other salmonids require streambed deposits of gravel-size particles in which to prepare nests (redds) for their eggs where there is substantial flow of water and dissolved oxygen. Until the fry emerge after two to six months, the redds are vulnerable to scour and deposition of other sediments that could block flow of water through the redd (Lisle 1989). When sediment inputs to a stream exceed the transport capacity of the channel, fine sediments (clays, silts, and sands) tend to accumulate on the bed surface (Lisle and Hilton 1992). Fine sediments have been found to fill substantial fractions of pools in streams on the Sierra National Forest that were known to have high sediment yields, such as Miami Creek (Hagberg 1993). These fine sediments often smother invertebrates, reduce permeability of streambed gravels and fish-egg nests (redds), impede emergence of fish fry, and cause poor health or mortality of fry at emergence because of reduced levels of dissolved oxygen (Burns 1970). Sedimentation also adversely impacts invertebrate habitat (Erman 1995). In many streams in the Sierra Nevada, suitable gravels for spawning are found only in isolated pockets and lower-gradient reaches (Kondolf et al. 1991; Barta et al. 1994). The limited extent of such areas increases their importance for fisheries maintenance. Fortunately, scour and deposition processes are highly variable within and between streams, so that some spawning areas are almost always available (Lisle 1989). Sediment transport processes in streams of the Sierra Nevada have been the subject of few studies (e.g., Andrews and Erman 1986), and even basic information is scarce. Much of the sediment in mountain streams consists of large particles known as bedload. In fourteen streams of the eastern Sierra Nevada, the proportion of bedload varied between 0% and 65% of the total sediment load (Skau et al. 1980).

Natural Sediment Yields

Natural surface erosion is generally regarded as small in the Sierra Nevada because of high infiltration capacity of the soils, predominance of snowmelt as a water input to soils, rarity of overland flow, predominance of subsurface flow, and relatively continuous vegetation cover. The sources and pathways of sediments supplied to stream channels are not completely understood. The channel system itself is an obvious candidate as a source for most of the sediment (King 1993). During persistent rainfall and peak snowmelt, the network of very small channels becomes rather extensive, mobilizing sediment from a large fraction of a watershed. Such sediment probably does not move very far but may be made available for transport by a high-magnitude runoff event. The sequence of events of different magnitudes can determine the net sediment transport over long time periods (Beven 1981). In the Sierra Nevada, the greatest potential for overland flow to occur appears to be below the snow zone in woodland-grassland communities between 300 and 900 m (1,000 and 3,000 ft) (Helley 1966). The maximum rates of sediment production have been observed in this same altitude range (Janda 1966). The woodland zone also was the primary sediment source in part of the American River Basin with annual erosion of about 150 m^3/km^2 (0.3 AF/mi^2) (Soil Conservation Service 1979).

Accelerated Erosion

Human activities often disrupt the natural geomorphic processes and accelerate erosion or destabilize hill slopes. Modeling erosion in the Camp and Clear Creek Basins suggests that disturbance, especially roads, can increase erosion many times above natural rates (McGurk et al. 1996). When soil loss and sediment transport occur at unusually high rates in response to some human disturbance, erosion and sedimentation become issues of concern. Accelerated soil loss is primarily a problem in terms of losing productivity for growing vegetation (Poff 1996). Excessive sedimentation can damage terrestrial plants and aquatic organisms. High levels of sediment deposition can also reduce the utility of facilities for water storage and diversion and hydroelectric production. At the extreme, hydraulic mining for gold on the west slope of the Sierra Nevada intentionally eroded entire hillsides. The resulting sedimentation in downstream river channels left deposits tens of meters thick. Sediment yield in the Yuba River was up to twenty-five times greater than natural rates (Gilbert 1917) and led to a legal decision effectively halting hydraulic mining. Activities that purposefully move soil, such as construction of roads and structures, have the greatest potential for increasing erosion. Activities that reduce vegetative cover and root strength can also increase erosion rates. Activities in and near stream channels have the greatest potential for altering sediment delivery and storage as well as channel form. For example, destruction of riparian vegetation can lead to massive streambank erosion, or dams can trap sediment from upstream while causing channel incision or narrowing downstream.

Processes involving movement of large units of soil or rock rather than individual particles are collectively known as mass wasting. Landslide activity is a typical mass failure in which a portion of a slope fails all at once. Movement may be catastrophic in seconds or progressive over years. Mass wasting may be important in providing a material supply to channels slowly through soil creep or suddenly when a debris flow reaches a stream, but it is not regarded as a major erosive agent in most of the Sierra Nevada (Seidelman et al. 1986). Mass movement typically occurs when most of the pores in the material become filled with water. The positive pressure of the pore water and its added mass may exceed the strength of the material, and failure of part of the slope may occur. Unusually high rates of water input to previously wet soils can lead to large numbers of landslides in the Sierra Nevada (De Graff et al. 1984). Disturbance of slopes accelerates the natural occurrence of landslides (Sidle et al. 1985). Excavations across slopes for roads intercept water flowing downslope through the soil and increase pore water pressure at the exposed seepage face. In granitic portions of the Sierra Nevada, ground-water flow is often at a maximum at the interface between the porous coarse-grained soils and underlying relatively impermeable bedrock (De Graff 1985). Exposure of this layer can bring large quantities of water to the surface (Seidelman et al. 1986). Such excavations also reduce the mechanical support for adjacent parts of the slope. Tree roots are often important in maintaining the integrity of a slope. Minimum strength occurs about ten years after fire or timber harvesting when roots from young trees have not yet compensated for the progressive loss of old roots (Ziemer 1981). Most opportunities to minimize mass wasting as a consequence of road construction and forest harvesting involve commonsense approaches to avoiding accumulation of subsurface water on steep slopes (Sidle 1980; McCashion and Rice 1983).

In years of high precipitation with large individual storms, the number, extent, and size of mass movements increase well above those of years with modest precipitation. Landslides were particularly active during the wet years of 1982 and 1983. In both those years, springs and seeps appeared in places they had not been noticed before, including many road cuts and fills. More than $2 million in damage occurred to roads on national forests in the Sierra Nevada during 1982, and additional damage estimated at more than $1 million occurred in 1983 (De Graff 1987). A landslide in the American River canyon blocked U.S. 50 for April, May, and June of 1983. Sustained high levels of soil moisture and ground water occurred throughout the winter and spring of each year. Additional water input from rainfall, combined rainfall and snowmelt, and snowmelt alone triggered the unusual number of failures (Bergman 1987; De Graff 1987). However, there seems to be relatively little interaction between high flows and initiation of landslides within the inner gorges of Sierra Nevada streams (Seidelman et al. 1986). Landslides can also be initiated by earthquakes (Harp et al. 1984) and summer thunderstorms (Glancy 1969). An extraordinarily intense storm occurred in the headwaters of the South Fork of the American River on June 18, 1982 (Kuehn 1987). About 100 mm (4 in) of rain fell in 30 minutes and produced a peak flow of about 200 m³/s/km² (19,000 ft³/s/mi²). These values of precipitation intensity and specific runoff are records for the Sierra Nevada and well above values assumed to be the maximum possible in the range (Kuehn 1987). The event also caused a large debris torrent in a small basin that had been burned the previous year.

Roads

Roads are considered the principal cause of accelerated erosion in forests throughout the western United States (California Division of Soil Conservation 1971a; California Division of Forestry 1972; Reid and Dunne 1984; McCashion and Rice 1983; Furniss et al. 1991; Harr and Nichols 1993). Roads destroy all vegetation and surface organic matter, minimize infiltration and maximize overland flow, oversteepen adjacent cut-and-fill slopes to compensate for the flat roadbed, and intercept subsurface flow, directing more water across the compacted surface (Megahan 1992). Stream crossings by roads are particularly effective at increasing sediment yields because of their direct impact on the channel. Stream banks are excavated for bridges and filled for culverts. Failure of inadequately designed and constructed culverts adds large amounts of sediment to streams. Increases in fine sediment and decreases in fish populations were associated with the number of culverts and roads near streams on the Medicine Bow National Forest in Wyoming (Eaglin and Hubert 1993). A classic study in the granitic batholith of Idaho found that sediment yields relative to an undisturbed forest increased by 60% as a result of logging and by 220 times (22,000%) from road construction (Megahan and Kidd 1972). A compilation of studies in the Oregon Coast Range showed that the quantity of mass movements associated with roads was 30 to 300 times greater than in undisturbed forest and was more than 10 times greater than that associated with large clear-cuts (Sidle et al. 1985). Large highway projects also produce significant amounts of sediment, with fill slopes often providing the most easily transported material (Howell et al. 1979). During major storms, highways are often damaged and provide much sediment to streams. For example, during February 1986, four serious debris flows in the Truckee River canyon closed Interstate 80, and sixty-three road failures occurred along 55 km (35 miles) of the Feather River Highway 70 (McCauley 1986; Keller and King 1986).

Land Development

Construction activities also have the potential to increase erosion rates (California Division of Soil Conservation 1971a). Residential construction around Lake Tahoe has been a major contributing factor in accelerating erosion and increasing nutrient inputs to the lake (Tahoe Regional Planning Agency 1988). In Nevada County, even by 1970, more than 35% of the length of streams in the county had been damaged by silta-tion and stream-bank erosion resulting from subdivision development (Gerstung 1970). Only a few examples of major erosion are well documented. For example, erosion from a single storm on freshly cleared land for a new subdivision in Plumas County killed 80% of the aquatic life in Big Grizzly Creek (California Division of Soil Conservation 1971b). Sediment from a failure of a channelization project for a new golf course largely filled Hunter's Reservoir on Mill Creek (California Division of Soil Conservation 1971b).

Logging

Timber harvesting itself seems to have relatively little effect on soil erosion compared with the construction of roads used for log removal (see McGurk et al. 1996; Poff 1996). Although soil disturbance associated with cutting trees and skidding logs exposes mineral soil to raindrop splash as well as to rill development where soils are compacted, in practice, comparatively little soil leaves harvested areas. The California Division of Forestry (1972) has asserted that "timber harvesting, when done carefully with provisions made for future crops, has little adverse effect upon soil erosion, sedimentation, or water quality." During his evaluation of sedimentation from hydraulic mining, Gilbert (1917) noted that erosional effects of timber harvesting were minor compared to other, non-mining effects such as overgrazing and roads. Several factors appear to mitigate potential adverse effects of harvesting: only small and discontinuous areas are compacted to an appreciable extent; infiltration capacity is generally maintained over large areas; a lot of slash is left behind; and some type of vegetation usually reoccupies the cutover land quickly. Another important factor to date has been the concentration of harvests in the most productive sites and most accessible areas, which tend to be on relatively gentle slopes. As harvesting moves to less desirable and steeper ground, risk of erosion and mass failure will increase. Avoidance of lands sensitive to disturbance, such as slopes greater than 60%, streams with soil-covered inner gorges, riparian areas, meadows, and known landslides, will minimize erosion associated with timber harvest (Seidelman et al. 1986).

Despite mitigating factors that can reduce logging-related erosion, some harvest units lose large amounts of soil. Such areas appear to be a minority, although their local effects can be quite significant. The degree of soil compaction seems to be a controlling influence on subsequent erosion (Adams and Froehlich 1981). Severe sedimentation in the West Fork of the Chowchilla was noted after upstream areas were virtually denuded of vegetation to supply fuel for a smelter at the Mariposa Mine about 1900 (Helley 1966). The headwaters of Last Chance Creek on the Plumas National Forest had erosion rates from 150 to more than 300 m³/km² (0.15 to 0.66 AF/mi²) during a severe thunderstorm following a salvage sale in the Clark Fire area (Cawley 1991). A series of studies of northern California streams, including some in the Sierra Nevada, found significantly greater amounts of fine sediments and altered benthic invertebrate communities downstream of logged slopes (Erman et al. 1977; Newbold et al. 1980; Erman and Mahoney 1983; Mahoney and Erman 1984). Some effects of logging on streams were persistent for more than a decade (Erman and Mahoney 1983; O'Connor 1986; Fong 1991). A study of erosion rates from small plots recently started by Robert Powers of the Redding office of the Pacific Southwest Research Station of the U.S. Forest Service should improve our understanding of erosion processes and rates in the Sierra Nevada.

Christmas tree plantations have been found to have very high rates of erosion (Soil Conservation Service 1979). Management for Christmas trees typically attempts to minimize other ground cover that would compete for water and, therefore, makes the plantations more vulnerable to erosion.

Measured Sediment Yields

Compared to other parts of California and the United States, the Sierra Nevada overall has relatively low sediment yields (Brown and Thorp 1947). A map of soil erodibility for California shows the absence of "very severe" ratings throughout the Sierra Nevada except for areas of western Plumas and eastern Butte Counties and in part of Yuba County, whereas such ratings are common in the Coast Range (California Division of Soil Conservation 1971a). General estimates shown on another statewide map show that the Sierra Nevada has the lowest sediment yield in California (generally less than $100 \text{ m}^3/\text{km}^2/\text{yr}$ [$0.2 \text{ AF}/\text{mi}^2/\text{yr}$]) (California Division of Forestry 1972). Sediment transport measurements in a variety of streams in the eastern Sierra Nevada were generally less than $10 \text{ m}^3/\text{km}^2$ ($0.02 \text{ AF}/\text{mi}^2$), but there were exceptions of up to $450 \text{ m}^3/\text{km}^2$ ($0.9 \text{ AF}/\text{mi}^2$) (Skau and Brown 1990). An estimate of annual sediment yield for the San Joaquin Basin above the San Joaquin valley based on a comprehensive geological investigation was about $38 \text{ m}^3/\text{km}^2$ ($0.08 \text{ AF}/\text{mi}^2$) (Janda 1966). For comparison, an average value for the entire United States is $76 \text{ m}^3/\text{km}^2$ ($0.16 \text{ AF}/\text{mi}^2$) (Schumm 1963). The Colorado River Basin produces about $300 \text{ m}^3/\text{km}^2/\text{yr}$ ($0.6 \text{ AF}/\text{mi}^2/\text{yr}$) and the Columbia River yields about $30 \text{ m}^3/\text{km}^2/\text{yr}$ ($0.06 \text{ AF}/\text{mi}^2/\text{yr}$) (Holeman 1968). A compilation of sediment studies from forested regions provided an average rate of about $30 \text{ m}^3/\text{km}^2/\text{yr}$ ($0.06 \text{ AF}/\text{mi}^2/\text{yr}$) from forest land in the United States excluding the Pacific Coast Ranges (Patric et al. 1984). A Soil Conservation Service report classified sediment yields below $150 \text{ m}^3/\text{km}^2$ as "low" with respect to nationwide rates (Terrell and Perfetti 1989).

The best means of determining sediment yields over long time periods is with repeated bathimetric surveys of reservoirs (Dunne and Leopold 1978; Hewlett 1982; Rausch and Heinemann 1984). Comparison of the bottom topography after a span of a few years allows calculation of the change in volume of sediment over the time interval (Rausch and Heinemann 1984; Jobson 1985; Mahmood 1987). Most of the information for the Sierra Nevada came from a Soil Conser-

vation Service study in the 1940s (Brown and Thorp 1947). This same data set has been republished many times (e.g., Dendy and Champion 1978; U.S. Army Corps of Engineers 1990; Kondolf and Matthews 1993), but there have been few additions to it. Until 1975, the Committee on Sedimentation of the Water Resources Council compiled data for reservoir surveys throughout the United States (Dendy and Champion 1978). Records of suspended sediment at water quality monitoring stations reported by the U.S. Geological Survey were also examined but did not prove to be useful. Almost all stations are downstream of dams, and uncertainty resulting from the assumptions required to estimate annual totals would mask any trends over time.

Estimates of average annual sediment yields in the Sierra Nevada were compiled from all available sources (tables 30.3 and 30.4). These values provide order-of-magnitude approximations of sediment yield. The numbers should be considered uncertain and may contain some serious errors resulting from the original measurements, assumption of inappropriate densities if reported as mass rather than volume, and conversion from some unusual units. The period of measurement varies greatly between basins, resulting in different sediment delivery regimes depending on the inclusion of floods. Some of the values in tables 30.3 and 30.4 were based on total basin area above the reservoir or measurement site, and others were based only on the sediment contributing area not regulated by upstream reservoirs and lakes. Tables 30.3 and 30.4 illustrate that sediment yields vary considerably between river basins but that the generalizations mentioned above seem appropriate. Most reported values are less than $100 \text{ m}^3/\text{km}^2/\text{yr}$ ($0.2 \text{ AF}/\text{mi}^2/\text{yr}$), which is the simple average of table 30.4. This value can be visualized as a tenth of a millimeter in depth over the entire contributing area, which is not how sediment is produced, but the conversion is useful for illustration. The relatively high sediment yields of the Kaweah and Tule are somewhat surprising, especially in the Kaweah Basin, which is largely in Sequoia National Park. However, this short period (1960–67) includes the massive floods of February 1963 and December 1964, which would tend to bias the annual sedimentation rate.

Unfortunately, very few measurements of reservoir sedimentation have been reported in the past two decades. The one-time measurements in isolation do not provide sufficient information or provide much confidence in using the values to infer differences between basins or over time. Comparison of modern sedimentation rates with those summarized by Brown and Thorp (1947) would be very useful in determining whether more intensive land management has altered sediment yields at the basin scale. A highly detailed bathimetric survey of Slab Creek Reservoir (in South Fork American River Basin) in 1993 revealed less than 0.5 m of accumulation on the bed of the reservoir since 1968 but did not estimate the volume of the deposit (Sea Surveyor, Inc. 1993). Crude estimates based on information provided in the report suggest an annual sediment yield less than $10 \text{ m}^3/\text{km}^2$ ($0.02 \text{ AF}/\text{mi}^2$).

TABLE 30.3

Sediment yields from reservoir surveys.

Site	Drainage Area (km^2)	Elevation of Dam (m)	Interval (years)	Annual Sediment Yield (m^3/km^2)	(AF/mi^2)	Source
Sacramento Tributaries						
Magalia	21	681	18–46	150	0.3	Brown and Thorp 1947
Yuba						
Bullards Bar	1,226	488	19–39	130	0.2	Brown and Thorp 1947
Bear						
Combie	330	488	28–35	360	0.8	Brown and Thorp 1947
American						
Ralston	1,095	362	66–89	80	0.2	EA 1990
Folsom	6,955	146	55–91	250	0.5	California Department of Water Resources 1992 in Kondolf and Matthews 1993
Cosumnes						
Big Canyon	14	232	34–45	30	0.1	Brown and Thorp 1947
Blodgett	8	48	40–45	80	0.2	Brown and Thorp 1947
Calaveras						
Davis	19	34	17–45	120	0.3	Brown and Thorp 1947
Gilmore	13	69	17–45	60	0.1	Brown and Thorp 1947
McCarty	1	350	37–45	140	0.3	Brown and Thorp 1947
Salt Spring Valley	47	357	82–45	100	0.2	Brown and Thorp 1947
Stanislaus						
Copperopolis	5	297	15–45	20	0.03	Brown and Thorp 1947
Lyons	102	1,287	30–46	50	0.1	Brown and Thorp 1947
Mokelumne						
Pardee	980	173	29–43	70	0.2	Brown and Thorp 1947
Pardee	980	173	29–95	150	0.3	EBMUD 1995
Upper Bear	72	1,791	00–46	10	0.2	Brown and Thorp 1947
Schadd's	72	886	40–90?	100	0.2	Euphrat 1992
Tuolumne						
Don Pedro	2,550	186	23–46	100	0.2	Brown and Thorp 1947
La Grange	3,842	92	95–05	40	0.1	Brown and Thorp 1947
Merced						
Exchequer	2,616	216	26–46	80	0.2	Brown and Thorp 1947
San Joaquin						
Crane Valley	135	1,026	01–46	80	0.2	Brown and Thorp 1947
Kerckhoff	3,031	296	20–39	80	0.2	Brown and Thorp 1947
Mammoth Pool	2,550	1,026	59–72	90	0.2	Anderson 1974
Kings						
Hume	62	1,616	09–46	10	0.03	Brown and Thorp 1947
Pine Flat	3,948	296	54–56	90	0.2	Dendy and Champion 1978
Pine Flat	3,948	296	54–56	30	0.1	Anderson 1974
Pine Flat	3,948	296	56–73	80	0.2	Dendy and Champion 1978
Wishon	445	2,000	58–71	10	0.03	Anderson 1974
Kaweah						
Terminus	1,453	212	61–67	360	0.8	Dendy and Champion 1978
Tule						
Success	1,006	190	60–67	400	0.9	Dendy and Champion 1978
Kern						
Isabella	5,309	776	53–56	35	0.1	Dendy and Champion 1978
Isabella	5,309	776	56–68	90	0.2	Dendy and Champion 1978
Walker						
Weber	6,241	1,284	35–39	10	0.02	Dendy and Champion 1978
Weber	6,241	1,284	?	30	0.05	Soil Conservation Service 1984

TABLE 30.4

Sediment yields from suspended sediment records and other estimates.

Site	Drainage Area (km^2)	Annual Sediment Yield (m^3/km^2)	(AF/mi^2)	Source
Feather				
Oroville	9,244	90	0.2	Jansen 1956
Oroville	9,244	100	0.2	U.S. Army Corps of Engineers 1990
Oroville	9,244	120	0.3	Soil Conservation Service 1989
East Branch North Fork	3,131	270	0.6	Soil Conservation Service 1989
Yuba				
Nonmining		160	0.3	Gilbert 1917
Hydraulic mining		3,300	7	Gilbert 1917
Castle Creek	10	70	0.1	Anderson 1979
Castle Creek (logged)	10	220	0.5	Anderson 1979
American				
Auburn dam site	2,485	130	0.3	U.S. Army Corps of Engineers 1990
Cameron Park		70	0.2	Soil Conservation Service 1985
Onion Creek	4	30	0.06	Dendy and Champion 1978
Cosumnes				
Michigan Bar	1,098	30	0.06	Anderson 1979
Stanislaus				
New Melones	2,314	60	0.1	U.S. Army Corps of Engineers 1990
Merced				
Happy Isles	463	3	0.01	Anderson 1979
Chowchilla				
Buchanan		40	0.1	Helley 1966
San Joaquin				
Kerckhoff		40	0.1	Janda 1966
Kings				
Teakettle	7	10	0.02	Dendy and Champion 1978
Kern				
????	2,613	150	0.3	Anderson 1979
Truckee				
Tahoe Basin	839	30–60	0.05–0.1	Tahoe Regional Planning Agency 1988
Tahoe (in 1850)	839	3	0.01	Tahoe Regional Planning Agency 1988
Upper Truckee	142	21	0.04	Hill and Nolan 1990
General Creek	19	13	0.03	Hill and Nolan 1990
Blackwood Creek	29	65	0.14	Hill and Nolan 1990
Ward Creek	25	63	0.13	Hill and Nolan 1990
Snow Creek	11	3	0.005	Hill and Nolan 1990
Third Creek	16	20	0.04	Hill and Nolan 1990
Trout Creek	95	12	0.03	Hill and Nolan 1990
Squaw Creek	21	12, 93	0.03, 0.2	Woyshner and Hecht 1989
Sagehen	28	2	0.005	Anderson 1979

Recent bathimetric surveys of Pardee Reservoir on the Mokelumne River suggest that the average annual rate of sediment deposition has more than doubled since the last survey in 1943 (150 m^3/km^2 [0.3 AF/mi^2]) (EBMUD 1995). Parts of the Mokelumne River Basin have been extensively roaded and logged in the past few decades, and there has been much concern about apparent increases in sediment yield from some of the erodible soils (e.g., Euphrat 1992). These new results offer evidence of a sedimentation response to large-scale disturbance of a forested basin. Much greater sedimentation rates are apparent in Camanche Reservoir, downstream of Pardee. Additional studies are needed to determine the sources of sediment trapped in Camanche. At rates of deposition suggested by the recent sediment surveys, half the original storage volume of Camanche would be lost in 380 years, and half of the original storage volume of Pardee would be lost in 600 years.

An Example of Disturbance Effects

The North Fork Feather River has perhaps the worst erosion and sediment problem of any large basin in the Sierra Nevada. Conditions were certainly much worse in several drainages during the hydraulic mining era and for following decades until most of the debris was flushed into the lower reaches of the river systems. Nevertheless, sediment production under current conditions in the North Fork Feather River can be considered high compared with natural background rates (Plumas National Forest 1988). A comprehensive evaluation of sediment sources in the basin found that about 90% of the erosion and about 80% of the sediment yield is accelerated (induced by human activities) (Soil Conservation Service 1989). That estimate and the current sediment yield of about 270 m^3/km^2 (0.6 AF/mi^2) imply that under natural conditions, sediment yield would be about 50 m^3/km^2 (0.1 AF/mi^2). The difference is caused mainly by bank erosion where riparian vegetation has been eliminated by overgrazing and erosion from road cut-and-fill slopes (Soil Conservation Service 1989; Clifton 1992, 1994). Mining, logging, and overgrazing before 1900 initiated widespread changes in hydrologic conditions of the land surface and channels. Gullying and channel erosion were noted by the 1930s (Hughes 1934). After about 1940, stream channels widened rapidly with little reestablishment of riparian vegetation along new channel banks. More than 75% of the stream length in the Spanish Creek and Last Chance Creek watersheds was found to be unstable and eroding (Clifton 1992). Bank erosion contributes sediment directly into the streams, which in turn transport it to lower elevations. About one-third of the forest roads are eroding rapidly as well and often contribute sediment directly into streams where roads cross or run parallel (Clifton 1992). By contrast, sheet and rill erosion appear to produce very little (less than 2% of the total) of the sediment in the basin because the nearly continuous vegetation cover protects the soil (Soil Conservation Service 1989). A cooperative effort among local landowners, public agencies, Pacific Gas and Electric Company, and private individuals is attempting to reduce erosion throughout the basin (Wills and Sheehan 1994; Clifton 1994). The Pacific Gas and Electric Company is involved because it operates two small reservoirs in the canyon of the North Fork Feather River as part of its hydroelectric network. Sediment is rapidly filling the reservoirs, interfering with operation of the control gates on the dams and accelerating turbine wear (Harrison 1992). A costly program of dredging and reconstruction of the dams to allow pass-through sluicing of sediment during high flows is being planned in addition to participation in the upstream erosion control program (Pacific Gas and Electric Company 1994). A few other reservoirs in the Sierra Nevada have filled with sediment and had to be dredged. This topic is discussed further in the section on dams and diversions.

GROUND WATER

Ground-water storage is generally limited throughout the Sierra Nevada compared with surface water resources. However, ground water is significant in providing small amounts of high-quality water for widely scattered uses, such as rural residences and businesses, campgrounds, and livestock watering. Without ground water, the pattern of rural development in the Sierra Nevada would be quite different. The geology of the mountain range is not conducive to storage of large quantities of subsurface water. Ground water occurs in four general settings: large alluvial valleys; small deposits of alluvium, colluvium, and glacial till; porous geologic formations; and fractured rocks. The shallow aquifers tend to be highly responsive to recharge and withdrawals. The effects of low precipitation in the recent drought cannot be readily separated from effects of increased pumping on declining water levels in some areas. Tens of thousands of wells tap ground water throughout the Sierra Nevada for local and distant municipal supply, individual residences, and recreational developments. Nearly one-quarter of all homes in the Sierra Nevada are supplied by private, on-site wells (Duane 1996a). More than 8,000 residents of Tuolumne County alone depend on wells for water supply. In 1982, there were about 5,800 wells in Placer County, 6,100 in El Dorado County, 3,400 in Amador County, and 2,200 in Calaveras County (California Department of Water Resources 1983a). Some of these wells were west of the SNEP study area. Contamination appears to be minimal overall (California State Water Resources Control Board 1992a).

Ground-Water Resources

A few ground-water basins in the Sierra Nevada store vast quantities of water, but they have limited recharge compared with some proposed exploitation plans. Honey Lake/Long Valley has a capacity of about 20 billion m^3 (16 million AF) in alluvial and lake sediments up to 230 m (750 ft) thick. The quality is poor in some areas, with high concentrations of boron, fluoride, sulfate, sodium, arsenic, and iron. Sierra Valley stores about 9 billion m^3 (7.5 million AF) of water in sediments up to 370 m (1,200 ft) deep. Hot springs occur in the center and southern part of the valley, and excessive amounts of boron, fluoride, and chloride have been found in some wells. Several schemes have been proposed for mining ground water from both Sierra Valley and Long Valley for export to Reno. Martis Valley contains about 1 billion m^3 (1 million AF) of water, is an important water source for the Truckee area, and has the lowest concentration of total dissolved solids (60–140 mg/l) of any large ground-water basin in the state. By contrast, ground water in the 4 billion m^3 (3.4 million AF) volume Mono basin is highly mineralized. The Owens Valley is the largest ground-water basin partially within the SNEP study area, with a storage capacity of about 47 billion m^3 (38 million AF). Export of ground water from the Owens Valley to southern California began in 1970. This ground-water development led to declines of ground water dependent vegetation (e.g., Groeneveld and Or 1994) and continues (as of 1995) to be the subject of negotiations between Inyo County and the City of Los Angeles. Smaller alluvial valleys include Indian and American Valleys of the Feather River Basin, Tahoe Valley in the upper Truckee River Basin, Slinkard and Bridgeport Valleys of the Walker River Basin, and Long Valley in the Owens River Basin.

Many wells in the Sierra Nevada are located in shallow deposits of glacial till, alluvium, and colluvium. These surficial deposits, which are often only a few tens of meters deep (Page et al. 1984; Akers 1986), are fairly porous and convey water to streams. Deeper deposits are capable of serving the needs of small communities but may be sensitive to recharge conditions. Placer County (1994) has determined that ground water in the foothills is not a reliable source of water for future growth.

Some rocks and other geologic formations, like buried river channels, are relatively porous and transmissive. Hydrogeologic properties of these formations are highly variable, as are well yields. Locating a well is often hit-or-miss, but drillers familiar with an area can usually find sources of water adequate for residential use. Mixed results have been obtained in recent drilling through the complex layers of till, volcanic ash, and basalt found in the Mammoth Lakes area. Some wells have been highly productive, and others have quickly gone dry.

Granitic and metamorphic rocks of the Sierra Nevada are essentially impermeable except where fractured. In some locations, the joint and fracture systems can transmit significant quantities of water. A recent study in the Wawona area of Yosemite National Park investigated fracture systems and the regional movement of deep ground water (Borchers et al. 1993). Most wells in southwestern Nevada County, and presumably in other parts of the foothills with similar geology, are located in areas of fractured rock (Page et al. 1984). Of some 13,000 wells drilled in Placer, El Dorado, Amador, and Calaveras Counties between 1960 and 1982, more than 90% were located in hard rock (California Department of Water Resources 1983). The size and frequency of fractures decline with depth away from the surface, so the more productive wells in Nevada County have been less than 60 m (200 ft) deep. Mean yield in that study area was less than 70 l/min (18 gal/min) (Page et al. 1984), with about half the wells yielding less than 38 l/min (10 gal/min) (California Department of Water Resources 1974). Average well yields determined from drillers' logs were less than 80 l/min (20 gal/min) in both Nevada and Amador Counties (Harland Bartholomew and Associates et al. 1992; California Department of Water Resources 1990a). Wells in Tuolumne County are often more than 90 m (300 ft) deep and are adequate for domestic use. The drought between 1987 and 1992 limited recharge throughout the Sierra Nevada, and yields of many wells declined through the period. There is insufficient information available to determine whether the proliferation of wells throughout the foothills in the past decade has had a pronounced effect on preexisting wells.

Pumping of water for industrial uses has lowered water tables in western Tuolumne County (comments by Tuolumne Utility District in DEIS on Yosemite Estates). Ground-water pumping can also impact local stream flow. Interactions between ground water and streams are very complex in some areas of the Sierra Nevada where glacial till is interlayered with volcanic mudflows and ash and is dissected by old stream courses and faults (Kondolf and Vorster 1992). Drilling of supplemental water-supply wells for Mammoth Lakes raised concerns that pumping could further reduce flows in Mammoth Creek, which is already diverted as the principal water source for the town (Kattelmann and Dawson 1994).

In a small lake basin in the alpine zone of Sequoia National Park, water released from short-term subsurface storage accounted for less than 15% of the annual stream-flow volume, but it controlled the chemistry of stream and lake water for more than two-thirds of the year (Kattelmann 1989b).

Springs are an important water source for small demands that require minimal development. Because springs are often fed by shallow aquifers, they are more susceptible to contamination than deep sources and often require protection of their contributing areas. Dense vegetation resulting from decades of fire suppression may maximize transpiration losses from hill slopes above springs, thereby reducing spring flow. Developing springs as a water source usually alters or even eliminates riparian and aquatic habitat in the immediate area. Springs are one of the most threatened habitats in the Sierra Nevada (see Erman 1996). Springs as well as pumped water are commercially developed for packaging as mineral water. Bottled water operations are present in the northern and southern Owens Valley.

Ground-Water Quality

The mineral content of ground water is generally much higher than that of surface water. The long residence time of water in the ground allows it to dissolve minerals and accumulate ions. Nevertheless, total dissolved solids in ground water in the Sierra Nevada are usually not an impediment for use. Deeper ground water in parts of the Honey Lake/Long Valley Basin and the Mono basin and below Mammoth Lakes contain substantial concentrations of various ions. Concentrations of naturally occurring iron are sometimes too high for domestic uses (Thornton 1992; Placer County 1994). Some wells in Kern Valley have very high levels of fluoride. Shallow ground water may be contaminated with nutrients from septic and sewage disposal systems, livestock, and chemicals applied to farms and gardens. Nutrients found in ground water in the Lake Tahoe Basin were relatively low in an absolute sense, but they still contributed to enrichment of the lake waters (Loeb and Goldman 1979). Water quality problems of the larger ground-water basins in the Sierra Nevada identified in the biennial state water quality assessment included drinking water impairment from heavy metals, fuel leaks, volatile organic compounds, naturally occurring radioactivity, pesticides, and wastewater (California State Water Resources Control Board 1992a). Some wells in the foothills of the southern west slope of the Sierra Nevada have been found to contain concentrations of uranium, radon, and radium above state health standards (California Department of Water Resources 1990b). Water in certain hot springs has high levels of natural radioisotopes. High levels of radionuclides have also been found in wells of the Lake Tahoe Basin (California Department of Water Resources 1994).

The leaking underground storage tank problem has probably introduced fuels to ground water in isolated spots throughout the Sierra Nevada. Gasoline contamination has been documented in Bishop, Mammoth Lakes, Bridgeport, and Placer County. Tetrachloroethylene, a solvent used in dry cleaning, was found in two municipal wells in East Sonora. The wells were removed from service, and an expensive extension to surface water supply was installed. Old landfills are another potential source of contamination.

MINING

Historically, mining had the most intense impact on rivers of the Sierra Nevada. As discussed in the history section, hydraulic mining for gold until 1884 truly wreaked havoc throughout the Gold Country. Affected streams and hill slopes have been recovering ever since. In most cases, the degree of recovery is remarkable. Much of the region appears to have healed over the past century. In terms of their more obvious hydrologic and biologic characteristics, the streams have improved dramatically compared to photographs and descriptions of the nineteenth century. Stream channels are now largely free of mining sediment, although large deposits remain as terraces (James 1988). Riparian vegetation has become reestablished. Aquatic biota have returned to the streams, at least partially. Some fish species that would be expected are not present in rivers heavily impacted by mining (Gard 1994). We can assume that the present form of the ecosystems is simplified compared to the pre–gold rush situation, but we really do not know what the west slope of the Sierra Nevada might have looked like had gold not existed in the range. Unfortunately, the Gold Country was so heavily mined that "natural" streams are not available for comparison. Portions of streams that were lightly impacted could be compared to those that were heavily impacted, but doubt would remain about what constitutes "natural" conditions. The water projects initiated during the mining period and other associated land uses have further modified the hydrologic system.

Legacy of Hydraulic Mining

After the 1884 *Sawyer* decision and the 1893 Caminetti Act, hydraulic mining continued on only a sporadic basis where the debris could be kept on-site. Mines in three old hydraulic

pits in the Yuba River Basin were active in the late 1960s: near French Corral, Birchville, and North Columbia (Yeend 1974). A few such mines continue operation today. When the original mines closed, there was no attempt at site reclamation, and the mines were simply abandoned. A variety of dams were constructed in attempts to prevent further movement of mining debris downstream (Rollins 1931). Only the larger, better-engineered structures did not fail. Dams such as Combie on the Bear, Englebright on the Yuba, and North Fork on the American have restrained vast amounts of mining debris from washing downstream to the Sacramento valley. An attempt to destroy a debris dam on Slate Creek with explosives was made in the 1960s by miners desiring another opportunity to recover gold. The initial bombing failed, but the structure is damaged and loses sediment during floods (Kondolf and Matthews 1993). Also along Slate Creek, a wooden wall retaining a large volume of mining debris appeared ready to fail in 1994. The hydraulic mine pits are slowly becoming revegetated, but they continue to release unnaturally high volumes of sediment as their walls continue to collapse until a stable slope angle is attained (Senter 1987). The unnaturally high sediment loads continue to affect aquatic biota (Marchetti 1994). A large open-pit gold mine that was operated at Jamestown until 1994 offers the first major opportunity for modern reclamation technology to be applied to a recently closed mine in the Sierra Nevada. The pit may also be used as a garbage dump. Current mineral potentials are discussed in Diggles et al. 1996.

Dredging

Massive riverbed dredging operations at the lower margins of the foothills persisted until 1967 (Clark 1970). The spoil piles may remain as a peculiar landscape feature for centuries. Some of the tailings in the Feather River were used in construction of the Oroville Dam, and other uses of the material and the land may be found. Small-scale suction dredging continues in many streams of the Gold Country. This activity has become widespread wherever there is easy access to the streams (McCleneghan and Johnson 1983). Powerful vacuums mounted on rafts remove stream gravels from the bed for separation of any gold particles, and the waste slurry is returned to the river, where the plume of sediment stratifies in the flowing stream. Turbidity obviously increases, and the structure of the bed is rearranged. The morphology of small tributaries can be dramatically altered by suction dredging (Harvey 1986; Harvey et al. 1995). Where stream banks are illegally excavated, the potential for damage is much greater. A study of effects of suction dredging on benthic macroinvertebrates showed local declines in abundances and species richness, but biota rapidly recolonized the disturbed sites after dredging stopped (Harvey 1986). Although dredging seems to have relatively little impact on adult fish, eggs and yolk-sac fry and amphibians within the gravel are usually killed by dredging (Johnston 1994). Dredging also has the potential to reintroduce mercury stored in sediments contaminated by early mining (Harvey et al. 1995; Slotten et al. 1995).

Underground Mining

Hard-rock mining often releases hazardous materials to ground water and streams. The nature and impacts of some of the typical mine effluents are reviewed by Nelson et al. (1991). Excavation of hard-rock mines exposes tunnel walls and tailings to water and oxygen and vastly increases the reactive surface area of minerals, allowing chemical reactions to occur at much faster rates than if undisturbed. If the mines or their waste piles contain sulfide minerals, oxidation in the flowing water can release sulfuric acid and metals into the drainage water. Exposure as a result of mining also allows reaction products to be leached from tailings piles or abandoned mines. Contaminated water can be flushed into streams in sudden pulses during storm runoff or slowly during base flow. In some cases, these products are highly toxic, and the runoff is acidic. The downstream extent of impacts along streams seems to depend on interactions between source concentrations, hydrologic characteristics of the mine or waste rock, storm characteristics, chemical behavior of the particular constituents, bacterial influences, presence of other substances as complexing agents, and dilution potential of the receiving waters. Fortunately, the mineralogy and geochemistry of most mines in the Sierra Nevada have resulted in relatively few serious surface-water problems (Montoya and Pan 1992). However, exceptions such as the Leviathan, Walker, and Penn mines have seriously degraded downstream areas. The substrate of a housing development built on tailings of the Central Eureka mine near Sutter Creek contains arsenic levels about seventy-five times greater than average values for soils in California. Discharge from mine dewatering and from rejuvenation of closed mines probably released toxic materials into nearby streams. Abandoned pits often fill with water and attract waterfowl and other wildlife. If the water contains toxic materials, these substances can enter the food chain.

Water Quality Impacts

An inventory of mines causing water quality problems has been developed by the Central Valley Regional Water Quality Control Board (1975). Mines in the Sierra Nevada included on that list appear in table 30.5. All except two are underground mines. The list is evenly split between gold mines and mines for other minerals, chiefly copper.

A more recent survey by the Central Valley Regional Water Quality Control Board (Montoya and Pan 1992) limited to the Sacramento valley investigated thirty-nine inactive mines from Butte Creek to the American River. Water quality of the drainage from these mines and waste piles was highly variable between mines and over time. For example, copper concentrations below the Spenceville mine on Dry Creek

TABLE 30.5

Mines cited by the Central Valley Regional Water Quality Control Board (1975) as degrading local water quality.

Mine	Receiving Stream
Cherokee	Sawmill Ravine / Dry Creek / Butte Creek
Mineral Slide	Little Butte Creek / Butte Creek
China Gulch	Lights Creek / Wolf Creek / North Fork Feather River
Engel	Lights Creek / Wolf Creek / North Fork Feather River
Iron Dyke	Taylor Creek / Indian Creek / Wolf Creek / North Fork Feather River
Walker	Little Grizzly Creek / Indian Creek / Wolf Creek / North Fork Feather River
Kenton	Kanska Creek / Middle Yuba River
Malakoff Diggings	Humbug Creek / North Fork Yuba River
Plumbago	Buckeye Ravine / Middle Yuba River
Sixteen to One	Kanska Creek / Middle Yuba River
Dairy Farm	Camp Far West Reservoir / Bear River
Lava Cap–Banner	Little Clipper Creek / Greenhorn Creek / Rollins Reservoir / Bear River
Alhambra Shumway	Rock Creek / South Fork American River
Copper Hill	Cosumnes River
Newton	Copper Creek / Sutter Creek / Dry Creek / Mokelumne River
Argonaut	Jackson Creek / Dry Creek /Mokelumne River
Penn	Mokelumne River
Empire	Copper Creek / Black Creek / Tulloch Reservoir / Stanislaus River
Keystone	Penny Creek / Sawmill Creek / Black Creek / Tulloch Reservoir / Stanislaus River

southwest of Grass Valley were up to eight times higher than EPA standards in the first hours of a rainfall-runoff event and then decreased with time. Such sudden spikes in concentrations may be harmful to aquatic life but are rarely captured in water quality sampling. Many of the adits of the different mines were dry when visited and were not releasing contaminants. Most of the mines studied in the Yuba River Basin were releasing high levels of arsenic because the gold in this region is associated with arsenopyrite minerals. Otherwise, mine runoff in this area was typically clear and was not acidic. Gold mines in the Bear River Basin were similar to those in the Yuba, but copper mines had acidic discharge with high levels of copper, zinc, cadmium, and other metals. Mines in the lower American River Basin near Folsom Lake were dry and did not appear to have serious water quality problems. The study demonstrated that surface-water quality problems associated with mines are highly site specific. Insufficient ground-water monitoring has been done in the vicinity of mines in the Sierra Nevada to identify potential problems.

The amount of mercury used in gold extraction in the Sierra Nevada and largely lost to soils and streams has been estimated at 3.4 million kg (7.6 million lb) (Central Valley Regional Water Quality Control Board 1987). Mercury is known to exist in streams below gold-ore processing sites; however, the bioavailability of mercury in the Sierra Nevada is not well understood. A survey found elevated concentrations of mercury in the upper tributaries of the Yuba, Bear, Middle Fork Feather, and North Fork Cosumnes Rivers

(Slotten et al. 1995). The heavy metal is readily trapped in reservoir sediments, and lower concentrations have been measured below reservoirs than above (Slotten et al. 1995). Mercury concentrations exceeded 0.5 mg/kg in sediment samples obtained from Camp Far West Reservoir, Lake Wildwood, Lake Amador, and Moccasin Reservoir (Central Valley Regional Water Quality Control Board 1987). Certain bacteria can convert metallic mercury to a methylated form that can be incorporated in tissue. Mercury tends to accumulate in the food chain. Although the opportunity for bacterial mercury methylation is minimized in cold, swift streams, the process can occur in the calm waters of reservoirs (Slotten et al. 1995). However, the reservoirs do not appear to be net exporters of bioavailable mercury. Instead, they seem to be sinks for both bioavailable and inorganic mercury (Slotten et al. 1995). Tissue samples of fish caught in the Yuba River contained more than 1 mg/kg, and samples exceeding 0.5 mg/kg were found in fish caught in Pardee, Don Pedro, and McClure Reservoirs (Central Valley Regional Water Quality Control Board 1987). A National Academy of Sciences report suggests that mercury amounts in tissue exceeding 0.5 mg/kg may be injurious to animals.

The Penn mine near the lower Mokelumne River has been considered one of the worst abandoned-mine problems in the Sierra Nevada. The mine was opened in 1861 and operated continuously until 1919 and then sporadically until the 1950s. Copper and zinc were the primary products of the mine (Heyl et al. 1948). More than 16,000 m (55,000 ft) of tunnels and the associated spoil provide the opportunity for percolating ground water to become acidic and leach zinc, copper, and cadmium from the mine. Flushing of some of the mine shafts in 1937 killed fish for 100 km (60 mi) downstream. A series of retention ponds were constructed and other attempts were made to restrict movement of the contaminants into the river in the 1980s, but they have had limited effectiveness (California State Lands Commission 1993). Until 1929, water draining from the mine into the Mokelumne River was diluted by the large volume of discharge. However, after the construction of Pardee Dam by the East Bay Municipal Utility District and export of up to one-third of the annual volume of the river upstream of the mine, concentrations of contaminants in the Mokelumne increased (Slotten et al. 1994). The dam for Camanche Reservoir just downstream of the mine was completed in 1964. Toxic materials leached from the mine are stored in sediments trapped by the dam. The potential for resuspension of the metals is minimal as long as water levels are kept relatively high (Slotten et al. 1994). In December 1993, the Environmental Protection Agency ordered the East Bay Municipal Utility District to control pollution from the mine. However, the utility contends that it is not responsible for the mine, which was last operated by the federal government during the Korean War.

The Leviathan mine provides another example of a water quality problem resulting from an abandoned operation. A copper and sulfur mine on Leviathan Creek in the Carson

River Basin near Monitor Pass was started by the Anaconda Copper Company in 1953. Overburden was dumped in the stream channel, causing the water to percolate through the material. Below the stream blockage, the water is highly acidic and polluted with toxic materials. The stream was sterile below the mine during the 1950s. In 1969, an isolated population of rainbow trout still existed in the unpolluted portion of Leviathan Creek above the mine. Below the mine, fish and macroinvertebrates were absent from 18 km (11 mi) of stream affected by the mine drainage. The effects of the pollution even extend for 3 km (2 mi) in the East Fork of the Carson River below the confluence with the contaminated creek (Davis 1969; Hammermeister and Walmsley 1985). Attempts at revegetating the spoils began in the 1970s (Everett et al. 1980).

Reclamation

California's Surface Mining and Reclamation Act of 1975 and amendments should prevent future disasters (Pomby 1987), but remediation of past problems requires massive investments. Even ascertaining the location of abandoned mines remains problematic (Desmarais 1977). Sealing of much of the Walker mine, a notorious problem in Plumas County, in 1987 significantly lowered copper concentrations in receiving waters and allowed partial recolonization of formerly sterile reaches by macroinvertebrates (Bastin et al. 1992). There is also a major question of liability in cleanup efforts. Current law holds those attempting remediation to be liable for any damage caused by their activities or, presumably, failure of the project to solve the problem. Therefore, under the cloud of legal liability, little action is undertaken by private or public agencies (California State Lands Commission 1993). Scores of small mines have been established under the terms of the antiquated 1872 Mining Act. In many cases, the properties are sources of sediment and toxic chemicals. Reform of portions of the Mining Act could finally alleviate some major land and water management problems associated with mining. Conversely, legislation has been introduced in California to weaken the state's regulations regarding reclamation of mined land.

Future Prospects

Changes in mineral economics and technology and new discoveries may lead to new mines. Reactivation of a large underground gold mine near Grass Valley has been proposed. Water pumped out of that mine would probably require thorough treatment before it could be discharged. In the eastern Sierra Nevada, the tungsten mine in Mt. Morgan on Pine Creek has been maintained on a standby basis awaiting an increase in the price of the metal. Reactivation of gold mines at Bodie and Independence Creek have been explored in recent years. A disseminated gold deposit in Long Valley near Mammoth Lakes has been identified through exploratory drilling in

1989–94. About one part per million of the ore is gold, which could be recovered through massive excavation and cyanide heap-leach processing.

Aggregate Mining

Sand and gravel are the most economically important nonfuel minerals mined in California. The $560 million value of sand and gravel produced in California in 1992 far surpassed the combined total value of all metallic minerals mined in the state (McWilliams and Goldman 1994). More aggregate is used per capita in California than in any other state, and the State Department of Transportation is the largest single consumer (California State Lands Commission 1993). Because aggregates are fundamental to most types of modern construction, they are used in almost every building and roadway project. The widespread demand and high transport costs of sand and gravel make aggregate production a highly dispersed mining activity (Poulin et al. 1994). Each 40 km (25 mi) of transport doubles the cost as delivered (California State Lands Commission 1993), so sources near the construction site are highly desirable. Materials excavated from stream deposits tend to be durable and have relatively few impurities and, therefore, are favored over hill slope deposits (Bull and Scott 1974).

Excavation within stream channels will obviously have direct effects on the fluvial system (Sandecki 1989; Kondolf and Matthews 1993). Removal of part of the streambed alters the hydraulic characteristics of the channel and interrupts the natural transport of bedload through the stream. The most immediate consequence is degradation of the bed both upstream and downstream. Creation of a hole in the streambed makes the channel locally steeper and thereby increases the shear stress on the bed. Erosion of the bed will propagate upstream as additional sections become steeper and erode progressively (Collins and Dunne 1990). The initial pit also serves as a bedload trap and relieves the stream of part of its load. The flowing water will then have greater availability to erode the bed in the downstream direction (Kondolf and Matthews 1993). The downcutting reduces the proportion of smaller sediments and can produce a bed composed of cobbles and boulders. Some stream reaches can lose their deposits of gravels that are suitable for fish spawning. The deeper channel can lower the local water table and kill riparian vegetation as the former floodplain dries out. Loss of the vegetation in turn makes the banks more susceptible to erosion. Incision of the channel limits the opportunity for overbank flooding to deposit sediments on the floodplain. These combined effects can result in dramatic changes in the overall form and structure of the channel and dependent aquatic and riparian habitat (Collins and Dunne 1990; Kondolf and Matthews 1993). Human structures in the channel such as bridges, culverts, pipelines, and revetments may be damaged by the geomorphic changes.

Gravel is also mined from streams by skimming a shallow layer off of gravel bars. Depending on the flow regime, distribution of particle sizes, and opportunities for establishment of riparian vegetation, a variety of complex channel and vegetation responses may occur (Kondolf and Matthews 1993). Mining of terrace deposits and abandoned channels can be problematic if the channel shifts enough to reoccupy the excavated areas. A swiftly flowing stream can be converted into a series of giant ponds if the floodplain and terraces are extensively excavated and then captured by the stream. Part of the lower Merced River suffered such a conversion in 1986 with serious impacts on a salmon population (California State Lands Commission 1993). Abandoned gravel pits and quarries well above the stream channel can also act as major sources of sediment input to streams (California Division of Soil Conservation 1971a).

The number and location of in-stream gravel operations in the Sierra Nevada is unknown, but large mines have been identified on the East Branch of the North Fork of the Feather, Middle Feather, North Yuba, Yuba near Camp Far West Reservoir, Bear, lower American, and Calaveras below New Hogan Dam (California State Lands Commission 1993). A major gravel mine operating on Blackwood Creek in the Tahoe basin increased sediment yield from the watershed about fourfold (Todd 1990). Smaller operations are assumed to be widespread throughout the Sierra Nevada. Reservoir deltas appear to be an environmentally benign source of aggregate, and removal would extend reservoir capacity. The delta of the Combie Reservoir on the Bear River has been mined for sand and gravel since 1946 (Dupras and Chevreaux 1984). Mining has occurred on the delta of Rollins Reservoir upstream from Combie (James 1988). Gold was recovered from sand and gravel operations during the construction of Friant Dam on the San Joaquin River in 1940–42 (Clark 1970).

Geothermal Resources

Geothermal energy is another subsurface resource that has potential adverse impacts on water resources when developed. Heat can be extracted from portions of the earth's crust that are unusually warm and close to the surface by pumping out hot water and using its heat to vaporize another fluid that drives turbines, which generate electricity. During the 1970s, many parts of the Sierra Nevada were explored for geothermal potential. Monache Meadows on the Kern Plateau was proposed for large-scale development. Geothermal energy in the Sierra Nevada has been developed most extensively in Long Valley near Mammoth Lakes. The large complex of geothermal power plants, located at Casa Diablo near the junction of Highways 395 and 203, had a capacity of more than 30 megawatts in 1991. The power plants operate as a completely closed system, reinjecting the water after some of its heat has been removed. After several years of operation, changes in nearby hot springs have been observed, and effects are suspected but not proven at springs feeding a fish hatchery downgradient. Additional geothermal development is being considered near Casa Diablo, Mono Craters, and Bridgeport Valley in Mono County.

DAMS AND DIVERSIONS

Impounding and diverting of streams are the principal impacts on the hydrologic system of the Sierra Nevada. While other resource management activities cause environmental alterations that, in turn, may affect stream flow, water management activities avoid the intermediate steps and intentionally and directly alter the hydrologic regime. The thoroughness of the hydraulic engineering in the Sierra Nevada that has been developed over a century and a half is probably underestimated by most water users in California. However, one simple fact stands out: no rivers reach the valley floor unaltered. Only three Sierra Nevada rivers greater than 65 km (40 mi) long flow freely without a major dam or diversion: Clavey, Middle Fork Cosumnes, and South Fork Merced Rivers (figures 30.2–30.4). Selected segments of the North Fork American, Middle Fork Feather, Kern, Kings, Merced, and Tuolumne Rivers receive some protection from additional dams under the National Wild and Scenic River System (Palmer 1993). Few streams get very far from their source before meeting some kind of structure. In the Mono Lake and Owens River Basins, about 730 km (460 mi) out of 850 km (530 mi) of streams are affected by water diversions (Inyo National Forest 1987). In California's Mediterranean climate, water is most available in winter and spring. Dams are built to reduce the peak flows of winter, provide irrigation water during the growing season, provide domestic and industrial water on a semiconstant basis, allow optimum hydroelectric generation, and secure some interannual storage for protection against drought. With so many uses of water, attempts to manage it are found throughout the Sierra Nevada.

Structures

The total number of water management structures in the Sierra Nevada is unknown but must be in the thousands. The storage capacity of all dams in the range is about 28 billion m^3 (23 million AF), which is about the average annual stream flow produced in the range. The dozen largest reservoirs (each with capacity greater than 500 million m^3 [400,000 AF]) account for about three-fourths of the rangewide storage capacity. The smallest dams in the Sierra Nevada are those built for minor domestic water supply on small creeks and may impound only a few cubic meters of water. Somewhat larger dams have augmented natural lakes and were often built for fisheries management purposes. Most of these dams were constructed before World War II, but a few continued to be built up to the 1960s. Their main purpose was to store water for releases in late summer to maintain some stream flow for fish

survival. About thirty such dams were built on the Eldorado National Forest (1980).

Dams are constructed in a great range of sizes for various purposes. Dams of a few meters' height are found throughout the Sierra Nevada to improve hydraulic conditions for tunnel intakes diverting water for municipal supply or irrigation or toward powerhouses. Such dams or weirs are not intended to have any effect on the seasonal pattern of stream flow. Dozens of small dams for small-scale hydroelectric production were proposed throughout the Sierra Nevada under the favorable climate created by the Public Utility Regulatory Policy Act of 1978 (California Energy Commission 1981). Many of these projects were ill conceived and were based on unrealistically high projections of future energy prices. Only a small proportion of those proposed were ever built. Several existing and proposed hydroelectric projects are being reconsidered by their owners or proponents because of currently low prices for electricity. The larger diversion dams can store stream flow accumulated over a few days. Dams intended to redistribute water over time have storage capacities equivalent to the stream flow of at least several weeks. A few of the megaprojects can hold more water than is produced in an average year. These massive structures account for most of the storage in an entire river basin. For example, the New Melones Reservoir has 84% of the total storage capacity in the Stanislaus River Basin, while the next largest forty dams in the basin represent only the remaining 16% (Kondolf and Matthews 1993). The dam at Lake Tahoe controls only 1.8 m (6 ft) of storage, but the vast area of the lake makes its storage volume the ninth largest in the Sierra Nevada. The big dams in the Sierra Nevada cause many of the same problems as other large dams in the western United States (e.g., Hagan and Roberts 1973). These dams prevent the further migration of anadromous fish and completely change the water and sediment regimes downstream. The combined effects of all the large dams on rivers tributary to the Sacramento River have significantly modified the annual hydrograph of the largest river in the state (Shelton 1994).

The other critical structures in water management are the conduits and canals for transferring water between rivers or to powerhouses or users. The vast network of artificial waterways redistributes water over short and long distances. A water molecule can take a very circuitous journey from the mountainside to the valley through several pieces of the plumbing system. Many of the old ditches and canals originally constructed during the mining era and that still supply water for hydroelectric generation, municipal use, or irrigation have become a secondary channel system. They both collect water from and discharge water to soils and slopes. In a 160 km (100 mi) long canal network in El Dorado County, about half of the initial water plus any gains en route are lost to seepage (Soil Conservation Service 1984). Water-supply agencies have sought to increase the efficiency of their antique delivery systems by reducing seepage from the old ditches. Replacement of the open ditches with pipes avoids

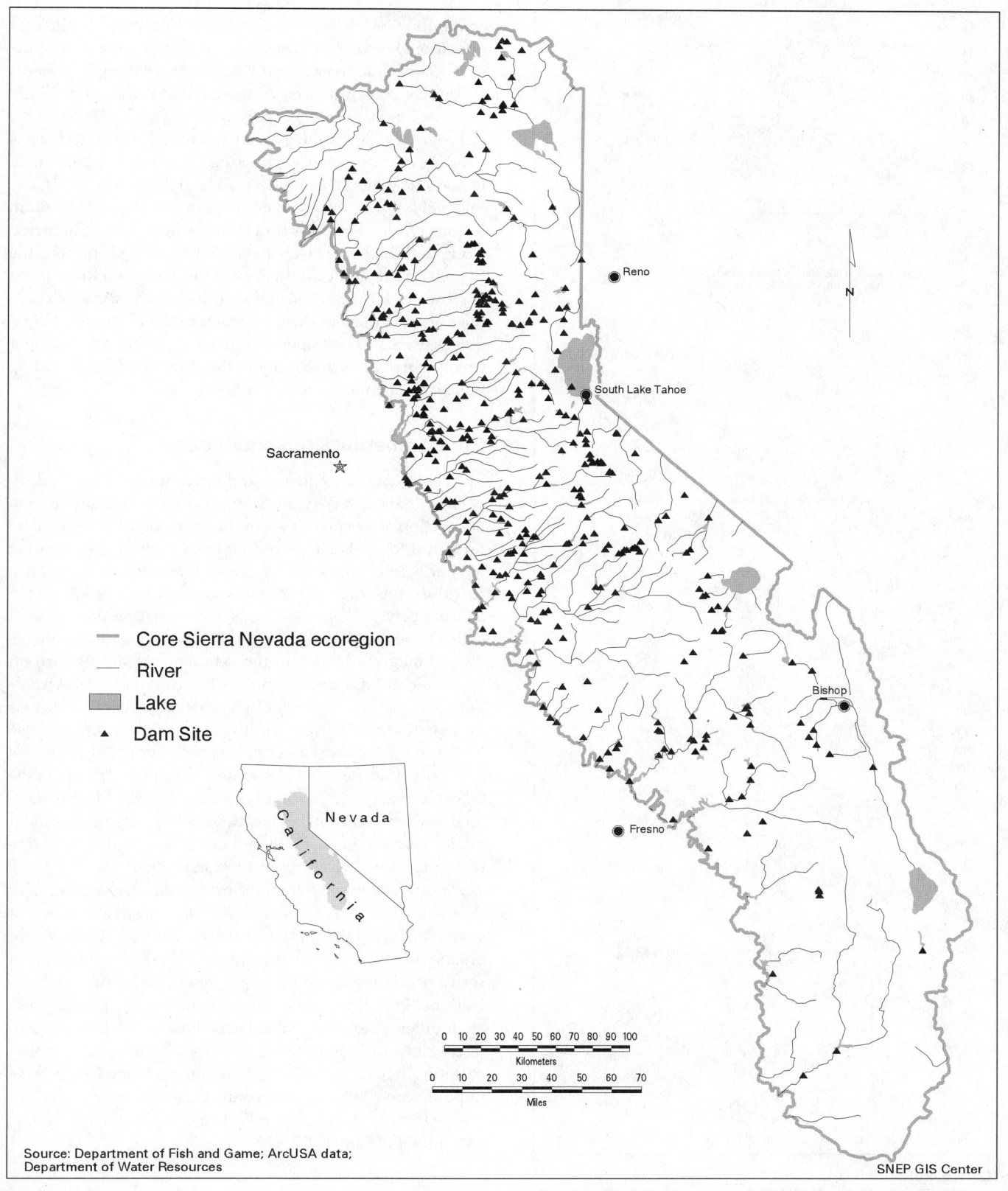

FIGURE 30.4

Larger dams that are regulated by the California Division of Dam Safety are found on almost all major streams of the Sierra Nevada.

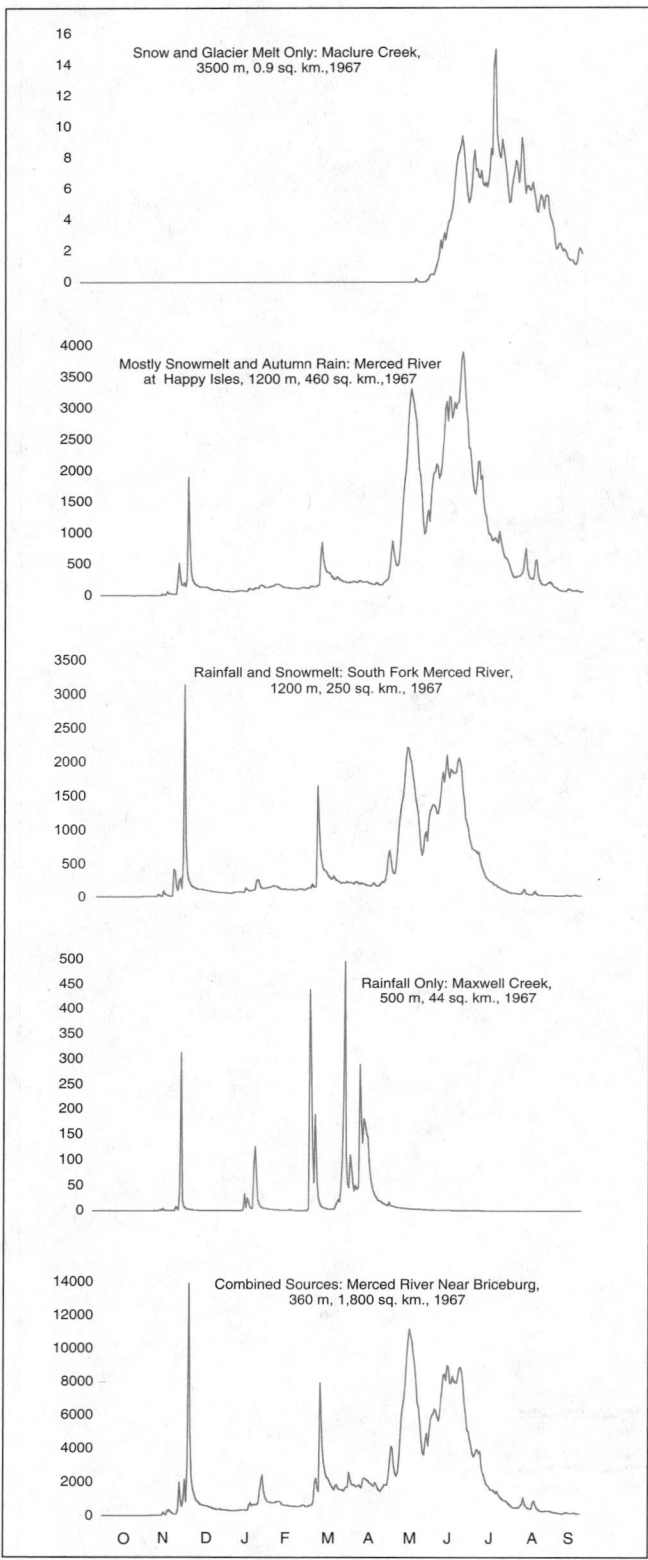

FIGURE 30.5

Watersheds with different elevation ranges and sources of runoff have different patterns of stream flow over a water year (October to September).

contamination of the enclosed water and provides greater operational flexibility. However, a finding by the staff of the State Water Resources Control Board held that improvements effectively constitute a new diversion that the ditch owner does not hold rights to. Leakage currently provides water for improvement of wildlife habitat and other uses. A decision is pending on this case involving the Crawford Ditch of the El Dorado Irrigation District (Borcalli and Associates 1993). Occasional failures of these (and more modern) canals result in serious erosion or debris flows. Four flume failures occurred along the Tule River just between 1962 and 1965. In 1992, the Cleveland fire in the South Fork of the American River Basin destroyed a large portion of the El Dorado Canal, which supplies about a third of the total water to the El Dorado Irrigation District. In November 1994, a fallen oak blocked the Tiger Creek Canal, diverting water to the slope below and eroding hundreds of cubic meters of soil.

Environmental Consequences

The construction, existence, and operation of dams and diversions have a variety of environmental effects. Inundation of a section of stream is the most basic impact. A river is transformed into a lake. The continuity of riverine and riparian habitat is interrupted. To creatures that migrate along such corridors, this fragmentation has consequences ranging from altering behavior of individuals to devastating populations. Dams have the potential to alters downstream flows by orders of magnitude and, at the extreme, can simply turn off the water and dry up a channel. Changing the natural transport of water and sediment fundamentally alters conditions for aquatic and riparian species. Changing stream flow also has dramatic impacts on chemical and thermal attributes of downstream water. The abundance of impoundments in the Sierra Nevada is impressive when one realizes that virtually all flat water at the lower elevations of the west slope is man-made. The terrain is simply not conducive to the formation of natural lakes below about 1,500 m (5,000 ft).

An obvious impact of water management is alteration of the natural hydrograph (temporal pattern of stream flow). For example, during the snowmelt season, the daily cycle of runoff and recession may be transformed into a constant flow. A series of hydrographs from streams in and near Yosemite National Park illustrate natural stream flow patterns generated under various watershed and climatic conditions at different elevations (figure 30.5). Dams are built to change those patterns (figure 30.6). Diversions not associated with large impoundments change the volume without much effect on timing (figure 30.7). Large projects usually alter both volume and timing (figure 30.8).

High Flows

The most obvious alterations in formerly natural hydrographs are decreases in peak flows. The size of an impoundment and

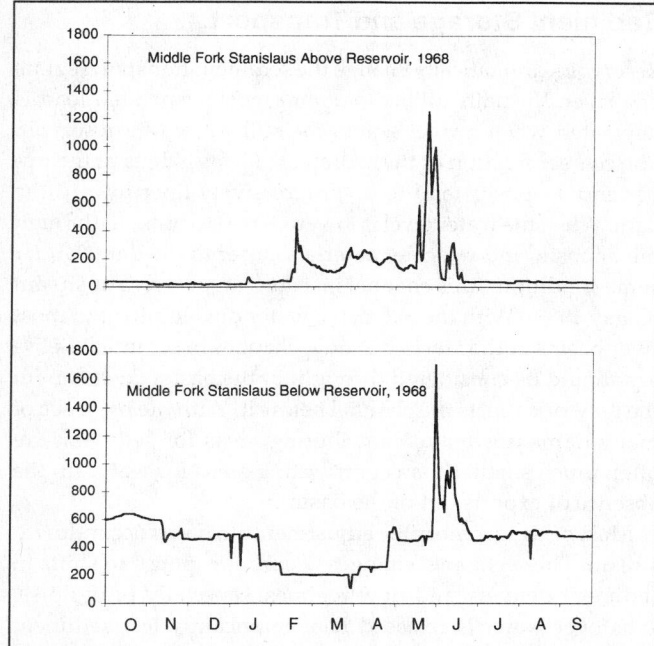

FIGURE 30.6

Storage reservoirs without diversions can greatly modify the natural hydrograph without reducing the annual volume.

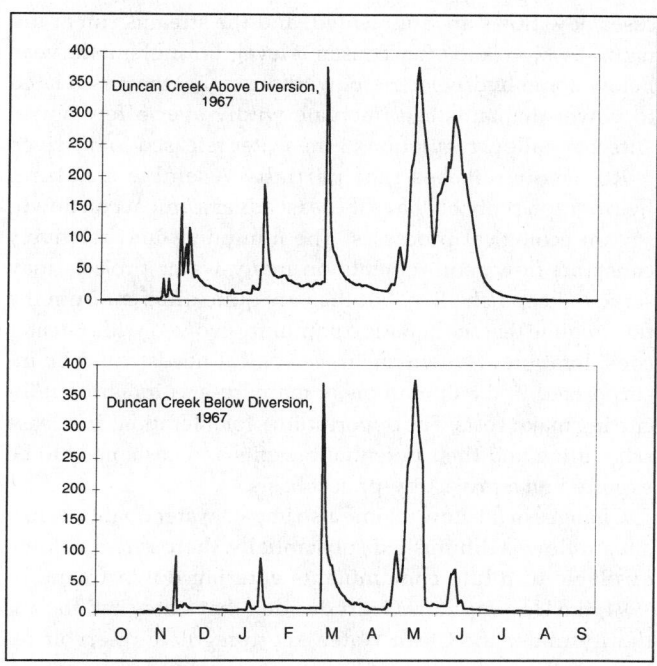

FIGURE 30.7

Diversions at small dams with minimal storage reduce the volume of stream flow without eliminating the natural pattern of fluctuations.

FIGURE 30.8

The largest reservoirs and associated diversions completely change the availability of water downstream. Note the extreme difference in scale (thousands of cubic meters per second in 1904 versus a constant 11 m^3/sec in 1979) after the Oroville Dam was completed.

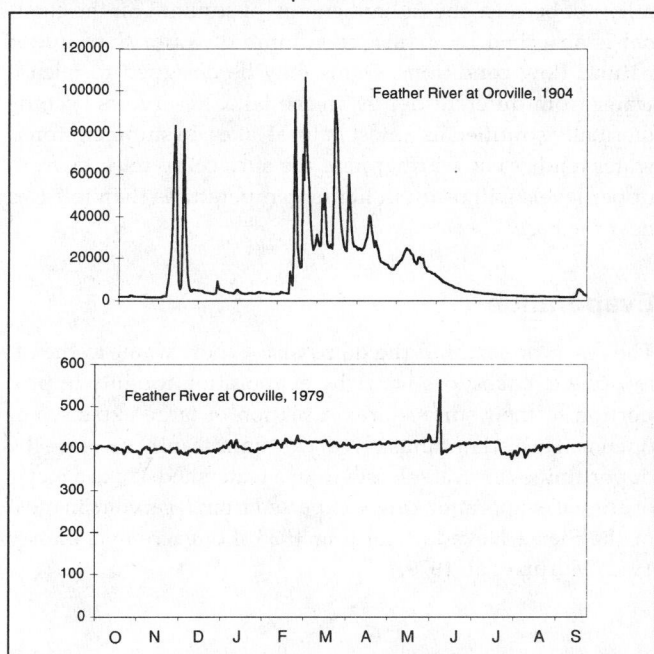

its flood reservation (management rules to keep a portion of the reservoir unfilled depending on the risk of floods at different times of the year) determine its ability to capture floodwaters and release them at a controlled rate. Small structures must pass the bulk of a flood without much influence. Large reservoirs can absorb large inflows by increasing the amount of water stored. Peak flows below some major reservoirs are reduced to essentially nothing as the dams perform their flood control functions. In a simplistic sense, all dams have a threshold for flood control. They can eliminate floods immediately downstream up to the point at which their storage capacity is exceeded. After they are filled, they exert no further control on stream flow. Of course, few reservoirs are operated in a static mode except small recreational impoundments such as Hume Lake. Most large reservoirs in the Sierra Nevada are multipurpose facilities whose releases are carefully controlled depending on inflows that are forecast, consequences of releases downstream, irrigation and power demands, and probability of additional precipitation.

Low Flows

Reservoir management also determines the releases under nonflood conditions. In the most severe cases, no water is allowed to flow in the natural channel; the entire natural flow is diverted elsewhere. Many streams in the Sierra Nevada, as in the classic example of inflows to Mono Lake, were completely dewatered below the points of diversion. In other cases, low flows are augmented, and the streams run at unnaturally high (and often constant) levels throughout the year. Below some hydroelectric powerhouses, discharges related to power demands can fluctuate wildly over a few hours. Afterbays allow regulation of the water released to the river.

Reservoir releases that partially resemble a natural hydrograph probably have the least adverse impact on downstream ecological processes. The minimum (and relatively constant) flow requirements on many water projects may serve to keep fish alive, but they are quite different from the flow regime that the aquatic community evolved with. Greater consideration of downstream ecological needs could be incorporated in the operations of many reservoirs without incurring major costs. The opportunities for alterations in release scheduling and their potential benefits and costs need to be explored on a project-by-project basis.

Changes in the flow regime also impact water quality. When stream flow is diminished substantially, there is less volume available to dilute contaminants entering downstream. In 1994, the U.S. Supreme Court decided that states had the authority under the Clean Water Act to regulate reservoir releases in the context of managing water quality. In some cases, reservoirs can improve riverine quality by allowing contaminants adsorbed on particles to settle out of the water column. However, this same process may be converting some reservoir beds into storage deposits of heavy metals (Slotten et al. 1994, 1995).

Water Temperature

Temperatures of streams below dams are affected by the volume and temperature of reservoir releases. If little water is released from a dam in summer, streams can become unnaturally hot because the radiant energy of sunlight on the channel is absorbed by a smaller volume of water than under natural flow conditions. Dams may be designed to release water from different depths in the lake. Reservoirs become thermally stratified like most natural lakes. In summer, stored water tends to be warmer near the surface, so releases from upper levels will result in higher temperatures than releases near the base.

Evaporation

The reservoirs behind the dams also export water to the atmosphere. Lakes lose water by evaporation roughly in proportion to their surface area. Creation of large expanses of open water by damming a river can significantly increase the opportunities for water losses from a watershed. Up to a meter of annual evaporation can be expected from reservoirs in most of the Sierra Nevada (Harding 1935; Longacre and Blaney 1962; Myrup et al. 1979).

Sediment Storage and Transport

Reservoirs dramatically change the sediment transport regime of a river. Virtually all bedload and most suspended load is deposited when a river enters the still water of a reservoir. The coarser fraction of the sediments forms a delta at the upper end. Deposits tend to be progressively finer toward the dam. When the water level is lowered, the streams cut through the deposits and relocate materials closer to the dam. Under some conditions, this channel incision can progress upstream (Galay 1983). With the extensive water development in most river basins of the Sierra Nevada, changes in sediment delivery should be considered throughout the basin. Each dam in the network affects the channel below it. With the presence of many dams upstream, contributing areas for sediment are often much smaller than contributing areas for water (in the absence of exports out of the basin).

Most of the geomorphic adjustments to dams occur downstream. These channel changes occur in response to shifts in sediment delivery and flow regimes, especially peak flows. Whatever water is released has significantly less sediment than when it entered the reservoir. Unless releases are minimal, the sediment-free discharge has the capacity to entrain and transport particles from the bed and banks of the downstream channel. Progressive lowering or degradation of the riverbed may occur after dam completion. Typical consequences of degradation include lowering of ground-water levels and consequent loss of riparian vegetation, reduction in overbank flooding and deposition of sediments and nutrients, bank erosion and loss of land, exposure of bridge foundations, and abandonment of diversion intakes (Galay 1983). The severity of channel incision depends on the size distribution of particles in the bed, characteristics of the channel, how the reservoir is operated and the sequence of flood events following construction (Williams and Wolman 1984). Downcutting seems to be greatest in rivers with fine-grained bed materials and where flood peaks are not greatly reduced by the dam. However, larger dams usually reduce flood peaks substantially and thereby limit the rate of degradation (Milhous 1982). Where channel incision occurs below dams, the finer particles are removed, and the larger cobbles and boulders are left behind. As the bed becomes coarser or "armored," it is more resistant to erosion and interferes with salmonid spawning. Also, downcutting decreases the channel gradient slightly, and degradation becomes somewhat self-limiting. The bed of the Yuba River below Englebright Dam has become armored with large cobbles and boulders but is still susceptible to incision during the largest floods (Kondolf and Matthews 1993). Conversely, flood control is so effective below both Pardee and Camanche Dams that channel degradation has not occurred and the gravels are immobile (BioSystems Analysis 1990, cited by Kondolf and Matthews 1993). Unfortunately, we lack any information about the condition of channels before placer mining and dam construction. Therefore, we are unable to make definitive statements about what constitutes natural channel conditions in most of the Sierra Nevada, although channels were unlikely to have been as armored as many are currently.

Where streambeds are not armored with large materials and are not actively degrading, fine sediments can interfere with fish spawning. Salmonids require gravels with sufficient pore space to allow interstitial flow to bring oxygen to eggs. Fine particles may be deposited between the gravels and limit the flow of water. Higher discharges are necessary on occasion to cleanse the gravels. Control of high flows by dams eliminates the opportunity to flush the fine sediments out of the spawning gravels. Many studies have been conducted in the past decade to define how much water is needed for this flushing function, and many rules-of-thumb have been suggested. However, variability in fluvial processes among streams illustrates the need to actually observe flows that begin to entrain particles of a particular size rather than depend on generalized procedures to estimate flow releases necessary to remove fine sediments from spawning gravels (Kondolf et al. 1987).

Limiting the size and frequency of floods below dams has also altered conditions for riparian vegetation. As total discharge and scouring flows decrease, riparian vegetation is able to become established in the former active channel (Williams and Wolman 1984). Roots stabilize the bank materials, and the plants slow overbank flows, which allows deposition of additional sediment. Gradually, the channel becomes narrower, and large trees occupy former parts of the channel. If allowed to become well established, mature riparian vegetation can resist significant flows. Confining the stream to a narrower channel can increase hydraulic forces on the bed and lead to incision and loss of riparian vegetation. To some degree, dams mimic the effects of long-term droughts on vegetation-channel interactions (Mount 1995). Depending on characteristics of the channel and plants, establishment of riparian vegetation can be enhanced by either higher or lower summer flows than occurred before dam construction. Encroachment of vegetation into river channels has been noted below Tulloch, Don Pedro, La Grange, and McClure Reservoirs (Pelzman 1973). Augmentation of flows at the receiving end of trans-basin diversion has widened channels and has pushed back riparian vegetation, as in the case of the upper Owens River.

Although larger dams seem to have sufficient space to store sediment for hundreds of years, at least at rates determined in the 1940s, smaller structures can become overwhelmed with sediment in just a few years. Unusually large floods can completely fill smaller diversion works, as occurred at Log Cabin Dam on Oregon Creek and Hour House Dam on the Middle Yuba in 1986 (Kondolf and Matthews 1993). Assuming that the dam is to remain in operation, the accumulated sediment must be removed. How that removal is accomplished can have an assortment of impacts. Dredging, trucking, and disposal

of the sediments in a stable location has been a costly approach to the problem. Ralston Afterbay on the Middle Fork American River has had sediment removed on six occasions between its completion in 1966 and 1986 (Georgetown Ranger District 1992). The average annual rate of filling of about 80 m^3/km^2 (0.2 AF/mi^2) is not excessive compared with that of other basins, but the Ralston Afterbay has a capacity of only 3.4 million m^3 (2,782 AF) with 530 km^2 (205 mi^2) of unregulated contributing area above it (EA Engineering, Science, and Technology 1990). Location of suitable sites for long-term storage of removed sediments within a short distance from the reservoir has been difficult (Georgetown Ranger District 1992). The small forebays on Southern California Edison's Bishop Creek system have also required dredging of accumulated sediments. Estimates of the costs of dredging and transportation depend on access and distance to a disposal site and have ranged from $26/m^3$ ($20/yd^3$) (EA Engineering, Science, and Technology 1990) to about $3,500/m^3$ ($2,700/yd^3$) (Kondolf and Matthews 1993).

Another option for removal of accumulated sediments is sluicing. Opening sluice gates or an outlet tunnel allows water levels to fall and sediment to be resuspended and flushed out with the water. This action creates a sudden pulse of sediment downstream. Problems have arisen when sluicing has been conducted during summer months, at times when flows are inadequate to disperse the redeposited sediment. Sluicing of Forbestown Reservoir on the South Fork Feather River in 1986 left a thin layer of sand over the entire channel well downstream of the dam. Another example was Democrat Dam on the Kern River in 1986. In the years following sluicing, high flows did not occur, and sand remained within the channel until scouring flows occurred in 1992 (Kondolf and Matthews 1993). Accidental releases of sediment occurred on the Middle Yuba River from Hour House Reservoir in 1986 and from Poe Dam on the North Fork of the Feather River in 1988. More than $1 million was spent excavating sand out of the channel below Hour House Dam, but a flood during the early stages of the North Fork Feather cleanup conveniently flushed all the excess sediments out of the channel (Ramey and Beck 1990; Kondolf and Matthews 1993).

When sediment is flushed out of reservoirs at low flows, it will be redeposited close to the dam; however, when it is introduced at higher flows, it will usually be carried downstream and dispersed. Engineering approaches to letting sediments pass through dams during high flows are being considered at several sites. The Pacific Gas and Electric Company (1994) is designing pass-through systems to retrofit two of its dams on the North Fork Feather River. Sediment is rapidly filling the reservoirs, complicating operation of the dams, and accelerating turbine wear (Harrison 1992). Such pass-through systems could allow reservoir operations to interfere less with natural sediment transport and could have geomorphic benefits with regard to channel degradation below dams (Kondolf and Matthews 1993).

Failure

Catastrophic failure of impoundments is always a concern of those living below dams. Sudden releases of water also have great potential for dramatic environmental change. During the gold-mining era, dam failures were fairly common, both because of design flaws and because of intentional releases to rearrange gold-bearing sediments in the practice known as booming. Early debris dams were also intentionally destroyed to allow fresh access to impounded gravels and to create new storage space. Unintentional collapse of the English Dam on the Middle Fork of the Yuba (Ellis 1939; McPhee 1993) in June 1883 released almost 18 million m³ (15,000 AF) of water suddenly and cleaned out much of the stored mining debris in that channel (James 1994). Excessive water releases from an upstream dam washed out a small dam on Bishop Creek in June 1909. Following failure of the Saint Francis Dam in the Ventura River Basin in 1929, the Division of Dam Safety of the Department of Water Resources has regulated larger dams and inspected them at least annually. Dams that are either more than 7.6 m (25 ft) tall and store more than 62,000 m³ (50 AF) or, alternatively, more than 1.8 m (6 ft) tall regardless of capacity or impound more than 19,000 m³ (15 AF) regardless of height are regulated by the Department of Water Resources (1988). Modern dams have little risk of failure; however, failures are not unknown. The best-known dam collapse in the Sierra Nevada in recent decades was that of the Hell Hole Dam on the Rubicon in December 1964 (Scott and Gravlee 1968). Failure of the North Lake Dam during a storm in September 1982 produced the largest flood of record on Bishop Creek and severely damaged one of the powerhouses. During the massive floods of February 1986, the coffer dam at the Auburn Dam site failed when diversion tunnels became clogged and the dam was overtopped. Structural failure of a penstock during high-pressure testing at the Helms Creek pumped storage facility in 1982 resulted in massive scouring of Lost Canyon (Chan and Wong 1989). Even partial failures, such as the gate damage on Folsom Dam in July 1995, can result in large releases of water and prolonged difficulties in project operation.

Eventually, some larger dams will become filled with sediment and no longer worth operating. The Federal Energy Regulatory Commission now has the authority to take dams out of service when they come up for relicensing. We have no real experience with what to do about a dam filled with sediment. Early debris dams on the Yuba and Bear Rivers just failed or were intentionally destroyed, and the sediments eventually moved downstream or became semistable terraces. However, that option probably won't be acceptable in the future. Plans are being made to decommission a dam on the Elwha River in Olympic National Park in Washington. Initial estimates suggest that removal of the dam could cost $60–80 million and sediment removal could cost $150–300 million. If estimates of reservoir sedimentation rates made during the 1940s turn out to be conservative, society will have a long time to think about what to do with the dams of the Sierra Nevada.

ROADS

Roads provide the most intensive modification of land surface properties relevant to the hydrology of common land-management practices. All vegetation is removed and prevented from reestablishment. Dirt-surfaced roads are compacted to a near-impervious state, and sealed and paved roads are completely impervious. Runoff from the surface is collected and discharged as potentially erosive flows at points below the road. Roads that are cut into slopes intercept subsurface water flow and bring it to the surface. Fill materials cover additional portions of the slope and often contribute to sediment yields slowly over time or catastrophically if they become saturated from subsurface water entry and then fail. Erosion from the actual roadbed of unpaved roads may be significant as well (Garland 1993; Adams 1993). Unauthorized use during wet surface conditions adds to the erosion of the road. A principal side effect of an extensive road network is the access that is provided to allow additional alterations. Few adverse impacts occur in the absence of roads. Avoidance of new road construction can minimize other potential impacts in currently unroaded areas.

Stream Crossings

The most serious impacts of roads occur where roads are in close proximity to streams or wetlands. Stream crossings by ford, culvert, or bridge have direct effects on the channel and local sediment regime. Although virtually any stream crossing will have some impact on the channel, careful engineering, construction, and maintenance can limit the severity. The basic problem just comes down to disturbing the bed, banks, floodplain, and terraces. Because the crossing is coincident with the channel, there is little opportunity to buffer the inadequacies of design or construction. Also, roadside ditches near the crossing drain directly into the stream, often contributing sediment to the stream. In past decades, very little attention was paid to stream crossings, and the cheapest alternative was usually chosen. Often, that choice was merely pushing a stack of cull logs into the channel and covering them with dirt. Installation of culverts sized only for summer flow, with anticipated reconstruction, was often a more cost-effective choice than a properly engineered crossing. Fortunately, engineering and construction practices have improved dramatically since crossings have become widely accepted as a potential problem (Furniss et al. 1991).

Forest Road Network

As with other disturbances, the proportion of a catchment that roads occupy greatly influences their net downstream impact. Sediment yield associated with roads has even been claimed to increase exponentially with their density in a watershed (California Division of Soil Conservation 1971a). Within national forests of the Sierra Nevada, gross road densities range from 0.6 km/km^2 (1.0 mi/mi^2) on the Inyo to 2.3 km/km^2 (3.6 mi/mi^2) on the Eldorado (U.S. Forest Service 1995a). There are approximately 28,000 km (18,000 mi) of roads on national forests of the Sierra Nevada. Construction of new forest roads has declined markedly in recent years, and reconstruction and obliteration have varied among years (table 30.6).

Sediment Production

A variety of studies have examined sediment production and mass movement occurrence from forest roads. As usual, there is little information from the Sierra Nevada. Studies of road impacts in northwestern California (e.g., Burns 1972; McCashion and Rice 1983), Oregon (e.g., Beschta 1978), Washington (e.g., Reid and Dunne 1984), Idaho (e.g., Megahan and Kidd 1972), and elsewhere have demonstrated increases in local erosion rates hundreds of times greater than natural rates as well as severalfold increases in sediment yield at the catchment scale. Sediment yield from roads is usually greatest in the first year following construction. Road construction and some timber harvesting in the 10 km^2 (4 mi^2) Castle Creek Basin near Donner Summit resulted in a fivefold increase in suspended sediment during the first year. Sediment yields decreased to twice the preconstruction levels during the second year (Rice and Wallis 1962; Anderson 1979). In a rapidly urbanizing part of the Lake Tahoe Basin, roadways were found to generate about half of the total sediment (California Division of Soil Conservation 1969). The presence of roads can increase the frequency of slope failures compared with the rate for undisturbed forest by up to hundreds of times (Sidle et al. 1985). Road location seems to be the most important single factor because it determines the opportunity of most other controlling influences to contribute to failure (Furniss et al. 1991; Rice and Lewis 1991). Road placement in topographic hollows caused ground-water flow to be impeded,

leading to several major failures along a principal road in the Tahoe National Forest (McKean 1987). In the past, there was little rational planning or design for inslope versus outslope road surfaces and associated drainage works as a means of minimizing erosion.

Landslides and surface erosion can often be traced to haphazard road design, location, and construction (McCashion and Rice 1983). Forest roads constructed as part of a carefully planned system usually disturb much less ground, produce less sediment, and have lower construction and maintenance costs (Brown 1980). Road stability is often jeopardized by infrequent maintenance. A looming problem for the Forest Service is how to maintain some 28,000 km (18,000 mi) of roads in the Sierra Nevada with budgets inadequate even at present. Declining budgets have decreased maintenance activities overall and placed roads in lower maintenance categories than specified in the original design (Clifton 1992). If maintenance is not improved, quality of both transportation and streams will suffer. Lack of maintenance is often used as an excuse for failures resulting from poor design or construction (Seidelman et al. 1986). The road network must be acknowledged as both an investment and a liability for the long term.

Rehabilitation

Casual examination of Watershed Improvement Needs Inventories on many of the national forests of the Sierra Nevada illustrated that fixing road problems is an overwhelming priority. The same engineering and construction skills needed to build roads can be used to repair, relocate, and obliterate roads that cause excessive water quality problems. Modern concepts of road location and design that are currently used to build new roads with minimal problems (Larse 1971) can be applied to reducing the adverse effects of existing roads (Clifton 1992). Reshaping road cuts, pulling back side-cast material, ripping compacted surfaces, and removing stream crossings were successfully employed in a watershed in northwestern Washington (Harr and Nichols 1993). The decommissioned roads survived with little damage two major storms that caused widespread failures of active roads. Sources of funding must be identified to maintain and stabilize the road network; otherwise, forests will be left with an analog to toxic waste dumps that get increasingly difficult to treat and cause additional impacts with the passage of time. Public education is also necessary to build acceptance for closing roads that damage public resources. Closure of unsurfaced roads during the wet season can also help to reduce erosion.

Streets and Highways

Although unsurfaced forest and rural roads have received most of the attention, urban streets and major highways can also create severe problems of slope instability and water quality (Scheidt 1967; Parizek 1971). Beyond sharing most of the impacts associated with forest roads, paved roads of higher

TABLE 30.6

Kilometers of road activities in national forests in the Sierra Nevada by fiscal year (U.S. Forest Service, Region 5, Engineering Section).

Activity	1990	1991	1992	1993	1994
Construction	113	83	75	53	11
Reconstruction	620	326	323	453	307
Obliteration	NA	136	86	111	180

standard have additional effects. Primarily, they are simply wider and affect more area per unit of length. A four-lane highway can occupy a substantial fraction of a small catchment. Their impervious surface can create overland flow over large areas where it was nonexistent before construction. They are designed for more traffic at higher speeds and so tend to be forced through the landscape, minimizing curvature and changes in grade instead of following the topography more closely. In partial compensation for the greater hill-slope alteration, highways are better engineered than lightly used roads. Large investments are made in adequate drainage structures, slope reinforcement, and revegetation. Nevertheless, mitigation for the sheer size and location of the highway projects is difficult at best. Major highways are immediately adjacent to portions of the Feather, North Yuba, South Yuba, Truckee, South Fork American, Merced, Walker, Kaweah, Tule, and Kern Rivers. Within cities and towns, the storm water drainage system for the entire road grid is often inadequate during large storms because communities tend to develop in a piecemeal fashion, rather than having a complete road and drainage network planned from the start. Contaminants from tire wear, fluid leaks, pet waste, and exhaust that accumulate on the roadway are washed off into the nearest waterway. Oils used for road dust abatement can also be problematic. For example, contamination of Ponderosa Reservoir on the South Fork Feather River with polychlorinated biphenyls (PCBs) was traced to the use of transformer oil on forest roads (Plumas National Forest 1988).

Deicing Agents

Chemicals used to remove snow and ice from roadways in winter can affect local water quality and roadside vegetation (Hawkins and Judd 1972; Scharf and Srago 1975; Goldman and Malyj 1990). During a heavy winter (1982/83), rock salt (sodium chloride) was applied to Interstate 80 near Donner Summit in an average quantity of about 45 metric tons per km (80 tons per mi) of roadway (Berg and Bergman 1984). Stream samples obtained about 0.5 km (0.3 mi) downstream from the last highway crossing of the channel contained up to 100 times more chloride and 10 times more sodium than water obtained just upstream from the highway (Berg and Bergman 1984).

FIRES, FIRE SUPPRESSION, AND POSTFIRE TREATMENTS

Catastrophic fire can produce some of the most intensive and extensive changes in watershed conditions of any disturbance. Within areas of intense fire, most vegetation is killed and stops transpiring, allowing soil moisture levels to remain high. Organic matter in the litter layer is volatilized and often forms a layer within the soil that reduces infiltration of water into the soil (see Poff 1996). Riparian zones that would not be harvested under current forest practices are often partially burned in intense fires. The combined effect of these changes is to increase total water yield and overland flow. As the proportion of overland flow increases, streams receive more water in less time than under prefire conditions, and peak flows may be increased. If a nearly continuous water-repellent (hydrophobic) layer is a few centimeters below the surface, the soil above that layer may become saturated and form shallow debris flows. With bare soil, increased overland flow, and lack of vegetation and litter, soil particles are more easily detached and transported. As with other impacts, the proportion of a catchment that is modified by fire and the location of the burned area with respect to the channel largely determine the effects on streams. A stream draining a watershed burned over 90% of its area will show much greater effects than a stream emanating from a similar watershed in which only the upper slopes and ridgetops were burned. Fire intensity is often highly variable over the landscape, and patches of unburned or lightly burned vegetation (especially near streams) can reduce the adverse effects of upslope areas that were intensely burned.

Water Yield

Fires affect water yield primarily by killing vegetation. Interception loss is decreased because of the loss of leaves, low vegetation, and litter. Transpiration is virtually eliminated wherever fire is intense. A daily cycle in stream flow reflecting transpiration demand during daylight hours in a catchment in Washington came to an abrupt halt following a catastrophic fire (Helvey 1980). Annual runoff in this completely burned watershed increased by 10–47 cm during the first seven years after the fire. Water yields in a small catchment in British Columbia that was burned over about 60% of its area increased by 25% on average for four years following the fire (Cheng 1980). Dramatic increases in flow of a small spring and a creek in the Sierra Nevada were observed following burning of riparian vegetation (Biswell 1989). A detailed modeling study for Pacific Northwest forests has suggested that a reduction in leaf area or basal area of about 50% is necessary before annual water-yield increases exceed about 50 mm (2 in) (Potts et al. 1989). Snow accumulation and melt rates might be expected to increase from opening a forest canopy by fire, and such effects have been observed in Washington and British Columbia (Helvey 1980; Cheng 1980) but not in Idaho (Megahan 1983).

Peak Flows

Peak flows can be expected to increase following significant fires because of higher soil moisture resulting from reduction of transpiration, decreased infiltration, and higher rates of snowmelt. Infiltration is usually the most important influence, and it is decreased in two ways. Removal of vegetation and the litter layer exposes bare mineral soil to raindrop impacts, which can physically force the soil particles closer together and disperse soil aggregates into surface pores, thereby reducing the infiltration capacity. Fires also vaporize organic compounds in the litter layer, some of which move into the soil until the vapor condenses and forms a layer that is water repellent, or hydrophobic (De Bano 1981). These layers tend to be more coherent in coarse-textured soils (e.g., decomposed granitics), under very hot fires, and where a thick litter layer and/or organic horizon was present (De Bano 1981; Poff 1989b). The continuity of such layers, which may be a function of fire intensity and litter distribution, determines their overall impact on hill-slope water movement. Additionally, larger macropores from roots and animals allow some water movement through the hydrophobic layers (Booker et al. 1993). Although the water-repellent layers tend to break down within a year or two, those formed in soils that are somewhat hydrophobic even without burning may be more persistent (Poff 1989b). Under some conditions, a hydrophobic layer forms on the surface of the soil and acts as a binder and sealant, maximizing overland flow while minimizing erosion (see Poff 1996). As usual, there is a lack of measured hydrologic response to fire in the Sierra Nevada. A variety of studies elsewhere in the western United States have demonstrated dramatic increases in peak flows following wildfire (Tiedemann et al. 1979).

Sediment Yield

In general, sediment yields increase markedly after fires, particularly if riparian vegetation was burned. Most of the sediment response seems to be from the channels themselves. In the absence of streamside vegetation, soil particles move into the channels from dry ravel erosion, and the banks become less stable. Increases in total discharge and peak flows result in channel erosion. Debris torrents may scour streams if extreme climatic events follow the fire (Helvey 1980; Kuehn 1987). If the fire is particularly hot, woody debris that helped stabilize the channel may be destroyed. Erosion from the general land surface usually increases, but it may not always be as important a delivery mechanism as has been assumed (Booker et al. 1993). Erosion from plots in brushland near North Fork in the San Joaquin River Basin increased by 200 to 400 times after repeated burning (Lowdermilk and Rowe 1934). In Dog Valley in the eastern Sierra Nevada near Reno, a single storm produced about 600 m^3/km^2 (1.3 AF/mi^2) of sediment from a burned catchment while an adjacent unburned area yielded only a trace of sediment (Copeland 1965). Under extraordinary rainfall, gully erosion, sheet erosion, and a debris torrent removed more than 19,000 m^3 (15 AF) of material from a burned catchment of about 0.8 km^2 (0.3 mi^2) in the headwaters of the South Fork of the American River in 1982 (Kuehn 1987).

Nutrient Yield

Fires provide an opportunity for nutrients that have been stored in vegetation and soils to move into streams. Materials that are not volatilized and lost to the atmosphere are left in ash on and near the soil surface in forms that are readily mobile. A variety of studies throughout the West have demonstrated that concentrations of nitrates and other ions in streams usually increase dramatically after fires (Tiedemann et al. 1979). However, the background concentrations of these constituents in streams draining healthy forests are typically so low that the relative increases following fires appear to be huge even though the absolute amounts often remain almost negligible or at least below water quality standards. Nevertheless, there is potential for a nutrient flush to dramatically increase algae in streams, which can have additional consequences. There is also the potential for large nutrient losses associated with physical erosion of soil particles that often carry nutrients with them (Tiedemann et al. 1979). A study of the chemistry of Sagehen Creek north of Truckee following the Donner Burn in 1960 did not detect any change in the ionic composition of the stream relating to the fire, which did not burn the riparian zone (Johnson and Needham 1966). The inevitable fires in urban intermix zones have the potential to release a variety of chemicals and combustion products into the aquatic environment. Reconstruction can keep soils bare and disturbed for years.

Aquatic Effects

Studies of the aquatic effects of a fire on the Plumas National Forest demonstrate how both physical and biological features of the stream change over time (Roby 1989; Roby and Azuma 1995). The lower two-thirds of this catchment, including riparian vegetation, was thoroughly burned. Initially, the channel widened in response to presumed higher flows of water and sediment. However, as vegetation became established and the watershed recovered, the cross sections of the channel returned to their prefire areas within six years of the burn. Partial recovery of the invertebrate community seemed to have occurred relatively quickly. No differences in community similarity were noted between burned and unburned reaches one year after the fire, and density and taxa richness were comparable within three years. However, significant (though declining) differences in a species-diversity index between the burned and unburned reaches remained throughout eleven years of monitoring (Roby and Azuma 1995).

Fire Suppression

Fire suppression during this century has created forests with greater density of vegetation than in the past (Chang 1996; Skinner and Chang 1996; Weatherspoon 1996). This forest structure has current and potential hydrologic consequences.

The present situation may decrease yields of water and sediment somewhat compared to a natural fire regime (if impacts of other activities, such as residential development and road construction, are ignored). Although these changes cannot be quantified, transpiration from the dense forests should be at or near maximum, and a more open forest structure resulting from more frequent fire could be assumed to use less water. The dense vegetation also increases the opportunity for intense conflagrations (Chang 1996; McKelvey et al. 1996; Skinner and Chang 1996) that could produce major increases in water and sediment yields. There is a basic contrast between moderately higher stream flow and sedimentation on a semiconstant basis with relatively frequent low-intensity fires and the other extreme of lower stream flow with less sediment for now with the looming possibility of damaging floods and sediment loads from less-than-perfect fire suppression.

Actual on-the-ground fire-fighting activities, as opposed to the general policy of fire suppression mentioned above, also have impacts on water resources (California Division of Forestry 1972). In general, the net effect of such actions is probably less than doing nothing, given current fuel loads. The principal impacts of the past, which presumably are rare under current practices, involved operation of heavy equipment in streams and riparian zones and down the fall line of slopes. In some large fires, an extensive network of fire breaks may be bulldozed and require rehabilitation. Aerial application of retardants can also have adverse aquatic impacts (Norris and Webb 1989).

Postburn Activities

Following fires, there is usually a strong desire by landowners, agencies, and the public to react quickly. Hastily constructed fire lines often require obliteration or drainage; otherwise, allowing natural recovery processes to function may often be the best policy (Beschta et al. 1995). For example, natural regrowth on north-facing slopes virtually stopped erosion within three years of a fire in Idaho that initially produced more than 1,000 m^3/km^2 (2.3 AF/mi^2) of sediment (Megahan and Molitor 1975). Unfortunately, the state of the art in postfire rehabilitation remains poorly developed. Despite vigorous implementation of various actions over vast areas of the western United States, there has been minimal monitoring of the effectiveness of those actions. We therefore have very little collective experience or documentation of what works and what doesn't work. There are a lot of different treatments recommended in rehabilitation handbooks, but there is little apparent basis for the recommendations measured as success or failure in years following the prescription. There is active debate among fire specialists, soil scientists, hydrologists, and ecologists about some very basic issues. For example, rye grass seeding has been encouraged for years as a means of getting some vegetation cover in place as quickly as possible. Despite evidence compiled over thirty years that it inhibits establishment of native vegetation, re-

sults in less total cover after a couple of years than in non-seeded areas, and may even enhance net soil loss or have other adverse effects (Krammes and Hill 1963; Booker et al. 1993; Roby and Azuma 1995), many people seem to view it as a panacea. Contour felling of logs and straw-bale check dams to trap sediment are other widely accepted practices that may be less effective than generally assumed. These techniques appear to meet certain objectives in some situations (e.g., De Graff 1982), but their indiscriminate use is ineffective at best and may be counterproductive. One of the lessons of the catastrophic fire in Oakland in 1991 was that we simply didn't know what erosion control measures would be appropriate.

In the past decade, salvage logging of dead and dying trees has become quite controversial, with some people feeling it is just an excuse to cut trees while others feel it is the only thing maintaining their business. Postfire salvage operations influence aquatic recovery in a variety of ways. Perhaps most important in the present economic climate, salvage sales have potential to generate revenue for watershed rehabilitation, which unfortunately seems to be underutilized. Culverts are often replaced with larger structures in anticipation of larger flows; the soil disturbance from reconstruction is much less than what would occur if the road were to fail. The replacements are probably better designed than the original and should be more stable in the long term. Some roads may be decommissioned. Logging slash can be used to provide some physical protection for soils. Cutting shallow-rooted trees avoids the displacement of their root masses if they were to be blown over. On the negative side, logging operations disturb soils when the soil is particularly sensitive to compaction and erosion in the absence of cover and organic matter. Significant ground disturbance during a salvage sale of the Clark Burn in the Last Chance Creek watershed of Plumas County led to severe erosion during a thunderstorm (Cawley 1991). Where strong hydrophobic layers have developed, such disturbance might be valuable in promoting infiltration (Poff 1989b). However, on slopes subject to deeper mass failure, hydrophobic layers may be desirable as a means of limiting accumulation of water in the soil. If postfire treatments of salvage logging and site preparation prevent rapid reestablishment of low vegetation, resulting erosion can be greater than that directly produced by the fire. Timing of major storms relative to the amount of bare soil is a dominant influence. A fire in the Tuolumne River Basin in 1973 was not immediately followed by any major erosion-producing events (Frazier 1984). However, widespread ground disturbance associated with salvage operations prolonged susceptibility of soils to erosion. Eventually, substantial rain-on-snow events provided the energy for serious rill, gully, and bank erosion, which resulted in significant soil losses (Frazier 1984).

The principal objectives of postfire rehabilitation work should be to avoid making things worse; repair potential problems from fire-fighting activities (e.g., bulldozed fire breaks); enhance establishment of native vegetation to provide soil cover, organic matter, stream-bank stability, and shade as quickly as possible; attempt to stabilize channels by non-structural means; minimize removal of large woody debris from streams; minimize adverse effects from the existing road network; schedule operations to minimize exposure of bare soil; and allow natural processes to heal the landscape.

Fuels Reduction

A major program of fuels reduction could increase gross water yields, peak flows, and sediment yields, depending on how extensively particular treatments are applied. As part of the investment in such a program, a team of soil scientists, hydrologists, and aquatic ecologists must actively participate to minimize the adverse effects on soil productivity and the aquatic environment. Although we have created forests that carry a high risk of damage to aquatic resources, pursuit of quick fixes in an atmosphere of crisis carries substantial risks as well (Beschta et al. 1995).

TIMBER HARVESTING

Harvesting of trees, especially in large clear-cut blocks, is commonly perceived as a major impact on the hydrology of river basins. Although timber removal has dramatic effects on the water balance of the immediate site, consequences at the catchment scale are not so obvious. As with many of the land management activities discussed in this chapter, the proportion of the catchment that is treated and the proximity of the treatment to water courses are critical in determining the impacts on water quantity, timing, and quality. In addition, associated activities such as road construction, yarding, slash treatment, and site preparation usually have much greater impacts than just the cutting of the trees. Hydrologic effects of selection harvests are generally considered to be less problematic than those of clear-cutting because the remaining trees remove soil moisture and provide some protection to the soil surface (Anderson et al. 1976). Harvest effects must also be considered with respect to time. Fortunately, trees and other plants quickly reoccupy most harvested areas, reestablishing protection from raindrop impact, uptake of soil moisture, deposition of organic matter to the soil, and support of soil masses by roots. Slopes are most vulnerable to surface erosion and generation of excess water immediately after harvest or site preparation, but they have minimal root strength about a decade after harvest (Ziemer 1981).

Water Yield

Harvesting timber has the potential to increase annual water yields via several mechanisms. Removal of all trees removes the possibility of any interception losses over the former area of the canopy. However, evaporative loss from rain and snow

detained in tree canopies may be a relatively small component of the water balance of forests in the Sierra Nevada and has been estimated at about 30 mm (1.2 in) per year (Kattelmann et al. 1983). Removing trees also terminates transpiration in rough proportion to the extent of removal and the ability of remaining plants to use the water. The depth and moisture storage capacity of forest soils largely control the amount of reduction in evapotranspiration from harvesting (Zinke 1987). When trees are harvested at the base of a slope near a stream, a large fraction of the soil water formerly used by those trees will enter the stream. If trees are cut near the top of a slope, residual trees below the area harvested may use much of the "excess" water not transpired in the harvest unit, and relatively little of this water may reach the stream. More than one hundred studies of stream-flow response to forest harvesting have been conducted around the world. These studies have been reviewed by many authors (e.g., Anderson et al. 1976; Bosch and Hewlett 1982; Ponce 1983; Kattelmann 1987; Reid 1993; Marvin 1995, 1996). In almost all cases, stream flow increases as basal area (and evapotranspiration) declines. As vegetation regrows on the site, evapotranspiration increases and stream flow declines correspondingly (e.g., Troendle and King 1985). Intensive timber harvesting under the usual constraints of national forest management could increase stream flow in most Sierra Nevada rivers by 1%–1.5% (6–9 mm [0.24–0.35 in]) (Kattelmann et al. 1983; Rector and MacDonald 1987).

Peak Flows

Although most work to date has been done on changes in the seasonal water balance, short-term changes with respect to storm response and flood augmentation are also important. Timber harvesting can affect peak flows through two principal mechanisms: maintenance of high soil moisture in the absence of evapotranspiration and higher rates of snowmelt during rain events. In a simplistic sense, less rainfall is required before runoff is produced if trees are not using stored soil moisture than if trees occupy the site. Creation of openings in the forest alters energy exchange and snow storage. During warm storms, most snowmelt occurs through turbulent exchange processes (condensation and convection), which are more effective at higher wind speeds. The greater wind speeds in forest clearings compared with dense forests increase the rate of snowmelt in the clearings relative to that under tree cover (Harr 1981; Berris and Harr 1987). Considerably more snow is found in forest openings than under forest canopies because wind deposition of snow is favored in openings and much of the canopy-intercepted snow drips off as liquid water and enters the soil. In the intermittent snowpack zone, this difference in deposition can result in an absence of snow under trees while several centimeters of snow water equivalence is available in openings to add water to storm runoff.

Potential effects of land management on flood generation

are most pronounced during small and moderate storm events and in small catchments (Hewlett 1982). During rare, intense storms, the differences in soil moisture storage or snow available for melt are almost incidental compared to tens of centimeters of rainfall (Ziemer 1981). At the river basin scale, flood peaks in the main river depend on synchronization of flood peaks from tributaries, which could be affected by drastic changes in land cover either positively or negatively.

Sediment Yield

Mass movements can be enhanced by timber harvesting by maintaining higher levels of soil moisture in the absence of evapotranspiration and loss of the reinforcement of the soil mass provided by roots (Sidle et al. 1985). On the average, the structural integrity of hill slopes is at a minimum about nine years after harvest, when the decay of old roots is not yet compensated for by the growth of new roots (Ziemer 1981). Roads tend to cause far more problems with respect to mass movement than does timber harvesting, and documentation of logging as a direct cause of mass failure in the Sierra Nevada has not been found.

Brushland Management

Conversion of brush fields to grass could increase stream flow and probably sediment yields at lower elevations (Anderson and Gleason 1960; Turner 1991). A proposal to manage about 130 km^2 (50 mi^2) of chaparral in the lower Feather River Basin with prescribed burning and some conversion to grass estimated that annual stream flow would increase by more than 3 million m^3 (2,500 AF) (California Department of Water Resources 1983b). Risks of increased erosion from such a program (e.g., Pitt et al. 1978) would need to be balanced against those from catastrophic fire. Conversion of brush fields to coniferous forest at higher elevations could delay snowmelt (Anderson 1963).

Observed Impacts

From a mechanistic point of view, forest harvesting in the past couple of decades has had limited opportunity to cause major changes in stream-flow volume, peak discharges, or sediment yield. In most river basins, the fraction of the basin area harvested per decade does not seem sufficient to cause major hydrologic responses. Nevertheless, peak flows in the South Fork Tule River appear to have increased in recent decades coincident with extensive road building and logging (see Marvin 1996). The level of harvesting since World War II has probably increased water yield somewhat in smaller catchments, but any increase may have been partially compensated for by increases in total vegetation density resulting from effective fire suppression over the same period. Unfortunately, neither influence can be quantified with any confidence. Similarly, we lack the appropriate data to observe

whether modest changes have actually occurred. Except in the Tule River case (see Marvin 1996), no changes in the stream-flow record that clearly exceed natural variability have been noticed. With respect to sediment yield, data are not currently available to show any change over the past few decades at larger scales except in the Mokelumne River, where sediment yield has increased dramatically (EBMUD 1995). Appropriate baseline data in reservoir surveys could be used to determine if sediment yields have increased as a cumulative result of all types of land disturbance. Carefully performed follow-up surveys are needed to find out if land-management activities have made a significant difference.

At the smaller watershed scale, there is at least some observational evidence to suggest that land management affects the hydrology of small streams in the Sierra Nevada. Again, the impacts are the result of all activities associated with harvesting, such as road construction, skidding, and site preparation. Landslides that begin at roads and that are the only occurrences of mass movement in a catchment can be attributed to management activities. Similarly, when the only pools in a channel reach that are filled with silt are those immediately below clear-cuts that included the riparian zone, we can infer some cause and effect. However, even when impacts are overwhelming at the local scale, they are quickly masked downstream because few other contributing catchments were treated in the same way. A few studies in the Sierra Nevada have indicated impacts at the small watershed scale. Suspended sediment increased in the 10 km^2 (4 mi^2) Castle Creek Basin near Donner Summit during the first year following road construction and timber harvesting (Rice and Wallis 1962). Sediment captured in weir ponds below a catchment 1.2 km^2 (0.5 mi^2) in area on the Sequoia National Forest increased severalfold after road construction and harvesting (McCammon 1977). Stream reaches in twenty-four small streams in the Sierra Nevada and Klamath Mountains had significantly higher indices of stored sediment than corresponding control reaches (Mahoney and Erman 1984). Water yields from Berry Creek (20 km^2 [7.5 mi^2]) near Yuba Pass seemed to have increased substantially following harvests on less than half the basin (Kattelmann 1982). Peak flows may have increased in part of the Mokelumne River Basin as a result of extensive harvesting (Euphrat 1992). Unfortunately, long-term paired-catchment studies have never been performed in the Sierra Nevada, so we are left attempting to infer impacts from experiments elsewhere.

Aquatic Effects

Studies that began in the 1970s on several streams in the northern Sierra Nevada and Klamath Mountains demonstrated that communities of aquatic invertebrates changed significantly in response to upstream logging (Erman et al. 1977; Newbold et al. 1980; Erman and Mahoney 1983; O'Connor 1986; Fong 1991). Some of the aquatic effects have persisted for two decades (Fong 1991). The aquatic communities are particularly sensitive to logging-related disturbance within 30 m of the channel (Erman and Mahoney 1983) and perhaps within 100 m (McGurk and Fong 1995). In a recent study of forest management effects on aquatic habitat in the Sierra Nevada, data were collected on channel characteristics, aquatic habitat, fish abundance and health, aquatic invertebrate abundance, large woody debris, water chemistry, and management history in twenty-eight different basins in the Sierra Nevada (Hawkins et al. 1994). In general, natural variability in the measured attributes masked effects of management activity. Response of aquatic organisms to disturbance in the watersheds tended to be small compared with response to natural factors. Observed increases in nutrient loading and temperature appeared to enhance abundance of some taxa without any noticeable adverse impacts on others. Most of the communities in the observed streams appeared to be limited by food resources. The study noted, "We cannot at this time either measure or predict with any degree of reliability or confidence the cumulative effects most types of land use practices will have on natural ecosystems" (Hawkins et al. 1994, 1).

GRAZING

Grazing of domestic livestock has probably affected more area in the Sierra Nevada than any other management practice (Menke et al. 1996). Over the past century and a half, cattle and sheep have been virtually everywhere in the mountain range that provides forage. The near-ubiquitous presence of grazing animals has left few reference sites that we can be certain were never used by livestock. The best approximations to ungrazed conditions are those areas that have been rested for a few decades. Even Sequoia National Park was grazed until 1930 (Dilsaver and Tweed 1990). The absence of reference sites leaves us uncertain about what an ungrazed stream looks like and how it functions. This uncertainty is not merely an academic concern. Major questions of grazing management depend on our confidence in our understanding of how natural systems function without human-induced perturbations. For example, we can hypothesize that overgrazing on the Kern Plateau in the 1800s contributed to the widespread arroyo development and conversion of wet meadows to dry terraces. There are several lines of evidence that support that hypothesis. However, we would have more confidence if one of the early shepherds had invested a couple of summers in fencing off an entire watershed and preventing entry just to satisfy the curiosity of future generations. This problem of uncertainty exists to some degree with all impacts, but there are many areas that were not mined, dammed, logged, roaded, or urbanized. There just are not many that were not grazed.

In 1924, Aldo Leopold wrote, "Grazing is the prime factor in destroying watershed values," in reference to an overgrazed site in Arizona. Since then, debate has continued about the validity of similar statements applied to watersheds throughout the West. The impacts of grazing that relate to hydrology depend primarily on the behavior of the animals: feeding, drinking, producing waste, and traveling. If the animals remain in one place too long and consume much more than about half the available forage, vegetative recovery may be impaired and an excessive amount of bare soil may be exposed to erosive rainfall (Fleischner 1994; Committee on Rangeland Classification 1994). Although the amount of consumption that constitutes "overgrazing" depends on vegetation and site characteristics (Menke et al. 1996), half of the initial forage is a useful, though admittedly crude, rule of thumb (California Division of Forestry 1972). When insufficient vegetation remains after grazing, raindrop impact can change surface conditions and consequently reduce infiltration and increase erosion (Ellison 1945). Soil can become compacted by the repeated pressure of moving animals, especially if the soil is wet. The combination of soil exposure and compaction can decrease infiltration and increase surface runoff. If infiltration capacity is severely limited on a large fraction of a catchment, the extra runoff can quickly enter streams and generate higher peak flows (e.g., Davis 1977).

Surface and Channel Erosion

Exposure of mineral soil and enhanced overland flow also accelerate erosion. A variety of studies around the West have found dramatic increases in sheet erosion and gullying in overgrazed sites compared with ungrazed areas (Fleischner 1994). Severe gully erosion in the uplands of the North Fork Feather River has been caused by decades of overgrazing (Soil Conservation Service 1989). Nevertheless, the worst erosion problems associated with grazing typically occur near streams. Cattle tend to congregate in riparian areas for obvious reasons: abundant food, water, shade, and lower temperatures. Consequently, riparian vegetation is overgrazed, banks are trampled and eroded back, and bed deposits are disturbed. All this activity adds significant amounts of sediment directly to the stream. Dislocation of sediments in the streambed by moving animals augments suspended sediment. Degradation of riparian vegetation permits bank erosion to accelerate under the more frequent peak flows that are caused by the decrease in infiltration capacity. About half of the channels in the Meiss allotment in the Upper Truckee River watershed were identified as being in fair or poor condition as a result of overgrazing (Lake Tahoe Basin Management Unit 1993). Changes in channel morphology have been related to overgrazing in headwater streams tributary to the Carson River (Overton et al. 1994). Elimination of riparian vegetation by overgrazing in the broad alluvial valleys of the North Fork Feather River has led to rapid channel widening and massive sediment loads (Hughes 1934; Soil Conservation Service 1989). In other areas, such as meadows of the Kern Plateau and San Joaquin River Basin, downcutting has followed overgrazing (e.g., Hagberg 1995). Development of these deep arroyos has lowered the local ground-water table and transformed wet meadows into dry terraces supporting sagebrush. The possibly compensatory effects of less bank storage and less transpiration by vegetation determine whether low flows in summer are decreased or increased by the downcutting. A recent study of channel characteristics between pairs of currently grazed areas on national forests and long-rested areas in national parks in the Sierra Nevada found significant differences in bank angle, unstable banks, bed particle size, and pool frequency (U.S. Forest Service 1995b). Significant differences in undercut and unstable banks were also observed between grazed areas and adjacent fenced exclosures with a few years of rest.

Water Temperature

Removal of riparian vegetation and channel widening by grazing expose the stream to much more sunlight. Therefore, stream temperatures in summer may be several degrees higher than if shade remained. In winter, the absence of riparian vegetation may allow wind scour of snow in exposed creeks in high-elevation meadows. With less snow serving as insulation, ice formation may be greater than in creeks with vegetation capable of trapping more snow, which provides insulation itself. These artificial changes in temperature impact aquatic organisms that rely on a more natural temperature regime.

Water Pollution

Congregation of cattle in and around streams provides a direct pathway for nutrients and pathogens to degrade water quality (Springer and Gifford 1980; Kunkle 1970). High nutrient loads promote the growth of aquatic algae, which can virtually clog streams at low flow. An example of proliferation of aquatic plants apparently augmented by cattle is found in the Owens River above the Benton Crossing road. High levels of coliform and other bacteria have been found in streams heavily used by livestock (Lake Tahoe Basin Management Unit 1993; Central Valley Regional Water Quality Control Board 1995). Cattle grazing in backcountry areas provides a source of Giardia cysts (Suk et al. 1985).

Associated Impacts

There are a variety of ancillary effects of grazing. Road construction to provide access to range improvements has similar impacts to those of roads in general, depending on the location and design. Springs are extensively developed for stock watering, at the expense of native biota. Irrigated pasture consumes immense quantities of water for a low-value product (Romm et al. 1988).

Improved Practices

Improved grazing practices as applied to the Sierra Nevada have the potential for limiting many of the possible adverse impacts (Albin-Smith and Raguse 1984). Major changes in grazing practices in some parts of the eastern Sierra Nevada have recently occurred. A study is currently under way monitoring the response of macroinvertebrates and fish to riparian fencing and rest-rotation management (Herbst and Knapp 1995a, 1995b). Degraded channels in the Meiss allotment at the south end of the Lake Tahoe Basin led to a decision by the Forest Service to rest the allotment for five to fifteen years until stream-bank vegetation has recovered (Lake Tahoe Basin Management Unit 1993). However, this decision was overturned on appeal by the regional forester in 1995.

URBAN, SUBURBAN, AND EXURBAN DEVELOPMENT

The population of the Sierra Nevada foothills is expected to increase rapidly in the next few decades (see Duane 1996a). Conversion of forests and woodlands to residential and commercial land uses has several serious hydrologic effects on local streams (Lull and Sopper 1969). Such conversions dramatically alter the disposition of rainfall or snowmelt on the landscape by reducing infiltration capacity of the surface to zero or near zero. Land that was formerly well vegetated and rarely, if ever, produced overland flow is converted to an impervious zone where virtually all precipitation becomes immediate runoff. The extent of such changes and the ability of adjacent land to absorb the additional runoff determines the response of streams. Gutters, ditches, drains, channels, culverts, and storm sewers are intended and designed to convey runoff as rapidly as possible to streams. The combination of greater volume of runoff, faster generation of runoff, and greater channel efficiency moves more water downstream faster than under natural conditions. This convergence of large volumes of water in short periods of time produces frequent floods downstream from even modest rainfall (Leopold 1968). The more frequent floods lead to channel enlargement by erosional processes (Dunne and Leopold 1978; Booth 1990). Unfortunately, stream gauging stations have not been placed in strategic locations to actually record changes in stream-flow regimes in response to development in the Sierra Nevada.

Impervious Surfaces

The proportion of impervious area created by residential construction is a rapidly decreasing function of lot size. Small urban lots can be effectively sealed over three-quarters of their surface area, while lots of 0.4 ha (1 acre) might be impervious on only 10%–15% of the area. So-called low-density residential lots can be 20%–30% impervious surface (California Division of Soil Conservation 1971b). Larger parcels would have much smaller proportions rendered impermeable by construction. Although impervious area is a small fraction of dispersed "ranchette" development, the amount and intensity of conversion of natural vegetation to other uses, such as orchards, vineyards, pasture, ostrich ranches, and Christmas tree farms, will determine the hydrologic impacts. The area occupied by roads is closely associated with the density of structures. In addition, the quality of road location and construction will influence the potential for adverse effects. Poorly designed roads for subdivisions are a principal source of sediment in Nevada County (Gerstung 1970).

Channelization

Channelization (forcing streams into engineered waterways) has been practiced in the Sierra Nevada since the first miners' ditches for water supply were constructed and rivers were confined in wooden flumes while their gravels were excavated. During the mining era and subsequent development of water resources for hydropower, municipal, and agricultural uses, streams were put into artificial channels to get the water to another place where it was wanted. Around roads and towns, the usual objective of channelization is to get water away from a place where it is *not* wanted. Creeks of all types and sizes have been relocated, smoothed, and straightened to get water away from roads and homes as quickly as possible. These ditches, canals, and storm sewers enhance the flood-producing effects of general land conversion by routing the extra runoff away from the town or road much more quickly than under natural conditions. Peak flows are augmented downstream, but that is typically beyond the concern of the local channelization project. Flooding in Roseville during January 1995 was a classic example of this phenomenon. Failure of artificial drainageways and streets to perform as expected can also cause damage within the community attempting to control the runoff, as occurred in Cameron Park in 1982 and 1983 (Soil Conservation Service 1985). At higher elevations, runoff rates from snowmelt may also be accelerated where infiltration is limited in significant fractions of a watershed (Buttle and Xu 1988).

Vegetation Removal

Other hydrologic impacts of conversion to residential and commercial land uses include reduction of interception and transpiration functions of trees and other vegetation via their removal. All plants intercept and store some proportion of the precipitation received. Water retained in the canopy eventually evaporates. Continuous vegetation cover can reduce the amount of water reaching the ground substantially, depending on storm amounts and frequency. Removing the vegetation largely eliminates this function. Whatever replaces the plants usually has some interception capacity. However, roofs,

latticework, and other structures do not transpire. So, vegetation conversion eliminates the active removal of soil moisture and its transfer to the atmosphere. Soil moisture would remain higher in the absence of transpiration if soil moisture recharge could take place through whatever covers the soil instead of vegetation. Where impervious areas are constructed, recharge of shallow and deep ground water is minimized. If the total area of limited infiltration is a significant fraction of a catchment, ground-water levels will decline. Stream flow during nonstorm periods that was formerly generated by seepage from ground water will also decline. Ground-water pumping for domestic and irrigation supply can exacerbate the problems of restricted recharge. In some cases, irrigation return flows may augment summer stream flow.

Water Pollution

The changes in runoff are closely related to declines in water quality associated with urban development. Enhanced runoff washes various contaminants off roofs, streets, parking lots, gutters, horse corrals, and golf courses and into streams. Diminished base flow increases the concentration of residual pollution entering after the floods. Urban pollutants include soil particles, nutrients, heavy metals, toxic organic chemicals such as pesticides, oil and grease, fertilizers, oxygen-demanding materials such as yard waste, and bacteria and other pathogens (Terrene Institute 1994). The diversity of sources makes control difficult, but best management practices are being developed and applied to control urban runoff. Development of riparian areas limits opportunities for filtering, uptake, and assimilation of contaminants. The combined effects of changes in runoff regime, water quality, and channel structure resulting from urbanization have profound effects on aquatic life. Eliminating infiltration on as little as a tenth of the catchment area led to declines in population of fish and amphibians near Seattle (Booth and Reinelt 1993).

Accelerated Erosion

Removal of vegetation, grading, and exposure of bare ground allows erosion to increase dramatically, especially during construction. Freshly cleared land for a new subdivision in Plumas County produced enough sediment in a single intense storm to kill 80% of the aquatic life in Big Grizzly Creek (California Division of Soil Conservation 1971b). In Nevada County, more than a third of the total length of streams has been damaged by siltation and stream-bank erosion resulting from subdivision development (Gerstung 1970). Erosion rates in the Middle Creek watershed near Shasta City increased more than twentyfold following urban development (Soil Conservation Service 1993). Residential construction around Lake Tahoe has been a major contributing factor in accelerating erosion and increasing nutrient inputs to the lake (Tahoe Regional Planning Agency 1988).

Sewage

Effluents from wastewater treatment facilities and leachates from dispersed septic systems add nutrients to ground water and streams. Breakdowns and spills from sewage facilities can introduce pathogens to receiving waters. Leaks from sewer lines have been recognized as an important source of nutrients in the Lake Tahoe Basin (Tahoe Regional Planning Agency 1988). Wastes from domestic animals and pets can also contaminate streams. Organic wastes can deplete dissolved oxygen in streams as well. Water quality was considered impaired in streams receiving wastewater from Nevada City, Grass Valley, Placerville, Jackson, and the Columbia-Sonora area (Central Valley Regional Water Quality Control Board 1991). Small sewage treatment plants serving recreational developments often suffer from inadequate financing, technology, and management (Duane 1996a). A facility at California Hot Springs alternately released excessive amounts of barely treated sewage or chlorine for several years. Giant Forest Village in Sequoia National Park was largely closed during the winter of 1994/95 because of poor performance of the wastewater treatment facility. Disposal of solid waste also has the potential to contaminate ground water and streams. Older landfills were probably not carefully located or designed and may be producing hazardous leachates. Location of new landfill sites is so difficult that Tuolumne County is planning to export its garbage to Lockwood, Nevada.

Water Supply

Supplying water for new development is problematic in many parts of the Sierra Nevada, which is ironic given the high runoff production of the mountain range. Coping with the seasonal distribution of runoff usually involves construction of storage reservoirs when large numbers of users are involved. Beyond the seasonal availability, most of the difficulties are legal and financial rather than physical. Surface waters are already overappropriated in many watersheds. Newcomers may find that all the local water is already claimed. Also, there seems to be a widespread belief that water should be supplied free or for a minimal charge—that somebody else (i.e., "the Government") should subsidize water supplies. Communities are often in favor of augmenting their water supplies for new development until they find that they are expected to share the cost. The Calaveras County Water District has financed its water development through the sale of hydroelectricity. Tuolumne County is faced with large costs and potentially high water rates from its redevelopment of the Lyons Reservoir system, acquired from the Pacific Gas and Electric Company.

The California Department of Water Resources (1990a) identified several generic problems facing rural water supply in the Sierra Nevada, which remain pertinent today:

- Rapid growth and development will burden existing water supplies and sewage treatment.

- Ground-water sources are not reliable in terms of quantity and quality.

- Water distribution systems are inefficient.

- Communities located on ridges are gravitationally disadvantaged.

- The best locations for impoundments have already been exploited by others.

- The revenue base is not sufficient to support water facilities at low rates per customer.

- Local funding sources are limited.

- Developing new water projects is economically and environmentally costly.

- Construction of new conveyance systems is expensive because of dispersed users and terrain.

Water companies and water-supply service districts in the Sierra Nevada vary in size from a few dozen customers to tens of thousands (Department of Water Resources 1983a; Harland Bartholomew et al. 1992). The nature of their sources, delivery and treatment systems, and demands are highly variable as well (e.g., Thornton 1992; Borcalli and Associates 1993). There are more than 160 separate water purveyors in Placer County alone (Placer County 1994). Water demands for individual households in the foothills have been estimated at 600–1,200 m^3/yr (0.5–1 AF/yr) (Page et al. 1984; Harland Bartholomew et al. 1992). Supplying new customers with existing water supplies may place current consumers at risk of shortfall during dry periods. Legislation was passed in California in 1995 (SB 901) to limit the ability of cities and counties to allow new developments unless local water purveyors certify that adequate water supplies exist for both present and expected residents. Sources of water for large proposed developments in the foothills, such as Yosemite Estates near Sonora, Las Mariposas near Mariposa, and Promontory near Placerville, are uncertain. Excessive water withdrawals from local streams can threaten recreational fisheries that form part of the economic base supporting the communities seeking the extra water for more development (Kattelmann and Dawson 1994). Development of additional water supplies is likely to become increasingly costly in both financial and environmental terms.

The projected demand for additional water in the foothills in the next few decades is staggering (see Duane 1996a). The Georgetown Divide Public Utilities District expects a 50% increase in water use in the next thirty years, and the El Dorado Irrigation District anticipates demand to double in the same period (Borcalli and Associates 1993). In Amador County, domestic water use was forecast to rise by between 2.6 times and 3.6 times between 1983 and 2020, depending on which population projections were used (Department of Water Resources 1990a). Tuolumne County expects to add 14,000 people in the next twenty years, who will need about 8.6 million m³ (7,000 AF) of water each year to support them. The largest expected increase has been postulated for the service area of the Nevada Irrigation District, where annual domestic use could go from about 15 million m³ (12,000 AF) to about 40 million m³ (33,000 AF) between 1992 and 2010 (Harland Bartholomew et al. 1992). Nevada County is in perhaps the best position to meet the expected demands. The Nevada Irrigation District currently has rights to more water than is used and has a vast base of agricultural use that is expected to decline. Calaveras County also appears to have a relatively secure water supply. Some streams that have been dewatered below diversions may receive some flow for instream needs as older contracts expire and are reviewed. Other regions must find new sources of water and presumably will want to build new storage facilities or acquire existing hydroelectric projects.

SKI AREAS

As the major industrial/commercial development in the higher-elevation parts of the Sierra Nevada, ski areas have generated public concern about impacts of the resorts on water resources. Because of their extensive marketing campaigns and their location along the major access roads of the range, one could get the impression that there are ski areas all over the Sierra Nevada. However, the twenty-five alpine resorts occupy a tiny fraction of the land area in the mountain range. Only a few of the larger resorts, such as Squaw Valley, Alpine Meadows, Heavenly Valley, and Kirkwood, occupy a major proportion of the immediate watershed they are situated in. Most of the more significant impacts of ski resort development are associated with base facilities, roads, and parking lots. Such facilities are usually located in a valley bottom and impact streams and wetlands. For example, the parking lot of Boreal Ridge converted a large subalpine meadow into an expanse of impermeable asphalt and channelized Castle Creek. Because of the need for flat ground at the base of ski areas, many streams have been rerouted or even put underground. Access roads are often located in riparian zones. Runoff from roads and parking lots is usually polluted. The base facilities, lodging, and recreational residences generate substantial amounts of wastewater, which has usually required a local sewage treatment plant. Sewage system failures occasionally occur under harsh winter conditions. Most of the impacts of small urban areas can be applied to resort development. Ensuring an adequate water supply for all uses (residential, commercial, snow making, landscaping, erosion control plantings, and golf courses) for a major resort community can be problematic even in prolific source areas of snowmelt runoff (Kattelmann and Dawson 1994).

Vegetation Conversion

Impacts related to the ski slopes themselves begin with tree removal for runs and lift access. Such clearing constitutes a permanent conversion of vegetation type, as opposed to forest harvesting, which implies hydrologic recovery. When runs are cleared, deep-rooted trees are replaced with shallow-rooted grasses, greatly reducing evapotranspiration and increasing soil moisture storage. Because ski runs are typically oriented down the fall line, there is little opportunity for trees downslope to use the extra soil water in transit. Type conversion on ski runs can generate at least 7–15 cm (3–6 in) per unit area harvested of additional stream flow (Hornbeck and Stuart 1976; Huntley 1992). In some situations, subsurface drainage pipes and new surface channels may need to be installed to accommodate the additional water and avoid saturated conditions that could lead to mass movement. In some areas, there is extensive excavation and shaping of the natural terrain. Maintenance of sufficient ground coverage for adequate erosion control may require artificial irrigation and fertilizers in summer. Excessive use of fertilizers can contribute to high nitrate levels in local ground water (Goldman et al. 1984). Erosion from ski areas seems to be fairly well controlled and largely in compliance with rules from the regional water quality control boards, especially in the Lahontan Region. General construction always has the potential for accelerating erosion, but best management practices for minimizing soil loss at ski areas are becoming fairly thorough (i.e., Calaveras Ranger District 1991). In general, ski areas can afford to invest in erosion control and slope stability techniques that are not possible outside of major engineering projects. Somewhat analogous to abandoned mines, abandoned ski areas have potential for severe erosion problems, as occurred at Pla-Vada, where high sediment loads from gullies on the ski slopes damaged fish habitat in the nearby South Yuba River (California Division of Soil Conservation 1971a).

Snow Compaction

Grooming operations, avalanche control, and skiing compact the snow and move some of it downhill. A study of effects of compacted snow near Donner Summit found that snow water equivalent on the narrow ski runs was up to 50% greater than that on adjacent uncompacted slopes and that ski runs remained snow covered for up to two weeks longer than adjacent uncompacted slopes (Kattelmann 1985). Chemicals, such as ammonium nitrate, sodium chloride, and calcium chloride, have been used at a few ski areas to prepare race courses and improve skiing conditions in spring and summer. In general, only small areas are treated with relatively small quantities of chemicals. Degradation of water quality is a concern and has been reported in Europe.

Artificial Snow

Snow making has become widespread among the ski areas of the Sierra Nevada because skier demand seems to be greatest in November and December, when natural snow cover may be marginal. Artificial snow is produced by mixing water and air under high pressure through a nozzle. The sudden expansion cools the water and forms ice particles, which provide a reasonably good skiing surface and base for natural snow. Typical depths of applied water range from 20 to 50 cm (8–20 in), so the area covered is the main determinant of the total volume of water used. Most of the water used is returned to the stream it was originally withdrawn from, but delayed by 5 to 8 months. Evaporative losses of 2%–5% occur at the nozzle, and sublimation losses from artificial snow on the ground (and not covered by natural snow) range from 10 to 50 mm depending on how long the snow is exposed (Eisel et al. 1988; Huntley 1992). If water diversions for snow making will seriously deplete stream flows during the low-flow part of the year, off-channel storage capturing one season's snowmelt runoff to artificially initiate the following season's snow cover, such as is practiced at Mammoth Mountain, may be warranted.

Prospects for Expansion

Despite seemingly flat skier demand and the failure of about a quarter of the nation's smaller ski areas in the past decade, future prospects appear good enough to the ski industry to add additional capacity (see Duane 1996b). For example, revenues at Northstar-at-Tahoe and Sierra-at-Tahoe grew by 4% in the first quarter of 1995. Major expansion occurred at Sugar Bowl in 1994 and is planned at Kirkwood. Squaw Valley has proposed construction of a new base complex and has plans for year-round skiing. An entirely new ski area has been approved by the Forest Service in the Mammoth Lakes area, but $50 million in financing for construction may be difficult to obtain. Various large-scale development schemes have been proposed in the Royal Gorge/Devil's Peak region near Soda Springs.

INTERPRETATIONS

Historic and Current Conditions

The most significant impacts to the hydrologic system of the Sierra Nevada started almost immediately with the boom in Euro-American entry into the mountains during the gold rush. The effects of riverbed and hydraulic mining were devastating to the rivers of the western slope. Substantial recovery of the obvious features of channel morphology and riparian vegetation provides the appearance of natural rivers, but the aquatic and riparian ecosystems may remain quite simplified

compared to the pre-1848 conditions. However, we will never know. As mining subsided, water development quickly took its place as an overwhelming, though less intensively destructive, impact. Although the severity of overgrazing may have peaked between about 1890 and 1930, continued grazing pressure has prevented thorough recovery of many degraded streams, and some (e.g., North Fork Feather River) continue to deteriorate. Early logging probably denuded larger expanses of the Sierra Nevada but may have applied less intense hydrologic disturbance to the soil than the road-building and tractor-skidding era that began after World War II. The various impacts of residential development have accelerated in the past decade. Impacts of fire and the legacy of fire suppression have yet to play out.

Water resources of the Sierra Nevada are highly controlled for various social purposes. That management causes the greatest current impacts to other social and ecological uses. The degree of alteration of natural stream flows generally increases in the downstream direction where the water passes through or is withdrawn by successive projects. However, the amount of unregulated flow from hydrologically intact tributaries also increases downstream, helping to "dilute" the effects of river engineering, at least until the big dams on the main rivers are reached. The dilution effect is also important in ameliorating changes in land use, which are most obvious close to their point of occurrence. The addition of water from relatively unimpacted watersheds helps offset the adverse cumulative impacts of assorted disturbances. Downstream of points of diversion, streams may lack the capacity to transport natural and accelerated sediment yields. Overall, water quality remains high compared with other rivers of the United States, but many problems exist locally. Alterations in the flow regime may be the most widespread degradation of water quality.

The primary trends related to water resource conditions at the scale of the entire mountain range have been recovery from gold mining and increasing regulation of stream flow via water developments. Both trends have diminished through time after the main geomorphic adjustments to mining debris and early dams occurred and the optimum dam sites and water rights were acquired and developed. Impacts from forest road building may have peaked as most of the potential road network would seem to be in place (see McGurk and Davis 1996); however, the high road density in some catchments ensures continued sediment yields at high rates. Impacts from residential road building and associated activities seem to keep increasing as the development of the foothills continues. An important question for planners is how to meet growing water demand while minimizing the environmental impacts of additional water development.

Some Implications

The overwhelming impacts on the water-resource system of the Sierra Nevada are those that directly modify the flow regime and the channel. Landscape impacts are secondary in those river basins of the Sierra Nevada with substantial water development. Even among land-use activities, those adjacent to or near a channel have far greater impacts than activities distant from water. Improvement of land-management practices and restoration should focus on issues closest to the streams if amelioration of aquatic impacts is a primary goal. Similarly, stream health will not suffer so much if disturbances are positioned well away from streams. Throughout this discussion of stream health, there is a presumption that fully functional aquatic ecosystems are inherently valuable and that attributes of streams beneficial to aquatic life (natural flow regime, low sediment transport, stable channel, good chemical quality, etc.) are also beneficial to the human uses of water. Aquatic ecosystems in headwater catchments are at greatest risk of damage from land disturbance. Combined effects of water engineering and land management may be particularly harmful to some aquatic communities.

Time Significance

Although there is no absolute urgency in changing the way society treats streams, the sooner damaging practices are improved or avoided, the sooner streams will benefit. Taking care of existing problems sooner rather than later and avoiding new mistakes will reduce the total impact (e.g., less sediment into stream pools or reservoirs) and may cost far less in the long run. Lake Tahoe is the best example of a system that needs urgent attention to slow the rate of deterioration of a particular resource or value (i.e., lake clarity). At Lake Tahoe, human activities have clearly altered a critical component of the ecosystem (nutrient cycling), and because the lake is extraordinarily sensitive in that regard, ecological responses are obvious and rapid. Although there are few real parallels to the Tahoe situation in terms of urgency, there are many other important problems to address. For example, reducing streambank erosion in the North Fork Feather River is clearly an important goal. As long as comprehensive action is delayed, productive alluvial land will continue to be lost, streams will continue to carry high sediment loads, and downstream reservoirs will continue to fill with sediment at unnaturally high rates. The most urgent problems are those where continued degradation could be irreversible or extremely expensive to mitigate if allowed to persist.

Perceptions

The adverse impacts of water management are probably overlooked by the public at large because of the obvious, personal benefits of that management. Perception of water-related impacts from residential development in the foothills is probably mixed depending on whether the individual has lived in a foothill community for decades, is a newcomer, relies on continued growth for personal income, or does not live in the area. Water-related problems associated with land management may be perceived by some people as more serious than they really are because of the visual impacts and media attention. The degree of destruction and subsequent recovery from placer and hydraulic mining are not widely recognized.

Gaps

Obviously, the operational difficulties lie in site-specific details. The assessment in this chapter is a very broad treatment of an entire mountain range. The problems are in particular streams. Every watershed has a story that is critical to its own stream. Management is conducted at that scale. The broad generalizations made here only provide the regional context for individual catchments and streams. The absence of information on recent sediment yields and limited stream-flow records from unregulated streams prevent any quantitative conclusions about how much land management has altered yields of water and sediments in the Sierra Nevada. Impacts on aquatic biota are also difficult to quantify because of scarce baseline data (see Erman 1996; Moyle 1996; Moyle and Randall 1996; Moyle et al. 1996).

Ecosystem Sustainability and Management

The physical recovery of streams from the gold mining era demonstrates the resiliency of rivers. Although recovery is probably to a more simplified state, with some lingering attributes of the original disturbance, this recovery illustrates an inherent long-term sustainability of the fluvial and aquatic systems, even in response to catastrophic impacts. On land, vegetation seems to reclaim favorable sites (i.e., riparian areas, north-facing slopes) very quickly after fire or logging. Drier sites and areas with special problems may require active intervention to reestablish ground cover in the short term or the avoidance of such sites in the first place. Recovery of other ecosystem properties and processes following disturbance requires much more time and possibly some management if we are impatient with nature's schedule. Ecosystem management must avoid impeding natural recovery processes after a fire or other disturbance, incorporate such processes in planning management programs, and augment them when necessary to accelerate ecological change in a desired direction, especially on difficult sites.

Remaining Questions

Among many important questions about hydrologic impacts of land management, three stand out:

1. How much has sediment yield been altered by human activities?

2. How much has the stream-flow regime (annual water yield, peak flows, low flows) been altered?

3. How do changes in water quantity and quality relate to declines in aquatic biota?

Reservoir sediment surveys on a 10- to 20-year cycle could be very informative about the first unknown. Establishment of a long-term network of stream gauges in strategic locations, such as actively managed headwater catchments, could be very informative about the magnitude of changes in hydrologic processes resulting from changes in land use. The present network informs us about water management and needs to be supplemented to inform us about land management. An aquatic research program, such as that suggested by Naiman et al. (1995), would help address many of the gaps in knowledge regarding streams of the Sierra Nevada.

CONCLUSIONS

From a hydrologic perspective, the Sierra Nevada seems to be functioning adequately as the preeminent water source for California society, agriculture, and industry. However, the hydrotechnical structures that facilitate exploitation of streams for social uses create the greatest impacts to those very uses as well as to aquatic ecosystems. This highly managed water system has created artificial patterns of stream flow in the lower reaches of most rivers and their principal tributaries. There are not many opportunities for further development of water resources in the mountain range, given existing infrastructure and water rights. Financing additions to community water supplies without subsidies from hydroelectric generation will be difficult at best. Existing ground-water development near foothill communities limits the availability of subsurface water as a dependable supply for future growth. The managed flows and physical barriers to movement of water, sediment, and biota have substantially altered aquatic and riparian ecosystems to something other than natural.

Compared with the intentional alteration of stream flow through water management, hydrologic side effects of changes in land use are difficult to measure but are still believed to be significant. Major changes in water and sediment regimes have not been observed in the main rivers and their larger tributaries as a result of shifts in land use. There may be a signal, but it is not obvious or well quantified. Hydrologic changes resulting from land management are most likely to be found in headwater areas, where a large fraction of the catchment has been affected. Diversion of water from a stream will limit transport of excess sediment loads and thereby compound the impacts of land disturbance. Roads are believed to have increased sediment yields substantially, but the inferred changes have not been measured in the Sierra Nevada. Overgrazing has probably altered channel conditions extensively, but the scarcity of ungrazed reference sites limits research-

ers' ability to quantify impacts. Rapid expansion of foothill communities has theoretically altered runoff and erosion processes enough to cause noticeable impacts in downstream channels, but quantitative and documentary evidence outside the Tahoe basin is lacking. Conversion of forestlands to roads associated with timber harvesting may have increased annual water yields and peak flows somewhat at the small watershed scale. However, decades of successful fire suppression may have increased evapotranspiration relative to a pre-1850 fire regime and partially compensated for the flow increases attributed to roads and harvests. The offsetting magnitudes of either impact cannot be quantified at this time. The legacy of fire suppression creates substantial risks of serious hydrologic impacts from potential conflagrations.

Overall, chemical water quality remains high, but water cannot be considered pristine. Because of widespread biological contamination, surface waters throughout the range cannot be assumed to be drinkable. A few local problems are very serious: Lake Tahoe, some abandoned mines, and some communities. Quality of receiving waters from the larger cities in the foothills has been degraded. These aquatic systems are not as sensitive to nutrient loading as Lake Tahoe. Excessive sediment production is the most widespread non-point-source problem, but its extent and severity are unknown. Studies in other areas suggest that roads are the overwhelming source of sediments that end up in wildland streams. Disturbance in and near stream channels generates the vast majority of sediment transported by the streams. Existing information about sediment yields in Sierra Nevada rivers is largely obsolete, and new reservoir sediment surveys are necessary to determine whether changing land use has accelerated sedimentation in the past few decades. Because of the importance of flowing water in diluting and dispersing pollution, alteration of stream flow by storage and diversion may be the fundamental water quality problem in the Sierra Nevada.

MANAGEMENT IMPLICATIONS

The ecological health of a stream is affected by all activities in its watershed. Those activities that directly control the flow regime or occur within the riparian zone usually have the greatest potential impacts. Changes in reservoir management practices may offer the best hope for improving aquatic ecosystems where they are known to be influenced by artificial flow regimes. In general terms, some shifts back toward a natural hydrograph, such as seasonally fluctuating flows, occasional flushing flows, maintenance of adequate low flows, or whatever is appropriate to a particular situation, will be beneficial to the local biota. Simply maintaining constant minimum flows is rarely sufficient. Stream habitat conditions and aquatic biota have developed in response to a highly variable natural flow regime. Restoring some aspects of that variability in managed streams should have ecological benefits in most cases. In some cases, changes in reservoir releases to benefit downstream organisms and water quality may have few adverse impacts on economics of the project. In other cases, there may be substantial costs, which may not be justified for the intended benefits. The tradeoffs between in-stream impacts and operational impacts must be carefully evaluated in the context of each water project, the watershed it is located in, and the ultimate downstream uses. There could be continued realignments in water rights as a result of application of the public trust doctrine, hydropower relicensing requirements, and regulation of reservoir releases by the regional water quality control boards for water quality management. The State Water Resources Control Board could ease legal and administrative matters by improving their water-rights database. An efficient, geographically referenced database for water rights could allow examination of in-stream flow conditions in a cumulative context for each stream and river system. Designation of additional wild and scenic rivers could help maintain ecological values of selected segments.

Major reconstruction of smaller dams to allow sediment pass-through under high-flow conditions could help restore some semblance of a natural sediment regime to many streams. Such work is a serious challenge in hydraulic engineering and reservoir management, but it would be an important contribution of technology to restoring natural processes in the managed rivers of the Sierra Nevada. Provision for flushing flows is particularly important where land disturbance may have augmented natural sedimentation and regulated flows encourage sediment deposition.

Recent actions by the State of California and the U.S. Forest Service to use watersheds as a geographic basis for planning and management are encouraging. As local agencies and citizens begin to incorporate a watershed basis into their own activities, overall conservation of aquatic resources should greatly improve. Continued public education about basic watershed concepts can only help. Application of watershed analysis methodologies developed in the Pacific Northwest (e.g., Montgomery et al. 1995) to the Sierra Nevada would be a worthwhile step toward improved management of wildlands at the landscape scale (see Berg et al. 1996). Watershed analysis can provide managers with better information about resource capabilities, existing problems, and sensitive areas before plans are made and projects are proposed. This analysis develops a logical foundation for decision making.

Reform of the 1872 Mining Act and greater application of California's Surface Mining and Reclamation Act to smaller claims could improve many isolated problems associated with mining and prevent future adverse impacts. Laws relating to liability that prevent rehabilitation of abandoned mines need to be modified, and funding must be generated to clean up problem mines. Mining of sand and gravel in streams of the Sierra Nevada should be directed toward reservoir deltas, despite the increase in transportation cost. Public agencies should set an example by using reservoir sediments as a source of aggregate whenever possible and avoiding chemical use in surface waters. The Department of Fish and Game (which administers streambed alteration agreements) might be able to negotiate agreements between reservoir operators and aggregate miners and users.

Modern information on sediment yields is needed to determine whether sedimentation has increased as a result of land-management activities. The California Department of Water Resources' Division of Dam Safety, the U.S. Geological Survey, and the U.S. Natural Resources Conservation Service might be the appropriate agencies to cooperatively administer a program of routine reservoir sediment surveys.

More efficient means of monitoring hydrologic impacts from land-management activities need to be explored at the operational field level and at the institutional level. Existing programs do not seem to provide the information necessary to evaluate how stream-flow regimes or water quality attributes are changing as a result of changes in land use. Current monitoring is also not adequate to determine whether restoration activities, including postfire treatments, are effective and appropriate. Maintenance and improvement of the snow survey, snow sensor, and climate station network is essential to management of water resources and detection of climate trends. Basic data collection programs to generate stream-flow, water quality, and climate information need to have long-term support to be worthwhile.

Now that the forest road network is largely complete, more attention should be focused on maintenance, relocation, upgrading, and decommissioning of roads by the engineering staffs of the national forests. Resource staffs have already identified many of the specific problems in the road network that need attention. Road construction budgets have been high in the past, and adequate funding is necessary to maintain, improve, and reduce the existing road system to minimize its aquatic impacts.

As foothill communities continue to grow, conversion from individual septic systems (and individual wells, in some cases) to community systems will be necessary to avoid cumulative impacts on local water quality. Construction of treatment facilities and collector systems is extremely expensive, especially where houses are far apart. The issue of who pays for such improvements is problematic. The community systems would not be necessary if not for the growth in potential pollution sources. At the same time, a community system is necessary because the capacity of the soil and ground-water system to treat household sewage is at or near its limit. Except for the service area of the Nevada Irrigation District, Calaveras County Water District, and a few others, foothill communities will need to develop major new sources of water or drastically reduce existing demand if they wish to continue their growth. Unless hydroelectric generating capacity is added when developing new sources of supply, project financing and end-user water rates may be serious constraints on new projects. Purchase of existing facilities (now largely owned by the Pacific Gas and Electric Company) by small communities and water agencies may be an increasing trend in attempts to augment community water supplies.

With all changes in land use and other disturbances, proximity to streams is a critical influence on the aquatic impacts of the activity. Simply minimizing disturbance of vegetation and soils near streams and conscientious application of best management practices for erosion control have the potential for reducing sediment problems. This locational emphasis is especially important with respect to grazing. Overgrazed riparian areas need substantial rest to adequately recover from past problems. Allowing such recovery means minimizing the presence of livestock and other disturbances in riparian zones on a continuing basis.

Management of forest fuels to reduce the risk of catastrophic fire must include thorough consideration of aquatic impacts and mitigation measures. If a major program of fuels treatment is started, a dedicated team of soil scientists, hydrologists, and aquatic ecologists should be involved in the planning and execution of such a program on local administrative units. A team of specialists, on either a zone or regional level, is also needed to monitor and evaluate the long-term effects of postfire treatments. Their experience could develop a rational set of best management practices for dealing with burned landscapes.

Prevention of further degradation and correction of existing water-related problems is expensive, as the Lake Tahoe experience has demonstrated. Rehabilitation of forest roads and restoration of degraded streams will require substantial investment. The forests of the Sierra Nevada contain three resources of substantial economic value to society: water, timber, and recreational opportunities. Some of their value in the marketplace could be returned to their sources and used to improve the conditions favorable to their production. Because the benefits of water from the Sierra Nevada contribute to so many aspects of California's economy, creative means of reinvesting a portion of those benefits into the watersheds need to be explored.

ACKNOWLEDGMENTS

Most members and associates of the SNEP team contributed to this chapter through comments, discussions, suggestions, contacts, reviews of drafts, and material in their own chapters. In particular, I wish to thank Peter Moyle, Don Erman, Bruce McGurk, Neil Berg, Jeff Dozier, Susan Ustin, Larry Costick, Debbie Elliott-Fisk, Rowan Rowntree, Bill Stewart, Jeff Romm, Roger Poff, Mike Diggles, Hap Dunning, Matt Kondolf, Roland Knapp, Doug Leisz, and Connie Millar for their help and insights. Karen Gabriel, Lian Duan, Steve Beckwitt, Paul Randall, and Russ Jones provided much analytical assistance with geographic information. Jen Lucas, Erin Fleming, Mike Oliver, Cindy Seaman, and Sue Enos kept the project going with their behind-the-scenes logistics. Mignon Moskowitz, Virginia Rich, and Zipporah Collins greatly improved the readability of the manuscript. Sarah Marvin and Maureen Davis obtained data that I was unable to access.

Beyond the greater SNEP team, hydrologists and other resource professionals on all the national forests of the Sierra Nevada were very helpful in providing suggestions, information, and access to their files. Ken Roby, Jim Frazier, Luci McKee, Terry Kaplan-Henry, and Bob Gecy were particularly helpful in this regard. Many other people involved with water issues in the Sierra Nevada provided information and assistance. The following list includes only a sample of those who were most helpful: Jane Baxter, Clay Brandow, Bob Curry, Gayle Dana, Gary Freeman, George Ice, Donna Lindquist, Brett Matzke, Sally Miller, John Munn, Robert Nuzum, Randall Osterhuber, Doug Powell, Terry Russi, Tom Suk, Darrell Wong, and Sue Yee.

REFERENCES

Adams, P. W. 1993. *Maintaining woodland roads*. Extension Circular 1137. Corvallis: Oregon State University, Extension Service.

Adams, P. W., and H. A. Froehlich. 1981. *Compaction of forest soils*. Pacific Northwest Extension Publication 217. Corvallis: Oregon State University.

Aguado, E. D., D. Cayan, L. Riddle, and M. Roos. 1992. Climatic fluctuations and the timing of west coast streamflow. *Journal of Climate* 5:1468–83.

Akers, J. P. 1986. *Ground water in the Long Meadow area and its relation with that in the General Sherman Tree area, Sequoia National Park, California*. Water-Resources Investigations Report 85-4178. Sacramento, CA: U.S. Geological Survey.

Albin-Smith, T., and C. A. Raguse. 1984. *Environmental effects of land use and intensive range management: A northern California example*. Contribution 187 ISSN 0575-4941. Davis: University of California, Water Resources Center.

Allen, H. L. 1987. Forest fertilizers. *Journal of Forestry* 85:37–46.

American Meteorological Society. 1992. Planned and inadvertent weather modification—a policy statement. *Bulletin of the American Meteorological Society* 73 (3): 1–4.

Anderson, H. W. 1963. *Managing California's snow zone lands for water*. Research Paper PSW-6. Berkeley, CA: U.S. Forest Service, Pacific Southwest Forest and Range Experiment Station.

———. 1974. Sediment deposition in reservoirs associated with rural roads, forest fires, and catchment attributes. In *Proceedings of the symposium on man's effect on erosion and sedimentation*, 87–95. Publication 113. Wallingford, England: International Association of Hydrological Sciences.

———. 1979. Sources of sediment induced reduction in water quality appraised from catchment attributes and land use. In *Third World Congress on Water Resources*. Urbana, IL: International Water Resource Association.

Anderson, H. W., and C. H. Gleason. 1960. *Logging and brush removal effects on runoff from snow cover*, 478–89. Publication 51. Wallingford, England: International Association of Scientific Hydrology.

Anderson, H. W., M. D. Hoover, and K. G. Reinhart. 1976. *Forests and water: Effects of forest management on floods, sedimentation, and water supply*. General Technical Report PSW-18. Berkeley, CA: U.S. Forest Service, Pacific Southwest Forest and Range Experiment Station.

Andrews, E. D., and D. C. Erman. 1986. Persistence in the size distribution of surficial bed material during an extreme snowmelt flood. *Water Resources Research* 22:191–97.

Armstrong, C. F., and C. K. Stidd. 1967. A moisture balance profile on the Sierra Nevada. *Journal of Hydrology* 5:258–68.

Aubury, L. E. 1910. *Gold dredging in California.* Bulletin 57. San Francisco: California State Mining Bureau.

Averill, C. V. 1946. *Placer mining for gold in California.* Bulletin 135. San Francisco: California Division of Mines.

Bailey, P. 1927. Summary report of the water resources of California and a coordinated plan for their development. Bulletin 12. Sacramento: California Division of Irrigation and Engineering.

Barta, A. F., P. R. Wilcock, and C. C. C. Shea. 1994. The transport of gravels in boulder-bed streams. In *Hydraulic engineering '94,* edited by G. V. Cotroneo and R. R. Rumer, vol. 2, 780–84. New York: American Society of Civil Engineers.

Bastin, E. W., K. J. Maier, and A. W. Knight. 1992. The recovery of a stream benthic community due to the Walker mine remediation project. In *Proceedings of the fourth biennial conference on watershed management,* 71. Davis: University of California, Water Resources Center and Watershed Management Council.

Bender, M. D. 1994. *Summary of information for estimating water quality conditions downstream of Lake Tahoe for the TROA DEIS/DEIR.* Denver: U.S. Bureau of Reclamation, Earth Sciences Division, Water Quality Section.

Beniston, M., ed. 1994. *Mountain environments in changing climates.* London: Routledge.

Berg, N. H., and J. A. Bergman. 1984. Roadway salting effects on snowmelt water quality. In *Water: Today and tomorrow,* 237–46. New York: American Society of Civil Engineers, Irrigation and Drainage Division.

Berg, N. H., and J. L. Smith. 1980. *An overview of societal and environmental responses to weather modification.* Vol. 5 of *The Sierra ecology project.* Denver: U.S. Bureau of Reclamation, Office of Atmospheric Water Resources Research.

Berg, N. H., K. B. Roby, and B. J. McGurk. 1996. Cumulative watershed effects: Applicability of available methodologies to the Sierra Nevada. In *Sierra Nevada Ecosystem Project: Final report to Congress,* vol. III. Davis: University of California, Centers for Water and Wildland Resources.

Bergman, J. A. 1987. Rain-on-snow and soil mass failure in the Sierra Nevada of California. In *Landslide activity in the Sierra Nevada during 1982 and 1983,* edited by J. V. De Graff, 15–26. San Francisco: U.S. Forest Service.

Berris, S. N., and R. D. Harr. 1987. Comparative snow accumulation and melt during rainfall in forested and clear-cut plots in the western Cascades of Oregon. *Water Resources Research* 23:135–42.

Beschta, R. L. 1978. Long-term patterns of sediment production following road construction and logging in the Oregon coast range. *Water Resources Research* 14:1011–16.

Beschta, R. L., et al. 1995. *Wildfire and salvage logging: Recommendations for ecologically sound post-fire salvage logging and other post-fire treatment on federal lands in the West.* Corvallis: Oregon State University.

Beven, K. 1981. The effects of ordering on the geomorphic effectiveness of hydrologic events. In *Erosion and sediment transport in Pacific Rim steeplands,* 510–15. Wallingford, England: International Association of Hydrological Sciences.

BioSystems Analysis, Inc. 1990. *Base-line analysis of riparian vegetation: Mill Creek, Mono County, California.* Rosemead: Southern California Edison Company.

Biswell, H. H. 1989. *Prescribed burning in California wildlands vegetation management.* Berkeley and Los Angeles: University of California Press.

Booker, F. A., W. E. Dietrich, and L. M. Collins. 1993. Runoff and erosion after the Oakland firestorm: Expectations and observations. *California Geology* 46 (6): 159–73.

Booth, D. B. 1990. Stream-channel incision following drainage-basin urbanization. *Water Resources Bulletin* 26:407–17.

Booth, D. B., and L. E. Reinelt. 1993. Consequences of urbanization on aquatic systems—measured effects, degradation thresholds, and corrective strategies. In *Watershed '93: A national conference on watershed management,* 545–50. Report EPA 840-R-94-002. Washington, DC: U.S. Environmental Protection Agency.

Borcalli and Associates. 1993. *Draft county water resources development and management plan.* Placerville, CA: El Dorado County Water Agency.

Borchers, J. W., S. Hickman, and G. J. Nimz. 1993. In-situ stress and ground-water flow in fractured granite at Wawona, Yosemite National Park, California: A model for the west-central Sierra Nevada. *EOS, Transactions American Geophysical Union* 74 (43): 581.

Bosch, J. M., and J. D. Hewlett. 1982. A review of catchment experiments to determine the effect of vegetation changes on water yield and evapotranspiration. *Journal of Hydrology* 55:3–23.

Bowen, E. 1972. *The High Sierra.* New York: Time-Life Books.

Broecker, W. S. 1995. Chaotic climate. *Scientific American* 267:62–68.

Brown, C. B. 1945. *Rates of sediment production in the southwestern United States.* Report SCS-TP-S8. Washington, DC: Soil Conservation Service.

Brown, C. B., and E. M. Thorp. 1947. *Reservoir sedimentation in the Sacramento–San Joaquin drainage basins, California.* Special Report 10. Washington, DC: Soil Conservation Service, Sedimentation Section, Office of Research.

Brown, D. L., C. M. Skau, J. Rhodes, and W. Melgin. 1990. Nitrate cycling in subalpine watershed near Lake Tahoe. In *Proceedings, international mountain watershed symposium: Subalpine processes and water quality,* edited by I. G. Poppoff, C. R. Goldman, S. L. Loeb, and L. B. Leopold, 147–59. South Lake Tahoe, CA: Tahoe Resource Conservation District.

Brown, G. W. 1980. *Forestry and water quality.* Corvallis, OR: OSU Book Stores.

Brown, G. W., A. R. Gahler, and R. B. Marston. 1973. Nutrient losses after clear-cut logging and slash burning in the Oregon coast range. *Water Resources Research* 9:1450–53.

Brown, T. C., and D. Binkley. 1994. *Effect of management on water quality in North American forests.* General Technical Report RM-248. Fort Collins, CO: U.S. Forest Service, Rocky Mountain Research Station.

Browne, J. R. 1868. *Mineral resources of the West.* Cited by Coleman 1952.

Bull, W. B., and K. M. Scott. 1974. Impact of gravel mining from urban streambeds in the southwestern United States. *Geology* 2:171–78.

Burgess, S. O. 1992. *The water king Anthony Chalfant: His life and times.* Davis, CA: Panorama West Publications.

Burns, J. W. 1970. Spawning bed sedimentation surveys in northern California streams. *California Fish and Game* 56 (4): 253–270.

———. 1972. Some effects of logging and associated road construction on northern California streams. *Transactions of the American Fisheries Society* 101 (1): 1–17.

Buttle, J. M., and F. Xu. 1988. Snowmelt runoff in suburban environments. *Nordic Hydrology* 19:19–40.

Cahill T. A., J. J. Carroll, D. Campbell, and T. E. Gill, 1996. Air quality. In *Sierra Nevada Ecosystem Project: Final report to Congress*, vol. II, chap. 48. Davis: University of California, Centers for Water and Wildland Resources.

Calaveras Ranger District. 1991. *Environmental assessment for snowmaking system development, Bear Valley Ski Area*. Sonora, CA: U.S. Forest Service, Stanislaus National Forest.

California Department of Public Works. 1923. Flow in California streams. Sacramento: California Department of Public Works.

California Department of Water Resources. 1974. *Water quality investigation of western Nevada County*. Sacramento: California Department of Water Resources, Central District.

———. 1983a. *Status of Sierra foothills water management studies*. Sacramento: California Department of Water Resources, Central District.

———. 1983b. *Potential for salvaging water through vegetation management in the lower Feather River watershed*. Sacramento: California Department of Water Resources.

———. 1989. *Surface water quality data evaluation for selected streams in Central District*. Sacramento: California Department of Water Resources.

———. 1990a. *Mountain counties water management studies: Amador County*. Sacramento: California Department of Water Resources, Central District.

———. 1990b. *Natural radioactivity in groundwater of the western Sierra Nevada*. Fresno: California Department of Water Resources, San Joaquin District.

———. 1994. *California water plan update*. Bulletin 160-93. Sacramento: California Department of Water Resources.

California Division of Forestry. 1972. *Wildland soils, vegetation, and activities affecting water quality*. Report for the State Water Resources Control Board. Sacramento: California Division of Forestry.

California Division of Soil Conservation. 1969. *Sedimentation and erosion in the upper Truckee River and Trout Creek watersheds, Lake Tahoe, California*. Sacramento: California Resources Agency, Division of Soil Conservation.

———. 1971a. *Problems of the soil mantle and vegetative cover of the state of California*. Sacramento: California Resources Agency, Division of Soil Conservation.

———. 1971b. *Environmental impact of urbanization of the foothill and mountainous land of California*. Sacramento: California Resources Agency, Division of Soil Conservation.

California Division of Water Resources. 1955. *Survey of mountainous areas*. Bulletin 56. Sacramento: California Resources Agency, Division of Water Resources.

California Energy Commission. 1981. *Small scale hydro: Environmental assessment of small hydroelectric development at existing sites in California*. Sacramento: California Energy Commission.

California Rivers Assessment. 1995. *Rivers assessment progress report*. Davis: University of California.

California State Lands Commission. 1993. *California's rivers: A public trust report*. Sacramento: California State Lands Commission.

California State Water Resources Control Board. 1984. *Water quality assessment for water years 1982 and 1983*. Water Quality Monitoring Report 84-3T5. Sacramento: California State Water Resources Control Board.

———. 1992a. *Water quality assessment*. Sacramento: California State Water Resources Control Board.

———. 1992b. *California report on water quality*. Sacramento: California State Water Resources Control Board.

Cawley, K. 1991. *Cumulative watershed effects in the Last Chance Creek watershed*. Milford, CA: U.S. Forest Service, Plumas National Forest, Milford Ranger District.

Cayan, D. R., and L. G. Riddle. 1993. A multi-basin seasonal streamflow model for the Sierra Nevada. In *Proceedings of the ninth Pacific climate workshop*, edited by K. T. Redmond and V. L. Tharp, 141–52. Sacramento: California Department of Water Resources, Interagency Ecological Studies Program.

Central Valley Regional Water Pollution Control Board. 1957. *Water pollution study, San Joaquin River watershed*. Sacramento, CA: Central Valley Regional Water Pollution Control Board.

Central Valley Regional Water Quality Control Board. 1975. *Central Valley basin plan*. Sacramento, CA: Central Valley Regional Water Quality Control Board.

———. 1987. *Regional mercury assessment*. Sacramento, CA: Central Valley Regional Water Quality Control Board.

———. 1991. *The water quality control plan (basin plan) for the Central Valley region*. Sacramento, CA: Central Valley Regional Water Quality Control Board.

———. 1995. Water quality monitoring of livestock grazing areas, summer 1994. Internal report. Central Valley Regional Water Quality Control Board, Sacramento, CA.

Chalfant, W. A. 1922. *The story of Inyo*. Bishop, CA: Chalfant Press.

Chan, F. J., and R. M. Wong. 1989. Reestablishment of native riparian species at an altered high elevation site. In *Proceedings of the California riparian systems conference*, edited by D. L. Abell, 428–35. General Technical Report PSW-110. Berkeley, CA: U.S. Forest Service, Pacific Southwest Forest and Range Experiment Station.

Chang, C. 1996. Ecosystem responses to fire and variations in fire regimes. In *Sierra Nevada Ecosystem Project: Final report to Congress*, vol. II, chap. 39. Davis: University of California, Centers for Water and Wildland Resources.

Chang, C. C. Y., J. S. Kuwabara, and S. P. Pasilis. 1990. Trace metal concentrations of three tributaries to Lake Tahoe, California and Nevada. In *Proceedings, international mountain watershed symposium: Subalpine processes and water quality*, edited by I. G. Poppoff, C. R. Goldman, S. L. Loeb, and L. B. Leopold, 103–15. South Lake Tahoe, CA: Tahoe Resource Conservation District.

Chang, H. H., L. L. Harrison, W. Lee, and S. Tu. 1994. Numerical modeling for sediment-pass-through operations of reservoirs. In *Hydraulic engineering '94*, edited by G. V. Cotroneo and R. R. Rumer, 1014–19. New York: American Society of Civil Engineers.

Cheng, J. D. 1980. Hydrologic effects of a severe forest fire. In *Symposium on watershed management, 1980*, 240–51. New York: American Society of Civil Engineers.

Chisholm, G. 1994. Building consensus for environmental restoration: The case of the Truckee-Carson Settlement Act. In *Effects of human induced changes on hydrologic systems*, edited by R. Marston and V. R. Hasfurther, 299–304. Bethesda, MD: American Water Resources Association.

Clark, W. B. 1970. *Gold districts of California*. Bulletin 193. San Francisco: California Division of Mines and Geology.

Clark, W. B., and P. A. Lydon. 1962. *Mines and mineral resources of Calaveras County, California*. County Report 2. San Francisco: California Division of Mines and Geology.

Clifton, C. 1992. *Stream classification and channel condition survey, with an inventory of sediment sources from roads and stream crossings conducted in the Last Chance and Spanish Creek watersheds.* Quincy, CA: U.S. Forest Service, Plumas National Forest.

———. 1994. *East Branch North Fork Feather River erosion control strategy.* Quincy, CA: East Branch North Fork Feather River Coordinated Resource Management Group.

Coats, R., and C. Goldman. 1993. Nitrate transport in subalpine streams, Lake Tahoe Basin, California-Nevada, U.S.A. *Applied Geochemistry,* Supplemental Issue 2, 17–21.

Coats, R. N., R. L. Leonard, and C. R. Goldman. 1976. Nitrogen uptake and release in a forested watershed, Lake Tahoe Basin, California. *Ecology* 57:995–1004.

Coleman, C. M. 1952. *Pacific Gas and Electric of California: The centennial story of Pacific Gas and Electric Company 1852–1952.* New York: McGraw-Hill.

Collins, B., and T. Dunne. 1990. *Fluvial geomorphology and river-gravel mining: A guide for planners.* Special Publication 98. Sacramento: California Division of Mines and Geology.

Colman, E. A. 1955. Operation wet blanket: Proposed research in snowpack management in California. Paper presented at the meeting of the American Geophysical Union, San Francisco, 5 February.

Committee on Rangeland Classification. 1994. *Rangeland health: New methods to classify, inventory, and monitor rangelands.* Washington, DC: National Research Council, National Academy of Sciences.

Conservation Commission. 1913. *Report of the Conservation Commission of the State of California.* Sacramento: California Conservation Commission.

Copeland, O. L. 1965. Land use and ecological factors in relation to sediment yields. In *Proceedings, federal interagency sedimentation conference, 1963,* 72–84. Washington, DC: U.S. Department of Agriculture.

Davis, G. A. 1977. Management alternatives for the riparian habitat in the Southwest. In *Importance, preservation, and management of the riparian habitat: A symposium,* edited by R. R. Johnson and D. A. Jones, 59–67. General Technical Report RM-43. Fort Collins, CO: U.S. Forest Service, Rocky Mountain Forest and Range Experiment Station.

Davis, M. L. 1993. *Rivers in the desert: William Mulholland and the inventing of Los Angeles.* New York: Harper-Collins.

Davis, S. P. 1969. *Pollution of Leviathan and Bryant Creeks, and East Fork Carson River, caused by Leviathan mine, Alpine County.* Sacramento: California Department of Fish and Game.

De Bano, L. F. 1981. *Water repellent soils: A state-of-the-art.* General Technical Report PSW-46. Berkeley, CA: U.S. Forest Service, Pacific Southwest Forest and Range Experiment Station.

De Decker, M. 1966. *Mines of the eastern Sierra.* Glendale, CA: La Siesta Press.

De Graff, J. V. 1982. *Final evaluation of felled trees as a sediment retaining measure, Rock Creek Burn, Kings River RD.* Fresno, CA: U.S. Forest Service, Sierra National Forest.

———. 1985. Using isopleth maps of landslide deposits as a tool in timber sale planning. *Bulletin of Association of Engineering Geologists* 22:445–53.

———. 1987. An overview of landslide hazard on national forests in the Sierra Nevada. In *Landslide activity in the Sierra Nevada during 1982 and 1983,* edited by J. V. De Graff, 1–14. San Francisco: U.S. Forest Service.

De Graff, J. V., J. McKean, P. E. Watanabe, and W. F. McCaffrey. 1984. Landslide activity and groundwater conditions: Insights from a road in the central Sierra Nevada, California. *Transportation Research Record* 965:32–37.

Dendy, F. E., and W. A. Champion. 1978. *Sediment deposition in U.S. reservoirs, summary of data reported through 1975.* Miscellaneous publication 1362. Oxford, MS: U.S. Department of Agriculture, Agriculture Research Service.

Dendy, F. E., W. A. Champion, and R. B. Wilson. 1973. Reservoir sedimentation surveys in the United States. In *Man-made lakes: Their problems and environmental effects,* edited by W. C. Ackermann, G. F. White, E. B. Worthington, and J. L. Ivens, 349–57. Washington, DC: American Geophysical Union.

Desmarais, S. 1977. *Locational research inventory of historic and inactive mines.* Placerville, CA: Sierra Planning/Sierra Economic Development District.

Diggles, M. F. , J. R. Rytuba, B. C. Moring, C. T. Wrucke, D. P. Cox, S. Ludington, R. P. Ashley, W. J. Pickthorn, C. T. Hillman, and R. J. Miller. 1996. Geology and minerals issues. In *Sierra Nevada Ecosystem Project: Final report to Congress,* vol. II, chap. 18. Davis: University of California, Centers for Water and Wildland Resources.

Dilsaver, L. M., and W. C. Tweed. 1990. *Challenge of the big trees.* Three Rivers, CA: Sequoia Natural History Association.

Dissmeyer, G. E. 1993. The economics of silvicultural best management practices. In *Watersheds '93: A national conference on watershed management,* 319–23. Report EPA 840-R-94-002. Washington, DC: U.S. Environmental Protection Agency.

Doppelt, B., M. Scurlock, C. Frissell, and J. Karr. 1993. *Entering the watershed: A new approach to save America's river ecosystems.* Washington, DC: Pacific Rivers Council/Island Press.

Duane, T. P. 1996a. Human settlement, 1850–2040. In *Sierra Nevada Ecosystem Project: Final report to Congress,* vol. II, chap. 11. Davis: University of California, Centers for Water and Wildland Resources.

———. 1996b. Recreation in the Sierra. In *Sierra Nevada Ecosystem Project: Final report to Congress,* vol. II, chap. 19. Davis: University of California, Centers for Water and Wildland Resources.

Dunne, T., and L. B. Leopold. 1978. *Water in environmental planning.* San Francisco: W. H. Freeman.

Dupras, D., and J. Chevreaux. 1984. Lake Combie specialty sands and gravels. *California Geology* 37:255–57.

EA Engineering, Science, and Technology. 1990. *Preliminary feasibility analysis of alternative sediment management option for Ralston Afterbay Reservoir.* Report for Placer County Water Agency. Lafayette, CA: EA Engineering, Science, and Technology.

Eaglin, G. S., and W. A. Hubert. 1993. Effects of logging and roads on substrate and trout in streams of the Medicine Bow National Forest, Wyoming. *North American Journal of Fisheries Management* 13: 844–46.

East Bay Municipal Utility District (EBMUD). 1995. Letter from R. C. Nuzum to Mokelumne River Joint Ownership Protocol Group, October 10. East Bay Municipal Utility District, Oakland, California.

Eisel, L. M., K. D. Mills, and C. F. Leaf. 1988. Estimated consumptive loss from man-made snow. *Water Resources Bulletin* 24:815–20.

Eldorado National Forest. 1980. *Status report on the stream flow maintenance dams.* Placerville, CA: U.S. Forest Service, Eldorado National Forest.

Elliott, R. D., D. A. Griffith, J. F. Hannaford, and J. A. Flueck. 1978. *Special report on background and supporting material for the Sierra cooperative pilot project.* Report 78-22, vol. 2. Salt Lake City: North American Weather Consultants.

Ellis, W. T. 1939. *Memories: My seventy-two years in the romantic county of Yuba.* Eugene: University of Oregon.

Ellison, W. D. 1945. Some effects of raindrops and surface flow on soil erosion and infiltration. *Transactions American Geophysical Union* 26:415–30.

Erman, D. C., E. D. Andrews, and M. Yodler-Williams. 1988. Effects of winter floods on fishes in the Sierra Nevada. *Canadian Journal of Fisheries and Aquatic Sciences* 45:2195–200.

Erman, D. C., and V. M. Hawthorne. 1976. The quantitative importance of an intermittent stream in the spawning of rainbow trout. *Transactions of the American Fisheries Society* 105:675–81.

Erman, D. C., and D. Mahoney. 1983. *Recovery after logging in streams with and without bufferstrips in northern California.* Contribution 186. Berkeley: University of California, Water Resources Center.

Erman, D. C., J. D. Newbold, and K. B. Roby. 1977. *Evaluation of streamside bufferstrips for protecting aquatic organisms.* Contribution 165. Davis: University of California, Water Resources Center.

Erman, N. A. 1996. Status of aquatic invertebrates. In *Sierra Nevada Ecosystem Project: Final report to Congress,* vol. II, chap. 35. Davis: University of California, Centers for Water and Wildland Resources.

Euphrat, F. D. 1992. Cumulative impact assessment and mitigation for the Middle Fork of the Mokelumne River, Calaveras County, California. Ph.D. diss., University of California, Berkeley.

Everett, R. L., R. O. Meeuwig, and R. I. Buttefield. 1980. Revegetation of untreated acid spoils, Leviathan mine, Alpine County, California. *California Geology* 32 (1): 8–10.

Farquhar, F. 1965. *History of the Sierra Nevada.* Berkeley and Los Angeles: University of California Press.

Federal Energy Regulatory Commission. 1990. *Environmental assessment of potential cumulative impacts associated with hydropower development in the Mono Lake Basin, California.* Washington, DC: Federal Energy Regulatory Commission, Public Reference Branch.

Fleischner, T. L. 1994. Ecological costs of livestock grazing in western North America. *Conservation Biology* 8:629–44.

Fong, D. R. 1991. Long-term logging-related influences on stream habitat and macroinvertebrate communities in northern California. Master's thesis, Department of Forestry and Resource Management, University of California, Berkeley.

Fowler, F. H. 1923. *Hydroelectric systems of California and their extensions into Oregon and Nevada.* Water Supply Paper 493. Washington, DC: U.S. Geological Survey.

Franklin, J. F., W. H. Moir, G. W. Douglas, and C. Wiberg. 1971. Invasion of subalpine meadows by trees in the Cascade Range, Washington and Oregon. *Arctic and Alpine Research* 3 (3): 215–24.

Frazier, J. W. 1984. *The Granite Burn: The fires and the years following, a watershed history, 1974–1984.* Sonora, CA: U.S. Forest Service, Stanislaus National Forest.

Frazier, J., and J. Carlson. 1991. *Surface and ground water quality monitoring protocol for herbicide application.* Appendix G, Groveland conifer release environmental assessment. Sonora, CA: U.S. Forest Service, Stanislaus National Forest.

Fredriksen, R. L. 1971. Comparative chemical water quality—natural and disturbed streams following logging and slash burning. In *Proceedings of a symposium on forest land uses and stream environment,* 125–37. Corvallis: Oregon State University.

Freeman, J. R. 1912. *On the proposed use of a portion of the Hetch Hetchy, Eleanor and Cherry Valleys.* San Francisco: Board of Supervisors.

Furniss, M. J., T. D. Roelofs, and C. S. Yee. 1991. Road construction and maintenance. In *Influences of forest and rangeland management on salmonid fishes and their habitats,* edited by W. R. Meehan, 297–323. Bethesda, MD: American Fisheries Society.

Galay, V. 1983. Causes of river bed degradation. *Water Resources Research* 19:1057–90.

Gard, M. F. 1994. Biotic and abiotic factors affecting native fishes in the South Yuba River, Nevada County, California. Ph.D. diss., University of California, Davis.

Garland, J. J. 1993. *Designing woodland roads.* Extension Circular 1137. Corvallis: Oregon State University, Extension Service.

Georgetown Ranger District. 1992. *Placer County Water Agency sediment storage site environmental assessment.* Placerville, CA: U.S. Forest Service, Eldorado National Forest.

Gerstung, E. 1970. *A brief survey of the impact of subdivision activity on the fish and wildlife resources of Nevada County.* Sacramento: California Department of Fish and Game.

Gilbert, G. K. 1917. *Hydraulic-mining debris in the Sierra Nevada.* Professional Paper 105. Washington, DC: U.S. Geological Survey.

Glancy, P. A. 1969. *A mudflow in the Second Creek drainage, Lake Tahoe Basin, Nevada, and its relation to sedimentation and urbanization.* Professional Paper 650-C. Carson City, NV: U.S. Geological Survey.

Gleick, P. H., P. Loh, S. V. Gomez, and J. Morrison. 1995. *California water 2020: A sustainable vision.* Oakland, CA: Pacific Institute.

Goldman, C. R. 1974. *Eutrophication of Lake Tahoe emphasizing water quality.* Report EPA–660/3-74-034. Washington, DC: U.S. Government Printing Office.

———. 1990. Long-term limnological research at Lake Tahoe. In *Proceedings, international mountain watershed symposium: Subalpine processes and water quality,* edited by I. G. Poppoff, C. R. Goldman, S. L. Loeb, and L. B. Leopold, 464–77. South Lake Tahoe, CA: Tahoe Resource Conservation District.

Goldman, C. R., R. Gersberg, and S. Loeb. 1984. *Water quality conditions of Kirkwood Basin.* Davis, CA: Ecological Research Associates.

Goldman, C. R., and G. J. Malyj. 1990. *The environmental impact of highway deicing.* Davis: University of California, Institute of Ecology.

Goldman, H. B. 1968. *Sand and gravel in California.* Bulletin 180. Sacramento: California Division of Mines and Geology.

Granger, O. 1979. Increasing variability in California precipitation. *Annals of the Association of American Geographers* 69:533–43.

Graumlich, L. 1993. A 1000-year record of temperature and precipitation in the Sierra Nevada. *Quaternary Research* 39:249–55.

Groeneveld, D. P., and D. Or. 1994. Water table induced shrub-herbaceous ecotone: Hydrologic management implications. *Water Resources Bulletin* 30:911–20.

Hagan, R. M., and E. B. Roberts. 1973. Ecological impacts of water storage and diversion projects. In *Environmental quality and water development,* edited by C. R. Goldman, J. McEvoy, and P. J. Richerson, 196–215. San Francisco: W. H. Freeman.

Hagberg, T. 1993. *Sierra National Forest preliminary V* findings.* Fresno, CA: U.S. Forest Service, Sierra National Forest, Kings River Ranger District.

———. 1995. Relationships between hydrology, vegetation, and gullies in montane meadows of the Sierra Nevada. Master's thesis, Humboldt State University, Arcata, CA.

Hall, W. H. 1881. *The irrigation question in California, report of the State Engineer.* Sacramento, CA: Office of the State Engineer.

Hammermeister, O. P., and S. J. Walmsley. 1985. *Hydrologic data for*

Leviathan mine and vicinity, Alpine County, California, 1981–83. Open File Report 85-160. Carson City, NV: U.S. Geological Survey.

Harding, S. T. 1935. Evaporation from large water-surfaces based on records in California and Nevada. *Transactions American Geophysical Union* 16:507–11.

———. 1960. *Water in California.* Palo Alto: N-P Publications.

Harland Bartholomew and Associates, Engineering-Science, and James R. Jones and Associates. 1992. *Public and private water systems and groundwater for the Nevada County General Plan.* Nevada City, CA: Nevada County.

Harp, E. L., K. Tanako, J. Sarmiento, and D. K. Keefer. 1984. *Landslides from the May 25–27, 1980, Mammoth Lakes, California, earthquake sequence.* Miscellaneous Investigations Map I-1612. Menlo Park, CA: U.S. Geological Survey.

Harr, R. D. 1981. Some characteristics and consequences of snowmelt during rainfall in western Oregon. *Journal of Hydrology* 53:277–304.

Harr, R. D., and R. A. Nichols. 1993. Stabilizing forest roads to help restore fish habitats: A northwest Washington example. *Fisheries* 18 (4): 18–22.

Harrison, L. L. 1992. *Rock Creek–Cresta sediment management project.* San Francisco: Pacific Gas and Electric Company.

Harvey, B. C. 1986. Effects of suction dredging on fish and invertebrates in two California streams. *American Journal of Fisheries Management* 6:401–9.

Harvey, B. C., T. E. Lisle, T. Vallier, and D. C. Fredley. 1995. *Effects of suction dredging on streams: A review and evaluation strategy.* Washington, DC: U.S. Forest Service.

Hawkins, C. P., et al. 1994. *Cumulative watershed effects: An extensive analysis of responses by stream biota to watershed management.* Final report on cooperative agreement PSW-88-0011CA. Albany, CA: U.S. Forest Service, Pacific Southwest Research Station.

Hawkins, R. H., and J. H. Judd. 1972. Water pollution as affected by street salting. *Water Resources Bulletin* 17:1246–52.

Helley, E. J. 1966. Sediment transport in the Chowchilla River Basin: Mariposa, Madera, and Merced Counties, California. Ph.D. diss., University of California, Berkeley.

Helms, J. A., and J. C. Tappeiner. 1996. Silviculture in the Sierra. In *Sierra Nevada Ecosystem Project: Final report to Congress,* vol. II, chap. 15. Davis: University of California, Centers for Water and Wildland Resources.

Helvey, J. D. 1980. Effects of a north central Washington wildfire on runoff and sediment production. *Water Resources Bulletin* 16: 627–34.

Henderson, T. J. 1995. Benefits from historic cloud seeding programs in California. *Proceedings of the western snow conference* 63:88–97.

Herbst, D., and R. Knapp. 1995a. Biomonitoring of rangeland streams under differing livestock grazing practices. *Bulletin of the North American Benthological Society* 12 (1): 176.

———. 1995b. *Evaluation of rangeland stream condition and recovery using physical and biological assessments of nonpoint source pollution.* Technical Completion Report UCAL-WRC-W-818. Davis: University of California, Water Resources Center.

Hermann, C., and F. R. McGregor, eds. 1973. *Water quality and recreation in the Mammoth Lakes Sierra.* Los Angeles: University of California, Department of Environmental Science and Engineering.

Herschy, R. W. 1985. *Streamflow measurement.* London: Elsevier.

Hewlett, J. D. 1982. *Principles of forest hydrology.* Athens: University of Georgia Press.

Heyl, G. R., M. W. Cox, and J. H. Eric. 1948. Penn zinc-copper mine, Calaveras County, California. *California Division of Mines Bulletin* 144:61–84.

Hill, B. R., and K. M. Nolan. 1990. Suspended-sediment factors: Lake Tahoe Basin, California-Nevada. In *Proceedings, international mountain watershed symposium: Subalpine processes and water quality,* edited by I. G. Poppoff, C. R. Goldman, S. L. Loeb, and L. B. Leopold, 179–89. South Lake Tahoe, CA: Tahoe Resource Conservation District.

Hinkle, G., and B. Hinkle. 1949. *Sierra Nevada lakes.* Indianapolis: Bobbs-Merrill.

Hoffman, A. 1981. *Vision or villainy: Origins of the Owens Valley–Los Angeles water controversy.* College Station: Texas A and M University Press.

Holeman, J. E. 1968. The sediment yield of major rivers of the world. *Water Resources Research* 4:737–47.

Hollett, K. J., W. R. Danskin, W. F. McCaffrey, and C. L. Walti. 1991. *Geology and water resources of Owens Valley, California.* Water Supply Paper 2370-B. Washington, DC: U.S. Geological Survey.

Hornbeck, J., and G. Stuart. 1976. When ski trails are cut through forest land, what happens to streamflow? *Ski Area Management,* fall, 34–36.

Howell, R. B., D. I. Nakao, and J. L. Gidley. 1979. *Analysis of short and long-term effects on water quality for selected highway projects.* Report FHWA/CA/TL-79/17. Sacramento: California Department of Transportation.

Hughes, J. E. 1934. *Erosion control progress report.* Quincy, CA: U.S. Forest Service, Plumas National Forest, Milford Ranger District.

Hundley, N., Jr. 1992. *The great thirst: Californians and water, 1770s–1990s.* Berkeley and Los Angeles: University of California Press.

Huntley, D. 1992. *Update of estimates of snowmaking water use/loss, Lake Tahoe and Truckee River Basins.* Davis, CA: West Yost and Associates.

Inyo National Forest. 1987. *Land and resource management plan.* Bishop, CA: U.S. Forest Service, Inyo National Forest.

———. 1993. *Interagency motor vehicle use plan revision—draft environmental impact statement.* Bishop, CA: U.S. Forest Service and Bureau of Land Management.

Jackson, W. T., and D. J. Pisani. 1973. *A core study in interstate resource management: The California-Nevada water controversy.* Contribution 142. Berkeley: University of California, Water Resources Center.

James, L. A. 1988. Historical transport and storage of hydraulic mining sediment in the Bear River, California. Ph.D. diss., University of Wisconsin, Madison.

———. 1994. Channel changes wrought by gold mining: Northern Sierra Nevada, California. In *Effects of human-induced changes on hydrologic systems,* edited by R. Marston and V. R. Hasfurther, 629–38. Bethesda, MD: American Water Resources Association.

Janda, R. J. 1966. Pleistocene history and hydrology of the upper San Joaquin River. Ph.D. diss., University of California, Berkeley.

Jansen, R. N. 1956. Sedimentation problems related to the design of the Feather River Project. In *Proceedings, conference on sediment problems in California,* edited by H. A. Einstein and J. W. Johnson, 44–46. Berkeley: University of California, Water Resources Center, Committee on Research on Water Resources.

Jenkins, O. P. 1948. *Geologic guidebook along Highway 49—Sierran gold belt, the Mother Lode country.* Bulletin 141. San Francisco: California Division of Mines.

Jennings, M. R. 1996. Status of amphibians. In *Sierra Nevada Ecosystem Project: Final report to Congress,* vol. II, chap. 31. Davis: University of California, Centers for Water and Wildland Resources.

Jobson, H. E. 1985. Field measurements. In *Methods of computing sedimentation in lakes and reservoirs*, edited by S. Bruk, 41–64. Paris: UNESCO.

Johnson, C. M., and P. R. Needham. 1966. Ionic composition of Sagehen Creek, California, following an adjacent fire. *Ecology* 47:636–39.

Johnson, P. C. 1971. *Sierra album*. Garden City: Doubleday.

Johnston, D. 1994. Wildlife and suction dredging. *Outdoor California*, July-August, 11–13.

Johnston, V. R. 1970. *Sierra Nevada*. Boston: Houghton Mifflin.

Jones, H. R. 1965. *John Muir and the Sierra Club: The battle for Yosemite*. San Francisco: Sierra Club.

Jones, J. 1991. *Truckee River atlas*. Sacramento: California Department of Water Resources.

Kahrl, W., ed. 1978. *California water atlas*. Sacramento: California Governor's Office of Planning and Research.

———. 1982. *Water and power*. Berkeley and Los Angeles: University of California Press.

———. 1993. Acquisitions and aqueducts: How California's water system evolved. *Pacific Discovery* 46 (1): 21–6.

Kattelmann, R. 1982. Water yield improvement in the Sierra Nevada snow zone: 1912–1982. *Proceedings, western snow conference* 50: 39–48.

———. 1985. Snow management at ski areas: Hydrological effects. In *Watershed management in the eighties*, 264–72. New York: American Society of Civil Engineers.

———. 1987. *Feasibility of more water from Sierra Nevada forests*. Report 16. Berkeley: University of California, Wildland Resources Center.

———. 1989a. Hydrology of four headwater basins in the Sierra Nevada. In *Proceedings of the headwaters hydrology symposium*, 141–47. Bethesda, MD: American Water Resources Association.

———. 1989b. Groundwater contributions in an alpine basin in the Sierra Nevada. In *Proceedings of the headwater hydrology symposium*, 361–69. Bethesda, MD: American Water Resources Association.

———. 1990. Floods in the high Sierra Nevada, California, USA. In *Artificial reservoirs: Water and slopes*, edited by R. O. Sinniger and M. Monbaron, 311–18. Vol. 2 of *Hydrology in mountainous regions*. Publication 194. Wallingford, England: International Association of Hydrological Sciences.

———. 1992. Historical floods in the eastern Sierra Nevada. In *The history of water in the eastern Sierra Nevada, Owens Valley and White Mountains*, edited by C. A. Hall, V. Doyle-Jones, and B. Widawski, 74–86. Berkeley and Los Angeles: University of California Press.

Kattelmann, R., and N. Berg. 1987. Water yields from high elevation basins in California. In *Proceedings, California watershed management conference*, edited by R. Z. Callaham and J. J. DeVries, 79–85. Report 11. Berkeley: University of California, Water Resources Center.

Kattelmann, R., N. Berg, and B. McGurk. 1991. A history of rain-on-snow floods in the Sierra Nevada. *Proceedings, western snow conference* 59:138–41.

Kattelmann, R., N. Berg, and J. Rector. 1983. The potential for increasing streamflow from Sierra Nevada watersheds. *Water Resources Bulletin* 19:395–402.

Kattelmann, R., and D. Dawson. 1994. Water diversions and withdrawal for municipal supply in the eastern Sierra Nevada. In *Effects of human-induced changes on hydrologic systems*, edited by R. A. Marston and V. R. Hasfurther, 475–83. Bethesda, MD: American Water Resources Association.

Kattelmann, R., and J. Dozier. 1991. Environmental hydrology of the Sierra Nevada, California, USA. *Hydrological Science and Technology* 7 (1–4): 73–80.

Kattelmann, R., and K. Elder. 1991. Hydrologic characteristics and water balance of an alpine basin in the Sierra Nevada. *Water Resources Research* 27:1553–62.

Keller, G., and A. King. 1986. 1986 storm impacts on the Feather River country. *Association of Engineering Geologists Newsletter* 29:21–22.

Kelley, R. 1959. *Gold vs. grain: The hydraulic mining controversy in California's Central Valley—a chapter in the decline of laissez-faire*. Glendale, CA: Arthur H. Clark.

———. 1989. *Battling the inland sea: American political culture, public policy, and the Sacramento Valley, 1850–1986*. Berkeley and Los Angeles: University of California Press.

King, J. G. 1993. Sediment production and transport in forested watersheds in the northern Rocky Mountains. In *Proceedings of the technical workshop on sediments*, 13–18. Washington, DC: Terrene Institute.

Kinney, W. C. 1996. Conditions of rangelands before 1905. In *Sierra Nevada Ecosystem Project: Final report to Congress*, vol. II, chap. 3. Davis: University of California, Centers for Water and Wildland Resources.

Kondolf, G. M. 1989. Stream-groundwater interactions along streams of the eastern Sierra Nevada, California: Implications for assessing potential impacts of flow diversions. In *Proceedings of the California riparian systems conference: Protection, management, and restoration for the 1990s*, edited by D. L. Abell, 352–59. Berkeley, CA: U.S. Forest Service.

———. 1993. Lag in stream channel adjustment to livestock exclosure, White Mountains, California. *Restoration Ecology* 4:226–30.

Kondolf, G. M., G. F. Cada, and M. J. Sale. 1987. Assessing flushing-flow requirement for brown trout spawning gravels in steep streams. *Water Resources Bulletin* 23:927–35.

Kondolf, G. M., G. F. Cada, M. J. Sale, and T. Felando. 1991. Distribution and stability of potential salmonid spawning gravels in steep boulder-bed streams of the eastern Sierra Nevada. *Transactions of the American Fisheries Society* 120:177–86.

Kondolf, G. M., R. Kattelmann, M. Embury, and D. C. Erman. Status of riparian habitat. In *Sierra Nevada Ecosystem Project: Final report to Congress*, vol. II, chap. 36. Davis: University of California, Centers for Water and Wildland Resources.

Kondolf, G. M., and W. V. G. Matthews. 1993. *Management of coarse sediment on regulated rivers*. Report 80. Davis: University of California, Water Resources Center.

Kondolf, G. M., and P. Vorster. 1992. Management implications of stream/groundwater interactions in the eastern Sierra Nevada, California. In *The history of water in the eastern Sierra Nevada, Owens Valley and White Mountains*, edited by C. A. Hall, V. Doyle-Jones, and B. Widawski, 324–38. Berkeley and Los Angeles: University of California Press.

Kraebel, C. J., and A. F. Pillsbury. 1934. *Handbook of erosion control in mountain meadows in the California region*. Berkeley: U.S. Forest Service California Forest and Range Experiment Station.

Krammes, J. S., and L. W. Hill. 1963. *"First aid" for burned watersheds*. Research Note PSW-29. Berkeley, CA: U.S. Forest Service, Pacific Southwest Forest and Range Experiment Station.

Kratz, K. W., S. D. Cooper, and J. M. Melack. 1994. Effects of single and repeated experimental acid pulses on invertebrates in a high altitude Sierra Nevada stream. *Freshwater Biology* 32:161–83.

Kuehn, M. H. 1987. The effects of exceeding "probable maximum precipitation" on a severely burned watershed in the Sierra Nevada of California. In *Landslide activity in the Sierra Nevada during 1982 and 1983,* edited by J. V. De Graff, 27–40. San Francisco: U.S. Forest Service.

Kunkle, S. H. 1970. Sources and transport of bacterial indicators in rural streams. In *Proceedings of the symposium on interdisciplinary aspects of watershed management,* 105–32. New York: American Society of Civil Engineers.

Lahontan Regional Water Quality Control Board. 1993. *Draft water quality control plan for the Lahontan Region.* South Lake Tahoe, CA: California Regional Water Quality Control Board, Lahontan Region.

Lake Tahoe Basin Management Unit. 1993. *Environmental assessment for the Meiss grazing allotment.* South Lake Tahoe, CA: U.S. Forest Service, Lake Tahoe Basin Management Unit.

Landers, D. H., et al. 1987. *Western lake survey phase 1, characteristics of lakes in the western United States.* EPA/600/3-86-054a. Washington, DC: Environmental Protection Agency.

Lardner, W. B., and M. J. Brock. 1924. *History of Placer and Nevada Counties, California, with biographical sketches.* Los Angeles: Historical Records Co. Cited by James (1994).

Larse, R. W. 1971. Prevention and control of erosion and stream sedimentation from forest roads. In *Proceedings of a symposium on forest land uses and stream environment,* 76–83. Corvallis: Oregon State University.

Lawton, H. W., Wilke, P. J., and W. M. Mason. 1976. Agriculture among the Paiute of Owens Valley. *Journal of California Anthropology* 3: 13–50.

Lee, W. S. 1962. *The Sierra.* New York: G. P. Putnam's Sons.

Leiberg, J. B. 1902. *Forest conditions in the northern Sierra Nevada, California.* Professional Paper 8, Series H, Forestry 5. Washington, DC: U.S. Geological Survey.

Leopold, A. 1924. Grass, brush, timber, and fire in southern Arizona. *Journal of Forestry* 22:1–10.

Leopold, L. B. 1968. *Hydrology for urban land planning: A guide book.* Circular 554. Washington, DC: U.S. Geological Survey.

———. 1994. *A view of the river.* Cambridge: Harvard University Press.

Leopold, L. B., and D. L. Rosgen. 1990. Natural morphology—key to stream channel stability. In *Proceedings, international mountain watershed symposium: Subalpine processes and water quality,* edited by I. G. Poppoff, C. R. Goldman, S. L. Loeb, and L. B. Leopold, 42–49. South Lake Tahoe, CA: Tahoe Resource Conservation District.

Likens, G. E., F. H. Bormann, N. M. Johnson, D. W. Fisher, and R. S. Pierce. 1970. Effects of forest cutting and herbicide treatment on nutrient budgets in the Hubbard Brook watershed-ecosystem. *Ecological Monographs* 40 (1): 23–47.

Lindgren, W. 1911. *Tertiary gravels of the Sierra Nevada, California.* Professional Paper 73. Washington, DC: U.S. Geological Survey.

Lindstrom, S. G. 1994. Great basin fisherfolk: Optimal diet breadth modeling to Truckee River aboriginal subsistence fishery. Ph.D. diss., Department of Anthropology, University of California, Davis.

Lisle, T. E. 1989. Sediment transport and resulting deposition in spawning gravels, north coastal California. *Water Resources Research* 25:1303–19.

Lisle, T. E., and S. Hilton. 1992. The volume of fine sediment in pools: An index of sediment supply in gravel-bed streams. *Water Resources Bulletin* 28:371–83.

Littleworth, A. L., and E. L. Garner. 1995. *California water.* Point Arena, CA: Solano Press Books.

Loeb, S. L., and C. R. Goldman. 1979. Water and nutrient transport via groundwater from Ward Valley into Lake Tahoe. *Limnology and Oceanography* 24:1146–54.

Logan, C. A. 1948. History of mining and milling methods in California. In *Geologic guidebook along Highway 49—Sierran gold belt, the Mother Lode country,* edited by O. P. Jenkins, 31–34. San Francisco: California Division of Mines.

Longacre, L. L., and H. F. Blaney. 1962. Evaporation at high elevations in California. *Journal of the Irrigation and Drainage Division of the American Society of Civil Engineers* 88:33–54.

Lowdermilk, W. C., and P. B. Rowe. 1934. Still further studies on absorption of rainfall in its relation to surficial runoff and erosion. *Transactions American Geophysical Union,* part II:509–15.

Lull, H. W., and W. E. Sopper. 1969. *Hydrologic effects from urbanization of forested watersheds.* Research Paper NE-146. Radnor, PA: U.S. Forest Service, Northeastern Forest and Range Experiment Station.

MacDonald, L. H., A. W. Smart, and R. L. Wissmar. 1991. *Monitoring guidelines to evaluate effects of forestry activities on streams in the Pacific Northwest and Alaska.* EPA/910/9-91-001. Washington, DC: Environmental Protection Agency.

Madej, M. A., W. E. Weaver, and D. K. Hagans. 1994. Analysis of bank erosion on the Merced River, Yosemite Valley, Yosemite National Park, California, USA. *Environmental Management* 18:235–50.

Mahmood, K. 1987. *Reservoir sedimentation: Impact, extent and mitigation.* Technical Paper 71. Washington, DC: World Bank.

Mahoney, D. L., and D. C. Erman. 1984. An index of stored fine sediment in gravel bedded streams. *Water Resources Bulletin* 20: 343–48.

Marchetti, M. P. 1994. Suspended sediment effects on the stream fauna of Humbug Creek. Master's thesis, University of California, Davis.

Marvin, S. 1996. Possible changes in water yield and peak flows in response to forest management. In *Sierra Nevada Ecosystem Project: Final report to Congress,* vol. III. Davis: University of California, Centers for Water and Wildland Resources.

Marvin, S. J. 1995. *Water yield in forest planning: A case study of the Sequoia National Forest.* Master's thesis, Department of Landscape Architecture, University of California, Berkeley.

May, P. R. 1970. *Origins of hydraulic mining in California.* Oakland, CA: Holmes Book Company.

McCammon, B. P. 1977. *Salmon Creek administrative study.* Porterville, CA: U.S. Forest Service, Sequoia National Forest.

McCashion, J. D., and R. M. Rice. 1983. Erosion on logging roads in northwestern California: How much is avoidable? *Journal of Forestry* 81 (1): 23–26.

McCauley, M. L. 1986. The effect of February 1986 storms on highways in central and northern California. *Association of Engineering Geologists Newsletter* 29 (4): 24–25.

McCleneghan, K., and R. E. Johnson. 1983. *Suction dredge gold mining in the Mother Lode region of California.* Administrative Report 83-1. Sacramento: California Department of Fish and Game, Environmental Services Branch.

McGurk, B. J., N. H. Berg, and M. L. Davis. 1996. Camp and Clear Creeks, El Dorado County: Predicted sediment production from forest management and residential development. In *Sierra Nevada Ecosystem Project: Final report to Congress,* vol. II, chap. 53. Davis: University of California, Centers for Water and Wildland Resources.

McGurk, B. J., and M. L. Davis. 1996. Camp and Clear Creeks, El Dorado County: Chronology and hydrologic effects of land-use change. In *Sierra Nevada Ecosystem Project: Final report to Congress,* vol. II, chap. 52. Davis: University of California, Centers for Water and Wildland Resources.

McGurk, B. J., and D. R. Fong. 1995. Equivalent roaded area as a measure of cumulative effect of logging. *Environmental Management* 19:609–21.

McKean, J. 1987. Landslide damage to the Mosquito Ridge road, Tahoe National Forest. In *Landslide activity in the Sierra Nevada during 1982 and 1983,* edited by J. V. De Graff, 56–62. San Francisco: U.S. Forest Service.

McKelvey, K. S., C. N. Skinner, C. Chang, D. C. Erman, S. J. Husari, D. Parsons, J. W. van Wagtendonk, and C. P. Weatherspoon. 1996. An overview of fire in the Sierra Nevada. In *Sierra Nevada Ecosystem Project: Final report to Congress,* vol. II, chap. 37. Davis: University of California, Centers for Water and Wildland Resources.

McPhee, J. 1993. *Assembling California.* New York: Farrar, Straus, and Giroux.

McWilliams, B., and G. Goldman. 1994. *The mineral industries in California: Their impact on the state economy.* Publication CNR 003. University of California, Division of Agriculture and Natural Resources.

Megahan, W. F. 1983. Hydrologic effects of clearcutting and wildfire on steep granitic slopes in Idaho. *Water Resources Research* 19:811–19.

———. 1992. An overview of erosion and sedimentation processes on granitic soils. In *Decomposed granitic soils: Problems and solutions,* edited by S. Sommarstrom. Davis: University of California, University Extension.

Megahan, W. F., and W. J. Kidd. 1972. Effects of logging and logging roads on erosion and sediment deposition from steep terrain. *Journal of Forestry* 7:136–41.

Megahan, W. F., and P. N. King. 1985. Identification of critical areas on forest lands for control of nonpoint sources of pollution. *Environmental Management* 9:7–18.

Megahan, W. F., and D. C. Molitor. 1975. Erosional effects of wildfire and logging in Idaho. In *Watershed management symposium,* 423–44. New York: American Society of Civil Engineers.

Melack, J. M., J. Dozier, C. R. Goldman, D. Greenland, A. M. Milner, and R. J. Naiman. In press. Effects of climate change on inland waters of the Pacific coastal mountains and western Great Basin of North America. *Hydrological Processes.*

Melack, J., and J. Sickman. 1995. Snowmelt induced chemical changes in seven streams in the Sierra Nevada, California. In *Biogeochemistry of seasonally snow-covered catchments,* edited by K. A. Tonnessen, M. W. Williams, and M. Tranter, 221–34. Publication 228. Wallingford, England: International Association of Hydrological Sciences.

Melack, J., J. Sickman, F. Setaro, and D. Dawson. 1995. Monitoring of wet deposition in alpine areas of the Sierra Nevada. Final report, contract A932-081. Sacramento: California Air Resources Board.

Melack, J., J. Sickman, F. Setaro, and D. Engle. 1993. Long-term studies of lakes and watersheds in the Sierra Nevada: Patterns and processes of surface-water acidification. Final report, contract A932-060. Sacramento: California Air Resources Board.

Melack, J. M., J. L. Stoddard, and C. A. Ochs. 1985. Major ion chemistry and sensitivity to acid precipitation of Sierra Nevada lakes. *Water Resources Research* 21 (1): 27–32.

Menke, J. W., ed. 1977. *Proceedings of the workshop on livestock and wildlife-fisheries relationships in the Great Basin.* Davis: University of California.

Menke, J. W., C. Davis, and P. Beesley. 1996. Rangeland assessment.

In *Sierra Nevada Ecosystem Project: Final report to Congress,* vol. III. Davis: University of California, Centers for Water and Wildland Resources.

Michaelson, J., L. Haston, and F. W. Davis. 1987. 400 years of central California precipitation variability reconstructed from tree rings. *Water Resources Bulletin* 23:809–18.

Milhous, R. T. 1982. Effect of sediment transport and flow regulation on the ecology of gravel-bed rivers. In *Gravel-bed rivers,* edited by R. D. Hey, J. C. Bathurst, and C. L. Thorne, 819–42. New York: John Wiley.

Millar, C. I. 1996. Tertiary vegetation history. In *Sierra Nevada Ecosystem Project: Final report to Congress,* vol. II, chap. 5. Davis: University of California, Centers for Water and Wildland Resources.

Momsen, J. H. 1996. Agriculture in the Sierra. In *Sierra Nevada Ecosystem Project: Final report to Congress,* vol. II, chap. 17. Davis: University of California, Centers for Water and Wildland Resources.

Montgomery, D. R., G. E. Grant, and K. Sullivan. 1995. Watershed analysis as a framework for implementing ecosystem management. *Water Resources Bulletin* 31:369–86.

Montoya, B. L., and X. Pan. 1992. *Inactive mine drainage in the Sacramento valley, California.* Sacramento: California Regional Water Quality Control Board, Central Valley Region.

Mount, J. F. 1995. *California rivers and streams: The conflict between fluvial processes and land use.* Berkeley and Los Angeles: University of California Press.

Moyle, P. B. 1996. Status of aquatic habitat types. In *Sierra Nevada Ecosystem Project: Final report to Congress,* vol. II, chap. 32. Davis: University of California, Centers for Water and Wildland Resources.

Moyle, P. B., and P. J. Randall. 1996. Biotic integrity of watersheds. In *Sierra Nevada Ecosystem Project: Final report to Congress,* vol. II, chap. 34. Davis: University of California, Centers for Water and Wildland Resources.

Moyle, P. B., R. M. Yoshiyama, and R. Knapp. 1996. Status of fish and fisheries. In *Sierra Nevada Ecosystem Project: Final report to Congress,* vol. II, chap. 33. Davis: University of California, Centers for Water and Wildland Resources.

Myrup, L. O., T. M. Powell, D. A. Godden, and C. R. Goldman. 1979. Climatological estimate of the average monthly energy and water budgets of Lake Tahoe, California-Nevada. *Water Resources Research* 15:1499–508.

Nadeau, R. A. 1950. *The water seekers.* Bishop, CA: Chalfant Press.

Naiman, R. J., D. G. Lonzarich, T. J. Beechie, and S. C. Ralph. 1992. General principles of classification and the assessment of conservation potential in rivers. In *River conservation and management,* edited by P. J. Boon, P. Calow, and G. E. Petts, 93–123. Chichester, England: John Wiley and Sons.

Naiman, R. J., J. J. Magnuson, D. M. McKnight, and J. A. Stanford. 1995. *The freshwater imperative: A research agenda.* Washington, DC: Island Press.

Nelson, R. L., M. L. McHenry, and W. S. Platts. 1991. Mining. In *Influences of forest and rangeland management on salmonid fishes and their habitats,* edited by W. R. Meehan, 425–57. Bethesda, MD: American Fisheries Society.

Newbold, J. D., D. C. Erman, and K. B. Roby. 1980. Effects of logging on macroinvertebrates in streams with and without buffer strips. *Canadian Journal of Fisheries and Aquatic Sciences* 37:1076–85.

Newton, M., and J. A. Norgren. 1977. *Silvicultural chemicals and protection of water quality.* EPA Report 910/9-77-036. Seattle: Environmental Protection Agency, Region X.

Norris, L. A., C. L. Hawkes, W. L. Webb, D. G. Moore, W. B. Bollen, and E. Holcombe. 1978. *A report of research on the behavior and impact of chemical fire retardants in forest streams.* Corvallis, OR: U.S. Forest Service, Forestry Sciences Laboratory.

Norris, L. A., H. W. Lorz, and S. V. Gregory. 1991. Forest chemicals. In *Influences of forest and rangeland management on salmonid fishes and their habitats,* edited by W. R. Meehan, 207–96. Bethesda, MD: American Fisheries Society.

Norris, L. A., and W. L. Webb. 1989. Effects of fire retardant on water quality. In *Proceedings of the symposium on fire and watershed management,* edited by N. H. Berg, 79–86. General Technical Report PSW-109. Berkeley, CA: U.S. Forest Service, Pacific Southwest Forest and Range Experiment Station.

O'Connor, M. D. 1986. Effects of logging on organic debris dams in first order streams in northern California. Master's thesis, University of California, Berkeley.

Overton, K. C., G. L. Chandler, and J. A. Pisano. 1994. *Northern/intermountain regions' fish habitat inventory: Grazed, rested, and ungrazed reference stream reaches, Silver King Creek, California.* General Technical Report INT-311. Ogden, UT: U.S. Forest Service, intermountain Research Station.

Pacific Gas and Electric Company. 1911. *Properties owned and operated, territories served.* San Francisco: Pacific Gas and Electric Company.

———. 1994. *Project description: Rock Creek–Cresta sediment management project.* San Francisco: Pacific Gas and Electric Company.

Page, R. W., P. W. Antilla, K. L. Johnson, and M. J. Pierce. 1984. *Groundwater conditions and well yields in fractured rocks, southwestern Nevada County, California.* Water-Resources Investigations Report 83-4262. Sacramento: U.S. Geological Survey.

Pagenhart, T. H. 1969. Water use in the Yuba and Bear River Basins, California. Ph.D. diss., University of California, Berkeley.

Palmer, T. 1988. *The Sierra Nevada.* Washington, DC: Island Press.

———. 1993. *The wild and scenic rivers of America.* Washington, DC: Island Press.

———. 1994. *Lifelines: The case for river conservation.* Washington, DC: Island Press.

Parizek, R. R. 1971. Impacts of highways on the hydrogeologic environment. In *Environmental geomorphology,* edited by D. R. Coates, 151–99. Binghamton: State University of New York.

Parsons Engineering Science. 1995. Environmental assessment for the Pacific Gas and Electric Company Mokelumne weather modification project. Parsons Engineering Science, Alameda, California.

Patric, J. H., J. O. Evans, and J. D. Helvey. 1984. Summary of sediment yield data from forested land in the United States. *Journal of Forestry* 82:101–4.

Peattie, R., ed. 1947. *The Sierra Nevada: The range of light.* New York: Vanguard Press.

Pelzman, R. J. 1973. Causes and possible prevention of riparian plant encroachment on anadromous fish habitat. Administrative Report 73-1. Sacramento: California Department of Fish and Game, Environmental Services Branch.

Perkins, M. A., C. R. Goldman, and R. L. Leonard. 1975. Residual nutrient discharge in stream waters influenced by sewage effluent spraying. *Ecology* 56:453–60.

Pisani, D. J. 1977. The polluted Truckee: A study in interstate water quality 1870–1934. *Nevada Historical Society Quarterly* 20 (3): 151–66.

———. 1984. *From the family farm to agribusiness.* Berkeley and Los Angeles: University of California Press.

Pitt, M. D., Burgy R. H., and H. F. Heady. 1978. Influences of brush conversion and weather patterns on runoff from a northern California watershed. *Journal of Range Management* 31 (1): 23–27.

Placer County. 1994. Placer County General Plan, background report, vol. 2. Auburn, CA: Placer County.

Plumas National Forest. 1988. Land and resources management plan. U.S. Forest Service, Plumas National Forest, Quincy, CA.

Poff, R. J. 1989a. Compatibility of timber salvage operations with watershed values. In *Proceedings of the symposium on fire and watershed management,* edited by N. H. Berg, 137–40. General Technical Report PSW-109. Berkeley, CA: U.S. Forest Service, Pacific Southwest Forest and Range Experiment Station.

———. 1989b. Distribution and persistence of hydrophobic soil layers on the Indian Burn. In *Proceedings of the symposium on fire and watershed management,* edited by N. H. Berg, 153. General Technical Report PSW-109. Berkeley, CA: U.S. Forest Service, Pacific Southwest Forest and Range Experiment Station.

———. 1996. Effects of silvicultural practices and wildfire on productivity of forest soils. In *Sierra Nevada Ecosystem Project: Final report to Congress,* vol. II, chap. 16. Davis: University of California, Centers for Water and Wildland Resources.

Pomby, J. 1987. Mined land reclamation program. *California Geology* 40 (1): 3–6.

Ponce, S. L., ed. 1983. *The potential for water yield augmentation through forest and range management.* Bethesda, MD: American Water Resources Association.

Postel, S. 1992. *Last oasis: Facing water scarcity.* New York: World Watch Institute/W. W. Norton.

Potts, D. F., D. L. Peterson, and H. R. Zuuring. 1989. *Estimating post fire water production in the Pacific Northwest.* Research Paper PSW-197. Berkeley, CA: U.S. Forest Service, Pacific Southwest Forest and Range Experiment Station.

Poulin, R., R. C. Pakalris, and K. Sindling. 1994. Aggregate resources: Production and environmental constraints. *Environmental Geology* 23:221–27.

Ramey, M. P., and S. M. Beck. 1990. *Flushing flow evaluation: The North Fork of the Feather River below Poe Dam.* Environment, Health, and Safety Report 009.4-89.9. San Ramon, CA: Pacific Gas and Electric Company, Research and Development.

Rausch, D. L., and H. G. Heinemann. 1984. Measurement of reservoir sedimentation. In *Erosion and sediment yield: Some methods of measurement and modeling,* edited by D. F. Hadley and D. E. Walling, 179–200. Norwich, England: Geo Books.

Rector, J. R., and L. H. MacDonald. 1987. Water yield opportunities on national forest lands in the Pacific Southwest Region. In *Proceedings of the California watershed management conference,* edited by R. Z. Callaham and J. J. DeVries, 68–73. Report 11. Berkeley: University of California, Wildland Resources Center.

Redinger, D. H. 1949. *The story of Big Creek.* Los Angeles: Angeles Press.

Reid, L. M. 1993. *Research and cumulative watershed effects.* General Technical Report PSW-141. Albany, CA: U.S. Forest Service, Pacific Southwest Research Station.

Reid, L. M., and T. Dunne. 1984. Sediment production from forest road surfaces. *Water Resources Research* 20:1753–61.

Reisner, M. 1986. *Cadillac desert.* New York: Viking Press.

Rice, R. M., and J. Lewis. 1991. Estimating erosion risks associated with logging and forest roads in northwestern California. *Water Resources Bulletin* 27:809–18.

Rice, R. M., and J. R. Wallis. 1962. How a logging operation can affect streamflow. *Forest Industries* 89 (11): 38–40.

Roby, K. B. 1989. Watershed response and recovery from the Will Fire: Ten years of observation. In *Proceedings of the symposium on fire and watershed management,* edited by N. H. Berg, 131–36. General Technical Report PSW-109. Berkeley: U.S. Forest Service, Pacific Southwest Forest and Range Experiment Station.

Roby, K. B., and D. L. Azuma. 1995. Changes in a reach of a northern California stream following wildfire. *Environmental Management* 19:591–600.

Rollins, J. L. 1931. Shall the hydraulic mines within the Bear River Basin, Placer and Nevada Counties, California, be rehabilitated or shall they be abandoned? Paper presented to California Debris Commission, Sacramento.

Romm, J., R. Z. Callaham, and R. Kattelmann. 1988. *Toward managing Sierra Nevada forests for water supply.* Report 17. Berkeley: University of California, Wildland Resources Center.

Roos, M. 1987. Possible changes in California snowmelt runoff patterns. In *Proceedings of the fourth Pacific climate workshop,* 22–31.

Roth, P., C. Blanchard, J. Harte, H. Michaels, and M. T. El-Ashry. 1985. *The American West's acid rain test.* Washington, DC: World Resources Institute.

Sandecki, M. 1989. Aggregate mining in river systems. *California Geology* 42 (4): 88–94.

Sauder, R. A. 1994. *The last frontier: Water diversion in the growth and destruction of Owens Valley agriculture.* Tucson: University of Arizona Press.

Scharf, R. F., and M. Srago. 1975. *Conifer damage and death associated with the use of deicing salt in the Lake Tahoe Basin of California and Nevada.* Forest Pest Control Report 1. San Francisco: U.S. Forest Service, California Region.

Scheidt, M. E. 1967. Environmental effects of highways. *Journal of the Sanitary Engineering Division, Proceedings of the American Society of Civil Engineers* 93 (SA5): 17–25.

Schindler, D. W. 1988. Effects of acid rain on freshwater ecosystems. *Science* 239:149–57.

Schumm, S. A. 1963. *The disparity between present rate of denudation and orogeny.* Professional Paper 4544. Washington, DC: U.S. Geological Survey.

Scott, K. M., and G. C. Gravlee. 1968. *Flood surge on the Rubicon River, California—hydrology, hydraulics, and boulder transport.* Professional Paper 422-M. Washington, DC: U.S. Geological Survey.

Sea Surveyor, Inc. 1993. *Results from a bathymetric/geophysical survey of Slab Creek Reservoir performed for the Sacramento Municipal Utility District.* Benicia, CA: Sacramento Municipal Utility District.

Seckler, D., ed. 1971. *California water: A study in resource management.* Berkeley and Los Angeles: University of California Press.

Seidelman, P., J. Borum, R. Coats, and L. Collins. 1986. *Land disturbance and watershed processes in Sierrian [sic] granitic terrain.* Earth Resources Monograph 9. San Francisco: U.S. Forest Service, Region 5.

Senter, E. 1987. *Erosion control at Malakoff Diggings State Historical Park.* Sacramento: California Department of Water Resources, Central District.

Shapiro, A. M. 1996. Status of butterflies. In *Sierra Nevada Ecosystem Project: Final report to Congress,* vol. II, chap. 27. Davis: University of California, Centers for Water and Wildland Resources.

Shelton, M. L. 1994. Reservoir storage and irrigation influences on the hydrologic system in the Sacramento River Basin, California.

In *Effects of human-induced changes on hydrologic systems,* edited by R. A. Marston and V. R. Hasfurther, 679–88. Bethesda, MD: American Water Resources Association.

Shoup, L. H. 1988. Historical overview of the New Melones locality, 1806–1942. In *Culture change in the central Sierra Nevada 8000 B.C.–A.D. 1950,* edited by M. J. Moratto, J. D. Tordoff, and L. H. Shoup, Fresno: Infotec Research, Inc.

Sidle, R. C. 1980. *Slope stability on forest land.* Publication 209. Corvallis: Oregon State University, Pacific Northwest Extension.

Sidle, R. C., A. J. Pearce, and C. L. O'Loughlin. 1985. *Hillslope stability and land use.* Water Resources Monograph Series II. Washington, DC: American Geophysical Union.

Silva, M. 1986. *Placer gold recovery methods.* Special Publication 87. Sacramento: California Division of Mines and Geology.

Skau, C. M., and J. C. Brown. 1990. A synoptic view of nutrients and suspended sediments in natural waters of forested watershed in east-central Sierra Nevada. In *Proceedings international mountain watershed symposium: Subalpine processes and water quality,* edited by I. G. Poppoff, C. R. Goldman, S. L. Loeb, and L. B. Leopold, South Lake Tahoe, CA: Tahoe Resource Conservation District.

Skau, C. M., J. C. Brown, and J. A. Nadolski. 1980. Snowmelt sediment from Sierra Nevada headwaters. In *Symposium on watershed management,* 418–29. New York: American Society of Civil Engineers.

Skinner, C. N., and C. Chang. 1996. Fire regimes, past and present. In *Sierra Nevada Ecosystem Project: Final report to Congress,* vol. II, chap. 38. Davis: University of California, Centers for Water and Wildland Resources.

Slotten, D. G., S. M. Ayers, J. E. Reuter, and C. R. Goldman. 1995. *Gold mining impacts on food chain mercury in northwestern Sierra Nevada streams.* Technical Completion Report W-816. Davis: University of California, Water Resources Center .

Slotten, D. G., J. E. Reuter, C. R. Goldman, R. Jepson, and W. Lick. 1994. *Camanche Reservoir bottom sediment study: Heavy metal distribution and resuspension characteristics.* Contribution No. 40. Davis: University of California, Institute of Ecology.

Smith, J. L. 1982. *The historical climatic regime and projected impact of weather modification upon precipitation and temperature at Central Sierra Snow Laboratory.* The Sierra Ecology Project, vol. 3. Denver: U.S. Bureau of Reclamation, Office of Atmospheric Resources Research.

Soil Conservation Service. 1979. *Sources of sediment: Georgetown, Camino-Fruitridge pilot study area, El Dorado County, California.* Placerville, CA: U.S. Soil Conservation Service.

———. 1984. *Foothills watershed area study, El Dorado unit.* Placerville, CA: U.S. Soil Conservation Service.

———. 1985. *Cameron Park watershed area study, El Dorado County, California.* Davis, CA: U.S. Soil Conservation Service, River Basin Planning Staff.

———. 1989. *East Branch North Fork Feather River erosion inventory report, Plumas County, California.* Davis, CA: U.S. Soil Conservation Service, River Basin Planning Staff.

———. 1993. Erosion and sediment control study, Middle Creek watershed, Shasta County, California. Redding, CA: U.S. Soil Conservation Service.

Springer, E. P., and G. F. Gifford. 1980. Unconfined grazing and bacterial water pollution: A review. In *Symposium on watershed management 1980,* 578–87. New York: American Society of Civil Engineers.

Stanislaus National Forest. 1993. *Domingo reforestation environmental assessment*. Sonora, CA: U.S. Forest Service, Stanislaus National Forest.

Stanley, O. G. 1965. Brief history of hydraulic mining, gold dredging, creation of the California Debris Commission, and birth of the Sacramento District of the Corps of Engineers. Manuscript on file at Water Resources Center Archives, University of California, Berkeley.

Stednick, J. D. 1991. *Wildland water quality sampling and analysis*. San Diego: Academic Press.

Steward, J. 1934. Ethnography of the Owens Valley Paiute. *American Archaeology and Ethnology* 33:233–324.

Stewart, W. C. 1996. Economic assessment of the ecosystem. In *Sierra Nevada Ecosystem Project: Final report to Congress,* vol. III. Davis: University of California, Centers for Water and Wildland Resources.

Stine, S. 1994. Extreme and persistent drought in California and Patagonia during mediaeval time. *Nature* 369:546–49.

———. 1996. Climate, 1650–1850. *In Sierra Nevada Ecosystem Project: Final report to Congress,* vol. II, chap. 2. Davis: University of California, Centers for Water and Wildland Resources.

Strong, D. H. 1984. *Tahoe, an environmental history*. Lincoln: University of Nebraska Press.

Sudworth, G. B. 1900. Stanislaus and Lake Tahoe Forest Reserves, California, and adjacent territory. In *Annual reports of the Department of the Interior, 21st annual report of the U.S. Geological Survey,* 505–61. Washington, DC: Government Printing Office.

Suk, T. J., J. L. Riggs, and B. C. Nelson. 1985. Water contamination with *Giardia* in back-country areas. In *Proceedings of the national wilderness research conference,* 237–40. Fort Collins, CO: n.p.

Suk, T. J., S. K. Sorenson, and P. D. Dileanis. 1986. *Map showing the number of Giardia cysts in water samples from 69 stream sites in the Sierra Nevada, California*. Open File Report 86-404. Menlo Park, CA: U.S. Geological Survey.

———. 1987. The relation between human presence and occurrence of Giardia cysts in streams of the Sierra Nevada, California. *Journal of Freshwater Ecology* 4 (1): 71–75.

Tahoe Regional Planning Agency. 1988. *Water quality management plan for the Lake Tahoe Region*. Zephyr Cove, NV: Tahoe Regional Planning Agency.

Taylor, T. P., and D. C. Erman. 1979. The response of benthic plants to past levels of human use in high mountain lakes in Kings Canyon National Park, California, U.S.A. *Journal of Environmental Management* 9:271–78.

Terrell, C. R., and P. B. Perfetti. 1989. *Water quality indicators guide: Surface waters*. Washington, DC: U.S. Soil Conservation Service.

Terrene Institute. 1994. *Urbanization and water quality*. Washington, DC: Terrene Institute.

Thornton, M. V. 1992. *A history of the Groveland Community Services District*. Groveland, CA: Groveland Community Services District.

Tiedemann, A. R., et al. 1979. *Effects of fire on water: A state-of-knowledge review*. General Technical Report WO-14. Washington, DC: U.S. Forest Service.

Todd, A. H. 1990. Watershed restoration and erosion control: Making it work in subalpine areas. In *Proceedings, international mountain watershed symposium: Subalpine processes and water quality,* edited by I. G. Poppoff, C. R. Goldman, S. L. Loeb, and L. B. Leopold, 290–99. South Lake Tahoe, CA: Tahoe Resource Conservation District.

Tonnessen, K. 1984. Potential for aquatic ecosystem acidification in the Sierra Nevada, California. In *Early biotic responses to advancing lake acidification,* edited by G. Hendrey, 147–69. Boston: Butterworth.

———. 1991. The Emerald Lake watershed study. *Water Resources Research* 27:1537–39.

Troendle, C. A., and R. M. King. 1985. The effect of timber harvest on the Fool Creek watershed, 30 years later. *Water Resources Research* 21:1915–22.

Troxel, B. W., and P. K. Morton. 1962. *Mines and mineral resources of Kern County, California*. County Report 1. San Francisco: California Division of Mines and Geology.

Turner, K. M. 1991. Annual evapotranspiration of native vegetation in a Mediterranean-type climate. *Water Resources Bulletin* 27:1–6.

U.S. Army Corps of Engineers. 1990. *American River and Sacramento metropolitan investigations, California*. Appendix K. Sacramento, CA: U.S. Army Corps of Engineers.

U.S. Environmental Protection Agency. 1989. *Rapid bioassessment protocol for use in streams and rivers: Benthic macroinvertebrates and fish*. EPA/444/4-89-001. Washington, DC: Environmental Protection Agency, Office of Water Regulations and Standards.

———. 1993. *Monitoring protocols to evaluate water quality effects of grazing management on western rangeland streams*. EPA 910/R-93-017. Washington, DC: Environmental Protection Agency, Water Division, Surface Water Branch.

U.S. Forest Service. 1988. *Record of decision on the California region final environmental impact statement for vegetation management and reforestation*. San Francisco: U.S. Forest Service.

———. 1992. *Investigating water quality in the Pacific Southwest Region, best management practices evaluation program: A user's guide*. San Francisco: U.S. Forest Service, Region 5.

———. 1995a. *Draft environmental impact statement: Managing California spotted owl habitat in the Sierra Nevada national forests of California, an ecosystem approach*. San Francisco: U.S. Forest Service, Pacific Southwest Region.

———. 1995b. Results of stream condition inventory of grazed and ungrazed meadow streams. Unpublished report, U.S. Forest Service, Pacific Southwest Region, San Francisco.

U.S. Geological Survey. 1984. *National water summary 1983—hydrologic events and issues*. Water Supply Paper 2250. Reston, VA: U.S. Geological Survey.

Wagner, J. R. 1970. *Gold mines of California*. Berkeley, CA: Howell-North Books.

Wagoner, L. 1886. Report on forests of the counties of Amador, Calaveras, Tuolumne, and Mariposa. In *First biennial report of the California State Board of Forestry for the years 1885–1886,* 39–44. Sacramento: California State Board of Forestry.

Wahl, K. L. 1991. Is April to July runoff really decreasing in the western United States? *Proceedings of the western snow conference* 59:67–78.

Walton, J. 1992. *Western times and water wars: State, culture, and rebellion in California*. Berkeley and Los Angeles: University of California Press.

Ward, R. C., J. C. Loftis, and G. B. McBride. 1986. The "data rich but information poor" syndrome in water quality monitoring. *Environmental Management* 10:291–97.

Weatherspoon, C. P. 1996. Fire-silviculture relationships in Sierra forests. In *Sierra Nevada Ecosystem Project: Final report to Congress,* vol. II, chap. 44. Davis: University of California, Centers for Water and Wildland Resources.

Webster, P. 1972. *The mighty Sierra: Portrait of a mountain world*. Palo Alto, CA: American West Publishing.

Whitney, J. D. 1880. *The auriferous gravels of the Sierra Nevada*. Vol. 6 of *Memoir of the Museum of Comparative Zoology*. Cambridge: Harvard University, Museum of Comparative Zoology.

Williams, G. P., and M. G. Wolman. 1984. *Downstream effects of dams and alluvial rivers*. Professional Paper 1286. Reston, VA: U.S. Geological Survey.

Williams, M. W., and J. M Melack. 1991. Solute chemistry of snowmelt and runoff in an alpine basin, Sierra Nevada. *Water Resources Research* 27:1575–88.

Wills, L., and J. C. Sheehan. 1994. *East Branch North Fork Feather River, Spanish Creek, and Lost Chance Creek non-point source water pollution study*. Quincy, CA: Plumas Corporation.

Woolfenden, W. B. 1996. Quaternary vegetation history. In *Sierra Nevada Ecosystem Project: Final report to Congress*, vol. II, chap. 4. Davis: University of California, Centers for Water and Wildland Resources.

Woyshner, M., and B. Hecht. 1990. Sediment, solute, and nutrient transport from Squaw Creek, Truckee River Basin, California. In *Proceedings, international mountain watershed symposium: Subalpine processes and water quality*, edited by I. G. Poppoff, C. R. Goldman, S. L. Loeb, and L. B. Leopold, 190–219. South Lake Tahoe, CA: Tahoe Resource Conservation District.

Yeend, W. 1974. *Gold-bearing gravel of the ancestral Yuba River, Sierra Nevada, California*. Professional Paper 772. Menlo Park, CA: U.S. Geological Survey.

Ziemer, R. R. 1981. Storm flow response to road building and partial cutting in small streams of northern California. *Water Resources Research* 17:907–17.

Zinke, P. 1987. Soil moisture budget: An example of effects of timber harvest and regrowth. In *Proceedings of the California watershed management conference*, edited by R. Z. Callaham and J. J. DeVries, 86–88. Report 11. Berkeley: University of California, Wildland Resources Center.

MARK R. JENNINGS
Research Associate
Department of Herpetology
California Academy of Sciences
Golden Gate Park
San Francisco, California

31

Status of Amphibians

ABSTRACT

The status of thirty-two amphibian taxa currently found in the Sierra Nevada region of California was reviewed. Of this number, thirty are native species or subspecies, one is an introduced species (bullfrog [*Rana catesbeiana*]), and one is of uncertain origin (tiger salamander [*Ambystoma tigrinum*]). Of the thirty definite native species or subspecies, nine are frogs and toads and twenty-one are salamanders. Fourteen (47%) of these taxa are native to the Sierra Nevada, and of this total, twelve taxa (86%) are in need of some form of protection, including six taxa (43%) that are either extinct or threatened with extinction in the near future. The most imperiled amphibians are the true toads (*Bufo* spp.) and true frogs (*Rana* spp.)—which make up 23% of the fauna—because of their widespread declines in the region over the past twenty-five years. For the salamanders, nine (43%) species or subspecies are at risk. As a whole, amphibians occurring in aquatic habitats are at greatest risk, because these habitats are being threatened by alteration of their physical or biotic structure by several types of human use of water and adjacent land. The uses that most severely affect aquatic habitats and their contained species are overgrazing by livestock; stream channelization; construction of hydroelectric, recreational, or water storage reservoirs of significant size; removal of ground and surface water near or beyond recharge or volume capacities; placer mining; and the introduction of a suite of exotic species (especially fishes) with which the native aquatic amphibian fauna frequently cannot coexist. The most imperiled aquatic habitats in the Sierra Nevada that harbor one or more of the taxa recommended for listing are springs, seeps, and bogs; rain (or vernal) pools; marshes; and small headwater streams. In the Sierra Nevada, taxa occurring in terrestrial habitats are generally less imperiled, because most terrestrial habitats in the region have a much greater total area than all aquatic habitats combined. Yet, aside from outright destruction and development, several widespread activities and land uses continue to alter the structure and vegetation of most terrestrial habitats in a manner unfavorable to the survival of their contained taxa. Among such uses the most significant are the impacts of off-road vehicles, overgrazing by livestock, timber harvest, mining, and urbanization.

INTRODUCTION

Over the past four years, there has been a heightened concern about the decline of a number of amphibian species in various parts of the world (see reviews in Blaustein 1994). In the Sierra Nevada region covered in this chapter (figure 31.1), such concern is borne out by the fact that all species of native true toads (*Bufo* spp.) and true frogs (*Rana* spp.) inhabiting the area have disappeared from significant portions of their ranges during the past twenty-five years, despite having large portions of habitat protected in wilderness areas and national parks (Jennings 1995). Because of the present uncertainty regarding the status of the amphibians of the Sierra Nevada region, this study was conducted to provide a benchmark or snapshot of the current status of each amphibian taxon. Although the information presented comes from data gathered from other studies, recommendations are made for taxa in need of active management by resource agencies.

METHODS

Much of the information contained in this chapter comes from Jennings and Hayes 1994, which reviews the status of the entire herpetofauna of California. An effort was made to include information on the status of taxa not already covered

Sierra Nevada Ecosystem Project: Final report to Congress, vol. II, *Assessments and scientific basis for management options*. Davis: University of California, Centers for Water and Wildland Resources, 1996.

FIGURE 31.1

Geographic area of the
Sierra Nevada as defined by
the SNEP Team. Amphibians
found only within the shaded
area are covered in the text.

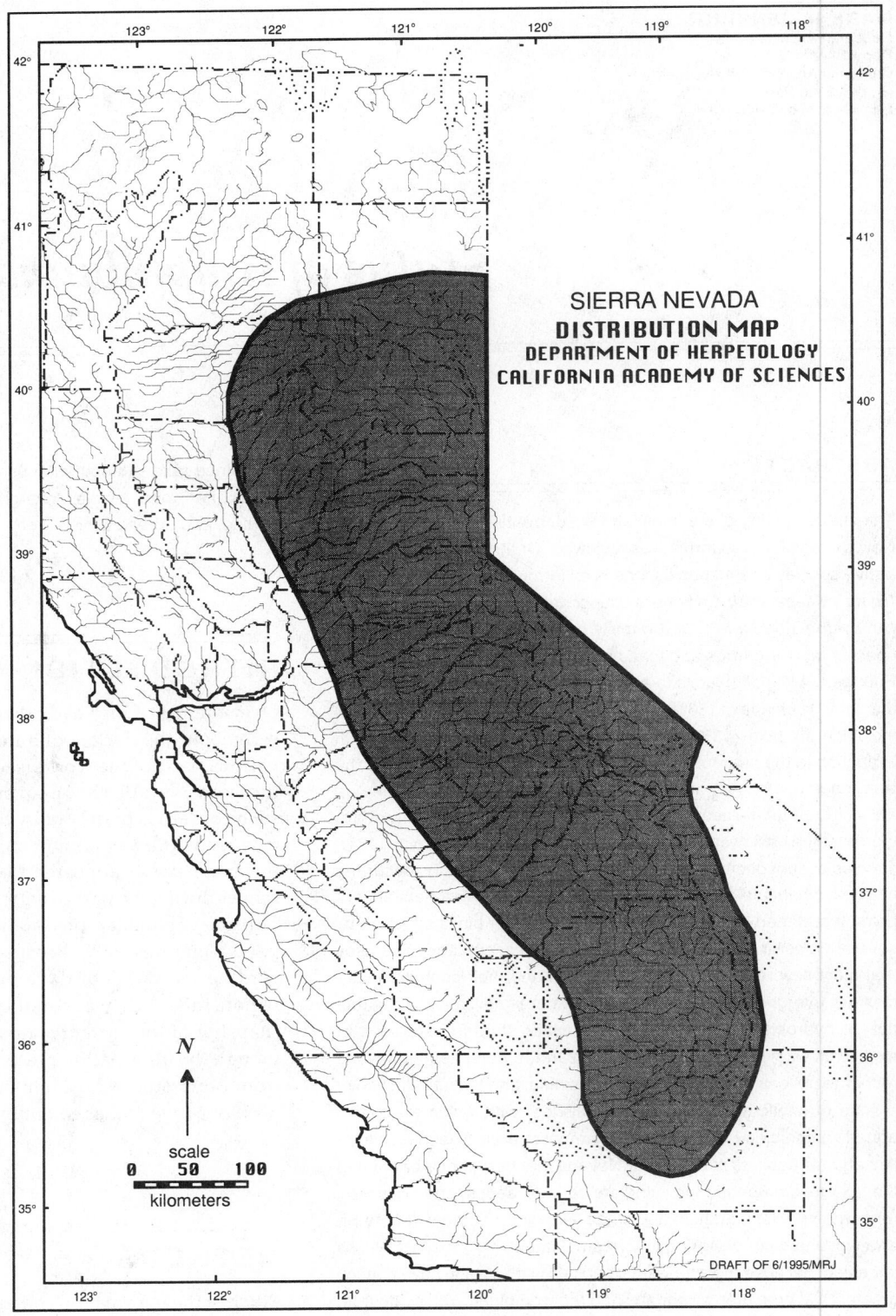

in that report, especially if it relates to amphibian populations in the Sierra Nevada. Distribution maps were compiled
with the aid of verified museum records and field reconnaissance in specific regions of California to help assess the presence or absence of taxa within their known range. Historical
assessments of the past distributions of each taxon were made
from a combination of verified museum specimens and the
field notes of current and former naturalists. A full descrip-

tion of the methodology used is provided in Jennings and Hayes 1994.

Data from the aforementioned sources were organized into generalized accounts for mole salamanders (Ambystomatidae), lungless salamanders (Plethodontidae), newts (Salamandridae), tree frogs (Hylidae), spadefoot toads (Pelobatidae), true toads (Bufonidae), and true frogs (Ranidae). Current distribution maps are provided for each taxon present within the study region.

For determining the status of each amphibian taxon in the Sierra Nevada, I followed Jennings and Hayes 1994 and assigned one of four categories:

1. Taxa for which endangered status is justified (i.e., those animals that are in serious danger of becoming extinct throughout all or a significant portion of their range due to one or more causes, including loss of habitat, change in habitat, overexploitation, predation, competition, or disease)

2. Taxa for which threatened status is justified (i.e., those animals that are not currently threatened with extinction but that are likely to become an endangered species in the foreseeable future in the absence of special protection and management efforts required by state and governmental agencies)

3. Taxa for which special concern status is justified (i.e., those animals that may become listed as threatened or endangered in the near future due to habitat modification or destruction, overcollecting, or disease or that are threatened in any way by introduced species)

4. Taxa for which no status is justified (i.e., those animals that are currently common throughout their range. Most amphibians fall under this category.)

I based my determination of whether endangered or threatened status was justified on the state-level definitions published in the California Fish and Game Code (California Administrative Code, title 14, sec. 670.5). For determining special concern status, I followed the criteria indicated in Williams 1986 and Moyle et al. 1989.

RESULTS

Of the thirty-two taxa reviewed, thirty are native species (or subspecies), one is an introduced species, and one species is of uncertain origin (table 31.1). Of the thirty definite species or subspecies, nine are frogs and toads, and twenty-one are salamanders. Fourteen (47%) of these taxa are endemic to the Sierra Nevada. At present, fourteen of the native species or subspecies (47%) do not have declining populations, although

five of these taxa (17%) have very localized distributions and as such are vulnerable to localized disturbances. Of the remaining sixteen species or subspecies, one species (3%) is apparently extinct, five species (17%) are formally listed (or proposed for listing) as threatened or endangered, five species (17%) clearly merit such listing, and five species (17%) are declining and so are of special concern (table 31.1). Of the fourteen endemics, only the Sierra Nevada salamander (*Ensatina eschscholtzii platensis*) and the Sierra newt (*Taricha torosa sierrae*) can be regarded as secure; the rest (86%) fit into one of the other three categories, including six taxa (43%) that are extinct or threatened with extinction in the near future. Of the twenty-one species or subspecies of salamanders, nine taxa (43%) are at risk, while eight (89%) of the nine frogs and toads are at risk. Accounts of all taxa follow.

Mole Salamanders (Ambystomatidae)

Mole salamanders are represented by three species in the Sierra Nevada: the California tiger salamander (*Ambystoma californiense*), the southern long-toed salamander (*A. macrodactylum sigillatum*), and the tiger salamander (*A. tigrinum* ssp.) (figure 31.2). The tiger salamander population in the eastern Sierra is of uncertain origin and may be the result of animals originally brought in for live fish bait at reservoirs such as Lake Crowley (Jennings and Hayes 1994).

All of these salamanders are long-lived (up to twenty years or more) (Bowler 1977; M. Allaback, Biosearch Wildlife Surveys, letter to the author, October 12, 1995). They have a three- to six-month aquatic larval stage, followed by metamorphosis into a terrestrial juvenile stage. After one or more years as juveniles, the salamanders then mature into the terrestrial adult form. They breed in temporary ponds at low to middle elevations (southern long-toed salamanders also breed at high elevations) and often use the same breeding ponds year after year. Juveniles and adults spend most of the year underground in small mammal burrows except during the winter months, when sufficient rainfall allows for surface activity and breeding. The California tiger salamander is a low-elevation species that is currently threatened by the destruction of its breeding ponds and the introduction of predatory fish (especially mosquito fish [*Gambusia affinis*]), Louisiana red swamp crayfish (*Procambarus clarkii*), and bullfrogs (*Rana catesbeiana*) into its habitat. It is considered threatened (Jennings and Hayes 1994) and has been found warranted for listing under the Endangered Species Act (ESA) (Sorenson 1994).

The southern long-toed salamander is currently recognized as a subspecies of the widely distributed long-toed salamander (*A. macrodactylum*) in northwestern North America (Stebbins 1985). Its Sierran populations are currently believed to be stable, although they are depleted in some areas due to the introduction of trout (*Oncorhynchus* spp.) and charr (*Salvelinus* spp.) into high-elevation lakes formerly used by salamanders for breeding purposes (e.g., see Liss and Larson 1991).

TABLE 31.1

Native amphibians of the Sierra Nevada (based largely on Stebbins 1985). Status levels based on Jennings and Hayes (1994).

Taxon	Drainage	Habitat	Status
Salamanders			
Mole Salamanders			
California tiger salamander, *Ambystoma californiense*	Sacramento–San Joaquin Rivers	Lowlands, foothills	Threatened
Southern long-toed salamander, *Ambystoma macrodactylum sigillatum*	Eagle Lake, Lahontan, Sacramento–San Joaquin Rivers	High elevations	Stable or expanding
Tiger salamander, *Ambystoma tigrinum* sp.	Owens Valley	High elevations	Stable or expanding; introduced?
Lungless Salamanders			
Arboreal salamander, *Aneides lugubris*	Sacramento–San Joaquin Rivers	Foothills	Stable or expanding
California slender salamander, *Batrachoseps attenuatus*	Sacramento–San Joaquin Rivers	Lowlands, foothills	Stable or expanding
Black-bellied slender salamander, *Batrachoseps nigriventris*	Sacramento–San Joaquin Rivers, Tulare Lake	Foothills	Stable or expanding
Pacific slender salamander, *Batrachoseps pacificus*	Sacramento–San Joaquin Rivers, Tulare Lake	Foothills	Special concern
Relictual slender salamander, *Batrachoseps relictus*	Sacramento–San Joaquin Rivers, Tulare Lake	Foothills	Threatened[a]
Kern Canyon slender salamander, *Batrachoseps simatus*	Tulare Lake	Foothills	Threatened[a]
Tehachapi slender salamander, *Batrachoseps stebbinsi*	Tulare Lake	High elevations	Endangered[b]
Breckenridge Mountain slender salamander, *Batrachoseps* sp.	Tulare Lake	Foothills	Stable or expanding
Fairview slender salamander, *Batrachoseps* sp.	Owens Valley, Tulare Lake	High elevations	Stable or expanding
Kern Plateau slender salamander, *Batrachoseps* sp.	Sacramento–San Joaquin Rivers	Foothills	Special concern
Hell Hollow slender salamander, *Batrachoseps* sp.	Tulare Lake	Foothills	Stable or expanding
Yellow-blotched salamander, *Ensatina eschscholtzii croceater*	Sacramento–San Joaquin Rivers, Tulare Lake	Foothills	Stable or expanding
Sierra Nevada salamander, *Ensatina eschscholtzii platensis*	Sacramento–San Joaquin Rivers	Foothills	Stable or expanding
Yellow-eyed salamander, *Ensatina eschscholtzii xanthoptica*	Sacramento–San Joaquin Rivers	Foothills	Threatened[a]
Limestone salamander, *Hydromantes brunus*	Sacramento–San Joaquin Rivers	High elevations	Special concern
Mount Lyell salamander, *Hydromantes platycephalus*	Lahontan, Owens Valley, Sacramento–San Joaquin Rivers, Tulare Lake		
Owens Valley web-toed salamander, *Hydromantes* sp.	Owens Valley	High elevations	Special concern
Newts			
Northern rough-skinned newt, *Taricha granulosa granulosa*	Sacramento–San Joaquin Rivers	Foothills	Stable or expanding
Sierra newt, *Taricha torosa sierrae*	Sacramento–San Joaquin Rivers, Tulare Lake	Foothills	Stable or expanding
Frogs and Toads			
True Toads			
California toad, *Bufo boreas halophilus*	All drainages	Lowlands, foothills, high elevations	Stable or expanding
Yosemite toad, *Bufo canorus*	Lahontan, Owens Valley, Sacramento–San Joaquin Rivers	High elevations	Endangered
Tree Frogs			
Pacific tree frog, *Hyla regilla*	All drainages	Lowlands, foothills, high elevations	Stable or expanding
True Frogs			
California red-legged frog, *Rana aurora draytonii*	Sacramento–San Joaquin Rivers, Tulare Lake	Lowlands, foothills	Endangered
Foothill yellow-legged frog, *Rana boylii*	Sacramento–San Joaquin Rivers, Tulare Lake	Foothills	Threatened
Cascade frog, *Rana cascadae*	Sacramento–San Joaquin Rivers	Foothills, high elevations	Endangered
Bullfrog, *Rana catesbeiana*	All drainages	Lowlands, foothills	Stable or expanding; introduced
Mountain yellow-legged frog, *Rana muscosa*	Lahontan, Owens Valley, Sacramento–San Joaquin Rivers, Tulare Lake	High elevations	Threatened
Northern leopard frog, *Rana pipiens*	Lahontan, Owens Valley, Sacramento–San Joaquin Rivers	High elevations	Threatened
Spadefoot Toads			
Western spadefoot, *Scaphiopus hammondii*	Sacramento–San Joaquin Rivers, Tulare Lake	Lowlands, foothills	Special concern

[a]Currently listed as threatened by the State of California (Jennings 1987).
[b]Probably extinct.

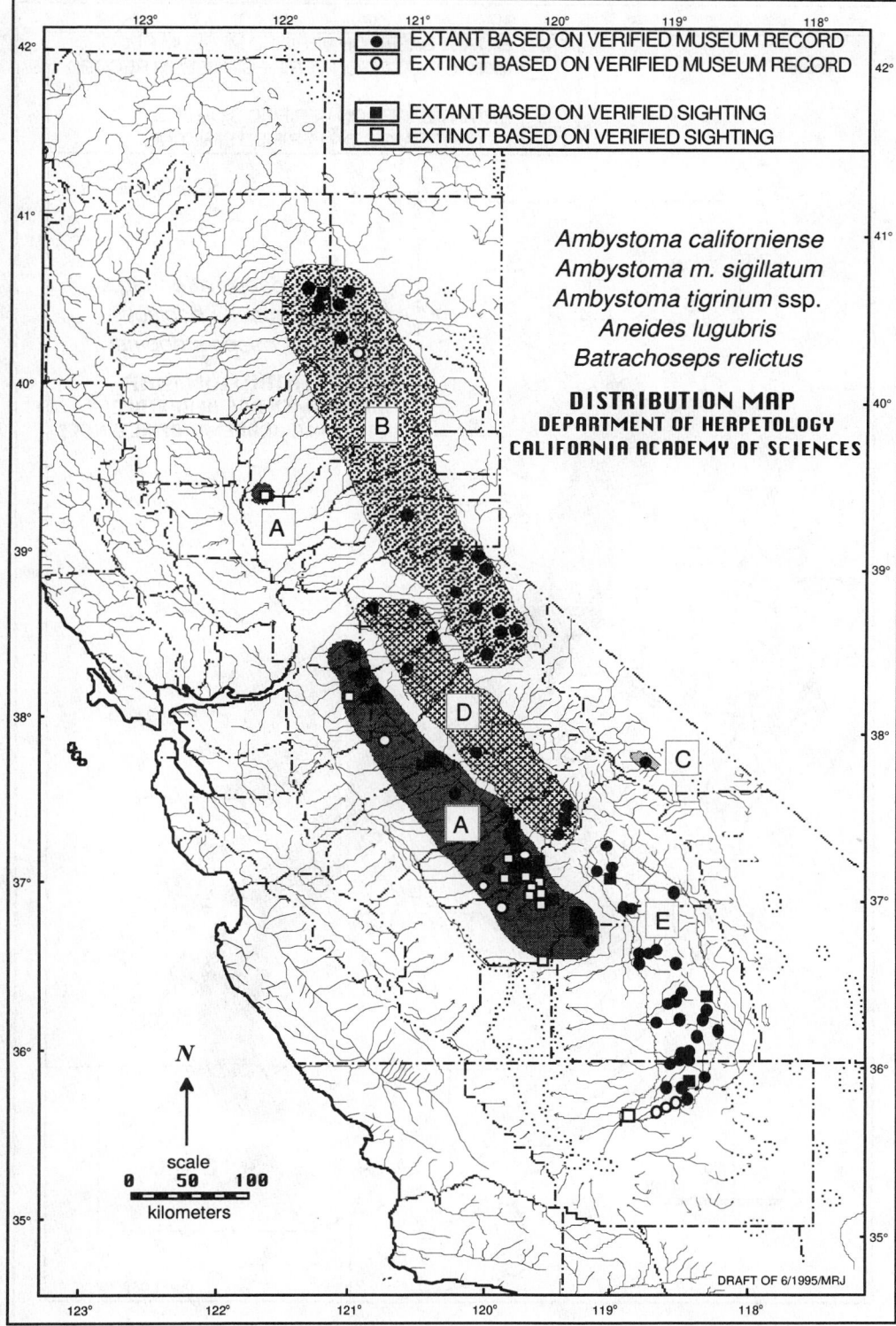

FIGURE 31.2

Historic and current distribution of (A) California tiger salamander *(Ambystoma californiense)*, (B) southern long-toed salamander *(Ambystoma macrodactylum sigillatum)*, (C) tiger salamander *(Ambystoma tigrinum* ssp.*)*, (D) arboreal salamander *(Aneides lugubris)*, and (E) relictual slender salamander *(Batrachoseps relictus)* in the Sierra Nevada.

Lungless Salamanders (Plethodontidae)

Lungless salamanders are represented by seventeen taxa (made up of fifteen species) in the Sierra Nevada (table 31.1; figures 31.2–31.7). They are among the most terrestrial of all amphibians in the Sierra Nevada because they do not require ponded surface water to breed. Instead, these salamanders lay their eggs in clusters in close contact with damp earth or wet rocky substrate, often hiding them in downed logs, un-

FIGURE 31.3

Historic and current distribution of (A) California slender salamander *(Batrachoseps attenuatus)*, (B) Pacific slender salamander *(Batrachoseps pacificus)*, and (C) Tehachapi slender salamander *(Batrachoseps stebbinsi)* in the Sierra Nevada.

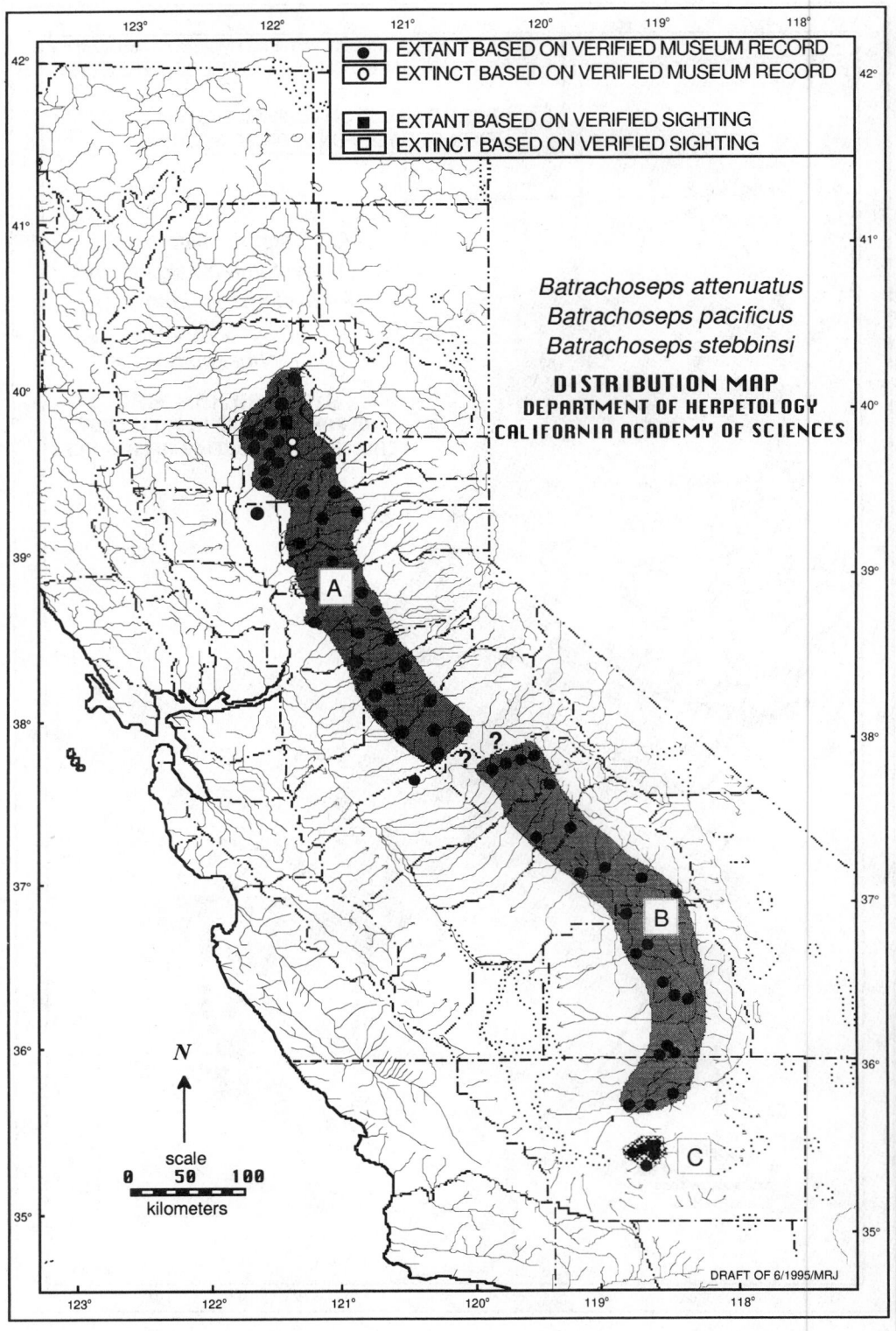

der talus slopes, or in natural rock caves (Stebbins 1951; Gorman 1956). Many species show parental care of eggs (Stebbins 1985). Development is direct. Because most of these salamanders are small in size and of ancient origin, they have undergone a high degree of isolation in the Sierra Nevada— so much so that twelve of the seventeen (71%) taxa are endemic to the region. Such endemism has only recently been recognized, and at least five species are in the process of be-

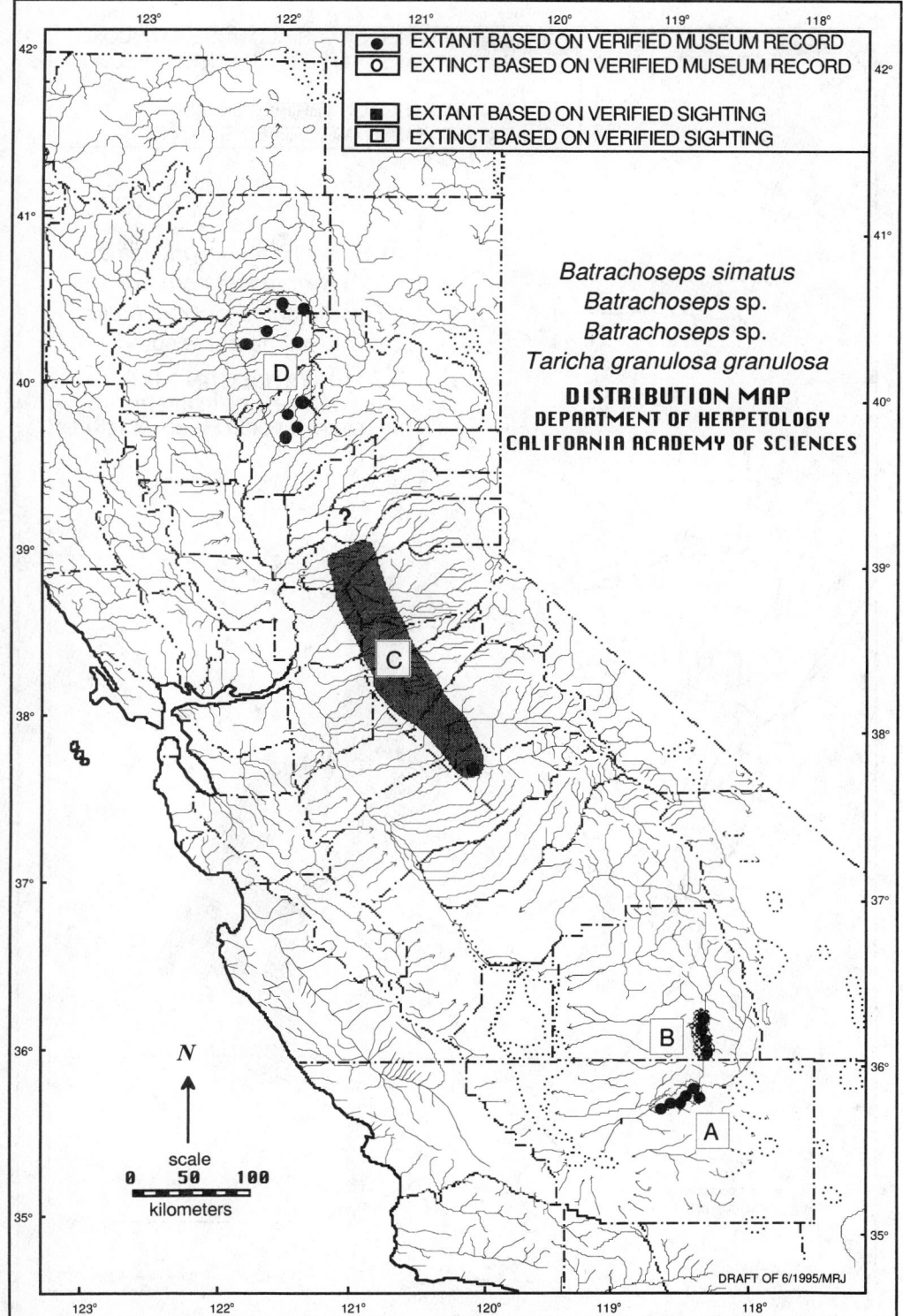

FIGURE 31.4

Historic and current distribution of (A) Kern Canyon slender salamander (*Batrachoseps simatus*), (B) Fairview slender salamander (*Batrachoseps* sp.), (C) Hell Hollow slender salamander (*Batrachoseps* sp.), and (D) northern rough-skinned newt (*Taricha granulosa granulosa*) in the Sierra Nevada.

ing formally described after genetic studies demonstrated their uniqueness (R. Hansen, editor, *Herpetological Review,* letter to the author, December 1, 1988; D. Wake, director, Museum of Zoology, University of California, Berkeley, letter to the author, July 19, 1994, conversation with the author, June 19, 1995). As a whole, lungless salamanders are generally restricted to small home ranges characterized by small patches of suitable habitat. Since much of the surrounding region is

FIGURE 31.5

Historic and current
distribution of (A) black-
bellied slender salamander
(Batrachoseps nigriventris),
(B) limestone salamander
(Hydromantes brunus), (C)
Mount Lyell salamander
*(Hydromantes
platycephalus),* and (D)
Owens Valley web-toed
salamander *(Hydromantes*
sp.) in the Sierra Nevada.

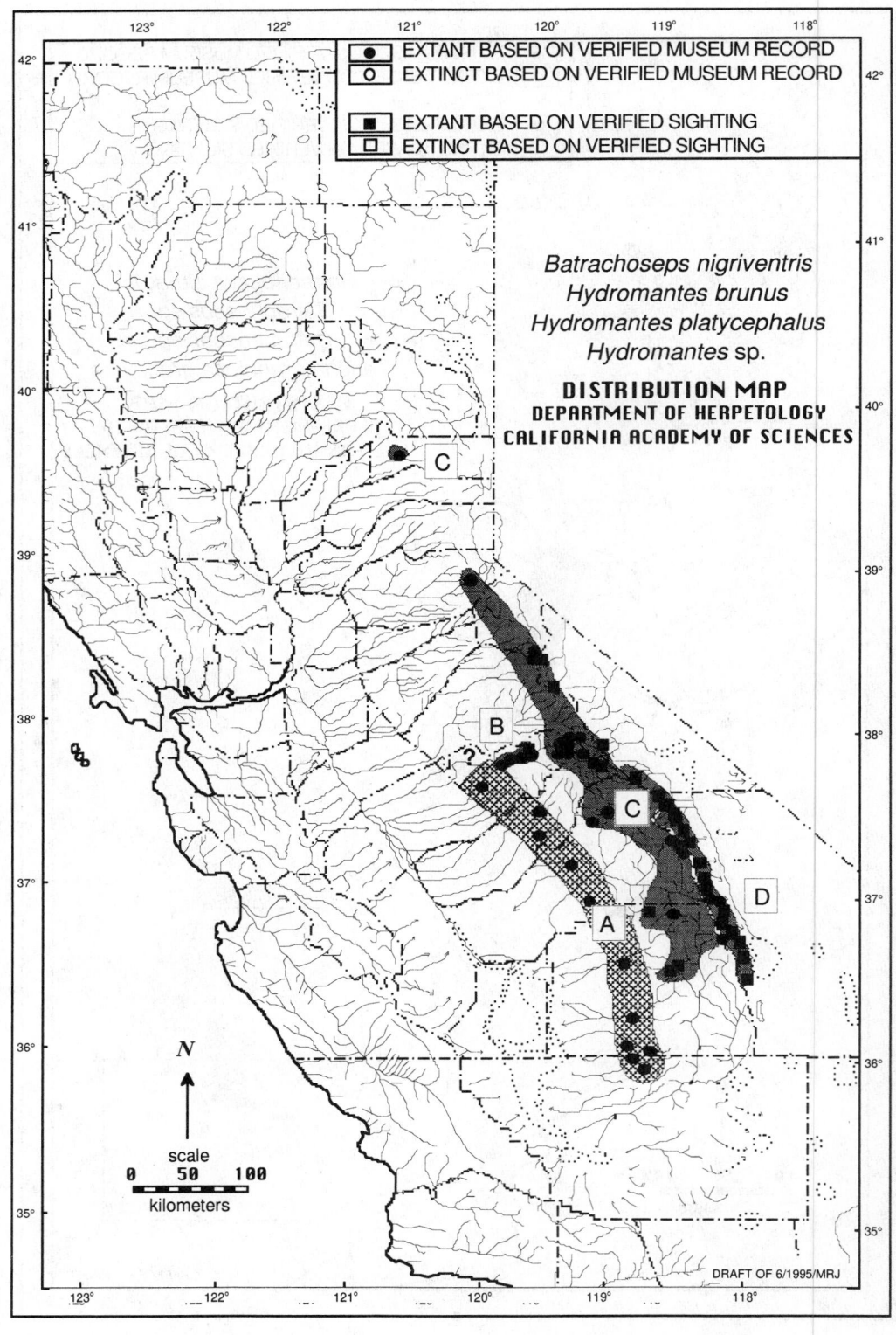

Batrachoseps nigriventris
Hydromantes brunus
Hydromantes platycephalus
Hydromantes sp.

DISTRIBUTION MAP
DEPARTMENT OF HERPETOLOGY
CALIFORNIA ACADEMY OF SCIENCES

● EXTANT BASED ON VERIFIED MUSEUM RECORD
○ EXTINCT BASED ON VERIFIED MUSEUM RECORD
■ EXTANT BASED ON VERIFIED SIGHTING
□ EXTINCT BASED ON VERIFIED SIGHTING

DRAFT OF 6/1995/MRJ

often composed of dry (or otherwise unsuitable) habitats, these salamanders are often vulnerable to activities that disrupt the hydrology of riparian canyons, the forest floor, and other mesic habitats (Jennings and Hayes 1994). Such negative activities include road building, mining, dam construction, and logging. Thus, at least seven species—the relictual slender salamander *(Batrachoseps relictus)* (special concern), the Kern Canyon slender salamander *(B. simatus)* (threatened),

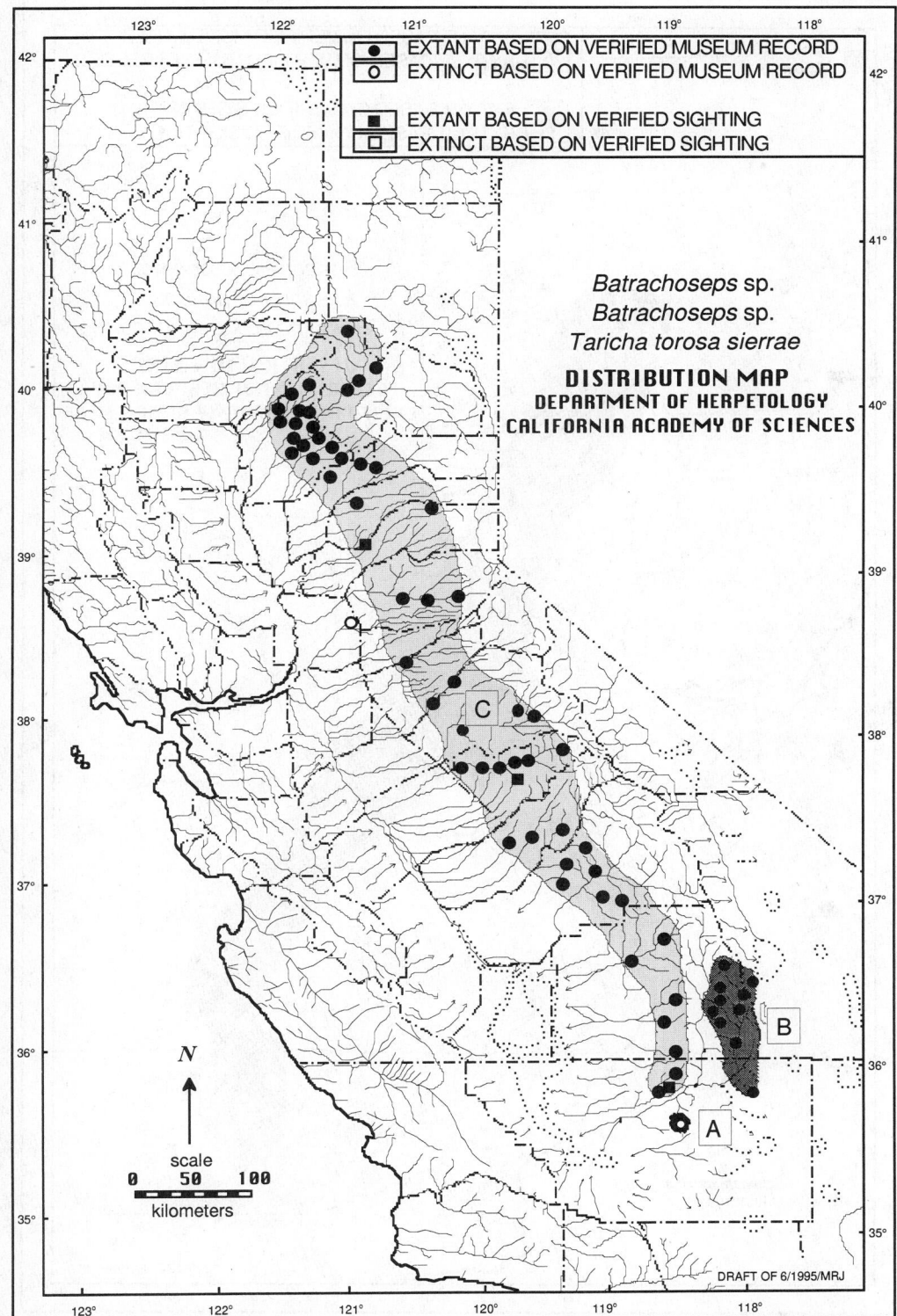

FIGURE 31.6

Historic and current distribution of (A) Breckenridge Mountain slender salamander (*Batrachoseps* sp.), (B) Kern Plateau slender salamander (*Batrachoseps* sp.), and (C) Sierra newt *(Taricha torosa sierrae)* in the Sierra Nevada.

the Tehachapi slender salamander *(B. stebbinsi)* (threatened), the yellow-blotched salamander *(Ensatina eschscholtzii croceater)* (special concern), the limestone salamander *(Hydromantes brunus)* (threatened), the Mount Lyell salamander *(H. platycephalus)* (special concern), and the Owens Valley web-toed salamander *(Hydromantes* sp.) (special concern)—are at risk because of these hydrology-disrupting activities in their habitats (e.g., see Steinhart 1990 and Jennings and Hayes 1994),

FIGURE 31.7

Historic and current
distribution of (A) yellow-
blotched salamander
*(Ensatina eschscholtzii
croceater)*, (B) Sierra Nevada
salamander *(Ensatina
eschscholtzii platensis)*, and
(C) yellow-eyed salamander
*(Ensatina eschscholtzii
xanthoptica)* in the Sierra
Nevada.

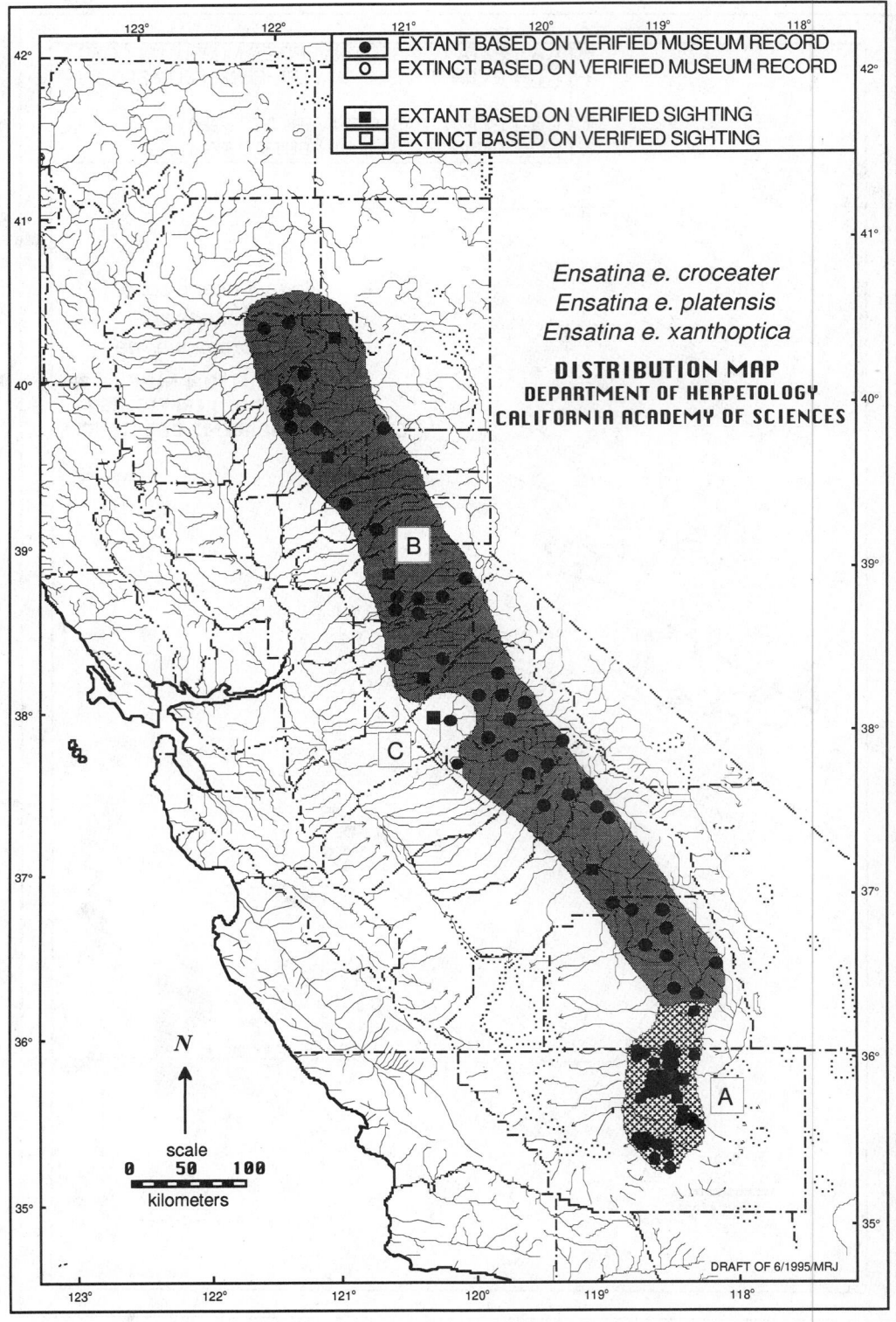

and the Breckenridge Mountain slender salamander *(Batrachoseps* sp.) apparently became extinct after a Forest Service road was rerouted above the seep that was its only known habitat (R. Hansen, conversation with the author, Oc-

tober 8, 1988). Without appropriate management actions by the responsible agencies or landowners, the same factors that have caused the decline of these seven species can also affect the five other endemic species with restricted distributions that so far have not had major disturbances within their known habitats.

Newts (Salamandridae)

Newts are represented by two species in the Sierra Nevada: the northern rough-skinned newt *(Taricha granulosa granulosa)* (figure 31.4) and the Sierra newt *(T. torosa sierrae)* (figure 31.6). Both newts have a life cycle like that of mole salamanders except that they also breed in streams as well as temporary pools, and adults are often found on the surface throughout much of the year. Both newts are secure within their current ranges, probably because the adults breed in small, often temporary streams at low to middle elevations (Stebbins 1951; observations by the author, 1988–95). They also seem better adapted to fluctuating conditions in streams than other aquatic salamanders. However, there is some recent evidence to indicate that aquatic newt larvae are highly susceptible to predation by introduced fishes (Liss and Larson 1991). Additionally, introduced bullfrogs are known to successfully consume juvenile and adult newts when given the opportunity (observations by the author). Recent ongoing genetic studies by students at the University of California, Berkeley, indicate that there are at least two taxa (probably species) within the subspecies currently recognized as *T. t. sierrae* (D. Wake, conversation with the author, June 19, 1995). Thus, further evaluation may be needed regarding these new endemic species with restricted distributions.

True Toads (Bufonidae)

True toads are represented by two species in the Sierra Nevada: the California toad *(Bufo boreas halophilus)* and the Yosemite toad *(B. canorus)* (figure 31.8). Both species require standing water (either in slow-moving streams or in ponds) for reproduction. The aquatic larval period for the true toads is short, usually only about two months (Storer 1925). After metamorphosis, the juveniles disperse into riparian habitats or other areas to mature (Stebbins 1951). Adults may live for ten years or more (Bowler 1977; Kagarise Sherman and Morton 1993).

The California toad is a subspecies of the widely distributed western toad *(Bufo boreas)*, a species that has undergone a substantial range reduction in the Rocky Mountain region and the Pacific Northwest during the past two decades (Carey 1993). In California, populations of the California toad seem to have been reduced as a result of urbanization, changing farming practices, and the use of pesticides, but the levels noted are not critical, as larvae, juveniles, and adults continue to be found in all known habitats (observations by the author, 1988–95). The Yosemite toad, on the other hand, is endemic to isolated high-mountain meadows in the central part of the Sierra Nevada. Its populations are declining so rapidly that the toad merits being listed as endangered (Jennings and Hayes 1994). For example, both Drost and Fellers (1994) and

Jennings and Hayes (1994) found this species to have disappeared from about half of its known historic localities. The causes for this decline are apparently similar to those for the decline of native true frogs (discussed later).

Tree Frogs (Hylidae)

Tree frogs are represented by a single species, the Pacific tree frog *(Hyla regilla)*, in the Sierra Nevada (figure 31.9). Like all frogs and toads, it has an aquatic larval stage that metamorphoses into a terrestrial juvenile. Pacific tree frogs reach maturity within one to two years after metamorphosis and are often found in terrestrial situations that may be more than 0.8 km (0.5 mi) from the nearest water source (Storer 1925; observations by the author, 1988–95). This tree frog is widely distributed throughout the American West and is found in good numbers at most Sierra Nevada localities (Bradford 1989; observations by the author, 1988–95). Bradford (1989) attributed this to the ability of Pacific tree frogs to breed in shallow water habitats or temporary ponds that are free of fish predators. However, Drost and Fellers (1994) note that while Pacific tree frogs are still widely distributed in the central Sierra Nevada, their numbers seem to be reduced at high elevations compared with historic observations. This reduction may be due to natural population fluctuations, as North American tree frogs (Hylidae) are known to undergo population fluctuations as much as thirtyfold or more (see Pechmann et al. 1991).

True Frogs (Ranidae)

True frogs are represented by six species in the Sierra Nevada: the foothill yellow-legged frog *(Rana boylii)* (figure 31.10), the northern leopard frog *(R. pipiens)* (figure 31.10), the bullfrog *(R. catesbeiana)* (figure 31.11), the California red-legged frog *(R. aurora draytonii)* (figure 31.12), the mountain yellow-legged frog *(R. muscosa)* (figure 31.12), and the Cascade frog *(R. cascadae)* (figure 31.13). The bullfrog is an introduced species originating in the United States east of the Rocky Mountains. It was first released in the Sierra Nevada about 1915 (Storer 1922) and has become well established in most perennial streams and ponds below 1,829 m (6,000 ft). Although currently considered a game species by the California Department of Fish and Game, it has been implicated in the decline of a number of native frog species (Moyle 1973; Hayes and Jennings 1986), and its game status is now under review by the California Fish and Game Commission (J. Brode, senior fisheries biologist, Inland Fisheries Division, California Department of Fish and Game, letter to the author, March 29, 1995).

All native true frogs have life histories like those of true toads and tree frogs. Their aquatic larval stage normally requires three to six months of development, and terrestrial juveniles require two to three years to reach adulthood (Zweifel 1955; Jennings and Hayes 1985), except for the mountain yel-

FIGURE 31.8

Historic and current distribution of (A) California toad *(Bufo boreas halophilus)* and (B) Yosemite toad *(Bufo canorus)* in the Sierra Nevada.

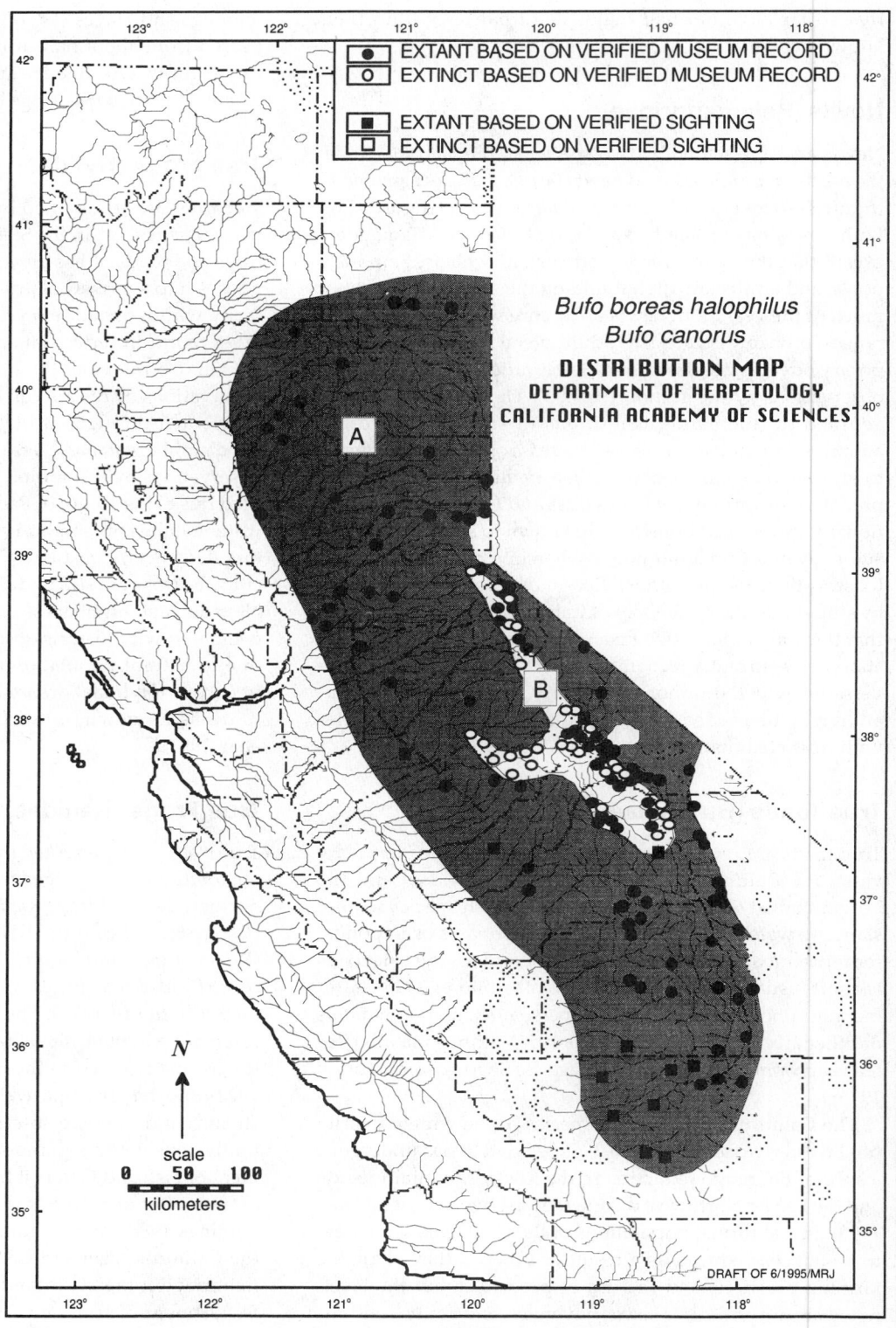

● EXTANT BASED ON VERIFIED MUSEUM RECORD
○ EXTINCT BASED ON VERIFIED MUSEUM RECORD

■ EXTANT BASED ON VERIFIED SIGHTING
□ EXTINCT BASED ON VERIFIED SIGHTING

Bufo boreas halophilus
Bufo canorus
DISTRIBUTION MAP
DEPARTMENT OF HERPETOLOGY
CALIFORNIA ACADEMY OF SCIENCES

scale
0   50   100
kilometers

DRAFT OF 6/1995/MRJ

low-legged frog, which has a considerably longer larval period of one to two and a half years (Bradford 1983). Such a long larval stage for the mountain yellow-legged frog makes it extremely vulnerable to predation by introduced aquatic predators such as trout, charr, and crayfish (Bradford 1989; Bradford et al. 1993).

The true frogs have shown the most dramatic declines of all groups of amphibians in the Sierra Nevada. They have

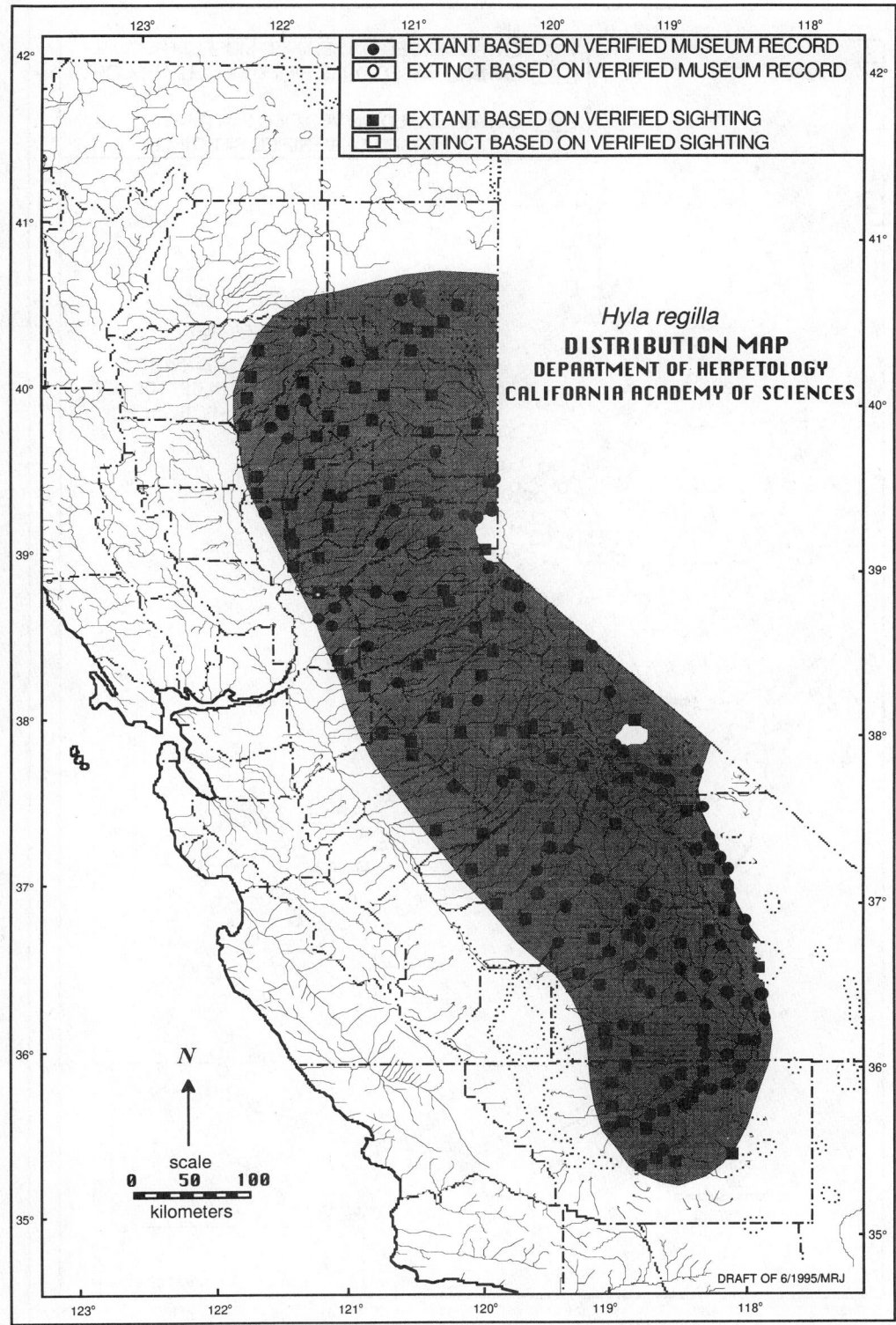

FIGURE 31.9

Historic and current
distribution of the Pacific tree
frog *(Hyla regilla)* in the
Sierra Nevada.

Legend on map:
- ● EXTANT BASED ON VERIFIED MUSEUM RECORD
- ○ EXTINCT BASED ON VERIFIED MUSEUM RECORD
- ■ EXTANT BASED ON VERIFIED SIGHTING
- □ EXTINCT BASED ON VERIFIED SIGHTING

Hyla regilla
DISTRIBUTION MAP
DEPARTMENT OF HERPETOLOGY
CALIFORNIA ACADEMY OF SCIENCES

scale
0 50 100
kilometers

DRAFT OF 6/1995/MRJ

disappeared from significant portions of their historic range over the past twenty-five years (Jennings 1995). In the Sierra Nevada, the California red-legged frog has disappeared from 99% of its historic range and has been proposed for listing as endangered under the ESA (Miller 1994). Similar observations of extensive frog declines in the Sierra Nevada continue to be made by many other biologists (see Hayes and Jennings 1986; Fellers and Drost 1993; Drost and Fellers 1994). For example,

FIGURE 31.10

Historic and current
distribution of (A) foothill
yellow-legged frog *(Rana
boylii)* and (B) northern
leopard frog *(Rana pipiens)*
in the Sierra Nevada.

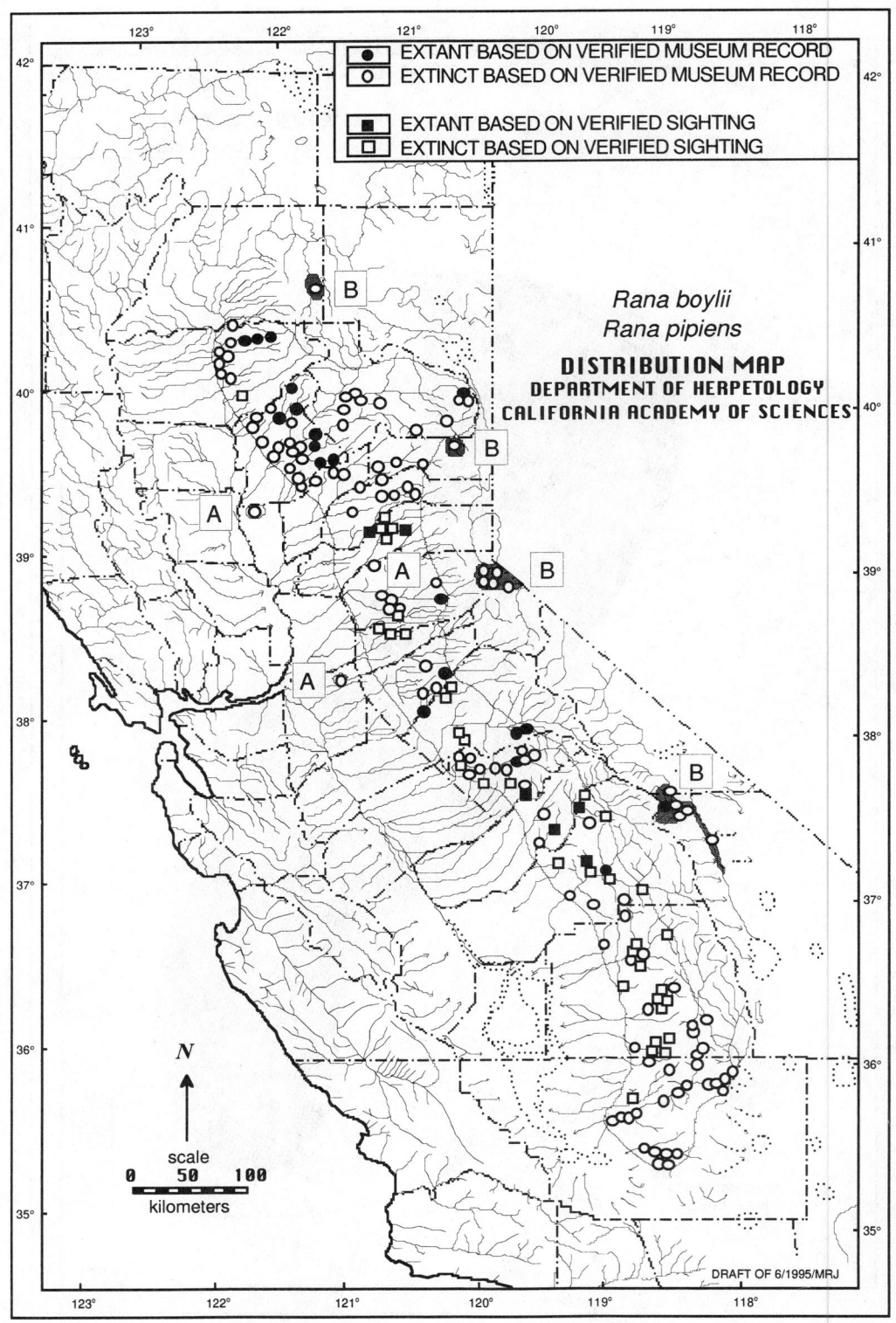

in 1987–88 foothill yellow-legged frogs were absent from all locations in the San Joaquin valley foothills where in 1970 they had been widespread and abundant (Moyle 1973; P. B. Moyle, Department of Wildlife, Fisheries, and Conservation Biology, University of California, Davis, conversation with the author, November 20, 1995). Currently, the foothill yellow-legged frog, Cascade frog, mountain yellow-legged frog, and northern leopard frog seem to have disappeared from

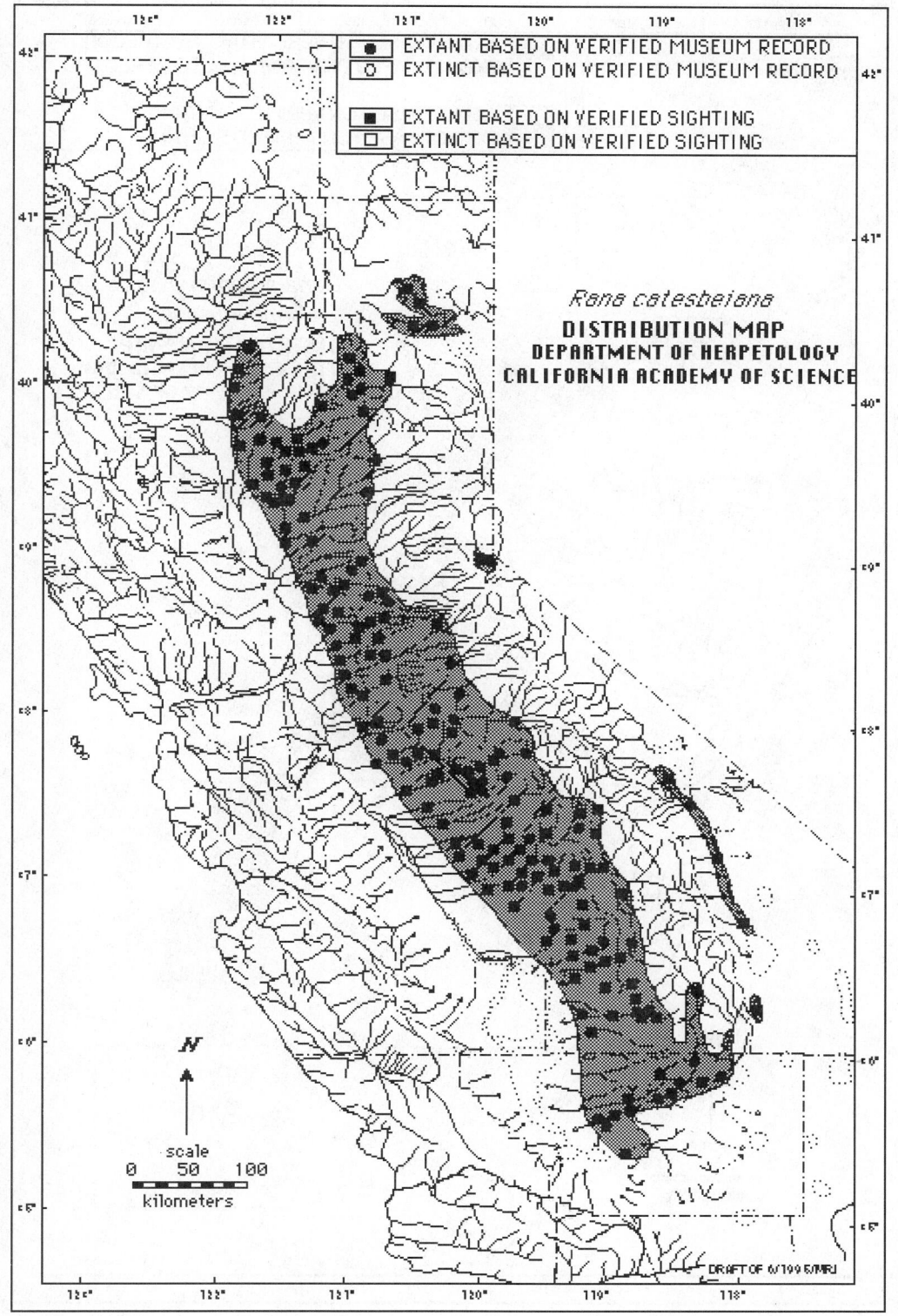

FIGURE 31.11

Current distribution of the introduced bullfrog (*Rana catesbeiana*) in the Sierra Nevada.

about 45%, 50%, 50%, and 95% of their historic ranges in California and from about 66%, 99%, 50%, and 99% of their historic ranges in the Sierra Nevada, respectively (Jennings and Hayes 1994). All of these frogs can now be considered to be threatened in the Sierra Nevada, except for the Cascade frog, which is considered to be endangered in this same region (Jennings and Hayes 1994).

FIGURE 31.12

Historic and current
distribution of (A) California
red-legged frog (*Rana aurora
draytonii*) and (B) mountain
yellow-legged frog (*Rana
muscosa*) in the Sierra
Nevada.

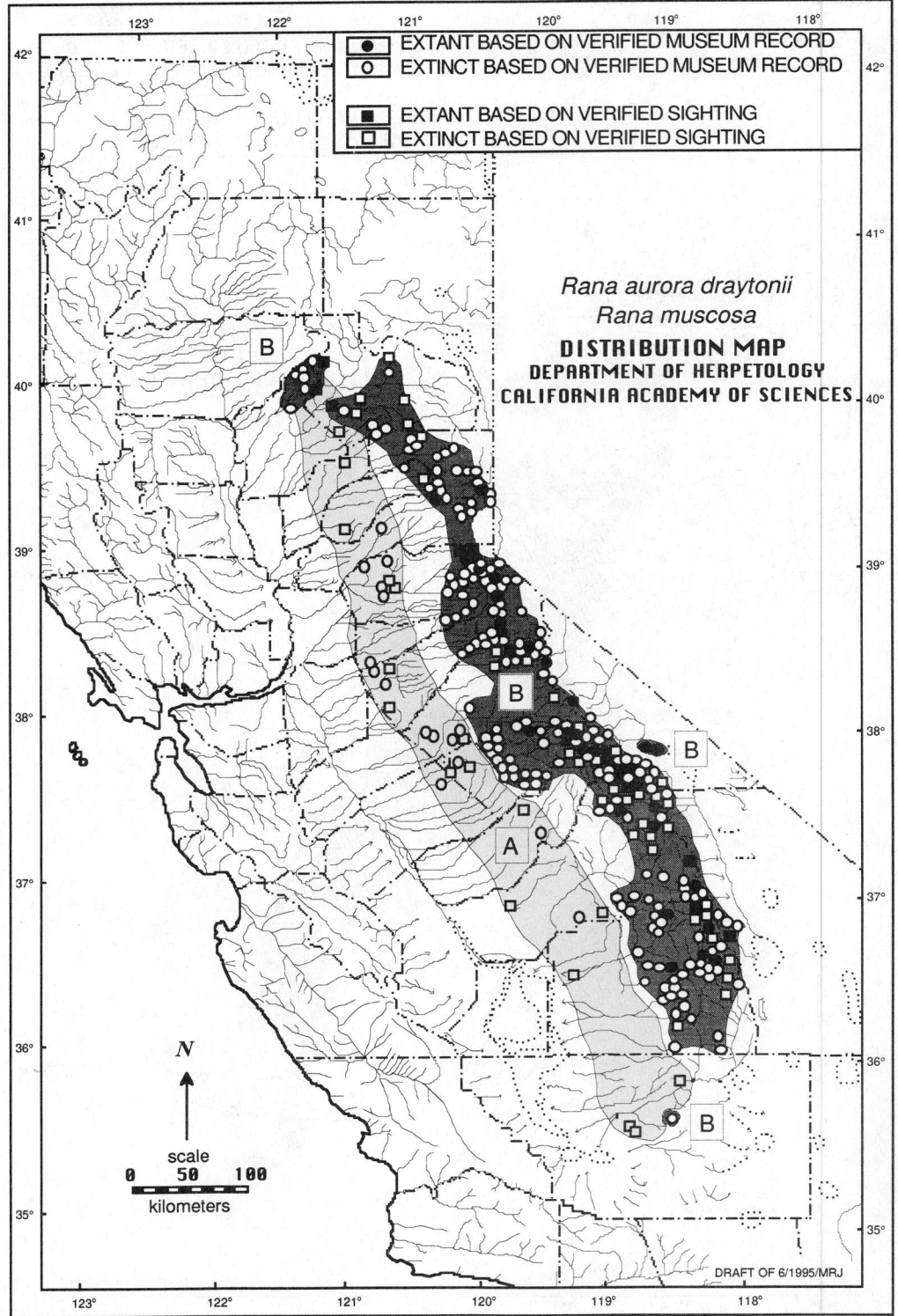

Spadefoot Toads (Pelobatidae)

Spadefoot toads are represented by a single species, the western spadefoot (*Scaphiopus hammondii*) in the Sierra Nevada (figure 31.13). It is a lowland species that has adapted to dry environments by breeding in temporary ponds and slow-moving streams (Stebbins 1985). Adults and juveniles burrow into suitable substrates near breeding sites or use small mammal burrows to avoid desiccation throughout most of

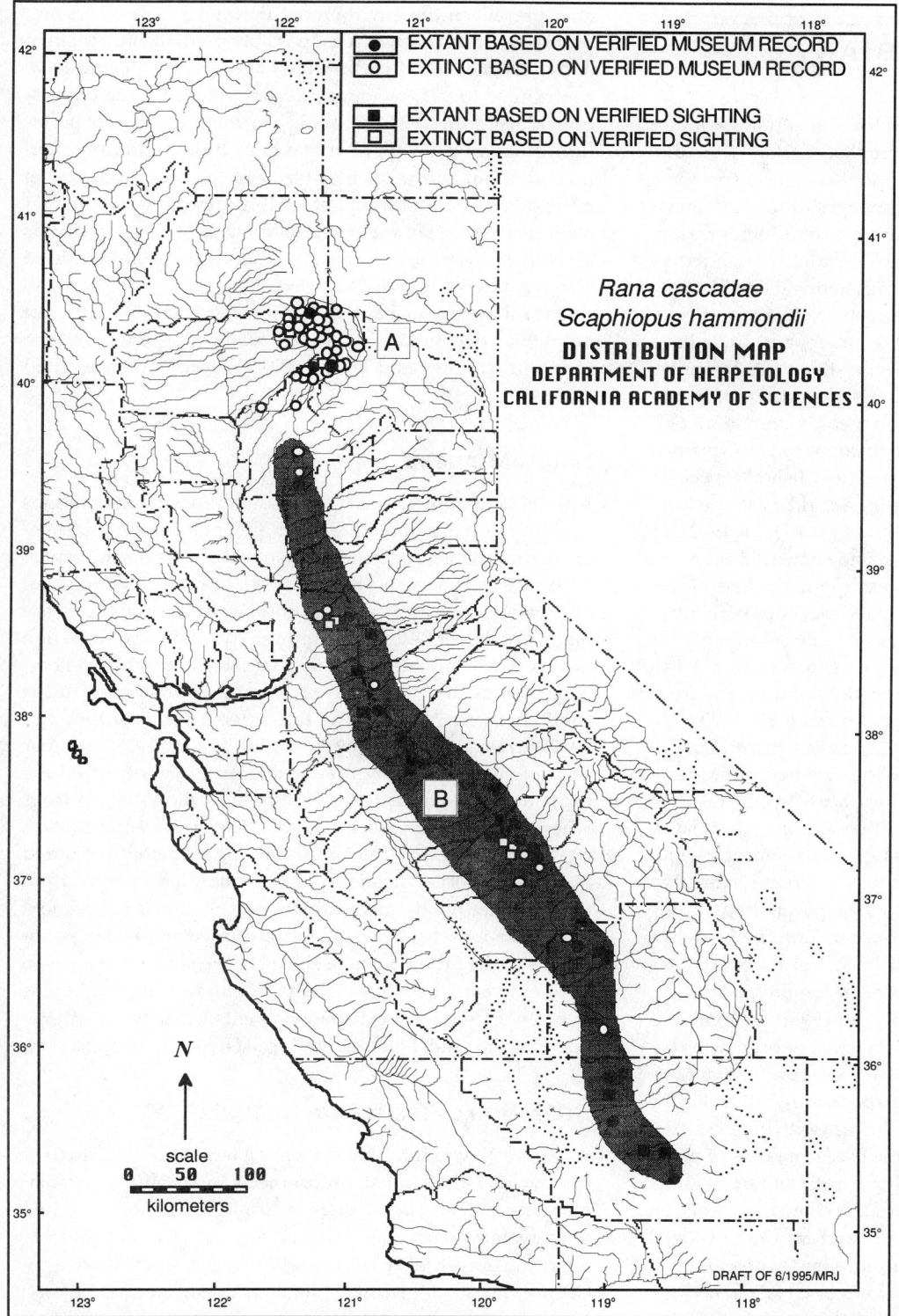

FIGURE 31.13

Historic and current distribution of (A) Cascade frog *(Rana cascadae)* and (B) western spadefoot *(Scaphiopus hammondii)* in the Sierra Nevada.

the year (Storer 1925). Adults are active on the surface only during short periods of time (such as the winter months) when conditions are suitable. The species is largely endemic to California and is found along the western edge of the Sierra Nevada foothills. Because of habitat loss due to agriculture and urbanization, this organism is considered a species of special concern (Jennings and Hayes 1994).

CAUSES OF AMPHIBIAN DECLINES

The reasons for the precipitous declines in certain amphibians (especially native frogs) are complex. Certainly the disappearance of all of the middle- to low-elevation species is due largely to habitat alteration from agriculture, urbanization, water development, placer mining, livestock grazing, drought, and the introduction of a wide variety of non-native predatory fishes, crayfish, and bullfrogs (Jennings 1995). However, in many localities none of these activities have occurred, yet amphibians such as native salamanders and frogs have still disappeared within the past twenty-five years (Jennings and Hayes 1994). The widespread disappearance of native frogs from middle- to high-elevation areas is even more perplexing. Some of these population extinctions can be explained by the widespread introduction of predatory fishes (especially trout and charr) in the Sierra Nevada over the past one hundred years (Bradford et al. 1993; Fellers and Drost 1993), as well as extensive livestock grazing and increased levels of recreation in sensitive breeding areas (Jennings and Hayes 1994). However, in literally hundreds of localities both introduced trout and native salamanders and frogs seemingly co-occurred together for at least fifty years, based on old fish planting records and common observations of trout and frogs in the same aquatic habitats in the 1960s and early 1970s (observations by the author, 1960–95). Apparently, a number of different factors are contributing to the declines in these amphibian species. Declines such as these are often the result of long-term, cumulative effects of multiple factors, where natural low points in amphibian population cycles synergize with widespread environmental alterations to create extinction events (e.g., many of the Yosemite toad populations in the Sierra Nevada underwent dramatic population crashes when they were unable to reproduce at historical breeding sites during extended periods of drought in the Sierra Nevada, 1986–90 [Kagarise Sherman and Morton 1993]). Amphibians seem to be in worse condition than most other organisms because they are uniquely vulnerable to these cumulative environmental effects. This is because species are either highly localized in their distribution (as is the case with lungless salamanders) or because they fit into the classic models of metapopulation dynamics (as is true of true frogs and true toads). Some of the best metapopulation studies come from work on true frogs in Scandinavia, where researchers find localized frog populations undergoing continuous cycles of extinction and recolonization from nearby sources (see Harrison 1991). Such recolonization events for native frog populations in the Sierra Nevada are now impossible for many areas because of the widespread extinction of many local source frog populations and the presence of introduced predators in most formerly suitable habitats (Bradford et al. 1993).

Whatever the problems are that are causing the decline among amphibians, there is no doubt that many stressors are now present in the environment that negatively affect amphibians such as native frogs, possibly predisposing them to native or introduced pathogens. These stressors could have contributed to the precipitous declines in many frog populations during the 1970s. Possible stressors include air pollution, increased levels of ultraviolet light radiation, acid precipitation, and pesticides (each of these is discussed in more detail later in this chapter). It is important to note that all of these stressors are linked to human needs, especially as a result of ever-increasing population growth in the Sierra Nevada and the rest of California.

The following is an annotated list of the possible causes of amphibian declines in the Sierra Nevada. More details can be found in Jennings and Hayes 1994 and the references cited therein.

Natural Causes

Amphibian populations naturally undergo wide fluctuations in abundance in response to environmental conditions, especially droughts, floods, and epizootic diseases (Pechmann et al. 1991; Pechmann and Wilbur 1994). Local or even regional extirpations are apparently common, but populations are maintained over wide areas through dynamic recolonization events. For example, some populations of foothill yellow-legged frogs disappeared from a number of streams in the southern Sierra Nevada after the extreme floods of 1968 and 1969 (personal observations by the author, 1970–95). Another example is provided by Bradford (1991), who observed the extinction of a population of mountain yellow-legged frogs when a flock of Brewer's blackbirds (*Euphagus cyanocephalus*), not normally regarded as significant frog predators, devoured the entire cohort of metamorphosing frogs that emerged from the breeding pond. Such extinction events (and subsequent recolonizations) have presumably governed amphibian populations in the Sierra Nevada since the Pleistocene. However, in recent years the natural ability of amphibian populations to recover from local extirpation events has been greatly reduced as the result of human-induced environmental changes.

Alteration of Terrestrial Habitats

All Sierra Nevada amphibians have a terrestrial stage to their life cycle. This is most pronounced in the lungless salamanders, which spend their entire lives without needing open-water environments. Thus, any activity that severely alters the terrestrial environment, such as urbanization, agriculture, livestock grazing, timber harvest, mining, or road building, is likely to result in the reduction and occasional extirpation of amphibian populations. For example, the release of domestic livestock in high-mountain meadows utilized by Yosemite toads (for reproduction) has resulted in the pollution of breeding ponds as well as the trampling of toad larvae and juveniles (Jennings and Hayes 1994). Except for unusual circumstances, such as the construction of the road that modified the only known habitat for the Breckenridge Mountain slender salamander (Jennings and Hayes 1994), single actions do not eliminate species. Nonetheless, changes to terrestrial habitats are often cumulative (as is the case with long-term livestock grazing, which tends to eliminate certain plants that provide important cover for many amphibian species) and may occur too frequently for population recovery following events that reduce resident amphibian populations.

Alteration of Riparian Habitats

Since most Sierra Nevada amphibians spend significant portions of their life cycles either in or moving through riparian habitats, these areas are important to their overall survival. For instance, foothill yellow-legged frogs and California red-legged frogs seem to require riparian areas that are well developed structurally (for cover and estivation as well as the production of food resources) but that also contain open areas for basking (Hayes and Jennings 1988). Thus, the degradation of riparian areas can lead to habitat fragmentation, loss of corridors necessary for recolonization, and the ultimate loss of local amphibian populations. Specific examples of factors contributing to this degradation are livestock grazing, road building, reservoir construction, and recreation (Jennings and Hayes 1994). The most obvious reasons for the demise of native amphibians due to these factors are (1) increased dehydration and increased predation due to the loss of vegetative cover; (2) changes in the structure and composition of the flora (thus affecting important food resources); and (3) the crushing or removal of small or cryptic individuals due to trampling, vehicles, or the results of human activities. Specific examples include (1) increased dehydration rates for slender salamanders in habitats where the riparian cover was removed (see Ray 1958); (2) the loss of riparian willows (*Salix* spp.), which resulted in increased predation on California red-legged frogs by raccoons (*Procyon lotor*) (Miller 1994); (3) the loss of important food resources that are critical for the growth and survival of juvenile frogs and toads, due to the removal of vegetation upon which invertebrates feed (Jennings and Hayes 1994); and (4) the crushing of individuals by livestock grazing in alpine meadows, which resulted in trampled larval and juvenile Yosemite toads (D. Martin, Martin, Canorus Ltd., letter to the author, May 12, 1991), or by motorcycle use in riparian zones, which crushed juvenile and adult foothill yellow-legged frogs and garter snakes (personal observations by the author, 1986–90).

Alteration of Aquatic Habitats

As is widely stated in the literature (e.g., see Moyle 1976), aquatic habitats of the Sierra Nevada have been greatly altered through dams, diversions, channelizations, siltation, livestock grazing, timber harvest, placer mining, and many other factors. The same factors that have made these habitats less suitable for native fishes have also made them less suit-

able for native amphibians. Reservoirs, found on most larger Sierra Nevada streams, disrupt native aquatic amphibians because most of these organisms cannot live in, or move through, the exposed shorelines, nor can they successfully reproduce in such fluctuating environments containing introduced predatory fish, crayfish, and bullfrogs. For example, mountain yellow-legged frogs seem unable to successfully produce a cohort of young in artificial reservoirs (with predatory fish) unless shallow side channels or disjunct pools are present that are separated from the main body of water (thus excluding the fish) (D. Bradford, U.S. Environmental Protection Agency, conversation with the author, February 4, 1992). Additionally, there are a number of observations of native adult frogs (*R. a. draytonii*, *R. cascadae*, and *R. muscosa*) being consumed by large introduced trout after the frogs were accidentally scared into the water by humans (Drost and Fellers 1994; L. Simons, graduate student, Department of Evolution and Ecology, University of California, Davis, letter to the author, September 9, 1994; observations by the author, 1989–90).

Besides the above, alteration of the natural hydrological regime often creates habitat conditions unfavorable to native amphibians, and the dams and their associated structures may create serious barriers to movements by dispersing juveniles and migrating adults. For example, open pipelines and canals have been found to catch and kill migrating adult California tiger salamanders that fall into them (Sorenson 1994). An additional example is the placement of reservoirs at middle to lower elevations in the Sierra Nevada, which has resulted in the creation of many year-round cold-water streams below dams. These reservoirs (along with unseasonal releases of water) resulted in unsuitable breeding habitats for foothill yellow-legged frogs and the scouring out of their egg masses downstream during the spring (Jennings and Hayes 1994). Thus, it is rare to find open-water-dependent native amphibians immediately below reservoirs, especially large reservoirs.

Introduction of Aquatic Predators

Hayes and Jennings (1986), Bradford (1989), Bradford et al. (1993), and Jennings and Hayes (1994), along with many others, have noted the generally negative correlation between the presence of introduced predators (especially fishes and bullfrogs) and the abundance of native amphibians in streams and lakes of California. Introduced bullfrogs, fishes, and crayfishes seem to be a particular problem for many species (such as California tiger salamanders and mountain yellow-legged frogs), probably because these organisms did not coevolve with a suite of aquatic predators (Hayes and Jennings 1986). Limited field and laboratory observations on how bullfrogs, fishes, and crayfishes feed on native amphibians indicate that although all life stages of the latter are eaten, it is the larvae of mole salamanders, newts, and true frogs and toads that are most susceptible (observations by the author, 1989–91). This is because most native amphibian larvae have traits that predispose them to introduced predators—especially sight predators such as trout or tactile predators such as crayfish—during periods of darkness (e.g., the larvae lack toxic skin secretions, lie on top of the benthos at night, have poor swimming escape tactics) (Hayes and Jennings 1986). Mountain yellow-legged frogs are probably the most obvious example of a species that is predisposed to predation because they have such a long larval period (one to two and a half years) that there is a relatively good chance of being exposed at some point to introduced aquatic predators if the latter are present. Overall, there is strong evidence that introduced fishes continue to limit the distribution and abundance of certain native amphibians in parts of the Sierra Nevada (Bradford et al. 1993; Jennings and Hayes 1994).

Disease

The presence of a wide variety of pathogens (some of which are native and others introduced) in salamanders and frogs has long been noted as a cause of local amphibian declines (Bradford 1991). The role of disease in the decline of certain frog and toad species in the American West has recently received more attention (see Carey 1993 and Scott 1993). Some of the more plausible hypotheses are that stressors, such as increased levels of UV-B radiation or air pollutants, cause a weakening of the immune system, which could cause an increased susceptibility to natural diseases (Blaustein and Wake 1995). Another hypothesis, supported by limited observations, is that diseases carried by planted trout may attack and kill amphibian eggs and larvae (Blaustein et al. 1994b). The overall importance of diseases as a cause of death among native amphibians is hard to assess, but it is probably the most important source of mortality for individuals stressed by other factors (Scott 1993).

Acid Precipitation

The widespread acidification of mountain streams and lakes in the Northeast, Rocky Mountains, and Europe has been associated with amphibian declines (Haines 1981). While unbuffered waters of the Sierra Nevada are subject to acidification from air pollution (Nikolaidis et al. 1991), Bradford et al. (1994) could not find any evidence that anthropogenic acidification is a major problem there, except for highly localized spots that receive acid runoff from a point source (such as a mine). However, the potentially negative effects of acidification were demonstrated by an examination of naturally acidic lakes in the Sierra Nevada. No lake with a pH value less than 6 supported amphibian populations (Bradford et al. 1994).

Pesticides

Like acid precipitation, pesticides have been suspected of affecting amphibian abundance, especially agricultural pesticides drifting upward from the San Joaquin valley. Previously, DDT was found in significant quantities in mountain yellow-legged frogs throughout the Sierra Nevada (Cory et al. 1970). More recently, the finding that pesticides mimic estrogen in vertebrates has been proposed as a hypothesis for amphibian declines (see the discussion in Stebbins and Cohen 1995). Pesticide deposition has increased in recent years in the San Joaquin foothills because of the rise of mega-agriculture on the valley floor (T. Cahill, Crocker Nuclear Laboratory, University of California, Davis, conversation with the author, April 26, 1995). However, none of these pesticide hypotheses have been tested, and their overall effects on Sierra Nevada amphibians are unknown.

Automobile Emissions

Recent studies by the Crocker Nuclear Laboratory, Air Quality Group, University of California, Davis, have noticed that the pattern of recent frog extinctions in the southern Sierra Nevada corresponds with the pattern of highest concentrations of air pollutants from automobile exhaust (T. Cahill, conversation with the author, April 26, 1995). It is possible that the increased nitrification (or other changes) in streams and lakes by these chemicals may be affecting frog reproduction and survival. Air pollution seems to be seriously weakening the coniferous trees of the Sierra Nevada (California Air Resources Board 1987) and may be having negative effects on other parts of the ecosystem as well.

Ultraviolet Light

As the ozone layer of the upper atmosphere thins due to some forms of air pollution, the earth has been bombarded by increased ultraviolet (UV) radiation. For amphibians that sun themselves (and amphibian eggs that develop in unshaded, shallow water habitats), exposure to increased levels of UV-B radiation may increase mortality rates, especially for those species that are unable to repair DNA damaged by UV-B radiation (see Blaustein et al. 1994a and Blaustein et al. 1995). The hypothesis that UV radiation is related to amphibian declines is favored by some herpetologists (see especially Blaustein and Wake 1995) because it could help explain (1) global amphibian declines, (2) the coincidence of rapid declines of several different species in many areas in recent years, and (3) the severe declines at high elevations (Wake 1991). However, this hypothesis has come under increasing attack by a number of scientists because "of its apparent lack of scientific rigor with regard to observed field situations" (e.g., see Roush 1995; but see also Blaustein 1995; Formanowicz 1995; Halliday 1995; and Reznick 1995). As Drost and Fellers (1994) state, "The evidence for an influence from ultraviolet radiation remains speculative and circumstantial, but until compelling evidence is brought forth for some other cause, this hypothesis must be considered an important possibility" (Drost and Fellers 1994, 31). Closer examination of the subject reveals three facts that make the hypothesis suspect. The first is that observed die-offs of native frogs and toads in the

American West occur among adults and juveniles (e.g., see Carey 1993 and Scott 1993), not among developing embryos, as shown in Blaustein et al. 1994a. Second, UV-B would negatively affect all organisms sensitive to this factor. However, there have been no documented die-offs of likely sensitive plants and insects due to "increased" UV-B in the Sierra Nevada. Finally, measurements by the Crocker Nuclear Laboratory, Air Quality Group, University of California, Davis, indicate that UV levels at high elevations in the Sierra Nevada have increased by no more than 5% over the past several decades (T. Cahill, conversation with the author, April 26, 1995). Thus, it is unlikely that increased UV-B levels are a major cause of amphibian declines in the Sierra Nevada.

CONCLUSIONS

It is apparent that a significant percentage of the native amphibian species inhabiting the Sierra Nevada have shown dramatic declines in abundance, distribution, and diversity in recent years. A total of 53% of the thirty native taxa now require some sort of protection. That these declines have something to do with the life history traits of the taxa and the disruption of aquatic environments is made evident by examining the status of reptiles in the Sierra Nevada (Jennings and Hayes 1994). Of the twenty-six Sierra Nevada species (twenty-four of which are terrestrial) within the study region—excluding another twenty desert species that occur on the periphery of the mountains—twenty (77%) are secure, four (15%) are listed or merit listing, and two (8%) are of special concern. Only one of the four threatened species is in serious decline—the western pond turtle (*Clemmys marmorata*), a highly aquatic species.

It is certain that amphibian declines and extinctions have been caused by a number of interacting factors, with each taxon being affected in different ways (table 31.2). Such factors can range from global to local, but the most important ones in the Sierra Nevada appear to be the alteration of terrestrial and aquatic habitat, habitat fragmentation, and the introduction of aquatic predators (table 31.2). Fortunately, there are still some watersheds where native amphibians thrive in sufficient numbers to ensure survival for the time being. In the foothills, these locations tend to be small streams that have a heavy riparian canopy, that are free of introduced predators, and that have been relatively undisturbed by livestock grazing, timber harvest, water development, and placer mining. At high elevations, such habitats tend to be in clusters of fishless lakes and streams in remote areas. These observations suggest that localized restoration of amphibian habitats, such as the creation of fishless basins (or watersheds) in wilderness areas, is possible. It is essential that the watersheds listed in table 31.3 (which have especially high values for amphibian conservation) be considered for protecting

important amphibian resources. Further, it is also important to note that native amphibians (especially true frogs) in the Sierra Nevada can no longer exist as metapopulations but rather must be seen as fragmented, individual populations that are highly vulnerable to extirpation. This fragmentation and likely extinction (without hope of recolonization) is certain to lead to local, then regional, then Sierra-wide extinctions of selected amphibian species if current trends continue.

Finally, it should be noted that there is a hopeful sign for the potential recovery of certain lower-elevation species if the habitat is restored and introduced aquatic predators are reduced or eliminated. For example, the South Fork of the Yuba River was badly sluiced by placer gold mining activities from the 1850s to the 1870s. With the recovery of the riparian zone, this stream currently has a good population of foothill yellow-legged frogs in suitable patches of habitat. There are a number of other large Sierran streams that fit this category (especially in the northern half of the Sierra Nevada), and efforts should be made to restore riparian and aquatic habitats and protect any sensitive native amphibians that are extant.

ACKNOWLEDGMENTS

My thanks to Mark L. Allaback, John M. Brode, David F. Bradford, Thomas A. Cahill, Robert W. Hansen, Marc P. Hayes, David L. Martin, Peter B. Moyle, Lee H. Simons, and David B. Wake for providing information used in this report. Museum curators and collection managers (acknowledged in Jennings and Hayes 1994) graciously provided the museum records used to produce the distribution maps of amphibians in the Sierra Nevada.

REFERENCES

Blaustein, A. R. 1994. Chicken Little or Nero's fiddle? A perspective on declining amphibian populations. *Herpetologica* 50 (1): 85–97.

———. 1995. Letters; ecological research. *Science* 269 (5228): 1201–2.

Blaustein, A. R., B. Edmond, J. M. Kiesecker, J. J. Beatty, and D. G. Hokit. 1995. Ambient ultraviolet radiation causes mortality in salamander eggs. *Ecological Applications* 5 (3): 740–43.

Blaustein, A. R., P. D. Hoffman, D. G. Hokit, J. M. Kiesecker, S. C. Walls, and H. B. Hays. 1994a. UV repair and resistance to solar UV-B in amphibian eggs: A link to population declines? *Proceedings of the National Academy of Sciences* 91 (5): 1791–95.

Blaustein, A. R., D. G. Hokit, R. K. O'Hara, and R. A. Holt. 1994b. Pathogenic fungus contributes to amphibian losses in the Pacific Northwest. *Biological Conservation* 67 (3): 251–54.

Blaustein, A. R., and D. B. Wake. 1995. The puzzle of declining amphibian populations. *Scientific American* 272 (4): 52–57.

Bowler, J. K. 1977. Longevity of reptiles and amphibians in North American collections. *Herpetological Circular* no. 6: 1–32. Lawrence: University of Kansas, Society for the Study of Amphibians and Reptiles.

TABLE 31.2

Relative importance of various factors in the decline of Sierra Nevada amphibians.

Species	Natural Causes	Terrestrial Alteration	Fragmen- tation	Riparian Changes	Aquatic Changes	Introduced Predators	Acid Rain	Pesticides/ Pollutants	UV Radiation	Disease
Salamanders										
Mole Salamanders										
California tiger salamander	1	3	2	3	1	2	0	1	?	1
Lungless Salamanders										
Relictual slender salamander	2	3	2	1	1	0	0	0	0	0
Kern Canyon slender salamander	0	3	2	1	0	0	0	0	0	0
Tehachapi slender salamander	0	3	2	0	0	0	0	0	0	0
Breckenridge Mountain slender salamander	1	3	3	3	0	0	0	0	0	0
Yellow-blotched salamander	0	2	1	1	0	0	0	0	0	0
Limestone salamander	0	2	2	1	0	0	0	0	0	0
Mount Lyell salamander	0	1	1	0	0	0	0	0	0	0
Owens Valley web-toed salamander	0	2	1	2	0	0	0	0	0	0
Frogs and Toads										
True Toads										
Yosemite toad	2	1	2	1	1	2	1	1	?	?
True Frogs										
California red-legged frog	2	2	3	2	1	3	0	1	0	1
Foothill yellow-legged frog	2	1	2	2	1	3	0	1	0	2
Cascade frog	1	0	2	1	0	2	0	0	0	2
Mountain yellow-legged frog	1	0	2	1	0	3	1	1	?	2
Northern leopard frog	1	1	1	1	1	1	0	0	0	2
Spadefoot Toads										
Western spadefoot	1	3	1	2	1	2	0	1	0	0
Totals	14	30	29	22	7	18	2	6	0	10

0 indicates the factor was of no importance.
1 indicates the factor was a minor contributor.
2 indicates the factor was an important contributor.
3 indicates the factor was a major contributor.
? indicates the importance of the factor is unknown.

TABLE 31.3

Watersheds with especially high values for amphibian conservation that should be protected.

County	Watershed	Species
Alpine	North Fork Mokelumne River (all tributaries)	Yosemite toad, mountain yellow-legged frog
Alpine	North Fork Stanislaus River (above Union Reservoir)	Yosemite toad, mountain yellow-legged frog
Amador	North Fork Mokelumne River (all tributaries)	Mountain yellow-legged frog
Butte	Big Chico Creek	Foothill yellow-legged frog
El Dorado	Alder Creek	Mountain yellow-legged frog
El Dorado	Camp Creek	Foothill yellow-legged frog
El Dorado	Caples Creek	Mountain yellow-legged frog
El Dorado	Silver Fork American River (all tributaries)	Mountain yellow-legged frog
Fresno	Big Creek	Yosemite toad
Fresno	Jose Creek	Foothill yellow-legged frog
Fresno	North Fork Kings River (all tributaries)	Slender salamander complex, Yosemite toad, mountain yellow-legged frog
Fresno	Piute Creek	Mountain yellow-legged frog
Fresno	South Fork Kings River (all tributaries)	Slender salamander complex, mountain yellow-legged frog
Fresno	South Fork San Joaquin River (all tributaries)	Slender salamander complex, Yosemite toad, mountain yellow-legged frog
Inyo	All eastern Sierra tributaries	Slender salamander complex, Owens Valley web-toed salamander, Yosemite toad, mountain yellow-legged frog, northern leopard frog
Kern	Breckenridge Mountain (all tributaries)	Slender salamander complex, yellow-blotched salamander, mountain yellow-legged frog
Kern	Caliente Creek	Slender salamander complex, yellow-blotched salamander
Kern/Tulare	Middle Kern River (all tributaries)	Slender salamander complex, yellow-blotched salamander
Kern/Tulare	South Fork Kern River (all tributaries)	Slender salamander complex
Mariposa	Bull Creek	Foothill yellow-legged frog
Mariposa	Middle and Upper Merced River (all tributaries)	Slender salamander complex, limestone salamander, Mount Lyell salamander, Yosemite toad, mountain yellow-legged frog
Mono	All eastern Sierra tributaries	Slender salamander complex, Owens Valley web-toed salamander, Yosemite toad, mountain yellow-legged frog, northern leopard frog
Plumas	Boulder Creek	Mountain yellow-legged frog
Plumas	Butt Creek	Cascade frog, mountain yellow-legged frog
Plumas	Canyon Creek	California red-legged frog
Plumas	Middle Fork Feather River (all tributaries)	Mountain yellow-legged frog
Plumas	North Fork Feather River (all tributaries)	Mountain yellow-legged frog
Tehama	Antelope Creek	Cascade frog, foothill yellow-legged frog
Tehama	Deer Creek	Cascade frog, foothill yellow-legged frog
Tehama	Mill Creek	Cascade frog, foothill yellow-legged frog
Tulare	Blossom Lakes	Mountain yellow-legged frog, Mount Lyell salamander
Tulare	Marble Fork Kaweah River (all tributaries)	Mountain yellow-legged frog, Mount Lyell salamander
Tulare	Upper Kern River (all tributaries)	Mountain yellow-legged frog, slender salamander complex
Tuolumne	Cherry Creek	Mountain yellow-legged frog
Tuolumne	Clavey River	Foothill yellow-legged frog, mountain yellow-legged frog
Tuolumne	Coyote Creek	Foothill yellow-legged frog
Tuolumne	Middle Fork Stanislaus River (all tributaries)	Yosemite toad, Mount Lyell salamander, foothill yellow-legged frog, mountain yellow-legged frog
Tuolumne	Rose Creek	Foothill yellow-legged frog
Tuolumne	South Fork Stanislaus River (all tributaries)	Yosemite toad, Mount Lyell salamander, foothill yellow-legged frog, mountain yellow-legged frog

Bradford, D. F. 1983. Winterkill, oxygen relations, and energy metabolism of a submerged dormant amphibian, *Rana muscosa*. *Ecology* 64 (5): 1171–83.

———. 1989. Allotopic distribution of native frogs and introduced fishes in the high Sierra Nevada lakes of California: Implication of the negative effects of fish introductions. *Copeia* 1989 (3): 966–76.

———. 1991. Mass mortality and extinction in a high elevation population of *Rana muscosa*. *Journal of Herpetology* 25 (2): 174–77.

Bradford, D. F., M. S. Gordon, D. F. Johnson, R. D. Andrews, and W. B. Jennings. 1994. Acidic deposition as an unlikely cause for amphibian population declines in the Sierra Nevada, California. *Biological Conservation* 69 (2): 155–61.

Bradford, D. F., D. M. Graber, and F. Tabatabai. 1993. Isolation of remaining populations of the native frog, *Rana muscosa*, by introduced fish in Sequoia and Kings Canyon National Parks, California. *Conservation Biology* 7 (4): 882–88.

California Air Resources Board. 1987. *Effect of ozone on vegetation and possible alternative ambient air quality standards*. Sacramento: California Air Resources Board.

Carey, C. 1993. Hypothesis concerning the causes of the disappearance of boreal toads from the mountains of Colorado. *Conservation Biology* 7 (2): 355–62.

Cory, L., P. Fjeld, and W. Serat. 1970. Distribution patterns of DDT residues in the Sierra Nevada Mountains. *Pesticides Monitoring Journal* 3 (4): 204–11.

Drost, C. A., and G. M. Fellers. 1994. *Decline of frog species in the Yosemite section of the Sierra Nevada*. Technical Report NPS/WRUC/NRTR-94-02. Davis: University of California, California National Park Resources Studies Unit.

Fellers, G. M., and C. A. Drost. 1993. Disappearance of the Cascades frog *Rana cascadae* at the southern end of its range, California, USA. *Biological Conservation* 65 (2): 177–81.

Formanowicz, D. R. 1995. Letters; ecological research. *Science* 269 (5228): 1203.

Gorman, J. 1956. Reproduction in plethodont salamanders of the genus *Hydromantes*. *Herpetologica* 12 (4): 249–59.

Haines, T. A. 1981. Acidic precipitation and its consequences for aquatic organisms: A review. *Transactions of the American Fisheries Society* 110 (5): 669–707.

Halliday, T. 1995. Letters; ecological research. *Science* 269 (5228): 1202–3.

Harrison, S. 1991. Local extinction in a metapopulation context: An empirical evaluation. *Biological Journal of the Linnean Society* 42 (1 & 2): 73–88.

Hayes, M. P., and M. R. Jennings. 1986. Decline of native frog species in western North America: Are bullfrogs *(Rana catesbeiana)* responsible? *Journal of Herpetology* 20 (4): 490–509.

———. 1988. Habitat correlates of distribution of the California red-legged frog *(Rana aurora draytonii)* and the foothill yellow-legged frog *(Rana boylii):* Implications for management. In *Proceedings of the symposium on the management of amphibians, reptiles, and small mammals in North America,* technical coordination by R. C. Szaro, K. E. Severson, and D. R. Patton, 144–58. General Technical Report RM-166. Rocky Mountain Range and Experiment Station, U.S. Forest Service, Fort Collins, CO.

Jennings, M. R. 1987. *Annotated check list of the amphibians and reptiles of California.* Special Publication 3. Van Nuys, CA: Southwestern Herpetologists Society.

———. 1995. Native ranid frogs in California. In *Our living resources: A report to the nation on the distribution, abundance, and health of U.S. plants, animals, and ecosystems,* edited by E. T. LaRoe, G. S. Farris, C. E. Puckett, P. D. Doran, and M. J. Mac, 131–34. Washington, DC: National Biological Service.

Jennings, M. R., and M. P. Hayes. 1985. Pre-1900 overharvest of the California red-legged frog *(Rana aurora draytonii):* The inducement for bullfrog *(Rana catesbeiana)* introduction. *Herpetologica* 41 (1) : 94–103.

———. 1994. *Amphibian and reptile species of special concern in California.* Rancho Cordova: California Department of Fish and Game, Inland Fisheries Division.

Kagarise Sherman, C., and M. L. Morton. 1993. Population declines of Yosemite toads in the eastern Sierra Nevada of California. *Journal of Herpetology* 27 (2): 186–98.

Liss, W. J., and G. L. Larson. 1991. Ecological effects of stocked trout on North Cascades naturally fishless lakes. *Park Science* 11 (3): 22–23.

Miller, K. J. 1994. Endangered and threatened wildlife and plants: Proposed endangered status for the California red-legged frog. *Federal Register* 59, no. 22 (2 February): 4888–95.

Moyle, P. B. 1973. Effects of introduced bullfrogs, *Rana catesbeiana,* on the native frogs of the San Joaquin valley, California. *Copeia* 1973 (1): 18–22.

———. 1976. *Inland fishes of California.* Berkeley and Los Angeles: University of California Press.

Moyle, P. B., J. E. Williams, and E. D. Wikramanayake. 1989. *Fish species of special concern of California.* Rancho Cordova: California Department of Fish and Game, Inland Fisheries Division.

Nikolaidis, N. P., V. S. Nikolaidis, and J. L. Schnoor. 1991. Assessment of episodic acidification in Sierra Nevada, California. *Aquatic Sciences* 53 (4): 330–45.

Pechmann, J. H. K., D. E. Scott, R. D. Semlitsch, J. P. Caldwell, L. J. Vitt, and J. W. Gibbons. 1991. Declining amphibian populations: The problem of separating human impacts from natural fluctuations. *Science* 253 (5022): 892–95.

Pechmann, J. H. K., and H. M. Wilbur. 1994. Putting declining amphibian populations in perspective: Natural fluctuations and human impacts. *Herpetologica* 50 (1): 65–84.

Ray, G. C. 1958. Vital limits and rates of desiccation in salamanders. *Ecology* 39 (1): 75–83.

Reznick, D. 1995. Letters; ecological research. *Science* 269 (5228): 1202.

Roush, W. 1995. When rigor meets reality. *Science* 269 (5222): 313–15.

Scott, N. J., Jr. 1993. Postmetamorphic death syndrome. *Froglog* no. 7: 1–2.

Sorenson, P. C. 1994. Endangered and threatened wildlife and plants: Twelve-month petition finding for the California tiger salamander. *Federal Register* 59, no. 74 (18 April): 18353–54.

Stebbins, R. C. 1951. *Amphibians of western North America.* Berkeley and Los Angeles: University of California Press.

———. 1985. *A field guide to western reptiles and amphibians.* 2nd ed. Boston: Houghton Mifflin.

Stebbins, R. C., and N. W. Cohen. 1995. *Natural history of the amphibians.* Princeton, NJ: Princeton University Press.

Steinhart, P. 1990. *California's wild heritage: Threatened and endangered animals in the Golden State.* San Francisco: California Department of Fish and Game, California Academy of Sciences, and Sierra Club Books.

Storer, T. I. 1922. The eastern bullfrog in California. *California Fish and Game* 8 (4): 219–24.

———. 1925. A synopsis of the amphibia of California. *University of California Publications in Zoology* 27:1–342.

Wake, D. B. 1991. Declining amphibian populations. *Science* 253 (5022): 860.

Williams, D. F. 1986. *Mammalian species of special concern in California.* Administrative Report no. 86-1. Sacramento: California Department of Fish and Game, Wildlife Management Division.

Zweifel, R. G. 1955. Ecology, distribution, and systematics of frogs of the *Rana boylei* group. *University of California Publications in Zoology* 54 (4): 207–92.

PETER B. MOYLE
Department of Wildlife, Fish, and
    Conservation Biology
University of California
Davis, California

32

Status of Aquatic Habitat Types

ABSTRACT

Sixty-six aquatic habitat types were described for the Sierra Nevada, based on the system of Moyle and Ellison (1991). Three aspects of each habitat type were rated on a scale of 1 to 5: rarity, amount of disturbance, and amount of protection it currently enjoys. The ratings were added to provide a measure of the status of each habitat type. Eighteen (27%) were rated as secure, thirty-three (50%) as of special concern, fourteen (21%) as threatened, and one (2%) as extirpated. Most of the secure habitat types were characteristic of high-mountain areas, while most of the threatened habitat types were found in lowland areas. The decline and loss of habitat types, especially rare or unusual habitat types, is one of the principal reasons that so many species of invertebrates, fish, and amphibians in the Sierra Nevada are in decline.

INTRODUCTION

The decline of native fishes, amphibians, and aquatic invertebrates in the Sierra Nevada (Erman 1996; Jennings 1996; Moyle et al. 1996) reflects, to a large extent, the deterioration in the quality of the range's aquatic habitats. Factors contributing to this deterioration are multiple, cumulative, and synergistic. They include changes in the amount and timing of stream flows, changes in water quality, reduction in structural complexity (from loss of riparian trees, channelization, and other factors), changes in stream channels, siltation, and invasions of non-native species (Meehan 1991). Not all Sierra Nevada habitats, of course, are affected equally by human influences. The unique alkaline lake habitats that once existed in the Owens Lake basin disappeared completely once the lake became dry as the result of diversion of inflowing water.

In contrast, small, fishless alpine ponds and streams exist by the hundreds, many little changed from pristine conditions.

In order to evaluate the relative state of aquatic habitats around California, Moyle and Ellison (1991) devised a classification system for aquatic habitat types. The term *habitat type* means a readily recognizable set of habitats or environmental conditions that is home to a distinctive assemblage of organisms. Moyle and Ellison used only the word *habitat* for this definition, rather than *habitat type*. Although their use of *habitat* in a broad context is consistent with the deliberately vague definitions in standard ecology texts (e.g., Rickleffs 1993), many of the "habitats" of Moyle and Ellison 1991 contain multiple habitats under more conventional classification schemes. For example, Lake Tahoe is usually considered to contain multiple habitats (deep water, shallow water, open water, etc.), rather than being only one habitat as defined by Moyle and Ellison. Habitats are also usually defined largely on the basis of their physical and chemical characteristics, including vegetation types (e.g., Cowardin et al. 1979). While the habitat types described here have distinct physical and chemical characteristics, animal assemblages, especially those containing endemic fishes and amphibians, are key parts of their descriptions. The habitat types have fairly recognizable boundaries, although stream habitat types often tend to blend into one another.

In devising their classification system, Moyle and Ellison (1991) tried to make sure that it

- covered all aquatic habitats

- was easy to use without being either too general or too specific

- took into account patterns of endemism in aquatic organisms

Sierra Nevada Ecosystem Project: Final report to Congress, vol. II, *Assessments and scientific basis for management options.* Davis: University of California, Centers for Water and Wildland Resources, 1996.

- was expandable, so that new categories could easily be added

- was based on a combination of physical, chemical, and biological aspects of each habitat type

- was predictive, so that once a site had been classified users could have a reasonably good idea as to what organisms were likely to be present

Fish and amphibian distribution patterns were used as the basis for the classification system, mainly because vertebrates are the best-known aquatic organisms. However, the system includes many habitat types that contain mainly invertebrates and can be expanded to include others. Thus, the first level in the classification system consists of the five ichthyological provinces (regions of endemism) found in California. The second level (within each province) is standing versus flowing waters. Within these waters, the first division is ephemeral versus permanent, followed by waters with and without fish. Further subdivisions are usually based on the size of the body of water and on the species present.

This chapter presents a classification system of Sierra Nevada aquatic habitat types, modified from Moyle and Ellison (1991), and then evaluates how much protection and management each habitat type (and its associated flora and fauna) needs if examples are to persist in the Sierra Nevada in the near future.

METHODS

The classification system of Moyle and Ellison (1991) was expanded and revised according to new knowledge obtained from personal observations, various forest management plans, consultation with other biologists, and other sources. Each habitat type was then rated by the author in three categories: rarity, degree of disturbance, and existing protection (table 32.1).

Rarity is essentially a rating of the frequency of the habitat type in the Sierra Nevada. Some habitat types, such as Mono Lake, are one of a kind; others are naturally rare (e.g., sphagnum bogs); and others are widespread (e.g., alpine lakes).

Disturbance is a subjective rating of the degree to which the habitat type as a whole has been disturbed by human activity, including fish introductions. The most disturbed habitat types are at low elevations such as rivers in which chinook salmon spawn, which are now largely modified or cut off by dams. Examples of habitat types having the lowest disturbance overall included small

TABLE 32.1

A rating system for aquatic habitat types of the Sierra Nevada to determine how much special protection and management is likely to be needed in the immediate future for persistence.

Rating	Description
Rarity	
1	Unique: only one or two examples exist/remain in the Sierra Nevada
2	Rare: probably only 2-5 examples exist in Sierra Nevada or a formerly common habitat type in which most examples have been irreversibly altered
3	Unusual: scattered or infrequent examples in the Sierra Nevada
4	Common: examples easy to find
5	Widespread: a major existing habitat type
Disturbance	
0	All known examples highly disturbed, not recoverable
1	All known examples highly disturbed/altered but some are recoverable to a defined desirable state
2	All known examples moderately to highly disturbed or altered but most are recoverable
3	Most examples disturbed but some relatively undisturbed examples exist or all known examples moderately to lightly disturbed (recoverable with minimal effort)
4	Fairly even mixture of disturbed and relatively undisturbed areas OR all known examples lightly disturbed
5	Most examples in good condition (relatively undisturbed)
Existing Protection	
1	No known examples in protected areas (national park, wilderness area, research natural area, etc.)
2	No known examples in protected areas but mostly on public land or just one or two protected examples
3	3-5 protected examples exist but most unprotected or a rare habitat type with partial protection
4	Moderately secure: several protected examples, many with de facto protection because of location, etc., or a rare habitat type with de facto protection
5	Secure: many examples in protected areas or with de facto protection or a rare habitat type in protected area
Status (Rarity + Disturbance + Protection)	
12–15	Secure: widespread, with many examples in good condition
8–11	Special concern: declining in abundance and quality but many examples still exist or a habitat type with only one or two examples in existence
4–7	Threatened: being lost/degraded rapidly
<4	Extirpated or likely to disappear soon if protective action is not taken

rainbow trout streams, found in many areas, and small, fishless glacial streams or ponds.

Existing protection was determined by examining maps and estimating the extent to which each habitat type is found in national parks, wilderness areas, and other categories of formal protection. Streams or lakes on national forest land were rated as having only moderate protection. The lowest protection ratings were generally for habitat types that occur mostly on private land.

Each factor was rated on a scale of 0 to 5 or 1 to 5 (table 32.2), and the scores were added. Habitat types with the lowest total scores are most likely to need special protection and management in the immediate future if they are to persist.

TABLE 32.2

Status of sixty-six habitat types in the Sierra Nevada, determined by subjectively rating rarity, degree of disturbance, and amount of formal protection. See table 32.1 for rating system. The confidence rating indicates the confidence the author has in his subjective ratings.

Number	Type	Rarity	Disturbance	Protection	Status Score	Rating	Confidence
A0000	*Sacramento-San Joaquin Province*						
A1151	Outcrop pool	4	4	5	13	Secure	Moderate
A1152	Mountain pond	5	5	5	15	Secure	Moderate
A1210	Alpine lake	5	3	5	13	Secure	High
A1220	Northeast volcanic lake	3	3	4	10	Special concern	Moderate
A1240	Dystrophic ponds/lake	3	4	4	11	Special concern	Low
A1260	Valley marsh	2	1	3	6	Threatened	Low
A1280	Sphagnum bog	2	3	4	9	Special concern	Moderate
A1290	Fen	3	4	3	10	Special concern	Low
A2110	Alpine snowmelt stream	5	5	5	15	Secure	High
A2120	Conifer forest snowmelt stream	5	4	5	14	Secure	High
A2130	Foothill/valley ephemeral stream	3	2	3	8	Special concern	High
A2140	Foothill canyon ephemeral stream	4	4	3	11	Special concern	Moderate
A2411	Alpine stream	5	3	5	13	Secure	High
A2412	Forest stream	5	3	5	13	Secure	High
A2413	Spring	4	3	3	10	Special concern	Moderate
A2414	Meadow stream	5	2	3	10	Special concern	High
A2415	Glacial melt stream	3	5	5	13	Secure	High
A2416	Hot springs outflow	2	4	5	11	Special concern	High
A2421	Resident rainbow trout stream	5	5	4	14	Secure	High
A2422	Rainbow trout/cyprinid stream	4	4	4	12	Secure	High
A2423	Kern golden trout stream	2	3	3	8	Special concern	High
A2431	Spring chinook stream	2	3	4	9	Special concern	High
A2441	Valley floor river	2	1	1	4	Threatened	High
A2442	Fall chinook salmon spawning stream	4	1	1	6	Threatened	High
A2443	Hardhead/squawfish stream	4	3	3	10	Special concern	High
A2444	Hitch stream	3	2	1	6	Threatened	High
A2445	California roach stream	4	3	2	9	Special concern	High
A2446	Squawfish-sucker stream	4	3	3	10	Special concern	Moderate
C0000	*Great Basin Province*						
C1110	Alkalai playa lake	2	4	2	8	Special concern	Moderate
C1120	Mountain pond	5	5	5	15	Secure	Moderate
C1130	Great Basin scrub pool	4	3	4	11	Special concern	Moderate
C1140	Rock pool	3	4	4	11	Special concern	Low
C1210	Alpine lake/pond	5	3	5	13	Secure	High
C1221	Great Basin scrub perennial pool	3	3	3	9	Special concern	Low
C1222	Spring pool	2	2	2	6	Threatened	Low
C1232	Mono Lake	1	3	3	7	Threatened	High
C1233	Owens Lake	1	0	1	2	Extirpated	High
C1241	Fen	3	3	3	9	Special concern	Moderate
C1242	Sphagnum bog	1	4	4	9	Special concern	Moderate
C1311	Alpine lake/pond	3	3	2	8	Special concern	Moderate
C1312	Lake Tahoe	1	3	3	7	Threatened	High
C1313	Caldera lake	2	1	3	6	Threatened	Moderate
C1320	Eagle Lake	1	5	2	8	Special concern	High
C1330	Honey Lake	1	3	2	6	Threatened	Moderate
C1341	Lahonton desert spring	2	2	1	5	Threatened	Moderate
C1343	Owens desert spring	1	1	4	6	Threatened	Moderate
C2110	Alpine snowmelt stream	5	5	5	15	Secure	High
C2120	Conifer forest snowmelt stream	5	4	5	14	Secure	High
C2130	Great Basin scrub snowmelt stream	4	2	3	9	Special concern	Low
C2140	Desert wash	4	3	3	10	Special concern	Low
C2211	Glacial melt stream	3	5	5	13	Secure	High
C2212	Exposed alpine stream	5	3	5	13	Secure	High
C2213	Spring	4	3	3	10	Special concern	Moderate
C2214	Conifer forest stream	5	3	5	13	Secure	High
C2215	Meadow stream	5	2	3	10	Special concern	High
C2216	Hot springs outflow	2	3	3	8	Special concern	Moderate
C2221	Desert scrub stream	3	3	4	10	Special concern	Low
C2310	Trout headwater	4	3	4	11	Special concern	Moderate
C2320	Trout/sculpin stream	3	3	4	10	Special concern	Low
C2331	Sucker/dace/redside stream, with trout	4	3	2	9	Special concern	Moderate
C2332	Sucker/dace/redside stream, without trout	3	3	2	8	Special concern	High
C2333	Pine Creek (Lassen County)	1	2	3	6	Threatened	High
C2340	Speckled dace stream	3	3	2	8	Special concern	Moderate
C2350	Whitefish/trout/sucker stream	5	3	2	10	Special concern	Moderate
C2360	Tui chub stream	1	1	2	4	Threatened	Moderate
C2374	Owens River	1	1	2	4	Threatened	Moderate

RESULTS

Sixty-six major habitat types were identified as occurring within the Sierra Nevada (appendix 32.1). Eighteen (27%) were found to be secure, thirty-three (50%) were rated as being of special concern, fourteen (21%) were rated as threatened (in rapid decline, with more extreme cases likely to disappear soon), and one (2%) was gone (table 32.2; figure 32.1). Of the eighteen habitat types designated as secure, fifteen were high-elevation habitat types, while eleven of the thirteen habitat types rated as being threatened were characteristic of lowland areas (e.g., Owens Valley, Central Valley). Most (79%) of the midmountain (foothill) habitat types were found to be in moderate decline (special concern). Although high-elevation habitat types were generally regarded as secure, this was mainly because of their wide distribution and abundance and their presence in national parks. In fact, most are degraded to a greater or lesser degree, especially by the introduction of trout and the grazing of livestock (both of which are allowed in wilderness areas).

Nine of the aquatic habitat types were unique or extremely rare, which automatically gave them at least special concern status. This is appropriate because such habitat types tend to contain endemic organisms and to be subject to degradation. Examples include large lakes such as Eagle Lake, Lake Tahoe, Mono Lake, and Owens Lake.

CONCLUSIONS

The diversity of natural aquatic habitat types in the Sierra Nevada is in the process of being diminished. As distinctive habitat types disappear, endemic or unusual native aquatic organisms disappear along with them. Habitat types in lowland regions, many of which just touch on the Sierra Nevada proper (e.g., valley lowland rivers), seem to be particularly degraded, presumably because they largely occur outside areas with formal protection, such as national parks, or are downstream of major dams. It is likely that the condition of aquatic habitat types in the Sierra Nevada is even worse than projected here, especially for fishless habitat types. This is because

- The classification system is based largely on vertebrates and so probably misses important habitat types dominated by invertebrates.

- Some of the habitat types are rather broad, especially the stream habitat types, and may therefore include special habitats (side channels, seeps, etc.) that are more vulnerable to degradation than the habitat type overall.

- The widespread disappearance of frogs and other amphibians may reflect subtle but widespread changes to common habitat types as well as increased fragmentation of aquatic habitats.

FIGURE 32.1

Status of aquatic habitats in the Sierra Nevada.

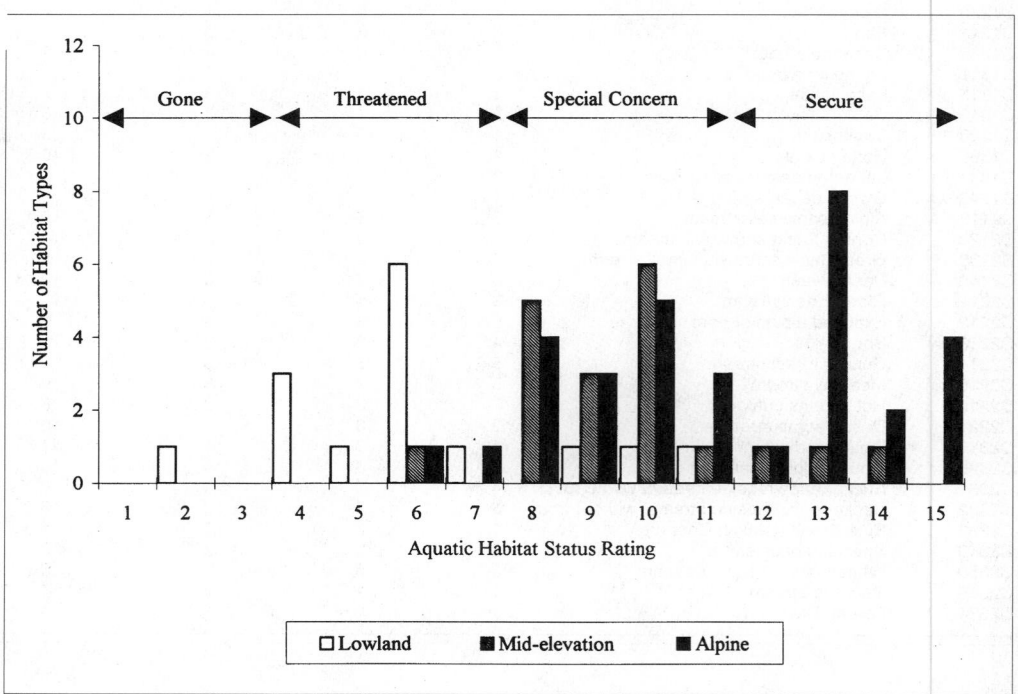

Although it is clear that many aquatic habitat types in the Sierra Nevada are in decline, the habitat type classification system presented here should be used with caution and the results obtained from the analysis regarded as conservative. The reason for this is that the habitat type classification system often makes arbitrary subdivisions of continuous habitat types that change with elevation or other factors. Also, not all habitat types have been rated by the author with high confidence (table 32.2), reflecting limited knowledge of the particular habitat types.

REFERENCES

Cowardin, L. M., V. Carter, F. C. Golet, and E. T. LaRoe. 1979. *Classification of wetlands and deepwater habitats of the United States.* Report FWS/OBS-79/31. Washington, DC: U.S. Fish and Wildlife Service.

Erman, N. A. 1996. Status of aquatic invertebrates. In *Sierra Nevada Ecosystem Project: Final report to Congress,* vol. II, chap. 35. Davis: University of California, Centers for Water and Wildland Resources.

Jennings, M. R. 1996. Status of amphibians. In *Sierra Nevada Ecosystem Project: Final report to Congress,* vol. II, chap. 31. Davis: University of California, Centers for Water and Wildland Resources.

Meehan, W. R., ed. 1991. *Influences of forest and rangeland management on salmonid fishes and their habitats.* Bethesda, MD: American Fisheries Society.

Moyle, P. B., and J. P. Ellison. 1991. A conservation-oriented classification system for the inland waters of California. *California Fish and Game* 77:161–80.

Moyle P. B., R. M. Yoshiyama, and R. Knapp. 1996. Status of fish and fisheries. In *Sierra Nevada Ecosystem Project: Final report to Congress,* vol. II, chap. 33. Davis: University of California, Centers for Water and Wildland Resources.

Rickleffs, R. F. 1993. *The economy of nature.* 3rd ed. New York: W. H. Freeman Co.

APPENDIX 32.1

A Classification System for
Aquatic Habitat Types of the
Sierra Nevada Region

This classification system is based on Moyle and Ellison (1991). Only habitat types within the Sierra Nevada proper were evaluated in this study (i.e., habitat types in the Modoc Plateau and some peripheral areas were not evaluated, although they are presented in the classification system).

A0000 SACRAMENTO–SAN JOAQUIN PROVINCE

A1000 Standing Waters

A1100   Ephemeral Waters
A1100-A1140 Not in Sierra Nevada
A1150 Alpine pond
A1151 Outcrop pool
Clear, oligotrophic pools found in shallow depressions on granitic outcrops at high elevations in which both freezing and drying are limiting factors; seasonally filled with snowmelt or rain water. Support communities of seasonal organisms such as fairy shrimp (*Brachinecta* sp., *Streptocephalus seali*) and larvae of long-toed salamanders (*Ambystoma macrodactylum*).
A1152 Mountain pond
Shallow (<1.5 m deep) ponds or small (<1 ha) lakes in alpine areas that periodically dry up, freeze solid, or become deoxygenated in winter; often associated with meadows and/or cirques.
A1200 Permanent Fishless Waters
A1210 Alpine lake
Clear, oligotrophic lakes found in cirques and other depressions carved out by glaciers. Historically, virtually all of these lakes were without fish and were dominated by aquatic insects, fairy shrimp and other crustaceans, and the larvae of frogs, principally *Rana muscosa*. Most of these lakes today contain one or more introduced species of salmonid fishes which have altered the native biotic communities considerably.
A1220 Volcanic lake
Permanent, isolated ponds and lakes created by old lava flows and landslides, especially in Lassen National Park area. Most now contain introduced fishes and original fauna is poorly known.
A1230 Caldera lake—no examples in drainage
A1240 Dystrophic pond/lake
Shallow alpine waters with boggy edges, presumably in the natural successional process of becoming bogs. Acidic and fishless.
A1250 Saline pond/lake—not in Sierra Nevada
A1260 Valley marsh
The floor of the Central Valley once supported extensive tule and cattail marshes that flooded seasonally and were permanently wet. Primarily fishless, but seasonally important for spawning and major habitat types for aquatic birds, including migratory waterfowl.
A1270 Northern volcanic pool—not in Sierra Nevada

A1280 Sphagnum bog
True bog containing marshy vegetation, including carnivorous plants, and ranid frogs. Example: Mt. Pleasant RNA, Plumas NF.
A1290 Fen
Minerotrophic, spongy, spring-fed peatlands located on hillsides and dominated by non-sphagnum mosses and sedges.
A1300 Permanent Lakes with Fish—none in drainage

A2000 Flowing Waters

A2100 Ephemeral Streams
A2110 Alpine snowmelt stream
Small, exposed, high gradient streams mainly above the timberline that exist only while snow is melting.
A2120 Conifer forest snowmelt stream
Small intermittent streams in conifer forest areas that also exist primarily while snow is still melting but have flows enhanced by seepage from bogs and meadows. Occasionally important as spawning areas for trout (*Oncorhynchus* spp.).
A2130 Foothill/valley ephemeral stream
Low elevation streams in oak woodland/valley grassland areas that flow primarily in response to winter and spring rainfall, although some water may be semi-permanent in bedrock pools. Have a distinctive succession of invertebrates and may be important for spawning of fishes from more permanent streams.
A2140 Foothill canyon ephemeral stream
High gradient seasonal tributary plunging down sides of steep canyons of foothill streams e.g., unnamed tributaries of lower Mill Creek, Tehama County.
A2200 Permanent Streams, Goose Lake Drainage
A2210 Fishless alpine stream
Small high-gradient streams in the Warner Mountains that are too steep or inaccessible to be colonized by native trout. Dominant fauna is aquatic insects.
A2220 Redband trout/lamprey spawning stream
Mid-elevation reaches of larger tributary streams (e.g., Willow and Lassen Creeks, Modoc County) to Goose Lake that contain enough gravel and spring flows to support spawning runs of redband trout and Goose Lake lamprey from the lake.
A2230 Resident redband trout stream
Small tributary streams (including tributaries that form A2220 streams) that support self-sustaining populations of redband trout.
A2240 Goose Lake sucker/speckled dace stream
Lower reaches of tributaries used for spawning by suckers and dace but are frequently dry in summer.
A2250 Valley tui chub stream
Streams reaches with low enough gradients to support Goose Lake tui chubs and other lake fishes. Typically warm and slightly turbid in late summer.
A2300 Permanent Streams, Pit River Drainage
A2310 Fishless streams
A2311 Glacial melt stream
Streams that drain melting glaciers on Mt. Shasta. Color is typically a milky brown from "rock flour" created by the grinding action of the glaciers. Biotic diversity low.

A2312 Alpine stream
Most streams above 3000 m elevation in the Sacramento-San Joaquin Basin contained no fish until various salmonids were introduced starting in the late 19th century. Originally dominated by aquatic insects and amphibian larvae.
A2313 Spring stream
Outflows of small springs too small or with too high gradients to be colonized by fish.
A2314 Forest stream
Small streams in forested areas with high gradients.
A2320 Low order trout streams
    A2321 Pit River rainbow/redband trout stream
    Typically, second, third, or fourth order tributaries to the Pit River with high enough gradients to exclude all fish but rainbow trout.
    A2322 McCloud River redband trout stream
    The upper McCloud River (above Upper Falls) and tributaries; the endemic McCloud River redband trout (*Oncorhynchus mykiss* ssp.) is the sole native fish.
A2330 Pit River tributaries
    A2331 Speckled dace/Pit sculpin stream
    Low elevation tributaries to the Pit River characterized by rocky substrates and large populations of speckled dace and Pit sculpin *(Cottus pitensis)*. Juveniles of the large cyprinids and catostomids characteristic of A2350 are often found here as well as they may use these streams for spawning.
    A2332 Squawfish/sucker stream
    The Pit River and the lower reaches of tributary streams (e.g., Ash Creek) in Big Valley, Modoc/Lassen/Shasta Counties. Gradient is low, water muddy and warm; dominant fishes are Sacramento squawfish *(Ptychocheilus grandis)* and Sacramento sucker *(Catostomus occidentalis)*.
    A2333 Modoc sucker stream
    Small, moderate gradient streams in Modoc County containing Modoc sucker *(C. microps)* but dominated numerically by speckled dace.
    A2334 Rough sculpin/Shasta crayfish spring stream
    Cold, clear, spring waters in lava areas that support a highly endemic fauna, including rough sculpin *(Cottus asperrimus)* and Shasta crayfish *(Pascifasticus fortis)*. Biggest examples are Fall River and its spring tributaries and lower Hat Creek.
A2340 Canyon rivers
    A2341 Lower Pit River (Hardhead/tule perch river)
    The Pit River proper as it flows through its canyon from Pit Falls to its confluence with the Sacramento River. Characterized by deep rocky pools containing hardhead *(Mylopharodon conocephalus)* and tule perch. Deep, swift riffles and runs contain rainbow trout.
    A2342 Lower McCloud River
    The McCloud River below Lower Falls was a cold, slightly milky river flowing through a deep canyon and characterized by deep pools that housed winter run chinook salmon *(Oncorhynchus tshawystscha)* and bull trout *(Salvelinus confluentus)*; both are now extinct in the river.
A2400 Permanent Streams, Central Valley Drainage
A2410 Fishless low-order tributaries
    A2411 Alpine stream
    Most streams above 3000 m elevation in the Sacramento-San Joaquin Basin contained no fish until various salmonids were introduced starting in the late 19th century. Originally dominated by aquatic insects and amphibian larvae.
    A2412 Forest stream
    Second or third order streams in fir, pine, or deciduous forest areas that are too small or too high in gradient to support fish.
    A2413 Spring
    Springs with constant temperature and flows, fine substrates, and clear water; can support unusual/endemic invertebrates. Several can unite to form a meadow stream (A2414).
    A2414 Meadow stream
    First or second order stream through subalpine meadows, low gradient with sinuous braided channel. Where not heavily grazed, abundant frogs. May have introduced trout populations.
    A2415G lacial melt stream
    Streams that drain melting glaciers; water is typically milky brown in color and biotic diversity is low.
    A2416 Hot Springs Outflow
    Streams created by outflows of large hot springs, containing no or highly specialized life forms in the high-temperature sections. Example: Mill Creek below Bumpass Hell, Lassen NP.
A2420 Resident trout stream
    A2421 Resident rainbow trout stream
    Low order, cold, high gradient streams, dominated by rainbow trout and, often, riffle sculpin.

    A2422 Rainbow trout/cyprinid stream
    Small streams of moderate gradient supporting rainbow trout and one or two species of cyprinids (mostly California roach, *Lavinia symmetricus*) and/or Sacramento sucker.
    A2423 Kern golden trout stream
    The upper Kern River (Kern County) and its branches and tributaries that support golden trout *(Oncorhynchus mykiss aquabonita; O. m. whitei; O. m. gilberti)*.
A2430 Salmon-steelhead streams
    A2431 Spring chinook stream
    Third to fifth order streams at elevations of 500-1500 m with deep canyons containing deep, cold pools that can sustain spring chinook salmon through the summer. Examples: upper San Joaquin River, Fresno County (formerly); Deer and Mill Creeks, Tehama County.
A2440 Low elevation streams
    A2441 Valley floor river
    The main channels of the Sacramento and San Joaquin rivers, plus the lower reaches of their tributaries. Much of the water sluggish in summer and considerable cover is provided by logs etc. from riparian forests. Floods seasonally. Fauna complex mixture of resident deep-bodied fishes, warmwater stream fishes, and anadromous fishes.
    A2442 Fall chinook salmon spawning stream
    Low elevation, low gradient tributaries to major rivers that dry up in summer but are used for spawning by both anadromous species and resident stream fishes in spring. Example: lower Deer Creek, tributary to Sacramento River (Tehama County).
    A2443 Hardhead/squawfish stream
    Low- to mid-elevation streams characterized by deep, bedrock pools, clear water, and cool temperatures (<25°C); characteristic fishes are hardhead, Sacramento squawfish, and Sacramento sucker, although typically 5–6 species are present.
    A2444 Hitch stream
    Warm, low-elevation streams with low to moderate current and long reaches with sandy bottoms. Typical fishes are hitch *(Lavinia exilicauda)*, Sacramento squawfish, Sacramento sucker. Example: Fresno River
    A2445 California roach stream
    Small, clear, mid-elevation second, third, or fourth order tributaries that typically contain deep pools in canyons and are often intermittent in flow by late summer. Dominant fish numerically are California roach, but juveniles of Sacramento squawfish and Sacramento sucker are often present.
    A2446 Squawfish-sucker stream
    Small low to mid-elevation streams with few deep pools that are dominated by Sacramento sucker, Sacramento squawfish and, often, California roach. Example: Deer Creek, Tulare County.

B0000 KLAMATH PROVINCE

C0000 GREAT BASIN PROVINCE

C1000 Standing Waters

C1100 Ephemeral Waters
    C1110 Alkali playa lake
    Shallow lakes in isolated desert basins that dry up annually (except during exceptionally wet years).
    C1120 Mountain pond
    See A1152
    C1130 Great Basin scrub pool
    Pools that form from seasonal rainfall or snowmelt in hardpan areas of the desert and rarely last more than a month or two.
    C1140 Rock pool
    Natural holes in rocks (often in washes) that fill with water seasonally and may be semipermanent if deep enough. Important sources of water for desert bighorn and other animals.
C1200 Permanent Fishless Waters
    C1210 Alpine lake/pond
    Small, usually isolated, oligotrophic lakes in high mountain areas formed by the action of glaciers or by cones of volcanos.
    C1220 Desert pool and pond
        C1221 Great Basin scrub perennial pool
        Small isolated ponds in lowland or sub-alpine areas formed by the

damming action of lava flows or landslides and dominated by predatory insects and amphibian larvae.

C1222 Spring pool
Isolated small springs in desert or scrub areas.

C1230 Desert lake

C1231 Playa lake
Terminal lakes, often large, that occupy desert basins, are too alkaline to support fish life, and may dry up during severe drought periods. Example: upper and lower Alkaline Lakes in Surprise Valley (Modoc County).

C1232 Mono Lake
A distinctive, permanent alkaline lake in Mono County with an endemic invertebrate fauna (e.g., *Artemia mona*).

C1233 Owens Lake
A large lake at the terminus of the Owens River that probably was similar in many of its characteristics to Mono Lake (C1232) but now dry due to diversion of inflowing water.

C1240 Fen and bogs

C1241 Fens
See A1290. Example: Mason Fen, Nevada County.

C1242 Sphagnum bog
See A1280. Example: Grass Lake

C1300 Permanent Waters with Fish

C1310 Alpine lake

C1311 Alpine lake/pond
Oligotrophic, permanent alpine lakes with connections to streams with fish. Example: Independence Lake (Sierra and Nevada Counties).

C1312 Lake Tahoe
A large, deep, extraordinarily clear alpine lake containing a complex fish fauna and unusual deepwater invertebrates.

C1313 Caldera lake
Lakes occupying calderas of extinct volcanos Example: Crater Lake, Lassen County.

C1320 Eagle Lake
An alkaline, permanent terminal lake in Lassen County that is productive of fish and fish-eating birds; contains Eagle Lake rainbow trout (*Oncorhynchus mykiss aquilarum*) and tui chubs.

C1330 Honey Lake
A large, shallow, terminal alkaline lake in Lassen County that fluctuates greatly in size, even drying up occasionally, but supports abundant fish life in whatever water it contains.

C1340 Desert Springs

C1341 Lahontan desert spring
Isolated desert springs and associated pools containing fish, usually tui chubs. Example: High Rock Springs, Lassen County.

C1342 Amargosa desert spring (not in Sierra Nevada)

C1343 Owens desert spring
Spring fed pools containing Owens pupfish (*C. radiosus*).

C2000 Flowing Waters

C2100 Ephemeral Streams

C2110 Alpine snowmelt stream
See A2110

C2120 Conifer forest snowmelt stream
See A2120

C2130 Great Basin scrub snowmelt stream
Small streams flowing seasonally through desert scrub carrying local snowmelt as well as that from higher elevations to permanent streams or terminal lakes.

C2140 Desert wash
Moderate-to-high gradient desert stream courses that mainly carry flood flows from usual rain or snow melting events.

C2200 Permanent Fishless Streams

C2210 Alpine stream

C2211 Glacial melt stream
See A2415

C2212 Exposed alpine stream
See A2411

C2213 Spring
See A2413

C2214 Conifer forest stream
See A2412

C2215 Meadow stream
See A2414

C2216 Hot spring outflow
See A2416

C2220 Desert stream

C2221 Desert scrub stream
Small streams in lowland areas, fed by mountain run-off.

C2300 Permanent Streams with Fish

C2310 Trout headwater
Small alpine streams with meadow systems; originally containing Lahontan (*Oncorhynchus clarki henshawi*) or Paiute cutthroat trout (*O. c. seleneris*) but now usually containing non-native trout. Example: By-Day Creek (Mono County).

C2320 Trout/sculpin stream
Alpine streams of sufficient size and low enough gradient to contain both cutthroat trout and Paiute sculpin (*Cottus beldingi*).

C2330 Sucker/dace/redside stream

C2331 With trout
Coldwater streams containing the typical Lahontan drainage stream fish community (5–6 species, including Lahontan cutthroat trout).

C2332 Without trout
Lower gradient reaches of C2231 that are too warm in summer to support cutthroat trout. Example: Willow Creek, Lassen County.

C2333 Pine Creek (Lassen County)
This is the only large tributary to Eagle Lake and the principal spawning stream of Eagle Lake trout, Tahoe sucker (*Catostomus tahoensis*), and Lahontan redside (*Richardsonius egregius*); it contains a community dominated by the juveniles of these three species, plus speckled dace.

C2340 Speckled dace stream
Small meadow streams, usually spring fed, that contain mainly speckled dace but occasionally Tahoe suckers and cutthroat trout. Example: Papoose Creek (Lassen County).

C2350 Whitefish/trout/sucker stream
Mainstem rivers (e.g., Truckee River, Walker River) and their larger tributaries that contain the complete Lahontan fish fauna including mountain whitefish (*Prosopium williamsoni*) as well as large adults of cutthroat trout and Tahoe sucker. Cutthroat trout now replaced by non-native trout species.

C2360 Tui chub stream
Low gradient streams, usually close to their confluence with lakes, that contain large populations of tui chubs and speckled dace but little else. Examples: Cowhead Lake Slough (Modoc County).

C2370 Desert streams

C2371-C2373 Not in Sierra Nevada

C2374 Owens River
The Owens River and the lower reaches of its tributary streams originally contained an endemic community of Owens tui chub (*Gila bicolor snyderi*), Owens sucker (*Catostomus fumeiventris*), Owens speckled dace, and Owens pupfish (*Cyprinodon radiosus*).

PETER B. MOYLE
Department of Wildlife, Fish, and
    Conservation Biology
University of California
Davis, California

RONALD M. YOSHIYAMA
Department of Wildlife, Fish, and
    Conservation Biology
University of California
Davis, California

ROLAND A. KNAPP
Sierra Nevada Aquatic Research
    Laboratory
University of California
Mammoth Lakes, California

33

Status of Fish and Fisheries

ABSTRACT

Forty kinds of fish are native to the Sierra Nevada; eleven of these taxa are found only in the range. The fish fauna and fisheries of the Sierra Nevada have changed dramatically since the massive influx of Euro-Americans began in 1850. Four broad patterns are evident: (1) anadromous fishes, especially chinook salmon, have been excluded from most of the riverine habitat they once used on the west side of the range; (2) most resident native fishes have declined in abundance, and the aquatic communities of which they are part have become fragmented, although a few species have had their ranges greatly expanded; (3) thirty species of non-native fishes have been introduced into or have invaded most waters of the range, including extensive areas that were once fishless, mainly at high elevations; and (4) Sierra Nevada fisheries have largely shifted from native fishes, especially salmon and other migratory fishes, to introduced fishes. One reflection of these patterns is that of the forty fishes native to the Sierra Nevada, six (15%) are formally listed by the federal and/or state government as threatened or endangered species, twelve (30%) are considered to be species of special concern because they are in trouble statewide and are potential candidates for listing or because they have limited distributions, four (10%) are in decline in the Sierra Nevada but are probably in less trouble than elsewhere, and eighteen (45%) seem to have stable or expanding populations. Among the species that have largely disappeared from the range are chinook salmon, steelhead, and five kinds of native trout. Fisheries for these species have been replaced, in part, by stream fisheries for non-native trout, often of hatchery origin, and by reservoir fisheries. The introduction of trout into several thousand originally fishless lakes at high elevations has greatly expanded fishing opportunities but has also caused declines of native invertebrates and amphibians. Introduction of non-native fish species has also been the single biggest factor associated with fish declines in the Sierra Nevada. However,

this factor is intimately tied to major habitat changes and other effects of dams and diversions, as well as habitat changes caused by grazing, channelization, and other streamside activities.

INTRODUCTION

The native fish fauna of the Sierra Nevada consists of forty native taxa. Eleven (28%) of these taxa, including five kinds of trout, are found only in (are endemic to) the Sierra Nevada as defined by SNEP, and most (85%) are endemic to the Californian region of which the Sierra Nevada forms the core. These fish were widespread, abundant, and an important source of food for the Native Americans of the region (Moyle 1976a; Lindstrom 1993). The fish fauna and fisheries of the Sierra Nevada have changed dramatically since the massive influx of Euro-Americans began in 1850 (Moyle 1995). Four broad patterns are evident:

1. Anadromous fishes have been excluded from most of the riverine habitat they once used on the west side of the range.

2. Most resident native fishes have declined in abundance, and the aquatic communities of which they are part have become fragmented; a few have had their ranges greatly expanded as the result of introductions.

3. Thirty species of non-native fishes have been introduced into or have invaded most waters of the range, including extensive areas that were once fishless, mainly at high elevations.

Sierra Nevada Ecosystem Project: Final report to Congress, vol. II, *Assessments and scientific basis for management options.* Davis: University of California, Centers for Water and Wildland Resources, 1996.

4. Sierra Nevada fisheries have largely shifted from native fishes, especially salmon and other migratory fishes, to introduced fishes.

This chapter examines these patterns by documenting (1) the original distribution patterns of the native fishes, (2) the current status of native fishes, (3) changes in the distribution and abundance of chinook salmon, (4) the causes of native fish declines, (5) the expansion of populations of non-native fishes, (6) the effect of the changing fish fauna on fisheries, and (7) the conservation implications of the changes.

ORIGINAL FISH DISTRIBUTION PATTERNS

The native fishes of the Sierra Nevada were found in four distinct zoogeographic regions, which shared surprisingly few species among them: (1) the Sacramento–San Joaquin drainage; (2) the Lahontan drainage, consisting of the Susan, Truckee, Carson, and Walker Rivers; (3) the Eagle Lake drainage; and (4) the Owens drainage. Each of these regions had assemblages (communities) of fish species that characterized different environments within the drainage (Moyle 1976a).

The *Sacramento–San Joaquin drainage,* which includes all watersheds on the west side of the range, had by far the richest native fish fauna, with twenty-two taxa found in the Sierra Nevada (table 33.1). This fauna included three abundant anadromous fishes—chinook salmon, steelhead rainbow trout, and Pacific lamprey—that were important in Native American fisheries. Chinook salmon, with four discrete runs, were particularly abundant and supported large commercial fisheries in the nineteenth and early twentieth centuries. The *Lahontan drainage* supported only ten native fish species in the Sierra Nevada, but these fish were also widespread and abundant in the low- to middle-elevation rivers and lakes, and· were major sources of food for the Native Americans (Lindstrom 1993). Lahontan cutthroat trout were abundant enough to support commercial fisheries in the nineteenth century, especially in Lake Tahoe and Pyramid Lake, Nevada.

Eagle Lake could be regarded as the northernmost part of the Lahontan drainage in the Sierra Nevada because it shares three fish species with the drainage, but it is an independent watershed that also supports an endemic subspecies of rainbow trout. The *Owens drainage,* in contrast, although also an eastern Sierra Nevada watershed, has its own distinct fish fauna of four endemic species, mostly confined to the Owens River itself. It was separated from the Lahontan drainage by the fishless Mono Lake basin.

All four of the major fish faunal regions shared a common trait with the Mono Lake basin: they were fishless at high elevations. The high-elevation regions were largely fishless (figure 33.1) because of the combination of extensive glacia-

tion during the Pleistocene (which created most of the lakes) and steep topography (which created many barriers to natural fish invasions). In streams, the highest elevations reached naturally by fish (ca. 3,000 m [9,800 ft]) occur either in unglaciated areas in the southern portion of the range (Kern River) or in the more accessible mountain streams on the east side. Only about twenty lakes naturally contained fish (e.g., Eagle, Tahoe, Donner, Fallen Leaf, Independence, Weber, Convict), which is considerably less than 1% of the total. All such lakes were closely associated with streams containing fish and had no barriers to invasion.

In the western Sierra Nevada, the fish reaching the highest elevations were trout, but in some circumstances other species were also found at elevations above 1,500 m (4,900 ft). Coastal rainbow trout, the trout native to most west-side watersheds, were mostly found below 1,500 m. For example, in the Merced River they reached only Yosemite Valley (1,400 m [4,400 ft]), and in the Tuolumne River they did not reach Hetch Hetchy Valley (1,100 m [3,600 ft]). In the Middle Fork of the Kings River, however, trout may have reached elevations higher than 2,200 m (7,200 ft). In the Kern River drainage, Little Kern golden trout reached about 2,400 m (7,900 ft), Kern River rainbow about 2,500 m (8,200 ft), and California golden trout about 3,000 m (9,800 ft). The only native nontrout species found at high elevations on the west side is the Sacramento sucker, which occurred naturally as high as 2,500 m (8,200 ft) in the Kern River.

In the eastern Sierra Nevada, the highest elevations were reached by Lahontan cutthroat trout (more than 3,000 m [9,800 ft]) and Paiute cutthroat trout (2,500 m [8,200 ft]). However, in the Carson, Walker, and Truckee drainages it was not unusual to find nontrout species (Paiute sculpin, Tahoe sucker, speckled dace, Lahontan redside) above 2,000 m (6,600 ft). These fishes also colonized Lake Tahoe (1,900 m [6,200 ft]), Independence Lake (2,118 m [6,950 ft]), Weber Lake (2,065 m [6,775 ft]), and a few other similar lakes. Fish were completely absent from the Mono Lake basin (including all streams), and the Owens River watershed did not historically contain trout. Of the four fishes native to the Owens River basin, only the Owens sucker was found above 1,500 m (4,900 ft), reaching Convict Lake (2,300 m [7,550 ft]), the only lake in the southeastern Sierra Nevada that naturally contained fish.

CURRENT STATUS OF NATIVE FISHES

The forty kinds of fish found in the Sierra Nevada represent twenty-four species. Six of the species can be divided into two to six forms (subspecies or runs of salmon) that can be recognized by their distinctive morphology and life history patterns. Many of the subspecies were originally described as distinct species (e.g., the golden trouts), and most are en-

TABLE 33.1

Native fishes of the Sierra Nevada.

Name	Drainage	Habitat	Status
Kern brook lamprey, *Lampetra hubbsi*[a]	Sacramento–San Joaquin	Lowlands	Special concern
Pacific lamprey, *Lampetra tridentata*	Sacramento–San Joaquin	Anadromous, foothills, lowlands	Declining
Mountain whitefish, *Prosopium williamsoni*	Lahontan	Foothills, high elevations	Stable
Chinook salmon, *Oncorhynchus tshawytscha*			
Spring run	Sacramento–San Joaquin	Anadromus, foothills, lowlands	Special concern
Winter run	Sacramento–San Joaquin	Anadromus, foothills, lowlands	Endangered
Fall run	Sacramento–San Joaquin	Anadromus, lowlands	Declining
Late fall run	Sacramento–San Joaquin	Anadromus, foothills, lowlands	Special concern
Rainbow trout, *Oncorhynchus mykiss*			
Resident rainbow, *O. m. irideus*	Sacramento–San Joaquin	Foothills, high elevations	Stable or expanding; introduced outside native range
Winter steelhead, *O. m. irideus*	Sacramento–San Joaquin	Anadromus, foothills, lowlands	Declining
Eagle Lake rainbow, *O. m. aguilarum*[a]	Eagle Lake	Foothills, high elevations	Special concern
Kern River rainbow, *O. m. gilberti*[a]	Sacramento–San Joaquin	High elevations	Special concern
Little Kern golden, *O. m. whitei*[a]	Sacramento–San Joaquin	High elevations	Endangered
California golden, *O. m. aquabonita*[a]	Sacramento–San Joaquin	High elevations	Special concern; introduced outside native range
Cutthroat trout, *Oncorhnychus clarki*			
Lahontan cutthroat, *O. c. henshawi*	Lahontan	Foothills, high elevations	Threatened; introduced outside native range
Paiute cutthroat, *O. c. seleneris*[a]	Lahontan	High elevations	Threatened; introduced outside native range
Tui chub, *Gila bicolor*			
Lahontan lake tui chub, *G. b. pectinifer*	Lahontan	Lowlands, foothills, high elevations	Special concern
Lahontan creek tui chub, *G. b. obesa*	Lahontan	Lowlands, foothills, high elevations	Stable or expanding
Owens tui chub, *G. b. snyderi*[a]	Owens River	Lowlands, foothills	Endangered
Eagle Lake tui chub, *G. b. ssp.*[a]	Eagle Lake	Foothills	Special concern
Lahontan redside, *Richardsonius egregius*	Lahontan	Lowlands, foothills, high elevations	Stable or expanding
Sacramento hitch, *Lavinia exilicauda exilicauda*	Sacramento–San Joaquin	Lowlands, foothills	Declining
California roach, *Lavinia symmetricus*			
Sacramento roach, *L. s. symmetricus*	Sacramento–San Joaquin	Foothills	Stable
San Joaquin roach, *L. s. ssp.*	Sacramento–San Joaquin	Foothills	Special concern
Red Hills roach, *L. s. ssp.*[a]	Sacramento–San Joaquin	Foothills	Special concern
Sacramento blackfish, *Orthodon microlepidotus*	Sacramento–San Joaquin	Lowlands	Stable or expanding
Hardhead, *Mylopharodon conocephalus*	Sacramento–San Joaquin	Lowlands, foothills	Special concern
Sacramento squawfish, *Ptychocheilus grandis*	Sacramento–San Joaquin	Lowlands, foothills	Stable or expanding
Speckled dace, *Rhinichthys osculus*			
Lahontan speckled dace, *R. o. robustus*	Lahontan, Eagle Lake	Lowlands, foothills, high elevations	Stable
Owens speckled dace, *R. o. ssp.*	Owens River	Lowlands	Special concern
Sacramento speckled dace, *R. o. ssp.*	Sacramento–San Joaquin	Lowlands, foothills	Stable
Sacramento sucker, *Catostomus o. occidentalis*	Sacramento–San Joaquin	Lowlands, foothills, high elevations	Stable or expanding
Tahoe sucker, *Catostomus tahoensis*	Lahontan, Eagle Lake	Lowlands, foothills, high elevations	Stable or expanding
Owens sucker, *Catostomus fumeiventris*[a]	Owens River	Lowlands, foothills, high elevations	Stable or expanding; introduced outside native range
Mountain sucker, *Catostomus platyrhynchus*	Lahontan	Lowlands, foothills, high elevations	Special concern
Owens pupfish, *Cyprinodon radiosus*[a]	Owens River	Lowlands	Threatened or endangered
Threespine stickleback, *Gasterosteus aculeatus*	Sacramento–San Joaquin	Lowlands	Stable or expanding; introduced outside native range
Sacramento tule perch, *Hysterocarpus t. traski*	Sacramento–San Joaquin	Lowlands, foothills	Stable
Prickly sculpin, *Cottus asper*	Sacramento–San Joaquin	Lowlands, foothills	Stable or expanding
Riffle sculpin, *Cottus gulosus*	Sacramento–San Joaquin	Foothills, high elevations	Stable
Paiute sculpin, *Cottus beldingi*	Lahontan	Lowlands, foothills, high elevations	Stable

[a]Taxa endemic to the region.

demic to the Sierra Nevada. This section briefly summarizes the status of each taxon to justify the status ratings (table 33.1). The overall causes of species declines are then discussed.

Species Accounts

The information in the following accounts is derived from California Department of Fish and Game (CDFG) 1992 for threatened or endangered species; Moyle et al. 1996 for species of special concern; and Moyle 1976a, Lee et al. 1980, and Sigler and Sigler 1987 for these and other species. Broad dis-

tribution, habitat, and status information is presented in table 33.1.

Kern brook lamprey: This small, nonpredatory lamprey is endemic to Sierran streams of the San Joaquin drainage. Today there are only four to five known populations, mostly below dams and isolated from one another (Brown and Moyle 1993).

Pacific lamprey: This species is anadromous, with a long (four- to seven-year) freshwater larval stage. Large runs

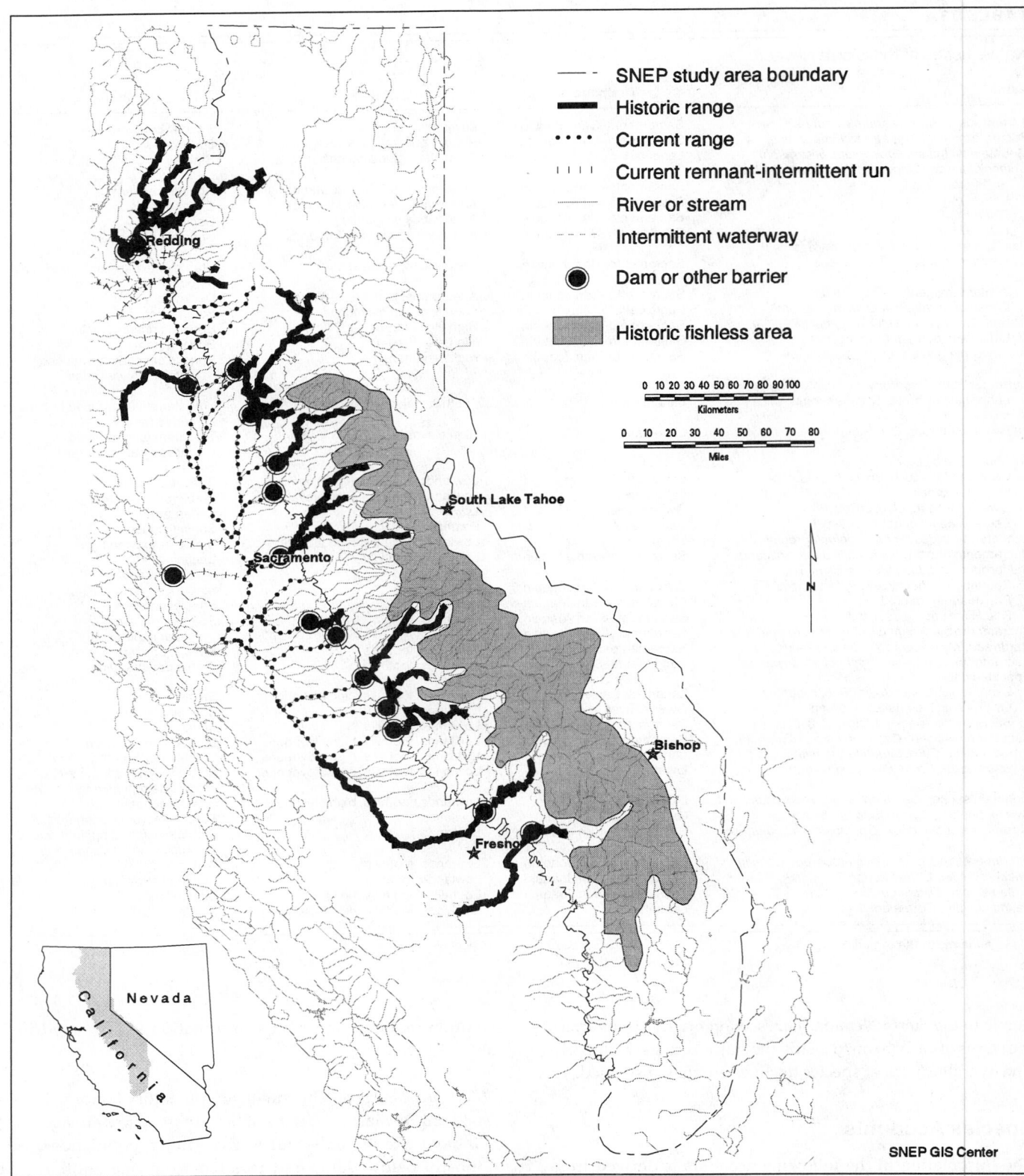

FIGURE 33.1

Two major changes in Sierra Nevada fish distribution. The shaded area shows streams and lakes that historically were without fish but that now mostly contain them. The dotted and heavy lines show the current and historic distribution of chinook salmon, respectively.

once spawned in most of the same places as chinook salmon, and lamprey populations appear to have declined for reasons similar to those for the salmon decline (e.g., dams). The decline in major prey species, salmon and steelhead, may have been an additional contributing factor. Lampreys still occur in reaches of west-side streams below major dams, and the extent of their decline is poorly documented.

Mountain whitefish: Whitefish are common inhabitants of the larger streams in the Lahontan drainage and Lake Tahoe. There is no evidence of major population declines, and they are subject to a sport fishery.

Chinook salmon: The four runs of chinook salmon were once extraordinarily abundant on the west side, but now the winter run is listed as endangered (state and federal), the spring and late fall runs both qualify for listing, and the fall run is largely supported by hatcheries. The historic distribution and abundance of salmon is treated in a separate section.

Resident rainbow trout: Rainbow trout were, and still are, the most widely distributed fish in the western Sierra. The resident populations were derived from steelhead and occupied all streams up to the highest barrier. Their range was greatly expanded by the transplanting of fish above barriers and the widespread stocking of hatchery fish both into fishless areas and throughout the eastern Sierra Nevada. If distinctive strains of rainbow trout existed in the western Sierra, they seem to have been genetically swamped by interbreeding with non-native strains.

Steelhead: It is likely that anadromous rainbow trout once inhabited most of the streams used by chinook salmon for spawning but ascended higher in the basins and into smaller tributaries. Unfortunately, their historic distribution is poorly documented. Today they are absent from the San Joaquin basin, and their distribution is limited by dams in the Sacramento basin, where the population(s?) are largely maintained by hatcheries. Fish of wild origin that have actually gone out to sea (rather than staying in the river) appear to be less than 10% of the population. It is estimated that 35,000 steelhead return to the Sacramento drainage each year, mostly to the Coleman, Feather River, and Nimbus hatcheries, but the trend is downward (CDFG 1990). In Mill and Deer Creeks, Tehama County, for example, runs of more than 1,000 fish have dwindled to 30 to 40 fish in each stream (Harvey 1995). All runs of steelhead in California, Oregon, and Washington have been proposed to the National Marine Fisheries Service for listing as threatened or endangered.

Eagle Lake rainbow trout: This subspecies is the only rainbow trout population native to the east side. It is unusu-

ally long-lived, late maturing, and capable of living in alkaline waters. By 1950, it was nearly extinct, largely as the result of the destruction of its spawning habitat by logging, grazing, and associated activities. It was saved at the last minute by the California Department of Fish and Game through a hatchery rearing program. All fish in the lake are now reared initially in a hatchery and are abundant enough to support a trophy trout fishery. A petition to list the Eagle Lake rainbow trout as a threatened species was denied by the U.S. Fish and Wildlife Service in 1995 because of ongoing efforts to restore its principal spawning stream, Pine Creek.

Kern River rainbow trout: This is a heavily spotted, brightly colored native of the mainstem Kern River. It is one of three subspecies in the complex of distinct "golden trout" forms that evolved in the upper Kern River basin, an isolated region that was mostly not glaciated. Until recently, this form was thought to be extinct as the result of interbreeding with hatchery rainbow trout that had been planted in the river, but it has managed to persist in small numbers. Active attempts to restore its populations are underway.

Little Kern golden trout: This subspecies is native to the Little Kern River basin and at one time was in danger of extinction as the result of invasions by non-native trouts and habitat degradation, especially from grazing. The downward trend in its populations has been reversed as the result of active programs to restore its habitats and exclude non-native trout (outlined in Christenson 1978). It was listed as an endangered species by the U.S. Fish and Wildlife Service in 1970.

California golden trout: The native range of the "classic" golden trout (also known as the Volcano Creek golden trout) was Golden Trout Creek and the South Fork Kern River, but this range has been reduced as the result of competition and predation from introduced trout, especially brown trout. Populations have been further reduced as the result of livestock grazing in the fragile meadow systems through which the streams flow. While the subspecies is in no danger of extinction, because many populations have been established through planting of lakes and streams outside its native range, maintenance of populations in the native range requires active management, including elimination of non-native trout within its native range and elimination of grazing along the streams.

Lahontan cutthroat trout: This was once the dominant trout and predator in streams of the Lahontan drainage as well as Lake Tahoe and other large lakes. It has been replaced throughout its range by non-native trout, except in a few scattered localities, where grazing by livestock often degrades the remaining habitat. It is a federally listed threatened species for which a recovery plan has been developed (Coffin and Cowan 1995).

Paiute cutthroat trout: This federally listed threatened species (upgraded from endangered in 1975) is endemic to the Silver King Creek drainage, Alpine County. It has been extirpated from its native range by non-native trout but survives in several transplanted populations, including one in upper Silver King Creek (Busack and Gall 1981). Grazing, however, continues to have negative effects on the streams in which it lives, and full recovery of populations will require the exclusion of grazing from the riparian zones and meadows (Kondolf 1994; Overton et al. 1994).

Lahontan lake tui chub: The principal native habitat of this plankton-feeding subspecies in the Sierra Nevada is Lake Tahoe, where its populations have presumably been depleted as the result of introductions of plankton-feeding competitors, especially mysid shrimp. However, populations in three reservoirs on the Little Truckee River may also belong to this subspecies.

Lahontan creek tui chub: This bottom-feeding form is abundant in many lowland streams and reservoirs on the east side of the Sierra Nevada and has apparently been introduced into other reservoirs outside its native range.

Owens tui chub: The Owens tui chub is listed as an endangered species by both state and federal governments. Its populations were depleted as the result of diversion of the Owens River, alteration of habitat, and displacement by introduced Lahontan creek tui chubs. Recovery efforts have resulted in populations being established in a number of isolated refugia, although the refugia have to be continually monitored for illegal introductions of predatory game fishes (mainly largemouth bass).

Eagle Lake tui chub: This form is confined to Eagle Lake, Lassen County, where it is a principal prey of Eagle Lake trout. It is extraordinarily long-lived (thirty or more years), an adaptation that has presumably allowed it to survive long droughts in the past, when the lake may have become too alkaline for successful reproduction. A tunnel constructed in the 1920s connecting the lake to Willow Creek keeps lake levels lower than they normally would be and has increased the possibility of an extended drought having a severe impact on the chubs.

Lahontan redside: This small minnow is abundant in streams and lakes in the Lahontan drainage, as well as in Eagle Lake, and it has successfully colonized a number of reservoirs.

Sacramento hitch: The hitch is a large cyprinid species adapted for lowland environments, including low-gradient, sandy-bottomed streams. There are still scattered populations, including one in a Sierra Nevada reservoir (Beardsley Reservoir), but they appear to be gradually disappearing. Brown and Moyle (1993) noted that hitch populations in the foothills in the southwestern Sierra Nevada were few and scattered and that several populations had disappeared in a fifteeen- to twenty-year period.

Sacramento roach: The California roach has numerous isolated and distinctive populations that are poorly described (Brown et al. 1992). The numerous isolated populations in tributaries to the Sacramento River are all considered to be one widely distributed form that is still fairly common and locally abundant.

San Joaquin roach: Like the Sacramento roach, this form is widely distributed in the Sierra Nevada foothills, but the populations in each tributary system have been demonstrated to be distinctive morphologically (Brown et al. 1992). Many of the small populations of this form have disappeared in recent years, a trend that seems to be ongoing (Moyle and Nichols 1974; Brown and Moyle 1993).

Red Hills roach: This undescribed subspecies is one of the most distinctive populations of California roach known (Brown et al. 1992) and inhabits the harsh environment of a few exposed streams in the Red Hills of Tuolumne County. The heavy use of the countryside around its streams for recreation and mining makes this form a possible candidate for endangered species status.

Sacramento blackfish: This blackfish is a lowland species in the Sacramento–San Joaquin drainage that is locally abundant and barely gets up into the streams of the Sierra Nevada.

Hardhead: The hardhead is one of the most specialized species of the Sacramento–San Joaquin fauna and requires clear, cool water in deep pools for its long-term survival. The principal habitats of these fish are in the same stream reaches that are favored for building dams, so their populations have become fragmented. They have become abundant in a few reservoirs (but are absent from most) and will thrive in regulated streams under certain conditions. However, many populations seem to have disappeared or declined in recent years, especially where smallmouth bass have invaded altered habitats, such as in the Kings River and South Yuba River (Brown and Moyle 1993).

Sacramento squawfish: Squawfish are predatory cyprinids that have managed to adapt to the altered conditions of California's rivers and are abundant in many westside streams. Their importance as a predator on salmonids is less than it seems (Brown and Moyle 1981).

Lahontan speckled dace: Speckled dace are the most widely distributed native fish in California and the only species native to both sides of the Sierra Nevada. The robust Lahontan form is abundant and widely distributed.

Owens speckled dace: In contrast to the Lahontan form, the diminutive Owens dace is in danger of extinction as the result of alteration of its small-stream and spring habitats and predation from introduced species. This taxon may actually represent two distinct subspecies.

Sacramento speckled dace: These dace are abundant and widely distributed in the Sacramento Valley, although the southernmost populations (in the Cosumnes River) are very limited in extent and so are in danger of extinction.

Sacramento sucker: The only sucker native to the Sacramento–San Joaquin drainage, this species is widespread and abundant, including in altered habitats. Although it is frequently accused of competing with trout for food and space, there is little evidence that this is true (Christenson 1978; Baltz and Moyle 1984).

Tahoe sucker: This is the common sucker of the Lahontan and Eagle Lake drainages, where it is abundant.

Owens sucker: Another Owens Valley endemic, the Owens sucker is still fairly abundant in its native range and has been introduced into reservoirs in the Mono Lake drainage. Its dependence on altered habitats that contain introduced predatory fishes, however, is a cause for concern.

Mountain sucker: This species frequently co-occurs with Tahoe sucker but is much less abundant. It has disappeared from or declined in much of its native range in the Sierra Nevada in recent decades (Decker 1989). The reasons for this are not clear but may be related to its inability to survive in reservoirs, which occur on most of the rivers to which it is native.

Owens pupfish: This small Owens endemic was once abundant in the sloughs and springs along the Owens River but became endangered when its habitats were drained and altered and exotic predators introduced into them. It is listed as endangered by state and federal governments and persists in only a few small refugia, whose abilities to protect the fish are continually threatened by illegal introductions of game fishes.

Threespine stickleback: This widespread native species is naturally found only in the San Joaquin River in the Sierra Nevada but has been introduced (as a contaminant in plantings of trout) into streams of the Mono Lake basin.

Sacramento tule perch: An unusual, live-bearing species endemic to the Sacramento–San Joaquin drainage, this fish has been largely extirpated from the San Joaquin basin, but it is abundant in much of its original range in the Sacramento basin, including in regulated streams.

Prickly sculpin: Prickly sculpin are primarily a low-elevation species and unusual for a sculpin (Cottidae) in that they can tolerate moderately warm water and become abundant in reservoirs. They are widespread and abundant in the Sacramento–San Joaquin drainage.

Riffle sculpin: This small fish, endemic to California, requires cold water of high quality and is found in many middle-elevation "trout" streams from the Kaweah River north. It recolonizes very slowly the areas from which it has been extirpated. It is missing from a number of streams where it might be expected on the basis of habitat (e.g., the South Yuba River) and so has probably been extirpated and been unable to recolonize.

Paiute sculpin: This is the ecological equivalent of the riffle sculpin in the Lahontan basin. It is widespread and abundant, although it may be locally extirpated as the result of dams and diversions.

Status of Native Fishes

Of the forty fishes native to the Sierra Nevada,

- Six (15%) are formally listed by the federal and/or state government as threatened or endangered species.

- Twelve (30%) are listed as species of special concern by Moyle et al. (1996) because they are in trouble statewide and are potential candidates for listing or because they have limited distributions.

- Four (10%) are in severe decline in the Sierra Nevada but are probably (but not necessarily) in less trouble elsewhere.

- Eighteen (45%) seem to have stable or expanding populations.

Three of the species of special concern (the Owens sucker, Eagle Lake tui chub, and Lahontan lake tui chub) are arguably reasonably secure in their populations, despite their vulnerability to a major drought. But even omitting these species from the concern list leaves nineteen species (48%) with significantly reduced populations and limited distributions within the Sierra Nevada, including all runs of three once-abundant anadromous species (the Pacific lamprey, chinook salmon, and steelhead rainbow trout) and six of the seven taxa of resident native trout. It is clear that the biggest declines in native fish abundance took place between 1850 and 1950, following intensive hydraulic mining, construction of hundreds of dams, and widespread introduction of non-native species. Although conservation efforts or reductions in the rate of habitat change have halted or reversed some declines, many species are still declining. The species in decline are found in all habitats and major drainages of the Sierra Nevada (table 33.1). Increasingly, the causes of continuing decline seem to be related to alteration of stream habitats (through new diversions, grazing, and urbanization) combined with the continued expansion of populations of non-native fishes. Even where declines seem to have been halted (e.g., golden trout in the Kern River basin), only management to prevent the reinvasion of non-native fishes (after eradication) and to restore habitats can prevent declines from starting once again.

Considering that so many of the fish taxa in the Sierra Nevada are threatened or in decline and that thirty species of non-native fish (see table 33.5 later) are established in the waters of the range, it is not surprising to find that the native fish assemblages have also been disrupted or have disappeared from many waters. In streams on the west side of the Sierra Nevada, most fish assemblages lost major components, mainly chinook salmon and other anadromous fishes, following the construction of dams in the nineteenth and early twentieth centuries (see the next section). The disruption of these communities is continuing. For example, in the San Joaquin drainage, the California roach assemblage has disappeared from many areas, including the upper San Joaquin River (Moyle and Nichols 1974), and the squawfish-sucker-hardhead assemblage is increasingly disrupted by reservoirs and introduced species (Brown and Moyle 1993). The most extreme example of loss of fish assemblages, however, occurs on the southeastern side of the range. In the Owens Valley, the original fish assemblage exists only in some tiny refuges especially created for it (Minckley and Deacon 1991).

CHINOOK SALMON DISTRIBUTION AND ABUNDANCE

At one time, millions of chinook salmon spawned in the streams of the western Sierra Nevada, from the Kings River in the south to Battle Creek and other tributaries to the north. In terms of numbers and biomass, they were among the most abundant fish in the streams. They were consequently a major source of energy for stream ecosystems, a major food for the Native Americans, and, after the Euro-American invasion in the nineteenth century, a mainstay of commercial fisheries. In recent years, their continuing decline has been a source of major conflict among various interest groups because their recovery will require major changes in the allocation of scarce resources, especially water. The importance of chinook salmon justifies a separate analysis of the changes in their distribution and abundance through time. This section is a summary of the more detailed analysis by Yoshiyama et al. (1996).

Sierra Nevada Chinook Salmon

The rivers draining the Sierra Nevada were renowned for their chinook salmon production in the nineteenth century, with annual runs of one to three million fish (Clark 1929; Skinner 1962). Even today these rivers are the source of most of the chinook salmon produced in California waters. Between 1980 and 1990, an average of 365,000 Central Valley chinook salmon were harvested annually by commercial fisheries (CDFG 1990). This catch is a fraction of the historic catch (Skinner 1962). Despite occasional years of high catches (e.g., 1995), the catch continues to decline even though the fishery is supported in good part by salmon reared in hatcheries (table 33.2) (Fisher 1994). Equally as dramatic as the decline in numbers of salmon has been the change in their distribution. Dams now block access to most upstream areas that were once major spawning grounds so that virtually all spawning now takes place just above the valley floor.

The main reason that Central Valley rivers, which include those that drain the Sierra Nevada, produced so many chinook salmon is that the salmon showed remarkable adaptations to the local conditions. There are four distinct runs of chinook salmon, more than in any other major river system on the Pacific Coast. Each run takes advantage of conditions that exist in different places and at different times in the drainage. The runs are defined by the times of adult freshwater migration to the spawning areas, the spawning periods, and juvenile residency and downstream migration periods (Fisher 1994). The runs are named on the basis of the season of the upstream spawning migration. The fish making the fall and late fall runs spawn soon after entering the natal streams, while those making the spring and winter runs, in their original natural circumstances, typically held in their streams for two to four months before spawning. Formerly, the runs could also be differentiated on the basis of their typical spawning habitats—spring-fed headwaters for the winter run, high-elevation streams for the spring run, upper mainstem rivers for the late fall run, and lower rivers and tributaries for the fall run. Different runs often occurred in the same stream—temporally staggered but broadly overlapping (Fisher 1994), with each run utilizing the appropriate seasonal stream-flow regime to which it had adapted.

The largest of the four runs were the fall run and the spring run, which were found in most of the major rivers. Fall-run fish historically spawned mainly in the valley floor and foothill reaches (less than 175 m [575 ft] elevation), where they still spawn today. The spring run, in contrast, ascended as high as 1,800 m (5,900 ft), the highest elevation known for any spawning salmon (in Mill Creek, Tehama County). The spring run was originally concentrated in the San Joaquin system, where the fish ascended and used high-elevation streams fed by snowmelt for over-summer holding until the fall spawning season (Fry 1961; F. Fisher, CDFG, conversations with the author, 1995). The winter run was present in the Sierra Nevada region only in Battle Creek (Tehama

County), which has the large cold springs that produce the habitat conditions required by this unique run. These habitat conditions otherwise existed mainly in the upper Sacramento River system (the Little Sacramento, Pit, McCloud, and Fall Rivers). Today winter-run salmon spawn only in the main Sacramento River below Keswick Dam. The late fall run probably was also most abundant in the upper Sacramento River system. However, late-fall-run fish spawned as well in the upper mainstem reaches of the larger rivers such as the American River and the San Joaquin River (Clark 1929; Fisher 1994).

Historic Distribution and Abundance of Chinook Salmon

Early distributions of salmon populations in the Sierra Nevada are not known exactly, due to a lack of scientific or historical records prior to 1850. However, the upstream limits of salmon distributions can often be inferred from the location of natural barriers to migration (e.g., major waterfalls) that exist or formerly existed. It was not until the late 1920s that reliable scientific surveys of salmon distribution in Central Valley drainages were conducted. Reports by Clark (1929) and Hatton (1939) give information on the accessibility of various streams to salmon, and they identify the human-made barriers present at those times. They also give limited qualitative information on salmon abundance. Fry (1961) provided the earliest comprehensive synopsis of chinook stock abundances in Central Valley streams, covering the period 1940–59. Since then, fairly regular surveys of spawning runs in the various streams have been carried out by the California Department of Fish and Game (CDFG 1990, 1993).

Of the four runs of chinook salmon, only the fall run still exists in any numbers (table 33.2). The winter run is listed by both state and federal governments as an endangered species, and the spring and late fall runs are considered to be species of special concern, with threatened species listing proposed for the spring run (Moyle et al. 1996). The principal cause of the decline of these runs has been the elimination of, or lack of access to, suitable habitat for holding, spawning, and rearing. This habitat loss started as soon as Euro-Americans arrived in large numbers to mine gold in Sierra Nevada streams in the 1850s. Numerous hydropower projects appeared in the 1890s and early 1900s, and collectively they eliminated the major portion of spawning and holding habi-

TABLE 33.2

Spawning stock estimates for the four seasonal runs of Central Valley chinook salmon during the period 1967–92, including hatchery returns. Stock estimates of the fall run are given separately for the Sacramento and San Joaquin drainages. The late fall, winter, and spring runs occurred only in the Sacramento drainage during this period, so the values listed for those runs pertain equally to that drainage and to the entire Central Valley. (Modified from Fisher 1994.)

Year	Sacramento Fall Run	San Joaquin Fall Run	Central Valley Late Fall Run	Central Valley Winter Run	Central Valley Spring Run	Central Valley Total
1967	157,643	22,785	37,208	57,306	23,840	301,182
1968	191,472	18,742	34,733	84,414	15,360	345,878
1969	268,178	52,212	38,752	117,808	27,447	506,482
1970	201,048	38,097	25,310	40,409	7,672	317,536
1971	193,762	42,996	16,741	63,089	9,274	331,062
1972	138,315	14,748	32,651	37,133	8,652	233,101
1973	263,385	7,895	23,010	24,079	11,967	332,936
1974	229,199	5,607	7,855	21,897	8,281	274,261
1975	187,564	7,825	19,659	23,430	24,044	264,922
1976	190,543	4,673	16,198	35,096	26,786	274,269
1977	184,090	1,050	10,602	17,214	13,951	227,157
1978	153,801	3,161	12,586	24,862	8,358	204,004
1979	222,549	5,087	10,398	2,364	2,960	244,865
1980	165, 041	7,098	9,481	1,156	11,937	197,944
1981	230,176	30,622	6,807	20,041	21,784	314,384
1982	210,975	19,761	4,913	1,242	28,082	274,345
1983	155,145	49,645	15,190	1,831	6,193	243,865
1984	198,517	58,820	7,163	2,663	9,923	284,237
1985	283,622	77,618	8,436	3,962	13,055	394,395
1986	264,212	24,268	8,286	2,464	20,329	324,478
1987	248,440	26,546	16,049	1,997	12,720	307,402
1988	252,542	22,522	11,597	2,094	18,486	307,753
1989	168,925	3,653	11,639	533	12,266	197,216
1990	118,309	1,092	7,305	441	6,630	134,208
1991	126,385	925	7,089	191	5,944	140,343
1992	109,218	3,098	10,370	1,180	2,997	128,495

tats for spring-run salmon well before the completion of the major dams constructed for water supply in later decades. By 1928, Clark (1929) estimated that the amount of salmon-spawning stream habitat had been reduced to about 820 km (510 mi) of river, which he considered to represent a loss of at least 80% of the spawning grounds. The obstructions to the spawners included eleven dams in the San Joaquin system and thirty-five dams in the Sacramento system.

The extent to which salmon (and other anadromous fish) habitat has been lost in the Sierra Nevada can be seen by examining the past and present distributions of salmon in each major river system (figure 33.1; table 33.3). In 1993, the CDFG estimated that the amount of spawning habitat left for salmon and steelhead in the Central Valley system totaled less than 480 km (300 mi) (CDFG 1993). Little of this is in the Sierra Nevada proper. Our estimates, based on the information presented in the stream-by-stream analysis in Yoshiyama et al. (1996), is that only 1,082 km (676 mi) of mainstream habitat remains of the 2,838 km (1,774 mi) originally available to chinook salmon for spawning, a loss of 62%. The actual percentage of spawning habitat lost is higher because in the San Joaquin drainage less than a third of the riverine habitat still accessible is suitable for spawning, and probably less than half of the accessible habitat is suitable in the Sacramento drainage. In addition, many of the smaller tributaries now located above dams were not added into the total of formerly accessible habitat, because it is likely that only small numbers of salmon used them for spawning. Thus, the estimate by CDFG that more than 90% of chinook salmon spawning habitat in the Central Valley drainage has been lost (CDFG 1993) seems reasonable, although the oft-cited estimate that more than 9,600 km (6,000 mi) of habitat were once available for chinook salmon spawning (Clark 1929) is probably high by a factor of three.

Conclusions

Chinook salmon and other anadromous fishes are largely gone from the Sierra Nevada, except where flows are provided for them below major dams at low elevations and in Butte, Deer, and Mill Creeks. This represents a major change in the riverine ecosystems of which they were once part. Not only are

TABLE 33.3

Estimated changes in lengths of streams available to chinook salmon in the major salmon-supporting drainages of the Central Valley.[a]

Drainage	Length (mi) of Stream Historically Available[b]	Length (mi) of Stream Presently Accessible[c]	Length (mi) of Stream Lost (or Gained)[d]	Percentage Lost (or Gained)
Sacramento Valley				
Pit River	93	0	93	100
McCloud River	43	0	43	100
Upper (Little) Sacramento River	53	0	53	100
Battle Creek	35	6	29	83
Antelope Creek	32	32	0	0
Mill Creek	44	44	0	0
Deer Creek	34	38	(4)	(12)
Big Chico Creek	21	21	0	0
Butte Creek	53+	53	>0	>0
Feather River	211	64	147	70
Yuba River	77	21	56	73
Bear River	16	16	0	0
American River	159	28	131	84
Clear Creek	25	16	4	16
Cottonwood Creek	79	79	0	0
Stony Creek	54	~3	51	94
San Joaquin Valley				
Cosumnes River	34	38	0	0
Mokelumne River	69	46	23	33
Calaveras River	~38	38	0?	0?
Stanislaus River	151	46	105	70
Tuolumne River	99	47	52	53
Merced River	99	43	56	57
Upper San Joaquin River	171	0	171	100
Kings River	84	0	84	100
Total	1,774	676	1,098	62

[a]The values for stream lengths originally available and subsequently lost are in most cases minimum estimates, because the full extent of the former salmon distributions is incompletely known. Additional, minor streams such as Thomes, Paynes, Cache, and Putah Creeks and perhaps a dozen others in the Sacramento Valley historically supported salmon runs (Fry 1961)—probably only the fall run and only during wet years when stream flows were adequate. The upstream distribution of salmon in those streams is too poorly known to allow inclusion in this table. Furthermore, current salmon production in those streams is limited because of a number of factors, including low stream flows, habitat degradation, and obstruction by irrigation canal crossings (CDFG 1993).
[b]Lengths of all stream reaches known or presumed to have been traversed or utilized by salmon in the drainage were summed.
[c]Length between the mouth of the stream and the current upstream limit.
[d]Length of stream gained is given in parentheses; this situation applies only to Deer Creek.

TABLE 33.4

Factors contributing to declines of native fishes of the Sierra Nevada region.

Species	Introduced Species	Dams and Diversions	Change in Aquatic Habitat	Watershed Disturbance	Other Factors
Kern brook lamprey	1	3	2	2	1
Pacific lamprey	0	3	2	1	1
Chinook salmon	1	3	2	2	2
Winter steelhead	1	3	2	2	2
Eagle Lake rainbow trout	0	1	1	3	1
Kern River rainbow trout	3	2	1	1	1
Little Kern golden trout	3	2	2	1	1
California golden trout	3	0	2	1	1
Lahontan cutthroat trout	3	3	2	1	1
Paiute cutthroat trout	3	0	2	0	1
Lahontan lake tui chub	3	0	0	0	1
Owens tui chub	3	1	0	0	1
Eagle Lake tui chub	2	0	1	1	1
Sacramento hitch	1	3	1	2	1
San Joaquin roach	2	2	2	2	2
Red Hills roach	2	1	2	3	2
Hardhead	3	2	2	1	1
Owens speckled dace	2	2	3	2	1
Mountain sucker	1	2	1	2	1
Owens pupfish	3	3	3	2	2
Total	40	36	33	29	25

0 indicates that the factor had no known effect.
1 indicates that the factor was of minor importance.
2 indicates that the factor was moderately important.
3 indicates that the factor was of major importance.

the once-abundant juvenile salmon and lampreys no longer part of local food webs, but the disappearance of adult fish has caused a loss of the annual influx of nutrients provided by the decaying carcasses. Attempts to replace the lost fish through the use of hatcheries have been only partially successful; total salmon numbers continue to decline. Achieving the officially stated goal of "doubling" salmon numbers (CDFG 1993) will presumably involve better management of flows in regulated rivers, habitat restoration where possible, and restoring salmon to some areas from which they are now gone (e.g., the American River above Folsom Dam).

CAUSES OF NATIVE FISH DECLINES

The causes of fish declines are multiple and interactive (table 33.4). They also can be quite different for different species and can change over time. In addition, what may be devastating for one species may favor another. Thus, Sacramento suckers and tui chubs do quite well in reservoirs in which most other native fishes cannot survive. The causes of decline can be broken into five broad categories: (1) introduced species, (2) dams and diversions, (3) changes in aquatic habitat, (4) watershed disturbance, and (5) other factors.

Introduced Species

Introduced species of fish have had strong negative effects on the abundance of ten of the twenty species in decline. Of the thirty introduced fish species, ten (33%) are abundant and widespread, eight (27%) are common with somewhat more restricted distributions than the abundant species, and the rest (40%) are rare or peripheral in the range (table 33.5). Introduced species particularly appear to be a problem at high elevations. The reason for this is that seven of the native species are trout characteristic of high-elevation habitats, and six of these trout have been negatively affected by competition, predation, and hybridization by non-native trout, especially brown trout. At lower elevations, predation by non-native fishes, especially centrarchid basses, has also been an important factor in the decline of native species. Hardhead, for example, have declined in response to expanding smallmouth bass populations (Brown and Moyle 1993), while the introduction of largemouth bass into the spring refuges of Owens pupfish, tui chub, and speckled dace continues to be a factor in their decline. Unfortunately, the introduction of new species of predatory fish into Sierra Nevada waters is continuing. Most recently, northern pike (*Esox lucius*) and white bass have been introduced illegally into reservoirs on the west side. If efforts by the CDFG to eradicate the populations have failed, it is likely that both of these predators will cause further changes to native fish assemblages.

Dams and Diversions

Although introduced species have been identified as a major cause of native fish declines, they often are as much a symptom of the decline as a cause. As a general rule, the more altered a stream or lake is by human disturbance, the more likely it is to become dominated by non-native species (Baltz and Moyle 1993). In many instances, the invasion of introduced fishes has followed habitat changes, especially those created by dams and diversions. Because of the importance of the Sierra Nevada as a supplier of water for California, virtually every stream of any size has at least one dam or diversion on it (Kattelmann 1996). The changes caused by such dams and diversions have been identified as a major cause of the declines of seven of the twenty declining species and as a contributing factor in most of the rest. Reservoirs generally favor exotic fishes, which can then invade both upstream and downstream. Dams and diversions also contribute to declines by flooding habitats, removing water, changing flow regimes, blocking movements and migrations, isolating populations, and causing increased human use of the watersheds. Dams on major rivers have blocked access by spring-run chinook salmon to more than 95% of its spawning and holding areas and have greatly reduced access to spawning grounds of other runs of salmon, steelhead, and Pacific lamprey.

Although Moyle and Williams (1990) identified dams and diversions as the single biggest cause of fish declines in California overall, it is important to recognize that the greatest impacts of dams occur immediately after they are built, when the changes they cause are fully in place for the first time. Thus, most runs of spring-run chinook salmon were eliminated before 1950, following the construction of dams on the major tributaries. However, dams, diversions, and reservoirs have a continued negative effect on native fishes through changes in flow regime and in the physical environment downstream because they block migrations to upstream areas and provide a continuous source of introduced species as predators and competitors to both upstream and downstream reaches. Upstream, these impacts on the isolated remnant populations of native fish are usually less than the effects of other activities in the watershed that alter stream habitats or water quality, such as grazing, road building, and mining.

TABLE 36.5

Introduced fishes of the Sierra Nevada.

Name	Habitat	Elevation[a]	Status
American shad, *Alosa sapidissima*	Anadromous; mainstem rivers; reservoirs and ponds	Low	Uncommon
Threadfin shad, *Dorosoma petenense*	Reservoirs and ponds	Low to middle	Abundant and widespread
Wakasagi, *Hypomesus nipponensis*	Reservoirs and ponds	Low to middle	Uncommon
Kokanee, *Oncorhynchus nerka*	Reservoirs and ponds; lakes	High	Common
Colorado cutthroat trout, *Oncorhynchus clarki pleuriticus*	Lakes	High	Rare
Brook trout, *Salvelinus fontinalis*	Cold-water streams; lakes	High	Abundant and widespread
Lake trout, *Salvelinus namaycush*	Lakes	High	Localized
Brown trout, *Salmo trutta*	Cold-water streams; lakes; reservoirs and ponds; mainstem rivers	Middle to high	Abundant and widespread
Arctic grayling, *Thymallus arcticus*	Reservoirs and ponds	High	Rare
Common carp, *Cyprinus carpio*	Warm-water streams; mainstem rivers; reservoirs and ponds	Low	Common
Goldfish, *Carassius auratus*	Reservoirs and ponds	Low	Uncommon
Golden shiner, *Notemigonus chrysoleucas*	Warm-water streams; lakes; reservoirs and ponds	Low to high	Common
Fathead minnow, *Pimephales promelas*	Warm-water streams; reservoirs and ponds	Low	Common
Channel catfish, *Ictalurus punctatus*	Mainstem rivers; reservoirs and ponds	Low to middle	Common
White catfish, *Ameiurus catus*	Reservoirs and ponds	Low to middle	Common
Brown bullhead, *Ameiurus nebulosus*	Warm-water streams; lakes; reservoirs and ponds	Low to high	Uncommon
Black bullhead, *Ameiurus melas*	Warm-water streams; lakes; reservoirs and ponds	Low to high	Abundant and widespread
Western mosquitofish, *Gambusia affinis*	Warm-water streams; reservoirs and ponds	Low to middle	Abundant and widespread
Striped bass, *Morone saxatilis*	Reservoirs and ponds	Low	Uncommon
White bass, *Morone chrysops*	Reservoirs and ponds	Low to middle	Localized
Sacramento perch, *Archoplites interruptus*	Reservoirs and ponds	Low to middle	Localized
Black crappie, *Pomoxis nigromaculatus*	Reservoirs and ponds	Low to middle	Abundant and widespread
White crappie, *Pomoxis annularis*	Reservoirs and ponds	Low to middle	Common
Green sunfish, *Lepomis cyanellus*	Warm-water streams; lakes; reservoirs and ponds	Low to high	Abundant and widespread
Bluegill, *Lepomis macrochirus*	Warm-water streams; lakes; reservoirs and ponds	Low to high	Abundant and widespread
Pumpkinseed, *Lepomis gibbosus*	Warm-water streams; reservoirs and ponds	Low to middle	Uncommon
Largemouth bass, *Micropterus salmoides*	Warm-water streams; reservoirs and ponds	Low to middle	Abundant and widespread
Spotted bass, *Micropterus punctualatus*	Warm-water streams; mainstem rivers; reservoirs and ponds	Low to middle	Common
Smallmouth bass, *Micropterus dolomieui*	Warm-water streams; mainstem rivers; reservoirs and ponds	Low to middle	Abundant and widespread
Redeye bass, *Micropterus coosae*	Warm-water streams; reservoirs and ponds	Low to middle	Rare

[a]Low elevation is less than 200 m (650 ft). Middle elevation is 200–1,500 m (650–4,900 ft). High elevation is more than 1,500 m (4,900 ft).

Alteration of Aquatic Habitats

Among the many factors affecting aquatic habitats, the most significant in the Sierra Nevada are road building, channelization, grazing, and mining. Road building and channelization are interrelated because hundreds of miles of Sierran streams have been channelized to support roads on their banks. The major transportation corridors in the Sierra Nevada follow streams and are often located in the riparian zones. The most noticeably altered streams are those that are sandwiched between highways and railroads (e.g., North Fork Feather River). However, smaller roads associated with logging, mining, and recreation can also alter streams, especially where they cross; there is a negative correlation between the abundance of roads in a watershed and the integrity of the native stream biota (Moyle and Randall 1996).

The effects of livestock grazing are pervasive throughout the Sierra Nevada, resulting in degraded stream habitats through loss of habitat complexity (by stream-bank alteration and removal of riparian vegetation), siltation, and other effects (Chaney et al. 1990; Menke et al. 1996). The loss of habitat quality and quantity associated with grazing contributes not only to the decline of native fishes but also to the reduction in populations of trout important in stream fisheries (Platts 1991; Dudley and Embury 1995). Mining also contributes to the presence of low fish populations in some areas, mainly through the residual effects (siltation, streambank alteration) of hydraulic mining (Gard 1994), the roads and tailing piles associated with hardrock mines, and the direct and indirect effects of suction dredge mining.

Watershed Disturbance

Cumulative watershed disturbances, as the result of urbanization, logging, grazing, mining, and other factors, have affected most species of fish to some extent, through changes in flow patterns, reductions in flows, and removal of riparian vegetation. These changes can be seen dramatically in Pine Creek (Lassen County), once the principal spawning stream of Eagle Lake trout. Heavy logging and grazing in the drainage, coupled with construction of road and railroad beds across key flowage areas, resulted in the lower reaches of the stream becoming dry much sooner and much more frequently than they had historically, denying adult trout access to spawning grounds and juvenile trout access to the lake (Moyle et al. 1996). Wissmar et al. (1994) found such activities to have caused long-term, cumulative degradation of stream habitats throughout eastern Oregon and Washington, resulting in degradation of fish communities. Similar problems are present throughout the West (e.g., Chaney et al. 1990; Meehan 1991) and the Sierra Nevada (Kattelmann 1996; Menke et al. 1996).

Other Factors

Other factors affecting fish populations include pollution, exploitation, and disturbance. Pollution has played a relatively minor role in fish declines in recent decades because many major sources (e.g., sewage from towns) were cleaned up as the result of the federal Clean Water Act and other regulations. However, pollution may play an increasing role in the future, as atmospheric deposition changes water chemistry and adds toxic materials to the water, especially in the southern portion of the range (Cahill et al. 1996), and as the effects of acid mine drainage and unregulated agricultural pollution (including livestock wastes) accumulate. Exploitation has affected salmon and steelhead populations. It was probably a bigger factor in the past (when salmon canneries were operating) than it is today, although existing commercial and sport fisheries may help to keep anadromous fish populations suppressed, making recovery more difficult. Likewise, the presence of salmon and trout of hatchery origin in streams may interfere with the recovery of wild populations, through behavioral interactions, genetic swamping, and introduced diseases (Steward and Bjornn 1990). Disturbance is a particular problem for anadromous fishes, especially spring-run chinook salmon. Heavy use of streams by rafters, anglers, or dredge miners may disturb fish that are holding or spawning, reducing the success of spawning (Moyle et al. 1996).

Conclusions

The native fishes and fish assemblages of the Sierra Nevada have declined largely as the result of water diversion, introduction of non-native species, and habitat alteration. A number of the species, and consequently the assemblages of which they are part, are likely to become extinct within the next fifty years if present trends continue. Increasingly, the native fish assemblages are found in streams that are isolated from one another by dams or other barriers. As a result, natural recolonization is not possible if a local extinction event occurs. The streams in which native fish assemblages still occur are mostly unregulated and flow through watersheds that are in relatively good condition. Such streams are also likely to support populations of native amphibians and invertebrates. Conservation of the native fish fauna ultimately will require active management of streams and lakes throughout the Sierra Nevada and probably the creation of an aquatic refuge system.

INTRODUCED FISHES IN THE SIERRA NEVADA

At least thirty introduced species of fish have reproducing populations in Sierra Nevada waters; twenty-seven of them are at least regionally abundant and eighteen are widely distributed (table 33.5). In addition, at least seven native species (rainbow trout, California golden trout, Lahontan cutthroat trout, Paiute cutthroat trout, tui chub, Owens sucker, threespine stickleback) have had their ranges expanded through introductions. Most (twenty-two) of the common non-native species are associated with reservoirs or highly altered streams at low to middle elevations, while two (kokanee, lake trout) are associated with large, high-elevation lakes and reservoirs. The two remaining non-native fishes (brook trout, brown trout) are widely distributed in high-elevation lakes and streams, along with introduced populations of rainbow trout and golden trout. The rainbow trout in particular has been widely introduced into streams, lakes, and reservoirs on the east side of the Sierra Nevada, where it (along with other trout) has largely displaced the native cutthroat trout. The other five native species occur in only a relatively few non-native waters.

As was indicated previously, these introduced fishes have had a major negative impact on native fishes in their native ranges, especially at high elevations. These impacts have been well documented in a general sense (e.g., Moyle 1976b, 1986; Li and Moyle 1993), although often only anecdotally in the Sierra Nevada. Ironically, the introduction of the two cutthroat trout subspecies into fishless waters was done because they had been displaced from native waters by introduced trout species. Less well known is the extent and impact of trout introductions into the vast areas of the range that were originally fishless. This section will therefore focus on these introductions and their impacts.

Because the high-elevation regions of the Sierra Nevada are largely within national parks, wilderness areas, and national forests, the hundreds of kilometers of clear-flowing streams and the more than 4,000 transparent lakes they contain are generally considered to be pristine. In fact, these waters are arguably among the most altered ecosystems in the Sierra Nevada. Historically, the single biggest factor altering these systems has been the introduction of trout. Most lakes and streams above 1,800 m (6,000 ft) were fishless until fish planting programs began in the nineteenth century. This section reviews (1) the history of trout stocking, (2) the current distribution of trout, and (3) the impacts of non-native trout on aquatic ecosystems. For more details, see Knapp (1996).

History of Trout Stocking

Although the indigenous peoples of the Sierra Nevada often lived at high elevations, there is no evidence that they moved fish into fishless waters. The upstream limits of fish were determined by natural barriers. This situation changed dramatically with the influx of Euro-Americans in the mid-nineteenth century, who brought with them a love of angling, especially for trout. The first introductions appear to have been transfers of native trout (Lahontan cutthroat trout, coastal rainbow trout, California golden trout) above waterfalls or into neighboring drainages by miners, sheepherders, and other people living in the mountains. Soon, however, non-native salmonids were being planted in the Sierra Nevada: brook trout (1872), brown trout (1872), lake trout (1889), kokanee (1941), Colorado cutthroat trout (1931), and arctic grayling (1930) (Moyle 1976a). Once established, exotic trout were moved to new locations both by private individuals or clubs and by state and federal agencies. Extensive trout planting in Yosemite, Sequoia, and Kings Canyon National Parks was accomplished by the U.S. Army (Christenson 1977). By the 1940s, official stocking of fish was done almost entirely by the CDFG, and today this agency has sole responsibility for stocking trout in the Sierra Nevada. In part because the vast majority of lakes and streams capable of supporting trout already had trout populations, the emphasis of CDFG stocking programs has been to supplement or maintain existing trout fisheries.

Although the fisheries supported in part by stocking programs included those in the national parks, the National Park Service (NPS) began to phase out fish stocking in 1969. This change in policy was a response to the Leopold (1963) report that recommended that "the natural biotic associations within each park should be maintained, or where necessary recreated, as nearly as possible in the condition that prevailed when the area was first visited by white man." A ban on fish stocking became official NPS policy in 1975, but limited stocking occurred in Sequoia, Kings Canyon, and Yosemite National Parks until 1991. Stocking is still permitted in all other waters on federal lands, with the exception of a few waters within wilderness areas that were not stocked prior to each area's designation as wilderness (Bahls 1992). In general, stocking trout in lakes and streams has been based on historic precedents, with little consideration given to the effects of the stocked trout on the native biota (Bahls 1992). Most waters stocked with fish are not regularly evaluated for their fish populations, angler use, or trends in their native biota (Bahls 1992).

It is worth noting that one of the side effects of indiscriminate planting of trout throughout the Sierra Nevada was the introduction of other species of fish either as "contaminants" in the water used for transporting the trout or as bait released by anglers. As a result, threespine stickleback, Owens sucker, and tui chub are present in the Mono Lake basin. Similar anomalous distributions can be found elsewhere in the Sierra Nevada, especially immediately above and below reservoirs or in lakes with easy road access. The exotic fish most commonly established through bait-bucket introductions is probably the golden shiner, which can survive in high-elevation lakes with deepwater refuges in winter and warm, shallow areas in summer.

Current Fish Distribution

All major watersheds of the Sierra Nevada contain introduced populations of trout, but records of exactly which waters contain trout are scattered and incomplete. A guide for anglers, for example, lists about 1,700 lakes with fish in them (Cutter 1984). More than half of these lakes are below 2,400 m (7,900 ft), yet Jenkins et al. (1994) estimate that there are more than 1,000 lakes above 2,400 m alone with fish in them. Existing estimates of the number of waters containing trout are based on interviews with fishery managers (Bahls 1992) or on extrapolations to the entire range from surveys of a small number of waters (Jenkins et al. 1994). Bahls (1992) estimated that 63% of all lakes above 800 m (2,600 ft) contained introduced fish and 52% were regularly stocked. Fishless lakes were generally too small (less than 2 ha [5 acres]) and shallow (less than 3 m [10 ft] deep) to support trout. Such lakes either freeze to the bottom, become depleted of oxygen in the winter, or become too warm in the summer. Jenkins and colleagues (1994) randomly selected thirty high-elevation (more than 2,400 m [7,900 ft]; more than 1 ha [2.5 acres]) lakes from throughout the Sierra Nevada for sampling. They then estimated that, of the 1,404 similar lakes in the range, trout existed in 63%, and the rest were fishless. Golden trout were estimated to occur in 36% of the lakes, rainbow trout in 33%, brook trout in 16%, brown trout in 8%, and cutthroat trout in less than 1%. Although these studies indicate that 60% to 65% of all Sierra Nevada lakes contain trout, the percentage of the larger, deeper lakes containing trout is much higher. This is echoed in the national forests and national parks, where larger, deeper lakes are also more likely to contain fish.

Christenson (1977) estimated that 95% of naturally fishless lakes that were large enough to support trout populations contained fish. To test this hypothesis, a detailed analysis was performed on the only large database available, that maintained by CDFG Region 5 (Knapp 1996). This database contains records of 649 lakes in portions of the Inyo, Sierra, and Toiyabe National Forests, more than 90% of the lakes in the region. Eighty-four percent of the lakes for which there are records lie within wilderness areas, 2% are in a U.S. Forest Service (USFS) Research Natural Area, and 14% are in other areas. The majority of these lakes are at elevations between 3,250 and 3,500 m (10,660 and 11,480 ft), and nearly all are less than 10 ha (25 acres) in area. All the lakes were originally without fish, but today 85% contain trout, 7% are fishless, and 8% are of unknown status. It seems reasonable to conclude that more than 90% of these lakes contain trout. Both fish-containing and fishless lakes occur at all elevations, although fishless lakes are most common at either high (more than 3,500 m [11,480 ft]) or low elevations (less than 2,500 m [8,200 ft]). Lakes containing fish are significantly larger than fishless lakes, which are mostly less than 1 ha (2.5 acres) in area. Of the lakes in this study with fish, 60% contained brook trout, 36% contained rainbow trout, 32% contained golden trout, 5% contained brown trout, and less than 1% contained other salmonids.

Of the 649 lakes, 299 (46%) are still planted on a regular basis by the CDFG, mostly with drops of juvenile fish from airplanes, either annually (35%) or every two years (65%). The principal trout planted are rainbow trout (49%) and golden trout (48%). Despite the regular plants of fish, the 649 lakes in the CDFG Region 5 database are only infrequently surveyed; only 32% were formally sampled for fish in the last ten years, 56% in the past twenty years. These surveys have been only for fish, with no effort made to determine the status of frogs and other native species.

Although lake surveys on non-national-park lands are limited, they are considerably more frequent than stream surveys. Presumably, most streams large enough to support trout contain them, especially if they are downstream of lakes containing trout or immediately upstream of such lakes. In a 1992 survey of 20 km (12.5 mi) of streams in the upper Lee Vining and Mill Creek watersheds (Mono Lake basin), Knapp (1996) found only 2 km (1.25 mi) without fish. It is likely that, as in the case of lakes, more than 90% of stream habitat suitable for trout now supports populations of them.

In the national parks, the percentages of lakes with fish are considerably lower than in areas outside the parks (Knapp 1996). Surveys of lakes greater than 1 ha (2.5 acres) in area in Yosemite National Park in the 1950s indicated that 35% were fishless (although about a third of the fishless lakes had been planted with fish at one time or another), 38% contained self-sustaining populations of trout, 24% had trout populations maintained by stocking, and 4% were of unknown status (Wallis 1952). A survey of 102 lakes in Yosemite National Park with a recent history of fish stocking (Botti 1977) indicated that trout were no longer present in 22% of the lakes and would probably disappear from 22% more following the cessation of stocking. Thus, it can be expected that at the present time about half the lakes in Yosemite National Park greater than 1 ha in area are fishless. This figure is roughly comparable to the frequency of fishless lakes (54%) recorded by Bradford and colleagues (1993) for Sequoia and Kings Canyon National Parks. However, a survey of 104 lakes in the most remote portions of Kings Canyon National Park indicated that only 17% contained trout (Bradford et al. 1994). In general, lakes in the national parks that have self-sustaining populations of fish are likely to be greater than 4 ha (10 acres) in area and likely to be in the most accessible parts of the parks.

For streams, it is likely that conditions in the national parks have not changed much since Wallis (1952) surveyed streams in Yosemite National Park and showed that 22% of 157 streams were fishless (including 2% that had been stocked with fish at one time or another), 58% contained trout populations, and for 20% there was no information. Limited recent data from Yosemite National Park indicate that, overall, at least 60% of all streams still contain trout (Knapp 1996).

Effects of Trout Introductions on Native Aquatic Biota

Trout are generalist predators, consuming whatever prey is available, from invertebrates to fish to amphibians (Moyle

1976a). For stream and lake biotic communities of invertebrates and amphibians that developed in the absence of a top predator like trout, the effects of trout introductions are potentially devastating. In the Sierra Nevada, introduced trout have had negative effects on native trout, amphibians, and invertebrates.

Native Trout

With the exception of rainbow trout and California golden trout, the native trout of the Sierra Nevada have declined in the face of competition, predation, and hybridization from non-native trout. Ironically, one endemic trout, the Paiute cutthroat trout, was saved from extinction because a sheepherder, in 1922, introduced it above an impassable falls on Silver King Creek. The population below the falls subsequently became extensively hybridized with rainbow trout (Busack and Gall 1981).

Amphibians

Amphibians are in decline throughout the Sierra Nevada (Jennings 1996), and introduced trout are one of a number of interacting factors responsible for the decline, especially at high elevations. In general, ranid frogs, bufonid toads, and ambystomid salamanders are less abundant than formerly in or near waters that contain introduced species of fish. It appears that the main mechanism by which trout affect amphibians is through predation on tadpoles, although diseases brought in with the fish may also play a role. The species apparently most affected by fish is the mountain yellow-legged frog (*Rana muscosa*), which is now found at fewer than 15% of the high-elevation sites at which it was present in 1915 (Drost and Fellers 1994). While the decline in this species is the result of many interacting factors (Jennings 1996), its long-term survival will apparently depend on predator-free lakes deep enough (more than 1.5 m [5 ft]) for the over-winter survival of adults and tadpoles, as well as a predator-free environment for the tadpoles, which take two to three years to mature. In addition, it is likely that the dispersal of this species depends in part on having predator-free streams that can be used as corridors between lakes.

Invertebrates

Introduced trout can affect the composition of zooplankton and benthic invertebrate communities in lakes and of benthic invertebrate communities in streams (Erman 1996). In a survey of seventy-five Sierra Nevada lakes, Stoddard (1987) found that the presence of trout was the best predictor of the zooplankton species present (or absent). As researchers have found in lakes elsewhere, large zooplankton species tend to disappear in the presence of fish because of their vulnerability to predation. The survey by Stoddard (1987) and other surveys indicate that the phantom midge, *Chaoborus americanus*, may have been extirpated from the Sierra Nevada by trout. This midge has planktonic larvae specialized for living in the larger, low- to middle-elevation lakes, which now universally contain trout. Trout probably have had similar effects on benthic insects, but these effects are much more poorly documented, although it is often noted that the abundance of benthic invertebrates, especially of larger species (caddisflies, mayflies, etc.), greatly declines following trout introductions (e.g., Reimers 1958). Whether or not invertebrate species eliminated from lakes by trout can recolonize a lake in which the trout have disappeared depends on the proximity of the lake to fishless lakes that contain the missing species.

Trout seem to have a less dramatic effect on stream invertebrates than they do on lake invertebrates. However, large, diurnal species may be eliminated from formerly fishless streams once trout are introduced, and the behavior of other species may be altered (Erman 1996).

Conclusions

The introduction of predatory trout into formerly fishless lakes and streams of the Sierra Nevada has caused major changes in the aquatic biota. As a result, relatively few lakes and streams have aquatic communities that are in near-pristine condition. Some invertebrate species may have been eliminated altogether, and a number of native fish, amphibians, and invertebrates have become endangered. The cessation of stocking of hatchery trout in lakes in the national parks has resulted in the partial reestablishment of the assemblages of aquatic organisms native to fishless lakes. Thus, it appears that it is not too late to restore some high-elevation watersheds to a fishless condition in order to restore populations of species sensitive to fish predation. If current trends continue, further extirpations of native organisms from the Sierra Nevada are likely as the result of trout predation in combination with other factors.

SIERRA NEVADA FISHERIES

The main reason that trout and other fish have been widely introduced in the Sierra Nevada is to support sport fisheries. The sport fisheries, including those for native fish, in turn are a major source of support for the recreational industry of the Sierra Nevada, bringing thousands of anglers into the range each summer. For many other people seeking recreation in the Sierra Nevada, catching fish is an important part of their total experience. Five major fisheries can be arbitrarily recognized in the range: high-elevation trout fisheries, wild trout fisheries, catchable trout fisheries, warm-water stream fisheries, and reservoir fisheries. Stock ponds at low elevations are also presumably important in fisheries, but they are largely on private land, and the extent to which they contribute to fisheries is poorly known. Fisheries for anadromous salmon, steelhead, and shad are now largely gone from the Sierra

Nevada and are present primarily in the main river channels at low elevations, mostly outside the SNEP area.

High-Elevation Lake Fisheries

There are about 4,000 lakes in the Sierra Nevada, probably 75% of which are large and deep enough to support trout. Some are accessible by road, but most can be reached only by hiking or similar means. Because the lakes have short ice-free seasons and are mostly in granitic, glacier-scoured basins, they are not very productive. They can support only relatively low densities of trout, and the trout present are usually rather slow growing, rarely exceeding 30 cm (12 in) in length. Aside from roadside lakes stocked with catchable trout, trout in the lakes come from two sources: natural reproduction and plants of small (fingerling) trout, usually by airplane. The majority of naturally reproducing trout populations in the lakes are brook trout, because they can spawn in the lakes. The problem with brook trout is that many populations are made up mainly of stunted individuals, mostly less than 15 cm (6 in) in length (Moyle 1976a). This occurs in part because the lakes are low in productivity, and so the trout have slow growth rates (Reimers 1958). There is presumably intense intraspecific competition among the trout for the limited food resources available in the lakes. In addition, brook trout are fall spawners, and post-spawning adults often have a hard time surviving the winters. Rainbow trout and brown trout often have difficulty maintaining populations in lakes because they require streams for spawning. For this reason, about half of all Sierra Nevada lakes are planted every one to two years with juvenile rainbow trout (Bahls 1992). However, lakes in national parks are no longer planted with trout, a policy decision that was highly controversial at the time it was made (Pister 1977).

A special case of lake fishery is found in Lake Tahoe, where the primary focus is on naturally produced lake trout, although rainbow trout, brown trout, and kokanee salmon are also caught (Cordone and Frantz 1966). Because of the low harvest of the lake's trout fishery (0.27 kg/ha/yr [0.24 lb/acre/yr]), opossum shrimp were introduced into the lake as additional forage. There is no evidence that this introduction improved the fishery, although it did dramatically change the ecology of the lake by eliminating most of the large zooplankton species (Morgan et al. 1978).

Bahls (1992), following a survey of fisheries managers responsible for high-mountain fisheries, characterized the management of high-mountain lakes, including those in the Sierra Nevada, as follows:

> Management . . . can best be summarized as intensive, on-going, and largely indiscriminate stocking. . . . Most regions stock mountain lakes with non-native trout species and with limited or nonexistent survey data upon which to make basic stocking decisions. . . . [There appears to be] little concern for protection of native fish species in lakes or downstream systems, no evident concern for maintaining representative pristine lakes, and no consideration for the effects of trout stocking on the indigenous fauna, aquatic ecosystems, and lake shore. . . . Furthermore, most regions appear to manage fisheries with little understanding of the high lake anglers whom they serve. (P. 191)

Wild Trout Fisheries

Fishing for trout produced naturally in California streams has always been an important recreational activity. California Trout, an angler organization, estimates that about 60% of the more than 150,000 licensed anglers in California fish primarily for trout, with most of the fishing effort concentrated on naturally produced trout (California Trout, unpublished studies). Wild trout are especially important in Sierra Nevada fisheries. Sierra Nevada streams have been estimated to have standing crops of trout 75 mm (3 in) long and longer (mean, 46 kg per ha [41 lb per acre]) that were typical of California streams but lower than those in Rocky Mountain streams (mean, 67 kg per ha [60 lb per acre]) (Gerstung 1973). Wild, catchable-size (longer than 15 cm [6 in]) trout in Sierra Nevada streams average 139 fish per km (224 per mi), with a range of 60–500 per km (100–800 per mi) (Gerstung 1973).

Recognizing that the harvest of trout in California waters was approaching or exceeding the maximum harvest rate (Gerstung 1973) and that a growing segment of the angling community preferred to release most of the fish they caught, the CDFG initiated a wild trout program in 1971. This program, authorized by the state legislature in 1979, allows the CDFG to designate streams and lakes as wild trout waters, in which no catchable-size trout are planted and which have restrictive angling regulations. Fifty such streams and lakes have been designated (Deinstadt et al. 1993), mostly in the Sierra Nevada. Other streams are added to the program on a regular basis, based on their ability to support wild trout fisheries. For example, the Upper Middle Fork of the San Joaquin River has been recommended for addition to the program based on the fact that it supports an average of 964 catchable trout per km (1,606 per mi), one of the highest densities of trout in the Sierra Nevada (Deinstadt et al. 1995). Although most wild trout waters allow a small number of trout to be kept by the anglers, some have catch-and-release fishing only. In either case, large numbers of fish are caught repeatedly over a season, "recycling" the trout. This program is very popular with trout anglers (Deinstadt et al. 1993).

Catchable Trout Fisheries

As fishing pressure increased on roadside streams and lakes in the 1940s and 1950s, the CDFG developed an extensive hatchery system to raise trout to catchable size (15–20 cm [6–8 in]). These fish are planted with the expectation that most will be caught within two weeks of release; at least 50% must be caught for a planting program to be considered successful (Butler and Borgeson 1965). Today, the CDFG supports ten production hatcheries to raise trout, which annually produce about 13 million catchable trout, 1.2 million "subcatchable" trout, and 12.3 million fingerling trout, mostly rainbow trout (Hashagen 1988). The catchable trout are planted in both streams and lakes, while the subcatchable trout (10–15 cm [4–6 in]) are planted mostly in reservoirs (because of higher survival rates). The fingerling trout are planted mainly in high-elevation lakes in the Sierra Nevada. A fairly typical hatchery serving Sierra Nevada streams is the Moccasin Creek hatchery on the Tuolumne River. This hatchery raises more than 1 million catchable rainbow trout each year, which are planted in forty heavily fished lakes and streams in the region, as well as more than 1 million fingerlings for aerial planting in alpine lakes (Groh 1990). The trout produced in such hatcheries presumably account for the bulk of the trout kept by anglers in the Sierra Nevada each year and contribute substantially to the recreational economy of the region.

Warm-Water Stream Fisheries

Warm-water stream fisheries occur in low-elevation streams, especially those with reduced flows due to diversions, and focus on various introduced black basses, sunfishes, and catfishes. In addition, Asian anglers capture common carp as well as various native minnows and suckers for consumption. However, compared to trout fisheries, these fisheries are relatively small, and little information exists regarding them.

Reservoir Fisheries

The creation of hundreds of reservoirs by damming streams throughout the Sierra Nevada has created many "new" habitats for fish and additional fishing opportunities. At high elevations, such reservoirs support mainly trout fisheries, and, because of their accessibility, they are heavily planted with hatchery trout. At lower elevations, warm-water fishes predominate, and these are largely sustained through natural reproduction. As in warm-water streams, the principal fish in angler catches in reservoirs are various bass, sunfish, and catfish species. Because reservoir volumes fluctuate considerably in response to water demands and the amount of water flowing into each reservoir, the fish populations show considerable fluctuation in size. While efforts were made by the CDFG in the 1960s and 1970s to evaluate reservoir fisheries (Calhoun 1966; Horton and Lee 1982), evaluation of fisheries in recent years has been confined to a few reservoirs of special interest (e.g., D. Lindstrom, Pacific Gas and Electric Company, conversation with the author, 1995). However, it is safe to assume that such fisheries are of major importance in the recreational economy of the area.

Conclusions

Recreational fisheries are clearly important in Sierra Nevada lakes and streams, but the intensity of fishing effort tends to diminish with distance from roads. Stream and reservoir fisheries for both hatchery trout and wild trout are important in the Sierra Nevada economy, representing a large chunk of the more than $3 billion contributed annually to the California economy by trout anglers (California Trout, unpublished studies). Although about half of the natural lakes in the Sierra Nevada are regularly stocked with fingerling trout, fishing intensity is not well known for most lakes. Nevertheless, lake fishing is an important part of the backcountry experience for many people. Clearly, an evaluation is needed that balances the economic and recreational benefits of the stocking of backcountry lakes with the biological costs to the local ecosystems.

CONSERVATION IMPLICATIONS

A number of patterns are apparent from the analysis of the status of fish and fisheries in the Sierra Nevada:

- Native fish communities have been disrupted, and the populations of a number of native fish species have been seriously depleted or are in decline.

- Anadromous chinook salmon, steelhead, and lampreys, which were once abundant and widely distributed in the western Sierra Nevada, are no longer important components of most riverine ecosystems in the Sierra Nevada.

- Dams and diversions have greatly altered fish habitats and blocked fish movements throughout the range, with the greatest effects probably occurring before 1950; the habitats they create favor mostly non-native fishes.

- A large percentage of the stream reaches in the Sierra Nevada have been altered to a greater or lesser degree by roads, railroads, grazing, mining, and other factors (Kattelmann 1996); this habitat change has depressed fish populations and continues to do so, but much of it is reversible.

- Trout have been introduced into most high-elevation lakes and streams capable of supporting them and have changed the nature of aquatic ecosystems in the high Sierra.

- Fisheries in the Sierra Nevada are predominantly for introduced fishes, including trout originating in hatcheries.

Obviously, the dramatic changes that have taken place in the fish fauna of the Sierra Nevada reflect dramatic changes in the aquatic ecosystems of which they are part, although our understanding of these changes is limited. It is equally obvious that while many of these changes are likely to be permanent, others are probably reversible, at least in limited areas. If conservation of the remaining native aquatic biota is to be accomplished, protection of the best remaining aquatic habitats will be necessary, as will restoration of the native biota in at least some areas in which it is now reduced or absent. Such protection must be systematic (Moyle and Yoshiyama 1994) and must recognize that there is no time to be lost.

REFERENCES

Bahls, P. 1992. The status of fish populations and management of high mountain lakes in the western United States. *Northwest Science* 66:183–93.

Baltz, D. M., and P. B. Moyle. 1984. Segregation by species and size classes of rainbow trout, *Salmo gairdneri*, and Sacramento sucker, *Catostomus occidentalis*, in three California streams. *Environmental Biology of Fishes* 10:101–10.

———. 1993. Invasion resistance to introduced species by a native assemblage of California stream fishes. *Ecological Applications* 3:246–55.

Bradford, D. F., S. D. Cooper, and A. D. Brown. 1994. *Distribution of aquatic animals relative to naturally acidic waters in the Sierra Nevada.* Final Report, Contract A132-173. Sacramento: California Air Resources Board.

Bradford, D. F., M. F. Tabatabai, and D. M. Graber. 1993. Isolation of remaining populations of the native frog, *Rana muscosa*, by introduced fishes in Sequoia and Kings Canyon National Parks, California. *Conservation Biology* 7:882–88.

Brown, L. R., and P. B. Moyle. 1981. The impact of squawfish on salmonid populations: A review. *North American Journal of Fisheries Management* 1:104–11.

———. 1993. Distribution, ecology, and status of the fishes of the San Joaquin River drainage, California. *California Fish and Game* 79:96–113.

Brown, L. R., P. B. Moyle, W. A. Bennett, and B. D. Quelvog. 1992. Implications of morphological variation among populations of California roach (Cyprinidae: *Lavinia symmetricus*) for conservation policy. *Biological Conservation* 62:1–10.

Busack, C. A., and G. A. E. Gall. 1981. Introgressive hybridization in populations of Paiute cutthroat trout (*Salmo clarki seleneris*). *Canadian Journal of Fisheries and Aquatic Science* 38:939–51.

Butler, R. L., and D. P. Borgeson. 1965. California "catchable" trout fisheries. *California Department of Fish and Game Fish Bulletin* 127: 1–47.

Cahill T. A., J. J. Carroll, D. Campbell, and T. E. Gill. 1996. Air quality. In *Sierra Nevada Ecosystem Project: Final report to Congress*, vol. II, chap. 48. Davis: University of California, Centers for Water and Wildland Resources.

Calhoun, A. 1966. *Inland fisheries management.* Sacramento: California Department of Fish and Game.

California Department of Fish and Game (CDFG). 1990. *Central Valley salmon and steelhead restoration and enhancement plan.* Sacramento: CDFG.

———. 1992. *Annual report on the status of California state listed threatened and endangered animals and plants.* Sacramento: CDFG.

———. 1993. *Restoring Central Valley streams: A plan for action.* Sacramento: CDFG.

Campbell, E. A., and P. B. Moyle. 1991. Historical and recent population sizes of spring-run chinook salmon in California. In *Proceedings, Northeast Pacific chinook and coho workshop*, edited by T. J. Hassler, 155–216. Arcata, CA: American Fisheries Society.

Chaney, E., W. Elmore, and W. S. Platts. 1990. *Livestock grazing on western riparian areas.* U.S. Environmental Protection Agency Report. Eagle, ID: Northwest Resource Information Service.

Christenson, D. P. 1977. History of trout introductions in California high mountain lakes. In *A symposium on the management of high mountain lakes in California's national parks*, edited by A. Hall and R. May, 9–16. San Francisco: California Trout, Inc.

———. 1978. *A fishery management plan for the Little Kern golden trout.* Special Publication 78-1. Sacramento: California Department of Fish and Game, Inland Fisheries Endangered Species Program.

Clark, G. H. 1929. Sacramento–San Joaquin salmon (*Oncorhynchus tschawytscha*) fishery of California. *California Department of Fish and Game Fish Bulletin* 17:38–39.

Coffin, P. D., and W. F. Cowan. 1995. *Lahontan cutthroat trout (Oncorhynchus clarki henshawi) recovery plan.* Portland, OR: U.S. Fish and Wildlife Service.

Cordone, A. J., and T. C. Franz. 1966. The Lake Tahoe sport fishery. *California Fish and Game* 52:240–74.

Cutter, R. 1984. *Sierra trout guide.* Portland, OR: Frank Amato Publications.

Decker, L. M. 1989. Coexistence of two species of sucker, *Catostomus*, in Sagehen Creek, California, and notes on their status in the western Lahontan basin. *Great Basin Naturalist* 49:540–51.

Deinstadt, J. M., D. C. Lentz, and S. Parmenter. 1995. *Assessment of the wild trout fishery in the Upper Middle Fork San Joaquin River, 1986–1993.* Inland Fisheries Administrative Report 95-1. Sacramento: California Department of Fish and Game.

Deinstadt, J. M., D. C. Lentz, G. F. Sibbald, and K. D. Murphy. 1993. *Fishing success on California wild trout waters in 1990-1991: Reports from angler box surveys.* Inland Fisheries Administrative Report 93-1. Sacramento: California Department of Fish and Game.

Drost, C. A., and G. M. Fellers. 1994. *Decline of frog species in the Yosemite section of the Sierra Nevada.* U.S. Department of the Interior Technical Report NPS/WRUC/NRTR-94-02. Washington, DC: Government Printing Office.

Dudley, T., and M. Embury. 1995. *Non-indigenous species in wilderness areas: The status and impact of livestock and game species in designated wilderness areas of California.* Oakland, CA: Pacific Institute for Studies in Development, Environment, and Security.

Erman, N. A. 1996. Status of aquatic invertebrates. In *Sierra Nevada Ecosystem Project: Final report to Congress*, vol. II, chap. 35. Davis: University of California, Centers for Water and Wildland Resources.

Fisher, F. 1994. Past and present status of Central Valley chinook salmon. *Conservation Biology* 8:870–73.

Fry, D. H. 1961. King salmon spawning stocks of the California Central Valley, 1940–1959. *California Fish and Game* 47:55–71.

Gard, M. 1994. Factors affecting fish populations in South Yuba River, Nevada County, California. Ph.D. diss., University of California, Davis.

Gerstung, E. R. 1973. Fish population and yield estimates from California trout streams. *Cal-Neva Wildlife* 1973:9–19.

———. 1989. Fishes and fishing in the forks of the American River: Then and now. In *The American River, north, middle, and south forks,* edited by S. Mandel et al., 302–5. Auburn, CA: Protect American River Canyons.

Groh, J. H. 1990. Moccasin Creek fish hatchery. *Outdoor California* 51 (2): 9–10.

Harvey, C. 1995. *Adult steelhead counts in Mill and Deer Creeks, Tehama County, October 1993–June 1994.* Inland Fisheries Administrative Report 95-3. Sacramento: California Department of Fish and Game.

Hashagen, K. 1988. California's hatchery system. *Outdoor California* 49 (2): 5–8.

Hatton, S. R. 1939. Progress report on the Central Valley fisheries investigations, 1939. *California Fish and Game* 26:334–73.

Hatton, S. R., and G. H. Clark. 1942. A second progess report on the Central Valley fisheries investigations. *California Fish and Game* 28:116–23.

Horton, J. L., and D. P. Lee. 1982. *Harvest and mortality of tournament caught and released largemouth bass in Don Pedro Reservoir, California.* Inland Fisheries Administrative Report 82-3. Sacramento: California Department of Fish and Game.

Jenkins, T. M. J., R. A. Knapp, K. W. Kratz, S. D. Cooper, J. M. Melack, A. D. Brown, and J. Stoddard. 1994. Aquatic biota in the Sierra Nevada: Current status and potential effects of acid deposition on populations. Final Report, Contract A932-138. Sacramento: California Air Resources Board.

Jennings, M. R. 1996. Status of amphibians. In *Sierra Nevada Ecosystem Project: Final report to Congress,* vol. II, chap. 31. Davis: University of California, Centers for Water and Wildland Resources.

Jennings, M. R., and M. Hayes. 1994. *Amphibian and reptile species of special concern in California.* Sacramento: California Department of Fish and Game.

Kattelmann, R. 1996. Hydrology and water resources. In *Sierra Nevada Ecosystem Project: Final report to Congress,* vol. II, chap. 30. Davis: University of California, Centers for Water and Wildland Resources.

Knapp, R. A. 1996. Non-native trout in natural lakes of the Sierra Nevada: An analysis of their distribution and impacts on native aquatic biota. In *Sierra Nevada Ecosystem Project: Final report to Congress,* vol. III. Davis: University of California, Centers for Water and Wildland Resources.

Kondolf, G. M. 1994. Livestock grazing and habitat for a threatened species: Land-use decisions under scientific uncertainty in the White Mountains, California. *Environmental Management* 18: 501–9.

Lee, D. S., C. R. Gilbert, C. H. Hocutt, R. E. Jenkins, D. E. McAllister, and J. R. Stauffer. 1980. *Atlas of North American freshwater fishes.* Raleigh: North Carolina State Museum of Natural History.

Leopold, A. S. 1963. *Wildlife management in the national parks.* Report to the Secretary of the Interior. Washington, DC: Department of the Interior.

Li, H. W., and P. B. Moyle. 1993. Management of introduced fishes. In *Inland fisheries management in North America,* edited by C. Koehler, 287–307. Bethesda, MD: American Fisheries Society.

Lindstrom, S. G. 1993. Great Basin fisherfolk: Optimal diet breadth modelling the Truckee River aboriginal subsistence fishery. Ph.D. diss., University of California, Davis.

Meehan, W. R., ed. 1991. *Influences of forest and rangeland management on salmonid fishes and their habitats.* Bethesda, MD: American Fisheries Society.

Menke, J., C. Davis, and P. Beesley 1996. Rangeland assessment. In *Sierra Nevada Ecosystem Project: Final report to Congress,* vol. III. Davis: University of California, Centers for Water and Wildland Resources.

Minckley, W. L., and J. E. Deacon, eds. 1991. *Battle against extinction: Native fish management in the American West.* Tucson: University of Arizona Press.

Morgan, M. D., S. T. Threlkeld, and C. R. Goldman. 1978. Impact of the introduction of kokanee (*Oncorhynchus nerka*) and opossum shrimp (*Mysis relicta*) on a subalpine lake. *Journal of the Fisheries Resources Board, Canada* 35:1572–79.

Moyle, P. B. 1976a. *Inland fishes of California.* Berkeley and Los Angeles: University of California Press.

———. 1976b. Fish introductions in California: History and impact on native fishes. *Biological Conservation* 9:101–18.

———. 1986. Fish introductions into North America: Patterns and ecological impact. In *Ecology of biological invasions of North America and Hawaii,* edited by H. A. Mooney and J. A. Drake, 27–43. New York: Springer-Verlag.

———. 1995. Conservation of native freshwater fishes in the mediterranean-type climate of California, U.S.A.: A review. *Biological Conservation* 72:271–79.

Moyle, P. B., and R. Nichols. 1974. Decline of the native fish fauna of the Sierra Nevada foothills, central California. *The American Midland Naturalist* 92 (1): 72–83.

Moyle, P. B., and P. J. Randall. 1996. Biotic integrity of watersheds. In *Sierra Nevada Ecosystem Project: Final report to Congress,* vol. II, chap. 34. Davis: University of California, Centers for Water and Wildland Resources.

Moyle, P. B., and J. E. Williams. 1990. Biodiversity loss in the temperate zone: Decline of the native fish fauna of California. *Conservation Biology* 4:275–84.

Moyle, P. B., and R. M. Yoshiyama. 1994. Protection of aquatic biodiversity in California: A five-tiered approach. *Fisheries* 19: 6–18.

Moyle, P. B., R. M. Yoshiyama, J. E. Williams, and E. Wikramanayake. 1996. *Fish species of special concern in California.* Sacramento: California Department of Fish and Game.

Overton, C. K., G. L. Chandler, and J. A. Pisano. 1994. *Northern/ intermountain regions' fish habitat inventory: Grazed, rested, and ungrazed reference stream reaches, Silver King Creek, California.* Technical Report INT-GTR-311. Boise: U.S. Forest Service Intermountain Research Station.

Pister, E. P. 1977. The management of high Sierra lakes. In *A symposium on the management of high mountain lakes in California's national parks,* edited by A. Hall and R. May, 27–33. San Francisco: California Trout, Inc.

Platts, W. S. 1991. Livestock grazing. In *Influences of forest and rangeland management on salmonid fishes and their habitats,* edited by W. R. Meehan, 389–424. Bethesda, MD: American Fisheries Society.

Reimers, N. 1958. Conditions of existence, growth, and longevity of brook trout in a small, high altitude lake of the eastern Sierra Nevada. *California Fish and Game* 44:319–33.

Sigler, W. F., and J. W. Sigler. 1987. *Fishes of the Great Basin: A natural history.* Reno: University of Nevada Press.

Skinner, J. E. 1962. *An historical review of the fish and wildlife resources of the San Francisco Bay area.* Sacramento: California Department of Fish and Game.

Steward, C. R., and T. C. Bjornn. 1990. *Supplementation of salmon and steelhead stocks with hatchery fish: A synthesis of published literature.* Project Report 88-100, part 2. Portland, OR: Bonneville Power Administration.

Stoddard, J. L. 1987. Microcrustacean communities of high-elevation lakes in the Sierra Nevada, California. *Journal of Plankton Research* 9:631–50.

Vogel, D. A., and K. R. Marine. 1991. *Guide to upper Sacramento River chinook salmon life history.* Report to U.S. Bureau of Reclamation, Central Valley Project. Redding, CA: CH2M Hill, Inc.

Wallis, E. O. 1952. Comprehensive review of trout fishery problems of Yosemite National Park: A report of Yosemite trout investigations, 1951–1953. Unpublished report. National Park Service, Yosemite National Park.

Wissmar, R. C., J. E. Smith, B. A. McIntosh, H. W. Li, G. H. Reeves, and J. R. Sedell. 1994. A history of resource use and disturbance in riverine basins of eastern Oregon and Washington (early 1800s–1990s). *Northwest Science* 68:1–35.

Yoshiyama, R. M., E. R. Gerstung, F. W. Fisher, and P. B. Moyle. 1996. Historical and present distribution of chinook salmon in the Central Valley drainage of California. In *Sierra Nevada Ecosystem Project: Final report to Congress,* vol. III. Davis: University of California, Centers for Water and Wildland Resources.

PETER B. MOYLE
Department of Wildlife, Fish, and
    Conservation Biology
University of California
Davis, California

PAUL J. RANDALL
Department of Wildlife, Fish, and
    Conservation Biology
University of California
Davis, California

34

Biotic Integrity of Watersheds

ABSTRACT

The biological health of one hundred Sierra Nevada watersheds was evaluated using an Index of Biotic Integrity (IBI). The IBI scores indicated that the biological communities of seven of the watersheds were in excellent condition, thirty-six were in good condition, forty-eight were in fair condition, and nine were in poor condition. The biggest factors contributing to low IBI scores were large dams and introduced fishes, although factors affecting local stream habitats, especially roads and activities associated with roads, were also important. All watersheds in the Sierra Nevada have experienced at least some loss of biotic integrity through the loss or decline of native organisms, but many have considerable potential for recovery.

INTRODUCTION

The Sierra Nevada can be divided into hundreds of small watersheds, which in turn are subdivisions of larger watersheds. All streams on the west side of the range are ultimately part of the Sacramento–San Joaquin watershed, while on the east side, all streams ultimately flow into the Great Basin, in three discrete drainages (Lahontan, Mono, and Owens). In many respects, watersheds are good units on which to base conservation efforts, especially for aquatic organisms, because they are relatively easy to define and because they can contain a wide variety of habitats and species, depending on the watershed's size. For aquatic organisms, watersheds are often the landscape unit in which evolution of distinct taxa takes place, because of the difficulty many aquatic organisms have in moving from one watershed to another (Moyle 1976a; Moyle et al. 1996). This chapter identifies watersheds in the Sierra Nevada that are still dominated by native aquatic species and communities and that contain a wide variety of habitats, rare habitats, or both. The watersheds with high scores

for biotic integrity may be logical places to focus large-scale conservation efforts.

INDEX OF BIOTIC INTEGRITY

The biological health of Sierra Nevada watersheds can be measured using a broad-scale Index of Biotic Integrity (IBI). Indices of biotic integrity are measures of the health of streams and have been developed as an alternative to physical and chemical measures of water quality (Karr 1981; Karr et al. 1986; Regier 1993). The early work on IBIs was largely funded by the U.S. Environmental Protection Agency (EPA) with the purpose of developing a rapid-assessment tool to help the EPA carry out the mandates of the Clean Water Act. The basic idea is to combine a number of measures of the structure and function of fish communities into an index, on the assumption that the responses of an integrated community of fishes to changes in the environment would reflect both major environmental insults (e.g., a pesticide spill) and more subtle long-term effects, such as chronic non-point-source pollution and changes in land use.

Biotic integrity is defined as "the ability to support and maintain a balanced, integrated, adaptive community of organisms having a species composition, diversity, and functional organization comparable to that of the natural habitat of the region" (Karr and Dudley 1981). An IBI is a method of measuring this complex idea, and IBIs can be developed independently for different regions or streams. IBIs are now widely used in the eastern United States, where fish communities are complex and largely made up of native species (Miller et al. 1988). For eastern streams it is possible to develop an IBI that uses ten to twelve different measures (metrics) in the creation of the final index (Karr et al. 1986). In California, the small number of native fishes in most

Sierra Nevada Ecosystem Project: Final report to Congress, vol. II, *Assessments and scientific basis for management options.* Davis: University of California, Centers for Water and Wildland Resources, 1996.

streams makes development of complex IBIs with numerous metrics (independent measures of the nature of the fish assemblage) difficult (Miller et al. 1986). In fact, two measures, number of native fish species and abundance of native fishes, provided much of the information needed to determine biotic integrity as defined previously (Moyle et al. 1986). In Sierra Nevada streams, if the fish communities are intact, the stream is likely to have a fairly natural hydrograph and the watershed is likely to be in reasonably good condition (Baltz and Moyle 1993). Native fishes, however, are only part of the biotic integrity picture, especially in relation to water quality, so we developed an IBI for Sierra Nevada watersheds that takes into account not only native fish assemblages but also the abundance of native frogs, the presence of anadromous fish, and the effects of the widespread introductions of trout into high-elevation streams. Ideally, this IBI should also include metrics based on invertebrates, but our knowledge of their distribution and abundance is too poor at this time to use them. It is worth noting that the IBI that we present here is designed to cover bigger watersheds than those for which most IBIs are designed. IBIs tend to be designed to evaluate specific types of streams or stream habitats. We are currently developing such specific IBIs for Sierra Nevada streams.

METHODS

The first problem to be resolved for this analysis was which watershed scale to use. The Calwater numbering system for watersheds, for example, breaks each major drainage basin (e.g., the Central Valley) into major tributary systems, labeled Hydrologic Units (HUs). Each HU is divided into Hydrologic Areas (HAs), which are divided into Hydrologic Subareas (HSAs), which in turn are divided successively into Super-Planning Watersheds and Planning Watersheds. There are thousands of watersheds in the latter two categories, so using them as the unit of analysis would both be difficult and have a high degree of redundancy. We chose as the basic unit of analysis, therefore, the HSA, using HAs or even HUs if the watersheds were too small to subdivide further. This choice resulted in one hundred watersheds being used in the evaluation, covering the entire mountain range (figure 34.1). The watersheds range in area from 4,816 ha (11,895 acres) (a partial drainage on the California-Nevada border) to 382,669 ha (945,192 acres) (the Upper Owens drainage). However, most (62%) of the analysis watersheds are between 15,000 and 90,000 ha (37,050 and 222,300 acres) in area; 28% are larger than 90,000 ha and 10% are smaller than 15,000 ha. Typical watersheds within these categories were the forks of large rivers (e.g., the South Yuba River) or independent drainages of modest size (e.g., Deer Creeks in Tehama, Placer, and Tulare Counties). An additional thirty-four watersheds were not evaluated because of inadequate information on their aquatic

biota. These watersheds are all at low elevations, most are small in size, and most seem to lack permanent water (figure 34.2). Nine of these watersheds mark the southern end of the SNEP area, twenty-two are in the foothills along the western edge of the boundary, and three are along the California-Nevada border.

The IBI developed for this analysis includes six metrics (table 34.1), each rated on a scale of 1 to 5, where 1 is low (poor) and 5 is high (good). The six metrics were added and standardized to a 100-point scale, because not all metrics could be used in all drainages. The following is an explanation of each metric.

Native ranid frogs: The foothill yellow-legged frog, mountain yellow-legged frog, and Cascade frog appear to be the amphibians most sensitive to environmental change. Their disappearance from much of their native habitat in the Sierra Nevada is a cause for concern, and their presence in a watershed is an indication that high-quality aquatic and riparian habitats still exist. We scored watersheds for this metric using information presented in Jennings 1996, Jennings and Hayes 1994, personal communications with M. R. Jennings, and observations by Moyle and his graduate students.

Native fishes: The native fishes of the Sierra Nevada are highly adapted to the natural flow regimes, and they tend to become depleted if the regimes are changed, especially by dams. Scores for this metric are based on field notes, University of California, Davis, stream surveys (Moyle et al. 1996), and studies such as Moyle and Nichols 1974 and Brown and Moyle 1993. Another important source of information was the data sheets of the Wild Trout Program of the California Department of Fish and Game. In many instances, agency biologists familiar with the watershed were consulted as well.

Native fish assemblages: One of the best indications of high-quality aquatic environments is the presence not only of native species but also of groups of species co-occurring in their natural assemblages of three to six species. Some of the native fishes can persist indefinitely in altered habitats and in the presence of exotic fishes, while others cannot. We scored this metric largely from information from the same sources as were used for the previous metric.

Anadromous fishes: Salmon, steelhead, and lamprey were important parts of the aquatic ecosystems at low to middle elevations in west-side Sierra Nevada streams, from the Kings River (Fresno County) north. Their exclusion by dams from much of their former habitat has significantly altered the stream communities of which they were once part. We scored this metric based on estimates of past and present distribution and abundance as presented in Yoshiyama et al. 1996.

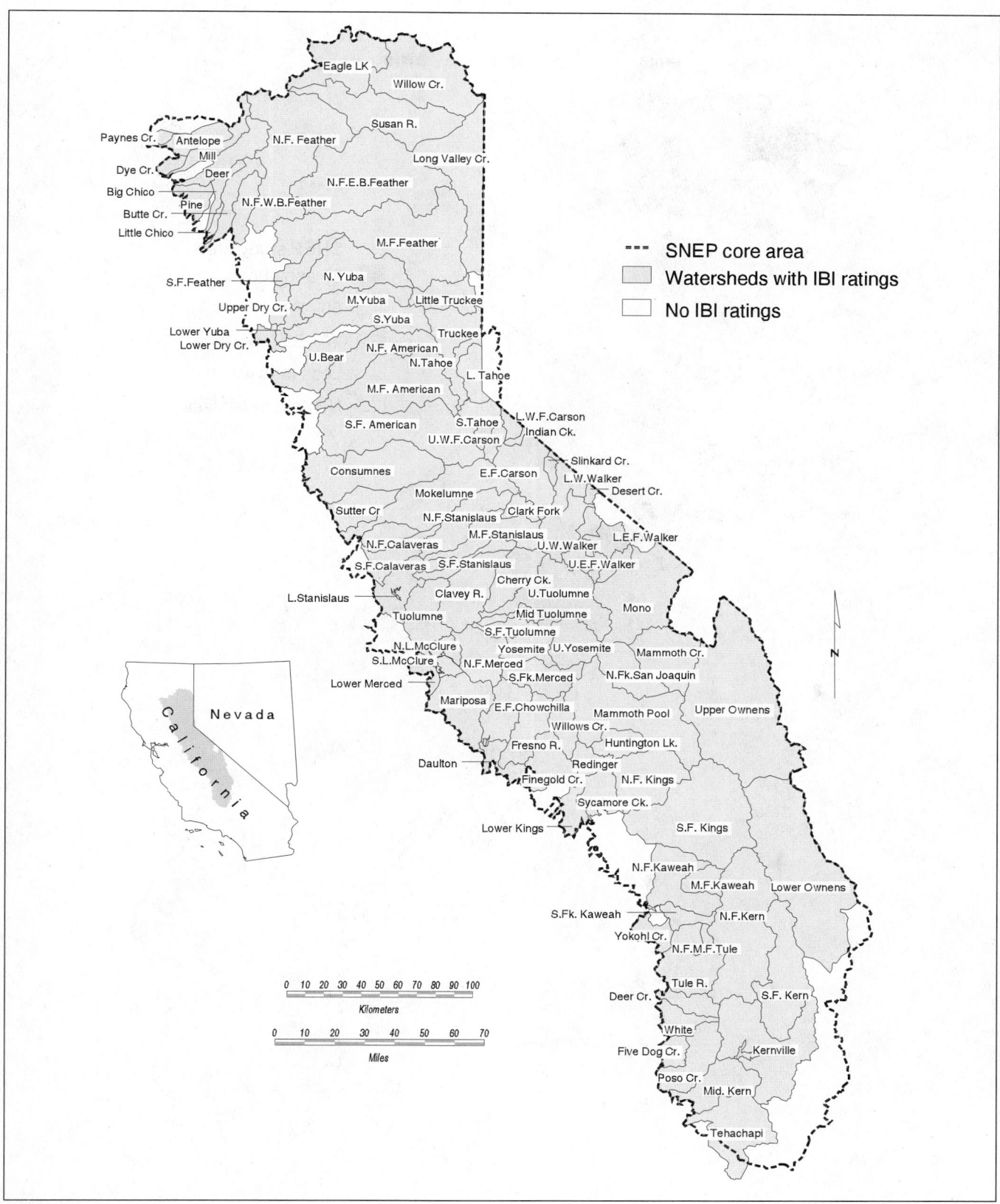

FIGURE 34.1

Watersheds selected for IBI analysis in the SNEP core area.

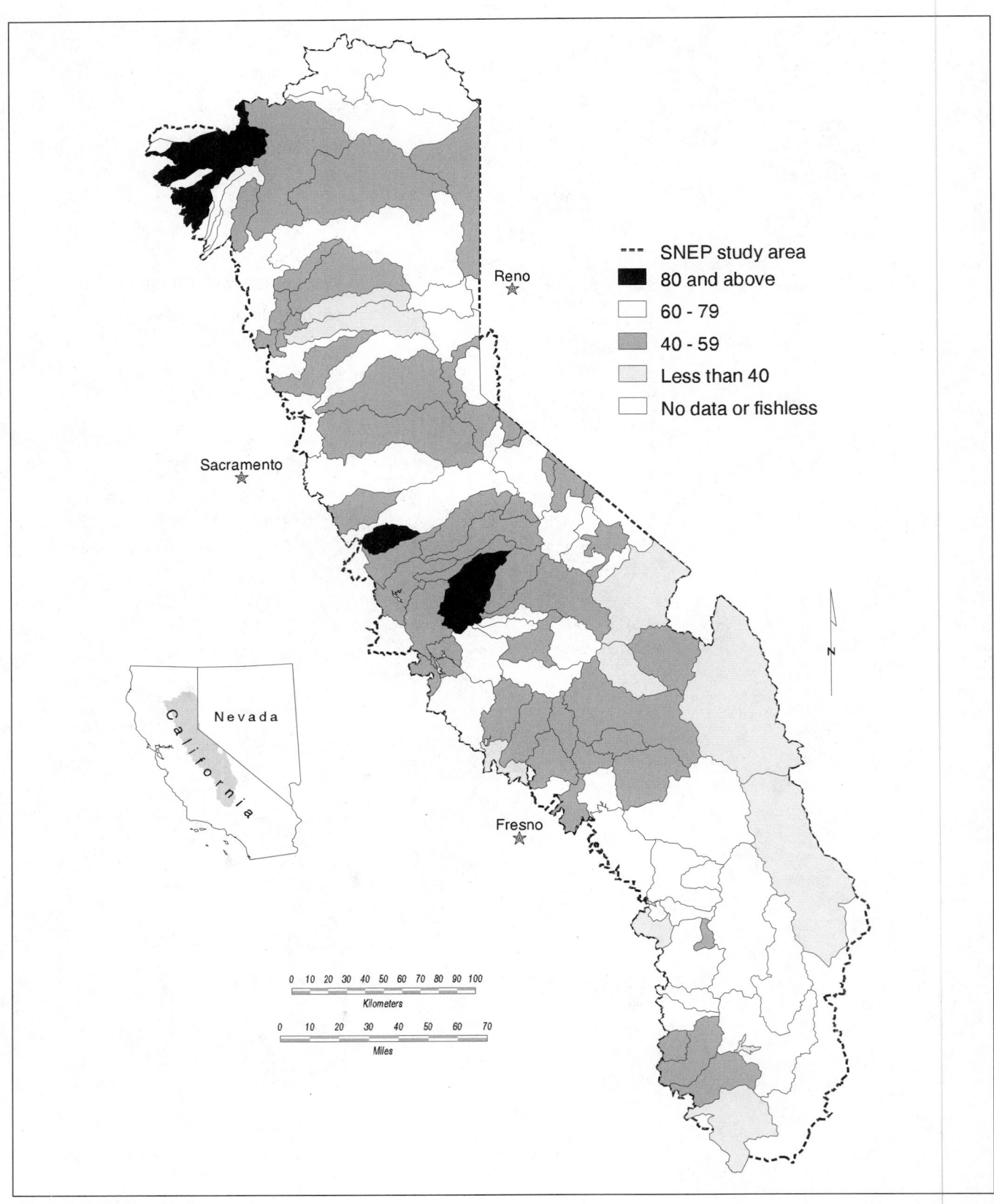

FIGURE 34.2

IBI ratings for Sierra Nevada watersheds.

TABLE 34.1

Metrics and scoring system for an Index of Biotic Integrity for Sierra Nevada watersheds.

Aquatic Community	Metrics
I. Native ranid frogs	1. Absent or rare 3. Present 5. Abundant and widely distributed
II. Native fishes	1. Absent or rare *or* introduced where not native 3. Present in much of native range 5. Abundant in most of native range
III. Native fish assemblages (excluding trout-only assemblage)	1. Largely disrupted 3. Present but scattered or containing exotic species 5. Largely intact
IV. Anadromous fishes (if historically present)	1. Absent or rare 3. Present mainly below dams or uncommon 5. Found in original range
V. Trout	1. Range greatly expanded, mixture of non-native and native species *or* range greatly reduced 3. Range expanded but includes native species *or* range about the same but native populations reduced, exotics present 5. Mostly native species in original range
VI. Stream fish abundance	1. Substantially lower than presumed historic levels *or* widespread and abundant in originally fishless areas 3. Somewhat lower overall than historic levels *or* present in fishless areas 5. About the same as or higher than historic levels

IBI score = [Total points possible/number of metrics] x 20
80–100   Aquatic communities in very good to excellent condition
60–79   Aquatic communities in good condition
40–59   Aquatic communities in fair condition
<40   Aquatic communities in poor condition

Trout: Rainbow and cutthroat trout were native to the Sierra Nevada, generally at elevations below 1,600 m (5,250 ft). However, a large region at high elevations was fishless until trout were introduced there by Euro-Americans. In addition, many of the trout introduced were not native to California. Because trout are now the dominant predators in the streams and lakes in which they were introduced, it is assumed that their introduction has had a significant negative effect on aquatic biodiversity. We scored this metric based on information sources similar to those used for the native fish metrics and on Knapp 1996.

Stream fish abundance: Often water projects and watershed alterations not only change the species composition of streams but also reduce the total biomass and abundance of fish, including non-native species. This metric is based on the same sources of information as the native fish metric.

Other analyses: To look for factors associated with high or low IBI scores, we determined the following variables for each watershed, based on a geographic information system (GIS) analysis of square landscape units (pixels) 1 ha (10,000 m^2 [2.47 acres]) in area:

- Dams: percentage of total hectares in each watershed that contain a dam of any size.

- Reservoirs: total capacity of reservoirs in the watershed, in acre-feet.

- Diversions: percentage of total hectares in each watershed containing a water diversion of any size. This figure is based on water rights filings and thus includes many small diversions and diversions that may not be active.

- Roads: percentage of hectares containing at least one road.

- Roads and streams: percentage of hectares containing both a road and a stream.

- Roadless area: percentage of watershed in areas that contain no roads and that are also at least 1,000 ha (2,470 acres) in area and are 0.2 km (0.125 mi) from a road.

- Fishless area: percentage of watershed that was presumably without fish historically, based on the map drawn for this chapter.

- Mean elevation: average elevation of hectares within the watershed.

The complete data set developed is presented in appendix 34.1. Once the data had been gathered, they were analyzed using principal components analysis. The purpose of the analysis was to determine the degree to which each of the eight variables, or a combination of them, seemed to influence IBI scores.

RESULTS

The IBI scores indicated that seven of the one hundred watersheds had aquatic communities in excellent condition (IBI values of 80–100) (figure 34.2; appendix 34.1). Another thirty-six had aquatic communities in good condition (IBI values of 60–79), while forty-eight had aquatic communities in fair condition (IBI values of 40–59) and nine had aquatic communities in poor condition (IBI values less than 40). Of the seven watersheds with the highest scores, three stand out with scores greater than 90: Deer Creek and Mill Creek (Tehama County) and the Clavey River (appendix 34.1). These watersheds contain intact native fish and amphibian faunas, and the biotic communities are still largely governed by natural processes.

Deer and Mill Creeks are highly unusual in that they both support runs of spring-run chinook salmon. There are three clusters of watersheds with high IBI scores: (1) the Deer–Mill–Antelope Creek and associated small watersheds in Tehama County, (2) the North Fork Calaveras and Clavey Rivers in the western central Sierra Nevada, and (3) the upper Kings and Kern River watersheds in the southern part of the range. Streams in the Tehama cluster flow through rugged volcanic terrain with low accessibility until recently; the streams were also too small to make large dams viable, generally. The western central cluster consists of medium-sized tributaries to larger, highly developed rivers that have managed to maintain much of their native fish fauna. The upper Kings and Kern watersheds are high-elevation watersheds with steep terrain and low accessibility. Most of their area is in either national parks or wilderness areas. Despite their high IBI scores, all of these watersheds have been altered by human activity, but less so than other watersheds in the Sierra Nevada, as indicated by their moderate scores for variables related to diversions and roads (table 34.2). None, however, contain large dams, so the natural hydrologic regimes are still intact.

Watersheds that received low scores are (1) low- to middle-elevation drainages that have been dammed and diverted and so tend to be dominated by introduced fishes and frogs and/or to have greatly diminished native fish and amphibian populations; (2) high-elevation watersheds that have lost most of their frogs and that are dominated by non-native trout; or (3) small, low-elevation watersheds that have been highly altered by human activity (urbanization, agriculture, mining, etc.), as indicated by high scores for variables related to dams, diversions, and roads (appendix 34.1).

Correlation analysis indicated that the IBI score was negatively correlated ($p < 0.05$) with the percentage of hectares containing dams (–0.22), reservoir capacity (–0.27), the percentage of hectares containing roads associated with streams (–0.22), and the percentage of the watershed that was historically fishless. This is not surprising, given that a low IBI score at high elevations would be strongly influenced by the presence of trout in naturally fishless areas, while a low score at low elevations would be related to the presence of major dams or road systems. This dichotomy is reflected in the results of the principal components analysis, which produced two factors with eigenvalues greater than 1.00 (table 34.2). Factor 1, explaining 42% of the variance, had only a moderate negative loading on the IBI score but was strongly positively loaded on the two road variables and strongly negatively loaded on mean elevation and the percentage of the watershed that was historically fishless. In factor 2, explaining 17% of the variance, the IBI score had a high negative loading while the percentage of hectares containing dams, reservoir capacity, and the percentage of the watershed that was historically fishless had high positive loadings.

DISCUSSION

The analysis of the IBI rating indicates that major dams at low to middle elevations and the introduction of fish at high elevations have had the greatest negative effects on lowering biotic integrity. These two factors are so dominant that they tend to obscure the effects of watershed degradation, as reflected in the variables related to the abundance of roads. For example, the historically fishless areas are also mostly wilderness areas and national parks today, and so have low numbers of roads, yet the presence of introduced fish greatly reduces the biotic integrity of the waters within these areas. In general, the watersheds with the highest IBI scores are at intermediate elevations, are without major dams, and have low to intermediate scores for variables related to human disturbance (roads, diversions).

The importance of dams and introduced species in reducing biotic integrity does not mean that other factors are not important, especially for smaller watersheds or for individual situations. Streams that are subject to high levels of sedimentation from numerous or poorly constructed roads, from mining, or from logging on steep hillsides will have reduced diversity of aquatic organisms, as will streams that have had their channels heavily modified for flood control or other purposes (e.g., Moyle 1976b). Streams heavily polluted by acidic water leaching out of an abandoned mine can have a very low diversity of organisms. Most of these factors, however, are likely to be more localized in their effects and reversible, often just by a cessation of the problem-causing activity. The native fish populations in particular have a high capacity to bounce back from being decimated (Moyle et al. 1983). For example, many small tributaries to the South Yuba River were devastated by hydraulic mining in the nineteenth century yet today show a high degree of recovery of their

TABLE 34.2

Factors created by the principal components analysis of variables related to the biotic integrity of Sierra Nevada watersheds.

Variable	Factor 1	Factor 2
Index of Biotic Integrity	–0.2242	–0.6065
Percentage of hectares containing dams	0.4245	0.5541
Reservoir capacity	0.2841	0.5414
Percentage of hectares containing diversions	0.5876	–0.2625
Percentage of hectares containing one or more roads	0.8606	0.1116
Percentage of hectares containing a road and a stream	0.8598	0.1644
Percentage of watershed that is roadless	–0.8997	0.0293
Percentage of watershed that is historically fishless	–0.5394	0.5541
Mean elevation	–0.7340	0.3865
Eigenvalue	3.7754	1.5434
Percentage of variance	42%	17%

native fish and amphibian faunas (Gard 1994; P. Randall, unpublished data). Species of fish that are missing from the local fauna appear to have been unable to reinvade because a combination of dams and introduced predators has made the movement of native fish in the main river difficult or impossible (Gard 1994). Reintroduction of the missing native fishes into some streams is now being considered (W. Frizzel, State Parks and Recreation, conversation with the authors, 1995).

Although the results of this analysis fit with other, even more subjective indicators of watershed health, they should nevertheless be treated with caution for a number of reasons.

- The information available to create an IBI score was limited for some watersheds, and the scoring was done by just one person, although many people, field notes, and references were consulted during the scoring process.

- The IBI scores essentially compare the present fish and amphibian assemblages to the presumed pre-Euro-American assemblages; the systems most resembling the original systems obtained the highest IBI values. All aquatic ecosystems in the Sierra Nevada have been altered to one degree or another, so even the highest-rated watersheds are far from pristine. Thus, a different value system, one that was more accepting of the changes, would result in different scores. For example, if it was assumed that the streams at high elevations should be rated positively on the basis of their ability to support large, fishable populations of wild trout, a number of high-elevation watersheds would receive higher IBI scores than they did under the scoring system used here. From the point of view of biotic integrity as defined in the introduction to this chapter, originally fishless streams and lakes that are now dominated by introduced trout must be considered as highly altered ecosystems. The presence of fish eliminates most of the large invertebrates and amphibians that once dominated these waters.

- A major factor lowering many of the scores was the scarcity or absence of native frogs from the watershed. The causes of frog declines (e.g., introduced diseases) are controversial and may have had little effect on the rest of the native biota. Nevertheless, frogs *were* once important parts of all aquatic ecosystems of the Sierra Nevada, and their absence lowers biotic integrity.

- The IBI does not consider aquatic invertebrates that may have disappeared from some areas where native fish and amphibians still exist. Invertebrates are likely to be particularly sensitive to land-use practices (road building, enclosure of springs, logging, grazing, etc.) that can cause extinctions of highly specialized endemic species that live in limited habitats (Erman 1996).

- Many of the one hundred watersheds analyzed here are very large in area and may have smaller watersheds within them that would score significantly higher or lower if

treated individually. For example, the North Fork of the Kings River received a mediocre IBI score (52) because it has been dammed for hydroelectric production, has been highly roaded from logging and recreational use, and has had its high-elevation waters filled with non-native trout. Within this drainage, however, is Rancheria Creek, a relatively inaccessible watershed that is one of the most undisturbed in the Sierra Nevada (E. Beckwitt, conversation with the author, 1995). At the opposite extreme is the Clavey River watershed (IBI score = 92), which has a number of small diversions in the upper watershed, has been heavily grazed and logged in places, and is only 26% in roadless areas. The stream nonetheless retains an abundant native fish fauna, with no exotic species, especially in the rugged lower canyon (S. Matern and M. Marchetti, unpublished field notes, 1993). Access to much of the Clavey River itself is limited because its north-south orientation means that few roads cross it, but none run parallel to it for a long distance (unlike most other major Sierra Nevada streams).

CONCLUSIONS

All aquatic ecosystems in the Sierra Nevada have lost biotic integrity to a greater or lesser degree. More than half (58%) of the watersheds, however, have been rated as having their native aquatic biota in poor to fair condition. Many of the processes that have contributed to the loss of biotic integrity have slowed down (e.g., the planting of trout, dam construction), and a number of waters are receiving special protection in national parks, as wild and scenic rivers, or through other actions (e.g., coordinated resource-management programs). There are still a few watersheds that are in remarkably good condition and many others that retain a good share of their original aquatic biota. However, there is no evidence that the overall trend in loss of biotic integrity that the waters of the Sierra Nevada have experienced over the past 150 years has been reversed, although it may have slowed down somewhat. There is every reason to suspect that the loss is continuing as new environmental problems related to human population growth are substituted for the old problems related to heavy exploitation of the landscape and as exploitation (e.g., grazing) continues, even if at reduced levels compared to those of twenty-five or fifty years ago.

REFERENCES

Baltz, D. M., and P. B. Moyle. 1993. Invasion resistance to introduced species by a native assemblage of California stream fishes. *Ecological Applications* 3:246–55.

Brown, L. R., and P. B. Moyle. 1993. Distribution, ecology, and status of the fishes of the San Joaquin River drainage, California. *California Fish and Game* 79:96–113.

Erman, N. 1996. Status of aquatic invertebrates. In *Sierra Nevada Ecosystem Project: Final report to Congress,* vol. II, chap. 35. Davis: University of California, Centers for Water and Wildland Resources.

Gard, M. 1994. Factors affecting the abundance of fishes in the South Yuba River, Nevada County, California. Ph.D. diss., University of California, Davis.

Jennings, M. R. 1996. Status of amphibians. In *Sierra Nevada Ecosystem Project: Final report to Congress,* vol. II, chap. 31. Davis: University of California, Centers for Water and Wildland Resources.

Karr, J. R. 1981. Assessment of biotic integrity using fish communities. *Fisheries* 6:21-27.

———. 1993. Measuring biological integrity: Lessons from streams. In *Ecological integrity and the management of ecosystems,* edited by S. Woodley, J. Kay, and G. Francis, 83–104. Ottawa: St. Lucie Press.

Karr, J. R., and D. R. Dudley. 1981. Ecological perspective on water quality goals. *Environmental Management* 11:249–56.

Karr, J. R., K. D. Fausch, P. L. Angermeier, P. R. Yant, and I. J. Schlosser. 1986. *Assessing biological integrity in running waters: A method and its rationale.* Special Publication 5. N.p.: Illinois Natural Historical Survey.

Knapp, R. A. 1996. Non-native trout in natural lakes of the Sierra Nevada: An analysis of their distribution and impacts on native aquatic biota. In *Sierra Nevada Ecosystem Project: Final report to Congress,* vol. III. Davis: University of California, Centers for Water and Wildland Resources.

Miller, D. L., et al. 1988. Regional applications of an index of biotic integrity for use in water resource management. *Fisheries* 13:12–20.

Moyle, P. B. 1976a. *Inland fishes of California.* Berkeley and Los Angeles: University of California Press.

———. 1976b. Some effects of channelization on the fishes and invertebrates of Rush Creek, Modoc County, California. *California Fish and Game* 62:179–86.

———. 1995. Conservation of native freshwater fishes in California, U.S.A.: A review. *Biological Conservation* 72:271–79.

Moyle, P. B., L. R. Brown, and B. Herbold. 1986. Final report on development of indices of biotic integrity for California. Unpublished report. U.S. Environmental Protection Agency, Washington, DC.

Moyle, P. B., and R. Nichols. 1974. Decline of the native fish fauna of the Sierra Nevada foothills, central California. *American Midland Naturalist* 92 (1): 72–83.

Moyle, P. B., B. Vondracek, and G. D. Grossman. 1983. Response of fish populations in the North Fork of the Feather River, California, to treatment with fish toxicants. *North American Journal of Fisheries Management* 3:48–60.

Moyle P. B., R. M. Yoshiyama, and R. A. Knapp. 1996. Status of fish and fisheries. In *Sierra Nevada Ecosystem Project: Final report to Congress,* vol. II, chap. 33. Davis: University of California, Centers for Water and Wildland Resources.

Regier, H. A. 1993. The notion of natural and cultural integrity. In *Ecological integrity and the management of ecosystems,* edited by S. Woodley, J. Kay, and G. Francis, 3–18. Ottawa: St. Lucie Press.

Yoshiyama, R. M., E. R. Gerstung, F. W. Fisher, and P. B. Moyle. 1996. Historical and present distribution of chinook salmon in the Central Valley drainage of California. In *Sierra Nevada Ecosystem Project: Final report to Congress,* vol. III. Davis: University of California, Centers for Water and Wildland Resources.

Variables Used in Analyzing Factors Affecting the Biotic Integrity of Watersheds (Arranged from Lowest IBI to Highest)

Name	Cal Water No.	Area (ha)	Mean Elevation	IBI	% Dams	Reservoirs (acre-ft)	% Diversions	% Roads	% Roads and Streams	% Roadless	% Fishless
Yokohl Cr.	553.50	27372	469.3	25	0.00	0	0.95	6.95	3.14	43.7	0.0
S. Yuba	517.30	92737	1492.4	30	3.13	227282	1.88	15.30	6.50	39.8	63.9
Daulton	539.20	28633	244.0	30	1.05	240182	1.64	8.75	5.48	42.3	0.0
Tehachapi	556.10	114518	1230.7	32	0.09	764	0.40	12.33	10.31	52.1	0.0
M. Yuba	517.40	54625	1416.2	33	0.92	54534	1.59	17.83	6.04	28.3	46.0
N.Fk.San Joaquin	540.60	65231	2766.9	36	0.00	0	0.08	0.57	0.21	96.9	100.0
Mono	601.00	174723	2428.4	36	0.57	87670	0.67	4.57	2.08	67.5	86.1
Upper Owens	603.20	382669	2121.7	36	0.21	42842	0.61	5.33	3.43	76.0	30.1
Lower Owens	603.30	331606	1761.3	36	0.03	46600	0.34	5.71	3.28	76.1	29.3
S.Fk. American	514.30	207832	1372.9	40	1.40	328037	2.58	19.90	9.76	28.9	38.9
N.Fk. Feather	518.40	238915	1598.7	40	0.59	1506087	0.80	14.36	5.73	33.4	0.0
L. Stanislaus	534.22	65305	560.4	40	1.69	2401459	1.06	15.53	10.12	21.1	0.0
Huntington Lk.	540.50	34647	2388.3	40	1.16	91016	1.41	11.96	5.31	52.1	82.4
Five Dog Cr.	555.40	22557	625.1	40	0.00	0	2.35	12.64	7.67	11.9	0.0
Mammoth Cr.	603.10	98451	2496.3	40	0.20	183570	0.94	13.59	6.10	46.7	47.4
Long Valley Cr.	637.10	154134	1473.1	40	0.07	140	0.58	11.29	3.32	40.1	0.0
N.Fk.E.Br.Feather	518.50	265413	1631.1	43	0.23	30396	0.76	14.27	8.22	31.9	76.3
N.Fk.Stanislaus	534.50	71098	1850.7	43	1.41	200833	0.66	16.66	8.93	41.0	0.0
Lower Merced	537.10	17016	214.9	43	0.59	9730	1.00	12.06	5.76	33.0	0.0
Redinger	540.30	26008	1375.2	43	1.15	171276	2.34	17.76	6.00	13.0	7.4
Mammoth Pool	540.40	209525	2502.2	43	0.38	313286	0.29	5.11	2.12	79.5	82.1
Clark Fork	534.60	17684	2487.0	44	0.00	0	0.17	1.64	1.36	95.4	100.0
Mid. Kern	554.10	71187	1249.1	44	0.14	15	2.66	9.83	5.48	67.1	0.0
M.Fk. American	514.40	159523	1498.3	47	0.82	448335	1.23	15.67	5.02	45.2	69.2
U. Bear	516.30	73125	705.3	47	1.23	77885	1.61	30.03	17.15	3.1	17.2
Lower Dry Cr.	517.12	13256	222.9	47	0.76	57000	1.74	16.16	15.47	12.1	0.0
Upper Dry Cr.	517.13	18643	659.2	47	2.68	1994	1.88	18.42	10.30	15.5	0.0
Lower Yuba	517.14	14374	510.0	47	2.09	1041505	1.74	16.76	11.14	21.1	0.0
Willow Cr.-SJ	540.20	33734	1500.3	47	0.89	45598	1.22	17.95	5.63	36.9	50.3
Sutter Cr.	532.40	54874	522.9	48	1.46	24517	2.75	15.70	14.25	16.0	0.0
U. Tuolumne	536.60	127114	2612.1	48	0.16	360115	0.02	0.82	0.36	96.1	95.6
N. Lk. McClure	537.21	19628	585.2	48	0.51	520	1.32	13.43	8.10	24.2	8.9
S. Lk. McClure	537.21	10640	585.2	48	0.00	0	0.66	8.56	3.47	62.4	0.0
Fresno R.	539.30	61064	777.7	48	0.49	966	2.52	17.58	5.09	23.6	9.0
Finegold Cr.	540.10	55058	588.8	48	0.55	524769	1.74	15.25	7.19	29.0	0.6
S.Fk. Feather	518.20	32813	1233.8	50	2.13	169012	0.61	22.27	7.50	18.1	51.6
N.Fk.W.Br.Feather	518.60	36769	1230.5	50	0.82	14927	2.80	14.86	7.04	29.6	0.0
Yosemite	537.50	49320	1935.2	52	0.00	1041505	0.26	6.32	3.93	76.9	45.6
N. Fk. Kings	552.33	100089	2376.6	52	0.40	242913	0.42	7.18	2.08	72.1	89.1
Poso Cr.	555.50	68691	1037.4	52	0.00	0	2.43	10.89	5.29	44.7	0.0
L.E.Fk. Walker	630.30	36162	2317.9	52	0.00	0	1.66	5.12	3.82	78.1	12.7
Slinkard Cr.	631.20	7836	2056.1	52	0.00	0	2.93	5.00	1.78	80.9	0.0
Desert Cr.	631.30	5762	2855.7	52	1.74	640	0.52	1.84	0.70	93.4	0.0
L.W.Fk. Carson	633.10	11638	2031.2	52	0.00	0	0.95	7.98	4.39	65.5	0.0
N. Tahoe	634.20	26642	2193.2	52	0.75	732070	2.48	21.27	7.70	41.6	33.9
N. Yuba	517.50	125798	1492.2	53	0.72	2568	1.50	17.97	6.64	35.7	53.0
S.Fk.Calaveras	533.30	46275	655.3	53	1.08	1715	1.73	16.33	11.65	32.2	14.6
S.Fk.Stanislaus	534.30	27733	1700.2	53	1.08	24660	0.76	15.33	9.96	42.1	69.7
M.Fk.Stanislaus	534.40	73638	2068.1	53	0.82	150334	0.49	10.55	5.39	61.0	86.8
Lower Kings	552.20	31276	398.6	53	0.32	900	0.83	10.34	6.71	54.2	0.0
L. Tuolumne	536.30	72867	576.5	56	1.37	17633	1.34	15.50	9.06	24.4	0.9
Cherry Ck.	536.50	60670	2181.8	56	1.81	305358	0.23	2.91	1.15	87.2	94.9
N.Fk.M.Fk.Tule	555.11	10463	1986.4	56	0.00	0	0.57	5.93	1.34	83.4	80.0
L.W. Walker	631.10	32886	2074.2	56	0.00	0	0.94	5.38	3.74	71.8	2.2
Indian Ck.	632.20	4816	1834.4	56	4.14	6860	0.00	6.73	5.59	59.3	0.0
S. Tahoe	634.10	34166	2303.7	56	0.88	2384	2.43	16.44	6.35	60.3	13.9
E.Fk. Chowchilla	539.11	60118	662.8	58	0.17	130	3.43	10.77	3.99	47.2	0.0
Big Chico Cr.	509.14	18689	943.6	60	0.00	0	1.87	13.63	5.67	36.7	0.0
Paynes Cr.	509.65	7707	373.8	60	0.00	0	0.00	8.43	2.33	59.5	0.0
Cosumnes	532.20	163766	824.4	60	0.98	48556	2.20	21.20	12.05	11.0	9.7

Name	Cal Water No.	Area (ha)	Mean Elevation	IBI	% Dams	Reservoirs (acre-ft)	% Diversions	% Roads	% Roads and Streams	% Roadless	% Fishless
Little Truckee	536.00	49215	2033.2	60	0.41	245000	0.67	15.62	7.30	23.6	10.4
Mid.Tuolumne	536.70	18750	1879.7	60	0.00	0	0.64	11.02	3.47	61.3	90.9
M. Fk. Kaweah	553.43	26633	2274.7	60	0.00	0	0.04	0.65	0.30	97.4	99.3
Tule R.	555.12	90865	1119.0	60	0.11	325	1.69	10.09	4.88	57.0	27.5
L. Tahoe	634.30	34792	1949.0	60	0.29	6800	1.78	0.09	0.17	0.0	0.0
Susan R.	637.20	147323	1525.8	60	0.75	35129	0.38	12.43	5.35	30.5	0.0
M.Fk. Feather	518.30	291916	1652.7	63	0.27	140096	0.53	13.98	6.64	39.3	12.3
S.Fk.Merced	537.40	62332	1859.2	63	0.00	0	0.45	6.75	2.39	71.4	44.9
Upper Mokelumne	532.60	150067	1611.4	64	0.93	227077	0.89	15.43	8.61	42.8	64.4
U. Yosemite	537.60	59320	2697.1	64	0.00	0	0.00	0.77	0.40	96.5	92.7
Mariposa	538.00	91822	390.9	64	0.65	26955	1.44	9.85	5.03	40.8	0.0
N. Fk. Kaweah	553.41	87928	1683.5	64	0.34	943	1.21	6.20	2.09	78.9	70.7
U.E.Fk. Walker	630.40	40652	2659.5	64	0.49	3500	1.11	2.27	2.02	92.1	23.4
U.E.Fk. Walker	630.40	15867	2659.5	64	0.00	0	2.14	5.39	2.77	78.4	46.2
Upper W.Walker	631.40	58923	2637.9	64	0.34	1385	0.51	4.32	3.22	83.0	47.0
E.Fk. Carson	632.10	84528	2351.4	64	0.95	7058	0.52	2.77	2.83	89.5	26.7
U.W.Fk. Carson	633.20	16448	2457.4	64	3.04	2630	0.49	8.16	4.98	71.3	8.2
Truckee	635.20	56456	2031.4	64	1.06	102570	0.97	20.36	9.69	31.4	15.2
Little Chico	521.20	7191	496.0	67	1.39	26	1.39	10.33	1.39	21.3	0.0
Butte Cr.	521.30	39986	1090.8	67	0.75	14680	2.33	16.81	7.00	30.6	0.0
N.Fk.Merced	537.30	65098	908.0	67	0.31	315	1.08	14.69	8.69	36.8	32.8
Sycamore Ck.	552.31	43047	922.9	68	0.00	0	1.77	11.35	5.71	53.9	10.2
Deer Cr.-SJ	555.20	29676	995.7	68	0.00	0	3.27	8.24	3.84	66.0	0.0
White R.	555.30	34026	643.3	68	0.00	0	0.71	13.02	5.38	28.7	0.0
Willow Cr.	637.40	163216	1589.4	68	0.43	10863	1.01	5.67	2.17	75.3	0.0
S.Fk.Tuolumne	536.80	23338	1658.9	72	0.00	0	0.39	14.59	3.51	48.0	94.6
S. Fk. Kaweah	553.42	22411	1520.1	72	0.00	0	1.65	4.06	1.52	83.3	45.7
S. Fk. Kern	554.23	137361	2347.5	72	0.00	0	0.81	3.11	1.30	88.5	2.2
Eagle Lk	637.30	111352	1788.9	72	0.00	0	0.48	14.87	2.38	16.4	0.0
N.Fk. American	514.50	89989	1302.6	73	1.44	33067	0.93	17.04	7.66	47.8	42.1
S. Fk. Kings	552.34	247075	2601.5	76	0.08	2780	0.14	2.84	1.03	88.8	74.9
Kernville	554.22	177969	1570.6	76	0.00	0	1.45	9.21	5.25	69.5	0.0
N. Fk. Kern	554.24	217331	2625.7	76	0.00	0	0.15	3.10	0.82	87.7	35.4
Pine Cr.	509.16	28057	491.7	80	0.00	0	0.57	3.21	1.60	78.0	0.0
Dye Cr	509.62	10529	364.5	80	0.00	0	0.10	12.20	1.43	31.5	0.0
Antelope Cr.	509.63	37439	950.5	80	0.27	70	1.44	11.85	4.65	51.4	0.0
N.Fk.Calaveras	533.20	31729	637.8	80	0.95	918	1.95	12.37	9.27	49.8	4.0
Clavey R.	536.40	92978	1408.9	92	0.43	8180	0.85	20.32	8.87	26.2	60.9
Deer Cr.	509.20	53975	1273.0	93	0.00	0	0.65	14.72	4.48	46.2	0.0
Mill Cr.	509.42	33863	1204.1	93	0.00	0	0.38	7.89	1.77	71.4	0.0

NANCY A. ERMAN
Department of Wildlife, Fish, and
    Conservation Biology
University of California
Davis, California

35

Status of Aquatic Invertebrates

ABSTRACT

The aquatic invertebrate fauna of the Sierra Nevada is diverse and extensive, with many endemic species throughout the range. Aquatic systems differ widely in the Sierra because of such natural factors as elevation, climate patterns, geology, substrate type, water source, water volume, slope, exposure, and riparian vegetation. These differences are reflected in the aquatic invertebrate fauna. Small, isolated aquatic habitats such as springs, seeps, peatlands, and small permanent and temporary streams have a high probability of containing rare or endemic invertebrates. Aquatic invertebrates are a major source of food for birds, mammals, amphibians, reptiles, fish, and other invertebrates in both aquatic and terrestrial habitats. Changes in a food source of such importance as aquatic invertebrates can have repercussions in many parts of the food web. The life cycles of aquatic invertebrates are intricately connected to land as well as water, and the majority of aquatic invertebrates spend part of their life cycle in terrestrial habitats. Aquatic invertebrates are affected by human-caused activities on land as well as activities in the water. Land and water uses and impacts are reflected in species assemblages in streams and lakes. Changes in aquatic invertebrate assemblages have been used for many decades to monitor impacts on land and in water. However, the level of detail of most monitoring is not sufficient to track species losses in aquatic invertebrates. Aquatic invertebrates have not been inventoried or well-studied at the species level in most of the Sierra. Aquatic invertebrates are rarely considered or evaluated in environmental impact assessments in the Sierra. Major changes have occurred in aquatic and terrestrial habitats in the Sierra over the last 200 years: we must logically assume that corresponding changes have occurred in aquatic invertebrate assemblages.

INTRODUCTION

To assess the status of aquatic invertebrates in the Sierra Nevada, we must first consider the status of aquatic habitats. Aquatic invertebrates have complicated life cycles that are inextricably connected to both aquatic and terrestrial environments (Erman 1984b). The impacts of human use of land and water are reflected in species assemblages in streams and lakes. As Gregory and colleagues (1987) noted, "The landscapes and biotic communities of terrestrial and aquatic ecosystems are intricately linked, and effective management must acknowledge and incorporate such complexity." Changes in aquatic invertebrate assemblages are measurable and have been used as a monitoring tool for more than eighty years (e.g. Cairns and Pratt 1993); thus, we know that invertebrate assemblages change with habitat changes. Major changes have occurred in aquatic habitats in the Sierra Nevada over the last 200 years (Beesley 1996; Kattelmann 1996; Kondolf et al. 1996; Mount 1995). We must logically assume that, as land and water are altered in the Sierra Nevada, aquatic invertebrate assemblages are changing; populations (e.g., Taylor 1981) and perhaps species are being lost. But most of these changes are occurring at unknown and undocumented rates.

In California, we do not have inventory data on aquatic invertebrates from 200 years ago. But neither do we have adequate inventory data on aquatic invertebrates at present. We have surveys of specific invertebrate groups in a few geographic areas of the Sierra, but a surprisingly small amount of survey information at the species level exists. There are not adequate systematic invertebrate inventories or surveys for even the national parks (Stohlgren and Quinn 1992). On the other hand, the responses of aquatic invertebrate assemblages to land and water alterations are well-known and have

Sierra Nevada Ecosystem Project: Final report to Congress, vol. II, *Assessments and scientific basis for management options.* Davis: University of California, Centers for Water and Wildland Resources, 1996.

been studied for decades in many parts of the world and, to some extent, in California. Therefore, we can predict generally how invertebrate assemblages will change in response to such environmental impacts as logging, grazing, mining, water development, construction, human settlement, and the introduction of exotic species. Habitat loss and degradation and the spread of "exotic" (non-native or nonindigenous) species are the greatest threats to biodiversity in running-water systems (Allan and Flecker 1993; Wilcove and Bean 1994). The extent of change in California river systems has recently been documented (California State Lands Commission 1993; Mount 1995). California may be unsurpassed for the extensive geographic scale and short time scale on which these basic changes have occurred.

Questions about invertebrate status in Sierra Nevada aquatic habitats are as follows:

- Are species disappearing?

- Are species assemblages changing or becoming simplified in response to changes in habitats?

- What is causing these changes?

- What can be done to reverse these changes?

Perhaps a fifth question we should be asking is

- Why have aquatic invertebrates been so little studied and so little considered in management in the Sierra Nevada and in California, in general?

This assessment can only begin to answer these questions.

Many aquatic invertebrates have specific and narrow habitat requirements and are restricted, therefore, to places that vary little from year to year. Others are generalists and can survive over a wide range of habitat types (Thorp and Covich 1991). The differences between these two groups and all the gradations between them are crucial to our understanding of what has been happening to aquatic invertebrate species and assemblages of species in the Sierra Nevada over the past 200 years, especially since the gold rush, when major alterations of aquatic systems began in the Sierra.

A knowledge of aquatic invertebrates at the species level is essential to assessing the status of biodiversity in the Sierra. Monitoring of invertebrates at a higher taxonomic level (genus, family, order) can be useful in indicating changes in invertebrate assemblages in response to some impact if proper controls are established, but such monitoring usually cannot determine loss of species. The term "species" has the same meaning for aquatic invertebrates as it has for any other group of living things; aquatic invertebrate species are not interchangeable. Just as the common pigeon (rock dove; Columbidae: *Columba livia*) is not the same bird as the band-tailed pigeon (Columbidae: *Columba fasciata*), nor a white fir the same as a giant sequoia, neither is one species (or genus,

family, or order) of aquatic invertebrate the same as another. Each species has different habitat requirements and different tolerances to environmental variables.

Endemic species of aquatic invertebrates in the Sierra Nevada (and in mountains in general) are often isolated at all elevations in small first- and second-order stream systems and can be limited in distribution to such habitats as springs, peatlands, and small headwater streams (Erman and Erman 1975; Hampton 1988; Stewart and Stark 1988; Erman and Erman 1990; Wiggins 1990; Erman and Nagano 1992; Hershler 1994). Some groups of aquatic invertebrates (e.g., some families of stoneflies, caddisflies, flatworms, and snails) exhibit high species endemism and great diversity in the Sierra Nevada.

Fish assemblages are not indicators or surrogates for aquatic invertebrate communities in much of the Sierra. Fish communities are not diverse in the Sierra; game fish have been introduced and moved throughout the range by humans, and some (e.g., rainbow, brown, and golden trout) are more tolerant of degraded habitats and/or a broad spectrum of conditions than are many invertebrate species and invertebrate assemblages. Historic distributions of fish were very different from current distributions, and much of the Sierra was originally fishless (see Knapp 1996; Moyle et al. 1996). Further, many small aquatic habitats rich in endemic invertebrates are lacking fish species.

Aquatic invertebrates are an important source of food for birds, mammals, amphibians, reptiles, fish, and other invertebrates. Changes in terrestrial and aquatic habitats lead to changes in invertebrate assemblages, which in turn increase, decrease, or change food supplies for other animals. As impacts occur in a stream, species (or taxa) richness (number of species) decreases but the population size of some species may increase. Further, large-sized species are usually replaced by small species (e.g., Wallace and Gurtz 1986). Conversely, when the stream condition improves, larger invertebrate species replace small species (Grubaugh and Wallace 1995). Such changes can have critical impacts on species that depend on invertebrates for a food supply.

Aquatic systems differ widely throughout the Sierra because of such natural factors as elevation, climate patterns, geology, substrate type, water source, water volume, slope, exposure, and riparian vegetation. For these reasons it is not possible to describe a typical Sierran stream, lake, spring, peatland, and so on, or a typical invertebrate assemblage. The natural variability among aquatic habitats must be understood when the effects on invertebrates of anthropogenic disturbance are studied.

The waters of the Sierra are the responsibility of many federal, state, and local agencies and are subject, through these agencies, to many laws and regulations. How these agencies work together and how they apply and enforce these laws determine the fate of the aquatic biota. Making connections among the aquatic biota, aquatic habitats, and institutional responsibility and performance is necessary to understand the present state of and future possibilities for Sierra waters.

PROCEDURES AND METHODS

To assess the extent of aquatic invertebrate work in the Sierra Nevada, we searched several standard library databases, using an extensive list of invertebrate names and aquatic habitat keywords. This method, while not complete, gave a reasonable indication of research on aquatic invertebrates over approximately the last twenty years (the general period covered by the databases). Key researchers were contacted to fill in some gaps in the list of studies. These contacts revealed that several papers had been missed in the databases, but also that the technique had given a fairly thorough indication of the topics being studied and of primary researchers or groups of researchers doing the work. For purposes of analysis, studies were grouped into a few general categories by geographical area or type of study. These groups were (1) taxonomic studies, (2) impact studies, (3) geographic surveys of certain taxonomic groups, (4) behavioral studies, (5) studies pf Mono Lake, (6) studies of Lake Tahoe, (7) other lake studies, and (8) studies on mosquitoes.

With such arbitrary groupings, there was much overlap. For example, many of the studies of Lake Tahoe could be considered impact studies or behavioral studies. But the groupings were made to provide an understanding of distinct aquatic systems or problems and to discover the studies' relevance (or lack of it) to the SNEP objective of assessing status. Much money has been spent on mosquito research and there were many papers on this group of organisms, but mosquitoes were not evaluated for this chapter. The reasons for this will be discussed later.

In addition to the general search of databases, we contacted agencies through a letter asking for information and made individual contacts with people known to have specific information on invertebrate work. This step revealed unpublished, nonrefereed reports and studies for which data sheets and notebooks, but no reports, existed.

A third step was to contact experts from North America known to be working on certain groups of invertebrates in an attempt to compile up-to-date species lists for the Sierra and to get some idea of the percentage of endemism among Sierra aquatic invertebrates. Most of these efforts are ongoing and incomplete. Recent published information for some groups (e.g., stoneflies, caddisflies, alderflies, dobsonflies, snails, and clams) was sufficient for estimates. Large gaps in our understanding and knowledge of aquatic invertebrates in California and in the Sierra are evident and will be discussed in a later section.

A fourth source of available information is museum collections. However, the short time allowed for this project did not permit us to explore these. Such collections as the California Academy of Sciences; the Bohart Entomology Museum at University of California, Davis; the Los Angeles County Museum; and the University of California, Berkeley, entomology museum have material from the Sierra, as do many other museums in North America (e.g., the Smithsonian and the Royal Ontario Museum in Toronto). Museum material is known and up-to-date for invertebrate groups being actively investigated by experts. But much other material has not been studied, and information is undoubtedly contained in these sources. To have meaning, this material requires examination by experts who are currently studying systematics in their respective fields. Taxonomy has changed rapidly and significantly in many invertebrate groups over the last twenty-five years. Hence, each specimen must be examined to determine its classification.

This chapter deals largely with aquatic macroinvertebrates (those that can be seen with the naked eye), not with the microinvertebrates (those that require a microscope to be seen). Such microinvertebrates as protozoans, tardigrades, and rotifers, for example, have not been assessed. The emphasis in this chapter is on running-water habitats. Some examples, however, are from standing-water habitats.

HISTORIC CONDITIONS AND AGENTS OF CHANGE

By describing conditions that existed in the Sierra Nevada prior to the immigration of Europeans and Asians, that is, conditions of 200 or 300 years ago, we can understand better what has happened to aquatic habitats and what the implications of those changes are for aquatic invertebrates. The numeric assessment of change to aquatic habitats (the numbers of dams, diversions, roads, grazing allotments, etc.) is described elsewhere (for example, see Kattelmann 1996, Menke et al. 1996, and Kondolf et al. 1996); therefore, this section gives a general description only, for the purpose of demonstrating habitat under which aquatic invertebrate species and species assemblages evolved in the Sierra Nevada over thousands of years and how that habitat has changed. It is not a complete listing of all of the changes and impacts that have occurred in Sierra aquatic invertebrate habitats.

Two hundred years ago Sierra Nevada streams were continuous running-water systems: there were no dams, reservoirs, water diversions, or interbasin transfers of water. There is no, or almost no, similarity between invertebrate species assemblages in running water and those in standing water. The major taxa of many invertebrate groups are found in both general habitat types, and in gradations between them, but the species that live in these two habitats are usually different. For example, true flies, in the order Diptera (a major insect taxon) are found in both reservoirs and in rapidly flowing water, but the species, and in many cases the genera and families, are different in the two habitat types. To continue this example, a family of true flies called the net-winged midges, Blephariceridae, is found exclusively in rushing mountain streams. It has suction-cup-like attachments on the underside of its larval body and lives only in the strongest currents. Widespread construction of dams throughout the mountainous areas of California has probably changed the distribution and possibly decreased the number of species of this family of flies.

In general, burrowing Chironomidae larvae (another type of midge fly in the order Diptera) and oligochaetes (aquatic segmented worms) predominate in habitats where sediments accumulate (Johnson et al. 1993), and their numbers rise where streams have been converted to reservoirs. Stoneflies (Plecoptera), found primarily in running water, are eliminated in reservoirs (Stewart and Stark 1988).

To illustrate the scope of change, figure 35.1 shows the locations of dams that are more than 7.6 m (25 ft) high or that have a capacity greater than 61,674 m3 (50 acre-ft). Smaller dams and water diversions exist on many other small Sierra Nevada streams but are not shown in this figure. Prior to the construction of reservoirs, natural hydrologic cycles existed on all streams and rivers. Water was high in the winter and spring and low in the summer and fall. Invertebrate life cycles evolved over thousands of years in response to such hydrologic cycles. Invertebrate biomass in the water was highest during the high water period and lowest in the summer and fall. Aquatic insects are the largest component of the aquatic invertebrate community, and most of them emerge as terrestrial adults in summer and fall in the Sierra Nevada, with fewer species emerging in spring and a small minority emerging in the winter (e.g., Erman 1989). Thus, invertebrate biomass is low when the water is low because many insects are in the terrestrial stage or are in the egg or small larval stage.

Invertebrates can accommodate the natural rise and fall of floodwater by moving up with the water and outside the stream banks, by burrowing into the substrate, or by taking refuge in root wads and debris along stream edges. They return to the stream channel as the water recedes. Natural floods perform the function of flushing sediment from the stream system, which, in turn, increases pore spaces within the stream-bottom substrate and provides surface area for invertebrates to inhabit.

The suddenly fluctuating water caused by some dams and water diversions has a different impact on invertebrate populations than does a natural flood. Invertebrates are stranded as water volume is lowered suddenly and stream channels dry up. Also, invertebrates drift downstream when water is rapidly lowered or raised (Minshall and Winger 1968; Bovee 1985). Year-round constant flow, a condition found in some artificially managed streams, is also abnormal to invertebrate communities of the Sierra Nevada. Under constant flow, sediment is not flushed from streams, and other poorly understood triggers to life cycle changes and in-stream migrations may not be present (Reiser et al. 1989).

Sediment from mining, logging, cattle grazing, roads, and construction had not entered Sierran streams 200 years ago. Natural sources of sediment, such as landslides from heavy rains and fires, were present, of course, prior to our recent history, as they are today. We can assume, therefore, that the quantity of sediment entering the aquatic systems of the Sierra today is far greater than it was. Much of this sediment is trapped behind dams at present (where it causes problems in water storage operations) (Kattelmann 1996) and is thereby removed from the stream system below dams. One example is the Mokelumne River watershed basin, where erosion rates estimated over the last twenty-five years are more than eight times higher than they were in 1944 (Robert C. Nuzum, Director of Natural Resources, East Bay Municipal Utility District, letter to Don C. Erman, September 25, 1995). The primary land use in the basin is timber harvesting (consisting of 98.5% of the land base).

The effects of sediment on aquatic macroinvertebrates have been amply demonstrated and known for many years (e. g., Cordone and Kelley 1961; Buscemi 1966; Chutter 1969; Brusven and Prather 1974; Luedtke et al. 1976; Waters 1995). In streams, sediment accumulation depletes available habitat for invertebrates, as pore spaces in the rocky substrate are filled with sand and silt. Over time, continued sedimentation can create a cemented stream bottom with no substrate pore spaces available for invertebrate colonization. Only a few highly tolerant invertebrate species will persist in these conditions. The gold mining areas contain examples of sediment accumulation that are even more dramatic. There, certain streams have become so filled with sediment that surface flow no longer exists. Where 200 years ago rocky-bottomed streams flowed, today sediment-filled, seasonally dry stream channels are evident many feet above the original channel (Mount 1995). A striking example of this impact is Shady Creek on Highway 49 near North Columbia. As sediment increases, species richness, density, and biomass decrease. Sediment obstructs respiration, interferes with feeding, causes loss of habitat and habitat stability, and may alter production of invertebrate food sources (Johnson et al. 1993).

In the past, streams were more shaded and were lower in temperature because there was more riparian cover. Headwater streams were deeper and narrower, in meadows and wetlands. They had rocky bottoms and were covered by either willows or alders, or by sedges and grasses. Today, small first- and second-order streams (small streams in the headwaters of river basins and also in river branches at all elevations of the Sierra) of this description are found largely in national parks. Livestock grazing has decreased or eliminated riparian vegetation, broken stream banks, widened stream bottoms, increased sediment, decreased shade, and increased water temperature (Platts 1978; California State Lands Commission 1993; Fleischner 1994; Li et al. 1994; Menke et al. 1996).

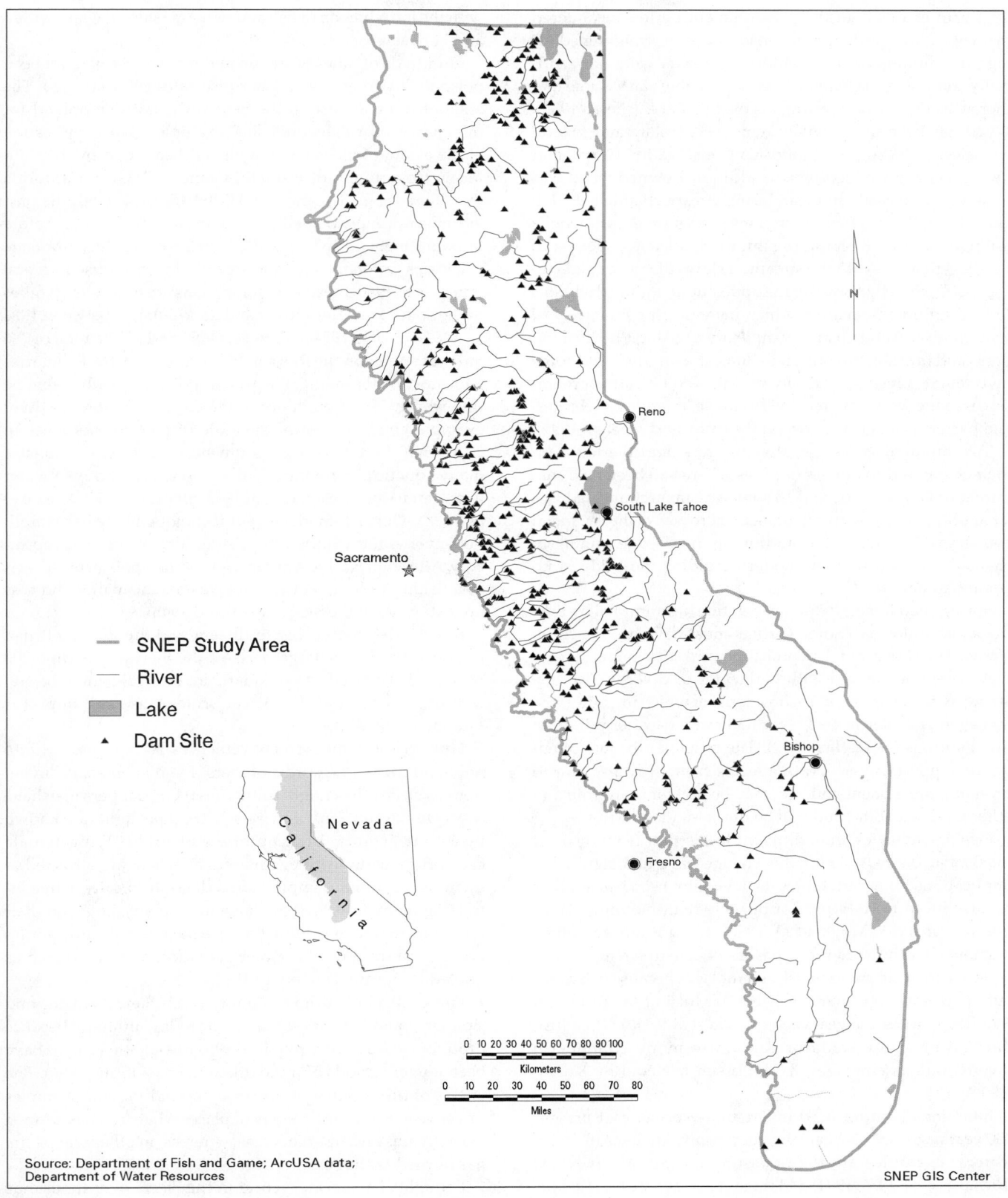

SNEP Study Area
River
Lake
Dam Site

Nevada

California

Reno

South Lake Tahoe

Sacramento

Bishop

Fresno

0 10 20 30 40 50 60 70 80 90 100
Kilometers

0 10 20 30 40 50 60 70 80
Miles

Source: Department of Fish and Game; ArcUSA data;
Department of Water Resources

SNEP GIS Center

FIGURE 35.1

Dams in California that are at least 7.62 m (25 ft) high or that have a capacity greater than 61,674 m³ (50 acre-ft) (from California State Lands Commission 1993).

Stream channels in the presettlement period meandered in areas of low gradient; they had not been straightened for logging, mining, or road building. Streams were not artificially confined to a channel. We can probably can assume that there was more wood in streams (Sedell and Luchessa 1982). Wood has been intentionally removed by state and federal agencies, including the California Conservation Corps, and by loggers and woodcutters. In addition, downed wood was retained more easily in meandering stream channels (Sedell and Maser 1994). Wood in streams serves several functions for invertebrates. It retains organic matter (leaves, sticks, and needles) that falls into the stream. It slows the water and creates pools, thereby allowing the opportunity for invertebrates to feed on organic matter, which increases the efficiency of nutrient use. Wood creates complexity of habitat by forming pools and breaking up otherwise long stream runs. And some invertebrates feed specifically on the wood or attach themselves to the wood and feed on the algae, microinvertebrates, and bacteria that grow on wood (Murphy and Meehan 1991).

Two hundred years ago, streams were not diverted from their channels into ditches or pipes. There had been no dynamiting of fish barriers, and so some stream sections had isolated populations of invertebrates. There was no heavy metal contamination of water from mining, no dredge mining in channels, no concrete, no modern building, no bridges, no riprapped banks.

Springs had intact riparian vegetation, untrampled by livestock and unlogged. Some springs must have been used by Native Americans, but they probably were not channeled. And they were not sprayed with herbicides or diverted, as they are today for a variety of reasons, some of which include game management (Bleich 1992). Nor had ground-water pumping dried springs (DeDecker 1992). The potential for loss of endemic aquatic invertebrate species from springs, due to present management and use, may be greater than from any other aquatic habitat and will be discussed later.

Fish assemblages were different in the past from those of today, and the reasons for this change have had significant implications for invertebrates. Fish had not been transported or introduced from Europe or put into high-mountain lakes (see Knapp 1996; Moyle et al. 1996). In addition, the introduction of fish into reservoirs has resulted in upstream as well as downstream changes in fish assemblages because fish move out of reservoirs (Erman 1973, 1986). Much of the upper Sierra was fishless 200 years ago (Moyle et al. 1996). The introduction of a top predator can cause many changes in invertebrate assemblages, as discussed in detail by Knapp (1996).

Intentionally introduced invertebrate species, not present 200 years ago, are also causing community changes in Sierra waters. Examples are the opossum shrimp, *Mysis relicta* (Richards et al. 1975, 1991) and the signal crayfish, *Pascifasticus leniusculus* (Flint and Goldman 1975; Elser et al. 1994). Other invertebrates likely have been introduced unintentionally

with the introduction of fish and with transfers of water within and between basins.

Hundreds of miles of stream and many lakes had not been poisoned by rotenone or other piscicides 200 years ago. The scale of rotenone use in the Sierra was not determined for this chapter, but a few published examples give an indication of the extent of its use. Rotenone has been used by the California Department of Fish and Game (CDFG) in California "for more than 45 years" (CDFG 1994a, 1994b). "In the past we have routinely treated the streams in a drainage . . . before impoundment" (Hashagen 1975). In the Kern River, drainage piscicides (rotenone and antimycin) have been used several times since 1960. Present ongoing plans are to poison 37 miles of the South Fork Kern River and its tributaries between 1994 and 1996 (CDFG 1994c). Between 1952 and 1954 a total of 286 miles of stream in the Russian River drainage (the tributaries and most of the main river) were poisoned with rotenone (Johnson 1975). Though not in the Sierra, the Russian River example gives an idea of the scale of past rotenone use in California. Rotenone and antimycin reduce populations of many aquatic invertebrates when applied to a body of water (Cook and Moore 1969; Degan 1973; Stefferud 1977; Maslin et al. 1988). This fish-management technique likely has simplified invertebrate communities, especially where used repeatedly. Although Native Americans used fish poison in streams as a fishing technique, the scale was much smaller and was not extensive in the Sierra (Rostlund 1948).

Insecticides, herbicides, fertilizers, and fire retardants had not been used over large parts of the Sierra landscape 200 years ago. Their effect on aquatic invertebrates may be significant (Norris et al. 1991). The scale of use and impact is unknown.

Humans with modern conveniences had not moved into wildland areas in record numbers. Even such small inventions as electric blacklight (ultraviolet) bug zappers may have a local impact on aquatic insects. Ultraviolet lights are known insect attractants; high numbers of night-flying female caddisflies, many with egg masses attached, are attracted to them (N. A. Erman, unpublished data). Ironically, while attracting many insects, they have little, if any, effect on their target insect, the mosquito, because most female mosquitoes are not attracted to ultraviolet light, according to Turpin as quoted in Purdue University 1993.

Fire probably had no more impact on Sierra streams and riparian zones in the past than it has today, although because drought cycles play a role in fire frequency there may have been longer periods of more fire in the past (Stine 1996). The effects of fire on streams are local and individual. Examples can be seen in the Sierra today of places where fire has jumped over streams and riparian areas, whereas, in other places, fire has burned to the stream edge.

Droughts have been cyclical in the Sierra over thousands of years, and some were longer and more severe than our recent eight-year drought (Fritts and Gordon 1980; Erman and

Erman 1995; Stine 1996). Many springs and wetlands (meadows and fens or minerotrophic peatlands) must have dried out or disappeared during those periods. Evidence indicates that the past 100 years or so has been a period of high moisture (Stine 1996).

Flooding of stream channels, on the other hand, probably was not as severe 200 years ago as it is under our present land use. We can assume that with more vegetative cover and fewer perturbations on the land, especially those due to road systems (Kondolf et al. 1996), water soaked into the soil in greater amounts than it does today.

In summary, invertebrate assemblages in the past were probably richer and species diversity was probably higher; that is, there were more species and their relative abundances were more evenly distributed in many habitats than they are today. Many cumulative impacts are present in Sierra Nevada aquatic systems that were not present 200 years ago. These impacts have a combined and often synergistic effect on stream systems. It is reasonable to assume that some species have probably disappeared from small, unique, isolated habitats (spring systems, small upper watershed streams, peatlands, and perhaps, high-mountain lakes and ephemeral ponds) that have been substantially altered or eliminated.

CURRENT CONDITIONS

Aquatic Invertebrate Resource

Endemic or Unusual Species

Many endemic species of aquatic invertebrates, known nowhere else in the world, are present in the Sierra Nevada. A wealth of evolutionary, ecological, and biogeographical information is contained in Sierra aquatic invertebrates. Among the more notable examples is the stonefly *Capnia lacustra*, present only in Lake Tahoe (Jewett 1963), the only stonefly in the world known to be fully aquatic in the adult stage (Nelson and Baumann 1989). Another unusual stonefly, *Cosumnoperla hypocrena*, is known from one intermittent spring in the Cosumnes River Basin (Szczytko and Bottorff 1987). Extensive searching has failed to produce this species from any other site. Worldwide, few stoneflies are known from intermittent habitats. A caddisfly, *Desmona bethula*, has been studied in sites at about 1,970 m (6,500 ft) elevation, where it leaves the stream on warm summer nights as an aquatic larva to feed on terrestrial vegetation, a behavior undescribed in the world prior to its being studied in the Sierra (Erman 1981). This species, too, is known from a small number of Sierra Nevada sites. Another caddisfly, known only from small streams in the Sierra, the Siskiyous, and the Cascades, *Yphria californica*, is possibly the most primitive living species of the tube-case-making caddisflies in the world (Wiggins 1962; Anderson 1976). A species of brine shrimp, *Artemia monica*, is endemic to Mono Lake (Belk and Brtek 1995). In the Sierra Nevada the symbiotic relationship between the midge larva *Cricotopus* and the algae *Nostoc* was discovered (Brock 1960). Endemic species of flatworms (Kenk 1970, 1972; Kenk and Hampton 1982; Hampton 1988), of amphipods (Holsinger 1974), and of hydrobiid snails (Hershler and Pratt 1990; Hershler 1994) have all been found in the Sierra. These are but a few of many such examples.

The percentage of endemism in aquatic invertebrates in the Sierra is apparently much higher than in terrestrial invertebrates (Kimsey 1996; Shapiro 1996). The reason is the discrete and isolated nature of small aquatic habitats in mountainous areas. The pattern for endemic amphibians occurring in aquatic habitats is similar to that of endemic aquatic invertebrates (Jennings 1996).

Aquatic Invertebrates as a Food Source

While the foregoing examples are of unusual species that are of great evolutionary interest to scientists, they may not be understood by most people as being of value. We are rarely taught the connections between small, seemingly obscure species and the larger species with which we are all familiar (Kellert 1993). And yet, in the details of small, unknown animals lies the fate of the animal world (Wilson 1987).

For example, the brine shrimp and alkali fly of Mono Lake provide food for thousands of migrating waterfowl from North and South America (Vale 1980; Lenz et al. 1986). Decreasing fresh water and increasing salinity in Mono Lake led to decreases in the alkali fly *Ephydra hians* prior to restoration of inflows to the lake and was the subject of concern and study of this critical invertebrate species (Herbst 1990, 1992; Herbst and Bradley 1993).

In Lake Tahoe, introductions of exotic fishes and an exotic invertebrate, the opossum shrimp, *Mysis relicta*, have led to periodic decreases in and disappearances of native zooplankton, species of *Bosmina* and *Daphnia*, which in turn have caused food shortages for fish (e.g., Goldman et al. 1979; Morgan 1979, 1980; Morgan et al. 1978; Richards et al. 1975, 1991) and possibly have caused increases in algae and decreases in water clarity. This story continues and is not completely understood. Suffice it to say that Lake Tahoe is now far from being a natural biotic community, and poorly considered introductions of exotic species, both fish and invertebrates, have played a major role in some of the changes that have occurred in the lake.

Aquatic invertebrates, in general, provide food to a vast array of birds, reptiles, amphibians, mammals, fish, and other invertebrates. Aquatic insects live in an aquatic habitat for only part of their lives and are terrestrial during other life stages, where they live primarily in riparian areas (Erman 1984b). They are a food source in all life stages for invertebrate-feeding animals in aquatic and riparian areas. Adult insects constitute a substantial percentage of the arthropod biomass and numbers near streams (Jackson and Fisher 1986;

Jackson and Resh 1989). Their numbers decline but are still significant 150 m (492 ft) or more from streams (Jackson and Resh 1989). This contribution of aquatic insects to the total arthropod assemblage near water makes riparian areas rich in food. Any vertebrate that is found in wet areas in the Sierra and that is known to eat invertebrates is likely feeding on aquatic invertebrates at some time. Many of these interrelationships have not been studied and are unknown in their specific details (but see, for example, Busby and Sealy 1979) but nevertheless are understood generally (Zeiner et al. 1988, 1990a, 1990b). For example, many of the bat species present in the Sierra forage over water for insects (Ron Cole, University of California, Davis, Department of Wildlife, Fish, and Conservation Biology, communication with the author, 1995; Zeiner et al. 1996). The water shrew eats aquatic insects, as may other shrews found in riparian areas (such as the vagrant shrew and the dusky shrew). Some part of the diet of the western jumping mouse, confined to wet areas, is likely aquatic insects, as are portions of the diets of the river otter, gray and red fox, mink, raccoon, marten, and western spotted skunk (Zeiner et al. 1990b).

A large number of bird species are dependent on aquatic invertebrates, either during all life stages (e.g. the American dipper, pied-billed grebe, and eared grebe); or during critical stages of breeding or the early life of the young (e.g., the gadwall, wood duck, tundra swan, American wigeon, belted kingfisher, red-winged blackbird, and yellow-headed blackbird); or during parts of the year when other food is unavailable. The list of birds that feed on invertebrates in and over wetlands, lakes, or streams is extensive (e.g., the hooded merganser, common merganser, spotted sandpiper, Forster's tern, black tern, tree swallow, violet-green swallow, bank swallow, barn swallow, willow flycatcher, black phoebe, Swainson's thrush, and yellow warbler). The few examples given here show the diversity of bird life in the Sierra and the northeastern plateau of California that depends on this food source (Zeiner et al. 1990a).

Several threatened amphibian species in the Sierra are highly dependent on aquatic invertebrates during the adult stage of their lives; these include the red-legged frog, foothill yellow-legged frog, mountain yellow-legged frog, and spotted frog (not in the Sierra but in Modoc County). Further, locally distributed salamanders that are present in springs and seeps—the Inyo Mountains salamander, and Mount Lyell salamander—likely feed on aquatic invertebrates. Also, the long-toed salamander, rough-skinned newt, California newt, and Yosemite toad, and the western pond turtle, a reptile, are all known to eat aquatic invertebrates (Zeiner et al. 1988). But this list is far from complete, and interested readers are referred to the three-volume work California's Wildlife (Zeiner et al. 1988, 1990a, 1990b) for specific details and references on Sierran vertebrate species.

Of course, many fish also depend on aquatic invertebrates, as do all animals that feed on those fish (e.g., the great blue heron, belted kingfisher, bald eagle, marten, mink, and river otter).

When a food source of such importance and magnitude as aquatic invertebrates is changed or extinguished in an area, even temporarily, it can have repercussions in many parts of the food web.

State of Knowledge of Aquatic Invertebrates

California has never undertaken the task of systematically and thoroughly surveying the invertebrates of its aquatic habitats, and in this regard, we lag behind the eastern United States, Europe, and much of Canada. Lack of expertise in California universities and state and federal resource agencies is a reason for this paucity of inventory data on aquatic invertebrates (see also Kimsey 1996). A shortage of aquatic invertebrate taxonomists and systematists worldwide is an obstacle to developing an understanding of issues of changing biodiversity and the impacts of environmental degradation (Disney 1989; Ehrenfeld 1989; Wiggins 1990; Erman 1992a).

Obsolete taxonomic keys and species lists are a problem for students of California aquatic invertebrates. When, *Aquatic Insects of California* was first published by Usinger in 1956 (it was reissued in 1963 and 1968) (Usinger 1956), it was considered a landmark work for a state and continued to be praised years later (Hynes 1984). But even at the time it was published, it was written from somewhat idiosyncratic collections of insects, not from systematic inventories, in most cases. Although it included species known then in California, it was far from complete. For example, 47 additional species of stoneflies alone have been found in California since the Usinger book was first published. (At present 167 species of stoneflies are known for California [R. L. Bottorff, R. Baumann, B. P. Stark, and N. Erman, unpublished list]). In addition, revisions of systematics for nearly all groups in the book have made many changes in species names and evolutionary relationships. Further, insects, though they constitute the largest taxon of freshwater invertebrates (that is, the taxon containing the greatest number of species), are not the only invertebrates present in Sierra waters. Examples of other groups not included in the Usinger book are flatworms, nematodes, segmented worms, snails, clams, and crustaceans (fairy shrimp, crayfish, isopods, etc.).

Species Inventories and Endemism in the Sierra Nevada

We can ascertain the extent and nature of aquatic invertebrate diversity by examining taxa in a few geographic areas in the Sierra where extensive survey work has been conducted on some groups. This effort is woefully incomplete because we do not have a Sierra-wide inventory, but it is a beginning and indicates the percentage of endemism and numbers of species in some basins (table 35.1). Endemism in the context of

TABLE 35.1

Estimated number of species of selected aquatic invertebrate taxa in some areas of the Sierra Nevada (see text for references).

Taxon, by Area	Number of Species Present	Number of Species Endemic to Sierra	Percentage of Species Endemic to Sierra
Sagehen Creek Basin			
Stoneflies	38	6	16
Caddisflies	77	11	14
Mayflies	22	?	?
Cosumnes River Basin			
Stoneflies	79	16	20
Fresno, Kern, Madera, and Tulare Counties			
Caddisflies	128	11	9

this chapter means species that are found only in the Sierra Nevada. Many other species are present in the Sierra in only one or a few other places but are not strictly endemic.

Specific Studies by Location

Sagehen Creek Basin. One of the better-studied stream basins is the Sagehen Creek basin on the east side of the Sierra, north of Truckee, where the University of California has operated a field station in the Tahoe National Forest since 1951 (see also Kimsey 1996). Even here, however, some groups, namely stoneflies (Plecoptera) and caddisflies (Trichoptera), have been well surveyed, and others, for example, true flies (Diptera) and mayflies (Ephemeroptera), are still incompletely known. Stoneflies were comprehensively surveyed in 1967 (Sheldon and Jewett 1967), and the list was revised and updated for the first North American Plecoptera Conference in 1985 (R. Baumann, W. Shepard, B. Stark, and S. Szczytko, unpublished data, available from N. A. Erman). Thirty-eight species of stoneflies are known from the Sagehen Creek basin. Seventy-seven species of caddisflies have been found in the basin (Erman 1989). Twenty-two species of mayflies have been identified in the basin, but this collection has not been verified by experts on Ephemeroptera, and the actual number is probably higher (D. C. Erman and N. A. Erman, unpublished data). Aquatic habitats surveyed included Sagehen Creek (a second-order stream), springs, spring streams, temporary streams, temporary ponds, and peatlands.

For the two well-studied groups in this stream basin we have an estimate of endemism: 11 of the 77 species of caddisflies are probably endemic to the Sierra (14%), and 6 of the 38 stonefly species (16%) are endemic, based on present information.

Cosumnes River Basin. Another study of stoneflies was conducted on the west side of the Sierra throughout the Cosumnes River basin (Bottorff 1990) where seventy-nine species were found over six stream orders, from headwater streams to the major river at the lower part of the watershed (R. Bottorff, telephone conversation with the author, October 11, 1995). Sixteen of these species are endemic to the Sierra; seventeen are endemic to California. Some species endemic to the Sierra are also found in Nevada and are therefore not considered California endemics but are, nevertheless, Sierra endemics. Twenty-six of the species found in the Cosumnes River basin were also present in the Sagehen Creek basin, and four of these are Sierra endemics.

Fresno, Kern, Madera, and Tulare Counties. Extensive black-light collections of caddisflies have been made in Fresno, Kern, Madera, and Tulare Counties, in the San Joaquin–Tulare basins, by D. Burdick and R. Gill (as reported in Brown 1993). Some species were collected by other methods. Species are reported by elevation from 30 m to 2,652 m (100 ft to 8,700 ft) above sea level. We eliminated species found only below 213 m (700 ft), synonymous species (species described more than once in the literature), and species listed as new species but for which no description exists in the literature. With these criteria, the number of caddisfly species reported by Burdick and Gill for these four Sierra Nevada counties was 128. Eleven of these species are endemic to the Sierra.

Black lights are known attractants of some insects and thus sample an unknown area. Some species are more attracted to them than are others, and day-flying species may not be collected with black lights. The results from blacklight collecting are difficult to interpret in terms of estimating the species richness of a given area or habitat. Therefore, this number is subject to revision, but it gives a general idea of west-side Sierra caddisfly species in these basins.

A comparison of this list with the east-side Sierra study of caddisflies in the Sagehen Creek basin (Erman 1989) shows that 50 species were collected in both areas and that 27 species were present in the Sagehen Creek basin that were not found in the San Joaquin–Tulare basins. Jaccard's index of similarity (Pielou 1984) showed a 32% similarity between the two areas.

With only these few comprehensive surveys and collections of Plecoptera and Trichoptera species in a given area of the Sierra, we can say little about relative diversity. Species-level surveys in other parts of the Sierra are greatly needed.

Selected Taxa of Invertebrates

A few taxa were selected for more in-depth analysis for this chapter to determine percentages of endemism in the greater SNEP study area. Taxa were selected because they have been studied recently, because databases existed and were being kept up-to-date by experts in that taxonomic group, or because a reasonably recent (since 1970) monograph had been published.

The difference between collections and surveys is important here. The total number of species for the state is based on collections. These usually consist of one-time visits to a site, with the collectors using one or two types of collecting methods. The numbers of species for the Sagehen Creek basin and the Cosumnes River basin, in the previous section, are based on surveys. A survey uses some kind of systematic sampling scheme. The sampling methods used by Sheldon and Jewett (1967), Bottorff (1990), and Erman (1989) were different, as would be expected when different groups of organisms are being sampled for different reasons, but all were year-round samplings of all habitat types. Surveys of species done in other parts of the Sierra would greatly enhance our knowledge of invertebrates and undoubtedly would reveal new species. Unless they are specifically designated as surveys, however, species numbers in this chapter are all based on collections (table 35.2).

Plecoptera (Stoneflies, Insecta). Plecoptera is one of the better known orders of freshwater invertebrates in California. It is also a small group (based on number of species) compared with the Trichoptera or Diptera. At present, 167 species are known in the state; 122 of these are present in the Sierra and 31 are endemic to the Sierra (R. L. Bottorff, R. Baumann, B. P. Stark, and N. A. Erman, unpublished list). The Sierra-Cascade system and the Appalachian system are considered the "two great centers of endemicity" for the North American Plecoptera. About 25 genera are thought to have evolved in each area (Stewart and Stark 1988). Most stoneflies are dependent on lotic habitats (running water) of high oxygen and low temperature, and so it is not surprising that their distribution would be concentrated in the Sierra.

Megaloptera (Alderflies and Dobsonflies, Insecta). Alderflies (Sialidae) are a small group of aquatic insects with only 24 North American species; 9 of these are present in the western United States (Whiting 1991a, 1991b, 1994). Six species are known from California, and one is endemic to California. Four

species are present in the Sierra as well as in other parts of the state.

Eleven species of dobsonflies (Corydalidae) are known in California, and seven of these are in the Sierra (Usinger 1968; Flint 1965; Evans 1984). Sierran endemism of species in this family was not determined.

Trichoptera (Caddisflies, Insecta). The caddisflies are a large and diverse group of aquatic insects, and species are found in nearly all freshwater habitats. At present, 308 species are known in the state; 199 of these are present in the Sierra, and 37 are endemic to the Sierra (Morse 1993; J. C. Morse, personal database of published literature; N. A. Erman, personal database). The largest family of caddisflies in the state is Limnephilidae (63 species), and the second largest is Rhyacophilidae (59 species). These are also the largest and second-largest families in the Sierra. At lower elevations in warmer water, the family Hydroptilidae, the microcaddisflies, is diverse and poorly known. New species of Trichoptera will be discovered with more extensive surveys.

Diptera (True Flies, Insecta). Diptera is the most diverse order of all freshwater invertebrates. Within the Diptera, the family Chironomidae (midges) is the largest (Allan and Flecker 1993). These taxa are some of the most difficult to identify and are greatly understudied in the Sierra. Many species of Diptera are semiaquatic and spend most of their lives in the riparian area at the land-water interface.

Two small and unusual families of Diptera are discussed here, but it should not be assumed that these in any way represent the vast spectrum and diversity of Sierra Nevada aquatic Diptera. A third family of Diptera, the mosquitoes (Culicidae), is briefly discussed.

Blephariceridae (Net-Winged Midges, Insecta: Diptera). As was mentioned earlier, a family of true flies called the net-winged midges, Blephariceridae, is found exclusively in rush-

TABLE 35.2

Species estimates of selected aquatic invertebrate taxa in California and the Sierra Nevada. (Includes greater SNEP study area. See text for sources.)

Taxon	Total in California	Total in Sierra	Number Endemic to Sierra	Percentage Endemic to Sierra
Stoneflies (Plecoptera)	167	122	31	25
Alderflies (Megaloptera: Sialidae)	6	4	0	0
Dobsonflies (Megaloptera: Corydalidae)	11	7	?	?
Caddisflies (Trichoptera)	308	199	37	19
Net-winged midges (Diptera: Blephariceridae)	16	11	1	9
Mountain midges (Diptera: Deuterophlebeiidae)	6	4	1	25
Snails, clams (Mollusca)	?	40	8	20
Fairy shrimp, brine shrimp (Anostraca)	23	10	1	10

ing mountain streams, primarily in the western United States. The larvae have suction-cup-like attachments on their abdominal segments. Sixteen species (in five genera) of these flies exist in California (to our present knowledge), more than in any other state (Hogue 1973, 1987). Seven are endemic to California. Eleven are present in the Sierra Nevada (including Modoc County). All sixteen are present primarily in the Sierra and/or Coast Ranges. One species, however, is known from only one area in the northeastern corner of California (and the northwestern corner of Nevada).

Deuterophlebiidae (Mountain Midges, Insecta: Diptera). Another family of Diptera, the Deuterophlebiidae, or mountain midges, lives in much the same habitat as the net-winged midges and is present in the western mountains of North America and in eastern and central Asia. The larvae have rings of hooks on the abdominal prolegs to attach to rocks in the strongest currents. Only six species have been described in North America; four are present in the Sierra, and one is endemic to the Sierra (Courtney 1990).

Culicidae (Mosquitoes, Insecta: Diptera). Mosquito research was not analyzed for this chapter and is mentioned here only because mosquitoes have probably been studied more than any other aquatic invertebrate in California. Much mosquito habitat in the Sierra is in tree holes in the lower elevations and in snowmelt pools at higher elevations. Mosquito researchers think that reservoirs have not had a significant impact on mosquito distribution (B. Eldridge, Entomology Department, University of California, Davis, telephone conversation with the author, 1994). Mosquitoes prefer shallow water, often with aquatic vegetation, which is not the general condition of reservoirs. But discussion of possible changes in mosquito distribution caused by reservoirs is speculative because studies on this issue were not found for the Sierra. It is known that in other parts of the world reservoirs have caused epidemics of invertebrate-borne diseases and the spread of invertebrates undesirable to humans (Petts 1989).

Mollusca (Snails, Clams). Our information about mollusks is incomplete, but what is known is instructive. In 1981, thirty-two species of mollusks were known from the Sierra and northeastern California (Taylor 1981). None were endemic to the Sierra, but several had only one to a few populations in California and those were in the Sierra or northeastern California. Some of these populations were known to be extinct.

In recent years several new species of snails have been described in the Sierra. Eight recently described species in the genus *Pyrgulopsis* are considered endemic to the greater SNEP study area and are present in springs (Hershler 1994, 1995). *Pyrgulopsis* is the second most diverse genus of freshwater snails. *Pyrgulopsis* are widespread in the United States, and their range extends into southern Canada and northern Mexico. Seventy-two species were known and considered

valid as of 1995, and eight of those were found only in a few spring systems in the Sierra Nevada study area (Hershler 1994, 1995).

Future work on mollusks will likely reveal new species of aquatic snails in the Sierra as thorough and systematic surveys are conducted, particularly on the west side of the Sierra, and as "modern" taxonomic study is used (R. Hershler, letter to the author, March 16, 1995).

Anostraca (Fairy Shrimps and Brine Shrimps). At present there are twenty-three species (six genera) of Anostraca known in California (Belk and Brtek 1995; B. Helm, personal database and conversations with the author, November 1995, January 19, 1996). Ten species are in the greater SNEP study area. Of the nine species endemic to California, three are in the SNEP study area. One is *Artemia monica*, a brine shrimp endemic to Mono Lake.

Fairy shrimp are generally restricted to small, fishless ponds and especially to temporary systems (Dodson and Frey 1991). Habitats of species in the foothills are probably the most threatened. These are the areas under greatest pressure from human development (Duane 1996).

Unique, Small, and Unusual Aquatic Habitats

Permanent Habitats

Some Sierra Nevada habitats, such as springs, seeps, peatlands, and small first- and second-order streams, have such a high probability of containing rare and/or endemic invertebrates and have received so little attention and protection from resource agencies that they deserve special mention. These habitats are also most likely to contain imperiled amphibians, according to Jennings (1996). Spring streams are first-order streams (though the reverse is not necessarily true) and are often isolated in mountainous watersheds. Second-order streams (formed when two first-order streams join) can also be small and isolated in steep terrain. Both stream types are found at all elevations in the Sierra; thus they are not necessarily synonymous with headwater streams.

Springs, because of their near-constant temperatures, are refuges for species from previous climate regimes. Invertebrates of both warmer and colder periods are present in springs. Thus, species living in springs are often isolated populations, far out of their present geographic range, either at much higher or much lower elevations or latitudes. They may undergo further evolution in isolated habitats, leading to new species. The more stable and long-lasting the spring, the greater the species richness and the greater the likelihood of its containing endemic species (Erman and Erman 1995). Isolated upper watershed streams have a similar probability, as do peatlands connected to these systems. Many endemic and unusual species have been found in Sierra spring systems where such systems have been studied (Erman 1981, Erman 1984a; Erman and Erman 1990; Hampton 1988;

Hershler and Pratt 1990; Hershler 1994, 1995; Holsinger 1974; Kenk 1970; Szczytko and Bottorff 1987; Wiggins 1973; Wiggins and Erman 1987).

Important to the understanding and management of these systems is that they are different from one another and are in close contact with the surrounding land. In a study of fourteen springs within one second-order (upper watershed) stream basin we found a similarity of only 25% among caddisfly assemblages in the various springs (Erman and Erman 1990). The springs differed widely in species richness and species composition. Some endemic species were present in only one or a few springs. The management implications of these findings are that spring "types" cannot be identified and set aside to protect or preserve species. In other words, all Sierran springs need some protection or consideration in land management.

The greatest threat to spring species in the Sierra today is probably livestock grazing because it is so all-pervasive and invasive to small, wet areas (Erman, unpublished information, S. Mastrup, California Department of Fish and Game, telephone conversation with the author; D. Sada, private consultant, telephone conversation with the author). But logging, road building, water development, dynamiting, wildfire, and other impacts in the vicinity of springs can affect riparian vegetation, water volume, timing of flow, chemical concentrations, solar radiation, and temperature regimes, making springs and spring streams uninhabitable to species restricted to them (Erman and Erman 1990).

One of the more ironic uses of headwaters is a multiagency (California Department of Fish and Game, Nevada Department of Wildlife, U.S. Fish and Wildlife Service, Pacific Southwest Region of the U.S. Forest Service [USFS] and Intermountain Region of the USFS) plan for spring streams to serve as safe holding areas for endangered fish. Headwater areas are poisoned prior to becoming repositories for fish that, in some cases, were not historically present (Gerstung 1986). Threats to unusual and endemic invertebrates under such a fish management scheme are apparent. This plan is an example of the fallacy of single-species management.

Temporary Habitats

Temporary aquatic habitats have been largely unprotected in management plans. While not rare in the Sierra, these habitats can have unique assemblages of invertebrate species. Temporary streams, ponds, and springs are not always recognized as aquatic habitats during dry seasons or periods, another reason they may be overlooked. Invertebrates that use temporary habitats have been studied somewhat in the Sierra (Abell 1957; Erman 1987, 1989; Szczytko and Bottorff 1987) and in western Oregon (Anderson and Dieterich 1992; Dieterich and Anderson 1995). Some species are confined to temporary habitats and require a drying phase to complete their life cycles. Such invertebrates are often widespread because of dispersal mechanisms evolved in response to variable habitats but not always (e.g., Szczytko and Bottorff 1987).

In addition to their importance for unusual invertebrates, temporary habitats can be areas of high invertebrate biomass and important spawning areas for fish. In the Sagehen Creek basin (on the east side of the Sierra), an intermittent stream that dries completely in most summers is the spawning ground for one-third to one-half of the rainbow trout population of the stream system (Erman and Hawthorne 1976). During the dry season, this streambed is grass covered and unrecognizable as an aquatic habitat.

The greatest threats to temporary aquatic habitats at present are logging operations and roads. These habitats should be treated as if they were permanent in terms of management protection: they are the habitat for species restricted to temporary water. Furthermore, intermittent or ephemeral streams connected to a permanent stream system are just as capable of transporting sediment downstream into larger streams as are permanent streams.

Aquatic Invertebrates as Monitoring Tools and Habitat Indicators

Values of Broad Taxa Invertebrate Monitoring

Invertebrates have been used as monitoring tools to assess water conditions for more than eighty years. Much of our knowledge about aquatic invertebrates in the Sierra and in California has come from this use. When using invertebrates as indicators of aquatic conditions, ecologists study a large assemblage of species at a site but identify and group species only at a broader taxonomic level (genus, family, order, or class). As water conditions change, some groups rise in numbers, others fall, and some may disappear or appear. As was discussed earlier, detrimental change due to some impact is usually in the direction of a decrease in organism size and a decrease in taxa richness (higher numbers of small species, fewer large-sized species, and perhaps fewer species overall, depending on the degree of impact [Wallace and Gurtz 1986; Grubaugh and Wallace 1995]).

One continuing, long-term study of logging impacts on invertebrate assemblages has been conducted in California since 1973 (Erman et al. 1977; Newbold 1977; Roby et al. 1977, 1978; Newbold et al. 1980; Erman and Mahoney 1983; Mahoney 1984; Mahoney and Erman 1984a, 1984b; O'Connor 1986; Fong, 1991). This study was conducted on 62 stream sites initially; 22 were in the Sierra Nevada. Logged and control (reference) streams were blocked into groups of three or four by geographical location, stream size, vegetation, stream morphology, and geology. Aquatic invertebrate assemblages were used to determine the effects of logging on streams and the effectiveness of wide and narrow buffered areas in protecting streams. The measurements used were diversity indices that examined invertebrate taxa richness and evenness of numbers within taxa. The study has been continued for two decades to assess stream recovery.

Major findings of the study were that the numbers of midge larvae (Chironomidae), the small mayfly *Baetis* spp., and small nemourid stoneflies rose significantly following logging in streamside zones without buffer strips, a result of increased sediments, increased light from loss of riparian vegetation, and increased algal growth (Erman et al. 1977).

Discrete, local disturbance from failed road crossings associated with logging caused a decline in the number of taxa downstream. Where wide buffer areas (strips) were left unlogged along the streams, invertebrate communities showed little difference from those of unlogged streams. However, narrow buffers incompletely protected streams, and the narrower the buffer, the greater the impact of logging on stream invertebrate communities. High levels of stored sediment remained in the set of streams logged without buffers when the streams were resampled five to six years later and compared to control streams. Full recovery of invertebrate communities required nearly twenty years after the initial disturbance from logging. A significant footnote to these long-term studies is that the control streams gradually were lost as controls because of further logging in the watersheds. New controls for research were established where possible.

Other impact studies in the Sierra that have used aquatic invertebrates cover a broad spectrum of subjects. A few examples are (1) the potential effects of copper in water (Leland et al. 1989); (2) the potential effects of acid deposition (Jenkins et al. 1994) (note that increased acidity in Sierra waters is not currently considered a problem according to Cahill and colleagues (1996); (3) the effects of suction dredge gold mining (Harvey 1982; Somer and Hasler 1992); (4) the effects of channelization (Moyle 1976); (5) the effects of fish introductions (Reimers 1979); (6) the effects of visitor use on high-mountain lakes (Taylor and Erman 1980); and (7) the effects of wildfire (Roby and Azuma 1995).

Limitations and Cautions of Broad Taxa Invertebrate Monitoring

A great deal of time, effort, and money have been spent on sampling and analyzing invertebrates from stream-bottom substrates in the Sierra Nevada. Much of this work has been conducted or funded by state and federal resource agencies (the U.S. Forest Service, California State Fish and Game, Bureau of Land Management, etc.). Most if it is in unpublished reports in agency files. Examples of such studies have been examined for this report. Problems of incomplete understanding of invertebrate sampling and what it can currently tell us in California are evident in the conclusions drawn from some of these efforts. Nevertheless, such sampling, if conducted with care and adequate controls, can serve as baseline work for future studies and should not be abandoned but rather expanded and conducted at more sophisticated levels in the Sierra.

In California several entities (e.g., timber companies, state agencies, citizen groups) are beginning programs in invertebrate monitoring to assess watershed condition. Therefore, a few points seem worth reviewing in regard to future studies in the Sierra Nevada and what such studies can reveal.

The natural variability of invertebrate assemblages in streams is poorly known in the Sierra. One-time or one-season invertebrate sampling cannot reveal the "health" of a stream or the extent of cumulative impacts in a stream basin at present. Changes over time in taxa richness and other various indices can show the direction of effect (i. e., are conditions worsening or improving?). Invertebrate sampling is a useful tool in stream monitoring if controls (references) in time and/or space (depending on the objectives of the study) are established, and if the limitations of stream-bottom substrate sampling are understood. Many papers and several books have been written on this subject (e.g., Plafkin, et al. 1989; Rosenberg and Resh 1993; Loeb and Spacie 1994), and a complete airing of the issues is beyond the scope of this chapter.

Stream-bottom samples, for the most part, cannot tell us what species are in streams. Species can be determined only from sexually mature forms for most aquatic invertebrates. The large majority of aquatic invertebrates (both biomass and species) in Sierran streams are insects, and the sexually mature form of most aquatic insects is the flying terrestrial adult. These adults are not collected in stream-bottom samples. Furthermore, the large majority of species descriptions for aquatic insects are based on males only, and so male adults are needed to determine species (Wiggins 1990).

Species identification becomes critical when invertebrates are being sampled to determine if a project or if cumulative impacts from many uses are having a permanent effect on species composition. The current interest in biodiversity and curbing the loss of species demands more rigorous analysis of invertebrates than is presently being conducted by resource agencies, universities, and consultants in California.

Recent examples of unproven conclusions based on invertebrate sampling are found in state documents for the continued use of rotenone in streams (CDFG 1994a, 1994b, 1994c). These documents and their use of supporting studies reveal a confusion between species and overall aquatic invertebrate assemblages. Invertebrate studies cited in the first two documents (1994a, 1994b) monitored not changes in numbers of species but rather changes in the numbers of larger taxonomic categories (e.g., order, family, or genus) of invertebrates. And the following statement is made in the Kern River negative declaration (CDFG 1994c): "Aquatic invertebrate populations will become reestablished in a few months. The species composition may be different initially and may require several years to return to the pre-project status." However, pre-project species composition was not determined for invertebrates. (It is not possible for species to "return to the pre-project status" if they have become extinct, and with the study that was conducted, we have no way of determining that.) These comments are not meant as a criticism of the original studies, but rather are meant to serve as a cautionary note about drawing conclusions that were not tested and then applying those conclusions to management policies.

Sampling must be appropriate for the question being asked. Many studies use sampling and analysis techniques merely because they provide numbers, with little regard for what is being sampled. An example is the use of invertebrate drift sampling as a general monitoring method in streams, without an understanding of the many natural causes of invertebrate drift. Another example is the rather arbitrary assignment of taxa (usually genus or family) to a functional feeding group. (Functional feeding groups are broad categories based on how invertebrates feed and can indicate broad trends in energy inputs to a stream system.) But what such categories indicate about stream conditions is questionable in the Sierra without controls and baseline data. A second problem with functional feeding groups in California is that they are based on general and incompletely researched tables from textbooks (e.g., Merritt and Cummins 1984) rather than on actual food-habit studies of the species (usually unknown) in question. Functional feeding groups may vary among species within a genus (that is, different species within a genus may feed in different ways) and with larval instar (stage) (Thorp and Covich 1991).

An example, for purposes of illustration, is the genus, *Dicosmoecus*, perhaps the most studied caddisfly genus in the western United States probably because of the large size of the individuals. Three species of the genus are present in the Sierra, sometimes in the same stream system (Erman 1989), but they live in different habitats. *Dicosmoecus gilvipes* feeds predominantly on diatoms and fine particulate organic matter by scraping substrate material. *D. atripes* and *D. pallicornis* feed largely on vascular plants and animal materials and are considered generalized predator-shredders (Wiggins and Richardson 1982). *D. gilvipes* is tolerant of warm temperatures, unshaded streams, and sedimentation. The other two species are present in cooler, shaded, undisturbed areas. The larvae of the three species are difficult to separate and are probably often confused. Conclusions about habitats or impacts based on larval identification or presumed functional feeding group could be quite misleading in this case.

A second example is of two caddisfly species with overlapping distributions in small Sierran streams, *Farula praelonga* and *Neothremma genella* (Trichoptera:Uenoidae). These two species, though in different genera, are difficult to separate as larvae except where they have already been studied (Wiggins and Erman 1987). *F. praelonga* reaches larval maturity in the winter and emerges as an adult in early spring, while *N. genella* matures through the summer and emerges in the autumn. Both feed by scraping diatoms from rocks. *F. praelonga*, the more rare of the two species, is most abundant in shaded areas with constant temperatures near spring stream sources, whereas *N. genella* reaches larger population numbers somewhat farther downstream from the source and in more open areas (Erman and Mahoney 1983; Wiggins and Erman 1987). Therefore, these two species, though in the same functional feeding group, would be affected differently by land management that, for example, opened the riparian canopy or changed water temperature.

Correct identification even of genera or families of invertebrates requires expertise, a knowledge of invertebrates, good up-to-date taxonomic keys, and knowledge of how to use the keys. Reference collections (sometimes called voucher specimens) are necessary to confirm identities of invertebrates from past studies. Taxonomy changes as groups of organisms are revised by experts. Without preserved specimens, studies from ten or twenty years ago become questionable, and there is no way to verify whether or not the taxa were correctly identified initially. These taxonomic changes or misidentifications probably would not affect the results of impact studies but would affect our knowledge of changes in species diversity, that is, of whether a species has disappeared over time.

Many habitat factors affect natural variability. Invertebrate assemblages can change rapidly over rather short distances if there are changes in light, temperature, substrate, water chemistry, elevation, and so on. An example from a study of Sierran spring streams illustrates this point. In one undisturbed spring stream, caddisfly (Trichoptera) species similarity between the spring source and a point 270 m (886 ft) downstream was only 38% (using Jaccard's index of similarity [Pielou 1984]). At 450 m (1,476 ft) downstream, species similarity with the spring source dropped to 20%. In another nearby spring with more water, less light, and lower temperature, caddisfly species were replaced less rapidly: similarity was 40% at 1000 m (0.6 mi) downstream and fell to 22% at 1,800 m (1.1 mi) downstream. Results were based on adult emergence traps, operated for a year on the first spring stream and for nineteen weeks through summer and autumn on the second spring stream. These two springs are near each other (about 1,600 m [1 mi] apart) in the same second-order stream basin and emerge from the same hillside, and yet the species similarity between the two spring systems was only 28% (Erman 1992b). Both were relatively protected from anthropogenic influence and had been for many years prior to the study.

This example may be somewhat dramatic because of rapid physical changes in small stream systems, but it nevertheless illustrates natural species replacement over a stream gradient and natural differences among nearby small, upper watershed streams.

If natural variation over a sampling gradient is not determined or accounted for, it can result in a study either underestimating or overestimating impacts to the invertebrate assemblage. Habitats already undergoing significant impacts may be selected as controls (references) and, thereby, underestimate the effects of some activity or project on the aquatic biota. Or, conversely, habitats naturally low in species (snowmelt streams, variable springs) may be considered degraded when they are not.

A recent survey of thirty-one Sierra Nevada streams from the north to the south Sierra was unable to detect the effect of cumulative impacts on invertebrate assemblages (based on one-time sampling at all but two sites), because natural variation over so broad an environmental gradient masked the effects within stream basins (Hawkins et al. 1994). An earlier study (Erman et al. 1977; Newbold 1977) had also concluded that natural variation rather than logging or disturbance effects accounted for variation in invertebrate assemblages when data were analyzed using multivariate analysis. However, in the earlier study, the effects of logging and buffer strips were clearly evident when streams were grouped by treatments and controls in the same geographic area. Further analysis of the Hawkins team's work must wait until the invertebrate data are published.

In conclusion, many levels of aquatic invertebrate monitoring are available for assessing environmental changes. The impacts on invertebrate assemblages of many land and water uses are known, and major changes have likely occurred in invertebrate assemblages in Sierra waters. But at present, we have little baseline information in the Sierra to know whether aquatic invertebrate species are being lost.

General Status of Aquatic Habitats

Currently, aquatic habitats are the most altered and threatened biotic communities in the state (Jensen et al. 1990). Recent forest plans contain reviews of conditions in the national forests. A few summaries are given here as examples of conditions, but no attempt has been made to review the aquatic analysis of all the forest plans. The plan for the Plumas National Forest (USFS 1988) found that one-third of the running-water fish habitat in the forest was in poor condition, 78% of it in small streams. Nearly half, 47.6%, of the small stream acreage was in poor condition. Only 20% of all running-water fish habitat in the Plumas National Forest was in good condition, according to the Forest Service's own assessment.

In the plan for the Stanislaus National Forest (USFS 1992) only two of sixteen watersheds were in very good condition. "Fair" and "poor" watersheds were lumped together, perhaps suggesting that there were more poor than fair watersheds. Analysis of aquatic habitat was less thorough in the Stanislaus plan than in the Plumas plan. Water projects were omitted from the discussion of cumulative watershed effects. Impacts of livestock grazing on streams were not analyzed in the plan, although 82% of the forest was grazed.

In the Modoc National Forest, 78% of riparian areas were in fair or poor condition in 1988 (U.S. General Accounting Office 1988).

The focus of resource management on game fish production poses a significant conflict with invertebrate diversity in Sierra waters (see also Knapp 1996; Moyle et al. 1996). Environmental assessments for projects of many types (e.g., hydroelectric projects, rotenone projects, proposed timber harvest operations, hydraulic mining regulations, Board of Forestry rules) have been reviewed for this chapter. None have contained adequate or realistic assessments of impacts to aquatic invertebrate communities; in most, there were no assessments. Projects are analyzed based on whether or not game fish (usually brown or rainbow trout) will be affected. Money and resources are directed toward that analysis objective. Little effort is made by state and federal agencies to protect species of no known economic value or species with few human defenders. More significant, however, is the apparent lack of understanding of the complex physical, chemical, and ecological processes and cycles that interact to determine the fate of biotic communities in Sierra aquatic habitats.

These assessments are not encouraging with regard to present trends in Sierra Nevada aquatic environments, but by admitting the problems and analyzing how they occurred, we can move on to restore degraded habitats and prevent the same problems in the habitats that are still in good condition.

Institutional Responsibilities

An assessment of aquatic habitats in the Sierra must include an assessment of the institutional management of those habitats. Many state and federal agencies have jurisdiction over the streams and wetlands of the Sierra, whether the land is privately or publicly owned. Such agencies as the California State Water Resources Control Board (the state water board) and its regional water quality boards, the Water Resources Agency, the California Department of Forestry, the Board of Forestry, the California Department of Fish and Game, Caltrans, the Fish and Wildlife Service, the Federal Energy Regulatory Commission, the U.S. Forest Service, the Bureau of Land Management, the Bureau of Reclamation, the Army Corps of Engineers, and many local county and city planning agencies all have authority and responsibility, regulations and laws governing Sierra waters and riparian areas. Evaluating the performance and effectiveness of these agencies is essential to improved watershed protection. How well are our present extensive regulations and laws working? How well are they obeyed and enforced? Do agencies communicate with one another and coordinate their efforts? Are agency decisions based on current scientific knowledge? Is continuing education encouraged within the agency? Do agencies recognize and admit resource problems and have the will to change? Do agencies evaluate their own past performance and effectiveness? How do they view the California Environmental Quality Act and the National Environmental Policies Act? Are agencies following the spirit as well as the letter of the law? Do they make decisions in an open and democratic manner? Do they welcome public input?

The answers to these questions are connected to and are crucial to the present and future status of the biota in Sierra waters.

FUTURE NEEDS

Future needs for the study and protection of aquatic and riparian habitats and, by implication, the status of aquatic invertebrates in the Sierra are in the areas of research, management, and institutional evaluation. The following recommendations are derived largely from the issues discussed earlier in the chapter.

Establish Reference Streams and Baseline Data

Aquatic habitat research for the Sierra should include the establishment of undisturbed reference streams and other aquatic sites to be monitored over time (controls). Streams in national parks may be the only reference sites remaining in the Sierra that are close to "natural" or undisturbed conditions, but they do not represent the full range of vegetation, elevation, and other Sierra Nevada stream conditions. We need to establish control sites in many parts of the Sierra, and this goal will require cooperation among many agencies.

We need complete aquatic invertebrate inventories and surveys, especially in undisturbed (or nearly undisturbed) sites to compare with disturbed (managed?) areas. University field stations could contribute substantially to this effort if they would make the commitment to undertake surveys at the species level. Few Sierra field stations have attempted to inventory aquatic invertebrate species. Where inventories have been conducted, they have been the specific interest of individual researchers and not a concerted university field station effort.

Improve Monitoring

Monitoring of stream invertebrates could be conducted at a more knowledgeable and meaningful level. Sampling to determine species and verification by taxonomic experts could make a significant contribution to our baseline knowledge of freshwater invertebrate diversity in California. Studies must be reproducible for long-term biomonitoring. We may not have the institutional organization in California at present to accomplish this level of research. California needs a natural history survey modeled after those of some eastern states (i.e., the Illinois Natural History Survey, and the Ohio Biological Survey). In some states these organizations are supported by private funding. An expanded role for the California Academy of Sciences could be explored in this regard.

Consider All Biota and All Impacts

We need better, more credible analysis of impacts to the entire biota. There are dangers in single species or even single-taxon management. There are also dangers in single-project review. The impacts of small hydroelectric projects or rotenone poisoning or grazing allotments or logging operations must be assessed in their entirety throughout the Sierra and for their cumulative and interactive impacts with one another. We need waterscape as well as landscape analysis.

Mechanical, species-specific means of removing unwanted exotic species should be encouraged. Chemical and mechanical methods that indiscriminately kill many species should be discouraged.

Value Citizen Groups

Citizens groups interested in watershed monitoring could be (and are) involved in identifying cumulative impacts; resource agencies should welcome this enormous source of energy and local expertise.

Recognize Problems with Reserves

We probably cannot protect aquatic diversity by setting aside reserves or key watersheds. We do not have the information at this time to determine what areas could serve as reserves for aquatic invertebrates, and it is unlikely that we ever will. We do not know the minimum habitat required to maintain genetic viability of aquatic invertebrate species. Rare and endemic aquatic invertebrates likely occur in every watershed in the Sierra, making every watershed "key" for some species. Endemic aquatic invertebrates are isolated in smaller streams and other small aquatic habitats throughout the Sierra. River basins are continuous systems; what happens upstream affects the downstream biota. Setting aside a piece of a stream or watershed for protection is not a long-term solution, though it may have some immediate benefits. Influences outside the boundaries of reserves (ground-water pumping, air pollution, changes in the ozone layer, exotic species, diseases, burgeoning human population) require us to consider issues far beyond the boundaries of reserves or watersheds and even beyond the Sierra. In short, reserves or key watersheds give a false sense of security about species conservation. Our best hope is to improve analysis, monitoring, and management; protect unspoiled areas; and work toward protection and restoration of all watersheds.

Concentrate on What We Can Change

We must concentrate on what we can change and what we know is having a negative impact on aquatic systems. Sedimentation from logging, roads, livestock grazing, construction of many kinds (housing, ski resorts, hydroelectric projects), and mining is a large problem in the Sierra and causes significant changes in invertebrate assemblages, as was discussed earlier. Reducing sediment in aquatic systems should become a major objective of resource agencies. "Sediment load and deposition constitutes one of the most serious water quality problems throughout the world" (Osborne and Kovacic 1993).

Evaluate Pulse and Press Disturbances

There is a need to recognize the difference between "pulse" disturbances (limited and definable duration) and longer-induration "press" disturbances (Bender et al. 1984; Niemi et al. 1990). Not surprisingly, streams recover more rapidly from pulse disturbances. But recognizing when pulse disturbances may become a continuous press on the system is important. For example, a logging operation that temporarily increases stream sediment or light is a pulse impact from which the system likely will recover within a few years. But continuing logging operations throughout a basin, or a broad network of roads, or a reservoir, or an old mining scar carrying sediment into a stream year after year, or continuous livestock grazing, or all of these together in a watershed become press disturbances and may irrevocably alter habitat and the biota.

Protect Upper Watersheds and Small Aquatic Habitats

Upper watershed streams and small aquatic habitats need far more protection than they are currently receiving. Buffer areas should be increased and should be dependent on landscape factors (slope, soils, geology) as well as stream size. In other words, the steeper the slope, the greater the buffer area should be regardless of stream size. The smaller the water body, the closer its connection to the surrounding land, and the more likely it is to be damaged by activities on the land. Present logging buffers required on private land for small streams are woefully inadequate, but also inadequate are buffers for small streams on public land. To protect the watershed, we must protect the headwaters.

Riparian areas are critical to aquatic habitats and the aquatic biota, and conversely the aquatic biota is a critical food supply to terrestrial animals that inhabit riparian areas. Riparian areas should not be abrupt and isolated zones, as they are presently in many logged areas, but should grade gradually into upland areas.

Protect Temporary Aquatic Habitats

Temporary water should receive the same protective safeguards and buffer areas as permanent water. There is as much biological justification for protecting temporary water as there is for protecting permanent water.

Reduce Total Roaded Area

Total roaded area should be reduced. Road construction around wetland and riparian areas needs more careful scrutiny.

Reduce, Eliminate, or Change Livestock Grazing

Cattle should be eliminated or greatly reduced in riparian and aquatic areas. Sheep should be moved rapidly through wet meadows and spring areas. No grazing of livestock should occur in peatland fens.

Restore Habitat Where Possible

Restoration should focus on eliminating the source of a problem. Some impacts to the Sierran waterscape are beyond complete restoration, but partial restoration may be possible. For example, large reservoirs are permanent and have an enormous impact on river systems; nevertheless, hydrologic regimes can be altered to more normal flows. Some streams, buried by mining spoils, may be beyond recovery.

Do Not Manage Riparian Areas for Fire Protection

Wildfire cannot be anticipated in riparian areas. Therefore, measures taken to prevent wildfire in riparian habitats may cause more harm than good by adding road systems and sediment, by decreasing wood and downed snags, and by opening riparian areas and changing the moist microclimate. Aquatic habitats will be better protected by preventing the known damage caused by known and predictable human activities than by trying to fire proof riparian areas.

ACKNOWLEDGMENTS

Many people have contributed to this chapter by supplying information and lists of species and studies. I thank Rich Bottorff for compiling the first draft of the Plecoptera of California and the Sierra Nevada and Richard Baumann, Brigham Young University, and Bill Stark, Mississippi College, for verification and additions to the list. John Morse, Clemson University, graciously contributed a list of Trichoptera of California from his database, from which estimates of endemic species were made. Robert Hershler, National Museum of Natural History, Smithsonian Institution, provided information on Hydrobiidae snails. Brent Helm, of Jones and Stokes, provided distribution of Anostraca from his database. I am grateful to Darrell Wong, California Department of Fish and Game, Region 5, for contributing studies and data from agency files, and to Ken Roby, U.S. Forest Service, for a list of types and locations of aquatic invertebrate studies. Dave Herbst, Sierra Nevada Aquatic Research Laboratory (SNARL), provided manuscripts and lists of studies from SNARL. Charles Goldman, University of California, Davis, gave us recent Lake Tahoe publications of invertebrate work. Theo Light conducted the library database search of invertebrate work in the Sierra Nevada.

Conversations with Lynn Decker, Sonke Mastrup, Don Sada, Judy Li, Rich Bottorff, and Ron Cole were helpful in assessing specific resource, habitat, and taxonomic issues.

I thank the following people for their helpful comments on the first draft of this paper: David Graber, John Hopkins, Roland Knapp, John Menke, Peter Moyle, Eric Gerstung, and Robert Motroni.

REFERENCES

Abell, D. L. 1957. *An ecological study of intermittency in foothill streams of central California.* Ph.D. diss., University of California, Berkeley.

Allan, J. D., and A. S. Flecker. 1993. Biodiversity conservation in running waters. *Bioscience* 43 (1): 32–43.

Anderson, N. H. 1976. *The distribution and biology of the Oregon Trichoptera.* Technical Bulletin 134. Corvallis: Oregon State University, Agricultural Experiment Station.

Anderson, N. H., and M. Dieterich. 1992. The Trichoptera fauna of temporary headwater streams in western Oregon, U. S. A. In *Proceedings of the Seventh International Symposium on Trichoptera,* edited by C. Otto, 233–37. Backhuys Publishers, Leiden, Netherlands.

Beesley, D. 1996. Reconstructing the landscape: An environmental history, 1820-1960. In *Sierra Nevada Ecosystem Project: Final report to Congress,* vol. II, chap. 1. Davis: University of California, Centers for Water and Wildland Resources.

Belk, D., and J. Brtek. 1995. Checklist of the Anostraca. *Hydrobiologia* 298:315–353.

Bender, E. A., T. J. Case, and M. E. Gilpin. 1984. Perturbation experiments in community ecology. *Ecology* 65(1) : 1–13.

Bleich, V. C. 1992. History of wildlife water development , Inyo County, California. In *The History of Water: Eastern Sierra Nevada, Owens Valley, White-Inyo Mountains,* edited by C. A. Hall, V. Doyle-Jones, and B. Widawski, 100–106. *Symposium* vol. 4. Los Angeles: University of California, White Mountain Research Station.

Bottorff, R. L. 1990. *Macroinvertebrate functional organization, diversity, and life history variation along a Sierra Nevada river continuum,* California. Ph.D. diss., University of California, Davis.

Bovee, K. D. 1985. Evaluation of the effects of hydropeaking on aquatic macroinvertebrates using PHABSIM. In *Proceedings of the Symposium on Small Hydropower and Fisheries,* edited by F. W. Olson, R. G. White, and R. H. Hamre, 236–41. Bethesda, MD: American Fisheries Society.

Brock, E. W. 1960. Mutualism between the midge *Cricotopus* and the alga *Nostoc. Ecology* 41:474–83.

Brown, L. R. 1993. *Water-quality assessment of the San Joaquin-Tulare Basins, California: Analysis of available information on aquatic biology, December 1992.* Open File Report 93 (preliminary). Sacramento, CA: U.S. Geological Survey, National Water Quality Assessment Program.

Brusven, M. A., and K. V. Prather. 1974. Influence of stream sediments on distribution of macrobenthos. *Journal of the Entomological Society of British Columbia* 71:25–32.

Busby, D. G., and S. G. Sealy. 1979. Feeding ecology of a population of nesting yellow warblers. *Canadian Journal of Zoology* 57: 1670–81.

Buscemi, P. A. 1966. The importance of sedimentary organics in the distribution of benthic organisms. In *Organism-substrate relationships in streams,* edited by K.W. Cummins, C.A. Tryon, and R. T. Hartman, 79–86. Special Publication 4. Pittsburgh: University of Pittsburgh, Pymatuning Laboratory of Ecology.

Cahill T. A., J. J. Carroll, D. Campbell, and T. E. Gill. 1996. Air quality. In *Sierra Nevada Ecosystem Project: Final report to Congress,* vol. II, chap. 48. Davis: University of California, Centers for Water and Wildland Resources.

Cairns, J. Jr. and J.R. Pratt. 1993. The history of biological monitoring using benthic macroinvertebrates. In *Freshwater biomonitoring and benthic macroinvertebrates.* Edited D. Rosenberg, and V. H. Resh, 10–27. New York: Chapman and Hall.

California Department of Fish and Game (CDFG). 1994a. *Draft programmatic environmental impact report (subsequent) for rotenone use for fisheries management, California.* Sacramento: California Department of Fish and Game

———. July 1994b. *Final programmatic environmental impact report (subsequent) for rotenone use for fisheries management, California.* Sacramento: California Department of Fish and Game.

———. 1994c. *Proposed negative declaration for South Fork Kern River golden trout restoration, South Fork Kern River, Inyo, California.* Sacramento: California Department of Fish and Game.

California State Lands Commission. 1993. *California's rivers: A public trust report.* Sacramento: California State Lands Commission.

Chutter, F. M. 1969. The effects of silt and sand on the invertebrate fauna of streams and rivers. *Hydrobiologia* 34:57–76.

Cook, S. K., and R. L. Moore. 1969. The effects of a rotenone treatment on the insect fauna of a California stream. *Transactions of the American Fisheries Society* 98 (3): 539–44.

Cordone, A. J., and D. W. Kelley. 1961. The influences of inorganic sediment on the aquatic life of streams. California Fish and Game 47 (2): 189–228.

Courtney, G. W. 1990. Revision of Nearctic mountain midges (Diptera: Deuterophlebiidae). *Journal of Natural History* 24:81–118.

DeDecker, M. 1992. The death of a spring. In *The History of Water: Eastern Sierra Nevada, Owens Valley, White-Inyo Mountains,* edited by C. A. Hall, V. Doyle-Jones, and B. Widawski, 223–26. Symposium vol. 4. Los Angeles: University of California, White Mountain Research Station.

Degan, D. J. 1973. Observations on aquatic macroinvertebrates in a trout stream before, during, and after treatment with antimycin. Master's thesis, University of Wisconsin, Stevens Point.

Dieterich, M., and N. H. Anderson. 1995. Life cycles and food habits of mayflies and stoneflies from temporary streams in western Oregon. *Freshwater Biology* 34: 47–60.

Disney, R. H. L. 1989. Does anyone care? *Conservation Biology* 3 (4): 414.

Dodson, S. I., and D. G. Frey. 1991. Cladocera and other Branciopoda. In *Ecology and classification of North American freshwater invertebrates,* edited by J. Thorp and A. Covich. Academic Press, Inc. San Diego: Harcourt Brace Jovanovich.

Duane, T. P. 1996. Human settlement, 1850-2040. In *Sierra Nevada Ecosystem Project: Final report to Congress,* vol. II, chap. 11. Davis: University of California, Centers for Water and Wildland Resources.

Ehrenfeld, D. 1989. Is anyone listening? *Conservation Biology* 3 (4): 415.

Elser, J. J., C. Junge, and C. R. Goldman. 1994. Population structure and ecological effects of the crayfish *Pacifastacus leniusculus* in Castle lake, California. *Great Basin Naturalist* 54 (2): 162–69.

Erman, D. C. 1973. Upstream changes in fish populations following impoundment of Sagehen Creek, California. *Transactions of the American Fisheries Society* 102: 626–29.

———. 1986. Long-term structure of fish populations in Sagehen Creek, California. *Transactions of the American Fisheries Society* 115:682–92.

Erman, D. C., and N. A. Erman. 1975. Macroinvertebrate composition and production in some Sierra Nevada minerotrophic peatlands. *Ecology* 56:591–603.

Erman, D. C., and V. M. Hawthorne. 1976. The quantitative importance of an intermittent stream in the spawning of rainbow trout. *Transactions of the North American Fisheries Society* 105 (6): 675–81.

Erman, D. C. and D. Mahoney. 1983. Recovery after logging with and without bufferstrips in northern California. Contribution 186. Davis: University of California, California Water Resources Center.

Erman, D. C., J. D. Newbold, and K. B. Roby. 1977. Evaluation of streamside bufferstrips for protecting aquatic organisms. Contribution 165. Davis: University of California, California Water Resources Center.

Erman, N. A. 1981. Terrestrial feeding migration and life history of the stream-dwelling caddisfly, *Desmona bethula* (Trichoptera: Limnephilidae). *Canadian Journal of Zoology* 59 (9): 1658–65.

———. 1984a. The use of riparian systems by aquatic insects. In *California Riparian Systems: Ecology, Conservation, and Productive Management*, edited by R. E. Warner and K. M. Hendrix, 177–82. Berkeley and Los Angeles: University of California Press.

———. 1984b. The mating behavior of *Parthina linea* (Trichoptera: Odontoceridae), a caddisfly of springs and seeps, 131–136. In *Proceedings Fourth International Symposium on Trichoptera*, edited by J. C. Morse, 131–36. Series Entomologica 30. The Hague, Netherlands: Dr W. Junk.

———. 1987. Caddisfly adaptations to the variable habitats at the land-water interface, 275–279. In *Proceedings of the Fifth International Symposium on Trichoptera*, edited by M. Bournaud and H. Tachet, 275–79. Dordrecht, Netherlands: Dr W. Junk.

———. 1989. Species composition, emergence, and habitat preferences of Trichoptera of the Sagehen Creek basin, California, U.S.A. *Great Basin Naturalist* 49 (2): 186–97.

———. 1992a. Aquatic invertebrates as indicators of biological diversity. In *Proceedings of Symposium on Biodiversity of Northwestern California*, technical coordination by R. R. Harris and D. C. Erman, edited by H. M. Kerner, 72–78. Report 29. Berkeley: Cooperative Extension and Wildland Resources Center.

———. 1992b. Factors determining biodiversity in Sierra Nevada cold spring systems. In *The History of Water: Eastern Sierra Nevada, Owens Valley, White-Inyo Mountains*, edited by C. A. Hall, V. Doyle-Jones, and B. Widawski, 119–27. Symposium vol. 4. Los Angeles: University of California, White Mountain Research Station.

Erman N. A., and D. C. Erman. 1990. Biogeography of caddisfly (Trichoptera) assemblages in cold springs of the Sierra Nevada (California, USA). Contribution 200. Davis: University of California, California Water Resources Center.

———. 1995. Spring permanence, Trichoptera species richness, and the role of drought. Biodiversity of Aquatic Insects and Other Invertebrates in Springs, L. C. Ferrington, Jr. (ed.). *Journal of the Kansas Entomological Society* 68 (2, supplement): 50–64.

Erman, N. A., and C. D. Nagano. 1992. A review of the California caddisflies (Trichoptera) listed as candidate species on the 1989 Federal "Endangered and threatened wildlife and plants: Animal notice of review." *California Fish and Game* 78 (2): 45–56.

Evans, E. D. 1984. A new genus and a new species of a dobsonfly from the far western United States (Megaloptera: Corydalidae). *Pan-Pacific Entomology* 60: 1–3.

Fleischner, T. L. 1994. Ecological costs of livestock grazing in western North America. *Conservation Biology* 8 (3): 629–44.

Flint, O. S. 1965. The genus *Neohermes*. *Psyche* 72: 255–63.

Flint, R. W. and C. R. Goldman. 1975. The effects of a benthic grazer on the primary productivity of the littoral zone of Lake Tahoe. *Limnology and Oceanography* 20:935–44.

Fong, D. R. 1991. Logging-related influences on stream habitat and macroinvertebrate community characteristics in northern California. Master's thesis, University of California, Berkeley.

Fritts, H. C., and G. A. Gordon. 1980. Annual precipitation for California since 1600 reconstructed from western North American tree rings. Agreement B53367. Sacramento: California Department of Water Resources.

Gerstung, E. R. 1986. Fishery management plan for Lahontan cutthroat trout (Salmo clarki henshawi) in California and western Nevada waters. Administrative Report. Sacramento: California Department of Fish and Game, Inland Fisheries.

Goldman, C. R., M. D. Morgan, S. T.Threlkeld, and N. Angeli. 1979. A population dynamics analysis of the cladoceran disappearance from Lake Tahoe, California-Nevada. *Limnology and Oceanography*. 24 (2): 289–97.

Gregory, S. V., G. A. Lamberti, D. C. Erman, K. V. Koski, M. L. Murphy, and J. R. Sedell. 1987. Influences of forest practices on aquatic production. In *Streamside management. Forestry and fishery interactions*, edited by E. O. Salo, and T. W. Cundy, 233–55. Contribution No. 57. Seattle: University of Washington, Institute of Forest Resources

Grubaugh, J. W., and J. B. Wallace. 1995. Functional structure and production of the benthic community in a Piedmont River: 1956–1957 and 1991–1992. *Limnology and Oceanography* 40 (3): 490–501.

Hampton, A. M. 1988. Altitudinal range and habitat of triclads in streams of the Lake Tahoe Basin. *American Midland Naturalist* 120 (2): 302–12.

Harvey, B. C. 1982. Effects of suction dredge mining on fish and invertebrates in California foothill streams. Master's thesis, University of California, Davis.

Hashagen, K. 1975. Non-game fish control: One view. In *Symposium on trout / non-gamefish relationships in streams*, edited by P. B. Moyle and D. L. Koch, 73–74. Reno: University of Nevada System, Center for Water Resources Research, Desert Research Institute.

Hawkins, C. P., J. P. Dobrowolski, L. M. Decker, J. N. Hogue, J. W. Feminella, T. Hougaard, and D. Glatter. 1994. *Cumulative watershed effects: An extensive analysis of responses by stream biota to watershed management*. Albany, CA: U.D. Forest Service, Pacific Southwest Forest and Range Experiment Station.

Herbst, D. B. 1990. Distribution and abundance of the alkali fly (*Ephydra hians*) Say at Mono Lake, California (USA) in relation to physical habitat. *Hydrobiologia* 197:193–205.

———. 1992. Changing lake level and salinity at Mono Lake: habitat conservation problems for the benthic alkali fly. In *The History of Water: Eastern Sierra Nevada, Owens Valley, White-Inyo Mountains*, edited by C. A. Hall, V. Doyle-Jones, and B. Widawski, 198–210. Symposium vol. 4. Los Angeles: University of California, White Mountain Research Station.

Herbst D. B., and T. J. Bradley. 1993. A population model for the alkali fly at Mono Lake: depth distribution and changing habitat availability. *Hydrobiologia* 267:191–201.

Hershler, R. 1994. *A review of the North American freshwater snail genus* Pyrgulopsis *(Hydrobiidae)*. Smithsonian Contributions to Zoology 554. Washington, DC: Smithsonian Institution.

———. 1995. New freshwater snails of the genus *Pyrgulopsis* (Rissooidea: Hydrobiidae) from California. *The Veliger* 38 (4): 343–73.

Hershler, R., and W. L. Pratt. 1990. A new *Pyrgulopsis* (Gastropoda: Hydrobiidae) from southeastern California, with a model for historical development of the Death Valley hydrographic system. *Proceedings of the Biological Society of Washington* 103 (2): 279–99.

Hogue, C. L. 1973. The net-winged midges or Blephariceridae of California. *Bulletin of the California Insect Survey* 15.

———. 1987. Blephariceridae. In *Flies of the Nearctic Region: Archaeodiptera and Oligoneura*, edited by G. C. D. Griffiths, 1–172. Vol. II, part 4. Stuttgart, Germany: E. Schweizerbart'sche Verlagsbuchhandlung (Nägele u. Obermiller).

Holsinger, J. R. 1974. *Systematics of the subterranean amphipod genus* Stygobromus *(Gammaridae), Part I: Species of the western United States*. Smithsonian Contributions in Zoology 160.

Hynes, H. B. N. 1984. The relationship between the taxonomy and ecology of aquatic insects. In *The ecology of aquatic insects*, edited by V. H. Resh and D. M. Rosenberg, 9–23. New York: Praeger.

Jackson, J. K., and S. G. Fisher. 1986. Secondary production, emergence, and export of aquatic insects of a Sonoran Desert stream. *Ecology* 67:629–38.

Jackson, J. K., and V. H. Resh. 1989. Distribution and abundance of adult aquatic insects in the forest adjacent to a northern California stream. *Environmental Entomology* 18 (2): 278–83.

Jenkins, T. M., Jr., R. A. Knapp, K. W. Kratz, S. D. Cooper, J. M. Melack, A. D. Brown, and J. Stoddard. 1994. *Aquatic biota in the Sierra Nevada: current status and potential effects of acid deposition on populations*. Final report, contract A932-138. Sacramento: California Environmental Protection Agency, Air Resources Board.

Jennings, M. R. 1996. Status of amphibians. In *Sierra Nevada Ecosystem Project: Final report to Congress*, vol. II, chap. 31. Davis: University of California, Centers for Water and Wildland Resources.

Jensen, D. B., M. Torn, and J. Harte. 1990. *In our own hands: A strategy for conserving biological diversity in California*. California Policy Seminar Research Report. Berkeley: University of California, Berkeley.

Jewett, S. G. 1963. A stonefly aquatic in the adult stage. *Science* 139:484–85.

Johnson, R. K., T. Wiederholm, and D. M. Rosenberg. 1993. Freshwater biomonitoring using individual organisms, populations, and species assemblages of benthic macroinvertebrates. In *Freshwater biomonitoring and benthic macroinvertebrates*, edited by D. Rosenberg and V. H. Resh, 40–158. New York: Chapman and Hall.

Johnson, W. C. 1975. Chemical control of non-game fish in the Russian River drainage, California. In *Symposium on trout/non-gamefish relationships in streams*, edited by P. B. Moyle and D. L. Koch, 47–61. Reno: University of Nevada System, Center for Water Resources Research, Desert Research Institute.

Kattelmann, R. 1996. Hydrology and water resources. In *Sierra Nevada Ecosystem Project: Final report to Congress*, vol. II, chap. 30. Davis: University of California, Centers for Water and Wildland Resources.

Kellert, S. R. 1993. Values and perceptions of invertebrates. *Conservation Biology* 7 (4): 845–55.

Kenk, R. 1970. Freshwater triclads (Turbellaria) of North America. II. New or little known species of *Phagocata*. *Proceedings of the Biological Society of Washington* 83 (2): 13–22.

———. 1972. *Freshwater planarians of North America. Biota of freshwater ecosystems, identification manual 1*. Washington, DC: Environmental Protection Agency, Water Pollution Control Research Series 18050 ELDO 2/72.

Kenk, R., and A. M. Hampton. 1982. Freshwater triclads (Turbellaria) of North America XIII. *Polycelis monticola*, new species from the Sierra Nevada Range in California. *Proceedings of the Biological Society of Washington* 95:567–70.

Kimsey, L. S. 1996. Status of terrestrial insects. In *Sierra Nevada Ecosystem Project: Final report to Congress*, vol. II, chap. 26. Davis: University of California, Centers for Water and Wildland Resources.

Knapp, R. A. 1996. Non-native trout in natural lakes of the Sierra Nevada: An analysis of their distribution and impacts on native aquatic biota. In *Sierra Nevada Ecosystem Project: Final report to Congress*, vol. III. Davis: University of California, Centers for Water and Wildland Resources.

Kondolf, G. M., R. Kattelmann, M. Embury, and D. C. Erman. 1996. Status of riparian habitat. In *Sierra Nevada Ecosystem Project: Final report to Congress*, vol. II, chap. 36. Davis: University of California, Centers for Water and Wildland Resources.

Leland, H. V., S. V. Fend, T. L. Dudley, J. L. Carter. 1989. Effects of copper on species composition of benthic insects in a Sierra Nevada, California, stream. 1989. *Freshwater Biology* 21 (2): 163–79.

Lenz, P. H., S. D. Cooper, J. M. Melack, D. W. Winkler. 1986. Spatial and temporal distribution patterns of three trophic levels in a saline lake. *Journal of Plankton Research* 8 (6): 1051–64.

Li, H., G. A. Lamberti, T. N. Pearsons, C. K. Tait, J. Li, and J. Buckhouse. 1994. Cumulative effects of riparian disturbances along high desert trout streams of the John Day Basin, Oregon. *Transactions of the American Fisheries Society* 123: 627–40.

Loeb, S. L., and A. Spacie (eds.). 1994. *Biological monitoring of aquatic systems*. Ann Arbor: Lewis Publishers, Ann Arbor.

Luedtke, R. J., M. A. Brusven, and F. J. Watts. 1976. Benthic insect community changes in relation to in-stream alterations of a sediment-polluted stream. *Melanderia* 23:21–39

Mahoney, D. L. 1984. Recovery of streams in northern California after logging with and without buffers. Ph.D. diss., University of California, Berkeley.

Mahoney, D., and D. C. Erman. 1984a. The role of streamside bufferstrips in the ecology of aquatic organisms. In *Proceedings California Riparian Systems*, edited by R.E. Warner and K. M. Hendrix, 168–76. Berkeley: University of California Press.

Mahoney, D., and D. C. Erman. 1984b An index of stored fine sediment in gravel bedded streams. *Water Resources Bulletin* 20:343–48.

Maslin, P., C. Ottinger, L. Travanti, and B Woodmansee. 1988. *A critical evaluation of the rotenone treatment of Big Chico Creek*. Report to California Department of Fish and Game. Chico: California State University, Department of Biological Sciences.

Menke, J., C. Davis, and P. Beesley. 1996. Rangeland assessment. In *Sierra Nevada Ecosystem Project: Final report to Congress*, vol. III. Davis: University of California, Centers for Water and Wildland Resources.

Merritt, R. W., and K. W. Cummins. 1984. *An introduction to the aquatic insects of North America*. Dubuque, IA: Kendall/Hunt.

Minshall, G. W., and P. V. Winger. 1968. The effect of reduction of streamflow on invertebrate drift. *Ecology* 49: 380–382.

Morgan, M. D. 1979. The dynamics of an introduced population of *Mysis relicta* (Loven) in Emerald Bay and Lake Tahoe, California-Nevada. Ph.D. diss., University of California, Davis.

———. 1980. Life history characteristics of two introduced populations of *Mysis relicta*. *Ecology* 61 (3): 551-61.

Morgan, M. D., S. T. Threlkeld, and C. R. Goldman. 1978. Impact of the introduction of kokanee *(Oncorhynchus nerka)* and opossum shrimp *(Mysis relicta)* on a subalpine lake. *Journal Fisheries Research Board Canada* 35 (12): 1572–79.

Morse, J. C. 1993. A checklist of the Trichoptera of North America, including Greenland and Mexico. *Transactions of the American Entomological Society* 119 (1): 47–93.

Mount, J. F. 1995. *California rivers and streams: The conflict between fluvial process and land use.* Berkeley and Los Angeles: University of California Press.

Moyle, P. B. 1976. Some effects of channelization on the fishes and invertebrates of Rush Creek, Modoc County, California. *California Fish and Game* 62 (3): 179–86.

Moyle P. B., R. M. Yoshiyama, and R. A. Knapp. 1996. Status of fish and fisheries. In *Sierra Nevada Ecosystem Project: Final report to Congress*, vol. II, chap. 33. Davis: University of California, Centers for Water and Wildland Resources.

Murphy, M. L., and W. R. Meehan. 1991. Stream ecosystems. In *Influences of Forest and Rangeland Management on salmonid fishes and their habitats*, edited by W. R. Meehan, 17–46. Special Publication 19. Bethesda, MD: American Fisheries Society.

Nelson, C. R., and R. W. Baumann. 1989. Systematics and distribution of the winter stonefly genus *Capnia* (Plecoptera: Capniidae) in North America. The *Great Basin Naturalist* 49 (3): 289–366.

Newbold, J. D. 1977. The use of benthic macroinvertebrates as indicators of logging impact on streams with an evaluation of buffer strip effectiveness. Ph.D. diss., University of California, Berkeley.

Newbold, J. D., D. C. Erman, and K. B. Roby. 1980. Effects of logging on macroinvertebrates in streams with and without buffer strips. *Canadian Journal of Fisheries and Aquatic Sciences* 37 (7): 1076–85.

Niemi, G. J., P. DeVore, N. Detenbeck, D. Taylor, A. Lima, J. Pastor, J. D. Yount, and R. J. Naiman. 1990. Overview of case studies on recovery of aquatic ecosystems from disturbance. *Environmental Management* 14: 571–88.

Norris, L. A., H. W. Lorz, S. V. Gregory. 1991. Forest chemicals. In *Influences of forest and rangeland management on salmonid fishes and their habitats*, edited by W. R. Meehan, 207–296. Special Publication 19. Bethesda, MD: American Fisheries Society.

O'Connor, M. D. 1986. Effects of logging on organic debris dams in first order streams in northern California. Master's thesis, University of California, Berkeley.

Osborne, L. L. and D. A. Kovacic. 1993. Riparian vegetated buffer strips in water quality restoration and stream management. *Freshwater Biology* 29: 243–58.

Petts, G. E. 1989. Perspectives for ecological management of regulated rivers. In *Alternatives in Regulated River Management*, edited by J. A. Gore and G. E. Petts, 3–24. Boca Raton, FL: CRC Press.

Pielou, E. C. 1984. *The interpretation of ecological data: a primer on classification and ordination.* New York: John Wiley.

Plafkin, J. L., M. T. Barbour, K. D. Porter, S. K. Gross, and R. M. Hughes. 1989. *Rapid bioassessment protocols for use in streams and rivers: Benthic macroinvertebrates and fish.* Report EPA/444/4-89-001.Washington, DC: U. S. Environmental Protection Agency, Office of Water.

Platts, W. S. 1978. Livestock interactions with fish and aquatic environments: Problems in evaluation. *Transactions of the 43rd North American Wildlife and Natural Resources Conference,* 498–504.

Purdue University. 1993. Bug zappers burn humans as much as insects. *Perspective* (Spring) 5.

Reimers, N. 1979. A history of a stunted brook trout population in an alpine lake: a lifespan of 24 years. *California Fish and Game* 65: 196–215.

Reiser D. W., M. P. Ramey, and T. A. Wesche. 1989. Flushing flows. In *Alternatives in Regulated River Management*, edited by J. A. Gore and G. E. Petts, 91–135. Boca Raton, FL: CRC Press.

Richards, R. C., C. R. Goldman, T. C. Frantz, and R. Wickwire. 1975. Where have all the *Daphnia* gone: The decline of a major cladoceran in Lake Tahoe, California-Nevada. *Verhandlungen International Verein, Limnologie* 19:835–42.

Richards, R., C. Goldman, E. Byron, and C. Levitan. 1991. The mysids and lake trout of Lake Tahoe: A 25-year history of changes in the fertility, plankton, and fishery of an alpine lake. *American Fisheries Society Symposium* 9:30–38.

Roby, K. B., D. C. Erman, and J. D. Newbold. 1977. *Biological assessment of timber management activity impacts and buffer strip effectiveness on National Forest streams in northern California.* Earth Resources Monograph 1. San Francisco: U.S. Forest Service.

Roby, K. B., and D. L. Azuma. 1995. Changes in a reach of a northern California stream following wildfire. *Environmental Management* 19 (4): 591–600.

Roby, K. B., J. D. Newbold and D. C. Erman. 1978. Effectiveness of an artificial substrate for sampling macroinvertebrates in small streams. *Freshwater Biology* 8:1–8.

Rosenberg, D., and V. H. Resh, eds. 1993. *Freshwater biomonitoring and benthic macroinvertebrates.* New York: Chapman and Hall.

Rostlund, E. 1948. Fishing among primitive peoples: A theme in cultural geography. *Yearbook of the Association of Pacific Coast Geographers* 10:26–32.

Sedell, J. and K. J. Luchessa. 1982. Using the historical record as an aid to salmonid habitat enhancement. In *Symposium on acquisition and utilization of aquatic habitat inventory information,* edited by N. B. Armantrout, 210–223. Bethesda, MD: American Fisheries Society, Western Division.

Sedell, J., and C. Maser. 1994. *From the forest to the sea: The ecology of wood in streams, rivers, estuaries, and oceans.* Delray Beach, FL: St. Lucie Press.

Shapiro, A. 1996. Status of butterflies. In *Sierra Nevada Ecosystem Project: Final report to Congress*, vol. II, chap. 27. Davis: University of California, Centers for Water and Wildland Resources.

Sheldon, A. L., and S. G. Jewett, Jr. 1967. Stonefly (Plecoptera) emergence in a Sierra Nevada stream. *Pan-Pacific Entomologist* 43:1–8.

Somer, W. L., and T. J. Hassler. 1992. Effects of suction-dredge gold mining on benthic invertebrates in a northern California stream. *North American Journal of Fisheries Management.* 12:244–52.

Stefferud, S. E. 1977. *Aquatic invertebrate monitoring, brown trout control program, South Fork Kern River.* Sacramento: California Department of Fish and Game.

Stewart, K. W., and B. P. Stark. 1988. *Nymphs of North American Stonefly Genera (Plecoptera).* N.p.: The Thomas Say Foundation, Entomological Society of America.

Stine, S. 1996. Climate, 1650–1850. In *Sierra Nevada Ecosystem Project: Final report to Congress,* vol. II, chap. 2. Davis: University of California, Centers for Water and Wildland Resources.

Stohlgren, T.J., and J.F. Quinn. 1992. An assessment of biotic inventories in western U.S. national parks. *Natural Areas Journal* 12 (3): 145–54.

Szczytko S. W., and R. L. Bottorff. 1987. *Cosumnoperla hypocrena,* a new genus and species of western Nearctic Isoperlinae (Plecoptera: Perlodidae). *Pan-Pacific Entomologist* 63 (1): 65–74.

Taylor, D. W. 1981. Freshwater mollusks of California: a distributional checklist. *California Fish and Game* 67: 140–163.

Taylor, T. P., and D. C. Erman. 1980. The littoral bottom fauna of high elevation lakes in Kings Canyon National Park. *California Fish and Game* 66 (2): 112–19.

Thorp, J., and A. Covich. 1991. *Ecology and Classification of North American Freshwater Invertebrates.* San Diego: Harcourt Brace Jovanovich.

U.S. Forest Service (USFS). 1988. *Plumas National Forest land and resource management plan: Final environmental impact statement.* San Francisco: U.S. Forest Service, Regional Office.

———. 1992. *Stanislaus National Forest land and resource management plan: Final environmental impact statement.* San Francisco: U.S. Forest Service, Regional Office.

U.S. General Accounting Office. 1988. *Public rangelands: Some riparian areas restored but widespread improvement will be slow.* GAO/RCED-88-105. Washington, DC: U.S. General Accounting Office.

Usinger, R.L. ed. 1956, 1963, 1968. *Aquatic insects of California.* Berkeley and Los Angeles: University of California Press.

Vale, T. R. 1980. Mono Lake, California: saving a lake or serving a city. *Environmental Conservation* 7 (3):190–92.

Wallace, J. B., and M. E. Gurtz. 1986. Response of *Baetis* mayflies (Ephemeroptera) to catchment logging, *American Midland Naturalist* 115 (1): 25–41.

Waters, T. F. 1995. *Sediment in streams: Sources, biological effects, and control.* Monograph 7. Bethesda, MD: American Fisheries Society.

Whiting, M. F. 1991a. A distributional study of *Sialis* (Megaloptera: Sialidae) in *North America. Entomological News* 102 (1): 50–56.

———. 1991b. New species of *Sialis* from southern California (Megaloptera: Sialidae). *Great Basin Naturalist* 51 (4): 411–13.

———. 1994. Cladistic analysis of alderflies of America north of Mexico (Megaloptera: Sialidae). *Systematic Entomology* 19:77–91.

Wiggins, G. B. 1962. A new subfamily of phryganeid caddisflies from western North America (Trichoptera: Phryganeidae). *Canadian Journal of Zoology* 40:879–91.

———. 1973. *New systematic data for the North American caddisfly genera* Lepania, Goeracea, and Goerita *(Trichoptera: Limnephilidae).* Life Science Contribution of the Royal Ontario Museum 91, Toronto: Royal Ontario Museum.

———.1990. Systematics of North American Trichoptera: present status and future prospect. In *Systematics of the North American insects and arachnids: status and needs,* edited by M. Kosztarab and C. W. Schaefer, 203–10. Information Series 90-1. Blacksburg: Virginia Polytechnic Institute and State University, Agricultural Experiment Station.

Wiggins, G. B., and J. S. Richardson. 1982. Revision and synopsis of the caddisfly genus Dicosmoecus (Trichoptera: Limnephilidae; Dicosmoecinae) *Aquatic Insects* 4 (4): 181–217.

Wiggins, G. B., and N. A. Erman. 1987. Additions to the systematics and biology of the caddisfly family Uenoidae (Trichoptera). *Canadian Entomologist* 119:867–72.

Wilcove, D. S., and M. J. Bean, eds. 1994. *The big kill: Declining biodiversity in America's lakes and rivers.* Washington, DC: The Environmental Defense Fund.

Wilson, E. O. 1987. The little things that run the world. (The importance and conservation of invertebrates). *Conservation Biology* 1 (4): 344–46.

Zeiner, D. C., W. F. Laudenslayer Jr., K. E. Mayer, and M. White, eds. 1988. *Amphibians and Reptiles.* Vol. I of *California's Wildlife.* Sacramento: Department of Fish and Game, California Statewide Wildlife Habitat Relationships System.

———. 1990a. *Birds.* Vol. II of *California's Wildlife.* Sacramento: Department of Fish and Game, California Statewide Wildlife Habitat Relationships System.

———. 1990b. *Mammals.* Vol. III of *California's Wildlife.* Sacramento: Department of Fish and Game, California Statewide Wildlife Habitat Relationships System.

G. MATHIAS KONDOLF
Department of Landscape Architecture
University of California, Berkeley

RICHARD KATTELMANN
Sierra Nevada Aquatic Research
 Laboratory
University of California, Santa Barbara

MICHAEL EMBURY
Sierra Nevada Aquatic Research
 Laboratory
University of California, Santa Barbara

DON C. ERMAN
Department of Wildlife, Fish, and
 Conservation Biology
University of California, Davis

36

Status of Riparian Habitat

ABSTRACT

Despite their ecological importance, riparian areas in the Sierra Nevada have been the subject of very little research and no systematic data collection at a scale adequate to directly evaluate the status of the resource over the entire range. In this chapter, we review the functioning and ecological importance of riparian areas, the effects of human activities on riparian areas, and the extent of these effects in the Sierra Nevada. Riparian areas in the Sierra Nevada have been directly removed or have had their functions impaired by gold mining, gravel mining, hydroelectric development, land clearance and diversions of water for irrigation, land drainage, phreatophyte removal programs, timber harvest, construction of roads and railroads, urbanization, livestock grazing, and groundwater abstraction. From a GIS (Geographic Information Systems) analysis of road influence on streams, we calculated the percentage of 100 by 100 m pixels containing streams that also contained a road, which we designated as the Road Influence Index (RII). RII values, a measure of stream length with a road within 100 m, range from 2% to 33%, with a median value of 14% for the Sierra Nevada. Aerial photographic analysis indicated that 121 of 130 study watersheds displayed obvious gaps in the riparian corridor, primarily from road and railroad crossings, timber harvesting, clearing of private lots, dewatering by dams and diversions, and livestock grazing. Examination of 1:100,000-scale topographic maps for the entire Sierra Nevada showed more than 150 gaps over 0.5 km long created by reservoirs and at least 1,000 km of riparian corridor eliminated by reservoir inundation. Management strategies to minimize effects on the riparian zone include buffer strips, flushing flows, and restoration of riparian habitat. Streamside management zones or land-use buffers may be used to filter pollutants and sediment from upland runoff and to provide adequate recruitment of organic matter to the channel. Deliberate release of high flows from reservoirs (flushing flows) may be used to mimic the effects of natural floods in maintaining bed substrate and active channel width. Riparian vegetation can also be replanted in sites from which it has been cleared.

INTRODUCTION

Riparian habitats are among the most ecologically productive and diverse terrestrial environments, by virtue of an extensive land-water ecotone, the diversity of physical environments resulting from moisture gradients, and a mosaic of habitats created by dynamic river changes (Naiman et al. 1993). Moreover, the importance of the physical and biological interchanges between aquatic and riparian habitats is increasingly recognized, so any consideration of aquatic habitat quality must account for the riparian conditions so influential upon the channel itself (Gregory et al. 1991). Riparian habitats are especially important in semiarid regions, where the availability of moisture and a cool, shaded microclimate gives these habitats an ecological importance disproportionate to their areal extent. For example, in the Inyo National Forest, riparian areas constitute less than 0.4% of the land area but are essential for at least one phase of life for about 75% of local wildlife species (Kondolf et al. 1987b). In this forest, many recreational activities for its annual seven million visitors are also concentrated in riparian zones (Federal Energy Regulatory Commission 1986). Of the total 401 Sierran species of mammals, birds, reptiles, and amphibians combined, 21% (84 species) depend on the riparian area near water, and of course many more use it occasionally or regularly to find food, water, and shelter (Graber 1996). Nearly one-quarter (24%) of

Sierra Nevada Ecosystem Project: Final report to Congress, vol. II, *Assessments and scientific basis for management options.* Davis: University of California, Centers for Water and Wildland Resources, 1996.

those dependent on the riparian community area are at risk of extinction (Graber 1996).

Until the 1960s, the ecological importance of riparian areas in mountain regions was largely undescribed in the scientific literature (Kauffman 1988). In the Sierra Nevada, little has been published on riparian areas per se, although the importance of riparian areas is implied by the habitat descriptions for many species. Interest in riparian areas in California has been growing over the past two decades, but most research has focused on the Central Valley or Coast Ranges. For example, in the proceedings of a conference held in 1988 on riparian habitat in California (Abell 1989), 46 papers concerned the Central Valley or Coast Range riparian systems and only 17 concerned Sierra Nevada (mostly eastern or southern) riparian systems.

In the 1980s, a proliferation of proposals for small hydroelectric developments generated a number of mostly sitespecific studies on the environmental effects of proposed hydroelectric projects (e.g., Taylor 1983; Harris et al. 1987; Kondolf et al. 1987b; Jones and Stokes Associates 1989; Smith et al. 1989; Nachlinger et al. 1989; Leighton and Risser 1989; Hicks 1995).

Land-management agencies have conducted studies of riparian areas as a component of other assessments or planning studies. Mono County is conducting detailed mapping of wetlands, including riparian areas (R. Curry, University of California, Santa Cruz, communication with R. Kattelmann, 1995). Riparian areas along streams tributary to Mono Lake have been studied by a National Academy of Sciences committee (National Research Council 1987), on behalf of parties to litigation over flow requirements for resident trout (Stromberg and Patten 1990), in support of a water rights adjudication (Stine 1991; State Water Resources Control Board 1994), and in related studies (Kondolf and Vorster 1993). The California Tahoe Conservancy is attempting to evaluate the health of riparian vegetation along streams tributary to Lake Tahoe using remotely sensed data and field observations (Manley 1995). In many cases, these site-specific studies have been sufficiently well funded and implemented that they provide valuable insights into the physical and ecological processes controlling the distribution and functioning of riparian vegetation. However, most have been concentrated in the Mono Basin, Lake Tahoe, or Kern River regions.

Attempts at a broader scale assessment of riparian conditions have been undertaken by the Bureau of Land Management (Myers 1987) and the U.S. Forest Service (e.g., U.S. Forest Service 1995). Unfortunately, inconsistencies in data collection, analysis, and reporting have inhibited the compilation of these various data into a coherent assessment of riparian conditions across the entire range. Moreover, many assessments, such as those undertaken by the national forests, have been conducted without the benefit of peer review of procedures or results, and some are based largely on subjective judgments of channel stability by nongeomorphologists and

thus contribute little to a scientifically based understanding of the status of riparian systems in the Sierra Nevada.

METHODS AND SCOPE

This chapter provides an overview of the functioning and ecological importance of riparian areas, the effects of human activities on riparian areas, and the extent of these impacts in the Sierra Nevada, based largely on a more detailed report (Kattelmann and Embury 1996). Although the continuity of riparian corridors was assessed from aerial photography of a sample of river systems over the entire Sierra Nevada, the effects of human activities upon riparian areas are merely inferred from the extent of human activities known to affect the extent or functioning of riparian vegetation. Direct measurement of riparian condition over a region as large as the Sierra Nevada was beyond the scope of this study.

A literature review was conducted on the ecological role of riparian areas, the physical conditions on which they depend, and the effect of human activities on riparian areas in general. These extensive references are summarized in tables (reviewed in more detail in Kattelmann and Embury 1996), and are generally not repeated in the text. Literature documenting physical and biological aspects of riparian areas in the Sierra Nevada was also reviewed, but this literature is relatively modest and does not reflect the full range of conditions found in the Sierra Nevada.

The courses of all rivers and streams appearing on U.S. Geological Survey (USGS) 1:100,000 topographic maps were examined to identify gaps greater than 0.5 km (0.3 mi) in the riparian corridor produced by reservoirs. The total length of inundated channel was estimated by assuming straight-line distances for channels under existing reservoirs.

More than 9,500 km (5,900 mi) of river and stream channel was examined on aerial photographs to identify gaps in the riparian corridor. Out of 694 Sierra Nevada Calwater Super-Planning Watersheds, as designated by the California Department of Forestry and Fire Protection (1996), 130 were selected for aerial photographic study. All blue-line channels appearing on USGS 1:100,000 scale maps were examined. Aerial photographs for most national forests were taken in 1991–93, but coverage for other watersheds dates back as far as 1981. Details of coverage, scale, and methods of assessment are presented in Kattelmann and Embury 1996. Aerial photographs provide little or no information on the condition of riparian vegetation below the canopy, and the small scale of the photos used limited the utility of this analysis to an assessment of canopy continuity (riparian fragmentation).

A more systematic analysis was conducted using a geographic information system (GIS) developed for the Sierra Nevada Ecosystem Project. For 141 Calwater Hydrologic Subareas (California Department of Forestry and Fire Protection

1996) in the Sierra Nevada study region (four subareas were omitted because they were reservoirs), the number of pixels (each 100 m by 100 m, or 1 ha [2.47 acres]) in which a road occurs was counted, and the number of pixels with a stream was counted. Sources of the digital road information were the U.S. Forest Service road layer of "system roads" for areas inside proclaimed boundaries of national forests and the Teale data center (1:100,000) for areas outside the proclaimed boundaries. Of the total number of pixels with streams, the percentage that also had roads was calculated, a statistic that can be restated as the percentage of stream length with a road within 100 m (328 ft) of the channel—a gross measure of the potential impact of roads upon streams. These percentages for each watershed were compiled for the entire Sierra Nevada and for the northern (north of Interstate 80), central (Merced River basin to I-80), southern (south of Merced River basin), and eastern (east of the divide, excluding Lake Tahoe) Sierra Nevada. For each data set, percentile values (10th, 25th, 50th, 75th, and 90th percentiles) were determined and box-and-whisker plots (modified from Tukey 1977) were generated to display the spread of values among individual watersheds.

We also convened a group of scientists familiar with riparian management issues in the Sierra Nevada to review an early draft of Kattelmann and Embury 1996, to discuss the topic, to contribute ideas and other published research to the review, and to consider how best to approach this broad subject. The comments received from this group were extremely helpful in preparing this chapter.

FINDINGS

Ecological Role of Riparian Areas

Riparian vegetation is vegetation associated with rivers, streams, and other aquatic systems (lakes, springs, seeps, wet meadows). The term has been variously defined, from meanings that restrict the term to vegetation occurring on the river banks (as implied by the Latin root *ripa*, bank) to more inclusive definitions that encompass floodplain and terrace vegetation as well. *Bottomland vegetation* is another term for the latter, more inclusive definition of riparian vegetation (e.g., Hupp 1986). Water-dependent vegetation found at springs and seeps is often referred to as riparian despite the lack of association with a stream or river. Our review and assessment has concentrated on the riparian areas associated with running water. Riparian vegetation is also distinguished as *obligate*, for species found only in riparian areas, and *facultative*, for species that commonly occur in riparian areas but that also occur in upland environments.

Individual riparian species are adapted to a range of conditions within the riparian zone, along gradients of water table depth, soil moisture, and frequency of disturbance. Characteristics typical of obligate riparian vegetation are dependence

on a high water table, tolerance to inundation and soil anoxia, tolerance to physical damage from floods, tolerance to burial by sediment, ability to colonize flood-scoured surfaces or fresh deposits, and ability to colonize and grow in substrates with few soil nutrients. The relative importance of these characteristics varies with the river system. In the Sierra Nevada, dependence on high water tables and ability to survive physical damage from high-velocity flood flows are important characteristics of riparian vegetation, whereas along the coastal plain rivers of southeastern North America, tolerance of prolonged inundation is more important.

The ecological importance of riparian areas derives from a range of attributes, such as moisture availability, structural complexity, linear continuity (for migration corridors), distinct microclimate (cooler in summer, protected in winter), diverse food resources (terrestrial and aquatic), and influence on aquatic habitat (table 36.1). Riparian vegetation has a greater influence on channel processes and aquatic habitat in smaller channels than in larger ones. The effect of roots in stabilizing banks, the role of large woody debris in channel processes, the importance of terrestrial food sources as opposed to autochthonous (within channel) food production, and the shading effect of bank vegetation are all relatively more important in small channels (Vannote et al. 1980).

Geomorphic and hydrologic processes and conditions important to riparian ecology include flood inundation, the physical effects of high-velocity flood flows, stream-groundwater interactions, and the extent and texture of alluvium and adjacent hill-slope soils (table 36.2). The relative importance of these physical controls, like that of vegetation, differs among riparian systems. Altering these controls can be expected to alter the distribution and structure of riparian vegetation.

Human Impacts on Riparian Areas

A wide range of human activities can affect riparian areas, either by direct removal of riparian vegetation or by altering the factors controlling the distribution and structure of riparian vegetation (table 36.3). The following paragraphs briefly review these impacts and consider their relative importance in the Sierra Nevada.

Gold mining has numerous effects on riparian vegetation, including the destruction of riparian bottomland forests for gold dredging, the damming and diversion of rivers, and increased sediment yield from hydraulic mining. Gold mining was extensive in the Sierra Nevada beginning in about 1850 (see Beesley 1996). To provide water and pressure for hydraulic mining, ambitious water diversion projects were undertaken, resulting in the dewatering of some reaches and the creation of new riparian habitats along artificial canals. The mining itself released more than 42 million m^3 (46 million yards^3) of sediment into steep canyons, burying existing vegetation before being flushed downstream for deposition in channels and on floodplains of rivers in the Central Valley

TABLE 36.1

Ecological attributes of riparian areas.

General Attribute	Specific Attributes	References
Moisture availability	Shallow water table supports phreatophytes	California State Lands Commission 1993
	Evapotranspiration, shading increase humidity	
	Moist environments for amphibians, reptiles	Reynolds et al. 1993; Jennings 1996
Structural complexity	Vegetation provides cover for wildlife, birds	
	Multiple plant canopies create multiple niches	Krzysik 1990
	Seasonal changes in deciduous vegetation	Reynolds et al. 1993
Periodic disturbance	Floods disrupt existing organisms, providing opportunities for pioneer species	Resh et al. 1988; Sparks et al. 1990; Junk et al.1989
Linear nature	Edge effect: terrestrial-aquatic ecotone	Schimer and Zalewski 1992
	Riparian zones serve as wildlife migration corridors	Thomas et al. 1979
Food resources	Diverse vegetation yields diverse foods	Cross 1988
	Diverse habitat harbors diverse prey	Raedeke et al. 1988
	Open water available for wildlife	
Microclimate	Shaded, cool, moist in summer	Raedeke et al. 1988
	Protected in winter: overwintering habitat	
Influences on aquatic habitat	Shading moderates water temperatures	Brown 1969
	Shading moderates algal growth	
	Plant materials and insects fall into stream, adding chemical energy and nitrogen	Cummins et al. 1989; Knight and Bottorff 1984
	Riparian zone "buffers" stream from upland	Erman and Mahoney 1983; Mahoney and Erman 1984
	Riparian vegetation stabilizes stream banks	Kondolf and Curry 1986

and San Francisco Bay (Gilbert 1917; Mount 1995). Along the Yuba River above Marysville, the "debris plain" built of these sediments exceeds 64 km^2 (40 mi^2) in area.

A later phase of mining involved dredgers. These reworked the natural floodplains or hydraulic mining debris and left behind elongated mounds of tailings, which are still largely unvegetated because their surfaces consist of open cobbles in which plants cannot become established. The dredgers required extensive, deep, relatively flat deposits to work, so they were concentrated in the lower Central Valley reaches of western Sierra Nevada rivers.

Gravel mining for construction aggregate from river channels and floodplains results in the direct removal of riparian vegetation for the creation of process yards, haul roads, and pits. Indirect effects of in-channel extraction typically include channel incision, which propagates both upstream and downstream, lowering the alluvial water table and inducing channel instability.

Gravel mining for construction aggregate is the largest mining industry in the state (see Diggles et al. 1996). More than 100 million metric tons are produced annually, virtually all from river channels and floodplains. Large gravel depos-

TABLE 36.2

Selected geomorphic and hydrologic processes in riparian areas.

Process	Physical Effect	Ecological Consequence	Reference
Flooding			
Inundation	Soil anoxia	Selects for plants tolerant of anoxia	Walters et al. 1980; Gill 1970
	Saturation of soil	Increases soil moisture	
High-velocity flow	Scour of seedlings	Prevents establishment of woody vegetation in channel	
	Physical damage to plants	Selects for tolerant plants	Sigafoos 1964
	Bank erosion and undercutting of mature vegetation	Creates new habitats for colonization	
Deposition	Burial of plants	Selects for tolerant plants	Sigafoos 1964
	Sand-gravel bar deposition	Selects for plants capable of colonizing sandy substrates	
	Fine-grained overbank deposition	Provides silty substrates	
Stream-Groundwater Interactions			
Drainage from hill slope	Maintains high water table	Supports vegetation independent of streamflow	
Bank storage	Recharges alluvial water table	Supports vegetation	Kondolf et al. 1987a
	Maintains base flow	Provides water downstream	

TABLE 36.3

Human activities, physical effects, and ecological consequences in riparian areas.

Activity and Potential Direct Physical Effects	Potential Ecological Consequences	References
Gold Mining		
Former floodplain forests reworked by placer mining into unvegetated dredger tailings	Riparian vegetation removed and replaced with unvegetated gravel	Clark 1970
Rivers and streams dammed and diverted through canals	Water stress in dewatered reaches, riparian vegetation established along canals and ditches	Averill 1946; Pagenhart 1969
Increased sediment from hydraulic mining debris leads to aggradation of sand and gravel in valley bottoms	Burial of existing vegetation	Gilbert 1917
Continued erosion from hydraulic mine sites	Elevated fine sediment loads affect aquatic biota	Marchetti 1994
Gravel Mining		
Direct removal of vegetation for gravel yards, processing plants, haul roads, pits	Riparian vegetation replaced by roads and industrial land use	Poulin et al. 1994
Mining-induced channel incision lowers alluvial water table	Increased mortality, decreased growth rate and crown volume in woody riparian vegetation	Kondolf 1994b; Scott et al. in press
Mining-induced channel instability results in increased bank erosion	Erosion of banks supporting riparian vegetation	Todd 1989
Mining tops of gravel bars ("skimming") lowers ground surface relative to water table	Riparian vegetation established in channel where water table was formerly too deep	Kondolf and Matthews 1993
Dams		
Reduced flood flows lead to reduced rate of channel migration	Reduced diversity of riparian habitats	Johnson 1992, 1994; Ligon et al. 1995; Hesse and Sheets 1993
Reduced flood flows eliminate frequent scour of active channel	Riparian vegetation encroaches into active channel	Williams and Wolman 1984; Bergman and Sullivan 1963; Brothers 1984
Increased base flows and raised alluvial water table	Waterlogging of vegetation	Parrish and Matthews 1993
Base flows reduced or eliminated, stream dries up	Riparian vegetation severely stressed or dies	Kondolf and Vorster 1993; Stine et al. 1984
Trapping of bedload sediments behind dam, release of sediment-starved water, channel incision	Alluvial water table drops and overbank flooding is less frequent due to channel incision	Williams and Wolman 1984
Reservoirs drown existing vegetation, fluctuating water levels may limit establishment of new vegetation along margins	Longitudinal continuity of riparian corridor interrupted	Hagan and Roberts 1973
Hydroelectric Generation		
Rivers and streams dammed and diverted through canals	Water stress in dewatered reaches, riparian vegetation established along canals and ditches	Harris et al. 1987; Kondolf et al. 1987b
Hydroelectric dams and associated canals, penstocks, power-houses, and access roads constructed within riparian zone	Riparian vegetation removed and replaced with roads and structures	Federal Energy Regulatory Commission 1986
Flow fluctuates rapidly to generate peak hydroelectric power	Rapid stage changes can lead to increased bank erosion	
Irrigation		
Water diverted from streams	Water stress in dewatered reaches, riparian vegetation established along canals and ditches	Erman 1992
Irrigation water may infiltrate, recharging groundwater	Excess irrigation water may support vegetation	Kondolf and Vorster 1993
Land Drainage		
Alluvial water table lowered by land drainage	Riparian plants desiccated	Hughes 1934
Land Clearance for Agriculture		
Removal of floodplain forest	Riparian vegetation removed and replaced with agricultural land	Katibah et al. 1984
Phreatophyte Removal		
Removal of riparian vegetation	Riparian vegetation removed, may require herbicides to prevent regrowth	Dunford and Fletcher 1947; Biswell 1989
Navigation		
Channel dredged, resulting in incision	Alluvial water table drops and overbank flooding is less frequent due to channel incision	Brookes 1988
Channel straightened and stabilized	Length, complexity, and dynamic nature of channel reduced	Brookes 1988
Timber Harvest		
Harvest of timber in riparian areas, removal of trees for logging road construction	Direct loss of large trees in riparian areas, reduction in structural complexity, elimination of supply of large woody debris to channel	Gregory et al. 1991; Maser and Sedell 1994
Log transport on rivers erodes banks, simplifies channel geometry	Habitat complexity reduced	Sedell and Luchessa 1981
Removal of timber on hill slopes, resulting in increased peak runoff and erosion	Bank erosion, conversion of vegetated bottomland into open gravel-bed channel	Lyons and Beschta 1983; Grant 1988

continued

TABLE 36.3 (continued)

Activity and Potential Direct Physical Effects	Potential Ecological Consequences	References
Road and Railroad Construction		
Railroads and highways often follow rivers, built along banks of river	Riparian habitat replaced by railroad or highway for long distances along one bank	Scheidt 1967
Railroads and highways cross rivers	Continuity of riparian corridor interrupted by gaps at crossings	Furniss et al. 1991
Failure of roads and culverts delivers sediment to channel	Sediment reduces invertebrate habitat and populations	Erman et al. 1977
Urbanization		
Settlement along riverbanks and on bottomlands	Riparian habitat replaced by urban infrastructure	Medina 1990
Increased impervious surface upstream increases peak runoff, induces channel widening, incision	Water table may fall with incising channel, resulting in moisture stress to vegetation	Dunne and Leopold 1978; Booth 1990
Land drainage to make land suitable for development	Desiccation of riparian vegetation	National Research Council 1992
Channel relocation or channelization for flood control	Engineered channel margins rarely provide suitable conditions for establishment of riparian vegetation	Brookes 1988
Grazing		
Livestock trample and compact banks	Prevent establishment of vegetation, crush amphibians	Armour et al. 1991; Chaney et al 1990; Jennings 1996
Livestock hooves chisel banks	Destroy existing vegetation, destroy undercut banks, contribute to channel widening	USFS 1995; Overton et al. 1994; Kondolf 1994c
Livestock browse seedlings	Recruitment of young woody riparian plants prevented	Platts 1991
Removal of vegetation and compaction in watershed leads increased peak runoff and erosion, possibly to decreased to base flow	Erosion of banks supporting riparian vegetation	Behnke and Raleigh 1979; Platts 1991; Dudley and Dietrich 1995
Previously listed factors lead to incision of channels, especially in meadows	Water table drops, desiccating wetland species	Odion et al. 1990
Lack of bank vegetation and undercut banks, channel widening, and higher water temperatures	Reduced fish populations, reduced invertebrate populations	Behnke and Raleigh 1979; Armour et al. 1991; Herbst and Knapp 1995
Groundwater Abstraction		
Groundwater pumping lowers alluvial water table	Water table may fall below root zone of riparian plants, inducing moisture stress or death	Kondolf and Curry 1986
Recreation		
Heavy foot traffic tramples vegetation, compacts soil, and physically damages bank	Loss of riparian vegetation, creation of bare banks prone to erosion	Liddle 1975; Madej et al. 1994
Trails (foot, horse, bicycle, motorcycle) often follow streams	Riparian vegetation removed and replaced by trail; continuity of riparian corridor interrupted at crossings	Holmes 1979; Lemons 1979

its tend to occur in wider alluvial reaches, and thus mining is concentrated in foothill and valley reaches of Sierran rivers, although mines are also active along the upper reaches of the Feather and Yuba Rivers (California State Lands Commission 1993) and the American River (Kondolf and Matthews 1993).

Dams have direct effects from the permanently removal of riparian habitat to construct roads, penstocks, powerhouses, canals, and dams. Reservoirs drown existing riparian vegetation, and fluctuating water levels usually prevent the establishment of comparable new vegetation stands along reservoir margins. Thus, reservoirs constitute significant gaps in the riparian corridors. The largest reservoirs are located in the foothills, but reservoirs large enough to constitute significant gaps occur at virtually all elevations, as reflected in plots of reservoirs by elevation for the Sacramento and San Joaquin River basins (figures 36.1 and 36.2). Maps of reservoir numbers and capacity by watershed reflect the widespread occurrence of reservoirs throughout the range, with greater capacity in the central Sierra Nevada (figure 36.3).

Indirect effects of dams derive from changes in the flow regime and sediment load on downstream channels. Reduc-

tion in floods leads to reduced rates of channel migration (which in turn reduces the diversity of riparian habitats) and to the encroachment of vegetation into (and thus the narrowing of) the active channel. Most vegetation encroachment and channel narrowing in Sierran rivers has been reported below large reservoirs in the foothills (Pelzman 1973), whose storage is adequate to substantially reduce flood flows. Large reservoirs are less common but do occur at higher elevations. Most have not been studied, but many would likely evince encroachment and narrowing downstream, as observed on the North Fork Kings River (Taylor and Davilla 1985).

By storing water during winter and spring for subsequent release, reservoirs can increase base flows, which can, in turn, waterlog riparian vegetation accustomed to well-drained conditions in late summer and fall. Summer base flows on the North Fork Stanislaus River have increased tenfold as a result of storage in a hydroelectric project, and mortality of many riparian trees has been predicted (Parrish and Matthews 1993). Where reservoir water is exported from the basin, base flows can be reduced. On Rush Creek, the principal tributary to Mono Lake, no regular base flow releases were made from

Grant Lake Reservoir from 1941 to 1981, and a massive die-off of woody riparian vegetation ensued (Stine et al. 1984).

Reservoirs also trap the coarser (sand and gravel) portion of the sediment load and some fraction of the suspended load (depending upon the capacity of the reservoir relative to inflow). As a result, reservoir releases are typically *sediment starved*—they have the energy to transport sediment but are deprived of this load. As a result they tend to erode their bed and banks (Williams and Wolman 1984). If the channel incises, the alluvial water table will probably drop, resulting in moisture stress for the riparian vegetation adapted to the previous water table.

Hydroelectric generation entails most of the effects of dams where storage is involved, but has a somewhat different suite of effects if the project involves diversion but no storage—a *run-of-the-river* project (figure 36.4). Small diversions are common in the Sierra Nevada, either for small run-of-the-river projects, or for seasonal diversion via tunnels into storage reservoirs in adjacent drainages (see Kattelmann 1996).

Irrigation usually involves storage reservoirs so that water is available during the growing season. Thus, irrigation projects typically involve many of the same effects as those described for dams, and because they cause a net decrease in river flow, irrigation projects dewater river reaches. Small fish can be pulled into unscreened diversions and killed when they are discharged onto agricultural fields. Excess irrigation water can support riparian vegetation in artificially created wetlands, fed either by surface flows or groundwater recharged by excess irrigation waters. Along Rush Creek in Mono Basin, excess irrigation water infiltrated into permeable bedrock and reemerged downstream as springs. This process maintained high water tables, reestablished perennial flow, and thereby supported riparian vegetation even when diversion had completely dried the channel upstream (Kondolf and Vorster 1993). The combination of dams and diversions results in impacts in the majority of watersheds of the Sierra Nevada (figure 36.3d). Although few large dams are found in the northern region of the study area, this region has in general a higher density of diversions than other regions.

Most large irrigation storage reservoirs on Sierran rivers are in the foothills; irrigation diversion from Friant Dam and downstream diversions completely dries up the San Joaquin River annually at Gravelly Ford. Seasonal diversions without storage were used to irrigate farmland on the Bishop Creek alluvial fan by Native Americans and subsequent European settlers. These irrigation canals now support lush riparian vegetation (Federal Energy Regulatory Commission 1986). A seasonal irrigation diversion on the Little Truckee River reduced flows and resulted in channel widening downstream (Erman 1992).

Land drainage (usually for agriculture or urbanization) results in desiccation of wetland plants. Drainage of former meadows has been common around Lake Tahoe, resulting in loss of many riparian plants and invasion by upland species. Probably the most widespread land drainage in the Sierra

Nevada has been in wet meadows, which have been drained deliberately (documented as early as the 1870s, as in the dynamiting of a moraine in Yosemite Valley by Galen Clark to drain upstream meadows) (Greene 1987) and inadvertently because of channel incision (generally attributed to effects of livestock grazing) (e.g., Odion et al. 1990).

Land clearance for agriculture has been most common in wide alluvial reaches in the foothills and Central Valley, where formerly extensive bottomland forests were cleared, leaving only a narrow band of riparian vegetation (if any) along the bank.

Removal of vegetation was undertaken in the southwestern United States, mostly on an experimental basis, to reduce water "losses" to evapotranspiration by phreatophytes. Although some phreatophytes were eliminated in the Sierra Nevada (Biswell 1989), the environmental impacts of this practice (Campbell 1970) are generally acknowledged to be too great to justify it. Nonetheless, an increased water yield anticipated as a result of forest harvesting (mostly upland) has been factored into the national forest planning process in California. The Sequoia National Forest attributed 30% of the "benefits" from its preferred alternative of the latest forest plan to the supposed value of increased water expected as a by-product of timber harvest (U.S. Forest Service 1988).

Navigation by large ships commonly requires channel dredging and straightening, mostly undertaken in lower, valley reaches of rivers, downstream of the study area.

Timber harvest affects riparian vegetation directly and indirectly. Riparian vegetation has been removed in harvests of bottomland forests, and the construction of logging roads along bottomlands replaces riparian vegetation with road surface. Past log drives down rivers resulted in extensive battering of banks, reducing habitat complexity along the water's edge. Removal of timber on hill slopes, along with road construction and skid trail compaction, typically results in increased peak runoff and increased erosion. These so-called *cumulative effects* can degrade aquatic habitat and can potentially lead to the erosion of banks supporting riparian vegetation and the conversion of well-vegetated valley bottoms into wide, open, gravel bed channels (Grant 1988; Lyons and Beschta 1983).

Timber harvest has been extensive in the Sierra Nevada. Riparian trees, notably giant sequoia and other old-growth stands on bottomlands of Sierran rivers have been harvested, directly affecting bank vegetation and aquatic habitat. Franklin and Fites-Kaufmann (1996) found that 95% of the 1,200 ha (3,000 acres) separately mapped as riparian hardwood forest type had no late successional/old growth characteristics left, although deep, inaccessible river canyons with other forest types contained some of best remaining examples of old growth. Given that average angular canopy densities (canopy measured at an angle that effectively blocks summer sun) of 75% were observed on unlogged first- and second-order channels in the northern Sierra Nevada (Erman et al. 1977), removal of riparian trees has a tremendous effect on aquatic habitat.

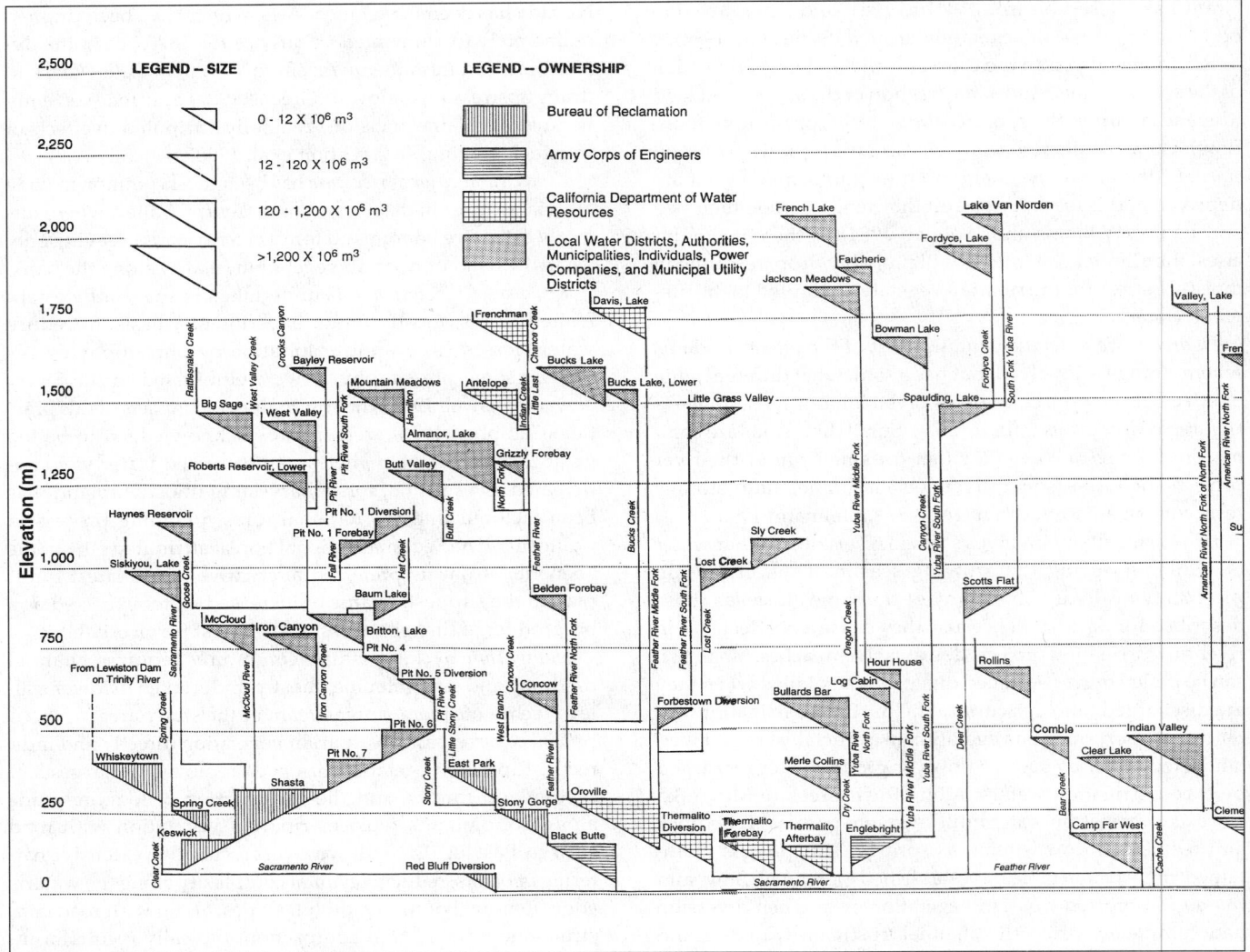

FIGURE 36.1

Schematic diagram of reservoirs in the Sacramento River basin, plotted by elevation. Reservoirs are included from two Coast Range drainages: Stony Creek (East Park, Stony Gorge, and Black Butte) and Putah Creek (Lake Berryessa and Solano Lake) and from the upper Sacramento River drainage that lies north of the Sierra Nevada. Otherwise, all reservoirs shown are in the Sierra Nevada or its foothills. (Adapted from a plot prepared by the California State Water Resources Control Board, Graphic Unit.)

The cumulative effects of timber harvest are widespread but poorly documented in the Sierra Nevada. Most of the timberlands in the Sierra Nevada lie within national forests. Despite this single ownership of large areas, and despite the mandate for the Forest Service (and other agencies) to analyze cumulative impacts of forest management activities, very little basic data collection on peak stream flow and sediment yield (the variables likely to be affected by timber harvest) is undertaken on the forests. Most field data collection and office analyses are apparently devoted to cumulative effects "assessment methods" (see Berg et al. 1996) that primarily involve office-based computations of such variables as area

of road surface and timber harvest within a watershed to predict cumulative impacts. These computations of effects are not verified by actual field measurements of peak flow or sediment yield, and in some cases, the results of these "methods" have been contradicted by field observations of Forest Service biologists (Kondolf 1994a).

Railroads and roads commonly follow rivers, taking advantage of flat bottomland and linking riverside settlements. These railroads and roads (and the additional settlement generated along them) displace riparian vegetation on the floodplain. In narrow canyons with limited bottomland, roads and railroads are commonly located along the riverbank itself,

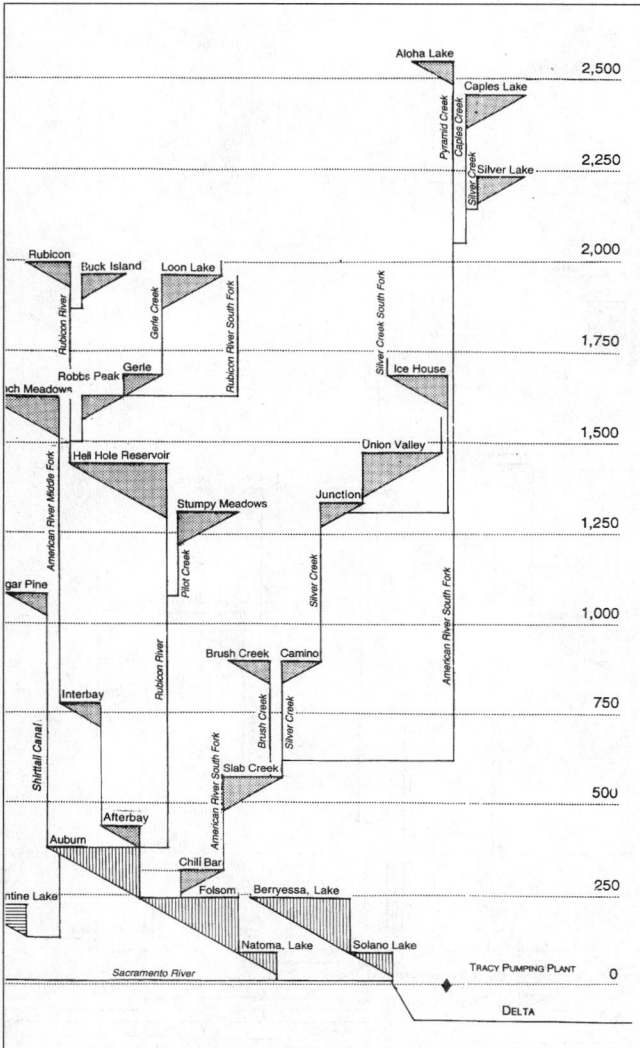

FIGURE 36.1 (continued)

Urbanization has occurred historically along bottomlands because of the flat land, proximity to water, and connection to communication and trade routes that often followed rivers. When such urbanization occurs, buildings, streets, parking lots, and other urban infrastructure directly displace riparian vegetation. The impervious surfaces of rooftops and pavement result in greater surface runoff per unit of precipitation, increasing peak flows and commonly inducing channel incision, bank erosion, and a drop in the water table (which may desiccate riparian plants) (Dunne and Leopold 1978). Sites with shallow water tables may be deliberately drained to permit development, resulting in desiccation of riparian vegetation. As floodplains are urbanized, flood damages increase by virtue of the increased value of the flood-prone land (whether the floods be naturally occurring or exacerbated by land-use change). Thus, urbanization commonly creates a demand for flood control, which involves structural measures such as channelization or levee construction, in turn reducing or eliminating riparian habitat.

Since the 1940s, California has experienced tremendous population increases and corresponding urbanization. From 1980 to 1990, the state's population increased from 24 million to 30 million (California Department of Finance 1990). From 1984 to 1990, urban land area increased by 123,000 ha (303,810 acres) in the 42-county state Office of Land Conservation farmland mapping area (California Office of Land Conservation 1988, 1990, 1992). In the last two decades an increasing proportion of the population increase has been accommodated by dispersed "ranchette" settlement in rural counties of the Sierra Nevada (see Duane 1996). This increased urbanization pressure has effects on riparian areas ranging from direct urbanization (riparian areas are often preferred sites for ranchettes), to fragmentation by roads and other infrastructure to support urbanization in uplands, to hydrologic changes induced by urbanization in the watersheds, to increased use of riparian areas by humans and domestic pets.

Grazing by livestock results in the trampling and compaction of riparian areas, the direct destruction of bank vegetation by bank through the chiseling of banks by hooves, and the elimination of recruitment of young woody riparian plants through browsing (Armour et al. 1991; Platts 1991; Menke et al. 1996). The lack of bank vegetation eliminates shading and terrestrial food sources for the channel, and reduces the stability of the bank. Grazing throughout a watershed can increase peak runoff and erosion rates, leading to channel incision (and thus lowered alluvial water tables and desiccation of riparian plants), bank erosion, and increasing fine sediment content in channels (Behnke and Raleigh 1979). Grazing is commonly concentrated in riparian areas because of vegetation supported by the greater moisture availability and because the stream provides drinking water.

Grazing by livestock was virtually ubiquitous in the Sierra Nevada from the nineteenth century through 1930 (Vankat and Major 1978; McKelvey and Johnston 1992; Kinney 1996),

replacing overhanging bank vegetation with riprap, which tends to narrow the channel with artificial fill. Even in the absence of these longitudinal impacts, the continuity of the riparian corridor is interrupted at each bridge crossing. Concentrated road runoff commonly carves gullies, and unpaved logging roads and their culvert crossings may wash out during storms, delivering pulses of sediment to the channel and degrading aquatic habitat and water quality.

Roads and railroads cross most of the Sierra Nevada, as indicated by the results of the GIS analysis of road influences on streams and by the aerial photographic analysis of riparian corridor gaps discussed later.

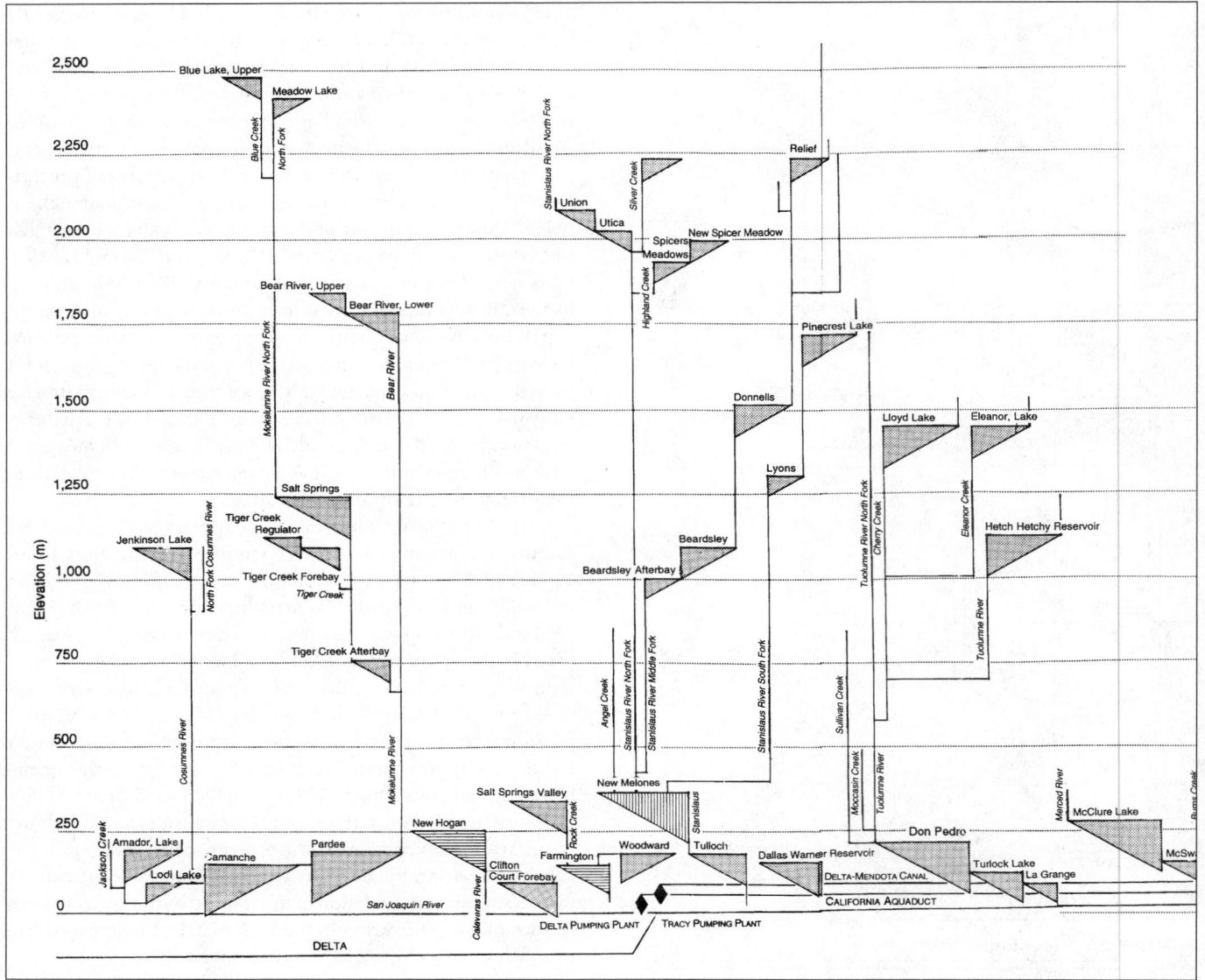

FIGURE 36.2

Schematic diagram of reservoirs in the San Joaquin River basin, plotted by elevation. Clifton Court forebay and four storage reservoirs in the Coast Ranges (San Luis, Little Panoche, O'Neil, and Los Banos) are included. Otherwise, all reservoirs shown are in the Sierra Nevada or its foothills. (Adapted from a plot prepared by the California State Water Resources Control Board, Graphic Unit.)

with heavy grazing even in many high-elevation meadows that remain inaccessible to vehicles today (Dudley and Embury 1995). As a result, channel incision and desiccation of meadow vegetation has been widespread in the Sierra Nevada. Grazing and its effects have been so pervasive and ubiquitous throughout the American West that virtually no unaffected "control" conditions exist for comparison, and what most people would regard as "natural" conditions are in fact influenced by historical (if not current) grazing (Elmore and Beschta 1987). Our best comparisons are derived from

studies of vegetation and channel recovery when streams are excluded from grazing, but channel conditions may be slow to recover from grazing effects (Kondolf 1993).

Groundwater abstraction for municipal or agricultural use can reduce alluvial water tables, stressing or killing riparian vegetation (Kondolf and Curry 1986; Wright and Berrie 1987). Groundwater pumping in the Owens Valley has had the greatest documented effects on vegetation known in the Sierra Nevada region (Perkins et al. 1984; Groeneveld and Or 1994).

Recreation can affect riparian corridors through the concen-

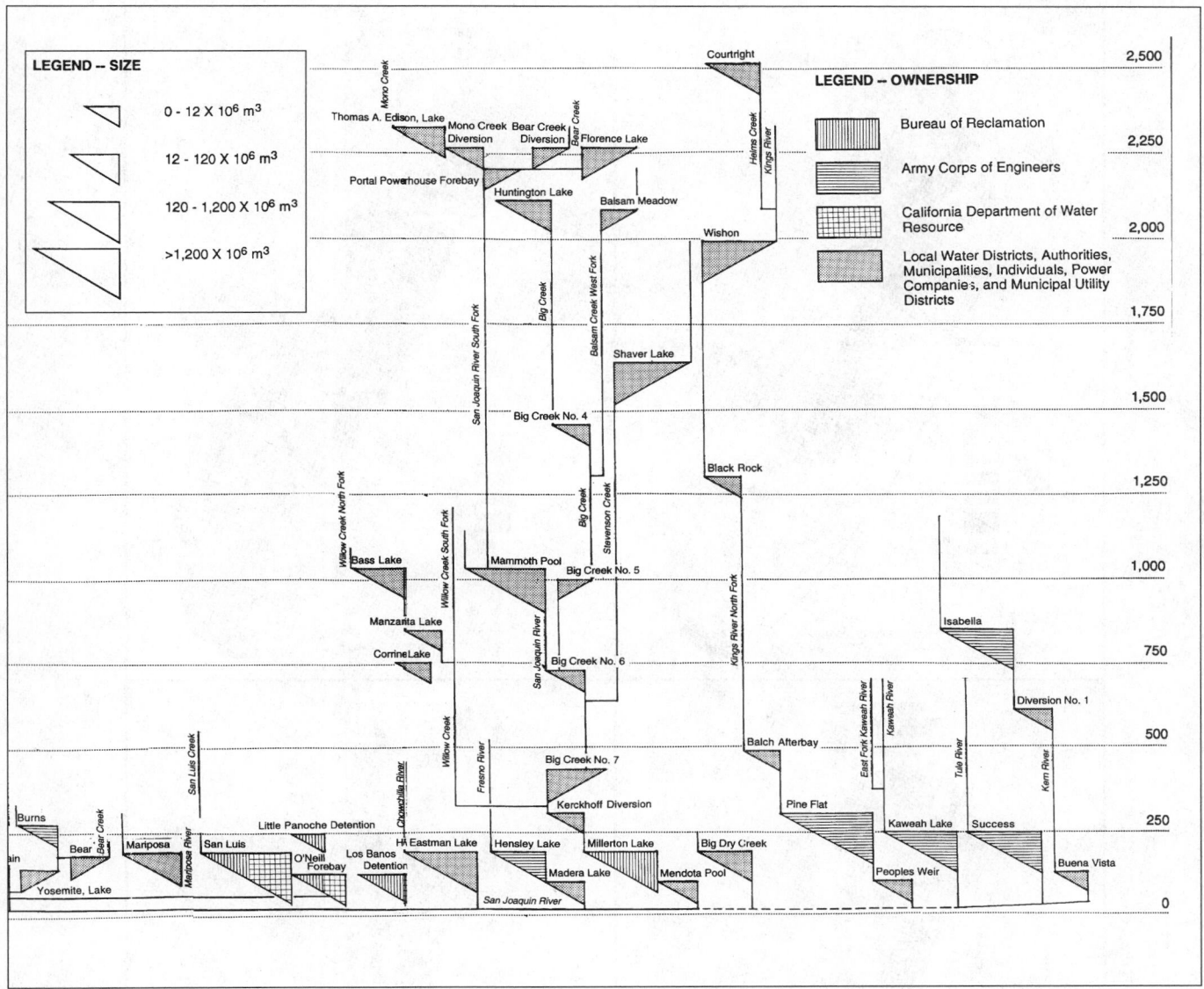

FIGURE 36.2 (continued)

tration of people along riverbanks: heavy foot traffic tramples vegetation, compacts soils, and can physically damage banks (Liddle 1975). Trails (foot, horse, bicycle, or motorcycle) replace riparian vegetation with pavement or bare, compacted earth and bring people into the riparian zone where they are then more likely to concentrate on banks, with the effects just described. Heavy concentration of anglers on the banks may have similar effects. These effects have been documented along the Merced River in Yosemite National Park (Madej et al. 1994) and are probably concentrated near popular camp-

grounds throughout the Sierra Nevada, but their overall extent has not been documented. Most national forest campgrounds in the Sierra Nevada are located in or near riparian areas.

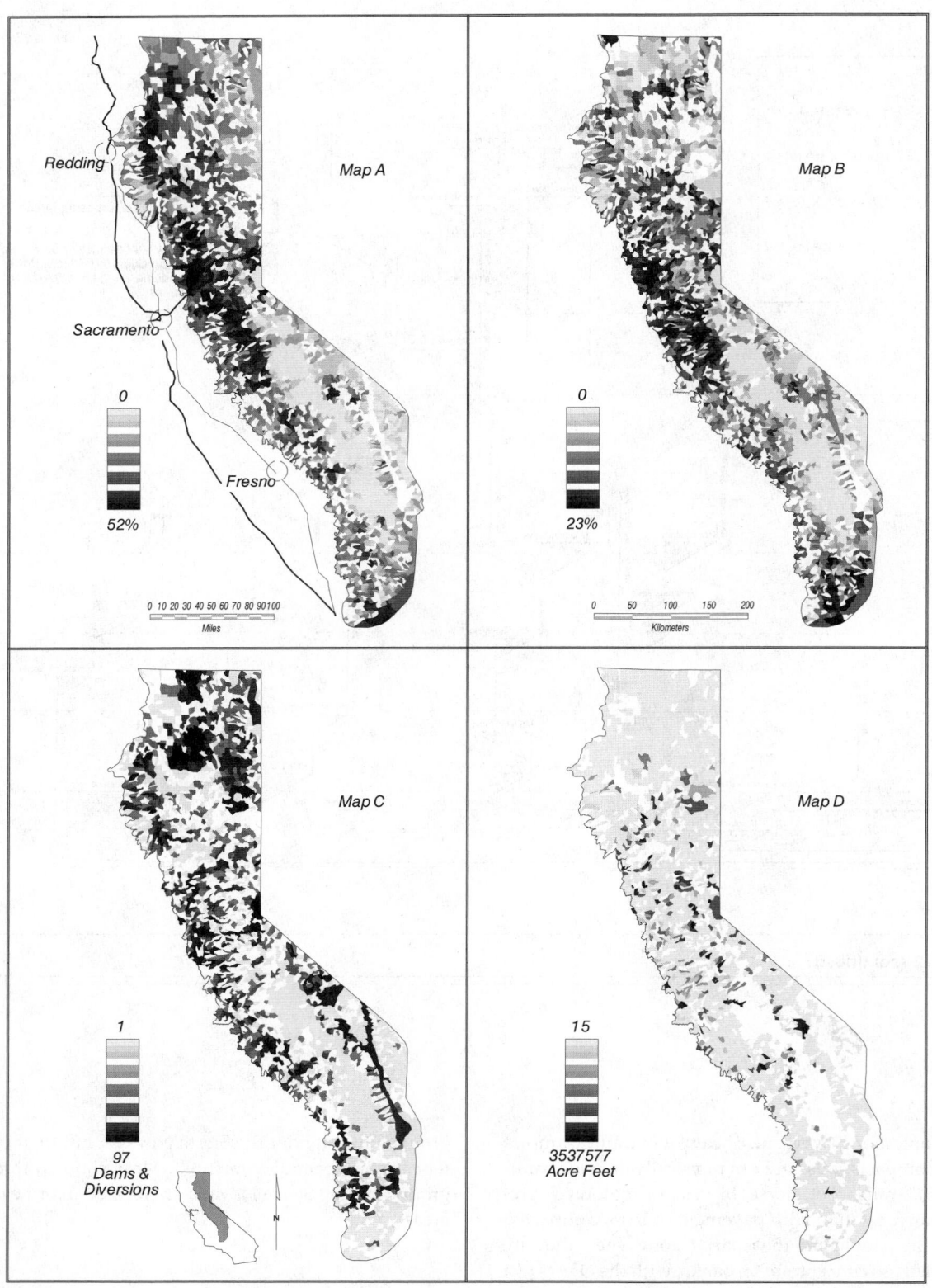

FIGURE 36.3

GIS plots of percentages of pixels (100 m x 100 m blocks) in each watershed that contain a road (map *a*), contain a road near a stream (map *b*), and contain a dam or diversion (map *c*). Map *d* shows dam capacity by watershed.

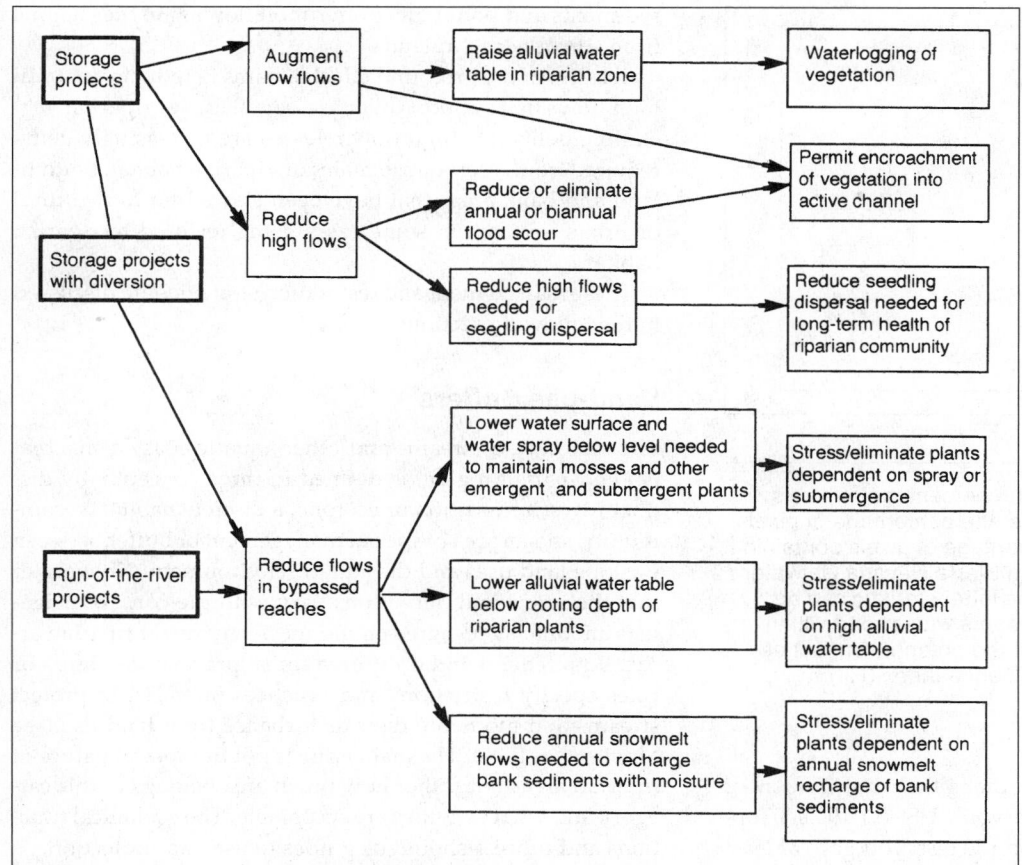

FIGURE 36.4

Effects of hydroelectric dams and diversions on riparian vegetation.

GIS ANALYSIS OF ROAD INFLUENCE ON STREAMS

An indication of the pervasiveness of road influence on Sierran rivers and streams is provided by the GIS analysis of 100 m by 100 m pixels in 141 watersheds (Calwater Hydrologic Subareas). In each watershed, the percentage of pixels with a road ranged from less than 0.6% to 31%, and the percentage with a stream ranged from 4% to 19% (figure 36.5). The results for roads are displayed for each watershed in figure 36.3a. When these patterns are overlaid, the more interesting result is obtained: the percentage of pixels with a stream that also contain a road, which we designate here as the Road Influence Index (RII) (figure 36.3b) The RII is a measure of the percentage of stream length with a road within 100 m. The RII ranges from 2% to 33%, with a median value of 14.1% (figure 36.5). The central 50% of the distribution (i.e., the 71 watersheds that fall in the center of the RII) have RII values between 10.8 and 17.4, and the central 80% have RII values between 8.7 and 21.3 (figure 36.5). Thus, in the vast majority (80%) of Sierra Nevada watersheds, 8% to 21% of stream reaches are potentially influenced by a road within 100 m. Additional detail, including values for this index, for thirty-three watersheds in the Eldorado National Forest is given in Costick 1996. He refers to this index as the percentage of roaded area inside a 100 m stream buffer.

The RIIs for watersheds in the northern Sierra Nevada (north of Interstate 80) are lower (median value 10) than those for watersheds in the central (median value 14), southern (median value 14), and eastern (median value 16) Sierra Nevada (figure 36.6).

The true values of RII are certainly higher than indicated here because the data sets used for roads were derived in large part from road maps, which do not show all roads. The total stream length would also be greater if smaller-scale maps (e.g., 1:24,000) were used to identify streams, as only larger streams are shown on the 1:100,000 scale maps.

Aerial Photograph and Map Analysis of Gaps in Riparian Corridors

Of the 130 Calwater Super-Planning Watersheds selected for assessment by aerial photography, 121 displayed obvious gaps in the riparian corridor. These gaps were caused primarily by road and railroad crossings, timber harvesting, clearing of private lots, dewatering by diversions and dams, and grazing.

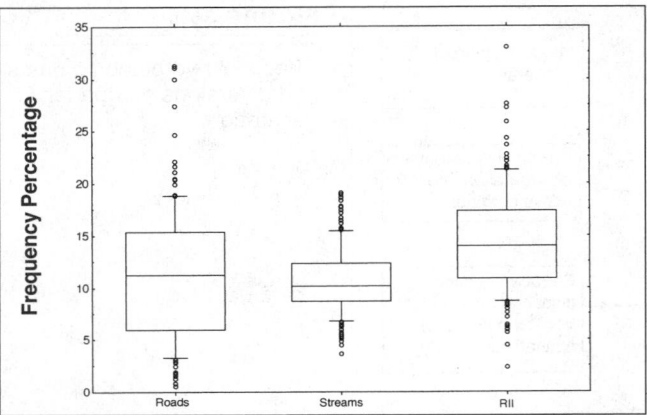

FIGURE 36.5

Box-and-whisker plots showing the percentage of pixels (100 m x 100 m) containing roads, the percentage of pixels containing streams, and the percentage of pixels containing streams that also contain roads in Sierra Nevada Calwater planning watersheds (*n*=141). The latter statistic can be restated as the percentage of streams with a road within 100 m of the channel, an index of the potential impact of roads upon streams, or Road Influence Index (RII).

The longest gaps (and thus perhaps the most influential ecologically) were created by reservoirs. USGS 1:100,000 topographic maps showed more than 150 reservoir gaps at least 0.5 km (0.3 mi) long. Highly developed basins such as the Feather and American Rivers had more than 20 reservoirs exceeding 0.5 km (0.3 mi) in length. The total length of riparian corridors inundated by reservoirs exceeds 1,000 km (600 mi).

MANAGEMENT IMPLICATIONS
Management Strategies

Management strategies can be used to minimize the impact of human activities on riparian areas or to restore ecological values of riparian areas. As described in preceding sections, human impacts to riparian systems have occurred by the direct removal or replacement of riparian vegetation or by the alteration of the physical conditions supporting riparian vegetation.

The most commonly applied, most straightforward, and probably most effective strategy is to define a *riparian management zone* or *riparian buffer strip* within which vegetation cannot be disturbed and ground compaction is avoided. This strategy serves not only to protect riparian vegetation for its own sake but also to maintain the beneficial influence of riparian vegetation upon aquatic habitat through shading, contribution of terrestrial food and nutrients, and filtering of

sediments and pollutants from runoff flowing to the channel from surrounding uplands.

Because of the profound effect of dams in reducing natural high flows that support diverse assemblages of riparian vegetation, deliberate high flow releases are increasingly being required from reservoirs to maintain riparian habitat. Bottomland and bank areas that have been cleared for agricultural or urban uses are in some cases being restored to riparian habitat.

These management and restoration strategies are discussed in the following sections.

Land-use Buffers

The region near streams and other aquatic ecosystems, that is, the riparian region, is defined in three conceptually distinct ways: a transition or ecotone, a discrete habitat or community, and an area of special management or buffer between upslope land uses and the aquatic environment. No wonder that the terms and definitions vary with the context. Scientists and managers agree on the special nature of riparian areas. Both federal and California forest practice standards or rules specify restrictions and practices intended to protect streams and moderate their disturbance from land use (see Moyle et al. 1996). The main issue is not the special nature of riparian areas but rather how much area belongs in this category and what activities are acceptable. The ecological functions and process should be guides to use and protection.

Riparian ecological functions and physical processes take

FIGURE 36.6

Box-and-whisker plots showing the percentage of pixels (100 m x 100 m) with streams that also contain roads (Road Influence Index) in northern (north of Interstate 80, *n*=38), central (from Interstate 80 south through the Merced River Basin, *n*=29), southern (south of the Merced River Basin, *n*=35), and eastern (east of the divide, *n*=35) Sierra Nevada watersheds. This statistic can be restated as the percentage of streams with a road within 100 m of the channel, an index of the potential impact of roads upon streams.

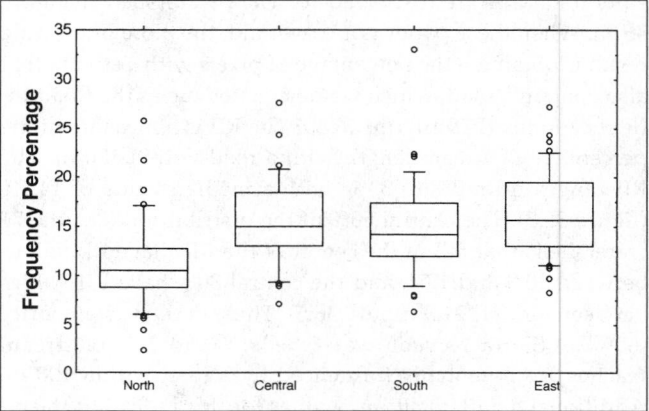

place in three areas at varying distances from the aquatic system: a community area, an energy area, and a land-use influence area. The size of these areas depends on the local characteristics that define them. Any one of the areas may be larger than the others; in other words the three areas are nested within each other, but the order is determined by the characteristics that define them rather than an arbitrary hierarchy. One other fact is important in understanding the dimensions of the entire riparian area: it is not proportional to the size of the aquatic system. Ephemeral ponds, intermittent streams, and small springs are as important to the suite of species that depend upon them as large rivers are to another suite of species (see Erman 1996). Smaller aquatic systems in forested environments are dominated by the land system. Consequently, the impacts from changes in riparian forest structure and composition and from land disturbance result in major changes in the aquatic system (Erman et al. 1977; Minshall 1994).

The direction of state and federal protection of riparian areas has been based on broad classification of the aquatic system—presence of a life-form (fish-bearing vs. non-fish-bearing, for example), size (rivers vs. spring runs), or permanence (year-round stream flow in most years vs. temporary flow in most years). Classification of aquatic habitats for management in this way does not recognize the connected nature of aquatic systems (upstream-downstream), does not recognize the needs of riparian-dependent species, and cannot work for the protection of aquatic biodiversity (which is particular to the type of system), or properly assist in the management of interconnected land-water systems. Shifting to a recognition of the community, energy, and buffering requirements of riparian areas will aid in the protection and management of the entire riparian system.

The Community Area

For any aquatic habitat there is a suite of species that depend on the combination of land and water. Some spend most of their life in the water, some on the land. Most aquatic insects, for example, develop in water but spend a portion of the life cycle on land—feeding, mating, and resting (see Erman 1996). Alder and cottonwood trees are always associated with nearby water—a spring, a lake, a stream, or groundwater near the surface. From a knowledge of the habitat requirements and life connections of the dependent species, we should be able to define the general dimensions of this community area in the various regions and elevation zones of the Sierra. However, the exact requirements and hence the dimensions for many species are unknown. The water shrew (*Sorex palustris*) is likely confined to the virtual stream bank. Beavers (*Castor canadensis*) may move tens of meters from water to cut aspen or other trees, as well as cottonwood on relatively flat floodplains that extend more than 100 m from low-water channels. The California tiger salamander (*Ambystoma californiense*), which occurs in the foothills zone (see Jennings 1996), lives in terrestrial habitats near temporary and permanent water

used for breeding. Adults migrate up to 129 m (423 ft) (average 36 m [118 ft]) and juveniles up to 57 m (187 ft) (average 26 m [85 ft]) between their breeding site and terrestrial burrows (Loredo et al. in press). Studies elsewhere on amphibians have found some species that live only in the cool, damp conditions near streams and up to several hundred meters from surface flow (Welsh 1993). Dramatic changes in riparian conditions due to the logging of forests near headwater streams have greatly reduced populations of riparian-dependent and terrestrial salamanders in the Appalachians (Petranka et al. 1994). Thus, to provide for the living requirements of those organisms dependent for their survival on the special conditions of the riparian area, the primary management should be maintenance of these conditions. Even the natural role of disturbance, documented in this chapter and others (see also Kattelmann and Embury 1996) does not require, in most situations, active restoration of the landscape in order to secure the habitat conditions necessary for the area.

The Energy Area

Major scientific understanding of the energy linkages between upstream and downstream (e.g., the river continuum concept, Vannote et al. 1980) and exchanges between the land area and aquatic systems has emerged in the last two decades (see reviews by Cummins et al. 1989; Carlson et al. 1991; Murphy and Meehan 1991). Riparian energy areas contribute a year-round supply of organic material that ranges from nearly the total supply of food at the base of the food chain (small forested streams and springs) to critical quality food (organic matter transported into larger streams from smaller upstream sources). Wind-blown seeds and leaves are a significant source of material entering meadow reaches with little forest canopy. The type of organic material is also important. Easily decomposed plant material (e.g., parts with a relatively low carbon-to-nitrogen ratio such as alder leaves), material that is slow to decompose (such as Douglas fir), as well as terrestrial insects carried in are needed to support an aquatic food web throughout the year. Flows of energy from the aquatic to surrounding terrestrial system (especially emerging insects) is also substantial (see Erman 1996) The surrounding riparian area also blocks energy from the sun and reradiation from the water (thus reducing temperature changes). And the role of large organic matter (trees, root-wads, debris dams) is of major importance to the structure and complexity of stream channels, to the routing of sediment, to the retention of nutrient supplies, and to the diversity of aquatic habitats. The dimensions of this region vary by the season (leaf fall of deciduous plants), by the hydrologic conditions (out-of-channel floods, size of stream), by the contributing area (large wood that can fall into the channel, plant parts and insects that blow in), and by the species mix (organic material breaks down and is useful as aquatic food at different times). A useful summary index of this area is the slope distance around the aquatic system equivalent to the height of the site potential tree (i.e., the height a mature tree can attain given the soil and other conditions at its location) (Chapel et al. 1992). For the Sierra Nevada, that height in many forest types is approximately 46 m (150 ft). However, the incorporation of wood and other organic material into streams will occur also during inundation of the floodplain. For larger streams in regions of gentle gradient, the width of a stream during major floods may extend much beyond 46 m.

The Riparian Buffer Area

The effects of land-use disturbance are reduced by keeping such activities at a distance from the aquatic system and by maintaining a buffer area capable of absorbing disturbance. The likelihood of disturbance to a stream from most land uses increases as a function of proximity to a stream, the steepness of surrounding hillsides, and the erodibility of soils. These relationships, as in many risk factors, are probably multiplicative and therefore a doubling of slope has more than twice the risk of disturbance to the stream (i.e., an exponential change). Current practice for designing buffer systems based on risk rely on classification of the aquatic system (as was mentioned earlier) and the creation of three or four categories of slope. As a consequence, a fixed width is chosen even though conditions on the land and requirements of the community would suggest a variable width (Bisson et al. 1987). We propose a more direct system for estimating a variable-width buffer system based on the community and energy area in combination with slope and other measurable risk factors.

For example, let us assume that a stream is in the mixed conifer zone. The determination of hillside slope can be made from topographic maps or from GIS. The SNEP GIS team has prepared a program that will calculate slope at 30 m (98.5 ft) increments along a stream channel. At each point, slope from five successive 30 m segments out from a channel are computed from the 30 m Digital Elevation Model. Slopes are then weighted 5, 4, 3, 2, 1, from closest to farthest away, and divided by 5 to produce a weighted average slope over the 150 m (slopes closest to the stream have the greatest effect on the average). Let's also assume that the stream has a community area defined by species as 110 ft (33.5 m) and an energy area that is 150 ft (46 m). Thus, a minimum region with maintenance of forest structure and minimal land disturbance is 150 ft for these two areas. This distance is then multiplied by the base of natural logs (e) raised to a power equal to 1+slope (in decimal form). If, for example, the slope were 25%, the equation would be

$$\text{Buffer width (ft)} = 150 * e^{(1+0.25)}$$

giving a value of 524 ft (160 m). If the average slope were 50%, the buffer would be 672 ft (205 m). In the first case, an additional 374 ft (114 m) of buffer would be needed. Soil erodibility, also available from soil maps and GIS, can be incorporated as the detachability value (see Costick 1996), and the exponent would be expanded to 1+slope+detachability –

(slope ∗ detachability). For example, if detachability were 0.30, the equation would be

$$\text{Buffer width (ft)} = 150 * e^{(1+0.25+0.30-0.075)}$$

giving a value of 656 ft (200 m). Extreme cases, when slope and detachability are both high, would result in even larger buffer zones, and as slope and detachability approach zero, buffer zones would become smaller—exactly the outcome common sense would indicate is appropriate. This additional area beyond 150 ft would not have the same land-use restrictions as the community and energy areas. Its purpose is to highlight a region in which probability of disturbance may affect the community or energy areas and the aquatic system. Silvicultural procedures should minimize soil disturbance and in general retain sufficient forest structure to ameliorate microclimate change within the community area and minimize the abrupt transition from the area upslope to the community area. Describing the buffer zone as a "probability of disturbance region" places the responsibility on managers for designing practices that have higher standards and are more carefully matched to conditions where mistakes will matter more.

Current information and computer-aided analytic methods are sufficient for layout of such a buffer system for many regions of the Sierra. An example is shown in figure 36.7 that illustrates a fixed buffer representing the energy area (150 ft) and the wider variable buffer area computed from the equation given earlier. Notice in the selected region along the North Yuba River near the town of Downieville that State Highway 49 lies within both areas for nearly all the distance illustrated. Stream channels in this case represent those modeled by GIS because existing USGS maps omit many actual streams. Refinements in scale of Digital Elevation Models from 30 m to 10 m are underway, and soil mapping (for estimating soil detachability and other factors) continues to expand and be incorporated into GIS layers. Most forest and land managers today could determine first approximations based on habitat requirements, energy inputs, and hillside slope calculations to produce a logical, ecologically based riparian management-protection system along these lines. It would lead to better protection of riparian-dependent organisms and of energy linkages between the land-water systems, and would assist managers in tailoring land-use activities to regions of greater need than is presently the case.

Riparian Maintenance Flows

The interrelations between physical channel processes and riparian vegetation are only now becoming better understood, and in any event the precise nature of these interrelations varies from river to river. Thus, specifying riparian maintenance flows (or "channel maintenance" or "flushing") can be viewed as essentially experimental at present.

To evaluate the effectiveness of riparian maintenance flows

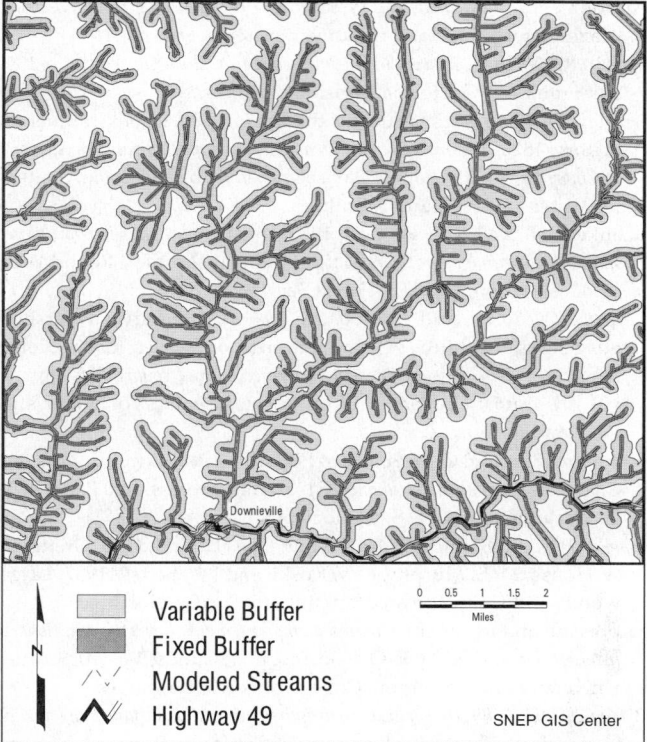

Variable Buffer
Fixed Buffer
Modeled Streams
Highway 49

0 0.5 1 1.5 2
Miles

SNEP GIS Center

FIGURE 36.7

Fixed-width buffer (150 ft) and variable-width buffer computed from an equation for slope adjacent to stream channels for a region of the North Yuba River. Channel locations are determined from geographic information system models.

requires that the broad goal of maintaining riparian vegetation be restated as specific objectives from which flows can be specified and actual effects observed (Kondolf and Wilcock in press). For example, to maintain diversity of riparian habitat may require continued lateral migration of a meandering alluvial channel, which in turn requires adequate flows to erode banks and deposit point bars. Similarly, to prevent invasion of xeric plants onto bottomlands may require periodic flooding and high river stages that maintain seasonally high water tables. The magnitude of the flows required to achieve these objectives can be determined from stage-discharge relations by reach.

Hill and colleagues (1991) suggested that floods with a return period of 25 years under natural, predam conditions might be needed to maintain valley form and riparian habitat.

Restoration of Riparian Vegetation

Riparian revegetation projects may be limited to revegetation of banks to increase bank stability, channel shading, and overhanging vegetation. Artificial floodplains (essentially the sec-

ond stage of two-stage flood channels), designed to be inundated every one to two years, are ideal sites to establish native riparian vegetation species (e.g. Matthews 1990). Much riparian revegetation has been undertaken to mitigate losses in riparian habitat elsewhere (Munro 1991), with mixed results in California. In general, riparian revegetation has been most successful along the banks of the low flow channel, less on higher surfaces. This difference is probably because of the nearly ubiquitous effect of reservoirs in reducing natural flood flows, eliminating hydrologic conditions needed for riparian vegetation on higher surfaces.

Probably the most ambitious riparian revegetation projects in the Sierra Nevada are being undertaken by the Nature Conservancy along the Cosumnes and South Fork Kern Rivers. Both rivers were chosen because their flood regimes are relatively natural, thus potentially maintaining near-natural hydrologic conditions on bottomlands.

Along the Cosumnes River, 80 ha (200 acres) have been replanted in valley oak (*Quercus lobata*) and other areas have been permitted to naturally revegetate in cottonwood (*Populus fremonti*) and various species of willow (*Salix* spp.). The suitability of various sites for different species was determined from flood inundation regime, soil type, and historical evidence of riparian vegetation present before these woodlands were cleared for agriculture (Griggs et al. 1994).

Along the South Fork Kern River, 130 ha (340 acres) were replanted, from 1987 to 1993 primarily, in Fremont cottonwood and red willow (*S. lavigata*) on floodplain sites from which these species had been cleared for agriculture. In large measure, the project was undertaken to create habitat for the yellow-billed cuckoo (*Coccyzus americanus*) and other avian species. Survival rates for plantings from 1991 to present have exceeded 90% (R. Tollefson, the Nature Conservancy, conversation with G. M. Kondolf, 1996).

CONCLUSIONS

Riparian areas are sites of exceptional ecological importance, typically having greater species diversity (floral and faunal) than surrounding uplands and providing essential food sources or habitat at certain life stages for upland wildlife species. Riparian areas also play a key role in maintaining water quality and aquatic habitat in streams and rivers, and because of their linear nature, riparian corridors are important routes for wildlife migration.

The riparian areas of the Sierra Nevada have been extensively affected by direct removal or inundation of riparian vegetation and by alterations to the conditions on which the riparian vegetation depends. Unfortunately, the field data base necessary to properly assess the health of riparian areas throughout the Sierra Nevada does not exist. However, from the extent of human activities known to affect riparian areas, we can infer substantial impacts. Moreover, map and aerial photograph analyses of a large sample of Sierran watersheds show that virtually all riparian corridors are interrupted by gaps caused by such human activities such as construction of road or railroad crossings, human settlements, dewatering of streams, grazing, timber harvest, and mining. The largest gaps are caused by reservoirs, many of which exceed 0.5 km (0.3 mi) in length, and which occur at a wide range of elevations in the Sierra Nevada.

Establishing riparian management zones (or "buffer strips") of adequate width is probably the single most effective strategy for protection and maintenance of the ecological values of riparian areas. Vegetation removal and ground disturbance should be prohibited in these zones, both to preserve the riparian habitat itself and for its beneficial influence upon aquatic habitat. Although the width of these zones has most commonly been set arbitrarily, variable-width buffer strips (based on attributes of the river itself, the riparian community, and hill-slope gradients) can be established to better protect riparian resources.

For channels below reservoirs, deliberate high flow releases can be made to mimic the hydrologic effects of natural floods in maintaining riparian vegetation. Restoration of riparian habitat, if based on careful analysis and on experience, can re-create many lost values to riparian and aquatic habitats.

ACKNOWLEDGMENTS

We are indebted to Lynn Decker, Nancy Erman, Diana Jacobs, and Peter Moyle for their constructive criticism of an early draft of Kattelmann and Embury (1996) and for their insightful suggestions on how to approach the broad, complex, yet very important topic of riparian areas in the Sierra Nevada. Many others contributed substantially to the literature review or provided useful background information. The Geometrics Division of the Engineering Department of the U.S. Forest Service Region V office in San Francisco contributed to the analysis of riparian corridors on aerial photographs, providing access to aerial photographs and use of office space and equipment. The Bureau of Land Management resource offices in Folsom, and the district offices in Bakersfield and Bishop, also provided access to aerial photographs. Sarah Marvin reviewed videotapes of riparian corridors, held by the Water Resources Center Archives, University of California, Berkeley. Steve Beckwitt conducted the GIS analysis of watersheds.

REFERENCES

Abell, D. L., ed. 1989. *Proceedings of the California riparian systems conference: Protection, management, and restoration for the 1990s.* General Technical Report PSW-110. Berkeley: U.S. Forest Service, Pacific Southwest Forest and Range Experiment Station.

Armour, C. L., D. A. Duff, and W. Elmore. 1991. The effects of livestock grazing on riparian and stream ecosystems. *Fisheries* 16 (1): 7–11.

Averill, C. V. 1946. *Placer mining for gold in California.* Bulletin 135. San Francisco: California Division of Mines.

Beesley, D. 1996. Reconstructing the landscape: An environmental history, 1820–1960. In *Sierra Nevada Ecosystem Project: Final report to Congress,* vol. II, chap. 1. Davis: University of California, Centers for Water and Wildland Resources.

Behnke, R. J., and R. F. Raleigh. 1979. *Grazing and the riparian zone: Impact and management perspectives.* General Technical Report WO-12. Washington, DC: U.S. Forest Service.

Berg, N., K. Roby, and B. McGurk. 1996. Cumulative watershed effects: Applicability of available methodologies to the Sierra Nevada. *Sierra Nevada Ecosystem Project: Final report to Congress,* vol. III. Davis: University of California, Centers for Water and Wildland Resources.

Bergman, D. L., and C. W. Sullivan. 1963. *Channel changes on Sandstone Creek near Cheyenne, Oklahoma.* C145-C148. Professional Paper 475-C. Washington, DC: U.S. Geological Survey.

Bisson, P. A., R. E. Bilby, M. D. Bryant, C. A. Dolloff, G. B. Grette, R. A. House, M. L. Murphy, K. V. Koski, and J. R. Sedell. 1987. Large woody debris in forested streams in Pacific Northwest: Past, present, and future. In *Streamside management: Forestry and fishery interactions,* edited by E. O. Salo and T. W. Cundy, 143–90. Seattle: University of Washington, College of Forest Resources.

Biswell, H. H. 1989. *Prescribed burning in California wildlands vegetation management.* Berkeley: University of California Press.

Booth, D. B. 1990. Stream-channel incision following drainage-basin urbanization. *Water Resources Bulletin* 26: 407–17.

Brookes, A. 1988. *Channelized rivers.* Chichester, England: John Wiley.

Brothers, T. S. 1984. Historical vegetation change in the Owens River riparian woodland. In *California riparian systems: Ecology, conservation, and productive management,* edited by R. E. Warner and K. M. Hendrix, 75–84. Berkeley and Los Angeles: University of California Press.

Brown, G. W. 1969. Predicting temperatures on small streams. *Water Resources Research* 5: 68–75.

California Department of Finance. 1990. *Ranking of California cities and counties.* Report E-8. Sacramento: Department of Finance, Demographic Research Unit.

California Department of Forestry and Fire Protection. 1996. *Calwater.* Database online. Sacramento: California Environmental Resource Evaluation System. Available from http://resources.agency.ca.gov.

California Office of Land Conservation. 1988. *Farmland mapping and monitoring program: Farmland conversion report 1984–1986.* Publication FM 88-01. Sacramento: California Department of Conservation, Office of Land Conservation.

———. 1990. *Farmland mapping and monitoring program: Farmland conversion report 1986–1988.* Publication FM 90-01. Sacramento: California Department of Conservation, Office of Land Conservation.

———. 1992. *Farmland mapping and monitoring program: Farmland conversion report 1988–1990.* Publication FM 92-01. Sacramento: California Department of Conservation, Office of Land Conservation.

California State Lands Commission. 1993. *California's rivers: A public trust report.* Sacramento: California State Lands Commission.

Campbell, C. J. 1970. Ecological implications of riparian vegetation management. *Journal of Soil and Water Conservation* 25:49–52.

Carlson, A., M. Chapel, A. Colborn, D. Craig, T. Flaherty, C. Marshall, D. Pratt, M. Reynolds, S. Tanguay, W. Thompson, and S. Underwood. 1991. Old growth and riparian habitat planning project: Review of literature addressing wildlife and fish habitat relationships in riparian and stream habitats. Unpublished report. U.S. Forest Service, Tahoe National Forest, Grass Valley, California.

Chaney, E., W. Elmore, and W. S. Platts. 1993. *Livestock grazing on western riparian areas*. Eagle, ID: Information Center, Inc.

Chapel, M., A. Carlson, D. Craig, T. Flaherty, C. Marshall, M. Reynolds, D. Pratt, L. Pyshora, S. Tanguay, and W. Thompson. 1992. Recommendations for managing late-seral-stage forest and riparian habitats on the Tahoe National Forest. Unpublished report. U.S. Forest Service, Tahoe National Forest, Nevada City, California.

Clark, W. B. 1970. *Gold districts of California*. Bulletin 193. San Francisco: California Division of Mines and Geology.

Costick, L. 1996. Indexing current watershed conditions using remote sensing and GIS. In *Sierra Nevada Ecosystem Project: Final report to Congress*, vol. II, chap. 54. Davis: University of California, Centers for Water and Wildland Resources.

Cross, S. D. 1988. Riparian systems and small mammals and bats. In *Streamside management: Riparian wildlife and forestry interactions*, edited by K. Raedeke. Seattle: University of Washington, Center for Streamside Studies.

Cummins, K. W., M. A. Wilzbach, D. M. Gates, J. B. Perry, and W. B. Taliaferro. 1989. Shredders and riparian vegetation. *BioScience* 39 (1): 24–30.

Diggles, M. F., J. R. Rytuba, B. C. Moring, C. T. Wrucke, D. P. Cox, S. Ludington, R. P. Ashley, W. J. Pickthorn, C. T. Hillman, and R. J. Miller. 1996. Geology and minerals issues. In *Sierra Nevada Ecosystem Project: Final report to Congress*, vol. II, chap. 18. Davis: University of California, Centers for Water and Wildland Resources.

Duane, T. P. 1996. Human settlement, 1850–2040. In *Sierra Nevada Ecosystem Project: Final report to Congress*, vol. II, chap. 11. Davis: University of California, Centers for Water and Wildland Resources.

Dudley, T. L., and W. E. Dietrich. 1995. Effects of cattle grazing exclosures on the recovery of riparian ecosystems in the southern Sierra Nevada. Technical Completion Report UCAL-WRC-W-831. Davis: University of California, Water Resources Center.

Dudley, T., and M. Embury. 1995. Non-indigenous species in wilderness areas: The status and impacts of livestock and game species in designated wilderness in California. Oakland: Pacific Institute for Studies in Development, Environment, and Security.

Dunford, E. G., and P. W. Fletcher. 1947. Effect of removal of streambank vegetation upon water yield. *Transactions of the American Geophysical Union* 28: 105–10.

Dunne, T., and L. B. Leopold. 1978. *Water in environmental planning*. San Francisco: W. H. Freeman.

Elmore, W., and R. L. Beschta. 1987. Riparian areas: Perceptions in management. *Rangelands* 9 (6): 260–65.

Erman. D. C. 1992. Historical background of long-term diversion of the Little Truckee River. In *The History of Water: Eastern Sierra Nevada, Owens Valley, and White Mountains*, edited by C. A. Hall, V. Doyle-Jones, and B. Widawski, 415–27. Los Angeles: University of California, White Mountain Research Station.

Erman, D. C., and D. Mahoney. 1983. *Recovery after logging in streams with and without bufferstrips in northern California*. Contribution 186. Berkeley: University of California, Water Resources Center.

Erman, D. C., J. D. Newbold, and K. B. Roby. 1977. *Evaluation of streamside bufferstrips for protecting aquatic organisms*. Contribution 165. Davis: University of California, Water Resources Center.

Erman, N. 1984. The use of riparian systems by aquatic insects. In *California riparian systems: Ecology, conservation, and productive management*, edited by R. E. Warner and K. Hendrix, 177–82. Berkeley and Los Angeles: University of California Press.

———. 1996. Status of aquatic invertebrates. In *Sierra Nevada Ecosystem Project: Final report to Congress*, vol. II, chap. 35. Davis: University of California, Centers for Water and Wildland Resources.

Federal Energy Regulatory Commission. 1986. *Final environmental impact statement, Owens River Basin: Seven hydroelectric projects, California*. Docket no. EL85-19-102. Washington DC: Federal Energy Regulatory Commission, Office of Hydropower Licensing.

Franklin, J. F., and J. A. Fites-Kaufmann. 1996. Analysis of late successional forests. In *Sierra Nevada Ecosystem Project: Final report to Congress*, vol. II, chap. 21. Davis: University of California, Centers for Water and Wildland Resources.

Furniss, M. J., T. D. Roelefs, and C. S. Yee. 1991. Road construction and maintenance. In *Influences of forest and rangeland management on salmonid fishes and their habitats*, edited by W. R. Meehan, 297–323. Special Publication 19. Bethesda, MD: American Fisheries Society.

Gilbert, G. K. 1917. *Hydraulic-mining debris in the Sierra Nevada*. Professional Paper 105. Washington, DC: U.S. Geological Survey.

Gill, C. J. 1970. The flooding tolerance of woody species—A review. *Forestry Abstracts* 31: 671–88.

Graber, D. 1996. Status of terrestrial vertebrates. In *Sierra Nevada Ecosystem Project: Final report to Congress*, vol. II, chap. 25. Davis: University of California, Centers for Water and Wildland Resources.

Grant, G. 1988. *The RAPID technique: A new method for evaluating downstream effects of forest practices on riparian zones*. General Technical Report PNW-220. Corvallis, OR: U.S. Forest Service, Pacific Northwest Forest and Range Experiment Station.

Greene, L. W., 1987. *Yosemite: The park and its resources; A history of the discovery, management and physical development of Yosemite National Park, California*. San Francisco: National Park Service.

Gregory, S. V., F. J. Swanson, W. A. McKee, and K. W. Cummins. 1991. An ecosystem perspective of riparian zones. *BioScience* 41: 540–51.

Griggs, F. T., V. Morris, and E. Denny. 1994. Five years of valley oak riparian forest restoration. *Fremontia* 22:13–17.

Groeneveld, D. P., and D. Or. 1994. Water table induced shrub-herbaceous ecotone: hydrologic management implications. *Water Resources Bulletin* 30 (5): 911–20.

Hagan, R. M., and E. B. Roberts. 1973. Ecological impacts of water storage and diversion projects. In *Environmental quality and water development*, edited by C. R. Goldman, J. McEvoy, and P. J. Richerson, 196–215. San Francisco: W. H. Freeman.

Harris, R. R., C. A. Fox, and R. Risser. 1987. Impacts of hydroelectric development on riparian vegetation in the Sierra Nevada region, California, USA. *Environmental Management* 11 (4): 519–27.

Herbst, D., and R. Knapp. 1995. Biomonitoring of rangeland streams under differing livestock grazing practices. *Bulletin of the North American Benthological Society* 14 (1): 176.

Hesse, L. W., and W. Sheets. 1993. The Missouri River hydrosystem. *Fisheries* 18 (5): 5–14.

Hicks, T. 1995. Riparian monitoring plan for hydropower projects. Bishop, CA: U.S. Forest Service, Inyo National Forest.

Hill, M. T., W. S. Platts, and R. L. Beschta. 1991. Ecological and geomorphological concepts for instream and out-of-channel flow requirements. *Rivers* 2: 198–210.

Holmes, D. O. 1979. Cultural influences on subalpine and alpine meadow vegetation in Yosemite National Park. In *First conference on scientific research in the national parks*, edited by R. M. Linn, 1267–72. Washington, DC: National Park Service.

Hughes, J. E. 1934. *Erosion control progress report*. Quincy, CA: U.S. Forest Service, Plumas National Forest, Milford Ranger District.

Hupp, C. R. 1986. The headward extent of fluvial landforms and associated vegetation on Massanutten Mountain, Virginia. *Environmental Management* 11: 545–55.

Jennings, M. R. 1996. Status of amphibians. In *Sierra Nevada Ecosystem Project: Final report to Congress*, vol. II, chap. 31. Davis: University of California, Centers for Water and Wildland Resources.

Johnson, W. C. 1992. Dams and riparian forests: Case study from the upper Missouri River. *Rivers* 3: 229–42.

———. 1994. Woodland expansion in the Platte River, Nebraska: Patterns and causes. *Ecological Monographs* 64 (1): 45–84.

Jones and Stokes Associates. 1989. Downstream effects of hydroelectric development on riparian vegetation: A joint PG&E/ SCE research project. Sacramento: Jones and Stokes Associates.

Junk, W. J., P. B. Bayley, and R. E. Sparks. 1989. The flood pulse concept in river-floodplain systems. In *Proceedings, International Large Rivers Symposium*, edited by D. P. Dodge, 110–27. Ottawa: Canadian Ministries of Fisheries and Oceans.

Katibah, E. F., K. J. Dummer, and N. E. Nedeff. 1984. Current condition of riparian resources in the Central Valley of California. In *California riparian systems: Ecology, conservation, and productive management*, edited by R. E. Warner and K. M. Hendrix, 314–21. Berkeley and Los Angeles: University of California Press.

Kattelmann, R. 1996. Hydrology and water resources. In *Sierra Nevada Ecosystem Project: Final report to Congress*, vol. II, chap. 30. Davis: University of California, Centers for Water and Wildland Resources.

Kattelmann, R., and M. Embury. 1996. Riparian areas and wetlands. In *Sierra Nevada Ecosystem Project: Final report to Congress*, vol. III. Davis: University of California, Centers for Water and Wildland Resources.

Kauffman, J. B. 1988. The status of riparian habitats in Pacific Northwest forests. In *Streamside management: Riparian wildlife and forestry interactions*, edited by K. J. Raedeke, 45–55. Seattle: University of Washington, Institute of Forest Resources.

Kinney, W. C. 1996. Conditions of rangelands before 1905. In *Sierra Nevada Ecosystem Project: Final report to Congress*, vol. II, chap. 3. Davis: University of California, Centers for Water and Wildland Resources.

Knight, A. W., and R. L. Bottorff. 1984. The importance of riparian vegetation to stream ecosystems. In *California riparian systems: Ecology, conservation, and productive management*, edited by R. E. Warner and K. M. Hendrix, 160–67. Berkeley and Los Angeles: University of California Press.

Kondolf, G. M. 1993. Lag in stream channel adjustment to livestock exclosure in the White Mountains of California. *Restoration Ecology* 1: 226–30.

———. 1994a. Cumulative watershed effects in the Sequoia National Forest: Summary of oral testimony before the House Agriculture Committee, 9 March 1994. *Hearing before the Subcommittee on Specialty Crops and Natural Resources on H.R. 2153, the Giant Sequoia Preservation Act of 1993*. 103rd Congress, 2nd session, 9 March. House of Representatives Serial No. 103–58.

———. 1994b. Geomorphic and environmental effects of instream gravel mining. *Landscape and Urban Planning* 28: 225–43.

———. 1994c. Livestock grazing and habitat for a threatened species: Land-use decisions under scientific uncertainty in the White Mountains of California. *Environmental Management* 18 (4): 501–9.

Kondolf, G. M., and R. R. Curry. 1986. Channel erosion along the Carmel River, Monterey County, California. *Earth Surface Processes and Landforms* 11: 307–19.

Kondolf, G. M., L. M. Maloney, and J. G. Williams. 1987a. Effect of bank storage, and well pumping on base flow, Carmel River, California. *Journal of Hydrology* 91: 351–69.

Kondolf, G. M., and W. V. G. Matthews. 1993. *Management of coarse sediment on regulated rivers*. Report 80. Davis: University of California, Water Resources Center.

Kondolf, G. M., and P. Vorster. 1993. Changing water balance over time in Rush Creek, eastern California, 1860–1992. *Water Resources Bulletin* 29: 823–32.

Kondolf, G. M., J. W. Webb, M. J. Sale, and T. Felando. 1987b. Basic hydrologic studies for assessing impacts of flow diversions on riparian vegetation: Examples from streams of the eastern Sierra Nevada, California. *Environmental Management* 11: 757–69.

Kondolf, G. M., and P. R. Wilcock. In press. The flushing flow problem: Defining and evaluating objectives. *Water Resources Research*.

Krzysik, A. J. 1990. Biodiversity in riparian communities and watershed management. In *Watershed planning and analysis in action*, edited by R. E. Riggins, E. B. Jones, R. Singh, and P. A. Rechard, 533–48. New York: American Society of Civil Engineers.

Leighton, J. P., and R. J. Risser. 1989. A riparian vegetation ecophysiological response model. In *Proceedings of the California riparian system conference: Protection, management, and restoration for the 1990s*, edited by D. L. Abell, 370–80. General Technical Report PSW-110. Berkeley: U.S. Forest Service, Pacific Southwest Forest and Range Experiment Station.

Lemons, J. 1979. Visitor use impact in a subalpine meadow, Yosemite National Park, California. In *First conference on scientific research in the national parks*, edited by R. M. Linn, 1287–92. Washington, DC: National Park Service.

Liddle, M. J. 1975. A selective review of the ecological effects of human trampling on natural ecosystems. *Biological Conservation* 7: 17–36.

Ligon, F. K., W. E. Dietrich, and W. J. Trush. 1995. Downstream ecological effects of dams: A geomorphic perspective. *BioScience* 45 (3): 183–92.

Loredo, I., D. Van Vuren, and M. L. Morrison. In press. Habitat use and migration behavior of the California tiger salamander. *Journal of Herpetology*.

Lyons, J. K., and R. L. Beschta. 1983. Land use, floods, and river channel changes: Upper Middle Fork Willamette River, Oregon (1936–1980). *Water Resources Research* 19 (2): 463–71.

Madej, M. A., W. E. Weaver, and D. K. Hagans. 1994. Analysis of bank erosion on the Merced River, Yosemite Valley, Yosemite National Park, California, USA. *Environmental Management* 18 (2): 235–50.

Mahoney, D. L., and D. C. Erman. 1984. The role of streamside bufferstrips in the ecology of aquatic biota. In *California riparian systems: Ecology, conservation, and productive management*, edited by R. E. Warner and K. M. Hendrix, 168–74. Berkeley and Los Angeles: University of California Press.

Manley, P. N. 1995. Biological diversity and its measure: An assessment in lotic riparian ecosystems of the Lake Tahoe basin. San Francisco: U.S. Forest Service, Pacific Southwest Region.

Marchetti, M. P. 1994. Suspended sediment effects on the stream fauna of Humbug Creek. Master's thesis, University of California, Davis.

Martin, K. E. 1984. Recreation planning as a tool to restore and protect riparian systems. In *California riparian systems: Ecology, conservation, and productive management*, edited by R. E. Warner and K. M. Hendrix, 748–57. Berkeley and Los Angeles: University of California Press.

Matthews, W. V. G. 1990. Design of restoration projects on gravel bed rivers emphasizing the reestablishment of riparian vegetation. Master's thesis, University of California, Santa Cruz.

Maser, C., and J. Sedell. 1994. *From the forest to the sea: The ecology of wood in streams*. Delray Beach, FL: St. Lucie Press.

McKelvey, K. S., and J. D. Johnston. 1992. Historical perspectives on forests of the Sierra Nevada and the Transverse Ranges of southern California: Forest conditions at the turn of the century. In *The California spotted owl: A technical assessment of its current status*, edited by J. Verner, K. S. McKelvey, B. R. Noon, R. J. Gutierrez, G. I. Gould Jr., and T. W. Beck, 225–46. Albany, CA: U.S. Forest Service, Pacific Southwest Research Station.

Medina, A. L. 1990. Possible effects of residential development on streamflow, riparian plant communities, and fisheries in small mountain streams in central Arizona. *Forest Ecology and Management* 33/34: 351–61.

Menke, J., C. Davis, and P. Beesley. 1996. Rangeland assessment. In *Sierra Nevada Ecosystem Project: Final report to Congress*, vol. III. Davis: University of California, Centers for Water and Wildland Resources.

Minshall, G. W. 1994. Stream-riparian ecosystems: Rationale and methods for basin-level assessments of management effects. In *Ecosystem management: Principles and applications: eastside forest ecosystem health assessment*, edited by M. E. Jensen and P. S. Bourgeron, 149–73. General Technical Report PNW-318. Portland, OR: U.S. Forest Service, Pacific Northwest Research Station.

Mount, J. F. 1995. *California rivers and streams: The conflict between fluvial processes and land use*. Berkeley and Los Angeles: University of California Press.

Moyle, P. B., R. Zomer, R. Kattelmann, and P. J. Randall. 1996. Watershed management: A summary of agency planning. *Sierra Nevada Ecosystem Project: Final report to Congress*, vol. III. Davis: University of California, Centers for Water and Wildland Resources.

Munro, J. W. 1991. Wetland restoration in the mitigation context. *Restoration and Management Notes* 9 (2): 80–86.

Murphy, M. L., and W. R. Meehan. 1991. Stream ecosystems. In *Influences of forest and rangeland management on salmonid fishes and their habitats*, edited by W. R. Meehan, 17–46. Special Publication 19. Bethesda, MD: American Fisheries Society.

Myers, L. H. 1987. *Riparian area management: Inventory and monitoring riparian areas*. Technical Reference 1737–3. Denver, CO: U.S. Bureau of Land Management.

Nachlinger, J. L., S. D. Smith, and R. J. Risser. 1989. Riparian plant water relations along the North Fork Kings River, California. In *Proceedings of the California riparian system conference: protection, management, and restoration for the 1990s*, edited by D. L. Abell, 366–69. General Technical Report PSW-110. Berkeley, CA: U.S. Forest Service, Pacific Southwest Forest and Range Experiment Station.

Naiman, R. J., H. DeCamps, and M. Pollock. 1993. The role of riparian corridors in maintaining regional biodiversity. *Ecological Applications* 3: 209–12.

National Research Council. 1987. *The Mono Basin ecosystem: Effects of changing lake level*. Washington, DC: National Academy Press.

———. 1992. *Restoration of aquatic ecosystems: Science, technology, and public policy*. Washington, DC: National Academy Press.

Odion, D. C., T. L. Dudley, and C. M. D'Antonio. 1990. Cattle grazing in southeastern Sierra meadows: Ecosystem change and prospects for recovery. In *Plant biology of eastern California*, edited by C. A. Hall and V. Doyle-Jones, 277–92. Los Angeles: University of California, White Mountain Research Station.

Overton, K. C., G. L. Chandler, and J. A. Pisano. 1994. *Northern/intermountain regions' fish habitat inventory: Grazed, rested, and ungrazed reference stream reaches, Silver King Creek, California*. General Technical Report INT-311. Ogden, UT: U.S. Forest Service, Intermountain Research Station.

Pagenhart, T. H. 1969. Water use in the Yuba and Bear River basins, California. Ph.D. diss., University of California, Berkeley.

Parrish, J. L., and W. V. G. Matthews. 1993. *Changes to Calaveras Big Trees State Park resulting from dam enlargement and flow changes at New Spicer Meadows Reservoir, North Fork Stanislaus River*. Berkeley: University of California, Center for Environmental Design Research.

Pelzman, R. J. 1973. Causes and possible prevention of riparian plant encroachment on anadromous fish habitat. Administrative Report 73-1. Sacramento: California Department of Fish and Game, Environmental Services Branch.

Perkins, D. J., B. N. Carlsen, M. Fredstrom, R. H. Miller, C. M. Rofer, G. T. Ruggerone, and C. S. Zimmerman. 1984. The effects of groundwater pumping on natural spring communities in Owens Valley. In *California riparian systems: Ecology, conservation, and productive management*, edited by R. E. Warner and K. M. Hendrix, 515–26. Berkeley and Los Angeles: University of California Press.

Petranka, J. W., M. P. Brannon, M. E. Hopey, and C. K. Smith. 1994. Effects of timber harvesting on low elevation populations of southern Appalachian salamanders. *Forest Ecology and Management* 67: 135–47.

Platts, W. S. 1991. Livestock grazing. In *Influences of forest and rangeland management on salmonid fishes and their habitats*, edited by W. R. Meehan, 389–423. Special Publication 19. Bethesda, MD: American Fisheries Society.

Poulin, R., R. C. Pakalnis, and K. Sinding. 1994. Aggregate resources: Production and environmental constraints. *Environmental Geology* 23: 221–27.

Raedeke, K. J., R. D. Taber, and D. K. Paige. 1988. Ecology of large mammals in riparian systems of Pacific Northwest forests. In *Streamside management: Riparian wildlife and forestry interactions*, edited by K. Raedeke, 113–32. Contribution 59. Seattle: University of Washington, Institute of Forest Resources.

Resh, V. H., A. V. Brown, A. P. Covich, M. E. Gurtz, H. W. Li, G. W. Minshall, S. R. Reice, A. L. Sheldon, J. B. Wallace, and R. C. Wissmar. 1988. The role of disturbance in stream ecology. *Journal of the North American Benthological Society* 7:433–55.

Reynolds, F. L., T. J. Mills, R. Benthin, and A. Low. 1993. *Restoring Central Valley streams: A plan for action*. Sacramento: California Department of Fish and Game.

Schiedt, M. E. 1967. Environmental effects of highways. *Journal of the Sanitary Engineering Division, Proceedings of the American Society of Civil Engineers* 93 (SA5): 17–25.

Schimer, F., and M. Zalewski. 1992. The importance of riparian ecotones for diversity and productivity of riverine fish communities. *Netherlands Journal of Zoology* 42 (2–3): 323–35.

Scott, M. L., E. D. Eggleston, G. T. Auble, J. M. Friedman, and L. S. Ischinger. In press. Effects of gravel mining on natural cottonwood stands. In *Proceedings of the sixth annual Colorado Riparian Conference.*

Sedell, J. S., and K. L. Luchessa. 1981. Using the historical record as an aid to salmonid habitat enhancement. In *Proceedings of the symposium on acquisition and utilization of aquatic habitat inventory information,* edited by N. B. Armantrout, 210–23. Bethesda, MD: American Fisheries Society.

Sigafoos, R. S. 1964. *Botanical evidence of floods and floodplain deposition.* Professional Paper 485-A. Washington, DC: U.S. Geological Society.

Smith, S. D., J. L. Nachlinger, A. B. Wellington, and C. A. Fox. 1989. Water relations of obligate riparian plants as a function of streamflow diversion on the Bishop Creek watershed. In *Proceedings of the California riparian system conference: Protection, management, and restoration for the 1990s,* edited by D. L. Abell. 360–65. General Technical Report PSW-110. Berkeley, CA: U.S. Forest Service, Pacific Southwest Forest and Range Experiment Station.

Sparks, R. E., P. B. Bayley, S. L. Kohler, and L. L. Osborne. 1990. Disturbance and recovery of large floodplain rivers. *Environmental Management* 14: 699–709.

State Water Resources Control Board. 1994. *Mono Lake Basin water right decision 1631.* Sacramento: State Water Resources Control Board.

Stine, S. 1991. *Extent of riparian vegetation on stream tributary to Mono Lake, 1930–1940: An assessment of the streamside woodlands and wetlands, and the environmental conditions that supported them.* Sacramento: California State Water Resources Control Board and Jones and Stokes Associates.

Stine, S., D. Gaines, and P. Vorster. 1984. Destruction of riparian systems due to water development in the Mono Lake watershed. In *California riparian systems: Ecology, conservation, and productive management,* edited by R. E. Warner and K. M. Hendrix, 528–33. Berkeley and Los Angeles: University of California Press.

Stromberg, J. C., and D. T. Patten. 1990. Riparian vegetation instream flow requirements: A case study from a diverted stream in the eastern Sierra Nevada, California. *Environmental Management* 14: 185–94.

Taylor, D. W. 1983. Assessing potential environmental impacts of small-hydro on riparian vegetation. Paper presented at Inyo County workshop on small-hydro projects, Bishop, California.

Taylor, D. W., and W. B. Davilla. 1985. *Riparian vegetation in the Crane Valley project.* San Ramon, CA: Pacific Gas and Electric Company.

Thomas, J. W., C. Maser, and J. E. Rodiek. 1979. Riparian zones. In *Wildlife habitats in managed forests: The Blue Mountains of Oregon and Washington,* edited by J. W. Thomas, 40–47. Agriculture Handbook 553. Washington, DC: U.S. Forest Service.

Todd, A. H. 1989. The decline and recovery of Blackwood Canyon, Lake Tahoe, California. In *Proceedings, International Erosion Control Association conference.* Vancouver, BC: International Erosion Control Association.

Tukey, J. W. 1977. *Exploratory data analysis.* Reading, PA: Addison-Wesley.

U.S. Forest Service (USFS). 1988. *Sequoia National Forest land and resource management plan.* Porterville, CA: Sequoia National Forest.

———. 1995. Results of stream condition inventory of grazed and ungrazed meadow streams. Unpublished report. U.S. Forest Service, Pacific Southwest Region, San Francisco.

Vankat, J. L., and J. Major. 1978. Vegetation changes in Sequoia National Park, California. *Journal of Biogeography* 5: 377–402.

Vannote, R. L., G. W. Minshall, K. W. Cummins, J. R. Sedell, and C. E. Cushing. 1980. The river continuum concept. *Canadian Journal of Fisheries and Aquatic Sciences* 37: 130–37.

Walters, M. A., R. O. Teskey, and T. M. Hinckley. 1980. *Impact of water level changes on woody riparian and wetland communities.* Vol. VII of *Mediterranean region, western arid and semi-arid region.* Washington, DC: U.S. Fish and Wildlife Service.

Ward, J. V., and J. A. Stanford. 1995. Ecological connectivity in alluvial river ecosystems and its disruption by flow regulation. *Regulated Rivers: Research and Management* 11:105–19.

Welsh, H. H. 1993. A hierarchical analysis of the niche relationships of four amphibians from forested habitats in northwestern California. Ph.D. diss., University of California, Berkeley.

Williams, G. P., and M. G. Wolman. 1984. *Downstream effects of dams on alluvial rivers.* Professional Paper 1286. Washington, DC: U.S. Geological Survey.

Wright, J. F., and A. D. Berrie. 1987. Ecological effects of groundwater pumping and a natural drought on the upper reaches of a chalk stream. *Regulated Rivers* 1: 145–60.

SECTION IV

Agents of Change in the Sierra Nevada

KEVIN S. McKELVEY AND
  SEVEN OTHER AUTHORS
Carl N. Skinner
Chi-ru Chang
Don C. Erman
Susan J. Husari
David J. Parsons
Jan W. van Wagtendonk
C. Phillip Weatherspoon

37

An Overview of Fire in the Sierra Nevada

ABSTRACT

Fire, ignited by lightning and Native Americans, was common in the Sierra Nevada prior to 20th century suppression efforts. Presettlement fire return intervals were generally less than 20 years throughout a broad zone extending from the foothills through the mixed conifer forests. In the 20th century, the areal extent of fire was greatly reduced. This reduction in fire activity, coupled with the selective harvest of many large pines, produced forests which today are denser, with generally smaller trees, and have higher proportions of white fir and incense cedar than were present historically. These changes have almost certainly increased the levels of fuel, both on the forest floor and "ladder fuels"—small trees and brush which carry the fire into the forest canopy. Increases in fuel, coupled with efficient suppression of low and moderate intensity fires, has led to an increase in general fire severity.

We suggest extensive modification of forest structure will be necessary to minimize severe fires in the future. In high-risk areas, landscapes should be modified both to reduce fire severity and to increase suppression effectiveness. We recommend thinning and underburning to reduce fire-related tree mortality coupled with strategically placed defensible fuel profile zones (DFPZs). DFPZs are areas in which forest structure and fuels have been modified to reduce flame length and "spotting", allowing effective suppression.

This chapter is an overview of work by the fire-subgroup of the Sierra Nevada Ecosystem Project. Details concerning these findings are found in Skinner and Chang 1996; Chang 1996; Husari and McKelvey 1996; McKelvey and Busse 1996; Erman and Jones 1996; van Wagtendonk 1996; and Weatherspoon 1996.

THE ECOLOGICAL ROLE OF FIRE

"The most potent factor in shaping the forest of the region has been, and still is, fire."
Leiberg 1902, 40

For thousands of years, the periodic recurrence of fire has shaped the ecosystems of the Sierra Nevada (Skinner and Chang 1996). Because fire was so prevalent in the centuries before extensive Euro-American settlement (presettlement), many common plants exhibit specific fire-adapted traits such as thick bark and fire-stimulated flowering, sprouting, seed release and/or germination (Chang 1996). In addition, fire affected the dynamics of biomass accumulation and nutrient cycling, and generated vegetation mosaics at a variety of spatial scales (Chang 1996). Because fire influenced the dynamics of nearly all ecological processes, reduction of the influence of fire through 20th century fire suppression efforts in these ecosystems has had widespread (though not yet completely understood) effects.

PATTERNS OF FIRE: PAST AND PRESENT

Estimates of presettlement median fire return intervals (length of time between fires), as recorded in fire scars, are typically less than 50 years (figure 37.1). More specifically, records from the foothill zone, mixed conifer zone, and east side pine showed median fire return intervals consistently less than 20 years.

Sierra Nevada Ecosystem Project: Final report to Congress, vol. II, *Assessments and scientific basis for management options.* Davis: University of California, Centers for Water and Wildland Resources, 1996.

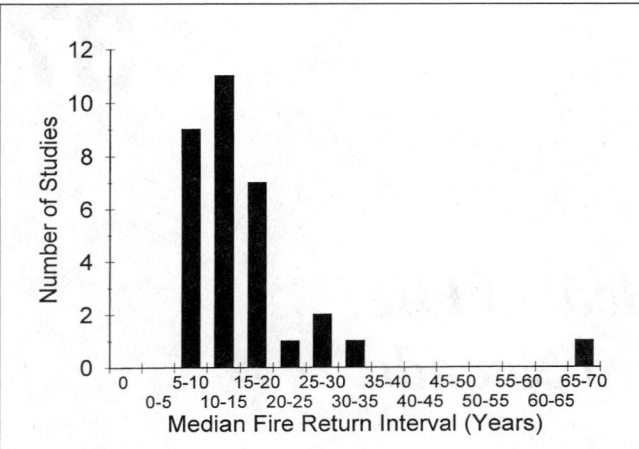

FIGURE 37.1

A histogram showing the number of tree ring studies (Skinner and Chang 1996) by median fire return interval. Only one study found a median return interval greater than 50 years.

Fire affects landscapes at a variety of scales, yet there have been few landscape-scale fire-history studies (Skinner and Chang 1996). Most fire history studies describe fire return intervals for local sites of generally 250 acres or less. The few existing landscape-scale fire-history studies found significant variation of fire occurrence from place to place (e.g., Kilgore and Taylor 1979; Caprio and Swetnam 1995). Fire return intervals varied in response to site and environmental factors such as ignition source, fuel accumulation, fuel moisture, and burning conditions (Kilgore and Taylor 1979). Furthermore, climate variation over many centuries is reflected in more frequent, less extensive fires during warmer periods and less frequent, more extensive fires during cooler periods (Swetnam 1993). Despite variability in fire occurrence patterns and uncertainties in how fire affected landscapes, we believe that the composite picture provided by these studies is compelling: presettlement fires were common enough to significantly affect forest structure and composition over much of the Sierra Nevada.

Knowledge of presettlement fire patterns helps to explain reports by early observers that describe the forests and woodlands as generally open (Sudworth 1900; Leiberg 1902). Sierra Nevada forests and woodlands at the turn of the century were altered by intense grazing by sheep and associated burning patterns in the late 1800s (McKelvey and Johnston 1992). However, these disturbances would have had a modest impact on the canopy trees that, according to Sudworth (1900), were large and on average 250 to 350 years old.

The general structure of forests and woodlands shaped by frequent fires may appear stable for decades or sometimes centuries, at least when viewed at the landscape level (1000s of acres). The pattern of change is usually limited in spatial extent (Skinner and Chang 1996), with patterns of tree mortality leading to patches of different sizes and ages of trees frequently confined to individuals or small groups. This pattern of mortality is most often caused by endemic insect activity, stem breakage due to weakening of trees from fire or physical scarring, or localized severe fire conditions sufficient to kill the canopy trees directly.

Fire regimes characteristic of presettlement conditions in the Sierra Nevada have been disrupted in the 20th century. The Forest Service has mapped fires since 1908, and these maps can be used to estimate the acreage burned in each forest type (McKelvey and Busse 1996). Fire rotation times were calculated from these data and compared with fire return intervals derived from fire scars (table 37.1). Fire rotation is defined as the number of years necessary to burn an area equal in size to an area of interest, in this case a forest type. Fire return intervals are based on the number of times a point (or small area) burned over a period of time. To compare the two metrics, we need to make the assumption that the fire return intervals generated from fire-scar studies represent the forest types in which they exist and therefore approximate fire rotations. Given this assumption, the deviation between current rates and presettlement fire return intervals is one to two orders of magnitude (table 37.1). We believe that this change is far too great to be accounted for by differences in measurement, climate, or potential biases associated with fire-scar data. Fire suppression has been extremely effective in the 20th century.

Another, and perhaps more subtle, difference between current and presettlement fire patterns lies in the recent decrease in fire frequency with increasing elevation. Throughout the 20th century, little of the higher elevation zones have burned (McKelvey and Busse 1996). This is in contrast to presettlement median fire return intervals that differed relatively little from the foothills through the upper mixed-conifer zone (table 37.1). The distribution of fires in the 20th century is closely associated with droughty conditions (McKelvey and Busse 1996) and probably is due to the effective suppression of low-to-

TABLE 37.1

Historic fire return intervals compared with 20th century patterns. Historical data are extracted from various sources (Skinner and Chang 1996) and are the average median return intervals for each forest type. Recent fire data are fire rotations based on area burned during the 20th century (McKelvey and Busse 1996).

Forest Type	Fire Return Period	
	20th Century	Pre-1900
Red fir	1,644	26
Mixed conifer-fir	644	12
Mixed conifer-pine	185	15
Ponderosa pine	192	11
Blue oak	78	8

moderate intensity fires. Before settlement, 10 times as much area in the foothills burned when compared with the 20th century, whereas 60 times as much burned in the red fir zone (table 37.1).

It is notable that 4,232 ha of the 38,828 ha red fir zone of Yosemite National Park has burned since a prescribed natural fire (PNF) program was initiated in 1972. Under this PNF program the calculated fire rotation is 163 years (van Wagtendonk 1995) for a much greater proportion than has burned outside the park (table 37.1). The PNF program has begun to approach presettlement fire rotation levels after 24 years.

CHANGES IN FUELS

The dramatic reduction in area burned in the 20th century, combined with the effects of forest management practices and generally warmer-moister climatic conditions (Graumlich 1993; Stine 1996), has almost certainly led to substantial increases in quantity and changes in arrangement of live and dead fuels. While data from the early 20th century are not available to test this assertion rigorously, it is based on comparisons with early conditions inferred from numerous historical accounts, documented fire histories, and structures of uncut stands (Kilgore and Sando 1975; Parsons and DeBenedetti 1979; Bonnicksen and Stone 1982; van Wagtendonk 1985; Biswell 1989; Weatherspoon et al. 1992; Chang 1996; Skinner and Chang 1996; Weatherspoon and Skinner 1996).

Live and dead fuels increased along with the development of denser conifer forests. These increases in stand density were concentrated mainly in small and medium size classes of shade-tolerant and fire-sensitive tree species. Lacking fire, the thinning that has occurred has been due to competition (primarily for water and light), diseases, and insects. The result has been a large increase in the amount and continuity of live forest fuels near the forest floor that provide a link between surface fuels and upper canopy layers. The lack of fire has also caused dead fuels on the forest floor to accumulate in excess of their presettlement levels.

The impact of forest management in the 20th century has primarily been to accelerate these trends. Logging on Forest Service and private lands has been primarily of the large overstory trees—accelerating growth in the dense understory and increasing landscape-level homogeneity of fuel structure (Weatherspoon 1996; McKelvey and Johnston 1992). Such cuttings, especially when combined with no treatment of slash (harvest-created fuel), increase the vulnerability of stands to damage from wildfires (Weatherspoon 1996). The national parks, while still maintaining extensive areas of large, old trees, have also experienced increased density of shade tolerant understory trees.

Therefore, compared with presettlement conditions, the current Sierra Nevada forests are generally younger, denser, smaller in diameter, and more homogeneous. Almost certainly there is increased dead biomass and on many sites increased live biomass as well (Weatherspoon and Skinner 1996). Due to high productivity and various forest management activities, the lower and middle-elevation mixed conifer forests have likely experienced greater change in structure and fuels conditions than have either higher elevation forests or foothill vegetation (Weatherspoon and Skinner 1996).

CHANGES IN FIRE INTENSITY

Frequently in the following chapters the assertion is made that fires today are more intense than either presettlement fires, or fires in the early 20th century. More precisely, the assertion is that current fires burn much larger contiguous areas at high intensities, resulting in a larger proportion of the burned area suffering severe fire effects. We have no direct data to support these assertions, but, as with the increase in fuels, such a conclusion is consistent with information available from fire history studies and other sources. The frequency and extensiveness of fires that occurred in the presettlement era were simply too high to allow the accumulation of dead fuel and live "ladder" fuels that support extensive crown fires.

Accounts of early surveyors explicitly state that crown fires were uncommon. In 1899, when George Sudworth was surveying the central Sierra Nevada, fires were so routinely encountered that "travel through a large part of the territory was at times difficult on account of dense smoke" (Sudworth 1900, p. 560). Nevertheless, Sudworth states of these, and previous fires:

Fires of the present time are peculiarly of a surface nature, and with rare exception there is no reason to believe that any other type of fire has occurred here. (Sudworth 1900, P. 557)

and

The incidences in this region where large timber has been killed outright by surface fires are comparatively rare. Two cases only were found. . . . One of these burns involved less than an acre, and the other included several hundred acres. (Sudworth 1900, P. 558)

Statements such as these by Sudworth would be absurd if they referred to today's forests. It is not likely that Sudworth could spend a year traversing the central Sierra Nevada today without at least noting the large burned patches created by stand replacing fire, sometimes covering tens of thousands of acres.

Leiberg surveyed an area north of where Sudworth worked, much of which is currently in the Tahoe and Plumas National Forests. Of the 2.8 million acres that he judged had burned at least once in the preceding 100 years, he concluded that total tree mortality had occurred on 214 thousand acres (Leiberg 1902). In making this determination, however, Leiberg assumed that most meadows and chaparral fields were fire-generated (Leiberg 1902). Even if this assumption was correct (which is doubtful), total tree mortality had occurred on less than 8% of the area burned.

THE ROLE OF SUPPRESSION

Many human uses and management activities have influenced patterns in Sierra Nevada ecosystems over the last century (e.g., grazing, logging, mining, recreation, settlement, fire management). However, only fire suppression has been applied throughout the Sierra Nevada landscape. Until recently, whatever the vegetation, if staff was available, fires were actively and vigorously suppressed. As a result, the fire suppression policy of the 20th century has played a primary role in the human induced changes in many Sierra Nevada vegetation types.

Though stand conditions and fire suppression methodologies and goals have changed during the 20th century (Husari and McKelvey 1996), the effect of suppression on a number of fire attributes has remained remarkably constant. An analysis of fires occurring on the national forests of the Sierra Nevada suggest there is no general time trend in total area burned (McKelvey and Busse 1996), and patterns of location have also remained stable. The relationships between fire occurrence and elevation have remained essentially constant (McKelvey and Busse 1996), as have the distributions of fire sizes for most national forests, the Eldorado and Stanislaus Forests being exceptions (Erman and Jones 1996). Of fires reaching a size of 100 acres or more, human caused fires have exceeded lightning fires in numbers and total area burned throughout this century. Only in the last two decades have the largest fires been caused by lightning (McKelvey and Busse 1996).

Recent Changes in Fire Suppression Resources

There is no question that we have the most mobile and highly organized fire suppression force ever assembled. However, the overall pool of available work force and equipment has recently declined (Husari and McKelvey 1996).

The number of fire suppression resources available for initial attack peaked during the 1970s and early 1980s and was declining by the late 1980s. Both the California Department of Forestry and Fire Protection and the U.S. Forest Service have seen an approximate 10% reduction in the numbers of

engines since the peak (Husari and McKelvey 1996). Although these declines are somewhat offset by increases in the numbers of state hand crews, the latter are generally used as reinforcements and not for initial assault.

The New Role of Lightning

In recent decades, the proportion of burned area contributed by lightning-caused fires has increased while the proportion burned by human-caused fires has decreased (McKelvey and Busse 1996; Weatherspoon and Skinner 1996). Additionally, both the number and size of human-caused fires have decreased. In contrast, the size of lightning fires has increased in the past three decades, particularly in the late 1980s and early 1990s (McKelvey and Busse 1996; Weatherspoon and Skinner 1996). If increased difficulty in suppressing individual fires was the only factor, one might expect an increase in the size of both lightning and human-caused fires.

A potential explanation for the recent increase in the proportion of area burned by lightning fires could lie in the temporal concentration of lightning strikes—lightning-caused ignitions often occur simultaneously during thunderstorms, overwhelming suppression resources.

A fire suppression organization that is efficient, highly organized and extremely mobile can usually control solitary ignitions even when staffing is limiting. When multiple ignitions occur, however, work force availability can become pressing: suppression resources allocated to one fire are not available for additional fires. For multiple simultaneous ignitions, the densities of resources available for initial attack are therefore critically important (Husari and McKelvey 1996).

While declines in suppression resources coincided with an increase in large lightning-caused fires in the Sierra Nevada, the extent to which they contributed to these lightning events is unknown. California experienced many large lightning-caused fires in 1977 (Biswell 1989) during the period of peak regional levels of suppression resources. Though none of the largest of these were in the Sierra Nevada, many occurred in the extended SNEP study area (i.e., Cascade Range and Modoc Plateau).

The increase in proportion of total burned area accounted for by lightning fires may be influenced by changing fuel conditions associated with fire suppression and other management activities (e.g., increasing stand densities, accumulating woody debris) discussed previously and in the next section. It is reasonable to expect these changes in fuel conditions to contribute to fires that are more difficult to control. Furthermore, it is likely that the effects of increasing suppression difficulty, despite suppression resource levels, would first become apparent under conditions of numerous, widespread, simultaneous ignitions (Weatherspoon and Skinner 1996).

THE ROLE OF FUEL TREATMENTS

The Rationale for Fuels Management

Fires in Sierra Nevada forests and woodlands occur less frequently and cover much less area than they did in the presettlement era; however, they are more likely to be more uniformly severe. These large, severe fires, in aggregate, are well outside the range of sizes and severity expected for the presettlement era and thus may be detrimental to the integrity and sustainability of Sierran ecosystems. Furthermore, the current prevalence of such fires is socially unacceptable.

The continuing accumulation of large quantities of forest biomass that fuel severe wildfires points to a need to increase the treatment of fuels substantially. To reduce the total area and average size burned by severe fires, designing treatments in landscape-level patterns that are strategically logical for fire management would be preferable. Concurrently, restoring more of the ecosystem functions associated with frequent low- to moderate-severity fires that previously characterized most Sierra Nevada forest ecosystems would be desirable (Weatherspoon and Skinner 1996).

The foothills and lower elevation mixed conifer zones have experienced rapid population growth in recent decades and this is unlikely to cease soon. The prevalence of private lands divided among many landowners, often with houses scattered through the landscape, makes it essential that cooperation among landowners, local entities, and fire agencies take place on a broad scale to effectively deal with the threat to human life and property.

Changing Stand Structure and Fire Behavior

The recent changes in forest structure in the Sierra Nevada probably have affected fire behavior in various ways (Weatherspoon 1996). Current forests are generally much denser than under presettlement conditions, and they contain more surface and ladder fuels (intermediate layers of smaller trees and shrubs). These changes create forests that are more likely to support large, severe fires.

Before the 20th century, forests and woodlands were generally more open. Such forests and woodlands have higher surface temperatures, lower relative humidities, dry more quickly, ignite more easily, and burn more rapidly than the dense forests of today. However, though the flame lengths can be high, fires in the more open forests and woodlands are more likely to be predominately surface fires. The more open conditions are less likely than multilayered, closed-canopy forests to support crown fires or have extensive areas of the overstory trees killed.

Potential Effects of Fuels Treatments on Fire Behavior

The recent accumulations of biomass (both living and dead) that fuel wildfires necessitate the development of strategies to manage fuels to reduce the extent of area burned by severe fire and to help ease the reintroduction of fire as an ecological process. Many fuels treatments involve thinning the smaller diameter trees or biomass removals (Weatherspoon 1996), in essence producing stands structurally similar to what are thought to have been presettlement conditions. Resulting forest structures will be more open, less likely to support crown fire, and less likely to exhibit extensive areas of severe fire effects. The post-treatment fire behavior will be strongly affected by the quantity of surface fuels left on site. Removal of trees necessary to open the stand (increasing drying and wind at the forest floor) usually produces much more severe fire behavior if slash is left untreated on site (van Wagtendonk 1996). Fuel treatments will need to be applied periodically to maintain effectiveness over time. If we fail to maintain treated areas, we will again be faced with hazardous fuel conditions.

Strategic Planning of Fuels Treatments

Given the massive scope of the fuels problem and budget constraints, a carefully-considered, landscape-level strategy is required. On public lands, treatments conducted to reduce the hazard of severe wildfires should be compatible with overall desired conditions for sustainable ecosystems. For private lands, creative processes will be needed that can balance society's desire to reduce the threat to lives, property, and resource values with land owners' individual goals and property rights. Treatments need to begin in the most logical, efficient, and cost-effective places. Additionally, the rate of treatment needs to be carefully planned: In the short term, rates of biomass removal may need to exceed rates of production, to return Sierra Nevada forests to a more sustainable, fire-resilient condition (Weatherspoon and Skinner 1996).

We believe a successful fuels management strategy can be based on three components: (1) a series of broad "defensible fuel profile zones" (DFPZs) to be described later; (2) use of fire for restoring natural processes while meeting fuels management goals; and (3) expansion of fuels treatments to other appropriate areas of the landscape, consistent with desired ecosystem conditions. In the short- to midterm (at least the first decade), installation and maintenance of DFPZ networks probably offer the greatest potential for reducing the area and average size burned by large, high-severity wildfires in the Sierra Nevada, and consequent losses of lives, property, and resource values. Increased use of prescribed fire should take place concurrently.

Development of DFPZs involves thinning and otherwise treating fuels as needed to reduce fire hazards. However, areas to be treated should be contiguous and reflect a planned strategy. DFPZs provide a zone of reduced fire intensity and reduced spotting potential (van Wagtendonk 1996) where suppression forces have a reasonable chance of stopping a fire.

We see DFPZs only as the first step in landscape-wide fuels treatments, not as the final solution. In this context, they should provide a foundation from which to extend subsequent prescribed burning or other treatments to broader areas of the landscape. A DFPZ network will help to achieve improved forest health, greater landscape diversity, increase availability of open forest habitats dominated by large trees, and, thus, probably a greater approximation of presettlement conditions along the ridges and upper southerly slopes where they would be concentrated. Periodic maintenance of DFPZs will be essential to their continued effectiveness, as with any fuel treatment. However, their contiguity, usually easy access (providing for easier, cheaper re-treatments), intended value for staging suppression forces, and protecting property and resources, increases the probability that they will be maintained (Weatherspoon and Skinner 1996).

Evaluating Fire Risk

One important factor that should be considered for selecting fuel treatment strategies is risk of fire occurrence based on historical patterns. An evaluation of fire-occurrence risk in high-value areas (e.g., wildland-urban interface areas, national and state parks, productive resource lands, and ecologically significant areas) should help prioritize the location of DFPZs and other fuel treatment areas. An assessment of risk of fire occurrence was developed from Forest Service records of fires mapped for the 20th century (McKelvey and Busse 1996), and the zones of highest risk were found in the foothills. During periods of drought, the area of high fire risk extends into the lower portions of the mixed conifer zone.

Use of Prescribed Fire

Prescribed fire is frequently advocated as a tool that can be used for landscape level fuel reduction while simultaneously restoring fire as an ecosystem process. We recognize the important role that prescribed fire can, and should, play in managing Sierra Nevada ecosystems; however, we must caution against reliance on prescribed fire as the only solution.

Practical and political considerations restrain reliance on prescribed fire, even for the more restricted objectives of restoration and maintenance of natural processes in parks and wilderness (Parsons and Botti in press). Both management ignited and prescribed natural fires (PNFs) ignited by lightning will occasionally escape prescriptions and boundaries, potentially resulting in unacceptable impacts and the ultimate threat of additional restrictions on fire use (such as those following the 1988 Yellowstone fires)(Botti and Nichols 1995). Additional constraints on prescribed fire programs include inadequate funding, inadequate number of personnel (due to competition for trained personnel during active wildfire

seasons), and air quality restrictions. The difficulties of carrying out a prescribed fire program are illustrated by the failure of the fire management program at Sequoia and Kings Canyon National Parks to approach the presettlement fire return frequency for the giant sequoia groves in spite of a well funded, aggressive program (Parsons 1995).

Modifying Fire Suppression Strategy

Although permitting low- and moderate-intensity wildfires to burn can provide benefits, the vast majority of ignitions in the Sierra Nevada are suppressed. Fire managers have been required to select the most economically efficient suppression option without considering the potential resource benefits of wildfires. Fires that would contribute to achieving presettlement vegetation conditions are regularly suppressed while small, because they are easy and inexpensive to put out. However, flexibility in present federal fire management policy exists that is rarely exercised outside the National Parks and a few wilderness areas in the Sierra Nevada. Fire managers may use less than full control strategies for fire suppression provided the strategy chosen is projected to incur the least cost of suppression plus loss of resource values. Use of appropriate suppression responses, expanded use of prescribed fire and use of PNF both inside and outside wilderness should be evaluated based on fire regime, expected fire behavior and weather regime throughout the Sierra Nevada. Agencies should seriously consider using managed wildfires to meet resource objectives in combination with prescribed fire and other forms of fuels treatment. Indeed, the proposed fire policies for the U.S. Departments of Agriculture and Interior would allow land managers the flexibility to use wildland fires to meet management objectives (USDI/USDA 1995).

CONCLUSION

Fire has been, and will continue to be, a major influence on Sierra Nevada landscapes. Each summer conditions occur where fires can easily ignite and spread due to the Mediterranean climate characterized by cool, wet winters and warm, dry summers. Under this prevailing climate, fire has served as a frequent, potent influence on Sierra Nevada ecosystems for millennia.

The combination of human uses and management activities over the last century and a half has profoundly altered fire regimes. The area influenced by low- and moderate-intensity fires in the Sierra Nevada landscapes has been greatly reduced. This has resulted in changes in forest structures and landscape patterns. Today many Sierra Nevada landscapes will more readily support more uniformly severe large fires than were characteristic of presettlement conditions.

It is likely that occurrence of large, severe fires will continue in the Sierra Nevada into the indefinite future. Fuel treatments may provide a long-term solution by reducing the likelihood of tree mortality and crown fires and producing defensible zones in which fires can be more successfully controlled. Fuel treatments must be periodically maintained to avoid a return to hazardous conditions. Initial treatment costs are often high and there will be continuing costs for maintenance. It is crucial that fuel treatment areas are carefully located to increase effectiveness and minimize costs. We must devise ways to provide the necessary financing, and where practical use the harvest of biomass to help pay for the needed treatments. The foothill zone will present the greatest challenge to manage fire effectively to protect human life and property, whereas, the mixed conifer and upper montane zones may present the greatest challenge from an ecological and resource perspective.

Even an extremely aggressive fuels treatment program will take more than a decade to accomplish. After more than a century of changing forest structures and fuel conditions, realizing significant regional shifts in fire behavior will take time, determination, and good financing. Judicious planning will help to achieve ecological goals while reducing the spatial extent and effects of severe fires.

Future Needs

Each of the following chapters has developed recommendations and assessed future needs that are best understood in the details of the separate reports. Together they lay the framework for action, better management, and better understanding of the critical role fire plays in the human and ecological landscape of the Sierra Nevada.

ACKNOWLEDGMENTS

This chapter was written by Kevin S. McKelvey, Carl N. Skinner, and C. Phillip Weatherspoon, all of the U.S. Forest Service, Pacific Southwest Research Station, Redding, California; David J. Parsons, of the U.S. Forest Service, Aldo Leopold Wilderness Research Institute, Missoula, Montana; Don C. Erman, of the Centers for Water and Wildland Resources, University of California, Davis; Jan W. van Wagtendonk, of the U.S. National Biological Survey, Yosemite Field Station, El Portal, California; Susan J. Husari, of the U.S. Forest Service, Region 5, San Francisco; and Chi-ru Chang, of the School of the Environment, Duke University, Durham, North Carolina.

We greatly appreciate the leadership, direction, and coordination for the Agents of Change working group provided by Joan Brenchley-Jackson. The helpful comments of William F. Laudenslayer Jr. and two anonymous reviewers improved the manuscript.

REFERENCES

Albini, F. A. 1976. *Estimating wildfire behavior and effects*. General Technical Report INT-30. Ogden, UT: U.S. Forest Service, Intermountain Research Station.

Biswell, H. H. 1989. *Prescribed burning in California wildlands vegetation management*. Berkeley and Los Angeles: University of California Press.

Bonnicksen, T. M., and E. C. Stone. 1982. Managing vegetation within U.S. national parks: A policy analysis. *Environmental Management* 6:101–2, 109–22.

Botti, S. J., and H. T. Nichols. 1995. Availability of fire resources and funding for prescribed natural fire programs in the National Park Service. In *Proceedings: Symposium on fire in wilderness and park management*, technical coordination by J. K. Brown, R. W. Mutch, C. W. Spoon, and R. H. Wakimoto, 94–104. General Technical Report INT-GTR-320. Ogden, UT: U.S. Forest Service, Intermountain Research Station.

Caprio, A. C., and T. W. Swetnam. 1995. Historic fire regimes along an elevational gradient on the west slope of the Sierra Nevada, California. In *Proceedings: Symposium on fire in wilderness and park management*, technical coordination by J. K. Brown, R. W. Mutch, C. W. Spoon, and R. H. Wakimoto, 173–79. General Technical Report INT-GTR-320. Ogden, UT: U.S. Forest Service, Intermountain Research Station.

Chang, C. 1996. Ecosystem responses to fire and variations in fire regimes. In *Sierra Nevada Ecosystem Project: Final report to Congress*, vol. II, chap. 39. Davis: University of California, Centers for Water and Wildland Resources.

Erman, D. C., and R. Jones. 1996. Fire frequency analysis of Sierra forests. In *Sierra Nevada Ecosystem Project: Final report to Congress*, vol. II, chap. 42. Davis: University of California, Centers for Water and Wildland Resources.

Graumlich, L. J. 1993. A 1000-year record of temperature and precipitation in the Sierra-Nevada. *Quaternary Research* 39:249–55.

Husari, S. J., and K. S. McKelvey. 1996. Fire management policies and programs. In *Sierra Nevada Ecosystem Project: Final report to Congress*, vol. II, chap. 40. Davis: University of California, Centers for Water and Wildland Resources.

Kilgore, B. M., and R. W. Sando. 1975. Crown-fire potential in a sequoia forest after prescribed burning. *Forest Science* 21:83–87.

Kilgore, B. M., and D. Taylor. 1979. Fire history of a sequoia–mixed conifer forest. *Ecology* 60:129–42.

Leiberg, J. B. 1902. *Forest conditions in the northern Sierra Nevada, California*. Professional Paper 8, Series H, Forestry, 5. Washington, DC: U.S. Geological Survey, Government Printing Office.

McKelvey, K. S., and K. K. Busse. 1996. Twentieth-century fire patterns on Forest Service lands. In *Sierra Nevada Ecosystem Project: Final report to Congress*, vol. II, chap. 41. Davis: University of California, Centers for Water and Wildland Resources.

McKelvey, K. S., and J. D. Johnston. 1992. Historical perspectives on forests of the Sierra Nevada and the Transverse Ranges of Southern California: Forest conditions at the turn of the century. In *The California spotted owl: a technical assessment of its current status*, technical coordination by J. Verner, K. S. McKelvey, B. R. Noon, R. J. Gutierrez, G. I. Gould Jr., and T. W. Beck, 225–46. General Technical Report GTR-PSW-133. Albany, CA: U.S. Forest Service, Pacific Southwest Research Station.

Parsons, D. J. 1995. Restoring fire to giant sequoia groves: What have we learned in 25 years? In *Proceedings: Symposium on fire in wilder-*

ness and park management, technical coordination by J. K. Brown, R. W. Mutch, C. W. Spoon, and R. H. Wakimoto, 256–58. General Technical Report INT-GTR-320. Ogden, UT: U.S. Forest Service, Intermountain Research Station.

Parsons, D. J., and S. J. Botti. In press. Restoration of fire in national parks. In *Proceedings of the international conference of the Society of Ecological Restoration. Session: The use of fire in forest restoration.* Ogden, UT: U.S. Forest Service, Intermountain Research Station.

Parsons, D. J., and S. H. DeBenedetti. 1979. Impact of fire suppression on a mixed-conifer forest. *Forest Ecology and Management* 2:21–33.

Skinner, C. N., and C. Chang. 1996. Fire regimes, past and present. In *Sierra Nevada Ecosystem Project: Final report to Congress*, vol. II, chap. 38. Davis: University of California, Centers for Water and Wildland Resources.

Stine, S. 1996. Climate, 1650–1850. In *Sierra Nevada Ecosystem Project: Final report to Congress*, vol. II, chap. 2. Davis: University of California, Centers for Water and Wildland Resources.

Sudworth, G. B. 1900. Stanislaus and Lake Tahoe Forest Reserves, California, and adjacent territories. In *Annual reports of the Department of Interior, 21st annual report of the U.S. Geological Survey.* Part 5, 505–61. Washington, DC: Government Printing Office.

Swetnam, T. W. 1993. Fire history and climate change in giant sequoia groves. *Science* 262:885–89.

U. S. Department of the Interior and U.S. Department of Agriculture (USDI/USDA). 1995. *Federal wildland fire management policy and program review: Final report.* Washington, DC: U.S. Department of the Interior and U.S. Department of Agriculture.

van Wagtendonk, J. W. 1985. Fire suppression effects on fuels and succession in short-fire-interval wilderness ecosystems. In *Proceed-*

ings, Symposium and workshop on wilderness fire, technical coordination by J. E. Lotan, B. M. Kilgore, W. C. Fischer, and R. W. Mutch, 119–26. General Technical Report INT-182. Ogden, UT: U.S. Forest Service, Intermountain Research Station.

———. 1995. Large fires in wilderness areas. In *Proceedings: Symposium on fire in wilderness and park management*, technical coordination by J. K. Brown, R. W. Mutch, C. W. Spoon, and R. H. Wakimoto, 113–16. General Technical Report INT-GTR-320. Ogden, UT: U.S. Forest Service, Intermountain Research Station.

———. 1996. Use of a deterministic fire growth model to test fuel treatments. In *Sierra Nevada Ecosystem Project: Final report to Congress*, vol. II, chap. 43. Davis: University of California, Centers for Water and Wildland Resources.

Weatherspoon, C. P. 1996. Fire-silviculture relationships in Sierra forests. In *Sierra Nevada Ecosystem Project: Final report to Congress*, vol. II, chap. 44. Davis: University of California, Centers for Water and Wildland Resources.

Weatherspoon, C. P., S. J. Husari, and J. W. van Wagtendonk. 1992. Fire and fuels management in relation to owl habitat in forests of the Sierra Nevada and Southern California. In *The California spotted owl: a technical assessment of its current status*, technical coordination by J. Verner, K. S. McKelvey, B. R. Noon, R. J. Gutierrez, G. I. Gould Jr., and T. W. Beck, 247–60. General Technical Report PSW-133. Albany, CA: U.S. Forest Service, Pacific Southwest Research Station.

Weatherspoon, C. P., and C. N. Skinner. 1996. Landscape-level strategies for forest fuel management. In *Sierra Nevada Ecosystem Project: Final report to Congress*, vol. II, chap. 56. Davis: University of California, Centers for Water and Wildland Resources.

CARL N. SKINNER
U.S. Forest Service
Pacific Southwest Research Station
Redding, California

CHI-RU CHANG
Duke University
School of the Environment
West Durham, North Carolina

38

Fire Regimes, Past and Present

ABSTRACT

Fire has been an important ecosystem process in the Sierra Nevada for thousands of years. Before the area was settled in the 1850s, fires were generally frequent throughout much of the range. The frequency and severity of these fires varied spatially and temporally depending upon climate, elevation, topography, vegetation, edaphic conditions, and human cultural practices.

Current management strategies and those of the immediate past have contributed to forest conditions that encourage high-severity fires. The policy of excluding all fires has been successful in generally eliminating fires of low to moderate severity as a significant ecological process. However, current technology is not capable of eliminating the high-severity fires. Thus, the fires that affect significant portions of the landscape, which once varied considerably in severity, are now almost exclusively high-severity, large, stand-replacing fires. The resulting landscape patterns are much coarser in grain.

Many gaps still exist in our knowledge of fire as an ecological process in the Sierra Nevada.

INTRODUCTION

Fire has been an important ecological force in Sierra Nevada ecosystems for thousands of years. In only a few vegetated areas of the Sierra would fire not be considered an important element of the ecosystem. The Mediterranean climate, with cool, wet winters and warm, dry summers, predisposes much of the Sierra Nevada to conditions that would carry fire annually. As a result, prior to the mid-1800s, many of the plant communities experienced fire at least once, and often a number of times, during the life spans of the dominant plant species. Appropriately, much of the vegetation of the Sierra Nevada exhibits traits that allow survival and/or reproduction in this environment of regular fire (e.g., Chang 1996).

The patterns in which fires occur in an area, known as the fire regime for that area, influence the nature of the vegetation mosaic to be found within a particular landscape. Knowing the history of fire in a landscape is therefore important in gaining an understanding of the role of fire in ecosystems. An understanding of fire ecology and fire history may provide managers, decision makers, and policy makers with information that will help them avoid being shocked by unanticipated situations. When one is planning for resource management, for ecosystem management, and for community well-being, a knowledge of fire history and fire ecology provides a reference for assessing how much deviation is developing from long-term past patterns and conditions (e.g., Swanson et al. 1994; Manley et al. 1995). Evaluating the probability of success of future long-term alternatives for resources or communities is problematic without considering the physical and biological potential for fire and its function in Sierran environments.

Substantial research and documentation regarding fire ecology and regimes have demonstrated the importance of fire in Sierra Nevada environments. It is not possible to summarize all of the relevant information here. This chapter provides only sufficient background to establish for the reader the general nature of, and thereby the importance of, fire in the ecosystems of the Sierra Nevada. For the reader wishing more on the topic, the references should serve as a good starting place. Kilgore 1973, Biswell 1989, and Arno in press provide excellent summaries. Additionally, though primarily focused on fire ecology in vegetation north of California, Agee 1993 and Agee 1994 summarize fire ecology information for vegetation types (i.e., mixed conifer, red fir, subalpine, and east-side vegetation types) found in the northern Sierra as well as in the

Sierra Nevada Ecosystem Project: Final report to Congress, vol. II, *Assessments and scientific basis for management options*. Davis: University of California, Centers for Water and Wildland Resources, 1996.

extended SNEP study areas of the southern Cascades and northeastern California.

Unless otherwise indicated, the descriptions of past fire regimes presented in this chapter pertain to conditions before 1850. We have done this because many changes in fire regimes have occurred since the influx of Euro-American culture during and following the gold rush of the mid-1800s.

patterns have been found to correspond well with variations in the frequency and apparent severity of fires (Swetnam 1993). Swetnam 1993 demonstrated that thirty years was the longest period without fire in any of five sequoia groves for more than 2,000 years. These fire scar records also show that many of these same groves have now experienced more than 100 years without fire under modern management policies (Swetnam 1993).

THE PALEOECOLOGICAL RECORD

Paleoecological studies show that Sierra Nevada fire regimes are dynamic in space and time on many scales. The long-term importance of fire in Sierran ecosystems is suggested by the common occurrence of charcoal in the paleoecological record of the Holocene (e.g., Smith and Anderson 1992; Davis and Moratto 1988). Analyses of fossil pollen suggest that climate and vegetation have varied considerably over this period (Woolfenden 1996). Vegetation and fire appear to have varied, sometimes greatly, in concert with the variation in climates (Davis et al. 1985). Fire may serve as a catalyst for the reorganization of vegetation during periods of rapid climate change (e.g., Whitlock 1992; Wigand et al. 1995). It is noteworthy that large charcoal peaks from the early Holocene were followed by vegetation that was considerably different from that found before this period of heightened fire activity (Edlund and Byrne 1990). However, the resolution of temporal data available for the Sierra Nevada is insufficient to define the role of fire in reorganizing vegetation at various times in the past.

The evidence from interpretation of long-term trends in sediment cores has shown that fire has been an important component of the Sierran environment since before current vegetation assemblages became established. Charcoal concentrations at one site were greatest during the warm period that followed the end of the Pleistocene, approximately 10,000 years ago (Smith and Anderson 1992). These concentrations of charcoal appear to coincide with the end of the Pleistocene vegetation typical of subalpine forests today. The charcoal concentrations were followed by species assemblages more similar to the mixed conifer forests found today at middle and lower elevations (Anderson 1990). Following this warm period there was a general cooling trend until approximately 3,000 years ago, when a relatively cooler, more moist climate regime appears to have become established. Charcoal, though varying over time with changes in climate and vegetation, is routinely present in sediment core samples (Smith and Anderson 1992).

More-detailed reconstructions of climate variations for the last few millennia have recently been developed using tree-ring analysis (e.g., Graumlich 1993; Hughes and Brown 1992; Stine 1994). These variations in temperature and moisture

THE ETHNOGRAPHIC RECORD

Ethnographic accounts show that Native Californians commonly used fire as a management tool in the Sierra Nevada (Reynolds 1959; Wickstrom 1987; Blackburn and Anderson 1993). Fire was used to provide many important foodstuffs and materials (Lewis 1993). The spatial extent of the influence of burning on the landscape is not known and has been subject to some debate (Barrett 1935; Wickstrom 1987). However, accounts of the frequency of fire necessary to maintain specific resources in conditions required by the various cultures suggest that extensive and very intensive burning would have been common in important vegetation types (Anderson and Moratto 1996).

Enhancing the production of foodstuffs was one important reason for burning (Wickstrom 1987). For example, acorns were a major staple in the diet of the Native Californians, and burning was reported to enhance the production of acorns. Acorn crops are described to have been improved by burning in two important ways: (1) by reducing the losses to insects and (2) by encouraging larger, more productive trees (Anderson 1993b).

A second important reason for burning was to encourage the production of basketry materials (Anderson 1993a; Lewis 1993). The better materials for making baskets were young, straight shoots of many sprouting species. As the shoots matured, they would become unsuitable due to side branching and lack of flexibility.

A third reason for burning usually given by Native Californians was to reduce the hazard of large, severe fires (Lewis 1993). The native cultures were reliant upon local resources for their livelihood. A large, severe fire could change the local plant communities in a way that affected the ability of the communities to survive in the area. For example, a large, severe fire could top-kill the old oaks that provided acorns, the main staple. These trees could not produce sufficient supplies for many years following a fire of this type.

FIRE REGIMES

Fire ecologists refer to the general characteristics of fires found within any specified area of interest as the fire regime. Fire regimes can vary considerably by vegetation and landscape. Thus, they offer a convenient way to categorize areas for study and management purposes. Fire regimes are described by the following characteristics: frequency, rotation, spatial extent, magnitude, and seasonality (White and Pickett 1985; Agee 1994). These terms are defined as follows:

Frequency: The frequency describes how often fires occur within a given time period. This characteristic is often described in terms of return intervals rather than frequency. The return interval is the length of time between fires.

Rotation: The fire rotation is the length of time necessary to burn an area equal to the area or landscape of interest. For example, if one is working with a landscape of 100,000 acres and it takes fifty years for fires to burn 100,000 acres within that landscape, the fire rotation would be fifty years. Keep in mind that all 100,000 acres need not burn if some acres are burned more than once. The only requirement for this term is a total accumulated burn area equal to the original area of interest.

Spatial extent: The spatial extent refers to the size or area covered by a fire and the spatial patterns created.

Magnitude: The magnitude of a fire refers to both its intensity and its severity. *Intensity* is a technical term used to describe the amount of energy released from a fire. Intensity may or may not be directly related to fire effects. *Severity* is related to the change in the ecosystem caused by the fire and can be either quantitatively or qualitatively related to fire effects. Fires that burn only surface fuels and in which most of the woody vegetation survives are usually considered low-severity fires. Fires that kill large trees over more than a few acres by burning their crowns are usually considered high-severity fires.

Seasonality: The seasonality, or timing, of a fire is important in relation to the moisture content of fuels, the phenology of the vegetation, and the resulting fire effects. The vegetation found within a particular ecosystem has adapted over time to the season or seasons in which the fires generally occur.

Few fire-history studies have attempted to describe all of the fire-regime characteristics just defined. Most describe the fire frequencies for points (a single tree) or small sites. These data are the easiest and least costly to obtain. Some have also included seasonality as interpreted from the location of the scar in the rings (i.e., latewood or early wood) of the year of the fire. Few studies have attempted to describe the rotation, spatial extent, or magnitude of past fires, because acquiring these data requires intensive sampling of many sites over a landscape. These latter studies are quite costly due to the time and labor involved in field sampling and laboratory analysis.

Each of the fire-regime characteristics, when used, is usually described in terms of the mean or median and sometimes in terms of a measure of variability. The median is used in this chapter because of the variability in fire-return intervals associated with vegetation types that do not have very regular, frequent fires. The median is less affected by erratic extremes than is the mean (Snedecor and Cochran 1980). The mean is often interpreted and applied in a way that would assume that the data come from a normal distribution with little variation, but fire-return intervals for many sites are often not represented by a normal distribution. Instead, they are often multimodal (Johnson and Gutsell 1994) or strongly skewed, with many shorter intervals and a few longer, extreme intervals. The pattern of fire-return intervals often varies from period to period, and a simple mean is not representative of longer records (Swetnam 1993).

FIRE HISTORY

Fire history can be reconstructed from a variety of data sources: written records, historical accounts, dendrochronology (tree-ring analysis), and the analysis of charcoal in sediments (Patterson and Backman 1988). Each of these data sources has its own limitations regarding spatial and temporal detail and accuracy. Within forested ecosystems, detailed reconstruction of fire histories before written records is possible through fire scar analysis using dendrochronology techniques (Agee 1993; Arno and Sneck 1977). Fire history, in contrast to human history, is not limited to written records or accounts.

Fire histories from fire scar analysis generally fall into one of three categories: (1) single-tree samples; (2) composites of multiple trees for specified areas (Dieterich 1980a, 1980b); and (3) composites of multiple sites for landscapes (Taylor 1993a). The single-tree sample is usually considered the most conservative estimate of past fire history for many areas (although Minnich et al. [in press] have some data to suggest this may not always be the case). This is because all fires passing a point may not have been of sufficient intensity to have scarred the single sample. Composites of multiple trees will usually provide a more comprehensive record of past fires for the site in question (Agee 1993). However, describing in spatial and temporal terms the influence of fire on landscape dynamics (age-class distributions, species composition patterns, stand structures, patch patterns, etc.) requires detailed landscape-level sampling (e.g., Teensma 1987; Morrison and Swanson

1990; Caprio and Swetnam 1995; Minnich et al. in press; Solem 1995). Because of the time and costs involved in this type of sampling, few studies of this nature have been undertaken. Instead, most fire-history studies have been site-specific, fire-return-interval studies.

In each type of fire history, the fire dates can be determined either by cross-dating (Fritts and Swetnam 1989) or by estimating correspondence among years (Arno and Sneck 1977). The cross-dating method is precise and can determine the calendar year of fires hundreds or thousands of years ago (Swetnam 1993). The second method is not as accurate in determining the actual calendar year of a fire and often underestimates the number of fires within a period of interest. However, it still provides valuable, though less detailed, fire-interval information (Madany et al. 1982) that can be useful in describing the fire regime, especially at the level of detail required by most natural resource managers.

Fire histories based on tree-ring analysis rely on interpretation of scars that formed in response to fire-caused damage in the tree ring of the year of the fire. A number of factors can influence the way in which fires are recorded as scars. Trees are the best recorders, since they are long-lived and are large enough to be able to survive fires of low to moderate intensity. Little information is usually left following fires in herbaceous or shrub communities because of heavy consumption and the fact that the parts of the plants that are above ground are often killed. In addition, the various tree species vary considerably in their susceptibility to damage or mortality by fire. In areas consisting solely of species of trees that are usually killed even by low-intensity fires, there may not be a record of fire prior to the last fire that initiated the current stand.

Fire-severity classes used in this chapter are

- Low severity: light surface fire; some small trees may be killed.

- Moderate severity: most small trees killed; some subcanopy trees killed or heavily damaged. Charring on bark of live trees. Overstory trees may occasionally be killed.

- High severity: small and subcanopy trees killed; many to most overstory trees killed.

Although none of the fire-history studies described in the next section meets the exacting standards of the randomized sampling design described by Johnson and Gutsell (1994), taken together they provide valuable information about the past temporal patterns of fires within forest stands on a localized scale. The sampling design suggested by Johnson and Gutsell was developed in forests characterized by infrequent, large, stand-replacing fires. These fires result in very coarse-grained landscape patterns. The objective of fire-history studies in such forests is primarily to describe when the last fire occurred in each patch by dating the age of the trees that regenerated after the burns. The spatial scales on which topography and vegetation vary, along with the spatial variability

in fire behavior and the effects from the frequent fires in the forests of the Sierra Nevada, create very complex, fine-grained spatial patterns. Attempting to carry out a landscape-scale study based upon the design suggested by Johnson and Gutsell under these latter conditions would be very difficult and costly.

Several fire-history studies have been completed within the Sierra Nevada and adjacent geographical areas. Most of these studies have been limited to providing information on fire-return intervals (FRIs) for a small area. A few have developed fire history at the landscape scale in the Sierra (e.g., Caprio and Swetnam 1995; Kilgore and Taylor 1979). For studies of areas with vegetation similar to that in portions of the Sierra Nevada, see McNeil and Zobel 1980, Taylor and Halpern 1991, Taylor 1993a, Solem 1995, and Minnich et al. in press. Each of these will be discussed in more detail later in relation to appropriate vegetation types.

Table 38.1 summarizes fire-history information from various published and unpublished sources for the Sierra Nevada and other areas that have similar vegetation and climate. The spatial context of the return intervals is given to facilitate comparisons among the areas. There exist a number of other fire-history studies within the Sierra Nevada (e.g., Rice 1990, 1992). These were not included in table 38.1 primarily because the spatial reference for the sampling was not clear and we were unsure of the spatial comparability of the reported fire frequencies. Other studies were not included because the data were presented in a fashion that was difficult to present within the structure of table 38.1 (e.g., Mensing 1988, 1992). These latter studies are referenced in the text where appropriate.

The FRIs as presented for a small, localized place do not necessarily provide information on how fire would have influenced the landscape-scale patterns. Periods of more frequent fires may have many small, low-severity fires scattered throughout the landscape, while periods of longer FRIs may be associated with larger, more severe fires (Swetnam 1993). The spatiotemporal variation in fire frequency and severity may be important in influencing stand structure and regeneration patterns over time (e.g., Minnich et al. in press; Stephenson et al. 1991), leading to the complex spatial patterns of the vegetation that are so characteristic of the Sierra.

Landscape-scale fire-history studies are especially important to our understanding of the role of fire in Sierran ecosystems. The continued lack of such studies for the Sierra Nevada leaves important questions unresolved concerning the spatial and temporal influences of fire on vegetation dynamics, aquatic and riparian environments, wildlife habitat, coarse woody debris accumulations, and so on. Although many fire-history studies have been conducted in the Sierra, there is a considerable need to expand the geographical coverage and to conduct landscape-scale studies of fire history tied to vegetation-related dynamics.

TABLE 38.1

Fire-return intervals (FRIs) from the Sierra Nevada and areas of similar vegetation and climate.

Area and Vegetation	Median FRIs[a]	Minimum FRI[a]	Maximum FRI[a]	Years Since Last Fire	Method	Sample Area	Years of Record[b]	Location	Source
West-Side Areas									
Foothill Zone									
Blue oak–gray pine	8 (29)	2 (8)	49 (49)	14–34	Composites of multiple trees	5 ha	78–267	Northern Sierra	McClaran and Bartolome 1989
Black oak– ponderosa pine	8	2	18	82–102	Composites of multiple trees	<2 ha	Not reported	Central Sierra	S. Stephens, e-mail communications with the author, 21 April and 9 and 30 May 1995[c]
	6–9[e]	2	23	Not reported	Composites of multiple trees	1 ha	175	Southern Sierra	Kilgore and Taylor 1979
Mixed Conifer Zone									
Mixed evergreen– tan oak	15 (13)	3 (5)	50 (41)	43–71	Composites of multiple trees	5–8 ha	235–245	Klamath Mountains	Wills and Stuart 1994
Canyon live oak– mixed conifer	13 (11)	7 (7)	39 (33)	5–75	Composites of multiple trees	<1 ha	112–116	Klamath Mountains	Taylor and Skinner in preparation[c]
Ponderosa pine– mixed conifer	8–10[e]	3	14	Not reported	Composites of multiple trees	1 ha	175	Southern Sierra	Kilgore and Taylor 1979
	5–11[e]	Not reported	Not reported	~3–135	Composites of multiple trees	<100 ha	~125–340	Southern Sierra	Caprio and Swetnam 1995[d]
	11 (11)	3 (5)	55 (46)	35–90	Composites of multiple trees	<2 ha	151–306	Klamath Mountains	Skinner in preparation[c]
Giant sequoia– mixed conifer	Not reported	1	15	Not reported	Composites of multiple trees	<100 ha	1,050	Southern Sierra	Swetnam et al. 1991[d]
	Not reported	1	30	Not reported	Composites of multiple trees	10–100 ha	1,350	Southern Sierra	Swetnam 1993[d]
	15–18[e]	4	35	Not reported	Composites of multiple trees	1 ha	175	Southern Sierra	Kilgore and Taylor 1979
	14–32[e]	Not reported	Not reported	~10–195	Composites of multiple trees	<100 ha	~195–380	Southern Sierra	Caprio and Swetnam 1995[d]
Douglas fir– mixed conifer	17 (16)	12 (12)	59 (18)	69	Composites of multiple trees	2 ha	169	Klamath Mountains	Agee 1991[d]
	13 (14)	3 (3)	57 (52)	26–93	Composites of multiple trees	2 ha	125–396	Klamath Mountains	Skinner in preparation[c]
	12 (15)	3 (3)	59 (59)	5–92	Composites of multiple trees	<1 ha	248–379	Klamath Mountains	Taylor and Skinner in preparation[c]
White fir–mixed conifer	12 (11)	3 (3)	33 (29)	72–82	Composites of multiple trees	1 ha	133–154	Southern Cascades	McNeil and Zobel 1980
	10 (13)	3 (5)	24 (24)	36–56	Composites of multiple trees	2 ha	192–289	Southern Cascades	Skinner unpublished data
	9 (10)	3 (4)	26 (26)	34–47	Composites of multiple trees	2 ha	214–268	Southern Cascades	Skinner unpublished data
	8 (13)	3 (3)	35 (35)	38–67	Composites of multiple trees	2 ha	103–252	Northern Sierra	Skinner unpublished data

[a]Values in parentheses are specifically pre-1850 where available. Other values are for the entire period of record.
[b]The number of years from the earliest fire scar to the latest fire scar. A range indicates multiple sample sites.
[c]Unpublished study.
[d]Cross-dated samples.
[e]These are means for the sites sampled, rounded to the nearest integer. This study reported only means and did not present data in a way to develop medians or ranges.

continued

TABLE 38.1 (continued)

Area and Vegetation	Median FRIs[a]	Minimum FRI[a]	Maximum FRI[a]	Years Since Last Fire	Method	Sample Area	Years of Record[b]	Location	Source
	12 (11)	4 (4)	32 (32)	47–101	Composites of multiple trees	2 ha	53–196	Central Sierra	Skinner unpublished data
	18 (18)	7 (7)	27 (27)	100	Single-tree sample		113	Central Sierra	Taylor 1993b[c,d]
	11–17[e]	2	39	Not reported	Composites of multiple trees	1 ha	175	Southern Sierra	Kilgore and Taylor 1979
Mixed conifer (Lake Tahoe Basin, Emerald Bay State Park)	12 (11)	3 (3)	114 (44)	0–109	Single-tree sample		97–373	Central Sierra	Rice 1988[c]
White fir	17 (16)	4 (4)	61 (39)	72–128	Composites of multiple trees	1 ha	84–154	Southern Cascades	McNeil and Zobel 1980
	9 (11)	4 (4)	56 (56)	35	Composites of multiple trees	2 ha	214	Southern Cascades	Skinner unpublished data
Riparian areas	31 (36)	7 (7)	71 (71)	49–102	Composites of multiple trees	1 ha	175–265	Klamath Mountains	Skinner in preparation[c]
Upper Montane Zone Jeffrey pine	16 (13)	4 (4)	157 (157)	93–139	Single-tree sample		81–345	Southern Cascades	A. H. Taylor, telephone conversation with the author, 17 May 1995[c,d]
Jeffrey pine–white fir	12 (12)	4 (4)	96 (96)	42	Composites of multiple trees	2 ha	480	Klamath Mountains	Skinner in preparation[c]
	29 (29)	4 (10)	100 (93)	93–156	Single-tree sample		88–326	Southern Cascades	A. H. Taylor, telephone conversation with the author, 17 May 1995[c,d]
Red fir–white fir	12 (11)	5 (5)	69 (69)	77–98	Composites of multiple trees	<4 ha	97–240	Central Sierra	Bahro 1993[c]
Red fir–white pine	69 (57)	14 (14)	109 (109)	102–211	Single-tree sample		31–167	Southern Cascades	A. H. Taylor, telephone conversation with the author, 17 May 1995[c,d]
Red fir	20 (20)	8 (8)	35 (35)	128	Composites of multiple trees	1 ha	63–98	Southern Cascades	McNeil and Zobel 1980
	11 (16)	1 (7)	47 (35)	45–52	Composites of multiple trees	3 ha	141–205	Southern Cascades	Taylor 1993[d]
East-Side Areas Ponderosa pine	16	8	32	34–75	Composites of multiple trees	<10 ha	70–169	Southern Cascades	Olson 1994[c]
	8	6	15	51	Single-tree sample		169	Southern Cascades	Olson 1994[c]
Mixed conifer (ponderosa pine–white fir)	9 (10)	3 (3)	71 (71)	40	Composites of multiple trees	<10 ha	143–362	Southern Cascades	Olson 1994[c]
Red fir	27 (21)	9 (19)	91 (22)	63–90	Composites of multiple trees	<10 ha	113–135	Southern Sierra	Hawkins 1994[c]
Red fir–Jeffrey pine	14 (14)	6 (6)	64 (23)	36–143	Single-tree sample		135–227	Southern Sierra	Hawkins 1994[c]
	17 (15)	5 (7)	56 (31)	41–43	Composites of multiple trees	<10 ha	161–243	Southern Sierra	Hawkins 1994[c]

[a]Values in parentheses are specifically pre-1850 where available. Other values are for the entire period of record.
[b]The number of years from the earliest fire scar to the latest fire scar. A range indicates multiple sample sites.
[c]Unpublished study.
[d]Cross-dated samples.
[e]These are means for the sites sampled, rounded to the nearest integer. This study reported only means and did not present data in a way to develop medians or ranges.

FIRE EFFECTS

There is generally more literature on the effects of fire than there is concerning fire history. The effects of fire on the mixed conifer forests and chaparral of the west side of the Sierra Nevada are the subject of an abundant literature, whereas the literature on fire effects for the higher elevations and the eastside forests is rather sparse (Chang 1996).

It would have been rare in most vegetation types under the Mediterranean climate and pre-1850 fire regimes for individuals of most woody species to have escaped fire during their life span. As a result, much of the Sierra Nevada vegetation exhibits traits that have allowed the various species to persist with periodic fire. Whether these plant communities are the result of community adaptations to fire (e.g., Mutch 1970) or are coincidental assemblages of species that individually developed fire-adaptive traits over long periods (e.g., Davis 1986) is subject to debate. Regardless, many of the more common Sierran vegetative communities are generally considered adapted to recurring fire (Chang 1996). This section discuss the general adaptive traits and responses of this vegetation to fire. For more detail on fire effects, including effects on soils and fauna, please refer to Chang 1996.

Conditions for successful reproduction of many plant species are most favorable immediately following a fire, owing to increased fertility, removal of potential allelochemicals (chemicals released by plants that are toxic to other plants), reduced competition, and so on (Canham and Marks 1985; Christensen 1993). Consequently, many plants have evolved adaptive traits that help them survive or reproduce after fire. Such traits include the following (Sweeney 1968; Christensen 1985; Agee 1993):

1. fire-stimulated seed germination, as in deer brush (*Ceanothus integerrimus*)

2. rapid growth and development that allows a complete life cycle between fires, as in many herbaceous species

3. fire-resistant buds and twigs, as in ponderosa pine (*Pinus ponderosa*) and Douglas fir (*Psuedotsuga menziesii*)

4. fire-resistant bark, as in ponderosa pine and Douglas fir

5. adventitious or latent axillary buds, as in oaks (*Quercus* spp.)

6. sprouting, as in chamise (*Adenostoma fasciculatum*)

7. serotinous cones and fire-stimulated seed release, as in knobcone pine (*Pinus attenuata*) and giant sequoia (*Sequoiadendron giganteum*)

8. fire-stimulated flowering, as in soap plant (*Chlorogalum pomeridianum*)

Additionally, these varied vegetative responses are often influenced by a complex interaction of external factors such as temperature, moisture conditions, heat duration (e.g., Rogers et al. 1989), and season of burn (e.g., Parker and Kelly 1989; Weatherspoon 1988).

Changes in Fuel Structure Related to Fire-Return Intervals

Because photosynthesis produces organic matter on a regular basis, vegetative biomass (fuel) accumulates with time and adds to total fuel accumulations. However, not all biomass is available fuel at any given moment. Available forest fuel is organic matter that could burn under the prevailing conditions if ignited. The amount of biomass available as fuel at any one time depends on factors such as the ratio of dead plant material to live material and fuel moisture content.

Fire plays an important role in regulating fuel accumulations. Fire also influences the horizontal and vertical continuities of fuels. The importance of fire in regulating fuel accumulations is amplified where fire occurs frequently. Under the regimes in the era before fire suppression, frequent fires would consume surface fuels, maintaining them at minimal levels. Periodic low-to-moderate-intensity fires also maintained gaps in vertical fuel continuity, inhibiting fires from moving into the crowns (e.g., Sudworth 1900; Leiberg 1902). Fire suppression in forests that previously experienced frequent fires has allowed fuels to build up both vertically and horizontally, increasing the chance of stand-replacement fires (Brown 1985; van Wagtendonk 1985; Kilgore 1987; Arno in press).

Landscape Patterns Resulting from Fire

Landscapes can be viewed as a dynamic mosaic of patches (White and Pickett 1985). The frequency, intensity, and spatial extent of successive fires, along with the vegetative response, have influenced and will continue to influence the grain and pattern of the landscape mosaic.

Fire regimes help to define the pattern or mosaic of age classes, successional stages, and vegetation types on the landscape (Turner et al. 1993). For example, periodic, low-intensity surface burning has been found to cause development of an uneven-aged stand, made up of even-aged groups of trees of various age classes (e.g., Bonnicksen and Stone 1981; Weaver 1967). Conversely, infrequent, high-intensity, stand-replacement fires result in larger patches of stands of more even age (e.g., Heinselman 1973; Hemstrom and Franklin 1982).

Because the dynamics determining landscape patterns are affected by fire regimes, characteristics of landscapes also respond to variations in fire regimes. Skinner (1995a), for example, found that forest openings have disappeared or become smaller in a remote area of the Klamath Mountains during the period of effective fire suppression.

Within a given landscape, fire behavior will vary on a variety of spatial scales, influenced by local microclimate, topography, and fuel conditions. This varying behavior will interact with the postfire climate to induce ecosystem responses that result in varying landscape mosaics (Christensen et al. 1989).

Much of the literature that discusses landscape patterns in relation to fire and changing fire regimes has concentrated on ecosystems in which the fire regimes are characterized by infrequent or long-return-interval, severe fires (e.g., Heinselman 1973; Hemstrom and Franklin 1982; Romme 1982; Baker 1989). Because of the differences in scale of the effects of such fires, most studies that discuss the spatial patterns associated with fire regimes consisting of frequent, low-to-moderate-intensity fires have concentrated more on stand-level patterns than on landscape-level patterns (e.g., Bonnicksen and Stone 1982). Generally, the relative extent of fires within a particular vegetation type appears to increase as the interval between fires lengthens (Swetnam 1993; Husari and Hawk 1994).

FIRE REGIMES OF MAJOR VEGETATION TYPES

Due to the past frequency of fires and their influences on species composition, stand structure, and spatiotemporal patterns, fire is generally considered an important ecological process throughout much of the Sierra Nevada. Some fire-regime characteristics (e.g., frequency [FRI] and seasonality) of specific vegetation types are described quite well in the Sierra Nevada. In the southern Sierra, especially in the national parks, the FRIs in ponderosa pine and mixed conifer forests (especially those with giant sequoia) are relatively well documented. Less-detailed data exist for some upper montane areas and for the foothill areas of the southern Sierra. However, for many other fire-regime characteristics, vegetation types, and geographical portions of the Sierra Nevada little data exist.

Most of the information about fire regimes in the Sierra Nevada is from studies that used tree-ring analyses to detect when and how often past fires occurred in a particular place. In the following discussions, for cases where no descriptions of the fire regimes for the Sierra Nevada were found, we have extrapolated information from studies of similar vegetation in other geographical areas, used anecdotal information from historical sources, or made inferences based upon our knowledge of fire behavior and effects. The type of source(s) used is noted in the discussions.

The northern Sierra Nevada is especially lacking in published research designed to describe the long-term fire regimes. Only a few sites, in the works of Show and Kotok (1924) and Wagener (1961), have been studied. The discussions of fire regimes for the northern Sierra Nevada therefore rely on information extrapolated from studies of similar vegetation types in the southern Cascades, the Klamath Mountains, and the southern Sierra Nevada.

The FRIs given in the following discussions are for the entire period of record unless otherwise stated. The period of record for FRIs from fire scars is the time from the earliest scar recorded for the site to the last scar recorded for the site. Refer to table 38.1 for more detail concerning the FRIs for the presettlement and postsettlement period, the length of the record, and the sizes of the sampling areas represented by the FRIs.

Foothill Zone

The foothill zone is generally below the main belt of the conifer forests and above the Sacramento and San Joaquin valleys. The blue oak–gray pine woodlands, chaparral, mixed evergreen woodlands, and black oak–ponderosa pine forests are the more common vegetation types found in the foothills (Parsons 1981). Little is known about fire history in these vegetation types. Woodlands usually promote fast-moving fires, generally of low severity, in herbaceous fuels that may not leave a record as scars in trees. Chaparral, on the other hand, usually supports severe fires that kill the above-ground parts of the plants. For the latter, only the time elapsed since the last fire can be reconstructed. The discussions that follow for these vegetation types should be viewed in this light.

Considerable evidence exists that the Native Californians burned frequently, usually in the late summer or fall months, within the foothill areas (Anderson 1993a; Lewis 1993). The spatial extent of this burning is unknown but appears to have been substantial, at least near communities, considering the amount of postfire resources required (Anderson and Moratto 1996). This burning increased the number of fires that would otherwise have been expected from lightning alone. Generally, fewer lightning strikes occur, and fewer resulting fires are ignited, in the lower elevations than farther upslope (Komarek 1967; van Wagtendonk 1991b). The relative proportion of the area burned by human-caused fires to that burned by lightning-caused fires could vary considerably under different intensities of management.

Blue Oak–Gray Pine

The blue oak–gray pine woodland is common throughout the lower elevations of the Sierra Nevada. Common trees are blue oak (*Quercus douglasii*), gray pine (*Pinus sabiniana*), interior live oak (*Q. wislizenii*), and California buckeye (*Aesculus californica*).

Fire History. Two fire-history studies are available for blue oak–gray pine woodlands, only one of which is from the Sierra Nevada proper. The one study in the Sierra Nevada is McClaran and Bartolome 1989, from the University of California Sierra Foothill Range Field Station east of Marysville. The median FRIs on two sites were 7 and 9 years. The study found shortened FRIs during and following the settlement period (post-1848). The pre-1848 FRIs ranged from 8 to 49 years, with a median of 28.5, whereas post-1848 FRIs ranged from 2 to 17 years with medians of 7 to 8 years.

Mensing (1988) found that fire frequency recorded as fire scars in blue oaks on three sites in the Tehachapi Mountains had changed considerably since European settlement. He found mean FRIs for the presettlement period to range from 9.6 to 13.6 years. During the settlement period (1843–1865), the mean FRIs were 3.3 to 5.8 years, and post-1865 FRIs ranged from 13.5 to 20.3. Interestingly, he found a period of more than 60 years (the 1860s to the 1920s) that lacked any evidence of fire scars. He found this period to coincide with the introduction of livestock grazing, suggesting a reduction in available fuels. A similar coincidence of grazing with fire scar reduction in the Sierra Nevada has been noted by Vankat and Major (1978).

Fire Effects. The blue oak–gray pine woodlands are well adapted to frequent, quick-moving, low-intensity surface fires (Arno in press). Fuels are usually light, and the primary carrier of fire is the surface herbaceous vegetation. Notably, the surface vegetation has changed from consisting largely of perennials in presettlement times to being dominated by introduced annuals (Heady 1977). Historically, the perennials may have limited the season of burning. Annuals, on the other hand, may promote an earlier onset to the burning season because they dry and cure earlier than the perennials.

The trees usually survive these surface fires except where increased fire intensity is created by fallen, dead trees or an increased density of understory shrubs. The oaks and buckeyes are strong sprouters when occasional fires do kill the above-ground portions of the plants. Most of the understory shrubs are scattered individuals or groups of species usually associated with chaparral. The frequency of fire in this vegetation type usually keeps the shrub cover limited.

Shrubs

Most shrub communities in the Sierra are considered chaparral. Chaparral is a term applied to communities of predominantly evergreen shrubs adapted to hot, dry summers and periodic fire typical of Mediterranean climates (Hanes 1977; Kilgore 1981; Barro and Conard 1991). Chaparral is common in the lower elevations of the Sierra Nevada, usually between the oak woodlands and the conifer forests and on steep, often rocky, south-facing slopes of canyons. Species composition varies considerably both locally and regionally in the chaparral. Some important common species are chamise, scrub oak *(Quercus dumosa),* interior live oak, manzanitas *(Arctostaphylos* spp.), ceanothus *(Ceanothus* spp.), toyon *(Heteromeles arbutifolia),* yerba santa *(Eriodictyon californicus),* and California buckeye. Locally important nonchaparral shrub communities are found where Brewer's oak *(Q. garryanna* var. *brewerii)* is an important component.

The severe nature of the fires and the intermingling of many

rural communities with areas of chaparral present a considerable challenge to natural resource managers regarding the need to ensure public safety as well as to manage wildlife habitat and watersheds (Sparks and Oechel 1984).

Fire History. Despite the common occurrence of chaparral and its importance to management, fire history for chaparral in the Sierra Nevada is lacking. Historical information is limited to fire records from this century and previous anecdotal accounts (Parsons 1981). Fire-history information about chaparral in California is generally confined to studies in the Coast and Transverse Ranges. These longer-term studies are from charcoal in oceanic sediment deposits (Byrne et al. 1977; Mensing 1993). Because of differences in lightning frequency and burning conditions, these studies may present conservative estimates of fire frequency for inland areas (Keeley 1982).

FRIs in chaparral appear to be quite variable, depending upon local site conditions, proximity to areas of aboriginal human use, and elevation. Chaparral FRIs generally have been estimated to be twenty to fifty years with ranges of approximately ten to more than a hundred years (Keeley 1982; Kilgore 1987; Barro and Conard 1991). FRIs in chaparral types have been limited to estimates because the severe nature of the fires in chaparral renders the areas unsuited to the reconstruction of fire history from dendrochronological techniques (Minnich and Howard 1984). However, Johnson and Gutsell (1994) suggest that FRIs in chaparral may be estimated by using a randomized spatial sampling design similar to that used in boreal forest types to determine the years since the last fire for various portions of the landscape. It is unlikely that this approach will work well in landscapes that have been affected by the extremely large fires of the last few decades, because the large fires will probably have destroyed the previous age-class patterns.

Fire Effects. Due to the dense growing habit of shrubland vegetation and the long dry season, fires in this type of community are usually severe and kill most above-ground portions of the vegetation (Christensen 1985; Barro and Conard 1991). Many of the shrubs found in foothills respond to fire by resprouting, germinating from seeds stored in soil seed banks, or both (Sweeney 1956; Keeley 1977; Parker and Kelly 1989). There is often a flush of herbaceous growth in the first few years following a fire that diminishes as the shrubs regain dominance (Sampson 1944; Sweeney 1956). Variations in FRIs differentially favor the various species, depending upon their method of response to fire. Variations in FRIs and species responses over time can lead to diverse patterns of vegetative communities, whereas short FRIs with little variation may lead to a reduction in vegetative diversity (Keeley 1991).

Closed-Cone Conifers

The closed-cone conifers are pines and cypresses that have adapted to fire by storing seeds in cones on the trees for many years. The resin melts from the heat of a crown fire and releases the seeds into a prepared seedbed. These trees are not as common in the Sierra Nevada as in the Klamath Mountains or the Coast Ranges, but they do occur in widely scattered areas, usually associated with chaparral, mostly in the central and northern Sierra (Griffin and Critchfield 1976). The more common species are knobcone pine and McNab cypress (*Cupressus macnabiana*). Knobcone pine appears to do well on poor soils in a fire regime similar to that of chaparral and is often found growing within and among chaparral stands (Vogl et al. 1977). The McNab cypress is more limited in distribution and is generally confined to areas in which the soils are derived from ultrabasics (Griffin and Stone 1967), where fuels are often limited. This may provide a longer minimum FRI than in the surrounding vegetation, due to the slower buildup of fuels on these sites (e.g., Vogl et al. 1977).

Fire History. We were unable to find any fire-history work related to the closed-cone conifers of the Sierra. FRIs for stands of closed-cone conifers are probably similar to, if not slightly longer than, those of the surrounding chaparral stands (Minnich and Howard 1984).

Fire Effects. As we stated earlier, the heat from fire opens the cones of closed-cone conifers and allows the seeds to disperse. Knobcone pine is a short-lived tree. It is found on sites where severe, stand-replacement fires usually occur within the life span of the tree. These species often regenerate dense stands following stand-replacement fires. The young trees can begin to produce cones by ten years of age. A loss of these species could be the result in areas of successful fire suppression. Once the trees die and fall over, a subsequent fire will either kill the seeds through the intense heat in the heavy fuel or consume the cones outright (Vogl et al. 1977; Howard 1992; Esser 1994).

Black Oak–Ponderosa Pine

The black oak–ponderosa pine forests and woodlands burned quite frequently with fires generally of low to moderate severity. Two factors contributed to this general fire regime: the ease of ignition and fire spread due to the relatively loose fuel beds of long needles and oak leaves (e.g., Rothermel 1983) and the regular use of fire to manage this forest type by the native tribes of the Sierra Nevada (Lewis 1993).

Fire History. Historical FRIs in these forests were generally from two to twenty-three years (Kilgore and Taylor 1979; S. Stephens, U.S. Forest Service, Pacific Southwest Research Station, e-mail communication with the author, April 29 and May 9 and 30, 1995), commonly being less than ten years (Swetnam et al. 1991; Swetnam 1993). Once it has been scarred, ponderosa pine usually becomes a good recorder of fire. The open wounds often do not heal rapidly and are easily scarred subsequently by even light fires (McBride 1983). This characteristic of ponderosa pine may allow for the development of

more comprehensive fire histories than in areas where the species is absent or sparse.

Fire Effects. The primary carrier of fire in black oak–ponderosa pine communities historically was probably grass and herbaceous vegetation with some needle and leaf litter. However, as was discussed previously for the blue oak–gray pine woodlands, the surface vegetation has changed from consisting largely of native perennials in presettlement times to being made up primarily of introduced annuals (Heady 1977). Again, historically the predominance of perennials may have narrowed the season of burning, whereas the annuals may promote an earlier onset to the burning season because they dry and cure earlier than the perennials.

Landscape Patterns

Little data exist to describe the historical landscape patterns of the foothill zone. Much of the information in this regard is from anecdotal accounts from the early to mid-1800s. The patterns were probably spatially complex in some areas and more simple in others, depending upon topography, soils, and past fire history. Areas dominated by grasses and herbaceous vegetation, with or without a tree component, would likely have supported frequent, low-severity fires. These areas could have remained for long periods as grasslands, savannas, or open woodlands.

Areas dominated by shrubs would probably show greater temporal variation due to the nature of the severe burns. These burns would then be followed by various stages of succession until a subsequent fire. The rates of fuel accumulation could vary both temporally and spatially over the landscape and could potentially lead to diverse patterns of age classes (Minnich 1983) and species composition (Keeley 1991).

Mixed Conifer Zone

The mixed conifer zone is the main middle-elevation zone of Sierran forest. The mixed conifer type varies from potentially being dominated by ponderosa pine to consisting largely of white fir (*Abies concolor*), with sugar pine (*Pinus lambertiana*) being an important component in many areas. This variation is generally associated with elevation, site quality, and topographic moisture effects (Rundel et al. 1977). Other tree species of importance in this zone are incense cedar (*Calocedrus decurrens*), black oak, and Douglas fir. A variety of hardwoods, shrubs, and herbaceous plants are also associated with the mixed conifer forests.

The lower portion of this zone, where ponderosa pine is often dominant, is commonly used to describe the characteristic fire regime of the Sierra Nevada. Generally, it is associated with frequent fires of low to moderate severity (Kilgore 1973). However, the fire regime can vary considerably in both frequency and pattern of severity by topographic position, site quality, vegetation, and other local factors.

Most published fire-history information for the Sierra comes from the southern Sierra. The fire histories are generally associated with the Yosemite, Sequoia, and Kings Canyon National Parks, with limited data from elsewhere. Often the fire histories were developed using giant sequoia samples because the long-lived trees have distinct rings that are easily cross-dated, have clear scars, and preserve a long record of fires and climate variation in the tree rings.

In contrast to the southern Sierra, the northern Sierra has very little published fire-history information available. The climate, while still Mediterranean, is more mesic than that of the southern Sierra. The northern Sierra Nevada receives precipitation in greater and more consistent amounts than the southern Sierra at equivalent elevations (Major 1977). Vegetation changes along this moisture gradient. Vegetation assemblages in the northern Sierra are often similar in species composition to those found in the southern Cascades and the Klamath Mountains. However, many of these species are missing or rare in the southern Sierra (Rundel et al. 1977). Due to these differences, the discussions of fire regimes in the northern Sierra Nevada draw on data available for similar vegetation from the southern Cascades and the Klamath Mountains.

Most of the divisions of the mixed conifer zone into vegetation types in the discussion that follows are based on Fites 1993 or on discussions with J. A. Fites at various SNEP Science Team meetings during the winter and spring of 1995.

Mixed Conifer–Ponderosa Pine

Ponderosa pine is found throughout the mixed conifer belt of the Sierra Nevada. The characteristic fire regime of much of the Sierra Nevada (frequent fires of low to moderate severity) favored the development of ponderosa pine–dominated forests on many different types of sites where the species is seral to other conifers (Wright 1978; Agee 1993; Arno in press). Ponderosa pine, being a shade-intolerant species (Oliver and Ryker 1990), is rarely a late-successional dominant. Exceptions exist where the sites are continually disturbed (usually by fire) or are warmer, are dryer, or have limited soil development compared with other mixed conifer sites (Agee 1993; Fites 1993).

Some of the earlier attempts to reconstruct FRIs from fire scars in the Sierra Nevada were in mixed conifer forests dominated by ponderosa pine (Show and Kotok 1924; Wagener 1961). Wagener found median FRIs of five to seven years (with a range of two to thirty years) for five sites ranging in size from 15 to 35 ha (37 to 86 acres). However, the fire scar portion of these studies essentially ignored west-side mixed conifer in the northern Sierra Nevada (Wagener 1961), as have more recent studies. Except in areas where fuel accumulates slowly due to local site conditions (e.g., Arno in press), there is no reason to believe that FRIs for ponderosa pine–dominated sites in the northern Sierra Nevada would be greatly different from those in other parts of its range (e.g., the southern Sierra Nevada, the southern Cascades, or the Klamath Mountains).

Median FRIs for seven sites from the Klamath Mountains, where the mixed conifer–ponderosa pine forest types are similar to those in the northern Sierra, were seven to fifteen years (with a range of three to fifty-five years) (Skinner in preparation).

Mixed Conifer–Canyon Live Oak

Canyon live oak (*Quercus chrysolepis*) is a widespread species that is found from the upper foothills into the mixed conifer belt (Myatt 1980). Where canyon live oak is an important component of the mixed conifer forests, it is typically associated with steep slopes and shallow, often rocky soils. Canyon live oak is often found with ponderosa pine on the harsher sites and with Douglas fir and/or sugar pine on the more mesic sites (Fites 1993).

Fire History. There appear to be no published fire histories of mixed conifer forests dominated by canyon live oak in the Sierra Nevada. The fire frequencies were probably similar to, if not longer and more variable than, those of other mixed conifer areas, due to lower fuel accumulations and less-continuous fuels because of site conditions (e.g., Minnich 1980; Fites 1993; Skinner 1995b). However, this is not always the case. Where conifers are well represented in the stands, a more consistent fuel bed can accumulate (e.g., Skinner 1978).

The characteristic fire regime of this type of forest was probably one of relatively frequent, spatially variable fires of low to moderate severity. In a fire-history study from the Klamath Mountains, median FRIs of eleven years (with a range of three to fifty-five years) were found on three sites by Taylor and Skinner (in preparation).

Fire Effects. Following the 1987 wildfires, Weatherspoon and Skinner (1995) found only small, widely dispersed patches of apparently stand-replacing fire effects in a large, roadless area near Hayfork in the Klamath Mountains. Much of the fire had apparently been a surface fire of low to moderate severity. The study assessed only sites considered commercial forestlands. However, much of the area is marginal or noncommercial forestland with a major component of canyon live oak. The large proportion of canyon live oak, often associated with generally sparse, discontinuous surface fuels, along with the strong atmospheric temperature inversions that are characteristic of the region, may have contributed to the minimal damage observed in the intermixed commercial forestlands.

Mixed Conifer–White Fir

Within the mixed conifer zone are broad areas where white fir is considered a major climax component and, depending upon the disturbance history of the sites, can make up a considerable portion of the stand, especially on more mesic sites (Laacke 1990). Many of these sites have supported frequent fires in the past (table 38.1), as evidenced by the occurrence of ponderosa and sugar pine (e.g., Agee 1993). It is in the upper elevations of the mixed conifer zone, where white fir often makes up a large component of the stands, that the greatest density of lightning fires has been found (van Wagtendonk 1986).

Fire History. In the southern Sierra, the fire regime was generally one of frequent fires of mostly low to moderate severity, with occasional, typically small, patches of high-severity fires (Kilgore 1973). The local FRIs vary in a pattern similar to the variation in potential species mixtures over the landscape (Kilgore and Taylor 1979; Caprio and Swetnam 1995). FRIs generally increase with increasing elevations (McNeil and Zobel 1980; Caprio and Swetnam 1995). In areas where white fir is well represented by large, old trees, the FRIs were likely to have been longer and more variable than those in areas where larger, older white fir are found only occasionally (Agee 1993).

The median FRIs for mixed conifer–white fir forests appear to have ranged from approximately seven to twenty years (with a range of three to forty years) for the southern Sierra Nevada (Kilgore and Taylor 1979; Caprio and Swetnam 1995). In the northern Sierra and the southern Cascades, the median FRIs ranged from eight to twelve years (with a range of three to thirty-five years) (table 38.1).

Fire Effects. The variation in actual species dominance is probably related to the local consistency of past fires. Those areas where fires were frequent, with little variation in the frequency, would tend to favor ponderosa pine. In areas where the fires were somewhat less frequent, especially where there was more variation in the frequency and severity, more sugar pine, Douglas fir, and white fir would tend to be found (e.g., Agee 1994).

Mixed Conifer–Giant Sequoia

The fire history of giant sequoia groves has been studied more than that of any other forest type of the Sierra Nevada. The giant sequoias are particularly interesting for studies involving tree-ring analysis because of the longevity of the species.

Fire History. In a study of five sequoia groves along a north-south 160 km (93 mi) transect, Swetnam (1993) found that during the last 1,500 years or so the longest fire-free period in any grove was thirty years before the 1860s. Generally, prior to 1860, the maximum FRIs were less than fifteen years, with mean FRIs of approximately three to eight years. Importantly, most of these fire scars (63%–92%) were found in latewood or between rings, suggesting that the fires occurred in either late summer or fall (Swetnam et al. 1992).

Fire Effects. The fire regime of mixed conifer forests dominated by giant sequoia has been described as being characterized by frequent fires of low to moderate severity (e.g., Kilgore and Taylor 1979; Kilgore 1981; Swetnam et al. 1991; Swetnam 1993; Caprio and Swetnam 1995), with occasional

areas of locally high-severity fires, where small patches of reproduction (young trees) and individuals occasionally burn more intensely (Stephenson et al. 1991).

Giant sequoias have closed cones (Harvey et al. 1980) that release seeds following relatively intense fires and regenerate best where seeds are scattered onto bare soil in open conditions (Stephenson et al. 1991). Fire appears to be a necessary ecological process to provide for adequate long-term reproduction in giant sequoia forests (Stephenson 1994).

Mixed Conifer–Douglas Fir

Douglas fir can be an important component of the mixed conifer forest in the northern Sierra. These areas are usually associated with more mesic conditions than those where ponderosa pine is more important (Fites 1993). Because of the moister, cooler conditions of these areas and the relatively compact fuel beds of short needles, these sites probably burned somewhat less frequently and less regularly than areas where the longer-needled pines are more dominant. Since we know of no published fire-history data from the Sierra Nevada for this forest type, we must rely on data from similar forests found in the Klamath Mountains and the southern Cascades.

Fire History. Presettlement median FRIs for areas of mixed conifers dominated by Douglas fir in the Klamath Mountains were found by Agee (1991) to have been sixteen years (with a range of twelve to fifty-nine years), by Skinner (in preparation) to have been ten to nineteen years (with a range of three to fifty-seven years) for seven sites, and by Taylor and Skinner (in preparation) to have been eleven to eighteen years (with a range of three to fifty-nine years) for six sites. The sites represented by these studies are geographically distributed from near Oregon Caves National Monument (southern Oregon) to near Castle Crags State Park (northern California). They cover a variety of elevations and topographic positions. Since they are so consistent for these geographically dispersed sites, we suggest that they may be representative of the type in the northern Sierra Nevada. Confirmation of this hypothesis will require fire-history studies in the Sierran mixed conifer–Douglas fir forests.

Fire Effects. It is important to note the range of intervals in the FRIs in this type. Most sites show infrequent longer periods (more than twenty-five years) without fire scars. Since the median intervals are not greatly different from those for the ponderosa pine–dominated areas, an important difference between these vegetation types may be the range of variability. This variability would periodically allow young trees to survive to reach a size and condition to become resistant to low-severity fire. Once Douglas fir is established on a site, the compact litter bed composed of short needles would also help reduce the intensity of subsequent surface fires (e.g., Rothermel 1983). As a mature tree, Douglas fir is quite resistant to fires of low to moderate intensity. The nonresinous, thick bark of Douglas fir does not appear to allow the tree to scar as readily as other species (e.g., ponderosa pine, incense cedar, and sugar pine), and it may heal more rapidly (McBride 1983; Skinner and Taylor in preparation). Due to the susceptibility of Douglas fir to fire damage as a seedling or sapling, it may be better suited to fire regimes where generally frequent (ten to twenty years) but variable FRIs allow the occasional survival of younger trees (e.g., Agee 1994).

Mixed Conifer–Tan Oak

The mixed conifer–tan oak (*Lithocarpus densiflorus*) forests of the northern Sierra are similar to those referred to as mixed evergreen forests in the Klamath and north Coast Ranges (Gudmunds and Barbour 1987). These forests are generally found within the mixed conifer zone at lower elevations associated with relatively high annual precipitation (Fites 1993).

Fire History. Fire histories for tan oak–dominated forests are from the Klamath Mountains of northwestern California and southwestern Oregon. Wills and Stuart (1994), in a study conducted near the Forks of the Salmon, found median FRIs to have been fifteen years (with a range of three to fifty years) on three sites. Agee (1993) reports a mean return interval of eighteen years for the type near Oregon Caves National Monument. The fire regime in mixed evergreen forests generally consisted of frequent fires of low to moderate severity, with occasional fires of locally high severity (Agee 1993). The FRIs were more variable than those of the mixed conifer–ponderosa pine forests.

Fire Effects. The tan oak, madrone (*Arbutus menziesii*), and other hardwoods of these forests are easily top-killed by fires of moderate or high intensity, but sprout vigorously following fire. The Douglas fir often associated with these forests can survive moderate fires when mature but may be killed in severe fires. Consequently, following a severe fire, sites can be dominated for extended periods by tan oak and other hardwoods, since recurring fires often kill the Douglas fir seedlings while the hardwoods continually resprout (e.g., Agee 1993).

Montane Chaparral

Throughout the mixed conifer zone and the upper montane are tracts dominated by shrubs often called montane chaparral. Some common species associated with these shrub fields are greenleaf manzanita (*Arctostaphylos patula*), deerbrush, snowbrush (*Ceanothus velutinus*), mountain whitethorn (*C. cordulatus*), bitter cherry (*Prunus emarginata*), and bush chinkapin (*Castanopsis sempervirens*) (Sampson and Jespersen 1963). Huckleberry oak (*Quercus vaccinifolia*) can be important on relatively poor sites (Rundel et al. 1977; Fites 1993).

Fire History. We were unable to find any fire-history studies that would shed light on FRIs for these vegetation types. The FRIs were probably quite variable due to the influence of poor growing conditions. The FRIs would often likely have been longer and more variable than those for the adjacent forest types within the mixed conifer zone. These areas sometimes have widely scattered individuals or clumps of old trees associated with the shrubs (Fites 1993). The large, old trees in these latter areas could potentially be used to determine the fire history of various sites of interest.

Fire Effects. These montane shrub fields can be rather stable communities on soils associated with poor growing conditions (Bolsinger 1989; Sampson and Jespersen 1963). However, many shrub fields of montane chaparral are the result of secondary succession initiated by stand-replacing fires, logging, or other disturbance (e.g., Leiberg 1902; Bock and Bock 1977; Bolsinger 1989). Most montane chaparral shrub species are disturbance adapted and can resprout or germinate from seeds stored in the soil following a fire (Kauffman 1990). Once a shrub field is established, the shrubs can maintain dominance for long periods. Fires that recur during the life of the shrubs and prior to the establishment of the succeeding forest will tend to maintain the shrub fields (Wilken 1967).

Landscape Patterns

For the mixed conifer zone there exist only anecdotal accounts of landscape patterns for most of the area prior to the 1900s. It is likely that by this time much of the mixed conifer had been affected by various activities associated with the settlement period (Cermak and Lague 1993). Many of the written accounts from the beginning of the twentieth century do not clearly indicate whether they describe presettlement conditions or conditions that reflect the effects of the settlement period (e.g., Sudworth 1900; Leiberg 1902).

Due to the physical structure of the landscapes (e.g., topography, geomorphology, etc.), it is likely that the landscapes of the mixed conifer zone varied considerably in their spatiotemporal patterns of species composition, age classes, and stand densities at a variety of scales. Cermak and Lague (1993) relate numerous accounts of vegetation that describe everything from open, parklike stands of large trees to thick stands of trees to dense stands of shrubs. However, there are little data to describe the extent of these conditions or to quantify what is meant by *open, thick, dense,* or other such descriptive terms. Recognizing the lack of such data, the California Spotted Owl EIS Team made a concerted effort to attempt landscape-scale characterizations for the Sierra, using the knowledge of specialists from a variety of disciplines (Toth et al. 1994).

It is impossible at this time, due to the lack of data, to conclusively describe the pre-1850s landscape characteristics and how they changed prior to the 1900s. However, based on available knowledge of fire history, fire effects, fire behavior, and the accounts noted previously, we surmise that the landscape patterns in the mixed conifer zone were of a relatively fine scale (e.g., Bonnicksen and Stone 1982; Stephenson et al. 1991). Large, old trees appear to have been characteristic of many forested areas. However, this certainly does not imply that varying sized patches of shrubs or younger trees were not present in the landscape. Variation in tree size and species composition was likely to be greater horizontally (across the landscape) than vertically (within a single stand). It appears that many forested areas were generally more open than they are today, due mostly to the frequency of fires. This may have promoted more grasses and herbs than are associated with most forest stands today. It is likely that riparian areas often served as barriers to low-intensity and some moderate-intensity fire movement, thus contributing to landscape diversity (see the discussion of riparian areas later in this chapter). The northerly aspects most likely had different species compositions from and greater densities of trees than the southerly aspects, as well as different scales of group, aggregation, or stand patterns. Fires were probably more variable in their severity and frequency on moist sites than on dryer sites. See Toth et al. 1994 for more detail concerning characteristics of landscape patterns.

In addition to site characteristics and landscape structure, past patterns of fire occurrence are likely to have influenced the patterns of species composition and dominance within the mixed conifer zone. The differences among the various species in traits that affect their survival of low-to-moderate-intensity fire (e.g., bark thickness, longevity, susceptibility to rots, etc.) may help suggest the characteristics of past fire regimes. An example would be bark thickness. The five widely spread conifers (ponderosa pine, sugar pine, Douglas fir, incense cedar, and white fir) develop thick bark when mature and are generally resistant to low-intensity fires (e.g., Starker 1934; Minore 1979; Wright and Bailey 1982). However, they vary in the relative ages at which they develop bark thick enough to withstand low-intensity fires. Thick bark generally develops rapidly in ponderosa pine and more slowly in white fir and incense cedar (Weaver 1974), with sugar pine and Douglas fir developing it at an intermediate rate. These differences among the species suggest that the spatiotemporal variability of fires may differentially influence the survival of young trees. More regular FRIs were likely to have been found in areas dominated originally by ponderosa pine. Where sugar pine and Douglas fir were a significant portion of the dominant trees, the frequency and intensity of fire may have been more variable, though fires were still generally frequent. Greater variability in both frequency and intensity would probably have occurred where white fir constituted a major proportion of the dominant trees. Thus, the variation in spatial and temporal fire patterns is likely to influence the variation in species composition of the mixed conifer zone over the landscape.

Upper Montane Zone

The upper montane zone includes the red fir (*Abies magnifica*), Jeffrey pine (*Pinus jeffreyi*), lodgepole pine (*P. contorta* var. *latifolia*), western white pine (*P. monticola*), aspen (*Populus*

tremuloides), and vegetation found in the higher elevations of white fir forests and woodlands (Rundel et al. 1977; Potter 1994). These forest types are found at altitudes and in topographic areas of high lightning frequency when compared with the mixed conifer or foothill zones (van Wagtendonk 1991a). The upper montane zone can be found on both the west and east sides of the Sierra Nevada. However, most research on fire history is from the west side, except in the southern Cascades.

In Sequoia National Park, Vankat (1983) found that the upper montane conifer types accounted for approximately 53% of the area of the park and 76% of the lightning ignitions. Although fires occur frequently in the upper montane zone, they are not likely to spread readily over the landscape except under unusual conditions. This is due to the shortness of the fire season, the compactness of the fuel beds, and the relatively common natural fuel breaks (meadows, rock outcrops, etc.).

The fire regimes in these upper montane areas are likely to be more variable in frequency and in severity than are those from the lower elevations (Agee 1993). The number of fires for a 1 ha (2.5 acre) upper montane Jeffrey pine–white fir site in the Klamath Mountains was found to vary considerably from one century to the next (Skinner in preparation). Six fires occurred in the 1500s, one in the 1600s, nine in the 1700s, four in the 1800s, and three between 1900 and 1944 (the year of the last fire). Mixed conifer sites at lower elevations in the same watershed were found to have had less temporal variation in the numbers of fires recorded in fire scars. Similar variation has been found on Jeffrey pine–white fir and red fir–white pine sites in upper montane areas of Lassen Volcanic National Park (A. H. Taylor, Department of Geography, The Pennsylvania State University, telephone conversation with the author, May 17, 1995).

Table 38.1 summarizes the high degree of variation in FRIs in upper montane forest types. Local variation in fuel continuity may contribute considerably to the variability in the FRIs from site to site.

Red Fir

The fire regimes of red fir forests appear to vary considerably from landscape to landscape. The surface litter is often sparse and compact (Parker 1984) and is usually not conducive to rapid fire spread. In landscapes broken up by many rock outcrops and meadow systems, such as those characteristic of the central and southern Sierra (e.g., Vale 1987), the fire regimes are characterized by longer FRIs (Pitcher 1987). Longer FRIs in landscapes of Lassen Volcanic National Park have been found where fuel accumulations are low and the fuel bed is broken by outcrops of volcanic rock (A. H. Taylor, telephone conversation with the author, May 17, 1995). However, in areas of more continuous litter beds, as are often found in the northern Sierra and the southern Cascades, the FRIs appear to have been shorter (Taylor 1993a). Fires in these areas, although patchy, appear to spread more easily over larger areas. Pre-suppression-period fire rotations in red fir landscapes within the Caribou Wilderness in the southern Cascades were found to have been approximately seventy years (Taylor 1995a).

Stand-replacing fires appear to have occurred infrequently (Taylor and Halpern 1991; Agee 1993). Leiberg (1902) reported sizable areas in the red fir zone of the Feather River where brush fields appeared to be the result of severe burns.

Some fires in red fir forests have been observed to spread primarily through branch wood and large woody debris, since the compact needle beds do not readily spread fire (Toth et al. 1994). This pattern of spotty fire spread helps contribute to the patchy nature of the burns. These patterns of fire occupance in red fir appear to hold for both the west side and the east side of the range (e.g., Hawkins 1994).

In the lower elevations of the upper montane, white fir can be a major component (Potter 1994), often mixing with red fir. Taylor (1993a) and Taylor and Halpern (1991) have reported on disturbance regimes and stand dynamics in forests of mixed red fir and white fir in the southern Cascades of northern California. They found that fire and wind had been major disturbance factors contributing to spatial patterns of age and tree sizes. White fir, generally more resistant to damage at a younger age than red fir (C. P. Weatherspoon, U.S. Forest Service, Pacific Southwest Research Station, personal conversation with the author, May 18, 1995), may occupy areas where FRIs are more regular than those where red fir is found without white fir.

Jeffrey Pine

The Jeffrey pine forests of the upper montane appear to have had more variable FRIs than the lower-elevation pine forests. Jeffrey pine is found from the upper mixed conifer zone through the upper montane on the west slope and extends onto the east slope of the Sierra. In the upper montane zone, Jeffrey pine is often associated with white fir, red fir, white pine, lodgepole pine, and other species (Rundel et al. 1977). Jeffrey pine is similar to ponderosa pine in its fire-associated characteristics.

Jeffrey pine is often found on more extreme sites (sites that are colder, drier, or more nutrient deficient) than many of its associates (Jenkinson 1990). Because of this, fire histories of Jeffrey pine forests may show much greater variability than those of its close relative, ponderosa pine. The increased variability is related to a limited burn season, slow fuel accumulations, and, often, landscapes broken up by rock outcrops. The fire regimes for Jeffrey pine forests in the southern Cascades (A. H. Taylor, telephone conversation with the author, May 17, 1995) and in the Sierra San Pedro Martir (Minnich et al. in press) appear to have followed the pattern of less frequent and more variable FRIs than would be expected of lower-elevation ponderosa pine forests of the Sierra Nevada.

Mean fire frequency prior to this century in forests dominated by Jeffrey pine and ponderosa pine in the Caribou Wilderness in the southern Cascade Range was found to be 18.8

years (the FRIs ranged from 5 to 39 years) for an area of 127 ha (314 acres). Fire rotation was determined to be 70 years (Taylor 1995a).

Aspen

We are aware of no published studies concerning aspen fire regimes in the Sierra Nevada. Elsewhere throughout the western United States it is recognized that stand-replacement fire has often played a major, yet infrequent, role in the development and maintenance of aspen stands (Kilgore 1981; Jones and DeByle 1985). However, because of the types of sites that aspen generally occupies in the central and southern Sierra (e.g., around moist meadows, near rock piles at the base of cliffs, etc.), many of these stands may be relatively stable and unrelated to fire (Rundel et al. 1977). In the northern Sierra Nevada and the southern Cascades, aspen may often be successional to more tolerant conifers such as white fir or red fir (Potter 1994). FRIs in these locations are likely to be quite variable and long. Fires have been shown to kill competing conifers and regenerate otherwise declining aspen stands (Brown and DeByle 1989; Bartos et al. 1991). Where aspen has become established, it may be able to survive more frequent fires in a shrub state similar to that described by Leiberg (1902).

Lodgepole Pine

Fire has long been recognized as an important ecological process in lodgepole pine forests (e.g., Clements 1910). Lodgepole pine is commonly thought of as a closed-cone conifer requiring the heat of fires to open the cones. However, this feature is absent from the species in the Sierra Nevada (Lotan and Critchfield 1990). The spatial patterns of age classes within stands of lodgepole pine in the central Sierra have been reported to be of a fine grain usually associated with small gaps rather than the large gaps created by the extensive crown fires characteristic of the type in the Rocky Mountains. Mature stands are often open and have sparse surface fuels (Parker 1986). These conditions do not easily promote ignition and fire spread (van Wagtendonk 1991b).

Fire History. No published information exists concerning fire history in the lodgepole pine forests of the upper montane in the Sierra. The type may have a fire regime that is intermediate between the red fir–white fir or Jeffrey pine–red fir forests and the subalpine areas.

Data from landscape-level studies in Lassen Volcanic National Park (A. H. Taylor, telephone conversation with the author, May 17, 1995) and the adjoining Caribou Wilderness area (Solem 1995; Taylor 1995a) in the southern Cascades suggest a disturbance regime similar to that reported by Stuart et al. (1989) for south-central Oregon. The primary difference was that there were more frequent fires in the Caribou Wilderness (Solem 1995). The mean FRI for nine-point samples of trees with multiple scars in stands dominated by lodgepole pine was 34.5 yrs (with a range of 28 to 41 years). The fire rotation prior to this century in two lodgepole pine–dominated areas of the Caribou Wilderness was calculated to be 57 years (162 ha [400 acres]) and 104 years (92 ha [228 acres]) (Taylor 1995a).

Fire Effects. The lodgepole pine–dominated forests of the Caribou Wilderness are multiaged but show the influence of larger-scale episodic events through the dominance of age classes by one or a few even-aged cohorts. These even-aged cohorts appear to be related to past fire events (Solem 1995).

Landscape Patterns

A greater variability in the spatial and temporal pattern may have developed in the upper montane zone than in the lower mixed conifer zone. The variability in landscape patterns is likely a result of a number of factors, such as heavy snow packs that can linger late into the year, influencing fire probability; patterns of soils and exposed rock; and compact litter beds, as well as other factors discussed in more detail previously.

Variation in the spatial extent and severity of individual fires helps lead to a dynamic, complex pattern of dominance by age classes and species over the landscape. Fire sizes estimated from fire scars, age classes, and other patterns of tree-ring variation in the Caribou Wilderness ranged from 22 ha to 1,067 ha (55 acres to 2,635 acres), with a median size of 101 ha (250 acres). Small fires appear to have been mostly of low severity, whereas larger burns had considerable portions affected by moderate-to-high-severity fire (Taylor 1995a).

Subalpine Zone

Subalpine areas of the Sierra Nevada are characterized by forests of widely spaced trees of short stature that often straddle the crest of the range (Rundel et al. 1977). These types generally have limited, usually discontinuous fuel accumulations (Kilgore and Briggs 1973; USFS 1983). Characteristic trees are mountain hemlock (*Tsuga mertensiana*), white bark pine (*Pinus albicaulis*), western white pine (*P. monticola*), foxtail pine (*P. balfouriana*), limber pine (*P. flexilis*), and western juniper (*Juniperus occidentalis*), with lodgepole pine in the lower portions. Subalpine areas receive a greater proportion of lightning strikes than do lower-elevation forests (van Wagtendonk 1991a). However, the number of ignitions is disproportionately low (Vankat 1983; van Wagtendonk 1991b).

Overall, fires are infrequent and of low severity within the subalpine types (Kilgore 1981). Only occasionally, and usually on relatively small areas, do the fires become more severe. Because of the nature of fire in the upper montane and subalpine forests, the National Park Service initiated a prescribed natural fire program in these areas more than two decades ago (van Wagtendonk 1986).

Keifer (1991) reports a study in the subalpine zone of Sequoia–Kings Canyon National Parks, where lodgepole pine and foxtail pine are found. She found that monospecific lodgepole stands always had evidence of past fires, whereas evidence of past fires was found only occasionally in the foxtail pine stands. The areas where the two species intermingled were intermediate in evidence of past fire. She noted that lodgepole pine recruitment appears to be pulsed with the age classes associated with past fires. Conversely, where foxtail pine stands showed evidence of fires the recruitment appeared to be more sporadic and not necessarily associated with fires. She suggests that the response to fire of lodgepole pine (a thin-barked tree) is regeneration in gaps created when the thin-barked trees are killed by fire. Foxtail pine, on the other hand, exhibits thicker bark, which may better protect the trees from the low-intensity fires characteristic of the zone. Climate variation and factors other than fire influencing mortality may account for foxtail pine recruitment patterns.

East-Side Ecosystems

We were unable to find published information concerning fire for the east side of the Sierra Nevada or the east side of the southern Cascades in California. Limited data from the southern Cascades in Lassen National Forest were supplied by Olson (1994) for three east-side mixed conifer stands dominated by ponderosa and Jeffrey pine with white fir and incense cedar present. He found median FRIs of eight to sixteen years (with a range of six to thirty-two years).

Agee (1993, 1994), individually and as part of the Eastside Forest Ecosystem Health Assessment, recently published reviews of the fire regimes for most of the vegetation types that would be found on much of the east-side SNEP assessment area. Finding only limited fire-history information specific to the SNEP assessment area, we refer the reader to these recent summaries as well as to Chang 1996.

Riparian Areas

Riparian areas are generally zones of transition from the terrestrial uplands to aquatic habitats. Riparian zones can be identified by vegetation that requires large amounts of free or unbound soil water. Because of available water and many other vegetative characteristics associated with riparian areas, these zones are disproportionately more important to many wildlife species than their limited extent on the landscape would indicate (Thomas et al. 1979).

We are not aware of any published fire-history studies that would shed light specifically on riparian fire regimes in the Sierra Nevada, southern Cascades, Klamath Mountains, or the east side of the Sierra-Cascade crest. Agee (1993) has conceptually described the probable relationships of fire with riparian areas of various forms. Agee suggests that narrower riparian zones will be more likely to have been more frequently disturbed by fire than will wider riparian zones, and that riparian zones in dryer areas will probably burn more frequently than those in wetter areas.

Skinner (in preparation) gathered fire-history data from four riparian areas along the east side of the Shasta-Trinity

Divide in the Klamath Mountains within the Sacramento River watershed north of Lake Shasta. These data were gathered from within the riparian zone and will be summarized here because of the lack of other data.

The forest type adjacent to all four sites would generally be described as the Klamath enriched mixed conifer type (e.g., Sawyer and Thornburgh 1977). Species common to all four sites were willows (*Salix* spp.), western azalea (*Rhododendron occidentale*), Port Orford cedar (*Cupressus lawsoniana*), and various grasses, sedges, and forbs associated with wet meadow systems. Other common species on these sites were spiraea (*Spiraea douglasii*), Sierra laurel (*Leucothoe davisiae*), thimbleberry (*Rubus parviflorus*), mountain alder (*Alnus incana* sp. *tenuifolia*), and California pitcher plant (*Darlingtonia californica*). Elevations ranged from 1,400 to 1,900 m (4,600 to 6,300 ft). Two sites were on north-trending, gently sloped swales, and two were on steeper south-facing slopes. The sites were all less than 1 ha (2.5 acres) each.

Fire History

The fire histories, which dated from the mid-1600s, suggest that riparian areas generally have longer and more variable FRIs than nearby upland sites. The pre-1850 median FRIs for north-facing swales were 31 and 36 years (with a range of 9 to 71 years). The time since the last fire scar formed on these sites was 49 and 95 years. The pre-1850 median FRIs for the south-facing slopes were 26 and 52 years (with a range of 7 to 65 years). No fires had been recorded in the stumps for 58 and 102 years. Nearby (less than 500 m [1,650 ft] away) upland sites had pre-1850 median FRIs of 12 and 15 years (with a range of 6 to 44 years).

On the west side of the Shasta-Trinity Divide, in the watershed of the East Fork of the Trinity River, data were collected for two 1 ha (2.5 acre) sites that were separated by a small creek with a well-developed riparian zone. The forest types were Klamath enriched mixed conifer with riparian species similar to those described previously. These sites were in the middle third of a long north-facing slope at an elevation of approximately 1,450 m (4,750 ft).

This fire history, dated from the mid-1500s, suggests that a narrow riparian zone only a few meters wide may have longer and more variable FRIs than adjacent upland sites. The pre-1850 median FRIs were 13 and 14 years (with a range of 5 to 47 years). Sixty-one years had passed since the last fire scar was formed. A total of nineteen fires was recorded for the period. Of these nineteen fires, only ten (53%) were recorded on both sides of the riparian zone. The median FRI for the fires recorded on both sides of the riparian zone was 29 years (with a range of 7 to 47 years).

Fire Effects

Many species associated with riparian areas are angiosperms that often can respond to fire by sprouting. Although the available moisture on these sites produces vegetation (potential

fuel) readily, FRIs may be longer than in the surrounding stands. Consequently, fires may tend to burn more severely, at least locally, when they do occur. However, the severity of fires may often be restrained by higher fuel moistures associated with riparian zones.

Localized severe burns in riparian areas may not greatly affect aquatic habitat at the landscape scale, depending upon the proportion of the riparian habitat that is burned severely (Amaranthus et al. 1989).

Landscape Patterns

Riparian areas may serve as effective barriers to many low-intensity and some moderate-intensity fires and thus influence landscape patterns beyond their immediate vicinity.

TWENTIETH-CENTURY FIRE REGIMES

The twentieth-century fire regimes of the Sierra Nevada are generally quite different from those prior to Euro-American settlement. Many factors occurring in the 1800s and early 1900s combined to induce drastic changes in fire regimes. Additionally, these factors have contributed to landscape patterns that are still evident today. We will first describe the major factors that contributed to the change in the fire regimes and then will discuss the nature of current Sierra Nevada fire regimes.

Factors in the Nineteenth Century That Helped Influence Changes in Fire Regimes

The following factors, mostly occurring in the 1800s, have combined to dramatically alter fire regimes in the Sierra. However, the effects of these factors were not spatially or temporally universal.

Population Decline among the Native Peoples

The populations of native peoples were declining throughout the nineteenth century. Initially the decline was due to diseases introduced by Europeans, and later it was augmented by systematic extermination and forced relocation (Beesley 1996; Cook 1971). These events caused considerable disruption of traditional land-use patterns and cultural practices (Moratto et al. 1988), probably including a reduction in the use of fire. Reductions in fire frequencies in the early 1800s have been noted on some higher-elevation mixed conifer sites. It has been hypothesized that these reductions may have been related to the decline in the native populations (Caprio and Swetnam 1995). Kilgore and Taylor (1979) suggest that the decrease in fires by the late 1800s for their study area may have been related to the decline in burning by natives.

Influx of Miners

Fire was used to aid in general land clearing during the settlement period and was associated with vegetation type conversions in some areas (Barrett 1935). There was a great influx of miners into the Sierra Nevada following the discovery of gold in 1848 (Beesley 1996). Leiberg (1902) noted that many areas that appeared to have been forested at one time had been converted to brush fields by severe fires, many of which appeared to be related to the mining locations.

Extraction of Wood Material during the Settlement Period

Extensive logging to provide materials to support mining and other settlement activities took place in many locations, for example, Nevada City, Placerville, and Lake Tahoe Basin (Beesley 1996; McKelvey and Johnston 1992). Shakes from sugar pines were more valuable than lumber and could often be produced economically where general logging did not take place because of distance to markets and lack of economical transportation (McKelvey and Johnston 1992). The extraction of shakes, lumber, and firewood left great quantities of residues behind, since the portions of the trees with limbs attached were often not used (Beesley 1996). The residues then fueled subsequent high-severity wildfires that would kill extensive tracts of residual and second-growth trees. These fires were notably more severe than fires that burned in forested areas that had not been subjected to the extraction of the wood materials (Leiberg 1902).

Sheepherding

Heavy grazing in the late 1800s appears to have reduced the landscape effects of fires in many areas. This alteration of fire regimes appears to have been effected in two basic ways by the large-scale sheepherding of the late 1800s and early 1900s. First, sheepherders burned extensive areas in higher elevations and more mesic sites to promote more forage for their herds (Barrett 1935; McKelvey and Johnston 1992). This burning was aimed at reducing the number of downed logs and patches of seedlings and saplings (Sudworth 1900). Second, the intensive grazing was also associated with nearly barren or very lightly covered ground (Sudworth 1900; Leiberg 1902). The combined effect of these practices appears to have been a significant reduction in fuel continuity that limited the actual spread of the fires by the late 1800s, so that many areas show an actual decrease in fire frequency as recorded in fire scars at that time (Vankat and Major 1978). This reduction in fire frequency associated with periods of heavy livestock grazing has been recorded in other areas of the western United States as well (e.g., Savage and Swetnam 1990; Mensing 1992).

Fire-Exclusion Policy

The move toward fire exclusion began early in California. The first law against starting fires was issued under Spanish rule in 1793 (Barrett 1935). This was aimed at halting Indian burning of grasslands, because it deprived the Spanish-owned horses and other livestock of forage. In the late 1800s, foresters were making strong arguments to persuade the public to support fire exclusion so that wood production would be higher in the future (California State Board of Forestry 1888). By the turn of the century, fire exclusion was becoming a general policy among government agencies, although it was not yet fully accepted by the public (Husari and McKelvey 1996; Office of the State Forester 1912).

Twentieth-Century Fire Regime Changes

Due to the initiation of fire-exclusion policies in the late nineteenth and early twentieth centuries, as well as to the suite of factors just discussed, the characteristic fire regimes of many forests of the Sierra Nevada appear to have changed dramatically since the mid-1800s. Before the nineteenth century, the characteristic fires affecting large portions of the landscape would most likely have been of low or low to moderate severity, with patches of higher severity. By the late twentieth century, the characteristic fire was generally of high severity, with only small portions of low to moderate severity. Those forests that have experienced the greatest changes are most likely those on productive sites where fires were more frequent in the past (Weatherspoon et al. 1992), for example, ponderosa pine, black oak, and mixed conifer stands.

The justification for eliminating fires was based primarily on the perceived damage done to the forests. Damage in this sense was related to two factors. First, the surface fires would often cause fire wounds to the bases of trees that reduced the value of these trees and, usually over many centuries, would contribute to the demise of the individual trees. Second, the frequent surface fires were noted to maintain the stands in open conditions by killing most seedlings and saplings in the understories and leaving the forests with low stocking levels (Sudworth 1900; Leiberg 1902; Show and Kotok 1924).

Despite the initial reluctance of local human populations to accept fire exclusion, there appeared to be the beginnings of successful reduction of fires by the end of the first decade of the twentieth century (Office of the State Forester 1912). Nationally, the disastrous fires of 1910 in the northern Rockies helped coalesce political support for exclusion of fire (Agee 1993). Data from the records of the national forests show a steady decline, especially in numbers and acres of human-caused fires, over most of this century (McKelvey and Busse 1996; Weatherspoon and Skinner 1996). However, numbers and acres of fires do not tell the whole story of the change in fire regimes. The major changes in fire regimes are related to the type of fire behavior and the spatial patterns associated with the fires. Fire behavior associated with most forest types in California has changed considerably over this century.

Sudworth (1900), Leiberg (1902), and Show and Kotok (1924) all remark that crown fires and extensive areas of mortality (except in previously logged areas) were unusual at the time of their studies. Show and Kotok (1929) describe the characteristic fires associated with major vegetation types in the late 1920s. Only chaparral and brush types were generally associated with crown fires. Forests composed of ponderosa pine or mixed conifers were associated with surface, litter fires, with crown fires being uncommon. They note that places where crown fires occurred were associated with logging slash or dense, young, second-growth stands. Fires in the upper montane forests were generally described as ground fires that moved primarily through duff. This is a very different picture from that of today, where most wildfires, if not immediately suppressed, quickly become at least severe surface fires capable of killing very large trees. A few examples of recent large, stand-replacement fires in the SNEP study area are the Scarface (1977), Indian (1987), Stanislaus Complex (1987), Stormy (1990), Cleveland (1992), Fountain (1992), and Cottonwood (1994) fires.

Early in the fire-exclusion era, Benedict (1930) indicated that the fire hazard was increasing even as the policy was leading to the achievement of the goal of increasing regeneration survival and ensuring greater stocking levels of trees. He noted that fire-suppression costs were increasing dramatically with the change in stand structures and fuel conditions. Please see Arno (in press) and Weatherspoon and Skinner 1996 for further discussion of fuels and fuel buildup during the fire-suppression era.

As most fires are now suppressed when they are quite small, the frequency of the fires that affect the landscape now appears to be related primarily to the occurrence of burning conditions that are outside the range that modern firefighting technology can deal with. Warm, dry summers guarantee that severe burning conditions will occur each year at some point. These conditions occur most often at the lower elevations and are less frequent as elevation increases. The significant fires now are more likely to occur during severe burning conditions of the inevitable drier years (McKelvey and Busse 1996).

Before the fire-exclusion policy, many fires probably burned for weeks or months. Those ignited in midsummer would have been able to burn until the fall rains or snows extinguished them (e.g., Agee 1993). These fires would have burned under a variety of weather conditions, ranging from hot, dry, and windy to relatively benign. This temporal variation in weather would influence fire behavior such that the resulting spatial patterns could be quite variable over the landscape. Even today, when a fire has burned for an extended period the result has been considerable spatial variation in the fire's severity. Recent examples are prescribed natural fires in the national parks of the Sierra (e.g., Kilgore and Briggs 1973); the 1987 wildfires in the Klamath Mountains (e.g., Weatherspoon and Skinner 1995); and the 1994 Dillon lightning fires (USFS 1995), also in the Klamath Mountains.

A characteristic of twentieth-century fire regimes that is different from those prior to the fire-exclusion policy is the spatial extent and pattern of severe burns. Most fires today, including the large, severe fires, usually burn for only a few days. These fires generally burn under severe conditions that

exceed the capabilities of suppression forces. When burning conditions moderate, the fires are quickly contained. The result is a more uniform spatial pattern within the burned area and a more coarse grain to the landscape mosaic as a whole.

HISTORICAL RANGE OF VARIABILITY

The historical range of variability (also called the reference variability, the natural range of variability, etc.) has recently been recognized as an important consideration in natural resources management (Swanson et al. 1994; Manley et al. 1995). Ecosystems are dynamic and constantly change in response to various environmental factors (Sprugel 1991; Johnson et al. 1994). Within the appropriate spatial and temporal context, the historical range of variability can provide a reference for assessing the status and possible trends of ecosystems (Laudenslayer and Skinner 1995).

The review of fire regimes given in this chapter, especially regarding the accumulating evidence from fire-history studies, reveals that many Sierran fire regimes (and associated vegetative characteristics) today may be outside their historical range of variability. The attendees at the paleoecology workshop held by SNEP in October of 1994 arrived at the same consensus: Sierran forest ecosystems, viewed at the scale of the Sierra Nevada, are outside the historical range of variability as to fire frequency and severity and associated stand structures and landscape mosaics.

The magnitude of the deviation from the historical range depends upon the spatial and temporal scale one considers. A small, localized area of less than a few hundred acres may not be outside conditions that existed sometime in the past. However, as we look at larger and larger areas, the conditions today are less and less likely to have existed during the last few hundreds of years. Large landscape patterns of relatively homogeneous multilayered forest stands, generally broken only by large changes in site conditions (rocky outcrops, thin soils, etc.) were probably uncommon before the twentieth century.

Many historical factors have contributed to the change in fire regimes. Yet it should be noted that only one of these factors, the implementation of a fire-exclusion policy, has been applied universally in the Sierran landscape. The effects of Euro-American settlement on the native populations and cultures were certainly pervasive. Nevertheless, we lack knowledge of the spatial extent of the native cultural influence on the fire regime (Anderson and Moratto 1996). However, we do know that the application of the fire-exclusion policy has been universal in the Sierra for much of this century (though it has recently been modified in the national parks).

Table 38.1 shows that the time since the last fire is generally equal to or greater than the longest FRI recorded for most of the sites. These fire-history studies suggest that the fire-exclusion strategy has been very successful in eliminating low-severity fires and most moderate-severity fires (those characteristic of the pre-1850s) from the Sierra. However, the attempt to exclude all fires from the environment has been only partially successful. Since it is not within current technological capabilities to suppress many fires burning under extreme conditions, the current management strategy has shifted the characteristic fire regime to one of infrequent, severe, large fires. This shift means that severe fires, rather than being rare events, have become the rule.

Both Leiberg (1902) and Sudworth (1900) comment on how open the forests in their areas of examination were. They each described the landscapes as characterized by large trees (except around previously logged and mined areas) that were widely spaced with sparse undergrowth. Of course, there were exceptions in local areas, such as the South Fork of the Feather River (Leiberg 1902) and others (Cermak 1988), but according to most accounts these latter areas were the exception. Both Sudworth (1900) and Leiberg (1902) also indicate that the fires in their time generally stayed on the surface. Only in unusual cases were extensive areas of larger trees killed, except where there had been considerable logging slash. These descriptions of landscapes and fire behavior are quite different from today's typical escaped fires and resulting landscape patterns, where the patches with high proportions of tree mortality are much larger (it is not unusual for a fire to kill thousands of acres of trees) and are continuous over the landscape. What was described as typical fire behavior in the forests (mostly low-severity, surface fire) is now atypical.

These human-induced changes in the characteristics of Sierran fire regimes have taken place during a climatic period that appears to have been unusually warm and moist when compared to previous centuries (Stine 1996; Woolfenden 1996; Hughes and Brown 1992; Graumlich 1993). A warmer and moister climate would likely have induced a variety of complex responses in Sierran ecosystems. For example, there is evidence that subalpine tree ecosystems have responded to this climatic variation through expansion near the upper tree line (Taylor 1995b). We think it is likely that these climatic variations may have affected the fire regimes even in the absence of the modern human influences described earlier. However, at this time we can only speculate on the direction and magnitude of change in the responses of the fire regimes to the anomalous warm, moist period. Nevertheless, the Mediterranean climate of warm, dry summers and cool, wet winters has remained a dominant feature.

We suggest that it is improbable that the overall effects of the recent variation in climate on the FRIs and fire regimes would have approached the direction and magnitude of the changes brought about by the various modern human policies and activities. The warm, dry summers would likely have continued to support fires in most years at all but the highest elevations (the historical fire records certainly support this). Fire frequencies would likely have varied somewhat from past

patterns, but we reason that fires would have remained frequent due to the warm, dry summers. Swetnam (1993), in describing long-term trends in fire frequency and associated climate patterns, indicates that fire frequency appears to be more strongly related to temperature than to moisture trends. Based on Swetnam 1993, Stine (1996) reasons that fire frequency, in the absence of modern fire-exclusion policies, would feasibly have increased over the past century in response to the increase in temperatures. Of course, the higher elevations, where snowpacks would remain for extended periods, and the less-exposed aspects would likely have experienced greater variation in FRIs due to increased moisture.

The rapidity of the changes in fire regimes over the last century appears to be remarkably unprecedented, especially considering the current climatic regime and the vegetation assemblages that can easily support frequent fire. Thus, if current management strategies are continued indefinitely, it is difficult to predict where this extraordinary, rapid change in fire regimes will ultimately lead, especially with the potential of future warmer and drier climate patterns. However, if warm, dry years become more common, as many suggest is likely, we would expect the recent paradigm of large, severe fires to continue.

RESEARCH NEEDS

There is much we do not know about the ecological role of fire in Sierra Nevada ecosystems. We have some idea of the progression of change in Sierran ecosystems since Euro-American settlement and the direction of change that would be necessary to develop vegetation conditions to restore more fire-resilient landscapes. However, we have only limited knowledge of fire as a continuing, ecological process. Much of our knowledge of how fire influences ecosystems is only in regards to fire as a single event (usually following many years of fire exclusion), not as a continuing, ongoing process. The resolution, or at least the partial resolution, of a number of poorly understood subjects would help greatly to formulate appropriate management goals and strategies and to determine appropriate methods for fire and ecosystem management. The following are some of the research areas that we believe are important.

Spatial-Temporal Dynamics

The spatial and temporal interactions of fire and Sierran ecosystems have not been extensively researched. Two general categories of research in this regard need to be addressed. First, there is need for information regarding the effects of frequent fires of low to moderate severity. This will be discussed in more detail later. Second, fire-history studies designed to describe patterns of fire and landscape dynamics over time are needed in order to resolve many management-oriented questions. These questions concern the interactions of fire and spatiotemporal patterns of landscapes. The ability to model these interactions (e.g., Miller 1994) will be extremely important as managers attempt to project the potential effects of various alternative management strategies.

There are two areas of need regarding spatiotemporal dynamics. First, fire-history information is lacking for the northern Sierra Nevada. The southern Sierra Nevada (primarily the national parks) is the source for most published fire-history studies. There are sufficient differences in moisture regimes and vegetation between the northern and southern Sierra to suggest that fire-history studies for the northern Sierra would be very valuable for long-term management. It should be emphasized that the fire-history record is being lost. Much of the logging over the past few decades has removed many of the old trees that contained the fire scar record. As time progresses, less and less of this record will be recoverable due to decomposition of the remaining material (e.g., stumps, logs, snags, etc.).

Second, as has been recounted throughout this chapter, very little published research has been designed to describe the influence of fire on spatial and temporal patterns of landscapes. The long-term spatiotemporal dynamics of landscapes appear to be related to climate variations, but the relationships are poorly understood. Until studies are undertaken specifically to address the role of fire in landscape dynamics, many questions will remain unresolved, for example, (1) how to evaluate appropriate long-term fire-management strategies, (2) how to project the effects of the spatiotemporal patterns of fire on wildlife habitat (e.g., food and cover patterns), and (3) how to model fire in landscapes as an ecosystem process.

Influence of Frequent Fire on Accumulations of Coarse Woody Debris

Setting appropriate standards and guidelines for coarse woody debris (CWD) in forests having fire regimes of frequent, low-to-moderate-severity fires requires research to specifically address the relationship between CWD and fire frequency. Available information describing the accumulations and function of CWD is generally not based on work within forests of functioning frequent, low-to-moderate-severity fire regimes. Much of the information is from ecosystems where fire was much more infrequent than that originally found in the Sierra Nevada (e.g., Maser and Trappe 1984; Harmon et al. 1986; Harmon et al. 1987). Associated information from the Sierra Nevada represents ecosystems affected by years of fire suppression. Continual suppression of the fires in many of these forests has probably increased CWD accumulations above that in pre-suppression-era forests.

Characterization of Old-Growth Forests under the Influence of Frequent Fires

Developing appropriate descriptions and guidelines for old-growth forests characterized by frequent fires of low to moderate severity will remain problematic until research designed to address the relationship between old-growth characteristics and fire is undertaken. The definition of old-growth forests has become standardized based upon work done in climates and forests of the Pacific Northwest that are quite different from those found in the Sierra Nevada (e.g., Franklin et al. 1981; Franklin and Spies 1991a, 1991b). In addition, definitions of old-growth forests in the Sierra Nevada were based upon describing sites that met the Pacific Northwest definitions and that had not been significantly disturbed by fire in recent years (Fites et al. 1992). These descriptions, while representing current conditions of old growth, are not necessarily representative of stands dominated by large, old trees that existed under a functioning presettlement fire regime. It is likely that stand structure, species composition, and understory conditions were very different in many presettlement old-growth stands from those in the old-growth stands found today. The conditions found in many old-growth stands today are at least in part the result of years of fire suppression and may not represent conditions characteristic of presettlement old growth. Research that includes landscape-level fire-history studies and large-scale prescribed fire programs is necessary to adequately develop Sierra Nevada descriptions of sustainable old-growth forests.

Smoke as an Ecosystem Process

The role that smoke management ultimately plays in achieving air-quality objectives may be a determining factor in the amount of prescribed fire eventually used for ecosystem management (e.g., Sandberg 1987). An estimate of smoke production from pre-suppression-era fire regimes based upon our understanding of those fire regimes would help build a baseline description of long-term patterns of air quality. Fahnestock and Agee (1983) have done something similar in regards to the Olympic Mountains of Washington. An assessment of the background levels of smoke that were characteristic of the pre-suppression-era fire regimes will provide policy makers and managers with information that will help them make more-informed choices concerning the long-term programmatic use of fire. Such information will help minimize the imposition of unnecessary restrictions on the use of prescribed fire (e.g., Cahill et al. 1996).

Fire Effects

Although much has been written about the effects of fire, much of the existing information is based upon describing the effects of unplanned wildfires. Few studies exist on the effects of recurring, low-to-moderate-severity fires that would have been more characteristic of the pre-suppression-era fire regimes. There are also few studies that display the effects of fire suppression on potential ecosystem responses to fire (e.g., Swezy and Agee 1991).

Interactions among Disturbance Agents

The interactions among various disturbance agents are poorly understood. Ferrell (1996) suggests that many of the factors that contribute to hazardous fire conditions in forests also increase the vulnerability of the forest to large-scale disturbance from insects and pathogens. Interdisciplinary studies designed to describe the complex interactions among the multiple agents of change (e.g., Gara et al. 1985; USFS 1994) will be necessary to gain a more comprehensive understanding of and to project the potential results of various management strategies.

MANAGEMENT IMPLICATIONS AND CONCLUSION

Fire was an important, regular ecological process in most vegetative communities of the Sierra Nevada for thousands of years before the last century. Euro-American settlement and management activities in Sierran ecosystems over the last 150 years or so have caused many changes in Sierran fire regimes and in the vegetation associated with those regimes. These changes include the significant reduction of fire occurrence, accompanied by a general increase in the density of woody vegetation and an accumulation of associated fuels over broad landscapes. Consequently, although most fires are kept quite small through fire-suppression activities, escaped fires often become large, severely burned patches.

For most of this century, fire has been regarded as a nuisance, as a destructive agent, or occasionally as a tool. In spite of the fact that a number of works on fire in natural resources management have been available for decades (e.g., Weaver 1943; Sampson 1944; Shantz 1947; Biswell et al. 1952), the ecological function of fire has been ignored, denied, or treated as an interesting but inconsequential, academic curiosity by most managers and policy makers (Mount 1969). Only recently, in response to attempts to define and carry out more comprehensive ecosystem management, has the ecological role of fire been generally acknowledged (e.g., Agee 1974; Williams 1993; Manley et al. 1995).

It is often said that an important first step to resolving a problem is to recognize or admit that the problem exists. Apparently, society in general is beginning to recognize that the failure to appreciate the role of fire in western North American ecosystems has contributed greatly to what has been characterized as a general forest health problem (e.g., Knudson 1994; Sampson and Adams 1994; Phillips 1995).

It is unlikely that fire will ever be as unrestrained as it was in past eras. Fire suppression will always play an important role in managing the Sierra Nevada. There are too many cultural values at risk to disallow it (Agee 1994). However, we know that fires are inevitable given modern climate and vegetation. Developing forest structures and landscape patterns that are comparable to those that developed under the more frequent fire regimes of the past will plausibly help ameliorate the ecosystem disruptions caused by the severe fires that are beyond fire-suppression capabilities. The chapters that follow in this section address various ways of analyzing and approaching strategies to deal with the ecological and cultural problems associated with the current condition of Sierra Nevada ecosystems.

ACKNOWLEDGMENTS

We would like to thank the following individuals for graciously supplying unpublished data and/or reports: B. Bahro, J. Fites, S. Gethen, R. Hawkins, R. Olson, S. Stephens, T. Swetnam, and A. Taylor. We would also like to thank the following people for valuable comments on an earlier version of the manuscript: M. Barbour, D. Erman, J. Fites, G. Greenwood, D. Leisz, K. McKelvey, C. Millar, D. Parsons, P. Weatherspoon, and four anonymous reviewers.

REFERENCES

Agee, J. K. 1974. *Environmental impacts from fire management alternatives.* San Francisco: National Park Service, Western Regional Office.

———. 1991. Fire history along an elevational gradient in the Siskiyou Mountains, Oregon. *Northwest Science* 65:188–99.

———. 1993. *Fire ecology of Pacific Northwest forests.* Washington, DC: Island Press.

———. 1994. Fire and weather disturbances in terrestrial ecosystems of the eastern Cascades. In *Assessment,* edited by P. F. Hessburg. Vol. 3 of *Eastside forest ecosystem health assessment.* General Technical Report PNW-GTR-320. Portland, OR: U.S. Forest Service, Pacific Northwest Research Station.

Amaranthus, M., H. Jubas, and D. Arthur. 1989. Stream shading, summer streamflow, and maximum water temperature following intense wildfire in headwater streams. In *Proceedings of the symposium on fire and watershed management,* technical coordination by N. H. Berg, 75–78. General Technical Report PSW-109. Berkeley, CA: U.S. Forest Service, Pacific Southwest Research Station.

Anderson, K. 1993a. Native Californians as ancient and contemporary cultivators. In *Before the wilderness: Environmental management by Native Californians,* edited by T. C. Blackburn and K. Anderson, 151–74. Menlo Park, CA: Ballena Press.

———. 1993b. The mountains smell like fire. *Fremontia* 21 (4): 15–20.

Anderson, M. K., and M. J. Moratto. 1996. Native American land-use practices and ecological impacts. In *Sierra Nevada Ecosystem Project: Final report to Congress,* vol. II, chap. 9. Davis: University of California, Centers for Water and Wildland Resources.

Anderson, R. S. 1990. Holocene forest development and paleoclimates within the central Sierra Nevada, California. *Journal of Ecology* 78:470–89.

Arno, S. F. In press. Fire regimes in western forest ecosystems. In *Fire effects on vegetation and fuels.* Ogden, UT: U.S. Forest Service, Intermountain Research Station.

Arno, S. F., and K. M. Sneck. 1977. *A method for determining fire history in coniferous forests of the mountain west.* General Technical Report INT-42. Ogden, UT: U.S. Forest Service, Intermountain Research Station.

Bahro, B. 1993. California spotted owl habitat area ED-79: An assessment of existing fuels and alternative fuel treatments. Unpublished report for Technical Fire Management VI. U.S. Forest Service, Eldorado National Forest, Placerville Ranger District, Placerville, CA.

Baker, W. L. 1989. Effect of scale and spatial heterogeneity on fire-interval distributions. *Canadian Journal of Forest Research* 19:700–706.

Barrett, L. A. 1935. A record of forest and field fires in California from the days of the early explorers to the creation of the forest reserves. Unpublished report to the Regional Forester. U.S. Forest Service, Pacific Southwest Research Station, Redding, CA.

Barro, S. C., and S. G. Conard. 1991. Fire effects on California chaparral systems: An overview. *Environment International* 17:135-49.

Bartos, D. L., W. F. Mueggler, and R. B. Campbell Jr. 1991. *Regeneration of aspen by suckering on burned sites in western Wyoming.* Research Paper INT-448. Ogden, UT: U.S. Forest Service, Intermountain Research Station.

Beesley, D. 1996. Reconstructing the landscape: An environmental history, 1820–1960. In *Sierra Nevada Ecosystem Project: Final report to Congress,* vol. II, chap. 1. Davis: University of California, Centers for Water and Wildland Resources.

Benedict, M. A. 1930. Twenty-one years of fire protection in the national forests of California. *Journal of Forestry* 28:707–10.

Biswell, H. H. 1989. *Prescribed burning in California wildlands vegetation management.* Berkeley and Los Angeles: University of California Press.

Biswell, H. H., R. D. Taber, D. W. Hedrick, and A. M. Schultz. 1952. Management of chamise brushlands for game in the north coast region of California. *California Fish and Game* 38:453–84.

Blackburn, T. C., and K. Anderson, eds. 1993. *Before the wilderness: Environmental management by Native Californians.* Menlo Park, CA: Ballena Press.

Bock, C. E., and J. H. Bock. 1977. Patterns of post-fire succession on the Donner Ridge Burn, Sierra Nevada. In *Proceedings of the symposium on the environmental consequences of fire and fuel management in Mediterranean ecosystems,* technical coordination by H. A. Mooney and C. E. Conrad, 464–69. General Technical Report WO-3. Washington, DC: U.S. Forest Service.

Bolsinger, C. L. 1989. *Shrubs of California's chaparral, timberland, and woodland: Area ownership and stand characteristics.* Resource Bulletin PNW-RB-160. Portland, OR: U.S. Forest Service, Pacific Northwest Research Station.

Bonnicksen, T. M., and E. P. Stone. 1981. The giant sequoia–mixed conifer forest community characterized through pattern analysis as a mosaic of aggregations. *Forest Ecology and Management* 3: 307–28.

———. 1982. Reconstruction of a presettlement giant sequoia–mixed conifer forest community using the aggregation approach. *Ecology* 63:1134–48.

Brown, J. K. 1985. The "unnatural fuel buildup" issue. In *Proceedings of the symposium and workshop on wilderness fire,* technical coordination by J. E. Lotan, B. M. Kilgore, W. C. Fischer, and R. W. Mutch, 127–28. General Technical Report INT-182. Ogden, UT: U.S. Forest Service, Intermountain Research Station.

Brown, J. K., and N. V. DeByle. 1989. *Effects of prescribed fire on biomass and plant succession in western aspen.* Research Paper INT-412. Ogden, UT: U.S. Forest Service, Intermountain Research Station.

Byrne, R., J. Michaelsen, and A. Soutar. 1977. Fossil charcoal as a measure of wildfire frequency in southern California. In *Proceedings of the symposium on the environmental consequences of fire and fuel management in Mediterranean ecosystems,* technical coordination by H. A. Mooney and C. E. Conrad, 361–67. General Technical Report WO-3. Washington, DC: U.S. Forest Service.

Cahill T. A., J. J. Carroll, D. Campbell, and T. E. Gill. 1996. Air quality. In *Sierra Nevada Ecosystem Project: Final report to Congress,* vol. II, chap. 48. Davis: University of California, Centers for Water and Wildland Resources.

California State Board of Forestry, ed. 1888. *Second biennial report of the California State Board of Forestry for 1887–1888.* Sacramento: State Printing Office.

Canham, C. D., and P. L. Marks. 1985. The response of woody plants to disturbance: Patterns of establishment and growth. In *The ecology of natural disturbance and patch dynamics,* edited by S. T. A. Pickett and P. S. White, 197–216. San Diego: Academic Press.

Caprio, A. C., and T. W. Swetnam. 1995. Historic fire regimes along an elevational gradient on the west slope of the Sierra Nevada, California. In *Proceedings of the symposium on fire in wilderness and park management,* technical coordination by J. K. Brown, R. W. Mutch, C. W. Spoon, and R. H. Wakimoto, 173–79. General Technical Report INT-GTR-320. Ogden, UT: U.S. Forest Service, Intermountain Research Station.

Cermak, R. W. 1988. Fire in the forest: Fire control in the California national forests, 1898–1955. Unpublished report. U.S. Forest Service, Eldorado National Forest, Placerville, CA.

Cermak, R. W., and J. H. Lague. 1993. Range of light—range of darkness: The Sierra Nevada, 1841–1905. Unpublished report. U.S. Forest Service, Pacific Southwest Regional Office, San Francisco.

Chang, C. 1996. Ecosystem responses to fire and variations in fire regimes. In *Sierra Nevada Ecosystem Project: Final report to Congress,* vol. II, chap. 39. Davis: University of California, Centers for Water and Wildland Resources.

Christensen, N. L. 1985. Shrubland fire regimes and their evolutionary consequences. In *The ecology of natural disturbance and patch dynamics,* edited by S. T. A. Pickett and P. S. White, 86–99. San Diego: Academic Press.

———. 1993. Fire regimes and ecosystem dynamics. In *Fire in the environment: The ecological, atmospherical, and climatic importance of vegetation fires,* edited by P. J. Crutzen and J. G. Goldammer, 233–44. New York: John Wiley and Sons.

Christensen, N. L., J. K. Agee, P. F. Brussard, J. Hughes, D. H. Knight, G. W. Minshall, J. M. Peek, S. J. Pyne, F. J. Swanson, J. W. Thomas, S. Wells, S. E. Williams, and H. A. Wright. 1989. Interpreting the Yellowstone fires of 1988: Ecosystem responses and management implications. *BioScience* 39:678–85.

Clements, F. E. 1910. *The life history of lodgepole burn forests.* Bulletin 79. Washington, DC: U.S. Forest Service.

Cook, S. F. 1971. Conflict between the California Indian and white civilization. In *The California Indians: A source book.* 2nd ed., edited by R. F. Heizer and M. A. Whipple, 562–71. Berkeley and Los Angeles: University of California Press.

Davis, M. B. 1986. Climatic instability, time lags, and community disequilibrium. In *Community Ecology*, edited by J. Diamond and T. S. Case, 269–84. New York: Harper and Row.

Davis, O. K., R. S. Anderson, P. L. Fall, M. K. O'Rourke, and R. S. Thompson. 1985. Palynological evidence for early Holocene aridity in the southern Sierra Nevada, California. *Quaternary Research* 24:322–32.

Davis, O. K., and M. J. Moratto. 1988. Evidence for a warm dry early Holocene in the western Sierra Nevada of California: Pollen and plant macrofossil analysis of Dinkey and Exchequer Meadows. *Madroño* 35:132–49.

Dieterich, J. H. 1980a. *Chimney Springs forest fire history*. Research Paper RM-220. Fort Collins, CO: U.S. Forest Service, Rocky Mountain Research Station.

———. 1980b. The composite fire interval: A tool for more accurate interpretation of fire history. In *Proceedings of the fire history workshop*, technical coordination by M. A. Stokes and J. H. Dieterich, 8–14. General Technical Report RM-81. Fort Collins, CO: U.S. Forest Service, Rocky Mountain Research Station.

Edlund, E. G., and R. Byrne. 1990. Climate, fire, and late Quaternary vegetation change in the central Sierra Nevada. In *Fire and the environment: Ecological and cultural perspectives*, technical coordination by S. C. Nodvin and T. A. Waldrop, 390–96. General Technical Report SE-69. Asheville, NC: U.S. Forest Service, Southeastern Research Station.

Esser, L. L. 1994. *Cupressus macnabiana*. In *The fire effects information system*, compiled W. C. Fischer. Database on magnetic tape. Missoula, MT: U.S. Forest Service, Intermountain Research Station, Intermountain Fire Sciences Laboratory.

Fahnestock, G. R., and J. K. Agee. 1983. Biomass consumption and smoke production by prehistoric and modern forest fires in western Washington. *Journal of Forestry* 81:653–57.

Ferrell, G. T. 1996. The influence of insect pests and pathogens on Sierra forests. In *Sierra Nevada Ecosystem Project: Final report to Congress*, vol. II, chap. 45. Davis: University of California, Centers for Water and Wildland Resources.

Fites, J. A. 1993. *Ecological guide to mixed conifer plant associations*. R5-ECOL-TP-001. San Francisco: U.S. Forest Service, Pacific Southwest Regional Office.

Fites, J. A., M. Chappel, B. Corbin, M. Newman, T. Ratcliff, and D. Thomas. 1992. *Preliminary ecological old-growth definitions for mixed conifer (SAF TYPE 243) in California*. San Francisco: U.S. Forest Service, Pacific Southwest Regional Office.

Franklin, J. F., K. Cromack Jr., W. Denison, A. McKee, C. Maser, J. Sedell, F. Swanson, and G. Juday. 1981. *Ecological characteristics of old-growth Douglas-fir forests*. General Technical Report PNW-118. Portland, OR: U.S. Forest Service, Pacific Northwest Research Station.

Franklin, J. F., and T. A. Spies. 1991a. Composition, function, and structure of old-growth Douglas-fir forests. In *Wildlife and vegetation of unmanaged Douglas-fir forests*, technical coordination by L. F. Ruggiero, K. B. Aubry, A. B. Carey, and M. H. Huff, 71–80. General Technical Report PNW-GTR-285. Portland, OR: U.S. Forest Service, Pacific Northwest Research Station.

———. 1991b. Ecological definitions of old-growth Douglas-fir forests. In *Wildlife and vegetation of unmanaged Douglas-fir forests*, technical coordination by L. F. Ruggiero, K. B. Aubry, A. B. Carey, and M. H. Huff, 61–69. General Technical Report PNW-GTR-285.

Portland, OR: U.S. Forest Service, Pacific Northwest Research Station.

Fritts, H. C., and T. W. Swetnam. 1989. Dendroecology: A tool for evaluating variations in past and present forest environments. *Advances in Ecological Research* 19:111–88.

Gara, R. I., W. R. Littke, J. K. Agee, D. R. Geiszler, J. D. Stuart, and C. H. Driver. 1985. Influence of fires, fungi, and mountain pine beetles on development of a lodgepole pine forest in south-central Oregon. In *Lodgepole pine: The species and its management*, edited by D. M. Baumgartner, R. G. Krebill, J. T. Arnott, and G. F. Weetman, 153–62. Pullman: Washington State University.

Graumlich, L. J. 1993. A 1,000-year record of temperature and precipitation in the Sierra Nevada. *Quaternary Research* 39:249–55.

Griffin, J. R., and W. B. Critchfield. 1976. *The distribution of forest trees in California, with supplement*. Research Paper PSW-82. Berkeley, CA: U.S. Forest Service, Pacific Southwest Research Station.

Griffin, J. R., and C. O. Stone. 1967. MacNab cypress in northern California: A geographical review. *Madroño* 19:19–27.

Gudmunds, K. N., and M. G. Barbour. 1987. Mixed evergreen forest stands in the northern Sierra Nevada. In *Proceedings of the symposium on multiple-use management of California's hardwood resources*, technical coordination by T. R. Plumb and N. H. Pillsbury, 32–37. General Technical Report PSW-100. Berkeley, CA: U.S. Forest Service, Pacific Southwest Research Station.

Hanes, T. L. 1977. California chaparral. In *Terrestrial vegetation of California*, edited by M. G. Barbour and J. Major, 417–69. New York: John Wiley and Sons.

Harmon, M. E., K. Cromack Jr., and B. Smith. 1987. Coarse woody debris in mixed-conifer forests, Sequoia National Park, California. *Canadian Journal of Forest Research* 17:1265–72.

Harmon, M. E., J. F. Franklin, F. J. Swanson, P. Sollins, S. V. Gregory, J. D. Lattin, N. H. Anderson, S. P. Cline, N. G. Aumen, J. R. Sedell, G. W. Lienkaemper, K. Cromack Jr., and K. W. Cummins. 1986. Ecology of coarse woody debris in temperate ecosystems. *Advances in Ecological Research* 15:133–302.

Harvey, H. T., H. S. Shellhammer, and R. E. Stecker. 1980. *Giant sequoia ecology: Fire and reproduction*. Scientific Monograph Series No. 12. Washington, DC: National Park Service.

Hawkins, R. 1994. Fire intervals in an eastern Sierra mixed conifer forest, preliminary results. Unpublished report. U.S. Forest Service, Inyo National Forest, Bishop, CA.

Heady, H. F. 1977. Valley grassland. In *Terrestrial vegetation of California*, edited by M. G. Barbour and J. Major, 491–514. New York: John Wiley and Sons.

Heinselman, M. L. 1973. Fire in the virgin forests of the Boundary Waters Canoe Area, Minnesota. *Quaternary Research* 3:329–82.

Hemstrom, M. A., and J. F. Franklin. 1982. Fire and other disturbances of the forests in Mount Rainier National Park. *Quaternary Research* 18:32–51.

Howard, J. L. 1992. *Pinus attenuata*. In *The fire effects information system*, compiled by W. C. Fischer. Database on magnetic tape. Missoula, MT: U.S. Forest Service, Intermountain Research Station, Intermountain Fire Sciences Laboratory.

Hughes, M. K., and P. M. Brown. 1992. Drought frequency in central California since 101 B.C. recorded in giant sequoia tree rings. *Climate Dynamics* 6:161–67.

Husari, S. J., and K. S. Hawk. 1994. The role of past and present disturbance in California ecosystems. In *Appendices*, IC1–IC56. Vol. 2 of *Draft Region 5 ecosystem management guidebook*, edited by P. N. Manley, P. Aune, C. Cook, M. E. Flores, D. G. Fullmer, S. J. Husari,

T. M. Jimerson, M. E. McCain, G. Schmitt, J. Schuyler, W. Bertrand, and K. S. Hawk. San Francisco: U.S. Forest Service, Pacific Southwest Regional Office.

Husari, S. J., and K. S. McKelvey. 1996. Fire management policies and programs. In *Sierra Nevada Ecosystem Project: Final report to Congress,* vol. II, chap. 40. Davis: University of California, Centers for Water and Wildland Resources.

Jenkinson, J. L. 1990. *Pinus jeffreyi* Grev. & Balf. Jeffrey pine. In *Conifers,* 359–69. Vol. 1 of *Silvics of North America,* technical coordination by R. M. Burns and B. H. Honkala. Agricultural Handbook 654. Washington, DC: U.S. Forest Service.

Johnson, C. G., Jr., R. R. Clausnitzer, P. J. Mehringer, and C. D. Oliver. 1994. Biotic and abiotic processes of eastside ecosystems: The effects of management on plant and community ecology, and on stand and landscape vegetation dynamics. In *Assessment,* edited by P. F. Hessburg, 66. Vol. 3 of *Eastside forest ecosystem health assessment.* General Technical Report PNW-GTR-322. Portland, OR: U.S. Forest Service, Pacific Northwest Research Station.

Johnson, E. A., and S. L. Gutsell. 1994. Fire frequency models: Methods and interpretations. *Advances in Ecological Research* 24:239–87.

Jones, J. R., and N. V. DeByle. 1985. Fire. In *Aspen: Ecology and management in the western United States,* edited by N. V. DeByle and R. P. Winokur, 77–81. General Technical Report RM-119. Fort Collins, CO: U.S. Forest Service, Rocky Mountain Research Station.

Kauffman, J. B. 1990. Ecological relationships of vegetation and fire in the Pacific Northwest. In *Natural and prescribed fire in Pacific Northwest forests,* edited by J. D. Walsted, S. R. Radosevich, and D. V. Sandberg, 39–52. Corvallis: Oregon State University Press.

Keeley, J. E. 1977. Fire-dependent reproductive strategies in *Arctostaphylos* and *Ceanothus.* In *Proceedings of the symposium on the environmental consequences of fire and fuel management in Mediterranean ecosystems,* technical coordination by H. A. Mooney and C. E. Conrad, 391–96. General Technical Report WO-3. Washington, DC: U.S. Forest Service.

———. 1982. Distribution of lightning- and man-caused wildfires in California. In *Proceedings of the symposium on dynamics and management of Mediterranean-type ecosystems,* technical coordination by C. E. Conrad and W. C. Oechel, 431–37. General Technical Report PSW-58. Berkeley, CA: U.S. Forest Service, Pacific Southwest Research Station.

———. 1991. Fire management for maximum biodiversity of California chaparral. In *Fire and the environment: Ecological and cultural perspectives: Proceedings of an international symposium,* technical coordination by S. C. Nodvin and T. A. Waldrop, 11–14. General Technical Report SE-69. Asheville, NC: U.S. Forest Service, Southeastern Forest Research Station.

Keifer, M. 1991. Forest age structure, species composition, and fire disturbance in the southern Sierra Nevada subalpine zone. Unpublished report submitted to the Sequoia Natural History Association. University of Arizona, Laboratory of Tree-Ring Research, Tucson.

Kilgore, B. M. 1973. The ecological role of fire in Sierran conifer forests: Its application to national park management. *Quaternary Research* 3:496–513.

———. 1981. Fire in ecosystem distribution and structure: Western forests and scrublands. In *Proceedings of the conference on fire regimes and ecosystem properties,* technical coordination by H. A. Mooney, T. M. Bonnicksen, N. L. Christensen, J. E. Lotan, and W. A. Reiners, 58–89. General Technical Report WO-26. Washington, DC: U.S. Forest Service.

———. 1987. The role of fire in wilderness: A state-of-knowledge review. In *Proceedings of the national wilderness research conference: Issues, state-of-knowledge, future directions,* compiled by R. C. Lucas, 70–103. General Technical Report INT-220. Ogden, UT: U.S. Forest Service, Intermountain Research Station.

Kilgore, B. M., and G. S. Briggs. 1973. Restoring fire to high elevation forests in California. *Journal of Forestry* 70:266–71.

Kilgore, B. M., and D. Taylor. 1979. Fire history of a sequoia–mixed conifer forest. *Ecology* 60:129–42.

Knudson, T. 1994. Smokey might be fall guy for fires. *Redding Record Searchlight,* 27 November.

Komarek, E. V., Sr. 1967. The nature of lightning fires. *Tall Timbers Fire Ecology Conference* 7:5–41.

Laacke, R. J. 1990. *Abies magnifica* A. Murr. California red fir. In *Conifers,* 71–79. Vol. 1 of *Silvics of North America,* technical coordination by R. M. Burns and B. H. Honkala. Agricultural Handbook 654. Washington, DC: U.S. Forest Service.

Laudenslayer, W. F., Jr., and C. N. Skinner. 1995. Historical California climate, forests, and disturbance: Understanding the past to manage for the future. *Transactions of the Western Section of the Wildlife Society* 31:19–26.

Leiberg, J. B. 1902. *Forest conditions in the northern Sierra Nevada, California.* Professional Paper 8, Series H, Forestry, 5. Washington, DC: U.S. Geological Survey.

Lewis, H. T. 1993. Patterns of Indian burning in California: Ecology and ethnohistory. In *Before the wilderness: Environmental management by Native Californians,* edited by T. C. Blackburn and K. Anderson, 55–116. Menlo Park, CA: Ballena Press.

Lotan, J. E., and W. B. Critchfield. 1990. *Pinus contorta* Dougl. ex. Loud. Lodgepole Pine. In *Conifers,* 302–15. Vol. 1 of *Silvics of North America,* technical coordination by R. M. Burns and B. H. Honkala. Agricultural Handbook 654. Washington, DC: U.S. Forest Service.

Madany, M. H., T. W. Swetnam, and N. E. West. 1982. Comparison of two approaches for determining fire dates from tree scars. *Forest Science* 28:856–61.

Major, J. 1977. California climate in relation to vegetation. In *Terrestrial vegetation of California,* edited by M. G. Barbour and J. Major, 75–108. New York: John Wiley and Sons.

Manley, P. N., G. E. Brogan, C. Cook, M. E. Flores, D. G. Fullmer, S. Husari, T. M. Jimerson, L. M. Lux, M. E. McCain, J. A. Rose, G. Schmitt, J. C. Schuyler, and M. J. Skinner. 1995. *Sustaining ecosystems: A conceptual framework.* R5-EM-TP-001. San Francisco: U.S. Forest Service, Pacific Southwest Regional Office.

Maser, C., and J. M. Trappe, eds. 1984. *The seen and unseen world of the fallen tree.* General Technical Report PNW-164. Portland, OR: U.S. Forest Service, Pacific Northwest Research Station.

McBride, J. R. 1983. Analysis of tree rings and fire scars to establish fire history. *Tree-Ring Bulletin* 43:51–67.

McClaran, M. P., and J. W. Bartolome. 1989. Fire-related recruitment in stagnant *Quercus douglasii* populations. *Canadian Journal of Forest Research* 19:580–85.

McKelvey, K. S., and K. K. Busse. 1996. Twentieth-century fire patterns on Forest Service lands. In *Sierra Nevada Ecosystem Project: Final report to Congress,* vol. II, chap. 38. Davis: University of California, Centers for Water and Wildland Resources.

McKelvey, K. S., and J. D. Johnston. 1992. Historical perspectives on forests of the Sierra Nevada and the Transverse Ranges of Southern California: Forest conditions at the turn of the century. In *The California spotted owl: A technical assessment of its current status,* technical coordination by J. Verner, K. S. McKelvey, B. R. Noon, R. J.

Gutierrez, G. I. Gould Jr., and T. W. Beck, 225–46. General Technical Report GTR-PSW-133. Albany, CA: U.S. Forest Service, Pacific Southwest Research Station.

McNeil, R. C., and D. B. Zobel. 1980. Vegetation and fire history of a ponderosa pine–white fir forest in Crater Lake National Park. *Northwest Science* 54:30–46.

Mensing, S. A. 1988. Blue oak (*Quercus douglasii*) regeneration in the Tehachapi Mountains, Kern County, California. Master's thesis, University of California, Berkeley.

———. 1992. The impact of European settlement on blue oak (*Quercus douglasii*) regeneration and recruitment in the Tehachapi Mountains, California. *Madroño* 39:36–46.

———. 1993. The impact of European settlement on oak woodlands and fire: Pollen and charcoal evidence from the Transverse Ranges, California. Ph.D. diss., University of California, Berkeley.

Miller, C. 1994. A model of the interactions among climate, fire, and forest pattern in the Sierra Nevada. Master's thesis, Colorado State University.

Minnich, R. A. 1980. Wildfire and the geographic relationships between canyon live oak, coulter pine, and bigcone Douglas-fir forests. In *Proceedings of the symposium on the ecology, management, and utilization of California oaks*, technical coordination by T. R. Plumb, 55–61. General Technical Report PSW-44. Berkeley, CA: U.S. Forest Service, Pacific Southwest Research Station.

———. 1983. Fire mosaics in Southern California and northern Baja California. *Science* 219:1287–94.

Minnich, R., and L. Howard. 1984. Biogeography and prehistory of shrublands. In *Shrublands in California: Literature review research needed for management*, edited by J. J. DeVries, 8–24. Contribution 191. Davis: University of California, California Water Resources Center.

Minnich, R. A., M. G. Barbour, J. H. Burk, and J. Sosa-Ramirez. In press. California conifer forests under unmanaged fire regimes in the Sierra San Pedro Martir, Baja California, Mexico. *Journal of Biogeography*.

Minore, D. 1979. *Comparative autecological characteristics of Northwestern tree species. A literature review.* General Technical Report PNW-87. Portland, OR: U.S. Forest Service, Pacific Northwest Research Station.

Moratto, M. J., J. D. Tordoff, and L. H. Shoup, eds. 1988. *Culture change in the central Sierra Nevada, 8000 B.C.–A.D. 1950.* Vol. 9 of *Final report of the New Melones archaeological project.* National Park Service Contract CX-0001-1-0053. Sonora, CA: Infotec Research Inc.

Morrison, P. H., and F. J. Swanson. 1990. *Fire history and pattern in a Cascade Range landscape.* General Technical Report PNW-254. Portland, OR: U.S. Forest Service, Pacific Northwest Research Station.

Mount, A. B. 1969. An Australian's impression of North American attitudes to fire. *Tall Timbers Fire Ecology Conference* 9:109–17.

Mutch, R. W. 1970. Wildland fires and ecosystems: A hypothesis. *Ecology* 51:1046–51.

Myatt, R. G. 1980. Canyon live oak vegetation in the Sierra Nevada. In *Proceedings of the symposium on the ecology, management, and utilization of California oaks*, technical coordination by T. R. Plumb, 86–91. General Technical Report PSW-44. Berkeley, CA: U.S. Forest Service, Pacific Southwest Research Station.

Office of the State Forester. 1912. *Fourth biennial report of the state forester of the state of California.* Sacramento: State of California.

Oliver, W. W., and R. A. Ryker. 1990. *Pinus ponderosa* Dougl. ex Laws. Ponderosa pine. In *Conifers*, 413–24. Vol. 1 of *Silvics of North America*, technical coordination by R. M. Burns and B. H. Honkala. Agricultural Handbook 654. Washington, DC: U.S. Forest Service.

Olson, R. D. 1994. Lassen National Forest fire history. Unpublished data. U.S. Forest Service, Lassen National Forest, Susanville, CA.

Parker, A. J. 1984. Mixed forests of red fir and white fir in Yosemite National Park, California. *American Midland Naturalist* 112:15–23.

———. 1986. Persistence of lodgepole pine forests in the central Sierra Nevada. *Ecology* 67:1560–67.

Parker, V. T., and V. R. Kelly. 1989. Seed banks in chaparral and other mediterranean climate shrublands. In *Ecology of soil seed banks*, edited by M. A. Leck, V. T. Parker, and R. L. Simpson, 231–55. New York: Academic Press.

Parsons, D. J. 1981. The historical role of fire in the foothill communities of Sequoia National Park. *Madroño* 28:111–20.

Patterson, W. A., III, and A. E. Backman. 1988. Fire and disease history of forests. In *Vegetation history,* edited by B. Huntley and T. Webb III, 603–32. Dordrecht, The Netherlands: Kluwer Academic Publishers.

Phillips, J. 1995. The crisis in our forests. *Sunset,* July, 87–92.

Pitcher, D. C. 1987. Fire history and age structure in red fir forests of Sequoia National Park, California. *Canadian Journal of Forest Research* 17:582–87.

Potter, D. A. 1994. *Guide to forested communities of the upper montane in the central and southern Sierra Nevada.* R5-ECOL-TP-003. San Francisco: U.S. Forest Service, Pacific Southwest Regional Office.

Reynolds, R. D. 1959. Effect of natural fires and aboriginal burning upon the forests of the central Sierra Nevada. Master's thesis, University of California, Berkeley.

Rice, C. L. 1988. Fire history of Emerald Bay State Park, California Department of Parks and Recreation. Unpublished report. Wildland Resources Management, Walnut Creek, CA.

———. 1990. Fire history of state parks of the Sierra District of the California Department of Parks and Recreation. Unpublished report. Wildland Resources Management, Walnut Creek, CA.

———. 1992. A fire history of the Station Creek Natural Area. Unpublished preliminary report, 14 May. U.S. Forest Service, Eldorado National Forest, Placerville, CA.

Rogers, C., V. T. Parker, V. R. Kelly, and M. K. Wood. 1989. Maximizing chaparral vegetation response to prescribed burns: Experimental considerations. In *Proceedings of the symposium on fire and watershed management*, technical coordination by N. H. Berg, 158. General Technical Report PSW-109. Berkeley, CA: U.S. Forest Service, Pacific Southwest Research Station.

Romme, W. H. 1982. Fire and landscape diversity in subalpine forests of Yellowstone National Park. *Ecological Monographs* 52:199–221.

Rothermel, R. C. 1983. *How to predict the spread and intensity of forest and range fires.* General Technical Report INT-143. Ogden, UT: U.S. Forest Service, Intermountain Research Station.

Rundel, P. W., D. J. Parsons, and D. T. Gordon. 1977. Montane and subalpine vegetation of the Sierra Nevada and Cascade Ranges. In *Terrestrial vegetation of California*, edited by M. G. Barbour and J. Major, 559–99. New York: John Wiley and Sons.

Sampson, A. W. 1944. *Plant succession on burned chaparral lands in northern California.* Bulletin 685. Berkeley: University of California, College of Agriculture, Agricultural Experiment Station.

Sampson, A. W., and B. S. Jespersen. 1963. *California range brushlands and browse plants.* Extension Service Manual 33. Davis: University of California, Division of Agricultural Sciences, California Agricultural Experiment Station.

Sampson, R. N., and D. L. Adams, eds. 1994. *Assessing forest ecosystem health in the inland West: Papers from the American Forests Workshop, November 14th–20th, 1993, Sun Valley, Idaho.* New York: Food Products Press.

Sandberg, D. V. 1987. Prescribed fire versus air quality in 2000 in the Pacific Northwest. In *Proceedings of the symposium on wildland fire 2000,* technical coordination by J. B. Davis and R. E. Martin, 92–95. General Technical Report PSW-101. Berkeley, CA: U.S. Forest Service, Pacific Southwest Research Station.

Savage, M., and T. W. Swetnam. 1990. Early and persistent fire decline in a Navajo ponderosa pine forest. *Ecology* 70:2374–78.

Sawyer, J. O., and D. A. Thornburgh. 1977. Montane and subalpine vegetation of the Klamath Mountains. In *Terrestrial vegetation of California,* edited by M. G. Barbour and J. Major, 699–732. New York: John Wiley and Sons.

Shantz, H. L. 1947. *The use of fire as a tool in the management of the brush ranges of California.* Sacramento: Division of Forestry, Department of Natural Resources.

Show, S. B., and E. I. Kotok. 1924. *The role of fire in the California pine forests.* U.S. Department of Agriculture Bulletin 1294. Washington, DC: Government Printing Office.

———. 1929. *Cover type and fire control in the national forests of northern California.* Department Bulletin 1495. Washington, DC: U.S. Department of Agriculture.

Skinner, C. N. 1978. An experiment in classifying fire environments in Sawpit Gulch, Shasta County, California. Master's thesis, California State University, Chico, CA.

———. 1995a. Change in spatial characteristics of forest openings in the Klamath Mountains of northwestern California, U.S.A. *Landscape Ecology* 10:219–28.

———. 1995b. Using prescribed fire to improve wildlife habitat near Shasta Lake. Unpublished report. U.S. Forest Service, Shasta-Trinity National Forest, Shasta Lake Ranger District, Redding, CA.

———. In preparation. Fire return intervals for the Shasta-Trinity Divide, Klamath Mountains, California.

Skinner, C. N., and A. H. Taylor. In preparation. Patterns of fire scarring in Douglas-fir, Klamath Mountains, California.

Smith, S. J., and R. S. Anderson. 1992. Late Wisconsin paleoecologic record from Swamp Lake, Yosemite National Park, California. *Quaternary Research* 38:91–102.

Snedecor, G. W., and W. G. Cochran. 1980. *Statistical methods.* 7th ed. Ames: Iowa State University Press.

Solem, M. N. 1995. Fire history of the Caribou Wilderness, Lassen National Forest, California, U.S.A. Master's thesis, Pennsylvania State University, University Park.

Sparks, S., and W. Oechel. 1984. General shrubland management procedures. In *Shrublands in California: Literature review and research needed for management,* edited by J. J. DeVries, 1–7. Contribution 191. Davis: University of California, Water Resources Center.

Sprugel, D. G. 1991. Disturbance, equilibrium, and environmental variability: What is "natural" vegetation in a changing environment? *Biological Conservation* 58:1–18.

Starker, T. J. 1934. Fire resistance in the forest. *Journal of Forestry* 32:462–67.

Stephenson, N. L. 1994. Long-term dynamics of giant sequoia populations: Implications for managing a pioneer species. In *Proceedings of the symposium on giant sequoias: Their place in the ecosystem and society,* technical coordination by P. S. Aune, 56–63. General Technical Report PSW-GTR-151. Albany, CA: U.S. Forest Service, Pacific Southwest Research Station.

Stephenson, N. L., D. J. Parsons, and T. W. Swetnam. 1991. Restoring fire to the sequoia–mixed conifer forest: Should intense fire play a role? *Tall Timbers Fire Ecology Conference* 17:321–37.

Stine, S. 1994. Extreme and persistent drought in California and Patagonia during medieval time. *Nature* 369:546–49.

———. 1996. Climate, 1650-1850. In *Sierra Nevada Ecosystem Project: Final report to Congress,* vol. II, chap. 2. Davis: University of California, Centers for Water and Wildland Resources.

Stuart, J. D., J. K. Agee, and R. I. Gara. 1989. Lodgepole pine regeneration in an old, self-perpetuating forest in south central Oregon. *Canadian Journal of Forest Research* 19:1096–1104.

Sudworth, G. B. 1900. Stanislaus and Lake Tahoe Forest Reserves, California, and adjacent territories. In *Annual reports of the Department of Interior, twenty-first annual report of the U.S. Geological Survey, Part 5,* 505–61. Washington, DC: Government Printing Office.

Swanson, F. J., J. A. Jones, D. O. Wallin, and J. H. Cissel. 1994. Natural variability—implications for ecosystem management. In *Ecosystem management: Principles and applications,* edited by M. E. Jensen and P. S. Bourgeron, 80–94. Vol. 2 of *Eastside forest ecosystem health assessment.* General Technical Report PNW-GTR-318. Portland, OR: U.S. Forest Service, Pacific Northwest Research Station.

Sweeney, J. R. 1956. Responses of vegetation to fire: A study of the herbaceous vegetation following chaparral fires. *University of California Publications in Botany* 28:143–250.

———. 1968. Ecology of some "fire type" vegetation in northern California. *Tall Timbers Fire Ecology Conference* 7:110–25.

Swetnam, T. W. 1993. Fire history and climate change in giant sequoia groves. *Science* 262:885–89.

Swetnam, T. W., C. H. Baisan, A. C. Caprio, R. Touchan, and P. M. Brown. 1992. *Tree-ring reconstruction of giant sequoia fire regimes.* Unpublished final report to Sequoia, Kings Canyon, and Yosemite National Parks, Cooperative Agreement DOI 8018-1-1002, Tucson: University of Arizona, Laboratory of Tree Ring Research.

Swetnam, T. W., R. Touchan, C. H. Baisan, A. C. Caprio, and P. M. Browns. 1991. Giant sequoia fire history in Mariposa Grove, Yosemite National Park. In *Proceedings of the Yosemite centennial symposium,* 249–55. NPS D-374. Denver, CO: National Park Service.

Swezy, D. M., and J. K. Agee. 1991. Prescribed fire effects on fine root and tree mortality in old growth ponderosa pine. *Canadian Journal of Forest Research* 21:626–34.

Taylor, A. H. 1993a. Fire history and structure of red fir *(Abies magnifica)* forests, Swain Mountain Experimental Forest, Cascade Range, northeastern California. *Canadian Journal of Forest Research* 23:1672–78.

———. 1993b. Letter to John Swanson dated 11 June. On file at the U.S. Forest Service, Lake Tahoe Basin Management Unit, South Lake Tahoe, CA.

———. 1995a. Fire history of the Caribou Wilderness, Lassen National Forest, California. Final report for cooperative agreement PSW-0006CA. U.S. Forest Service, Pacific Southwest Research Station, PSW Silviculture Lab, Redding, CA.

———. 1995b. Forest expansion and climate change in the mountain hemlock *(Tsuga mertensiana)* zone, Lassen Volcanic National Park, California, U.S.A. *Arctic and Alpine Research* 207–16.

Taylor, A. H., and C. B. Halpern. 1991. The structure and dynamics of *Abies magnifica* forests in the southern Cascade Range, U.S.A. *Journal of Vegetation Science* 2:189–200.

Taylor, A. H., and C. N. Skinner. In preparation. Fire regimes and landscape dynamics in the Klamath Mountains.

Teensma, P. D. A. 1987. Fire history and fire regimes of the central western Cascades of Oregon. Ph.D. diss., University of Oregon.

Thomas, J. W., C. Maser, and J. E. Rodiek. 1979. Riparian zones. In *Wildlife habitats in managed forests: The Blue Mountains of Oregon and Washington*, edited by J. W. Thomas, 40–47. Agricultural Handbook 553. Washington, DC: U.S. Forest Service.

Toth, E., J. Laboa, D. Nelson, R. Hermit, and R. S. Andrews, eds. 1994. *Ecological support team workshop proceedings for the California Spotted Owl Environmental Impact Statement.* San Francisco: U.S. Forest Service, Pacific Southwest Regional Office.

Turner, M. G., W. H. Romme, R. H. Gardner, R. V. O'Neill, and T. K. Kratz. 1993. A revised concept of landscape equilibrium: Disturbance and stability on scaled landscapes. *Landscape Ecology* 8: 213–27.

U.S. Forest Service (USFS). 1983. Emigrant Wilderness fire management area plan. Unpublished management plan. U.S. Forest Service, Stanislaus National Forest, Sonora, CA.

———. 1994. Interdisciplinary research program at Blacks Mountain Experimental Forest, California. Unpublished research prospectus. U.S. Forest Service, Pacific Southwest Research Station, Redding, CA.

———. 1995. *Dillon Creek watershed analysis.* Yreka, CA: U.S. Forest Service, Klamath National Forest, Happy Camp and Ukonom Ranger Districts.

Vale, T. R. 1987. Vegetation change and park purposes in the high elevations of Yosemite National Park, California. *Annals of the Association of American Geographers* 77:1–18.

Vankat, J. L. 1983. General patterns of lightning ignitions in Sequoia National Park, California. In *Proceedings of the symposium and workshop on wilderness fire*, technical coordination by J. E. Lotan, B. M. Kilgore, W. C. Fischer, and R. W. Mutch, 408–11. General Technical Report INT-182. Ogden, UT: U.S. Forest Service, Intermountain Research Station.

Vankat, J. L., and J. Major. 1978. Vegetation changes in Sequoia National Park, California. *Journal of Biogeography* 5:377–402.

van Wagtendonk, J. W. 1985. Fire suppression effects on fuels and succession in short-fire-interval wilderness ecosystems. In *Proceedings of the symposium and workshop on wilderness fire*, technical coordination by J. E. Lotan, B. M. Kilgore, W. C. Fischer, and R. W. Mutch, 119–26. General Technical Report INT-182. Ogden, UT: U.S. Forest Service, Intermountain Research Station.

———. 1986. The role of fire in the Yosemite wilderness. In *Proceedings of the national wilderness research conference: Current research*, compiled by R. C. Lucas, 2–9. General Technical Report INT-212. Ogden, UT: U.S. Forest Service, Intermountain Research Station.

———. 1991a. GIS applications in fire management research. In *Fire and the environment: Ecological and cultural perspectives: Proceedings of an international symposium*, technical coordination by S. C. Nodvin and T. A. Waldrop, 212–14. General Technical Report SE-69. Asheville, NC: U.S. Forest Service, Southeastern Forest Experiment Station.

———. 1991b. Spatial analysis of lightning strikes in Yosemite National Park. In *Proceedings of the eleventh conference on fire and forest meteorology*, edited by P. Andrews and D. F. Potts, 605–11. Bethesda, MD: Society of American Foresters.

Vogl, R. J., W. P. Armstrong, K. L. White, and K. L. Cole. 1977. The closed-cone pines and cypresses. In *Terrestrial vegetation of California*, edited by M. G. Barbour and J. Major, 295–358. New York: John Wiley and Sons.

Wagener, W. W. 1961. Past fire incidence in Sierra Nevada forests. *Journal of Forestry* 59:739–48.

Weatherspoon, C. P. 1988. Preharvest prescribed burning for vegetation management: Effects on *Ceanothus velutinus* seeds in duff and soil. In *Ninth annual forest vegetation management conference*, edited by J. H. Tomascheski, D. Coombes, F. Burch, R. Stewart, D. Thomas, and B. Heald, 125–41. Redding, CA: Forest Vegetation Management Conference.

Weatherspoon, C. P., S. J. Husari, and J. W. van Wagtendonk. 1992. Fire and fuels management in relation to owl habitat in forests of the Sierra Nevada and Southern California. In *The California spotted owl: A technical assessment of its current status*, technical coordination by J. Verner, K. S. McKelvey, B. R. Noon, R. J. Gutierrez, G. I. Gould Jr., and T. W. Beck, 247–60. General Technical Report PSW-133. Albany, CA: U.S. Forest Service, Pacific Southwest Research Station.

Weatherspoon, C. P., and C. N. Skinner. 1995. An assessment of factors associated with damage to tree crowns from the 1987 wildfires in northern California. *Forest Science* 41:430–51.

———. 1996. Landscape-level strategies for forest fuel management. In *Sierra Nevada Ecosystem Project: Final report to Congress*, vol. II, chap. 56. Davis: University of California, Centers for Water and Wildland Resources.

Weaver, H. 1943. Fire as an ecological and silvicultural factor in the ponderosa pine region of the Pacific Slope. *Journal of Forestry* 41:7–14.

———. 1967. Fire and its relationship to ponderosa pine. *Tall Timbers Fire Ecology Conference* 7:127–49.

———. 1974. Effects of fire on temperate forests: Western United States. In *Fire and ecosystems*, edited by T. T. Kozlowski and C. E. Ahlgren, 279–320. New York: Academic Press.

White, P. S., and S. T. A. Pickett. 1985. Natural disturbance and patch dynamics: An introduction. In *The ecology of natural disturbance and patch dynamics*, edited by S. T. A. Pickett and P. S. White, 3–13. San Diego: Academic Press.

Whitlock, C. 1992. Vegetational and climatic history of the Pacific Northwest during the last 20,000 years: Implications for understanding present-day biodiversity. *Northwest Environmental Journal* 8:5–28.

Wickstrom, C. K. R. 1987. *Issues concerning Native American use of fire: A literature review.* Publications in Anthropology 6. Yosemite, CA: National Park Service, Yosemite National Park, Yosemite Research Center.

Wigand, P. E., M. L. Hemphill, S. Sharpe, and S. Patra [Manna]. 1995. *Eagle Lake Basin, northern California, paleoecological study: Semi-arid woodland and montane forest dynamics during the late Quaternary in the northern Great Basin and adjacent Sierras.* Reno: University and Community College System of Nevada, Quaternary Sciences Center, Desert Research Institute.

Wilken, G. C. 1967. History and fire record of a timberland brush field in the Sierra Nevada of California. *Ecology* 48:302–4.

Williams, J. T. 1993. Fire related considerations and strategies in support of ecosystem management. Unpublished staffing paper. U.S. Forest Service, Fire and Aviation Management, Washington, DC.

Wills, R. D., and J. D. Stuart. 1994. Fire history and stand development of a Douglas-fir/hardwood forest in northern California. *Northwest Science* 68:205–12.

Woolfenden, W. B. 1996. Quaternary vegetation history. In *Sierra Nevada Ecosystem Project: Final report to Congress*, vol. II, chap. 4.

Davis: University of California, Centers for Water and Wildland Resources.

Wright, H. A. 1978. *The effect of fire on vegetation in ponderosa pine forests: A state-of-the-art review.* Range and Wildlife Information Series 2, College of Agricultural Sciences Publication T-9-199.

Lubbock: Texas Tech University, Department of Range and Wildlife Management.

Wright, H. A., and A. W. Bailey. 1982. *Fire ecology: United States and Canada.* New York: John Wiley and Sons.

CHI-RU CHANG
School of the Environment
Duke University
Durham, North Carolina

39

Ecosystem Responses to Fire and Variations in Fire Regimes

ABSTRACT

This chapter summarizes the literature available on the effects of fire on Sierra Nevada ecosystems. A general theme that will emerge from the discussion is the intimate, even circular, relationship between fire and postfire ecosystem processes. Fire affects individual species through direct mortality and postfire changes in nutrient, food, and habitat availability. The diversity of species' responses to fire as well as the variety of fire intervals and fire intensities contribute to the overall biodiversity of the Sierra Nevada. The diversity of plants leads to the accumulation of different quantities and quality of fuel. Animals and fire change the quantity and the horizontal and vertical continuity of these fuels, which in turn generate variations in fire behavior. Fire also interacts with other ecosystem processes to create heterogeneity across the landscape. The heterogeneity in the amount, structure, and continuity of fuels across the landscape in turn generates variations in fire regimes. Such diversity, variation, and changes are important components of Sierra Nevada ecosystems.

INTRODUCTION

This chapter compiles and summarizes the literature available on the effects of fire on Sierra Nevada ecosystems. Key elements in the approach were to (1) understand how ecosystem structure and processes respond to variations in fire regimes, for example, fire intensity, frequency, and extent, and (2) understand the role of such responses in generating variations in fire regimes. A general theme that will emerge is the intimate, even circular relationship between fire and postfire ecosystem processes.

The chapter is organized into two major parts: the first part presents general ecosystem responses to fire and variations in fire regimes; the second part deals specifically with the various types of ecosystems of the Sierra Nevada. Successional responses, biodiversity and community-structure responses, fuel-structure changes, landscape-pattern responses, and biogeochemical and soil changes are among the ecosystem structures and processes discussed.

OVERVIEW

Classification of Fire Regimes

The following is a simplified classification of fire regimes (Kilgore 1987; Skinner and Chang 1996), with some examples from the Sierra Nevada. The types of regimes are described in relative terms and would differ if observed from a different temporal or spatial scale. The categories also overlap and have high variations within each type. Nevertheless, they provide a useful structure for discussion.

1. Short-interval, stand-replacement fires: Fire burns frequently and intensively, allowing only fire-adapted species to dominate. One example is the chaparral, but Keeley and Zedler (1978) noted that chaparral is adapted to both short and long fire-free intervals, reflecting how unpredictable fire is in that environment. They suggest a model in which a short fire cycle favors sprouting shrubs over those reproducing entirely from seed, and a longer fire cycle in which "sprouters" and "seeders" coexist.

Sierra Nevada Ecosystem Project: Final report to Congress, vol. II, *Assessments and scientific basis for management options.* Davis: University of California, Centers for Water and Wildland Resources, 1996.

2. Short-interval, low-intensity surface fires: Fire burns regularly and frequently and, as such, rarely allows organic fuels to accumulate to a point where higher-intensity fires may develop (van Wagtendonk 1972). Examples of such regime types include ponderosa pine forest, mixed conifer forest, and sequoia groves. In such regime types the following effects occur:

   • fire controls species composition by favoring species that require sunlight (such as pines and sequoia) over shade-tolerant forms (such as white fir and incense cedar), and by favoring fire-resistant and fire-dependent types over non-fire-dependent forms.

   • fire recycles understory vegetation without damaging the overstory canopy.

   • crown fires are rare if not nonexistent (Kilgore and Taylor 1979).

   • small patches of intense surface burning often result in small openings and consequent fine-grained landscape.

3. Variable-interval, variable-intensity surface fires: Fire usually spreads slowly and rarely crowns. An example of this regime type in the Sierra Nevada is the upper montane red fir forests: the occasional longer fire-free periods provide the chance for red fir seedlings to survive their fire-susceptible stage and therefore establish themselves.

4. Long-interval, low-intensity surface fires: Fire usually spreads slowly or not at all and rarely burns the crowns or kills stands of overstory trees (Kilgore and Briggs 1972). Examples of this regime type in the Sierra Nevada are the subalpine forests of whitebark pine (*Pinus albicaulis*) and lodgepole pine (*Pinus contorta* var. *murrayana*). The effects of fire on such vegetation vary with species, stand age, and burning intensity.

5. Long-interval, high-intensity surface fires: Fire burns rarely, but whenever it happens, it becomes a high-intensity possibly stand-replacing fire. Kilgore (1987) cited coastal redwood (*Sequoia sempervirens*) as an example of this regime type, but see Brown and Swetnam (1994) for a different observation. For the Sierra Nevada, piñon pine and juniper in the eastern Sierra might fit this category (C. Millar personal communication).

6. Very long interval, stand-replacement fires: Stand-replacing fires burn at mean intervals longer than three hundred years. Such types are typical of very damp forests, for example, spruce fir forests, cedar-hemlock forests, and true fir forests (Hemstrom and Franklin 1982), and are not typical in the Sierra. Even if found in the Sierra, they would be sparse and local. In general, in the long absence of fire, succession shifts dominance to shade-tolerant "climax" species.

7. Variable regime: Both short-interval, low-intensity surface fires and long-interval, stand-replacement fires occur on the landscape. Examples include boreal forests, Great Lakes forests, Pacific Northwest Douglas fir (*Pseudotsuga menziesii*) forests, and Rocky Mountain lodgepole pine (*Pinus contorta* var. *latifolia*) forests. Such variable regimes allow complex local variables to give complex results.

The vegetation in these fire-regime types responds differently to fire and consequently responds differently to variations in fire regimes.

Changes in Sierra Nevada Fire Regimes and Their Probable Causes

It is generally agreed that fires in the forest and woodland areas of the Sierra Nevada have become less frequent (e.g., longer return intervals) and generally more severe since about 1850 (Swetnam 1993; Skinner and Chang 1996). Nevertheless, there is considerable disagreement regarding the causes behind these changes. First, climate changes have always had a prominent impact on fire regimes (Swetnam and Betancourt 1990), and because the climate in the region shifted markedly circa 1850 (Stine 1996), some have argued that climate drove changes in fire regimes. However, a closer look at how the climate has changed suggests that such changes would have shifted fire regimes in the opposite direction (Swetnam 1993; Stine 1996).

A second group of explanations for the changes in fire regimes is anthropogenic based on the fact that European-American settlement in the Sierra Nevada also occurred around 1850. Examples include the alteration of fuel loads by sheep grazing and the decrease in ignition sources by the diminished presence of Native Americans. Although many researchers doubt the magnitude of the impact of these two arguments and therefore discount them as important factors, one should still be aware of their impacts on applicable localities. (More detailed discussion of the probable driving forces of fire regimes can be found in Skinner and Chang 1996.)

The most discussed probable cause of the changes in fire regime is the effect of fire-suppression policies. Even so, many question the effectiveness of fire suppression and whether longer fire-return intervals due to fire suppression have dramatically altered ecosystems. Recent research on fire ecology suggests that the influence of historical fire-suppression activities most likely varies depending on ecosystem fire regime: the effects of fire suppression on ecosystem structure and processes is far less important in the longer-interval types than in the short-interval types because more cycles of fire and associated fire effects would have been excluded in the short-interval regimes (Brown 1985; Habeck 1985; van Wagtendonk 1985; Weatherspoon et al. 1992). However, not all short-interval types are equally affected by the longer fire intervals. For example, Keeley (1995) used examples from Hedrick (1954), Keeley (1992a), and Keeley and Zedler (1978) to show that a hundred years of fire-free conditions do not seem to pose a threat to the persistence of any chaparral species, and the century-old chaparral stands' recovery from fire does not seem to differ from that of younger stands.

The effectiveness of fire suppression also varies across Sierran landscapes. Fire suppression has been more successful in areas where access is easier and less so in remote areas. Fire suppression has also been more successful in areas closer to developments because protecting private property has always been given higher priority. There has been greater success in suppressing low- to medium-intensity fires than in suppressing fast-spreading fires induced by extreme weather. Finally, fire suppression has had greater impact in the middle-elevation zones: the small-area burns typical of the higher elevations usually went out on their own; the fast-spreading fires typical of chaparral sites were often beyond the control of humans and were less successfully suppressed; but the low- to medium-intensity surface fires typical of the middle-elevation zones were more controllable and therefore more successfully suppressed. The variation in the effectiveness and impact of fire-suppression activities has added more intangibles to the understanding of the already complex and variable Sierra Nevada ecosystems.

The Issue of Natural Vegetation and Natural Fire Regimes

In trying to define management goals for the Sierran landscape, people have tried to study past patterns in the hopes of finding what was "natural" (Kilgore 1985). However, it may not be useful to worry about whether the present fire regimes and vegetation are natural, for several reasons. First of all, Native Americans have been influencing the fire regimes for centuries through active burning (Blackburn and Anderson 1993; Skinner and Chang 1996), and thus it would be almost impossible to tell what the vegetation and fire regimes might have been without human influence. Second, because human beings are a part of the environment now, avoiding the influence of people is not realistic and may not even be a desirable management goal (Parsons 1981; Christensen et al. 1989). Third, climate changes often occur on a temporal scale that is within the life span of trees (Delcourt et al. 1982; Davis 1984), and thus the forests often lag behind what would be the equilibrium under such climatic conditions. So even without the influence of people, ecosystems are constantly changing (Sprugel 1991), and a natural state of vegetation or fire regimes throughout time is nonexistent.

Variation and Scale Matter

To extract a take-home message from the literature on fire effects, "Variation and scale matter." The effects of fire and the consequences of different fire regimes vary among ecosystems and among their constituent species, from time to time and from location to location (Christensen 1985). The

magnitude of the variation depends on the temporal and spatial scale of observation. Such variability in fire regimes and its consequences result in a dynamic mosaic of shifting patches that may or may not achieve a higher level of "equilibrium," making fire ecology a complex and fascinating subject.

GENERAL ECOSYSTEM RESPONSES TO FIRE AND VARIATIONS IN FIRE REGIMES

Plant Species Responses to Fire

Conditions for successful reproduction of many fire-adapted plant species are often most favorable immediately following fire, owing to increased fertility, removal of potential allelochemicals, reduced competition, and so on (Canham and Marks 1985; Christensen 1993). Subsequently, many plant species have evolved adaptive traits that help them survive fires or reproduce after fire. Such traits include (1) fire-stimulated germination (e.g., California lilac [Ceanothus spp.], manzanita [Arctostaphylos spp.], and sumac [Rhus spp.]); (2) rapid growth and development that allow a complete life cycle between fires (e.g., many annual species and knobcone pine [Pinus attenuata]); (3) fire-resistant foliage (e.g., ponderosa pine [Pinus ponderosa], Douglas fir [Pseudotsuga menziesii]); (4) fire-resistant bark (e.g., Douglas fir, ponderosa pine); (5) adventitious or latent axillary buds (e.g., oaks [Quercus spp.], California lilac [Ceanothus spp.]); (6) lignotubers (e.g., manzanita and chamise [Adenostoma fasciculatum]); (7) serotinous cones and fire-stimulated seed release (e.g., knobcone pine, giant sequoia [Sequoiadendron giganteum]); and (8) fire-stimulated flowering (e.g., soap plant [Chlorogalum pomeridianum]) (Sweeney 1968; Christensen 1985; Agee 1993; nomenclature following Munz and Keck 1973).

Plant responses to fire vary and are often determined by a complex interaction among external factors such as temperature, soil moisture, and heat duration (Rogers et al. 1989) and season of burn (Parker and Kelly 1989; Weatherspoon 1988b). More information on plant nutritional responses to fires can be found in Chapin and Van Cleve 1981.

Because fire has been a potent selective force for individual plant species in fire-prone areas (Sweeney 1968; Parker and Kelly 1989), some researchers have suggested that fire and fire-adapted species may have coevolved. Because many of the fire-adapted species are also more flammable and encourage burn, which would in turn facilitate regeneration of these fire-adapted species (Loucks 1970), it has been suggested that higher flammability of these fire-adapted species may also be selected for (Mutch 1970). However, to argue that fire selects for increased flammability of an individual, it would be necessary to show that greater flammability of an individual significantly enhances the likelihood of that individual's being burned, thus resulting in enhanced survival or reproduction. However, that is not the case. Because the likelihood of de novo ignition (e.g., by lightning) of a particular plant is infinitely small, the likelihood of an individual's being burned is less dependent on its own flammability than on the flammability of its immediate surroundings and the overall continuity of flammable individuals in the landscape. Therefore, increased flammability of an individual is unlikely to be selected for (Christensen 1993). The observed higher flammability of fire-adapted species is more likely a secondary effect of herbivory, or other selective forces, selecting for higher contents of secondary compounds that at the same time increase flammability, deter insects, and serve many other purposes (Mooney and Dunn 1970; Christensen 1985).

Community-Level Responses to Fire

Fire, like many disturbances, often resets the successional sequence. However, not all fires are equal: some are stand-replacing fires that completely reset succession, whereas others are surface fires that clear only the understory but leave the canopy intact. Thus, the fire intensities and fire intervals characteristic of a site determine the seral stage most observed on the landscape.

Following the previously discussed observation of higher flammability of fire-adapted species, Mutch (1970) proposed that natural selection might have favored the evolution of flammable characteristics in fire-dependent plant communities. Given the probability that burn is a character expressed at the level of a stand, and not an individual, this hypothesis may seem more likely to hold. Furthermore, there is evidence that fuels in fire-prone areas are more flammable than those in less fire-prone areas (Rundel 1981). However, whether this fire-driven natural selection, if it exists, operates at a community level is very questionable (Agee 1993). Natural selection operates at the level of an individual plant, not the whole plant community: if there are traits less adapted to an environment, it is the individuals with those traits, not the whole community, that are out-competed. Conversely, if there are traits that enhance the chance of survival, these genetic traits are maintained within a species and probably are not exchanged among different species. This means that similar adaptations of different species have most likely evolved independently within each species and not as a whole community. Moreover, paleological studies have shown that individual species have migrated at unique rates over time (Davis 1981; Brubaker 1986) and likely changed associated species (i.e., "belonged" to different communities) throughout time. Therefore, even if there were such a thing as an evolving community, each community (the assembly of migrating "accidental tourists") may not have stayed around long enough to evolve and respond to selective pressures together. Therefore, the community-level aspect of the Mutch hypothesis may not hold either. Nevertheless, the hypothesis may still have some intriguing aspects worth testing if restated for kin selection in pure species stands.

Another aspect of the community-level response to fire is the notion of "direct succession" (Romme and Knight 1981) and "accelerated succession" (Abrams and Scott 1989). Direct succession asserts that where fire is less frequent, the disturbance-free period is more likely to be long enough to allow the establishment of later seral species. Therefore, when fire burns through such areas, these sites would have a higher abundance of seed source for the later seral species than would the more frequently burned sites. Consequently, the later seral species would be established much faster in these sites, and thus the term *direct succession*. Accelerated succession asserts that disturbance, including fire, kills off the overstory seral species and releases the understory climax species, thus accelerating the successional process. Although both notions seem to describe the same observed results, their hypothesized mechanisms are very different. The mechanisms suggested for direct succession seem more probable for fire and may invite more investigation, whereas the scenarios described by accelerated succession (killing overstory trees and releasing understory trees) are seldom observed in burns and seem more likely a result of other disturbances such as windthrow, insects, and disease.

Biodiversity: Floral Community Structure Responses to Fire Regimes

Because fire has been shown to select for fire-adaptive traits in individual species, different fire regimes may also determine the types of plant characteristics dominant in the community. For example, fire regimes have historically varied widely in their interval between occurrences, dimensions, and fire characteristics, and even varied, though on a lesser scale, in the seasons of occurrence. These variable regimes form a diverse set of environmental characteristics and offer a mechanism for promoting and maintaining biodiversity (Keeley 1991a; Martin and Sapsis 1992). Quoting Martin and Sapsis (1992), "Pyrodiversity promotes biodiversity."

If pyrodiversity promotes biodiversity, would fire suppression reduce pyrodiversity and consequently reduce biodiversity? Ledig (1992) and Husari and Hawk (1994) are among those who answer affirmatively, but, as previously discussed, the extent of fire suppression's influence on pyrodiversity and biodiversity will be determined by the effectiveness of fire-suppression activities.

The effect of fire on ecosystem stability, however, is less agreed upon by ecologists, and this disagreement results mostly from different interpretations of stability and scales of observation. One interpretation of stability has been "the ability to resist change" (Vogl 1970). Vegetative cycles maintained and driven by fires were considered stable from this perspective. Another interpretation of stability has been "less prone to disturbance and significant change" (Christensen 1991), and thus many fire-prone ecosystems were thought to become increasingly unstable over the course of succession. Another emerging line of thought asserts that equilibrium may not always be possible, and for some ecosystems, disequilibrium is the norm (Sprugel 1991). So for these ecosystems, the important question will not be how fire affects ecosystem stability, but how fire affects ecosystem dynamics (see also "Landscape-Pattern Responses to Fire Regimes" later in this chapter).

Biodiversity: Faunal Responses to Fire

The effects of fire on wildlife vary widely depending on fire intensity, duration, frequency, location, shape and size, season of year, fuel types and amount, soils and other site characteristics, as well as the animal species involved (Bendell 1974; Chandler et al. 1983).

Immediate Responses to Fire

Changes in animal populations immediately after fire derive from emigration and mortality during fire and, in some cases, immigration during and immediately after fire. Immigration and emigration behaviors during fire are largely dependent on the type of animal; animal mortality during fire is largely dependent on the type of fire, the type of animal, and the animal's corresponding behavioral responses to fire.

Invertebrate populations decrease immediately after fire, because the animals or their eggs are killed by the flames or heat and their food supply and shelter are diminished. This is true for both soil and surface insects. In some instances, flying insects are attracted by heat, smoke, or dead or damaged trees; thus the population of such species may increase during and after a fire (Lyon et al. 1978).

Amphibians are relatively vulnerable to fire kill because they are less mobile than other animals and often seek shelter in leaf litter or old logs that are consumed by fire. Typically, however, California's amphibians prefer moist habitats, which generally remain unburned except in the most intense wildfires, and thus show better survivorship (Nichols and Menke 1984).

Reptiles generally exhibit low mortality from fire, because they retreat into burrows or rocks to escape the flames. However, some reptile species habitually hide in litter or under logs and thus are vulnerable to fire-induced mortality (Nichols and Menke 1984).

Birds' responses to fire vary, but they generally show no fear of fire. Some fly ahead of a burn or escape into unburned refuges, some ignore fires, and many insectivorous birds, birds of prey, and other birds are attracted to smoke where insects are abundant (Lyon et al. 1978; Nichols and Menke 1984).

Most small mammals react to fire by hiding or seeking shelter. Small nonburrowing mammals live and hide in flammable shelters above ground and thus suffer heavy direct losses in wildfires (Quinn 1979); small burrowing mammals are much less affected directly by wildfires because of the insulation of the soil (Sampson 1944; Lyon et al. 1978; Quinn 1979). Small rodents are also more likely to exhibit panic behavior and have been observed to run in circles or even back into fires, making them more susceptible to mortality (Nichols and Menke 1984).

Larger, more mobile animals usually move calmly and can escape fires more easily. This is especially true for predatory animals that have exceptional mobility. Large, rapidly moving wildfires may still trap and kill escaping animals, although recorded incidence is rare (Lyon et al. 1978; Nichols and Menke 1984).

To summarize, large, intense, rapid-moving wildfires are known to trap and kill animals through heat and suffocation, whereas smaller, less intense fires, such as those typical of the presettlement era fire regimes of the Sierra (Skinner and Chang 1996) and present-day prescribed burns, are easier for wildlife to escape and survive (Sampson 1944; Quinn 1979). Animals that either have greater mobility or find shelter in refuges, such as burrows, unburned islands of vegetation and rocky patches, and riparian zones, have better chances of surviving a fire, whereas animals that exhibit panic or find shelter in tinderbox-type shelters, such as woody litter, logs, and brush piles, are most vulnerable to fire mortality. Although there is certainly some vertebrate mortality during fire, the most common opinion is that direct deaths from fires are rare, and such mortality does not have significant long-term impacts on species' populations (Vogl 1977; Lyon et al. 1978).

Postfire Influences

The postfire influence on wildlife is largely related to fire's role in (1) stimulating germination or sprouting of shrubs, herbaceous plants, or trees that are useful to mammals or birds as food or shelter; (2) making openings in the forest understory or canopy that favor wildlife such as deer; or (3) creating snags or hollow trees that provide shelter (Kilgore 1973; Andrews 1994).

Short-term changes following fire are often detrimental to wildlife, especially after intense wildfires, because fires clear large areas of vegetation and litter, removing food and shelter. However, these detrimental effects last only a few years. Vegetative biomass increases shortly after fire, leading to a greater abundance of food, cover, and structural heterogeneity. In fact, some communities, such as the chaparral, attain the highest levels of productivity, diversity, and carrying capacity for consumer organisms during the early stages of succession (Mooney and Parsons 1973; Lillywhite 1977). As with the case of pyrodiversity's maintaining floral biodiversity, optimal wildlife habitat is created when fire maintains a mosaic of different vegetative age classes. Such mosaics lead to higher spatial diversity of food and habitat and create a maximum amount of ecotone or edge areas, both of which help maintain higher faunal biodiversity (Leopold 1932; Wright and Bailey 1982).

Influence of Animals on Fire Behavior

Animals may influence the probability and intensity of fires through alteration of fuel quantity and structure (Chandler et al. 1983). Insects and beavers can increase the amount of dead fuel on the ground and therefore increase fire hazards (Flieger 1970; Geiszler et al. 1980; Chandler et al. 1983), whereas browsers and grazers such as deer and livestock may reduce the fuel amount (Campbell 1954). Animals that create nests on trees may also increase the flammability of trees (Rowe 1970; Chandler et al. 1983).

Fuel-Structure Changes in Response to Variations in Fire Regimes

Because photosynthesis produces organic matter on a regular basis, vegetative biomass accumulates with time in ecosystems where net primary production is positive. Different types of biomass fall along a gradient of flammability, and the amount available as fuel depends on the prevailing weather conditions and the intensity of the burn. For example, living biomass (especially that in tree boles) does not dry out easily and does not burn in most lower-intensity forest fires. However, in cases of extended drought, high-intensity fires, or fires that burn for long periods of time, part of this biomass may dry out and contribute to the fuel load. This usually nonfuel biomass also becomes fuel when it dies and is added to the fuel complex on the ground. Empirical evidence suggests that fuels are often high immediately following a stand-replacing or intense fire, then decline, and finally build back up again (Romme 1980; Agee and Huff 1987). However, the stochastic natures of ignition, fire weather, and the causes of tree mortality complicate this tendency (Baker 1989). As a result, fuel buildup is not related to stand chronology in a simple way (Paysen and Cohen 1990).

Fire plays an important role in regulating fuel accumulations; fire can decrease dead fuels by consuming them or increase them by killing live vegetation. Fire also affects the horizontal and vertical continuity of fuels. The importance of fire regulating fuel accumulation is especially significant in short-fire-interval types, where natural decay rates are usually very slow. In such types, frequent fires consume and maintain fuels at lower levels. Periodic surface fires also maintain gaps in vertical fuel continuity and prevent fires from moving up to the crown. In cases where fire suppression has effectively lengthened the fire-return intervals in these surface fire types, fuel buildup is more abundant and more vertically continuous, thus increasing the chance of stand-replacement fires (Brown 1985; van Wagtendonk 1985; Kilgore 1987).

In long-fire-interval types, decay, rather than fire, recycles much of the dry matter. In these ecosystems, heavy fuel accumulations are more commonly found historically (Hemstrom and Franklin 1982), and fire suppression does not affect fuel conditions as much as in the short-interval types. Quite the contrary; there has been a concern that fires under a no-suppression regime in these ecosystem types may increase fuels and lead to higher levels of flammability for longer periods of time than under a suppression regime (Brown 1985; Habeck 1985). Still, studies of the Yellowstone fires of 1988 suggest that although the fuel conditions within any individual stand may not be significantly altered by suppression efforts, the extent and continuity of flammable old-growth stands may have been greater in 1988 than they would have been with no previous fire suppression (Romme and Despain 1989).

In the Sierra Nevada, where both short and slightly longer interval types exist, the relationship between fuel structure and fire regimes may be even more complex. In places where fuel has become more continuous in the shorter-interval, low- and middle-elevation zones, the chances that fires originating in these types could spread uphill into the longer-interval, upper-elevation zones are increased. However, the implications of this possibility are still unknown.

Landscape-Pattern Responses to Fire Regimes

Landscapes can be viewed as a collection of patches undergoing successional change (Pickett and White 1985). The character and pattern of landscapes are determined by (1) the underlying physical template, including larger-scale factors such as climate, elevation, and aspect, and smaller-scale factors such as local topographic features and soil characteristics; (2) the frequency, intensity, and spatial extent of disturbances; and (3) the rate and nature of biotic processes (Cooper 1961; Levin 1978; Urban et al. 1987). The importance of disturbances in shaping landscape patterns is determined by the temporal frequency and spatial scale of the disturbance relative to the biotic processes on the landscape.

Patches formed by biotic processes (e.g., succession) are usually at the scale of a tree-fall gap. As individual trees shed branches or die at different times across the landscape, gaps form and undergo successional processes (Runkle 1981; Brokaw 1985). Hence, if the landscape is less prone to disturbance and therefore dominated by biotic processes, the landscape is usually composed of gap-sized patches that are each at different successional stages.

On the other hand, patches formed by stand-replacing fires could have a size ranging from a few trees to thousands of hectares, depending on the characteristic fire size of the landscape. Within each burn, heterogeneity may result from the local variations in the intensity of the burn (Turner and Romme 1994), but each patch created by similar-intensity burns would be relatively homogeneous in regard to stand characteristics such as stand age and community structure. Therefore, if the landscape is dominated by large, stand-replacing burns, the landscape would consist of coarser grains than would sites dominated by smaller stand-replacing burns, other disturbances (e.g., localized insects and diseases), or biotic processes.

In landscapes characterized by surface fires, however, burning does not create distinct patches as stand-replacing fires do. Instead, the patterns of these landscapes are shaped by an interaction among biotic processes, other localized disturbances, and the surface burns: the biotic processes and localized disturbances create heterogeneity in fuel levels and fuel conditions within the landscape, which in turn create a higher level of variation in the local burn conditions. Therefore, the patches shaped by surface burns are usually smaller than those created by stand-replacing fires, but larger than those created by biotic processes. There is also higher heterogeneity in both size and stand structure among these patches created by surface fire.

The surface burns also have a secondary effect on the undercanopy composition and vertical structure of these patches (see "Fuel-Structure Changes in Response to Variations in Fire Regimes," earlier in this chapter). The interesting consequence is that a longer fire-free period in these landscapes would increase the probability of a stand-replacing fire, which would in turn create a coarser-grained landscape.

Because chance factors play more important roles in determining stand composition in early seral stages than in later seral stages (Margalef 1968; Christensen and Peet 1984), if fire recurs at an interval shorter than the time needed for the ecosystem to reach a later seral stage, the seed source may be more variable and thus may result in a wider range of outcomes. It follows that the relative rate of fire return versus the successional process determines ecosystem stability, which in this case means "keeping the same successional trajectory" (Turner et al. 1993).

The size of fire relative to the size of the landscape may also have an effect on landscape stability. The concept of a "shifting mosaic" stability, or dynamic equilibrium, has been asserted by many for ecosystems undergoing frequent disturbances (Heinselman 1978; Bormann and Likens 1979; Kilgore 1987; Clark 1991). Such a concept asserts that whereas individual patches on the landscape may undergo various changes throughout time and the patches may differ greatly from each other at one time, the characteristics of the landscape as a whole change little over time. However, such an argument comes under scrutiny when the sizes of the disturbance and thus the patches become relatively large in comparison to the whole landscape. In this case, the landscape cannot buffer the impacts of the disturbance and shifts out of equilibrium (Christensen et al. 1989; Sprugel 1991). Could this be the case for the present or future Sierra Nevada? We do not have enough evidence to state either way and need more research on this issue.

An interesting aspect of fire is that there is an interaction between its temporal frequency and spatial scale. Because longer fire-free periods allow an increase in both fuel volume and continuity, the relative size of fires generally increases as the average interval between fires lengthens (Heinselman 1981; Baisan and Swetnam 1990). Findings from tree-ring analysis in giant sequoia groves of the Sierra Nevada support this general rule: fires were smaller in size during the higher-fire-frequency period of the Medieval Warm epoch (1000 to 1300 A.D.), whereas fires were more widespread dur-

ing longer interval periods from 500–1000 A.D. and after 1300 A.D. (Swetnam 1993). It has therefore been argued that decreased fire frequency in the Sierra Nevada since 1850 has increased fire size and consequently increased the size of landscape patches (Bonnicksen and Stone 1982; Skinner 1995a; USFS 1995).

Other than the asserted increase in size of landscape patches, landscape-pattern responses to changes in fire regimes are not clear in the Sierra Nevada. Nevertheless, studies from other regions have suggested the complexity of the issue. Baker (1993), for example, studied the responses of landscapes to fire suppression in Minnesota (of variable regime, according to the classification of Kilgore 1987) across several spatial scales. His findings once again supported the importance of variation and scale dependence in fire effects:

- Whereas some characteristics of landscapes respond immediately to fire (e.g., Shannon diversity of patch types), other landscape characteristics (e.g., mean patch size) are slow to respond.

- Variations in the disturbance regime may produce a spatially heterogeneous response, with some parts of the landscape responding immediately after the disturbance, and other parts responding after some time lag.

- This difference in response time becomes more obvious as the scale of observation becomes finer.

- The condition of the landscape at the time of the change in disturbance regime (in this case, fire suppression) can affect the location and timing of a response.

- Even when the disturbance regime is uniform across the landscape (e.g., all fires suppressed), there will still be spatially heterogeneous effects.

To summarize, although Baker was convinced that fire suppression affected the Minnesotan landscape, the spatially heterogeneous character of the responses led him to conclude that it may be difficult to definitively attribute certain vegetation changes to fire suppression (or in general terms, changes in fire regimes).

By shifting down on the temporal scale and looking at the effect of daily burn area on landscape heterogeneity, we further realize the complexity of the issue. Turner and colleagues (1994) found that when burned area exceeded 1,250 ha (3,100 acres) per day in the 1988 Yellowstone fires, the proportions of burned area in different burn severity classes were more fixed, whereas when the area burned per day was smaller, the proportion of burned area in different burn severity classes varied widely.

As an aside, other characteristics of fire besides fire regimes can alter landscape patterns. For example, local variations in microclimate, landform, and fuel beds also influence fire behavior on small spatial scales and contribute significantly to the resulting mosaic of burn intensities, causing a wide variation in patterns of plant mortality, ash deposition, and soil heating. In addition to variation during fire, postfire climatic patterns may also influence the trajectories of many ecosystem processes. These variations in ecosystem process will eventually determine the ecosystem responses at specific localities, resulting in a complex mosaic of variable patterns (Christensen et al. 1989).

Biogeochemical and Soil Changes after Fire

Although there are studies that suggest prescribed fires had limited effects on soils, nutrient cycling, and hydrologic systems for some ecosystems (e.g., Richter and Ralston 1982), it is commonly believed that fire can have significant effects on soil properties because organic matter located on or close to the soil surface is rapidly combusted. This is particularly true for the studies performed in the Sierra Nevada. Fire has been suggested to alter mineral soil-nutrient concentrations by means of five important mechanisms: (1) Direct volatilization of nutrients reduces the total amount of nutrients in the soil; (2) Mineralization induced by heating increases nutrient availability; (3) Ash deposition and subsequent leaching further add nutrients to the mineral soil; (4) Soil erosion following fires decreases total nutrient amount; and (5) Transportation of nutrients due to the differences in the relative availability of nutrients in the ash versus the mineral soil further influences the relative abundance of nutrients in the mineral soil (Behan 1970; DeBano 1991; Rice 1993). Because intense fires deposit greater quantities of ash than do lighter fires, variations in burning intensities may contribute to the variations in soil-nutrient concentrations following fire.

Because the threshold temperatures for volatilization of nitrogen (N), potassium (K), and sulfur (S) are lower than the glowing combustion temperature (650°C), and all of these nutrients, plus phosphorus (P), have volatilization temperatures lower than the flaming temperatures of woody fuels (1,100°C), these nutrients are readily volatilized and lost from organic matter during combustion (DeBano 1991). However, the amount of nutrient volatilized is related to the intensity of fires, and thus in cooler soil-heating regimes, less nutrient is volatilized (Wells et al. 1979; DeBano 1991). The responses of the different nutrients to heating also indicate that little change is likely to occur more than 4 to 5 cm (about 2 in) below the soil surface, unless a very intense, long-duration fire occurs. Therefore, although volatilization is the most direct response of nutrients to fire, its effects are limited.

Despite the loss of nutrients from volatilization plus the loss due to soil erosion following fires (Behan 1970; Christensen 1995), most nutrients, including phosphorus, potassium, calcium (Ca), and magnesium (Mg), are made more available for use by vegetation by the rapid mineralization induced by fire (Hare 1961; St. John and Rundel 1976; Boerner 1982; Kilgore 1987; DeBano 1991). Such fire-induced mineralization releases nutrients much faster than the decomposition processes, which may require years, or, in some cases,

decades. The increased growth and increased nutrient content in surviving trees following fire may be a direct result of this increased availability of nutrients (Weaver 1947; Hartesveldt 1964; Rundel and Parsons 1980), but it may also be a result of reduced competition due to fire-caused mortality.

Nutrient availability (particularly nitrogen) in the soil can also be increased by the translocation of nutrients downward into the soil during a fire (Wells 1971). For example, although total nitrogen (TN) decreases immediately after burning, available ammonium nitrogen (NH_4-N) is usually higher in the underlying soil following a fire, because of this transfer mechanism (DeBano 1991). Phosphorus, however, does not appear to be translocated downward in the soil profile as readily as nitrogen compounds. Thus, phosphorus increases mainly in the ash, on or near the soil surface (DeBano 1991).

All these increases in nutrient availability are usually short term. Plant production may increase in the first or second growing season after fire, but soil-nutrient concentration declines to preburn levels with time (Christensen 1995).

Fire may also affect the nitrogen cycle indirectly through vegetation change. For example, Clark (1990) studied fire and its relationship with soil nutrients in a Minnesota forest and suggested that the tendency for species that are more nitrogen-efficient to colonize more mesic sites that burned rarely in the past may have increased the nitrogen content in the litter accordingly. In the Sierra, some chaparral species that follow fire (e.g., California lilac [*Ceanothus* spp.]) are known to fix nitrogen (Riggan et al. 1988). Therefore, given that the plants are not phosphorus limited, they may dominate the site and increase the rate of nitrogen cycling (Delwiche et al. 1965; Kilgore 1973; Christensen 1995).

Fire may also affect nutrient cycling through its effect on microbes. Nitrifying bacteria and endo- and ectomycorrhizae appear to be particularly sensitive to soil heating (Dunn et al. 1985; DeBano 1991). This aspect may be especially important in systems where light burns are typical and ash deposition is lighter (Christensen 1995).

Overall, low-intensity surface fires increase soil pH, stimulate mineralization, facilitate cycling of nutrients, and generally do not increase soil erosion, whereas high-intensity fires may volatilize large amounts of nitrogen and other volatile nutrients, disrupt soil structure, and induce water repellency and erosion (Wells et al. 1979). The trend of increasing intensity in fires, therefore, may have adverse consequences on soil nutrients (Kilgore 1987).

ECOSYSTEM RESPONSES TO FIRE AND VARIATIONS IN FIRE REGIMES

In the Sierra Nevada region, not all ecosystems have been equally studied. For example, most publications concentrate on the west slope, especially the chaparral and mixed conifer forests, whereas very little of the literature examines the higher elevations and the east slopes. The following is an attempt to synthesize the published materials to date and to provide a more comprehensive picture of what we know today.

Shrublands

Primary Vegetation and Its Historical Relationship with Fire

The major shrubland type found throughout the Sierra is the chaparral, which is characterized by sclerophyllic evergreen woody shrubs. The major types of chaparral found in the Sierra Nevada (Hanes 1977) are the foothill chaparral and the montane chaparral.

The foothill chaparral (chamise chaparral, *Ceanothus* chaparral, scrub oak chaparral, mixed chaparral, and so on) typically occur in the elevation range of 450–1,700 m (1,480–5,580 ft) in the southern Sierra Nevada (Vankat and Major 1978), with different types occurring under different environmental conditions. For the northern Sierra, foothill chaparrals are more widely scattered and are generally restricted to the drier slopes (Hanes 1977). Chamise chaparral is the dominant chaparral type throughout California. It is dominated by chamise (*Adenostoma fasciculatum*) and is associated with hot, xeric sites (south- and west-facing slopes and ridges). *Ceanothus* chaparral is a successional form of chaparral in the southern Sierra but is a climax form in the northern Sierra in the more mesic sites. Its dominant species is buck brush (*Ceanothus cuneatus*). Scrub oak chaparral is also a mesic type, occurring in north-facing slopes below 900 m (3,000 ft) and all slope aspects above 900 m (Hanes 1971) in the southern Sierra, and lies above chamise chaparral in the northern Sierra. Scrub oak (*Quercus dumosa*) is the dominant species, but in the northern parts, interior live oak (*Q. wislizenii* var. *frutescens*) is also abundant. Mixed chaparral consists mainly of chamise, plus tree species such as buckeye (*Aesculus californica*), interior live oak, and canyon oak (*Q. chrysolepis*). It occupies the shady slopes above 900 m in the southern Sierra (Hanes 1977).

Montane chaparral occurs in higher elevations than the foothill chaparral, generally within the same elevation zone as mixed conifers. The dominant species are deer brush (*Ceanothus integerrimus*), manzanita (*Arctostaphylos* spp.), and bush chinquapin (*Castanopsis sempervirens*).

Another type of shrubland found mainly in the northern Sierra is that dominated by Brewer's oak. It generally falls in the upper foothill to lower mixed conifer ecotone and consists mainly of Brewer's oak (*Quercus garryana* var. *breweri*) and deer brush (*Ceanothus integerrimus*).

Although fire-return intervals in chaparral vary among different sites (Skinner and Chang 1996), all the above-mentioned shrublands are believed to be adapted to and intimately related to fire and are suggested to persist where there are recurring fires (Show and Kotok 1924; Skinner 1995b).

Vegetation Responses to Fire

Chaparral is highly flammable and, at the same time, possesses various adaptations to fire. It can sprout massively and quickly from thickened root bases after even severe burning or produce a heavy crop of fire-resistant and fire-stimulated seeds that germinate following fire (Baker et al. 1982; Biswell 1974; Christensen 1985). Vegetative responses on chaparral sites vary and are determined by a complex interaction of temperature, soil moisture, heat duration, depth of burn, and season of burn (Baker et al. 1982; Weatherspoon 1988b; Rogers et al. 1989). Some chaparral shrubs have been shown, through a population model based on size-specific demographic characteristics, to be able to survive more than twenty-three fires (Stohlgren and Rundel 1986). Chamise chaparral regrows more slowly after fire compared to other chaparral types, because of the poor site conditions; montane chaparral responds to fire in a very variable manner (Hanes 1977).

Keeley (1991b, 1992a, 1992b) emphasized the difference between the two types of life-history adaptation to fires: "fire-recruiters" and "fire-persisters." Fire-recruiters, or seeders, establish their seedlings during the first rainy season after fire. Examples of this type are chamise, manzanita, and buck brush. Fire-persisters, or sprouters, are resilient to frequent fires (mostly by vegetative resprouting) but require fire-free periods for recruiting new seedlings. Examples include scrub oak, Christmas berries (*Heteromeles arbutifolia*), holly-leafed cherry (*Prunus ilicifolia*), mountain mahogany (*Cercocarpus betuloides*), and buckthorn (*Rhamnus* spp.). Of the seeders, some *Ceanothus* species are obligate seeders because of their inability to resprout after fire. Zedler and colleagues (1983) found that a seeded chaparral site exposed to a second reburn after one year resulted in the exclusion of buck brush (*Ceanothus* spp.) from the site, providing evidence of the dependence of these species on fire-free intervals.

Keeley (1992a, 1992b) found that chamise chaparral stands older than sixty years often are declining in vitality. Old stands were characterized by a high proportion of deadwood, little annual growth, and no new seedling development. Various phytotoxic substances and changes in dormancy related to aging (Hadley 1961, as cited by Parker and Kelly 1989) were suggested to account for the loss of vitality and lack of regeneration, and maintenance of vigorous chamise chaparral was suggested to be dependent on fire (Hanes 1971). Although this "decadence" idea has never really been demonstrated, it is partially supported by Rundel and Parsons (1979), who observed that with chamise chaparral, "between 16 and 37 years, shrub senescence increases with no increase in aboveground biomass and a sharp reduction of available photosynthetic surface area." In contrast, species that survive fire solely by vegetative regeneration from the rootcrown (mainly oaks and Christmas berries) were found to be capable, in the absence of fire, of continuously regenerating their canopy with basal sprouts, had very little mortality even in the century-old stands, and did not show signs of decadence. Obligate seeding shrubs that did not initiate new stems from the rootcrown, on the other hand, had 100% of the stems date back to the years immediately following the last fire and suffered greatest mortality (Keeley 1992a, 1992b).

Differences in vegetation structure and composition may influence fire behavior in ways that maintain different chaparral types within the same ecosystem. Fires (e.g., the Tower fire in 1980, which occurred west of Redding near Whiskeytown Lake) have been observed to burn rapidly across south-facing slopes dominated by either chamise or annual grasses and to skip or burn slowly through the oak/manzanita-dominated shrub fields on the north slopes (Skinner, personal observations). Such microscale variation in fire regimes allowed for the coexistence of these species of very different life-history strategies (Christensen 1985).

Succession after Fire

For the foothill chaparral, postfire community composition closely resembles that of the prefire community (Hanes 1971), suggesting that chaparral is the norm, if not the climax, of the system. Such communities recover quickly from fire and regain dominance soon after (Biswell 1974). However, at the shrub-herb interface in the more open chaparral types, seeders and sprouters seem to follow slightly different successional paths. After fire, herbaceous plants usually appear in abundance, and the shrub seedlings usually have to invade these pioneer herb communities to regain dominance. The seeder seedlings have higher survival rates under competition from the herbaceous species and stand a better chance of invading the areas covered by herb. Sprouter (e.g., chamise) seedlings, on the other hand, do not compete well with the herbs and require intense fire to clear the site of competitors to survive. Such intense fires usually do not occur on the sites originally occupied by herbs but can occur on the sites occupied by seeder chaparral that has successfully excluded the herbs. Once sprouters are established, they are capable of invading neighboring herb-covered sites by vegetative reproduction. Therefore, it is suggested that at the shrub-herb interface, the seeders "pave the way for the sprouting species" (Biswell 1974).

For the montane chaparral, the shrubs face competition not with the herbaceous species, but with the tree species. Therefore, at the forest-shrub interface, it takes intense fires to remove the forests and allow the establishment of chaparral. After initial establishment, frequent fires help maintain the chaparral by killing the trees before they shade out the shrubs, whereas less-rigorous fire regimes allow the trees to regain dominance.

Biodiversity: Floral Community Structure Responses to Fire and Fire Regimes

During postfire chamise chaparral succession, total woody cover, canopy height, litter, and dominance of chamise increased, while species diversity and herbaceous cover decreased in postfire succession (Parsons 1976). Thus, the observed increase in cover and biomass of chamise in Sequoia

National Park appeared to result from reduced fire frequency (Vankat 1970; Vankat and Major 1978).

Ceanothus chaparral that reproduces only from seed has been known to suffer under rare, low-intensity fire regimes. Black sage (*Salvia mellifera*), California buckwheat (*Eriogonum fasciculatum*), and manzanitas (*Arctostaphylos* spp.) can occupy the resulting openings in the canopy, and their abundant deadwood and compact biomass readily allow the spread of low-intensity fires, thereby further limiting buck brush (*Ceanothus* spp.) establishment (Riggan et al. 1988).

For the mixed chaparral, little information about postfire succession is available, but it is believed that reduced fire frequency has increased stand density and decreased species diversity. In some areas near stands of ponderosa pine forest, observed increases in density and cover of forest species are also thought to result from reduced fire frequency (Vankat and Major 1978).

For the montane chaparral, an increase in both shrub and tree cover and density (thus perhaps invasion of tree species) has been observed for some stands in Sequoia National Park. This increase may have resulted from either reduced fire frequency or recovery from sheep grazing (Vankat and Major 1978). However, Wilken (1967) suggested that for the northern Sierra Nevada, such increases in tree cover and density were temporary and occurred only until the fire-susceptible montane chaparral was burned again.

Fire has been associated with the maintenance of species diversity in nearly all shrubland types (Christensen 1985). In general, species richness and equability tend to be highest immediately following fire. Shortly after, both species richness and equability decline (Christensen et al. 1981). However, when the seed pool is included, species richness changes very little during the fire cycle in many shrublands. In a study of chaparral stands that have been unburned for 56 to 120 years, Keeley (1992a) found that although successional changes in community composition were evident, there was no indication of a decline in species diversity. In addition, he also identified two reproductive modes in chaparral shrubs in response to fire (see preceding discussion) that required very different durations of fire-free intervals for seedling establishment and suggested that both regimes were necessary for the maintenance of both adaptive traits. Also, chaparral communities, unlike other vegetation types, have been found to consist of some fire-adapted herbaceous species that occur only in the burned patches (Sweeney 1968). Therefore, no single, constant fire regime will meet habitat needs of all species. It is thus suggested that to maintain maximum biodiversity, a mosaic of variable fire regimes is required (Keeley 1991a; Husari and Hawk 1994).

Biodiversity: Faunal Responses to Fire

Few quantitative studies of chaparral fire effects on wildlife have been of sufficient duration to assess fully the long-term consequences. General patterns of faunal response to fire have been observed, and these were discussed earlier. Specific species responses to fire in the chaparral are noted as follows:

Very few data concerning direct effects of chaparral fire on invertebrates are available, except that, generally, mortality of soil invertebrates is low and that certain insects, such as the smoke fly (*Microsamia occidentalis*), are known to be attracted to the smoke of fire in other localities (Nichols and Menke 1984).

Some vertebrate animal species decrease whereas others increase following a burn, but no species is totally eliminated, nor is there any apparent diminution of total life on a burn after plant growth resumes (Lawrence 1966).

Mortality of reptiles in chaparral fires is generally low. Several snake tracks and a live king snake have been found on a fresh chaparral wildfire burn (Tratz and Vogl 1977). Legless lizards and other reptile species that habitually hide in litter or under logs are more directly affected by fire (Nichols and Menke 1984).

Birds seem to be less directly affected by fire in the chaparral, as there are no reported instances of bird deaths attributed directly to controlled burning in this vegetation type (Nichols and Menke 1984). However, birds that normally exhibit a strong preference for chaparral habitat were observed to decrease in numbers in the years immediately following the burn. Conversely, some birds that normally prefer grassland or oak woodland increased in number (Lawrence 1966). Such changes in relative species abundance are less likely a result of fire-induced mortality but more likely a result of migration due to habitat preference. Overall, fire resulted in an increase in density of nesting birds.

Most small mammals are not found in the chaparral immediately after a wildfire. Wood rats and brush rabbits are vulnerable to fire because they often exhibit panicked behavior described previously, whereas dusky-footed wood rats (*Neotoma fuscipes*) and some other small mammals are also vulnerable to fire-caused mortality because they retreat to their woody shelters when frightened (Nichols and Menke 1984). Kangaroo rats are the only abundant rodent species in chaparral immediately after a wildfire. In the second and third years after a fire, the number of species and population densities of rodents increase, and such trends continue for at least five years, and probably much longer (Quinn 1979).

Larger mammals such as mule deer (*Odocoileus hemionus*) are more mobile and generally have little trouble fleeing chaparral fires, but large, rapidly moving wildfires have been known to trap and kill deer. Predatory, fur-bearing animals such as foxes (*Vulpes* spp. and *Urocyon* spp.) and coyotes (*Canis latrans*) are especially mobile animals that can flee from most fires, and there are no reports of direct kills by fire for these species (Sampson 1944). Howard and colleagues (1959) have even observed a bobcat (*Lynx rufus*) trotting away from a controlled burn in a leisurely fashion.

As with birds, mammals that normally exhibit a strong preference for chaparral habitats substantially decreased in numbers in the years following the burn. None of the small mammals increased in numbers after the burn, but some of the larger predators, such as the coyote and badger, immigrated into the stand during the months following the fire (Lawrence 1966).

As is the case in pyrodiversity maintaining plant biodiversity, it has been suggested that the number of species and the population density of small mammals would be maximized by breaking up chaparral into small areas of different ages, maximizing ecotones, emphasizing physical heterogeneity, and leaving a few areas of unburned brush (Quinn 1979).

Landscape-Pattern Responses to Fire Regimes

Fire in itself is known to cause fragmentation and clumpiness of chaparral vegetation cover at the population level. For example, fire-induced mortality tended to move a chaparral stand toward a more clumped distribution, after which seedling establishment in the patchy empty spots and subsequent self-thinning eventually restored the pattern toward regular spacing (Stohlgren et al. 1984; Keeley 1992a).

Regular fire regimes also increase the clumpiness in shrubland landscape by forming smaller patches than would be formed during longer fire-free periods. Analyzing historical accounts of some chaparral burns in the San Gabriel Mountains, Minnich (1987) suggested that because of frequent burning, chaparral in southern California watersheds before 1900 comprised patches smaller than those seen today. Because these fire-created patches would constrain later burns, the patchiness has tended to persist over time. Irregular fire behavior and development of secondary burns from smoldering or from embers moving long distances beyond the fire zone were also suggested to have added to the complexity of vegetation fragmentation (Minnich 1987, 1989). In Baja California, where regular fire regimes still persist, fires are also recorded as being generally medium-sized fires that form an interlocking mosaic and burn at varied intensities (Minnich 1983).

Conversely, longer fire intervals are expected to push the chaparral landscape toward a more uniform distribution. Minnich (1987), for example, noted that shrubland fuels in the San Gabriel Mountains are more continuous than they were decades ago. Similar findings have been recorded for the shrublands of Sequoia National Park, where instead of a mosaic of different successional stands, old-age stands were found to be dominant across the landscape, and reduced fire frequency was suggested to be the cause (Vankat 1970; Parsons 1976; Vankat and Major 1978).

As a consequence of these continuous fuels, the few fires escaping control under the most extreme weather conditions (usually strong Santa Ana winds) have turned into enormous, high-intensity conflagrations and have resulted in even more spatially continuous landscapes (Minnich 1983, 1987, 1989). A study on fire history of chaparral in the Los Padres National Forests also indicated that fires are getting larger and less scattered (Radtke et al. 1981).

Biogeochemical and Soil Changes after Fire

In general, nutrient availability in the chaparral has been found to be higher immediately following fire (Christensen and Muller 1975; Westman et al. 1981). In a study on nutrient changes after a prescribed chaparral burn, only two (N and K) of the six nutrients studied (N, P, K, Na, Mg, and Ca) showed measurable losses (DeBano and Conrad 1978).

In response to this increased availability of nutrients, chaparral species not only grew more quickly (chamise chaparral, Christensen and Muller 1975), but also increased their consumption levels of nitrogen, phosphorus, and potassium beyond their immediate metabolic requirements (chamise and *Ceanothus* chaparral, Rundel and Parsons 1980). This ability to take up extra amounts of nutrients and store them for future use is an important fire adaptation in plants on low-nutrient soil (Rundel and Parsons 1980).

In addition to being available after fire, soil nutrient concentrations in the chaparral have also been found to be considerably more variable after than before fire (Christensen and Muller 1975). This variation arises because of the previously discussed local variations in fire intensity and uneven distribution of ash.

During the course of chaparral succession after fire, nutrient availability continued to change. Nitrogen availability increased with increasing stand age up to fifty to sixty years. Beyond sixty years, nitrogen availability declined. Phosphorus availability decreased logarithmically with increasing age. Declining nitrogen availability in older stands was attributed to declining total soil nitrogen and a decline in the fraction available, whereas decline in the phosphorus availability was attributed to a decline in the fraction available (Marion and Black 1988).

Nutrient concentrations in the plants also changed over this course. Chamise and buck brush (*Ceanothus* spp.) stands studied in Sequoia National Park showed a rapid decline in foliage concentrations of nitrogen during the first six years after fire, followed by a more gradual decline over succeeding years. Phosphorus concentrations showed a similar early decline but increased in older-age stands. A sharp increase in above-ground nutrients per unit of chamise canopy was found for the first sixteen years of growth, before a plateau was reached. Frequent chaparral fires, such as those observed for the presettlement era, were suggested to promote fire cycling of nutrients at intervals consistent with periods when nutrient availability became limiting (Rundel and Parsons 1980).

The aspects of fire-biogeochemical-plant interactions in chaparral were also discussed by Riggan and colleagues (1988). They suggested that copious nitrogen volatilization during burning is promoted by high concentration of nitrogen in the foliage and fine woody biomass of buck brush and heavy leaf litter of scrub oak (*Quercus dumosa*). They accordingly concluded that communities most prone to severe fires also accumulated and cycled nitrogen and phosphorus rapidly.

Foothill Woodlands

Primary Vegetation and Its Historical Relationship with Fire

Foothill woodlands can be viewed as a group of variable communities geographically placed between grasslands or shrublands and the montane forests (Griffin 1977). In the Sierra, the oak foothill woodlands (especially the interior live oak woodlands) are closely related to the chaparral (Griffin 1977). A detailed description of the ecological relationships between foothill woodland and chaparral communities can be found in Rundel (1981).

Three major types of foothill woodlands are found in the Sierra Nevada: blue oak, live oak, and black oak.

The blue oak foothill woodland is scattered throughout the western portion of the Sierra Nevada at 150–910 m (500–3,000 ft) (McDonald 1990a). Blue oak (*Quercus douglasii*) is dominant, with manzanita (*Arctostaphylos viscida*) and buckthorn (*Rhamnus crocea*) commonly seen. Soil beneath blue oak stands is found to be significantly lower in total nitrogen, total phosphorus, and organic matter content than adjacent sites with mixed evergreen woodland (Vankat and Major 1978).

The live oak foothill woodland usually occurs above the blue oak woodlands, especially on the north-facing slopes. Interior live oak (*Quercus wislizenii*) is the dominant species, and gray pine (*Pinus sabiniana*) is common. A highly variable mixture of shrubs (e.g., mountain mahogany [*Cercocarpus betuloides*]) and trees (e.g., buckeye [*Aesculus californica*], canyon oak [*Quercus chrysolepis*]) is found primarily on mesic north slopes and at higher elevations (Parsons 1981).

Some live oaks occur in riparian areas, and fires frequently go out or reduce their intensity dramatically when they reach these wetter areas. Most associated plant species have aboveground parts that are susceptible to damage by fire. However, most species quickly resprout after fire (Andrews 1994).

The California black oak woodlands occur on the west slopes of the Sierra from near Lassen Peak to Kings Canyon. Their elevation zones range from 460–1,980 m (1,500–6,500 ft) in the northern Sierra to 1,220–2,380 m (4,000–7,800 ft) in the southern Sierra (McDonald 1990b). California black oak (*Quercus kelloggii*) is the dominant species and is commonly associated with ponderosa pine (*Pinus ponderosa*). Burning by Native Americans has been considered a primary factor in maintenance of black oak stands (Anderson 1993). Without such disturbance, it has been suggested that black oak will eventually be crowded out of most suitable sites and will retreat to scattered remnants in mixed conifer forests (McDonald 1990b).

Vegetation Responses to Fire

Oaks are highly variable in their response and resistance to fire because of differences in their bark thickness, tree structure, and sprouting response. Individual survival is also influenced by understory composition and the degree of fire intensity (Husari and Hawk 1994). Plumb (1980) and Plumb and Gomez (1983) offer more detailed accounts of the different responses of various oak species to fire.

Blue oak is thought to benefit from fires. Although acorn survival and germination are thought to be negatively affected by fire, the positive association between blue oak ages and fire dates suggests a temporal concentration of postfire sprouting. The low rate of recruitment since the 1940s may be partly due to fire suppression (McClaran and Bartolome 1989).

In contrast, fires are damaging to live oak vegetation, because most associated species are susceptible to fire damage. In particular, canyon oak (*Q. chrysolepis*), interior live oak (*Q. wislizenii*), sycamore (*Platanus* spp.), and cottonwood (*Populus* spp.) have fairly thin bark and are easily top killed by fire. Fire is also known to cause basal wounds and development of hollow trees, which structurally weaken trees, leaving them susceptible to windthrow. However, most species resprout quickly after fire (Andrews 1994).

Succession after Fire

There is very little information about succession after fire in oak stands. In lowland live oak stands, light surface fires may trigger succession through sprouting from resident vegetation in the lower-canopy layers. For succession to proceed from outside seed sources, less-frequent, high-intensity fires are required (Andrews 1994).

Biodiversity and Community-Structure Responses to Fire

For the blue oak stands, livestock grazing has been proposed as the cause of their increase in density. Livestock grazing removes herbaceous competition for blue oak seedlings and decreases fuel levels, so that fires are less intense and thus less detrimental (Vankat and Major 1978).

In lowland live oak woodlands, stand density is controlled by fire frequency, because live oaks are found at their highest density in areas without recent fire (Davis et al. 1988). It is thus believed that interior live oaks and some other woody species have increased cover and density as a result of reduced fire frequency. Before European settlement, some of the stands may have been as open as today's blue oak woodland (Vankat and Major 1978).

Fire also influences faunal diversity in the lowland live oak stands. For example, the basal wounds and hollow trees that can result from fire are thought to provide important habitats for some animals. On the other hand, because riparian live oak stands are less intensely burned, they may also serve as a critical refuge during wildfire (Andrews 1994).

Fuel-Structure Changes in Response to Different Fire Regimes

The Sequoia National Park fire atlas, which maps all fires that burned more than 4 ha (10 acres) since 1920, shows that much of the foothill zone of Sequoia National Park has not burned in at least sixty years. Buildup in both live and dead fuel is thought to indicate a serious overabundance of "dense, overmature, highly flammable brush" (Parsons 1981). Loss of distinct age-class boundaries as a result of this longer fire-free period also deprives the system of effective firebreaks and adds to the continuity of the fuel (Parsons 1981).

Biogeochemical and Soil Changes after Fire

Fire running upslope from lowland oak stands is thought to affect runoff and sedimentation. In particular, high-intensity fires are thought to cause water-repellent soil upslope that reduces permeability and increases runoff and erosion (Andrews 1994).

Ponderosa Pine Forests

Primary Vegetation and Its Historical Relationship with Fire

Ponderosa pine forests dominate the xeric sites on the lower-elevation west slopes of the Sierra Nevada, from about 300 to 610 m (1,000 to 2,000 ft) in the north, and from 1,615 to 2,225 m (5,300 to 7,300 ft) in the south (Burns 1983). Their range of occurrence also differs on different aspects, starting below 1,220 m (4,000 ft) on north-facing slopes, and 1,830–2,440 m (6,000–8,000 ft) on south-facing slopes. Ponderosa pines (*Pinus ponderosa*) are dominant, and sugar pine (*P. lambertiana*) and incense cedar (*Libocedrus decurrens*) are commonly found associates. At the upper margin, ponderosa pines are often replaced by Jeffrey pines (*P. jeffreyi*) (Rundel et al. 1977).

Historically, surface fires were most common in this type, with occasional flare-ups occurring in brush patches. Crown fires were unlikely in most stands (Husari 1980). Long, loosely packed pine needles and herbaceous species maintained frequent and mild surface burns. These fires created openings for pine seedling establishment, thus maintaining its persistence. These fires also thinned saplings and maintained the relatively open understories documented by early settlers (Muir 1894; Sudworth 1900; Leiberg 1902; Cooper 1961). The recent increase in understory density has been attributed to the current longer fire intervals, which in turn have been suggested to result largely from fire suppression.

Vegetation Responses to Fire

Ponderosa pine is fire-adapted in all stages of its life history and is especially well adapted to light, regular surface fires (Rundel et al. 1977). The seeds prefer openings with mineral soil usually prepared by fire for seedbed. For the seedlings and young trees, early development of insulative bark, shielded meristems, high moisture content in living needles, and rapid extension of taproots reduce their mortality from fire (Husari 1980). For the mature trees, thick bark, deep roots, and low-flammability crown structures help them survive most fires (Starker 1934).

Generally, well-spaced ponderosa pine seedlings and saplings are able to survive low-severity fires, as are pole-sized and mature trees. Moderate- to high-severity fires, however, kill pole-sized and smaller trees, and crown fires kill mature trees. The main cause of their mortality following fire is crown scorch rather than damage to the cambium or roots. Crown scorching has also been noted to make pines more vulnerable to bark-beetle infestation, thus adding to the indirect mortality caused by fire (Andrews 1994). Fortunately for the pine stands, self-pruning of lower branches and open crown structure reduce the chance of crown fires, and such incidences rarely have been observed in this vegetation type historically (Husari 1980).

Succession after Fire

Following a stand-replacement fire, the successional sequence in these types proceeds from herbaceous species to shrub and hardwood, and finally ponderosa pine stages. Occasional sugar pines and incense cedars are found with the ponderosa pines when seed sources are available. The lower-intensity surface fire typical of this type generally does not completely reset succession but, rather, thins the understory and allows pine establishment in slightly more intensely burned small patches where fuel loads are locally higher.

Biodiversity: Floral Community Structure Responses to Fire and Fire Regimes

Under the presettlement fire regime of frequent, low-intensity fires, the ponderosa pine forest canopies were kept open with spaced trees of an uneven-aged structure (Weaver 1943, 1967; Husari 1980; Andrews 1994). Herbaceous species were much more common, and except for brush fields, shrubs in the forest were rarer and younger. Multilayered stands existed but were less extensive than today (Andrews 1994).

At present, the understories of ponderosa pine forests are dense in many places with unthinned pine seedlings and increased hardwood and shrub cover. The resulting high-intensity crown fires shift species dominance to hardwoods, because these stand-replacing fires kill all conifers and the above-ground portion of hardwoods, but hardwoods sprout vigorously after fire and capture the site. It has been suggested that such early seral vegetation will dominate the site for some time after fire (Andrews 1994).

Following large, severe fires, shrubs may also occupy sites for very long periods before the pine can again attain a superior position (Husari 1980; Andrews 1994). But instead of giving way to pine, these brush fields may also maintain a continual cycle of fire, therefore maintaining itself (Andrews 1994). The exclusion of fire has also been noted to allow the establishment of brush, usually manzanita (Husari 1980).

Biodiversity: Faunal Responses to Fire

Bock and Bock (1983) studied bird and deer mouse populations in response to a prescribed burn in a ponderosa pine forest. They found that total breeding birds were more abundant on burned than on unburned sites during the first postfire summer. In the second summer, however, one of the four burned sites had fewer birds than its control, whereas the remaining three paired sites did not differ. Although species composition of burned versus control areas remained almost the same through both years, seven species were more abundant on the burns during the first postfire nesting season, whereas none was more common on control plots. In the second summer, however, only one species was more abundant on the burns, whereas two were more common on the controls. Deer mice were also more abundant on the burned sites during the first summer, whereas there was no difference between burned and unburned sites during the second summer. It seems that for the birds and rodents, the population change after fire is short term.

Fuel-Structure Changes in Response to Different Fire Regimes

Fuel loads from the ponderosa pines are relatively light (Blonski 1980, as cited by Husari 1980) and are known to have carried frequent, light surface fires under the presettlement regimes. But with fire exclusion, large amounts of pine needles and small branches have accumulated over time and created a bed of fine fuels with large surface-to-volume ratios that often lead to fast-moving, intense fires. The dense understories also increase the chance of crown fires.

There has also been an increase in the accumulation of downed logs and snags in these forests as a result of the increased mortality from recent, severe fires, from insects, and from stressed, overcrowded pine stands. Such mortality increases the debris and the vertical and horizontal structural development of fuels in ponderosa pine stands. As a consequence, large (4,000 ha [10,000 acres] or more), high-severity fires, once rare, have become commonplace in recent years, as have many small (less than 4 ha [10 acres]), high-intensity fires (Andrews 1994).

Landscape-Pattern Responses to Fire Regimes

Ponderosa pine stands are a classic example of how gap-sized local variation shapes the heterogeneity of the landscape. Within a stand, single mature trees or groups of trees are killed by insects, disease, lightning, or windthrow. These dead trees form gap-sized patches of concentrated fuel in the landscape. When a characteristic low-intensity surface fire burns through the area, these "tinderboxes" result in patches of higher-intensity burns, open up the ground, and allow young pines to germinate. In the remaining areas where mature trees still stand, heavy accumulations of flammable needles, cones, and bark scales build up, and when the next fire comes through, it burns more intensely under these canopies than in the openings, thereby killing seedlings and saplings under their canopies. The young pines in the gap-sized openings, however, survive the event, because the small accumulation of needles in the openings will not support a surface fire. Hence, until the pines are large enough to build up fuels under themselves, fires are not intense enough to kill them, and by the time the pines do create such heavy fuels, many of them are also large enough to withstand the surface fires (Kilgore 1973; Husari and Hawk 1994). The periodic fires typical of the area in the presettlement era are thought to support this process and thereby maintain a mosaic of different-aged patches.

At present, because of the increased continuity of fuels, high-intensity fires typically kill all vegetation over large to very large areas, increasing the homogeneity and patch sizes in these forests (Andrews 1994).

Mixed Conifer Forests

Primary Vegetation and Its Historical Relationship with Fire

Mixed conifer forests are largely restricted to the west slope of the Sierra Nevada at middle elevation, from about 760 to 1,400 m (2,500 to 4,600 ft) in the north, and from 915 to 3,050 m (3,000 to 10,000 ft) in the south (Eyre 1980). Dominant species include ponderosa pine *(Pinus ponderosa)*, sugar pine *(P. lambertiana)*, incense cedar *(Libocedrus decurrens)*, white fir *(Abies concolor)*, California black oak *(Quercus kelloggii)*, and Douglas fir *(Pseudotsuga menziesii)*. Jeffrey pine *(Pinus jeffreyi)*, red fir *(A. magnifica)*, lodgepole pine *(P. contorta)*, and patches of giant sequoia *(Sequoiadendron giganteum)* are also found (Eyre 1980; Tappeiner 1980). The proportion of each species in different stands is highly variable (Eyre 1980) and is thought to be determined by elevation, precipitation, and fire frequency: more frequent fire favors a higher percentage of fire-adapted ponderosa and Jeffrey pines; less frequent fire favors the less fire-tolerant white fir and incense cedar. Fire scar analysis indicates that the majority of this forest type was historically subject to low- to moderate-severity fires (Show and Kotok 1925; Wagener 1961). There was a great deal of variation in fire intensity and effect within similar sites, even within a single fire (Stephenson et al. 1991).

Vegetation Responses to Fire

Weatherspoon (1988a) discussed the relative fire tolerance of the various species in the mixed conifer forests. Sugar pine, Douglas fir, red fir, white fir, and incense cedar typically survive less crown scorch as a percentage of crown volume than do ponderosa and Jeffrey pines. A study of mortality in four species in the mixed conifer forest showed that ponderosa pine is able to survive longer flame lengths than sugar pine, incense cedar, and white fir (van Wagtendonk 1983).

The various species in the mixed conifer forests also responded differently to the various burning conditions in terms of seedling establishment. Ponderosa pine and Jeffrey pine seedlings benefit from high-intensity surface burns that open up the understory (Vlamis et al. 1956; Bock and Bock 1969), whereas sugar pine and giant sequoia are known to require patches of stand-replacing burns that create canopy gaps (Stephenson et al. 1991). Kilgore and Biswell (1971), for example, found the largest number of sequoia seedlings in places that experienced the most intense fire, and vice versa. Both white fir and red fir germination also seem to be favored by

fire (Agee and Biswell 1969; Laacke and Fiske 1983), although both firs also do well without fire (Kilgore 1973; Taylor 1990a).

Although the fewest numbers of shrubs were found in patches that had gone through intense fires (Kilgore and Biswell 1971), deer brush (*Ceanothus* spp.) and manzanita (*Arctostaphylos*) species are known to be stimulated to germinate by fire and can persist only within canopy gaps (Andrews 1994). Although there is no totally fire-dependent herbaceous species associated with the conifer forests in California (Sweeney 1969), most herbs are known to prefer more open habitats and may be affected by pine litter. Several species of herbs have been found to increase in coverage or frequency following burns in the giant sequoia–mixed conifer forest (Kilgore 1973).

Denser stands that have resulted from the exclusion of fire cause more competition for available water and, therefore, greater moisture stress. Extensive mortality thus results from droughts, either directly from drought stress or from stress-induced bark-beetle outbreaks (Weatherspoon et al. 1992).

Succession after Fire

During the first few years after fire, a herbaceous layer develops but is eventually shaded out by the growing woody layer. Among the many tree species in the mixed conifer zone, the pines and sequoia benefit most from postfire conditions and dominate the site in the years following fire. Although white fir and incense cedar also benefit from improved seedbeds after fire, their seedlings do not grow as fast as pine and sequoia seedlings do and thus are suppressed in the early seral stages. However, these two species are more shade tolerant and are known to be released at the chance of a gap formation, even after many years, and thus benefit from longer fire-free periods. Under presettlement conditions, forest succession was influenced by generally low-intensity fire with inclusions of localized patches of vegetation that were either completely burned or unburned (McBride and Sugihara 1990).

Biodiversity: Floral Community Structure Responses to Fire and Fire Regimes

Fire affects the character of the mixed conifer forests by effectively decreasing woody plant density and influencing species composition (Rundel et al. 1977). The frequency of fire determines the percentage of fir in these forests (Agee et al. 1978; Husari and Hawk 1994): frequent surface fire eliminates the less fire-resistant white fir and incense cedar seedlings and favors the more fire-resistant but shade-intolerant black oak and pine seedlings, whereas longer fire-free periods allow the shade-tolerant fir to dominate the canopy with time (Sellers 1970; Lyon et al. 1978). White firs in Sequoia National Park have been observed to increase in density during dates that correspond to the onset of reduced fire frequency (Vankat and Major 1978). Other observations of present white fir dominance in the understory are abundant (Parsons and De-Benedetti 1979; Bonnicksen and Stone 1982; van Wagtendonk 1985; Weatherspoon et al. 1992). In addition to pine, sequoias also may be severely affected by fire-suppression policies (Parsons and DeBenedetti 1979). Periodic fire has been thought to maintain a mix of fire-tolerant and fire-intolerant conifers (Husari 1980), whereas fire exclusion is thought to be leading toward fir dominance. A prolonged period with intense, stand-destroying fires is thought to convert the habitat to a montane chaparral type (Husari 1980).

Before European settlement, these ecosystems were probably more open than they are today (Weaver 1974) and sufficiently open to support a well-developed herbaceous-layer community of forbs, perennial bunchgrasses, and dispersed shrubs. These understory components are now lost from the system and impossible to describe in detail from available historical data. Shrubs are also thought to have been present in the understory but were patchy and variable in the percentage of ground covered (Andrews 1994). Vankat and Major (1978), for example, documented a decrease in some shrub species, especially manzanita (Arctostaphylos viscida), in Sequoia National Park following European settlement. They suggested that this decrease may be a result of decreased germination of their fire-stimulated seeds.

Fire also affects the forest structure of the mixed conifer zone. Periodic fire thins the trees, and thus fire suppression has been thought to be the reason for the observed increase in the density of small trees (Parsons and DeBenedetti 1979). Periodic fires before European settlement were also thought to maintain uneven-aged stands (Weaver 1974), which were dominated by larger, older trees as compared with today's predominant smaller, younger trees (Andrews 1994). Fire also changes forest structure by altering the species composition. For example, the fire-sensitive tree species, especially white fir, have been observed to have increased dramatically in abundance, particularly in small to medium size classes. The resulting multiple-canopy stands consisting largely of these shade-tolerant species are now common but are thought to have been much less common previously, except in the cool and moist extremes of the type (Parsons and DeBenedetti 1979; Bonnicksen and Stone 1982; van Wagtendonk 1985). Consequently, stands have become more complex vertically but less complex and more homogeneous in terms of aerial arrangement (Weatherspoon et al. 1992).

Biodiversity: Faunal Responses to Fire

The findings of the few studies on faunal responses to fire in the mixed conifer forests show that deer populations have increased after burns (Lawrence and Biswell 1972); bird populations also have increased in numbers or biomass following fire (Lawrence 1966; Bock and Lynch 1970), but elimination of saplings less than 3.5 m (11.5 ft) tall did not make major changes in species composition of a breeding-bird population (Kilgore 1971a).

There is no evidence to suggest that the overall wildlife species richness before European settlement was markedly different from that currently found in the mixed forest. However, it is likely that the species' relative abundance has changed (Andrews 1994). Also, the shift from open- to closed-canopy forests with a more complex vertical structure has probably benefited wildlife associated with closed forests over those associated with open forests, such as chickadees. Spotted owls may have been affected by this change in forest structure too, because the increased vertical stand structure could close their flyways and affect their hunting success (Andrews 1994).

Fuel-Structure Changes in Response to Different Fire Regimes

The mixed conifer forests before European settlement were thought to be uneven-aged, patchy, broken, and varied in cover type. This discontinuous structure made the forest fairly immune to extensive, stand-replacing crown fires. Although local crown fires may have occurred, they probably extended at most over a few hundred acres (Show and Kotok 1924).

Because of the lack of periodic fires, fuels on the forest floor (including coarse, woody debris) have accumulated over time (Parsons and DeBenedetti 1979). Increased mortality due to drought stress (see earlier discussion) has also added greatly to fuel loads. The increase in snags and large woody fuels is likely to increase fire spotting and make fires harder to suppress. Because pine fuels are easier to burn and reduce, the species-composition shift from ponderosa and sugar pines to a white fir–incense cedar mix makes the system less readily and thus less frequently burned (Agee et al. 1978), therefore further adding to fuel accumulation. Opening of the canopy as a result of tree mortality also results in warmer and drier fuels (Countryman 1955), which ignite more easily and support faster-spreading fires (Weatherspoon et al. 1992).

In addition to changes in fuel amount, fuel structure has also changed following the change in species composition. Increased prevalence of white fir in the understory has created multilayered structure and fuel ladders, linking surface fuels to upper canopy layers. In addition to having increased the vertical continuity of fuels, the lack of periodic fire has also resulted in a more homogeneous landscape and horizontally continuous fuel. This increase in both the horizontal and vertical continuity of fuels, combined with the greater quantity of fuels, has substantially augmented the probability of large-scale, catastrophic fires (Kilgore 1973; Kilgore and Sando 1975; van Wagtendonk 1985).

Landscape-Pattern Responses to Fire Regimes

Present fire regimes within the mixed conifer forests are characterized by infrequent, high-intensity surface fires or infrequent, stand-replacement fires. These fires range from medium to very high severity. With present fire recurrence rates, it is thought that stands will generally convert to an "infrequent, very high intensity fire regime with uniformly severe effects" (Husari and Hawk 1994).

Before European settlement, fire regimes were characterized by periodic low-intensity surface fires. Because these low-intensity fires were very much influenced by local variables, they varied in intensity and effect even within similar sites (Lindenmuth 1960; Sweeney and Biswell 1961). These fires also had large fluctuations in the timing of fire frequency and set various scenarios for vegetation establishment, composition, and survival. As a consequence, each patch bore a distinct aggregation of trees that developed successively as an independent entity, creating a complex mosaic of aggregations with patch sizes typically 0.2 to 2 ha (.5 to 5 acres) (Bonnicksen and Stone 1982; McBride and Sugihara 1990; Stephenson et al. 1991). Fire patterns similar to those ascribed to the presettlement period are currently found in the ponderosa pine and mixed conifer forests of the Sierra San Pedro de Martir National Park, Baja Norte, Mexico, where historical fire patterns still persist (Barbour et al. 1994; Minnich et al. 1995).

Toth and others (in Andrews 1994) agreed with Bonnicksen and Minnich's assertion that the presettlement mixed conifer landscapes had a fine-grained patch structure, in which small patches of single-storied, even-aged, and uniformly sized stands shifted in time across the landscape. They noted that these small patches were typical on flat terrain and most aspects. However, they asserted that the presettlement mixed conifer landscapes also had coarser-grained seral stage patterns (larger patches). These larger patches, though scarce, were thought to be found in areas of catastrophic fires or with longer fire-return intervals, particularly on moister north and east aspects. Such big patch patterns were also thought to be more uniform in nature and to have initially formed a more coarse-grained mosaic. However, such homogeneity in the landscape would have persisted only until both age and localized disturbance events, such as insects, diseases, blowdowns, and other disturbances, blended these areas into the overall landscape pattern of finer grains. Because of the general big tree character, they also asserted that age-class distinctions would have blurred at the landscape level, so that the fire-induced vegetation pattern in the presettlement landscape appeared "relatively uniform on a broad scale."

At present, the landscape pattern is changing from the fine-grained, open, presettlement forest, which generally contained large numbers of large-diameter, older trees, toward a patchy, coarse-grained mosaic of openings and more closed-canopy forest, with a much larger proportion of the landscape in younger stands (Andrews 1994).

Heterogeneity is a relative term, and it is useful to note that earlier discussions of landscape heterogeneity are based on the comparison of presettlement and post-settlement Sierran landscapes. Even the relatively heterogeneous presettlement landscape could be seen as relatively homogeneous: Parker (1984), for example, compared Yosemite forests to those in Glacier National Park and concluded that the surface fire regime promoted "uniformity of structure and compositional dynamics" in Yosemite forests, whereas a broader spectrum of disturbance regimes in Glacier forests promoted "heterogeneity of structure and complex patterns of compositional dynamics."

Biogeochemical and Soil Changes after Fire

Fire plays a significant role in recycling various mineral nutrients back to the soil in all Sierran conifer forests. Light burns typical of the presettlement mixed conifer forests are thought to increase soil pH, stimulate nitrification, and increase available phosphorus, potassium, calcium, and magnesium through ash deposition (Hare 1961). The effects of burning on soil nitrogen, though, are more complex: some studies show a loss of nitrogen from the forest floor (Knight 1966), while others report a net gain (Klemmedson et al. 1963).

However, because the exclusion of fire has resulted in a change in fuel structure, when a wildfire occurs in dry weather, it is much more likely to develop crown fire behavior. This type of fire behavior has been thought to "seriously disrupt energy and nutrient cycle stability" (Agee et al. 1978). The increased amount of large materials (downed logs and snags) has also raised the concern of "greater heating damage to soils" because of these localized concentrations of fuels (Weatherspoon et al. 1992).

Upper Montane Fir Forests

Primary Vegetation and Its Historical Relationship with Fire

Two major species of firs are present in this type: red fir (*Abies magnifica*) and white fir (*A. concolor*). In the northern Sierra, the elevation range of these firs runs from about 1,520 to 2,440 m (5,000 to 8,000 ft). From Mount Lassen to as far south as Sonora Pass, where peaks extend above the 2,440 m (8,000 ft) level, red fir extends to (or near) timberline. An exception is the Warner Mountains, where there is no red fir (Skinner personal communication). To the south and on the east side of the Sierra crest, the elevational limits gradually shift upward to about 2,130–3,040 m (7,000–10,000 ft) (Andrews 1994).

In the northern Sierra Nevada, there is a gradual transition from mixed conifer forests, to more or less pure white fir stands (up to 1,830 m, or 6,000 ft), to a mixed red and white fir zone, and finally to red fir forests (above 2,130 m, or 7,000 ft). White fir is frequently the dominant species on mesic sites between 1,500 and 2,000 m (5,000 and 6,500 ft) (Conard and Radosevich 1982).

In the southern Sierra Nevada, the transition goes directly from mixed conifer forests to red fir forests, although the lower portions of these red fir forests often contain white fir, and upper portions often contain lodgepole pine (Vankat and Major 1978).

Although red fir appears to be a climax species in many stands at high elevations, it is rarely found in extensive, pure stands. Common associates include white fir, lodgepole pine (*Pinus contorta* var. *murrayana*), incense cedar (*Libocedrus decurrens*), sugar pine (*P. lambertiana*), Jeffrey pine (*P. jeffreyi*), western white pine (*P. monticola*), western juniper (*Juniperus occidentalis*), mountain hemlock (*Tsuga mertensiana*), and quaking aspen (*Populus tremuloides*) (Eyre 1980).

Although lightning-ignition frequencies are much higher in the upper montane zone (Vankat 1983), fires spread less readily because (1) biomass accumulates more slowly; (2) the fuel is more compact; (3) weather conditions that will support a fire occur less often; and (4) the high-elevation areas of the Sierra contain many natural fuel breaks, including sharp or sparsely vegetated ridges, barren rocky areas, streams and draws with relatively fire-resistant riparian vegetation, and large areas of sparsely spaced vegetation, that do not support fire well (Kilgore and Briggs 1972; Husari 1980; Weatherspoon et al. 1992). It has been observed, for example, that although red fir type constitutes only 8% of Yosemite National Park, 16% of the fires recorded in the park between 1930 and 1983 were in this zone. The majority of those were of single trees, and larger fires occurred when red fir was mixed with montane chaparral (van Wagtendonk 1986). Crown fire is unusual in this type except under rare high winds, partly because of the sparse understory vegetation (Kilgore and Briggs 1972; Rundel et al. 1977). The potential for stand-destroying fire exists on steep, south-facing slopes with heavy fuels and dense regeneration (Husari 1980). As a consequence, fires in the red fir zone were historically less frequent and usually far less intense than fires in the lower elevations, but the fire regimes varied from landscape to landscape (Kilgore and Briggs 1972; Husari 1980; Skinner and Chang 1996).

Fire-suppression actions are thought to have had less effect on the red fir type, both because suppression activities began later (1920s to 1930s) in red fir forests and because fewer fires would have burned there without suppression anyway.

Vegetation Responses to Fire

The relationship between red fir and fire is paradoxical. On the one hand, red firs seem to do well without fire and may even suffer under prolonged periods of frequent fires: they can germinate and grow in light litter and thus do not require fire-generated openings to establish themselves (Laacke and Fiske 1983; Taylor 1990a); they are shade tolerant and do well under canopies (Fowells 1965; Helms and Standiford 1985); and they are highly susceptible to fire when young (Kilgore 1971b). On the other hand, red firs seem to be at the same time fairly resistant to fire and may even benefit from fire: they develop thicker bark as they mature and become moderately resistant to the low- and medium-intensity fire typical of the vegetation type (Kilgore 1971b; Husari 1980); they establish best in postfire bare mineral soils (Laacke and Fiske 1983; Parker 1986); and episodes of their regeneration seem to be associated with fire events (Weatherspoon et al. 1992; Andrews 1994). It has been argued that the decrease in fire frequencies in this century may have led to a decrease in red fir establishment (Pitcher 1987).

Thus, we can conclude that red fir requires some fire for regeneration but does not do well during prolonged periods of frequent fires. Frequent fire eliminates fir seedlings and favors pine (Husari 1980). Firs benefit most from a variable-interval, low-intensity fire regime with occasional longer fire-free periods that allow young fir trees to survive through the fire-susceptible age. The short growing season and heavy snowpack may be important in limiting the frequency of fire, thus benefiting fir in this vegetation zone (Husari 1980).

Succession after Fire

Fir forest is the climax community over much of the coniferous type in the upper montane zone. Small fires produce openings where lodgepole seedlings may become established. Stand-destroying fires replace red fir stands with montane chaparral type, but subsequent seral changes will shift the dominance back to red fir (Husari 1980; Conard and Radosevich 1982). On older sites, openings in the canopy caused by tree mortality allowed patchy understory development (Conard and Radosevich 1982).

Biodiversity and Community-Structure Responses to Fire and Fire Regimes

Because of its ability to regenerate in both shade and sunlight, it has been suggested that red fir will be able to reach dominance in this vegetation zone even under different fire regimes. Given the present seed source in the area, fire suppression may not alter species composition in this vegetation zone by shifting the dominance of species. However, fire-dependent associates of red fir (e.g., lodgepole pine, Jeffrey pine, western white pine, and quaking aspen) may decrease with continued fire suppression (Andrews 1994).

As in other conifer types, fires are thought to have historically thinned young fir patches, leaving scattered trees that then developed into scattered, mature individuals. Therefore, although fire suppression may not have had as significant an effect in the red fir forests as in the lower-elevation forests, decreased fire frequency still may have resulted in denser red fir forests (Vankat and Major 1978).

Stand age structure is also influenced by fire events. A study of the relation between stand age structure of red fir and disturbances in the southern Cascade Range revealed that the red fir forest patches may be even-aged or multiaged, depending on the disturbance history of the site. Even-aged patches were a result of synchronous postdisturbance establishment. Gaps in the study area were created mostly by frequent wildfire and by windstorms. Severe fire cleared larger openings, initiated mass establishment of red fir and white fir, and resulted in larger even-aged patches, whereas smaller gaps created by windthrow released already established individuals, resulting in smaller, even-aged patches. Low-intensity fire stimulated little recruitment and probably caused thinning but did not affect the age structure of patches. Multiaged patches, on the other hand, were a result of continuous recruitment of seedlings moderately tolerant of shade during disturbance-free periods. Thus, both episodic and continuous recruitments, as determined by the type and severity of natural disturbances, were thought to be the driving forces shaping the complex age and structures of these red fir forests (Taylor 1991, 1993).

Because the patchy distribution of red fir forests in the Sierra Nevada originates not only from fire and windthrow, but also from a wide variety of other agents (including insect attacks, diseases, avalanches, or landslides), and because fires are not as regular as in the lower-elevation zones, changes in stand structure due to fire suppression are much less marked than in lower-elevation mixed conifer forests. Nevertheless, continuous establishment of red fir due to fire suppression will likely lead to changes in the vertical structure of forests, especially along the red fir–white fir ecotone (Andrews 1994).

Regarding fire influence on faunal biodiversity in red fir forests, no changes in deer or bird numbers were noted after a prescribed surface fire (Kilgore 1971b).

Fuel-Structure Changes in Response to Different Fire Regimes

Fir types tend to accumulate fuels of larger size classes (Husari 1980). The finer fuels that accumulate are made up of short needles that form a dense litter layer, which is further compressed by a heavy snowpack (Weatherspoon et al. 1992). Such fuel structure is capable of supporting only slow burns and thus does not promote rapid fire spread. The historical low- to medium-intensity fires in the red fir forest tend to spread to and from large, downed logs, burning the areas around the logs lightly, often spotting from log to log. Distribution of the large, decaying logs, therefore, tends to be spotty and concentrated in areas that missed being burned for a variety of reasons. One potentially important change in the ecosystem is the slight increase in the amount of large, downed woody material, which may now be more uniformly distributed throughout the forest (Andrews 1994). However, because this trend has not been commonly reported, whether it is representative of the whole Sierra Nevada is unknown.

Landscape-Pattern Responses to Fire Regimes

The level of heterogeneity in the Sierra Nevada fir forests changes from north to south: the mix of species becomes more variable and diverse, and spatial heterogeneity increases from north to south. A range of severity and frequencies of fire, in addition to a variety of other disturbances, has shaped the landscape into a complex pattern of various patch sizes and tree ages (Agee 1989; Weatherspoon et al. 1992; Andrews 1994). These fires burned irregular-shaped areas with varying spread rates (Kilgore 1971b) and were thought to behave so because of the variations in fuel-type pattern and fuel-moisture content (Kourtz and O'Regan 1971; van Wagtendonk 1972) and local topographic variations and weather conditions (Kilgore 1973).

Stand-replacing fires occasionally occur in these zones, and they initiate large cohorts of red and white firs (example from Cascade Range, Taylor 1993), but lower-severity fires that create smaller canopy openings and consequent smaller patches of thinned saplings are most common (Kilgore 1971b; van Wagtendonk 1986; Taylor 1993).

Biogeochemical and Soil Changes after Fire

Few studies of the biogeochemical and soil responses to fire have been done in the fir zone, but it has been suggested that fire may be an important factor in nutrient cycling by releasing nutrients from the accumulated litter on the forest floor, which would otherwise decompose very slowly (Andrews 1994). Water quality of creeks in this vegetation zone was not altered by prescribed surface burn (Kilgore 1971b).

Lodgepole Pine Forests

Primary Vegetation and Its Historical Relationship with Fire

The elevation range for lodgepole pines is about 2,000–3,000 m (6,560–9,840 ft) for the central Sierra Nevada, and slightly higher, 2,150–3,400 m (7,050–11,160 ft) for Sequoia National Park in the southern Sierra.

Lodgepole pine (*Pinus contorta* var. *murrayana*) is the dominant tree, whereas western white pine (*P. monticola*), ponderosa pine (*P. ponderosa*), and Jeffrey pine (*P. jeffreyi*) are common associates. Gooseberry (*Ribes montigenum*) is the most common shrub.

Small or moderate-intensity fires are thought to favor perpetuation of the lodgepole habitat in the north (Husari 1980), although Vankat and Major (1978) have suggested that the elimination of sheep grazing could have triggered a major pulse of lodgepole pine reproduction in the south.

Vegetation Responses to Fire

The mature lodgepole trees have very thin bark, deep roots, and medium to low foliage flammability and are considered intermediate in fire tolerance (Starker 1934; Husari 1980; Kilgore 1971b). Mature lodgepole trees are most commonly killed by scorching cambium or crowning (Starker 1934). Lodgepole seedlings are shade intolerant and thus require openings for growth. These seeds germinate best in openings and sterilized mineral seedbeds produced by high-intensity fire, and they seem to be stimulated by fires (Kilgore 1971b, Husari 1980).

In many regions, lodgepole pine owes its prominence to repeated fires, particularly in the Rocky Mountains (e.g., Brown 1975). In the Sierra, this dependency varies with location. For example, lodgepole pines seem to be fire dependent in the north (Husari 1980), but not quite so further south (Parker 1986).

Sierran lodgepoles do not have serotinous cones as the Rocky Mountain lodgepoles do, but they produce seeds in abundance each year and can set seed at an early age. The combination of heavy, early seed production and the ability to reseed in any openings that occur gives lodgepole pine a competitive advantage over its associates in the presence of fire, whether the intensity is low, medium, or high (Husari 1980).

In the central Sierra Nevada, an observed small-scale clumping of lodgepole seedlings suggests that tree falls may often create regeneration sites for these trees. Despite the virtual absence of crown fires in these areas, lodgepole pines apparently persist in these stands by continuous successful establishment in these tree-fall gaps (Parker 1986). Thus, for the central Sierra, lodgepole pines do not appear to be fire dependent.

Succession after Fire

Lodgepole has long been considered mainly a seral stage, to be replaced by fir in the absence of fire in most locations. Some localities seem to support a lodgepole climax, however, and Husari (1980) has suggested that for the northern Sierra and the Cascades, frost pockets and poorly drained soils may be the edaphic factors that define a lodgepole climax.

Biodiversity, Community-Structure, and Landscape Responses to Fire and Fire Regimes

Not much information is available on the compositional changes of lodgepole forests after fire, except that many areas in the north are found with developing understory of white and red firs (*Abies concolor, A. magnifica*), especially white fir, and that the suspected cause of this encroachment of firs is the exclusion of fires (Husari 1980).

Lodgepole pines in the Rocky Mountains are known to have a relatively even-aged structure and unimodal or bimodal diameter distribution, which have been associated with the crown fires typical of that area. For the northern Sierra, however, uneven-aged stands are more commonly seen, and low-intensity, small fires are suggested to be the process maintaining this structure. For the central Sierra, negative-exponential distribution of age classes, random dispersal of trees, and aggregated seedlings in small clumps are typically observed, and tree-fall gaps are thought to have had a major influence on the forest structure. Differences in community structure and landscape heterogeneity are due to the regional variations in site productivity and in the scale and frequency of disturbance: tree-fall or small-fire gaps in the Sierra Nevada lead to a more heterogeneous landscape, while larger crown fires in the Rocky Mountains create a more homogeneous landscape (Parker 1986).

Fuel-Structure Changes in Response to Different Fire Regimes

Not much is known about the effect of fire on fuel structure in the Sierran lodgepole pine forests, except that the recent development of understory fir in the northern Sierra creates fuel ladders that increase crown fire potential (Husari 1980).

Subalpine Meadows

Very little information exists for fire effects on subalpine meadows, like lodgepole pine ecosystems. But fortunately a case study on a lightning-ignited burn that ran through Ellis Meadow, a subalpine meadow in the Kings Canyon National Park, in 1977 (DeBenedetti and Parsons 1979, 1984) recorded detailed descriptions on the immediate postfire responses and a follow-up of four years of succession. Although the Ellis Meadow study may not be representative of all meadow burns, it contains valuable details. Most of the following discussion derives from this study.

Primary Vegetation and Its Historical Relationship with Fire

Subalpine meadows consist mainly of grasses and sedges, with some lodgepole pines invading from the borders. Ratliff (1982) presents a description of the herbaceous species in these meadows.

Rather than focusing on "what maintains meadows," the literature available on meadow ecology is most concerned with "what leads to tree invasion of meadows."

Three major factors have been suggested to define forest-meadow boundary dynamics: climatic change, livestock grazing, and fire (Husari 1980; Taylor 1990b). Although climate has been identified as an important force behind forest-meadow dynamics in the Sierra Nevada over thousands of years (Wood 1975), the effect of climate over shorter temporal scales is less clear. In the Pacific Northwest, warm, dry weather has been associated with a period of tree establishment in subalpine meadows (Franklin and Dyrness 1973; Agee and Smith 1984). In the Cascade Range, tree establishment was associated with wetter periods (Vale 1981; Taylor 1990b). For the Sierra Nevada, tree invasion has been associated with drier periods (Boche 1974; Helms 1987). The paradoxical relationship between climate and tree invasion along the Pacific Coast suggests that climate alone cannot explain the forest-meadow boundary dynamics observed over the past few decades.

The most commonly held hypothesis on such tree invasion is that past grazing activities resulted in increased precipitation runoff, soil erosion, and stream entrenchment and that these in turn lowered the water table, dried the meadows, and improved conditions for tree seedling establishment (Vankat and Major 1978). However, because sheep ate the tree seedlings and kept them off, tree invasion did not start until grazing activities ceased (Magee 1885; Bradley 1911). Large fires were also an important factor determining the meadow-forest boundary (Vankat and Major 1978), and there has been a concern that meadows are being lost to lodgepole invasion because of the exclusion of fire (Husari 1980).

Charcoal layering indicates that meadows in Yosemite National Park and Sequoia National Park burned once every 250 to 300 years (Leonard et al. 1968; Botti 1979). However, because charcoal layers are less indicative of less severely burned periods, these subalpine meadows historically may or may not have burned less frequently than the surrounding forests. An observation that some fires burning in the nearby forests may stop at the edge of meadows (e.g., Sugarloaf Meadow in 1974, as described by DeBenedetti and Parsons 1979) provides a mechanism that could result in a reduced fire frequency. Meadows characterized by high productivity and biomass are most likely to carry fires through, and prolonged drought greatly facilitates fire spread (DeBenedetti and Parsons 1979).

Vegetation Responses to Fire

Generally, fire does little damage to grasses and sedges (Husari 1980). Hot fires may kill well-established trees and may greatly damage meadows, but light fires do little harm to trees or meadows (Ratliff 1985). In the case of the Ellis Meadow burn, the fire killed many lodgepole seedlings but only a few of the numerous sapling or pole-sized pines found around the perimeter of the meadow (DeBenedetti and Parsons 1979).

Succession after Fire

In Ellis Meadow, grass and graminoid species recovered quickly after fire, increasing from an 8% ground cover one year after fire to a 75% ground cover four years after fire. In areas that burned less severely, above-ground biomass and cover of grass and graminoid appeared comparable to those of unburned sites the following year. Broadleaf species also increased, but at a much slower rate, from a 28% ground cover one year after fire to 49% four years after fire. Absolute cover by annuals declined steadily during the second through fourth years after fire. Postfire succession brought the vegetation back toward the characteristics of the preburn state (DeBenedetti and Parsons 1984).

Many of the lodgepole pine seedlings and saplings located around the boundaries of Ellis Meadow were killed by fire. Lodgepole pine seedlings were not found in burned portions of the meadow during any of the four years following fire (DeBenedetti and Parsons 1984).

Biodiversity and Community-Structure Responses to Fire and Fire Regimes

In Ellis Meadow, the total number of vascular plant species increased from the first year after fire to the second year after fire and remained essentially constant through the fourth year after fire. Although the number of species remained relatively constant, the composition fluctuated considerably from year to year. Species that appeared in the first year through the fourth year following the fire constituted the most important plant group for every postfire year, in terms of their contribution both to the number of species present and to total cover. No species was found only during the first year following fire (DeBenedetti and Parsons 1984).

Landscape-Pattern Responses to Fire Regimes

In Ellis Meadow, the burn was more intense in flat places dominated by wideleaf sedge. In these areas, subsurface fires did not carry well through the above-ground fuels and smoldered around them. In hummocky microtopography, patterns of burns varied: in places where burns were of low intensity, fires were largely confined to the trough between hummocks, and green vegetation was rarely totally consumed, leaving the landscape patchy with surviving vegetation on top of hummocks, whereas in places where fire intensity was high, the surfaces of both troughs and hummocks were continuously burned, leaving the landscape relatively homogeneous.

The same pattern occurred for lodgepole seedlings: those growing on flat topography were mostly killed by the fires, whereas those growing on top of the hummocks, where the fire was usually not hot enough to burn through the bases of the trees, were rarely killed (DeBenedetti and Parsons 1979).

Biogeochemical and Soil Changes after Fire

In Ellis Meadow, where the topography was flat, the fires smoldered for some time, consumed the organic layer, and left an ash layer 10 to 30.5 cm (4 to 12 in) deep. Where the fire was most intense, the surface of the meadow was lowered between 10 and 25 cm (4 and 10 in) relative to adjacent vegetation. Soil puddling occurred in most of these areas, and initial stages of channelization in the form of shallow rills were sometimes present, probably a result of a heavy rain that occurred one year after fire. DeBenedetti and Parsons hypothesized that this would alter overall meadow drainage patterns and change the distribution of major plant communities, but subsequent observation has not supported this hypothesis (DeBenedetti and Parsons 1984).

East-Slope Vegetation Types

The literature on east-slope Sierran ecosystems is sparse. There are several vegetation types on the east slope, including piñon-juniper, sage-bitterbrush, and east-side lodgepole (which is slightly different from the west-side lodgepoles discussed earlier), and east-side pine and mixed conifers. But of these various types, only the east-side pine forests have received enough study to be comprehensively reviewed here. Also, it is important to keep in mind that these following statements have relatively little scientific support and are mostly from the workshop supporting the California spotted owl environmental impact statement (Andrews 1994), which consists of ad hoc summarization by scientists and managers from the area.

East-Side Pine Forests

Primary Vegetation and Its Historical Relationship with Fire

Generally, the east-side pine forest is roughly defined by the region dominated by various pine species east of the Sierra Nevada crest (McDonald 1983). Because the east-side pines have not been widely studied, they are poorly described both geographically and ecologically. In the northeastern California region, they fall in the elevation range of 1,220–1,980 m (4,000–6,500 ft) (McDonald 1983). Ponderosa and Jeffrey pines (*Pinus ponderosa, P. jeffreyi*) are the dominant species, and white fir (*Abies concolor*), incense cedar (*Libocedrus decurrens*), and, on poorer sites, juniper (*Juniper* spp.) are commonly found associates (Andrews 1994). Fire is an important factor in maintaining this vegetation type (Sweeney 1968), but because the rate of the biotic processes is slow, the system is less resilient and requires a longer recovery period after fire than the west-side vegetation types (Andrews 1994).

Succession after Fire

The plant succession sequence following severe fires in this type generally proceeds from herbs, to shrubs, to pine, to fir (Sweeney 1968). For the Inyo area, the postfire sequence proceeds from herbs to a shrub and pine mix (Millar personal communications). Succession of western juniper is usually a function of disturbance: after severe fire, western juniper is usually reduced greatly in abundance, sometimes almost to elimination, and perennial grasses increase in both abundance and productivity (Andrews 1994).

Biodiversity and Community-Structure Responses to Fire and Fire Regimes

Concerning biodiversity, east-side pine forests have low species diversity when compared to their west-side counterparts (Andrews 1994). Understory grasses and herbaceous vegetation were thought to be generally abundant historically, because of the frequent fire and open canopy conditions in the past. Although many junipers are believed to germinate better after fire (Millar personal communications), western junipers are thought to be susceptible to fire and would decrease with fire (Andrews 1994). A major regeneration pulse of white fir (*A. concolor*), shrubs, western juniper (*Juniperus occidentalis*), whitebark pine (*P. albicaulis*), ponderosa pine (*P. ponderosa*), and Jeffrey pine (*P. jeffreyi*) was observed for the early 1900s (before 1930), but the driving forces behind these changes are largely unknown. Suggested candidates include the cessation of sheep grazing, logging of large trees, fire suppression, and a concurrent wet climatic period.

Fire exclusion and selective logging have been suggested to have caused the observed shift toward shade-tolerant conifers, especially white fir and, in some places, incense cedar. Andrews (1994) has suggested that high-severity fires may benefit species adapted to such fires (e.g., deer brush [*Ceanothus* spp.] and manzanita [*Arctostaphylos* spp.]), introduced herbaceous species (e.g., cheat grass [*Bromus secalinus*]), and persistent herbaceous species (e.g., mule ear [*Wyethia glabra*]).

Historically, tree canopies in this area were characteristically open and exhibited a high degree of horizontal diversity but relatively low vertical diversity. There was a diverse mosaic of seral stages and slow-growing, long-lived tree species. In the recent past, vertical diversity in this area has increased, and horizontal diversity has decreased. It has also been observed that the small patches of older, large trees have been lost from the system extensively (Andrews 1994).

Fuel-Structure Changes in Response to Different Fire Regimes

Because of the dry climate, the rates of both fuel accumulation and decay are slower on the east slope than on the west. Before European settlement, fuel structure was thought to consist mostly of low levels of small, woody fuels, litter, and duff. Coarse, woody debris was thought to have been patchy. Snags were thought to have stood longer than they do now, but for exactly how long is unknown (Andrews 1994).

Fire suppression is thought to have greatly changed the fuel complex to more small, surface fuels, more vertical fuel distribution favoring crown fires, and greater fuel loading overall. Together, these changes increase the probability of large, high-severity fires (Andrews 1994).

Landscape-Pattern Responses to Fire Regimes

Fires in this region were primarily of low severity, with patches of high severity corresponding mostly to areas with heavy fuel accumulations or dense patches of small trees. More mesic sites burned less often, but fires were somewhat more severe when they did occur. Such fire patterns resulted in a mosaic of diverse, small, even (or similar) aged (or sized) patches, which exhibited little vertical diversity. Such "fine-grained" forest mosaic was occasionally fragmented into a more "coarse-grained" mosaic by a number of landscape elements common to the east-side pine type, including sagebrush flats, low sites, rock outcrops and scarps, meadows, springs, cold air pockets, brush fields, lava flows, and occasional large, high-severity burns. Large, intense fires create large patches that remain for a long time in early- and mid-seral stages (Andrews 1994).

Biogeochemical and Soil Changes after Fire

East-side pine forests are characterized by low levels of stocking, productivity, and growth rates, and nutrient cycling and decomposition are slow. Fire seems to help increase the rate of nutrient cycling. There has been a significant loss of soil productivity, and Andrews (1994) has suggested that the large, high-severity fires and earlier impacts from logging and harsh mechanical site preparation are the major causes.

CONCLUSION

Because of variations in fire tolerance, the time needed to reach maturity, and reproductive strategies, different species respond differently to the various fire intervals and fire intensities. Some species (e.g., obligate seeding chaparral shrubs and ponderosa pine) respond well to frequent fires, but others (e.g., sprouting chaparral shrubs, incense cedar, and firs) are favored by slightly longer fire-free periods. The successional process depends on the seed source at the time of burn and the consequent competition among the seedlings.

Different plant species also produce different amounts and types of fuel accumulation and, therefore, have the potential of supporting different fire regimes. Some (e.g., chaparral and mixed conifer types) accumulate fuels more rapidly than others (e.g., montane fir forests) and thus are capable of supporting fires at shorter intervals. Some (e.g., ponderosa pine) create loosely packed fuel that favors fast-spreading surface fires; others (e.g., firs and incense cedar) create ladder fuels that facilitate crown fires.

Depending on the relative temporal frequency and spatial scale of fire, other local disturbances, and the biotic processes on the landscape, landscape structure also responds differently to the various fire regimes. Landscapes characterized by stand-replacing fires (e.g., chaparral) are composed of coarser grains with patch sizes reflecting the typical range of fire sizes in the system. Landscapes characterized by surface fires (e.g., mixed conifer forests) have higher levels of variation in their patch sizes and stand structures. Patches created by stand-replacing fires often define the boundary of the next burn; patches shaped by surface fires often create local heterogeneity in burning intensities during the next burn. Because both horizontal and vertical fuel continuities increase as the fire-free period lengthens, longer fire-free periods result in larger burns. In ecosystems characteristically visited by surface burns, longer fire-free periods also increase the probability of stand-replacement fires. Therefore, a longer fire-free period often results in an increase in patch size, and the landscape becomes coarser grained compared to the landscapes that experience more frequent burns.

The complex geography of the region also has a significant effect on fire spread. The rocky exposures near the mountain ridges serve as natural firebreaks, limiting the spread of fire in and across that region, whereas the continuous sheet of fuel bed from the foothills to the mixed conifer zones often results in a larger area burned, especially when the fuel within each type is continuous.

Today, many of the issues in the Sierra Nevada surround debates in land use and forest management: Active or passive management? To burn or not to burn? If to burn, how often, and how intense? If not to burn, what to do instead?

From the ecology of the Sierra Nevada, we have learned that change is constant and that ecosystems have a wide range of variation in their responses to fire. We should not study past patterns of change in the hopes of recreating them. Rather, we should seek to understand how change will determine the patterns of the future. We have to decide what it is we want and work with the natural processes to achieve those goals. If we seek biodiversity, a diversity of fire regimes may be critical. If we seek to control fires in vegetation close to residential areas, lower fuel loads and lower fuel continuity may help us gain that control. But if it is stability in terms of persistence of some "steady state" that we are after, we are bound to fail; change is the very heart of nature.

ACKNOWLEDGMENTS

Thanks are due to N. Christensen, G. Greenwood, C. Millar, C. Skinner, D. Urban, and two anonymous reviewers for valuable comments on earlier versions of the manuscript.

REFERENCES

Abrams, M. D., and M. L. Scott. 1989. Disturbance-mediated accelerated succession in two Michigan forest types. *Forest Science* 35 (1): 42–49.

Agee, J. K. 1977. Fire and fuel dynamics of Sierra Nevada conifers. *Forest Ecology and Management* 1:255–65.

———. 1989. Wildfire in the Pacific west: A brief history and implications for the future. In *Proceedings, symposium on fire and watershed management*, 11–16. General Technical Report PSW-109. Berkeley, CA: U.S. Forest Service .

———. 1993. *Fire ecology of Pacific Northwest forests*. Washington, DC: Island Press.

Agee, J. K., and H. H. Biswell. 1969. Seedling survival in a giant sequoia forest. *California Agriculture* 23 (4): 18–19.

Agee, J. K., and M. H. Huff. 1987. Fuel succession in a western hemlock/Douglas-fir forest. *Canadian Journal of Forest Research* 17:697–704.

Agee, J. K., and L. Smith. 1984. Subalpine tree establishment after fire in the Olympic Mountains, Washington. *Ecology* 65 (3): 810–19.

Agee, J. K., R. H. Wakimoto, and H. H. Biswell. 1978. Fire and fuel dynamics of Sierra Nevada conifers. *Forest Ecology and Management* 1:255–65.

Anderson, M. K. 1993. The mountains smell like fire. *Fremontia* 21 (4): 15–20.

Andrews, R. S., ed. 1994. *Ecological support team workshop proceedings for the California Spotted Owl Environmental Impact Statement*, August 1993. San Francisco: U.S. Forest Service, Pacific Southwest Region.

Baisan, C. H., and T. W. Swetnam. 1990. Fire history of a desert mountain range: Rincon Mountain Wilderness, USA. *Canadian Journal of Forest Research* 20:1559–69.

Baker, G. A. 1989. Effect of scale and spatial heterogeneity on fire-interval distributions. *Canadian Journal of Forest Research* 19: 700–706.

———. 1993. Spatially heterogeneous multi–scale response of landscapes to fire suppression. *Oikos* 66:66–71.

Baker, G. A., P. W. Rundel, and D. J. Parsons. 1982. Post-fire recovery of chamise chaparral in Sequoia National Park, California. In *Proceedings of the symposium on dynamics and management of Mediterranean-type ecosystems*, 584. General Technical Report PSW-58. Berkeley, CA: U.S. Forest Service, Pacific Southwest Forest and Range Experiment Station.

Barbour, M. G., R. A. Minnich, and R. F. Fernau. 1994. Sixty years of change in Californian mixed conifer forests: Reconstruction of the pre-fire-suppression landscape. Annual meeting of the Botanical Society of America, Knoxville, TN, 7–11 August. *American Journal of Botany* 81 (6 Suppl.): 67–68.

Behan, M. J. 1970. The cycle of minerals in forest ecosystems. In *Role of fire in the Intermountain West: Proceedings of a symposium*, 27–29, October, 11–29. Missoula, MT: Intermountain Fire Research Council.

Bendell, J. F. 1974. Effects of fire on birds and mammals. In *Fire and ecosystems*, edited by T. T. Kozlowski and C. E. Ahlgren, 73–138. New York: Academic Press.

Biswell, H. H. 1974. Effects of fire on chaparral. In *Fire and ecosystems*, edited by T. T. Kozlowski and C. E. Ahlgren, 321–64. New York: Academic Press.

Blackburn, T. C., and K. Anderson, eds. 1993. *Before the wilderness: Environmental management by Native Californians*. San Francisco: Ballena Press.

Boche, K. E. 1974. Factors affecting meadow–forest borders in Yosemite National Park, California. Master's thesis, University of California, Los Angeles.

Bock, C. E., and J. H. Bock. 1983. Responses of birds and deer mice to prescribed burning in ponderosa pine. *Journal of Wildland Management* 47 (3): 836–40.

Bock, C. E., and J. F. Lynch. 1970. Breeding bird populations of burned and unburned conifer forest in the Sierra Nevada. *The Condor* 72:182–89.

Bock, J. H., and C. E. Bock. 1969. Natural reforestation in the northern Sierra Nevada-Ridge Burn. *Tall Timber Ecological Conference* 9:119–26.

Boerner, R. E. J. 1982. Fire and nutrient cycling in temperate ecosystems. *BioScience* 32 (3): 187–92.

Bonnicksen, T. M., and E. C. Stone. 1982. Reconstruction of a presettlement giant sequoia–mixed conifer forest community using the aggregation approach. *Ecology* 63 (4): 1134–48.

Bormann, F. H., and G. E. Likens. 1979. *Pattern and process in a forested ecosystem*. New York: Springer-Verlag.

Botti, S. 1979. Natural, conditional, and prescribed fire management plan. Yosemite National Park, CA: National Park Service.

Bradley, H. C. 1911. The passing of our mountain meadows. *Sierra Club Bulletin* 8:39–42.

Brokaw, N. V. L. 1985. Treefalls, regrowth, and community structure in tropical forests. In *The ecology of natural disturbance and patch dynamics*, edited by S. T. A. Pickett and P. S. White, 53–69. Orlando, FL: Academic Press.

Brown, J. K. 1975. Fire cycles and community dynamics in lodgepole pine forests. In *Management of lodgepole pine ecosystems: Proceedings of the symposium*, edited by D. M. Baumgartner, 429–56. Pullman: Washington State University, Cooperative Extension Service.

———. 1985. The "unnatural fuel buildup" issue. In *Proceedings, symposium and workshop on wilderness fire*, 15-18 November, 1983, Missoula, MT, technical coordination by J. E. Lotan, B. M. Kilgore, W. C. Fischer, and R. W. Mutch, 127–28. General Technical Report INT-182. Ogden, UT: U.S. Forest Service, Intermountain Forest and Range Experiment Station.

Brown, P. M., and T. W. Swetnam. 1994. A crossdated fire history from coast redwood near Redwood National Park, California. *Canadian Journal of Forest Research* 24:21–31.

Brubaker, L. B. 1986. Responses of tree populations to climatic change. *Vegetatio* 67:119–30.

Burns, R. M., ed. 1983. *Silvicultural systems for the major forest types of the United States*. Agricultural Handbook 445. Washington, DC: U.S. Forest Service.

Campbell, R. S. 1954. Fire in relation to forest grazing. *Unasylva* 8:154–58.

Canham, C. D., and P. L. Marks. 1985. The response of woody plants to disturbance: Patterns of establishment and growth. In *The ecology of natural disturbance and patch dynamics*, edited by S. T. A. Pickett and P. S. White. Orlando, FL: Academic Press.

Chandler, C., P. Cheney, P. Thomas, L. Trabaud, and D. Williams. 1983. *Fire in forestry*. Vol. I of *Forest fire behavior and effects*. New York: Wiley.

Chapin, F. S., III, and K. Van Cleve. 1981. Plant nutrient absorption and retention under differing fire regimes. In *Proceedings of conference, fire regimes and ecosystem properties*, technical coordination by H. A. Mooney, T. M. Bonnicksen, N. L. Christensen, J. E. Lotan, and W. A. Reiners, 301-21. General Technical Report WO-26. Washington, DC: U.S. Forest Service.

Christensen, N. L. 1985. Shrubland fire regimes and their evolutionary consequences. In *The ecology of natural disturbance and patch dynamics*, edited by S. T. A. Pickett and P. S. White, 85-100. Orlando, FL: Academic Press.

———. 1991. Variable fire regimes on complex landscapes: Ecological consequences, policy implications, and management strategies. In *Fire and the environment: Ecological and cultural perspectives*, technical coordination by S. C. Nodvin and T. A. Waldrop, ix–xiii. General Technical Report SE-69. Asheville, NC: U.S. Forest Service, Southeastern Research Station.

———. 1993. Fire regimes and ecosystem dynamics. In *Fire in the environment: The ecological, atmospherical, and climatic importance of vegetation fires*, edited by P. J. Crutzen and J. G. Goldammer, 233–44. Chichester, UK: John Wiley and Sons.

———. 1995. *Fire ecology*. Vol. 2 of *Encyclopedia of environmental biology*. Orlando, FL: Academic Press.

Christensen, N. L., J. K. Agee, P. F. Brussard, J. Hughes, D. H. Knight, G. W. Minshall, J. M. Peek, S. J. Pyne, F. J. Swanson, J. W. Thomas, S. Wells, S. E. Williams, and H. A. Wright. 1989. Interpreting the Yellowstone fires of 1988: Ecosystem responses and management implications. *BioScience* 39 (10): 678–85.

Christensen, N. L., R. B. Burchell, A. Liggett, and E. L. Simms. 1981. The structure and development of pocosin vegetation. In *Pocosin wetlands*, edited by C. J. Richardson, 43–61. Stroudsburg, PA: Dowden, Hutchinson and Ross.

Christensen, N. L., and C. H. Muller. 1975. Effects of fire on factors controlling plant growth in Adenostoma chaparral. *Ecological Monograph* 45:29–55.

Christensen, N. L., and R. K. Peet. 1984. Convergence during secondary forest succession. *Journal of Ecology* 72:25–36.

Clark, J. S. 1990. Landscape interactions among nitrogen mineralization, species composition, and long-term fire frequency. *Biogeochemistry* 11:1–22.

———. 1991. Disturbance and population structure on the shifting mosaic landscape. *Ecology* 72:1119–37.

Conard, S. G., and S. R. Radosevich. 1982. Post-fire succession in white fir *(Abies concolor)* vegetation of the northern Sierra Nevada. *Madroño* 29 (1): 42–56.

Cooper, C. F. 1961. The ecology of fire. *Scientific American* 204:150–60.

Countryman, C. M. 1955. Old-growth conversion also converts fireclimate. In *Proceedings of Society of American Foresters meeting*, 16–21 October, Portland, OR, 158–60. Washington, DC: Society of American Foresters.

Davis, F. W., D. E. Hickson, and D. C. Odion. 1988. Composition of maritime chaparral related to fire history and soil, Burton Mesa, Santa Barbara County, California. *Madroño* 35 (3): 169–95.

Davis, M. B. 1981. Quaternary history and the stability of forest communities. In *Forest succession: Concepts and application*, edited by D. C. West, 132–53. New York: Springer-Verlag.

———. 1984. Climatic instability, time lags, and community disequilibrium. In *Community ecology,* edited by J. Diamond and T. J. Case, 269–84. New York: Harper and Row.

DeBano, L. F. 1991. The effect of fire on soil properties. In *Proceedings of the symposium on management and productivity of western-montane forest soils,* Boise, ID, 10–12 April, compiled by A. E. Harvey, and L. F. Neuenschwarder, 151–56. Ogden, UT: USFS Intermountain Forest and Range Experiment Station.

DeBano, L. F., and C. E. Conrad. 1978. The effect of fire on nutrients in a chaparral ecosystem. *Ecology* 59 (3): 489–97.

DeBenedetti, S. H., and D. J. Parsons. 1979. Natural fire in subalpine meadows: A case description from the Sierra Nevada. *Journal of Forestry* 77:477–79.

———. 1984. Post-fire succession in a Sierran subalpine meadow. *The American Midland Naturalist* 111 (1): 118–25.

Delcourt, H. R., P. A. Delcourt, and T. Webb III. 1982. Dynamic plant ecology: The spectrum of vegetational change in space and time. *Quaternary Science Review* 1:153–75.

Delwiche, C. C., P. J. Zinke, and C. M. Johnson. 1965. Nitrogen fixation by ceanothus. *Plant Physiology* 40 (6): 1045–47.

Dunn, P. H., S. C. Barro, and M. Poth. 1985. Soil moisture affects survival of microorganisms in heated chaparral soil. *Soil Biology and Biochemistry* 17:143–48.

Eyre, F. H., ed. 1980. *Forest cover types of the United States and Canada.* Washington, DC: Society of American Foresters.

Flieger, B. W. 1970. Forest fires and insects: The relation of fire to insect outbreaks. *Tall Timbers Fire Ecology Conference Proceedings.* 10:107–16.

Fowells, H. A. 1965. *Silvics of North American trees.* Agricultural Handbook 271. Washington, DC: U.S. Department of Agriculture.

Franklin, J. F., and C. T. Dyrness. 1973. *Natural vegetation of Oregon and Washington.* General Technical Report PNW-8. Portland, OR: U.S. Forest Service, Pacific Northwest Research Station.

Geiszler, D. R., R. I. Gara, C. H. Driver, V. F. Gallucci, and R. E. Martin. 1980. Fire, fungi, and beetle influences on a lodgepole pine ecosystem of south-central Oregon. *Oecologia* 46:239-43.

Griffin, J. R. 1977. Oak woodland. In *Terrestrial vegetation of California,* edited by M. G. Barbour and J. Major, 383–415. New York: John Wiley and Sons.

Habeck, J. R. 1985. Impact of fire suppression on forest succession and fuel accumulations in long-fire-interval wilderness habitat types. In *Proceedings, symposium and workshop on wilderness fire,* 15–18 November 1983, Missoula, MT, technical coordination by J. E. Lotan, B. M. Kilgore, W. C. Fischer, and R. W. Mutch, 110–18. General Technical Report INT-182. Ogden, UT: U.S. Forest Service, Intermountain Forest and Range Experiment Station.

Hanes, T. L. 1971. Succession after fire in the chaparral of southern California. *Ecological Monographs* 41 (1): 27–52.

———. 1977. California chaparral. In *Terrestrial vegetation of California,* edited by M. G. Barbour and J. Major, 417–69. New York: John Wiley and Sons.

Hare, R. C. 1961. *Heat effects on living plants.* Occasional Paper 183. New Orleans, LA: U.S. Forest Service, Southern Forest Experiment Station.

Hartesveldt, R. J. 1964. Fire ecology of the giant sequoias: Controlled fires may be one solution to survival of the species. *Natural History Magazine* 73 (10): 12–19.

Hedrick, D. W. 1954. Studies on the succession and manipulation of chamise brushlands in California. Ph.D. dissertation, Texas A & M College.

Heinselman, M. L. 1978. Fire in wilderness ecosystems. In *Wilderness management,* edited by J. C. Hendee, G. H. Stankey, and R. C. Lucas, 249–78. U.S. Forest Service Miscellaneous Publication 1365. Washington, DC: U.S. Forest Service.

———. 1981. Fire intensity and frequency as factors in the distribution and structure of northern ecosystems. In *Proceedings of conference, fire regimes and ecosystem properties,* technical coordination by H. A. Mooney, T. M. Bonnickson, N. L. Christensen, J. E. Lotan, and W. A. Reiners, 7–57. General Technical Report WO-26. Washington, DC: U.S. Forest Service.

Helms, J.A. 1987. Invasion of *Pinus contorta* var. *murrayana* (Pineaceae) into mountain meadows at Yosemite National Park, California. *Madroño* 34:77–90.

Helms, J. A., and R. B. Standiford. 1985. Predicting release of advance reproduction of mixed conifer forest species in California following overstory removal. *Forest Science* 31:3–15.

Hemstrom, M. A., and J. F. Franklin. 1982. Fire and other disturbances of the forests in Mount Rainier National Park. *Quaternary Research* 18:32–51.

Howard, W. E., R. L. Fenner, and H. E. Childs Jr. 1959. Wildlife survival on brush burns. *Journal of Range Management* 12:230–34.

Husari, S. J., 1980. Fire ecology of the vegetative habitat types in the Lassen Fire Management Planning Area (Caribou Wilderness and Lassen Volcanic National Park). In *Fire management plan: Lassen Fire Management Planning Area park, Caribou unit.* Lassen Volcanic National Park and Lassen National Forest.

Husari, S. J., and K. S. Hawk. 1994. The role of past and present disturbance in California ecosystems. Draft Region 5 Ecosystem Management Guidebook, Vol. 2, Appendices I–C. San Francisco: U.S. Forest Service, Pacific Southwest Region.

Keeley, J. E. 1991a. Fire management for maximum biodiversity of California chaparral. In *Fire and the environment: Ecological and cultural perspectives: Proceedings of an international symposium,* 20–24 March 1990, Knoxville, TN, edited by S. C. Nodvin and T. A. Waldrop, 11–14. General Technical Report SE-69. Ashville, NC: U.S. Forest Service, Southeastern Forest Experiment Station.

———. 1991b. Seed germination and life history syndromes in the California chaparral. *Botanical Review* 57 (2): 81–116.

———. 1992a. Demographic structure of California chaparral in the long-term absence of fire. *Journal of Vegetation Science* 3:79–90.

———. 1992b. Recruitment of seedlings and vegetative sprouts in unburned chaparral. *Ecology* 73 (4): 1194–1208.

———. 1995. Future of California floristics and systematics: Wildfire threats to the California flora. *Madroño* 42 (2): 175–79.

Keeley J. E., and P. H. Zedler. 1978. Reproduction of chaparral shrubs after fire: A comparison of sprouting and seeding strategies. *American Midland Naturalist* 99 (1): 142–61.

Kilgore, B. M. 1971a. Response of breeding bird populations to habitat changes in a giant sequoia forest. *American Midland Naturalist* 85 (1): 135–52.

———. 1971b. The role of fire in managing red fir forests. In *Transactions of the 36th North American Wildlife and Natural Resource Conference,* 7–10 March, 405–16. Washington, DC: Wildlife Management Institute.

———. 1973. The ecological role of fire in Sierran conifer forests: Its application to national park management. *Quaternary Research* 3:496–513.

———. 1985. What is "natural" in wilderness fire management? In *Proceedings, symposium and workshop on wilderness fire,* November 15–18, 1983, Missoula, MT, technical coodination by J. E. Lotan,

B. M. Kilgore, W. C. Fischer, and R. W. Mutch, 57–67. General Technical Report INT-182. Ogden, UT: U.S. Forest Service, Intermountain Forest and Range Experiment Station.

———. 1987. The role of fire in wilderness: A state-of-knowledge review. In *Proceedings, national wilderness research conference: Issues, state-of-knowledge, future directions.* General Technical Report INT-220. Ogden, UT: U.S. Forest Service, Intermountain Research Station.

Kilgore, B. M., and H. H. Biswell. 1971. Seedling germination following fire in a giant sequoia forest. *California Agriculture* 25 (2): 8–10.

Kilgore, B. M., and G. S. Briggs. 1972. Restoring fire to high elevation forests in California. *Journal of Forestry* 70:266–71.

Kilgore, B. M., and R. W. Sando. 1975. Crown-fire potential in a sequoia forest after prescribed burning. *Forest Science* 21:83–87.

Kilgore, B. M., and D. Taylor. 1979. Fire history of a sequoia–mixed conifer forest. *Ecology* 60:129–42.

Klemmedson, J. O., A. M. Schultz, H. Jenny, and H. H. Biswell. 1963. Effect of prescribed burning of forest litter on total soil nitrogen. *Soil Science Society of America Proceedings* 26 (2): 200–202.

Knight, H. 1966. Loss of nitrogen from the forest floor by burning. *Forestry Chronicle* 42 (2): 149–52.

Kourtz, P. H., and W. G. O'Regan. 1971. A model for a small forest fire . . . to simulate burned and burning areas for use in a detection model. *Forest Science* 17 (2): 163–69.

Laacke, R. J., and J. N. Fiske. 1983. Red fir and white fir. In *Silvicultural systems for the major forest types of the United States,* edited by R. M. Burns, 41–43. U.S. Forest Service Agricultural Handbook 445. Washington, DC: U.S. Forest Service.

Lawrence, G. E. 1966. Ecology of vertebrate animals in relation to chaparral fire in the Sierra Nevada foothills. *Ecology* 47:278–91.

Lawrence, G. E., and H. H. Biswell. 1972. Effects of forest manipulation on deer habitat in giant sequoia. *Journal of Wildlife Management* 36 (2): 595–605.

Ledig, F. T. 1992. Human impacts on genetic diversity in forest ecosystems. *Oikos* 63:87–108.

Leiberg, J. G. 1902. *Forest conditions in the northern Sierra Nevada, California.* U.S. Geological Survey, Professional Paper 8, Series H, Forestry 5. Washington, DC: Government Printing Office.

Leonard, R., C. M. Johnson, P. Zinke, and A. Schultz. 1968. Ecological study of meadows in lower Rock Creek, Sequoia National Park. Progress report on file at Sequoia National Parks, Three Rivers, CA.

Leopold, A. S. 1932. *Game management.* New York: Scribner's.

Levin, S. A. 1978. Pattern formation in ecological communities. In *Spatial pattern in plankton communities,* edited by J. S. Steele, 433–65. New York: Plenum Press.

Lillywhite, H. B. 1977. Animal responses to fire and fuel management in chaparral. In *Proceedings of the symposium on the environmental consequences of fire and fuel management in Mediterranean ecosystems,* 368–730. General Technical Report WO-3. Washington, DC: U.S. Forest Service.

Lindenmuth, A. W., Jr. 1960. *Effects of intentional burning on fuels and timber stands of ponderosa pine in Arizona.* Paper 54. Fort Collins, CO: U.S. Forest Service, Rocky Mountain Forest and Range Experiment Station.

Loucks, O. L. 1970. Evolution of diversity, efficiency, and community stability. *American Zoologist* 10:17–25.

Lyon, L. J., H. S. Crawford, E. Czuhai, R. L. Fredriksen, R. F. Harlow, L. J. Metz, and H. A. Pearson, eds. 1978. *Effects of fire on fauna: A state-of-knowledge review.* National fire effects workshop, Denver, CO, 10–14 April. General Technical Report WO-6. Washington, DC: U.S. Forest Service.

Magee, T. 1885. *Ascent of Mt. Whitney.* In Tulare County Scrapbook. California State Library, Sacramento.

Margalef, R. 1968. *Perspectives in ecological theory.* Chicago: University of Chicago Press.

Marion, G. M., and C. H. Black. 1988. Potentially available nitrogen and phosphorus along a chaparral fire cycle chronosequence. *Soil Science Society of America Journal* 52 (4): 1155–62.

Martin, R. E., and D. B. Sapsis. 1992. Fires as agents of biodiversity: Pyrodiversity promotes biodiversity. In *Proceedings of the symposium on biodiversity of northwestern California,* 28–30 October 1991, Santa Rosa, CA, edited by H. M. Kerner, 150–57. Report 29. Berkeley, CA: Wildland Resources Center, Division of Agriculture and Natural Resources, University of California.

McBride, J. R., and N. G. Sugihara. 1990. Forest succession of the upper mixed conifer zone. Department of Forestry and Resource and Management, University of California, Berkeley.

McClaran, M. P., and J. W. Bartolome. 1989. Fire-related recruitment in stagnant *Quercus Douglasii* populations. *Canadian Journal of Forest Research* 19 (5): 580–85.

McDonald, P. M. 1983. Climate, history, and vegetation of the eastside pine type in California. In *Management of the eastside pine type in northeastern California: Proceedings of a symposium,* edited by T. F. Robson and R. B. Standiford, 1–16. Arcata, CA: Northern California Society of American Foresters.

———. 1990a. Quercus douglasii Hook. & Arn. In *Hardwoods,* Vol. 2 of *Silvics of North America,* technical coordination by R. M. Burns and B. H. Honkala, 631–39. U.S. Forest Service Agriculture Handbook 654. Washington, DC: U.S. Forest Service.

———. 1990b. Quercus kelloggii Newb. In *Hardwoods,* Vol. 2 of *Silvics of North America,* technical coordination by R. M. Burns and B. H. Honkala, 661–71. U.S. Forest Service Agriculture Handbook 654. Washington, DC: U.S. Forest Service.

Minnich, R. A. 1983. Fire mosaics in Southern California and northern Baja California. *Science* 219:1287–94.

———. 1987. Fire behavior in southern California chaparral before fire control: The Mount Wilson burns at the turn of the century. *Annals of the Association of American Geographers* 77 (4): 599–618.

———. 1989. Chaparral fire history in San Diego county and adjacent northern Baja California: An evaluation of natural fire regimes and the effects of suppression management. In *The California chaparral: Paradigms reexamined,* edited by S. C. Keeley, 37–47. Natural History Museum of Los Angeles County, Science Series 34.

Minnich, R. A., M. G. Barbour, J. H. Burk, and J. Sosa-Ramirez. 1995. California conifer forests under unmanaged fire regimes in the Sierra San Pedro Martir, Baja California, Mexico.

Mooney, H. A., and E. L. Dunn. 1970. Convergent evolution of Mediterranean climate evergreen sclerophyll shrubs. *Evolution* 24:292–303.

Mooney, H. A., and D. J. Parsons. 1973. Structure and function of the California chaparral: An example from San Dimas. In *Mediterranean type ecosystems,* edited by F. Di Castri and H. A. Mooney, 83–112. New York: Springer-Verlag.

Muir, J. 1894. *The mountains of California.* Boston: Houghton.

Munz, P. A., and D. D. Keck. 1973. *A California flora and supplement.* Berkeley and Los Angeles: University of California Press.

Mutch, R. W. 1970. Wildland fires and ecosystems—a hypothesis. *Ecology* 51 (6): 1046–51.

Nichols, R., and J. Menke. 1984. Effects of chaparral shrubland fire on terrestrial wildlife. In *Shrublands in California: Literature review and research needed for management*, edited by J. J. DeVries, 74–96. Contribution 191. Davis: University of California, Davis, Water Resources Center.

Oosting, H. J., and W. D. Billings. 1943. The red fir forest of the Sierra Nevada: *Abietum magnificae. Ecological Monographs* 13:261–74.

Parker, A. J. 1984. A comparison of structural properties and compositional trends in conifer forests of Yosemite and Glacier National Parks, USA. *Northwest Science* 58 (2): 131–41.

———. 1986. Persistence of lodgepole pine *(Pinus contorta* ssp. *murrayana)* forests in the central Sierra Nevada, California, USA. *Ecology* 67 (6): 1560–67.

Parker, V. T., and V. R. Kelly. 1989. Seed banks in California chaparral and other Mediterranean climate shrublands. In *Ecology of soil seed banks*, edited by M. A. Leck, V. T. Parker, and R. L. Simpson, 231–55. San Diego: Academic Press.

Parsons, D. J. 1976. The role of fire in natural communities: An example from the Southern Sierra Nevada, California. *Environmental Conservation* 3 (2): 91–99.

———. 1981. The historical role of fire in the foothill communities of Sequoia National Park. *Madroño* 28 (3): 111–20.

Parsons, D. J., and S. H. DeBenedetti. 1979. Impact of fire suppression on a mixed-conifer forest. *Forest Ecology and Management* 2:21–33.

Paysen, T. E., and J. D. Cohen. 1990. Chamise chaparral dead fuel fraction is not reliably predicted by age. *West Journal of Applied Forestry* 5 (4): 127–31.

Pickett, S. T. A., and P. S. White. 1985. *The ecology of natural disturbance and patch dynamics*. Orlando, FL: Academic Press.

Pitcher, D. C. 1987. Fire history and age structure in red fir forests of Sequoia National Park, California. *Canadian Journal of Forest Research* 17:582–87.

Plumb, T. R. 1980. Response of oaks to fire. In *Proceedings of the symposium on the ecology, management, and utilization of California oaks*, 26–28 June 1979, Claremont, CA, 202–15. General Technical Report PSW-44. Berkeley, CA: U.S. Forest Service, Pacific Southwest Forest and Range Experiment Station.

Plumb, T. R., and A. P. Gomez. 1983. *Five southern California oaks: Identification and postfire management*. U.S. Forest Service General Technical Report PSW-71. Berkeley, CA: U.S. Forest Service.

Quinn, R. D. 1979. Effects of fire on small mammals in the chaparral. *Cal-Neva Wildlife Transactions* 1979:125–33.

Radtke, K., A. Arnde, and R. H. Wakimoto. 1981. Fire history of the Santa Monica Mountain. In *Proceedings, symposium on dynamics and management of Mediterranean-type ecosystems*, 22–26 June, San Diego, CA, technical coordination by C. Conrad, C. Eugene, and W. C. Oechel, 438–43. General Technical Report PSW-58. U.S. Forest Service, Pacific Southwest Forest and Range Experiment Station.

Ratliff, R. D. 1982. *A meadow site classification for the Sierra Nevada, California*. General Technical Report PSW-60. Berkeley, CA: U.S. Forest Service, Pacific Southwest Forest and Range Experiment Station.

———. 1985. *Meadows of the Sierra Nevada of California: State of knowledge*. General Technical Report PSW-84. Berkeley, CA: U.S. Forest Service.

Rice, S. K. 1993. Vegetation establishment in post-fire *Adenostoma* chaparral in relation to fine-scale pattern in fire intensity and soil nutrients. *Journal of Vegetation Science* 4:115–24.

Richter, D. D., and C. W. Ralston. 1982. Prescribed fire: Effects on water quality and forest nutrient cycling. *Science* 215:661–63.

Riggan, P. J., S. Goode, P. M. Jacks, and R. N. Lockwood. 1988. Interaction of fire and community development in chaparral of Southern California (USA). *Ecological Monographs* 58 (3): 155–76.

Rogers, C., V. T. Parker, V. R. Kelly, and M. K. Wood. 1989. *Proceedings of the symposium of fire and watershed management*, 26–28 October 1988, Sacramento, CA, technical coordination by N. H. Berg. General Technical Report PSW-109:158. Berkeley, CA: U.S. Forest Service, Pacific Southwest Range and Experiment Station.

Romme, W. H. 1980. Fire frequency in subalpine forests of Yellowstone National Park. In *Proceedings of the fire history workshop, Tucson, Arizona*, 20–24 October, edited by M. A. Strokes and J. H. Dieterich, 27–30. General Technical Report RM-81. Fort Collins, CO: U.S. Forest Service.

Romme, W. H., and D. G. Despain. 1989. Historical perspective on the Yellowstone fires of 1988. *BioScience* 39 (10): 695–99.

Romme, W. H., and D. H. Knight. 1981. Fire frequency and subalpine forest succession along a topographic gradient in Wyoming. *Ecology* 62 (2): 319–26.

Rowe, J. S. 1970. Spruce and fire in northwest Canada and Alaska. *Tall Timbers Fire Ecology Conference Proceedings* 10:245–54.

Rundel, P. W. 1981. Ecological relationships of foothill woodland and chaparral communities in the southern Sierra Nevada, California. *Ecologia Mediterranea* T8:461–71.

Rundel, P. W., and D. J. Parsons. 1979. Structural changes in Chamise *(Adenostoma fasciculatum)* along a fire-induced age gradient. *Journal of Range Management* 32 (6): 462–66.

———. 1980. Nutrient changes in two chaparral shrubs along a fire-induced age gradient. *American Journal of Botany* 67 (1): 51–58.

Rundel, P. W., D. J. Parsons, and D. T. Gordon. 1977. Montane and subalpine vegetation of the Sierra Nevada and Cascade Ranges. In *Terrestrial vegetation of California*, edited by M. G. Barbour and J. Major, 559–99. New York: John Wiley and Sons.

Runkle, J. R. 1981. Gap regeneration in some old-growth forests of the eastern United States. *Ecology* 62:1041–51.

Sampson, A. W. 1944. Plant succession on burned chaparral lands in northern California. *U. of California Agriculture Experiment Station Bulletin 685*. Berkeley: University of California.

Sellers, J. A. 1970. Mixed conifer forest ecology: A quantitative study in Kings Canyon National Park, Fresno County, California. Master's thesis, Fresno State College.

Show, S. B., and E. I. Kotok. 1924. *The role of fire in the California pine forests*. U.S. Department of Agriculture Bulletin 1924. Washington, DC: Government Printing Office.

———. 1925. *Fire and the forest: California pine region*. Circular 358. Washington, DC: U.S. Department of Agriculture.

Skinner, C. N. 1995a. Change in spatial characteristics of forest openings in the Klamath Mountains of northwestern California. *Landscape Ecology* 10 (4): 219–28.

———. 1995b. Using prescribed fire to improve wildlife habitat near Shasta Lake. Unpublished manuscript. Redding, CA: Pacific Southwest Research Station.

Skinner, C. N., and C. Chang. 1996. Fire regimes, past and present. In *Sierra Nevada Ecosystem Project: Final report to Congress*, vol. II, chap. 38. Davis: University of California, Centers for Water and Wildland Resources.

Sprugel, D. G. 1991. Disturbance, equilibrium, and environmental variability: What is "natural" vegetation in a changing environment? *Biological Conservation* 58:1–18.

Starker, T. J. 1934. Fire resistance in the forest. *Journal of Forestry* 32 (4): 462–67.

Stephenson, N. L., D. J. Parsons, and T. W. Swetnam. 1991. Restoring natural fire to the sequoia–mixed conifer forest: Should intense fire play a role? *Proceedings, Tall Timbers fire ecology conference* 17:321–37.

Stine, S. 1996. Climate, 1650–1850. In *Sierra Nevada Ecosystem Project: Final report to Congress,* vol. II, chap. 2. Davis: University of California, Centers for Water and Wildland Resources.

St. John, T. V., and P. W. Rundel. 1976. The role of fire as a mineralizing agent in a Sierran coniferous forest. *Oecologia* 25:35–45.

Stohlgren, T. J., D. J. Parsons, and P. W. Rundel. 1984. Population structure of *Adenostoma fasciculatum* in mature stands of chamise chaparral in the southern Sierra Nevada, California (USA). *Oecologia* 64 (1): 87–91.

Stohlgren, T. J., and P. W. Rundel. 1986. A population model for a long-lived, resprouting chaparral shrub: *Adenostoma fasciculatum. Ecological Modelling* 34 (3–4): 245–58.

Sudworth, G. B. 1900. Stanislaus and Lake Tahoe Forest Reserves, California, and adjacent territories. In *Annual reports of the Department of Interior,* 505–61. 21st Annual Report of the U.S. Geological Survey, Part 5. Washington, DC: Government Printing Office.

Sweeney, J. R. 1968. Ecology of some "fire type" vegetation in northern California. *Proceedings, Tall Timbers fire ecology conference* 7:110–25.

———. 1969. The effects of wildfire on plant distribution in the Southwest. In *Proceedings of the symposium on fire ecology and the control and use of fire in wild land management, Journal of the Arizona Academy of Science:* 23–29.

Sweeney, J. R., and H. H. Biswell. 1961. Quantitative studies of the removal of litter and duff by fire under controlled conditions. *Ecology* 42:572–75.

Swetnam, T. W. 1993. Fire history and climate change in giant sequoia groves. *Science* 262:885–89.

Swetnam, T. W., and J. L. Bentancourt. 1990. Fire–southern oscillation relations in the southwestern United States. *Science* 249:1017–20.

Tappeiner, J. C., II. 1980. Sierra Nevada mixed conifer. In *Forest cover types of the United States and Canada,* edited by F. H. Eyre, 118–19. Washington, DC: Society of American Foresters.

Taylor, A. H. 1990a. Habitat segregation and regeneration patterns of red fir and mountain hemlock in ecotonal forests, Lassen Volcanic National Park, California. *Physical Geography* 11 (1): 36–48.

———. 1990b. Tree invasion in meadows of Lassen Volcanic National Park, California. *Professional Geographer* 42 (4): 457–70.

———. 1991. The structure and dynamics of *Abies magnifica* forests in the southern Cascade Range, USA. *Journal of Vegetation Science* 2 (2): 189–200.

———. 1993. Fire history and structure of red fir *(Abies magnifica)* forests, Swain Mountain Experimental Forests, Cascade Range, northeastern California. *Canadian Journal of Forest Research* 23 (8): 1672–78.

Tratz, W. M., and R. J. Vogl. 1977. Post-fire vegetation recovery and herbivore utilization of a chaparral-desert ecotone. In *Proceedings of the symposium on the environmental consequences of fire and fuel management in Mediterranean ecosystems,* 426–30. General Technical Report WO-3. Washington, DC: U.S. Forest Service.

Turner, M. G., W. W. Hargrove, R. H. Gardner, and W. H. Romme. 1994. Effects of fire on landscape heterogeneity in Yellowstone National Park, Wyoming. *Journal of Vegetation Science* 5:731–42.

Turner, M. G., and W. H. Romme. 1994. Landscape dynamics in crown fire ecosystems. *Landscape Ecology* 9 (1): 59–77.

Turner, M. G., W. H. Romme, R. H. Gardner, R. V. O'Neill, and T. K. Kratz. 1993. A revised concept of landscape equilibrium: Disturbance and stability on scaled landscapes. *Landscape Ecology* 8 (3): 213–27.

Urban, D. L., R. V. O'Neill, and H. H. Shugart Jr. 1987. Landscape ecology: A hierarchical perspective can help scientists understand spatial patterns. *BioScience* 37 (2): 119–27.

U.S. Forest Service (USFS). 1995. Draft environmental impact statement: Managing California spotted owl habitat in the Sierra Nevada national forests of California (An ecosystem approach). San Francisco, CA: USFS Pacific Southwest Region.

Vale, T. R. 1981. Ages of invasive trees in Dana Meadows, Yosemite National Park, California. *Madroño* 28:45–47.

Vankat, J. L. 1970. Vegetation change in Sequoia National Park, California. Ph.D. dissertation, University of California, Davis.

———. 1983. General patterns of lightning ignitions in Sequoia National Park, California. In *Proceedings—symposium and workshop on wilderness fire,* edited by J. E. Lotan, B. M. Kilgore, C. William, and R. W. Mutch, 408–11. General Technical Report INT-182. Ogden, UT: U.S. Forest Service, Intermountain Research Station.

Vankat, J. L., and J. Major. 1978. Vegetation changes in Sequoia National Park, California. *Journal of Biogeography* 5:377–402.

van Wagtendonk, J. W. 1972. Fire and fuel relationships in mixed conifer ecosystems of Yosemite National Park. Ph.D. dissertation, University of California, Berkeley.

———. 1983. Prescribed fire effects on understory mortality. In *Proceedings, American Meteorological Society 7th Conference on Fire and Forest Meteorology* 7:136–38.

———. 1985. Fire suppression effects on fuels and succession in short-fire-interval wilderness ecosystems. In *Proceedings, symposium and workshop on wilderness fire,* 15–18 November 1983, Missoula, MT, technical coordination by J. E. Lotan, B. M. Kilgore, W. K. Fischer, and R. W. Mutch, 119–26. General Technical Report INT-182. Ogden, UT: U.S. Forest Service, Intermountain Forest and Range Experiment Station.

———. 1986. The role of fire in the Yosemite Wilderness. In *Proceedings, national wilderness research conference: Current research,* 23–26 July 1985, Fort Collins, CO, compiled by R. C. Lucas, 2–9. General Technical Report INT-212. Ogden, UT: U.S. Forest Service, Intermountain Forest and Range Experiment Station.

Vlamis, J., H. H. Biswell, and A. M. Schultz. 1956. Seedling growth on burned soils. *California Agriculture* 10 (9): 13.

Vogl, R. J. 1970. Fire and plant succession. In *Role of fire in the Intermountain West,* 65–75. Missoula, MT: Intermountain Fire Research Council.

———. 1977. Fire: A destructive menace or a natural process. In *Recovery and restoration of damaged ecosystems: Proceedings of the international symposium on the recovery of damaged ecosystems,* 23–25 March 1975. Virginia Polytechnic Institute and State University, Blacksburg, edited by J. Cairns Jr., K. L. Dickson, and F. F. Herrides, 261–89. Charlottesville: University Press of Virginia.

Wagener, W. W. 1961. Past fire incidence in Sierra Nevada forests. *Journal of Forestry* 59:739–48.

Weatherspoon, C. P. 1988a. Evaluating fire damage to trees. In *Proceedings, ninth annual forest vegetation management conference,* 4–5 November 1987, 106–10. Redding, CA: Forest Vegetation Management Conference.

———. 1988b. Preharvest prescribed burning for vegetation management: Effects on *Ceanothus velutinus* seeds in duff and soil. In *Proceedings, ninth annual forest vegetation management conference,* 4–

5 November 1987, 125–41. Redding, CA: Forest Vegetation Management Conference.

Weatherspoon, C. P., S. J. Husari, and J. W. van Wagtendonk. 1992. Fire and fuels management in relation to owl habitat in forests of the Sierra Nevada and southern California. In *The California spotted owl: A technical assessment of its current status*, technical coordination by J. Verner, K. S. McKelvey, B. R. Noon, R. J. Gutíerrez, G. I. Gould Jr., and T. W. Beck, 247–60. General Technical Report PSW-133. Albany, CA: U.S. Forest Service, Pacific Southwest Research Station.

Weaver, H. 1943. Fire as an ecological and silvicultural factor in the ponderosa pine region of the Pacific slope. *Journal of Forestry* 41:7–15.

———. 1947. Fire, nature's thinning agent in ponderosa pine stands. *Journal of Forestry* 45:437–44.

———. 1967. Fire and its relationship to ponderosa pine. *Proceedings, Tall Timbers fire ecology conference* 7:127–49.

———. 1974. Effects of fire on temperate forests: Western United States. In *Fire and ecosystems*, edited by T. T. Kozlowski and C. E. Ahlgren, 279–319. New York: Academic Press.

Wells, C. G. 1971. Effects of prescribed burning on soil chemical properties and nutrient availability. In *Prescribed burning symposium proceedings*, 14–16 April 1971, Charleston, SC, 86–97. Ashville, NC: U.S. Forest Service, Southeast Forest Experiment Station.

Wells, C. G., R. E. Campbell, L. F. DeBano, E. L. Fredriksen, R. C. Froelich, and P. H. Dunn. 1979. *Effects of fire on soil: A state-of-knowledge review.* General Technical Report WO-7. Washington, DC: U.S. Forest Service.

Westman, W. E., J. F. O'Leary, and G. P. Malanson. 1981. The effects of fire intensity, aspect, and substrate on post-fire growth of Californian coastal sage scrub. In *Components of productivity of Mediterranean-climate regions—basic and applied,* edited by M. S. Margaris and H. A. Mooney, 151–79. The Hague: Junk.

Wilbur, R. B., and N. L. Christensen. 1983. Effects of fire on nutrient availability on a North Carolina coastal plain pocosin. *American Midland Naturalist* 110:54–63.

Wilken, G. C. 1967. History and fire record of a timberland brush field in the Sierra Nevada of California. *Ecology* 48:302–4.

Wood, S. H. 1975. Holocene stratigraphy and chronology of mountain meadows, Sierra Nevada, California. Ph.D. dissertation, California Institute of Technology.

Wright, H. A., and A. W. Bailey. 1982. *Fire ecology: United States and southern Canada.* New York: John Wiley and Sons.

Zedler, P. H., C. R. Gautier, and G. S. McMaster. 1983. Vegetation change in response to extreme events: The effect of a short interval between fires in California chaparral and coastal scrub. *Ecology* 64:809–18.

SUSAN J. HUSARI
U.S. Forest Service
Pacific Southwest Region
San Francisco, California

KEVIN S. McKELVEY
Redwood Sciences Laboratory
Pacific Southwest Research Station
Arcata, California

40

Fire-Management Policies and Programs

ABSTRACT

For most of this century the goal of fire management in the Sierra was to control fire. The policy was aggressively and successfully applied, substantially reducing annual acres burned. This goal was based on a fire policy that emphasized keeping wildland fires as small and inexpensive as possible. As the role of fire in maintaining Sierran ecosystems has been recognized, fire has been reintroduced through the application of planned prescribed fire and prescribed natural fire. Despite changes in fire-management policy that have allowed expanded use of fire, relatively few acres have been managed using fire in the Sierra Nevada. This chapter explores options for expanding the role for fire in the Sierra through more liberal application of current fire policy and through changes in existing fire policy. These recommendations are tempered by the knowledge that the number of available fire-fighting resources has been steadily declining since the mid-1970s and that social, economic, and biological factors are making all aspects of fire management more costly and difficult.

INTRODUCTION

This chapter describes the history of fire management in the Sierra and discusses present programs in the context of changing public expectations of fire organizations and evolving management objectives.

For most of this century, the goal of fire management in the Sierra was to control fire. The policy was aggressively and successfully applied, substantially reducing annual acres burned. Fire-suppression programs, although effective in achieving this goal, are very expensive. The cost of the U.S. Forest Service (USFS) presuppression program in the Sierra,

for example, was $30,000,000 in fiscal year 1995, and this amount does not include aircraft contracts and the money spent actually suppressing fires. National fire-suppression costs are increasing at a rate higher than that of inflation. Fire-fighting costs are rising at an even faster rate in the Pacific West than in the rest of the country (USFS 1995b; Schmidt 1995). A USFS study aimed at determining the reasons for increasing fire-suppression costs concluded that the explosive fuel types that have developed across the West have made traditional fire-suppression tactics very expensive and sometimes ineffective, and this expense was a major contributor to the record-breaking fire expenditures during the 1994 fire season (USFS 1995c). A series of reports has highlighted cost increases due to emphasis on protection of private property (USFS 1995d). Rising costs, increasing numbers of firefighter injuries and fatalities, and concerns about the ecological effects of excluding natural disturbance from fire-adapted ecosystems have prompted national review of fire programs and policies. These issues are magnified in the Sierra, where fire suppression has been highly successful in reducing the annual acres burned by wildfire, fuel treatments have not affected enough acres to influence fire regimes, and more and more people are moving into vegetated wildlands adjacent to or mixed with federal and state lands.

Fire-management organizations are more than fire trucks and helicopters. Fire-management programs encompass presuppression activities aimed at reducing the land area burned by wildfire, as well as fire-suppression activities aimed at putting out fires and repairing the damage caused by wildfires that escape initial attack. Presuppression includes reducing the flammability of fuels through removal or rearrangement; engaging in fire-prevention and public-education activities; training fire personnel to fight fires; detecting fires; and operating fire stations, air tanker bases, and other facili-

Sierra Nevada Ecosystem Project: Final report to Congress, vol. II, *Assessments and scientific basis for management options.* Davis: University of California, Centers for Water and Wildland Resources, 1996.

ties during the fire season each year. Fire-suppression includes fire-fighting activities and emergency rehabilitation of burned areas.

Five agencies have fire-management responsibilities in the Sierra Nevada: the California Department of Forestry and Fire Protection (CDF), the USFS, the Bureau of Land Management (BLM), the National Park Service (NPS), and several Native American tribes. All the fire agencies cooperate closely. Many dispatch or coordination centers in the Sierra dispatch resources from more than one agency in the vicinity. Wildland fire fighting in the Sierra is conducted using the "closest forces" concept, where the fire-fighting resources closest to the fire are dispatched, regardless of agency. Actual protection boundaries between the larger agencies were set through a process called *balancing of acres* in 1990. These boundaries redistribute protection responsibilities to ensure that fire-suppression resources are used most efficiently. The balancing of acres also reorganized responsibilities to avoid the need for reimbursement among agencies for initial-attack fire protection. As a result, each agency provides fire protection on lands in the other agencies' jurisdictions. Each agency has responsibility for prescribed burning in its own jurisdiction. Local government, in the form of fire districts and through CDF contracts, is responsible for structural fire protection within their areas within the State Responsibility Area. Many local fire departments also participate in suppression of wildland fires.

Inherent differences in the missions of fire-fighting agencies affect their fire-management programs. The California Department of Forestry and Fire Protection provides fire protection primarily for private lands with roads. CDF has the highest percentage of wildlands mixed with structures (urban intermix or interface lands) in its protection area. CDF protects much of the lower-elevation lands in the Sierra foothills as well as large areas of private timberlands. These lands dry earliest and have the longest fire season (McKelvey and Busse 1996). CDF also protects state parks and other state-owned lands. CDF works closely with the Office of Emergency Services and rural fire departments. Fire-suppression strategies, tactics, and activities are influenced by state statutes, the types of vegetation in the CDF protection area, access, and the need to protect lives and private property. CDF conducts prescribed burns cooperatively with landowners through the vegetation-management program.

The national forests in the Sierra Nevada range from the foothills through the high-elevation zone. The USFS manages most of the publicly owned timber-producing belt in the Sierra Nevada. Fire-management activities are conducted to meet the objectives outlined for the various management areas in each forest's land- and resource-management plan. The forests are managed with many objectives in mind, from recreation, cattle grazing, scenic values, and water quality to late successional forests, wilderness, timber harvest, and wildlife habitat. The varied land uses and management objectives result in a variety of fire-management strategies for each forest.

Fire-suppression strategies, tactics, and activities are influenced by vegetation type, management objectives, proximity to development, private/public ownership patterns, elevation, and other factors. The forests have large fire-management programs that include fire-suppression, fuels management, and a small amount of prescribed natural fire.

Four national parks fall within the Sierra Nevada Ecosystem Project (SNEP) analysis area. These four areas—Yosemite National Park, Sequoia National Park, Kings Canyon National Park, and Lassen Volcanic National Park—are managed primarily for their wilderness, ecological, and recreational values. Most of the park acreage is inaccessible by road. The national parks put great emphasis on restoring natural processes, including fire. The parks have complex fire-management programs that include fire suppression, prescribed burning, and prescribed natural fire.

The BLM protects lands on the southern end and the east side of the Sierra Nevada range, outside the core SNEP area. The agency has protection responsibilities east of the Sierra in the Susanville area. Most of the protection area is in Great Basin vegetation types. The BLM has a complex fire-management program that includes fire suppression and prescribed burning.

Native American lands are protected by either the USFS or the BLM through agreements or contracts. None of the tribes in the Sierran area maintain separate fire-fighting organizations. Activities include fire suppression and vegetation management.

EVOLUTION OF FIRE-SUPPRESSION POLICY

One of the fundamental purposes for establishing forest reserves (the original name given to the national forests) and national parks was to provide organized fire protection for public lands. The *Forest Reserve Use Book* issued in 1905 listed protection of reserves from fire as one of the three duties of forest officers. Disastrous fires in 1910 claimed eighty-five lives and burned 1,011,750 ha (2.5 million acres) in the northern Rocky Mountains (Cermak 1988). The 1910 fires focused emphasis on fire control nationally. During the same time period, California was the site of a nearly two-decade debate over the application of "light burning" as a management tool in forests and rangeland. This debate was resolved in favor of aggressive fire control. The USFS quantified its fire-protection mission in 1926 by adopting the objective of controlling all fires at 4 ha (10 acres) or less. Wildfires were to be suppressed, minimizing the costs of fire suppression and resource loss. These concepts were the basis of fire suppression in the National Park Service as well. The USFS sought to strengthen its fire-protection policy by adopting the "10 AM Policy" in 1935. If aggressive initial attack did not control a fire, then enough fire-fighting resources would be assigned to control it by 10 A.M. the next day. The policy was simple, was easy to understand, and provided clear direction. The developers of the 10 A.M. policy considered it consistent with the objective of minimizing fire-suppression costs and resource damage because they expected suppression costs to decrease if all fires were attacked aggressively.

In 1971 the USFS adopted a 10-acre control plan for 90% of all fires as a planning objective. Rising fire presuppression and suppression costs and the need to link fire protection with land-management planning led to the replacement of the 10 A.M. policy in 1978. Terminology changed from fire control to fire management. The new fire-management policy directed fire managers to minimize fire-suppression costs and damage consistent with land and resource objectives. It defined appropriate suppression response (ASR) as a range of suppression strategies. These strategies—called contain, confine, and control—were to be employed to accomplish a cost-effective response to fires that escaped initial attack. ASR implies that the most cost-effective response might deviate from a suppression philosophy that emphasized keeping *all* fires small.

Starting in 1983, ASR was expanded to allow the federal agencies to use confine, contain, or control strategies during initial-attack fire fighting. The NPS requires a rationale for the use of a strategy other than control during initial attack. The USFS requires justification (completion of a fire situation analysis, or FSA) if a fire is managed for more than a single burning period without being considered to have escaped. At a minimum the FSA must include a decision analysis that considers expected suppression cost, damage, and the probability of success or failure. If it is determined that the initial action response does not meet or is anticipated not to meet established fire-management direction minimizing fire-suppression cost and damage from fire, the fire is declared an escaped fire.

PRESENT FIRE-SUPPRESSION POLICY

The fire-suppression programs pursued by fire-management agencies have limited the number of fires that escape initial attack. Nationally, only 2% of all fires in USFS jurisdiction required large-scale suppression efforts in 1994. Ninety-four percent of the total burned acres resulted from 2% of the fires (USFS 1995a). The California Department of Forestry and Fire Protection estimates a similar success rate in suppressing wildfires in the CDF protection area.

The National Park Service, Bureau of Land Management, and national forests have similar fire policies. These policies are likely to be further standardized in response to the recent federal wildland fire policy review recommendations. The ob-jective of fire suppression in the NPS is to "suppress wildfires at minimum cost consistent with values at risk while minimizing the impacts from suppression activities" (NPS 1990b). The BLM policy states that "wildfire losses will be held to the minimum through timely and effective suppression action consistent with the values at risk." The USFS manual states that "the objective of fire suppression is to safely suppress wildfires at minimum cost consistent with land and resource management objectives and fire management direction as stated in fire management action plans" (USFS 1994b). The goal for fire control on CDF lands is "to detect, respond to and control each fire occurring in or threatening State Responsibility Area (SRA) at a size that will hold net damages to resources and exposed life and property to a minimum" (CDF 1986). All four agencies recognize confine, contain, and control strategies as appropriate suppression strategies for managing escaped fires. The NPS and USFS define the strategies slightly differently. ASR is a continuum of fire strategies from monitoring through control. Figure 40.1 contrasts the NPS and USFS definitions of confine, contain, and control. BLM and CDF policy manuals do not include definitions.

Present NPS and USFS directions specifically prohibit the use of wildfire to meet resource-management objectives. This interpretation is based on the philosophy of economic efficiency adopted in 1928, which directed that fires must be suppressed using the alternative that cost the least and most effectively reduced resource loss. Fires are managed to minimize cost and damage without considering their benefits to the resource.

FIGURE 40.1

Definitions for confine, contain, and control in the NPS and USFS.

Confine:
NPS: To restrict the wildfire within determined boundaries established either prior to, or during the fire. These identified boundaries will confine the fire, with no action being taken to put the fire out.

USFS: To limit fire spread within a predetermined area principally by use of natural or preconstructed barriers or environmental conditions. Suppression actions may be minimal and limited to surveillance under appropriate conditions.

Contain:
NPS: To restrict a wildfire to a defined area, using a combination of natural and constructed barriers that will stop the spread of the fire under the prevailing and forecasted weather conditions, until out.

USFS: To surround a fire, and any spot therefrom, with a control line as needed, which can reasonably be expected to check the fire's spread under prevailing and predicted conditions.

Control:
NPS: To aggressively fight a wildfire through the skillful use of personnel, equipment and aircraft to establish fire lines around a fire to halt the spread and to extinguish all hot spots, until out.

USFS: To complete the control line around a fire, any spot fires therefrom, and any interior islands to be saved; to burn out any unburned area adjacent to the control line, until the line can reasonably be expected to hold under foreseeable conditions.

IMPLEMENTATION OF FIRE-SUPPRESSION POLICY IN THE SIERRA NEVADA

The four national parks and the nine national forests in the Sierra Nevada have had the option of applying appropriate suppression response since 1978. The degree to which the flexibility inherent in ASR is exercised is highly variable in the Sierra Nevada, both within and among agencies. The application of fire-management policy by the various agencies in the Sierra Nevada could be summarized as follows: CDF has a rigid fire-suppression policy that is applied flexibly. The USFS and BLM have flexible suppression policies that are applied conservatively. The NPS has a flexible fire-suppression policy applied liberally.

The 1986 fire-management plan for the California Department of Forestry and Fire Protection establishes an objective of controlling all fires during initial attack on CDF's jurisdiction. Appropriate suppression response is allowed on fires that have escaped initial attack.

The forest plans for the Inyo, Tahoe, and Lassen National Forests and the Lake Tahoe Basin Management Unit allow use of ASR on all fires on forest land. The Eldorado, Sierra, Sequoia, and Stanislaus National Forests allow the use of ASR in wilderness and in high-elevation areas of the forests but specify control in other portions of the forests. The Modoc National Forest has used ASR since 1971 in the Big Sage Management Unit. The Plumas forest plan specifically prohibits the use of any suppression strategy other than control anywhere on forest land and at any stage of fire suppression.

The fire-management plans for Sequoia and Kings Canyon National Parks, Lassen Volcanic National Park, and Yosemite National Park allow use of ASR for any fire in any location.

Reading plans and policies alone does not give an accurate picture of how and where fires are suppressed in the Sierra Nevada. Forest plans, fire-management plans, and other documents describe the options available to the fire manager but do not explain how often each strategy is applied. The way in which the plans are carried out varies from place to place. The differences in application of initial-attack strategies are displayed in table 40.1. As can be seen, there seems to be little relationship between what is written in the plans and what is applied. Although confine and contain strategies are allowed, they are not frequently employed, since the manager generally opts for the control strategy.

In discussing the application of policy in the Sierra Nevada, fire managers listed the following reasons for selecting confine and/or contain initial-attack strategy on federal lands in the Sierra:

Confine or contain is used to reduce fire-suppression impacts and costs, particularly in wilderness. This also reduces rehabilitation costs.

TABLE 40.1

Estimated use of confine, contain, and control strategies for fire suppression during initial attack in Sierran forests and parks, through 1994, listed by percentage of total wildfires (survey of Fire Management Officers of parks and forests conducted for this chapter).

Unit	Percentage Confine	Percentage Contain	Percentage Control
Eldorado National Forest	5	5	90
Inyo National Forest	35[a]		65
Lake Tahoe Basin	1[a]		99
Lassen Volcanic National Park	10	20	70
Lassen National Forest	1[a]		99
Modoc National Forest	23	1	76
Plumas National Forest	0	0	100
Sequoia and Kings Canyon National Parks	17[a]		83
Sequoia National Forest	0	0	100
Sierra National Forest	2[a]		98
Stanislaus National Forest	0	0	100[b]
Tahoe National Forest	1[a]		99
Yosemite National Park	5[a]		95

[a]Confine or contain.
[b]All fires are controlled except lightning fires in the Emigrant Wilderness. However, an amendment to the Stanislaus National Forest forest plan allowed use of confine and contain strategies in other areas starting in 1995.

Confine or contain may be selected because of firefighter safety concerns. Fires may be confined or contained when inaccessible to firefighters, such as those located on cliffs or in steep drainages.

Confine or contain may be selected if the fire is confined by natural barriers to a small area of continuous fuels that will burn and go out.

Confine or contain may be selected for some fires when resources are needed for higher-priority fires.

Confine or contain may be selected when no resource damage is expected.

Confine or contain may be selected when fewer fire-fighters can accomplish the job of suppression over more time. The fire gets larger, but fewer firefighters are committed, though they may be on the fire for a longer period. For example, a single crew may take several days to suppress a fire at a larger final size using ASR, as compared to several crews controlling the fire at a small area. This may be chosen either because fire-fighting resources are scarce or to minimize suppression costs.

Federal fire managers listed these limitations to applications of appropriate suppression response in the Sierra Nevada:

Mixed ownership patterns occur in many areas of the Sierra Nevada. For example, much of the Tahoe National

Forest is a checkerboard pattern of sections in public and private ownership. Aggressive initial-attack and control strategies are used because of risk to private land.

Many areas of the Sierra Nevada have continuous, homogeneous fuels with few of the natural barriers or fuel type changes that provide opportunities for application of contain or confine strategies.

Most of the area protected by CDF and much of that protected by the Forest Service is intermixed with or adjacent to homes, communities, and other development. Even a remote chance of an escaped fire is unacceptable because of the dire consequences.

Many areas of the Sierra Nevada are subject to frequent and unpredictable severe fire weather patterns.

The Sierra's Mediterranean-type climate (wet winters and long, dry summers) results in a lengthy fire season with few breaks in the fire danger.

Managers and firefighters do not want to take on additional risk associated with some fire-management strategies.

The concept of appropriate suppression response is poorly understood. Most fire managers have not received training in its application or in matching tactics to any strategy except control.

There is no incentive or encouragement to apply the full range of appropriate suppression response.

Long-term management of wildfires is discouraged because it ties up fire-fighting resources that could be used on other incidents.

Control strategies are generally viewed as the least costly suppression response for fires in the Sierra Nevada, given the restricted definition of cost that fire managers use in selecting fire-suppression alternatives. Many lightning fires start under low to moderate burning conditions and spread slowly. It is consistently less expensive to assign a fire crew to put a fire out when it is confined to a single tree than it is to pay to monitor the same fire for a longer period in a containment or confinement mode. Fires in red fir, lodgepole pine, or upper-elevation mixed conifer forests spread slowly at first. The litter is tightly packed and burns slowly, at low intensity. Dense canopies shelter the fire from the wind and from the direct rays of the sun. These vegetation types are also under snow for a longer period and at higher elevation where the fuels dry slowly and the fire season is shortened. Such fires can be extinguished easily and inexpensively when they are small. Fire managers recognize the lower risk associated with these fires but cannot justify allowing them to get larger because of the requirement to select the least-cost-plus-loss suppression

alternative. As a result, most wildfires that burn in locations and under conditions that would produce results most similar to those that occurred under historic conditions are suppressed at small size.

Fires initiate and spread rapidly in fuel types with light, quick-drying fuels or with more-open canopies that allow wind and sunlight to reach the surface litter. In the Sierra Nevada these types include ponderosa pine, eastside pine, grassland, oak savanna, deciduous oak stands, lower-elevation mixed conifer, sagebrush, and chaparral. These are the same types in closest proximity to structures and other development. The risk and suppression cost of managing fires in these types limit suppression action to rapid, aggressive control.

In practice a combination of several fire-suppression strategies may be applied to a single fire. Fire managers and members of specialized Incident Management Teams agree that a single fire-suppression strategy is rarely applied on a large fire. One flank may be allowed to run into rocks, another may be contained by a river, and a third, adjacent to structures, may be controlled by direct or indirect methods. There are, however, only three examples of large fires in which the contain or confine strategy has been selected in the Escaped Fire Situation Analysis (document that describes the selected suppression alternative) on Sierran forests in the last ten years. Control strategies have been used to suppress all large fires on the Eldorado, Lassen, Sequoia, Plumas, Tahoe, and Stanislaus National Forests in this time period. Confine or contain has been used regularly on escaped fires in national parks, especially when prescribed natural fires have been declared wildfires because of national fire emergencies or because of smoke-management concerns.

FIRE-SUPPRESSION TACTICS

Once a fire strategy is selected, it can be accomplished using a variety of fire-suppression tactics. Minimum impact suppression tactics (MIST) are those fire-suppression techniques that use the minimum tool needed to do the job. They also accomplish fire suppression using methods that produce the least visual impact. Techniques include flush cutting of stumps, use of natural barriers or roads as firelines, retention of snags, narrow firelines, and other techniques that minimize the impacts of fire suppression.

Minimum impact suppression tactics are used in all four parks whenever it is safe to do so. All forests use MIST in wilderness areas. In addition, the Eldorado and Inyo National Forests and the Lake Tahoe Basin apply MIST whenever possible outside wilderness areas. Several fire managers mentioned the cost savings in reduced rehabilitation through implementing these tactics.

FIRE-MANAGEMENT RESOURCES IN THE SIERRA

There are fewer fire-management resources in the Sierra today than in past decades. For example, in 1963, the Lassen National Forest had sixteen engines, two helicopters, twelve lookouts, nine prevention units, and two air tankers. By 1995 the number of resources had been reduced by nearly half, to nine engines, one helicopter, six lookouts, three prevention units, and one air tanker. Table 40.2 displays the number of USFS resources in California from 1982 to 1995. The table lists all USFS wildland fire-fighting resources in California. Approximately a third of these USFS resources are located in the Sierra. The table illustrates the gradual decrease in the numbers of wildland fire-fighting resources during the fourteen-year period. The number of fire engines, for example, has been reduced by 12%.

Table 40.3 illustrates the number of CDF wildland fire-fighting resources in California. CDF has experienced reductions in some types of fire-fighting resources in the Sierra similar to those displayed for the USFS. The number of CDF fire engines in California has been reduced by 12% since 1970. The number of hand crews available to the CDF has increased substantially in this same period. These fire crews, from the California Department of Corrections, California Conservation Corps, and other sources, are generally not dispatched as initial-attack forces except during high fire-danger periods. They take thirty minutes to an hour to arrive at a fire.

The decrease in numbers of USFS and CDF fire-fighting resources cannot be directly linked to a decrease in the amount of presuppression funds. The presuppression budget for the Pacific Southwest region was $96,200,000 in FY82 and $97,800,000 in FY95, expressed in constant FY95 dollars. Nationally, USFS presuppression funding peaked in 1977 and has not increased or decreased in real dollars during the past twenty years (USFS 1995b). Adjusted for inflation the annual CDF fire-protection base budget has been relatively constant during the period FY84/85 through FY93/94, with an average of $310,551,384 in 1994 dollars. There are a number of reasons for the decline in available fire-fighting resources in both agencies, given the reasonably stable presuppression budget. In the USFS, there has been a decrease in the availability of project funds to pay portions of the base salaries of fire crews when they are not fighting fires. In the past, portions of fire crews' salaries and basic costs were paid to improve wildlife and fisheries habitat, build fences, thin plantations, construct fuel breaks, and clean up slash resulting from timber sales. The impact of declining project funds has been greatest on forests that had large timber-sale programs, where collections for brush disposal (dollars collected to clean up slash resulting from timber harvest) have dropped dramatically. The portion of national presuppression funding used to treat natural fuels accumulations has decreased steadily since the mid-1970s.

Table 40.2 does not fully illustrate the decrease in numbers of fire-suppression resources in the USFS because it does not include brush disposal crews. Districts formerly employed hand crews to complete slash clean up. These crews were also available to fight fires. Most of the Sierran forests had approximately one ten-person crew per district at the height of the brush disposal program in the early 1980s. Fire-fighting ability has also been impacted by the decrease in the number of USFS employees from outside the fire-fighting organization (foresters, administrators, biologists, and others) who are available or willing to fight fires. Currently, only 53% of USFS employees hold red-card qualifications, which certify them to participate on wildfires. In 1994 25% of the red-carded employees accounted for 75% of the fire-fighting efforts (USFS 1995a).

Administrative support costs have absorbed an increasing portion of fire funds because the fire program has become a

TABLE 40.2

USFS wildland fire-fighting resources in California during fiscal years 1982–95 (summarized from records on file in the regional office of the Pacific Southwest Region of the USFS).

Fiscal Year	Air Tankers	Helicopters	Air Attack	Hotshot Crews[a]	Engines	Prevention Units
FY82	13	19	8	17	240	282
FY83	13	19	8	17	228	245
FY84	13	17	6	17	241	251
FY85	13	18	6	16	254	229
FY86	12	18	6	16	237	215
FY87	11	18	6	16	231	228
FY88	11	18	6	18	236	240
FY89	11	18	6	18	228	222
FY90	11	18	6	18	228	222
FY91	11	18	6	18	228	222
FY92	13	18	6	18	228	222
FY93	11	18	6	18	217	205
FY94	11	18	6	18	221	182
FY95	11	18	6	18	219	176

[a]Hotshots are organized, twenty-person fire crews.

TABLE 40.3

CDF wildland fire-fighting resources available in California from 1970 through 1994 (summarized from statistics compiled by CDF).

Year	Air Tankers	Helicopters	Crews[a]	Engines	Dozers	Lookouts
1970	23	2	116	367	58	78
1971	23	2	114	374	58	79
1972	23	7	110	370	58	82
1973	21	7	113	370	67	82
1974	21	7	113	368	71	83
1975	21	7	113	367	70	80
1976	21	7	113	362	68	79
1977	21	7	113	362	67	78
1978	21	8	114	362	67	78
1979	21	9	114	355	55	78
1980	21	8	132	352	63	76
1981	21	8	132	352	63	75
1982	21	8	150	344	63	72
1983	21	8	148	344	63	72
1984	21	8	153	344	63	72
1985	21	8	157	344	63	72
1986	21	8	177	344	63	71
1987	21	9	188	344	63	71
1988	21	9	206	344	63	64
1989	21	9	217	344	63	64
1990	21	9	230	344	63	64
1991	21	9	231	338	58	33
1992	15	9	184	338	58	24
1993	19	9	173	336	58	24
1994	19	9	173	334	58	32

[a]California Department of Corrections hand crews, California Conservation Corps, and other crews.

larger percentage of the forests' organization as other parts of the organization have shrunk. Unemployment claims have risen dramatically because many temporary firefighters cannot find jobs during the off-season.

Both CDF and the USFS have experienced increasing module costs (cost to staff and operate individual pieces of fire-fighting equipment). Within CDF, labor costs have risen dramatically as a result of court decisions regarding the Fair Labor Standards Act. Changes in overtime pay policies have also increased costs. USFS module costs have increased because of changes in job classification and grade structure that have resulted in more highly paid employees on modules.

A discussion of the declining availability of fire-suppression resources would be incomplete without focusing on the impact of structure protection on fire-fighting resources. A recent USFS publication (1995a) states, "Forest Service manual direction for planning wildfire suppression strategies prioritizes the protection of life and private property above protecting natural resources. Suppression forces therefore protect urban values at the expense and detriment of forest ecosystem values. The result is even greater acreage of burned wildfires." This statement is echoed in the draft Federal Wildland Fire Management Policy and Program Review (1995) and the Strategic Assessment of Fire Management in the U.S. Forest Service (USFS 1995d). None of these reports have included quantitative estimates of the increased costs or the drain on wildland fire-fighting resources created by increasing demands to protect private property interspersed with wildland.

However, the California Department of Forestry and Fire Protection has defined three fire-management environments in California: undeveloped, developed, and mixed interface (CDF 1995). These categories can be used to display the degree to which development affects fire-management programs and decisions. Undeveloped lands are defined as those areas with less than one house per 160 acres located more than five kilometers from areas with a housing density greater than one house per 160 acres and arranged in contiguous blocks of 50,000 acres (20,000 ha). Developed lands include all areas of the state with a housing density greater than one house per five acres plus all areas within two kilometers of such developed areas. Mixed-interface areas are those between the developed and wildland areas. When this classification is applied to the Sierra, approximately 39.2 million acres (15.9 million ha) are undeveloped wildlands, 9.7 million acres (3.9 million ha) are developed, and 34.9 million acres (15.6 million ha) are mixed interface. The three categories are distributed across all ownerships and jurisdictions. Sierran forest fire managers estimate that the efficiency (speed at which fireline is constructed and held) of fire-fighting resources decreases by 20% to 25% in portions of the forests where demands to protect private property are high.

California has one of the most mobile, highly organized fire-suppression forces ever assembled. The pool of available manpower and equipment has, however, declined. An organization that increases its response efficiency but decreases overall manpower would exhibit the patterns we see in the

Sierra Nevada: more-effective average response coupled with exhaustion under extreme circumstances (McKelvey and Busse 1996). To understand why simultaneous ignitions can create problems for fire suppression and how an organization can be effective at controlling single ignitions but fail when faced with multiple ignitions, consult the conceptual model in appendix 40.1.

FIRE-MANAGEMENT PLANNING

The National Park Service emphasizes understanding fire regimes in developing fire-management plans. Restoring fire to its natural role in park ecosystems is one of the highest resource-management priorities in all four Sierran parks (NPS 1990a). The three plans divide the parks into zones: a high-elevation zone where lightning fires are managed as prescribed natural fire under all but the most extreme conditions; a middle-elevation, conditional fire-management zone where prescribed fire is used to restore fuel conditions to natural range of variability and then prescribed natural fire is employed; and a suppression zone where only fire suppression or prescribed fire is employed. Full suppression zones are found around the perimeter of parks, at low elevations, and around improvements within parks. The use of prescribed natural fire is influenced, in all zones, by the national fire situation, availability of fire-fighting resources to manage a lightning-caused ignition as a prescribed natural fire, the current drought situation, and funds.

Fire is not a central issue in the current forest plans for the national forests in the Sierra Nevada. It is discussed in the context of protection of resources in the various management areas described in the plans. Although acre objectives for wildfire control were superseded by ASR in 1978 and 1984, the USFS has continued to use acre objectives (maximum fire-size objectives) as a convenient method of relating forest-plan objectives for individual Management Areas to Standards and Guides for fire management, as required in forest plans. Most of the Sierran forest plans set different maximum fire-size objectives for different fire-management zones depending on fire-intensity level. For example, the Stanislaus Forest land and resource management plan may have a maximum fire-size objective of 40 ha (100 acres) if the fire intensity is low but a maximum fire-size objective of 4 ha (10 acres) if intensity is high. Maximum fire-size limits of 4 ha (10 acres) are the upper limit for most of the other forests. The size limit is negotiated in the planning process through discussion of fire effects on resources and is based on the objectives of the unit, such as watershed management, timber management, or wilderness management.

The maximum fire-size objective does not exempt the fire manager from selecting a least-cost-plus-loss alternative. This brings up a fundamental point of confusion in USFS fire planning and policy: both planning and future budget requests

for presuppression (National Fire Management Analysis System) are based on suppression cost plus the net value change in the resource. Net value change includes consideration of both the benefits and detriments of wildfire. The combination of cost of fire suppression plus the net value change in resources (timber value, watershed values, recreation values, forage, wildlife habitat, and others) is used to justify a level of protection on each national forest, defined by the most efficient level of fire suppression. The future funding tool encourages high valuation of resources to maximize presuppression funding.

There is no mechanism within the current USFS planning system to display the effects of excluding fire from the ecosystem. Fuel management can be considered beneficial only in the sense that a reduction in suppression costs can be demonstrated. Currently, the USFS does not organize fire-management planning units around similar fire behavior types, and the fire-planning model does not allow planning for multiple ignitions.

The statewide CDF fire plan is currently being revised. The new fire plan will be based on a damage-plus-cost analysis of fire-protection performance similar to that used by the USFS. The purpose of the analysis is to provide a fire-protection system that equally protects lands of similar type. The analysis will define a level of service rating that can be used to compare, on a relative basis, the level of fire protection provided for wildland areas in California. The level of service rating will be used to set program priorities and provides a means to integrate various program elements like fire prevention, vegetation management, and engineering. Public input will be used to adjust the level of service acceptable to California residents.

PRESCRIBED NATURAL FIRE POLICIES AND PROGRAMS

In 1964 the Wilderness Act recommended that fire be allowed, as much as possible, to play its natural role in wilderness. In 1968 Sequoia and Kings Canyon National Parks began a prescribed-burning program that used prescribed natural fire and management ignition. Yosemite National Park started a prescribed natural fire program in 1972. Lassen Volcanic National Park began a prescribed natural fire program in cooperation with the Lassen National Forest in 1983. All three programs were suspended for revisions called for by the Interagency Fire Policy Review Team in 1988 (Fire Management Policy Review Team 1989). The Yosemite and Sequoia Kings Canyon programs were restarted in 1990, and the Lassen program was restarted in 1994.

USFS fire-management policy was amended to allow use of prescribed natural fire in wilderness in 1971. In 1985 it was again revised to allow use of planned-ignition prescription burning in wilderness, under a limited set of conditions and objectives. Before 1988 there were approved prescribed natural fire programs in the Lassen National Forest's Caribou Wilderness, the Stanislaus National Forest's Emigrant Wilderness, and the Jennie Lakes Wilderness on the Sequoia National Forest. The programs were suspended after the 1988 fire season for a review of the prescribed natural fire program and fire policy. The prescribed natural fire programs in the Caribou and Emigrant wilderness areas were restarted in 1993 and 1994, respectively.

Planned prescribed-burning programs are permissible in eighteen USFS wilderness areas in the Sierra. A combination of planned-ignition prescribed-burning programs and prescribed natural fire programs is called for in thirteen wilderness areas. Prescribed natural fire alone is called for in one wilderness area. The Lake Tahoe Basin plan allows only fire suppression in the Desolation Wilderness. The prescribed natural fire program has been taken from the planning to implementation stage in two Forest Service wilderness areas in the Sierra Nevada.

The NPS and USFS have similar wilderness fire-management policies. The fire programs for the two agencies differ in the degree to which the policies have been applied locally. Sequoia and Kings Canyon and Yosemite National Parks have had active prescribed natural fire programs for most of the last twenty-five years. Both parks have used extensive prescribed burning to restore fuel loadings and forest structure to levels within the natural range of variability. Lassen Volcanic National Park has had an active program for almost fifteen years. However, analysis of even the most successful prescribed natural fire programs (Botti and Nichols 1995; Parsons 1995) indicates that these programs fall far short of duplicating the role of natural process in Sierran ecosystems. Acres burned are much fewer than the number of acres burned under historic fire regimes. Smoke-management constraints, risk to adjacent jurisdictions, or improvements and limitations on programs during periods of high wildfire activity are among the factors that have limited accomplishments. The national and state interagency preparedness plans have required that no new prescribed natural fires be managed during periods of high activity and may require that ongoing prescribed natural fires be suppressed during extremely high activity.

The plans for the Sierran national forests authorize the use of prescribed natural fire and, in most cases, management-ignited fire, as shown in table 40.4. Despite program authorization in nearly every plan, only two USFS wilderness areas have prescribed natural fire programs, with a total Sierra-wide burned area of less than 40 ha (100 acres) in the entire period that the programs have been in place. No management-ignited prescribed burns have been conducted in USFS wilderness areas in the Sierra Nevada.

When surveyed, fire managers in both agencies gave the following reasons for the differences in implementation of wilderness fire-management programs between agencies:

The National Park Service has provided consistent funding for the planning and implementation of prescribed natural fire programs. Forests must use scarce project dollars for both planning and implementation. These same funds are in demand for prescribed burning outside wilderness and for other recreation and wilderness activities. There is no indication that additional dollars will be made available for managing prescribed natural fires. As a result, there is little or no incentive to develop programs.

Before 1988, the National Park Service used emergency dollars to manage prescribed natural fires. This practice was suspended in 1988 but is once again in place throughout the Department of the Interior. This mechanism provides the flexibility to allow changes in the size of the prescribed natural fire program from year to year in response to variation in the number of lightning fires. The U.S. Department of Agriculture does not use emergency fire dollars to manage prescribed natural fires, because of the department's interpretation of fiscal policy and allowable uses of emergency funds.

Yosemite and Sequoia and Kings National Parks have developed a strong research basis for implementation of fire-management strategies within the parks. The southern Sierra Nevada have been the focus of most of the dendrochronology/fire history studies conducted in the range. A fire-history study has also been conducted in Lassen Volcanic National Park and the adjacent Caribou Wilderness.

The Forest Service and National Park Service differ in philosophical basis of fire programs in wilderness. The National Park Service focus is on management of fire as a disturbance within its natural range of variability. In Forest Service wilderness, on the other hand, the emphasis has been on allowing natural processes to operate freely, without making judgments about whether the effects of these processes are good or bad. There is a subtle but important difference between managing fire freely and managing it as a process that has a distinct ecological role. The fire-management plans for the NPS areas in the Sierra include use of planned prescribed fire to reduce fuel loadings, prior to reintroduction of prescribed natural fire.

National Park Service wilderness areas are substantially larger than Forest Service wilderness areas. When comparing two areas of similar fuel and fire behavior characteristics, the risk of long-duration fires leaving the prescribed natural fire zones is reduced in a larger area. It is notable, however, that the Emigrant Wilderness, Hoover Wilderness, Yosemite National Park, Ansel Adams Wilderness, John Muir Wilderness, Sequoia and

TABLE 40.4

Summary of wilderness fire direction in forest plans and NPS fire-management plans.

Wilderness Area	Acres	Forest/Park	Fire-Management Options[a]
Ansel Adams	228,669	Inyo National Forest (INF), Sequoia National Forest (SQF)	INF 1; SQF 2
Bucks Lake	21,000	Plumas National Forest (PNF)	3
Caribou	20,625	Lassen National Forest (LNF)	2, approved interagency plan with Lassen Volcanic National Park (LAVO)
Carson-Iceberg	160,000	Stanislaus National Forest (STF)	2
Desolation	63,475	Eldorado National Forest (ENF)	2
Dinkey Lakes	30,000	Sierra National Forest (SNF)	2
Domeland	94,686	SQF	2
Emigrant	112,191	STF	2, approved plan
Golden Trout	303,287	INF, SQF	INF 1; SQF 2
Granite Chief	25,000	Tahoe National Forest (TNF)	2
Hoover	48,601	INF, Toiyabe National Forest	1
Ishi	41,600	LNF	1
Jennie Lakes	10,500	SQF	2
John Muir	580,675	INF, SNF	INF 1; SNF 2
Kaiser	22,700	SNF	2
Lassen Volcanic	79,000	LNF, LAVO	2, approved plan
Mokelumne	104,461	ENF, STF, Toiyabe National Forest	ENF 2; STF 2; Toiyabe 2
Monarch	45,000	SQF, SNF	SQF 2; SNF 2
Sequoia and Kings Canyon Parks (SEKI)	736,584	SEKI	2, approved plan
South Sierra	63,000	SQF, INF	SQF 1; INF 1
South Warner	70,385	Modoc National Forest	2
Thousand Lakes	16,335	LNF	2
Yosemite National Park (YOSE)	677,600	YOSE	2, approved plan

[a]1 indicates planned ignition only; 2 indicates planned ignition and prescribed natural fire; 3 indicates prescribed natural fire only.

Kings Canyon National Parks, Monarch Wilderness, Jennie Lakes Wilderness, and Golden Trout Wilderness, when grouped, form a 3 million acre unit where consistent prescribed-fire programs could be developed. At the present time, agreements are in place to allow prescribed natural fires to cross agency boundaries between Lassen National Park and the adjoining Caribou Wilderness on the Lassen National Forest and between Yosemite National Park and the Emigrant Wilderness. These agreements have not been used to date.

PRESCRIBED-FIRE PROGRAM AND POLICY

The objectives for application of management-ignited prescribed fire vary between agencies, but the policies, planning requirements, and implementations are very similar. The Federal Wildland Fire Management Policy and Program Review Team (1995) recommends that policy concerning prescribed fire be standardized for all federal agencies. The revised policy statement reads, "Wildland fire will be used to protect, maintain, and enhance resources, and be allowed to function, as nearly as possible, in its natural ecological role."

Each agency requires the completion of a prescribed-burn plan for each prescribed burn. The plans describe quantifi-

able objectives for the burn, the burning prescription designed to meet the objective, the organization that will accomplish the burn, the ignition plan, the holding plan, the mop-up plan, and the contingency plan should the burn escape. The burn plan also describes smoke-management requirements, monitoring requirements, and values at risk.

The effectiveness of the prescribed-fire program in the Sierra is limited chiefly by the scale at which it is currently applied. As an example, table 40.5 shows the extent of recent and planned burning in the Sierra Nevada forests. The extent of burning is negligible when compared to the historic fire regimes. Currently, 20,235 ha (50,000 acres) are burned in the Sierra each year using prescribed fire. Evidence suggests that a much greater area burned yearly under historic fire regimes (Skinner and Chang 1996). Further discussion of the prescribed-burning program and fuels-management strategies is included in Weatherspoon and Skinner 1996.

CDF's Vegetation-Management Program

In 1981 the California Department of Forestry and Fire Protection implemented a vegetation-management program (VMP) on private lands in California. The goal of the program is to reduce large, damaging wildfires by reducing fire hazards on wildlands.

CDF's intent is to realize the best mix of natural resource benefits from these lands, consistent with environmental protection and landowner/steward objectives.

The VMP identifies three broad goals:

1. Reduce conflagration fires.

2. Optimize soil and water productivity.

3. Protect and improve intrinsic floral and faunal values.

The VMP identifies twelve subgoals:

1. Reduce the number and intensity of large, damaging wildfires with corresponding savings of suppression costs.

2. Increase public safety.

3. Increase water quantity and maintain water quality from managed watersheds.

4. Decrease the potential for damage from flooding and siltation.

5. Protect and improve soil productivity, and decrease erosion over the long term.

6. Improve wildlife and fisheries habitat.

7. Improve oak woodlands through fire management and regeneration.

8. Establish and maintain desired plant communities.

9. Propagate rare and endangered plant species that are fire dependent.

10. Improve air quality over the long term.

11. Improve forage and browse for livestock.

12. Increase opportunities for recreation and improve scenic vistas.

The VMP was originally established to reduce fire hazard by treating standing brush. Since its inception in 1981, there have been 61,919 ha (153,400 acres) burned in the Sierra, an average of 4,775 ha (11,800 acres) per year.

The VMP was never intended to replace landowner burning; however, this has been a consequence in some areas. Some private landowners no longer burn vegetation because they would rather let the state assume the liability.

Currently, the VMP is being reviewed with the intent of expanding the program to include fuel types other than standing brush, for example, understory burning. Such expansion would add areas to the program that have not historically been treated. The program may also expand to include methods other than burning to accomplish its goals.

Costs of Prescribed Burning versus Wildfire Suppression

Table 40.6 displays some examples of costs per acre for implementing planned prescribed burns in forests and parks. A discussion and comparison of the costs of various fire-management activities are beyond the scope of this chapter. Prescribed burning, however, is much cheaper than fire suppression, when the two are compared on a per-acre basis. For example, on the Stanislaus National Forest current fire-suppression costs range from $6,400 per acre for fires up to 1 acre in size to a low of $1,000 per acre for fires 5,000 acres or larger. The cost per acre for underburning is $50 per acre. Average cost per acre for suppression of wildfires in Yosemite National Park between 1970 and 1994 was $216 to $358 per acre compared to $19 per acre for prescribed burning and prescribed natural fire during the same period.

Prescribed-burning costs are difficult to quantify because information collection is not standardized. Costs for differ-

TABLE 40.5

Number of acres burned using prescribed fire in 1993 and 1994 compared to planned future acreage per year.

Unit	Acres Burned in 1993	Acres Burned in 1994	Future Acres/Year
Eldorado National Forest	4,267	3,235	7,000
Inyo National Forest	165	365	800
Lassen National Forest	9,193	6,772	not available
Modoc National Forest	2,527	2,781	40,000
Plumas National Forest	5,099	4,443	10,000
Sequoia National Forest	2,452	2,280	11,000
Sierra National Forest	1,035	3,794	6,000
Stanislaus National Forest	8,353	11,587	13,000
Tahoe National Forest	2,725	not available	5,000
Lake Tahoe Basin Management Unit	355	355	1,100
Sequoia and Kings Canyon National Parks	2,851[a]	1,294[a]	16,000–18,000
Yosemite National Park	1,075[a]	3,490[a]	not available
CDF	11,800[b]	11,800[b]	not available
Total	51,897	52,196	

[a]Includes both prescribed fire and prescribed natural fire.
[b]Average figure per year for all CDF areas in the Sierra combined.

TABLE 40.6

Estimated prescribed-burning costs in dollars per acre for 1995.

Unit	Underburning, Dollars per Acre	Burning Piles (Hand Piles and Machine Piles), Dollars per Acre	Broadcast Burning of Slash, Dollars per Acre	Brush Burning, Dollars per Acre
Eldorado National Forest		40–100		
Inyo National Forest		53–111		
Lassen National Forest	205–559	42–124	169–509	50–86
Modoc National Forest	80–180	30–75	170–420	
Sequoia National Forest	229	45		107
Stanislaus National Forest	50	40–110		
Tahoe National Forest	450	60–100	650	
Sequoia and Kings Canyon National Parks	22–356			2.50–52
Yosemite National Park	19[a]			

[a]Average value for all planned prescribed burns for 1982–88.

ent units are not necessarily comparable because they include different things. In particular, planning and prefire survey costs for endangered species or archaeological values can increase costs. Each forest, park, or ranger unit differs in the amount and degree of planning and public involvement needed for the individual project.

FUTURE FIRE-MANAGEMENT POLICY AND PROGRAM OPTIONS

Land managers are struggling to reconcile ecosystem management, which emphasizes the role of natural processes in maintaining healthy ecosystems, with the tremendous success of fire suppression, which has all but eliminated the influence of fire on ecosystems. The National Park Service began its program of natural-process management in 1968 after reassessing its policy of suppressing all fires, at least partially in response to the Leopold report (Leopold et al. 1963). Both the USFS and the BLM are reassessing the role of fire in California's ecosystems through ecosystem-management efforts. Manley and her colleagues (1995) have recommended that fire frequency, intensity, size, and seasonality be used as key environmental indicators of ecosystem health in the national forests of California. Two recent reports have dealt with this emerging dilemma on a national, interagency scale. The USFS recently issued a strategic assessment of its fire-management programs (USFS 1995d) recommending a shift from the traditional focus on fire suppression and control to true fire management. A review of the federal wildland fire-management policy and program, undertaken in light of the severe 1994 fire season, highlighted needed changes in federal fire policy. The report recommends that federal agencies standardize their fire-management policies, taking into consideration the role of fire as an essential ecological process and

natural change agent (Fire Management Policy and Program Review Team 1995).

Possible changes in fire-management programs in the Sierra Nevada fall into two categories: those possible under current policy, especially if additional funding were made available, and changes possible if policy were altered.

Changes Possible under Current Policy

The agencies responsible for fire management in the Sierra Nevada must cooperate to take full advantage of the present flexibility in fire-management policy. Under current policy the prescribed natural fire program could be expanded to all suitable wilderness areas and to many high-elevation areas outside wilderness. Consistent prescriptions and programs across jurisdictions for both prescribed natural fire and planned prescribed fire would reduce perceived risk and cost, because fires would not be suppressed along some jurisdictional boundaries.

The four agencies in the Sierra Nevada have the complementary skills in all areas of fire management needed to implement a more effective overall program. For example, the NPS has the most experience managing and restoring natural processes in Sierran ecosystems. The USFS has the greatest experience using mechanical methods to reduce fuels. CDF has experience protecting private lands and structures. BLM has specialized in rangeland burning. They must work together more closely, especially in the planning phase.

The agencies must also consider the organizational structures best suited to the changing role of fire management. Several recent documents have emphasized the difficulty of linking fire-management objectives to resource- or ecosystem-management objectives, if the fire-management organizations specialize in fire suppression and emergency response at the expense of vegetation management, fuels management, or fire planning.

Forests, BLM areas, and parks could reexamine the opportunities to fully exercise appropriate suppression response. It is essential that fire-management planning be organized in ecological units, which emphasize similar fire regimes. McKelvey and Busse (1996) have displayed the relative increase in fire-suppression effectiveness with elevation in the Sierra. Current presuppression planning takes into account differences in fire regimes by dispatching fewer initial-attack resources to fires in dispatch units with lower fire potential and during periods of low to moderate fire danger. However, this approach has not been extended to application of different strategies with increasing elevation, variation in fire behavior, or different values at risk, except in a few areas.

Changes Possible under Revised Fire Policy

Frustrating for a number of fire managers surveyed is their inability to use wildfire to meet resource-management objectives. The cost-efficiency requirement makes it impossible to allow low- to moderate-intensity fires to burn to significant size, as wildfires. The Federal Wildland Policy and Program Review (1995) approaches this issue by suggesting that "Planning should consider all wildland fires, regardless of ignition source, as opportunities to meet management objectives." Planning documents for all agencies could be revised to prescribe conditions under which wildfires could be used to meet resource objectives, even if fire-suppression costs increased. The basis for applying the proposed policy change is already present in the fire-management plans for the national parks, which contain natural fire prescriptions for most areas. The forests would need to determine the relationship between fire characteristics and resource objectives through landscape-level analysis. One vehicle for such analysis is watershed analysis. The use of wildfire to meet resource objectives is not recommended on private lands, unless the landowner supports the proposal.

Several managers suggested changes in planning methods to take into account the cost of repeatedly suppressing lightning fires in the same watershed, when it could be burned by a single low- to moderate-intensity wildfire at lower cost over time. Small fires on the Stanislaus National Forest cost an average of $6,000 per acre to suppress. Current policy requires that the cost effectiveness of each wildfire be analyzed individually. Again, up-front planning would be needed to contrast the long-term costs and benefits of fire suppression in a watershed.

The risks associated with widespread use of fire throughout the Sierra are daunting, especially given the risks to developed areas. It is essential, however, that fire-management programs are realigned to match suppression strategies and prescribed-burning applications with the known burning characteristics of the different fuel types. The fire-management agencies simultaneously pursue two fire-management objectives, one with the goal of eliminating fire from the ecosystem (fire suppression), and the other with the goal of reintroducing fire in areas from which it has been intentionally eliminated (prescribed fire). In the Sierra Nevada 20,235 ha

(50,000 acres) are burned each year using prescribed fire, at a cost of approximately $5,000,000. The average cost of twenty-six large fires that burned in California in 1994 was $2,920,989 each (USFS 1995c). Five of these fires—the Cottonwood, Hirschdale, Crystal, Big Creek, and Doyle—burned in the Sierra, cost an estimated $27,000,000 (charges do not include costs for mobilization and transport and do not include resource damage), and burned approximately 25,496 ha (63,000 acres) of federal, state, and private land.

To begin to influence fire regimes in the Sierra Nevada, prescribed burning and fuel treatments must be increased by at least five to ten times their current levels. It is essential that the costs of the prescribed-burning and fuels-treatment program be put in clear perspective by assessing their value to Sierran ecosystems and contrasting them to the considerable costs and effects of wildfires that do occur.

REFERENCES

Botti, S. J., and H. T. Nichols. 1995. Availability of fire resources and funding for prescribed natural fire programs in the National Park Service. In *Proceedings: Symposium on fire in wilderness and park management, 30 March–1 April 1993, Missoula, MT*, technical coordination by J. K. Brown, R. W. Mutch, C. W. Spoon, and R. H. Wakimoto, 74–103. General Technical Report INT-GTR-320. Ogden, UT: U.S. Forest Service, Intermountain Research Station.

California Department of Forestry (CDF). 1986. California Department of Forestry and Fire Protection fire plan. Sacramento: CDF.

———. 1995. Fire management for California ecosystems. Unpublished report. Sacramento: CDF.

Cermak, R. W. 1988. Fire control in the California national forests, 1898–1955. Unpublished report. Tahoe National Forest: National Park Service.

Fire Management Policy and Program Review Team. 1995. Federal wildland fire management policy and program review. Washington, DC: U.S. Department of the Interior and U.S. Department of Agriculture.

Fire Management Policy Review Team. 1989. Final report on fire management policy. Washington, DC: U.S. Department of Agriculture and U.S. Department of the Interior.

Leopold, A. S., S. A. Cain, C. M. Cottam, I. N. Gabrielson, and T. L. Kimbal. 1963. Wildlife management in the national parks. *Trans America Wildlife Natural Resources Conference* 28:1–18.

Manley, P. N., G. E. Brogan, C. Cook, M. E. Flores, D. G. Fullmer, S. Husari, T. M. Jimerson, L. M. Lux, M. E. McCain, J. A. Rose, G. Schmitt, J. C. Schuyler, and M. J. Skinner. 1995. *Sustaining ecosystems: A conceptual framework*. San Francisco: U.S. Forest Service, Pacific Southwest Region and Station.

McKelvey, K. S., and K. K. Busse. 1996. Twentieth-century fire patterns on Forest Service lands. In *Sierra Nevada Ecosystem Project: Final report to Congress*, vol. II, chap. 41. Davis: University of California, Centers for Water and Wildland Resources.

Mills, T. J., and F. W. Bratten. 1982. *FEES: Design of a fire economics evaluation system*. General Technical Report PSW-65. Berkeley, CA: U.S. Forest Service, Pacific Southwest Forest and Range Experiment Station.

———. 1988. *Economic efficiency and risk character of fire management programs, northern Rocky Mountains.* Berkeley, CA: U.S. Forest Service, Pacific Southwest Forest and Range Experiment Station.

National Park Service (NPS). 1990a. Fire management plan. Unpublished report. Yosemite National Park: NPS.

———. 1990b. Wildland fire management guideline. NPS-18. Washington, DC: NPS.

Parsons, D. J. 1995. Restoring fire to giant sequoia groves: What have we learned in 25 years? In *Proceedings: Symposium of fire in wilderness and park management, 30 March–1 April 1993, Missoula, MT,* technical coordination by J. K. Brown, R. W. Mutch, C. W. Spoon, and R. H. Wakimoto, 256–58. General Technical Report INT-GTR-320. Ogden, UT: U.S. Forest Service, Intermountain Research Station.

Pyne, S. J. 1982. *Fire in America: A cultural history of wildland and rural fire.* Princeton, NJ: Princeton University Press.

Schmidt, G. 1995. Emergency fire suppression expenditure trends in the Forest Service. In Fire suppression costs on large fires: A review of the 1994 fire season, Appendix A. Unpublished report. Washington, DC: USFS.

Skinner, C. N., and C. Chang. 1996. Fire regimes, past and present. In *Sierra Nevada Ecosystem Project: Final report to Congress,* vol. II, chap. 38. Davis: University of California, Centers for Water and Wildland Resources.

U.S. Forest Service (USFS). 1994a. Fire related considerations and strategies in support of ecosystem management. Unpublished report. Washington, DC: USFS.

———. 1994b. Fire management. In *USFS manual 5100.* Washington, DC: USFS.

———. 1995a. Course to the future: Positioning fire and aviation management. Unpublished report. Washington, DC: USFS.

———. 1995b. Fire economic assessment report. Unpublished report. Washington, DC: USFS.

———. 1995c. Fire suppression costs on large fires: A review of the 1994 fire season. Unpublished report. Washington, DC: USFS.

———. 1995d. Strategic assessment of fire management in the USDA Forest Service. Unpublished report. Washington, DC: USFS.

Weatherspoon, C. P., and C. N. Skinner. 1996. Landscape-level strategies for forest fuel management. In *Sierra Nevada Ecosystem Project: Final report to Congress,* vol. II, chap. 56. Davis: University of California, Centers for Water and Wildland Resources.

APPENDIX 40.1

A Conceptual Model
for Fire Suppression

To understand why simultaneous ignitions can create problems for fire suppression and how a suppression organization can be effective at controlling single ignitions but fail when faced with multiple ignitions, a simple conceptual model is useful.

Fire-resource scheduling models are designed to evaluate the effectiveness of suppression response for specific geographic zones. They are, therefore, complex, with detailed descriptions of resource capabilities and travel times (Mills and Bratten 1982, 1988). They have, however, common structural features:

For input they require

- a list of available resources

- the travel time for each resource to each potential fire location

- the rate at which each unit resource creates fireline in various fuel types

- the assumed fuel type in which fires occur

- weather

When fires occur, in the models,

- They spread at constant rates based on fuels and weather, and the fire perimeter forms an ellipse.

- There are rules controlling the suppression response—which resources are dispatched to the fire.

- The fire is contained when the total length of line created exceeds the fire perimeter.

To be useful for evaluating the effectiveness of a suppression organization, each of these inputs needs to be as accurate as possible, and hence these models are complex and extremely data-bound. For purposes of developing a simple conceptual model, however, it is possible to simplify each of these requirements without altering the basic model form.

SIMPLIFYING ASSUMPTIONS

The model can be greatly simplified by assuming that there is only one type of suppression resource, evenly distributed across the landscape. Travel time is simply the straight-line distance between the resource and the fire multiplied by the rate of speed at which the resource can travel. In addition it is assumed that there is only one fuel type and one weather condition. We will also assume that fire spreads at a constant rate and forms the simplest possible ellipse, a circle. Last, we will assume that all resources are dispatched to the nearest ignition and that resources continue to be dispatched until the fire is contained.

SINGLE IGNITION

Think of this model as a parking lot with people scattered on it. Suddenly a light turns on (a fire) somewhere in the lot, and everyone runs toward it as quickly as they can (they all run at the same speed and don't get in each other's way).

At any time t after the ignition, all resources within a distance r from the ignition will have arrived (figure 40.A1). r is simply the speed (s) at which the resources can travel $* t$, the time elapsed since the fire started. For instance, if the resources can travel toward the fire at 30 mph, then at time $t = 1$ hr, all resources from up to 30 miles away will be at the fire. At $t = 2$ hr, all the resources from up to 60 miles away will arrive, and so on.

If the resources are uniformly distributed on the landscape (one of our simplifying assumptions), then the forces available to suppress the fire at any time $t(S_t)$ will be:

$$S_t = \pi(s*t)^2 d \tag{1}$$

that is, the area of a circle of radius $r = s*t$ times the density (d) of resources per unit area. If $s = 30$ mph and $d = 2$ firefighters per square mile, then at time $t = 1$ hr, 5,654 firefighters could be on the scene.

Because the radius of the response circle gets larger at a

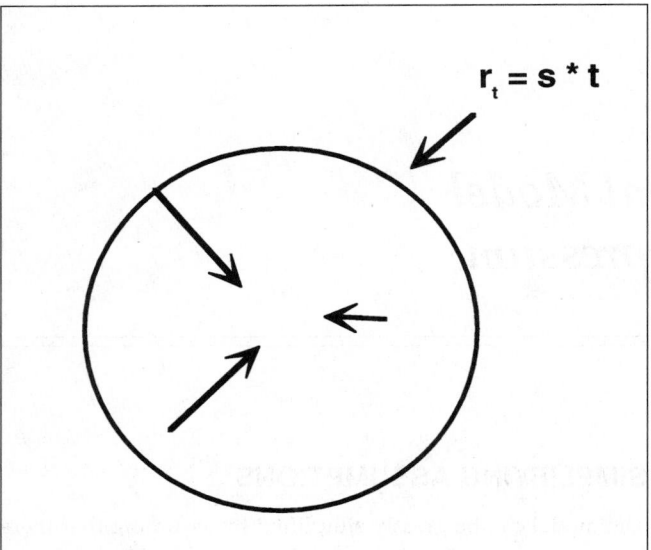

FIGURE 40.A1

The suppression response circle at time *t*. All resources within the area of the circle will have arrived at the fire.

steady rate over time, the number of forces arriving per unit time increases quadratically, leading to a "power curve" in suppression resources at the fire over time (figure 40.A2).

While resources are streaming toward the fire, the fire is spreading according to our simplified rules—rate of spread (ROS) is constant, and the fire expands in a circular manner. Its radius at any time *t*, therefore, is *t*ROS*. The key to suppression is that the resources don't fight the area of the fire, only its perimeter:

$$P_t = 2\pi(ROS*t) \tag{2}$$

where P_t is the perimeter at time *t*. Because this is a linear function of time, whereas our response function is quadratic, if we can maintain our power function in suppression resources, we will eventually control the fire (figure 40.A3). The point at which the suppression-resources curve crosses the fire-perimeter curve is the time at which the fire is controlled (*tc*). The average acreage associated with a fire in a suppression environment is directly related to *tc*. The important variables controlling when *tc* is achieved are, on the suppression side, the speed of response and the density of resources. On the fire side they are the rate of spread and the resistance to control—that is, how many resources are required to control a unit distance of the fire perimeter.

Assume that we are dissatisfied with *tc* and want to shorten the time necessary to achieve it. Should we increase the speed of response (*s*) or increase our resources (*d*)? Looking at equation 1, the answer will always be to increase the speed of response. Increases in *s* are squared, while increases in *d* are not (figure 40.A4).

This model, while simple, captures the basic dynamics of the suppression process. In reality, resources come in clumps—and some are more mobile than others—they have different suppression capabilities, and they are unevenly distributed. But this doesn't change the basic power-curve structure of suppression response. Fire ROS is also not constant, but that doesn't change the basics of fire perimeter growth. And there will be a *tc*. When the fire calms down because of a change in weather, if sufficient resources have been gathered, the fire will be contained.

FIGURE 40.A2

The "power curve" for suppression response to a single ignition.

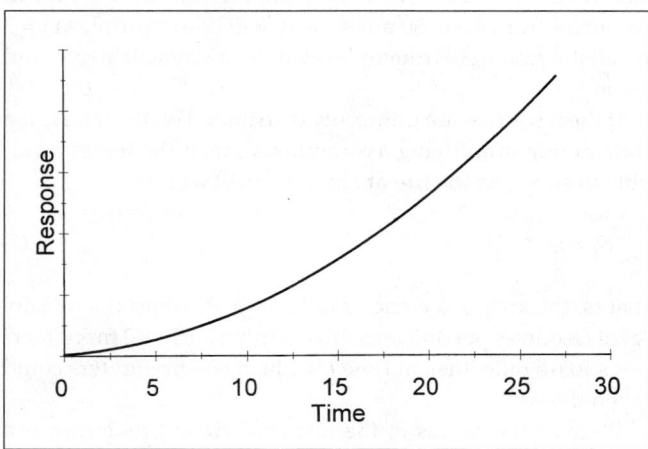

FIGURE 40.A3

Where the suppression-response curve (curved line) crosses the fire-perimeter curve (straight line), the fire will be contained, at time *tc*.

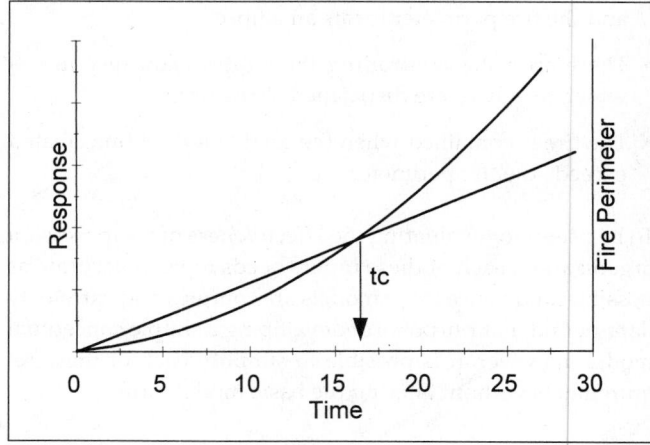

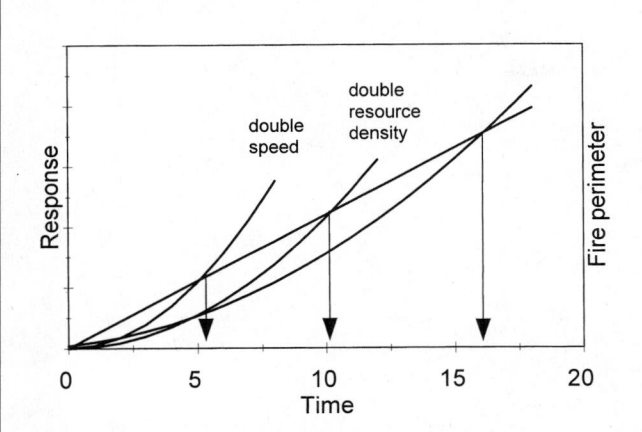

FIGURE 40.A4

Curves showing the change in suppression response associated with doubling the rate of speed at which resources can converge on a fire *(s)* or the number of resources per unit area *(d)*. Arrows point to the *tc*, or time at which the fire is controlled, associated with changes in resource availability.

MULTIPLE IGNITIONS: WHY SUPPRESSION FAILS

If individual ignitions are the expectation, then the optimization will be heavily weighted toward speed of response. Not only is this tactic more effective, but it is generally less expensive than large increases in the resource pool. In many cases (such as by keeping crews fire ready), response time can be shortened at no cost. This happy world begins to come undone, however, when there are multiple ignitions in the same area. Returning to our basic model, assume that *tc* is known, that is, the resource density, speed of response, and ROS of the fire are all fixed. For a single ignition, *tc* occurs when all resources in a circle of radius r_{tc} around the ignition are at the fire. Figure 40.A5 shows the problem. Figure 40.A5 shows a snapshot of four closely spaced ignitions at time *t*. The small, black-outlined circles are the areas whose resources have responded to each strike by time *t*, and at this time there is no conflict: the responses to all fires are still following their power curves. Unfortunately, *tc* hasn't been reached—r_{tc} will require resources from a larger area, and there will be a resource conflict. Resources necessary to achieve suppression will already have been dispatched to the nearest fires. This conflict fundamentally alters the power curve of suppression response. In the worst case, there will be a gap during which

no new resources arrive at the fire. During this period the fire perimeter will continue to grow unchecked.

So, for multiple ignitions, the key to avoiding breakdown is to avoid competition for resources, and to do so the radius of the resource area associated with control should be as small as possible. In figure 40.A5, for instance, if r_c was achieved at time *t*, when the snapshot was taken, there would be no conflict and hence no breakdown. So the optimization for dealing with multiple strikes is very different from the single-strike model. In the single-strike model a small quantity of resources can be very effective if they are mobile enough. For multiple strikes, the density of resources is much more important. It should also be noted that for any suppression organization, regardless of its structure, there will be a point of resource exhaustion. No suppression agency can guarantee that extreme fire events characterized by multiple strikes will not get out of hand.

FIGURE 40.A5

In this scenario there are four fires close together. Each fire requires resources from an area r_{tc} in size to be contained. The resource demands of these fires will overlap and the "power function" cannot be maintained.

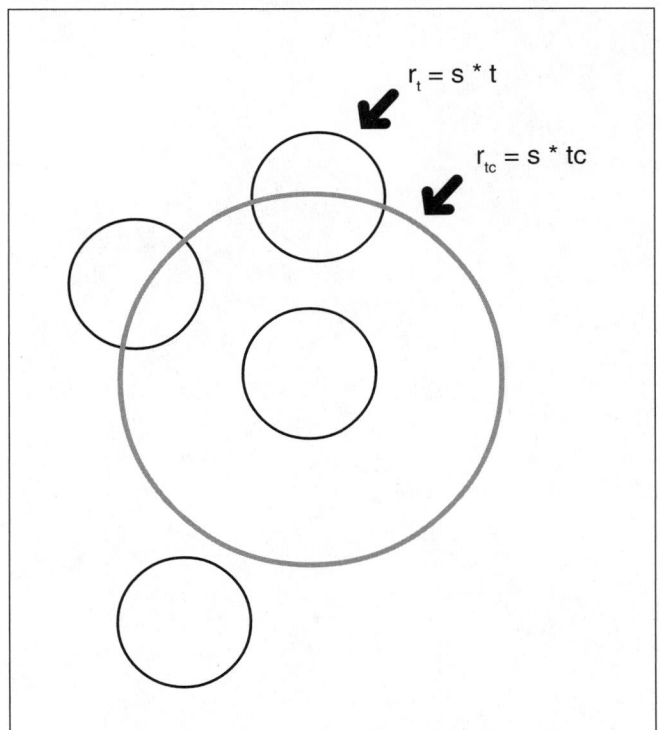

KEVIN S. McKELVEY
U.S. Forest Service
Redwood Sciences Laboratory
Arcata, California

KELLY K. BUSSE
U.S. Forest Service
Redwood Sciences Laboratory
Arcata, California

41

Twentieth-Century Fire Patterns on Forest Service Lands

ABSTRACT

Maps of twentieth-century fires on Forest Service lands were analyzed. Time trends showed no overall trend in acreage, but human-caused fires decreased and lightning fires increased. The increase in lightning fire was dominated by two recent years (1987 and 1990), but more subtle trends prior to 1987 indicated that lightning fires were following a trajectory separate from that of human-caused fires. Landscape-level analysis indicated a strong and stable elevation gradient in burn frequency, and this allowed the development of an accurate descriptive model. An analysis relating fire frequency to vegetation type showed that certain types of vegetation burn more than expected given their elevation, but that burning within these types followed the general trend, with higher-elevation types burning less frequently.

An analysis of reburn patterns showed that, given a particular risk zone, fire location is nearly random. Acreage that burned more than three times had a greater burn frequency than would be expected if the fires were random, but the total area with multiple burns was tiny. The location of multiple-burn sites indicated that they were associated with special features such as busy roads.

Fire correlations with general weather indices were weak, but more area burned in hot, dry years. Perhaps more importantly, all of the extreme fire years occurred when it was hot and dry. Short-term (1979–89) analyses of drought patterns indicated that drought decreased with increasing elevation, paralleling the decrease in fire frequency. Periods of drought were highly synchronized between weather stations, increasing the window during which extreme fire events could occur at all elevations.

The strong and stable elevation trends in fire frequency indicated that future risk could be inferred from twentieth-century fire patterns. The largely random location of fires within each risk zone indicated that a general zonal strategy for fire control would be most effective. The fire acreage patterns over time indicate that while suppression and possibly education can reduce human-caused fires, large lightning fires will continue to occur. Overall risk will vary with weather, but we can expect that large fires will occur during future droughts. If the weather in the twenty-first century is similar to weather in the twentieth century, we might reasonably expect 40% to 60% of the foothills zone to see fire at least once in the next 100 years.

INTRODUCTION

Determining the nature of fire patterns on the landscape is fundamental to both fire ecology and the assessment of fire risk. Critical questions include how often an area can be expected to burn, the size distribution of fires, the impacts of weather and changes in vegetation on fire patterns, and likely fire effects.

In 1994, maps of historical fires on U.S. Forest Service lands were collated to support the development of the California spotted owl environmental impact statement (CALOWL EIS) (USFS 1996). Mapped fires covered a period from 1900 to 1993, with reasonably complete coverage from 1908 to 1992. One forest, the Inyo, didn't report any fires prior to 1960 (Erman and Jones 1996). In all, 2,536 fires, ranging in size from 1 ha (2.5 acres) to more than 50,000 ha (123,500 acres), were individually mapped, digitized, and properly georeferenced. Small fires (less than 1 ha) were not mapped, and therefore these data represent a subset of all fires and only part of the acreage. However, the mapped fires are probably those that are most important biologically, financially, and socially. These are the fires that escaped initial containment and were, at least for a time, uncontrolled.

Sierra Nevada Ecosystem Project: Final report to Congress, vol. II, *Assessments and scientific basis for management options*. Davis: University of California, Centers for Water and Wildland Resources, 1996.

While these maps are not a complete set of all fires, they probably include most of the acreage burned (Strauss et al. 1989), and, because they were made available in digital format, they represent a remarkable and rich information source concerning the nature of fire in the twentieth-century Sierra Nevada.

Utilizing these maps, we have been able to address several important questions concerning fire. In this chapter, we first analyze the general patterns in acreage over time. We then examine the location and frequency of fire on the landscape, including reburn patterns, and lastly we explore the relationships between fire patterns and weather. Because of the large number of analyses associated with these topics, each is presented separately, with the major themes summarized at the end.

Because of data availability, the time period varied with each analysis. For analyses of general fire patterns, all mapped fires (1900–1993) were utilized. If yearly data were used, the time series was narrowed to 1908–92, the years in which coverage appeared to be reasonably complete. Representative weather data were not available prior to 1933, and the daily weather database we utilized extends only to 1989. Most of the analyses involving weather, therefore, extended from 1933–89. Lastly, because the number of reporting stations in the mountains has increased in recent years, analyses of weather changes with elevation utilized the years 1979–89.

EVALUATING ACREAGE BURNED OVER TIME

The number of mapped fires decreased with fire size, following a linear pattern when graphed on a log-log scale (figure 41.1). This pattern, which is similar to fire patterns developed for Southern California (Minnich 1983), means that there are many small fires and a few large fires. The fire acreage was quite variable from year to year and showed no time trend in yearly area burned (figure 41.2; $r^2 = 0.01$, $p = 0.33$). When viewed as a cumulative distribution, total acres burned appeared to increase in spurts in the 1920s, late 1950s, and late 1980s with slower, but rather constant, increases in between (figure 41.3).

While there are no significant trends in overall acreage burned, a number of weak, but significant, patterns were detected. The total number of fires between 1908 and 1992 decreased ($r^2 = 0.09$, $p = 0.005$), and the proportion of the total yearly acreage contributed by the largest fire increased ($r^2 = 0.11$, $p = 0.002$). Neither of these trends was very strong, and neither appears to be linear (figure 41.4). The divergence in fire size, for instance, was strongly influenced by three large fires that occurred since 1940 (figure 41.5). Due to their large size, these fires caused a significant shift in the acreage distribution (figure 41.6).

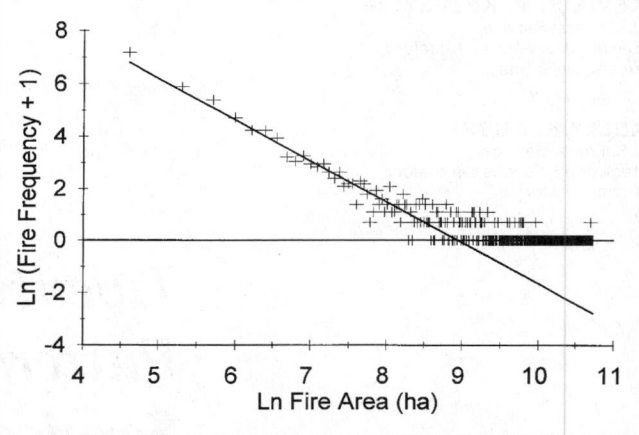

FIGURE 41.1

Number of fires as a function of size on U.S. Forest Service lands. All mapped fires (1900–1993) are included. Both axes are ln transformed.

The relative importance of people and lightning as ignition sources has also changed over time. Over the course of the twentieth century, human-caused fires dominated, both in numbers and in area. Of the 2,536 fires mapped by the Forest Service, 2,046 were caused by humans. This same pattern was reflected in the acreage statistics, with 1,164,439 ha (2,876,164 acres) attributable to human causes and 277,110 ha (684,462 acres) attributable to lightning. Although the number of human-caused fires decreased over time ($r^2 = 0.17$, $p < 0.001$), the dominance of humans as a fire source remained constant for most of the century.

In the late 1980s, however, this pattern changed radically (figure 41.7). In 1987 and 1990, lightning-caused fires domi-

FIGURE 41.2

Yearly area burned between 1908 and 1992. The years 1900–1907 and 1993 are excluded due to incomplete fire records.

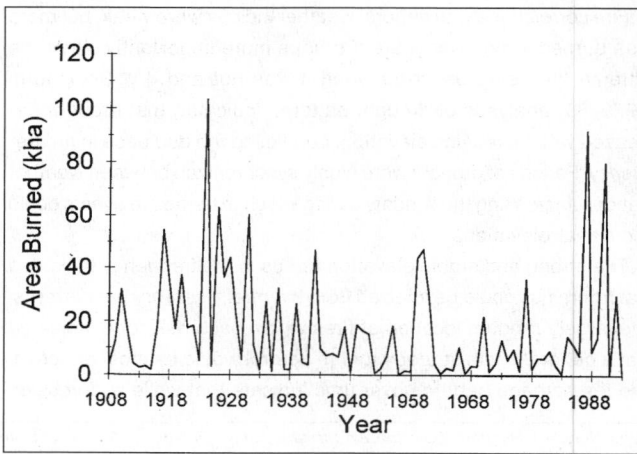

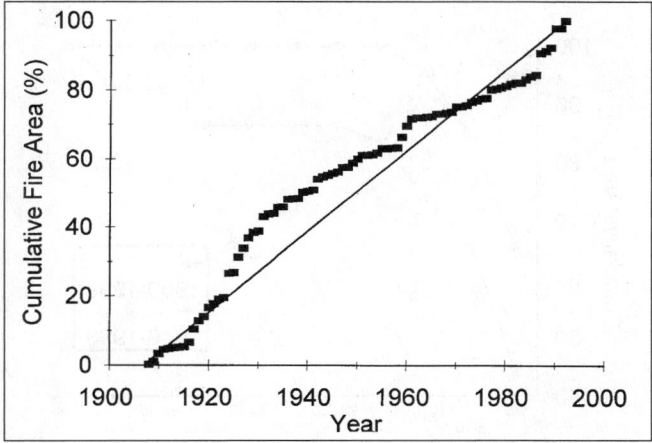

FIGURE 41.3

Cumulative fire area burned between 1908 and 1992. The line represents the rate of increase, assuming a constant area burned per year.

nated the fire acreage. These lightning fires were also unusual due to their size (table 41.1). Of the ten largest fires on record, two occurred in 1987, and both were caused by lightning.

The years 1987 and 1990 were extreme fire years. The fires during these years were caused primarily by lightning (more than 90% of the area burned due to lightning fires) and were enormous in size (table 41.1). Studying a phenomenon in the immediate aftermath of an extreme event, it is difficult to ascertain the significance of that event in a more global sense. One cannot determine directly whether the event was due to random chance or whether it was a portent of things to come.

In the case of the recent lightning fires, however, clues can be found in the longer-term patterns of human- and lightning-caused fires. While the arithmetic means for the areas burned by human- and lightning-caused fires virtually identical (569 ha and 564 ha [1,405 acres and 1,393 acres], respectively) their distributions are different, as is reflected in their median values (105 ha and 39 ha [259 acres and 96 acres], respectively). Most lightning fires were small, but occasionally they were huge (table 41.1). The trends in the size and occurrence of human- and lightning-caused fires over time were also different. When we compared pre-1940 fire patterns (1908–39) with post-1940 patterns (1940–92), we found that the median size of the largest annual human-caused fire decreased by a factor of 2 in the post-1940 period (table 41.2), as did the median number of fires per year (table 41.2). Conversely, the number of lightning-caused fires, while much smaller, remained constant. Interestingly, this statement is still valid if we eliminate the recent large lightning fires (table 41.2).

It appears, therefore, that fire suppression and possibly public education have had a measurable impact on the characteristics of human-caused fires, but not on those of lightning-caused fires. This may be due to differences in ignition patterns

and the necessary response of the fire-suppression organization to lightning and human ignitions. Human-caused fires generally occur as singular events. This allows the fire-suppression organization to respond to individual fires with a large body of fire-suppression resources. Lightning fires, on the other hand, often occur as multiple simultaneous ignitions. In years that are drier than the norm, the amount of resources necessary to deal with simultaneous multiple ignitions can quickly exceed what is available.

These relationships between lightning fires and increased use of resources have been understood for a long time. Because of the 1917 fire season, Show and Kotok (1923) recognized that lightning events have the potential to strain the fire-suppression organization severely. They noted that this was especially true when lightning events were general storms starting multiple fires across the region.

FIGURE 41.4

a, Trends in the number of fires and the acreage burned by the largest fire: the number of fires recorded each year between 1908 and 1992; b, the average proportion of the total yearly acreage burned that can be attributed to the largest fire that occurred that year.

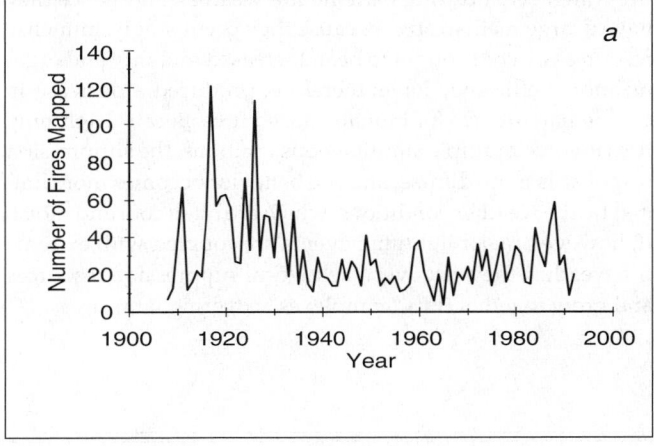

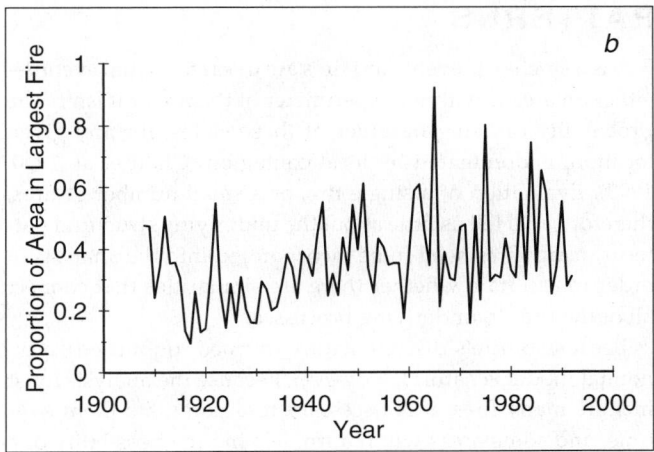

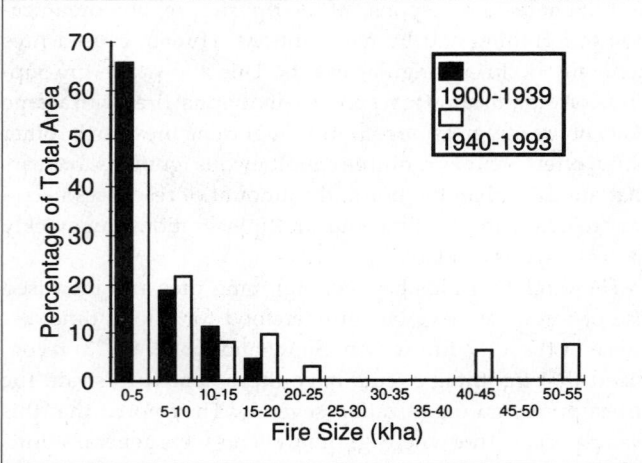

FIGURE 41.5

Proportion of total area burned by fires in different size classes. No fires larger than 20 kha were mapped prior to 1940.

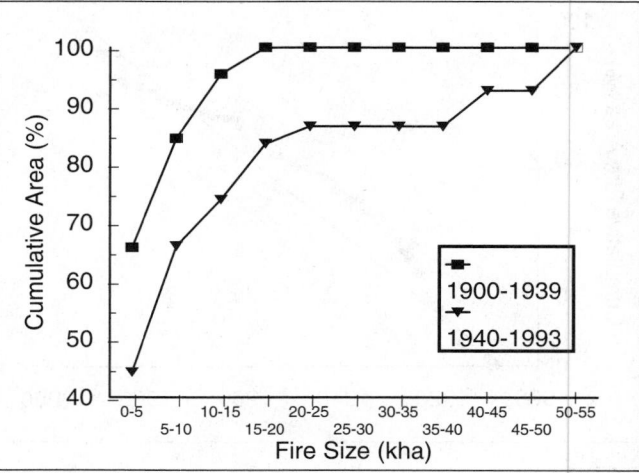

FIGURE 41.6

Cumulative fire area as a function of fire size.

FIGURE 41.7

Acres burned by fires, 1908–92: *a*, caused by humans; *b*, caused by lightning.

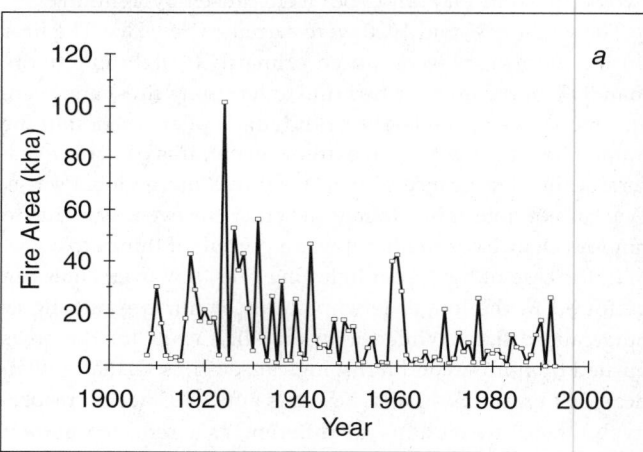

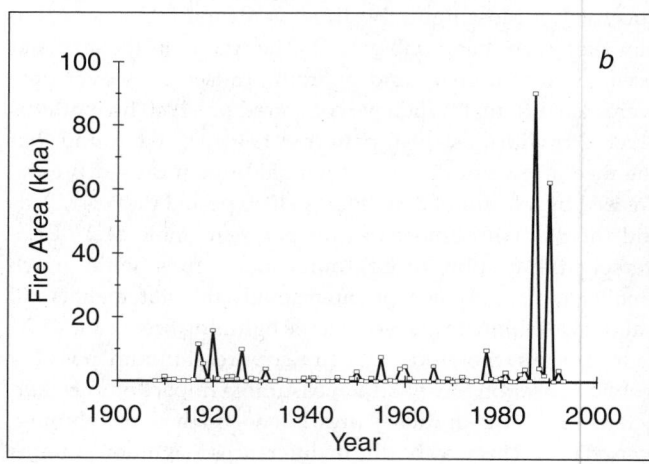

This difference in ignition patterns may explain the observed patterns in human- and lightning-caused fires. Human-caused fires often occur during extreme fire weather, and hence they have a large median size. Because they occur singly, immense resources can be brought to bear. Increased mobility and organizational efficiency have, therefore, produced a decrease in the median fire size for human-caused fires. Because lightning fires involve multiple simultaneous ignitions, the suppression response is more diffuse, and fire behavior becomes more limited by the weather conditions, which are often cool and moist. If, however, major lightning events are coupled with extreme fire weather, they can overwhelm local suppression resources and grow together into "complexes" covering large areas.

EVALUATING LANDSCAPE PATTERNS

Fire is a stochastic event, and fires are discrete spatial events—either an area is within the perimeter of the fire or it isn't. The probability of being in either of these states, for any given location, is dominated by local contagion (Chou et al. 1990, 1993). Evaluation of a single fire, or a small number of fires, therefore, will tell us little about the underlying structural patterns; many fires need to be incorporated into the analysis in order to ascertain whether there are similarities that connect all of the fires to underlying processes.

Because of fire's discrete nature (burned/unburned), a binomial model is natural. However, because the analysis must include many fires, it is necessary to accumulate them over time, and some areas will reburn, leaving the possibility of a multinomial structure (unburned, burned once, burned twice, etc.). For simplicity, we suppressed the repeated burns and categorized the landscape as either unburned or burned at least once. Using this description, we utilized logistic regression to explore fire patterns.

The need to include many fires, covering extensive domains both in space and time, imposed additional constraints. The analyses required independent variables that were spatially extensive, as well as being reasonably accurate and consistent over time.

Obtaining the Available Data

Because the CALOWL EIS fire maps (USFS 1996) were collated for fires on or adjacent to national forest lands in the Sierra Nevada, our initial data set was confined to the area described by these maps (figures 41.8 and 41.9).

When the mapped fires were overlaid on national forest ownership boundaries, however, approximately 40% of the fire area was exterior to U.S. Forest Service (USFS) land. To include acreage exterior to USFS lands, a new map boundary was required. This boundary needed to incorporate as much of the fire area as possible without extending so far beyond USFS boundaries that fire frequency would be controlled by a lack of reporting rather than a lack of fire. To achieve this we arbitrarily buffered USFS land boundaries by 2,000 m (6,600 ft), enclosing approximately 50% of the exterior fire area.

To describe the fire patterns in terms of other descriptive variables, we needed spatially referenced data that covered this same region. Topographic data were available through a 100 m (330 ft) Digital Elevation Map (DEM) compiled by the SNEP Geographic Information System (GIS) center (plate 41.1). Slope, aspect, and elevation were, therefore, available at 1 ha (2.5 acre) resolution across the entire region.

For these analyses, the DEM-derived data were modestly manipulated. Examination of the DEM showed that it was "banded" in the north-south direction (banding is a data artifact common to DEMs). Because both slope and aspect are sen-

TABLE 41.1

The ten largest mapped fires, 1900–1993. All of the large lightning-caused fires occurred in 1987 and 1990. Two of these are the largest fires on record and are more than twice the size of other large fires.

Year	Size (ha)	Cause	Rank
1924	15,055	Human	9
1931	17,715	Human	5
1942	21,234	Human	3
1960	18,100	Human	4
1960	17,057	Human	7
1961	17,459	Human	6
1987	53,011	Lightning	1
1987	16,152	Lightning	8
1990	44,272	Lightning	2
1990	14,508	Human	10

TABLE 41.2

Median values for annual fire data. The period 1940–86 is included here to remove the impact of recent large lightning-caused fires.

	Time Period			
	1908–92	**1908–39**	**1940–92**	**1940–86**
Maximum Annual Fire Size (ha)				
All causes	2,588	4,406	2,199	2,154
Lightning	217	215	217	172
Human	2,525	4,406	1,881	1,881
Number of Fires per Year				
All causes	24	33	21	20
Lightning	4	4	5	4
Human	18	31	16	16

sitive to banding, we applied a 3x3 focal mean function to the DEM to remove these artifacts (Brown and Bara 1994). Aspect and slope were then generated using standard Arc/Info (Environmental Systems Research Institute Inc., Redlands, California) GRID functions.

Aspect is undefined for flat areas (and unreliable in areas of minimal slope) and is a circular statistic. For fire, aspect should be related to potential evapotranspiration (PET), which is a function of both temperature and radiant energy (Campbell 1977). South-facing slopes and flat areas receive the most sun, but the temperature maximum occurs after solar noon. Hence, the peak PET will occur on flat areas and southwest (SW) slopes. Fire studies have found that fire occurrence conforms to these understandings (Agee et al. 1990), with the highest fire frequencies on SW aspects. We therefore used these understandings to develop a simple transform of aspect, using angular distance from south-southwest (SSW) (203°) as a metric. SSW aspects, therefore, were given a value of 0, and north-northeast (NNE) aspects had a value of 180. Flat areas (defined as areas with slopes of less than 10%) also received a value of 0.

In addition to slope, aspect, and elevation, there are significant differences in rainfall between the northern and southern Sierra. To capture these differences, we obtained a digital copy of an isohyetal map developed by the U.S. Geological Survey (plate 41.2) (Rantz 1969). This map covers only the state of California and therefore precluded analysis on those areas where USFS lands, or our buffer, extended into Nevada.

Lastly, we removed all large bodies of water from the map, using the CALVEG (Matyas and Parker 1979) water layer.

Developing a Logistic Function to Describe Burn Patterns

To relate fires to topography and vegetation, Chou et al. (1990, 1993) divided the landscape into irregular polygons based on topography and utilized values associated with each poly-

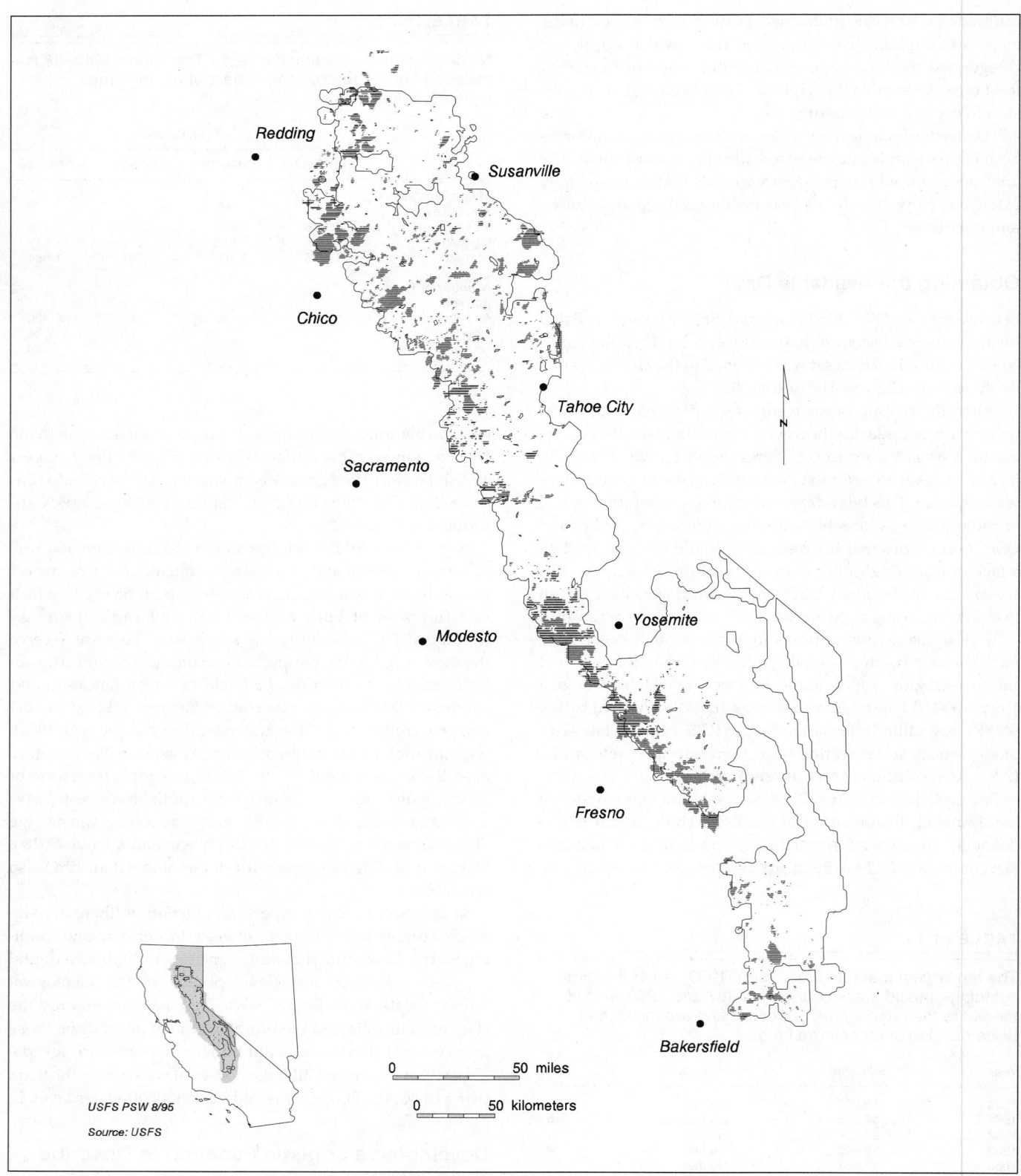

FIGURE 41.8

Map of fires (1900–1939) on and around U.S. Forest Service lands. The analysis area is overlaid on the mapped fires, and any fire area exterior to the study area was excluded from the landscape analyses (adapted from USFS 1996).

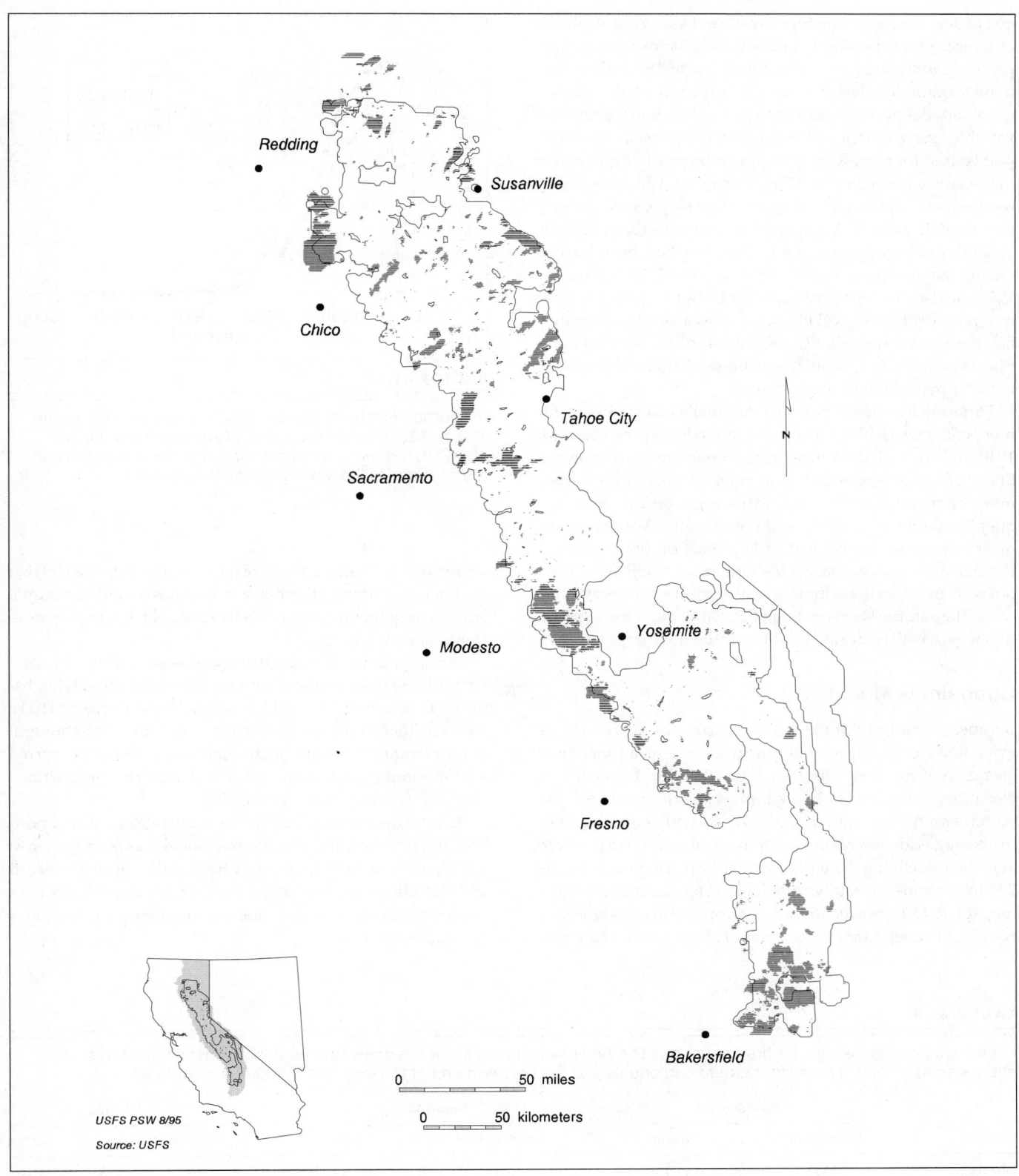

FIGURE 41.9

Map of fires (1940–93) on and around Forest Service lands. The analysis area is overlaid on the mapped fires, and any fire area exterior to the study area was excluded from landscape analyses (adapted from USFS 1996).

gon to develop regression functions. We chose to use attributes of sample points randomly located on the landscape or, more precisely, individual grid cells (1 ha [2.5 acres]) as if they were point locations on the landscape. This approach has a number of advantages over the use of polygons, obviating the need for variables such as area and perimeter. The number of sample points used for regression was quite arbitrary (we had access to the entire landscape of 5.7 million cells). We chose 32,000 random cell locations and obtained their slope, aspect, elevation, rainfall, location (Universal Transverse Mercator [UTM]), and information regarding whether they had been burned during the time periods 1900–1939 (pre-1940), 1940–93 (post-1940), or at all during the time period 1900–1993 (all). For these analyses, 1940 is a logical break point because approximately half the area burned prior to 1940 (figure 41.3). We used the S-Plus (MathSoft, Inc., Seattle, Washington) statistical package to perform the logistic regressions.

Three models were developed, originally using all available independent variables. Fires were grouped into pre-1940, post-1940, and all. In all three models, the dominant factors controlling fire frequency were elevation, slope (steeper slopes burned more frequently), and rainfall (drier areas burned more frequently) (table 41.3). Of these three, elevation was by far the most important. Aspect had little impact on fire frequency. Formal significance tests on the regression coefficients were not warranted, as the sample size was arbitrary. However, because the sample size was large (32,000 points), the expectation was that all relevant variables would have large t-values.

Choosing a Model

Because elevation dominated the regression equation, we evaluated fire-frequency patterns over time as a function of elevation. When the proportion burned within 100 m (330 ft) elevation bands was compared, pre-1940 and post-1940 fire patterns were very similar (figure 41.10), with burn frequency increasing with decreasing elevation to about 500 m (1,650 ft) and then declining. The decrease in burn frequency below 500 m is almost certainly unreliable. An examination of the map (plate 41.1) showed that areas below 500 m elevation were confined to deep inner gorge areas of major rivers and repre-

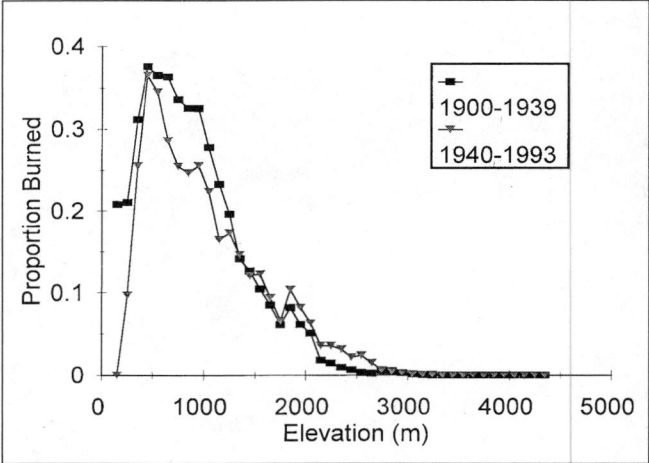

FIGURE 41.10

Proportion of 1 ha (2.5 acre) cells that burned, binned into 100 m (330 ft) elevation zones. Elevations below 500 m (1,650 ft) represent a tiny proportion of the study area and are confined to the inner gorges of major rivers.

sented only 0.7% of the map area. We therefore do not feel that the frequency data for these elevation zones should be extrapolated to the general lower foothills zone, which is largely exterior to our study area.

The cumulative acreage distributions were also very similar, with the median elevation for each distribution lying between 1,000 m and 1,500 m (3,300 ft and 4,950 ft) (figure 41.11). Based on these data, it is reasonable to state that fires throughout the twentieth century in the Sierra Nevada have occurred within similar elevation zones and that the proportions burned have not varied appreciably.

Because there is very little difference between pre- and post-1940 fire patterns, and because the model was stronger when all fires were included, a model utilizing the entire fire record and including elevation, slope, and rainfall was chosen as an appropriate model to describe fire frequency in the Sierra Nevada (table 41.4).

TABLE 41.3

Logistic regression models for fire frequency. The All Fires columns show the acres that burned at least once across the entire period. Row is a proxy for latitude and equals 1 at the northern end of the map and 6,400 at the southern end.

	Pre-1940		Post-1940		All Fires	
	Coefficient	t-value	Coefficient	t-value	Coefficient	t-value
Intercept	7.045E-1	7.6	7.467E-1	8.4	1.785	23.8
Elevation	−1.971E-3	−49.8	−1.560E-3	−44.2	−1.922E-3	−61.1
Slope	1.470E-2	5.4	2.987E-2	11.7	2.605E-2	11.8
Rainfall	5.769E-4	1.2	−7.581E-3	−16.1	−4.250E-3	−11.5
Aspect	2.200E-4	0.6	−1.055E-3	−2.8	−6.719E-4	−2.1
Row	−2.026E-5	−1.5	2.644E-5	2.1	8.021E-6	0.8

Testing Model Fit

We tested the fit of the model by binning the landscape into deciles of risk based on model predictions (Hosmer and Lemeshow 1989, 142). Hosmer and Lemeshow (1989) present a statistic closely related to X^2 to evaluate the quality of this fit, but, as we mentioned before, our sample size is arbitrary and significance tests are artificial. However, because we had access to the entire population of cells, we directly assessed the quality of fit by using the following method:

1. For each grid cell on the map (approximately 5.7 million locations), we computed the model estimate of fire probability and assigned the cell to a decile based on this estimate.

2. For each decile, we accumulated the model estimates, the number of cells which burned, and the total number of grid cells.

We were then able to compute the ratio of burned to unburned cells (the actual frequency for that decile) and compare it with the model estimate,

$$M_j = \frac{\sum_{i=1}^{t_j} E\left(y \mid e_{i,j}, s_{i,j}, r_{i,j}\right)}{t_j} \qquad (1)$$

where M_j is the model estimate for decile j, t_j is the total number of cells in the decile, and $e_{i,j}$, $s_{i,j}$, and $r_{i,j}$ are the elevation, slope, and rainfall evaluated for cell i in decile j.

This test is a strong validation of the quality of the model as a descriptor of historic fire frequency because the results

TABLE 41.4

Logistic regression model based on all fires and excluding aspect and row. This model, with the provision that model estimates would be truncated at 60%, was used to generate a fire-frequency map for the Sierra Nevada.

	All Fires	
	Coefficient	**t-value**
Intercept	1.803	27.8
Elevation	−1.925E-3	−61.2
Slope	2.436E-2	13.0
Rainfall	−4.375E-3	−13.1

are an exact statement of the relationship between the model and the surface it is supposed to represent. For purposes of developing a fire-risk map, the test is particularly appropriate: For a particular area of the map, if the model predicted a particular fire frequency, how close to this estimate was the actual historical fire frequency evaluated within that same area?

This test was performed for all fires, using the final model (table 41.4). The match was very close for deciles 0–10 through 50–60 (figure 41.12), but the model overestimated fire in the higher fire-frequency deciles. The total acreage in these higher deciles, however, was tiny (figure 41.12), and the model was therefore insensitive to these divergences. To correct this problem, we simply truncated the model estimates at 60% (in essence creating a 60+ bin). When this was done, the model estimates conformed to the measured frequencies very closely (plus or minus 2.5%) for all deciles (figures 41.13 and 41.14).

Testing model fit in logistic regression is generally done at the decile level, probably because small sample size precludes

FIGURE 41.11

Cumulative acreage burned, by elevation. The arrows point to the median elevation (half the fire area is below, half above this point) for each curve.

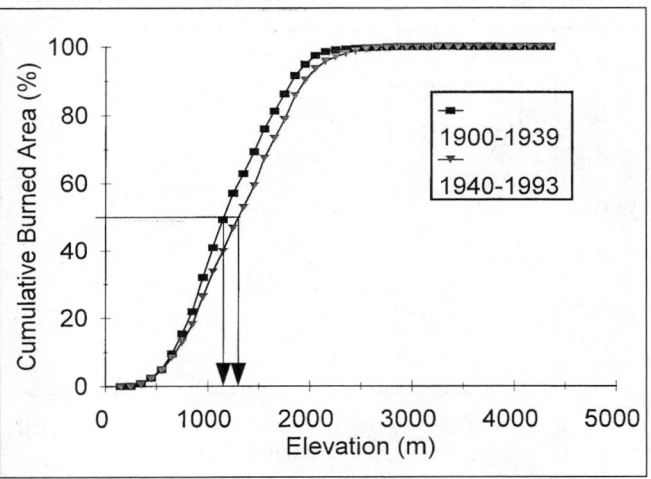

FIGURE 41.12

A test of model fit to actual burn patterns based on risk deciles (10% bins). The model overestimates fire probability for high-risk zones.

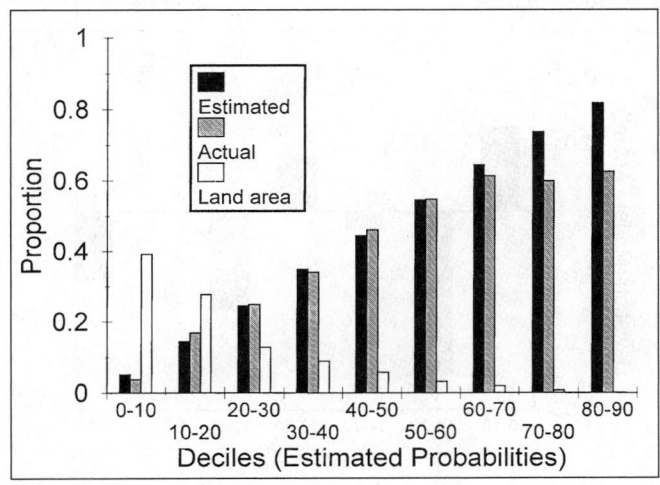

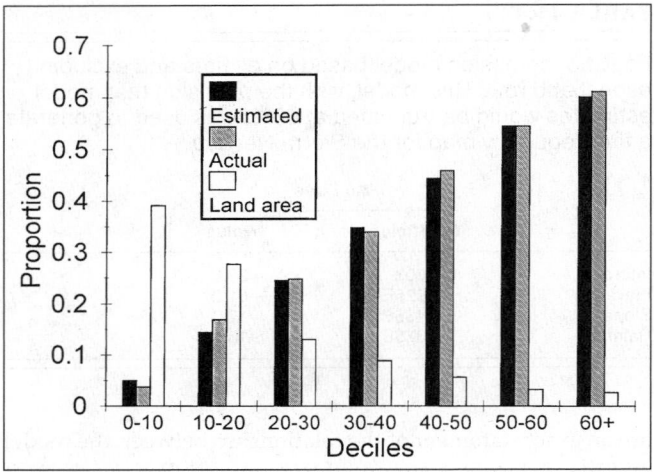

FIGURE 41.13

A test of model fit to actual burn patterns based on risk deciles (10% bins), but with model estimates greater than 0.6 combined into a single bin.

finer groupings. Here, however, we were able to test the model at a finer scale. When binned into groups of 1% (centiles), the model fit was still reasonable (figure 41.15). Deviation between model estimates and measured fire frequencies was within 4% for all centiles.

Developing a Risk Map

Because the model accurately described the probability of fire, we generated a generalized fire-probability map by comput-

ing the model estimates for each grid cell (plate 41.3). Because the fit to historical patterns was good, and because these burn patterns remained remarkably constant over time, we feel that it is reasonable to use this map to evaluate relative fire risk in the near future for specific locations.

Estimating Fire Frequency Based on Vegetation Types

Vegetation was excluded from the logistic regression analysis just described for a number of reasons, the most important of which is that vegetation is not static on the landscape. An area currently mapped to a vegetation type may not have been that same type when a fire occurred. In fact, the area may be in its current condition because of its past fire history. Vegetation may, however, be less abstract than a multidimensional topographic model, and the burn patterns linked to vegetation types may give insights that are missed in the topographic analysis.

The vegetation map covering the analysis area available to us was the CALVEG map (Matyas and Parker 1979). When we looked at fire patterns within each CALVEG vegetation stratum, we found that the strata with the highest proportions burned were generally associated with the west-side foothills (table 41.5). The highest rates of burning occurred in the interior live oak and blue oak types, with the next two highest levels in chaparral types. The fire frequencies associated with these types were at or above (mostly above) the fire frequencies associated with the arithmetic-mean elevation of each type (figure 41.16). The elevation trend for these fire-prone strata, however, followed the general elevation trend—strata with higher mean elevations burned less frequently (figure 41.16).

FIGURE 41.14

Deviation between the model and the actual fire distribution for each risk decile (10% bins). This is not a statistical measure but rather an exact statement of the divergence between model expectations and the patterns upon which the model is based.

FIGURE 41.15

Deviation between the model and actual fire distribution for risk centiles (1% bins). The actual proportion burned is compared with weighted model estimates for each centile.

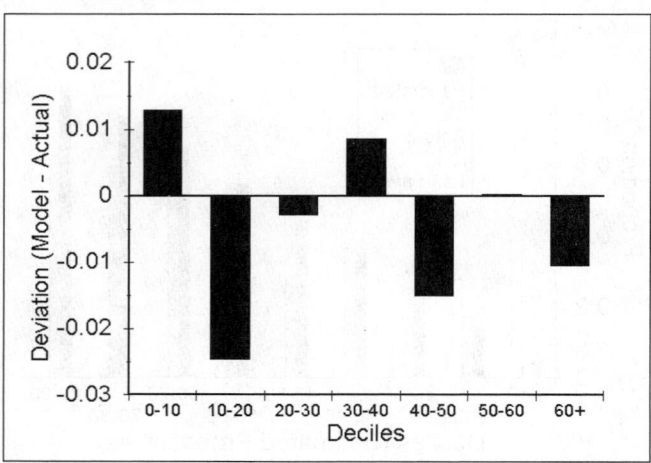

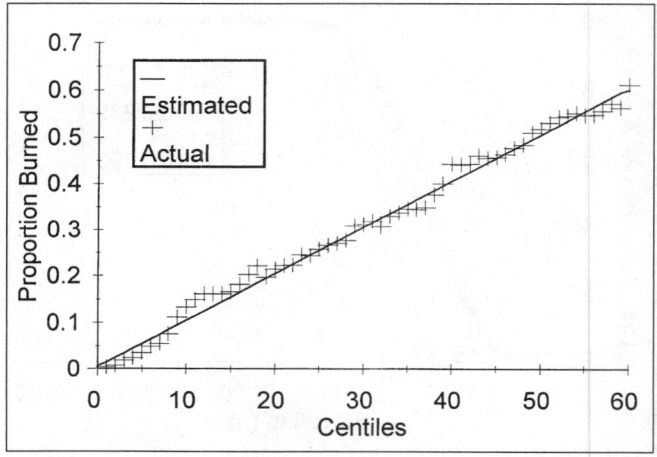

TABLE 41.5

Fire frequencies for selected CALVEG vegetation strata in the Sierra Nevada. Strata larger than 10,000 ha and with more than 30% of the strata burned at least once were selected. If the estimated fire rotation exceeds the measured one, the strata reburn more frequently than would be expected based on random fire patterns.

Strata name	Area (ha)	Area Burned[a] (ha)	Proportion Burned[b]	Measured Fire Rotation[c] (Years)	Estimated Fire Rotation[d] (Years)
Interior live oak	21,673	16,509	0.76	57	59
Blue oak	84,417	59,092	0.70	78	70
Chamise	111,307	64,389	0.58	114	98
Whiteleaf manzanita	91,031	42,464	0.47	106	133
Ponderosa pine	332,974	112,869	0.34	192	205
Mixed conifer–pine	699,740	235,251	0.34	185	208
Canyon live oak	102,445	32,222	0.31	189	223

[a]Area burned at least once based on the CALOWL EIS fire map.
[b]Proportion burned at least once based on the CALOWL EIS fire map.
[c]Based on total area burned/total strata area, for the period 1908–92.
[d]Based on the Poisson expectation, λ/t, where $t = 85$ years.

EVALUATING REBURN PATTERNS

The Poisson Model

If planar shapes (disks, for instance) are randomly scattered on a surface, the areas of overlap will approximate a Poisson distribution (see Horn 1971, chapter 5):

$$P(k,\lambda) = e^{-\lambda}\frac{\lambda^k}{k!} \qquad (2)$$

where k is the number of disks that overlap on a specific location and λ is combined projection area of all of the disks. Given $P(0,\lambda)$, the probability that a site is not covered by any disks, equation 2 simplifies to

$$P(0,\lambda) = e^{-\lambda} \qquad (3)$$

and can be solved for λ by taking logarithms.

This approach can be used to test for the random placement of fires (Agee 1993). If fires occur randomly on the landscape, the reburn patterns will follow Poisson expectations.

We know, based on the logistic regression described previously, that in a global sense fires are not randomly distributed on the landscape. Slope, elevation, and rainfall determine their likelihood. Hence, globally we can reject the expectation of randomness. Of more interest, given that there is a zone of equal fire probability (that is, $P(0,\lambda)$ is constant), is the question, Is the reburn pattern random within that zone? This question is important both for ecological reasons and for controlling fire. If reburn patterns deviate significantly from a random pattern, we can use the specific fire history for a site to modify local risk assessment.

The logistic regression described earlier developed fire-expectation functions based on whether a site had never burned in the twentieth century or had burned at least once. The frequency-of-burning statistics produced by the regression are, therefore, equivalent to $(1-P(0,\lambda))$. Because the model does a good job of predicting $(1-P(0,\lambda))$, we can use the model to bin

the map into equal risk zones (the deciles in figure 41.13) and compute $P(0,\lambda)$ directly for each bin as fraction of cells in the bin which did not burn (based on the "actual" bars in figure 41.13). Solving for λ, given $P(0,\lambda)$, we can then use equation 2 to estimate $P(1,\lambda) \ldots P(n,\lambda)$ where $(1..n)$ is the number of times that a site reburned during the time period and assuming a random spatial location for all burned cells. We can then compare the actual distributions of reburns and determine whether they conform to Poisson expectations.

If sites burn more frequently than expected, estimates for $P(1,\lambda)$, and possibly $P(2,\lambda)$ in high-risk areas, will be lower than the measured frequencies, and $P(3,\lambda) \ldots P(n,\lambda)$ will be higher. If sites reburn less frequently than the random model, this pattern will be reversed.

FIGURE 41.16

All of the CALVEG classifications for which more than 30% of the area burned at least once, based on all mapped fires. The bars are placed at the arithmetic mean elevation for each vegetation type, rounded to the nearest 100 m, and the line is the proportion of the landscape that burned at that elevation.

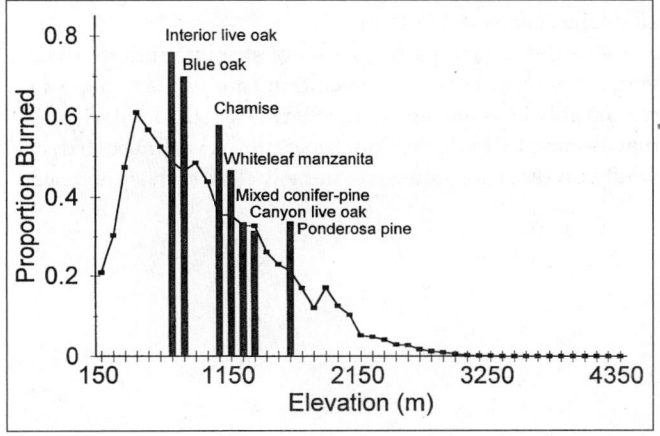

Evaluating Reburn Patterns

Based on this logic, we broke the landscape into deciles of fire risk based on the logistic regression described previously (figure 41.13), counted the proportion of cells in each bin that had burned 0..n times, computed λ and the expected proportions (1..n) based on the Poisson distribution, and compared the measured reburn patterns with Poisson expectations. A X^2 test is often used to evaluate fit to the Poisson distribution (Feller 1968). In this case, however, this test is inappropriate due to the large and arbitrary size of the sample (see Wonnacott and Wonnacott 1977, pgs. 506–507). The actual n is related to the number of fires (2,536) and their acreage distribution rather than to the number of sample points. However, determining the relationship of these fires to individual decile bins was difficult because individual fires crossed multiple bins. We therefore present these data without statistical measures of fit. As was the case with the validation of the logistic regression, these relationships are exact statements of the relationship of a predictive model (in this case the Poisson distribution) to the measured landscape patterns.

Figure 41.17 presents the results of these comparisons. $P(0,\lambda)$ is directly computed from the map frequency and, therefore, is always exact. In general, the distributions conform reasonably closely to the Poisson distribution. The Poisson model overestimates $P(1,\lambda)$ and underestimates $P(3,\lambda)$ to $P(8,\lambda)$ for all deciles—areas reburn more frequently than would be expected if the fires were random. For the highest risk decile, $P(2,\lambda)$ is also overestimated. A reasonable way to simplify these patterns is to look at the fire rotation within each decile. This can be computed directly as

$$R_d = \frac{t}{\sum_{k=1}^{n_d} k p_{k,d}} \qquad (4)$$

where R_d is the fire rotation for decile d, k is the number of times a site was burned, $p_{k,d}$ is the proportion of decile d that burned k times, and t is the elapsed time. This statistic is sometimes referred to as the natural fire rotation (Agee 1993, 100). R_d can also be estimated from the Poisson distribution as λ/t (Feller 1968, 152–53). λ/t overestimates fire-return periods for all decile classes (table 41.6).

While the reburn patterns are not strictly random, the divergences from Poisson expectations are modest and affect remarkably little acreage in the Sierra (figure 41.18). A model that assumed simple random fire patterns within each decile would produce fire patterns extremely close to those observed.

Possible Explanations for the Observed Reburn Patterns

The consistency of divergence from Poisson expectations for all reburn classes and all bins makes it unlikely that this deviation is due purely to chance. This suggests that areas that have a record of frequent fires in the past are somewhat more likely to burn again than areas of similar elevation, slope, and rainfall that have not burned. A number of potential processes could account for these patterns. Previous fire could make a site more flammable through changes in vegetation and fuel loadings. An alternate explanation is that flammability remains constant and that some locations have higher ignition rates. Due to the small areas involved (of the 1.4 million ha that burned, 3 ha [7.5 acres] burned nine times), it is difficult to determine which explanation might be more correct, but an examination of the maps suggests the second. Areas that burn repeatedly appear to be ones that we would identify as high-risk sites (low elevation, steep slopes) that are adjacent to major highways. The canyon wall on the north side of the Merced River (plate 41.4) is a good example. Not only is this area identified as being in a high-risk zone, but a major highway into Yosemite National Park (State Highway 140) runs along the base of the high reburn zone.

Interestingly, there is no evidence that areas that have burned in the past are less likely to burn in the future, as might be the case in a fuel-limited system (Minnich 1983). All of the calculated reburn intervals are, however, longer than eighty years (table 41.6), and it is probable that, for fuels to limit fire acreage, fires would need to burn far more frequently than the observed pattern. This is an important understanding, because the area burned in the twentieth century is not insignificant. The total fire acreage in the CALOWL EIS (USFS 1996) fire map is 1,441,549 ha (3,562,068 acres). In the two zones with the highest fire frequency (more than 50% burned this century), 177,050 ha (437,313 acres) have burned during the period for which we have records, producing 3 ha (7.5 acres) that burned nine times and 116 ha (286 acres) that burned eight times. If we had systematically burned this zone on a ten-year cycle in order to reduce fuels, we would have burned approximately 2.6 million ha (308,299 ha × 8.5) (6.4 million acres) over the same time period—more than fifteen times the amount burned in wildfires.

Another interesting aspect of these analyses is that while, in a broad sense, fire patterns are described by elevation and climate, the patterns of fire overlap are very complex on any smaller scale (plate 41.4). Because the overlap patterns appear to conform reasonably well to a random model, we conclude that most of the overlap within a specific risk zone is due to chance. This provides a cautionary note to descriptions of "average" fire return evaluated either on a site-specific basis or on a limited and nonrandom sample. The expectation is that many sites will have burned more frequently (as well as less frequently) than expectation due simply to chance.

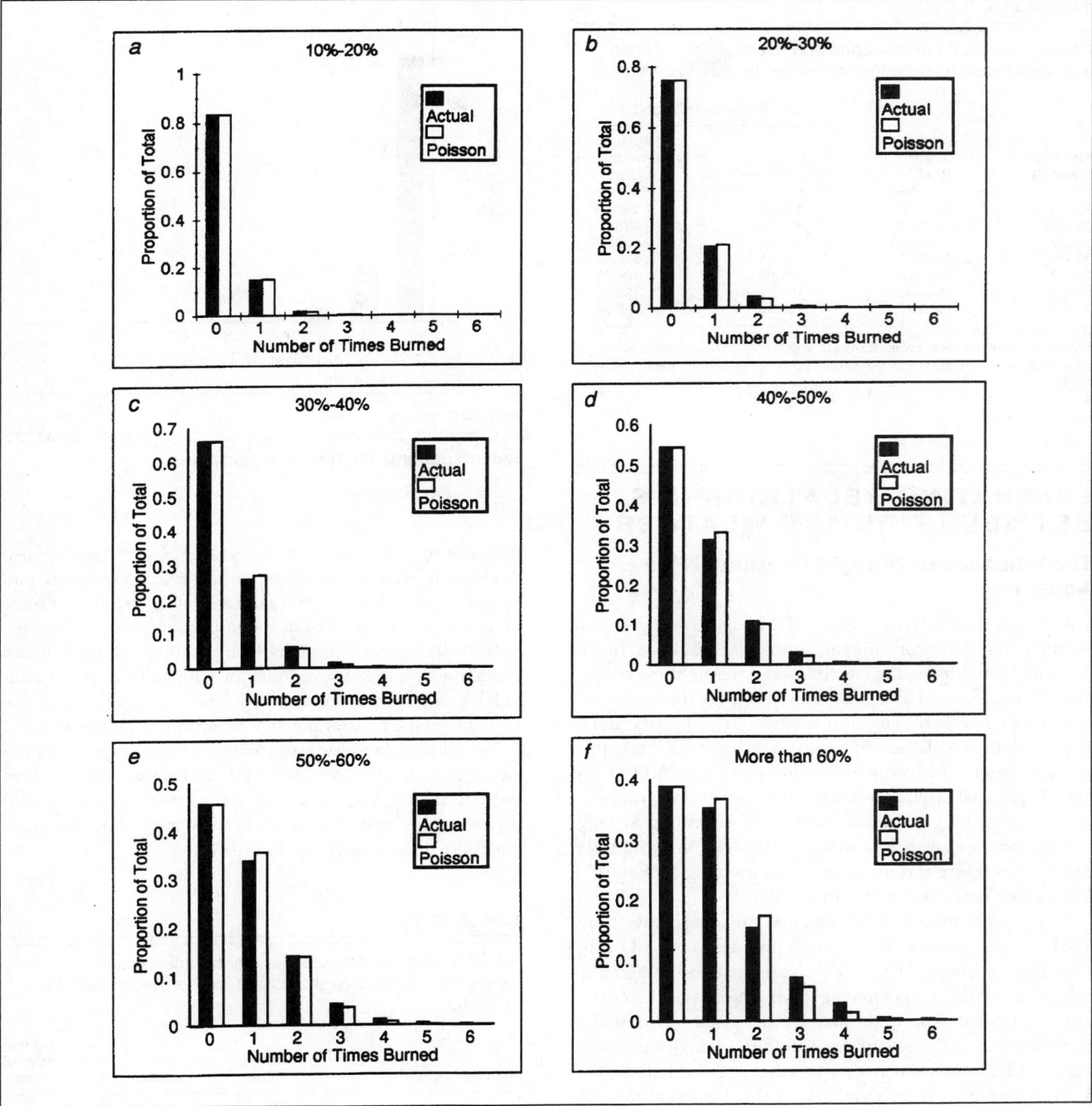

FIGURE 41.17

Reburn patterns for risk decile zones identified through logistic regression (figure 41.13). Poisson expectations for burning 1..*n* times during the time period (1900–1993) are derived directly from fitting the I value to the proportion that did not burn. The "0" bars therefore always match exactly.

TABLE 41.6

Hectares in each fire-risk zone and average fire-rotation intervals based on mapped fires in the twentieth century.

Fire-Risk Class (%)	Area (ha)	Fire-Return Period	
		Actual Frequencies	Poisson Approximation (t/λ)[a]
0–10	1,913,476	2,429.28	2,498.55
10–20	1,440,833	451.21	469.00
20–30	678,572	288.90	303.54
30–40	454,085	194.54	205.81
40–50	295,206	131.92	139.72
50–60	167,508	101.33	108.57
60+	140,791	82.80	89.88

[a] t is the amount of time elapsed, 85 years; λ defines a Poisson distribution (see equation 2, earlier) where in this case k is the number of times a site was burned.

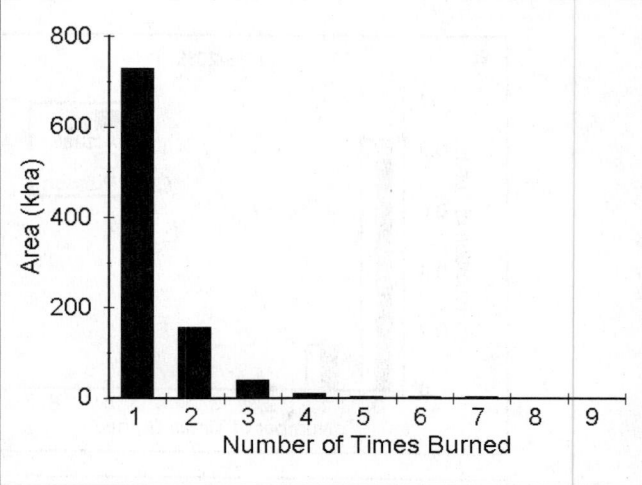

FIGURE 41.18

Reburn patterns for the entire study area.

EVALUATING RELATIONSHIPS BETWEEN FIRE AND WEATHER

The Influence of Drought on Annual Fire Acreage

The overall fire patterns (figures 41.2 and 41.3) suggest that weather may have an influence on acreage burned. In the twenties, and thirties, the late fifties and the eighties—all periods that we would associate with drought—there were increases in fire acreage (figure 41.3). To quantify these relationships we utilized historical weather data (daily precipitation and temperature) from areas near the SNEP study area. The available data set was weather station data compiled by EarthInfo, Inc., of Boulder, Colorado, extending to 1989. Fourteen stations were reasonably close to the Sierra and had daily temperature and moisture data extending back to 1933 (fifty-seven years) (table 41.7; figure 41.19).

To compare these data with annual fire acreage, we averaged the precipitation and temperature data over all of the sites. Because several of the stations are very close geographically, some of them were averaged separately and their average was treated as a single station entry. This was done for Sacramento and Davis, and for Hanford, Orange Cove, and Visalia. Many daily records are missing in this database. For a given station, if more than 10% of the data were missing from the weather record, it was not included in the average for that year or season. To compute yearly or seasonal estimates of rainfall and temperature, we summed daily precipitation and high-temperature readings for each station.

Attempts to correlate fire with annual patterns of precipitation and temperature produced no significant patterns. Because annual precipitation is dominated by winter storms that may not affect fires directly, we repeated the analyses with weather data restricted to the months March through October. Using these data, we found significant relationships between temperature and precipitation (table 41.8). Fire acreage was positively correlated with cumulative seasonal temperature and negatively correlated with seasonal rainfall. To build a simple index comparing both of these understandings, we normalized the values for cumulative temperature and rainfall and created a simple drought index of temperature minus rainfall. We regressed this index on the natural log of fire acreage and obtained a modest positive relationship (figure 41.20).

The relationship was not particularly strong, but yearly fire patterns probably have a strong random component. All of the extreme fire years occurred during years that we would classify as hot and dry, but there were many hot, dry years with very little fire activity (figure 41.21).

TABLE 41.7

Stations used to compute seasonal precipitation, temperature, and drought indices for the Sierra Nevada, 1933–89.

Stations Chosen	Starting Year	Elevation (m)	Annual Rainfall (cm)
Chico University Farm	1906	58	64
Davis Experimental Farm	1917	18	45
Hanford	1927	76	21
Hetch Hetchy	1931	1,180	89
Independence	1927	1,204	13
Modesto	1931	27	31
Nevada City	1931	847	140
Orange Cove	1931	131	32
Red Bluff	1933	103	58
Sacramento	1877	24	47
Sonora	1931	533	80
Susanville	1931	1,265	37
Tahoe City	1931	1,898	81
Visalia	1927	101	26

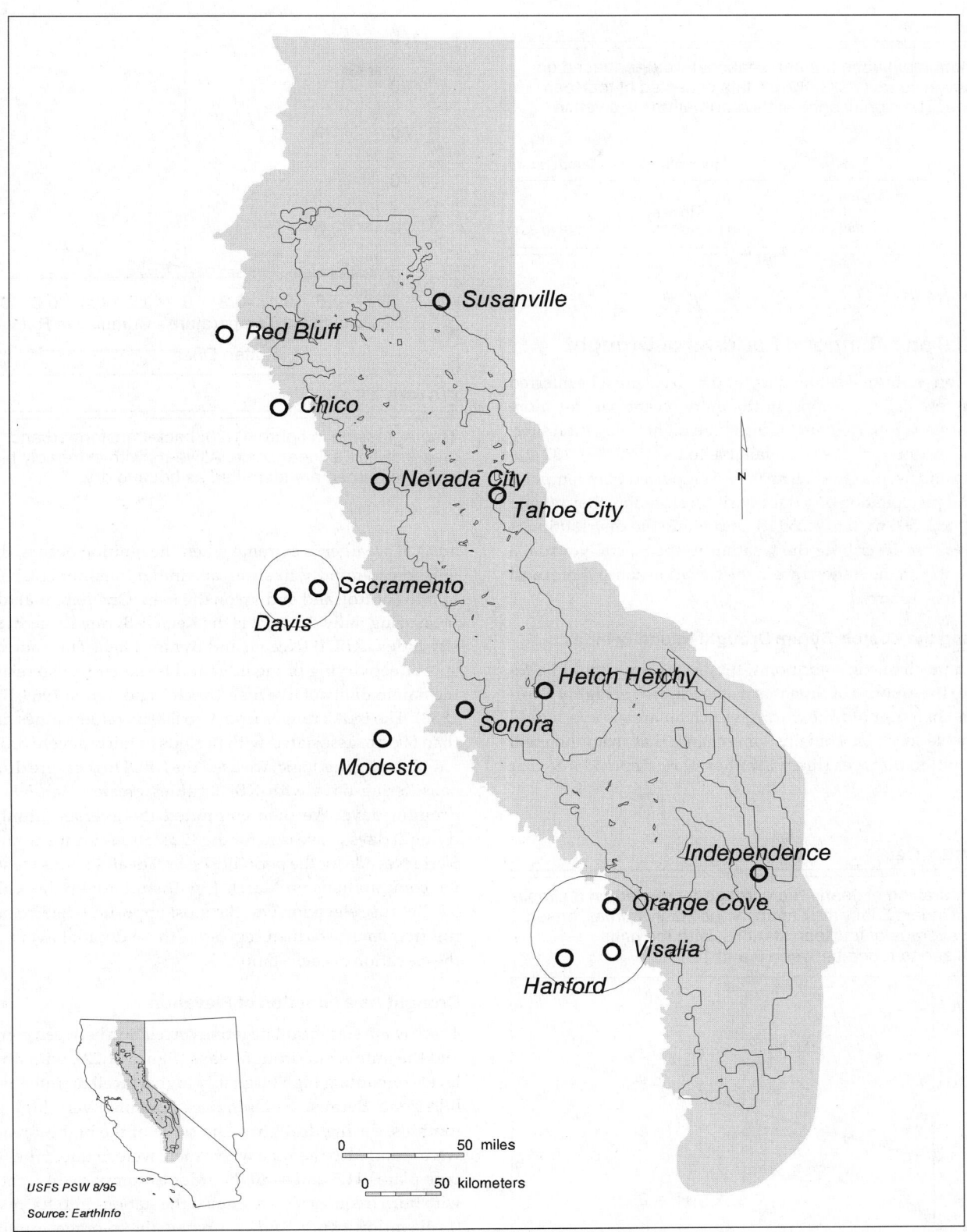

FIGURE 41.19

Locations of fourteen stations for which daily weather data (temperature and precipitation) were available for the period 1933–89. Data from circled stations were averaged and treated as a single station when developing yearly averages.

TABLE 41.8

Correlation between fire and seasonal weather based on fifty-seven years (1933–89) of data collected at fourteen stations. The significance of the correlations is given in parentheses.

	Fire	Rainfall	Temperature
Rainfall	−0.360 (0.006)		
Temperature	0.350 (0.008)	−0.255 (0.056)	
Year	0.042 (0.754)	0.041 (0.759)	−0.126 (0.351)

Spatial and Temporal Patterns of Drought

While longer-term weather data for the Sierra must be inferred from a few valley stations, in the more recent past far more stations have reported, including some higher-elevation sites. For the most recent decade available to us (1979–89), 132 stations in and around the Sierra Nevada reported both temperature and precipitation on a daily basis. In elevation, they ranged from 0 to 2,500 m (0 to 8,250 ft), and while the distribution is not ideal for describing the weather in the Sierra Nevada, a fair number of the stations are in the elevation zone where most of the fires occurred.

Utilizing the Keetch-Byram Drought Intensity Index

From a mechanistic standpoint, fire frequency should be related to the amount of time that a specific area is highly flammable. Each unit of time during which an area is in a highly flammable condition contains some chance that an ignition will occur within that area (fire will either start or spread into that

FIGURE 41.20

The natural log of yearly fire acreage regressed on a simple drought index: Σdaily max temp minus Σdaily rainfall, based on the average of fourteen stations, with the units normalized to range between 0 and 1.

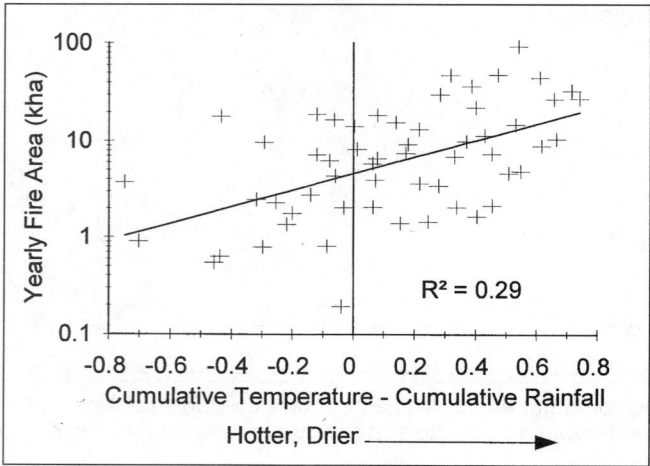

FIGURE 41.21

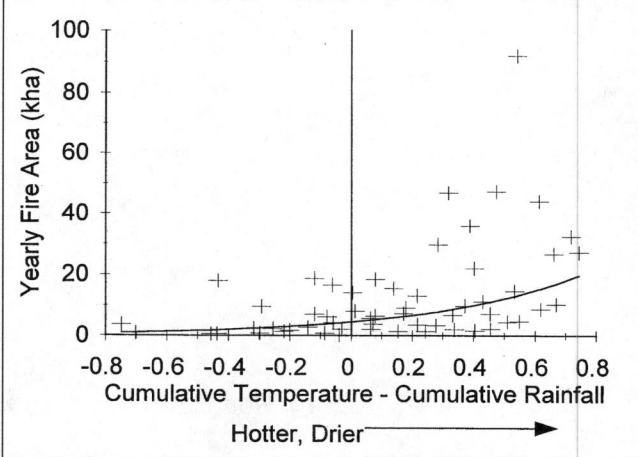

The regression in figure 41.20, back-transformed and presented on a linear scale. All years with extremely high acreage burned are identified as hot and dry.

area). If weather is extreme when the ignition occurs, the fire will spread rapidly, likely escape initial containment, be resistant to control, and end up on the map. One potential tool for measuring daily drought is the Keetch-Byram drought intensity index (KBDI) (Keetch and Byram 1968). This index measures deep drying of the litter and is therefore also related to the flammability of live fuels (Keetch and Byram 1968; Burgan 1988). The index ranges from 0 to 800 in value; values greater than 500 are associated with periods of fairly severe drought.

Following this logic, we used the KBDI to measure drought, considering days with KBDI values greater than 500 to be drought days. We then computed the average number of drought days per season for the 132 stations in and around the Sierra Nevada for the period 1979–89. For all stations we started the computations on March 1, with an initial index value of 0.0. Because elevation was the most important determinant of fire frequency, we then regressed these decadal averages on the elevation of each station.

Drought As a Function of Elevation

There was a significant negative correlation between elevation and the number of drought days (figure 41.22), with drought levels remaining high (arguably higher) well up into the foothills zone. Because KBDI indices remained very high in the foothills, during drought years some of the highest readings were reported in the zone where there were frequent fires (compare plates 41.5 and 41.6). In order to compare drought days with burn frequencies, we binned the stations into 100 m (3,330 ft) elevation bands and compared the average number of drought days with the fire frequency for those same elevation zones. When the numbers of drought days were scaled to the same maximum value as the burn-frequency data, the patterns were similar (figure 41.23). The elevation at which fire activity vanished in the Sierra was approximately the same elevation at which there were, on average, no days with KBDI levels greater than 500.

Drought As a Temporal Phenomenon

During the 1980s, there was a wide range in seasonal temperatures and precipitation: 1987 was an extreme fire year, 1982 was one of the least droughty years in the last fifty-seven years, and 1979 was about average. Binning the stations into 100 m elevation groups and comparing these three years reveals parallel patterns, with 1987 having more drought days at all elevations and 1982 having fewer (figure 41.24). When compared to the decade average, stations averaged twenty-two more drought days in 1987 and twenty fewer in 1982. Although the sample was small, at least for the decade of the 1980s, drought in the Sierra Nevada appears to have been highly synchronized. While each station appeared to have had unique weather (hence the scatter observed in figure 41.22), the effect of drought was to increase the number of drought days for all stations. Hence, drought simultaneously enlarged the window of opportunity for extreme fire to occur for all sites and provided a window of opportunity for higher-elevation sites that would normally be too cool and wet.

SUMMARY

Throughout these analyses, we have assumed that the fire maps developed by the U.S. Forest Service were reasonably accu-

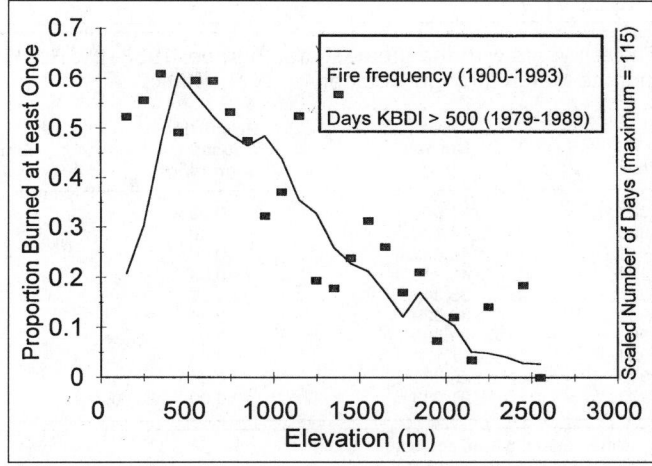

FIGURE 41.23

Average number of drought days for 132 stations (1979–89) grouped into 100 m elevation bins and compared with the fire/frequency pattern based on all mapped fires. The number of drought days has been scaled so that its maximum is 0.61, the maximum fire-frequency level.

rate and the record reasonably complete. Although we have verified their accuracy for the most recent years (1980–92), we have no direct means to test these assumptions throughout the twentieth century. If there are omissions, they would most likely be concentrated in the smaller fires early in the twentieth century. Most of the results presented in this chapter, however, would be either unaffected or strengthened by these

FIGURE 41.22

Average drought days, defined as days with KBDI levels greater than 500, for the period 1979–89. Each point represents an individual recording station (*n* = 132). Drought remains high up to about 750 m elevation and then declines. No stations higher than 2,600 m reported weather data during this period.

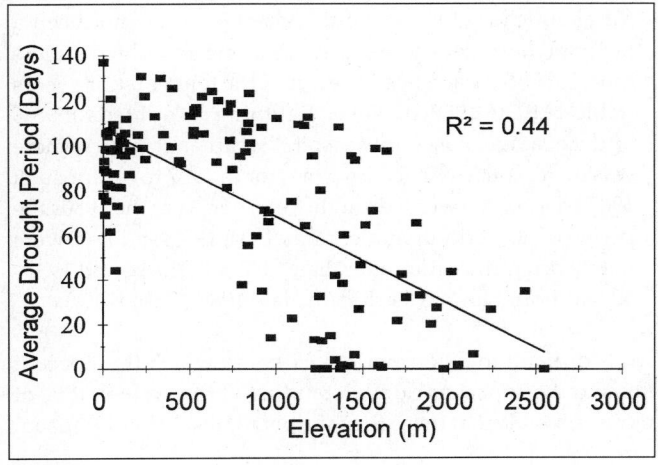

FIGURE 41.24

Average number of days with KBDI values greater that 500 on elevation for the years 1979, 1982, and 1987. Weather station data (approximately 100 stations had complete weather data for each year) was grouped into 100 m elevation bins and averaged.

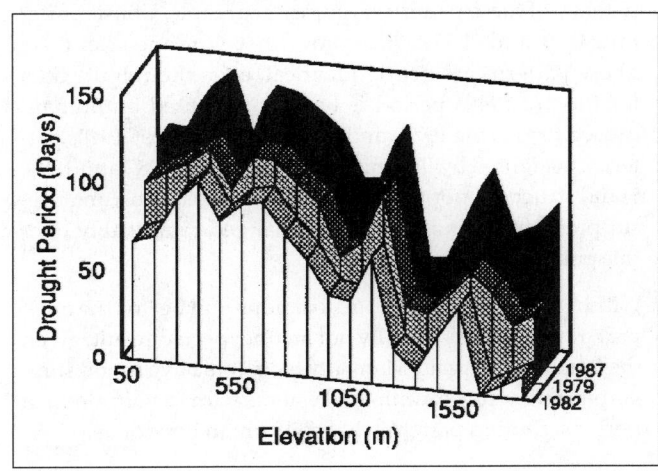

TABLE 41.9

The ten years with the greatest area burned, 1933–89. Rankings are based on fourteen weather stations in the Sierra for the months March through October.

Year	Area Burned (ha)	Proportion Caused by Lightning	Temperature Rank[a]	Rainfall Rank[b]	Cumulative KBDI Rank[c]	Maximum KBDI Rank[d]
1987	91,869	0.98	8	13	11	27
1960	47,204	0.10	13	9	5	1
1942	46,868	0.00	18	38	44	12
1959	43,950	0.09	7	2	7	8
1977	35,841	0.27	19	5	37	37
1936	32,329	0.00	1	18	10	7
1961	29,695	0.04	22	23	17	13
1934	27,172	0.02	2	4	2	17
1939	26,581	0.04	3	30	12	11
1970	21,983	0.00	17	7	6	10

[a] 1 is the hottest, 57 the coolest.
[b] 1 is the driest, 57 the wettest.
[c] Cumulative KBDI levels summed for each season; 1 is the highest cumulative value, 57 the lowest.
[d] Daily KBDI values are averaged for the fourteen stations, and the highest daily value for each season is recorded. 1 is the highest, 57 the lowest.

potential omissions. Acreage statistics and patterns would be largely unaffected, because the larger fires contribute most of the acreage. In particular, the patterns of fire risk with elevation and overlap would be robust, because these statistics were based on all twentieth-century fires. Some of the time-trend data could be changed if significant early fire acreage was omitted. We might have a slight downward trend in total acreage, and human-caused fires would be down even more dramatically, but acreage in lightning fires would still almost certainly be up. Data on the number of fires per year are probably the most sensitive to the potential omission of small fires (most fires are small), but these patterns remain similar and the inferences unchanged even when small fires are excluded from the analyses (Weatherspoon and Skinner 1996). We therefore believe that analyses based on the mapped fires present a reliable picture of fire patterns in the Sierra Nevada.

Based on these analyses, we feel that the following points have been adequately demonstrated or can reasonably be inferred:

- Fires will occur in the high-risk zones. We feel that the consistency of these patterns is quite remarkable. Consider figures 41.10 and 41.11, which show fire before and after 1940. These patterns are nearly identical, even though the data for the pre-1940s period is heavily weighted by human-caused fires in the 1920s and the data for the post-1940s pattern is weighted by lightning-caused fires in 1987 and 1990. Stand structure, population and land use patterns, and fire-suppression technologies have changed remarkably over this period.

- Within a given risk zone, fire locations will be, for the most part, random. This is really not an unexpected result, given the long fire intervals. Many investigators have found similar patterns when examining systems with long fire-return periods, finding patterns that conform to Poisson expecta-

tions (Henstrom and Franklin 1982; Agee 1991) or age structures that fit exponential decline models (Van Wagner 1978, but see Baker 1989; Pitcher 1987; Johnson and Larsen 1991). In addition, these fires all occurred in the context of suppression, and hence probably burned under fairly extreme conditions. Extreme fires are much less sensitive to fuels and topography than are more moderate fires (Turner and Romme 1994).

- If current stand conditions are not altered, current levels of suppression cannot prevent large, stand-replacing fires.

- Periods of drought of the magnitude that has occurred several times in the twentieth century will cause a significant increase in the likelihood of extreme fire events and will likely be accompanied by large wildfires.

- The last drought, though not the most extreme in terms of weather, caused the most extreme fire behavior. Several recent large fires have been caused by lightning, which is also unusual. The degree to which these years are unusual can be seen by examining the "worst" fire years on record (table 41.9). In general, while lightning has frequently accounted for as much as 40% of the fire acreage, it has not been a factor in the extreme fire years. Lightning accounted for less than 10% of the acreage in seven of the ten worst fire years (table 41.9). In 1987, however, lightning accounted for 98% of the acreage. Based on seasonal weather statistics, there was no reason to expect fire behavior in 1987 to be unusual: 1987 was not the worst drought year, nor were the 1980s the period of worst drought. Certainly 1934, 1959, and 1960 were hotter, drier, and droughtier than 1987, and the period 1933–40 was hotter and drier than the late 1980s (table 41.9).

It is difficult to determine the extent to which these recent fires should impact our understanding of the system. It is, of course, possible that these fires simply represent very unusual weather. It should be noted that the late 1980s saw many extreme fires across the western United States, including the 1988 Yellowstone fires. While fuel buildups may be implicated in these events, patterns of fuel accumulation and the impacts of fire suppression in other locations may have little to do with the condition of the lower-elevation Sierra Nevada. Fire patterns in the Greater Yellowstone Ecosystem, for instance, appear to be following very long natural cycles (Romme 1982; Romme and Despain 1989).

There are, however, subtle signs that the situation in the forest may well have changed. These signs are found in the decoupling of human-caused and lightning-caused fire patterns. Human-caused fires have been declining steadily—in total acreage, in number, and in average size. This pattern is probably due to a combination of increased efficiency in fire suppression and public education. Lightning fires have remained constant or have increased over this same time period, and this statement was true prior to the 1987 fires.

Lightning fires are grouped both in space and time. Because of this grouping, the resources required to suppress each ignition become critical. If changes in forest structure and fuel loading are increasing the suppression fire costs per acre, then it will take fewer simultaneous ignitions to exhaust the local resources.

While there are unknowns concerning recent extreme fire behavior, we believe that, taken as a whole, these points reasonably lead to the following conclusions concerning land-management policies:

- We can accurately describe risk zones in the Sierra; hence we can approach fire control efficiently. In the twentieth century 50% of the fire acreage has been below 1,300 m, and 92% has been below 2,000 m (figure 41.25). Fire control policies must recognize this reality.

- If conditions present in the twentieth century (fuels, weather, and suppression capabilities) continue, we can expect the patterns developed in these analyses to be stable into the near future. Absolute acreage will go up and down with shifts in weather, but fires will be probably be concentrated in the higher-risk (lower-elevation) zones.

- The high frequency of fire in lower-elevation areas, coupled with a high degree of randomness, suggests a zonal strategy. The goal must be to reduce the probability of destructive fires within these areas, either by increasing suppression capabilities to the extent that few fires escape initial containment (assuming that this is possible) or by reducing fuels to the point that canopy trees and structures are not threatened by fires when they occur.

- Either increasing suppression efforts or reducing fuels sufficiently to have a significant impact on fire behavior will be extremely costly. It is essential to evolve control plans that are cost-effective on a per-acre basis and that have a strong strategic emphasis so as to efficiently treat the most

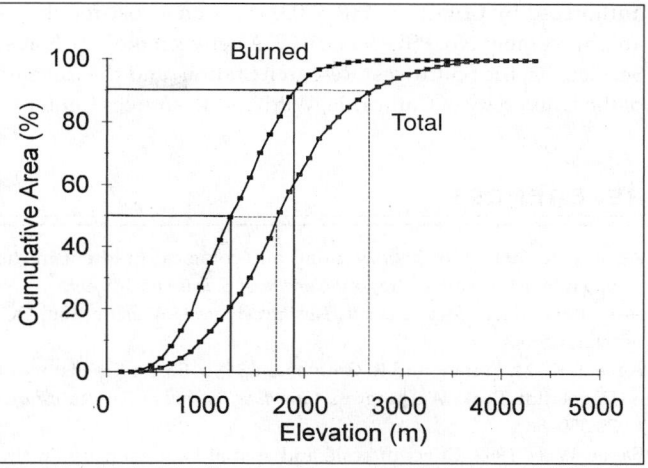

FIGURE 41.25

Fire area and total land in the study area. Lines mark the 50th and 90th percentiles for both distributions.

critical areas. The acreage distribution of the risk categories, which shows relatively small acreages in the highest risk categories (table 41.6; figure 41.13), suggests that strategies that concentrate treatment in high-risk zones may be cost-effective.

- Because within a risk zone overlap patterns of fire are largely random, we believe that fire-risk management would be best served by a policy that first allocated resources to control the destructive effects of fire in high-risk topographic zones. Within these zones, suballocations would appropriately be based on many factors, such as the consequences of fire and the costs of either treatment or suppression. We believe that in this suballocation process it would be prudent also to consider the number of times that the site has burned in the past as a risk factor. We do not believe, however, that localized history of burning should dominate resource allocation decisions.

ACKNOWLEDGMENTS

We would like to acknowledge Carl N. Skinner for providing analyses demonstrating the divergence between human- and lightning-caused fires in the Sierra. In addition, we would like to acknowledge the national forest employees past and present who created the fire maps, without which none of these analyses could have been accomplished; the CALOWL EIS team; the SNEP GIS team, and the members of the SNEP disturbance subgroup. In addition we would like to thank two anonymous reviewers for their suggestions, as well as many others who provided additional reviews and comments. This study was supported (in part) by the Sierra Nevada Ecosystem Project as authorized by Congress (HR 5503) through a cost-reimbursable agreement No. PSW-93-001-CRA between the U.S. Forest Service, Pacific Southwest Research Station, and the Regents of the University of California, Wildland Resources Center.

REFERENCES

Agee, J. K. 1991. Fire history along an ecological gradient in the Siskiyou Mountains, Oregon. *Northwest Science* 65:188–99.

——. 1993. *Fire ecology of Pacific Northwest forests.* Washington, DC: Island Press.

Agee, J. K., M. Finney, and R. Gouvenain. 1990. Forest fire history of Desplation Peak, Washington. *Canadian Journal of Forest Research* 20:350–56.

Baker, W. L. 1989. Effect of scale and spatial heterogeneity on fire interval distributions. *Canadian Journal of Forest Research* 19: 700–706.

Brown, D. G., and T. J. Bara. 1994. Recognition and reduction of systematic error in elevation and derivative surfaces from 7 1/2-minute DEMs. *Photogrammetric Engineering and Remote Sensing* 60:189–94.

Burgan, R. E. 1988. *Revisions to the 1978 national fire-danger rating system.* Ashville, NC: U.S. Forest Service, Southeastern Forest Experiment Station.

Campbell, G. S. 1977. *An introduction to environmental biophysics.* New York: Springer-Verlag.

Chou, Y. H., R. A. Minnich, and R. A. Chase. 1993. Mapping probability of fire occurrence in San Jacinto Mountains, California, USA. *Environmental Management* 17:129–40.

Chou, Y. H., R. A. Minnich, L. A. Salazar, J. D. Power, and R. J. Dezzani. 1990. Spatial autocorrelation of wildfire distribution in the Idyllwild Quadrange, San Jacinto Mountain, California. *Photogrammetric Engineering and Remote Sensing* 56:1507–13.

Erman, D. C., and R. Jones. 1996. Fire frequency analysis of Sierra forests. In *Sierra Nevada Ecosystem Project: Final report to Congress,* vol. II, chap. 42. Davis: University of California, Centers for Water and Wildland Resources.

Feller, W. 1968. *An introduction to probability theory and its applications.* New York: John Wiley and Sons.

Henstrom, M. A., and J. F. Franklin. 1982. Fire and other disturbances in the forests in Mount Rainier National Park. *Quaternary Research* 18:32–51.

Horn, H. 1971. *The adaptive geometry of trees.* Monographs in Population Biology, no. 3. Princeton, NJ: Princeton University Press.

Hosmer, D. W., and S. Lemeshow. 1989. *Applied logistic regression.* New York: John Wiley and Sons.

Johnson, E. A., and C. P. S. Larsen. 1991. Climatically induced change in fire frequency in the southern Canadian Rockies. *Ecology* 72: 194–201.

Keetch, J. J., and G. M. Byram. 1968. *A drought index for forest fire control.* Research Paper, SE-38. Asheville, NC: U.S. Forest Service, Southeast Forest Experiment Station.

Matyas, W. J., and I. Parker. 1979. *CALVEG: Mosaic of the existing vegetation of California.* San Francisco: U.S. Forest Service, Regional Ecology Group.

Minnich, R. A. 1983. Fire mosaics in southern California and northern Baja California. *Science* 219:1287–94.

Pitcher, D. C. 1987. Fire history and age structure in red fir forests of Sequoia National Park, California. *Canadian Journal of Forest Research* 17:582–87.

Rantz, S. E. 1969. *Mean annual precipitation in the California region.* Menlo Park, CA: U.S. Geological Survey.

Romme, W. H. 1982. Fire and landscape diversity in subalpine forests of Yellowstone National Park. *Ecological Monographs* 52:199–221.

Romme, W. H., and D. G. Despain. 1989. Historical perspective on the Yellowstone fires of 1988. *Bioscience* 39:695–99.

Show, S. B., and E. I. Kotok. 1923. *Forest fires in California, 1911–1920: An analytical study.* Department Circular 243. Washington, DC: U.S. Department of Agriculture.

Strauss, D., L. Bednar, and R. Mees. 1989. Do one percent of forest fires cause ninety-nine percent of the damage? *Forest Science* 35:319–28.

Turner, M. G., and W. H. Romme. 1994. Landscape dynamics in crown fire ecosystems. *Landscape Ecology* 9:59–77.

U.S. Forest Service (USFS). 1996. *Final environmental impact statement: Managing California spotted owl habitat in the Sierra Nevada National Forests of California—An ecosystem approach (CALOWL EIS).* 2 vols. Berkeley, CA: U.S. Forest Service, Pacific Southwest Research Station.

Van Wagner, C. E. 1978. Age class distribution and the forest fire cycle. *Canadian Journal of Forest Research* 8:220–27.

Weatherspoon, C. P., and C. N. Skinner. 1996. Landscape-level strategies for forest fuel management. In *Sierra Nevada Ecosystem Project: Final report to Congress,* vol. II, chap. 56. Davis: University of California, Centers for Water and Wildland Resources.

Wonnacott, T. H., and R. J. Wonnacott. 1977. *Introductory Statistics.* New York: John Wiley and Sons.

DON C. ERMAN
Department of Wildlife, Fish, and
  Conservation Biology
and
Centers for Water and Wildland
  Resources
University of California
Davis, California

RUSSELL JONES
Sierra Nevada Ecosystem Project
Centers for Water and Wildland
  Resources
University of California
Davis, California
Current address:
P.O. Box 4226
Evergreen, CO 80437

Fire Frequency Analysis of Sierra Forests

ABSTRACT

The pattern and frequency of fire size reported for seven national forests and Sequoia–Kings Canyon National Parks were assessed by frequency analysis, a method commonly used in hydrology for establishing probabilities of future events from historical data. The common generality that throughout the Sierra frequently occurring fires have become smaller and infrequently occurring large fires are becoming larger is not supported by the data. In the Plumas and Sequoia National Forests size and frequency of fires have not changed significantly during this century; in the Sierra National Forest all fires were larger before 1950; in Sequoia–Kings Canyon National Parks frequent small fires have been larger since 1950, and infrequent large fires have been smaller since 1950. The pattern of smaller, frequent fires and larger, infrequent fires since 1950 was true, however, in the central-western Sierra, in the Eldorado and Tahoe National Forests, and—to a lesser extent—the Stanislaus National Forest and the Lassen National Forest. Since 1950 in the Eldorado National Forest, frequent fires (those with recurrence intervals from 2 to 5 years) were 70%–80% smaller than before 1950, and infrequent large fires (those with recurrence intervals from 10 to 40 years) were 250%–500% larger than before 1950. The average fire for all forests (taken as approximately the size at a 2-year recurrence interval) ranged from about 350 acres to 1,500 acres (142 to 607 hectares), illustrating the variation throughout the Sierra. There was large variation in diligence of fire data collection among the national forests and national parks, ranging from forty-eight years (Lassen National Forest) to seventy-nine years (Stanislaus and Sequoia National Forests) of record. With additional data other applications of frequency analysis may be useful in examining the patterns of fire on the landscape. By recognizing that fire patterns differ throughout the Sierra, we can begin to dispassionately examine causes of fire.

INTRODUCTION

The current popular and oft-repeated hypothesis about fires in the Sierra Nevada is that large fires are occurring more frequently and are larger and more intense than they were in the past (Knudson 1994; USFS 1995). Some reasons given to explain this hypothesis are drought, fire suppression, fuel buildup, silvicultural methods, livestock grazing, and higher human population. But the hypothesis of a change in fire size and frequency must first be examined. If it is not true, it may lead to inappropriate forest management, wasted resources, and even increased fire. Currently, there are no data to examine changes in degree of fire intensity. We have the data, however, to examine questions of frequency and fire size. We also have the tool—frequency analysis—adapted from hydrology, where it is used to estimate probabilities of floods, droughts, large waves, low oxygen concentration, and other water-quality conditions (Chow 1964; Dunne and Leopold 1978; Greb and Graczyk 1995).

With this tool we can answer some critical questions:

1. Do the forests in different Sierra regions have similar fire size at the same frequencies?

2. Have the fire frequencies changed over time?

3. If fire frequency is different, is the difference large or small?

4. Do all the forests exhibit a similar pattern in fire frequency?

5. Is there any regional pattern in fire frequency?

In the Sierra, all national forests and, to a somewhat lesser degree, national parks extend over a substantial elevational range and hence include several forest types. Ignition frequency and fuel loadings are therefore not uniformly distrib-

Sierra Nevada Ecosystem Project: Final report to Congress, vol. II, *Assessments and scientific basis for management options.* Davis: University of California, Centers for Water and Wildland Resources, 1996.

uted. Forests also extend north to south along the axis of the Sierra, and east and west of the crest, over broad gradients of climate and change in species composition. Because of these and other differences, we must be cautious about making generalizations among forests.

The major objective of this study was to analyze the frequency and size of fires in this century using the method of frequency analysis borrowed from hydrology.

METHODS

Fire perimeter maps showing location, year, and cause of the fires were prepared for the California Spotted Owl Environmental Impact Statement in 1994 and stored in digital form for all Sierran national forests (USFS 1995). These data represent all fires greater than 1 hectare (ha) that were mapped from 1908 to 1993. Completeness of the record varied among the forests (for example, some years were missed, perimeters of fires were roughly estimated, some smaller fires were not recorded). In addition, fires were mapped at varying distances outside the national forest administrative boundaries. Digital maps of fire perimeters were also obtained for Yosemite and Sequoia–Kings Canyon National Parks. Data for Yosemite National Park consisted only of lightning fires from 1931 to 1993 (USGS 1994), whereas for Sequoia–Kings Canyon National Parks data for all fires from 1921 to 1993 were obtained (USGS 1993).

For this analysis, fires were included for a national forest if the majority of the burn area fell within the respective forest's administrative boundary. Data for analysis of fires classified as outside national forests are incomplete, and so our analysis is restricted to the regions of public land described earlier. Fire area within a perimeter was determined from the Arc/Info software program for geographic information systems (GIS). Care was taken to eliminate duplicate counting by examining the shape of the mapped fires as well as the date and cause of the fire. For all study areas, the great majority of fires were of human origin.

The earliest records began in 1908 but were not consistent for the study areas or reporting units; that is, many years are missing in agency records. Data from seven national forests—Lassen, Plumas, Tahoe, Eldorado, Stanislaus, Sierra, Sequoia—and Sequoia–Kings Canyon National Parks were used. The records for the Inyo National Forest and Yosemite National Park were too incomplete to use.

Frequency analysis is a statistical procedure for interpreting past events in terms of their chances of occurring in the future (Chow 1964). We used the extreme-event series (Chow 1964) or the annual maximum fire to determine the recurrence interval for fires of different sizes. The term *recurrence interval* here does not refer to the return of fires in a specific place as it does in other studies of fire (see McKelvey and Busse 1996;

Skinner and Chang 1996) but, rather, to a probability of an event of a given size occurring somewhere in the unit.

The following methods are described briefly for professionals and others who may want to follow the procedure. The annual maximum fire for each unit was ranked from largest to smallest (largest is rank 1), and the recurrence interval was calculated as the number of years of record plus 1, divided by the rank of each annual maximum fire in the series. The resulting recurrence intervals (in years) were then plotted against the size of the fire (on logarithmic scales). The resulting plot shows the average interval of time within which the size of any given fire will be equaled or exceeded at least once. The inverse of the recurrence interval is the probability of the fire size in any year. Thus a fire size having a recurrence interval of 5 years has a 0.2, or 20%, chance of occurring in any year.

The scatter of points in the frequency analysis is used to determine a best fit (theoretical) distribution. (See Interagency Advisory Committee on Water Data 1981, the standard guide for fitting a distribution to the data points, for details of this procedure.) For some graphs, however, the data did not follow a simple distribution. In these cases, hand-smoothed curves (using a French curve) were used to fit portions of the distribution. All fitted curves differ somewhat from actual data points. To simplify comparisons, we selected the 2-, 5-, 10-, 20-, and 40-year recurrence intervals to highlight patterns. Properties of the frequency analysis also yield estimation of the overall average annual event (see Chow 1964), which is the fire with a 2.33 recurrence interval. For simplicity in our analysis, we have used the 2-year recurrence interval fire as a close approximation of the average fire.

We divided the fire records into those before 1950 and those from 1950 to 1993 to determine if fire frequencies have changed over time. This division gave two periods of approximately forty years, and we could test the assumption that fire frequency has changed in the latter half of the century. It also reduced the magnitude of recurrence interval estimation and explains why we used forty years as a maximum. To more closely examine the question of relative change in fire size after 1950 we subtracted the post-1950 fire size from the pre-1950 fire size, divided by the pre-1950 fire size for the five recurrence intervals, and expressed differences as a percentage.

RESULTS

The eighty-six-year period of record (1908 to 1993) for the study areas was incomplete in all cases (figure 42.1) and ranged from twenty-five years (Inyo National Forest) to seventy-nine years (Stanislaus and Sequoia National Forests). For particular years, the number of units with recorded fires ranged from two to ten (all units); however, all ten units re-

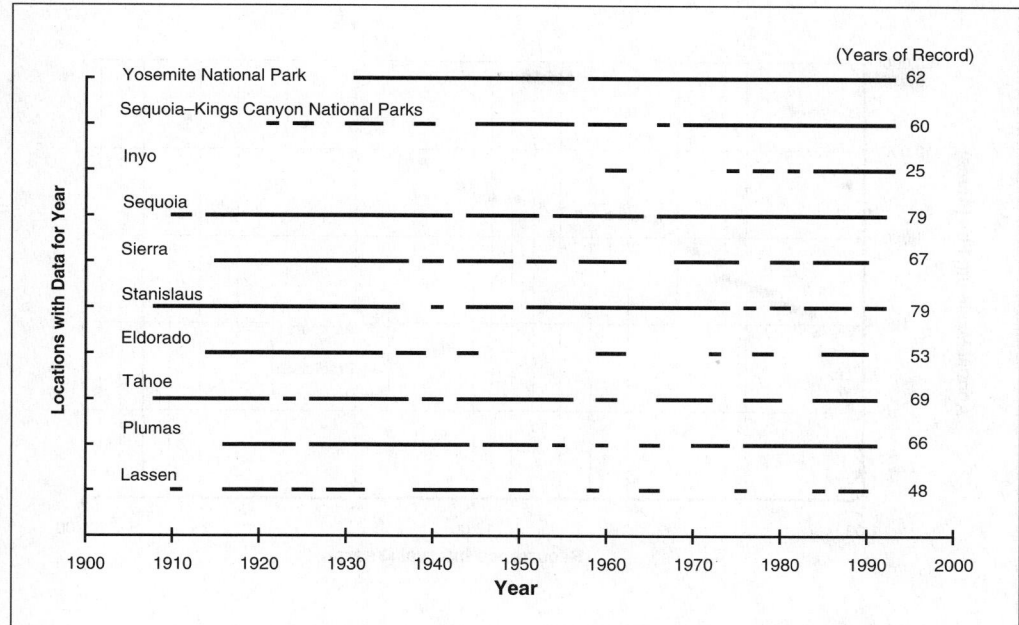

FIGURE 42.1

Years of record for fires in Sierra Nevada locations (national forests and national parks). Gaps in horizontal lines represent missing data.

ported fires from only six years in common. All the units had missing data for periods of two years in a row or more. Both missing data and years with no (zero) fires are likely explanations but were not discriminated in the records. Fires from the Inyo National Forest and Yosemite National Park were not analyzed further because of brief and incomplete records. Also, major gaps from the Eldorado and Lassen National Forests after 1950 weaken the analysis for those forests (figure 42.1).

Six national forests (Lassen, Plumas, Tahoe, Eldorado, Stanislaus, and Sierra) had smaller fires after 1950 at the 2-year recurrence interval, and all but Plumas had smaller fires at the 5-year recurrence interval. For illustration, fire frequency plots are shown for the Eldorado National Forest (figure 42.2) for all years of record (upper plot) and by pre- and post-1950 (lower plot); plots for other units are given in appendix 42.1. In the Eldorado National Forest the 2-year fire (the approximate average fire) was reduced from 700 acres (283 ha) before 1950 to 140 acres (57 ha) after 1950. Looked at in another way, the chance of a 700-acre fire occurring within a given year changed from 50% (chance of one year out of two) before 1950 to about 20% (chance of one year out of five) after 1950 (figure 42.2, lower plot). By contrast, frequent (2- to 5-year recurrence interval) fires were about the same or larger after 1950 than before in Plumas National Forest (for greater than 2-year fires), Sequoia–Kings Canyon National Parks, and Sequoia National Forest. The average fire (2-year recurrence interval) among the units ranged from 700 to 2,500 acres (283–1,012 ha) before 1950 and from 140 to 1,000 acres (57–405 ha) after 1950.

A comparison of fire size at recurrence intervals of 2, 5, 10,

20, and 40 years for all locations is summarized in figures 42.3–42.7. These results clearly show that the general hypothesis of large fires occurring more frequently throughout the Sierra is false. Frequency of large fires is highly variable among the regions. Only the Tahoe, Eldorado, and Lassen National Forests have a clear and consistent pattern of large fires occurring more frequently since 1950 (figures 42.5–42.7). In the Stanislaus National Forest, the 10- and 20-year recurrence interval fires (figures 42.5–42.6) were smaller or no different after 1950, but the 40-year recurrence interval fire was substantially larger after 1950. In the Plumas, Sierra, and Sequoia National Forests and in the Sequoia–Kings Canyon National Parks frequence of large fires either has not changed or has declined.

For forest units showing the pattern of smaller, more-frequent fires after 1950 and larger, less-frequent fires, the crossover point for frequency distributions from the two periods occurred from the 5-year to about the 11-year recurrence interval (see appendix 42.1).

The relative changes in fire size before and after 1950 are shown in figure 42.8. In the Lassen, Tahoe, Eldorado, and Stanislaus National Forests frequent fires (2- to 5-year recurrence interval) were 30%–85% larger before 1950, but infrequent fires (10- to 40-year recurrence interval) were generally larger after 1950 (30%–500%). However, the Stanislaus National Forest (about 85% larger 2- and 5-year fires before 1950) had no increase in the size of infrequent fires until the 40-year recurrence interval, when a very large fire, the result of a complex of fires in 1987, burned more than 100,000 acres (40,469 ha). The Eldorado National Forest differed dramatically from all others in the large increase in fire size for the infrequent fires. After 1950, fires in the 10- to 40-year recur-

FIGURE 42.2

Fire frequency plots of the annual maximum fires for the Eldorado National Forest. The upper plot shows the actual and predicted fires for the full fifty-three years with data; the lower plot shows the same data divided into two periods, before and after 1950 with fitted points and lines.

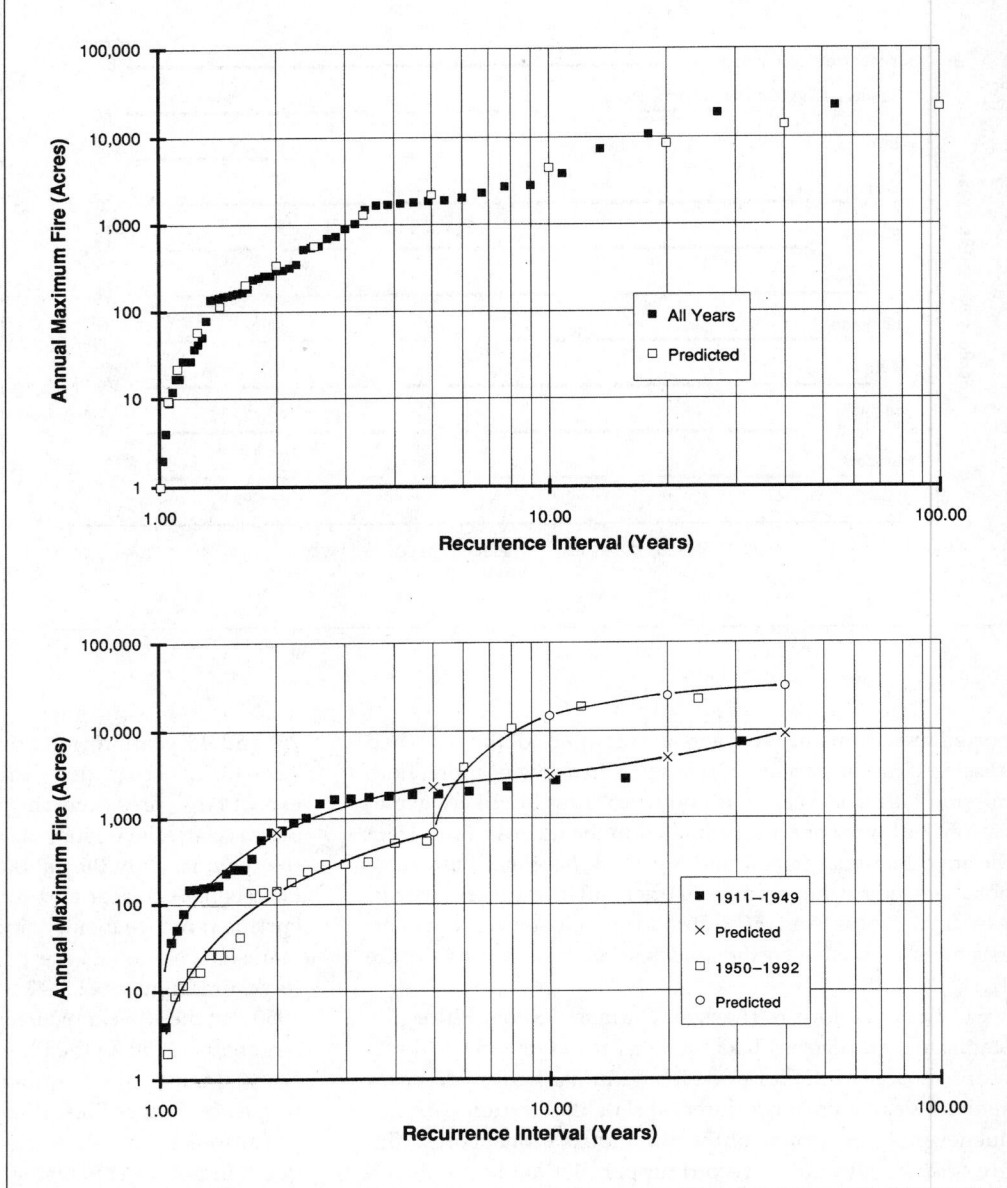

rence interval increased in size from 250% to 500%. The patchy record of years with reported fires (figure 42.1) in the Eldorado National Forest (fifty-three years) and the Lassen National Forest (forty-eight years) is unfortunate and leaves some uncertainty about the pattern.

No clear pattern emerged for relative difference in fire size before and after 1950 in the Plumas and Sequoia National Forests (figure 42.8). In the Sierra National Forest all fires were larger before 1950. And in Sequoia–Kings Canyon National Parks the pattern of fire frequency is different from that in all the national forests: since 1950 frequent fires have been larger and infrequent fires have been smaller.

DISCUSSION

The five questions posed in the introduction have been answered by fire frequency analysis:

1. The forests in the different Sierra regions do not have similar fire size at the same frequencies. For example, the 2-year recurrence interval fire since 1950 ranged from 140 acres (57 ha) to 1,000 acres (405 ha), whereas the 40-year fire was from 12,000 acres (4,856 ha) to 99,000 acres (40,064 ha).

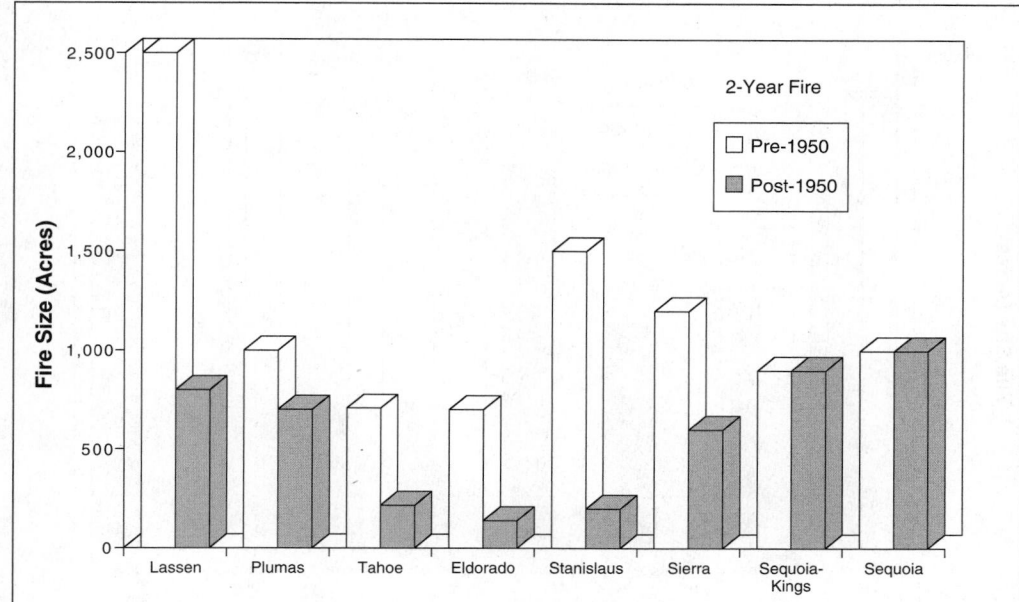

FIGURE 42.3

Estimated fire size for each forest location at the 2-year recurrence interval for the periods before and after 1950. Locations are arranged in order from the northern (Lassen National Forest) to the southern (Sequoia National Forest) Sierra Nevada.

2. The fire frequencies have changed over time for most but not all forests in this century.

3. The magnitude of the change in fire frequencies among the forests depends on the forest. Those in the central-western Sierra showed the greatest change.

4. Not all forests exhibit a similar pattern of fire frequency.

5. There is a regional pattern in fire frequency. The trend in the central-western Sierra, particularly the Eldorado National Forest, is for small, frequent fires to be smaller since 1950 and large, infrequent fires to be larger. A nearly opposite pattern occurs in the southern Sierra, particularly Sequoia–Kings Canyon National Parks. These opposite patterns lead to some questions about causes and management and indicate a new area for exploration. For example, has the prescribed burning program been responsible for the approximately 80% reduction in the size of the 40-year fire since 1950 in Sequoia–Kings Canyon National Parks?

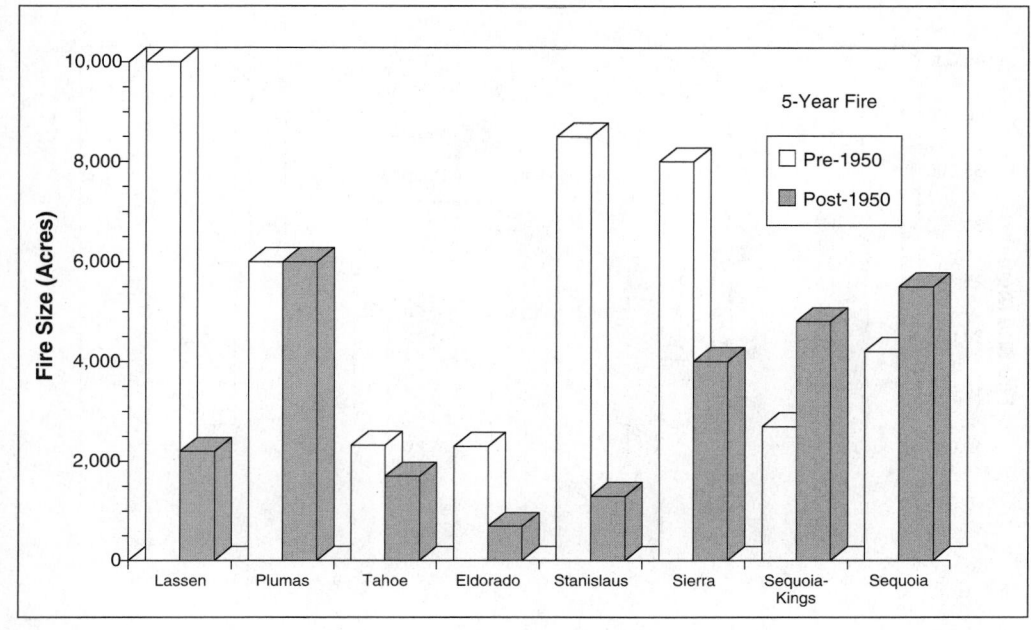

FIGURE 42.4

Estimated fire size for each forest location at the 5-year recurrence interval for the periods before and after 1950. Locations are arranged in order from the northern (Lassen National Forest) to the southern (Sequoia National Forest) Sierra Nevada.

FIGURE 42.5

Estimated fire size for each forest location at the 10-year recurrence interval for the periods before and after 1950. Locations are arranged in order from the northern (Lassen National Forest) to the southern (Sequoia National Forest) Sierra Nevada.

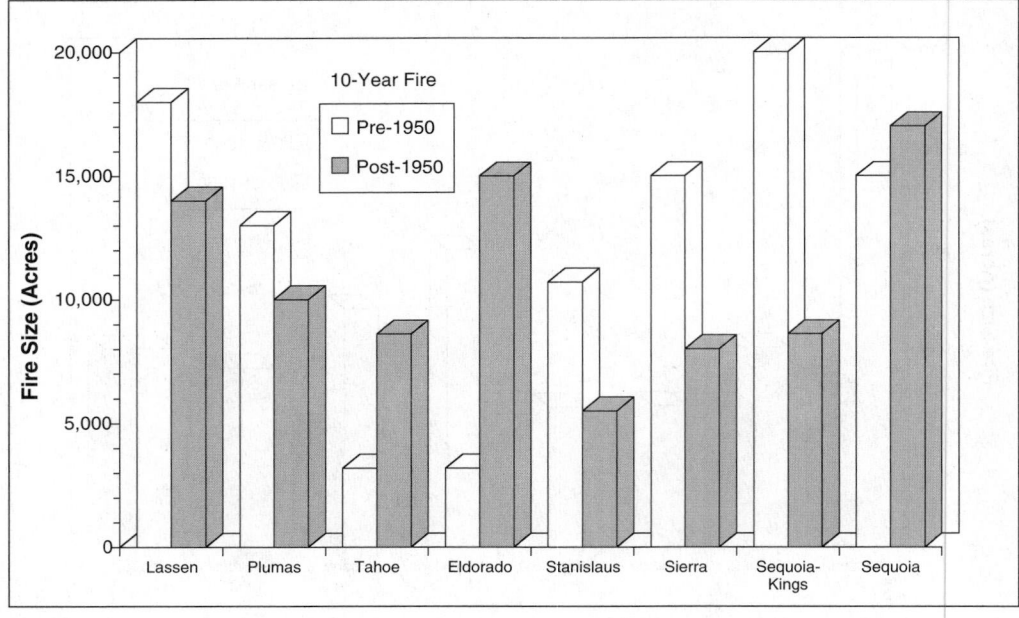

Many of the popular generalizations about increasing probability of large fires seem to be borne out for the Eldorado National Forest and to a lesser extent for the Tahoe and Stanislaus National Forests. The reasons for this pattern—whether natural factors (weather, for example) or human factors (fuel modifications, fire-fighting strategies, human activity)—must be explored through other means.

The patterns of fire frequency also indicate regions where apparent risk of large fires has not changed substantially during the period of record. These regions may provide useful comparisons of institutions, climate, fire behavior, and other factors with regions that have changed.

In some forests, such as the Plumas National Forest, recent large fires that may seem unusual are, in our analysis, not out of the ordinary and are similar in frequency and size to those in the first half of the century. Thus, the entire record of fire frequency since 1910 can be used to examine fires with recurrence intervals greater than 40 years (appendix 42.1). For the Plumas National Forest the 100-year recurrence interval fire is on the order of 60,000 acres (24,281 ha).

FIGURE 42.6

Estimated fire size for each forest location at the 20-year recurrence interval for the periods before and after 1950. Locations are arranged in order from the northern (Lassen National Forest) to the southern (Sequoia National Forest) Sierra Nevada.

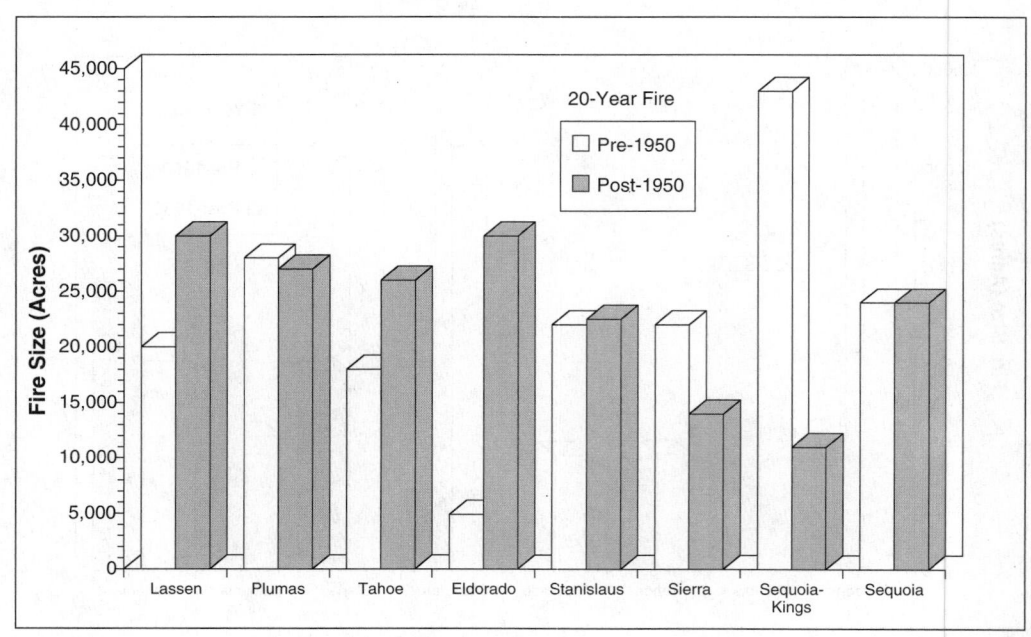

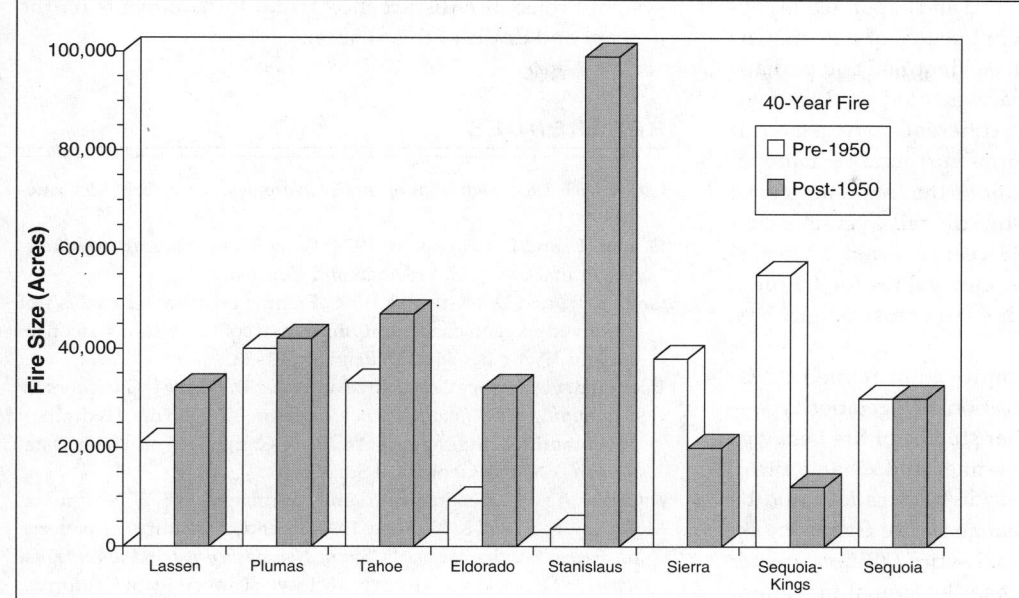

FIGURE 42.7

Estimated fire size for each forest location at the 40-year recurrence interval for the periods before and after 1950. Locations are arranged in order from the northern (Lassen National Forest) to the southern (Sequoia National Forest) Sierra Nevada.

What is not explained in our frequency analysis of fire size is possible changes in intensity. If data eventually become available or are collected in the future, fires could be sorted by intensity class and examined separately. However, in the absence of a robust method of testing, generalizations about the pattern of fires, as in the case of fire size, may be heavily influenced either by a few locations (such as the Eldorado National Forest) or by the vagaries of memory and human judgment about individual events.

With these questions about fire size answered, a new set of questions emerges about the reasons for these differences. Many possible causes for differences in fire frequency or changes over time have been offered, but the wide range in pattern among the forests analyzed suggests a variety or interaction of causes. Our results also suggest that fires in the 5- to 11-year recurrence interval range (where the lines cross over for the two distributions) may represent some limitation of human control or some natural process that deserves further investigation.

The method of extreme-event analysis that we applied to

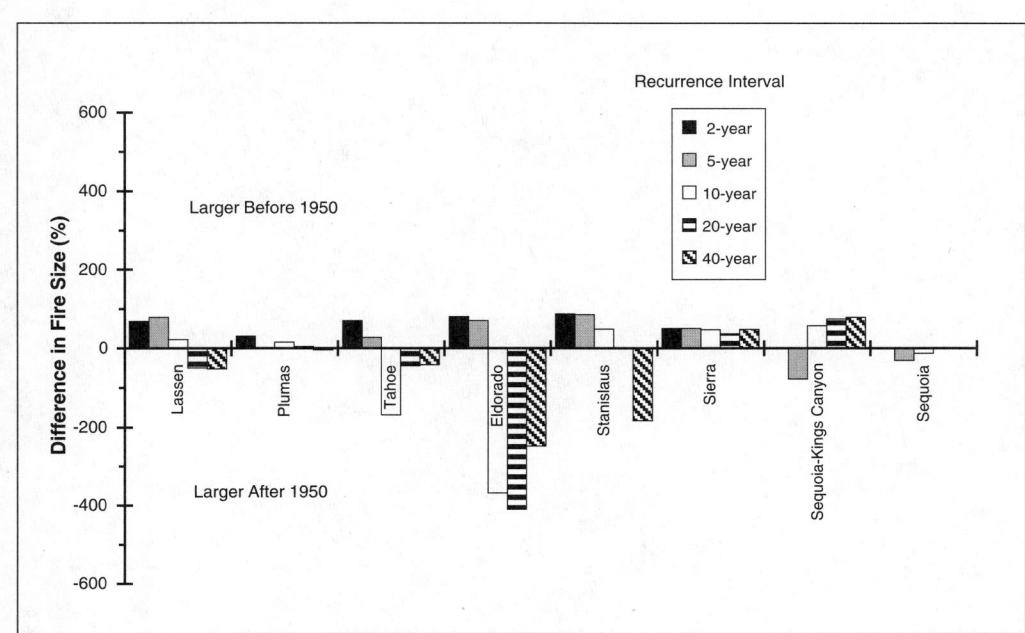

FIGURE 42.8

Relative change in fire size before and after 1950 for the five comparison recurrence intervals. The percentage difference in fire size for each recurrence interval is the fire size before 1950 minus the size after 1950 divided by the size before 1950 times 100. Locations are arranged horizontally from north to south. (Apparent data gaps represent zero difference between periods.)

the problem of fires could be improved and expanded in several ways. First, the problem of completeness of data may be rectified if true years of no fires can be identified and perhaps gaps in the record filled. Similar analyses could produce comparisons with private land under different management if sufficient records exist for the regions surrounding national parks and forests. In other applications the frequency curve provides a simple extraction of the overall average-sized event, which is the size of the 2.33-year recurrence interval event (Dunne and Leopold 1978). Our values for frequent events are probably overestimated in size because smaller fires may be less well recorded.

The analysis could be further improved or refined by assembling fire frequency data by elevation or vegetation type—critical elements considered in other studies of fire behavior (see McKelvey and Busse 1996; Skinner and Chang 1996). Different time periods, such as decade by decade, could be examined for other patterns of change in fire frequency by adopting the procedure of the partial series (all fires greater than some minimum size) rather than the annual maximum event (Chow 1964).

Nevertheless, the important conclusion from our analysis is that a Sierra-wide generalization about the pattern of fire size and change in frequency in this century is not supported by the data.

ACKNOWLEDGMENTS

We thank Nancy A. Erman for help in reviewing and revising this chapter. We also thank Carl Skinner, Gary Biehl, David Parsons, William Stewart, John Helms, Linda Blum, John Buckley, Michael Fry, and members of the Sierra Nevada Ecosystem Project fire-disturbance group for comments on the concept and drafts of this chapter.

REFERENCES

Chow, V. T. 1964. *Handbook of applied hydrology.* New York: McGraw-Hill.

Dunne, T., and L. B. Leopold. 1978. *Water in environmental planning.* San Francisco: W. H. Freeman and Company.

Greb, S. R., and D. J. Graczyk. 1995. Frequency-duration analysis of dissolved-oxygen concentrations in two southwestern Wisconsin streams. *Water Resources Bulletin* 31:431–38.

Interagency Advisory Committee on Water Data. 1981. *Guidelines for determining flood flow frequency.* Bulletin 17 B of the Hydrology Subcommittee, Interagency Advisory Committee on Water Data. Reston, VA: U.S. Geological Survey.

Knudson, T. 1994. Feeding the flames. *Sacramento Bee,* 27 November.

McKelvey, K. S., and K. K. Busse. 1996. Twentieth-century fire patterns on Forest Service lands. In *Sierra Nevada Ecosystem Project: Final report to Congress,* vol. II, chap. 41. Davis: University of California, Centers for Water and Wildland Resources.

Skinner, C. N., and C. Chang. 1996. Fire regimes, past and present. In *Sierra Nevada Ecosystem Project: Final report to Congress,* vol. II, chap. 38. Davis: University of California, Centers for Water and Wildland Resources.

U.S. Forest Service (USFS). 1995. *Draft environmental impact statement: Managing California spotted owl habitat in the Sierra Nevada national forests of California, an ecosystem approach.* Albany, CA: USFS.

U.S. Geological Survey (USGS). 1993. *GIS fire atlas data for Sequoia and King's Canyon National Park: All fires 1921–1993.* Yosemite National Park, CA: National Park Service.

———. 1994. Spatial patterns of lightning strikes and fires in Yosemite National Park. *Proceedings of the 12th Conference on Fire and Forest Meteorology.* Jekyll Island, GA.

APPENDIX 42.1

Fire Frequency Analysis of Sierra Forests

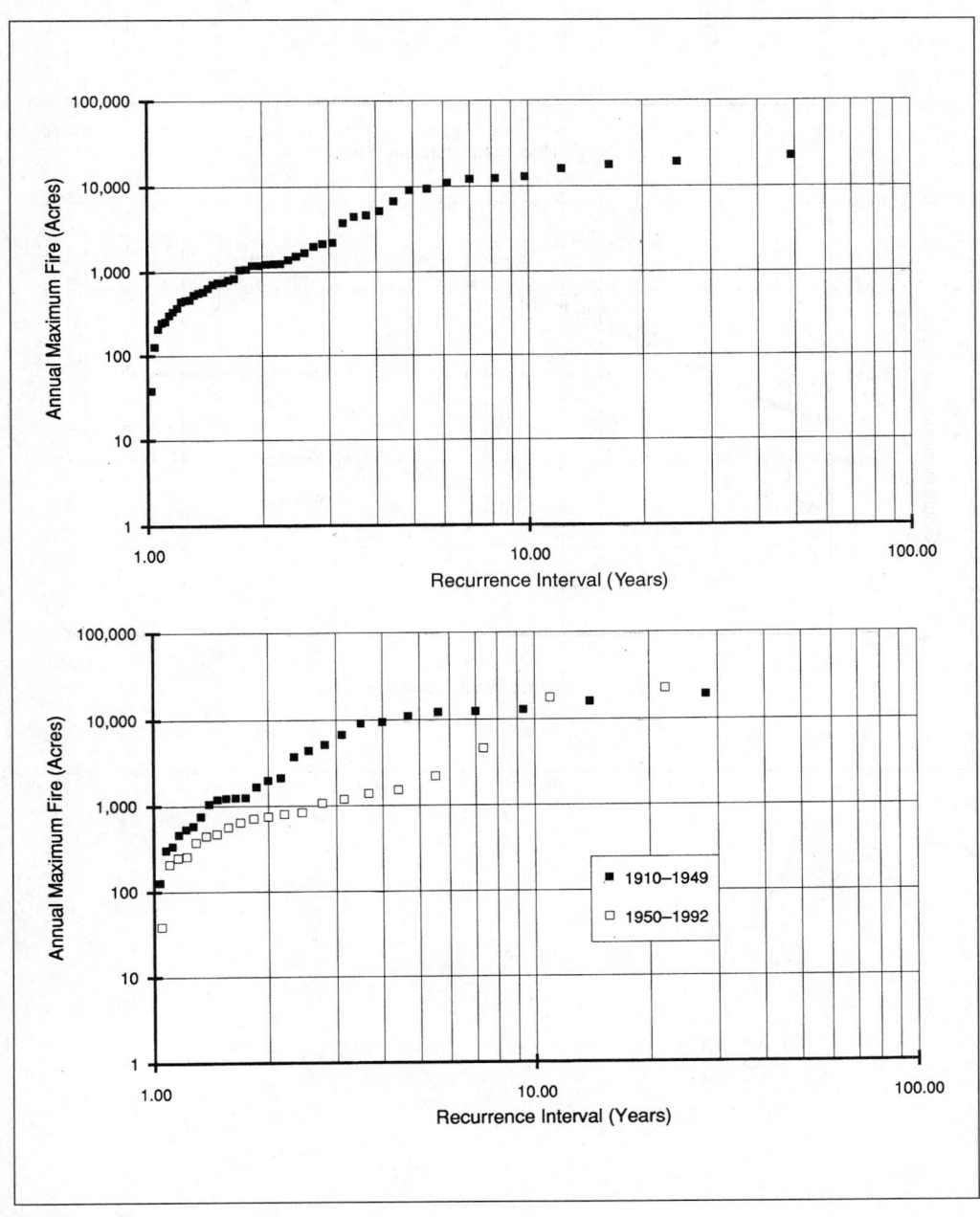

FIGURE 42.A1

Fire-frequency plots based on the annual maximum fire for all years of record (upper plot) and for the periods before and after 1950 (lower plot) for the Lassen National Forest.

FIGURE 42.A2

Fire-frequency plots based
on the annual maximum fire
for all years of record (upper
plot) and for the periods
before and after 1950 (lower
plot) for the Plumas National
Forest.

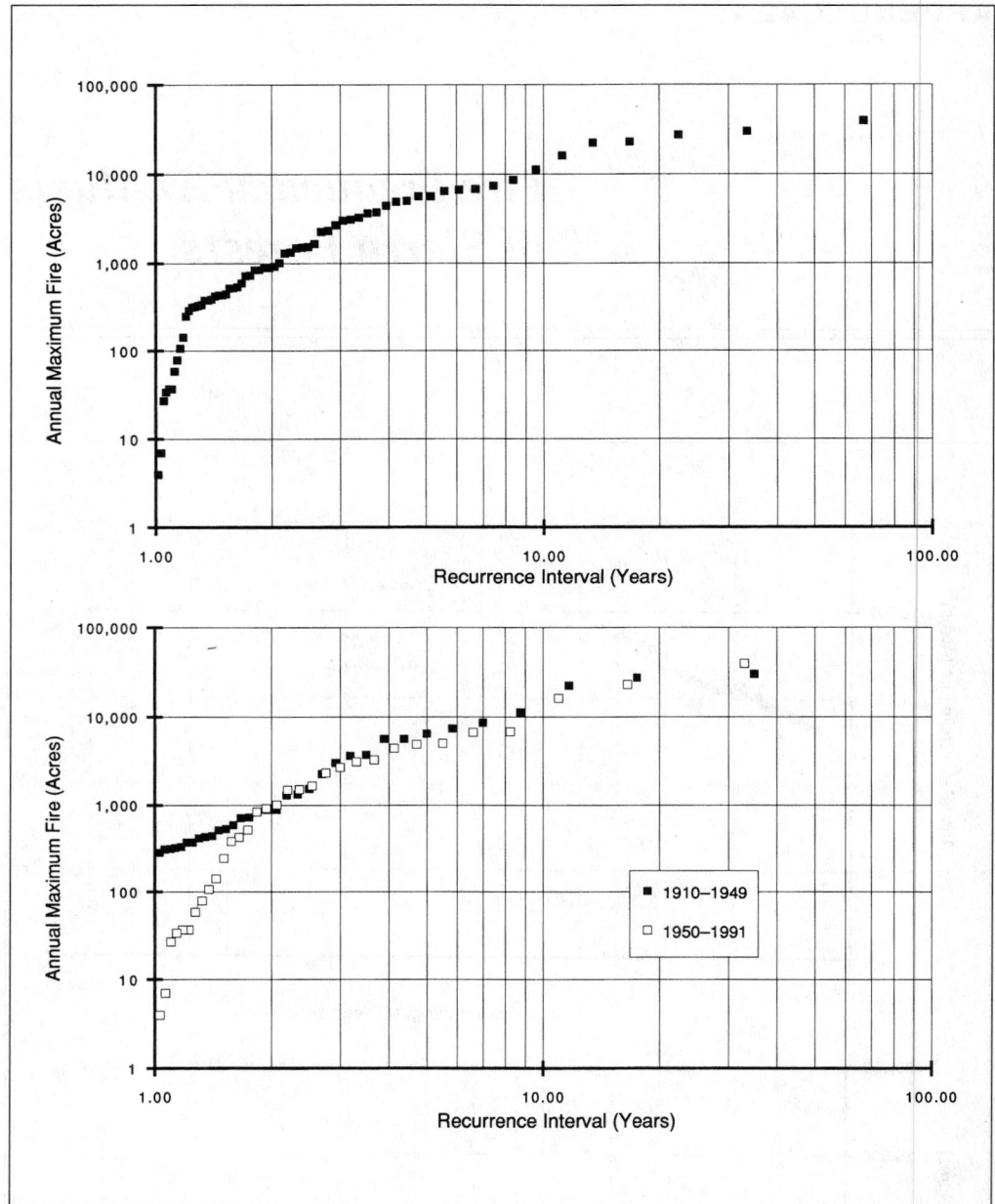

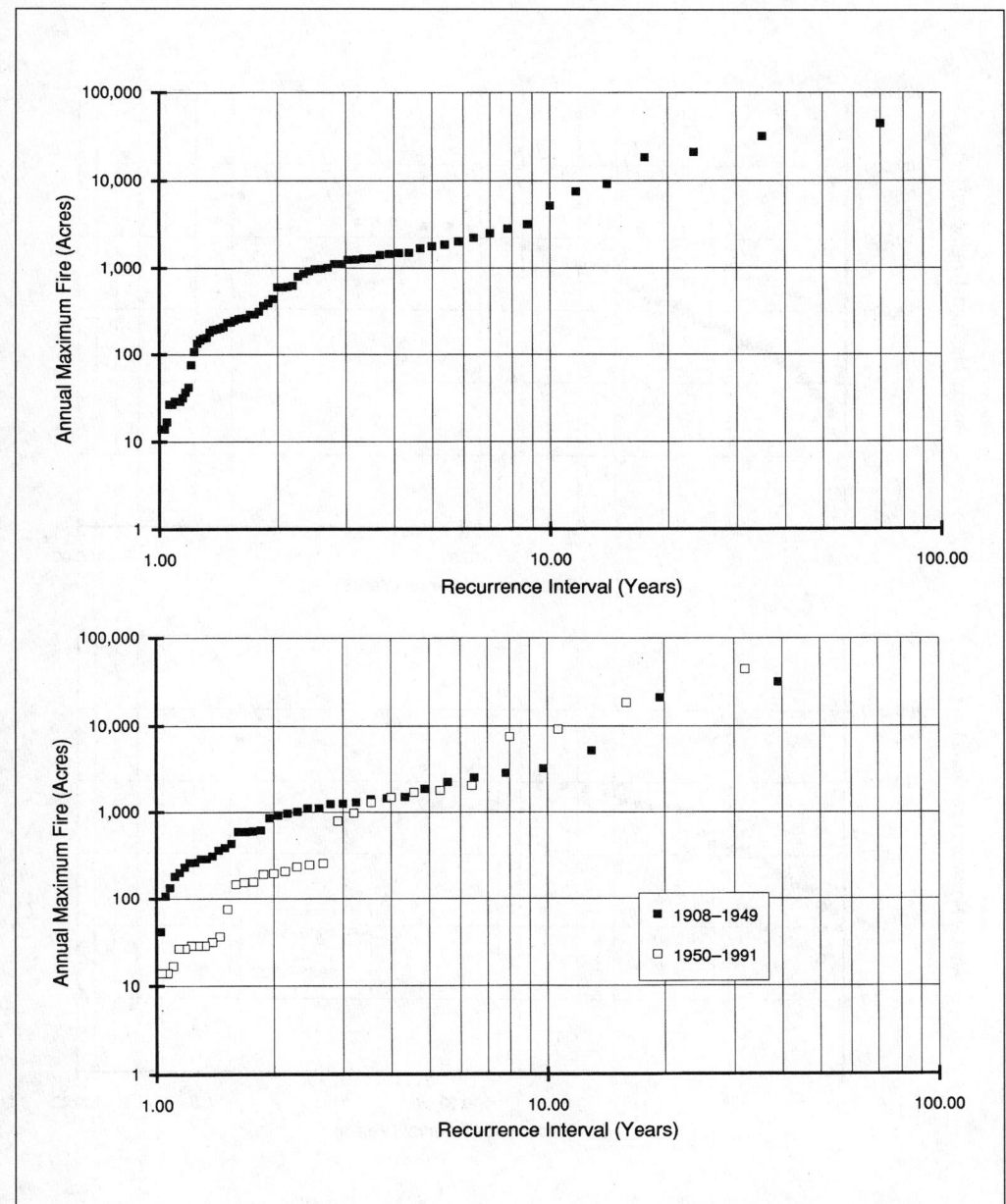

FIGURE 42.A3

Fire-frequency plots based on the annual maximum fire for all years of record (upper plot) and for the periods before and after 1950 (lower plot) for the Tahoe National Forest.

FIGURE 42.A4

Fire-frequency plots based on the annual maximum fire for all years of record (upper plot) and for the periods before and after 1950 (lower plot) for the Stanislaus National Forest.

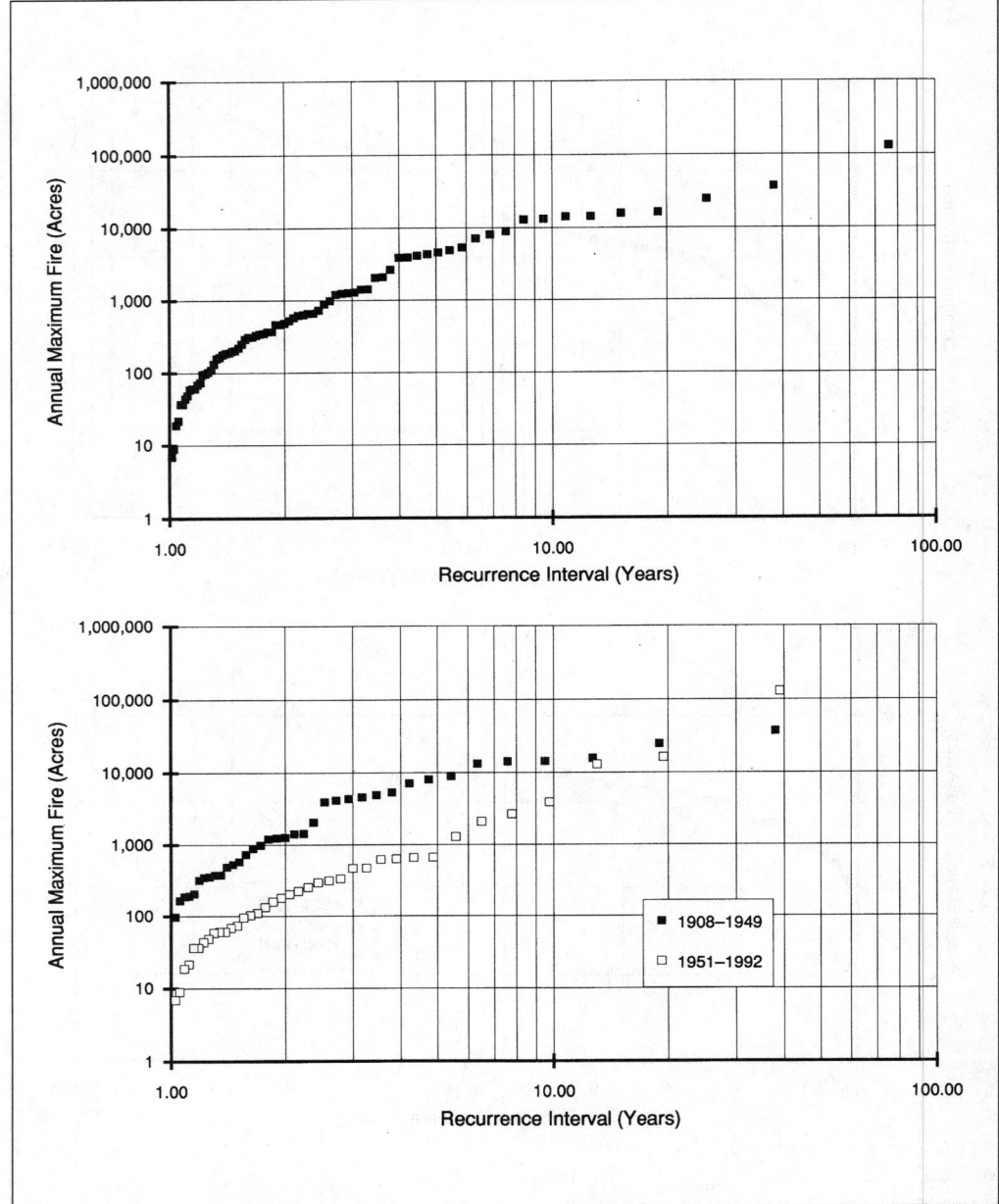

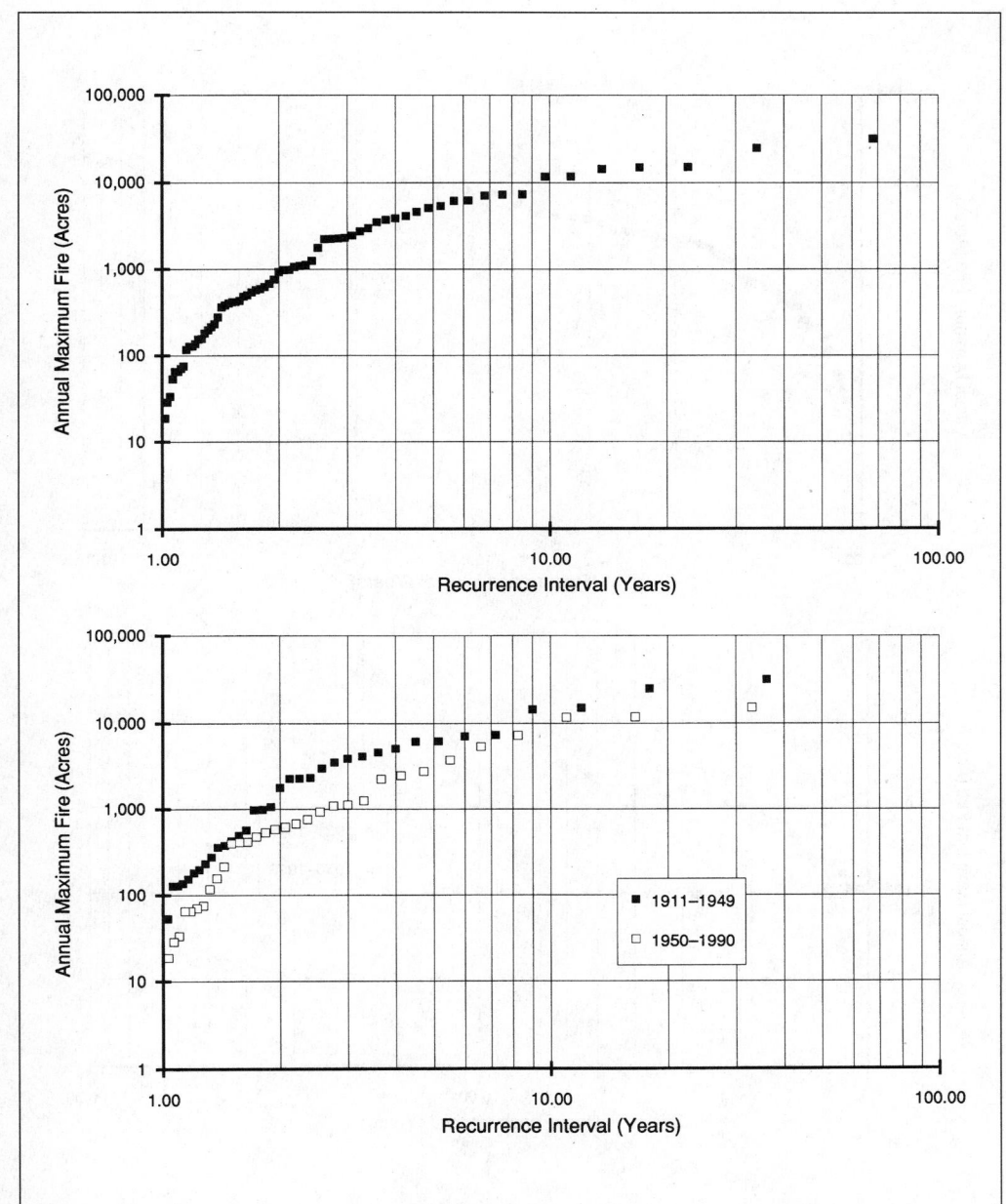

FIGURE 42.A5

Fire-frequency plots based on the annual maximum fire for all years of record (upper plot) and for the periods before and after 1950 (lower plot) for the Sierra National Forest.

FIGURE 42.A6

Fire-frequency plots based on the annual maximum fire for all years of record (upper plot) and for the periods before and after 1950 (lower plot) for the Sequoia–Kings Canyon National Parks.

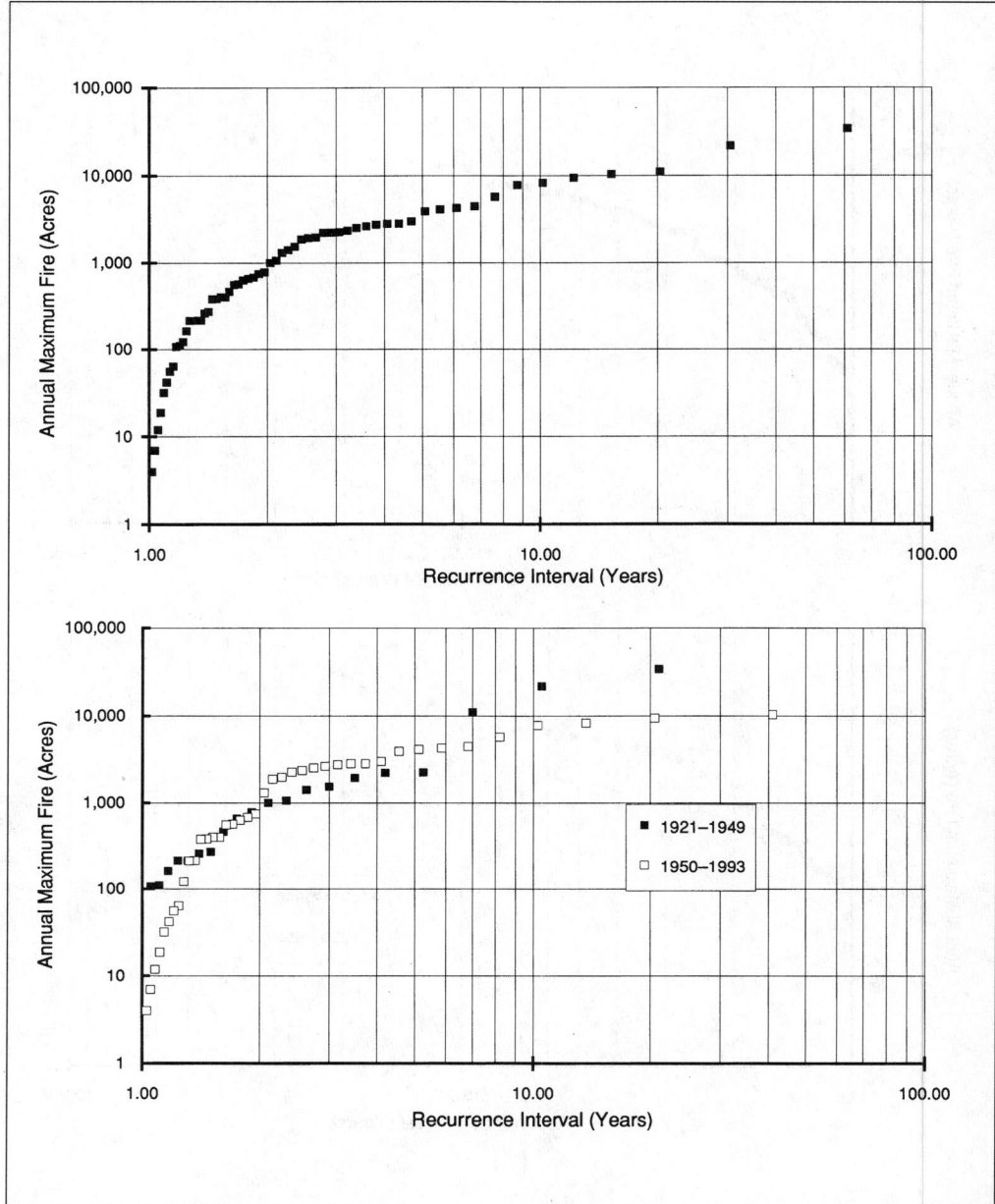

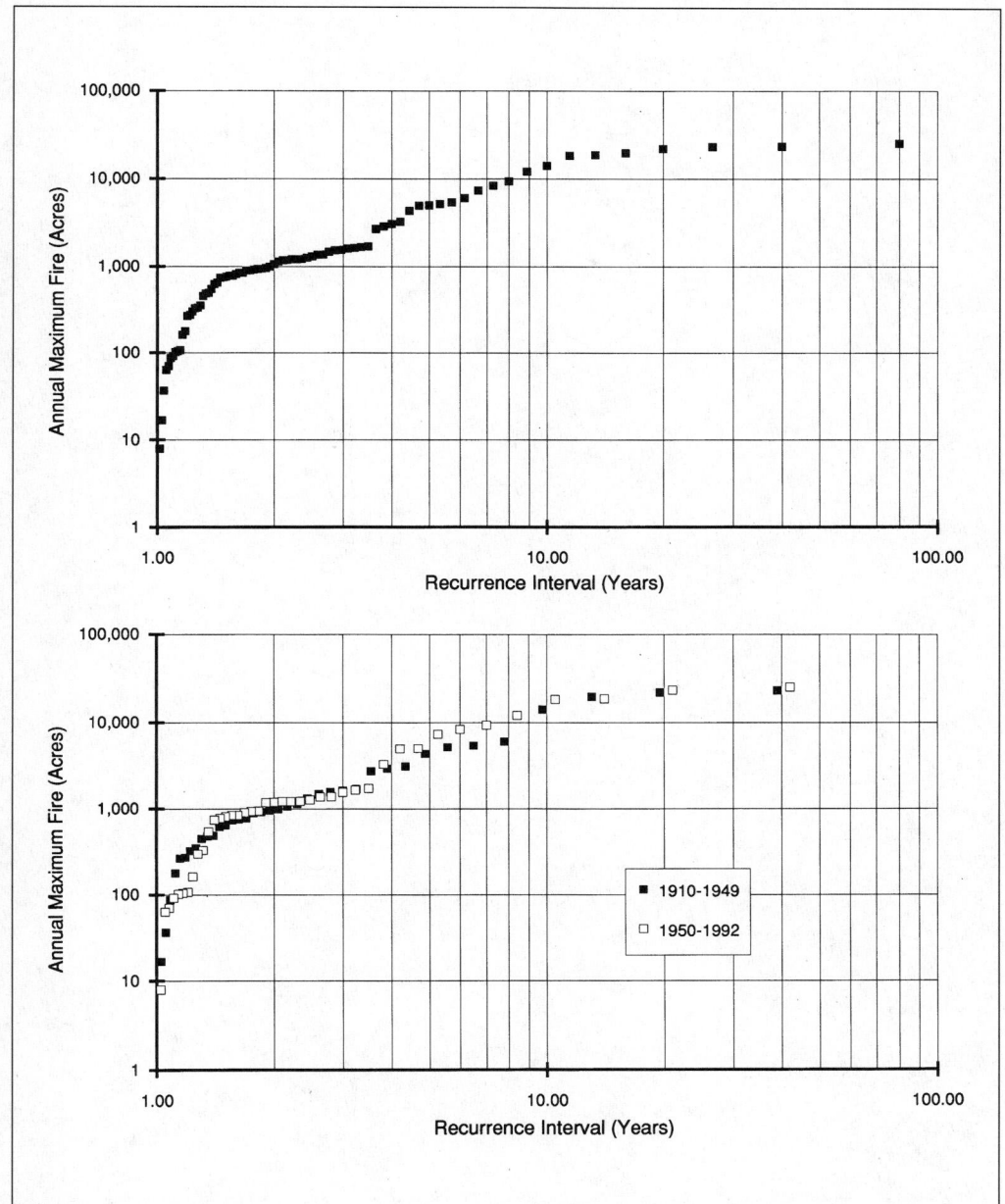

FIGURE 42.A7

Fire-frequency plots based on the annual maximum fire for all years of record (upper plot) and for the periods before and after 1950 (lower plot) for the Sequoia National Forest.

JAN W. VAN WAGTENDONK
National Biological Service
Yosemite Field Station
El Portal, California

43

Use of a Deterministic Fire Growth Model to Test Fuel Treatments

ABSTRACT

Fuel treatments are necessary in many vegetated areas of the Sierra Nevada to mitigate the effects of decades of fire suppression and land-management activities on fuel accumulations and understory canopies. Treating fuels will reduce the severity of wildfires and, as a result, the threat to human lives, the destruction of property and valuable resources, and the alteration of natural fire regimes. This chapter describes the use of a deterministic fire-modeling approach to obtain information about the relative effectiveness of fuel treatments, including fuel breaks, prescribed burning, biomassing, piling and burning, and cutting and scattering. Wildfire spread was simulated under idealized conditions to see how specific fuel and stand treatments affect fire behavior. It was obvious from the simulations that fuel breaks alone do not halt the spread of wildfire. Prescribed burning appears to be the most effective treatment for reducing a fire's rate of spread, fireline intensity, flame length, and heat per unit of area. A management scheme that includes a combination of fuel treatments in conjunction with other land-management scenarios should be successful in reducing the size and intensity of wildfires.

INTRODUCTION

It is evident that it will be necessary to reduce the amount of accumulated fuel in many vegetated areas of the Sierra Nevada to mitigate the effects of decades of fire suppression and management activities on fuel accumulations and understory canopies. Treating fuels should reduce the severity of wildfires and, as a result, should reduce the threat to human lives, the destruction of property and valuable resources, and the

alteration of natural fire regimes and ecosystem processes such as succession and nutrient flows. Several fuel-treatment options have been suggested, including the creation of fuel breaks of various widths, the prescribed burning of understory and surface fuels (the duff, live fuels, and dead woody fuels lying on the forest floor), the use of biomassing (thinning and chipping of trees up to a specific size class), and the removal of understory trees and branches to reduce ladder fuels. The efficacy of these treatments is largely unknown, and some means of evaluating them is necessary. Few field examples exist that can provide definitive data on the comparative value of the various treatments, alone or in combination. Information about the effectiveness of fuel treatments is critical for selecting alternatives and setting priorities.

Fuel treatment must be an integral part of any management scenario for the Sierra Nevada. This fact is beginning to be accepted by Congress, land-managing agencies, commodity interests, fire-fighting organizations, and the public. Less agreement exists on the best methods to use to achieve fuel-management objectives. Fuel breaks are the preferred option in the California spotted owl draft environmental statement (U.S. Forest Service [USFS] 1995) and are also mentioned in proposals submitted by the public. Acceptance of any large-scale fuel-treatment program will depend upon costs and threats to perceived values, including landscapes and personal property.

Although fuel breaks may prove useful, their value alone and in conjunction with fuel treatments within areas bounded by breaks must be evaluated. Most land-management agencies generally prefer zones where fuels are reduced with prescribed fires. Industry groups favor salvage logging and biomassing as alternatives to burning. This chapter describes the use of a deterministic fire-modeling approach to test the

Sierra Nevada Ecosystem Project: Final report to Congress, vol. II, *Assessments and scientific basis for management options.* Davis: University of California, Centers for Water and Wildland Resources, 1996.

relative effectiveness of fuel treatments, including fuel breaks, prescribed burning, biomassing, piling and burning, and cutting and scattering. In this approach, wildfire spread under idealized conditions is simulated to see how specific fuel and stand treatments affect fire behavior.

The key question addressed by this chapter is, what is the effect on fire behavior of various fuel-treatment alternatives?

BACKGROUND

Fuel treatments have been universally suggested as a means to limit the size and intensity of wildfires, yet little evidence of their effectiveness exists. Fuel breaks have been used in California since 1914 and have included the Ponderosa Way, which linked the Sierra and Sequoia National Forests, and more recent efforts in southern California. Most have failed because of the costs of maintenance. Some of the southern California programs have had examples of successes with fuel breaks, but data on their effectiveness are usually buried in lengthy fire reports.

Some anecdotal information exists about the effectiveness of prescribed burning. Biswell (1963) indicates that when a wildfire burned into an area of the Coast Range in California that had previously been burned under a program of prescribed fires, it was easily controlled. In the treated area scarcely any needles on the trees were scorched, while a majority of the trees outside of the area of the prescribed burn were killed. Similarly, wildfires have burned into Sierra Nevada park areas that have previously been burned by prescribed fires. In 1987, the Pierce fire crowned uphill into the Redwood Mountain Grove in Kings Canyon National Park, where it dropped to the ground in an area that had been burned five years before (Stephenson et al. 1991). The eventual control of the A-Rock fire in Yosemite in 1990 was attributed, in part, to the prescribed burns that had greatly reduced surface and understory fuels (Clark 1990).

No known experiments have tested the effectiveness of various fuel treatments on subsequent wildfire behavior. Although it would be important to include these sorts of experiments in fuel-treatment efforts, they are very difficult and costly to conduct. An alternative method is to use computer simulation, but until recently the tools for such experiments were not available. Quantitative models can now describe surface fuel arrays and canopy characteristics (Albini 1976; Van Wagner 1977). In addition, models for predicting fire behavior, including fire growth, spotting, and crowning, have been developed (Albini 1983; Rothermel 1983, 1991; Van Wagner 1993). The BEHAVE fire behavior prediction system combines the fuel and fire behavior models to make predictions of fire spread and intensity from a point source (Andrews 1986; Burgan and Rothermel 1984). Predictions from BEHAVE are adjusted in the field to account for the coarse temporal and spatial scale of the data used for the calculations (Rothermel and Reinhart 1983).

Finney (1994) developed a fire area simulator called FARSITE as a deterministic model for simulating the spatial and temporal spread and behavior of fires under conditions of heterogeneous terrain, fuels, and weather. Since it also includes spotting and crowning, the FARSITE simulator is an ideal tool to use to evaluate fuel treatments. The simulator has been verified in the field, using prescribed natural fires in Yosemite and Glacier National Parks (Finney and Ryan 1995).

The limitations of the FARSITE model include those of Rothermel's (1972) original fire-spread equation. His model describes a fire consisting of a flaming front advancing steadily in uniform and continuous surface fuels within 2 m (6 ft) of, and contiguous to, the ground (Rothermel 1983). Fuel models also simplify the array of burnable material on the ground into a set of parameters that are measurable and repeatable. FARSITE uses simplified weather and wind inputs, and assumes that fire spread is elliptical and independent of the shape of the fire front (Finney 1995). The uniform fuel constraint is limited only by the resolution of the fuel-model map.

Since simulation models are simplifications of reality and are based on numerous assumptions, their results are often in question. Models can serve as one source of information for decision making, but their primary usefulness is to gain understanding of complex systems. Deterministic models suffer from having enormous data requirements and practically infinite combinations of input variables. Researchers can overcome this limitation somewhat by simplifying the conditions under which the models operate. Simulations of fire processes are subject to all of these limitations but are often the only way, short of actual tests on the ground, of analyzing proposed scenarios.

METHODS

The FARSITE model was used to test the various fuel treatments in mixed conifer vegetation and fuels. One of the assumptions made in the simulations performed for this chapter is that the fires are unconstrained. This was necessary in order to isolate the effect of fuel treatments from the effects that might result from any number of suppression actions. Obviously, fire suppression during the simulation period would affect the results, making extrapolations to future conditions problematic. The simulation will, however, indicate the fire behavior that could be expected when suppression forces reached the fire.

Uniform terrain and weather were used to simplify the conditions in order to isolate the treatment effects. The simulation surface was a 3,000 m by 6,000 m (9,843 ft by 19,686 ft) area, with the long axis in a north-south direction. The first 3,000 m (9,843 ft) of the surface was at a 20% slope facing south, and the remaining 3,000 m (9,843 ft) was flat. Although some areas of the Sierra Nevada are steeper, 20% was selected as representing the majority of the areas subject to treatment. For instance, in Yosemite National Park, more than two-thirds of the park is on slopes of less than 20% (van Wagtendonk 1991). The mean elevation of the simulation surface was 1,500 m (4,921 ft), and the latitude was 38°. The forest was mixed conifer–pine typical of the Sierra Nevada, with an average tree height of 20 m (66 ft).

Eight different fuel-treatment scenarios were run with both 95th percentile and 75th percentile weather. These weather conditions are exceeded only 5% and 25% of the time, respectively, during the fire season. Fuel treatments were confined to the sloped portion of the simulation surface. The effectiveness of 90 m and 390 m (295 ft and 1,280 ft) fuel breaks was tested. The simulation was set to begin on August 1 and run for a twenty-four-hour period, until noon the next day. A single fire was started at the center of the simulation surface at a point 500 m (1,640 ft) from the bottom of the slope. Crown fires, embers from torching trees, and spot fire growth were all enabled. The trees most likely to be engaged in torching are ponderosa pines, which have a low tolerance for shade.

Fuel-Treatment Scenarios

The simulation model differentiates among fuel treatments by changing the fuel-model values for load and depth (table 43.1); the canopy characteristics for canopy cover, crown base height, and crown density; and the wind reduction factors (table 43.2). Fuels are categorized according to the time it takes for a fuel particle to reach 63% of its equilibrium moisture content (Lancaster 1970). Fuels with a 1-hour time lag consist of dead herbaceous plants and branchwood less than 0.64 cm (0.25 in) in diameter, as well as the uppermost litter layer. Dead branchwood fuels from 0.64 cm to 2.54 cm (0.25 in to 1 in) in diameter have a 10-hour time lag. Fuels with a 100-hour time lag include dead branchwood from 2.54 cm to 7.62 cm (1

in to 3 in) in diameter. Live fuels consist of forbs, grasses, and understory foliage within 1 m (3 ft) of the surface.

The fuels and canopies of the eight treatments and fuel breaks were defined by custom fuel models (Burgan and Rothermel 1984) and canopy characteristics. The control custom model (model 14) is identical to Albini's (1976) fuel model 10 (timber with litter and understory), although the 100-hour fuel load has been reduced to 2 tons per acre and the depth of the fuel bed has been increased to 1 foot to more accurately depict Sierra Nevada conditions. In the prescribed-burn model (model 15), the load in each fuel class and the fuel-bed depth are reduced by 50% compared to those of the control model. The cut-and-scatter model (model 16) increases loads and depth by 50% compared to the control model. The fuel-break model (model 17) keeps a fuel load of only 1 ton per acre of 1-hour fuels and half a ton each in the two larger classes.

Crown densities for each treatment were based on values derived by Brown (1978). Since crown base height is the only measurement of understory fuels, treatments were assumed to remove all of these fuels up to the specified height. Changes in surface fuels as a result of each treatment were represented in the custom fuel models by increases or decreases in fuel load and depth (table 43.1). Adjustment factors were used to tune the simulation to actual fire-spread patterns (van Wagtendonk and Botti 1984).

Control

The control scenario assumes that the simulation area has been subjected to effective fire suppression for at least fifty years. Surface fuels have accumulated over that period to 9 tons per acre and are 0.3 m (1 ft) deep. The understory is crowded with small trees, and crown bases are within 1 m (3 ft) of the ground, providing numerous fuel ladders. Canopy cover ranges from 81% to 100% closure, and crown densities are high. Although many stands in the Sierra Nevada that have been subjected to logging might have heavier and deeper accumulations than those modeled here, the results from this scenario will serve as a minimum example. More accumulated fuel will only exacerbate the resulting fire behavior.

Prescribed Burn

The prescribed-burn treatment reduces surface fuels by 50% in both load and depth compared to the control model (table 43.1). Duff and small branchwood up to 0.64 cm (0.25 in) in diameter are reduced from 3 tons per acre to 1.5 tons, while woody fuels up to 7.62 cm (3 in) are reduced to 2 tons per acre. Fuel-bed depth is decreased from 0.3 m (1 ft) to 0.15 m (0.5 ft). This approximates the effects of a safe and effective prescribed burn for this type (van Wagtendonk 1974). Two tons per acre of understory trees, brush, and branches up to 2 m (6 ft) in height are removed by this treatment, but canopies are not opened up or thinned. Complete removal of the understory had to be assumed because of limitations in the model. As a result, simulated subsequent fires are slightly less intense than might occur under actual conditions.

TABLE 43.1

Custom fuel-model values used in the simulations.

Fuel Variable	Model 14	Model 15	Model 16	Model 17
1-hour load (tons/acre)	3.0	1.5	4.5	1.0
10-hour load (tons/acre)	2.0	1.0	3.0	0.5
100-hour load (tons/acre)	2.0	1.0	3.0	0.5
Live load (tons/acre)	2.0	0.0	2.0	0.0
1-hour surface:volume ratio	2,000	2,000	2,000	3,000
Live surface:volume ratio	1,500	1,500	1,500	1,500
Depth (feet)	1.0	0.5	1.5	0.5
Moisture of extinction (%)	35	35	35	12
Dead heat content (BTU/lb)	9,000	9,000	9,000	9,000
Live heat content (BTU/lb)	8,000	8,000	8,000	8,000
Adjustment factor	0.5	0.5	0.5	0.5

TABLE 43.2

Fuel models and canopy characteristics for fuel treatments and fuel breaks.

Scenario	Custom Fuel Model	Canopy Cover (Percentage)	Crown Density (kg/m^3)	Crown Base Height (m)	Wind Reduction Factor
Control	14	81–100	0.30	1	0.22
Prescribed burn	15	81–100	0.30	2	0.22
Pile and burn	14	81–100	0.30	2	0.22
Cut and scatter	16	81–100	0.30	2	0.22
Biomassing	14	50–80	0.15	1	0.32
Biomassing and prescribed burn	15	50–80	0.15	2	0.32
Biomassing and pile and burn	14	50–80	0.15	2	0.32
Biomassing and cut and scatter	16	50–80	0.15	2	0.32
Fuel breaks	17	1–20	0.05	3	0.69

Pile and Burn

The pile-and-burn model assumes that hand crews are used to remove and pile all of the small understory trees, brush, and branches up to 2 m (6 ft) in height. This removal results in a reduction of 2 tons per acre of live fine fuels. The vertical and horizontal continuity of the fuels that constitute "ladder" fuels is eliminated. The piles are then covered with a tarp until after the fall rains have soaked through the surface fuels, and then they are ignited. This is a common practice in areas where it is thought to be too risky to use prescribed fire. Since the piles are burned during conditions when surface fuels will not ignite, there is no reduction in surface fuel load or depth. Areas underneath the piles are assumed not to burn, since the piles are smaller than the 30 m (98 ft) resolution of the input maps used for the simulation. This scenario does not include cutting or thinning in the upper canopy.

Cut and Scatter

The cut-and-scatter treatment is similar to the pile-and-burn treatment except that the understory trees and branches are cut, lopped, and scattered over the treatment unit. This situation often occurs when there are insufficient funds to remove the material resulting from an understory cutting operation. Surface fuel loads in each size class and fuel depth are increased from a total of 7 tons per acre in the control model to 10.5 tons per acre. Some of the increase in the fine dead fuel loads comes from the cut live fuels. Crown bases are raised to 2 m (6 ft) and no upper canopy thinning occurs.

Biomassing

The biomassing model includes the cutting, chipping, and hauling away of overstory trees up to a certain size class. This treatment has been proposed for many areas of the Sierra Nevada. This scenario assumes that 50% of the overstory trees are removed. The associated live and dead crown fuels are also removed from the site, but not the surface fuels. Although biomassing sometimes results in the crushing of surface fuels and the removal of some understory trees, fuel depth and crown base height are assumed not to change.

Biomassing and Prescribed Burning

Biomassing and prescribed burning involve thinning the overstory trees and canopies 50% through biomassing and treating the remaining surface fuels with prescribed fire. Surface fuel loads and depths are decreased by 50% compared to the control model, and crown base height is raised to 2 m (6 ft).

Biomassing and Piling and Burning

Biomassing and piling and burning treatment combines overstory biomassing with the cutting, piling, and burning of understory trees and branches during moist conditions. This treatment increases crown base height to 2 m (6 ft) but does not reduce surface fuels.

Biomassing and Cutting and Scattering

When biomassing is combined with the cut-and-scatter treatment, 50% of the overstory canopy is removed, and the remaining understory trees and branches are cut, lopped, and scattered. Surface fuel load and depth are not changed.

Fuel-Break Alternatives

Fuel breaks are often seen as an option when time or money is limited and surface fuel treatments are not considered feasible. Although fuel breaks are not intended to stand alone and should be integrated with other fuel treatments on adjacent lands, very often there is only enough money to construct the break. The fuel-break alternatives were designed to test their efficacy against the range of fuel-treatment scenarios, including one involving no treatment. The first alternative is a 90 m (295 ft) wide fuel break at the crest of the slope. This width corresponds to that suggested by Green and Schimke (1971) for fuel breaks in the Sierra Nevada. The second alternative is a fuel break 390 m (0.24 mi) in width that is also located at the top of the slope. These breaks were designed to approximate the widths proposed by a public group and the California spotted owl draft environmental statement (USFS 1995). The breaks have sparse crowns (1% to 20%) that have been pruned to a height of 3 m (10 ft) and shelter grass understories.

TABLE 43.3

Weather scenarios.

Variable	95th Percentile	75th Percentile
Maximum temperature (°F)	90	65
Minimum temperature (°F)	60	45
Maximum humidity (%)	40	60
Minimum humidity (%)	10	20
Wind speed (mph)	18	6
Wind direction (degrees from north)	180	180
1-hour fuel moisture (%)	4	6
10-hour fuel moisture (%)	6	8
100-hour fuel moisture (%)	8	10
Live herbaceous fuel moisture (%)	90	110
Live woody fuel moisture (%)	90	110
Foliar moisture (%)	80	100

Weather Scenarios

The values for the two weather scenarios are listed in table 43.3. These are based on readings taken at the Crane Flat weather station in Yosemite National Park. The percentiles are based on the normal fire season, which runs from May through October.

Output Measures

After the simulation runs had been completed, their output was compared. Maps and tables were created that display the amount of the simulation surface within treated and fuel-break areas that had been exposed to various levels of fire-behavior parameters. These parameters include rate of spread, fireline intensity, flame length, and heat per unit of area. In addition, severe fire behavior, such as torching, spotting, and crowning, was listed for each scenario. The fuel-break alternatives were evaluated based on whether or not they were sufficiently wide to prevent spot fires from occurring beyond the break.

RESULTS

Treatment of surface and understory fuels affected the behavior of fires within the treated areas and, to a lesser extent, initial fire behavior within the fuel breaks. Treatments also affected severe fire behavior such as spotting and crowning, which spread the fire beyond the fuel breaks.

Fire Behavior within Treated Areas with 95th Percentile Weather

Fires in the treated areas burning with 95th percentile weather showed considerable variation in their behavior (table 43.4). The prescribed-burn treatment without any overstory thinning produced the lowest average values for rate of spread,

fireline intensity, and flame length. The prescribed-burning treatment and the biomassing and prescribed-burning treatment had the lowest average values for heat per unit of area because surface and understory fuels had been treated and no crowns were involved. Biomassing combined with cutting and scattering of the understory fuels had the highest values for all parameters except heat per unit area and exceeded the behavior in the control area. The additional surface fuel load and depth resulting from cutting and scattering contributed to the more extreme behavior. The four biomassing treatments produced more intense fires than the equivalent treatments without overstory thinning. The sparser overstory left fuels more exposed to the sun, resulting in lower fuel moisture. In addition, wind speed was not reduced as much as in denser canopies, resulting in higher midflame winds.

Table 43.5 lists the total area burned within the treatments during the first twenty-four hours, as well as whether or not those fires torched, spotted, or crowned. The biomassing, cutting, and scattering treatment resulted in the largest burned area, while the prescribed-burning treatment without biomassing had the smallest burned area. Torching and spotting occurred in all scenarios except the two prescribed-burning treatments, while crowning was present only in the control, pile-and-burn, and cut-and-scatter treatments.

Control Simulation

The fire that burned in the area that was not treated spread quickly upslope, aided by torching trees, spot fires, and crowning, finally covering a total of 414 ha (1,023 acres). The fire had an average rate of spread of 1.88 m/min (9.84 ft/min) and a maximum spread rate of 11.65 m/min (38.22 ft/min). Fireline intensities averaged 490.83 kW/m and ranged up to 4,854 kW/m (figure 43.1). Flame length varied throughout the burn, reaching a maximum of 3.85 m (12.63 ft) in the area of fastest spread. Fire behavior was greatly influenced by heavy surface fuels, low crown base heights, and dense canopies. Sufficient heat was generated by the surface fuels to create spot fires and to initiate crowning. The addition of crown fuels contributed to the high maximum heat per unit of area of 43,549 kJ/m^2.

Plate 43.1 displays flame length for the eight scenarios with 95th percentile weather. The inclined portion of the simulation surface is the treated area for each scenario, and the level portion is untreated. The ignition point is indicated by the yellow pointer. Head fires burn upslope from that point, while backing fires burn downslope, and flanking fires spread laterally. The red flames are from 0 to 1 m (0 to 3 ft) in length, while the yellow flames are from 4 to 5 m (12 to 15 ft). The high "ridge" of yellow and orange flames in the control, pile-and-burn, and cut-and-scatter scenarios indicates when crowning occurred. The lower flame lengths in the treated areas are a result of flanking and backing fires and, in the untreated areas, are the result of the flat terrain. Rates of spread for flanking and backing fires are determined by the model, and flame lengths are calculated based on the amount of fuel

TABLE 43.4

Average fire behavior for fires within fuel-treatment areas with 95th percentile weather.

Scenario	Rate of Spread (m/min)	Fireline Intensity (kW/m)	Flame Length (m)	Heat/ Unit of Area (kJ/m^2)
Control	1.88	490.83	1.27	14,629
Prescribed burn	1.74	117.80	0.68	4,015
Pile and burn	1.86	457.78	1.25	14,389
Cut and scatter	2.86	964.53	1.75	19,360
Biomassing	2.15	516.26	1.34	14,266
Biomassing and prescribed burn	2.10	142.17	0.74	4,007
Biomassing and pile and burn	2.15	515.71	1.34	14,268
Biomassing and cut and scatter	3.28	1,070.74	1.85	19,243

present. The effect of each treatment can be seen by comparing the lower flame lengths in the prescribed-burn area and the higher flames in the cut-and-scatter area to the flame lengths in the control and pile-and-burn areas.

Prescribed-Burning Simulation

The 50% reduction in surface fuels and the complete removal of understory fuels up to 2 m (6 ft) using prescribed fire reduced the magnitude of subsequent fire behavior. No torching, spotting, or crowning occurred. The average rate of spread dropped to 1.74 m/min (5.71 ft/min), the maximum rate of spread dropped to 3.32 m/min (10.89 ft/min), and the size after twenty-four hours of burning was 2,608 ha (6,444.5 acres). The average fireline intensity was reduced by 76 percent to 117.80 kW/m (figure 43.1), while the maximum intensity was reduced by 95 percent to 272 kW/m. This drop in intensity can be attributed to the fact that crown fuels were not involved. The flame lengths depicted in plate 43.1 show a typical pattern for fires burning under uniform conditions without spotting and crowning. The lack of crown fuel involvement is also evident in the low average heat per unit of area of 4,015 kJ/m^2.

Pile-and-Burn Simulation

When a scenario in which understory fuels are cut, piled, and then burned at a time when surface fuels will not ignite was simulated, the results were very similar to those of the control treatment. Only slight decreases in each of the fire behavior parameters were observed (table 43.4; figure 43.1). Torching, spotting, and crowning did occur, since understory ladder fuel removal without treatment of the surface woody and duff fuels is not sufficient to prevent severe fire behavior. Flames were long enough to reach into the upper canopy (plate 43.1), reaching a maximum of 3.11 m (10.20 ft). Heat per unit of area (table 43.4) and fire size after twenty-four hours (table 43.5) were also similar to the control simulation.

Cut-and-Scatter Simulation

Although the understory trees are removed up to 2 m (6 ft) in the cut-and-scatter scenario, the 50% increase in surface fuel loads and depth resulted in significant increases in fire be-

havior. The average rate of spread increased to 2.86 m/min (9.38 ft/min), and maximums exceeded 15 m/min (49 ft/min). Fireline intensities were also nearly double those of the control simulation (table 43.4). Flame lengths averaged 1.75 m (5.74 ft) and reached a maximum of 4.83 m (15.85 ft). Plate 43.1 depicts the flame lengths over the simulation surface. The average heat per unit of area exceeded that of the control simulation by nearly 5,000 kJ/m^2.

Biomassing Simulation

The biomassing scenarios all have 50% of the overstory canopy removed, both in density and in cover. This removal has the effect of slightly increasing fire behavior parameters because surface fuels are less shaded and are therefore drier (table 43.4). Crowning does not occur, since fire is not able to spread through the less dense canopies (table 43.5). In the biomassing scenario without surface fuel treatment, the rate of spread averaged 2.15 m/min (7.05 ft/min), and fireline intensity averaged 516.26 kW/m (figure 43.2). Flame length reached a maximum of 2.01 m (6.59 ft) and averaged 1.34 m (4.40 ft). Plate 43.1 shows the flame lengths for this scenario. Although a limited amount of torching and spotting occurred, the average heat per unit of area was slightly less than that of the control simulation, since crowns were not involved.

TABLE 43.5

Area burned in treatment areas and severe fire behavior for fires with 95th percentile weather.

Scenario	Area Burned (ha)	Torching	Spotting	Crowning
Control	414.0	Yes	Yes	Yes
Prescribed burn	260.8	No	No	No
Pile and burn	404.6	Yes	Yes	Yes
Cut and scatter	708.1	Yes	Yes	Yes
Biomassing	457.2	Yes	Yes	No
Biomassing and prescribed burn	348.3	No	No	No
Biomassing and pile and burn	455.4	Yes	Yes	No
Biomassing and cut and scatter	730.9	Yes	Yes	No

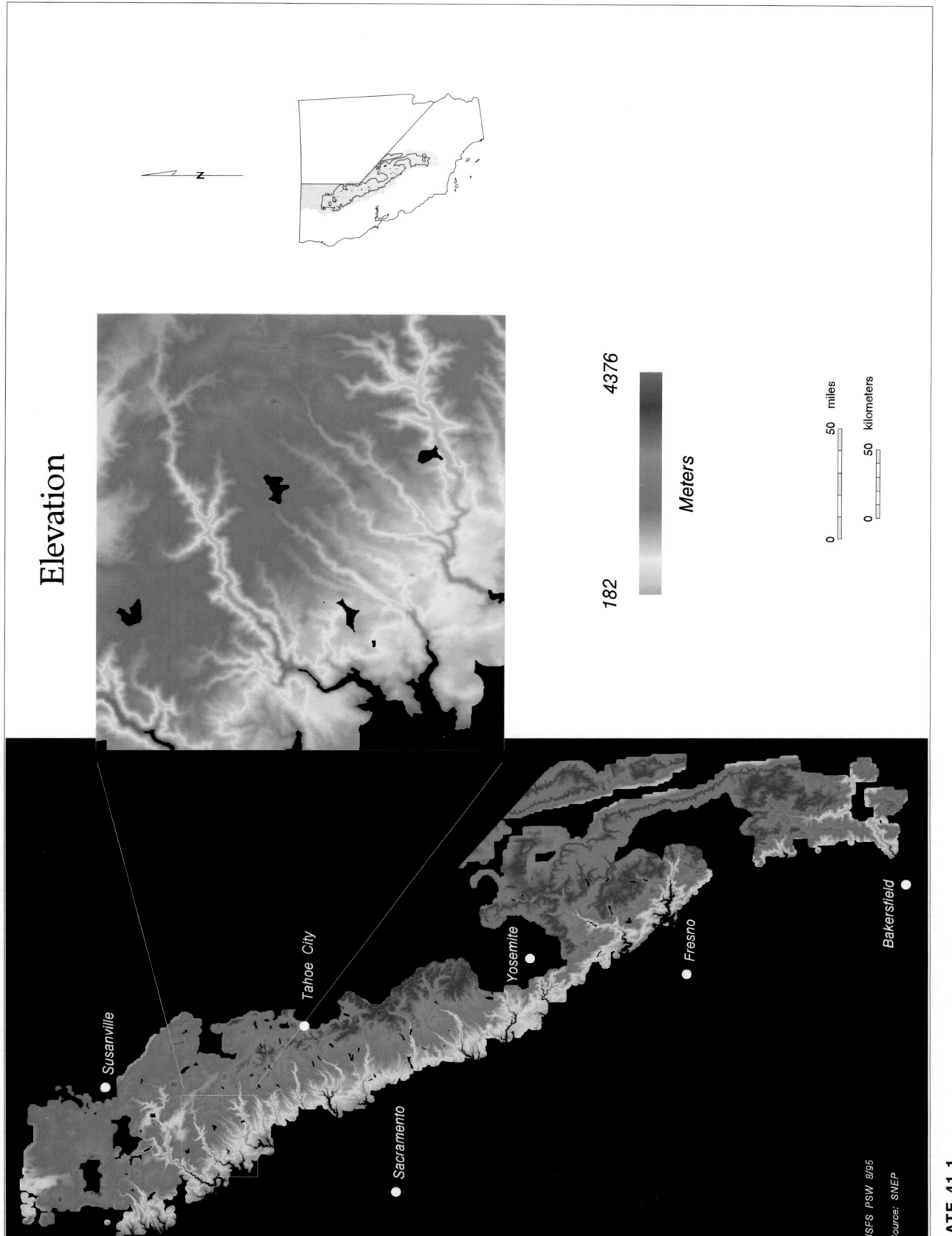

Elevation

182

4376

Meters

USFS PSW 8/95

Source: SNEP

Susanville

Sacramento

Tahoe City

Yosemite

Fresno

Bakersfield

PLATE 41.1

The elevation surface used to generate slope, aspect, and elevation for analysis of fire patterns. Elevation data are spaced 100 m (330 ft) apart, giving the map 1 ha (2.5 acre) resolution.

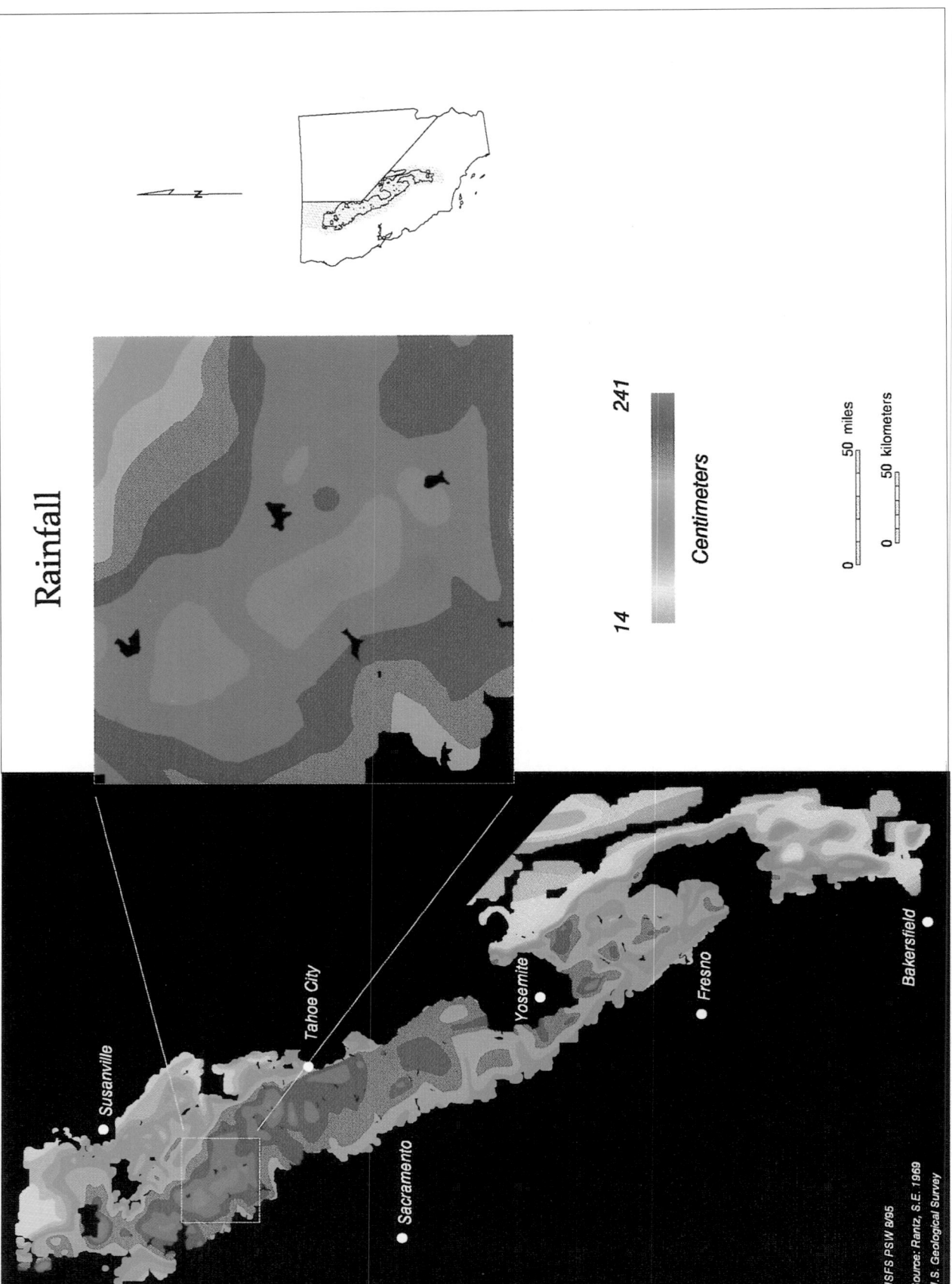

PLATE 41.2

Isohyetal map of precipitation patterns. Rainfall increases in the northern Sierra, whereas the highest elevations are in the south (adapted from Rantz 1969).

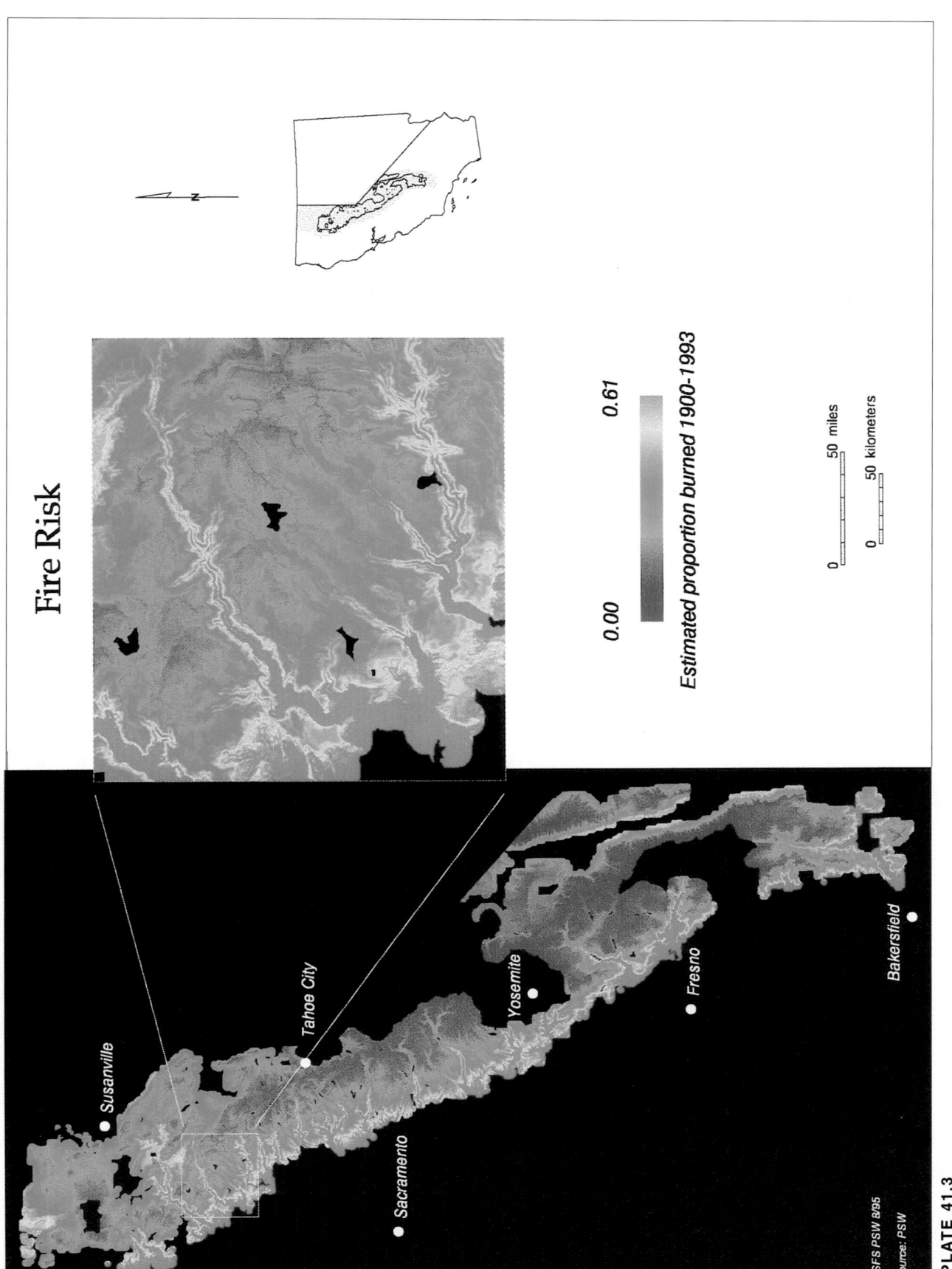

Fire Risk

Estimated proportion burned 1900-1993

0.00   0.61

0   50 miles

0   50 kilometers

Susanville

Sacramento

Tahoe City

Yosemite

Fresno

Bakersfield

N

USFS PSW 8/95

Source: PSW

PLATE 41.3

Map of model expectations for fire frequency for the entire study area. Risk levels were based on fire frequency for the period 1900–1993, using elevation, slope, and rainfall as independent variables.

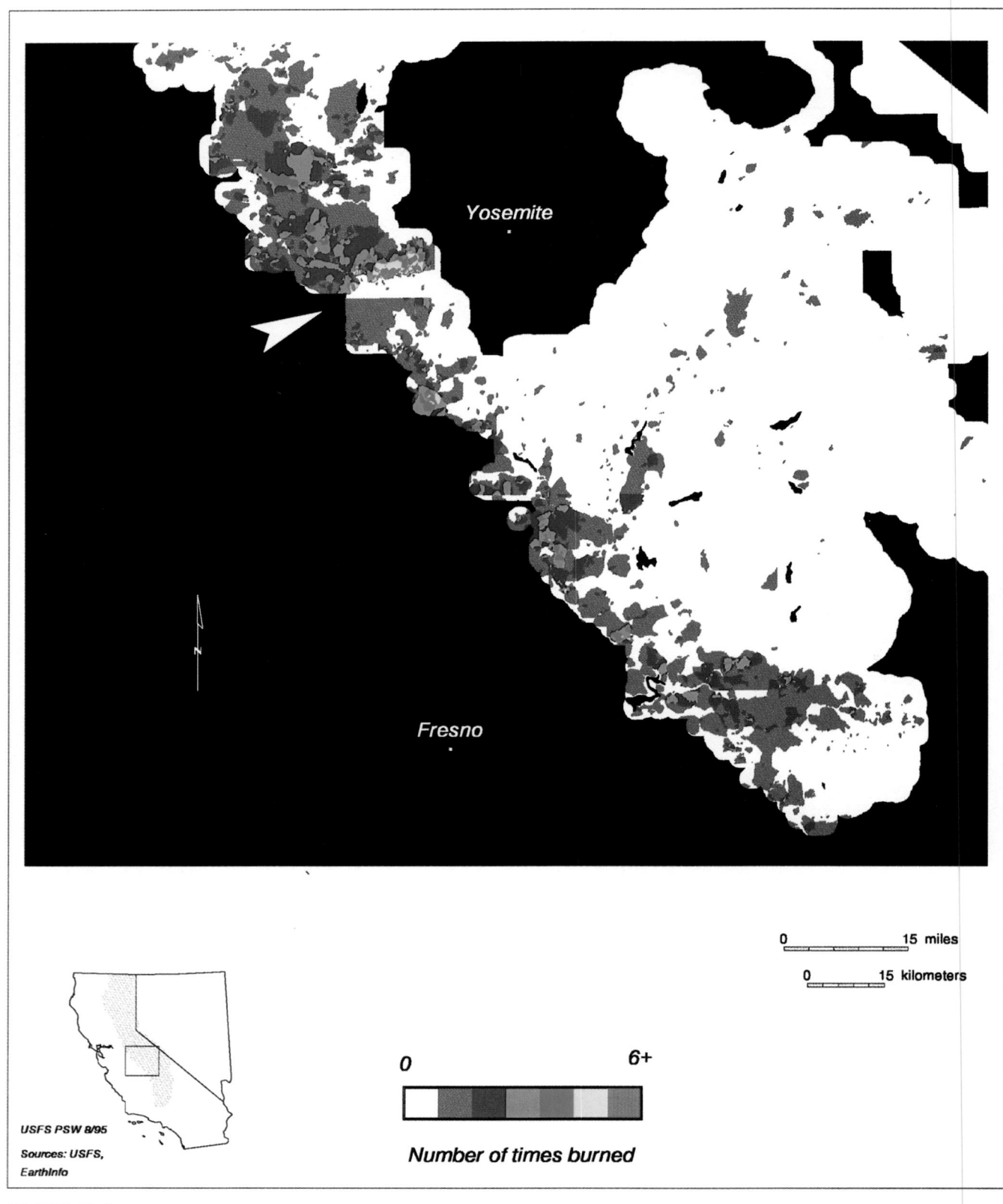

Yosemite

Fresno

0   15 miles

0   15 kilometers

USFS PSW 8/95

Sources: USFS,
EarthInfo

0   6+

Number of times burned

PLATE 41.4

An area of high fire activity in the southern Sierra Nevada. The area, which reburned more than 5 times, is confined to a small section on the north side of the Merced River, immediately adjacent to State Highway 140, a major route into Yosemite National Park (see arrow).

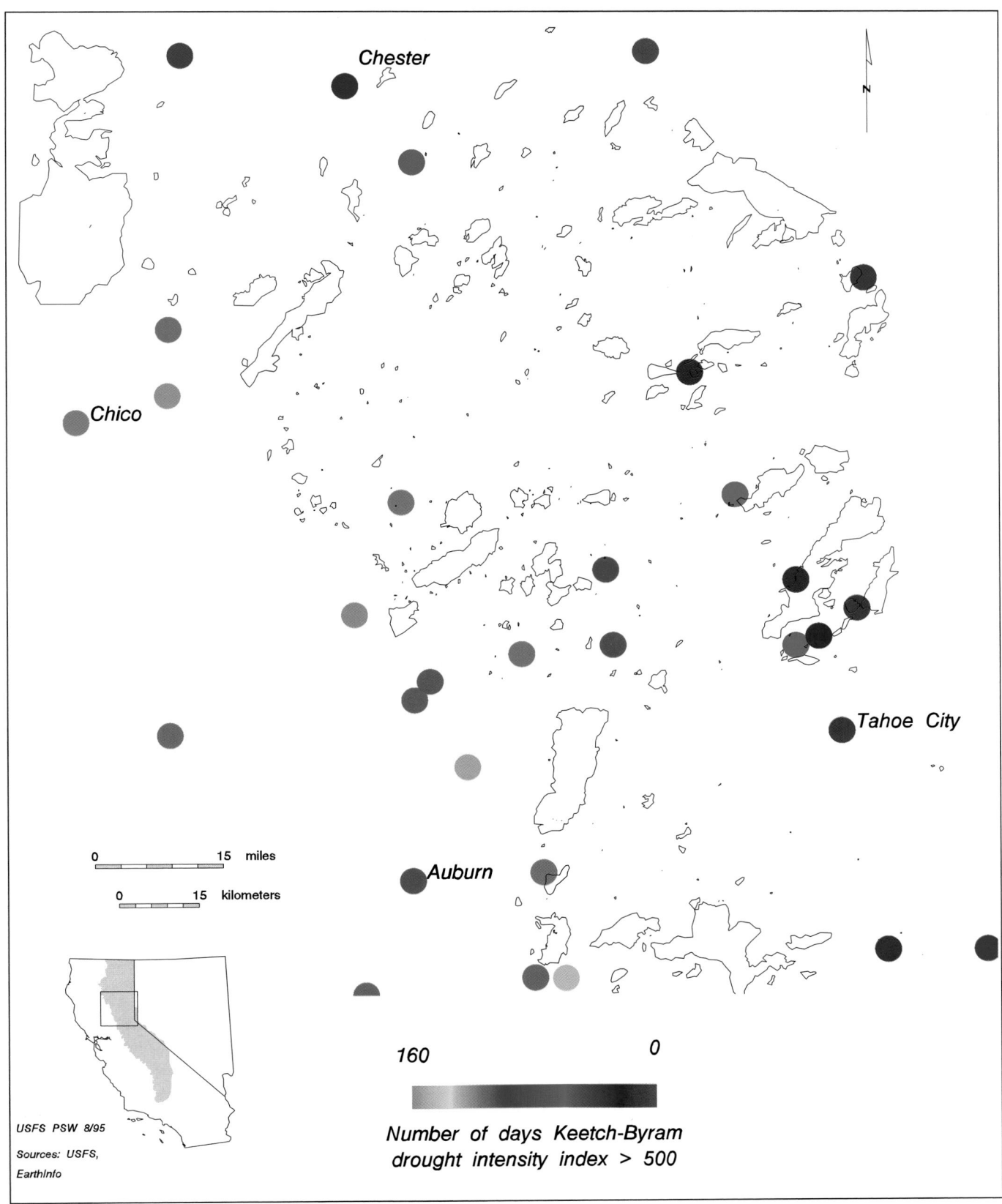

Chester

Chico

Tahoe City

Auburn

0   15   miles

0   15   kilometers

160   0

Number of days Keetch-Byram
drought intensity index > 500

USFS PSW 8/95

Sources: USFS,
EarthInfo

PLATE 41.5

Average number of drought days for the period 1979–89 for an area of the northern Sierra chosen due to the high density of recording stations.

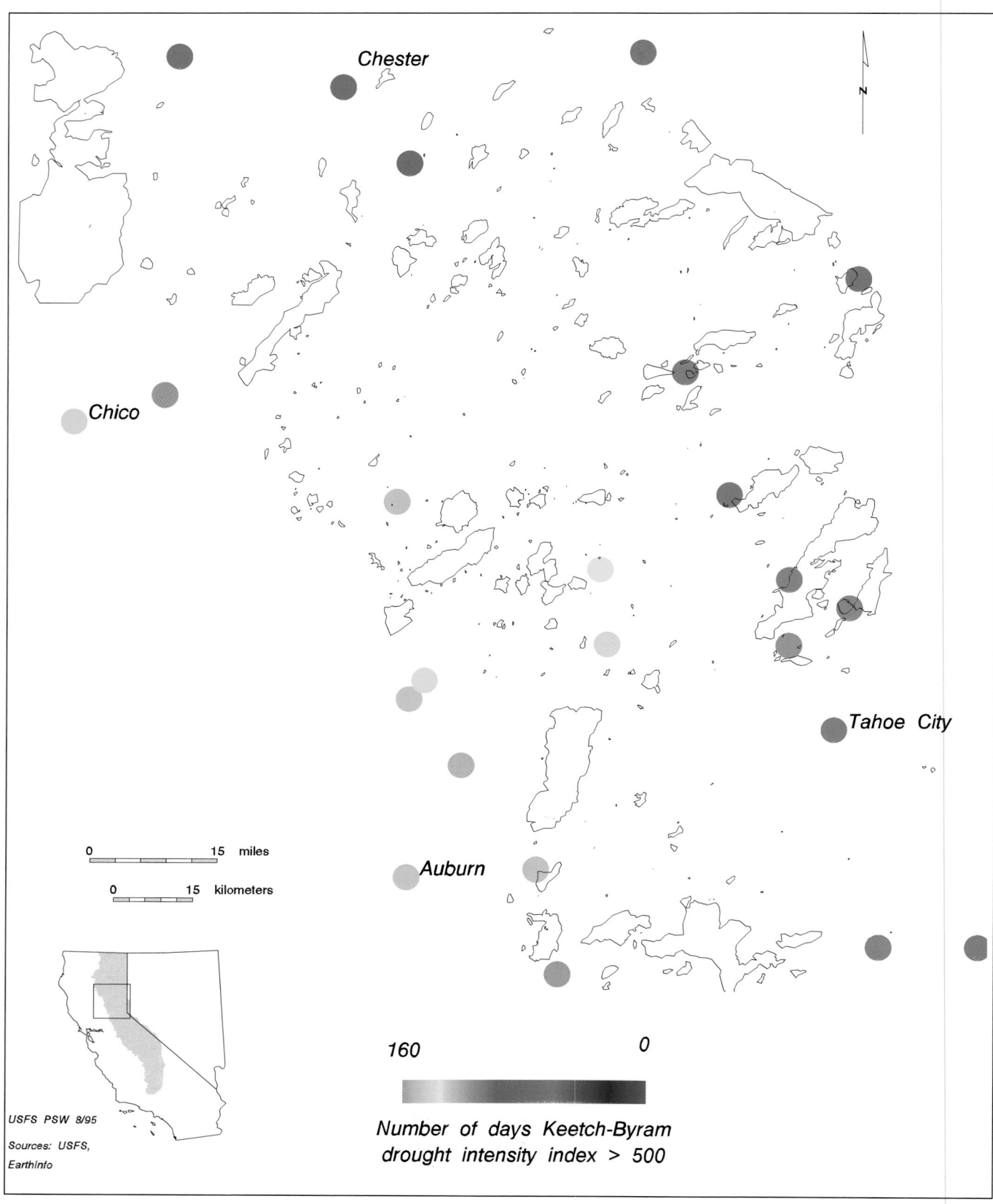

Chester

Chico

Tahoe City

Auburn

0   15 miles

0   15 kilometers

USFS PSW 8/95

Sources: USFS,
EarthInfo

160   0

Number of days Keetch-Byram
drought intensity index > 500

PLATE 41.6

Number of drought days in 1987 for the area shown in plate 41.5. More area burned in 1987 than in any year since the
1920s.

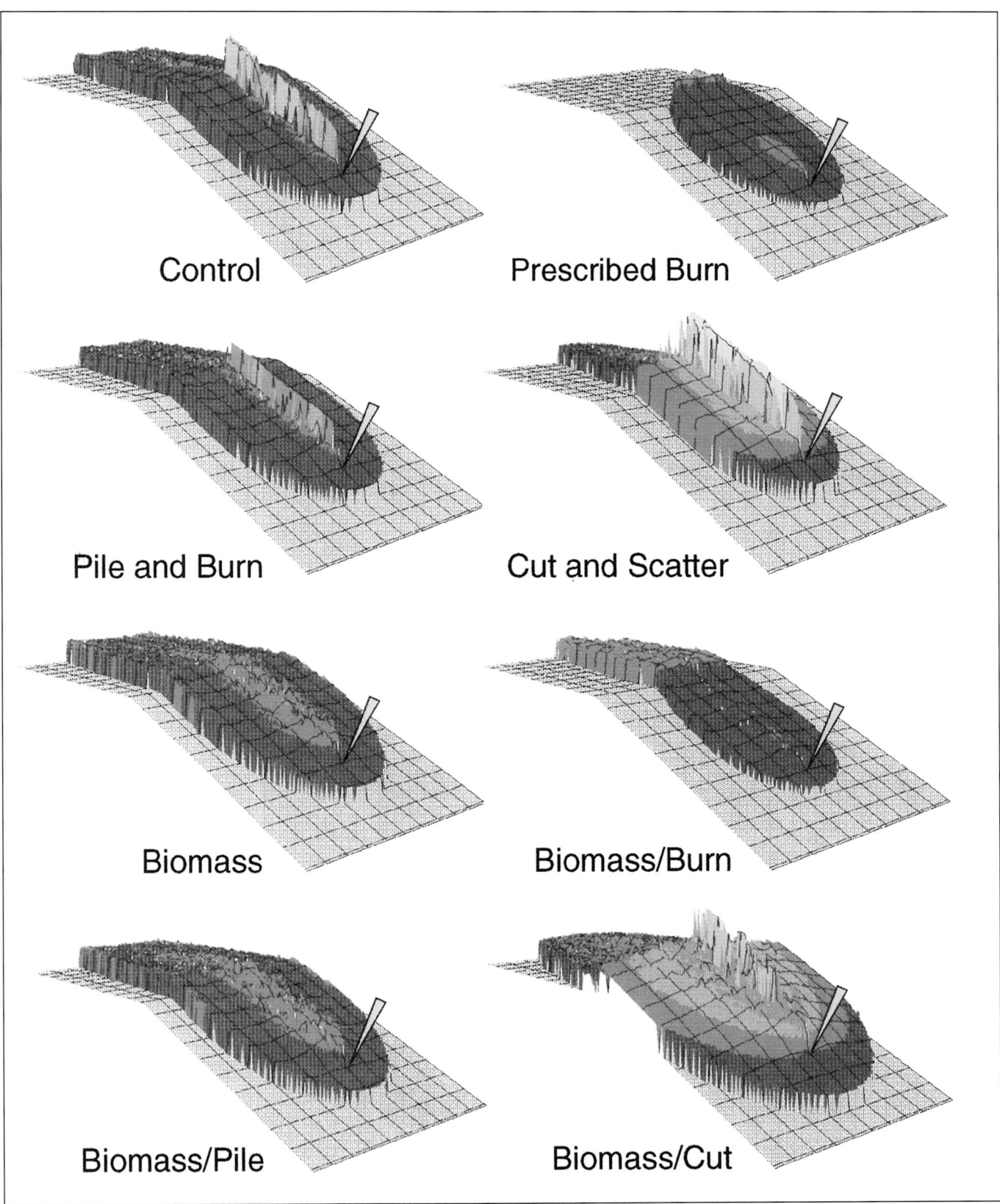

PLATE 43.1

Flame length for fuel-treatment simulations with 95th percentile weather.

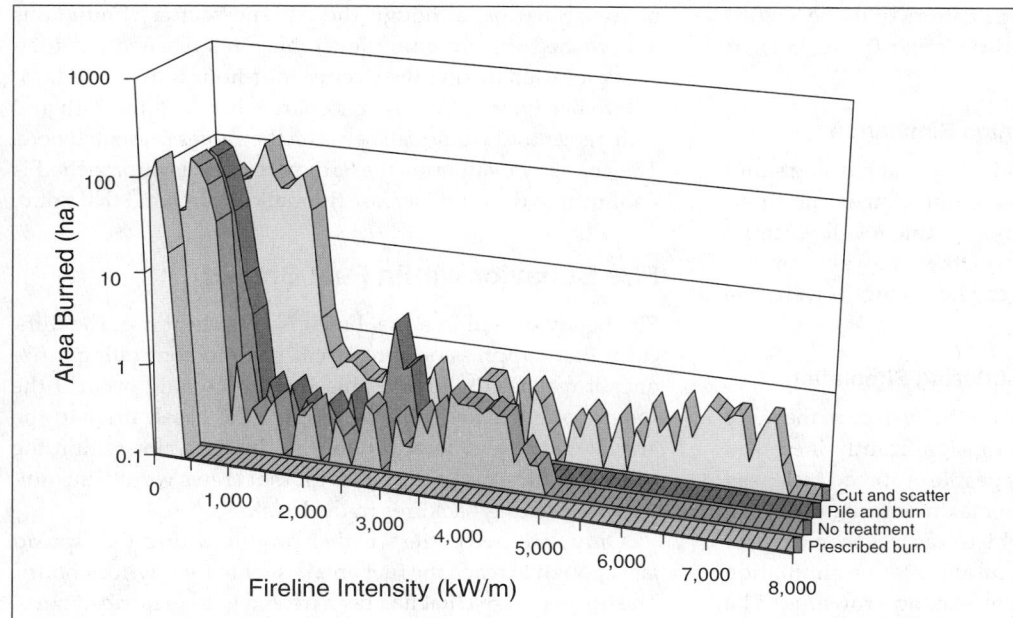

FIGURE 43.1

Fireline intensity for surface fuel treatments with no crown thinning under 95th percentile weather.

Although crowning did not occur with any of the biomassing treatments, because of the thinned canopies, flames approached the crowns when fuels were cut and scattered. The biomassing with prescribed-burning treatment had the smallest flame lengths. Since the surface fuels were the same between the biomassing treatment and the biomassing with piling and burning treatments, their flame lengths were similar. Biomassing combined with cutting and scattering the fuels resulted in the highest flames and the largest area burned.

Biomassing and Prescribed-Burning Simulation

Adding burning of surface fuels to the biomassing scenario resulted in reduced fire behavior values for subsequent fires. The average rate of spread, fireline intensity, and flame length were slightly greater than in the prescribed-burning scenario

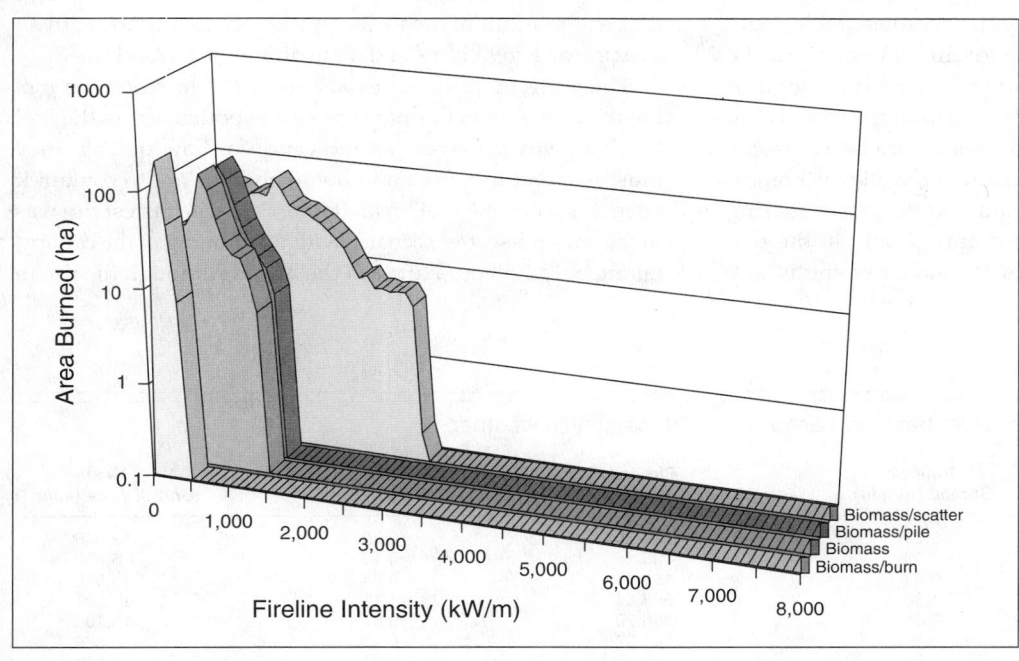

FIGURE 43.2

Fireline intensity for surface fuel treatments with crown thinning under 95th percentile weather.

without biomassing (table 43.4). The pattern of flame lengths over the simulation surface was also similar (plate 43.1), as was the heat per unit of area.

Biomassing and Piling and Burning Simulation

The results from the biomassing and piling and burning simulation were nearly identical to those of the biomassing simulation without surface fuel treatment (table 43.4). Since no crowning occurred in either scenario, the removal of understory fuels had no appreciable effect. Flame lengths were also virtually the same (plate 43.1).

Biomassing and Cutting and Scattering Simulation

Scattering the cut and lopped fuels on the surface in the more open conditions created by thinning significantly increased subsequent fire behavior. The average rate of spread increased to 3.28 m/min (10.76 ft/min). The maximum rate of spread was 7.75 m/min (25.43 ft/min), which was less than those of the control simulation and the cut-and-scatter simulation without biomassing, because there was no crowning. The average fireline intensity, however, at 1,070.74 kW/m was the highest of any scenario (table 43.4). The average heat per unit of area was only slightly less than that of the cut-and-scatter simulation. Flame length patterns approached those of the cut-and-scatter scenario, although the maximum flame lengths were shorter as a result of the lack of crowning in the thinned canopies (plate 43.1).

Fire Behavior within Treated Areas with 75th Percentile Weather

As would be expected, fire behavior for all scenarios was reduced when run with 75th percentile weather (table 43.6). Reductions ranged from 72% for fireline intensity in the biomassing and cutting and scattering scenario to 8% for heat per unit of area in the biomassing and piling and burning scenario. The lowest values for all behavior characteristics were for the prescribed-burning treatment, while the highest values were for the biomassing and cutting and scattering treatment. Torching and spotting occurred only in the two cut-and-scatter simulations. There was no crowning in any

of the scenarios, although the cut-and-scatter simulations approached the crowning level. No fires reached the fuel-break location during the twenty-four-hour burn period.

Fire behavior for weather scenarios between the 75th and 95th percentiles would fall between the values presented here. The change in values as the 95th percentile is approached is nonlinear and would increase dramatically beyond that point.

Fire Behavior within Fuel Breaks

Fire behavior within a fuel break is an indicator of the difficulty that suppression forces will have in controlling a fire once it reaches the break. This situation would occur if the forces have not had time to prepare the break prior to the arrival of the fire. In addition, the fire behavior within the break is indicative of the behavior that crews would encounter when setting backfires in the break.

Only 95th percentile weather produced fires that spread fast enough to reach the fuel breaks within twenty-four hours. The fire in the area that had been treated with prescribed burning, however, did not reach the fuel break within that time. Although fire behavior within fuel breaks was initially influenced by the rate of spread and fireline intensity of the fire when it reached the breaks, the behavior quickly adjusted, becoming determined by conditions within the break. The grass fuels with a sparse overstory of trees burned with an average rate of spread of 3.35 m/min (10.99 ft/min). The maximum rate of spread of 7.35 m/min (24.11 ft/min) within the breaks occurred on the sloped portion of the breaks. Fireline intensity averaged 99.80 kW/m and reached a maximum of 267.27 kW/m. The average flame length in the fuel breaks was 0.63 m (2.07 ft), with a maximum of 1.01 m (3.31 ft). Heat per unit of area was typical for grass fuels, with an average of 1,759 kJ/m² and a maximum of 2,195 kJ/m².

Knowing the time necessary for a fire to reach the fuel breaks, and whether or not those fires spotted across the fuel breaks, gives managers an indication of how quickly they must respond to a fire and how likely they are to contain it once it reaches the fuel break (table 43.7). The fastest fire was in the cut-and-scatter scenario with no thinning of the canopy, taking only 1.5 hours to reach the break. Although the fire in

TABLE 43.6

Average fire behavior for fires within fuel-treatment areas with 75th percentile weather.

Scenario	Rate of Spread (m/min)	Fireline Intensity (kW/m)	Flame Length (m)	Heat/ Unit of Area (kJ/m²)
Control	0.81	177.20	0.83	13,049
Prescribed burn	0.59	36.44	0.40	3,638
Pile and burn	0.83	181.84	0.84	13,095
Cut and scatter	1.36	393.84	1.19	17,321
Biomassing	0.89	196.28	0.87	13,106
Biomassing and prescribed burn	0.65	40.37	0.42	3,666
Biomassing and pile and burn	0.90	199.28	0.87	13,147
Biomassing and cut and scatter	1.49	435.56	1.25	17,434

TABLE 43.7

Time to reach fuel breaks and spot fire occurrence beyond fuel breaks for fires burning with 95th percentile weather.

Scenario	Hours to Reach Fuel Break	Spot Fires beyond 90 m Fuel Break	Spot Fires beyond 390 m Fuel Break
Control	2.5	Yes	No
Prescribed burn	—	No	No
Pile and burn	3.0	Yes	No
Cut and scatter	1.5	Yes	No
Biomassing	4.0	Yes	No
Biomassing and prescribed burn	20.5	No	No
Biomassing and pile and burn	4.0	Yes	No
Biomassing and cut and scatter	3.5	Yes	No

the prescribed-burn treatment area did not reach the break in 24 hours, when the canopy was thinned, it took 20.5 hours for a fire to burn into the break.

If it is assumed that control efforts make the fuel breaks impermeable to fire, their effectiveness in stopping fires is directly related to the ability of the fire to loft embers over the break to start spot fires. Table 43.7 shows that all scenarios except the prescribed-burn treatments produced fires that spotted across the 90 m (295 ft) breaks and that no fires spotted across the 390 m (0.24 mi) breaks.

DISCUSSION

Like any model, FARSITE is based on simplifying assumptions and has its limitations. Inherent in the model are the assumptions of homogeneous fuels within the map resolution, surface fire spread in an elliptical shape, simplified weather and wind inputs, and no extreme behavior such as fire whirlwinds, plume-dominated fires, or fire-induced weather. Albini (1976) points out, however, that the internal consistency of a well-disciplined model allows it to be used to assess the impacts of changes in important variables.

A serious limitation in applying models such as FARSITE to actual situations is the need for spatially accurate fuels data (Finney and Ryan 1995). Efforts are underway in Yosemite to use satellite imagery from the Thematic Mapper to provide high-resolution fuels data for research and management purposes. The techniques, when developed, will enable managers throughout the Sierra Nevada to acquire the needed data in a relatively simple and inexpensive manner. Currently, weather and wind data are provided from the nearest station, but these need to be supplemented with on-site observations. Given these limitations, however, the results from this study can be applied to similar situations in the Sierra

Nevada. In fact, if information is available to accurately model fuels for the various treatment scenarios, there is no reason why these results would not hold for other areas as well.

The results described in this chapter amplify the proposal by Weatherspoon and Skinner (1996) for a landscape-level strategy for fuels management in Sierra Nevada forests. They recommend a strategy of establishing a network of defensible fuel profile zones, enhancing the use of prescribed fire for restoring natural processes and meeting other ecosystem goals, and expanding fuel treatments to other appropriate areas of the landscape consistent with management goals. The fuel zones are envisioned to be similar to the 390 m (0.24 mi) wide fuel breaks tested here, except that crown cover would be reduced to between 20% and 50% rather than to between 1% and 20%. Weatherspoon and Skinner (1996) emphasize that construction of the fuel zones must be combined with treating fuels within areas bounded by the zones. This point is reinforced by this chapter, which indicates that fuel breaks alone will not be sufficient to stop all wildfires without some internal fuel treatment and active fire suppression.

The use of FARSITE to test fuel-treatment scenarios on simulated terrain has been extended to actual conditions in Yosemite National Park and Eldorado National Forest. Stephens (1995) used the FARSITE model to test fuel treatments for protecting the Tuolumne Grove of giant sequoias at the head of North Crane Creek in Yosemite. He tested prescribed burns of moderate intensity as well as the mechanical removal of ladder fuels and salvage logging with and without slash treatments. His results complement those of this chapter, reiterating the importance of fuel treatments such as prescribed burning and defensible fuel profile zones in areas requiring protection.

The policy scenarios proposed by Johnson et al. (1996) for managing late successional old-growth forests in the Sierra Nevada include four of the fuel treatments tested with FARSITE in this chapter. These are (1) no active management of fuels, (2) prescribed burning, (3) reduction of stand density with prescribed burning, and (4) fuel breaks with prescribed burning. They ran their analyses with and without budget constraints to learn which approaches would be most effective in protecting and restoring late successional forests. Their budget constraints were to require each treatment to pay for itself at the stand level and that the total budget spent on treatments must be less than a specified amount.

Recent work on mapping fire risk and fire hazard will make it possible to set priorities for treating areas. Using twentieth-century fire data from Sierra Nevada national forests, McKelvey and Busse (1996) found a strong correlation between elevation and fire frequency, with low elevations burning more frequently. Areas that burned more than three times were associated with special features such as roads. Greenwood (Sapsis et al. 1996) developed a fire hazard map of the Sierra Nevada by relating forest stand areas to fuel models and expected fire behavior. He found that low-elevation forests had the highest hazard and, because of their proximity to developed areas, the highest risk as well. These analyses indicate that the effective fuel treatments determined by this study would be most proficiently applied to low-elevation forests near high-risk areas.

CONCLUSION

The key mechanisms at work that affected the results of the simulations were the amount of surface fuels and the presence of low crowns or ladder fuels. If there is insufficient fuel on the ground either to cause the fire to spread quickly or to generate enough heat to move it into the crowns, sufficient time will be available either to suppress the fire or to use a fuel break ahead of the fire. Scenarios that did not treat surface fuels, such as biomassing only the overstory or piling and burning, did not appreciably change fire behavior. Adding the additional fuels resulting from cutting, lopping, and scattering understory trees and branches exacerbated fire behavior.

In those scenarios where surface fuels were not treated or were increased, fires spread rapidly, were very intense, spotted ahead of the main fire, and moved into the crowns. High flame lengths and large values for heat per unit of area were associated with this behavior. This extreme fire behavior occurred when large accumulations of woody and duff fuels burned uphill, with the wind producing flames that reached the low crown bases. Not only are these fires difficult to suppress, they also do not provide adequate time for treating fuel breaks ahead of the fire.

An obvious next step is to assess the costs and benefits of the various treatments. If fuel breaks are not effective until they reach a certain width, the additional costs of widening and maintaining the breaks must be compared to the cost of treating fuels within areas bounded by them, with and without the appropriate use of fuel breaks. Future applications of FARSITE should include the testing of various control strategies, using various combinations of fuel breaks and fuel treatments. The model can simulate the setting of backfires and the creation of fire breaks by hand crews, mechanized equipment, and aerial retardants.

It is obvious from this simulation project and from actual experience that fuel breaks alone will not alleviate the spread of wildfire. Although fuel breaks can form effective barriers to a fire during a suppression action if they are cleared of all flammable fuels and they are wide enough, the time available to defend them is critical to their success. This time can be greatly increased if adjacent fuel treatments are accomplished beforehand. Prescribed burning appears to be the most effective treatment for reducing a fire's rate of spread, fireline intensity, flame length, and heat per unit of area. Not only are surface fuels reduced by this treatment, but understory and ladder fuels are also reduced to the point where spotting and crowning are not a serious threat. Removing a portion of the canopy has the obvious effect of reducing the chance of a crown fire with or without surface fuel treatment. A management scheme that includes a combination of fuel treatments in combination with other land-management scenarios is critical for successfully reducing the size and intensity of wildfires. Land-management agencies and private landowners must cooperate to take the necessary steps on their lands to reduce the risk of catastrophic fire. Prescribed fire, in conjunction with fuel profile zones, appears to be the most effective strategy to accomplish that goal.

ACKNOWLEDGMENTS

This project would not have been possible without the FARSITE model that Mark Finney has so tirelessly developed. It is a true labor of love, and I have enjoyed working with him to test it to its limits. Rob Anderson deserves thanks for running all of the scenarios and plotting the figures, and thanks to Les Chow for producing all of the data for the tables and figures. Peggy Moore helped by handling all of my administrative matters while I was immersed in the project.

REFERENCES

Albini, F. A. 1976. Estimating wildfire behavior and effects. General Technical Report INT-30. Ogden, UT: U.S. Forest Service.

———. 1983. Potential spotting distance from wind-driven surface fires. Research Paper INT-309. Ogden, UT: U.S. Forest Service.

Andrews, P. L. 1986. BEHAVE: Fire behavior prediction and fuel modeling system—BURN subsystem, part 1. General Technical Report INT-194. Ogden, UT: U.S. Forest Service.

Biswell, H. H. 1963. Research in wildland fire ecology in California. In *Proceedings first annual tall timbers fire ecology conference*, edited by E. V. Komarek, 63–97. Tallahassee, FL: Tall Timbers Research Station.

Brown, J. K. 1978. Weight and density of crowns of Rocky Mountain conifers. Research Paper INT-197. Ogden, UT: U.S. Forest Service.

Burgan, R. E., and R. C. Rothermel. 1984. BEHAVE: Fire behavior prediction and fuel modeling system—Fuel modeling subsystem. General Technical Report INT-167. Ogden, UT: U.S. Forest Service.

Clark, W. 1990. Fire behavior in relation to prescribed fires. Unpublished report. Fire Management Office, Yosemite National Park.

Finney, M. A. 1994. Modeling the spread and behavior of prescribed natural fires. In *Proceedings of the twelfth conference on fire and forest meteorology*, edited by J. D. Cohen, J. M. Saveland, and D. D. Wade, 138–43. Bethesda, MD: Society of American Foresters.

———. 1995. *FARSITE fire area simulator.* Missoula, MT: Systems for Environmental Management.

Finney, M. A., and K. C. Ryan. 1995. Use of the FARSITE fire growth model for fire prediction in the U.S. national parks. In *International emergency management and engineering conference*, edited by J. D. Sullivan, J. L. Wybo, and L. Buisson. Paris, France: International Emergency Management and Engineering Society.

Green, L. R., and H. E. Schimke. 1971. Guides for fuel breaks and the Sierra Nevada mixed-conifer type. Station Report. Berkeley, CA: U.S. Forest Service.

Johnson, K. N., J. Sessions, and J. F. Franklin. 1996. Some ecological and economic implications of alternative forest management policies. In *Sierra Nevada Ecosystem Project: Final report to Congress.* Davis: University of California, Centers for Water and Wildland Resources.

Lancaster, J. W. 1970. Timelag useful in fire danger rating. *Fire Control Notes* 31 (3): 6–8.

McKelvey, K. S., and K. K. Busse. 1996. Twentieth-century fire patterns on Forest Service lands. In *Sierra Nevada Ecosystem Project: Final report to Congress,* vol. II, chap. 41. Davis: University of California, Centers for Water and Wildland Resources.

Rothermel, R. C. 1972. A mathematical model for predicting fire spread in wildland fuels. Research Paper INT-115. Ogden, UT: U.S. Forest Service.

———. 1983. How to predict the spread and intensity of forest and range fires. Technical Report INT-143. Ogden, UT: U.S. Forest Service.

———. 1991. Predicting behavior and size of crown fires in the northern Rocky Mountains. Research Report INT-438. Ogden, UT: U.S. Forest Service.

Rothermel, R. C., and G. C. Reinhart. 1983. Field procedures for verification and adjustment of fire behavior predictions. Technical Report INT-142. Ogden, UT: U.S. Forest Service.

Sapsis, D., B. Bahro, J. Gabriel, R. Jones, and G. Greenwood. 1996. An assessment of current risks, fuels, and potential fire behavior in the Sierra Nevada. In *Sierra Nevada Ecosystem Project: Final report to Congress,* vol. III. Davis: University of California, Centers for Water and Wildland Resources.

Stephens, S. L. 1995. Effects of prescribed and simulated fire and forest history of giant sequoia (*Sequoiadendron gigantem* [Lindley] Buchholz.)–mixed conifer ecosystems of the Sierra Nevada, California. Unpublished Ph.D. diss., University of California, Berkeley.

Stephenson, N. L., D. J. Parsons, and T. W. Swetnam. 1991. Restoring natural fire to the sequoia–mixed conifer forest: Should intense fire play a role? In *Proceedings seventeenth tall timbers fire ecology conference,* edited by S. Hermann, 321–37. Tallahassee, FL: Tall Timbers Research Station.

U.S. Forest Service (USFS). 1995. Draft environmental impact statement: Managing California spotted owl habitat in the Sierra Nevada national forests of California (an ecosystem approach). San Francisco: U.S. Forest Service, Pacific Southwest Regional Office.

Van Wagner, C. W. 1977. Conditions for the start and spread of crown fires. *Canadian Journal of Forest Research* 7:23–24.

———. 1993. Prediction of crown fire behavior in two stands of jack pine. *Canadian Journal of Forest Research* 23:442–49.

van Wagtendonk, J. W. 1974. Refined burning prescriptions for Yosemite National Park. Occasional Paper 2. Washington, DC: National Park Service.

———. 1991. Spatial analysis of lightning strikes in Yosemite National Park. In *Proceedings eleventh conference on fire and forest meteorology,* edited by P. L. Andrews and D. F. Potts, 605–11. Bethesda, MD: Society of American Foresters.

van Wagtendonk, J. W., and S. J. Botti. 1984. Modeling behavior of prescribed fires in Yosemite National Park. *Journal of Forestry* 82 (8): 479–84.

Weatherspoon, C. P., and C. N. Skinner. 1996. Landscape-level strategies for forest fuel management. In *Sierra Nevada Ecosystem Project: Final report to Congress,* vol. II, chap. 56. Davis: University of California, Centers for Water and Wildland Resources.

C. PHILLIP WEATHERSPOON
Pacific Southwest Research Station
U.S. Forest Service
Redding, California

44

Fire-Silviculture Relationships in Sierra Forests

ABSTRACT

Many of the tools available for managing forested ecosystems lie within the disciplines of silviculture and fire management. These two sets of management practices, in fact, are commonly used in concert. Understanding the relationships between these two disciplines, therefore, can contribute to more intelligent ecosystem management. Silvicultural techniques mimic to varying degrees some of the disturbance functions—such as facilitating establishment of regeneration and influencing forest structure and composition—performed naturally by fire. This chapter provides a brief overview of some of these relationships for a range of stand structures and fire regimes. Effects of partial cuttings on fire hazard also are discussed. Research is needed to clarify basic relationships between fire regimes and the dynamics and structures of stands and landscapes. Adaptive management experiments also should be undertaken to determine the practicability and long-term ecological consequences of a range of silvicultural and fire treatments.

INTRODUCTION

Before Euro-American settlement, relatively frequent fires strongly influenced the composition, structure, and dynamics of most forest ecosystems in the Sierra Nevada, in concert with other disturbance factors (Ferrell 1996; Skinner and Chang 1996). These fires, mostly low to moderate in severity, caused changes by damaging or killing plants and setting the stage for regeneration (including sprouting of top-killed plants) and vegetation succession. They maintained surface fuels at fairly low levels, and in most areas kept forest understories relatively free of trees and other vegetation. In addition, fires influenced many processes in the soil and forest floor, including the organisms therein, by consuming organic matter, affecting nutrient cycling, and inducing other thermal and chemical changes (Agee 1993; Chang 1996). These fire effects in turn resulted in a wide array of effects on other ecosystem components and processes, including wildlife communities and watershed properties.

Human activities since the mid-1800s have greatly changed the occurrence, nature, and effects of fire in the Sierra Nevada (Husari and McKelvey 1996; McKelvey and Johnston 1992; Skinner and Chang 1996; Weatherspoon et al. 1992). Organized fire suppression, which began early in the twentieth century, has been extremely effective in limiting the area burned by wildfires (Husari and McKelvey 1996; McKelvey and Busse 1996). The resulting virtual exclusion of low- and moderate-severity fire has profoundly affected the structure and composition of most Sierra Nevada vegetation, especially in low- to middle-elevation forests. Conifer stands have become denser, mainly in small and medium size classes of shade-tolerant and fire-sensitive tree species. Stands have also become more complex when viewed vertically, but less complex and more homogeneous in terms of areal arrangement (Weatherspoon et al. 1992). "Selective" cutting of large overstory trees (McKelvey and Johnston 1992) and the relatively warm and moist climate that has characterized most of the twentieth century (Graumlich 1993) have probably reinforced these trends. Excessively dense stands have led to drought stress and bark beetle outbreaks, resulting in widespread mortality of trees in many areas and the potential for extensive additional mortality (Ferrell 1996). One consequence of

Sierra Nevada Ecosystem Project: Final report to Congress, vol. II, *Assessments and scientific basis for management options*. Davis: University of California, Centers for Water and Wildland Resources, 1996.

these changes has been a large increase in the amount and continuity of both live and dead forest fuels, resulting in a substantial increase in the probability of large, severe wildfires (Weatherspoon and Skinner 1996). In many areas, ecosystem diversity and sustainability appear jeopardized by these changes, even without the threat of severe fires.

The necessity of restoring and sustaining these at-risk ecosystems is emerging as a major challenge confronting those responsible for managing Sierra Nevada forests. The means to accomplish these goals is the subject of some controversy. Some would advocate a hands-off philosophy of forest management, one of "letting nature take its course." Such a philosophy may well be appropriate for some upper montane and subalpine forests that have been affected relatively little by past management activities and in which wilderness values and/or restoration of natural processes are primary management emphases, especially if lightning fires are permitted to resume their natural role. This approach, however, is very unlikely to be successful in most lower- and middle-elevation Sierran forests, whose presettlement disturbance regimes were dominated by frequent low- to moderate-severity fires (Skinner and Chang 1996). Given the excessive quantities of fuels present in most of these forests, continued fire suppression (which certainly is not a hands-off approach) at a minimum will be required to avoid wildfire losses that are completely unacceptable ecologically and socially. Suppression alone, however, will only exacerbate the growing problems of overly-dense stands and excessive fuels. In addition to fire suppression, therefore, some form of active management, designed to replace critical missing elements of the largely defunct historic disturbance regimes, is probably essential to begin to reverse these problems and to ensure the diversity and sustainable productivity of these forests into the future.

Many of the tools available for managing forested ecosystems, and thereby mimicking to various degrees the functions of historic fire regimes (Skinner and Chang 1996) or other disturbance processes, lie within the disciplines of silviculture and fire management. These two sets of management practices are commonly used in concert, and in fact the line between silviculture and fire management can be quite blurry. For example, cuttings can be effective in breaking up the horizontal and vertical continuity of live fuels in lower canopy layers or in pretreating a stand to facilitate the introduction of prescribed fire. Alternatively, cuttings can add fuels and otherwise increase wildfire hazard. Prescribed fire and other techniques are often used for the dual purpose of reducing hazardous fuels and preparing a site for successful establishment of tree regeneration. Silvicultural techniques are used to emulate some of the historic effects of fire on forest structure. In fact, prescribed fire itself is considered by some to be a silvicultural technique.

The many ecosystem functions of frequent low- to moderate-severity fire can be restored fully only through the use of fire. Silvicultural cuttings and other fire "surrogates" can substitute only partially for fire. As is described in the following section, silvicultural techniques can mimic to varying degrees some of the functions performed naturally by fire, including facilitating the establishment of regeneration and influencing forest structure and composition. A wide array of thermal and chemical effects of fire (Agee 1993; Kilgore 1973; Chang 1996), however, are not mimicked by other methods. Fire and fire "surrogates" also differ markedly in terms of other factors, including potential for soil compaction and components of biomass removed from a site (i.e., fire tends to consume greater proportions of smaller size classes of biomass, whereas larger size classes typically are removed by cuttings). Accordingly, it seems desirable for low- to moderate-severity fire—both prescribed fire and "managed wildfire" (Husari and McKelvey 1996; Weatherspoon and Skinner 1996)—to assume a considerably expanded role in the management of Sierra Nevada forests. In those areas from which such fire continues to be excluded, for whatever reasons, managers should recognize that some ecosystem components and processes will depart significantly from their historical ranges of variability (Manley et al. 1995), with mostly unknown consequences for long-term ecosystem viability.

Nevertheless, it is important to understand that the reintroduction of fire alone cannot restore millions of acres of degraded Sierra Nevada forests. Silvicultural techniques are needed in addition to or in lieu of fire in many areas to move conditions away from dense forests dominated by small trees and containing excessive fuels toward more open forests dominated by large trees. Given the realities of modern civilization, it is inconceivable that fire in its presettlement extent, frequencies, and severities could be restored fully to the Sierra Nevada. Even at a reduced scale, a number of factors constrain the use of both management-ignited prescribed fires and prescribed natural (lightning) fires (Husari and McKelvey 1996; Weatherspoon and Skinner 1996). Furthermore, like nonfire methods, prescribed fire cannot fully mimic the ecosystem functions of presettlement fire, at least in the short term. The effects of newly reintroduced fire are likely to be quite different from those of presettlement fires because the forests (including fuels) have changed so greatly. If fire alone were used, several sequential entries with prescribed fire would probably be necessary, especially in densely stocked stands with heavy fuel concentrations, before the desired forest conditions would be approached. Early prescribed burns in such stands would tend to be expensive and have a relatively high risk both of escapes and of undesirable fire effects. In contrast, where feasible and compatible with management objectives, appropriate silvicultural cuttings preceding prescribed burns may significantly speed the movement toward desired forest structure and composition and in turn could hasten the use of prescribed fire in a way that more nearly mimics the natural ecosystem functions of frequent low- to moderate-severity fire. Of course, cuttings also provide opportunities to meet human needs for jobs and utilization of wood fiber.

Understanding the relationships between silviculture and fire can contribute to more intelligent ecosystem management. This chapter provides a brief overview of some of these relationships, discussed in two general categories: (1) silvicultural cutting methods as approximations of stand and landscape structural effects of fire in different fire regimes, and the compatibility of fuel-management techniques with these cutting methods; and (2) effects of partial cuttings on wildfire hazard. Although even-aged cutting methods are discussed briefly, this chapter emphasizes methods other than even-aged ones because (1) they more closely mimic the natural disturbance regimes prevailing in most Sierra Nevada forests, and (2) any landscape-level needs for large, even-aged stands are likely to be met by severe wildfires and subsequent plantation establishment for the foreseeable future.

SILVICULTURE, FIRE REGIMES, AND FUEL-MANAGEMENT TECHNIQUES

Silviculture was originally developed to produce timber efficiently and sustainably (Smith 1962), and in fact timber production has been the principal focus of the discipline during most of its existence. In the minds of many, silviculture is still the handmaiden of timber management. Over the years, however, silviculturists and others have come to recognize that silviculture employs a powerful and flexible set of techniques for meeting a wide array of resource management objectives and desired values (Daniel et al. 1979; Helms and Tappeiner 1996). These techniques need to be used with intelligence and discrimination. In prescribing and implementing management treatments, the silviculturist should consider site capabilities, species requirements, and key ecological processes, including the natural disturbance (mainly fire) regimes that prevailed in the area. The extent to which these factors, especially fire regimes, have been considered in the past has varied considerably.

One key function that both silviculture and natural disturbance have in common is facilitating the establishment of regeneration. The long-term sustainability of any desired forest condition in the Sierra Nevada depends in part on adequate establishment of regeneration at suitable intervals. Silvicultural systems are designed to promote the establishment of regeneration and in fact are classified by the methods they use to achieve this goal and the types of structures they create (Ford-Robertson 1971). In most Sierran forest types, fire historically was the primary agent that set the stage for regeneration of conifers and many other plants. Fires typically produced at least two conditions that promoted conifer regeneration: they provided the mineral soil seedbed favored by many species for seed germination and seedling survival,

and they created openings ranging from a fraction of an acre to perhaps hundreds of acres—needed for survival and subsequent growth of shade-intolerant species. Other effects of fire that often influenced regeneration establishment included increased nutrient availability, reduced density of potentially competing vegetation, and reduced populations of soil microorganisms pathogenic to tree seedlings.

In many cases, regeneration was not established after a fire of low to moderate intensity burned through the understory. Such fires, however, influenced stand structure and species composition in other ways. A disproportionate percentage of smaller trees were killed by fire, thereby tending to keep the understory relatively open. In addition, fire discriminated against thin-barked or otherwise fire-sensitive species. Silvicultural counterparts exist for these nonregeneration functions of fire: thinning from below (removing smaller trees and leaving larger trees) and thinning to modify species composition. In fact, the short- to medium-term need most apparent in many Sierran forests is not the establishment of new regeneration but rather the removal, or thinning, of excessive numbers of small understory trees. This is a high priority, both to reduce the hazard of severe wildfire and to begin to restore forests to a healthier, more sustainable condition (Weatherspoon and Skinner 1996).

A Range of Fire Regimes and Their Associated Stand Structures

Fire as a disturbance event, and the variability in the way fire functions as reflected in various fire regimes, is largely responsible for the range of natural stand structures found in forests of the western United States. Stephenson and colleagues (1991, 322–23) defined five "fire types" representing points along a continuum of increasing dominance by intense fire (and decreasing survival by main canopy trees), in order to account for the patchy nature of fires:

(1) uniform low intensity, in which all or most canopy trees survive; (2) low intensity with patchy high intensity . . . in which groups of canopy trees are killed locally within a matrix of surviving trees; (3) mixed intensity, in which roughly equal areas of canopy trees are killed and survive, with neither obviously predominating; (4) high intensity with patchy low intensity, in which groups of canopy trees survive within a matrix of killed trees; and (5) uniform high intensity, in which all or most canopy trees are killed.

These fire types provide useful reference points in the sections that follow.

The natural fire regime of most Sierra Nevada forests is generally characterized as one of comparatively frequent fires of low to moderate severity, with small patches of high severity (Skinner and Chang 1996). This fire regime, which corresponds to fire type 2 (Stephenson et al. 1991), prevailed historically in most ponderosa pine and mixed conifer forests both west and east of the Sierran crest and in portions of the upper montane forests as well. Greater variability in fire regimes occurred in more mesic sites within the mixed conifer forest type, especially those dominated by white fir, and in significant portions of the red fir and other upper montane types (Skinner and Chang 1996). This greater variability in fire regimes probably translated to greater variability in fire types as well, so that significant, albeit probably small, proportions of these cooler and/or more mesic types may have been characterized by fire types 3, 4, or 5 (Stephenson et al. 1991).

It is noteworthy that the extensive changes in Sierran forests brought about largely by fire suppression and other human activities over the past 150 years have included a virtual reversal of fire types (Stephenson et al. 1991). Fire type 2, historically the dominant fire type in Sierra Nevada forests, has now been virtually eliminated. Conversely, fire types 4 and 5, relatively rare historically, now account for a large proportion of wildfire acreage in the Sierra Nevada.

As was noted earlier, fire type 2 (Stephenson et al. 1991) corresponds to the presettlement fire regime that evidently dominated most Sierra Nevada forests, especially those low- to middle-elevation forests now in greatest need of restorative management. The corresponding stand structure type (a mosaic of small, even-sized groups) and its silvicultural counterpart (the group selection cutting method) are therefore of special interest in the discussion that follows. Three additional basic stand structures are discussed, however, in the interest of providing information on a more complete range of silvicultural and fire techniques to accommodate varied current stand conditions and to help meet management objectives for achieving structural diversity across the landscape. The extent to which it is desirable to mimic with management the kinds of stand and landscape structures associated with presettlement fire regimes (as best we can reconstruct those structures) is a subject of debate. At a minimum, however, we need to recognize and understand those historic structures as a frame of reference, so we know what we are departing from and can better assess the significance and sustainability of such departures.

The sections that follow contrast even-aged stands with three other basic types of stand structures that may be found in more of our managed forests in the future. These are simplified representations of stand structure; the real world is more complex. Nevertheless, they should provide useful reference points for illustrating silvicultural alternatives. One could probably approximate any realistic stand structure by varying the arrangement and stocking of particular canopy levels, using one of these four structures as a starting point. A desired stand structure could also be viewed as a point on the multidimensional continuum connecting the four basic types of structures. For example, as the structure created by the retention shelterwood cutting method becomes clumpier, it begins to approximate the structure created by the group selection cutting method; as openings created by group selection cuttings become larger, they grade into small clearcuts; as the openings become smaller, the structure approximates that created by the individual tree selection cutting method. Stand components other than live trees—such as snags, downed logs, and nontree vegetation—are also important parts of stand structure for many purposes, and within limits they can be manipulated silviculturally. For simplicity, however, the live tree component is emphasized here.

The discussion that follows, which is adapted in part from McKelvey and Weatherspoon 1992, deals with generalized stand structures and associated management practices primarily at the stand level. Just as numerous stand-level variations in structure are possible, as was indicated earlier, it is important to emphasize that great flexibility also exists for distributing variations and combinations of these structures across the landscape and through time. This provides opportunities to arrange landscape-level vegetation structures to meet varying management objectives.

The sections that follow are organized around stand structures associated with different regeneration cutting methods. For each of these structures, however, nonregeneration, or intermediate, cutting methods such as thinnings are integral components of the overall silvicultural system, and, like regeneration cutting methods, mimic natural disturbance functions to various degrees.

Standard silvicultural terminology is used (Daniel et al. 1979; Ford-Robertson 1971; Smith 1962). As was indicated earlier, these silvicultural systems and the associated terminology were developed in the context of timber management. The terms, however, are descriptive of cuttings that result in a broad range of stand conditions—clearly of interest to many resource areas—and are widely used and recognized.

A short consideration of fuel-treatment options relevant to each of the basic stand structures is included. It is assumed that, to the extent practicable, fuels are removed from the site to promote utilization as well as to reduce wildfire hazard. In the case of partial cuttings (cuttings other than clear-cuts), this includes the removal of small understory trees that form hazardous fuel ladders. Historically, effective fuel management has not always been a strong emphasis, due largely to short-term economic considerations. However, it is becoming an increasingly important concern in treatments prescribed today.

With all of the cutting methods, the use of tractors or other ground-based machines for yarding logs or for piling or otherwise manipulating harvest residues is limited to relatively moderate slopes. Treatment options are much more limited on steep slopes.

Even-Aged Stands

In an even-aged stand, the ages of all of the trees in the stand are similar. Natural even-aged stands originate mostly from high-severity fires that kill the great majority of trees in the stand (fire type 5) (Stephenson et al. 1991). With natural fire regimes, such fires in coniferous forests normally are separated by fairly long intervals (usually more than 100 years) and typically occur in forest types found in moist or cold regions.

Even-aged forest stands in the Sierra Nevada were probably relatively uncommon in the presettlement era. Such stands may have been represented best in portions of the upper montane forests—for example, in some red fir areas—and in widely-scattered stands of knobcone pine (Skinner and Chang 1996). In contrast, fire type 5 characterizes a large proportion of current wildfire acreage in the Sierra Nevada because of increased fuel quantities and continuity.

Silvicultural regeneration cutting methods that produce even-aged stands include clear-cutting, seed-tree, and shelterwood cutting. In a complete cycle of practices in the even-aged silvicultural system, such a regeneration cutting would normally be followed by establishment of a plantation or natural regeneration, removal of seed trees or shelterwood trees (retained initially to provide seed and/or protection for regeneration) where present, appropriate tending of the young stand, a series of intermediate cuttings (precommercial and commercial thinnings and possible "improvement" cuttings), and, at rotation age, another regeneration cutting to begin the cycle again. Either broadcast burning or machine piling and burning is commonly used to prepare the site for regeneration (including reducing competing vegetation and physical obstacles to planting) following the regeneration cutting. Underburning or other fuel treatments may take place at subsequent times during the life of the stand, especially after any intermediate cuttings. Prescribed burning is relatively straightforward in even-aged stands except when the trees are very young.

Even-aged stands resulting from even-aged silvicultural systems and from infrequent severe fires may be similar in terms of the general structure and arrangement of live trees. Other stand components, however, including large woody material such as snags and downed logs, and their ecological functions in the new stand, can be quite different in the two kinds of stands.

Two-Storied Stands

As the name suggests, two-storied stands consist of trees of two quite different ages and sizes. These stands are, in a sense, intermediate in structure between even-aged and uneven-aged stands. Natural two-storied stands tend to be associated with a moderate- to high-severity fire regime, in which only scattered live trees or clumps of trees (generally the larger trees and those of fire-resistant species) survive a fire within a matrix of killed trees (fire type 4) (Stephenson et al. 1991). The fire also promotes the establishment of a new age class of trees in the understory. Climates tend to be fairly moist but somewhat drier than those of the high-severity, long-interval fire regimes.

The presettlement occurrence of fire type 4 and two-storied stands in the Sierra Nevada was probably somewhat more frequent than fire type 5 and even-aged stands, although direct evidence of this is very limited. Some upper montane forests, along with the more mesic mixed conifer sites, such as those dominated by Douglas fir or white fir, may have accounted for much of this stand structure type (Skinner and Chang 1996).

The silvicultural technique associated with this kind of stand structure is retention shelterwood (also sometimes called irregular shelterwood or shelterwood without removal). Typically beginning with a shelterwood seed cutting, shelterwood trees (and trees reserved for other reasons) are left in place after regeneration has become established, instead of being removed. These trees may remain in the stand through much or all of the following rotation. Some will become snags, and some may be removed at the end of the next rotation (at which time a new set of overstory shelterwood trees will be selected for retention).

Other conditions could be used as starting points for creating a two-storied stand structure. Understocked stands, traditionally a high priority for clear-cutting, could instead be underplanted, leaving most of the overstory in place. This kind of structure could also be initiated in an older plantation by having a heavy commercial thinning double as a shelterwood-type regeneration cutting. The cut could be followed by site preparation/fuel treatment and underplanting with the desired mix of species. Throughout the "rotation" of such a stand, thinnings could be conducted as needed to maintain desired size classes and species. These should be followed by prescribed burning or other fuel treatments such as mastication or chipping. Snags could be created as needed. Once created, the stand would never be devoid of large trees: each regeneration cutting would be accompanied by the retention of some overstory trees.

Fuel treatments, including prescribed burning, should not be particularly difficult for a two-storied stand. Initial site preparation/fuel treatment before establishment of the understory would be the same as for a shelterwood cut. Subsequent treatments would be comparable to those for an even-aged plantation. Separation of canopy layers would normally be sufficient to keep wildfires from torching into overstory crowns.

Uneven-Aged Stands Consisting of a Mosaic of Small, Even-Aged or Even-Sized Groups

In an uneven-aged stand of small, even-aged or even-sized groups, each of several age or size classes occurs in a number of small (mostly from 1/4 acre to about 2 acres in size) groups or aggregations distributed throughout the stand. For the most part, age or size classes are separated horizontally rather than vertically. Natural stand structures of this type originate primarily in fire regimes in which fires burn relatively frequently but generally at low to moderate severity. Most areas are underburned, with many small trees being killed but most large trees surviving. Scattered individuals and groups of main canopy trees, however, are killed where the fire locally flares up or burns more severely (or groups of trees previously killed by other agents such as bark beetles are consumed to varying degrees by the fire), leaving scattered small openings within a matrix of surviving trees (fire type 2) (Stephenson et al. 1991). The locally intense fire exposes mineral soil (a favorable substrate for seedling establishment) and temporarily reduces competing vegetation (including reserves of dormant seeds stored in duff and soil). Given good cone crops and favorable soil moisture and other conditions, tree seedlings become established. Seedlings in an opening may be even-aged—originating from a single cone crop—or they may become established over a number of years. This fire regime and this stand structure were common during the presettlement era in the Sierra Nevada, especially in the ponderosa pine and mixed conifer forest types (Skinner and Chang 1996).

Silviculturally this kind of stand structure is approximated with the group selection cutting method. Group sizes should be large enough to permit successful regeneration of shade-intolerant tree species. In a sense, each group can be regarded as a small even-aged stand, which can be carried through the full cycle of regeneration cutting, regeneration establishment and tending, intermediate cuttings, and regeneration cutting once again. So within a stand that contains many of these small, even-aged groups, the group (regeneration) cuttings can be accompanied by concurrent intermediate cuttings in the other groups within the stand (mimicking small, high-severity burn areas within a matrix of low- to moderate-severity fire). Keeping track of numerous small openings and groups for management purposes, long considered a major obstacle to the use of group selection, should be significantly easier with the advent of geographic information systems and satellite-based global positioning systems.

In groups to be regenerated, all trees could be removed, or, especially in larger groups, scattered live trees and/or snags could be retained. To facilitate fuel treatment and reduce damage to the surrounding stand, cut trees should be felled as much as possible into the newly created opening.

Openings could be regenerated, either naturally or artificially and with or without vegetation management (reduction of competing vegetation). Even with planting and vegetation management, growth of tree seedlings would be less in an opening typical of group selection than in a large opening because of competition for site resources from large trees surrounding the opening. (The degree of competition will depend on the density or stocking level of the surrounding stand as well as the distance from the edge of the opening.) Without planting and some control of nonconifer vegetation, however, the development of conifers could be delayed for several decades. Under such conditions, fuel treatment would be complicated as well.

The development of a mosaic of small groups could be initiated in a wide range of stand conditions—for example, in an older plantation, an uneven-aged young-mature stand, or an old stand with patchy, uneven distributions of size classes or species.

Harvesting and other treatments are more difficult and expensive in an uneven-aged stand with a mosaic of even-aged or even-sized groups than in an even-aged stand. Implementing group selection cuttings on steep slopes, however, is especially problematical. Helicopters can be used but are very expensive. This area is ripe for some good logging engineering research and development. Hopefully, practical and economically viable methods will be developed for using skyline systems to yard group selection cuttings while keeping damage to the residual stand within acceptable limits. This could also provide opportunities for cable yarding of residues or for the use of other means of reducing fuel loads, such as removing tree tops (which contain considerable potential fuel) together with adjacent merchantable logs.

Fuels should be treated not only in the regeneration openings but also in the rest of the stand. On machine-operable slopes, the whole range of mechanical fuel-management techniques would be available. These could include tractor piling and burning of slash in regeneration openings, mastication, and removal (with or without utilization). Residual stand damage and soil impacts, however, must be kept within acceptable levels. Machine size and capabilities and operator skill are all critical factors.

Prescribed understory burning is an option on steep as well as moderate slopes. Prescribed burning would be more difficult than in even-aged or two-storied stands, simply because a variety of conditions and tree sizes occur within the stand. However, the fact that these size or age classes are separated horizontally rather than vertically, if combined with proper temporal spacing of treatments (McKelvey and Weatherspoon 1992), should alleviate many of the potential problems. Two-stage burning (sequential burns under different conditions) or jackpot burning (burning of residue concentrations under conditions that impede fire spread into adjacent areas) may be applicable in some situations. One could broadcast-burn regeneration cut areas after harvest, and then underburn the rest of the stand at the same time or perhaps at a later stage, when understory fuels have dried a little more. Depending on stand conditions, some preburn treatment may be necessary prior to the first fire entry to reduce fuel ladders and overall flammability to acceptable levels. This could be expensive and might include biomass harvest, cutting and hand piling, or other methods. If litter from ponderosa pine is available, prescribed burns can be conducted under moister conditions and therefore in more difficult situations. Again depending on stand conditions, a first burn might create substantial additional fuel by scorching or killing (mostly small) trees, necessitating a second and possibly a third burn to get the fire hazard down to an acceptable level.

Uneven-Aged Stands Consisting of a Fine Mosaic of Individual Trees

In an uneven-aged stand containing a fine mosaic of individual trees, three or more sizes and ages of all tree species present are distributed more or less uniformly throughout the stand. Openings are very small, the size of individual large trees. This occurs in nature (at least in a sustainable mode) only in forest types composed entirely of shade-tolerant species and in fire regimes having very long fire-return intervals. It develops long after a stand-replacement fire, as the overstory begins to break up and a full range of understory canopy layers has had a chance to develop. This stand type is incompatible with frequent periodic fires. (Some observers have considered certain open-growing ponderosa pine stands with short fire-return intervals to have this kind of stand structure. In such cases, the distinction between stands of uneven-aged individual trees and stands of uneven-aged groups of trees becomes largely one of semantics.)

This stand condition is produced and maintained silviculturally using the individual-tree selection cutting method. Unless a definition of individual-tree selection is used that includes openings up to 1/4 acre or so (or involves very open stands), this method will not allow for adequate regeneration and development of shade-intolerant species on most sites. If the stand does not already consist of shade-tolerant conifers, it will move in that direction under this cutting method as long as such species are present in the area. Retention of the smallest size classes of trees well distributed through the stand—a necessity for sustaining this stand structure through time—creates dangerous fuel ladders and makes prescribed understory burning essentially impracticable.

On gentle terrain, various machine treatment methods are available, at least theoretically, for accomplishing individual-tree selection cuttings. Residues remaining after harvesting could be machine piled, chipped, or masticated. But skillful operators and tight controls over fuel-treatment activities would be necessary to avoid unacceptable damage to the residual stand.

Other alternatives include jackpot burning of slash concentrations and the much more costly option of hand piling and burning—either applied preferably at a time when surrounding fuels are too moist to carry fire. Both of these methods would also be available on steep slopes. Implementation of individual-tree selection on steep slopes may be feasible only with expensive helicopter logging systems.

At higher elevations or other mesic sites where the probability of severe wildfire is not great, some combination of lopping, bucking, and scattering of slash, or no fuel treatment at all, may be acceptable. If individual-tree selection is to be used at all, it will be on such mesic sites that it probably makes the most sense anyway because it is more nearly compatible with presettlement fire regimes and stand and landscape structures.

EFFECTS OF PARTIAL CUTTINGS ON WILDFIRE HAZARD

The effects of partial cuttings on wildfire hazard in the residual stand result from combinations and interactions of two general factors: effects on fuels, and effects on microclimate.

Effects of Partial Cuttings on Fuels

Thinnings, insect sanitation and salvage cuts, and other partial cuttings add slash, or activity-generated fuels, to the stand unless all parts of the tree above the stump are removed from the forest. Small trees damaged by harvest activities but not removed from the forest often add to the fuel load. To the extent that it is not treated adequately, this component of the total fuel complex tends to increase the probability of a more intense, more damaging, and perhaps more extensive wildfire.

Foliage and small branches of live forest vegetation also contribute to the total amount of available fuel. The position and continuity of these fuels are important. Dense understory trees, for example, can provide both the horizontal and the vertical continuity of live fuels needed to move a fire from the surface into the main forest canopy and sustain it as a crown fire. This kind of stand condition is currently widespread in Sierra Nevada forests. Cutting and removing a large proportion of such a dense understory, thus interrupting much of the live fuel continuity, can substantially reduce the probability of a crown fire.

Partial cuttings also have longer-term, more indirect effects on fuels. Thinning or not thinning overly dense stands, for example, influences overall levels of competition for limiting resources (water, nutrients, and sunlight) in the stand and consequent levels of stress-induced mortality (including but not limited to that caused by insects). Dead trees obviously add to the total dead fuel load and may increase both the severity of a future wildfire and its spread rate via spotting. Thinning also influences the subsequent regeneration and development of understory vegetation—trees, shrubs, and herbs—which becomes part of the live fuel component.

Effects of Partial Cuttings on Microclimate

A related but separate kind of concern has to do with changes in microclimate brought about by stand opening. Thinning or otherwise opening a stand allows more solar radiation and wind to reach the forest floor. The net effect, at least during periods of significant fire danger, is usually reduced fuel moisture and increased flammability (Countryman 1955). The greater the stand opening, the more pronounced the change in microclimate is likely to be.

Interactions of Changed Fuels and Microclimate

The ways in which changes in these two sets of factors—fuels and microclimate—as a result of a management activity interact to affect wildfire hazard can be quite complex. The net effect, in terms of the direction of change in hazard, may be obvious in many cases, however. For example, removing most of the large trees from a stand, leaving most of the understory in place, and doing little or no slash treatment—a situation all too familiar in the past—will certainly increase the overall hazard and expected damage to the stand in the event of a wildfire. Everything points in the same direction: removing most of the fire-tolerant large trees; retaining most of the easily damaged small trees; increasing the loading (quantity) and depth of the surface fuel bed; and creating a warmer, drier, windier environment near the forest floor during times of significant fire danger. In contrast, heavily thinning an overstocked stand from below and using whole-tree removal (or chipping and spreading the limbs and tops), followed by a prescribed understory burn to reduce natural fuels, will almost certainly reduce the wildfire hazard of the stand. Computer simulations of the effects of such treatments on fire behavior (van Wagtendonk 1996), along with anecdotal reports of how such stands have fared during a wildfire in comparison with surrounding untreated stands, provide strong support for this conclusion. In this case, the "negative" effects on microclimate of opening the stand are outweighed by the reduction in live and dead fuel loading and continuity. Past cuttings in the Sierra Nevada (Helms and Tappeiner 1996) have spanned the range represented by these two contrasting situations but have tended generally, like the first situation, to create a net increase in fire hazard.

An example of a more complex relationship was reported by Weatherspoon and Skinner (1995) as part of a large retrospective study of factors—including prior management activities—that affected the degree of tree damage resulting from the extensive 1987 wildfires in northern California. Among three categories of uncut or partial-cut stands, they found that uncut stands (with no treatment of natural fuels) suffered the least fire damage, followed by partial-cut stands with some fuel treatment; partial-cut stands with no treatment had the most damage. The fact that partial-cut stands with no fuel treatment experienced more damage than partial-cut stands with some fuel treatment is no surprise. One might wonder, however, why the uncut stands experienced less damage than the partial-cut and treated stands. The explanation probably lies in a combination of the following factors:

- The partial cuttings created a warmer, drier microclimate compared with that of the uncut stands—an inevitable effect of cuttings, as was explained earlier.

- The partial cuttings were typical of many past cuttings that removed big trees and left small ones. The more readily scorched small trees thus constituted a higher percentage of the residual stand. Furthermore, the live fuel ladder component of fire hazard in the uncut stand was not reduced in the partial-cut stand.

- Fuel treatments may have been only partially effective. Two types of fuel treatments—lop and scatter and underburning—were combined in the analysis (their separate effects on fire damage were indistinguishable). Lop-and-scatter treatments reduced slash depth (and so presumably reduced flammability compared with no treatment) but did not change the fact that total downed dead fuel loading in those partial-cut stands (consisting of natural plus activity-generated fuels) was greater than downed dead fuel loading in uncut stands (consisting of natural fuels only). The underburns were not planned treatments but rather were burns that were allowed to creep around between clear-cut units that had been broadcast-burned or to move away from burned roadside piles. Thus, fuel consumption may have been spotty in these areas. More intensive treatment of surface fuels might well have reduced fire damage further.

- When only the management compartments containing fuel-treated stands (a small subset of the total number of compartments in the study) were analyzed separately, differences in fire damage between uncut and partial-cut and treated stands virtually disappeared. Evidently, lower average levels of damage in uncut stands in the remaining compartments changed the relationship in the overall analysis.

CONCLUSIONS AND RESEARCH NEEDS

Restoration and maintenance of Sierra Nevada forests in productive, sustainable conditions will almost certainly require combinations of silvicultural and fire-management techniques. Understanding the ecological and operational linkages between these two disciplines will facilitate this task.

It is generally recognized that recurring fires historically played a key role in influencing the species composition, stand structure, and landscape mosaic of most forest types in the Sierra Nevada as well as elsewhere in western North America. But the basic relationships between fire regimes and stand and landscape dynamics are poorly understood for many forest types, including those in the Sierra Nevada. Clarifying these relationships through research should help managers as they seek to define desired forest conditions and processes.

We also have little information about the long-term consequences of various forest conditions on a range of ecosystem components. The long-term nature of these questions and the need to find answers on a landscape scale means that the necessary studies will need to be done in the context of adaptive management, an organized process of learning by doing (Everett et al. 1994; Walters and Holling 1990). Managers and scientists should cooperate in long-term adaptive management experiments to (1) devise silvicultural and fire treatments that mimic historical or other desired conditions in certain key respects; (2) define treatments representing reasonable management alternatives that "bracket" those conditions; and (3) incorporate these treatments into long-term, interdisciplinary studies of the consequences of alternative management strategies in terms of ecosystem productivity, diversity, and sustainability. Because of the key role of fire historically and the broad range of fire effects on forest ecosystems, it is important that the suite of treatments include comparable stand structures produced and maintained by prescribed fire alone (requiring multiple burns), through silvicultural cuttings and mechanical fuel treatments alone (i.e., without fire), and through combinations of cuttings, mechanical fuel treatments, and prescribed fire. Only in this way will it be possible to determine which ecosystem functions of fire can be emulated satisfactorily by other means, which may be irreplaceable, and the implications of these findings for management.

Although the basic theory of silvicultural systems has been well established, actual application of systems other than even-aged ones in California is quite limited. Practical methods for implementing such treatments, especially on steep ground and in conjunction with a variety of fuel-treatment methods, will require considerable applied research as part of the adaptive management efforts discussed previously.

At least in the short to medium term, much of the needed silviculture in Sierran forests will involve thinning of small trees. To make such operations economically sustainable, cooperative research and development efforts are needed to develop more efficient technology for harvesting and processing of small material and new markets for utilizing it (Lambert 1994).

While we have much to learn, it is important to note that we do not have to have all the answers before beginning needed restoration work. We know enough at this point to recognize that current conditions in most low- to middle-elevation forests of the Sierra Nevada are unacceptable in terms of wildfire hazard, diversity, and sustainability. Regardless of the extent to which presettlement conditions are used as a guide to desired conditions, most informed people would agree that these forests generally should be less dense, have less fuels, and have more large trees. Even if we have not precisely identified target conditions, we certainly know the direction in which we should begin moving. That beginning alone will require a large measure of commitment and hard work. We can adjust along the way as we learn more and become better able to define desired conditions for Sierran forests.

ACKNOWLEDGMENTS

I would like to thank the following individuals for valuable comments on earlier versions of the manuscript: L. Blum, J. Buckley, J. Fites, D. Fullmer, M. Landram, D. Leisz, K. McKelvey, D. Parsons, C. Skinner, J. Tappeiner, J. Woods, and two anonymous reviewers.

REFERENCES

Agee, J. K. 1993. *Fire ecology of Pacific Northwest forests.* Washington, DC: Island Press.

Chang, C. 1996. Ecosystem responses to fire and variations in fire regimes. In *Sierra Nevada Ecosystem Project: Final report to Congress,* vol. II, chap. 39. Davis: University of California, Centers for Water and Wildland Resources.

Countryman, C. C. 1955. Old-growth conversion also converts fireclimate. In *Proceedings of Society of American Foresters Annual Meeting,* 158–60. Portland, OR: Society of American Foresters.

Daniel, T. W., J. A. Helms, and F. S. Baker. 1979. *Principles of silviculture.* 2nd ed. New York: McGraw-Hill.

Everett, R., C. Oliver, J. Saveland, J. R. Boeder, and J. E. Means. 1994. Adaptive ecosystem management. In *Ecosystem management: Principles and applications,* edited by M. E. Jensen and P. S. Bourgeron, 340–54. Vol. 2 of *Eastside forest ecosystem health assessment.* General Technical Report PNW-GTR-318. Portland, OR: U.S. Forest Service, Pacific Northwest Research Station.

Ferrell, G. T. 1996. The influence of insect pests and pathogens on Sierra forests. In *Sierra Nevada Ecosystem Project: Final report to Congress,* vol. II, chap. 45. Davis: University of California, Centers for Water and Wildland Resources.

Ford-Robertson, F. C., ed. 1971. *Terminology of forest science, technology, practice, and products.* Washington, DC: Society of American Foresters.

Graumlich, L. J. 1993. A 1,000-year record of temperature and precipitation in the Sierra Nevada. *Quaternary Research* 39:249–55.

Helms, J. A., and J. C. Tappeiner. 1996. Silviculture in the Sierra. In *Sierra Nevada Ecosystem Project: Final report to Congress,* vol. II, chap. 15. Davis: University of California, Centers for Water and Wildland Resources.

Husari, S. J., and K. S. McKelvey. 1996. Fire management policies and programs. In *Sierra Nevada Ecosystem Project: Final report to Congress,* vol. II, chap. 40. Davis: University of California, Centers for Water and Wildland Resources.

Kilgore, B. M. 1973. The ecological role of fire in Sierran conifer forests: Its application to national park management. *Quaternary Research* 3:496–513.

Lambert, M. B. 1994. Establish stable stand structures and increase tree growth: New technologies in silviculture. In *Restoration of stressed sites, and processes,* compiled by R. L. Everett, 93–96. Vol. 4 of *Eastside forest ecosystem health assessment.* General Technical Report PNW-GTR-330. Portland, OR: U.S. Forest Service, Pacific Northwest Research Station.

Manley, P. N., G. E. Brogan, C. Cook, M. E. Flores, D. G. Fullmer, S. Husari, T. M. Jimerson, L. M. Lux, M. E. McCain, J. A. Rose, G. Schmitt, J. C. Schuyler, and M. J. Skinner. 1995. *Sustaining ecosystems: A conceptual framework.* R5-EM-TP-001. San Francisco: U.S. Forest Service, Pacific Southwest Region.

McKelvey, K. S., and K. K. Busse. 1996. Twentieth-century fire patterns on Forest Service lands. In *Sierra Nevada Ecosystem Project: Final report to Congress*, vol. II, chap. 41. Davis: University of California, Centers for Water and Wildland Resources.

McKelvey, K. S., and J. D. Johnston. 1992. Historical perspectives on forests of the Sierra Nevada and the Transverse Ranges of Southern California: Forest conditions at the turn of the century. In *The California spotted owl: A technical assessment of its current status*, technical coordination by J. Verner, K. S. McKelvey, B. R. Noon, R. J. Gutierrez, G. I. Gould Jr., and T. W. Beck, 225–46. General Technical Report PSW-133. Albany, CA: U.S. Forest Service, Pacific Southwest Research Station.

McKelvey, K. S., and C. P. Weatherspoon. 1992. Projected trends in owl habitat. In *The California spotted owl: A technical assessment of its current status*, technical coordination by J. Verner, K. S. McKelvey, B. R. Noon, R. J. Gutierrez, G. I. Gould Jr., and T. Beck, 261–73. General Technical Report PSW-133. Albany, CA: U.S. Forest Service, Pacific Southwest Research Station.

Skinner, C. N., and C. Chang. 1996. Fire regimes, past and present. In *Sierra Nevada Ecosystem Project: Final report to Congress*, vol. II, chap. 38. Davis: University of California, Centers for Water and Wildland Resources.

Smith, D. M. 1962. *The practice of silviculture.* 7th ed. New York: John Wiley.

Stephenson, N. L., D. J. Parsons, and T. W. Swetnam. 1991. Restoring natural fire to the sequoia–mixed conifer forest: Should intense fire play a role? *Tall Timbers Fire Ecology Conference* 17:321–37.

U.S. Forest Service (USFS). 1995. *Draft environmental impact statement: Managing California spotted owl habitat in the Sierra Nevada national forests of California (an ecosystem approach).* San Francisco: U.S. Forest Service, Pacific Southwest Region.

van Wagtendonk, J. W. 1996. Use of a deterministic fire growth model to test fuel treatments. In *Sierra Nevada Ecosystem Project: Final report to Congress*, vol. II, chap. 43. Davis: University of California, Centers for Water and Wildland Resources.

Walters, C. J., and C. S. Holling. 1990. Large-scale management experiments and learning by doing. *Ecology* 71:2060–68.

Weatherspoon, C. P., S. J. Husari, and J. W. van Wagtendonk. 1992. Fire and fuels management in relation to owl habitat in forests of the Sierra Nevada and Southern California. In *The California spotted owl: A technical assessment of its current status*, technical coordination by J. Verner, K. S. McKelvey, B. R. Noon, R. J. Gutierrez, G. I. Gould Jr., and T. W. Beck, 247–60. General Technical Report PSW-133. Albany, CA: U.S. Forest Service, Pacific Southwest Research Station.

Weatherspoon, C. P., and C. N. Skinner. 1995. An assessment of factors associated with damage to tree crowns from the 1987 wildfires in northern California. *Forest Science* 41:430–51.

———. 1996. Landscape-level strategies for forest fuel management. In *Sierra Nevada Ecosystem Project: Final report to Congress*, vol. II, chap. 56. Davis: University of California, Centers for Water and Wildland Resources.

GEORGE T. FERRELL
U.S. Forest Service
Pacific Southwest Research Station
Redding, California

45

The Influence of Insect Pests and Pathogens on Sierra Forests

ABSTRACT

Currently, Sierra Nevada forests have high levels of mortality caused by bark beetles infesting trees stressed by drought, fire, overly dense stands, and pathogens. Fuel loads and fire hazard are high. Past logging and fire exclusion practices are partially responsible for this situation. Mitigative restoration requires thinning overly dense stands, primarily by controlled burning in parks and wilderness areas, combined with mechanical thinning and other selective tree-cutting practices elsewhere. Care will have to be taken to avoid creating more pest problems than are remedied, and some insect and disease activity should be tolerated as part of the restorative process. Failure to act may lead to continuation and perhaps worsening of present conditions. Widespread restoration should lead to sustenance of the biodiversity and productivity of these forests.

INTRODUCTION

Over the last decade, Sierra Nevada forests have sustained widespread, severe levels of tree mortality associated with a protracted drought. Most of the common conifer species and forest types were affected on both the western and eastern slopes of the range. Outbreaks of bark beetles were the proximate cause of tree mortality, affecting trees already stressed by drought and chronically weakened by pathogens and other agents (U.S. Forest Service [USFS] 1994). In California's Mediterranean climate, such outbreaks are not rare. Forest pest condition reports indicate that they have occurred somewhere in the Sierra Nevada and contiguous ranges in nearly every decade of this century (USFS 1917–49; California Forest Pest

Council 1951–93). Climatological and tree-ring studies have established that recurrent droughts have been a long-standing feature of the Sierra Nevada climate (Fritts and Gordon 1980; Fritts et al. 1979; Graumlich 1993).

Large and diverse communities of insects and microbes influence the structure, composition, and function of Sierra Nevada forests. Most of these insects and microbes are indigenous, and their effects are considered beneficial and necessary for forest health (e.g., pollination, nutrient cycling). A few, however, are considered key pest species because they can cause widespread and severe injury or mortality to the conifer species that are major components of Sierra Nevada forests and thus can adversely affect management goals. Several alien species are also considered key pests, either currently or potentially, based on the same criteria. Taken together, these key pest species are omnipresent in the Sierra Nevada, affecting virtually every major tree species in every forest type (Furniss and Carolin 1977; Scharpf 1993). Their effects have commonly been perceived as negative, and mitigation methods have been developed and applied. Mitigation has sometimes not been successful or cost-effective, depending to some degree on whether it was undertaken before (preventive) or during (remedial) a pest outbreak. As our understanding of forest ecosystem functioning increases, however, we are reassessing the roles of key pest species in forest health. Some of their effects may be viewed in the future as positive and even necessary for forest sustainability over the long term. Indeed, because many insects and pathogens can seriously injure only those trees and forests already under some form of environmental stress, epidemics of these agents are increasingly recognized as symptoms, rather than causes, of poor forest health (Wickman 1992).

Although bark beetle outbreaks in the Sierra Nevada are

Sierra Nevada Ecosystem Project: Final report to Congress, vol. II, *Assessments and scientific basis for management options*. Davis: University of California, Centers for Water and Wildland Resources, 1996.

commonly viewed as primarily drought-related, the severity of recent tree mortality has raised the question of whether there might be other important contributing factors. Attention has focused on whether the health of Sierra Nevada forests has declined as a result of human influences in the present century. In particular, it is believed that fire suppression and past logging practices have resulted in overly dense understories of the more shade-tolerant conifers, such as incense cedar, and the drought-susceptible white fir. In such stands, even species that are normally more drought-resistant, such as ponderosa pine, may suffer water stress and become susceptible to bark beetles during droughts. With their heavy complement of dead and dying trees and "fuel ladders" of highly flammable foliage, such stands are also highly susceptible to stand-destroying wildfires. Other factors may be contributing as well to this environmental stress and decline in forest health. Air pollution, urbanization, past logging practices, and invasion of alien pests (including insects and pathogens) are prominent among the factors being examined in this context. Nor has the possibility been ruled out that natural or anthropogenic (human-related) climatic shifts are causing changes in forest composition and structure.

Widespread concerns over the perceived decline in the health of Sierra Nevada forests have led to informal proposals of practices to mitigate and reverse current trends. Most of these proposals have the stated or unstated goal of restoring forest composition and structure to approximate those before settlement of the country by people of European origin. Inherent in this is the belief that restoration and maintenance of presettlement forests provide the best chance of sustaining these forests and their biodiversity in a reasonably intact and healthy state. It has been hypothesized that presettlement forests contained more old-growth (late-seral-stage) stands and that under extant fire regimes these stands were not choked with understory but rather more open and parklike. Unfortunately, because of the paucity of historical descriptions there is considerable uncertainty over just what these forests were like. There is also concern that some restoration and maintenance practices, if not judiciously applied, may have unforeseen side effects and cause more problems, especially with insects and pathogens, than they solve.

This chapter focuses on key insect pests and pathogens, both indigenous and alien, affecting major conifer components of Sierra Nevada forests. This focus was selected because conifers are the dominant vegetation in Sierran forests and therefore have the largest influence on the composition and functioning of these ecosystems. Moreover, many of the conifers are commercially valuable. Consequently, much more is known about the insects and pathogens affecting them than about pests with nonconifer hosts. Within this focus, the chapter describes current and past effects of key insects and pathogens on forest composition, structure, and functioning, as well as interactions with other environmental factors such as weather, fire, logging, urbanization, and air pollution. Criti-

cal knowledge gaps and research opportunities are identified. Finally, mitigation methods are described and evaluated in light of known or potential effects, and the potential results of taking no action or continuing present practices are evaluated. Particular attention is paid to potential effects on development and maintenance of old-growth (late-seral-stage) stands.

KEY QUESTIONS

1. *What are the current conditions and trends in Sierra Nevada forests?* It is necessary to describe current patterns and trends in forest conditions and effects of insects and pathogens as the starting point for assessing the effects of any changes in biotic and environmental factors, including human impacts, that may occur in the future.

2. *What are the key insect pests and pathogens affecting Sierra Nevada forests?* To understand the role of these agents in Sierra Nevada forests it is necessary first to identify which species have historically had major impacts and why (host tree species attacked, potential for spread, effects on the forest, history of outbreaks).

3. *What are the interactions of key insect pests and pathogens with other biotic and environmental factors known to be important influences on these forests?* Drought, fire, logging, air pollution, and urbanization have current or potential major influences on Sierra Nevada forests. Existing and potential interactions with these factors thus must be described in the context of the historical record in order to understand and predict the role of these agents.

4. *What are the effects of key insect pests and pathogens on the composition and structure of these forests?* To understand the role of these agents in influencing current conditions and to predict their effects in the future it is necessary to describe their effects in the past. Because Sierra forests fall into distinct and diverse forest types, it is necessary to describe the effects of these agents by forest type.

5. *What mitigation methods are available?* To explore how to mitigate current and future effects of key insect pests and pathogens it is necessary to examine mitigation methods currently available and assess their efficacy in the past.

6. *What major knowledge gaps exist?* It is necessary to identify major knowledge gaps in order to understand the uncertainties inherent in present assessments and predictions and to indicate the kinds of research and monitoring needed to fill these gaps.

7. *How will key insect pests and pathogens respond to various management scenarios?* Because past, current, and future environmental, biotic, and human conditions will constrain future management, certain management scenarios are more likely to be implemented in the future. It is therefore useful to explore probable responses of key insect pests and pathogens to these scenarios.

METHODS

The scientific and forest management literature, published and unpublished reports on pest conditions and on control projects compiled by federal and state agencies, and national forest management plans were used to identify key insects and diseases, their effects and interactions with other environmental factors, and their past and present effects on forest conditions. Key issues and relevant references were identified by means of a workshop convened in Davis, California, January 19–20, 1995, in which approximately twenty scientists offered their research findings and expert opinions on the subject. The same sources were consulted to identify mitigation methods and to assess their efficacy. Primary emphasis was given to reports from the Sierra Nevada, with reports from the contiguous southern Cascades and Modoc Plateau playing a supplementary role. All of this information was collated to develop models predicting the response of these agents to various environmental factors and management scenarios and the resulting impact on the effects of the agents. It was at this stage that major data and knowledge gaps became fully apparent. Except for a few research projects, quantitative data on trees killed by insects and pathogens were limited in time and space to outbreaks, while pest populations and tree mortality in other situations were reported merely as "endemic," "normal," or "balanced." Formal quantitative modeling was therefore abandoned, and primary reliance had to be placed on the development of informal conceptual models.

Assumptions inherent in these methods are all subsidiary to the main one that past interactions among insect populations, pathogen populations, and natural and human-caused environmental factors, and their past effects on forest attributes, are applicable to understanding the future. Observations, however, are limited to the present century. If future trends and ranges of variability in any of these components exceed those of the present century, future forest responses may well be unpredictable. As it was, particular care had to be taken to ensure that qualitative terms used in the past such as *endemic* or *normal* were comparable with current definitions. Also, it was necessary to be mindful of biases in the information from earlier decades of the century when forest conditions, management, and utilization differed from those of today. For example, previously only the larger and more commercially valuable conifers (pines) received much detailed attention, while smaller or less commercially valuable species (firs, cedars) were largely ignored.

Sierra Nevada forests and their associated insect pests and pathogens are highly variable because of the variety in their growing sites and histories. It was difficult if not impossible to present this wealth of variety in a report of reasonable length. To achieve this and also provide focus, much variation is covered only briefly, and emphasis is placed on key insect pests and pathogens affecting the important conifers of the major forest types. Key pests were those that have had continuing, widespread, major impacts on Sierra Nevada forests. In general discussions and summaries, the mixed conifer forest type, because of its wide areal extent and because it exemplifies the major problems perceived to be facing Sierra Nevada forests, was emphasized. To be optimally effective, however, management prescriptions must be site-specific and take into account the full range of variation both in these forests and in the insect pests and pathogens affecting them.

CURRENT CONDITIONS AND TRENDS

Currently, Sierra Nevada forests are in the aftermath of the 1987–92 drought. Over the last few years, these forests have sustained catastrophic levels of tree mortality due to drought, fire, disease, and bark beetles. In Sierra Nevada national forests alone, this mortality totaled over two billion board feet of timber by 1993 (California Forest Pest Council 1951–93, for the years 1990 and 1993). Although these losses have occurred throughout the Sierra Nevada, they have been particularly high on the east side of the range, where mortality, mainly of pines and firs, has exceeded 80% of the standing volume in some stands (U.S. Forest Service [USFS] 1994). Mortality has been greatest in overly dense stands, especially those where past logging and/or fire-exclusion practices have promoted tree species susceptible to insects, pathogens, fire, and drought. Wildfires also occurred during the drought, leaving many scorched trees susceptible to insects. Exacerbating these losses are the extreme fire hazards resulting from the dead and dying trees. Although levels of insect-caused tree mortality are expected to subside with the cessation of the drought, the accumulation of fuel in the dead trees and overly dense stands will cause the fire hazard to remain critical for some years. Adding to this threat are potential negative impacts from increased logging, inappropriate management practices, urbanization, air pollution, and invasion of alien insects and pathogens.

KEY INSECT PESTS AND PATHOGENS

Among the large and diverse communities of insects and microbes occurring in Sierra Nevada forests, relatively few are considered key pests and pathogens that have had major impacts on current as well as past forest conditions (USFS 1994). Virtually all of the key insect pests and pathogens are indigenous, but the threat from alien species is increasing, and some of these are considered potential key pests.

Indigenous Species

Insect pests and pathogens currently and historically having major impacts on Sierra Nevada forests are listed by host tree and forest type in table 45.1. All of these agents are indigenous except for white pine blister rust.

Bark beetles have the largest impacts, with sporadic outbreaks causing widespread tree mortality in virtually all major conifers and forest types. These beetles are a continuing source of mortality, but at lower levels, during nonoutbreak periods. Western pine beetles and mountain pine beetles are the major killers of ponderosa pine, and the latter are the major killer of sugar, lodgepole, and most other pines in the Sierra Nevada except for Jeffrey pine, whose major killer is the Jeffrey pine beetle. To a lesser extent, several species of pine engraver beetles also damage pines, but mortality is usually limited to smaller trees or tops of larger trees. The fir engraver beetle is the major killer of white fir and red fir, as the Douglas fir beetle is of Douglas fir, although outbreaks of the latter are rare in the Sierra. Trees killed by bark beetles are also often infested with other bark and timber beetles, but these are usually not considered the primary cause of death.

Having lesser impacts are defoliator insects, outbreaks of which can reduce tree growth and sometimes cause tree mortality if they are persistent or if the defoliated trees are rendered susceptible to bark beetles; noteworthy in this regard are Douglas fir tussock moth (*Orygia pseudotsugata*), pandora moth (*Coloradia pandora*), white fir needle-miner (*Epinotia meritana*), and lodgepole needle-miner (*Coletechnites milleri*). Effects of these defoliators have been more limited temporally and spatially than those of the key insect pests as defined here, and none of these defoliators is currently reported to be in outbreak status in the Sierra Nevada. Thus they were not considered key insect pests for the purposes of this report. However, their effects on Sierra Nevada forests will require reassessment should their outbreaks become more widespread and protracted, as they have in forests of eastern Oregon and Washington (Wickman 1992). Similarly, a variety of insects infest regeneration (seedlings and saplings) but do not usually cause serious impacts except in limited situations, such as plantations. If more large plantations are established in the aftermath of large forest fires, however, these insects may become key pests as defined in this report.

Pathogens having major impacts by killing trees on their own and by predisposing them to bark beetles are the root disease fungi, mistletoes, and white pine blister rust. Of the fungus-caused root diseases, the most important are annosus root disease, armillaria root disease, and black-stain root disease. Taken together, these root disease fungi infect most of the major conifers in Sierra Nevada forests. Spread is by aerial spores, subterranean root contacts, or, in the case of black-stain, bark beetle vectors. Other major pathogens are the dwarf and true or leafy mistletoes, various species of which weaken and sometimes kill, either by themselves or in concert with bark beetles, most of the conifer species of the Sierra Nevada. The sticky seeds are spread either by hydraulic bursting of fruits (dwarf mistletoes) or by birds (true mistletoes).

Alien Species

One introduced pathogen, white pine blister rust, is having a major impact on sugar pine throughout most of its range in the Sierra Nevada and is now spreading to other white pines (western white and whitebark pines) in the subalpine forests

TABLE 45.1

Key insect pests and pathogens and major conifer hosts in Sierra Nevada forests (USFS 1994).

Species	Host Trees	Forest Types
Western pine beetle (*Dendroctonus brevicomis*)	Ponderosa pine	Ponderosa pine, mixed conifer
Mountain pine beetle (*Dendroctonus ponderosae*)	Sugar pine, lodgepole pine, ponderosa pine, western white pine	Ponderosa pine, mixed conifer, red fir, subalpine conifer
Jeffrey pine beetle (*Dendroctonus jeffreyi*)	Jeffrey pine	Jeffrey pine, mixed conifer, red fir
Douglas fir beetle (*Dendroctonus pseudotsugae*)	Douglas fir	Mixed conifer
Pine engravers (*Ips* species)	All pines	All
Fir engraver beetle (*Scolytus ventralis*)	White fir, red fir	Mixed conifer, red fir
Annosus root disease (*Heterobasidion annosum*)	All conifers	All
Armillaria root disease (*Armillaria ostoyae*)	All conifers	All
Black-stain root disease (*Leptographium wageneri*)	Ponderosa pine, Jeffrey pine, Douglas fir, single-leaf piñon pine	Mixed conifer, Jeffrey pine, piñon-juniper
White pine blister rust (*Cronartium ribicola*)	Sugar pine, western white pine	Mixed conifer, red fir, subalpine conifer
Dwarf mistletoes (*Arceuthobium* species)	All conifers	All
True mistletoes (*Phoradendron* species)	White fir, incense cedar, western juniper	Mixed conifer, piñon-juniper

of the Sierra. First reported in California in 1929, this rust fungus infects stems of hosts, weakening them and predisposing them to bark beetles. Damaging infections of sugar pine are now so common and widespread that there is concern that sugar pine might not survive as a significant component of Sierra Nevada forests in the future.

Other alien species have been identified as major potential pests should they become established in the Sierra Nevada. Pine pitch canker, caused by an introduced pathogenic fungus (*Fusarium subglutinans* f. sp. *pini*), infects Monterey and bishop pines on the California coast. It kills twigs, branches, and tops, weakening trees and making them susceptible to bark beetles. Spread is primarily by bark and cone beetles. Except for pruning and disposing of infected material, methods of mitigation have not yet been developed. In coastal California, this disease has been found in ponderosa pine, a major component of Sierra Nevada forests. Although it has not yet been reported in the Sierra Nevada, there is concern that the disease may spread inland by transport of infected Christmas trees, firewood, or logs.

Originally from Europe, the gypsy moth (*Porthetria dispar*) is an introduced defoliator of oaks and other hardwoods in the northeastern United States, where outbreaks have caused widespread defoliation. Established infestations have not yet been reported in the Sierra Nevada, but several newly established infestations have been eradicated by insecticidal spraying elsewhere in California. Transcontinental spread is mostly by cocoons and egg masses adhering to vehicles or cargo. Feeding trials indicate that oaks and other hardwoods indigenous to the Sierra Nevada are suitable hosts for the gypsy moth caterpillar. Introduction of the Asian form of this moth, not yet reported as established in North America, might be particularly damaging to Sierra Nevada forests, as the female moths fly (the European females are flightless) and feed on some conifers as well as hardwoods. Assessments have indicated that the risk of introducing the Asian form is particularly high if proposed large-scale importations of Siberian logs occur (USFS 1991). Forests of western North America, including those of the Sierra Nevada, are considered potentially susceptible to a number of serious forest pest insects and pathogens from eastern Siberia because the Bering land bridge connection was recent in geological time, and this suggests close affinities with Siberian forests. All such risk assessments are hampered, however, by the impossibility of assessing all potential pests, and historically it has often been an unsuspected agent, such as chestnut blight devastating indigenous chestnuts in eastern North America, that has proved most destructive. This is particularly sobering in light of contemplated increased importations of raw logs and wood into the United States from other parts of the world.

PEST INTERACTIONS WITH BIOTIC AND OTHER ENVIRONMENTAL FACTORS

Biotic Complexes

The key insect pests and pathogens affecting Sierra Nevada forests usually function as members of biotic complexes in which the members are highly interactive (USFS 1994). Frequently, but not always, infection by one or more pathogens weakens the host tree, making it susceptible to bark beetles. Common complexes are root disease/bark beetles, mistletoe/bark beetles, and root disease/mistletoe/bark beetles. Membership in the complexes varies with changing environmental conditions. In years during and immediately after drought, increasing percentages of trees are killed by insects (primarily bark beetles) only, while percentages of trees killed by insects and pathogens are initially high and fall steadily (table 45.2). This indicates that trees already weakened by pathogens tend to be killed early in the drought, whereas later in the drought most trees are killed by bark beetles only. Populations of all of these insects and pathogens are primarily controlled by the number of susceptible host trees. Any factor that increases the latter will usually produce population increases or even epidemics of the former. In California's Mediterranean climate, drought is probably the most important predisposing factor (Ferrell and Hall 1975; Ferrell et al. 1994; Miller and Keen 1960). But overly dense stands, fire, logging, urbanization, air pollution, snow breakage, windthrow, and flooding can also weaken trees and cause them to become susceptible to pathogens and insects (Furniss and Carolin 1977; Scharpf 1993). Like biotic complexes, environmental factors can be highly interactive. Fire, for example, can exacerbate the effects of drought in increasing susceptibility of trees to insects and pathogens.

Site and Stand Factors

Overly dense forest stands are an important cause of tree susceptibility to insects and pathogens (Ferrell et al. 1994; Oliver

TABLE 45.2

Agents causing tree mortality during and after the 1975–77 drought in twelve northern California national forests (USFS 1994).[a]

| Agents | Annual Percentage of Trees with Agents | | | |
	1976	1977	1978	1979
Insects only	24	43	59	82
Pathogens only	1	2	1	1
Insects and pathogens	69	53	40	14
Other	6	2	0	3

[a]The national forests studied include those in the Klamath Province as well as those in the Sierra Nevada.

1995; Slaughter and Parmeter 1989; Scharpf and Parmeter 1976). Intense tree-to-tree competition in overly dense stands tends to slow growth and decrease resistance of trees. Spread of insects, disease, and fire is also enhanced in dense stands. Overly dense stands are a major cause of tree mortality in Sierra Nevada forests during both drought and nondrought periods. This is evident if average annual mortality over a decade or more is expressed as a percentage either of levels of growing stock (total stem volume) or of growth. Expressed either way, tree mortality tends to be higher in forests with higher levels of growing stock (stand densities), especially if growth, expressed as a percentage of growing stock, is also low (table 45.3).

Stand composition and quality of the growing site interact with stand density in their effects on forest susceptibility to insects and pathogens. Sites have specific capabilities for the amount of forest biomass they can grow, and, when that level is exceeded, mortality of the trees will eventually occur, especially during periods of drought. There is not an absolute level for any site, but general ranges do exist for a particular quality of site. Poorer sites are usually drought-prone, because they occupy steep slopes with hot, dry exposures and thin, rocky or coarse-textured soil. Typical examples of such sites are steep canyon walls and ridgetops. Over the long term, these sites are able to support only sparse forest stands and have a higher risk of tree mortality, especially from mistletoe/root disease/bark beetle complexes, during both drought and nondrought periods. These effects are particularly pronounced in lower-to-middle-elevation forests on both the western and (especially) the eastern slopes of the Sierra, which tend to be hot and dry (Ferrell 1986; USFS 1994). Effects of stand density and site quality are compounded by the influence of stand composition. For example, during the recent (1987–92) drought, mortality was particularly high in mixed conifer and east-side pine stands that had been invaded by drought-sensitive white firs (Ferrell et al. 1994; USFS 1994).

Drought

Recurrent droughts are characteristic of the Sierra Nevada climate. Summers are usually hot and dry, with the bulk of the precipitation occurring in winter, much of it as snow. But in addition to the dry summers, there have been droughts of one or more years' duration in nearly every decade of this century (Graumlich 1993). Tree-ring studies have established that, compared to weather over the previous two centuries, however, weather during the present century has been relatively moist, without the decades-long droughts that occurred earlier (Fritts and Gordon 1980; Graumlich 1993). Records in this century indicate that increases in tree mortality caused by bark beetles are often triggered by droughts (table 45.4). Increased tree mortality may continue for one to several years after the return of normal precipitation (table 45.5). Increased mortality usually occurs first at the lower and middle elevations on both western and eastern slopes of the range and spreads to the upper elevations only if the drought is protracted (California Forest Pest Council 1951–93). During droughts, lack of spring precipitation has a particularly large influence, not only by increasing the susceptibility of the trees, as indicated by their rates of growth and beetle-caused mortality (figures 45.1–45.3), but also probably by aiding dispersal of and host selection by the flying beetles.

Fire

Recurrent fires, particularly during droughts, are also characteristic of the Sierra Nevada (Swetnam 1993), although their frequency and effects have varied temporally and spatially (Martin 1982; Pitcher 1987; Parsons and DeBenedetti 1979; Taylor and Halpern 1991). Trees with scorched trunks or crowns can die directly from the injury (Wagener 1961), or they can become susceptible to tree-killing bark beetles and suitable for beetle reproduction, but this is highly dependent on the severity and pattern of scorching (table 45.6). Trees badly scorched are readily infested by bark beetles but may not be optimal for bark beetle reproduction, and those only lightly scorched may not be rendered susceptible to beetle attack. If the fire is not during a drought, usually only a sharp spike of tree mortality, confined largely to scorched trees, occurs, and no protracted bark beetle outbreak ensues (Miller and Keen 1960). During a drought, however, beetles emerging from scorched trees can spread to surrounding stands and

TABLE 45.3

Endemic levels of tree mortality in relation to growing stock and growth in six national forests in the Sierra Nevada (Bolsinger 1980).[a]

National Forest	Growing Stock (ft³/acre)	Annual Growth as a Percentage of Growing Stock	Annual Mortality As a Percentage of Growing Stock	As a Percentage of Growth
Eldorado	3,670	2.2	0.1	4.8
Inyo	1,830	1.4	0.1	7.6
Plumas	3,690	1.6	0.2	10.2
Sequoia	3,860	1.1	0.2	14.3
Stanislaus	4,460	2.0	0.2	7.4
Tahoe	4,380	1.7	0.2	12.2

[a]The Sierra National Forest is excluded because it was inventoried during the 1975–77 drought, when epidemic mortality occurred.

TABLE 45.4

Increased tree mortality in relation to subaverage annual precipitation in seven national forests in the Sierra Nevada, 1917–49 (USFS 1917–49).

Year	Subaverage Precipitation[a]	Increased Tree Mortality (More Than 0.2 Trees/Acre Killed) by National Forest[b]						
		Eldorado	Inyo	Plumas	Sequoia	Sierra	Stanislaus	Tahoe
1917	√			√				
1918	√			√				
1919	√	√	√		√	√	√	√
1920	√	√	√		√		√	√
1921				√				
1922	√		√	√		√	√	
1923	√	√	√	√		√	√	√
1924	√			√	√			√
1925	√	√			√	√	√	
1926	√		√			√	√	√
1927	√		√					
1928	√							
1929								
1930	√							
1931								
1932	√	√				√		√
1933	√	√	√			√		√
1934	√	√				√		√
1935	√	√						√
1936	√	√						
1937	√	√		√				
1938	√			√				
1939								
1940								
1941								
1942								
1943								
1944								
1945								
1946								
1947								
1948								
1949								

[a]From the statewide precipitation index by Fritts and Gordon (1980). A blank indicates that precipitation was at or above average that year.
[b]A blank indicates that mortality was at endemic levels that year.

contribute to the developing beetle outbreak. The beetle aftermath of forest fires, then, tends to be highly variable, depending on the intensity and pattern of the fire and the local levels of bark beetle populations. There is concern, however, that bark beetles, especially the red turpentine beetle, which is attracted to and readily reproduces in scorched pines, may become more of a problem if prescribed burning becomes widely used. This beetle usually does not kill trees on its own but rather weakens them and renders them more susceptible to other, tree-killing, bark beetles. Bark beetle infestations may be a serious problem especially following the initial reintroduction of fire into stands with high fuel levels due to long-term fire exclusion. In such stands, the number of fire-scorched, beetle-susceptible trees is expected to be greatest. Some damage to trees may be avoided by reducing fuel, for example, by pruning fuel ladders before the prescribed burning is started. Then scorching of trees should be lessened, and less beetle-susceptible host material will be produced.

Logging

Logging operations have been conducted in the Sierra Nevada ever since settlement began in the mid-nineteenth century. Levels and patterns of logging have, however, changed greatly over time with changing methods, technology, markets, and forest management (Laudenslayer and Darr 1990). Early logging concentrated mainly on the largest trees of the most commercially valuable species (a procedure called *high-*

TABLE 45.5

Tree mortality in relation to precipitation during and after the 1975–77 drought in twelve northern California national forests (USFS 1994; U.S. Weather Bureau 1977–80).

	1977	1978	1979	1980
Number of dead trees (millions)	4.5	5.8	1.1	0.9
Trees per acre killed	.71	.92	.17	.06
Percentage of normal precipitation	20–38	93–205	79–138	90–148

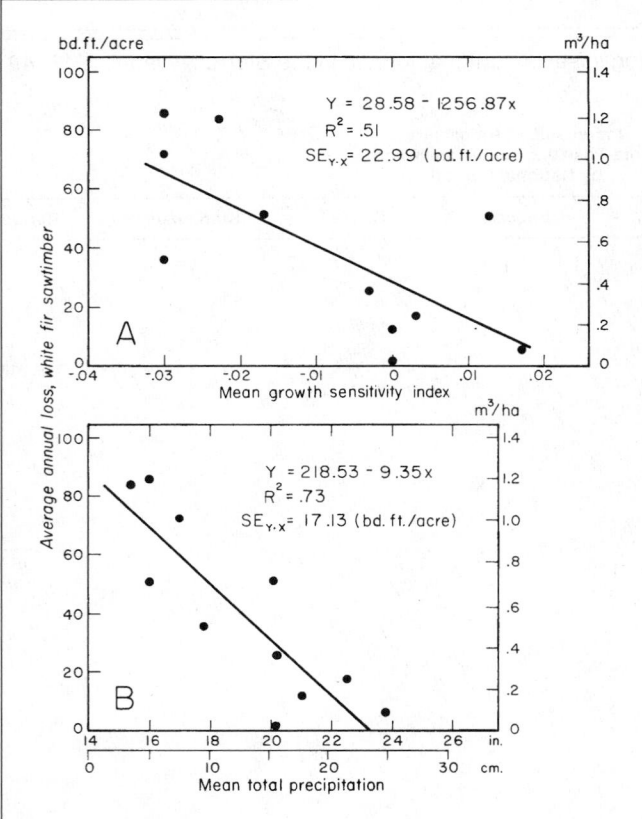

FIGURE 45.1

Relationship between white fir mortality in California caused by the fir engraver beetle and *a*, tree growth sensitivity index, which measures acceleration (>0) or deceleration (<0) of tree growth over the past two years; and *b*, average annual precipitation for the years 1944–54 (Ferrell and Hall 1975). The curve is the average of the levels shown.

grading). As readily accessible stocks of the most desirable trees became less available, foresters began to look toward husbanding this resource. They perceived, however, that this was not possible under prevailing mortality rates of the highest-value trees because many were old, diseased, and highly susceptible to bark beetles. To try to harvest these trees before they were killed by beetles, forest entomologists developed risk-rating systems for identifying the trees most likely to die soon (Keen 1936; Salman and Bongberg 1942; Ferrell 1989). The object was to cover the forest quickly with a light cut, removing the high-risk trees before they could be killed by beetles and thereby reducing tree mortality rates and maybe lowering beetle population levels as well. This risk-cutting was successful, particularly in east-side pine stands. Harvesting trees before they became infested by bark beetles reduced tree mortality rates for periods up to twenty-two years (Wickman and Eaton 1962), but it did not produce large volumes of timber for the market. With the increased

market demand and improved harvesting and replanting technology of the 1950s, foresters began to implement intensive or high-yield management, based on heavier overstory removal cuts or clear-cuts to convert the old growth to younger, even-aged stands capable of producing higher yields of timber.

Each phase of logging had both positive and negative effects on forest health as influenced by insects and pathogens. High-grading, being essentially a single-tree selective cut, tended to favor shade-tolerant species such as firs and cedars at the expense of the more shade-intolerant and drought-resistant pines. Many of the high-graded trees were high-risk trees but others were still vigorous and among the best seed trees. Risk-cutting, also being a single-tree selection system, also tended to favor the drought-sensitive species, but to a lesser extent than high-grading, because the high-risk trees that were cut were not usually the best seed trees. Both of these single-tree selective cutting practices tended to lower the mortality rates of large, valuable pines. They may also have lowered the rates at which understory trees became infected by dwarf mistletoes by reducing the number of infected overstory trees, which are an important infection source (Hawksworth 1961). However, the promotion of a dense, drought-sensitive understory has increased the susceptibil-

FIGURE 45.2

The annual level of *a*, fir mortality caused by the fir engraver and round-headed fir borer, varied inversely with *b*, the levels of total precipitation, and *c*, the levels of spring precipitation and mean air temperature (Ferrell and Hall 1975).

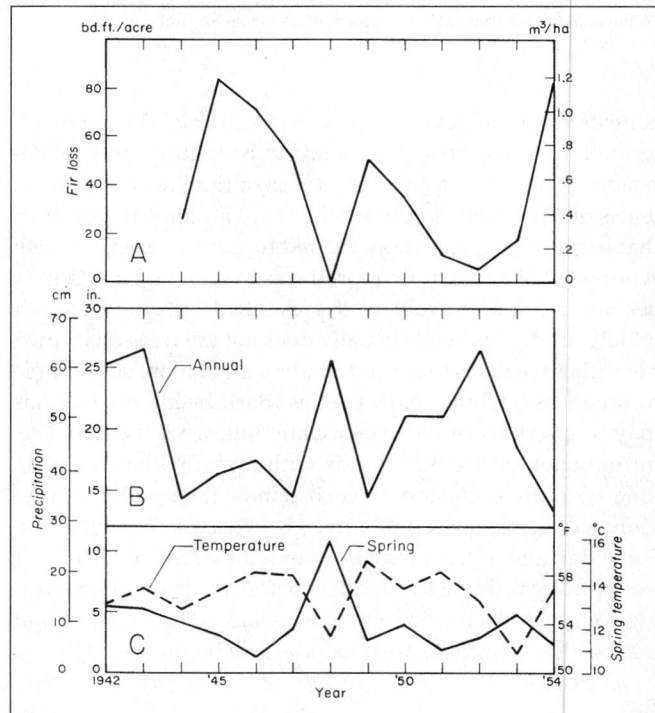

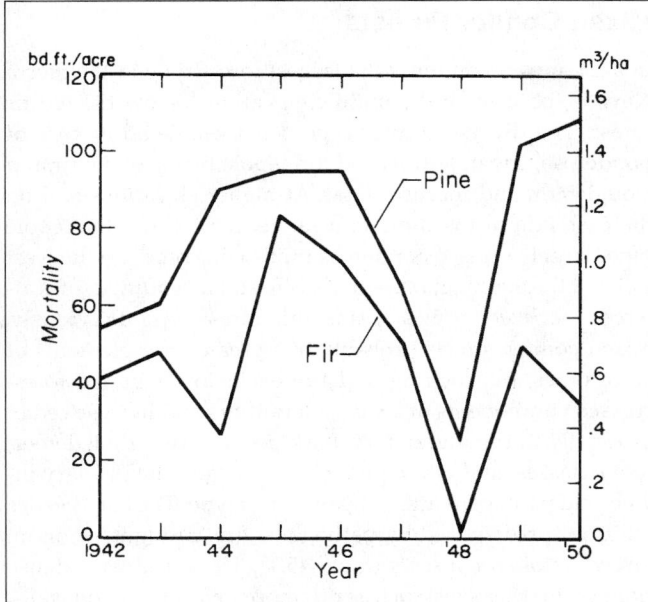

FIGURE 45.3

Mean annual mortality caused by subcortical insects was similar for white fir and pine during the years 1942 to 1950 (Ferrell and Hall 1975).

ity of these stands to bark beetles during droughts (Ferrell et al. 1994). Moreover, repeated logging of stands by these two selective methods, if not carefully done to avoid damage to the residual trees, led to increased levels of root disease by repeatedly providing stumps and trunk or root wounds, which are infection sites for root pathogens (Gast et al. 1991). Potentially, intensive forestry, involving conversion of old-growth to younger, even-aged stands, should mitigate most of these problems, but in practice this has not always been the case. Too often in the past the valuable overstory was removed, leaving a battered, pest-prone understory, which was left unthinned and unsanitized. Even if intensive forestry is well planned and executed, some critics are concerned that the resulting young, even-aged stands may be more pest-prone in the future. At present the evidence for this is at best unclear.

Urbanization

Urban development is increasing in the Sierra Nevada, although it is regulated by public ownership of lands and by the effects of the county land-use planning process on private lands (USFS 1995). Despite these constraints, urbanization, with attendant traffic and infrastructure development, is increasingly affecting Sierra Nevada forests and their management. Commonly occurring problems are construction, trenching, paving, sewage effluent disposal, insecticidal spraying of trees for mosquito abatement, and applications

of highway deicing salt. All of these can injure trees or alter their environments and render them susceptible to insects and pathogens, with root diseases and bark beetles again the most serious. Transport and improper storage of bark-beetle-infested firewood and failure to thin overly dense stands of trees, or improper disposal of the thinnings, can cause problems at times. Fire risk is also increased at the urban/forest interface. Many of these problems can be avoided or mitigated by good planning, regulation, and education, and by adequate public and private expenditures for infrastructure.

Air Pollution

Forest injury from air pollution has been found in some parts of the Sierra Nevada (Cahill et al. 1996; Urban and Miller 1996; Miller and Millecan 1971; Peterson et al. 1992; Pronos and Vogler 1981). This pollution is generated by urban and agricultural sources in the San Joaquin and Sacramento valleys and transported to the Sierra Nevada by the prevailing winds, with the west-side forests thus receiving the highest concentrations. Towns, highways, and major resort areas such as Lake Tahoe can be sources of considerable locally produced air pollution at times. The effects of air pollution are now monitored closely in those parts of the Sierra forests suspected to be at highest risk for damage. Effects of air pollution on forests very similar to those in the Sierra Nevada have been studied intensively in the mountains surrounding Los Angeles. Tree species differ somewhat in their sensitivity, but affected trees display chlorotic, sparse foliage, reduced exudation of defensive resin in response to bark beetle attack, and therefore increased susceptibility to bark beetles (Miller et al. 1963; Stark et al. 1968). Ozone is known to be an important source of the damage because it destroys chlorophyll in conifer foliage. Incipient symptoms have been noticed in west-side forests of the southern Sierra Nevada, but, as yet, no pronounced increases in tree mortality or bark beetle populations have been clearly attributable to this source, probably because air pollution levels have not yet been high enough for long enough periods. Indeed, recent monitoring in the western slope of the Sierra Nevada indicates that levels of injury from

TABLE 45.6

Trees killed by bark beetles in relation to fire injury to crown and cambium (Salman 1934).[a]

	Percentage of Trees Killed	
Percentage of Crown Defoliation	Slight Cambium Injury	Moderate–Severe Cambium Injury
0–25	8.7	10.0
25–50	3.6	13.3
50–75	18.2	37.5
75–100	19.2	72.2

[a]Data are from the Sugar Hill fire, Modoc National Forest.

air pollution are not increasing as expected. In the future, however, rising levels of air pollution could interact with effects of fire and drought to produce widespread outbreaks of tree-killing bark beetles.

EFFECTS ON FOREST COMPOSITION AND STRUCTURE

Insects and pathogens, working in concert with other environmental factors such as drought and fire, strongly influence the composition and structure of Sierra Nevada forests. Specific effects, however, differ by forest type.

Ponderosa Pine Forests

Lying at the lower elevational edge of the commercial forest zone, the ponderosa pine forest type intergrades at its lower elevational limit with chaparral and at its upper elevational limit with the mixed conifer forest type. Ponderosa pine is the primary tree species on most sites, with an admixture of scattered California black oak. Because of the relatively low elevation, water availability, not temperature, is the strongest factor limiting forest growth. Primary insects and pathogens of concern here are complexes of bark beetles, fungal root diseases, and dwarf mistletoe killing ponderosa pines (table 45.1). Western pine beetle kills mainly mature pines weakened by root disease, dwarf mistletoe, or drought, while pine engravers kill mainly pole-size pines in overly dense stands, especially during droughts. Logging slash is also readily infested. Red turpentine beetle acts as a predisposing agent for the other pine bark beetles and may become even more important if controlled burning is reintroduced. Pines are usually killed singly or in small clumps during nondrought periods. But during droughts both mature and pole-size pines may be killed in large groups, and the openings thus created, as well as larger openings created by intense fires, are strongly invaded by manzanita and ceanothus shrubs. These shrubs may dominate such openings for many years, retarding reforestation. Oaks, too, are favored by fire, because they are able to sprout from stumps. Unless mitigated, logging increases the incidence of root disease by providing freshly cut stumps, which serve as entrance points for fungal infection. Because of the accessibility and relatively moderate climate of ponderosa pine forest, it has been greatly affected by previous logging, widespread urbanization, and fire exclusion. Many of the large pines have been logged; current stands are thus younger and denser, with a higher shrub component, than presettlement stands. Continued intense disturbances could cause more of this type to revert to brush fields, especially at the lower elevations.

Mixed Conifer Forests

In area, mixed conifer is the largest forest type in the Sierra Nevada, occupying the middle elevations below the red fir forest. The diverse conifer type is a variable admixture of ponderosa, sugar, Jeffrey, and lodgepole pines, white fir and Douglas fir, and incense cedar. At higher elevations and on the east side of the mountain range, Jeffrey and lodgepole pines largely replace ponderosa pine in this type. On the west slope of the Sierra, giant sequoia is found in a number of scattered, localized groves within this forest type. Previously, mixed conifer forests probably had greater complements of large pines, but logging and fire exclusion have led to increased components of shade-tolerant firs and incense cedar, especially in the understory. Bark beetles, dwarf mistletoes, root diseases, and white pine blister rust are the primary insects and pathogens affecting this forest type. The bark beetles kill trees predisposed by pathogens, fire, drought, or, more rarely, defoliator insects (table 45.1). Trees in overly dense stands and drier sites, such as ridgetops or steep canyon walls, are particularly affected. Windthrows, snow breakage, and logging slash also provide readily infested host materials. Western pine beetle and pine engraver are the primary insects affecting ponderosa pine, while mountain pine beetle is most important in killing sugar pine. The fir engraver is the most serious killer of white fir, the Douglas fir beetle is for Douglas fir, and the Jeffrey pine beetle is for Jeffrey pine. Major predisposing root diseases, caused by somewhat host-specific strains of fungi, are annosus root disease, black-stain root disease, and armillaria root disease. Host-specific dwarf and true mistletoes are also important predisposing agents. White pine blister rust, caused by an introduced stem rust fungus, has spread to sugar pine throughout this forest type, weakening and killing trees and predisposing them to bark beetles. Red turpentine beetle may become a more important predisposing agent for pines if controlled burning is reintroduced in this forest because the beetle readily infests fire-scorched trees. Neither incense cedar nor giant sequoia is host to any serious insect pests, but both are afflicted to some degree by annosus root disease.

Along with climatic fluctuations and fire, insects and pathogens have been postulated to contribute to maintaining the mix of conifer species in this forest type. In this view, when some combination of these agents increases densities and competitional stress in some of the component conifer species, host-specific bark beetles and predisposing pathogens combine to kill those species, reducing their proportion in the stand composition. Ecologists term such a process, in which the organisms with the highest population densities suffer the highest mortality rates, *density-dependent mortality*. Later on, when some other combination of environmental factors increases the density of another component and thus predisposes it to bark beetles, that species in turn is reduced, so that, although the stand composition fluctuates like a pendulum, the mixed composition is maintained over time. Such a process, involving pathogens and bark beetles, appears to have played a role in maintaining a mixed forest of Jeffrey pine and white fir near South Lake Tahoe in the Sierra Nevada (Ferrell et al. 1994; Scharpf and Bega 1981). There are indications that postsettlement human influences have led to an increase in the amplitude of these compositional variations and thus are a destabilizing influence. A key question for future management of mixed conifer forests is how to use natural processes, including the effects of insects and pathogens, to increase rather than decrease the stability of the composition and structure of the forests.

Red Fir Forests

Composed mainly of red fir, this forest type also includes other conifer species such as white fir, mountain hemlock, and western white, lodgepole, and Jeffrey pines as lesser components. The forests lie at the upper elevations in the Sierra Nevada, where winter snowpacks are usually deep and persist well into summer. Tree growth is thus normally limited by low temperatures; in fact, growth may increase during droughts, with their lesser snowpacks and longer growing seasons, provided the drought is not protracted. Development and dynamics of insects and pathogens are also temperature-controlled and usually slower at higher elevations. During nondrought periods, bark beetles kill scattered trees weakened by root diseases and dwarf mistletoes (table 45.1). Trees damaged by avalanches, windthrows, and snow breakage are also readily infested, as is logging slash. But fire damage is seldom a major source of beetle-breeding material, because large, intense fires are less common than at lower elevations. Not only is the fire season shorter, but also the rockiness of the subalpine terrain causes a disjunct distribution or patchiness of this forest. Increased mortality of trees occurs in response to droughts, but usually not in large groups and only if the drought lasts longer than two or three years. Because of their shallow root systems, lodgepole pines that have invaded wet meadows may be particularly susceptible to mountain pine beetle when water table levels change. The fir engraver is responsible for most of the mortality of red fir and white fir, killing both mature and pole-size trees; mountain pine beetle plays a similar role in lodgepole and western white pines. Jeffrey pine beetle is the primary bark beetle infesting Jeffrey pine. Important root diseases are annosus and armillaria root diseases, both of which are especially abundant in dense stands having red fir or white fir as the major component. Red fir in some stands is heavily infected by dwarf mistletoe, which weakens the trees and renders them susceptible to the fir engraver. Insects and pathogens normally kill individual trees or small clumps, rather than large groups of trees, in this forest type and thus normally cause few major shifts in the composition and structure of the forest.

Subalpine Conifer Forests

At the upper limit of tree growth in the Sierra Nevada occurs a variable mixture of several pine species (western white, whitebark, limber, foxtail, and lodgepole) and mountain hemlock, red fir, and Sierra juniper. Mountain pine beetle is the primary bark beetle infesting all of these pines, but outbreaks are rare because of low temperatures and the sparsity and disjunction of the stands due to the extreme rockiness of the growing sites. Usually only single trees or small clumps are killed by the beetle. All of the pines except lodgepole are, however, susceptible to white pine blister rust, and there is currently concern that this virulent pathogen may spread more widely into this forest type in the future. As susceptible pines are virtually the only trees in some high-elevation forests, they could be considered keystone species, because many animals at that elevation are almost totally dependent on them.

Jeffrey Pine Forests

Located at middle elevations on the steep eastern slope of the Sierra Nevada, Jeffrey pine forests are found on somewhat dry sites. They are composed almost wholly of Jeffrey pine with an understory of bitterbrush, sagebrush, and scattered mountain mahogany. Key insect pests and pathogens are Jeffrey pine beetle, annosus root disease, and dwarf mistletoe, operating in pest complexes as described for the other forest types (table 45.1). Located in the rain shadow of the Sierra Nevada, this forest type is well adapted to dry conditions, but, during droughts, outbreaks of Jeffrey pine beetle cause widespread mortality of trees, especially those weakened by root disease or mistletoe.

Piñon-Juniper Forests

Found at the lowest, driest elevations on the east flank of the Sierra Nevada, the piñon-juniper forest type is composed almost wholly of single-leaf piñon pine and western juniper, growing singly or in combination, with a mainly sagebrush shrub layer. Key insects and pathogens are pine engravers killing piñon pines, often those weakened by black-stain root disease or annosus root disease, with the latter also infecting western juniper. Usually single trees or small groups are killed, but where stands are dense, large clumps can be killed.

MITIGATION METHODS

Mitigation methods exist for most of the key insects and pathogens affecting Sierra Nevada forests, but their efficacy is highly dependent on the underlying condition of the forest, which is influenced by environmental and human factors (Furniss and Carolin 1977; Parmeter 1978; Otrosina and Cobb 1989; Kinloch in press). Consequently, the efficacy of past mitigation has been highly variable. Mitigation falls into two categories, based on the approach or strategy employed. Remedial or suppressive methods, sometimes termed *direct control,* are intended to remedy existing situations by directly suppressing pest populations and thus mitigating the damage they cause. Examples are pesticide applications and removal of infested trees from the forest. Experience has shown that beneficial results tend to be only temporary, and retreatments may be needed unless the forest, environmental, or human conditions that provoked the situation in the first place are altered.

In contrast, preventive methods aim to avert the development of future pest problems. These are usually the most effective methods, in terms of both cost and long-term efficacy, because their beneficial results tend to last, barring changes in the conditions governing stand susceptibility. In forest pest management they usually involve what is termed *indirect control,* in that they seek to decrease pest damage by decreasing the susceptibility of the forest rather than attacking the pests and thus are usually silvicultural in nature. Because patterns of forest susceptibility can change over time, and endemic populations of insects and pathogens are virtually omnipresent, however, even the mitigation provided by silvicultural treatments may not be lasting, and retreatment may still be required. In situations where the key insects and pathogens tend to operate as biotic complexes, direct and indirect control may not be strictly separable in practice, because direct control of, say, a pathogen may have its greatest effect as an indirect control of an insect because it reduces the susceptibility of the host tree to the insect.

As preventive methods, particularly those that decrease stand susceptibility to pests, are potentially the most efficacious and cost-effective in the long run, the following discussion focuses on them as the preferred approach to management of the disturbances caused by insects and pathogens in Sierra Nevada forests, although suppressive methods are mentioned where appropriate.

Mistletoes

Mitigation methods developed for the mistletoes infecting Sierra Nevada conifers usually include either direct reduction of sources of infection or increasing the growth and vigor of target trees so that they can outgrow the infection or tolerate it better or both (Parmeter 1978). Reduction of infection sources usually involves removal of infected overstory trees so that mistletoe seeds cannot drop from them and infect the understory. Source reduction can also entail thinning dense stands to reduce competition among residual trees and increase their height growth, thereby slowing upward spread of mistletoe in tree crowns. These approaches have the added benefit of reducing competition within the stand, thus increas-

ing the growth and vigor of residual trees and increasing their ability to outgrow or tolerate the mistletoe. Pruning of infected branches is sometimes used to achieve the same results; it often increases tree growth but, of course, does not suffice if the trunk is also infected. Since mistletoes are host-specific, selective cutting can be used to alter the species composition of stands in the direction of increased resistance to a particularly prevalent and damaging form of mistletoe. Genetic resistance to mistletoes is known, and programs to develop resistant trees may in future provide resistant planting stock. Such stock may be needed if uneven-aged management (single tree or group selection) is used in stands with one major tree species in both the understory and the overstory. In such stands, unless all infected overstory is removed, it will seriously infect the understory, including naturally regenerated seedlings (Parmeter 1978; Scharpf and Parmeter 1976). Prevention or merely reduction of mistletoe infection by any of these methods often leads to increased resistance of the stand to bark beetles.

Blister Rust

Mitigation methods available and effective for white pine blister rust are few (Kinloch in press). Several approaches, including eradication of wild currant or gooseberry (*Ribes* species), the rust's required alternate host, have been tried and found to be costly and insufficiently effective. Infected branches of the pines can be pruned away and disposed of, but this is, of course, not effective if the infection has already spread to the stem. Maintenance of a mixed species stand composition may help slow buildup of the rust. More promising, perhaps, is the development of rust-resistant host trees. Genetically resistant sugar pines have been found, but there is evidence that even resistant trees might become infected under certain environmental conditions (Kinloch and Byler 1981). The current effort is to speed the development of resistance through selection and breeding, but, over the long term, resistance is also expected to develop in natural host tree populations. Meanwhile, cutting practices should avoid large reductions in host inventories so as to maintain sufficiently diverse gene pools to ensure host adaptability in the future.

Root Diseases

Effective mitigation methods for root diseases are mostly preventive (Otrosina and Cobb 1989). Reliable remedial treatments for established infection centers have not yet been developed for any of the major root diseases afflicting Sierra Nevada conifers despite much effort to do so. The difficulty arises from the ability of the root pathogens to spread to surrounding trees through root-to-root contacts and to survive for decades in infected or dead root systems. The only approach currently effective is to keep infections from occurring in the first place. Methods to do so involve borate treatment of freshly cut stumps to prevent their infection by annosus root disease. Additionally, care is taken during logging to avoid creating basal trunk wounds or open root wounds, which could serve as entry points for fungal infections. White fir is particularly susceptible to annosus infection through basal stem wounds. Avoiding such injuries has been difficult, given the heavy logging equipment used in the past, but lighter, more maneuverable equipment is rapidly coming into use. Greater understanding of the host-specific strains of root disease fungi is providing insights about how to manipulate stand composition silviculturally so as to reduce future infections.

Bark Beetles

For bark beetles, both suppressive and preventive methods are available, although lasting beneficial effects of the former are infrequent if the underlying susceptibility of the stand remains undiminished (Furniss and Carolin 1977). Suppressive methods include insecticidal sprays or "fell, peel, and burn" applied to infested trees or logs. Fell, peel, and burn involves felling the infested tree, peeling the bark, and burning, or otherwise disposing of, the bark, so no beetles can emerge from it to attack other trees. The trap-tree method, in which trees are felled to induce attack and, once infested, are destroyed or removed from the forest, also can be used. Prompt salvage logging, in which recently dead and perhaps still infested trees are removed from the woods, is another way to reduce bark beetle populations. Experience has shown, however, that unless virtually all infested trees are located and removed from the woods before the beetles have emerged, tree mortality rates are not greatly affected (Miller and Keen 1960). Preventive methods, most of which aim at reducing the number of susceptible trees in the forest, are far more likely to result in a lasting reduction in tree mortality rates. Sanitation logging, in which diseased or otherwise weakened (high-risk) trees are removed from the forest, has historically been efficacious, particularly in east-side pine stands (Wickman and Eaton 1962). A similar risk-cutting approach is currently being tested for sanitation logging of red fir and white fir in both west- and east-side forests (Ferrell 1989).

Thinning of overly dense stands is also a very important preventive method for bark beetles, although site-specific goals must be established, specifying the density of trees to leave. There have been insufficient studies to establish site-specific goals for the Sierra Nevada, but results so far (Oliver 1995) indicate that, at the least, "catastrophic" (extremely high) tree mortality from bark beetles can be prevented by reducing stand densities below about 14 m² (150 ft²) per acre in basal area. Basal area is a frequently used measure of stand density. It is the cross-sectional area of all tree trunks, usually at 1.4 m (4.5 ft) above ground, and thus expresses total tree occupancy of the site. Developed only in young ponderosa pine stands, this 14 m² standard appears to apply to such stands on sites varying considerably in quality on both the west and the east side of the Sierra, but the standard has not yet been validated for other forest types (Oliver 1995).

With any of these preventive silvicultural treatments, opportunity exists to alter the species composition of the forest from one that is susceptible to one that is less so. However, care must be taken in any treatment involving logging to avoid leaving untreated stumps and injured trees susceptible to bark beetles and pathogens. All logging slash must be promptly treated or disposed of to prevent it from being infested by the beetles. Such considerations are particularly important in old-growth (late-seral-stage) stands, which can be maintained only if much care is taken to avoid injury to the residual stand following any forest treatments.

KNOWLEDGE GAPS

Critical gaps in knowledge about insect pests and pathogens in Sierra Nevada forests include the following:

- Insufficient understanding of conifer resistance to insects and pathogens and of the population levels of these agents that are required to kill trees under various conditions of stress (the so-called *threshold of susceptibility* of trees). Thresholds vary with changing forest and environmental conditions, and inability to specify them has been a major barrier in developing models predicting the forest damage that will be caused by these agents.

- Inadequate knowledge of how, and to what extent, various sources of stress, such as drought, air pollution, and fire, interact to determine tree and stand susceptibility. Air pollution and fire increase during droughts but their interactive effect on forest stress and susceptibility to insects and pathogens is not well understood.

- Paucity of systems for rating the risk of trees or stands to damage from insects and pathogens during both outbreak and nonoutbreak periods. Risk and hazard rating systems exist, but not for all major tree species and forest types in the Sierra Nevada.

- Dearth of site-specific guidelines to determine the reserve or residual stand densities to maintain after logging or thinning so as to maintain forest resistance during both drought and nondrought periods. Rough guidelines are used, but most of these were derived from a limited range of stands and forest types.

- Inadequate knowledge and ability to predict interactions between fire and insects or pathogens. Present knowledge derives largely from case studies of the aftermath of relatively few wildfires, with little characterization of affected trees and stands and their susceptibility to these agents.

- Unavailability of predictive models linking population levels of these agents to environmental factors, mitigative treatments, and forest dynamics. Population models are available for only some of these agents, and the models have not been linked to environmental variables and stand dynamics models, largely because thresholds of susceptibility are not known. Existing population models usually do not have a spatial component able to predict the spatial population patterns produced by these contagiously spreading agents.

- Insufficient ability to predict results of mitigative treatments for these agents, especially results for new methods, such as bark beetle attractants and repellents, on levels of forest damage. These results are needed both to judge whether mitigation is worthwhile and to add to predictive models.

- Unavailability of stand dynamics models that adequately predict spatial patterns of forest growth and survival in quantitative terms suitable for multiple-resource planning. Spatial patterns are critically relevant for contagiously spreading agents such as insects and pathogens, and quantitative predictions of forest dynamics must include nontimber as well as timber resources to facilitate the planning process. Most existing models have no spatial component and some do not produce outputs in units relevant to nontimber, as well as timber, resource planning.

- Inadequate knowledge of the role played by insects and pathogens in forest ecosystem function, especially those functions necessary for the long-term sustenance of forest biodiversity and productivity.

PEST RESPONSES TO FUTURE MANAGEMENT SCENARIOS

If present trends and management practices continue, Sierra Nevada forests will experience outbreaks of bark beetles and wildfires in response to the recurrent droughts characteristic of the California climate. It is not possible to predict the extent of the damage produced by these outbreaks, because they will depend partly on the intensity and duration of the droughts. But, because highly susceptible stands have developed in some forest types, in part as a result of past management practices and the synergistic effects of drought, pests, and fire, it is likely that the high levels of forest damage caused by these agents will increase in the future over much of the Sierra Nevada. Under such conditions, experience has shown that mitigation will not be adequate. Even previously treated stands and young plantations may suffer damage, because they will be surrounded by a large backlog of untreated stands with high fuel loads and populations of insects and pathogens. Moreover, the mortality of desirable remnant old-growth

trees will increase due to their age-related higher susceptibility to insects and pathogens. Compounding these losses will be those arising from likely increased disturbances, such as frequent salvage logging, urbanization, and its many side effects, including air pollution and construction. Under these conditions, endemic or nonoutbreak populations of insects and pathogens may well increase, leading to quicker outbreaks at the onset of droughts, and the health of Sierra Nevada forests is likely to continue to decline.

If, on the other hand, management strategies are altered in the direction of restoring presettlement forest composition and structure, current forest conditions, including damage from insects and pathogens, would probably be mitigated. Likely scenarios include thinning the overly dense understory of firs and cedars to reduce its susceptibility to fire, insects, and pathogens. Reduction of this understory not only reduces fuel loads and the spread of insects and pathogens from tree to tree but also reduces competitive stress on the overstory and thus lowers its susceptibility to insects and pathogens. Controlled burning may be an ecologically sound method of thinning this understory, but mechanical thinning may have to be used, at least initially, in commercial forests where present fuel loads are too high to use burning. Fire alone may have to be used in parks and wilderness areas or other areas where logging is not an option.

Although understory reduction may reduce mortality rates of the remaining large overstory trees, mortality of the overstory trees may still be too high to meet the inventory goals of ecosystem management. Important among these goals is maintaining sufficient numbers of large trees to meet wildlife needs. For example, some owl species prefer to roost in such trees; some cavity-nesters such as woodpeckers prefer to excavate nest holes and to forage in such trees after they become snags. Snags do not remain standing forever, however. Survival and replacement of large trees must therefore provide adequate numbers of standing snags in future decades. Removal of competing smaller trees around large trees may prolong survival of the latter. The same treatment around rapidly growing midsize trees may speed their replacement of larger old trees as the latter die.

After wildlife and other ecosystem needs are met, there may be sufficient inventories of large trees to permit some selective logging. If so, this should be done on a sanitation basis. Removing diseased or otherwise high-risk trees decreases endemic populations of bark beetles and pathogens, thereby protecting the remaining stand and retarding outbreaks during droughts. This sanitation logging could be done on both a single-tree and a group basis, depending on the spatial distribution of high-risk trees. Indeed, this distribution is usually clumped because of the contagious distribution of root diseases and mistletoes. Group selection or patch-cutting should therefore be used to ensure that entire clumps of pathogen-infected trees are removed.

All restoration management has to be done carefully, especially because it may require repeated reentry into the stand,

and with every entry there is a risk of injuring the residual stand. In particular, scorching and mechanical injury to the residual stand should be minimized. Appropriate equipment and methods should be used to minimize soil compaction and root damage. Large stumps should be treated to avoid initiating root disease infection centers. Treatments should prevent not only root diseases but also entrance of heart-rot and other decay fungi, which can structurally weaken other tree trunks, leading to wind and snow breakage. Sufficient slash should be left for wildlife but treated to prevent bark beetle infestation. If, because of wildlife requirements, logging slash cannot be piled and burned or scattered and dried in full sunlight (proven beetle prevention methods), it should be lightly burned without piling, to dry it and render it unsuitable for bark beetles. Unless such precautions are taken, forest susceptibility to fire, insects, and pathogens may be higher after treatment than before.

A variety of thinnings, overstory sanitation cuttings, controlled burns, and other treatments will have to be used, with the specific combination dependent on the particular site and the goals for managing it.

If this program is undertaken, losses caused by insects and pathogens will still occur, at least initially, but as treatments become more widespread, the damage caused by these agents should decline over the long term.

The guiding strategy should be to restore the presettlement composition and structure of the forests by using the natural processes (or reasonable substitutes) under which these forests evolved. In this way chances of inadvertent extinctions are minimized, and chances of keeping the entire biota reasonably intact are maximized. Some treatments, such as mechanical thinning and sanitation logging, are largely precluded in forests within established parks and wilderness areas, where human influences are minimized. Instead, natural or prescribed fires will have to be relied upon largely to restore presettlement conditions. In stands with unnaturally high fuel loads due to previous fire suppression, fire injury to the forest overstory may, at times, occur, and insect- and pathogen-caused damage may also be extensive. If viewed as natural processes, however, the effects of these agents can be considered beneficial and necessary to restore presettlement conditions.

Accomplishing this restoration will not be quick and inexpensive. The current situation took many decades to develop, and due to its scope and magnitude it will require many decades to rectify. With widespread implementation of mitigative restoration, however, these forests can be sustained reasonably intact for future generations.

REFERENCES

Bolsinger, C. L. 1980. *California forests: Trends, problems, and opportunities.* Resource Bulletin PNW-89. Portland, OR: U.S. Forest Service, Pacific Northwest Forest and Range Experiment Station.

Cahill T. A., J. J. Carroll, D. Campbell, and T. E. Gill. 1996. Air quality. In *Sierra Nevada Ecosystem Project: Final report to Congress,* vol. II, chap. 48. Davis: University of California, Centers for Water and Wildland Resources.

California Forest Pest Council. 1951–93. *Forest pest conditions in California—1951–93.* Sacramento: California Forest Pest Council.

———. 1993. *Forest pest conditions in California—1993.* Sacramento: California Forest Pest Council.

Ferrell, G. T. 1986. *Using indicator plants to assess susceptibility of California red fir and white fir to the fir engraver beetle.* Research Note PSW-388. Berkeley, CA: U.S. Forest Service.

———. 1989. *Ten-year risk-rating systems for California red fir and white fir: Development and use.* General Technical Report PSW-115. Berkeley, CA: U.S. Forest Service.

Ferrell, G. T., and R. C. Hall. 1975. *Weather and tree growth associated with white fir mortality caused by fir engraver and roundheaded fir borer.* Research Paper PSW-109. Berkeley, CA: U.S. Forest Service.

Ferrell, G. T., W. J. Otrosina, and C. J. DeMars Jr. 1994. Predicting susceptibility of white fir during a drought-associated outbreak of the fir engraver, *Scolytus ventralis,* in California. *Canadian Journal of Forest Research* 24:302–5.

Fritts, H. C., and G. A. Gordon. 1980. *Annual precipitation in California since 1600 reconstructed from western North America tree rings.* Report under California Department of Water Resources Agreement B53367. Tucson: University of Arizona.

Fritts, H. C., G. R. Lofgren, and G. A. Gordon. 1979. Variations in climate since 1602 as reconstructed from tree rings. *Quaternary Research* 12:18–46.

Furniss, R. L., and V. M. Carolin. 1977. *Western forest insects.* Miscellaneous Publication 1339. Washington, DC: U.S. Forest Service.

Gast, W. R., D. W. Scott, C. Schmitt, and C. G. Johnson Jr. 1991. *Blue Mountains forest health report: New perspectives in forest health.* Portland, OR: U.S. Forest Service, Pacific Northwest Region.

Graumlich, L. J. 1993. A 1000-year record of temperature and precipitation in the Sierra Nevada. *Quaternary Research* 39:249–55.

Hawksworth, F. G. 1961. Dwarfmistletoes of ponderosa pine. *Recent Advances in Botany* 2:1537–41.

Keen, F. P. 1936. Relative susceptibility of ponderosa pines to bark beetle attack. *Journal of Forestry* 34:919–27.

Kinloch, B., Jr. In press. Mechanisms and inheritance of resistance to blister rust in sugar pine. In *Sugar pine: Status, values, and roles in ecosystems: A proceedings,* edited by B. Kinloch Jr., M. Marosy, and M. Huddleston. Agriculture and Natural Resources Publication 3362. Davis: University of California.

Kinloch, B. B., and J. W. Byler. 1981. Relative effectiveness and stability of different resistance mechanisms to white pine blister rust in sugar pine. *Phytopathology* 71:386–91.

Laudenslayer, W. F., and H. H. Darr. 1990. Historical effects of logging on the forests of the Cascade and Sierra Nevada Ranges of California. *Transactions of the Western Section of the Wildlife Society* 26:12–23.

Martin, R. E. 1982. Fire history and its role in succession. In *Forest succession and stand development research in the Northwest,* edited by J. E. Means, 92–99. Corvallis: Oregon State University, Forest Research Laboratory.

Miller, J. M., and F. P. Keen. 1960. *Biology and control of the western pine beetle.* Miscellaneous Publication 800. Washington, DC: U.S. Forest Service.

Miller, P. R., and A. A. Millecan. 1971. Extent of oxidant air pollution damage to some pines and other conifers in California. *Plant Disease Reporter* 55:555–59.

Miller, P. R., J. R. Parmeter Jr., O. C. Taylor, and E. A. Cardiff. 1963. Ozone injury to foliage of *Pinus ponderosa*. *Phytopathology* 53:1072.

Oliver, W. W. 1995. Is self-thinning in ponderosa pine ruled by *Dendroctonus* bark beetles? *Proceedings of the 1995 National Silviculture Workshop, May 18–21, Mescalero, New Mexico*, 213–18. General Technical Report RM-GTR-267. Fort Collins, CO: U.S. Forest Service.

Otrosina, W. J., and F. W. Cobb Jr. 1989. Biology, ecology, and control of *Heterobasidium annosum*. In *Proceedings of the symposium on research and management of annosus root disease in western North America*, technical coordination by W. J. Otrosina and R. F. Scharpf, 26–34. General Technical Report PSW-116. Berkeley, CA: U.S. Forest Service.

Parmeter, J. R., Jr. 1978. Forest stand dynamics and ecological factors in relation to dwarf mistletoe spread, impact, and control. In *Proceedings of the symposium on dwarf mistletoe control through forest management, April 11–13, 1978, Berkeley, California*, technical coordination by R. F. Scharpf and J. R. Parmeter Jr., 16–30. General Technical Report PSW-31. Berkeley, CA: U.S. Forest Service.

Parsons, D. J., and S. H. DeBenedetti. 1979. Impact of fire suppression on a mixed-conifer forest. *Forest Ecology and Management* 2:21–23.

Peterson, D. L., R. D. Doty, D. L. Schmoldt, J. M. Eilers, and R. W. Fisher. 1992. *Guidelines for evaluating air pollution impacts on Class I wilderness areas in California*. General Technical Report PSW-GTR-136. Berkeley, CA: U.S. Forest Service.

Pitcher, D. C. 1987. Fire history and age structure in red fir forest of Sequoia National Park, California. *Canadian Journal of Forest Research* 17:582–87.

Pronos, J., and D. R. Vogler. 1981. *Assessment of ozone injury to pines in the southern Sierra Nevada, 1979/80*. Forest Pest Management Report C90-14. San Francisco: U.S. Forest Service.

Salman, K. A. 1934. Entomological factors affecting salvage of fire-injured trees. *Journal of Forestry* 32:1016–17.

Salman, K. A., and J. W. Bongberg. 1942. Logging high-risk trees to control insects in the pine stands of northeastern California. *Journal of Forestry* 40:533–39.

Sartwell, C., and R. E. Stevens. Mountain pine beetle—prospects for silvicultural control in second-growth stands. *Journal of Forestry* 73:136–40.

Scharpf, R. F. 1993. *Diseases of Pacific Coast conifers*. Agriculture Handbook 521. Albany, CA: U.S. Forest Service.

Scharpf, R. F., and R. V. Bega. 1981. *Elytroderma disease reduces growth and vigor, increases mortality of Jeffrey pines at Lake Tahoe Basin, California*. Research Paper PSW-155. Berkeley, CA: U.S. Forest Service.

Scharpf, R. F., and J. R. Parmeter Jr. 1976. *Population buildup and vertical spread of dwarf mistletoe on young red and white firs in California*. Research Paper PSW-122. Berkeley, CA: U.S. Forest Service.

Slaughter, G. W., and J. R. Parmeter Jr. 1989. Annosus root disease in true firs in northern and central California forests. In *Proceedings of the symposium on research and management of annosus root disease in western North America*, technical coordination by W. J. Otrosina and R. F. Scharpf, 70–77. General Technical Report PSW-116. Berkeley, CA: U.S. Forest Service.

Stark, R. W., P. R. Miller, F. W. Cobb Jr., D. L. Wood, and J. R. Parmeter Jr. 1968. Photochemical oxidant injury and bark beetle (*Coleoptera: Scolytidae*) infestation in injured trees. *Hilgardia* 39:121–26.

Swetnam, T. W. 1993. Fire history and climate change in giant sequoia groves. *Science* 262:885–89.

Taylor, A. H., and C. B. Halpern. 1991. The structure and dynamics of *Abies magnifica* forests in the southern Cascade range, USA. *Journal of Vegetation Science* 2:189–200.

Urban D., and C. Miller. 1996. Modeling Sierran forests: Capabilities and prospectus for gap models. *Sierra Nevada Ecosystem Project: Final report to Congress*, vol. III. Davis: University of California, Centers for Water and Wildland Resources.

U.S. Forest Service (USFS). 1917–49. Insect conditions reports for Sierra Nevada national forests. Unpublished file reports. USFS, San Francisco.

———. 1991. *Pest risk assessment of the importation of larch from Siberia and the Soviet Far East*. Miscellaneous Publication 1495. Washington, DC: USFS.

USFS, Pacific Southwest Region. 1995. *Managing California spotted owl habitat in the Sierra Nevada national forests of California*. Draft Environmental Impact Statement. San Francisco: USFS.

USFS, Pacific Southwest Region, Forest Pest Management. 1994. *California forest health*. Publication R5-FPM-PR-001. San Francisco: USFS.

U.S. Weather Bureau. 1977–80. *Annual Summary of California Weather*. Washington, DC: Government Printing Office.

Wagener, W. W. 1961. *Guidelines for estimating the survival of fire-damaged trees in California*. Miscellaneous Paper 60. Washington, DC: U.S. Forest Service.

Wickman, B. E. 1992. *Forest health in the Blue Mountains: The influence of insects and diseases*. General Technical Report PNW-GTR-295. Portland, OR: U.S. Forest Service, Pacific Northwest Research Station.

Wickman, B. E., and C. B. Eaton. 1962. *The effects of sanitation-salvage cutting on insect-caused mortality at Blacks Mountain Experimental Forest 1938–1959*. Technical Paper 66. Berkeley, CA: U.S. Forest Service, Pacific Southwest Experiment Station.

JOE R. McBRIDE
Environmental Science, Policy, and
    Management
University of California
Berkeley, California

WILLIAM RUSSELL
Environmental Science, Policy, and
    Management
University of California
Berkeley, California

SUE KLOSS
Environmental Science, Policy, and
    Management
University of California
Berkeley, California

Impact of Human Settlement on Forest Composition and Structure

ABSTRACT

Human settlement in the Sierra Nevada has resulted in a decrease in crown canopy cover, a reduction in tree density, and an introduction of exotic tree species. The decrease in proportion of crown canopy cover, about 30%, was found to be fairly uniform in all forest types. Changes in tree density and species richness were more variable among the forest types. The greatest decrease in tree density was observed in the mixed conifer forest (70%), and the greatest increase in species richness was found in the ponderosa pine forest, where average species richness of trees increased from four to thirty-eight species. Human settlement increases the amount of impervious surface as a result of the building of roads and houses. This effect was greatest in the foothill woodland, where an average of 41% of the ground surface was made impervious on lots less than 1 acre. A detailed analysis of land use was conducted along Highway 49 in the foothill woodland vegetation type. This analysis provides insights into landscape-scale changes that have occurred as a result of human settlement. The significance of these changes for fire hazard, forest hydrology, and wildlife habitat are discussed.

INTRODUCTION

Small populations of Native Americans lived in the forests of the Sierra Nevada for at least 6,000 years. The management practices used by these populations relied on a sophisticated understanding of plant ecology and horticulture. These practices maintained a character of forest composition and structure adjacent to settlements that many twentieth-century Californians have come to regard as the natural condition. Ground fires were used in these areas to maintain visibility into adjacent forests for security, to enhance the collection of plant materials for food and fiber, and to reduce fire hazards (Lewis 1973). This use of fire resulted in somewhat lower tree density, reduced crown canopy cover, and the favoring of fire-resistant species such as ponderosa pine. Because of the absence of any written history during this period, one can only speculate on the magnitude of the change in forest characteristics associated with Native American settlement of the Sierra Nevada.

Modern human settlement began in the Sierra Nevada with the California gold rush in 1849. The population of gold seekers entering the mountains required materials for housing and mining activities. Many of the towns and highways in the Sierra Nevada today date from the gold rush period. The initial population of these towns declined as local reserves of gold were depleted, but the rapid growth of California's population following World War II resulted in renewed population growth in the Sierra Nevada. This growth involved the expansion of existing towns, suburban development, and new town development. The impact of modern human development on the Jeffrey pine–dominated forest at the south end of Lake Tahoe was reported in studies by McBride and Jacobs (1979, 1986). Their studies showed a 66% decrease in tree canopy cover, a 51% decrease in tree density, and an increase in the number of tree species from one to six as a result of the development of suburban areas. The even-aged structure of the presettlement forest at South Lake Tahoe was modified into an uneven-aged structure by tree planting and the natural establishment of Jeffrey pines within the urban forest. McBride and Jacobs compared their findings at South Lake

Sierra Nevada Ecosystem Project: Final report to Congress, vol. II, *Assessments and scientific basis for management options.* Davis: University of California, Centers for Water and Wildland Resources, 1996.

Tahoe with similarly collected data from Menlo Park, California, where the presettlement vegetation was a mosaic of oak savanna and oak woodland. The comparison showed a similar trend in the effects of human settlement (decreased tree canopy and density, increased species richness) but significant quantitative differences between the two areas. This comparison indicated that the degree of reduction in tree canopy cover and tree density and the increase in species richness associated with human settlement varies between forest types.

The purpose of this study is to quantify changes in tree canopy cover, tree density, and species richness as a result of human settlement in the forests of the Sierra Nevada. Knowledge of the quantitative nature of these changes can be used to project the effects of various development scenarios on fire hazard, forest hydrology, and wildlife habitat quality. The four forest types selected for the study were the foothill woodland, ponderosa pine forest, mixed conifer forest, and red fir–lodgepole pine forest. Three classes of lot sizes (less than 1 acre, 3–5 acres, 10–20 acres) were studied in the foothill woodland, while only one size class (less than 1 acre) was studied in the other vegetation types because of the lack of available maps showing boundaries of larger properties. A detailed analysis of land use was conducted along Highway 49 in the foothill woodland vegetation type in order to better understand the distribution and juxtaposition of land use associated with human settlement.

STUDY AREA

This study was limited to woodlands and forests occurring in portions of Sacramento, El Dorado, Amador, Nevada, and Calaveras Counties. These counties were selected because of the focus on the central Sierra Nevada by other Sierra Nevada Ecosystem Project researchers interested in the impacts of human settlement. The characteristics of the woodlands and forests in these central Sierra Nevada counties are typical of the woodlands and forests farther north and south. The major human settlements studied were El Dorado Hills, Cameron Park, and Shingle Springs (foothill woodland); Camino, Nevada City, and Forest Springs (ponderosa pine forest); Arnold, Dorrington, and Sly Park (mixed conifer forest), and Bear Valley, Kirkwood, and Echo Lake (red fir–lodgepole pine forest).

The foothill woodland in the study area occurs from elevations of 500 ft to 2,500 ft and is dominated by blue oak (*Quercus douglasii*). Other common tree species include maul oak (*Quercus chrysolepis*), interior live oak (*Quercus wislizenii*), and foothill pine (*Pinus sabiniana*). Much of the foothill woodland has been grazed by cattle and usually supports an understory of annual grasses. On moist sites the understory contains poison oak (*Toxicodendron diversiloba*), toyon (*Heteromeles*

arbutifolia), and snowberry (*Symphoricarpos rivularis*). Average annual precipitation in the foothill woodland ranges from 15 in to 25 in. Winters are relatively mild, with temperatures seldom dropping below 25°F. Summers are very hot, with daytime temperatures often exceeding 105°F. Summaries of the ecology of the foothill woodland are presented by Griffin (1977), Holland and Keil (1986), Pavlik et al. (1991), Barbour et al. (1993), and Johnston (1994).

The ponderosa pine forest occurs from elevations of about 2,000 ft to 2,500 ft in the central Sierra Nevada. The ponderosa pine forest extended to lower elevations (about 1,500 ft) in portions of the central Sierra Nevada in pre–gold rush times but was largely removed from these lower elevations during the early mining period. The dominant species in this type is ponderosa pine (*Pinus ponderosa*). California black oak (*Quercus kelloggii*) is commonly found in the ponderosa pine forest. At higher elevations in the type, incense cedar (*Calocedrus decurrens*) occurs. Understory shrubs include various species of manzanita (*Arctostaphylos* spp.) and ceanothus (*Ceanothus* spp.). Rainfall averages 20 in to 50 in annually. Snow occurs in the winter months, but accumulations seldom exceed 1 foot. Winters are cool, with temperature minimums down to 10–15°F. The ecology of the ponderosa pine forest has been reported by Rundel et al. (1977) and Holland and Keil (1986).

At elevations of about 2,500 ft, one first encounters the mixed conifer forest type of the central Sierra Nevada. This type extends up to elevations of about 6,000 ft in the central portion of its range. Precipitation averages from 30 in to 60 in, with the bulk falling in the form of snow. Temperatures commonly drop to near or below 0°F in the winter. Summers are warm, with temperatures often reaching into the 90°F range. The mixed conifer forest type supports five conifer species: ponderosa pine (*Pinus ponderosa*), incense cedar (*Calocedrus decurrens*), white fir (*Abies concolor*), Douglas fir (*Pseudotsuga menziesii*), and sugar pine (*Pinus lambertiana*). California black oak (*Quercus kelloggii*) is also common. A large variety of shrubs and herbaceous species occur in the understory of less dense stands in the mixed conifer forest. Dense stands support fewer shrubs and herbs. The ecology of the mixed conifer forest has been reviewed by Rundel et al. (1977), Bonnicksen and Stone (1981), Holland and Keil (1986), Barbour et al. (1993), and Johnston (1994).

The red fir–lodgepole pine forest occurs at elevations above 6,000 ft in the central Sierra Nevada. The dominant species of this forest type (red fir [*Abies magnifica*] and lodgepole pine [*Pinus contorta* var. *murrayana*]) extend to elevations above 9,000 ft, but the forest type is generally restricted to elevations of about 8,000 ft. At these high elevations, summers are cool, with day temperatures in the low 70s. Winters are cold, with nighttime temperature readings below zero. Within the red fir–lodgepole forest zone, one encounters pure stands of either species and stands supporting mixtures of both species. Red fir dominates on deeper, more nutrient-rich soils, while lodgepole pine is found on areas either too wet or too dry for red fir. Understory species are few in number. Snow accumulation and dense tree canopy cover inhibits the development of a diverse understory flora. Common shrubs may include huckleberry oak (*Quercus vaccinifolia*), sticky currant (*Ribes viscosissimum*), mountain snowberry (*Symphoricarpos vacciniodes*), and tobacco brush (*Ceanothus velutinus*). These commonly occur in openings in the forest or at the margins of stands. The ecology of these higher-elevation forests has been reviewed by Oosting and Billings (1943), Rundel et al. (1977), Holland and Keil (1986), Barbour et al. (1993), and Johnston (1994).

METHODS

Three methods were used to sample the characteristics of the woodlands and forests in and adjacent to human settlements in the Sierra Nevada. These included (1) estimation of cover on aerial photographs, (2) measurement of tree density and determination of tree species richness on plots, and (3) point sampling along Highway 49 to determine the frequency and adjacency of different types of land use associated with human settlement.

Cover Estimation on Aerial Photographs

Black-and-white aerial photography (1991, scale 1:35,000), available in the Map Library of the University of California at Berkeley, was used to estimate the percentage of the ground covered by (1) tree canopy, (2) structures, (3) paved surfaces (roads, driveways, sidewalks, patios), and (4) plants other than trees (lawns, shrubs, vegetable gardens, hedges). Dot grids on transparent acetate were dropped at random onto the area of each photograph where cover estimates were needed, and the dots superimposed on each category of cover were counted. The percentage of dots "falling" on each cover category was then calculated. The number of samples taken to estimate cover varied with woodland or forest type because of the frequency of human settlements and the availability of maps showing individual lot boundaries. From three to ten

settlements (areas of urban development), referred to as sites, were selected for each lot (individual home owner's property) size in each vegetation type. For each site, three lots were sampled (table 46.1) with the dot-grid sampling encompassing an area greater than the selected lot. An area of 3 to 5 acres was sampled for lots less than 1 acre (samples ranged from twelve to twenty adjacent lots), 20 acres for lots from 3 to 5 acres (samples ranged from three to four adjacent lots), and 10 to 20 acres for lot sizes of 10 to 20 acres (samples ranged from a single to two adjacent lots).

For the largest lot size class (10 to 20 acres), samples were taken for the entire lot and for the "developed areas" immediately around structures on the lot. For each site, three adjacent control areas were chosen that occurred on a similar aspect and slope and supported a similar woodland or forest type as the developed area. Control areas ranged from 5 to 20 acres in each vegetation type. Cover estimates were made using the dot grid on these control areas.

Measurement of Tree Density and Species Richness

Ten ground plots were located in developed areas with lot size less than 1 acre, and three control plots were located in adjacent control (undeveloped) areas at three sites (three separate human settlements) in each vegetation type. In developed areas, each plot was located at an intersection of two streets. The initial intersection at each site was chosen at random, and subsequent intersections were selected at points on a three-by-three block grid, moving out from the initial intersection. Plots at intersections were used to allow observation of trees in the backyards of lots. It was not possible to accurately observe backyard trees from the street in front of a house. All observations were made without entering private property, with the exception of the control plots and on occasions when the field crew was invited by a home owner to go into a backyard. Some corner lots initially chosen for sampling were rejected when it was not possible to see from the side street into the backyard. Some bias was no doubt involved in only sampling corner lots. These sites may be more open to wind-dispersed seeds, and they may have been planted to provide visual privacy from adjacent streets. No evidence was

TABLE 46.1

Number of sites (human settlements) and samples used in various woodland and forest types to estimate canopy cover.

Type	Less than 1 Acre		1–3 Acres		10–20 Acres	
	Number of Settlements	Number of Samples	Number of Settlements	Number of Samples	Number of Settlements	Number of Samples
Foothill woodland	10	30	10	30	10	30
Ponderosa pine forest	3	9	0	0	0	0
Mixed conifer forest	6	18	0	0	0	0
Red fir–lodgepole pine forest	3	9	0	0	0	0

observed of a bias toward wind-dispersed species on these corner lots, however. Corner lots that were densely planted with screening trees were rejected as sample lots. All trees over 4 in in diameter at breast height occurring on the sampled lots were recorded by species. The dimensions of each lot were determined by pacing the property lines along the two streets. All control plots were circular and 1/10 acre in area. As with the developed lots sampled, all trees over 4 in in diameter at breast height were recorded by species. Tree density was calculated on a per acre basis.

Species richness was calculated by summing the number of tree species in the total sample of developed lots. In this study, species richness is defined as the total number of tree species found in a sample. Species richness for control plots was also calculated by summing the number of species on the three sample plots. Comparison of the species richness on developed lots and control plots is seen as a measure of the effect of human settlement on the diversity of trees.

Point Sampling Along Highway 49

A point sample was taken every mile along Highway 49 from Sonora to Grass Valley, California. This procedure was designed to provide more information on the character of land use and vegetation within the foothill woodland type. This concentration of sampling was deemed necessary because of the potential for development in this vegetation type. At each point the land use adjacent to the highway on each side was recorded, and notes were made concerning the condition of the vegetation associated with each type of land use. In all, 116 points were sampled, giving 116 paired samples of land use for the area between the two towns. The data was summarized by calculating the relative frequency of occurrence of each land-use type and the relative frequency of pairing of any two land-use types.

RESULTS AND DISCUSSION

Tree Canopy Cover, Density, and Species Richness

Tree canopy cover and tree density were reduced by human settlement in all four vegetation types, while species richness was increased. Changes in cover associated with human settlement are summarized in table 46.2. Figure 46.1 compares the tree canopy cover on developed and control plots in each of the four vegetation types for lots less than 1 acre.

The differences in tree canopy cover between developed and control plots in all lot size classes were significant at the 0.01% level (ANOVA [analysis of variance]; F-test). No significant difference was found in the percentage of reduction in crown canopy cover among the four vegetation types. The percentage of the surface covered by impervious materials

(structures, roads, sidewalks) was significantly greater in the less-than-1-acre lot size class in foothill woodlands (41%) than in the larger lot size classes (7.5%). No significant difference in impervious surface cover occurred between the 3-to-5 and the 10-to-20-acre lot size classes. The category "other" refers to ground covered by plants other than trees. The area covered by "other" was not significantly different among the lot sizes or vegetation types. The character of this vegetation was, however, different when developed areas were compared with control sites and when developed areas in different vegetation types were compared. In the less-than-1-acre lot size class, the "other" cover consisted primarily of irrigated lawn in the foothill woodland and ponderosa pine forest, while the control sites in these vegetation types were primarily annual grassland. In the mixed conifer and red fir–lodgepole pine types, the "other" cover was primarily native shrubs or montane meadows, for both developed and control sites. Little evidence was seen of lawn installation and irrigation in the developed areas of this higher-elevation zone.

Tree density was decreased by human settlement in all four vegetation types, as shown in table 46.3 and figure 46.2. Average tree density on developed sites versus control sites was significantly different in all vegetation types except the ponderosa pine forest (ANOVA; F-test; 0.01%).

Species richness increased as a result of human settlement in all vegetation types on less-than-1-acre lots. Species rich-

TABLE 46.2

Percentage cover on developed and nondeveloped property associated with human settlement.

Property	Percentage Cover			
	Tree	Other	Structure[a]	Road
Foothill Woodland				
Less than 1 acre				
Developed	43	16	25	16
Control	69	31	0	0
3-5 acres				
Developed	70	22	3	5
Control	90	10	0	0
10-20 acres				
Structures[a]	48	41	7	4
Lot[b]	56	37	4	3
Control	74	24	0	0
Ponderosa Pine				
Less than 1 acre				
Developed	62	9	13	16
Control	90	10	0	0
Mixed Conifer				
Less than 1 acre				
Developed	64	11	9	16
Control	92	8	0	0
Red Fir–Lodgepole Pine				
Less than 1 acre				
Developed	59	22	6	13
Control	79	21	0	0

[a]Area immediately around structures (houses, farm buildings, sheds, etc.).
[b]Portion of property not adjacent to structures.

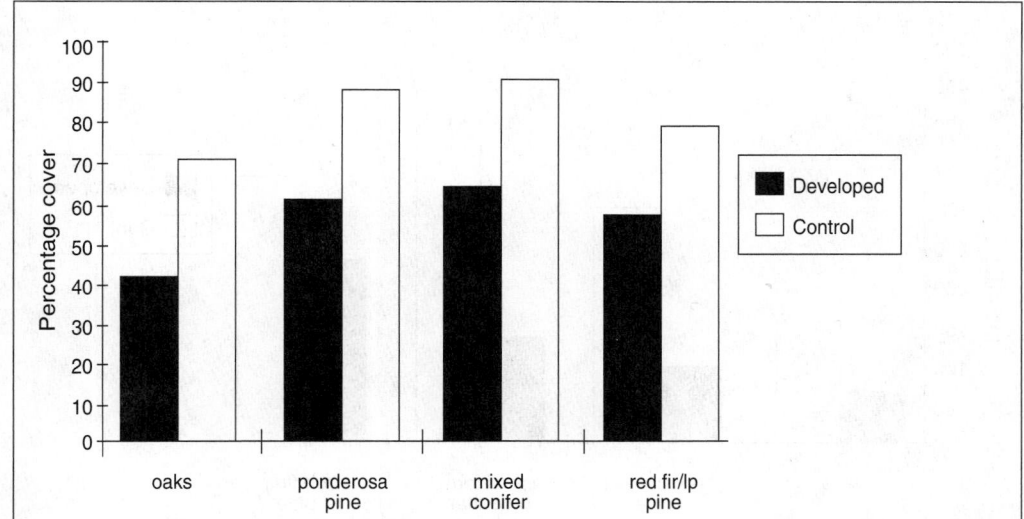

FIGURE 46.1

Forest canopy cover on developed and control sites in the less-than-1-acre lot size class.

ness is summarized in table 46.4 and figure 46.3. The increase in species richness was statistically significant (ANOVA; F-test; 0.01%) in developed areas in the foothill woodland and ponderosa pine forest but not in the mixed conifer and red fir–lodgepole pine forests. A large number of exotic tree species contributed to the increase in species richness in the foothill woodland and ponderosa pine forest. The increase in species richness in the red fir–lodgepole pine forest was due entirely to the planting of native species.

The changes in cover, tree density, and species richness observed in this study are similar in direction to those reported by McBride and Jacobs (1979, 1986) for areas of human settlement in Jeffrey pine forests in the Lake Tahoe Basin. Construction of structures, roads, and other infrastructure elements in forests often necessitates the removal of trees and results in reduction of canopy cover and tree density. Trees may also be removed to facilitate access to sunlight, especially in more densely wooded areas. Conversion of tree cover to lawn also contributes to the decrease in tree canopy cover and density. The increase in species richness in developed areas is primarily due to tree planting by home owners. Some of the increase in species richness may be due to the invasion of exotics. The greater number of exotic tree species observed in developed areas in the foothill woodland and ponderosa pine forest may be related to the more moderate winter temperatures, use of lawn irrigation around structures, and the year-round nature of residency in communities in these lower-elevation zones. Moderate winter temperatures allow a greater number of species to be used without fear of frost damage; lawn irrigation provides summer moisture to many exotic tree species that could not be established or survive without supplemental water; year-round residency encourages the planting of trees and other gardening activities. Year-round residents in the Lake Tahoe area studied by McBride and Jacobs (1986) indicated that they had time for and interest in gardening.

Seasonal residents who were using their homes in the mountains for recreational activities, often for short stays in mid-summer and winter, did not wish to devote time to gardening. Lots in the higher-elevation forests are typically steeper than lots in the lower-elevation forest types. These steeper lots tend to be less maintained and stay in something closer to a "wild state," which accounts in part for the lower species richness. In addition, nursery supplies of trees are more limited in species variety in the high-elevation zones compared with the lower-elevation zone. Tree establishment in human settlements in the higher-elevation forest often involved transplanting or invasion of native trees from adjacent or nearby forests.

The decrease in crown canopy cover in developed areas has implications for fire hazard, hydrology, and wildlife habitat value. Most fire models involve canopy cover as a measure of vegetation associated with increased fire hazard (Andrews and Burgan 1985). These models predict an increase

TABLE 46.3

Tree density on less-than-1-acre lots and control plots.

Type of Plot	Trees per Acre
Foothill Woodland	
Developed	78
Control	156
Ponderosa Pine	
Developed	100
Control	278
Mixed Conifer	
Developed	134
Control	454
Red Fir–Ponderosa Pine	
Developed	234
Control	361

FIGURE 46.2

Tree density on less-than-1-acre lots and control plots.

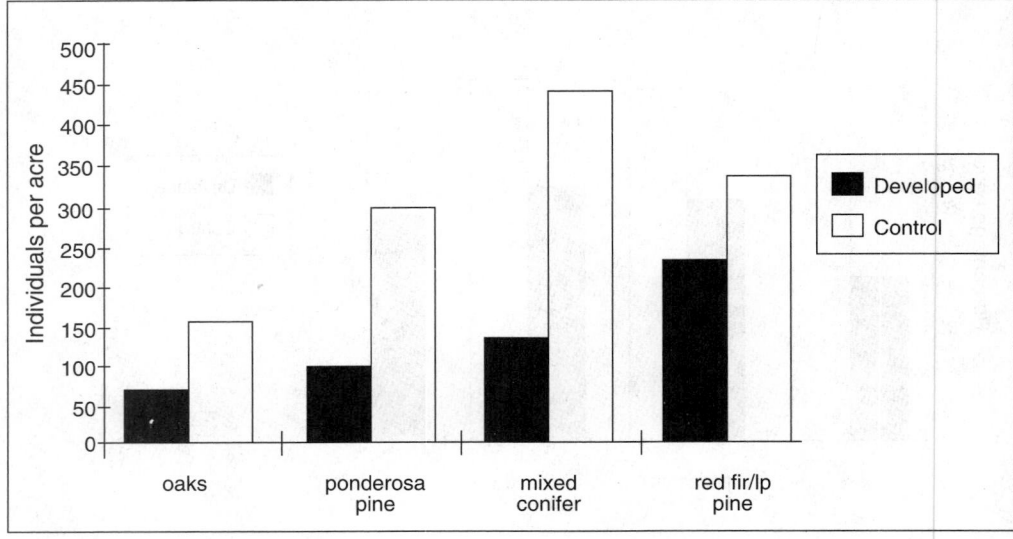

in fire hazard with increasing crown canopy cover. The fragmentation that is occurring in the forest canopy of the Sierra Nevada as a result of human settlement could lead to a reduction in fire hazard, however, if development extended over sizable areas. This reduction is augmented in the lower-elevation woodland and forest types by the maintenance of higher fuel-moisture levels in trees due to lawn irrigation. Fuel ladders are often eliminated as a result of development. Large, woody ground fuels have also disappeared following development. These changes in the quantity and structure of natural fuels must be balanced against the overall change in the fire hazard of a landscape when structures and people become a part of the fuel/ignition complex. It should be noted that, along with the changes in the reduction of fire hazard, the increased human population increases fire risk associated with arson and accidental fires.

The change in cover from native forest to human settlement has important ramifications for the hydrology of the woodlands and forests of the Sierra Nevada. This change will affect the runoff of precipitation from the landscape. Values of the rational runoff coefficient for impervious surfaces in urban areas are on the order of 0.70 to 0.95, while woodlands and forests have coefficients of 0.30 to 0.40 (American Society of Civil Engineers 1969; Rantz 1971). The reduction in tree canopy cover influences the runoff coefficients by reducing the interception of rainfall. The increase in impervious surface found on less-than-1-acre lots decreases the infiltration of precipitation and contributes to greater and more rapid runoff, thus higher runoff coefficients. Lawn irrigation also contributes to greater quantities of runoff from developed areas by maintaining a higher level of soil moisture storage.

The changes in canopy cover, tree density, and species richness associated with development will mean a change in the characteristics of the food and cover available to wildlife species. Studies of the effects of urbanization on wildlife have generally shown a shift in the species using an area following development (Noyes and Porgulske 1973; Adams and Dove 1989; Mills et al. 1989). In general, wildlife species diversity declines along gradients of increasing urbanization, while the population density of some well-adapted urban species increases. A significant factor in this relationship is the proximity of the developed area to undeveloped woodlands and forests.

Pattern of Land Use in the Foothill Woodland Along Highway 49

Thirty-two vegetation and land-use types were observed in the survey of land use along Highway 49. Of these, twenty-nine were tallied at the point samples taken every mile along the 116-mile survey from Sonora to Grass Valley. The three additional types were observed along the highway but did

TABLE 46.4

Species richness on less-than-1-acre lots and control plots.

Type of Plot	Native	Exotic	Total
Foothill Woodland			
Developed	5	24	29
Control	3	0	3
Ponderosa Pine			
Developed	8	32	40
Control	4	0	4
Mixed Conifer			
Developed	6	3	9
Control	5	0	5
Red Fir–Lodgepole Pine			
Developed	4	0	4
Control	2	0	2

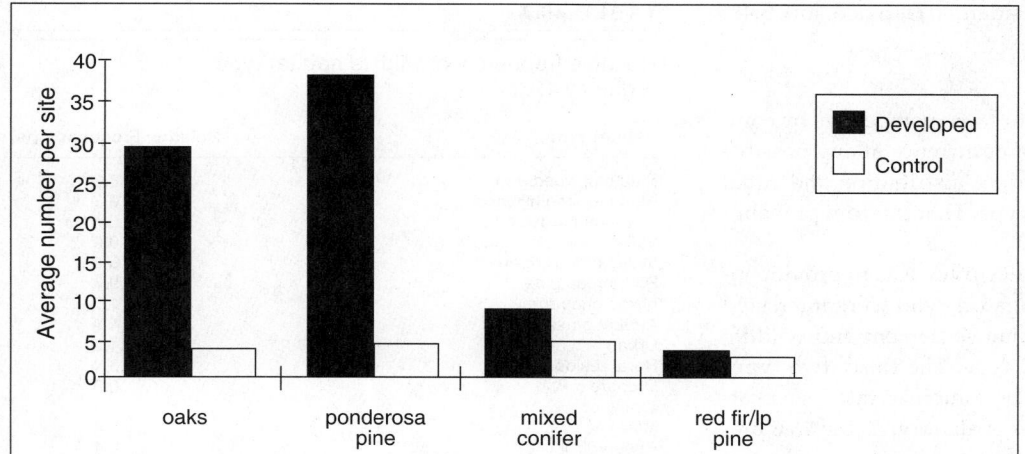

FIGURE 46.3

Species richness on less-than-1-acre lots and control plots.

not occur at any of the 116 sample points. These vegetation and land-use types are listed in table 46.5 with their relative frequency of occurrence. The relative frequency of natural vegetation types along Highway 49 was 59.2%. Agriculture, rural residential, urban, and other land uses had frequencies of 11.6%, 15.9%, 10.7%, and 2.5%, respectively. The five most frequently encountered types were blue oak woodland (11.2%), pasture (10.3%), residential (1–3-acre lot, 9.9%), blue oak–foothill pine (9.5%), and urban (commercial, 9.0%). This method has probably overestimated the relative frequency of urban (commercial) development on the landscape because this land use is situated along roads for the purpose of commerce; while other land uses are to be found along roads, they do not need to be adjacent to a road or highway. Similarly, urban (quarter-acre lot), trailer park, and residential (less-than-1-acre lot) land uses are probably overrepresented in the sample.

With a total of thirty-two vegetation and land-use types, there would be 512 possible pairs. At the 116 sample points along Highway 49, only 70 of the 512 potential pairs were recorded (at each sample point the "pair" consisted of the two types, one on each side of the road). The small number of pairs recorded compared with the 512 potential pairs is due to the lack of randomness in the distribution of vegetation and land-use types, the difference in the relative frequency of occurrence of each type, and the compatibility of different land uses. The majority of the vegetation and land-use pairs were encountered only one time in the 116-point sample. The most frequently encountered pairs were

- urban commercial:urban commercial, 6.8% of the 188 pairs

- blue oak savanna:pasture, 5.1%

- black oak–ponderosa pine forest:black oak–ponderosa pine forest, 4.2%

TABLE 46.5

Relative frequencies of vegetation and land-use types along Highway 49.

Type	Relative Frequency (%)
Natural Vegetation	**59.2**
Blue oak savanna	11.2
Blue oak–foothill pine woodland	10.6
Blue oak woodland	6.9
Black oak–ponderosa pine woodland	6.5
Riparian woodland	5.1
Maul oak–foothill pine woodland	4.3
Maul oak woodland	3.9
Grassland	3.4
Valley oak savanna	2.1
Maul oak savanna	1.7
Foothill pine–chaparral	0.9
Maul oak–ponderosa pine woodland	0.9
Ponderosa pine forest	0.9
Foothill pine woodland	0.4
Mixed broadleaf forest	0.4
Agriculture	**11.6**
Pasture	10.3
Orchard	0.9
Vineyard	0.4
Christmas tree farm[a]	0.0
Hay field[a]	0.0
Rural Residential	**15.9**
Residential (1–3-acre lots)	9.9
Residential (1-acre lots)	4.3
Trailer park	1.7
Urban	**10.7**
Commercial	9.0
Residential (quarter-acre lots)	1.3
Institutional property (school, fire station, etc.)	0.4
Industrial[a]	0.0
Other	**2.5**
State park	0.9
Mine	0.4
Reservoir	0.4
Himalayan blackberry (erosion-control planting)	0.4
Scotch broom (invasion on disturbed land)	0.4

[a]Land-use types observed along Highway 49 that did not occur at sample points.

- blue oak–foothill pine:rural residential (1–5-acre lot), 3.4%

- Pasture:residential, 3.4%.

These frequencies may be put into some perspective by comparing them to the probability of occurrence of any possible pair at any point, assuming random distribution and equal frequency of occurrence of each type. That random probability is 0.19%.

The purpose of the Highway 49 survey was to provide information to other SNEP investigators who were interested in the relationship between human settlement and wildlife habitat in the foothill woodland type. The thirty-two types observed along Highway 49 can be combined into fewer categories for the purpose of habitat evaluation. Table 46.6 presents combinations of the thirty-two types that are considered to have similar structural characteristics and, therefore, may be considered similar wildlife habitats. This table reduces the thirty-two vegetation and land-use types to fourteen habitat types, many of which are habitats recognized in the Califor-

TABLE 46.6

Wildlife habitats and vegetation and land-use types along Highway 49.

Wildlife Habitat	Vegetation or Land-Use Type
Blue oak woodland[a]	Blue oak woodland
	Blue oak savanna
	Maul oak woodland
	Maul oak savanna
	Regenerating blue oak woodland
Blue oak–foothill pine[a]	Blue oak–foothill pine woodland
	Blue oak–foothill pine woodland (recently burned)
	Black oak–foothill pine woodland
	Foothill pine woodland
	Maul oak–foothill pine woodland
	State historic park
Montane hardwood[a]	Mixed broadleaf forest
Valley oak woodland[a]	Valley oak savanna
Valley foothill riparian[a]	Riparian woodland
Ponderosa pine[a]	Ponderosa pine forest
	Ponderosa pine forest (recently burned)
	Maul oak–ponderosa pine woodland
	Black oak–ponderosa pine woodland
Mixed chaparral[a]	Foothill pine–chaparral
	Himalayan blackberry erosion-control planting
	Scotch broom scrub
	Christmas tree farm
Annual grassland[a]	Grassland
	Pasture
	Hayfield
Urban[a]	Industrial
	Urban residential (quarter-acre lot)
	Urban commercial
	Trailer park
	Institutional property
Rural residential	Rural residential (less-than-1-acre lot)
	Rural residential (1–3-acre lot)
Orchard	Orchard
Vineyard	Vineyard
Mine	Mine
Reservoir	Reservoir

[a]California Wildlife Habitat Relationship Program type.

TABLE 46.7

Relative frequency of wildlife habitat types along Highway 49.

Habitat Type	Relative Frequency (%)
Blue oak woodland	28.8
Blue oak–foothill pine	13.7
Montane hardwood	0.4
Valley oak woodland	0.9
Valley foothill riparian	3.4
Ponderosa pine	9.7
Mixed chaparral	1.7
Annual grassland	14.3
Urban	11.5
Rural residential	13.4
Orchard	0.9
Vineyard	0.4
Mine	0.4
Reservoir	0.4

nia Wildlife Habitat Relationship Program (Mayer and Laudenslayer 1988). The relative frequency of these habitat types is shown in table 46.7. The habitat types with the highest relative frequencies were blue oak woodland (28.8%), annual grassland (14.3%), blue oak–foothill pine (13.7%), rural residential (13.4%), and urban (11.5%). The fourteen habitat types could potentially be combined into 105 pairs. Only 33 of these potential pairs occurred at the sampling points along Highway 49. The number of these pairs tallied at the sampling points, and their relative frequencies are shown in table 46.8. Knowledge of the frequency of pairing between various land-use types and vegetation types may provide some insights into the character of future wildlife habitats in the foothill woodland. For example, pairing of rural residential development with blue oak woodland would produce a landscape that would support a different mosaic of habitats than the pairing of rural residential with annual grassland.

Table 46.9 conveys a sense of the fragmentation of the natural cover as a result of human settlement along Highway 49. Of the 116 pairs of land use sampled, 46% could be categorized as natural cover, suggesting slightly less than one-half of the area of natural vegetation along the highway has suffered fragmentation (excluding the fragmentation caused directly by the highway). The other 54% of the pairs were combinations of natural cover and land uses resulting from human settlement. Just under 16% of the pairs were the result of juxtaposition of natural cover and rural development for housing. The other major source of fragmentation was agriculture. Urban development accounted for almost 13% of the 116 pairs of samples along Highway 49. This direct loss of wildlife habitat, combined with habitat fragmentation along a major state route through the foothill woodland vegetation type, may be representative of the potential impact of human settlement patterns along new highways through the foothill woodland type. The impacts of human settlements along highways through other forest types in the Sierra Nevada are not expected to be similar to that along Highway 49 because of the limited use of land for agricultural purposes in other forest types. In addition, the common siting of highways along ridges at higher elevations in the Sierra Nevada may tend to concentrate development along the highway corridor because of slope limitations. Analyses of development patterns along highways in other Sierra Nevada forest types would be appropriate to better understand their patterns of human settlement in relation to highways.

FUTURE RESEARCH DIRECTION

Additional data are needed to round out our understanding of the changes in forest conditions on larger-sized lots in the ponderosa pine, mixed conifer, and red fir–lodgepole pine forests. In order to obtain these data, it would be necessary to have access to maps showing private property boundaries (lot boundaries) in these forest types. It would also be valuable to extend data collection into both the northern and the southern Sierra Nevada and to the east-side vegetation and land-use types. If the technique used along Highway 49 has

TABLE 46.8

Relative frequencies of habitat pairs along Highway 49.

Habitat Pair	Relative Frequency (%)
Blue oak woodland:blue oak–foothill pine woodland	11.9
Blue oak woodland:annual grassland	10.2
Blue oak woodland:blue oak woodland	9.3
Urban:urban	9.3
Ponderosa pine forest:ponderosa pine forest	5.1
Rural residential:blue oak woodland	5.1
Annual grassland:annual grassland	4.2
Rural residential:annual grassland	4.2
Blue oak woodland:ponderosa pine forest	3.4
Rural residential:blue oak–foothill pine woodland	3.4
Blue oak woodland:valley foothill riparian woodland	2.5
Rural residential:ponderosa pine forest	2.5
Rural residential:rural residential	2.5
Blue oak–foothill pine woodland:valley foothill riparian woodland	2.5
Urban:rural residential	2.5
Blue oak–foothill pine woodland:blue oak–foothill pine woodland	1.7
Blue oak–foothill pine woodland:annual grassland	1.7
Valley oak woodland:valley foothill riparian woodland	1.7
Rural residential:valley foothill riparian woodland	1.7
Orchard:annual grassland	1.7
Blue oak–foothill pine woodland:ponderosa pine forest	0.8
Valley oak woodland:valley oak woodland	0.8
Valley foothill riparian woodland:ponderosa pine forest	0.8
Ponderosa pine forest:mixed chaparral	0.8
Ponderosa pine forest:annual grassland	0.8
Mixed chaparral:mixed chaparral	0.8
Mixed chaparral:annual grassland	0.8
Urban:blue oak woodland	0.8
Vineyard:blue oak–foothill pine woodland	0.8
Mine:valley foothill riparian woodland	0.8
Reservoir:blue oak woodland	0.8
Reservoir:blue oak–foothill pine woodland	0.8

TABLE 46.9

Relative frequencies of land-use pairs along Highway 49.

Land-Use Pair	Relative Frequency (%)
Natural cover:natural cover (e.g., blue oak woodland:blue oak woodland)	46.0
Natural cover:rural development (e.g., blue oak woodland:rural residential)	15.7
Urban:urban (e.g., commercial:urban residential)	12.6
Natural cover:agriculture (e.g., blue oak woodland:pasture)	12.4
Agriculture:agriculture (e.g., pasture:orchard)	4.2
Agriculture:rural development (e.g., pasture:rural residential)	4.2
Natural cover:other (e.g., blue oak woodland:reservoir)	2.4
Rural development:rural development (e.g., rural residential:trailer park)	1.7
Other:other (e.g., mine:reservoir)	0.8

provided useful data to the wildlife experts associated with the SNEP project, this technique could be applied to other forest types in the Sierra Nevada.

ACKNOWLEDGMENTS

The authors wish to express their appreciation for the assistance and counsel of Rowan Rowntree (USFS Pacific Southwest, Albany, California), Tim Duane (Departments of Landscape Architecture and City and Regional Planning, University of California, Berkeley), Karl Goldstein (Department of Landscape Architecture, University of California, Berkeley), Greg Greenwood (Strategic Planning, California Department of Forestry and Fire Protection, Sacramento), Dave Graber (U.S. Biological Service, Yosemite National Park), and the librarians in the Map Library at the University of California.

Many home owners in the Sierra Nevada graciously invited us into their backyards to see trees they had planted and loved. We appreciate their generosity in allowing us onto their property and their insights into the urbanization process in the Sierra Nevada.

REFERENCES

Adams, L. W., and L. E. Dove. 1989. *Wildlife reserves and corridors in the urban environment*. Columbia, MD: National Institute for Urban Wildlife.

American Society of Civil Engineers. 1969. *Design and construction of sanitary and storm sewers*. ASCE Manuals and Reports on Engineering Practices 37. New York: ASCE.

Andrews, P. L., and R. E. Burgan. 1985. "BEHAVE" in the wilderness. In *Proceedings, symposium and workshop on wilderness fire*. General Technical Report INT-43. Missoula, MT: U.S. Forest Service.

Barbour, M., B. Pavlik, F. Drysdale, and S. Lindstrom. 1993. *California's changing landscape.* Sacramento, CA: California Native Plant Society.

Bonnicksen, T. M,. and E. C. Stone. 1981. The giant sequoia–mixed conifer forest .cmmunity characterized through pattern analysis as a mosaic of aggregations. *Forest Ecology and Management* 3: 307–28.

Griffin, J. 1977. Oak woodland. In *Terrestrial vegetation of California,* edited by M. Barbour and J. Major, 383–415. New York: J. Wiley and Sons.

Holland, V. L., and D. J. Keil. 1986. *California vegetation.* San Luis Obispo, CA: El Corral Publications.

Johnston, V. R. 1994. *California forests and woodlands.* Berkeley and Los Angeles: University of California Press.

Lewis, H. T. 1973. *Patterns of Indian burning in California: Ecology and ethnohistory.* Anthropological Paper 1. Santa Barbara, CA: Ballena Press.

Mayer, K. E., and W. F. Laudenslayer, Jr. 1988. *A guide to wildlife habitats of California.* Sacramento: California Department of Forestry and Fire Protection.

McBride, J. R., and D. F. Jacobs. 1979. Urban forest structure: A key to urban forest planning. *California Agriculture* 33:24.

———. 1986. Presettlement forest structure as a factor in urban forest development. *Urban Ecology* 9:245–66.

Mills, G. S., J. B. Dunning, Jr., and J. M. Bates. 1989. Effects of urbanization on breeding bird community structure in southwestern desert habitats. *Condor* 91:416–28.

Noyes, J. H., and D. R. Porgulske, eds. 1973. *A symposium on wildlife in an urbanizing environment.* Amherst, MA: University of Massachusetts, Cooperative Extension Service.

Oosting, H. J., and W. D. Billings. 1943. The red fir forest of the Sierra Nevada: *Abietum magnificae. Ecological Monographs* 13:259–74.

Pavlik, B. M., P. C. Muick, S. Johnson, and M. Popper. 1991. *Oaks of California.* Los Olivos, CA: Cachuma Press.

Rantz, S. E. 1971. Suggested criteria for hydrologic design of storm-drainage facilities in the San Francisco Bay Region, California. Open File Report. Menlo Park, CA: U.S. Geological Survey.

Rundel, P. W., D. J. Parsons, and D. T. Gordon. 1977. Montane and sub-alpine vegetation of the Sierra Nevada and Cascade Ranges. In *Terrestrial vegetation of California,* edited by M. Barbour and J. Major, 559–99. New York: J. Wiley and Sons.

MARK W. SCHWARTZ
Center for Population Biology
University of California
Davis, California

DANIEL J. PORTER
Center for Population Biology
University of California
Davis, California

JOHN M. RANDALL
The Nature Conservancy
Davis, California

KELLY E. LYONS
Section of Plant Biology
University of California
Davis, California

47

Impact of Nonindigenous Plants

ABSTRACT

Nonindigenous species play a dominant role in the vegetation of many ecosystems that adjoin or overlap the Sierra Nevada. Many of these habitats, such as the valley and foothills grasslands, appear to be saturated with these species. In contrast, the high elevations of the Sierra Nevada, like most high alpine regions, are not as heavily impacted by nonindigenous species. In between these extremes is a gradient of impact that is heavily influenced by the amount and extent of human disturbance of natural ecosystems. The most heavily affected regions within the Sierra Nevada are the foothill grassland and oak savanna habitats, infested with a diversity of Mediterranean annual grasses as well as herbaceous dicots; the riparian zones, infested by woody plants; and the eastern slope, which is strongly dominated by cheat grass. Infestation at middle elevations is most closely linked to disturbances such as clear-cuts and roadsides. Non-native species such as cheat grass, yellow star thistle, salt cedars, Russian olive, ailanthus, Himalayan blackberry, and Scotch broom affect ecosystem attributes such as grazing potential, forest regeneration, and water availability along stream courses. At the present time there are few restrictions on the importation of species that may, in the future, pose additional threats to the integrity and utility of the Sierra Nevada. This chapter recommends a series of actions, beginning with educational programs that may limit the sale of non-native species that are known problems. Programs from other states are profiled as models for potential importation restrictions that would be useful in protecting the integrity of the Sierra Nevada.

INTRODUCTION

Introduced weedy species present a disturbance to both natural ecosystems and managed habitats in ways that most citizens do not fully recognize. When one pauses to think about introduced species and how they affect our lives, from commerce to recreation, the evidence is abundant. Food that crosses international borders is inspected for potential pests and pathogens; the agrochemical industry has developed around the effort to reduce weedy pests (plants and animals) in agricultural systems; teams of scientists monitor the spread of plant pests such as the gypsy moth, killer bees, and disease agents that cause Dutch elm disease, oak wilt and white pine blister rust, to name a few.

Biological invasions have persistent and far-reaching consequences. Although plant invasions have often reduced biological diversity and the aesthetic value of natural lands, to humans these costs are often intangible, vary among individuals, or are difficult to measure. In contrast, the direct economic effects to agricultural systems are more easily assessed. A conservative estimate of the cumulative losses to the United States from selected harmful nonindigenous species from 1906 to 1991 is \$97 billion (U.S. Congress, Office of Technology Assessment [OTA] 1993). This estimate includes only losses to agriculture, industry, and human health, and not the undocumented costs of the loss of native biological diversity, the disturbance of healthy forest ecosystems, and the loss of the recreation potential of habitats compromised by the presence of weedy pests (OTA 1993).

California, with respect to vascular plants, is one of the most biologically diverse regions of North America. Similarly, California has a greater diversity of problems regarding intro-

Sierra Nevada Ecosystem Project: Final report to Congress, vol. II, *Assessments and scientific basis for management options.* Davis: University of California, Centers for Water and Wildland Resources, 1996.

duced species than most places. With its rich agricultural industry, California is arguably one of the most extreme examples of the costs associated with living with introduced species, including the Mediterranean fruit fly, Argentinean ants, killer bees, cheat grass, and star thistle. The list seems endless. Yet these problems are relatively recent. By definition, no nonindigenous plants had been introduced prior to European contact in 1769. By 1860 there were at least 134 established alien plant species (Raven 1990). By 1993 this number has grown to more than 1,000, or 15% of the entire flora (Hickman 1993). This century has seen exponential growth in the number of nonindigenous plant species (figure 47.1) as well as exponential growth in their impacts. Because problems with non-native species are typically less severe in the mountains, the Sierras are burdened with less of a problem with non-native species than is most of California. Nonetheless, the Sierras have a greater problem than most mountainous regions.

KEY QUESTIONS

This chapter addresses four key questions regarding non-native species:

1. Why are nonindigenous species problematic?

2. What habitats within the Sierra are most heavily affected by non-native species, and which species are responsible for these effects?

3. What disturbances, human generated or natural, exacerbate the non-native species problem, and in what ways do they do so?

4. What mechanisms can be implemented to minimize the risk associated with non-native plant species?

To answer these questions, we have divided this chapter into four parts. First, we define and describe nonindigenous weedy species and address attributes that contribute to the ability of certain species, once introduced, to spread uncontrollably and become management problems. Second, we describe habitats of the Sierra Nevada that are heavily disturbed as a result of non-native plants and discuss the specific ecology of some of the most problematic of these species. Third, we link the disturbances created by nonindigenous plants to other disturbance factors in the Sierra. Finally, we propose mechanisms to limit the potential for new damage by nonindigenous species and to remediate damage already inflicted upon the Sierra Nevada by the species that have already been introduced.

AN INTRODUCTION TO NONINDIGENOUS PLANTS

Terminology for defining non-native species has been somewhat unclear (Lukens 1994). Plants and animals have often been dispersed beyond their historic distribution limits through human activities. These species with human-enhanced distributions that become established (that is, that grow and reproduce without further human intervention) in their new environments are variously called "exotic," "non-native," "introduced," "alien," and "nonindigenous." A small but significant subset of invasive nonindigenous species experiences rapid and uncontrolled population growth. This chapter considers all plant species known to have been introduced, intentionally or unintentionally, to the flora of the Sierra Nevada since the time of European settlement to be nonindigenous.

The natural environment is a complex combination of biotic and abiotic interactions. Each species has numerous relationships with other organisms and its environment, some subset of which typically regulates population size. Predation, competition, disease, and unpredictable, harsh weather conditions can all limit the number of individuals in a population. Without regulation, populations of all species have the ability to grow exponentially (Silvertown and Doust 1993). When organisms are transported to a new habitat, they may experience a release from factors that typically limit their population growth. Such population expansions are frequently at the expense of native species that would otherwise occupy the space, but whose populations are regulated by their environment.

Researchers have tried to predict successful plant invaders, using life histories, genetics, and ecological traits. Baker (1974, 1986) addresses characteristics possessed by the hypo-

FIGURE 47.1

Number of nonindigenous plant species by date as reported in botanical treatments of the California flora.

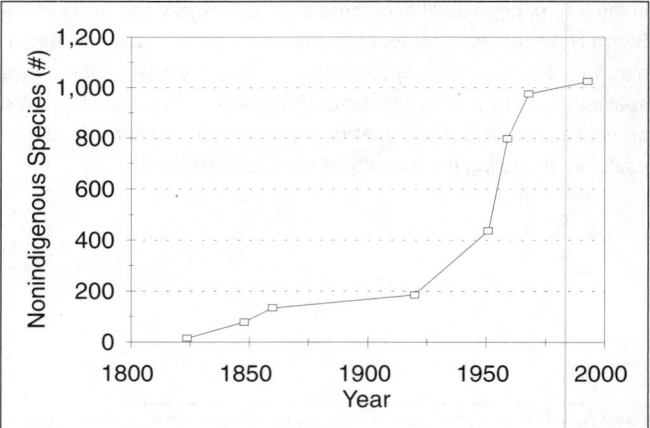

thetical ideal weed (e.g., few germination requirements, rapid growth, continuous seed production, and strong competitive ability). Similarly, polyploidy (containing more than one full set of chromosomes) (Bazzaz 1986) and preadaptation to climate, nutrient levels, and the disturbance regime of the new area (Kruger et al. 1989) may also increase invasiveness. While this is interesting speculation, the value of such characterizations is questionable because biological systems are inherently complex (Noble 1989). The success of a plant species in a novel environment depends on the outcome of an undefined number of interactions with the abiotic and biotic environment. Hence, a plant possessing few "weedy" characteristics may succeed because one environmental factor may drive the ability of its population growth. Rejmanek (1989) has compiled a list of more than fifty species that invade "undisturbed" communities. After reviewing these species and their invaded communities, Rejmanek concludes, "While it seems to be possible to make some generalizations about successful invaders in disturbed and successionally young communities (Baker 1965; Heywood 1989), there is apparently nothing unifying for invaders of 'undisturbed' natural communities." Alternatively, Reed (1977) has generated a list, based on the plants' performance elsewhere, of plants not currently in the United States that, if introduced, could become invasive. Similarly, Rejmanek (1995) has identified potentially invasive pines by examining regeneration characteristics and past performance as an invader. Such lists could be successful in preventing the import of invasive species. Fewer than one-third of import protocols, however, require information on an organism's foreign performance (Ruesink et al. 1995).

The impact of nonindigenous plant species generally decreases with increasing elevation. While there are species (e.g., Kentucky bluegrass [*Poa pratensis*]) that invade high-elevation Sierra Nevada meadows, the number of these is few relative to lower elevations in the Sierras. The relative paucity of nonindigenous plants at higher elevations has led some to suggest that there is an elevational barrier to species introductions. The question remains as to whether the lower impact is due to a scarcity of invasive nonindigenous species that survive at higher elevations or to a lower frequency of introductions of high-elevation species. Evolution of high-elevation ecotypes from low-elevation nonindigenous species may eventually decrease this gradient. Populations of yellow star thistle (*Centaurea solstitialis*) in California have begun to differentiate as a result of differences in climatic conditions between the coast and inland areas (Maddox and Mayfield 1985).

In general, nonindigenous species are not randomly distributed with respect to biogeographic regions. Historically, regions with a mediterranean climate have been among those most affected by biological invasions. These regions, such as California, have cool, wet winters followed by hot, dry summers. A combination of a long potential growing season and a long human history of agriculture in the Mediterranean has resulted in numerous species that seem to be well adapted to invading human-disturbed habitats in such climates (Kruger et al. 1989; Groves 1986). The majority of successful invasives in other regions seem to come from Eurasia. The abundance of biological invasions in regions with a mediterranean climate is staggering. To cite a few examples,

- The invasive plant *Mimosa pigra* threatens to decimate waterbird populations and to reduce reptilian and mammalian diversity in Australia (Braithwaite et al. 1989).

- By 1977, more than 9 million hectares (22 million acres) of California grassland and woodland had been severely invaded by non-native plants (Heady 1977).

- Recent estimates show that 60% of South Africa's fynbos community (brushy, chaparrallike habitat) has been replaced by non-native woody invasives (Heywood 1989).

- Several dozen nonindigenous species have been deemed so serious as to be officially considered "plagues of agriculture" in Argentina (Mack 1989, citing Marcoza 1984).

In addition, a multitude of human-induced disturbances make communities more invasible. For example, agriculture (Mack 1989), clear-cutting (Heywood 1989), fire (Heywood 1989; Baird 1977; Groves 1986), road construction (Frenkel 1970), and urban expansion and grazing (Kruger et al. 1989; Orians 1986; Milchunas et al. 1988; Milchunas et al. 1989) have all historically been implicated or experimentally shown to facilitate plant invasion.

BIOLOGICAL INVASION OF THE SIERRA NEVADA

An examination of the nonindigenous plant species in the Sierra Nevada highlights three high-impact areas: valley grasslands and foothill oak woodland, riparian zones, and the eastern slope (table 47.1). Our treatment of nonindigenous plants focuses on highlighting general problems within these more heavily affected habitats, followed by a discussion of the autecology of key species. A separate, smaller section will follow, highlighting other problematic species for which we have less information. This survey of heavily affected sites and highly problematic species is not exhaustive (see, for example, table 47.1), but represents the status of a significant portion of the Sierra Nevada and is suggestive of the potential future of the region.

Sierra Nevada Valley Grassland and Foothill Oak Woodland

California has several extensively invaded plant communities. One of the most heavily affected is the Central Valley,

TABLE 47.1

A partial list of invasive nonindigenous plants of the Sierra Nevada, listed by habitat.

Valley Grasslands, Foothill Oak Woodlands	Riparian Zones	Eastern Slope Desert	General to Many Disturbed Habitats
Bull thistle (Cirsium vulgare)	Salt cedar (Tamarix spp.)	Klamath weed (Hypericum perforatum)	Kentucky bluegrass (Poa pratensis)
Cheat grass (Bromus tectorum)	Ailanthus (Ailanthus altissima)	Wild oats (Avena spp.)	Spanish broom (Spartium junceum)
Scotch broom (Cytisus scoparius)	Russian olive (Eleagnus angustifolia)	Fescue (Festuca spp.)	Tansy ragwort (Senecio jacobaea)
Yellow star thistle (Centaurea solstitialis)	Giant reed (Arundo donax)	Brome grass (Bromus spp.)	Spotted knapweed (Centaurea maculosa)
Wild oats (Avena spp.)	Harding grass (Phalaris aquatica)	Woolly mullein (Verbascum thapsus)	Leafy spurge (Euphorbia esula)
Fescue (Festuca spp.)	Eucalyptus (Eucalyptus spp.)	Camel thorn (Alhagi camelorum)	Canada thistle (Cirsium arvense)
Smooth brome (Bromus inermis)	English ivy (Hedera helix)		Bull thistle (Cirsium vulgare)
Artichoke thistle (Cynara cardunculus)	Himalayan blackberry (Rubus discolor)		Foxglove (Digitalis purpurea)
Wild fennel, anise (Foeniculum vulgare)	Periwinkle (Vinca major)		Woolly mullein (Verbascum thapsus)
Eucalyptus (Eucalyptus spp.)	Canada thistle (Cirsium arvense)		Halogeton (Halogeton glomeratus)
French broom (Cytisus monspessulanus)	Bull thistle (Cirsium vulgare)		Castor bean (Ricinus communis)
English ivy (Hedera helix)	Edible fig (Ficus carica)		
Fountain grass (Pennisetum setaceum)	Black locust (Robinia pseudoacacia)		
Himalayan blackberry (Rubus discolor)	Purple loosestrife (Lythrum salicaria)		
Periwinkle (Vinca major)	Lippia (Phyla nodiflora)		
Purple starthistle (Centaurea calcitrapa)			
Tansy ragwort (Senecio jacobaea)			
Spanish broom (Spartium junceum)			
Gorse (Ulex europaeus)			
Pennyroyal (Mentha pulegium)			
Harding grass (Phalaris aquatica)			
Smilo grass (Piptatherum milaceium)			
Castor bean (Ricinus communis)			
Camel thorn (Alhagi camelorum)			
Purple loosestrife (Lythrum salicaria)			
Giant plumeless thistle (Carduus acanthoides)			
Lippia (Phyla nodiflora)			
Tree tobacco (Nicotiana glauca)			

which has one of the worst problems with nonindigenous species in the world. The Central Valley, measuring 100 km (60 mi) wide and 650 km (400 mi) long, includes 15% of the state's total area. The original California grassland (pre-European settlement) was eliminated by the invasion of introduced annual species during the nineteenth century. Historical accounts of the area's vegetation are vague, but it is likely that two species of perennial bunchgrasses (Nassella cernua and N. pulchra) dominated in many areas (Barbour et al. 1993). In contrast, the present-day grassland is dominated by annual grasses, most of which are introduced. The first introductions began with the Spanish missionaries in about 1769. The effect of these earliest introductions is thought to have been minimal compared to the introductions by migrants moving to the West following the discovery of gold in 1848, who brought with them a huge number of livestock accompanied by contaminated seed lots, imported forage, and packing materials. The cattle and sheep also significantly intensified grazing pressure on native vegetation (figures 47.2 and 47.3). The perennial bunchgrasses, with their flowers and seeds high above the ground, were not well adapted for in-

tense, year-around grazing pressure. It is thought that the native grasses disappeared rapidly with intense grazing and were replaced by grazing-tolerant non-native annual grasses (Mack 1989). In addition, widespread agriculture provided the disturbance and nutrients needed for many introduced plants to outcompete native plants (Mack 1989). Finally, woodland and chaparral communities were burned to make room for grazing and mining, thus changing the fire regime to one that favored annual grasses (Mack 1989).

The condition of the central valley grassland, which lies immediately to the west of the Sierra, has profound effects on the Sierra Nevada. Given the frequency with which land is cleared in the Central Valley, newly introduced plants have a good chance of becoming established and spreading. The Central Valley currently serves as a launching platform for nonindigenous plants, allowing them to ascend to higher elevations (figure 47.4). It may be only a matter of time before species introduced to the Central Valley move upslope into the Sierras. Species like Scotch broom (Cytisus scoparius) and yellow star thistle (Centaurea solstitialis) already exist in the middle-elevation ranges. Riparian areas transecting both the

Central Valley and the Sierra Nevada also provide corridors for possible invasion. The construction of dams and the associated disturbance of natural river flows also make riparian areas more susceptible to invasion (see the section on riparian areas, later in this chapter, for examples).

Oak woodlands form a large ellipse around the Central Valley, covering several million hectares. Three endemic trees dominate the community (blue oak [*Quercus douglasii*], valley oak [*Q. lobata*], and gray pine [*Pinus sabiniana*] [Griffin 1990]), along with interior live oak (*Q. wislizenii* var. *wislizenii*) and California buckeye (*Aesculus californica*). The oak woodlands support more than a hundred species of birds during the breeding season and sixty species of mammals (Barbour et al. 1993). Both black-tailed deer and the acorn woodpecker rely heavily on the acorns from oak trees (Barbour et al. 1993). In most sites, the understory of this community, as is the case in the Central Valley grassland, is dominated by non-native grasses. Within this matrix, additional weedy species are invading, making the habitat less suitable as forage for cattle (Menke et al. 1996). Further, the habitat itself is potentially threatened by the nonindigenous annual grasses. The characteristic overstory dominant blue oak is failing to regenerate within California oak woodlands (Momen et al. 1994). One hypothesis for the failure of blue oak is that it cannot compete with the annual grasses (Gordon et al. 1989; Welker et al. 1991; Gordon and Rice 1993; Rice et al. 1993).

Scotch Broom

Scotch broom (*Cytisus scoparius*) presents one of the most severe threats to the oak woodlands of the Sierra foothills. Since its introduction in the 1850s, Scotch broom has spread in coastal and foothill regions and now covers more than 250,000 hectares (618,000 acres). Scotch broom has expanded up out of the oak woodlands to an elevation of 1,200 m (4,000 ft) (C. C. Bossard, telephone conversation with D. J. Porter, spring

FIGURE 47.2

Percentage of California land area used for grazing (redrawn from Mooney et al. 1989).

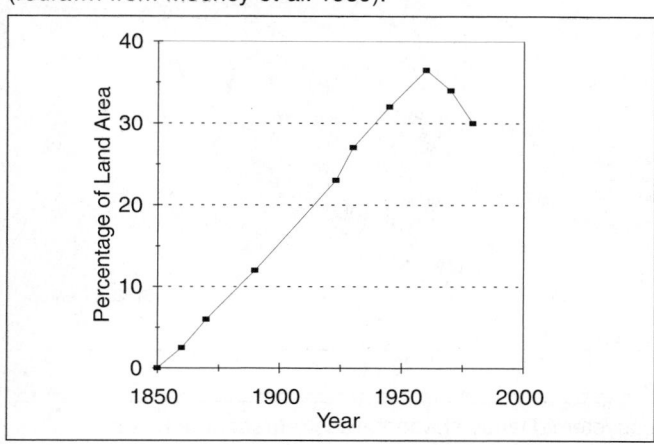

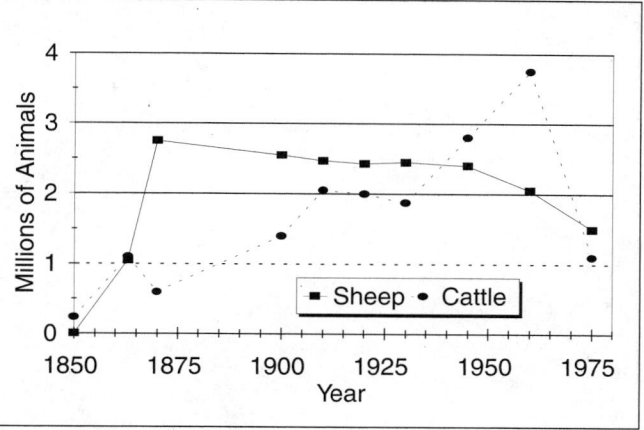

FIGURE 47.3

Sheep and cattle in California (redrawn from Mooney et al. 1989).

1995). As an invasive nonindigenous species, Scotch broom aggressively displaces native vegetation (Andres 1979), making reforestation efforts more difficult (Bossard and Rejmanek 1994). Within the Sierra Nevada, Scotch broom is found in Plumas, Yuba, Butte, Sierra, Nevada, Placer, El Dorado, and Calaveras Counties, primarily around the ponderosa pine forest–chaparral transition, but also along roadsides across both of these habitat types.

Much of Scotch broom's success can be attributed to its weedy life history attributes and to continued habitat disturbance. Scotch broom lives up to seventeen years (M. Rejmanek, unpublished data, 1993) and produces a large number of seeds. It produces two types of seeds, which vary in average distance of dispersal, and maintains a long-lasting seed bank (Bossard 1990a). Scotch broom also has the ability to resprout after being cut or burned (Bossard and Rejmanek 1994), thus defying commonly used eradication methods. Scotch broom has the ability to colonize nitrogen-poor, seasonally dry, frequently disturbed soils (Bossard 1991; Williams 1981; Johnson 1982; Simandl and Kletecka 1989), a combination of edaphic characteristics that is commonly found along roadsides and other places. Frenkel (1970) conducted a survey of California vegetation along roadways and concluded that roads provided both the disturbance needed for establishment of Scotch broom and the corridors for it to spread. Bossard and Rejmanek (1994) found that in the Sierra Nevada, cutting of native vegetation and subsequent soil disturbance promoted the invasion of Scotch broom. Road construction, similarly, facilitates the spread of this plant.

Given the increasing rate of rural development and concomitant construction of roads, it seems likely that Scotch broom will continue to expand its range and zone of heavy impact. Further spread may have a number of ecological and economic impacts. Scotch broom, like many introduced species, provides little value for wildlife, and thus invasion rep-

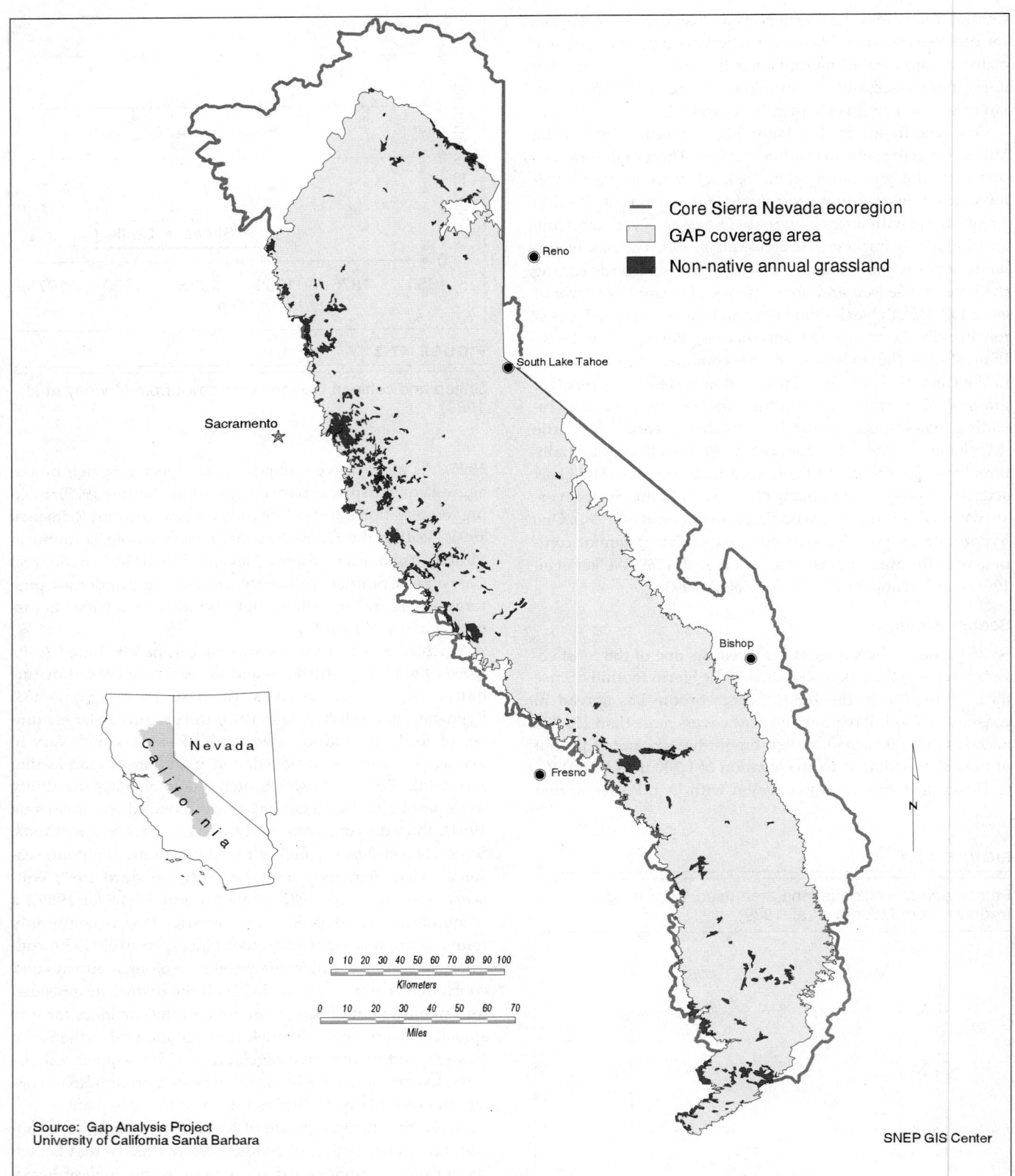

Source: Gap Analysis Project
University of California Santa Barbara

SNEP GIS Center

FIGURE 47.4

Non-native annual grassland coverage within the Sierra Nevada ecosystem. Darkly shaded areas represent habitats dominated by non-native grasslands. The *occurrence* of non-native grasses is much broader.

resents a loss of wildlife food resources. No bird or rodent has been observed eating Scotch broom seed, and experiments have shown that there is no significant granivory by vertebrates (Bossard 1990b; 1991). Scotch broom can also affect the survival of pine seedlings. For example, at one site 70% of ponderosa pine seedlings planted were killed following the invasion of Scotch broom (Bossard and Rejmanek 1994). A clear-cut recently conducted without the removal of the adjacent population of Scotch broom resulted in subsequent invasion of the cleared area, forming a dense monoculture and little regeneration of trees (C. C. Bossard, telephone conversation with the author, May 1995). A wide variety of landowners, from environmentalists to timber managers, must now include Scotch broom removal in their management plans.

Yellow Star Thistle

The genus *Centaurea* is noted for its invasive abilities, with twelve species introduced to the state, three of which are listed as noxious weeds (Hickman 1993). The genus includes purple star thistle (*C. calcitrapa*), Iberian star thistle (*C. iberica*) and spotted knapweed (*C. maculosa*), all found within the Sierra Nevada region. Also within the genus, yellow star thistle (*C. solstitialis*) is a highly invasive plant now found in most parts of the world (Maddox and Mayfield 1985). Yellow star thistle typically invades grasslands, orchards, agricultural fields, and oak woodlands. Since its introduction in the mid to late 1800s, yellow star thistle has run rampant across U.S. rangelands, displacing native vegetation and reducing forage quality. Surveys were conducted between 1958 and 1985 to assess changes in the coverage of yellow star thistle in California (figure 47.5), with the foothills of the Sierra Nevada among the most heavily invaded regions. Maddox and Mayfield (1985) estimated that nearly 8 million acres were infested in

California, with more than 75% of this acreage located in drainage regions of the Sierra Nevada.

Yellow star thistle has several characteristics conducive to its behavior as an invasive weed. It has broad ecological tolerance and can be found (1) from sea level to 2,500 m (8,200 ft); (2) in deep, well-drained soils or shallow, rocky soils; and (3) in areas that receive between 25 and 100 cm (10 and 40 in) annual precipitation (Maddox et al. 1985). Once established, yellow star thistle populations may expand quickly. The plant produces abundant seeds in many heads, each with 50 to 100 seeds. The seed is dimorphic, with the inner seeds being smaller and bearing a pappus that facilitates wind dispersal as well as secondary dispersal by being caught on animals (Roche 1991). The outer seeds are larger and bear no pappus; they stay in the heads longer and tend not to be dispersed as far. Remaining in the seed heads may allow these larger seeds to have different germination requirements and seed longevity, diversifying the life history strategies and hence enhancing the potential success of this species. The plants also appear to produce toxins that exclude the establishment of other species (Maddox et al. 1985). Sharp spines around the flower heads deter grazers (Roche 1991). Intense grazing provides the disturbance and removal of competing vegetation needed for yellow star thistle to colonize, spread, and dominate an area (Maddox and Mayfield 1985).

The impact of yellow star thistle on ranching operations and recreational use of lands is substantial. Nearly all of the semiarid to subhumid rangeland in the western United States is susceptible to yellow star thistle invasion (Callihan et al. 1982). Although cattle will graze on yellow star thistle in early spring, when the spines are not well developed, the nutritional value is low, and cattle lose weight on a sole diet of yellow star thistle (Maddox et al. 1985). Later in the year cattle will not normally use the plant for forage. The thistle also causes a neurological disorder in horses commonly referred to as "chewing disease" in which loss of mouth control prevents horses from eating, eventually causing them to starve. Further, yellow star thistle invades many additional habitats in the Sierra: grainfields, orchards, vineyards, cultivated crops, pastures, roadsides, trails, and wastelands (Maddox et al. 1985). The thistle spines are as unpleasant to humans as they are to cattle, resulting in a loss of recreational value in infested land. Methods of control are numerous, ranging from mechanical to chemical. Recent studies indicate that carefully timed grazing of yellow star thistle during the bolting, prespiny stage, followed by three subsequent grazing events can significantly reduce populations (Thomsen et al. 1993).

Riparian Areas of the Sierra Nevada

Riparian communities contain more plant and animal species than any other California community type (Schoenherr 1992). Riparian communities extend from high-elevation snowmelts down to the ocean, spanning a wide range of environmental conditions. A long walk along a river or stream

FIGURE 47.5

Total acreage estimates of yellow star thistle infestation in California (redrawn from Maddox and Mayfield 1985).

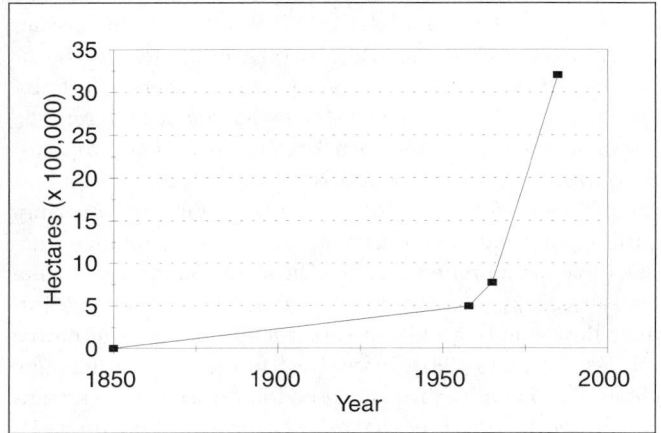

from the Sierran crest to the Pacific would reveal a number of unique vegetation types described as riparian communities. High-elevation riparian zones are dominated by quaking aspen, willows, and deciduous shrubs (Schoenherr 1992). Valley oak dominates in the foothills, followed by cottonwoods, box elders, and willows, with an understory of coyote bush, California grape, and poison oak toward the Central Valley (Barbour et al. 1993). The low-elevation riparian communities are threatened primarily by habitat loss because their fertile soil and relatively level profile make valuable agricultural land. Some of California's most expensive agricultural property is former riparian woodland (Barbour et al. 1993). Currently, less than 10% of the original 900,000 acres of Central Valley riparian woodland exists, and more than half of this is degraded (Barbour et al. 1993). Much of this degradation is due to invasion by nonindigenous plants.

Salt Cedar

The salt cedars (*Tamarix chinesis, T. ramosissima, T. parviflora*) are invasive trees that can transform arid riparian communities into monocultures with diminished ecological and agricultural value. Because of its preference for arid environments, salt cedar is invading mostly in the southern and eastern portions of the Sierra Nevada. Salt cedars have a number of properties that make spread likely and eradication difficult:

- Seed production is prolific: a single tree can produce half a million seeds a year.

- Seed production is prolonged: salt cedars mature seed long into the dry season after co-occurring native species have stopped (Brothers 1981).

- The seeds are small and are distributed widely by wind.

- The seeds germinate where soil remains moist (Neill 1993).

- The plants are fast growing, highly competitive, and drought tolerant, the roots often penetrating deeper to obtain water where native species do not (Griffin et al. 1989).

Salt cedars will often outgrow native species before the drought season and then continue to grow while natives are dormant (Neill 1993). The salt cedars are so invasive that, under optimal conditions, a desert riparian area containing only a few trees can be converted to an impenetrable thicket in less than ten years (Neill 1993).

Salt cedars may also alter the natural flood regime of riparian areas. Native plants are typically swept out of floodplains by seasonal floods, allowing natural regeneration. In contrast, salt cedars grow quickly enough to withstand the seasonal floods. In the Green River of Utah, *Tamarix ramosissima* has been observed to stabilize riparian soils and induce increased sedimentation, causing a 13% to 57% reduction in channel width (Macdonald et al. 1989, citing Graf 1978). Similarly, artificial changes in the flood regime may promote the establishment of salt cedars. The construction of the Los Angeles

aqueduct changed the flow of the Owens River (Inyo County) and the frequency of floods such that the river level was high for most of the year; both flood magnitude and frequency were decreased. Native species have had difficulty becoming established in such conditions, while *T. chinesis* has established itself and continues to spread (Brothers 1981). Salt cedars may also consume water (via transpiration) faster than other native riparian species, representing a significant pathway for water loss from reservoirs (Brothers 1981). The invasion of salt cedars into Death Valley's Eagle Borax Spring was followed by the complete disappearance of water from what used to be a large marsh. Following the removal of the salt cedars, the water returned (Vitousek 1986; Neill 1983).

Salt cedars continue to spread through Sierra Nevada, particularly at low elevations along stream courses in the northern Sacramento valley, southern San Joaquin valley, the south fork of the Kern River, and near the mouths of streams draining from the east side of the Sierra into the Owens Valley. The potential impact of salt cedars on the Sierra Nevada is substantial. Once established, salt cedars are among the most difficult plants to eradicate. Fire, ground-level cutting, and application of herbicide are ineffective in isolation; all result in vigorous resprouting of the crown (Neill 1993). Successful removal involves combinations of cutting and herbicide application in a labor-intensive program that has proved impractical over large infected areas (Neill 1993).

Russian Olive

Introduced during the colonial period, Russian olive (*Elaeagnus angustifolia*) has spread to all of the western states. It is an introduced tree with a dense growth form and edible fruit (Olson and Knopf 1986). It can germinate and survive in a wide range of temperature, soil, and moisture conditions (Olson and Knopf 1986; Borell 1976; Elias 1980). Within a ten-year period (1964–74), Russian olive became one of the dominant trees in Arizona's Canyon de Chelly National Monument (Harlan and Dennis 1976). Currier (1982) reports that Russian olive has the ability to decrease site suitability for native cottonwood germination, thus slowing the recovery of degraded riparian areas. Currier (1982) also postulates that Russian olive may increase overbank deposition, eventually transforming a site to a relatively dry upland with Russian olive as the climax species. Russian olive interferes with farming operations and wildlife refuge management efforts by forming dense, monospecific stands that are hard to remove (Olson and Knopf 1986). Once established, it is also difficult to control and virtually impossible to eradicate (Olson and Knopf 1986). Mowing seedlings, cutting, burning, spraying, girdling, and bulldozing have all been used to remove Russian olive, with limited success (Olson and Knopf 1986). Like the salt cedars, Russian olive may also tolerate changes in river flow due to aqueduct construction better than native species (Brothers 1981). In contrast to most nonindigenous plants, Russian olive provides good food and cover for a number of species such as sharp-tailed grouse, fox squirrels, whitetailed deer, and non-game birds like northern flickers and black-headed grosbeaks. Knopf and Olson (1984) report that the spread of Russian olive may widen some riparian woodlands and benefit the avian species that depend on tall shrubs.

Russian olive currently presents a threat to the Sierra Nevada ecosystem in three areas: the Mojave River bed near Victorville, a small area near Antioch, and along Lone Pine Creek in Inyo County (Olson and Knopf 1986). Detailed accounts from Olson and colleagues note that Inyo County is the only area in California where Russian olive infestations are extensive; however, the species has been reported near Oroville (northern California) and along the Cosumnes and Mokelumne Rivers (near Sacramento) (Olson and Knopf 1986, citing unpublished reports by D. Barbe, B. Bartholomew, O. Clark, T. Combs, M. DeDecker, G. Levi, and W. Wisura). Given the lack of information on community effects and the persistent nature of this species, we recommend that Russian olive be watched carefully and eradicated where resources are available.

Ailanthus, or Tree of Heaven

Ailanthus (*Ailanthus altissima*) was first introduced to the United States in 1784 from Asia (Heisey 1990). Since its early introduction, ailanthus has invaded most of North America, quickly gaining a reputation for its persistence in harsh, urban environments, where it withstands drought, poor soil, and pollution. Even under such poor conditions, ailanthus can grow to 15 to 18 m (50 to 60 ft) tall with a dense canopy. The potential to transform desolate and barren city streets into shaded urban forests was immediately recognized, and ailanthus has frequently been used as a street tree (Newton 1986). Ailanthus vegetatively reproduces well from long lateral roots, sending up sprouts as far as 15 m (45 ft) away from the original tree (Hunter et al. 1993). The cutting of unwanted stems often results in resprouting. It is likely that ailanthus was brought to California's Sierra Nevada foothills by Chinese immigrants working in the gold mines in about 1850 (McClintock 1981). A number of trees were planted near the Placer mining camps and from there spread to the surrounding areas (McClintock 1981). By 1936, ailanthus had spread throughout much of California, so much so that it has been referred to as the only tree in California that is "aggressively spontaneous" (McClintock 1981).

Ailanthus is a fast-growing tree, with seedling growths of 1 to 2 m (3 to 7 ft) in the first year (Miller 1965) and vegetative sprout growths of 3 to 4 m (10 to 13 ft) in a year (Miller 1965). Ailanthus competes well above ground by overtopping many of the surrounding trees and below ground with an extensive and adaptable root system. It has the ability to send out fibrous lateral roots in search of water, and can form a long taproot to extract water from deeper sources (Miller 1965; Newton 1986; J. Hunter, University of California, Davis, conversation with the author, May 1995). Further, ailanthus is allelopathic (produces toxins that exclude the establishment of other species) on more than thirty-five species of hardwoods and thirty-four species of conifers (Miller 1965). The seeds (samaras) of ailanthus are small, wind dispersed, and numerous (more than 1 million per tree) (Hunter 1995). The seeds often remain on the tree and are dispersed throughout the winter (J. Hunter, conversation with the author, May 1995). The expansion of ailanthus appears to be limited by a closed forest canopy (ailanthus is shade intolerant) (Miller 1965). Further, wind, snow, and hard freezes seem to kill the tops of seedlings, possibly excluding ailanthus from higher elevations. It has, however, been known to sprout vegetatively in shaded areas and to resprout following wind, snow, or freeze damage (Miller 1965).

Data describing the current distribution of ailanthus are sparse, but large populations are known to exist in several regions, for example, at Angels Camp in San Andreas County (Robbins et al. 1951). Ailanthus is known to invade riparian zones of the Sierra Nevada foothills (McClintock 1981; J. Hunter, conversation with the author, May 1995; M. Rejmanek, University of California, Davis, conversation with the author, May 1995). It does not seem to invade much above 1,000 m (3,280 ft) (M. Rejmanek, conversation with the author, May 1995). Scientists and land managers generally agree that the problems associated with ailanthus invasion have just begun. Although very little is known about the effects of ailanthus on community diversity, most people familiar with the tree agree that without action it will expand in riparian systems.

The Eastern Slope of the Sierra Nevada

Owing to the rain shadow created by the Sierra Nevada, there is a sharp environmental gradient from the relatively moist crest of the Sierra down the eastern slope to some of the most arid environments in North America. Desert regions in California are, in general, common and cover close to 28 million acres, or approximately 28% of the state (Barbour et al. 1993). The desert communities of California, often severely degraded by mineral extraction, water diversion, military training, suburb expansion, and motorized recreation, recover very slowly (on the order of hundreds of years) (Barbour et al. 1993). Exacerbating the current pressures on desert communities are a number of aggressive nonindigenous plants introduced by early European settlers.

The vegetation of the eastern slope varies considerably with altitude and latitude (Barbour and Major 1988) . The slope vegetation of the Great Basin desert is dominated by a mixture of woody shrubs such as Great Basin sagebrush, rabbitbrush, and bitterbrush. In pristine, ungrazed sites, native perennial grasses make up the understory of this two-layer landscape, but in most places non-native grasses have replaced the native species. In the southern Sierra, the Mojave slope vegetation is dominated by an overstory of evergreen creosote bush, burro bush, and brittle bush. The understory of native annual and perennial species has also been largely replaced by nonindigenous plants. The problems caused by introduced plants in this habitat are exemplified by one single species that is the most widespread and pervasive of all weeds in these arid grasslands: cheat grass.

Cheat Grass

Cheat grass *(Bromus tectorum)*, being indigenous to Central Asia, has a long association with human occupation and disturbance. Cheat grass is well adapted to frequent burning, intense grazing, and agriculture, and so it spreads rapidly in disturbance-dominated landscapes. In its native range, cheat grass thrives in chronically disturbed grasslands (Pierson and Mack 1990a, citing Hess et al. 1967). Like with many of the early introductions, cheat grass probably came to the western United States via contaminated seed lots in the mid to late 1800s (Mack 1989). When introduced to western North America, cheat grass encountered an equitable climate, ample disturbance, and a landscape free of its associated pests and pathogens. Its spread was rapid, filling more than 200,000 km^2 (80,000 mi^2) of the intermountain west in just forty years (Mack 1989). Cheat grass now dominates much of the arid western United States and the eastern slope of the Sierra Nevada, having both negative ecological and negative economic impacts. It and a related species, *B. rubens,* now dominate much of the annual flora in California and the Mojave desert (Hunter 1991).

In presettlement times native ungulates browsed on overstory shrubs, which resprout following herbivory, leaving the native understory grasses relatively untouched. Cattle and sheep, in contrast, prefer understory grasses. The mass introduction of livestock completely changed the local grassland grazing regime. Large areas of ground were cleared and disturbed. Successful pioneer species, such as cheat grass, do well in these situations because they exploit the available resources quickly. Cheat grass has an efficient seed dispersal mechanism, making colonization of recently grazed sites more probable (Ellner and Shmida 1981; Levin et al. 1984). Further, cheat grass is an efficient competitor for several potentially limiting resources in arid environments: (1) it rapidly sequesters nitrogen, which is typically seasonally depleted to the point that it limits plant growth in arid desert environments; (2) it appears to grow its root system faster than native species (a proposed mechanism by which it competitively displaces the native *Agropyron spicatum* [Harris 1967]); and (3) it has the ability to grow under shrubs, where soil is relatively fertile (Hunter 1991, citing Soholt and Irwin 1976).

One of the principal negative impacts of cheat grass has been to promote wildfire (Macdonald et al. 1989; Macdonald et al. 1988). Most perennial grasses of North America mature slowly and do not dry out until after the fire season has passed (Macdonald et al. 1989). In contrast, cheat grass matures in early June and dries one to two weeks after maturity (Macdonald et al. 1989). Earlier drying and high biomass production increase fuel loads and increase fire hazard. After a fire, cheat grass quickly germinates, recolonizes, and expands, resulting in stand domination (Macdonald et al. 1989). The

ubiquitous nature of cheat grass makes revegetation with native species impractical (Barbour et al. 1993). From an economic standpoint, cheat grass has several disadvantages as a forage grass: it grows slowly in the spring, has a short "green feed" period, is highly flammable in the summer, and has a highly erratic yield from year to year (Melgoza et al. 1990). Poor-quality food means that ranchers require more land to support the same number of cattle and sheep than if they had native grass pastures available. Cheat grass has also reduced agricultural yields in the Great Basin desert (Hunter 1991, citing Morrow 1984).

While cheat grass peaks in abundance in grassland communities, it is also found in forest communities, even in subalpine zones (Hess et al. 1967). Sheep and cattle bring cheat grass seeds into Sierran forest communities, although these forest populations do not persist (Pierson and Mack 1990a). Deep forest litter appears to inhibit the germination and growth of cheat grass (Pierson and Mack 1990b). The species also seems to be more apparent in forest communities, making it highly susceptible to grazing by small native mammals (Pierson and Mack 1990b). Finally, shading limits populations of cheat grass (Pierson et al. 1990). In relatively sparse forests, such as those dominated by ponderosa pine and Douglas fir, cheat grass can establish populations with understory disturbance, but in denser forests, such as those dominated by fir and cedar, simultaneous disturbance of both the understory and the overstory is required for establishment. Current fire-suppression policies have minimized the amount of understory disturbance, resulting in a dense understory. Crown wildfires, as well as prescribed burning, however, could promote the invasion of grass species.

Other Problematic Invasive Plants of the Sierra Nevada

A large number of additional species also pose problems in the Sierra Nevada. Himalayan blackberry *(Rubus discolor)* infests riparian and moist areas below 1,600 m (5,280 ft). When present, Himalayan blackberry forms dense thickets. It is favored by rats for food and shelter, providing additional ecological problems (Hickman 1993). Mullein *(Verbascum thapsus),* Canada thistle *(Cirsium arvense)* and bull thistle *(Cirsium vulgare)* infest middle-elevation Sierran meadows, including those in Yosemite Valley, as well as clear-cuts in the Eldorado National Forest. Canada thistle is a rhizomatous plant that has become very difficult to control; stems are easily killed, but individuals are not. Foothill grasslands are literally inundated with additional Mediterranean annual grasses *(Avena fatua, A. barbata, Bromus diandrus, B. mollis, Hordeum* spp., *Lolium multiflorum,* and *Taneatherum caput medusae,* to name a few), in addition to cheat grass and a wide range of herbaceous species (e.g., *Elodea canadensis, Lythrum salicaria,* etc.). This overview has merely given a synopsis of the worst species in the most heavily infested habitats, and is by no means an exhaustive discussion of the problem.

RELATIONSHIPS BETWEEN PLANT INVASIONS AND ANTHROPOGENIC DISTURBANCE

Human Land Use

The single most influential factor allowing one to predict problems with nonindigenous plant species in the Sierras is estimating development pressure. Development pressure creates potential invasion problems through increases in: (1) the extent of habitat disturbance through road and other types of construction; (2) the amount of plant material sold through horticulturalists; (3) the quantity of nonnatural habitats (yards, pastures, etc.) that support populations of potentially problematic species; (4) the importation of landfill that may carry seeds of invasive nonindigenous species; and (5) the movement of humans and their domesticated animals that may be bearing seeds of invasive weedy species. This list provides just a few of the examples by which human presence directly increases the risk of creating an ecologically harmful and potentially expensive nonindigenous plant problem. Quite simply, increased human presence means increased risk of plant invasion. Beyond these obvious mechanisms, however, there are several less obvious ways in which human alteration of the natural landscape has made it possible for nonindigenous species to invade natural habitats.

Fire

Fire can promote the invasion of nonindigenous plants by creating open space for new seed germination. Humans influence the amount and types of fire that occur on a landscape. In the early twentieth century, the United States adopted a stringent policy of fire suppression to prevent the destruction of forests and property. This policy did not recognize the importance of forest ground fires. Prior to fire suppression, frequent ground fires maintained an open understory and an intact overstory. With fire suppression, the plant density of the forest understory has increased, making catastrophic fire more likely (Skinner and Chang 1996). In addition, vegetation changes such as the increased dominance of white fir increase the likelihood of ground fires being transported to the canopy and turning into stand-clearing crown fires. Fire frequency in chaparral communities has been reduced to protect property, allowing fuel loads to build. Crown fires and intense chaparral fires frequently leave behind expanses of bare ground, which favor the invasion and establishment of new species, including nonindigenous plants. Nonindigenous plants are typically characterized by efficient seed-dispersal mechanisms that allow them rapidly to colonize newly opened sites. Aggressive species, such as those described earlier, once established, competitively exclude native understory plants and hinder the establishment of trees and other native species.

Nitrogen Deposition Pollutants

The nitrogen cycle, upon which plants and animals depend, is being augmented by human activities. Nitrogen is an essential building block of all proteins, as well as the predominant atmospheric gas. (Approximately 80% of the atmosphere is nitrogen.) Animals acquire nitrogen, directly or indirectly, through plants. Plants, however, can utilize nitrogen only when it is found in the soil in one of two compounds, ammonium or nitrate. These compounds are seasonally depleted in most soils, and nitrogen availability often limits plant growth in natural habitats. Currently, society produces mass quantities of fixed nitrogen (mostly for fertilizer, but also as a pollutant emitted by cars and as industrial waste), effectively fertilizing the earth's surface. The added nitrogen disrupts the natural cycle and changes the relationships between plants and animals. Anthropogenic nitrogen fixation has risen steadily throughout the twentieth century and currently exceeds the combined natural nitrogen fixation of plants and microorganisms. Vitousek (1994) estimates that approximately 50% of all industrial nitrogen fertilizer used in human history through 1992 has been applied since 1982.

Nonindigenous plants often compete well in artificially enriched sites (witness the preponderance of pest plants in agricultural settings). A number of experimental studies have shown that artificial augmentation of soil nitrogen can promote the invasion of introduced plants. Huennecke et al. (1990) confirm that increased nutrient availability, without physical disturbance of soil or native vegetation, can favor the invasion and success of introduced weeds in an ecosystem where natural levels of resources are low. Heil et al. (1987) have shown that small increases in soil ammonium were sufficient to change the competitive relationships among species in favor of the faster-growing, nonindigenous weeds. While the relationship between nitrogen deposition and plant invasion has not been studied specifically in the Sierra Nevada, there are observations and data that may help guide further research.

Cahill et al. (1996) summarizes the biological effects of air pollution, including nitrification, in the Sierra Nevada. The Sierra Nevada serves as a sink for atmospheric nitrogen. (Miller 1995). Over time, increased deposition rates could lead to altered nutrient cycling and eventually to nitrogen saturation. Already water from the San Gabriel Mountains exceeds federal standards for nitrogen content in drinking water, and nitrate levels in the southern Sierra Nevada and San Bernardino Mountains are among the highest in the nation. Cahill et al. (1996) also notes that nitrogen accumulation appears to be a distinct characteristic of California wildlands exposed to photochemical smog. Nitrogen deposition in the Sierra is an ongoing problem and, while more chronic in close proximity to urban areas, is likely to have long-term negative effects on native species in the Sierra by increasing the importance of nonindigenous species.

Carbon Dioxide

Increased levels of CO_2 associated with human combustion of fossil fuels is well documented. Like nitrogen, CO_2 is often a limiting resource for plants, and its enhancement increases biomass production (Melillo et al. 1990). Increases in CO_2 are also likely to change the competitive abilities of plants (Melillo et al. 1990). Plants are divided into three broad categories, based on their photosynthetic pathway. The three pathways are similar but differ physiologically in ways that are important to the plants' survival. C_3 plants are the largest group of plants and include many introduced pest plants. These plants, in general, respond more to increased CO_2 levels than the other two groups (C_4 and CAM) of plants (Bazzaz 1990; Poorter 1993, Melillo et al. 1990). In addition, species with rapid growth rates (a common characteristic of nonindigenous invasive species) may be more responsive to added CO_2 than species with slower growth rates (Hunt et al. 1990; Poorter 1993).

MECHANISMS TO LIMIT NONINDIGENOUS SPECIES AND REMEDIATE CURRENT DAMAGE

Overview of the Current Status of Regulations

The regulation of introduced pest species is a loose patchwork of federal, state, and local laws and ordinances. Regulation efforts focus on economically important industries such as agriculture, aquaculture, and forestry, giving minimal attention to the protection and restoration of natural areas. For example, in 1992, more than $100 million was spent on agricultural quarantine and port inspection for introduced species, compared to $3 million spent on port inspection for species posing threats to natural areas (OTA 1993). Most regulatory organizations, like the Animal and Plant Health Inspection Service (APHIS), lack the funding and technical expertise to handle the problems and research associated with introduced species. In fact, the Office of Technology Assessment's survey of state fish and wildlife agencies found that a clear majority (63%) favored an increased federal role in addressing problems with introduced species (OTA 1993). The decentralized approach to introduced species, coupled with the lack of funding, has both short-term and long-term consequences. Currently, nonindigenous species that are known to be problematic are still legally imported. Education on the nonindigenous species problem is typically ranked as a low priority in most state and federal agencies and private organizations involved with natural resources, receiving less than 1% of their budgets (OTA 1993). The California Exotic Pest Plant Council (EPPC) is using voluntary restrictions and education as its primary tools to limit the potential hazard of additional problems with nonindigenous species in California.

An in-depth discussion of all the laws, ordinances, and local efforts is beyond the scope of this report. The Office of Technology Assessment's report (OTA 1993) provides a thorough treatment of this subject. A few of the federal laws pertaining to nonindigenous plants, some model state ordinances and organizations, and the regulatory bodies of California are briefly highlighted in the paragraphs that follow.

APHIS does much of the current risk assessment for introduced species primarily using two federal laws (the Federal Noxious Weed Act [FNWA] and the Federal Seed Act) to prevent the import of potentially invasive plants. The FNWA prohibits the importation of listed noxious weeds and provides the authority to quarantine species entering the country. There are, however, a number of problems with the implementation of the act. The major problem is the cumbersome nature of the listing process. In eight years, APHIS has placed 93 species on the current list of federal noxious weeds, even though more than 750 weeds meeting the act's definition remain unlisted (OTA 1993). There is no emergency mechanism in the FNWA to allow rapid action in cases of unlisted species known to have large negative effects, despite a recognized backlog. Another major problem associated with this act is APHIS's narrow interpretation of the interstate transport sections of the FNWA. Section 4 of the FNWA requires a permit for moving listed species between states. APHIS interprets this section as applying only to species for which a specific quarantine has been issued under Section 5 and has issued only one such quarantine in eighteen years. As a result, at least nine known noxious weeds were sold in interstate commerce as of 1990 (OTA 1993). The proposed 1990 Farm Bill, which did not become law, included several amendments to the FNWA that would have required each Federal and land management agency to establish and fund an "undesirable plant management program" for lands under their jurisdiction (OTA 1993). A proposed 1995 Farm Bill contains many of the same amendments.

The Federal Seed Act provides for accurate labeling and purity standards (or impurity tolerance standards) for seeds in commerce (OTA 1993). By 1993, only twelve species had been listed under this act, only one of which is also among the ninety-three noxious weeds listed by the FNWA. Clearly, the issue of whether seeds of FNWA-listed species should be banned is not resolved at this time. Both FNWA and the Federal Seed Act are barely effective in preventing the introduction and transport of agricultural weeds, and neither mentions nonagricultural nonindigenous plants. Natural communities are also not recognized in most of the local protocols regarding unplanned introductions of nonindigenous species (Ruesink et al. 1995).

A thorough review of state regulations by the Office of Technology Assessment (OTA 1993) highlights some exemplary state efforts. Georgia has addressed what may be the most ill-founded principle in the species introduction issue, burden of proof of safety. Currently, APHIS takes full responsibility for assessing the risk of incoming species. The agency generally treats unregulated imports under the presumption "that everything is enterable until we (APHIS) determine it should not be" (OTA 1993). Given that we have yet to successfully eradicate *any* nonindigenous species other than the smallpox virus, the burden of proof would, logically, lie in proving a species safe prior to importation. Georgia treats the importation and release of nonindigenous plants as a privilege to be granted only upon "clear demonstration" that review criteria are met (OTA 1993).

State regulation, like overall regulation, is a patchwork of groups with significant gaps. Some states, like Hawaii, have conducted thorough reviews of their organizations to find such gaps. After its review, Hawaii found that no organization was addressing the problem of weeds entering forest communities. As a result, Hawaii has written an interagency agreement to research the biological control of forest weeds (OTA 1993). Other state groups have filled similar gaps. The Exotic Pest Plant Council (EPPC) of Florida is a collection of agency officials, botanists, and other environmentalists who focus on nonagricultural introduced plants that threaten biodiversity. This council has succeeded in passing the only model local law that addresses introduced plant species. The law combines the eradication of nonindigenous pest plants with land development. Predevelopment removal of introduced plants and tax reductions for property owners who remove them are two ways this law promotes responsible development. In California, a similar organization based on the Florida model was formed in 1992.

Proposed Responses within the Sierra Nevada

Seven steps can be identified that would help limit the flow of nonindigenous species into the Sierra Nevada (table 47.2). The least stringent of these responses would be to seek voluntary compliance from nurseries and horticulturalists in stopping the sale and planting of recognized problem species within the Sierra. A proposed list has been provided by the California Exotic Plant Pest Council and is summarized in table 47.1. Lacking sufficient state and federal legislation, California's EPPC has adopted this strategy.

In assembling this report we conducted a survey of nursery operators and their customers in El Dorado County. Appendix 47.1 contains a full description of the methods, questions asked, and frequencies of responses to the survey. Eighteen nursery owners and thirty-eight customers participated in the survey. Although it does not represent a random sample of merchants and residents of the Sierra, and despite the small sample size, the survey is informative because it suggests that the implementation of simple educational programs could substantially reduce the risk of nonindigenous species in the Sierra. We found that most nursery owners are fairly knowledgeable regarding problematic nonindigenous plants. The typical nursery owner would not be willing to restrict the product lines sold to only native species, but would

shift away from selling nonindigenous species that invade natural habitats and would be willing not to sell them. Owners typically do not track the proportion of their sales of native species to non-native ones.

The second recommended step is to increase educational programs and horticultural incentives to plant native species. In our survey we found that most nursery customers are substantially less well informed than the nursery owners but just as well intentioned. That is, they would forgo purchasing a plant product that was viewed as problematic, but for the most part they lack the information they need to act as environmentally friendly consumers.

These two steps alone would help alleviate many problems, because the intentional planting of problematic exotic species is one of the major driving forces of our expanding problems with exotic pest plants. The third step is to create a legislative constraint similar to the Georgia model that places the burden of proof for the introduction of any additional nonindigenous species upon the importer. Those who wish to market new plant products would need to demonstrate clearly that a species proposed to be introduced is not invasive in Sierran habitats. These first three recommendations could be implemented with relatively little direct cost and little impact on commerce.

Our fourth recommended step is to adequately fund eradication programs for species whose spread and impact we have a chance to contain (e.g., ailanthus and Scotch broom). This process of attacking the vanguard of spreading populations is most likely to be effective in controlling problem plants, but is somewhat counterintuitive. Our impulse is to fund a massive program to eradicate a clearly widespread problem like cheat grass. It may be more effective and more economical, however, to eradicate new and currently sparse populations of species that, if let go, may become the next cheat grass.

Fifth, we recommend funding specific research programs in biological control, or natural herbicides, for those species such as cheat grass and yellow star thistle that are already chronic problems. Sixth, land managers can identify sites that are currently relatively free from exotic pest plants and maintain them as "clean" sites. These latter three steps require expenditures.

Our seventh recommendation is to restrict the sale of nonnative plant species that are recognized as problems. This seems like an obvious and necessary first step in the control of nonindigenous plant species, but it has proven very difficult to implement because of resistance from nurserymen's associations.

CODA

Despite a recognized threat and seemingly adequate restrictions, the Asian gypsy moth was recently accidentally im-

TABLE 47.2

Seven steps proposed to limit the impact of nonindigenous plants in the Sierra Nevada.

Cost and Impact of Program	Steps
A. Programs for which there would be little direct cost or direct impact on commerce.	1. Seek voluntary compliance in stopping the sale and planting of recognized exotic weed problem species within the Sierra. 2. Increase educational programs regarding the cost of exotic species introductions, and provide incentives to plant native species. 3. Legislate constraints that place the burden of proof for the introduction of species not currently propagated or sold in the Sierra upon the importer, who would need to demonstrate clearly that the proposed species is not invasive.
B. Programs for which there would be a substantial direct cost but no negative (and possibly a positive) impact on commerce.	4. Adequately fund eradication programs. 5. Fund research into biological control, or natural herbicides, for chronic pest plants such as yellow star-thistle and cheat grass. 6. Identify "safe site" regions and maintain them free from exotic pest plants.
C. Programs for which there may or may not be a direct negative impact on commerce.	7. Cease and desist the sale of recognized problematic non-native pest plant species.

ported to the Pacific Northwest, resulting in an emergency eradication effort costing $14 million to $20 million. Further, while habitat management stops at political boundaries, fluxes of water, particulates, and organisms do not (Saunders et al. 1991). Likewise, legislative boundaries are not biological boundaries. Restrictions can not safeguard against the next yellow star thistle or cheat grass to invade the Sierra. Luckily, relatively few high-elevation species have been introduced into California. However, the expansion of human populations into higher elevations, bringing with them a desire for new horticultural varieties, is likely to result in additional species that the Sierra will be saddled with forever. While there can be no absolute safeguard from nonindigenous pest plants, there are several constraints that could easily be invoked that would limit the likelihood of introducing a particularly severe problem. Lest we think that by this point in the twentieth century the Sierra is saturated with exotic pest plants, we should bear in mind that zebra mussels, the Asian gypsy moth, and tiger mosquitoes have all been introduced to the United States within the past decade, and all already cost millions of dollars in control efforts. The curve of increasing numbers of introduced species in the California flora appears exponential and shows no tendency toward leveling off. With the current level of control, it is only a matter of time before we will be paying, in real dollars, for the impact of yet another new nonindigenous pest plant.

ACKNOWLEDGMENTS

The authors would like to thank the SNEP team members who provided supporting information for our report. This study was supported by the Sierra Nevada Ecosystem Project as authorized by Congress (HR 5503) through a cost-reimbursement agreement No. PSW-93-001-CRA between the U.S. Forest Service, Pacific Southwest Research Station, and the Regents of the University of California, Wildlands Resources Center.

REFERENCES

Andres, L. 1979. Biological control: Will it solve the broom problem? *Fremontia*, 7(3): 9–11.

Baird, A. M. 1977. Regeneration after fire in King's Park, Perth, Western Australia. *Journal of the Royal Society of Western Australia*, 60:1–22

Baker, H. G. 1965. Characteristics and modes of origin of weeds. In *The genetics of colonizing species*, edited by H. G. Baker and G. L. Stebbins, 147–69. New York: Academic Press.

———. 1974. The evolution of weeds. *Annual Review of Ecology and Systmantics* 5:1–24.

———. 1986. Patterns of plant invasion in North America. In *Ecology of biological invasions of North America and Hawaii*, edited by H. A. Mooney, 44–57. New York: Springer-Verlag.

Barbour, M. G., and J. Major, eds. 1988. *Terrestrial vegetation of California*. Expanded ed. Sacramento: California Native Plant Society.

Barbour, M., B. Pavlik, F. Drysdale, and S. Lindstrom. 1993. *California's changing landscapes: diversity and conservation of California vegetation*. Sacramento: California Native Plant Society.

Bazzaz, F. A. 1986. Life history of colonizing plants: Some demographic, genetic, and physiological features. In *Ecology of biological invasions of North America and Hawaii*, edited by H. A. Mooney, 96–110. New York: Springer-Verlag.

———. 1990. The response of natural ecosystems to rising global CO_2 levels. *Annual Review of Ecology and Systematics* 21:167–96.

Borell, A. E. 1976. *Russian-olive for wildlife and other reservation uses*. Washington, DC: U.S. Department of Agriculture.

Bossard, C. C. 1990a. Secrets of an ecological interloper: Studies on *Cytisus scoparius* (Scotch broom) in California. Ph.D. diss., University of California, Davis.

———. 1990b. Tracing of ant-dispersed seeds: A new technique. *Ecology* 51:2370–71.

———. 1991. The role of habitat disturbance, seed predation, and ant dispersal on establishment of the exotic shrub *Cytisus scoparius* in California. *The American Midland Naturalist* 126:1–13.

Bossard, C. C. and M. Rejmanek. 1994. Herbivory, growth, seed production, and resprouting of an exotic invasive shrub *Cytisus scoparius*. *Biological Conservation* 67:193–200.

Braithwaite, R. W., W. M. Lonsdale, and J. A. Estbergs. 1989. Alien vegetation and native boita in tropical Australia: The impact of *Mimosa pigra*. *Biological Conservation* 48:189–210.

Brothers, T. S. 1981. Historical vegetation change in the Owens River riparian woodland. Paper presented at the California Riparian Systems Conference (University of California, Davis, September 17–19, 1981).

Cahill T., J. J. Carroll, D. Campbell, T. E. Gill, and P. R. Miller. 1996. Air quality. In *Sierra Nevada Ecosystem Project: Final report to Congress*, vol. II, chap. 48. Davis: University of California, Centers for Water and Wildland Resources.

Callihan, R. H., R. L. Sheley, and D. C. Thill. 1982. Yellow starthistle identification and control. Current Information Series 634. Moscow: University of Idaho.

Currier, P. J. 1982. The floodplain vegetation of the Plate River: Phytosociology, forest development, and seedling establishment. Ph.D. diss., Iowa State University.

Elias, T. S. 1980. *The complete trees of North America: Field guide and natural history*. New York: Van Nostrand Reinhold.

Ellner, S., and A. Shmida. 1981. Why are adaptations for long-range seed dispersal rare in desert plants? *Oecologia* 51:133–44.

Frenkel, R. E. 1970. *Ruderal vegetation along some California roadsides*. Berkeley and Los Angeles: University of California Press.

Gordon D. R., and K. J. Rice. 1993. Competitive effects of grassland annuals on soil water and blue oak (*Quercus douglasii*) seedlings. *Ecology* 74:68–82.

Gordon D. R., J. M. Welker, J. W. Menke, and K. J. Rice. 1989. Competition for soil water between annual plants and blue oak (*Quercus douglasii*) seedlings. *Oecologia* 79:533–41.

Graf, W. L. 1978. Fluvial adjustments to the spread of tamarisk in the Colorado Plateau region. *Geological Society of America Bulletin* 89:1491–1501.

Griffin, G. E., D. M. Stafford Smith, S. R. Morton, G. E. Allan, K. A. Masters, and N. Preece. 1989. Status and implication of the invasions of tamarisk (*Tamarix aphylla*) on the Finke River, Northern Territory, Australia. *Journal of Environmental Management* 29:297–315.

Griffin, J. R. 1990. Oak woodland. In *Terrestrial vegetation of California*, edited by M. G. Barbour and J. Major, 383–416. Sacramento: California Native Plant Society.

Groves, R. H. 1986. Invasion of mediterranean ecosystems by weeds. In *Resilience in mediterranean-type ecosystems*, edited by B. Dell, A. J. M. Hopkins, and B. B. Lamont, 129–46. Dordrecht, The Netherlands: Dr W. Junk Publishers.

Harlan, A., and A. E. Dennis. 1976. A preliminary plant geography of Canyon de Chelly National monument. *Journal of the Arizona-Nevada Academy of Sciences* 11:69–78.

Harris, G. A. 1967. Some competitive relationships between *Agropyron spicatum* and *Bromus tectorum*. *Ecological Monographs* 37:89–111.

Heady, H. F. 1977. Valley grassland. In *Terrestrial vegetation of California*, edited by M. G. Barbour and J. Major, 491–514. Sacramento: California Native Plant Society.

Heil, G. W., M. J. A. Werger, W. deMol, D. Van Dam, and B. Heijne. 1987. Capture of atmospheric ammonium by grassland canopies. *Science* 239:764–65.

Heisey, R. M. 1990. Evidence for allelopathy by tree-of-heaven (*Ailanthus altissima*). *Journal of Chemical Ecology* 16:2039–56.

Hess, H. E., E. Landolt, and R. Hirzel. 1967. Flora der Schweiz. Basel: Band I. Birhauser Verlag.

Heywood, V. H. 1989. Patterns, extends, and modes of invasions by terrestrial plants. In *Biological invasions: A global perspective*, edited

by J. A. Drake, H. A. Mooney, F. DiCastri, R. H. Groves, F. J. Kruger, M. Rejmanek, and M. Williams, 31–60. New York: John Wiley.

Hickman, J. C., ed. 1993. *The Jepson manual: Higher plants of California*. Berkeley and Los Angeles: University of California Press.

Huenneke, L. F., S. P. Hamburg, R. Kloide, H. A. Mooney, and P. M. Vitousek. 1990. Effects of soil resources on plant invasion and community structure in Californian serpentine grassland. *Ecology* 7:478–91.

Hunt, R., D. W. Hand, M. A. Hannah, and A. M. Neal. 1991. Response to CO_2 enrichment in twenty-seven herbaceous species. *Functional Ecology* 5:410–21.

Hunter, J. C. 1995. *Ailanthus altissima* (Miller) Swingle: Its biology and recent history. *CalEPPC News* 3(4): 4–5.

Hunter, J. C., S. D. Viers, and P. Reeburg. 1993. *Vegetation of Mt. Wanda, John Muir National Monument, Martinez, California*. Davis, CA: National Park Service, Co-operative Studies Unit.

Hunter, R. 1991. *Bromus* invasion on the Nevada test site: Present status of *B. rubens* and *B. tectorum* with notes on their relationship to disturbance and altitude. *Great Basin Naturalist* 51:176–82.

Johnson, P. N. 1982. Naturalised plants in south-west South Island, New Zealand. *New Zealand Journal of Botany* 20:131–44.

Knopf, F. L. and T. E. Olson. 1984. Naturalization of Russian-olive: Implications to Rocky Mountain wildlife. *Wildlife Society Bulletin* 12:289–98.

Kruger, F. J., G. J. Breytenbach, I. A. W. Macdonald, and D. M. Richardson. 1989. The characteristics of invaded mediterranean-climate regions. In *Biological invasions: A global perspective*, edited by J. A. Drake, H. A. Mooney, F. DiCastri, R. H. Groves, F. J. Kruger, M. Rejmanek, and M. Williams, 181–214. New York: John Wiley.

Levin, S. A., D. Cohen, and A. Hastings. 1984. Dispersal strategies in patchy environments. *Theoretical Population Biology* 26:165–91.

Lukens, J. O. 1994. Valuing plants in natural areas. *Natural Areas Journal* 14:295–99.

MacDonald, I. A. W., D. M. Graber, S. DeBenedetti, R. H. Groves, and R. R. Fuentes. 1988. Introduced species in nature reserves in mediterranean-type climatic regions of the world. *Biological Conservation* 44:37–66.

MacDonald, I. A. W., L. L. Loope, M. B. Usher, and O. Hamann. 1989. Wildlife conservation and the invasion of nature by introduced species: A global perspective. In *Biological invasions: A global perspective*, edited by J. A. Drake, H. A. Mooney, F. DiCastri, R. H. Groves, F. J. Kruger, M. Rejmanek, and M. Williams, 215–56. New York: John Wiley.

Mack, R. N. 1989. Temperate grasslands vulnerable to plant invasions: Characteristics and consequences. In *Biological invasions: A global perspective*, edited by J. A. Drake, H. A. Mooney, F. DiCastri, R. H. Groves, F. J. Kruger, M. Rejmanek, and M. Williams, 155–80. New York: John Wiley.

Maddox, D. M., and A. Mayfield. 1985. Yellow starthistle infestations are on the increase. *California Agriculture* 39:10–12.

Maddox, D. M., A. Mayfield, and N. H. Poritz. 1985. Distribution of yellow starthistle (*Centuarea solstitialis*) and Russian Knapweed (*Centaurea repens*). *Weed Science* 33:315–27.

Marcoza, A. 1984. Manual de malezas., 3rd ed. Buenos Aires: Editorial Hemisferio Sur

McClintock E. 1981. Trees of Golden Gate Park 19: Tree of heaven, *Ailanthus altissima*. *Pacific Horticulture* 42(2): 16–18.

Melgoza, G., R. S. Snowak, and R. J. Tausch. 1990. Soil water exploitation after fire: Competition between *Bromus tectorum* (cheat grass) and two native species. *Oecologia* 83:7–13.

Melillo, J. M., T. V. Callaghan, F. I. Woodward, E. Salati, and S. K. Sinha. 1990. Effects on ecosystems. In *Climate change: the IPCC scientific assessment*, edited by J.T. Houghton, G. J. Jenkins and J. J. Ephraums, 283–10. Cambridge: Cambridge University Press.

Menke, J., C. Davis, and P. Beesley. 1996. Rangeland assessment. In *Sierra Nevada Ecosystem Project: Final report to Congress*, vol. III. Davis: University of California, Centers for Water and Wildland Resources.

Milchunas, D. G., W. K. Lauenroth, P. L. Chapman, and M. K. Kazempour. 1989. Effects of grazing, topography, and precipitation on the structure of a semiarid grassland. *Vegetatio* 80:11–23.

Milchunas, D. G., O. E. Sala, and W. K. Lauenroth. 1988. A generalized model of the effects of grazing by large herbivores on grassland community structure. *American Naturalist* 132: 87–106.

Miller, J. H. 1965. *Ailanthus altissima* (Mill) Swingle. In *Hardwoods*, 101–104, vol. 2 of *Silvics of North America*, technical coordination by R. M. Burns and B. H. Honkala. Agriculture Handbook 654. Washington, DC: U.S. Forest Service.

Momen, B., J. W. Menke, J. M. Welker, and K. J. Rice. 1994. Blue-oak regeneration and seedling water relations in four sites within a California oak savanna. *International Journal of Plant Sciences* 155:744–49.

Mooney, H. A., S. P. Hamburg, and J. A. Drake. 1989. The invasions of plants and animals into California. In *Biological invasions: A global perspective*, edited by J. A. Drake, H. A. Mooney, F. DiCastri, R. H. Groves, F. J. Kruger, M. Rejmanek, and M. Williamson, 250–72. New York: John Wiley.

Morrow, L. A., and P. W. Stahlman. 1984. The history and distribution of downy brome (*Bromus tectorum*) in North America. *Weed Science* 32 (suppl. 1):2–7.

Neill, W. M. 1983. *The tamarisk invasion of desert riparian areas*. Educational Bulletin 83-4. Spring Valley, CA: Desert Protective Council.

Neill, W. M. 1993. The tamarisk invasion of desert riparian areas. *CalEPPC News: Newsletter of the California Exotic Pest Plant Council*, 1(1): 6–7.

Newton, E. 1986. Arboreal riffraff or ultimate tree? *Audubon* 88: 12–19.

Noble, I.R. 1989. Attributes of invaders and the invading process: terrestrial and vascular plants. In *Biological invasions: A global perspective*, edited by J. A. Drake, H. A. Mooney, F. DiCastri, R. H. Groves, F. J. Kruger, M. Rejmanek, and M. Williamson, 301–14. New York: John Wiley.

Olson, T. E., and F. L. Knopf. 1986. Naturalization of Russian-olive in the western United States. *Western Journal of Applied Forestry* 1:65–69.

Orians, G. H. 1986. *Site characteristics favoring invasions*. In *Ecology of biological invasions of North America and Hawaii*, edited by H. A. Mooney and J. A. Drake, 133–48. New York: Springer-Verlag.

Pierson, E. A., and R. N. Mack. 1990a. The population biology of *Bromus tectorum* in forest: Distinguishing the opportunity for dispersal from environmental restriction. *Oecologia* 84:519–25.

———. 1990b. The population biology of *Bromus tectorum* in forest: Effect of the disturbance, grazing, and litter on seedling establishment and reproduction. *Oecologia* 84:526–33.

Pierson, E. A., R. N. Mack, and R. A. Black. 1990. The effect of shading on photosynthesis, growth, and regrowth following defoliation for *Bromus tectorum*. *Oecologia* 84:534–43.

Poorter, H. 1993. Interspecific variation in the growth response of plants to an elevated ambient CO_2 concentration. *Vegetatio* 104/105:77–97.

Raven P. H. 1990. The California flora. *Terrestrial vegetation of California*, edited by M. G. Barbour and J. Major, 109–37. Sacramento: California Native Plant Society.

Reed, C. F. 1977. *Economically important foreign weeds: Potential problems in the United States*. Agricultural Handbook 498, Washington, DC: U.S. Department of Agriculture.

Rejmanek, M. 1989. Invasibility of plant communities. In *Biological invasions: A global perspective*, edited by J. A. Drake, H. A. Mooney, F. DiCastri, R. H. Groves, F. J. Kruger, M. Rejmanek, and M. Williamson, 369–88. New York: John Wiley.

———. 1995. What makes a species invasive? In *Plant invasions*, edited by P. Pysek, K. Prach, M. Rejmanek, and P. M. Wade, 3–13. The Hague, The Netherlands: SPB Academic.

Rice K. J., D. R. Gordon, J. L. Hardison, and J. M. Welker. 1993. Phenotypic variation in seedlings of a keystone tree species (*Quercus douglasii*): The interactive effects of acorn source and competitive environment. *Oecologia* 96:537–47.

Robbins, W. W., M. K. Bellue, and W. S. Ball. 1951. *Weeds of California*. Sacramento: State Printing Division.

Roche, B. F. Jr. 1991. Achene dispersal in yellow starthistle (*Centuarea solstitialis L.) Northwest Science* 66:62–65.

Ruesink, J. L., I. M. Parker, M. J. Groom and P. M. Kareiva. 1995. Reducing the risks of nonindigenous species introductions: Guilty until proven innocent. *BioScience* 45:465–77.

Saunders, D. A., R. J. Hobbs, and C. R. Margules. 1991. Biological consequences of ecosystem fragmentation: A review. *Conservation Biology* 5:18–27.

Schoenherr, A. A. 1992. *A natural history of California*. Berkeley and Los Angeles: University of California Press.

Silvertown, J. W. and J. L. Doust. 1993. *Introduction to plant population biology*. Oxford: Blackwell Scientific Publications.

Simandl, J., and Z. Kletecka. 1989. Drevokazni brouci a jejich sukcese na janovci metlaten. *Ziva* 4:169–70.

Skinner, C. N., and C. Chang. 1996. Fire regimes, past and present. In *Sierra Nevada Ecosystem Project: Final report to Congress*, vol. II, chap. 38. Davis: University of California, Centers for Water and Wildland Resources.

Soholt, L. F., and W. K. Irwin. 1976. The influence of digging rodents on primary production in Rock Valley. US/IBP Desert Biome Research Memo 76-78. Logan: Utah State University.

Thomsen, C. D., W. A. Williams, M. Vayssieres, F. L. Bell, and M. R. George. 1993. Controlled grazing on annual grassland decreases yellow starthistle. *California Agriculture* 47:36–40.

U.S. Congress, Office of Technology Assessment. 1993. *Harmful nonindigenous species in the United States*. OTA-F-565. Washington, DC: U.S. Government Printing Office.

Vitousek, P. M. 1986. Biological invasions and ecosystem properties: Can species make a difference? In *Ecology of Biological Invasions of North America and Hawaii*, edited by H. A. Mooney and J. A. Drake, 163–76. New York: Springer-Verlag.

———. 1994. Beyond global warming: Ecology and global change. *Ecology* 75:1861–76.

Welker J. M., D. R. Gordon, K. J. Rice. 1991. Capture and allocation of nitrogen by *Quercus douglasii* seedlings in competition with annual and perennial grasses. *Oecologia* 87:459–66.

Williams, P. 1981. Aspect of the ecology of broom (*Citisus scoparius*) in Canterbury, New Zealand. *New Zealand Journal of Botany* 19:31–43.

Survey of the Opinions of Nursery People and Their Customers Concerning the Use and Spread of Nonindigenous Plant Species in El Dorado County, California

Introduction

<u>Purpose:</u> These surveys were conducted to learn about the opinions and awareness of professional nursery people and their customers on the issues surrounding the use and spread of non-native and invasive species. El Dorado county was chosen as a geographical location for the survey as it is within the SNEP project boundaries, spans diverse rural and urban areas and is easily accessed from Davis, CA.

<u>General Methods:</u> The survey questions were designed to gain descriptive information about public opinions; in other words, to elucidate what people thought, not why. For this reason no demographic questions were included in the surveys. In addition, questions of a personal nature were avoided so that people felt comfortable when administering and completing the survey. Guidelines developed by deVaus (1990) were used in the survey design.

Nursery Surveys

<u>Methods:</u> Managers or owners of 18 nurseries (Table 1) were surveyed by phone. The surveys consisted of 11 questions (Figure 1). Nurseries were chosen from either the directory of the California Association of Nurserymen or the El Dorado county phone book.

Table 1

Participating Businesses (Business and Customer [c] Surveys)	
Add Growers (Rescue)	Frontyard Nursery & Landscape (Placerville)
Anderson Backhoe & Trucking (Camino)	Georgetown Divide Supply, Inc. (Greenwood) **(c)**
Blue Oak Growers (Rescue)	Gold Hill Nursery (Placerville)
Camino Garden Center (Camino)	Green Valley Nursery & Landscape (Rescue) **(c)**
Carter's Garden & Pet Supply (Placerville)	Hall's Pleasant Valley Supermarket (Rescue)
Clifton & Warren (Placerville)	Homebuilder's Outlet (Placerville)
Divide Nursery (Cool) **(c)**	Maple Leaf Nursery (Placerville)
Dusty Creek Lumber (El Dorado Hills)	Pollock Pine True Value Hardware (Pollock Pines)
Foothill Nursery (Shingle Springs) **(c)**	Red Bed Farms (Georgetown)

Results: The one nursery owner in the Tahoe region with whom we were in contact consistently failed to return our calls. Therefore, all nurseries surveyed are in the western foothills of the Sierra Nevadas near Placerville and Auburn. Although the owner of the Lotus Valley Nursery & Gardens

Figure 1

Nursery Survey with Results	
(n=18, % response to question found in parenthesis)	
1. Are you familiar with any of the following plants that are considered to be invasive in your area? (check boxes)	☐ Scotch broom (*Cystisus scoparius*) (94%) ☐ Cheat grass (*Bromus tectorum L.*) (22%) ☐ Tree of heaven (*Ailanthus altissima*) (61%) ☐ Gum tree (*Eucalyptus*) (89%) ☐ Russian olive (*Elaeagnum angustifolius L.*) (56%) ☐ Bull thistle (*Cirsium vulgare*) (61%) ☐ Kentucky Blue Grass (*Poa pratensis*) (94%) ☐ Yellow star-thistle (*Centaurea solistitialis L.*) (100%) ☐ Salt cedar (*Tamarix*) (56%) ☐ None (0%)
2. Approximately what percentage of your inventory is made of these plants? (Circle your answer from the following choices)	0% (56%) 20-39% (0%) 60-79% (0%) 1-19% (44%) 40-59% (0%) 80-100% (0%) DNK (0%)
3. Would your business suffer if you could not sell one or more of these species?	YES (0%) NO (100%) DNK (0%)
4. Can you name any plants in your inventory, not mentioned above, that you suspect to be invasive out of a garden setting? (Please list)	
5. Do you maintain records of the number of native and non-native species in your inventory?	YES (17%) NO (83%)
6. If not, have you considered maintaining these sorts of records?	YES (17%) NO (67%) N/A(17%)
7. Approximately what percentage of your inventory is in native species?	0% (11%) 20-39% (17%) 60-79% (0%) 1-19% (39%) 40-59% (6%) 80-100% (6%) DNK (22%)
8. Do some of your customers distinguish between native and non-native species?	YES/occasionally (61%) If no, go to #10
9. Which do they request more often?	NATIVE (6%) NON-NATIVE (56%) DNK (0%)
10. If you had access to the appropriate information would you sell only non-invasive species?	YES (78%) NO (11%) DNK (11%)
11. If you do not already, would you consider selling only native species?	YES (6%) NO (83%) DNK/NA (11%)

refused to participate in the survey, all others were very cooperative.

Question #1: All 18 business people surveyed were familiar with yellow star-thistle and 17 (94%) were familiar with scotch broom and Kentucky Blue Grass (Figure 2). Familiarity with Eucalyptus was also noteworthy at 89%.

Figure 2

Question #1 Customer and Business Surveys		
plant name	% familiar (customers) (n=38)	% familiar (businesses) (n=18)
Scotch broom	82	94
Cheat grass	03	22
Tree of heaven	37	61
Gum tree	26	89
Russian olive	13	56
Bull thistle	37	61
Kentucky Blue Grass	32	94
Yellow star-thistle	82	100
Salt cedar	08	56

Question #2: Fifty-six percent sold none of the species listed in question #1. The remainder either sold one of the brooms, Eucalyptus, russian olive or a grass as part of a grass mix. The manager at the Green Valley Nursery and Landscape added that they were sure to warn their customers about potential problems with those species that were notoriously invasive. However, invasiveness seems to be more troublesome to nursery people and their customers within a gardening context.

Question #3: One-hundred percent of those surveyed answered that their business would not suffer if they could not sell the species listed in question #1.

Question #4: Species stated to be potential invasives outside of garden settings were:

Oenothera (spp?)   Black Locust
Ivy   Vinca (spp?)
Ceanothus (spp?)   Blackberries
Baccharis (spp?)   Bamboos
other Brooms   Convolvulus (spp?)
Euphorbia (spp?)   Honey Suckle
Poplar

Question #5: Eighty-three percent answered that they did not maintain records of the number of natives and non-natives in their inventory. One of those that answered "yes" also stated that they "maintained records on everything".

Question #6: Ten of the 15 that answered no to #5 (that they do not maintain records on the number of natives in their inventory) have not considered maintaining these sorts of records. Five said that they had considered it.

Question #7: Thirty-nine percent said that their inventory consisted of between 1 and 19% natives. Eleven and 17% said that their inventory consisted of 0% and 20-39%, respectively. Hall's Pleasant Valley Supermarket/Nursery said their inventory consisted of between 80-100% natives. Twenty-two percent did not know how much of their inventory was in natives.

Question #8: Sixty-one percent of those surveyed said that their customers occasionally distinguished between natives and non-native species. The remainder said that their customers did not.

Question #9: Of the eleven that answered "yes" to #8, ten answered that their customers request non-natives most often. The one business that answered "natives" to the question is Blue Oak Growers, a nursery specializing in natives.

Question #10: Seventy-eight percent answered that they would sell only non-invasive species if they had access to the appropriate information. Eleven percent (or two businesses) answered that they would not sell only non-invasives and another two that they did not know.

Question #11: Eighty-three percent said that they would not consider selling only native species. Six percent (1 business) answered "yes" and 11% did not know.

Discussion: Most of those surveyed conveyed a feeling of concern for the problem with invasive species in their area. All, save Lotus Valley Nursery and Garden, responded favorably to our request for their time and were willing to discuss some of their feelings after the survey was conducted. Some were amused when asked if they were familiar with species such as yellow star-thistle and scotch broom. Lotus Valley Nursery and Garden felt very strongly about their right to import and sell exotic species and gave us a leaflet, written by seedsman J.L. Hudson, to further express their views.

Question #5 on occasion seemed to elicited default responses. Some assumed that if they maintained records on all species they automatically had records on the native/non-native status of those species. No explanation of the meaning of the question was offered so that standardization could be maintained. Most did not want to bother keeping these types of records.

Blue Oak Growers was the only business in the survey that specialized in native species. On question #7 they answered that their inventory consisted of between 40-59% natives. The owner mentioned that she would like to sell more but that people generally did not use only natives and that to stay in business she had to sell some non-natives. On the other hand, Hall's Pleasant Valley Supermarket and Nursery answered 80-100% on this question. While this may be true, it is rather dubious in light of the substantially lower percentage sold by a native specialty nursery. It is interesting and instructive that Blue Oak Growers answered that they would not consider carrying an entirely native inventory.

The definition of non-native did become vague at times in both the nursery and customer surveys. Those surveyed sometimes assumed that if the plant grew vigorously in the area and was deer proof it was therefore native. This added confusion to the surveys that was not anticipated.

As mentioned before, most of the owners and managers were concerned about the invasive species in their area. Although not asked directly it seems that they would be responsive to educating themselves and their customers on this issue and a campaign to eliminate the sale or propagation of invasive exotic species; however, they seem generally skeptical of legislation and scientists

Customers Surveys

Methods: Customer survey questions were constructed as were the business survey questions. Surveys were hand delivered with explicit verbal and written instructions to ten cooperating nurseries. A contact person was established by phone prior to our arrival. Each nursery was given fifty surveys consisting of four questions (Figure 3). Pencils were provided. The contact was instructed to place the surveys on the counter for two weeks in a place where the customers could complete them leisurely. A SASE was left with each nursery.

Results: Of the 500 surveys delivered, 38 were completed and returned by 4 businesses ([c] Table 1). Therefore, 40% of the businesses were compliant and 13% of the surveys delivered were returned to us completed by customers. All of the contact persons that did not return the survey were called until they either sent the surveys or were convinced that the surveys were lost.

Question #1. Thirty-four percent of the people surveyed were familiar with three of the plants in the list. Sixteen and 18% were familiar with 1 and 2 plants respectively. Eighty-two percent were familiar with scotch broom and yellow star-thistle, thirty-seven with tree of heaven and bull thistle, thirty-two with Kentucky Blue Grass and 26% with Eucalyptus.

Question #2 Eighty-two percent would buy native (versus non-native) if the plants were clearly designated. Sixteen percent would not and three did not know.

Question #3 Eighty-two percent would not buy a plant if it were considered to be invasive.

Thirteen would and five did not know.

Question #4 Eleven percent of the customers want rural development to increase in their area. Sixty-one percent favor no change and 24% would like to see development decrease.

Figure 3

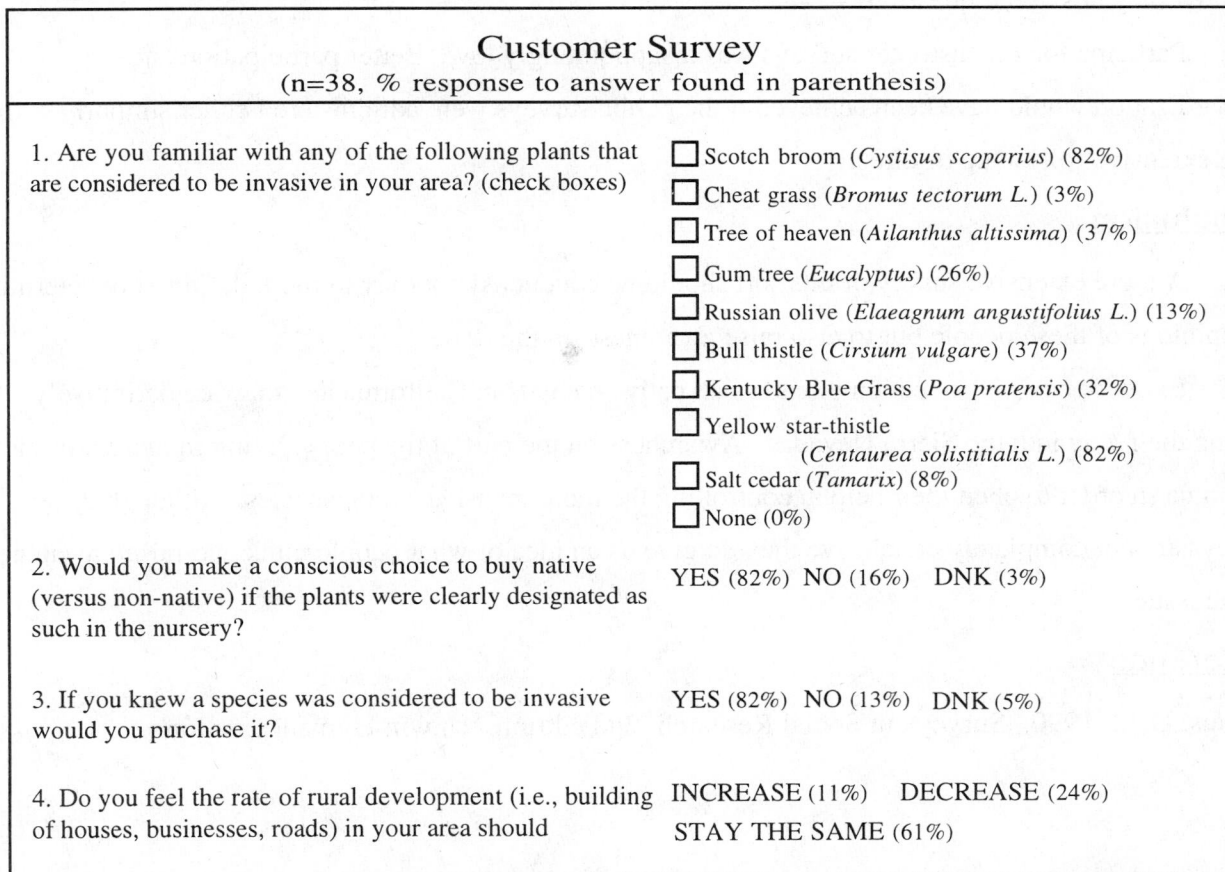

Customer Survey
(n=38, % response to answer found in parenthesis)

1. Are you familiar with any of the following plants that are considered to be invasive in your area? (check boxes)

☐ Scotch broom (*Cystisus scoparius*) (82%)
☐ Cheat grass (*Bromus tectorum L.*) (3%)
☐ Tree of heaven (*Ailanthus altissima*) (37%)
☐ Gum tree (*Eucalyptus*) (26%)
☐ Russian olive (*Elaeagnum angustifolius L.*) (13%)
☐ Bull thistle (*Cirsium vulgare*) (37%)
☐ Kentucky Blue Grass (*Poa pratensis*) (32%)
☐ Yellow star-thistle
        (*Centaurea solistitialis L.*) (82%)
☐ Salt cedar (*Tamarix*) (8%)
☐ None (0%)

2. Would you make a conscious choice to buy native (versus non-native) if the plants were clearly designated as such in the nursery?

YES (82%) NO (16%) DNK (3%)

3. If you knew a species was considered to be invasive would you purchase it?

YES (82%) NO (13%) DNK (5%)

4. Do you feel the rate of rural development (i.e., building of houses, businesses, roads) in your area should

INCREASE (11%) DECREASE (24%)
STAY THE SAME (61%)

Discussion: A positive correlation was found between questions #2 and 3. Seventy-one percent of those customers surveyed answered that they would buy native (Q#2) but not buy invasive (Q#3). Seventy-seven percent of those (or 55% of the total) said they preferred that development either decrease or stay the same (Q#4). Half of the six that would not buy natives said that they would buy invasive species.

Perhaps the question concerning the purchase of natives should have clearly indicated that the available native plants would serve the same purpose in the garden as the non-natives. According to many business owners (even the one that specialized in natives), customers prefer the highly horticultural non-native species to the natives and it would be difficult to maintain a completely native inventory.

It is surprising, considering its leading nature, that anyone would answer "yes" to question #3. In order to understand better what the customer was thinking it would have been more informative to word the question to suggest that we were referring to plants that would be invasive in habitats outside of the garden. There was no trend with regard to the number or type of species with which people were familiar and how they answered questions #2-4.

Participation on customer surveys was disappointingly low. Better participation and standardization would have been achieved if the public surveys were administered either similarly with more extensive follow-up or directly.

Conclusion

A more extensive survey of this sort should be conducted not only to more definitively determine the opinions of these people but to also raise awareness on the issue.

Exotic species introduction threatens all native habitats in California and they are definitively finding their way into the Sierra Nevadas. Awareness on the part of the people living in these areas is important in order to elicit their help in controlling the more invasive exotic species. Although these surveys are not completely conclusive they do give us an idea of what people think and raised awareness on the issue.

Reference

deVaus, D.A. 1990. Surveys in Social Research. 2nd edition. Unwin Hyman, London.

THOMAS A. CAHILL
Atmospheric Sciences
Department of Land, Air,
    and Water Resources
and
Air Quality Group
Crocker Nuclear Laboratory
University of California
Davis, California

JOHN J. CARROLL
Atmospheric Sciences
Department of Land, Air,
    and Water Resources
University of California
Davis, California

DAVE CAMPBELL
Air Quality Group
Crocker Nuclear Laboratory
University of California
Davis, California

THOMAS E. GILL
Atmospheric Sciences
Department of Land, Air,
    and Water Resources
and
Air Quality Group
Crocker Nuclear Laboratory
University of California
Davis, California

Air Quality

ABSTRACT

Air quality in the Sierra Nevada is highly variable in quality: excellent much of the time and in many places, seriously degraded at other times and places. In this chapter, we summarize air quality in the Sierra Nevada both in terms of state and federal ambient air quality standards (ozone, particulate mass, visibility reduction, and so on) which are periodically violated in the Sierra Nevada, and in terms of air quality impacts generally of more ecological import (acid deposition, transport of air toxics, eutrophication of Lake Tahoe, and so on). The emphasis is on well-documented rangewide impacts of human origin. Many other impacts, actually or potentially of no less importance, must be omitted in this short summary because of their more local or limited scope and/or weaker documentation.

At present, the most important deleterious air quality impacts are closely tied to the efficient transport of air pollutants from the Central Valley of California into the western slopes of the Sierra Nevada, up to elevations 6,000 feet or more. This transport is strong in summer and weak or absent in winter, severe in the southern reaches and more modest north of Sacramento, where mountain slopes are more gentle. Of these pollutants, ozone has the best documented and most important effects, especially in its connection to serious injury to the economically important Jeffrey and ponderosa pines. Fine particulate sulfates, nitrates, and smoke are also transported on the same winds, especially between April and October, sharply reducing visibility. Other components in valley air, including nitrates, pesticides, herbicides, and so on, are also efficiently transported into the mountains and deposited on vegetation and in watersheds.

In terms of air pollution sources within the Sierra Nevada, degradation of air quality is one of the difficult questions raised by the potentially increased use of prescribed fire in controlling the high levels of fuel in present Sierras Nevada forests. There is good documentation on degradation of air quality in massive, uncontrolled fires, but other than local data and visual smoke, smoke from prescribed fires is low enough that it is difficult to detect in the rangewide fine particulate mass records since 1988. While smoke from prescribed fires is usually much smaller than that from wildfires, it can, under exceptionally unfavorable conditions, also approximate similar levels.

High-elevation towns of modest population can generate very high levels of fine particles in winter smoke, with concentration levels larger than typically seen even in the largest urban areas of California. Rather surprisingly, there is a rough equality between the maximum mass of fine particles seen in winter urbanized areas, that seen near downwind of massive forest fires, and that from prescribed fires under the most unfavorable conditions. All of these can exceed state and even federal 24-hour particulate mass (PM-10) standards. Lake Tahoe has sharply reduced water clarity and increased algae, some of which is tied to local urban and/or transported air pollutants such as nitrates. Other typically urban air pollutants, such as carbon mon-

Sierra Nevada Ecosystem Project: Final report to Congress, vol. II, *Assessments and scientific basis for management options*. Davis: University of California, Centers for Water and Wildland Resources, 1996.

oxide, have been high enough to warrant special air standards to protect respiration at these high-altitude sites.

The rapid desiccation of the eastern Sierra Nevada lakes, Mono and Owens lakes, has resulted in dust storms that in most years generate the highest 24-hour fine dust levels in the United States. Much of this dust is transported into the Sierra Nevada and the White-Inyo Mountains, the latter site being the home of the ancient bristle-cone pines. The dust levels near Mono Lake should improve greatly following recent legal and regulatory decisions limiting water export.

On the other hand, acid rain and snow are not as much a problem in the Sierra as in the eastern United States. No permanently acidified lakes or streams occur in the Sierra Nevada, although pulses of acidity can occur during spring snowmelt and during occasional thunderstorms from the southern California desert. In the winter, over much of the nonurbanized Sierra Nevada, levels of some characteristic human pollutants such as sulfates are extremely low, mimicking those at the high-altitude world baseline station on Mauna Loa in Hawaii, which helps explain the modest concentrations of sulfates and nitrates in the snow. Transport of Sierra Nevada smoke downwind into the Colorado plateau national parks appears minor except in catastrophic wildfires.

INTRODUCTION

The Sierra Nevada has always existed in a dynamic communion with the Earth's atmosphere, responding to changes global, regional, and local, and in turn changing the atmosphere itself. The pace of these dynamic changes has accelerated with man's involvement, modestly in the period of native Americans, but more rapidly at present, threatening responses from the litho-, hydro-, and bio-spheres that may seriously alter the social and economic values of the Sierra Nevada to the state, nation, and world. However, because of its difficult topography, severe weather, relative lack of mineral resources, and low population density, the Sierra Nevada still retains at many times and most places some of the best air quality in the state and the nation. Yet at other times, air pollutants transported into the range, or generated within the range itself, can result in such severe degradation of air quality that at some times and some places, air quality may be as bad or worse than any place in the state or the nation. For example, the highest dust levels seen anywhere in California were near Mono Lake in 1993. Winter smoke levels at Mammoth Lakes resulted in fine particle masses 1.7 times the worst seen all year in downtown Los Angeles. The early morning ozone level (5–10 AM) at Sequoia National Park is usually 2 to 3 times greater than that seen in Fresno. In order to clarify the contradictory nature of air quality in the Sierra Nevada, this report considers three spatial and three temporal scales. The spatial scales are (1) global, (2) regional or directly upwind, and (3) local, within the Sierra Nevada. The effect of the Sierra Nevada itself on air quality downwind of the range is also ad-

dressed. The temporal scales are (1) recent historical, past 200 years, (2) present, and (3) near future, next few decades.

This report has as its primary goal the broadest overview of air quality in the Sierra Nevada. Yet in order to make this report useful and readable, it is severely constrained in length and detail. This report has focused on air quality data based on widely dispersed monitoring sites within the Sierra Nevada, largely generated by state and federal air quality programs with multi-year duration. While this decision was based on both statistical relevance and quality assurance considerations, it also represents an effort to make these data more widely available to a research community that tends to emphasize a refereed literature that generally favors studies of more limited scope. In addition, the brevity of this report means that it cannot adequately consider the numerous air quality studies relevant to the Sierra Nevada, the mere listing of which alone would represent approximately many pages of text. Those that are listed and cited must represent many other such studies not cited in this report.

Some important information that will be used repeatedly is given in appendix 48.1, which includes state and federal ambient air quality standards relevant to the Sierra Nevada and a list of California and federal air monitoring sites within the SNEP study region (California Air Resources Board 1972–94). The same source is used for data from sites outside of the study area, such as in the Central Valley of California. The annual reports of the Interagency Monitoring of Protected Visual Environments (IMPROVE 1995) and its NPS/EPA predecessors are the source for much of the federal aerosol data, with individual parks adding specific sites and species. Also important are the annual summaries of the ARB's California Acid Deposition Monitoring Program (CADMP 1995) and the Tahoe Regional Planning Agency.

At many points, choices had to be made regarding what materials could be included. Choices were made that favored well documented, range wide, and significant impacts, (e.g., ozone damage to Jeffrey pines, degraded visibility, ...) while neglecting or only mentioning in passing more local impacts (ozone injury to deciduous vegetation in a specific watershed, geothermal injury near Mammoth Lakes, ...). Detail is often lacking, references truncated to the most important, and judgments made regarding relevance of data to the overall picture. Further, any efforts to even partially achieve such ambitious goals, immediately confront a critical paucity of air quality data in the Sierra Nevada. At most places and most times, no data whatsoever of any type are available. For example, all high altitude (greater than 2,700 meters or 9,000 feet) ozone data are based upon one summer's sampling at one site in Sequoia NP. The paucity of data often demands extrapolations from limited measurements to predict air quality away from the sampling site or sampling times. This is in reality a form of modeling that, of course, becomes even more suspect when one looks into past air quality or tries to predict future air quality (e.g., global climate change, regulatory changes to anthropogenic emissions or ambient levels of criteria pollutants, and so on). Any serious errors or omissions caused by such extrapolations are the sole responsibility of the original author.

Changes in global climate are already occurring, with large (+ 25% for CO_2, + 100% for methane, order of magnitude in chlorofluorocarbons, ...) changes in atmospheric chemistry. These may well produce both positive and negative consequences of uncertain magnitude and timing. Some estimate that the most important consequence for the Sierra Nevada is likely to be a shift in the hydrological cycle towards more frequent and intense rain events and away from historical snow patterns. Observed increases in temperature and carbon dioxide, and predictions of increased moisture, could lead to increased bio-productivity. However, recent decreases in the worldwide rate of rise in carbon dioxide, methane, and other "greenhouse gasses" may be harbingers of somewhat lower peak values in the 21st century than some models have predicted, thus limiting changes in climate. While there are many other subtle impacts on the biosphere, it appears that decreases in the northern latitude ozone shield probably are not responsible for the decline in Sierran amphibians. Other, more local, non-atmospheric causes are implicated.

The impacts on Sierran air quality from regional, upwind sources of air pollution are dramatic and easily measurable, from the persistent hazes in summer to ozone damage to Jeffrey and ponderosa pines. The ozone damage is both serious and persistent, and poses both social and economic costs to the Sierra Nevada. Despite massive and costly efforts, the decline in peak ozone values in the Central Valley source regions is slow (unlike the dramatic decreases in the Los Angeles basin), so that relief is not imminent. The persistent hazes have been definitively linked to California sources, and great improvements in visibility could be achieved by a number of proven methods including suppression of summer and fall agricultural burning and controls of sulfur and nitrogen emissions, especially in the Bay Area. The hydrological cycle is dominated by winter snowfall, and the impacts of upwind sources of sulfates and nitrates on mean Sierran snow composition is modest. That does not rule out pulses of moderate acidity at snow melt. It does reflect that winter storm processes do not have the same local connection to California emissions as summer aerosols and ozone.

The impacts on Sierran air quality from local sources, within the study area, are highly variable in magnitude and timing, resulting in major degradation of air quality to levels among the worst in the state and nation superimposed on typical air quality that is so clean as to be the envy of the state and the nation. We consider three major impacts: smoke from fires, influence of urbanized enclaves, and the desiccation of eastern Sierra lakes.

Smoke from major wildfires can be seen for hundreds of miles downwind of the Sierra Nevada, filling valleys (even on occasion the Central Valley) and clearly causing the most obvious and extensive air pollution impact from any local source. This is enhanced by the growing intensity of wildfires caused by fuel build-up over the past decades. Yet, perhaps surprisingly, the impacts of wildfires on the one particulate pollutant subject to state and federal regulation, respirable particulate mass, are not major for several reasons. First, these events are infrequent, so that they have only a modest impact on long-term averages. But perhaps more surprisingly, the maximum smoke impacts of major wildfires fires are generally less in magnitude, and far less in frequency, than smoke impacts in urbanized enclaves such as Mammoth Lakes, South Lake Tahoe, Truckee, and others. The situation is even more favorable for controlled burns designed to limit fuel loading for the major wildfires. First, a great deal of the smoke in the Sierra Nevada during the summer comes from the Central Valley. This smoke is more extensive than that developed by most controlled burns, partially through careful planning of burn periods and burning procedures. Thus, it is our opinion that limits on controlled burning could be relaxed significantly without danger to public health, and with major benefits to public welfare including increased human safety as a result of reduced wildfire events.

The urbanized enclaves referred to above can generate local air pollution that mimics and even surpasses that present in major urban areas of California, but on a much more local spatial scale. Summer levels for standard gaseous pollutants may be significant, while winter urban smoke in small Sierran towns can result in the highest winter particulate mass loading of any site in California, higher even than in the South Coast Air Basin, Bay Area, and San Joaquin Valley. Mass loading at these winter sites may not, however, be directly comparable to those at other warmer, drier sites at times, since measurements have shown that about one-third of the mass can be driven off easily by modestly elevated temperatures. One suspects that trapped water of combustion is retained in very cold climates. The question of other pollutants, such as polyaromatic hydrocarbons (PAHs), is much more important to questions of potential health impacts of wood smoke. The impacts of smoke on local winter visibility are on occasions extreme. Other influences from air pollution in urbanized enclaves include accelerated nutrient input to Lake Tahoe and other pure bodies of water, causing algal growth and lack of clarity. It is our opinion that atmospheric nitrate, a major and occasionally limiting nutrient, from transported, upwind sources is not as important as local nitrate sources around Lake Tahoe, but this is still controversial.

Finally, the influence on local air pollution from the artificially desiccated beds of Mono and Owens lakes is severe, causing in most years the highest respirable dust loading in the entire United States, although for relatively few days per year. The recent Water Resources Control Board ruling on Mono Lake used this air quality information as a component in setting the lake level to a value that should make such events a thing of the past. No such near-term improvements are imminent for the even more severe problems at Owens (dry) Lake.

The next sections will be organized as follows:

1. The Impacts of Global Climate Change

2. The Impacts of Upwind Air Pollutants, generated largely within California

3. The Impacts of Local Air Pollutants, generated within the Sierra Nevada

In addition, we must consider impacts caused by the Sierra Nevada:

4. The Impact of Sierra Nevada Sources on Downwind Areas

THE IMPACTS OF GLOBAL CLIMATE CHANGE

This section will not specifically treat questions of the hydrologic cycle, as that is handled elsewhere. It will deal with possible effects on the Sierra Nevada by factors loosely called "global climate change," or more specifically, "anthropogenic forcing of the global climate." While it is not our intent to delve deeply into global climate debate, it is important to separate those aspects that are certain from those that involve modeling that can verge on pure speculation. The former include changes in global atmospheric chemistry including the 25% rise in CO_2, a key nutrient to plants, and a doubling of methane since the late 19th century, massive increases in chlorofluorocarbons after World War II, the Antarctic ozone hole, and other such well documented changes. In an intermediate category are changes that are at the edge of statistical significance but likely to be true, such as the 2% to 3% decrease in the stratospheric ozone shield above the Sierra Nevada, a 0.5°C rise in global temperature (now that the cooling effect of man made aerosols is included in the models), and the transition from the stable, warm climate of 1900 to 1960 into a more highly variable pattern of weather. In the final category, the effect of global climate on local meteorology such as rainfall-snowfall amounts is highly uncertain and while predictions of the models probably contain some guidance, the details are changing rapidly. For a discussion of the topic in depth, a University of California analysis (Knox 1989) discussed many important factors that may be very significant in terms of the amount and timing of rainfall and/or snowfall. In summary, the documented rise in CO_2 (carbon dioxide), CH_4 (methane), and other "greenhouse gases" raises global temperature and, more importantly, results in greater energy input to the equatorial Pacific, strengthening storms and adding more energy to the southern branch of the jet stream. The "El Nino–Southern Oscillation" (ENSO) events could become more frequent, with more heavy tropical rains in some parts of the cycle and droughts in others. While research on this phenomenon is still in its infancy, the most general and least contested conclusions predict an increase of rain, especially in the southern part of the range, and a decrease in snow in the northern sections. Quantitative uncertainties in such predictions are high, however, and are made more so by recent indications of a slowing down or leveling off of the rise of CH_4 and CO_2, which if continued, would limit the ultimate temperature rise (NOAA 1995).

A second aspect of global climate change is the direct effect of the observed increases in CO_2 levels on the growth of vegetation. In areas in which CO_2 is the limiting nutrient, the 25% increase in CO_2 since the late 19th century would increase bio-productivity. Research in this area also has not yet delivered a clear and convincing answer, especially since there are many parts of the Sierra Nevada where nitrogen, water, or other factors limit plant growth. Furthermore, local anthropogenic influences can also change growth rates, including pollutants such as ozone, acidic deposition, and alkaline salts from Mono Lake and Owens (dry) Lake. The latter source has, for most of this century, dusted the bristlecone pines (*Pinus longaeva*) of the White-Inyo Range with alkaline and saline salts. These same trees are used for studies of the effect of globally-enhanced CO_2 on growth rates.

A further aspect of global climate change is the observed thinning of the stratospheric ozone layer and the subsequent increase in ground-level ultraviolet radiation. This change has a potential impact on a number of areas, one of which is the global decline of amphibians. In this section, we will discuss this topic in some greater detail as it is needed for discussions of amphibians elsewhere in this report.

Our analysis indicates that increased ultraviolet radiation may not be as big a factor as some predict (Blaustein and Wake 1995). As seen in figure 48.1b, the mean change in the stratospheric ozone shield in the 25° to 35° north latitudes is only about 2% to 3% since 1978 (NOAA 1995), much less than natural variations caused by sunspot cycles, volcanic eruptions like those of Mt. Pinatubo, and even some weather patterns. Further, in figure 48.1a, we see that the recent patterns in the fluorocarbon concentrations that thin the ozone layer show that the rapid growth of the past decades has leveled off, limiting ozone reduction and hence ultraviolet increases in the future if it persists.

The modest decreases in ozone, a few percent at most, would increase in ultraviolet (B) also by only a few percent. When combined with mortality data from recent studies (Blaustein and Wake 1995), this predicts an almost negligible (and certainly undocumented) change in the survival of the eggs of certain amphibian species. Figure 48.1c shows such amphibian extinction rates in the Sierra Nevada. Survival rates of amphibians are closely tied to altitude in the range, with the species with the highest altitudinal range (and thus receiving the highest ultraviolet fluxes) having the highest survival. The survival pattern of the yellow-legged frog (*Rana muscosa*) is similar to that of other Sierran species, as summa-

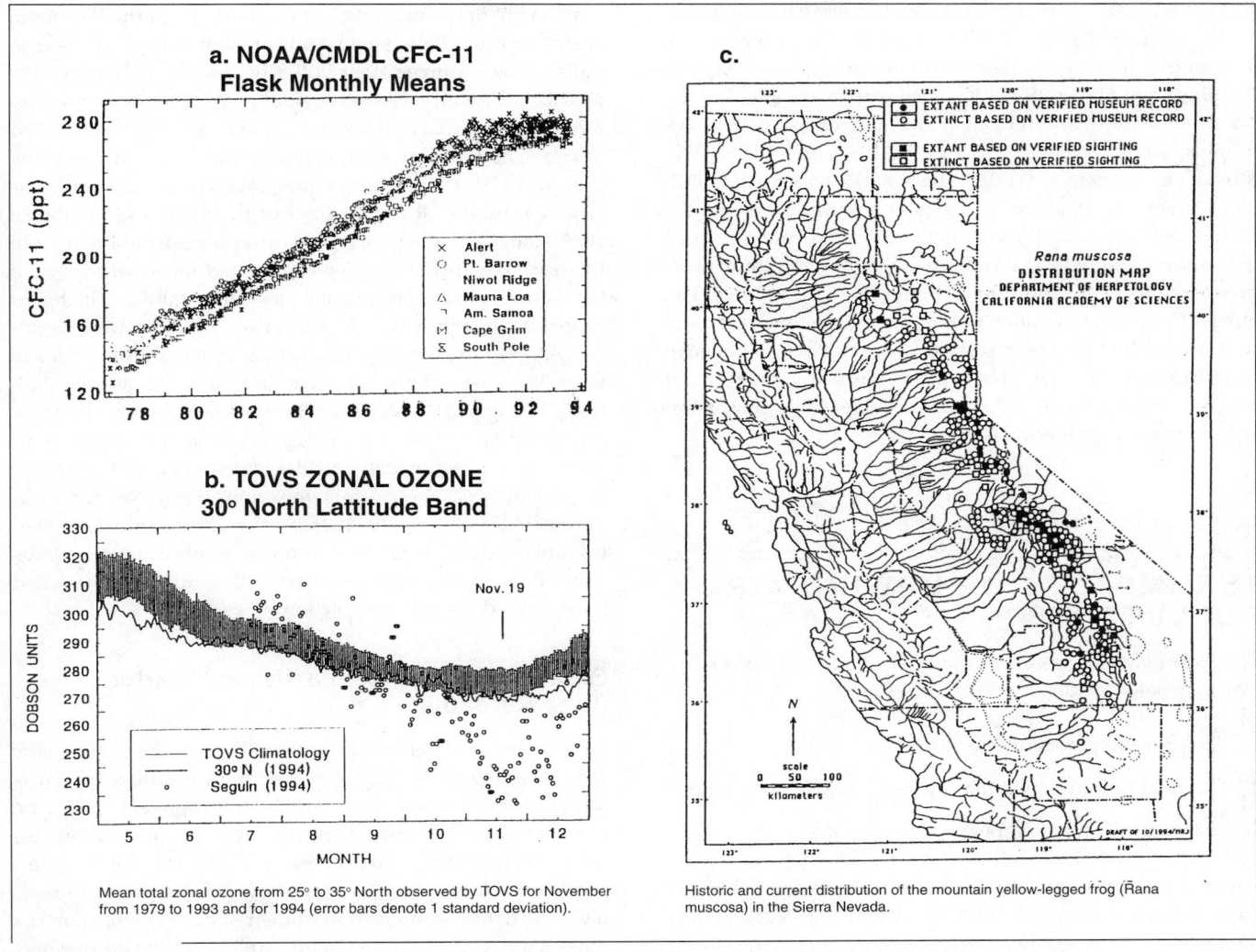

a. NOAA/CMDL CFC-11 Flask Monthly Means

Legend:
- × Alert
- ○ Pt. Barrow
- Niwot Ridge
- △ Mauna Loa
- ⊓ Am. Samoa
- Cape Grim
- South Pole

b. TOVS ZONAL OZONE 30° North Lattitude Band

Nov. 19

Legend:
- —— TOVS Climatology
- —— 30° N (1994)
- ○ Seguin (1994)

Mean total zonal ozone from 25° to 35° North observed by TOVS for November from 1979 to 1993 and for 1994 (error bars denote 1 standard deviation).

c.

EXTANT BASED ON VERIFIED MUSEUM RECORD
EXTINCT BASED ON VERIFIED MUSEUM RECORD
EXTANT BASED ON VERIFIED SIGHTING
EXTINCT BASED ON VERIFIED SIGHTING

Rana muscosa
DISTRIBUTION MAP
DEPARTMENT OF HERPETOLOGY
CALIFORNIA ACADEMY OF SCIENCES

N

scale 0 50 100 kilometers

DRAFT OF 10/1994/MRJ

Historic and current distribution of the mountain yellow-legged frog (*Rana muscosa*) in the Sierra Nevada.

FIGURE 48.1

Changes in *a,* ozone precursors, and *b,* the ozone shield, 1978–present, along with *c,* historic and current distribution of the mountain yellow-legged frog *(Rana muscosa)* in the Sierra Nevada region.

TABLE 48.1

Status of some Sierra Nevada amphibians. (For comparison, the status of the California red-legged frog *(Rana aurora)* at some non-Sierra sites is also shown.)

Common Name (Scientific Name)	Elevation of Sites (ft)		Survival at Sites		
	Range	Average[a]	Extant	Extinct	% Survival
California yellow-legged frog *(Rana aurora draytonii)*[b]	0–5,000	2,500	1	33	3
Foothill yellow-legged frog *(Rana boylii)*	1,000–6,300	3,650	0	Many	0
Cascade frog *(Rana cascadae)*	760–8,250	4,500	3	41	7
Mountain yellow-legged frog *(Rana muscosa)*	4,500–12,000	8,250	44	220	17
Yosemite toad *(Bufo canorus)*	6,400–11,400	8,900	25	29	46
California Red-Legged Frog (Rana aurora)					
Los Angeles south (south coastal)	0–5,000	2,500	4	66	6
North of Los Angeles (central coastal)	0–5,000	2,500[c]	135	Approx. 65	Approx. 50

[a]Average of elevational extremes of historical range.
[b]Only in the Mt. Lassen–Mt. Shasta region; now extinct in the Sierra Nevada proper.
[c]Most extant sites near sea level.

rized in table 48.1. The amphibian survival rate is closely tied to the absence of planted trout, to ponds that freeze to the bottom in winter, and other local non-atmospheric impacts (Jennings and Hayes 1994). It is also worthy of note that the rapid extinction of amphibians at low- and mid-elevation sites in the southern Sierra Nevada in the 1960s and 1970s (P. B. Moyle, Department of Wildlife, Fish, and Conservation Biology, University of California, Davis, communication with the author, 1995) occurred immediately after the completion of the California aqueducts and vast changes in agricultural practices in the southern San Joaquin Valley (Scheuring 1983). The efficient transport of valley pollutants into the mountains is well established (see next section). In summary, we believe that ultraviolet increases due to global climate change are not a significant factor in the massive decline of some amphibian populations in the Sierra Nevada.

THE IMPACTS OF UPWIND AIR POLLUTANTS

The Sierra Nevada is closely coupled to upwind air quality by three major factors:

1. The prevailing northwesterly winds

2. The local terrain-generated (upslope-downslope) winds

3. The Central Valley temperature inversions

There is also influence from other directions, probably the most important of which from air quality considerations is transport from the southeast to the eastern slope of the Sierra Nevada. This becomes important for both acidic rain episodes in the summer and alkaline-saline dust from Owens and Mono lakes in the spring and fall. Such events are relatively rare, however, and total impacts are low compared to the close coupling to upwind sources west of the range. In terms of the winds from the west, the prevailing winds dominate local terrain winds in winter, while the local terrain-generated winds dominate in summer. The presence of strong winter inversions in the Central Valley is a major factor in trapping local pollutants near the ground and reducing transport into the range in winter.

In the rest of this section, we will summarize what little is known about air quality upwind of and in the Sierra Nevada before the European settlement, and then consider the present-day air quality of the region.

Natural Status of Air Quality— Pre–European Immigration

Information on the pre-European air quality upwind of the Sierra Nevada is difficult to ascertain at present. No matter which way the wind blows, to the south from the Bay Area, north from the Bakersfield area, or west across the Central Valley, major sources of air pollution now exist that severely modify air quality. However, some information may be inferred from spatial and temporal patterns of both source and effects. The northern Sacramento Valley has a low population and industrial density and receives its incoming air largely from the North Pacific over the Klamath Mountains. With some exceptions, we may extrapolate from the data in this area to natural conditions before European settlement of the Sierra Nevada. The major exception is smoke, which was probably quite prevalent in pre-European times due to lightning-started fires and deliberate fires by native Americans to favor black oak plantations (e.g., the floor of Yosemite Valley). During early decades of European settlement of the Sierra Nevada, anecdotal reports at the time indicated that air quality in the region was considered extremely pure, pristine and therapeutic, and doctors prescribed Sierra Nevada air as a curative (Thompson 1972). However, other reports and photographs indicate significant summer smoke from numerous small fires "of the surface variety" that burned unchecked for extended periods (see section on fires).

Present Air Quality—Particulate Matter (Fine Aerosols)

The first upwind pollutant that we address is the atmospheric aerosol, fine particles that persist in the atmosphere for hours to days, which range from largely anthropogenic (sulfates, nitrates) to largely natural (soil dusts) in composition. We consider these before ozone, which is biologically more important, because particles retain much information regarding their nature and location of their sources. Ozone, on the other hand, is confounded by multiple sources and complex photo chemistry. Thus, data on fine particles can be used to clarify ozone behavior.

Regional patterns of the two most important constituents of fine (D_p < 2.5 microns diameter) aerosols, ammonium sulfate and ammonium nitrate, are shown in figure 48.2. The data are derived from IMPROVE sites (IMPROVE 1995), mostly in national parks, monuments, and wilderness areas, or other sites that represent regional as opposed to local air quality. The nitrate levels in the southern Sierra Nevada and San Bernardino Mountain sites are higher than at any other site in the nation (example: Shenandoah NP, 0.4 $\mu g/m^3$) while even the highest sulfate levels are 20% of those in the east (example: Shenandoah NP, 11.8 $\mu g/m^3$). Conversely, the northern Sierra Nevada values for sulfate and nitrate average only about 40% higher than those at Denali NP in Alaska and the Mauna Loa world baseline observatory in Hawaii. The sulfate levels at coastal California sites have not been corrected for sea spray sulfate, which will sharply reduce most of these values. Note that the Sierra Nevada seems to act as a barrier, limiting transport of Californian aerosols into the largely pristine intermountain area.

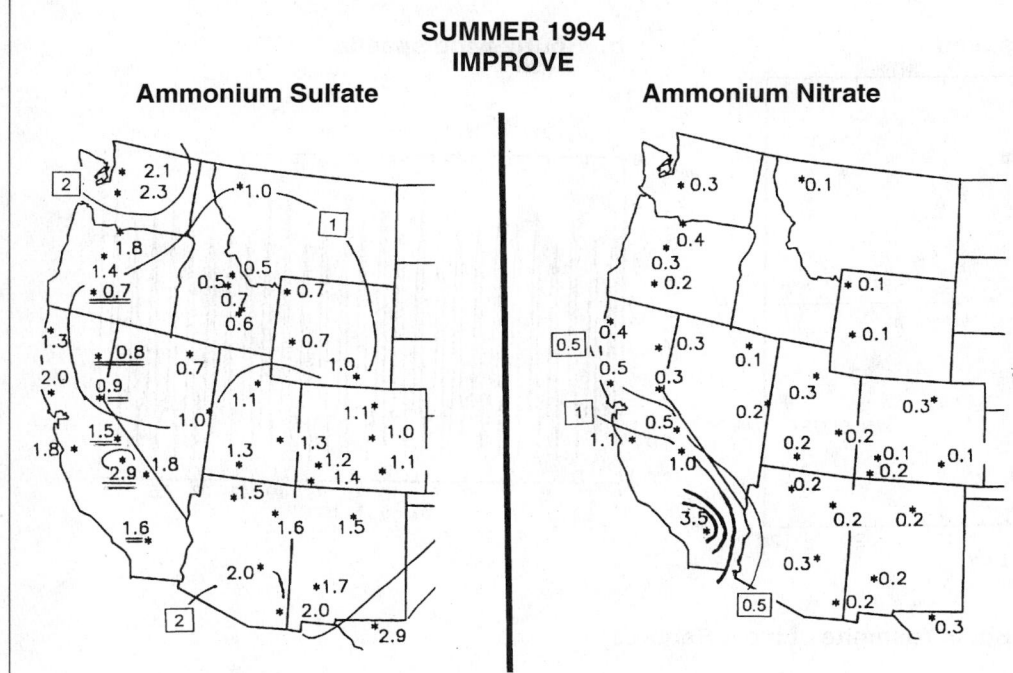

FIGURE 48.2

Spatial patterns of fine particulate ammonium sulfate ($\mu g/m^3$) and ammonium nitrate ($\mu g/m^3$) in summer 1994. The doubly underlined sites all occur at elevations above 1,700 m (4,000 feet) and will later be used for north to south spatial trends.

The summer pattern of strong, terrain-driven diurnal winds in the Sierra Nevada is shown in figures 48.3a and b. The regularity of wind direction and velocity are extraordinary, resulting in efficient transport of valley sulfates into the mountains as shown in figure 48.3c. Most other valley pollutants, particles and ozone, show similar behavior.

The diurnal wind patterns are both strong and regular, resulting in strong upslope winds each day and weak downslope winds each night. The concentration of sulfate pollutants in the valley, measured by local air monitoring stations, is relatively constant throughout the year. If anything, it is the inverse of that in the mountains—higher in winter, lower in summer. Clearly, upslope transport from the Central Valley dominates air quality in the mountains, since there are no major sources of sulfate particles within the western slope of the Sierra Nevada. The dominant effect of the Central Valley upon summer air quality on the western slopes of the Sierra Nevada is probably the most important lesson learned from air quality research at mountain sites in the past 20 years. The eastern slopes, on the other hand, can be influenced by air masses moving up from Southern California and Arizona. Summer storms from this direction have shown high acidity.

Further information can be gained through analysis of particulate pollutant patterns as a function of distance along the Sierra Nevada. Figure 48.4 shows the upwind anthropogenic sources of the precursor gasses, sulfur dioxide (SO2) and oxides of nitrogen (NO$_x$), roughly integrated from the Sierra Nevada to the coast along an east-to-west strip.

Figure 48.5 shows annual fine sulfate and nitrate aerosols at mountain sites between 4000 and 7000 feet (1200 to 2100 m) from the San Bernardino Mountains (San Gorgonio/Barton Flats) through the Sierra Nevada into the Cascade Range (Crater Lake National Park). All sites except Sequoia NP lie between 4,000 and 7,000 foot elevation, (1200 to 2100 m). The Sequoia site in 1992 was at elevation 2200 feet (660 m), but using the 1987 particle transects, a 10% correction has been applied to make the data from the present site equivalent to that at Giant Forest at 6,000 feet (1800 m). While there are some similarities between the sources and the resultant fine aerosols, there are also major differences. First, NO$_x$ sources dominate SO$_2$ sources, but sulfate particles dominate nitrate particles. Both types are aerosols in neutralized forms, ammonium sulfate and ammonium nitrate respectively, due to abundant ammonia sources in the Central Valley. Note how rapidly nitrates drop from the very high levels at San Gorgonio (Barton) to low ones at Bliss, Lassen, and Crater Lake. The drop off is much more rapid than for sulfates. San Gorgonio is influenced by the strong and chemically complex upwind sources of nitrogen from the Los Angeles area, but even so, nitrates fall off more sharply than sulfates from Sequoia NP north to Crater lake. This probably reflects that the northern Sacramento Valley is approaching the global background for sulfate aerosols, typically around 0.3 to 0.5 $\mu g/m^3$, seen at sites like the NOAA world baseline Mauna Loa Observatory (MLO), Hawaii (NOAA 1995). Global baseline values for nitrates are not well established.

Figure 48.6 shows fine particulate mass, which is dominated by sulfates, nitrates and organic matter, and fine soils. Mass has been divided by 10 to show how fine soils are only about 10% of mass at mountain sites.

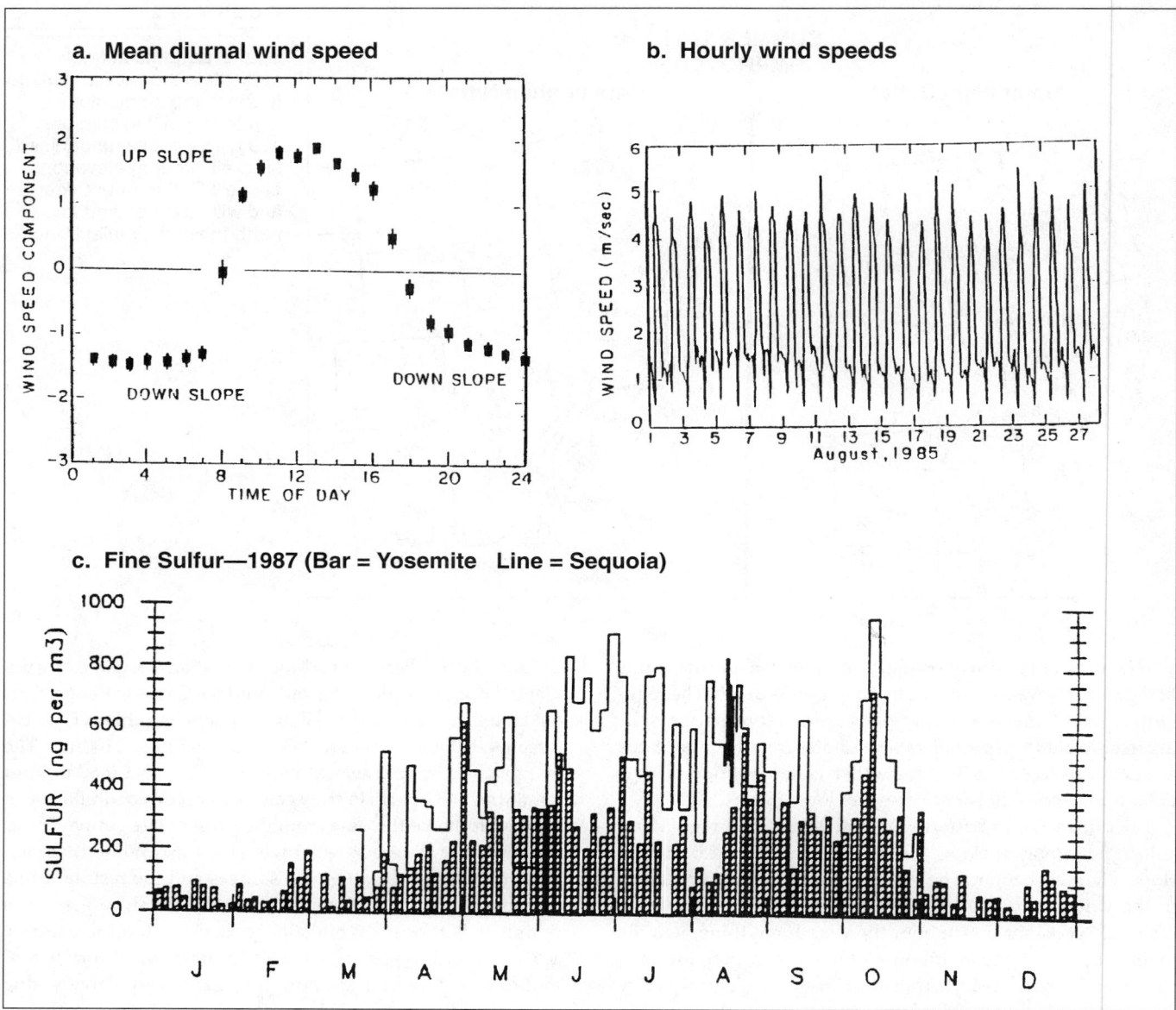

FIGURE 48.3

Terrain-driven diurnal wind patterns *(a, b)* and sulfur aerosols *(c)*, Sequoia National Park, during 1987. A direct comparison is shown for Yosemite National Park, since both sites were near 6,000 feet (1,800 m) elevation.

Table 48.2 contains these data, as well as trace element data, along with comparisons to two urbanized areas, South Lake Tahoe, and, for comparison purposes only, Washington, D.C. All were collected with identical IMPROVE instrumentation and identical analytical and quality assurance protocols (IMPROVE 1995).

The ecological impact of these fine particles is not yet clear. They are acidic and hygroscopic, precursors to acid fogs (generally rare in the Sierra Nevada) and acid rain (generally weak in the Sierra Nevada), and directly involved in foliar dry deposition. However, they are neutralized by the abundant am-

monia sources in the valley to ammonium sulfate and ammonium nitrate—common fertilizers.

One impact of these fine aerosols is clear to every visitor; fine particles are the major cause of haze, and severely degrade visibility along and into the western slope of the range (Cahill et al. 1989; Malm et al. 1994). This is especially true downwind of the San Joaquin Valley, a U.S. Environmental Protection Agency "non attainment area" for PM_{10} (particulate matter smaller than 10 μm diameter). The impact of fine particulate matter includes an aesthetic component and may ultimately impact tourism. Such impacts have already oc-

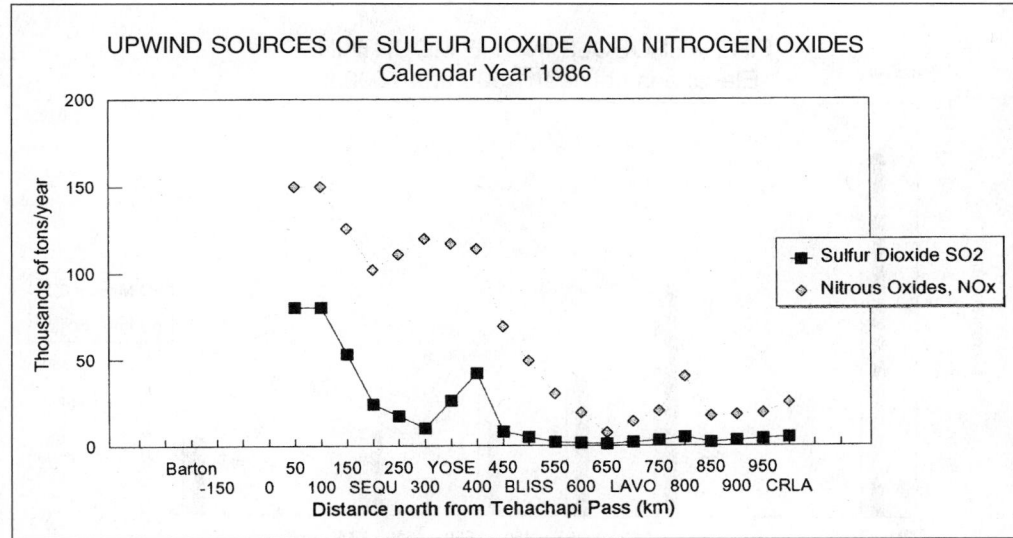

FIGURE 48.4

Anthropogenic emissions of sulfur dioxide and nitrogen oxides, gaseous precursors to sulfates and nitrates, on a strip across California roughly 50 km wide, west from the south-to-north transect starting at Tehachapi Pass. Also noted are the approximate locations of Sequoia NP (SEQU), Yosemite NP (YOSE), Bliss State Park at Lake Tahoe (BLISS), Lassen Volcanic NP (LAVO), and Crater Lake NP (CRLA).

curred in several national parks in the eastern U.S.A. (Shenandoah and Great Smoky Mountains), where summer visitation rates have declined in part because tourists literally can't see very much.

An analysis of the impact of fine particles on visibility was performed at 36 national parks, monuments, and wilderness areas as part of the IMPROVE program (Malm et al. 1994). The results of these analyses for western mountain sites is given in table 48.3, with an additional comparison to Shenandoah NP. These results must be interpreted carefully, however, as mountain sites often have extreme summer-winter aerosol differences (figure 48.3c). Thus, an annual average may dilute very poor summertime conditions by adding in almost pristine winter conditions, as happens in the Sierra

Nevada and even at Appalachian sites such as Shenandoah NP. Thus, mean summer visibility at Yosemite NP is actually closer to 60 km (38 mi) than to 117 km (73 mi), coinciding with the highest visitor use. At most other types of sites, no such extreme seasonal gradient exists.

Some points are evident. First, visibility is sharply reduced at many Sierran sites, again following a north to south gradient seen in aerosols. Again, Lassen Volcanic (and, it turns out, Bliss State Park) has almost the same visibility as Denali NP in Alaska, one of the best sites in the network. Organic matter is the most important single component, derived at least in part from biomass (mostly agricultural) burning in the Sacramento and San Joaquin Valleys. Sulfates are about a quarter, but are more important than this in summer, about one-third

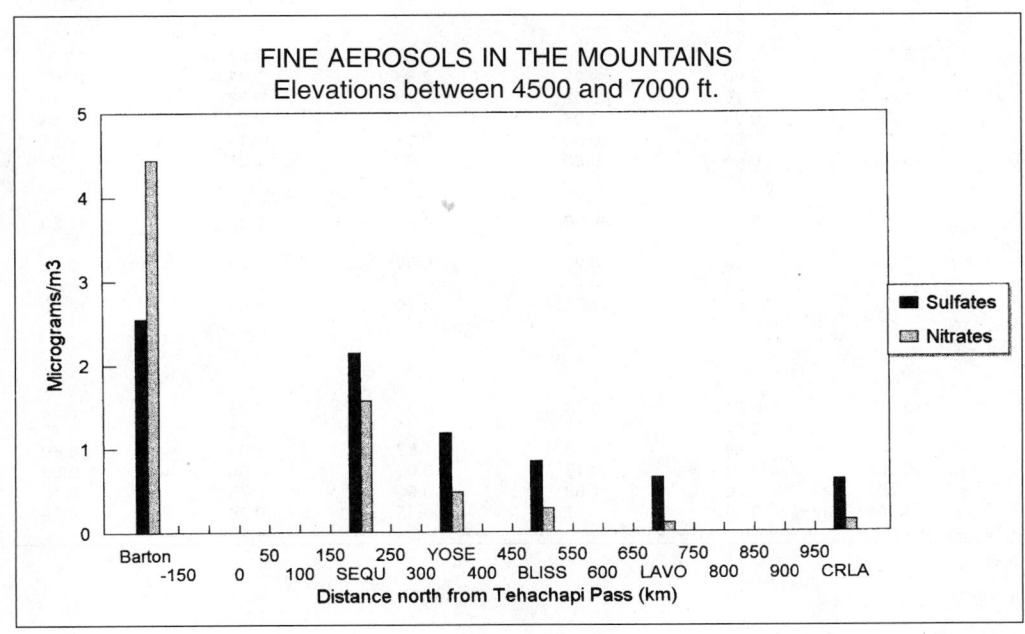

FIGURE 48.5

Concentrations of fine sulfate and nitrate aerosols, from San Bernadino NF (Barton Flats/San Gorgonio) to Crater Lake NP, Oregon, for 1992–93. Distances are marked from Tehachapi Pass.

FIGURE 48.6

Fine particulate mass, divided by 10, versus fine soils. Soils provide approximately 10% of the fine mass at all sites.

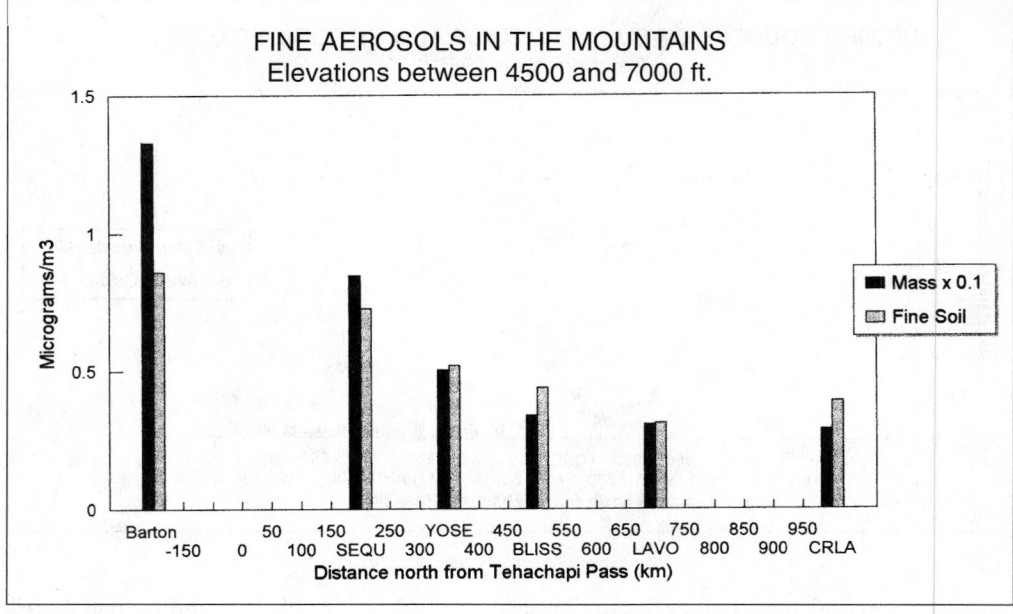

of the haze. Fine soils, nitrates, and soot are the remaining fractions, with the latter also tied in part to biomass burning. San Gorgonio responds to the Los Angeles area with its intense nitrate sources, but more recent data show that Sequoia National Park is also nitrate-impacted.

Figure 48.7 shows three components or tracers of smoke—organic matter, optical absorption, and excess non-soil fine potassium, K-NON. The smoke signature of the Central Valley is very evident even at these high altitude sites.

Note from figure 48.7 the large smoke signature in the Sierra

TABLE 48.2

Comparison of Sierra-Cascade aerosol concentrations at high elevations, 1992–93.

	Sequoia NP (x.09)[a]	Yosemite NP	Bliss SP	South Lake Tahoe[b]	Lassen NP	Crater Lake NP	Washington DC[b]
Major Constituents (Micrograms/m³)							
Coarse Mass							
PM-10	N.A.	11.00	5.85	18.20	7.05	6.21	23.10
Fine Mass							
PM-2.5	8.49	5.04	3.39	9.65	3.05	2.87	19.70
Estimated sum	6.86	4.17	2.90	9.21	2.55	2.41	18.50
Organics	2.94	1.84	1.16	5.52	1.18	0.98	5.12
Sulfates	2.14	1.19	0.84	1.03	0.65	0.62	9.54
Nitrates	1.58	0.48	0.29	0.53	0.11	0.14	2.41
Soil	0.73	0.52	0.44	0.88	0.03	0.39	1.03
Smoke Tracers							
b(abs) (10-8m-1)	13.50	7.83	5.66	29.30	4.99	5.25	41.80
Estimated mass (micrograms/m³)	0.14	0.08	0.06	0.29	0.04	0.05	0.42
KNON (ng/m³)	37.89	23.80	14.80	41.60	12.40	6.58	16.40
Trace Elements[c] (Nanograms/m³)							
Nickel	0.19	0.09	0.07	0.13	0.07	0.08	3.53
Copper	1.58	0.47	0.44	1.41	2.53	0.84	4.76
Zinc	3.12	1.55	1.29	5.13	1.52	3.33	18.60
Selenium	0.28	0.19	0.12	0.12	0.09	0.06	2.08
Bromine	2.65	1.56	1.16	1.63	0.91	0.75	5.12
Lead	1.29	0.78	0.68	1.71	0.53	0.96	6.99

[a]Corrects to Giant Forest elevation, 6,000 feet.
[b]Urbanized sites.
[c]Ni, As, and Se often below detectable limit.

TABLE 48.3

Causes of haze and visibility loss in the Sierra Nevada, with comparisons to other IMPROVE sites.

Site Name	Mean Annual Visibility	Soil	Sulfates	Organic Matter	Nitrates	Soot
Denali NP	154 km (96 mi)	18%	43%	30%	4%	4%
Crater Lake NP	139 km (87 mi)	13%	26%	45%	6%	10%
Lassen Volcanic NP	**140 km (88 mi)**	**17%**	**23%**	**41%**	**10%**	**10%**
Yosemite NP	**117 km (73 mi)**	**15%**	**25%**	**35%**	**15%**	**11%**
Sequoia NP	**74 km (46 mi)**	—	—	—	—	—
San Gorgonio Wilderness Area	61 km (38 mi)	14%	14%	18%	44%	9%
Shenandoah NP	35 km (22 mi)	3%	69%	15%	8%	4%

Nevada is larger than that of either the San Bernardino or Cascade Mountains. This is caused mostly by transport of smoke into the Sierra Nevada from agricultural burning and other anthropogenic sources on the Central Valley floor, not wildfires or controlled burns in mountain forests. This is indicated both by air samples taken on the valley floor, showing persistent smoke, and by the almost total lack of prescribed natural fire or other controlled burns during summer months, June through mid-September. We must be very careful in interpreting these results, for as we investigate pollution sources within the Sierra Nevada, we often find highly polluted mini-urban sites that often serve as the sites for air pollution monitoring stations. The sites chosen for the data above, generally part of the national IMPROVE network, were sited so as to avoid this problem.

Future Air Quality—Aerosols

Reductions in concentrations of fine particulate matter in the Sierra Nevada would be dramatically reflected in improved visibility throughout the range. This visibility degradation is both serious and anthropogenic (Malm et al. 1994). The Clean Air Act Amendments of 1977, extended by the amendments of 1990, mandate that one must mitigate human sources of haze in mandatory Class I areas. This could be accomplished by limitations of upwind sulfur emissions in the Bay Area, especially by the oil refineries and chemical plants near the Carquinez Strait which dominate SO_2 inventories. Continued efforts to control oxides of nitrogen, and tighter controls or elimination of all agricultural burning during summer months would result in sharply improved visibility. This latter policy is in effect in Oregon. On the other hand, increased use of prescribed fire would increase smoke in the mountains, especially during late spring and fall, preferred times for controlled burns.

FIGURE 48.7

Fine aerosol concentrations for aerosol components that reflect combustion sources.

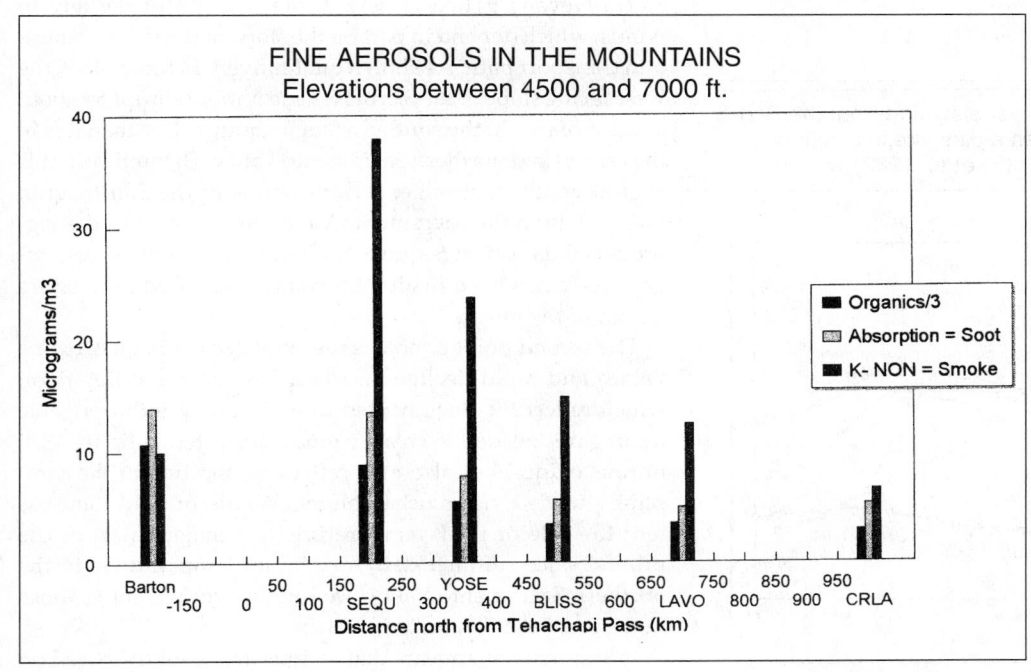

Ozone

Of all anthropogenic pollutants, ozone has the best documented impact upon the Sierran biosphere. Injury to trees, especially the Jeffrey pine (*Pinus jeffreyi*), has been investigated extensively in the past decades, leading to an association with ozone. Due to its importance, a companion chapter by Paul Miller (Miller 1996) deals exclusively with our knowledge of ozone injury to vegetation.

Ozone is generated from emissions of hydrocarbons and oxides of nitrogen that, given sunlight and time, form ozone. The O_3 molecule is a very strong oxidizer that, when in contact with biological material, can break up biological molecules, including DNA, destroying their function. The removal of ozone from the air comes from its destruction on surfaces, incorporation into clouds, and scavenging by oxides of nitrogen that destroy it at night. The patterns of ozone are therefore tied to sunlight. Figure 48.8 shows patterns of ozone as a function of time of day, from the valley floor (Visalia) into Sequoia National Park, including Ash Mountain at 2,200 feet (670 m), Giant Forest at 6000 feet (1,800 m) to, at almost 10,000 feet (3,000 m), Emerald Lake.

Clearly, peak daily ozone values at Giant Forest are almost the same as on the valley floor, while ozone values in the key early morning hours, when the stomata of pines open up, are actually much higher at Giant Forest than at Visalia.

Recently, two extensive sets of ozone measurements and analysis have become available, the first covering all western forests (Bohm et al. 1995a, 1995b), and the second an extensive set of ozone measurements at sites on the western slope of the Sierra Nevada (Van Ooy and Carroll 1995). The sites used in the Van Ooy and Carroll study are shown in figure 48.9. Statistical summaries of the data from these sites are shown in figure 48.10.

FIGURE 48.8

Mean diurnal variations of ozone vs. elevation from the San Joaquin Valley floor (Visalia) to an alpine site in Sequoia National Park (Emerald Lake) (Cahill et al. 1989).

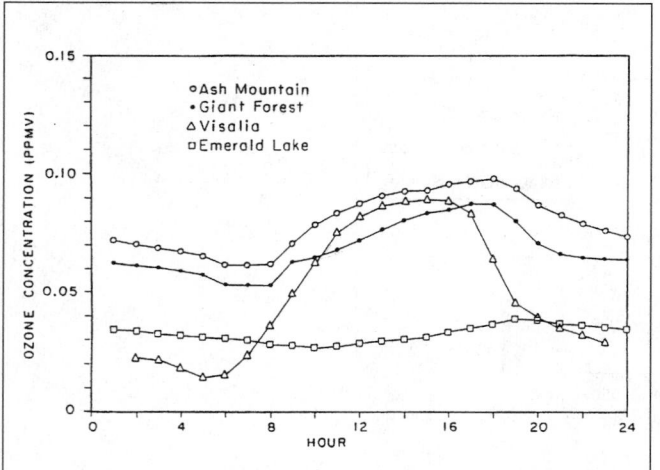

The patterns are complex, reflecting the presence of urban plumes of increased ozone and efficient transport at some sites. The diurnal patterns are also complex, but fall generally into two categories similar to the several categories of Bohm et al. 1995b. One type was seen before at Sequoia NP, with a sharp daytime peak (figure 48.8), while the other is much flatter, resulting in the narrower of the probability distributions seen at Jerseydale and White Cloud. The value 0.09 ppm (90 ppb), the California standard for peak daily hour, has been highlighted, since it will later be used in exposure and dose calculations.

Miller 1996 has documented ozone injury to forests of the Sierra Nevada, expressed by an Ozone Injury Index (OII). Since the dominant ozone sources are in the Central Valley, one can attempt to match averaged ozone peak values to forest injury. This is done in figure 48.11, with poor results.

The Ozone Injury Index (OII) tends to fall dramatically from Barton Flats in the San Bernardino Mountains, with peaks in the urban plumes of Fresno and Sacramento, until low levels of injury are seen at Lassen Volcanic NP (LAVO). However, the averaged peak hours, typically the 10 worst hours per year averaged in a three-year rolling average (California Air Resources Board 1991; see below), only fall slowly in this profile. The resolution of the problem comes through consideration of exposure and dose.

Now the match between ozone and injury is much more satisfactory, using only hours above 90 pphm (0.09 ppm) for ozone and multiplying the hours by the concentration (figure 48.12). This reflects the fact that the northern Sacramento Valley can have high ozone peaks, but they are of shorter duration than those in the San Joaquin Valley.

There is also the question of transport of ozone into the mountains. The efficient transport of valley ozone into the Sierra Nevada is tied closely to the strength of the terrain winds, which depend in part on the slope of the valley-mountain transport path. As shown qualitatively in figure 48.9, the west-facing slope of the Sierra Nevada is more abrupt by about a factor of two in the southern San Joaquin Valley than it is in the central and northern Sacramento Valley. Theoretically, this should result in stronger terrain winds in the San Joaquin Valley than in the Sacramento Valley, as reflected in the vigorous winds seen at Sequoia NP (figure 48.3), but no systematic study has been made of terrain winds along the entire length of the range.

The second point to note is the slow decline in peak ozone values and rapid decline in ozone dose on the valley floor, which reflect the documented diminution of anthropogenic input gases needed to create ozone (illustrated in figure 48.4, nitrous oxides) but also may reflect the location of the sampling site vis à vis an urban plume. We discount to some extent the role of peak temperature as a major cause of the profile, since summer daily maximum temperatures in the northern Sacramento Valley are roughly equivalent to those in the southern San Joaquin Valley.

In summary, it appears that serious ozone injury, based on

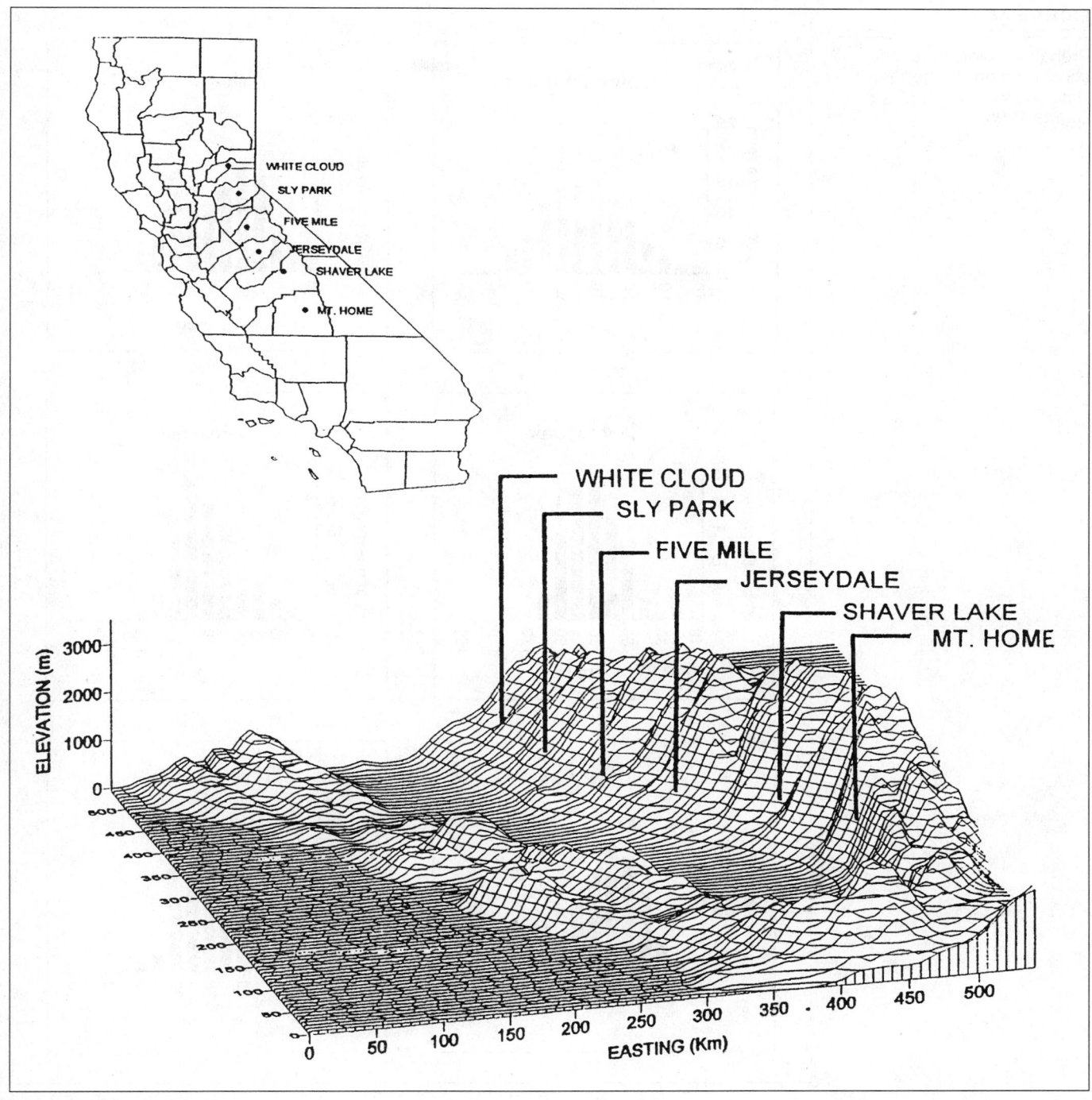

FIGURE 48.9

Perspective view of central California, showing approximate site locations. Horizontal area shown is 530 km on a side. Vertical exaggeration is a factor of 3.

the ozone injury index (OII) occurs on the most sensitive but also economically important species, ponderosa pine (*Pinus ponderosa*) and Jeffrey pine, when the average peak hour valley ozone concentration exceeds 0.09 ppm, which is the California state standard. Injury severity is best reflected by a dose

calculation, based upon the product of the concentration times all hours during which the concentration exceeds 0.09 ppm.

Note that neither dose nor average peak ozone values reflect the individual daily-hour peak values upon which health standards are based. Figure 48.13 shows the daily peak ozone

FIGURE 48.10

Probability plots for ozone concentrations in the Sierra Nevada (Van Ooy and Carroll 1995).

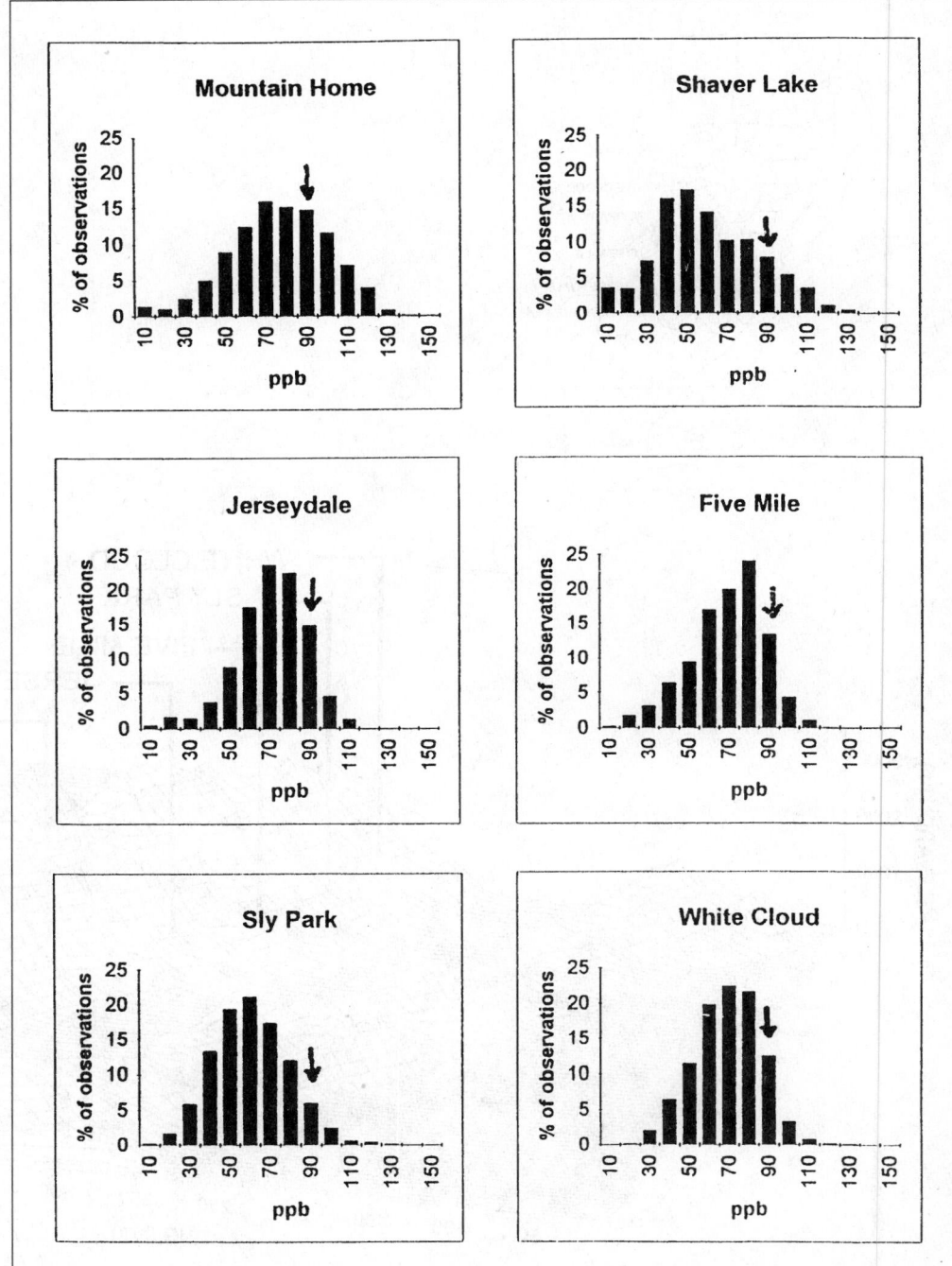

values for Sacramento in summer 1992. Average peak hourly values were 0.069 ppm, while peak hour levels as high as 0.12 ppm occurred. The peak hour values may now be compared with the 0.12 ppm national standard and the 0.09 ppm California standard. The state standard was equaled or exceeded on about one-quarter of all days in summer 1992. It is therefore plausible that by achieving the state health standard of 0.09 ppm ozone at valley floor sites for peak hours, we would also achieve a value that would largely protect an important Sierran bio-resource, the ponderosa and Jeffrey pine forests of the western slope.

The ozone profiles in the Central Valley also bring us back to the question with which we started. What was the ozone value in pre-settlement times, and what will it be in the future? A general consensus appears to be developing that average daily peak summer ozone levels at pristine, mid-northern latitude sites are about 0.03 to 0.04 ppm, based upon remote sites used in global monitoring studies such as

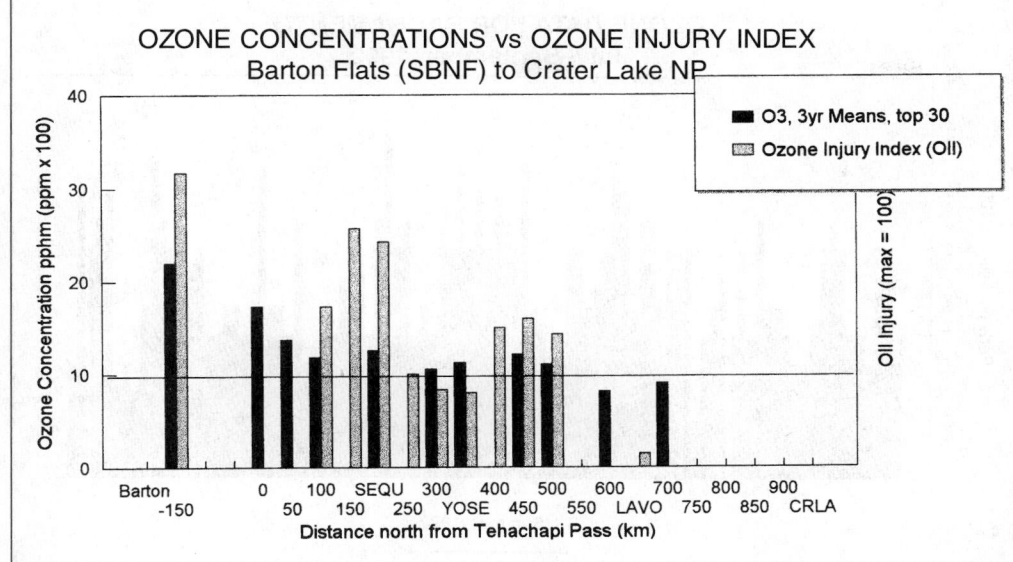

FIGURE 48.11

Comparison of averaged peak ozone hours versus ozone injury. The California standard, 0.09 ppm, is 9 on this scale (pphm).

NOAA's Mauna Loa Observatory, Hawaii. If true, we can infer that since summer ozone levels in the northern Sacramento Valley are approaching those levels (0.059 ppm, Redding; 0.052 ppm, Chico; 0.069, Sacramento) and anthropogenic influences are modest (figures 48.3 and 48.4), then the pre–European settlement ozone levels in the central and southern Sacramento Valley were somewhere between 0.035 and 0.05 ppm. We are thus seeing roughly a doubling to tripling of ozone upwind of the Sierra Nevada from historical levels.

Future Air Quality—Ozone

We can also infer something about the future ozone values from this information and current ozone trends on the valley floor. Figure 48.14 shows recent trends in ozone in the San Joaquin Valley and Sacramento Valley air basins, 1981–90 (California Air Resources Board 1991). These were the data used in figure 48.11.

What is evident is that the dramatic decline of peak ozone

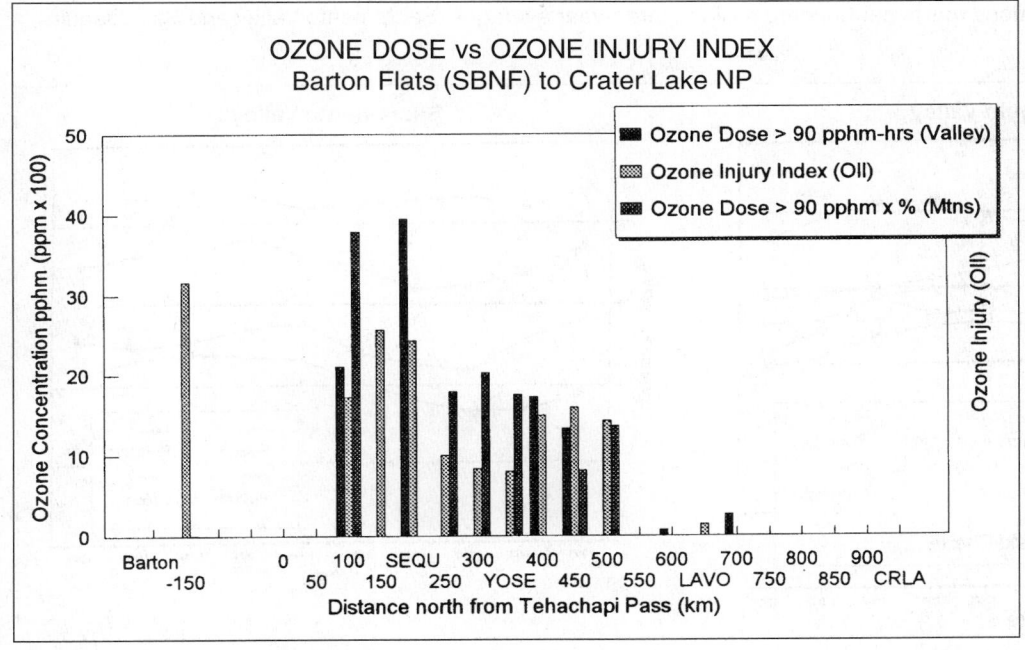

FIGURE 48.12

Comparison of ozone exposure (valley sites), ozone exposure (mountain sites), and the ozone injury index (OII) taken from mountain sites.

FIGURE 48.13

Peak daily hour ozone concentration for Sacramento (Citrus Heights–Sunrise site), July–September 1992.

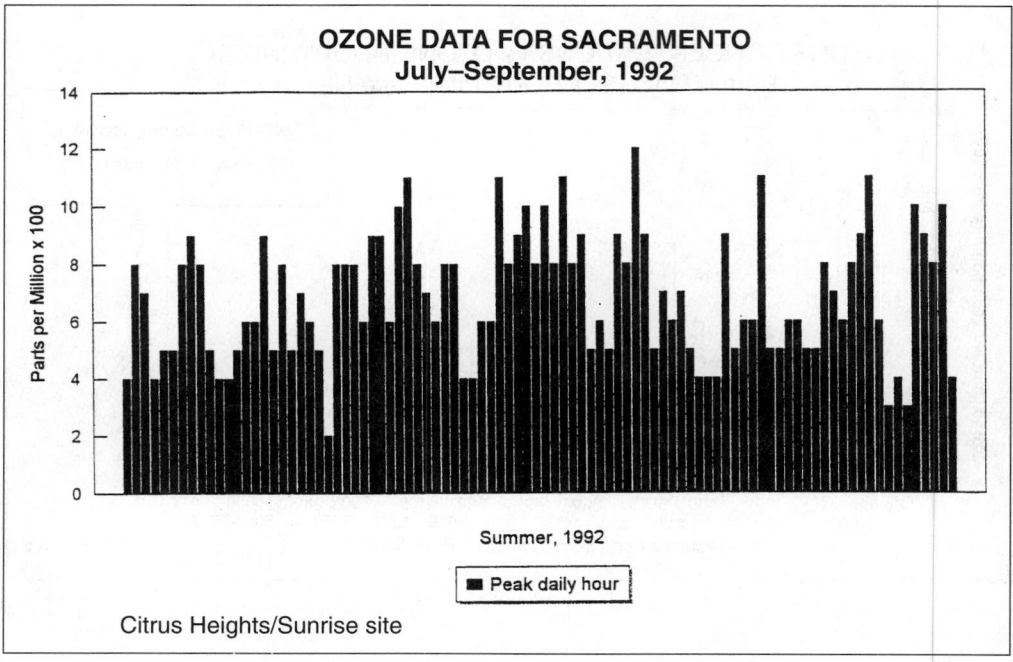

concentration seen in places such as the Los Angeles Basin (1981–90) is not seen in the Central Valley. Since both areas share much of the same controls on vehicular emissions, and since there has been roughly comparable (fractional) growth in vehicular miles per year in both areas during this time, this probably indicates that mobile source emissions are somewhat less of a factor in the Central Valley than in Los Angeles. Certainly, the intense biological activity of the valley floor,

prehistoric to a degree and agriculturally enhanced in the present, introduces biological sources of hydrocarbons and ozone precursor gases such as nitrous oxide that are much less important in the anthropogenically dominated Los Angeles Basin. Certainly, peak summer temperatures are somewhat higher and persist longer, and pollutant retention times are longer, in the Central Valley, allowing pollutants to more fully convert to ozone. It is important to note that ozone ex-

FIGURE 48.14

Mean of top 30 ozone concentrations (parts per hundred million), three-year averages, Sacramento Valley and San Joaquin Valley air basins, 1981–90.

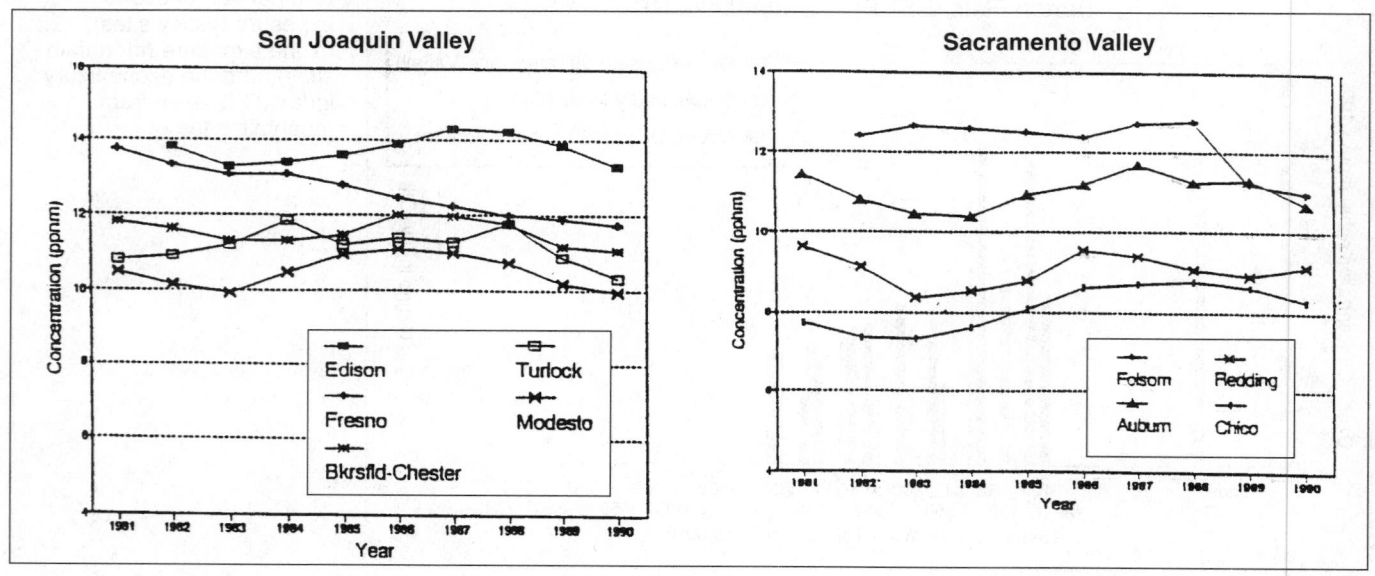

posure values have decreased in the Central Valley since 1980, more so than peak values. The reasons are complex and still under intense investigation. What is clear is that there is a serious ozone problem in the Central Valley, and the state and federal air quality agencies will continue to pursue ozone control measures. Based on the past decade, however, all indications are that progress will be slow.

The situation is not without hope. Recent studies confirm the impact of a small number of "grossly emitting" vehicles (~10%) that generate perhaps two-thirds of all automotive emission. New technologies allow these vehicles to be identified on the road, allowing very effective reductions of emissions (Beaton et al. 1995). Until these gross emitters have been removed from the highway, little improvements in automotive emissions could be achieved by much more expensive technologies like "zero emission vehicles" (electric) that pose new types of environmental problems (Lave et al. 1995).

In winter, ozone levels in the Sierra Nevada are, on average, higher than those on the Central Valley floor, which fall to near-zero levels due to fog, lack of sun, and intense scavenging by anthropogenic gases and particles (CARB 1994; South Lake Tahoe site). The ozone patterns at high elevations do not fit typical summertime patterns, with peaks occurring at strange times and places (i.e., at night). More and more evidence indicates that these winter ozone levels, though moderate, may be partially caused by subsidence of stratospheric ozone (the "good" ozone that stops ultraviolet) down to high elevation sites such as Lake Tahoe. Since these ozone concentrations are not high, and since the biosphere is largely quiescent, the effects are assumed to be modest.

Other Upwind Pollutants

The same conditions that efficiently transport particulate matter and ozone from the Central Valley into the Sierra Nevada also transport other valley pollutants including gaseous pollutants (NO_x [nitrogen oxides], etc.), other photochemical compounds (PAN [peroxyacetyl nitrate], etc.), herbicides, pesticides, and other air pollutants. Gaseous pollutants can be converted into more damaging substances such as PAN, which could be a significant factor in declining forest health. Only limited data are available on these other valley floor pollutants, but they are surely present. The importance of their presence is unknown, but should be the focus of future studies.

Winter Conditions

In winter transport from the valley is cut off by persistent valley inversions, photochemistry is weak, and local sources are quiescent. As shown in figure 48.2, aerosols are at very low values in remote (non-urban) areas. But it is this time that most of the annual water input to the Sierra Nevada occurs in the form of snow. The best information on pollutants in precipitation is probably derived from snow surveys.

Figure 48.15 shows snow survey results for sulfates and nitrates, after Laird et al. 1986. We have plotted using the spatial scale of figure 48.4, which shows the upwind sources. Clearly, a very different pattern is shown in the sulfate and nitrate content of snow, measured near the end of the snowfall season in February and March. The pattern is now relatively flat from north to south, and even flatter if one factors in the amount of snowfall at each site (higher in the north). Thus, net deposition of sulfates and nitrates in the snow is relatively constant, as expected from the nature of the synoptic winter storms which respond to nitrate and sulfate sources, natural and anthropogenic, over a very wide area upwind of California.

The introduction of nitrates and sulfates into the Sierra Nevada hydrological cycle leads to the possibility of perma-

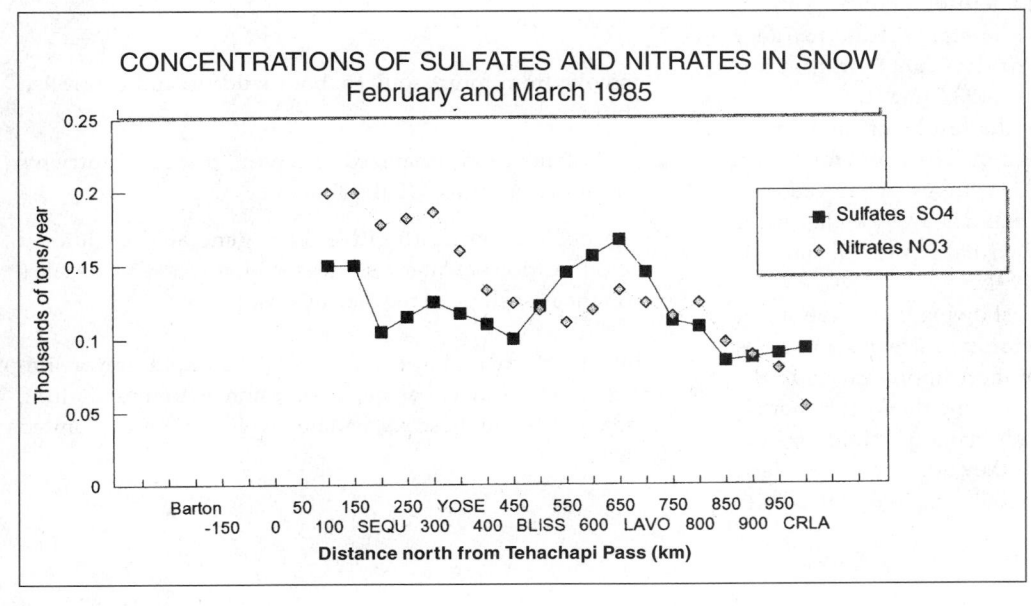

FIGURE 48.15

Sulfate and nitrate concentrations in snow on a south-to-north transect along the Sierra Nevada (after Laird et al. 1986).

nent or ephemeral acidification of lakes and streams, exacerbated by the generally low buffering capacity of high Sierra granite watersheds. Extensive work has been done on this problem in the past decade, most notably by and for the California Air Resources Board's Research Division. The results of the California Acid Deposition Monitoring Program (CADMP) are summarized annually (CADMP 1995) and provide an excellent basis for evaluating the impact of acidic deposition. The results show that "In contrast to the Eastern U.S., no permanently acidified lakes and streams have been found in California" despite the observation that ". . . many California lakes exhibit a very low buffering capacity" (CADMP 1995). This was attributed to the relatively un-acidified winter storms that dominate total annual precipitation in the Sierra Nevada. Highly acidified storms have been observed, such as those in Sequoia N.P. (Cahill et al. 1989), originating over the southern California desert, some carrying effluents attributed to the Arizona copper smelters. However, total precipitation from such events is, on the average, only a modest contribution to total precipitation.

Averaging over the hydrological cycle for the years 1984–88, CADMP found that pH in the Sierra Nevada was 5.26 +/- 0.04. This can be compared to the cleanest California sites on the Northwest coast, 5.33 +/- 0.02, close to the weak acidity of pristine rainfall in equilibrium with atmospheric CO_2. Another measure of human impact is nitrate and non–sea salt sulfate. The sulfate deposition was 1.4 times higher at the southern sites, Lake Isabella through Sequoia N.P., than at the northern sites, Lake Tahoe through Quincy, while nitrates were higher by a factor of 1.8 at the southern sites than the northern sites. Recall the equivalent ratios for aerosols were, respectively, about 2.5 for sulfates, and 8.0 for nitrates. One striking pattern did emerge—a strong dependence upon the elevation at which the precipitation was collected. The well documented Sequoia N.P. transect from Ash Mountain (2200 ft) through the Giant Forest (6,000 ft) to Emerald lake (10,000 ft) used previously for ozone and aerosols showed a strong gradient in acidic deposition; Ash Mountain, 7.1 and 16.6 μeg/L for sulfates and nitrates, respectively; Giant Forest, 4.5 and 8.4 μeg/L, and Emerald Lake, 3.0 and 3.5 μeg/L.

These data are in accord with the results of the national Acid Deposition Assessment Program, which also found that only one western lake out of 10,393 surveyed showed even marginal acidity, below 6.0, and that lake was geothermal in origin. "It appears that there are virtually no acidic lakes in these (western) areas" (NAPAP 1991).

None of this contradicts results showing that in the spring melt, pulses of acidity can surge through streams and lakes. But it does put into context the far more important acidic dry deposition of gasses and particles during the long, generally dry summers of the Sierra Nevada. Highly acidic fogs can occur in California, but generally they are rare in the Sierra Nevada region since they are generally trapped in the valley floor inversions.

Summary of Present-Day Effects of Upwind Air on Sierra Nevada Air Quality

The present status of air quality in the Sierra Nevada can be conveniently separated into summer and winter conditions. In summer, the southern parts of the western slope of the range are highly impacted, as terrain winds pull into the mountains, to altitudes of at least 6,000 feet, all the air pollutants of the San Joaquin and Sacramento valleys. In the winter, the Sierra Nevada's non-urban air is very clean. This occurs because in clear air conditions under high pressure, a strong inversion closes off the Central Valley floor and prevents transport into the range. In winter storm conditions, the inversion is broken, but air motion and mixing are extremely vigorous while precipitation rapidly strips the air of pollutants. Such storms come off the Pacific, and thus have a very limited time over land to pick up anthropogenic pollutants.

Thus, in terms of regional impacts of upwind sources:

Summer: highly impacted air quality in the south (for ozone, among the worst in the nation), but fair to good air quality, especially north of Sacramento

Winter: very clean air, close to pristine quality, both in terms of wet deposition (snow) and aerosols (sulfates, nitrates, etc.)

THE IMPACTS OF LOCAL AIR POLLUTANTS

There are significant sources of air pollution generated within the Sierra Nevada, strongly affecting local air quality, and more weakly affecting downwind sites in the Great Basin deserts of Nevada and California. These will be considered in the following order:

1. smoke from forest sources, both wildfire and controlled burns

2. pollutants from urbanized enclaves, including nutrients from urban sources (Lake Tahoe)

3. particulate matter (fugitive dust) generated by human modification of water resources at Mono and Owens lakes on the east slope of the Sierra Nevada

While the global and upwind sources of air pollution are important, local sources can be, on occasion, extreme, resulting in some of the highest particulate levels seen in the United States.

Prehistoric Air Quality

There is a paucity of data on factors within the Sierra Nevada that affected air quality in prehistoric times. Clearly, there were no urban centers, so that component was absent. Mono and Owens lakes were in slow decline after the Pleistocene "ice ages," and without exposed playas, were generally a minor source of dust to air quality that was generally very good. Deep Springs Lake and other lakes at the western edge of the Great Basin that had larger natural cycles probably had some blowing dust. However, the situation is very different for smoke. From historical data on fires and the natural timing of lightning-induced fires, we can infer that there was two to four times as much area burned on an average summer day than in present times, with removal of timber plus present-day catastrophic fires redressing the biomass balance (see the following section on fires). Thus, we expect that there was much more summer smoke in the past, and less smoke in spring and fall, the time where prescribed natural fires and controlled burns are now encouraged (California Department of Fish and Game 1994). An example can be found in the Blue Mountains of northeastern Oregon, which were noted and named in the past for having much fire smoke commonly present in the summer. Another example is in historical pictures of the Lake Tahoe area, many of which show smoke. However, at the time of these pictures (late 19th century) massive human impact was evident as this was the peak of the Comstock lumbering period. There would also be changes in the nature of the smoke due to changes in the temperature of the fires.

The extent of transport of smoke into the Sierra Nevada from the Central Valley, a major factor at present (figure 48.7), is not known from presettlement times. All one has are a few logs from the first Spanish explorers. It may have been less than at present, based upon statements on what the explorers could see. However, native Americans were known to encourage fires to favor generation of oaks and for other purposes.

Present-Day Air Quality: Effect of Forest Fire Smoke within the Sierra Nevada on Air Quality

This section derives much of its information from the previous sections on fire, and they should be used for the more detailed and definitive discussion of the role of fire. Here we examine what is known about present and proposed air quality impacts of forest burning in the Sierra Nevada based on data taken in or near the Sierra Nevada. There is also a very large and important state and federal effort to model smoke emissions from fires, and this will not be discussed here. However, the modeling, if successful, will be constrained by the actual measurements that are reported herein.

There are many ways to categorize forest fires, but the following classifications appear the most widely used at present. Fires are categorized as:

Wildfires: These are large, sometimes catastrophic crowning wildfires that burn all before them. They may be initiated by lightning, accidents, and arson. The 24,500 acre Cleveland fire mostly in the El Dorado NF in 1992 is an example, used several times in this section. Such fires are predominantly in late summer, with dry fuel, high temperatures, but relatively vigorous smoke dispersal.

Prescribed fires: Prescribed fires, sometimes called prescribed burns or controlled burns, are human-initiated fires that burn under some prescription that involves weather, terrain, location, fuel moisture, fuel configuration (piles) and especially meteorology. These fires are used in national forests, usually in spring and fall, and are often keyed to an approaching rainstorm that will both disperse smoke and prevent loss of control.

Prescribed natural fires: These are lightning-initiated forest fires that are allowed to burn under some prescription. Lightning-started fires are most common in summer, and such fires are allowed to burn in national parks and wilderness areas. Sometimes the terms 'control and contain' are used for such fires.

The second and third types are also generally "surface burns" that spread only occasionally to tree crowns, and thus represent an approximation of historical patterns of fire that existed before human impacts.

Other terms have recently arisen that categorize the types of prescribed fires, namely 'ecological burns' and 'activity burns' (WESTAR Council 1995). There was some suggestion that the former may be considered differently from the latter in terms of federal air quality regulations since the former merely returns the forest to a more natural situation that avoids smoke from the otherwise inevitable and much more damaging wildfires.

There are two other source of smoke in the Sierra Nevada—the transport of smoke from the central valleys into the mountains, generally in summer and early fall, generally from biomass (agricultural, levee maintenance, and so on) burning (see earlier), and the heavy but localized smoke in late fall and winter in urban enclaves, especially those in valleys, derived largely from wood stoves and fireplaces (see later).

It is surprisingly difficult to establish the effect of each of these smoke sources on air quality in the Sierra Nevada. Smoke has a visual impact all out of proportion with the mass of smoke present, so that smoke levels must be extreme before the record of particulate mass reflects a major impact. Yet the only 24-hour federal particulate standard is for particle mass below 10 micrometers in diameter (PM-10), which is not violated until visibility drops to about 2 miles. Most of the air particulate air sampling in the Sierra Nevada measures only PM-10 mass, and thus is of limited use in identifying small and moderate smoke impacts. These sites only operate on a one-day-in-six cycle, and due to urban locations, are of little use to establish non-urban smoke levels. Further, the data on how many acres are burned each day from either wild-fires or prescribed burns are often difficult to access. Meteo-rological measurements in the mountains are scarce, and terrain effects major.

The IMPROVE data base is somewhat better in several re-gards. The measurements are PM-2.5, a better match to the size of smoke. The sites operate Wednesday and Saturday, in non-urban , non-valley locations, and have full meteorology, chemical, and optical analysis. However, as can be seen in appendix 48.1, there are only two such sites in the Sierra Ne-vada—Sequoia and Yosemite N.P. Fortunately, the paired sta-tions at Lake Tahoe, operated for the Tahoe Regional Planning Agency (TRPA) using full IMPROVE protocols, provide a very important third site, as well as an invaluable non-urban to urban comparison. Finally, data are extended by using simi-lar sites in the Cascade and San Bernardino Mountains. It is this data set that we must use for long-term data on Sierran smoke, supplemented by local studies.

Analysis of aerosol data from several sites in the Sierra Nevada indicates that the most severe impacts on air quality occur from large wildfires, but shows little effect of controlled fires at remote locations. Using figure 48.16 (data from the IMPROVE air sampler at Turtleback Dome, Yosemite N P) as an example, it is evident that the highest levels of particulate pollution occurred during a prescribed natural fire that burned in the park from July 3 to August 18, 1994. On only one occa-sion, however, did the pollution exceed the state air quality standard of 50 $\mu g/m^3$ for PM_{10}. The presence of smoke at the site during these episodes is evident from the unusually high peaks in non-soil potassium (K-non), a tracer of biomass smoke, and from human observations. Relatively low levels of particulate matter were seen (figure 48.16) during the sub-sequent fall season when the majority of agricultural waste burning is occurring in the San Joaquin Valley as well as con-trolled burning in nearby forests for fire suppression and sil-viculture.

In contrast, high levels of PM_{10} occur frequently in the heavily developed Yosemite Village (in Yosemite Valley) dur-ing the same period, even when no large fires are burning in the area, as shown in figure 48.17. The presence of many small local sources (campfires, fireplaces, and vehicles) and the micro meteorology of the valley, which tends to trap air un-der a nighttime inversion, result in a high background level of pollution.

Another comparison of local, anthropogenic sources ver-sus wildfires and controlled burns occurs in the Tahoe Basin. Air quality data taken near the relatively urbanized High-way 50 corridor in South Lake Tahoe show high levels of aero-sol pollution in the winter. Large peaks occur in both organic matter and in K-non (non-soil potassium) indicating wood smoke as the source (figure 48.18). At D. L. Bliss State Park, located in a largely undeveloped area on the west shore of

FIGURE 48.16

Some components of air quality (concentrations of non-soil potassium and particulate matter) at Turtleback Dome, Yosemite National Park, June through November 1994, showing impact of fires.

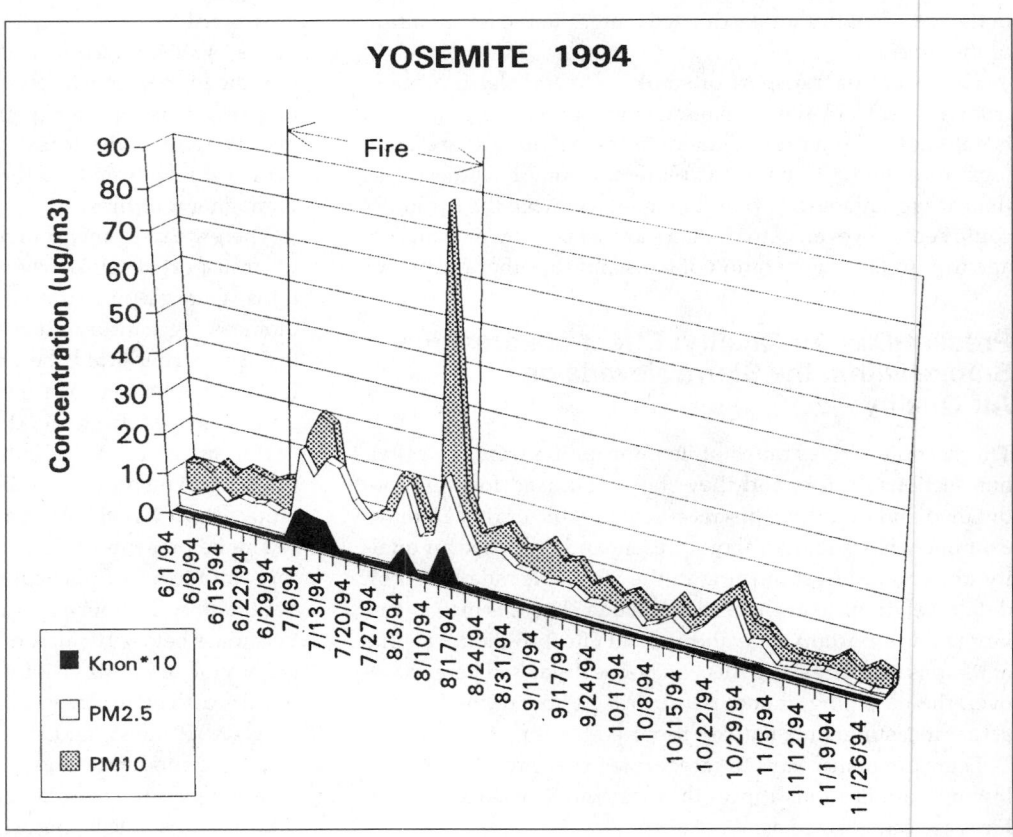

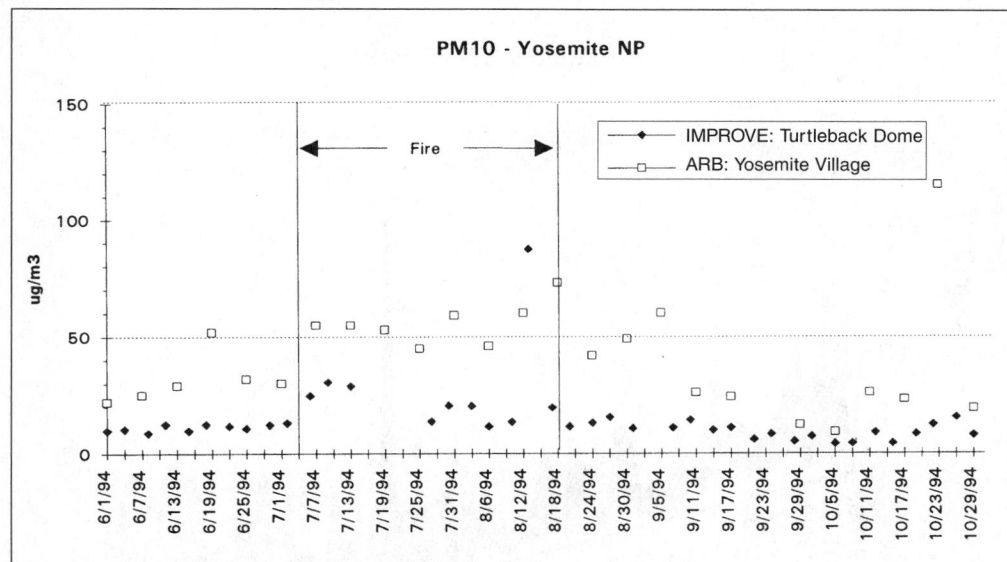

FIGURE 48.17

PM10 concentrations in Yosemite Village, Yosemite National Park, June through November 1994, as measured by California Air Resources Board (ARB) and UC Davis IMPROVE sampler (IMPROVE).

the lake, the winter is the cleanest season (figure 48.19). This suggests that residential wood combustion is the primary source at South Lake Tahoe. The only period in which occasional elevated levels of smoke are detected at both sites, indicating a source outside the basin, is the late fall when large amounts of cropland are being burned in the Sacramento Val-

ley and controlled burning in the surrounding national forests is at its peak. But the smoke levels even in these conditions are far less than the winter peaks at South Lake Tahoe roughly 20%, and of much shorter duration.

A final direct comparison between wildfires and residential wood burning is shown in figure 48.20 for Truckee, Cali-

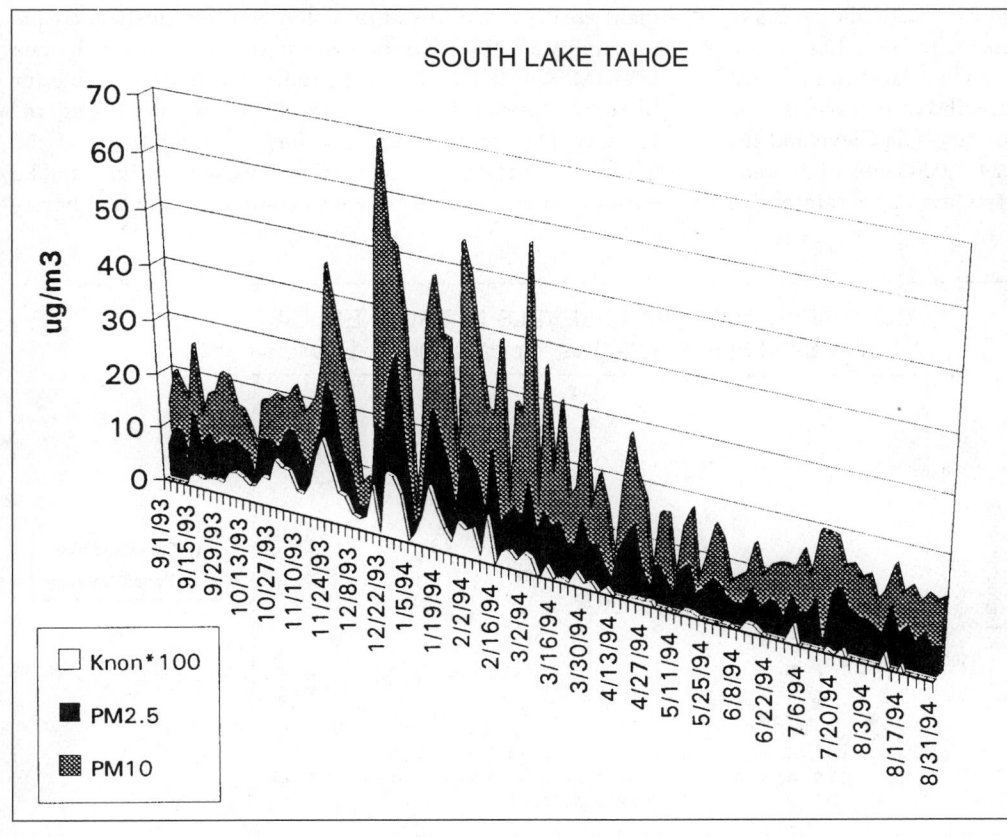

FIGURE 48.18

Some components of air quality (concentrations of non-soil potassium and particulate matter) at South Lake Tahoe, September 1993 through August 1994, showing impact of smoke.

FIGURE 48.19

Some components of air quality (concentrations of non-soil potassium and particulate matter) at D. L. Bliss State Park, September 1993 through August 1994, showing impact of smoke.

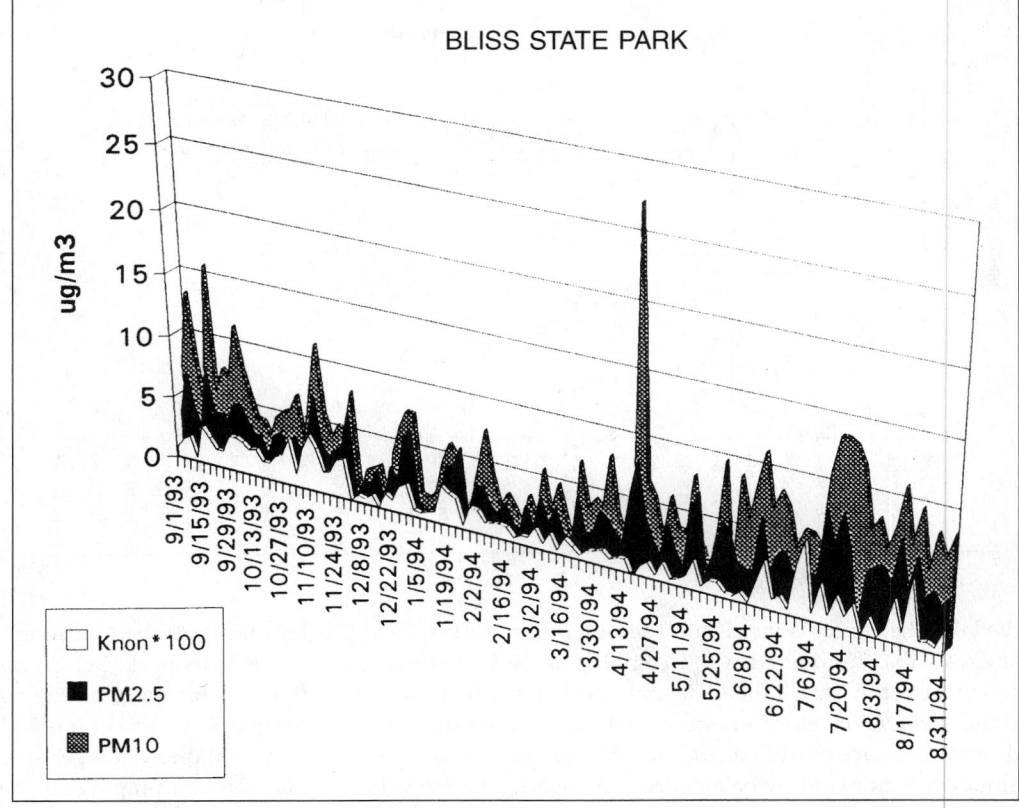

fornia. The availability of a new type of particulate measuring unit, a TEOM, allows hourly measurements to be made of mass. In figure 48.20, the intense Cleveland fire of 1992, located south of Interstate 80 and upwind of Truckee, is compared to winter smoke levels in the city. The Cleveland fire burned 5,500 acres on September 29, 7,000 acres of September 30, and 7,500 acres on October 1, when a light rainfall (0.1

inch) greatly aided fire suppression, limiting further acreage to roughly 3,500 acres until it was declared out about 1 week later (McKey 1995). Not only are the levels comparable for these two cases, 121 μg/m^3 for the Cleveland fire, 124 μg/m^3 for a typical January day, one has to remember that the Cleveland fire lasted for only a few days, while winter smoke episodes at Truckee are extremely common under the charac-

FIGURE 48.20

Comparison of hourly PM$_{10}$ levels at Truckee, CA, for the peak day of the Cleveland wildfire (9/30/92) and a typical January day (1/16/92).

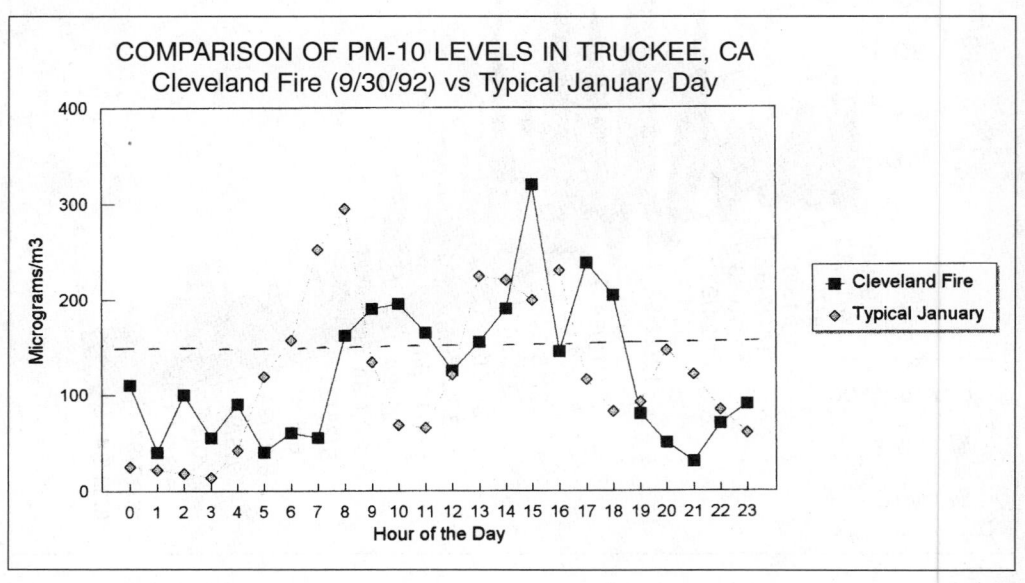

teristic winter subsidence inversion common to almost all high mountain valleys. The peak winter day in Truckee, January 1992, was 179 $\mu g/m^3$, measured by TEOM. But the TEOM probably understates the equivalent filter mass by 30% since the TEOM filter is heated, which drives off some of the water. Thus, in terms of person-dose, a typical winter is at least a factor of 10 or greater more important to health than a major local forest fire.

On the same day as the Truckee data, September 30, 1992, samples were taken at Bliss State Park as part of the regular Wednesday-Saturday IMPROVE-compatible protocol for TRPA. Bliss was not nearly as impacted as Truckee, since the wind was driving the smoke northeast. However, it recorded 13 $\mu g/m^3$ of PM-10, versus the 121 $\mu g/m^3$ for Truckee during the Cleveland fire. Even so, it was the highest PM-10 recorded on 16 sampling days between September 2 and November 30, 1992, more than double the annual average of 5.85 $\mu g/m^3$ recorded 1992–1993. The corresponding fine mass was 5.35 $\mu g/m^3$, versus an annual average of 3.39 $\mu g/m^3$. A strong nitrate signal was received, in fact the highest level seen all year, 1.45 $\mu g/m^3$, versus an annual average of 0.29 $\mu g/m^3$. This raises the question of whether the strong nitrate peak seen in South Lake Tahoe in winter could have a significant component from residential combustion of pine wood. Another more likely possibility is that the nitrate resulted from the volatilization of dry deposited nitrate on pine needles. Other species reached the highest level during the fire, including trace amount of chloride, arsenic, selenium, and bromine. The nonsoil potassium smoke tracer, K-NON, reached its second highest level on June 30, supporting the heavy transport of valley smoke into the mountains since grass smoke has a K-NON/mass ratio of at least ten times that of pine.

The results of the Cleveland fire help put into context the smoke from controlled burns, which for an entire season might total 7,000 acres, roughly as much was consumed per day in the Cleveland fire. In addition, the Cleveland fire occurred at a dry, hot period of the summer, without the meteorological mitigation built into controlled burns. Hence, the absence of any obvious signature due to controlled burns at Bliss, along with only one day of moderate impact at Yosemite, can now be readily understood since so little fuel is burned per day as compared to the uncontrolled Cleveland fire.

The relative importance to human health of local wood burning, as compared to forest fires, can be explained by the (by definition) higher population densities in urban areas, the regular pattern of residential wood fires, and the penchant for these urbanized areas to be in valleys rather than ridges, and the common nighttime inversions that trap smoke close to the ground. Wildfires, by their very nature, generate lots of heat, and tend to loft much of their pollutant load into the sky.

The smoke produced by biomass combustion is composed of water vapor, other gases, and particles less than 2.5 μm in diameter, but a significant amount of larger particles may also be produced by large, intense fires due to entrainment of soil and partially combusted matter in the strong updrafts. Significantly larger particles present little threat to health or visibility, and typically do not persist in the atmosphere for more than a few hours before they settle out due to gravity. Fine particles (smaller than 2.5 μm), however, are very effective in reducing visibility because they scatter light and aid the condensation of water vapor in the air. These smaller particles also contain a significant quantity of organic compounds known collectively as polycyclic aromatic hydrocarbons (PAH) which include a number of toxic and potentially carcinogenic substances. Since the fine particles are readily inhaled and retained in the lungs, and may settle onto the surface of vegetation, increased concentrations of smoke represent a potential hazard to both human health (Larson and Koenig 1994) and the environment. These concerns are not limited to emissions from forest fires. Research data indicate that burning of grasses, agricultural wastes, and other types of wood produce even higher concentrations of PAH (Jenkins et al. 1995b).

Woodburning emits a variety of gaseous pollutants (Jenkins et al. 1995a). These are composed primarily of CO_2 and H_2O, with the remainder dominated by CO (carbon monoxide) and a variety of hydrocarbons, including PAHs. Since carbon monoxide is relatively inert and disperses readily, it should not have any significant impact on air quality beyond the immediate area of the fire. Hydrocarbons, on the other hand, can be transported over large areas and contribute to ozone formation in the presence of other pollutants. NO_x is also produced, as in all combustion, but in relatively small concentrations in comparison to their emissions from vehicles and industrial sources.

Finally, there is evidence that part of the water of combustion of wood smoke may be trapped in the smoke, especially in cold, humid, winter conditions (Molenar et al. 1996), and seen also in the 30% difference between TEOM and standard PM_{10} filters (above). If even a small fraction of the water is trapped, it can greatly raise the smoke mass. More detailed analyses (see above) are needed before this can be resolved. Nevertheless, a certain caution should be retained about ways to reduce wood smoke by reducing temperature of combustion and air flow, as opposed to an oxygen-rich open flame. Low temperature smoke is far more chemically complicated than high temperature smoke, retaining compounds that are known mutagenic (and perhaps carcinogenic) agents.

Fall 1995 saw a good deal of activity in the area of fire pollution. First, the fall was exceptionally dry, with the first significant rain occurring in early December. The meteorology was stable, with weak winds and strong inversions forming in the Central Valley. Several prescribed natural fires and controlled burns persisted into periods of poor ventilation, with major smoke impacts on local communities. This occurred for fires in and near Sequoia N.P., which totaled about 9,000 acres by early December. Prescribed fires were ignited near Mineral King and in a chaparral zone about 10 miles upslope of Three Rivers. Heavy smoke was recorded in local communities, resulting in four violations of the 150 µg/m³ federal 24-hour PM-10 regulations, with the maximum value of 194 µg/m³ (D. Ewell, Sequoia National Park, communication with the author, 1995). Another fire burned for about a month in the Lake Tahoe basin, in Bliss State Park near the TRPA aerosol site. Smoke impacts were regularly reported (B. Mahern, Tahoe Regional Planning Agency, communication with the author, 1995). Both of these fires represent patterns of prescribed and controlled burns that may become more likely in the future, and the experience gathered in these events will be useful in avoiding such impacts. Clearly, the concentration of so much burn acreage in a single watershed of air basin at times of poor ventilation resulted in unacceptable levels of smoke, although the anomalous weather of fall 1995 was a major factor in these episodes.

Finally, there was a major wildfire/prescribed fire workshop sponsored by the WESTAR Council, an association of air resource agencies from western states, Alaska to the Dakotas, San Francisco, November 27–29, 1995. While the reports and recommendations of this meeting are not yet released, minutes of the presentations have several points of interest (WESTAR Council 1995). One of these was the conceptual separation by a speaker from the U.S. Environmental Protection Agency on various options, including separation of smoke from "ecological burns" and "activity burns" and possible trade-offs against wildfire smoke. The logic is that the ecological burns are really a way of avoiding future smoke from the much more serious crowning wildfires, as well as a way to maintain a healthy forest. The consensus also was reached that the nuisance effects of smoke, including visibility reduction, will become more important as a constraint on burning than possible violations of federal fine particulate air quality standards.

Overall, current data suggest that controlled forest burns are not as major a source of particulate mass in populated areas of the Sierra Nevada as residential wood combustion and campfires. Large wildfires produce severe short-term impacts on air quality, but because they are rare, average smoke dose to individuals is generally limited. Prescribed or controlled burns are more common, but the amount of materials burned is more modest, and the measures to limit human smoke impacts are generally quite effective, leading to very low contributions to PM_{10} particulate loading in inhabited areas. Thus it would appear that prescribed fires are usually performed in such a way as not to cause a significant threat to regional air quality as measured by fine particulate mass. The obvious exception is for some local visibility reduction, but this must be compared to improved air quality by decreasing the impacts of major wildfires. Given that fire is a natural part of the Sierra Nevada ecosystem (Phillips 1995), the beneficial effects on the Sierra Nevada ecosystem of increased fire use should not result in widespread violations of state and/or federal fine particle health standards.

The very real problems of perceived smoke and visibility reduction must be addressed, however. One way is to couple the presence of modest summer smoke with the overall health of the forest and the reduced chances of major wildfires, which cause drastic reductions of visibility and direct and indirect health effects. The other is to ascertain the relationship between visibility reduction and smoke mass, showing that even in visibly dense smoke, mass loadings are modest. Using results of studies of Oregon and Washington fires (Radke et al. 1990), a relationship was measured. Visibility due to smoke must be reduced to 3.0 +/- 1.8 km (1.9 +/- 1.1 mi) before one reaches the federal particulate air quality standard of 150 µg/m³ six miles before one reaches the California standard of 50 µg/m³. The same relationship is found for IMPROVE's fine (Dp< 2.5 mm) particulate mass. A "best fit" between visibility and mean annual mass at 44 sites gives 3.0 kilometers (1.9 miles) for the federal standard of 150 µg/m³, assuming no contribution from particles greater 2.5 µm diameter (S. Copeland, U.S. Forest Service, Fort Collins, Colora-do, communication with the author, 1995). The corresponding visibility at the 50 µg/m³ California standard is 9.1 km (5.7 miles). The problem of visibility is compounded for fires that occur in scenic areas by the fact that people are used to seeing many miles. Thus, visibility reductions are obvious. The plumes tend to be well above the ground, which makes them more visible as it reduces ground level mass concentrations. The same effects do not occur for the even greater smoke densities in towns like Truckee during the winter, for example, since the densely populated core of the town is less than one mile long.

There are also indirect effects of fires, in which they act as a means of transporting materials from one location to another. An example is agricultural burning in the central valleys of California. The mass of smoke by itself may not be a serious factor in terms of particulate mass, but health effects are reliably reported when smoke impacts cities, such as visits to doctors by asthmatics (Betty Turner, American Lung Association, communication with the author, 1995). The answer appears to be in the reactions of sensitive populations to all the other materials lofted into the atmosphere with the smoke, which in the valleys include pollens, fungal spores, partially pyrolyzed pesticides and herbicides, and other components.

Effect of Urbanization within the Sierra Nevada on Air Quality

The ecological and touristic values of the Sierra Nevada have naturally generated areas of moderate population density in small cities, towns, and other areas. These areas in turn modify the local environment in many ways, including impacts on air quality associated with increased traffic, changes in land use, heating, and other activities. It was not anticipated, however, that the focusing of developmental pressures on areas of especially high scenic value would then generate quasi-urban areas with traffic and population densities similar to other, larger cities in California. Examples include the Lake Tahoe literal, Yosemite Valley, and Mammoth Lakes, but there are others. These quasi-urban areas then in turn degrade to a greater or lesser degree the values that drew the population in the first place, and even lead to levels of air pollution that result in violations of state and federal air quality standards. Visitors became clearly aware of other impacts of urbanization, including traffic jams, parking problems, smoke from fires, etc. But most casual observers would be startled to realize that some of the highest particulate mass loadings in California occur in the Sierra Nevada.

Health and Regulatory Air Quality Issues

Summer, Lake Tahoe Sites. In response to heavy and congested traffic levels at South Lake Tahoe, and a few air samples that showed high particulate lead levels, a study was mounted in summer 1973 by the California Air Resources Board at sites all around the Lake Tahoe Basin. While the results showed a wide variation in air quality at sites around the lake, it was clear that sites near the Nevada state line, the locus for casinos and the target of much of the daily traffic, had levels of gaseous and particulate pollution that were typical of other, much larger areas in California.

Table 48.4 shows a representation of the ARB data, placed into a comparison with other California sites (Goldman and Cahill 1975). One entry in the table is incorrect, although that was not known at the time. The lead value for Los Angeles, submitted to the ARB, had been arbitrarily divided by a factor of 2.8 in order to maintain continuity with earlier (erroneous) measurements. Thus, the true Los Angeles lead value is actually 2.7 $\mu g/m^3$. Nevertheless, the fact that even some air pollution levels at some Tahoe sites were worse than in downtown Los Angeles was a cause of considerable comment, leading to a designation of a special Lake Tahoe Air Basin, special, stricter standards for both visibility and carbon monoxide, establishment of a permanent air pollution site, and several state and federal air quality studies that continue to this day,

including major efforts by the Tahoe Regional Planning Agency (TRPA).

The key question that immediately arose was the attribution of air pollution to anthropogenic sources within the basin, potentially amenable to mitigation, versus either natural sources or anthropogenic sources upwind of the Lake Tahoe area. Figure 48.21 shows the results of recent TRPA studies that address this question.

Comparisons of a site at Bliss State Park that responds only to pollutants coming across the mountains from upwind sites, versus at site at South Lake Tahoe, give a convincing answer to the question. Since the sulfur (sulfate) particles are essentially the same at both sites, this pollutant comes from upwind sources. Organic matter (smoke, ...) and nitrates appear to be largely from upwind sources in the summer, but of strongly local sources in the winter. A similar analysis using ARB data shows that, in the summer, ozone behaves like sulfur and responds to upwind sources, while most other gaseous pollutants are largely local in origin. Methane has a significant (roughly 50%) natural source, as do coarse particles that respond to pollen, bio-debris, etc. Soils are largely local, especially in winter and spring when road sanding debris is present.

Since the early 1970s, the air pollution at South Lake Tahoe has been reduced due to improved auto emissions, control of road surfaces, and other efforts. Lead is essentially absent, while NO, NO_2, and CO have been cut roughly in half. No such improvement has been seen for ozone, which has actually risen slightly in this period, though still representing moderate levels.

Nevertheless, visibility at Lake Tahoe has degraded since the base year of 1982 as both transported and local sources of fine particles have increased. This is a serious source of concern and closely tied to the enjoyment of the extraordinary vistas for which Lake Tahoe is justly famous. Studies are un-

TABLE 48.4

Air quality at two Lake Tahoe sites, with comparison to other California cities. Gaseous data were collected over the month of July 1973.

Pollutant[a]	Incline	Stateline	For Comparison Purposes		
			Monterey	Sacramento	Los Angeles
Oxidant (ppm)	0.063	0.049	0.04	0.09	0.11
Carbon monoxide (ppm)	1.5	6.4	1.0	2.0	6.0
Sulfur dioxide (ppm)	—	—	—	—	0.04
Nitrogen dioxide (ppm)	0.009	0.051	0.00	0.040	0.11
Nitric oxide (ppm)	0.003	0.024	0.02	0.020	0.10
Oxides of nitrogen (ppm)	0.012	0.068	0.03	0.060	0.17
Hydrocarbons (ppm)	2.5	5.2	—	2.0	3.0
Hydrocarbons (ppm) (non-methane)	0.17	1.97	—	—	—
Lead particulate (micrograms/m³)	0.203	1.72	0.23[b]	0.49	0.95
Suspended particulates (micrograms/m³)	95	87	36	78	116

[a]The values for the gaseous pollutants (the first 8) are in parts per million of air, maximum hour averaged over the month, while the particulate values are 24-hour averages taken at random times throughout the month and averaged, expressed as micrograms of material per cubic meter of air. (—) indicates pollutant not measured at that site.
[b]Data averaged from two nearby sites, since lead was not measured at Monterey.

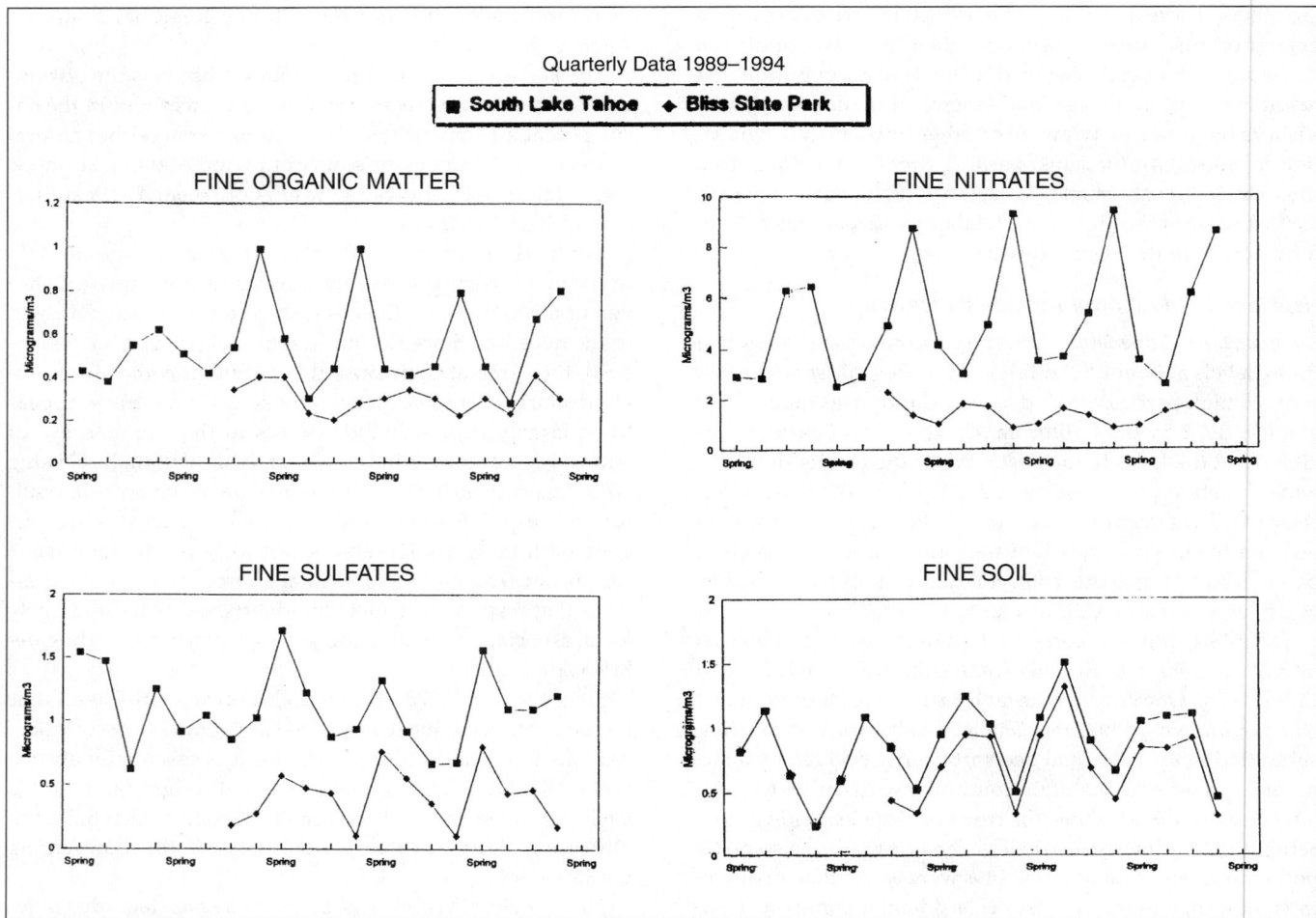

FIGURE 48.21

Aerosols at two Lake Tahoe sites.

derway to identify exactly which factors are dominant in this problem (Molenar et al. 1996). Note that similar problems of urban haze occur in Yosemite Valley from a combination of local smoke and transported San Joaquin Valley air pollution.

Winter, Lake Tahoe, Mammoth Lakes, and Yosemite Valley. The question of the relative impact of urban versus non-urban wood smoke has been largely addressed above, in the section on fires. These data also show, however, that there are additional sources of smoke in the urbanized areas that are not wood smoke. This is shown by analysis of the optical opacity of the smoke as compared to the known smoke tracer, K-NON, at South Lake Tahoe and Bliss, suggesting some fossil fuel/diesel source contributions. But wood smoke dominates the mass of smoke particles.

The mass concentrations in smoke levels in winter are elevated, often exceeding state and federal air quality standards as well as causing intense haze. This is a fact also realized at other winter resorts such as Aspen and Vail, Colorado, which

have initiated vigorous controls on residential woodburning. Examples were shown above at Yosemite Valley and Truckee. As a further example, the city of Mammoth Lakes, California, achieved levels of particulate matter that ranked among the highest urban values in California, and gross violation of state and federal standards. Figure 48.22 shows that not only are such levels high in the peak days, but they are also high on average, unlike the intense but infrequent episodes of wildfires. Since 1990, serious efforts at smoke suppression have been in place, with some success.

Ecological Impacts: The Case of Lake Tahoe

As seen in figure 48.21, there are significant concentrations of airborne particulate nitrates at Lake Tahoe sites, along with much smaller levels of phosphorous, both limiting nutrients in the nutrient-poor lake. Some of this material will enter this lake through dry and wet deposition, thus fertilizing the lake and contributing to the roughly 30% degradation in water quality observed since 1958 (Goldman 1994). Clearly, the question of local versus transported sources becomes critical, as does the ratio of these nutrients to those contributed to run-off from urbanized areas and soil disturbance from development. This is the subject of active investigation at this time, and a clear consensus has yet to be achieved.

The location of the sources of particulate nitrates has largely been resolved (figure 48.21). During spring, summer, and fall, most particulate nitrate is transported from upwind sources. During this time, most gaseous nitrogen, NO_x, is of local, motor vehicle origin (Cahill et al. 1977). In winter, both the particulate nitrate and the gaseous nitrate is local in origin, with heavy transportation sources but also including other forms of combustion.

One opinion is that atmospheric deposition is a major factor in nitrate input to the lake, resulting in a tons/year prediction of nitrate input to the lake. In comparison to the nitrate input in streams and run-off, this gives it an atmospheric source that dominates nitrate input to the lake. However, dry deposition measurements are notoriously difficult to do accurately, and questions remain on input pathways of nitrogen (Jassby et al. 1994).

From the atmospheric data given in figure 48.21, we made calculations of dry deposition from the measured nitrate concentrations (Sehmel 1980). Using a mean transported nitrate concentration of 0.3 $\mu g/m^3$, from the Bliss site but averaged over the entire lake surface, yields deposition values between 0.4 and 1.0 ton/year, well below those inferred from the TRG measurements (Jassby et al. 1994). Adding in the local anthropogenic particulate nitrate, 0.3 $\mu g/m^3$, inferred from the South Lake Tahoe site after subtracting the transported component, assuming a somewhat larger particle size for humid, winter conditions, and averaging over that portion of the lake near urbanized areas, yields an additional 0.1 to 0.3 tons/year. In contrast, using local gaseous NO_x concentrations from ve-

hicles, and the same type of calculations but this time over a 1 km wide band around the lake, yields a mean NO_x concentration of 22.6 $\mu g/m^3$, roughly 75 times the concentration of transported particulate nitrate. If only 10% is scavenged onto trees and surfaces to eventually reach the lake in spring snow melt, this yields on the order of 20 (or more) tons/year into the lake, with a spatial pattern that closely matches observed maxima in algal growth. Since there are major uncertainties in making sub-surface nitrate measurements from urban run-off, direct observation of this effect is difficult.

Even these factors do not appear to explain the increasing turbidity of Lake Tahoe; since NO_x levels have been steadily decreasing over the past 20 years while the lake is getting steadily worse. A good match is seen, however, when one compares development around the lake, with soil disturbance and mobilization of phosphorous, to algal growth. Local traffic will also be driven in part by this development.

Dust Storms Caused by the Desiccation of Mono and Owens Lakes

At the interface of the Sierra Nevada and the Great Basin lie several saline lakes or playas (dry lake beds), remnants of large "pluvial lakes" that stored glacial melt water and run-off from the Sierra Nevada during the Pleistocene ice ages. As stated previously, a certain amount of airborne dust is generated from some of these playas (such as Deep Springs and Honey Lakes), but this material has a limited impact, and, as it is of natural occurrence, is not an air pollutant *per se*. However, when saline lakes are desiccated by human action, any dust generated is considered "fugitive" and subject to the National Ambient Air Quality Standards (NAAQS). Wind-driven sand moving across unvegetated, recently exposed playa surfaces kicks up dust plumes composed of silicate

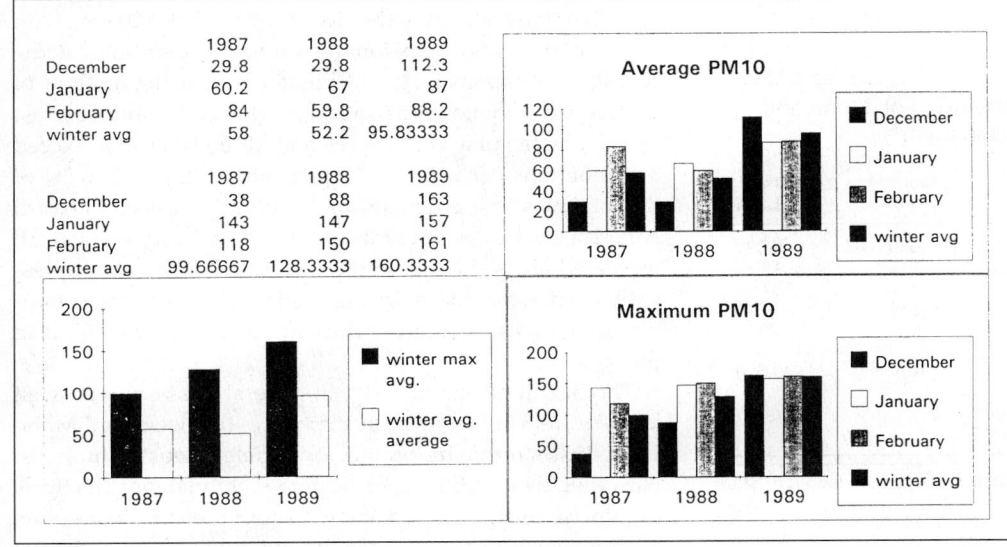

FIGURE 48.22

Particulate matter (PM_{10}) at Mammoth Lakes, California, winter, 1987–89. All values are in micrograms/m^3. The California 24-hour standard is 50 micrograms/m^3, while the federal 24-hour standard is 150 micrograms/m^3.

minerals and salts. About half of the mass of playa dust aerosols is contained in particles of diameter less than 10mm (PM_{10}), small enough to be transported long distances and inhaled deeply into the human respiratory tract. Fugitive dust storms from playas are a problem in several areas around the globe, of which the Sierra Nevada region possesses several significant examples (Gill in press). Water diversions from the Truckee and Walker rivers flowing out of the Sierra Nevada have resulted in minor blowing dust at Pyramid and Walker lakes in Nevada. However, serious PM_{10} problems exist at Mono and Owens (Dry) lakes at the Sierra Nevada's eastern base (table 48.5). These areas are two of the three "non-attainment areas" for PM_{10} formally designated by the U.S. EPA within the Sierra Nevada region; the third is the community of Mammoth Lakes in Mono County, which is impacted by wood smoke.

All significant dust storms from the playas of Mono and Owens lakes are dependent on one major factor external to the Sierra Nevada—sustained winds caused by synoptic (large-scale) weather systems affecting the region. A few dust events, generally minor and short-lived and especially at Owens Lake, can be caused by mesoscale (regional) atmospheric circulation (upslope-downslope winds and convective storms) caused or enhanced by the steeply-sloping topography of the Sierra Nevada itself (Cahill et al. 1994).

As much as 65 km^2 of playa has been exposed along the shore of Mono Lake (directly east of Yosemite National Park) since water diversions by the Los Angeles Department of Water and Power (LADWP) began in 1940. When no dust is observed (in recent years, more than 90% of all days), the air in the Mono Basin is among the "cleanest" in California. But when the lake was near its historical low, average dust concentrations on the remaining days exceeded the then-existing California standard for particulate matter by a factor of six (Kusko and Cahill, 1984). Mono dust storms can violate the California airborne sulfate standard, and may contain suf-

ficient arsenic to elevate cancer risk in humans (Cahill and Gill 1988). The occurrence and significance of dust storms from Mono Lake's northeastern playa has been a major factor in the legal and environmental battle over LADWP's water rights and protection of the Mono Lake ecosystem.

Although the level of Owens Lake (in the shadow of Mt. Whitney) was already slowly receding due to Owens River water withdrawals for Owens Valley agriculture, LADWP diversions into the Los Angeles Aqueduct caused the lake's complete desiccation. The water transfer began in 1913, and dried Owens into a 280 km^2 playa within fifteen years. The outer third of the playa, a zone of crystalline salts, clays, and fine silts, is vulnerable to severe wind erosion by the abrasion of blowing sand; Owens Lake dust events represent the highest estimated PM_{10} levels recorded to date in the U.S.A., a 24-hour PM_{10} average of 4,184 $\mu g/m^3$ at the town of Keeler and a 2-hour PM_{10} concentration on the lakebed exceeding 40,600 $\mu g/m^3$ (Cahill et al. 1994, in press). For comparison, the U.S. EPA 24-hour limit for PM_{10} is 150 $\mu g/m^3$; this standard is exceeded at least 48 days per year downwind of Owens Lake (Great Basin Unified Air Pollution Control District [GBUAPCD] 1994). Just as in the Mono Basin, on dust-free days air quality in the Owens Valley is generally very good, with PM_{10} levels on the order of 10 $\mu g/m^3$ or less.

Dust plumes from Owens Lake tend to blow north or south and hug the eastern slope of the Sierra Nevada, blocking scenic views of the mountains with white dust haze, occasionally disrupting traffic on U.S. Highway 395, and constituting a general nuisance to local residents. Saline, alkaline dust from Owens Lake is known to encrust the needles of pines and leaves of other plants in the White-Inyo Range, and is deposited within the borders of Death Valley National Park. Owens dust is transported onto the eastern slope of the Sierra Nevada, impacting the John Muir, Golden Trout, South Sierra and Dome Lands Wilderness Areas and adjacent parts of Inyo National Forest before spilling over the crest of the range. To the south, the dust clouds enter the Indian Wells Valley east of Walker Pass, affecting the city of Ridgecrest (120 km south of the playa), and occasionally suspend operations at the Naval Air Weapons Station, China Lake, causing millions of dollars in economic losses each year. The total amount of dust emitted by the playas of Owens and Mono lakes may exceed 8 million tons per year. This represents perhaps 3% to 5% of the total mass of particulate air pollution produced in North America (and is several times greater than the sum total of all regulated air pollutants in the Los Angeles air basin), presently placing these two dry lake beds among the largest individual sources of fugitive dust in North America (Gill in press).

The health effects of PM_{10} in general are becoming well known, and chronic or acute exposures to Owens and Mono Lake dust storms are bound to be deleterious to humans. However, there is little specific data on human health effects of mineral dust, even less known about the effects of saline, alkaline particles from lake beds, and only anecdotal data at

TABLE 48.5

Air quality impact of dust storms downwind of Mono and Owens Lakes, 1979–83[a] (all values in $\mu g/m^3$).

	Mono Lake[b]	Owens Lake
Maximum 24-hour PM_{10} concentration	1,650	2,092
Worst 1.3% of all days	912	1,098
Worst 5% of all days	416	599
Worst 11% of all days	265	315
Remaining 89% of all days (non-dust storm days)	9	14
Maximum short-term PM_{10} concentration, $\mu g/m^3$ from UC Davis field measurements	N.A.	40,620[c]
U.S. EPA 24-hour standard for PM_{10}, 150 $\mu g/m^3$		
California 24 hour standard for PM_{10}, 50 $\mu g/m^3$		

[a]Based on GBUAPCD data, in Kusko and Cahill (1984) converted from Total Suspended Particulates to PM_{10}.
[b]Level of Mono Lake, 6,373 +/- 1 feet above MSL, 1979–82.
[c]Based on calculations in Cahill et al. 1994, 1995.

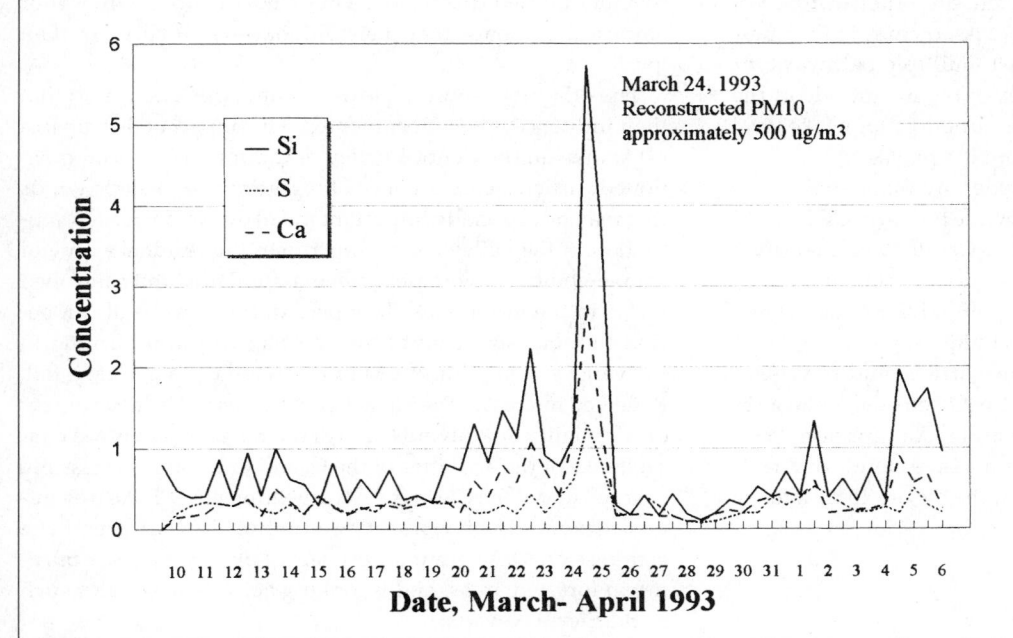

FIGURE 48.23

Elemental chemistry of dust for silicon, sulfur, and calcium at Schulman Grove, Ancient Bristlecone Pine Forest, March–April 1993, showing impact of March 24 Owens Lake event; concentrations in $\mu g/m^3$, data for particles 2.5 μm or smaller collected with UC Davis SMART sampler (after Cahill et al. 1994).

best on specific health effects of Mono-Owens aerosols. The effects of this dust on ecosystems are also not well known, though we can make inferences from other studies. Prolonged deposition of alkaline dust causes chemical, physical and biological changes in soil profiles and eventually changes vegetation communities and ecosystem structure; there is anecdotal evidence that such changes have started to occur in the Mono Basin (Cahill and Gill 1988). Alkaline, saline dust coating needles or leaves limits plant germination, growth, respiration, transpiration, and photosynthesis; blocks the stomata; exacerbates secondary stresses such as drought, insects and pathogens; modulates the uptake of toxic metals and other air pollutants; and may cause visible injury and even cell death to needles, leaves and bark (Farmer 1993). No detailed monitoring for these problems has been undertaken in the Inyo National Forest, but dry deposition of PM_{10} from Mono and Owens lakes is known to occur on its slopes. Since the most damaging effects of dust take place on arctic-alpine vegetation (Farmer 1993), it may well have some of the aforementioned effects on high-altitude ecosystems of the Sierra Nevada.

A significant fraction of the soil in alpine environments in the White-Inyo Range, including the Ancient Bristlecone Pine Forest Area of Critical Ecological Concern, was created by the fallout of fine airborne sediments originating in the Owens and/or Mono basins in the geological past (Marchand 1970). The enhanced input of PM_{10} to these areas from dust storms (figure 48.23) simulates these soil-building episodes of the geologic past (Gill in press), and should have some effect on the health and growth of bristlecone pines. Since bristlecone pine growth in the White Mountains is being used to evalu-

ate global climate change, and prevailing wind trajectories into the groves pass through the Mono Basin and Owens Lake areas, the impact of Owens Lake and Mono Basin dust could provide a "false signal" to this system.

Ruling D-1631 of the State Water Resources Control Board in 1994 provided that water exports from the Mono Basin must be restricted in a manner to "result in the water level of Mono Lake rising to a level of 6,391 feet in approximately 20 years." When this occurs, blowing dust from the Mono Lake playa will be significantly reduced and will be unlikely to have a serious environmental impact.

Figure 48.24 shows all measured values of air pollution downwind of Mono Lake, 1979–1993, with corrections to convert to the maximum PM_{10} values on the land around Mono Lake. These corrections include conversion from total suspended particulates (TSD), measured for 1979 through 1986 to PM_{10}, measured at present using a multiplicative factor of 0.47. It also includes corrections for drought, 1987–1993, using Owens Lake as a model (about a factor of 2) and conversion for the Simis Ranch site to the maximum site, generally near Warm Spring, using (1) the Fugitive Dust Model (FDM) used by Jones and Stokes, the WRCB contractor, (2) the Industrial Source ISC2 model used by the GBUAPCD's contractor, and (3) the Mono-Owens Davis Dust Model (MODDM) based on the Davis work (Cahill and Gill 1988).

The latter model (figure 48.24), based as it is on a linear fetch hypothesis closely tied to sand motion, and calibrated against observed dust levels 1979–1983, is not as sensitive to the source strength assumptions inherent in the first two models, neither of which was designed for the two-step dust resuspension process that dominates dust loadings at Mono (and Owens) Lake. In addition, there are other factors not considered in the models of PM_{10} mass, including the questions of interior and persistent dust, multiple pathways for arsenic input from the dust, and even the magnitude of the extraordinary dust events. All these support a high lake level for elimination of blowing dust from the playas.

At Owens Lake, even though federally mandated, PM_{10} mitigation may take much longer. While the early UCD/ARB work (Barone et al 1979; Kusko and Cahill 1984) identified the cause, and mitigation studies were undertaken as early as 1982 (State Lands Commission WESTEC Report, 1984), progress has been slow. Several techniques, including flood irrigation, deployment of sand fence arrays, and re vegetation, are presently being tested on the Owens Lake playa for dust storm suppression, (State Lands Commission 1991; GBUAPCD 1995), but control of the massive clouds of dust is deemed not feasible in the short term by the local agency (GBUAPCD 1995).

EFFECT OF THE SIERRA NEVADA ON DOWNWIND AIR QUALITY

The three major sources of air pollutants within the Sierra Nevada are forest smoke (wildfires, prescribed natural fires, controlled burns), urban sources (again mainly smoke, some vehicular) and the partially/completely desiccated lake beds of Mono Lake and Owens Lake (alkaline/saline dusts). The urban sources are, however, minor in total emissions (tons/year), and their high winter concentrations are due mainly to

severely limited dispersion. Thus, there is no theoretical or empirical evidence that their influence is much more than local.

The other two sources, however, are large enough so that their influence is well documented. The impact of Sierran forest smoke on the Central Valley of California has been mentioned earlier, a consequence of nighttime downslope winds that may be especially important in fall due to decreased ventilation in the valleys, which increases the residence time of smoke, combined with prescribed natural fires and controlled burns in the mountains. This period, however, is also a period that has significant acreage of agricultural burning in the valleys, many hundreds of thousands of acres each fall. Renewed interest in the impact of these Sierra Nevada sources on air quality downwind (east) of the range is partially a consequence of the activities of the Grand Canyon Commission, charged under the Clean Air Act amendments of 1990 to evaluate all sources of visibility reduction in that area. The commission's task groups are aware of plans to increase burning in forested areas, and is looking actively at sources such as the Sierra Nevada.

Data from the Sierra Nevada can place this evaluation in perspective. The results at lake Tahoe show efficient transport of smoke aerosols (and other components such as ozone) from the California Central Valley into the Tahoe Basin, across passes at roughly 7,000 feet and around mountains that rise to 10,000 feet. This occurs during much of each spring, summer, and early fall. These pollutants certainly influence the Great Basin air quality, although levels are modest. The results of the Cleveland fire of 1992 show massive transport of smoke downwind of the range, but such events are infrequent. Conversely, the valley to mountain transects in Sequoia NP,

FIGURE 48.24

Maximum PM_{10} near Mono Lake, extrapolated from Simis, drought-corrected.

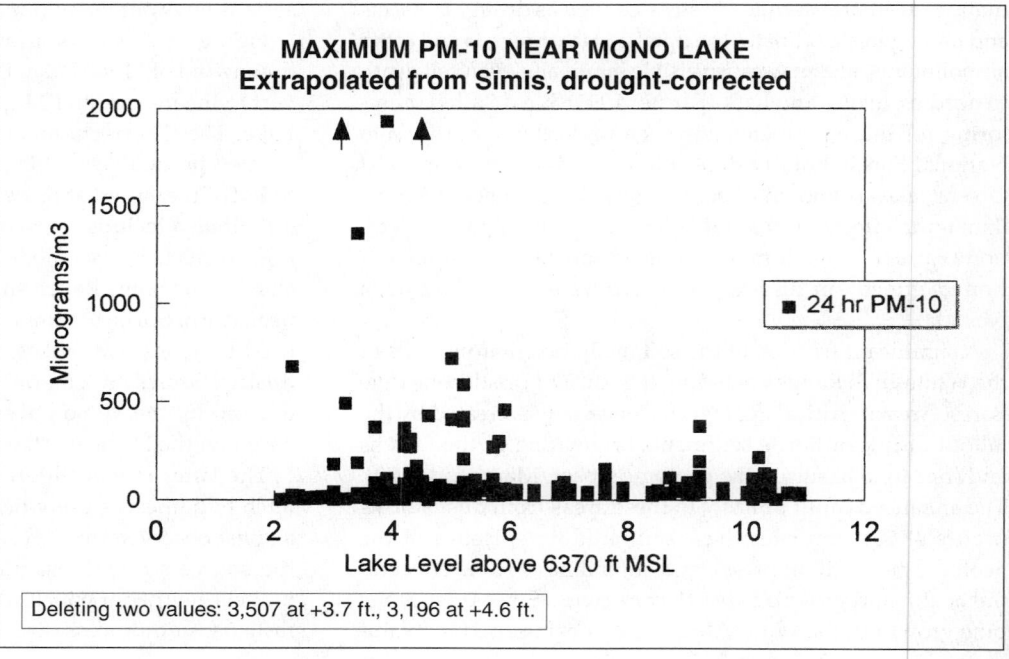

1987 (Cahill et al. 1989) show a sharp reduction in ozone and aerosols between Giant Forest at 6,000 feet and Emerald lake at 10,000 feet. The Emerald Lake site is west of the Great Western Divide, and the peaks to the east of it rise to over 14,000 feet. This supports a very limited pollutant transport efficiency over the mountains to downwind sites in the central and southern Sierra Nevada, both for local forest smoke and valley smoke. Finally, there is well-documented transport across Tehachapi Pass into the Mojave Desert (Pitchford et al, 1984) as the elevation drops to around 4,000 feet. As temperatures drop each fall, all mountain transport processes weaken and smoke of all kinds tends to stay in the Central Valley. This was certainly the experience in the dry fall of 1995, when smoke from the Sequoia N.P. prescribed fires drifted downslope into the valley. It is highly unlikely that any significant amount of this smoke ever made it to the Grand Canyon. In summary, fires that burn under summertime conditions can contribute smoke downwind of the range, while spring and fall fires tend to have greatly reduced transport east of the mountains and, conversely, the greatest local impact. The wintertime inversions in the Central Valley, and lack of fires in the Sierra Nevada, indicate little Sierra Nevada influence at downwind sites.

A final piece of evidence concerning transport into the intermountain area can be gathered by comparing aerosols at Bliss State Park, Lake Tahoe, with the Great Basin and Grand Canyon National Parks. The characteristic signatures of wood smoke are, in order of uniqueness, excess fine potassium (K-NON), organic carbon from carbon (C) and hydrogen (H), optical absorption, and elemental carbon. The mean values for each are shown in table 48.6 (IMPROVE 1995).

Thus, there is no convincing evidence that there is major transport from the Sierra Nevada into the great basin region or the Grand Canyon N.P., since such long scale transport would cause values to decrease as particle are lost during transit. Further, since the highest values at Bliss occur in summer, and these come from the Sacramento valley floor, a better case could be made for the impact of agricultural burning in California on air quality in the great basin region.

Finally, transport of aerosols from Mono and Owens Lake into the mountains and then downwind into the Great Basin

reported photographically by aerosol measurements and by satellite (Cahill et al. in press). While these events are infrequent, occurring about 11% of all days (Kusko and Cahill 1984) they are intense and carry a great deal of fine alkaline/saline dust into this region. The source at Owens Lake alone is estimated to represent on the order of 5% of all the fine dust generated in the United States each year (Gill and Gillette 1991). They tend to occur preferentially in spring and fall, but they can occur at any time in the year. The Mono Lake source is being effectively mitigated, and efforts are underway to control Owens Lake, which, at the very least, will not be getting any worse. The particles are also coarser than smoke, typically around 5 µm diameter, as opposed to smoke at 0.3 µm, and thus get removed more readily during transport. Thus, interest in this source is not as keen as for forest smoke, which at least in some situations may be increasing in future years.

CONCLUSIONS

Air quality in the Sierra Nevada is, at times, as good as that found anywhere in the world, and, at times, as bad as that found anywhere in the world. Fortunately, good air quality is much more common than bad air quality, but the present impacts are important and future threats serious.

Changes in air quality from past values have accelerated with man's involvement, modestly in the period of native Americans, but more rapidly at present, threatening responses from the litho-, hydro-, and biospheres that may seriously alter the social and economic values of the Sierra Nevada to the state, nation, and world.

Changes in global climate are already occurring, with both positive and negative consequences of uncertain magnitude and timing. The most important is most likely to be a shift in the hydrological cycle towards intense rain events and away from historical snow patterns. Observed increases in temperature and carbon dioxide (CO_2), and predictions of increased moisture, could lead to increased bioproductivity. However, recent decreases in the worldwide rate of increase in CO_2 and methane (CH_4) may be harbingers of somewhat lower peak values in the 21st century than some models have predicted, and thus limiting changes. While there are many other subtle impacts on the biosphere, it appears that decreases in the northern latitude ozone shield are probably not responsible for the decline in Sierran amphibians. Other, more local, non-atmospheric causes are implicated.

The impacts on Sierran air quality from upwind sources of air pollution are dramatic and easily measurable, from the persistent hazes in summer to ozone damage to Jeffrey and ponderosa pines. The ozone damage is both serious and persistent, and poses both social and economic costs to the Sierra Nevada. Despite massive and costly efforts, the decline in peak

TABLE 48.6

Mean values of aerosols at Bliss State Park, Great Basin NP, and Grand Canyon NP.

	Bliss State Park, CA	Great Basin NP	Grand Canyon NP
KNON	8.77 ng/m³	6.70 ng/m³	7.83 ng/m³
Organic carbon (C)	1.12 µg/m³	0.79 µg/m³	0.63 µg/m³
Organic carbon (H)	1.17 µg/m³	1.01 µg/m³	0.87 µg/m³
Optical absorption	5.42 Mm⁻¹	4.18 Mm⁻¹	4.26 Mm⁻¹
Elemental carbon[a]	0.15 µg/m³ (?)	0.12 µg/m³ (?)	0.09 µg/m³ (?)

[a]The (?) indicate values close to the detectable limit and thus statistically weak.

ozone values in the Central Valley source regions is slow (unlike the dramatic decreases in the Los Angeles basin), so that relief is not imminent. The persistent hazes have been definitively linked to California sources, and great improvements in visibility could be achieved by a number of proven methods including suppression of summer and fall agricultural burning, further controls of sulfur emissions in the Bay Area, and increased efforts to reduce NO_x emissions, including nonvehicular sources. The hydrological cycle is dominated by winter snowfall, and the impacts of upwind sources of sulfates and nitrates on mean Sierran snow composition is modest and no acidified lakes and streams are found. That does not rule out pulses of moderate acidity at snow melt. It does reflect that winter storm processes do not have the same local connection to California emissions as summer aerosols and ozone.

The impacts on Sierran air quality from local sources are highly variable in magnitude and timing, resulting in major degradation of particulate air quality to levels among the worst in the state and nation superimposed on typical air quality that is so clean as to be the envy of the state and the nation. We consider three major areas: smoke from fires, influence of urbanized enclaves, and the desiccation of eastern Sierra lakes.

Smoke from major wildfires can be seen for hundreds of miles downwind of the Sierra Nevada, filling valleys (even on occasion the Central Valley) and clearly causing the most obvious and extensive air pollution impact from any local source. This is enhanced by the growing intensity of wildfires caused by fuel build-up over the past decades. Yet, perhaps surprisingly, the air pollution impacts of wildfires on state and federal fine particulate mass standards is generally not major for several reasons. First, these events are infrequent, so that they have only a modest impact on long-term averages. But perhaps more surprisingly, the maximum smoke impacts of major fires are generally less in magnitude, and far less in frequency, than smoke impacts in urbanized enclaves such as Mammoth Lakes, California, South Lake Tahoe, Truckee, and others. The situation is even more favorable for controlled burns designed to limit fuel loading for the major wildfires. First, there is a great deal of smoke in the Sierra Nevada range from the Central Valley. This is in fact more extensive than that developed by most controlled burns, partially through careful planning of burn periods and burning procedures. Thus, it is our opinion that limits on controlled burning could be significantly relaxed without danger to public health, and with major benefits to public welfare including increased human safety from reduced wildfire events.

The urbanized enclaves referred to above can generate local air pollution that mimics and even surpasses that present in major areas of California, but on a much more local spatial scale. Winter urban smoke can result in the highest winter particulate mass loading of any site in California. Yet we believe that using mass loading alone may be misleading, since there is growing evidence that the abundant water of combustion in low temperature burning of wood, especially pine wood, becomes trapped in the smoke in cold conditions and gives misleading values for mass that may not have equivalent health impacts to equal mass loading in other urban areas of California. The question of other pollutants, such as polyaromatic hydrocarbons (PAHs), is much more important to questions of potential health impacts of wood smoke. The impacts of smoke on local winter visibility are on occasions extreme.

Other influences from air pollution in urbanized enclaves include accelerated nutrient input to Lake Tahoe and other pure bodies of water, causing algal growth and lack of clarity. It is our opinion that atmospheric nitrates, a major and occasionally limiting nutrient, from transported, upwind sources, are not as important as local nitrate sources in Lake Tahoe, but this is still controversial.

Finally, the influence on local air pollution from the artificially desiccated beds of Mono and Owens lakes is severe, causing in most years the highest respirable dust loading in the entire United States, although for relatively few days per year. The recent Water Resources Control Board ruling (D-1631, 1994) on Mono Lake used this air quality information as a component in setting the lake level to a value that should make such events a thing of the past. No such near-term improvements are imminent at the even more severe problems at Owens (dry) Lake.

Returning to the very beginning of this report, one final conclusion must be proposed. Any future studies of air quality in the Sierra Nevada would be improved immeasurably by rectifying deficiencies in the air quality data set. These should be based on the importance of the ecological effects and impacts on the mountains, as well as human health considerations. For example, ozone transects from valley floor to high elevation should be routinely done at three or four sites (perhaps Visalia through Sequoia National Park, Merced through Yosemite, Sacramento through Lake Tahoe, and Chico east, plus Redding through Lassen Volcanic National Park in the Cascades) in order to document ozone dose for comparisons to ozone injury. These same sites might well allow for measurements of other valley pollutants, including herbicides and pesticides from agricultural operations. Much more information is needed on smoke from fires in the forest, especially the smoke from the historic/prescribed surface based fires proposed for increased use in fuel control. Some effort should be expended to study health effect in winter urban smoke episodes and blowing alkaline dust from Owens (dry) Lake. Ultraviolet measurements of all kinds are almost totally lacking. Other examples come to mind.

ACKNOWLEDGMENTS

Some of the conclusions in this report are based on data collected as part of various contracts funded by the U.S. Department of Interior / National Park Service (IMPROVE program), USDA Forest Service, California Department of Forestry, and the California Air Resources Board. We acknowledge the assistance of M. Corey, M. Brown, D. Dutcher, G. Torres and L. Ashbaugh in the preparation of this report.

REFERENCES

Barone, J.B., L. Ashbaugh, R. Eldred, and T. Cahill. 1979. Further investigation of air quality in the Lake Tahoe Basin. Final Report to the California Air Resources Board.

Beaton, S. P., et al. 1995. On-road vehicle emissions: Regulations, costs, and benefits. *Science* 268:991–93.

Blaustein, A. R., and D. B. Wake. 1995. The puzzle of declining amphibian populations. *Scientific American* 272 (4): 52–57.

Bohm, M., B. McCune, and T. Vandetta. 1995a. Ozone regimes in or near forests of the western United States. Part 2. Factors influencing regional patterns. *Journal of the Air and Waste Management Association* 45:477–89.

Bohm, M, B. McCune, T. Vandetta, and M. Flores. 1995b. Ozone regimes in or near forests of the western United States. *Journal of the Air and Waste Management Association* 45:235–46.

Cahill, T.A. 1991. Mitigation of Wind Blown dust from Owens (dry) Lake. Report to the California State Lands Commission.

Cahill, T.A., L. Ashbaugh, J. Barone. 1977. Sources of Visibility Degradation in the Lake Tahoe Air Basin. Air Resources Board contract #A-5-005-87.

Cahill, T. A., and T. E. Gill. 1988. Air quality at Mono Lake. Appendix D5. in *The future of Mono Lake*, edited by D. Botkin et al. Report 68. (Often referred to as the CORI report for the organization that organized the study as part under Senate Bill 270.) Davis: University of California, Water Resources Center.

Cahill, T. A., et al. 1989. *Monitoring of atmospheric particles and ozone in Sequoia National Park: 1985–1987*. Final report on contract A5-180-32. Sacramento: California Air Resources Board.

Cahill, T. A., et al. 1994. *Generation, characterization, and transport of Owens (dry) Lake dusts*. Final report on contract A132-105. Sacramento: California Air Resources Board.

Cahill, T. A., et al. In press. Saltating particles, playa crusts, and dust aerosols from Owens (Dry) Lake, California. *Earth Surface Processes and Landforms.*

California Acid Deposition Monitoring Program (CADMP). 1995. *Wet deposition data summary, July 1987 through June 1990*. Sacramento, CA: Air Resources Board, Research Division.

California Air Resources Board (CARB). 1972–94. *California air quality data*. Published quarterly. Sacramento: California Air Resources Board.

California Air Resources Board. 1991. *Ozone trends in California, 1981–1990*. Sacramento: California Air Resources Board.

California Department of Fish and Game. 1994. *California spotted owl draft environmental impact report*. Sacramento: California Department of Fish and Game.

Eldred, R. A., T. A. Cahill, M. Pitchford, and W. C. Malm. 1988, IMPROVE—A new remote area particulate monitoring system for visibility studies. *Proceedings of the Air Pollution Control Association 81st annual meeting*. Paper 88—54.3:1—16.

Farmer, A. M. 1993. The effects of dust on vegetation: A review. *Environmental Pollution* 79 (1): 63–75.

Gill, T. E. In press. Eolian sediments generated by anthropogenic disturbance of playas; Human impacts on the geomorphic system, geomorphic impacts on the human system. *Geomorphology.*

Gill, T. E., and D. L. Gillette. 1991. Owens Lake: A natural laboratory for aridification, playa desiccation, and desert dust. *Annual meeting of the Geological Society of America abstracts and programs* 23 (5): 426.

Goldman, C. R. 1994. *Annual report of the Tahoe Research Group*. Davis: University of California.

Goldman, C. R., and T. A. Cahill. 1975. Danger signs for Tahoe's future. *Cry California: The Journal of California Tomorrow* 10:30–35.

Great Basin Unified Air Pollution Control District (GBUAPCD). 1994. *Report: Owens Valley PM$_{10}$ planning area best available control measures state implementation plan (SIP)*. Bishop, CA: Great Basin Unified Air Pollution Control District.

———. 1995. Mono Basin planning area PM$_{10}$ state implementation plan. Draft report. Bishop, CA: Great Basin Unified Air Pollution Control District.

IMPROVE. 1995: *Data base and quarterly summary of interagency monitoring of protected visual environments (IMPROVE), 1988–1995*. Davis: University of California, Air Quality Group.

Jassby, A. D., J. Reuter, R. Axler, C. Goldman, and S. Hackley. 1994. Atmospheric deposition of nitrogen and phosphorus in the annual nutrient load of Lake Tahoe (California-Nevada). *Water Resources Research* 30 (7): 2207–16.

Jenkins, B., A. Jones, S. Turn, and R. Williams. 1995a. Emissions of polycyclic aromatic hydrocarbons (PAH) from biomass burning. Paper presented at the 209th annual meeting of the American Chemical Society, Anaheim, California, April.

Jenkins, B. M., et al. 1995b. *Atmospheric pollutant emission factors from open burning of agricultural and forest biomass by wind tunnel simulations*. Draft final report. CARB project A932-126. Sacramento: California Air Resources Board.

Jennings, M. R., and M. P. Hayes. 1994. *Amphibian and reptile species of special concern in California*. Report to the California Department of Fish and Game, contract # 8023. San Francisco: California Academy of Sciences, Department. of Herpetology.

Jones and Stokes Associates. 1993. *Draft environmental impact report for the review of Mono Basin water rights of the City of Los Angeles*. Prepared for California State Water Resources Control Board, Sacramento.

Knox, J. 1989. *Proceedings of the University of California task force on the effects of global change on California*. Berkeley: University of California.

Kusko, B. H., and T. A. Cahill. 1984. Study of particle episodes at Mono Lake. Unpublished final report on contract A9-147-31, California Air Resources Board, Sacramento.

Laird, L. B., H. E. Taylor, and V. C. Kennedy. 1986. Snow chemistry of the Cascade–Sierra Nevada Mountains. *Environmental Science and Technology* 20 (3): 275–90.

Larson, T. V., and J. Q. Koenig. 1994. Wood smoke: emissions and noncancer respiratory effects. *Annual Review of Public Health* 15:133–56.

Lave, L. B., C. T. Hendrickson, F. C. McMichael. 1995. Environmental implications of electric cars. *Science* 268:993–95.

Malm, W. C., J. F. Sisler, D. Huffman, R. A. Eldred, and T. A. Cahill. 1994. Spatial and seasonal trends in particle concentration and optical extinction in the United States. *Journal of Geophysical Research* 99 (D1): 1347–70.

Marchand, D. E. 1970. Soil contamination in the White Mountains, eastern California. *Geological Society of America Bulletin* 81 (8): 2497–505.

McKey, M. 1995. Unpublished U. S. Forest Service records on the Cleveland Fire.

Miller, P. 1996. Biological effects of air pollution in the Sierra Nevada. *Sierra Nevada Ecosystem Project: Final report to Congress,* vol. III. Davis: University of California, Centers for Water and Wildland Resources.

Molenar, J. V., D. Dietrich, D. Cismoski, T. Cahill, and P. Wakabayashi. 1996. Aerosols and visibility at Lake Tahoe. Submitted for publication. *Atmospheric Environment.*

National Acid Precipitation Assessment Program (NAPAP). 1991. *Acidic deposition: State of science and technology, summary report.* Edited by P. Irving.

National Oceanic and Atmospheric Administration (NOAA). 1995: *Annual report of the Climate Monitoring and Diagnostic Laboratory.* Boulder, CO: National Oceanic and Atmospheric Administration.

Phillips, J. 1995. The crisis in our forests. *Sunset,* July, 87–92.

Pitchford et al. 1984. *Report on the RESOLVE project in the Mohave Desert.* Washington, DC: U.S. Environmental Protection Agency.

Radke, L. F., et al. 1990. *Airborne monitoring and smoke characterization of prescribed fires on forest lands in Western Washington and Oregon: Final report.* General Technical Report PNW-GTR-251. Portland, OR: U.S. Forest Service, Pacific Northwest Research Station.

Richmond, T. 1994. Report on PM_{10} modelling of dust events at Mono Lake, CA. Submitted to the Water Resources Control Board. Great Basin Unified Air Pollution Control District, Bishop, California.

Scheuring, A. F. 1983. *A guidebook to California agriculture.* Berkeley and Los Angeles: University of California Press.

Sehmel, G. A. 1980. Particle and gas dry deposition. *Atmospheric Environment* 14:983–1011.

Thompson, K. 1972. The notion of air purity in early California. *Southern California Quarterly* 54 (3): 203–10.

Van Ooy, D. J., and J. J. Carroll. 1995. The spatial variation of ozone climatology on the western slope of the Sierra Nevada. *Atmospheric Environment* 29 (11): 1319–30.

WESTAR Council. 1995. Preliminary notes, Wildfire/Prescribed Fire Workshop. Edited by G. W. Gause. San Francisco, 27–29 November.

WESTEC, Inc. 1984. Dust mitigation at Owens (dry) Lake. Final Report to the California State Land Commission.

Air Quality Standards and Monitoring Stations

SUMMARY OF AMBIENT AIR QUALITY STANDARDS ESPECIALLY RELEVANT TO THE SIERRA NEVADA

Species	Averaging Period	California	Federal, Primary	Comment
Ozone	1 hour	0.09 ppm	0.12 ppm	
Carbon Monoxide	8 hour	9.0 ppm	9.0 ppm	Lake Tahoe, 6.0 ppm
Carbon Monoxide	1 hour	20.0 ppm	35.0 ppm	
Nitrogen Dioxide	1 hour	0.25 ppm	0.053 ppm (annual average)	
Suspended Particulate Matter (PM-10)	Annual	30 $\mu g/m^3$	50 $\mu g/m^3$	
Suspended Particulate Matter (PM-10)	24 hour	50 $\mu g/m^3$	150 $\mu g/m^3$	
Visibility Reducing miles	8 hour (day)	~10 miles		Lake Tahoe, ~35

(also sulfur dioxide, lead (30 day average, 2.5 $\mu g/m^3$), hydrogen sulfide, and vinyl chloride)

AIR MONITORING STATIONS

Air monitoring stations in the Sierra Nevada Ecosystem Project (SNEP) study region, the southern Cascade Mountains, and San Bernardino Mountains, in operation for all or part of 1993.

Site Name	California		Federal		Other	Comments
	Gases	Particles PM$_{10}$	Particles, PM$_{10}$	Particles, PM$_{2.5}$		
Cascade Mountains						
Burney, Shasta County	Yes	Yes				closed 3/93
Lassen Volcanic National Park	Yes	No			Yes, IMPROVE	
SNEP Region						
Chester, Plumas County	No	Yes				
Quincy, Plumas County	Yes	Yes				
Graeagle, Plumas County	No	Yes				closed 9/93
Loyalton, Sierra County	No	Yes				
Nevada City, Nevada County	Yes	Yes				closed 6/93
Grass Valley, Nevada County	Yes	Yes				4 sites
Truckee, Nevada County		Yes	Yes			2 sites
Colfax, Placer County	Yes	Yes				
Lake Tahoe, Placer/El Dorado County	Yes	Yes			Yes, TRPA	4 sites
Placerville, El Dorado County	Yes	Yes				
Jackson, Amador County	Yes	No				
Sonora, Tuolumne County			Yes	No		
Yosemite, Camp Mather, Tuolumne County			Yes	No		
Yosemite National Park	Yes	Yes			Yes, IMPROVE	3 sites
Mono Lake, Mono County			No	Yes		2 sites
Mammoth Lakes, Mono County	Yes	Yes				
Wilsonia, Tulare County	Yes	No				
Sequoia National Park	Yes	No			Yes, IMPROVE	2 sites
San Bernardino Mountains						
Lake Gregory, San Bernardino County	Yes	Yes				
San Gorgonio Wilderness	No	No			Yes, IMPROVE	

RICHARD KATTELMANN
University of California Sierra Nevada
  Aquatic Research Lab
Mammoth Lakes, California

49

Impacts of Floods and Avalanches

ABSTRACT

Floods in the Sierra Nevada are produced by snowmelt, winter rainfall and rain-on-snow events, summer thunderstorms, and failure of impoundments. Floods routinely modify channel conditions and therefore affect aquatic and riparian communities. Riparian vegetation has a variety of interactions with peak flows and sediment transport. Floods function as a disturbance mechanism primarily as they damage or remove riparian plants and alter riparian habitat. Land management or disturbance alters flood processes mainly if changes in land cover are dramatic and extend over a large fraction of a river basin.

Avalanches are a natural process that occasionally alters forests at higher elevations. The location of a forest stand with respect to avalanche-prone terrain is the primary risk factor. Stands in vulnerable locations are subject to destruction on an irregular basis. Weather and snow conditions determine the timing and extent of damage. When trees located in potential avalanche-starting zones die off because of fire or disease, avalanche activity may be enhanced, with downslope forests subject to damage. Humans alter avalanche size and frequency in a few limited locations above highways and ski areas for safety reasons. However, there is little that humans can or should do about the forest alterations caused by avalanches. In the context of the Sierra Nevada Ecosystem Project, avalanches are a forest influence that must be considered, but they are not a management or policy concern.

FLOODS

Floods are merely events of higher than average stream flow in response to storms or other large inputs of water. Generally, high-flow events that rise above stream banks are the phenomena of concern. Considering the overwhelming role of snow in the hydrologic cycle of the Sierra Nevada, snowmelt floods are the most obvious source of peak flows. Snowmelt floods are an annual event each spring of sustained high flow, long duration, and large volume. However, they usually do not produce the highest instantaneous peaks. The Sierra Nevada snowpack at the maximum of winter accumulation represents an enormous reservoir of potential runoff. The sustained input of water into reservoirs and canals can overwhelm storage and conveyance capabilities and can cause substantial leakage through levees (Dean 1975). Particularly large snowmelt floods in Sierra Nevada rivers have been documented in 1906, 1938, 1952, 1969, 1983, and 1995. Although their peak discharges were generally less than twice the mean annual snowmelt flood and only one-tenth to one-half as great as the largest rain-on-snow floods, their total volumes were two to four times larger than average. In all cases, snow deposition was more than twice average amounts and persisted into April or May. Thus, snow cover was still extensive in late spring when energy available for melt was much greater than in early spring (Kattelmann 1990). There was substantial potential for serious snowmelt floods in 1995 with snow water equivalence almost twice average amounts at many sites. However, cloudy conditions during the spring and early summer limited the rate of snowmelt runoff generation so that instantaneous peaks were not exceptional. Nevertheless, the duration of moderately high water and the total volume of runoff were extraordinary.

Midwinter rainfall on snow cover has produced all the highest flows in major Sierra Nevada rivers during this century. In the past sixty years, six floods of large magnitude have occurred in almost all rivers draining the snow zone. Rainfall has occurred up to the highest elevations of the Sierra Nevada during winter, but the freezing level of winter storms generally fluctuates between about 1,000 m (3,300 ft) and 2,500 m (8,200 ft). Even during the warmest storms, snowpacks

Sierra Nevada Ecosystem Project: Final report to Congress, vol. II, *Assessments and scientific basis for management options.* Davis: University of California, Centers for Water and Wildland Resources, 1996.

above 2,500 m (8,200 ft) rarely melt much because temperatures are close to freezing. The interaction of precipitation amount, freezing level, energy availability, and basin characteristics determines the relative response of rivers at different elevation zones. Large-magnitude warm storms do not seem to occur during spring in the Sierra Nevada. There are only a few moderate rain-on-snow events superimposed on spring snowmelt floods in the stream-flow record. Storms in April and May generally do not incorporate the warm air masses from low latitudes that lead to the warm storms that occasionally occur in the winter months (Kattelmann et al. 1991). In basins that are largely above 2,000 m (6,600 ft), the highest peaks also tend to be caused by rain-on-snow events, even though almost all the other floods in the annual series are of snowmelt origin.

Although the summer and early autumn seasons in the Sierra Nevada tend to be dry, a few minor storms or brief showers occur in most years (Hannaford and Williams 1967). In general, summer rainfall is much less of a flooding concern in the Sierra Nevada than in the Rocky Mountains (e.g., Jarrett and Costa 1982). However, subtropical storms occasionally move into the southern Sierra Nevada in late summer. Intense thundershowers occurring over a period of three or four days can generate local flooding, cause extensive surface erosion, and destabilize hill slopes. These storms may generate the greatest floods in some alpine basins that are sufficiently high to avoid midwinter rain-on-snow events and are oriented so that snowmelt rates are kept low because of northern exposure over much of the basin. In August 1989, a flood and debris flow generated by a thunderstorm in the 2,000 to 3,000 m (6,600 to 9,800 ft) headwaters of Olancha Creek in the southeast part of the Sierra Nevada damaged the Los Angeles Aqueduct several kilometers downstream at 1,200 m (3,900 ft).

The sudden release of water from storage generates the most extreme floods (Costa and O'Connor 1995) but occurs under a limited set of conditions in a small fraction of the Sierra Nevada. Although this type of flooding is localized, it may produce flood peaks that are at least several times greater than those caused by any other process, and it is likely to entrain large quantities of bed and bank material. Sierra Nevada lakes tend to be stable, with little risk of failure of their impoundments of bedrock or broad moraines. Failures of artificial dams were almost common during the hydraulic mining era. Recent dam failures in the Sierra Nevada include Hell Hole Dam on the Rubicon River in December 1964 (Scott and Gravlee 1968), North Lake Dam on a tributary to Bishop Creek in September 1982, and the coffer dam near Auburn on the American River in February 1986. A gigantic gate on Folsom Dam broke in July 1995 and allowed a large volume of water to be released but did not produce a high flood wave. The failure of landslide and snow-avalanche dams that temporarily impound streams undoubtedly occurs at a variety of scales in the Sierra Nevada, but large events of this type are not known to have been documented. Displacement of lake water by snow avalanches is yet another flood generation

process in high-elevation streams of the Sierra Nevada (Kattelmann 1990, 1992). The impact of an avalanche on the ice cover of a lake can force large volumes of water into the outlet channel and affect aquatic organisms (Williams et al. 1993). These events may be relatively common and are the only means (other than earthquakes) of generating high flow immediately below lakes, which otherwise tend to attenuate floods.

These various flood-generation mechanisms modify stream channels to various extents. Although debate continues about the relative effectiveness of common events (e.g., annual snowmelt floods) versus catastrophic events (e.g., rain-on-snow events) in shaping the landscape (e.g., Wolman and Gerson 1978; Beven 1981; Costa and O'Connor 1995), large floods would seem to be particularly important in mountain streams because of the high proportion of material transported as bedload. In mountain rivers, rare high-magnitude floods are generally required to significantly alter the channel because material composing the bed and banks tends to be large and resistant to entrainment (Lisle 1987). However, the sequence of events of different magnitudes also determines the geomorphic effectiveness of particular floods (Beven 1981). Large floods that destabilize a channel can lead to enhanced sediment transport from low-magnitude events over several decades (Lisle 1987). Such effects have been documented in the Lake Tahoe Basin following extreme rain-on-snow or thunderstorm events (Nolan and Hill 1987; Glancy 1988). Similarly, two large rain-on-snow events in 1982 may have created channel conditions favorable for the high bedload transport measured in the snowmelt flood of 1983 (Andrews and Erman 1986). These interactions of different flood processes may be a critical influence on channel form and sediment transport in the Sierra Nevada.

Changes in channels as a result of floods have major impacts on aquatic and riparian communities (e.g., Swanson et al. 1982; Erman et al. 1988; Lisle 1989). Floods both flush fine sediments out of spawning gravels and deposit these fine sediments elsewhere depending on hydraulic factors and sediment supply. Riparian vegetation protects banks against erosion and aids in bank construction by enhancing deposition of sediment during overbank flows. Floods are often required for dispersal of propagules of riparian plants. However, shear stresses imposed by high flows often destroy riparian vegetation directly. Erosion of stream banks and excessive deposition of sediment also kill riparian plants. In meadows where the sod has been cut by vehicles or cattle, high flows can erode deep gullies, which consequently lower the local ground-water table and completely change plant composition. Catastrophic floods can initiate landslides directly above the channel that remove upland vegetation. These various actions cast floods in the role of a disturbance mechanism with regard to terrestrial communities. Fortunately, riparian vegetation tends to become reestablished quickly if adequate soil water is available (e.g., Gregory et al. 1991). Riparian vegetation tends to survive routine flooding (magnitudes that occur up to once in five or ten years on the average). Rare, high-magnitude events have the potential to alter large portions of riparian communities.

Modest changes in forest cover tend to have little measurable effect on flood generation (Hewlett 1982). Where trees are harvested, transpiration is reduced, and there is less soil moisture deficit that could otherwise store potential runoff from storms. Therefore, streams can rise more quickly and receive greater volumes of water in areas devoid of trees than in forests. However, such effects tend to be local under conventional forest practices. If forest vegetation is converted (long-term change with little opportunity to recover to its original state) to sparse and/or shallow-rooted vegetation (or pavement at the extreme) over a large fraction of a watershed, then there is potential for greater effects. Increases in peak flows resulting from forest harvesting tend to be most noticeable in the early part of the rainy season and during small storms. During major storms, almost all available soil moisture storage is filled under all vegetation types, and rates of runoff production from all lands become similar (Hewlett 1982). In larger rivers, floods are a product of water volumes received from tributaries. The synchronization of incoming flows determines the flood level, and the original runoff generation processes on the landscape become irrelevant. There is no evidence to suggest that peak flows in larger streams have changed in the Sierra Nevada as a result of forest management activities. Although such increases may have occurred, we lack the data to demonstrate a change. In creeks influenced by urban and suburban development, flood magnitudes have probably increased as a result of increases in impermeable surface, but flow records have not been located to quantify such impacts. Channels generally increase their cross-sectional area to accommodate persistent increases in flood size such as can be expected following urbanization (Dunne and Leopold 1978). Water management activities, particularly the construction of large dams, have dramatically reduced flood magnitudes throughout the range.

If forests are replaced by shallow-rooted vegetation over a large proportion of a basin, then floods can be markedly increased. The greatest danger of such a widespread change would be from catastrophic fire. Intense fires can also create hydrophobic layers within the soil, which dramatically increase runoff (Anderson et al. 1976). In the snow zone, widespread reductions in forest density and/or forested area would tend to increase the local rate of snowmelt and advance the local timing of snowmelt runoff. The effect of such changes on spring peaks in stream flow would depend on the relative timing and synchronization of tributary peaks under present conditions (Anderson 1963). In smaller basins within the forested zone, the current slow rate of snowmelt runoff from forested areas tends to spread the seasonal hydrograph over several weeks. Changing forest to clearings would compress the snowmelt season, and, if enough area were cleared, flood peaks could be expected to increase. In larger basins, the earlier melting of snow in the former forest might lower water levels during late spring when the alpine snowmelt contributions would be at a maximum. The snowmelt runoff regime of the Sierra Nevada could be further affected by interactions of changes in both vegetation and climate.

Floods are commonly described in terms of their magnitude and frequency—how big they are and how often a flood of a particular size has been observed or is likely to occur. Flood magnitude at a particular point is expressed as discharge—volume of water over a time interval (usually cubic feet per second)—or as stage—height of the river surface on a fixed rule (or distance below a bridge). Estimating the frequency of a flood with a particular magnitude depends on availability of records of floods over time. If we have 100 years of recorded stream flow at a point on a river, we can identify the ten largest floods in that century, for example. The magnitude of the smallest of those ten floods was exceeded ten times in that century, or once every ten years on the average. We can call the flood of that size a "ten-year flood" and expect that a flood at least that big has a 1 in 10 chance of occurring in any particular year. The likelihood of floods of other sizes can be estimated in a similar way. Some relatively simple statistical procedures are used to refine these estimates, especially for rare floods, for which the observed record is generally too short. Flood frequency must be considered over a long time span. In general, floods should be considered as independent of one another. The qualifying phrase *on the average* is critical. Floods exceeding the ten-year level could occur twice in the same year or perhaps be thirty-five years apart, but 1,000 of them would be expected in 10,000 years.

Floods become natural hazards when they interact with people and our structures and activities. These natural hazards occur on floodplains. The hazard posed by floods could be avoided entirely by avoiding floodplains during floods. Floodplains are essentially parts of rivers that are occupied by flowing water on the occasions of floods. If our society incorporated that definition in our collective development plans, we would experience much less trouble during those occasions. An individual considering construction of a house "on a floodplain" would be much less likely to want to build a house "in a river." However, modern society has often ignored such considerations and extensively developed floodplains instead. In years like 1995, we are reminded that some people have built in a river. In 1995, 53 of California's 58 counties qualified for "disaster assistance" after the floods of January and March. Communities within the Sierra Nevada did not sustain damage comparable to those in the Sacramento area or the Coast Range, but several were threatened by high water. All lands adjacent to streams that have been inundated before are at risk in the future.

Floodplain occupants have long set public policies that effectively subsidize that occupancy. Structural attempts at flood control such as dams and levees and broad financial compensation for flood damages are generally paid for by the vast majority of taxpayers who do not live anywhere near a stream.

Development on floodplains creates political pressure to build more flood control reservoirs upstream at the expense of nature and the nation. The new flood-control structures inundate additional river channels, riparian corridors, and deep canyons. Despite its enormous environmental and financial costs, the Auburn Dam was again being promoted as a means of protecting occupants of the American River floodplain after the storms of 1995 raised concerns. Nevertheless, after the 1993 floods on the Mississippi and the 1995 floods in California, there are signs that flood policies may be changing. Agencies at various levels of government are beginning to purchase land on floodplains as a cheaper alternative to paying for recurring damages or giant new dams. Such expenditures were rarely questioned until society began to view flooding as a natural, normal process of rivers that occurs in locations that are easy to recognize and can be avoided instead of a disaster that strikes the unlucky. Ideally, any construction on floodplains must be designed with the risk of flood damage clearly in mind—design the structure to avoid or withstand floods of a particular magnitude, anticipate and accept the eventuality of damage or loss, or relocate upslope. Often the risk of damage involves more than just damage to the structure itself. Failure of inadequate culverts or bridges can lead to massive amounts of bank erosion. Toxic chemicals stored in structures on the floodplain can be released into the stream. Pieces of structures destroyed by the flood and transported downstream can damage other structures and vegetation. It is to be hoped that the floods of 1995 will provide incentives for individuals, communities, and agencies to begin some real floodplain management.

AVALANCHES

Snow and avalanches are important influences on forests of the Sierra Nevada. In the forested snowpack zone (above about 1,500 m [4,900 ft]), snow insulates the soil against freezing and extends availability of soil water for weeks beyond the winter precipitation season. Snow is also responsible for mechanical damage to trees by overloading branches intercepting snowfall, trimming off limbs caught within the snowpack as it settles, bending and breaking trees as the snowpack slowly creeps (deforms) or glides downslope, and snapping limbs and trunks during avalanches (Salm 1979; Wakabayashi 1979).

Avalanches may be defined as rapid downslope movements of snow. They can range from a few snow grains rolling a few centimeters to immense volumes of snow falling thousands of meters with tremendous impact pressures. After snow crystals precipitate from the sky, they tend to lose their complex shapes and become semirounded grains, bonding with other grains in the process. Snow layers from individual storms constitute a snowpack that evolves over time. Grains within

a storm layer tend to form stronger bonds with one another than those at the interface between layers, thus forming a cohesive slab of snow that acts as a structural unit with possibly poor bonding to the underlying layer. The force of gravity imposes mechanical stresses within a snowpack. If these stresses exceed strength at some point, local failure occurs. After such a failure, adjacent areas receive additional stress and are either strong enough to withstand it or fail in turn, possibly leading to propagation of the failure over a large area and release of an avalanche. The balance between stress and strength within the snowpack is extremely complex. In general, snowpack strength increases as bonds grow between grains, reducing the risk of an avalanche soon after a storm ends (Perla and Martinelli 1976). Avalanches can also occur when strength decreases in the presence of liquid water. When high-elevation snowpacks initially get wet in spring, bonds between layers can weaken and avalanches occur without an increase in stress (Kattelmann 1985). Rainfall on a snowpack can both increase stress by adding weight and decrease strength. Wet snow avalanches can occur on shallower slopes than dry snow avalanches.

In the Sierra Nevada, the vast majority of avalanches occur during and shortly after storms. If loading of new snow increases stress at a rate faster than strength develops, the slope will fail. Intense snowfall (greater than 25 mm [1 in] per hour) and high winds redepositing snow increase the load much faster than typical storms, so avalanche activity is enhanced during severe storms. Critical stresses develop more quickly on steeper slopes and where deposition of wind-transported snow is common. Consequently, certain slopes are prone to avalanching during almost every storm while most terrain simply is not steep enough or never accumulates enough snow to fail. Between these extremes is a continuum of terrain conditions that require increasingly severe (and rare) weather and snow conditions to produce an avalanche. Therefore, while some avalanche paths consistently run several times each winter, other areas may slide only under unusual sets of conditions that occur perhaps once in a hundred or a thousand years. Such extreme situations produce massive avalanches over much of the Sierra Nevada.

Our short and geographically limited records (and even shorter memories) of weather in the mountains provide little basis for anticipating the potential of major avalanche cycles. For example, storms in late March 1982 resulted in a very destructive and tragic avalanche cycle. Press accounts called it the "storm of the century" even though greater snowfall quantities for various time intervals had occurred at least four times in the previous two decades (Stetham 1992; Osterhuber 1993). Less than four years later, precipitation totals for one, two, three, and four days during a series of severe storms were more than 1.7 times greater than previous records. Nevertheless, avalanches in the Alpine Meadows area near Lake Tahoe did not even approach the size of those generated in 1982, while elsewhere in the Sierra Nevada, damage was extreme (Wilson 1986). The winters of 1993 and 1995 left exceedingly deep snowpacks but did not produce catastrophic avalanches.

Avalanches occur throughout the snow zone of the Sierra Nevada but become more common with increasing elevation and steeper slopes. Because prevailing wind direction during storms is from the southwest, snow is scoured from south- and southwest-facing slopes and redeposited on north- and northeast-facing slopes, with consequent differences in avalanche occurrence. The greater solar radiation input and higher temperatures on south-facing slopes tend to stabilize those slopes faster than shaded slopes. The influence of avalanches on forests generally increases with increasing elevation up to local timberline. Starting zones of avalanches (places where avalanches begin) are usually above timberline, but the slides continue into the trees below. Red fir and lodgepole pine forests are probably impacted the most, but some avalanche paths extend into the upper mixed conifer zone. Subalpine trees such as mountain hemlock, foxtail pine, and whitebark pine generally occur outside avalanche paths, which run more frequently at higher elevations and do not allow the trees to become established. Avalanche paths in the Sierra Nevada have been mapped only in existing and proposed ski areas, highway and rail corridors, and mountain communities where the forest, terrain, and avalanche hazard might be managed. For example, avalanche path mapping identified forty-nine paths in a 21 km^2 (8 mi^2) area of the Galena Basin, the site of a proposed ski area northeast of Lake Tahoe (Frutiger 1990).

Avalanches can be a dominant influence on plant community structure and create a fragmented vegetation mosaic (Patten and Knight 1994). In the forest zone, avalanche paths are easily recognized as strips oriented straight down the hill containing a different age or type of vegetation than that adjacent to the strip (Martinelli 1974; Mears 1992). These vertical paths through the forest are particularly obvious when the strips are devoid of vegetation or contain deciduous trees. Aspen and other fast-growing, light-tolerant trees often colonize avalanche tracks. A series of avalanches may progressively force a path through a forest stand. After a clear path is established, a major avalanche can break through and continue into a previously untouched forest (Perla and Martinelli 1976). A thick jumble of debris can remain in the runout zone for decades if undisturbed. Such debris could potentially influence other disturbance factors such as fire, insects, or disease. Conversely, fire and insect kill can allow avalanche paths to develop that would not had the forest remained alive (Fohn 1979).

Vegetation in the avalanche path can be used to infer the size and frequency of avalanches (Perla and Martinelli 1976; Wakabayashi 1979; Mears 1992). If an avalanche occurs at least every decade, its path will be free of trees or include a few large individuals with obvious damage. Shrubs and flexible trees up to a couple of meters in height may be present. Where an avalanche has not occurred for up to thirty years, aspen and small conifers may occupy the path. Larger conifers of

uniform age but younger than the adjacent forest may be found where avalanches have not occurred for several decades. Branches of the older trees along the borders of the path are usually missing. If paths above timberline avalanche infrequently, a forest can recover between major avalanches. These extreme events can occur on a timescale similar to the growth of a mature stand (deQuervain 1979). A single avalanche in Switzerland in 1962 destroyed about 100 ha (250 acres) of mature forest (Fohn 1979). Loss of productivity of forest land is considered an economic cost of avalanches (Voight 1990). However, little commercial forest is known to be impacted by avalanches in the Sierra Nevada because the most productive forests are generally found in lower-elevation terrain not particularly prone to avalanches. At higher elevations, several hundred hectares of forest were destroyed by avalanches in 1986 (Wilson 1986). A large proportion of these trees were 125–150 years old. Some trees destroyed near Sonora Pass were 350 years old.

Forests offer a substantial protection role with respect to avalanche hazard. This function was formally recognized in 1876 when Switzerland enacted a forest protection law to maintain forests above inhabited areas and to reforest places that might provide protection from avalanches (Armstrong and Williams 1986). Forests influence snow in a variety of ways. Canopies intercept and retain snowfall. Some of this snow sublimates, some melts and drips into the snowpack below, and some just falls off as clumps that are often wet. Besides reducing the amount of snow compared with adjacent open areas, interception ultimately leads to strengthening the snowpack around the tree. The drip and snow falling from branches form a rim around the vertical projection of the crown, which significantly increases the overall strength of the snow in the forest compared to the stems alone (Gubler and Rychetnik 1991). Forests also tend to disturb stratification of the snowpack, break up weak layers, increase density, and minimize surface hoar (which can become an extremely weak layer within the snowpack if buried by snowfall). Under extreme conditions, avalanches can start in openings within the forest as small as 30 m (100 ft) long and 15 m (50 ft) wide (Gubler and Rychetnik 1991).

Avalanches also produce a variety of geomorphic effects, such as scouring of vegetation and soils from hill slopes, maintenance of vertical troughs, accumulation of debris in the runout zone, and creation of impact and scour pits (Davis 1962). When avalanches dam streams, serious floods and channel damage may occur following eventual failure of the snow dam (Perla and Martinelli 1976; McClung and Schaerer 1993). Avalanches also generate floods by suddenly displacing water from lakes. Avalanches can even affect fisheries. For example, formation of a plunge pool in a lake by avalanches provided a high-quality spawning area for brook trout by transporting gravel and removing floculent organic matter from the hatching area (Williams et al. 1992).

Avalanches are defined as a hazard when they influence human activities. In the Sierra Nevada, avalanches did not have as great an effect on nineteenth-century mining as in the Wasatch and the Rockies, where avalanche tragedies were common (Armstrong and Williams 1986). Until after World War II, very few people occupied the higher portions of the Sierra Nevada in winter. Rapid growth of winter recreation put many people at risk. Ski areas remain the principal foci of avalanche hazard where the steepest runs are avalanche paths that require artificial control. Control usually implies an explosion to trigger release of an avalanche at a time when the path is empty. The force of an explosion usually propagates only a short distance and ruptures critical bonds under a slab. As more roads were maintained in winter, more travelers were exposed to avalanches. Avalanche paths cross Highways 4, 50, 80, 88, 89, 158, and 395 as well as many local roads in mountain communities. Highway closure during periods of avalanche danger is a major indirect cost of avalanches (Voight 1990). Control via hand charges, artillery, and, recently, propane-fueled exploders (GazEx) on California Highways 50, 88, and 158 and Nevada Highway 431 allows roads to be open sooner during storms. Rapid expansion of mountain communities led to construction of vacation and year-round residences in avalanche paths. In recent years, homes and other structures have been damaged or destroyed at Virginia Lakes, Twin Lakes, and Long Valley, and near Tahoe City. A fatal avalanche occurred within a residential area of Mammoth Lakes in 1993. The large avalanche cycles of 1982 and 1986 led several Sierra Nevada counties to consider zoning and other land-use restrictions to reduce avalanche hazards (Penniman 1992). However, property owners and real-estate interests vigorously fought such restrictions, and the counties concerned have settled on some form of a "fair warning" to owners and renters within avalanche zones. Placer County requires that new construction in avalanche areas be designed to resist avalanche forces (Placer County 1994). Washoe County has ignored the recommendations of its consultants and has taken no action, raising liability concerns when damage eventually occurs (Penniman 1992). The lack of agreement between avalanche consultants in defining hazards on the ground has impeded avalanche zoning efforts (Penniman 1992) and is another example of the perils of scientific uncertainty in developing public policy.

REFERENCES

Anderson, H. W. 1963. *Managing California snow zone lands for water.* Research Paper PSW-6. Berkeley, CA: U.S. Forest Service, Pacific Southwest Forest and Range Experiment Station.

Anderson, H. W., M. D. Hoover, and K. G. Reinhart. 1976. *Forests and water: Effects of forest management on floods, sedimentation, and water supply.* General Technical Report PSW-18. Berkeley, CA: U.S. Forest Service, Pacific Southwest Forest and Range Experiment Station.

Andrews, E. D., and Erman, D. C. 1986. Persistence in the size distribution of surficial bed material during an extreme snowmelt flood. *Water Resources Research* 22 (2): 191–97.

Armstrong, B. R., and K. Williams. 1986. *The avalanche book.* Golden, CO: Fulcrum.

Beven, K. 1981. The effect of ordering on the geomorphic effectiveness of hydrologic events. In *Erosion and sediment transport in Pacific Rim steeplands*, 510–25. Publication 132. Wallingford, England: International Association of Hydrological Sciences.

Costa, J. E., and J. E. O'Connor. 1995. Geomorphically effective floods. In *Natural and anthropogenic influences in fluvial geomorphology, the Wolman volume*, edited by J. E. Costa, A. J. Miller, K. W. Potter, and P. R. Wilcock, 45–56. Geophysical Monograph 89. Washington, DC: American Geophysical Union.

Davis, G. 1962. *Erosional features of snow avalanches, Middle Fork Kings River, California.* Professional Paper 450-D. Washington, DC: U.S. Geological Survey.

Dean, W. W. 1975. Snowmelt floods of April–July 1969 in the Buena Vista Lake, Tulare Lake, and San Joaquin River Basins. In *Summary of floods in the United States during 1969*, 77–87. Water-Supply Paper 2030. Washington, DC: U.S. Geological Survey.

deQuervain, M. 1979. Wald und lawinen. In *Mountain forests and avalanches*, 219–39. N.p.: International Union of Forestry Research Organizations.

Dunne, T., and L. B. Leopold. 1978. *Water in environmental planning.* San Francisco: W. H. Freeman.

Erman, D. C., E. D. Andrews, and M. Yoder-Williams. 1988. Effects of winter floods on fishes in the Sierra Nevada. *Canadian Journal of Fisheries and Aquatic Sciences* 45:2195–2200.

Fohn, P. M. B. 1979. Avalanche frequency and risk estimation in forest sites. In *Mountain forests and avalanches*, 241–54. N.p.: International Union of Forestry Research Organizations.

Frutiger, H. 1990. Maximum avalanche runout mapping: A case study from the central Sierra Nevada. In *Proceedings of the International Snow Science Workshop, Bigfork*, 245–51.

Glancy, P. A. 1988. *Streamflow, sediment transport, and nutrient transport at Incline Village, Lake Tahoe, Nevada, 1970–73.* Water-Supply Paper 2313. Washington, DC: U.S. Geological Survey.

Gregory, S. V., F. J. Swanson, W. A. McKee, and K. W. Cummins. 1991. An ecosystem perspective on riparian zones. *BioScience* 41: 540–51.

Gubler, H., and J. Rychetnik. 1991. Effects of forests near the timberline on avalanche formation. In *Snow, hydrology, and forests in high alpine areas*, edited by H. Bergmann, H. Lang, W. Frey, D. Issler, and B. Salm, 19–38. Publication 205. Wallingford, England: International Association of Hydrological Sciences.

Hannaford, J. F., and M. C. Williams. 1967. Summer hydrology of the high Sierra. *Proceedings of the Western Snow Conference* 35:73–84.

Hewlett, J. D. 1982. Forests and floods in the light of recent investigation. In *Proceedings of the Canadian Hydrology Symposium*, 545–60. Ottawa: Natural Research Council of Canada.

Jarrett, R. D., and J. E. Costa. 1982. Multidisciplinary approach to the flood hydrology of foothill streams in Colorado. In *International Symposium on Hydrometeorology*, 565–69. Bethesda, MD: American Water Resources Association.

Kattelmann, R. 1985. Wet slab instability. In *Proceedings of the International Snow Science Workshop, Aspen*, 102–8.

———. 1990. Floods in the high Sierra Nevada, California, USA. In *Hydrology in mountainous areas. II—Artificial reservoirs; water and slopes*, edited by R. O. Sinniger and M. Monbaron, 311–17. Publication 194. Wallingford, England: International Association of Hydrological Sciences.

———. 1992. Historical floods in the eastern Sierra Nevada. In *The history of water in the eastern Sierra Nevada, Owens Valley, and White Mountains*, edited by C. A. Hall, V. Doyle-Jones, and B. Widawski, 74–86. Berkeley and Los Angeles: University of California Press.

Kattelmann, R., N. Berg, and B. McGurk. 1991. A history of rain-on-snow floods in the Sierra Nevada. *Proceedings of the Western Snow Conference* 59:138–41.

Lisle, T. E. 1987. Overview: Channel morphology and sediment transport in steepland streams. In *Erosion and sedimentation in the Pacific Rim*, 287–97. Publication 165. Wallingford, England: International Association of Hydrological Sciences.

———. 1989. Sediment transport and resulting deposition in spawning gravels, north coastal California. *Water Resources Research* 25 (6): 1303–19.

Martinelli, M., Jr. 1974. *Snow avalanche sites, their identification and evaluation.* Agriculture Information Bulletin 360. Washington, DC: U.S. Forest Service

McClung, D., and P. Schaerer. 1993. *Avalanche handbook.* Seattle: The Mountaineers.

Mears, A. I. 1992. *Snow-avalanche hazard analysis for land-use planning and engineering.* Bulletin 49. Denver: Colorado Geological Survey.

Nolan, K. M., and B. R. Hill. 1987. Sediment budget and storm effects in a drainage basin tributary to Lake Tahoe. *Eos, Transactions American Geophysical Union* 68 (16): 305.

Osterhuber, R. S. 1993. *Climatic summary of Donner Summit, California.* Soda Springs, CA: U.S. Forest Service.

Patten, R. S., and D. H. Knight. 1994. Snow avalanches and vegetation pattern in Cascade Canyon, Grand Teton National Park, Wyoming, U.S.A. *Arctic and Alpine Research* 26 (1): 35–41.

Penniman, D. 1992. The political dilemma of avalanche hazard zoning: A comparative analysis of four Sierra Nevada counties. In *Proceedings of the International Snow Science Workshop, Breckenridge*, 236–45. Denver: Colorado Avalanche Information Center.

Perla, R., and M. Martinelli Jr. 1976. *Avalanche handbook.* Agriculture Handbook 489. Washington, DC: U.S. Forest Service.

Placer County. 1994. *Placer County general plan background report.* Vol. 2. Auburn, CA: Placer County.

Salm, B. 1979. Snow forces on forest plants. In *Mountain forests and avalanches*, 157–81. N.p.: International Union of Forestry Research Organizations.

Scott, K. M., and G. C. Gravlee Jr. 1968. *Flood surge on the Rubicon River, California—hydrology, hydraulics, and boulder transport.* M1–M38. Professional Paper 422-M. Reston, VA: U.S. Geological Survey.

Stetham, C. 1992. The Alpine Meadows avalanche: One expert's recollection. *The Avalanche Review* 10 (5): 11.

Swanson, F. J., S. V. Gregory, J. R. Sedell, and A. G. Campbell. 1982. Land-water interactions: The riparian zone. In *Analysis of coniferous forest ecosystems in the western United States*, edited by R. L. Edmonds, 267–91. Stroudsburg, PA: Hutchinson Ross.

Voight, B. 1990. *Snow avalanche hazard and mitigation in the United States.* Washington, DC: National Research Council, National Academy Press.

Wakabayashi, R. 1979. Deformation and damage to forest plants by snow forces. In *Mountain forests and avalanches*, 205–9. N.p.: International Union of Forestry Research Organizations.

Williams, M., K. Elder, C. Soiseth, and R. Kattelmann. 1992. Aquatic ecology as a function of avalanche runout into an alpine lake. In *Proceedings of the International Snow Science Workshop, Breckenridge*, 47–56. Denver: Colorado Avalanche Information Center.

Wilson, N. 1986. A widespread cycle of unusual avalanche events. In *Proceedings of the International Snow Science Workshop, Squaw Valley*, 153–54.

Wolman, M. G., and R. Gerson. 1978. Relative scales of time and effectiveness of climate in watershed geomorphology. *Earth Surface Processes* 3:189–208.

SECTION V

SNEP Case Studies

CONSTANCE I. MILLAR
Institute of Forest Genetics
U.S. Forest Service
Pacific Southwest Research Station
Albany, California

50

The Mammoth-June Ecosystem Management Project, Inyo National Forest

ABSTRACT

The Sierra Nevada Ecosystem Project (SNEP) case-study assessment of the Mammoth-June Ecosystem Management Project (MJEMP) was undertaken to review and analyze the efficacy of a local landscape analysis in achieving ecosystem-management objectives in the Sierra Nevada. Of primary interest to SNEP was application of the new U.S. Forest Service (USFS) regional process for landscape analysis, especially use of historic and natural range of variability. An underlying assumption in current USFS approaches is that managing lands within historical and natural ranges of variability will promote ecological sustainability. Another assumption of interest to SNEP is that social goals can be incorporated into ecological goals to arrive at integrated management objectives. Success in describing historical condition varied considerably by ecological indicator. A few quantitative measures were developed for short- (decade) to medium-term (several centuries) periods, but many descriptions were qualitative, highly inferential, and based on very short-term studies. If the intent were to develop desired conditions from scientifically defensible, quantitative descriptions of historical variabilities, the MJEMP analyses would be inadequate; the team found that it was difficult to take a science-based approach when there was not time, budget, or qualifications to do the science. For the MJEMP team, however, the value of historic data was not to develop a desired condition that mimicked past structural conditions, but to be informed about natural processes and how they can be severely disrupted by human activities (present and past). Thus, the information obtained by the MJEMP was useful for describing the status, trends, and apparent changes in successional pathways caused by humans. Without detailed information about historic ranges of conditions, however, the team had difficulty describing desired future conditions, finding it oversimplified to say they wanted to maintain natural or current conditions.

Public involvement in the MJEMP was at first low to moderate, but built to strong participation and interest. However, a segment of the local public expressed dissatisfaction with the general USFS approach to landscape analysis and the specific implementation in the Mammoth-June area, and began to mount legal action against it. The main concern of this group is that the landscape analysis is actually a decision process, yet it has been considered exempt from (or outside of) National Environmental Policy Act procedures. The outcome of these discussions could have implications for landscape analysis on national forests throughout the Sierra Nevada.

INTRODUCTION

The Sierra Nevada Ecosystem Project (SNEP) is primarily an assessment study. In addition to assessing ecological and sociological conditions and trends in the Sierra Nevada, SNEP is charged with assessing relevant methods, approaches, and policies. This direction includes both methods that SNEP itself uses and also policies and approaches to ecosystem management potentially or actually employed by others in the Sierra Nevada. For this reason, five SNEP case studies were chosen as ongoing examples of ecosystem management in the Sierra Nevada.

The case studies illustrate diverse conditions in the Sierra Nevada and do not parallel one another in intent, histories, magnitude, funding, or other attributes. Each exemplifies a particular approach to common institutional problems encountered in ecosystem management of the Sierra Nevada. Collectively they sample many significant situations encountered in ecosystem management. SNEP will evaluate the effi-

Sierra Nevada Ecosystem Project: Final report to Congress, vol. II, *Assessments and scientific basis for management options.* Davis: University of California, Centers for Water and Wildland Resources, 1996.

cacy of these approaches to the physical and biological communities each represents, to the human communities involved, for value to SNEP in its analyses, and for their value in wider application of these approaches in the Sierra Nevada.

Case-Study Objectives

Each of the three SNEP assessment questions pertains to analysis of SNEP case studies. In addition, because most have involved some form of projecting and evaluating land-management alternatives, they also represent approaches to SNEP's questions about policy scenarios. These issues are woven into five questions that pertain directly to each case study:

1. What conditions does this case study represent for ecosystem management in the Sierra Nevada? Conditions of interest include natural and social environment, land-ownership patterns, current land-management objective, historical use and policies, nature of public involvement, and policy context.

2. What are the specific ecosystem management methods, approaches, or policies being applied? These include intended, planned, actual, and implemented methods, as well as biological and social aspects.

3. How effective have these specific methods been in reaching goals? Effectiveness is assessed relative to the natural (physical conditions, biodiversity) and social (local communities, interest groups, common good) environments.

4. How representative of other situations in the Sierra Nevada is the case study?

5. What can be learned from the case study? Specifically, what are the implications for local conditions (both the local natural environment and local human communities), for SNEP, and for broader application in the Sierra Nevada?

Mammoth-June Case-Study Objectives

The Mammoth-June Ecosystem Management Project (MJEMP) of the Inyo National Forest was selected by SNEP because it meets the preceeding conditions and exemplifies a set of representative issues in Sierra Nevada ecosystem management. The MJEMP

- Represents eastern Sierra landscape and management conditions in

  - patterns of land ownership (almost exclusively federal)

  - focus on recreation and habitat protection with diverse but low intensity commodity values

  - forest structure and composition with associated physical and biotic environment

- competing and conflicting desires for management of parts of the area

- active public involvement

- relatively strong scientific information base

- Applies new U.S. Forest Service (USFS) guidelines for ecosystem management, both national policy (Forest Plan Implementation, USFS 1992a), and Pacific Southwest regional approach (Manley et al. 1995). These guidelines contain the conceptual thinking and procedural models that are to be adopted by and guide land-management planning on the national forests across the country and throughout the Sierra Nevada in the future.

- Relies on comparisons of current conditions to inferred historical conditions (especially natural ranges of variability and ecological indicators) to arrive at ecological management goals. It assumes (explicit in Manley et al. 1995) that landscapes managed within relevant natural range of variabilities are sustainable.

- Assumes that social desires can be accommodated by modifying ecological goals to arrive at integrated management objectives for the landscape (desired conditions).

METHODS AND ASSUMPTIONS

SNEP's approach to assessing the MJEMP was primarily observation by participation, interview, and review of secondary sources. SNEP scientists involved have ongoing experience independent of SNEP in the eastern Sierra, the Inyo National Forest, and especially the Owens River headwaters region. Each has a history of research and management interest in the area and has participated to some degree in management processes for the area in recent years. By participating in the meetings and field trips of the MJEMP team, through interviews and informal discussion with team members and members of the public, and through working in residence in the eastern Sierra, SNEP scientists were directly involved (although to varying degrees) from the beginning of the current Mammoth-June project.

Several explicit assumptions are accepted:

1. MJEMP reflects general approaches (e.g., Grumbine 1994) being taken in land management.

2. MJEMP is a serious attempt to adopt the specific steps outlined in the national Forest Plan Implementation (USFS 1992a) and the new regional ecosystem management manual (Manley et al. 1995) and thus reflects a process that may be repeated commonly throughout Sierra Nevada national forests.

3. MJEMP is one of the first landscape analyses in the Sierra Nevada to implement these specific national and regional guidelines at the landscape or watershed scale.

4. The conditions local to the Mammoth-June landscape are not unique nor so unusual as to limit application of lessons learned there for landscape analyses elsewhere in the Sierra Nevada.

5. Knowledge and experience gained at the scale of Mammoth-June (14,750 ha [36,000 acres]) are relevant to landscape analyses and ecosystem management analyses at other scales in the Sierra Nevada.

6. By choosing to evaluate historical conditions, natural ranges of variation, and their application to desired conditions, SNEP does not necessarily endorse the concept. Rather, because this concept is so widely promoted and discussed in conservation biology and restoration ecology communities, SNEP felt it important to evaluate its application in a Sierran case study.

BACKGROUND: ECOSYSTEM MANAGEMENT

General Context

Within only a few years ecosystem management has taken on almost symbolic meaning in social, political, management, and scientific communities. Although scientists and analysts alike debate the nature of ecosystem management, some elements that are common to its philosophical and conceptual bases can be summarized as follows (Grumbine 1994, references in Duane 1994):

- maintain viable populations of all native species in situ

- represent, within protected areas, all native ecosystem types across their native range of variation

- maintain evolutionary and ecological processes (e.g., disturbance regimes, hydrological processes, nutrient cycles)

- manage over periods of time long enough to maintain the evolutionary potential of species and ecosystems

- accommodate human use and occupancy within these constraints

Forest Service Interpretations

The USFS has evolved its own lexicon and interpretations for ecosystem management. Because landscapes in the Sierra Nevada are predominantly influenced by management decisions of the national forests, interpretations specific to the USFS are widely relevant to the Sierra, and hence, to SNEP.

Ecosystem management is the current theme guiding USFS land management. In June 1992, the chief of the USFS instituted ecosystem management throughout the national forests of the United States and defined it as "the skillful, integrated use of ecological knowledge at various scales to produce desired resource values, products, services, and conditions in ways that also sustain the diversity and productivity of ecosystems" (Robertson 1992). Further emphasized have been sustainability of resilient ecosystems, restoration and maintenance of ecological conditions; production of desired resource uses within the capabilities of ecosystems; aesthetic, cultural, and spiritual values; and collaboration internally, among agencies, and with diverse publics. Goals are healthy ecosystems, vital human communities, and organizational effectiveness (Robertson 1994).

The USFS recognizes ecosystem management as a "means not an end, scientifically credible, legally defensible, and socially accepted" (Manley et al. 1995) . The focus for land management is changed from *output*-driven project planning to *outcome*-driven planning.

Forest Plan Implementation

Although the USFS emphasizes that ecosystem management, as a means to sustaining healthy ecosystems, cannot be prescriptively assigned, there have been attempts to standardize general approaches and develop manuals and guidelines. Relevant national guidelines, although not explicitly under the ecosystem management banner, have been widely taught under the title of *Forest Plan Implementation* (USFS 1992a). This approach has an implicit ecosystem management philosophy that is developed through a three-phased approach to implementing national forest land and resource management plans (figure 50.1). An initial landscape analysis phase (National Forest Management Act [NFMA] component) is described at length. This phase involves evaluating existing conditions of the landscape, determining desired conditions relative to the present, and outlining management opportunities, practices, and projects to achieve the desired conditions. The emphasis in these analyses is on ecosystem capacities, limitations, and thresholds, and on determining ecologically and socially acceptable environmental outcomes (not projects). Because these steps constitute analysis and are not decisions affecting land allocations or land disturbance, there is no involvement of National Environmental Policy Act (NEPA 1970) analysis. This interpretation, however, is being challenged within the MJEMP (discussed later in this chapter), with potentially significant implications for the planning process in general.

The second phase occurs when a USFS official selects one or more site-specific management practices to implement within the landscape analyzed as a project. Once an actual project is proposed, it becomes subject to NEPA, calling into play the standard phases of public scoping, issue identification, development of alternatives, environmental effects, significance of impacts, and decision notification. The NEPA

FIGURE 50.1

Three phases of the USFS forest plan implementation guidelines (USFS 1992a).

phase focuses just on the actual project activities proposed as a result of the landscape analysis, not the landscape analysis itself (except for some cumulative effects analyses).

The final phase, adaptive management, uses monitoring to provide feedback between expected results based on the initial NFMA analysis and actual results based on implementation of the NEPA project (figure 50.1). Unexpected or undesired results indicate how the landscape analysis may need to be adjusted.

Public participation is encouraged during all phases of forest plan implementation. Significantly, it is only during the NEPA phase, however, that it is legally required, a formal process is prescribed, and an appeal procedure outlined for some decisions. No prescriptive steps for public participation are outlined for the other phases, although the emphasis is on iterative dialogue with interested members of the public.

The significant change that forest plan implementation effected was the emphasis on an independent, interdisciplinary landscape (ecosystem) analysis prior to determining specific projects. In effect such emphasis implies that managers think first about the whole landscape and only then act on site-specific projects. Traditionally much of the thinking (i.e., landscape analysis) has been forced into evaluations within specific NEPA projects to meet administrative targets (either proposed within the agency or from outside), with the result that analyses have often been hurriedly conducted in response to specific projects. The geographic and scientific scope of these analyses has been constrained by activities and management projects rather than by comprehensive ecological analysis. In effect, forest plan implementation sought to link NFMA analysis with NEPA evaluations to promote more responsible, scientifically defensible, and proactive management planning.

Pacific Southwest Regional Ecosystem Management

Responding to national imperatives to develop regionally appropriate guidelines for ecosystem management, in February 1994 the Pacific Southwest (PSW) Region of the USFS distributed a three-volume draft ecosystem management guidebook (USFS 1994) . The goals of this effort were to (1) develop clear objectives for ecosystem management in the region, (2) define the major ecosystems in the region, as well as their components and functions, and (3) develop a process by which these objectives could be incorporated into planning. The draft guidelines, a subsequent regionwide workshop, and the final version, *Sustaining Ecosystems: A Conceptual Framework* (Manley et al. 1995) were intended to disseminate and catalyze an approach to ecosystem management that would be implemented in national forests throughout the California (Pacific Southwest) region.

The Pacific Southwest conceptual framework (Manley et al. 1995) builds on the basic three-phase outline of forest plan implementation, but adds scientific (especially ecological) rationale and detail. It embraces the notion that ecosystems are dynamic and evolving in time and over space, and that resilience to disturbance and adaptability to environmental change characterize natural ecosystems. Further, it firmly endorses the notion that the landscapes that will be favored mimic conditions within and across watersheds that have occurred over evolutionary time. The underlying assumption here, widely supported within other USFS regions, is that "restoring and maintaining landscape conditions within distributions that organisms have adapted to over evolutionary time is the management approach most likely to produce sustainable ecosystems" (Manley et al. 1995) .

The basic steps outlined for landscape analysis (figure 50.2) are:

1. Determine which ecosystem elements (components, structures, processes) are key for the landscape under analysis. An element is key if it reflects ecosystem integrity.

2. Identify environmental indicators (previously called ecological indicators) that measure (directly or indirectly) the key ecosystem elements (table 50.1). Selected indicators should take into account both coarse-filter (habitat condition; ecosystem processes) and fine-filter (specific needs of unique elements) aspects.

3. Determine natural ranges of variabilities (in the final version of the guidelines, these are called reference variabilities; also sometimes called historic ranges of variation) for the environmental indicators (figure 50.2). The ranges of variabilities are developed from inferences about historic conditions and/or spatial variability of the ecological indicators at one time.

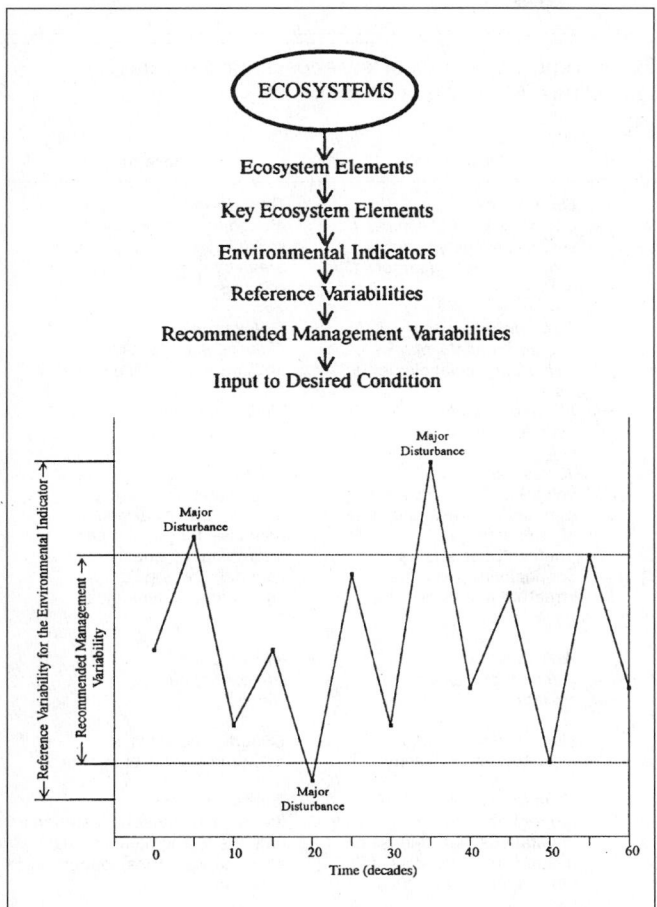

FIGURE 50.2

Steps in the ecosystem management guidelines of the USFS regional guidelines (top) and illustration of reference variability and recommended management variability (bottom). (Manley et al. 1995).

4. Determine recommended management variability (previously called recommended range of variation) for the environmental indicators. This range is the subset of natural (later called reference) range of variability that would set the thresholds for management of each indicator (figure 50.2) by eliminating the extreme events that management would not want to replicate.

5. Develop (or, extract from the forest plan, should it contain an adequate description) the desired condition for the landscape. Desired condition is the portrayal of how the landscape would be if management goals are achieved. Desired conditions reflect an integration of physical, biological, and cultural/social considerations. The desired condition is stated quantitatively for the environmental indicators and ideally lies within the recommended management variability for each indicator.

6. Inventory existing conditions of environmental indicators in the landscape (physical, biological, and cultural/social). Develop a baseline database.

7. Compare desired condition with existing conditions and develop a sequence of potential projects (opportunities) to move the landscape toward the desired condition. (Steps 1-7 approximately encompass phase 1 of forest plan implementation.)

8. Select projects, using NEPA procedures as appropriate.

9. Implement projects according to NEPA procedures. (Steps 8–9 are approximately phase 2 of forest plan implementation.)

10. Invoke adaptive management through monitoring and feedback, to adjust details of analysis, including variabilities, desired conditions, and proposed projects (approximately phase 3 of forest plan implementation).

Through the background material in the conceptual framework, foundations of dynamic ecosystems, the role of historic change, scales of temporal and spatial hierarchies, and discussions about inferring ranges of variabilities are developed. Similarly, an extensive section of the conceptual framework develops and describes sets of key ecosystem elements and indicators for the cultural/social, the hydrological, and the terrestrial hierarchies.

The Mammoth-June Ecosystem Management Project adopted both the forest plan implementation general approach and the landscape analysis of the draft regional guidelines.

THE MAMMOTH-JUNE ECOSYSTEM MANAGEMENT PROJECT

Background and assessment of the MJEMP are discussed here relative to the five questions posed for SNEP case studies.

Question 1. What Conditions Does the MJEMP Represent for Ecosystem Management in the Sierra Nevada?

Physical Setting

Lying entirely within the administrative boundaries of the Inyo National Forest, the 14,570 ha (36,000 acre) land area included in the MJEMP extends from the town of Mammoth Lakes and Highway 203 on the south to June Lake on the north (figure 50.3). The western boundary lies adjacent to the Sierra Nevada crest (San Joaquin Ridge, highest elevation, 3,515 m [11,600 ft]) and the eastern boundary traverses U.S. Highway 395 (lowest elevation, about 2,240 m [7,400 ft]) (figure

TABLE 50.1

Example of environmental indicators (formerly called ecological indicators) showing matrix of key ecosystem elements, terrestrial hierarchy (from USFS regional ecosystem management guidelines, Manley et al. 1995).

Ecosystem Element	Land Unit	Landscape	Subregion	Ecoregion
Vegetation mosaic (structure)	**Description** vegetation patch(s) identified by similar compositions and structure, each defined as a "series"	**Description** vegetation patches identified by similar compositions and structure, each defined as a "series"	**Description** an aggregation of similiar series to simplify complex areas by creating larger patches	**Description** an aggregation of similar series to simplify complex areas by creating larger patches
	Relevance *Cultural/Social Links* all cultural/social elements	**Relevance** *Cultural/Social Links* all cultural/social elements	**Relevance** *Cultural Social Links* all cultural/social elements	**Relevance** *Cultural Social Links* all cultural/social elements
	Hydrologic Links hydrologic cycle	*Hydrologic Links* hydrologic cycle	*Hydrologic Links* hydrologic cycle	*Hydrologic Links* hydrologic cycle
	Terrestrial Links animal and plant species (amount of habitat for individuals, fragmentation effects, species composition); erosion	*Terrestrial Links* animal and plan species (amount and arrangement of habitat for individual or populations, fragmentation effects, connectivity, population structure and dynamics); erosion	*Terrestrial Links* animal and plant species (amount and arrangement of habitat for populations, fragmentation effects, connectivity, population structure and dynamics)	*Terrestrial Links* animal and plant species (amount and arrangement of habitat for populations, fragmentation effects, connectivity, population structure and dynamics)
	Affected By *Atmospheric Links* climate	**Affected By** *Atmospheric Links* climate	**Affected By** *Atmospheric Links* climate	**Affected By** *Atmospheric Links* climate
	Cultural/Social Links all cultural/social elements	*Cultural/Social Links* all cultural/social elements	*Cultural/Social Links* all cultural/social elements	*Cultural/Social Links* all cultural/social elements
	Terrestrial Links erosion; fire regimes; insect infestations; pathogens and disease; plant species (structural heterogeneity—	*Terrestrial Links* erosion; fire regimes; insect infestations; pathogens and disease; plant species; soil porductivity; topography	*Terrestrial Links* fire regimes; insect infestations; pathogens and disease; plant species; soil types; typography	*Terrestrial Links* fire regimes; insect infestations; pathogens and disease; plant species; soil series; topographic variation
	Environmental Indicators suitale habitat area (including cover by species); patch size; shape indices; landscape location; pattern analysis (re: gaps); nearest neighbor analysis	**Environmental Indicators** total habitat area; suitable habitat area; habitat arrangement; frequency of patch sizes; shape indices; connectivity; fragmentation; subregion location; pattern analysis; frequency of occurrence across the range of potential substrates (re: plant communities)	**Environmental Indicators** total bahitat area; suitable habitat area; habitat arrangement; frequency of patch sizes; fragmentation; connectivity; ecoregion location; pattern analysis; frequency of occurrence accross the range of potential substrates (re: plant communities)	**Environmental Indicators** total habitat area; suitable habitat area; habitat arrangement; fragmentation; connectivity; pattern analysis frequency of occurrence across the range of potential substrates (re: plant communities)

50.3). The area is about 120 km (45 mi) north of Bishop and about 25 km (10 mi) south of the Mono Basin.

Included within the boundaries of the Mammoth-June area (MJ area) are the entire upper watersheds of Glass Creek and Deadman Creek, and all but the uppermost portion of the upper Dry Creek watershed. Lands at the south extend slightly into the Mammoth Creek watershed, where the landscape abuts the town of Mammoth Lakes. These permanent streams have significant value as the headwaters of the Owens River, the dominant watershed of the eastern Sierra.

Geologically and topographically the area differs from adjacent eastern Sierran escarpment environments. The Sierra crest is a relatively low ridge in this region and does not dominate as a weather or migration barrier as it does north and

south of the area. The lands east of the crest within the MJ area have relatively gentle topography, although several mountains and low buttes lie east of the crest within the area. Whitewing Mountain (3,035 m [10,014 ft]), situated in the middle of the MJ area, is the largest. The southernmost six Inyo Craters form a north-south chain within the area (Deer Mountain has a maximum elevation of 2,665 m, [8,796 ft]). Repeated explosions from the craters (as recently as 530–650 years ago) dispersed volcanic ash, tephra, and lava over portions of the area, and have been important periodic forest and landscape disturbance agents (Miller 1985; Sieh and Bursik 1986; Wood 1977).

About three-quarters of the MJ area is forested. Significant are the extensive red fir (*Abies magnifica*) forests that cover

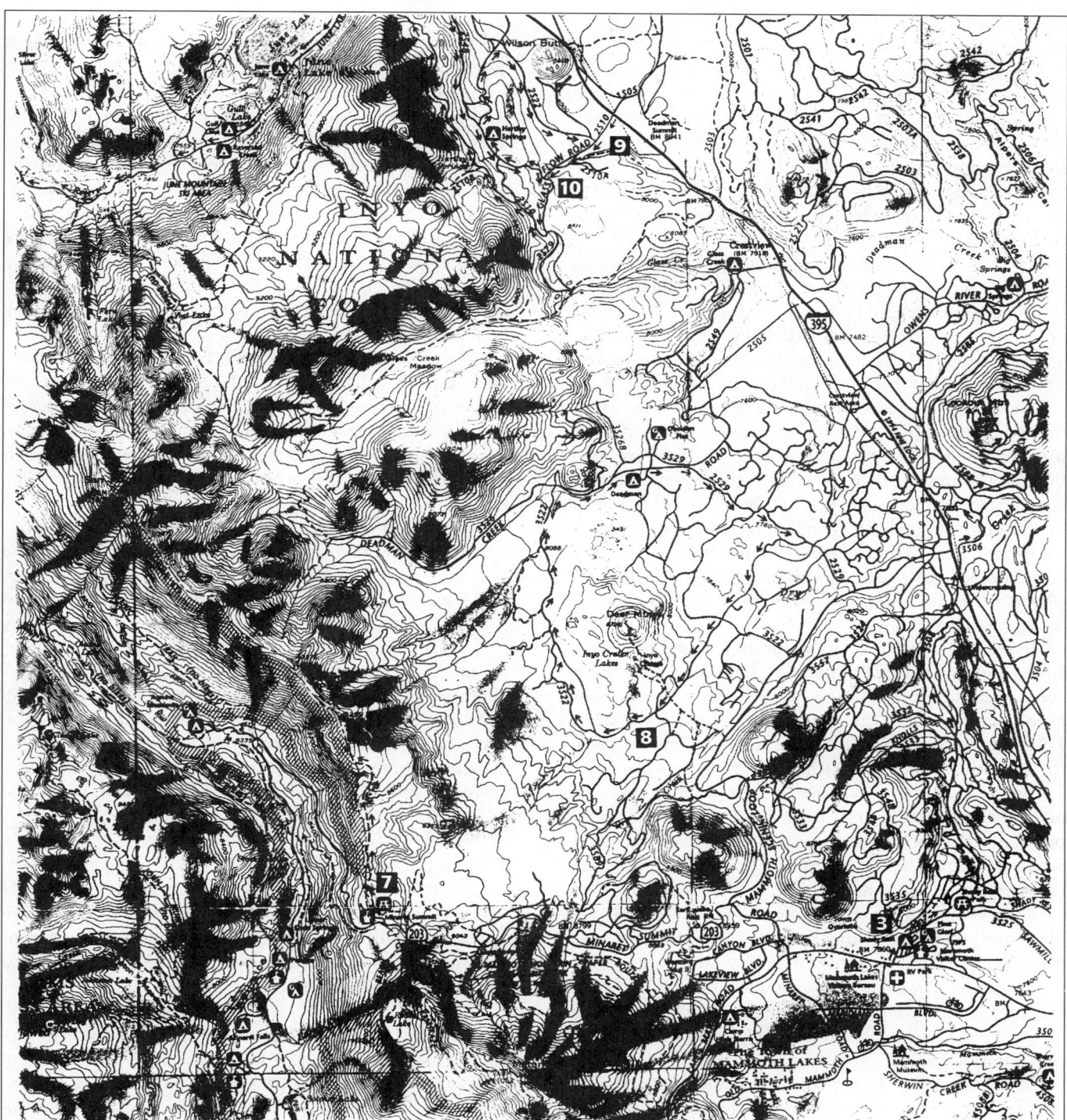

FIGURE 50.3

Map of the Mammoth-June landscape.

much of the western portions of the area. Red fir is infrequent east of the Sierran crest, occurring elsewhere only as scattered trees or small groves south of the Tahoe area. The red fir forest within the MJ area, however, exists not only in mixed as-

sociations with other montane conifers (predominantly Jeffrey pine [*Pinus jeffreyi*] and lodgepole pine [*P. contorta*]), but also in pure red fir associations, which are very rare in the eastern Sierra. Many of these forest stands have late successional characteristics and are valued from many standpoints for their old-growth conditions.

Below the pure red fir forests are mixed montane forests (red fir, Jeffrey pine, lodgepole pine, white fir [*Abies concolor*], western juniper [*Juniperus occidentalis*]) locally dominated by single species where soils and other conditions limit diversity (e.g., pure lodgepole pine stands in some areas). Toward the eastern edges of the area, especially in the Dry Creek drainage, extensive stands of Jeffrey pine (locally mixed with other conifers) dominate.

Above the red fir forests are montane and subalpine forests of western white pine (*Pinus monticola*), whitebark pine (*P. albicaulis*), mountain hemlock (*Tsuga mertensiana*), lodgepole pine, western juniper, and possibly limber pine (*Pinus flexilis*). Scattered subalpine forests, alpine meadows, scree slopes, and barren rock occur in the higher reaches.

Openings in the forested and nonforested areas occur throughout the MJ area. The most important meadow complex are the Glass Creek Meadows in the northern portion of the area. Comprising both wet and dry components, this area is rich in shrubs, forbs, grasses, and sedges. Sand flats occur in some areas around the crater, and influence vegetative development. In the eastern part of the area, the forest meets an ecotone with sagebrush (*Artemesia tridentata*)/bitterbrush (*Purshia tridentata*) associations, which extend eastward through Long Valley.

Understory plant diversity in the area is rich. Species occur in this area that have affinities to at least seven floristic zones, including areas of the western Sierra and assemblages in low abundance in the xeric eastern Sierra (Constantine 1994). The low elevation of the Sierran crest appears to contribute to this diversity as a corridor for plant migration.

The unusual occurrences and mixes of forest types and nonforest associations (especially mesic types) in this area provide important habitat for wildlife in the semiarid eastern Sierran bioregion. As it did for plants, the low elevation of the Sierran crest in this region appears to have served historically and at present as a trans-Sierran corridor for some wildlife. Major vertebrates of importance are furbearers, especially marten and possibly wolverine, whose prime habitats are the red fir forests. The MJ area is a significant mule deer migration area, offering important summer habitat and fawning areas. Raptors are important, especially goshawks. Although no California spotted owls have been found in the MJ area (or anywhere else on the Inyo National Forest east of the Sierran crest), they have been sighted closely adjacent in the upper San Joaquin drainage, and the MJ area may serve as occasional foraging habitat. Willow flycatcher and other neotropical migrants may use Glass Creek Meadows and other meadows in the area.

The aquatic fauna is also rich by eastern Sierran standards. Glass Creek Meadows supports a diverse herptofauna, including the endangered Yosemite toad. Although no natural salmonids occur in the Upper Owens River Basin, Glass Creek and Deadman Creek support permanent populations of ex-

otic trout. Glass Creek is part of the California Department of Fish and Game's Lahontan trout restoration plan, even though the trout is not known to be native to the stream.

Landscape disturbance in the forested areas historically was most frequently caused by fire, insects, and disease. Fire influence and effect in the pre-suppression era varied by forest type (discussed later in this chapter). Fire appears to have been most common in the Jeffrey and lodgepole pine forests. Fire intervals were much longer in the red fir forests, where the range of fire-return times could be quite long and where stand-replacing fires did occur, although not exclusively. The role of fire in the sagebrush/bitterbrush types is less clear. Fires were probably uncommon in the meadows and only very local in the high elevations.

Insects and disease contribute to forest structure in most of the forest types, acting alone and interacting with other disturbances. Bark beetles have caused significant mortality recently in both pine and fir types and most likely played an important role in thinning stands and creating regeneration gaps and forest mosaics before fires were anthropogenically suppressed.

Windfall and avalanches are important secondary contributors to forest structure and mosaic in this area. An extreme avalanche cycle in February 1986 opened or expanded several avalanches in the forest along the east side of the San Joaquin Ridge.

Regular blasts from the Inyo and Mono Craters throughout the last 30,000 years (Wood 1977) have been steady, low-frequency disturbance events to forest, associated wildlife, and aquatic biota. Their role in initiating primary succession in blast zones may be significant in determining the course of modern vegetation and development of aquatic faunal compositions, and in influencing forest age and stand dynamics.

Land Use and Management Context

The area now included in the MJEMP has long been the subject of public interest and policy focus. Appendix 50.1 details land-use and management history of the area. Included here is a summary of historical trends relevant to evaluation of the current ecosystem management project. Use and policy trends fall roughly into five historical periods.

Pre-1950. The western portion of the MJ area was designated as part of the Sierra Timber Reserve in 1890, with the remainder added in 1905. This area was transferred to the Inyo National Forest in 1908. Heavy sheep and cattle grazing dominated use of this area from the mid-1800s into the early years of national forest administration, and early records indicate that huge numbers of sheep foraged throughout the area. Although the creation of the national forest provided an opportunity to regulate grazing, it was not until the mid-1940s that numbers of animals were actually brought in line with thresholds based on range capabilities. By 1950, the Animal Unit Months had been reduced on the forest as a whole by 40%. From the standpoint of current conditions and management, however, many areas, especially the meadows and grasslands, probably still show evidence from the early days of unrestricted grazing.

In addition to bringing use of the range for grazing under control, the early orders of the national forest rangers were to extinguish wildfires. By the early 1900s, fire suppression had begun in the MJ area, although undoubtedly with variable success because of a small workforce, limited access, and simple equipment.

The first recorded timber sale on the Inyo National Forest occurred in 1908, near Mammoth Lakes, in the extreme southern end of the MJ area. The first timber-planning efforts for the national forest began about 1920, resulting in several small timber sales that supplied early construction in Mammoth Lakes and the agricultural communities of the Owens Valley. There is no indication of harvest in the area during the 1930s and only a few sales in the 1940s. All were focused in the southern end of the area, and most were overstory removal of large trees (probably Jeffrey pine), with the exception of undocumented firewood cutting.

Recreational interest has focused on the MJ area since the early 1900s with the development of resorts in the Mammoth Lakes area. The Civilian Conservation Corps (CCC) was active in the MJ area during the 1930s building roads; campgrounds at Hartley Springs, Shady Rest, and Glass Creek (still in use); trails; and ranger facilities. Early recreational uses of the MJ area were fishing, hunting, and camping.

1950–79. The Integrated Use Plan of 1950 is the earliest planning document of the Inyo National Forest to systematically outline and coordinate management objectives for the MJ area. The MJ area was in the Mammoth Zone, with all but the southern boundary coinciding with the present MJEMP boundaries. The dominant objective for the MJ area under this plan was to manage for recreation, water quality, and wildlife protection. It is important to underscore that the explicit management directions in planning documents and action of this period clearly emphasized the priority of recreation in the western portion of the area. Timber harvest was not allowed unless it was considered to have an effect on recreation and scenic values.

The area subsequently fell under the Multiple Use Plans for the Mono Lake and Mammoth Ranger Districts, both of 1970. In these plans, objectives for the MJ area continued to emphasize recreation, especially protecting the scenic beauty of the area by constraining timber harvest and other extractive uses if they affected recreation values. The plans allowed timber harvest in portions of the MJ area that were considered low in scenic and recreation value.

Shortly after the development of the Multiple Use Plans, the Inyo National Forest and Mono County signed a "community forest" agreement to produce a coordinated land-management plan for the region of Mammoth and upper Owens River. The resulting document, the Mono Plan of 1976, involved primarily private land but some national forest lands in the MJ area.

Eight timber sales, with about 60 thousand board feet (MMBF) harvested, occurred in the far north (Hartley), Deadman Creek, Dry Creek, and far southern portions of the area. Timber was not harvested, or harvested in limited quantities, in the western red fir portion of the MJ area. Continuing in the earlier pattern, harvest consisted of removing old, large, high-value trees, with most areas having only 30%–40% of the overstory removed, while a few areas had up to 70% overstory removed. Small, younger trees were left, no areas were clear-cut, and no plantations appear to have been established. Lack of clear-cuts was probably due to the fact that large trees were abundant and highly valued, rendering clearcutting unnecessary. By the late 1960s, most of the eastern half of the area (considered to have low or no recreation value) had been entered for harvest.

By 1950, grazing in the MJ area was contained in two allotments which are still used today: the June Lake and the Sherwin-Deadman allotments. Records to present indicate that total head (1,800) for each allotment remained quite stable once grazing had been regulated, whereas the permitted number of days has been sharply reduced over the years. In the 1950s, the season of use seems to have extended from early June through late October.

Recreational development with an emphasis on intensive use escalated in the eastern Sierra between 1950 and 1970, especially in the mountains and resorts surrounding the MJ area. Since the early 1900s, the area has been linked economically to southern California and especially Los Angeles (Kennedy 1995). The resort potential of the Mammoth region was recognized and actively developed by an increasingly mobile southern California populace. The Mammoth Mountain ski area was established in 1949 and continued to expand extensively throughout the 1960s (now the largest ski area in the United States). Emphasis in the planning documents was on preparing development plans for high-class winter-sports facilities. Campgrounds were added at Pine Glen and Deadman Creek and a "vista site" added at Minaret Summit. Interpretive sites were built, including trails and a new visitor center at the edge of the area in 1969. This period saw a rapid acceleration of developed recreation along the edges of the MJ area, adjacent to the booming resort communities.

1979–88. The Land Management Plan for the Mammoth-Mono planning unit of the Inyo National Forest was developed in 1979 to meet requirements from both the Mono Plan of 1976 and the National Forest Management Act of 1976. Significantly, this was the first time the MJ area was segregated into many discrete management zones, each with different primary management objectives and permissible activities. Thus, integrity of the whole landscape was diminished from a planning perspective, coinciding with several major changes to management in the MJ area as a whole.

Whereas in the previous planning periods, timber harvest and extractive commodities had always ranked below water, recreation, and wildlife, the 1979 plan listed timber second only to watersheds in management emphasis. Language in the plan explicitly emphasized the timber value and left open for the first time much more intensive overstory removal and clear-cutting. Although public comment in response to this plan was encouraged and recorded, there were no comments on harvest, and the public apparently was more concerned about developed recreation at that time.

During this period, seven timber sales occurred, with a total of about 30 MMBF removed. For the first time, in the late 1970s, serious and comprehensive planning efforts began for major timber harvests in the mostly roadless red fir forests of the western part of the MJ area, along the base of the San Joaquin Ridge. Although there had been light harvest in some of these areas in the 1940s, the forests had retained much of their old-growth character. Plans called for major opening of this area to development, including new multipurpose roads, recreation sites, and proposed harvest of 11.5 MMBF.

Through the middle to late 1970s and early 1980s, much of this western portion of the MJ area, including the proposed timber-sale areas and lands west in the San Joaquin drainage, was included in the national Roadless Area Review and Evaluations (RARE I and RARE II). This process kept the proposed timber sales in the red fir zones on hold. In 1984, wilderness legislation allocated lands west of the MJ area into the Ansel Adams Wilderness but excluded from wilderness designation any of the roadless portions of the MJ area (called the San Joaquin roadless area). Thus, this area was considered released from mandatory roadless condition and available to be reconsidered for new management directions. For reasons considered later, however, timber plans proposed for the area were never implemented.

In addition to the major change in emphasis toward timber harvest, the 1979 Land Management Plan for the Mammoth-Mono planning unit also proposed for the first time intensive recreation development in the MJ area. By the mid-1970s, growth in Mammoth Lakes had surpassed earlier expectations. In 1971, the Inyo National forest plan reported that Mammoth Lakes was the "fastest growing community in the country." Growth and recreation demands were expected to continue to explode. The MJ area was considered a major national/international recreation destination, capable of being developed to accommodate the expanding resort population. Plans for a major trans-Sierran highway (which had first been considered in the 1950s–1960s) to complete a gap in the national interstate system were developed through the MJ area over Mammoth Pass. Attention was focused on expanding winter sports facilities (alpine skiing) in the area. Included among alternatives discussed in the 1979 plan were various ski developments that would connect existing Mammoth Mountain and June Mountain ski resorts along San Joaquin Ridge and Whitewing Mountain. This development would have affected 5,665 ha (14,000 acres) of roadless area. Although the allocation of the San Joaquin drainage to Ansel Adams Wilderness finally terminated the idea of a trans-Sierran highway, the 1984 California Wilderness Act provision that released the San Joaquin roadless area for management reconsideration left proposals for ski area and timber harvest open.

Geothermal development issues were raised for the first time in the 1979 plan, and a Geothermal Management Zone (primarily in the southeastern portion of the area) was considered. This zone included areas suitable for leasing and further development, pending exploration and study. Geothermal development became a subject of attention during this period because of the successful establishment of the Casa Diablo Geothermal Plant at Mammoth Lakes in 1990. The USFS and Bureau of Land Management prepared an environmental assessment to determine lands (including the eastern two-thirds of the MJ area) suitable for leasing. In 1984, a lease was approved.

Independently of USFS activities, the San Joaquin Ridge was nominated as a candidate area for an inventory of National Natural Landmarks of the Sierra Nevada, commissioned by the National Park Service (Burke et al. 1982). The planned evaluation phases for landmark designation have never been undertaken by the National Park Service.

1988–93. Planning began in the mid-1980s for the Inyo National Forest Land and Resource Management Plan (LMP), and was published in 1988 (USFS 1988). The growing incompatibilities and conflicts among uses desired and proposed in the MJ area led to considerable controversy during and after the LMP planning era. The primary conflicts related to development and intensive use (alpine ski area expansion, timber harvest, geothermal development, grazing, road building) versus non-manipulative, non-intrusive uses (wilderness, wildlife habitat, old growth, biodiversity protection, water quality, Nordic skiing, and backcountry hiking).

In preparing for the LMP, a "common ground" work group including a cross-section of participants from the USFS and the public was convened to evaluate the issues and determine a concensus management objective for the MJ area. The work group found that the detailed information it needed for evaluation was not available and that desires for the area were so mutually exclusive that consensus could not be reached for the LMP. The group agreed that further analysis was needed before any significant development could take place and that such development would likely trigger the preparation of an environmental impact statement. This recommendation led to the direction in the LMP that defined the so-called Mammoth to June Study Area, which was the foundation for the current MJEMP and differed from the current area only slightly.

The 1988 LMP allocated the MJ area to seven management prescriptions, each with specific management objectives and direction (figure 50.4). The issue of geothermal development was not allocated a specific prescription, although the lease conditions were made part of the plan's overall direction. Zones were allocated for concentrated recreation use (e.g., Inyo Craters), dispersed recreation (e.g., Deadman Creek),

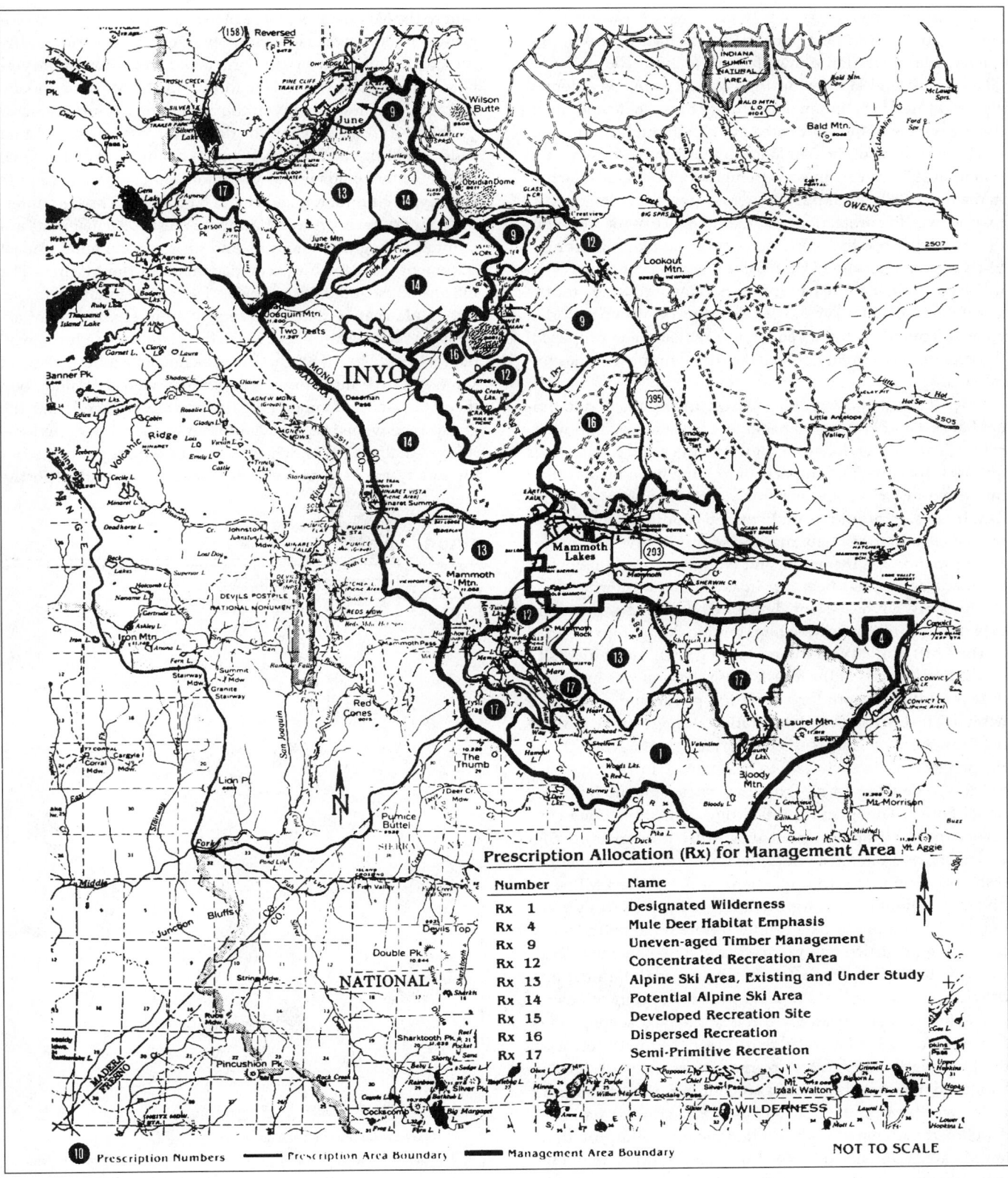

Prescription Allocation (Rx) for Management Area

Number	Name
Rx 1	Designated Wilderness
Rx 4	Mule Deer Habitat Emphasis
Rx 9	Uneven-aged Timber Management
Rx 12	Concentrated Recreation Area
Rx 13	Alpine Ski Area, Existing and Under Study
Rx 14	Potential Alpine Ski Area
Rx 15	Developed Recreation Site
Rx 16	Dispersed Recreation
Rx 17	Semi-Primitive Recreation

Prescription Numbers — Prescription Area Boundary — Management Area Boundary

NOT TO SCALE

FIGURE 50.4

Management prescriptions for the Mammoth-June study area as developed in the 1988 Inyo National Forest land-management plan (USFS 1988).

semi-primitive recreation (e.g., Glass Creek Meadows), uneven-aged timber management (the eastern portion), and range. A large section along the San Joaquin Ridge, upper Glass Creek, and Hartley Springs was allocated as potential alpine ski area to maintain opportunity for such development during review of the entire area after LMP publication.

The allocation of the San Joaquin released roadless area to potential ski area remained extremely controversial. Starting in the mid-1980s, local public interest grew for primitive recreation, wildlife protection, and maintenance of roadless conditions in the red fir forest. As part of the planning for the LMP, it was determined that no timber would be harvested in the red fir belt at the base of the San Joaquin Ridge for the life of the LMP (10–15 years). Because this provision included areas in which the major red fir harvest had been proposed, those plans were finally canceled and the proposed sales terminated.

Since the mid-1980s, public interest and opposition to timber harvest and grazing management has increased dramatically. Clear-cutting, loss of old-growth trees and habitat, loss of forest diversity, and impacts of grazing have increasingly interested an active and organized environmental community. Interest in primitive, undisturbed conditions, protection of wildlife and forest habitats, and semiprimitive and primitive recreation has increased, and wilderness designation for the roadless area has been a primary goal. Countering these interests are advocates for developed recreation, primarily expanding winter sports facilities (alpine skiing) into the area.

The LMP of 1988 called for retaining seral forest diversity on the timberlands of the forest. This retention was indicated in several places, both for vegetation per se and as wildlife habitat. The LMP left implementation strategy open, indicating only the seral classes (seedling through old growth) and the percentages in each class to be retained. In 1990, a group formed to develop an old-growth strategy for the Inyo National Forest. The intent of this group, which included concerned public (including local wildlife biologists), USFS biologists, and biologists from the California Department of Fish and Game, was to implement in detail the LMP seral diversity guidelines. Old-growth forests in the Jeffrey pine, lodgepole pine, and mixed conifer timber types on the forest were to be identified and mapped, and a management strategy was to be developed for enhancing, maintaining, and providing adequate acres to meet the LMP specifications.

The group met repeatedly through the next several years and mapped 2,935 ha (7,250 acres) of old-growth forest on the Inyo National Forest. A prime concern was to develop habitat in adequate amounts and configurations to support viable wildlife populations. By late 1992 a strategy had been developed and maps produced that included old-growth retention areas connected by corridors that could be moved as they became scheduled for silvicultural treatment (USFS 1992b). Old-growth recruitment areas were also identified, because inadequate acreages were available in several forest types to meet the LMP guidelines on seral diversity. Signifi-

cant to the MJ area were large blocks of land included in old-growth retention areas, recruitment areas, and corridors (figure 50.5). This plan in effect removed from timber-harvest potential most of the red fir forests in the MJ area and restricted activities in many of the remaining MJ forest areas.

The Record of Decision for the Inyo LMP had stated that "additional significant development of any kind on National Forest System lands in the Mammoth/June area will require a study of culmmative effects" (USFS 1988). Thus, an interdisciplinary team of Inyo National Forest specialists was appointed to begin an analysis and an environmental impact statement (EIS) for future management of the MJ area. The EIS and subsequent Record of Decision were intended to "select an alternative that will identify an integrated set of actions that will be implemented in the Mammoth to June area" (USFS 1988). The MJ Study Group began in October 1990, and some resource inventories and studies were initiated soon thereafter. In September 1992, leadership changed, and the group reconvened as the Mammoth to June Integrated Resource Analysis (IRA), which was to be a pilot project under the newly adopted ecosystem management. Although the

FIGURE 50.5

Old-growth management strategy.

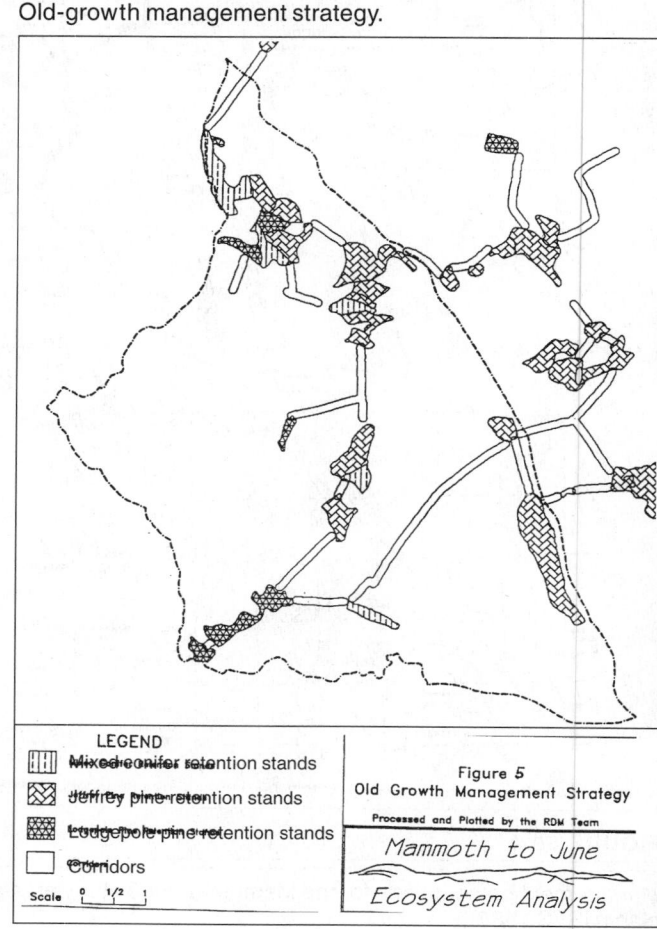

LEGEND
Mixed conifer retention stands
Jeffrey pine retention stands
Lodgepole pine retention stands
Corridors

Scale 0 1/2 1

Figure 5
Old Growth Management Strategy
Processed and Plotted by the RDM Team

Mammoth to June

Ecosystem Analysis

team met several times, other forest priorities stalled progress, and the IRA never crystallized.

Also in 1992 the Research Natural Areas Program of the USFS Pacific Southwest Research Station and Region proposed a research natural area (RNA) in middle of the MJ area (USFS 1992c). This proposal targeted Whitewing Mountain because of the unusual presence of an ancient downed forest on its summit (figure 50.6). The logs appear to have been blown down by blasts from one of the nearby Inyo Craters 530–650 years ago (radiometry on the downed logs coincides with the blast date). The downed trees remained intact for six centuries on the high summit of this peak because of the arid environment. They are valued by the research community for their age and their role in understanding vegetation response to climate change. Several of the logs have been identified as species that no longer live in the eastern Sierra (e.g., sugar pine *Pinus lambertiana)* or as local species that grow in ecologically very different sites today (Millar 1995) . The growth and form of these logs are different from those of the vegetation on this arid and barren summit today, and their presence indicates the conditions that led to the dramatic vegetation change that occurred. Research on these logs continues. In 1993 the Inyo Forest Supervisor requested of the Regional RNA Committee that a decision on the designation or rejection of the proposed Whitewing RNA be delayed pending analysis of the entire MJ area. In the meantime, the area is managed for protection of scientific values.

Extended drought through the late 1980s and early 1990s and the occurrence of several local and/or large fires within and closely adjacent to the MJ area (Laurel Fire 1987, Mammoth Fire 1987, Rainbow Fire 1992, and Bald Mountain Fire 1993) increased the fear of catastrophic fire in the eastern Sierra. The community of Mammoth Lakes in particular has grown increasingly concerned that actions be taken by the USFS to reduce risks of catastrophic fires starting in the wildlands and burning into town (Kennedy 1995). In November 1993, the Inyo National Forest proposed a fuel reduction project in the MJ area, which would have included salvage timber harvest and prescribed fires of 2,670 ha (6,600 acres) of the MJ area. Red fir and mixed conifer forests within the released roadless area were included in this proposal. The local environmental community reacted angrily against this or any ground-disturbing activities in the released roadless area before completion of the Mammoth-June cumulative effects study. In February 1994, the Inyo National Forest rescinded the proposal for salvage or prescribed burning in the released roadless area pending completion of the MJ study but retained plans for fuel reduction in the zones with roads near Mammoth Lakes.

1993–Present. In January 1993, Inyo National Forest line officers and staff were trained in the national forest plan implementation (USFS 1992b), one of the first forests in the Pacific Southwest region in which this training was done. In February 1994, the Draft Region 5 Ecosystem Management Guidebook was issued (USFS 1994). Because staff on the Inyo National Forest had been involved in both teaching others about the forest plan implementation and developing the regional ecosystem management guidebook, they had a high degree of knowledge about the underlying concepts and the intent of the guidelines as well as the motivation to apply them.

FIGURE 50.6

Ancient downed logs on Whitewing Mountain, blown down 530–650 years ago by a blast from the nearby Inyo Crater, are included in the proposed Whitewing Research Natural Area that would be maintained for the study of vegetation response to climate change.

With the release of the draft regional ecosystem management guidelines in February 1994 the Mammoth-June IRA evolved into the Mammoth-June Ecosystem Management Project. The intent was to follow the new guidelines for landscape analysis rather than proceed with the EIS process. At the same time, the project was adopted as a case study of ecosystem management by the Sierra Nevada Ecosystem Project.

One reason that the MJ Integrated Resource Analysis effort did not get under way was its low priority among other Inyo National Forest projects. In 1994 the priority of the MJEMP rose and has remained a relatively high priority since. In addition to the LMP imperative, two situations prompted this recent urgency. One was the fire hazard issue. Environmentalist reaction notwithstanding, the communities of Mammoth Lakes and June Lake remained concerned about the risk of fire burning through the MJ area and insistent that the Forest Service reduce the hazard. An evaluation of environmental conditions at MJ had to be done before measures could be taken to address these fuel situations adequately.

Another pressing issue was geothermal leasing. Renewed interest has been expressed in developing the geothermal leases near the Dry Creek Basin, originally explored by Unocal. The 1981 lease gave contract opportunities to the lessee throughout the entire southeastern half of the MJ area. The so-called diligence requirements meant that the lessee must take exploration actions to maintain the lease. Exploration does not imply development of geothermal power sources, but it does entail action on the part of the lessee, public involvement, and, potentially, environmental analysis. If geothermal development were to proceed, a power plant similar to that at Casa Diablo might be built in the MJ area. It was appropriate to conduct the MJ landscape analysis before any project-specific actions such as this were taken, but lease requirements imposed deadlines to this preference.

Furthermore, a major water development project was proposed for the MJ area. The Mammoth County Water District has sought a special-use permit from the Inyo National Forest for four ground-water wells and a pipeline along Dry Creek to augment its other water supplies for the town of Mammoth Lakes, especially in dry years. Exploratory wells drilled and tested in 1988–1990 demonstrated the feasibility of a large-scale pumping project. Annual production could be as high as 2.4 million m^3 (2,000 acre-feet). Concern exists that ground-water pumping could alter the discharge of Big Springs, to which the Dry Creek Basin is assumed to contribute water. However, the proposed maximum volume to be extracted is less than 15% of the estimated annual recharge within the upper basin and less than 1% of the annual flow of Big Springs. The California Department of Fish and Game is also concerned about the impact of pumping on Mammoth Creek and the Hot Creek Fish Hatchery.

Summary of Management History

From this brief history, it is evident that the MJ area has been the focus of land-management attention for decades. The groups most actively involved have been the land administrators (USFS), the traditional local public (longtime resident communities), the "resort public" (Mammoth Lakes and connected urban communities of southern California), local scientists, and the new local public (recent in-migrants, who may have different values and interests). Over time the international reputation of Mammoth Mountain Ski Resort and an increasingly sophisticated environmental community brought the values and controversies of the MJ area to the attention of nonresident and distant populaces. Complex interactions between national and local policy, societal trends, and incidental local situations over the years resulted in the patterns of land management summarized earlier.

The history of land use and public interest in the MJ area shows repeating cycles over time. Since the early 1900s, prevailing attitudes about the MJ area fluctuated between pro-development and pro-protection (or manipulative versus non-manipulative uses). After the earliest days when grazing was rampant and timber harvest—though low-level—was unrestricted, the first swing toward strong protection of recreation and scenic values began in the 1930s. A long period followed when grazing was brought into regulation, and harvest was low priority, allowed only if it benefited recreational values. The dominant value during this time was based on primitive camping, fishing, and hunting. As the southern California population became more affluent and mobile, and as resort development in Mammoth Lakes grew, so did the interest in development of the adjacent MJ area. This interest triggered a swing toward aggressive development. By the 1960s, alpine ski areas were being proposed in the MJ area, roads were built and paved, a trans-Sierran highway was designed, and much more active and extensive timber programs were planned and conducted. Subsequently, a strong environmental faction once again opposed the pro-development designs, which nevertheless have remained to the present. The environmental group favored protection not just for scenic and recreational opportunities, but also for undisturbed wilderness and inherent ecological values.

Some of these cycles seem to mirror national trends in land management, as reflected first in the early "presence" era of the USFS, and the early fire-suppression and CCC recreation decades. Increasing affluence brought urban travelers interested in rugged outdoor scenery and intensive outdoor play and provided money to develop mountain areas for these ends. The spirit of the Multiple Use Act of 1960 was reflected locally in the MJ area at that time. Backlash to this intensive development period was felt in the eastern Sierra and in attitudes toward the MJ area, as the environmental laws of the 1970s (especially those that greatly affected the USFS, such as NEPA and NFMA) became widely used and advocated by environmentalists. The swing toward ecosystem management in the early 1990s has been perceived by the pro-protection public as yet another turn back toward human intervention in the MJ area, because the focus on sustainable ecosystem conditions—in ways that seem paradoxical at first—appears to favor manipulative or interventionist actions to restore natural processes and structures. This reaction appears in part related to overall distrust of the USFS as an agency, resulting from past actions in local and nonlocal contexts.

Concomitant with the cycles in types of management emphases have been cycles in the way the MJ area has been geographically zoned in management units. During the early agency days and the primitive recreation period, the MJ area was considered primarily a single management unit. During the more intensive development era, the area became increasingly fragmented into several distinct management units, each under different jurisdictions and with different management objectives. The highest degree of fragmentation occurred in the LMP of 1988, when the area was divided into seven management prescriptions, each with different management objectives. The ecosystem management era brought a return in focus to the MJ landscape as a single holistic unit for cumulative planning and management. The periods of protection (for either recreation or ecosystems) favored management of the area with the least fragmentation.

Linkages of the MJ area and adjacent areas to various human communities have influenced the prevailing attitudes toward management and land use. Since the early 1900s, the relatively rare presence in the eastern Sierra of ready access to extensive forests and lush meadows with relatively gentle topography attracted local and distant recreationists to the MJ area. The increasing links between southern California and Mammoth Lakes have tended to bring urban money tinged with development and intensive recreation interests. By contrast, the links of Mono Lake activists to environmental communities in northern and southern California (and elsewhere) have provided educated resident and distant populations concerned about and well-versed in environmental protection. An increasingly active local community of environmental scientists, who bring urban-educated values and insights, has focused on the MJ area's important physical and biological resources. Local communities have favored maintenance of traditional dispersed recreation (hunting, fishing, woodcutting, off-road vehicles) as well as activities that bring economic prosperity to the small towns (skiing, hiking, nature study).

Question 2. What Specific Ecosystem Management Methods, Approaches, or Policies Are Being Applied in the Mammoth-June Ecosystem Management Project?

Intended Goals and Process

The ultimate goal of the MJEMP was to resolve the issue of resource thresholds for the Mammoth-June area left unresolved by the 1988 Land Management Plan. Since the MJ IRA had lapsed, a new team (though with many of the same players) was composed for the MJEMP in early 1994. This reconstituted team met for the first time in April 1994. Represented

on the team are the following resource specialties (all Inyo National Forest staff): team leader (forest ecosystem management coordinator; recreation/fire), geology, soils, air quality, insects and disease, fisheries, range, recreation, vegetation ecology, wildlife biology, fire and fuels, hydrology, landscape architecture/visual-quality management, archaeology/historical ecology, and land-management planning (appendix 50.2). Specialists from both the forest level and the two ranger districts that the MJ area spans are involved. About fifteen staff members have primary responsibilities to conduct analyses, while many more attend meetings and participate in technical aspects.

Whereas the previous MJ IRA, like all environmental analyses, had focused on identifying issues and developing alternative projects to resolve conflicts, the intended goals for the MJEMP were much different. Following the guidelines of forest plan implementation (USFS 1992a) and the draft Regional Ecosystem Management Guidelines (USFS 1994), the goals of the MJEMP were to develop a desired condition for the landscape (management objectives) and generate potential management practices that would allow the desired condition to be achieved over time. The desired condition was to be consistent with the LMP as much as possible, within its inherent flexibilities. Analysis would be based on physical and ecological capabilities, thresholds, and health conditions of the landscape; social goals and public conflicts would be incorporated subsequently into a final desired condition. The process identified at the first meeting involved the following seven intended or planned steps:

1. Define the Analysis Area. The team considered five alternatives for adjusting the boundary that had been used in previous planning efforts. These ranged from expanding the boundary to include entire watersheds (uppermost Dry Creek had been excluded previously because of the presence of Mammoth Mountain Ski Area and other developments) or expanding to include the adjacent communities of Mammoth Lakes and June Lake. The final decision was to keep the original (LMP) boundaries (figure 50.4) even though they do not adhere to current ecosystem management guidelines to avoid fragmenting watersheds. The existing boundaries focus on potential land-use issues and constrain further development until the analysis is complete. The team decided to retain the original boundaries but allow boundaries to be fluid for the purposes of data collection and responsiveness to issues relevant in individual analyses. Thus, data could be collected outside the area, and adjacent landscapes would be brought into analysis, but the intent of determining a desired condition for the MJ area would be as defined in the LMP.

2. Describe the Existing Condition. Much attention was given to approaches for describing current conditions in the MJ area. Ecological indicators (later called environmental indicators) were chosen as a basis for evaluating ecosystem health and sustainability. Specific indicators were chosen that represented, in the team's view, the key compositional, structural, and process elements in the MJ landscape. The ecological indicators would be used initially to focus analysis of existing conditions and subsequently to describe measures of desired condition. Thirty-nine ecological indicators in seven categories were tentatively chosen and assigned to team members for analysis (table 50.2).

The team chose ecological indicators that could be measured and managed practically. Thus, although avalanches, earthquakes, volcanic eruptions, contemporary weather and climate change potentially greatly affect natural dynamics in the MJ area, the team decided to study these for their influence on the ecosystem but not to include them as practical descriptors of desired condition.

Social indicators for existing and historic conditions were not developed initially. The team hoped to have assistance in developing creative approaches to social analysis, but when this did not occur, the team settled on using traditional visual-opportunity-spectrum ratings and visitor recreation measures as indicators for assessing and developing a desired condition.

3. Describe the Historic Condition. The team agreed that although the ability to successfully infer and describe historic condition varies by ecological indicator, the value of historic understanding in analysis made the effort to obtain historic information worthwhile. The team chose to understand historical information not for creating a desired condition that mimicked the past, but rather, for assessing viabilities and health of current conditions and making recommendations about (not targets for) future management. The team accepted that each ecological element would require a slightly different approach to historical analysis, some quantitative, some qualitative, some entirely inferential and even anecdotal or speculative. The group further acknowledged that the relevant time depth for understanding historic ranges of conditions varied with the attribute, because of both scientific and practical considerations. The most appropriate time period for

TABLE 50.2

Ecological indicators initially chosen to describe existing and desired conditions in the Mammoth-June Ecosystem Management Project.

Key Resource Area	Ecological Indicator	Unit of Measure
Air quality	Visibility	Miles
	PM-10	Microgram/m^3
	Ozone	ppm
Watershed	Stream-flow duration	cfs
	Stream-flow timing	cfs
	Stream-flow magnitude	cfs
	Springs	Number of springs
	Channel stability	Channel stability ratings
	Soil erosion	Tons/acre
	Soil productivity	
	Water quality	Temperature Degrees F/C
	Turbidity	jtu
	Conductivity	mv
	pH	pH
	Total suspended solids	mg/l
	Total dissolved solids	ppm
Biodiversity	Key species habitat available	Acres
	Key species habitat distribution	
	Key species population	Number of individuals/pairs
	Key species distribution	
	Vegetation composition	Acres
	Vegetation structure	Seral stage/strata
Fisheries	Pool habitat	Number of pools/mile
	Biomass	Pounds/acre
	Woody debris	Number of pieces/mile
	Species distribution	
	Trophic status	
	Macroinvertebrates	
Fire	Size	Acres
	Intensity	Flame length, btu ft^2
	Frequency	Recurrence interval
	Distribution	
	Fuel loading	Tons/acre
	Fuel model	NFFL fuel-model type
	Fuel structure	
Insects/pathogens	Severity of epidemics	Percentage mortality
	Size	Acres
	Distribution	
	Species affected	

analysis would be one that was responsive to the temporal sensitivity of the resource and that embraced basically modern conditions. For understanding many attributes, several hundred to several thousand years was considered an appropriate theoretical length of time. Practically, information was available for most attributes only for far shorter time periods.

4. Describe the Desired Condition. In USFS terms, echoed by the MJEMP, desired conditions are an expression first of future land conditions that are within bounds of ecosystem sustainability. The analyses of environmental conditions and descriptions of existing and historic conditions and variabilities would provide a background for assessing the condition of the present environment relative to a sustainable one. Conditions determined to be unsustainable, artificial, unnaturally unstable, anthropogenically vulnerable, or significantly outside natural ranges of variabilities would be identified. These factors would lead to conceptualization of desired environmental conditions that would be within a window of ecological sustainability.

Desired condition is not recognized by the USFS nationally or regionally, however, as a statement solely of ecological sustainability. Ecological analysis forms the first part of the

process, followed by incorporation of socially desired goals and conditions for the landscape (figure 50.7). When these are incorporated with ecological goals (with various potentials for conflict among goals), desired conditions are determined. Thus arriving at a final desired condition itself takes several steps. The team agreed that the desired condition for the MJ landscape was an integrative statement of the future of the MJ landscape, resulting from ecological analysis and integrating public input and analysis of socially desirable conditions.

The team recognized the following locations and issues to have high public interest. Although the desired condition would not indicate management categories or land designations, the following public interest issues would be considered in development of the final desired condition in the MJ area:

- Alpine versus Nordic skiing
- Deer migration/fawning areas
- Geothermal leasing
- Glass Creek Meadows
- Managed fires
- Marten/goshawk habitat
- Mortality/fuels/fire hazards
- Old-growth forests
- Potential wilderness
- Proposed research natural area
- Recreation: dispersed versus developed
- Red fir forest
- Released roadless area
- Timber harvest
- Trout recovery
- Water development

FIGURE 50.7

The Pacific Southwest Region approach used in MJEMP of integrating biological, physical, and cultural/social considerations to arrive at a desired condition for a landscape (Manley et al. 1995).

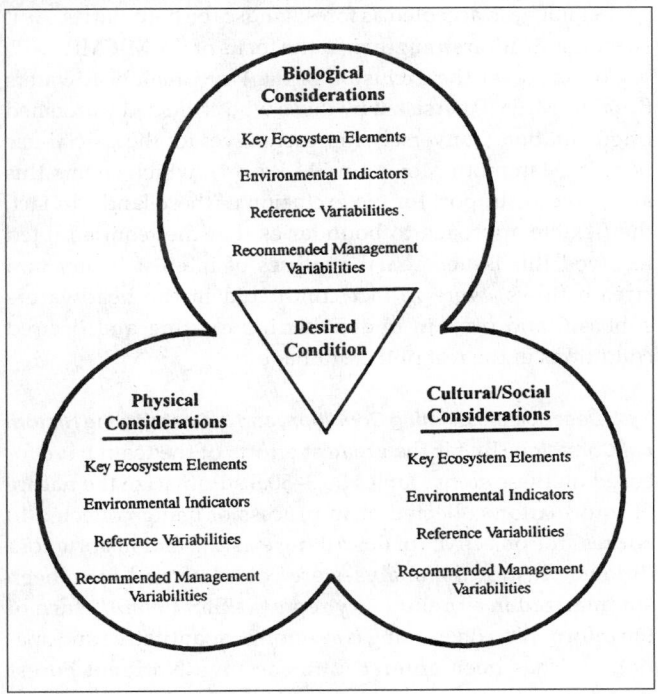

5. Identify Opportunities. Following both the forest plan implementation and draft regional guidelines, the MJEMP team intended to compare existing condition to the desired condition as a starting point for recommending various strategies by which the USFS could potentially bring about the desired conditions.

6. Identify Management Practices. Once the opportunities for action are identified, specific management activities and practices would be proposed that would steer the MJ landscape toward the desired future. Alternative management practices and strategies, compatible with ecological and social constraints determined in earlier steps, would be identified.

7. Select, Schedule, and Prioritize Practices. Scheduling management projects for NEPA analysis was the final step identified for the MJEMP. The team acknowledged that the work would not end there but would become part of the analysis, documentation, decision, and monitoring feedback loop for adaptive management. The MJEMP asserts that its final step ends before any NEPA analysis begins. Thus it attempts to stay in a non-decision-making phase of the forest plan implementation process (figure 50.1) (USFS 1992a).

From a planning perspective, the MJEMP process was viewed as partial LMP implementation. The LMP's formal management prescriptions were binding during the analysis period. However, some of the management directions in the LMP are forestwide, ambiguously described, and geographically unfocused. Thus they do not integrate individual ecosystems (landscapes, watersheds) into functional wholes but view resources piecemeal over the landscape. The MJEMP was intended to focus the LMP's general direction on the specific needs of the MJ landscape and to consider cumulative effects of potential projects on the landscape. The ecological and physical analyses, however, were to be "blind" to LMP prescriptions, focusing primarily on landscape capabilities. If the landscape analysis brought to light information and conclusions that indicated the LMP to be in error, or if the refined desired conditions significantly deviated from the LMP, then a LMP amendment would be implicated.

Public participation in the MJEMP was intended to occur in several ways. Informally, each team member was responsible for collecting and using research information from specialists who work in the MJ area or have expertise pertinent to the analyses. For the MJ area, there is considerable published work on some aspects of the landscape and a moderate amount of informal research, research interest, and local scientific focus.

More general public participation was planned for formal meetings in which information would be shared during the course of steps 1 through 3. At the time the desired condition was being developed, more active public participation was planned, with iterative meetings to identify conflicts, propose modifications, and develop an integrated desired condition. A consensus process was not intended, although input and evaluation was encouraged.

Initially the MJEMP team considered a formal collaboration with the Coalition for Unified Recreation of the Eastern Sierra (CURES) to provide iterative public feedback for developing the desired condition. Although CURES members represent diverse backgrounds, their focus on recreation meant that they would not adequately represent all public sectors likely to be interested in the management of the MJ area. Thus the team decided to use open public meetings to share information and get public feedback.

The timetable for completion of the work was initially two years for steps 1 through 5, at which point a report would be written and NEPA analyses could begin if implicated.

Data from analysis of both existing and historic conditions would be entered into a geographic information system (GIS) wherever possible. At the beginning of the MJEMP, there was no existing GIS for the forest, and team members took responsibility for developing a system that used preexisting electronic information and could provide the capacity for analyses needed during the MJEMP. Data not appropriately handled by GIS would be maintained in tabular or narrative form.

Funding for the MJEMP was on an ad hoc basis. Individual Inyo National Forest resource departments would pay from their budgets, a situation that became defensible only when the MJEMP was identified as high priority in forest planning for 1994 and 1995. Available funding played a role in determining the level of analysis of the ecological indicators and the nature of the analyses included.

Actual Goals, Process, and Progress

Between February 1994 and September 1995, the full MJEMP team held many office meetings and two field sessions. Subgroups (e.g., vegetation working group) met more often. These meetings were all working meetings of the MJEMP team, and the public was not specifically invited. Four public meetings and a field trip focused specifically on the MJEMP. The actual goals, process, and progress of the MJEMP through October 1995 are described under each of the seven steps intended to guide the process.

1. Define the Analysis Area. As indicated earlier, five alternative boundaries were discussed by the team, with those from the original LMP chosen for the MJEMP (figure 50.4). The reorganization of the Inyo National Forest (discussed under Question 3 later in this chapter), which took place at approximately the same time as the MJEMP (1994–1995), realigned management areas on the Forest into eight key landscapes (figure 50.8). The MJEMP is entirely contained within the upper Owens Watershed and coincides with that landscape on all but the eastern edge. Thus, because management philosophy has changed drastically throughout the forest as a result of reorganization, the designation of these landscape boundaries might have affected the MJEMP had the boundaries of the project not been accepted as forest landscape boundaries. As it is, reorganization strengthened the focus of the MJEMP.

Concern over the exclusion of the Dry Creek headwaters from the MJEMP persisted by those who criticized watershed fragmentation. Conversely, representatives for the special-use permit (Mammoth Mountain Ski Resort), which covers this area, voiced support for the exclusion of these lands. In fact, the flexible approach to boundaries that the team adopted resolved this issue. Actual analyses of interest (water and stream flows) were in fact conducted in the headwaters subbasin and used in understanding existing and desired conditions in the rest of the drainage.

2. Describe the Existing Condition, and 3. Describe the Historical Condition. By far the greatest efforts of the team have focused on these steps. Tables 50.3–50.5 summarize the nature of information collected or in process of being collected to address the objectives of describing existing and historic conditions. Most of the analyses are complete and have been summarized in a preliminary report (USFS 1995b). Much of the information on existing condition is quantitative and spatial, and has been entered into the Inyo National Forest geographic information system (USFS 1995c) (table 50.5).

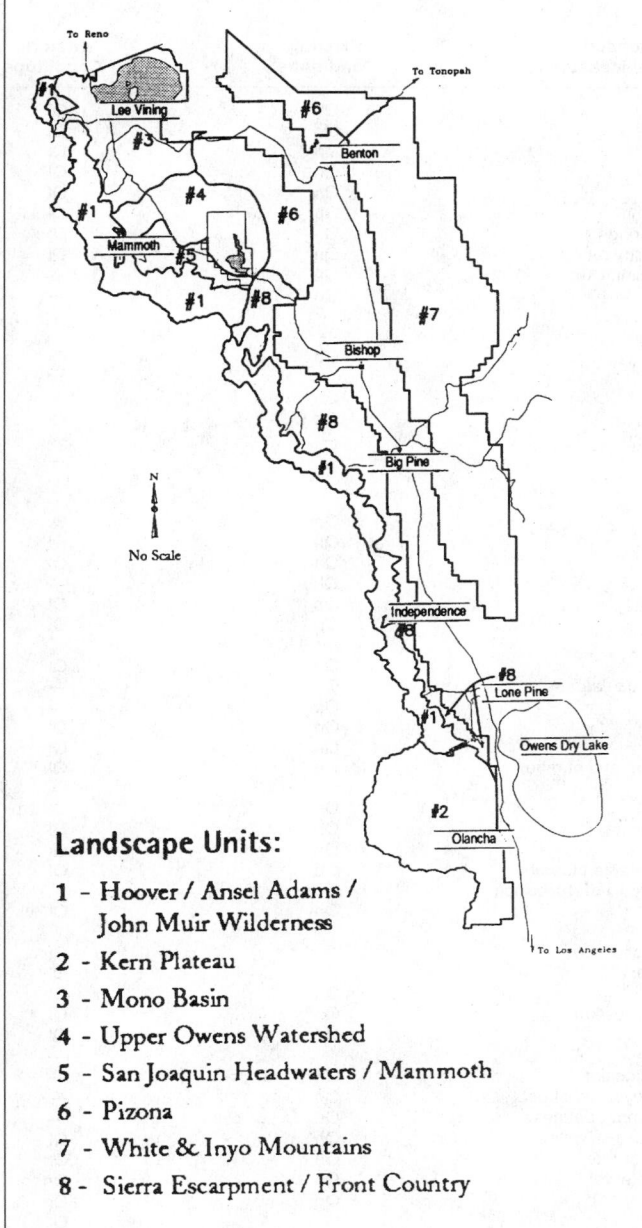

Landscape Units:

1 - Hoover / Ansel Adams / John Muir Wilderness

2 - Kern Plateau

3 - Mono Basin

4 - Upper Owens Watershed

5 - San Joaquin Headwaters / Mammoth

6 - Pizona

7 - White & Inyo Mountains

8 - Sierra Escarpment / Front Country

FIGURE 50.8

Inyo National Forest landscapes resulting from the forest reorganization of 1994–95. The MJEMP is fully contained in the Owens Watershed (landscape 4), and its boundaries coincide with that planning unit on all but the eastern edge (USFS 1995a).

Coverage, intensity, and validity of data vary by indicator. Some features have been described over the entire area, others logically included only parts of the area (e.g., aquatic conditions), and others were geographically limited by time and available funds. Depth of coverage and intensity of analyses were determined by the importance of the ecological indica-

tors and practical factors. Because they were considered critical elements, emphasis was given to forest vegetation structure and composition and to water. Key questions considered and use of data in the analyses are summarized by ecosystem elements as follows (further detail in tables 50.3–50.5, appendix 50.2, and USFS 1995b).

Water. Key questions address the nature and abundance of water flow, streams and springs, and annual/seasonal fluctuations, the impact of human activities on water quality, and the effects of diversions on stream flow. Significant natural variations have been documented for only five years. Overall, the watersheds in the MJ area, regionally very important as the headwaters of the Owens River, are in good condition. Impacts are localized, and water quality is excellent. Many springs in the area provide important scattered wetland habitat.

Vegetation structure and composition. Maps of dominant/codominant vegetation based on several sources have been compiled in the forest GIS. Because information about forest structure and composition was considered one of the most important indicators, considerable fieldwork was done to complete inventories on overstory and understory vegetation. Field mapping and polygon verification of dominant vegetation have been completed for about one-third of the forested MJ area. This mapping was augmented by information from the USFS ecology program work on red fir classification (Potter 1994). A floristic study is partially completed for much of the area outside the meadows (Constantine 1994). Based on results from fieldwork and previous data, models of species replacement with elevation and slope are being developed. The model will be used to extrapolate vegetation inventory to unsurveyed parts of the landscape.

Fire/fuels. Primary questions are, What are the existing fuel loads, stocking levels, current and historic fire intervals (spatial and temporal ranges)? What did the area look like before fire exclusion? Sampling for fuels has focused in the pine and mixed-conifer timber areas. Fire-history analyses have been done for much of the area, but sampling has been low. History studies from fire scars go back only to the 1700s; charcoal analyses from pollen cores in Glass Creek Meadows will extend these fire data when the study is complete.

Analyses indicate highly variable pre-suppression fire intervals in pine and mixed conifer types, with an average of 10–20 years, and longer in red fir (up to 30 years or more). Fires appear to have been low intensity in both pine and fir types, although stand structure in the fir types suggests that stand-replacing events occurred during certain past climatic periods. No evidence exists, however, for large fires in the fir forest in recent times. Sampling suggests high variability in the site and inten-

TABLE 50.3

Ecological indicators included in analyses of existing and desired conditions in the Mammoth-June Ecosystem Management Project compared to intended indicators.

Key Resource Area	Intended Ecological Indicator	Intended Unit of Measure	Existing Conditions[a]	Historic Conditions
Air Quality	Visibility	Miles	Ql	Ql
	PM-10	Microgram/m^3	Qn	Ql
	Ozone	ppm	Qn	Ql
Watershed	Stream-flow duration	cfs	Qn	Ql[b]
	Stream-flow timing	cfs	Qn	Ql
	Stream-flow magnitude	cfs	Qn	Ql
	Springs	Number of springs	Qn	Ql
	Channel stability	Channel stability ratings	Qn	Ql
	Stream crossings[c]	Number crossings/km	Qn	
	Soil erosion	Tons/acre	Qn	—
	Soil productivity	Qn	—	
	Snowpack water content	Qn	—	
Water Quality	Temperature	Degrees F/C	Qn	Ql
	Turbidity	jtu		
	Conductivity	mv	Qn	—
	pH	pH	Qn	Ql
	Total suspended solids	mg/l	—	—
	Total dissolved solids	ppm	—	—
Biodiversity				
Vertebrate	Key species habitat available	Acres		
	Spotted owl		Qn	Ql
	Goshawk		Qn	Ql
	Willow flycatcher		Qn	Ql
	Breeding birds		Ql	—
	Marten		Qn	Ql
	Mule deer		Qn	Ql
Vertebrate	key species habitat distribution for species listed above		Qn	Ql
	Key species population sizes	Number individuals/pairs	Qn	—
	Key species distribution		Qn	—
Amphibians	**Habitat availability**	Acres	Qn	Ql
	Habitat distribution		Qn	Ql
	Metapopulation structure	Population size and number	Qn	Ql
Plant	Vegetation composition (species mix, veg types)	Acres	Qn	Ql/Qn
	Forest structure	Seral stage/strata	Qn	Ql
		Density	Qn	Ql
		Range of age/size classes	Qn	Ql
		Susceptibility to disturbance	Qn	
	Down woody debris	Tons/acre	Qn	Ql
	Cover of: total vegetation, duff, litter, bare ground	Percentage basal cover	Qn	Ql
Fisheries	Pool habitat	Number pools/100 m	Qn	Ql
	Biomass	kg/ha	Qn	Ql
	Woody debris	Number of pieces/km	Qn	Ql
	Species distribution (trout)		Qn	Ql
	Trophic status		—	—
	Macroinvertebrates	Species composition	Ql	—
	Pool sedimentation	Percentage coverage of pool	Qn	—
	Cobble condition	Percentage embeddedness	Qn	—
Fire	Size	Acres	Qn	Ql
	Intensity	Flame length/btu ft^2	Qn	Qn
	Frequency	Recurrence interval	Qn	Qn
	Distribution		Qn	Qn
	Fuel loading	Tons/acre	Qn	Ql
	Fuel model	NFFL fuel-model type	Qn	Ql
	Fuel structure		Qn	Ql
	Cause of fire	Human, lightning, etc.	Qn	Ql
	Vegetation type	Percentage of fire in type, species, age of trees	Qn	Ql
	Fire risk	Percentage of fire type	Qn	
Insects/Pathogens	Severity of epidemics	Percentage mortality	Qn	—
	Size	Acres	Qn	—
	Distribution		Qn	—
	Species affected		Qn	—
Human Use	Prehistoric culture	Number of sites	Qn	
Recreation	**Recreation facilities**	Number facilities	Qn	Qn
	Recreation use levels	Recreation visitor days	Qn	Ql
	Visual quality	ROS/VQO ratings	Qn	Ql
		Acres in seen area	Qn	—
	Sensitivity levels	Ratings	Qn	—
	Variety class	Ratings	Qn	—
	Visual absorption capability	Ratings	Qn	—

[a]Ql indicates qualitative information; Qn indicates quanititative information.
[b]Historic conditions inferred only for Glass Creek watershed.
[c]Bold indicates a new indicator or unit of measure beyond those initially intended.

TABLE 50.4

Additional information available or being collected by the MJEMP team for use in analysis of existing and historic conditions and for development of desired condition.

Ecosystem Element[a]	Information Available or Being Collected[b]
Geology	Bedrock geology (Qn)
	Seismic, volcanic, landslide hazards (Qn)
	Fault locations (Qn)
	Geothermal areas (Qn)
	Glacial conditions (existing/historic) (Qn)
	Topography
	Stream locations
Climate	Climate regimes (existing/historic) (Qn/Ql)
	Fire weather (current) (Qn)

[a]These elements are not considered environmental indicators in analysis.
[b]Qn indicates quantitative information; Ql indicates qualitative information.

sity of fires within the MJ area. Evidence confirms a change in fire pattern (size, frequency, distribution) concomitant with suppression era. Fire suppression has considerably changed forest structure and composition, primarily in the pine types. Sampling is adequate to reach tentative and generalized conclusions about the natural fire regimes in the vegetation types within the MJ area.

Wildlife. For critical vertebrates, direct studies are assessing presence, population size, and habitat-use parameters. Determining the value of the MJ area to wildlife is especially critical because of its proximity to transmontane migration corridors. East-side habitat may be used in ways very different from habitat on the west side, where most knowledge about vertebrates derives, and habitat requirements and population viabilities may be unique. Critical species include spotted owls (first confirmed sightings reported recently on the Inyo National Forest in red fir forests; the question of interest is the extent of easterly ecotone along the San Joaquin Ridge), marten and other furbearers, goshawks, willow flycatchers, and breeding birds. The red fir forests, meadows, and riparian zones are the focus of wildlife study in both disturbed and undisturbed areas throughout the MJ area.

Insects/pathogens. Key questions are, What is the temporal and spatial pattern of mortality caused by insects and pathogens in different vegetation types? What is the nature of disturbance from insects and pathogens under normal fire regimes? To what extent has fire suppression or drought created conditions of abnormal insect and pathogen epidemic? Mortality in the MJ forests is considerably higher than background levels for the pine type, and is attributed to drought, fire suppression, and silvicultural activities.

Fish/aquatic biodiversity. Key questions pertain to the composition of fish and amphibian fauna, conditions of the stream habitat, viability of fish and amphibian popula-

tions relative to known healthy or undisturbed populations, and extent to which human activities have affected habitats and populations of the aquatic fauna. Comparisons are being made of habitat characteristics in reaches above and below disturbed areas in an attempt to assess health of populations in disturbed streams. Surveys are being conducted on springs and streams in the MJ area for fish and amphibians. Intensive surveys have been done for Yosemite toad, mountain yellow-legged frog (*Rana muscosa*), and exotic naturalized Lahontan trout. For remaining species, inventories have focused on available habitat. The fish biologist is working closely with the hydrologist to develop an understanding about the relationship of water quality and stream/spring flows to species' requirements and to evaluate the history of fish planting and management. Glass Creek is dramatic in its large population of trout, highly productive for other aquatic species, and unusual in low amounts of woody debris, all of which appear to be natural conditions. Effects of fish on frog populations are being investigated.

TABLE 50.5

Data layers in the Inyo National Forest geographic information system developed by the MJEMP team and other available sources for analyses of existing and desired conditions in the Mammoth-June landscape.

Administered Lands
Administrative Withdrawn Lands
Air Pollution Control Districts
Air Quality Non-Attainment Areas
Archaeology Sites
Bald Eagle Habitat
California Airshed Class Designation
County Boundaries
Fire History
Forest Plan Management Areas
Forest Plan Prescriptions
Forest Service Unit Boundaries
Geothermal Lease Boundaries
Goshawk Habitat
Inyo Plantations
Lakes
Land Ownership Data
Local Government Boundaries
Mammoth to June Updated Vegetation
Management Areas
Old-Growth Management Strategy (Areas)
Old-Growth Management Strategy (Corridors)
Outside Boundary of Administered Lands
Proposed Wild and Scenic Areas
Public Land Survey
Roadless Area Review and Evaluation
Soil
Springs
Stream-Flow
Streams
Trails
Unsuitable Lands
Vegetation from Landsat
Visual Quality Inventory
Visual Quality Objectives
Watershed Designation (CalWater)
Wild and Scenic River Study Areas
Wild and Scenic RiverAreas
Wilderness Areas

Historical ecology. Key questions focus on reconstructing late Holocene forests, and from this developing an un–derstanding of the pattern of vegetation dynamics through time, describing trends and periodicities in vegetation composition, rates of change, successional pathways, variation in disturbance regimes, and the nature and rate of response of vegetation to climate change. Studies include floristic and climate analysis over several thousand years as documented in pollen cores from Glass Creek Meadows and analysis of ancient downed woods atop Whitewing Mountain.

Archaeological studies are addressing the patterns and trends of human land use since the initial occupation; the history of human adaptation in terms of settlement, technologies, and resource bases; the response of human populations to environmental variation; and, conversely, the nature and extent to which human populations have affected the environment.

Together, these analyses focus directly on the historic condition of the landscape, especially on the dynamics of climate and physical changes as they relate to biotic responses.

Geology. Questions asked and data collected provide important background information on the physical landscape. These data in turn provide a baseline for analyzing impacts from and imposing constraints on future uses in the MJ area. This information will be especially important for evaluating anticipated geothermal proposals.

Recreation/visual resources. Potentials in the MJ area have been developed by surveying its visual quality, condition, sensitivity, and absorption capability. Inventory maps exist for visual data, showing Visual Quality Opportunity (VQO) classes, seen areas, sensitivity classes, and existing visual conditions. From a recreational standpoint, the presence of major montane forests—the most extensive, accessible forests of their kind on the eastern slope of the Sierra Nevada between Los Angeles and Lake Tahoe—is of prime importance. The combination of old-growth forests and expansive mountain views provides diverse, desirable scenic quality and recreational opportunities. The deep forest also serves to screen use; thus the area can accommodate a fairly large number of users without decreasing its value. Facility development has remained low and stable. Most sites are within the roaded natural recreation opportunity class. Use of the area is relatively heavy in both winter and summer. Summer use is associated with specific sites; winter use is more dispersed. Demand for mountain biking, Nordic track trails, and winter snow play areas has greatly increased in the recent period.

In sum, information about existing conditions, including variabilities, will be extensive, quantitative, and spatially based. In contrast, information about historical conditions and variabilities is highly inferential and much of it qualitative. In many cases, historical condition is based on inference about inferences (e.g., speculating on historic animal distributions based on inferred historical distribution of habitat). Thus for perhaps only one or two ecosystem elements is it possible for the team to follow the detailed, prescriptive, and quantitative approach outlined in the regional ecosystem management guidelines (Manley et al. 1995) . There are some differences of opinion on how historical information would be used to develop the desired condition. To describe this disagreement requires explaining the team's interpretations of sustainability, which is an underlying goal for developing the ecological desired condition of the MJ area (details in appendix 50.2). Several of the team members, especially vegetation, physical, and historic specialists, regard sustainability as the ability to maintain a dynamic and shifting mosaic of vegetation types, seral stages, water flows, and aquatic patterns across the landscape into an uncertain but variable future. Plants and animals would be able to cope with most disturbances without irreplaceable loss of biodiversity. Process is most important, although diversity of composition and structures is desired such that ecosystems can continue to evolve concordant with physical changes (environment/climate/human impacts). In sum, natural processes are favored because they provide resilience and adaptability to change. From a physical perspective, a sustainable watershed allows the natural fluvial processes of the stream channel to determine the habitat condition and populations of aquatic organisms. Thus, carrying capacities of terrestrial and aquatic systems would be set by natural potential of the stream ecosystem. Many team members emphasize this coarse, habitat focus, in which not all conditions and species are managed in intricate matrices of species and areas. Instead, mixes of structures and processes would be favored in levels and intensities that are the consequences of allowing natural process to predominate. The view of these team members tends to be that management for single species should be subordinate to or nested within maintenance of holistic ecosystem properties.

Further, many team members feel that information about the historic condition should not provide a target for the MJ landscape; the goal should not be to replicate specific historic conditions (appendix 50.2). For these team members historic information informs their assessment of present conditions and guides their thinking about the way elements of the ecosystem interact and respond to disturbance. Knowledge of historic vegetation pattern, for instance, provides more understanding of interactions among seral classes, successional pathways in the different forest types, and disturbance/regeneration processes. These in turn allow assessments of the effect (e.g., increased susceptibility to insects and pathogens, changed forest structure, altered species abundances in meadow) of human actions (e.g., grazing, fire suppression).

The MJEMP team is further using information on historic conditions and variabilities as a basis for determining if the rationale behind choice of ecological indicators is valid and for evaluating whether obvious structural diversity or ecological processes are missing or grossly deviant in MJ systems. In thinking about natural and historic variability, most accept that extreme and catastrophic events (e.g., catastrophic fire and flood) are within natural ranges. These events are not, however, within the ranges that would necessarily be included in the desired condition, although they undoubtedly would still occur and must be anticipated. Understanding historic conditions informs these team members about the nature of processes and disturbances that molded the landscape and thus informs about how ecosystem elements might respond to future management or environmental changes (appendix 50.2).

Knowledge of the historic fluctuations and behaviors informs the managers about constraints on levels of accepted human use (e.g., water diversions should be conservatively far within known natural ranges of fluctuation). Team members who adopt the conceptual approach of managing within known natural variabilities, however, have generally considered the concept of assigning ecologically acceptable, quantitative thresholds risky. Most have not felt that their knowledge was good enough to allow assignment of discrete, absolute thresholds, preferring instead to operate far within bounds of variability and approaching values distant from estimated "means" are only to be approached with great caution. Management proposals near the extremes of known natural variabilities would signal a need for further study before such projects were considered.

Not all team members characterize sustainability in this way, nor do they use historic information as described. For some, preservation of certain indicator species and specific habitat is the critical focus of this part in the MJEMP. These members take a single-species or single-indicator approach, advocating that priority for determining forest structure and composition should be given to requirements for viability and persistence of key species. Preserving these habitat structures in conditions desirable to indicators forms their ideas about sustainable management and appropriate goals for desired condition at MJ.

Regarding the intensity of analysis conducted to determine existing and historic conditions, priority was given to geographic areas of ecological importance or vulnerability (e.g., Glass Creek Meadows, red fir forests, streams, and springs), to keystone processes (e.g., fire, forest and meadow seral development, stream-channel morphogenesis), to critical species (e.g., marten, owl, amphibians), and to conditions that seemed already greatly outside natural variability (fire, forest structure, meadow diversity, aquatic diversity).

Public meetings during this phase emphasized information sharing in informal demonstrations. Team members set up tables where members of the public could review information, discuss topics with the specialists, and submit comments.

Plenary discussions took place before and after each session. In general, the public supported the ecological indicators chosen, although they seemed unclear (as did the team) in how social values were being measured.

4. Describe the Desired Condition. From a vegetation perspective, the ecologically desired condition is described with reference to seral diversity, stem densities, overstory species composition, canopy closure, presence of fire, patch size, cohort structure, insects and disease, regeneration, and primary understory composition (appendix 50.3). Based on inferred assessments, the present condition of the MJ forests (especially the pine types) has been variably affected by humans. Several processes (fire, insects and disease, successional patterns) and landscape characteristics inferred to have been in the historical landscape are included in the desired condition with reference to historic patterns and abundances, although the historical condition per se will not become a strict target. The emphasis is on reintroducing processes that have been significantly influenced by humans (fire, forest structure, understory composition, patch size, regeneration dynamics) and restoring more natural dynamics.

Glass Creek Meadows is another area that appears to have been significantly affected by humans, primarily through heavy grazing at the turn of the century. Early indications suggest that grazing led directly and indirectly to a change in plant species composition in the meadow, favoring forb diversity at the expense of native grasses. Grazing impacts seem to have acted in concert with climatic changes over the decades toward drier conditions, especially in meadow soils, leading to changes in species composition, gradual succession of pines into the meadow, and to head-cutting and stream-channel incision (though relatively minor compared to effects on the Kern Plateau and in Plumas County). An ecological desired condition, if described in historical terms, would suggest meadow conditions wetter than current conditions, boundaries of the forest maintained back from Glass Creek, grass species dominating forbs, and a normal but stable riparian strip blending the meadow with the stream channel. Because achieving such conditions would require heavy manipulation in the meadow, and because such a scenario is unlikely to be socially accepted (for reasons described later), the final desired condition is not written in terms of rapid return to historical conditions and processes. Rather, the desired condition emphasizes basic ecological conditions similar to those of the present, with focus on improving meadow and aquatic habitat. These goals would be achievable by allowing natural successional processes to unfold with little human intervention.

Aquatic diversity is also greatly outside the range of inferred historic variability. Introduced trout are thriving with very high productivity at least in Glass Creek. The natural condition of the stream is assumed to be fishless. Changes in invertebrate diversity due to the presence of exotic fish are assumed to be great. Habitat exists for mountain yellow-legged frog, although no extant populations are known, and Yosemite toad populations are estimated to be smaller than in the past. Introduced fish and grazing are implicated in amphibian decline.

Despite this significant deviation of current condition in aquatic diversity, this is another area where the MJ team does not favor restoration of historic conditions. Removing introduced trout from the creeks would be difficult and likely to be strongly disfavored by some public groups, especially because Glass Creek has been promoted as a site for recovery of exotic Lahontan trout. More natural (historic) stream-channel conditions—and improved water quality—is favored to restore habitat for other native aquatic species. Restoring populations of amphibians is also favored.

Several ecosystem elements in the MJ area seem to be in good to excellent condition and deviate little from inferred historical conditions and processes. These include water quality and quantity (streams and springs) and air quality. For these, the desired condition will most likely suggest that conditions continue within the present trajectories and variabilities.

Several other ecosystem elements have been difficult to assess for current trends or viability. These are elements for which inventory data of present conditions are scarce or where prediction of historic conditions or future habitat requirements is highly speculative. Examples include spotted owl, marten, mule deer, and willow flycatcher. In the case of the spotted owl, for instance, it isn't clear whether spotted owls use (or would or ever have used) the MJ area as a stable breeding habitat, whether the area was incidentally used historically or at present, or if it has become an expansion area because of changing habitat conditions (anthropogenic or natural) on the west side of the Sierra. It is difficult to assess viability of the owls in the MJ area, or their desired condition, without this knowledge. Thus, the desired condition would not be described using inferences about historical conditions, but would favor the status quo for these species and recommend further study.

Public opinion, although relatively accepting of ecological indicators, has been divided over aspects of the process that relate to desired conditions. During public meetings and in letters to the MJEMP from members of the resident local as well as transient public, strong and opposing desires were voiced about management of specific areas (USFS 1995d). There is a well-organized environmental advocacy group in the eastern Sierra that is dedicated to establishment of wilderness in the released roadless area, opposes development of alpine skiing, and favors minimal development and manipulation in other parts of the MJ area. In contrast, some members of the public who promote developed recreation support the expansion of Mammoth Mountain and development of recreation and geothermal facilities elsewhere in the MJ area. Many members of the nonlocal public travel to the MJ area for winter and summer dispersed recreation and want

parking facilities, campgrounds, trailheads, snow-play areas, roads for sightseeing, and Nordic ski trails.

With these social preferences as input, the MJEMP team will evaluate public goals relative to the ecological desired conditions that the team developed. In some cases, the goals will be compatible, and a combined desired condition can easily be written. In other cases, certain public interests will be very different from the ecological desired conditions. In other cases, public interests will conflict among themselves. These will take case-by-case analysis to arrive at specific, integrated desired conditions.

The team's general approach to integrate and resolve these public and ecological goals is to address issues at two geographic scales (appendix 50.3). At the broader landscape scale, desired conditions will be stated predominantly in ecological terms for each vegetation cover type. These will mostly be determined as described earlier, emphasizing landscape conditions that include (but do not mimic) historic processes and dynamic structures as well as reintroduction of processes that have been significantly altered by humans. Deviations as noted occur.

At a finer geographic scale, within vegetation cover types, detailed desired conditions have been developed for specific sites where public interest or resource condition dictate. Desired conditions for these sites are subordinate or nested within the desired condition for relevant vegetation types. For instance, the draft desired condition for the Inyo Craters area is stated in terms of recreation conditions as well as ecological and physical conditions (appendix 50.3). The deviations, however, would be accepted only if they do not have a significant effect on the ecological desired conditions dictated by the vegetation type in which the Inyo Craters lie.

Some social goals will also limit certain extreme conditions from being defined as management objectives, even if they are clearly natural from a historic perspective. For instance, severe, stand-replacing fire is most likely within the range of natural variation for most of the MJ vegetation types. Because of the proximity of Mammoth Lakes and June Lake communities, and because public interest for old-growth forests in the MJ area is high, large stand-replacing fires are not included in any of the desired condition statements. To the contrary, aggressive fire protection is part of the desired condition around these communities, with homogeneous forest conditions that have low risk of severe fire.

Steps 5–7. These steps of the landscape analysis have not been completed or informally addressed, and thus cannot be evaluated at this time.

Question 3: How Effective Have These Methods Been in Reaching the Goals of the Mammoth-June Ecosystem Management Project?

The goal of the MJEMP was to conduct an analysis that would result in the description of future desired conditions (management objectives) for the MJ area. These conditions would, first, provide ecological sustainability of MJ ecosystems and, second, integrate social desires with ecological objectives. The MJEMP process can be assessed for its effectiveness in

- scientific logic and feasibility (especially historical condition and natural variabilities)

- integration of ecological capacities and social values

- logistics and feasibility of the team process

- institutional effectiveness

Together with an understanding of how representative MJEMP is of other Sierran conditions (Question 4), we can assess how valuable this process, as prescribed by the USFS regional guides (Manley et al. 1995) and applied on the Inyo National Forest, would be to other situations in the Sierra Nevada and to SNEP (Question 5).

Scientific Logic and Feasibility

The premise of the USFS regional guidelines (Manley et al. 1995), adopted with minor modification for the MJEMP, relies on fundamental assumptions about ecological sustainability. In the first place is the underlying assumption that sustainability is ecologically meaningful, recognizable, attainable, and practical as a management objective. This assumption is widely debated in ecological, conservation, and political communities. For it to be useful as a management tool, sustainability requires a definition that can be translated into measurable or descriptive (not necessarily quantitative) terms, so that conditions (current, past, future) can be evaluated and monitored relative to attaining goals.

As described earlier, the MJEMP team members differed somewhat in their view of sustainability. Most took a dynamic view of structure, emphasizing the importance of natural processes, recognizing that ecosystem elements shift and change not only in cyclic, recognizable patterns of succession, but also along unique trajectories in response to novel environments and climates. In this view, what is sustained is not a static landscape structure, but resilience and adaptability to change. Change in ecological elements per se is not viewed as contrary to sustainability; indeed, it is considered part of it. A minority of the team emphasized, instead, preserving structural aspects of forest conditions (especially as habitat) over time (e.g., 100 years). In this case, what is sustained is ecosystem structure (current or restored). This view is much less accepting of change in the ecosystems, except in the direction of

restoration. In both cases, however, the assumption is that a naturally functioning ecosystem is a sustainable one.

Corollary logic was that conditions significantly modified by human activities might not be sustainable and thus should be avoided or treated with special concern. The question of whether prehistoric humans were different from modern humans in their influence on MJ ecosystems is not central in this context. Rather, actions taken by any humans, prehistoric or modern, that significantly modify natural processes would be considered potentially nonsustainable. Where human actions do not cause significant changes in ecosystem evolution, impacts are considered nonsignificant and in line with ecological sustainability. Thus, the question remains focused on defining the natural, variable (or its proxy, historic) condition and recognizing whether a system is within it or far from it. To address these questions, the MJEMP attempted to practice the logic of comparing existing conditions to inferred historical conditions. The main challenge has come in reliably describing the historic condition of each ecological indicator and the ecosystem as a whole.

Success in this description varied considerably by ecological indicator (table 50.3). An initial factor that differed by ecological indicator was the time depth used to infer historic condition. The team had earlier decided that several thousand years was a time span theoretically appropriate for understanding changes in the MJ ecosystem elements. During that time, climates have been basically modern, yet have fluctuated through warm, dry, cold, and wet intervals, in intensities and durations that could occur in the future. Strong evidence for climate changes and vegetation response in the local region is available from geomorphological studies at Mono Lake (Stine 1994) and paleoecological studies in the eastern Sierra (references and details in Millar [1996] and Woolfenden [1996]). These investigations give detailed information on the exact centuries and duration of dry and wet periods in the recent millennia. This time depth was acknowledged as important for getting beyond the climate (and hence vegetation response) of the last several centuries, which, relative to variability in the last several thousand years, has been at the coolest, wettest extreme.

The ability in fact to accomplish this time depth will be possible only in the case of the Glass Creek Meadows pollen core analysis, which is incomplete. This analysis will provide information on species compositions during the last several thousand years, fire intervals and intensities from charcoal, and the relationship of the volcanic eruptions to forest and meadow succession. Analyses of the downed forest on Whitewing Mountain, secondarily, are giving information on tree species composition in the last 500–1,000 years.

For direct measures of fire (fire scars) and species composition (using ages from stumps to grow the forest back in time) the understanding of historical condition was measurable directly and quantitatively over only several hundred years, from stem cores and stump analyses. Historic photographs (from the late 1800s) and early air photographs were used to estimate conditions 50–100 years ago. Even here, statistically the sample sizes were small, the spatial coverage limited, the investigations minimal, and the baseline post-Euro-American settlement. Primary emphasis is on understanding the historic structure of the fir and pine vegetation types, because they are then subsequently used to infer current and historic wildlife species habitat, abundance, and distribution. Glass Creek Meadows were also a primary focus, because the question of grazing versus natural succession and climate change was important. The fact that these forests and meadows grew during the cool, wet interval, which may not reflect forest conditions in the future, means that the information from the last hundred years may be skewed.

Potentially confounding the analyses further is the uncertain effect that volcanic eruptions have had on forest succession in the past 500–800 years. Tephra, ash, lava, and pyroclastic flows from volcanic vents in the area undoubtedly created conditions for primary succession in some forest and meadow areas. The extent to which current forests reflect response to those conditions or are still under the influence of these effects is unknown. If they are, forest conditions of the past several hundred years may be poor indicators of future conditions.

For some other indicators, time depth used in analysis was extremely short and clearly inadequate for assessing anything near a natural range of variation. For instance, variation in water flows of streams and springs can be known only from direct measurement; such variations have been monitored for ten years at most in the MJ area. No direct information is attainable to extend these data to other periods, and only indirect evidence from general climate can inform estimates of a truly historic range of variation. These limitations were underlined by the team.

For most variables, inferences of historic structure from data on current conditions in the MJ area are based circumstantial evidence left in the forest, estimates of conditions in "healthy" forests, and information gleaned from the literature about the expected behavior of the elements elsewhere. Most of these variables are described in broad, general terms, emphasizing landscape structure and process (forest gaps, species compositions, age class mixtures, regeneration processes, nature of senescence, and disturbance). In only a few cases has the natural range of variation been displayed quantitatively in anything like a metric distribution. From the standpoint of scientific investigation of the historic condition and of statistically validated, quantitative measurement of natural variation, these attempts appear shallow, poorly documented, and limited in time depth, and incapable of even coming close to describing realistic ranges of variation.

If the attempt were to document historic condition and natural variability at the level suggested by the regional guidelines, and to use these measures to define quantitatively precise ecological desired conditions, the MJEMP effort could be judged inadequate. However, because the MJEMP team used historic inferences to inform its understanding of the ecosystem rather than as a target for future management, obtaining precise and quantitative information about the past becomes less important. For reasons described earlier, most of the team members feel that attempting to return current ecosystems to historic architecture is both inappropriate and unattainable. This is an important deviation from interpretations of ecosystem management processes elsewhere. Rather, they argue that a more logical and achievable goal is to favor or reintroduce to the landscape some of the structures and processes that had been present in the natural or historic condition but are missing or greatly deviant from the present. This approach should be more effective, flexible, and open to adaptive management than one that attempts to go back to a specific historic condition.

Further, the team tried to emphasize historic variability most representative of current and anticipated climate variability (i.e., unknown specifics but increasing fluctuations with more extremes). Thus the goal would not be to achieve an exact fire interval of any historic period in the fir forest (which is unknowable for the past and probably inadequately indicates the current or future natural state) nor to restore exact landscape pattern and structure. Rather, the goals are to reintroduce fire per se into the fir and pine forests and to favor the general ecological pattern of regeneration and forest structure that might have occurred in the historic condition and are defensible under present and anticipated climates.

The question arises whether the desired conditions that the team describes, which are general and qualitative, are of value in guiding land management. In other words, is a science-based approach appropriate if the science cannot be done adequately? The MJEMP has cautiously answered this question in the affirmative by modifying the goals for the project. In its use of information, science-based thinking is appropriate. The team is taking an important step by not attempting to set specific quantitative goals when detailed information is not available to support them. Rather, the team acknowledges the need for general statements based on scientific insight from the studies done, which allows for a flexible approach to long-term management. Casting goals in general terms (e.g., reintroduce a missing process such as fire disturbance) should not result in management ambiguity or inability to evaluate specific projects. Rather, casting management objectives in general terms, with emphasis on important process and grossly deviant structural diversity, acknowledges the real situation in which knowledge and understanding, as well as the ability to manipulate forests precisely or prescriptively over long periods, are limited and it's impossible to predict future events (fire, drought). The criticism that historic conditions for many elements in the MJ area could not be analyzed in depth becomes less important. Detailed knowledge is not essential for the team to achieve its modified goals.

Further, the concept of recommended management range as described in the regional guidelines was not quantitatively expressed by the MJ team. Instead, these concepts are discussed in narrative as constraints or supplements to the discussion of ecological desired condition. For instance as described in the case of fire, large, severe fires may be within the historic range in most of the forest types (although rare), but are described as undesirable in the MJ area. As another example, relative species abundances in Glass Creek Meadows at present may likely have been different before heavy grazing in the late 1800s. Wildflower diversity is highly valued by the public, however, and restoration to a historic condition would not only conflict with the probable natural climate/successional trend in the meadows even without grazing but would require heavy manipulation. The desired condition is written to favor a slow and natural succession (not equal to restoration) as is occurring unassisted now since grazing and other impacts have been reduced over the last fifty years.

As a final example, the concept of recommended management range at a recreation site in the MJ area (Inyo Craters) has been described in terms of the modifications needed to promote low-impact recreation (appendix 50.3). These modifications are nested within the ecological goals for the vegetation type and watershed in which Inyo Craters occurs. This use of recommended management range seems appropriate, in that the accepted recreation activities would not cause the system to deviate from natural trends, and is considered to be consistent with the goal of achieving or maintaining ecological sustainability.

In the cases of ecosystem elements for which the current condition itself is not adequately understood and the literature for the area is inadequate (e.g., use of the MJ area by spotted owl), the team is not attempting to develop a historic condition. For these cases, the team will not rely on assumptions about historic proxies or natural range of variation. Instead, the team will favor conditions that maintain or enhance the status quo for these elements, at least until more is understood about their behavior in these east-side environments. This position is defensible and consistent in light of limited information.

The concept of thresholds has been debated energetically by the MJEMP. Although management leadership on the national forest wanted thresholds as guides for later management, and to make defensible decisions, the team was unwilling to set quantitative values to thresholds. The team has felt that (1) information was inadequate to determine defensible ecological thresholds within useful orders of magnitude, (2) ecological thresholds would become management targets, where decisions would be made to "manage to the thresholds," and (3) even with excellent knowledge, thresholds are not static and absolute values. They vary with space, time, disturbance, climate, and so on. Adding these variabilities to threshold setting made it an even more impossible task.

Instead, the team felt more confident making guesstimates for ranges of important conditions. For example, the desired condition for stream flow and spring volume is expressed widely around the range and fluctuation that have been measured in recent years (and is assumed to be an undisturbed and adequate condition). This short measured span would serve as a very general benchmark or reference for any future water diversion projects, recognizing that much wider ranges of climate may occur in the future. Those projects that would not cause deviations more significant than the natural range might be considered acceptable (from the water perspective). Any projects, however, that would cause deviations near the extremes in natural fluctuation would trigger more intensive investigation into water flow and natural variation. Projects that clearly result in conditions outside the natural fluctuations would be considered inappropriate. This represents a realistic approach to the use of historic and existing knowledge for setting management objectives and evaluating proposed projects. It emphasizes the uncertainty at the extremes of variability and would allow for more intensive study and analysis when it was needed.

Significant in the MJEMP process, and clearly implied in the regional ecosystem management guidelines, has been the resistance to propose land designations or to make recommendations for or against any of the existing projects that have been proposed for the area, such as alpine ski development, Research Natural Area, fuel reduction, or wilderness designation. In this regard, the MJEMP successfully separated the concepts of ecosystem capacities and conditions from administrative land classifications. This approach keeps the emphasis where it should be for this phase of analysis: on the primary aspects of ecosystem element viability and sustainability, rather than on the indirect aspect of land allocation. The analysts on the team have focused on what their expertise allows them to do: understand and assess the biological and physical relationships among the ecosystem parts, and infer the requirements and sufficiencies for natural ecosystem functioning. They further do not assume that a particular desired condition has a direct correlation with a land-designation category or management practice. For instance, if maintaining the ancient downed forest on Whitewing Mountain is desired, it may be achieved through several management paths, not necessarily a Research Natural Area. They leave to the next phase in the process—which will involve significantly more public input and normally would invoke the NEPA process—decisions about how best to achieve or maintain the desired landscape conditions. At that time, and only then, might land designations be proposed as a mechanism to aid in achieving these conditions.

Integration of Ecological Capacities and Social Values

MJEMP's primary focus was on analysis of existing ecological conditions and inferring (directly or indirectly) historical conditions. From the earliest meeting, the team recognized the need to develop and analyze social indicators along with ecological indicators. Quantitative measures of these attributes had already been surveyed and mapped for the MJ area (table 50.5). Further, the years of public involvement over the MJ area prior to the MJEMP had yielded enormous amounts of scoping information, which the team took as background for understanding the range of social interests in the MJ area.

Public meetings held during the end of the information-gathering period fielded information about current social desires (USFS 1995d). The public participated much more actively during workshops held later in the process, by contributing ideas about desired conditions.

Public reaction to the MJEMP process in general has been varied. Some groups that have traditionally been involved in eastern Sierra and USFS issues have been quiet or have reserved judgment. Others have expressed concern that options might be foreclosed by the emphasis on ecological sustainability. Many groups feel that the resolution of the MJ area has dragged on far too long and that over the years the USFS has made too many new attempts (new projects) to address management objectives in this area. The prevailing mood is of guarded suspicion about the procedures and success of yet another new project in the MJ area.

The most organized and vocal opposition came midway in the process from members of the local environmental community who reacted both individually and under the auspices of "Friends of the Inyo." Their primary objection was to the landscape-analysis process per se. In letters to the Inyo National Forest (Miller 1994), personal communications (conversations between B. Hawkins and various members of the public, 1994), and finally a letter from the organization's lawyer (Emerson 1995), the group argued that the entire MJEMP process should be subject to NEPA. The group stresses that the development of a desired condition for the MJ area constitutes a decision about how the land will be managed in the future and potentially permits or forecloses certain kinds of activities and land designations. Such a process, they argue, must legally be done within an environmental impact statement analysis, with full public scoping, input, reaction to alternatives, and opportunity to appeal.

Clearly, the group considered the opportunities to meet with the MJ team in the public meetings (to that date) to be inadequate. More importantly, the Friends of the Inyo group felt that the MJEMP's entire landscape-analysis process illegally weakened the public's role in determining management objectives. In the traditional NEPA process, which the LMP anticipated as a cumulative-effects analysis for the MJ area (USFS 1988), members of the public are given a powerful role in determining the fate of land management. Much of the group's concern came from its long-standing desire to establish a wilderness area in the former San Joaquin roadless area and its worry that an alternate fate might result. This sector of the public expresses the opinion that the USFS has reneged on its obligation to conduct an EIS analysis and has chosen a path that purposefully and illegally limits public involvement.

This reaction seems traceable to three situations: lack of full understanding of the landscape-analysis process, opposition to some of the intents of landscape-analysis, and distrust of the USFS to conduct any analyses that might affect land management without full public input. All three might have been mitigated if the team had conducted more public meetings early in the process. Regarding the first, the confu-

sion (and thus suspicion) about the intent and process of landscape analysis is understandable. Ecosystem management as an overall approach is itself new; its implementation as a guiding philosophy is almost without precedent. The underpinnings of forest plan implementation have not been widely described to the public. More importantly, the actual process of landscape analysis is new even to Californian national forests; the MJEMP was following a draft version of the guidelines. USFS team members themselves are learning by participating in the process, so it is not surprising that the public, steeped in the traditional NEPA process, does not clearly understand the intent of landscape analysis.

Opposition to the intent of landscape analysis, as understood by some members of the public, is based on the perceived preoccupation on identifying management projects (e.g., steps 5-7 earlier in the chapter). The public stresses that the objective of landscape analysis should not be to propose actions and projects. This concern stems from opposition to the USFS goal to "do something" on the land and the local public's desire for wilderness allocation in the MJ area, a "do-nothing" alternative. Current ecosystem management thinking in the USFS, however, considers "do-nothing" alternatives as valid (Manley et al. 1995), but this purported support has not convinced some of the eastern Sierra public.

Related to this concern is the unease which some members of the local public feel about the way landscape analysis takes the focus off administrative land designation. Land designations determine with certain finality the range of permitted or non-permitted activities (e.g., Wilderness or RNA) for a piece of land. The focus of landscape analysis, instead, on desired physical and ecological conditions purposely leaves the issue of actual land management open and subject to change.

Finally, any approach that appears to withdraw opportunities for public involvement and power in decision making is opposed. This perception extends to a distrust of the USFS's ability to carry out unbiased and adequate scientific analyses, to include the range of opinions held by special interest groups, and to manage the land as promised.

The integration of public opinion in developing the desired condition has yet to evolve. Members of the public disagreed among themselves on most of the MJ issues. The MJ team is treating public input in much the same way SNEP has: it will be informed by the input, use good ideas that surface, and try to find solutions that satisfy public desires without compromising ecological requirements. It will not try to solve conflicting public opinion independently of ecological needs.

In sum, the effectiveness of the MJEMP in reaching its goals was hindered by inadequately informing and involving the public in the early stages. More meetings, open dialogue, earnest partnership attempts, and use of public information went a long way toward engendering collaborative attitudes. By late summer 1995, public opinion seemed more supportive and less suspicious. Legal resistance, should it continue from Friends of the Inyo, may force a complete change in the planned strategy and a return to the NEPA process, thus rendering the MJEMP process inefficacious. Should this happen, the opportunity to use this process for other projects on the forest, and potentially throughout the Pacific Southwest region, could be challenged.

Logistics and Feasibility of the Process

Many idiosyncratic, or situation-specific, factors influence whether a planned strategy is effectively conducted. In a team process such as the MJEMP, there are very significant issues of personal leadership, team synergy, incentive, and support; institutional aspects such as agency culture, policy and management history, budget, staff resources, time, and supervisory support; and social factors such as local public issues, political climate, and background societal movements. Only a sampling of these is considered here.

A primary condition enabling the MJEMP team to work as successfully as it has are the individual and collective attitudes of the team leader, members, and supervisors. Although the institutional backdrop began and remained highly pessimistic—local and national budgets were inadequate, forest GIS was unavailable when the project started, forest- and national-level reorganization and downsizing were draining enthusiasms, workforces were being cut, and public pressure was heated—the team maintained a determined, "can-do" attitude. The team members worked on inner initiative. When there was a void, they filled it, creating their own interpretations, GIS, work plans, schedules, and priorities. Diversity in backgrounds of team members led generally not to conflict but to collaboration. Diverging positions, on scientific or technical aspects or in views on project orientation (e.g., management for general forest structure versus single species) did not polarize team members or create barriers to work. A challenge for the team will come in resolving the final ecological desired condition. However, the process itself, with its focus on analyzing landscape conditions and not on a course of management action, relieves the traditional pressures felt in NEPA interdisciplinary teams where specialists often end up polarized in defense of particular disciplinary views.

This is not to say that the team always worked as a unit. Much of the work was done independently or in subgroups, and there was, at times, a fair amount of confusion about the process in general and individual assignments and schedules in particular. Significant aspects limiting effectiveness were the inability to meet intended schedules, conduct planned social analyses, or meet full goals of public involvement.

A major detriment to achieving these particular goals was the lack of dedicated time and priority available to staff to accomplish the work. The routine course of national forest business requires that team members often carry up to twenty or thirty projects (most of them "urgent") at one time. Forest priorities are established annually but rarely strictly supported or enforced, so staff feel pushed and pulled on a daily basis to reprioritize from one project to the next. Time is grossly inadequate to accomplish within proposed schedules even a frac-

tion of work each specialist is assigned, with the result that all projects are compromised to some extent. With members drawn in different directions by their individual obligations, the ability to focus a large team on any one project such as the MJEMP is an enormous challenge. By contrast, the model many team members hold up as preferable is one in which they could all work on just the MJEMP together for a dedicated time.

The MJEMP team has faced additional challenges to time and work structure because of reorganization of the Inyo National Forest (USFS 1995a), a process that coincided almost entirely with the MJEMP. Reorganization has instituted major changes in the way staff work and has been enormously disruptive and time consuming. Although in the long run the reorganization should make team projects and landscape analyses such as MJEMP more effective, in the short run, the transition has greatly diminished the working capacity and incentive of many staff members.

In sum, the effectiveness of the MJEMP has depended primarily on personal staff commitment and interest and on general forest priority given to the project, and secondarily on available budgets. Primary factors limiting the logistic effectiveness of the team have been the inability to focus dedicated time on the project, staff overload, inability to foster needed public participation and the roadblocks that resulted, and the fact that forest reorganization coincided with the MJEMP.

Institutional Effectiveness

Several conditions of the MJEMP suggest at first that the landscape analysis might logically be done internally, as a USFS staff effort. The lands under analysis in the MJ area are administered by the Inyo National Forest, technical agency staff representing the major areas under study were available, funding was primarily internal, no land allocations or management prescriptions were to be made, and no environmental analysis (NEPA) was involved. As conceived nationally by the USFS and described regionally in the California handbook, landscape analysis is a technical exercise intended to identify resource capacities, limits, trends, and future conditions. Public participation is encouraged, but no formal process is outlined or required. Projects and treatments, should they be proposed, would come later in an independent process, within traditional NEPA scope.

Under closer scrutiny, the MJEMP actually had several components, some of which might not be appropriately confined to analysis by a single-agency technical team. The MJ area has a large and diverse constituency, consisting of both people interested in the area itself and those concerned about implications for adjacent lands and communities. Further, the role of the MJEMP as a flagship ecosystem management project of the Inyo National Forest meant that it received attention as a pilot process per se beyond the implications for a particular area. Public understanding of what ecosystem management actually entails, or how it will be implemented locally, was poor. The relationship of the Land Management Plan to the

MJEMP, and especially to land allocations or decisions about the future of the landscape, was unclear. Suspicion of the new process was high.

The challenge to the agency in such situations is how to co-ordinate an interactive, adaptive-management process with stakeholders prior to, and concomitant with, the technical analysis. Information needs to be brought out early, between the agency and constituents and among the different interest groups themselves, about changes in intent since the Land Management Plan, about elements of ecosystem management, and about how and why a landscape analysis would be conducted. Public views on the current and future condition of the area need to be heard early in the process, so that they can be incorporated as needed before the technical team begins.

The actual science work of the technical team belongs to specialists and resource professionals. However, this too is best conducted as an open process with vigorous input and review from experts outside the team and outside the agency. Because the analysis and interpretation of historic variability are not straightforward, significantly more scientific involvement is needed than if a routine resource inventory were being done. Opportunities for the public to learn from the specialists about technical findings in meetings and workshops, as the MJEMP team held occasionally, are important throughout the process.

The appropriate role for the various stakeholders in developing a future condition is less clear. If sustainability could be robustly described with high confidence and little variability by specialists, then the technical team would properly be the primary author. As it is, however, in situations like the MJ analysis, there is such limited understanding of what conditions (averages, ranges, and temporal variabilities) result in long-term ecological sustainability, such disparity in fact about what is socially or ecologically implied by sustainability, and such low accuracy in quantitative estimates, that the process extends beyond science and data collection. More appropriately, during the development of a desired condition the technical team would prepare technical information and analyses, including its best interpretations of long-term capacities and sustainability. The final development of a desired condition, however, is best handled as a mutually interactive, iterative, discursive process among agency staff (decision makers, planners, and specialists) and diverse constituencies (scientists, interest groups, other agencies). This approach will challenge all involved to communicate openly, and will require conscious commitment to a continuing dialectic.

Question 4. How Representative Is the MJEMP of Other Situations in the Sierra Nevada?

Biophysical Aspects

By eastern Sierran standards, the natural environment of MJEMP is unusually diverse for a small landscape and the type of diversity does not directly apply to adjacent landscapes. However, this diversity means that within a small area many of the plant, animal, and physical conditions occur that exist elsewhere in the eastern Sierra (especially Jeffrey pine, sagebrush, east-side meadows and riparian corridors), and in cismontane (red fir, montane mixed conifer), subalpine (whitebark pine/mountain hemlock), and alpine zones. Thus experience with these elements within the MJEMP will apply more broadly to these types and situations elsewhere.

The management condition of the natural environment is relatively representative for these elevations both in the eastern Sierra and elsewhere. Limited roadless areas, large blocks of harvest, and forest structure altered because of fire suppression are typical. East-side pine stands with long histories of partial overstory removal are typical for the Inyo National Forest but atypical for east-side pine in the northern Sierra where clear-cutting has been more common. Recreation development is representative for these elevations and landscapes in the eastern Sierra, as are grazing effects from past and current use.

Management Context

The dominance of USFS administration and ownership is representative of the eastern Sierra but less so for other subregions in the Sierra Nevada. This pattern of ownership makes management issues and strategies within the eastern Sierra unique in the Sierra Nevada. The MJEMP, thus, represents an eastern Sierra subregional model in this regard.

Within the context of the eastern Sierra, the management history of the MJ area samples many of the dominant issues past and present—again with great diversity in a small area—and has probably received more concentrated attention than that of other areas. The current mix of public interest and management issues captures many of the primary concerns in the eastern Sierra. Involvement, participation, and reactions of the public are representative of the eastern Sierra.

Ecosystem Management Model

The MJEMP process is widely representative of approaches to ecosystem management by many agencies and groups, especially those that favor an approach based on historic conditions, ecological sustainability, and natural variation. In particular, it reflects the most recent and specific interpretations and guidelines developed at the national and regional USFS levels. As such, the approaches adopted by MJEMP are intended to be repeated for most of the lands under USFS jurisdiction throughout portions of the Sierra Nevada within the Pacific Southwest region.

Question 5. What Can Be Learned from the MJEMP Case Study?

Local Natural Environments and Local Social Issues

The approach of comparing existing conditions to inferred historic conditions appears appropriate for most ecosystems within the MJ area at the level of analysis intended for the MJEMP. The approach is applied without obvious problems at a more qualitative, inferential level than implied by the USFS regional guidelines (Manley et al. 1995). This application is allowed by the MJEMP team's modified approach, which uses inferred historic condition to inform the understanding of current and future ecosystem relationships rather than as a target to mimic. The consequences of these situations are that (1) quantitative targets, specific targets, and detailed descriptions of desired conditions will not be developed, and thus a more flexible approach will be enabled, (2) historic conditions (and range of conditions) are not used to set targets for a future that mimics the past, and (3) important ecological processes will be favored for reintroduction where practical and implied by scientific analysis. These are realistic advances in ecosystem management thinking, acknowledging inevitable dynamism of the ecological future in the MJ area and the fact that the present and future are different enough from the past that there is little reason to consider "going back" even if it were practical.

For some attributes it has been extremely difficult even to understand current conditions, and thus it is not yet possible to estimate historic condition with an acceptable level of confidence. For these attributes, the team recommends that the status quo be maintained or that, within best professional judgment, changes to improve conditions be made where they are judged to be degraded.

To the MJ ecosystem as a whole, the approach taken has the benefit over traditional NEPA analyses in that all elements are considered together, thus providing for understanding of ecosystem interactions and cumulative effects. Because the process is proactive and holistic (albeit at a relatively superficial level), it provides a broad baseline for understanding effects of specific projects that might be proposed in the future and avoids reactive project management.

With several very important exceptions, from the standpoint of the local community, many of the social vales that both local and adjacent communities have expressed are being incorporated into the desired condition for the MJ area. Important exceptions include the desires to retain legal public participation and appeal privileges at all stages of the analysis and to set management designations on certain areas within the landscape.

Sierra Nevada Ecosystem Project

In many ways, the MJEMP represents a mini-version of SNEP although it does not represent the full diversity of conditions SNEP must contend with in the Sierra Nevada. However, SNEP may be guided in the value and limitations of the approach for assessing conditions using a historic perspective, in the choice and rationale of ecological indicators, and in the specific application of the method of using historic conditions and natural range of variation in projecting future management objectives. SNEP could take the lesson from MJEMP that because historic conditions can never be known precisely, and because current and future natural trends (even without human presence) may be very different from the past anyway,

the goal is to be informed by inferences about the past. This knowledge would best be used in making broad assessments of future conditions and choices for the future.

SNEP should be guided by the lessons learned at MJEMP of the importance of early, dedicated, and sincere involvement and communication with the public. Distrust of top-down approaches, institutional control, academic advice, and holistic solutions to problems will apply also to SNEP. Early disclosure and communication mean a greater likelihood that the results might be understood, accepted, and used.

Many of the procedural difficulties encountered on a minor scale with MJEMP involving team participation, focus, leadership, decision making, internal conflict, networking, and communication with communities of peers also have challenged SNEP. Clearly, large, interdisciplinary teams require working relationships, personal behaviors, and ground rules with which scientists and natural resource managers alike have as yet limited experience.

Situations Elsewhere in the Sierra Nevada

As mentioned earlier, the MJEMP is representative of anticipated future planning throughout the national forests of the Sierra Nevada, and thus the procedural lessons learned here apply broadly. Results from the MJEMP apply to conditions on private lands, lands with checkerboard ownership, and other federal agency lands only to the extent that management or institutional approaches resemble those of the USFS Pacific Southwest region.

Significantly, the questions raised by the attorney for Friends of the Inyo regarding the legality of the landscape process in regard to NEPA analysis pertain much more broadly than to the MJ area or the Inyo National Forest alone. Should this issue be pursued, it could cause major revisions of the nascent ecosystem management guidelines and could thus affect the way USFS landscape analysis is conducted throughout the Pacific Southwest region. Concerns expressed by the letter might have been alleviated by earlier and more extensive public involvement. Detailed analysis of alternative approaches that might result as a consequence of such public reaction are beyond the scope of this report.

ACKNOWLEDGMENTS

Many thanks go to the Inyo National Forest MJEMP team members for their openness to SNEP participation, specifically their willingness to share ideas, data, and analysis, frankness in discussions and interviews; and continued candor in describing the process and its implications. They are model ecosystem analysts and ecosystem managers. Specific thanks to Robert Hawkins and Dale Johnson of the Inyo National Forest for developing the environmental history of the MJ area and for their critical reviews of this manuscript; and to SNEP team member Tim Duane for his substantial participation in the project and with manuscript preparation. We also thank Hap

Dunning, Debbie Elliott-Fisk, Rick Kattelmann, Jonathan Kusel, and Doug Leisz (SNEP); John Schuyler and Tom Higley (Inyo National Forest) for critical and careful review comments.

REFERENCES

Burke, M. T., R. Curry, J. Major, and D. W. Taylor. 1982. *Natural landmarks of the Sierra Nevada.* Davis, CA: National Park Service.

Constantine, H. 1994. Floristic affinities of the San Joaquin Roadless Area, Inyo National Forest, Mono County, CA, Progress Report 1. Arcata, CA: Humboldt State University.

Duane, T. P. 1994. Ecosystem management and bioregional planning in the Sierra Nevada: Integrating top-down funding and analysis with bottom-up solutions. Paper presented at Annual Conference of the Association of Collegiate Schools of Planning.

Emerson, L. 1995. Letter to Inyo National Forest, representing Robbins and Livingston, Attorneys at Law.

Grumbine, R. E. 1994. What is ecosystem management? *Conservation Biology* 8 (1):27–38.

Kennedy, C. B. 1995. Development, environmental concerns, and constraints in Mono County, California. Unpublished paper, Department of Geology, State University of California, Hayward.

Manley, P. N., G. E. Brogan, C. Cook, M. E. Flores, D. G. Fullmer, S. Husari, T. M. Jimerson, L. M. Lux, M. E. McCain, J. A. Rose, G. Schmitt, J. C. Schuyler, and M. J. Skinner. 1995. *Sustaining Ecosystems: A Conceptual Framework.* San Francisco: USFS, Pacific Southwest Region and Station.

Millar, C. I. 1995. Identification and dating of downed logs on Whitewing Mountain, Inyo National Forest. Internal research report. Unpublished report. San Francisco: USFS Pacific Southwest Region and Station.

Millar, C. I. 1996. Tertiary vegetation history. In *Sierra Nevada Ecosystem Project: Final report to Congress*, vol. II, chap. 5. Davis: University of California, Centers for Water and Wildland Resources.

Miller, C. 1985. *Geology* 13:14–17.

Miller, S. 1994. Letter to Inyo National Forest.

National Environmental Policy Act (NEPA). 1970. National Environmental Policy Act. Public Law 91-90. *Federal Register*, 1 January.

Potter, D. A. 1994. *Guide to forested communities of the upper montane in the central and southern Sierra Nevada.* San Francisco: USFS Pacific Southwest Region and Station, Ecology Program.

Robertson, D. 1992. Ecosystem management of the national forests and grasslands. Washington, DC: USFS, Office of the Chief.

———. 1994. Mission, vision, and guiding principles of the U.S. Forest Service. Washington, DC: USFS.

Sieh, K., and M. Bursik. 1986. *Journal of Geophysical Research* 91:12, 539–71.

Stine, S. 1994. Extreme and persistent drought in California and Patagonia during the mediaeval time. *Nature* 369:546–49.

U.S. Forest Service (USFS). 1988. *Inyo National Forest Land and Resource Management Plan.* Inyo National Forest: USFS, Pacific Southwest Region.

———. 1992a. forest plan implementation training manual: Washington, DC: USFS.

———. 1992b. *Old growth management strategy for the Inyo National Forest.* Inyo National Forest: USFS.

———. 1992c. *Reconnaissance report for the proposed Whitewing Research Natural Area, Inyo National Forest.* San Francisco: USFS Pacific Southwest Research Station and Region.

———. 1994. *Draft Region 5 Ecosystem Management Guidebook.* San Francisco: Pacific Southwest Region.

———. 1995a. *1995 pocket guide to the new Inyo.* Inyo National Forest: USFS.

———. 1995b. Existing and historic conditions of the Mammoth-June Area: Preliminary results of the Mammoth-June Ecosystem Analysis. Bishop, CA: USFS.

———. 1995c. Inyo National Forest Geographic Information Systems Data Index (Yellow Pages). Inyo National Forest: USFS.

———. 1995d. Mammoth-June Ecosystem Management Project. Summary of notes from the 17 August 1995 meeting and copies of letters received by 21 August 1995. Bishop, CA: USFS.

Wood, S.H. 1977. Distribution, correlation, and radiocarbon dating of late Holocene tephra, Mono and Inyo Craters, eastern California. *Geological Society of America Bulletin* 88:89–95.

Woolfenden, W. B. 1996. Quaternary vegetation history. In *Sierra Nevada Ecosystem Project: Final report to Congress*, vol. II, chap. 4. Davis: University of California, Centers for Water and Wildland Resources.

Summary of Management and Land-Use History of the Mammoth-June Study Area

INTRODUCTION

The Mammoth to June Study Area is located on the east side of the Sierra Crest, between the Town of Mammoth Lakes and the Community of June Lake. The 36,000 acres of land within this area are part of the Inyo National Forest's "Mammoth to June Ecosystem Management Project". This report summarizes the management history of the Mammoth to June Study Area from the time it was designated as part of the National Forest System until the present date. It is organized into sections that address management plans, logging history, grazing history, access development, and recreation. This information was compiled from records at the various offices of the Inyo National Forest, so the focus of the report is on those resource areas of interest to the Forest Service. This report does not include information about the use of the area by native americans or by the early settlers prior to the reservation of the land from the public domain.

OVERVIEW

The western portion of the study area was designated as part of the Sierra Timber Reserve on October 1, 1890. The remainder of the area was added on July 25, 1905, to the then renamed Sierra National Forest. This area was transferred to the Inyo National Forest in 1908. Grazing dominated the use of the study area at the turn of the century, often with little or no restrictions. Timber harvest within the study area was limited to several hundred acres in the southern part of the area. Grazing use continued and timber harvest increased after the area became part of the Inyo National Forest. Recreational use increased in the 1920s, particularly in the Crestview area. Both management and use of the area since the 1920s has been a balance of recreation, timber harvest, and grazing. Roaded access to the area increased slowly, primarily associated with timber harvest. Approximately one third of the area remains unroaded.

MANAGEMENT PLANS

The general orders for Rangers at the turn of the century were to put out fires and to keep trespassing sheep out of the Reserve. There were no Rangers east of the crest until 1903, and when they did arrive they had their hands full chasing sheep and putting out fires until more help arrived in 1905. As the staffing of the Inyo National Forest increased through 1910, resource management became more focused and policy was established. Grazing administration remained the focus of planning efforts through 1920. Recreation and timber management plans were started in the 1920s, and grazing plans were revised. Copies of those plans have not been located at this time.

Integrated Use Plan

The earliest planning document found in the records is the Integrated Use Plan prepared by Forest Supervisor Neal M. Rahm in 1950. The purpose of the plan was to provide a coordinating key between all uses, facilitate administration decisions, resolve conflicts between uses, and assure continuity and consistency of Forest Administration. The plan divided the Forest into eight (8) Forest Units. Each unit was defined by conflicts that were occurring within the area, and boundaries were drawn without regard to administrative boundaries. The Forest Units could be further divided into Zones, which were based on a combination of natural features, resources, and uses.

The Mammoth to June area was within the Mammoth-Mono Unit, Mammoth Zone. The Mammoth Zone included the area from June Mountain south to Rock Creek. The north, west, and east boundaries of the Mammoth Zone are the same as the boundaries set for the Mammoth to June area. Recreation was the dominant use of the zone, particularly within the Mammoth Lakes and Deadman Creek areas. The plan called for managing the area for Rec-Wildlife, with other uses allowed if they didn't conflict with the primary use. This plan was intended to cover a ten year period, but in fact guided management until 1970. The following section summarizes the pertinent management direction:

Timber: No timber cutting will be permitted unless non-detrimental to recreation; That an adequate highway strip be preserved along Highway 395 for scenic values.

Recreation: Acquire lands in Deadman Area; Develop Deadman-Glass Creek area to relieve local pressure

Lands: Reserve government land for public service sites and encourage commercial development on private land.

Winter Sports: Prepare development plans and issue prospectus for a high class winter sports facility.

Multiple Use Plan

The next generation of comprehensive plans were the Multiple Use Plans. These plans were developed for each District, with land classification based on a standard framework of descriptions developed for the Northern California Subregion of the Pacific Southwest Region of the National Forest System. The standard guide contained definitions, characteristics, management direction, and coordinating requirements for each land classification. District Plans were attached to the subregion Management Guide, and provided management direction and coordination requirements specific to an area. The Mammoth to June Study Area is located in two districts, so management under this planning effort was directed by the Mono Lake District for the north half, and the Mammoth District for the south half. The Multiple Use Plan for the Mono Lake District was prepared by District Ranger Harold Cahill on 3/23/70, and approved by Forest Supervisor John Radel on 3/24/70. The Mammoth District Multiple Use Plan was prepared by District Ranger Richard Austin on 2/11/71, and approved by Forest Supervisor Radel on 2/16/71.

The Mammoth to June area was classified into several zones. The San Joaquin Ridge was part of the Mammoth Crest Zone. Management Direction for this zone was to safeguard the natural environment, and protect interesting and unusual features. Specific direction for this zone included direction to avoid expanding the transportation system in the zone.

The majority of the Mammoth to June study area was classified as general forest zone, further divided into three units. The northern end was part of the Hartley Springs-June Mtn-Glass Creek & Deadman unit (GF-1)on the Mono Lake District. The southeast quarter was part of the Sawmill unit (GF-1) on the Mammoth District, and the southwest quarter was part of the Mammoth Fringe unit (GF-2), also on the Mammoth District. All three units recognized the importance of recreation in this area. Although the subregional direction for this zone was to emphasize sustained yield of timber, the District plans focused on the recreational use of these areas.

Timber harvest was allowed on a limited basis, and the practices needed to enhance the recreation value of the forest. The Mammoth Fringe and Sawmill units also provided direction for expanding overnight camping areas, as well as exchanging land to allow for expansion of the community of Mammoth Lakes.

Both districts also identified Travel Influence zones that were located within the Mammoth to June area. The direction for these zones was to maintain or enhance beauty and attractiveness, and to develop suitable recreational sites. The travel influence zones included Mammoth Mountain, the Mammoth Fringe, Minaret Summit road, Inyo Craters road, Sawmill Road, Highway 395, and the Hartley Springs road.

Mono Plan

Shortly after approving the Multiple Use Plans, the Inyo National Forest and Mono County signed a "community-forest" agreement to produce a land management plan for 300 square miles of private and National Forest System land. The Mono County Board of Supervisors approved the Mono Plan for the private land around Mammoth Lakes in 1976. The sections of the Mono Plan that applied to the National Forest needed additional work to comply with the National Environmental Policy Act, the National Forest Management Act, and the Geothermal Steam Act. As a result, the Mono Plan was reviewed and converted into the Land Management Plan for the Mammoth-Mono Planning Unit.

Land Management Plan for the Mammoth-Mono Planning Unit

The Mammoth-Mono Planning Unit covered 695 square miles of National Forest Land from Mono Lake to Crowley Lake. The goal of the planning effort was to develop a plan that met the requirements of the National Forest Management Act while responding to the Mono Plan that was jointly developed by the Forest Service and Mono County.

This planning document was the first land management plan to address the "build-out" of winter sports facilities and also the development of geothermal resources. Both issues related directly to resources within the Mammoth to June Study Area. The ski area development issue was identified as a key issue in the Record of Decision for the plan. Some alternatives would have proposed development to support 71,000 skiers-at-one-time (SAOT). This level of development would have required the connection of the Mammoth and June ski areas, as well as development of Sherwin Bowl. The connection of Mammoth and June would have been along San Joaquin ridge, affecting approximately 14,000 acres of roadless area. The issue of allocating the "Roadless Areas" was deferred to the Roadless Area Review and Evaluation process that was being completed at the national level, but the other issues were addressed in the plan.

The Land Management Plan, which was approved on May 23, 1979 by Regional Forester Zane G. Smith, allocated National Forest Lands into various management zones. The San

Joaquin ridge and Glass Creek area was allocated to Zone C. The management objective for Zone C was to emphasize watershed, visual quality, recreation, and fisheries. Policies provided for activities that maintained visual quality, forest health, and habitat productivity.

The upper Dry Creek watershed was allocated to Zone D. This was the winter sports allocation, and the emphasis was on watershed, recreation, and visual quality. Policies provided for a cap on ski area development, set at 31,000 SAOT for the forest. This cap would be shared by the Mammoth and June ski areas. Policies also reserved the area connecting the Mammoth and June ski areas, as well as Sherwin Bowl, to allow for further evaluation and development of winter sports sites if development at Mammoth and June could not meet the 31,000 SAOT capacity.

The Glass Creek, Deadman Creek, and the Crestview rest stop area was allocated to Zone E, which was the developed recreation allocation. The emphasis was on watershed, visual quality, recreation, and fisheries. Other activities were allowed as long as they supported the emphasis items. This allocation applied to all the major recreation centers in the planning unit, and the plan did not provide any specific direction for the Mammoth to June area.

The majority of the study area was allocated to Zone G. This allocation emphasized watershed, timber, visual quality, and wildlife habitat. Recreation, timber, and grazing were all recognized uses. Policies provided direction for maintaining a visual quality objective of retention around Highway 395, the Inyo Craters Road, the Hartley Springs Loop Road, the Deadman Creek Road, and the Sawmill Road. Protection and enhancement of important wildlife habitats located in the vicinity of Wilson Butte and Inyo Craters were also recognized.

The issue of geothermal development was considered and a Geothermal Management Zone was identified and overlayed on top of the other Management Zone allocations. The purpose of the overlay was to identify areas suitable for leasing, but no leasing action would occur without additional studies. The southeast portion of the study area was included in this geothermal overlay.

Roadless Area Review and Evaluation

The Roadless Area Review and Evaluation process was completed for a second time at the national level in the late seventies. This process, referred to as RARE II, reviewed the existing roadless areas and evaluated their suitability for inclusion into the National Wilderness System. The western third of the Mammoth to June Study Area was part of the larger San Joaquin unit that extended into the Sierra National Forest. Two issues factored into the eventual recommendation, one was the potential ski area development along San Joaquin ridge, and the other was the construction of the trans-Sierra highway from Mammoth to the central valley along the San Joaquin drainage. In the end, San Joaquin ridge was left out of the proposed wilderness to maintain the poten-

tial for ski area development, and the San Joaquin drainage was added to the adjacent wilderness areas, eliminating the route for the trans-Sierra highway.

Geothermal Leasing for Lease Block II

The Forest Service and BLM jointly prepared an Environmental Assessment to determine what National Forest Lands in Lease Block II of the Mono-Long Valley Known Geothermal Resource Area (KGRA) were suitable for leasing. The eastern two thirds of the Mammoth to June study area were part of Lease Block II. The result of that assessment, approved by Forest Supervisor Eugene Murphy and District Manager Robert Rheiner on May 14, 1984, was to approve leasing with restrictions. A lease was issued in 198?. One of the key restrictions was a limitation on surface occupancy in key areas. This restriction applies to much of the lease in the Mammoth to June Study Area.

Inyo National Forest Land and Resource Management Plan

The Land and Resource Plan, otherwise known as the Land Management Plan or LMP, is the current planning document for the Inyo National Forest. The LMP was structured around Forest-wide Standards and Guidelines (S&G's), Management Prescriptions, and Management Areas. The S&G's are broad direction for all the resources on the Forest. The Management Prescriptions are focused direction for particular resources and are applied to a specific area of the forest, however, one Management Prescription can be applied to several different areas. The Management Areas are designated geographic areas defined by issues, opportunities, uses, or topography. Direction is more specific at the Management Area level.

The LMP was primarily a land allocation or "zoning" document, in that certain activities are allowed by the LMP depending on the Management Prescription. Considerable controversy developed over the allocation of the area between Mammoth Lakes and June Mountain, particularly as it related to ski area expansion, geothermal development, timber harvest, and recreation development. The Forest assembled a "common ground" workgroup from a cross-section of interests to evaluate these issues. The workgroup found that the detailed resource information necessary to fully evaluate the issues, as well as the future of the area, was not available. The individual values and desires of group members were also very different. Faced with the combination of different opinions and lack of information, the group was unable to reach consensus on all the issues. They did agree that further analysis was needed before any significant development took place in the area. That recommendation lead to direction in the LMP that defined the Mammoth to June Study Area, as well as creating the foundation for the Mammoth to June Ecosystem Management Project.

The LMP, which was approved by Regional Forester Paul F. Barker on August 12, 1988, allocated the Mammoth to June area to seven different Management Prescriptions (MP). The

Mammoth Mountain Ski Area was allocated to MP 13, Existing Alpine Ski Area. The purpose of this prescription is to manage the existing downhill ski areas for public use. San Joaquin ridge, upper Glass Creek, and portions of the Hartley Springs area were allocated to MP 14, Potential Alpine Ski Area. The purpose of this prescription is to maintain the potential for alpine ski development, and to retain the value as potential downhill ski developments. The Inyo Craters, Glass Creek, and Deadman Creek areas were allocated to MP 12, Concentrated Recreation Area. The purpose of this allocation is to manage the areas to maintain or enhance major recreational values and opportunities. The center section of the area was allocated to MP 16, Dispersed Recreation. The purpose of this prescription is to maintain the potential for both winter and summer high quality dispersed recreation opportunities. Glass Creek Meadows was allocated to MP 17, Semi-Primitive Recreation. The purpose of this prescription is to limit vehicular access to existing routes to protect and maintain recreation and wildlife values. The eastern section of the area was allocated to MP 9, Uneven-aged Timber Management. The purpose of this prescription is to manage suitable timberlands for the production of wood products using silvicultural treatments that maintain options for other resource emphases during the planning period. The final allocation of MP 11, Range, was applied to a small section of the southeast corner of the study area. The purpose of this prescription is to maintain or increase forage production and achieve uniform livestock distribution through maintenance or expansion of structural and nonstructural range improvements.

The allocation of the released roadless area to MP 14, and Glass Creek Meadow to MP 17, remain extremely controversial. There is strong support for opposite positions regarding future use of this area. One side would like to see the area added to the adjacent Ansel Adams Wilderness, while the other side supports increased recreational use of the area and leaving the option of ski area development open.

The issue of geothermal development leasing had been settled prior to the LMP, so the existing lease conditions were made part of the plan. All the allocations within the lease block are subject to the pre-existing rights of the leasee.

TIMBER HARVEST

The first recorded timber sale on the Inyo NF occurred in 1908, near Mammoth and in the extreme southern portion of the Study Area. The records do not indicate a purchaser, but since Mammoth boasted several small mills at this time, it no doubt went to a local mill for local use. A 1907 vintage map located at the USFS office in Lee Vining gives the location of "Home Lumber Company," as very near the present day Shady Rest Park and ballfields. It is presumed this is why the "Sawmill Cutoff Road" and the "Sawmill Timber Compartment" are named as they are.

The first mention of any timber planning effort occurs around 1920, the previous decade apparently being spent primarily on grazing issues. A timber map prepared in the mid-1960s shows several small timber sale areas, located immediately north of Mammoth, which were sold and cut in the years 1923 thru 1930. Volumes from these cuttings were probably relatively small and went to feed the small mills in Mammoth, which in turn supplied the agriculturally based communities in the Owens Valley.

There is no indication of any formal timber harvesting in the Study Area during the 1930s. The Inyo NF Integrated Use Plan of 1949-50 includes a history of past uses and trends, but gives no report on timber activities during the 1930s. The creation of the Civilian Conservation Corps (CCC) in 1933, apparently had the greatest influence on the Inyo NF during the 1930s. Roads, trails, campgrounds and Ranger Stations were all constructed by the CCC during the 1930s.

Timber planning and harvesting records improve markedly, beginning in the 1940s and continuing on up to the present day. The following is a list of all known sawlog timber sales in the Mammoth to June Study Area. While a fair amount of commercial fuelwood sales have occurred in the Study Area, their volume is relatively small in comparison and the records on these sales prior to 1970, are somewhat sketchy and incomplete.

Sale Name	Year	Volume	Location
Sawmill(?)	1923-30	?	Near Shady Rest Park
Inyo	1944-45	?	Sawmill Timber Compartment
West Crater	1946-47	?	Dry Creek area
Hartley Springs	1952-53	10.0 MMBF	Hartley Timber Compartment
Deadman Creek	1957-58	7.5 MMBF	Dry Creek/Glass Timber Compartments
Sawmill(?)	1958-59	?	East side of Sawmill Comp.
Upper Deadman	1962-63	8.5 MMBF	North side, upper Deadman Road
Shady Rest	1963-64	6.9 MMBF	South end of Sawmill Comp.
Sawmill	1964	10.6 MMBF	Sawmill Compartment
Hartley Springs	1967	?	Hartley Timber Comp.
Middle	1967-71	8.2 MMBF	North end of Dry Crk. Comp.
Mammoth Fir	1970-75	3.5 MMBF	Sawmill Timber Compartment
Glass	1972-74	4.7 MMBF	Glass Timber Compartment
Dry Creek	1978-82	9.9 MMBF	Dry Creek Timber

Prior to 1980, all harvesting in the Study Area was most likely overstory removal in nature. The old, high value trees were cut, leaving the generally smaller, younger trees to continue growing. The Study Area contains no large, old clearcuts or plantations from the past. Several factors contributed to this lack of clearcutting. In general, the Study Area was well stocked with younger trees, which were not yet large enough to be valuable for lumber and so were left behind after cutting. Also, management direction for the Study Area in particular and Forest Service land in general, favored cutting methods other than clearcuts.

In the 1949-50 Inyo NF Integrated Use Plan, the Mammoth to June Study Area falls into the Mammoth and June Lake Loop Zones. In both of these zones, the Use Plan directs water, recreation and wildlife to take precedence over timber when conflicts arise. Specifically, no timber harvesting was to occur in these zones unless "non-detrimental to recreation." The only exception was to be an unspecified area of Dry Creek where "no recreational value exists." This direction obviously was not meant to exclude timber harvesting, as the type of harvesting that began in the 1940s continued in the 1950s, but rather was meant to prevent wholesale losses in recreational opportunities that might occur if recreation potential did not exist.

Even by 1950, the growth potential for recreation in the Mammoth area was recognized. The population and economy of Los Angeles was growing, roads and automobiles were improving and plans for a ski resort at Mammoth Mountain were in the works. So, in the 1950s and early 1960s, Mammoth grew slowly and timber harvesting continued, with respect to future recreation needs. By the late 1960s, most of the eastern half of the Study Area had been harvested at least once. Typically, 30 to 40 % of the overstory had been removed, but in some areas, 60 to 70% of the overstory had been removed by 1970. The harvesting was concentrated in the eastern half of the Study Area, as this was where the more valuable pine was located. Some harvesting did occur in the fir-dominated regions of the Study Area, but the level of harvest was much lighter than in the pine-dominated areas.

By the late 1960s and early 1970s, growth in Mammoth had reached and begun to surpass its earlier expectations. At the time of publication in 1971, the Inyo NF Multiple Use Plan reported that Mammoth was the "fastest growing community in the country." The direction in this plan for the area immediately north of Mammoth, was to maintain it for recreation needs. The direction did not preclude timber harvesting from occurring, but generally supported past direction, which gave precedence to possible future recreation needs, namely campgrounds. The growth figures for Mammoth seemed to indicate the needs would come sooner, rather than later. Further to the north in the Study Area, potential recreation needs also influenced timber activities. The Multiple Use Plan recommends timber roads be coordinated with recreation needs and that a continuous green cover be maintained, even if this means only partial overstory removal cuts and multiple entries. In general, the direction was to maintain a quality recreation environment, by harvesting timber on a selection or very small group basis.

Timber harvesting in the 1970s, in the Mammoth to June Study Area more or less followed the selection harvesting direction given in the Multiple Use Plan. Despite Mammoth's growth in this decade, no new campgrounds were constructed in the Study Area. Additionally, the much discussed northward expansion of Mammoth Mountain did not occur. In 1979, the Land Management Plan for the Mammoth - Mono Planning Unit was released. This plan marked a significant change

for timber management in the Mammoth to June Study Area. Previously, timber had been ranked behind water, recreation and wildlife when conflicts arose. The Mammoth - Mono Plan of 1979, lists timber second only to watersheds in order of management emphasis. The plan lists as a goal for the Mammoth to June area, as well as much of the rest of the timbered lands on the forest, the following: "Irregular size structured stands of healthy, vigorous trees within and adjacent to existing or potential recreation development sites, scenic roads and key wildlife habitat: generally even size structured stands of healthy, vigorous trees on all other productive forest land." This seems to open the door for more intensive harvesting of large, old trees via overstory removal and clearcutting. Previous plans used more restrictive language when describing allowable harvesting in the Mammoth to June Study Area.

The 1979 Mammoth - Mono Plan solicited public input, but public comment was directed toward ski area expansion and geothermal development in the Mammoth area and no mention was made of any comments on the timber program. Apparently, old growth trees and old growth habitat were not issues yet. This new plan was the guide for the early and mid 1980s, and under this plan, there was indeed a change in timber harvest techniques. Where understory stocking was deemed inadequate, a few, small clearcut units were established in the Sawmill and Dry Creek Compartments. Overstory removal became somewhat more intensive, leaving fewer large trees standing.

Also in the late 1970s, serious planning efforts were begun for timber harvesting in the Earthquake and Deer Mtn. Timber Compartments. These compartments sit at the base of the San Joaquin Ridge and are comprised mainly of stands of large diameter, pure red fir and red fir-pine mix. Some areas of these compartments had seen harvesting many years earlier, but the harvesting was relatively light and so in large part, the Earthquake and Deer Mtn. compartments had retained most of their old growth forest characteristics.

By 1980, the data collection and environmental analysis work were complete. Plans called for harvesting of 11.5 MMBF of red fir and Jeffrey pine sawlogs, from scattered areas within the Earthquake and Deer Mtn. compartments. A forest road was to be constructed from just west of the Mammoth Mtn. Main Lodge, northward, passing on the west side of Crater Flat and ending at the Deadman Creek Road, at a point west of the Deadman Campgrounds. The road was to be multi-functional in design, serving both immediate timber access needs, as well as providing long-term recreational access to the area.

For a number of reasons, the Earthquake - Deer Mtn. Timber Sale EA was never signed by the Forest Supervisor, and hence, never implemented. From early on in the process, there was internal Forest Service opposition to the proposal. Letters and other documents included in the EA indicate that recreation and wildlife issues were the primary reasons for opposition to the proposal. Wildlife staff felt the ability to regenerate red fir by planting was still unproven, that the wildlife input was completely inadequate, especially for mule deer and that the proposed road construction would fragment the wildlife habitat. Recreation staff felt the Inyo NF, and this project area in particular, are more suited for recreation use, rather than timber procurement. They were concerned that Nordic skiing potential for the area would be compromised if the harvesting were allowed to occur.

The Earthquake - Deer Mtn. issue dragged on into the early and finally mid 1980s before final resolution was reached. In the early 80s, much of the Earthquake and Deer Mtn. Compartments were included in the Roadless Area Review and Evaluation II (RARE II), with potential for wilderness designation by the U.S. Congress. In 1984, wilderness legislation did pass, but the San Joaquin Roadless Area, which included the project area, was not included. By this time, the EA was getting somewhat out-of-date, EA guidelines were changing and the new forest plan for the Inyo NF was on the horizon. Additionally, local environmentalists were becoming organized and opposed the timber harvest plans on the grounds that it would preclude the area from any future, possible wilderness designation. In 1986, as part of the work on the new Inyo NF Land Management Plan (LMP), it was determined that no timber harvesting would occur in the red fir belt at the base of San Joaquin Ridge, for the life of the LMP (ten years). This quite clearly included the Earthquake - Deer Mtn. project area and so with removal of the red fir from the timber base, the issue was finally resolved.

Technically, current direction for timber is to be taken from the Inyo NF LMP, which was approved in 1988. More practically, current timber direction is taken from the guidelines put forth from the Washington Office, in the form of Ecosystem Management. The LMP has effectively carved up the Mammoth to June Study Area into a wide variety of land management prescription areas. Most of the Study Area is classified as either "Potential Alpine Ski Area" or "Dispersed Recreation." Roughly the eastern third of the Study Area (portions of the Hartley, Glass, Dry Creek and Sawmill Compartments) is classified as "Uneven-aged Timber Management." Aside from the mostly pure stands of red fir in the Earthquake and Deer Mtn. compartments, timber harvesting could continue, as long as uneven-aged management methods were used, openings were kept small and harvesting in general was not detrimental to higher value resources in the area.

In reality, aside from salvage harvesting and small fuelwood sales for local consumption, no timber harvesting has occurred in the Study Area since implementation of the LMP. In-house and public concerns about old growth trees and old growth habitat, and more recently the need for furbearer population surveys has curtailed the timber activities in the area. The concern over old growth gave rise to the Inyo NF Old Growth Management Strategy, which has resulted in a mapping out of old growth retention and recruitment areas throughout the Forest and a series of corridors connecting these areas. The development of this old growth issue slowed or stalled work on timber harvesting plans during the late 80s and early 90s.

The furbearer issue is ongoing and has currently stopped progress on the Dry Creek Timber Sale EA. This EA calls primarily for harvesting to be accomplished by thinning, with retention of nearly all old trees (180 yrs or older) within the sale area. However, wildlife staff have felt furbearer data are incomplete, with respect to population numbers and locations at various levels of canopy closure.

Since the mid 1980s, public interest and public comment on timber management on the Inyo NF has greatly increased. Clearcutting, loss of old growth trees and old growth habitat, loss of forest diversity, deer cover issues, excessive road systems and conflicts with numerous animal species and recreation uses have all been cited as concerns by the public. A relatively small, but sophisticated and well educated group of citizens have consistently commented on timber harvesting proposals over the past ten years. They are well connected to larger environmental groups and have been a rather effective voice for change, with respect to timber management on the Inyo NF.

A more recent and important change for timber management on the Inyo NF, has been the 1994 Ecosystem Management (EM) direction out of the Washington Office. EM, along with the more recent Forest Health initiatives, have provided support for a timber management program which looks at current forest conditions and desired future conditions, with an eye toward improved ecosystem health. On the Inyo NF and within the Mammoth to June Study Area, this has given rise to a timber program which focuses on reducing the over-stocked stands to former, healthier, more sustainable levels. By and large, this will be accomplished by thinning cuts. The timber sale in the Hartley Compartment, scheduled for sale in 1995, is an on-the-ground example of this new timber management direction. The EA for this compartment was appealed to the Regional and Washington levels, but the appeals were turned down. Interestingly, some members of the public traditionally opposed to timber harvesting have expressed some positive comments on the new timber direction on the Inyo NF. Many others, however, remain steadfastly opposed to virtually all timber harvesting, regardless of harvesting motive.

GRAZING HISTORY

The grazing history in the eastern Sierra, and presumably the Mammoth to June Study Area, dates back to the middle of the 19th century. Reports indicate that huge numbers of livestock formerly grazed on what is now Inyo NF land. Unrestricted by regulation, these bands of sheep and herds of cattle roamed throughout the area, in search of good forage.

The creation of the Inyo NF in 1907, provided the opportunity to put grazing under regulation. A shortage of personnel made the regulating process difficult, at best. The 1949-50 Inyo NF Integrated Use Plan reports that initial efforts at control consisted of nothing more than placing the livestock under permit. Actual management plans did not begin to appear until the 1920s, but numbers of animals allowed appear to reflect demand, rather than a carrying capacity dictated by the range condition. Beginning in 1944, an aggressive adjustment plan was initiated to bring permitted AUMs in line with range capabilities and to solve present and/or potential conflicts with higher ranking resources. By 1950, the AUMs on the forest as a whole, for both cattle and sheep, had been reduced by over 40 percent. Interestingly, this reduction was achieved not so much by a reduction in actual animal numbers, but by a reduction in the number of days animals were allowed on a given range allotment.

Within the Mammoth to June Study Area, there are currently two allotments. The June Lake allotment runs from the south June Lake Junction, down Highway 395 to Glass Flow Road, then west to San Joaquin Mtn, including Glass Creek Meadow, then north to include Yost Meadow and then northeast, back to south June Lake Junction, excluding the town of June Lake. Within the Study Area, 1800 sheep are allowed on this allotment, with a use period of July 1st thru August 31st. In actuality, the current sheep use of this portion of the allotment is closer to one month, rather than two months. Typically, the sheep head up toward the Hartley Springs area and spend two or three days in this area, then move up to Glass Creek Meadow for three to five days and finally over to Yost Meadow for a week to ten days. They then return out by the same route they followed in, stopping as they did on their way in. Actual days in each location vary, depending on the availability of forage.

The Sherwin-Deadman allotment covers much of the Study Area south of Deadman Creek, west of highway 395, north of highway 203 and roughly east of a line from Lower Deadman Campground, south to Deer Mtn. and southeast to near the Mammoth Ranger Station, staying out of the Mammoth Knolls and Inyo Craters areas. Within the Study Area, 1500 sheep are allowed on this allotment, with a use period of July 1st thru September 30th. In recent years, the permitee has come onto the range later, left earlier and has had only around 1000 sheep, rather than the 1500 he is allowed. Typically, the sheep enter the allotment from the south, crossing highway 203 near the junction with highway 395. The sheep are herded north and west throughout the allotment, avoiding areas of high public use and stopping short of Deadman Creek, where no grazing is allowed. Since no water is available, the permitee trucks all water into the sheep. After grazing is complete, the sheep exit the allotment to the south, again crossing highway 203.

Old records for both allotments indicate relatively stable actual numbers of sheep on the allotments over many years. Records for the June Lake allotment date back to 1914 and records for the Sherwin - Deadman allotment go back to the 1940s. While the actual numbers have remained consistent with today's figures, the permitted number of days on the allotments have been sharply reduced. Formerly, these allotments were eligible for grazing from early June thru late Oc-

tober. The current grazing periods reflect a reduction of roughly 50 percent from the allowable periods of the past.

The treatment of range issues in the various Inyo NF planning documents has remained relatively consistent for the areas within the Mammoth to June Study Area. Grazing has always ranked lower than other resource values in the Study Area, when conflicts between resources would occur. In general, these conflicts have been avoided by range management staff, or unrecognized by other resources.

Very recently, however, the issue of sheep grazing in Glass Creek Meadow has become important due to the presence of Yosemite toads in the meadow. This toad species is in the midst of a presumably unprecedented and rather spectacular decline, with causes for the decline still uncertain. Sheep grazing is probably not beneficial, and more likely harmful, to the toads and informed members of the public and wildlife personnel are pressing for the elimination of domestic livestock from Glass Creek Meadow. Range management staff have responded by urging the permitee to spend more time in Yost Meadow and less time in Glass Creek Meadow. This has been a somewhat effective stopgap measure, while a more permanent solution is worked out.

By comparison with other issues in the Mammoth to June Study Area, grazing has received scant attention by both forest planning efforts and the interested public. It seems likely that issues such as the Yosemite toad/sheep grazing conflict in Glass Creek Meadow will stimulate interest in grazing in future planning efforts by the Inyo NF and in the public at large.

RECREATION HISTORY

Early recreation use of the study area was typically camping associated with hunting and fishing. Mammoth served as base area for many early excursions as early as 1904, when Langille noted the areas popularity as a summer resort in his report on potential additions to the Sierra Forest Reserves. The first developed areas were designated in the late 1920s Plat maps for the Crestview Resort and Glass Creek summer homes are dated 1929. The CCC's were based in the Shady Rest area during the 1930s, and development of campgrounds in the Hartley Springs and Glass Creek area are probably associated with CCC projects.

Recreation use and development was fairly static through the 1950s, although management plans recognized the recreation values of the area. Recreational use on other areas of the forest grew rapidly as roads and cars improved in quality.

In most instances the study area was considered an expansion zone for Mammoth Lakes. The picture changed rapidly with the introduction of winter sports into the area. Mammoth Mountain ski area was established in 1949 in the upper Dry Creek watershed with a portable rope tow, and has grown to be the largest single ski area in the country. The Forest Service approved the first warming hut in 1952, and the first chair lift in 1955.

The forest also added recreation facilities in the late 1950s, constructing Minaret Vista and the Deadman Creek campground. By the 1960s, recreation developments were expanding in several areas. Mammoth Mountain continued a steady expansion of lifts and runs, including the addition of a gondola. The Mammoth Mountain Inn was built in 1960, and the Mammoth Chalets were constructed through 1965. The forest built the Earthquake Fault Interpretive site, as well as expanding Shady Rest Campground and constructing Pine Glen Campground. The decade of the 60s was capped with the dedication of a new visitor center next to Shadey Rest in 1969.

Recreation development since the 1960s has been limited to Mammoth Mountain and the addition of a campground south of Glass Creek. Recreation use has changed significantly. The pressure for dispersed use has shifted from hunting and fishing to camping and driving for pleasure. Nordic skiing increased in popularity in the 70s, and has held steady since then. Snowmobiling and snowplay are growing uses in the 90s, and capacity does not meet the demand for either activity. Mountain Biking is the growing summer activity, and there is an increasing demand for single track trails.

ACKNOWLEDGMENTS

This appendix was written by: Dale Johnson, Silviculture, and Robert Hawkins, Ecosystem Management/Recreation, Inyo National Forest, Bishop, California.

Summary of Records of Information, Winter 1995, Mammoth-June Landscape Analysis, Inyo National Forest

Team Member:
Date of Record: 3/23/95
Subject Area for Team: Fuels, Fire History

Key Questions for your Subject Area (i.e., what questions are guiding the studies/inventories you are doing, and how do they support the overall analysis?):Finding fire intervals (fire history), existing fuels loads, stocking levels of conifer and brush species, finding the historical range of variability, what did this area look like before fire exclusion. By finding out fire intervals and all other items address, this information directs us towards our desired condition.

Information Available about M-J Area for your Subject Area (include citation and a brief description of each; if there are many references, give general description, and where the info can be found):
Fire history- Sample slabs from stumps and trees, historical records of fire occurrence.
Fuel loading- Using Brown's inventory to find out levels of fuel loading by class. May use photo series for some areas.
    Maps? Maps showing fires of 10 acres or more. Not sure of dates. May have some of the fire history on GIS.
Vegetation
    Documents?
    Ongoing studies or surveys in the area? Only the fuels inventory.

TEAM ASSIGNMENT AND PROGRESS (i.e., your assignment & your progress)

I. Existing Conditions

What "ecological indicators" or other variables are you inventorying or mapping? Fire occurrence and size , fuel loadings, weather, topographic features, and riparian areas.

Progress? Expected completion date (be real):
Fire history is partially done, should be completed this summer.
Fuel inventories have been started and completed this summer.

What portions of M-J landscape are you focusing on? Why?
Sawmill bench area- Existing timber sale in this area that fuel treatment has started. Broadcast, jackpot, handpiles, understory burning. The rest of the landscape. We do not have good data at this time of fuel loadings, fire history etc. for the rest of this landscape.

BRIEF summary of what you've found to date:
The areas that have been sampled for fuel loadings have come up with loading of 12 ton/acre.

Found post activity fuels in decaying condition for treatment of lop and scatter. Date of activity fuels was sometime in the late 70's.

In some areas we have found encroachment of white fir in the Jeffrey Pine stands and white fir encroachment in the Red Fir stands.

Found no sign of large fire occurrence in the Red Fir area in recent years.

II. Historical Conditions

What historical conditions are you attempting to analyze?
Fire occurrence intervals, intensities, and sizes. Type of fire- was it a person caused fire or natural. Condition, age, and species of vegation. Role that fire played in this area.

Progress? Expected completion date:
Late summer or early fall. Have some info gathered but need to still do some field verification.

What portions of M-J are you focusing on? Why?
We are looking at the whole area. This give us better data for future projects

BRIEF summary of what you've found to date:
Found samples of fire history in the Sawmill bench area, fuel loadings in the same area. These samples are showing us that frequent low intensity fires at close intervals in the past and less frequent high intensity fire in present time.

III. Analysis

How do you think the info or data you are collecting on existing conditions will or should be used in the overall analysis?
Yes. It would lead to finding the desired condition.

How do you think the info or data you are collecting on historical conditions will or should be used in the overall analysis?
Yes. It will help establish the range of variability.

How do you anticipate describing natural ranges of variability for the data you are collecting? Is this concept meaningful to your part of the M-J analysis?
Intervals and intensities
Yes

What does "ecological sustainability" mean in the context of your contribution to the M-J landscape analysis?
Strategies to obtain desired conditions.

RECORD OF INFORMATION, WINTER 1995
MAMMOTH-JUNE LANDSCAPE ANALYSIS
Inyo National Forest

Team Member:
Date of Record: 3/23
Subject Area for Team: Fire History, Recreation

Key Questions for your Subject Area (i.e., what questions are guiding the studies/inventories you are doing, and how do they support the overall analysis?):

What is the historic fire frequency and intensity? Fire was a primary disturbance factor and had a significant role in shaping vegetation composition and structure. Fire history will help define the historic landscape conditions as well as contributing to a description of desired condition.

What recreational opportunities exist in the area? What potential exists? What are the demands for recreation? These will support the development of a desired condition.

Information Available about M-J Area for your Subject Area (include citation and a brief description of each; if there are many references, give general description, and where the info can be found):

Fire History - Field survey in the fall of 94 in the Sawmill Salvage timber sale area. Forest inventory of fires from the 1960's to present, with maps of larger fires. Research findings for similar vegetation cover types available in my office.

Recreation - Several maps and reports that document recreation opportunities in the area, including potential developed sites. The Forest Plan also documents recreation opportunities based on the "ROS" system.

TEAM ASSIGNMENT AND PROGRESS (i.e., your assignment & your progress)

I. Existing Conditions
What "ecological indicators" or other variables are you inventorying or mapping?

Recreation facilities and opportunities, recreation use levels

Progress? Expected completion date (be real):

Field work is complete, GIS input is in progress. Target date for completion is by mid May

What portions of M-J landscape are you focusing on? Why?

For this resource we are covering all the area

BRIEF summary of what you've found to date:

Facility development is relatively low. Most sites fit the roaded natural ROS class, however some private sector sites fit the urban ROS. Use of the area is heavy both winter and summer. Summer use is associated more with sites, winter use is more dispersed. Demand for mountain biking in the area has increased, and demand for single track trails is not being met by the available supply. Demand for winter snow play areas is also growing, and while terrain is available, safe parking or other services is not.

II. Historical Conditions
What historical conditions are you attempting to analyze?

Historic fire frequency and intensity

Progress? Expected completion date (be real):

Field work is complete, results will be combined with research findings to describe fire history by cover type. Descriptions will be done by the end of April.

What portions of M-J landscape are you focusing on? Why?

For the fire history, focus is on the Jeffrey pine and mixed conifer cover types, because fire suppression has altered the fire regime the most in these areas. The change in fire regime, combined with timber harvest, has resulted in considerable change in vegetation composition and structure. The consequences of these changes need to be considered in the discussion of desired condition.

Fire history in the sage brush is documented in several papers, and we don't have much in the study area. We have enough info to discuss desired condition.

Fire history is also well documented for the red fir cover type, and our area seems to fit the research findings. Fires occurred less frequently, so suppression of fires hasn't had the time to alter vegetation composition and structure. Most of our red fir has also not been logged. We have enough info now to determine the desired condition from a fire regime standpoint.

BRIEF summary of what you've found to date:

For the mixed conifer stands, fire occurred quite regularly until the turn of the century, and then less extensively through the 1950's. The fire interval on a stand basis (50-100 acres) ranges from 10 to 20 years.

Stands show a marked break in age classes, and preliminary work indicates that the break is tied to the last observed fire in the stand. One area examined had fires scars that indicated the last fire occurred around 1900, younger pines were 70-80 years old and fir were 60-70 years old. Stand basal area was in excess of 300 sq ft.

III. Analysis

How do you think the info or data you are collecting on existing conditions will or should be used in the overall analysis?

Recreation - The existing recreation information should be used to help shape the desired condition after the resource thresholds have been estimated. Evaluating and incorporating the use of the area by people is critical to the desired condition.

How do you think the info or data you are collecting on historical conditions will or should be used in the overall analysis?

Fire History - The fire history information is going to provide useful background on how the cover types were formed and how disturbance processes operated in the past. It will help us estimate how much the existing condition varies from the historic condition. It should provide an indication of what forest types are sustainable under different disturbance regimes.

How do you anticipate describing natural ranges of variability for the data you are collecting? Is this concept meaningful to your part of the M-J analysis?

The historic range of variability applies to this information, but the data from the survey only goes back to the early 1700's, so anything beyond that is extrapolation. The historic range of variability will be expressed as the average fire interval on a stand basis by cover type for various time periods. The information we have for two stands will be compared to the literature to extend the comparison.

What does "ecological sustainability" mean in the context of your contribution to the M-J landscape analysis?

Sustainability would be the ability to maintain a dynamic mosaic of cover types and seral stages across the landscape. The vegetation would have to be able to cope with most disturbances without irreplaceable loss of a particular component. The mosaic is defined in a social context based on a combination of social and resource values. Human influence is integrated into the process, so that uses do not impair the ability to maintain the desired mix of ecosystem components.

RECORD OF INFORMATION, WINTER 1995
MAMMOTH-JUNE LANDSCAPE ANALYSIS
Inyo National Forest

Team Member:
Date of Record: 3/24/95
Subject Area for Team: Plant Ecology

Key Questions for your Subject Area (i.e., what questions are guiding the studies/inventories you are doing, and how do they support the overall analysis?):

"What is the current existing vegetation and its spatial distribution?"

Analysis of the information answering this question will be used in determining the desired future condition.

Information Available about M-J Area for your Subject Area (include citation and a brief description of each; if there are many references, give general description, and where the info can be found):

Maps?
Currently there are maps of the dominant/co-dominant vegetation developed from orthophoto quads, LMP data, timber data, that are housed in the M-J GIS data base. There is some wildlife habitat data that the Mammoth district has and the timber compartment data set housed at Lee Vining.
Documents?
Ongoing studies or surveys in the area?
For the above two questions contact Bob Hawkins and/or Connie Millar. They have been involved with fire history, paleoecology, and a Glass Creek watershed floristic study.

TEAM ASSIGNMENT AND PROGRESS (i.e., your assignment & your progress)

I. Existing Conditions
What "ecological indicators" or other variables are you inventorying or mapping?

Percent aerial cover of existing vegetation by lifeform and species.
Slopes
Aspect
Percent basal cover of total vegetation, duff, litter, bare
        ground/gravel/rock
Down woody debris and snags

Progress? Expected completion date (be real):

We have completed the field mapping and polygon verification of dominant vegetation for approximately a third of the M-J landscape area. We will be completing the analysis on that acreage and attempting to extrapolate to the other portions of the landscape using environmental data and the existing vegetation data sources mentioned above. 5/1

What portions of M-J landscape are you focusing on? Why?

The whole landscape.

BRIEF summary of what you've found to date:

Observations from the field notes:
PIJE and PICO seem not occur together with ABMA except as traces. They seem to follow the following gradient:

```
East-----------------------------------------West
Lower elevation-----------------------------------Higher elevations
PIJE---------------------PIJE/ABMA------------ABMA
PICO---------------------PICO/ABMA------------ABMA
```

At approximately 2600 meters ABMA became dominant over PIJE or PICO.

PICO occurs on a wider amplitude of moisture conditions then does PIJE appearing to be more abundant in the higher elevations and the pumice soil types.

ABMA was more dominant than ABCO.

PIJE-ABCO has a significant shrub diversity as compared to PIJE-ABMA (soil type differences?)

PIJE-PICO tree overstory had a low shrub cover component.

ABMA does not have a shrub component except as inclusions.

Only pure sizable stand of ABCO was found on the north facing slopes above June Lake. Mammoth Mtn area was the next area of high occurrence but mixed with PIJE.

ABCO shrub understory was found to be very diverse.

PIJE seemed to be the dominant in areas of ABMA harvest.

Mainly seeing PUGL2 in understory of PIJE stands but PUTR was associated with ARTR2 stands.

PIJE- Jeffrey Pine; PICO- Lodgepole; ABMA- Red Fir; ABCO- White Fir; PUGL2/PUTR- Bitterbrush varieties; ARTR2- Sagebrush.

II. Historical Conditions
What historical conditions are you attempting to analyze?

I will be using input from other team members in the future, especially harvest information which has played a role in determining the existing vegetation.

Progress? Expected completion date:
What portions of M-J are you focusing on? Why?
BRIEF summary of what you've found to date:

III. Analysis
How do you think the info or data you are collecting on existing conditions will or should be used in the overall analysis?

In developing the desired future condition.

How do you think the info or data you are collecting on historical conditions will or should be used in the overall analysis?

In developing the historic site potentials.

How do you anticipate describing natural ranges of variability for the data you are collecting? Is this concept meaningful to your part of the M-J analysis?

I anticipate, if the information is there, to describe the natural range of variability for vegetation on the basis of current site potentials.

What does "ecological sustainability" mean in the context of your contribution to the M-J landscape analysis?

A synergistic complex of plants, animals and cultural components whose structures and processes can be maintained indefinitely.

RECORD OF INFORMATION, WINTER 1995
MAMMOTH-JUNE LANDSCAPE ANALYSIS
Inyo National Forest

Team Member:
Date of Record: MARCH 2, 1995
Subject Area for Team: SILVICULTURE

Key Questions for your Subject Area (i.e., what questions are guiding the studies/inventories you are doing, and how do they support the overall analysis?):
1. What was the historical condition of the conifer forest?
2. What is the current condition of the conifer forest?
3. What were the process that shaped the historical condition and what processes or lack of processes have shaped the current condition?
4. What is the natural range of variability for the Forest types?

Knowledge of Forest structure both historic and current as well as what structures (ecosystems) are potential and sustainable integrates directly with all other resources and tends to dictate long term sustainable desired conditions for all other resources.

Information Available about M-J Area for your Subject Area (include citation and a brief description of each; if there are many references, give general description, and where the info can be found):

Information available to me includes Don Potter's work on red fir and Jeffrey pine. CIA data for the earthquake and deer mountain compartments, old sale data and recent stand exam data for Dry Creek, Hartley and Glass compartments.
Some of this info is on GIS also Ralph Warbington's remote sensing data is on GIS.
Ongoing studies include pine marten surveys and analysis of deer hiding cover.

TEAM ASSIGNMENT AND PROGRESS (i.e., your assignment & your progress)
I. Existing Conditions
What "ecological indicators" or other variables are you inventorying or mapping?
Ecological indicators include species mix, stand structure, density, range of age/size classes, susceptibility to minor and major disturbances.

Progress? Expected completion date (be real):
I have completed the current and historic condition of Jeffrey pine, working on Red fir and lodgepole pine. Expect to finish in sometime in June.

What portions of M-J landscape are you focusing on? Why?
I have been focusing on the Jeffrey pine forest and the true fir belt below San Joaquin ridge since I have data on those areas.

BRIEF summary of what you've found to date:
Due to years of timber harvesting and fire suppression the Jeffrey pine forest no longer resembles its historic condition and is more susceptible to major disturbance from fire, pathogens and insects than ever before. However, additional thins and prescribe fire could return it to its natural range of variability within 10 to 20 years. The red fir/lodgepole pine ecosystem is eliciting some interesting info and potential hypothesis in my mind. Frequently the lodgepole is relatively young and has scattered fir and/or pine overstory. The overstory is frequently 200 to 400 years of age. This suggests to me that stand replace fires (or some other disturbance has occurred) and the lodgepole has invaded and now fir has seeded in under the lodgepole. In another 100 years may convert to fir. Also there are fewer large trees per acre in the red fir than I would have expected.

II. Historical Conditions
What historical conditions are you attempting to analyze?
I am attempting to describe historical conditions for the major cover types ie Jeffrey pine, mixed conifer, red fir and lodgepole pine.

Progress? Expected completion date: See above.

What portions of M-J are you focusing on? Why? See above.

BRIEF summary of what you've found to date: See above.

III. Analysis

How do you think the info or data you are collecting on existing conditions will or should be used in the overall analysis?

Existing conditions should provide the land manager a benchmark from which to determine if the landscape is within its desired condition, and provide a basis from which to generate projects that push the landscape to or maintain it at it desired condition.

How do you think the info or data you are collecting on historical conditions will or should be used in the overall analysis?

Information on historic conditions should provide us a basis for determining if the rationale behind our ecological indicators is valid and a basis for determining if current and proposed desired conditions are sustainable.

How do you anticipate describing natural ranges of variability for the data you are collecting? Is this concept meaningful to your part of the M-J analysis?

I expect to describe the natural range of variability in terms of forest structure, species mix etc, as well as the processes and disturbances that have molded the landscape into a sustainable ecosystem.

What does "ecological sustainability" mean in the context of your contribution to the M-J landscape analysis?

Ecological sustainability is the ability of the desired condition to maintain and replicated itself over a long period of time, ie 100s of years. It does not mean that the historic condition is replicated nor are all species and or conditions managed for.

RECORD OF INFORMATION, WINTER 1995
MAMMOTH-JUNE LANDSCAPE ANALYSIS
Inyo National Forest

Team Member:
Date of Record: 3/14/95
Subject Area for Team: Fisheries/aquatics

Key Questions for your Subject Area (i.e., what questions are guiding the studies/inventories you are doing, and how do they support the overall analysis?):

What is the makeup of the fish population, the condition of the stream habitat, and what factors (human and other) brought the existing condition to its present state?? How will this information help in the analysis of determining the DFC?? Where is the potential amphibian habitat, how do present populations compare to other populations regarded as healthy, and how much has existing occupied habitat and its populations been affected by recent, historical management activities?? Not sure how it will assist the overall M-J analysis yet.

Information Available about M-J Area for your Subject Area (include citation and a brief description of each; if there are many references, give general description, and where the info can be found):

Maps?

I have found no maps pertaining to aquatic species for the M-J area yet. The springs in the area have been mapped by the hydrologic technician in the Bishop Office, and I plan to have the amphibian crew use this as part of the M-J frog/toad survey this summer.

Documents? None located yet.

Ongoing studies or surveys in the area? The Forest has stream survey data for Glass and Deadman Creeks, and some cursory data concerning Yosemite toads in Glass Creek Meadows. I plan to survey most or all likely habitat for Yosemite toads and mountain yellow-legged frogs in the M-June Analysis Area this summer. I plan to have the stream survey data entered into our PC database this spring, and analyze the data shortly thereafter to see if/how it can be used for development of DFC's.

TEAM ASSIGNMENT AND PROGRESS (i.e., your assignment & your progress)

Literature search and review is complete to the best of our knowledge. Stream habitat data has been collected, but not completely analyzed. Fish distribution surveys are essentially complete for Glass Creek, but stage of completion is unknown for Deadman Creek; however, there may be enough data for a meaningful analysis. I will not know if amphibians can or should be used in the analysis until July-August (after some field work is done).

I. Existing Conditions

What "ecological indicators" or other variables are you inventorying or mapping?

Fisheries
% coverage of pool area by fine sediment
% cobble embeddedness
amount of pool habitat (pools/100 meters)
large woody debris (pieces/1000 meters)
biomass (kg/hectare)
distribution of different trout species throughout Glass Creek and possibly Deadman Creek
any available macroinvertebrate data

Amphibians
Habitat availability
Habitat distribution
Metapopulation pattern

I plan to work with the hydrologist on water quality and hydrology DFC's

Progress? Expected completion date (be real): 6/15/95 or soon after our Range Allotment EA's are done. If amphibians are used in the analysis, it will be late August-early Sept. If this is too late for use by the team, amphibians will have to be dropped.

What portions of M-J landscape are you focusing on? Why?

I am focusing mostly on Glass Creek and surrounding meadow areas because that is where I have the most data and it is where the only known amphibian populations exist in M-J.

BRIEF summary of what you've found to date:

Glass Creek supports a tremendous population of wild brook and brown trout and displays other signs indicative of a highly productive stream. The stream contains very low amounts of large woody debris that appears to be a natural condition. A complete barrier to upstream fish movement exists approximately 1200 meters downstream of Glass Creek Meadows. Large numbers of Yosemite toadlets were found in Glass Creek Meadows in 1993. Roads were built too close to the stream channel on parts of Deadman Creek and may be limiting the natural hydrologic conditions to some degree.

II. Historical Conditions
What historical conditions are you attempting to analyze?
Fish distribution and possible affects to amphibians

Progress? Expected completion date:
6/15/95

What portions of M-J are you focusing on? Why?
Glass Creek due to data availability and the highest amount of amphibian habitat found in the analysis are to date.

BRIEF summary of what you've found to date:
Have just started investigating historical conditions and availability of this type of data and have no concrete findings yet.

III. Analysis
How do you think the info or data you are collecting on existing conditions will or should be used in the overall analysis?
I think it should be used in a comparative role to approximate how much the ecosystem is being controlled or influenced by primarily human processes vs natural processes. In other words, are past or present human activities (management or general public use) the dominant forces driving the M-J ecosystem, or are natural ecosystem functions the dominant forces.

How do you think the info or data you are collecting on historical conditions will or should be used in the overall analysis?
I don't have enough information yet to answer this.

How do you anticipate describing natural ranges of variability for the data you are collecting? Is this concept meaningful to your part of the M-J analysis?
It would be meaningful, but I'm not sure enough data can be collected with the current deadlines to be able to determine natural ranges of variability for some of the aquatic species and their habitat.

What does "ecological sustainability" mean in the context of your contribution to the M-J landscape analysis?

A watershed in a condition that allows the natural fluvial processes of the stream channel determine the habitat condition and populations of aquatic organisms. In this way, the carrying capacity of aquatic organisms would be set by the natural potential of the stream ecosystem.

I would like to compare habitat characteristics in reaches above and below disturbed areas and attempt to get an idea of the health of the systems (Glass and Deadman Creeks). If the data allow this, I can probably approximate DFC's for various EI's in the disturbed and undisturbed reaches. I have so little data on amphibians and their habitat in M-J, I'm not sure I can develop anything accurate for amphibians beyond an approximation of unoccupied habitat, and a comparison of unoccupied habitat available now vs historically.

RECORD OF INFORMATION, WINTER 1995
MAMMOTH-JUNE LANDSCAPE ANALYSIS
Inyo National Forest

Team Member:
Date of Record: 3/23/95
Subject Area for Team: Water (air)

Key Questions for your Subject Area (i.e., what questions are guiding the studies/inventories you are doing, and how do they support the overall analysis?):

How much water is available for diversion in the project area?
Define existing conditions, past and current human-caused impacts.

Information Available about M-J Area for your Subject Area (include citation and a brief description of each; if there are many references, give general description, and where the info can be found):

Maps? Geological maps, maps created for project
Documents? Ken Heim's 1993 survey
Ongoing studies? None, except those related to geothermal developments, and our own studies.

TEAM ASSIGNMENT AND PROGRESS (i.e., your assignment & your progress)

I. Existing Conditions
What "ecological indicators" or other variables are you inventorying or mapping?

--water quality parameters: pH, conductivity, temperature, DO
--snowpack water content, stream flow
--spring locations, flow
--channel morphology with cross sections
--watershed improvement needs and conditions

Progress? Expected completion date (be real):
Ongoing

What portions of M-J landscape are you focusing on? Why?

Deadman and Glass Creeks, focusing on these as the primary flows in the watershed (ignoring Mammoth Creek watershed)

BRIEF summary of what you've found to date:

Overall, the watershed is in good condition. Impacts are localized (campsites, roads). The water quality in the project area is excellent with the exception of brief episodes of elevated sediments during the snowmelt period. There are many springs in the area providing high quality wetland habitat.

II. Historical Conditions
What historical conditions are you attempting to analyze?

None, with the exception of Glass Creek Meadow watershed condition.

Air quality: qualitative discussion of standard air quality parameters.

Progress? Expected completion date (be real):
What portions of M-J landscape are you focusing on? Why?

BRIEF summary of what you've found to date:

III. Analysis

How do you think the info or data you are collecting on existing conditions will or should be used in the overall analysis?

Existing and historic conditions should help us develop DFC. Specifically, a general water balance will help us determine the reference (NRV) variability to provide the recommended management variability.

How do you think the info or data you are collecting on historical conditions will or should be used in the overall analysis?

How do you anticipate describing natural ranges of variability for the data you are collecting? Is this concept meaningful to your part of the M-J analysis?

Water flows: quantitative
Air quality: qualitative

Could also include water quality and channel morphology.

The concept is meaningful here, because future development relies on water availability, and NRVs assist to determine how much water is available.

What does "ecological sustainability" mean in the context of your contribution to the M-J landscape analysis?

Critical. We need to insure that the land uses allowed in the project area sustain the current aquatic, riparian, and groundwater dependent ecosystem components, structures, and processes.

RECORD OF INFORMATION, WINTER 1995
MAMMOTH-JUNE LANDSCAPE ANALYSIS
Inyo National Forest

Team Member:
Date of Record: April 3, 1995
Subject Area for Team: Geology

Key Questions for your Subject Area (i.e., what questions are guiding the studies/inventories you are doing, and how do they support the overall analysis?):

Uncertain how to answer this question. The question implies that I am doing studies or inventories. To this point all I have done is input existing data. I believe this data will identify the existing geologic condition in the area and provide a baseline for analyzing impacts from and imposing constraints to future uses proposed within the study area.

Information Available about M-J Area for your Subject Area (include citation and a brief description of each; if there are many references, give general description, and where the info can be found):

Maps?
Geologic map of the Long Valley Caldera, Mono-Inyo Craters Volcanic Chain, and Vicinity, Eastern California, Bailey, Roy A., U.S. Geological Survey, Miscellaneous Investigations Series, 1989.
(Principal source of bedrock geology).

Documents?
Geologic Resource Inventory, and Evaluation of Landslide, Seismic, and Volcanic Hazards in the Inyo National Forest, Merrill and Seeley, Inc. Contract No. 53-9JC9-0-50, For the Inyo National Forest, May 1981. (Source of geologic hazard information).

Basement Structure Implication for Hydrothermal Circulation Patterns in the Western Moat of Long Valley Caldera, California, Sumnicht, Gene A. and Varga, Robery J. Journal of Geophysical Research, Vol. 93, No. B11, Pages 13,191-13,207, November 10, 1988. (I used one of the maps as a source of fault locations in the central portion of the M-J area).

Geothermal Leases issued during 1980 and 1984 contain the lease boundaries. The Environmental documents written to document the affects from leasing contain the surface and seasonal occupancy restrictions on those leases. (Sources for the geothermal lease area boundaries and use restrictions).

Hundreds of other article and publications which include all or portions of the Mammoth to June area are on file at the Inyo National Forest office in Bishop and are available for review but were not used as primary sources of information.

Ongoing studies or surveys in the area?

Ongoing monitoring of seismicity and ground deformation within the Long Valley Caldera by U.S.G.S personnel from Menlo Park, California.

TEAM ASSIGNMENT AND PROGRESS (i.e., your assignment & your progress)

I. Existing Conditions

What "ecological indicators" or other variables are you inventorying or mapping? None. Current data input entirely from existing data sources. These consist of bedrock geology, seismic, volcanic, and landslide hazards, fault locations, and existing geothermal lease areas and surface use restrictions associated with each lease.

Progress? Expected completion date (be real): April 30, 1995. Currently 90% complete.

What portions of M-J landscape are you focusing on? Why? Not focusing on any specific portion of the M-J landscape.

BRIEF summary of what you've found to date: No answer to question.

II. Historical Conditions
What historical conditions are you attempting to analyze? None

Progress? Expected completion date:
What portions of M-J are you focusing on? Why? None
BRIEF summary of what you've found to date:

III. Analysis
How do you think the info or data you are collecting on existing conditions will or should be used in the overall analysis?
Unknown

How do you think the info or data you are collecting on historical conditions will or should be used in the overall analysis?
Unknown

How do you anticipate describing natural ranges of variability for the data you are collecting? Is this concept meaningful to your part of the M-J analysis?
Unknown

What does "ecological sustainability" mean in the context of your contribution to the M-J landscape analysis? Little meaning

RECORD OF INFORMATION, WINTER 1995
MAMMOTH-JUNE LANDSCAPE ANALYSIS
Inyo National Forest

Team Member:
Date of Record: 3/15/95
Subject Area for Team: Wildlife Biologist

Key Questions for your Subject Area (i.e., what questions are guiding the studies/inventories you are doing, and how do they support the overall analysis?):1)Marten and other forest carnivores habitat use parameters
2)Spotted Owl-determine occupancy based on habitat interface between east-side pine and red fir (easterly expansion "line")
3)Goshawk-population density/estimate
4)willow flycatcher and 5)Yosemite toad-Glass Creek inventories
6)Breeding bird inventory

This biological information will provide baseline data from which to measure our progress toward a desired condition. Establish habitat parameters for sustaining viability of key species (mature habitat components, riparian integrity, bird communities, etc.)

Information Available about M-J Area for your Subject Area (include citation and a brief description of each; if there are many references, give general description, and where the info can be found):

Maps? Survey strategy maps
Documents? forest carnivore study proposal
Ongoing studies or surveys in the area? multiple

TEAM ASSIGNMENT AND PROGRESS (i.e., your assignment & your progress)

I. Existing Conditions
What "ecological indicators" or other variables are you inventorying or mapping?
See above

Progress? Expected completion date (be real):
August 31, 1995

What portions of M-J landscape are you focusing on? Why?
Timbered and riparian
Habitat for indicator species
BRIEF summary of what you've found to date:

Forest carnivores- marten in red fir, Jeffrey Pine/white fir (habitat parameters in progress)

Spotted Owl- confirmed sightings in red fir; potential habitat mapped and surveys ongoing

Goshawk- nest sites in red fir and east-side pine

Willow flycatcher- habitat survey completed

Yosemite toad- occupancy established, population estimate to be completed

Breeding bird survey-to be developed

II. Historical Conditions
What historical conditions are you attempting to analyze?

Level of sustainability for indicators species

Progress? Expected completion date:
Unknown

What portions of M-J are you focusing on? Why?

Red fir mature habitat
Riparian areas
1) Provides greatest potential for historical conditions analysis
2) Habitat is high quality, therefore provides greater opportunity for data collection and resulting development of habitat parameters

BRIEF summary of what you've found to date:
In progress

III. Analysis
How do you think the info or data you are collecting on existing conditions will or should be used in the overall analysis?

Provide for goals and objectives in order to provide for species viability

How do you think the info or data you are collecting on historical conditions will or should be used in the overall analysis?

Future planning

How do you anticipate describing natural ranges of variability for the data you are collecting? Is this concept meaningful to your part of the M-J analysis?

What does "ecological sustainability" mean in the context of your contribution to the M-J landscape analysis?

Providing for continued existence of indicator species and habitat

RECORD OF INFORMATION, WINTER 1995
MAMMOTH-JUNE LANDSCAPE ANALYSIS
Inyo National Forest

Team Member:
Date of Record: 3/20/95
Subject Area for Team: Visual Resources

Key Questions for your Subject Area (i.e., what questions are guiding the studies/inventories you are doing, and how do they support the overall analysis?):

What is the visual quality, condition, sensitivity, and absorption capability for the M-J area. From where and to what degree is the M-J area seen from key viewing platforms. What will be the visual impacts of various management alternatives and resource modifications. Will the selected management prescription meet the visual objectives identified in the Inyo Land and Resource Management Plan.

Information Available about M-J Area for your Subject Area (include citation and a brief description of each; if there are many references, give general description, and where the info can be found):

Maps? Inventory Maps for all inventoried visual data is located in the flat files in the Xerox room of the SO. Maps showing seen area, variety class, sensitivity levels, inventoried VQO's, visual absorption capability, and existing visual condition are available. This info is being entered into the GIS format for M-J.

Documents? Visual direction is located in the Forest Plan and the Mono Basin CMP. The old Visual resource handbook (big-eye book, etc.) is being replaced by a new Scenic Resource handbook this fiscal year. A draft is in my office.

Ongoing studies or surveys in the area? Not at the current time.

TEAM ASSIGNMENT AND PROGRESS (i.e., your assignment & your progress)

I. Existing Conditions
What "ecological indicators" or other variables are you inventorying or mapping?

At this time I am not inventorying any other information other than what we already have. We make occasional corrections from time to time, as conditions change, particularly viewpoints as roads or trails change or are added.

Progress? Expected completion date (be real):

Inventory and information basically completed. Terminology and language may change when the new handbook is finished.

What portions of M-J landscape are you focusing on? Why?

The visual resources are focused on the whole area. However any impacts relating to the steep escarpment lands, such as the Knolls, San Joaquin Ridge, and Whitwing will be especially critical because of their high visibility from everywhere.

BRIEF summary of what you've found to date:
Being the only major forest area between LA and Reno the M-J area provides a landscape that receives close scrutiny. The combination of timber and broad views and landscapes with high mountains and ridges provides for exciting scenery. The timber in this area also provides excellent screening ability for a large variety of impacts. Susceptible to visual disturbances are the steep slope areas and ridges mentioned above.

II. Historical Conditions
What historical conditions are you attempting to analyze?
At this time none other than existing visual condition which maps areas of disturbance and the degree of disturbance.

Progress? Expected completion date:
done.

What portions of M-J are you focusing on? Why?
The total area. Every acre has visual significance and every acre is managed with an assigned VQO through the Forest Plan.

BRIEF summary of what you've found to date:
Major Disturbances have been created by the Hwy 395 corridor and are related to this corridor. Most other areas within the M-J area have received relatively minor disturbances that are visible or none at all. Most of the area is seen as a natural appearing landscape to most viewers.

III. Analysis
How do you think the info or data you are collecting on existing conditions will or should be used in the overall analysis?
The existing visual condition can be used to determine the amount of visual change will take place under various management alternatives and with the selected alternative, if any.

How do you think the info or data you are collecting on historical conditions will or should be used in the overall analysis?
See previous answer.

How do you anticipate describing natural ranges of variability for the data you are collecting? Is this concept meaningful to your part of the M-J analysis?
   Don't think so.

What does "ecological sustainability" mean in the context of your contribution to the M-J landscape analysis?
I'm not sure yet. Any scenario that is ecologically sustainable will probably not change the quality of the scenic resource to any noticeable degree.

RECORD OF INFORMATION, WINTER 1995
MAMMOTH-JUNE LANDSCAPE ANALYSIS
Inyo National Forest

Team Member:
Date of Record: Mar. 17, 1995
Subject Area for Team: Paleoecology, archaeology

Key Questions for your Subject Area (i.e., what questions are guiding the studies/inventories you are doing, and how do they support the overall analysis?):

Paleoecology:

1. To what extent can a Late Glacial/Holocene landscape history be compiled from existing and planned data bases, for both the analysis area and the regional context?

2. Assuming the vegetation record will provide the most complete and continuous chronology and because vegetation integrates many ecological variables (i.e., hydrologic variation, soil development, substrate control, geomorphic processes) and can act as a proxy for environmental conditions, a series of related questions can be asked: what has been the pattern of vegetation dynamics through time; what were the trends and periodicities in vegetation composition; what were the rates of change; what have been the autogenic and allogenic successional trends and pathways; what have been the major disturbance regimes and at what frequency have they occurred; how has vegetation responded to climate change, and have the responses been in equilibrium with climate or not?

Archaeology

1. In general, what have been the patterns and trends of human land use since initial occupation of the analysis area; what is the history of human adaptation in terms of settlement/transhumance patterns, economic systems, technologies, and resource bases?

2. How have exchange networks evolved and what have been their relationships to local economic systems and broader interaction spheres?

3. What has been the response of human populations to environmental variation and, conversely, how and to what extent have human populations affected the environment?

Relation to the analysis:
Paleoecological and archaeological information will contribute to the description of the historic condition. It will define historic trends and ranges of variability and help to assess the stability and resilience of ecosystems and the status of the existing condition. This is essential for predicting the success of achieving the desired condition.

Information Available about M-J Area for your Subject Area (include citation and a brief description of each; if there are many references, give general description, and where the info can be found):

Maps? Archaeological sites are plotted on Heritage Resource Atlases (7.5' and 15' Quads) filed in the Lee Vining and Bishop offices. Site locations within the M/J analysis area have been entered in the Forest GIS system.

Documents? The nearest paleoecological sites (pollen and macrofossils), Starkweather Pond and Barrett Lake, along with other, more distant Sierra Nevada sites, have been published by Scott Anderson in Journal of Ecology 78:470-489 (1990). Pleistocene glacial geology of the Mammoth Lakes area has been summarized by R. R. Curry in University of Montana Department of Geology, Geological Serial Publication 11 (1971). Late Pleistocene glacial geology of Mono Basin (including the north end of the M/J analysis area) is summarized by M. Bursik and A. Gillespie in Quaternary Research 39:24-35 (1993). Holocene climatic and glacial history is presented by R. R. Curry in Geological Society of America Special Paper 123 (1969). Holocene volcanism of the Inyo Craters is discussed by C. Miller in Geology 13:14-17 (1985) and by D. E. Sampson and K. L. Cameron in Journal of Geophysical Research 92(B10) (1987). Research on the Holocene volcanism of the Mono Craters is presented by K. Sieh and M. Bursik in Journal of Geophysical Research 91:12,539-12,571 (1986). The overall geology of the Mount Morrison Quadrangle is summarized by C. D. Rinehart and D. C. Ross in U.S. Geological Survey Professional Paper 385:1-106 (1964). There are numerous published and unpublished reports on the archaeology of the area, both surveys and data recovery projects. This gray literature has been generated under contract with the Forest Service, Bureau of Land Management, CalTrans, and private developers. All reports are on file at the Bishop and Lee Vining Offices, Inyo National Forest.

Ongoing studies or surveys in the area? None that I know of.

TEAM ASSIGNMENT AND PROGRESS (i.e., your assignment & your progress)

My assignment is, with other team members, to inventory and analyze the historic condition. To date, all archaeological and paleoenvironmental documents have been compiled and read. A short sediment core for pollen, macrofossil and charcoal analysis has been retrieved from Glass Creek Meadow and is in cold storage, awaiting processing. Cross-sections of logs from the Whitewing Mountain "ghost forest" are undergoing analysis, in collaboration with Connie Millar of PSW; included is taxon identification, tree-ring dating, and radiocarbon dating. Cross-dating using skeleton plots was unsuccessful and so direct measurement with computer plotting is planned.

I. Existing Conditions

What "ecological indicators" or other variables are you inventorying or mapping? The only indicator relevant to my analysis is fire frequency (see historic condition for details).

Progress? Expected completion date (be real):

What portions of M-J landscape are you focusing on? Why?

BRIEF summary of what you've found to date:

II. Historical Conditions

What historical conditions are you attempting to analyze? Prehistoric vegetation, volcanic, tectonic, glacial, fire, and cultural history since the last glacial maximum (18,000 years). See above for details.

Progress? Expected completion date: Sept. 1995.

What portions of M-J are you focusing on? Why? The entire project area for cultural, volcanic, tectonic, and glacial history since archaeological sites, tephra, faulting, and glacial till and moraines are located throughout; Glass Creek Meadow for the pollen and charcoal (fire) record although the record (among the other pollen and macrofossil records) ranges from local to regional in extent.

BRIEF summary of what you've found to date: Even a summary of the data base to date is too extensive for this questionnaire, but I'll send it if you want.

III. Analysis

How do you think the info or data you are collecting on historic conditions will or should be used in the overall analysis?

The prehistoric data should help define the temporal range of variability of ecosystem processes in the study area and place the present condition of the ecosystems along their temporal trajectories. Data on prehistoric conditions and trends will provide comparative information to assess the extent of ecosystem modification by historic and present management practices. Reconstruction of past disturbance regimes (fire and volcanism) will contribute to knowledge about the resilience and thresholds of M/J ecosystems as reflected in the vegetation (pollen) record.

How do you anticipate describing natural ranges of variability for the data you are collecting? Is this concept meaningful to your part of the M-J analysis?

1. Temporal ranges of variability in vegetation patterns (abundances of taxa and composition of plant communities relative to disturbances, climate change, and internal biotic dynamics) will be described by analyzing pollen profiles using numerical methods (such as Spearman's rank correlation coefficient, principal components analysis, and cluster analysis) for comparing stratigraphic sequences. Baseline pollen representation of past vegetation will be estimated using a dissimilarity coefficient (probably squared chord distance) to identify modern analogs for pollen spectra.

2. Rates of vegetation change will be determined by the analysis of pollen percentage data using the ordination technique of Detrended Correspondance Analysis.

3. An attempt to relate pollen spectra to climatic variables for a paleoclimatic chronology will be made using ecological response surfaces as a transfer function.

4. Large and small particle charcoal accumulation rates, correlated with fire scar chronologies, will be used to estimate variation in fire frequencies (intervals).

The concept of natural ranges of variability has a central place in my analysis.

What does "ecological sustainability" mean in the context of your contribution to the M-J landscape analysis?

Sustainability is relative to the management objective for an ecosystem. If the objective is a complete transformation or destruction of an ecosystem, then sustainability is meaningless. If minimal disturbance to or protection of an ecosystem is a management objective, then paleoecological information is necessary for determining the thresholds within which ecological processes can be maintained, not to sustain the function or structure of an ecosystem indefinitely, since that is contrary to the nature of ecosystem dynamics, but to sustain the operation of those processes to allow its evolution relative to ongoing contextual trends (including external human influences).

APPENDIX 50.3

Draft Desired Condition for the Mammoth to June Analysis Area

INTRODUCTION

This document is the second of three reports that will be developed by the Mammoth to June analysis team. It describes the proposed Desired Condition for the Mammoth to June analysis area. This description was prepared by the Inyo National Forest line officers and analysis team members. This proposed Desired Condition will be distributed to the public for review and comment. The analysis team and line officers will review the comments before releasing the final version of the Desired Condition.

The Desired Condition is an integrated description of how we want our analysis area to exist, now and into the future. We have developed desired conditions for the landscape as a whole, and for specific geographic areas within the landscape. At each scale, the desired condition will address the blend of social, physical, and biological conditions that we would like to see in the area.

DEVELOPMENT OF THE DESIRED CONDITION

The desired condition is based on several sources of information, including the LMP, our understanding of the resource capabilities, and the comments from the public. The LMP provides the basic framework for minimum resource conditions, as well as providing direction for land allocations. Included with the LMP are current decisions that have committed the Forest Service to specific desired conditions within the analysis area. Some key decisions include the geothermal leases in the area, existing permits, public utilities and roads, the approved Master Development Plan for Mammoth and June Ski Resorts, the Shady Rest Community Park development plan, and the Glass Creek campground rehab project. Even with these sideboards, the LMP provides enough inherent flexibility to develop a desired condition that is based on the principles of ecosystem management.

The Existing and Historic Conditions report summarized our understanding of landscape components, structure and process. While we recognize that we did not document every facet of the analysis area, the report provides specific information for many key elements. This information was used to identify the variability within each element, as well as processes that currently operate or have operated within the landscape. Past and present structural components were also identified. Considered together, this information helped the team identify resource conditions that could be sustained over time.

The public comments provided proposed desired conditions from groups and individuals, as well as indicating key areas of interest. We collected comments from over 40 individuals at our August 17 public meeting, as well as receiving over 90 letters about the desired condition. The desired conditions expressed by the public covered a wide range of social settings, ecosystem processes, and structural elements. Some common areas of interest included the red fir vegetation series and the area around Glass Creek meadow.

The team took this information and worked with the line officers to develop an integrated description of the Desired Condition. The process was conducted in a series of meetings over several days. The focus was on identifying ranges of ecosystem elements that were compatible. The results are presented in the following sections. Some aspects of the proposed Desired Condition represent a change from the condition identified in the LMP. When this is the case, the LMP will need to be amended before we can implement that aspect of the desired condition. The amendment process will follow the requirements of the National Forest Management Act as well as the National Environmental Policy Act.

The Desired Condition is written in the present tense to represent landscape conditions after implementation of the LMP.

DESIRED CONDITION FOR THE MAMMOTH TO JUNE ANALYSIS AREA

The Mammoth to June landscape encompasses 36,000 acres of National Forest land between the Town of Mammoth Lakes and the community of June Lake. The area is bounded to the east by Highway 395, and to the west by the Ansel Adams wilderness. This area is considered the headwaters of the Owens River.

The Desired Condition is organized in two sections, the first is the overall Desired Condition at the landscape scale, and the second is a description of Desired Conditions for specific, smaller areas within the landscape. Although the analysis team recognized the interdependence between many of the elements, the landscape descriptions are divided into physical, biological, and social elements for organizational purposes and clarity. The Desired Condition for vegetation provided the best opportunity to consider the interaction for several elements, and as a result, the Desired Condition for fire, soils, and wildlife habitat will be discussed in that section.

Physical Elements

Geology. The scientific integrity of the volcanic features in the White Wing and Obsidian flow area is maintained.

Watershed: Overall watershed condition is very good. Riparian vegetation functions to filter sediment and provide bank stability; the uplands, wetlands, and streambanks function to allow water storage during high flow with releases during the rest of the year to sustain stream flow; and the channel and associated floodplains transport the high flows without accelerated erosion or accelerated alterations of channel morphology. Streamflow timing and magnitude for Glass Creek and Deadman Creek, as well as the springs in those watersheds, reflect climatic input from precipitation and snowmelt. Dry Creek flow regimes fluctuate in response to erosion and runoff control practices at the Mammoth Mountain Ski Area. Water releases from the ski area are regulated to avoid accelerated erosion of the Dry Creek stream channel.

The interaction between surface flows and groundwater is recognized. Groundwater resources are used to support consumptive uses of water, but extraction does not adversely impact beneficial uses that depend on base flow levels that are sustained by groundwater.

Water quality parameters are within the legal limits as defined by the Regional Water Quality Control Board and described in the Basin Plan for the Owens River. Stream turbidity, temperature, conductivity, and pH falls within the bounds described in the following table:

Component	Upper Owens River	Deadman and Glass Creeks	Dry Creek
Turbidity (JTU's)	0–15	0–15	0–100 ave.=15
Temp C	0–20	0–20	0–20
Conductivity (umh)	30–180	20–60	20–250
pH	6.2–7.5	6.2–7.5	6.2–7.5

There is no evidence of the introduction of human waste into the surface waters of the area.

Soils. Soil loss does not exceed the rate of soil formation (approximately the long-term average of 1 ton/acre/year). Areas with a high or very high erosion hazard are managed to minimize soil loss. Infiltration rates are high.

Volcanic soils have low bulk density, are readily permeable, and have high infiltration rates. Soil porosity (measured as bulk density) in vegetated areas is at least 90 percent of the total porosity found under undisturbed conditions.

Surface bulk densities of upland pumice soils within the area range from approximately .85 gm/cm3 to 1.10 gm/cm3. Subsoil bulk densities range from approximately 1.25 gm/cm3 to 1.58 gm/cm3. Soils forming in granitic or metamorphic parent materials tend to have slightly higher bulk densities. Soil compaction areas are limited to those sites designated for intensive use, such as roads, trails, and developed recreation sites. Temporary use sites that result in soil compaction are treated to reduce compaction.

The amount of organic matter in the soil varies depending on the vegetation series. The Desired Condition for vegetation describes this component of soils.

Air Quality. Air quality is within all legal standards, and activities that affect air quality are in compliance with the State Implementation Plans for the Mammoth Lakes and Mono Basin nonattainment areas. Air quality and visibility are usually outstanding. Smoke from vegetation fires may be present during certain times of the year, reducing visibility.

Fire and Fuels. Fire operates as a process throughout the landscape to provide nutrient cycling, fuel reduction, and vegetation succession. The degree to which fire occurs will vary by vegetation type, as detailed in the Desired Condition for vegetation.

The Town of Mammoth Lakes, Mammoth and June Mountain ski areas, the community of June Lake, and the Glass Creek recreation area are protected from high intensity fires that might occur within the area, and the landscape is protected form fires that might originate in these areas of concentrated use. The risk of fire spread is reduced by decreasing ladder fuels around these areas, by reducing fuel loads to less than 20 tons per acre, and by evenly spacing tree crowns in adjacent area so that contact between crowns does not occur.

The risk of large, high intensity fires within the landscape is low. Lightning caused fires may be managed using Prescribed Natural Fire plans, or one of the three fire suppression strategies (confine, contain, control). Human caused wildfires will be suppressed in the most cost effective manner, with cost being defined as suppression cost plus net value change.

Biological Environment

Vegetation. The vegetation across the landscape represents a complex mosaic of vegetation series. The relative occurrence or distribution of vegetation series changes little over time, and any large scale changes are in response to climatic changes and not human activities. The structure, species distribution, and size of individual plants will change over time as plant communities respond to processes operating within the landscape.

The Desired Condition for vegetation includes descriptions for species composition, vegetative cover, size class, snags (standing dead trees >12" diameter and >20' tall), logs (downed trees >12 diameter and >20' long), fire frequency, fire intensity, fuel loading, and surface cover. These descriptions apply to the overall condition within the vegetation series. Vegetation conditions for a specific area may vary from the range of conditions described for the series as a whole so that other desired conditions can be achieved.

Subalpine Series

This series is characterized by dispersed stands and scattered individual trees over rocky and often steeply sloping terrain. The common species are: whitebark pine, lodgepole, limber pine, western white pine, mountain hemlock, and western juniper. Shrubby thickets of aspen may be found on moist talus slopes. Shrubs and herbaceous vegetation are found throughout the high elevation rocky slopes. A significant number of springs and associated vegetation are also found throughout this zone and are important contributors to the biodiversity of this area.

Fire occurs relatively infrequently in this zone, and fire size is small, sometimes limited to a single tree.

Red Fir Series

The Red Fir Ecosystem comprises much of the western 1/3 to 1/2 of the study area. While red fir is the dominant tree species throughout this area, it is commonly associated with lodgepole pine, Jeffrey pine, western white pine, whitebark pine, white fir and mountain hemlock.

The desired vegetation community types can be grouped into two Red Fir Subseries: Red Fir with a Lodgepole component and Red Fir with a Jeffrey Pine Component. The red fir/lodgepole component is the dominant community type and is found throughout the Red Fir ecosystem. The red fir/Jeffrey pine community is typically found on southern exposures and lower elevations within the Red Fir Ecosystem.

Red Fir/Lodgepole Subseries. The area will display a mix of pure red fir community types with tree canopy closures of 60% or greater and red fir/lodgepole community types with tree canopy closures of 30-60%. While red fir is expected to be the dominate tree species, disturbance processes will create small to large openings (greater than 10 acres) in which lodgepole pine will dominate for periods of time. When lodgepole is dominate average tree canopy closure is expected to be lower and the larger size classes will account for a smaller percentage of the total cover. Without major disturbances, lodgepole pine might drop out entirely leaving communities comprised of red fir and mountain hemlock. High density, multi-layer stands will be the norm in those areas.

Common associates are western white pine and white fir on southern exposures and lower elevations and mountain hemlock and white bark pine at higher elevations and/or northern exposures.

The shrub and herbaceous understory will typically be less than 1%.

The tree canopy cover for these red fir community types is expected to be 60% or greater, unless lodgepole pine dominates in which case the canopy cover may be 30% or less. The mix of size classes over red fir community types will be:

Size Class DBH	Percent of Total Tree Cover
1-6"	3–7%
6-11"	3–7%
11-18"	5–9%
18-25"	5–9%
25-30"	5–9%
30-40"	15–35%
40"+	20–40%

The snag component is typically 3 to 8 snags per acre, 10 to 15 logs per acre contribute to the down woody debris, and the duff thickness is approximately 3" covering greater than 85% of the ground. Bare ground is less than 5%.

Fire operates as a process throughout this vegetation series. The range of fire indicators are:

Size	Intensity	Frequency	Fuel Load
0–2 Acres	Low	25–50 years	30–60 tons/acre
0–50 Acres	Mod–High	75–100 years	30–60 tons/acre

Red Fir/Jeffrey Pine Subseries. This subseries was divided into two groups as a result of site quality. The desired condition is described for areas of high site quality and areas of low site quality.

The sites of higher quality typically have tree canopy closures of 30-60% and lower site quality typically have tree canopy closures of 10 to 20%, with Jeffrey Pine comprising approximately 10-20% of the stand on both sites. A trace of lodgepole (less than 1%) may occur. The shrub and herbaceous components are typically less than 5%. The mix of size classes will be similar on both high and low sites with only the tree canopy cover differing and will be as follows:

Size Class DBH	Percent of Total Tree Cover
1–6"	3–7%
6–11"	3–7%
11–18"	5–9%
18–25"	5–9%
25–30"	5–9%
30–40"	15–35%
40"+	20–40%

Canopy cover will reach its maximum in areas dominated by the larger size classes. Areas occupied by smaller size classes will have correspondingly lower tree canopy closure.

The snag component is typically 3 to 5 snags per acre, 8 to 12 logs per acre contribute to the down woody debris, and the duff thickness is approximately 2-3" covering 50-85% of the ground. Bare ground is less than 5%.

Fire occurrence is similar to the Red Fir Lodgepole series, however, fire frequency is higher and fuel loadings slightly lower due to the southern aspects or lower elevations that this series occupies.

Size	Intensity	Frequency	Fuel Load
0–5 acres	Low	20–30 years	20–50 tons/acre
0–50 acres	Mod-High	75–100 years	30–60 tons/acre

Mixed Conifer Series

This series was divided into three groups as a result of site quality. The Desired Condition is described for areas of high site quality, areas of high site quality but modified by higher elevations, and areas of low site quality.

The sites of higher quality typically have tree canopy closures of 30-45%. Jeffrey Pine is dominant and white fir comprises approximately 5-15% of the stand. The shrub and herbaceous components are typically less than 10%. The mix of size classes over these mixed conifer community types and percent of the 30-45% tree canopy closure will be:

Size Class DBH	Percent of Total Tree Cover
1–6"	3–7%
6–11"	3–7%
11–18"	5–9%
18–25"	5–9%
25–30"	5–9%
30–40"	15–35%
40"+	20–40%

Forest structure is a combination of small, single-layered, even-aged groups of trees and small, uneven-aged, multi-layer, multi-species groups of trees. When viewed as a whole, these groups combine to provide continuous forest cover with a wide range of structural diversity.

The snag component is typically 4 to 6 snags per acre, 8 to 10 logs per acre contribute to the down woody debris, and the duff thickness is approximately 1-3" covering 30-70% of the ground. Bare ground is less than 10%.

The sites of higher quality/higher elevation typically have tree canopy closures of 30-40%. Jeffrey Pine is dominant and white fir comprises approximately 5-15% of the stand. The primary difference between this group and the previous group is the stature. The trees are much shorter and the size classes as a result are smaller. The shrub and herbaceous components are typically less than 10%. The mix of size classes over these mixed conifer community types and percent of the 30-40% tree canopy closure will be:

Size Class DBH	Percent of Total Tree Cover
1–6"	5–15%
6–11"	5–15%
11–18"	15–40%
18–25"	15–40%
25–30"	trace
30–40"	trace
40"+	absent

Forest structure is a combination of small, single-layered, even-aged groups of trees and small, uneven-aged, multi-layer, multi-species groups of trees. When viewed as a whole, these groups combine to provide continuous forest cover with a wide range of structural diversity.

The snag component is typically 4 to 6 snags per acre, 8 to 10 logs per acre contribute to the down woody debris, and the duff thickness is approximately 1-3" covering 30-70% of the ground. Bare ground is less than 10%.

The sites of lower quality typically have tree canopy closures of 10-30% and a total vegetative cover of 20-45%. Jeffrey Pine is dominant and white fir comprises approximately 5-15% of the stand. The mix of size classes over these mixed conifer community types and percent of the 10-30% tree canopy closure will be:

Size Class DBH	Percent of Total Tree Cover
1–6"	3–7%
6–11"	3–7%
11–18"	5–9%
18–25"	5–9%
25–30"	5–9%
30–40"	15–35%
40"+	20–40%

The shrub and herbaceous species are important components contributing 15-35% to the total vegetative cover within this third mixed conifer group. The shrub species will vary depending on the environmental parameters of the site. Manzanita, California lilac, or snowberry will be the dominant shrubs on the more mesic, cooler sites. Sagebrush and/or bitterbrush will be the dominant shrubs on the hotter or drier sites. The herbaceous vegetation will contribute 5-10% of the total vegetative cover and will be comprised of at least 10 different species such as squirrel tail, stipas, Ross' sedge, buckwheats, mustards and gayophytums.

Forest structure is more open, but still dominated by a mix of even-aged and uneven-aged groups of trees. The snag component is typically 4 to 6 snags per acre, 8 to 10 logs per acre contribute to the down woody debris, and the duff thickness is approximately 1" covering 15-25% of the ground. Bare ground is 20-35%.

Fires occur frequently in this vegetation series, and serve as the primary process influencing vegetation composition, species mix, and down woody debris.

Size	Intensity	Frequency	Fuel Load
0–200 acres	Low–Mod	10–25 years	20–40 tons/acre

Lodgepole Series

Lodgepole Riparian Subseries. This subseries was divided into three groups based on elevation and year round availability of surface water. The *higher elevation group* is within the same elevation zone as the red fir/lodgepole subseries and located along perennial surface water. The desired condition for this lodgepole riparian group will be to have a total vegetative cover of 70-85% comprised of trees, shrubs, and herbaceous vegetation. The mix of tree size classes over this lodgepole riparian community will be:

Size Class DBH	Percent of 25–35% Tree Canopy Closure
1–6"	5–15%
6–11"	5–15%
11–18"	5–15%
18–25"	20–30%
25–30"	20–30%
30–40"	5–15%
40"+	5–15%

Lodgepole is the dominant tree in this stand and is well represented in all size classes. Aspen is the next dominant and is well represented in the first four size classes. Red fir may be found contributing less than 2% to the tree canopy closure.

The shrubs are a key biodiversity component of this lodgepole riparian group and contribute 15-25% to the total vegetative cover. Willow species of various age classes are the dominate shrubs. The herbaceous component is made up of mesophytes and hydrophytes contributing 15-25% to the total vegetative cover.

The snag component is typically 4 to 6 snags per acre and 8 to 10 down logs per acre contribute to the down woody debris. Bare ground is less than 10%.

The *lower elevation group* is found outside the red fir/lodgepole subseries zone and is located along perennial surface water. The desired condition for this lodgepole riparian group will be to have a total vegetative cover of 70-85% comprised of trees, shrubs, and herbaceous vegetation. The mix of tree size classes over this lodgepole riparian community will be:

Size Class DBH	Percent of 15–25% Tree Canopy Closure
1–6"	5–10%
6–11"	5–15%
11–18"	5–15%
18–25"	10–20%
25–30"	10–20%
30–40"	5–10%
40"+	5–10%

Lodgepole is the dominant tree in this stand and is well represented in all size classes. Aspen is the next dominant and is well represented in the first four size classes.

The shrubs are a key biodiversity component of this lodgepole riparian group and contribute 20-35% to the total vegetative cover. Willow species of various age classes are the dominate shrubs. The herbaceous component is made up of mesophytes and hydrophytes contributing 20-35% to the total vegetative cover.

The snag component is typically 4 to 6 snags per acre and 8 to 10 down logs per acre contribute to the down woody debris. Bare ground is less than 10%.

The *third lodgepole riparian group* is found along ephemeral stream corridors. Dry Creek would be an example. The desired condition for this lodgepole riparian group will be to have a tree canopy closure of 25-35%. Lodgepole is the dominant species. The shrub and herbaceous species are typically less than 5%. The mix of tree size classes over this lodgepole riparian community will be:

Size Class DBH	Percent of 25–35% Tree Canopy Closure
1–6"	3–7%
6–11"	3–7%
11–18"	5–9%
18–25"	5–9%
25–30"	5–9%
30–40"	15–35%
40"+	20–40%

The snag component is typically 4 to 6 snags per acre and 8 to 10 down logs per acre contribute to the down woody debris. Duff thickness is approximately 1.5" covering 60-80% of the ground. Bare ground is less than 10%.

Fire occurrence in the riparian area is generally low, and would normally be associated with fires burning into the riparian area from surrounding areas. Due to the generally moister conditions associated with the riparian zone, fire intensity is usually low, and riparian areas frequently stop the spread of fires that do occur in adjacent areas. An infrequent, high intensity fire could occur within these areas during extended droughts.

Lodgepole Non-Riparian Subseries. This subseries was divided into two groups as a result of site quality. The desired condition is described for areas of high site quality and areas of low site quality.

The sites of higher quality typically have tree canopy closures of 15-30%, however the higher canopy cover would only be reached in the larger size classes. The shrub and herbaceous components are typically less than 5%. The mix of size classes over these lodgepole pine community types and percent of the 15-30% tree canopy closure will be:

Size Class DBH	Percent of 15–30% Tree Canopy Closure
1–6"	3–7%
6–11"	3–7%
11–18"	5–9%
18–25"	5–9%
25–30"	15–30%
30–40"	20–40%
40"+	Trace

The snag component is typically 3 to 5 snags per acre, 8 to 10 logs per acre contribute to the down woody debris, and the duff thickness is approximately 1-3" covering 40-70% of the ground.

The sites of lower quality typically have tree canopy closures of 5-15%. The shrub and herbaceous components are typically less than 5%. Trees are short, less than 60 feet in height. The mix of size classes over these lodgepole pine community types and percent of the 5-15% tree canopy closure will be:

Size Class DBH	Percent of 5–15% Tree Canopy Closure
1–6"	3–7%
6–11"	3–7%
11–18"	5–9%
18–25"	15–35%
25–30"	20–40%
30–40"	Trace
40"+	Absent

The snag component is typically 2 to 4 snags per acre, 8 to 12 logs per acre contribute to the down woody debris, and the duff thickness is approximately 1/2" covering 15-25% of the ground.

Fires occur with moderate frequency in this series. Fire intensity is generally low due to the lower fuel loads and discontinuous nature of the fuel. Although the overall fire intensity is low, there may small areas of high intensity fires associated with concentrations of fuels.

Size	Intensity	Frequency	Fuel Load
0–50 acres	Low	20–30 years	10–30 tons/acre

Jeffrey Pine Series

The desired condition for the series is divided into two components based on forest structure and stand density. The first component is characterized by small, single-layer, even-aged groups of trees. There would be a distribution of age and size classes between groups of trees. Stand density would be 15 to 30%, however the higher densities would only be reached in the larger size classes. Openings will be less than two acres in size. The forest would appear to be dominated by open stands of large trees mixed with patches of smaller, younger trees. The distribution of size classes within this Jeffrey pine community will be:

Size Class DBH	Percent of Series
1–6"	3–7%
6–11"	3–7%
11–18"	7–9%
18–25"	7–9%
25–30"	10–20%
30–40"	15–30%
40"+	20–40%

The snag component is typically 2 to 4 snags per acre and 3 to 5 down logs per acre contribute to the down woody debris, and duff thickness is approximately 1" covering 20% of the ground. Bare ground is less than 20%.

The second component would apply to 15% of the Jeffrey pine series. This component has a greater area covered by 6 to 11 inch size class trees, and the stand density of these areas is greater, ranging from 30 to 40%. The area covered by larger size classes will be less. The purpose of this is to provide greater wildlife habitat diversity through increased stand diversity. The mix of size classes over this Jeffrey pine community will be:

Size Class DBH	Percent of Series
1–6"	3–7%
6–11"	10–15%
11–18"	7–9%
18–25"	7–9%
25–30"	10–15%
30–40"	15–30%
40"+	10–20%

The snag component is typically 2 to 4 snags per acre and 3 to 5 down logs per acre contribute to the down woody debris,

and duff thickness is approximately 1" covering 20% of the ground. Bare ground is less than 20%.

The total vegetation canopy within the two scenarios will be 30-55% with shrubs contributing 20-40% to the total vegetation canopy closure. The shrub species will vary depending on the environmental parameters of the site. Manzanita and California lilac will be the dominant shrubs on the more mesic, cooler sites. Mountain Mahogany, sagebrush and/or bitterbrush will be the dominant shrubs on the hotter or drier sites. The herbaceous vegetation will contribute 5-10% of the total vegetative cover and will be comprised of at least 10 different species such as squirrel tail, stipas, Ross' sedge, buckwheats, mustards and gayophytums.

Fires occur very frequently in this vegetation series, and serve as one of the processes influencing vegetation composition, species mix, and down woody debris. Although the overall fire intensity is low, there may small areas of high intensity fires.

Size	Intensity	Frequency	Fuel Load
0–500 acres	Low	5–15 years	20–40 tons/acre

Aspen Series

The total vegetation cover is 70-85%. Stands will be managed to provide the following mix of size classes:

Size Class DBH	Percent of 70–85% Tree Canopy Closure
1–6"	5–15%
6–11"	20–30%
11–18"	20–30%
18–25"	5–15%
25–30"	5–15%
30–40"	trace
40"+	trace

Other tree species may contribute no more than 5% to the total tree canopy closure.

Shrubs provide 10-20% cover of various age classes to the total vegetation cover and are an important biodiversity component. Snowberry is the dominant shrub with rabbitbrush, currents, sagebrush and bitterbrush contributing less than 5% cover. The herbaceous composition will contain a large number of species (greater than 10), and contribute 15-25% cover to the total.

The snag component is typically 10 to 12 snags per acre and 10 to 12 down logs per acre contribute to the down woody debris. Duff/litter thickness is approximately 1" covering 35-55% of the ground. Bare ground is less than 10%.

Fire occurs with moderate frequency and low intensity in the vegetation series, and is the primary process affecting species composition.

Wet Meadow Series

Wet meadow systems are not predominant throughout the study area; however, three distinct locations contain hydric vegetation that are characteristic of wet meadow systems. The headwaters of upper Deadman Creek and upper Glass Creek along with the more notable Glass Creek Meadow are characterized by a vegetative cover of at least 90%, consisting of primarily sedges and willows. These areas are well vegetated although they contain areas that are naturally unstable. The Glass Creek Meadow area is highly diverse in its vegetative associations, with more than 150 species previously identified. Some species typically occurring with greater frequency in this vegetation series include: *Carex* sp., *Trifolium* sp., *Deschampsia caespitosa, Juncus* sp., *Ranunculus alismifolius, Hordeum brachyanthrum* and *phleum alpinum* among others.

Both the headwaters of Deadman Creek and Glass Creek are smaller riparian systems containing dense stands of willow and aspen canopy intermingled with Jeffrey pine. The area of hydric influence within these small riparian areas does not provide for extensive openings within the woody vegetation. The understory is comprised of 40% *carex* sp. with the remaining vegetation being equally divided between 40% grasses/forbs and 20% litter/duff.

Montane Chaparral Series

Montane Chaparral occurs in two places within the landscape. The first area is within openings created by disturbance. In these areas, montane chaparral will eventually be replaced by the next successional vegetation species, resulting in spatial shifts in distribution. The second area is comprised of rocky sites or steep slopes. Shrubs will provide 40 to 50% vegetative cover for these areas. Fire occurrence is tied to the fire occurrence in the surrounding vegetation series, and will vary greatly within this series.

Great Basin Sagebrush Series

The distribution of this series is fairly constant within the area. Vegetation cover ranges from 40 to 60%, dominated by sagebrush. Grass and forbes provide cover between 5 to 15%. There is an even distribution of size classes throughout the series, but the spatial distribution is characterized by large (30-100 acre) patches of even-aged vegetation. Fire occurs with moderate frequency but high intensity, and is the primary process affecting age class distribution. Fires frequently burn into the sagebrush from the surrounding forest.

Wildlife. The vegetation patterns within the landscape provide diverse, connected, habitat components for a wide range of species. Migration corridors for deer are free of barriers.

Fisheries. Aquatic habitat consists of cold water streams and springs free of fish, cold water streams with self-sustaining populations of fish that colonized the area from past stocking, and cold water streams with fish populations supplemented with stocking. The extent of colonized reaches varies in response to changes in stream channels due to flooding, landslides, avalanches, and debris jams. Fish stocking will only occur near Glass Creek and Deadman Creek campgrounds. Stocking levels are low.

Substrate embeddedness ranges from 20 to 58% at designated sample stations. Until more data is available, the desired condition is to not increase embeddedness levels beyond the current levels at the sample stations. Percent pool area coverage by fine sediment varies from 5 to 40% at designated sample sites. Until more data is available, the desired condition is to not increase fine sediment coverage in pools beyond the current levels at the sample stations.

Amphibians. Habitat with known populations of amphibians is maintained. Potential habitat is surveyed to determine the presence of amphibians. Reaches of Deadman Creek above fish migration barriers are monitored for trout absence, so that potential impacts to amphibian habitat from trout invasion will be detected as early as possible.

Social Components

Heritage. All historic, prehistoric, and traditional properties are identified and evaluated. Significant properties are listed on the National Register of Historic Places and protected from damage.

Visual Resources. Activities within the area meet the Visual Quality Objective of Retention or Partial Retention as shown in figure 50.A1. Special emphasis is given to the visual landscape as seen from Highway 395 and the Scenic Loop Road.

Recreation Opportunity Spectrum. The Recreation Opportunity Spectrum (ROS) System is used to describe the social setting that visitors and users will encounter within the landscape. The description will also include specific activities and opportunities that will occur.

Visitors to the area find recreation opportunities that range from semi-primitive non-motorized to roaded natural. Rural settings are limited to the Mammoth Mountain Ski Area, the Glass Creek Recreation Area, and the Crestview Station area. Urban settings are limited to the area around the Mammoth Mountain Inn, Shady Rest Park, and private lands within the Town of Mammoth Lakes. The distribution of these opportunities are shown in figure 50.A2.

The following section describes the setting found in each of the areas.

Semi-Primitive Non-Motorized—The area is characterized as by predominantly natural-appearing landscapes. Visitors have a strong feeling of remoteness from more heavily used areas. Motorized vehicles are not allowed. Access is provided by trails, but much of the area can only be accessed by off-trail travel. Mechanized vehicles are allowed along designated trails. Mechanized or motorized equipment is allowed for resource management, although the use of motorized vehicles is limited. There are no permanent roads. Facilities are provided for resource protection.

This area will provide opportunities for hiking, hunting, fishing, off-track Nordic skiing, back-country alpine skiing, nature viewing/study, and mountain biking. Fa-

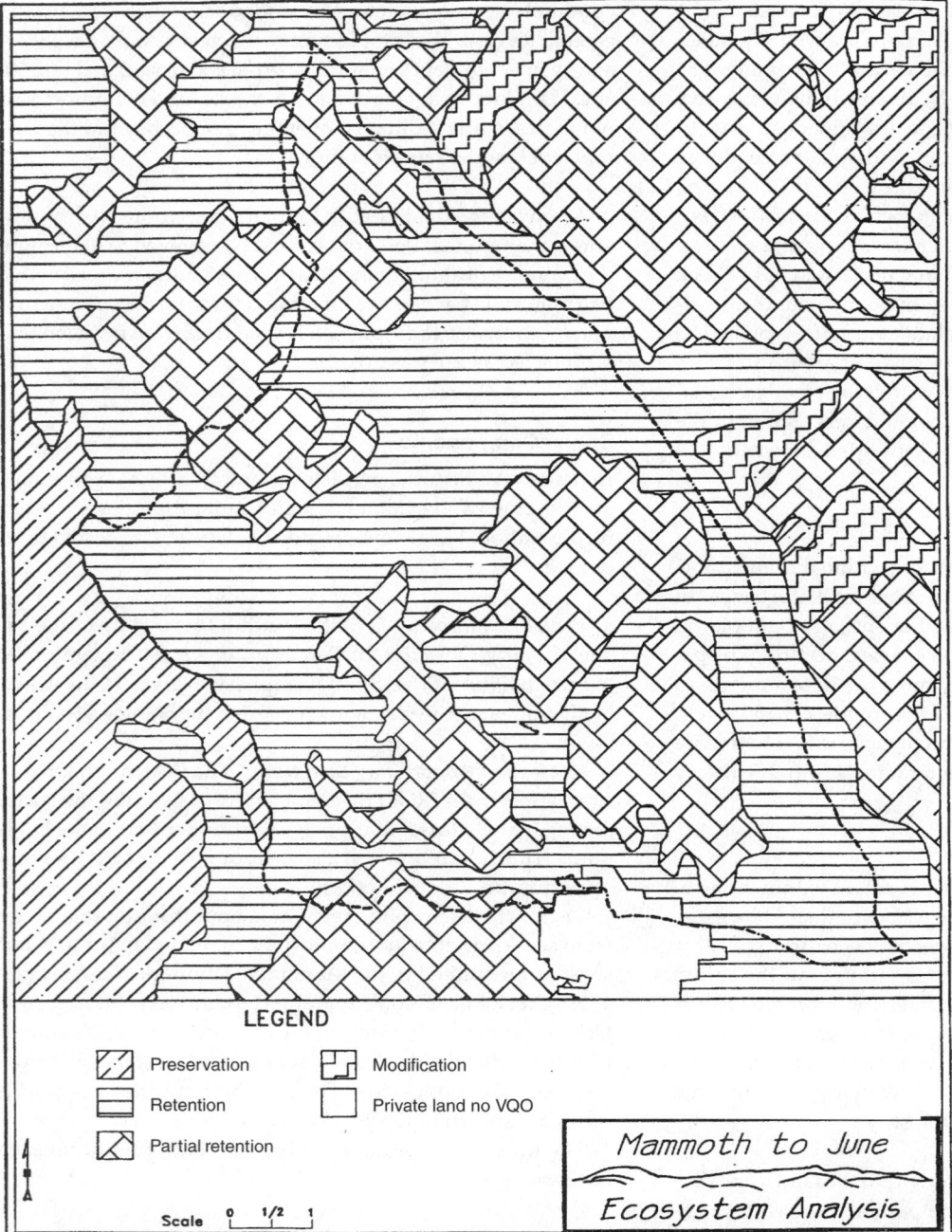

FIGURE 50.A1

Visual quality objectives.

LEGEND

▨	Preservation	⊡	Modification
▤	Retention	☐	Private land no VQO
▧	Partial retention		

Scale 0 1/2 1

Mammoth to June
Ecosystem Analysis

cilities to support these activities, such as parking and accessible rest rooms, will be outside of the area.

Semi-Primitive Motorized—The area is characterized as by predominantly natural-appearing landscapes. Visitors have a strong feeling of remoteness from more heavily used areas. Motorized vehicles are allowed along designated routes, but route density is low, usually less than 2 miles of routes per square mile of area. Facilities are provided for user safety and resource protection.

This area will provide opportunities for hiking, hunting, fishing, off-track and track Nordic skiing, back-country alpine skiing, nature viewing/study, mountain biking, snowmobiling, and off-highway or 4WD vehicle use. Facilities to support these activities, such as parking and accessible rest rooms, will be provided at concentrated use areas. Roads and parking areas are generally surfaced with native materials if surfaced at all.

FIGURE 50.A2

Desired ROS designations.

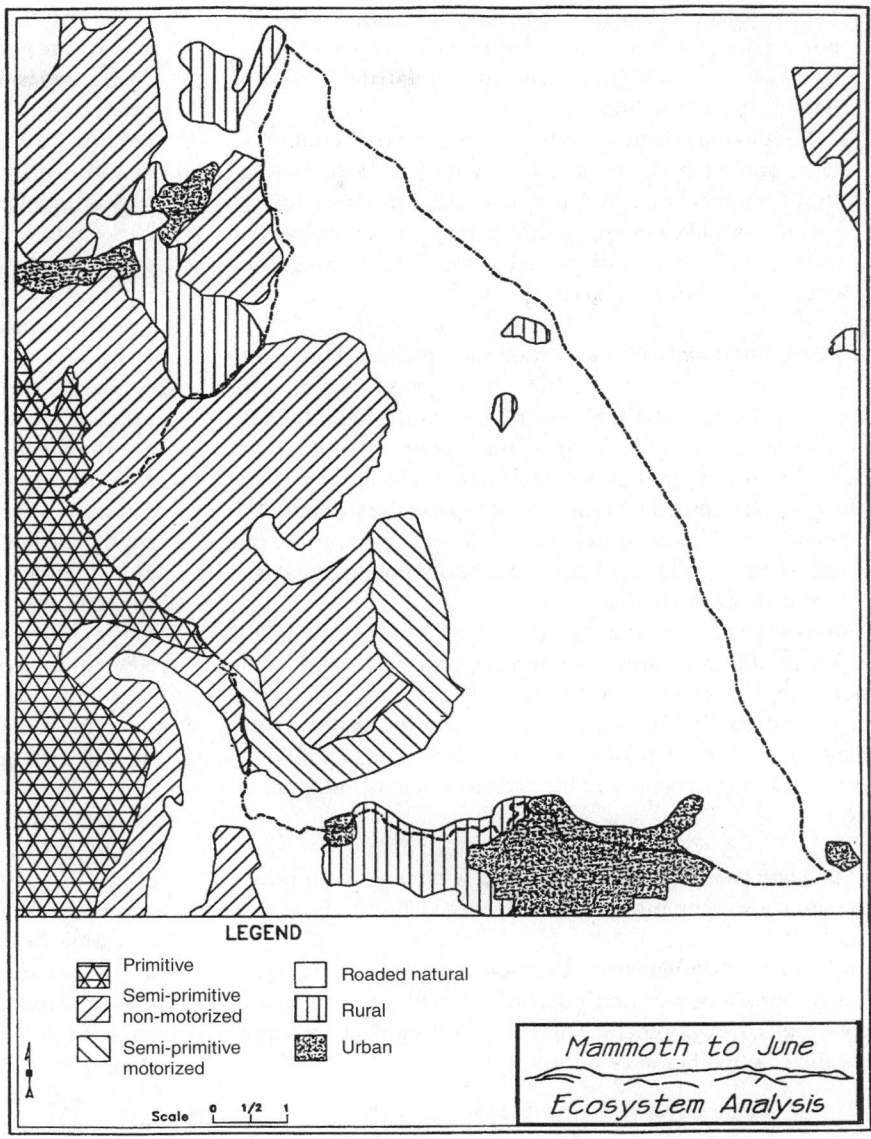

LEGEND

Primitive

Semi-primitive non-motorized

Semi-primitive motorized

Roaded natural

Rural

Urban

Scale 0 1/2 1

Mammoth to June Ecosystem Analysis

Roaded Natural—The area is characterized by a naturally appearing area with moderate evidence of the sights and sounds of humans. Roads and motorized vehicles are common to the area, with road density ranging from 3 to 6 miles of roads per square mile of area. Users may see evidence of a wide range of activities. Facilities are provided for user safety, convenience, and resource protection.

This area provides opportunities for the widest range of activities, including hiking, hunting, fishing, dispersed and developed site camping, viewing interpretive exhibits, OHV and 4WD vehicle use, mountain biking, Nordic and alpine skiing, snowmobiling, snowshoeing, dog sledding, snowplay and guided activities. Facilities such as

parking and accessible rest rooms are provided to support these activities. Roads and parking areas can be surfaced for resource protection.

Rural—The sights and sounds of human activity are readily evident. Use levels are moderate. Highly developed facilities are provided for user safety, convenience, and resource protection.

Rural areas are limited to the area around Glass Creek Campgrounds and the Crestview Administrative site. The Glass Creek site is described in greater detail later in this document. Opportunities for other activities are generally associated with the camping use.

Urban—Urban areas are characterized by high levels of human activity and by concentrated development. Use levels can be very high. Developed sites are highly modified for specific activities.

Besides the private lands within the Town of Mammoth Lakes, only the Mammoth Mountain Inn and the Shady Rest Area are considered urban. In addition to providing specific uses like lodging or urban sports, both areas serve as starting points for dispersed recreation activities that occur throughout the landscape.

Roads and Trails. Access to the area is provided by an integrated system of roads and trails that accommodate motorized vehicles, mountain bikes, and foot traffic in both winter and summer. Travel loop opportunities are identified, and dead-end routes are limited. Motorized vehicles and mountain bikes are limited to designated routes in the summer, with off-road travel allowed in designated areas for specific purposes, such as fuel wood gathering. Snowmobile travel is allowed in designated areas.

Access routes are well signed, and information is available at key locations to inform visitors of opportunities and restrictions. Local partners take an active role in managing the access system, including assistance with signing, route maintenance, trail grooming, and use monitoring. Facilities are provided to support use of the access system throughout the year.

Existing Uses. The area supports a wide range of approved uses that are compatible with the Desired Condition.

Geothermal Development. Proposals to develop the geothermal resource are evaluated in accordance with geothermal lease conditions. Development proposals are designed to minimize conflicts with other uses in the area.

UPPER GLASS CREEK WATERSHED

Introduction

The Upper Glass Creek watershed area is located in the north central section of the Mammoth to June analysis area (figure 50.A3). It is bounded to the north by June Mountain, and to the south by White Wing. Glass Creek Meadow lies in the center of the area. Vegetation series within the basin include wet meadow, red fir/lodgepole, aspen, and sub alpine. Glass Creek is one of the primary tributaries to Deadman Creek and the Upper Owens River.

Desired Condition

The Upper Glass Creek watershed provides quality wildlife habitat appropriate for represented vegetation series as well as quality aquatic habitat associated with the streams and springs. Ecosystem processes such as fire, avalanches, and vegetation succession operate with little interference from hu-

mans or human activities. The sights and sounds of human activity are generally absent. Access to the area is provided by designated trails open to hikers or equestrians. These trails are connected to the integrated road and trail system accessing the area. Motorized vehicles or mountain bikes are not allowed in the majority of the basin, but transition points are provided so that travelers using motorized or mechanized transport can park and walk to Glass Creek Meadow. Facilities are constructed only for resource protection. Market resources (grazing, timber) are not expected as a byproduct of management activities, since the focus is on non-market resources such as solitude, wildlife habitat, and aquatic habitat.

Glass Creek Meadow is the keystone of the basin. The meadow is comprised of numerous springs and seeps throughout the meadow system with an associated stream channel supporting overhanging streambank vegetation in those locations where natural sloughing is limited. Vegetation varies slightly across the meadow zone from heavily hydric to marginally xeric near the meadow margins. The willow riparian vegetation occupies 80% of its natural streamside habitat. In the more hydric areas adjacent to springs there is an 80%/20% composition of sedges (*carex* sp.) to grasses and forbs. Within the more xeric sites, the vegetation is typically 40% *carex* sp., 20% grasses, 20% forbs and 20% bare ground. Of the 20% bare ground, 10% is undisturbed with litter in place and 10% contains some level of natural disturbance (sidehill erosion, gopher activity).

The stream channel supports self-sustaining populations of fish that were planted in the past. Plans to use Glass Creek as a recovery reach for Lahontain cutthroat trout have been modified to remove this area from the recovery plan. The upper reaches of Glass Creek provide aquatic habitat for Yosemite toads that includes slow moving and standing water with adjacent willow and aspen vegetation and talus slopes.

The upland areas around the meadow provide a diverse mosaic of red fir, lodgepole, aspen, and sub alpine forests. Fire, insect mortality, windthrow, and avalanches are the primary processes that influence vegetation composition and forest structure.

LOWER GLASS CREEK RECREATION AREA

Introduction

The Lower Glass Creek Recreation Area is located along Glass Creek, just above the confluence with Deadman Creek. It includes both sections of Glass Creek Campground, as well as the Glass Creek summer home tract. Vegetation is predominantly Jeffrey pine with some lodgepole.

Desired Condition

The Glass Creek Recreation area is divided into three clearly designated areas. Glass Creek #1 is located east of Glass Creek and north of the access road, and consists of a developed camp-

FIGURE 50.A3

Landscape units.

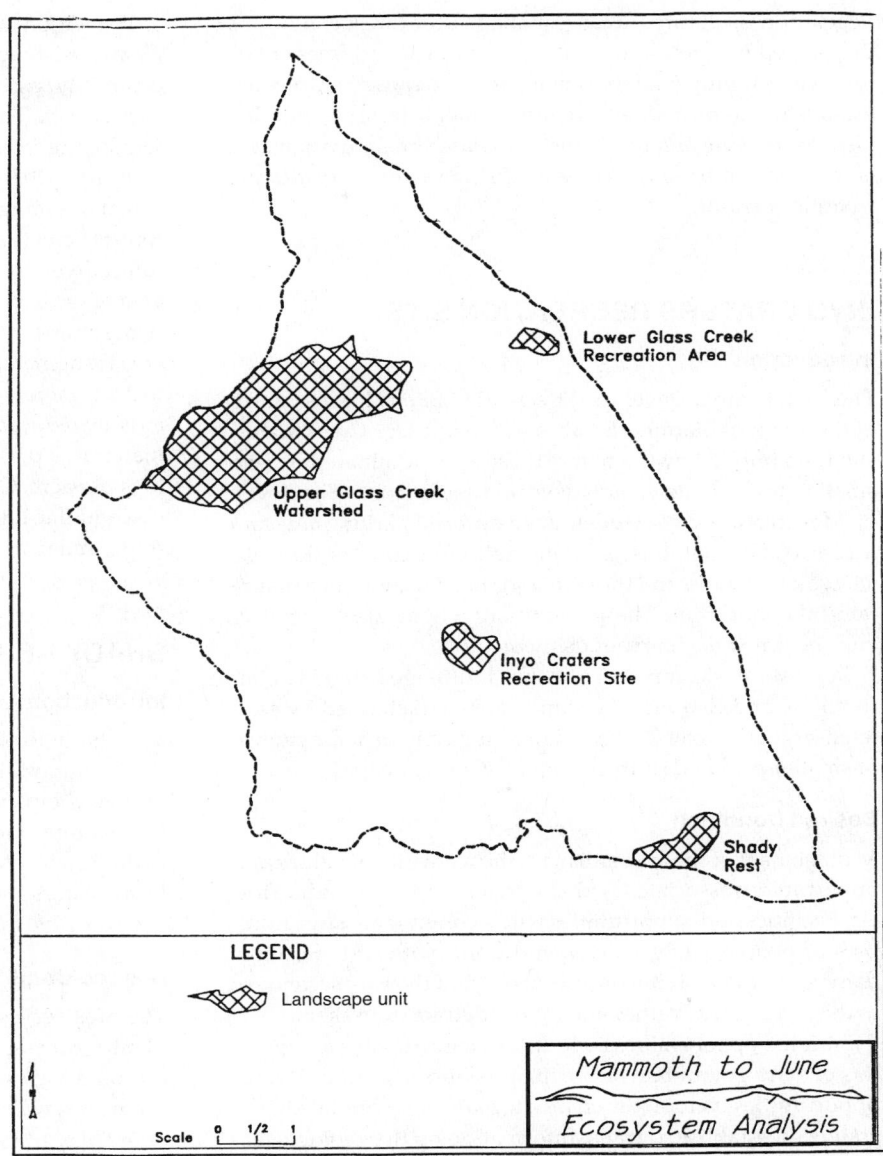

LEGEND

Landscape unit

Mammoth to June
Ecosystem Analysis

Scale 0   1/2   1

ground. Glass Creek #2 is west of Glass Creek and south of the access road, and consists of a developed campground. Glass Creek #3 is west of Glass Creek and north of the access road, and includes the recreation residence tract.

Glass Creek #1—The campground provides short term tent, car, or motorhome camping opportunities in designated spots. Access roads are paved and campsites clearly designated. Potable water is provided throughout the campground. Toilet facilities have flush toilets. Campsites are set back 100 feet from Glass Creek. The campground has between 75 and 125 designated sites. Numerous sites are designed to be universally accessible. All toilets are accessible.

Glass Creek #2—The campground provides long term tent, car, or motorhome camping opportunities. Access roads and campsites are clearly defined. Campsites are designed to accommodate a mix of individual and group camping arrangements. Campsites are set back 100' from Glass Creek. Vault toilets meeting the "Sweet Smelling Toilet" standard are accessible to all users. Potable water is provided at the toilet building for maintenance and for campers.

Glass Creek #3—The summer homes provide private recreation opportunities to families with permits. There are no changes proposed for this area.

Vegetation in all three areas is managed for forest conditions dominated by open stands of large trees. Hazard trees are be removed for public safety. Small patches of younger trees are established to provide future replacement trees and provide visual screening and diversity. The Glass Creek riparian area is managed to balance access for fishing with a functioning riparian corridor.

INYO CRATERS RECREATION SITE

Introduction

The Inyo Craters Developed Recreation Site is located north of the Town of Mammoth Lakes within the Dry Creek watershed. As implied by the name, the area is dominated by two small volcanic craters and a volcanic cinder cone. The developed recreation site includes an unsurfaced parking area and access road, an outhouse, a hiking trail, cross country ski trails, snowmobile trails, and the start and finish sections of a mountain bike loop trail. The primary attractions are the craters and the forest that surrounds them.

Vegetation is composed of dense multi-aged stands of Jeffrey Pine and Red Fir. The stands are characterized by scattered large trees over 250 years in age, mixed with dense clumps of smaller pine and fir trees under 100 years of age.

Desired Condition

Management of the area enhances the recreational values and opportunities associated with the craters and surrounding forest. Facilities and opportunities will accommodate large numbers of people safely, conveniently, and with little resource damage. Other activities are occurring but they are secondary to the recreational values and do not detract from them.

Summer opportunities include the chance to hike to the craters or start a mountain bike trip on established trails. Winter opportunities include the chance to ski or snowmobile to the craters on established trails. Information on the geologic features and the surrounding forest is available to visitors during both seasons. Parking is provided in the summer and accessible toilets that meet the "Sweet Smelling Toilets" standards are available all year. A safe viewing area is provided for visitors at the rim of the craters. The access road and parking area have stable surfaces that will accommodate light trucks and passenger cars. The parking area includes adequate controls to confine vehicles to the parking area and avoid adverse effects from uncontrolled vehicle use. Established trails have stable surfaces that provide easy access to the crater rim; however, people with disabilities will find the trail access challenging. Trails are constructed and maintained in a manner that discourages off trail use thus minimizing the formation of multiple user trails. All facilities meet the Recreational Opportunity Spectrum guidelines for roaded natural. Interpretative signing will focus on vegetation management along the trail and geologic processes at the crater rim.

Vegetation management promotes forest conditions that increase resistance to insect and disease attacks, reduce the susceptibility to catastrophic fire, and provide an aesthetic visual appearance. The majority of the area is characterized by scattered large trees with open understories. Tree density ranges from 10 to 20 large (>32") trees per acre. Small patches (1 to 2 acres) of younger, smaller trees are scattered throughout the forest to create a fine textured matrix. This matrix provides for forest diversity as well as visual complexity and variety. Basal area of most stands will range between 120 to 160 sq.ft./acre, with some isolated pockets of higher density stands with basal areas ranging up to 210 sq.ft./acre. Snag density ranges from 2 to 3 snags per acre, with snag location managed so that hazards are reduced along trails and near facilities. Dead and down material is present at low density, with 1 to 2 large downed logs per acre. Ground cover is composed primarily of needle cast and fine litter, and exposed soil is only found along roads, trails, and in association with rock outcrops or pumice flats.

SHADY REST

Introduction

The Shady Rest area includes the Mammoth Ranger Station, the Mammoth FS administrative site, the Shady Rest Campground complex, the Sawmill road parking area, and the Shady Rest Community Park. These areas are immediately adjacent to the Town of Mammoth Lakes, and receive considerable use from visitors and the community. The areas are located within a Jeffrey pine forest.

Desired Condition

The area serves large numbers of people safely and conveniently during all seasons. Summer opportunities include camping, viewing interpretive displays, attending interpretive programs, walking, hiking, and biking. The community park offers playground activities for all visitors and team sport activities for local residents. The park also supports special events. All activities at the park are limited to daylight hours. The limits of the campground, administrative site, and community park have not expanded.

Winter opportunities include Nordic skiing on groomed trails, snowmobiling, dog sledding, snowshoeing, other snowplay activities, and interpretive activities. Because the area is fairly confined, some use restrictions are enforced to avoid user conflicts. Parking and toilets are provided in the winter to support these users.

ACKNOWLEDGMENTS

This appendix was written by Robert Hawkins, team leader, and the Inyo National Forest Mammoth-June Analysis Team.

MARK BAKER
Postdoctoral Fellow
Department of Political Science
University of California
Berkeley, California

WILLIAM STEWART
Senior Research Associate
Pacific Institute for SIDES
Oakland, California

51

Ecosystems under Four Different Public Institutions: A Comparative Analysis

ABSTRACT

Two-thirds of the Sierra Nevada lies within the jurisdiction of public land-based resource management institutions. Public land-based institutions operate within a context of increasingly complex political and social environments and of ecological independence between reserves and adjacent non-reserve resource systems. The core analysis is based on an institutional comparison of four adjacent yet different public institutions managing forests in the southern and central Sierra Nevada. Sequoia National Forest, Mountain Home Demonstration Forest, Sequoia and Kings Canyon National Parks, and the Tule River Indian Reservation all manage areas with similar ecological characteristics.

The present landscape pattern associated with each institution and the probable direction of these landscape patterns can be best accounted for by the interaction between internal organizational characteristics and institutional mandates, rather than by bio-physical endowments or scientific principles of land, timber, forest, or ecosystem management. The key organizational factors which emerge from this case study are the degree of institutional centralization, criteria used for budget allocations, means for ensuring public accountability, and lastly, degree of local-level planning and management flexibility. Maintaining ecosystem integrity based on the "island-in-time" self-contained reserve model is inadequate to ensure resource preservation or conservation because significant impacts on areas within a reserve arise from outside of it and management regimes within a reserve affect those aspects of the ecosystem that lie outside of it. The three factors found to positively affect institutional performance are tight feedback loops between responsible research and resource management, high levels of institutional legitimacy and public trust, and active inter-organizational coordination.

INTRODUCTION

The current condition of the Sierra Nevada ecosystem is a product of the historical interaction between spatially variant ecological processes and social and institutional dynamics. Approximately one third of the Sierra Nevada is privately owned. The remaining two thirds of the Sierra Nevada lies within the jurisdiction of public land-based resource management institutions which include federal, state and county agencies and Native American authorities. These public institutions differ in terms of their purposes, mandates, organizational characteristics, histories, and planning procedures. In addition to these land-based public institutions, a marble cake of federal, state and county authority extends across and through the whole Sierra Nevada, in some cases providing the basis for cooperative exchange, and in other cases for conflict. Within this marble cake context of overlapping authority, public resource management institutions have managed the resources within their jurisdictions in different ways to produce unique combinations of public and private, and commodity and non-commodity benefits.

The ecological and institutional characteristics of the southern Sierra Nevada provide an opportunity to examine how and why public institutions differentially shape the landscapes they manage. Within close proximity to each other are four different land-based public institutions which manage areas with similar ecological characteristics. The institutions are a national forest (Sequoia National Forest), a state forest (Mountain Home Demonstration Forest), a national park (Sequoia and Kings Canyon National Parks), and a Native Ameri-

Sierra Nevada Ecosystem Project: Final report to Congress, vol. II, *Assessments and scientific basis for management options*. Davis: University of California, Centers for Water and Wildland Resources, 1996.

can Indian reservation (Tule River Indian Reservation). The national forest and national park cover western sierra ecosystems ranging from lower elevation oak and grass woodlands up through the mixed conifer and true fir belts to areas above timberline. The Native American reservation extends from oak and grass woodlands up through the mixed conifer and true fir belt and the state forest falls totally within the mixed conifer belt. Giant sequoia (*Sequoiadendron giganteum*) groves are located within the boundaries of all four of these institutions (see figures 51.1 and 51.2). Table 51.1 summarizes the elevation range, area and vegetation types of the four institutions.

Although the four institutions examined in this study manage some comparable ecosystems, the unique characteristics of each institution in combination with their different mandates, have produced different patterns on the landscape, different mixes of benefit flows, and different conflicts. The first two sections of this report address the question of how the different purposes, mandates, organizational characteristics, histories and operating rules of each institution account for the observable patterns on the landscape and the particular mix of benefits that each institution provides. We briefly examine each institution in terms of its original purpose, current operating mandate, and its key organizational characteristics. We then trace the linkages between these factors and institutional outcomes in terms of landscape patterns, benefit flows and the degree and nature of conflicts it is engaged in. We show that in some cases institutions with similar legislative mandates can produce different landscape patterns and conversely that institutions with different mandates can sometimes produce similar landscape patterns.

These four institutions were originally endowed with similar biological and physical resources. Each has experienced the changing social values concerning forests and other natural resources. Our premise is that the present landscape pattern associated with each institution and the probable direction of these landscape patterns, can be best accounted for by the interaction between internal organizational characteristics and institutional mandates, rather than by biophysical endowments or scientific principles of land, timber, forest or ecosystem management. The key organizational factors which emerge from this case study are the degree of in-

stitutional centralization, the criteria used for budget allocations, the means for ensuring public accountability, and lastly, the degree of planning and management flexibility.

The perception that current social pressures for timber, grazing, water diversions, recreation opportunities and development on the Sierra Nevada ecosystem threaten the integrity of ecosystem structure and function was a key driving force behind the creation of the Sierra Nevada Ecosystem Project. Therefore, it is prescient to also analyze some of the elements which comprise effective resource management strategies and policies under conditions of social conflict over the "proper" goals of public institutions charged with managing forest ecosystems. Accordingly, the last section of this report examines factors which contribute to the ability of public land management institutions to respond to increasingly sophisticated, differentiated and numerous public(s) while simultaneously maintaining their legitimacy within society and the integrity of the ecosystems within their jurisdiction. Indications that the context of resource management on public lands has become more uncertain and complex include increasing legislative oversight, the extent of public controversy concerning federal public lands management especially regarding the ecological and social consequences of past and present timber harvesting, grazing practices and fire suppression activities and policies, current efforts to use "adaptive management" principles in resource management policy and planning, shifts away from the preservation of objects to the management of ecosystems within the National Park Service and from single resource to "multiple use" to ecosystem management within the National Forest Service. Perhaps most significant is the recognition that resource management and stewardship efforts based on the "island-in-time" self contained reserve model are decreasingly effective strategies for resource preservation or conservation because significant influences on areas within a reserve arise from outside of it and management regimes within a reserve impact those aspects of the ecosystem which lie outside of it. Examples of the permeability of reserve boundaries, which we call porosity in this report, include air pollution, fire, and in some cases sedimentation and changes in hydrologic regimes resulting from upstream management activities.

To mitigate against the bias that ensues when only one case

TABLE 51.1

Elevation, area, and vegetation types of the four institutions.

Characteristic	Sequoia National Forest	Sequoia–Kings Canyon National Park	Mountain Home State Demonstration Forest	Tule River Indian Reservation
Lowest elev. (ft)	928	1,443	4,903	918
Highest elev. (ft)	12,218	14,494	7,583	7,334
Total acres	1,118,241	863,372	5,048	53,907
Hardwoods and shrubs	28.6%	9.2%	1.3%	69.4%
Mixed conifer	68.8%	52.1%	98.7%	30.6%
Bare rock	2.6%	38.7%	0%	0%

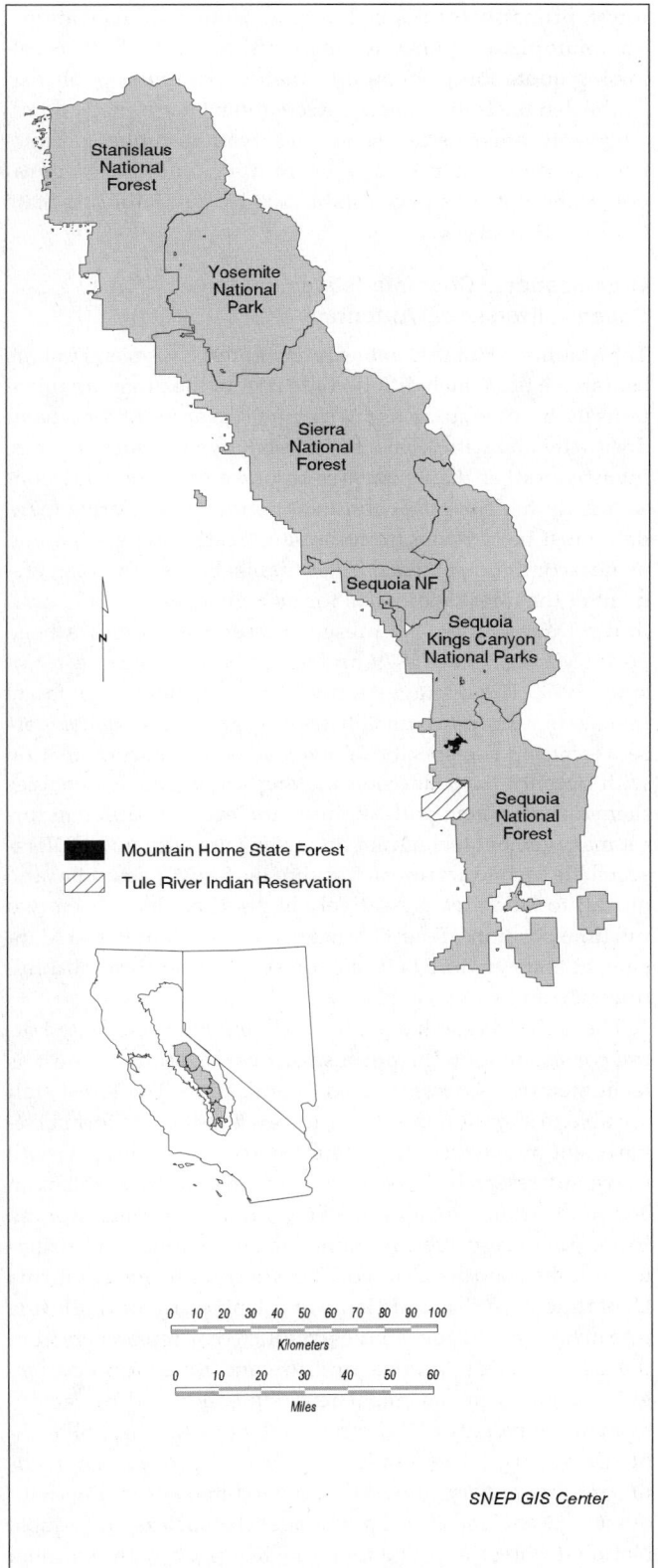

FIGURE 51.1

Land-based public institutions in the southern Sierra Nevada.

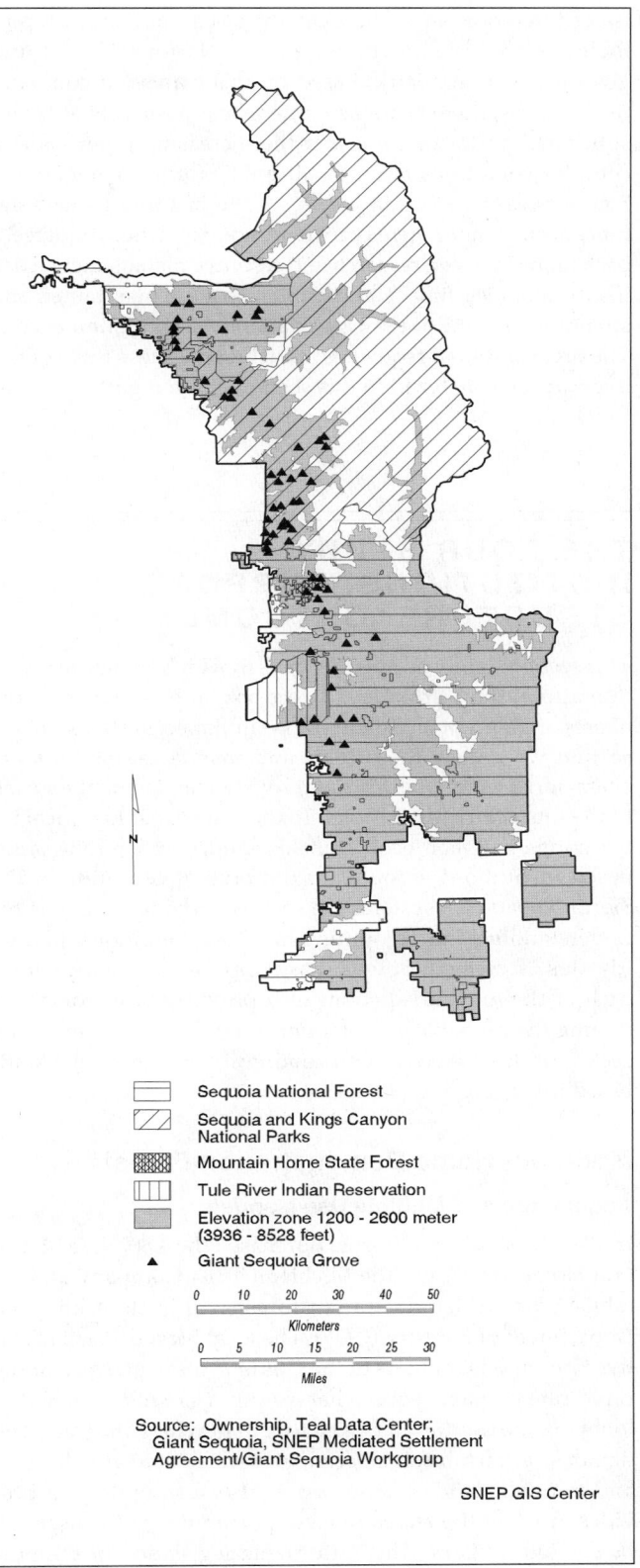

FIGURE 51.2

Detailed ownership and ecological map of the four public institutions.

is used to represent a whole set, the last section of the report includes analyses of the Stanislaus National Forest and Yosemite National Park. Based on this expanded comparative analysis, three factors emerge which positively affect institutional performance under the increasingly porous and complex conditions faced by all public forest owners in the Southern Sierra Nevada. While all the institutions face significant challenges there is greater optimism where tight feedback loops between research and resource management exist, institutional legitimacy and public trust are maintained and strengthened, and inter-organizational coordination occurs. The organizational requisites for achieving these three objectives are also analyzed in this section of the report.

THE FOUR STUDY INSTITUTIONS: PURPOSE, STRUCTURE, OUTCOMES

This section demonstrates the ways in which agency jurisdiction affects the Sierra Nevada ecosystem by comparing the effects of different jurisdictions on similar ecosystems. In this section we review the origins and mandates, institutional characteristics, and benefits of resource management for each of the four study institutions. To the extent that historical interactions between ecosystem dynamics and public land-based institutions account for the present condition of the Sierra Nevada ecosystem, the section provides one lens for understanding and accounting for those conditions. Implicitly, this assessment of what has happened and why, within each of the four jurisdictions also provides basis for determining the probable outcomes of different policies that may seek to influence ecosystem condition through public land-based institutions.

Mountain Home Demonstration Forest

Acquisition and Multiple Use Mandate

In 1946 the state of California purchased the 4,807 acre Mountain Home Tract from the Michigan Trust Company and established as a demonstration forest within the California Department of Forestry (CDF). The local Fresno-Visalia "Native Sons and Daughters of the Golden West," alarmed at the rapid rate of giant sequoia harvesting, were instrumental in lobbying the California Legislature to purchase the tract. The demonstration forest is adjacent to private parcels on the west and to public lands on the east. It also surrounds, and provides much of the recreational opportunities to the users of, Balch County Park. The authorizing legislation of the purchase (section 4426, chapter 1496 of the Statutes of the State of California) clarifies the purpose of the forest, "The Mountain Home Tract Forest in Tulare County shall be developed and maintained, pursuant to this chapter as a multiple use

forest, primarily for public hunting, fishing and recreation." The multiple use policy for the forest is clarified in the following quotation from an information pamphlet published by the demonstration forest, "Recreational use is made dominant, with other uses—water conservation, timber production, forage and mining—secondary, and the general governing policy is to be established by the California State Board of Forestry."

Organizational Characteristics: Decentralized Local Autonomy

The Mountain Home Demonstration Forest has several unique characteristics which differentiate it from the other organizations in this case study and which help to account for what it does, why, and with what effects. Its relatively small size forces intensive rather than extensive resource management. Consequently multiple uses consistent with its legislative mandate must be satisfied from the same, rather than adjoining or non-contiguous, areas. The organization of the management of the forest is unusual for its high levels of staff continuity: Dave Dulitz, the present Forest Manager, has been Forest Manager since 1979, and he has been on the forest staff since 1974. The average tenure of the previous four forest managers was approximately eight years. The staff also possess localized site-specific knowledge which, in combination with decentralized decision making arrangements, enables them to experiment with, monitor and evaluate different forest management techniques and to engage with non-CDF researchers to conduct research within the forest. Lastly, the ratio of staff to land area is relatively high—there have been two full time positions, Forest Manager and Assistant Forest Manager, to manage the 5,000 acre forest and recently a third full time position was created.

The active research agenda at Mountain Home State Forest, consistent with the purpose of a demonstration forest, is facilitated by its decentralized organization. The forest staff are able to submit requests for research to the California Department of Forestry, or sometimes to contract directly with university researchers for specific research projects which can be funded from the California Forest Improvement Program. Examples of research programs include wildlife and fisheries studies conducted in collaboration with the California Department of Fish and Game and University of California researchers, fire history studies conducted by researchers from the University of Arizona, and fire and forest (particularly giant sequoia) management research conducted by faculty from the University of California and the California Polytechnic University. Several on-going research projects, i.e. wildlife and fire history, are also conducted in coordination with other adjacent land-based public agencies such as the Sequoia National Forest and Sequoia and Kings Canyon National Parks. Mountain Home Demonstration Forest also provides most of the seed stock for California Department of Forestry nurseries. Last year 150 sacks of giant sequoia cones were collected for this purpose. Individual trees resistant to blister rust have been identified and their cones are collected in order to propagate blister rust resistant seedlings.

Landscape Patterns and Benefit Flows: Balanced Multiple Use

The Mountain Home Demonstration Forest is an intensively managed, multiple use demonstration forest. Grazing permits are not issued because of incompatibility with recreation and to allow historically overgrazed areas to regenerate. Timber harvesting is tempered by the proximity and density of recreation use within the forest. Harvests are planned to minimize visual impacts by using only single tree and small group selection harvests. Clear cutting is not practiced for aesthetic reasons, and harvest entry intervals have been increased from 15-20 to 30 years to minimize entry-related forest damage and the associated negative visual impacts. The forest has been under a sustained yield management plan consistent with its multiple use mandate that was implied in the 1946 authorizing legislation. The volume of timber on the forest has increased from 92,454 mbf at the time of its purchase to 105,458 mbf in 1990. During this period (1946-1993) 96,028 mbf of timber was harvested (Mountain Home Demonstration Forest information pamphlet). The forest is managed as two overlapping forests. The sequoia groves and the camping facilities constitute a preserve/recreational forest while the non-sequoia forest is managed as an uneven-aged production forest. Unlike the national forests where these two uses are practiced on widely separated areas, at the Mountain Home Demonstration Forest adjacent areas of production forest and preserve/recreational forest create a mosaic of different management regimes within a concentrated area. Recreation uses not consistent with this mosaic pattern of land management, e.g. wilderness backpacking, are not feasible at Mountain Home Demonstration Forest. However, the demonstration forest borders on, and provides direct access to, the Golden Trout Wilderness of the Sequoia National Forest.

Annual recreational use has increased from 3,000 visitors in 1963 to close to 40,000 in recent years. Managing day and overnight visitors occupies an increasing proportion of the staff's time and energy and has lead to the creation of a third full time position primarily to carry out recreation related and interpretive work. Partly due to the increasing recreational use of the forest, and because it is a demonstration forest, the forest managers have developed several on-site interpretative programs and disseminate information through the CDF series, "California Forest Notes". Their public education and outreach efforts include the financing and construction of an interpretative center in Balch County Park which is surrounded by Mountain Home Demonstration Forest, the closure of a campground located on a Native American archeological site and its conversion into a self-guided archeological trail, and a public education campaign about the effects of white pine blister rust and pine bark beetle damage and the importance of salvage logging of diseased trees. A combination of tempered harvesting practices, outreach and

education efforts, and the short two week public comment period required under the Timber Harvest Plan (THP) planning process limited public controversy and conflict over Mountain Home Demonstration Forest harvesting operations. The primary source of conflict relates not to controversy over forest management but to unruly visitors, especially during the major holiday weekends.

The original purpose for which the Mountain Home Demonstration Forest was purchased, i.e. to preserve old growth giant sequoia groves, combined with its mandated emphasis on recreation and the subsequent evolution of maximum sustained yield production forestry in non-sequoia areas, has produced a mosaic of differently managed and used patches within a relatively constrained geographical area. The forest is extensively roaded and is intensively used as a production and recreation forest. Consequently there are no large intact landscape units. Riparian areas and meadows are in better condition than they would be otherwise due to the ban on stock grazing. The decentralized organization of the forest administration has provided the local decision making autonomy necessary for establishing and maintaining feedback loops between resource science and resource management. For example, in response to research findings which suggested that larger openings were required for successful sequoia regeneration, selective harvesting methods were shifted from single tree to small group selections which created open patches from .5 to 1 acre large. Although the forest administration staff do not have the capacity to conduct research themselves, they successfully compensate for this by contracting with other agencies and universities which do have research capacity.

Sequoia National Forest

Reservation for Multiple Use

The 1.1 million acres which comprise the Sequoia National Forest (SNF) were originally reserved from the public domain as part of the 4 million acre Sierra Forest Reserve in 1893. Local lobbying efforts, spearheaded by George Stewart, editor of the *Visalia Delta*, and other Tulare County residents were instrumental in influencing President's Harrison's decision to reserve the southern portion of the Sierra Nevada range. While local support was also an important factor in the California State legislature's decision to authorize the purchase of the Mountain Home Demonstration Forest, in this case local concerns focused primarily on threats to the San Joaquin Valley's water supply posed by uncontrolled upstream mining, grazing, fire and lumbering, as well as concerns, shared perhaps by fewer individuals, about the negative impacts of these activities on the natural beauty of the area and especially the large giant sequoia trees (Dilsaver and Tweed, 1990). In 1908 the Sequoia National Forest was created from that portion of the Sierra National Forest south of the watershed of the Middle fork of the Kings River which by this time had been transferred along with the other forest reserves from the Division of Forestry of the Department of Interior to the newly formed U.S. Forest Service headed by Gifford Pinchot within the Department of Agriculture. The original purpose of the national forest reserves are described in the 1897 Organic Act which established the purposes for which forest reserves could be withdrawn from the public domain, and provided the primary statutory authority for the administration of the forest reserves by what was to become the U.S. Forest Service. The Organic Act states that the purposes of national forests are to "preserve and protect the forest within the reservation" in order to secure "favorable conditions of water flow" and "to furnish a continuous supply of timber". The 1960 Multiple Use Sustained Yield Act expanded the purposes for which the National Forests were to be managed to include outdoor recreation, range, wildlife and fish, in addition to those purposes set forth in the 1897 legislation (Dana and Fairfax, 1980).

Organizational Characteristics: Centralized Hierarchy

In his organizational (and now historical) study of the Forest Service, Kaufman analyzes how the Forest Service is able to ensure that widely dispersed field officers operating under a wide variety of social and ecological conditions will do what is asked of them in a manner which achieves the centrally determined goals of the organization. Kaufman (1981) argues that the Forest Service counteracts the centrifugal tendencies towards fragmentation through hierarchical organization and specialization, the development of centrally controlled "preformed decisions" and concomitant means to detect deviation, and the "homogenization" of staff through in-service training and indoctrination and through frequent personnel transfers.

The Multiple Use Sustained Yield Act of 1960 and new planning procedures mandated in the National Environmental Policy Act of 1969 (NEPA), the Forest and Rangeland Renewable Resources Planning Act of 1974 (RPA) and the National Forest Management Act of 1976 (NFMA) have broadened the mission of the Forest Service and transformed the means for accomplishing that mission since Kaufman's study was completed. However, considerable debate exists as to the extent to which the Forest Service has institutionalized its diversified mission. For example Twight and Lyden (1989) show that the attitudes, preferences and values of 394 district rangers are similar to those of the Forest Service's "resource user constituency" and differ strongly from the service's "environmentalist constituency". They attribute this to the Forest Services "institutionalized socialization process, the most important features of which have apparently changed little since Kaufman's 1960 analysis" (Briggs 1982, cited in Twight and Lyden). Culhane (1981), on the other hand, rejects the claim of isomorphism between forest ranger's attitudes and those of their industry clients. Instead he posits a model of interest group politics and multiple clientelism to explain public lands politics and policies. Consistent with Culhane, Tipple and Wellman (1991) argue that the Forest Service has transformed from an agency which emphasizes efficiency and economy to

one which now also embraces "responsiveness and representativeness."

While the degree of change within the Forest Service is debatable and probably varies from region to region, significant structural continuities have persisted which are relevant for this study's focus on the Sequoia National Forest. First, the Sequoia National Forest is still part of a hierarchical organization which follows centrally mandated and externally legislated standardized planning procedures. Second, its staff are relatively frequently transferred (although less often than previously), relative to the other three institutions in this study. Third, at the forest level, administrators and staff have minimal control over funding for research and the generation of research questions. Lastly, the majority of funding for forest management activities, excluding fire protection funds, is tied to commodity production targets.

Examples of the effects of centralization on decision making autonomy on the Sequoia National Forest can be taken from three arenas: planning, fire management, and research. Resource planning on the forest must comply with the NEPA, RPA and NFMA federal planning requirements, the 1988 Land Management Plan for Sequoia National Forest as modified by the locally negotiated Mediated Settlement Agreement, and the 1993 interim guidelines for the California Spotted Owl. The combined prescriptive effects of these multiple layers of internally mandated and externally legislated planning procedures leave little opportunity for local level planning innovation. They also decrease incentives for intensive monitoring and evaluation of the impacts of resource management plans other than to ensure that legal stipulations are fulfilled because the flexibility does not exist to incorporate the new information monitoring and evaluation generates into subsequent management plans. Finally, the plethora of requirements each of these documents contains, combined with the uncertainty of possible future changes in planning requirements and procedures, and the possibility of successful legal challenges, generates considerable uncertainty about what resource management activities will be possible in the future. This uncertainty mitigates against effective long range planning and fosters a more ad hoc approach which approximates what Lindblom (1959) termed "muddling through."

The Sequoia National Forest is also constrained in its ability to use fire as a management tool. Internal fire policies severely limit the opportunity to use a prescribed natural fire program that would allow some lightning fires to burn. Funding for the planning and implementation of prescribed burns is tied to funds allocated for timber management and harvesting (Aaron Gelobter, Deputy Fire Management Officer pers. comm.). Most of the funds for prescribed burns and fire suppression not related to timber management are available only in high risk contexts, i.e. situations where either structures or urban interface areas are threatened. Although mechanisms exist to enable fire managers to plan and conduct prescribed burns to promote non-commodity ecosystem values, they are generally not well funded. The commodity orientation of fire programs and funding priorities constrains the ability of the managers of the forest to effectively use fire as a resource management tool within an "adaptive management" framework.

In addition to constraints on local level planning autonomy and ability to use fire to achieve non-commodity purposes, there is little research capacity at the national forest level. An exception to this are "administrative studies" which focus on applied management issues and are carried out by forest service staff. In an attempt to insulate research from "administrative evangelism" the 1915 internal restructuring of the Forest Service established a separate and parallel research branch comprised of experiment stations accountable directly to the Chief Forester rather than to the Regional Forester (Schiff 1962). Schiff shows that despite the organizational separation of science and administration, research establishing the important role of fire in enabling longleaf pine (*Pinus palustris*) regeneration in the southern United States was deliberately suppressed throughout the first half of this century because it ran counter to the anti-fire sentiment prevailing within the administrative branch of the Forest Service. Although concentrating the agency's research capacities in a separate research branch did not insulate scientific inquiry from administrative concerns, it did separate resource management from resource science. Without effective use of the institutionalized channels of communication and feedback between managers at the forest level and scientists at the experimental stations, it was inevitable that questions of concern to managers of forests relatively distant from experiment stations, such as the Sequoia National Forest, would remain unanswered. For example, pending research questions include the effects of forest management activities on sensitive furbearers such as the red fox and pine marten, the effects of different kinds of fire and other management regimes on giant sequoia regeneration and tree failure rates, the effects of grazing on range and riparian ecosystems, and the hydrological and aquatic resource effects of alternative management practices.

Landscape Patterns and Benefit Flows: Multiple Use and Sustain(able) Yields?

Up until the 1950s, the primary uses of the Sequoia National Forest had been low levels of hydroelectric development and mining, some logging on the western slopes of the forest, and recreational use particularly in the Mineral King and Kern Plateau areas. During this time primary resource management activities consisted of fire suppression and livestock control. Beginning in the 1950s the Forest Service began an extensive timber harvesting program which focused on achieving maximum sustained yield timber production in some areas, and in other areas such as the Kern Plateau, sought to integrate timber production with other multiple use land management objectives. Previously unroaded areas were roaded and where roads already existed, particularly on the west side, they were

widened and rebuilt to satisfy both intensified timber harvesting and recreational use pressures. During this period timber was harvested using salvage and selective cutting methods. During the 1970s the Forest Service, aware of the ecological importance of giant sequoia groves, "pursued an aggressive grove acquisition and protection program" which excluded groves from timber management goals, and involved creating four grove classes and prescribing acceptable management activities per class (Doug Leisz, pers. comm.).

By the late 1970s some stands had become understocked. In response to these forest conditions, to pressure from private timber interests to increase harvest levels and to "the allowable cut effect" which linked allowable cut levels to future anticipated growth rates, management of the forest shifted to extensive clearcutting and a shortened cutting cycle (from 150 to 70 years). The shift to clearcutting and short rotations maximized the present net value of commodity outputs, provided opportunities to quickly restock harvested areas with desirable species and hence maximize long term timber yields, and was an effective response to the non-declining even flow constraint which required sustained timber harvest levels. This accelerated short rotation timber harvesting program continued through the mid-1980s. Although the logic of the shift to short rotation clearcutting was silviculturally sound, inadequate investment in post-harvest site preparation and reforestation as well as harsh sites created other problems. By the mid-1980s public concern about clearcutting and other environmental consequences of the timber harvesting program, and the threat that harvesting in and adjacent to giant sequoia groves posed to that species, lead to 22 administrative review appeals challenging the 1988 Forest Land Management Plan and the supporting Environmental Impact Statement. The Forest Service's attempts to respond to the appeals lead to a series of mediated negotiations which produced the Mediated Settlement Agreement. The MSA is the product of a political process of negotiation, not the result of consensus based decisions grounded in resource science. It addresses management, and the monitoring and evaluation of the effects of management, of the full range of ecological processes concerning both the sequoia and non-sequoia forests, meadows, and riparian areas within Sequoia National Forest.

Grazing is regulated by annual permits for specific allotments and has continued since the forest was reserved from the public domain. Because of the steep slopes in the southern portion of the range, some riparian areas are steep, rocky and inaccessible to livestock. However high elevation meadows and lower elevation riparian zones and the blue oak savannah are areas of current concern in terms of range and riparian condition, aquatic habitat, and blue oak regeneration. The Forest Service is now under pressure to revise its grazing policies due to concern about possible range deterioration, the adequacy of existing range condition monitoring efforts, and the timing of grazing permits. Current policy allows livestock grazing on the forest in the spring when soil moisture levels make the range particularly vulnerable to erosion, rather than later in the year when the range ecosystem is more resilient to grazing effects. An interagency study team recently evaluated the Sequoia National Forest's range condition and management program. The team's report suggested changes in grazing policy such as improved monitoring and evaluation of the range condition, and delaying the grazing season by several weeks to lessen the negative effects of grazing.

Sequoia National Forest receives more visitors than the more famous but less developed Sequoia-Kings Canyon National Park (figure 51.3). The majority of users come from the Central Valley and the Los Angeles area, a smaller percentage come from the San Francisco Bay area and other countries. The permanent communities and resorts surrounding Lake Isabella are technically within the National Forest. Differences in how these establishments are treated may account for the large variations in visitor days in the 1970s. In order to accommodate the growing recreation activity within the forest, campground management and other recreational activities are contracted to private firms through special use permits. In recent years the number of law enforcement officers on the staff has increased from 2-3 to 10-12 to respond to the law and order problems associated with increased visitor use. As at the Mountain Home Demonstration Forest, altercations between visitors, and visitors who do not observe Forest Service rules are the most common forms of conflict resulting from increased recreational use of the forest.

In summary, the management of Sequoia National Forest has been characterized by intensive and extensive timber harvesting and associated road construction, continued grazing, and high levels of recreational use. These uses reflect the multiple use mandate of the Forest Service embodied in the 1960 Multiple Use Sustained Yield Act. However the landscape effects of the Forest Service's mandate have also been shaped by the centralized organization of the service, the budget priority given to commodity resource production activities, the lack of adequate reinvestment in reforestation and other non-commodity resource values and the lack of effective integration of research with resource management. Together these factors have mitigated against innovation in forest management and grazing policy, have restricted the use of fire as a means to restore pre-fire suppression policy ecosystem structure and function, and have made it exceedingly difficult to sustain a feedback loop based on intensive monitoring and evaluation between resource science and resource management. In 1983 these factors led to logging operations in and around giant sequoia groves on the basis of slender evidence that logging in groves would promote giant sequoia regeneration. The controversy that resulted when the logging was "discovered" by the public was a key factor in precipitating the Mediated Settlement Agreement in 1990.

Sequoia and Kings Canyon National Parks

Reservation, Expansion and Preservationist Mandate

Sequoia and Kings Canyon National Parks comprise approximately 864,000 acres of land primarily in the upper watersheds of the Kaweah, Kern, and middle and south forks of the Kings Rivers. This area was acquired in stages, beginning in September 1890 with the withdrawal of two townships and four sections from the public domain to establish the beginning of the Sequoia National Park, and culminating in the creation of Kings Canyon National Park and its transfer from the Sequoia National Forest to the National Park Service in 1940. The Mineral King area was officially included in the park in 1978 (Dilsaver and Tweed, 1990).

FIGURE 51.3

Sequoia and Kings Canyon National Park and Sequoia National Forest visitor trends.

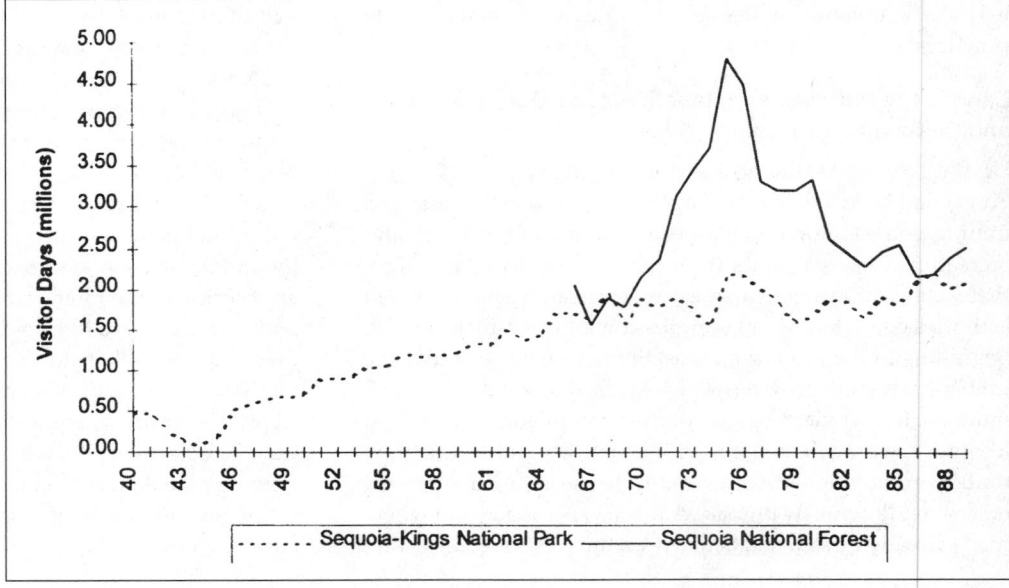

The legislative history of the park's early expansion exemplifies the interrelationship between reserved and non-reserved areas. Less than one week after the initial legislation was passed which established Sequoia National Park (H.R. 11570), a second bill (H.R. 12187) was passed without debate by the House and the Senate and was signed by President Harrison. This bill called for a large federal reservation of public lands surrounding the existing state reservations of Yosemite Valley and the Mariposa Grove of giant sequoias, the addition of five townships to the initial one township comprising Sequoia National Park, and the permanent reservation of the Grant Grove. The bill is interesting because its origin illustrates that preservationist goals and corporate interests are sometimes complementary. H.R. 12187 apparently was quietly substituted for another bill (HR 8350) which called for a much smaller reservation around Yosemite Valley and no extension of Sequoia National Park. Runte (1990) argues that in order to gain congressional support for a larger reservation than called for in H.R. 8350, John Muir and his preservationist friend, editor Robert Johnson, sought and received the support of Southern Pacific Railroad executives for a larger reservation. Soon afterwards H.R. 12187 was introduced. It authorized a reservation five times larger than the alternative bill called for. Inspired by Daniel K. Zumwalt who was a land agent for Southern Pacific Railroad, the bill passed the House and Senate "virtually without debate" September 29 and 30, and on October 1, 1890, President Harrison signed it into law (Runte 1990).

Dilsaver and Tweed suggest that the Southern Pacific Railroad stood to gain from a large federal reservation in three ways. It would benefit from the increased passenger travel and the associated tourist enterprises which the national parks were expected to attract. The railroad's substantial agricultural land holdings in the San Joaquin Valley would have an assured water supply. And the reservation would increase the value of sequoia groves on private timberlands to the north and south of the park which the railroad was involved in, and would eliminate the Kaweah Colony of loggers operating within the reserve who may have otherwise have competed with the railroad for a share of the timber market (Dilsaver and Tweed, 1990). Runte draws an analogy between the support Southern Pacific gave for the reservation of Yosemite National Park and the active promotion of Yellowstone National Park by the Northern Pacific Railroad. He also demonstrates that this was not the only time preservationist and capitalist interests coincided. In 1905 and again in 1906 Muir appealed directly to railroad magnate Edward H. Harriman, whose empire included not only Southern Pacific but also the Union Pacific and Illinois Central Railroads, to solicit his support for the retrocession of Yosemite from the custodial authority of California to the federal government. The railroad's interest in the transfer was based on anticipated increased tourism and passenger rail travel following federally funded improvements in visitor facilities. John Muir and the Sierra Club had been long standing critics of the Yosemite

Commission's management of the park. They felt federal control would be more efficient and effective at removing hunters, herders and other trespassers and at providing facilities for increasing visitor use. Runte attributes the speedy passage of California's transfer bill and of its acceptance by the U.S. House and Senate to Harriman's lobbying efforts. This case represents one early example of the interdependence between preservationist interests and economic interests. Because the areas inside and outside of the newly reserved areas are part of a larger, shared system of ecological and economic relations, management policies taken within reserves influence the management decisions (and consequently the landscape patterns) on private and public lands outside the reserve.

The initial land reservation which created Sequoia National Park differed from the reservation of the Sierra Forest Reserve in 1893 in one significant way—the legislation establishing Sequoia National Park called for the protection of the natural features within its boundaries while the presidential proclamation establishing the Sierra Forest Reserve only required that land sales within the reserve be stopped. Consequently, military protection was provided to protect the park from illegal activities such as grazing, logging and trapping, while in the Sierra Forest Reserve mining, grazing and logging were allowed to continue (Dilsaver and Tweed, 1990). The difference between prohibiting and allowing extractive uses in these two reserves presaged the rancorous debates between preservationists and conservationists whose differences were crystallized in the legislative mandates of the National Park Service and U.S. Forest Service.

The protectionist land management philosophy within national parks was codified in the 1916 legislation establishing the National Park Service. The Park Service's organic act states that the purpose of the Park Service is, "to conserve the scenery and the natural and historic objects and the wildlife therein and to provide for the enjoyment of same in such manner and by such means as will leave them unimpaired for the enjoyment of future generations" (Dana and Fairfax, 1980). The tension between the potentially contradictory goals of preserving "natural and historical objects" and "providing for the enjoyment of same," especially within the current context of shifts in park management philosophy from preserving objects to managing ecosystems, presents a considerable challenge to the management of Sequoia and Kings Canyon National Parks (SEKI).

Organizational Characteristics: Moderately Centralized Flexibility

The organization of the management of SEKI represents a mid-range alternative to the examples of decentralized and centralized organizations which the Mountain Home Demonstration Forest and the Sequoia National Forest represent. One important consequence of SEKI's mid-range position is the potential this provides for scientific research and for integrating research with resource management. Until recently, this capacity was institutionalized within the Division of Natural Science which was created in the wake of the influential 1963 Leopold Report. By 1971, the Division of Natural Science was comprised of a chief scientist, a research botanist, several wildlife rangers and other permanent and temporary research positions (Dilsaver and Tweed, 1990). Prior to the creation of this division, scientific research within SEKI waxed and waned with fiscal conditions, the support of park superintendents, and the interests of individual researchers. The decade and a half preceding the Second World War was a period of active research primarily focused on giant sequoia reproduction and vegetation and wildlife management. Examples of this early research included Emilio Meinecke's 1926 study of human impacts on giant sequoia, George Wright's system-wide study (1929-31) of wildlife policy which led to the creation of a wildlife division, and Lowell Sumner's research at SEKI, begun in 1935, which lead to the eventual development of wildlife, vegetation and backcountry management programs (Dilsaver and Tweed, 1990). This research was actively supported by the then Park Superintendent Colonel White who was one of the earliest administrators to advocate prioritizing "atmosphere preservation", e.g. ecosystem level values, over enhancement of visitor experience. The park's research program (and that of the whole park service) withered during the war and throughout the 1950s, until at the behest of preservationist groups, several studies were commissioned which reviewed the status of ecosystem management within national parks. The most influential of these was the Leopold Report which "provided a framework for the organized expansion of science as a management tool" and was instrumental in shifting the goals of park management from the "protection of objects ... to an aggressive attempt to reestablish ecosystems" (Dilsaver and Tweed, 1990). At SEKI the Leopold Report gave renewed impetus and support to the park's various wildlife management programs and to prescribed burning as a management tool. This included research on prescribed burning as a management tool for giant sequoia and other conifer species by R. J. Hartesveldt, H.T. Harvey, H.S. Shellhammer, R.E. Stecker, B.M. Kilgore, and H.H. Biswell.

SEKI's location in the mid-range of the centralized-decentralized spectrum is a necessary but not always sufficient condition for it to use the knowledge gained through research, monitoring and evaluation to improve the scientific basis and reduce the unanticipated consequences of subsequent resource management plans. SEKI has the decision making autonomy, planning authority and staff resources necessary to use the information research and experience generate to modify, amend, and tailor their resource management plans. However, financial constraints and political and constituency pressures sometimes challenge the ability of park resource managers to implement the management plans developed in consultation with park resource scientists. For example, due in part to funding constraints and conflicting attitudes towards fire within the public and other resource management agencies, only about 10% of areas which should be burned "to protect park resources" are prescribed burned each year (Jeff Manly and William Tweed, pers. comm.). Similarly, under conditions of fiscal retrenchment, competition for funding often emerges between resource managers who feel that an adequate knowledge base already exists to implement more resource management programs, and research scientists who often feel that further research is necessary and should be funded in tandem with resource management programs. The creation of the National Biological Service (NBS) and subsequent transfer of all the research scientists from SEKI to the NBS is the most recent threat to SEKI's research capability. It appears therefore, that effectively integrating science with resource management requires not only local level autonomy and capacity to generate research questions and conduct research, but also institutionalized mechanisms which ensure that the knowledge gained through research will be incorporated into subsequent management plans.

In addition to coupling science with resource management, SEKI must also maintain and cultivate the support of its public constituencies, even as it implements resource management programs, for example controlled burning in giant sequoia groves, which run counter to prevailing and historical norms and attitudes about what should and should not be allowed in forests and what forests should look like. However, the park's relatively narrow preservation mandate restricts the number of special interest groups it must be responsive to, especially in comparison to the Sequoia National Forest whose multiple use mandate ensures that there will be multiple and conflicting special interest groups.

Unlike the Sequoia National Forest which, prior to the 1970s legislation requiring public involvement in resource planning, did not depend on public support in order to fulfill and justify its mission, the early superintendents of SEKI depended on "visitor days" to legitimize the park's purpose and budget and to help justify its expansion. The low number of visitors to Sequoia National Park during the first thirty years of its existence lead to concerns among park administrators that without adequate public support it, and the National Park Service, might not survive. In order to generate more public support for the park, radio and magazine publicity was encouraged and Park Superintendent White initiated the campfire programs and guided nature walks for park visitors which have become the hallmark of the National Park Service's onsite interpretive program (Dilsaver and Tweed, 1990). Consequently, SEKI has not had the organizational autonomy to proceed along a course of action against which substantial public opposition existed. As a result the organization has always funded extensive outreach and extension efforts designed to create a supportive public constituency, and more recently, to generate public support for controversial management programs such as increased prescribed burning in sequoia groves and elsewhere within the parks. When a resource management plan generates controversy, as prescribed burning did following the Yellowstone fire, and when a prescribed burn within the park scorched several large giant sequoia trees, the park management generally responds, in this case by curtailing the burning program, until public support can be regained or perhaps more realistically, the opposition simply wanes.

Landscape Patterns: From Protectionism to Ecosystem Management

Consistent with its legislative mandates, and in contrast to the multiple use mandates of Mountain Home Demonstration Forest and Sequoia National Forest, SEKI has followed a preservationist strategy of land management in combination with efforts to initially encourage visitors and then, when their increasing numbers threatened the natural features the park was mandated to protect, to control and restrict their activities. Some early park management programs would be considered inappropriate within the current interpretation of the park's legislative mandate. These included allocating grazing permits for 2,675 cattle within the park from 1918-1931, a predator control policy which depended on steel traps and poison, indiscriminate killing of problem black bears (15 between 1922 and 1931), and the construction of bleachers to enable tourists to view black bears pawing through the garbage dump (Dilsaver and Tweed, 1990). Fortunately, these programs were relatively short lived, had relatively low level landscape impacts, and provided the impetus for developing management programs based on principles of ecosystem management.

The factors most significant in producing SEKI's current ecological landscape are the historical institutionalization of total fire suppression, the park's preservation mandate which prohibits predominantly commercial uses of the park's natural resources, the historically high visitor use rates and concentration of visitors in some areas, especially Giant Forest, and the commitment among park administrators to block proposed highways into the park's backcountry and across the Sierra crest to Owens Valley (Dilsaver and Tweed, 1990). The absence of commercial timber harvesting (significant numbers of trees have been removed to reduce hazards), mining and grazing, combined with a commitment to minimize road construction, has preserved the integrity of larger landscape blocks than has occurred on the other landscapes in this study managed for multiple use. However, the long standing policy of total fire suppression has interrupted ecological processes, transformed forest structure, and halted the regeneration of some conifer species, notably giant sequoia. Therefore, while landscape blocks may have been preserved relatively intact, ecosystem structure and function has been less successfully maintained. In addition to this, but on a smaller landscape scale, areas of high visitor use such as campgrounds, Giant Forest, and other areas where concessionaires facilities are concentrated, have been disturbed to the extent that the very objects of preservation, e.g. the giant sequoias, have become threatened. Current research agendas and resource management programs address restoration of ecosystem structure and function and explore ways to reduce the negative ecological impacts of visitor congestion. For example, the prescribed burn program in Mineral King with funding secured for five years, represents the most ambitious attempt so far to reduce fuel buildup and restore forest structure to pre-fire suppression conditions across a relatively large landscape block. Similarly, public hearings regarding current SEKI proposals to shift concessions out of the Giant Forest area may actually achieve that end. That park superintendents since the 1940s have unsuccessfully attempted to either reduce the number of accommodations at Giant Forest or to relocate the facilities elsewhere speaks to the extent to which SEKI's management decisions have been tempered, and at times driven, by the organizational necessity of maintaining public support for its activities. This is perhaps analogous to the manner in which timber harvesting on the Mountain Home State Forest has been tempered by the exigency of promoting and managing for recreation.

Tule River Indian Reservation

Establishment—Sovereignty as Mandate

The 55,356 acre Tule River Indian Reservation located in southern Tulare County was established in 1873. More than nine Californian tribes speaking different languages were relocated here from an area extending from the Kings River south to the desert beyond and to the southeast of the Tehachapi range. Consequently only a few of the culturally significant areas for the tribes are located within the reservation. Most areas of cultural significance are scattered across a much broader region encompassing their former seasonal migration areas. The reservation contains a full west side Sierra transect including grassland, blue oak woodland and chaparral below 4,000 feet, black oak and ponderosa pine between 4,000 and 5,000 feet, mixed conifer forest extending to 7,000, and true fir above 7,000 feet (Rueger, 1992). Giant sequoia groves are located within the mixed conifer belt and extend into neighboring Sequoia National Forest.

Current Organization: High Public Accountability

The nine elected members of the Tribal Council set the objectives and policy which govern resource management on the reservation (Rueger, 1992). In addition to the elected council, the traditional elders council also provides considerable leadership. The USDI Bureau of Indian Affairs (BIA) has formal authority on the reservation but currently does not play an active role in management. In the 1950s and 1960s the BIA-sponsored timber harvest plans achieved high levels of production and supported a sawmill on the reservation. Currently, resource management programs are implemented by the Tribe's Natural Resource Department with assistance provided by their consulting forester Brian Rueger from Integrated Forest Management. The reservation's vegetation types have been mapped using aerial photographs obtained from the consulting firm Hammond Jenson and Wallen. This provided the basis for an initial resource inventory, and for the establishment and subsequent monitoring of growth plots in both the mixed conifer and oak woodland belts.

Landscape Patterns—Culturally Attuned Multiple Use

The resource management philosophy of the reservation closely approximates Mountain Home Demonstration Forest's multiple use mandate with the exception that the public(s) is/are on-site," i.e. they live on the reservation as opposed to the demonstration forest whose public owners are the citizens of California. As on the Demonstration Forest, timber sales have historically been a primary source of locally generated revenue for the reservation. Since the reservation assumed direct control of its natural resources from the Bureau of Indian Affairs, the reservation's timber management program has sought to balance the economic values of timber with recreational and aesthetic values and the socio-cultural benefits the forests provide the reservation's inhabitants. From a technical point of view, the forest is harvested at less than its sustained yield potential. Although giant sequoia trees are not harvested, whitewood species distributed throughout giant sequoia areas are intensively managed. Timber harvest levels and employment generation are sometimes reduced if planned timber harvests or other resource extraction activities would damage tribally defined ecological, cultural resources or other non-commodity resources. Unlike the nearby federal or state properties, the social review process does not involve complex reporting and legal analysis. Given the extreme attention paid to sequoia groves on adjacent federal and state ownerships, it is surprising how little attention giant sequoia groves on the reservation receive. Dead and down giant sequoias are harvested for forest products and the groves are used primarily for recreation and other cultural values. However, other flora such as red willow and riparian vegetation have greater cultural significance to many tribal members.

In addition to timber harvesting, grazing and firewood cutting are important consumptive uses of the reservation's resource base. Firewood cutting is important both for local use and off-reservation sale. Firewood cutters (only tribal members can cut firewood) are supposed to pay $5 per cord and harvest only in specified locations. However rules restricting cutting areas are difficult to enforce and there is evidence of over-harvesting of oaks similar to what can be seen on some private ranches throughout the southern Sierra Nevada. The resource management staff apparently feel that the social conflict that strict enforcement would generate does not warrant the slight improvement in resource management enforcement would provide. Grazing on the reservation follows 1983 guidelines established to promote long term range productivity and reduce some of the localized overgrazing problems. Stocking levels have decreased as some tribal members no longer graze stock and others have not increased their herd sizes. The oak woodland and grass lands appears to have more dry residual matter than adjacent ranches which suggests that

overgrazing is less serious than on many other lands in the region. The physical impacts of relatively loose policies towards both firewood harvesting and grazing are visible to both the resource management staff and interested tribal members. At the present time, the low-cost monitoring strategy appears sufficient. Stronger responses could be developed and implemented if needed but the staff clearly weighs this against the potential conflict among tribal members.

The Tule River Indian Reservation's approach to resource management, as shaped by the Tribal Council, exemplifies the key tenets of a multiple use management philosophy which balances commodity and non-commodity resource values. After historically fluctuating timber harvest levels, harvests are now planned to be compatible with non-commodity uses of the forest. In a manner analogous to the Mountain Home State Forest, timber harvest receipts subsidize other resource management activities and still produce a large financial surplus. Unlike the other public institutions, most of the beneficiaries live on the parcel. Daily contact between stakeholders holding a range of goals and the resource managers who report to the Tribal Council provides numerous avenues for these parties to discuss resource management without the formal reporting procedures used in most state and federal systems. While this can be considered a constraint for professional resource managers, it reduces the political uncertainty which arises for national forest and national park managers whose stakeholders are often situated outside the local area.

PAIRED INSTITUTIONAL COMPARISONS

The Many Meanings of Multiple Use

Merely knowing the legislative mandate of an institution is inadequate basis for anticipating what it will actually do and with what impacts. The case study descriptions suggest that an institution's internal organization, the criteria used for budget allocations, the relationship between research science and resource management, and relationships with the public(s) who have stake in it strongly influence the way an institution interprets and implements its mandate and with what ecological effects. Table 51.2 compares the purpose, organizational characteristics, levels of conflict and forest structure of the four study institutions. The structural diversity of the mixed conifer forests in the national forest, national park and state forest is described using the 'later seral and old growth' (LSOG) ranking system developed by Franklin and Fites-Kaufmann (1996) for the Sierra Nevada Ecosystem Project. The rankings are based on large landscape level units consisting of thousands of acres. Rankings of 1 and 2 represent young and relatively simple forests, a ranking of 3 represents mature forests with some late seral attributes, and forests

TABLE 51.2

The four study institutions by organizational characteristics and forest structure.

Institution	Mt. Home State Demonstration Forest	Sequoia National Forest	Sequoia and Kings Canyon National Parks	Tule River Indian Reservation
Purpose/mandate	Multiple use	Multiple use	Preservation	Sovereignty
Organization	Decentralized	Centralized hierarchy	Moderately centralized	Decentralized
Planning autonomy	High	Low	Medium	High
Budget allocation criteria	Floating	Linked to commodity output	Linked to visitor use	Floating
On-site research	High	Low	High	Low
Means for maintaining accountability	Formal	Formal	Formal	Informal
Conflict level	Low	Medium/high	Low	Low
Later seral and old growth (LSOG) ranking, by area				
1&2 (low)	1.2 thousand acres (24% of mixed conifer forest area)	488.7 (66%)	8.0 (9%)	Not surveyed (NS)
3 (med)	2.9 (57%)	185.2 (25%)	74.9 (38%)	NS
4&5 (high)	1.0 (19%)	75.7 (10%)	105.4 (54%)	NS

with rankings of 4 or 5 have considerable structural complexity with many large diameter trees, snags, and down logs. Figure 51.4 graphically represents the percentage distribution of the structural diversity categories for the conifer forests in the mixed conifer forest belt (1200-2600 meters). Bare rock and other areas unsuitable for forest growth in the mixed conifer belt were excluded from the analysis (see table 51.3 for these areas). The figure shows that the most structurally complex mixed conifer forests are in Sequoia and Kings Canyon National Parks, the least structurally complex forests are in Sequoia National Forest, and the forests in the Mountain Home Demonstration Forest are of intermediate structural complexity.

Two paired examples from table 51.2 illustrate how organizational characteristics influence the ways institutions interpret their mandates, and the consequent social and ecological outcomes: the Sequoia National Forest and the Mountain Home Demonstration Forest, and Mountain Home Demonstration Forest and the Tule River Indian Reservation. On paper, the mandates of the Sequoia National Forest and the Mountain Home Demonstration Forest both emphasize "multiple use", but they give different weights to the importance of those multiple uses. The original mandate of the Forest Service which focused solely on timber supply and water flow, was subsequently modified to include recreation and other multiple uses. The mandate for the demonstration forest emphasized recreation over other forest uses such as "water conservation, timber production, forage and mining". Based only on knowledge of their respective mandates, we would expect the Sequoia National Forest to resemble a multiple use forest and Mountain Home Demonstration Forest to be primarily oriented towards preserving giant sequoia and

FIGURE 51.4

Structural diversity of mixed conifer forests in different ownerships.

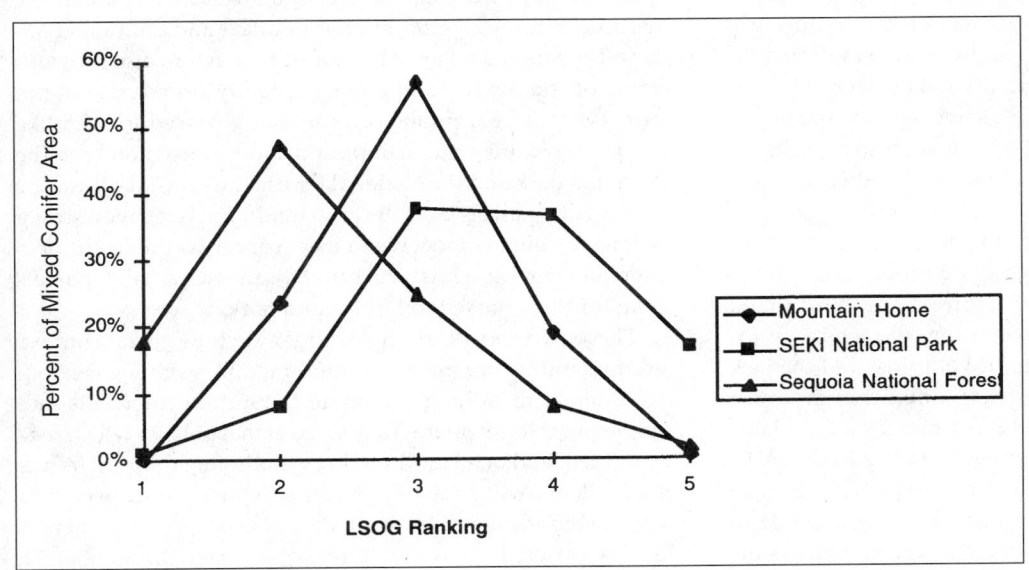

TABLE 51.3

Late seral/old growth structural diversity rankings of the surveyed institutions.

Characteristic	Sequoia National Forest	Sequoia–Kings Canyon National Park	Mountain Home State Demonstration Forest	Tule River Indian Reservation
Mixed conifer region acres 1,200–2,600 m (4,000–8,500 ft)	905,654	248,767	5,048	Not surveyed (NS)
Acres with forest cover in mixed conifer region	749,715	198,312	4,990	NS
Low structural diversity LSOG 1&2	66%	9%	24%	NS
Medium structural diversity LSOG 3	25%	38%	57%	NS
High structural diversity LSOG 4&5	10%	54%	19%	NS

providing recreational opportunities. Contrary to this, we have shown not only that both forests are managed as multiple use forests, but also that the ways in which conflicting resource use patterns are reconciled, the integrity of the feedback loop between research and resource management, and the degree of controversy over resource management activities significantly differs between them. We suggest that these differences can be accounted for by examining the differences in relative degree of centralization, constituency relations and the structure of funding between the national forest and the demonstration forest.

The Mountain Home Demonstration Forest emphasizes timber production to a greater extent than their mandate would lead one to expect to cross-subsidize the administration and management of the rest of the forest. However, due to its decentralized organization and local planning autonomy, Mountain Home forest managers can practice intensive forest management in small patches of mixed conifer forest while simultaneously enhancing recreational opportunities and preserving giant sequoia groves in adjacent areas. Freedom from the need to maximize commodity output targets enables the forest managers to temper timber harvesting to reduce potential conflicts with recreation use by using single tree or small group selection harvest methods and by decreasing the entry frequency by 50 percent. These same organizational and funding characteristics enable forest managers to experiment with, and monitor and evaluate, alternative timber management and fire regimes.

The Sequoia National Forest, on the other hand, also manages for multiple use but through quite different organizational, planning and funding structures. The centralized organization of the forest administration, the tendency for funding to be linked with commodity outputs, and the lack of local level research capacity, restrict the ability of the forest managers to develop innovative timber management plans. This resulted in accelerated timber harvest rates in the 1970s and early 1980s whose ecological effects eventually lead to the multiple appeals of the forest's Land Management Plan and the negotiated Mediated Settlement Agreement. We ar-

gue that more flexible funding arrangements which do not tend to prioritize commodity over non-commodity resource management, a more complete feedback loop between research and resource management, and a more vigorous set of outreach and interpretive programs, could have enabled managers to produce less controversial resource management plans. This paired example shows how and why the meaning of "multiple use" can vary according to organizational and social context.

The Mountain Home Demonstration Forest and the Tule River Indian Reservation illustrate an example in which the high public accountability of the latter and the mandate of the former produced roughly comparable landscape outcomes. Both these institutions follow intensive resource management programs which nevertheless are able to balance commodity and non-commodity resource values in ways which satisfy the diverse needs of the public(s) they are accountable to. The Tule River Indian Reservation is not mandated to follow any specific resource management approach. Its present culturally attuned multiple use management regime developed because of the high levels of accountability reinforced through a number of political and cultural channels. The Mountain Home Demonstration Forest also provides a mix of commodity and non-commodity resources, but not because of formal public accountability procedures. Unlike the more complex public input procedures used on both the National Park and the National Forest, the procedures for the two smaller parcels meet the legal minimum but have a strong record of being responsive to local concerns. Public input is more informal and less structured for the two smaller parcels than for the large federal forest and park.

These two sets of examples illustrate how social context, accountability, organization, funding and planning flexibility interact to influence how an institution interprets and implements its mandate. They suggest that policies which seek to influence what institutions do by modifying the legal framework alone, will probably produce as many unexpected as expected outcomes. However, they also suggest that there is a wide variety of possible sources of leverage through which policy can influence institutions, and that an effective policy will probably combine several.

Funding Effects on Accountability

Regardless of other factors, the criteria used to determine budget allocations always exert considerable influence on the course an organization steers. The funding for Sequoia and Kings Canyon National Parks and for Sequoia National Forest has been historically tied to visitor days and commodity targets, respectively. The national parks' dependence on visitor days renders constituency support essential for the organization's survival. The relatively high scientific research budget at the national parks was based in part on competitive bidding for funding outside of the normal Park Service appropriations. Sequoia National Forest's dependence on meeting internally defined target output levels retains decision making and planning control within the organization and makes it difficult to justify expending the resources required for maintaining external constituency support. The decision to pursue an externally mediated settlement in the late 1980s and the lingering difficulty of getting significant support from the signatories illustrate some of the long term problems which emerge when constituency support is not maintained.

The structure of funding for these two organizations has also helped to generate the central tensions that each faces. Within Sequoia and Kings Canyon National Parks this tension stems from the contradiction between the historical emphasis on preservation and current shifts towards active resource management which must not threaten the primarily non-local constituency support which the national parks depend on. Within the Sequoia National Forest, a central tension exists between meeting internally (internal to the National Forest Service but often determined above the national forest level in the regional or national offices) defined commodity targets and simultaneously satisfying increasing and often conflicting constituency demands for the protection and provision of non-commodity resource values.

EFFECTIVE MANAGEMENT RESPONSES IN COMPLEX ENVIRONMENTS

This section addresses elements of an effective strategy for managing public lands within a context of increasingly complex political and social environments and of growing ecological interdependence between reserves and adjacent resource use systems. Under these conditions we suggest that three elements necessary for effective resource management are applied research programs, the maintenance of public trust and institutional legitimacy, and inter-agency coordination. To reduce the potential biases which stem from studying only

a single national park or forest, in this section we expand the comparative analysis to include Yosemite National Park and the Stanislaus National Forest.

From "Islands in Time" to Porous Reserves and Complex Environments

Protection of natural resources, whether defined as preservation or conservation, has historically involved establishing boundaries around that which is to be protected and then developing the political capacity to secure the boundaries from external threats and the administrative capacity to control what occurs within the boundaries. The Sierra and Stanislaus National Forests and Sequoia and Kings Canyon and Yosemite National Parks were reserved from the public domain in order to protect the otherwise threatened natural resources which fell within their boundaries. While the initial motivations for reserving these lands were similar, e.g. to protect them from degradation by private interests, soon after their reservation they were imbued with different management philosophies and purposes, administrative structures, and political constituencies. Although purchased, instead of reserved from the public domain, Mountain Home Demonstration Forest was similarly established with strong local support to protect the large giant sequoias it contained from harvesting. Finally, the Tule River Indian Reservation established the boundaries of sovereignty of people who once claimed a much larger territory. Here again, it could be argued, one of the initial purposes of the reservation was to protect those inside it from an environment so hostile that there was little or no chance of surviving in it.

While the "island in time" approach to resource protection may have succeeded in the past, it is now ineffective. Mounting external pressures on reserve boundaries make them appear increasingly porous, and the increasing differentiation of the social and political environments of reserves heightens the tensions between alternative and sometimes mutually exclusive management objectives. The porosity of reserve boundaries refers to situations where influences which impact areas within an agency's jurisdiction arise from outside it. Examples include the effects of air pollution on southern Sierra conifers, wildfires which move without regard to jurisdiction, water claims from outside the reserve which affect the supply of water within the reserve, and habitat degradation and/or reduction on adjacent lands on which migratory wildlife are seasonally dependent. The threats these external pressures and influences pose are not insignificant. For example in 1980, in response to two private studies of the threats adjacent land use posed for public lands management (NPCA 1979) the National Park Service released a report on threats to the national parks which showed that more than fifty percent of the threats originated from sources or activities external to the park, and that air quality was endangered in more than forty-five per-cent of the parks.

In addition to increasingly porous boundaries, the social

and political environment of public lands management has grown more complex and uncertain. Long term planning on national forests is hindered by the uncertainty created by restrictive court injunctures and temporary guidelines which determine what management activities are allowed where. In many instances court orders and temporary guidelines such as for the California Spotted Owl and the Mediated Settlement Agreement on the Sequoia National Forest are manifestations of the difficulty of managing public lands for constituencies with conflicting values and attitudes about the prioritization and acceptability of different resource uses. In addition to being asked to satisfy increasing claims from diverse public(s) for different resource amenities, the agencies responsible for managing public lands are also asked to adopt "adaptive management" and "ecosystem management" approaches, which if they are to be effective, require a level of integration between resource science and resource management which has historically been difficult to sustain.

Research Capacity and Integration with Resource Management

Although little disagreement exists regarding the importance of research, there is considerable debate about the most effective way to organize research and to integrate it with resource management. Variables across which the organization of research can vary are: the degree to which it is centralized or decentralized; the extent to which research budgets constitute a separate line item or are subsumed within other budget categories; the degree to which research scientists are accountable to administrators or to other scientists; the degree of complementarity between research agendas and outcomes and the information needs of resource managers; the extent to which research agendas dovetail with a macro-level coordinated strategy or are tailored to meet site-specific objectives; and the extent to which research capacity is concentrated within an organization or is accessed from other institutions through contracts and coordinated agreements.

The formal organization of research within the National Park Service and National Forest Service exemplify alternative combinations of the above variables and therefore, opportunity to identify the strengths and weaknesses of different research organization structures. Additionally, differences in research activity and degree of linkage with resource management exist between units even when they share the same formal structure. For example substantial differences exist between Sequoia and Stanislaus National Forests in terms of research activity and its links with management. This section of the report comparatively analyzes the organization of research on these two national forests and then between them and Yosemite and Sequoia Kings Canyon National Parks.

Although both the Sequoia and Stanislaus National Forests share the same centralized model of research common throughout the Forest Service, there are unexpected differences in the level of research on the two forests. On the Stanislaus National Forest there is one experimental forest, the Stanislaus-Tuolumne Experimental Forest, as well as several Research Natural Areas (RNA's). Both the experimental forest and the RNA's have been set aside as research sites by and for Pacific Southwest (PSW) Research Station scientists who are involved in a wide variety of ongoing research projects. Examples of research on the forest include silvicultural experiments in the experimental forest, aspen and black oak research and other research projects in various of the RNA's, research on herbicide use and effects, and the interagency Mokelumne River Watershed Project (Henly, 1993). Following the Stanislaus Complex Fire in 1987 which burned approximately 145,500 acres, PSW scientists established a paired watershed study to measure the erosion and sedimentation associated with two different salvage logging methods. In addition to research carried out by scientists from off-site PSW research stations, administrative studies which approximate research but are often explicitly related to management goals, are conducted by on-site personnel. Recent examples of this type of research include studies of forest health by entomologists and pathologists on the forest staff, as well as studies of the advanced cut-to-length logging technology and the complementary relationship between it and prescribed burning for fuels reduction under conditions of high fuel loading. The Stanislaus National Forest also has the only funded prescribed natural fire management and research program in California.

The relatively high level of research activity on the Stanislaus National Forest contrasts sharply with that of the Sequoia, at least in recent years as described earlier in this report. Several factors account for this difference. Among them include the closer proximity of the Stanislaus to PSW research stations in Albany and Redding, possible ecological differences between the two forests which make the Stanislaus more attractive given the research agendas of PSW scientists, differences in receptivity to "outside" (PSW) researchers among the personnel on the two forests, and, on the Sequoia National Forest, a possible dearth of the initiative and enthusiasm required to carry out in-house administrative studies given the public controversies and confrontations the forest has been embroiled in since the late 1980s.

However, even on the Stanislaus National Forest the highly centralized research structure has generated shortcomings and criticism. Two recurring criticisms concern the research agenda setting process which forest managers feel they have little or no ability to significantly influence and contribute to, and the project based funding of many management programs whose target driven structure often leaves inadequate funds for on-site research and monitoring. Past and present forest managers stated that the research agendas of PSW scientists often do not address forest-level management concerns and information needs, and that the management implications of research are not explicitly stated in a manner which promotes communication between researchers and managers. Managers contrasted this with research and administrative studies conducted by forest entomologists and pathologists on the forest staff which are aimed at answering management related research questions. Because of the lack of perceived benefits from research conducted on the forest, some forest managers felt that the reductions in the area which could be managed for multiple use because of land allocations for research natural areas and experimental forests were not warranted.

Research within the National Park Service is organized very differently from the U.S. Forest Service. Important differences include the lack of a separate research branch within the Park Service, research budget allocations which are not clearly separated from resource management activities, and the assignment of research scientists to individual parks where they report either to the park superintendent or to the regional chief scientist. Interviews with research scientists and resource managers at Sequoia and Kings Canyon and Yosemite National Parks indicate that, at least in these two parks, this decentralized model of agency research has generated research with explicit and clearly communicated management objectives. The organizational proximity of researchers and managers facilitates communication and coordination between them and enables park scientists to address research questions of applied significance to managers. While this model makes it possible to integrate research and resource management and thereby avoid the problems associated with the Forest Service's centralized research organization, it also has its own particular shortcomings. These are detailed in the many reviews of the Park Service's research program beginning with the report of Leopold and colleagues (1963) and ending most recently with a report by the National Research Council (1992). Some of the most common criticisms these reports contain are the lack of an integrated and coordinated research agenda at the national and sometimes even regional level which can result in fragmentation and duplication of research effort, the low budget priority of research and competition with resource management activities for funding, the sacrifice of long-term research goals in the face of administrative pressures to provide guidance for shorter term resource management decisions, and a tendency for research scientists to leave the Park Service because of these and related constraints. To redress these problems these reports often suggest a centralized model of research organization within an independent research branch which would resemble the organization of research within the U.S. Forest Service. Yet it is unclear how these recommendations, if implemented, would avoid the weaknesses documented within the Forest Service research model.

The comparison of the advantages and disadvantages of the centralized Forest Service research program and the decentralized Park Service program suggests that only a hybrid research organization will be able to provide the local autonomy required for effective feedback between research and resource management while simultaneously providing the organizational resources and insulation from short term administrative imperatives necessary for the sustained ecological research which is needed to define and achieve conditions of ecosystem health. While an independent research branch is probably a necessary element of this hybrid approach, effective integration of research with resource management will only occur when research is organized at the local level either through "in-house" administrative studies, or through cooperative studies involving university researchers or scientists in other state and federal natural resource agencies. The extensive program of contracted research of this type described earlier in this report for the Mountain Home Demonstration Forest exemplifies an effective use of this strategy in a situation where there is little or no "in-house" research capacity within the forest management staff. Another example of how research can be integrated with resource management is the envisioned organization of future research within Yosemite National Park. In the proposed plan inventorying, monitoring and evaluation functions would be accomplished with park personnel and all other research needs would be met by contracting with scientists from other agencies such as the National Biological Survey as well as universities. Although administrative studies and cooperative studies with outside researchers are possible on national forests, they appear to be effectively utilized only by National Forest staff with a "can do" reputation. Only by changing the structure of funding, the incentives for investing time and energy in these types of research activities and providing the requisite local level staff autonomy will administrative and cooperative studies be conducted by other than "can do" forests.

Constituency Support: Its Importance, Maintenance and Restoration

The ability of public land-based resource management agencies to maintain their institutional legitimacy and the trust of the public(s) is especially important, and difficult to achieve, as their social and political environments become increasing complex and the tensions inherent in satisfying diverse and sometimes conflicting values grow stronger. The conflicts and administrative appeals associated with the Sequoia National Forest's land management plan which eventually lead to the negotiated Mediated Settlement Agreement (MSA) suggest that the procedures necessary for satisfying the legal provisions for public involvement in resource management planning mandated by the National Environmental Policy Act (NEPA), the Resources Planning Act (RPA), and the National Forest Management Act (NFMA) are not sufficient to ensure minimal public support for agency resource management plans. This section of the report identifies some of the components of agency-constituency relations which help to maintain institutional legitimacy and public trust in the agency. We do this by analyzing the factors which account for the variation over time in relations with the public on the Stanislaus National Forest and by briefly comparing this with Yosemite National Park.

Relations between the Stanislaus National Forest and the public(s) it serves and the degree of engagement of forest personnel with local communities, have waxed and waned over time. While it is difficult to fully explain these fluctuations, they are related to factors such as the degree of decision making autonomy at the ranger district and forest level, the extent of conflict the forest is embroiled in and the response of forest service personnel to conflict, e.g. whether they withdraw into a defensive position or not, and the extent to which leadership is willing and able to take the initiative with regards to outreach and provides support and incentives for staff to do the same. Interviews with retired Forest Service personnel from the Stanislaus National Forest revealed the extensive, proactive outreach efforts which existed on the forest during the 1970s and early 1980s. For example, at the district level forest service personnel would organize field trips one or more times a year to which local community members, county supervisors, personnel from other state and federal resource agencies including the local representative of the Regional Water Quality Control Board, and representatives of non-government organizations were invited. The purpose of the field trips was generally to observe and discuss a planned, in-process, or completed resource management plan such as a timber harvest, reforestation effort, watershed rehabilitation project, prescribed burn, or fuel break near an urban-wildland interface. These "show-me" trips were not simply geared for generating public support for pre-determined courses of action, but were an effort to solicit informed public input which was used to modify plans and projects where possible to better satisfy constituency needs and objectives. Modifications to resource management plans which at least partially resulted from the public input solicited during these fieldtrips included reductions in the size of clearcut blocks, not harvesting trees along travel corridors for aesthetic reasons, spatial harvest patterns which least hindered the movements of wildlife, and attempts through fencing and the control of stock numbers to reduce damage to sensitive areas from stock grazing.

At the forest level outreach and extension activities included maintaining regular contact with the staff and elected members of the state and federal legislatures, elected county officials, and local newspaper editors, and inviting local journalists to attend annual planning meetings. As one forest service retiree described, "It (was) a matter of getting out on your own and running the people down." One important component of this proactive outreach effort was the willingness to publicly admit errors, to explain how and why they were made, and to redress them and minimize the likelihood of a recurrence. This was crucial in maintaining the public's trust, the agency's legitimacy and local political support. Two factors which enabled these relatively high levels of outreach and extension were leadership support at the forest and region levels for staff to spend time in outreach and at least minimal levels of decision making autonomy at the district and forest levels within the Forest Service. Together these fac-

tors provided adequate incentives and rewards for Forest Service personnel to invest time and resources for maintaining public trust and institutional legitimacy.

During the latter half of the 1980s and the early 1990s, and for a variety of reasons including the perceived reduction of leadership support for outreach, declining decision making autonomy at the district and forest levels, and increasing conflicts over timber harvests and related resource management issues, the level of public trust in and the legitimacy of the Stanislaus National Forest waned. It appears that the informal modes of public participation described above were gradually supplanted by the formalized involvement methods mandated in the Resources Planning Act, and that possibly in a defensive move, the previously robust informal relationships with community and political leaders were no longer as actively cultivated. Concomitant with this process was a decline in the perceived willingness of the forest service personnel to admit mistakes and acknowledge when errors were made. Not surprisingly, this has generated friction and resentment in some local circles, one manifestation of which was a recent home-rule initiative that, although voted down, had adequate support to be placed on the county election ballot.

More recently, the Stanislaus National Forest's innovative and aggressive prescribed burn program, and the program's integration with other forest management activities, may be triggering a resurgence of the type of outreach which existed previously. Last June and during the recent 5,000 acre prescribed burn, fieldtrips were organized to which non-Forest Service researchers, environmental group and forest industry representatives, and congressional staff members were invited. While this and similar outreach efforts may help improve public awareness, public trust will probably not be regained without some devolution of planning authority to the district and forest level which allows substantive public involvement, and without an "error embracing" attitude and the concomitant organizational openness this attitude requires. Public confidence and trust, once lost, is hard to restore. However, and perhaps paradoxically, as the social and political environment within which the Forest Service operates becomes more complex and contentious, the importance of maintaining institutional legitimacy becomes increasingly important.

The National Park Service has, for the most part, not suffered the same loss of legitimacy and trust which some argue the Forest Service has. This is at least partly due to its narrower legislative mandate and because its activities have not aroused the same degree of public controversy and subsequent scrutiny that the Forest Service's have. One park administrator at Yosemite National Park said that the "white hat" image of the Park Service has sometimes enabled park managers to "skate on thin ice" regarding the scientific basis of their resource management programs and the extent to which NEPA and other planning laws are followed "in letter rather than in spirit." The difficulty of managing the tension inherent in the Park Service's mission between providing for visitor enjoyment and managing natural resources for future generations has in some cases generated ecological harm and/or conflicts with the public. Despite the relatively narrow mandate of the Park Service, the more complex social and political environment within which it operates presents challenges that did not exist even a decade previously. For example, in Yosemite National Park these challenges include poaching within park boundaries, increased gang activity in the park as well as drug-related crimes, and conflicts between public attitudes that national parks exist primarily for personal recreation and enjoyment and park research and resource management programs which seek to restore ecosystem processes and values. In an attempt to address these challenges in a proactive rather than reactive fashion, park administrators are planning an unusual off-site outreach program which will involve sending rangers to elementary schools in Fresno, Modesto, Merced and other local school districts to teach students basic ecosystem principles and concepts and to discuss the nature and purpose of national parks.

Investing resources in maintaining constituency relations and institutional legitimacy becomes increasingly important as the environments of resource management institutions grow more complex and contestatory. Under these conditions a public agency will likely be able to retain its institutional legitimacy and the trust of the public(s) for whom it manages the resources within its jurisdiction by following a proactive strategy of public outreach, on- and off-site interpretative programs, and extension work which involves all of the various and concerned interest groups. Accomplishing this probably requires minimal degrees of local level organizational autonomy, widening the envelope of acceptable planning outcomes in the interests of fostering substantive public involvement, providing leadership support and organizational incentives for personnel to invest time and energy in outreach efforts, and a non-defensive attitude which allows errors to be acknowledged and transformed into learning opportunities.

Inter-Agency Coordination

Coordination between public resource management institutions is increasingly important as reserve boundaries become more porous and social and political environments more complex. Examples of formal coordination among the agencies in this study include cooperative fire suppression agreements between California Department of Forestry and the U.S. Forest Service, the coordination of research activities on fire history studies and wildlife research between university researchers, Sequoia Kings Canyon National Parks, Mountain Home Demonstration Forest and Sequoia National Forest, and efforts to mesh wilderness use policies between U.S. Forest Service and the Park Service. One of the more long-lived examples of inter-agency coordination is the Sierra wilderness group. The wilderness group was begun in the

mid-1970s to coordinate research, resource use and condition monitoring and fire management within the wilderness areas distributed across different land-based public institutions. More recently the group has sought to establish uniform wilderness and backcountry regulations throughout all wilderness areas in the central and southern Sierra regardless of agency jurisdiction. Other examples of coordination include an inter-agency manager's group, wildlife and fisheries research and management group, a GIS group and annual meetings of the region's forest supervisors and park superintendents. Cost-share programs, using funds from tax receipts or provided by interested non-government organizations, also provide opportunities for mutually beneficial inter-agency resource management programs. For example, through cooperative agreements California Department of Fish and Game revenues from hunting licenses and other taxes fund wildlife habitat improvement programs on national forest lands such as meadow restoration and controlled burning.

One of the more recent examples of inter-agency coordination in the southern Sierra Nevada is the Giant Sequoia Ecology Cooperative which was formed soon after the 1992 symposium on Giant Sequoias held in Visalia, California. It emerged as an inter-agency response to public controversy about management and regeneration of large giant sequoias, and common agency recognition of the sparse scientific basis for giant sequoia management. The Giant Sequoia cooperative will facilitate the coordination and sharing of giant sequoia related research amongst the member institutions. Ideally the cooperative will combine the comparative strengths of each member institution in a manner which strengthens the linkage between resource scientists and resource managers, and improves the public accountability of the participating agencies vis a vis sequoia management. More informal forms of coordination include reciprocal road easements between the Tule River Indian Reservation and Sequoia National Forest, and between Sequoia National Forest and Mountain Home Demonstration Forest.

Coordination between agencies also sometimes includes county government and other local community organizations. Mariposa County has four or five memorandums of understanding (MOU) with Yosemite National Park addressing a wide range of activities from building permits to search and rescue coordination. One of the MOU's regarding transportation involves four other adjacent counties and the California Department of Transportation. MOU's can function to help maintain channels of communication between public agencies and local communities and government. They are one vehicle through which tensions and conflicts of interest between public agencies and local communities can be reduced and/or resolved. In this respect they perform a similar political function as do the forms of outreach and extension discussed above.

Inter-agency coordination emerges under conditions of porosity and complexity when the benefits outweigh the costs of coordination.(Romm and Baker, 1990). It capitalizes on the comparative advantages of different resource management agencies, for example coordination between Sequoia Kings Canyon National Parks and Sequoia National Forest, or Mountain Home Demonstration Forest and university researchers, compensates for the lack of on-site research capacity on the national and demonstration forests. Inter-agency coordination also provides local level arenas for resolving potential conflicts among agencies and between them and local communities and government, it enables more efficient utilization of scarce resources through coordinated project planning, and it helps bring policy and managerial coherence to ecosystems riven by jurisdictional boundaries.

CONCLUSION

Integrating resource science with resource management, maintaining constituency support and public accountability, and coordinating the activities of multiple state, federal, and county organizations and various non-government interests and organizations, are increasingly important as the political environment within which resource agencies work becomes more complex and contestatory. The research basis of resource management decisions can be presented and defended in public forums. This is crucial in order to maintain an institution's credibility and legitimacy. Although good science will not produce a consensus when differing values are at stake, it will provide the basis for establishing viable alternative policy options.

The Sierra Nevada ecosystem is likely to become more porous and complex in the future. In some cases the most serious threats to ecosystem health originate from outside agency jurisdictions. The organizational autonomy necessary in order to respond effectively to these and other threats to ecosystem health will only be granted by the public to those government agencies with legitimacy and accountability. Under these conditions, static legislated or rigid centrally planned policies are unlikely to produce lasting ecosystem protection. The range of organizational policy levers which can be used to affect the public land-based institutions in the Sierra Nevada include shifts in the funding and organization of research to create hybrid research organizations, relaxing the links between commodity outputs and budget levels, providing the local level flexibility, means and incentives necessary for maintaining institutional accountability and legitimacy, and facilitating formal and informal modes of interagency coordination at all levels. Policies which operate in these non-legislative arenas are often process rather than target oriented: instead of legislating outcomes they attempt to create institutional mechanisms for resolving conflict which incorporate scientific research and maintain institutional accountability.

ACKNOWLEDGMENTS

The research this report is based on would not have been possible without the generous cooperation of the staff of the four institutions we studied. Without exception they were willing to meet with us and to discuss the issues we were interested in. We would also like to thank the retired Forest Service personnel and other individuals with whom we met for their interest, insights and cooperation. Thanks are also due to the Sierra Nevada Ecosystem Project's internal reviewers whose comments and criticisms of an early draft of this report helped substantially to strengthen it. The views expressed in this report as well as any remaining shortcomings are those of the authors.

REFERENCES

Chase, A. 1986. *Playing God in Yellowstone: The destruction of America's first national park.* San Diego: Harcourt Brace Jovanovich.

Conservation Foundation. 1979. *Federal resource lands and their neighbors.* Washington, DC: Conservation Foundation.

Culhane, P. J. 1981. *Public lands politics: Interest group influence on the forest service and the Bureau of Land Management.* Baltimore: Johns Hopkins.

Dana, S. T., and S. K. Fairfax. 1980. *Forest and range policy.* 2nd ed. New York: McGraw-Hill.

Dilsaver, L. M., and W. C. Tweed. 1990. *Challenge of the big trees.* Three Rivers, CA: Sequoia Natural History Association.

Franklin, J., and J.A. Fites-Kaufmann. 1996. Analysis of late successional forests. In *Sierra Nevada Ecosystem Project: Final report to Congress,* vol. II, chap. 21. Davis: University of California, Centers for Water and Wildland Resources.

Henly, R. 1993 . Policy, legal, and institutional considerations in the control of cumulative environmental impacts on forested watersheds in California. Ph.D. dissertation. University of California, Berkeley.

Kaufman, H. 1981. *The forest ranger: A study in administrative behavior.* Baltimore: Johns Hopkins University Press.

———. 1994. *The paradox of excellence.* Manuscript, May 15.

Leopold, A., S. Cain, C. Cottam, I.Gabrielson, and T. Kimball. 1963. *Wildlife management in the national parks.* Transcript North American Wildlife and Natural Resources Conference 28:28–35.

Lindblom, C. 1959. The science of muddling through. *Public Administration Review* 19:79–88.

NPCA (National Parks and Conservation Association). 1979. *NPCA adjacent lands survey: No park is an island.* Washington, DC: National Parks and Conservation Association.

NRC (National Research Council). 1992. *Science and the national parks.* Washington, DC: National Academy Press.

Romm, J., and M. Baker. 1990. *Emerging institutional linkages between land and water.* Report for Forest and Rangeland Resources Assessment Program, California Department of Forestry and Fire Protection, Contract 8CA74717.

Romm, J., and S. K. Fairfax. 1985. The backwaters of federalism: Receding water rights and the management of national forests. *Policy Studies Review* 5 (2): 413–30.

Rueger, B. 1992. Giant sequoia management strategies on the Tule River Indian Reservation. Paper presented at the Symposium on giant sequoias: Their place in the ecosystem and society, Visalia, CA. PSW-GTR-151:5–7.

Runte, A. 1990. *Yosemite: The embattled wilderness.* Lincoln: University of Nebraska Press.

Schiff, A. 1962. *Fire and water: Scientific heresy in the forest service.* Cambridge, MA: Harvard University Press.

Setting forest ablaze to save it. *San Francisco Chronicle.* 25 November 1995.

Tipple, T. J., and J. D. Wellman. 1991. Herbert Kaufman's forest ranger thirty years later: From simplicity and homogeneity to complexity and diversity. *Public Administration Review* 51 (5): 421–28.

Tweed, W. C. 1992. Public perceptions of giant sequoia over time. Paper presented at the Symposium on giant sequoias: Their place in the ecosystem and society, Visalia, CA. PSW-GTR-151:5–7.

Twight, B. W., and F. J. Lyden. 1989. Measuring forest service bias. *Journal of Forestry* 87 (5): 35–41.

U.S. Forest Service. 1991. *Stanislaus National Forest land and resource management plan.* Sonora, CA: US Forest Service, Pacific Southwest Region.

Wildavsky, A. 1984. Federalism means inequality: Political geometry, political sociology, and political culture. In *The costs of federalism,* edited by R. Golembiewski and A. Wildavsky. 55–69. New Brunswick, NJ: Transaction Books.

BRUCE J. McGURK
Pacific Southwest Research Station
U.S. Forest Service
Albany, California

MAUREEN L. DAVIS
Pacific Southwest Research Station
U.S. Forest Service
Albany, California

52

Camp and Clear Creeks, El Dorado County: Chronology and Hydrologic Effects of Land-Use Change

ABSTRACT

As part of the Sierra Nevada Ecosystem Project's Camp Creek/Clear Creek case study, modeling was done to assess the relative hydrologic effect of fifty years of land management in two basins. The Camp Creek Basin encompasses 8,425 ha (20,821 acres), ranges in elevation from 975 to 2,316 m (3,200 to 7,600 ft), and is managed by the Eldorado National Forest to provide timber and other products. The Clear Creek Basin encompasses 3,068 ha (7,580 acres), ranges in elevation from 512 to 1,250 m (1,680 to 4,100 ft), and has been extensively developed with low-density housing as well as ranching. The goal of this project was to quantify changes in runoff timing and volume stemming from changes in land management over time. Hydrologic effects were quantified for representative climatic conditions occurring in high-, medium-, and low-magnitude water years. Aerial photographs and a geographic information system (GIS) were used to quantify changes in cover density, impervious area, and other information for 1940, 1952, 1966, 1976, 1986, and 1991.

Disturbance in the Camp Creek Basin has been primarily associated with logging and roads. Between 1940 and 1991, April runoff increased by about 18% for the medium- and high-magnitude water years. For these years, and associated with decreasing forest cover and an increasing road network, annual snowmelt and subsurface flows increased over time, and annual ground water and evapotranspiration amounts decreased. For the medium-magnitude water year on Camp Creek, runoff shifted to earlier in the melt season, and summer base flows were smaller when compared with outflows predicted with the 1940 land-use condition.

Disturbance in the Clear Creek Basin has been primarily associated with residential development and roads. Because Clear Creek is lower in elevation and does not accumulate a seasonal snowpack, its runoff pattern does not have an April peak. High-flow months are between December and April, depending on storm occurrence. For the medium- and high-magnitude water years, predicted stream-flow increases in February, March, and April ranged from 1% to 4%. For the low-magnitude water year, the increase was between 14% and 18%. The change in total runoff was due to a large increase in the surface runoff contribution to total flow, and it is distinctly different from the Camp Creek response to land-use change. Evapotranspiration losses, ground water, and subsurface flows all declined during the analysis interval. Runoff responses to storms occurred faster in Clear Creek by the end of the analysis period, a result associated with changes in land condition.

Because the basins are at such different elevations, their hydrologic responses are different even under similar inputs. For these two basins, however, fifty years of changes in hill slope condition due to forest management and suburbanization appear to produce changes in runoff timing and volume and to change the relative contribution of flow components.

INTRODUCTION

The Sierra Nevada Ecosystem Project (SNEP) is an assessment of the entire Sierra Nevada ecoregion. Late successional for-

Sierra Nevada Ecosystem Project: Final report to Congress, vol. II, *Assessments and scientific basis for management options.* Davis: University of California, Centers for Water and Wildland Resources, 1996.

ests, watersheds, and significant natural areas are critical concerns, but the assessment also includes the social, economic, and ecological components of the entire set of Sierra Nevada ecosystems. In areas of the Sierra Nevada such as the 243,000 ha (600,000 acre) Cosumnes River Basin (figure 52.1), humans have been modifying the landscape for at least 150 years. Landscape disturbances include fire, logging, mining, water resource development, residential and road construction, and grazing. By the 1940s, road building and land clearing were common in the middle elevations of the Cosumnes, and logging was common in the upper elevations. In the 1960s, however, the pace of resource extraction accelerated, and demands for housing and water began to increase in the foothills below the lands that had become national forests. By the 1980s, many of the other river basins in the Sierra Nevada had been developed to supply hydropower, irrigation, or municipal water supplies. Retirement, recreation, and vacation communities in the Sierran foothills became widespread as Sacramento and other Central Valley towns grew. This growth has led to concerns about fragmentation of ecosystems and wildlife habitat, the role of wildland fire, and the effect of land disturbance on the hydrologic regime and the associated riparian ecosystems.

The Cosumnes River Basin was selected by the SNEP Science Team as a case study area because it is typical of many basins in terms of development. It is atypical in that there are no major dams on the system, although diversions are common and a municipal water supply dam disrupts natural flow in the Camp Creek and Sly Park Creek tributaries to the North Fork of the Cosumnes.

Suburbanization and forest management are two major uses of the Sierra Nevada and the Cosumnes basin. The SNEP team selected two small catchments within the Cosumnes for intensive analysis of the effects of these two uses. The two catchments have an extensive soils, fire, and road-network database and are in close proximity. Only Camp Creek has a stream gauge, but few small basins such as Clear Creek are gauged anywhere in the Sierra Nevada.

This project uses a process model to assess changes in runoff volume and timing between 1940 and 1991. Trends in vegetative cover, road extent, fire, and other factors were quantified at approximately ten-year intervals, and the changes were incorporated into the land-use condition portion of the hydrologic model. A thirty-three-year record of precipitation and runoff was ranked, and low-, medium-, and high-magnitude water years were selected to represent

FIGURE 52.1

Location of study area and climate and discharge stations in the Cosumnes River Basin, California.

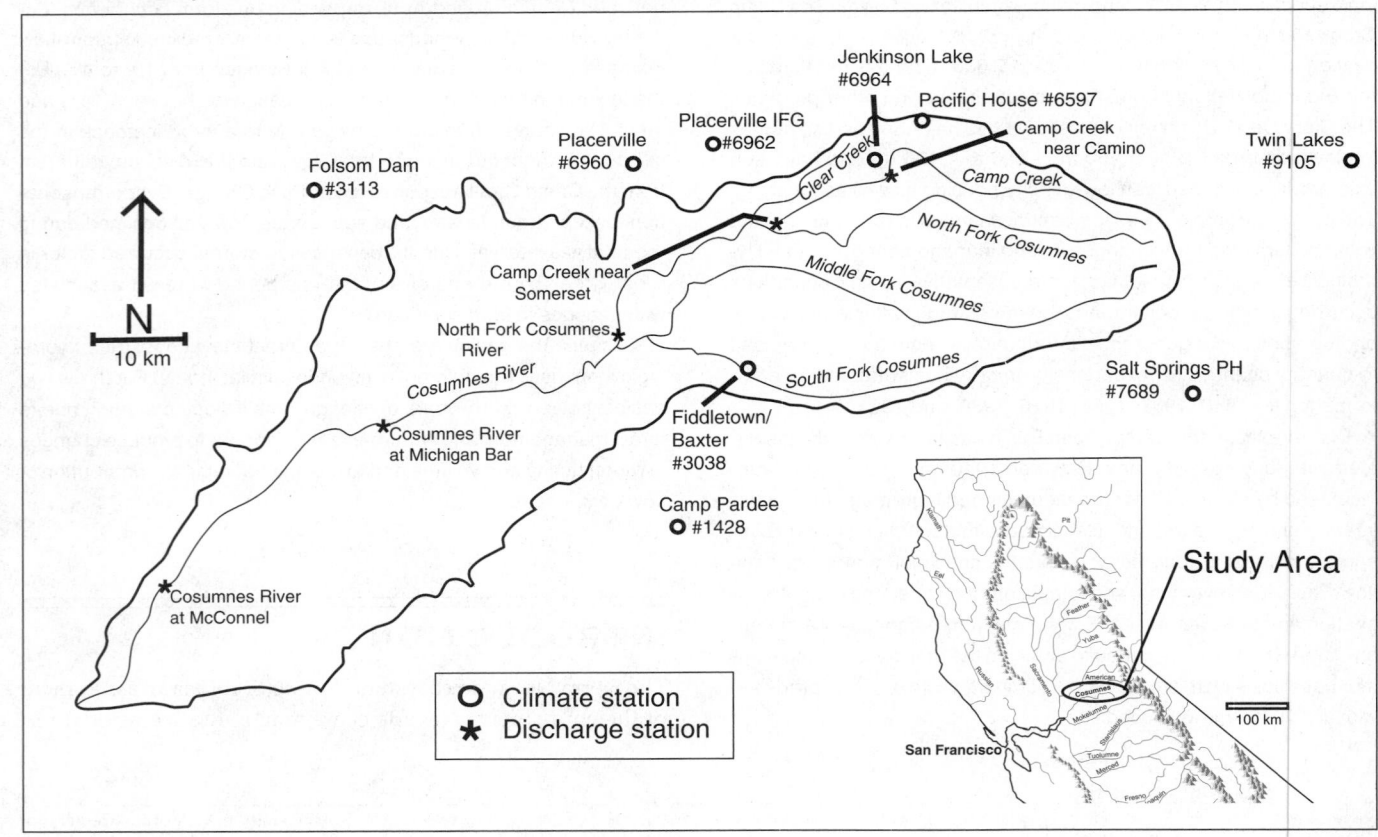

drought, normal, and wet years. The hydrologic model was configured for each of the land-use conditions, and analysis was done for each of the three types of water years. The results were compared to assess the effects of changes in land use and water-year magnitude. A companion report to this one is an analysis of changes in sediment yields associated with the same set of land-use changes (McGurk et al. 1996).

STUDY SITE DESCRIPTION

The Camp Creek Basin ranges from 975 to 2,316 m (3,200 to 7,600 ft) in elevation and is managed by the Eldorado National Forest (ENF) to provide timber and other products (figures 52.2 and 52.3). Camp Creek is about 20 km (12.4 mi) east of Placerville, has a west-facing aspect, and is a tributary to the North Fork of the Cosumnes. A 8,246 ha (20,821 acre) portion of the Camp Creek Basin was selected by the SNEP Hydrology Team and is the contributing area upstream of a U.S. Geological Survey (USGS) gauging station (#113315, Camp Creek near Camino). Because the portion of the Camp Creek Basin between the #113315 and the #113330 stations (Camp Creek near Somerset) is not all within the ENF boundaries and has not been extensively logged or roaded because of the steepness of the canyon, it was excluded from the analysis.

The Clear Creek Basin is 3,068 ha (7,580 acres) in size, ranges from 512 to 1,250 m (1,680 to 4,100 ft), and has been extensively developed with low-density housing (figure 52.4). The Clear Creek Basin has a southwest aspect and is predominantly private land.

Both basins are composed primarily of loam soils (Cohasset, Josephine, Mariposa, and McCarthy series) (Mitchell and Silverman n.d.; Rogers 1974), but Clear Creek has some areas with clay soils. Camp Creek soil depths range from 66 to 178 cm (26 to 70 in) and average 88 cm (35 in). Clear Creek soils range from 45 to 131 cm (18 to 52 in) and average 103 cm (41 in). Clear Creek has grass- and shrublands along with forested areas, but Camp Creek is almost entirely forested with mixed conifers. Clear Creek slopes range from 9% to 41% and average 24%. Camp Creek slopes range from 15% to 43% and average 31%.

METHODS
Scale of Analysis

This study is local in scale and assesses the land-use and hydrologic changes in two small basins in the central Sierra Nevada. It is likely that the sequence of land-use develop-

ments and the forest management activities that have taken place in the Camp and Clear Creek Basins are typical of activities that have occurred across a much wider scale. Suburbanization and logging are common uses of much of the western slope of the Sierra Nevada, so impacts documented here should have wide application.

A fifty-year period between 1940 and 1991 was selected for analysis for several reasons. One factor was data availability; aerial photographs are uncommon prior to 1940, and stream discharge and climate station data are less common as well. Both basins were relatively undeveloped in 1940. Logging was certainly being practiced in the Camp Creek Basin by 1940, but records that would allow a detailed compilation of practices, locations, and yield are too sparse to allow analysis. Further, logging and development had not progressed to the levels that they did in the post–World War II era.

Land-Use History

Land-use information was acquired from a variety of public sources. The ENF's Supervisor's Office and the Placerville Ranger District provided most of the logging, grazing, and road information. Aerial photographs supplied information on the date of construction of roads. The construction dates of new residences were obtained from the El Dorado County Assessor's Office records and parcel maps.

A considerable amount of land-use information was acquired from the raster-based geographic information system (GIS) used by the ENF, the Distributed Wildland Resource Information System. DWRIS was used extensively during this project, both as a source of basic land-use information layers (roads, plantations that resulted from clear-cuts, soils, fire extent, slope) and to determine area and distance of land uses derived from the photographs. DWRIS coverage included the Clear Creek Basin because a number of sections west of the national forest boundary are public land.

Aerial photographs were acquired for the following dates: 1940, 1952, 1965/66 (hereafter referred to as 1966), 1976, 1986, and 1991. In addition, 1988 orthophotoquads (7.5-minute mosaics of aerial photographs, rectified to a uniform scale) were obtained. Mosaics were created out of photocopies of each of the six sets of aerial photographs, and the mosaics were analyzed successively by date for clear-cuts and roads.

The grouping of land-use changes into photo intervals is not a perfect process. We recognize that an activity such as logging could occur in 1953 and be assigned to the 1966 photo series, thereby ignoring twelve years of recovery. Because aerial photos are often the only source of information, it is rarely possible to assign a more accurate date to a disturbance. However, the primary disturbances in the early photo intervals are roads and structures, which do not "recover" in terms of their effects on runoff; roads and structures create impervious areas that are essentially permanent. The dates of the fires were known, and a thirty-year recovery was presumed.

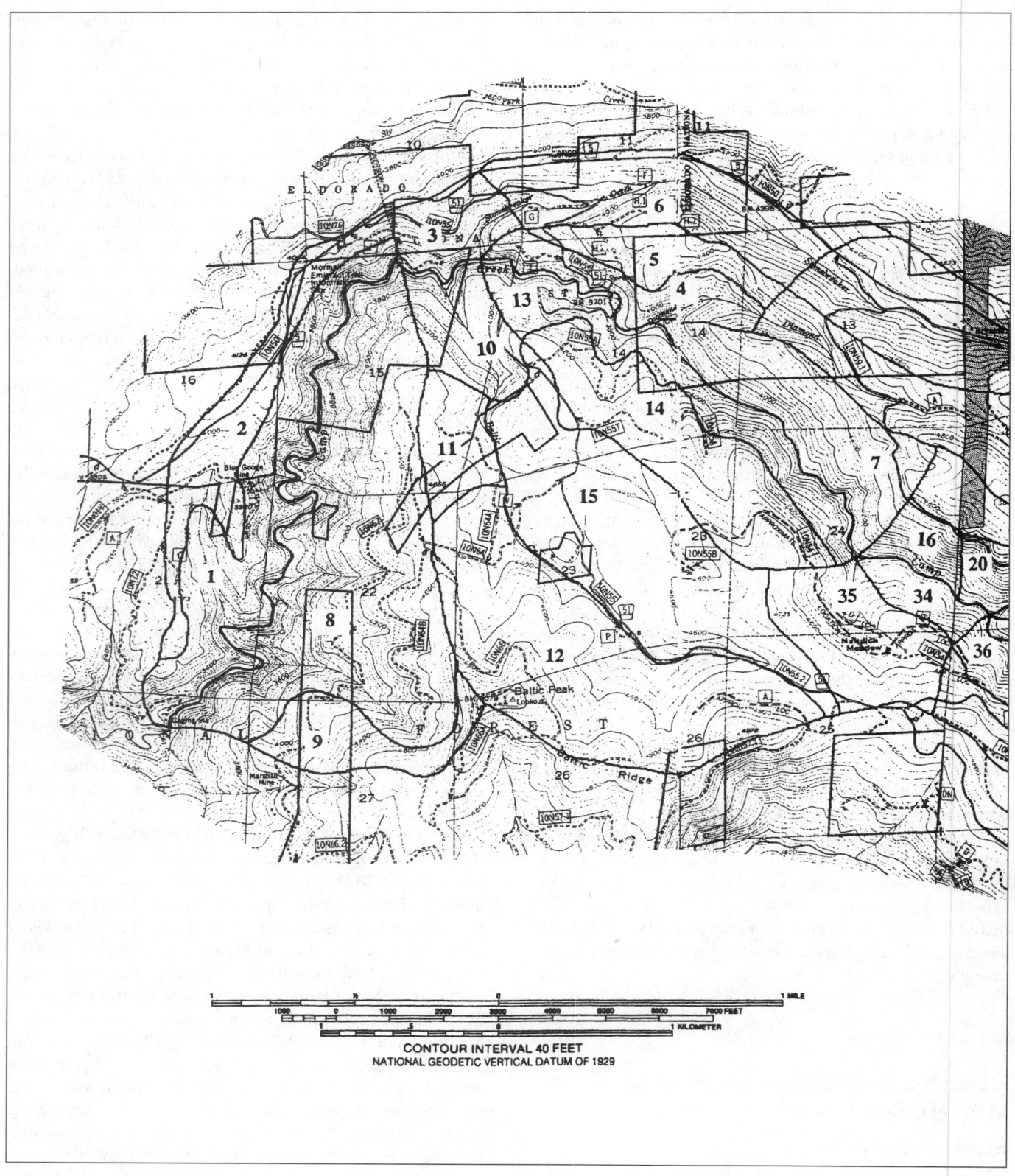

FIGURE 52.2

Map of lower Camp Creek, a tributary to the North Fork of the Cosumnes River, showing Hydrologic Response Units with labels, located in the Eldorado National Forest, California.

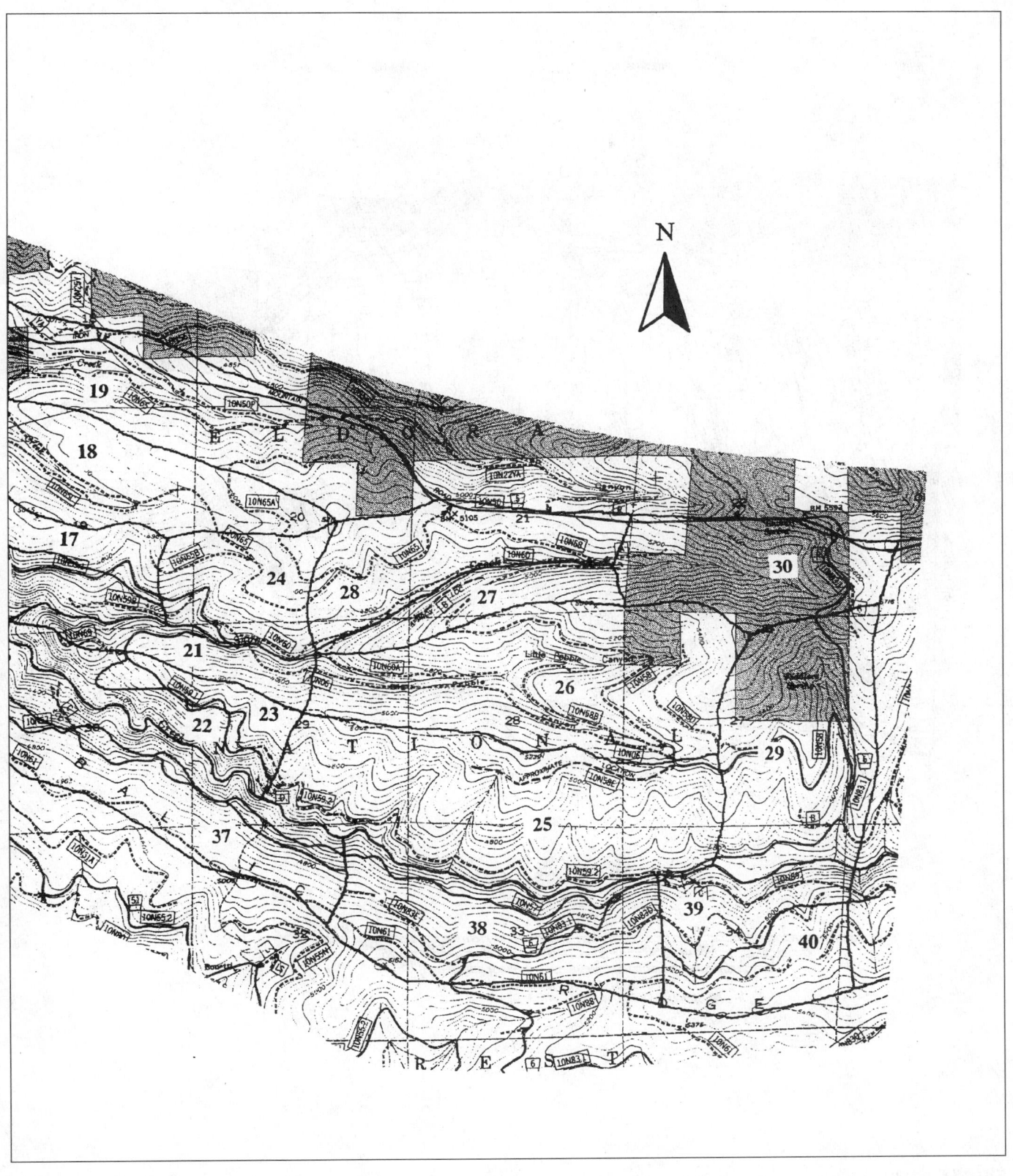

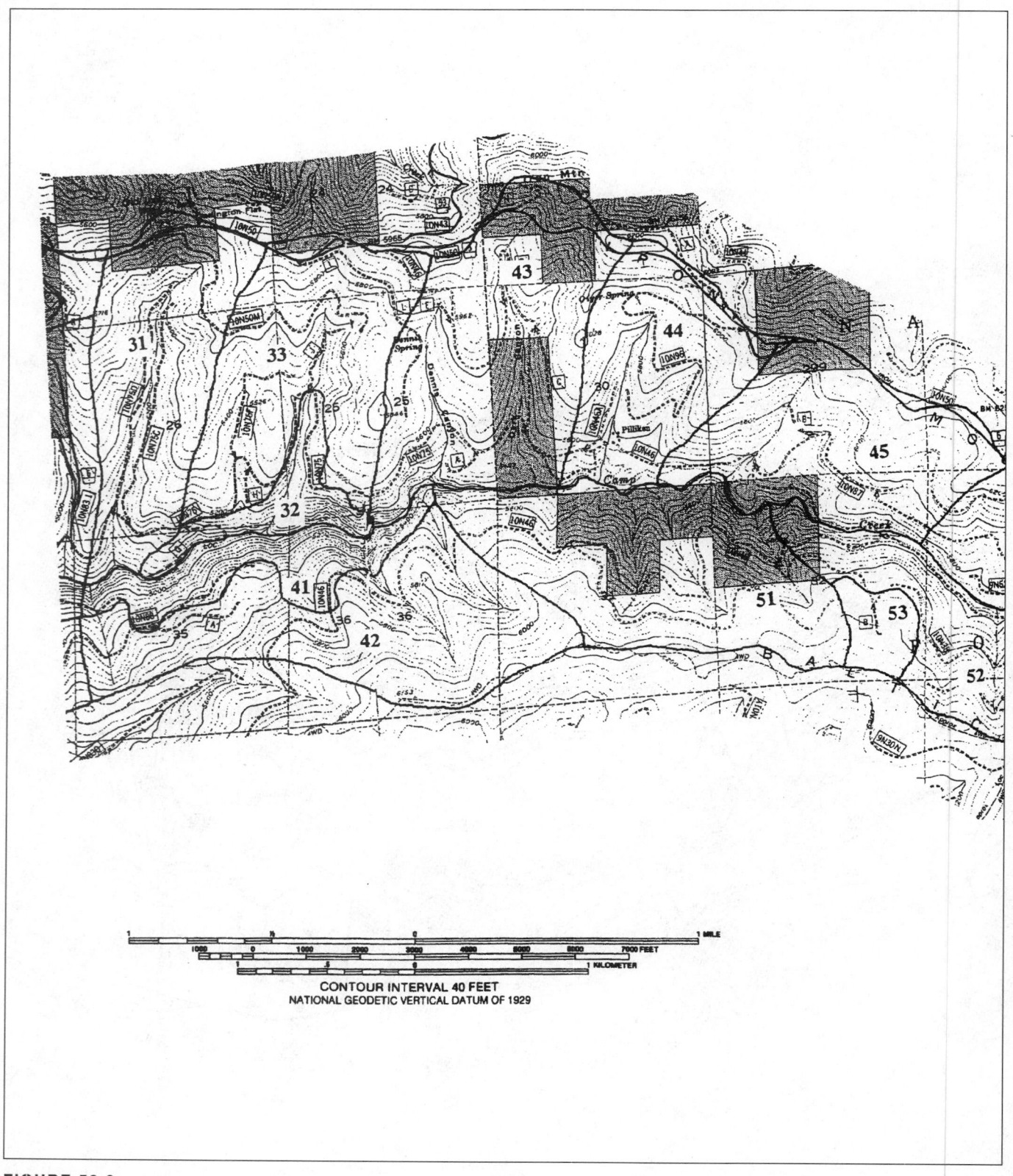

CONTOUR INTERVAL 40 FEET
NATIONAL GEODETIC VERTICAL DATUM OF 1929

FIGURE 52.3

Map of upper Camp Creek, a tributary to the North Fork of the Cosumnes River, showing Hydrologic Response Units with labels, located in the Eldorado National Forest.

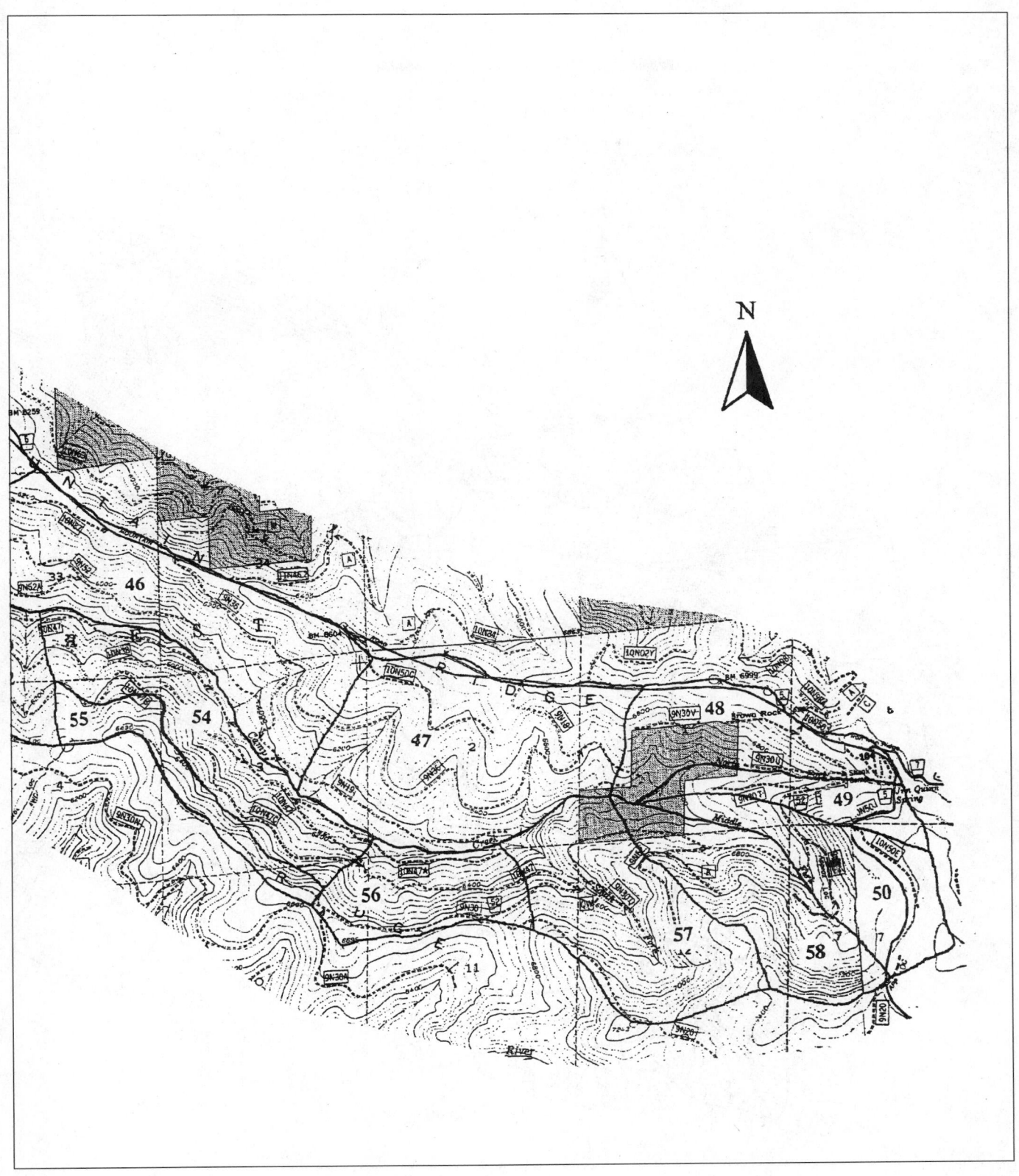

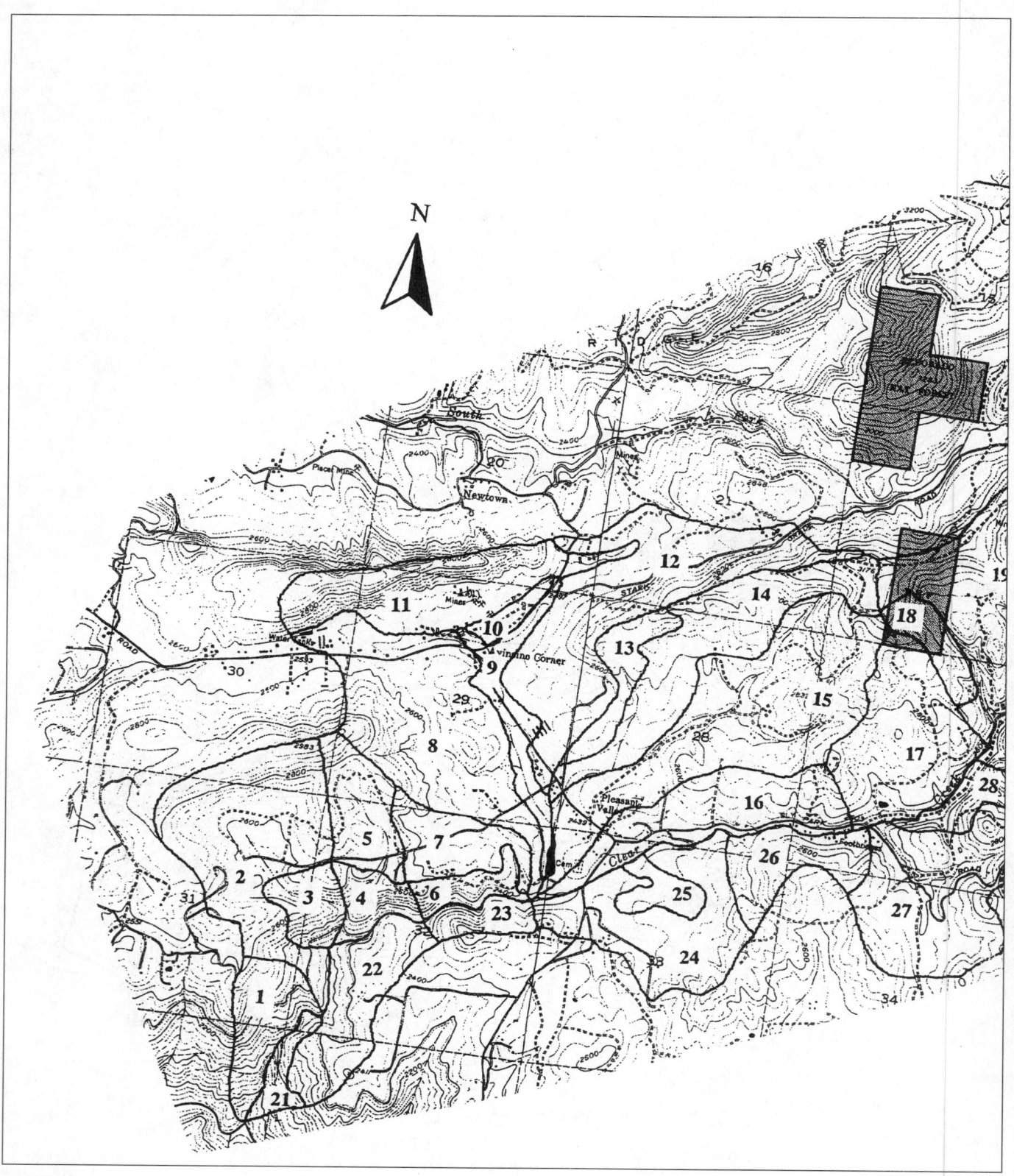

FIGURE 52.4

Map of Clear Creek, a tributary to the North Fork of the Cosumnes River, showing Hydrologic Response Units with labels, located in El Dorado County, California.

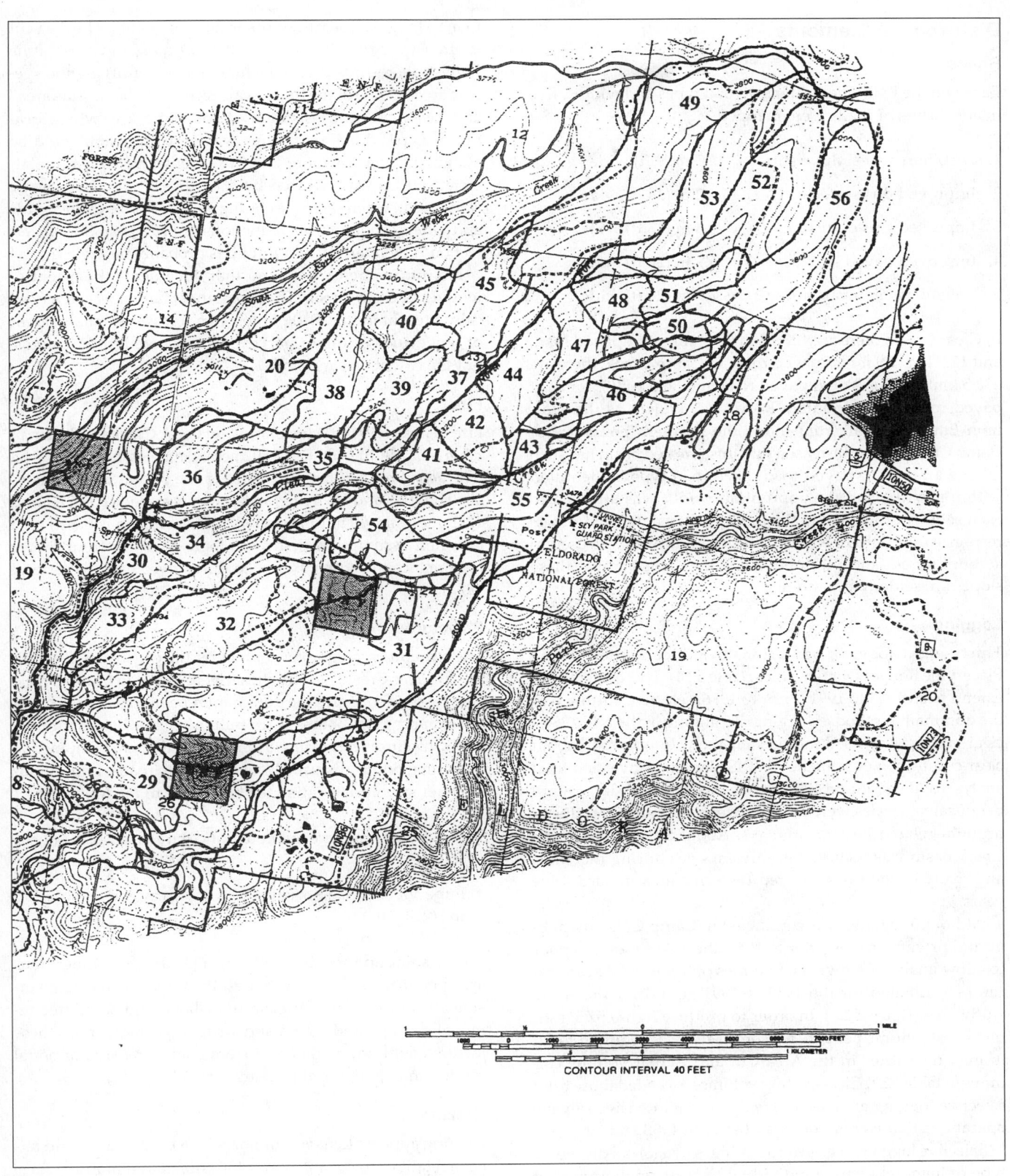

CONTOUR INTERVAL 40 FEET

Disturbance Elements

Roads

Based on the ENF's definitions of road types and widths, five major classes of roads were mapped:

1. Dirt/four-wheel drive surface: 2.7 m (9 ft) wide

2. Improved dirt surface: 4.2 m (14 ft) wide

3. Improved gravel surface: 5.5 m (18 ft) wide

4. Improved paved surface: 6.7 m (22 ft) wide

5. Secondary highway surface: 7.3 m (24 ft) wide

In Clear Creek, there are 128.1 km (79.7 mi) of unpaved road, and 12.7 km (7.9 mi) of paved road. In Camp Creek, there are 232.5 km (144.5 mi) of unpaved road and 56.6 km (35.2 mi) of paved road. Iron Mountain Road, also known as the Mormon Emigrant Trail, runs along the northern boundary of Camp Creek and is the reason that this forested basin has such a large amount of paved road. Each road type is identified within DWRIS, and the appropriate width was attached to each segment to generate the total road area per hydrologic response unit (HRU). No extra width multiplier was used to account for cut-and-fill slopes along the roads because these slopes are seldom impervious.

Logging

Thirty-eight logging operations are documented in the Placerville Ranger District's files (table 52.1). These operations extend from 1953 to 1991. Prior to 1976, selective cutting was the dominant method of logging. Beginning in the mid 1980s, clear-cutting was used more extensively. Information about clear-cutting comes from the DWRIS plantation layer and from analysis of the 1986 and 1991 photos. An unusually large number of salvage sales were done between 1988 and 1991 due to drought-related insect outbreaks. About 80% of the Camp Creek Basin was included in salvage sales during this time, and many of the sales covered the same area in successive years.

Although logging was conducted in Camp Creek by public and private groups prior to 1953, the records are too poor to allow analysis. However, volumes of timber and areas cut can be estimated for the 1953 to 1991 period for the entire study area (figure 52.5). In order to produce figure 52.5, estimation of volume per acre for some sales was required to fill in missing entries in the Area Cut and Actual Volume columns in table 52.1. Clear-cutting volumes exceeded those from selective logging after 1977, even with the intensive salvage operations that were under way between 1988 and 1991.

Selective logging (e.g., group cutting, salvage) is a dispersed type of logging wherein individual trees or small groups of trees are cut from within a much larger sale area. The sales in table 52.1 illustrate the problem associated with attempting to map logging operations other than clear-cuts. The area cut is often a very small fraction of the sale area, and most of the ENF's sale records do not include maps identifying the specific areas. Some of the timber sale records do have "sale area" and "acres cut" entries, but most do not. In cases where location was provided, or volume was very large and could be assigned to one or more HRUs, percentage cover in the affected HRUs was reduced by 5%–10%. From a hydrologic standpoint, when selective cutting is done from existing roads and when canopy cover reductions are less than 5%, the effect on flow quantity and timing is probably negligible.

No information was found that documents the extent of logging in Clear Creek. The grasslands there now were typically present in the 1940 photos. In a few cases, enlargement of rangelands or minor harvesting was noted in the photos, but the extent was minimal.

Fire

Fire history (1911–91) for Camp and Clear Creeks is shown in table 52.2. Of the 608 ha (1,502 acres) affected by fire, 257 ha (636 acres) burned by 1920, prior to the period of this analysis. The 1915 and 1920 fires are shown in appendixes 52.1 and 52.2 under the 1940 time period. This inclusion is based on a common assumption that hydrologic recovery from fire or logging occurs in twenty-five to thirty-five years (Satterlund 1972).

Structures

The structures category encompasses the construction activities in Clear Creek. No structures were found in Camp Creek. Private residences are the dominant type of structure in Clear Creek, and most homes were built after the 1966 photo period. Structures create impervious areas that are presumed to be permanently "disturbed," analogous to roads. Impervious area per structure is estimated according to the following:

- Impervious roof area of house and garage is 232 m^2 (2,500 ft^2).

- Impervious area associated with driveway and walks is 232 m^2 (2,500 ft^2).

This assumption results in 464 m^2 (5,000 ft^2) of impervious area per structure unit. Although there may also be conversions from rangeland to pasture or lawn and some tree removal during subdivision and structure construction, these factors could not be detected from county records or aerial photos and are thought to be minor.

Grazing

Grazing appears to have a minor role in Camp Creek and has been relatively constant (about 150 animal unit months) since 1900, according to records kept by the ENF. No information was obtained for Clear Creek.

Model Selection

Two types of models are suitable for analyzing hydrologic change associated with land-use change: conceptual process models and distributed-parameter numerical models. Statistical models were not considered because of the need to document the contribution of hydrologic flow components to total runoff.

A range of conceptual models have been formulated and are typified by the Stanford Watershed Model (Crawford and Linsley 1966) and the Sacramento soil moisture accounting model (Peck 1976). Most conceptual models are nonlinear and time-invariant and have lumped parameters that are representative of gross watershed characteristics. These models are generally accepted as being reliable in forecasting important features of the hydrograph (e.g., timing, shape, and volume) (Sorooshian 1983).

Numerical models such as TOPMODEL (Moore et al. 1988) are spatially distributed, time-variant, and subdivide the basin into small cells using high-resolution, digital elevation data and detailed soil data. Numerical models provide detailed analysis of flow paths and runoff using kinematic routing and other complex mathematical techniques. Numerical models require comprehensive topographic and soil data and have typically been restricted to application to small basins.

The goal of this project was to discern the effects of long-term land-use changes on the hydrologic response of two medium-sized basins with rather coarse physiographic, cli-

TABLE 52.1

Logging history in the Camp Creek Basin, Eldorado National Forest.

Sale	Date	Activity	Sale Area (ha)	Area Cut (ha)	Volume Estimated (MBF)	Volume Actual (MBF)
1953–66[a]						
Sly Park	5/64–11/64	Insect salvage	135	135	263	—[b]
Schenck	6/64–4/68	Group cut	1,004	630	46,400	47,983
1967–76						
Oiyer Spring	11/67–1/68	—	113	—	1,400	2,506
Pilliken	6/68–2/71	Group cut	402	171	13,600	—
Pebble Cyn	4/71–3/72	Fire salvage	56	56	1,470	—
Baltic Regen	4/71–11/71	Regen salvage	—	—	66	—
Lode	4/70–12/73	Group cut	53	30	2,400	—
Baltic	8/75–12/75	Salvage	35	35	296	—
Iron Park	5/75–3/76	Salvage	502	134	255	383
Brown Rock	7/76–8/76	Salvage	—	—	18	—
Matulich	7/76–3/77	Salvage	121	121	88	—
Brandon Cyn	8/76–2/77	Insect salvage	61	—	84	—
Dennis Cyn	9/76–8/78	Insect salvage	—	—	83	—
1977–86						
Corky	6/79–8/79	Insect salvage	24	4	141	—
Premat	6/80–12/81	Insect salvage	728	—	1,000	—
Iron Mtn.	7/81–11/82	Insect salvage	850	178	1,200	—
Diamond T	8/81–10/82	Insect salvage	111	24	249	465
Quinn	9/81–2/86	Various	1,803	601	25,000	—
Brandon	3/83–3/87	Clear-cut and other	411	80	5,720	—
Blue Gouge	5/79–10/84	—	13	13	643	—
Diamond	5/84–3/87	Clear-cut	389	63	5,270	5,252
Dennis	5/82–2/89	Clear-cut and regeneration	—	—	6,700	—
Pebble	11/85–5/89	Clear-cut and overstory removal	643	71	14,676	—
1987–91						
Diamond Jim	1988	Salvage	4,862	—	2,285	—
Bonetti	1989	Salvage	507	—	1,535	—
Iron	1989	Salvage	2,539	—	3,774	—
Morrison	1989	Salvage	23	—	45	—
Sleek	1989	Salvage	2,601	—	6,827	—
Jimbean	1990	Salvage	23	—	23	—
Quinn Addon	1990	Salvage	45	—	7,000	—
Rathole	1990	Salvage	814	—	3,641	—
Vancamp	1990	Salvage	1,630	—	2,358	—
Willow	1990	Salvage	27	—	51	—
Beetlebattle	1991	Salvage	20	—	12	—
Lost Larva	1991	Salvage	8	—	11	—
Peanut Bee	1991	Salvage	2,531	—	1,587	—
Pitchstream	1991	Salvage	1,649	—	759	—
Plum Dead	1991	Salvage	38	—	23	—

[a]No sale records for 1940–52.
[b]Dash indicates no data.

FIGURE 52.5

Logging extent by area and timber volume for Camp Creek, Eldorado National Forest, 1953–91 (select areas estimated).

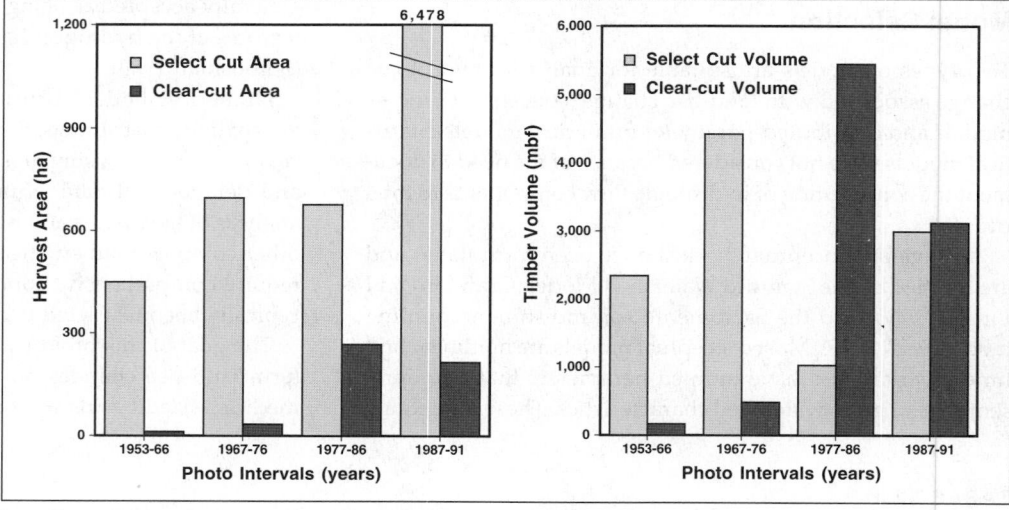

mate, and soil information. The sizes of the basins, the lack of data, and the long analysis period mandated the use of a conceptual hydrologic model that included algorithms associated with changes in impervious area and vegetation type and density. The U.S. Geological Survey's Modular Modeling System (MMS) was selected because it is well documented, has seen extensive use, incorporates land-use change factors, and is in the public domain.

Modular Modeling System

The watershed model contained within the MMS is the USGS's Precipitation-Runoff Modeling System (PRMS) (Leavesley et al. 1983). PRMS is a conceptual process modeling system. To reproduce the physical reality of the hydrologic system as closely as possible, each component of the hydrologic cycle is mathematically expressed via known physical laws. Where physical laws or information are lacking, empirical relationships are used that have some physical interpretation and may be based on measurable watershed characteristics.

The MMS or PRMS has been used in a number of studies to investigate the effects of global climate change in California (Jeton and Smith 1993) and Colorado (Leavesley et al. 1992; Hay et al. 1993), to simulate runoff in small basins in Colorado (Norris 1986), to simulate dry-season runoff in Guam (Nakama 1994), to evaluate hydrologic response to surface coal mining (Stannard and Kuhn 1989), and to evaluate the effect of forest management on hydrology (Grant et al. 1990).

The watershed system embodied in PRMS is schematically depicted in figure 52.6. System inputs are precipitation, air temperature, and optional snow accumulation and solar radiation data. Precipitation is classified as rain or snow or mixed, based on temperature, and is delivered to the watershed surface. The energy inputs of temperature and solar radiation drive the processes of evaporation, transpiration, sublimation, and snowmelt. The watershed system is concep-

tualized as a series of four reservoirs whose outputs combine to produce the total system response.

Stream flow is the sum of the surface, subsurface, and base flow outputs. No channel routing is done when the model is run on a daily time step, as it was in this study.

Impervious Zone

One reservoir is the impervious-zone reservoir, which has no infiltration capacity and represents areas such as roads. This reservoir has a maximum retention storage capacity that must be filled before surface runoff will occur. Retention storage is depleted by evaporation when the area is snow free.

Soil Zone

The soil-zone reservoir represents that part of the soil mantle that can lose water through evaporation and transpiration. Average rooting depth of the predominant vegetation cover-

TABLE 52.2

Fire history for Clear Creek and Camp Creek Basins, El Dorado County, California, by photo interval and watershed.

Photo Interval	Total Area (ha)	Subarea (ha)	Year (ha)	Watershed
1911–39	257.46	256.10	1915	Camp Creek
		1.36	1920	Camp Creek
1940–52	0			
1953–66	0			
1967–76	336.46	97.50	1969	Camp Creek
		84.98	1972	Clear Creek
		153.98	1973	Camp Creek
1977–86	0			
1987–91	14.20		1988	Camp Creek

Total 1911–39 = 257.46
Total 1940–91 = 350.66

Total Clear Creek = 84.98
Total Camp Creek = 523.14

Grand total = 608.12

ing the soil surface defines the depth of this zone. Water storage in the soil zone is increased by infiltration of rainfall and snowmelt and depleted by evapotranspiration (ET). The depth and water-storage characteristics of the upper layer of this reservoir, termed the *recharge zone,* are user defined and based on vegetation type. Losses from the recharge zone are assumed to occur from evaporation and transpiration. Losses from the lower zone occur only through transpiration.

Infiltration into the soil zone depends on whether the input source is rain or snowmelt. All snowmelt is assumed to infiltrate until field capacity is reached. At field capacity, the soil zone is assumed to have a maximum daily snowmelt infiltration capacity. Snowmelt in excess of field capacity contributes to surface runoff. Infiltration in excess of field capacity is first used to satisfy recharge to the ground water reservoir, up to a maximum daily amount. Excess infiltration after recharge to the ground water reservoir becomes recharge to the subsurface reservoir.

For rainfall with no snow cover, the volume infiltrating the soil zone is computed as a function of soil characteristics, antecedent soil-moisture conditions, and storm size. For daily-flow computations, the volume of rain that becomes surface runoff is computed using a contributing-area concept. Daily infiltration is computed as net precipitation less surface runoff.

Subsurface Zone

The subsurface reservoir routes the soil-water excess that percolates to shallow ground water zones near stream channels or that moves downslope from the point of infiltration to some point of discharge above the water table. Subsurface flow is considered to be water in the soil and ground water zones that is available for relatively rapid movement to a channel system.

Ground Water Zone

Recharge to the ground water reservoir can occur from the soil zone and the subsurface reservoir. Soil recharge has a daily upper limit and occurs only when the field capacity is exceeded in the soil zone. Subsurface recharge is computed daily

FIGURE 52.6

Schematic diagram of the mathematical watershed system and its inputs (after Leavesley et al. 1983).

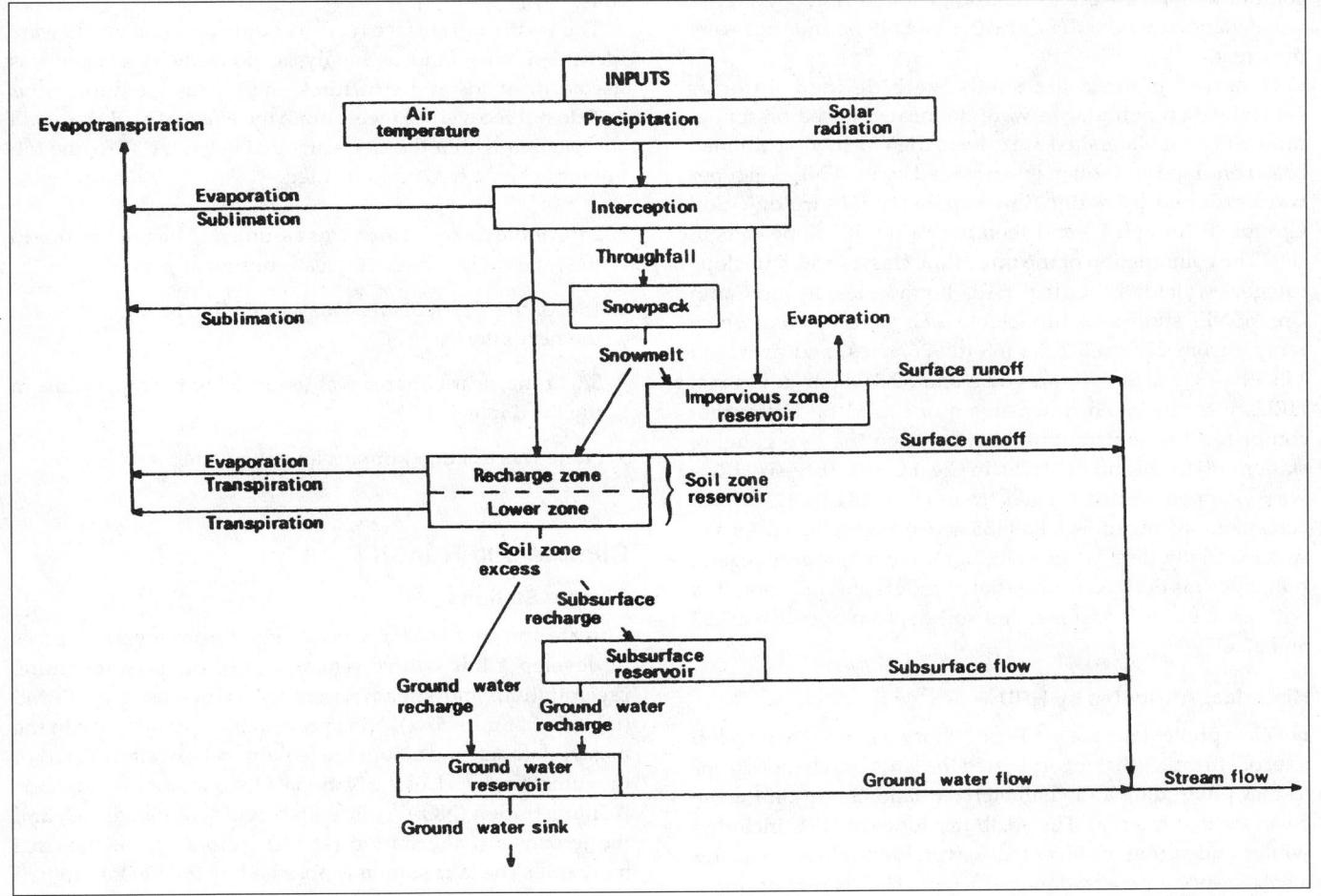

as a function of a recharge rate coefficient and the volume of water stored in the subsurface reservoir. Release of ground water is controlled by a linear equation, and this release is the source of all stream base flow.

Hydrologic Response Units

Hydrologic models partition a basin into "homogeneous" grid cells or polygons, and PRMS is based on polygons called Hydrologic Response Units. HRUs are delineated based on physiographic properties that affect runoff generation: slope, aspect, elevation, vegetation type, soil type, and precipitation distribution. The goal of the polygon (HRU) delineation process is to subdivide the basin into units that respond similarly to rainfall-runoff processes and also to allow separation of historical land-use changes. Clear Creek was divided into smaller HRUs than Camp Creek because suburbanization occurs at a smaller spatial scale than forest management activities. Two sources of boundaries for Camp and Clear Creeks were obtained: the Calwater boundaries (Brandow 1994) were mapped at the 7.5-minute scale by the SNEP Geographic Information Center, and the GIS laboratory at the Supervisor's Office of the ENF supplied watershed boundary maps at the same scale. All boundaries were manually compared to the contour lines on the USGS quadrangle maps, and boundaries were changed where the Calwater or ENF boundaries were incorrect.

Once the major watersheds were defined, interior subwatershed boundaries were delineated based on topography. The subwatershed areas were then further partitioned based on slope and soil maps provided by the ENF. Soil types were grouped by water flow capability (Hydrologic Soil Groups B through D) and then overlaid with slope categories. The combination of the three flow classes and four slope categories yielded the final HRU boundaries. In the Camp Creek SNEP study area, fifty-eight HRUs were mapped, which ranged from 27.5 to 375 ha (68 to 926 acres) and averaged 145.3 ha (359 acres) (figures 52.2 and 52.3). Four additional HRUs were included downstream of the SNEP HRUs that comprised the contributing area between the two gauging stations (#113330 and #113315). In Clear Creek, fifty-six HRUs were mapped, which ranged from 8.5 to 184 ha (21 to 455 acres) and averaged 54.8 ha (135 acres) (figure 52.4). Aspect was manually derived from topographic maps, and vegetation type was derived from aerial photographs and compiled with area, elevation, slope, and soil depth (appendixes 52.3 and 52.4).

Historical Attributes by HRU

For each photo date, the land-use history was used to develop a set of attributes that incorporated the land-use changes prior to that photo date and characterized land condition for the basin for that interval. The attributes for each HRU included winter and summer percentage cover, impervious area, and a solar-canopy-penetration coefficient. The six sets of photo

mosaics were overlaid by the HRU boundaries and percentage cover was subjectively determined. All photos were taken in the summer, and unless the vegetation cover type was conifer, percentage cover for winter was estimated based on the fraction of deciduous trees, shrubs, or grass. Impervious area was the percentage of the area of the HRU that was devoted to roads, residences and driveways, and portions of recently burned areas. The solar-canopy-penetration factor was estimated via a graphical procedure presented in the PRMS users manual (Leavesley et al. 1983).

Recovery from Disturbance

A complex recovery scheme was not incorporated in the analysis for several reasons. In most cases, the only source of information on activities was the aerial photos, so each road segment and logging activity could not be identified by exact date. Dates were known for fires, so a simple recovery scheme was implemented: 50% of the area of a fire in an HRU was set as impervious for the five years after a fire (Krammes and DeBano 1965; Dyrness 1976; Poff 1989), and thereafter the fire had no effect beyond a reduction in percentage cover that was estimated from the aerial photographs. A value of 50% was chosen to include compacted area created during fire suppression as well as to incorporate the effect of hydrophobic soils.

The width and surface type of county and ENF roads were identified in the land-use analysis. No recovery scheme was used with roads and structures; their areas are impervious and do not recover. To incorporate the effect of skid trails and compaction within the clear-cuts that began by 1976, the following scheme was implemented:

- 25% of the clear-cut area was assumed to be impervious in the interval in which the clear-cut first appeared

- 15% of the clear-cut area was assumed to be impervious in the next interval

- 5% of the clear-cut area was assumed to be impervious in the third interval

- None was assumed impervious thereafter

Climate and Runoff Data

Climate Stations

Climate and runoff data were obtained from several sources to develop a thirty-three-year record of daily temperature, precipitation, and stream discharge values used by PRMS (table 52.3; figure 52.1). Nine precipitation stations are in the master data set, and all but the Jenkinson Lake site are part of the climate network of the National Oceanic and Atmospheric Administration (NOAA). The Placerville station (#6960) and the Jenkinson Lake station (#6964) were used for the final modeling. The lake station is operated by the El Dorado Irrigation District (EID) and was given a NOAA-style site code of "#6964" for purposes of record keeping. Six temperature stations with maximum and minimum temperatures were used during varying steps in the modeling, but the Placerville station (#6960) was used in the final simulation runs.

Data quality at the NOAA climate stations is generally good, but all the temperature and precipitation stations had periods of varying lengths for which data were missing. Computer programs were written to identify the year, month, and duration of all missing data, and both manual and programmed patches were applied to the records. Nearby stations were used to allow interpolation of missing data for periods of 2–3 days. Month-long periods of missing data were filled by splicing records from neighboring stations after determining bias due to difference in elevation.

Discharge Stations

USGS discharge data for four stations were used during the analysis, but a modified Camp Creek gauge record was used in the final simulations. Water from Camp Creek is diverted to a municipal reservoir approximately 5 km (3 mi) upstream of the discontinued #113315 gauge. Jenkinson Lake (also known as Sly Park Reservoir) was built in 1955–56 in the Sly Park Creek Basin, just to the north of Camp Creek. Sly Park Creek flows into Camp Creek upstream of the #113330 Camp Creek gauge, so runoff from the Sly Park Basin is included with Camp Creek runoff in years with high runoff volumes. In normal or drought years, all the runoff from the Sly Park Basin above Jenkinson Lake and the diverted water from Camp Creek is exported via the Camino Conduit to the cities of Camino and Placerville by the EID. Except during exceptional winter storms or occasional spring runoff conditions, the only water flowing in Sly Park Creek is inflow from tributaries and hillslopes below the dam, and a release from the conduit or the dam of 0.057 cms (2 cfs) to satisfy water rights. The EID manages the reservoir and provided daily data for the Camp Creek diversion canal for the 1956–94 period. It also provided daily spill rates from the lake into Sly Park Creek. Spills are generally zero, but they can exceed 28 cms (1,000 cfs) during large storm events such as occurred in 1982 and 1983. Diversions from Camp Creek can exceed 25 cms (900 cfs), so the Camp Creek gauge record as measured by the USGS is not at all indicative of the natural flows.

To allow calibration of the hydrologic model, a more natural stream-flow record was required. To this end, daily diversion and reservoir spill flows were added or subtracted to the #113330 flow to create an "unimpaired" flow record, which was named #1133301 Camp Creek Unimpaired for record-keeping purposes. The flow correction algorithm is as follows:

1. Camp unimpaired equals observed Camp plus diversion

2. If Camp unimpaired is greater than spill, Camp unimpaired equals Camp unimpaired minus spill

3. If spill is greater than Camp unimpaired, Camp unimpaired equals zero

Diversions typically occurred in late winter and during spring runoff. Diversions are curtailed after June 1.

TABLE 52.3

Climate and stream discharge stations used in modeling Clear and Camp Creeks, El Dorado County, California.

Station	Index #	Basin	Elevation (m)	Data	Latitude	Longitude
Camp Pardee	1428	Mokelumne	201	P*,T,S	38°15'	120°51'
Fiddletown/Baxter	3038	Cosumnes	218	P,S	38°32'	120°42'
Folsom Dam	3113	American	107	P,T	38°42'	121°10'
Jenkinson Lake	6964	Cosumnes	1,058	P	38°43'	120°33'
Pacific House	6597	American	1,049	P,S	38°45'	120°30'
Placerville	6960	American	564	P,T,S	38°43'	120°49'
Placerville IFG	6962	American	840	P,T,S	38°44'	120°44'
Salt Springs PH	7689	Mokelumne	1,128	P*,T,S	38°30'	120°13'
Twin Lakes	9105	American	2,438	P,T	38°42'	120°02'

Station	Index #	Drainage Area (km²)	Period of Record	Latitude	Longitude
Camp Creek near Camino	113315	83	1949–56	38°42'	120°32'
Camp Creek near Somerset	113330	163	1955–90	38°39'	120°40'
Camp Creek near Somerset, Unimpaired	1133301	163	1955–90	38°39'	120°40'
North Fork Cosumnes River near El Dorado	113335	531	1912–87	38°33'	120°50'
Cosumnes River at Michigan Bar	113350	1,388	1966–90	38°30'	121°03'
Cosumnes River at McConnel	113360	1,875	1941–82	38°22'	121°20'

P* indicates precipitation data, hourly recording summed to daily.
P indicates daily precipitation.
T indicates maximum and minimum temperature.
S indicates new snowfall or snow on ground.

Water-Year Ranking

Because watersheds respond differently to different water-year magnitudes, the thirty-three-year record was analyzed to identify high-, medium-, and low-magnitude water years. The relative magnitude of a water year can be judged by either the precipitation depth or the volume of the runoff. In this analysis, both the precipitation at the Placerville station and the unimpaired runoff at Camp Creek were used. Because the pattern of precipitation and runoff can vary greatly between two years that might have the same annual total, an additional ranking was done that included the mean ranks of monthly values between December and June. A table was prepared that assigned the smallest three years in the list to the low-magnitude year class, the middle three to the medium-magnitude class, and the top three to the high-magnitude class (table 52.4). The magnitude increases among the three in each class from top to bottom.

For the high-magnitude water year, 1983 was consistently the middle value in all four rankings. Water year 1983 was selected as the example of a high-magnitude precipitation and runoff year. Water year 1979 appears in three of the four rankings in the medium-magnitude category and was selected as representative of the medium-magnitude water year.

The 1977 water year appears as the driest year on record in three of the four rankings. The 1976 water year appears in two of the four classes and is the driest ranked by mean monthly precipitation. Because 1977 was the second year of a two-year drought, 1976 was selected as being more representative of drought years in general.

Hydrologic Model Calibration

Physical process models such as MMS have numerous coefficients to allow adjustment of rates of energy or mass between reservoirs. They also have coefficients to allow extrapolation of the temperature and precipitation data from the climate station to the basin of interest. Calibration is the process of varying the coefficients to make the predicted daily hydrograph match the hydrograph for the unimpaired Camp Creek discharge station. In addition to matching the hydrographs, water input must match the sum of the outflows, must be physically reasonable, and must match hydrologic theory. Because MMS has a long history of use in many basins, each coefficient has a recommended range. In the calibration of the model to Camp Creek and Clear Creek, water balance was the first step. Correct identification of precipitation type was the second step. Adjustment between flow mechanisms to obtain matching peaks and recession curves was the third step.

Water Balance

The water balance in MMS can be examined at any time step, but monthly or yearly steps are the most effective. The MMS extracts sublimation of rain or snow from interception, so the

TABLE 52.4

Results of ranking water years from a 33-year record of Placerville Climate Station (#6960) and the Camp Creek Gauging Station (#1133301).

Basis for Ranking	Magnitude of Water Year		
	Low	Medium	High
Yearly Ranking			
By flow	1977	1957	1965
	1961	1972	1983
	1988	1979	1969
By precipitation	1977	1979	1958
	1987	1989	1983
	1976	1957	1982
Monthly Ranking			
By flow	1977	1957	1967
	1961	1979	1983
	1987	1972	1969
By precipitation	1976	1965	1982
	1977	1989	1983
	1985	1970	1958

residual, effective precipitation equals the sum of evapotranspiration (ET), surface runoff, subsurface runoff, ground water outflow, and changes in storage. The MMS estimates potential ET based on solar radiation (Jensen and Haise 1963), which is predicted by HRU from slope, aspect, and maximum and minimum temperatures (Leavesley et al. 1983). Actual ET is estimated based on soil water availability, vegetation type, and percentage cover. Monthly values and annual totals were referenced to data from an evaporation pan at Placerville (Farnsworth and Thompson 1982), and when the model's values ranged around those values for a number of years, the coefficients were judged to be correct. The sum of the annual ET and the runoff categories, plus any changes in storage, was less than 1.9% different than the total annual effective precipitation in all runs (table 52.5). Effective precipitation is the amount that is estimated to reach the soil surface, and it accounts for sublimation during interception.

Rain and Snow Discrimination

The MMS classifies precipitation as rain, snow, or a mixture of the two based on the maximum and minimum daily air temperatures for the HRU. Temperatures are predicted for each HRU from the temperature station based on the relative elevation of each and a specified lapse rate. Threshold temperature values are also declared above which all precipitation is rain and below which all precipitation is snowfall. A ten-year subset, water years 1973–83, was partitioned out of the thirty-three-year data set to speed the processing and simplify the calibration. Simulations of daily discharge were made for the ten-year period while adjusting temperature coefficients, lapse rate coefficients, and control values. Adjustments in model coefficients controlling discrimination between rain and snow were made so that the runoff following large storm events matched the observed runoff record.

Flow Component Adjustment

The MMS model sums surface runoff, subsurface outflow, and ground water outflow to create stream flow. In keeping with hydrologic theory, each reservoir releases water at a different rate and therefore stores precipitation for a different duration. Ground water storage is released the most slowly and comprises summer base flow, so adequate water must be moved into the ground water reservoir and released at a rate so that the post-snowmelt recession curve matches the observed recession curve. Surface runoff is thought to be rare in forest lands other than from near-channel saturated areas or from compacted surfaces (Dunne and Leopold 1978), but surface runoff occurs quickly and contributes to storm peaks. Subsurface runoff contributes to saturated areas and contributes water to the channel more slowly than surface runoff but faster than base flow. By adjusting the coefficients that control allocation of precipitation into these reservoirs and adjusting the coefficients that control release, the predicted hydrograph was matched to the observed hydrograph. This process was subjective, and several calibration strategies were tested. For Camp Creek, a runoff regime that allocated much of the precipitation into the subsurface reservoir appeared to best fit the observed hydrograph.

Clear Creek had no observed hydrograph with which to compare the predicted runoff, so the Camp Creek coefficients were initially applied except for the land-use parameters associated with land condition. Surface runoff was very large based on the Camp Creek coefficients, so the coefficients were adjusted to reduce the overwhelming surface flow component. In a basin with loam soils and reasonable amounts of vegetation, very large amounts of surface flow contradict hydrologic theory and experience (Dunne and Leopold 1978). Surface flow in Clear Creek remained, however, a larger component of total flow than in Camp Creek (table 52.5).

Other than mass balance, goodness-of-fit measures (e.g., root mean squared error) were not used during the calibration. The major reason was the poor quality of the observed

discharge record. The large variability, especially during the recession phase of the annual hydrograph, produced spurious results when an attempt was made to optimize the model using root mean squared error as the objective function.

Interpretation of the results of a simulation model assumes that the important hydrologic processes and linkages are represented in the model. Because of the extensive development and wide use of MMS, we assume that the outputs are repeatable, conserve mass, and vary according to climate and basin inputs. Limitations include the inability to affix error bands to the predictions. We acknowledge that calibration is a subjective process, and different combinations of rate coefficients might appear as reasonable as the ones selected, yet weight the major processes differently.

Simulation Design

The precipitation record for the years 1983, 1979, and 1976 became the input data to the MMS in the evaluation of the effects on runoff generation of changes in land use over time. Soil-moisture accounting models such as MMS begin with a default set of values in the various reservoirs. For a representative simulation of any given year, the simulation should begin at least one year prior to the year of analysis so that the reservoir values are set properly. To set the reservoir value, two years of climate data were modeled prior to each of the three water years (1983, 1979, and 1976) that were analyzed. Each set of HRU attributes linked to the six photo dates was analyzed by the model with each of the three water years, so eighteen simulations were done for each of the two basins. Simulation output files included monthly and annual water balance results, plot files of predicted versus observed hydrographs, and plot files of the flow components.

TABLE 52.5

Annual water balance from simulation runs of the three water years with the 1976 land condition coefficients (cm).

Water-Year Magnitude	Ground Water Outflow	Subsurface Runoff	Surface Runoff	Evapo-transpiration	Sum of Fluxes	Effective Precipitation
Camp Creek						
Low	6.6	2.3	0.7	42.3	51.9	52.6
Medium	32.4	17.5	1.0	39.9	90.8	90.1
High	76.7	59.2	3.5	59.5	198.9	195.2
Clear Creek						
Low	4.9	2.1	2.0	34.1	43.1	44.3
Medium	18.2	16.1	3.8	40.4	78.5	77.1
High	49.0	63.3	13.3	54.4	180.0	177.1

RESULTS

Land Condition Changes

The partitioning of the basins into polygons allowed historical land-use change to be allocated by date to a particular HRU. Through acquired GIS layers and analysis of six series of aerial photographs, sets of land-use data were compiled for each of the following years: 1940, 1952, 1966, 1976, 1986, and 1991. Summaries of disturbance area were compiled for the Camp and Clear Creek Basins (appendicxes 52.1 and 52.2). These tables list the disturbances by category for each of the 118 HRUs. The disturbance is summed within each photo interval to allow comparison of the totals.

A decrease in percentage cover over time was observed in both basins (figure 52.7). Percentage cover was found to decrease in Clear Creek from 59% to 57% during the fifty-year study interval. Percentage cover in Camp Creek decreased from 68% to 55% between 1952 and 1991.

Impervious area associated with road building increased in both basins throughout the fifty-year period, but it slowed markedly in the last five-year interval (figure 52.8). Impervious area associated with residential development increased in Clear Creek from about 0.3% to 3.6% during the study interval. The increase was especially rapid between 1966 and 1986. Disturbed area in Camp Creek increased from near 0% to 7.4% during the study interval, and the large increases were associated with clear-cuts after 1976. The hectares of disturbed land in figure 52.8 are cumulative, and roads and houses are assumed to be impervious surfaces. Clear-cuts are termed disturbed areas and are treated differently from roads or houses, as described later.

FIGURE 52.7

Changes in mean percentage cover across all HRUs over time and for summer and winter in Clear and Camp Creek Basins, El Dorado County.

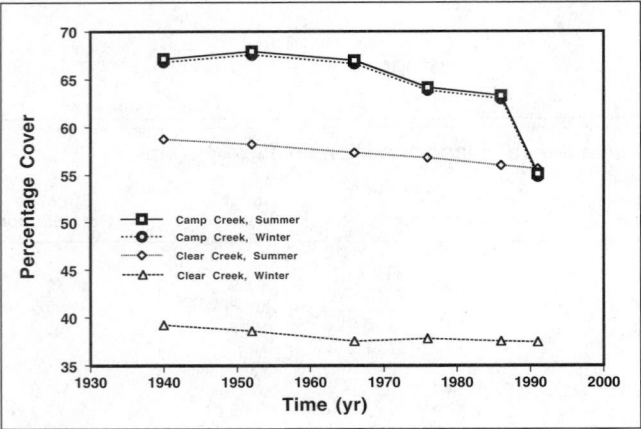

Clear Creek

The disturbance in the Clear Creek Basin is due entirely to roads, structures, and fire. Residential development in Clear Creek caused a 44 ha (109 acre) increase in road area and a 58 ha (143 acre) increase in area covered by structures and appurtenant impervious areas. Fire occurred in 1972 and affected 85 ha (210 acres) of Clear Creek. Road surface doubled between 1940 and 1952, and again between 1966 and 1976 (figure 52.8). The road network grew again between 1976 and 1986 but grew only a little between 1986 and 1991. This pattern of growth was associated with the platting of the subdivisions. Structures have had a more constant growth pattern as individual parcels within the subdivisions have been purchased and developed (figure 52.8). Structures have at least doubled in every photo interval except the most recent five-year interval. The slope of the housing curve is unchanged and very steep since 1976.

Camp Creek

The disturbance in the Camp Creek Basin is due to a combination of clear-cutting, roads, and fire. Forest management caused a decrease in percentage cover, an increase in road area from 11 to 107 ha (27 to 264 acres), and over 570 ha (1,412 acres) of clear-cuts. Fire occurred in 1969, 1973, and 1988 and affected 265 ha (655 acres) of Camp Creek. Roads show a steady increase in all intervals except 1991, supporting the supposition that the basin is fully roaded for logging activities (figure 52.8). Large increases in roads occurred in the intervals ending in 1976 and 1986. Clear-cuts were rare in the first three photo intervals but increased dramatically in the last three. The clear-cut area increased seven-fold in the 1976–86 interval and doubled during the 1986–91 interval. The 1991 clear-cutting occurred in addition to the salvage logging that covered 80% of the basin.

Although appendix 52.1 shows over 946 out of 8,426 ha (2,337 out of 20,821 acres) as disturbed by 1991, it is important to remember that this table does not recognize the recovery of forest vegetation. Although nearly 60% of the disturbance is due to clear-cutting, at least half that area was cut at least five years earlier. Nevertheless, it is evident that the rate of disturbance associated with forest management accelerated dramatically over the last three decades. The type of disturbance has changed, however, in that little new road construction appears to have been required during the most recent logging activities.

Hydrologic Simulation Results

Results from model simulations are grouped into monthly runoff trends over time and by water year, annual runoff trends over time and by water year, daily runoff values for the high- and medium-magnitude water years, annual flow component hydrographs for the medium-magnitude year, and

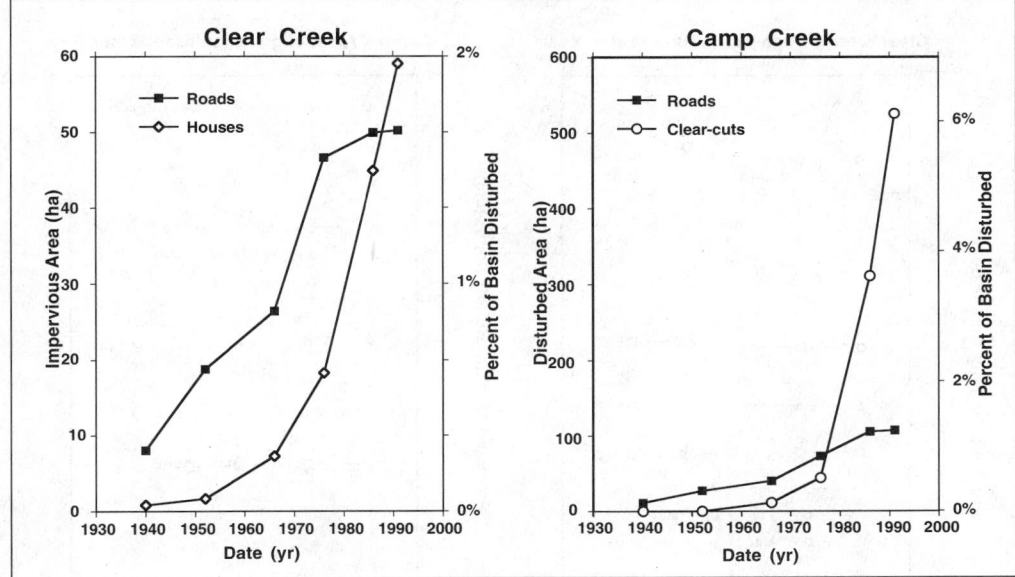

FIGURE 52.8

Trends in disturbed or impervious areas associated with roads, houses, and clear-cuts in Clear and Camp Creeks.

flow components over time and by water year. In subsequent sections, identification of trends is based on visual, not statistical, analysis of the plots. Statistical tests of the changes in predicted values over time are not possible, because no measures of variance can be calculated on each predicted value. In this report, runoff is discussed in terms of areal runoff depths rather than volume or flow rates. Runoff depths are analogous to precipitation depth across the basin and have the advantage of correcting runoff volume for size of the basin. This correction allows easy comparison between basins of different size.

Monthly Trends

Three water years (1983, high-; 1979, medium-; and 1976, low-magnitude) were analyzed to determine if hydrologic response changed based on water-year magnitude and over time because of land condition (figures 52.9 and 52.10). Monthly runoff depths for Clear Creek show minor increases after the 1966 photo period, concurrent with the increases in roads and in fire-affected area. The plot for the low-magnitude water year is not presented because the change in runoff depths over time is negligible. February, March, and April are the months with a slight change over time for the low-magnitude water year. For the medium-magnitude year, the increase over time is most notable in the major runoff months: January, February, and March (figure 52.10).

April and May runoff trends upward in the 1986 and 1991 years for both the high- and medium-magnitude water years in Camp Creek, associated with snowmelt (figures 52.8 and 52.9). For the low-magnitude water year in Camp Creek, the upward trend in runoff depth shifts to February and March, reflecting earlier melt of the snowpack. April runoff for the high- and medium-magnitude years shows the largest trend related to changes in land condition (figure 52.10). Although

there is some variability, a predicted change in monthly runoff between 1940 and 1991 of over 2 cm (0.8 in) is notable.

Annual Trends

Because daily values have wide variation and obscure trends, the annual hydrographs are plotted using mean monthly values rather than daily means (figures 52.11 and 52.12). The high-magnitude water year (October 1, 1982, through September 30, 1983) had in excess of 240 mm (9.5 in) of precipitation each month from November through March. March alone produced 380 mm (15.1 in) of precipitation, and the April precipitation was 180 mm (7.1 in). As a result, both Camp and Clear Creeks had large predicted monthly runoff depths of 32.9 cm (13.0 in) and 26.6 cm (10.5 in), respectively. Camp Creek has a larger monthly runoff depth because it is at a higher mean elevation than Clear Creek and therefore receives more precipitation. Camp Creek has a single runoff peak in March, while the largest of Clear Creek's two peaks was in December 1982. Because Camp Creek accumulates significant snow during cold storms, some storms may fail to produce runoff peaks. Clear Creek, however, only occasionally receives minor amounts of snow, and storm runoff is more immediate and larger in proportion to the basin size than Camp Creek's.

For the high-magnitude water year, no runoff change associated with land condition was found on Clear Creek (figure 52.11). The traces for the six photo intervals are almost identical. For Camp Creek, however, there is an apparent difference in the runoff regime during the last six months of the water year. The 1991 trace is elevated above the other years' traces in April, May, and June, coincides with the other traces in July, and then moves below the other traces in August and September. This pattern suggests that changes in land use extend the snowmelt runoff period and decrease late-summer base flow.

FIGURE 52.9

Predicted monthly runoff trends over time associated with changes in land condition for the high-magnitude water year in Clear and Camp Creek Basins.

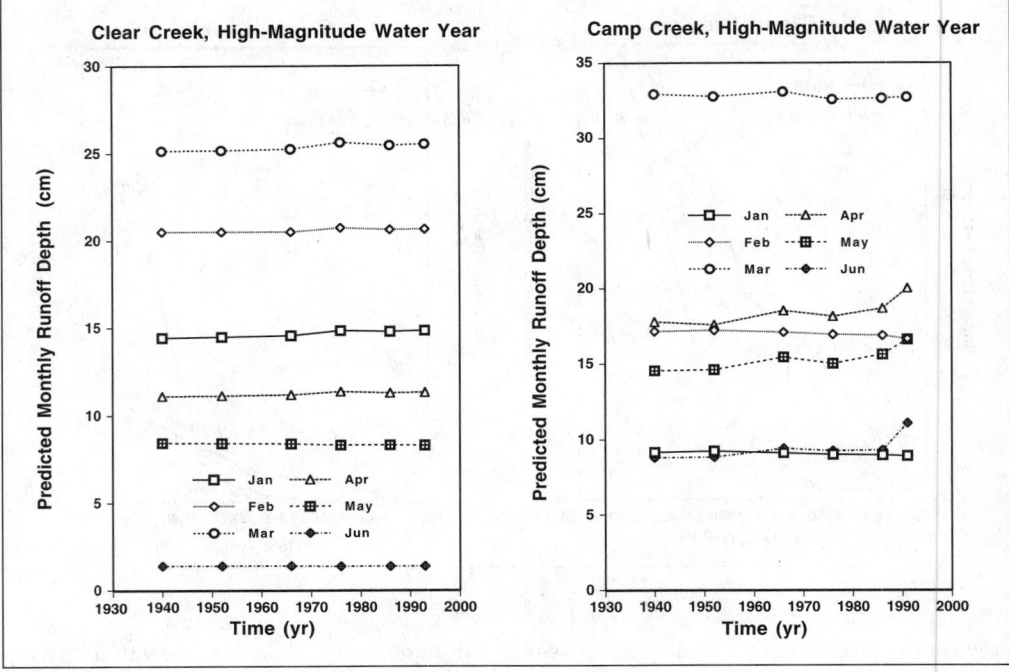

For the medium-magnitude water year on Clear Creek, an increase in February peak flows from 1940 to 1991 is visible, but the volume is minor (figure 52.12). The April peak on Camp Creek shows a larger increased runoff for 1966 and later years. There is also a slight decrease in summer base flow visible in July for the land condition associated with 1991. The difference in runoff timing between the low- and high-elevation basins is also apparent in figure 52.12. For the low-magnitude water year on Clear Creek, changes over time are negligible, so no figure is presented.

Daily Runoff

Predicted and observed daily discharges for Camp Creek for the high-magnitude water year illustrate the rapid increase

FIGURE 52.10

Predicted monthly runoff trends over time associated with changes in land condition for the medium-magnitude water year in Clear and Camp Creek Basins.

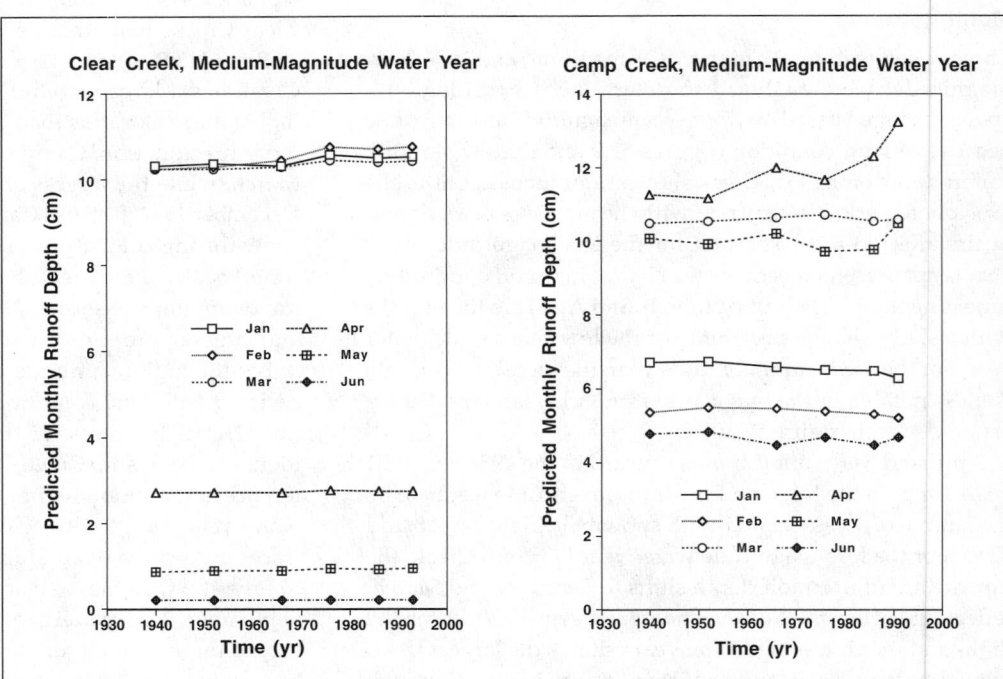

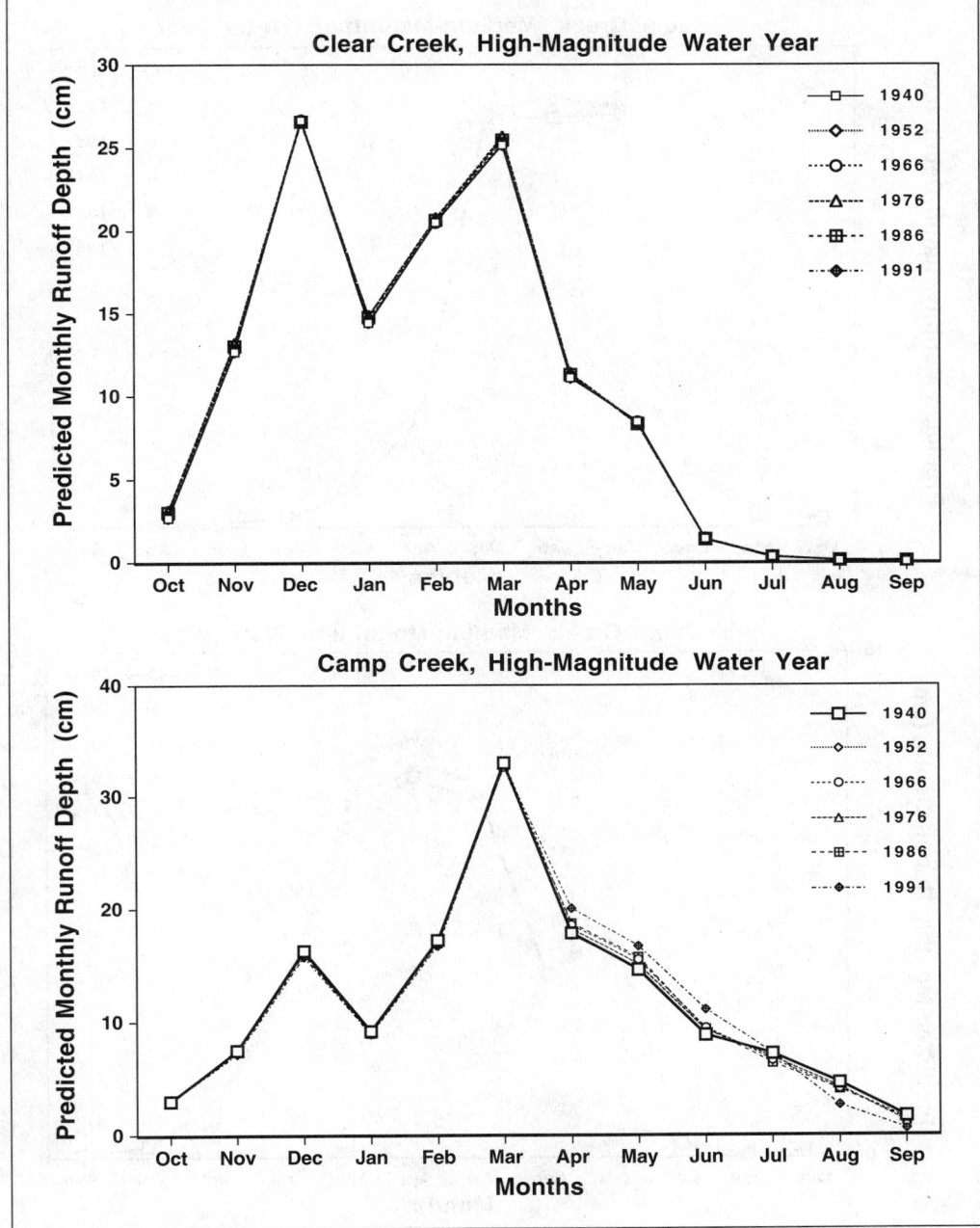

FIGURE 52.11

Predicted annual runoff patterns associated with six land condition descriptions between 1940 and 1991 for the high-magnitude water year in Clear and Camp Creek Basins.

in flow in the Camp Creek Basin and the moderate snowmelt runoff (figure 52.13). The Clear Creek hydrograph has a predicted trace and no observed trace because there is no stream gauge. Clear Creek simulations used only the Placerville precipitation station. The Camp Creek simulations used the Placerville precipitation station for the low-elevation HRUs and the Jenkinson Lake station for the high-elevation stations. Because of snow, the large peak flow near December 1 on the Clear Creek plot is much smaller on the Camp Creek plot.

The dashed line in the Camp Creek plot is the unimpaired outflow estimated from the USGS gauging station, Camp Creek near Somerset. In spite of efforts to compensate for the effects of the diversion from Camp Creek into Jenkinson Lake and the spill from the lake in times of high flow, the observed flow is not an accurate representation of flow from an unmodified basin. The EID record includes very large daily fluctuations in diversion flows, especially in April and May. There is no reason for EID to alternately open and close a very large discharge structure during a time when they are attempting to fill their reservoir, so the variations are believed to be er-

FIGURE 52.12

Predicted annual runoff patterns associated with six land condition descriptions between 1940 and 1991 for the medium-magnitude water year in Clear and Camp Creek Basins.

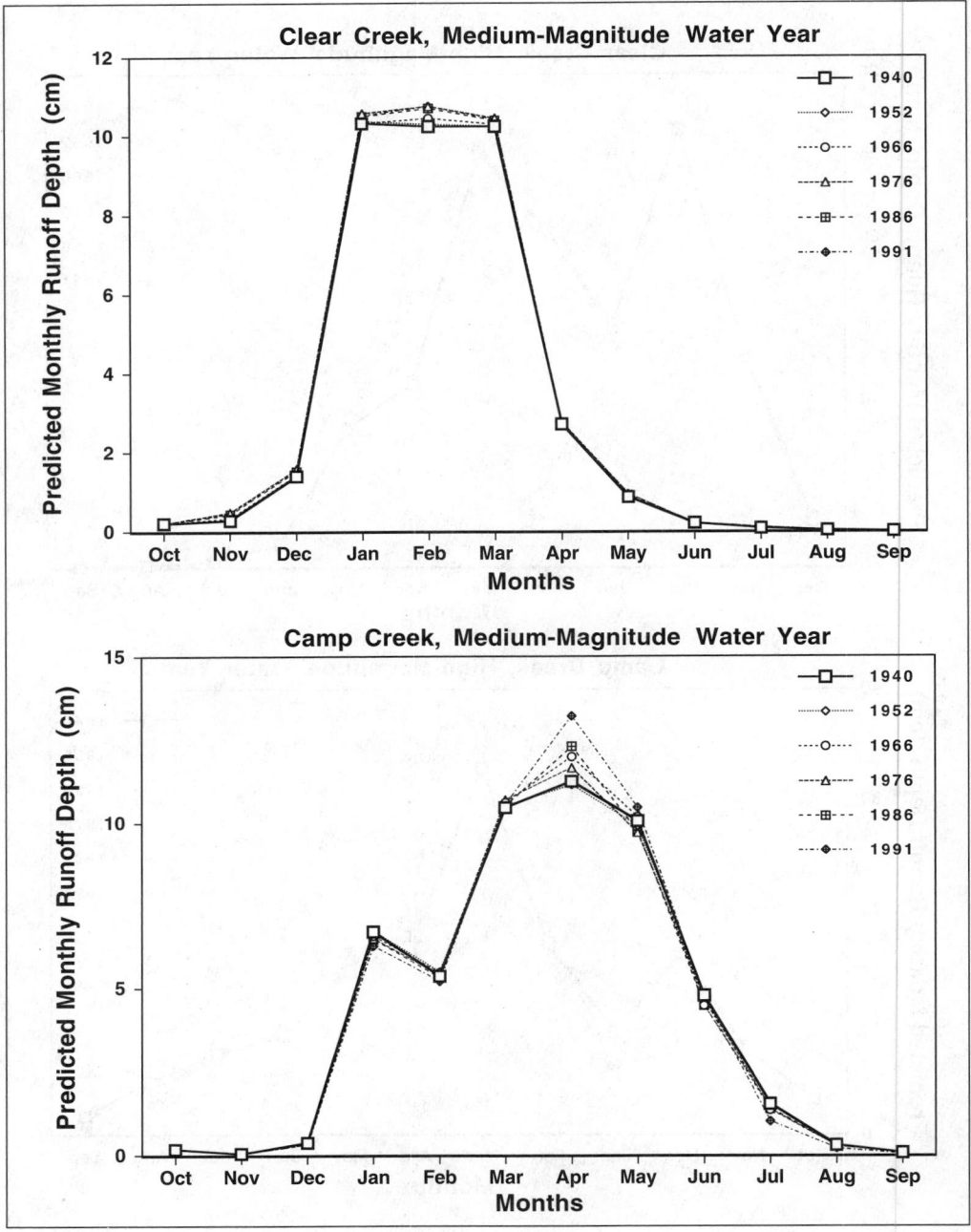

rors. These variations are reflected in the swings of the trace between near-zero flow and the estimated snowmelt discharge.

The predicted flow shown in figure 52.13 excludes flow generated from four HRUs between the discontinued upper Camp Creek gauge (#113315, drainage area of 83 km² [32 mi²]) and the present Camp Creek gauge (#113330, drainage area of 163 km² [63 mi²]). The drainage area shown for the present gauge includes the above-dam Sly Park Basin, an area that was not modeled in this analysis. In all but the most extreme years, water from that basin is trapped in Jenkinson Lake and

exported to Placerville. The flow reconstruction technique that was used also excluded Sly Park Basin's runoff, in that spills were subtracted from the USGS record. The four excluded HRUs below Jenkinson Lake total 3,366 ha (8,316 acres), and the addition of their predicted contribution to runoff increased the predicted flows so that the observed and predicted flow peaks match better than shown in figure 52.13.

The medium-magnitude water year has predicted runoff peaks of between a quarter and a third of the high-magnitude water year (figure 52.14). The unimpaired runoff trace further illustrates the difficulty in reconstructing the Camp

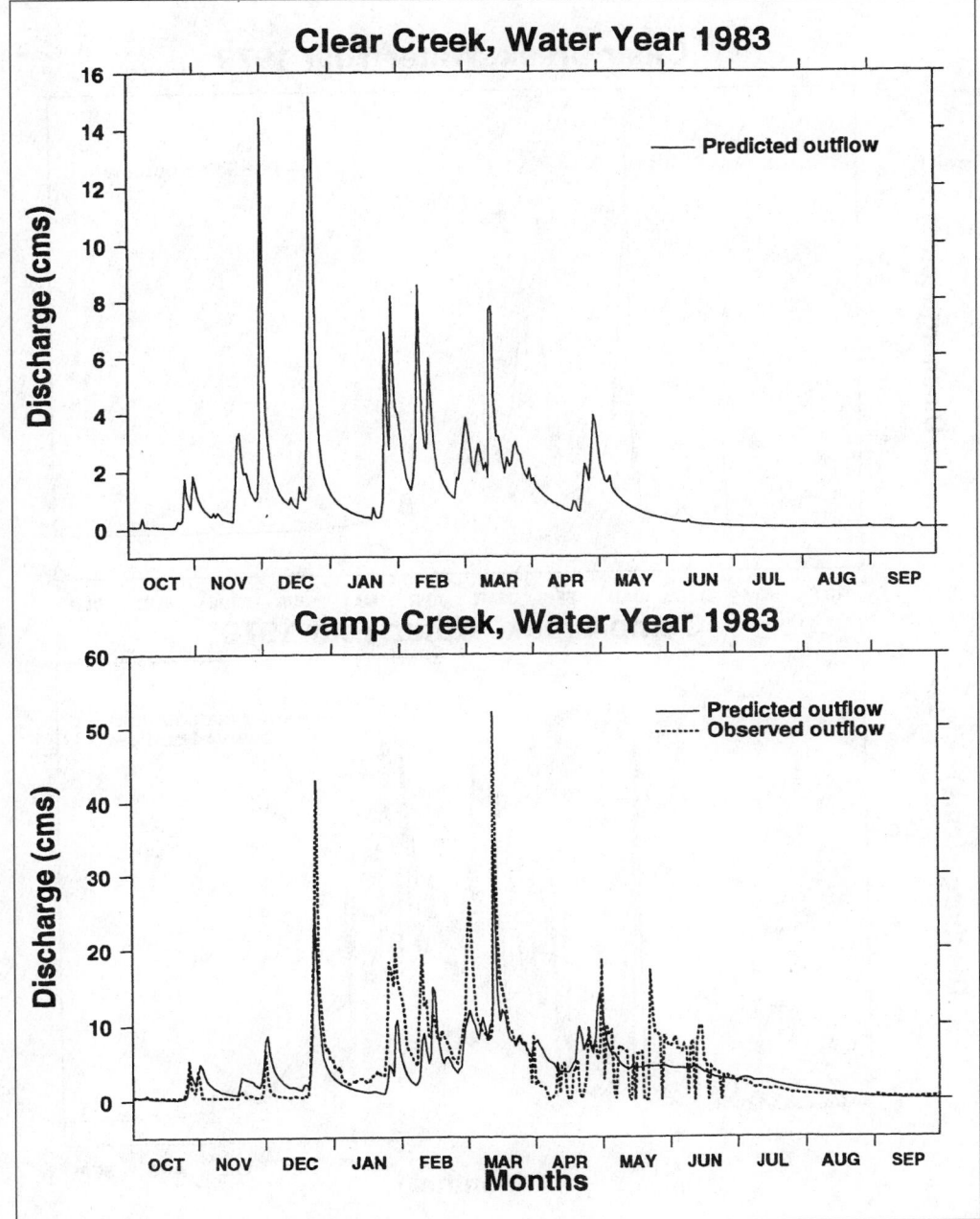

FIGURE 52.13

Predicted daily discharges for the high-magnitude water year for Clear and Camp Creek Basins, and observed discharges for Camp Creek Basin.

Creek record. The reconstruction of the record leads to slightly lower peaks in some years and higher average rates of flow in May and June. However, the unimpaired record is also quite variable in May and June because of large daily changes in the Camp Creek diversion rate and the slow decline in lake level when spill occurs. Current operating rules for Jenkinson Lake change on June 1, and the change from approximately 3.7 cms (130 cfs) to a near-zero flow is a product of the poor record from EID and the reconstruction algorithm, not natural processes. As with the high-magnitude water year, the peaks for both basins match in timing, and the results of snow-

pack accumulation and ablation (melting and evaporation) are evident in the Camp Creek plot.

Daily Flow Components

Total predicted runoff from the PRMS is composed of ground water outflow, subsurface outflow, and surface runoff, so the total flows for the medium-magnitude water year for Camp and Clear Creeks can be separated into their components (figure 52.15). The ground water outflow is proportional to the amount of water stored in the ground water reservoir, but the release rate is small compared to potential input. For the

Predicted daily discharges for the medium-magnitude water year for Clear and Camp Creek Basins, and observed discharges for Camp Creek Basin.

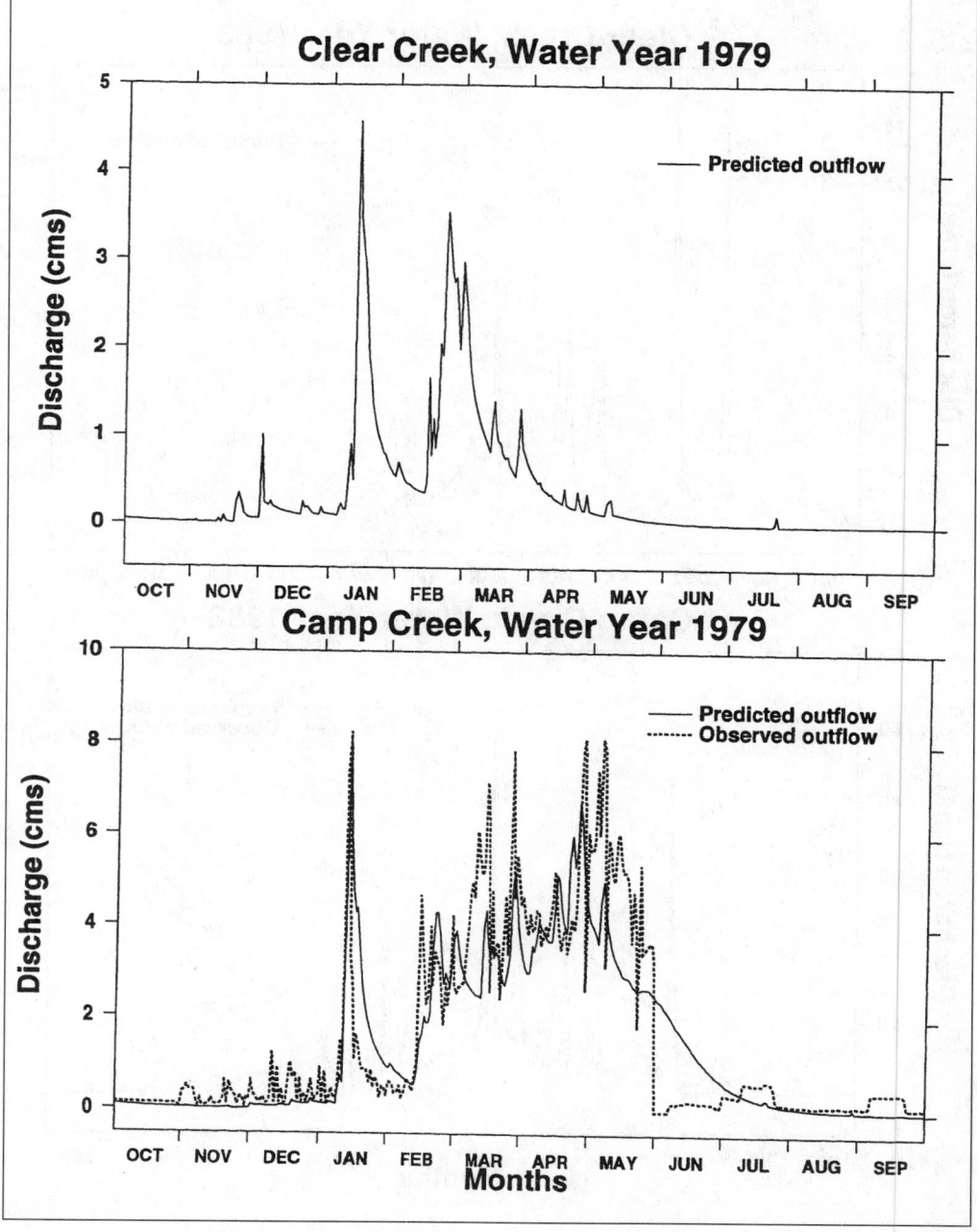

medium-magnitude water year on Camp Creek, ground water accounts for 64% of the total outflow (excluding ET) (table 52.5). For the medium-magnitude water year on Clear Creek, ground water accounts for 48% of the total outflow. Subsurface runoff percentages are 34% and 42% for Camp and Clear Creeks, respectively. Surface flow percentages are 2% and 10% for Camp and Clear Creeks, respectively. This pattern is shown in figure 52.15, and the surface runoff component is especially noticeable for Clear Creek. The large ground water component for Camp Creek is created by melting snow and provides stream flow through the summer.

The Clear Creek ground water outflow reaches zero by early July for the medium-magnitude year, and this pattern matches observations of streams at this elevation. Clear Creek flow does not actually cease, as the model predicts, but only because the EID releases 0.06 cms (2 cfs) from the Camino Conduit during the summer months. The model predicts that Camp Creek would be dry at the gauge site by the end of August during low- and medium-magnitude water years, and USGS records support this prediction. Base flow upstream of the upper gauge (#113315) has been observed in August of several years, but flows are quite small.

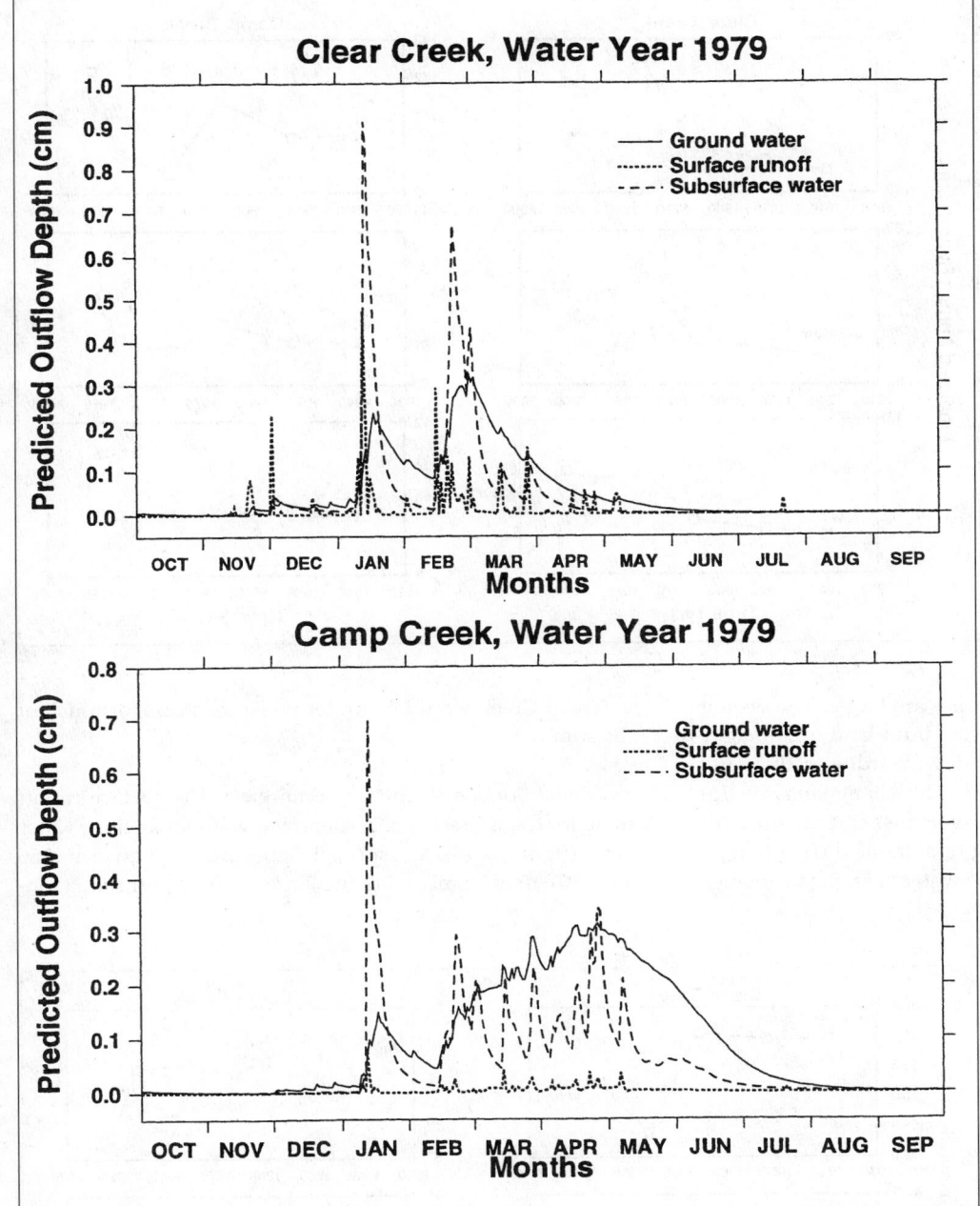

FIGURE 52.15

Predicted ground water, subsurface, and surface flows for the medium-magnitude water year for Clear and Camp Creek Basins.

Annual Flow Components

Trends in flow components can be analyzed over time to assess the effect of changes in land condition. Annual runoff depth for the three water years in both basins suggests the difference between them may be due to forest management and suburbanization (figure 52.16). Total flow increases consistently in Clear Creek over time, and the magnitude is greater with increasing precipitation magnitude. The large increase in 1976 may be attributed to a fire that burned 85 ha (210 acres) in 1972, a doubling of the road network, and a more than doubling of the area covered by structures.

The overall changes in Camp Creek are much less consistent. Runoff increases between 1976 and 1991 for the high- and medium-magnitude water years. In that total runoff is the sum of the other flow components, there may be compensatory changes in the components without overall changes.

Evapotranspiration. Except for the low-magnitude water year in Camp Creek, decreases in percentage cover caused decreases in ET over time (figure 52.17). The Clear Creek traces show a major effect of the 1972 fire mentioned above, but the overall trend of ET is down. Even when wetter years make

FIGURE 52.16

Predicted annual runoff for Clear and Camp Creek Basins, showing changes over time associated with changing land condition and water-year magnitude.

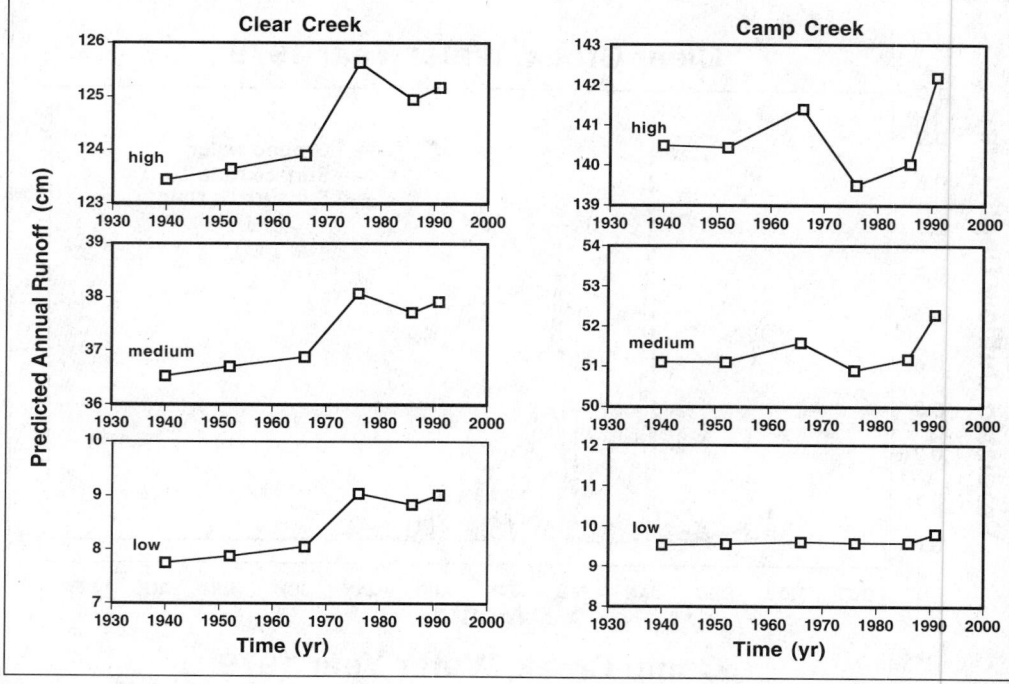

more moisture available, ET declines are larger. The decline over time in Camp Creek for the medium- and high-magnitude water years is consistently downward, and there is a large drop between 1986 and 1991. This change may be due to reductions in percentage cover associated with the aggressive insect salvage logging operations during that interval. During the low-magnitude water year, the vegetation

in Camp Creek would be under moisture stress for much of the summer.

Annual Surface Runoff and Snowmelt. The surface runoff trend for Clear Creek is consistently upward for all three water years (figure 52.18). An especially large increase between 1966 and 1976 may be related to the large increase in the area ren-

FIGURE 52.17

Predicted annual evapotranspiration for Clear and Camp Creek Basins, showing changes over time associated with changing land condition and water-year magnitude.

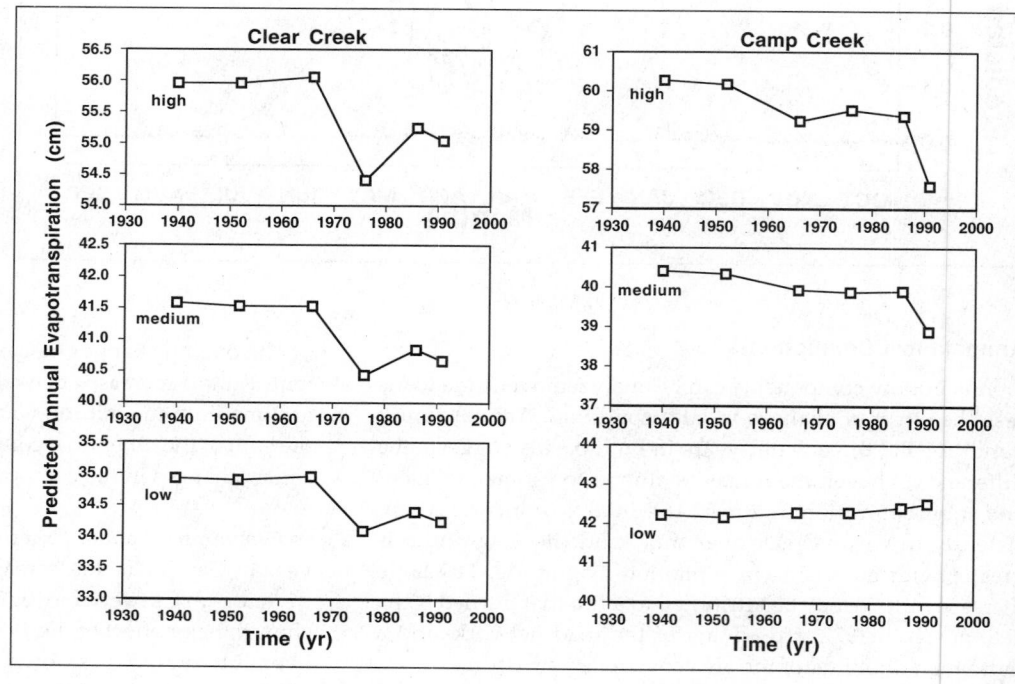

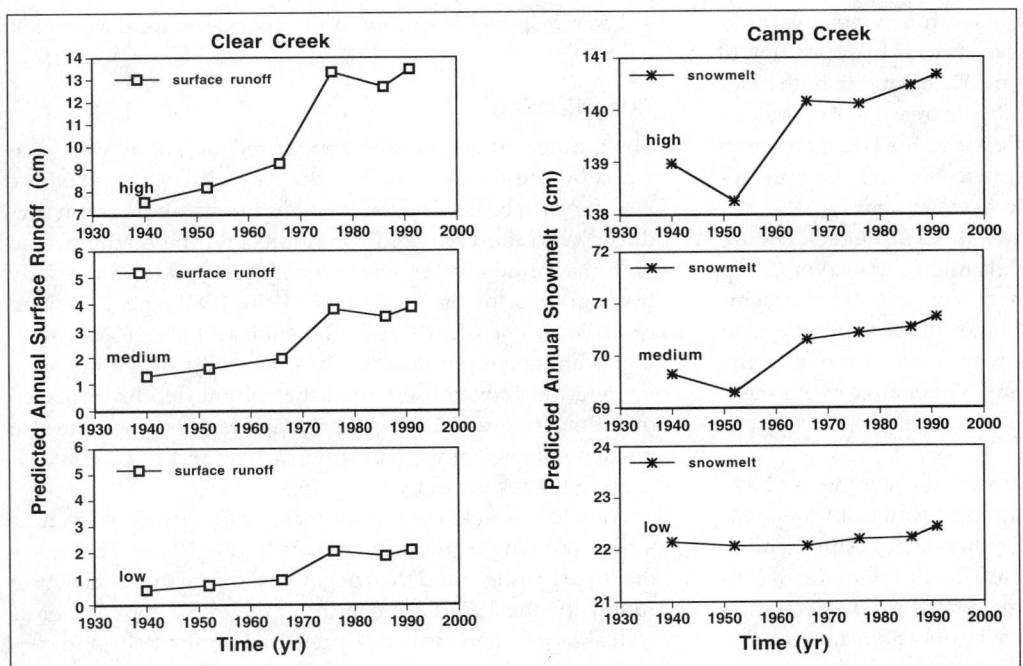

FIGURE 52.18

Predicted annual surface runoff for Clear Creek Basin and snowmelt for Camp Creek Basin, showing changes over time associated with changing land condition and water-year magnitude.

dered impervious by roads and structures. A slight decline occurs between 1976 and 1986 in spite of continued increases in the area covered by structures.

There was no change in surface runoff in Camp Creek between 1940 and 1991, but there was an interesting change in snowmelt depth (figure 52.18). After an initial decline between 1940 and 1952, snowmelt increased fairly consistently over

time. Decreased percentage cover is generally thought to reduce interception losses and thereby increase the snowpack (McGurk and Berg 1987), and the PRMS annual interception estimates do decline slightly over time.

Annual Subsurface Outflow. Subsurface outflow decreases slightly over time for all water years for Clear Creek, but the

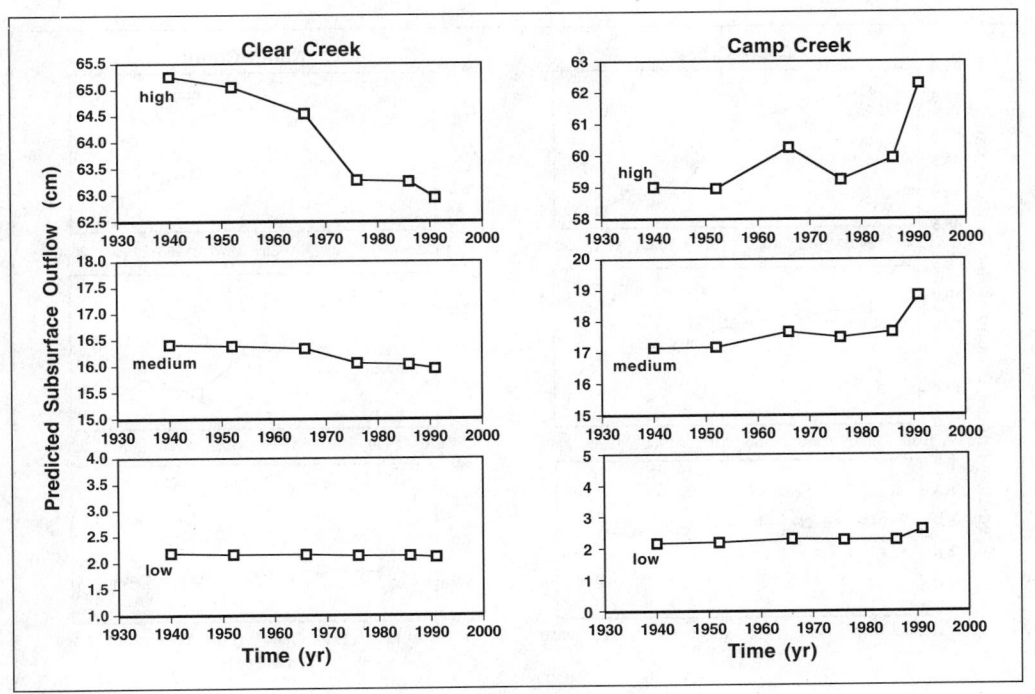

FIGURE 52.19

Predicted subsurface outflow for Clear and Camp Creek Basins, showing changes over time associated with changing land condition and water-year magnitude.

magnitude is largest for the high-magnitude water year (figure 52.19). This decline in subsurface flow is in opposition to the increase in surface runoff (figure 52.18), and in both cases the decreases are associated with the increase in impervious surface over time in the basin (figure 52.8). The increase in surface runoff, however, is about twice the decrease in subsurface flow for the high-magnitude water year.

The pattern of subsurface flows in Camp Creek (figure 52.19) closely resembles the annual runoff pattern for Camp Creek (figure 52.16). This is not surprising, in that the subsurface flow component makes up 34% of the annual total flow (table 52.5). Subsurface flow increases over time, and the magnitude is greater in wetter years. The change over time is negligible in the low-magnitude water year.

Ground Water Outflow. Both Camp and Clear Creeks have a consistent downward trend of ground water outflow over time as land condition changes (figure 52.20). This trend is certainly associated with the increase in impervious area (figure 52.8) and surface runoff (figure 52.18). The decreased ground water outflow has negative implications for summer base flow levels in the two basins. As stream flow decreases, aquatic fauna could be adversely affected by warmer water and lower concentrations of dissolved oxygen. To the extent that vegetation is using ground water, moisture stress might occur earlier in the year and become severe more often if ground water levels have declined and continued to do so. However, some experimental results outside the Sierra Nevada demonstrate an increase in base flow with decreased cover because of reduced ET and greater residual moisture storage (Kattelmann et al. 1983). Other authors have demonstrated decreased base flow with decreased cover (Harr 1980),

so basin response is variable and depends on local characteristics.

Runoff Timing

The timing and magnitude of predicted daily flow were analyzed by setting the runoff hydrograph in 1940 as the base year for each basin. The daily values for the hydrograph produced with the 1940 land conditions and the medium- and high-magnitude water years were subtracted from the daily hydrographs for the 1952, 1966, 1976, 1987, and 1991 land conditions. For Clear Creek, the subtraction yielded values that were generally positive and increased over time, confirming the above observation that runoff depths increased over time (figure 52.16). For Camp Creek, the subtraction also yielded values that were positive in 1966 and 1991, also confirming the results in figure 52.16.

For Clear Creek, the runoff peaks shifted forward in time as well as being larger in later years (figure 52.21). The spikes that trend positive and then negative indicate that early storm runoff for the 1976 land condition may be a day earlier as well as larger than the runoff predicted for the 1940 land condition. The shifts were evident for the 1976 and later storm responses for the high- and medium-magnitude years for Clear Creek. The pattern was more pronounced for the high-magnitude water year than for the medium-magnitude water year (figure 52.21) and was negligible in the low-magnitude year. In the results from Camp Creek, this characteristic sawtooth pattern was not observed. Camp Creek's snowmelt pattern was evident, however, and the analysis again demonstrated the increased snowmelt and decreased base flow illustrated in figure 52.11.

FIGURE 52.20

Predicted ground water outflow for Clear and Camp Creek Basins, showing changes over time associated with changing land condition and water-year magnitude.

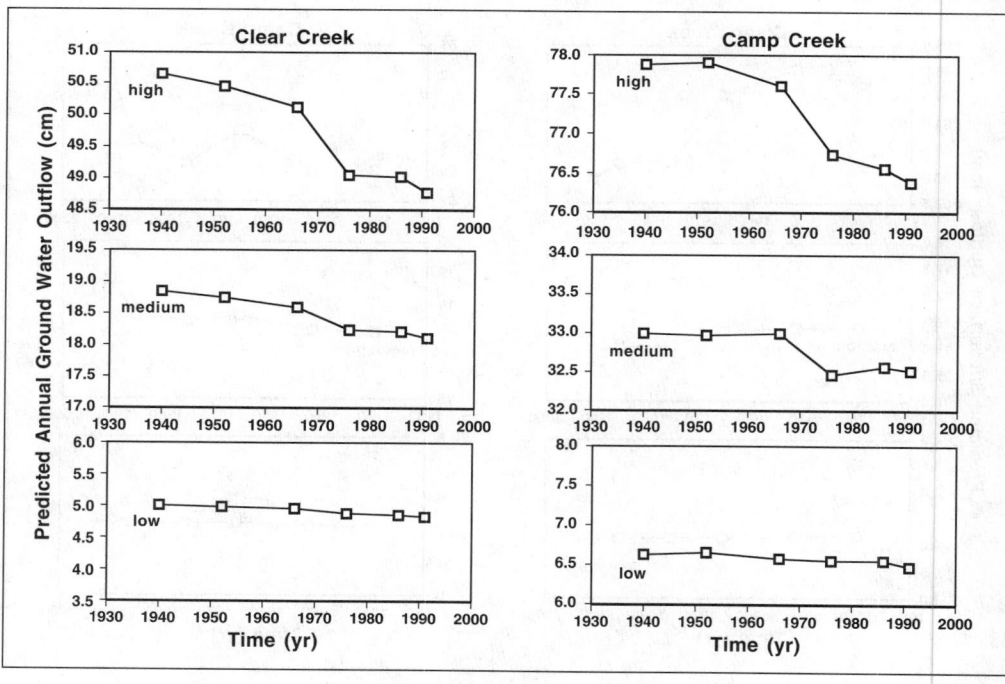

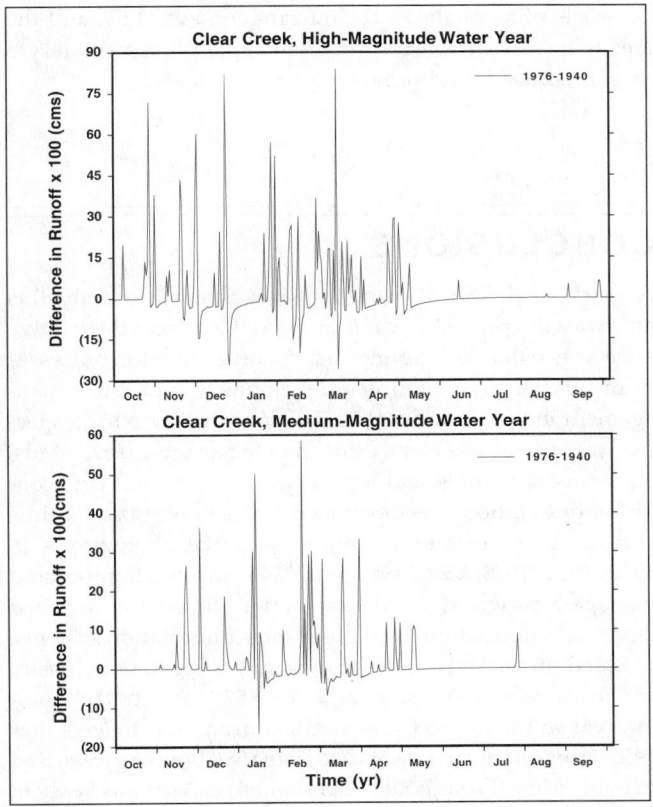

FIGURE 52.21

Sawtooth pattern indicating that changing land condition has caused earlier and greater runoff in 1976 than in 1940 for Clear Creek Basin.

DISCUSSION

Water-Year Trends

High-, medium-, and low-magnitude water years were included in this analysis so that the effect of the water year could be examined via modeling. In general, many changes were evident with the high- and medium-magnitude water years, and changes were negligible at the low-magnitude water year. Changes in monthly runoff trends (figures 52.9 and 52.10) were evident in both the medium- and high-magnitude water years, but the medium-magnitude curves revealed a major increase in April snowmelt runoff over time for Camp Creek. On Clear Creek, the medium-magnitude water year revealed runoff increases over time in January, February, and March (figure 52.10), illustrating the winter rainfall response of a low-elevation basin. For the annual runoff patterns (figures 52.11 and 52.12), the analysis of water years revealed the increased spring runoff and decreased summer base flow only in the high-magnitude water year for Camp Creek.

The analysis of flow components for the three water years also revealed trends linked to water-year magnitude. Clear Creek's increased flow over time was larger with larger-magnitude water years, but evident even in a drought year such as 1976 (figure 52.16). Camp Creek, however, shows virtually no change in runoff over time for the low-magnitude water year. Both basins showed declines in ET over time, except Camp Creek in the low-magnitude water year (figure 52.17). Clear Creek's surface runoff response to changing land condition was scaled to the magnitude of the water year (figure 52.18), but on Camp Creek surface runoff increased noticeably only for the high-magnitude water year (table 52.5). Camp Creek's snowmelt increased over time for the medium- and high-magnitude water years, but the increase was negligible in the low-magnitude water year (figure 52.18). Both basins showed larger changes in subsurface flow for the high-magnitude water year, but the change was slight for Clear Creek in the medium- and low-magnitude water years. Both basins showed similar trends in decreasing ground water outflow associated with changing land condition, and the trends are proportional to water-year magnitude (figure 52.20).

Land Condition Trends

Suburbanization and logging have been shown to cause changes in the hydrology of basins (Dunne and Leopold 1978). The changes are usually attributed to increases in impervious surface because of compaction and reduction in ET associated with removal of vegetation. By holding constant the PRMS coefficients other than percentage cover and impervious area and using the same three water years, this study isolated the effect of changes in percentage cover and impervious area associated with land-use change. The percentage cover declined over time for Clear Creek and declined after 1952 for Camp Creek (figure 52.7), and impervious or disturbed area increased over time (figure 52.8). For Camp Creek between 1940 and 1952, the increase in road area was small, no clear-cuts occurred, and percentage cover increased slightly. Because of this, there was generally little difference in the values of the flow components in figures 52.16–52.20 for the 1940 and 1952 photo periods. Road building, logging, and fire in Camp Creek caused increases in disturbed area and decreases in percentage cover after 1952, and there were concurrent changes in the predicted flow components. There is a large relative increase in runoff in 1991 (figure 52.16), concurrent with the largest decrease in percentage cover during the analysis period (figure 52.7).

Clear Creek's smooth decrease in percentage cover over time (figure 52.7) is paralleled by the steady increase in impervious area due to road building and home construction (figure 52.8). The trends in the plots of the flow components are similarly rather smooth except for the 1976 photo point (figures 52.16–52.20). As mentioned earlier, fire burned 85 ha (210 acres) in 1972, and this reduced percentage cover and increased impervious area for this photo interval alone. The fire caused exceptional increases in total flow and surface runoff and exceptional decreases in predicted ET, subsurface flow, and ground water flow.

Model Configuration and Application

A runoff model such as PRMS translates our conceptualization of the way hydrologic theory operates into a mathematical process model that predicts runoff from a basin. The model is based on the hydrologic processes that research has shown are important (Leavesley et al. 1983). Assumptions are incorporated in the application of the model, however. For example, because warm days occur during the winter in the Sierra Nevada, the model permits ET during winter. ET is estimated from temperature, and although ET rates are low in winter, conifers have been shown to transpire whenever air temperatures exceed 0°C (32°F) (Dunne and Leopold 1978).

Process models have shortcomings associated with the way they represent the basin. Overland flow from each HRU, for example, is summed with other outflows to become stream flow, regardless of the location of the HRU. It is possible that the overland flow in an HRU distant from a channel might be reabsorbed after flowing onto a neighboring HRU, but this form of model eliminates water fluxes in each HRU independently. The model does allow prediction of flow at any point along the "channel" because the contributing HRUs "upstream" can be grouped and fluxes reported for that group. This option was used to predict flow at the discontinued gauge site (#113315).

The inability to allocate disturbances to a particular location within an HRU is another shortcoming of process models. Roads or harvests that are near a channel are thought to be more likely to produce hydrologic effects than the same disturbance distant from a channel. This concern can be addressed to some degree by subdividing HRUs into smaller polygons with more homogeneous properties. If detailed physical and management data are not available, however, subdivision of the HRUs does not increase homogeneity, and modeling results do not improve. Subdivision of HRUs into smaller polygons also increases the computation time used to run the model. The PRMS can simulate a year of runoff for the Camp and Clear Creeks in less than a minute; the physical models that use small cells, short time steps, and mass flux and energy equations may take tens of hours on a similar computer to model a single year.

The results from the MMS type of process model are most useful when considered in a comparative manner, rather than as absolute measures of the effects of land-use change. The results from this analysis allow the comparison of two different land management strategies over a fifty-year period. Information on soils, vegetation, and other physiographic features is typically rather coarse in scale, whereas management activities are often fine-scale. Suburbanization occurs at an even finer scale than forest management. Depending on the goals of an analysis, the information available, and the area to be modeled, users can often apply process models to meet their needs and provide useful information.

CONCLUSIONS

As part of the SNEP Cosumnes River Basin case study, this study used a process hydrologic model to assess the relative effects on water flow, timing, and runoff generation processes from fifty years of suburban development and forest management in the Clear Creek and Camp Creek Basins, respectively. The land-use history documented an increase in roads, clear-cuts, structures, and burned areas over time. A land condition description was developed for each basin at six time periods, based on the availability of aerial photographs in 1940, 1952, 1966, 1976, 1986, and 1991. Information on land management was derived from aerial photos, unpublished data, and information from geographic information systems.

Based on analysis of the photographs, vegetation density in Clear Creek declined from 59% to 57% during the study interval and in Camp Creek declined from 68% to 55% after 1952. Impervious area associated with road building increased in both basins throughout the period but slowed markedly in the last five-year interval. In Clear Creek, impervious area associated with residential development increased from 0.3% to 3.6% throughout the study interval and increased dramatically after 1976. Disturbed area in Camp Creek associated with clear-cuts and roads increased from near 0% to 8.1%, and disturbance has increased dramatically since 1976, similar to the pattern in Clear Creek.

A thirty-three-year series of data from nine climate and four stream discharge stations was assembled and analyzed to identify high-, medium-, and low-magnitude water years, and three water years were selected to represent these classes. The hydrologic model was calibrated and percentage cover and impervious area information was incorporated so that six versions of the model existed for each basin, each representing the land conditions of a specific photo period. Simulations were done with each of the three water years, and the eighteen sets of output data were analyzed to determine the effect of water-year magnitude and land condition on the hydrology of the basins as predicted by the model.

The simulations showed that the changes in land condition were associated with changes in runoff depth in both the suburbanized and logged basins. The predicted mean monthly flow from Clear Creek increased over time, but the change was less evident for Camp Creek except in the high-magnitude water year. Surface runoff for both basins increased in the high-magnitude water year, but only in Clear Creek for the other years. Ground water flow and ET declined over time for both basins.

Changes in runoff timing and pattern were also found. Clear Creek's storm-based runoff was both larger and slightly earlier when the later years were compared to the 1940 year. Camp Creek's runoff, being at least partly based on snow-melt, showed an increased snowmelt flow during the runoff period and a decline in summer base flow for the high-magnitude water year. A predicted increase in the mean April flow was shown for Camp Creek for the 1986 and 1991 years.

The results obtained from this case study should have general application along the west slope of the Sierra Nevada. In many other basins, residential development is occurring at an elevation similar to Clear Creek's, and these basins would be likely to have a similar response to that demonstrated here. Forest management is similarly located both north and south of Camp Creek, so analogous results are to be expected. Because of these factors, this case study has wider application than to just these two basins.

Because the Clear and Camp Creek Basins are at different elevations and are of different size, direct comparisons of suburban development and forest management are somewhat difficult. Both basins showed definite changes in their hydrology over time, and those predicted changes are the result of changes in land condition, as simulated by the hydrologic model. Suburbanization in a rain-dominated basin appears to have the most distinct signature: increased runoff that is largely due to increases in surface runoff. The model predicts increases in snowmelt runoff when forest cover is reduced in a basin that receives a significant amount of snow.

ACKNOWLEDGMENTS

We are grateful to the individuals who contributed to this project. Thad Edens (PSW) assembled and perfected the climate and hydrologic data. Susan Rodman and Annette Parsons (Eldorado National Forest's Supervisor's Office) provided critical advice, invaluable map products and analysis, as well as historical information. Patricia Ferrell (Placerville Ranger District) provided access to and guidance in the use of district records and photo archives. George Leavesley, Linda Stannard, and Steve Markstrom (U.S. Geological Survey, Denver, Colorado) provided technical and professional support and advice about the hydrologic modeling system.

REFERENCES

Brandow, C. 1994. Calwater: A standardized set of California watersheds. Sacramento: California Department of Forestry and Fire Protection.

Crawford, N. H., and R. K. Linsley. 1966. *Digital simulation in hydrology: Stanford watershed model IV.* Technical Report 39. Stanford, CA: Stanford University, Department of Civil Engineering.

Dunne, T., and L. B. Leopold. 1978. *Water in environmental planning.* San Francisco: W. H. Freeman.

Dyrness, C. T. 1965. Soil surface condition following tractor and high-lead logging in the Oregon Cascades. *Journal of Forestry* 63:272–75.

———. 1972. *Soil surface conditions following balloon logging.* Research Note PNW-182. Portland, OR: U.S. Forest Service.

———. 1976. *Effects of wildfire on soil wettability in the high Cascades of Oregon.* Research Paper PNW-202. Portland, OR: U.S. Forest Service.

EarthInfo, Inc. 1993. *Hydrodata.* CD-ROM database. Boulder, CO: EarthInfo, Inc.

Farnsworth, R. K., and E. S. Thompson. 1982. *Mean monthly seasonal and annual pan evaporation for the United States.* Technical Report NWS 34. Washington, DC: National Oceanic and Atmospheric Administration.

Fogelman, R. P., T. C. Hunter, J. R. Mullen, R. G. Simpson, and D. A. Grillo. 1984. *Water resources data: California, water year 1984.* Vol. 3. Water-Data Report CA-84-3. Sacramento, CA: U.S. Geological Survey.

Grant, G. E., R. D. Harr, and G. Leavesley. 1990. Effects of forest land use on watershed hydrology—a modelling approach. *Northwest Environmental Journal* 6: 414–15.

Harr, R. D. 1980. *Streamflow after patch logging in small drainages within the Bull Run municipal watershed.* Research Paper PNW-268. Portland, OR: U.S. Forest Service.

Hay, L. E., W. A. Battaglin, R. S. Parker, and G. H. Leavesley. 1993. Modeling the effects of climate change on water resources in the Gunnison River Basin, Colorado. In *Environmental modeling with GIS,* edited by M. F. Goodchild, B. O. Parks, and L. T. Steyaert, 173–81. New York: Oxford University Press.

Jensen, M. E., and H. R. Haise. 1963. Estimating evapotranspiration from solar radiation. *Proceedings of the American Society of Civil Engineers, Journal of Irrigation and Drainage* 89 (IR4): 15–41.

Jeton, A. E., and J. L. Smith. 1993. Development of watershed models for two Sierra Nevada basins using a geographic information system. *Water Resources Bulletin* 29: 923–32.

Kattelmann, R. C., N. H. Berg, and J. Rector. 1983. The potential for increasing streamflow from Sierra Nevada watersheds. *Water Resources Bulletin* 19: 395–402.

Krammes, J. S., and L. F. DeBano. 1965. Soil wettability: A neglected factor in watershed management. *Water Resources Research* 1: 283–86.

Leavesley, G. H., R. W. Lichty, B. M. Troutman, and L. G. Saindon. 1983. *Precipitation-runoff modeling system—user's manual.* Water Resources Investigations Report 83-4238. Denver: U.S. Geological Survey.

Leavesley, G. H., P. Restrepo, L. G. Stannard, and M. Dixon. 1992. The modular hydrologic modeling system—MHMS. In *Managing water resources during global change,* edited by R. Herrmann, 263–64. Bethesda, MD: American Water Resources Association.

McGurk, B. J., and N. H. Berg. 1987. Snow redistribution: Strip cuts at Yuba Pass, California. In *Forest hydrology and watershed management,* edited by R. H. Swanson, P. Y. Bernier, and P. D. Woodward, 285–95. Vancouver, BC: International Association of Hydrological Sciences.

McGurk, B. J., N. H. Berg, and M. L. Davis. 1996. Camp and Clear Creeks, El Dorado County: Predicted sediment production from forest management and residential development. In *Sierra Nevada Ecosystem Project: Final report to Congress,* vol. II, chap. 53. Davis:

University of California, Centers for Water and Wildland Resources.

Mitchell, C. R., and K. J. Silverman. N.d. *Soil survey of Eldorado National Forest area, California*. Washington, DC: U.S. Forest Service.

Moore, I. D., E. M. O'Laughlin, and G. J. Buick. 1988. A contour-based topographic model for hydrologic and ecological applications. *Earth Surface, Processes, Landforms* 13:305–20.

Nakama, L. Y. 1994. *Application of the precipitation-runoff modeling system model to simulate dry season runoff for three watersheds in south-central Guam*. Water Resources Investigations Report 93-4116. Denver: U.S. Geological Survey.

National Oceanic and Atmospheric Administration. 1990. *Climatological data: California* 94(1). Asheville, NC: NOAA.

Norris, J. M. 1986. *Application of the precipitation-runoff modeling system to small basins in the Parachute Creek Basin, Colorado*. Water Resources Investigations Report 86-4115. Denver: U.S. Geological Survey.

Peck, E. L. 1976. *Catchment modeling and initial parameter measurement for National Weather Service River Forecasting System*. NOAA Technical Memo NWS HYDRO-31. Silver Spring, MD: National Oceanic and Atmospheric Administration.

Poff, R. J. 1989. Distribution and persistence of hydrophobic soil layers on the Indian Burn. In *Symposium on fire and watershed management*, edited by N. H. Berg. General Technical Report PSW-109. Berkeley, CA: U.S. Forest Service.

Rogers, J. H. 1974. *Soil survey of El Dorado area, California*. Washington, DC: U.S. Department of Agriculture, Soil Conservation Service and Forest Service.

Satterlund, D. R. 1972. *Wildland watershed management*. New York: Ronald Press.

Sorooshian, S. 1983. Surface water hydrology: On-line estimation. *Review of Geophysics* 21: 706–21.

Stannard, L., and G. Kuhn. 1989. Watershed modeling. In *Summary of the U.S. Geological Survey and U.S. Bureau of Land Management national coal-hydrology program, 1974–84*, edited by L. J. Anderson et al. Professional Paper 1464. Denver: U.S. Geological Survey.

APPENDIX 52.1

Areas Occupied by Three Disturbance Types for Six Photo Intervals in Camp Creek Basin

HRU	Area (ha)	1940 Road	1940 Clear-cut	1940 Fire	1952 Road	1952 Clear-cut	1952 Fire	1966 Road	1966 Clear-cut	1966 Fire	1976 Road	1976 Clear-cut	1976 Fire	1986 Road	1986 Clear-cut	1986 Fire	1991 Road	1991 Clear-cut	1991 Fire
1	162.6	0	0	0	0	0	0	0.13	2.54	0	0.42	2.54	0	0.42	3.68	0	0.42	3.68	0
2	103.4	0.1	0	0	1.1	0	0	1.35	0	0	2.74	3.82	0	2.74	4.62	0	2.74	4.62	0
3	46.6	0.07	0	15.95	0.07	0	0	0.3	0	0	1.03	0	0	1.03	0	0	1.03	0	0
4	60	0	0	59.01	0.16	0	0	0.23	0	0	0.4	0	0	0.79	0	0	0.79	0	0
5	51.6	0	0	45.65	0	0	0	0	0	0	0.27	0	0	0.27	0	0	0.27	0	0
6	137.4	0.94	0	7.38	0.94	0	0	0.94	0.77	0	3.93	2.21	0	3.93	2.21	0	3.93	2.21	0
7	126.1	0	0	27.2	0.4	0	0	0.48	0	0	0.76	0	0	1.47	0	0	1.47	0	0
8	375	0	0	15.26	0.77	0	0	0.77	0	0	1.8	0	0	1.9	0	0	1.9	0	0
9	62.2	0	0	0	0.24	0	0	0.46	1.32	0	0.46	1.32	0	0.46	1.32	0	0.46	1.32	0
10	60.8	0	0	18.05	0	0	0	0	0	0	0	0	0	0.51	0	0	0.51	0	0
11	47.6	0	0	0	0	0	0	0	0	0	0	0	0	0.38	0	0	0.38	0	0
12	288.9	0	0	0	0	0	0	0.56	2.85	0	1.37	2.85	0	2.15	2.85	0	3.02	2.85	0
13	102.7	0	0	48.48	0	0	0	0	0	0	0	0	0	0.95	0	0	0.95	0	0
14	152.8	0	0	15.95	0	0	0	0	0	0	0	0	0	1.5	0	0	1.5	0	0
15	250.6	0	0	0	0	0	0	0.04	0	0	0.63	8.72	0	3.97	8.72	0	3.97	8.72	0
16	27.9	0	0	0	0	0	0	0	0	0	0	0	0	0.36	0	0	0.36	0	0
17	70	0	0	0	0	0	0	0	0	0	0	0	0	0.49	13.54	0	0.49	13.54	0
18	229.9	0	0	4.53	0.26	0	0	0.38	0	0	0.38	0	0	2.6	53.17	0	2.6	53.17	0
19	339.2	2.81	0	0	3.61	0	0	3.62	0	0	5.08	0	0	6.84	13.53	0	6.84	34.96	0
20	54.1	0	0	0	0	0	0	0	0	0	0	0	0	0.99	4.56	0	0.99	4.56	0
21	63.2	0	0	0	0.72	0	0	0.72	0	0	0.74	0	0	0.99	0	0	0.99	2.62	0
22	60.4	0	0	0	0.24	0	0	0.24	0	0	0.24	0	7.65	0.31	0	7.65	0.31	0	7.65
23	45	0	0	0	0.71	0	0	0.71	0	0	0.75	0	12.79	1.16	0	12.79	1.16	4.47	12.79
24	98.5	0	0	0	0.04	0	0	0.04	0	0	0.04	0	0	1.74	12.56	0	1.74	12.56	0
25	327.4	0	0	0	1.4	0	0	1.56	0	0	3.14	0	48.73	3.27	0	48.73	3.27	0	53.66
26	299.1	1	0	0	1.73	0	0	1.73	0	0	2	0	26.1	3.49	14.63	26.1	3.49	38.41	32.21
27	67.5	0.43	0	0	0.43	0	0	0.43	0	0	0.43	0	0	0.83	0	0	0.83	19.17	0
28	141.6	1.01	0	0	1.04	0	0	1.04	0	0	1.46	0	0	2.77	0	0	2.77	0	0
29	215.1	0	0	0	1.35	0	0	1.99	0	0	1.99	0	0	3.64	7.47	0	3.64	7.47	3.16
30	142.3	0.51	0	0	0.62	0	0	0.62	0	0	0.96	0	0	2.22	13.48	0	2.22	46.07	0
31	225.4	0.41	0	0	0.41	0	0	1.25	0	0	2.47	0	0	2.47	31.08	0	2.47	31.08	0
32	60.3	0	0	0	0	0	0	0.25	0	0	0.25	0	0	0.69	3.05	0	0.69	3.05	0
33	247.4	0.37	0	0	0.53	0	0	1.43	0	0	1.77	0	0	2.95	47.5	0	2.95	47.5	0
34	27.5	0	0	0	0	0	0	0	0	0	0	0	0	0.21	0	0	0.21	0	0
35	102.5	0	0	0	0.02	0	0	0.23	0	0	0.23	0	0	1.73	0	0	1.73	0	0
36	80.1	0	0	0	0	0	0	0.18	0	0	1.05	0	2.23	1.3	0	2.23	1.3	0	2.23
37	142.6	0	0	0	0.73	0	0	1.21	0	0	2.4	0	0	2.52	0	0	2.52	0	0
38	198.7	0	0	0	0.59	0	0	0.74	2.08	0	2.5	2.08	0	4.36	2.08	0	4.36	2.08	0
39	72.7	0	0	0	0	0	0	0.5	0	0	0.5	0	0	0.95	0	0	0.99	0	0
40	78	0	0	0	0	0	0	0	0	0	0	0	0	0.6	0	0	0.67	0	0
41	95.3	0.12	0	0	0.12	0	0	0.22	0	0	0.22	0	0	0.52	0	0	0.52	0	0
42	279.8	0.32	0	0	0.75	0	0	1.74	0	0	2.35	0	0	2.74	0	0	2.74	0	0
43	305.3	0.56	0	0	1.72	0	0	2.72	0	0	3.41	0	128	3.51	22.19	127.97	3.51	22.19	127.97
44	180.8	0.64	0	0	1.21	0	0	1.37	0	0	1.93	0	25.98	2.9	15.51	25.98	2.9	15.51	25.98
45	222	0.29	0	0	0.29	0	0	1.6	0	0	1.96	0	0	2.26	25.78	0	2.26	25.78	0
46	270.1	0.89	0	0	0.9	0	0	1.23	0	0	3.79	0	0	3.79	0	0	3.97	23.93	0
47	270.3	0.35	0	0	0.73	0	0	0.73	0	0	3.17	0	0	2.78	0	0	2.78	5.68	0
48	112.8	0.7	0	0	0.84	0	0	0.84	0	0	2.38	0	0	2.61	0	0	2.61	10.23	0
49	58.4	0	0	0	0.25	0	0	0.25	0	0	0.58	0	0	0.81	0	0	0.81	2.94	0
50	116.1	0	0	0	0.6	0	0	0.6	0	0	1.3	0	0	1.54	0	0	1.87	7.76	0
51	291.2	0	0	0	0.27	0	0	2.76	0	0	2.88	0	0	2.88	0	0	2.88	4.4	0
52	132.4	0	0	0	0.46	0	0	0.88	0	0	1.64	0	0	1.64	0	0	1.64	20.11	0
53	33.6	0	0	0	0.04	0	0	0.07	0	0	0.27	0	0	0.27	0	0	0.27	11.33	0

HRU	Area (ha)	1940			1952			1966			1976			1986			1991		
		Road	Clear-cut	Fire	Road	Clear-cut	Fire	Road	Clear-cut	Fire	Road	Clear-cut	Fire	Road	Clear-cut	Fire	Road	Clear-cut	Fire
54	156.3	0	0	0	0.47	0	0	0.47	0	0	2.03	3.53	0	2.73	3.53	0	2.73	22.68	0
55	59.1	0	0	0	0	0	0	0	0	0	0.59	3.08	0	0.59	3.08	0	0.59	12.74	0
56	91.7	0	0	0	0	0	0	0	0	0	1	3.08	0	1.32	3.08	0	1.32	8.17	0
57	179.2	0	0	0	0	0	0	0	1.63	0	1.31	11.06	0	1.39	11.06	0	1.39	24.09	0
58	127	0	0	0	0.03	0	0	0.03	0	0	0.88	0	0	1.1	0	0	1.1	9.88	0
Totals	8426.3	11.3	0	257.4	26.8	0	0	39.6	11.2	0	73.9	44.29	251.4	107.3	324.3	251.4	108.8	571.5	265.6

APPENDIX 52.2

Areas Occupied by Three Disturbance Types for Six Photo Intervals in Clear Creek Basin

HRU	Area (ha)	1940 Road	1940 House	1940 Fire	1952 Road	1952 House	1952 Fire	1966 Road	1966 House	1966 Fire	1976 Road	1976 House	1976 Fire	1986 Road	1986 House	1986 Fire	1991 Road	1991 House	1991 Fire
1	36.5	0	0	0	0	0	0	0	0	0	0	0	0	0	0	0	0	0	0
2	117.2	0.45	0	0	1.17	0	0	1.17	0.09	0	1.19	0.19	0	1.19	0.84	0	1.19	0.98	0
3	21.8	0	0	0	0.15	0	0	0.15	0	0	0.15	0	0	0.15	0.04	0	0.15	0.09	0
4	18.6	0	0	0	0.04	0	0	0.04	0	0	0.04	0	0	0.04	0	0	0.04	0	0
5	38.3	0.23	0.04	0	0.66	0.04	0	0.66	0.04	0	0.67	0.19	0	0.67	0.51	0	0.67	0.74	0
6	16.8	0	0	0	0.28	0	0	0.33	0.04	0	0.33	0.04	0	0.33	0.19	0	0.33	0.19	0
7	36	0.28	0	0	0.59	0	0	0.59	0.04	0	0.59	0.09	0	0.59	0.42	0	0.59	0.61	0
8	133.4	0.62	0	0	0.97	0.09	0	1.13	0.19	0	1.78	0.32	0	1.78	0.74	0	1.78	1.17	0
9	31	0.51	0	0	0.57	0.04	0	0.57	0.04	0	0.93	0.04	0	0.93	0.04	0	0.93	0.09	0
10	8.5	0.28	0	0	0.55	0	0	0.55	0.09	0	0.76	0.09	0	0.76	0.14	0	0.76	0.19	0
11	81.7	0	0	0	0.21	0.09	0	0.21	0.09	0	0.27	0.14	0	0.27	0.23	0	0.27	0.51	0
12	137	0.97	0.04	0	1.93	0.09	0	2.11	0.23	0	2.88	0.42	0	2.88	1.44	0	2.88	1.81	0
13	24.2	0	0	0	0.03	0	0	0.07	0	0	0.07	0	0	0.07	0.04	0	0.07	0.04	0
14	90.3	0.11	0.08	0	0.18	0.09	0	0.45	0.14	0	0.84	0.19	0	0.84	0.19	0	0.84	0.19	0
15	156.6	0.56	0.23	0	0.83	0.28	0	1.43	0.7	0	2.01	0.93	0	2.21	1.72	0	2.21	2.05	0
16	73.4	1	0.09	0	1.29	0.14	0	1.79	0.56	0	2.58	1.11	0	2.58	2.83	0	2.58	3.12	0
17	84.9	0.04	0	0	0.2	0	0	0.86	0	0	1.16	0.56	0	1.7	0.98	0	1.7	1.21	0
18	19.9	0	0	0	0	0	0	0.07	0	0	0.2	0	0	0.35	0	0	0.35	0	0
19	121.6	0	0	0	0.79	0	0	0.88	0.09	0	1.32	0.09	3.04	1.32	0.23	3.04	1.32	0.6	3.04
20	117.2	0	0	0	0.01	0	0	0.18	0	0	2.55	0.28	64.91	2.59	1.11	64.91	2.59	1.53	64.91
21	12.6	0	0	0	0	0	0	0	0	0	0	0	0	0	0.14	0	0	0.19	0
22	54.3	0	0	0	0.08	0	0	0.08	0	0	0.08	0	0	0.08	0	0	0.08	0	0
23	15.6	0	0	0	0	0	0	0.05	0	0	0.05	0	0	0.05	0.09	0	0.05	0.14	0
24	80.3	0.15	0	0	0.29	0	0	0.81	0.32	0	0.92	0.37	0	0.92	0.6	0	0.92	0.74	0
25	14.5	0	0	0	0	0	0	0	0	0	0	0	0	0	0.14	0	0	0.19	0
26	37.9	0	0	0	0	0.14	0	0.06	0.47	0	0.06	0.74	0	0.06	1.02	0	0.06	1.4	0
27	122.6	1.29	0.04	0	1.94	0.23	0	2.29	0.56	0	3.28	0.84	0	3.28	1.49	0	3.28	2.04	0
28	23.5	0	0	0	0.21	0.04	0	0.37	0.32	0	0.48	0.37	0	0.48	0.74	0	0.48	0.84	0
29	117.8	0.32	0.04	0	1.66	0.04	0	2.24	0.51	0	2.85	1.07	0	3	1.72	0	3	2.27	0
30	31.8	0	0	0	0.31	0	0	0.31	0	0	0.31	0	0	0.31	0	0	0.31	0	0
31	184.1	0.58	0	0	1.35	0	0	2.52	0.65	0	4.26	2.37	0	4.28	6.41	0	4.28	8.36	0
32	94.6	0	0	0	0.42	0	0	0.48	0	0	1.95	0.93	0	1.99	3.53	0	1.99	4.83	0
33	40.6	0	0	0	0.28	0	0	0.61	0	0	0.61	0.09	0	0.61	0.42	0	0.67	0.7	0
34	32.3	0	0	0	0	0	0	0	0	0	0	0	2.75	0	0	2.75	0.21	0	2.75
35	25.7	0	0	0	0	0	0	0.53	0.23	0	0.56	0.32	5.42	0.56	0.47	5.42	0.56	0.65	5.42
36	41.4	0	0	0	0	0	0	0.1	0	0	1.22	0.04	8.86	1.22	0.47	8.86	1.22	0.89	8.86
37	28	0	0	0	0	0	0	0.06	0	0	0.45	0	0	0.51	0	0	0.51	0	0
38	42.9	0	0	0	0	0	0	0.13	0	0	0.47	0.04	0	0.59	0.19	0	0.59	0.23	0
39	23.1	0	0	0	0	0	0	0.06	0	0	0.35	0	0	0.48	0	0	0.48	0	0
40	17.4	0	0	0	0	0	0	0	0	0	0	0	0	0.14	0	0	0.14	0	0
41	20.2	0	0	0	0	0	0	0.11	0	0	0.78	0.09	0	0.78	0.37	0	0.78	0.65	0
42	20.2	0	0	0	0	0	0	0.06	0	0	0.17	0	0	0.17	0	0	0.17	0	0
43	11.5	0	0	0	0	0	0	0	0	0	0	0	0	0.48	0	0	0.48	0	0
44	24.5	0	0	0	0	0	0	0	0	0	0.01	0	0	0.2	0	0	0.2	0	0
45	39.6	0	0	0	0	0	0	0	0.04	0	0.51	0.09	0	0.69	0.28	0	0.69	0.42	0
46	38.1	0	0	0	0	0	0	0.4	0.23	0	0.91	1.72	0	0.93	2.93	0	0.93	3.85	0
47	20.9	0	0	0	0	0	0	0	0	0	0.25	0.42	0	0.25	1.11	0	0.25	1.53	0
48	16.1	0	0	0	0	0	0	0	0	0	0.12	0.19	0	0.14	0.37	0	0.14	0.42	0
49	112.5	0.51	0.04	0	0.66	0.14	0	0.66	0.19	0	1.7	0.19	0	1.7	0.32	0	1.7	0.42	0
50	10.4	0	0	0	0.01	0	0	0.09	0	0	0.69	0.19	0	0.69	0.84	0	0.69	1.02	0
51	14.5	0	0	0	0.06	0	0	0.06	0	0	0.06	0	0	0.06	0.09	0	0.5	0	0
52	66.5	0	0	0	0.24	0	0	0.24	0	0	0.44	0	0	0.44	0	0	0.35	0.42	0
53	52.3	0.03	0	0	0.16	0	0	0.16	0	0	0.17	0	0	0.35	0.19	0	0.35	0.42	0
54	39.2	0	0	0	0	0	0	0	0	0	1.72	0.65	0	1.72	4.32	0	1.72	5.35	0
55	94.5	0.1	0.23	0	0.1	0.23	0	1.25	0.74	0	2.08	1.76	0	2.19	3.39	0	2.19	4.46	0
56	80.8	0	0	0	0.48	0	0	0.68	0.65	0	1	1.11	0	1.03	1.53	0	1.03	1.81	0
Totals	3067.7	8.03	0.83	0	18.7	1.68	0	27.65	7.25	0	48.83	18.27	84.98	51.63	44.86	84.98	51.96	59.03	84.98

APPENDIX 52.3

Physical Attributes for the Hydrologic Response Units That Comprise Camp Creek Basin

HRU	Area (ha)	Soil Type	Slope (%)	Soil Depth (cm)	Aspect	Elevation (m)	Vegetation Type
1	162.7	Loam	43	72.9	SE	1112.5	Trees
2	103.4	Loam	23	92.2	SE	1188.7	Trees
3	46.6	Loam	38	109.7	S	1173.5	Trees
4	60	Loam	40	74.7	SW	1158.2	Trees
5	51.6	Loam	29	122.2	SW	1255.8	Trees
6	137.5	Loam	29	122.7	SW	1194.8	Trees
7	126.1	Loam	40	70.4	SW	1335	Trees
8	375	Loam	40	126.5	W	1219.2	Trees
9	62.2	Loam	33	118.1	NW	1341.1	Trees
10	60.8	Loam	43	69.6	NE	1204	Trees
11	47.6	Loam	26	69.6	NE	1341.1	Trees
12	288.9	Loam	25	124	NE	1432.6	Trees
13	102.7	Loam	43	105.9	NE	1188.7	Trees
14	152.8	Loam	27	114.6	NE	1341.1	Trees
15	250.6	Loam	15	97.8	W	1371.6	Trees
16	27.9	Loam	43	68.8	SW	1371.6	Trees
17	70	Loam	26	142.7	S	1493.5	Trees
18	229.8	Loam	20	119.6	W	1432.6	Trees
19	339.2	Loam	22	118.4	N	1432.6	Trees
20	54.2	Loam	43	177.8	SW	1371.6	Trees
21	63.2	Loam	34	126.5	N	1402.1	Trees
22	60.6	Loam	43	138.7	SW	1341.1	Trees
23	45	Loam	28	88.6	SW	1447.8	Trees
24	98.5	Loam	22	96.5	S	1487.4	Trees
25	327.4	Loam	28	77.7	S	1478.3	Trees
26	299.1	Loam	27	66	W	1554.5	Trees
27	67.5	Loam	33	66	NW	1511.8	Trees
28	141.6	Loam	25	66	S	1524	Trees
29	215.1	Loam	28	69.1	SE	1585	Trees
30	142.3	Loam	27	66	W	1630.7	Trees
31	225.4	Loam	30	68.1	S	1621.5	Trees
32	60.3	Loam	41	109.5	SW	1566.7	Trees
33	247.4	Loam	27	68.1	S	1694.7	Trees
34	27.5	Loam	44	127.3	N	1280.2	Trees
35	102.5	Loam	24	138.2	NE	1386.8	Trees
36	79.7	Loam	41	103.6	NE	1341.1	Trees
37	142.6	Loam	29	78.7	NE	1463	Trees
38	198.7	Loam	31	76.2	N	1499.6	Trees
39	72.7	Loam	39	80.8	N	1493.5	Trees
40	78	Loam	29	66	N	1597.2	Trees
41	95.3	Loam	43	109	NW	1603.2	Trees
42	279.8	Loam	27	67.6	NW	1691.6	Trees
43	305.3	Loam	26	75.4	S	1752.6	Trees
44	180.8	Loam	24	68.1	SW	1767.8	Trees
45	222	Loam	24	69.3	SW	1798.3	Trees
46	270.1	Loam	26	68.1	SW	1889.8	Trees
47	270.3	Loam	25	67.8	SW	1981.2	Trees
48	112.8	Loam	18	69.6	SW	2060.4	Trees
49	58.4	Loam	23	91.4	NW	2133.6	Trees
50	116.1	Loam	27	70.9	W	2164.1	Trees
51	291.2	Loam	28	72.9	N	1767.8	Trees
52	132.4	Loam	33	82.6	NE	1828.8	Trees
53	33.6	Loam	19	68.6	NE	1889.8	Trees
54	156.3	Loam	30	73.2	NE	1859.3	Trees

HRU	Area (ha)	Soil Type	Slope (%)	Soil Depth (cm)	Aspect	Elevation (m)	Vegetation Type
55	59.1	Loam	30	70.9	NE	1950.7	Trees
56	91.7	Loam	31	75.9	N	1950.7	Trees
57	179.3	Loam	30	78.7	NW	2072.6	Trees
58	127.0	Loam	33	91.9	N	2133.6	Trees
59	1226.6	Loam	19	102.6	S	1020.8	Trees
60	762.9	Loam	9	102.6	W	811.1	Trees
61	797.7	Loam	21	71.1	W	1077.5	Trees
62	578.3	Loam	15	66.0	SW	1058.9	Trees
SNEP Basin	8426.1						
Total	11791.6						

APPENDIX 52.4

Physical Attributes for the Hydrologic Response Units That Comprise Clear Creek Basin

HRU	Area (ha)	Soil Type	Slope (%)	Soil Depth (cm)	Aspect	Elevation (m)	Vegetation Type
1	36.5	Clay	35	124.2	SE	652.3	Shrubs
2	117.2	Loam	24	130.8	SE	816.9	Shrubs
3	21.8	Loam	33	77.2	SE	755.9	Trees
4	18.6	Loam	36	68.8	S	725.4	Trees
5	38.2	Loam	17	127.8	SE	841.2	Shrubs
6	16.7	Loam	27	74.2	SE	755.9	Shrubs
7	36	Loam	15	125.2	NE	737.6	Grass
8	133.4	Loam	16	121.9	NE	810.8	Grass
9	31	Loam	10	126.7	S	743.7	Grass
10	8.5	Loam	12	171.2	SE	755.9	Grass
11	81.7	Clay	29	120.1	SE	810.8	Shrubs
12	137	Loam	20	133.9	NW	819.9	Grass
13	24.2	Loam	17	127	S	792.5	Shrubs
14	90.3	Loam	18	75.2	S	853.4	Shrubs
15	156.6	Loam	17	126.7	SW	841.2	Trees
16	73.4	Loam	17	126.5	S	780.3	Grass
17	84.9	Loam	21	122.2	SE	853.4	Shrubs
18	19.9	Loam	29	110.5	E	883.9	Shrubs
19	121.6	Loam	30	70.6	SE	877.8	Shrubs
20	117.1	Loam	31	54.6	SE	893.1	Shrubs
21	12.6	Loam	41	137.2	W	609.6	Trees
22	54.3	Loam	23	149.9	W	682.8	Shrubs
23	15.6	Loam	30	76.2	N	762	Trees
24	80.3	Loam	14	128.8	NW	768.1	Grass
25	14.5	Loam	9	77.7	NW	755.9	Grass
26	37.9	Loam	24	122.4	N	804.7	Trees
27	122.6	Loam	22	124.2	W	841.2	Shrubs
28	23.5	Loam	40	100.6	W	826	Trees
29	117.8	Loam	26	93.7	N	841.2	Shrubs
30	31.8	Loam	40	71.6	W	865.6	Trees
31	184.1	Loam	21	120.9	W	969.3	Trees
32	94.6	Loam	21	106.9	W	944.9	Trees
33	40.6	Loam	33	130.3	NW	929.6	Trees
34	32.3	Loam	41	73.9	NW	914.4	Trees
35	25.7	Loam	40	66.8	S	914.4	Trees
36	41.4	Loam	25	90.7	NW	914.4	Shrubs
37	28	Loam	26	77.5	SE	975.4	Trees
38	42.9	Loam	25	103.1	W	960.1	Trees
39	23.2	Loam	23	49.5	SE	987.6	Shrubs
40	17.4	Loam	29	47.8	W	1036.3	Shrubs
41	20.2	Loam	24	82.6	SW	975.4	Trees
42	20.2	Loam	23	122.2	NW	999.7	Trees
43	11.5	Loam	16	42.7	S	1024.1	Grass
44	24.5	Loam	27	66.8	W	1018	Shrubs
45	39.6	Loam	23	45.5	SE	1036.3	Shrubs
46	38.1	Loam	19	116.8	S	1066.8	Trees
47	20.9	Loam	26	116.8	NW	1060.7	Trees
48	16.1	Loam	27	98.6	NW	1048.5	Trees
49	112.5	Loam	21	128.5	SE	1097.3	Trees
50	10.4	Loam	14	60.2	SW	1121.7	Shrubs
51	14.5	Loam	28	49.5	NW	1085.1	Trees
52	66.5	Loam	21	127.3	W	1121.7	Trees
53	52.3	Loam	21	119.6	W	1097.3	Trees
54	39.2	Loam	17	127	NW	1036.3	Trees
55	94.4	Loam	25	134.1	NW	1036.3	Trees
56	80.8	Loam	14	167.4	S	1170.4	Trees
Total	3067.6						

BRUCE J. MCGURK
Pacific Southwest Research Station
U.S. Forest Service
Albany, California

NEIL H. BERG
Pacific Southwest Research Station
U.S. Forest Service
Albany, California

MAUREEN L. DAVIS
Pacific Southwest Research Station
U.S. Forest Service
Albany, California

53

Camp and Clear Creeks, El Dorado County: Predicted Sediment Production from Forest Management and Residential Development

ABSTRACT

As part of the Sierra Nevada Ecosystem Project case study, we assessed the relative sediment production of fifty years of land use (1940–91) in the Camp Creek and Clear Creek Basins, two catchments in the Cosumnes River Basin. The Camp Creek Basin encompasses 8,245 ha (20,821 acres), ranges in elevation from 975 to 2,316 m (3,200 to 7,600 ft), and is managed by the Eldorado National Forest. The Clear Creek Basin encompasses 3,068 ha (7,580 acres), ranges in elevation from 512 to 1,250 m (1,680 to 4,100 ft), and has been extensively developed with low-density housing. We used the Water Yield and Sediment Model (WATSED) to predict the relative sediment production rates and volumes associated with forest management and residential development in these two basins. Although the best available data were used in this study, the predicted sediment rates should be used with caution.

Camp Creek: Predicted peak sediment production from roads was 4 mT/ha·yr (1.8 T/acre·yr). Two estimates of sediment production from logging were made. At the low rate (based on limited local data), peaks in sediment production for three periods of activity were 0.2 mT/ha·yr (0.09 T/acre·yr) or less. At the high rate (an order of magnitude greater than the low rate), peaks were less than 2 mT/ha·yr (0.9 T/acre·yr). Fires affected 256 ha (633 acres), and minimal quantities of sediment were predicted. At the low erosion rate for logging, relative sediment production for roads and logging was predicted at 98% and 1%, respectively; at the high erosion rate, 90% and 9%.

Clear Creek: Peak sediment production from roads was 2.6 mT/ ha·yr (1.2 T/acre·yr). To account for the effects of residential development, sediment yield predictions were made for the area attributed to unsurfaced driveways. Peak sediment production from driveways was 0.4 mT/ha·yr (0.2 T/acre·yr). A fire affected 85 ha (210 acres) in 1972, and a minimal quantity of sediment was predicted. Predicted sediment from roads was 83% of the total, and 16% of the total was from housing.

Predicted mean annual erosion rates due to both disturbance regimes range from 29 to 119 times the assumed natural rate of erosion during the sixty-two-year period. For residential development, the average ratio of mean annual erosion to the natural rate of erosion was 53:1, and for logging the mean ratio was 87:1. Mean total sediment production rates per hectare for Clear Creek were between 53% and 67% the magnitude of the Camp Creek Basin's rates over the period of analysis.

A second normalization procedure was devised to measure sediment from housing and logging by the area of road built to support the land use, rather than by hectare of land involved in the land use. The residential rate was 11 mT/yr per hectare of road (5 T/yr per acre of road), more than twice the rate of 5 mT/yr per hectare of road (2 T/yr per acre of road) for logging. Logging and residential development have different relative scales of impact depending on whether sediment production is normalized by road area or by land area.

Sierra Nevada Ecosystem Project: Final report to Congress, vol. II, *Assessments and scientific basis for management options*. Davis: University of California, Centers for Water and Wildland Resources, 1996.

INTRODUCTION

Accelerated (management-induced) erosion and sediment production are often the greatest effects of logging, road building, and residential development in forested areas (Hewlett 1982; Anderson et al. 1976). These potential threats to land and water resources can be exacerbated in regions of the Sierra Nevada where erodible, decomposing granitic rock may comprise the bulk of the soil parent material. The few available data from the Sierra Nevada suggest that sediment yield may increase significantly after logging (McCammon 1977). The potential for accelerated erosion and sedimentation is consistently identified by the public in reviews of environmental assessments and appeals of proposed timber sales. A recent overview of hydrology and water resources in the Sierra Nevada, prepared for the Sierra Nevada Ecosystem Project (SNEP), identified sediment yield as one of two preeminent hydrologic issues needing resolution for the Sierra Nevada (Kattelmann 1996). Sediment prediction systems from elsewhere, such as the Idaho Batholith (Megahan and Kidd 1972) and northern California (Lewis and Rice 1989), need to be modified prior to use in the Sierra Nevada because of differences between these areas in physiography, geomorphologic processes, and soils and in the distribution, timing, and type of precipitation.

As part of an analysis of conditions in the Sierra Nevada, members of the SNEP selected two small basins, Camp Creek and Clear Creek near Placerville on the west slope of the central Sierra Nevada, for detailed hydrologic and other analyses. The basins were selected because they are typical of much of the Sierra in their history and land-use patterns. This chapter is a companion to a case study of land-use changes and their hydrologic effects in these basins (McGurk and Davis 1996). The Camp and Clear Creek catchments contain examples of two common land uses in the Sierra Nevada, logging (Camp Creek) and residential development (Clear Creek). The goal of this analysis was to estimate with a simulation model accelerated sediment production from fifty years of land-use change in these basins and to compare the levels.

Few data exist on natural or accelerated sediment production in the Sierra Nevada. Some information on sediment accumulation in large, low-elevation reservoirs is available (Kattelmann 1996), but these data are not directly applicable to higher-elevation, managed landscapes. Reservoir sedimentation may also combine natural and induced hillslope erosion and in-channel sediment production. Soil erosion studies often do not address movement of the eroded soil downslope; the actual delivery of sediment between point of detachment and entry (if at all) into a stream channel has generally not been addressed in the Sierra Nevada.

Management activities that bare the soil surface or accelerate mass movement can lead to increased erosion. Depending on the distance of the hillslope activities from the channel and on barriers between the activity and the stream, eroded sediment may be delivered to the channel. Models attempt to simulate these processes either with "lumped" coefficients that aggregate several processes together or with more physically based models that simulate the mechanics of each geomorphologic process. Physically based models require extensive physical data and typically operate over short time intervals, such as an individual rainstorm.

With high-resolution digital elevation data and detailed soils data, numerical models such as TOPMODEL predict surface and channel flow that lead to particle detachment (Moore et al. 1988). Neither the high-resolution digital elevation nor the soils data were available for the Camp and Clear Creek Basins, so a distributed-parameter, numerical model could not be used. The Water Erosion Prediction Project model (WEPP) was also rejected because of input data requirements (e.g., thirty-three coefficients for vegetation, fifteen values per soil horizon, etc.) and the lack of a completed module for forest land (Agricultural Research Service 1994).

Because of the paucity of local data, the need to model long time periods and large basins, and the SNEP stipulation to use available technologies and information, the Water Yield and Sediment Model (WATSED) was selected. WATSED, an empirically based accounting model that is described in a later section, is currently used on several national forests in Montana (U.S. Forest Service 1991). Our analysis estimated the amount of sediment reaching channels but did not estimate potential in-channel sources of sediment or routing within the channel.

STUDY SITE DESCRIPTION AND HISTORICAL LAND USE

The Cosumnes River Basin, located on the west slope of the central Sierra Nevada (figure 53.1), has been modified by settlers for at least 150 years by logging, fire, mining, grazing, and diversion of water for agricultural use. Using aerial photographs taken between 1940 and 1991, McGurk and Davis (1996) quantified the land-use changes for the Camp and Clear Creek Basins, located in the middle- to upper-elevation range of the Cosumnes River watershed.

Study Site

The Camp Creek Basin is about 20 km (12 mi) east of Placerville. The basin has a west aspect and is a tributary to the North Fork of the Cosumnes River. The upper 8,246 ha (20,821 acres) of the Camp Creek Basin were selected for analysis by the SNEP Hydrology Team. The lower portion of the basin is in steep terrain and has not been extensively logged or roaded and was therefore excluded from the analysis. The analysis area in the middle and upper portions of the Camp Creek Basin ranges in elevation from 975 to 2,316 m (3,200 to

1409

Camp and Clear Creeks, El Dorado County: Predicted Sediment Production from Forest Management and Residential Development

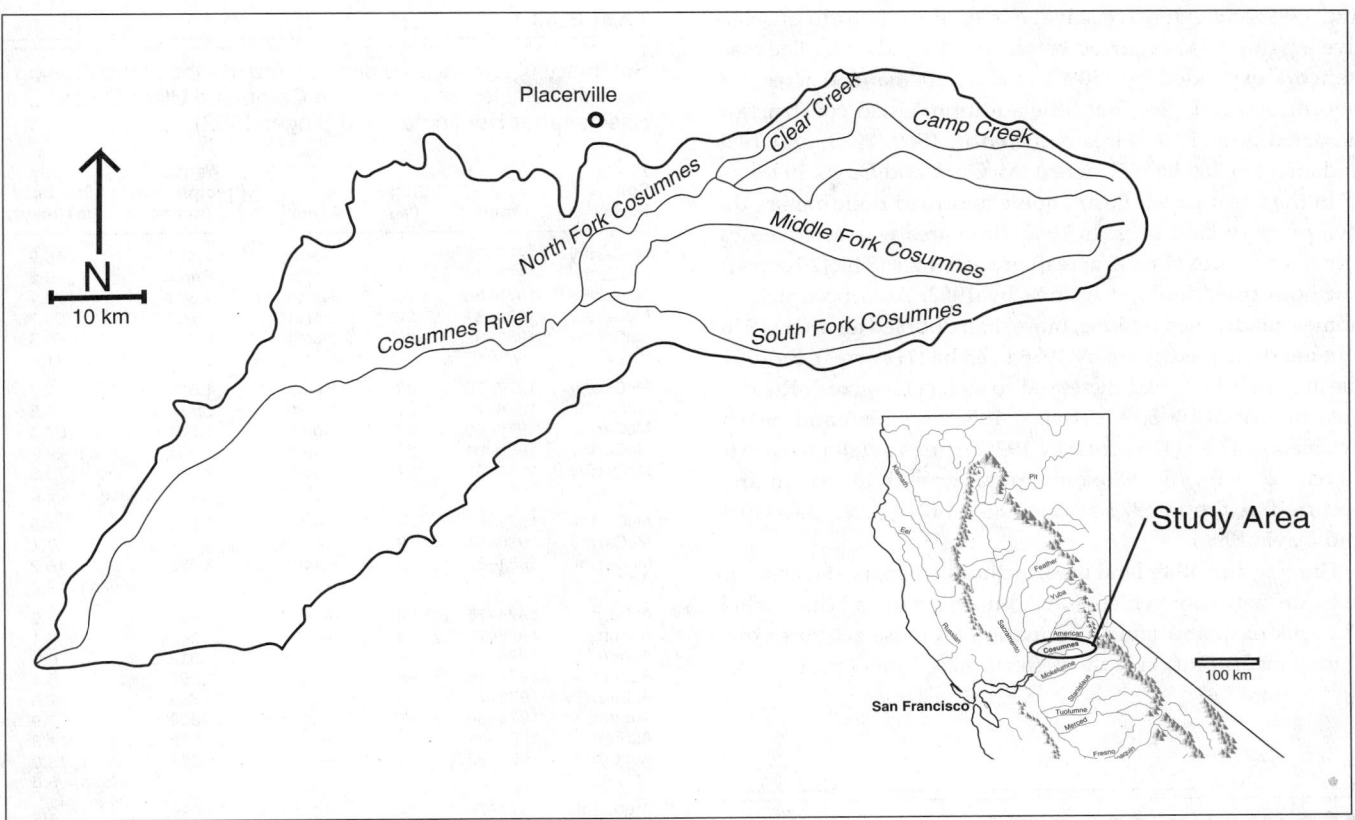

FIGURE 53.1

Location of study area, Placerville, and Clear and Camp Creeks in the Cosumnes River Basin, California.

7,600 ft) and is managed by the Eldorado National Forest (ENF). The main Camp Creek channel ranges from a moderately deep canyon (160 m [400 ft]) to a shallow canyon. The upper reaches are the shallowest, with many forks ending in open meadows.

The Clear Creek Basin is located 14 km (9 mi) east of Placerville. This basin encompasses 3,068 ha (7,580 acres) and is directly west of the Camp Creek Basin. The Clear Creek Basin is predominantly private land and ranges from 512 to 1,250 m (1,680 to 4,100 ft) in elevation; it has a southwest aspect (McGurk and Davis 1996).

Both basins are composed primarily of loam soils that have high infiltration rates (e.g., McCarthy, Josephine, and Cohasset series), although the Clear Creek Basin has some clay soils (Mitchell and Silverman n.d.; Rogers 1974). Soils in the Camp Creek Basin range from 66 to 178 cm (26 to 70 in) in depth and average 88 cm (35 in). The Clear Creek Basin's soils range from 45 to 131 cm (18 to 52 in) in depth and average 103 cm (41 in). Because of the depth and texture of the soils, surface runoff is likely to occur only on compacted zones such as roads.

Geologic parent materials in both basins are dominated by volcanic lahars, although approximately 20% of the soils were

formed in materials weathered from granitic rock. Bedrock channel sections are more evident in the middle portion of Camp Creek than the upper portion, where alluvial (volcanic and granitic) components are common (Fiore 1992). About 55% of the soil in the Camp Creek Basin has a high to very-high erosion hazard, and 75% of the soil in the Clear Creek Basin has a high to very-high erosion hazard. Slopes range from 9% to 41% and average 24% in the Clear Creek Basin and range from 15% to 43% and average 31% in the Camp Creek Basin (McGurk and Davis 1996). For Camp Creek, the steepest slopes are part of an inner gorge along the main stem or the mid-basin tributaries (Swanson et al. 1993). Grass, shrubs, and forest areas are found in the Clear Creek Basin, but the Camp Creek Basin is almost entirely mixed conifer forest.

Historical Land Use

In the Camp Creek Basin, logging and road building are the two major land uses. Selective logging (a technique in which small clumps or individual trees are removed from large areas) was performed almost exclusively until the mid-1980s. Over 527 ha (1,303 acres, or 6.4%) were clear-cut between 1984 and 1989, and extensive salvage operations (a form of selective logging) also occurred between 1988 and 1991. The road network expanded by 250% to nearly 100 ha (250 acres) between 1966 and 1986, but little additional road construction occurred after 1986. Fires occurred in 1969, 1973, and 1988 and affected 266 ha (657 acres) (McGurk and Davis 1996).

In the Clear Creek Basin, home and road building are the two primary land uses. In 1940, little area was occupied by either structures (1 ha [2 acres]) or roadways (8 ha [20 acres]), but both uses doubled in area by 1952. Area occupied by homes quadrupled by 1966, more than doubled again by 1976, and nearly tripled again by 1986 to 45 ha (111 acres). By 1991, the area in homes had increased to 59 ha (146 acres). Roaded area increased by 50% between 1952 and 1966 and nearly doubled to 47 ha (116 acres) by 1976. After a small increase to 50 ha (124 acres) in 1986, only minor growth in roaded area occurred. A fire in 1972 affected 85 ha (210 acres) (McGurk and Davis 1996).

Grazing and other land uses, including dispersed recreation (e.g., off-highway vehicle [OHV] use), occur in both basins. The lack of quantifiable information on these activities precluded incorporation of their effects on sediment production.

SEDIMENT STUDIES

Two primary concerns associated with erosion are the potential loss of soil productivity and the degradation of stream water quality (Poff 1996). Undisturbed upland forests, where mineral soil is fully carpeted by litter and humus, have low natural erosion rates and deliver little or no sediment to streams (Anderson et al. 1976). In this undisturbed situation, sediment in the channel often originates from within the stream zone, often as a result of soil creep, mass failure, or debris torrents. Alternatively, ground-disturbing activities such as logging and road building may become the major sediment sources.

No erosion, sediment transport, or sediment deposition data were available from plots within these two basins. Soil maps and erosion hazard potential maps exist for both basins and are discussed later in the chapter. In addition, results from research or administrative studies in nearby areas were available.

Erosion Plot Studies

Runoff and erosion data from several soils common in the Camp and Clear Creek Basins are available in unpublished studies. From 1976 through 1982, sheet and rill erosion were monitored under natural precipitation conditions at standard erosion plots for soil series that together make up 40% of the soil in the Camp Creek Basin (table 53.1—sediment yield is decribed in metric tons per hectare per year [mT/ha·yr])

TABLE 53.1

Summary of annual precipitation and soil loss from erosion plots on selected soils from the Camp and Clear Creek Basins (after Huntington and Singer 1982).

Soil Series	Year	Slope (%)	Aspect	Annual Precipitation (mm)	Soil Loss (mT/ha·yr)
McCarthy	1977–78	20	North	1,588	46.9
McCarthy	1978–79	20	North	Snow	0.2
McCarthy	1979–80	20	North	1,801	72.1
McCarthy	1980–81	20	North	607	35.7
McCarthy	1981–82	20	North	2,073	50.3
				Mean = 41.0	
McCarthy	1977–78	27	South	1,588	77.7
McCarthy	1978–79	27	South	Snow	4.5
McCarthy	1979–80	27	South	1,801	107.4
McCarthy	1980–81	27	South	607	29.9
McCarthy	1981–82	27	South	2,073	75.3
				Mean = 59.0	
McCarthy	1979–80	9	East	1,801	34.6
McCarthy	1980–81	9	East	607	7.4
McCarthy	1981–82	9	East	2,073	16.2
				Mean = 19.4	
Auburn	1974–75	8	—	625	4.0
Auburn	1975–76	8	—	389	0.1
Auburn	1976–77	8	—	312	0.1
Auburn	1977–78	8	—	1,090	5.4
Auburn	1978–79	8	—	754	7.6
Auburn	1979–80	8	—	800	7.9
Auburn	1980–81	8	—	378	1.3
Auburn	1981–82	8	—	1,331	18.0
				Mean = 5.6	
Argonaut	1977–78	14	—	996	2.9
Argonaut	1978–79	14	—	726	10.3
Argonaut	1979–80	14	—	772	26.5
Argonaut	1980–81	14	—	386	5.8
Argonaut	1981–82	14	—	1,265	90.3
				Mean = 27.2	
				Grand mean = 30.4	

(Huntington and Singer 1982). In addition, measurements of runoff and erosion during simulated rainfall were made for soil series comprising 60% of the soil in the Camp Creek Basin (tables 53.2 and 53.3) (Huntington et al. 1979; Baker et al. 1994).

Of these three data sets, the long-term information from the standard erosion plots most closely represents the type of data required by WATSED as first-year values after timber harvest (table 53.1). These values were derived under conditions of natural precipitation, with each plot weeded and raked every two weeks between precipitation events to retain a bare-ground condition. Because of the weeding and raking, the erosion rates from this study certainly represent upper limits for erosion rates from logging and site preparation. Annual soil loss rates are quite variable within each soil series. Although some of the variability may be attributable to variation in precipitation and slope, the range in soil loss rates is typically large, even within a single slope class (table 53.1). Based on the relative ranking of the loss rates in table 53.2, we calculated the relative ratios of soil loss for Holland, Josephine, Aiken, and Windy soils compared with the soil loss

1411

Camp and Clear Creeks, El Dorado County: Predicted Sediment Production from Forest Management and Residential Development

for the McCarthy soil. Nearly 55% of the basins are composed of McCarthy and Josephine soils, which have similar erosional characteristics.

Simulated rainfall intensities in the Baker et al. (1994) study were potentially greater than naturally possible (table 53.3). In their study, simulated rainfall was initially applied to dry ground for one hour at a rate of 5.3 cm/hr (2.1 in/hr). The 50-yr 15-min natural rainfall intensity for the study sites equated to approximately 5.3 cm/hr. The duration of the simulated rainfall was therefore four times the duration of the 50-yr rainfall for that rainfall intensity. Long-term soil losses as great as those listed in table 53.3 are impossible to sustain. However, the data from the studies using simulated rainfall were considered to be upper bounds for erosion in this study.

Road and Trail Erosion

Between 1990 and 1993, the ENF monitored erosion downslope of OHV trails and roads at a site in the American River Basin, immediately north of the Cosumnes River Basin (U.S. Forest Service 1995). Silt fences trapped sediment transported off the road or trail surface at three adjacent sites that covered a range of hillslope angles. The roads and trails were designed to conform to current standards for road and trail design. Soil parent material at these sites is metasedimentary rock, a rock type that is not as common and is somewhat more erodible than most of the soils in the Camp or Clear Creek Basins. Sediment yield rates shown in table 53.4 are based on the assumption that the sediment was produced under natural precipitation regimes by concentrated surface runoff. Except for the high-slope category, these soil loss rates are similar to those measured by Huntington and Singer (1982) (table 53.1). Disturbance conditions between the two studies differed in that compaction of the soil surface was incorporated into the OHV study only. The Huntington and Singer (1982) site more nearly matches typical disturbances in logging operations, and the OHV results specifically address road and trail erosion. Differences in parent material between the OHV study site and the Camp and Clear Creek Basins add uncertainty to the applicability of the OHV study results.

TABLE 53.2

Soil loss rate under simulated rainfall and relative erodibility of five forest soils found in Camp and Clear Creek Basins (after Huntington et al. 1979).

Soil Series	Soil in Basins (%)	Soil Loss Rate (mT/ha)	Relative Erodibility
Windy (Typic Xerumbrept)	2.6	3.5	0.63
Aiken (Xeric Haplohumult)	1.6	4.8	0.87
Josephine (Typic Haploxerult)	17.5	5.4	0.97
McCarthy (Umbric Vitrandept)	36.7	5.5	1.00
Holland (Ultic Haploxeralf)	0.7	7.0	1.25

TABLE 53.3

Extrapolated annual soil loss from sixty minutes of simulated rainfall on soils found in Camp and Clear Creek Basins (after Baker et al. 1994).

Soil Series	Soil Loss (mT/ha·yr)	Slope (%)	Aspect	Condition
Aiken/Sites 1a	289.8	24	South	Logged, cleared
Aiken/Sites 1b	68.2	25	South	Logged, cleared
Aiken/Sites 2a	933.0	21	North	Logged, cleared
Aiken/Sites 3a	51.1	22	South	Logged, cleared
Aiken/Sites 3b	717.4	21	South	Logged, cleared
Mean = 411.9				
McCarthy 1a	315.5	21	West	Logged, ripped
McCarthy 1b	537.2	26	West	Logged, ripped
McCarthy 2a	66.7	24	West	Logged, ripped
McCarthy 2b	865.6	27	West	Logged, ripped
McCarthy 3a	437.8	15	South	Logged, ripped
McCarthy 3b	478.3	14	South	Logged, ripped
Mean = 450.0				

Watershed Improvement Needs (WIN)

As used on the ENF, the WIN inventory identifies disturbance types (e.g., sheet, rill, channel, or gully erosion, mass wasting, soil compaction) and disturbance sources. Besides induced sources of disturbance (e.g., animal grazing, off-highway vehicles, silvicultural methods, roads, hydrologic diversions), natural mass wasting is listed as a potential disturbance source. Analysis of over forty WIN reports compiled in 1991 for the Camp Creek Basin identified four locations where mass wasting was found. In three of the four cases, roads were associated with the disturbance. WIN reports from 1993 for the upper Camp Creek Basin identified road-related mass wasting as a disturbance cause in two of nineteen reports. In both surveys, roads and silvicultural activities were the most common causes of sediment entering the channels.

Pfankuch Channel Ratings

Pfankuch stream channel ratings (Pfankuch 1978) were assessed during stream surveys conducted on Camp Creek by the ENF (U.S. Forest Service 1993). Pfankuch's channel ratings include a category on the amount of mass wasting on each channel reach and are a second source of information on mass wasting in the Camp Creek Basin. As used on the ENF, existing or potential mass wasting is classed as

excellent: no evidence of past or potential for future mass wasting into channels

good: infrequent and/or very small, mostly healed over, low future potential

fair: moderate frequency and size, with some raw spots eroded during high flows

poor: frequent or large, causing sediment nearly yearlong or imminent danger of same

TABLE 56.4

Sediment yields from off-highway-vehicle roads on the Eldorado National Forest (after U.S. Forest Service 1995).

Slope (%)	Sediment Yield (mT/ha·yr)
0–15	18
16–22	45
23–30	89
>30	135

Of the 200 sites that were assessed, approximately 75% of the mass wasting ratings in the Camp Creek Basin study area were in the "excellent" category, and fewer than 2% were in the "fair" or "poor" categories (U.S. Forest Service 1993). Taken together, the WIN and Pfankuch data suggest that although localized mass movement can occur naturally, especially on steeply sloping inner gorge landtypes, mass wasting is otherwise a minor natural source of sediment. For this reason, mass wasting estimates were not included in the modeling.

MODEL DESCRIPTION AND IMPLEMENTATION

Because of data availability, basin size, and the long study interval, we decided to model sediment production in the Camp and Clear Creek Basins with the WATSED accounting model (figure 53.2). WATSED is used by the Lolo and Kootenai National Forests and others in the Northern Region (U.S. Forest Service 1991). WATSED has comprehensive documentation and experienced users available for assistance and is designed to address the cumulative effects of multiple management activities over time in forested landscapes. WATSED models both water and sediment yield (figure 53.2), but the WATSED water component was not used in this analysis. A more detailed analysis was performed (McGurk and Davis 1996) using the U.S. Geological Survey's Modular Modeling System (MMS) (Leavesley et al. 1992).

Eleven independent data files (or databases) are required by WATSED (figure 53.2). Eight coefficient files describe physical characteristics of the basin, and the manual specifies that these files be calibrated to fit the conditions of the area using local data. The remaining three data files are supplied by the users and incorporate areal information and management history data for the analysis watershed(s).

WATSED requires information on natural and accelerated erosion rates. Unfortunately, field data for the central Sierra Nevada are not common, so we adapted the few locally available data sources to the study area. We used the erosion information and physical data from Camp and Clear Creek Basins to complete WATSED's databases as follows:

Landtypes

Each basin is segmented into landtypes delineated on the basis of morphology, parent material, soils, and vegetation type. Physical information for each landtype is incorporated into the Land Systems Inventory (LSI), and disturbances (entered into the Activities database) are located by landtype.

In a companion hydrologic analysis of these basins, McGurk and Davis (1996) used the MMS to examine hydrologic changes due to changing land use. This model uses many of the same landscape parameters and divides the basin into polygons called Hydrologic Response Units (HRU). We decided to equate HRUs with landtypes. There are fifty-six landtypes designated for the Clear Creek Basin and fifty-eight landtypes identified for the Camp Creek Basin. Landtype areas range from 8 to 184 ha (21 to 455 acres) and average 55 ha (135 acres) in the Clear Creek Basin and range from 28 to 375 ha (68 to 926 acres) and average 145 ha (359 acres) in the Camp Creek Basin.

Land Systems Inventory (LSI)

The LSI is the primary data source for the model. It contains information on natural sediment yield, surface and mass erosion hazards, average slope, soil depth, and sediment delivery ratios for each landtype. In addition, the LSI links the landtypes with other databases (e.g., Surface Erosion Curves).

Natural Rate of Erosion

The term *natural erosion* refers to the erosion that occurs without anthropogenic disturbances such as roads. Natural rates of erosion vary widely, dependent on geologic parent material, slope, climate, soil, and vegetative cover. Heede (1984) cites 0–0.013 mT/ha·yr (0–0.006 T/acre·yr) as the range of natural erosion for small drainages in Arizona with climate and slope characteristics similar to those of our study basins. Euphrat (1992) cites a rate of 0.07 mT/ha·yr (0.03 T/acre·yr) for the Mokelumne River Basin. Dunne and Leopold (1978) report a rate of 0.004 mT/ha·yr (0.002 T/acre·yr) for an undisturbed forest in North Carolina. Kattelmann (1996) referenced a 1972 California Division of Forestry report that lists a Sierran value of 0.014 mT/ha·yr (0.006 T/acre·yr). Based on these literature values and the types of soils in the Camp and Clear Creek Basins, a mean rate of 0.02 mT/ha·yr (0.009 T/acre·yr) was used in this study for comparison of natural versus accelerated sediment production.

Surface Erosion Hazards

The ENF has incorporated soil and slope data, as well as other factors, into an Erosion Hazard Rating (EHR) system. Percentages of the different categories (very high, high, moderate, and low hazard) by landtype were obtained from the ENF geographic information system, and landtypes were assigned a category based on the dominant percentage (figure 53.3). WATSED adjusts the basic surface erosion curves to reflect slope and natural erodibility. In this case, landtypes with low

1413

Camp and Clear Creeks, El Dorado County: Predicted Sediment Production from Forest Management and Residential Development

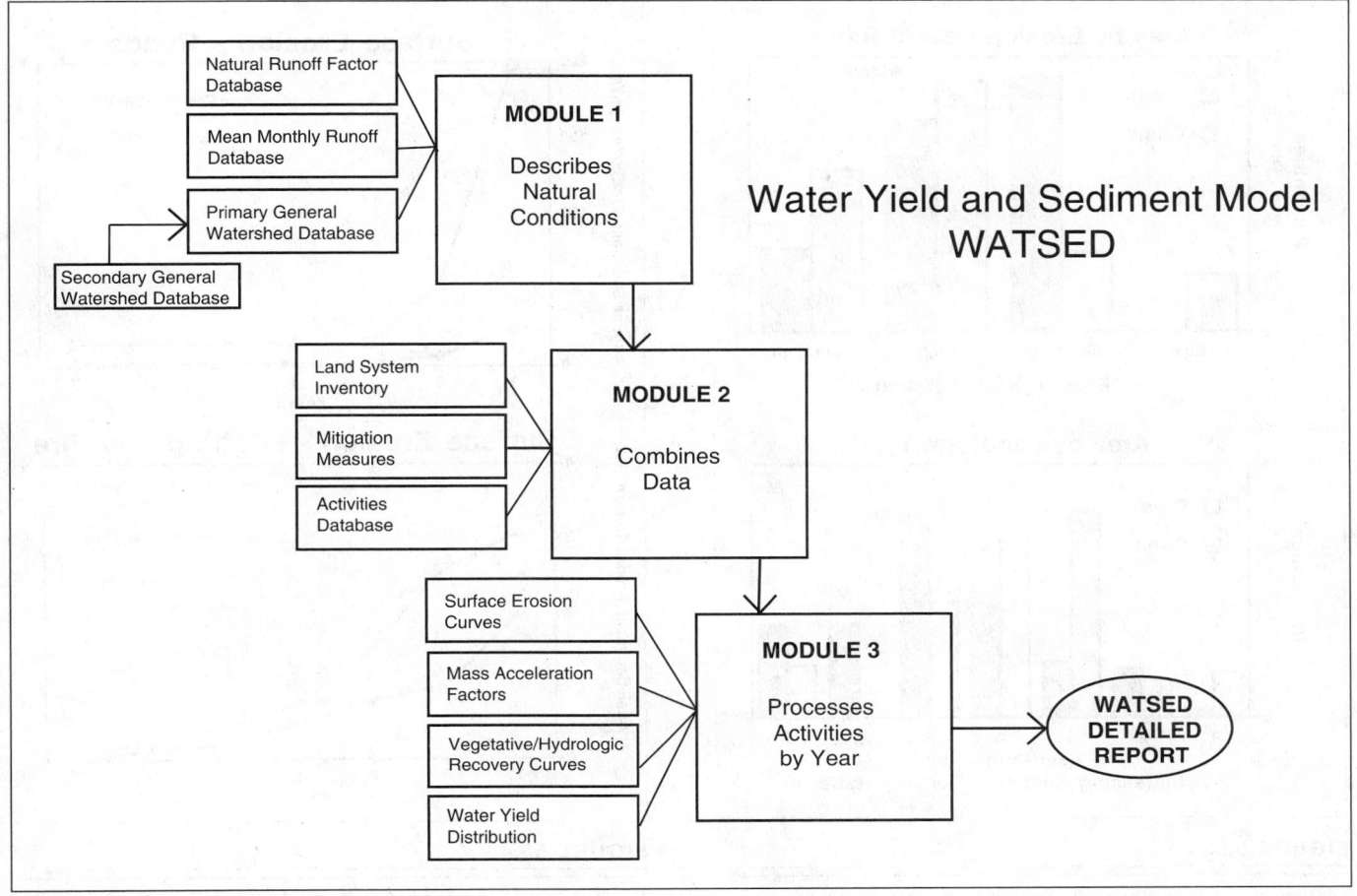

FIGURE 53.2

Schematic of Water Yield and Sediment Model used to predict sediment yields from Camp and Clear Creek Basins (after U.S. Forest Service 1991).

erosion hazard ratings were assigned a correction factor of 25%. This means that only 25% of the potential sediment from a disturbance in a landtype with a low erosion hazard would be generated. Moderate erosion hazard landtypes were assigned a correction factor of 50%, high-hazard landtypes 75%, and very high hazard landtypes 100%. These adjustment values were used for all disturbance types.

Landform Type and Sediment Delivery Ratios

WATSED incorporates hillslope position and presence or absence of ephemeral and first-order drainages by establishing landform types and assigning sediment delivery ratios to each. Using the landform types described in the WATSED manual as a guide, seven landform types were defined for the Camp and Clear Creek Basins (figure 53.3 and table 53.5). Delivery ratios incorporated advice from soils experts at the ENF and at the California Department of Forestry and Fire Protection. Based on analysis of topographic maps, a landform was assigned to each landtype, and the appropriate sediment delivery ratio for each landtype was entered into the LSI.

Primary and Secondary General Watershed Databases

Information for the entire analysis watershed(s) is entered into the Primary database. Water Resources Council code, name, total acres, natural sediment yield, and precipitation and runoff values for Camp and Clear Creek Basins were entered into their respective databases. Area and precipitation for each landtype were entered into the Secondary database.

Surface Erosion Curves

WATSED estimates sediment from each activity over time by applying the area of the disturbance to the appropriate erosion curve. Curves from the WATSED manual (U.S. Forest Service 1991) for roads, logging, and fire were modified to reflect local conditions.

The road curve (figure 53.4), based on observations in Idaho of roads on granitic parent material, predicts that the peak amount of sediment occurs during the year following road construction. Sediment is reduced to 27% of the original value

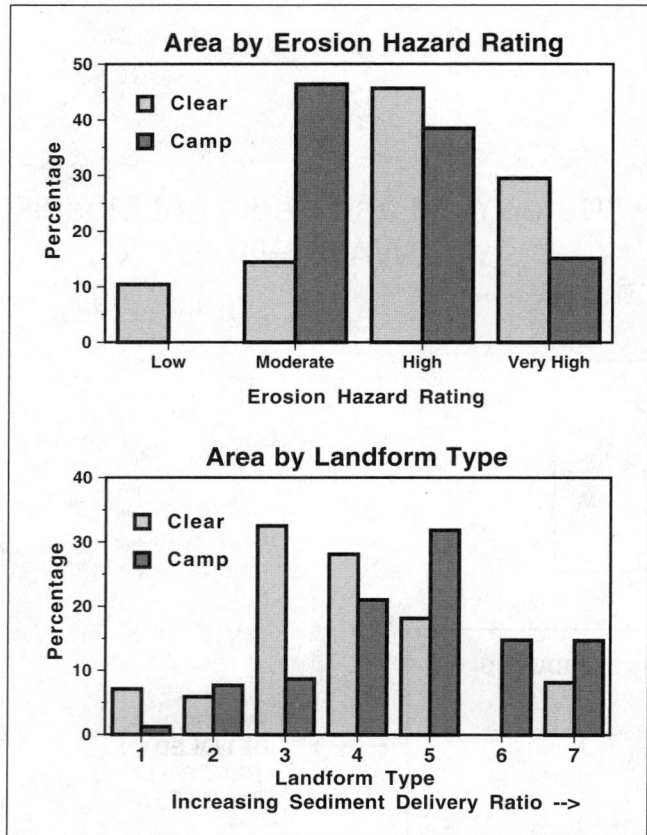

FIGURE 53.3

Distribution of area by soil erosion hazard rating and landform type in the Clear and Camp Creek Basins.

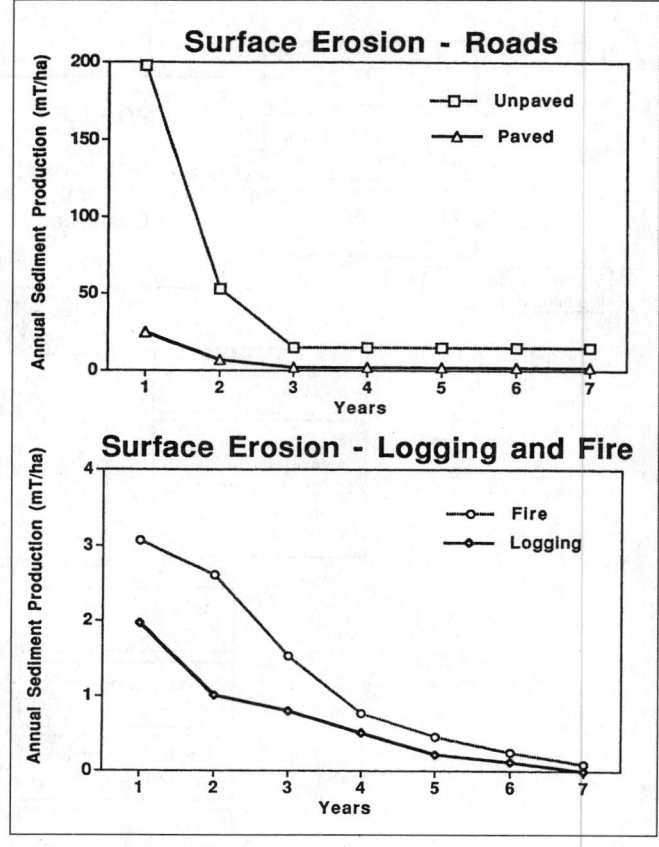

FIGURE 53.4

Surface erosion curves for roads, logging, and fire that were used for sediment predictions in Clear and Camp Creek Basins (modified from U.S. Forest Service 1991).

in the second year and falls to a base level of 7% in the third year and thereafter. For forest roads in Montana, WATSED uses a first-year sediment production value of 237 mT/ha·yr (105 T/acre·yr), but the manual recommends that this value be adjusted based on local data. Erosion data by slope class from the OHV study on the ENF were plotted, and a curve was fit to the data. The maximum average landtype slope was

TABLE 53.5

Definitions of landform types and the sediment delivery ratio values that were assigned to each landtype (after U.S. Forest Service 1991).

Landform Type (#)	Sediment Delivery Ratio (%)
Ridgetop (1)	2
Undissected slopes not adjoining a channel (2)	7
Dissected slopes not adjoining a channel (3)	10
Undissected slopes adjoining a channel (4)	15
Dissected slopes adjoining a channel (5)	20
Very dissected slopes adjoining a channel (6)	25
Inner gorge (7)	75

43%, and the erosion value for this slope was 198 mT/ha·yr (88 T/acre·yr). This value is appropriately higher than the erosion values on more gentle slopes (table 53.1) and lower than the values associated with intense simulated rainfall (tables 53.2 and 53.3). The 198 mT/ha·yr value became the initial value in the surface erosion curve for unpaved roads (figure 53.4), and the subsequent years were estimated based on the percentage changes in the WATSED road curve.

Unlike the watersheds on the national forests where WATSED was developed, the Camp and Clear Creek Basins include paved roads. We developed a new road curve to predict sediment yield generated from the paved roads. Paved roads in these basins are either 6.7 or 7.3 m (22 or 24 ft) wide, according to ENF records. We assumed the paved roads also include an additional 2.4 m (8 ft) of unpaved shoulder and ditch, for a total width of 9.1 or 9.7 m (30 or 32 ft). Although approximately one-quarter of the total surface is unpaved, we selected a paved to unpaved ratio of 1:8 for sediment yield because shoulders and ditches associated with paved roads are not disturbed as often as those associated with unpaved roads. Based on this ratio, 25 mT/ha·yr (11 T/acre·yr) became

1415

Camp and Clear Creeks, El Dorado County: Predicted Sediment Production from Forest Management and Residential Development

the initial value in the surface erosion curve for paved roads, 7 mT/ha·yr (3 T/acre·yr) the second-year value, and 2 mT/ha·yr (0.9 T/acre·yr) the value for the third and subsequent years (figure 53.4).

The dominant disturbances in the Clear Creek Basin are roads and residential development. McGurk and Davis (1996) assumed that each residential unit was composed of 762 m² (2,500 ft²) of impervious roof area and an equal amount of driveway and paths. Though sediment may be produced during the construction phase, no information exists on sediment production from house construction for this basin. Most houses are built during the summer season, so erosion during construction may be negligible. Roofs do not generate sediment, and it was assumed that the homeowners maintain ground cover across the rest of their properties. Based on personal inspection, most driveways in Clear Creek subdivisions are unpaved. To account for sediment yield generated from residential development, driveways were represented as roads that are 4.9 m (16 ft) wide.

Surface erosion curves for fire and logging were also established. A survey of the literature yielded relative sediment production factors from logging and roads (table 53.6). The ratios range from 0.005 to 0.11, and a value of 0.01 was initially selected as most appropriate based on soils and climate. This factor was applied to the established erosion value for roads to yield a first-year erosion value for logging of 1.98 mT/ha·yr (0.88 T/acre·yr). This value is between ten and thirty times less than the bare soil loss values for McCarthy soil in table 53.1. To establish a potential upper bound on erosion from logging, a second WATSED prediction was made using a value of 19.8 mT/ha·yr (8.8 T/acre·yr), approximately equal to the lowest McCarthy soil loss values in table 53.1 and an order of magnitude greater than the ratio-based value. Values for subsequent years were then generated using the percentage change values specified in the WATSED manual.

The ratio of the fire value to the Idaho road value used by WATSED was 1.56%, and use of this factor yields a sediment production rate of 3.07 mT/ha·yr (1.37 T/acre·yr) for wildfire in both the Camp and Clear Creek Basins (figure 53.4). This rate was within the bounds established using Sierra data, and Anderson et al. (1976) reported a 45-fold increase over the natural rate of erosion after fire in northern California. Haig (1938) reported a twofold to 239-fold increase over the natural rate after fire in the pine region of the Sierra Nevada. Wells et al. (1979) found that a pine area in Arizona produced 30 mT/ha·yr (13 T/acre·yr) for the first several years after an intense fire. Rates for subsequent years were generated using the percentage values specified in the WATSED manual (figure 53.4).

Mitigation

WATSED accounts for mitigation measures aimed at reducing the amount of eroded material delivered to the stream channel. Mitigation is treated as a linear reduction of sedi-

TABLE 53.6

Relative sediment production factors from tractor logging and road building.

Source	Location	Sediment Production (mT/ha·yr)		Ratio of Logging to Roads
		Logging	Roads	
Megahan 1975	Idaho	0.2	24	0.008
Fredriksen 1970	Oregon	1.1[a]	28	0.04
U.S. Forest Service 1991	Montana	1.2	237	0.005
Binkley and Brown 1993	Georgia	0.4	3.6	0.11

[a]Skyline yarding.

ment from surface or mass erosion for each management action. The WATSED manual provides standard mitigation values for several management activities. We applied our own mitigation measures to logging activities to reflect improved management practices over time. Mitigation values were as follows: for logging in 1966, 1%; logging in 1976, 10%; logging between 1984 and 1986, 25%; and logging between 1987 and 1991, 40%. For example, for logging in 1976 there was a 10% reduction in the amount of sediment estimated by the logging surface erosion curve.

Activities Database

Disturbances in the basins are represented by the Activities database. Land-use information (disturbances) developed by McGurk and Davis (1996) for use with the MMS hydrologic model (Leavesley et al. 1992) was modified for use in WATSED.

For MMS, disturbances were given a "photo date," that is, the date (year) of the aerial photograph in which they first appeared. Thus, disturbances were clustered on a small number of years, separated roughly by a decade. Our goal with WATSED was to predict annual sediment yield, so we needed disturbance information that was not clustered.

The actual dates for fires and logging were known through ENF records. We had no information on the actual year of construction for roads or driveways in the basins, so dates were assigned to roads and driveways based on their photo date. These disturbances were distributed, roughly equally, every other year, back into the photo interval in which they first appeared. For example, roads with the photo date of 1976 were distributed equally across the years 1968, 1970, 1972, 1974, and 1976.

Each disturbance (e.g., road segment, clear-cut, fire) was described appropriately by date, activity type code, location by landtype, area, logging method, crown removal percentage, road length and width, and mitigation method applied. Although WATSED can account for the varying effects of high- and low-intensity fires and tractor, cable, and aerial logging, all fires were assumed to be high-intensity and all logging was assumed to be tractor logging. ENF records show that very little cable logging has been done in Camp Creek.

SEDIMENT PREDICTION RESULTS AND DISCUSSION

A sixty-two-year period, 1940 to 2002, was analyzed with the model. The year 2002 was selected to examine the sediment produced during the ten-year period following the last activities in 1991. The model's detailed output was summarized to provide a year-by-year sediment output from roads, houses, logging, and fire (figures 53.5 and 53.6). The sawtooth pattern that is evident in both figures is because road building was spread out at two-year intervals during the photo period. Mean production rates per decade are also presented to further smooth the sediment production rates (figure 53.7).

Annual Sediment Outputs

Roads

Construction of roads produced the largest amount of sediment in both the Camp and Clear Creek Basins (figures 53.5 and 53.6). Camp Creek has over 100 ha (250 acres) of road surface, and Clear Creek has about 50 ha (124 acres) of road surface. Camp Creek (8,246 ha [20,821 acres]) is almost three times the size of Clear Creek (3,068 ha [7,580 acres]), but the Clear Creek Basin has a higher road density (1.7%) than the Camp Creek Basin (1.2%). The Camp Creek Basin has 55 km

(34 mi) of paved road because of the presence of Iron Mountain Road, about 4.3 times the paved roads in the Clear Creek Basin. Camp Creek has 256 km (159 mi) of roads overall compared with Clear Creek's 114 km (71 mi) of roads.

Sediment from roads was generated during the entire sixty-two-year history of the basins, and the peaks vary widely in their magnitude. Camp Creek sediment peaks range from 1 to 4 mT/ha·yr (0.4 to 1.8 T/acre·yr). The decade mean rate for Camp Creek climbs over the analysis interval to nearly 2.5 mT/ha·yr (1.1 T/acre·yr). Clear Creek sediment peaks range from 0.6 to 2.6 mT/ha·yr (0.3 to 1.2 T/acre·yr). The decade mean rate for Clear Creek reaches 1.5 mT/ha·yr (0.7 T/acre·yr) in the 1971–80 period and declines to 1.3 mT/ha·yr (0.6 T/acre·yr) by the end of the analysis period. Over the sixty-two-year period, roads in Camp Creek produced 883,665 mT (803,340 T) of sediment, and the Clear Creek roads produced 168,537 mT (153,217 T) of sediment. The Camp Creek sediment volume is over 5.2 times the Clear Creek volume because the annual rate is higher and the watershed is 2.7 times larger.

Housing

As a proxy for houses in the Clear Creek Basin, sediment was also produced from 29 ha (72 acres) of driveway. Rates exceeded 0.1 mT/ha·yr (0.04 T/acre·yr) after 1959, and peaks sometimes exceeded 0.4 mT/ha·yr (0.2 T/acre·yr) between 1979 and 1991. Over the sixty-two-year analysis period, 33,263 mT (30,239 T) of sediment was produced from driveways. Driveways accounted for 16% of the sediment from the Clear Creek Basin.

FIGURE 53.5

Sediment yields over time from roads, logging, and fire in the Camp Creek Basin.

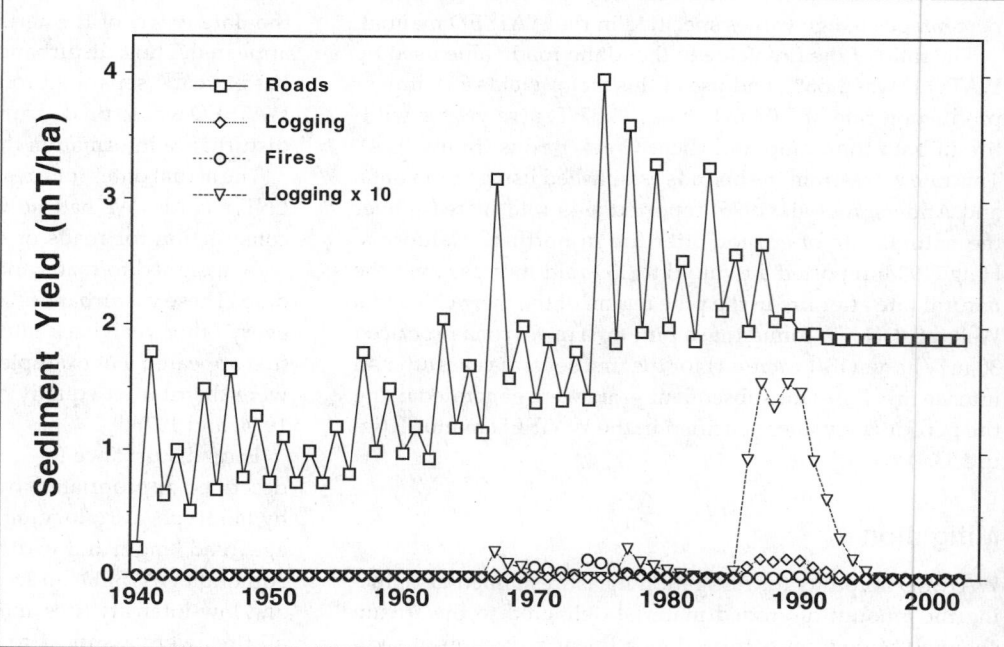

1417

Camp and Clear Creeks, El Dorado County: Predicted Sediment Production from Forest Management and Residential Development

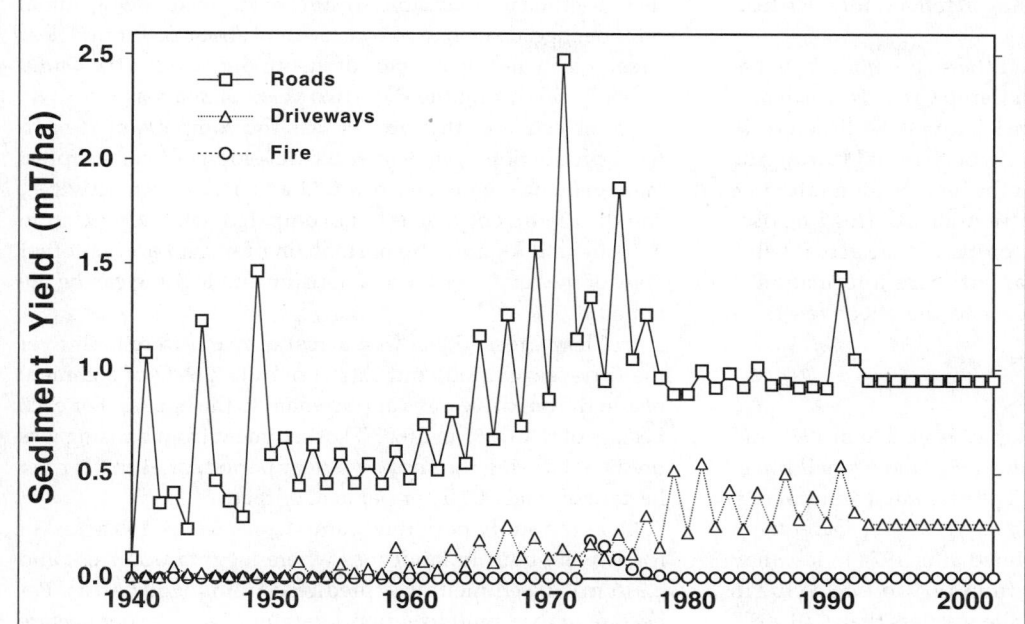

FIGURE 53.6

Sediment yields over time from roads, driveways, and fire in the Clear Creek Basin.

Logging

Clear-cut logging in Camp Creek occurred in 1966, in 1976, and between 1984 and 1989. Salvage logging that could be allocated to specific areas occurred between 1988 and 1991. Sediment was generated from logging in Camp Creek between 1967 and 1972, between 1977 and 1982, and between 1985 and 1997. Curves for two erosion rates for logging are presented, but for the low rate, only the last of the three time periods is visible in figure 53.5. All three periods are visible for the high erosion rate. Both the 1967 and 1977 periods produced 0.02 mT/ha·yr (0.01 T/acre·yr) or less in any year at the low rate, and the values for the high rate are larger by a factor of ten. The 1989 value of 0.2 mT/ha·yr (0.09 T/acre·yr) was the peak (1.6 mT/ha·yr [0.7 T/acre·yr] for the high-rate curve), and the values declined to zero by 1997. The total sediment yield from logging was 8,316 mT (7,560 T) between 1967 and 1997 for the low erosion rate and 83,161 mT (75,602 T) for the high erosion rate.

An analysis was performed to determine if more of the highly erodible sites were logged in the later time intervals. This issue pertains to an old nostrum that suggests that the "best sites" would be cut before the "poorer sites." We plotted, by decade, the hectares of roads built (to support logging) in each of the four erosion hazard classes. No shift toward the high or very-high classes was evident. In all but the earliest interval, the largest area of roads was built in the medium hazard zone. In all but the 1970 decade, which had none, two or fewer hectares of roads were built in the very-high class.

About 108 million board feet of timber have been removed from Camp Creek by selective logging between 1952 and 1991

(McGurk and Davis 1996). Because we were unable to allocate most of the selective logging by landtype, we were unable to develop erosion information for selective logging for the entire study interval. However, we have included in our predictions all of the roads built in support of selective logging. In that roads are consistently identified as the primary sediment producer compared with logging (Fredriksen 1970; Binkley and Brown 1993), we feel that the lack of sediment

FIGURE 53.7

Average annual sediment yield from 1940 to 2002, by decade, from roads, driveways, logging, and fire in the Clear and Camp Creek Basins.

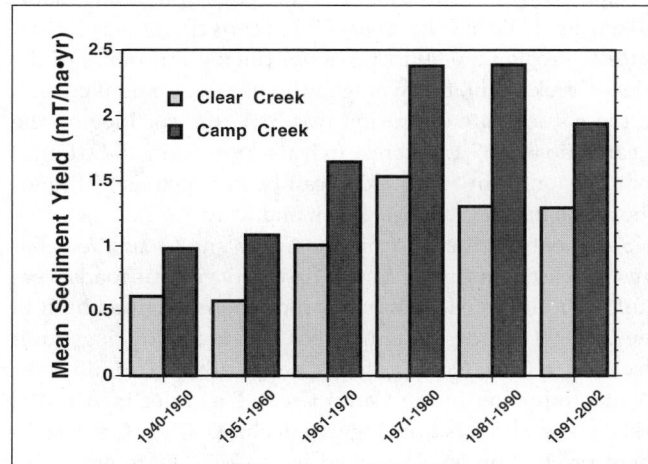

projections from selective logging itself is not a critical omission.

Although roads produce most of the sediment in both basins, the relative role of logging in Camp Creek is worth noting. At the low erosion rate (1.98 mT/ha·yr [0.88 T/acre·yr]), sediment from logging is about 1% of the total during the sixty-two-year analysis period. At the high erosion rate (19.8 mT/ha·yr [8.8 T/acre·yr]), the contribution from logging rises to 8.5%. We believe this range encompasses the actual value for logging in the Sierra Nevada, but more information is needed on logging-related erosion and the effectiveness of mitigation.

Fire

The fire curve on the Camp Creek plot is visible in 1970 and 1974 in response to fires in 1969 and 1973, and a small fire in 1988 is undetectable (figure 53.5). The estimated annual sediment production reaches 0.1 mT/ha·yr (0.04 T/acre·yr) in 1974. Sediment from that fire declined after 1974 to less than 0.006 mT/ha·yr (0.003 T/acre·yr) from 1979 to 1988. A fire in 1988 caused a small rise, and the rate was less than 0.01 mT/ha·yr (0.004 T/acre·yr). The total sediment yield from land affected by fire was 6,654 mT (6,049 T) between 1970 and 2002.

Clear Creek was affected by fire in 1972, but the 1973 level was only 0.2 mT/ha·yr (0.08 T/acre·yr). After a six-year decline (controlled by the fire erosion curve, figure 53.4), sediment production remains below 0.01 mT/ha·yr (0.004 T/acre·yr) for the rest of the analysis period. The total sediment yield for land affected by fire was 1,994 mT (1,813 T) between 1973 and 2002.

Mean Sediment Yield

To smooth the sawtooth pattern and to combine the yield from all activities in figures 53.5 and 53.6, total mean annual rates were calculated for six intervals between 1940 and 2002 (figure 53.7). The bars in figure 53.7 are the mean of the values for approximately ten-year intervals. The means include sediment for roads, driveways, logging, and fire for each basin. Compared with the midpoint of the range of assumed natural erosion (0.02 mT/ha·yr [0.009 T/acre·yr]), the accelerated rates of erosion are 29 to 119 times the natural rate. For the Clear Creek Basin, the average ratio of mean annual erosion to the natural rate of erosion was 53:1, and for logging the mean ratio was 87:1. Past research at a location in Idaho found that erosion from roads alone can be as much as 750 times above the natural rate (Megahan and Kidd 1972).

Sediment production from the Camp and Clear Creek Basins increased uniformly for the first forty years as roads were built in both basins. Sediment production remained high in the 1981–90 decade in Camp Creek because of logging, but it declined in Clear Creek as fewer new roads were built. Sediment production in the Camp Creek Basin declined in the last decade as yields from logging declined. Clear Creek sediment production levels peaked in the 1971–80 interval and then declined as road building almost stopped. Mean annual rates per decade increased by a factor of about 2.5 for the Clear Creek Basin and by a factor of just under two for the Camp Creek Basin during the sixty-two years of analysis.

An objective of this project was the comparison of sediment production from residential development versus forest management. Roads are included as part of both activities, and their sediment yield is large compared with the yield from housing and logging. By normalizing for road area, another view of the relative effects of housing and logging can be obtained.

In Clear Creek, 50 ha (124 acres) of roads were built over the fifty-year period, and 33,263 mT (30,239 T) of sediment resulted from driveways appurtenant to the houses. For each hectare of road, 665 mT (605 T) of sediment from housing was predicted during the sixty-two-year period, or 11 mT/yr per hectare of road (4.9 T/yr per acre of road).

Over the study period in Camp Creek 68 ha (168 acres) of roads were built in landtypes where logging occurred, and 8,316 mT of sediment was predicted from logged areas. Per hectare of road built to support logging, 5 mT/yr per hectare of road (2 T/yr per acre of road) of sediment resulted. The residential rate is more than twice the logging rate, because sediment yield from logging is of short duration compared with the permanent effect of driveways, coupled with the higher road density in the Clear Creek (1.7%) versus Camp Creek (1.2%) Basins.

Prediction Accuracy

The state-of-the-art in large-scale, long-term sediment modeling does not permit prediction of precise amounts of sediment. Models such as WATSED allow comparisons of the relative differences between various disturbances and disturbed and undisturbed conditions. They are accurate enough for planning and to support management decisions. We recognize, however, that the results from an analysis such as this are very sensitive to the erosion values selected from the literature to represent the processes that were analyzed. The analysis reported here must be viewed from this perspective. Because of the lack of basin data, no attempt was made to validate that model's prediction. Although absolute values (e.g., mT/ha·yr of sediment) are reported, the confidence interval around these estimated values is unknown.

WATSED has many strengths because of its flexibility, but it also has limitations. Users can incorporate new erosion curves to meet their needs, and the magnitude and shape are set by the user. Polygon size is variable and can be quite small if information is available. The model is limited to yearly average values, however, so variability in the climatic regime is ignored. We did not use the hydrologic portion of the model, so we cannot comment on the accuracy of its predicted changes in water yield.

1419

Camp and Clear Creeks, El Dorado County: Predicted Sediment Production from Forest Management and Residential Development

CONCLUSIONS

A sixty-two-year period was analyzed with a sediment accounting model (WATSED), which included the 1940–91 period of changing land use plus a ten-year post-disturbance recovery period. The primary disturbances in Camp Creek are road building, logging, and fire. The primary disturbances in Clear Creek are road building, residential development, and fire. The natural sediment production rate for this area is thought to be 0.02 mT/ha·yr (0.009 T/acre·yr).

Roads produced the bulk of the sediment in both basins. As of 1991, Clear Creek had about 50 ha (124 acres) of roads, and Camp Creek had over 100 ha (250 acres) of roads. Mean annual sediment production increased in both basins from 1940 through the 1971–80 interval and declined thereafter in Clear Creek. Mean annual sediment production in the Camp Creek Basin stayed constant between the fourth and fifth decades because of logging between 1984 and 1991. Between 1987 and 1991, road construction was insignificant in both basins.

In Clear Creek, roads and then homes were built, and sediment yield was predicted to increase because of areas allocated to roads and unsurfaced driveways. About 29 ha (72 acres) of driveways were assumed to complement the 50 ha (124 acres) of roads, but only 16% of the total predicted sediment from the basin was due to driveways. The combined effect of roads and driveways accounted for 99% of the predicted sediment.

Logging occurred on Camp Creek only, and clear-cut logging became common after 1984. Selective logging has been practiced in the basin since prior to 1940, but selective logging was not analyzed in this study except for salvage logging between 1988 and 1991. Because roads were built in support of the selective logging, the major source of sediment from selective logging is included. The predicted sediment levels after 1984 were 0.2 mT/ha·yr (0.09 T/acre·yr) or less for the low rate of erosion. At the high rate of erosion from logging, peak sediment production level was 1.6 mT/ha·yr (0.7 T/acre·yr). The literature generally supports the lower value.

Fire affected both Camp Creek (256 ha) and Clear Creek (85 ha), but only minimal sediment was produced because the erosion rate is relatively small and the model assumes that erosion due to fire stops after seven years. The peak sediment production rates from fire were 0.1 and 0.2 mT/ha·yr (0.04 and 0.08 T/acre·yr) in the two basins, respectively.

Predicted mean annual erosion rates due to disturbance are 29 to 119 times the assumed natural rate of erosion during the sixty-two-year period. This increase is due to the nearly 700-fold difference between the long-term erosion rate from roads and the natural rate. Mean total sediment production rates per hectare for Clear Creek were between 53% and 67% the magnitude of the Camp Creek Basin's rates over the period of analysis. Camp Creek is steeper and has more area in

zones of higher sediment delivery than Clear Creek (figure 53.3), thereby explaining the higher sediment production rates from a lower density of roads.

The WATSED accounting model, once calibrated with local data, was relatively easy to use and could be applied to other parts of the Sierra. In each case, however, local erosion rates for common soils would be required. By proxy, an evaluation of relative road densities would provide a preliminary evaluation of sediment production from logging and residential development.

A normalization procedure was devised to attribute sediment from logging and housing by the area of road built to support the land use. The residential development rate was 11 mT/yr per hectare of road (5 T/yr per acre of road), more than twice the rate of 5 mT/yr per hectare of road (2 T/yr per acre of road) for logging.

Results and conclusions in this study are based on the best available data, but many assumptions had to be made before the sediment model could be implemented. Local erosion data for post-fire and -logging would improve the accuracy of the sediment predictions. The ten-fold range in erosion rates for logging changes the logging contribution from a negligible (relative to roads) to a rather serious, albeit short-term, quantity of sediment. Because roads produced the bulk of the sediment, additional information on erosion from roads is of the highest priority. Conclusions on relative sediment production levels between land uses may be made, but confidence intervals around predicted rates are unknown.

ACKNOWLEDGMENTS

We thank the individuals who provided data or technical advice during this project. Rosa Nygaard (U.S. Forest Service Regional Office, Northern Region) provided the WATSED program we used. John Casselli (Lolo National Forest) provided considerable technical advice on applying WATSED. Chuck Mitchell, Annette Parsons, Sue Rodman, Anne Boyd, and Christine Christiansen (Eldorado National Forest) provided erosion data and information on local erosional processes. John Munn (California Department of Forestry and Fire Protection) also provided erosion data and technical advice. Thad Edens (PSW Research Station) provided invaluable assistance during the model implementation and data analysis.

REFERENCES

Agricultural Research Service. 1994. *Prediction project: Erosion prediction model, version 94.3 user summary.* ARS National Soil Erosion Research Laboratory report no. 8. West Lafayette, IN: U.S. Department of Agriculture, Agricultural Research Service.

Anderson, H. W., M. D. Hoover, and K. G. Reinhart. 1976. *Forests and water: Effects of forest management on floods, sedimentation, and water supply.* General Technical Report PSW-18. Berkeley, CA: U.S. Forest Service.

Baker, M. B., L. F. DeBano, and S. Krammes. 1994. Runoff and erosion from selected forest and rangeland sites in California. Draft final report. Sacramento: California Department of Forestry and Fire Protection.

Binkley, D., and T. C. Brown. 1993. *Management impacts on water quality of forests and rangelands.* General Technical Report RM-239. Fort Collins, CO: U.S. Forest Service.

Dunne, T., and L. B. Leopold. 1978. *Water in environment planning.* San Francisco: W. H. Freeman.

Euphrat, F. D. 1992. Cumulative impact assessment and mitigation for the Middle Fork of the Mokelumne River, Calaveras County, California. Ph.D. diss., Forestry Department, University of California, Berkeley.

Fiore, H. 1992. Upper Camp Creek stream survey. Draft report. Placerville, CA: Supervisor's Office, Eldorado National Forest.

Fredriksen, R. L. 1970. *Erosion and sedimentation following road construction and timber harvest on unstable soils and three small western Oregon watersheds.* Research Paper PNW-104. Portland, OR: U.S. Forest Service.

Haig, I. T. 1938. Fire in modern forest management. *Journal of Forestry* 36:1045–49.

Heede, B. H. 1984. Sediment source area related to timber harvest on selected Arizona watersheds. In *Effect of land use on erosion and slope stability,* edited by C. L. O'Laughlin and A. J. Pierce, 123–30. Honolulu: IUFRO–East-West Center.

Hewlett, J. D. 1982. *Principles of forest hydrology.* Athens: University of Georgia Press.

Huntington, G. L., and M. J. Singer. 1982. Soil interpretations and socio-economic criteria for land use planning. Termination report for project W-125. Sacramento: California Department of Forestry and Fire Protection.

Huntington, G. L., M. J. Singer, and E. L. Begg. 1979. Soil interpretations and socio-economic criteria for land use planning. Annual report for project W-125. Sacramento: California Department of Forestry and Fire Protection.

Kattelmann, R. 1996. Hydrology and water resources. In *Sierra Nevada Ecosystem Project: Final report to Congress,* vol. II, chap. 30. Davis: University of California, Centers for Water and Wildland Resources.

Leavesley, G. H., P. Restrepo, L. G. Stannard, and M. Dixon. 1992. The modular hydrologic modeling system—MHMS. In *Managing water resources during global change,* edited by R. Herrmann, 263–64. Bethesda, MD: American Water Resources Association.

Lewis, J., and R. Rice. 1989. Site conditions related to erosion on private timberlands in northern California: Final report. In *Critical sites erosion study,* vol. 2. Sacramento: California Department of Forestry and Fire Protection.

McCammon, B. P. 1977. Salmon Creek administrative study. Porterville, CA: U.S. Forest Service, Sequoia National Forest.

McGurk B. J., and M. L. Davis. 1996. Camp and Clear Creeks, El Dorado County: Chronology and hydrologic effects of land-use change. In *Sierra Nevada Ecosystem Project: Final report to Congress,* vol. II, chap. 52. Davis: University of California, Centers for Water and Wildland Resources.

Megahan, W. F. 1975. Sedimentation in relation to logging activities in the mountains of central Idaho. In *Present and prospective technology for predicting sediment yields and sources.* ARS-S-40. Washington, DC: Department of Agriculture, Agricultural Research Service.

Megahan, W. F., and W. J. Kidd. 1972. *Effects of logging and logging roads on sediment production rates in the Idaho Batholith.* Research Paper INT-123. Odgen, UT: U.S. Forest Service.

Mitchell, C. R., and L. J. Silverman. N.d. *Soil survey of Eldorado National Forest area, California.* Washington DC: U.S. Forest Service.

Moore, I. D., E. M. O'Laughlin, and G. J. Buick. 1988. A contour-based topographic model for hydrological and ecological applications. *Earth Surface, Processes, Landforms* 13:305–20.

Pfankuch, D. J. 1978. *Stream reach inventory and channel stability evaluation.* Missoula, MT: U.S. Forest Service, Northern Region.

Poff, R. J. 1996. Effects of silvicultural practices and wildfire on productivity of forest soils. In *Sierra Nevada Ecosystem Project: Final report to Congress,* vol. II, chap. 16. Davis: University of California, Centers for Water and Wildland Resources.

Rogers, J. H. 1974. *Soil survey of El Dorado area, California.* Washington DC: U.S. Department of Agriculture Soil Conservation Service and Forest Service.

Swanson, M., B. Emery, D. de Clercq, J. Vollmar, and E. Bianci. 1993. Upper Camp Creek watershed restoration and monitoring plan. Draft final report prepared by Swanson and Associates/ Biosystems Analysis, Inc. Placerville, CA: Supervisor's Office, Eldorado National Forest.

U.S. Forest Service. 1991. *WATSED, water yield and sediment.* Missoula, MT: U.S. Forest Service, Northern Region.

———. 1993. Eldorado National Forest stream channel evaluation. Placerville, CA: Supervisor's Office, Eldorado National Forest.

———. 1995. Rock Creek OHV Area sediment delivery analysis. Placerville, CA: Supervisor's Office, Eldorado National Forest.

Wells, C. G., R. E. Campbell, L. F. DeBano, C. E. Lewis, R. L. Fredriksen, E. C. Franklin, R. C. Froelich, and P. H. Dunn. 1979. *Effects of fire on soil.* General Technical Report WO-7. Washington DC: U.S. Forest Service.

LARRY A. COSTICK
Graduate Group in Ecology
Department of Land, Air, and Water
  Resources
University of California
Davis, California

54

Indexing Current Watershed Conditions Using Remote Sensing and GIS

ABSTRACT

Two objectives of the Sierra Nevada Ecosystem Project (SNEP) were to evaluate the current condition of watersheds in the Sierra Nevada and to identify physical processes such as soil erosion that affect watershed health and sustainability. In response to this request for a resource inventory, an indexing or screening model has been developed that produces both a natural erosion potential (NEP) and sedimentation hazard index (SHI), which are indicators of the current cumulative condition in watersheds of the Sierra Nevada.

The goal of the study undertaken here is to design and test a methodology using geographic information systems (GIS) and remote sensing to rank watersheds prone to soil erosion and locate specific sites where stream sedimentation is likely to occur. One hundred and thirty-four watersheds on the Eldorado National Forest (ENF) were analyzed and ranked using a method that selects the parameters of slope, cover, and soil detachability, which were assumed to be the most significant contributors to soil erosion, given uniform climatic conditions. Threshold values established for these parameters provided the link to locations where there is a high probability of sediment reaching the watercourse.

Correlation with U.S. Forest Service equivalent roaded acres (ERA) and cumulative watershed effects (CWE) work previously completed and in progress on the ENF was positive when compared to NEP and SHI rankings created by this model. Additional correlation opportunities yet to be implemented using change detection techniques with Landsat TM imagery, spectral mixture analysis (SMA) with high resolution AVIRIS imagery, and the identification of large rock outcrops are expected to improve results. The model described here gives the resource manager a tool that can be used to quickly screen proposed CWE assessment areas and focus both human and financial resources on potential "hot spots." Once located, the cumulative effects benefit of a specific mitigation opportunity may be evaluated as to its cost and to the watershed improvement that it provides.

INTRODUCTION

Two objectives given to the Sierra Nevada Ecosystem Project were to evaluate the current condition of watersheds in the Sierra and to identify physical processes such as soil erosion that affect watershed health and sustainability. The goal of this project was to utilize geographic information systems and remote sensing as the basis of a watershed assessment model. This model ranks watersheds prone to soil erosion and locates specific sites where stream sedimentation is likely to occur.

A healthy watershed is defined here as an area of land having the structure and density of vegetative stands to support a diverse wildlife population and having the natural stability of geology and soils to maintain the contribution of eroded sediments reaching streams at a level where natural hydrologic processes balance the ability of the system to both store and transport these sediments without degrading aquatic habitats. One hundred and thirty-four Cal-Water planning watersheds on the Eldorado National Forest (ENF) were analyzed and ranked using a method that selects three physical landscape parameters most likely to contribute to soil erosion: slope, surface cover, and soil erosivity or detachability. Threshold values established for these parameters provided the link to locations with high probability of sediment reaching a watercourse.

Sierra Nevada Ecosystem Project: Final report to Congress, vol. II, *Assessments and scientific basis for management options.* Davis: University of California, Centers for Water and Wildland Resources, 1996.

Resource managers need practical tools for watershed assessment. Those tools should be based on simple concepts and built around readily available or easily acquired information. The method proposed here requires the user to have access to Landsat imagery and a limited knowledge of soils, geomorphology, and ecology. Doing hierarchical analysis, first using a screening tool followed by more data-intensive and quantitative procedures, allows managers to identify and prioritize both analytical and restoration activities. This model gives the resource manager a tool that may be used to quickly screen proposed cumulative watershed effects assessment areas and focus both human and financial resources on potential "hot spots." Once located, the cumulative effects benefit of a specific mitigation opportunity may be evaluated relative to its cost versus the environmental watershed improvement that it provides. Figure 54.1 locates the Eldorado National Forest relative to SNEP's regional study area (see inset) and identifies specific drainages such as Fry Creek and Camp Creek.

Regional Background

Years of grazing, mining, road building, home construction, and logging disturbances as well as fire, landslides, and plant disease have modified forest ecosystems in the Sierra Nevada. Present remote sensing technology provides for observing, measuring, and monitoring natural and management-induced changes such as soil loss, vegetative cover, and habitat disturbance. There are, however, very few predictive ecosystem models that use spatial and temporal remote sensing data to infer cumulative watershed condition or ecosystem health. Comparison of current condition on a watershed-by-watershed basis allows us to index ecosystems relative to each other. An accurate indexing methodology is a valuable tool when allocating resources for cumulative watershed effects (CWE), mitigation, or adjudicating disturbance rights among landowners in mixed ownership watersheds.

The methodology presented here assesses the ecosystem, as defined by watershed boundaries, for natural erosion potential and sedimentation hazards. It suggests physical parameters for ecosystem assessment and an accounting system to track and recalculate a watershed condition index. Data on these parameters—that is, amount of ground cover, bare soil, soil detachability, or sensitivity to erosion as well as slope—are used to quantify the ecosystem's sensitivity to accelerated erosion and sedimentation. Geographic information system (GIS) layers of slope, soil type, soil detachability, disturbance history data, and road and stream proximity are integrated to spatially display current relative watershed condition.

Both national and state environmental quality acts (NEPA and CEQA) require cumulative effects assessment for all land disturbance "projects" on private, state, and federal land. Definitions of cumulative effects vary, and there are no universally accepted techniques for their measurement or monitoring. Our inability to objectively quantify cumulative effects and the absence of standards for comparison have created difficulty for regulatory agencies. This model aids resource managers and agency regulators in objectively analyzing ecosystem complexities with particular regard for cumulative and synergistic impacts of human activity and natural processes.

With the advent of GIS technology, spatial analysis procedures are available to quantify both present and historic physical features and land-use practices on a landscape basis. From the rates of change in these features, as determined by GIS interpretations of aerial and space imagery, habitat improvement or degradation and habitat potential may be inferred. Simultaneous analysis of several GIS layers provides a more objective view of ecosystem condition.

METHODS

Model Description

The model produces a natural erosion potential (NEP) and sedimentation hazard index (SHI) as indicators of the "current cumulative condition" in the watershed. Watershed characteristics used to assess relative health include an estimate of natural sensitivity to erosion, and an analysis of the location and number of roads, to allow prediction of probable origins of sediment. Because this present model is a screening tool, it focuses on initial soil-forming and erosional processes. The susceptibility of soils and geology to mass failure and rill and gully erosion are part of this process.

GIS Methodology and Logic

The model is similar to pre-GIS geographic map overlaying techniques in which clear acetate sheets scribed with information at one spatial and temporal scale are overlaid by other maps with information from a different time period or a different spatial arrangement. In this system both of these sheets are fixed to a base map containing information such as topography, streams, and soils common to both overlays. In order to begin to understand the relationships between the aggregated information, we analyze the composite. New values may be constructed, or the data may be classified by forming statistically similar clusters, which are referred to as *polygons*, or if at single points, *cells*. Using a commercial grid GIS computer program, data are distributed over a matrix with 0.22-acre cells, with dimensions of 30 meters on each side. This fine scale allows large numbers of attribute variables such as soil, slope, and vegetation to be viewed individually or simultaneously in very rapid order for a single cell or a cluster of cells, speeding the analysis process. Every major attribute, soil, for instance, may have dozens or hundreds of variables that describe the soil at a specific location.

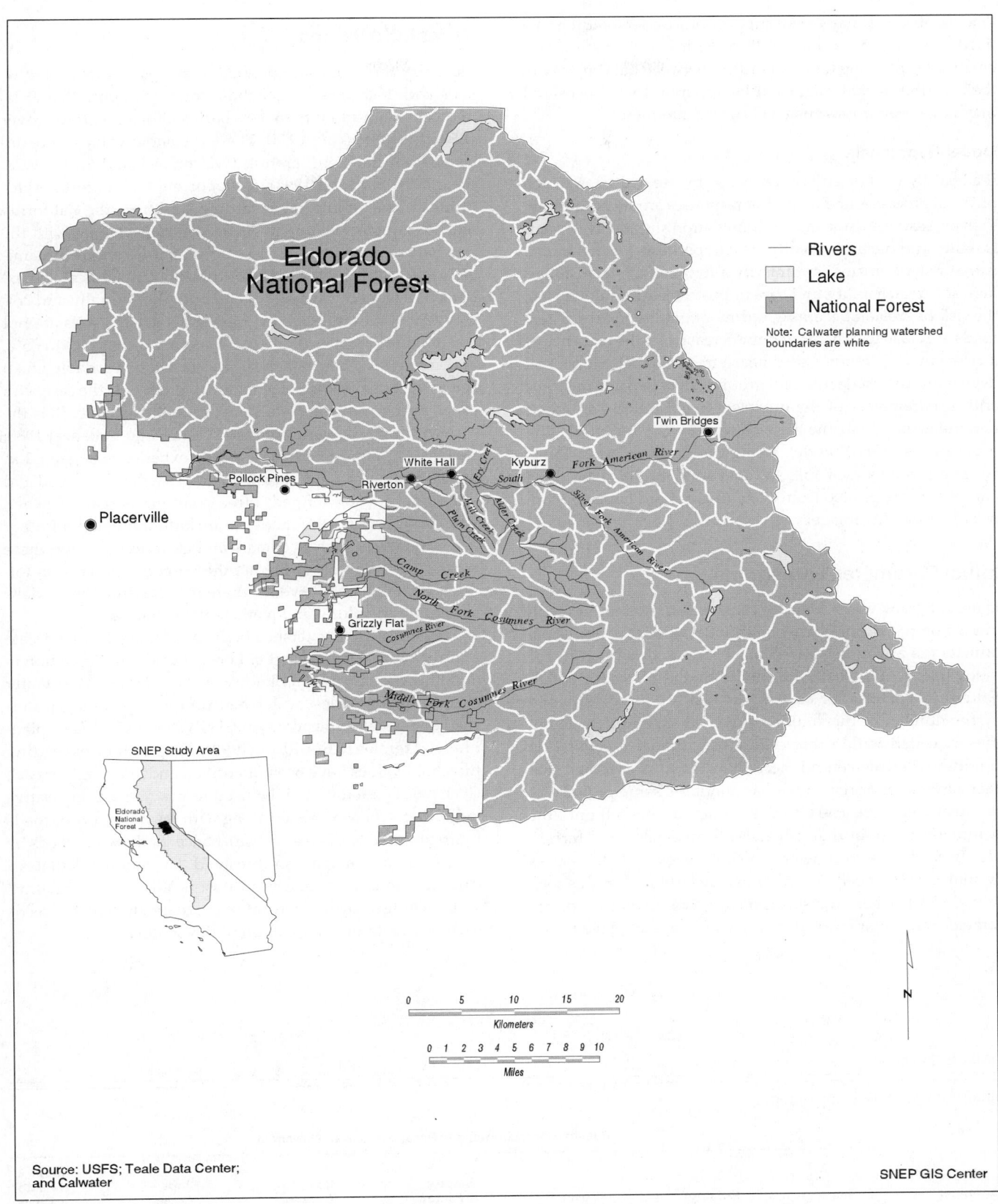

FIGURE 54.1

SNEP study area and Eldorado National Forest location map.

One database or many may support our understanding of a fixed point on the ground. GIS provides a means of mathematically searching for relationships between data layers and their attributes that might not be apparent to our eyes and may have been missed using earlier techniques.

Model Hypothesis

If a healthy watershed is determined by the degree to which physical processes and biological responses are at equilibrium, then excessive erosion and sedimentation suggest system instability and declining health. The hypothesis for this model is that risk of erosion is primarily a function of steep slopes, high soil detachability, and bare unprotected ground. Further, the risk of erosion becoming sedimentation increases where roads are close to streams and is decreased by the presence of a riparian vegetation buffer near stream banks. Slope; soil detachability, or K-factor; and ground cover become the three critical parameters of the models. Stream buffers are not a parameter but limit the area viewed by the sedimentation hazard index (SHI) model. Using the program ARC/INFO GRID as well as available soil and topography data, these parameters are plotted from GIS and Landsat Thematic Mapper (TM) satellite imagery.

Initial Parameter Thresholds

In order to rank watersheds for comparison, erosion and sedimentation hazard risks are quantified. Each of the above parameters is assigned a threshold value as described in the following section. These thresholds are indicators for potential erosion. Values for each watershed cell are determined by the number of thresholds—slope, cover, and detachability—exceeded within that cell. Given normal precipitation conditions for the central Sierra Nevada, it is assumed that each parameter or risk factor has about the same probability of causing erosion. The GIS does not count the cell until the parameter value in that cell exceeds an established threshold. Each time a parameter threshold is exceeded, a "1" is tabulated for that cell. A cell value may be 0, 1, 2, or 3, as seen in table 54.1, where the seven possible combinations of parameters and their corresponding values are displayed.

Threshold Values

Here risk thresholds are defined as slopes in excess of 40%, soils with K-factors (detachability ratings) higher than 0.28, and cells with more than 40% bare soil or no surface cover (Elwell and Stocking 1974). These threshold values were derived from the soil literature (Wischmeier and Smith 1978; Rose 1994; Stocking 1994), from current U.S. Forest Service limits for tractor and cable yarding, and from the California and Washington State Forest Practice rules. Along with the intensity of precipitation, these three parameters are dynamically interactive, with each contributing to "critical shear," detachment, and transport of soil both individually and collectively. For example, bare, highly detachable soils are not as erodable at slopes of 0% to 5% as they are at 15% to 35%; and conversely, bare, steep slopes are not as erodable when soil textures have low detachability values, as is the case with clays, as they are when soils are highly detachable, as is the case with very fine sandy loams (Mitchell and Bubenzer 1980; Kirkby and Morgan 1980). As more experience is gained in using this model, other threshold values will be explored and the model further refined into a continuous scale. Further, we plan to extend the model to include influences of other external factors such as climate and elevation. Adding these factors will allow us to predict the potential for erosion following rain-on-snow events, thereby increasing the model's sensitivity to natural and management changes.

This study used watershed boundaries mapped by the state of California's Department of Forestry and Fire Protection in a data dictionary project known as "Cal-Water." Cal-Water defines their smallest watershed unit as a planning watershed and gives it the acronym CWPWS, for Cal-Water planning watershed (Brandow 1995). Parameters exceeding threshold values have been quantified and analyzed for each CWPWS. These data will be used to provide a comparative index that, when examined along with the proximity of roads to streams and total area of disturbance, ranks watersheds by their areal percentage over threshold. The model calculates a current condition ranking on a "most-healthy to least-healthy" scale as judged by the percent of the watershed that exceeds each threshold or combination of thresholds.

TABLE 54.1

Maximum cell value calculation.

Parameter	Possible Combinations of Parameters over Threshold						
	Slope	K-Factor	Cover	Slope + K-Factor	Cover + K-Factor	Slope + Cover	Slope + K-Factor + Cover
Value	1	1	1	2	2	2	3

Soils Database and Derived Map Products

This analysis draws from three primary sources of information: slope is derived from a 30-meter digital elevation model (DEM) produced by the U.S. Geological Survey, bare ground is derived from a 1994 Landsat Thematic Mapper satellite image, and the soils information is found in four soil surveys from the Eldorado NF and the US Natural Resources Conservation Service (NRCS). Soils database attributes were derived from several sources of soil survey data such as engineering properties and physical/chemical properties. A number of products have resulted from this derived data. Plate 54.1 uses soil parent material and particular geologic formations to group soils that have similar erosion characteristics. This map provides foresters and resource managers with a ready reference of spatial information by basic geologic group and soil series.

Calculation of Natural Erosion Potential Percentage

Natural erosion potential (NEP) is an index of stability or resilience, predicting an unmanaged watershed's ability to withstand erosion causing events. As seen in plate 54.2 and table 54.2, this model operates on Boolean logic: when a cell's value exceeds any threshold, it is assigned an index value of 1; conversely, if the feature being assessed is less than the threshold, the cell value is assigned a value of 0. Cell value accuracy is a function of grid size. In the case of slope, using the best available information, which is the 30-by-30-meter DEMs, means that the angle formed using one cell's centroid elevation when compared to the centroid value of its neighbors either does or does not exceed the threshold. Each parameter has its own data layer in the GIS. Again referring to table 54.1, when two thresholds are exceeded for the same cell, the cell's value is the combination of those two data layers and has an index value of 2. Likewise, for the combination where all three thresholds are exceeded in the same cell, the value of that cell becomes 3. Therefore in the worst case (maximum NEP) every cell would have a value of 3. Multiplying three times the number of cells in a watershed yields the maximum potential NEP watershed value. The present watershed value is generated by counting the total number of cells over threshold in the composite GIS layers. The total values of cells exceeding thresholds, divided by the maximum potential for the watershed, times 100, becomes the relative watershed score or percentage NEP. The NEP for the whole national forest, graphically projected, is found in plate 54.2, where K-factor, bare ground, and slope have one column and the presence of a "1" indicates the parameter is over threshold. If a "1" is present in more than one column, it is interpreted as an increased erosion risk up to a value of 3. (See table 54.2 for further explanation.)

Using a computer monitor, plate 54.2 can be expanded so that individual 30-by-30-meter cells may be located and re-viewed for soil, slope, or bare ground attributes. Even the very small scale displayed in this map provides sufficient spatially explicit information to make reasonable visual watershed comparisons and guide additional assessment work. Each watershed is given an attribute table that provides the user with specific information about those parameters being evaluated. These attribute tables on either a watershed or parameter scale may be accessed to add or edit data.

Plate 54.3 is the type of map that is used for field assessment work. It is the basis for the tables used to calculate the ranking of every cell and for aggregating up to planning watershed or river basin. Roads and stream buffers are shown so that areas of special concern may be reviewed for possible mitigation opportunities—for example, the Fry Creek watershed, shown in red and located in the upper center of plate 54.3. (Also see figure 54.1.)

Table 54.2 is an example of one of the data tables built for each watershed. The Interpretation column has been added for reader assistance. Cells over threshold and their corresponding acreages are summed at the bottoms of the columns. Values are not duplicated when thresholds are combined. Maps similar to the one in plate 54.3 were used in the field to validate the location of the soil and slope parameters and the percentage of bare ground estimated by the bare ground threshold. Both individual cells and clusters of cells were targeted and found for examination using a global positioning system (GPS).

Calculation of Sedimentation Hazard Index Percentage

Although NEP reflects a watershed's natural stability, SHI focuses on the potential to upset that stability through road construction and maintenance practices. Stream sedimentation is often the result of a very small erosional failure becoming a very large CWE disturbance (Megahan et al. 1991; Lewis and Rice 1989; Rice 1993). SHI seeks to evaluate detailed patterns in a stream buffer zone by identifying areas at risk, predicting specific points most likely to fail, and reflecting SHI reductions as road segments are abandoned, rocked, or paved.

As defined earlier, a healthy watershed is one in which the natural stability of geology and soils maintains the contribution of sediments at a level where natural hydrologic processes balance the ability of a system to both store and transport these sediments without degrading aquatic habitats. Assuming vegetation and debris in stream buffer zones can trap and stabilize incoming sediments, an adequate width for these buffers must be determined in order to protect habitats of aquatic and terrestrial species while permitting access to lands for management. Erman and colleagues (1977, 1983) looked at stream buffer widths and the impacts on benthic organisms. They found the population count and species diversity of these organisms were indicators of the condition of the

TABLE 54.2

Fry Creek data interpretation.

Number of Cells	Total Acres	Percentage of Watershed	Soil K-Factor >0.28 (Acres)	Bare Ground >40% (Acres)	Steep Slopes >40% (Acres)	Stream Buffer (Acres)	Road in Stream Buffer (Acres)	Interpretation
28,535	6,345	100	0 *	0	0	1,123	134	6,345 acres is the watershed: 1,123 in stream buffers and 134 with roads in the buffers.
420	93	1.5	0	0	0	1	0	93 acres of stream buffer under threshold without roads.
81	18	0.3	0	0	0	1	1	18 acres of stream buffer under threshold with roads.
380	84	1.3	0	0	1 *	0	0	84 acres of slopes >40% outside of stream buffers and without roads.
17	4	0.1	0	0	1	1	0	4 acres of slopes >40% in the stream buffer.
1,363	303	4.8	0	1	0	0	0	303 acres of bare ground outside the stream buffer without roads.
14	3	0	0	1	0	1	0	3 acres of bare ground inside the stream buffer but without roads.
13	3	0	0	1	1	0	0	3 acres of bare, steep area outside the stream buffer and without roads.
8,916	1,982	31.2	1	0	0	0	0	1,982 acres of high K-factor soils outside of stream buffers or roads.
2,697	600	9.5	1	0	0	1	0	600 acres of high K-factor soils in stream buffers.
404	90	1.4	1	0	0	1	1	90 acres of high K-factor soils in stream buffers and beside roads.
5,131	1,141	18	1	0	1	0	0	1,141 acres of high K-factor and steep lands outside of stream buffers or roads.
765	170	2.7	1	0	1	1	0	170 acres of high K-factor and steep lands inside stream buffers without roads.
18	4	0.1	1	0	1	1	1	4 acres of high K-factor and steep lands inside stream buffers with roads.
2,334	519	8.2	1	1	0	0	0	519 acres of high K-factor and bare lands outside of stream buffers or roads.
420	93	1.5	1	1	0	1	0	93 acres of high K-factor and bare lands inside stream buffers without roads.
97	22	0.3	1	1	0	1	1	22 acres of high K-factor and bare lands inside stream buffers with roads.
1,358	302	4.8	1	1	1	0	0	302 acres of high K-factor, bare and steep lands outside stream buffers or roads.
112	25	0.4	1	1	1	1	0	25 acres of high K-factor, bare and steep lands inside of stream buffers without roads.
3	1	0	1	1	1	1	1	1 acre of high K-factor, bare and steep land inside of stream buffers with roads.
			4,949	1,271	1,734	1,123	134	

* "0" means no data in the "Number of Cells" column for this parameter. "1" means the number of cells shown in the "Number of Cells" column are the cells, acres, or percentage over threshold for this parameter or combination of parameters.

habitat, but only as it pertains to aquatic species. Buffers originally thought to be adequate to meet the needs of terrestrial invertebrates and to prevent or minimize sedimentation may not be adequate to maintain stream organic inputs or provide for the needs of mammals and riparian species (Kattelmann 1996). Because this study is a screening process attempting to focus the resource manager's attention on the most acute problem areas, roads that fall within 60 meters (197 feet) of a perennial stream become the target of GIS querying. Cells fully located in the buffer between a road and stream that exceed any of the index thresholds are tagged. Where multiple thresholds are exceeded in the same cell, the magnitude of severity ensures that management attention will be focused on that location. Parameters exceeding thresholds for cells within a 60-meter buffer zone along perennial streams and adjacent to roads are calculated in the same manner as for NEP, except that the maximum potential SHI value becomes three times the total number of stream buffer cells where roads are present. Actual SHI is composed of those cells over threshold within the stream buffer where roads are present. Dividing the actual by the potential maximum, yields the percentage SHI in the same manner as percentage NEP was generated. These new values are the most critical of the process because they reflect the increased probability that sedimentation will occur at a location under specified conditions. Potential problem cells are noted and are uniquely identifiable, thus facilitating monitoring and/or mitigation. Maps of roads, stream buffers, watershed boundaries, and parameters over threshold are produced along with the tables so that graphical comparisons can be made and checked in the field. Plate 54.4 emphasizes the fact that the occurrence of cells over threshold inside stream buffers is limited. This limitation points to locations where increased sedimentation should be expected and to critical areas that should be monitored.

Watershed Assessment Terminology

Because of data limitations for areas beyond the national forest boundaries, the application of the NEP and SHI methodology was limited to the Eldorado National Forest and to those CWPWS that were completely within the national forest. Table 54.3 features twenty-seven of 177 watersheds reviewed for this work. Differences in watershed boundaries selected by the Forest Service and Cal-Water were reconciled by consolidating Cal-Water watersheds in some cases and Forest Service watersheds in others. The consolidating process yielded 120 watersheds with adequate data for comparison. Only seventy-six of the USFS watersheds had complete data directly compatible with the model. However, all 120 watersheds had the USFS-generated natural sensitivity index (NSI). Designed by Kuehn in 1989 for cumulative watershed effects analysis on the ENF, this indexing system considers both hillslope and in-channel hydrologic and erosional processes. Soils, stream channel conditions, geomorphic instability, drainage density, and precipitation regimes are all part of the NSI calculation. NSI as seen in table 54.4 is used to generate a watershed's threshold of concern (TOC) (U.S. Forest Service, 1987). TOC relates to the percent of equivalent roaded acres (ERA), which is a watershed ranking by the amount and type of land disturbance within a watershed. TOC for a watershed is determined by the NSI number, where less than 15 is very low and greater than 65 is very high. For watersheds with very low NSI numbers, the TOC will range from 18% to 20% ERA, meaning that 20% of the watershed may be disturbed before significant cumulative effect occurs. Likewise, watersheds with very high NSI numbers have lower TOCs, and as little as 10% ERA may trigger significant CWE.

RESULTS AND DISCUSSIONS

Model Comparison with USFS Outputs

One of the highest Forest Service NSI and TOC rankings is that of the 6,346-acre Fry Creek watershed (see plate 54.3 and table 54.3). A tributary of the South Fork of the American River, much of its ground cover was burned in the 1993 Cleveland fire. It has steep slopes and highly detachable soils. The NSI is 183 and the percentage TOC is 138%. A TOC of 138% indicates that this watershed is significantly over the USFS threshold and that further unmitigated disturbance may result in considerable harm to the ecosystem. This model calculated Fry Creek as one of its highest risk watersheds, with NEP and SHI ratings of 41.7% and 35.5%, respectively. The USFS erosion hazard rating (EHR) risk number for this watershed, as seen in the seventh column of table 54.3, is 5: extreme.

Model Construction Time and Proposed Uses

The model yields a relative ranking for each watershed without extensive field surveys and could be used to guide future mitigation activity. Advantages of the model include lower dollar costs to produce, objective generation, capacity to be easily updated, responsiveness to changes in elevation and precipitation conditions, and reduced data corruption because minimal staff (one or two individuals) are required to process data.

After the soil database was constructed, 177 Cal-Water planning watersheds were reviewed and 134 analyzed for natural erosion potential and sedimentation hazards using approximately ten days of GIS and analysis time. Positive correlation with the Eldorado National Forest's natural sensitivity index and equivalent roaded acres methodology provides significant encouragement to continue refining this model and expanding its application to other portions of the Sierra Nevada Ecosystem Project study area. NEP and SHI rankings may be modified by testing mitigation alternatives, which include the surfacing and abandonment of road segments in areas over erosion parameter threshold.

Correlation Comparisons

As seen in table 54.5 the correlation coefficients (r) are positive when comparing the Forest Service indexes: USFS's NSI, ERA, and TOC and this model's NEP and SHI. Using all 120 comparable watersheds, the correlation between NSI and NEP is 0.54. Between TOC and SHI, with data from seventy-six watersheds, it is 0.44, and between ERA and SHI it is 0.34. The model has only been run once; hence ground truthing to better calibrate its predictions will continue. Percentage roaded acres (RdAcs) in this model is not the equivalent of ERA because ERA includes all cumulative logging disturbance and RdAcs is only the 30 meters of the roadway. Likewise, stream buffer acres (StBufAc) include only the percentage of the watershed where roads are present inside the stream buffer. Finding that 54% of the variation in NSI ranking is explained by the variation in NEP is encouraging, considering the difference in these methodologies. Improving calibration for large areas of exposed bedrock, accounting for precipitation isohyets and their influence on areas of high rain-on-snow potential, and change detection analysis should improve the correlation between the USFS assessment method and this model. The ERA method of analysis is likewise an evolving technique requiring large commitments of personnel time in both field and disturbance history research. Greater opportunity for human bias has led to the objectivity of the ERA method to be questioned.

Model-Directed Mitigation

After reviewing the results of the screening analysis and making an on-the-ground inspection of potential hazards, one of

TABLE 54.3

Twenty-seven Eldorado National Forest watersheds with the highest sedimentation hazard indexes and their corresponding natural sensitivity index and threshold of concern rankings.

Cal-Water Planning Watershed ID Number	Cal-Water Watershed Name	Cal-Water Planning Watershed Acres	Natural Sensitivity Index	Percentage Threshold of Concern	Percentage Equivalent Roaded Acres	Erosion Hazard Rating Number	Percentage Natural Erosion Potential	Percentage Sedimentation Hazard Index	Percentage Roaded Acres	Percentage - Roaded Acres Inside Stream Buffers
514.33021	Peavine Creek	11,510	60	125.0	15	5	40.9	38.6	10.5	11.4
514.35021	Fry Creek	6,346	183	138.0	13.8	5	41.7	35.5	9.1	11.9
514.32010	Gaddis Creek	8,684	81	106.0	10.6	5	35.2	34.6	8.2	9
532.60051	Beaver Creek	2,464	95	100.0	10	5	31.7	34.6	8.2	9.5
514.33035	Camp Seven	4,248	291	70.0	7	3	32.4	34.4	6.1	5.3
514.32012	Brush Creek	5,132	37	36.4	5.1	2	36.6	34.0	10.2	7.7
532.23043	Clear Creek	2,896	34	61.3	9.8	3	28.1	32.2	10.3	14.1
514.33030	Little Silver Creek	8,604	28	68.1	10.9	3	30.6	32.0	9.6	11.3
514.35050	Twenty-Five Mile Cyn	10,972	138	129.0	12.9	5	33.6	31.6	9.6	10.8
532.60061	W Panther Creek	5,853	79	104.0	10.4	5	26.1	30.2	11.3	9.9
532.23042	Middle Butte	2,925	160	53.0	5.3	2	31.1	29.8	6.2	3.2
514.36033	Middle Creek	4,735	119	50.0	5	3	24.5	29.8	7.0	6.9
514.32022	Whaler Creek	10,209	91.3	62.0	6.2	3	29.7	29.4	11.5	11
532.23033	North Canyon	3,541	25	23.1	3.7	1	29.5	28.8	10.0	15.3
514.32011	Slab Creek	5,493	114	43.0	4.3	2	32.3	27.9	11.0	8.6
532.23062	Clear Creek	6,840	28	50.0	8	2	28.0	27.7	13.1	14.5
514.32031	Bear Creek	5,358	59	68.3	8.2	3	28.1	27.4	12.6	16
514.32013	Slab Creek Res	5,723	174	51.0	5.1	2	28.7	26.8	9.8	6.1
514.35022	Mill Creek	2,178	61	117.5	14.1	5	11.5	24.9	8.6	8.1
514.35051	Grays Canyon	8,308	173	51.0	5.1	2	31.2	24.6	9.0	5.7
514.43033	Zero Spring	8,212	220	30.0	3	2	34.8	24.5	6.0	4.2
514.32015	Iowa Canyon	5,107	41	95.0	13.3	4	18.2	24.5	14.2	10.9
514.32021A	AWS1	13,502	94	34.0	3.7	2	25.1	24.4	10.3	0
532.24012	Cat Creek	5,655	93	138.0	13.8	5	14.4	23.9	10.7	13.8
532.23032	Van Horn Creek	7,516	77	64.0	6.4	3	26.5	23.8	10.2	12.3
514.35052	Soldier Creek	3,414	52	103.3	12.4	5	17.6	23.5	9.3	13.4
532.23051	Camp Creek	10,140	92	66.0	6.6	3	29.9	23.3	7.4	3.9

TABLE 54.4

Relationship of natural sensitivity index to equivalent roaded acres and threshold of concern (from Carlson and Christiansen 1993).

NSI	Sensitivity	TOC
<15	Very low	18–20% ERA
16–35	Low	16–18% ERA
36–50	Moderate	14–16% ERA
51–65	High	12–14% ERA
>65	Very high	10–12% ERA

the first questions to be asked is "What are the mitigation opportunities?" It is not possible to change a soil's K-factor, but one can consider abandoning or surfacing roads when they are located adjacent to streams on highly detachable soils, especially where they are combined with steep slopes and bare ground. In the Fry Creek example (table 54.2), 603 cells, totaling 134 acres or sixty-one hectors of roads, are present within stream buffers and represent opportunities for possible CWE mitigation. Being able to locate these cells using a GPS, portable computer, and GIS programs provides a means for immediate optimization of mitigation alternatives based on the recalculation of SHI. Some high-risk cells will become candidates for road abandonment, road surfacing, culvert replacement, or fill-slope ripraping. The current cumulative condition of the watershed can be evaluated and improved as soon as mitigation has been completed to reduce risks. Abandoning a portion of road within a stream buffer or on steep bare ground where the soils are highly detachable reduces the denominator in the formula equation, thereby reducing the percentage SHI. Reducing the risk factors will also reduce the percentage of the watershed exceeding thresholds, and, if it does not improve the watershed's ranking relative to others, it will at least allow for additional management to take place without excessive risk.

Other opportunities to influence the NEP or SHI are available through planting, seeding, and/or mulching of bare areas. The Fry Creek watershed has 5,716 cells, 1,271 acres, or 578 hectors that could be considered for this treatment. The number includes all those cells or combinations of cells that are bare and exceed other thresholds. It includes many areas

TABLE 54.5

Correlation coefficients, r, for the NEP and SHI model output.

	NSI	ERA	TOC
NEP	0.54	0.19	0.33
SHI	0.43	0.34	0.44
RdAcs	−0.12	0.47	0.42
StBufAc	−0.29	0.37	0.24

that are already planted in trees but are not yet tall enough to provide a closed canopy. Bare rock outcrops and heavily grazed meadows also give the spectral signature of bare ground or bare ground covered with nongreen vegetation such as logging slash or litter. Some of these conditions cannot be mitigated or may not need treatment. After mitigation, however, treated cells are deducted from the list, and the NEP and SHI indexes are recalculated. Resource managers may choose to optimize both environmental and economic investment strategies by locating those areas that have the greatest impact on cumulative watershed effects and selecting the mitigation that is most cost and environmentally effective. Thinking of this as an environmental accounting system permits one to allocate resources to those projects that have the most immediate impact on the net reduction of cumulative effects.

CONCLUSIONS

Need for Public-Private Cooperation

One of the most important assessment findings of the SNEP Hydrology Group is the absence of standardized tools that provide resource managers with the information they need for sound economic and environmental decision making. To make the Sierra Nevada ecosystem sustainable, both public and private landowners must be able to exchange information about their individual activities accurately, quickly, and in similar formats. Sustainability of forest ecosystems in both the eastern and western United States depends on understanding the "current cumulative condition." In order to gain this understanding at a regional scale, one must have information on what resource elements are present and how are they distributed, regardless of ownership.

Thirty-six percent of the nearly twenty-nine million acres in the SNEP study area are privately owned. These private lands are relatively evenly dispersed in "mixed ownership watersheds." The natural boundaries of many mixed ownership watersheds often exceed the administrative boundaries of the national forests and divide the watershed for analysis purposes. As an example, there might be one-third of the watershed inside the national forest, two-thirds outside, half of which is held by large landowners and half held in small lots for residential use or investment. While each landholding group may have different management plans, all agencies and private operators need a standardized database in order to calculate the combined impacts of their land use histories and from which to project their combined future activities.

Model Expansion and Improvement

As standardized and integrated soils databases are completed for other portions of the Sierra Nevada, and as DEMs of higher resolution become available, a Sierra-wide NEP and SHI analysis could be completed and reanalyzed periodically to evaluate the impacts of residential development, fire, timber harvest, and other regionally important phenomena that can be observed from space. Although this model will continue to be evaluated and validated, effects of additional elements such as climate, rain on snow, and geology will be tested to improve model performance. High-elevation basins are important and sensitive even if unmanaged; however, their contribution to the sediment load is not potentially as high as that of lower-elevation areas that are heavily managed. Therefore the problem of separating bare rock outcrops from bare exposed soil will be an important element of future NEP models.

Adjudication of "Disturbance Rights" in CWE Limited Watersheds

An accurate indexing methodology is a valuable tool when allocating resources for watershed improvement or adjudicating "disturbance rights" between landowners. Predicting the erosional potential for a given unit area of land is the objective of this methodology; it is intended to index current cumulative condition for individual planning watersheds relative to their neighbors.

The adjudication of logging rights has not yet been implemented in mixed ownership watersheds. As watersheds become "cumulative effects limited," or over the threshold of concern, to the extent that management operation must be modified, this model will provide the basis for selection of mitigation projects as well as locate areas to be avoided or managed with more informed sensitivity.

With these tools, decisions about road location, road abandonment, skid trail layout, recreation, and grazing practices may be reviewed on local or regional scales and provide better information for balancing ecosystem health, cumulative effects, and human need in order to maintain all systems in sustainable condition.

REFERENCES

Brandow, C. 1995. *Calwater 1.0—California planning watersheds data dictionary.* Sacramento, California Department of Forestry and Fire Protection.

Carlson, J., and C. Christiansen. 1993. *Eldorado National Forest cumulative off-site watershed effects (CWE) analysis process.* Placerville, CA: U.S. Department of Agriculture, Forest Service, Eldorado National Forest.

DaCosta, L. M. 1979. *Surface soil color and reflectance as related to physiochemical and mineralogical soil properties.* Columbia: University of Missouri.

Elwell, H. A., and M. A. Stocking. 1974. Rainfall parameters and a cover model to predict runoff and soil loss from grazing trials in the Rhodesian sandveld. *Proceedings of the Grassland Society of South Africa* 9:157–64.

Erman, D. C., and D. Mahoney. 1983. *Recovery after logging in streams with and without bufferstrips in northern California.* Davis: University of California, Water Resources Center.

Erman, D. C., J. D. Newbold, and K. B. Roby. 1977. *Evaluation of streamside bufferstrips for protecting aquatic organisms.* Berkeley: University of California, Department of Forestry and Conservation.

Kattelmann, R. 1996. Hydrology and water resources. In *Sierra Nevada Ecosystem Project: Final report to Congress,* vol. II, chap. 30. Davis: University of California, Centers for Water and Wildland Resources.

Kirkby, M. J., and R. P. C. Morgan, eds. 1980. Modelling water erosion processes. In *Soil erosion,* edited by M. J. Kirkby, 183–212. Chichester, England: John Wiley.

Lewis, J., and R. M. Rice. 1989. Critical Sites Erosion Study. Vol. II: Site conditions related to erosion on private timber lands in northern California. In final report submitted to the California Department of Forestry and Fire Protection, May.

Megahan, W. F., S. B. Monsen, and M. D. Wilson. 1991. Probability of sediment yields from surface erosion on granitic roadfills in Idaho. *Journal of Environmental Quality* 20 (1): 53–60.

Rice, R. M. 1993. A guide to data collection and analysis in support of an appraisal of cumulative watershed effects in California forests. Special report for the Georgia Pacific Corporation, Martell, California.

Rose, C. W. 1994. Research progress on soil erosion processes and a basis for soil conservation practices. In *Soil erosion research methods,* edited by R. Lal, 159–80. Delray Beach, FL: St. Lucie Press.

Stocking, M. A. 1994. Assessing vegetative cover and management effects. In *Soil erosion research methods,* edited by R. Lal, 211–34. Delray Beach, FL: St. Lucie Press.

U.S. Forest Service. 1987. *Cumulative off-site watershed effects analysis.* R-5 FSH 2509.22. U.S. Forest Service, Pacific Southwest Region.

Wischmeier, W. H., and D. D. Smith. 1978. *Predicting rainfall erosion losses—A guide to conservation planning.* No. 537. Vol. Agricultural Handbook. Washington, DC: U.S. Department of Agriculture.

NATHAN L. STEPHENSON
National Biological Service
Sequoia and Kings Canyon National Parks
Three Rivers, California

55

Ecology and Management of Giant Sequoia Groves

ABSTRACT

As a result of recent changes in U.S. Forest Service (USFS) policy, the two public agencies that collectively manage most giant sequoia groves—the USFS and the National Park Service—now share remarkably similar sequoia management goals: to protect, restore, and conserve giant sequoia ecosystems for their non-commodity values. The goal of greatest immediate importance is to protect sequoia groves from unusually severe wildfires; the hazard of such fires has increased with the accumulation of forest fuels during a century of fire exclusion. By reducing surface fuels, tree density, and the vertical continuity of aerial fuels, restoration of pre-Euroamerican grove conditions automatically confers a good deal of protection from extreme wildfires. Managers wishing to restore pre-Euroamerican grove conditions face at least four complex issues: (1) defining specific restoration goals (e.g. is the goal simply to restore low- to moderate-intensity fire as a natural process, letting it determine forest structure, or to mechanically restore a particular forest structure before reintroducing fire?), (2) describing the physical targets for restoration (what was the range of pre-Euroamerican grove conditions?), (3) evaluating the practicality and possibility of re-creating the target grove conditions (can we restore past conditions, given the limitations imposed by present grove conditions?), and (4) choosing specific restoration tools and approaches (what are the trade-offs among using prescribed fire, saws, or both as restoration tools?).

Once groves have been protected and restored a conservative approach to assuring their long-term sustainability is to maintain the processes that sustained them in the past, especially frequent low- to moderate-intensity surface fires. Undisturbed hydrology is also important, thus special management attention should focus on the local watershed above and adjacent to groves. There is no evidence that the long-term sustainability of giant sequoia ecosystems as a whole depends on adding to the public land base. Continuing and future threats to sequoia ecosystems include air pollution, unnatural effects of pathogens, and anthropogenic climatic change.

Present conditions in many sequoia groves demand immediate attention—particularly the ongoing failure of giant sequoia regeneration and the accumulation of hazardous fuels. Yet our present understanding of grove restoration and conservation is imperfect, meaning that management must move forward in spite of uncertainties. Success therefore depends on managers practicing adaptive management, which formalizes the common-sense process of trying something, seeing what happens, learning from the experience, then trying something new. Successful adaptive management depends on monitoring the results of different management actions, a step that is often ignored. Within certain bounds, there is no single clearly correct approach to grove restoration and conservation; thus, the different sequoia management agencies are likely to apply a variety of different management approaches. Knowledge will grow most rapidly if the various agencies cooperate in comparing the consequences of their different management approaches.

For the agencies managing giant sequoias, meeting obligations to protect, restore, and conserve sequoia ecosystems will be difficult, time-consuming, and expensive. Efforts seem sure to fail unless there is strong institutional support at all levels, including significant permanent base funding.

INTRODUCTION

The charge of the Sierra Nevada Ecosystem Project (SNEP) included conducting "[a]n examination of the Mediated Settlement Agreement [U.S. Forest Service 1990], Section B, Sequoia Groves for the Sequoia National Forest and recommendations for scientifically based mapping and management of Sequoia groves" (SNEP 1994). This chapter is limited to addressing the last part of this charge: providing an assessment for scientifically-based management of sequoia groves, with some

Sierra Nevada Ecosystem Project: Final report to Congress, vol. II, *Assessments and scientific basis for management options*. Davis: University of California, Centers for Water and Wildland Resources, 1996.

attention given to scientifically-based grove mapping. The remainder of the SNEP charge relating to giant sequoias (i.e., examining the Mediated Settlement Agreement between Sequoia National Forest and various appellants, and related institutional issues) is addressed by Elliott-Fisk et al. (1996).

To a large degree, this chapter is shaped by three premises. The first is that sequoia management policy, at its broadest, is an ethical decision reflecting human values (Croft 1994); the role of science is to inform and support the expression of those values. This chapter accepts as a given that the management goal for the majority of naturally-occurring sequoia groves, as determined by decades of public and political discourse, is to protect, restore, and conserve the natural character of the groves (see "Broad Goals of Sequoia Management," below). Science's most important role in sequoia management is to suggest different means to achieve this end, and to evaluate their possible consequences. This chapter therefore musters the best available scientific information to support a critical review and analysis of the complex policy, scientific, and practical issues related to the protection, restoration, and conservation of giant sequoia ecosystems for their amenity values. A handful of sequoia groves, both public and private, currently are managed for commodity production in addition to amenity values (e.g. see Dulitz 1986; Rueger 1994); however, a review of issues related to commodity production is beyond the scope of this chapter.

The second premise is that forest managers, policy-makers, and the public will best be served by a chapter that focuses on broad principles of sequoia ecology and management, not on site-specific management prescriptions or in-depth discussions of the mechanics of specific management tools. During my sixteen years of interactions with sequoia managers and the interested public, I have come to conclude that meaningful debate about sequoia management has been most hindered by people's differing assumptions as to the fundamental nature and dynamics of sequoia ecosystems. By focusing on general principles, then, this chapter helps lay a necessary foundation for informed discussion among scientists, policy-makers, managers, and the public. The critical review of principles will also help managers set justifiable, site-specific goals and objectives, and implement sequoia management practices that are based on sound science and consistent with policy. Additionally, by focusing on principles the chapter becomes relevant to sequoia grove management in general, not just the management of groves in the Sequoia National Forest (as emphasized in the SNEP charge).

The third underlying premise, consistent with the policies of the major sequoia land management agencies, is that the overarching goal of sequoia management is to restore and sustain the health of whole, functioning giant sequoia ecosystems. Sequoia ecosystems include the physical environment and all living organisms found where giant sequoias grow, including everything from bacteria to mice to the giant sequoias themselves. At times, approaches to managing whole sequoia ecosystems have seemed in conflict with the tremen-

dous social value placed on individual large sequoias, such as when prescribed fire has charred the trunks of some sequoias (Croft 1994; Parsons 1994; Tweed 1994). However, managing whole sequoia ecosystems and managing selected individual sequoias as objects of great social importance are not mutually exclusive, and the analysis I present here should not be taken to preclude the special status of selected big trees. Rather, managing whole sequoia ecosystems should be viewed as a conservative approach to maintaining the sequoias themselves, assuring their perpetuation for the enjoyment and benefit of future generations.

Even though the overarching goal of sequoia grove management is to sustain all the pieces of giant sequoia ecosystems, this chapter focuses almost exclusively on trees. This is because (1) social values are such that most past management conflicts have centered on trees, (2) trees are the components of sequoia ecosystems for which the best available scientific information is available, and (3) through their dominant influence on habitat structure, microclimate, and soil properties, trees exert tremendous influence on most other organisms within sequoia ecosystems.

The remainder of the chapter is divided into six sections. The first summarizes present conditions in giant sequoia groves throughout the Sierra Nevada, and is followed by a brief section summarizing the new, broad sequoia management goals adopted by the USFS. The next three sections sequentially assess sequoia grove protection, restoration, and conservation. By far most attention is given to the complex and difficult task of defining specific grove restoration goals and describing targets for restoration; of necessity, new syntheses of available scientific information are presented to support this analysis. The final section offers some general conclusions and summarizes some of the alternatives for implementing giant sequoia management.

PRESENT GROVE CONDITIONS

Giant sequoias are the largest trees on the planet and are among the oldest, sometimes living for 3,200 years or more. They often occur in stately groups which some people have likened to living cathedrals (figure 55.1). Few organisms on the planet, plant or animal, have inspired as much human admiration.

Sequoia groves are portions of Sierra Nevada mixed conifer forest that contain giant sequoias. Groves contain a mix of tree species in which sequoia is a numerically minor, but visually striking, component. Numerically, most groves are overwhelmingly dominated by white fir, with sugar pine commonly being the next most abundant species, followed by giant sequoia (Rundel 1971). Black oak, ponderosa pine, incense-cedar, Jeffrey pine, and red fir are often additional grove components.

FIGURE 55.1

Generations of Americans, as well as people from all over the world, have been awed and inspired by giant sequoias. Sequoias are the largest trees on the planet and are among the oldest, sometimes living for 3200 years or more. (Photograph by George Grant, courtesy of the National Park Service.)

FIGURE 55.2

The 75 naturally-occurring sequoia groves in the Sierra Nevada (indicated by dots) are small and scattered. Most are found south of the Kings River (which separates the Sierra and Sequoia National Forests) and are on national forest, national park, or other public land. Roughly 8% of all grove area is privately owned. (SNEP map by John Aubert.)

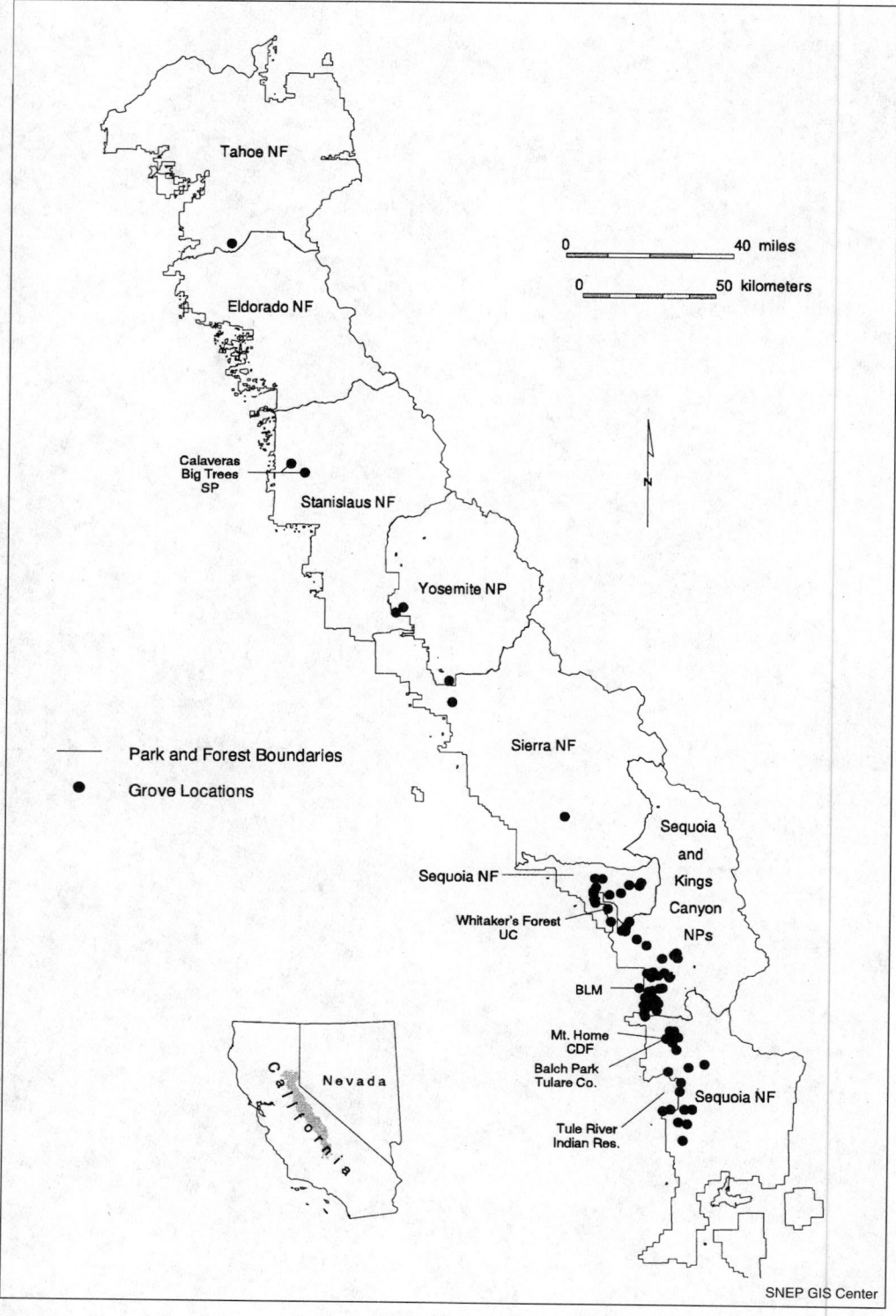

Most of the 75 naturally-occurring sequoia groves occur in the southern Sierra Nevada, south of the Kings River (Rundel 1972a; figure 55.2), collectively occupying about 14,600 ha (36,000 acres).[1] Most are under federal jurisdiction; about 49% of all grove area in the Sierra Nevada is managed by the U.S.

Forest Service (USFS), about 28% by the National Park Service (NPS), and less than one percent by the Bureau of Land Management.[2] (Percentages are of total Sierra Nevada grove area, not number of groves.) Other public ownership includes 11% of all grove area, variously managed by the California Department of Forestry and Fire Protection, California Department of Parks and Recreation, the University of California, and Tulare County. About 4% is managed by the Tule River Indian Reservation. The remaining approximately 8% of grove area in the Sierra Nevada is privately owned.

Grove areas can be classified simplistically into four broad categories according to past management, which strongly shapes future management objectives and possibilities: (1) grove areas which have been continuously protected from both fire and logging (presently about 53% of all grove area in the Sierra Nevada), (2) grove areas which have been protpected from logging, but treated with prescribed fire (about 18% of all grove area), (3) grove areas which were logged, by whatever method, before 1980 (i.e. before the most recent round of logging on the Sequoia National Forest—about 23% of all grove area), and (4) grove areas which were logged since 1980 (about 6% of all grove area). The latter two categories fail to distinguish between grove areas that have been logged more or less continuously over several decades (such as Mountain Home Grove and some private lands) from those that were logged, usually intensely, over a very short period (most other logged areas). The former represents only a relatively small portion of all grove area.

The following brief overviews describe current ecological conditions of groves in each of the four categories. However, it is important to recognize that broad variability in grove conditions occurs within each of these categories. Other sources of information on current grove conditions can be found in Hartesveldt et al. (1975), Harvey et al. (1980), Weatherspoon et al. (1986), Aune (1994), and Willard (1994b).

Grove Areas Protected from Both Fire and Logging (About 53% of All Grove Area)

For at least the two or three millennia preceding Euroamerican settlement, predominantly low- to moderate-intensity surface fires burned within individual sequoia groves on the order of every 2 to 10 years (Kilgore and Taylor 1979; Swetnam et al. 1992; Swetnam 1993). Because of the loss of Native American ignitions and suppression of lightning ignitions that followed Euroamerican settlement, most groves areas today have experienced a 100- to 130-year period without significant fire (figure 55.3)—a fire-free period that is unprecedented over at least the last two millennia (Swetnam et al. 1992). This lack of fire has resulted in important changes in grove conditions. Soil characteristics in unburned groves are more homogeneous than in burned groves (Gebauer 1992). Giant sequoia reproduction, which in the past depended on frequent fires, has effectively ceased in groves protected from fire and logging, and reproduction of other shade-intolerant species has been reduced (Harvey et al. 1980; Stephenson 1994 and in preparation). Today more area is dominated by dense intermediate-aged forest patches, and less by young patches, than in the past (Bonnicksen and Stone 1978, 1982a; Stephenson 1987). Forest conditions have become more closed in many areas (figure

55.4), and shrubs and herbaceous plants are probably less abundant than in the past (Kilgore and Biswell 1971; Harvey et al. 1980). Perhaps most significantly, dead material has accumulated, causing an unprecedented buildup of surface fuels (Agee et al. 1978; van Wagtendonk 1985; see figure 55.5). Additionally, "ladder fuels" capable of conducting fire into the crowns of mature trees have increased (Kilgore and Sando 1975; Parsons and DeBenedetti 1979; see figure 55.4).

One of the most immediate consequences of increased fuels is an increased hazard of wildfires sweeping through groves with a severity rarely encountered before Euroamerican settlement (cf. Stephens 1995; Chang 1996; McKelvey and Busse 1996; Skinner and Chang 1996; Weatherspoon and Skinner 1996; van Wagtendonk 1996). High-severity fires are those that kill many or most mature forest trees, sometimes even monarch sequoias. Though pre-Euroamerican fires usually consisted of small (on the order of 0.1 ha) patches of high-severity fire within a matrix of low-severity surface fire (Harvey et al. 1980; Stephenson et al. 1991), fires of more uniformly high severity occasionally burned large portions of individual groves (Swetnam et al. 1992; Caprio et al. 1994). Fuel conditions today are such that these formerly relatively rare, high-severity fires could become more common.

Grove Areas Treated with Prescribed Fire (About 18% of All Grove Area)

Prescribed fires for both fuels management and ecosystem management were first introduced in sequoia groves on a large scale in the late 1960s, mostly in Sequoia and Kings Canyon National Parks. Though prescribed fire has caused immediate and dramatic reductions of surface fuels (forest litter, duff, and all downed woody debris; figure 55.5), fuel re-accumulation has been relatively rapid. In Redwood Mountain Grove, Parsons (1978; see also Kilgore 1973a; Agee et al. 1978; Gebauer 1992) found that prescribed fires of the late 1960s to mid-1970s reduced the average surface fuel load to about 8% of its pre-burn value of 190 tonnes/ha (85 tons/acre). Within seven years of the fires, however, fuels had accumulated to 53% of pre-burn levels. In contrast, prescribed fires of the 1980s and 1990s, generally burning in other groves and under somewhat moister and cooler conditions, reduced the average fuel load to about 33% of its pre-burn value of 126 tonnes/ha (56 tons/acre) (Keifer 1995 and personal communication). After ten years, surface fuels within the older of these burns had nearly reached pre-burn values. Compared to pre-burn fuels, however, post-burn fuel accumulation was more heavily dominated by woody debris. Much of the rapid re-accumulation of fuel is due to fire-caused death of small trees in the abnormally dense thickets which have become established in the absence of frequent fires (Parsons 1978). It is therefore evident that a sustained reduction of fuel accumulation rates to their probable pre-Euroamerican levels will require at least two prescribed fires, the second of which removes the small

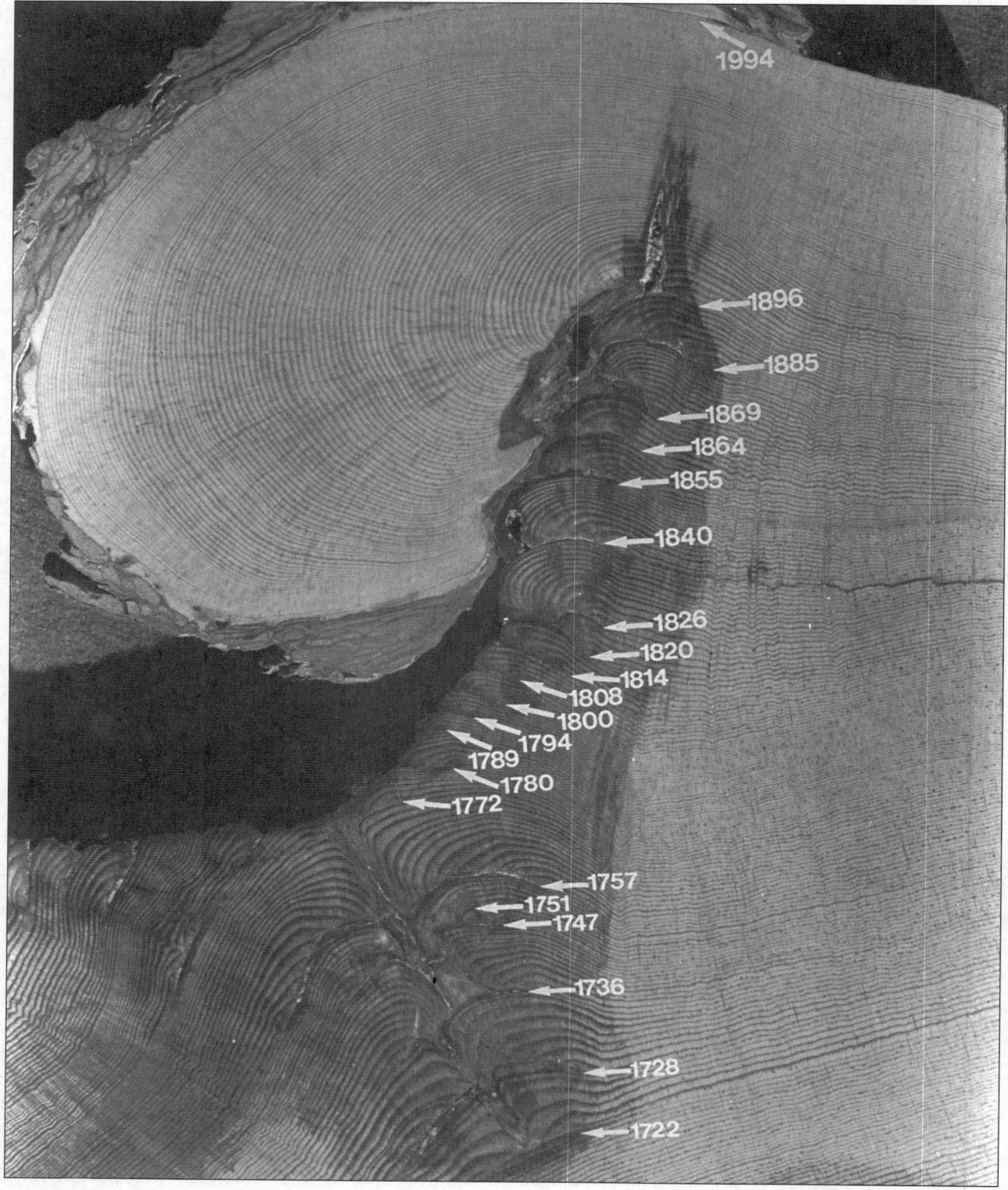

FIGURE 55.3

The dates of past fires are revealed by scars in the growth rings of this ponderosa pine from the edge of the Big Stump Grove, Kings Canyon National Park. Between 1722 and 1896, 20 fires burned at the base of this particular tree at intervals ranging from 4 to 16 years, averaging one fire every 8.7 years. In most groves, pre-Euroamerican fires were predominantly low- to moderate-intensity surface fires. Fire scarring in most groves ceased abruptly in the 1860s or 1870s, due to the loss of ignitions by Native Americans, suppression of lightning ignitions, and perhaps due to reduction of fine fuels by grazing. The last two fires revealed in this cross-section (1885 and 1896) were almost certainly related to logging activities in the heavily-logged Big Stump Grove (see figure 55.6). The recent 100- to 130-year fire-free interval in most sequoia groves is unprecedented during at least the last 2000 to 3000 years. (Photograph courtesy of C. Baisan, T. Swetnam, and M. Wilkenson, University of Arizona.)

FIGURE 55.4

Top: The Confederate Group of giant sequoias in Mariposa Grove, Yosemite National Park, was nearly free of understory trees in about 1890.

Bottom: By 1970, in the absence of frequent surface fires, a dense thicket of white firs grew at the base of the sequoias. Such thickets provide fuels that could conduct fire high into the sequoias. (Photographs courtesy of Bruce M. Kilgore, National Park Service.)

FIGURE 55.5

Top: Lack of frequent surface fires has led to heavy build-up of surface fuels, such as shown in this photo taken within the Redwood Mountain Grove, Kings Canyon National Park. Such fuels increase the hazard of wildfires that are generally more severe than those of pre-Euroamerican times.
Bottom: Low- to moderate-intensity prescribed fires can greatly reduce hazardous fuels, as in this view of the same spot in Redwood Mountain Grove following a prescribed fire. (Photographs by Dan Taylor, courtesy of Bruce M. Kilgore, National Park Service.)

trees killed by the first. Importantly, the death of small understory trees and the lower branches of larger trees has reduced "ladder fuels" otherwise capable of conducting wildfires into the crowns of the largest trees (Kilgore and Sando 1975).

Tree density is reduced in burned groves. After a prescribed fire in Redwood Mountain Grove, Kilgore (1973a) found that total tree density was reduced by 81%. Almost all of the dead trees were firs and pines less than 30 cm diameter, and especially less than 15 cm diameter; these small trees occurred in abnormally dense thickets which had become established during a century of fire exclusion. In a more extensive sample of 29 plots in several groves, small trees were less common in the pre-burn forest and average tree density was reduced by only 36% one year following prescribed fires, again with greatest mortality in the smallest trees (Keifer 1995 and personal communication; see also Keifer and Stanzler 1995). Because almost no sequoias of any size were killed by fire in these plots, the relative density of sequoia increased at the expense of white fir. (This is partly because most sequoias in modern groves are large and relatively fire-resistant.) The absolute density of sequoias greater than 1.4 m tall doubled within 10 years of the fires and is likely to keep increasing, due to the rapid ingrowth of seedlings following the fires. There has not yet been a corresponding increase in the density of firs greater than 1.4 m tall. In a separate study which followed the fate of 1135 giant sequoias for 23 years, Lambert and Stohlgren (1988) found that the death rate (not the proportion dying) of sequoias less than 30 cm in diameter increased by 65% in areas that had been burned. Their data also suggested that death rates increased for sequoias larger than 30 cm in diameter, but death rates of large sequoias in both the presence and absence of fire were so low as to be statistically indistinguishable.

Tree death following prescribed fires is spatially clumped (Kilgore 1973a, 1973b; Harvey et al. 1980; Stephenson et al. 1991; Demetry 1995; Demetry and Duriscoe 1996). Demetry (1995) found that 18 forest gaps created by prescribed fires in Giant Forest were of variable size (the author's non-random sample included gaps of 0.067 to 1.17 ha), with 0.1 ha being the approximate (to the nearest order of magnitude) modal gap size for a large portion of Giant Forest (A. Demetry, personal communication). These fire-created forest gaps are the site of abundant sequoia seedling establishment and rapid growth, and appear to be essential for successful sequoia regeneration in the absence of other gap-creating disturbances (Harvey et al. 1980; Stephenson et al. 1991; Mutch 1994; Stephenson 1994).

Shrubs, particularly *Ceanothus* and *Ribes,* are much more abundant in burned groves than in unburned and otherwise undisturbed groves (Kilgore and Biswell 1971; Harvey et al. 1980). Herbaceous cover also is greater in burned groves, though it generally begins to decline a few years following a fire (Harvey et al. 1980).

Grove Areas Logged before 1980 (About 23% of All Grove Area)

The heaviest logging of sequoia groves occurred south of the Kings River between about 1880 and 1920 (figure 55.6). Nearly all pines and many firs were removed from several groves, though trees of lesser value (particularly small trees) often were left, providing a seed source for regeneration. In many areas (particularly the groves in or near Converse Basin), nearly all old-growth sequoias were removed.

Today, these logged groves have regenerated as complex mosaics of forest patches of differing structures (tree diameter, height, and density) and species compositions (R. Rogers, personal communication). Some patches are densely stocked with nearly pure, century-old giant sequoias; this regeneration now typically ranges from 0.3 to 1 m (1 to 3 ft) in diameter and 30 to 50 m (100 to 150 ft) tall. There are few understory trees or shrubs in these dense patches. Other patches have similar structure but contain additional species, particularly white fir and sugar pine. In still other patches, trees are sparse and the once-forested lands are dominated by shrubs (on dry sites) or grasses, sedges, and forbs (on wetter sites), with scattered large sequoia stumps standing as reminders of past conditions. In some regenerating patches, fuels (and therefore fire danger) are generally high and are steadily increasing. Some heavily-logged sites have been re-disturbed more recently, such as by the 1955 McGee wildfire; many of these sites are now dominated by young-growth white fir and planted ponderosa pine rather than giant sequoia.

In the heavily-logged Big Stump Grove (figure 55.6), Stohlgren (1992) found that patches of regeneration that are presently dominated by sequoias do not necessarily grow in the same places within the grove that were dominated by sequoias in pre-Euroamerican times. Locations of slash burning and other factors may have influenced this spatial redistribution of sequoia dominance. Because some sequoia seedlings became established beyond the original grove boundary, there was a small net increase in grove area following logging. Stohlgren reported that in some respects the grove seemed resilient to heavy logging; 85 years after logging, it had already recovered nearly half of its pre-logging sequoia basal area. However, as Stohlgren noted, after 85 years the sequoia stumps he used to estimate pre-logging basal area had probably lost most of their bark and sapwood, meaning that he underestimated pre-logging sequoia basal area for the grove. If estimates of lost sapwood and bark thicknesses are added to the pre-logging basal area calculations (Stephenson, unpublished data), the grove seems somewhat less resilient; a revised estimate is that after 85 years the grove had recovered about one third of its pre-logging sequoia basal area. Much of this regeneration is dense and would have been thinned by recurring fires, had they not been excluded.

FIGURE 55.6

In the heavily-logged Big Stump Grove, a photographer captured that last moments of the Mark Twain Tree as it was felled in 1891 to provide cross-sections for museum exhibits. For scale, note the people standing on the edge of, and to the right of, the Twain Tree's 7 m (24 ft) diameter stump. About one-fourth of all naturally occurring sequoia grove area was logged between 1880 and 1980, mostly early in that period. The logging often including most or all of the largest sequoias. Most of these heavily-logged grove areas have regenerated and often are now dominated by dense growth of young trees or brush. Fuels have re-accumulated to the point that many logged areas are now at risk of unusually severe wildfires. (Negative 42130, courtesy of the Department of Library Services, American Museum of Natural History.)

A few larger sequoias, on the order of 200 or 300 years old, are scattered throughout areas that were heavily logged near the turn of the century. These sequoias are just beginning to show the rounded crowns characteristic of old growth sequoias; their crowns contrast sharply with the extremely pointed crowns of sequoias that became established since logging. These older and larger sequoias will help restore old-growth character to once-devastated groves.

Grove Areas Logged since 1980 (About 6% of All Grove Area)

Logging in USFS sequoia groves during the 1980s helped spur the events leading to the Mediated Settlement Agreement (MSA) between Sequoia National Forest and various appellants. Overall, about 490 ha (1,200 acres) of sequoia groves, mostly on the Sequoia National Forest, were logged (R. Rogers, personal communication). About one third of this area was selectively logged, with only occasional trees being cut; no large

FIGURE 55.7

These recently-logged blocks in the Black Mountain Grove, Sequoia National Forest, illustrate some of the effects of logging during the 1980s. Though large sequoias were not cut, logging such as this helped spark the controversies that ultimately led to the Mediated Settlement Agreement between Sequoia National Forest and various appellants. (Photograph by Nathan L. Stephenson, National Biological Service.)

sequoias were cut. There is presently little visual difference in forest structure between these areas and the surrounding uncut forest matrix (R. Rogers, personal communication).

The other two thirds of logged grove area was logged in distinct 2 to 10 ha (5 to 25 acre) patches within an otherwise intact grove matrix (figure 55.7). The patches presently are open stands occupied by scattered large sequoias (again, no large sequoias were cut) and occasionally other mixed-conifer trees left as seed sources or along riparian corridors. This approach

to logging was described as "modified clear-cutting" by critics (Cloer 1994), and helped spark the chain of events that led to changes in USFS policy regarding sequoia groves. Most of the logged forest openings today are dominated by shrubs and planted or naturally-seeded trees. Most gap area was planted with mixed ponderosa pine and giant sequoia, either in equal numbers or with ponderosa pine dominating. In some areas, sugar pine and white fir were also planted. Shrubs presently are about 0.5 to 1.2 m (2 to 4 ft) tall, with tree seedlings somewhat taller, though conditions are variable from place to place. There is little surface fuel accumulation, as these sites were cleared before planting.

BROAD GOALS OF SEQUOIA MANAGEMENT

The Mediated Settlement Agreement (MSA) (U.S. Forest Service 1990), President Bush's Presidential Proclamation (Bush 1992), and Regional Forester Stewart's policy directive (Stewart 1992) collectively provide a uniform suite of policy and management direction for all naturally-occurring sequoia groves in national forests. All USFS grove areas containing old-growth giant sequoias are to be managed in such a way that will "... protect, preserve, and restore the Groves for the benefit and enjoyment of present and future generations" (U.S. Forest Service 1990).[3] Additionally, "... groves shall be protected as natural areas with minimal development. ... [A]ny proposed development shall provide for aesthetic, recreational, ecological, and scientific value" (Bush 1992). Regional Forester Stewart (1992) further defined the new policy direction: "Naturally occurring groves, including an appropriate ecological 'buffer', shall be withdrawn from the land base considered suitable for the long term sustained (regulated) production of timber. Groves shall also be withdrawn from other forms of consumptive entry such as mineral and geothermal developments." These policy and management directions chart a new course for USFS management of the groves, and raise a broad spectrum of challenging issues.

As explained in more detail later, I interpret the intent of the policy statements and management goals expressed in the documents cited above to be:

- To protect sequoia ecosystems from commodity-driven uses (such as logging and associated road construction) and from other major disturbances (such severe wildfires) which could preempt future management and use options,

- To restore sequoia ecosystems to the range of conditions that existed before Euroamerican settlement

- To conserve sequoia ecosystems, assuring their long-term sustainability in the face of changes resulting from Euroamerican settlement and potential future threats such as air pollution, unnatural effects of pathogens, and anthropogenic climatic change.

These goals are remarkably similar to the sequoia management goals of the NPS (see Parsons 1994). Therefore the following assessments of the policy and practical issues surrounding protection, restoration, and conservation draw heavily on decades of NPS experience managing giant sequoia ecosystems.

GROVE PROTECTION

The term *protection* has had several different meanings with reference to sequoia groves. In the MSA (U.S. Forest Service 1990), Presidential Proclamation (Bush 1992), and Regional Forester's policy directive (Stewart 1992), grove protection mostly refers to protection from mechanical human disturbances (such as road construction and logging for commodity production) that are inconsistent with the amenity values of groves. As a consequence of the recent changes in USFS policy (Stewart 1992), this form of grove protection currently is in place for USFS groves (as it has been for decades in NPS groves).

For much of the history of sequoia management by Euroamericans, protection has also meant exclusion of all fire. This form of protection has allowed both surface and aerial fuels to accumulate within and surrounding groves, thereby increasing the hazard of unusually severe wildfires (cf. Stephens 1995; Chang 1996; McKelvey and Busse 1996; Skinner and Chang 1996; Weatherspoon and Skinner 1996; van Wagtendonk 1996). In the last 50 years, four groves (Case Mountain, Cherry Gap, Converse Basin, and Redwood Mountain) have been burned at least partly by large wildfires. Some areas within these groves were burned severely, particularly areas that had been logged before the fires but that had not been subjected to subsequent fuel reductions. The hazard of similar or more severe fires occurring in other groves is steadily increasing. Unlike protection from logging, groves cannot be protected from severe wildfires simply by a change in written policy.

Thus, the foremost immediate concern of all giant sequoia managers is to assure that future management and use options are not preempted by unusually severe wildfires; this is the form of grove protection that will receive the greatest attention in this chapter. Groves can be protected from wildfire by altering fuel conditions inside of groves, altering fuel conditions outside of groves, or both. Methods for altering fuel conditions are discussed elsewhere (Stephens 1995; van Wagtendonk 1996; Weatherspoon 1996; Weatherspoon and Skinner 1996). The effectiveness of grove protection conferred by fuels management within groves has been demonstrated twice within the last decade. In August of 1987, a lightning-ignited wildfire swept into the Redwood Mountain Grove (Sequoia National Forest and Kings Canyon National Park). The fire grew quickly in size and severity, in some places completely scorching or consuming the crowns of huge pines, firs, and even monarch giant sequoias, killing the trees (Stephenson et al. 1991). Fire crews were successful in containing the blaze only after it died down upon entering the portion of the grove that had been prescribed burned by NPS managers a few years before (Nichols 1989). In October of 1988, a wildfire caused by a carelessly discarded cigarette raced upslope through heavy chaparral toward the famous Giant Forest grove in Sequoia National Park. To contain the blaze and protect the grove, fire crews began to ignite backfires along the edge of Giant Forest. Recent prescribed burns had been so effective in reducing fuels that the fire crews could barely get their backfires to burn. This freed some firefighting resources to be focused elsewhere, and created a no-panic situation in which fire control efforts could go forward more deliberately. The fire was easily contained upon reaching the grove.

A fundamental premise of this chapter is that restoration of sequoia groves to pre-Euroamerican conditions automatically confers a large measure of protection from extreme wildfires (Fullmer et al. in press; Weatherspoon and Skinner 1996). As discussed elsewhere in this chapter, pre-Euroamerican sequoia groves were less dense, had lower average fuel loads, and had less continuous vertical fuels than typical unlogged and otherwise undisturbed groves today. Consequently, pre-Euroamerican groves supported fires of predominantly low to moderate severity (i.e., spatially variable fires that killed many seedlings and saplings and some subcanopy trees, with occasional patches of high severity which locally killed many or most trees of all ages; Stephenson et al. 1991). (There were a few notable exceptions, such as the predominantly high-severity fire that swept through Mountain Home Grove in A.D. 1297; Swetnam et al. 1992; Caprio et al. 1994.) It stands to reason that once grove structure (which broadly includes the spatial arrangement and sizes of forest patches and the diameters, heights, and densities of trees in the patches) and fuel characteristics are restored to pre-Euroamerican conditions, pre-Euroamerican fire behavior will follow, and groves will thus be less susceptible to severe wildfires (Weatherspoon and Skinner 1996).

This line of reasoning suggests that restoration and protection can proceed simultaneously, as a single action. In fact, this may be the only reasonable approach to grove management, since changes in grove structure automatically accompany fuel manipulations designed to protect groves from unusually severe wildfires. For these reasons many of the issues surrounding protection are addressed below, in the section on restoration.

GROVE RESTORATION

The MSA (U.S. Forest Service 1990) does not clearly define the term *restore*. It states that "[t]he objectives of regenerating cutover Giant Sequoia Groves will be to restore these areas, as nearly as possible, to the former natural forest condition" (p. 27). Unfortunately, it is unclear whether the "former natural forest condition" means what occurred immediately before logging (which might have been quite different from pre-Euroamerican conditions), or what occurred "naturally" in pre-Euroamerican times. Another statement hints at the latter: "The objective of fuel load reduction plans shall be to preserve, protect, restore, and regenerate the Giant Sequoia Groves ..." (pp.

10–11). The use of the term *restore* in this context implies that present conditions in many undisturbed groves, which often include unusually heavy fuel loads, are not the desired "natural forest condition." Regional Forester Stewart's 1992 sequoia management policy directive offers a specific interpretation of the intent of the MSA. Stewart directed that sequoia management activities "... shall generally be designed to recreate and maintain stand structure, including the long term recruitment of 'specimen' giant sequoia trees and understory vegetation, that would have occurred naturally prior to the settlement of California by European immigrants." This USFS policy statement, which is meant to comply with the intent of the MSA, declares that pre-Euroamerican grove structure is to be restored and maintained. NPS policy implies similar goals for restoration.

The goal of recreating pre-Euroamerican grove conditions is logical and defensible. As described earlier, groves returned to pre-Euroamerican conditions will be protected, to a large degree, from unusually severe wildfires because of reduced surface and aerial fuels (Weatherspoon and Skinner 1996). Additionally, re-creating the conditions that sustained the groves for millennia is the most conservative approach to assuring their continued long-term sustainability (Fullmer et al. in press).

Managers wishing to restore groves to pre-Euroamerican conditions face at least four complex issues:

1. *Defining specific restoration goals consistent with the overall goals and policies of the agencies or individuals managing the groves.* A specific restoration goal discussed in detail later, for example, is to restore grove structure, species composition, and function to the usual range of conditions that existed in the 1,000 years preceding Euroamerican settlement.

2. *Describing the targets for restoration.* For the preceding example, this would entail explicitly describing the range of grove conditions that occurred in the 1,000 years preceding Euroamerican settlement.

3. *Evaluating the practicality (and possibility) of meeting restoration objectives.* Having determined the range of grove conditions that existed in the 1,000 years preceding Euroamerican settlement, can we realistically expect to restore groves to this range, given that present grove structure and composition limit the possible changes we can make?

4. *Choosing restoration tools and approach.* Choice of restoration tools and approach depends on various trade-offs and on the starting grove structure.

As in past analyses of issues surrounding grove restoration, the following analysis focuses most strongly on unlogged, unburned groves—the most abundant grove type in the Sierra Nevada and among the most vulnerable to unusually severe wildfires. However, attention is also given to specific issues surrounding restoration of logged groves.

Defining Restoration Goals

Structural versus Process Restoration

The most lively and instructive debate over appropriate grove restoration goals has been between "structural restorationists" and "process restorationists" (Vale 1987). Simply stated, structural restorationists have argued that grove structure (the spatial arrangement and sizes of forest patches and the diameters, heights, and densities of trees in the patches) and species composition must be restored, by whatever means possible, before natural processes (particularly fire) are allowed to run a more natural course in determining grove dynamics (Bonnicksen and Stone 1978, 1982b, 1985). In contrast, process restorationists have argued that initial grove structure is unimportant; the goal of restoration is to restore the major processes (particularly fire) that shaped sequoia ecosystems in pre-Euroamerican times in such a way that "the interaction of those processes with other ecosystem elements ... [is] ... similar to that which would have occurred had modern humans not intervened" (Parsons et al. 1986; Parsons 1990a). The debate has largely centered on NPS grove management; the details of the debate therefore have been colored by the assumption (following NPS policy) that at some point during or after grove restoration, fire becomes the tool of choice in determining future grove structure and composition. Broadly, however, the debate is equally relevant to management approaches in which mechanical manipulation, not fire, is the management tool of choice.

Some of the disagreements between structural restorationists and process restorationists have hinged upon differing interpretations of NPS legislation, goals, and policies. No attempt is made here to assess or reconcile differences among these interpretations; excellent summaries can be found elsewhere (Bonnicksen and Stone 1978, 1982b; Bancroft et al. 1985; Parsons et al. 1986; Lemons 1987; Parsons 1990a). Instead, this section summarizes the philosophical and practical issues that have driven the debate—those issues most relevant to managers and policy makers establishing future restoration goals for sequoia groves.

Both the structural and process restoration viewpoints recognize the dynamic nature of forests and share a similar goal of restoring "natural" forest conditions lost during a century of fire exclusion (Vale 1987). However, the viewpoints differ in several important respects. According to Vale (1987), the structural restorationist viewpoint is characterized by (1) reestablishing a precise forest structure, (2) calibrating initial restoration to a particular point in time, (3) focusing only on the effects of fire suppression on forest trees, (4) emphasizing the vegetation structure needed to reestablish natural process (fire), and (5) believing that vegetation change is not easily reversible. In contrast, the process restorationist viewpoint is characterized by (1) reintroducing a general forest process

(fire), (2) calibrating restoration to a general period, not a precise point in time, (3) recognizing multiple causes of vegetation and ecosystem change, (4) believing that reestablishing process (fire) will eventually allow the forest to reestablish its natural structure, and (5) believing that vegetation change is easily reversible. Lemons (1987) has persuasively argued that many of these differences in viewpoints are based not on science, but on largely unarticulated human values.

Championing the structural restorationist viewpoint, Bonnicksen and Stone (1978, 1982b) argued for the necessity of structural restoration preceding the reintroduction of fire, largely based on their contention that fire suppression had led to more uniform fuel and vegetation conditions within sequoia groves, thus blurring the boundaries between formerly distinct forest patches of different ages and structures. This increased uniformity in forest conditions, they argued, would be perpetuated even after fire was reintroduced, thereby erasing the original character of the forest mosaic. As part of their analysis, Bonnicksen and Stone (1978, 1982b) evaluated the practicality and desirability of several potential vegetation restoration goals that at some point have been considered by land managers: (1) maintaining vegetation exactly as it is today, (2) restoring vegetation to its pre-Euroamerican state, then maintaining it exactly in that condition, (3) restoring pre-Euroamerican conditions, then allowing fire to determine future conditions, (4) reintroducing fire in the present vegetation without first restoring pre-Euroamerican structure, and (5) restoring conditions as they would be today, had Euroamericans never arrived, then reintroducing fire. Bonnicksen and Stone recognized that the first two goals are impossible to achieve; vegetation is dynamic and cannot be frozen in time. Of the latter three goals, they felt that the last would most closely fit NPS legislation and policies. They also recognized that the last goal could never be met perfectly due to physical limitations imposed by present grove conditions (see below).

There is much appeal to the structural restorationists' preferred goal of restoring conditions as they would be today, had Euroamericans never arrived, followed by allowing natural processes (especially fire) to play a major role in determining future forest conditions. However, some sequoia managers have argued that the goal is impractical. Expanding on Vale's (1987) summary, some of the reasons that NPS sequoia managers have adopted process rather than structural restoration goals are listed below. No attempt has been made to assure consistency among the reasons. As will be discussed later, the following reasons do not preclude the possibility of setting some broad structural goals.

It is difficult or impossible to quantitatively define precise structural targets for grove restoration. There are strong limitations on the accuracy and precision with which we can determine grove characteristics in the past, much less hypothetical grove conditions that would exist today if Euroamericans had never arrived.

Even if they could be quantified, it is impossible to approach some structural goals in less than several centuries (if ever). As elaborated later, this fact is a practical, not philosophical impediment.

The urgent need to protect groves from wildfire eliminates the option of waiting until quantitative structural goals can be defined and implemented with confidence.

At least within national parks (where policy generally prohibits timber sales), structural restoration by mechanical means is prohibitively expensive for all but small areas. A possible rebuttal to this argument is that prescribed fire is less expensive than mechanical manipulation, and can be used as a tool to restore at least some aspects of grove structure (see the next subsection). Outside of national parks, sale of trees removed during restoration might partially offset costs.

A large proportion of grove area is legally-defined wilderness, where mechanical tools for restoration are generally prohibited. Again, structural restoration by prescribed fire may be an option.

It is difficult to justify expending scarce funds on fine-tuning forest structure when other threats may unravel the restoration efforts. The interacting effects of air pollution, introduced diseases (such as white pine blister rust), and potential anthropogenic climatic change collectively threaten to force large changes on sequoia ecosystems; such changes could overwhelm efforts to restore forest structure altered by fire suppression. Of course, structural restoration (specifically, reduction in tree density within groves) could increase grove vigor by reducing competitive stresses, and therefore could diminish the effects of the other threats (Ferrell 1996).

The informed opinion of some managers is that the pre-Euroamerican range of variation in forest conditions will be restored if process (fire) is restored. At least one model of landscape change suggests that this belief may be true in at least some forest types (Baker 1994). However, the possibility has yet to be convincingly demonstrated for all aspects of sequoia grove structure (see the next subsection).

Some managers have suggested that present forest structure and composition may already fall within the pre-Euroamerican range of variability, therefore there is no need for structural restoration. The fossil pollen record (Anderson 1994; Anderson and Smith 1994) shows that large compositional changes (and presumably structural changes) occurred in sequoia groves over the last 10,000 years, sometimes including combinations of tree species that no longer exist (such as sequoia growing with lodgepole pine). Some managers have suggested that at some point during these wide variations in grove conditions, forest structure may have been similar to today's (which is partly a consequence of fire exclusion). However, if we limit ourselves to considering only past forest structures that existed under climatic conditions relatively similar to today's, this argument is unpersuasive.

Given how little we know about sequoia ecosystems (which include much more than just the trees focused on by structural restorationists), a conservative management approach is to (1)

avoid introducing new processes (such as mechanical restoration and its accompanying soil disturbance) which have unknown immediate or long-term effects on many ecosystem components, and (2) restore and maintain those processes that sustained the grove ecosystems in the past. To paraphrase Aldo Leopold, a wise tinkerer keeps all the pieces, including fire, and avoids adding new pieces.

Reconciling Structural and Process Restoration Goals

Nearly a decade of renewed sequoia ecosystem research, mostly funded by the NPS and the National Biological Service (Parsons 1990b; Stephenson and Parsons 1993), has provided a rich background of new findings to support a reassessment of the debate between structural and process restorationists. Partly in response to the results of this new research, NPS managers have begun cautiously to step back from pure process restoration and to also consider structural goals (Parsons 1995). Here I attempt to close the formerly enormous gap between the structural and process restorationist viewpoints by finding a balance between the idealistic view of structural restorationists, which recognizes that grove structure and process are inextricably intertwined, and the pragmatic view of process restorationists, which recognizes physical limitations to structural restoration. To reach this end, I muster the available scientific information to demonstrate that (1) process restoration alone, without a preceding mechanical treatment, can restore and sustain at least some aspects of pre-Euroamerican grove structure, and (2) broad structural restoration goals can be defined which bracket a wide range of possible outcomes, consistent with our limited knowledge of past grove conditions, physical limitations to grove restoration, and the intrinsic variability of giant sequoia ecosystems.

In support of their contention that reintroduction of fire without a preceding structural restoration would perpetuate unnatural grove changes, Bonnicksen and Stone (1981) cited spatial data from a single 80 m x 80 m plot established in a recently-burned portion of Redwood Mountain Grove. Bonnicksen and Stone found that trees 41 to 60 years old (i.e., a cohort that became established since fire suppression became effective) within the recently-burned plot were clumped in a hierarchical pattern. Since two separate unburned plots in the same grove showed similar hierarchical clumping within the same tree age class, they concluded that "the prescribed burn did not significantly alter the pattern for this age class." They additionally concluded that "[s]ince this [hierarchical] pattern was not characteristic of most older age classes [in the same three plots] it was probably not characteristic of the presettlement giant sequoia - mixed conifer forest community." By Bonnicksen and Stone's reasoning, these findings demonstrated that fire perpetuated a Euroamerican-induced change in the forest mosaic. They suggested that similar changes existed in, and would be perpetuated in, other sequoia groves in which prescribed fire is reintroduced without a preceding structural restoration (Bonnicksen and Stone 1978, 1981, 1982b).

For several reasons Bonnicksen and Stone's arguments are unpersuasive. First, their analysis was based on only a single burned plot. Second, they did not actually measure changes in forest pattern resulting from a fire; they inferred changes by comparison with two different unburned plots. Third, though they concluded that the present clumping of 41- to 60-year-old trees was unnatural because it differed from that of older trees, it has long been known that tree spatial pattern changes fundamentally with age (e.g. Laessle 1965). Finally, and most important, direct evidence from the studies outlined below demonstrates that high spatial heterogeneity in present-day fuels, prescribed fire behavior, and consequent forest response continue to result in a forest mosaic that, at least in gap and patch sizes, is similar to that of pre-Euroamerican times.

After a century of fire exclusion, average surface fuel loads (forest litter, duff, and all downed woody debris) within unlogged groves are high (128 tonnes/ha [57 tons/acre]; Keifer 1995 and personal communication) but are also highly variable. (Data are not available for logged groves.) Kilgore (1973a) found extreme fuel variability at scales of a few meters. Variability is high even at larger spatial scales; averaged fuel loads within each of 26 approximately 0.1 ha plots within several sequoia groves (stratified random sampling) ranged from 42 tonnes/ha (19 tons/acre) to 301 tonnes/ha (134 tons/acre), a seven-fold difference (Keifer 1995 and personal communication). This variability in fuels accentuates variability in prescribed fire behavior, which is already high due to differences in local topography and changes in daily and seasonal weather and fuel moisture (Kilgore 1973b; Harvey et al. 1980). Kilgore (1973a) found that total energy released during a prescribed fire in the Redwood Mountain Grove varied by several orders of magnitude over a distance of a few meters. During two prescribed fires in other groves, flame length (a measure related to fire intensity) at predesignated monitoring points varied from 0 (smoldering combustion) to more than 1 m (M. Keifer, personal communication). In a pocket of extremely heavy fuels during another prescribed fire, flame lengths were more than 12 m (Nichols 1977).

Such variability in fire behavior and intensity, in turn, contributes to variability in fire effects and forest response. For example, Gebauer (1992) showed that spatial heterogeneity in four of seven soil characteristics was significantly greater in recently-burned areas of Giant Forest than in areas that had not burned for more than a century (there was no significant trend in the remaining three soil characteristics). Kilgore (1973a) showed that in a study plot in the Redwood Mountain Grove (different from the plots studied by Bonnicksen and Stone), non-uniform fuels and fire behavior broke a uniform thicket of young white fir into a distinct gap (greater than 0.05 ha) and two smaller remaining thickets. Demetry (1995) found that 18 forest gaps created by a number of prescribed fires burning under different conditions in Giant Forest were of variable size (the author's non-random sample included gaps of 0.067 to 1.17 ha). (Gap size depends on how gaps are defined; Demetry defined gaps using a slight modification of the meth-

ods used by Spies et al. 1990.) The approximate modal gap size (to the nearest order of magnitude) was 0.1 ha for a large portion of Giant Forest (A. Demetry, personal communication). These gap sizes correspond to pre-Euroamerican gap sizes inferred from sequoia age structure analysis (Stephenson 1994); they also roughly correspond to the modern-day 0.0135 to 0.16 ha forest patch sizes found in Redwood Mountain Grove by Bonnicksen and Stone (1981, 1982a). Demetry additionally found that composition and structure of tree and shrub regeneration varied with gap size (Demetry 1995; Demetry and Duriscoe 1996).

Collectively, these data suggest that process restoration alone can restore or sustain at least one component of pre-Euroamerican forest structure: the relative abundances of different sizes of forest gaps and, presumably, resulting forest patches. We do not yet know, however, whether other aspects of grove structure ultimately will be restored, such as the relative proportions of forest patches in different age classes or with particular species compositions. However, it is worth examining the results of a landscape dynamics model developed by Baker (1994), which simulated the effects of different fire regimes on eight measures of forest mosaic pattern in the Boundary Waters Canoe Area, Minnesota. Baker's simulations suggested that by simply restoring the pre-Euroamerican fire regime to a forest that had been altered by 82 years of fire suppression, the forest mosaic would be restored to its pre-Euroamerican range of variability. Restoration of some aspects of the simulated forest mosaic occurred within 50 to 75 years; all eight measures of forest pattern were restored within 125 to 250 years after fire reintroduction. While Baker's results apply to a forest type that differs in many ways from sequoia groves, they suggest that it is not unreasonable to believe that grove structure might be restored by process reintroduction alone. The possibility should be tested more thoroughly with the aid of linked fire and forest models tailored specifically to sequoia groves, and by better monitoring of prescribed fire effects.

I now turn to evidence supporting part of the structural restorationists' viewpoint—that it is possible to define broad structural restoration goals that reflect practical limitations. As described in later subsections, physical constraints limit our ability to quantitatively describe grove conditions at a specific point in time, and especially grove conditions that would exist today if Euroamericans had never arrived. Even if we could describe these conditions, physical constraints limit our ability to recreate them. Of necessity, then, a reasonable goal for structural restoration must bracket a range of possible outcomes (often called "natural range of variability," or NRV; Manley et al. 1995). Conversation among sequoia managers and scientists has centered on two slightly different structural restoration goals that meet the latter criterion: (1) restore grove structure and composition within the range of variation that occurred during pre-Euroamerican periods in which the climate was similar to today's, or (2) restore structure and composition within the range of variation that occurred over a long but relatively recent pre-Euroamerican period. Describing targets appropriate to these goals is still an enormous challenge (see the next subsection), but is more realistic than calibrating restoration to a specific point in time.

Paleoecological records from pollen sediments and tree rings help set limits on what pre-Euroamerican time periods might provide appropriate targets for the two goals. Pollen records from meadow sediments demonstrate that within present grove boundaries, sequoias began to increase dramatically in importance relative to pines about 4,500 years ago, coincident with a slight global cooling (Anderson 1994; Anderson and Smith 1994). Though the pollen records suggest that changes in the relative proportions of tree species in groves have continued up to the present, most of the changes were completed by about 1,000 years ago. Even though climate and fire regimes have varied within groves during the last 1,000 years (Hughes and Brown 1992; Graumlich 1993; Swetnam 1993; Caprio et al. 1994), the variation has been relatively non-directional and the combined effect on giant sequoia demography at centennial time scales has been moderate; by far the largest deviation from equilibrium conditions (stationary age distribution) in giant sequoia populations over the last two to three millennia is due to the effects of fire suppression during the last century (Stephenson in preparation). The millennium preceding Euroamerican settlement therefore seems to be a good period for calibrating goals for structural restoration.

However, it seems unnecessarily restrictive and perhaps philosophically difficult to defend calibrating structural restoration only to those climatic periods during the last millennium that were similar to the present. Sequoia groves in the millennium preceding Euroamerican settlement experienced only a few brief periods of climate comparable to the warm, wet conditions of the last few decades (Graumlich 1993). Additionally, the natural tendency for vegetational change to lag behind climatic change ("vegetational inertia"; Cole 1985) suggests that forest structure and composition during these brief periods was to a large degree a legacy of preceding annual-, decadal-, and centennial-scale shifts in climate and fire regimes, not the climate of the moment. It therefore seems reasonable to conclude that a variety of different grove structures, not a single predictable grove structure, probably occurred during periods that shared similar climates. On the other hand, this does not imply that it is appropriate to calibrate restoration to any arbitrary time period. As described earlier, large directional changes in grove composition coincided with a general climatic cooling from about 4,500 to 1,000 years ago. I suggest that only the millennium preceding Euroamerican settlement is an appropriate period for calibration, because changes in the relative proportions of tree species slowed during this period, and grove composition was more similar to today's than at any time period for which we have pollen or other fossil records. Climatic changes during the last millennium have tended to be relatively non-directional when compared to the preceding several thousand years.

The preceding arguments lead me to suggest that a reasonable structural restoration goal is to come as close as is practical to restoring grove structure and composition to the usual range of conditions that existed during the 1,000 years preceding Euroamerican settlement. The term "usual range of conditions" is meant to exclude rare extremes that may have occurred over the last millennium, such as the large expanses of trees that were likely killed by the widespread, severe fire of A.D. 1297 in Mountain Home Grove (Swetnam et al. 1992; Caprio et al. 1994). While such extremes fall within the "natural" (pre-Euroamerican) range of variability, their deliberate creation is not likely to be tolerated by many managers or the public (though it is reassuring to know that, on the scale of centuries, groves seem resilient to such extremes).

The structural goal described above allows managers to step back from the unrealistic limitations imposed by trying to determine and replicate conditions calibrated to a specific year or other narrow time period. I now turn the discussion to an enormous challenge: describing structural restoration targets by determining the range of grove conditions that occurred in the millennium preceding Euroamerican settlement.

Describing Restoration Targets

Given the high quality and great length of paleoecological records of climatic and fire regimes in sequoia groves (Hughes and Brown 1992; Swetnam et al. 1992; Graumlich 1993; Swetnam 1993), explicit process restoration targets are relatively easy to describe. Discussion of several issues surrounding restoring fire is deferred to the section on grove conservation.

Tools for Describing Past Grove Structure

In contrast, describing targets for structural restoration (i.e., descriptions of the desired spatial arrangement and sizes of forest patches and the diameters, heights, and densities of trees in the patches) has proven difficult. Many constraints limit our ability to precisely and confidently determine pre-Euroamerican grove structure. Thus, before describing our best current understanding of past grove conditions, I present a brief overview of the capabilities and limitations of the various tools and approaches used to describe those conditions. Though conclusions drawn from any one tool or approach may be suspect, they can be powerful in combination.

Old written accounts (see especially the summaries in Bonnicksen 1975; Bonnicksen and Stone 1978). Old written accounts supply qualitative descriptions of conditions surrounding (and sometimes within) sequoia groves in the late 1800s and early 1900s. There is some disagreement among the written accounts and their interpretations (e.g., contrast the summaries in Otter [1963], who believed grove conditions of the late 1800s were artifacts of shepherds' fires, with those in Bonnicksen [1975] and Bonnicksen and Stone [1978]). Many written accounts were probably biased toward scenes that were particularly memorable to the chroniclers.

Reconstruction from repeat photography (e.g. Vankat 1969, 1970; Kilgore 1972; Vankat and Major 1978; Kauper et al. 1980). Photographs are one of the best windows on past grove conditions, sometimes showing dramatic changes (see figure 55.4). So far, photographic analyses of past grove conditions have been limited and qualitative. Some photographs undoubtedly were biased by photographers seeking attractive (but not necessarily representative) scenes. Most early photographs from sequoia groves date from the late 1800s, usually one or more decades after Euroamerican arrival.

Biological inference based on tree life-history traits and responses to fire (e.g. Harvey et al. 1980). (This might also be called the modern analog approach.) Modern studies of the shade tolerance, seed dispersal, and seedling germination and establishment traits of the various Sierran conifers, coupled with our understanding of the present effects of fires in groves and our knowledge that fires burned frequently through groves in pre-Euroamerican times, allow us to qualitatively infer past grove conditions relative to today. By themselves, they do not allow us to define a precise forest structure and composition for a specific location or time in the past.

Analysis of forest age structure (e.g. Vankat and Major 1978; Kilgore and Taylor 1979; Parsons and DeBenedetti 1979). Forest age structure alone is difficult to interpret. For example, without further information one cannot determine whether finding many more young trees than old—the usual condition in forests worldwide—indicates an increasing, steady-state, or declining tree population. However, obviously multi-modal age distributions reveal periods of high and low success in tree recruitment, thereby extending our ability to qualitatively infer conditions in centuries past.

Analysis of forest age structure coupled with demographic models (e.g. Stephenson in preparation). Age structure data coupled with demographic models gives clues as to forest trends (increasing, steady-state, decreasing, or fluctuating tree populations) over the last several hundred years (or thousands of years, in the case of giant sequoia). Such analysis can help define general, but not precise forest structures at a specific time in the past.

Forest dynamics models (e.g. Kercher and Axelrod 1984; Miller 1994; Miller and Urban in preparation; Urban et al. in preparation). Though the potential exists, no gap-phase forest dynamics model has yet been explicitly applied to estimate grove conditions at a specific time or time period in the past (but see the related approach listed in the next paragraph). Forest dynamics models depend heavily on the empirical data that drive them, and in many cases the data are weak, meaning that broad, untested assumptions sometimes must be made. However, this approach deserves more serious attention, especially in conjunction with the other approaches listed here.

Analysis of the physical legacies of past forest conditions (e.g. Bonnicksen and Stone 1978, 1982a, 1982b). Through analysis of logs, snags, and the sizes and ages of living trees, past forest conditions can be estimated by backward projection from present conditions. Unfortunately, white fir logs rot quite rap-

idly in the Sierra Nevada, having a half-life of only 14 years (Harmon et al. 1987). Thus, some of the material needed to accurately determine pre-Euroamerican conditions may be missing. Accurate reconstructions may therefore be limited to the postsettlement era (see Stephenson 1987). With consideration given to this caveat, this approach can still be used to help set limits on possible past grove conditions.

Analysis of old plot data (e.g. the data presented by Sudworth 1900, and summarized by Stephens 1995). The earliest available plot data from sequoia groves—Sudworth's 1900 data—are probably biased. His size structure data for every species are modal in the middle size classes, not the smallest size classes. This strongly suggests that his sampling was biased toward older forest patches, that he ignored small trees, or both. However, his data might help us understand conditions specifically in old-growth patches 30 to 40 years after Euroamerican settlement.

Inferring forest composition and structure from pollen records and macrofossils (e.g. Anderson 1994; Anderson and Smith 1994). Pollen and macrofossils from meadow or lake sediment can reveal changes in the relative abundances of different tree species over periods of 10,000 years or more. General forest aspect—open or closed—can be inferred from the relative abundances of pollen from shade-intolerant trees and understory plants. Pollen cannot reveal other aspects of forest structure, such as gap and patch sizes, relative proportions of trees in different age classes, and so on.

Thus, descriptions of past grove conditions often are limited in three ways: (1) most descriptions should be considered qualitative, not quantitative, (2) information is skewed toward describing grove conditions in the late 1800s or early 1900s, not earlier periods, and (3) results are usually specific to only a few locations.

Current Best Estimates of Past Grove Conditions

With consideration given to the preceding cautions, I provide a qualitative best estimate of average grove conditions in the late 1800s, followed by a best estimate of conditions in the millennium preceding the late 1800s. The descriptions are based on a synthesis of available studies and, when reasonable, assume that results apply to sequoia groves in general; however, potentially large within- and between-grove variation may be obscured. As discussed later, local research can be used to derive better descriptions for individual groves or portions of groves.

The description of grove conditions in the late 1800s is presented as a list of twenty-one brief statements supported by references. The references are not meant to be exhaustive; they were selected as being the most recent, most relevant, unique, or offering an entry into the broader literature (also see Harvey et al. 1980; Weatherspoon 1990; and Aune 1994 for other entries into the literature). Differences in past grove conditions are described relative to modern groves that have never been disturbed by logging, and have not burned since the late 1800s.

1. The dominant tree species in groves of the late 1800s were the same as today; no tree species has been lost or gained, though there have been some shifts in density and age structure (Bonnicksen and Stone 1978, 1982a; Stephens 1995).

2. Groves of the late 1800s (as well as today) could be described as a mosaic of generally small forest patches of differing ages, vegetation structures, and species composition (Bonnicksen and Stone 1981, 1982a; see figure 55.8).

3. These forest patches generally originated in forest gaps created or modified by locally severe fire; recently-created forest gaps were an integral part of the grove landscape (Bonnicksen and Stone 1978, 1982a; Stephenson et al. 1991; Stephenson 1994).

4. The gaps and patches comprising the forest mosaic often were characterized by diffuse boundaries, grading together without sharp edges; scattered trees (particularly large sequoias) often survived in gap interiors (Demetry and Duriscoe 1996; Stephenson unpublished data).

5. Gap sizes were variable, ranging from single tree gaps to gaps of several hectares; the modal (most common) gap size may have been near 0.1 ha, to the nearest order of magnitude (Stephenson et al. 1991; Stephenson 1994).

6. Rarely, large gaps of more than ten hectares were created by avalanches or single or repeated fires dominated by high intensities (Fry 1933; Stephenson et al. 1991; Caprio et al. 1994).

7. Forest patches of more-or-less uniform structure and composition were probably generally smaller than the gaps in which they became established, due to non-uniform regeneration within gaps (Demetry 1995; Demetry and Duriscoe 1996).

8. The structure and composition of forest regeneration partly depended on gap size (Demetry 1995; Demetry and Duriscoe 1996).

9. Virtually all successful sequoia regeneration occurred within recently-created forest gaps (Stephenson et al. 1991; Stephenson 1994; figure 55.9).

10. Giant sequoia seedlings often were by far the most abundant tree seedlings within gaps (Demetry 1995; Demetry and Duriscoe 1996), though they would not necessarily maintain their dominance as the new forest patch matured.

11. The largest sequoia seedlings in a given cohort (and presumably those most likely to survive to maturity; Harvey and Shellhammer 1991; Demetry 1995) occurred toward the center of gaps larger than about 0.1 ha (Demetry 1995; Demetry and Duriscoe 1996).

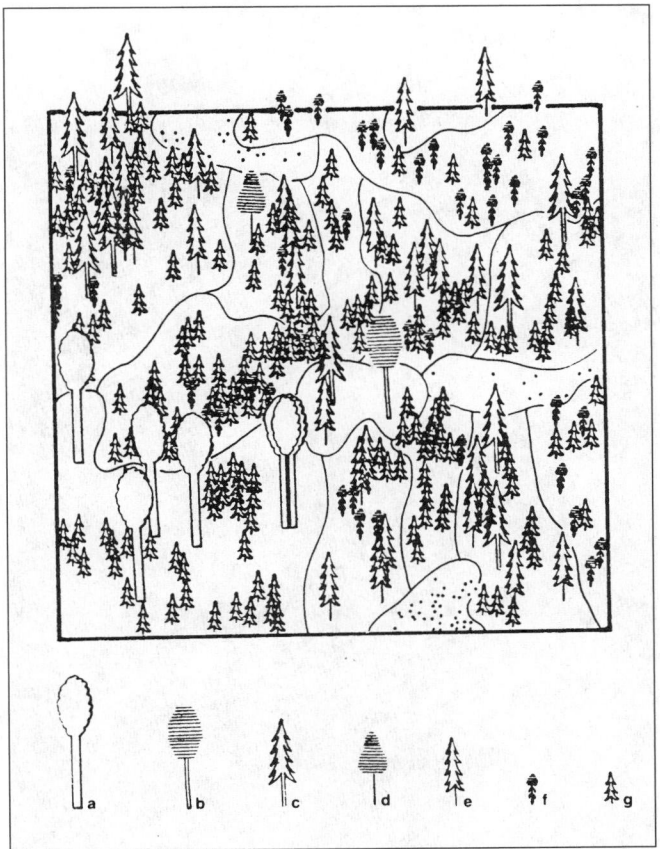

FIGURE 55.8

Sequoia grove structure and dynamics can be understood in terms of a mosaic of forest gaps and patches. This schematic diagram shows the locations of trees in a 50 m x 50 m (164 ft x 164 ft) section of the Redwood Mountain Grove, unburned for about a century. Lines are meant to accentuate the forest mosaic by delimiting patches of relatively uniform forest structure and composition, though it is clear that patch boundaries are not always distinct and their designation can be somewhat arbitrary. The tree symbols represent *a*, giant sequoias greater than 35 m (115 ft) tall, *b*, sugar pines greater than 35 m tall, *c*, white firs greater than 35 m tall, *d*, sugar pines 10 to 35 m (33 to 115 ft) tall, *e*, white firs 10 to 35 m tall, *f*, sugar pines 3 to 10 m (10 to 33 ft) tall, and *g*, white firs 3 to 10 m tall. For clarity, the tree symbols are reduced in size relative to the plot, lending a somewhat open appearance to the stand. (Reproduced from Bonnicksen and Stone [1982a], with permission of the Ecological Society of America.)

12. Young sequoias (1 to 100 years old) were orders of magnitude more abundant in the late 1800s than today (Stephenson 1994 and in preparation).

13. For trees less than about 30 cm diameter, the overall proportion of pine, oak, and sequoia relative to fir was greater than today (Kilgore 1973a, 1973b).

FIGURE 55.9

Sequoia is a pioneer species, requiring forest gaps for successful regeneration. This vigorous sequoia reproduction is in a forest gap created by locally intense fire during a prescribed burn in the Redwood Mountain Grove, Kings Canyon National Park. The white firs and incense-cedars in the background were killed by the fire, whereas the large sequoia at right was not. Most forest gaps created by fire in pre-Euroamerican groves were relatively small, probably covering fractions of hectares (a hectare is about 2.5 acres). Modern wildfires burning through heavy fuel accumulations are likely to be more severe, killing more trees and creating unusually large gaps. (Photograph by Nathan L. Stephenson, National Biological Service.)

14. Relative to today, more area was occupied by conifers less than about 10 m tall, and correspondingly less area was occupied by conifers 10 to 35 m tall; that is, more area was occupied by young forest patches than today (Bonnicksen and Stone 1978, 1982a).

15. More total area was occupied by young forest patches than old (Bonnicksen and Stone 1978, 1982a; Stephenson 1987).

16. More area was dominated by shrubs (particularly *Arctostaphylos, Ceanothus,* and *Prunus*) or open ground (de-

pending on time since last fire) than today (Bonnicksen and Stone 1978, 1982a).

17. More area was dominated by mature *Quercus kelloggii* (California black oak) than today (Bonnicksen and Stone 1978, 1982a).

18. Understory trees (and perhaps to a lesser extent overstory trees) within forest patches were generally more sparse than today, although thickets still existed (Kilgore 1973a).

19. Vertical fuels were less continuous than today; fires removed the lower branches of trees (Kilgore and Sando 1975).

20. The crowns of monarch giant sequoias and other trees were sometimes much more sparse than today (when comparing the same individual trees); local high-severity fires scorched some trees high into their crowns (photos in Vankat 1969, 1970).

21. Average surface fuel loads were much lower than today, with fuels distributed in a patchy mosaic ranging from light or absent to locally heavy (Kilgore 1973b; Bonnicksen 1975; van Wagtendonk 1985).

Some authors have suggested that grove conditions in the late 1800s were an artifact of abnormally frequent and severe wildfires ignited by shepherds and cattle ranchers (see especially Otter 1963). This is probably not the case; reconstructed fire histories suggest that fires of the 1800s were of the normal range of intensities, with a marked decline in fire frequency beginning soon after the arrival of shepherds and ranchers in the 1860s (Swetnam et al. 1992; Swetnam 1993; Caprio and Swetnam 1995). For the most part, then, the description of grove conditions given above is based on information from groves that had been fire-free for one to a few decades. This falls within the range of fire-free intervals recorded over the preceding two millennia (Swetnam et al. 1992), suggesting that any grove structural changes that had taken place between the arrival of Euroamericans in the 1860s and the photographs and descriptions of the 1870s, 1880s, and 1890s were within the normal range of variability, and were probably far too small to affect the broad, qualitative description of grove conditions listed above.

Interpreting their finding that more total grove area was occupied by young forest patches than old in the late 1800s, Bonnicksen and Stone (1978, 1982b) concluded that grove structure in the late 1800s, at least in the Redwood Mountain Grove, was far from equilibrium (stationary age distribution). It would logically follow that conditions in the late 1800s were not representative of earlier periods in the same century. However, Stephenson (1987) reevaluated Bonnicksen and Stone's conclusions using well-established demographic models, demonstrating that the presence of more forest area in young patches than old qualitatively fits what is expected for groves near (but probably fluctuating around) equilibrium. (This same conclusion was reached by Van Wagner [1978] for other forest types.) If conditions in the Redwood Mountain Grove were indeed fluctuating near equilibrium in the late 1800s, it follows that they were probably near equilibrium in the early 1800s, given the multi-century life spans of the mixed conifer tree species. We do not know whether this conclusion would apply to sequoia groves in general; however, it lends support to the notion that the qualitative description of grove conditions in the late 1800s (listed above) is broad enough to apply also to the early 1800s.

As described in the preceding subsection, even though climate and fire regimes have varied within sequoia groves during the last 1,000 years (Hughes and Brown 1992; Graumlich 1993; Swetnam 1993), their combined effect on giant sequoia demographics at centennial time scales has been moderate. By far the largest deviation from equilibrium conditions (stationary age distribution) in giant sequoia populations over the last two to three millennia is due to the effects of fire suppression during the last century (Stephenson 1994 and in preparation). Additionally, the pollen record indicates no major changes in species composition over the last millennium (Anderson 1994; Anderson and Smith 1994). Collectively, these data suggest that our broad, qualitative description of grove conditions in the 1800s may also apply to other portions of the millennium preceding Euroamerican settlement.

Tree-ring reconstructions of fire history and sequoia population age structure suggest that grove conditions during the millennium preceding Euroamerican settlement experienced some local (and perhaps sometimes widespread) deviations from grove conditions found in the late 1800s. In A.D. 1297, during a severe drought, Mountain Home Grove experienced a widespread, severe fire which probably killed many mature trees, including some large sequoias (Swetnam et al. 1992; Caprio et al. 1994). This fire apparently induced the establishment of a large cohort of new sequoia regeneration in the grove. Big Stump Grove may have experienced a similar event at about the same time; Huntington's (1914) sequoia age determinations demonstrate that a large cohort of sequoias became established at Big Stump (he called it "Comstock") in about the early 1300s.

Linked fire and forest dynamics computer models show good promise of helping us more precisely infer the possible range of past grove conditions. Such modeling efforts are presently being made by the National Biological Service's Sierra Nevada Global Change Research Program (Stephenson and Parsons 1993; Miller 1994; Miller and Urban in preparation; Urban et al. in preparation). The models will be driven with climatic and fire regimes reconstructed from tree rings for the millennium preceding Euroamerican settlement; model output will be spatially-explicit patterns of grove structure and composition through time.

Describing Structural Restoration Targets for Specific Groves

In pre-Euroamerican times, forest structure and composition undoubtedly varied from grove to grove in response to chance, local cultural practices of Native Americans, and local environmental conditions (elevation, slope aspect, slope steepness, soil characteristics, soil moisture, surrounding vegetation types, local ignitions and fire regime, and so on). It follows that managers should, to the extent possible, recognize the unique environment and history of each grove by setting grove-specific targets for restoration. Ideally, the targets would be set by using as many as possible of the tools and approaches listed earlier to determine past grove conditions in each grove. However, at least four problems interfere with our ability to define precise structural restoration targets for individual groves. I will first present the problems, then possible solutions.

First, some groves lack the information needed to directly determine past conditions. For example, many heavily logged groves have lost all signatures of their past structure, except for large sequoia stumps. If such groves also lack photographic and written records, there is no reliable direct way to determine their past structure and composition. Second, even if the needed information on past conditions could be gathered and analyzed, it may be too expensive and time consuming to attempt to describe past conditions unique to each grove unless significant new funds are made available. This is especially true during the present period of shrinking federal and state budgets. Third, even if both the funds and needed information are available, grove-specific descriptions of past conditions may still be so qualitative as to be nearly indistinguishable from the generic target listed earlier. For individual groves, targets will usually be quite broad, such as "reduce the area occupied by pole-sized white firs by 10 to 50%," leaving much room for overlap in target structures among groves. Finally, for reasons listed earlier, individual grove targets will usually be defined only by conditions near the turn of the century, which in some cases may have been extreme relative to local conditions in the preceding millennium. For example, if there is good evidence that a particular grove in the late 1800s experienced a predominantly high-severity fire which killed much of the forest (such as portions of the Redwood Meadow Grove, as indicated by present grove structure and the photographs and commentary by Sudworth [1900]), managers would probably be hard-pressed to justify re-creating such conditions, no matter how "natural" they might have been.

There are at least three ways of partly overcoming these obstacles to describing restoration targets for individual groves. First, the best available target might be defined by the pre-Euroamerican conditions of other groves found in similar environments (assuming that such analog groves exist, and that their past conditions can be determined). Second, linked fire and forest dynamics computer models using local environmental variables might be used to simulate the possible range of past grove conditions unique to a particular grove (see the preceding subsection). Finally, if all else fails, perhaps the only reasonable option would be to move forward with restoration using as a target the generic, qualitative description of past grove conditions listed in the preceding subsection.

Uncertainties are inevitable and are no reason to halt restoration. We do know many things with certainty, such as the existence of a continuing and unprecedented failure of sequoia regeneration (which is relatively easily reversed) in groves protected from fire and logging. Most uncertainty surrounds the details of restoration, not the big picture. Having considered issues related to describing targets for structural restoration, I now move the discussion to the step that managers must take after defining targets: evaluating the practicality of reaching restoration objectives.

Evaluating the Practicality of Reaching Restoration Objectives

Once restoration targets have been described, managers must determine how precisely and how quickly they can be reached. The fidelity with which restored grove structure matches target structure will depend both on the pre-restoration grove structure and the target structure. Today, pre-restoration structure varies widely among four broad categories of groves (as described earlier): (1) groves that have been protected from both fire and logging, (2) groves that have been treated with prescribed fires, (3) groves that were logged near the turn of the century, and (4) groves that were logged more recently.

Our ability to rapidly achieve structural restoration goals is severely limited by a simple fact: it is possible to remove trees of any size, but not to plant trees of any size. For example, research has demonstrated that groves protected from logging and fire during the last century have orders of magnitude fewer 20- to 100-year-old sequoias than they would if fires had not been excluded (Stephenson 1994 and in preparation). This is well outside of the range of pre-Euroamerican conditions. Even if enough sequoias in this age class could be found in plantations around the world (which is doubtful), it would be impossible to transplant them successfully into groves (not to mention that such an operation would be prohibitively expensive and would compromise the genetic integrity of groves). Realistically, only small sequoia seedlings can be planted. Broadly speaking, then, selective removal of trees and selective encouragement of seedling establishment (either by planting or natural seeding) can be used to relatively quickly bring grove structure closer, but not exactly, to a specific target.

Agee (1995) has suggested the possibility of thinning and even fertilization aimed at increasing the sizes, and therefore apparent ages, of younger trees in selected forest stands, thereby creating large trees to "replace" a missing cohort. Even with this admittedly labor-intensive approach, it may take cen-

turies (the time it takes for seedlings to become mature trees) of ongoing, hands-on management to increase the precision of structural restoration. Those grove areas recently logged of all but large sequoias will take longest to regain old-growth character. (On the other hand, in some ways the latter areas presently come close to mimicking the structural effects of rare but particularly severe fires of the past.)

It seems inevitable that, for centuries to come, most groves will bear at least some imprint of twentieth century fire suppression and logging. Managers must accept that even when they have a well-defined structural target, they may not be able to reach it. The best option may be to put in motion the structure and dynamics that someday may result in a structure that falls within the range of pre-Euroamerican variability.

Choosing Restoration Tools and Approach

Simplistically, structural restoration can been seen as consisting of two parts: taking things out (removing trees) and putting things in (sowing seeds or planting seedlings). Two major tools can be used to remove trees: fire and saws. In many people's minds fire is associated with process restoration, but fire can be (and is) used also as a tool for structural restoration (figure 55.10). Fire intensity and effects can be controlled with moderate precision by judiciously locating fire lines, burning under selected weather and fuel conditions, and using different ignition techniques. This gives managers moderate control over which trees are killed and which are not, conferring moderate control over final forest structure.

Managers seeking to restore grove structure must consider potential tradeoffs between the two major tools for removing trees (table 55.1). Of course, the tools are not mutually exclusive; either or both can be used, depending on objectives and practical considerations. For example, saws can be used to girdle or fall trees selected for removal, then fire used to consume them. By whatever means selected trees are removed, the resulting release from competition results in dramatic increases in the height and especially diameter growth of mature sequoias (Dulitz 1986; Gasser 1994; Mutch 1994; Mutch and Swetnam 1995).

By whatever means it is accomplished, opening the forest canopy and clearing litter and duff from the forest floor also creates conditions favorable to sequoia seedling establishment, growth, and survival (Harvey et al. 1980; Harvey and Shellhammer 1991). However, there are large differences in sequoia seed release (and therefore seedling establishment) after fire and cutting. Stephens (1995) found more than one million sequoia seedlings per hectare in forest gaps created by fire, compared to a maximum of 90 seedlings per hectare in gaps of similar sizes created by cutting, even if slash fires had burned in the cut gaps. Benson (1986) found greater, but still low, sequoia seedling establishment after logging (820 seedlings/ha), with the number of seedlings dropping off rapidly in the following years. Insignificant sequoia seed release in cut

FIGURE 55.10

This prescribed fire is burning in a century's worth of accumulated fuels in the Giant Forest sequoia grove, Sequoia National Park. Prescribed fire can be used both as a tool for fuel reduction (grove protection) and as a tool for restoring aspects of grove structure to pre-Euroamerican conditions, while simultaneously maintaining the ecosystem processes of the past (such as soil sterilization and nutrient cycling). (Photograph by Betty Knight, courtesy of the National Park Service.)

gaps is due to the absence of a large, cone-opening heat pulse delivered to the crowns of the mature sequoias within and surrounding the gaps. Simple demographic models show that, due both to the high natural death rates of sequoia seedlings and to centuries of compounding of low death rates in sapling and mature sequoias (Harvey and Shellhammer 1991, Lambert and Stohlgren 1988), 90 to 820 seedlings per hectare is simply not enough to ensure recruitment of old-growth sequoias at pre-Euroamerican rates (Stephenson in preparation). Thus, sequoia regeneration in forest gaps created by cutting must be encouraged by one or more of three possible ways: (1) carefully-positioned slash fires deliver a heat pulse to the crowns of some nearby mature sequoias, (2) sequoia seeds are collected

from manually-harvested cones, then scattered in the gaps, or (3) nursery-raised sequoia seedlings are planted. In either of the latter two cases, maintenance of the genetic integrity of local sequoia populations depends on the use of only local seed stock (Libby 1986).

The mechanics of restoration will be influenced heavily by individual grove histories (that is, initial grove structures). Currently, some unlogged grove areas that have been prescribed burned two or more times are probably closer to pre-Euroamerican conditions than any other grove areas. Among the other grove areas, those presently unlogged and unburned, with their dense mixture of trees in a broad spectrum of age classes, generally have the most potential for being restored to something approaching pre-Euroamerican conditions. The next greatest potential for restoration is in groves that were logged near the turn of the century, which now are often dominated by dense regeneration of all species. Where appropriate, dense stands can be thinned and new gaps created in the forest—a first step toward re-creating a forest mosaic of stands of many different ages (Weatherspoon 1996).

The most difficult grove areas in which to make rapid, significant progress in restoration are those that were logged of all trees except large sequoias in the 1980s. These 2 to 10 ha (5 to 25 acre) grove areas now lack intermediate ages classes, being dominated only by scattered old sequoias and relatively uniform expanses of seedlings of mixed species (figure 55.7). On the other hand, these forest gaps individually may resemble gaps created by uncommon, extremely severe fires of the past millennium. Thus, natural regeneration patterns in large fire-caused gaps could serve as a guiding analog for restoration of the cut gaps (see Demetry 1995; Demetry and Duriscoe 1996). Immediate attention should be given to restoring the species composition of tree seedlings in the gaps created by logging in the 1980s, most of which were planted with only ponderosa pine and giant sequoia. The findings of Demetry (1995) and Demetry and Duriscoe (1996), coupled with site-specific knowledge of forest composition around the gaps, will provide guidance for restoring the composition and spatial arrangement of tree seedlings within these gaps.

The mechanics of restoration will also be heavily influenced by the details of the desired target structures, such as the relative proportions of gaps and patches of different sizes, and their spatial relationships. Collectively, uncertainties about both the past structure of individual groves (and of groves in general) will make it difficult for managers to specify precise restoration actions on the ground. For example, if managers are confronted with ten white fir thickets and, to meet a broad objective of reducing white fire thickets by 10 to 50%, decide to mechanically remove some of the thickets, how do they choose precisely how many of the ten to remove? How do they choose which to remove? Should they remove only part of a given thicket or all of it? Should they expand the resulting

gap into adjacent forest patches? Should a few trees be left standing in the newly-created gap? How many, and where? Should the boundary of the gap be relatively sharp, or should there be a slow transition of increasing tree density into the intact forest matrix?

Such uncertainty need not halt restoration; managers confronted with broad structural restoration objectives but lacking the information needed to prescribe the details of the restoration still have reasonable options for moving forward. First, managers might use fire as the main tool for achieving broad structural objectives. (Earlier I presented evidence that fire is a reasonable tool for restoring at least some aspects of pre-Euroamerican grove structure.) Managers using fire can achieve at least some of their broad structural objectives by controlling the season of burns, fire line locations, and ignition patterns, while allowing the details of the structural restoration to be determined by the same process that shaped the forest in the past—fire. This approach presumes that fire can do a better job of restoring the details of pre-Euroamerican forest structure than humans with saws, who lack complete information on past fire effects and therefore must make some arbitrary choices. However, it remains to be seen whether fire alone can meet all broad structural objectives. For example, it may be difficult to reduce the abnormally large cohort of large firs that has grown since fire suppression became effective (Bonnicksen and Stone 1978, 1982a). However, Kilgore (1973a) reports success in using fire to reduce this cohort. The use of fire in restoration should err on the conservative; once a tree is killed, it cannot be brought back.

If fire is not used as the primary tool for meeting broad restoration targets, managers must then precisely define, on the ground, which trees are to be removed by cutting. Given that structural targets are likely to be very broad, choosing the details of the reconstruction will be somewhat arbitrary. In this case, the best approach is conservative, because once a tree is removed, it cannot be put back. Restoration using saws will proceed in ways unlike standard silvicultural treatments (though it will most closely resemble the group selection cutting method; Weatherspoon 1996). Restoration targets will be

TABLE 55.1

Some tradeoffs between the two major tools for tree removal.

Fire	Saws
Structural objectives achieved with moderate precision.	Structural objectives achieved with excellent precision.
Conservative: maintains the processes that sustained groves in the past, such as nutrient cycling and soil sterilization.	More likely to have unknown or unexpected short- and long-term ecosystem consequences.
No soil compaction, usually low erosion.	Potential for soil compaction and greater erosion, depending on approach.
Policy allows use in many designated wildernesses.	Would require special exemption for use in designated wildernesses.
Low or no potential for commodity production as an incidental byproduct of restoration.	High potential for commodity production as an incidental byproduct of restoration.
Smoke production and a chance of fire escape.	No smoke or chance of fire escape unless debris is removed by burning.
High potential for scarring trees, providing possible entry points for pathogens.	Lower potential for scarring trees, but high chance of entry of root pathogens through cut stumps.
Adequate natural seed release following treatment.	Some species (particularly sequoia) will require manual seeding or planting.
Relatively inexpensive to apply over large areas.	May be very expensive if costs are not partially offset by commodity production.

based on ecological principles aimed at restoring variable pre-Euroamerican conditions, not on commodity values, maximization of site production, or ease of silvicultural treatment. (Commodity production, however, still could be an incidental byproduct of restoration.) Restored groves will include suppressed trees, insects and pathogens, snags, logs, brush patches, small forest gaps of different sizes, and different levels of forest thinning which grade into one another—sometimes gradually, sometimes more abruptly. Some of the different cutting treatments that have been used, sometimes for decades, in different groves and plantations (e.g. Mountain Home Demonstration State Forest, Calaveras Big Trees State Park, the Tule River Indian Reservation, the University of California's Whitaker's Forest and Blodgett Forest Research Station, and on private lands) may help illustrate the consequences of different approaches (Benson 1986; Dulitz 1986; Harrison 1986; Heald 1986; Gasser 1994; Rueger 1994; Stephens 1995).

Prescribing sizes and spatial arrangement of forest gaps and patches may be difficult. I have emphasized that the modal gap size in pre-Euroamerican sequoia groves may have been near 0.1 ha, but was also highly variable. To complicate matters, patterns of clumping in trees are hierarchical, spanning scales from a few meters to whole groves, and can vary among groves (Bonnicksen and Stone 1981; Stohlgren 1993). To replicate the hierarchical clumping of pre-Euroamerican times will require thoughtful consideration of the spatial arrangement of newly-created gaps of different sizes.

A related problem is determining at which spatial scales restoration success is to be judged. Individual forest stands can deviate widely from the general pre-Euroamerican grove conditions listed earlier, yet collectively they might re-create the conditions. Clearly, criteria for judging restoration success need to be defined differently at different spatial scales: individual trees, stands, groves, and the whole population of naturally-occurring groves.

Restoration of Areas Affected by Roads and Foot Traffic

The preceding subsections were devoted to issues relevant to restoration of the structure and dynamics of grove vegetation altered by fire exclusion or logging. In contrast, some of the earliest and most urgent calls for grove restoration centered on counteracting the effects of roads and tourist foot traffic on the rooting zones of sequoias (Meinecke 1926; Hartesveldt 1962). These earlier concerns bear reexamination.

In contrast to Meinecke's (1926) expectation that sequoias would be harmed by foot traffic and by the placement of road fill and pavement over their rooting zones, Hartesveldt (1962, 1965) found that most mature sequoias actually showed a distinct increase in growth rate after these disturbances. Pavement over a sequoia's rooting zone eliminates competition for moisture and nutrients by other plants, reduces losses of soil moisture to evaporation, and causes substantial soil warming. At 30 cm (1 ft) beneath pavement in the summer-

time, Hartesveldt (1965) found soils to be 14°C (25°F) warmer than nearby soils not covered with pavement. These warm, moist soils result in accelerated growth and longer growing seasons for the affected sequoias. Similarly, soils compacted by foot traffic often retain more soil moisture than uncompacted soils. Consequently, mature sequoias with compacted rooting zones also tend to grow faster than sequoias in undisturbed areas (Hartesveldt 1965).

In no way do these findings mean that paving or trampling of sequoia rooting zones is desirable. First, it is possible that warmer and wetter soils also provide a better environment for root pathogens, potentially leading to accelerated toppling of infected sequoias (Hartesveldt 1962). While there is presently no evidence of increased failure rates among sequoias with paved or trampled rooting zones, such an effect, if it existed, could be subtle and could take decades or more to become evident. Second, heavy foot traffic around sequoias can result in erosion that exposes sequoia roots (Hartesveldt 1962), possibly opening corridors to root pathogens. Third, heavy trampling eliminates other plant species native to sequoia ecosystems, while compacting soils to the point that plant reestablishment is inhibited long after trampling ceases. Finally, many (if not most) people find trampled areas esthetically less pleasing than untrampled areas.

Hartesveldt (1962, 1965) found that the primary negative effect of roads on mature sequoia growth rates was from roots being cut during road construction. Some sequoias with cut roots showed signs of growth recovery over a period of decades. Hartesveldt further proposed that improper road drainage could damage sequoias and other species through the direct effects of accelerated erosion, which might undermine and topple trees, or by the indirect effects of extreme erosion or deposition that raises or lowers local water tables (Hartesveldt 1966).

Because roads and foot traffic tend to be localized, and because their negative effects on most mature sequoias seem to be small to insignificant, restoration of such areas seems less urgent than reducing fuel loads within groves. However, some restoration actions can be taken relatively easily. First, the most conservative approach to grove management would avoid new road construction and would minimize areas of concentrated off-trail foot traffic. Second, foot, stock, and off-highway vehicle trails showing signs of significant soil loss or exposure of roots might be moved away from the rooting zones of sequoias, which, to the best of our knowledge, generally extend 30 m (100 ft) or more from the bases of sequoias (Hartesveldt et al. 1975). Dirt roads experiencing or inducing significant erosion should be repaired, or closed and rehabilitated. Third, areas that were formerly heavily trampled might be re-planted with native plants and lightly mulched with forest litter and duff to encourage site recovery. In extreme cases of topsoil loss (see Hartesveldt 1962), a layer of topsoil taken from adjacent mixed-conifer forest might be added. Attempts to reduce soil compaction by tilling should probably be avoided; tilling might encourage pathogen entry through root wounds. Fourth, various efforts can be taken to help stabilize trees obviously weakened by road cuts or accelerated erosion. Finally, original drainage patterns could be restored in those apparently few areas where water tables have been significantly raised or lowered by erosion, deposition, or other causes.

GROVE CONSERVATION

Active management cannot end with protection and restoration; once protected and restored, groves must be maintained. Here I address the following pertinent issues: What approach or approaches should be taken to assure long-term grove sustainability? How much protected land adjacent to groves is needed, and how should it be managed? What are likely future threats to the long-term sustainability of sequoia ecosystems?

Maintaining Restored Groves

As is the case for restoration, groves can be sustained through the judicious use of fire, saws, or both. The most conservative approach to assuring the long-term sustainability of sequoia ecosystems is to maintain the processes that sustained them in the past. Specifically, prescribed fire is the most conservative tool for sustaining sequoia ecosystems. Some important ecosystem functions of fire which cannot be mimicked by other means are mobilization of nutrients locked in litter, duff, and woody fuels; killing of pathogens in the upper soil layers; changing soil structure and wettability without causing soil compaction; and inducing seed release from serotinous cones (Kilgore 1973b; Harvey et al. 1980; Chang 1996; Weatherspoon 1996). To paraphrase J. B. S. Haldane, sequoia ecosystems are not only more complex than we suppose, but more complex than we can suppose. Given this complexity, fire probably plays other important roles of which we are not yet aware, and may never be aware. To permanently remove fire from its former role would put sequoia ecosystems on a new, unknown track (Weatherspoon 1996).

It is clear that for a number of purposes Native Americans lit fires extensively in the foothills and mixed conifer forest of the Sierra Nevada (Reynolds 1959; Lewis 1973; Anderson and Moratto 1996). (Interestingly, one of the primary purposes of these generally low-intensity fires was to reduce fuels that could lead to catastrophic fires [Anderson and Moratto 1996]). A non-trivial policy question therefore accompanies the restoration of fire: should the restored fire regime mimic the pre-Euroamerican fire regime (which included fires ignited by Native Americans), or mimic a lightning-only fire regime appropriate to the present climate? (An additional non-trivial question is whether rare high-severity fires, even if natural, should be allowed to burn.) This choice between Native American and lightning-only fire regimes is partly philosophical,

driven by conflicting definitions of "natural" and by ethical choices as to the proper role of humans in ecosystems. For discussions of the philosophical aspects of the dilemma, I refer readers to other articles (e.g. Kilgore 1985, Graber 1985, 1995).

It is possible that the choice between Native American and lightning-only fire regimes is of little material importance. First, some fire ecologists intimately familiar with the Sierra Nevada have examined patterns of lightning fire ignitions and think that, in contrast to Kilgore and Taylor's (1979) conclusions, in some places fire frequency may have been limited mostly by weather, fuel quantity, and fuel quality—not by availability of ignitions (Swetnam et al. 1992; J. van Wagtendonk personal communication). This possibility should be examined with the aid of computer simulations by linked forest and fire dynamics models specifically tailored to sequoia ecosystems (e.g. Miller 1994; Finney 1995; Miller and Urban in preparation). Second, if differences exist between Native American and lightning-only fire regimes, they might not be large enough to have a major effect on grove structure and composition. Again, this possibility should be examined with the aid of computer simulations. However, until convincing evidence exists that the choice of fire regimes makes little difference, managers' choices should be based on clearly articulated policy justifying one approach or the other.

If the managers' choice is to mimic the Native American fire regime (which includes fires started by lightning), they should burn so as to mimic the size, frequency, season, and usual range of intensities of fires that burned during climatic periods similar to the present, and that occurred within the last few millennia (Kilgore 1985; Parsons 1990a). (Whether or not Native Americans significantly influenced fire regimes, we know that fire regimes generally tracked climatic change; Swetnam 1993.) Good quantitative targets are available; tree-ring analyses have produced excellent multi-millennial records of climate, fire frequency, and fire season in several sequoia groves (Hughes and Brown 1992; Swetnam et al. 1992; Graumlich 1993; Swetnam 1993; Caprio and Swetnam 1995). Mutch (1994) has demonstrated that tree rings can also be used to infer past fire severity, but this approach has yet to be applied broadly (see also Caprio et al. 1994; Mutch and Swetnam 1995).

In contrast, no quantitative targets are presently available to allow managers to mimic lightning-only fire regimes. It would not be enough for managers simply to avoid interfering with lightning ignitions within groves; such a fire regime would not include fires that started outside of groves and would have burned into the groves if they had not been suppressed, and if land-use changes had not created barriers to fire spread. Targets for lightning-only fire regimes might be simulated using available fire spread models (e.g. Finney 1995). Patterns of lightning strikes and ignitions could be examined and the resulting fires would be allowed to burn across a simulated landscape free of "unnatural" barriers to fire spread. The simulated fire regime, including fires that the models suggest would have burned into groves from the outside, would then become the target for management-ignited prescribed fires.

Though fire is the most conservative tool for sustaining sequoia ecosystems, there are potential limitations to its use. Protection of people and property from escaped prescribed fires is of primary importance; fortunately, protection is usually a straightforward task. Perhaps the greatest hurdle is meeting air quality standards, both locally (which affects local residents and tourists) and regionally in the San Joaquin Valley. Air quality issues related to prescribed fire are discussed by Cahill et al. (1996).

Cutting and planting alone might be used to sustain groves, but this approach has its own problems. To prescribe cutting and planting as the primary grove maintenance tool presumes that we understand most aspects of sequoia ecosystem dynamics, and that we can mimic them without fire. This is probably true for regenerating the major tree species (which we understand relatively well), but not for most other organisms in sequoia ecosystems. If cutting and planting are used to sustain groves, the ecosystem consequences should be closely monitored and compared with those of burning. Of course, an intermediate path would be to use saws to girdle or fall trees which are then burned in situ, followed by planting and/or natural seeding. Again, the ecosystem consequences of this approach should be monitored.

Special attention might be given to restoring and maintaining the genetic integrity of sequoia groves (Fins 1979; Libby 1986; Fins and Libby 1994). For example, the Placer County Grove is a tiny grove consisting of six naturally-occurring sequoia trees. The grove lies far to the north of all other naturally-occurring sequoia groves and shows some unique genetic traits (Fins 1979; Libby 1986). Dozens of sequoia seedlings, probably from Mountain Home Grove, were planted among the Placer County Grove sequoias in about 1951. Some of these introduced sequoias are reaching sexual maturity, and thus threaten to introduce foreign genes into the local population. Maintaining the genetic integrity of this grove would be simple: the 45-year-old sequoia seedlings would be cut. This course of action was recommended at least seventeen years ago by Fins (1979), but still no action has been taken. If the genetic integrity of this unusual grove is to be maintained, the introduced trees should be removed immediately.

Land Needs

Grove Influence Zones

The MSA (U.S. Forest Service 1990) specified that sequoia grove boundaries would be defined by "... an interim 500 foot buffer extending from a hypothetical perimeter line around the outermost known giant sequoias in the Grove[s]." There was to be no logging or other mechanical entry in this zone, except that with the specific purpose of reducing fuel loads. An additional 500-foot zone, called the grove influence zone, was to extend beyond the 500-foot administrative boundary; certain

restrictions were placed on logging within the grove influence zone. Many specific exceptions to these methods of defining groves and grove influence zones are listed in the MSA (U.S. Forest Service 1990). For example, several groves were to have 300-foot administrative boundaries surrounded by 300-foot grove influence zones. Additionally, topographic features such as ridges could take precedence in finalizing grove boundaries and influence zones, when such features logically and physically separated giant sequoias from the general forest. Rogers et al. (1995) describe the issues and mechanics that led to the final mapping of USFS groves and their influence zones.

The MSA's definition of grove influence zones has little ecological basis. USFS and other managers need an ecologically sound basis for defining grove influence zones, and must state clearly what land management practices are appropriate within these zones. The defining element of sequoia ecosystems is the giant sequoia itself; all known plant and animal species in sequoia groves (with the exception of a single species of beetle, *Callidium sequoiarum*, which is host-specific to sequoia) are also found within the much more extensive mixed-conifer forest surrounding groves (DeLeon 1952; Harvey et al. 1980). Thus, perhaps the most obvious measure of a sequoia ecosystem's sustainability is its ability to support sequoias themselves. To the best of our knowledge, high soil moisture availability in well-drained soils is the primary factor allowing sequoias to grow within present grove boundaries but not in adjacent mixed-conifer forest (Rundel 1969, 1972b). Thus, one of the primary needs for assuring sequoia ecosystem sustainability is undisturbed grove hydrology.

Until individual grove assessments suggest otherwise, the most conservative approach to restoring or maintaining grove hydrology (and therefore long-term sustainability) begins with defining a hydrologic influence zone—the local watershed above and adjacent to groves. Certain upslope areas falling within the same topographic watershed as a grove might be designated as outside of a grove's hydrologic influence zone, if it is convincingly demonstrated that there is no significant aboveground or belowground hydrologic connection with the grove. This might be true for the more distant portions of large watersheds. Such determinations will need to be made by qualified forest hydrologists working with sequoia ecologists.

An additional important component to the grove influence zone is defined by fire behavior. Fire influence zones should be added immediately adjacent to groves, and managed in such a way that fires entering the grove will behave as they would have in pre-Euroamerican times. Fire influence zones will usually be widest immediately below groves, but occasionally may extend beyond the hydrologic influence zones above groves, usually for groves that extend to ridgetops. The widths of fire influence zones will vary with local conditions, but typically might be the equivalent of two tree heights: 100 to 150 m (300 to 500 ft, which in this case is similar to the width of grove influence zone boundaries defined by the MSA). Individual fire influence zones should be determined by fuels and fire behavior specialists.

The boundary of the final grove influence zone would be defined by the wider of the hydrologic and fire influence zones at each point around the grove periphery. Conservatively, management practices within the grove influence zone would be limited to those identical to the management practices outlined for the groves themselves: protect and restore, then to the extent possible let natural processes (particularly fire) shape forest dynamics and hydrology.

Additional Land Needs

The USFS and NPS collectively manage more than three-fourths of all grove area in the Sierra Nevada; public agencies as a whole manage about 90%. There is no compelling evidence that the long-term sustainability of giant sequoia ecosystems as a whole depends on adding more to the public land base. For example, it is highly likely that the majority of genetic diversity among sequoias is already found on public land, especially considering that genetic variation within groves tends to be greater than variation between groves (Fins 1979; Fins and Libby 1994; however, not all groves have been genetically explored). Additionally, a diversity of grove ownerships promotes a diversity of management approaches. Those private landowners who take an active interest in sequoia stewardship are potentially valuable partners with public agencies in determining the consequences of different management approaches to sequoia ecosystems.

However, logical reasons for public purchase of groves from willing private owners might include providing additional public recreational opportunities, conserving specific ecological or genetic features unique to particular groves, and increasing the public agencies' ability to manage grove areas already in their protection. For example, about 500 acres of private land within the Alder Creek Grove, if added to the USFS land base, would include the largest sequoia outside of the national parks (which also happens to be the sixth largest tree in the world; Flint 1987) and the only known wild example of the unusual "weeping" variety of giant sequoia (R. Rogers, personal communication). As a further example, USFS presently manages all of Freeman Creek Grove (a USFS Botanical Area) except for about 10 privately-owned acres in the heart of the grove. USFS purchase of this small parcel could greatly facilitate the future use of prescribed fire as a tool for managing the grove.

Air Pollution

Some of the worst air pollution in the United States is found periodically along the western flank of the southern Sierra Nevada, the home of the vast majority of sequoia groves (Peterson and Arbaugh 1992; Cahill et al. 1996) (figure 55.11). Mature giant sequoias seem to be resistant to present levels of ozone, the most damaging component of Sierran air pollution. One hundred twenty-year-old sequoias exposed to ozone for two months, some at concentrations up to three times ambient, showed no visible foliar injury or detectable changes in photosynthetic rates (Miller et al. 1994). In contrast, newly-emerged sequoia seedlings were more vulnerable. Seedlings exposed to ozone over an entire summer showed very slight foliar injury at ambient ozone levels; however, those exposed to 1.5 times ambient levels showed extensive foliar injury and lowered photosynthetic efficiency (Miller et al. 1994; Miller 1996).

Some other tree species found in sequoia groves are more susceptible to ozone injury than giant sequoia—particularly ponderosa pine and Jeffrey pine. Ozone-sensitive individuals

FIGURE 55.11

As seen in this view from the edge of the Giant Forest sequoia grove, some of the worst air pollution in the United States is periodically found along the western flank of the southern Sierra Nevada, home of most of the world's naturally occurring sequoias. Sequoia seedlings, but not mature trees, show some damage at present levels of air pollution; ponderosa pine and Jeffrey pine are more strongly affected. Air pollution, unnatural effects of pathogens, and potential for climatic change all threaten giant sequoia ecosystems to varying degrees. (Photograph by Diane Ewell, courtesy of the National Park Service.)

of these pines show extensive foliar injury at present ozone levels in the southern Sierra Nevada (Peterson and Arbaugh 1992; Duriscoe and Stolte 1992; Patterson 1993; Miller 1996). Compared to ozone-resistant individuals, ozone-sensitive pines have lower photosynthetic rates, lose their needles earlier, and have diminished annual ring growth (Miller 1996). Smaller trees are the most severely affected. Pines in the Sierra Nevada east of Fresno, particularly in Grant Grove and Giant Forest of Sequoia and Kings Canyon National Parks, show some of the most severe ozone damage in the Sierra Nevada (Peterson and Arbaugh 1992; Stolte et al. 1992). Patterson (1993) found that nearly 90% of Jeffrey pines in or near the Giant Forest sequoia grove showed visible signs of ozone injury; however, he ranked only 10% of the pines as showing severe injury.

If ozone concentrations in the Sierra Nevada remain relatively constant into the future (as they have over the last decade, due to increasing pollution control efforts in the face of rapid population growth; Cahill et al. 1996), air pollution may have some limited effects on the genetic composition of sequoia seedling populations, while significantly contributing to increased death rates and decreased recruitment of ponderosa pine and Jeffrey pine within sequoia groves (Miller 1996). If pollution were to increase beyond present levels, adult pines stressed by air pollution (compounded by crowding caused by fire suppression) may become more susceptible to fatal insect attacks, as they have in the Los Angeles basin to the south (Miller 1973; Ferrell 1996; Miller 1996). Additionally, sequoia seedling establishment, survival, and recruitment might eventually be reduced (assuming that conditions for establishment are otherwise favorable). Options for counteracting the effects of air pollution include (1) reducing production of air pollution, (2) reducing competition among trees by thinning (whether by fire or saws), and (3) identifying, breeding, and planting pollution-resistant varieties of selected tree species. In the latter case, genetic diversity within groves may diminish.

Pathogens

Annosus root rot (*Heterobasidion annosum*), a native fungus, may be killing more sequoias now than in pre-Euroamerican times. Fire suppression has allowed white fir to grow more densely in sequoia groves than it did in the past, meaning that there are more opportunities for root rot to spread from infection centers and to be transmitted to sequoias through root contact (Piirto 1977; Piirto et al. 1984). Sequoias weakened by root rot are more susceptible to falling than those free of infection. Restoration of groves to their more open pre-Euroamerican conditions probably will reduce the occurrence of annosus root rot; the direct effects of fire on the pathogen are less certain (Piirto et al. 1992). Serious consideration should be given to chemically treating freshly-cut fir stumps that might be created during grove restoration, which otherwise can provide a major entry path for various root rots (Ferrell 1996).

Throughout its range sugar pine, generally the second or third most abundant tree species in sequoia groves, is succumbing to white pine blister rust (*Cronartium ribicola*), an epidemic disease introduced from Asia. Attempts to eradicate white pine blister rust have been unsuccessful (Ferrell 1996); most likely, groves will continue to lose sugar pine. Consequently, populations of small mammals and birds that depend on sugar pine seeds might also eventually be reduced. Over the range of most sequoia groves, roughly ten percent of sugar pines are resistant to the epidemic strains of blister rust. However, a more virulent strain has been identified and its spread is a distinct possibility (Kinloch and Comstock 1980; Kinloch and Dupper 1987). Even if more virulent strains do not spread, sugar pines of all sizes will survive in groves, but in greatly reduced numbers. The effects of this change on other ecosystem components are unknown.

A long-term strategy for maintaining sugar pines in sequoia groves will probably include planting seedlings of resistant varieties taken from local stock. USFS already has tested thousands of candidate sugar pines for resistance, is protecting resistant seed trees, and is growing and planting resistant seedlings. NPS efforts lag.

Climatic Change

There is no serious doubt that the average global temperature has been rising in this century (Houghton et al. 1990). Internationally, there is now a near-consensus among climatologists and atmospheric scientists that at least part of this warming is due to human activities (Kerr 1995). California, like the rest of the world, is vulnerable to climatic changes induced by the global increase in atmospheric greenhouse (heat-trapping) gases. Though available projections are crude, climatic models suggest that California and the Sierra Nevada may experience significant changes in temperature and the timing and amount of precipitation, leading to fundamental changes in climatic regime over the next several decades (Knox and Scheuring 1991; Westman and Malanson 1992). Snow melt, a major source of soil-water recharge in sequoia groves (Rundel 1972b; Stephenson 1988), is likely to come earlier in the spring than at present, potentially prolonging the summer drought characteristic of the Sierra's mediterranean-type climate. Depending on their magnitude, such climatic changes could have tremendous effects on giant sequoia ecosystems.

The paleoecological record is one of our best tools for understanding the possible magnitude of biotic changes resulting from climatic changes. Contrary to John Muir's glacial hypothesis (Muir 1876, quoted in Axelrod 1959), the fossil pollen record suggests that the present highly disjunct distribution of sequoias is due to the generally higher global summertime temperatures and prolonged summer drought in California of the early and middle Holocene (about 10,000 to 4,500 years ago) (Anderson 1994; Anderson and Smith 1994; this explanation was earlier proposed by Rundel 1972b and Axelrod 1986). During this period, sequoias were probably much rarer than today (at least in areas where they are presently found; Anderson 1994; Anderson and Smith 1994), existing only along creek and meadow edges where present groves exist. Pines were more abundant, firs less abundant. Only since cooling and shortening of summer droughts began about 4,500 years ago has sequoia been able to spread out and create today's groves, over a period of only two or three sequoia life spans (Anderson 1994; Anderson and Smith 1994).

This record of sequoia's response to past climatic changes offers an imperfect but instructive analog to the possible effects of future climatic changes. Projected increases in global temperature over the next several decades are of similar or greater magnitude than those that caused the dramatic increase in sequoia abundance during the last 4,500 years, but they are in the opposite direction (Houghton et al. 1990). It therefore seems reasonable to conclude that, if model projections are correct, increasing temperature over the next several decades, by inducing earlier snowmelt and prolonging summer droughts, may cause a return to conditions unfavorable to sequoias. An immediate effect probably would be a widespread and continuing failure in sequoia reproduction, even in the presence of prescribed fires; this would be a consequence of the high vulnerability of sequoia seedlings to prolonged drought (Harvey et al. 1980; Mutch 1994). Death rates might increase among adult sequoias and associated species as drought stress makes them more vulnerable to insects, pathogens, and air pollution. Of course, there may be other species in the giant sequoia community that would be equally or more severely affected by climatic change than sequoias.

Global warming might also increase the probability of destructive wildfires, particularly within groves that have not yet been restored. Models predict that global warming will be accompanied by increased lightning strike frequencies at the latitudes spanned by the Sierra Nevada (Price and Rind 1991). Compounding the possible increase in wildfire ignitions, extreme weather conditions are likely to make individual fires burn more total area, be more severe, and escape containment more frequently (Torn and Fried 1992). Ryan (1991) raises some of the questions faced by park and wilderness managers confronted with climatic change and the resulting changes in fire regimes and vegetation.

Managers have few, if any, viable options for counteracting the effects of climatic change. Mature sequoias cannot be transplanted upslope to cooler conditions, and even if seedlings are planted at higher elevations in an attempt to start new groves, soils there are less well developed and have generally lower water-holding capacities (Huntington et al. 1985). Selected areas within existing groves might be artificially irrigated to reduce drought stress, though increased competition with urban areas for water may limit the effectiveness of this admittedly desperate approach. More drastically, managers may choose to favor giant sequoias by severely thinning (whether with saws or with fire) non-sequoia trees within the groves, thereby reducing competition for water. Managers would also need to

focus more closely on reducing surface and ladder fuels within groves to reduce the chances of severe wildfires.

CONCLUSIONS AND STEPS FOR IMPLEMENTATION

General Conclusions

I wish to highlight four broad conclusions. First, inaction threatens the sustainability of giant sequoia ecosystems; the ongoing changes in forest succession and buildup of hazardous fuels in most groves cannot be ignored. To do nothing is to assure greater changes away from some of the very conditions that inspired protection of the groves, until such time that severe wildfires preempt options for the future. Protection, restoration, and conservation of giant sequoia ecosystems demand active, science-based management, starting today and continuing indefinitely.

Second, our present knowledge of grove restoration and conservation is imperfect, meaning that grove managers must have the flexibility to change (and must change) their practices as knowledge increases. Rephrased, managers must practice adaptive management. Simply put, adaptive management is the common-sense approach to management, in which managers formalize the process of trying something, seeing what happens, learning from the experience, then trying something new. All too often, however, the cycle is broken at the "seeing what happens" stage; that is, adequate monitoring does not parallel management. If adaptive management (with its indispensable monitoring step) is successfully implemented, specific management prescriptions aimed at grove protection, restoration, and conservation will change as knowledge increases. Within the bounds outlined in this chapter, there is no single clearly correct approach to grove restoration and conservation; rather there is a suite of reasonable and practical approaches. Thus, the different sequoia management agencies are likely to apply a diversity of management approaches.

Corollaries of the preceding two paragraphs are that there will be uncertainties in sequoia management, and that management must move forward in spite of these uncertainties. Additionally, even if there were no uncertainties, attempts to restore groves to pre-Euroamerican conditions will be imperfect due to physical constraints.

Third, the new knowledge needed to guide sequoia management will grow most rapidly if the various land management agencies cooperate in management planning, management actions, and in comparing the consequences of their different management approaches. Coordinated research and monitoring is especially important during these times of shrinking budgets, and would offer indispensable support to the ambitious restoration efforts outlined on previous pages. A step in the right direction has been made with the recent formation of the Giant Sequoia Ecology Cooperative, which includes representatives from the U.S. Forest Service (USFS), National Park Service (NPS), the National Biological Service, California Department of Forestry and Fire Protection, and the University of California. However, it is highly unlikely that the Cooperative will be effective unless new funds become available for coordinated research and monitoring.

Finally, for USFS to meet its new mandate, permanent new base funding must be earmarked for sequoia management, research, and monitoring. It will be expensive to meet the demands of the Mediated Settlement Agreement (MSA) (U.S. Forest Service 1990) and Stewart's (1992) policy directive; this reality cannot be avoided. Private funds in support of management, research, and monitoring can and should be sought, but are not likely to be adequate or provide needed program continuity. Even if public acceptance were to allow the sale of trees removed during grove restoration, such funds might only partly offset costs. And once groves are restored, their maintenance will require committed, active management indefinitely into the future, albeit at a lower level.

Nearly all research and monitoring in support of sequoia management has been funded by NPS or the National Biological Service. These agencies have never had a base-funded sequoia program; funding has been temporary and sporadic, often in response to crises (e.g. see Parsons 1990b). The latest flurry of research and monitoring began in 1987 and is likely to end when funded projects in the National Biological Service's Sierra Nevada Global Change Research Program expire in September of 1996.

Implementation

The following summary steps can help guide sequoia managers in their efforts to protect, restore, and conserve sequoia ecosystems:

1. *Prioritize groves for fire protection and restoration.* Sequoia National Forest personnel already are analyzing the data needed to prioritize groves for protection and restoration according to their vulnerability to wildfires. The assessments should consider fuel conditions within groves, fuel conditions adjacent to groves, and historic patterns of ignitions. This information can be translated into probability of groves experiencing damaging wildfire by using available fire behavior and spread models (e.g. Finney 1995). Protection is to a large degree automatically conferred by grove restoration; thus both goals might be met in one step. In fact, it can be reasonably argued that actions toward both goals (protection and restoration) must proceed simultaneously, since fire protection alters forest structure.

2. *Define broad goals for restoration and conservation.* The interagency Giant Sequoia Ecology Cooperative and interested publics, as required by the National Environmental Policy Act and the MSA (U.S. Forest Service 1990), should be con-

sulted before finalizing broad goals for restoration and conservation. I suggest that a reasonable restoration goal (subject to modification as knowledge increases or policy changes) is to come as close as is practical to restoring grove structure and function to the usual range of conditions that existed during the 1,000 years preceding Euroamerican settlement. The goal as it is stated is meant to allow managers to step back from the unrealistic limitations that would be imposed by trying to replicate conditions calibrated to a specific year or other narrow time period. The term "usual range of conditions" is meant to exclude socially unacceptable extremes that may have occurred during the last 1,000 years, though these extremes may have been important in shaping modern groves. If prescribed fire is chosen as the main tool for maintaining restored groves, a choice consistent with policy must be made between two possible fire regimes: mimic fire regimes from pre-Euroamerican periods with climates similar to today's (which includes both lightning and Native American ignitions), or mimic a lightning-only fire regime.

3. *Define targets for individual groves.* Ideally, defining targets for restoration of individual groves will be guided by an inventory of past and present ecological conditions in each grove (see the section "Grove Restoration"). However, given the immediate need for grove protection through fuel reduction, two actions should proceed simultaneously: determining targets for restoration for individual groves, and restoration itself, with the former staying at least one step (one grove) ahead of the latter. Inevitably, lack of information, time, or funds will mean that there are uncertainties in defining targets for restoration; however, present conditions in many (if not most) groves are such that it may be worse not to act at all than to go forward armed with limited knowledge. (It is important to emphasize that we do already know much about sequoia grove dynamics—particularly about the conditions needed for sequoia regeneration.) Restoration targets should err on the side of the conservative; once trees are removed, they cannot be put back. Individual grove targets should be developed and reviewed with the aid of knowledgeable sequoia experts, including members of the interagency Giant Sequoia Ecology Cooperative. As before, public participation should be nurtured, as required by the National Environmental Policy Act and the MSA.

4. *Choose restoration tool(s) and approach.* Once targets are defined, grove restoration can be accomplished using fire, saws, or some combination of the two. Each tool has advantages and disadvantages, as listed in table 55.1. If saws are used, consideration should be given to immediately treating freshly-cut stumps with borax or some other agent to reduce the possibility of annosus root rot establishment. Felled trees would either be burned on-site (with or without being chipped before burning) or removed. If they are

removed, the MSA (U.S. Forest Service 1990) specifies that it should be done with minimal site disturbance, such as by using helicopters or other low-impact means (cut-to-length systems might be a less expensive option on shallower slopes [O'Connor 1991; Hartsough and McNeel 1994]). Regardless of the tool used, restoration will likely proceed in ways unlike standard silvicultural treatments. Restoration targets will be based on ecological principles aimed at re-creating variable pre-Euroamerican conditions, not on commodity values, maximization of site production, or ease of silvicultural treatment. Commodity production, however, could be an incidental byproduct of restoration (see below). Restored groves will include suppressed trees, insects and pathogens, snags, logs, brush patches, small forest gaps of different sizes, and different levels of forest thinning which grade into one another—sometimes gradually, sometimes more abruptly.

If saws are used as the main tool for grove restoration, restoration costs might be partly offset by the incidental sale of the trees removed. Adoption of this choice would likely involve intense public participation, as required by the National Environmental Policy Act and the MSA. With inadequate public participation, some members of the public might suspect that managers are trying to make restoration pay for itself by adding high-valued trees to those being removed; public involvement would likely reduce the potential for misperceptions. Additionally, detailed grove restoration plans should be reviewed by the interagency Giant Sequoia Ecology Cooperative, which includes members from USFS, NPS, the National Biological Service, the California Department of Forestry and Fire Protection, and the University of California. This would be in addition to the public involvement required by the National Environmental Policy Act and the MSA. At their discretion, members of the Cooperative might ask for additional reviews by other sequoia managers and researchers.

5. *Implement adaptive management.* Adaptive management cannot go forward unless an active research and monitoring program is developed to determine the ecosystem consequences of the different agencies' approaches to sequoia grove management. Ideally, monitoring programs will involve close cooperation among sequoia researchers and managers, coordinated by the interagency Giant Sequoia Ecology Cooperative. The information gained by these efforts would be used to continually assess and refine management approaches.

For the agencies managing giant sequoias, meeting obligations to protect, restore, and conserve sequoia ecosystems will be difficult, time-consuming, and expensive. Grove management cannot go forward piecemeal, drawing only from resources ultimately dedicated to other tasks. Efforts seem sure to fail unless there is strong institutional support at all levels, including programmatic designation and significant permanent base

funding. Responsible stewardship therefore demands a deep and continuing commitment from the management agencies.

ACKNOWLEDGMENTS

For supplying information, input, and useful critical review, I thank L. Bancroft, A. Caprio, A. Demetry, D. Duriscoe, D. Elliott-Fisk, M. Keifer, D. Leisz, L. Mutch, D. Parsons, R. Rogers, S. Stephens, T. Swetnam, J. van Wagtendonk, and two of three anonymous reviewers. Additional thanks go to Bob Rogers for consistently responding, always on short notice, to an extraordinary number of requests for data, documents, and consultation.

NOTES

1. Willard (1994a) corrects some errors in Rundel's (1972a) grove list, and recognizes 65 groves (lumping several of Rundel's groves).
2. Percentages are derived from the latest estimates of grove areas, as compiled by R. Rogers (USFS), P. Lineback (NPS), J. Manley (NPS), and D. Willard (1994b). The new estimates of grove areas dramatically reverse Hartesveldt's estimates of 21% managed by USFS and 68% managed by NPS (Hartesveldt et al. 1975). Still earlier estimates by the California Department of Natural Resources (1952) had roughly equal areas managed by USFS and NPS (38% and 41%, respectively). It appears that earlier estimates were biased by both underestimated USFS grove acreages and overestimated NPS acreages; however, estimates of total grove area in the Sierra Nevada are virtually unchanged. Though more accurate than earlier estimates, the estimates presented in this chapter may change as information improves.
3. Although the USFS uses *preserve*, I prefer the term *conserve*. *Conserve* implies maintaining dynamic grove ecosystems within a range of desired conditions, whereas to some people *preserve* implies maintaining groves in an unchanging state—an impossible task.

REFERENCES

Agee, J. K. 1995. Alternatives for implementing fire policy. In *Proceedings: Symposium on fire in wilderness and park management,* technical coordination by J. K. Brown, R. W. Mutch, C. W. Spoon, and R. H. Wakimoto, 107–12. General Technical Report INT-GTR-320. Ogden, UT: U.S. Forest Service.

Agee, J. K., R. H. Wakimoto, and H. H. Biswell. 1978. Fire and fuel dynamics of Sierra Nevada conifers. *Forest Ecology and Management* 1:255–65.

Anderson, M. K., and M. J. Moratto. 1996. Native American land-use practices and ecological impacts. In *Sierra Nevada Ecosystem Project: Final report to Congress,* vol. II, chap. 9. Davis: University of California, Centers for Water and Wildland Resources.

Anderson, R. S. 1994. Paleohistory of a giant sequoia grove: The record from Log Meadow, Sequoia National Park. In *Proceedings of the symposium on giant sequoias: Their place in the ecosystem and society,* technical coordination by P. S. Aune, 49–55. General Technical Report PSW-151. Albany, CA: U.S. Forest Service.

Anderson, R. S., and S. J. Smith. 1994. Paleoclimatic interpretations of meadow sediment and pollen stratigraphies from California. *Geology* 22:723–26.

Aune, P. S., technical coordinator 1994. *Proceedings of the symposium on giant sequoias: Their place in the ecosystem and society.* General Technical Report PSW-151. Albany, CA: U.S. Forest Service.

Axelrod, D. I. 1959. Late Cenozoic evolution of the Sierran big tree forest. *Evolution* 13:9–23.

———. 1986. The Sierra redwood (Sequoiadendron) forest: End of a dynasty. *Geophytology* 16:25–36.

Baker, W. L. 1994. Restoration of landscape structure altered by fire suppression. *Conservation Biology* 8:763–69.

Bancroft, L., T. Nichols, D. Parsons, D. Graber, B. Evison, and J. van Wagtendonk. 1985. Evolution of the natural fire management program at Sequoia and Kings Canyon National Parks. In *Proceedings: Symposium and workshop on wilderness fire,* edited by J. E. Lotan, B. M. Kilgore, W. C. Fischer, and R. W. Mutch, 174–80. General Technical Report INT-182. Ogden, UT: U.S. Forest Service.

Benson, N. J. 1986. Management of giant sequoia on Mountain Home Demonstration State Forest. In *Proceedings of the workshop on management of giant sequoia,* technical coordination by C. P. Weatherspoon, Y. R. Iwamoto, and D. D. Piirto, 30–31. General Technical Report PSW-95. Albany, CA: U.S. Forest Service.

Bonnicksen, T. M. 1975. Spatial pattern and succession within a mixed conifer–giant sequoia forest ecosystem. M.S. thesis, University of California, Berkeley.

Bonnicksen, T. M., and E. C. Stone. 1978. An analysis of vegetation management to restore the structure and function of presettlement giant sequoia–mixed conifer forest mosaics. Unpublished report to the United States National Park Service, Sequoia and Kings Canyon National Parks, California.

———. 1981. The giant sequoia–mixed conifer forest community characterized through pattern analysis as a mosaic of aggregations. *Forest Ecology and Management* 3:307–28.

———. 1982a. Reconstruction of a presettlement giant sequoia–mixed conifer forest community using the aggregation approach. *Ecology* 63:1134–48.

———. 1982b. Managing vegetation within U.S. national parks: A policy analysis. *Environmental Management* 6:101–2, 109–22.

———. 1985. Restoring naturalness to national parks. *Environmental Management* 9:479–86.

Bush, G. 1992. Giant sequoias in national forests. A proclamation by the President of the United States of America, signed July 14, 1992.

Cahill T. A., J. J. Carroll, D. Campbell, and T. E. Gill. 1996. Air quality. In *Sierra Nevada Ecosystem Project: Final report to Congress,* vol. II, chap. 48. Davis: University of California, Centers for Water and Wildland Resources.

California Department of Natural Resources. 1952. The status of *Sequoia gigantea* in the Sierra Nevada. Report to the California Legislature, Sacramento.

Caprio, A. C., L. S. Mutch, T. W. Swetnam, and C. H. Baisan. 1994. Temporal and spatial patterns of giant sequoia radial growth response to a high severity fire in A.D. 1297. Contract report to the California Department of Forestry and Fire Protection, Mountain Home State Forest.

Caprio, A. C., and T. W. Swetnam. 1995. Historic fire regimes along an elevational gradient on the west slope of the Sierra Nevada, California. In *Proceedings: Symposium on fire in wilderness and park management,* technical coordination by J. K. Brown, R. W. Mutch, C. W. Spoon, and R. H. Wakimoto, 173–79. General Technical Report INT-GTR-320. Ogden, UT: U.S. Forest Service.

Chang, C. 1996. Ecosystem responses to fire and variations in fire regimes. In *Sierra Nevada Ecosystem Project: Final report to Congress,* vol. II, chap. 39. Davis: University of California, Centers for Water and Wildland Resources.

Cloer, C. A. 1994. Reflections on management strategies of the Sequoia National Forest: A grassroots view. In *Proceedings of the symposium on giant sequoias: Their place in the ecosystem and society,* technical coordination by P. S. Aune, 129–36. General Technical Report PSW-151. Albany, CA: U.S. Forest Service.

Cole, K. 1985. Past rates of change, species richness, and a model of vegetational inertia in the Grand Canyon, Arizona. *American Naturalist* 125:289–303.

Croft, W. 1994. Sequoia grove preservation: Natural or humanistic? In *Proceedings of the symposium on giant sequoias: Their place in the ecosystem and society,* technical coordination by P. S. Aune, 8–10. General Technical Report PSW-151. Albany, CA: U.S. Forest Service.

DeLeon, D. 1952. Insects associated with *Sequoia sempervirens* and *Sequoia gigantea* in California. *Pan-Pacific Entomologist* 23:75–91.

Demetry, A. 1995. Regeneration patterns within canopy gaps in a giant sequoia–mixed conifer forest: Implications for forest restoration. M.S. thesis, Northern Arizona University, Flagstaff.

Demetry, A., and D. Duriscoe. 1996. Fire-caused canopy gaps as a model for the ecological restoration of Giant Forest Village. Unpublished report to Sequoia and Kings Canyon National Parks, California.

Dulitz, D. J. 1986. Growth and yield of giant sequoia. In *Proceedings of the workshop on management of giant sequoia,* technical coordination by C. P. Weatherspoon, Y. R. Iwamoto, and D. D. Piirto, 14–16. General Technical Report PSW-95. Albany, CA: U.S. Forest Service.

Duriscoe, D. M., and K. W. Stolte. 1992. Decreased foliage production and longevity observed in ozone-injured Jeffrey and ponderosa pine in Sequoia National Park, California. In *Effects, modeling and control,* 663–80. Vol. 2 of *Tropospheric ozone and the environment.* Pittsburgh, PA: Air and Waste Management Association.

Elliott-Fisk, D. L., J. Aubert, D. Murphy, J. Schaber, and S. Stephens. 1996. Mediated settlement agreement, giant sequoia ecosystems. In *Sierra Nevada Ecosystem Project: Final report to Congress,* vol. III. Davis: University of California, Centers for Water and Wildland Resources.

Ferrell, G. T. 1996. The influence of insect pests and pathogens on Sierra forests. In *Sierra Nevada Ecosystem Project: Final report to Congress,* vol. II, chap. 45. Davis: University of California, Centers for Water and Wildland Resources.

Finney, M. A. 1995. Fire growth modeling in the Sierra Nevada of California. In *Proceedings: Symposium on fire in wilderness and park management,* technical coordination by J. K. Brown, R. W. Mutch, C. W. Spoon, and R. H. Wakimoto, 189–91. General Technical Report INT-GTR-320. Ogden, UT: U.S. Forest Service.

Fins, L. 1979. Genetic architecture of giant sequoia. Ph.D. diss., University of California, Berkeley.

Fins, L., and W. J. Libby. 1994. Genetics of giant sequoia. In *Proceedings of the symposium on giant sequoias: Their place in the ecosystem and society,* technical coordination by P. S. Aune, 65–68. General Technical Report PSW-151. Albany, CA: U.S. Forest Service.

Flint, W. D. 1987. To find the biggest tree. Sequoia Natural History Association, Three Rivers, California.

Fry, W. 1933. The great sequoia avalanche. *Sierra Club Bulletin* 18: 118–20.

Fullmer, D. G., R. R. Rogers, J. D. Manley, and N. L. Stephenson. In press. Restoration as a component of ecosystem management for giant sequoia groves in California. In *Proceedings of the annual meeting of the Society for Ecological Restoration*, Seattle, Washington, September 14–16, 1995.

Gasser, D. G. 1994. Young growth management of giant sequoia. In *Proceedings of the symposium on giant sequoias: Their place in the ecosystem and society*, technical coordination by P. S. Aune, 120–25. General Technical Report PSW-151. Albany, CA: U.S. Forest Service.

Gebauer, S. B. 1992. Changes in soil properties along a post-fire chronosequence in a sequoia–mixed conifer forest in Sequoia National Park, California. M.S. thesis, Duke University, Durham, North Carolina.

Graber, D. M. 1985. Coevolution of National Park Service fire policy and the role of national parks. In *Proceedings: Symposium and workshop on wilderness fire*, edited by J. E. Lotan, B. M. Kilgore, W. C. Fischer, and R. W. Mutch, 345–49. General Technical Report INT-182. Ogden, UT: U.S. Forest Service.

———. 1995. Resolute biocentrism: The dilemma of wilderness in national parks. In *Reinventing nature? Responses to postmodern deconstruction*, edited by M. E. Soulé and G. Lease, 123–35. Washington, DC: Island Press.

Graumlich, L. J. 1993. A 1000-year record of temperature and precipitation in the Sierra Nevada. *Quaternary Research* 39: 249–55.

Harmon, M. E., K. Cromack, Jr., and B. G. Smith. 1987. Coarse woody debris in mixed-conifer forests, Sequoia National Park, California. *Canadian Journal of Forest Research* 17:1265–72.

Harrison, W. 1986. Management of giant sequoia at Calaveras Big Trees State Park. In *Proceedings of the workshop on management of giant sequoia*, technical coordination by C. P. Weatherspoon, Y. R. Iwamoto, and D. D. Piirto, 40–42. General Technical Report PSW-95. Albaby, CA: U.S. Forest Service.

Hartesveldt, R. J. 1962. The effects of human impact upon *Sequoia gigantea* and its environment in the Mariposa Grove, Yosemite National Park, California. Ph.D. diss., University of Michigan, Ann Arbor.

———. 1965. An investigation of the effect of direct human impact and of advanced plant succession on *Sequoia gigantea* in Sequoia and Kings Canyon National Parks, California. Report on contract number 14-10-0434-1421. San Francisco: National Park Service.

———. 1966. Study of the possible changes in the ecology of sequoia groves in Sequoia National Park to be crossed by the new Mineral King highway. Report on contract number 6177-H. Fresno: California Division of Highways, District VI.

Hartesveldt, R. J., H. T. Harvey, H. S. Shellhammer, and R. E. Stecker. 1975. *The giant sequoia of the Sierra Nevada*. Washington, DC: National Park Service.

Hartsough, B. R., and J. R. McNeel. 1994. Comparison of mechanized systems for thinning ponderosa pine and mixed conifer stands. Report presented before the International Winter Meeting of the American Society of Agricultural Engineers, Atlanta, Georgia, 13–16 December.

Harvey, H. T., and H. S. Shellhammer. 1991. Survivorship and growth of giant sequoia (*Sequoiadendron giganteum* [Lindl.] Buchh.) seedlings after fire. *Madroño* 38:14–20.

Harvey, H. T., H. S. Shellhammer, and R. E. Stecker. 1980. *Giant sequoia ecology*. Washington, DC: National Park Service.

Heald, R. C. 1986. Management of giant sequoia at Blodgett Forest Research Station. In *Proceedings of the workshop on management of giant sequoia*, technical coordination by C. P. Weatherspoon, Y. R. Iwamoto, and D. D. Piirto, 37–39. General Technical Report PSW-95. Albany, CA: U.S. Forest Service.

Houghton, J. T., G. J. Jenkins, and J. J. Ephraums. 1990. *Climate change: The IPCC assessment*. New York: Cambridge University Press.

Hughes, M. K., and P. M. Brown. 1992. Drought frequency in central California since 101 B.C. recorded in giant sequoia tree rings. *Climate Dynamics* 6:161–67.

Huntington, E. 1914. *The climatic factor as illustrated in arid America*. Publication 192. Washington DC: Carnegie Institute of Washington.

Huntington, G. L., R. G. Burau, and L. D. Whittig. 1985. Pedologic investigations in support of acid rain studies, Sequoia National Park, California. Davis: University of California, Department of Land, Air, and Water Resources.

Kauper, D., N. Stephenson, and T. Warner. 1980. The effect of past management actions on the composition and structure of vegetation in the Grant tree portion of Grant Grove, Kings Canyon National Park, California. Unpublished report to Sequoia and Kings Canyon National Parks, California.

Keifer, M. 1995. Changes in stand density, species composition, and fuel load following prescribed fire in the southern Sierra Nevada mixed conifer forest. *Supplement to the Bulletin of the Ecological Society of America* 76:138–39.

Keifer, M., and P. M. Stanzler. 1995. Fire effects monitoring in Sequoia and Kings Canyon National Parks. In *Proceedings: Symposium on fire in wilderness and park management*, technical coordination by J. K. Brown, R. W. Mutch, C. W. Spoon, and R. H. Wakimoto, 215–18. General Technical Report INT-GTR-320. Ogden, UT: U.S. Forest Service.

Kercher, J. R., and M. C. Axelrod. 1984. A process model of fire ecology and succession in a mixed-conifer forest. *Ecology* 65:1725–42.

Kerr, R. A. 1995. It's official: First glimmer of greenhouse warming seen. *Science* 270:1565–67.

Kilgore, B. M. 1972. Fire's role in a sequoia forest. *Naturalist* 23:26–35.

———. 1973a. Impact of prescribed burning on a sequoia–mixed conifer forest. *Proceedings of the Tall Timbers Fire Ecology Conference* 12:345–75.

———. 1973b. The ecological role of fire in Sierran conifer forests. *Journal of Quaternary Research* 3:496–513.

———. 1985. What is "natural" in wilderness fire management? In *Proceedings: Symposium and workshop on wilderness fire*, edited by J. E. Lotan, B. M. Kilgore, W. C. Fischer, and R. W. Mutch, 57–67. General Technical Report INT-182. Ogden, UT: U.S. Forest Service.

Kilgore, B. M., and H. H. Biswell. 1971. Seedling germination following fire in a giant sequoia forest. *California Agriculture* 25:8–10.

Kilgore, B. M., and R. W. Sando. 1975. Crown-fire potential in a sequoia forest after prescribed burning. *Forest Science* 21:83–87.

Kilgore, B. M., and D. Taylor. 1979. Fire history of a sequoia-mixed conifer forest. *Ecology* 60:129–42.

Kinloch, B. B., and M. Comstock. 1980. Race of *Cronartium ribicola* virulent to major resistance in sugar pine. *Plant Disease* 65:604–5.

Kinloch, B. B., and G. E. Dupper. 1987. Restricted distribution of a virulent race of the white pine blister rust pathogen in the western United States. *Canadian Journal of Forest Research* 17:448–51.

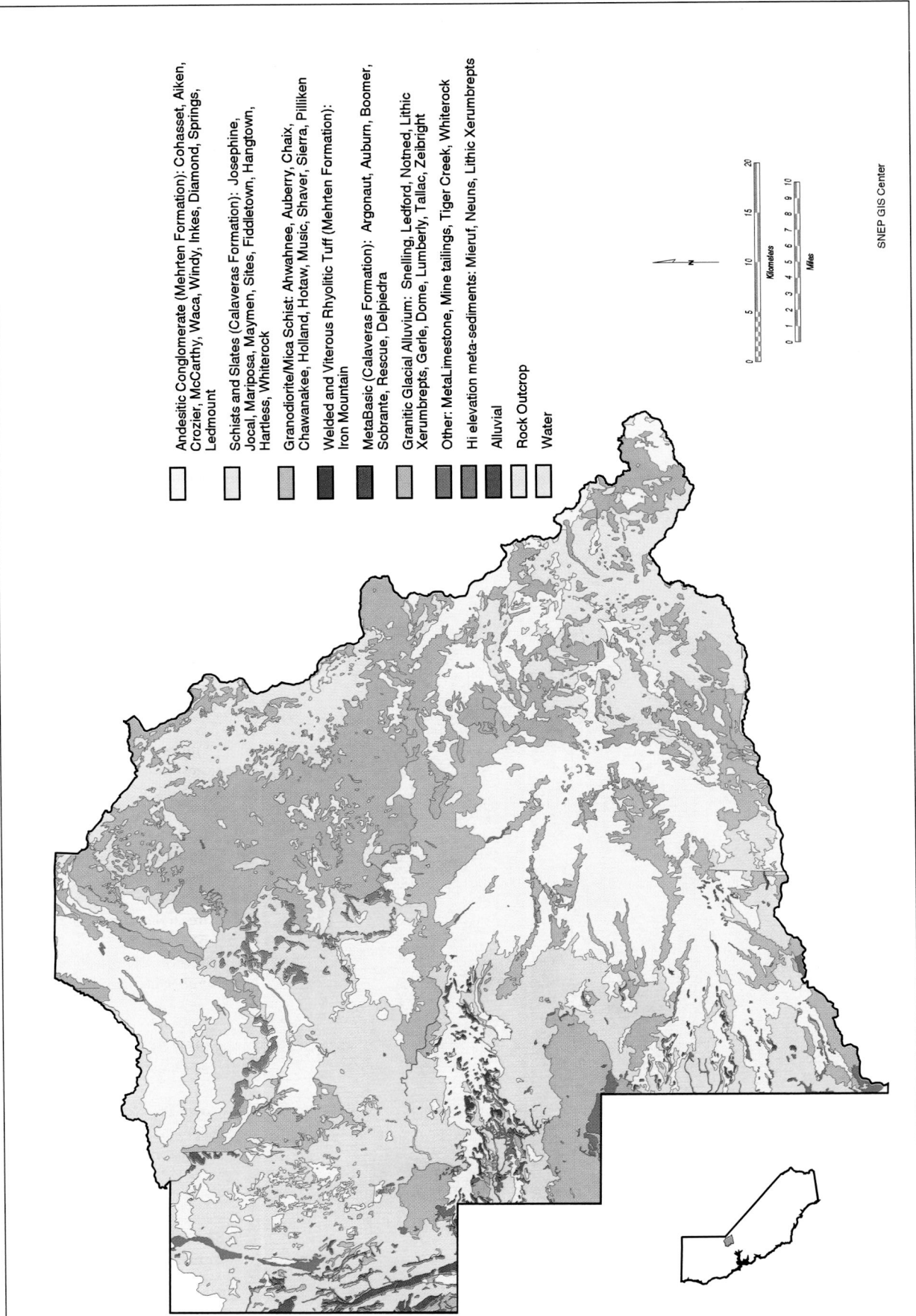

Andesitic Conglomerate (Mehrten Formation): Cohasset, Aiken, Crozier, McCarthy, Waca, Windy, Inkes, Diamond, Springs, Ledmount

Schists and Slates (Calaveras Formation): Josephine, Jocal, Mariposa, Maymen, Sites, Fiddletown, Hangtown, Hartless, Whiterock

Granodiorite/Mica Schist: Ahwahnee, Auberry, Chaix, Chawanakee, Holland, Hotaw, Music, Shaver, Sierra, Pilliken

Welded and Viterous Rhyolitic Tuff (Mehrten Formation): Iron Mountain

MetaBasic (Calaveras Formation): Argonaut, Auburn, Boomer, Sobrante, Rescue, Delpiedra

Granitic Glacial Alluvium: Snelling, Ledford, Notned, Lithic Xerumbrepts, Gerle, Dome, Lumberly, Tallac, Zeibright

Other: MetaLimestone, Mine tailings, Tiger Creek, Whiterock

Hi elevation meta-sediments: Mieruf, Neuns, Lithic Xerumbrepts

Alluvial

Rock Outcrop

Water

SNEP GIS Center

PLATE 54.1

Eldorado National Forest geology of parent material.

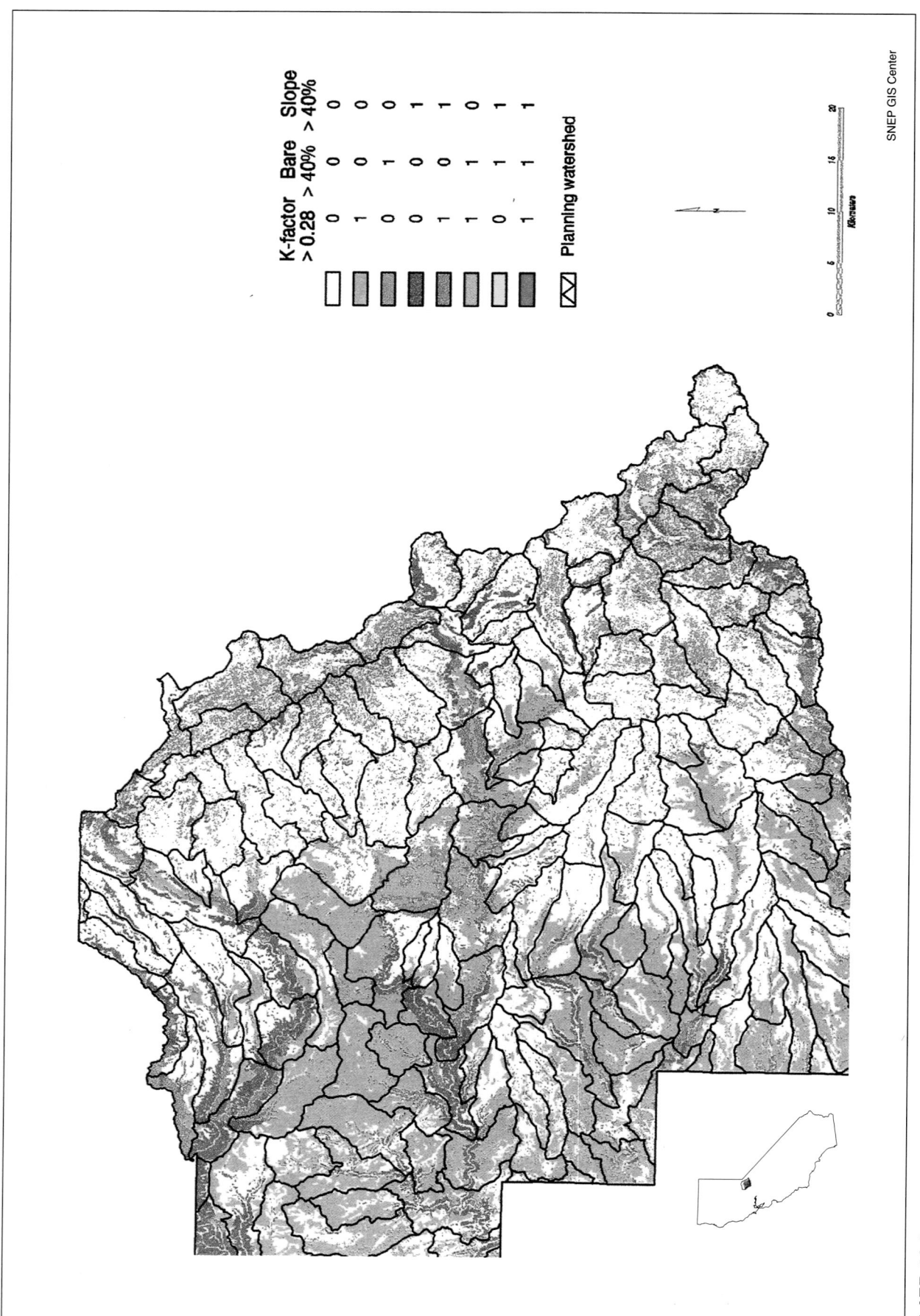

PLATE 54.2

Eldorado National Forest natural erosion potential.

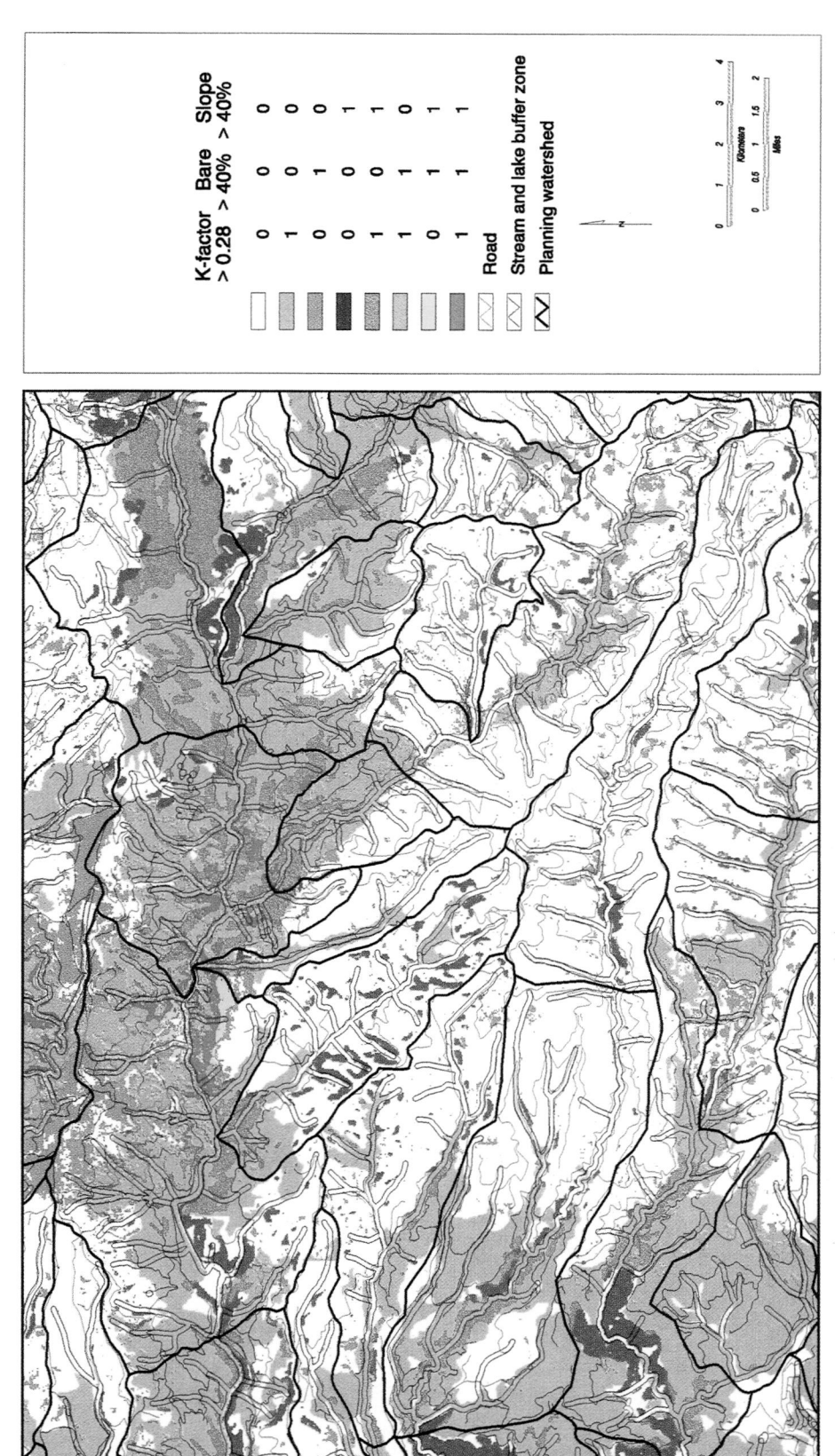

K-factor Bare   Slope
> 0.28 > 40%   > 40%

0	0	0
1	0	0
0	1	0
1	1	0
0	0	1
1	0	1
0	1	1
1	1	1
0	1	0
1	1	1

Road

Stream and lake buffer zone

Planning watershed

Kilometers
0   1   2   3   4

Miles
0 0.5 1 1.5 2

PLATE 54.3

Camp Creek area natural erosion potential.

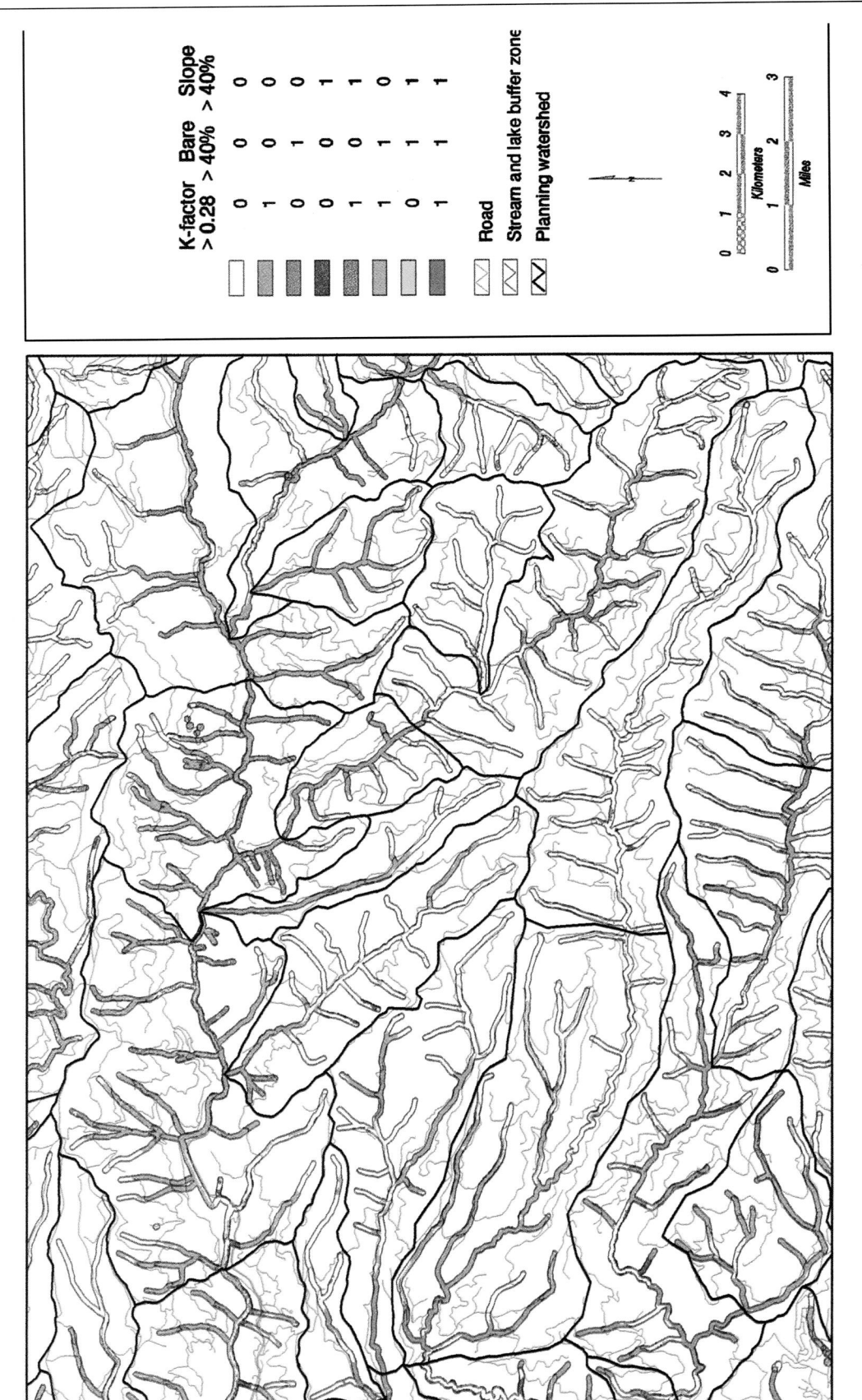

PLATE 54.4

Camp Creek area sedimentation hazard index.

Knox, J. B., and A. F. Scheuring, eds. 1991. *Global climate change and California*. Berkeley and Los Angeles: University of California Press.

Laessle, A. M. 1965. Spacing and competition in natural stands of sand pine. *Ecology* 46:65–72.

Lambert, S., and T. J. Stohlgren. 1988. Giant sequoia mortality in burned and unburned stands. *Journal of Forestry* 44:44–46.

Lemons, J. 1987. Unites States' national park management: Values, policies, and possible hints for others. *Environmental Conservation* 14:328-40.

Lewis, H. T. 1973. *Patterns of Indian burning in California: Ecology and ethnohistory*. Ballena Press Anthropological Papers 1. Ramona, CA: Ballena Press.

Libby, W. J. 1986. Genetic variation and early performance of giant sequoia in plantations. In *Proceedings of the workshop on management of giant sequoia*, technical coordination by C. P. Weatherspoon, Y. R. Iwamoto, and D. D. Piirto, 17–18. General Technical Report PSW-95. Albany, CA: U.S. Forest Service.

Manley, P. N., G. E. Brogan, C. Cook, M. E. Flores, D. G. Fullmer, S. Husari, T. M. Jimerson, L. M. Lux, M. E. McCain, J. A. Rose, G. Schmitt, J. C. Schuyler, and M. J. Skinner. 1995. *Sustaining ecosystems: A conceptual framework*. San Francisco, CA: U.S. Forest Service Pacific Southwest Region.

McKelvey, K. S., and K. K. Busse. 1996.Twentieth-century fire patterns on Forest Service lands. In *Sierra Nevada Ecosystem Project: Final report to Congress*, vol. II, chap. 41. Davis: University of California, Centers for Water and Wildland Resources.

Meinecke, E. P. 1926. Memorandum on the effects of tourist traffic on plant life, particularly big trees, Sequoia National Park, California. Unpublished memorandum to the Director of the National Park Service, Washington, D.C.

Miller, C. 1994. A model of the interactions among climate, fire, and forest pattern in the Sierra Nevada. M.S. thesis, Colorado State University, Fort Collins.

Miller, C., and D. L. Urban. In preparation. Fire, climate, and forest pattern in the Sierra Nevada. To be submitted to *Ecology*.

Miller, P. R. 1973. Oxidant-induced community change in a mixed conifer forest. In *Air pollution damage to vegetation*, edited by J. A. Naegele, 101–17. Advances in Chemistry Series 122. Washington, DC: American Chemical Society.

———. 1996. Biological effects of air pollution in the Sierra Nevada. *Sierra Nevada Ecosystem Project: Final report to Congress*, vol. III. Davis: University of California, Centers for Water and Wildland Resources.

Miller, P. R., N. E. Grulke, and K. W. Stolte. 1994. Air pollution effects on giant sequoia ecosystems. In *Proceedings of the symposium on giant sequoias: Their place in the ecosystem and society*, technical coordination by P. S. Aune, 90–98. General Technical Report PSW-151. Albany, CA: U.S. Forest Service.

Muir, J. 1876. On the post-glacial history of *Sequoia gigantea*. *Proceedings of the American Association for the Advancement of Science* 25:242–53.

Mutch, L. S. 1994. Growth responses of giant sequoia to fire and climate in Sequoia and Kings Canyon National Parks, California. M.S. thesis, University of Arizona, Tucson.

Mutch, L. S., and T. W. Swetnam. 1995. Effects of fire severity and climate on ring-width growth of giant sequoia after burning. In *Proceedings: Symposium on fire in wilderness and park management*, technical coordination by J. K. Brown, R. W. Mutch, C. W. Spoon, and R. H. Wakimoto, 241–46. General Technical Report INT-GTR-320. Ogden, UT: U.S. Forest Service.

Nichols, H. T. 1977. A quantification and analysis of the 1977 Redwood Mountain prescribed burn program. Unpublished report filed in the Resources Management office, Sequoia and Kings Canyon National Parks, California.

———. 1989. Managing fire in Sequoia and Kings Canyon National Parks. *Fremontia* 16:11–14.

O'Connor, P. R. 1991. Advantages and implications of forwarding cut-to-length wood. Report presented before the International Winter Meeting of the American Society of Agricultural Engineers, Chicago, 17–20 December.

Otter, F. L. 1963. *The men of Mammoth Forest*. Ann Arbor, MI: Edwards Brothers.

Parsons, D. J. 1978. Fire and fuel accumulation in a giant sequoia forest. *Journal of Forestry* 76:104–5.

———. 1990a. Restoring fire to the Sierra Nevada mixed conifer forest: Reconciling science, policy, and practicality. In *Proceedings of the First Annual Meeting of the Society for Ecological Restoration*, edited by H. G. Hughes and T. M. Bonnicksen, 271–79. Madison: University of Wisconsin.

———. 1990b. The giant sequoia fire controversy: The role of science in natural ecosystem management. In *Proceedings of the Third Biennial Conference on Research in California's National Parks*, edited by C. van Riper, T. Stohlgren, S. Veirs, and S. Hillyer, 257–68. Washington, DC: National Park Service. Transactions and Proceedings Series 8.

———. 1994. Objects or ecosystems? Giant sequoia management in national parks. In *Proceedings of the symposium on giant sequoias: Their place in the ecosystem and society*, technical coordination by P. S. Aune, 109–15. General Technical Report PSW-151. Albany, CA: U.S. Forest Service.

———. 1995. Restoring fire to giant sequoia groves: What have we learned in 25 years? In *Proceedings: Symposium on fire in wilderness and park management*, technical coordination by J. K. Brown, R. W. Mutch, C. W. Spoon, and R. H. Wakimoto, 256–58. General Technical Report INT-GTR-320. Ogden, UT: U.S. Forest Service.

Parsons, D. J., and S. H. DeBenedetti. 1979. Impact of fire suppression on a mixed-conifer forest. *Forest Ecology and Management* 2:21–33.

Parsons, D. J., D. M. Graber, J. K. Agee, and J. W. van Wagtendonk. 1986. Natural fire management in national parks. *Environmental Management* 10:21–24.

Patterson, M. T. 1993. The ecophysiology of ozone-stressed Jeffrey pine. Ph.D. diss., University of California, Los Angeles.

Peterson, D. L., and M. J. Arbaugh. 1992. Mixed conifer forests of the Sierra Nevada. In *The response of western forests to air pollution*, edited by R. K. Olson, D. Binkley, and M. Bohn, 433–59. Ecological Studies 97. New York: Springer-Verlag.

Piirto, D. D. 1977. Factors associated with tree failure of giant sequoia. Ph.D. diss., University of California, Berkeley.

Piirto, D. D., K. L. Piper, and J. R. Parmeter, Jr. 1992. Biological and management implications of fire/pathogen interactions in the giant sequoia ecosystem: Part I, Fire scar/pathogen studies. Final report. Cooperative Agreement 9000-8-0005. Three Rivers, CA: National Park Service.

Piirto, D. D., W. W. Wilcox, J. R. Parmeter, Jr., and D. L. Wood. 1984. Causes of uprooting and breakage of specimen giant sequoia trees. Bulletin 1909. University of California, Division of Agriculture and Natural Resources, Berkeley.

Price, C., and D. Rind. 1991. Lightning activity in a greenhouse world. In *Proceedings of the 11th Conference of Fire and Forest Meteorology*, 598–604. Bethesda, MD: Society of American Foresters.

Reynolds, R. D. 1959. Effect of natural fires and aboriginal burning upon the forests of the central Sierra Nevada. M.A. thesis, University of California, Berkeley.

Rogers, R. R., L. H. Jump, and C. E. Spencer. 1995. GPS among the giants: Mapping the Sequoia National Forest. *GPS World*, November, 24–33.

Rueger, B. 1994. Giant sequoia management strategies on the Tule River Indian Reservation. In *Proceedings of the symposium on giant sequoias: Their place in the ecosystem and society*, technical coordination by P. S. Aune, 116–17. General Technical Report PSW-151. Albany, CA: U.S. Forest Service.

Rundel, P. W. 1969. The distribution and ecology of the giant sequoia ecosystem in the Sierra Nevada, California. Ph.D. diss., Duke University, Durham, North Carolina.

———. 1971. Community structure and stability in the giant sequoia groves of the Sierra Nevada, California. *American Midland Naturalist* 85:478–92.

———. 1972a. An annotated check list of the groves of Sequoiadendron giganteum in the Sierra Nevada, California. *Madroño* 21:319–28.

———. 1972b. Habitat restriction in giant sequoia: The environmental control of grove boundaries. *American Midland Naturalist* 87: 81–99.

Ryan, K. C. 1991. Vegetation and wildland fire: Implications of global climate change. *Environment International* 17:169–78.

Skinner, C. N., and C. Chang. 1996. Fire regimes, past and present. In Sierra Nevada Ecosystem Project: Final report to Congress, vol. II, chap. 38. Davis: University of California, Centers for Water and Wildland Resources.

SNEP. 1994. Sierra Nevada Ecosystem Project progress report.

Spies, T. A., J. F. Franklin, and M. Klopsch. 1990. Canopy gaps in Douglas-fir forests of the Cascade Mountains. *Canadian Journal of Forest Research* 20:649–58.

Stephens, S. L. 1995. Effects of prescribed and simulated fire and forest history of giant sequoia (*Sequoiadendron giganteum* [Lindley] Buchholz)–mixed conifer ecosystems of the Sierra Nevada, California. Ph.D. diss., University of California, Berkeley.

Stephenson, N. L. 1987. Use of tree aggregations in forest ecology and management. *Environmental Management* 11:1–5.

———. 1988. Climatic control of vegetation distribution: The role of the water balance with examples from North America and Sequoia National Park, California. Ph.D. diss., Cornell University, Ithaca, New York.

———. 1994. Long-term dynamics of giant sequoia populations: Implications for managing a pioneer species. In *Proceedings of the symposium on giant sequoias: Their place in the ecosystem and society*, technical coordination by P. S. Aune, 56–63. General Technical Report PSW-151. Albany, CA: U.S. Forest Service.

———. In preparation. Age structure, survivorship, and long-term viability of giant sequoia populations with and without fire. To be submitted to *Ecological Applications*.

Stephenson, N. L., and D. J. Parsons. 1993. A research program for predicting the effects of climatic change on the Sierra Nevada. In *Proceedings of the Fourth Conference on Research in California's National Parks*, edited by S. D. Veirs, T. J. Stohlgren, and C. Schonewald-Cox, 93–109. Transactions and Proceedings Series 9. Denver: National Park Service.

Stephenson, N. L., D. J. Parsons, and T. W. Swetnam. 1991. Restoring natural fire to the sequoia–mixed conifer forest: Should intense

fire play a role? *Proceedings of the Tall Timbers Fire Ecology Conference* 17:321–37.

Stewart, R. E. 1992. Internal memorandum to Region 5 forest supervisors and staff directors. File designation #2470. 19 June.

Stohlgren, T. J. 1992. Resilience of a heavily logged grove of giant sequoia (*Sequoiadendron giganteum*) in Kings Canyon National Park, California. *Forest Ecology and Management* 54:115-40.

———. 1993. Spatial patterns of giant sequoia (*Sequoiadendron giganteum*) in two sequoia groves in Sequoia National Park, California. *Canadian Journal of Forest Research* 23:120–32.

Stolte, K. W., M. I. Flores, D. R. Mangis, and D. B. Joseph. 1992. Tropospheric ozone exposures and ozone injury on sensitive pine species in the Sierra Nevada of California. In *Effects, modeling and control*, 637–62. Vol. 2 of *Tropospheric ozone and the environment*. Pittsburgh, PA: Air and Waste Management Association.

Sudworth, G. 1900. Notes on big tree groves: Excerpts on fire, lumbering, range, and soil and water conditions. Compiled by A. M. Avakian, December 1939, California Forest and Range Experiment Station, Berkeley.

Swetnam, T. W. 1993. Fire history and climate change in giant sequoia groves. *Science* 262:885–89.

Swetnam, T. W., C. H. Baisan, A. C. Caprio, R. Touchan, and P. M. Brown. 1992. Tree-ring reconstruction of giant sequoia fire regimes. Final report on Cooperative Agreement DOI 8018-1-0002 to National Park Service, Sequoia and Kings Canyon National Parks, California.

Torn, M. S., and J. S. Fried. 1992. Predicting the impacts of global warming on wildland fire. *Climatic Change* 21:257–74.

Tweed, W. 1994. Public perception of giant sequoias over time. In *Proceedings of the symposium on giant sequoias: Their place in the ecosystem and society*, technical coordination by P. S. Aune, 5–7. General Technical Report PSW-151. Albany, CA: U.S. Forest Service.

Urban, D. L., C. Miller, and N. L. Stephenson. In preparation. Gradient response in the mixed conifer forest of the Sierra Nevada: The physical template. To be submitted to *Ecology*.

Urban, D. L., C. Miller, N. L. Stephenson, and D. M. Graber. In preparation. Gradient response in the mixed conifer forest of the Sierra Nevada: Biotic response. To be submitted to *Ecology*.

U.S. Forest Service. 1990. Sequoia National Forest mediated settlement agreement. Unpublished administrative document, Porterville, California.

Vale, T. R. 1987. Vegetation change and park purposes in the high elevations of Yosemite National Park, California. *Annals of the Association of American Geographers* 77:1–18.

Vankat, J. L. 1969. The pristine vegetation conditions of Sequoia National Park. Report on National Park Service contract 14-10-9-900-112, Sequoia and Kings Canyon National Parks, California.

———. 1970. Vegetation change in Sequoia National Park, California. Ph.D. diss., University of California, Davis.

Vankat, J. L., and J. Major. 1978. Vegetation changes in Sequoia National Park, California. *Journal of Biogeography* 5:377–402.

Van Wagner, C. E. 1978. Age-class distribution and the forest fire cycle. *Canadian Journal of Forest Research* 8:220–27.

van Wagtendonk, J. W. 1985. Fire suppression effects on fuels and succession in short-fire-interval wilderness ecosystems. In *Proceedings: Symposium and workshop on wilderness fire*, edited by J. E. Lotan, B. M. Kilgore, W. C. Fischer, and R. W. Mutch, 119–26. General Technical Report INT-182. Ogden, UT: U.S. Forest Service.

———. 1996. Use of a deterministic fire growth model to test fuel treatments. In *Sierra Nevada Ecosystem Project: Final report to Congress,* vol. II, chap. 43. Davis: University of California, Centers for Water and Wildland Resources.

Weatherspoon, C. P. 1990. *Sequoiadendron giganteum* (Lindl.) Buchholz. In *Conifers,* technical coordination by R. M. Burns and B. H. Honkala, 552–62. Vol. 1 of *Silvics of North America.* Agriculture Handbook 654. Washington, DC: U.S. Forest Service.

Weatherspoon, C. P. 1996. Fire-silviculture relationships in Sierra forests. In *Sierra Nevada Ecosystem Project: Final report to Congress,* vol. II, chap. 44. Davis: University of California, Centers for Water and Wildland Resources.

Weatherspoon, C. P., Y. R. Iwamoto, and D. D. Piirto (technical coordinators). 1986. Proceedings of the workshop on management of giant sequoia. General Technical Report PSW-95. Albany, CA: U.S. Forest Service.

Weatherspoon, C. P., and C. N. Skinner. 1996. Landscape-level strategies for forest fuel management. In *Sierra Nevada Ecosystem Project: Final report to Congress,* vol. II, chap. 56. Davis: University of California, Centers for Water and Wildland Resources.

Westman, W. E., and G. P. Malanson. 1992. Effects of climatic change on mediterranean-type ecosystems in California and Baja California. In *Global Warming and Biological Diversity,* edited by R. L. Peters and T. E. Lovejoy, 258–76. New Haven: Yale University Press.

Willard, D. 1994a. The natural giant sequoia *(Sequoiadendron giganteum)* groves of the Sierra Nevada, California—an updated annotated list. In *Proceedings of the symposium on giant sequoias: Their place in the ecosystem and society,* technical coordination by P. S. Aune, 159–64. General Technical Report PSW-151. Albany, CA: U.S. Forest Service.

———. 1994b. *Giant sequoia groves of the Sierra Nevada: A reference guide.* Privately published by D. Willard, P.O. Box 7304, Berkeley, CA 94707.

SECTION VI

Building Strategies for the Future Sierra Nevada

C. PHILLIP WEATHERSPOON
Pacific Southwest Research Station
U.S. Forest Service
Redding, California

CARL N. SKINNER
Pacific Southwest Research Station
U.S. Forest Service
Redding, California

56

Landscape-Level Strategies for Forest Fuel Management

ABSTRACT

As a result largely of human activities during the past 150 years, fires in Sierra Nevada forests occur less frequently and cover much less area than they did historically but are much more likely to be large and severe when they do occur. High-severity wildfires are considered by many to be the greatest single threat to the integrity and sustainability of Sierra Nevada forests. The continuing accumulation of large quantities of forest biomass that fuel wildfires points to a need to develop landscape-level strategies for managing fuels to reduce the area and average size burned by severe fires. Concurrently, more of the ecosystem functions of natural fire regimes—characterized in most areas by frequent low- to moderate-severity fires—need to be restored to Sierran forests. This chapter reviews past and current approaches to managing fuels on a landscape basis and, based on a synthesis of many of these approaches, proposes an outline for a potential fuel-management strategy for Sierra Nevada forests.

INTRODUCTION

Prior to concentrated Euro-American settlement in the middle to late 1800s, low- and middle-elevation forests in the Sierra Nevada were characterized by relatively frequent low- to moderate-severity fires (Skinner and Chang 1996). These frequent fires performed important ecological functions (Kilgore 1973). As a result largely of human activities during the past 150 years, including but not limited to fire suppression, fires now occur less frequently and cover much less area but are much more likely to be large and severe when they do occur (Husari and McKelvey 1996; McKelvey and Johnston 1992;

Skinner and Chang 1996; U.S. Forest Service 1995; Weatherspoon et al. 1992). In aggregate, such high-severity fires are well outside the natural range of variability for these ecosystems and are considered by many to be the greatest single threat to the integrity and sustainability of Sierra Nevada forests. In addition, related human-induced changes in forest structure, composition, and processes (including many of the functions once performed by frequent fires) are in many areas so profound that they jeopardize ecosystem diversity and viability even without reference to severe fire (Skinner and Chang 1996; U.S. Forest Service 1995).

These concerns are prominent among the issues confronting those interested in the well-being of the Sierra Nevada. This chapter addresses potential landscape-level strategies intended to reduce the extent of severe fires in Sierra Nevada forests and to restore more of the ecosystem functions of frequent low- to moderate-severity fires. As a byproduct, these strategies offer tools that could contribute significantly to improving the health, integrity, and sustainability of Sierra Nevada ecosystems.

To keep the scope of the chapter manageable, we focus on the low- to middle-elevation coniferous forests of the Sierra Nevada, on both west and east sides of the crest. Our reasons include the following:

- These forests rank at or near the top among Sierran vegetation zones in terms of overall richness and diversity of resources and values.

- Twentieth-century fire occurrence in these forests has been much greater than in higher-elevation forests (McKelvey and Busse 1996). High-severity wildfires are much less a concern in the higher-elevation forests.

Sierra Nevada Ecosystem Project: Final report to Congress, vol. II, *Assessments and scientific basis for management options*. Davis: University of California, Centers for Water and Wildland Resources, 1996.

- Based on records of twentieth-century fire occurrence, the probability of wildfire in low- to middle-elevation coniferous forests is somewhat less than in the lower elevation foothill woodland and chaparral vegetation types (McKelvey and Busse 1996). However, the negative effects of severe wildfire on the dominant vegetation—and by extension on numerous other resources—generally are more profound and more long lasting in the coniferous forests.

- The composition and structure of the dominant vegetation in low- to middle-elevation coniferous forests probably have been affected more adversely by removal of the natural fire regime (and thus potentially could benefit more from its partial restoration) than in higher or lower vegetation types.

We recognize the problems associated with the threat of wildfires to lives and property in the urban-wildland intermix areas in the Sierran foothills. Management of foothill vegetation is mentioned in our discussion of these intermix areas. Many of the same general principles and approaches for fuel-management strategies that we discuss for the coniferous forests apply also to the foothill vegetation types.

A CAUTIONARY TALE OF FOREST BIOMASS

A simplified, qualitative accounting of production and disposition of biomass may help to clarify the problem of fuel accumulation in many Sierra Nevada forests. As indicated earlier, low- and middle-elevation forest types—west-side pine, west-side pine–mixed conifer, and east-side pine—are emphasized. It is appropriate here to consider only aboveground biomass, both for simplicity and relevance to the topic at hand. While we recognize the importance to today's forests of events in the latter half of the nineteenth century (McKelvey and Johnston 1992), we focus here on contrasts between the periods before 1850 and after 1900.

Biomass Production

Sierra Nevada forests produce a great deal of biomass. While considerable variation exists in terms of the site and climatic variables that largely determine net primary productivity, in general terms Sierra Nevada forests are quite productive (Helms and Tappeiner 1996). For the forest types indicated earlier, the west-side types are substantially more productive than east-side pine. The average rate of biomass production during most of the twentieth century probably has exceeded that which occurred from, say, 1650 to 1850 because this century generally has been warmer and wetter than the earlier

period (Graumlich 1993). More complete site occupancy, in the form of denser forests in many areas (Gruell 1994), also may have contributed to greater production now than then. Allocation of total biomass production apparently has differed considerably between the two periods. A much greater percentage of biomass historically was stored in the boles of large trees and in herbaceous vegetation in relatively open stands, whereas now much more goes into small trees in dense stands.

Biomass Disposition

The main factors accounting for disposition or removal of forest biomass are decomposition (oxidation), fire (oxidation), and herbivores and humans (utilization).

Decomposition

In California's Mediterranean climate, decomposition rates generally are low, limited by low temperatures in the winter and inadequate moisture in the summer. In some portions of the Sierra Nevada mixed conifer forest type, however, sufficient moisture may be retained well into the summer to support fairly high rates of decay (Harmon et al. 1987). Decomposition rates in Sierra Nevada forests probably have been greater during this century than during the period 1650–1850 because (1) this century has been warmer and wetter (Graumlich 1993), (2) the generally denser stands during this century have provided more mesic microclimates that favor decomposition, and (3) more forest floor biomass has been available for decomposition because it has not been removed regularly by fire during the twentieth century. Neither historically nor now, however, has decomposition been the primary remover of biomass in Sierra Nevada forests.

Presettlement Fire

In presettlement forests most biomass ultimately was oxidized by frequent low- to moderate-severity fires. High-severity fires more than a few acres in size were unusual (Kilgore 1973; Skinner and Chang 1996; Weatherspoon et al. 1992). Across much of the landscape, dead biomass on the forest floor was kept at low levels, and most small understory trees were killed and subsequently consumed by fire. While small areas of high-severity fire killed patches of large trees (Stephenson et al. 1991), most large trees survived the fires and were consumed at some point after their death by subsequent fires. The longevity of large snags and downed logs under presettlement fire regimes is a subject of debate. It seems likely, however, that relatively few downed logs reached advanced stages of decay on xeric sites before being consumed by fire, whereas a greater proportion could last for longer periods (and also decay faster) on more mesic sites. Physical removal from the site was a minor component of total biomass disposition, although harvest of biomass by Native Americans, especially for firewood, may have been a significant factor locally (Anderson and Moratto 1996).

Twentieth-Century Fire

If we skip now to the twentieth century, the relative roles of fire and biomass removal have changed drastically. As fire suppression was initiated and took effect early in the century, the proportion of biomass consumed by fire dropped precipitously, as did annual burned area. During the course of the twentieth century, however, annual burned area for the Sierra Nevada has shown no overall time trend, even though it has fluctuated considerably from year to year. Large fires have composed an increasing proportion of that burned area as the century has progressed (McKelvey and Busse 1996). In recent years, large fires have become less controllable and more severe, evidently reflecting in part increased fuel loadings.

Another possible indicator of changing fuel conditions is a shift in the distribution of fires between human and lightning ignitions over the course of the twentieth century. We observed this shift as part of an evaluation of twentieth-century fire records for Sierran national forests. We used records for fires greater than 40 ha (100 acres) within the twenty-four core SNEP river basins. Because of the extraordinary extent of the 1987 and 1990 lightning fires, we present the summaries for two intervals of time so as to exclude and include these two years: 1910 through 1986 (table 56.1) and 1910 through 1993 (table 56.2). We arbitrarily split each interval into two time periods for these summaries.

These summaries suggest some conspicuous differences between human-caused fires and lightning fires. Whether the extraordinary years of 1987 and 1990 occurred simply by chance we cannot say based on these limited data. However, whereas the fire-suppression organization does appear to have reduced total area burned by, and number of, large human-caused fires, it has not been effective in reducing either the area burned by or the number of large lightning fires.

In table 56.3 we summarize fire characteristics for each of the three years of greatest burned area for each time period. All six of these years were quite dry. The summaries show that total area burned was similar in these years. However, lightning fires contributed only small proportions of total area burned for the first four years but very large proportions for the last two years—1987 and 1990. It is interesting to note that the total number of fires also differs considerably between the earlier years and 1987 and 1990. Those two years had fewer and much larger fires contributing most of the area burned.

The pattern of fire starts and the necessary response of the fire-suppression organization differ considerably between the two types of ignition. Human-caused fires generally occur as a singular event or occasionally a few simultaneous events. This allows the fire-suppression organization to respond to individual fires with a relatively large body of fire-suppression resources. Lightning fires, in contrast, usually occur as simultaneous multiple ignitions. In unusually dry years, resource requirements necessary to deal with simultaneous

TABLE 56.1

Summary of fire characteristics for 1910–47 compared with 1948–86.

Years	Total Annual Burned Area (ha)		Maximum Annual Fire Size (ha)		Total Annual Number of Fires	
	1910–47	1948–86	1910–47	1948–86	1910–47	1948–86
All Fires						
Greater than 40 ha						
Minimum	882	125	283	66	5	2
1st quartile	3,990	1,257	1,260	559	12	6
Median	14,483	4,295	3,324	2,026	19	9
3rd quartile	21,285	11,443	8,421	6,599	35	14
Maximum	95,126	43,330	21,234	18,100	82	23
Total for entire period	685,880	319,806			983	395
Human-Caused Fires						
Greater than 40 ha						
Minimum	882	125	283	66	5	2
1st quartile	3,732	1,182	1,022	553	11	5
Median	14,202	3,781	3,324	1,333	17	8
3rd quartile	20,708	8,690	8,421	6,599	33	11
Maximum	93,588	39,402	21,234	18,100	65	20
Total for entire period	651,801	273,526			890	318
Lightning-Caused Fires						
Greater than 40 ha						
Minimum	0	0	0	0	0	0
1st quartile	12	21	12	21	0	1
Median	324	217	175	197	2	1
3rd quartile	901	1,233	550	708	3	4
Maximum	9,738	7,356	5,748	7,238	18	6
Total for entire period	34,079	46,280			93	77

TABLE 56.2

Summary of fire characteristics for 1910–51 compared with 1952–93.

Years	Total Annual Burned Area (ha)		Maximum Annual Fire Size (ha)		Total Annual Number of Fires	
	1910–51	1952–93	1910–51	1952–93	1910–51	1952–93
All Fires						
Greater than 40 ha						
Minimum	828	44	283	44	5	1
1st quartile	3,990	1,178	1,260	526	12	6
Median	13,856	4,537	3,654	2,107	18	9
3rd quartile	20,110	12,125	8,880	6,144	34	13
Maximum	95,126	81,887	21,234	53,011	82	23
Total for entire period	730,131	454,861			1039	403
Human-Caused Fires						
Greater than 40 ha						
Minimum	828	0	283	0	5	0
1st quartile	3,732	1,120	1,022	481	11	4
Median	13,585	3,108	3,654	1,099	17	7
3rd quartile	19,585	6,993	8,880	4,434	32	9
Maximum	93,588	39,402	21,234	18,100	65	20
Total for entire period	692,170	267,879			934	306
Lightning-Caused Fires						
Greater than 40 ha						
Minimum	0	0	0	0	0	0
1st quartile	12	44	12	44	0	1
Median	324	347	175	272	2	1
3rd quartile	902	2,625	550	1,960	3	4
Maximum	9,738	80,704	5,748	53,011	18	13
Total for entire period	37,960	186,982			105	97

multiple ignitions can quickly exceed those available (e.g., 1977, 1987, 1990). Show and Kotok (1923) recognized early, on the basis of the 1917 fire season, that general regional lightning events have the potential to strain the fire-suppression organization severely.

The period of record is insufficient to conclude that there is a definite trend toward larger severe lightning fires or that a threshold has been crossed. However, we suggest that the potential influences of changing fuel mosaics, stand conditions, and landscape patterns on the fire environment logically would begin to show up first in dry years under lightning situations.

Utilization

In contrast to the changed role of fire in removing biomass, utilization of biomass has increased by orders of magnitude over the levels that prevailed before Euro-American settlement. The components of biomass removed by logging have changed dramatically from those that previously were removed by fire. Fire-resistant large trees have been harvested and replaced by much more fire-susceptible small trees. Dead biomass in the form of logging slash and natural (i.e., not produced by management activities) fuels has built up on the forest floor because of lack of fire and inadequate or nonex-

TABLE 56.3

Fire characteristics in the three major fire years (years of greatest burned area) during 1910–51 compared with those during 1952–93.

Year	1910–51			1952–93		
	1924	1926	1931	1959	1987	1990
Fire size (ha)						
1st quartile	95	101	119	155	182	120
Median	305	222	249	673	277	606
3rd quartile	1,307	572	1,095	3,268	785	3,405
Maximum	15,054	10,252	17,715	7,710	53,011	3,405
Total burned area	95,126	57,527	52,540	43,330	81,887	38,624
Lightning percentage[a]	2	17	2	9	99	57,099
Total number of fires	56	80	40	23	18	95
						11

[a]Percentage of total area burned.

istent fuel treatment. Total decomposition probably has accelerated, but at a rate not nearly sufficient to compensate for the increasing fuel load. Together, surface fuels and dense understories have greatly increased the risk of crown fires (Kilgore and Sando 1975; Parsons and DeBenedetti 1979). Heightened stress from overly dense stands, often dominated by shade-tolerant species no longer kept in check by frequent fires, also has increased mortality from insects (Ferrell 1996), further adding to dead biomass available as fuel.

Fuel Management

As managers began to see the consequences of increased fuel loads, they undertook a variety of fuel-management activities. These activities have included a range of treatments that mimic or facilitate the natural processes of biomass disposition: (1) burning on site (with or without prior piling or rearrangement), (2) accelerating decomposition (and reducing flammability) by rearranging the fuel bed closer to the ground, and (3) physical removal from the site. Adequacy of slash treatment following timber harvest or other vegetation management activity has varied from quite good to nonexistent.

For the Sierra Nevada as a whole, however, vegetation management activities have produced considerably more new fuels than they have eliminated. Furthermore, the increasing problem of live understory fuels has been addressed inadequately in silvicultural or fuel-management activities. Efforts to treat accumulating amounts of natural fuels, often with prescribed fire, also have fallen far behind rates of fuel accretion, due in large part to inadequate funding and various concerns about the use of prescribed fire. Even the active prescribed burning programs in Sierran national parks over the past twenty-five years, utilizing both natural and management ignitions, have restored fire to the forests at rates well below presettlement levels (Botti and Nichols 1995; Husari and McKelvey 1996; Parsons 1995). Consequently, these burns have been unable even to keep up with new biomass accumulation, let alone to consume all the excess biomass generated by decades of fire suppression. The basic problem is the same outside the parks: current quantities of flammable biomass—primarily small trees and surface fuels—in low- to middle-elevation Sierran forests are unprecedented during the past several thousand years and are continuing to accumulate at a much faster rate than they are being removed.

The Fuel Problem and the Need for a Strategy

Given current federal and state budget climates, increasing suppression costs, and attrition of skilled firefighters, reductions in suppression forces seem more likely than substantial increases (Husari and McKelvey 1996; U.S. Department of the Interior and U.S. Department of Agriculture 1995). According to a growing consensus among fire managers, more suppression capability is not the solution anyway. This idea is reinforced, we believe, by the data presented earlier on distributions of lightning and human ignitions. History tells us that periodic dry years are inevitable and that regional-scale lightning events that limit the effectiveness of suppression forces are not unusual.

If more suppression is not the answer, and if flammable biomass continues to accumulate at current rates, and if we do nothing substantive to arrest that accumulation, simple physics and common sense dictate that the area burned by high-severity fires will increase. Losses of life, property, and resources will escalate accordingly. This conclusion is strengthened by the fact that recent "drought" years, during which many large, severe fires burned (McKelvey and Busse 1996), appear to be relatively common when viewed on a time scale of centuries (Graumlich 1993).

Therein lies the rationale for large-scale fuel management. Given the massive scope of the problem and budget constraints, brute force is likely to be neither feasible nor adequate. A carefully considered strategy is required. Treatments need to begin in the most logical, efficient, cost-effective places. Specific components of biomass—mostly small trees and surface fuels—need to be targeted. We must devise ways to pay for the needed treatments. At least on public lands, treatments conducted to reduce the hazard of severe wildfires should be compatible with overall desired conditions for sustainable ecosystems. In general, conditions need to be moved away from dense, small-tree-dominated forests toward more open, large-tree-dominated forests. And the rate of treatment needs to be carefully planned: in the short term, rates of biomass removal may well need to exceed rates of production in order to return these forests to a more sustainable, fire-resilient condition. The remainder of this chapter displays and discusses various considerations for developing such a landscape-level fuel-management strategy.

A REVIEW OF FUEL-MANAGEMENT STRATEGIES

Our use of the term *fuel-management strategies* here refers to methods for prioritizing or locating fuel treatments on a landscape scale in such a way as to increase their overall effectiveness for reducing the extent of severe wildfires. Most past fuel management in the Sierra Nevada has taken place in the national forests. Most of that has not been characterized by strategic planning: management emphasis and funding have directed fuel management primarily toward treatment of activity fuels following timber sales, and sales usually were not located with strategic fuel considerations in mind. In fact, timber sales often were dispersed—thereby reducing overall effectiveness of fuel treatments—intentionally in an attempt to meet various management objectives, such as minimizing cumulative watershed impacts of harvest-related activities. In recent years, however, innovative fire and fuel managers have begun to think much more strategically and to collaborate with foresters and silviculturists to address landscape-level forest health concerns. This change has been stimulated and supported by the general move toward ecosystem management and by new capabilities for spatial, landscape-level planning provided by geographical information system (GIS) technology.

Some of these evolving ideas are included in the following sections, which provide a sampling of various types of fuel-management strategies that have been proposed and, to varying degrees, implemented. Also incorporated here are some of the ideas discussed by a group of experts in a Fuels Management Strategies Workshop sponsored by SNEP in March 1995 (Fleming 1996). Three somewhat distinct but certainly overlapping approaches have been used: (1) identifying fuel-management approaches appropriate within each of several landscape zones defined by general characteristics, uses, or emphases; (2) setting priorities based on various combinations of risk, hazard, values at risk, and suppression capabilities; and (3) employing a fuelbreak-type concept intended to interrupt fuel continuity on a landscape scale and to aid in limiting the size of fires by providing defensible zones for suppression forces. A fourth "approach" that has received explicit emphasis recently, although it is implicit to some degree in the other approaches, is rate or timing of implementation.

Strategies Based on Zones

Arno and Brown (1989) proposed three landscape zones. In Zone I, wilderness and natural areas, the emphasis would be on prescribed natural fire (PNF), augmented by management-ignited prescribed fires (MIPF) as necessary to restore much of the natural role of fire to these ecosystems. In Zone II, the general forest management zone, well-planned and well-implemented fuel management, both in conjunction with and in addition to proper timber harvests, would contribute significantly to good overall management. In Zone III, the residential forest, education of homeowners and local officials about the realities of fire hazards in the wildland-urban interface would go hand in hand with effective, esthetically pleasing manipulation of fuels. The authors suggested that shaded fuelbreaks around homes and developments could be an effective measure. They recommended concentrating most efforts in Zone III and adjacent portions of Zone II.

A somewhat different zone approach provides the basis for fire-management direction in Sequoia–Kings Canyon National Parks (Manley 1995). Zones are defined by estimated proximity of current conditions to the natural range of variability. In Zone 1, areas essentially unaffected by postsettlement activities (mostly higher elevations), natural processes, including PNF, are permitted to operate with little restriction. In Zone 2, areas significantly modified by postsettlement activities, corrective actions, including conservative use of PNF and MIPF, are required before permitting resumption of all natural processes. In Zone 3, built-up areas with highly flammable

fuel types near park boundaries, full suppression is combined with mechanical fuel treatments and conservative use of MIPF.

Greenwood (1995) described a land classification system based on structure density (presumably closely related to population density) plus appropriate fire-related buffers. While his analysis was done for the entire state of California, the subset of Sierra Nevada data could easily be analyzed separately, and most of his general conclusions probably would still apply. He labeled the classes wildland, intermix, and developed, corresponding to increasing structure densities, and noted the surprisingly high percentage of land in the intermix category, even on public lands. He emphasized that the presence of people and their structures constrains many of the options available for both fuel management and fire suppression. Approaches suggested ranged from reestablishment of presettlement conditions and processes in some wildland areas to reliance on fire-safe regulations, public education, aggressive initial attack, and only minimal vegetation manipulation in more densely settled developed areas.

Strategies Based on Risk, Hazard, Values at Risk, and Suppression Capabilities

To provide a common frame of understanding for the discussion that follows, definitions of "risk," "hazard," and "values at risk" (McPherson et al. 1990) are given here.

FIRE RISK: (1) The chance of fire starting, as affected by the nature and incidence of causative agents . . . (2) Any causative agent. (P. 45)

FIRE HAZARD: A fuel complex, defined by volume, type, condition, arrangement, and location, that determines the degree of ease of ignition and of resistance to control. (P. 42) "Resistance to control" is related both to fire behavior and resistance to line construction.

VALUES-AT-RISK: Any or all natural resources, improvements, or other values which may be jeopardized if a fire occurs. (P. 131)

A number of authors have reported the use of decision analysis to aid in fuel-management decision making (Anderson et al. 1991; Cohan et al. 1983; Radloff and Yancik 1983). Decision analysis became the cornerstone of the National Activity Fuel Appraisal Process (Hirsch et al. 1981; Radloff et al. 1982), which was intended to provide a consistent means of evaluating the important factors affecting fuel-treatment decisions. The Fuel Appraisal Process provided probabilities of various-sized fires by intensity class, based on information about topography, historical weather, historical fire occurrence (risk), suppression capability, and hazard (measured or projected based on alternative fuel treatments).

Biehl (1995) described an "all risk management" strategy in use on the Stanislaus National Forest. Fuel profiles, expected ignitions, and suppression resources are used in conjunction with management-defined acceptable resource loss to determine whether, where, and what kind of fuel treatment is needed. The Stanislaus National Forest is combining the most active prescribed burning program of all California national forests—concentrated mainly in natural (i.e., nonactivity) fuels—with considerable biomass thinning. Fuelbreaks are employed, but only as anchor lines to facilitate initiation of areawide fuel treatments using prescribed fire.

Perkins (1995) has devised a similar fire-analysis system for use on the Klamath National Forest as part of the forest's landscape-analysis system. Risk, fire behavior potential (based on fuel classification, slope class, and ninetieth-percentile summer wildfire weather conditions), and resource values (based on forest plan direction) are the primary factors used to determine fuel-management treatment priorities. Fuels information is derived from vegetation classification, modified by management history and large-fire history.

James (1994) developed a simple system for estimating a "catastrophic fire vulnerability rating," based on a point total derived from separate qualitative assessments of risk, hazard, value, and suppression capability. The system includes three sets of "fire/fuel treatment standards" corresponding to fire vulnerability ratings of high, moderate, or low. Finally, it provides a straightforward feedback mechanism for adjusting the posttreatment vulnerability rating. All vulnerability factors are weighted equally, but local managers should be able to modify weightings fairly easily to account for their assessment of the relative importance of various factors.

Strategies Based on Fuelbreaks or Similar Landscape-Level Interruptions of Fuel Continuity

FUELBREAKS: Generally wide (60–1,000 feet) strips of land on which native vegetation has been permanently modified so that fires burning into them can be more readily controlled. (McPherson et al. 1990, 56)

Early Experiences with Fuelbreaks

Green (1977) traced the long history of fuelbreaks and their predecessors, firebreaks (narrower strips usually cleared to mineral soil), in California. Perhaps surprisingly, a recommendation to the State Board of Forestry for blocking out the forest with strips of "waste" land wide enough to prevent fire from crossing was made as early as 1886. The Sierra Nevada was a part of early firebreak history. S. B. Show, District Forester, proposed in 1929 that a firebreak be constructed along the entire length of the western slope of the Sierra Nevada at the interface of the chaparral and the pine forest. Depression-related federal funding, especially for the Civilian Conservation Corps, permitted work to begin in 1933 on what came to be known as the "Ponderosa Way and Trucktrail." The intent of this strip, which when completed was about 1,050 km (650

mi) long and generally 45–60 m (150–200 ft) wide (Green 1977), was to help prevent fires from burning from the chaparral up into the more valuable Sierran timber (Green and Schimke 1971).

The transition from firebreaks to fuelbreaks came about as part of preattack planning in the early 1950s (Green 1977). Most early fuelbreak construction was in southern California chaparral. The Duckwall Conflagration Control Project on the Stanislaus National Forest, initiated in 1962, extended the fuelbreak concept into the Sierra Nevada mixed conifer forest type (Green and Schimke 1971). Green and Schimke (1971), Pierovich and colleagues (1975), and Green (1977) provided a number of guidelines for planning, constructing, and maintaining fuelbreak systems. Among their recommendations: The number and location of fuelbreaks, along with the size of blocks to be separated by the fuelbreak network (1,000 ha [2,500 ac] for the Duckwall program), should be determined by fire-control objectives as part of the preattack planning process. Needs for protecting populated areas or high resource values should be given high priority in fuelbreak location. Planned management projects—in range, wildlife, recreation, timber, watershed, and forest roads and trails—should be reviewed to see how they might contribute to the fuelbreak network. Ridges usually are preferred for locating fuelbreaks, although other locations can be used. Locating fuelbreaks along existing roads where possible was recommended to facilitate access by suppression forces. Suggested fuelbreak widths varied from about 60 to 120 m (200 to 400 ft). The necessity of maintaining reduced-fuel conditions on fuelbreaks, through a combination of appropriate vegetation (e.g., low volume and/or low flammability) and periodic treatments, was emphasized.

A number of anecdotal accounts of the effectiveness of fuelbreaks (or lack thereof) during wildfire incidents, mostly during the 1960s and early 1970s, were summarized by Pierovich and colleagues (1975) and Green (1977). Although experiences were mixed, fuelbreaks were found to be effective much of the time in stopping wildfires except under the most extreme conditions. Success was most likely when fuelbreaks were properly installed, properly maintained, and adequately staffed by suppression forces during wildfires.

The same authors (Pierovich et al. 1975; Green 1977) discussed existing economic analyses of fuelbreak effectiveness, which differed in their conclusions but for the most part found that a fuelbreak system could be justified economically as part of a well-integrated fire-management system. A subsequent study of fuelbreak investments in southern California, using a linear programming model, predicted that increasing fuelbreak widths could substantially reduce area burned and fire-related damages if initial investments were concentrated in a specific "damage-potential zone" (Omi 1979). Although potential corollary—i.e., nonfire—benefits of fuelbreaks have been recognized (Green 1977), such benefits generally have not been considered in evaluations of their efficacy or cost effectiveness. In a study of three forested fuelbreaks in the central Sierra Nevada, however, Grah and Long (1971) found that fuelbreak construction increased timber values within the fuelbreaks by reallocating site resources to larger, faster growing, and more valuable trees. A portion of fuelbreak costs, therefore, was offset by the benefit to the timber resource.

Recent Experiences and Recommendations for Using Fuelbreaks

Fuelbreak construction and maintenance have retained some emphasis in southern California. Salazar and Gonzalez-Caban (1987) found that in a large 1985 wildfire in chaparral on steep terrain, the fuelbreak system apparently influenced the location of the final fire perimeter. Except during the most extreme burning conditions, fuelbreaks functioned as intended.

In contrast, most forested areas in the state have seen little attention given to fuelbreaks over the past twenty years. Fuel management in Sierra Nevada national forests has been dominated by support of the timber management program during most of that period. Budgets for other fuel activities have been quite limited. Furthermore, many fire and fuel specialists have viewed fuelbreaks as being of little value for a variety of reasons, including the following: (1) to be effective for stopping fires, fuelbreaks need to be staffed by suppression forces, which often have been unavailable when needed, frequently because of demands for protecting structures in urban-wildland intermix areas; (2) in general, recommended fuelbreak widths of 60–120 m (200–400 ft) (Green and Schimke 1971; Green 1977) have been considered too narrow to be effective under many conditions, especially with extensive spotting (ignition of new fires outside the perimeter of the main fire by windborne sparks or embers); (3) fuelbreaks often have been viewed as standalone measures that competed with more effective areawide fuel treatments; and (4) fire control has been viewed as the sole beneficiary of fuelbreaks, with little thought given to other potential resource benefits.

Over the past ten years or so, a number of large, severe fires in California and elsewhere in the western United States have emphasized the seriousness and the enormity of the wildland fuel problem. Fuelbreaks have begun to receive renewed attention as one part of the solution. Arno and Brown (1989) suggested their use around homes and developments in the wildland-urban interface. In the recovery plan for the northern spotted owl, Agee and Edmonds (1992) recommended the use of fuelbreaks along with underburning to reduce the probability of catastrophic wildfires in "designated conservation areas" within the Klamath and East Cascades subregions. Weatherspoon and colleagues (1992) suggested a two-stage fuelbreak strategy to help reduce the occurrence of severe fires in California spotted owl habitat in Sierra Nevada mixed conifer forests. Known owl sites first would be "isolated" using a broad band of prescribed burns, followed by a more general program of breaking up fuel continuity on a landscape scale. Fites (1995) proposed a similar approach to help protect "areas of late-successional forest emphasis" and to restore more sustainable, fire-resilient conditions across

the landscape. Arno and Ottmar (1994, 19) pointed out the need for "an interconnected network of natural fire barriers and treated stands as zones of opportunity for controlling wildfires."

In the draft Environmental Impact Statement (EIS) for managing California spotted owl habitat in Sierra Nevada national forests (U.S. Forest Service 1995), Alternatives C and D included an upper slope/ridge zone that would be dominated by large, widely spaced shade-intolerant trees. These alternatives were viewed as creating conditions in this zone closer to those thought to have existed before Euro-American settlement. In addition, the zone would provide many of the fire-management benefits of a wide shaded fuelbreak. Alternative F incorporated some of the fuelbreak-related concepts of the Quincy Library Group (QLG) proposal (summarized later) for the northern Sierra Nevada.

LaBoa and Hermit (1995) presented a number of ideas for strategic fuel planning and treatment, based on their recent work as members of the California spotted owl EIS Team (sufficiently recent that these ideas were not included in the draft EIS). They included the use of fuelbreaks; however, they stressed the need not to stop with a fuelbreak network but to build from it to accomplish large-scale fuel modification on a landscape level.

The most detailed fuel-management strategies to date have been proposed for the northern end of the Sierra Nevada—the Lassen and Plumas National Forests and the Sierraville Ranger District of the Tahoe National Forest. The two strategies, which were developed semi-independently by the QLG and the U.S. Forest Service, have much in common and build on many of the ideas cited earlier. Rapid implementation of a network of broad fuelbreaks is key to both proposals.

QLG is a community-based group whose members represent a wide range of interests, including fisheries and environmental groups, timber industry, and county government. The group has made strategic fuel management a central focus of its land management proposal (Quincy Library Group 1994). QLG proposes that an intensive four-year effort be focused on installing a network of strips approximately 0.4 km (0.25 mi) in width, mostly along existing roads, that break up fuel continuity across the landscape and provide defensible zones for suppression forces. During this period, essentially all forest management activities, including biomass and other thinnings, salvage activities, and treatment of surface fuels, would be focused on implementing this fuelbreak network. Each year 1/32 of the total forest acreage would be treated, so that at the end of the four-year period 1/8 of the forest would be a part of these strips. The strips would have reductions in stand density, lower canopy ladder fuels, and surface fuels, and they would have relatively low levels of snags and large downed woody debris. After the initial period, a longer term fuel-management strategy would add some strips to isolate areas of high value and/or high risk, but the emphasis generally would shift to areawide treatments.

The Technical Fuels Report, prepared by fire/fuel specialists from the Lassen, Plumas, and Tahoe National Forests (Olson et al. 1995), is similar in several respects to the QLG proposal. The "defensible fuel profile zone" (DFPZ), a concept first described by Olson (1993), is central to the strategy outlined in the report. Much like a broad fuelbreak, a DFPZ is a low-density, low-fuel zone averaging 0.4 km (0.25 mi) in width, located mostly along roads, and designed to support suppression activities. Like the strips in the QLG proposal, DFPZs are intended to be installed over a period of just a few years. The authors point out that DFPZs are intended not to take the place of widespread fuel treatment but rather to increase the effectiveness of initial fuel treatment and to facilitate subsequent treatment of adjacent areas. Olson et al. (1995) describe the "community defense zone" (CDZ) as another component of their strategy concerned with urban interface areas within or near national forest boundaries. Similar in concept to a DFPZ, a CDZ is designed to reduce the threat of wildfire spreading onto national forest land from private land, or vice versa. Like DFPZs, CDZs would have a high priority for completion within a short period of time. The authors stress the importance of the involvement and cooperation of local communities in implementation of CDZs. A third type of zone, the "fuel reduction zone" (FRZ), refers to general area fuel treatment that would take place mainly after the high-priority system of DFPZs and CDZs is in place. The Technical Fuels Report (Olson et al. 1995) emphasizes the importance of site-specific considerations and local decision making in setting priorities and implementing the details of the broad fuel-management strategy outlined.

A POTENTIAL FUEL-MANAGEMENT STRATEGY FOR SIERRA NEVADA FORESTS

The approaches summarized in the previous section, along with the discussion at the SNEP Fuels Strategies Workshop (Fleming 1996), seem to point to some degree of convergence of thinking about the fuel problem and some components of a strategy to deal with it. In this section we attempt to synthesize many of the previously mentioned approaches into an outline for a potential fuel-management strategy for Sierra Nevada forests.

The ideas presented here are necessarily general in nature. The Sierra Nevada is enormously complex and diverse. Landowners and ownership objectives vary widely. While agencies and large landowners may choose to set some priorities on a regional or subregional scale, any attempt on our part to recommend or prescribe specific management practices rangewide would be naive, counterproductive, and contrary to the SNEP charter. Readers should view this "strategy" as a set of principles and ideas to consider as they develop their own landscape-specific strategic plans. (Additional ideas can

be found in cited references.) Such plans will be greatly facilitated and improved by developing and maintaining good GIS databases. Later in this chapter we discuss the nature and role of such databases for supporting fire and fuel-management decision making in the context of adaptive ecosystem management (Everett et al. 1994; Walters and Holling 1990).

Although landscape-specific planning is focused on a small portion of the entire Sierra Nevada, it nevertheless requires thinking on a much broader scale than often has occurred in the past. Making significant progress toward these goals will require long-term vision, commitment, and cooperation across a broad spectrum of land-management agencies and other entities. Dealing with fuels on only a local, piecemeal basis will be inadequate.

Goals of the Fuel-Management Strategy

The strategy has three general goals, ranging from short to long term and from relatively narrow to broad. Each goal can be viewed as nesting within the following one. The goals are consistent and complementary, as are the means to work toward their accomplishment. For example, the strategy provides that short-term approaches to reducing hazard be compatible with longer-term goals of ecosystem sustainability (Arno and Ottmar 1994).

The first goal—the immediate need from a fire-management standpoint—is to reduce substantially the area and average size burned by large, severe wildfires in the Sierra Nevada. Ideally this will be a short- to medium-term goal, whose urgency will lessen as the fuel-management strategy becomes increasingly effective. A second, longer-term goal should be to restore more of the ecosystem functions of frequent low- to moderate-severity fire. The two goals are closely linked. They could be met simultaneously by replacing most of the high-severity acreage with the same, or preferably much greater, acreage of low- to moderate-severity fire. A third, overarching goal is to improve the health, integrity, and sustainability of Sierra Nevada ecosystems. This goal certainly goes beyond fire considerations. Progress toward achieving the first two goals, however, is critical to the third.

Management actions to progress toward these three goals should be occurring concurrently. Often it will be possible for a single treatment or project to address all three goals simultaneously. In fact, opportunities for such congruence should be sought. In this chapter, however, we spend the most time addressing the first goal—not because it is most important in the long run but because it is the most urgent in the short run to reduce losses of lives, property, and resources, and to make it possible to work more effectively toward achieving the second and third goals. Stated in another way, the fuel-management strategy has joint themes of protection and restoration of ecosystems, and, in many portions of the Sierra Nevada, protection is a prerequisite to restoration. In a longer term context, strategies geared specifically toward reducing losses from large, severe wildfires should gradually become less important; restoration in turn should provide a more fundamental level of protection along with improved ecosystem health.

Goal 1: Reduce Substantially the Area and Average Size Burned by Large, High-Severity Wildfires

Large, high-severity fires were unusual historically in most Sierra Nevada forests. Fire regimes in the Sierra Nevada generally were characterized by relatively frequent, low- to moderate-severity fires (Skinner and Chang 1996). Changes in low- and middle-elevation forests and their associated fuel complexes, brought about largely by human activities since Euro-American settlement (including but not limited to fire suppression), have made these forests much more prone to large, severe fires (Chang 1996; Husari and McKelvey 1996; McKelvey and Johnston 1992; Skinner and Chang 1996; U.S. Forest Service 1995). Such fires, in aggregate, are well outside the natural range of variability and thus can be considered detrimental to Sierra Nevada ecosystems (Manley et al. 1995). Furthermore, the current prevalence of such fires is unacceptable socially. The rapidly increasing population of the Sierra Nevada increasingly places people's houses at risk of loss to severe wildfires and makes potential solutions to the problem much more difficult.

In pursuing goal 1, it is essential for the wildland fire agencies to continue support for suppression and prevention activities. These fire-management efforts alone, however, cannot resolve the problems of fire in the Sierra Nevada. Aggressive, strategically logical fuel-management programs, compatible with overall desired conditions for sustainable ecosystems, are necessary to address the basic problem of excessive fuel accumulation.

Goal 2: Restore More of the Ecosystem Functions of Frequent Low- to Moderate-Severity Fire

The frequent low- to moderate-severity fires that occurred throughout much of the Sierra Nevada until about 150 years ago performed many important ecological functions (Kilgore 1973; Chang 1996). Wildfires of this type, however, have been virtually eliminated from Sierra Nevada ecosystems (as measured by annual area burned by such fires), because these are the fires that are suppressed most easily. As a result, the ecological functions historically performed by such fires have been largely lost, with some known and many unknown consequences. It is highly unlikely that fires will ever burn as much area as often and with the same distribution of severities as they once did. Nevertheless, it makes sense to try to restore fire to a more nearly natural role in those parts of the landscape where it is practical to do so. Where fire alone cannot be used practically, fire surrogates such as silvicultural techniques and mechanical fuel reduction methods (Helms and Tappeiner 1996; Weatherspoon 1996) can be employed—either by themselves or in conjunction with prescribed fire—as appropriate to mimic some of the functions of fire and to move landscapes toward desired conditions (Manley et al.

1995). Over time, adaptive management (Everett et al. 1994; Walters and Holling 1990) should help us to determine which ecosystem functions of fire can be emulated satisfactorily by surrogates, which may be irreplaceable, and the implications for management.

Goal 3: Improve the Health, Integrity, and Sustainability of Sierra Nevada Ecosystems

The third goal is consistent with the first two and is central to overall SNEP goals. It should be achievable (1) by reducing the incidence of high-severity fires, which are detrimental to ecosystem sustainability in natural fire regimes characteristic of most of the Sierra Nevada; and (2) by moving ecosystems closer to pre-European-settlement conditions and processes, assumed by many to be a useful first approximation of sustainable ecosystems (e.g., Manley et al. 1995; Swanson et al. 1994), at least on public lands. We cannot define those presettlement conditions with any great precision, but we do know enough to be reasonably confident that this strategy would move us in the desired direction.

Components of the Strategy

The strategy we discuss here has three basic components: (1) networks of defensible fuel profile zones (DFPZs) (the term adopted from Olson 1993 and Olson et al. 1995) created and maintained in high-priority locations; (2) enhanced use of fire for restoring natural processes and meeting other ecosystem management goals; and (3) expansion of fuel treatments to other appropriate areas of the landscape, consistent with desired ecosystem conditions. We also discuss possible institutional changes that might increase the effectiveness of the strategy. This strategy builds upon and draws freely from the various strategies cited elsewhere in this chapter.

Defensible Fuel Profile Zones

Given the massive scope of the problem that goal 1 is intended to address, a carefully considered strategy is required for prioritizing fuel treatments. Such a strategy should permit managers to multiply the benefits of treatments in order to make the most rapid and most efficient progress toward achieving goal 1. We focus our discussion in this section on DFPZ networks. Multiple benefits of DFPZs may include (1) reducing severity of wildfires within treated areas (as with any fuel-management treatment), (2) providing broad zones within which firefighters can conduct suppression operations more safely and more efficiently, (3) effectively breaking up the continuity of hazardous fuels across a landscape, (4) providing "anchor" lines to facilitate subsequent areawide fuel treatments, and (5) providing various nonfire benefits. We are aware of no other strategy with as great a potential in the short term to progress reasonably rapidly toward achieving goal 1.

Rationale

The basic purposes of fuelbreaks were summarized earlier. These stated purposes generally do not include some of the potential benefits we envision for DFPZs, however. We offer an expanded rationale here, including the reasons for our choosing not to use the term fuelbreak as part of the strategy we describe.

Fuel-management activities in forested ecosystems normally involve some combination of (1) removing or modifying surface dead fuels to reduce their flammability; (2) removing or modifying live fuels to reduce their horizontal and/or vertical continuity, thereby reducing the probability of crown fire; and (3) felling excess snags that could be safety hazards and sources or receptors of firebrands.

The kind of protection afforded by fuel-management treatments depends not only on the localized nature of the treatments but also on their scale and spatial relationships. If you do a good job of treating fuels on a 1-acre (0.4 ha) patch of forest but do nothing in the surrounding forest, the edge effects probably will overwhelm the treatment in the event of a severe fire, and the small patch will be lost as well as everything around it. (There is a lesson here for group selection cuttings [Helms and Tappeiner 1996; Weatherspoon 1996]: it makes little sense to do fuel treatments in only the small regeneration openings and ignore the rest of the forest [Weatherspoon and Skinner 1995].) If you treat fuels to the same standard in a square 40-acre (16 ha) stand, edge effects are relatively much less important. Fire intensity will be much lower than in the surrounding (untreated) forest, and under most conditions the majority of the stand probably will survive. However, that 40-acre stand probably will have only a limited effect on fire damage in the untreated forest downwind. If you now treat the fuels on n 40-acre stands scattered randomly across the landscape, essentially the same result is expected, times n—i.e., the treated stands probably will not suffer excessive damage from a fire, but their intensity-reducing effect will not extend much beyond the treated areas. This last scenario, incidentally, approximates most of our past fuel treatments, which were not planned with strategic fuel management in mind.

If you take that same total treated acreage (40n) and string it together into a broad zone (DFPZ) that makes sense strategically, you have still protected those treated acres, with even less edge effect. In addition, however, you now have a reasonable chance of putting suppression forces into that zone and stopping the fire, thereby protecting areas on the downwind side of the DFPZ.

The term fuelbreak or shaded fuelbreak has been used to describe some of the same ideas. We do not use either term in describing this strategy, however, because they tend to carry some undesirable connotations:

- A shaded fuelbreak is often envisioned as a strip of land too narrow (60–120 m [200–400 ft] [Green and Schimke 1971;

Green 1977]) to be effective for stopping a fire under many conditions. In contrast, 0.4 km (0.25 mi) has been suggested as a nominal width for DFPZs (Olson et al. 1995; Quincy Library Group 1994). Use of the term zone (the Z in DFPZ) suggests a broader treated area than fuelbreak.

- A shaded fuelbreak is usually considered to have a single purpose—a relatively safe, accessible location in which suppression forces can initiate suppression actions. A DFPZ also serves this suppression function, almost certainly more effectively (because of its greater width) than a normal shaded fuelbreak. In addition, however, the DFPZ represents a substantial portion of the landscape—perhaps 10 to 25 percent for a completed network—within which fire damage is likely to be much reduced in the event of a wildfire. Furthermore, a DFPZ network may represent a number of potential additional benefits, including improved forest health, greater landscape diversity, increased availability of open forest habitat, and probably greater proximity to the historic range of variability and desired conditions.

- A shaded fuelbreak is often envisioned as "an alternative"— i.e., a standalone option for dealing with fuels. The DFPZ incorporates the notion that landscape treatment of fuels must start somewhere, so it makes sense to begin in strategically logical locations. The DFPZ is a place to start—a place from which to build out in treating other appropriate parts of the landscape—not an end in itself.

General Location, Description, Creation, and Maintenance

For the most part, DFPZs should be placed primarily on ridges and upper south and west slopes. All else being equal, DFPZs should be located along existing roads to simplify construction and maintenance and to facilitate use by suppression forces. Where roads do not follow ridges, road locations in relatively gentle terrain—e.g., along broad valley bottoms— are usually suitable for DFPZs. Roads that follow side slopes and canyon bottoms in steep terrain should be avoided except where they might facilitate stream crossings by DFPZs.

A network of DFPZs that define discrete blocks of land would require some DFPZ segments to cross drainages. Decisions about how best to deal with stream crossings should be based upon site-specific analyses. In most cases, however, we anticipate that the function of a DFPZ network would not be seriously jeopardized by limiting any treatments within the riparian zone portion of a DFPZ to those treatments (if any) deemed acceptable elsewhere in the riparian zone. Prescribed burning might be particularly appropriate as a treatment. Because of their relatively moist environment, untreated or minimally treated riparian zones normally should not present an undue risk of serving as a "fuse" to spread fire across a DFPZ adequately staffed with suppression forces.

A reasonable nominal width for DFPZs is probably 0.4 km (0.25 mi) (Olson et al. 1995; Quincy Library Group 1994) until experience indicates otherwise. It seems logical, however, to vary the width based on strategic importance, topography, or other conditions. For example, a broad, major ridge with a main road might warrant a considerably wider DFPZ than a spur ridge with steep side slopes. Using the fire-growth model FARSITE to model various fuel-treatment alternatives, van Wagtendonk (1996) found that fires burning under ninety-fifth-percentile weather conditions spotted across 90-m (300 ft) fuelbreaks under most fuel treatment scenarios but did not spot across 390-m (slightly less than 0.25 mile) fuelbreaks under any of the scenarios.

The Quincy Library Group (1994) proposed that DFPZs be used to break up the land into blocks averaging 4,000–5,000 ha (10,000–12,000 ac). We have no reason to argue with that as a first approximation, but the appropriate area certainly will vary among landscapes as a function of topography and the various factors discussed later. In many cases it may be logical to implement an initial high-priority "low-density" DFPZ network—e.g., along major ridges and main roads and in the vicinity of forest communities. Subsequent efforts would be a combination of maintaining existing DFPZs, constructing new ones to break up the landscape into smaller blocks, and broadening existing DFPZs in conjunction with areawide fuel treatments.

Treatment of DFPZs should result in a fairly open stand, dominated mostly by larger trees of fire-tolerant species. DFPZs need not be uniform, monotonous areas, however, but may encompass considerable diversity in ages, sizes, and distributions of trees. The key feature should be the general openness and discontinuity of crown fuels, both horizontally and vertically, producing a very low probability of sustained crown fire. Similarly, edges of DFPZs need not be abrupt but can be "feathered" into the adjacent forest. Posttreatment canopy closure usually should be no more than 40%, although adjustments in stand density based on local conditions certainly are appropriate. In some areas, for example, greater canopy closure may be desirable to slow encroachment by highly flammable shrubs or other understory vegetation, so long as tree crowns are high enough that a sustained crown fire in the denser canopy is very unlikely.

Available treatment techniques for DFPZs include silvicultural cutting methods, prescribed fire, mechanical fuel-reduction techniques, and combinations of these. In most cases, cuttings of various kinds will be the most effective initial treatments to accomplish needed adjustments in stand structure and composition (Helms and Tappeiner 1996; Weatherspoon 1996). Thinning from below often will be a desirable technique to move DFPZs from overly dense, small-tree-dominated stands toward more open, large-tree-dominated stands. Prescribed fire frequently will be the treatment of choice following a cutting. In some areas, prescribed fire alone may be the preferred approach because existing stand conditions are near desired conditions or because cuttings are precluded or otherwise inappropriate. Generally, however, prescribed fire is not likely to be a suitable standalone technique for bring-

ing about major changes in stand structure on the large scale necessary for timely implementation of DFPZ networks in Sierra Nevada coniferous forests. Factors that argue against massive and rapid increases in standalone prescribed burning include lack of adequate funding (initial burns in unthinned stands may be quite expensive), air-quality restrictions, competition for trained personnel during active wildfire seasons, and risk of escapes. Moreover, needed reductions in stand density using fire alone could require a number of successive burns spanning several decades. Failure to utilize biomass in the process would generate large quantities of smoke from consumption of excess biomass and would forgo opportunities to generate income to finance treatments. Opportunities for economic and social benefits would be forfeited as well. Furthermore, effects of initial burns probably would not closely approximate "natural" fire effects because of fuel complexes that differ greatly from those of the presettlement era (Skinner and Chang 1996; Weatherspoon 1996).

To ensure effectiveness of a DFPZ, basic adjustments in stand structure must be followed by reduction in surface fuels to a low-hazard condition using prescribed fire or mechanical methods, or both. In some cases, adequate mechanical "treatment" may result from crushing of fuels during harvest operations, especially where whole trees are removed from the stand. Prescribed fire was the best choice among van Wagtendonk's (1996) modeled scenarios from the standpoint of reducing surface fuels, and it also can raise the bases of live crowns (by killing lower branches) to increase vertical discontinuity of live fuels. Where feasible economically, removal and utilization of cut trees are preferable to treating them in place as fuels. Densities of snags and downed logs should be kept relatively low and compensated as appropriate by higher densities outside DFPZs.

From a fire standpoint, ridges and upper southerly slopes generally should benefit more than average from thinning and hazard reduction: they tend to dry out faster and without treatment would support severe fires a higher proportion of the time than other aspects and slope positions. The heavy thinning also would promote faster growth of trees into large size classes less susceptible to fire damage. Their low-fuel character, low density of snags, and resistance to sustained crown fires should make DFPZs substantially safer for suppression personnel than most other locations. Furthermore, the efficiency and productivity of suppression forces in building and holding firelines and in backfire operations should be significantly enhanced in DFPZs, especially in those containing roads. Aerial retardant drops should be considerably more effective in DFPZs as well because of the open canopy and relative ease of getting retardant to the forest floor.

To retain their effectiveness, DFPZs should be maintained in low-fuel conditions with periodic retreatments, targeting especially accumulated surface fuels and new growth of understory vegetation. Retreatment with prescribed burns should be relatively easy and inexpensive in the open environment of DFPZs. (It should be noted in this regard that

DFPZs are not unique in their need for maintenance. Fuel treatments anywhere require maintenance to retain their effectiveness. A DFPZ should cost less to maintain than an equal area of comparable fuel treatment elsewhere, however, because of its contiguity and relative accessibility.) Burns may be required about once every ten years or more often depending on rate of encroachment by shrubs and other understory fuels. DFPZ retreatment may be combined with broadened area treatment, using the DFPZ as an "anchor line." Appropriate vegetative ground covers, including perennial grasses and low-volume shrubs (e.g., bear clover), can reduce maintenance needs (Green 1977).

As main canopy trees grow and increase in crown area, they will need to be thinned periodically to maintain desired crown spacing. A few may be left to become snags, but snag density generally should be lower than elsewhere in the forest. In addition, long-term maintenance of a large-tree-dominated DFPZ will require periodic regeneration of portions of the zone. Long-rotation, low-density versions of group selection (Weatherspoon 1996) might be the best silvicultural method for this purpose, because it provides for regeneration of shade-intolerant (generally fire-tolerant) species and permits the maintenance of single canopy layers in any given location, thereby discouraging crown fires. With long rotations, a DFPZ could have sustainable age-class structures and still be occupied mostly by fire-resistant large trees.

Potential Nonfire Benefits

A range of benefits not directly related to fire would be expected to accrue from having more open stand conditions along ridges and upper southerly slopes. In general, such open conditions probably would be somewhat similar to those that dominated the same topographic positions in presettlement forests (Skinner and Chang 1996)—on average more open than other sites because of more xeric conditions and more frequent fires. A probable reduction in total evapotranspiration could lead to increased water yield from these sites. Probability of adverse watershed effects from harvesting and other management activities should be reduced because of greater-than-average distances from streams (Kattelmann 1996). These areas should contribute to overall habitat diversity and esthetic variety in landscapes that currently tend to be deficient in open, large-tree-dominated structures (Graber 1996; U.S. Forest Service 1995). Forage conditions should be improved in more open forest areas, especially with prescribed fire (Menke et al. 1996), and conceivably could help to reduce livestock grazing pressure in riparian areas. From a timber standpoint, total production of woody biomass might be reduced but would be concentrated in larger, more valuable trees (e.g., Grah and Long 1971). Lower stand density should reduce stress on trees and make them less susceptible to insect attack (Ferrell 1996). It is possible, though unproved, that broad zones of relatively low susceptibility to insects could reduce "contagion" effects of insect activity, thus perhaps slowing movement of outbreaks (Mason and Wickman 1994). If found

to be true, this idea would provide an interesting parallel to the effect of a low-hazard DFPZ on fire movement.

The concept that DFPZs may have multiple nonfire benefits emphasizes the point that strategic fuel management is an integral component of overall ecosystem management. It also argues for focusing a large proportion of overall management efforts in the short term on planning and implementing a sound DFPZ network.

Factors to Be Considered in Prioritizing DFPZ Locations

In the next sections we present a number of factors that should be considered in designing a DFPZ network. We do not attempt to set priorities among these factors—to presume, for example, that values should be weighted more heavily than historical fire occurrence or that one value is more important than another value. Such prioritization is best left to local managers using local fire planning and other information.

"Biggest Bang for the Buck." This concept says, in essence, "All else being equal, do the cheapest, easiest areas first."

Some stands already may be in an open, low-fuel condition because of recent management activities. Other areas, such as rocky outcrops and relatively bare ridges, may provide natural barriers to the spread of fire. Where it makes sense strategically to do so, such areas should be incorporated into a DFPZ network.

For areas requiring some degree of treatment to be suitable as a DFPZ, we suggest that those areas sometimes considered "most in need of treatment"—i.e., dense stands and heavy fuels—should not necessarily be given high priority. Their costs per unit area may be quite high. This subject can, and should, be debated. Our feeling, however, is that from a strategic standpoint, it seems advisable to treat first those areas that currently would not function effectively as a DFPZ but that could be brought to acceptable standards most quickly and inexpensively. Thus a greater total length of effective DFPZ could become functional for a given cost or in a given period of time. That larger treated area of DFPZ also would be more likely itself to survive in the event of a severe fire.

Some areas may be acceptably open but require surface fuel treatment. Prescribed burning may be the most desirable and cost-effective option. More often, some thinning is likely to be necessary. Except in areas where they are precluded for various reasons, cuttings (preferably with utilization of cut trees) generally provide a more efficient route to desired forest structures than prescribed burns. Where thinning is needed, the "biggest bang for the buck" principle may translate to giving priority to multiproduct sales that are economically self-sustaining by removing some sawtimber to pay for the removal of smaller trees.

Other examples of locations or conditions that might be given priority under this principle include (1) accessible areas with relatively gentle terrain and (2) areas with a significant component of relatively large pine or Douglas fir trees.

An additional benefit of the "biggest bang for the buck" principle may be in more quickly developing demonstration areas or other examples of successful implementation of DFPZs. Such areas may be valuable for building and sustaining trust and support for strategic fuel management.

Historical Fire Occurrence and Risk. A major consideration in locating DFPZs on the landscape should be the broad zones within the Sierra Nevada that have experienced the highest occurrence of large fires during this century—reflecting a combination of relatively high risk and high hazard. McKelvey and Busse (1996) found a strong elevational trend in the occurrence of twentieth-century fires in Sierran national forests. The frequency (percentage of area burned at least once) of large fires was highest below 1,000 m (3,300 ft) elevation and dropped fairly rapidly at higher elevations. This elevation zone corresponds generally with the foothill vegetation types and lower coniferous forests. It is consistent with observations by others that the highest twentieth-century fire occurrence in Sierra Nevada forests has been in the west-side pine and pine–mixed conifer types and in the east-side pine type (LaBoa and Hermit 1995; U.S. Forest Service 1995; Weatherspoon et al. 1992).

This information suggests a fairly simple guideline for accounting for historical fire occurrence: all else being equal, and in the absence of more site-specific fire-occurrence information, begin establishing a DFPZ network at the lowest elevations of ponderosa or Jeffrey pine forests and work upward into the mixed conifer type. In the general forest zone—i.e., away from settlements or other high-value areas—true fir and other upper montane types probably have low priority for a DFPZ network from the standpoint of wildfire control. Certainly other management objectives, however, may call for zones of more open forest conditions than those common in most locations today.

Where managers have good "landscape-specific" data on fire-occurrence, it of course should be weighed more heavily than regionwide trends. Local fire data also may indicate the direction of prevailing winds that accompany extreme weather events and/or large fires; this information should be used in planning DFPZ locations. Current and projected information on risk—i.e., ignition sources—should be considered as well. For example, DFPZs should have a role in isolating heavily traveled transportation corridors and other areas where ignitions historically have been high. This certainly applies to urban-wildland intermix areas, which are discussed next.

Urban-Wildland Intermix Areas. DFPZs have a potential benefit as protective buffers around high-value locations. Urban-wildland intermix areas are prominent in this regard. A protective buffer should help reduce the incidence of fires moving from wildlands into these high-value areas and (from the risk standpoint) also reduce the movement into wildland areas of fires initiating in intermix areas. These reasons, along with the fact that most populated areas in the Sierra Nevada lie within the elevation zone most frequently burned during the twentieth century (Greenwood 1995; McKelvey and Busse 1996), give a high overall priority to strategic fuel management in urban-wildland intermix areas.

As compared with DFPZs elsewhere, in forested intermix areas it may be desirable to focus more on nonfire silvicultural treatment methods in order to minimize concerns about smoke and potential escapes. In woodland and chaparral vegetation types, however, prescribed burning may be the most practical treatment approach except for limited areas of mechanical treatment. Opportunities may exist for the California Department of Forestry and Fire Protection's Vegetation Management Program (Husari and McKelvey 1996) to develop DFPZs near urban-wildland intermix areas in conjunction with some of its prescribed burning in foothill vegetation types.

The need to deal with fire and fuel issues in intermix areas is confounded by the considerable complexity of those issues. The physical problems associated with the juxtaposition of people, personal property, and wildlands are compounded by an array of problems linked to political and institutional conditions, multiple and diverse ownerships, and a wide range in understanding and attitude.

Any overall fuel strategy for urban-wildland intermix areas must begin with the use of appropriate fire-safe practices by individual property owners. Prominent among those practices are adequate clearance between structures and flammable vegetation and the use of fire-resistant roofing and other fire-safe construction practices (Davis 1990). Part of the process of achieving better compliance with fire-safe regulations is simply education of property owners—necessarily an ongoing task. Another part may involve stronger incentives, including significant fines for noncompliance, revision of insurance premiums and insurability requirements (Davis 1990), and possibly increased tax rates, to reflect more accurately the risk of fire loss in wildland settings as modified by personal fire-safe practices.

Cooperative efforts to reduce hazard within and around communities represent another critical component of fuel management in intermix areas. Partnerships that include local governments, local landowners, community groups, bioregional councils, and, as appropriate, state and federal agencies could be effective. Fostering such cooperative efforts is a high priority for the recently formed California Fire Strategies Committee. Sponsored by the California Resources Agency, the committee consists of representatives of a wide array of government and private entities with a common interest in dealing effectively with California's wildfire problems. Members have adopted an ambitious set of action items in support of the committee's mission "to reduce the risk of catastrophic fire for the protection of Californians and the natural environment."

Fuel-management activities in urban-wildland intermix areas should be coordinated with similar activities on nearby national forest or other public land and with activities of large private landowners. In a recent strategic assessment of fire management in the U.S. Forest Service, Bacon and colleagues (1995) proposed that priority for hazard mitigation on national forests in intermix areas be placed on areas where adjacent landowners agree to participate with the U.S. Forest Service in fuel management and other fire-safety projects. While designing and implementing an effective DFPZ network in and around complex intermix areas often will not be easy, it will be greatly facilitated by effective cross-ownership cooperative efforts.

Concerns about intermix areas do not stop with current conditions. Population in Sierran foothill areas is projected to continue rapid growth (Duane 1996). An important potential set of solutions related to fire issues rests with state and local officials, including legislators and county planning and zoning commissioners, who should implement appropriate limitations and disincentives for new construction in high-fire-hazard areas.

Fire-related connections between urbanized areas and nearby wildlands go beyond the potential spread of fire from one area to the other. Increasingly in recent years, federal wildland fire-control agencies have been put into the position of having to assume responsibility for structure protection during major wildfires (Bacon et al. 1995; Husari and McKelvey 1996). This imposes costs on other landowners and the general public in two ways: (1) Taxpayers at large pay for these fire-protection services, and (2) losses to natural resources on public lands increase when these forces are diverted to structure protection (Davis 1990). Bacon and colleagues (1995, 4) proposed a redefinition of responsibilities: "(1) fire protection on State and private lands is the responsibility of State and local governments, (2) homeowners have a personal responsibility to practice fire safety, (3) the role of the Forest Service is stewardship of adjacent National Forests, cooperative assistance to State and local fire organizations, and cooperative suppression during fire emergencies." They suggested two general approaches for the U.S. Forest Service in response to these responsibilities: (1) The U.S. Forest Service would phase out of responsibility for direct initial attack in urbanized areas. Existing protection agreements would be renegotiated to reflect this change. Cooperative fire-protection programs would be expanded to facilitate state efforts to take on the additional work. (2) Protection priorities would be changed from the present order of life first, property second, and resources third, to life first, followed by property and resources valued on a par. These recommendations are consistent with policy changes for federal agencies proposed in the Federal Wildland Fire Management Policy and Program Review (U.S. Department of the Interior and U.S. Department of Agriculture 1995). Bacon and colleagues (1995) also recommended that opportunities be sought for land exchanges that would improve the ability to manage fire in urban-wildland intermix areas.

Other High-Value Areas. A number of other kinds of high-value areas may warrant buffering with DFPZs—e.g., areas of late-successional emphasis (Franklin et al. 1996), biodiversity management areas (Davis et al. 1996), and plantations (Wilson 1977). Such protection may be particularly useful when fuel reduction within the high-value area itself is undesirable or infeasible because of the nature of the value being emphasized and/or high costs of treatment. It might be desirable to treat a high-value area with prescribed fire, for example, but appropriated funds might be inadequate, especially since initial reintroduction of fire without mechanical pretreatment can be rather expensive in some places. In contrast, a DFPZ outside the high-value area could be self-financing through removal of a product. It also could aid in the subsequent reintroduction of fire into the area.

DFPZs need not be placed immediately adjacent to a high-value area. In most cases it probably is desirable to back off to a location that makes sense for other reasons, as discussed earlier—e.g., a ridge or an upper south slope, along a road, relatively cheap to treat.

Using a DFPZ to provide a buffer between adjacent areas may also be useful where management emphases or intensities, rather than values per se, differ. For example, it might be desirable to provide such a separation between an area managed primarily for natural values, including use of PNF, and an adjacent area managed primarily for commodities. This might or might not be associated with an ownership boundary.

Fire Hazard. Hazard is another factor that needs to be considered in locating DFPZs. All else being equal, a landscape dominated by continuous heavy fuels is in greater need of zones of fuel discontinuity than one with light fuels. Insofar as possible, however, actual DFPZ location should favor relatively open, low-fuel sites in order to treat more area with the available funds. In other words, DFPZs should separate high-hazard areas but not necessarily be built through them.

It is reasonable to assume that high-hazard areas may be relatively more of a concern with respect to the potential for high-severity wildfires in drier years. In such years, a higher percentage of the total fuel profile (including live fuels) becomes readily available for combustion. Drier fuels and drier microclimate near the forest floor favor easier ignition and faster fire spread. The significance of such changes in dry years is increased by the preponderance of dry years in the past ten years and by the fact that such years may be more nearly the norm when viewed on a time scale of centuries (Graumlich 1993).

Professional and Public Support. Many forest-management activities are controversial, among resource professionals as well as various segments of the public. We believe that creating and maintaining DFPZs may offer multiple benefits, including reduced wildfire hazard, improved forest health, and utilization of excess forest biomass, which in most cases should outweigh potential ecosystem damage. Adequately explained and understood, therefore, DFPZs should be reasonably well supported. Nevertheless, some areas proposed for DFPZs may be controversial. All else being equal, we suggest that, at least initially, creation of DFPZ networks be concentrated in areas where professional and public support are relatively high and disagreement relatively low. In most cases, more than enough work will need to be done to permit activities to be focused in these areas and to defer more controversial work. Well-designed and properly implemented early DFPZs may generate additional support for further development of a strategic fuel-management program.

Rate of Implementation and Practicability

We believe that, in the short term, planning and implementing DFPZ networks should have a high priority for management of low- to middle-elevation Sierran forests and appropriate portions of foothill woodland and chaparral types. Ideally, these networks should be in place within ten years. Implementing these networks will require a great deal of concentrated and cooperative effort. It also may well require "departures" from nondeclining even flow of timber volume under the National Forest Management Act. Potential benefits could be substantial, however, in terms of strategic reduction of wildfire hazard, improvement in forest conditions, and increases in economic and social well-being in forest-based communities.

By any measure, implementing a rangewide system of DFPZs within ten (or even twenty) years is a formidable undertaking. Responsible managers must be concerned with the feasibility and potential value of such a task compared with alternative management actions. Given the high priority of fire-protection and restoration issues in Sierran forests and the multiple benefits (cited earlier) that might be anticipated from DFPZ networks, a number of managers may judge such networks to have a high overall priority for management.

To be achievable, implementation of a DFPZ system cannot be viewed simply as a fire function or goal. Rather, it should be considered a multiresource or ecosystem management goal, with much of the overall activity of the management unit in the short term being integrated with and focused on planning and implementing a sound DFPZ network. Similarly, multifunction funding would improve the feasibility of accomplishing this task.

How will we pay for all the silviculture and fuel management that will be necessary to implement DFPZ networks, given the large areas that need to be treated? Considering historical levels of funding and current directions of federal budgets, it seems highly unlikely that federal appropriated funds—even from multiple functions—will be adequate. And managers may decide that most of the limited appropriated funds for fuel treatment are best spent to support prescribed burning of natural fuels in areas with special emphases on reestablishing natural processes (see the following section). Thus, truly significant progress on DFPZs and other large-

scale fuel treatments will have to be the result of economically self-sustaining activities. Yet much of the needed treatment involves removal of small trees that often have marginal or negative market value. Part of the solution may come from multiproduct sales, in which sawtimber and other high-value products subsidize the removal of lower value material. One of the challenges for managers will be to locate and design multiproduct or other sales in ways that make them economically viable. In addition, however, it probably will be important to support the establishment of particleboard or other plants capable of generating value from small trees. Public land managers and private entrepreneurs need to discuss whether and how it may be possible to provide sufficient assurances of a continuing supply of biomass from public lands (e.g., for several decades) to warrant the capital investment in such plants. Research and development efforts also are needed to develop more efficient technology for harvesting and processing small material and new markets for utilizing it (Lambert 1994).

Most resource professionals would agree that fuel reduction and thinning of overly-dense stands are high-priority needs in most pine and mixed-conifer forests of the Sierra Nevada. These are precisely the kinds of activities envisioned for DFPZs, with the added proviso that they be placed in strategically logical locations. It is important to note, therefore, that the major barriers to DFPZ implementation—e.g., economic viability of small trees and maintenance of treated areas—are not unique to DFPZs: they apply much more widely. Thus, these barriers must be resolved in any case if large-scale thinning and fuel management are to be implemented. The contiguous nature and relative accessibility of DFPZs, however, may help to lessen the severity of these problems in DFPZs.

Enhanced Use of Fire

Restoring the many functions of fire as an ecosystem process can be accomplished fully only by using fire. Alternative and supplementary methods must play a large part in needed restoration, but they can substitute only partially for fire (Weatherspoon 1996). In the context of goal 2, therefore, we believe that a considerably expanded use of prescribed fire can and should play an important role in the management of Sierra Nevada ecosystems (Husari and McKelvey 1996; Mutch et al. 1993).

In some portions of the Sierra Nevada, especially higher elevation areas, large high-severity fires are not much of a concern. Thus neither goal 1 nor DFPZs are particularly applicable. Many such areas are located in national parks and wilderness areas, but substantial additional acreage of red fir and other high-elevation vegetation types fits in this category. Our suggestion in these areas would be to extend the use of prescribed natural fire (PNF) as much as possible (including appropriate areas outside parks and wildernesses) and to augment PNF with management-ignited prescribed fires

(MIPF) as needed to reestablish a near-natural distribution of fire frequencies.

MIPF also should become a key part of the management of other areas in which restoration of natural processes is a major management objective. Examples of such areas might include areas of late-successional emphasis (Franklin et al. 1996), biodiversity management areas (Davis et al. 1996), and research natural areas.

As indicated earlier, DFPZs require periodic maintenance to retain their effectiveness, and prescribed fire often will be the treatment of choice. Since the structure and composition of DFPZs are intended to be closer to presettlement conditions than most other areas of the landscape, it would seem logical for fire to assume a dual role there—maintenance of the low-fuel nature of DFPZs and restoration of natural processes.

A number of practical and political considerations constrain the use of both MIPF and PNF on a large scale. Constraints include risk of escapes, lack of adequate funding, competition for trained personnel during active wildfire seasons, and air quality restrictions (Husari and McKelvey 1996; Parsons 1995). The difficulties of applying prescribed fire on a significant scale are illustrated by the inability of the prescribed fire program at Sequoia and Kings Canyon National Parks—certainly among the most active in the Sierra Nevada—even to begin to approach the presettlement fire frequency for the giant sequoia groves. A National Interagency Fire Center study to be undertaken beginning in 1996 will test the feasibility of and constraints on landscape-scale application of prescribed fire in the Kaweah River drainage of Sequoia National Park.

In addition to prescribed burning, significant benefits related to goal 2 could be achieved by allowing low- and moderate-intensity wildfires to burn. Potentially, many more burned acres could be achieved by this means than with prescribed fire. The vast majority of ignitions in the Sierra Nevada are suppressed using fast, aggressive control. The flexibility already existing in present federal fire-management policy to use alternative suppression responses is rarely exercised outside the national parks and a few wilderness areas in the Sierra Nevada (Husari and McKelvey 1996). Fire managers currently are required to select the most economically efficient suppression option without considering potential resource benefits of wildfires. Fires that would produce results most similar to those that occurred under presettlement conditions are regularly suppressed while small, because they are easy and inexpensive to put out. Proposed new federal policies (U.S. Department of the Interior and U.S. Department of Agriculture 1995) would permit wildfires to be "managed" if they meet resource objectives.

More flexible use of appropriate suppression responses, possible use of managed wildfires to meet resource objectives, and expanded use of both MIPF and PNF jointly offer considerable opportunities for managers to restore more of the ecosystem functions of fire to the Sierra Nevada. All of these opportunities should be enhanced as forest and fuel conditions are improved over time. It should be recognized that in those areas from which fire continues to be excluded, for whatever reasons, some ecosystem components and processes will depart significantly from their natural range of variability, with unknown consequences.

Areawide Fuel Treatments

The development of DFPZs described in this chapter is a logical place to begin, but it is intended to be only a first step toward achieving the three goals of the fuel-management strategy discussed earlier. DFPZs should help to limit the spatial extent of severe fires (van Wagtendonk 1996; Sessions et al. 1996); however, they will not reduce the susceptibility of the intervening landscape areas to severe fire effects, nor will they improve forest health or restore more nearly natural processes in those intervening areas. Landscape mosaics and vegetative profiles will need to be managed on broader scales, using mainly silvicultural cuttings and fire, to achieve desired forest conditions and processes (Mutch et al. 1993).

The implementation of areawide landscape treatments should be significantly facilitated by using previously established DFPZ networks as anchor lines from which to build out. Factors considered in prioritizing DFPZ locations, discussed earlier, may also be useful as guides for prioritizing areawide treatments. From the standpoint of topography, for example, middle and upper south and west aspects on relatively gentle (machine-operable) slopes may be logical locations for early work.

RESEARCH AND ADAPTIVE MANAGEMENT NEEDS

The Role of Adaptive Management

Ecosystem management is increasingly espoused as a guiding concept for managing public lands (Jensen and Bourgeron 1994; Manley et al. 1995; Salwasser 1994). Managing for ecosystem integrity and sustainability, however, is more difficult and fraught with more uncertainties than managing for a set of specific outputs. We have much to learn. For many reasons, including the complexity and variability of forested ecosystems and the broad spatiotemporal scale that provides the context for ecosystem management, traditional research cannot provide all the answers. Scientists, managers, and interested members of the public must work together as partners in a process of learning by doing—i.e., adaptive ecosystem management (Everett et al. 1994; Mutch et al. 1993; Walters and Holling 1990).

A key concept of adaptive management is that we cannot wait for perfect information, because we will never have it. Despite the uncertainties, we must move forward with managing for sustainable ecosystems using the best information we have, knowing that with time we will learn more and be able to manage more intelligently.

The subject of landscape-level fuel-management strategies is certainly appropriate to address through adaptive management. For example, we can make educated assumptions about how a network of DFPZs might help to reduce high-severity fires and contribute to desired conditions and landscape diversity. Only through monitoring, experience, and time, however, will we know the validity of those assumptions. Only through adaptive management will we learn what locations, target conditions, and treatment schedules for implementing a DFPZ network will work for what kinds of landscapes—or whether a DFPZ network makes sense in the first place.

Similarly, we know that the ecosystem functions of frequent low- to moderate-severity fire have been largely lost from Sierran forests. Restoring these functions can be accomplished fully only by using fire. Yet in many areas silvicultural techniques and other fire "surrogates" are needed in addition to or in lieu of fire to accomplish needed restoration (Weatherspoon 1996). The extent to which natural fire regimes can or should be emulated, and the consequences for long-term ecosystem viability of alternative approaches to using fire versus fire surrogates on large scales, will become clear only through carefully designed research and adaptive management.

A GIS Database in Support of Fuel-Management Strategies and Adaptive Ecosystem Management

Good information is essential to intelligent planning of specific fuel-management strategies in the short term, and to assessing the effectiveness of those strategies (and adjusting subsequent management as appropriate) in the mid to long term. An integrated GIS database can provide a good focus for this information. The concept is quite simple and logical, given the increasingly GIS-oriented world in which we operate. Actually accomplishing the monitoring and other data collection necessary to make it fully functional may be another matter. From a fire standpoint, it probably makes sense to use the same general priorities for this data collection as discussed earlier for locating DFPZs.

In the following sections we indicate some thoughts about the directions in which we should be moving with GIS databases. We are not suggesting a standalone fire and fuel GIS. Rather, the following kinds of data needed to support fire and fuel decision making would be integrated into a larger database to inform overall land management.

Management Direction

Management objectives and guidelines, including those specific to fire and fuel management, should be indicated by area.

Vegetation and Fuels Data

The need for data on vegetation and fuels is basic and well recognized. (Much of the living vegetation is fuel, of course, but to simplify the discussion here we list vegetation and fuels separately.) Mapping should utilize the best sampling strategies combining remote sensing imagery (perhaps at several scales) and ground truthing. The reliability of existing vegetation maps should be verified before they are incorporated into the database. Fire-relevant attributes of vegetation (including understory composition and structure, and vertical and horizontal continuity) need to be characterized adequately. Similarly, surface fuels should be described, utilizing field-verified vegetation/fuels correlations to the extent feasible.

Since vegetation and fuels change over time, the dynamics occurring naturally through succession and growth must be dealt with using models combined with periodic field evaluations. Natural and human-caused disturbances also change vegetation and fuels, from a little to a lot. The database must be updated as needed to reflect these disturbance-induced changes. To account for these dynamics adequately, we need to go beyond traditional spatial GIS to incorporate new concepts in spatiotemporal GIS (Peuquet 1994; Skinner et al. 1992).

Management Activities and Other Disturbances

For our land management activities (including prescribed fire and fuel management) that significantly alter vegetation and fuels, monitoring must be carried out to determine the extent to which management objectives were met and the effects on vegetation, fuels, and other key ecosystem components. The GIS database should be updated to indicate the nature, date, spatial extent, and costs of the activity and the resulting spatially referenced vegetation and fuels. "Natural" or unplanned disturbances—especially wildfires—must also be incorporated into the database. Wildfires should be mapped by severity classes and key fire effects. To the extent allowed by available data, burning conditions at different times and places on a fire, along with suppression actions and costs, also should be entered. After postfire activities are completed, the new vegetation/fuel complex should become part of the database. To permit long-term evaluation of fires and management activities, however, it is important to maintain—not discard—prefire vegetation and fuel data. A spatiotemporal GIS would serve this purpose more efficiently than the systems generally available today (Peuquet and Niu 1995; Peuquet et al. 1992).

Other Fire-Related Data

Risk (historical fire occurrence and historical and projected ignition patterns), values at risk (for both populated and wildland areas), suppression capabilities, and any other spatially relevant fire-planning data should be included in the database. It may well be advisable for public and private landowners to cooperate in establishing data standards and protocols applicable to fire and fuels, thereby permitting data sharing, cross-ownership analyses, and the like when mutually desirable.

Benefits of the GIS Database

This kind of database, in even a rudimentary form, certainly will permit better planning for fuel-management strategies. As data are improved and accumulated over time, moreover, its value will increase. We will begin to have the data necessary to relate wildfire severity and effects to prior management activities (including fuel treatments), fuel conditions, and site and stand characteristics (e.g., Weatherspoon and Skinner 1995). Over time, as more wildfires are documented, our ability to assess the efficacy and cost-effectiveness of various fuel-management strategies in terms of both behavior and effects of subsequent wildfires and suppression costs will grow. We also will be able to evaluate trade-offs involving environmental effects of the treatments themselves. We will be much better able to learn by doing and monitoring—the essence of adaptive management (Everett et al. 1994; Mutch et al. 1993; Walters and Holling 1990).

Establishing and maintaining an accurate GIS database of this kind will require considerable effort and commitment on the part of managers and landowners. It will be a long-term, ongoing process. Many other resource benefits will accrue, however, and in fact it is difficult to see how real ecosystem management in a fire-prone region such as the Sierra Nevada will be feasible without such a database.

CONCLUSIONS

Fire has been an important component of most Sierran ecosystems for thousands of years (Skinner and Chang 1996). However, human activities since European settlement, along with variation in climate, have profoundly altered fire regimes, leading to anomalous vegetation and fuel conditions throughout much of the range. Two major fire-related "problems" have developed in the Sierra Nevada: (1) too much high-severity fire and the potential for much more of the same and (2) too little low- to moderate-severity fire, along with a variety of ecological changes attributable at least in part to this deficiency. Clearly, these are not just "fire problems." They influence virtually all resources and values in the Sierra Nevada and cut across all of SNEP's subject areas.

Given the realities of our modern civilization, we must recognize that the changes in ecosystem conditions and in the role of fire are only partially reversible. We can and should reduce the extent of large, severe wildfires. However, such fires will continue at an appreciable level (almost certainly at a higher level than in the presettlement period) into the foreseeable future. We can and should restore more of the ecosystem functions of low- and moderate-severity fire, utilizing such fire to the extent feasible. It is inconceivable, however, that fire in its presettlement extent and frequencies could be restored fully to the Sierra Nevada.

Nevertheless, a partial solution is far better than no solution at all or than a continuing deterioration of Sierran forests from a fire standpoint. There is much that we as land stewards can and should do. The two fire-related problems cited earlier can be translated into the three strategic goals that have been discussed in this chapter. Making significant progress toward these goals will require long-term vision, commitment, and cooperation across a broad spectrum of land-management agencies and other entities. The problems were created over a long period of time, and they certainly cannot be solved overnight. Progress also will require landscape-scale strategic thinking, planning, and implementation. This chapter has provided some ideas for managers to consider as they develop their own landscape-specific plans.

We have much to learn as we move more fully into an era of ecosystem management, including strategic fuel management. Adaptive management must be an integral part of our management activities, as discussed earlier. It is important to note in this regard that we do not have to have all the answers before beginning needed restoration work. We know enough at this point to recognize that current conditions in most low- to middle-elevation forests of the Sierra Nevada are unacceptable in terms of wildfire hazard, diversity, and sustainability. Regardless of the extent to which presettlement conditions are used as a guide to desired conditions, most informed people would agree that these forests generally should be less dense, have less fuels, and have more large trees. Even if we have not precisely identified target conditions, we certainly know the direction in which we should begin moving. That beginning alone will require a large measure of commitment and hard work. We can adjust along the way as we learn more and become better able to define desired conditions for Sierran forests.

ACKNOWLEDGMENTS

We would like to thank Russ Jones of the SNEP GIS staff for his considerable efforts in providing updated data related to human-caused and lightning fires on Sierra Nevada national forests during the twentieth century. We gratefully acknowledge the Cal Owl EIS team for providing the original fire data used by Russ. We also want to thank the following individuals for valuable comments on earlier versions of the manuscript: J. Brenchley-Jackson, J. Fites, D. Fullmer, S. Husari, J. LaBoa, D. Leisz, K. McKelvey, D. Parsons, R. Powers, J. Reiss, E. Roberson, L. Salazar, J. Tappeiner, G. Terhune, J. Wood, and two anonymous reviewers. Finally, we sincerely appreciate the efforts and contributions of all those who participated in the SNEP Fuels Management Strategies Workshop in March 1995 (Fleming 1996).

REFERENCES

Agee, J. K., and R. L. Edmonds. 1992. Forest protection guidelines for the northern spotted owl. In *Recovery plan for the northern spotted owl: Appendix F.* Washington, DC: U.S. Department of the Interior.

Anderson, M. K., and M. J. Moratto. 1996. Native American land-use practices and ecological impacts. In *Sierra Nevada Ecosystem Project: Final report to Congress,* vol. II, chap. 9. Davis: University of California, Centers for Water and Wildland Resources.

Anderson, P. J., R. E. Martin, and J. K. Gilless. 1991. Decision analysis in the evaluation of wildfire hazard reduction by prescribed burning in the wildland-urban interface. In *Proceedings of the 11th conference on fire and forest meteorology,* 291–98. Washington, DC: Society of American Foresters.

Arno, S. F., and J. K. Brown. 1989. Managing fire in our forests—Time for a new initiative. *Journal of Forestry* 87:44–46.

Arno, S. F., and R. D. Ottmar. 1994. Reducing hazard for catastrophic fire. In *Restoration of stressed sites, and processes,* compiled by R. L. Everett, 18–19. Vol. 4 of *Eastside forest ecosystem health assessment.* General Technical Report PNW-GTR-330. Portland, OR: U.S. Forest Service, Pacific Northwest Research Station.

Bacon, D., S. Conard, G. Ferry, P. Frey, S. Heywood, D. Lamb, D. MacCleery, D. Mangan, S. Pedigo, D. Radloff, J. Saveland, C. Swanson, and H. Tran. 1995. Strategic assessment of fire management in the U.S. Forest Service. Unpublished report. Washington, DC: U.S. Forest Service.

Biehl, G. 1995. Untitled presentation at SNEP Fuels Management Strategies Workshop, Davis, California, March.

Botti, S. J., and H. T. Nichols. 1995. Availability of fire resources and funding for prescribed natural fire programs in the National Park Service. In *Proceedings: Symposium on fire in wilderness and park management,* technical coordination by J. K. Brown, R. W. Mutch, C. W. Spoon, and R. H. Wakimoto, 94–103. General Technical Report INT-GTR-320. Ogden, UT: U.S. Forest Service, Intermountain Research Station.

Chang, C. 1996. Ecosystem responses to fire and variations in fire regimes. In *Sierra Nevada Ecosystem Project: Final report to Congress,* vol. II, chap. 39. Davis: University of California, Centers for Water and Wildland Resources.

Cohan, D., S. Haas, and P. J. Roussopoulos. 1983. Decision analysis of silvicultural prescriptions and fuel management practices on an intensively managed commercial forest. *Forest Science* 29: 858–70.

Davis, J. B. 1990. The wildland-urban interface: Paradise or battleground? *Journal of Forestry* 88:26–31.

Davis, F. W., D. M. Stoms, R. L. Church, W. J. Okin, and K. N. Johnson. 1996. Selecting biodiversity management areas. In *Sierra Nevada Ecosystem Project: Final report to Congress,* vol. II, chap. 58. Davis: University of California, Centers for Water and Wildland Resources.

Duane, T. P. 1996. Human settlement, 1850–2040. In *Sierra Nevada Ecosystem Project: Final report to Congress,* vol. II, chap. 11. Davis: University of California, Centers for Water and Wildland Resources.

Everett, R., C. Oliver, J. Saveland, P. Hessburg, N. Diaz, and L. Irwin. 1994. Adaptive ecosystem management. In *Ecosystem management: Principles and applications,* edited by M. E. Jensen and P. S. Bourgeron, 340–54. Vol. 2 of *Eastside forest ecosystem health assessment.* General Technical Report PNW-GTR-318. Portland, OR: U.S. Forest Service, Pacific Northwest Research Station.

Ferrell, G. T. 1996. The influence of insect pests and pathogens on Sierra forests. In *Sierra Nevada Ecosystem Project: Final report to Congress,* vol. II, chap. 45. Davis: University of California, Centers for Water and Wildland Resources.

Fites, J. A. 1995. Untitled presentation at SNEP Fuels Management Strategies Workshop, Davis, California, March.

Fleming, E. 1996. Compilation of workshops contributing to Sierra Nevada assessments. In *Sierra Nevada Ecosystem Project: Final report to Congress,* vol. III. Davis: University of California, Centers for Water and Wildland Resources.

Franklin, J. F., D. Graber, K. N. Johnson, J. A. Fites-Kaufmann, K. Menning, D. Parsons, J. Sessions, T. A. Spies, J. C. Tappeiner, and D. A. Thornburgh. 1996. Comparison of alternative late successional conservation strategies. In *Sierra Nevada Ecosystem Project: Final report to Congress.* Davis: University of California, Centers for Water and Wildland Resources.

Graber, D. 1996. Status of terrestrial vertebrates. In *Sierra Nevada Ecosystem Project: Final report to Congress,* vol. II, chap. 25. Davis: University of California, Centers for Water and Wildland Resources.

Grah, R. F., and A. Long. 1971. California fuelbreaks: Costs and benefits. *Journal of Forestry* 69:89–93.

Graumlich, L. J. 1993. A 1,000-year record of temperature and precipitation in the Sierra Nevada. *Quaternary Research* 39:249–55.

Green, L. R. 1977. *Fuelbreaks and other fuel modification for wildland fire control.* Agricultural Handbook 499. Washington, DC: U.S. Forest Service.

Green, L. R., and H. E. Schimke. 1971. *Guides for fuel-breaks in the Sierra Nevada mixed-conifer type.* Station Report. Berkeley, CA: U.S. Forest Service, Pacific Southwest Research Station.

Greenwood, G. 1995. How population defines fire-management environments and fire-planning options. Presentation at the conference on fire ecology and fuels management in California wildlands, 5–7 April, Sacramento, CA. Sponsored by the American Fisheries Society, California Chapter; Society for Range Management, California Section; Northern California Society of American Foresters; and Wildlife Society, Western Section.

Gruell, G. E. 1994. Perspectives on historical vegetation change in the Sierra Nevada. In *Proceedings, 15th annual forest vegetation management conference,* 10–24. Redding, CA: Forest Vegetation Management Conference.

Harmon, M. E., K. Cromack Jr., and B. Smith. 1987. Coarse woody debris in mixed-conifer forests, Sequoia National Park, California. *Canadian Journal of Forest Research* 17:1265–72.

Helms, J. A., and J. C. Tappeiner. 1996. Silviculture in the Sierra. In *Sierra Nevada Ecosystem Project: Final report to Congress,* vol. II, chap. 15. Davis: University of California, Centers for Water and Wildland Resources.

Hirsch, S. N., D. L. Radloff, W. C. Schopfer, M. L. Wolfe, and R. F. Yancik. 1981. *The activity fuel appraisal process: Instructions and examples.* General Technical Report RM-83. Fort Collins, CO: U.S. Forest Service, Rocky Mountain Research Station.

Husari, S. J., and K. S. McKelvey. 1996. Fire management policies and programs. In *Sierra Nevada Ecosystem Project: Final report to Congress,* vol. II, chap. 40. Davis: University of California, Centers for Water and Wildland Resources.

James, T. E. 1994. Catastrophic fire vulnerability rating system: "A process." Unpublished report, Groveland Ranger District, Stanislaus National Forest, 30 April.

Jensen, M. E., and P. S. Bourgeron, eds. 1994. *Ecosystem management:*

principles and applications. Vol. 2 of *Eastside forest ecosystem health assessment.* General Technical Report PNW-GTR-318. Portland, OR: U.S. Forest Service, Pacific Northwest Research Station.

Kattelmann, R. 1996. Hydrology and water resources. In *Sierra Nevada Ecosystem Project: Final report to Congress,* vol. II, chap. 30. Davis: University of California, Centers for Water and Wildland Resources.

Kilgore, B. M. 1973. The ecological role of fire in Sierran conifer forests: Its application to national park management. *Quaternary Research* 3:496–513.

Kilgore, B. M., and R. W. Sando. 1975. Crown-fire potential in a sequoia forest after prescribed burning. *Forest Science* 21:83–87.

LaBoa, J., and R. Hermit. 1995. Strategic fuels planning in the Sierra Nevada province. White paper prepared for SNEP Fuels Management Strategies Workshop, Davis, CA, March.

Lambert, M. B. 1994. Establish stable stand structures and increase tree growth: New technologies in silviculture. In *Restoration of stressed sites, and processes,* compiled by R. L. Everett, 93–96. Vol. 4 of *Eastside forest ecosystem health assessment.* General Technical Report PNW-GTR-330. Portland, OR: U.S. Forest Service, Pacific Northwest Research Station.

Manley, J. 1995. Untitled presentation at SNEP Fuels Management Strategies Workshop, Davis, California, March.

Manley, P. N., G. E. Brogan, C. Cook, M. E. Flores, D. G. Fullmer, S. Husari, T. M. Jimerson, L. M. Lux, M. E. McCain, J. A. Rose, G. Schmitt, J. C. Schuyler, and M. J. Skinner. 1995. *Sustaining ecosystems: A conceptual framework.* R5-EM-TP-001. San Francisco, CA: U.S. Forest Service, Pacific Southwest Region.

Mason, R. R., and B. E. Wickman. 1994. Procedures to reduce landscape hazard from insect outbreaks. In *Restoration of stressed sites, and processes,* compiled by R. L. Everett, 20–21. Vol. 4 of *Eastside forest ecosystem health assessment.* General Technical Report PNW-GTR-330. Portland, OR: U.S. Forest Service, Pacific Northwest Research Station.

McKelvey, K. S., and K. K. Busse. 1996. Twentieth-century fire patterns on Forest Service lands. In *Sierra Nevada Ecosystem Project: Final report to Congress,* vol. II, chap. 41. Davis: University of California, Centers for Water and Wildland Resources.

McKelvey, K. S., and J. D. Johnston. 1992. Historical perspectives on forests of the Sierra Nevada and the transverse ranges of Southern California: Forest conditions at the turn of the century. In *The California spotted owl: A technical assessment of its current status,* technical coordination by J. Verner, K. S. McKelvey, B. R. Noon, R. J. Gutierrez, G. I. Gould Jr., and T. W. Beck, 225–46. General Technical Report GTR-PSW-133. Albany, CA: U.S. Forest Service, Pacific Southwest Research Station.

McPherson, G. R., D. D. Wade, and C. B. Phillips. 1990. *Glossary of wildland fire management terms used in the United States.* Washington, DC: Society of American Foresters.

Menke, J., C. Davis, and P. Beesley. 1996. Rangeland assessment. In *Sierra Nevada Ecosystem Project: Final report to Congress,* vol. III. Davis: University of California, Centers for Water and Wildland Resources.

Mutch, R. W., S. F. Arno, J. K. Brown, C. E. Carlson, R. D. Ottmar, and J. L. Peterson. 1993. *Forest health in the Blue Mountains: A management strategy for fire-adapted ecosystems.* General Technical Report PNW-GTR-310. Portland, OR: U.S. Forest Service, Pacific Northwest Research Station.

Olson, R. D. 1993. Defensible fuel profile zones. Unpublished report. Lassen National Forest: U.S. Forest Service.

Olson, R., R. Heinbockel, and S. Abrams. 1995. Technical fuels report. Unpublished report. Lassen, Plumas, and Tahoe National Forests: U.S. Forest Service.

Omi, P. N. 1979. Planning future fuelbreak strategies using mathematical modeling techniques. *Environmental Management* 3:73–80.

Parsons, D. J. 1995. Restoring fire to giant sequoia groves: What have we learned in 25 years? In *Proceedings: Symposium on fire in wilderness and park management,* technical coordination by J. K. Brown, R. W. Mutch, C. W. Spoon, and R. H. Wakimoto, 256–58. General Technical Report INT-GTR-320. Ogden, UT: U.S. Forest Service, Intermountain Research Station.

Parsons, D. J., and S. H. DeBenedetti. 1979. Impact of fire suppression on a mixed-conifer forest. *Forest Ecology and Management* 2:21–33.

Perkins, J. 1995. Untitled presentation at SNEP Fuels Management Strategies Workshop, Davis, California, March.

Peuquet, D. J. 1994. It's about time: A conceptual framework for the representation of temporal dynamics in geographic information systems. *Annals of the Association of American Geographers* 84: 441–61.

Peuquet, D. J., R. J. Laacke, and C. N. Skinner. 1992. Representational issues involved in the development of a decision support system for spatial and temporal analysis of forest ecosystems. In *Resource Technology 92,* 381–87. Vol. 5 of *Proceedings of the ASPRS/ACSM/RT92 conference: Monitoring and mapping global change.* Bethesda, MD: American Society for Photogrammetry and Remote Sensing and American Congress on Surveying and Mapping.

Peuquet, D. J., and D. A. Niu. 1995. An event-based spatio-temporal data model (ESTDM) for temporal analysis of geographic data. *International Journal of Geographical Information Systems* 9:7–24.

Pierovich, J. M., E. H. Clarke, S. G. Pickford, and F. R. Ward. 1975. *Forest residues management guidelines for the Pacific Northwest.* General Technical Report PNW-33. Portland, OR: U.S. Forest Service, Pacific Northwest Research Station.

Quincy Library Group. 1994. Fuels management for fire protection. Unpublished report. Quincy Library Group position paper, Quincy, CA.

Radloff, D. L., and R. F. Yancik. 1983. Decision analysis of prescribed burning. In *Proceedings of the 7th conference on fire and forest meteorology,* 85–89. Washington, DC: Society of American Foresters.

Radloff, D. L., R. F. Yancik, and K. G. Walters. 1982. *User's guide to the national fuel appraisal process.* Fort Collins, CO: U.S. Forest Service, Rocky Mountain Research Station.

Salazar, L. A., and A. Gonzalez-Caban. 1987. Spatial relationships of a wildfire, fuelbreaks, and recently burned areas. *Western Journal of Applied Forestry* 2:55–58.

Salwasser, H. 1994. Ecosystem management: Can it sustain diversity and productivity. *Journal of Forestry* 92:6-7,9-10.

Sessions, J., K. N. Johnson, D. Sapsis, B. Bahro, and J. T. Gabriel. 1996. Methodology for simulating forest growth, fire effects, timber harvest, and watershed disturbance under different management regimes. In *Sierra Nevada Ecosystem Project: Final report to Congress.* Davis: University of California, Centers for Water and Wildland Resources.

Show, S. B., and E. I. Kotok. 1923. *Forest fires in California 1911–1920: An analytical study.* Department Circular 243. Washington, DC: US Department of Agriculture.

Skinner, C. N., and C. Chang. 1996. Fire regimes, past and present. In *Sierra Nevada Ecosystem Project: Final report to Congress,* vol. II, chap.

38. Davis: University of California, Centers for Water and Wildland Resources.

Skinner, C. N., R. J. Laacke, and D. J. Peuquet. 1992. Developing a user-oriented natural resource decision support system. In *Resource Technology 92*, 374–80. Vol. 5 of *Proceedings of the ASPRS/ACSM/ RT92 conference: Monitoring and mapping global change*. Bethesda, MD: American Society for Photogrammetry and Remote Sensing and American Congress on Surveying and Mapping.

Stephenson, N. L., D. J. Parsons, and T. W. Swetnam. 1991. Restoring fire to the sequoia–mixed conifer forest: Should intense fire play a role? *Tall Timbers Fire Ecology Conference* 17:321–37.

Swanson, F. J., J. A. Jones, D. O. Wallin, and J. H. Cissel. 1994. Natural variability—implications for ecosystem management. In *Ecosystem management: Principles and applications,* edited by M. E. Jensen and P. S. Bourgeron, 80–94. Vol. 2 of *Eastside forest ecosystem health assessment*. General Technical Report PNW-GTR-318. Portland, OR: U.S. Forest Service, Pacific Northwest Research Station.

U.S. Department of the Interior and U.S. Department of Agriculture. 1995. *Federal wildland fire management: Policy and program review*. Washington, DC: U.S. Department of the Interior and U.S. Department of Agriculture.

U.S. Forest Service. 1995. *Draft environmental impact statement: Managing California spotted owl habitat in the Sierra Nevada forests of California (an ecosystem approach)*. San Francisco: U.S. Forest Service, Pacific Southwest Region.

van Wagtendonk, J. W. 1996. Use of a deterministic fire growth model to test fuel treatments. In *Sierra Nevada Ecosystem Project: Final report to Congress*, vol. II, chap. 43. Davis: University of California, Centers for Water and Wildland Resources.

Walters, C. J., and C. S. Holling. 1990. Large-scale management experiments and learning by doing. *Ecology* 71:2060–68.

Weatherspoon, C. P. 1996. Fire-silviculture relationships in Sierra forests. In *Sierra Nevada Ecosystem Project: Final report to Congress*, vol. II, chap. 44. Davis: University of California, Centers for Water and Wildland Resources.

Weatherspoon, C. P., S. J. Husari, and J. W. van Wagtendonk. 1992. Fire and fuels management in relation to owl habitat in forests of the Sierra Nevada and Southern California. In *The California spotted owl: A technical assessment of its current status*, technical coordination by J. Verner, K. S. McKelvey, B. R. Noon, R. J. Gutierrez, G. I. Gould Jr., and T. W. Beck, 247–60. General Technical Report PSW-133. Albany, CA: U.S. Forest Service, Pacific Southwest Research Station.

Weatherspoon, C. P., and C. N. Skinner. 1995. An assessment of factors associated with damage to tree crowns from the 1987 wildfires in northern California. *Forest Science* 41:430–51.

Wilson, C. C. 1977. How to protect western conifer plantations against fire. Paper presented at the annual meeting of the Western Forest Fire Committee, Western Forestry and Conservation Association, Seattle.

PETER B. MOYLE
Department of Wildlife, Fish, and
    Conservation Biology
University of California
Davis, California

57

Potential Aquatic Diversity Management Areas

ABSTRACT

Aquatic ecosystems in the Sierra Nevada have been highly altered as the result of dams and diversions, watershed alterations, and introductions of non-native species. The native aquatic biota has declined in diversity and abundance as a result. Reversing this trend requires appropriate, systematic management of watersheds throughout the range. Assuming that maintenance of some basic set of the native biota is desirable, a number of options for watershed management are possible, ranging from biodiversity-oriented management of all watersheds to simply reacting to the need to keep species from becoming extinct. A middle series of options, presented here, focuses on designating forty-two watersheds as Aquatic Diversity Management Areas (ADMAs), whose first goal of management is the protection of aquatic biodiversity. The watersheds were chosen on the basis of size (greater than 50 km² [19 mi²]), natural hydrologic regime, presence of native fish and amphibians, and representativeness. To achieve more complete protection of aquatic biodiversity, a series of small reserves (Significant Natural Areas, or SNAs) could also be established to protect special or unique habitats. Management objectives for ADMAs and potential methods for achieving the objectives are presented in two example scenarios, one involving complete watersheds and one involving public lands only.

INTRODUCTION

Aquatic ecosystems are among the most highly altered ecosystems in the Sierra Nevada. This is the result of three broadly interacting factors: (1) extensive development of the rivers to supply water and to generate power; (2) watershed alterations through activities such as logging, grazing, road building, and mining; and (3) widespread introduction of non-native species into Sierran lakes and streams. As a result, the depleted native aquatic biota continues to decline, and populations of species are increasingly isolated from one another. To reverse this trend, or at least to ensure the survival of the diverse communities of aquatic organisms found in the Sierra, ecosystems and habitats have to be protected on a systematic basis. Moyle and Yoshiyama (1994) provide a general framework for accomplishing this.

The Moyle and Yoshiyama (1994) approach has five tiers, listed in order of the ease with which they can be accomplished and in reverse order of permanence of conservation:

1. Protect threatened and endangered species.

2. Protect clusters of co-occurring native species, centering on threatened and endangered species.

3. Create a system of Aquatic Diversity Management Areas (ADMAs), watersheds or other aquatic areas in which maintenance of aquatic biodiversity is the first goal of management.

4. Implement conservation plans for all watersheds, in which protection of aquatic biodiversity is an important goal.

5. Implement bioregional (landscape) plans for integrated use by humans and other organisms of natural landscape units, usually clusters of watersheds.

The first two approaches are the principal means by which state and federal agencies approach conservation of species today. Moyle et al. (1995) present recommendations for fish species to be added to the list of threatened species in California, as well as a list of clusters of species that could be comanaged for conservation. Jennings and Hayes (1994) present similar recommendations for amphibians in California. The problem with species-oriented management is that the number of endangered species is growing faster than the ability of management agencies to protect them. Species-ori-

Sierra Nevada Ecosystem Project: Final report to Congress, vol. II, *Assessments and scientific basis for management options.* Davis: University of California, Centers for Water and Wildland Resources, 1996.

ented management also often does not address the root causes of the declines, especially ecosystem-level changes to the environment. The all-watershed and bioregional approaches (tiers 4 and 5) are much discussed but not implemented, because of the enormous political difficulties of doing so. Therefore the creation of a system of waters that can be managed with biodiversity as a high priority, while allowing other beneficial activities to take place, seems like a practical solution to the problem of biodiversity conservation in the immediate future.

The purpose of this chapter is to identify ADMAs that could become part of a rangewide system of waters managed to favor native aquatic organisms and to suggest management guidelines appropriate for ADMAs. The ADMAs are arbitrarily divided into two groups: large watersheds (watersheds larger than 50 km² [19 mi²]) and Significant Natural Areas (SNAs). SNAs are smaller watersheds or fragments of watersheds that contain habitats or species of an exceptional nature (e.g., an endangered species, an unusually pristine stream, a rare habitat type) that are likely to need more intense management or protection than an ADMA watershed. SNAs are closer to the traditional idea of a preserve or reserve than are ADMA watersheds. Two sets of suggested guidelines are presented, one for a strategy involving entire watersheds and the other for a strategy involving only public lands.

METHODS

General Strategy

The first step in identifying potential ADMAs was to examine the available information on all watersheds. The idea was to identify watersheds that seemed to have the most potential for perpetuating native organisms in the future in all areas of the Sierra Nevada. Ideally, the ADMA system should contain a good representation of all aquatic habitat types found in the Sierra Nevada, either within ADMA watersheds or as SNAs. To achieve a balance between examining only very large watersheds (major tributary systems) and examining the thousands of Planning Watersheds in the Calwater Data Dictionary, I examined watersheds that were either Hydrologic Areas (HAs) or Hydrologic Subareas (HSAs) in the Calwater system. Potential ADMA watersheds and aquatic SNAs were identified from a variety of sources, including interviews with biologists familiar with the Sierra Nevada and field investigations. Once a potential ADMA watershed or SNA was identified, information on the area was reported in a standard fashion (table 57.1).

TABLE 57.1

Format for the ADMA watershed descriptions. Actual accounts of the forty-two watersheds selected as ADMAs are presented in Moyle et al. 1996.

ADMA Watershed
Name: Name of largest unit (stream, lake).
Drainage: Major drainage to which the watershed is tributary.
Calwater No.: Number assigned to the hydrologic unit(s) through the Calwater system.
County: All counties in which the watershed exists.
Location: Description of the location from headwaters to mouth.
Elevation Range: In meters.
Drainage Area: In square kilometers.
Description: Physical description of watershed: geomorphology, dominant vegetation, etc.
Aquatic Province: Ichthyological provinces of Moyle and Ellison 1991.
Habitat Types: Number and name of aquatic habitat types from updated version of Moyle and Ellison (1991).
Native Fishes: Common names only. If known, abundance is indicated by A (abundant), C (common), U (uncommon), or R (rare), where "abundant" indicates that the species is either widespread throughout the drainage or is found in large numbers in appropriate habitats, and "rare" indicates that only a few individuals have been observed.
Amphibians: As for native fishes.
Other Vertebrates: Species with a strong connection to aquatic or riparian habitats, especially rare, unusual, or high-visibility species (e.g., Pacific pond turtles, otters). Ubiquitous species such as black phoebes or aquatic garter snakes are not mentioned.
Invertebrates: Unusual or rare invertebrates; a general evaluation of aquatic insect abundance if possible.
Riparian Zone: Comments on the nature and condition of riparian habitats. Rare plants and animals listed if information available.
Human Impacts: Description of major anthropogenic factors that have altered the watershed or affected flows or water quality (e.g., diversions, clear-cutting, riparian grazing, hydraulic mining).
Ownership: List of principal public managers of land, in order of relative importance.
Existing Protection: Special designations already assigned to the water or watershed: Wild Trout Water, Wild and Scenic River, national park, Nature Conservancy preserve, etc.
Significant Natural Areas (Aquatic): Small areas or sub-basins within the watershed that have especially high value for the protection of aquatic biodiversity: spring systems, lakes, small tributaries, etc. These are listed separately in an SNA catalog.
Overall Quality Rating: A rating on a scale of 1 to 3, where 1 = near-pristine, native biota largely intact; 2 = altered, but in fair to excellent condition, potentially restorable to a rating of 1; 3 = natural appearing and important as a refuge for some native species, but probably irreversibly altered, usually because of a large dam or urban area. See Moyle and Yoshiyama 1994 for details. Because most ADMA watersheds have a rating of 2, this category is further scaled according to how close to pristine (2.1) or to nearly irreversibly altered (2.9) the watershed is perceived to be.
Reasons for Rating: Justification for the rating; a description of such aspects as condition of watershed, presence of rare species, and unusual abundance and diversity of native species.
IBI Score: Index of Biotic Integrity score for the watershed (range: 5–100). This is a standardized score based on six metrics (ranid frogs, native fishes, native fish assemblages, anadromous fishes, trout distribution, and stream fish abundance). A score of 80–100 indicates that the aquatic communities are in very good to excellent condition, 60–79 indicates that they are in good condition, 40–59 means that they are in fair condition, and 39 or less means that they are in poor condition.
Notes: Other information of potential use or interest to decision makers.
UC Davis survey? If yes, a survey by P. B. Moyle and/or coworkers was a major source of information.
Sources: Individuals who provided information, or references to published literature or reports.
Date: Date of latest major revision.
Compiler: PBM, Peter Moyle; PR, Paul Randall; RY, Ronald Yoshiyama.

Definitions

ADMA Watersheds

An ADMA watershed is one having a high value for aquatic biodiversity because it is rich in native aquatic species and communities and/or contains some particularly rare or unusual biotic element. In the Pacific Northwest, similar watersheds have been called key watersheds, but this term has been used mainly to refer to watersheds with exceptionally high value for the production of anadromous fishes, especially salmon and steelhead (Moyle and Yoshiyama 1994). The ADMA watershed concept is broader.

ADMA watersheds have the following six characteristics:

1. *They are greater than 50 km² (19 mi²) in area.* This is a fairly arbitrary figure, but it represents a watershed large enough to allow most natural processes to function indefinitely and also large enough for most aquatic species within the watershed to have a low probability of extinction from random demographic events. In general, however, the larger the watershed, the better it can serve to protect aquatic biodiversity. All large watersheds in the Sierra Nevada with an Index of Biotic Integrity (IBI) score of 60 or better (Moyle and Randall 1996) are considered to be candidates for ADMA watershed status.

2. *They have a natural hydrologic regime.* This means that the central watercourse does not have dams or diversions on it that significantly alter the way the system operates, such as eliminating flood flows in a stream or lowering the level of a lake. It also means that the watershed has not been so severely altered that runoff patterns have changed to increase the magnitude of high-flow events or to decrease flows during low-runoff periods.

3. *The waters within them are dominated by native species.* As a rule of thumb, 75% of the fish found within an ADMA watershed should belong to native species. A major exception to this rule occurs in the case of trout, because non-native trout tend to interact with other organisms in a fashion similar to native trout and because non-native trout are often widespread in watersheds that would otherwise fit the ADMA watershed definition well. A major problem from a biodiversity perspective is that many high-elevation lakes and streams did not contain *any* fish until trout (and other fishes) were introduced into them. Thus, fishless waters, or waters that can be reclaimed as fishless waters, are especially important as ADMA watersheds.

4. *The watersheds contain a wide representation of aquatic habitat types.* Moyle (1996) has identified sixty-six aquatic habitat types in the Sierra Nevada. Ideally, all these habitat types should be included within an ADMA system, and each ADMA watershed should contain most of the habitat types found in that particular region. Redundancy of habitat types between ADMAs is also important, to account for localized differences in the biota, especially aquatic invertebrates.

5. *The terrestrial and riparian ecosystems they contain are in reasonably good condition.* "Good condition" means that the watershed has a high degree of biological integrity as outlined by Regier (1993) and Karr (1993). Because anything that happens in a watershed (e.g., erosion, pollution) tends to be magnified in the low-lying waters within it, a stream or lake in a highly disturbed watershed is likely to have a highly altered biota. This criterion has to be applied flexibly because all watersheds in the Sierra Nevada are altered to a greater or lesser degree. An ADMA watershed must at least have the potential to be restored to a state that is fairly close to its original condition, especially in terms of ecosystem processes.

6. *They have other characteristics that make them special or unusual.* Potential special characteristics include representativeness, uniqueness, and scientific value, and these characteristics can partially override one or more of the first four characteristics. For example, Mariposa Creek (IBI score = 64) is recommended as an ADMA watershed, even though its fish fauna has a large non-native component, because the watershed is one of the best representatives of west-side foothill streams, which in general are highly altered. Likewise, the Mono Lake watershed is recommended as an ADMA watershed, despite the fact that the streams are regulated and dominated by exotic trout, because Mono Lake itself is a unique ecosystem. It also has high scientific value in that it has been studied intensively for years, and such long-term ecological studies can give us insights into what is happening to the Sierran environment on a much larger scale. Such scientific values are among the principal reasons for also singling out Sagehen Creek and Convict Creek for inclusion as ADMA watersheds, despite their relatively small drainage areas.

Significant Natural Areas

The term *Significant Natural Area* (SNA) is used by the California Department of Fish and Game (CDFG) to indicate areas with unusual biological value, usually as habitat for rare or endangered species or communities. Such areas are typically small and localized. SNAs designated by the CDFG have no formal protection but can form the basis for preserves. Here the term is used to designate aquatic habitats or ecosystems that contain unusual biotic elements but that are too small to be included as ADMA watersheds. Aquatic SNAs usually need special protection because they contain especially fragile species (e.g., spring-dwelling caddisflies) and/or because they are not contained in an ADMA watershed. Because of their small size and sensitivity to disturbance, aquatic SNAs will typically have to be treated as preserves if they are to continue to maintain their unusual elements; that is, they will have to be actively protected from heavy human use. A sys-

TABLE 57.2

Potential ADMA watersheds of the Sierra Nevada region. A full description of each ADMA watershed is provided in Moyle et al. 1996.

West-Side Drainages
Sacramento River Tributaries
1. Antelope Creek
2. Dye Creek
3. Mill Creek
4. Pine Creek
5. Deer Creek
6. Big Chico Creek

Feather River Drainage
7. Yellow Creek
8. Middle Fork Feather River

Yuba River Drainage
9. Lavezolla Creek/Downey River

American River Drainage
10. North Fork American River
11. Rubicon River above Hell Hole Reservoir
12. Jones Fork of Silver Fork (above Union Valley Reservoir)
13. Rock Creek

Cosumnes River Drainage
14. Entire drainage

Mokelumne River Drainage
15. North Fork Mokelumne River

Stanislaus River Drainage
16. North Fork Stanislaus River
17. South Fork Stanislaus River above Pinecrest Reservoir
18. Rose Creek

Tuolumne River Drainage
19. Clavey River
20. South Fork Tuolumne River

Merced River Drainage
21. Entire drainage above McClure Reservoir

Upper San Joaquin Drainage
22. Mariposa Creek above Mariposa Reservoir
23. East Fork Chowchilla River
24. Finegold Creek

Kings River Drainage
25. Rancheria Creek
26. South and Middle Forks Kings River

Kaweah River Drainage
27. South Fork Kaweah River

Tule River Drainage
28. North and Middle Forks Tule River

Tulare Lake Foothill Drainages
29. Deer Creek

Kern River Drainage
30. Kern River above Isabella Reservoir
31. South Fork Kern River
32. North Fork Kern River

East-Side Drainages
Eagle Lake Drainage
33. Entire drainage, including Pine Creek

Susan River/Honey Lake Drainage
34. Willow Creek

Truckee River Drainage
35. Upper Little Truckee River
36. Sagehen Creek

Carson River Drainage
37. East Fork Carson River

Walker River Drainage
38. Buckeye Creek
39. West Walker River drainage

Mono Lake Basin
40. Mono Lake

TABLE 57.2 (continued)

Owens River Drainage
41. Owens River drainage above Crowley Reservoir
42. Convict Creek

Modoc Region[a]
Pit River Drainage
43. Mill Creek (South Fork Pit River)
44. Cedar Creek above Tule Reservoir
45. Ash Creek
46. Turner Creek

Goose Lake Drainage
47. Goose Lake

Cowhead Lake
48. Cowhead Slough

[a]Potential ADMA watersheds for the Modoc Region are included here for the sake of completeness, although they will not be discussed further in this chapter.

tem of protected aquatic SNAs would supplement a system of ADMA watersheds, helping to ensure that all native species and natural communities in the Sierra Nevada can persist. Examples of aquatic SNAs include small, isolated streams that contain remnant populations of Lahontan cutthroat trout (e.g., By-Day Creek, Mono County) and spring systems with unusual invertebrate assemblages (e.g., Bendorf Spring, El Dorado County). Many areas designated as research natural areas by the U.S. Forest Service also fit the definition of aquatic SNAs. Aquatic SNAs are not considered systematically in this chapter or in Moyle et al. 1996. This is not, however, a reflection of their importance in an overall strategy to protect aquatic biodiversity in the Sierra Nevada.

RESULTS

Forty-two potential ADMA watersheds were identified (table 57.2). They are widely distributed over the Sierra Nevada (figure 57.1). A description of each ADMA watershed is presented in Moyle et al. 1996. These watersheds contain sixty of the sixty-six major aquatic habitat types identified for the Sierra Nevada, with forty-nine of them represented two or more times. The habitats not covered by ADMAs either are lowland habitats that have been strongly affected by water diversions (e.g., Valley Floor River, Owens Lake) or are limited habitats that will need to be protected in SNAs (e.g., sphagnum bogs, Lahontan desert springs). Table 57.3 presents examples of potential SNAs.

The ADMA watersheds include habitats for most of the native fish and amphibians of the range. How well the native aquatic invertebrates are represented in the forty-two ADMA watersheds is not known, although it is likely that a high percentage of them are covered, given the distribution and size

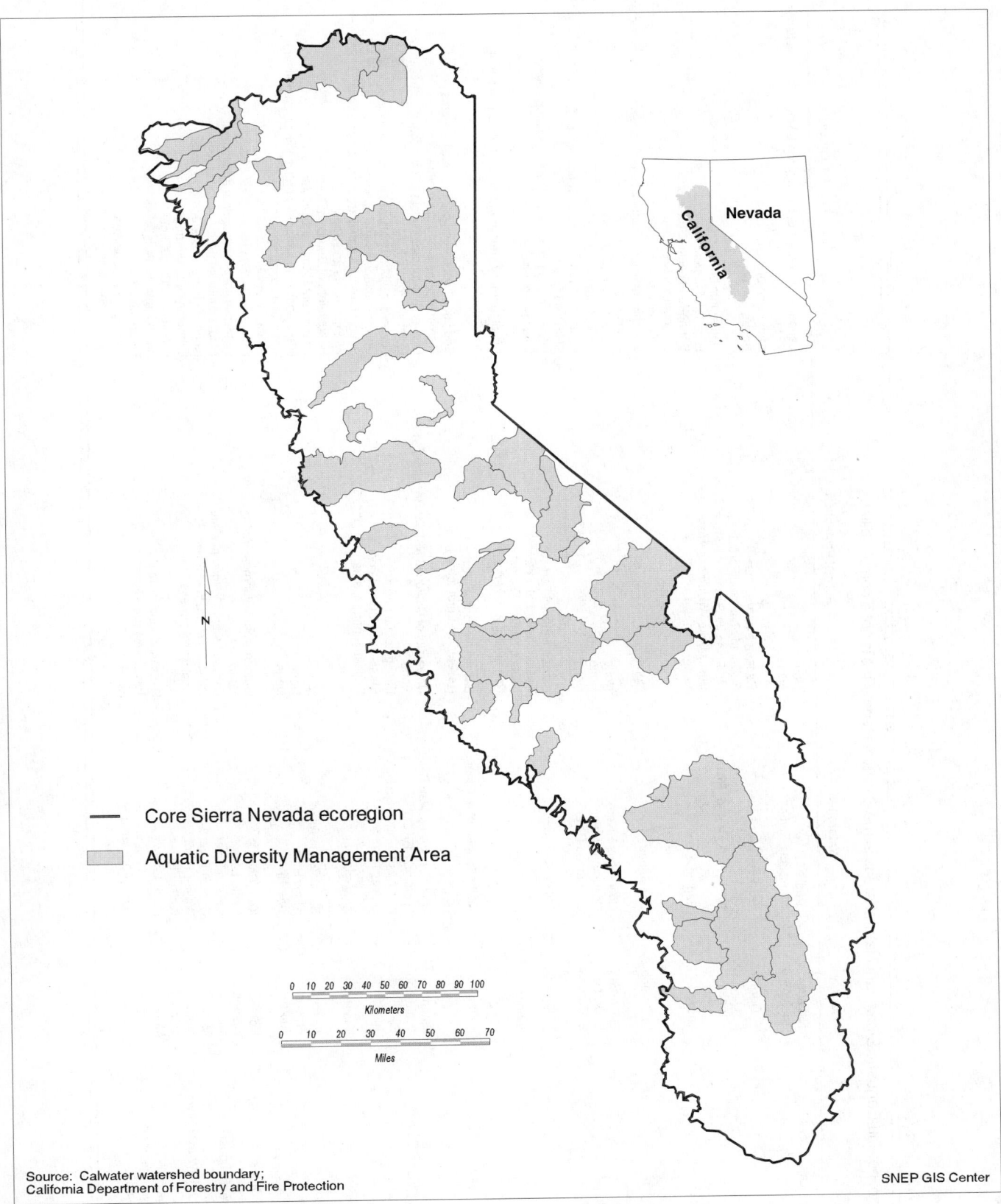

Core Sierra Nevada ecoregion

Aquatic Diversity Management Area

Nevada

California

0 10 20 30 40 50 60 70 80 90 100
Kilometers

0 10 20 30 40 50 60 70
Miles

Source: Calwater watershed boundary;
California Department of Forestry and Fire Protection

SNEP GIS Center

FIGURE 57.1

Potential ADMA watersheds of the Sierra Nevada region.

TABLE 57.3

Examples of potential aquatic Significant Natural Areas in the Sierra Nevada. This is not a complete list.

Name	County	Watershed	Formal Protection	Attributes
Independence Lake/Creek	Nevada	Truckee River	None	Lahontan cutthroat trout, one of few lakes with native fishes
Three Meadows	Tuolumne	Stanislaus River	None	Fishless meadow stream
Grass Lake	El Dorado	Truckee River	Research natural area	Amphibian breeding area
			El Dorado National Forest	Sphagnum bog
Mill Creek	Plumas	North Fork Feather River	Mt. Pleasant Research Natural Area	Sphagnum bog
Papoose Creek	Lassen	Eagle Lake	Bucks Lake Wilderness Area	High-rainfall creek
			None	Spring-fed speckled dace stream
Jackass Canyon	El Dorado	Cosumnes River	None	Large meadow system
				Amphibians and reptiles
				Intermittent fishless foothill canyon stream with endemic aquatic insects
Stump Spring	El Dorado	Cosumnes River	None	Spring with rare stoneflies
Bendorf Spring	El Dorado	Cosumnes River	None	Spring with rare stoneflies
Camp Creek	El Dorado	Cosumnes River	None	Foothill stream with native amphibians
				SNEP study site
Indian Creek	Tehama	Antelope Creek	Lassen National Forest	Intact community of native fishes and amphibians
			Research natural area	Intact riparian zone
Cub Creek	Tehama	Deer Creek	Lassen National Forest	
			Research natural area	Small tributary stream with intact riparian and aquatic communities
Green Island Lake	Plumas	North Fork Feather River	Lassen National Forest	Bog lakes, ephemeral ponds, small streams
			Research natural area	Rich invertebrate and plant biota
The Cedars	Placer	North Fork American River	University of California Natural Reserve System	Low-order tributaries in old-growth forest, native trout
Six Bit Gulch (Horton Creek)	Tuolumne	Tuolumne River	None	Red Hills roach (fish)
Onion Creek	Placer	North Fork American River	Experimental forest	Low-order tributary stream in managed forest
Bell Meadow	Tuolumne	Clavey River	Stanislaus National Forest	Unusual meadows and riparian areas
			Research natural area	Small creek with native trout and amphibians
Bourland Meadow	Tuolumne	Clavey River	None	Wet meadows and bogs
				Headwaters of Bourland Creek
Little Finegold Creek	Merced	San Joaquin River	None	Intermittent foothill stream with native fishes, including hitch
Mill Flat Creek	Fresno	Kings River	None	Major spawning stream for native fishes, amphibians
Doyle Springs	Tulare	Tule River	None	Travertine spring system
				Fauna unknown
Soda Springs Creek	Kern	Kern River	Golden Trout Wilderness Area	Little Kern golden trout
Mahogany Lake	Lassen	Eagle Lake	None	Fishless lake with abundant amphibians
Silver King Creek (upper)	Mono	Carson River	Carson-Iceberg Wilderness Area	Paiute cutthroat trout
White Cliff Lake	Mono	Carson River	Carson-Iceberg Wilderness Area	Isolated cirque lake with native frogs
Headwaters, Little Walker River	Mono	Walker River	Hoover Wilderness Area	Lahontan cutthroat trout
Big Dry Creek	Mono	Walker River	?	Lahontan cutthroat trout
Harvey Monroe Hall Research Natural Area	Mono	Mono Lake	Research natural area	Amphibians, fishless lakes
			Toiyabe National Forest	Research area
New York Ravine	Sierra	North Fork Yuba River	None	Endangered caddisflies
				Steep, fishless stream

of many of the recommended ADMAs. For example, multiple springs and their outflows probably exist in most of the ADMAs (prominent springs have been identified in twenty-one of them), a habitat type that supports a wide variety of unusual and endemic forms. However, it is likely that a number of rare or endemic invertebrates will need to be protected in SNAs or other intensively managed areas.

MANAGEMENT OF ADMA WATERSHEDS

The key for managing ADMA watersheds is initially to halt or reduce all activities on public lands that are contributing to habitat deterioration or loss of diversity. Examples of the kinds of restrictions that could apply are given in table 57.4. Such restrictions could be relaxed once a formal analysis of the watershed's biotic and abiotic characteristics had been completed and once mechanisms were in place to allow adaptive management strategies (Lee 1993) to be worked out with consensus from stakeholder groups. This process should allow for increased flexibility in the management of these watersheds and would allow many activities, such as recreation, logging, and mining, to continue in most areas. Presumably, such activities would follow guidelines that were compatible with maintaining biodiversity, and the most intense management efforts would be focused on areas in which substantial improvements in habitats could be attained.

In ADMA watersheds covering a mixture of public and private lands, landowner and other stakeholder involvement in management is essential. Stakeholder organizations (watershed associations) seem to be an effective way to achieve this involvement, and there are now a number of models to follow (e.g., the Deer Creek Watershed Conservancy in Tehama County). Such organizations can work with environmental organizations (e.g., the Nature Conservancy and Friends of the River), public agencies, and educational institutions to develop strategies that maintain their values while also maintaining natural systems. For example, the Deer Creek Watershed Conservancy has worked out measures to protect spring-run chinook salmon and other native fishes (including a no-new-dams agreement, now in state law) in exchange for measures to protect private property from regulations regarded as intrusive (e.g., those associated with Wild and Scenic River status).

For a large watershed, it may also be desirable to have a professional stream keeper, a person paid by agencies or stakeholders or both to monitor the aquatic and riparian habitats, organize restoration activities, and generally keep everyone informed of activities within the watershed. It is important, however, to have the stream keeper focus on just one watershed, rather than be an agency scientist who has stream-keeper duties assigned as part of a larger job.

DISCUSSION

A spectrum of alternatives is available to protect aquatic biodiversity in the Sierra Nevada, of which the ADMA approach is just one. One extreme alternative is to treat all watersheds without major dams or diversions as ADMA watersheds and to manage them under guidelines such as those in table 57.4. In addition, under this alternative, all rivers below dams and diversions would be provided with flow regimes and riparian protections that offer the most benefit to aquatic life. This option would be highly desirable from the perspective of protecting aquatic biodiversity but would presumably reduce the amount of water available for out-of-stream uses as well as greatly restricting other uses of the watersheds.

Another alternative is to keep using the piecemeal system of protection that now exists and wait for crises to develop before taking major steps to protect many endangered species or unusual assemblages of organisms (if they are to be protected at all). This system will inevitably lead to increased and irreversible loss of biodiversity on the local, regional, and Sierra-wide scales and will result in many painful conflicts among diverse interests. It is also likely to lead to declines in fisheries, losses in water quality, reductions in storage capacities of reservoirs, and other consequences with direct effects on humans. A more extreme version of this alternative would be to make biodiversity protection a low priority in watershed management in general. It is likely that this alternative would result in an accelerated loss of species and biotic communities, as well as a significant loss of ecosystem services that healthy watersheds provide (e.g., clean water, high aesthetic values).

If the aquatic biodiversity of the Sierra Nevada is going to be maintained at the present level or improved, avoiding widespread extinction of species, the creation of a system of ADMA watersheds and aquatic SNAs, or an approach similar to it, would seem to be necessary. Redundancy in ADMAs and SNAs is extremely important, because if one watershed or SNA is hit by a major disaster (e.g., a severe fire), recolonization or reintroduction of the biota should be possible. Redundancy is also insurance against random extinctions that occur in isolated populations (Moyle and Sato 1991). The system of ADMA watersheds presented in this chapter should be regarded as an example of the kind of system that can be developed, rather than the only approach of this type. Many permutations and combinations of watershed protection are possible. For example, less restrictive management activities in ADMA watersheds could take place on public lands, with private lands managed for biodiversity largely through voluntary means (table 57.5). An approach of this nature would probably result in the loss of some species and habitats and slower recovery of damaged systems, but would ultimately be beneficial to aquatic and riparian life.

TABLE 57.4

Management objectives and potential methods for achieving the objectives for ADMA watersheds.[a]

I. Monitoring and management
   A. Objective: to have continuous and responsible management and monitoring of ADMA watersheds
   B. Potential methods
      1. Hire professional stream keepers to monitor and lead management efforts in streams
      2. Encourage development of watershed associations made up of landowners and other interested parties to guide watershed management
      3. Develop watershed management plans for both public and private lands that encourage adaptive management and protect biodiversity while permitting other uses
      4. Provide educational/extension programs to assist landowners in developing management strategies for private lands

II. Flows
   A. Objective: to maintain natural flow regime and natural passage for fish movements
   B. Potential methods
      1. Allow no new dams; retire old dams wherever possible
      2. Prohibit any increase in diversions within the watershed
      3. Enhance flow within range of natural flow variation (if channel used for water conveyance)
      4. Manage watershed to reduce "flashiness" of runoff
      5. Remove or modify artificial barriers to fish movement

III. Riparian areas
   A. Objectives
      1. To maintain and enhance structure of in-stream, lake, and wetland habitat
      2. To maintain natural temperature regimes in streams
      3. To provide continuous habitat for riparian-dependent native plants and animals
      4. To maintain large riparian trees
      5. To maintain native riparian vegetation
   B. Potential methods
      1. Establish 100 m buffer (streamside protection) zones on 3+ order streams and 50 m zones on 1–2 order streams or to top of canyon until watershed evaluation done
      2. Establish a 50 m buffer zone along all lakes and wetlands
      3. Eliminate grazing from riparian buffer zones, except for small, fenced access points for watering
      4. Eliminate logging from riparian zones
      5. Limit the number of road crossings to one or fewer per 10 km
      6. Keep roads out of riparian zones or locate them to minimize effects on aquatic and riparian habitats; prohibit new roads in riparian areas; reduce riparian roads by 50% in ten years
      7. Prohibit planting of non-native fishes or hatchery trout in the watershed except in areas with high public access (e.g., roadside sections)
      8. Eliminate camping or other 24-hour uses of riparian areas; develop recreational trails that minimize negative effects on riparian areas
      9. Restrict number and size of in-stream dredge mining operations; enforce laws to keep mining (and related activity) out of riparian areas
      10. Eliminate dumping of all mine spoils into riparian zones
      11. Develop incentives to keep new buildings out of riparian zones/floodplains as well as for removal of existing structures

IV. Pollution
   A. Objective: to reduce pollution from toxic materials and sediments from local sources to levels within the presumed range of natural variation
   B. Potential methods
      1. Manage the watershed to reduce sediment runoff (e.g., require proper road construction, eliminate "bad" roads, minimize grazing and logging practices that increase erosion, provide better erosion control in ski areas)
      2. Eliminate or reduce toxic drainage from abandoned mines
      3. Allow only tertiary treated sewage, if any, to be dumped into streams
      4. Prevent septic tanks and leach fields from leaking into streams
      5. Limit use of pesticides to emergencies or to situations where short-term use is needed to assist recovery of native organisms

TABLE 57.4 (continued)

V. Land use
   A. Objective: to minimize or reduce human-caused disturbance of existing terrestrial systems in the watersheds, in order to reduce human impacts on aquatic systems
   B. Potential methods
      1. Allow no net increase in road kilometers; no roads in existing roadless areas
      2. Eliminate or restrict use of off-road vehicles to highly disturbed areas (old quarries, etc.)
      3. Construct trails to minimize impacts on sensitive areas
      4. Protect all remaining spring systems (fence, remove boxes from source, etc.)
      5. Develop incentives to keep human population levels at reduced levels compared to neighboring non-ADMA watersheds
      6. Discourage development of activities likely to degrade the watershed (e.g., new ski resorts or other intensive recreation sites)
      7. Maximize cover of late successional old-growth forest
      8. Develop fire management strategies that minimize the potential for large-scale, devastating fires
      9. Create educational/extension programs to enable private landowners to maximize income and benefits from land while minimizing the impact on aquatic systems

VI. Exotic species
   A. Objective: to reduce the influence of non-native species on aquatic and riparian ecosystems
   B. Potential methods
      1. Eliminate planting of fish in high-elevation lakes
      2. Systematically eradicate trout from selected stream and lake systems in areas that were originally fishless
      3. Develop techniques for bullfrog eradication
      4. Manage streams and riparian areas to favor native organisms
      5. Reintroduce native fishes and frogs into areas from which they were extirpated
      6. Develop programs to encourage use of native fishes and other organisms on habitats on private lands, including stock ponds

VII. Salmon restoration
   A. Objective: to increase spawning areas for chinook salmon in order to increase populations and improve habitats for all stages in lowland rivers
   B. Potential methods
      1. Restore salmon to selected areas from which they were extirpated, such as the North Fork American River (over Nimbus and Folsom Dams)
      2. Actively improve spawning gravels, riparian areas, holding pools, and other habitats in degraded sections of rivers

VIII. Recreation
   A. Objective: to reduce impacts of recreational activities on the native biota
   B. Potential methods
      1. Restrict take of wild fish by anglers
      2. Reduce recreational use of riparian and aquatic areas (trails, roads, etc.)
      3. Reduce or eliminate in-stream activities (gold dredging, rafting, etc.) that disturb anadromous or spawning fish or breeding amphibians
      4. Identify sensitive areas (spring systems, etc.) and protect them from human entry

IX. Significant Natural Areas
   A. Objective: to protect unique or sensitive habitats within the watersheds that are limited in area (e.g., large springs)
   B. Potential methods
      1. Inventory watersheds to locate SNAs
      2. Acquire title or conservation easements (or equivalent) if private land or special protective designations if public land
      3. Provide individualized protective measures (e.g., signing, fencing, road removal)
      4. Develop educational/extension programs to provide landowners with tools to provide voluntary protection of SNAs on private land

[a]This list is not comprehensive and is meant to suggest activities and actions that would be appropriate for the maintenance and/or enhancement of aquatic biodiversity in selected watersheds.

TABLE 57.5

Management objectives and potential methods for partially achieving the objectives for ADMA watersheds by focusing on management activities on public lands or on regulations within the normal purview of public agencies.[a]

I. Monitoring and management
   A. Objective: to have continuous and responsible management and monitoring of ADMA watersheds
   B. Potential methods
      1. Hire professional stream keepers to monitor and lead management efforts in streams on public lands
      2. Encourage development of watershed associations made up of landowners and other interested parties to guide watershed management
      3. Develop watershed management plans for public lands that encourage adaptive management and protect biodiversity while permitting other uses
      4. Provide educational/extension programs to assist landowners in developing management strategies for private lands

II. Flows
   A. Objective: to maintain natural flow regime and natural passage for fish movements
   B. Potential methods
      1. Allow no new dams to be constructed on public lands; retire old dams wherever possible
      2. Allow no increase in diversions on public lands
      3. Protect headwater areas to reduce "flashiness" of runoff
      4. Remove or modify artificial barriers to fish movement

III. Riparian areas
   A. Objectives
      1. To maintain and enhance structure of in-stream, lake, and wetland habitat
      2. To maintain natural temperature regimes in streams
      3. To provide continuous habitat for riparian-dependent native plants and animals
      4. To maintain large riparian trees
      5. To maintain native riparian vegetation
   B. Potential methods
      1. Follow riparian prescriptions in table 57.4 for federal and state land
      2. Develop incentives for improved riparian management on private lands
      3. Restrict number and size of in-stream dredge mining operations; enforce laws to keep mining activity out of riparian areas

IV. Pollution
   A. Objective: to reduce pollution from toxic materials and sediments from local sources to levels within the presumed range of natural variation
   B. Potential methods
      1. Manage portions of watersheds on federal and state lands to reduce sediment runoff (e.g., require proper road construction, eliminate "bad" roads, minimize grazing and logging practices that increase erosion, provide better management of ski areas)
      2. Develop incentives for reducing sediment and non-point-source pollution on private lands
      3. Eliminate or reduce toxic drainage from abandoned mines
      4. Allow only tertiary treated sewage, if any, from municipal plants to be dumped into streams

V. Land use
   A. Objective: to minimize or reduce human-caused disturbance of existing terrestrial systems in the watersheds, in order to reduce human impacts on aquatic systems
   B. Potential methods
      1. Develop incentives to keep human population levels at reduced levels compared to neighboring non-ADMA watersheds
      2. Institute fire management strategies that minimize the potential for large-scale, devastating fires
      3. Create educational/extension programs to enable private landowners to maximize income and benefits from land while minimizing the impact on aquatic systems

VI. Exotic species
   A. Objective: to reduce the influence of non-native species on aquatic and riparian ecosystems
   B. Potential methods

TABLE 57.5 (continued)

      1. Eliminate planting of fish in high-elevation lakes more than 4 km (2.5 mi) from a road or trailhead
      2. Allow systematic eradication of trout from selected stream and lake systems on public lands that were originally fishless
      3. Develop techniques for bullfrog eradication
      4. Manage streams and riparian areas to favor native organisms on public lands
      5. Reintroduce native fishes and frogs into areas on public lands from which they were extirpated
      6. Develop programs to encourage use of native fishes and other native organisms on private lands, including their use in stock ponds

VII. Salmon restoration
   A. Objective: to increase spawning areas for chinook salmon in order to increase populations and improve habitats for all stages in lowland rivers
   B. Potential methods
      1. Restore salmon to selected areas from which they were extirpated, such as the North Fork American River (over Nimbus and Folsom Dams)
      2. Actively improve spawning gravels, riparian areas, holding pools, and other habitats in degraded sections of rivers

VIII. Recreation
   A. Objective: to reduce impacts of recreational activities on the native biota
   B. Potential methods
      1. Reduce or eliminate in-stream activities (gold dredging, rafting, etc.) that disturb anadromous or spawning fish or breeding amphibians on waters flowing through public lands
      2. Identify sensitive areas (spring systems, etc.) on public land and protect them from human entry
      3. Develop educational programs and incentives to reduce the impact of recreation on aquatic systems on private land

IX. Significant Natural Areas
   A. Objective: to protect unique or sensitive habitats within the watersheds that are limited in area (e.g., large springs)
   B. Potential methods
      1. Inventory watersheds on public lands to locate SNAs
      2. Provide individualized protective measures (e.g., signing, fencing, road removal) for SNAs on public land
      3. Develop educational/extension programs to provide landowners with tools to provide voluntary protection of SNAs on private land

[a]This list is not comprehensive and is meant to suggest activities and actions that would be appropriate for the maintenance and/or enhancement of aquatic biodiversity in selected watersheds.

Regardless of the management details, the development of a system of ADMA watersheds that at least maintains aquatic biodiversity at current levels within the selected watersheds depends on a number of assumptions:

- It will be public policy to maintain self-sustaining populations of all aquatic and riparian-dependent species presently inhabiting the Sierra Nevada. This means that extinctions of species or of taxonomically distinct populations cannot occur and that there must be representatives of all aquatic and riparian habitat types under protective management.

- A major and continuing role of all watersheds is to supply high-quality water to the people of California either to consume or to leave in-stream for ecosystem purposes.

- The watersheds of the Sierra Nevada will continue to have increasing use by humans, and streams, lakes, and riparian areas will continue to be a focus of human activities, despite their high sensitivity to disturbance. Therefore, protection of aquatic biodiversity will require more restrictive regulation of human use in at least some watersheds.

- The aquatic and riparian systems in many areas have been severely and perhaps irreversibly altered. There is an ongoing trend toward continued degradation of the native biota, and it is highly desirable to halt and/or reverse this trend.

- Watershed management strategies must be individualized for each watershed. The major factors affecting biodiversity are often quite different at low elevations than at high elevations, as well as among different regions of the Sierra Nevada. However, until watershed-specific management strategies are developed, it is highly desirable to use broad-scale prescriptions for land and water use that err on the side of protection of habitats and biota. Such prescriptions can become more flexible (adaptive) once the condition of each watershed has been analyzed and areas and waters especially sensitive to human disturbance have been identified.

- A systematic inventory and monitoring program for the biota will be established, not only to keep track of trends in the aquatic organisms and habitats, but to identify SNAs for the protection of endemic invertebrates and other poorly known organisms.

It is obvious that for an ADMA system to work, the first goal of management for each ADMA watershed or SNA must be to protect its biotic integrity. This does not mean that each ADMA watershed should be locked away from human use, but rather that human use should be as gentle as possible. Roads, for example, should be minimized and constructed in such a way that little erosion occurs and contact with aquatic and riparian areas is minimal. Diversions of water within the watershed should be minimal so that the natural flow regime is not altered and so that reservoirs and other disruptive habitats are not created. Recreation should be limited to activities that do not significantly alter the landscape or biota (limited use by off-road vehicles, restricted camping in riparian zones, etc.). In contrast, SNAs will presumably need a much higher degree of protection, especially ones not included within an ADMA watershed.

A system of ADMA watersheds is not meant to signal that all other watersheds can be treated without respect for the natural biota and processes. Ultimately, a system of ADMA watersheds will work only if the drainages connecting them and the watersheds around each ADMA watershed are not highly degraded. A system of properly managed watersheds could demonstrate the considerable economic and social benefits of good watershed management. Such a system could also help avoid many of the problems created by the loss of biodiversity and ecosystem services while not seriously interfering with the ability of Sierran streams to provide continuing economic benefits to the people of California.

REFERENCES

Jennings, M. A., and M. Hayes. 1994. *Amphibian species of special concern in California.* Sacramento: California Department of Fish and Game.

Karr, J. R. 1993. Measuring biological integrity: Lessons from streams. In *Ecological integrity and the management of ecosystems,* edited by S. Woodley, J. Kay, and G. Francis, 83–104. Ottawa: St. Lucie Press.

Lee, K. N. 1993. *Compass and gyroscope: Integrating science and politics for the environment.* Washington, DC: Island Press.

Moyle, P. B. 1996. Status of aquatic habitat types. In *Sierra Nevada Ecosystem Project: Final report to Congress,* vol. II, chap. 32. Davis: University of California, Centers for Water and Wildland Resources.

Moyle, P. B., and J. P. Ellison. 1991. A conservation-oriented classification system for the inland waters of California. *California Fish and Game* 77:161–80.

Moyle, P. B., and P. J. Randall. 1996. Biotic integrity of watersheds. In *Sierra Nevada Ecosystem Project: Final report to Congress,* vol. II, chap. 34. Davis: University of California, Centers for Water and Wildland Resources.

Moyle, P. B., P. L. Randall, and R. M. Yoshiyama. 1996. A catalogue of potential ADMA watersheds for the Sierra Nevada. In *Sierra Nevada Ecosystem Project: Final report to Congress,* vol. III. Davis: University of California, Centers for Water and Wildland Resources.

Moyle, P. B., and G. M. Sato. 1991. On the design of preserves to protect native fishes. In *Battle against extinction: Native fish management in the American West,* edited by W. L. Minckley and J. E. Deacon, 155–69. Tucson: University of Arizona Press.

Moyle, P. B. and R. M. Yoshiyama. 1994. Protection of aquatic biodiversity in California: A five-tiered approach. *Fisheries* 19:6–18.

Moyle, P. B., R. M. Yoshiyama, E. Wikramanayake, and J. E. Williams. 1995. *Fish species of special concern in California.* Sacramento: California Department of Fish and Game.

Regier, H. A. 1993. The notion of natural and cultural integrity. In *Ecological integrity and the management of ecosystems,* edited by S. Woodley, J. Kay, and G. Francis, 3–18. Ottawa: St. Lucie Press.

FRANK W. DAVIS
Institute for Computational
  Earth System Science
University of California
Santa Barbara, California

DAVID M. STOMS
Institute for Computational
  Earth System Science
University of California
Santa Barbara, California

RICHARD L. CHURCH
Department of Geography
University of California
Santa Barbara, California

WILLIAM J. OKIN
Department of Geography
University of California
Santa Barbara, California

K. NORMAN JOHNSON
College of Forestry
Oregon State University
Corvallis, Oregon

Selecting Biodiversity Management Areas

58

ABSTRACT

Here we present and evaluate a conservation strategy whose objective is to represent all native plant communities in areas where the primary management goal is to sustain native biodiversity. We refer to these areas as Biodiversity Management Areas (BMAs), which we define as specially designated public or private lands with an active ecosystem management plan in operation whose purpose is to contribute to regional maintenance of native genetic, species and community levels of biodiversity, and the processes that maintain that biodiversity. Our purpose in this chapter is to explore opportunities for siting BMAs in the Sierra Nevada region. The strategic goal is to design a BMA system that represents all major Sierran plant community types, which we use as a coarse surrogate for ecosystems and their component species. We consider a community type to be represented if some pre-defined fraction of its mapped distribution occurs in one or more BMAs. We use a multi-objective computer model to allocate a minimum of new land to BMA status subject to the constraints that all community types must be represented, and that the new BMA areas should be located in areas of highest suitability for BMA status. Our purpose in this exercise is not to identify the optimal sites for a Sierran BMA system; instead it is to measure some of the likely dimensions of plausible, alternative BMA systems for the Sierra Nevada and to develop a rationale that would guide others in formulating such a system. Thus we examine a wide range of possible BMA systems based on different assumptions, constraints, target levels for representation, and priorities.

If one ignores current land ownership and management designations and sets out to represent plant communities in a BMA system based on Calwater planning watersheds (which average roughly 10,000 acres in size), an efficient BMA system requires land in direct proportion to the target level, at least over the range of target levels examined in this study. In other words, it takes roughly 10% of the region to meet a 10% goal, and 25% of the region to meet a 25% goal. The pattern of selected watersheds is very different from the current distribution of parks and wilderness areas, which are concentrated at middle and high elevations in the central and southern portion of the range.

Public lands alone are insufficient to create a BMA system that adequately represents all plant community types of the Sierra Nevada. Many of the foothill community types occur almost exclusively on private lands. Terrestrial vertebrates are reasonably well represented in a BMA system selected for plant communities. A BMA system selected for vertebrates alone, however, has little overlap with the one for plant communities.

Areas selected by the BMAS model show only a modest amount of overlap with areas selected by other SNEP working groups as focal areas for conserving aquatic biodiversity or late successional/old growth forests. However, the BMAS model can be formulated to favor these areas with little loss of efficiency, especially in the northern Sierra.

Sierra Nevada Ecosystem Project: Final report to Congress, vol. II, *Assessments and scientific basis for management options.* Davis: University of California, Centers for Water and Wildland Resources, 1996.

PROBLEM STATEMENT

The Sierra Nevada Ecosystem Project (SNEP) has highlighted some pervasive and resource-specific impacts of human activities on the region's aquatic and terrestrial biodiversity. We will not review these here, instead referring the reader to SNEP's key findings in Volume 1 of this report. Suffice it to say that virtually every ecosystem in the Sierra Nevada is impacted by one or more human activities such as impoundment and diversion of water, residential and agricultural development, logging, livestock grazing, suppression of wildfire, air pollution, introduction of non-native plant and animal species, mining, and recreation. To address these impacts, SNEP scientists have developed and evaluated a dozen or so alternative strategies for management and conservation of regional biodiversity.

Here we present one such strategy for conserving native biodiversity based on establishing a Sierra-wide system of Biodiversity Management Areas (BMAs), defined as specially designated public or private lands with an active ecosystem management plan in operation whose purpose is to contribute to regional maintenance of native genetic, species and community levels of biodiversity, and the processes that maintain that biodiversity. The BMA system is located to be representative of biodiversity but is not a comprehensive reserve strategy that in itself can guarantee the viability of the native biodiversity of the Sierra Nevada.

The assumption underlying the BMA strategy is that future land use activities as well as pressures on rural and wilderness areas will only increase the risk of losing native Sierran species and ecosystems. We also assume that the vulnerability of Sierran biodiversity to human activities can be reduced by increasing the amount of land devoted to conservation and management of the native biota. Finally, because managing an area as a BMA may conflict with other social or economic goals, we assume that BMA land should ideally be located as efficiently as possible both in terms of the amount and the suitability of the area that is allocated to this management objective.

The specific goal of the BMA strategy is to design a regional system of managed areas that represents all major Sierran plant community types, which we are using as a coarse surrogate for ecosystems and their component species. A community type is considered represented if some pre-defined fraction of its mapped distribution occurs in one or more BMAs. We use a computer model to produce BMA systems for the Sierra Nevada that represent all community types as efficiently as possible, that is, with minimal land allocation and located in areas with highest suitability for biodiversity conservation goals.

The purpose of this exercise is not to identify a specific set of sites that form the optimal design for Sierran BMA system. Instead, the purpose is to measure the likely dimensions of plausible, alternative BMA systems for the Sierra Nevada and to develop a rationale that would guide others in formulating such a system. Towards this end, we have examined a wide range of possible BMA systems based on different assumptions, constraints, target levels for representation, and priorities.

Specifically, we sought answers to the following questions:

1. What is the minimal area required to represent all Sierran plant community types in BMAs? How does an "optimal" BMA system compare to the existing set of parks, wilderness areas, and reserves in the region?

2. How does the location of BMAs relate to the distribution of areas of special interest that have been identified in other SNEP assessments and biodiversity strategies, in particular, Aquatic Diversity Management Areas, Significant Ecological Areas, and Areas of Late Successional Emphasis?

3. Can a representative BMA system be established on public lands only? If not, what area of private lands is required? How does the area requirement change if lands that are currently administratively withdrawn from grazing and timber harvest are classified as BMA lands?

4. How sensitive is the siting of BMAs to the way in which biodiversity is measured? Specifically, how do solutions to represent plant community types compare to solutions based on representing vertebrate species?

5. Do some general areas emerge from the analysis that appear especially well suited to serve as BMAs?

THE BIODIVERSITY MANAGEMENT AREA CONCEPT

Biological conservation strategies have traditionally centered on biological reserves, where a reserve is "an area with an active management plan in operation that is maintained in its natural state and within which natural disturbance events are either allowed to proceed without interference or are mimicked through management (e.g., most national parks, Nature Conservancy preserves, some USFWS National Wildlife Refuges, research natural areas)" (Scott et al. 1993, 34). Large (e.g., >10,000 ha) reserves are the most common strategy to maintain biotic communities over long time periods in areas undergoing large-scale conversion from wildlands to agricultural and urban systems (e.g., Shafer 1990). In areas of extensive habitat conversion, the design of reserve systems is typically based on a model of reserves as isolated islands of habitat for native species. The viability of a reserve system is gauged based on the size, shape, and connectedness of these remnant habitat areas.

A different situation prevails over much of the Sierra Nevada because a large portion of public and private lands is managed for renewable natural resources such as livestock forage, timber, and recreation. In contrast to largely agricultural or urban landscapes such as the Central Valley or Los Angeles Basin, the prevailing land cover types of the Sierra Nevada are managed forest, rangeland, and alpine ecosystems that sustain many if not most elements of native biodiversity while also supporting natural resource-based economies. In this setting, a BMA system could serve to provide "core habitat areas" of higher habitat quality for many species, sanctuaries for species and habitat types that are especially negatively impacted by human activities, and could possibly serve to buffer populations from unexpected environmental change or unintended consequences of extractive activities on remaining lands.

There is much debate among conservation biologists on the design of conservation land systems in regions such as the Sierra Nevada where the matrix lands (i.e., those outside of BMAs) contribute significantly to maintaining biodiversity. One view holds that nothing short of very large, well-connected wilderness areas can maintain native biodiversity over the long run, particularly if the biota includes wide ranging predators and migratory herds (e.g., Noss 1992). An opposing view is that smaller and more dispersed areas could suffice as long as the matrix lands are well managed for sustained yield of natural resources and the BMA lands are actively managed for native biodiversity (e.g., Alverson et al. 1994). In our opinion, there is not sufficient long term evidence to evaluate the relative merits of these opposing approaches to conserving Sierran biodiversity. The latter view forms the premise for the scenario presented here. That is, we will assume that Sierran biodiversity can be maintained by ecologically sound management of lands managed for renewable resource extraction, in combination with a rationally designed and located set of moderately sized areas managed specifically for native biodiversity. This assumption is most tenuous in areas such as the foothill zone of the western central Sierra Nevada where native habitats are being converted to urban, residential and agricultural purposes. The assumption also may not be sufficient to maintain some wide ranging predators such as the fisher that are especially sensitive to human activities.

Our concept of a BMA is similar to the Diversity Maintenance Areas concept proposed by Alverson and colleagues (1994). Diversity Maintenance Areas are envisioned as national forest lands that have biodiversity as the *primary* management priority. Alverson and colleagues (1994) propose that Diversity Maintenance Areas should be positioned to the degree possible to include existing forest reserve lands, should account for site history and biological legacy, and should be large and designed according to accepted principles of conservation biology. Economic activities including hunting, timber harvest and recreation are allowed on Diversity Maintenance Areas as long as they do not conflict with the primary management goal.

The BMA concept differs from that of Diversity Mainte-nance Areas in that it extends to both private and public lands and across both forest and non-forest habitats. Economic activities can go forward to the extent that they are compatible with the goal of maintaining native biodiversity. Most importantly, each BMA is managed as part of a system of BMAs that are themselves managed to limit the total risk to regional biodiversity through maintenance of a representative amount of all plant communities.

MODEL FORMULATION

The modeling approach is summarized here and presented in detail in appendix 58.1. First we divided the SNEP core area into six separate planning regions whose boundaries were defined by major river drainages (figure 58.1). This division was deemed necessary to capture latitudinal and longitudinal gradients in Sierran habitats, ecosystem processes, plant community composition, and population genetics that are not adequately reflected in current plant community classification systems (Davis and Stoms 1996; Shevock 1996; Millar et al. 1996). We have only analyzed the northern, central, and southern regions (figure 58.1).

In each region we defined a starting BMA system based on maps of land ownership and management. (For example, one alternative considers all parks, designated natures reserves, and ungrazed designated wilderness areas as BMA lands.) Next we established a target level for representing plant community types in BMAs. This level can be set for each individual element, but for simplicity we use the same target level, for example, 10% of the mapped distribution, for every plant community type. We then overlaid the map of existing BMAs on the map of plant community types to determine which types are not adequately represented and how much additional BMA land is needed for each type. This process of assessing vulnerability in relation to land management categories is the essence of gap analysis (Davis and Stoms 1996). We then used a multi-objective siting model to allocate a minimum of new land to BMA status subject to the constraints that all community types must be represented, and that the new BMA areas should be located in areas of highest suitability for BMA status.

We used Calwater planning watersheds as the land units for allocating new BMAs (figure 58.1). These watersheds were delineated by the California Department of Forestry and Fire Protection to support regional planning, and have a minimum size of 3,000 acres. There are 1,785 watersheds in the three regions averaging 3,024 ha (7,470 ac) in size. These watersheds make logical units for BMAs because they are readily located on the ground and are appropriate physiographic units within which to manage ecosystem and hydrologic processes. A single watershed might be sufficient to maintain viable populations of many plant and small animal species (e.g.,

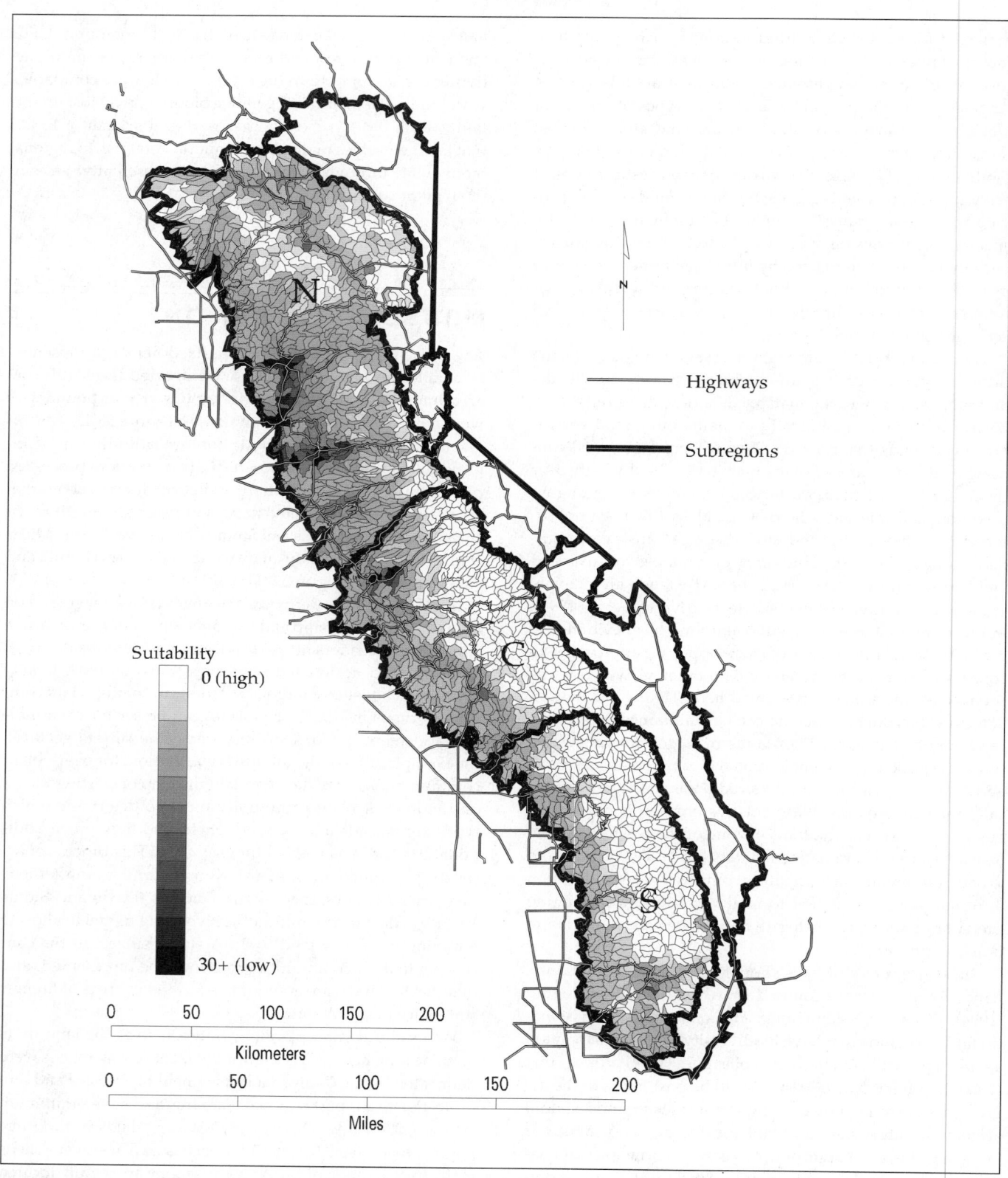

FIGURE 58.1

Outline of the SNEP core area showing the boundaries of the six hydrologic regions (bold lines) and Calwater planning watersheds for the northern, central, and southern hydrologic regions of the SNEP core area. The Watershed Suitability Index (WSI) is shown by a gray scale, where darker areas have a higher WSI (i.e., lower suitability).

Schonewald-Cox 1983), although our approach does not depend on this premise.

Only whole watersheds can be allocated to BMA status in the model. The area of different plant community types in each watershed are calculated by intersecting the watershed boundaries with a map of plant community types (Davis and Stoms 1996). The vegetation map was prepared at 1:100,000 scale for the gap analysis of the Sierra Nevada. In general, there are several vegetation units per watershed.

The suitability of each watershed for BMA management is estimated with a Watershed Suitability Index (WSI), which is the weighted sum of four factors: human population density, the fraction of the watershed affected by roads, the fraction of the watershed that is privately owned, and the degree to which public and private ownership are intermingled. These factors represent many of the known impacts on biodiversity, both in terms of habitat quality and management constraints. Certainly other factors also negatively affect biodiversity but spatial data covering the entire region were not readily available. The higher the value of this index, the less suitable a watershed is for BMA status (figure 58.1). This counter-intuitive scaling of WSI is needed to be consistent with the model's objective function, which seeks to minimize the area and WSI of the model solution. The development of the WSI is explained in greater detail in appendix 58.1.

Because we might consider hundreds of plant community types or species, and we can select from among hundreds of watersheds in a region, the number of potential BMA systems becomes quite large and the problem of selecting the optimal set of watersheds is relatively complex. We have formulated this decision problem as a robust heuristic model where the objectives are to minimize the total area and maximize the suitability of the regional BMA system, subject to the constraint that enough area is selected for all elements to be considered adequately represented (appendix 58.1). We call this model the Biodiversity Management Area Selection (BMAS) model. The output of BMAS is a map of a hypothetical BMA system.

Because the objective of the analysis is not to design a reserve system in the traditional sense, the current version of the BMAS model does not explicitly consider the spatial pattern of the selected watersheds. Based on general principles of conservation biology, one could argue that larger, better connected BMAs would tend to maintain biodiversity better than small, poorly connected systems (Reid and Murphy 1995). On the other hand, there is evidence that populations in several scattered sites are less vulnerable to large-scale environmental disturbances than populations in a single larger site (Harrison and Quinn 1989). Obviously, it would be useful to incorporate spatial considerations in the BMAS model in order to explore these issues more analytically. However, the BMAS model used here provides solutions that are the most efficient solutions only in terms of minimizing the area and WSI for a given set of parameters. Thus the solutions can be considered planning benchmarks in terms of the area re-

quirements for representative BMA systems. Any additional constraints such as spatial design would necessarily increase the area of the solution.

Trade-off Analysis of the BMAS Model

The BMAS model solves a multi-objective decision problem that balances selecting the least area versus minimizing WSI. In general, one would expect a solution weighted towards watersheds with a low WSI to require more area than a solution for which suitability was given less weight. Conversely, one would expect the least area for a solution that ignored WSI. We conducted a sensitivity analysis in the northern hydrologic region to explore the range of feasible solutions generated by varying the weights for the two objectives.

We ran 12 variations of the model that varied in the weighting of area and suitability, in target level for representation, and in the starting BMA system. More specifically, we solved the BMAS model for area only, for WSI only, and for a balanced weighting of area and WSI. Each set of weights was applied to models with a target goal for representation of 10% or 25% for all plant community types, and for an initial BMA system consisting of Class 1 lands or for Classes 1 and 2 lands as defined for SNEP's gap analysis. (See Davis and Stoms 1996; discussions in the next section and in appendix 58.1.)

Solutions based on different weights can be plotted as trade-off curves of total area versus total WSI, which is obtained by summing the WSI values for all selected watersheds. These curves indicate the limits of feasible solutions for the specified target levels and initial conditions (figure 58.2). No solutions will be feasible closer to the origin of the graph. The curves also show that solving for both objectives simultaneously makes a substantial improvement in one objective with only a slight reduction (or trade-off) in the other. For instance, for the model with an initial BMA system of Class 1 and Class 2 lands (C1+C2) and a target level of 25%, the minimum area that can be selected while still meeting the 25% target for all elements is 341,153 ha (figure 58.2). However, the total WSI in that solution is 725. The minimum possible WSI (i.e., the cumulative index of the most suitable set) of selected planning units is 369, but this requires a total area of 444,550 ha to meet the 25% target. The multi-objective solution in the middle of the curve reduces WSI by 41% while only increasing total area by 15,000 ha (4%) relative to the "area only" alternative. Similarly, for a 20% reduction in area, the multi-objective alternative only increases total WSI by 16% relative to the "WSI only" alternative.

Exploring all possible combinations of weights for each set of definitions was beyond the scope of this analysis. It may be possible to improve the multi-objective solution through further moderate adjustments in the weights. However, the trade-off curves in figure 58.2 suggest that the opportunity for improvement is slight, and we expect that any improvement in the objective function will not significantly change the configuration of BMAs that were selected. Based on the

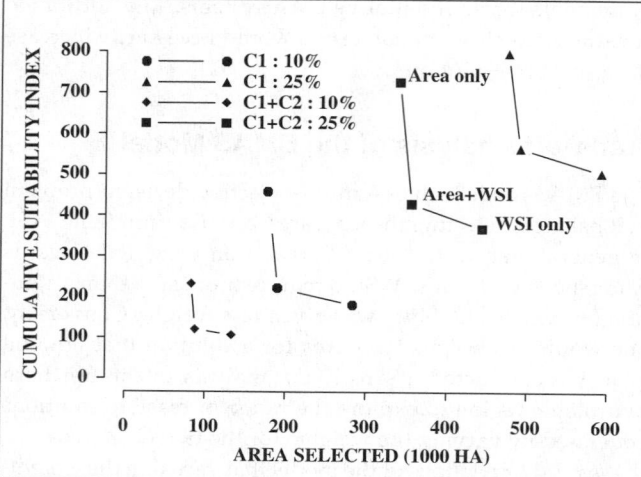

FIGURE 58.2

BMAS trade-off curves for the northern hydrologic region. Curves show the total area and summed Watershed Suitability Index (WSI) for selected watersheds for four different alternatives that vary in target levels (10% or 25%) and starting BMA system (Class 1 (C1) lands or Class 1 and Class 2 lands (C1+C2)). See text and appendix 58.1 for explanation of C1 and C2 lands. For each alternative, three sets of weights were applied: setting the area weight to 0 (WSI only), setting the suitability weight to 0 (Area only), and positive weighting of the two (Area + WSI, with weights scaled to their range of values). Thus the curves show the trade-off between solutions weighted towards area versus suitability for the four models.

sensitivity analysis for the northern region, we selected weights for multi-objective solutions for all of the planning regions. In the results section, all runs were based upon weights for area and WSI that roughly weighted them equally. These weights were determined by trial and error.

SUMMARY OF ALTERNATIVES AND FINDINGS

To answer the five questions listed above in the Problem Statement, we analyzed several dozen alternatives that varied in one or more model specifications. These alternatives can be grouped into six classes that vary in their starting BMA systems, inclusion of private lands, or measure of biodiversity. Within each classes of alternative, we ran the model with targets of 10% and/or 25% representation for two different levels of biodiversity emphasis. These alternatives and their solutions for each hydrologic region are summarized below. Our objective is not to provide a detailed analysis of the specific watersheds that are selected. Instead, we have focused on how much land is required for each alternative and how

that land is distributed among different ownership and management categories. Also, we do not analyze specific plant community types in any detail. Appendix 58.2 provides a full listing of the plant community types and indicates which types required additional area under each model alternative.

Alternative 1

Represent all plant community types in BMAs, ignoring the current distribution of designated conservation lands. Minimize the area and WSI of the solution.

Solving this model alternative reveals the areal requirements of a representative BMA system whose extent and distribution is not tightly constrained by existing land allocations. Essentially, we are posing the question: "If there were no parks, reserves, or other areas in the Sierra Nevada managed primarily for biodiversity, how would one allocate land to a new set of areas in order to most efficiently represent the region's biodiversity?" Private lands were considered available for BMA allocation, although private ownership increases the WSI and thus reduces the likelihood of selection.

We solved this alternative for the central and southern regions based on a 25% target, which is roughly the fraction of those regions that is currently in designated parks, ungrazed wilderness areas, and other conservation areas. This land management category is labeled Class 1 lands by Davis and Stoms (1996) and in appendix 58.1. We do not show the solution for the northern region, where the amount of Class 1 land is only 2% of the region, because ignoring Class 1 lands has a negligible effect on the solution.

For the central region the solution required a total of 348,898 ha (861,778 ac) for the central region, or roughly 21% of the total area (table 58.1; figure 58.3). In the southern region the solution required 361,219 ha (892,212 ac) or 23% of the region (table 58.1; figure 58.3). The reason that the required area is less than 25% of the entire region is because we did not try to represent land use or cover types such as orchards and cropland, water, and barren lands.

In minimizing WSI, the solution favors public lands over private lands, and favors less roaded areas such as parks and wilderness areas over others. Nevertheless, only 22% of the selected area is drawn from existing Class 1 lands, an amount that is in proportion to the extent of Class 1 lands in each region. Although private lands comprise one-third of the two regions, they also contribute around 22% of the selected area. Lands administratively withdrawn from commercial timber harvest on the national forests (Class 2) contribute a relatively small area of the solution (4%), again in proportion to their extent in the two regions.

The selected watersheds are distinctly clustered, notably in east-west lines that span elevational gradients (figure 58.3). This clustering appears to be related to clustering in the watershed suitability index at the scale of larger drainage basins that encompass several to many planning watersheds.

In summary, the results for Alternative 1 suggest that an

TABLE 58.1

Comparison of biodiversity management alternatives.

Alternative	Region	Initial BMA System	Target Level	Suitability Factors	Initial BMA Area (ha)	Vulnerable/Total Community Types	# Selected Watersheds/Total	Additional Watershed Area Selected	Management Class Selected[a]					Class 5 as % of Selected Area	Class 5 as % of Region	Total BMA Area (ha)	BMA as % of Region
									Class 1 Area	Class 2 Area	Class 3 Area	Class 4 Area	Class 5 Area				
1a	Central	None	25%	WSI	0	55/55	100/482	348,898	79,968	15,132	107,848	66,767	79,183	22.7%	33.1%	348,898	21.0%
1b	South	None	25%	WSI	0	65/65	110/527	361,219	81,514	14,256	127,024	60,135	78,290	21.7%	30.5%	361,219	22.7%
2a	North	C1	10%	WSI	43,572	54/59	55/776	189,138	866	15,319	36,119	58,864	77,970	41.2%	47.9%	231,844	10.8%
2b	North	C1	10%	WSI+AD	43,572	54/59	53/776	202,456	349	17,540	33,458	53,116	97,993	48.4%	47.9%	246,028	11.4%
2c	North	C1	25%	WSI	43,572	56/59	123/776	489,326	1,190	38,532	82,363	142,088	225,153	46.0%	47.9%	531,708	24.7%
2d	Central	C1	10%	WSI	385,791	28/55	19/482	67,765	3,882	3,570	24,399	10,998	24,916	36.8%	33.1%	449,674	27.1%
2e	Central	C1	10%	WSI+AD	385,791	28/55	18/482	74,811	3,757	2,577	30,798	5,810	31,869	42.6%	33.1%	460,602	27.8%
2f	Central	C1	10%	WSI+SA	385,971	28/55	18/482	67,298	3,741	3,514	25,103	7,878	27,062	40.2%	33.1%	453,269	27.3%
2g	Central	C1	25%	WSI	385,791	35/55	42/482	179,428	3,367	9,585	49,791	37,457	79,228	44.2%	33.1%	561,852	33.9%
2h	South	C1	10%	WSI	403,500	30/65	23/527	71,785	98	2,963	32,175	10,295	26,254	36.6%	30.5%	475,187	29.9%
2i	South	C1	10%	WSI+AD	403,500	30/65	22/527	83,268	805	2,513	33,676	10,194	36,080	43.3%	30.5%	485,963	30.6%
2j	South	C1	10%	WSI+SA	403,500	30/65	20/527	79,154	805	2,366	34,162	9,698	32,123	40.6%	30.5%	481,849	30.3%
2k	South	C1	25%	WSI	403,500	47/65	56/527	195,886	852	5,284	82,730	24,446	82,574	42.2%	30.5%	598,534	37.7%
2l—Superplan	North	C1	10%	WSI	43,572	54/59	29/251	240,328	837	17,692	46,578	69,319	104,955	43.7%	47.9%	282,116	13.1%
3a	North	C1+C2	10%	WSI	212,456	36/59	25/776	87,461	0	445	15,736	17,652	54,073	61.8%	47.9%	299,917	14.0%
3b	North	C1+C2	25%	WSI	212,456	50/59	86/776	350,999	107	3,659	64,281	87,786	198,932	56.7%	47.9%	563,455	26.2%
3c	Central	C1+C2	10%	WSI	441,484	21/55	15/482	50,818	2,614	1,629	21,000	6,182	23,636	46.5%	33.1%	492,302	29.7%
3d	Central	C1+C2	25%	WSI	441,484	32/55	38/482	154,009	2,740	3,737	42,359	25,601	76,962	50.0%	33.1%	586,406	35.3%
3e	South	C1+C2	10%	WSI	461,789	24/65	15/527	61,451	805	4	18,218	5,943	35,545	57.8%	30.5%	521,495	32.8%
3f	South	C1+C2	25%	WSI	461,789	41/65	47/527	163,167	1,392	670	60,727	16,478	83,750	51.3%	30.5%	622,744	39.2%
4	North	C1+AL	10%	WSI	335,036	36/59	25/776	84,768	0	4,569	10,650	14,135	55,414	65.4%	47.9%	419,804	19.5%
5—pub only	North	C1	10%	WSI	43,572	54/59	204/776	706,426	829	24,630	99,620	102,117	470,390	66.6%	47.9%	740,329	34.4%
6—verte- brates	North	C1	10%	WSI	43,572	300/375	50/776	175,100	94	8,019	28,470	65,432	73,085	41.7%	47.9%	218,672	10.2%

[a]The management classes are:
Class 1: Public or private land formally designated for conservation of native biodiversity.
Class 2: national forest land that is generally managed for its natural values but is not formally designated for conservation of native biodiversity.
Class 3: public land that is generally managed for its natural values, is treated in existing management plans as unsuitable for timber harvest, and may be grazed.
Class 4: Other public lands not included in Classes 1 through 3, mainly multiple-use federal lands.
Class 5: private lands other than those in Class 1.

FIGURE 58.3

Calwater planning watersheds in the central and southern regions showing selected watersheds (dark shading) in the BMAS solution for Alternative 1. The objective is to represent all plant communities over at least 25% of their distribution, ignoring current land ownership and administrative designation. Existing parks and ungrazed wilderness areas are shown for reference (light shading).

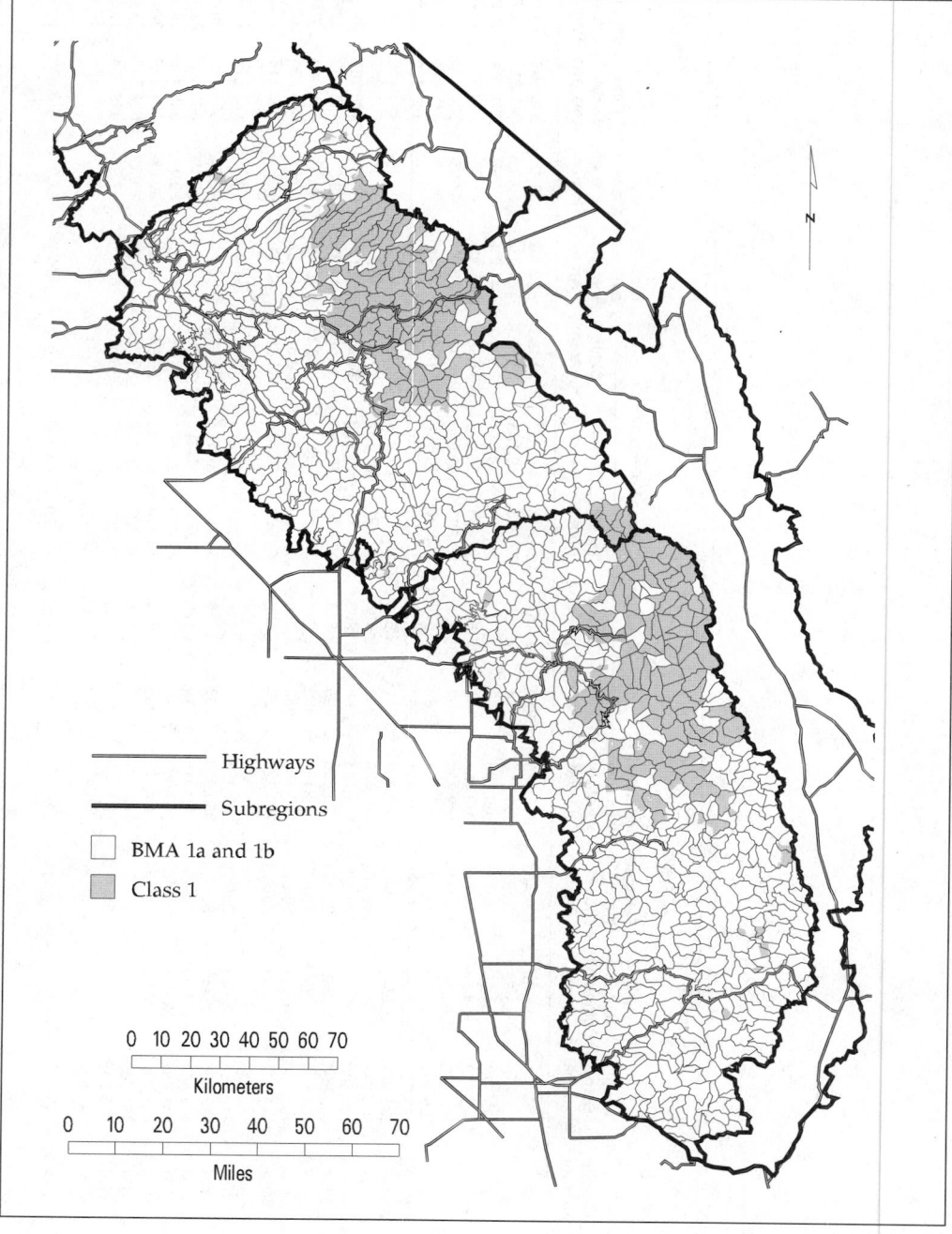

Highways

Subregions

BMA 1a and 1b

Class 1

0 10 20 30 40 50 60 70
Kilometers

0 10 20 30 40 50 60 70
Miles

optimal allocation of lands to BMA status would appear very different from the existing pattern of Class 1 lands in the central and southern regions, mainly in being more evenly distributed across elevation zones. Four-fifths of the BMA land would be selected from public lands. In spite of the fact that Class 1 lands generally have lower WSI values (i.e., are more suitable) than other public lands, only 20% of existing Class 1 lands would appear in the solution because, based on a 25% target, they currently overrepresent many plant community types and underrepresent many others.

Alternative 2

Represent all plant community types in BMAs but treat GAP Class 1 lands (parks, reserves, ungrazed wilderness areas) as the starting BMA system. Find additional area to meet target goals of 10% or 25% for all plant community types, balancing the objectives of minimal area and minimal WSI. Additional variations of the model with a 10% target level include 1) weighing towards Aquatic Diversity Management Areas (ADMAs), 2) weighing towards Significant Ecological Areas (SEAs), and 3) increasing the size of the sites from

Calwater planning watersheds to Calwater Superplanning watersheds (roughly a factor of three times larger).

Given a starting BMA system, this class of models begins with fewer underrepresented types and with smaller areal requirements than alternative 1. The model favors locating additional BMAs on public lands in watersheds with lower population density, road density, and fragmentation of public lands. The 10% and 25% target levels are arbitrary but in our view span the range from a relatively modest to a substantial allocation of lands to BMA status. Model variations 1) and 2) account for other SNEP biodiversity objectives and were run to evaluate the degree to which the objective of representing plant communities in BMAs can be met within areas identified by other SNEP scientists as of special biological significance. Model variation 3) tests the sensitivity of the result to the size of the planning unit, and provides a "large reserve" solution for comparison with the smaller watershed approach.

In the northern region, starting with Class 1 lands, 54 of 59 community types do not meet the 10% target for representation on BMA lands (figure 58.4; table 58.1, Alternative 2a). The 10% solution requires 55 watersheds and a total area that is roughly three-fourths the size of Yosemite National Park. This large areal requirement reflects the small amount of Class 1 lands in the northern region. Selected watersheds exhibit some clustering in the Plumas National Forest near Sierra Valley, in the Tahoe National Forest north of Highway 49, and on private lands in eastern Calaveras County. Despite weighting towards public lands, 41% of the selected BMA area falls on private lands, in order to capture foothill woodland, shrubland, grassland and meadow community types that are almost entirely in private ownership (Davis and Stoms 1996). Only 15,319 ha (18.6%) of the final BMA solution occurs on administratively withdrawn national forest lands.

Figure 58.5 indicates how the management profiles of individual plant community types in the northern region would change under Alternative 2a. Notice that 40% representation is exceeded for 4/59 types, and 25% representation is met or exceeded for 15/59 types. This "excess coverage" is an effect of selecting whole watersheds for BMAs, and applies especially to rare or widely scattered community types. For instance, the bar marked "KP" with nearly 50% representation indicates Knobcone Pine Forest, which occupies only 5 km^2 in the northern region. This is an indirect, and from a conservation perspective some might consider desirable, effect of the model: it tends to be most efficient for widespread types, and tends to provide excess coverage for rare or restricted types.

In assessing aquatic biodiversity in the Sierra Nevada, Moyle (1996a, 1996b; Moyle and Randall 1996; Moyle et al. 1996) identified forty-two watersheds that had unusually high value because they were rich in native aquatic vertebrate species and communities and/or contained particularly rare or unusual biotic elements. He referred to these watersheds as Aquatic Diversity Management Areas (ADMAs) and recom-mended that they be managed for their natural values. Solving the BMAS model with selection weighted towards ADMAs results in a large change in selected watersheds so that 36 of 53 occur in these larger basins that are of special interest in terms of aquatic biodiversity, notably the Cosumnes River Basin and the Middle Fork of the Feather River Basin. There is only a modest change in area (figure 58.4; table 58.1, Alternative 2b). This indicates that there is a good deal of flexibility in selecting BMAs to represent plant community types in this region, especially among publicly owned watersheds where WSI values do not vary greatly from one watershed to the next. Note however that, because ADMAs include private lands, a higher fraction of the solution occurs on private lands.

When the target representation level is raised from 10% to 25%, area requirements for the northern region increase by a factor of 2.56 to cover 56 community types (table 58.1, Alternative 2c). A slightly larger fraction of the solution must come from private lands (46% vs. 41%). The solution, which is comparable to solutions for the central and southern regions under Alternative 1, appears very efficient in the sense that it only requires 24.7% of the total region.

A very different set of solutions is obtained for the central and southern regions (Alternatives 2d through 2k). Because the national parks form a large starting BMA system, fewer plant community types are underrepresented. For example, with a 10% target only 28 of 55 communities are underrepresented in the central region (table 58.1, Alternative 2d), and 30 of 65 communities are underrepresented in the south (Alternative 2h). BMA systems to meet a 10% target require from 27% to 31% of the region. The large "excess" in BMA lands at this target level is due to the fact that a large portion of the distribution of many middle and higher elevation plant community types already falls on Class 1 lands (Davis and Stoms 1996).

The fraction of the solution coming from private lands is very consistent for the different variations on Alternative 2 in these regions (37–44%), and, unlike the northern region, is consistently higher than the private fraction of the total region (33.1% of the central region and 30.5% of the southern region). This reflects the fact that a disproportionate share of the types requiring additional acreage occur largely on private lands.

As in the northern region, the BMA solutions weighted towards ADMAs (Alternatives 2e and 2i) require slightly more land than those weighted only by WSI (table 58.1; figure 58.6). However, the BMA solution for plant communities is not as flexible as that for the northern region. In the central region, only 7/18 of the selected watersheds actually fall within ADMAs, notably within the lower Merced River Basin. Furthermore, 10/18 watersheds in Alternative 2a are selected in both the WSI-weighted and ADMA-weighted models. This is because the ADMAs for the central region are mainly middle and high elevation drainages that contain plant community types that are already well represented. The situation is even

FIGURE 58.4

Selected watersheds in the northern region for Alternatives 2a (hatched areas) and 2b (shaded areas). Both alternatives start with Class 1 lands, provide at least 10% representation, and account for both area and Watershed Suitability Index (WSI). Model 2b is weighted towards Aquatic Diversity Management Areas (ADMAs), which are outlined with heavy lines.

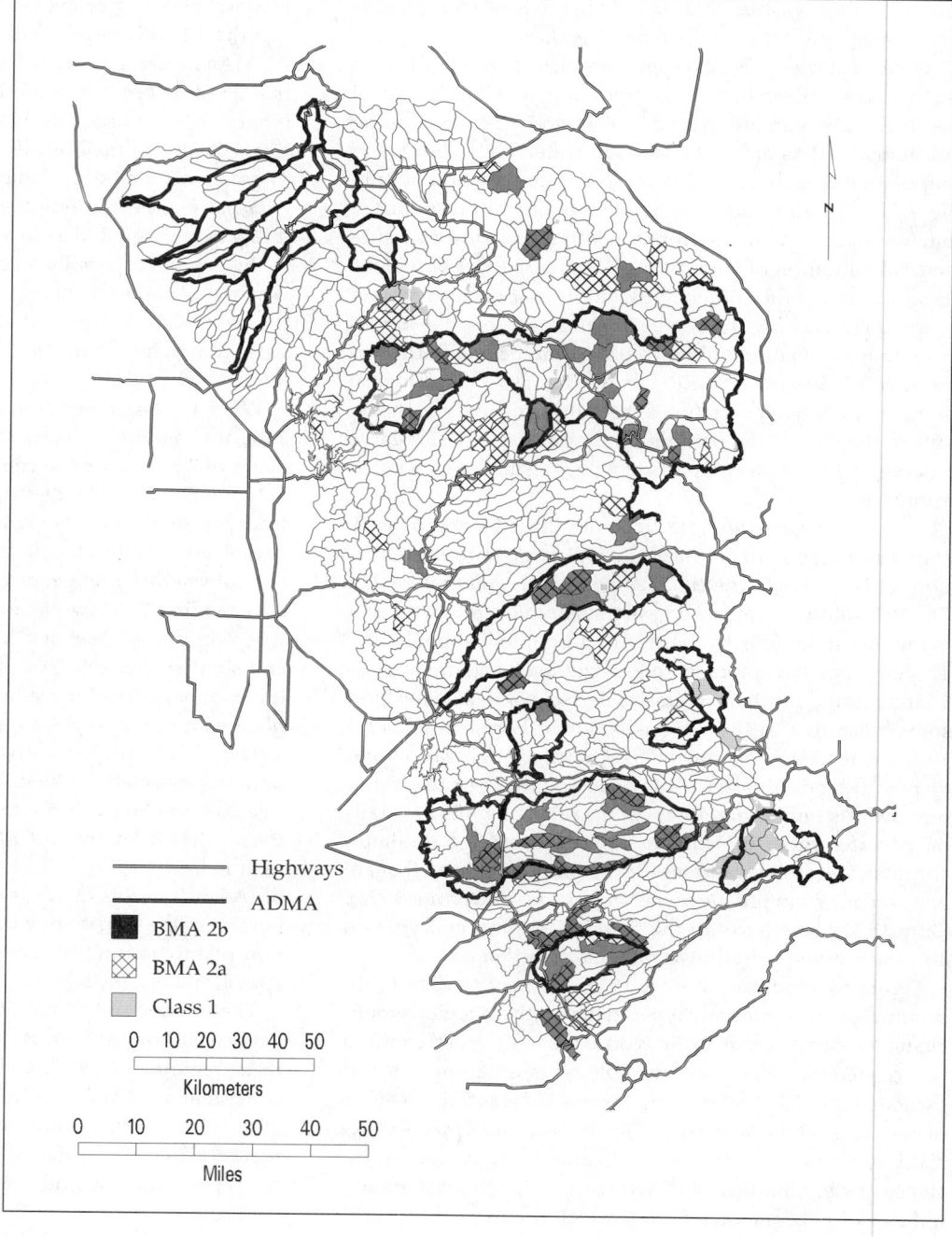

more extreme in the southern region, where only 6/23 selected watersheds fall within ADMAs (figure 58.6).

Millar et al. (1996) identified Significant Ecological Areas (SEAs) in the Sierra Nevada that were distinguished by containing unusually rare, diverse, or representative components of native biodiversity. Solving the BMAS model for the central and southern regions with selection weighted towards SEAs results in a somewhat different set of selected watersheds (figure 58.7, Alternatives 2f and 2j). The degree of co-location of BMAs and SEAs is limited, however, because SEAs were only mapped on public lands, whereas many of the underrepresented types occur mainly on private lands.

Meeting the 25% target requires 561,852 ha and 598,534 ha in the central and southern regions, respectively (table 58.1, Alternatives 2g and 2k). The additional area required increases by a factor of roughly 2.6 compared to the 10% solution, while the total area in BMA status only increases by a factor of 1.25. Unlike the northern region, the area requirement increases only by a factor of 1.25 because of the excess coverage at 10%.

We examined the effect of increasing the size of BMAs by

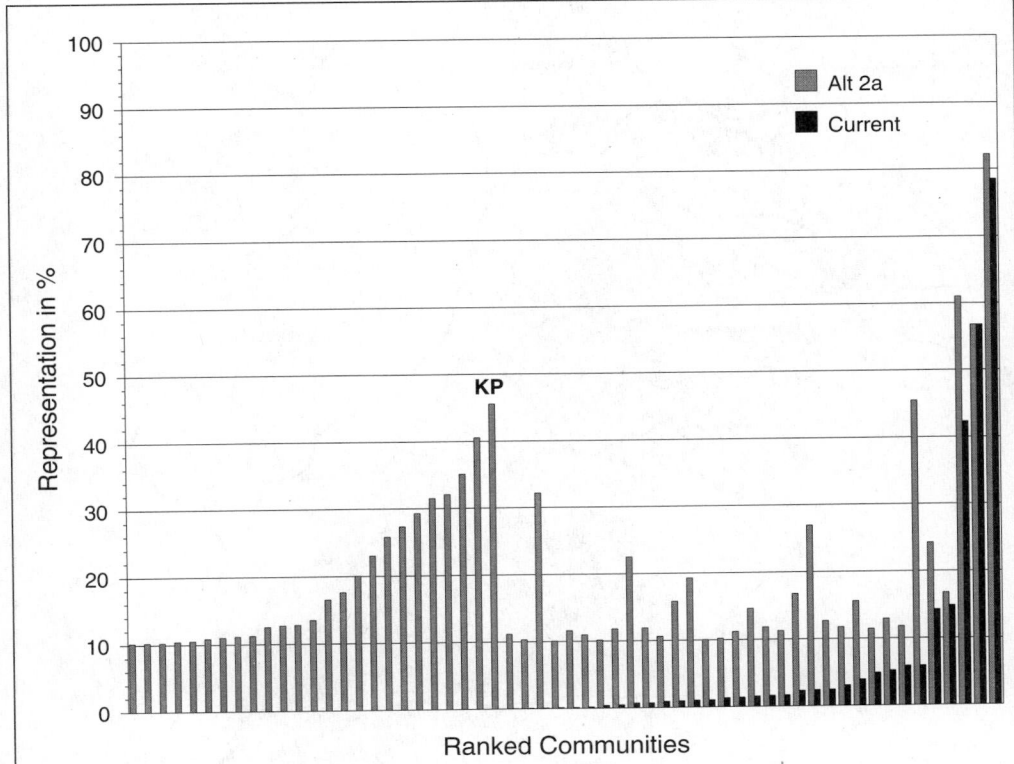

FIGURE 58.5

Representation of 59 plant community types in the northern region in existing BMAs (black bars) and in the BMAS solution to Alternative 2a (gray bars). Community types are ranked according to current representation on Class 1 lands. Ranks 1 through 25 have percentages close to zero, so only the BMAS solution is visible on the chart. "KP" indicates the Knobcone Pine Forest community, which occupies only 5 km^2 of the region.

using Calwater "superplanning watersheds" instead of planning watersheds as units of analysis. These superplanning watersheds are aggregates of the Calwater planning watersheds, which are the lowest level in the Calwater hierarchy. The average size of the superplanning watersheds is 8,565 ha (21,156 ac) or three times that of the average planning watershed.

The BMAS solution for the northern region changes considerably when superplanning watersheds are used as candidate BMA sites instead of the smaller planning watersheds. Twenty-nine of 251 watersheds were selected to meet the 10% target level (figure 58.8). Thus the number of new BMAs is reduced from 55 to 29 when the size of the sites is increased three-fold. While the spatial distribution of the two sets of sites is similar, only 19 of 55 planning watersheds selected in Alternative 2a are also selected in Alternative 2l. As expected, the area requirement increases as the size of the planning site increases. In this case, the total area increased by 51,190 ha (126,439 ac). The fraction of the solution from private lands increases slightly from 41.2% to 43.7%.

Alternative 3

Represent all plant community types in BMAs, but treat GAP Class 1 and Class 2 lands (forest service lands that are administratively withdrawn from timber harvest and grazing based on current allotment boundaries and mapped land suitability classes) as the start-

ing BMA system. Find additional area to meet target goals of 10% and 25%, balancing the objectives of minimal area and WSI.

The gap analysis of the Sierra Nevada (Davis and Stoms 1996) showed that lower to mid-elevation forest community types are not well represented in Class 1 lands, but that for a variety of reasons related to environmental and biological concerns, extensive tracts of these forest types are classified in current forest plans as unsuitable for intensive timber harvest and also fall outside of grazing allotments. These conservation lands are referred to as Class 2 lands by Davis and Stoms (1996) and in appendix 58.1. We considered one class of model alternatives in which these Class 2 lands were included with Class 1 lands as a starting BMA system for the region. This reduces considerably the number of underrepresented types and area requirements, and places greater emphasis in terms of area requirements on non-forest community types, especially on private lands at lowest elevations.

Starting with Class 1 and Class 2 lands in the northern region, 36/59 types do not meet the 10% target, compared to 54/59 when starting with only Class 1 lands (table 58.1, Alternative 3a). The solution requires 87,461 ha in 25/776 watersheds, which is less than half as much area as Alternative 2a. However, 62% of the area is on private lands, and the overall BMA system requires nearly 70,000 ha more than the solution that begins with Class 1 lands only. This indicates the fact that Class 2 lands in this region provide excess coverage

FIGURE 58.6

Selected watersheds in the central and southern regions for Alternatives 2d and 2h (hatched areas), and for Alternatives 2e and 2i (shaded areas). All of the alternatives start with Class 1 lands, provide at least 10% representation, and attempt to minimize both area and Watershed Suitability Index (WSI). Models 2e and 2i are weighted towards Aquatic Diversity Management Areas (ADMAs), which are outlined with heavy lines.

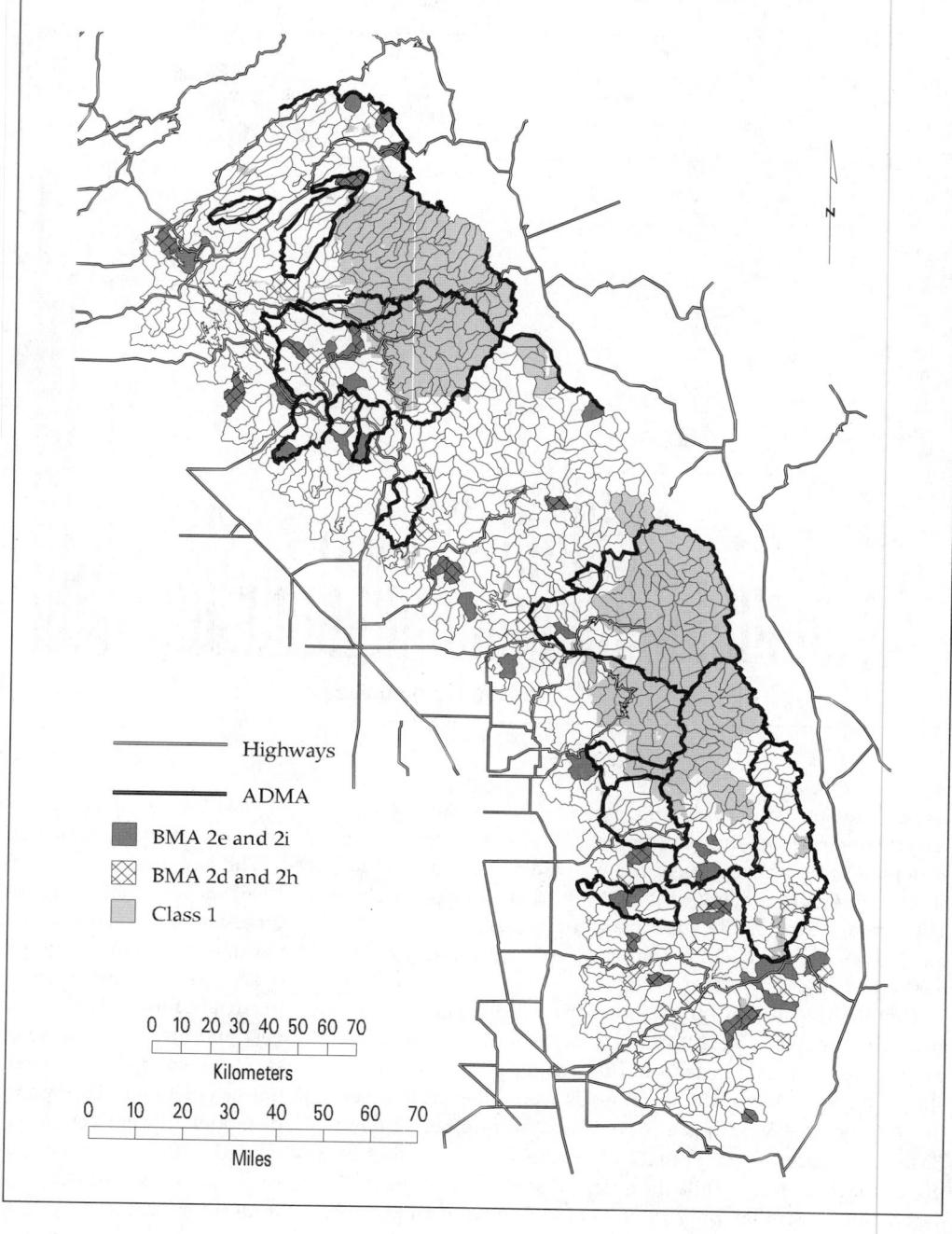

for a few community types, notably middle elevation forest types, but provide little or no representation for many other community types, especially hardwood forest, foothill woodland, and chaparral types. The solution becomes more efficient as the target level is raised (Alternative 3b). Thus the area required to meet a 25% target, while still predominantly falling on private lands, is only 1.5% more of the region than the model that starts with Class 1 lands only.

There is not much Class 2 land in the central and southern regions, so that the solutions for Alternative 3 are closer to those for Alternative 2 than in the northern region. In the central region (Alternative 3c), starting with Class 1 and Class 2 lands and a 10% target, the number of underrepresented types drops from 28/55 to 21/55, but the final area in BMA lands increases from 449,674 to 491,979, and an increasing fraction of new BMA land is located on private lands. The trend is similar for the 25% target (Alternative 3d). These findings reflect the fact that most Class 2 lands in the central region support plant community types that are already well represented in the national parks. Including Class 2 lands in the

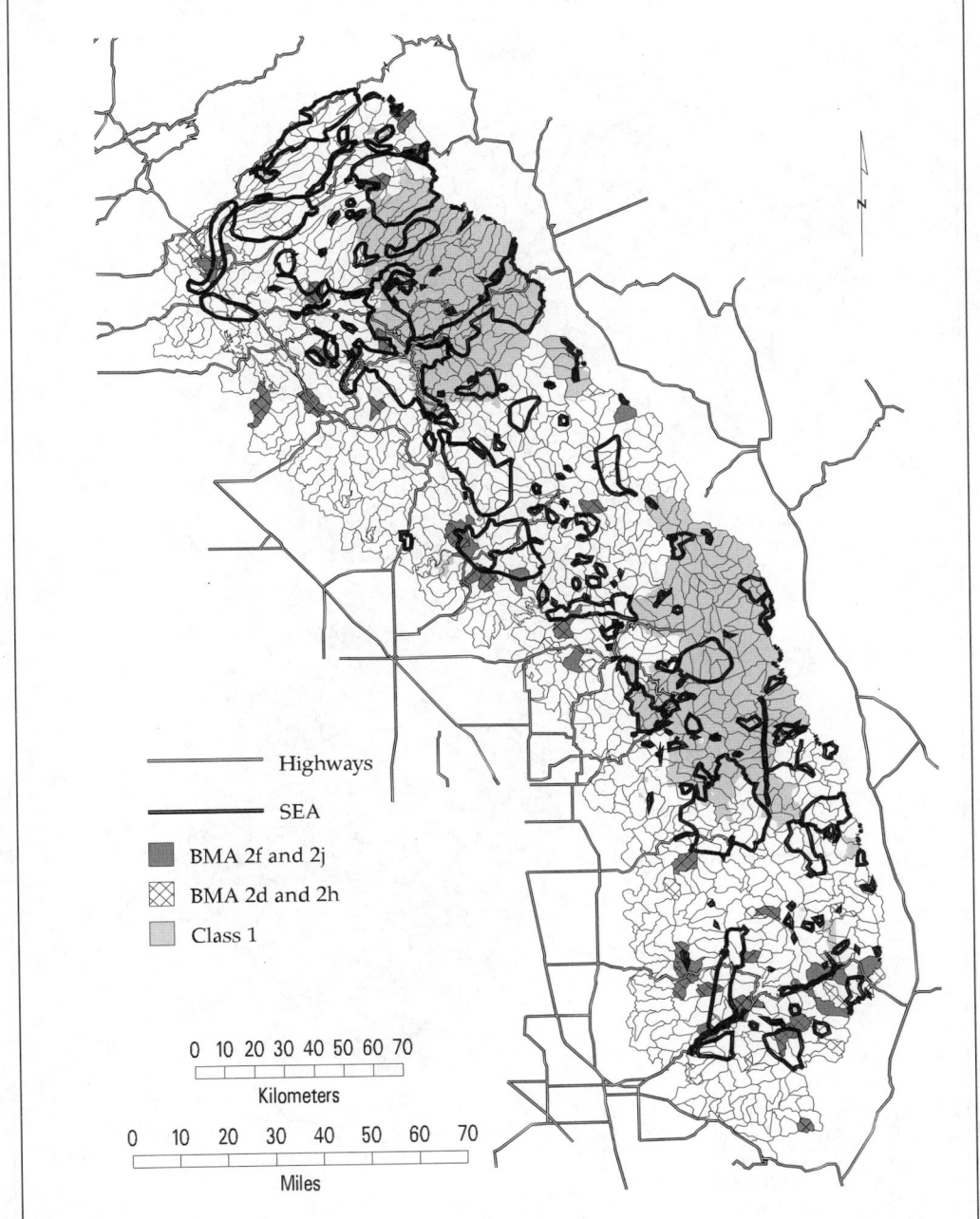

FIGURE 58.7

Selected watersheds in the central and southern region for Alternatives 2d and 2h (hatched areas), and for 2f and 2j (shaded areas). All of the alternatives start with Class 1 lands, provide at least 10% representation, and minimize area and Watershed Suitability Index (WSI). Models 2f and 2j are weighted towards Significant Ecological Areas (Millar et al. 1996).

Highways

SEA

BMA 2f and 2j

BMA 2d and 2h

Class 1

0 10 20 30 40 50 60 70
Kilometers

0 10 20 30 40 50 60 70
Miles

starting BMA system for the southern region reduces area requirements slightly (Alternatives 3e and 3f). Once again, however, including Class 2 lands has little effect on the amount of private land in the final solutions. In fact the contribution from private lands actually increases in solutions to Alternative 3 for the central and southern regions.

Alternative 4

Represent all plant community types in BMAs, but treat GAP Class 1 and ALSE lands as the starting BMA system. Find additional area to meet target goals of 10% and 25%, balancing the objectives of minimal area and WSI.

In their assessment of late seral/old growth forests of the Sierra Nevada, Franklin and Fites-Kaufmann (1996) delineated Areas of Late Successional Emphasis (ALSEs), which were large landscape units on the public lands that contained

FIGURE 58.8

Selected watersheds in the northern region for Alternatives 2a (hatched areas) and 2l (shaded areas). Both alternatives start with Class 1 lands, provide at least 10% representation, and account for both area and Watershed Suitability Index (WSI). The BMA sites for Model 2a are planning watersheds (thin lines), and for Model 2l are superplanning watersheds (heavy lines).

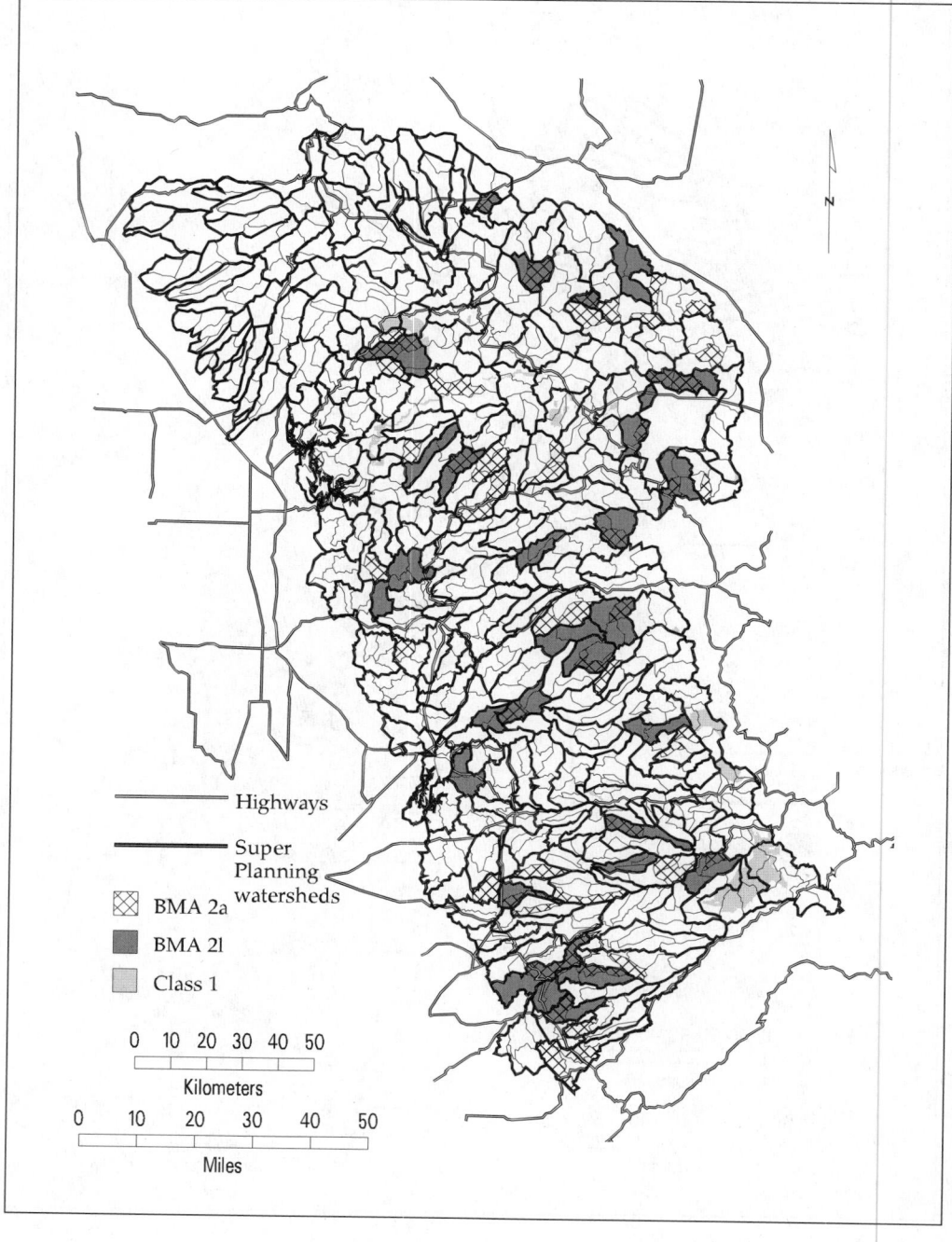

Highways

Super Planning watersheds

⊠ BMA 2a

■ BMA 2l

□ Class 1

0 10 20 30 40 50
Kilometers

0 10 20 30 40 50
Miles

one or more extensive tracts of late seral/old growth forests. One strategy that they considered for conserving late seral/old growth forests in the Sierra Nevada was active management of designated ALSEs for late seral conditions.

This BMAS alternative tests the effect of adding ALSE lands to a starting BMA system that also includes Class 1 lands. (We should note that ALSE boundaries were slightly modified subsequent to our analysis, which was based on draft maps as of 1 Oct, 1995.) Defining ALSEs as BMAs presumes that they could also be managed to maintain non-forest plant communities occurring in those areas, as well as to maintain the compositional and structural components of forest types that would be the main focus of management practices.

Only the northern region was analyzed. The total area mapped as ALSEs in this region is 291,464 ha (13.5% of the study region) compared to 43,572 ha in Class 1 lands. Given a 10% target, the ALSEs provide excess representation for selected forest types, notably Sierran mixed conifer and Red fir community types. Because ALSEs were not aimed at woodland, shrubland and herbaceous types, the coverage of these

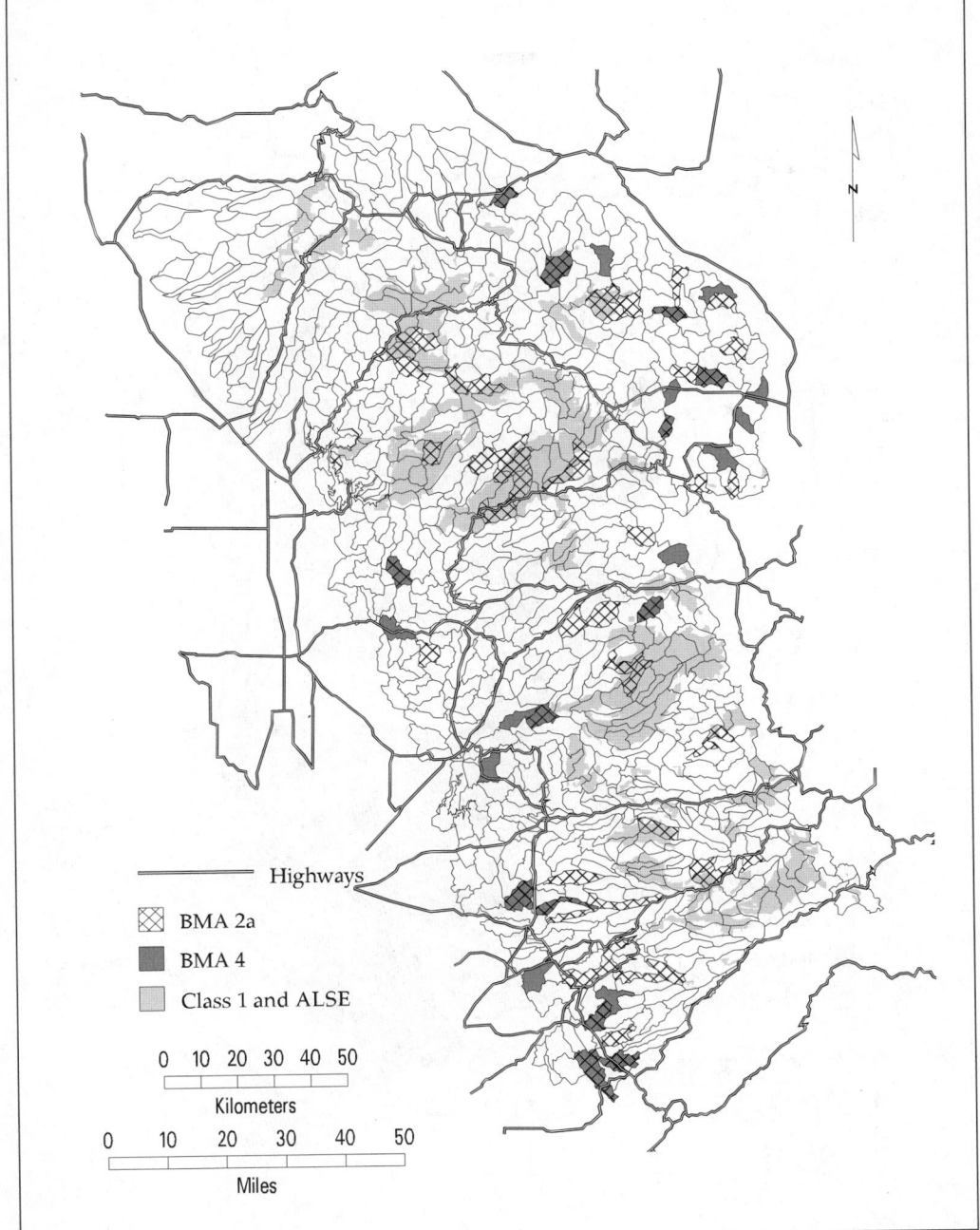

FIGURE 58.9

Selected watersheds (dark areas) in the northern region for Alternative 4. This alternative starts with Class 1 lands and ALSEs (light shaded areas), provides at least 10% representation, and accounts for both Watershed Suitability Index (WSI) and area. Selected watersheds for Alternative 2a (hatched areas) are shown for comparison.

Highways

⊠ BMA 2a

■ BMA 4

▨ Class 1 and ALSE

0 10 20 30 40 50
Kilometers

0 10 20 30 40 50
Miles

types is largely unaffected by adding ALSEs to the starting BMA system. Thus the total area required for a 10% target is greater by a factor of 1.8 for this Alternative (figure 58.9; table 58.1, Alternative 4a) than the area of the solution that starts with only Class 1 lands (Alternative 2a). The selected watersheds fall almost entirely at lower elevation in the northeastern and southwestern portions of the region. Two-thirds of the additional area is selected from private lands.

Alternative 5

Represent all plant community types in BMAs, treating GAP Class 1 areas as the starting BMA system, but with the added constraint that the solution be comprised entirely of public lands. Find additional area to meet target goals of 10%, or as close to 10% as possible, for all plant community types, balancing the objectives of minimal area and WSI.

This alternative aims to build a representative BMA system from the public land base. We know from the gap analy-

FIGURE 58.10

Selected watersheds (shaded areas) in the northern region for Alternative 5. This alternative starts with Class 1 lands, provides 10% representation or as close to that level as possible, accounts for both area and Watershed Suitability Index (WSI), but only public lands contribute toward representation targets. The solution when private lands are eligible (Alternative 2a) is shown as hatched areas.

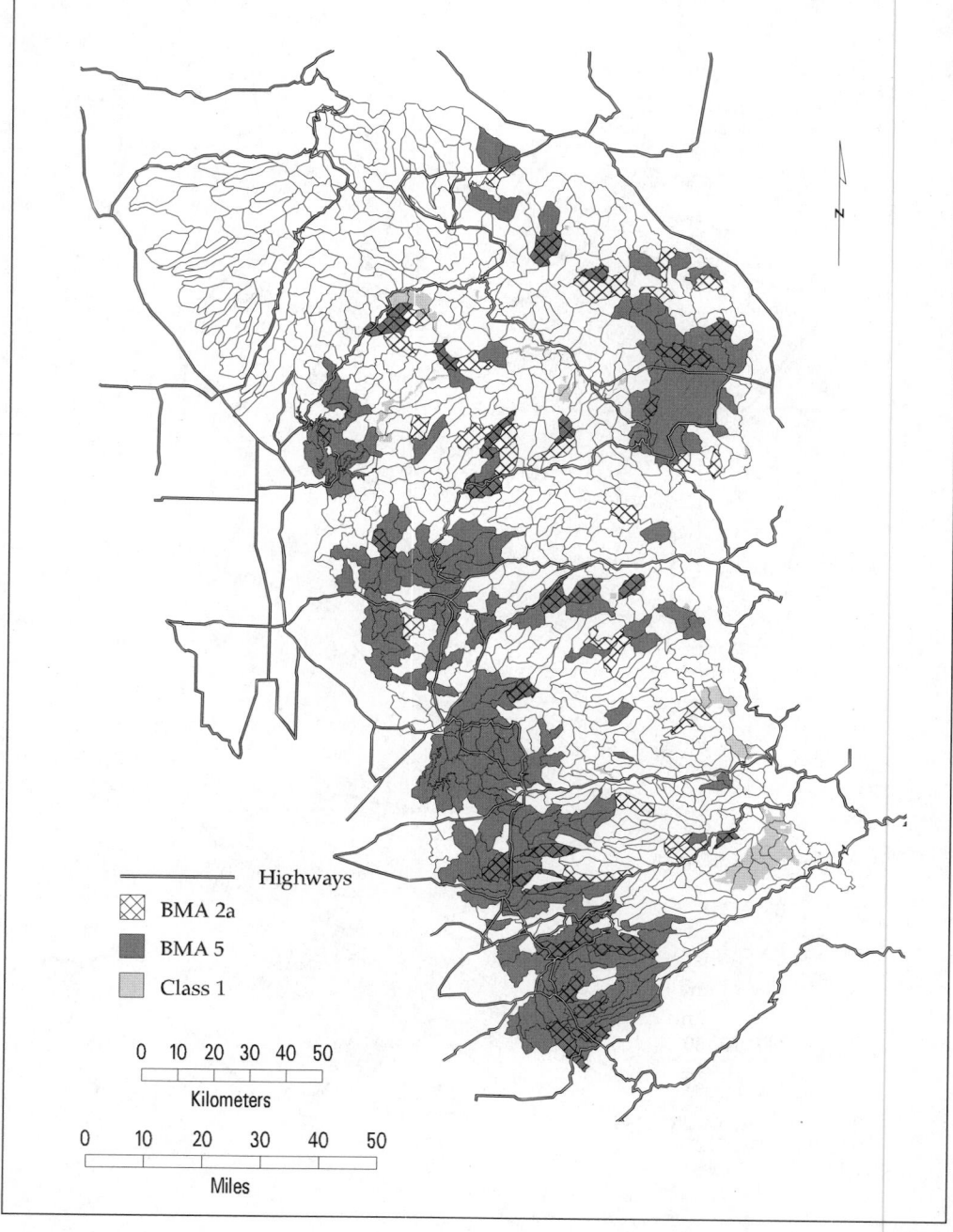

Highways

⊠ BMA 2a

■ BMA 5

▨ Class 1

0 10 20 30 40 50
Kilometers

0 10 20 30 40 50
Miles

sis data that several foothill types have less than 10% of their distribution on public lands. This model effectively allocates all of the public land in these types to BMAs, irrespective of the suitability of the watersheds, still minimizing area and WSI for the entire set of BMAs. Although private lands do not contribute towards representation of plant communities in this alternative, watersheds with private lands can still be selected if any public lands occurs within them.

This alternative results in selecting a very large number of watersheds at lower elevations on both western and north-

eastern sides of the northern region in order to accumulate public lands with underrepresented types (figure 58.10). Over 1/3 of the land area in the region would have to be allocated to BMAs. This result certainly highlights the extremes that would be required in order to focus Sierran biodiversity management and conservation entirely on public lands if entire planning watersheds were the basic management units.

Alternative 6

Represent all vertebrates in BMAs, treating GAP Class 1 lands as the starting BMA system. Based on equation 3) in appendix 58.1, find additional acreage to meet a target goal of 10%.

This alternative was developed to test how much the BMA solution might change if a different measure of biodiversity were applied. We selected terrestrial vertebrates because their distributions are better known and more readily modeled than most other organisms. The distributions used in this analysis were produced using the gap analysis method as described by Hollander et al. (1994). The method entails intersecting coarse range maps (presence or absence in 7.5-minute USGS quadrangles) with maps of suitable habitats. Habitat maps were derived by re-classifying the vegetation data in the gap analysis database into the habitat types used in the Wildlife Habitat Relationships (WHR) system (Mayer and Laudenslayer 1988). WHR rates the suitability of general habitat types as well as structural classes within habitat types for breeding, feeding, and resting activities of all native vertebrates.

We applied a regionalized version of WHR that was modified from the original version by David Sterner and David Graber to apply more specifically to the Sierra Nevada Region. Our predicted vertebrate distributions were based on habitat type, except for forest types where we also subdivided the habitat into general size classes. Structural information was obtained from USFS timber strata maps, SNEP's Late seral/Old growth forest database, or in the absence of other information, by interpretation of recent air photos. (As it turned out, at the scale of the GAP vegetation map, there was very little difference between distributions predicted with or without forest structural information, so we will forego a more detailed discussion of this aspect of the modeling.) A species was predicted to be present if the mapped habitat was within the range of the species and of at least moderate suitability for breeding.

The main differences between using plant community types and vertebrate distributions predicted from habitat types derived from the vegetation map are 1) the use of vertebrate range limits (thus different species may occupy the same habitat type but in different parts of the Sierra), and, 2) the use of WHR habitat types, which are generally aggregates of the plant community types (especially the shrubland types).

Before reporting on the BMAS model results for vertebrates, it is also useful to examine how vertebrates are represented in the model results for plant community types, i.e. a "sweep analysis." To do this we overlaid the hypothetical BMAS system from Alternative 2a on the vertebrate distributions and tallied the percent of each species' distribution that fell within existing Class 1 or new BMAS areas. Of 375 native vertebrates predicted to occur in the northern region, 302 species were represented at the 10% target level or higher, and all species were represented over at least 7.5% of their mapped distribution. The number of species represented compared to the to-

tal in the group is 168/216 for birds, 85/97 for mammals, 19/26 for amphibians, and 30/36 for reptiles.

The BMAS solution to represent 300 vertebrates in the northern region that do not meet a target level of 10% requires 218,672 ha, almost exactly the same area as the solution for plant communities (Alternative 6 versus Alternative 2a). However, the spatial pattern of the vertebrate solution overlaps only slightly with the plant community solution (figure 58.11). Only 12 watersheds are common to both solutions, and there appears to be more grouping of the watersheds selected for vertebrates, for example up canyons and across elevational gradients. Although there are many more vertebrates than plant communities, the predicted distributions of most vertebrates are broader than the distributions of plant community types. Thus watershed selection for vertebrates is driven more strongly by WSI than it is for plant community types.

DISCUSSION

Weaknesses and Limitations of the Approach

Our stated objectives for this scenario were to measure some likely dimensions of plausible BMA systems for the Sierra Nevada. To do this we formulated and applied an optimization model to produce BMA systems that represent all biodiversity elements in BMA sites using as little area as possible given the additional objective of selecting the most suitable sites. Following the logic of the gap analysis assessment, we have simply identified which types might be vulnerable given the geography of *permitted land use* in the region, rather than actual land use. We have not attempted to project possible future trends in regional biodiversity under existing or alternative land management systems.

Most previous applications of siting models to conservation planning have focused on designing a system to efficiently represent biodiversity and have not accounted for site suitability. Margules et al. (1991) argued that biological conservation should aim for sites of the highest biological importance, irrespective of other social or economic considerations. This argument may be compelling when the sites are very dissimilar from one another, for instance when the planning regions span biogeographic regions (e.g., national surveys) and when the planning sites are relatively large (e.g., hundreds of thousands of hectares). Our approach is tailored towards more regional or local scales where many sites may have very similar biota and differ mainly in their suitability for conservation management. The data that we used were tailored to a regional analysis and would not be appropriate for more detailed, local applications.

A BMAS solution that is optimal for minimizing area and WSI may not be optimal with respect to other design criteria, for example, political feasibility or economic cost. The model is useful for establishing minimal area requirements and in

FIGURE 58.11

Selected watersheds in the
northern region for
Alternatives 2a (hatched
areas) and 6 (shaded areas).
Both alternatives start with
Class 1 lands, provide at
least 10% representation,
and account for both area
and WSI. Alternative 2a
represents plant community
types, while Alternative 6
represents all terrestrial
vertebrate species.

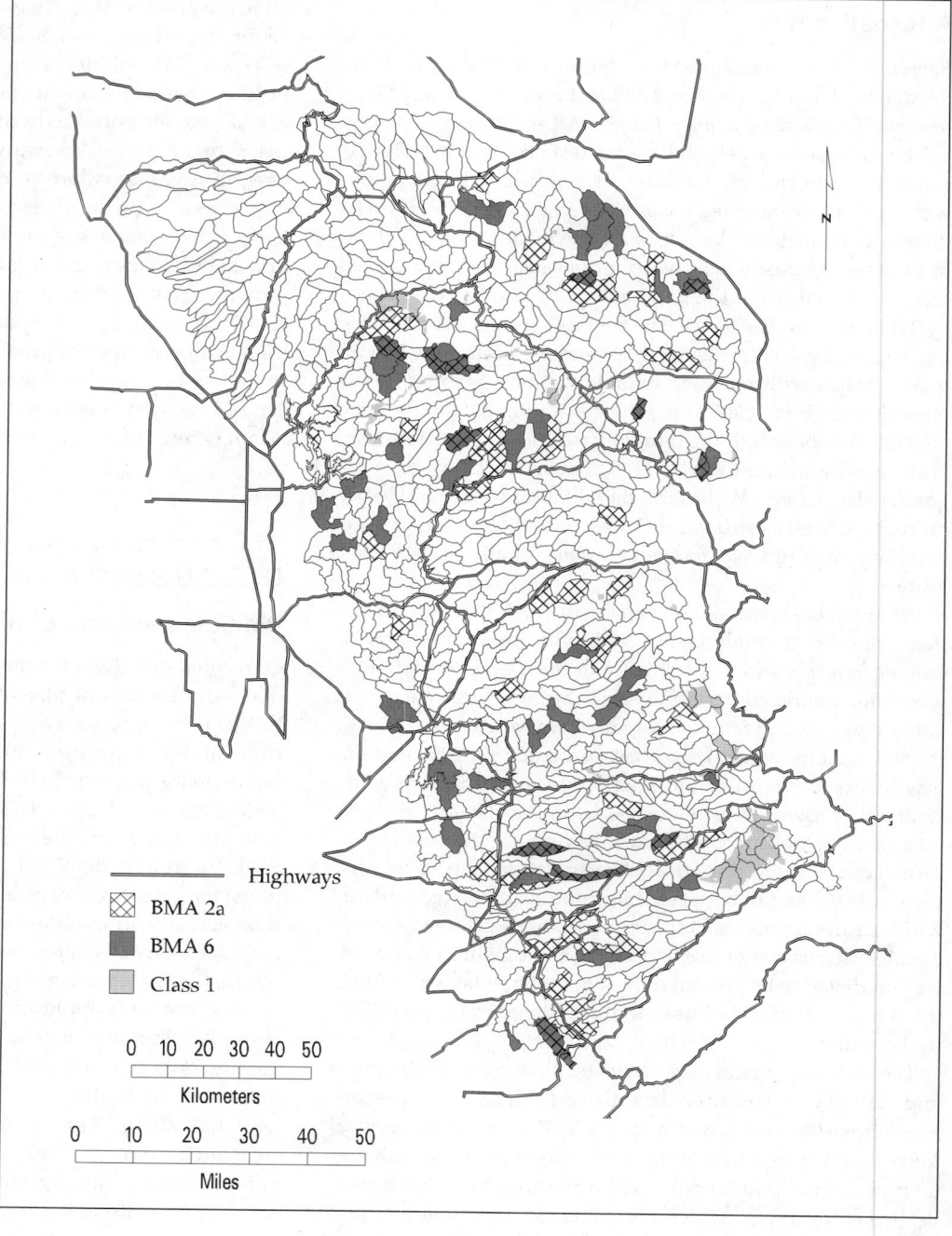

highlighting some areas that would appear to be good sites
for a regional system. However, it does not consider many
local political, economic, or biotic perspectives.

The BMAS model is relatively simple and straightforward,
but it still requires that the user specify a weight for the area
term and for each variable used to measure suitability, as well
as a target level for each biodiversity element. Solutions could
be very sensitive to weights applied to each term in the model,
although there appears to be only modest sensitivity in this
specific application. For example, the analysis of trade-offs

between area and WSI indicates that area requirements might
differ by only 20-30% between the model extremes where the
weight for WSI is set to zero and those where the weight for
area is set to zero.

Obviously, solutions will vary considerably depending on
how one sets the target levels for each biodiversity element.
Setting credible target levels is the most difficult aspect of the
BMAS model. Given that the ultimate purpose of a BMA system is to contribute to maintaining regional biodiversity, one
should probably relate a change in the level of an element's

representation in a BMA system to a change in the likelihood of maintaining that element in the region over some specified time period. We have not attempted to do this kind of viability analysis for several reasons. First, the concept of viability is more readily applied to single elements such as a species or a specific component of an ecosystem (e.g., forest structure or chaparral fire return interval) than it is to a whole assemblage of species or ecosystem types. Secondly, the relative merits of one target level versus another could only be measured based on modeling ecological processes over both BMA and non-BMA lands. Such modeling could be extremely informative but was beyond the scope of our analysis. Thirdly, we felt that a viability analysis would focus too much attention on the specific sets of watersheds that comprised the model solutions.

We used plant community types and wildlife habitat types as our elements of biodiversity, but other mapped variables could also be applied, for example, an alternative vegetation classification system, physical environmental types, or species localities. Similarly, other criteria could be used to measure suitability. Based on results that we obtained using different biodiversity and suitability measures, site selection could be very sensitive to the choice of measures, however the total area selected and general distribution of the solution among biotic zones will remain fairly constant.

We did not explore how sensitive our results could be to errors in the mapping of plant community types or predicted vertebrate distributions. Davis and Stoms (1996) discuss some of the sources of error in the vegetation map, but to date no quantitative map accuracy analysis has been conducted. We expect that map errors affect the set of watersheds selected in the different model alternatives more than the total area or general spatial pattern of the model solutions. We also expect the errors to be most significant for very rare or localized vegetation types whose distribution was not as reliably mapped at the coarse resolution used for SNEP's gap analysis.

A more sophisticated siting model would also account for the neighboring watersheds in the site selection process. For example, it may be desirable to cluster BMAs into larger blocks. Also, one may want to adjust each watershed's suitability to incorporate the suitability of adjacent watersheds. The latter could be readily accomplished. The former objective of spatial clustering is a more difficult problem that we are currently pursuing.

Our analysis is limited by the fact that we have treated input biotic and cultural factors as static when in fact they can be very dynamic, even at the relatively coarse spatial scale of the Calwater planning watersheds. Any local change in plant community distributions or in watershed suitability could lead to a different optimal configuration of BMAs in the region. Although beyond the scope of our analysis, it is certainly possible and desirable to incorporate expected changes in biological distributions and human activities into the model. Even accounting for such dynamics, however, does not address the more complex problem that implementing a

system of BMAs would necessarily be a staged, locally adaptive process. That staging could also affect how optimal any particular solution was. These concerns should serve to re-emphasize that the main value of the model is as a tool for ongoing analysis and evaluation of conservation strategies at the regional level.

Answers to Key Questions

Question 1

What is the minimal area required to represent all Sierran plant community types in BMAs? How does an "optimal" BMA system compare to the existing set of parks, wilderness areas and reserves in the region?

If one ignores current land ownership and management designations and sets out to represent biodiversity in a BMA system based on Calwater planning watersheds, an efficient system requires land in direct proportion to the target level, at least over the range of target levels examined in this study. In other words, it takes roughly 10% of the region to meet a 10% goal, and 25% of the region to meet a 25% goal. The pattern of selected watersheds is very different from the current distribution of parks and wilderness areas, which are concentrated at middle and high elevations in the central and southern portion of the range. An efficient BMA system to meet a 25% target for all community types would require only slightly more land than existing parks and reserves, but this system would require much more even dispersal of BMAs from north to south and across elevations and land ownerships than occurs in the existing situation.

Although Class 1 lands occupy 15% of the combined north, central and southern regions, only 5/59 plant community types exceed a 10% target level in the northern region. Starting with Class 1 lands has little effect in the northern region, where efficient BMA systems still require roughly 200,000 ha to represent all plant community types.

Despite their large size, Yosemite and Sequoia-Kings Canyon National Parks do not encompass the full suite of plant community types that occur in the central and southern Sierra. Roughly half of the native plant community types in these regions do not meet or exceed a 10% target. Meeting that target would require a minimum of roughly 150,000 ha of additional BMA land, 30% of which is currently privately owned. A similar proportion would come from Class 3 lands (mainly national forest lands in grazing allotments and outside of areas suitable for timber harvest).

Increasing the size of the BMA sites by a factor of three from planning watersheds to superplanning watersheds has a surprisingly large effect on the distribution and areal efficiency of the solution, for example, increasing the area for a 10% target by 27%. This illustrates both the sensitivity of the model results to the choice of planning sites and also the trade-off between increased BMA size and decreasing efficiency for representing regionally dispersed elements of biodiversity.

Question 2

How does the location of BMAs relate to the distribution of areas of special interest that have been identified in other SNEP assessments and biodiversity strategies, in particular, Aquatic Diversity Management Areas, Significant Ecological Areas, and Areas of Late Successional Emphasis?

Solutions to the BMAS model show only a modest amount of overlap with other SNEP biodiversity strategies unless the model weights are set to favor ADMAs, SEAs, and ALSEs. In the northern region, the pattern of biodiversity provides sufficient flexibility to find solutions of roughly similar area that also favor these areas. For example, 37% of the BMA area in alternative 2a occurs within ADMAs, but this doubles to 76% when the suitability index was weighted for ADMAs. In the central and south regions, many ADMAs are located at higher elevations and on public lands and thus do not supply a representative set of plant community occurrences to draw from in meeting BMAS objectives. Thus the proportions of overlap of BMAs and ADMAs in the central and southern regions are only 38% and 21%, respectively, for models weighted towards ADMAs. Similarly, even when the suitability index is weighted to favor SEAs, the overlap of BMAs with SEAs is only 27% in the central and 16% in the southern region.

ALSEs were developed on public lands to conserve late seral forest structure, especially in the mixed conifer and red fir types. ALSEs are oriented towards forested environments and do not cover many other types. Including ALSEs to the base level of currently protected areas provides a very high level of representation for forest types, but the total area required nearly doubles. It should be noted that there are several ALSE alternatives and that they are hypothetical. We used only one alternative configuration for BMAS modeling.

Question 3

Can a representative BMA system be established on public lands only? If not, what area of private lands is required? How does the area requirement change if lands that are currently administratively withdrawn from grazing and timber harvest are classified as BMA lands?

Many community types do not occur on public lands, or are present in insufficient extent to be adequately represented on public lands alone. To represent as much of these types that is on public lands requires over 1/3 of the land in the region. In our model, we allocated entire planning watersheds when selecting BMAs, even if the public land containing a type needing additional representation was only a small portion of it. Therefore, a large amount of private land was swept into this solution even though it did not count toward biodiversity representation. Of course, it would be possible to allocate individual parcels of public land for biodiversity management with much less total area required, but this would violate our premise that larger, ecologically-based units make superior BMAs.

Including Class 2 lands has a significant effect in the north but much less effect on solutions in the central and southern regions, because most plant communities that are widespread on Class 2 lands in those regions are also well represented in the national parks. Therefore the amount of private land required to satisfy the representation targets remains quite similar to that of Alternative 2.

Question 4

How sensitive is the siting of BMAs to the way in which biodiversity is measured? Specifically, how do solutions to represent plant community types compare to solutions based on representing vertebrate species?

The predicted distributions of most vertebrates are well represented by the solution for plant communities. Even the vertebrate species not fully represented in alternative 2a were nearly so. There was, however, considerable sensitivity in terms of the sites that were chosen in Alternative 2a compared to Alternative 6. Because vertebrates tend to be more widespread than plant communities, i.e., they occur in more planning units, the vertebrate alternative was more driven by suitability factors than was the corresponding plant community alternative 2a.

Question 5

Do some general areas emerge from the analysis that appear especially well suited to serve as BMAs?

As stated above, our purpose in this exercise was not to identify the optimal sites for a Sierran BMA system. Rather, we have attempted to scope out the dimensions of the decision space through evaluation of a set of plausible alternatives. Nevertheless, certain geographic areas were consistently identified in the alternatives based on the biological, efficiency, and suitability criteria and therefore were less sensitive to model assumptions and objectives. In the northern region, these general areas include the lower elevations in Calaveras County and portions of the Cosumnes River basin, mid-elevations of Sierra County north of Highway 49, and parts of Plumas County east of Highway 89 and south of Highway 70. Frequently selected watersheds in the central region are scattered along Highway 49, particularly in Mariposa County. Few watersheds are needed from higher elevation zones because Yosemite National Park provides coverage for most conifer and subalpine community types. Likewise in the southern region, higher elevation communities are generally well represented in the National Parks. The areas of BMAs from the alternatives tend to concentrate along the South Fork of the Kern River to Walker Pass and along the Greenhorn Mountains. These watersheds warrant more detailed study in any biodiversity management strategy for the Sierra Nevada region.

REFERENCES

Alverson, W. S., W. Kuhlmann, and D. A. Waller. 1994. *Wild forests: Conservation biology and public policy.* Washington, DC: Island Press.

Davis, F. W., and D. M. Stoms. 1996. Sierran vegetation: A gap analysis. In *Sierra Nevada Ecosystem Project: Final report to Congress,* vol. II, chap. 23. Davis: University of California, Centers for Water and Wildland Resources.

Franklin, J. F., and J. A. Fites-Kaufmann. 1996. Analysis of late successional forests. In *Sierra Nevada Ecosystem Project: Final report to Congress,* vol. II, chap. 21. Davis: University of California, Centers for Water and Wildland Resources.

Harrison, S., and J. F. Quinn. 1989. Correlated environments and the persistence of metapopulations. *Oikos* 56:293–98.

Holland, R. F. 1986. *Preliminary descriptions of the terrestrial natural communities of California.* Sacramento: California Department of Fish and Game.

Hollander, A. D., F. W. Davis, and D. M. Stoms. 1994. Hierarchical representation of species distributions using maps, images, and sighting data. In *Mapping the diversity of nature,* edited by R. I. Miller, 71–88. New York: Chapman and Hall.

Margules, C. R., R. L. Pressey, and A. O. Nicholls. 1991. Selecting nature reserves. In *Nature conservation: Cost effective biological surveys and data analysis,* edited by C. R. Margules and M. P. Austin, 90–97. Melbourne, Australia: CSIRO.

Mayer, K. E., and W. F. Laudenslayer Jr. 1988. *A guide to wildlife habitats of California.* Sacramento: California Department of Forestry and Fire Protection.

Millar, C. I., M. Barbour, D. L. Elliott-Fisk, J. R. Shevock, and W. B. Woolfenden. 1996. Significant natural areas. In *Sierra Nevada Ecosystem Project: Final report to Congress,* vol. II, chap. 29. Davis: University of California, Centers for Water and Wildland Resources.

Moyle, P. B. 1996a. Potential fisheries diversity management areas. In *Sierra Nevada Ecosystem Project: Final report to Congress,* vol. II, chap. 57. Davis: University of California, Centers for Water and Wildland Resources.

Moyle, P. B. 1996b. Status of aquatic habitat types. In *Sierra Nevada Ecosystem Project: Final report to Congress,* vol. II, chap. 32. Davis: University of California, Centers for Water and Wildland Resources.

Moyle, P. B., and P. J. Randall. 1996. Biotic integrity of watersheds. In *Sierra Nevada Ecosystem Project: Final report to Congress,* vol. II, chap. 34. Davis: University of California, Centers for Water and Wildland Resources.

Moyle P. B., R. M. Yoshiyama, and R. Knapp. 1996. Status of fish and fisheries. In *Sierra Nevada Ecosystem Project: Final report to Congress,* vol. II, chap. 33. Davis: University of California, Centers for Water and Wildland Resources.

Noss, R. 1992. The Wildlands Project: Land conservation strategy. *Wild Earth* special issue: 10–25.

Reid, T. S., and D. D. Murphy. 1995. Providing a regional context for local conservation action. *BioScience* supplement: S84–S90.

Schonewald-Cox, C. M. 1983. Conclusions: Guidelines to management, a beginning attempt. In *Genetics and conservation: A reference for managing wild animal and plant populations,* edited by C. M. Shonewald-Cox, S. M. Chambers, B. MacBryde, and W. L. Thomas, 414–45. Redwood City, CA: Benjamin/Cummings.

Scott, J. M., F. Davis, B. Csuti, R. Noss, B. Butterfield, C. Groves, H. Anderson, S. Caicco, F. D'Erchia, T. C. Edwards Jr., J. Ulliman, and R. G. Wright. 1993. Gap analysis: A geographic approach to protection of biological diversity. *Wildlife Monographs* 123:1–41.

Shafer, C. L. 1990. Nature reserves: Island theory and conservation practice. Washington, DC: Smithsonian Institution Press.

Shevock, J. R. 1996. Status of rare and endemic plants. In *Sierra Nevada Ecosystem Project: Final report to Congress,* vol. II, chap. 24. Davis: University of California, Centers for Water and Wildland Resources.

APPENDIX 58.1

The Biodiversity Management Areas Selection (BMAS) Model

Definitions and Units of analysis

Biodiversity element: physical or biological feature of an area that serves as a metric of biodiversity. In the analysis reported here the elements are plant community types as defined by Holland (1986). In a companion report, we described the distribution, ownership and management status of these community types in the Sierra Nevada (Davis and Stoms 1996).

Landscape: map unit (polygons) used to represent the spatial distribution of biodiversity elements. For this analysis we use the Gap Analysis database and map of plant community types. Each map unit contains information on the occurrence and extent of up to three plant community types.

Biodiversity Management Area: an area with an active ecosystem management plan in operation whose purpose is to contribute to regional maintenance of native genetic, species and community levels of biodiversity and the processes that maintain that biodiversity. Generally a BMA will be comprised of several *landscapes*. In this analysis we considered four different starting points for a BMA system for the Sierra Nevada: 1) ignore current land allocation and assume no existing BMAs, 2) public or private lands that are formally designated for conservation of native biodiversity and within which economic activities such as development, grazing and timber harvest are precluded (Class 1 lands as defined in SNEP's gap analysis of the Sierra Nevada (Davis and Stoms 1996), Class 1 lands plus national forest lands that are administratively withdrawn from grazing and intensive timber harvest (Class 2 lands as defined in the SNEP gap analysis, and 4) Class 1 lands plus other lands identified by SNEP as Areas of Late Successional Emphasis (ALSEs).

Planning unit: spatial aggregate of landscapes used to map Biodiversity Management Areas. Because they provide comprehensive coverage and form rational units for ecosystem management, we use the CalWater Planning watersheds (~ 2400 at 3-10k acres each) as planning units. The Calwater system is hierarchical, with planning watersheds being aggregated into superplanning watersheds. In the northern region, the superplanning watersheds average about three times the size of planning watersheds. These were used in alternative 2l to test the sensitivity of the model to the size of planning units.

Planning region: the spatial domain of the analysis, in this case three of the six hydrologic regions of the SNEP core area (figure 1). We consider these regions as renewable resource zones that are relatively homogeneous in terms of their biotic composition.

Suitability Elements: mapped indicators of human activities that affect the suitability of an area for BMA designation. Our present model includes the following elements:

Human Population Density: 1990 Census data were obtained as an ARC vector coverage by block group from Professor John Radke at the University of California, Berkeley. The coverage had attributes for population and population density for each block group (population on the order of 1,000 persons per block group which is the first level of aggregation from census blocks). Data by block group were resampled to watersheds. The vector data were converted to an ARC GRID using the population density values over a 100 m grid. Next this grid was combined with a grid of watersheds and the population density per 1 ha was summed over all grid cells in the watershed. Then the total population per watershed was converted to density by dividing by the area of the watershed in hectares. Values range from 0 to 559.8 people per km^2 (mean of 7.4), with high values indicating urbanized watersheds that would generally be unsuitable for protection of most forms of native biodiversity.

Road Density: USGS 100,000 scale Digital Line Graph (DLG) datasets were obtained from USGS and converted to ARC coverages. The road arcs were buffered with a buffer width related to the class of road according to table 1. This buffer operation was used to estimate the area of land actually impacted by the presence of each road, where freeways were assumed to affect a greater spatial extent than dirt roads. This operation also accounted for the spatial distribution of roads which a simple measure of road density (i.e., km of road length per km^2 of area) does not. For instance, urban streets could total a long length but because they are so closely spaced, they do not affect as large an area of habitat as a similar length of road spread uniformly across a watershed. The road density index was calculated by summing the total area of buffered roads per watershed and converting the area to a percentage of watershed area.

There are a number of issues with the DLG data and with the index. The DLG files were largely compiled between 1975 and 1985 and therefore do not include more recent residential development in the foothills and logging activity in the mixed conifer belt. The index itself does not consider the kind of habitat the roads are in. Clearly a road in an urban area has less impact than a road in a natural land cover type. Values for the road density index range from 0 to 98 percent (mean of 15.7), with high values indicating watersheds heavily disturbed by a variety of human activities, making them less suitable for biodiversity management areas.

Percent of Land in Private Ownership: The cost of changing land management to better protect the long-term viability of native biodiversity is partly a function of current land ownership and management. Therefore we have included an index of the proportion of land in a watershed that is in private ownership (either individuals or corporations), derived from the land ownership/management coverage developed for Gap Analysis (Davis and Stoms 1996). Values range from 0 to 100 percent (mean of 31.9), with high values indicating watersheds with high probable costs for management of biodiversity.

Public-Private Interface: Intermingling of public and private lands, or fragmentation of ownership, can greatly complicate BMA management. To capture this feature of an area's land ownership we summed the length of boundary separating public from private lands. The total length was divided by watershed area to derive an index of length per unit area. Values range from 0 to 1.94 km / km^2 (mean of 0.33), with high values of the index indicating complex ownership patterns.

Watershed Suitability Index: There are many ways these four elements (road density, human population density, percentage of private ownership, and public-private interface) could be combined to create an overall index representing suitability for biodiversity management areas. Factors are scaled differently and should be normalized in some way before combining them. Because the factor indices are quite skewed in their distributions, we chose not to weight them by the reciprocal of their maximum value but by the reciprocal of their mean value. This has the effect of contributing very high factor values for watersheds when the values approach their maximum, such as urban areas for the population density factor. In such cases, the factor would receive a weighted score well above 1, whereas a low value would only be a small fraction of 1. Values of WSI range from 0 to 82.6 (mean of 4.0), with high values indicating watersheds that are either extremely high in one of the factors or are moderately high in several factors.

The BMAS Model

In order to formulate this model, we use the following notation:

j, J	index and set of planning units (e.g., small watersheds)
k, K	index and set of biodiversity elements considered vulnerable
a_j	the area of planning unit j
a_{jk}	area in planning unit j which contains element k and is potentially impacted by planned management activities in j
Min_k	minimum area containing element k that needs to be brought under biodiversity management in order to remove element k from the list of vulnerable elements
Hd_j	human density measurement for planning unit j
Rd_j	the percent of the area in planning unit j that is impacted by roads

Pla_j	the percent of the area of unit j that is held in private ownership
PPI_j	the density of public-private land interface
WSI_j	watershed suitability index for planning unit j (index approaches 0 if most suitable)
w_l	weight attached to term l in the objective function
X_j	$\begin{cases} 1 \text{ if site } j \text{ is selected for a Biodiversity Management Area} \\ 0 \text{ if not} \end{cases}$
l	index of weights for terms in the objective function

We can formulate the general Biodiversity Management Area Selection (BMAS) Model in the following manner:

$$\text{Minimize } Z = \sum_j \left(w_1 a_j + w_2 WSI_j \right) X_j \tag{1}$$

Subject to the following conditions:

1) Element k is sufficiently represented in BMAs to be considered not vulnerable, that is,

$$\sum_j a_{jk} X_j \geq Min_k \quad \text{for each } k \,\varepsilon\, K \tag{2}$$

2) Integer requirements: $X_j = 0$ or 1 for each $j \,\varepsilon\, J$

This model, which we have termed the biodiversity management area selection (BMAS) model, involves selecting the most suitable planning units that contain underrepresented biodiversity elements. An element is considered vulnerable until a specified fraction of its mapped distribution occurs in areas designated as BMAs. This is established by condition 1). Either a planning unit is selected as a BMA or it is isn't. This is enforced by condition 2).

The objective function (Equation (1)) contains two terms. The first term is strictly an area term, and has the effect of minimizing the total area selected for biodiversity management options. The second term is a suitability term. As we define it , the less suitable an area is, the higher this term becomes. Thus this term operates to minimize the total "unsuitability" of the selected areas. The target levels for representation of plant communities can be set for each type individually. This formulation provides some flexibility in the way that rare or endemic elements are treated relative to more widespread types.

The Suitability index can be expanded to include a number of cultural or biological variables. For example, in Equation 3 the term has been expanded to four terms that contribute weighted values for the suitability elements in the jth watershed planning unit.

$$\text{Minimize } Z = \sum_j \left(w_1 a_j + w_2 Hd_j + w_3 Rd_j + w_4 Pla_j + w_5 PPl_j \right) X_j \tag{3}$$

Unless we state otherwise, we have used this last equation as the model formulation for the results described in the text. We use the term *Watershed Suitability Index* (WSI) to describe the sum of the four suitability elements in Equation 3. That is,

$$WSI_j = \sum \left(w_2 Hd_j + w_3 Rd_j + w_4 Pla_j + w_5 PPl_j \right)$$

where the weights w_l equal 1 / mean of the factor score.

Vulnerable Plant Community Types by Alternative

Holland Code	Holland Name	1a	1b	2a	2b	2c	2d	2e	2f	2g	2h	2i	2j	2k	2l	3a	3b	3c	3d	3e	3f	4	5
34200	Mojave mixed scrub and steppe		X								X	X	X	X						X	X		
34210	Mojave mixed woody scrub		X								X	X	X	X						X	X		
34300	Blackbush scrub		X								X	X	X	X						X	X		
35100	Great Basin mixed scrub	X	X	X	X	X	X	X	X	X						X	X	X	X			X	X
35210	Big sagebrush scrub	X	X	X	X	X					X	X	X	X		X	X	X		X	X	X	X
35211	Low sagebrush scrub[a]	X		X	X	X	X	X	X	X						X	X	X	X			X	X
35212	Silver sagebrush scrub[a]		X	X	X	X					X	X	X	X		X	X	X		X	X	X	X
35220	Subalpine sagebrush scrub					X																	
35300	Sagebrush steppe	X	X	X	X	X	X	X	X	X	X	X	X	X		X	X	X	X	X	X	X	X
35400	Rabbitbrush scrub			X	X	X										X	X	X				X	X
35500	Cercocarpus ledifolius woodland[a]		X	X	X	X					X	X	X	X		X	X	X		X	X	X	X
35600	Wyethia mollis[a]	X		X	X	X				X					X		X		X			X	X
37100	Upper Sonoran mixed chaparral			X	X	X									X								
37110	Northern mixed chaparral	X	X	X	X	X	X	X	X	X					X	X	X	X	X			X	X
37200	Chamise chaparral	X	X	X	X	X	X	X	X	X					X	X	X	X	X			X	X
37400	Semidesert chaparral		X																				
37510	Mixed montane chaparral	X	X	X	X					X			X	X	X	X		X			X		X
37520	Montane manzanita chaparral	X	X	X	X	X	X	X	X	X					X	X		X	X				X
37530	Montane ceanothus chaparral	X		X	X	X	X	X	X	X					X	X	X	X	X				X
37531	Deer brush chaparral			X	X	X									X	X	X						X
37541	Shin oak brush		X										X		X								
37542	Huckleberry oak chaparral	X	X	X	X	X				X			X		X		X		X			X	X
37550	Bush chinquapin chaparral	X		X	X	X									X	X	X						X
37810	Buck brush chaparral	X	X	X	X	X	X	X	X	X	X	X	X	X	X	X	X	X	X			X	X
37900	Scrub oak chaparral			X	X	X					X	X	X	X	X					X	X	X	X
37A00	Interior live oak chaparral	X		X	X	X	X	X	X	X					X	X	X	X				X	X
37B00	Upper Sonoran manzanita chaparral	X	X	X	X	X	X	X	X	X					X	X	X	X	X			X	X
37E00	Mesic north slope chaparral	X	X	X	X	X	X	X	X	X					X	X		X		X		X	X
39000	Upper Sonoran subshrub scrub		X								X	X	X	X	X						X		
42110	Valley needlegrass grassland	X		X	X	X										X	X	X				X	X
42200	Non-native grassland	X	X	X	X	X	X	X	X	X	X	X	X	X	X	X	X	X	X	X	X	X	X
45100	Montane meadow	X	X	X	X	X										X	X	X				X	X
45210	Wet subalpine or alpine meadow	X	X	X	X	X										X	X	X				X	X
45310	Alkali meadow		X								X	X	X	X						X	X		
52320	Transmontane alkali marsh		X								X	X	X	X						X	X		
61410	Great Valley cottonwood riparian forest		X	X	X	X								X		X	X	X			X	X	X
61420	Great Valley mixed riparian forest		X								X	X	X	X						X	X		
61430	Great Valley valley oak riparian forest	X	X				X	X	X	X	X	X	X	X				X		X	X	X	
61510	White alder riparian forest	X					X	X	X	X								X	X				
61530	Montane black cottonwood riparian forest		X								X	X	X	X						X	X		
63500	Montane riparian scrub	X	X	X	X	X										X	X	X		X	X	X	X
71110	Oregon oak woodland		X	X	X	X					X	X	X	X	X	X	X			X	X	X	X
71120	Black oak woodland	X	X	X	X	X	X	X	X	X	X	X	X	X	X	X	X	X	X		X	X	X
71130	Valley oak woodland	X	X	X	X	X	X	X	X	X	X	X	X	X	X	X	X	X	X	X		X	X
71140	Blue oak woodland	X	X	X	X	X	X	X	X	X	X	X	X	X	X	X	X	X	X	X	X	X	X

continued

[a]Addition to the standard Holland classification.
X indicates that the alternative was required to select additional area for that community type. Communities are not necessarily vulnerable in every region in which they occur.

Holland Code	Holland Name	Alternative																					
		1a	1b	2a	2b	2c	2d	2e	2f	2g	2h	2i	2j	2k	2l	3a	3b	3c	3d	3e	3f	4	5
71150	Interior live oak woodland	X	X	X	X	X	X	X	X	X	X	X	X	X	X	X	X	X	X		X	X	X
71310	Open foothill pine woodland	X	X	X	X	X	X	X	X	X	X	X	X	X	X	X	X	X	X	X	X	X	X
71322	Nonserpentine foothill pine woodland	X	X	X	X	X				X	X	X	X	X	X	X	X			X	X	X	X
71410	Foothill pine–oak woodland	X	X	X	X	X	X	X	X	X	X	X	X	X	X	X	X	X	X	X	X		X
71500	Cismontane juniper woodland[a]	X	X	X	X	X											X	X	X				X	X
71600	Oak–piñon woodland[a]		X											X							X		
72110	Northern juniper woodland	X		X	X	X											X	X	X			X	X
72121	Great Basin piñon–juniper woodland		X											X							X		
72122	Great Basin piñon woodland	X	X				X	X	X	X	X	X	X	X				X	X	X	X		
72123	Great Basin juniper woodland and scrub	X					X	X	X	X									X	X			
72220	Mojavean juniper woodland and scrub		X								X	X	X	X						X	X		
73000	Joshua tree woodland		X								X	X	X	X						X	X		
81320	Canyon live oak forest	X	X	X	X	X								X	X	X							X
81330	Interior live oak forest	X	X	X	X	X	X	X	X	X	X	X	X	X	X	X	X	X	X		X	X	X
81340	Black oak forest	X	X	X	X	X	X	X	X	X				X	X	X		X	X				X
81400	Tan oak forest			X	X	X									X		X						X
81B00	Aspen forest	X		X	X	X									X		X					X	X
83210	Knobcone pine forest	X		X	X	X	X	X	X	X					X			X	X			X	X
83330	Southern interior cypress forest		X								X	X	X	X						X	X		
84210	West-side ponderosa pine forest	X	X	X	X	X				X					X	X	X		X				X
84220	East-side ponderosa pine forest			X	X	X									X	X	X					X	X
84230	Sierran mixed conifer forest	X	X	X	X	X				X					X		X						X
84240	Sierran white fir forest	X	X	X	X	X				X					X		X			X			X
84250	Big tree forest		X																				
85100	Jeffrey pine forest	X	X	X	X	X								X	X		X				X	X	X
85120	Red fir–western white pine forest[a]	X	X			X																	
85210	Jeffrey pine–fir forest	X	X	X	X	X								X	X		X				X		X
85310	Red fir forest	X	X	X	X	X									X		X						X
86100	Lodgepole pine forest	X	X	X	X	X									X		X						X
86210	Whitebark pine–mountain hemlock forest	X																					
86220	Whitebark pine–lodgepole pine forest	X		X	X	X									X	X	X						X
86300	Foxtail pine forest		X																				
86600	Whitebark pine forest	X	X																				
86700	Limber pine forest	X	X				X	X	X	X								X	X				
87100	Lower cismontane mixed conifer–oak forest[a]	X	X	X	X	X	X	X	X	X					X	X	X	X		X			X
87200	Upper cismontane mixed conifer–oak forest[a]	X	X									X	X	X	X						X		
91120	Sierra Nevada fell field	X	X				X	X	X	X					X			X	X		X		
94000	Alpine dwarf scrub	X	X											X							X		

[a]Addition to the standard Holland classification.
X indicates that the alternative was required to select additional area for that community type. Communities are not necessarily vulnerable in every region in which they occur.